National Association of Broadcasters

ENGINEERING HANDBOOK

10th Edition

National Association of Broadcasters

ENGINEERING HANDBOOK

10th Edition

Editor-in-Chief

EDMUND A. WILLIAMS

Associate Editors

GRAHAM A. JONES
DAVID H. LAYER
THOMAS G. OSENKOWSKY

NEW YORK AND LONDON

First published 1935
by Focal Press

Published 2014
by Focal Press
70 Blanchard Road, Suite 402, Burlington, MA 01803

Published in the UK
By Focal Press
2 Park Square, Milton Park, Abingdon, Oxon OX14 4RN

Focal Press is an imprint of the Taylor & Francis Group, an informa business

Notices

Practitioners and researchers must always rely on their own experience and knowledge in evaluating and using any information, methods, compounds, or experiments described herein. In using such information or methods they should be mindful of their own safety and the safety of others, including parties for whom they have a professional responsibility.

To the fullest extent of the law, neither the Publisher nor the authors, contributors, or editors, assume any liability for any injury and/or damage to persons or property as a matter of products liability, negligence or otherwise, or from any use or operation of any methods, products, instructions, or ideas contained in the material herein.

Library of Congress Cataloging-in-Publication Data
Application submitted

British Library Cataloguing-in-Publication Data
A catalogue record for this book is available from the British Library.

ISBN 13: 978-0-240-80751-5 (hbk)
ISBN 13: 978-0-240-80982-3 (CD-ROM)

Contents

SECTION 2: TECHNICAL MANAGEMENT AND SAFETY

SECTION 3: AUDIO PRODUCTION AND STUDIO TECHNOLOGY

SECTION 4: RADIO TRANSMISSION

SECTION 5: VIDEO PRODUCTION AND STUDIO TECHNOLOGY

SECTION 6: TELEVISION TRANSMISSION

SECTION 7: BROADCAST TOWERS

SECTION 8: SIGNAL MEASUREMENT AND ANALYSIS

SECTION 9: ANCILLARY BROADCAST SYSTEMS

Foreword

In his introduction to the first edition of the *NAB Engineering Handbook*, published in 1935, NAB Managing Director James W. Baldwin wrote that its purpose was "for use as a guide in matters on which little or no authoritative collected information has heretofore been generally available in readily usable form."

Is this still a valid basis for publishing a new edition of this book over 70 years later? After all, with the huge quantity of information available on the Internet and the explosion of high speed broadband Internet access, it's quite likely that details on any desired subject can be found out there somewhere on the Net, given a powerful enough search tool, the time needed to conduct the search, and a way to verify the information found.

And that is the problem. So much information is on the Internet, finding just what is needed in a reasonable amount of time is a difficult (sometimes impossible) challenge, and the credibility of the information that is found may be unknown or questionable. For the broadcast engineer in today's fast-paced business and technological environment, the mere known existence of online information alone is not enough; there must also be certainty of timely access to the information and there must be confidence in its veracity.

In 1982, author and futurist John Naisbitt recognized the impending dilemma of the Information Age in his book *Megatrends* when he wrote: "We are drowning in information and starved for knowledge." With this perspective in mind, we are proud to present this new 10th edition of the *NAB Engineering Handbook* and conclude that James W. Baldwin's words from 1935 still apply today.

This new edition of the Handbook has been completely revised and updated from the 1999 9th edition with twenty-four new chapters that encompass the digital transition in broadcasting at all levels. There are new chapters that reflect how information technology has been adopted by broadcasters, how both radio and television operations benefit from the use of servers, LANs, the Internet, compression techniques, nonlinear editing, digital transmission, metadata, and many other important and timely subjects.

The goal of the editorial staff on this project is to ensure that the information in the 10th edition represents what broadcast engineers need to know to be comfortable, competent, and effective in the rapidly changing broadcast technology environment.

The great 18th century English literary figure Samuel Johnson once insightfully remarked: "Knowledge is of two kinds. We know a subject ourselves, or we know where we can find information on it." The *NAB Engineering Handbook* is where broadcast engineers know they can find the information needed to do their jobs. Please feel free to contact us with comments on how to continue to maintain this Handbook as a readily usable form of an authoritative collection of information on broadcast engineering topics.

Lynn D. Claudy
Senior Vice President, Science and Technology
National Association of Broadcasters

Notes and Acknowledgments

"In theory, there is no difference between theory and practice. But, in practice, there is." These words, attributed to Jan L.A. van de Snepscheut, computer scientist and author, ring true to broadcast engineers who make a living from implementing practical applications in real world operating environments using equipment designed "in theory" to do the job.

In this completely revised and substantially updated 10th edition of the *NAB Engineering Handbook*, we have used the phrase "what the engineer needs to know to do the job" as a motto for the authors when writing or revising a particular chapter.

While all the images in the handbook are in black and white many were originally in color and are reproduced in color in the compact disc (CD) that accompanies the book.

In this handbook of 104 chapters are 24 new chapters that bring the *NAB Engineering Handbook* fully into the digital era. Assisting the Editor-in-Chief, and working one-on-one with the 140 authors, were three Associate Editors who commissioned, assembled, edited, reviewed, formatted, and verified the material:

Graham A. Jones
Director, Communications Engineering,
Science & Technology
National Association of Broadcasters,
Washington, D.C.

David H. Layer
Director, Advanced Engineering,
Science & Technology
National Association of Broadcasters,
Washington, D.C.

Thomas G. Osenkowsky
Radio Engineering Consultant
Brookfield, Connecticut

The editors of the *NAB Engineering Handbook* thank the authors, co-authors, and contributors for the time and expertise that they have given to make this book an important part of a broadcast engineer's library and reference materials.

The preparation of this edition has taken nearly two years, a year longer than expected, but well worth the time and considerable effort to insure the material is accurate and up to date.

Of course, such an undertaking could not have been accomplished without the considerable support of other NAB Science and Technology staff members Lynn Claudy, Senior Vice President, Janet H. Elliott, Senior Director of Operations and Managing Editor for the 9th edition, and Dan Landrigan, NAB Publications Manager.

We hope you find this newest edition of the *NAB Engineering Handbook* useful in your work in broadcast engineering and operations.

Edmund A. Williams
Editor-in-Chief

The Editors

Edmund A. Williams
Editor-in-Chief

During his 50 year career in broadcasting, Mr. Williams has worked for Ohio State University Radio and Television, Ohio University Radio and Television, the Public Broadcasting Service, the National Association of Broadcasters, and the Advanced Television Test Center. He participated in the Emmy Award-winning PBS Captioning for the Deaf, Satellite Interconnection, and UHF Improvement Projects and developed a terrain-sensitive broadcast coverage prediction technique (AREAPOP). He conducted lab and field tests for AM Stereo, TV Stereo, Ghost Canceling, and managed the field testing of the Grand Alliance Digital Television Broadcast System. He has been a member of NAB, ATSC, and IEEE standards committees, and the FCC Advisory Committee for Advanced Television Systems. He authored numerous technical papers for industry symposia, and conducted demonstrations of high definition television broadcasting for the FCC, Congress, and broadcast industry groups. He participated as engineer and technical seminar presenter on the Harris/PBS DTV Express that toured over 40 cities in the U.S. in 1997-1999 with a mobile HDTV and digital television demonstration facility.

Mr. Williams is a Life Senior Member of the IEEE and serves on the Broadcast Technology Society AdCom as Technical Activities chair, is a Fellow and former Governor of the SMPTE, a Senior Member of SBE and SCTE, and Associate Member of the AFCCE. Mr. Williams retired from the Public Broadcasting Service in 2004. His amateur radio license is W8APE. He can be reached at ed.williams@ieee.org.

Graham A. Jones
Associate Editor

Graham Jones is Director, Communications Engineering with NAB Science and Technology, specializing in advanced television technology issues and standards. Previously with Harris Corporation, he was Engineering Director for the Harris/PBS DTV Express—the educational road show that introduced DTV to many U.S. broadcasters. He started his career with the BBC in London, and for nearly 20 years with International Broadcasting Consultants worked as a consultant to broadcasters in many parts of the world. He chairs the ATSC Planning Committee and is a member of the ATSC Technology and Standards Group. He is also the Director of Education for SMPTE and an active member of several SMPTE technology committees.

Mr. Jones is author of *A Broadcast Engineering Tutorial for Non-Engineers* (Focal Press, 2005). He has presented papers at SMPTE, NAB, and other conferences and was principal author of *Digital Television Station and Network Implementation*, published in the 2006 Proceedings of the IEEE on Global Digital Television, for which he served on the ATSC board of editors. Mr. Jones holds a degree in Physics from the University of Nottingham, England. He is a Chartered Engineer, a Fellow of SMPTE, and a member of the SBE, the Institution of Engineering and Technology, and the Royal Television Society. He is a Governor of SMPTE and a manager of the SMPTE Washington DC Section. In 2004, he was awarded the ATSC Bernard J. Lechner Outstanding Contributor award. He can be reached at GJones@nab.org.

David H. Layer
Associate Editor

David Layer is Director, Advanced Engineering in the Science & Technology Department of NAB. He has been with NAB since 1995, and has been very active in the radio standards setting area. He is also involved in NAB's technical conference planning and technical publication activities, and has been an author and contributing author for numerous technical publications, including IEEE Spectrum magazine and the McGraw-Hill Yearbook of Science and Technology. Mr. Layer currently serves as the Vice Chairman and Secretary of the ATSC Specialist Group on RF Transmission. He has also been active in the work of the ITU-R, and participates in Study Group 6, having previously been a member of the U.S. delegation to the Working party 10A and 10B (now 6E) meetings.

Mr. Layer is a frequent presenter at broadcasting industry events around the world, having made presentations to numerous state broadcasting associations, trade associations and technical societies as well as at conferences in Switzerland, Chile, Brazil, Mexico, Puerto Rico, and Uruguay. Prior to joining NAB, he was the Associate Manager of the Transmission and Channel Processing department of COMSAT Laboratories. Mr. Layer is an Associate Member of AFCCE where he serves on the association's board of directors, and is a Senior Member of the Institute of Electrical and Electronics Engineers (IEEE). He served as the Chairman of the annual IEEE Broadcast Symposium from 2002 to 2004. He received a BSEE degree from the University of Maryland, and an MSEE degree from Purdue University, where he was also a teaching assistant. Mr. Layer has also served on the faculty of American University and Frederick Community College. He can be reached at dlayer@nab.org.

Thomas G. Osenkowsky
Associate Editor

Thomas G. Osenkowsky began his career in broadcasting in 1973 while a senior in high school. He has held positions as announcer, Chief Engineer, Operations Manager, and General Manager at broadcast stations in Connecticut. He has written software for antenna system design and analysis, RFR, mapping and other broadcast engineering related tasks. He has served as a consultant to AM and FM stations in the United States and Caribbean islands. He is a Senior Member of the Society of Broadcast Engineers (SBE), Institute of Electrical and Electronics Engineers (IEEE) and National Association of Radio and Telecommunications Engineers (NARTE).

Mr. Osenkowsky earned his Professional Broadcast Engineer Life Certification (CPBE) from the SBE, a Certified Master Engineer with Master RF Radiating Endorsement from NARTE, and holds a First Class Radiotelephone license from the FCC. He has presented papers at Broadcast Engineering Conferences held at NAB Conventions, SBE Annual Conventions, and state broadcasters annual conventions. He is a regular contributor to Radio World magazine, a private pilot, and amateur radio operator – N1IXJ. Reach him at Engineer_Tom_2000@Yahoo.com.

Contributors to the 10th Edition

Jay C. Adrick (Chapter 6.2) is Vice President, Technology, Harris Corporation Broadcast Communications Division. As a 42+ year veteran of the broadcast industry, Jay has led the teams that designed and built many of the leading broadcast facilities, including The Golf Channel, The Weather Channel, The Discovery Channel, National Public Radio, The Voice of America, Georgia Public Broadcasting, Iraqi Media Network, and many other broadcast facilities. At Harris, he led the product development for the Harris FlexiCoder, MasterPlus, and MonitorPlus products during the U.S. digital rollout.

Prior to joining Harris, he was Executive Vice President of Midwest Communications Corporation. He also taught broadcast communications at Xavier University in Cincinnati, Ohio, for eight years, served as the University's Director of Radio and Television, and was a founder of public radio station WVXU-FM. His broadcast career began in commercial radio and later included work at several television stations.

Mr. Adrick holds a Bachelor of Science degree in communication arts and a Master's degree in educational communications from Xavier University. He is a member of the Society of Broadcast Engineers, Society of Motion Picture and Television Engineers, and the Society of Telecommunications Engineers.

Mr. Adrick served on the Board of Directors and as Vice Chairman of the Advanced Television Systems Committee. He is currently a board member of the ATSC Forum. In addition, he has served on both of the FCC Media Security and Reliability Councils.

W. C. (Cris) Alexander (Chapter 4.1) is Director of Engineering for Crawford Broadcasting Company. He began his career in broadcasting in Amarillo, Texas, in the mid-1970s. Obtaining a FCC First Class Radiotelephone License, he worked in the engineering departments of AM, FM, and TV stations in that market before moving to Dallas/Fort Worth, Texas, where he worked first in television and then in radio, landing in his present position with Crawford Broadcasting Company in 1984.

Cris is a graduate of the Cleveland Institute of Electronics Broadcast Engineering course. He is an SBE member, holds SBE Professional Broadcast Engineer (CPBE) and AM Directional Specialist (AMD) certifications and is Certification Chairman of the Denver SBE chapter. He is an associate member of the Association of Federal Communications Consulting Engineers (AFCCE). He is a partner in Au Contraire Software, Ltd., a provider of broadcast engineering software and database services.

Cris lives with his family in the Denver area.

José Alvarez (Chapter 9.2) is the owner and founder of Wavetech Associates Inc., a 15-year year-old company specializing in power quality solutions for business. He is a 1983 graduate of the New Jersey Institute of Technology (NJIT), with a Bachelor of Science degree in Electrical Engineering.

Mr. Alvarez has been quoted and written about in journals such as *Business News New Jersey* and *Power Quality Assurance*. His 22 years of experience has led him into numerous applications, including the broadcast field where he has successfully applied power conditioning solutions.

Hiroshi Asami (Chapter 1.14) is Director-General of Hokuriku Bureau of Telecommunications, Ministry of Internal Affairs and Communication, Japan.

Mr. Asami was born in Japan, and received an MS in electrical engineering from Kyoto University in 1980, and an MS in engineering economic systems from Stanford University in 1985.

Mr. Asami joined the Ministry of Posts and Telecommunication in 1980, and has been engaged in standardization and regulations of telecommunication and broadcasting systems since then. He has been working for channel planning of digital broadcasting, and contributed significantly to the start of digital terrestrial television in Japan in 2003. He was Director of Broadcasting Technology until 2005.

David Baden (Chapter 2.2) is currently Chief Technical Officer for Radio Free Asia where, in 1996, he was an organizational founding member of senior management. He began his career as a musician in 1974 and continued to perform until 1980. During that time he also worked as an engineer for various theatrical lighting, recording studios, and sound reinforcement companies in the Greater Washington, D.C. area. Over the course of his career he has been intimately involved with all aspects of technical documentation. Using CAD, which has replaced the drafting table, he has created and maintained technical documentation from the preliminary proposal to as-built stage for multiple projects worldwide.

In 1980, Mr. Baden was one of the founding principles at db Tech Inc., a company that designed, installed, and maintained professional recording studios and commercial sound reinforcement systems and manufactured specialized audio devices.

In 1984, Mr. Baden began his career in professional broadcasting as Technical and Production Services Supervisor for the Washington News Bureau of Radio Free Europe/Radio Liberty, Inc (RFE/RL). He was responsible for all broadcast systems as well as daily production operations. In 1989, as Deputy Director Broadcast Operations, USA, his responsibilities included all of RFE/RL's USA USA-based operations. In 1995, he relocated to Prague, Czech Republic as Manager of Technical Construction for the Prague Task Force. He was responsible for all technical systems design, integration, logistics, and physical facility construction for the RFE/RL Munich to Prague headquarters relocation.

Dave Bancroft (Chapter 5.23) is Manager, Advanced Technology, for divisions within the three Strategic Business Units of Thomson: Grass Valley, Technicolor, and Technology, and is based in Reading, United Kingdom. He coordinates the work of these business units in technology issues such as scanning and workflow for digital intermediate film production, color management in displays, digital cinematography, and HDTV broadcasting for Europe. He represents Thomson in the standards activities of the SMPTE and related organizations.

Mr. Bancroft began his career as a trainee engineer with the BBC, specializing in live television outside broadcasts. After leaving the UK, he held positions with RCA, Ampex Corporation, BTS, Philips, and Thomson, in Europe, Africa, and the United States, before returning to the UK.

Mr. Bancroft has a certification in broadcast engineering from the BBC. He is a Member of the Institution of Engineering and Technology, United Kingdom (IET), a Fellow of the SMPTE, a Fellow of the Royal Television Society, a Director of Council of the BKSTS Moving Image Society, UK, an International Governor of the SMPTE, and a Member of the Administrative Committee of the IEEE Broadcast Technology Society. He is currently Chairman of the Study Group on Display Technologies in SMPTE. Mr. Bancroft has presented papers at NAB, IBC, SMPTE, and other broadcast conventions. He received the 2004 IBC Presidents Award for his paper on Universal Content Production, and the 2000 SMPTE Journal Award for his paper, "Recent Advances in the Transfer and Manipulation of Film Images in the Data and HDTV Domains." He contributed the HDTV standards chapter to the 2003 *Broadcast Engineer's Reference Book,* and the "Digital Picture Exchange (DPX)" file format chapter to the 2004 *File Interchange Handbook,* both published by Focal Press.

J. Robert (Bob) Beach (Chapter 5.14), Radtec, Inc., began his technical career in the computing research group of a major aircraft manufacturer, developing mathematical algorithms and the associated software for the analysis of structural vibration in aircraft and space vehicles. Subsequently, he became a project manager for a major computer manufacturer, managing the design, integration, and installation of very large computer systems. A number of those systems were developed for weather applications and weather agencies, both domestically in the United States and worldwide. For the last 10 years, he has provided both technical and system-level expertise for the design and implementation of weather radar systems.

Bob is a graduate of Iowa State University (BSEE), and in addition to the *NAB Engineering Handbook* section on weather radar, has written *The Practical Physics of Airport Weather Radar* for Air Traffic Technology International, and *Doppler Weather Radar: Benefiting from Innovation* for the World Meteorological Organization Bulletin.

Michael Bergeron (Chapter 5.3) is HD Camera Product Engineer for Panasonic Broadcast in Secaucus, New Jersey, supporting the studio camera systems. He started in production as a film camera technician at Abel Cine Tech in New York, in 1991. He became Service Department Manager in 1995 and served as Director of Engineering from 1997 to 1999. After two years in research & development at Bell Laboratories and two more years with the NYU Physics Department, Michael returned to broadcast/production in 2003.

While at Panasonic, Michael has been responsible for product development and support for HD system cameras and camcorders. He has been involved in the development and introduction of film cameras including the Aaton 35, XTR Prod and A-Minima, as well as the development of Panasonic HDTV cameras including the AJ-HVX200, AJ-HDX900, AK-HC1500 HD Box camera.

Michael holds a BS in physics from the University of Massachusetts, an MS in electrical engineering from New Jersey Institute of Technology, and is a member of SMPTE.

Richard B. Bernhardt, P.E., (Chapter 9.3) is Founder, President, and Senior Consultant of John-Winston

Engineers and Consultants, Allenhurst, New Jersey, and has successfully guided the business to be recognized internationally as a high-value technical services consultancy.

Mr. Bernhardt led multidiscipline teams of up to 25 involved in power project planning and budget development, technical infrastructure system design, and construction. Sample engagements include traditional planning, design, construction, and maintenance of industrial and institutional electrical distribution systems.

Mr. Bernhardt has developed plans and designs identifying multiple alternative capital and operational requirements for hospitals, data centers, industrial plants, and high technology manufacturing operations. Systems designed included on-site generation systems and UPS equipment, specialty grounding and surge protection, and industrial electrical distribution.

Mr. Bernhandt's project management of industrial and institutional electrical distribution systems include managed construction activities for individual projects with construction values up to $15M, reviewed equipment submittals, conducted system performance testing, and commissioning, and he has developed operation and maintenance plans for electrical system operations.

Mr. Bernhardt holds a BS in engineering from New Mexico State University, 1981. His studies included civil and mechanical engineering coursework. His graduate studies in engineering economics at New Mexico State University in 1981 included the study of project economic evaluation. He received the Centennial Distinguished Alumni Award, New Mexico State University, 1996, for accomplishment in Utility Management Consulting.

Greg Best (Chapter 6.12) is President of Greg Best Consulting, Inc. His firm performs broadcast consulting services for a wide variety of customers and serves the RF communications industry in general. He earned his BSEE degree from the University of Missouri-Rolla and MSEE degree from Illinois Institute of Technology and is a Registered Professional Engineer.

Greg has 30 years of experience in the design, marketing, and product management of RF communications equipment. His corporate experience includes 16 years with the Broadcast Division of Harris where he was responsible for TV transmitter design and management. While there, he was responsible for coordinating the development of the Platinum Series VHF-TV transmitter, as well as many other VHF and UHF-TV transmitters. He also worked for Motorola on the original 800 MHz AMPS cellular phone system transmitter development and for IFR Systems Test and Measurement division developing 3 GHz spectrum analyzers.

Greg has published papers on TV transmitter multichannel sound and others on TV transmitter design architecture. He is a member of the Association of Federal Communications Consulting Engineers, IEEE, and SBE. Greg currently heads the IEEE DTV RF Measurement Standards activities and serves as an Associate Editor for the *IEEE Transactions on Broadcasting*.

Ralph S. Blackman (Chapter 5.1) has been with Rees Associates, Inc., in Dallas, Texas, since 1978. Currently Vice-President, his responsibilities at Rees have included serving as the Broadcast Market Segment Leader, Project Director, Project Programmer/Planner/Designer, and Project Manager. He has been involved with over 120 of the 300 300-plus broadcast projects Rees has completed.

Mr. Blackman is a licensed architect, member of the American Institute of Architects, and is NCARB Certified, with over 30 years of experience, 25 of which have been dedicated to serving the broadcast industry.

After receiving his Bachelor of Environmental Design and Bachelor of Architecture degrees from the University of Oklahoma, Mr. Blackman continued his education by attending Management of Design & Planning Firm at the Graduate School of Design at Harvard University, and Cox School of Business at Southern Methodist University to study Mid-Level Management. He maintains licenses in over 25 states.

Over the years, Mr. Blackman has been invited to present several papers at NAB Convention Engineering Conferences and PBS Technology Conferences.

Jim Boston (Chapters 3.9, 5.17) is a California California-based consultant specializing in system test and architecture. and he has a BSEE from Cleveland State University.

Jim has been in the television industry for over 30 years. He has been in the trenches and at upper levels of technical management. His experience spans the range of television technical equipment from the camera lens to the RF plumbing headed out towards the television transmission tower. He has also worked in such ancillary areas such as C and Ku band uplinking and has been Engineer-in-Charge of large production trucks.

Jim is a frequent contributor to *Broadcast Engineering* magazine. He has authored the book *"DTV Survival Guide,"* and co-authored the book *"TV on Wheels: The Story of Remote Television Production,"* with George Hoover.

Vyacheslav (Slava) Bulkin, Ph.D., (Chapter 4.9) is currently Antenna Engineer with Jampro Antennas, Inc., in Sacramento, California, where he develops broadband TV and FM broadcasting antennas and feed systems.

Dr. Bulkin earned his MS in physics from Moscow State University, Russia, and his PhD in EE from Radio Industry Institute in Moscow, Russia, working on developing radar antennas systems.

Dr. Bulkin worked as Visiting Scholar at University of Illinois at Urbana-Champaign, developing time-domain domain-based computational techniques in electromagnetics.

He also worked as Visiting Scholar at Penn State University, University Park, Pennsylvania, developing compositions of thin layered frequency selective surfaces materials.

Steve Campbell (Chapters 3.1, 3.5) began his career in commercial audio engineering in 1970 performing sound and projection engineering services and remodeling designs for movie theatres throughout the Rocky Mountain area. In 1977, while attending college, he worked in broadcasting as an assistant engineer at Montana State University's FM station and later as chief engineer for other stations in the state of Montana. In 1988, Steve became the Director of Engineering for Citadel Communications Corporation, and was principally concerned with studio construction and consolidation through the large growth period of Citadel.

Since 2002, he has been an independent broadcast consultant working on new broadcast development.

Steve holds a Bachelor's degree in electrical engineering, is a Registered Professional Engineer, and holds a General Class (previously First Class) Radiotelephone License from the FCC. His interests are commercial architecture and software development supporting broadcast engineering.

Ted N. Carnes, Ph.D., P.E., (Chapter 3.2) is Senior Acoustical Consultant for Pelton Marsh Kinsella, LLC, and is a registered Professional Engineer in the state of Texas.

Dr. Carnes has broad experience as an acoustical consultant in the fields of architectural acoustics which that includes the acoustics and noise control on all types of building systems, industrial noise control which that includes the acoustics and noise reduction in both production areas and office facilities, mechanical vibrations as it applies to HVAC and other types of equipment upon building structural systems, and environmental noise assessment of transportation systems on different types of buildings and developments. He also serves as an expert witness in the fields of acoustics and mechanical engineering.

Dr. Carnes has a B.S. in mechanical and aerospace engineering, and an M.S. in mechanical engineering from Oklahoma State University, and a Ph.D. in mechanical engineering from The University of Texas at Austin. He is a member of the American Society of Mechanical Engineers, the Audio Engineering Society, the National Council of Acoustical Consultants, and the American Society of Heating, Refrigeration and Air-conditioning Engineers professional organizations. He has written and presented a number of professional papers on acoustics and contributed to several books and standards on acoustics and noise control.

Tim Carroll (Chapters 1.15, 5.18) is President of Linear Acoustic Inc., a company he founded to offer innovative, customer-centric solutions for managing multichannel surround sound audio and loudness issues in digital broadcasting—a goal that has been supported by the development of several acclaimed products including the StreamStacker system, AutoNorm, upMAX 2251, OCTiMAX 5.1, and AEROMAX-TV.

Mr. Carroll was previously the Product Manager for the Professional Audio Division of Dolby Laboratories in San Francisco, where he helped define and develop Dolby Digital (AC-3), Dolby E, and Dolby Surround products for High Definition Television, DVD, Digital Radio, and Digital Cinema applications in the United States and abroad.

Mr. Carroll was honored by the Academy of Television Arts and Sciences with an Emmy® for his work on the Dolby E system. He is also an inventor of several patent-applied-for technologies.

Mr. Carroll remains actively involved in the creation of digital broadcast standards and practices. He is a member of IEEE, AES, SBE, SMPTE, and BKSTS and is an active participant in the work of the ATSC. He was also a member of the NRSC DAB Subcommittee and the Evaluation Working Group that formulated NRSC-5, the FM IBOC digital radio standard.

Ron Castro (Chapter 4.12) is the Chief Technical Officer and part owner of Results Radio, LLC, a chain of small-market FM stations in northern California.

Ron began his broadcast career as an air personality and engineer in Pennsylvania in 1966 and later joined the U.S. Navy as a Communications Technician. After working in Honolulu and San Francisco, he became the owner of a small FM station in Santa Rosa, California.

Ron is an active ham radio operator (N6AHA), holder of an FCC Lifetime General Class Radiotelephone License, and a member of the Society of Broadcast Engineers.

Garrison C. Cavell (Chapter 1.4) has been involved in the broadcast industry since the early 1970's. His experience in radio and television runs the gamut from announcer to free lance production engineer to staff engineer, Chief Engineer, and Radio Station General Manager. He even participated in the ownership of a small small-market radio station. Along the way, he has designed and built radio and television systems, supervised their construction, assisted facility acquisition efforts, and served as a consultant on the technical aspects of government regulation in the broadcast, microwave, and cellular telephone industries. He participates in industry committees and speaks at industry forums, training sessions, and panel discussions.

Mr. Cavell is a member or associate member in several professional organizations, including the National Society of Professional Engineers (NSPE), the Institute of Electrical and Electronic Engineers (IEEE), the Society of Broadcast Engineers (SBE), the Virginia Society of Professional Engineers, the Royal Television Society, and the Society of Motion Picture and Television Engineers (SMPTE). He was formerly the President of the IEEE Broadcast Technology Society, the New Orleans chapter of the IEEE Communications Society, and a local chapter of the Virginia Society of Professional Engineers. He holds an Extra Class Amateur Radio

License and is an instrument instrument-rated private pilot.

Dr. Richard Chernock (Chapter 5.22) is currently Director of Technology at Triveni Digital, Princeton, New Jersey. In that position, he is developing strategic directions for metadata management, content distribution, and monitoring in emerging digital television systems and infrastructures.

Previously, Dr. Chernock was a Research Staff Member at IBM Research, investigating digital broadcast technologies. He is active in a number of the ATSC standards committees, particularly in the area of metadata and data broadcast. He chairs a number of ad hoc committees within ATSC whose work relates to metadata and transport issues. He is Vice Chair of the ATSC Technology Group on Distribution (TSG).

In a previous position, Dr. Chernock used transmission electron microscopy to study materials characteristics for advanced ceramics packaging and semiconductor technology at IBM. His ScD was from MIT in the field of nuclear materials engineering.

Steve Church (Chapter 3.10) spent the first part of his career as a Chief Engineer at stations WFBQ in Indianapolis, IN Indiana, and WMMS in Cleveland, Ohio. He was also a part-time talk-show host. In this dual role, he determined that broadcast telephone interfacing systems needed to be improved. As a result, he developed the first digital audio telephone interfacing product for the radio broadcasting industry, the Telos-10 on-air phone interface. After selling a few to friends and getting a positive reaction, he decided it had commercial potential and founded Telos Systems to manufacture and market it.

Twenty years later, Mr. Church remains head of Telos Systems, which has grown to include Zephyr ISDN/IP codecs, Omnia processing, and Axia IP-based studio equipment. He was the first to use MP3 in a telephone interface product.

Mr. Church is a co-inventor of Livewire, an Ethernet/IP technology for the transport and routing of professional studio-grade audio.

Tim Claman (Chapter 5.11) is Director of Product Design for Avid Technology, Inc., the leading manufacturer of media content creation tools.

During his six-year tenure at Avid, Tim has contributed to a number of product and technology areas, including Avid's nonlinear editing products, as well as the Avid Unity family of collaborative storage and asset management solutions. His unique perspective on media creation solutions stems directly from his 10-year career in film and television post production.

An accomplished sound editor, designer, and mixer, Tim has worked in a variety of environments, from full-service post production facilities like Pacific Ocean Post in Santa Monica to small boutiques such as the Crescendo! Studios in San Francisco. He has contributed to countless productions, including feature films (Oliver Stone's *The Doors*), episodic and cable TV programs ("NYPD Blue"), and national commercials (Budweiser Frogs).

Kerry W. Cozad (Chapters 4.11, 6.6) joined Dielectric Communications in 1998 and presently holds the position of Senior Vice President, Broadcast Engineering.

Mr. Cozad is a native of Jonesboro, Georgia. He received his B.E.E. degree (with highest honors) from the Georgia Institute of Technology in 1981. He joined the Broadcast Division of Harris Corporation in June of 1981 and became the Lead Engineer for the Antenna Group in 1986. In this position, he was responsible for the electrical design of high power VHF, UHF, and multiplexed FM antennas. In 19881988, he joined Andrew Corporation. As Engineering Manager for Broadcast Products, he was responsible for the design of high and low power broadcast transmission systems including UHF and VHF antennas, transmitting antennas for ITFS/MMDS services, HELIAX, coaxial cable products, and rigid transmission lines for broadcast services.

Mr. Cozad is a member of the IEEE and served for six years on the Broadcast Technology Society Administrative Committee, as well as serving for four years on the board of the Association of Federal Communications Consulting Engineers. He also served on the Technical Advisory Committee for the Wireless Cable Association (WCA) during the transition from analog to digital transmission for the ITFS/MMDS services.

Mr. Cozad has written several articles published in trade magazines in Europe, Asia, and the United States, and gives presentations regularly at the NAB Engineering Conference and local SBE conferences. He authored Chapter 19, "Coax/Transmission Line" in *The Electronics Handbook* by CRC Press.

Aldo Cugnini (Chapters 1.13, 1.14) is a consultant in the digital television industry, with expertise in broadcast systems, consumer electronics product development, market research and analysis, intellectual property analysis and defense, and industry standardization. His clients include the Association for Maximum Service Television, major consumer electronics companies, and electronic and print media companies.

Aldo held technical and management positions at Philips Electronics' Research and Consumer Electronics Divisions and at ACTV, an interactive-television developer. He was a leader in the development of the ATSC system and its progenitor, the "Grand Alliance" digital HDTV system. Earlier, he developed audio and RF systems at Broadcast Technology Partners, the CBS Technology Center, and was an RF specialist at RCA Broadcast Systems.

Aldo received BS and MS degrees from Columbia University, holds an FCC First Class Commercial Radiotelephone Operator's License, and holds six patents in the fields of digital television and broadcasting. He served on the Board of Directors of the Advanced Television Technology Center, and writes numerous industry reports, technical papers, and publications.

Aldo received a 1997 Engineering Emmy and *R&D Magazine's* 1998 R&D 100 Award. He was a finalist in the 2005 IEEE-USA Congressional Fellowship program, and is a member of the Academy of Digital Television Pioneers, the Institute of Electrical and Electronics Engineers, and Eta Kappa Nu. He is a past member of the American Association for the Advancement of Science, the Audio Engineering Society, and is listed in *Who's Who in America*.

Aldo is a concert timpanist by avocation, and serves on the Board of Trustees of the Hanover Wind Symphony. He resides in New Jersey with his family.

Birney Dayton (Chapter 5.5) is currently the Chairman and CTO of NVISION, Inc., which he and two others founded in 1989.

Since 1968, Mr. Dayton has been active in the broadcast industry. He spent 4 years in television production and equipment maintenance while completing his BSEE at the University of Nevada, Reno. In 1973, he joined Grass Valley Group, Inc., and for the next 16 years he designed and managed the development of many products. From 1983 to 1989, he was VP of Engineering.

Over the last 30 years, Mr. Dayton has spent considerable time working on industry committees helping to advance the state of the art. He was involved in the development of SMPTE analog and digital component video standards, and was cochairman of the SMPTE High Definition Electronic Production working group. He also chaired the Systems Analysis Working Party of the Advisory Committee on Advanced Television Service.

Mr. Dayton has authored numerous industry papers, is a Fellow of the SMPTE, and holds 16 patents.

Jed Deame (Chapter 5.12) is Cofounder, Vice President, and General Manager of Teranex, a division of Silicon Optix. He developed the concept for the Teranex Video Computer, the first fully software-based real-time video supercomputer.

Mr. Deame began his career in television in 1982 while attending the University of Central Florida. He worked in a post production house as an engineer for four years while earning his Bachelor of Science degree in electrical engineering. He then worked at Lockheed Martin as a Video Systems Design Engineer, developing parallel processing architectures for real-time image processing applications.

Mr. Deame is a member of SMPTE, The Society of Information Display, and the Hollywood Post Alliance. He presents regularly at various conventions and technical symposiums around the world. He is also a contributing author for various home theater magazines.

Jeff R. Detweiler (Chapter 4.13) is Director of Broadcast Technology for iBiquity Digital Corporation, where he directs broadcast product development and the introduction and launch of its HD Radio™ brand of In-Band On-Channel (IBOC) technology to radio stations worldwide. He manages the technical relationships with broadcast equipment manufacturers and coordinates the transfer of technology to iBiquity's licensees.

Mr. Detweiler has 27 years of experience in the radio industry. Prior to joining iBiquity Digital, he spent 12 years in sales management at QEI Corporation, last serving as Worldwide Sales & Marketing Manager. Prior to joining QEI, he was the Northeast Sales Engineer for Allied Broadcast Equipment. He served as Chief Engineer at Lake Erie Radio (WWWE and WDOK) in Cleveland, Ohio, from 1985 to 1987 and in the same capacity at Nassau Broadcasting (WHWH and WPST) Princeton, New Jersey, from 1983 to 1985.

Mr. Detweiler is a frequent presenter at NAB, IEEE, and other industry conferences and events and a frequent contributor to broadcast publications.

Anne Devine (Chapter 2.6) is Manager of Public Relations for Medtronic Emergency Response Systems in Redmond, Washington, where she has worked for more than five years in marketing communications and public relations. Before Medtronic, she was Director of Communications at the University of Washington School of Nursing for six years. She provided communications consulting and writing and editing services for several high-tech startups, including drugstore.com, where she was pharmacy editor.

Ms. Devine holds a Master of Arts degree from the University of Washington and a BSN from the University of Missouri. She has served on the boards of Women in Communications and the Public Relations Society of America, and is a member of the American Marketing Association and Women in Digital Journalism. She is a freelance writer on a number of topics and enjoys hiking and traveling with her family and black standard poodle.

Medtronic's Emergency Response Systems business unit, located in Redmond, WA, pioneered defibrillation technology over 50 years ago, and with more than 600,000 LIFEPAK defibrillators distributed worldwide, it is the world's leading provider of external defibrillators for the treatment of sudden cardiac arrest. Go to www.medtronicers.com or call 1-800-442-1142 for more information.

Bruce Devlin, MA, CEng, MIEE, (Chapter 5.6) is Vice President, Technology, Snell & Wilcox.

Bruce graduated from Queens' College Cambridge in 1986 and has been working in the broadcast industry ever since. He joined the BBC Research Department, where he worked on Radio-Camera systems before moving to France, where he worked on sub-band and MPEG coding for Thomson. He joined Snell & Wilcox in 1993, where he started the company's work on compression coding.

Bruce holds several patents in the field of compression, has written international standards, and contributed to books on MPEG and File Formats. He is coauthor of the *MXF File Format Specification* and an active contributor to the work of SMPTE and the AAF association. Bruce is a Fellow of the SMPTE (Society of Motion Picture and Television Engineers).

Daniel L. Dickey (Chapter 4.2) is Vice President of Engineering at Continental Electronics Corp. based in Dallas, Texas. He began is career in broadcasting in 1976, working at a 500W daytime AM station. He has over 25 years of experience designing AM transmitters from 1 kW to over 500 kW. In addition to working with AM transmitters, he has also led design teams for numerous FM and DTV transmitters. At Continental Electronics, his engineering team is responsible for the research, design, and development of all products manufactured by the company.

Mr. Dickey has a Bachelor of Science degree from the University of Missouri in Rolla, Missouri. He is a member of the IEEE Broadcast Technology Society.

Michael A Dolan (Chapter 1.17) is founder and President of Television Broadcast Technology, providing specialized professional encoders, test tools, and technical consulting in the field of digital television.

Mr. Dolan received a BSEE degree from Virginia Tech in 1979 and has worked for and founded various leading-edge computer graphics and real-time systems companies, including early foundational work in W3C technology and analog data broadcasting.

Mr. Dolan has been involved in digital television engineering for the past eight years, including data broadcast system architecture and digital receiver design and compliance. He currently chairs the ATSC Data Broadcasting Specialist Group (TSG/S13), chairs various groups in SMPTE and CEA, and is active in other data-related television standards activities.

Mr. Dolan is an SMPTE Fellow and holds several patents in computer web technology.

Joan Dollarhite (Chapter 1.3) is Director, Legal Operations, for the National Association of Broadcasters in Washington, D.C.. She is responsible for fulfilling legal research requests from NAB attorneys and other staff members. This work includes document retrieval from the FCC, other federal agencies, and the courts. She maintains the NAB Legal Library comprised of books, periodicals, and online and reporter services.

Joan assists in preparation of the department's operating budget as well as the budgets for the department's participation in the NAB Show and the Radio Show. She is the designated conference planner for Legal Operations and takes care of the logistic and administrative responsibilities associated with the conventions.

Joan drafts monthly reports updating the Executive Committee and Board Members on NAB Legal activities and legal developments occurring within the industry. She is the lead staff person for the department on NAB's I-Team, tasked with designing, updating, and maintaining legal sections of NAB website.

Joan graduated with honors from the University of Maine with a BA in broadcasting and has a law degree from New York Law School. She joined NAB in 1998 after working in radio news, public access, and cable television. Her legal background includes litigation, family law, and legal recruiting, as well as a year spent on Capitol Hill monitoring telecommunications legislation.

Stephen P. Dulac (Chapter 6.15) is currently Director, Standards and Regulatory, in the Set-Top Box Engineering organization of DirecTV Inc., El Segundo, California. Since 1997, he has been with the company in system engineering roles supporting U.S. service launches including HDTV, local channel rebroadcasts, DVR, interactive services, and home networking.

Mr. Dulac received a BS degree in electrical engineering and an MSEE degree in telecommunications engineering from the University of California, Los Angeles (UCLA), in 1985 and 1987, respectively. Joining Hughes Aircraft Company in 1986 as a Masters Fellow, he contributed to many company projects as it evolved into Hughes Electronics and most recently into the DirecTV Group. At subsidiary Hughes Communications from 1986–1991, his responsibilities included the first digital satellite link budget analyses applicable to the DirecTV network. At Hughes Space and Communications from 1992–1994, he was Lead Payload Engineer for the Solidaridad system of communications satellites. From 1994–1997, he held the position of Director, Conditional Access, supporting the development and launch of the DirecTV Latin America service.

Mr. Dulac coauthored a paper, "Adjacent Satellite and Ground Station Interference," for the *Society of Motion Picture and Television Engineers (SMPTE) Journal* in December 1989. He led the authoring and adoption in 2000 of the CEA/EIA-805 standard for delivering data services across HD analog component interfaces. A Senior Member of the Institute of Electrical and Electronics Engineers (IEEE), he coauthored a paper, "Satellite Direct-to-Home," for the January 2006 *Proceedings of the IEEE Special Issue on Global Digital Television.* He holds three patents.

Thomas Edwards (Chapter 5.6, 5.21 Appendix) is the Senior Manager, Interconnection Engineering, for the PBS Interconnection Replacement Office. He is currently responsible for the engineering management of the PBS Next Generation Interconnection System (NGIS), including work on the development of an MXF Application Specification for video file distribution to public television stations, and development of the NGIS file transfer and station edge server systems. Thomas is also the author of Internet RFC 4539 "Media Type Registration for SMPTE MXF."

Before joining PBS in 2002, Mr. Edwards was the Streaming Media Product Manager at Cidera, where he developed a broadband desktop video channel for technical employees delivered using IP-over-satellite. He was also responsible for streaming media production and delivery at the Internet service provider DIGEX as well as his own streaming media company, The Sync.

Mr. Edwards holds a Master's degree in electrical engineering from the University of Maryland, and is a member of IEEE and SMPTE.

John Robert Emmett, PhD, MIBS, MRI, (Chapter 3.4) joined the Engineering Department of Thames Television in London after gaining a PhD at Durham University and BSc (Hons) in applied physics. He started the UK audio equipment manufacturer EMO Systems. While at Thames Television, he worked on subjects as diverse as film archive formats and psychoacoustics, and along the way gained six international patents as sole inventor. Jointly with Lee Lighting, he received a Technical Oscar for developing the flicker-free Lighting Ballast.

Dr. Emmett continued as R&D Manager with Pearson Television, and is currently Technical Director and Chief Executive of Broadcast Project Research, a new independent studio based research group. In conjunction with Channel Four Television in the UK, BPR was responsible for developing the "Gordon" Photoepilepsy Monitor.

Dr. Emmett, representing the UK as principle digital audio expert to the IEC, is perhaps best known for his work on digital audio and video standards. At present he chairs the EBU digital audio advisory group, which was responsible for the Broadcast Wave Format.

Dr Emmett also keeps closely in touch with the international academic community and he chaired the IEE Broadcast Summer School in the UK for many years.

Steve Epstein (Chapter 5.8) is the owner/editor of BroadcastBuyersGuide.com. He began his broadcasting career 1978 in Woodward, Oklahoma. A graduate of Broadcast Center in St. Louis, he has held positions as Announcer, Master Control Operator, Chief Engineer, and Director of Operations at broadcast and production facilities throughout the Midwest.

Steve served as Technical Editor of *Broadcast Engineering* magazine and several other Intertec (now Prism Business Media) publications, writing and editing many of the features and columns. He has also written software for switcher control, departmental workflow, and automation systems as well as other broadcast engineering related tasks. He has served as a consultant to a variety of facilities and manufacturers in broadcast and related industries.

Steve holds a Professional Broadcast Engineer Certification from the SBE and a First Class Radiotelephone License from the FCC. He has presented papers at Broadcast Engineering Conferences held at NAB Conventions, SBE Annual Conventions, and ghost-written numerous papers that are included in conference proceedings from conventions around the world. He is an instrument-rated private pilot.

Dane E. Ericksen, P.E., CSRTE, 8VSB, CBNT, (Chapters 1.6, 1.7) received a BSEE from California State University, Chico, in 1970. His first job was an inspector for the San Francisco FCC office. He rose to Senior FM/TV/CATV Specialist, operating one the FCC Enforcement Unit monitoring trucks.

In 1982, Mr. Ericksen joined Hammett & Edison, Inc., Consulting Engineers, and is a senior Consulting Engineer designing broadcast and broadcast auxiliary services stations licensed by the FCC, except standard broadcast.

Mr. Ericksen is a fifth-term SBE national director, has chaired the SBE FCC Liaison Committee since 1987, and has responsibility for national-level SBE filings with the FCC. He has served on the SBE Certification Committee for over 20 years, and was Chapter Chairman, Vice Chairman, and Secretary of SBE Chapter 40, San Francisco.

Mr. Ericksen is an SBE Fellow and recipient of the 1999 SBE Engineer of the Year award. He is a registered Professional Engineer (electrical) in the State of California.

Mr. Ericksen has written technical articles for broadcasting trade journals, and writes a column for the SBE *Signal* regarding regulatory issues affecting BAS spectrum. He served on the NAB/SBE Engineering Conference Committee for eight years and on the ANSI/IEEE C95.1 Committee for five years. He is a regular participant in NAB Broadcasting Engineering Conferences.

Mr. Ericksen has served as Chairman of ATSC Technology & Standards Group S3 Specialist Group on Digital ENG since November 2003, which deals with standards for the 2 GHz TV BAS data return link channels and for Digital ENG communications.

Mr. Ericksen resides in Sonoma, California.

Donald G. Everist (Chapters 4.15, 6.9) is a graduate electrical engineer from the University of Illinois. He is a Registered Professional Engineer in the District of Columbia and is President of the firm of Cohen, Dippell and Everist, P.C., located in Washington, D.C.

Mr. Everist has been in the consulting engineering business with this firm or its predecessors for over 45 years. During this time, he has performed numerous AM, FM, and DTV interference and allocation studies, including microwave and other communication systems and field strength surveys worldwide for AM, FM, and TV broadcast stations. He has prepared exhibits and documentations for submission before the Federal Communications Commission (FCC) and international forums. He has also served as a U.S. delegate to several international (ITU) conferences.

Mr. Everist is a member of the Institute of Electrical and Electronic Engineers (IEEE); the National Society of Professional Engineers (NSPE); the Illinois Society of Professional Engineers, and a member and past president of the Association of Federal Communications Consulting Engineers (AFCCE).

Rod Fairweather (Chapter 5.16), Harris Corporation, studied electronic engineering at Edinburgh University, and obtained his MBA at Imperial College, London.

After joining BBC TV as an engineer, Rod moved into studio directing, where he built up extensive live directing experience with the BSkyB, GMTV, and ITN facilities. He joined VH-1 as Senior Producer for the launch of the UK version of the channel before traveling extensively to launch channels and to train operational staff.

Rod joined Harris Corporation to set up and run a major multichannel broadcast operation before moving into product management of broadcast software.

Rod is the author of *Basic Studio Directing*, published by Focal Press.

Ty Ford (Chapter 3.3) formed his own company, Ty Ford Audio & Video, in 1986, and simultaneously began reviewing professional audio equipment for Radio World. He is recognized internationally as a specialist in the field of microphones and preamplifiers. He writes feature stories and equipment reviews for *Radio World, TV Technology*, and *Pro Audio Review*. He keeps an online review archive at www.tyford.com.

Mr. Ford passed his First Class Radio Telephone Operator's License test in 1969. He spent 17 years as a Chief Engineer and Production Director at top twenty market radio stations in Baltimore and Washington. He was recognized by the NAB as a significant contributor to NAB publications and conventions. He has given seminars about audio production and rewrote the *NAB Guide to Radio Copy Writing*.

Frank Foti (Chapter 3.8) is the Founder and President of Omnia Audio, an innovative signal processing company, located in Cleveland, Ohio. He leads a talented team that researches and designs audio processing for every form of broadcasting: FM/AM/TV, HD Radio, DAB/DRM, Music Mastering, and Netcasting. He is working diligently towards implementing discrete 5.1 surround audio for FM radio using the HD Radio system and the Internet. He is a frequent presenter at NAB, AES, SBE, and state broadcaster association conventions and has written for all of the broadcast trade publications.

Mr. Foti, in an earlier life, was "in the trenches" of day-to-day radio as Chief Engineer for a number of successful radio stations in the United States, including WMMS/WHK (Cleveland), KSAN/KNEW (San Francisco), and the well-known Z-100, WHTZ-FM (New York City), which he designed and built in 1983. Outside the world of broadcast audio, he is deeply involved in the world of 1/8th scale live steam railroading.

John D. Freberg (Chapter 6.4 Appendix) heads The Freberg Engineering Company in Homewood, Illinois, and has a broad and varied background in electronic media and communications technology. As Manager of Technical Operation at CBS's WBBM-TV in Chicago, he is actively integrating digital television and advanced networking as part of the construction of a new television facility from the ground up. He also directed the engineering departments at several AM and FM stations. As a Project Manager for Tribune Company, he was involved in the implementation of advanced network infrastructures and information systems.

John received his Bachelor's degree from Northern Illinois University in De Kalb, Illinois. In 1997, he completed his Master of Science in communication systems at Northwestern University in Evanston, Illinois.

In addition to pursuing a dual-track career in broadcast engineering and information systems, John has taught courses in advanced network management at Northwestern University, and is currently on the faculty of Lewis University in Romeoville, Illinois. His current course assignments include electronics, engineering, and communications technology.

John is active in industry developments as a member of several IEEE Societies, including the Broadcast Technology Society, and is currently serving on the G2.2 RF Standards Committee. He is a member of the Society of Broadcast Engineers and the Society of Cable Telecommunications Engineers. He has written articles and presented seminars for the NAB Engineering Conference, *Broadcast Engineering* magazine, and other organizations.

James "Brad" Gilmer (Chapter 5.6) is President of Gilmer & Associates, Inc, a management and technology consulting firm in Atlanta, Georgia, providing services to a wide variety of clients in the television industry.

Brad graduated Summa Cum Laude with a major in business management from Georgia State University, and is a member of Beta Gamma Sigma.

Brad started his broadcasting career in Albuquerque, New Mexico, first as Assistant Chief Engineer at KRKE Radio, then as a transmitter engineer at KGGM-TV. He was Director of Engineering and Operations at Turner Broadcasting Systems in Atlanta and was responsible for implementing digital facilities serving the U.S., Asian, and Latin American markets, for the first installations of broadcast automation systems at TBS, TNT, and Cartoon networks, and some of the industry's first large multichannel, multilanguage network launches in Atlanta, London, and Asia.

Brad is Executive Director of the Advanced Authoring Format Association, which developed the Advanced Authoring Format (AAF), and participated, along with the Pro-MPEG Forum and the SMPTE, in co-development of the Material eXchange Format (MXF). He served as cochair of the Pro-MPEG Forum's File Interchange working group. He also serves as Executive Director of the Video Services Forum (VSF), an association focused on delivering video over packetized networks.

Brad is Editor-in-Chief of the *File Interchange Handbook* (Focal Press), author of the monthly "Computer and Networks" column in *Broadcast Engineering* magazine. He is a frequent presenter at broadcast conventions including SMPTE, VidTrans, NAB, and IBC.

Brad is a Fellow of the SMPTE, has been an active participant within the SMPTE since 1984, and a Manager of the Atlanta Section.

John P. Godwin (Chapter 6.15) is currently with Gretna Green Associates, Los Angeles, California, a consultancy in electronic media and telecommunications. His personal research interests include exploring applications for the High Altitude Platform Service (HAPS). He received a BA degree in physics from DePauw University, Greencastle, Indiana, and MSE degrees in aerospace engineering and in computer, information, and control engineering from the University of Michigan, Ann Arbor.

Mr. Godwin was a Laboratory Manager with Hughes Communications, a satellite-based communications carrier, in the 1980s, and part of the DirecTV startup team formed in 1991. He contributed to system architecture and management including the satellites, broadcast centers, consumer electronics, and conditional access. He held the position of Senior Vice President, Engineering, from 1994 to 1997 during the DirecTV service launch. From 1997 to 2002, he was Senior Vice President, New Technologies, providing strategic planning and architecture enhancements. He led the workflow activities of DirecTV-Japan and provided oversight for a substantial company investment in XM Satellite Radio. From 2002 to 2004, he was Chief Technology Officer at Movielink, LLC, an online movie rental and distribution service owned by major movie studios.

Mr. Godwin was one of three DirecTV managers accepting the company's Emmy Award for Pioneering Achievement in Direct Broadcast Television. He is a member of the Space Technology Hall of Fame. He holds six patents.

Matthew Goldman (Chapter 5.22) has been Vice President of Technology, Compression Systems, for TANDBERG Television Inc. since 2004.

Mr. Goldman has been involved digital television (DTV) systems developments since 1992, particularly in the Moving Picture Experts Group (MPEG) where he helped create the MPEG-2 Systems standard, the baseline transport technology used in terrestrial broadcasting, digital cable, direct broadcast satellite, and DVD-video. He also served as project editor for the MPEG-2 DSM-CC standard, the control signaling used in video-on-demand systems and for DTV data downloads and carousels. He was a co-developer of the ATSC Program and System Information Protocol (PSIP) standard used by DTV receivers to navigate and tune to broadcast DTV services. He is an active member of the SMPTE, SCTE, the Consumer Electronics Association, and the ATIS IPTV Interoperability Forum.

Until 1996, Mr. Goldman was a Consulting Engineer at Digital Equipment Corporation, and from 1996 to 2000, he was Director of Engineering, Advanced Systems Development, at DiviCom Inc. From 2001to 2003, he was a technology consultant specializing in DTV system solutions, including the definition of a compressed-domain high definition program splicer for the FOX Broadcasting Company.

Mr. Goldman received his BS (high honors) and MS degrees in electrical engineering from Worcester Polytechnic Institute. He holds six patents related to digital video transport. He is a senior member of the IEEE, SMPTE, and the Academy of Digital Television Pioneers. He has presented papers at the NAB Convention, the SMPTE Technical Conference, the SCTE Cable-Tec Expo, Supercomm, TelecomNext, the International Broadcasting Convention (IBC), and the Hollywood Post Alliance Retreat. He has been published in the *IEEE Proceedings* and has been a contributing writer to trade magazines.

Dave Guerrero (Chapter 8.2) is Vice President and General Manager for Videotek® Test and Measurement, Broadcast Communications Division, Harris. He brings to the company more than 30 years of real-world experience in the broadcast industry and applies his substantial knowledge and expertise to new product development.

Dave's long and diverse career includes extensive hands-on experience in live and studio audio and video productions. Before joining Videotek in 2002, Dave held progressively responsible positions with various broadcast radio and television stations, culminating in the prestigious role of Engineer-in-Charge of network-level remote productions for all U.S. broadcast networks.

Dave is a highly respected authority in the area of television production and has supervised the engineering for coverage of numerous high-profile events, including the Macy's Thanksgiving Day Parade, the State of the Union Address, multiple Super Bowls, and the 1988, 1996, and 2002 Olympics.

Dave holds an FCC General (formally First) Class Radiotelephone License, is an active member of SMPTE, AES, SBE, and IEEE and has earned two Emmy® Awards for his contributions in the area of engineering expertise.

Linley Gumm (Chapter 8.4), now retired, spent nearly 40 years designing RF test equipment for Tektronix, Inc. A major effort was to lead the RFA300A (8-VSB signal analyzer) project, where he learned the intricacies of testing the 8-VSB signal.

Mr. Gumm is an inactive professional engineer in the state of Oregon and an IEEE member. He holds a BSEE from Washington State University, an MSEE from the University of Washington, and has 22 United States patents. He is currently active on the IEEE standards committee, creating RF measurement standards for DTV.

Harold Hallikainen (Chapter 9.5) was one of the founders of Hallikainen & Friends in 1974. The company started as a contract engineering firm, then quickly moved into manufacturing, specializing in broadcast transmitter telemetry and control. In 1995, he joined Dove Systems where he does embedded system design for Dove and several other manufacturers.

Mr. Hallikainen studied electronic engineering at Cal Poly, San Luis Obispo, California. He has written over 100 articles on FCC Rules and maintains a website devoted to FCC Rules. He has taught electronics part time at Cuesta College in San Luis Obispo for 25 years.

William F. Hammett, P.E., (Chapter 2.4) has over 20 years of experience measuring, calculating, and mitigating RF exposure conditions at radio and TV stations, satellite and radar facilities, cell sites, and industrial plants. He authored the book *RF Radiation—Issues and Standards*, published in 1997 by McGraw-Hill. He is a frequent speaker on the topic and has presented papers at three NAB Engineering Conferences.

Mr. Hammett graduated from Dartmouth College magna cum laude in 1977 and earned a Master of Science degree from the University of Illinois in 1978. He has worked at Standard Oil of California and Dean Witter Reynolds and is a principal with Hammett & Edison, Inc., Consulting Engineers.

Mr. Hammett is a Registered Professional Engineer in California (Mechanical and Electrical) and other states.

Peter J. Harman (Chapter 5.4) currently holds the position of Product Marketing Manager for the Lightweight Vinten brand worldwide. He began his career with Vinten Broadcast Limited in 1972, having served his mechanical apprenticeship with Corning Medical, then a UK subsidiary of the US Corning group. He has held positions within Vinten Broadcast Limited as Development Engineer, Design Engineer, Customer Support Manager, Quality Manager, and Marketing & Training Manager.

Mr. Harman has earned professional qualifications within Quality Management (Associate Member IQA), Marketing (Member IDM), and Training (Associate Member CIPD). He also holds national diplomas in Performance Coaching and Business Excellence and regularly contributes articles to many international magazines and professional publications.

James B. Hatfield, P.E., (Chapter 4.5) was a founding partner of Hatfield & Dawson Consulting Engineers in 1973. He is registered as a Professional Engineer in the states of Washington, Oregon, and Hawaii.

Mr. Hatfield has written numerous papers over the years on the analysis and adjustment of AM directional arrays using Method of Moment techniques. These papers have been presented at IEEE BCT, NAB, Applied Computational Electromagnetics Society meetings, and published in their journals. He has authored chapters on this subject in the *NAB Engineering Handbook* and *The Electronics Handbook*. He is a charter member of the Applied Computational Electromagnetics Society, ACES.

Mr. Hatfield serves on the IEEE International Committee on Electromagnetic Safety (ICES) Standards Coordinating Committee SCC28, Non-Ionizing Radiation, and its subcommittees SC-1, SC-2, SC-4, and SC-5 of SCC28. SC-4 was charged with revision of IEEE C95.1-2005, *Standard for Safety Levels with Respect to Human Exposure to Radio Frequency Electromagnetic Fields, 3 kHz to 300 GHz.* He is current chair of the SC4, C95.1-1991 Interpretations Working Group, and a member of the SC4 Engineering Working Group RFR Literature Review and Risk Assessment Working Group.

Dr. Paul Hearty (Chapter 6.14) is Associate Dean of the Faculty of Communication & Design and Director of the Rogers Communications Centre of Ryerson University in Toronto, Ontario, Canada. He joined Ryerson University in 2003. His role there is to enhance the Rogers Communication Centre's capacity and agenda for in-house and collaborative research.

Dr. Hearty received his PhD from Queen's University in 1981. He joined Industry Canada's Communications Research Centre in 1980. He founded several labs, including the Advanced Television Evaluation Laboratory, which carried out tests of High Definition Systems as part of the joint U.S.–Canada DTV standardization effort.

Dr. Hearty joined General Instrument (now Motorola Broadband) in 1994. He led the digital HDTV Grand Alliance group responsible for the video subsystem and served in the group overseeing development and standardization of the overall system. He spearheaded the deployment of General Instrument's technology for digital compression, multiplexing, and satellite transmission in Canada.

In 2001, Paul joined DemoGraFX, a company developing advanced video compression technology. In 2002, he started his own company, assisting clients in assessing, developing, and marketing advanced digital technologies.

Dr. Hearty has been active in many standards bodies, including the FCC Advisory Committee on Advanced Television Service, the Advanced Television Systems Committee, and the International Telecommunications Union. He chaired the Society for Cable Telecommunications Engineers' Digital Video Subcommittee, which sets standards for digital cable in North America since its inception in 1996, and currently is cochair of Canada's Digital Television Technology Group (DTV-TC), a government-industry-university advisory group.

Richard G. Hickey (Chapter 7.4) entered the field of technical sales in 1983. His entry into the world of Aviation Obstruction Lighting occurred in 1996 by joining the staff of Flash Technology in Franklin, Tennessee. In his tenure at Flash, he has held the positions of High Intensity Sales Manager, Product Manager, International Sales Manager, and Inside Sales Manager.

Richard has designed lighting systems within the parameters established by the FCC/FAA, ICAO, DGAC, Transport Canada, and the NEC for various high-profile sites around the world.

Richard has spoken at many broadcast industry events and SBE meetings in 36 states and Canada. His advice and input are sought for a wide-ranging assortment of lighting challenges, including the Egyptian Minister of Aviation, the Suez and Panama Canals, wind turbines in Austria and Germany, a U.S. observatory in the Antarctic, and the former World Trade Center Towers in New York. His favorite projects, however, have been the many broadcast tower configurations with which he has worked across the United States.

Thomas J. Hoenninger (Chapter 7.2) is a member of Stainless LLC and currently holds the positions of Vice President of Operations and Chief Engineer. He has been designing, analyzing and modifying towers for Stainless for 18 years. Prior to this, he worked for FMC and Symons Corporation designing complex structures and concrete forming systems.

Tom holds a BSCE and an MSCE from Drexel University and is a Licensed Professional Engineer in 26 states. He is actively involved in the TIA TR14.7 subcommittee, which maintains the TIA 222 structural standard for antenna support structures and antennas. He is also actively involved in the ASTM A01.02 subcommittee, which maintains the ASTM standards on structural steel for bridges, buildings, rolling stock, and ships.

Tom performs peer reviews of submitted articles to be published in the *AISC Engineering Journal* and has presented papers at the NAB convention and PBS conferences.

George Hoover (Chapter 5.17) serves as Senior Vice President and Director of Engineering for all divisions of NEP Broadcasting, Inc., including the mobile, studio, and video screens groups in the United States and England and has been a member of the NEP Broadcasting family since 1993. He possesses substantial television technical engineering expertise and design skills, and he plays a vital role in long-term strategic planning for NEP and its clients.

Prior to joining NEP, Mr. Hoover was General Manager of the Public Broadcasting Authority (New Jersey Network) of the state of New Jersey from 1982 to 1993. Prior to NJN, he was a partner in Video East; a Philadelphia-based mobile unit company. He was with RCA Broadcast Systems in the mid-seventies.

Mr. Hoover attended Florida State University. He earned a Sports Emmy® for the 1996 Summer Olympics in Atlanta as technical team supervisor and a second Emmy® as Producer of the PBS Super Band series with Amad Jamal. He published a book with coauthor Jim Boston in 2003 entitled *TV on Wheels, The Story of Remote Television Production.*

Mr. Hoover is a member of the Society of Motion Picture and Television Engineers, Omicron Delta Kappa, and the American Institute of Architects.

Jeff Hutchins (Chapter 5.24) retired in 2006 as Chairman of the Accessible Media Industry Coalition, a trade association of companies that provide services such as captioning and video description to make media programs accessible to people with hearing and/or vision impairments.

Prior to this, he was an owner and the Executive Vice President, Planning and Development, of VITAC, a Pittsburgh-based company providing complete captioning services nationwide. He also was Director of Systems Development at the National Captioning Institute (1980–86); and from 1973 to 1980 was producer of *The Captioned ABC News* and an executive for The Caption Center, WGBH-TV, in Boston.

He has been honored as one of the pioneers who helped implement closed captioning. He was the author of the closed-captioning specifications adopted by the FCC in 1992, and the principal author of the EIA-608 standard.

Mr. Hutchins currently is a member of the Boards of Trustees of the Western Pennsylvania School for the Deaf (in Pittsburgh, Pennsylvania) and the American Community School in Beirut, Lebanon.

He has a Bachelor of Science degree in broadcasting and film from Boston University.

Jim Jachetta (Chapter 6.10) is the Senior Vice President of Sales and Marketing at MultiDyne Video & Fiber Optic Systems. His responsibilities include sales, marketing, product management, and developing new markets for the expanding MultiDyne product line.

Jim began his career as a technician in 1982 while earning a Bachelor of Science degree in electrical engineering from Polytechnic University in Brooklyn, New York. After college, he worked for Micro Corp designing Hybrid Micro-Electronics for military video guidance systems. While working full-time at Micro Corp he earned a Master of Science degree in electrical engineering and communications from Polytechnic.

Jim took a position in 1986 at MultiDyne Video & Fiber Optic Systems as a Design Engineer. He was part of the team that developed a high-end, broadcast quality, RS-250C Short-haul fiber optic product line. He also contributed to the design of the TS12 Video Test Signal Generator; VDA8505 Field Video DA and Equalizer; AB200 Automatic Bypass; VLD-2 Video Loss, Black, White Detector, and DTV100/200 Series SDI Fiber Transport. In 1994, he became Vice President of Engineering and Product Development at MultiDyne.

Jim is a member of the Society of Motion Picture and Television Engineers, the National Association of Broadcasters, and The Alternative Board. On weekends he volunteers as a sound technician at his local church.

Brett Jenkins (Chapter 6.4) is the Vice President of Engineering for Thomson Broadcast & Multimedia, Inc., which is part of the Grass Valley business unit within Thomson. He heads the research, product development, and systems engineering groups for Thomson Broadcast & Multimedia's U.S. headquarters located in Southwick, Massachusetts.

Prior to his current position, Brett was responsible for modulator and exciter technology development for broadcast products for the company. He was the lead U.S. engineer in a global team responsible for the development of Digital Adaptive Precorrection technology. Thales received a technical and engineering excellence Emmy® award for pioneering this technology in 2003.

For the past several years, Brett has been active in many broadcast industry groups involving Digital Television. He has authored and presented technical papers and tutorials on various digital communications topics.

Brett received a BSEE with honors from the University of Massachusetts, Amherst, in 1992 and an MBA from Boston University in 2005. He is a member of the Institute of Electrical and Electronics Engineers (IEEE) and the Society of Motion Picture and Television Engineers (SMPTE).

J. Dane Jubera (Chapter 4.9) is presently Senior Antenna Engineer at Jampro Antennas, Inc., where he has spent over a decade designing antennas for FM and television broadcast applications.

Mr. Jubera earned his BS in electrical engineering from Wayne State University, Detroit, Michigan. He has worked in the broadcast and cable television industries throughout his career.

Thomas Kite (Chapter 8.1) is Principal Engineer at Audio Precision, Inc., in Beaverton, Oregon. He has written much of the signal processing code for the System Two Cascade/2700 and APx500 line of products. Before joining the company in 1999, he worked at Hewlett-Packard Labs in Palo Alto, California, and Xerox Labs in Webster, New York.

Mr. Kite received his Bachelor's degree in physics from Oxford University, England, in 1991, and his Master's and PhD degrees in electrical engineering from the University of Texas at Austin in 1993 and 1998, respectively.

Mr. Kite regularly gives technical seminars and presentations. He is a member of the Audio Engineering Society (AES), and sits on the AES digital audio measurement standards committee.

Stephen Kolvek (Chapter 6.7) joined MYAT, Inc. in September, 1991, as an Electrical Engineer for transmission line products. His broadcast experience career began with work on complex hybrid feed systems. Over the years at MYAT, he has been involved in directing new product development and design changes. He is currently responsible for developing and implementing MYAT Quality Control Systems, Engineering Design Control Procedure, Corrective Action and Preventive Action Procedure, Nonconforming Product Procedure, and Contract Control Procedure for Government Applications.

Mr. Kolvek is also responsible for directing and administration of the Drafting Department operations and infrastructure and the RF Test Laboratory and R&D Facilities, and interfaces closely with manufacturing, production, and assembly departments. His knowledge of transmission line theory, RF combiners, and feed systems have aided MYAT in developing 7-3/16" 75 Ohm Coaxial transmission line and related accessories, MYAT EStar N-Way power combiners, *SpectraLine* Broad Band Rigid Coaxial transmission line systems, *SpectraGuide* Rigid Elliptical Wave Guide, and custom components specific to customer applications.

Mr. Kolvek received a BSEET degree in 1987 from DeVry Institute of Technology in Chicago, Illinois, and an Advanced Electronics Engineering Diploma from DeVry Technical Institute, Woodbridge, New Jersey. He is coauthor of a patent for the MYAT EStar power combiner.

Geoffrey A. Krenkel (Chapter 9.4) is a Senior Associate at John-Winston Engineers and Consultants, Inc., for which he provides professional consulting services mainly in the field of electrical power engineering. He is a Licensed Professional Engineer in the state of New Jersey.

Mr. Krenkel received his Bachelor of Science degree from Rutgers College of Engineering in the curriculum of electrical and computer engineering. He is a Fellow of the IEEE. His memberships include The Industry Applications Society and The Power Engineering Society. His early career achievements include telecommunications design for the government/military and later broadcast system engineering design for major cable networks and traditional broadcast facilities.

Alan Lambshead, BEng, PEng, (Chapter 5.24) is Vice President of Engineering for Evertz Microsystems, and an acknowledged leader in the fields of HDTV, post production and production with over 25 years of experience. Since joining Evertz in 1979, he has led the design of post production and time code and closed captioning products.

In 1999, he pioneered encoding of Film Transfer metadata into 1080p/24 High Definition video with Sony and Laser Pacific Media Corporation. In 2003, he worked with Industrial Light and Magic and other manufacturers to advance this technology to 4:4:4 HDTV acquisition for the shooting of *Star Wars Episode III*. He led the Evertz team in the development of specialized HDTV products for James Cameron, Pace Technologies, and others in the high definition production and post production field.

Alan's list of industry standard production products include the Tracker Telecine Logging and Configuration system with HD9025TR-HD Film Footage Encoder and new 4:4:4 based HD9045TR. He designed the HD9155Q-AUD-HD Production Afterburner/Downconverter and worked with many of Evertz's other post-production accessories, including HD and SD Graticule Generators.

Alan is Lead Designer of Evertz's ECAS, High Definition Fiber-optic Enabled Camera Adapter System and downconverter accessory for HD production workflows for Sony and Panasonic HD and SD camcorders. He designed this system for use with current and future markets, including HDSDI with embedded audio and time code, NTSC/PAL, SDI, and IEEE1394A downconverted video.

Alan is active on SMPTE engineering committees and is author of several SMPTE recommended practices. He has presented papers at local chapter and national SMPTE conferences and is a widely recognized expert in the field of HDTV time code and sync issues, HDTV production, HDTV metadata and film to tape transfers.

Alan graduated from McMaster University in Hamilton, Ontario, with a BEng in electrical engineering in

1972, and was licensed as a Professional Engineer in Ontario in 1988.

Chris Lennon (Chapters 5.13, 5.22) works at Harris Corporation as Director of Integration and Standards in the Software Systems group and has worked in the broadcasting industry for over 20 years, the majority of that time on the software systems side. He has managed a wide array of products and led several large-scale projects at broadcast facilities around the world. He has also led the development of well over 100 interfacing/integration projects between broadcast systems.

Mr. Lennon is an Ontario Scholar, and earned his degree in commerce and computer science at McMaster University in Hamilton, Ontario, Canada. He resides in the United States as a dual citizen of both the United States and Canada.

Mr. Lennon is Chair of the SMPTE S22-10 Working Group, whose task is to standardize communications between Traffic, Automation, and Content Delivery Systems. He is also an active participant in the ATSC, where he was one of the early instigators of the effort that became PMCP, enabling PSIP-related data to flow between broadcast systems.

Mattew Lightner (Chapter 3.5) is the President of Lightner Electronics, Inc, a broadcast integration company that he founded in 1995. He also serves as the Broadcast Engineering consultant for the Pennsylvania Association of Broadcasters. At the PAB he is responsible for the Alternative Broadcast Inspection Program and he maintains the state Emergency Alert System.

Matt gives credit to his father for getting him interested in electronics. When he was only 7, he was helping in his father's part-time TV repair and satellite business. At the age of 15, a friend who worked as an announcer at a local radio station asked him to repair their cart machines. Curious as to what a cart machine was, he agreed to look at them and the rest is history. After he graduated from high school he was promoted to Chief Engineer of WVAM/WPRR radio in Altoona, Pennsylvania. Matt then engaged in contract engineering, which grew into Lightner Electronics, Inc., and has since designed and built over 60 radio studios and the radio broadcast facilities for AccuWeather.

When Matt gets a break from his busy schedule he enjoys spending time with his friends and family, traveling, and playing music.

Ronald E. Lile (Chapter 6.8) is currently the TV Antenna Engineering Manager for Electronics Research, Inc. He earned a Bachelor of Science degree in electrical engineering in 1979 from the University of Missouri at Rolla, Rolla, Missouri, after serving in the U.S. Air Force.

Mr. Lile's work experience includes radar, commercial flight simulators, and communications systems from HF to UHF. While with Texas Instruments, he participated in the development of one of the first microprocessor controlled automatic antenna tuning units for marine applications. Prior to joining Electronics Research, he worked for a telecommunications company designing UHF power amplifiers, wide area network control and monitoring systems, plus he served as Regulatory Engineer interfacing with the FCC, ICC, UL, and customers. While with this company, he held positions as Systems Engineering Manager, Regulatory Compliance Manager, and Director of Customer Support Services.

Mr. Lile is a Member of the IEEE Antennas and Propagation Society and Broadcast Technology Society, and a Fellow in the Radio Club of America. He has an FCC First Radiotelephone License, is an amateur radio operator and a Life Member of the ARRL. He also holds a membership in the Quarter Century Wireless Association and actively participates in the Amateur Radio Emergency Service and RACES for Southern Indiana.

Edward Lobnitz, PE, (Chapter 7.3) serves as Consulting Principal at TLC Engineering for Architecture in Orlando, Florida. A foremost authority on lightning protection, he was a principal organizer and Chairman of the Lightning Protection Institute's Professional Division and a speaker at numerous LPI and United Lightning Protection Association events. He is a principal member of the National Fire Protection Association's (NFPA 780) Lightning Protection Standards Committee and has served on several other NFPA committees as a principal member.

As a highly honored Fellow in the Florida Engineering Society, Ed is the recipient of the FES Outstanding Service Award, the Young Engineer of the Year Award, and the Central Florida Engineer of the Year Award. He served an eight-year stint on the State Board of Professional Engineers, including two years as chairman, and chaired the NCEES committee that wrote the National Electrical Examination for Professional Engineers. He is the recipient of the National Council of Examiners for Engineering and Surveying Distinguished Service Award. In 1987, he was recognized by the Florida Board of Professional Engineers for outstanding contributions to the profession. He is also a Senior Member of the National Society of Professional Engineers and the Institute of Electrical and Electronic Engineering.

A Lay Leader at the First Methodist Church of Orlando, he is cofounder of the Central Florida Chapter of Engineering Ministries International. Ed joined TLC, then called Tilden Lobnitz Cooper, in 1967, and was CEO from 1981–1991.

Peter Ludé (Chapter 5.25) is Senior Vice President of Engineering for the Broadcast and Business Solutions division of Sony Electronics, based in Silicon Valley, and has been involved in broadcast engineering and production for over 30 years. In this role, he is responsible for engineering and business development for Sony's media and business electronics groups. Previously, he served as Chief Technology Officer at iBlast, the pioneering datacasting network, and before that was a systems integration executive for cutting-edge broadcast projects, including DBS, digital cable, and DTV.

Mr. Ludé was the founder of Ludé Broadcast Engineering in San Francisco, which was eventually acquired by Sony. He is an active member of IEEE, SBE, and is an SMPTE Fellow. He is a frequent speaker and panelist on topics of digital communications systems and content distribution, and currently serves as Editorial Vice President for the SMPTE.

Mr. Ludé is a graduate of the College of San Mateo, and lives in San Francisco with his wife, Lani, and two young children. In his occasional spare time, he plays bluegrass music, makes Pinot Noir wine, and hunts mushrooms.

John A. Luff (Chapters 2.3, 5.21) is an independent consultant specializing in television technology, facility planning, and system design. He has 40 years of experience in broadcasting, post production, facilities management, remote production, project management, technical consulting, and system design. He is a consultant and lecturer on emerging media technologies, and a graduate of the Honors College of Ohio University. He is a Fellow of the Society of Motion Picture and Television Engineers.

Mr. Luff was founder and President of Synergistic Technologies Inc., purchased in 2000 by AZCAR USA, Inc. He served at AZCAR until 2006 as Senior Vice President Business Development.

Mr. Luff was responsible for the design, project management, and construction of the first progressive scan network origination center at ABC in New York and HDTV mobile unit used for Monday Night Football and the Super Bowl in 1999/2000. Since 1985, he has been involved in the planning and management of coverage of major news events around the world, including the Hong Kong Handover, the "Pro Democracy Movement" in Beijing in 1989, American Political Party Conventions from 1980 through 1996, and summits and elections across Eastern Europe.

Mr. Luff writes a regular column for *Broadcast Engineering* magazine, and is writing a book on Centralized Broadcast Operations for Focal Press.

Mr. Luff's clients include the ABC Television Network, the European Broadcasting Union, Sinclair Broadcasting, Corporation for Public Broadcasting, Panasonic Broadcast and Television Systems, NexStar Broadcasting, Turner Entertainment Networks, WQED, Pennsylvania Public Television Network, CNN, Gamecreek Video, Acme Television, North Carolina Public Television, Cornerstone TeleVision, and others.

He lives and works in Pittsburgh, Pennsylvania.

John M. Lyons (Chapters 4.1, 6.1) has been involved in the communications industry for over 40 years, from WNYE at Brooklyn Technical High School to his present position as Assistant Vice President and Director of Broadcast Communications at The Durst Organization.

Mr. Lyons has held engineering positions at New York City broadcast stations WRFM, WWRL, WOR, WAXQ, WLTW, WXLO, and WEBR-TV. He constructed ZDK Radio in St. John's, Antigua, BWI, where he also served as Director of Engineering. He also served as Chairman of the Master FM Broadcasters Committee at the Empire State Building for a total of 12 years.

Mr. Lyons joined DSI Communications where he was the Senior Project Manager responsible for communications and broadcast facility build-outs. In 1994, he assumed a consulting position at the Sony Worldwide Radio Networks, working to establish and set the standards for the nationwide satellite-programming network.

Mr. Lyons was responsible for the design of the Clear Channel backup FM transmitting site at 4 Times Square. This facility is also capable of point-to-point microwave, spread spectrum, broadband, two-way, STL/TSL, RPU TV broadcast, and ENG services. In the fall of 2005, Lyons was elevated to his present position and is responsible for the communications needs of the entire 10 million square foot Durst portfolio, including interfacing to the NYPD/NYFD/EMS systems.

Steve Mahrer (Chapter 5.3) is Director of Engineering for the Business Development Group of Panasonic Broadcast & Television Systems Company. His responsibilities include managing the technologies used in Panasonic products designed for the broadcast and production marketplaces. These technologies include imaging and display, video compression and recording, and file systems and networking. The recent move to high definition and the increasing pace of the convergence of the IT and broadcast industries have made this an "interesting" experience.

Prior to joining Panasonic in 1991, Mr. Mahrer was employed at NBC's Technical Development group at 30 Rockefeller Plaza, New York, where he worked on small-format tape integration, automation, and early high definition work. The high definition work included assistance with testing at the ATTC in the late 1980s. He was awarded two patents for embedded data signaling during his work at NBC.

Before joining NBC, Mr. Mahrer was employed in 1974 by the RCA Broadcast System Group in Gibbsboro, New Jersey, working on the development of CCD ENG and traditional studio camera design, including Triax system variants. In 1984 he was transferred to the United States from RCA (Jersey) Ltd., a European manufacturing facility located in the British Channel Isles. His work there included the custom design and reengineering of RCA equipment intended for the European, African, Australian, and Asian markets and included the many variants of PAL, SECAM, and NTSC.

George Maier (Chapter 9.1) started Orion Broadcast Solutions in 1997 to answer the need for outsourced transmission engineering and marketing support. The company specializes in system design, FCC licensing, and project management for ENG, SNG, terrestrial microwave, VHF and UHF, studio routing and terminal equipment, and IT-based studio systems. The Orion client base includes broadcast networks, major groups, and individual stations. In the marketing area, Orion has worked with numerous equipment manufacturers to develop new products and strengthen existing lines.

Mr. Maier has been involved in broadcast transmission systems for most of his career and has worked for a number of well-known companies, including Telco Systems, Inc., *Television Broadcast* magazine, Artel Video Systems, ADC Telecom, NORTEL Networks, Harris Corporation, M/A-COM Inc., and Lucent Technologies.

Mr. Maier is Senior Member of the SBE and holds a CSTE certification as well as FCC General Class Radiotelephone and Amateur Extra Licenses. He also serves on the SBE National Frequency Coordination Committee, and is the Director of Monthly Programs for local SBE Chapter 11

Donald L. Markley, P.E., (Chapter 8.3) is Senior Consultant to and President of D. L. Markley & Associates in Peoria, Illinois. He is a graduate of Bradley University in Peoria, Illinois, having earned Bachelor of Science and Master of Science degrees in electrical engineering. Prior to fulltime consulting, Don was a tenured Associate Professor at Bradley.

Don is the Technical Facilities Editor for *Broadcast Engineering* magazine with over 200 articles published since 1967. He has earned Honors from Tau Beta Pi and Sigma Tau in Engineering, Eta Kappa Nu for Electrical Engineering, and Phi Kappa Phi Graduate School Scholarship. He was honored as 1992 IEEE Illinois Valley Chapter Engineer of the Year.

Don is an amateur radio operator K9WFG and has spoken at many conferences sponsored by various organizations.

William G. Marshall (Chapter 5.20) is a Principal at Harvey Marshall Berling Associates. He has over 25 years of experience in the design of audio and video facilities for theatrical, broadcast, industrial, and educational television facilities and architectural lighting design services for hotels, amusement parks, arenas, restaurants, and military and government buildings and projects. Prior to forming Harvey Marshall Berling, he directed facilities design for Imero Fiorentino Associates, Caribiner International, and Jack Morton Worldwide.

Mr. Marshall has been involved in video and film broadcast, corporate, and educational production facilities that include General Electric, J.C. Penney, American Express, AT&T, Proctor & Gamble, Northwest Mutual Life, J.P. Morgan Trust, Merrill-Lynch, Philip Morris, State Farm, and Sony. Broadcast clients include WNET, NBC-Chicago, HBO, the Manhattan Studio Center, Unitel, Teletronics, the Dallas Communications Complex, and Chelsea Studio/All Mobile Video. Educational studio facilities include the University of Iowa, Northwestern University, and the City University of New York.

Mr. Marshall provides lectures on teleconference lighting to many trade groups and has been responsible for broadcast and teleconference lighting of the U.S. Senate Chamber, U.S. House of Representatives, the Military Airlift Command at Scott AFB, and the Joint Chiefs of Staff conference room at the Pentagon. He worked on teleconferencing facilities designs for Western Electric, the U.S. Department of Defense, AT&T Bell Labs, ISACOMM, and Andersen Consulting. He has also designed hybrid architecture and video lighting systems for Marble Collegiate and Trinity Church in New York.

Jim Martinolich (Chapter 5.15) is VP of Product Development at Chyron, where he helped move Chyron's focus from "big iron" to open-platform software-based solutions. He has worked in all aspects of television technology for over 20 years, from video games to missile guidance systems.

Jim earned a Bachelor of Science degree in electrical engineering from Rensselaer Polytechnic Institute in Troy, New York, and an MSEE from Polytechnic University in Brooklyn, New York. His first job and "graduate degree in television" was working at CBS Laboratories on early video disc development. From there, he worked for Atari on video games and home robotics, and for Chyron as an analog design engineer on the VP-2 and Scribe. Later, he designed video and infrared imaging systems for various military and space applications, including the B-52, Tomahawk cruise missile, and the Space Station. At California Microwave, he helped develop one of the first digital satellite news-gathering systems built on DVB standards.

Jim hold several patents, has presented papers at NAB Broadcast Engineering Conferences, the SBE, and other regional technical conferences. He has contributed articles on the subject of broadcast graphics to several magazines and journals in the United States and Europe.

David Mathew (Chapter 8.1) is Technical Publications Manager at Audio Precision, Inc. in Beaverton, Oregon. He has been with Audio Precision since 1999. An Emmy-awarded sound engineer, he has worked in audio recording studios and in film sound as chief engineer, recording engineer and production mixer.

Mr. Mathew has written technical articles for Recording Engineer/Producer, Studio Sound, Pro Sound News, and other trade organs, and has authored user manuals for Audio Precision, Mackie, Audio Control Industrial, and Abekas.

Mr. Mathew is a member of the Audio Engineering Society and holds an FCC General Radiotelephone license.

David P. Maxson (Chapter 2.5) is a founding partner of Broadcast Signal Lab and has been providing broadcasters with engineering services since the company was formed in 1982. He is the former Vice President, Director of Engineering of Charles River

Broadcasting Company, Boston's long-time classical music broadcaster, where he served for 20 years.

David's experience as a corporate executive and independent service provider has exposed him to a broad spectrum of engineering issues, including facility design, construction, operation, and safety management. With a Bachelor of Science degree in broadcasting, David's career has led to certification by the Society of Broadcast Engineers, and he is a member of the Institute of Electrical and Electronics Engineers. He holds an FCC First Class License with radar endorsement, and is also a licensed construction supervisor.

David is a regular presenter of technical papers at NAB conferences on topics ranging from datacasting technology, to RF safety signs and exposure management, to measurements of IBOC signals. He developed an authoritative, ANSI-compliant set of RF safety signs available on RFSigns.com. As his company's representative to the National Radio Systems Committee, he actively participated in the development of the NRSC-5 IBOC standard. He is coauthor of *The IBOC Handbook*, published by Focal Press.

Jeff Mazur (Chapter 5.10) is the Senior Director of Technology for ABC Entertainment, which includes the Los Angeles-based Primetime On-Air Promotions Department. He has written over 50 articles and several books on broadcasting, electronics, and computers.

Jeff attended the University of Michigan in Ann Arbor and earned a BS degree in physics from UCLA. He has several Emmy® awards for Technical Achievement and is also a member of the Academy of Television Arts & Sciences. He has FCC General Radiotelephone and Amateur Radio Extra Class Licenses. He can be reached at jeff.mazur@abc.com.

Gary L. McAuliffe (Chapter 3.2) is a Partner and Principal Consultant at Pelton Marsh Kinsella, LLC, and has been involved in the architectural and technical systems design and construction of facilities for broadcasters, industry, institutions, and government for over 30 years.

Mr. McAuliffe has broad experience in radio, television, and allied industries. His experience includes work at TV and radio stations, in mobile television, systems integration companies, and professional services firms. His work involves the design of studio facilities for government, education, and commercial users, television trucks, television planning for stadiums and arenas, private satellite networks, and cable television systems. His recent work includes microwave and television transmission facility planning for what will become the world's tallest building—the Burj Dubai in Dubai, United Arab Emirates.

Mr. McAuliffe is a member of the Society of Motion Picture and Television Engineers, the Audio Engineering Society, and the Stadium Managers Association, and is a regular contributor to *Stadia Magazine*.

William Meintel (Chapter 1.5) holds a degree in electrical engineering and has 37 years of experience in the communications field. After graduation, he was employed by the Federal Communications Commission, first as a Field Engineer and then in the Mass Media Bureau's Policy and Rules Division. While in Policy and Rules, Bill served as the division's computer expert, directed the development of several major computer modeling projects related to spectrum utilization and planning, and represented the United States at international spectrum planning conferences.

Bill entered private practice in 1989, and has been heavily involved in technical consulting, computer modeling, and spectrum planning for the broadcast industry. In April 2005, he merged his consulting practice into the firm Meintel, Sgrignoli, & Wallace.

In the spectrum planning area Bill has, among other things, coauthored a report for the NAB on spectrum requirements for Digital Audio Broadcasting (DAB), created a plan for independent television broadcasting for Romania, and has been extensively involved in spectrum planning for digital television (DTV) in the United States and internationally. He developed the software used by both the U.S. television industry and the FCC for DTV spectrum planning as well as that now used by the FCC for processing analog and digital television applications (OET-69 Longley-Rice analyses). He has published several articles and presented numerous papers related to spectrum planning.

Bill is currently Vice President of the IEEE Broadcast Technology Society, editor of its Newsletter, and a member of the Engineering Honor Society Tau Beta Pi.

Geoffrey N. Mendenhall, P.E., (Chapter 4.7) has spent most of his 42-year career in the broadcast industry. He started as a broadcast technician while attending high school, then worked for several Atlanta radio and TV stations while earning his electrical engineering degree from Georgia Tech. In 1973, he joined the Harris Broadcast Division as an electrical engineer, designing directional antenna phasing equipment followed by the MS-15 FM exciter. Later, he became Vice President of Engineering at Broadcast Electronics, where he led the development of the FX series of analog exciters and a full line of AM/FM radio transmitters.

Mr. Mendenhall rejoined Harris in 1993, as Vice President—Radio Product Line Manager, where his team successfully launched DIGIT, the world's first digital FM exciter, the first 2-megawatt, solid-state AM transmitter, and CD-Link, the first uncompressed, 950-MHz, digital STL. In 1995, Geoff assumed overall responsibility for the development of all Harris radio and television transmission products. He is now leading the teams designing next generation digital radio products including the new FlexStar HDx exciter and HT-HD+ common amplification transmitter.

Mr. Mendenhall has authored over 40 technical papers. In 1999, he received the NAB Radio Engineering Achievement Award recognizing his many innovations and contributions to the broadcast industry. He holds five U.S. patents for broadcast equipment.

Mr. Mendenhall is a Registered Professional Engineer. He is a member of the Association of Federal Communications Consulting Engineers (AFCCE) and a Senior Member of the IEEE.

Richard Miller (Chapter 9.2) currently is Manager of the Systems/Applications Engineering Group for Microwave Radio Communications, part of Vislink Communications LLC. Since joining MRC in 1991, he has been Manger of Field Services, System Test Manager, and Systems Engineer.

Mr. Miller attended Northeastern University in 1996, pursuing a degree in electrical engineering. He began his career with Microwave Associates in 1980 as a Field Service Technician, eventually graduating to Engineer. Prior to this, he worked in the communications industry, primarily in broadcast.

Mr. Miller gained practical and field experience in the RF field, eventually leading to management of field projects in the design, installation, and commissioning of microwave systems. With this experience he held positions at several communications companies, Microwave Associates, M/A-COM Inc., California Microwave, and Adaptive Broadband.

Steven E. Mitchel (Chapter 1.2) is Reference Librarian for the National Association of Broadcasters. He has previously held positions as Senior Researcher for ABC News, and Librarian and Legislative History Researcher for the law firm of Winston & Strawn. He has presented papers at conferences of the American Association of Law Libraries, and his articles have appeared in *Law Library Journal* and *Government Publications Review*. He is a recipient of the Donald G. Wing Bibliographic Award.

Tom Mock (Chapter 4.14) heads WYE Consulting and is a consultant to industries specializing in areas involving the application and development of consumer electronic products.

Tom was formerly with the Consumer Electronics Association, a sector of the EIA, from 1981 until 2000, where as Director, Technology and Standards, his duties included staffing of technical committees in the areas of product safety and compliance, vehicular electronics, and liaison with other associations and consortiums. From 1956 until 1981, he served in various engineering capacities in the Government and private sectors related to the military and Department of Defense.

While with the CEA/EIA, Tom was directly involved with the development of stereo broadcasting for TV for which the Association won an Emmy, initiation of the home automation program that became CEBus, and introduction and standardization of the Radio Data System (RDS) in the United States. He was also involved with the standardization of the Closed Captioning and the eXtended Data System (XDS) for TV, the development of the V-Chip parental control system and, coordination of CEA Intelligent Transportation System activities with ITSA and SAE.

Tom has served as U.S. delegate on several IEC committees and is a member of the Institute of Electrical and Electronics Engineers (IEEE), the National Radio Systems Committee (NRSC), and the RDS Forum. He is also a member of Underwriters Laboratories (UL) Standards Technical Panel (STP) for UL 6500 and UL 1678.

Jeff Moore (Chapter 5.10) is President of Ross Video in Iroquois, Canada. Jeff joined Ross Video in 1997, serving as the Director of Marketing and Sales before being promoted to Vice-President Marketing and Sales in 2002. In 2006, Jeff was promoted to President. He holds an MBA from the University of Ottawa.

Jeff was raised in Canada's north, Whitehorse, Yukon, across from an area known as Whiskey Flats. His father had an electrical contracting business and was responsible for sparking Jeff's interest in electronics when he brought home various pieces of equipment that Jeff began experimenting with.

After high school, Jeff spent a few years following a career in geology before realizing that it wasn't his calling. Rekindling his interest in electronics, he attended the Southern Alberta Institute of Technology where he studied broadcast electronics engineering technology, earning an honors diploma.

Jeff got his broadcast start working at CHUM Television in Toronto, at the time CITY TV, Much Music, in the engineering group assisting the Manager of Engineering with the complete redesign and move of the facility. Jeff spent 14 years in Toronto, where he also worked at Azcar Technologies managing broadcast design and installation projects; Sony as a Broadcast Account Manager, Broadcast Television Systems as Regional Sales Manager, Acura Technology Group as Regional Sales Manager, and Tektronix VND as National Business Manager.

When not working, Jeff enjoys cooking, canoeing and kayaking, reading, and spending time with family.

Andrew Morris (Chapter 5.19) is a consulting engineer with over 30 years of experience in the broadcast and cable industries. A four-time Emmy® award winner for his work designing and managing the broadcast communication systems for three of NBC's Olympic broadcasts, Mr. Morris has performed a number of engineering and operational roles in the broadcast industry.

A member of SMPTE, SBE, AES, and IEEE, Andrew resides in Denver, Colorado, and currently works with a variety of systems integrators and broadcasters on the upgrade to and creation of modern broadcast systems.

Richard Morris (Chapter 1.10) studied electronic engineering at Birmingham University in the UK and joined the British Broadcasting Corporation in 1987. Over the next few years, he worked on projects building monitoring and control systems for broadcast networks, including a year reengineering a BBC shortwave transmitter site in Cyprus. Following a short period working on digital studio to transmitter links, he joined the team that built the world's first fully engineered Eureka 147 DAB transmitter network in 1995. Richard was later the technical authority and team leader for digital radio head end systems with Crown Castle UK, and built DAB coding and multiplexing systems for three commercial Eureka 147 digital radio networks.

In 2003, Richard left the UK to join Commercial Radio Australia, the industry body that is currently working with the commercial radio networks to prepare for rolling out digital radio networks in Australia. He is currently the principal engineer for the DAB trial in Sydney.

Richard has an honors degree in electronics. He is a Chartered Engineer and a member of the Institute of Engineering and Technology (UK). He has presented papers at international digital radio conventions, and is a member of the WorldDAB technical committee.

John Norgard (Chapter 1.1) is a Professor Emeritus at the University of Colorado at Colorado Springs, the President and CEO of ElectroMagnetic Techniques, the Chief Scientist of zeeWAVES, and Senior Research Scientist in the Sensors Directorate at the Rome Research Site of the Air Force Research Laboratory. He was also a Distinguished Visiting Professor at the U.S. Air Force Academy. He has taught graduate and undergraduate courses in electromagnetics for over 35 years and is the Director of the Electromagnetics Laboratory at the University of Colorado. He was a Professor of Electrical Engineering at Georgia Tech and a Post-Doctoral Fellow at the Norwegian Defense Research Establishment in Oslo, Norway. He worked at the Jet Propulsion Laboratory while studying at Caltech (PhD/1969, MS/1967, applied physics) and was a co-op student at Georgia Tech (B.S./1966, ECE) while working at the Charleston Naval Shipyard. He was a Visiting Professor at Tel-Aviv University and a member of the technical staff of the Bell Telephone Laboratories.

He has worked on numerous computational electromagnetic problems, including plasmas (polar ionosphere), field-to-wire coupling (cross-talk, NEMP, lightning), EMI, EMC, EMV/S, HPM, GPR, RF tomography, IR metrology, and holography.

He is a Fellow of IEEE for IR measurements of EM fields, on the Board of Directors of the IEEE/EMC Society serving as the Vice President for Technical Services, on the Board of Physics and Astronomy for the National Academy of Sciences, Past Chairman for Commission A/Metrology of URSI, and an Associate Editor for the *IEEE/EMC Transactions* for antenna metrology.

Robert Orban (Chapter 3.8) received a BSEE degree from Princeton University in 1967 and an MSEE degree from Stanford University in 1968. In 1970, he founded Orban Associates, originally as a manufacturer of studio equipment. In 1975, Orban Associates introduced the original Optimod-FM 8000, the first in a long line of broadcast audio processors for AM, FM, TV, and digital broadcasting from the company.

Mr. Orban has been involved in professional recording for many years, and has mixed several records released on the Warner Bros. label, as well as on small independent labels. As a composer, his music has been heard on classical radio stations in New York and San Francisco, and his score for a short film, *Dead Pan,* was heard on PBS television in Chicago. He has designed studio reverberation systems, stereo synthesizers, compressors, parametric equalizers, enhancers, and de-essers under both the Orban and dbx brand names.

Mr. Orban was actively involved in NRSC committee AM improvement work. He is widely published in both the trade and refereed press (including *J. Audio Engineering Soc., Proc. Soc. Automotive Engineers,* and *J. SMPTE*). He is the author of the chapter on "Transmission Audio Processing" in the *NAB Engineering Handbook,* 9th edition. He currently holds over 20 U.S. patents.

Mr. Orban was elected a Fellow of the Audio Engineering Society in 1973. In 1993, he shared with Dolby Laboratories a Scientific and Engineering Award from the Academy of Motion Picture Arts and Sciences. In 1995, he received the NAB Radio Engineering Achievement Award.

Thomas G. Osenkowsky (Chapter 2.7) began his career in broadcasting in 1973 while a senior in high school. He has held positions as announcer, Chief Engineer, Operations Manager, and General Manager at broadcast stations in Connecticut.

Tom has written software for antenna system design and analysis, RFR, mapping, and other broadcast engineering related tasks. He has served as a consultant to AM and FM stations in the United States and Caribbean islands. He is a Senior Member of the Society of Broadcast Engineers (SBE), Institute of Electrical and Electronics Engineers (IEEE), and National Association of Radio and Telecommunications Engineers (NARTE).

Tom earned the Professional Broadcast Engineer Life Certification from the SBE, the Certified Master Engineer with Master RF Radiating Endorsement from NARTE, and holds a First Class Radiotelephone license from the FCC. He has presented papers at NAB Broadcast Engineering Conferences, SBE Annual Conventions, and state broadcasters annual conventions.

Tom is a regular contributor to *Radio World* magazine, a private pilot, and an amateur radio operator.

David Philip Otey (Chapter 1.6), AZCAR Technologies, began his broadcasting career in 1974, at age 15, as an announcer for radio station KEYE in Perryton, Texas. After college, he focused on television. In 1983, he was part of the engineering team that launched Channel 42 in Austin, Texas, then known as KBVO but now—by remarkable coincidence—carrying the call sign KEYE. He joined the engineering department of public TV station KLRU in 1985, serving as Chief Engineer, 1990–1996.

From 1996 to 2001, David worked for HSE Communications, a Colorado-based integrator of microwave systems, as Chief Engineer and later Operations Manager. During this time, he became involved in frequency coordination of Broadcast Auxiliary Services. From 2002 to 2005, he served on the staff of the Society of Broadcast Engineers as Frequency Coordination Director. His work on 2 GHz relocation led to his joining SignaSys in 2005 to lead the industrywide training effort in Digital Electronic News Gathering, and in 2007 to AZCAR Technologies.

David is an SBE-certified Professional Broadcast Engineer. His articles have appeared in *TV Technology, Radio World,* and *Radio Guide* magazines, and he has presented technical papers and training seminars at SBE national and regional conferences as well as the annual conventions of several state broadcasting associations. A 1981 graduate of Trinity University in San Antonio, David holds two master's degrees from the University of Texas at Austin. He lives with his family in the Denver area, where he also pursues a hobby in community theater.

Karl Paulsen (Chapter 5.9) is currently the Chief Technology Officer for AZCAR Technologies with offices in Markham, Ontario, Canada, and AZCAR USA, Inc., in Canonsburg, Pennsylvania. He has over 30 years of industry experience as a broadcast operator, engineering director, and consultant in the fields of broadcast, IPTV, mobile television, 3D graphics-and-animation, and systems integration.

Mr. Paulsen is a monthly columnist for *TV Technology* magazine, having contributed over 120 articles related to video servers, storage, and media management. He is the author of *Video and Media Servers: Technology and Applications*, 2nd edition, published by Focal Press.

Mr. Paulsen is an SBE Life Certified Professional Broadcast Engineer, an IEEE member, and a Fellow in the SMPTE.

Howard K. Pelton, P.E., (Chapter 3.2) began his career in noise control engineering and acoustical consulting in 1963. He has held positions with an industrial silencer manufacturer and several acoustical consulting firms, the latest of which is Pelton Marsh Kinsella, LLC. Mr. Pelton is a registered engineer in Texas and Louisiana. He is the author of Noise Control Management and many noise control articles. He also belongs to a number of professional organizations including ASA, ASHRAE, and ASME.

Mr. Pelton has extensive experience in the development of noise control management systems and programs for a wide variety of public and private sector clients. His experience includes plant surveys, feasibility studies, hearing conservation programs, training programs, seminars, noise control design, and construction management of detailed noise control programs. His architectural acoustics background includes design and development of noise and vibration control systems for high-rise office buildings, hotels, convention centers, hospitals, and schools, as well as remedial noise and vibration control evaluation, development of recommendations for corrective action, construction management, and partition testing.

Mr. Pelton has designed quality acoustical environments for spaces such as schools, churches, auditoriums, and meeting and lecture rooms. He has also performed assessments of environmental noise impacts from industrial plants, airports, freeways, and railroads for city, state, and federal regulations.

Skip Pizzi (Chapter 3.11) is Technical Policy Manager for Microsoft's Entertainment & Devices Division, a position in which he helps define the company's media-related public and business policies. In addition, he has represented Microsoft in digital broadcast standards organizations worldwide, and serves as a corporate liaison to the broadcast technology industry.

Skip is a contributing editor at *Radio World* magazine, where he writes "The Big Picture" column that appears in every issue. Previously, he served as Editor-in-Chief of *BE Radio* (now *Radio*) magazine, and earlier spent 13 years at National Public Radio where he served as technical director for numerous award-winning programs, and founded the company's technical training program. His book, *Digital Radio Basics,* the world's first on the subject of DAB, was published in 1992. He has contributed to several other technical texts, including McGraw Hill's *Digital Consumer Electronics Handbook* and the CRC/IEEE Press *Electronics Handbook.*

Skip is a frequent speaker at international conferences on broadcasting and audio, and currently cochairs the Surround Sound Audio Task Group of the National Radio Systems Committee. He also serves as a judge for the National Television Academy's "Technical Emmy" Awards. He is a graduate of Georgetown University in Washington, D.C.

Schuyler Quackenbush (Chapter 3.7) is an expert in digital audio technology. He is active in the area of standardization of audio coding algorithms and is currently the chair of the International Standards Organization Motion Picture Experts Group (ISO/MPEG) Audio subgroup. He was one of the authors of the ISO/IEC MPEG Advanced Audio Coding standard. He has worked on audio and speech coding algorithms, audio and speech quality assessment, audio error mitigation algorithms, and real-time signal processing algorithms and hardware. He holds 12 patents and is the author of more than 30 publications in these areas, including one book, Objective Measures of Speech Quality.

Dr. Quackenbush received a BS degree from Princeton University in 1975, an MS degree in electrical engineering in 1980, and a PhD degree in electrical engineering in 1985 from Georgia Institute of Technology. He was Member of Technical Staff at AT&T Bell Laboratories from 1986 until 1996, when he joined AT&T Laboratories as Principal Technical Staff Member. In 2002, he founded Audio Research Labs, an audio technology consulting company. He is a Fellow of the Audio Engineering Society (AES) and a senior member of the Institute of Electrical and Electronics Engineers (IEEE).

Ronald D. Rackley, P.E., (Chapter 4.3) is a Consulting Radio Engineer with the firm of du Treil, Lundin & Rackley, Inc. of Sarasota, Florida. He has been a specialist in the design, adjustment, and testing of the antenna systems that are employed by medium-wave (known as AM in North America) radio stations for more than 30 years and practices internationally as well as domestically.

Mr. Rackley developed several antenna design, analysis, and measurement techniques that are in common use today. He is a frequent author and lecturer on the subject of medium-wave antenna systems and participant in advisory committees for regulatory agencies. His work has included medium-wave directional antennas ranging in power to 2,000,000W and in complexity to 12 towers.

Mr. Rackley is a member and has served as President of the Association of Federal Communication Consulting Engineers. He is a member and has served as Vice President of the Institute of Electrical and Electronic Engineers Broadcast Technology Society, and is a member of the Institute of Electrical and Electronic Engineers Antennas and Propagation Society.

Mr. Rackley holds a BSEE degree from Clemson University in his native South Carolina and is a Registered Professional Engineer.

Phillip Rayson (Chapter 7.1) is currently Radian Communications Services Corporate Safety Officer. He is a Certified Hoisting Engineer-Mobile 1 with over 25 years of experience in the construction industry. For 16 years, he was involved with workplace safety, tower assembly, tower erection, and station and switch yard construction for a major utility company in Canada. Over the past six years he has been responsible for managing safety in both the United States and Canada for Radian Communication Services.

Mr. Rayson works with the Radian safety team conducting safety inspections, audits, safety training, and provides safety support to Aerial and Technical staff involved with broadcast, telecommunications, and wind turbine projects. He has been involved with developing and maintaining OSHA partnership agreements in regions 5 and 7.

Mr. Rayson holds the following certifications: OSHA 510, OSHA 500, OSHA Construction Outreach Trainer, Instructional Techniques Workshop, Propane CCR, Workplace Health and Safety Agency Core Certification Training Cert. ID: WHSC05395, TapRoot Incident Investigation & Root Cause Analysis, and Hoisting Engineer-Mobile 1 339A-171158.

Jeffrey Riedmiller (Chapter 5.18) is Senior Broadcast Product Manager at Dolby Laboratories in San Francisco, where he is responsible for the specification and development of Dolby's broadcast products and technologies including Dolby Digital, Dolby Digital Plus, Dolby E, and Metadata, with a strong focus on loudness estimation and control processes. He is also responsible for managing the specification and integration of Dolby technologies and systems into major cable networks and cable distribution facilities. He was responsible for creating and leading the development of the Dolby LM100 Broadcast Loudness Meter with Dialogue Intelligence, for which he received an Emmy® in 2004 for Outstanding Achievement in Engineering Development. He is coinventor of two patents in the area of speech-based loudness estimation and control, as well as methods to analyze, process, and correct audio metadata in the compressed domain.

Mr. Riedmiller has 20 years of experience in audio engineering and electronics as a Chief Engineer of a major upstate New York recording studio, a member of Time Warner Communications Studio Engineering Group, and an independent consultant on various facility, system, and custom electronic designs for recording studios, broadcast, and postproduction facilities throughout the United States.

Mr. Riedmiller is a member of the Mathematical Association of America, an active member of the Society of Cable Telecommunications Engineers and its Standards Committees, and serves as Associate Editor for *Transactions on Broadcasting* for the IEEE Broadcast Technology Society. Previously, he also served as cochairman for the National Cable & Telecommunications Association Engineering Committee Audio Quality Subcommittee.

Roy W. Rising (Chapter 3.1) is retired after 38+ years with ABC-TV, Hollywood. His career began in 1965 following graduation from U.C.L.A. with a Bachelor of Science degree in engineering. He has held supervisory positions in TV Engineering Facilities and TV Engineering Operations. His activities included design, construction, installation, maintenance, modification, and operation of broadcast audio facilities. His show mixing credits range from Music-Comedy-Variety with live orchestras, Sitcoms, Game Shows, News and Sports to the Oscar®, Emmy®, and Grammy® award shows.

Mr. Rising's writing credits include 15 years of the "Sound Ideas" monthly column in *Video Systems Magazine* and feature articles for *Broadcast Engineering* and *Recording Engineer–Producer* magazines. He also has contributed sections to *The Electronics Handbook* (CRC Press/IEEE Press) and the *NAB Handbook,* 9th edition.

Richard A. Rudman (Chapter 2.7) retired from KFWB Radio in Los Angeles (Infinity/CBS) as Director of Engineering on June 15, 2002, after 27 years in that position. He now owns and operates his own firm, Remote Possibilities, that consults on emergency public information, emergency preparedness for communications facilities, construction of 802.11 (b) (g) wireless nodes, and specialized broadcast spectrum issues. He holds Certification from the Society of Broadcast Engineers as Professional Broadcast Engineer (CPBE).

Richard has authored many papers on EAS, AMBER, emergency public information, emergency preparedness for communications facilities, and broadcast auxiliary spectrum. His paper, *Disaster Recovery for Broadcast Facilities,* was presented at the 1996 Spring NAB Convention and published in the Convention Proceedings. He is also the author of book chapters on broadcast auxiliary frequency coordination and disaster preparedness for communications facilities in the *NAB Engineering Handbook* and in two electronics textbooks published by CRC Press.

Richard was one of the 17 founding Trustees for the Partnership for Public Warning (founded in November, 2001) and contributed to several reports submitted to FEMA and the FCC on the subject of public warnings and the EAS.

Richard was elected National President of the SBE in 1985 and elected as a Fellow of the SBE in 1987. He still serves as a member of several national SBE committees. In November 2002, he received a lifetime achievement award from the SBE for his contributions to emergency public information, EBS and EAS, and for his contributions to broadcast auxiliary spectrum coordination.

Stanley Salek, P.E., (Chapters 1.9, 8.1) is a Senior Engineer in the consulting engineering firm of Hammett & Edison, Inc., San Francisco, California. Since joining the firm in 1991, he managed numerous projects related to broadcast radio and television technology, including analog and digital radio and television transmission analysis, signal coverage evaluation, and RF safety compliance. He is presently cochair of the AM Broadcasting Subcommittee of the National Radio Systems Committee, working toward the development and enhancement of industrywide technical standards.

Prior to his present position, Stan was Director of Radio Engineering at the National Association of Broadcasters, a design engineer with RF and audio broadcast equipment manufacturers, and has managed technical projects at several broadcast stations. He is a graduate of Florida Institute of Technology, earning a BS in electrical engineering, and is a Registered Professional Engineer in the state of California. He holds a U.S. patent for an RF modulator design. He also holds FCC commercial and amateur radio licenses, and is a member of IEEE, AES, SMPTE, and AFCCE. He has authored articles and book chapters on various topics related to broadcast engineering.

Gary Sgrignoli (Chapter 8.4) received Bachelor and Master of Science degrees in electrical engineering from the University of Illinois, Champaign-Urbana in 1975 and 1977, respectively. He joined Zenith Electronics Corporation in January 1977, where he worked as an engineer in the Research and Development department for 27 years. In March 2004, he set up Sgrignoli Consulting, a DTV-transmission consulting firm, and in April 2005, he merged his practice with those of Bill Meintel (Techware, Inc.) and Dennis Wallace (Wallace and Associates) to create Meintel, Sgrignoli, and Wallace (MSW). Further information can be found at www.MSWdtv.com.

Mr. Sgrignoli has worked in the R&D design area on television "ghost" canceling, cable TV scrambling, and cable TV two-way data systems before turning to digital television transmission systems. Since 1991, he has been extensively involved in the VSB transmission system design, its prototype implementation, the ATTC lab tests in Alexandria, Virginia, and both ACATS field tests in Charlotte, North Carolina. He holds 35 U.S. patents.

Mr. Sgrignoli was involved with the DTV Station Project in Washington, D.C., helping to develop DTV RF test plans. He has also been involved with numerous television broadcast stations around the country, training them for DTV field testing and data analysis, and participated in numerous DTV over-the-air demonstrations with the Grand Alliance and the ATSC, both in the United States and abroad.

In addition to publishing technical papers and giving presentations at various conferences, he has held numerous VSB transmission system seminars around the country.

Thomas B. Silliman, P.E., (Chapter 6.8) is President as well as the Chairman of the Board of Directors of Electronic Research Inc. (ERI). He attended Cornell University in 1964, where he received a Bachelor of Electrical Engineering degree in 1969 and a Master of Electrical Engineering degree in 1970.

After graduating from Cornell University, Mr. Silliman went to work for ERI, where he designed the ERI Rototiller antenna in 1975 for which he was granted a patent (assigned to ERI) on August 22, 1975. While working for ERI, he was also a partner in the broadcast consulting firm of Silliman and Silliman with Robert Silliman, his father.

Mr. Silliman is a Registered Professional Engineer in Indiana, Minnesota, and Maryland, and a Senior Member of the IEEE. He is currently serving as Secretary of the IEEE Broadcast Technology Society Administrative Committee. He is a full member of the Association of Federal Communications Consulting Engineers (AFCCE), and a past two-term President of that organization. He also serves as a Board Member of WNIN, an Evansville, Indiana, local public radio and television provider.

Wes Simpson (Chapter 6.13) is president and founder of Telecom Product Consulting, an independent consulting firm that focuses on helping companies develop and market video and telecommunications products. He holds a BSEE from Clarkson University and an MBA from the University of Rochester.

Wes has over 20 years of experience in the design, development, and marketing of products for telecommunication applications. Before founding Telecom Product Consulting, he was COO of VBrick Systems, Inc., a manufacturer of MPEG video equipment. Earlier, at ADC Telecommunications, he was the Director of Product Management for the DV6000, a market leading video transport system. He previously held a variety of marketing and engineering positions in the telecommunications industry. Wes was a founding member of the Video Services Forum, and was a member of its Board of Directors from 1997 to 2001.

Wes is a frequent speaker and analyst for the video transport marketplace, and is author of the book *Understanding Video Transport over IP Networks, A Practical User's Guide to Technologies and Applications*, published by Focal Press in 2005.

Sidney M. Skjei, P.E., (Chapter 6.11) is President of Skjei Telecom, a technical consulting company providing engineering and operational support for satellite and broadcasting applications. Since its founding in 1994, Skjei Telecom has provided consulting and support services to over 100 different companies or organizations in a broad spectrum of sectors involved in satellite communications and broadcasting.

Prior to co-founding Skjei Telecom, he held senior-level engineering management positions with GTE, Southern Pacific Satellite Company (SPRINT), and COMSAT World Systems. He has over 25 years of experience in engineering and developing a wide range of hardware and software telecommunications and broadcasting products, systems and services, and in providing engineering support to the procurement, operation, and marketing numerous geosynchronous communications satellites. His experience and expertise includes domestic U.S., international, and military communications satellites.

Mr. Skjei has taught numerous satellite and digital video courses for both private industry and government students. He is the author of numerous articles and a textbook on satellite telecommunications and has testified as Litigative Consultant and Expert Witness. He holds a BS degree from the U.S. Naval Academy and an MSEE degree With Distinction from the Naval Postgraduate School. He is a Registered Professional Engineer in the Commonwealth of Virginia and a member of the IEEE, Eta Kappa nu, SMPTE, and AFCCE.

Derek Small (Chapter 6.7) is Director of Filter Products for MYAT, Inc. After receiving his BS in electrical engineering from the University of Maine in Orono in 1986, he was immediately immersed in the design of passive microwave components and filter subsystems for military programs with M/A-Com and Continental Microwave.

From 1993 to 1999, Mr. Small was the primary developer of high power filter products and other passive components for Passive Power Products of Grey, Maine. Key among his activities at Passive Power Products was his work in passive components and the development of very stable, high power UHF cavities. In January 2000, he established Lowpass Prototype Inc. to develop and manufacture filter based products for the broadcast and wireless industries, and served as its President until its acquisition by MYAT, Inc. in October 2001.

Mr. Small has published numerous papers and articles, and is the holder of one patent, with others pending. His research interests are focused on filters and other products for high power applications.

Eric Small (Chapters 4.8, 6.3) is founder and Chief Technology Officer of Modulation Sciences, Inc., in Somerset, New Jersey. Modulation Sciences manufacturers products including the CP-803 Composite Processor, the Sidekick™ SCA Generator, the StereoMaxx™ Spatial Image Enlarger, and the FM ModMinder™ Modulation Monitor.

Mr. Small's career began at classical music station WNCN, New York, in 1964. In 1969, he joined A&R Recording under Phil Ramone. Later, as a consultant, he worked for most of the major broadcasters in North America. He was an aerospace hardware and software designer for the visual portion of the F/A-18 combat flight simulator in the 1980s. In 1974, he participated in the design and subsequent widespread use of the original Optimod™ 8000 FM processor. In 1975, he authored the technical chapter in the *CPB Handbook* for setting up SCA-based Radio Reading Services for the blind. He has remained active in radio reading services since.

When Multichannel Television Sound emerged, Mr. Small was a voting member of the BTSC, the group that wrote the standard for stereo TV sound. Modulation Sciences went on to design and manufacture a TV Stereo Generator and a TV Stereo Reference Decoder. It also designed and a manufactured NTSC vide/audio demodulator, and digital TV 8VSB analyzer.

Mr. Small continues to participate in international standards committees and represented and spoke on behalf of the United States as a member of several U.S. delegations to ITU-R, SG10B meetings in Europe.

Martin Stabbert (Chapter 3.5) is Director of Corporate Engineering for Citadel Broadcasting, a position he's held since 2003. When asked why he chose to enter broadcasting, he contends that the fit was so natural, it's almost as if radio chose him. He recalls working in his father's appliance business as early as the age of 6, repairing toasters and coffee makers. The inclination toward technology, all things mechanical, and a general interest in audio led him to KVIP/KVIP-FM in Redding, California, where he began his radio career as a board operator. He quickly realized that the engineering aspect of radio was for him.

Graduating with degrees in electronic technology and electronic engineering from Shasta College in Redding, California, Martin spent several years as a contract engineer with clients in Nevada as well as Northern and Central California before accepting a position with Citadel Broadcasting in 1992 as Market Engineer for their Reno, Nevada facility.

When not tuned in to radio, Martin enjoys snowmobiling, pistol marksmanship, and is occasionally found on the ham radio bands with call sign KC7FTK. Martin and his wife Monica currently reside in Reno with their three dogs.

Leon Stanger (Chapter 6.15) is an independent consulting engineer serving the television industry. He has served the television industry in a broad range of engineering design, development, and management capacities at KWCS, KOET, Harris Corp., The Grass Valley Group, Utah Scientific, and DIRECTV.

Mr. Stanger received a Bachelor's degree in electronic engineering in 1968 from Weber State University and a Master's degree in business administration from the University of Phoenix in 1995.

Mr. Stanger is an active member of the Institute of Electrical and Electronic Engineers (IEEE) and The Society of Motion Picture and Television Engineers (SMPTE). He has served on a number of engineering committees to develop standards and practices for the broadcast industry as well as private companies. He holds 10 patents relating to television production, distribution, and transmission. He can be reached via e-mail at stanger62@msn.com.

Mike Starling (Chapter 3.4) joined NPR in 1989 as Senior Engineer, was named Director of Engineering in 1991, Vice President for Engineering in 1998, and Vice President, Chief Technology Officer in 2005. He is also Executive Director of NPR Labs, NPR's broadcast technology research and development unit.

Mr. Starling's undergraduate degree is in broadcast journalism and radio-TV from the University of Maryland. He also earned a BSL and Doctor of Jurisprudence degree from National University School of Law in San Diego.

In the 1970s Mr. Starling founded, built, and managed commercial and noncommercial stations in Virginia (WKYY-AM, WUDZ-FM, and WWLC-FM) and was Chief Engineer for NPR Member Station KPBS-FM in San Diego in the 1980s. He is a Board Member of the Richardson Maritime Museum in Cambridge, Maryland, and a past Board Member of the International Association of Audio Information Services, and the North American Broadcasters Association. He consults for radio stations in the United States and Southern Africa, and has been a U.S. delegate to the ITU.

Mr. Starling received the IAAIS 2004 C. Stanley Potter Award for work on digital radio reading services, and was named one of *Radio Ink's* 35 "Most Admired Engineers for 2005," and *Radio World's* "Excellence in Engineering—Engineer of the Year 2005" for his work on digital radio multicasting.

Mr. Starling is Chair of the Radio Subcommittee for the North American Broadcasters Association, a founding member of the Association of Public Radio Engineers, a member of the IEEE and the Radio Club of America. He is also a lawyer and a member of the California and D.C. bars.

Robert A. Surette (Chapter 4.10) is Manager of RF Engineering with Shively Labs, a Division of Howell Laboratories. He has been directly involved with design and development of broadcast antennas, filter systems, and RF transmission components since 1974, and as an RF Engineer for six years with the original Shively Labs in Raymond, Maine, and for a short time with Dielectric Communications.

Mr. Surette graduated from Lowell Technological Institute, Lowell, Massachusetts, in 1973 with the degree of Bachelor of Science in electrical engineering.

Mr. Surette is currently an Associate Member of the Association of Federal Communications Consulting Engineers and a Senior Member of IEEE. He has participated in writing several articles for trade publications. He was a lecturer for the National Public Radio seminars, has spoken at many local SBE meetings, and presented papers at NAB national conventions and national SBE meetings. He has authored a chapter on filters and combining systems for the latest edition of *The Electronics Handbook* from CRC Press and for the *NAB Engineering Handbook*, 9th edition.

Dr. Norman M. (Sam) Swan (Chapter 2.1) began his broadcasting career in 1967 as a radio announcer at KFVS-AM in Cape Girardeau, Missouri, in radio news and programming. In 1969, he joined sister station KFVS-TV, where he served as host of the Breakfast Show and as a news anchor and reporter.

Mr. Swan graduated from Southeast Missouri State University and worked in Sri Lanka in 1970–71 developing radio quiz shows for Young Farmers clubs. Later, he was a radio and television reporter for the University of Missouri Extension Service in Columbia, producing radio and TV reports used by Missouri broadcast stations.

Mr. Swan completed a PhD at the University of Missouri in 1978 and became Electronic Media Leader at the University of Minnesota. There he led a group of writers and producers in the production of programs used by radio and TV stations throughout Minnesota.

In 1981, Mr. Swan became Head of the Department of Radio-Television at Southern Illinois University in Carbondale. He produced several programs for WSIU-FM and WSIU-TV. He later became Head of the Department of Broadcasting at the University of Tennessee in Knoxville. He also served as General Manager of WUTK-FM and WUTK-AM. He produces a weekly news magazine program for WBIR-TV in Knoxville.

Mr. Swan currently serves as a trainer and consultant to radio and television stations around the world. He has conducted over 100 workshops on broadcast management and broadcast journalism for the Voice of America, Radio Free Europe/Radio Liberty, IREX, U.S. State Department, and other agencies. He conducts annual management workshops for broadcast engineers as part of the USTTI program in Washington, D.C. He writes a monthly newsletter for radio and TV affiliates of Voice of America in Central and Eastern Europe and published a book on broadcast management for the International Broadcasting Bureau in Prague.

Peter Symes (Chapter 5.7) is Manager, Advanced Technology, for Grass Valley division of Thomson Broadcast, with responsibilities that include strategic planning, intellectual property, and technological liaison. In the latter role, he represents Grass Valley in many organizations, including the Society of Motion Picture and Television Engineers (SMPTE). He served two terms as Engineering Vice President of SMPTE, and is currently the SMPTE Financial Vice President.

Mr. Symes gained his Bachelor of Science degree with honors in 1967 and began his career in television in the engineering department of the British Broadcasting Corporation. He worked in product management for Philips and Central Dynamics before joining Grass Valley in 1983.

Mr. Symes holds patents and is the joint recipient of Emmy® awards for the architecture of the digital picture processor. He is a Senior Member of the Institution of Electrical Engineers (IEE), and a Fellow of the SMPTE.

Mr. Symes has written and presented numerous papers at industry conferences, and is the author of *Video Compression* (1998), *Video Compression Demystified* (2001), and *Digital Video Compression* (2003), all published by McGraw-Hill. He has also contributed to other books, including the *NAB Engineering Handbook* and *Understanding Digital Cinema* (Focal Press, 2004).

Rolf Taylor (Chapter 3.10) is currently Product Manager for telephony product lines at Telos Systems and works with the Research and Development Department as well as the Support Engineering Department. Telos Systems, with headquarters in Cleveland, Ohio, and offices in several states and Europe, makes digital network and telephone interface products for talk-show programming, call-in segments, teleconferencing, audio production, remote broadcasts, and intercom applications.

After joining Telos Systems in 1995, Rolf was involved with telephony in the form of ISDN and T1. As Senior Customer Support Engineer, he helped customers solve telephone system interface problems

Rolf uses his knowledge of "working with the phone company" originally gained during his analog program loop days and often takes support calls to keep in practice. In Research and Development, he is involved in user-interface design, as well as being an expert on matters pertaining to telephony.

Rolf learned telephone technology at age 10 by dismantling a telephone. Later, as Technical Director at WRUW-FM, he learned the "art" of dealing with the phone company, and understanding analog program loops and 1A2 key systems. At WRUW-FM, he managed a complete refit of the air studio and replacement of old program loops with a 950 MHz STL.

Rolf writes white papers and product manuals at Telos. He has presented a paper at the 2002 NAB Engineering Conference, and contributed several definitions for *Newton's Telecom Dictionary.*

Rolf's hobbies include collecting and restoring vintage telephones and reading books about telephone technology or science fiction.

Phil Tudor (Chapter 5.6) is a Senior Engineer at the BBC's Research Department at Kingswood Warren, Surrey, UK and is a Chartered Engineer. He studied electrical and information sciences at Cambridge University, graduating in 1990.

Over the last six years, Phil has been active in industry efforts to develop standard file formats for use in program making, in particular MXF and AAF. He represents the BBC on the board of the Advanced Authoring Format Association, and leads their engineering work. He is also a member of SMPTE W25 technology committee.

Phil's technical background includes MPEG-2 video standardization, video coding optimization, and MPEG bitstream manipulation. His current work areas include file format standardization, metadata interoperability, and technical architectures for program making.

David T. Turner (Chapter 3.6) is currently Executive Vice President of ENCO Systems, Inc., in Southfield, Michigan, where he directs development of software-based solutions for the broadcast industry.

Mr. Turner started in broadcasting in 1972 while a freshman in high school. He has an extensive background in radio and television broadcasting, including 20 years of top ten market experience. He has performed a variety of responsibilities including disc jockey, board operator, Transmitter Supervisor, Chief Engineer, Technical Director, Operations Supervisor, Facility Design Engineer, and has operated and serviced nearly every type of broadcast equipment.

Mr. Turner holds a Bachelor of Science degree in electrical engineering from the University of Michigan, is an FCC-licensed First Class Radio Telephone Operator, and holds an Extra Class Amateur Radio Operator License.

Mr. Turner is a contributing author to the *NAB Engineering Handbook* and has been a guest speaker and NAB, SBE, and SMPTE conferences.

Jay Tyler (Chapter 3.5) currently manages the U.S. distribution of Audioarts Engineering and works directly with the major industry groups on the Wheatstone product line. He has been a Sales Engineer for Wheatstone Corporation in New Bern, North Carolina since 1996.

Mr. Tyler was working his way through community college in his hometown of Syracuse, New York at an audio-video store when he remembers attending to a particular client well-versed in audio. That client turned out to be Wheatstone founder Gary Snow. After getting to know each other and touring the Wheatstone factory with Gary, Jay couldn't resist the opportunity to work in a technology field, where he was hooked.

Although Wheatstone is his first career position in the broadcast industry, Mr. Tyler started working with consumer audio at the age of 14 with basic 12-volt automotive audio. By the age of 18 he was designing, selling, and installing premier home AV systems throughout the Syracuse area.

When he's not traveling for Wheatstone, Jay enjoys boating, skiing, golfing, and of course, "Listening to the Radio."

Doug Vernier (Chapter 1.8) is President and Head Engineer at V-Soft Communications, a broadcast engineering and software development firm with television and radio clients across the United States. His propagation prediction software is used by a majority of the country's broadcast engineering consultants and station engineers. He is principal technical advisor for the Corporation for Public Broadcasting's digital conversion program.

Mr. Vernier retired in 2002 from a 30-year appointment as Director of Broadcasting Services at the University of Northern Iowa in Cedar Falls. He has served two terms on the National Public Radio Board of Directors. His bachelor and masters degrees are from the University of Michigan, where he studied engineering and telecommunications. Mr. Vernier is certified as an SBE Professional Broadcast Engineer. He has an FCC First Class Radiotelephone License and an amateur radio operator Extra Class (K0DV). He is a motorcycle rider and a collector of antique radios.

John Wahba, PhD., PEng., P.E., (Chapter 7.1) is President and Principal Engineer of Turris Corp. Dr. Wahba has over 16 years of experience with structural engineering and design of broadcast and telecommunication towers. He has research experience in that field in addition to his design activities. He has designed literally hundreds of towers including some 600m guyed towers with candelabras for multiple broadcast antennas and a 395m self-supporting structure. He has coauthored several publications in the field of dynamic and static analyses and design of towers.

Dr. Wahba has held several senior engineering positions with Radian Communication Services and also headed the safety group at Radian. He holds engineering licenses in over 40 jurisdictions across the United States and Canada. He has been active on several technical committees (CSA, TIA, and ASCE dynamics of latticed structures).

Dr. Wahba is a member of the Canadian Standards Association Technical Committee on Communication Towers and the workgroup on Wind Turbines.

John Warner (Chapter 4.6) is Vice President of AM Engineering for Clear Channel and is responsible for the maintenance of nearly 400 AM stations.

John attended the University of Maryland, and after over a decade in defense electronics and communications took a position at WBAL in Baltimore. During his almost 20 years there, he was involved in numerous upgrades, and he spent many weekends working on directional antennas in the area. He came to Jacor in 1999 and moved into Clear Channel shortly thereafter. At Clear Channel, he has been responsible for the construction and commissioning of numerous directional arrays, including several high power diplexes.

John is a member of the Antennas and Propagation and Broadcast Technology Societies of the IEEE and has been a presenter at several NAB conferences.

John holds an Amateur Radio Advanced License, is a licensed pilot, and enjoys fishing and cross-country skiing.

S. Merrill Weiss (Chapter 6.5) heads the Merrill Weiss Group LLC in Metuchen, New Jersey, and is a consultant in electronic media technology and technology management. He is a graduate of the Wharton School of the University of Pennsylvania.

Mr. Weiss' career is dedicated to designing and building systems for broadcast and related-industry entities. He also has worked on developing many of the technologies that underlie the digital television systems currently being implemented. He has led several development efforts and tests and chaired committees that have prepared standards for the Society of Motion Picture and Television Engineers (SMPTE), the FCC Advisory Committee on Advanced Television Service (ACATS), and the Advanced Television Systems Committee (ATSC).

Mr. Weiss has pursued, for over 15 years, use of multiple transmitters by television broadcasters to cover their service areas. Such service can be provided by distributed transmitters, distributed translators, and digital on-channel repeaters, or combinations of them. Use of multiple transmitters requires synchronized with one another. To make such uses practicable, he invented technology that allows synchronization of 8-VSB modulation and recently received a patent on the method. The current ATSC standard for transmitter synchronization (A/110) embodies his technology.

Mr. Weiss is a Fellow of SMPTE and received its 1995 Sarnoff Gold Medal Award and its 2005 Progress Medal. He was the 2006 recipient of the NAB Television Engineering Achievement Award.

Robert D. Weller, P.E., (Chapter 2.4) is a Senior Engineer at Hammett & Edison, Inc., Consulting Engineers, and has been involved with RF safety issues for over 20 years. He is a member of the International Committee on Electromagnetic Safety (Subcommittee IV, Radio Frequencies), the IEEE Committee on Man and Radiation (COMAR), an Associate Member of the Bioelectromagnetics Society, a Full Member of the Association of Federal Communications Consulting Engineers, and a Senior Member of the IEEE.

Mr. Weller is a graduate in electrical engineering and computer science of the University of California at Berkeley, and a Registered Professional Engineer in California and Colorado. He spent nine years at the U.S. Federal Communications Commission.

Mr. Weller's philosophical interests include the interaction of electromagnetic (EM) waves with matter, EM wave propagation, EM modeling, and measurement. He has published a number of papers in these areas, and serves as a peer reviewer for two IEEE publications.

Jerry Westberg (Chapter 4.4) has been at Broadcast Electronics for the past 16 years in the research and development department as a Principal Engineer. He earned Bachelor's and Master's degrees from Western Illinois University and an engineering degree from Southern Illinois University at Edwardsville, Illinois.

Jerry left the world of teaching high school math and physics in 1979. He is best known for his work in the AM phasing and matching design work, where he spent eight years at Harris Broadcast as a design engineer. He holds the patent on 4M Modulation. He also sells software for AM Phasor and Diplex designs.

Jerry Whitaker (Chapters 1.11, 1.16, 5.2, 9.2) is Vice President for standards development for the Advanced Television Systems Committee.

Mr. Whitaker supports the work of the various ATSC technology and planning committees and assists in the development of ATSC Standards and related documents. He currently serves as Secretary of the Technology and Standards Group and Secretary of the Planning Committee, and is closely involved in work relating to educational programs.

Mr. Whitaker is a Fellow the Society of Broadcast Engineers and a Fellow of the Society of Motion Picture and Television Engineers.

Mr. Whitaker is the author and editor of more than 30 books on technical topics, including *The Standard Handbook of Video and Television Engineering*, 4th edition, *NAB Engineering Handbook*, 9th edition, *DTV Handbook*, 3rd edition, and *The Electronics Handbook*, 2nd edition. Prior to joining the ATSC, Mr. Whitaker headed the publishing company Technical Press, based in Morgan Hill, California. He has served as a Board Member and Vice President of the Society of Broadcast Engineers.

Danny Wilson (Chapter 8.2) is the founder, CEO, and President of Pixelmetrix Corporation, a Singapore-based company that specializes in the design and production of management and telemetry systems for digital broadcasters. The company's award-winning products are deployed globally at numerous terrestrial, satellite, cable, and IPTV operators.

Mr. Wilson began his management career at Hewlett-Packard. Initially based in Canada as the Business Manager of HP's Communication Measurement Division, he was responsible for the introduction of the MPEGScope Transport Stream Analyzer as well as the world's first ATM/B-ISDN Test System, which accelerated the development and deployment of ATM technology worldwide. He later went on with HP to Kobe, Japan, where he established and managed HP's Asia Business Centre, leading the development of the world's first multi-port network monitoring system, which was accepted and implemented throughout Japan by telecom giant NTT.

Mr. Wilson is a frequent speaker at IPTV and television conferences in Europe, Asia, and North America. He currently chairs the IPTV Quality of Service working group within ITU-T/IPTV-FG. Born in Edmonton, Canada, he holds a computer engineering degree from the University of Alberta.

David Wilson (Chapter 1.18) is Director, Engineering and Standards at the Consumer Electronics Association in Arlington, Virginia, where he oversees CEA's work on various technical activities affecting consumer electronics products. He joined CEA in 2000 after six years at the National Association of Broadcasters, where he was Manager of Technical Regulatory Affairs.

Prior to NAB, Mr. Wilson spent four and a half years in the Office of Engineering and Technology at the Federal Communications Commission, providing engineering support on equipment authorization and spectrum allocation issues. Before that, he spent six years at WUVA–FM in Charlottesville, Virginia, serving for several years as the station's Chief Engineer before becoming its General Manager and President.

Mr. Wilson holds a BS degree in electrical engineering and an MS degree in accounting, both from the University of Virginia. He is also an SBE-certified Broadcast Engineer.

David Wood (Chapter 1.14) is Head of New Media for the European Broadcasting Union (EBU), Geneva, Switzerland.

The EBU is the collective organization for Europe's 65 national broadcasters, and has a further 60 associate members from the rest of the world. It acts in matters of common interest to its members in television, radio, and multimedia, in sports coverage, news coverage, music, drama, and documentary, and in legal and technical matters.

David has chaired new media coordination activities for the EU IST program for many years, and chaired several Working Groups in the ITU. He is particularly interested in the future success of radio as a media form in the new media environment, and in the evolution of audio components of television.

David was educated at Southampton University in the UK, the UNIIRT in Odessa, and the Harvard Business School in the United States. He worked for the BBC and the IBA in the UK before joining the EBU.

Authors whose material was used from the 9th Edition and not listed above are:

Edward J. Anthony (Chapter 4.2)

Lynn Claudy (Chapter 3.2)

James H. Cook, Jr. (Chapter 6.11)

Eric Dye (Chapter 4.9)

Clifford D. Ferris (Chapter 2.6)

J.J. Gibson (Chapter 1.13)

Earnest Hickin (Chapter 9.1)

Randall Hoffner (Chapter 1.15)

Chip Morgan (Chapter 1.15)

Peter K. Onnigian (Chapter 4.9)

D.H. Pritchard (Chapter 1.13)

James H. Rooney, III (Chapter 9.1)

Greg Silsby (Chapter 3.3)

Edmund A. Williams (Chapter 6.3)

Scott A. Wright (Chapter 4.14)

Fred Wylie (Chapter 3.7)

NAB Engineering Achievement Award Recipients

Since 1959 the National Association of Broadcasters annually recognizes broadcast industry engineers for outstanding achievements during their distinguished professional careers and for significant contributions that have advanced the state of the art of broadcast engineering.

In 1991 a second award was developed in order to separately recognize achievements in Radio and Television engineering and, on occasion, a third award is given to recognize individuals for their Service to Broadcast Engineering.

The awards are presented during the Technology Luncheon at the annual NAB Convention.

For more information on the award and how nominations may be made see: www.nab.org

1959 John T. Wilner, Vice President of Engineering, Hearst Corporation, Baltimore, Maryland

1960 Commissioner T. A. M. Craven, Federal Communications Commission, Washington, District of Columbia

1961 Raymond F. Guy, Consultant

1962 Ralph N. Harmon, Vice President for Engineering, Westinghouse Broadcasting Co., New York, New York

1963 Dr. George R. Town, Dean of Engineering, Iowa State University, Ames, Iowa

1964 John H. DeWitt, Jr., President, WSM, Inc., Nashville, Tennessee

1965 Edward W. Allen, Jr., Chief Engineer, Federal Communications Commission,Washington, District of Columbia

1966 Carl J. Meyers, Senior Vice President and Director of Engineering, WGN Continental Broadcasting Co., Chicago, Illinois

1967 Robert M. Morris, Staff Consultant, Engineering Department, American Broadcasting Company, New York, New York

1968 Howard A. Chinn, Director, General Engineering, CBS Television Network, New York, New York

1969 Jarrett L. Hathaway, Senior Project Engineer, NBC Television Network, New York, New York

1970 Philip Whitney, General Manager, WINC, Winchester, Virginia, and SupervisoryEngineer for Richard F. Lewis radio stations

1971 Benjamin Wolfe, Vice President Engineering, Post–Newsweek Stations, Washington, District of Columbia

1972 John M. Sherman, Director of Engineering, WCCO, Minneapolis, Minnesota

1973 A. James Ebel, President and General Manager, KOLN–TV, Lincoln, Nebraska

1974 Joseph B. Epperson, Vice President Engineering, Scripps Howard Broadcasting Co., Cleveland, Ohio

1975 John D. Silva, Director, Research and Development, Golden West Broadcasters, Los Angeles, California

1976 Dr. Frank G. Kear, Consulting Engineer, Washington, District of Columbia

1977 Daniel H. Smith, Senior Vice President for Engineering, Capital Cities Communications, Inc., Philadelphia, Pennsylvania

1978 John A. Moseley, President, Moseley Associates, Inc., Goleta, California

1979 Robert W. Flanders, Vice President and Director of Engineering, McGraw–HillBroadcasting Co., Inc., Indianapolis, Indiana

1980 James D. Parker, Staff Consultant, Telecommunications, CBS Television Network, New York, New York

1981 Wallace E. Johnson, Executive Director, Association for Broadcast EngineeringStandards, Washington, District of Columbia

1982 Julius Barnathan, President, Broadcast Operations and Engineering, AmericanBroadcasting Companies (ABC), Inc., New York, New York

1983 Joseph Flaherty, Vice President, Technology, CBS Inc., New York, New York

1984 Otis S. Freeman, Vice President and Director of Engineering, WPIX, Inc., TribuneBroadcasting, New York, New York

1985 Carl E. Smith, President, Smith Electronics, Cleveland, Ohio

1986 Dr. George Brown, RCA Laboratories, Princeton, New Jersey

1987 Renville H. McMann, CBS Technology Center, Stamford, Connecticut

1988 Jules Cohen, Jules Cohen and Associates, Washington, District of Columbia

1989 William Connolly, President, Sony Advanced Systems, Montvale, New Jersey

1990 Hilmer Swanson, Senior Staff Scientist, Harris Corporation, Broadcast Division, Quincy, Illinois

1991 George Marti, President and CEO, Marti Electronics, Cleburne, Texas (Radio)

Kerns Powers, David Sarnoff and NBC Consultant, Princeton, New Jersey (Television)

1992 Edward Edison and Robert L. Hammett, Hammett & Edison, San Francisco, California (Radio)

James C. McKinney, Chairman, Advanced Television Systems Committee, Washington, District of Columbia (Television)

1993 Robert M. Silliman, Silliman and Silliman, Silver Spring, Maryland (Radio)

Stanley N. Baron, Managing Director, Technical Development, NBC, New York, New York (Television)

Herb H. Schubarth, Vice President of Engineering, Gannett Broadcasting, Denver, Colorado (Service to Broadcasting Engineering)

1994 Charles T. Morgan, Vice President and Director of Engineering, Susquehanna Radio Corporation, York, Pennsylvania (Radio)

Thomas J. Vaughan, President, PESA Micro Communications, Inc., Manchester, New Hampshire (Television)

1995 Robert Orban, Chief Engineer, AKG Acoustics, Inc. San Leandro, California (Radio)

Carl G. Eilers, Manager of Electronic Systems R&D, Zenith Electronics, Glenview, Illinois (TV)

1996 Ogden Prestholdt, A.D. Ring & Associates, Nakomis, Florida (Radio)

Charles Rhodes, Advanced Television Test Center, Alexandria, Virginia (Television)

Gerald Robinson, Hearst Broadcasting, Milwaukee, Wisconsin (Service to Broadcast Engineering)

1997 George Jacobs, George Jacobs & Associates, Silver Spring, Maryland (Radio)

Michael Sherlock, NBC, New York, New York (Television)

1998 John Battison, P.E., John Battison, Consultant, Loudonville, Ohio (Radio)

Robert Hopkins, Sony Pictures High Definition Center, Culver City, California (Television)

1999 Geoffrey Mendenhall, P.E. Harris Corporation, Quincy, Illinois (Radio)

John Turner, Turner Engineering, Mountain Lakes, New Jersey (Television)

2000 Michael Dorrough, Dorrough Electronics, Woodland Hills, California (Radio)

Max Berry, Capital Cities/ABC Elkins Park, Pennsylvania (Television)

2001 Arno Meyer, Belar Electronics Laboratory, Devon, Pennsylvania (Radio)

Larry Thorpe, Sony Electronics, Inc., Park Ridge, New Jersey (Television)

2002 Paul Schafer, Schafer International, Bonita, California (Radio)

Bernard Lechner, Consultant, Princeton, New Jersey (Television)

2003 John W. Reiser, FCC, Mt. Vernon, Virginia (Radio)

Robert P. Eckert, FCC, Washington, District of Columbia (Television)

2004 Ira Goldstone, Tribune Broadcasting, Los Angeles, California (Television)

E. Glynn Walden, Infinity Broadcasting, New York, New York (Radio)

2005 Dr. Oded Bendov, TV Transmission Antenna Group, Inc. Cherry Hill, New Jersey (Television)

Milford K. Smith, Greater Media Inc. Braintree, Massachusetts (Radio)

2006 Benjamin F. Dawson, III, P.E., Hatfield & Dawson Consulting Engineers, Seattle, Washington (Radio)

Ronald D. Rackley, P.E., du Treil, Lundin & Rackley, Sarasota, Florida (Radio)

S. Merrill Weiss, Merrill Weiss Group, LLC, Metuchen, New Jersey (Television)

2007 Victor Tawil, Senior Vice President of the Association for Maximum Service Television (MSTV), Washington, D.C. (Television)

Louis A. King, Founder and Chairman, Kintronic Labs, Inc. Bristol, Tennessee (Radio)

SECTION

1

BROADCAST ADMINISTRATION, STANDARDS, AND TECHNOLOGIES

CHAPTER

1.1

The Electromagnetic Spectrum

JOHN NORGARD
Air Force Research Laboratory
Rome, New York

INTRODUCTION

The electromagnetic (EM) spectrum consists of all forms of EM radiation, from DC to light to gamma rays. A chart of the EM spectrum can be arranged in order of frequency or wavelength into a number of regions,[1] usually wide in extent, within which the EM waves have some specified common characteristics (*e.g.*, those characteristics relating to the production or detection of the radiation). A common example is the spectrum of the radiant energy in the region referred to as *white light*, which when dispersed by a prism will produce a rainbow of its constituent colors.

The EM spectrum is typically displayed as a function of frequency (or wavelength), as shown schematically in Figure 1.1-1. In air, frequency (f) and wavelength (λ) are inversely proportional ($f = c/\lambda$, where $c = 2.998 \times 10^8 \approx 3 \times 10^8$ m/s is the speed of light in a vacuum). The meter-kilogram-second (MKS) unit of frequency is the hertz (Hz, where 1 Hz = 1 cycle per second); the MKS unit of wavelength is the meter.

Frequency is also measured in the following subunits:

- Kilohertz (1 kHz = 10^3 Hz)
- Megahertz (1 MHz = 10^6 Hz)
- Gigahertz (1 GHz = 10^9 Hz)
- Terahertz (1 THz = 10^{12} Hz)
- Petahertz (1 PHz = 10^{15} Hz)
- Exahertz (1 EHz = 10^{18} Hz)

Electromagnetic energy at a particular frequency f has a photon energy (E) associated with it as follows:

$$E = hf \text{ (units are electron volts [eV])}$$

where h is Planck's constant ≈ 4.13567 μeV/GHz, and f is the frequency (in Hz).

Wavelength is also measured in the following subunits:

- Centimeters (1 cm = 10^{-2} m)
- Millimeters (1 mm = 10^{-3} m)
- Micrometers (microns) (1 μm = 10^{-6} m)
- Nanometers (1 nm = 10^{-9} m)
- Ångstroms (1 Å = 10^{-10} m)
- Picometers (1 pm = 10^{-12} m)
- Femtometers (1 fm = 10^{-15} m)
- Attometers (1 am = 10^{-18} m)

SPECTRAL SUBREGIONS

For convenience, in this chapter the overall EM spectrum is divided into three main subregions:

- *DC to light* spectrum
- *Optical* spectrum (this spectrum is treated first)
- *Light to gamma ray* spectrum

[1]Note that specific frequency ranges are often called *bands*; several contiguous frequency bands are usually called *spectrums*; and subfrequency ranges within a band are sometimes called *segments*.

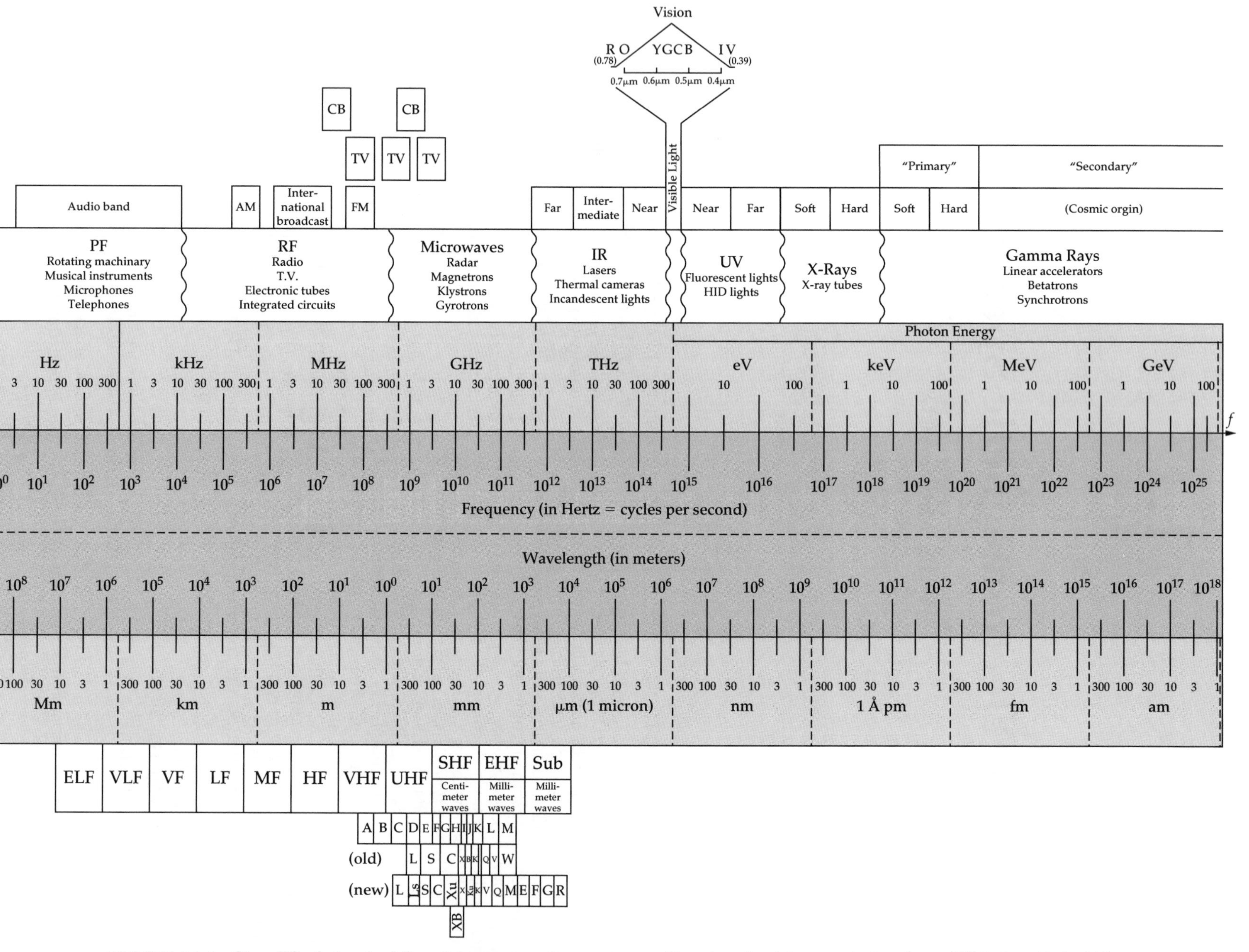

FIGURE 1.1-1 Simplified chart of the electromagnetic spectrum. (Reprinted with permission from Whitaker, J. C., Ed., *The Electronics Handbook*, CRC Press, Boca Raton, FL, 1996.)

Note that the boundaries between some of the spectral regions are somewhat arbitrary. Some spectral bands have no sharp edges and merge into each other, and some spectral segments overlap each other slightly.

Optical Spectrum

The optical spectrum is the middle frequency/wavelength region of the EM spectrum. It is defined here as the visible and near-visible regions of the EM spectrum and includes the *infrared*, *visible*, and *ultraviolet* bands (listed by wavelength/frequency):

- *Infrared* (IR) band, ≈300–0.7 μm/≈1–429 THz
- *Visible light* band, ≈0.7–0.4 μm/≈429–750 THz
- *Ultraviolet* (UV) band, ≈0.4 μm–≈10 nm/≈750 THz–≈30 PHz

Because frequencies in the optical spectrum are so high, these regions of the EM spectrum are usually described in terms of their wavelengths. Atomic and molecular radiation produce radiant light energy. Atomic radiation (outer shell electrons) and radiation from arcs and sparks produce EM waves in the UV band. Molecular radiation and radiation from hot bodies produce EM waves in the IR band.

Visible Light Band

The visible light band, in the middle of the optical spectrum, extends in wavelength from approximately 0.4 μm (violet) to 0.7 μm (red) and in frequency from approximately 750 THz (violet) to 429 THz (red). EM radiation in this region of the EM spectrum, when entering the eye, gives rise to visual sensations (colors), according to the spectral response of the eye, which responds only to radiant energy in the visible light band extending from the extreme long wavelength edge of red to the extreme short wavelength edge of violet.[2] This visible light band is further subdivided into the various colors of the rainbow (listed in decreasing wavelength/increasing frequency):

- Red—a primary color;[3] peak intensity at 700.0 nm (429 THz)
- Orange
- Yellow
- Green—a primary color; peak intensity at 546.1 nm (549 THz)
- Cyan
- Blue—a primary color; peak intensity at 435.8 nm (688 THz)
- Indigo
- Violet

[2]The spectral response of the eye is sometimes described as extending from 0.38 μm (violet) to 0.75 or 0.78 μm (red) (*i.e.*, from 789 THz to 400 or 385 THz).

[3]*Primary colors* are those that cannot be created by mixing other colors. The three *additive* primary colors are red, green, and blue, and the three *subtractive* primary colors are magenta, yellow, and cyan.

IR Band

The IR band is the region of the EM spectrum lying immediately below the visible light band in frequency. The IR band consists of EM radiation with wavelengths extending between the longest visible red, approximately 0.7 μm (429 THz), and the shortest microwaves, 300 μm–1 mm (1 THz–300 GHz). The IR band is further subdivided in wavelengths into the *near* (shortwave), *intermediate* (midwave), and *far* (longwave) IR segments (listed by wavelength/frequency):[4]

- *Near* IR segment, 0.7–3 μm/429–100 THz
- *Intermediate* IR segment, 3–7 μm/100–42.9 THz
- *Far* IR segment, 7–300 μm/42.9–1 THz

Note that the submillimeter region of wavelengths is sometimes included in the very far region of the IR band:

- *Submillimeter* band, 100 μm–1 mm/3 THz–300 GHz

In addition to emanating from electronic devices specifically designed for EM radiation purposes, EM radiation is produced in all matter by the oscillation and rotation of the molecules and atoms of which that matter is comprised; therefore, all objects at temperatures above absolute zero emit EM radiation by virtue of their thermal motion (warmth) alone. Objects near room temperature emit most of their radiation in the IR band; however, even relatively cool objects emit some IR radiation, and hot objects, such as incandescent filaments, emit strong IR radiation.

IR radiation is sometimes incorrectly referred to as *radiant heat*, because warm bodies emit IR radiation and bodies that absorb IR radiation are warmed; however, IR radiation is not itself heat. This radiant energy is more properly referred to as *black body radiation*. Such waves are emitted by all material objects; for example, background cosmic radiation (≈2.7 K) emits microwaves, room temperature objects (≈293 K) emit IR rays, the sun (≈6000 K) emits yellow light, and the solar corona (≈1 million K) emits X-rays.

IR astronomy uses the 1 μm to 1 mm portion of the IR band to study celestial objects by their IR emissions. IR detectors are used in night vision systems, intruder alarm systems, weather forecasting, and missile guidance systems. IR photography uses multilayered color film, with an IR-sensitive emulsion in the wavelengths between 700 and 900 nm, for medical and forensic applications and for aerial surveying.

UV Band

The UV band is the region of the EM spectrum lying immediately above the visible light band in frequency. The UV band consists of EM radiation with wavelengths extending between the shortest visible violet

[4]Some reference texts use 2.5 μm (120 THz) as the breakpoint between the near and the intermediate IR bands and 10 μm (30 THz) as the breakpoint between the intermediate and far IR bands. Also, 15 μm (20 THz) is sometimes considered as the long wavelength end of the far IR band.

(≈0.4 μm) and the longest x-rays (≈10 nm);[5] that is, from 750 THz (≈3 eV) to ≈30 PHz (≈100 eV). The UV band is further subdivided in frequency into the *near* and the *far*[6] UV segments (listed by wavelength/frequency/photon energy):

- *Near* UV segment, ≈0.4 μm–100 nm/≈750 THz–3 PHz/≈3–10 eV
- *Far* UV segment, 100–≈10 nm/3–≈30 PHz/≈10–100 eV

UV radiation is produced by electron transitions in atoms and molecules, as in a mercury discharge lamp. Radiation in the UV range is easily detected, can cause fluorescence in some substances, and can produce photographic and ionizing effects. In UV astronomy, the emissions of celestial bodies in the wavelength band between 50 and 320 nm are detected and analyzed to study the heavens. The hottest stars emit most of their radiation in the UV band.

DC to Light

Below the IR band are the lower frequency (longer wavelength) regions of the EM spectrum, subdivided generally into the *microwave*, *radiofrequency*, and *power* spectral bands (listed by frequency/wavelength):

- *Microwave* band,[7] 300 GHz–300 MHz/1 mm–1 m
- *Radiofrequency* (RF) band, 300 MHz–10 kHz/1 m–30 km
- *Power frequency* (PF)/*telephony* band, 10 kHz–DC/30 km–∞

These regions of the EM spectrum are usually described in terms of their frequencies. EM radiation for which the wavelengths are of the order of millimeters and/or centimeters are called *microwaves*, and those still longer are *radiofrequency* (RF) or *Hertzian* waves. Radiation from electronic devices produces EM waves in both the microwave and the RF bands. Power frequency energy is generated by rotating machinery. Direct current (DC) is produced by batteries or rectified alternating current (AC).

Microwave Band

The microwave band is the region of wavelengths between the far IR/submillimeter region and the conventional RF region. The boundaries of the microwave band have not been definitively fixed but are commonly regarded as the region of the EM spectrum extending from about 1 mm to 1 m (300 GHz–300 MHz). The microwave band is further subdivided into *centimeter* and *millimeter* segments (listed by frequency/wavelength):

- *Millimeter* waves,[8] 300–30 GHz/1 mm–1 cm (EHF band)
- *Centimeter* waves, 30–3 GHz/1–10 cm (SHF band)

The microwave band usually includes the UHF band from 3 GHz to 300 MHz (10 cm–1 m). Microwaves are used in radar, in communication links spanning moderate distances as radio carrier waves in radio broadcasting, and for mechanical heating and cooking (*e.g.*, in microwave ovens).

Radiofrequency (RF) Band

The RF range of the EM spectrum is the wavelength band suitable for utilization in radio communications from 10 kHz to 300 MHz (30 km–1 m).[9] Some radio waves serve as the carriers of low-frequency audio signals; other radio waves are modulated by video and digital information. The *amplitude modulated* (AM) broadcasting band uses waves with frequencies between 550 and 1705 kHz; the *frequency modulated* (FM) broadcasting band uses waves with frequencies between 88 and 108 MHz. In the United States, the Federal Communications Commission (FCC) is responsible for assigning a range of frequencies (*e.g.*, a frequency band in the RF spectrum) to a broadcasting station or service. The International Telecommunication Union (ITU) coordinates frequency band allocation and coordination on a worldwide basis. Radio astronomy uses radio telescopes to receive and study radio waves naturally emitted by objects in space. Radio waves are emitted from hot gases (thermal radiation), from charged particles spiraling in magnetic fields (synchrotron radiation), and from excited atoms and molecules in space (spectral lines), such as the 21 cm line emitted by hydrogen gas.

Power Frequency (PF)/Telephone Band

The PF range of the EM spectrum is the wavelength band suitable for generating, transmitting, and consuming low-frequency *prime* power (*e.g.*, electrical power provided by electric utility companies), extending from 10 kHz to DC (zero frequency; 30 km to ∞ wavelength). In the United States, most prime power is generated at 60 Hz (some military and computer applications use 400 Hz); in Europe, prime power is generated at 50 Hz.

FREQUENCY BAND DESIGNATIONS

The combined microwave, radiofrequency, and power/telephone spectra are subdivided into the specific bands shown in Table 1.1-1. The U.S. military triservice designations for radio communication bands are shown in Table 1.1-2. Another set of designations

[5] Some reference texts use 4, 5, or 6 nm as the upper edge of the UV band.

[6] The far UV band is also referred to as the *vacuum* UV band, because air is opaque to all UV radiation in this region.

[7] Some reference works define the lower edge of the microwave spectrum at 1 GHz.

[8] Some reference articles consider the top edge of the millimeter region to stop at 100 GHz.

[9] Some authors consider the RF band to extend from 10 kHz to 300 GHz, with the microwave band as a subset of the RF band from 300 MHz to 300 GHz.

TABLE 1.1-1
International Radio Frequency Band Designations and Numerical Designations

Band Name	Band Designation	Frequency Range	Wavelength Range
Extremely low frequency	ELF (1)	3 Hz–30 Hz	100,000 km–10,000 km
Super low frequency	SLF (2)	30 Hz–300 Hz	10,000 km–1000 km
Ultra low frequency	ULF (3)	300 Hz–3 kHz	1000 km–100 km
Very low frequency	VLF (4)	3 kHz–30 kHz	100 km–10 km
Low frequency	LF (5)	30 kHz–300 kHz	10 km–1 km
Medium frequency	MF (6)	300 kHz–3 MHz	1 km–100 m
High frequency	HF (7)	3 MHz–30 MHz	100 m–10 m
Very high frequency	VHF (8)	30 MHz–300 MHz	10 m–1 m
Ultra high frequency	UHF (9)	300 MHz–3 GHz	1 m–10 cm
Super high frequency	SHF (10)	3 GHz–30 GHz	10 cm–1 cm
Extremely high frequency	EHF (11)	30 GHz–300 GHz	1 cm–1 mm
Tremendously high frequency	THF*(12)	300 GHz–3 THz	1 mm–100 μm

*THF is not a universally accepted designation for this band.

that predate the U.S. tri-service designations are given in Table 1.1-3. The European Community (EC) radar band designations in prior use are listed in Table 1.1-4. An alternative and more detailed subdivision of the UHF (9), SHF (10), and EHF (11) bands is shown in Table 1.1-5. Several other frequency bands of interest (not exclusive) are shown in Tables 1.1-6 and 1.1-7.

A comprehensive and informative chart of the U.S. frequency allocations of the radio spectrum has been prepared by the Office of Spectrum Management (OSM) of the National Telecommunications and Information Administration (NTIA) of the U.S. Department of Commerce (DoC). The chart graphically partitions the radiofrequency spectrum (3 kHz–300 GHz) into

TABLE 1.1-2
Current U.S. Tri-Service Radar Band Designations

Band Designation	Frequency Range	Wavelength Range
A	0 Hz–250 MHz	∞–1.2 m
B	250 MHz–500 MHz	1.2 m–60 cm
C	500 MHz–1 GHz	60 cm–30 cm
D	1 GHz–2 GHz	30 cm–15 cm
E	2 GHz–3 GHz	15 cm–10 cm
F	3 GHz–4 GHz	10 cm–7.5 cm
G	4 GHz–6 GHz	7.5 cm–5 cm
H	6 GHz–8 GHz	5 cm–3.75 cm
I	8 GHz–10 GHz	3.75 cm–3 cm
J	10 GHz–20 GHz	3 cm–1.5 cm
K	20 GHz–40 GHz	1.5 cm–7.5 mm
L	40 GHz–60 GHz	7.5 mm–5 mm
M	60 GHz–100 GHz	5 mm–3 mm
N	100 GHz–200 GHz	3 mm–1.5 mm
O	200 GHz–300 GHz	1.5 mm–1 mm

TABLE 1.1-3
Designations Predating the U.S. Tri-Service Designations

Band Designation	Frequency Range	Wavelength Range
I	100 MHz–150 MHz	3 m–2 m
G	150 MHz up to 225 MHz	2 m–1.33 m
P	225 MHz–390 MHz	1.33 m–76.9 cm
L	390 MHz–1.5 GHz	76.9 cm–19.4 cm
S*	1.5 GHz–3.9 GHz	19.4 cm–7.69 cm
C*	3.9 GHz–6.2 GHz	7.69 cm–48.4 mm
X*	6.2 GHz–10.9 GHz	48.4 mm–27.5 mm
K	10.9 GHz–36 GHz	27.5 m–8.33 mm
Q	36 GHz–46 GHz	8.33 mm–6.52 mm
V	46 GHz–56 GHz	6.52 mm–5.36 mm
W	56 GHz–	5.36 mm–

*An alternative prior S band designation extended from 1.5 GHz to 5.85 GHz, and similarly a prior X band extended from 5.85 GHz to 10.9 GHz, eliminating the C band in that designation.

TABLE 1.1-4
European Community Radar Band Designations in Prior Use

Band Designation	Frequency Range	Wavelength Range
L	1–2 GHz	30 cm–15 cm
S	2–4 GHz	15 cm–7.5 cm
C	4–8.2 GHz	7.5 cm–3.66 cm
X	8.2–12.4 GHz	3.66 cm–2.42 cm
Ku	12.4–18 GHz	2.42 cm–1.67 cm
K*	18–26.5 GHz	1.67 cm–11.3 mm
Ka*	26.5–40 GHz	11.3 cm–7.5 mm
V	40–75 GHz	7.5 mm–4 mm
W	75–110 GHz	4 mm–2.73 mm
mm	110–300 GHz	2.73 mm–1 mm

*The prior K band sometimes included the Ka band and extended from 18 to 40 GHz.

TABLE 1.1-5
Alternate and More Detailed Subdivision of the UHF (9), SHF (10), and EHF (11) Bands

Band Designation	Frequency Range	Wavelength Range
L	1.12–1.7 GHz	26.8 cm–17.6 cm
LS	1.7–2.6 GHz	17.6 cm–11.5 cm
S	2.6–3.95 GHz	11.5 cm–7.59 cm
C(G)	3.95–5.85 GHz	7.59 cm–5.13 cm
XN(J, XC)	5.85–8.2 GHz	5.13 cm–3.66 cm
XB(H, BL)	7.05–10 GHz	4.26 cm–3 cm
X	8.2–12.4 GHz	3.66 cm–2.42 cm
Ku(P)	12.4–18 GHz	2.42 cm–1.67 cm
K	18–26.5 GHz	1.67 cm–1.13 cm
V(R, Ka)	26.5–40 GHz	1.13 cm–7.5 mm
Q(V)	33–50 GHz	9.09 mm–6 mm
M(W)	50–75 GHz	6 mm–4 mm
E(Y)	60–90 GHz	5 mm–3.33 mm
F(N)	90–140 GHz	3.33 mm–2.14 mm
G(A)	140–220 GHz	2.14 mm–1.36 mm
R	220–325 GHz	1.36 mm–0.923 mm

TABLE 1.1-6
Power Band

Band Designation	Frequency Range
Subsonic band	0 Hz–10 Hz
Audio band	10 Hz–10 kHz
Ultrasonic band	10 kHz and up

over 450 frequency bands and uses distinct colors to distinguish the allocations for 30 different radio services and for 3 different radio activities. The chart presents a graphical summary of the detailed allocations contained in the U.S. Table of Frequency Allocations found in

- The *NTIA Manual* (*Manual of Regulations and Procedures for Federal Radio Frequency Management*, Chapter 4)
- The FCC Rules (47 CFR, Part 2)

This radiofrequency allocation chart may be viewed online at the Web site (http://www.ntia.doc.gov/osmhome/allochrt.html), and printed copies of this chart are available for a nominal fee from the U.S. Government Printing Office (telephone: 202-512-1800; Stock No. 003-000-00691-3). In the chart, the radio spectrum is divided into the following three subareas according to radio activity:

- Government exclusive (NTIA)
- Government/nongovernment shared (NTIA/FCC)
- Nongovernment exclusive (FCC)

The chart is also subdivided into the following 30 subareas according to radio service:

- Aeronautical mobile
- Aeronautical mobile satellite

TABLE 1.1-7
RF Band

Band Designation	Frequency Range
Longwave broadcasting	150–290 kHz
AM broadcasting	550–1705 kHz
International broadcasting	3–30 MHz
Shortwave broadcasting (8 bands)	5.95–26.1 MHz
VHF TV (channels 2–4)	54–72 MHz
VHF TV (channels 5–6)	76–88 MHz
FM broadcasting	88–108 MHz
VHF TV (channels 7–13)	174–216 MHz
UHF TV (channels 14–51)*	512–698 MHz

*As part of the U.S. transition to digital television, TV channels 52 through 69 (698–806 MHz) were removed from the TV band, leaving a core TV band that contains channels 2 through 51.

- Aeronautical radionavigation
- Amateur
- Amateur satellite
- Broadcasting
- Broadcasting satellite
- Earth exploration satellite
- Fixed
- Fixed satellite
- Inter-satellite
- Land mobile
- Land mobile satellite
- Maritime mobile
- Maritime mobile satellite
- Maritime radionavigation
- Meteorological aids
- Meteorological satellite
- Mobile
- Mobile satellite
- Radio astronomy
- Radiodetermination satellite
- Radiolocation
- Radiolocation satellite
- Radionavigation
- Radionavigation satellite
- Space operations
- Space research
- Standard frequency and time signal
- Standard frequency and time signal satellite

Light to Gamma Rays

Above the UV spectrum are the higher frequency (shorter wavelength) regions of the EM spectrum, subdivided generally into the *x-ray* and *gamma ray* spectral bands (listed by electron voltage/wavelength/frequency):[10]

- *X-ray* band, ≈10 eV–1 MeV/≈10 nm–≈1 pm/ ≈3 PHz–≈300 EHz
- *Gamma ray* band, ≈1 keV–∞/≈300 pm–0 m/ ≈1 EHz–∞

[10]Note that *cosmic rays* (from astronomical sources) are not EM waves (rays) and, therefore, are not part of the EM spectrum. Cosmic rays are high-energy charged particles (electrons, protons, and ions) of extraterrestrial origin moving through space which may have energies as high as 10^{20} eV. Cosmic rays have been traced to cataclysmic astrophysical or cosmological events, such as exploding stars and black holes. Cosmic rays are emitted by supernova remnants, pulsars, quasars, and radio galaxies. Cosmic rays that collide with molecules in the Earth's upper atmosphere produce secondary cosmic rays and gamma rays of high energy which also contribute to the natural background radiation. These gamma rays are sometimes called *cosmic* or *secondary* gamma rays. Cosmic rays are a useful source of high-energy particles for experiments.

These regions of the EM spectrum are usually described in terms of their photon energies in electron volts. Note that the bottom of the gamma ray band overlaps the top of the X-ray band. Radiation from atomic inner shell excitations produces EM waves in the X-ray band. Radiation from naturally radioactive nuclei produces EM waves in the gamma ray band.

X-Ray Band

The X-ray band is further subdivided into *soft* and *hard* X-rays (listed by photon energy/wavelength/frequency):

- *Soft* X-rays, ≈10 eV–10 keV/≈ 10 nm–100 pm/ ≈3 PHz–3 EHz
- *Hard* X-rays, ≈10 keV–1MeV /100 pm–≈1 pm/ 3 EHz–≈300 EHz

Because the physical nature of these rays was at first unknown, this radiation was named *X-rays* by Wilhelm Roentgen, the German scientist who discovered them (with *X* being used as the symbol for an unknown quantity). The more powerful X-rays are the hard X-rays, which are of high frequencies and, therefore, more energetic; less powerful X-rays are the soft X-rays, which have lower energies.

X-rays are produced by transitions of electrons in the inner levels of excited atoms or by rapid deceleration of charged particles (*Brehmsstrahlung*, or "breaking radiation"). An important source of X-rays is synchrotron radiation. X-rays can also be produced when high-energy electrons from a heated filament cathode strike the surface of a target anode (usually tungsten) between which a high alternating voltage (approximately 100 kV) is applied.

X-rays are a highly penetrating form of EM radiation, and applications of X-rays are based on their short wavelengths and their ability to easily pass through matter. X-rays are very useful in crystallography for determining crystalline structure and in medicine for photographing the body. Because different parts of the body absorb X-rays to a different extent, X-rays passing through the body provide a visual image (negative) of its interior structure when striking a photographic plate. X-rays are dangerous and can destroy living tissue. They can also cause severe skin burns. X-rays are useful in the diagnosis and nondestructive testing of products for defects.

Gamma Ray Band

The gamma ray band is subdivided into *primary* and *secondary* gamma ray segments (listed by photon energy/wavelength/frequency):

- *Primary* gamma rays, ≈1 keV–1 MeV/≈ 300 pm–300 fm/≈1 EHz–1000 EHz
- *Secondary* gamma rays, ≈1 MeV–∞/≈300 fm–0 m/ 1000 EHz–∞

Secondary gamma rays are created from collisions of high-energy cosmic rays with particles in the Earth's upper atmosphere. The primary gamma rays are further subdivided into *soft* and *hard* gamma ray segments (listed by photon energy/wavelength/frequency):

- *Soft* gamma rays, ≈1 keV–≈300 keV/≈300 pm–≈3 pm/≈1 EHz–≈100 EHz
- *Hard* gamma rays, ≈300 keV–1 MeV/≈3 pm–300 fm/≈100 EHz–1000 EHz

Gamma rays are essentially very energetic X-rays. The distinction between the two is based on their origin. X-rays are emitted during atomic processes involving energetic electrons; gamma rays are emitted by excited nuclei or other processes involving subatomic particles.

Gamma rays are emitted by the nucleus of radioactive material during the process of natural radioactive decay as a result of transitions from high-energy excited states to low-energy states in atomic nuclei. Cobalt 90 is a common gamma ray source (with a half-life of 5.26 years). Gamma rays are also produced by the interaction of high-energy electrons with matter. Cosmic gamma rays cannot penetrate the Earth's atmosphere.

Applications of gamma rays are found both in medicine and in industry. In medicine, gamma rays are used for cancer treatment, diagnosis, and prevention. Gamma ray emitting radioisotopes are used as tracers. In industry, gamma rays are used in the inspection of castings, seams, and welds.

Defining Key Terms

- *Cosmic rays*—Highly penetrating particle rays from outer space. Primary cosmic rays (particles) that enter the Earth's upper atmosphere consist mainly of protons. Cosmic rays of low energy have their origin in the sun, those of high energy in galactic or extragalactic space, possibly as a result of supernova explosions. Collisions with atmospheric particles result in secondary cosmic rays (particles) and secondary gamma rays (EM waves).
- *Electromagnetic spectrum*—EM radiant energy arranged in order of frequency or wavelength and divided into regions within which the waves have some common specified characteristics (e.g., the waves are generated, received, detected, or recorded in a similar way).
- *Gamma rays*—Electromagnetic radiation of very high energy (<30 keV) emitted after nuclear reactions or by a radioactive atom when its nucleus is left in an excited state after emission of alpha or beta particles.
- *Infrared (IR) radiation*—Electromagnetic radiation with wavelengths in the range of 0.7 nm (the long wavelength limit of visible red light) to 1 mm (the shortest microwaves). A convenient subdivision is as follows: *near*, 0.7 µm to 2–5 µm; *intermediate*, 2–5 µm to 10 µm; *far*, 10 µm to 1 mm.
- *Light*—White light, when split into a spectrum of colors, is composed of a continuous range of merging colors: red, orange, yellow, green, cyan, blue, indigo, and violet.
- *Microwaves*—Electromagnetic waves that have wavelengths between approximately 0.3 cm (or 1 mm) and 30 (or 10) cm, corresponding to frequencies between 1 GHz and 100 GHz. Note that there are no well-defined boundaries distinguishing microwaves from infrared or radio waves.
- *Radio waves*—Electromagnetic radiation suitable for radio transmission in the range of frequencies from about 10 kHz to about 300 MHz.
- *Ultraviolet (UV) radiation*—Electromagnetic radiation with wavelengths in the range of 0.4 nm (the shortest wavelength limit of visible violet light) to 3 nm (the longest X-rays). A convenient subdivision is as follows: *near*, 0.4 µm to 100 nm; *far*, 100 nm to 3 nm.
- *X-rays*—Electromagnetic radiation of short wavelengths (3 nm to 30 pm) produced when cathode rays impinge on matter.

Bibliography

Collocott, T. C. and Dobson A. B., Eds., *Chambers Dictionary of Science and Technology*, rev. ed., HarperCollins, New York, 1983.

Condon, E. U. and Odishaw, H., Eds., *Handbook of Physics*, McGraw-Hill, New York, 1958.

Judd, D. B. and Wyszecki, G., *Color in Business, Science and Industry*, 3rd ed., John Wiley & Sons, New York, 1975.

Kaufman, J. E. and Christensen, J. F., Eds., *IES Illumination Handbook*, Illumination Engineering Society, New York, 1984.

Lapedes, D. N., Ed., *The McGraw-Hill Encyclopedia of Science and Technology*, 2nd ed., McGraw-Hill, New York.

Stimson, A., *Photometry and Radiometry for Engineers*, John Wiley & Sons, New York, 1974.

The Cambridge Encyclopedia, Cambridge University Press, London, 1990.

The Concise Columbia Encyclopedia, Columbia University Press, New York, 1993.

Webster's New World Encyclopedia, Prentice Hall, Englewood Cliffs, NJ, 1992.

Wyszecki, G. and Stiles, W. S., *Color Science: Concepts and Methods, Quantitative Data, and Formulae*, 2nd ed., John Wiley & Sons, New York, 1982.

CHAPTER

1.2

Broadcast-Related Organizations and Information

STEVEN MITCHEL
National Association of Broadcasters
Washington, D.C.

INTRODUCTION

This chapter provides a general overview of entities whose activities affect the work of broadcast engineers. It is divided into five categories:

- Federal government
- State and local government
- Trade associations
- Professional associations
- Related broadcast-oriented organizations

Web site references are included to direct broadcast engineers to more information on the listed entity, including publications, conferences, committees, and other resources. Information contained in this chapter was current as of June 2006. Much of the information contained in this chapter has been obtained from the Web sites of the entities described herein. The chapter concludes with a list of broadcast engineering and related periodicals.

FEDERAL GOVERNMENT

U.S. Department of Agriculture (USDA), Forest Service

14th and Independence Avenue, S.W.
Washington, D.C. 20250
Phone: (202) 205-1523
Web site: http://www.fs.fed.us

The U.S. Department of Agriculture (USDA) Forest Service oversees 190 million acres of public land under the domain of the National Forest System. Often, the use of a mountaintop transmitter site is administered by the Forest Service. Leases authorizing communications facilities are issued by the Forest Supervisor for the particular forest of interest. Specific information can be can be found on the Service's Web site under "Passes & Permits" and "Special Use Permits."

U.S. Department of Commerce, National Institute of Standards and Technology (NIST)

Quince Orchard and Clopper Roads
Gaithersburg, Maryland 20899
Phone: (301) 975-2000
Web site: http://www.nist.gov

The National Institute of Standards and Technology (NIST) is a non-regulatory agency within the U.S. Department of Commerce Technology Administration. NIST's mission is to promote U.S. innovation and industrial competitiveness by advancing measurement science, standards, and technology in ways that enhance economic security and improve our quality of life.

U.S. Department of Commerce, Time and Frequency Division, Boulder Laboratories

325 Broadway
Boulder, Colorado 80305
Phone: (303) 497-6461
Web site: http://tf.nist.gov

The functions of the Time and Frequency Division, part of the Physics Laboratory of NIST, include:

- Maintaining the primary frequency standard for the United States
- Developing and operating standards of time and frequency
- Coordinating U.S. time and frequency standards with other world standards
- Providing time and frequency services for U.S. clientele
- Performing basic and applied research in support of improved standards and services

Precise time and frequency information is required by radio and television stations, electric power companies, telephone companies, air traffic control systems, participants in space exploration, computer networks, scientists monitoring data of all kinds, and navigators of ships and planes. These users need to compare their own timing equipment to a reliable, legally traceable, internationally recognized standard. NIST provides this standard for civilian users in the United States. Broadcast services include shortwave radio signals from NIST radio stations WWV and WWVH, 60-kHz signals from WWVB, and time signals carried by satellites, over phone lines, and on the Internet using the NIST Internet Time Service (ITS). Precise frequency references are available from WWV, WWVH, WWVB, and Loran-C. Time synchronization can also be accomplished via global positioning system (GPS) satellites. The division offers a fee-supported Frequency Measurement and Analysis Service (FMAS) that allows users to make accurate, NIST-certified frequency calibrations at their site, rather than sending their oscillators to NIST or elsewhere for calibration.

National Telecommunications and Information Administration (NTIA)

14th Street and Constitution Avenue, N.W.
Washington, D.C. 20230
Phone: (202) 482-7002
Web site: http://www.ntia.doc.gov

The National Telecommunications and Information Administration (NTIA) is the President's principal adviser on telecommunications and information policy issues, and in this role frequently works with other Executive Branch agencies to develop and present the Administration's position on these issues. In addition to representing the Executive Branch in both domestic and international telecommunications and information policy activities, NTIA also manages the federal use of spectrum; performs telecommunications research and engineering, including resolving technical telecommunications issues for the federal government and private sector; and administers infrastructure and public telecommunications facilities grants. NTIA's mission is to promote the development of an advanced telecommunications and information infrastructure that efficiently serves the needs of all Americans, creates job opportunities for American workers, and enhances the competitiveness of U.S. industry in the global marketplace. NTIA addresses a broad range of telecommunications issues and concerns. Many efforts cross-cut across the agency, drawing on telecommunications policy expertise, radiowave propagation knowledge, spectrum engineering, and lessons learned from actual applications. Of particular interest to broadcast engineers is NTIA's Office of Spectrum Management (OSM), which is responsible for managing the federal government's use of the radiofrequency spectrum. To achieve this, OSM receives assistance and advice from the Interdepartment Radio Advisory Committee.

U.S. Department of the Interior, Bureau of Land Management (BLM)

1849 C Street, N.W.
Washington, D.C. 20240
Phone: (202) 452-5125
Web site: http://www.blm.gov

The Bureau of Land Management (BLM), an agency within the U.S. Department of the Interior, administers 261 million surface acres of America's public lands, located primarily in 12 western states.

U.S. Department of the Interior, U.S. Geological Survey (USGS) National Center

12201 Sunrise Valley Drive
Reston, Virginia 20192
Phone: (703) 648-4000
Web site: http://www.usgs.gov

The U.S. Geological Survey (USGS) serves the nation by providing reliable scientific information to describe and understand the Earth; minimize loss of life and property from natural disasters; manage water, biological, energy, and mineral resources; and enhance and protect our quality of life. In support of the U.S. Geological Survey's mission to provide information about the Earth and its physical resources, *The National Map* provides geographic, cartographic, and remote sensing information, maps, and technical assistance. The USGS also makes available maps, images, spatial data, remote sensing data, and related information; provides assistance in selecting, acquiring, and using geographic and cartographic products; and designs, prints, and distributes maps of the National Atlas. It coordinates federal topographic mapping and digital cartographic activities and provides leadership in the development and advancement of surveying and mapping technology. Topographic maps usually used in the prediction of coverage and other engineering studies that require accurate information about the position and elevation of terrain features may be obtained from the U.S. Geological Survey Web site.

Department of Labor (DOL), Occupational Safety and Health Administration (OSHA)

200 Constitution Avenue, N.W.
Washington, D.C. 20210
Phone: (202) 693-2000
Web site: http://www.osha.gov

Broadcast engineers should be aware that two federal laws—the Federal Labor Standards Act and the Occupational Safety and Health Act—empower the Department of Labor (DOL) to regulate workplace safety standards and the wages and hours of employment for broadcast employees. Occupational safety standards are enforced by the Occupational Safety and Health Administration (OSHA) of the DOL, and the minimum wage and hours of employment are enforced by the DOL's Wage and Hour Division (WHD). The applicability of OSHA regulations depends on hazards, but WHD regulations encompass many factors, such as the station's geographic location, the number of station employees, and the type of work they perform. Broadcast engineers should consult their station's attorney to determine what regulations apply to them and how to abide by them.

Department of Transportation, Federal Aviation Administration (FAA)

800 Independence Avenue, S.W.
Washington, D.C. 20590
Phone: (202) 267-3111
Web site: http://www.faa.gov

When construction or alteration of a broadcast tower is proposed, the Federal Aviation Administration (FAA) conducts an aeronautical study to determine the potential impact that the proposal may have on the navigable airspace. Broadcasters are required to notify the FAA of new proposals or alterations to existing towers. The obstructions standards in Title 14 of the Code of Federal Aviation Regulations (Part 77) are applied to determine the effect the proposal would have on aeronautical operations. It may be necessary for broadcasters to amend their proposal by modifying the tower height or location in order to eliminate a determination of hazard to air navigation. When an aeronautical study results in a determination of no hazard to air navigation, the FAA will recommend the marking and/or lighting for that structure, if appropriate. The Federal Communication Commission's license will contain the marking and/or lighting recommended by the FAA. Conspicuity is achieved only when all recommended lights are working; therefore, it is important that any outage be corrected as soon as possible. The FAA must receive notice immediately of any failure or malfunction that lasts more than 30 minutes and affects a top light or flashing obstruction light, regardless of its position.

Environmental Protection Agency (EPA)

401 M Street, S.W.
Washington, D.C. 20460
Phone: (202) 260-2090
Web site: http://www.epa.gov

The Environmental Protection Agency (EPA) protects public health and safeguards and improves the natural environment—air, water, and land—upon which human life depends. The EPA's purpose is to ensure that:

- Federal environmental laws are implemented and enforced fairly and effectively.
- Environmental protection is an integral consideration in U.S. policies concerning economic growth, energy, transportation, agriculture, industry, international trade, and natural resources.
- National efforts to reduce environmental risk are based on the best available scientific information.
- All parts of society—business, state and local governments, communities, and citizens—have full access to information so they can become full participants in preventing pollution and protecting human health and the environment.

When broadcasters construct station facilities such as antenna towers or large satellite Earth stations they should be aware that these activities may fall under the scope of the National Environmental Protection Act (NEPA), which grants the EPA authority to regulate activities that may affect the "quality of the human environment." The Federal Communications Commission cooperates with the EPA in enforcing provisions of NEPA that relate to telecommunication licensees.

Federal Communications Commission (FCC)

445 12th St., S.W.
Washington, D.C. 20554
Phone: (202) 418-0200
Web site: http//www.fcc.gov

For detailed information on the Federal Communications Commission (FCC), see Chapter 1.3, "FCC Organization and Administrative Practices," and Chapter 1.4, "FCC Rules Compliance." Every station should have a current copy of the FCC's broadcast rules.

National Labor Relations Board (NLRB)

1099 14th Street, N.W.
Washington, D.C. 20570
Phone: (202) 273-1000
Web site: http://www.nlrb.gov

Broadcast engineers should be aware that the National Labor Relations Act (NRLA) protects the right of station employees to bargain collectively with management over the "terms and condition of employment."

The National Labor Relations Board (NLRB), an independent federal agency, was established to enforce the right of workers to organize and engage in "concerted activity." As with the Department of Labor regulations, the extent of these rights depends upon many factors, such as the type and number of employees involved.

STATE AND LOCAL GOVERNMENT

Several aspects of a broadcast engineer's job are affected or controlled by state or local government agencies. It is important for engineers to have some familiarity with the laws, codes, and zoning ordinances governing such matters as building construction, electrical wiring, and fire safety. Regulations may vary from one community to another, even within the same state or county.

Although there are model national codes, these codes may or may not be adopted by a state or local government. If adopted, there may be some changes from the national model. The only way to determine this is to check with the local agency or agencies having jurisdiction over the matter in question. For more information on state or local regulations, see the following:

- The broadcast station's local lawyer
- The county or city business licensing office
- The county or city building inspector or fire marshal
- A licensed local contractor who performs the type of work in question

TRADE ASSOCIATIONS

Association for Maximum Service Television (MSTV)

P.O. Box 9897
4100 Wisconsin Avenue, N.W.
Washington, D.C. 20016
Phone: (202) 966-1956; *fax:* (202) 966-9617
Web site: http://www.mstv.org

The Association for Maximum Service Television (MSTV) is a national association of local television stations dedicated to preserving and improving the technical quality of free, universal, community-based television service to the American public.

Association of Public Television Stations (APTS)

666 11th Street, NW, Suite 1100
Washington, D.C. 20001
Phone: (202) 654-4200; *fax:* (202) 654-4236
Web site: http://www.apts.org

The Association of Public Television Stations (APTS) is a nonprofit membership organization established in 1980 to support the continued growth and development of a strong and financially sound noncommercial television service for the American public. APTS provides advocacy for public television interests at the national level, as well as consistent leadership and information in marshaling support for its members: the nation's public television stations. APTS' affiliated organization, APTS Action, Inc., provides legislative advocacy and seeks grassroots and congressional support.

Association of Radio Industries and Businesses of Japan (ARIB)

Nittochi Building
1-4-1 Kasumigaseki, Chiyodaku
Tokyo 100-0013, Japan
Phone: +81 (3) 5510-8590; *fax:* +81 (3) 3592-1103
Web site: http://www.arib.or.jp/english/index.html

The objectives of the Association of Radio Industries and Businesses of Japan (ARIB) are to investigate, conduct research and development, and provide consultation regarding the utilization of radiowaves from the view of developing radio industries and to promote the realization and popularization of new radio systems in the field of telecommunications and broadcasting.

Cable Television Laboratories, Inc. (CableLabs)

858 Coal Creek Circle
Louisville, Colorado 80027
Phone: (303) 661-9100; *fax:* (303) 661-9199
Web site: http://www.cablelabs.com

Cable Television Laboratories, Inc. (CableLabs), a nonprofit research and development consortium, is dedicated to helping its cable operator members integrate new cable telecommunications technologies into their business objectives. CableLabs serves the cable television industry by:

- Researching and identifying new broadband technologies
- Authoring specifications
- Certifying products
- Disseminating information

Consumer Electronics Association (CEA)

2500 Wilson Boulevard
Arlington, Virginia 22201
Phone: (703) 907-7600
Web site: http://www.ce.org

The Consumer Electronics Association (CEA) promotes growth in the consumer technology industry through technology policy, events, research, promo-

tion, and the fostering of business and strategic relationships. The CEA represents more than 2100 corporate members involved in the design, development, manufacturing, distribution, and integration of audio, video, mobile electronics, wireless and landline communications, information technology, home networking, multimedia, and accessory products, as well as related services that are sold through consumer channels. The CEA also sponsors and manages the annual International Consumer Electronics Show, the world's largest consumer technology tradeshow. The CEA develops standards in such areas as product safety, audio and video systems, antennas, mobile electronics, and cable compatibility. In addition, the CEA is the cosponsor, along with the National Association of Broadcasters, of the National Radio Systems Committee (described below).

Electronic Industries Alliance (EIA)

2500 Wilson Boulevard
Arlington, Virginia 22201
Phone: (703) 907-7500; *fax:* (703) 907-7501
Web site: http://www.eia.org

The Electronic Industries Alliance (EIA) is a national trade organization that includes the full spectrum of U.S. manufacturers. The EIA is a partnership of electronic and high-tech associations and companies whose mission is promoting the market development and competitiveness of the U.S. high-tech industry through domestic and international policy efforts. The EIA is composed of nearly 1300 member companies whose products and services range from the smallest electronic components to the most complex systems used by defense, space, and industry, including a full range of consumer electronic products.

European Broadcasting Union (EBU)

17A Ancienne Route
CH-1218 Grand-Saconnex, Switzerland
Phone: + 41 (0) 22-717-2111; *fax:* +41 (0) 22-747-4000
Web site: http://www.ebu.ch

The European Broadcasting Union (EBU) is the largest professional association of national broadcasters in the world, with 74 members in 54 countries of Europe, North Africa, and the Middle East, as well as 44 associate members in 25 additional countries. Working on behalf of its members in the European area, the EBU negotiates broadcasting rights for major sports events, operates the Eurovision and Euroradio networks, organizes program exchanges, stimulates and coordinates coproductions, and provides a full range of other operational, commercial, technical, legal, and strategic services. In the technical area, the EBU is involved in the research and development of new broadcast media, such as Radio Data System (RDS), digital audio broadcasting (DAB), digital video broadcasting (DVB), and high-definition television (HDTV).

National Association of Broadcasters (NAB)

1771 N Street, N.W.
Washington, D.C. 20036
Phone: (202) 429-5300; *fax:* (202) 775-4981
Web site: http://www.nab.org

The National Association of Broadcasters (NAB) represents the broadcasting industry before Congress, the courts, regulatory agencies, the White House, and the general public. The NAB Science and Technology (S&T) Department works with other NAB departments to represent the industry before the Federal Communications Commission and other agencies on issues affecting spectrum management and technical regulations. The S&T Department's mission is to preserve and improve the ability of radio and television stations to distribute services to consumers and businesses. The S&T Department conducts a series of technical conferences, including the NAB Broadcast Engineering Conference (which celebrated its 60th anniversary in 2006) at the annual NAB spring convention, as well as the technical program portion of the annual NAB Radio Show. The department also provides timely, useful, and accurate technical information in *Radio TechCheck* and *TV TechCheck*, which are weekly technical newsletters made available to the entire broadcasting industry, and through numerous other NAB technical publications including the annual *Broadcast Engineering Conference Proceedings.*

North American Broadcasters Association (NABA)

P.O. Box 500, Station A
Toronto, Ontario M5W 1E6
Canada
Phone: (416) 598-9877; *fax:* (416) 598-9774
Web site: http://www.nabanet.com

The North American Broadcasters Association (NABA) is a nonprofit union of broadcasting organizations throughout North America committed to advancing the interests of broadcasters at home and internationally. As a member of the World Broadcasting Unions, NABA creates the opportunity for North American broadcasters to share information, identify common interests, and reach consensus on issues of an international nature. NABA provides representation for North American broadcasters in global forums.

National Cable Television Association (NCTA)

1724 Massachusetts Avenue, N.W.
Washington, D.C. 20036
Phone: (202) 775-3550
Web site: http://www.ncta.com

The National Cable and Telecommunications Association (NCTA), formerly the National Cable Television Association, is the principal trade association of the cable television industry in the United States.

Founded in 1952, NCTA's primary mission is to provide its members with a strong national presence by providing a single, unified voice on issues affecting the cable and telecommunications industry. The NCTA represents cable operators serving more than 90% of the nation's cable television households and more than 200 cable program networks, as well as equipment suppliers and providers of other services to the cable industry. In addition to offering traditional video services, NCTA's members also provide broadband services such as high-speed Internet access and telecommunications services such as local exchange telephone service to customers across the United States. The NCTA also hosts the industry's annual trade show, which serves as a national showcase for the cable industry's services.

Japan Electronics and Information Technology Industries Association (JEITA)

3rd Floor, Mitsui Sumitomo Kaijo Building Annex
11, Kanda Surugadai 3-chome, Chiyoda-ku
Tokyo 101-0062, Japan
Web site: http://www.jeita.or.jp/english

The objective of the Japan Electronics and Information Technology Industries Association (JEITA) is to promote the healthy manufacturing, international trade, and consumption of electronics products and components to contribute to the overall development of the electronics and information technology (IT) industries and thereby further Japan's economic development and cultural prosperity. JEITA's mission is to foster a digital network society for the 21st century, in which IT advancement brings fulfillment and a higher quality of life to everyone. The association is also actively promoting environmental preservation countermeasures, including those to combat global warming.

World Broadcasting Unions (WBU)

P.O. Box 500, Station A
Toronto, Ontario M5W 1E6
Canada
Phone: (416) 598-9877; *fax:* (416) 598-977
Web site: http://www.worldbroadcastingunions.org/wbuarea/

The World Broadcasting Unions (WBU) is the coordinating body for broadcasting unions that represent broadcaster networks across the globe. It was established in 1992 as a coordinating body at the international broadcasting level. The North American Broadcasters Association (NABA) acts as secretariat for the WBU. The broadcasting unions that belong to the WBU include the Asia-Pacific Broadcasting Union (ABU), the Arab States Broadcasting Union (ASBU), the International Association of Broadcasting (IAB/AIR), the Caribbean Broadcasting Union (CBU), the European Broadcasting Union (EBU), the North American Broadcasters Association (NABA), the Organización de la Televisión Iberoamericana (OTI), and the Union des radiodiffusions et télévisions nationales d'Afrique (URTNA). The WBU has three working committees: the International Satellite Operations Group (WBU-ISOG), the Technical Committee (WBU-TC), and the Sports Committee.

PROFESSIONAL ASSOCIATIONS

Asia-Pacific Broadcasting Union (ABU)

P.O. Box 1164
59700 Kuala Lumpur, Malaysia
Fax: (60-3) 2282-5292
Web site: http://www.abu.org.my

The Asia-Pacific Broadcasting Union (ABU) is a nonprofit, nongovernment, professional association of broadcasting organizations, formed in 1964 to facilitate the development of broadcasting in the Asia-Pacific region and to organize cooperative activities among its members. It has over 150 members in 55 countries, and its broadcaster members reach a potential audience of about 3 billion people. The ABU provides a forum for promoting the collective interests of television and radio broadcasters and engages in activities to encourage regional and international cooperation among broadcasters.

Association of Federal Communications Consulting Engineers (AFCCE)

Web site: http://www.afcce.org

The Association of Federal Communications Consulting Engineers (AFCCE) was founded in 1948 as a professional association of communications engineers practicing before the Federal Communications Commission. Engineering for broadcast stations in the AM, FM, and television services; for microwave, cellular radio, PCS, paging, and cable systems; and for satellite facilities are some of the areas in which AFCCE members offer their professional services. Associate membership is offered to technical personnel and to other professionals sharing an interest in the technical aspects of communications. The purpose of the Association is to aid and promote the proper federal administration and regulation of those engineering and technical phases of communications that are regulated by the Federal Communications Commission.

Broadcast Education Association (BEA)

1771 N Street, N.W.
Washington, D.C. 20036
Phone: (202) 429-5354
Web site: http://www.beaweb.org

The Broadcast Education Association (BEA) provides professional development for people who teach and research electronic media and multimedia. Academics, media professionals, and students participate in

this organization, which has served the industry for more than 43 years. The BEA publishes the *Journal of Electronic Media*, *Feedback*, and the *Journal of Radio Studies*. BEA administers a variety of scholarships annually to honor broadcasters and the broadcast industry. The BEA 2-year scholarship is for study at schools offering only freshman and sophomore instruction or for study at 4-year institutions by graduates of BEA 2-year programs. All other scholarships are awarded to juniors, seniors, and graduate students at BEA member colleges and universities.

Institute of Electrical and Electronics Engineers (IEEE) Broadcast Technology Society

3 Park Avenue, 17th Floor
New York, New York 10016
Phone: (732) 562-3846
Web site: http://www.ieee.org/bts

The Institute of Electrical and Electronics Engineers (IEEE) has 365,000 members in 150 countries, including engineers and scientists in electrical engineering, electronics, computers, and allied fields as well as over 60,000 students. The IEEE Broadcast Technology Society is one of the Institute's 39 societies addressing member interests. The IEEE holds numerous meetings and special technical conferences (for example, the Broadcast Technology Society holds an annual Broadcast Symposium), conducts lecture courses at the local level on topics of current engineering and scientific interest, assists student groups, and awards medals, prizes, and scholarships for outstanding technical achievement. Publications of the IEEE include their *Proceedings* (monthly), *IEEE Spectrum* (monthly), and *Directory* (annually) and more than 700 industry standards. The societies and councils publish journals, magazines, and conference proceedings.

Society of Broadcast Engineers (SBE)

9102 North Meridian Street, Suite 150
Indianapolis, Indiana 46260
Phone: (317) 846-9000; *fax:* (317) 253-0418
Web site: http://www.sbe.org

With more than 5300 members, the Society of Broadcast Engineers (SBE), the largest national organization for broadcast engineers, is devoted to the professional development of its members and the field of broadcast engineering. The SBE promotes communication among SBE members and provides national representation for those members before federal and state regulatory agencies, manufacturers, and the general public. The SBE, through its 100-plus local chapters, holds monthly meetings and technical conferences. It administers a certification program that recognizes various levels of engineering achievement: six classes of engineering certifications, two operator certifications, and one broadcast networking certification, each valid for a period of 5 years. There are also two specialist certifications to establish a benchmark of individual strengths for those already certified at specific certification levels. The certification designations are as follows:

- Certified Radio Operator (CRO)
- Certified Television Operator (CTO)
- Certified Broadcast Technologist (CBT)
- Certified Broadcast Networking Technologist (CBNT)
- Certified Audio Engineer (CEA)
- Certified Video Engineer (CEV)
- Certified Broadcast Radio Engineer (CBRE)
- Certified Broadcast Television Engineer (CBTE)
- Certified Senior Broadcast Radio Engineer (CSRE)
- Certified Senior Broadcast Television Engineer (CSTE)
- Certified Professional Broadcast Engineer (CPBE)
- Certified 8-VSB Specialist
- Certified AM Directional Specialist (AMD)

The SBE National Frequency Coordinating Committee was established in 1982 when the Federal Communications Commission (FCC) asked the broadcast industry to identify local contacts for Part 74 frequency coordination. The SBE formed a national network of volunteers to develop local databases of frequencies and users to assist the FCC and Part 74 users. Voluntary frequency coordination is handled on a local basis by SBE coordinators. The Ennes Educational Foundation Trust was founded in 1981. Each year, this trust awards scholarships to qualified individuals pursuing an education in broadcast engineering or to those pursuing continuing education in broadcast technology. The trust also presents workshops around the country, providing in-depth instruction to members, and underwrites educational publications of a technical broadcast nature.

Society of Cable Telecommunications Engineers (SCTE)

140 Phillips Road
Exton, Pennsylvania 19341
Phone: (610) 363-6888; *fax:* (310) 363-5898
Web site: http://www.scte.org

The Society of Cable Telecommunications Engineers (SCTE), is a nonprofit, professional organization formed in 1969 to promote the sharing of operational and technical knowledge in the field of cable television and broadband communications. Through the efforts of both the national and local chapter levels of the organization, the SCTE provides training opportunities, standards development, and certification at the technician and engineer levels. Additional information and exchange opportunities are provided through the SCTE's national conferences, board of directors, committees, newsletter, and associated trade journals. The SCTE has 15,000 members and over 70 national and international chapters and meeting groups.

Society of Motion Picture and Television Engineers (SMPTE)

3 Barker Avenue
White Plains, New York 10601
Phone: (914) 761-1100
Web site: http://www.smpte.org

The membership of the Society of Motion Picture and Television Engineers (SMPTE) is comprised of professional engineers and technicians in motion pictures, television, and allied arts and sciences. SMPTE advances engineering technology, disseminates scientific information, and sponsors lectures, exhibitions, and conferences to advance the theory and practice of motion-picture and television engineering. As an accredited standards developer under the American National Standards Institute, SMPTE develops national standards for motion pictures, television, and sound associated with motion picture and television images. It also develops recommended practices and engineering guidelines. (For copies of SMPTE engineering documents, contact their Standards Department.) SMPTE also makes available picture and sound test films and video tapes for use as standardized measuring tools and serves as administrator of the Secretariat of International Organization for Standardization (ISO) Technical Committee 36 on Cinematography and of the U.S. Technical Advisory Groups for ISO/TC 36 and International Electrotechnical Commission Subcommittee (IEC/SC) 100 on audio, video, and multimedia systems and equipment. SMPTE sponsors technical courses at universities on such subjects as digital television, sound techniques, laboratory processing, special effects, and lighting for technicians and students. The Society presents nine annual awards for outstanding contributions to motion-picture and television engineering. SMPTE engineering committees include:

- Digital Cinema Technology
- Video Compression Technology
- Television Recording and Reproduction Technology
- Television Systems Technology
- Television Audio Technology
- Television Image Technology

SMPTE publications include a monthly journal, papers presented at their conferences, and other books on motion-picture and television technology.

RELATED BROADCAST-ORIENTED ORGANIZATIONS

Advanced Television Systems Committee (ATSC)

1750 K Street, N.W., Suite 1200
Washington, D.C. 20006
Phone: (202) 872-9160; *fax:* (202) 872-9161
Web site: http:\\www.atsc.org

The Advanced Television Systems Committee (ATSC) is an international, nonprofit organization developing voluntary standards for digital television. The ATSC member organizations represent the broadcast, broadcast equipment, motion picture, consumer electronics, computer, cable, satellite, and semiconductor industries. ATSC creates and fosters implementation of voluntary standards and recommended practices to advance terrestrial digital television broadcasting and to facilitate interoperability with other media. ATSC was formed in 1982 by the member organizations of the Joint Committee on Intersociety Coordination: the Electronic Industries Alliance, the Institute of Electrical and Electronics Engineers, the National Association of Broadcasters, the National Cable and Telecommunications Association, and the Society of Motion Picture and Television Engineers. ATSC has approximately 140 members representing the broadcast, broadcast equipment, motion picture, consumer electronics, computer, cable, satellite, and semiconductor industries. ATSC digital television standards include digital high-definition television (HDTV), standard-definition television (SDTV), data broadcasting, multichannel surround-sound audio, and interactive television. In 1996, the Federal Communications Commission adopted the major elements of the ATSC Digital Television (DTV) Standard (A/53).

Advanced Authoring Format (AAF) Association

2207 Ringsmith Drive
Atlanta, Georgia 30345
Phone: (770) 414-9952; *fax:* (925) 475-6700
Web site: http://www.aafassociation.org/

The Advanced Authoring Format (AAF) Association, Inc., is a broad-based trade association intended to promote the development and adoption of AAF technology, which enables content creators to easily exchange digital media and metadata across platforms and between systems and applications throughout the media industry. The Advanced Authoring Format simplifies project management, saves time, and preserves valuable metadata that was often lost when transferring media between applications in the past. With representatives from several major players in the industry, the AAF Association intends to help deliver the full benefits of digital media to content creators, including film, television, Internet, and post-production professionals.

American National Standards Institute (ANSI)

25 W. 43rd Street
New York, New York 10036
Phone: (212) 642-4900; *fax:* (212) 398-0023
Web site: http://www.ansi.org

The American National Standards Institute (ANSI) is a private, nonprofit organization that administers and coordinates the U.S. voluntary standardization and

conformity assessment system. The mission of ANSI is to enhance both the global competitiveness of U.S. business and the U.S. quality of life by promoting and facilitating voluntary consensus standards and conformity assessment systems and by safeguarding their integrity. ANSI serves as the clearinghouse for nationally coordinated voluntary safety, engineering, and industrial standards. One of ANSI's important functions is accreditation. ANSI accredits standards developers, certification bodies, and technical advisory groups (TAGs) to both the International Organization for Standardization and the International Electrotechnical Commission. The Society of Motion Picture and Television Engineers, Society of Cable Telecommunications Engineers, and Consumer Electronics Association are ANSI-accredited standards development organizations (SDOs).

American Radio Relay League (ARRL)

225 Main Street
Newington, CT 06111
Phone: (860) 594-0200; *fax:* (860) 594-0259
Web site: http://www.arrl.org

The American Radio Relay League (ARRL) is the largest organization of radio amateurs in the United States. The ARRL is a not-for-profit organization that promotes interest in amateur radio communications and experimentation, represents U.S. radio amateurs in legislative matters, and maintains fraternalism and a high standard of conduct among amateur radio operators. ARRL is also international secretariat for the International Amateur Radio Union, which is made up of similar societies in 150 countries around the world. ARRL publishes the monthly journal *QST*, as well as newsletters and many publications covering all aspects of amateur radio. Its headquarters station, W1AW, transmits bulletins of interest to radio amateurs and Morse code practice sessions. The ARRL also coordinates an extensive field organization, which includes volunteers who provide technical information for radio amateurs and public-service activities. In addition, ARRL represents U.S. amateurs with the Federal Communications Commission and other government agencies in the United States and abroad.

Association for Computing Machinery (ACM)

1515 Broadway
New York, New York 10036
Phone: (800) 342-6626
Web site: http://www.acm.org

The Association for Computing Machinery (ACM) delivers resources that advance computing as a science and a profession. ACM provides the computing field's premier digital library and serves its members and the computing profession with leading-edge publications, conferences, and career resources.

Audio Engineering Society (AES)

60 East 42nd Street, Room 2520
New York, New York 10165
Phone: (212) 661-8528
Web site: http://www.aes.org

The Audio Engineering Society (AES) is devoted exclusively to audio technology. Its membership includes engineers, scientists, and other authorities throughout the world. The AES serves its members, the industry, and the public by stimulating and facilitating advances in the constantly changing field of audio. It encourages and disseminates new developments through annual technical meetings and exhibitions of professional equipment, and through the *Journal of the Audio Engineering Society*, the professional archival publication in the audio industry.

British Kinematograph Sound and Television Society (BKSTS)

G104 Pinewood Studios
Iver Heath, Buckinghamshire SL0 0NH
United Kingdom
Phone: +44 (0) 175-365-6656; *fax:* +44 (0) 175-365-7016
Web site: http://www.bksts.com/

Originally called the British Kinematograph Society, it was founded in London, England, in 1931 to serve the growing film industry. The name was later changed to the British Kinematograph Sound and Television Society to reflect the wide range of interests of the membership and was then to become known as BKSTS—The Moving Image Society, as its membership and the industry became increasingly international. As well as meetings, presentations, seminars, international exhibitions, and conferences, BKSTS—The Moving Image Society also organizes an extensive program of training courses, lectures, workshops, and special events in addition to its regular publications *Image Technology* and the quarterly *Cinema Technology*.

Canadian Digital Television (CDTV)

c/o 2727 Russland Road
Vars, Ontario K0A 3H0
Canada
Web site: http://www.cdtv.ca/

Canadian Digital Television (CDTV) is a not-for-profit Canadian television industry organization dedicated to providing expert information on high-definition television implementation. CDTV is guiding the orderly migration to advanced digital television services by directing research and testing; identifying and advising on policy, regulation, and marketplace issues; and developing and monitoring the digital television transition plan for the industry in Canada.

Digital Cinema Society

P.O. Box 1973
Studio City, California 91614
Phone: (818) 762-2214; *fax:* (818) 763-8769
Web site: http://www.digitalcinemasociety.org/

The Digital Cinema Society is a nonprofit corporation dedicated to educating and informing the entertainment industry about digital motion picture production, post-production, delivery, and exhibition. The Society's purpose is to objectively examine all media, solutions, services, and technologies without favoring any one brand or service over another, simply looking for the best tools available.

Digital Television Group (DTG)

7 Old Lodge Place
St. Margarets, Twickenham MIDDX TW1 1RQ
United Kingdom
Phone: +44 (0) 870-242-7346; *fax:* +44 (0) 870-242-7347
Web site: http://www.dtg.org.uk/

The Digital Television Group (DTG) is the industry association for digital television in the United Kingdom. It is independent, platform neutral, and technology agnostic. The group is currently focused on digital switchover and the rich media services and products it will help enable. Emerging consumer devices and experiences include high-definition television, mobile television, video-on-demand, broadband television, and television metadata.

European Industry Association for Information Systems, Communication Technologies, and Consumer Electronics (EICTA)

20, Rue Joseph II
B-1000 Brussels, Belgium
Phone: +32 (2) 609-53-10; *fax:* +32 (2) 609-53-39
Web site: http://www.eicta.org/

The European Industry Association for Information Systems, Communication Technologies, and Consumer Electronics (EICTA) is dedicated to improving the business environment for the European information and communications technology (ICT) and consumer electronics (CE) sectors and to promoting the industry's contribution to economic growth and social progress in the European Union. EICTA promotes the collective European interests in the ICT and CE sectors, seeks to participate in the development and implementation of EU policies, and facilitates long-term business generation for the digital technology industry in Europe by supporting the diffusion and usage of ICT and CE technologies.

European Telecommunications Standards Institute (ETSI)

650, Route des Lucioles
06921 Sophia-Antipolis Cedex, France
Phone: +33 (0) 4-92-94-42-00; *fax:* +33 (0) 4-93-65-47-16
Web site: http://www.etsi.org/

The European Telecommunications Standards Institute (ETSI) is an independent, nonprofit organization responsible for the standardization of information and communication technologies within Europe. These technologies include telecommunications, broadcasting, and related areas such as intelligent transportation and medical electronics. ETSI unites around 700 members from over 50 countries inside and outside Europe, including manufacturers, network operators, administrations, service providers, research bodies, and users. ETSI's prime objective is to support global harmonization by providing a forum in which all the key players can contribute actively.

Guild of Television Cameramen (GTC)

Web site: http://www.gtc.org.uk/

The Guild of Television Cameramen (GTC) is an international nonprofit organization offering a way for camera manufacturers, when designing new equipment, to consult with working cameramen to produce outline specifications for such things as cameras, lenses, and mountings. The GTC is not a trade union and avoids any political involvement, but it is an authoritative source of advice and information on all matters concerning television cameramen. Its aim is to preserve the professional status of the television cameraman and to establish, uphold, and advance the standards of qualification and competence of the television cameraman.

Hollywood Post Alliance (HPA)

846 S. Broadway, Suite 601
Los Angeles, California 90014
Phone: (213) 614-0860; *fax:* (213) 614-0890
Web site: http://www.hpaonline.com/

The Hollywood Post Alliance (HPA) is the trade association representing the Southern California-based professional community of businesses and individuals who provide expertise, support, tools, and infrastructure for the creation and finishing of motion pictures, television, commercials, digital media, and other dynamic media content. The HPA:

- Provides a forum for the networking of colleagues and peers who have dedicated their careers to post-production, with a goal of representing a large and diverse community of interests and experience
- Facilitates information exchange on issues that relate to business, technology, skills training, and industry education

- Acts as an industry advocate and speaks with the power of the "larger voice" of the entire industry on common issues and topics such as government affairs, local community and business issues, and technology
- Provides a platform for industry events, seminars, electronic e-mail exchange, Web-based information outlets, and professional special interest groups dedicated to specific skills and areas of expertise

Institution of Engineering and Technology (IET)

Savoy Place
London, WC2R 0BL
United Kingdom
Phone: +44 (0) 207-240-1871; *fax:* +44 (0) 207-240-7735
Web site: http://www.theiet.org

The Institution of Engineering and Technology (IET) was formed by the Institution of Electrical Engineers (IEE) and the Institution of Incorporated Engineers (IIE) and now has more than 150,000 members worldwide. It is the largest professional engineering society in Europe and the second largest of its type in the world.

International Association of Broadcasting Manufacturers (IABM)

P.O. Box 2264
Reading, Berkshire RG31 6WA
United Kingdom
Phone: +44 (0) 118-941-8620
Web site: http://www.theiabm.org

The International Association of Broadcasting Manufacturers (IABM) is an international trade association that represents manufacturers and suppliers of products and services to the broadcasting and electronic media industries. The Association's aim is to identify and promote the interests of its members and to provide benefits and services that enhance their business performance.

International Broadcasting Convention (IBC)

Aldwych House
81 Aldwych
London, WC2B 4EL
United Kingdom
Phone: +44 (0) 207-611-7500; *fax:* +44 (0) 207-611-7530
Web site: http://www.ibc.org/

The International Broadcasting Convention (IBC) is a European trade show established in 1967 as a showcase for broadcasting technology. The IBC consists of conferences and exhibits; exhibitors on the show floor include every major supplier of technology for the creation, management, and delivery of entertainment content. The IBC is designed to serve everyone involved in the creation, management, and delivery of content for the entertainment industry. It is owned by a partnership of six sponsoring organizations: the International Association of Broadcasting Manufacturers, the Institution of Engineering and Technology, the IEEE Broadcast Technology Society, the Royal Television Society, the Society of Cable TV Engineers, and the Society of Motion Picture and Television Engineers.

International Electrotechnical Commission (IEC)

3, rue de Varembé
P.O. Box 131
CH-1211 Geneva, Switzerland
Phone: +41 (22) 919-0211; *fax:* +41 (22) 919-0300
Web site: http://www.iec.ch/

Founded in 1906 with British scientist Lord Kelvin as its first president, the International Electrotechnical Commission (IEC) is a global organization for the preparation and dissemination of international standards for all electrical, electronic, and related technologies. The IEC charter embraces all electrotechnologies, including electronics, magnetics and electromagnetics, electroacoustics, multimedia, telecommunications, and energy production and distribution, as well as associated general disciplines such as terminology and symbols, electromagnetic compatibility, measurement and performance, dependability, design and development, safety, and the environment.

International Standard Audiovisual Number (ISAN) International Agency

26, rue de Saint Jean
CH-1203 Geneva, Switzerland
Phone: +41 (22) 545-1000; *fax:* +41 (22) 545-1040
Web site: http://www.isan.org/

The International Standard Audiovisual Number (ISAN) International Agency has responsibility for the overall ISAN system maintenance and administration. The ISAN is a voluntary numbering system for the identification of audiovisual works. It provides a unique, internationally recognized, and permanent reference number that distinguishes one audiovisual work from all other audiovisual works registered in the ISAN system. Other methods of identifying audiovisual works, such as by title, can result in confusion about the specific work being referenced. The ISAN remains the same for an audiovisual work regardless of the various formats in which the work is distributed (*e.g.*, DVD, video recording) or the uses to which it is put.

International Organization for Standardization (ISO)

1, rue de Varembé
Case postale 56
CH-1211 Geneva, Switzerland
Phone: +41 (22) 749-0111; *fax:* +41 (22) 733-3430
Web site: http://www.iso.org/

The International Organization for Standardization (ISO) is a nongovernmental organization comprised of a network of the national standards institutes of 156 countries, on the basis of one member per country, with a central secretariat in Geneva, Switzerland, that coordinates the system. The ISO occupies a special position between the public and private sectors by bridging organizations to meet both the requirements of business and the broader needs of society. ISO standards contribute to making the development, manufacturing, and supply of products and services more efficient, safer, and cleaner. They make trade between countries easier and more equitable. They provide governments with a technical base for health, safety, and environmental legislation and aid in transferring technology to developing countries.

International Telecommunications Union (ITU)

Place des Nations
CH-1211 Geneva, Switzerland
Phone: +41 (22) 730-5111; *fax:* +41 (22) 733-7256
Web site: http://www.itu.int

The International Telecommunications Union (ITU) is an international organization within the United Nations system where governments and the private sector coordinate global telecommunications and radiocommunications networks and services. The Union's standardization activities, which have already helped foster the growth of new technologies such as mobile telephony and the Internet, are now being put to use in defining the building blocks of the emerging global information infrastructure and designing advanced multimedia systems that deftly handle a mix of voice, data, audio, and video signals. The ITU is organized into three sectors:

- The Radiocommunication Sector (ITU-R) is charged with determining the technical characteristics and operational procedures for wireless services and plays a role in the international management and use coordination of the radiofrequency spectrum.
- The Telecommunication Standardization Sector (ITU-T) embodies ITU's oldest activity—developing internationally agreed-upon technical and operating standards and defining tariff and accounting principles for international telecommunication services. The work of the ITU-T aims to foster seamless interconnection of the world's communication network and systems.
- The Telecommunication Development Sector (ITU-D) seeks to promote investment and foster the expansion of telecommunications infrastructures in developing nations throughout the world.

Internet Society (ISOC)

1775 Wiehle Avenue, Suite 102
Reston, Virginia 20190
Phone: (703) 326-9880; *fax:* (703) 326-9881
Web site: http://www.isoc.org

The Internet Society (ISOC) is a professional membership society composed of more than 100 organizations and over 20,000 individual members in over 180 countries. It provides leadership in addressing issues that confront the future of the Internet and is the organizational home for the groups responsible for Internet infrastructure standards, including the Internet Engineering Task Force and the Internet Architecture Board. Through its annual International Networking (INET) conference and other sponsored events, developing-country training workshops, tutorials, statistical and market research, publications, public policy and trade activities, regional and local chapters, standardization activities, committees, and an international secretariat, the Internet Society serves the needs of the growing global Internet community. From commerce to education to social issues, the Internet Society's goal is to enhance the availability and utility of the Internet on the widest possible scale.

Internet Engineering Task Force (IETF)

Web site: http://www.ietf.org/

The Internet Engineering Task Force (IETF), an organized activity of the Internet Society, is a large open international community of network designers, operators, vendors, and researchers concerned with the evolution of the Internet architecture and the smooth operation of the Internet. It is open to any interested individual. The actual technical work of the IETF is done in its working groups, which are organized by topic into several areas (*e.g.*, routing, transport, security). Much of the work is handled via mailing lists. The IETF holds meetings three times per year.

Illuminating Engineering Society of North America (IESNA)

120 Wall Street, 17th Floor
New York, New York 10005
Phone: (212) 248-5000
Web site: http://www.iesna.org

The Illuminating Engineering Society of North America (IESNA) is the recognized technical authority for the illumination field. For over 90 years, its objective has been to communicate information on all aspects of good lighting practice to its members, to the lighting community, and to consumers through a variety of programs, publications, and services. IESNA is a forum for the exchange of ideas and information and a vehicle for its members' professional development and recognition. Through its technical committees, with hundreds of members from the lighting community, the IESNA correlates vast amounts of research, investigations, and discussions to guide lighting experts and laymen on research- and consensus-based lighting recommendations. Complete lists of current and available recommendations may be obtained by writing to the IESNA Publication Office. In addition to the *IESNA Lighting Handbook*, the IESNA also publishes *Lighting Design + Application* (*LD+A*) and the *Journal of the Illuminating Engineering Society* (*LEUKOS*).

Media Communications Association International (MCA-I)

810 Crossroads Drive, Suite 3800
Madison, Wisconsin 53718
Phone: (608) 443-2464; *fax:* (608) 443-2474
Web site: http://www.mca-i.org/

The Media Communications Association International (MCA-I) is a not-for-profit, member-driven, global community that provides opportunities for networking, members-only benefits, forums for education, and the resources for information to media communications professionals. Through facilitating effective communication, MCA-I offers media professionals the connections needed to succeed in a highly competitive environment.

National Captioning Institute (NCI)

1900 Gallows Road, Suite 3000
Vienna, Virginia 22182
Phone: (703) 917-7600; *fax:* (703) 917-9878
Web site: http://www.ncicap.org/

The National Captioning Institute (NCI) was established in 1979 as a nonprofit corporation with the mission of ensuring that deaf and hard-of-hearing people, as well as others who can benefit from the service, have access to television's entertainment and news through the technology of closed captioning. With a highly skilled captioning staff and state-of-the-art facilities, NCI provides the highest quality captioning services for broadcast and cable television, home video and DVD, and government and corporate video programming. NCI also provides subtitling and language translation services in over 40 languages and dialects.

National Radio Systems Committee (NRSC)

1771 N Street, N.W.
Washington, D.C. 20036
Phone: (202) 429-5346, *fax:* (202) 775-4981
Web site: http://www.nrscstandards.org

The National Radio Systems Committee (NRSC) is jointly sponsored by the National Association of Broadcasters (NAB) and the Consumer Electronics Association (CEA). Its purpose is to study and make recommendations for technical standards that relate to radio broadcasting and the reception of radio broadcast signals. The NRSC is a vehicle by which broadcasters and receiver manufacturers can work together toward developing solutions to common problems in radio broadcast systems. Anyone who has a business interest in the technology being investigated by the NRSC is welcome to join and participate in its activities. Members of the NRSC are generally engineers, scientists, or technicians with in-depth knowledge of the subject being studied. To promote the free exchange of ideas during committee work, members of the press are not allowed to attend NRSC meetings; however, members of the press are free to contact committee chairpersons, the NAB, or the CEA with general questions about meetings. NRSC meetings are held on an as-needed basis, and NRSC members participate at their own expense.

National Center for Accessible Media (NCAM)

125 Western Avenue
Boston, Massachusetts 02134
Phone: (617) 300-3400; *fax:* (617) 300-1035
Web site: http://ncam.wgbh.org/

The Corporation for Public Broadcasting (CPB)/WGBH National Center for Accessible Media (NCAM) is a research and development facility dedicated to the issues of media and information technology for people with disabilities in their homes, schools, workplaces, and communities. NCAM acts as the research and development arm of WGBH's Media Access Group; its mission is to expand access to present and future media for people with disabilities; to explore how existing access technologies may benefit other populations; to represent its constituents in industry, policy, and legislative circles; and to provide access to educational and media technologies for special needs students.

Professional MPEG Forum (Pro-MPEG)

Web site: http://www.pro-mpeg.org/

The Professional MPEG Forum (Pro-MPEG) is an association within the broadcast industry open to all interested parties. It was established in 1998 to take forward the work of the EBU/SMPTE Task Force for Harmonized Standards for the Exchange of Program Material as Bitstreams, whose Final Report recommended MPEG-2 as one of two key digital compression technologies for professional use in the broadcast environment. The report also identified the need to develop advanced solutions for future file-based network environments. Manufacturers and broadcasters have joined together in the Pro-MPEG Forum to ensure that the needs of the industry are met with end- to-end interoperable products. The Forum strives to share its work through the publication of educational material, presentations, test results, recommendations, and demonstrations and to complement and work with other internationally recognized organizations interested in promoting interoperability.

Radio and Television News Directors Association (RTNDA)

1600 K Street, N.W., Suite 700
Washington, D.C. 20006-2838
Phone: (202) 659-6510; *fax:* (202) 223-4007
Web site: http://www.rtnda.org

The Radio and Television News Directors Association (RTNDA) exclusively serves the electronic news profession, made up of more than 3000 news directors, news associates, educators, and students. RTNDA's purpose is to set standards of news gathering and reporting. Although news techniques and technologies have changed since the early years of its founding, RTNDA's commitment to encouraging excellence in the electronic journalism industry remains the same. RTNDA represents journalists in radio, television, cable, and emerging forms of electronic journalism. RTNDA members benefit from publications, training, advocacy, and many opportunities to meet with and learn from colleagues.

Royal Television Society (RTS)

5th Floor, Kildare House
3 Dorset Rise
London EC4Y 8EN
United Kingdom
Phone: +44 (0) 207-822-2810; *fax:* +44 (0) 207-822-2811
Web site: http://www.rts.org.uk

The Royal Television Society (RTS) is a British forum for discussion and debate on all aspects of the television community. Formed in 1927 and originally a meeting place for television engineers—both amateur and professional—the Society's earliest publications chart the birth of television and document the pioneering work of, among others, John Logie Baird. In the late 1960s and early 1970s, the balance of the industry shifted from engineering to program production and broadcasting. The Society broadened to become an independent forum, hosting regular symposia on topical subjects at which members from all the different television companies could meet together. These symposia have developed into a wide range of different events that continue to be an essential focus for the television industry.

Special Interest Group on Graphics and Interactive Techniques (SIGGRAPH)

Web site: http://www.siggraph.org/

The Special Interest Group on Graphics and Interactive Techniques (SIGGRAPH) is dedicated to promoting the generation and dissemination of information on computer graphics and interactive techniques by fostering a membership community whose core values help them to catalyze the innovation and application of computer graphics and interactive techniques. Also known as ACM SIGGRAPH (because the Association for Computing Machinery is SIGGRAPH's parent organization), SIGGRAPH sponsors not only the annual SIGGRAPH conferences but also focused symposia, chapters in cities throughout the world, awards, grants, educational resources, online resources, a public policy program, a traveling art show, and the *SIGGRAPH Video Review.*

Society for Information Display (SID)

610 S. 2nd Street
San Jose, California 95112
Phone: (408) 977-1013; *fax:* (408) 977-1531
Web site: http://www.sid.org/

The Society for Information Display (SID) is composed of over 6000 professionals in the technical and business disciplines that relate to display research, design, manufacturing, applications, marketing, and sales. Each member belongs to the SID chapter of his or her choice. Chapters hold periodic meetings, conferences, and trade shows that have national as well as international appeal. SID's largest international gathering is the annual SID Symposium, Seminar, and Exposition, which attracts thousands of attendees, speakers, and exhibitors from around the world.

Society of Television Lighting Directors (STLD)

Web site: http://www.stld.org.uk/

The Society of Television Lighting Directors (STLD) provides a forum that stimulates a free exchange of ideas in all aspects of the television profession, including discussion of techniques and the use and design of equipment. Professional meetings are organized throughout the United Kingdom and abroad; also, technical information and news of members and their activities are published in the Society's magazine. The STLD has no union affiliations and is therefore recognized as a valuable platform for open discussion and demonstration by production, management, and industry.

Underwriters Laboratories (UL)

333 Pfingsten Road
Northbrook, Illinois 60062
Phone: (847) 272-8800
Web site: http://www.ul.com/

Underwriters Laboratories (UL) is an independent, not-for-profit, product-safety testing and certification organization. UL has been evaluating products in the

interest of public safety since 1894. UL seeks by scientific investigation, study, experiments, and tests to determine the relation of various materials, devices, products, equipment, constructions, methods, and systems to identified hazards and how their use affects life and property. It also seeks to ascertain, define, and publish standards, classifications, and specifications for materials, devices, products, equipment, construction, methods, and systems affecting such hazards, and other information tending to reduce loss of life and property from such hazards.

Video Services Forum (VSF)

Web site: http://www.videoservicesforum.org/

Video Services Forum (VSF) is an international not-for-profit association open to businesses, public sector organizations, and individuals worldwide and dedicated to video transport technologies, interoperability, quality metrics, and education. VSF is composed of service providers, users, and manufacturers. The organization's activities include providing forums to identify issues involving the development, engineering, installation, testing, and maintenance of audio and video services; exchanging nonproprietary information to promote the development of video transport service technology; fostering resolution of issues common to the video services industry; identification of video services applications and educational services utilizing video transport services; promoting interoperability; and encouraging technical standards for national and international standards bodies.

World Wide Web Consortium (W3C)

32 Vassar Street, Room 32-G515
Cambridge, Massachusetts 02139
Phone: (617) 253-2613; *fax:* (617) 258-5999
Web site: http://www.w3.org

The World Wide Web Consortium (W3C) is an international consortium where member organizations, a full-time staff, and the public work together to develop Web standards. W3C's mission is to lead the World Wide Web to its full potential by developing standards, protocols, and guidelines that ensure long-term growth for the Web. Since 1994, W3C has published more than 90 such standards, referred to as W3C Recommendations. W3C also engages in education and outreach, develops software, and serves as an open forum for discussion about the Web.

BROADCAST ENGINEERING AND RELATED PERIODICALS

ABU Technical Review

Asian Pacific Broadcast Union
P.O. Box 1164, 59700 Kuala Lumpur, Malaysia
Phone: (603) 2282-3592; *fax:* (603) 2282-5292
Web site: http://www.abu.org.my

AGL (Above Ground Level)

P.O. Box 284
Waterford, Virginia 20197
Phone: (540) 882-4290
Web site: http://www.agl-mag.com/

Broadcast Engineering

P.O. Box 2100
Skokie, Illinois 60076-7800
Phone: (866) 505-7173; *fax:* (847) 763-9682
Web site: http://www.broadcastengineering.com

Broadcasting & Cable

P.O. Box 5655
Harlan, Iowa 51593
Phone: (800) 554-5729 (U.S.);
(515) 247-2984 (international)
Web site: http://www.broadcastingcable.com

Digital Content Producer

9800 Metcalf Avenue
Overland Park, Kansas 66212
Phone: (913) 341-1300; *fax:* (913) 967-1898
Web site: http://www.digitalcontentproducer.com

EBU Technical Review

European Broadcast Union
Case Postale 67
CH-1218 Grand-Saconnex, Switzerland
Phone: +41 (22) 717-2111; *fax:* +41 (22) 798-5897
Web site: http://www.ebu.ch/en/technical/trev

ECN Electronic Component News

Reed Business Information
360 Park Avenue South
New York, New York 10014
Phone: (646) 746-6400
Web site: http://www.ecnmag.com

Electronic Design

Penton Media, Inc.
2 Greenwood Square, Suite 410
3331 Street Road
Bensalem, Pennsylvania 19020
Phone: (215) 245-4555; *fax:* (215) 245-4060
Web site: http://www.elecdesign.com

Electronic Engineering Times

CMP Publications, Inc.
600 Community Drive
Manhasset, New York 11030
Phone: (516) 562-5000; *fax:* (516) 562-5995
Web site: http://www.eetimes.com/

Electronic Supply & Manufacturing

CMP Publications, Inc.
600 Community Drive
Manhasset, New York 11030
Phone: (516) 562-5000; *fax:* (516) 562-5995
Web site: http://www.my-esm.com/

GPS World

201 Sandpointe Avenue, Suite 500
Santa Clara, California 92707
Web site: http://www.gpsworld.com

IEEE Transactions on Broadcasting

3 Park Avenue, 17th Floor
New York, New York 10016
Phone: (212) 419-7900; *fax:* (212) 752-4929
Web site: http://www.ieee.org/organizations/society/bt/public.html

IEEE Transactions on Consumer Electronics

3 Park Avenue, 17th Floor
New York, New York 10016
Phone: (212) 419-7900; *fax:* (212) 752-4929
Web site: http://www.ewh.ieee.org/soc/ces/publications.html

Journal of the Audio Engineering Society

60 East 42nd Street, Room 2520
New York, New York 10165
Phone: (212) 661-8528; *fax:* (212) 682-0477
Web site: http://www.aes.org/journal/

Mix

6400 Hollis Street, No. 12
Emeryville, California 94608
Phone: (510) 653-3307; *fax:* (510) 653-5142
Web site: http://www.mixonline.com

Popular Communications

25 Newbridge Road
Hicksville, New York 11801
Phone: (516) 681-2922; *fax:* (516) 681-2926
Web site: http://www.popular-communications.com

Pro Sound News

460 Park Avenue South, 9th Floor
New York, New York 10016
Phone: (212) 378-0400; *fax:* (212) 378-2160
Web site: http://www.prosoundnews.com

QST

American Radio Relay League
225 Main Street
Newington, Connecticut 06111
Phone: (860) 594-0200; *fax:* (860) 594-0303
Web site: http://www.arrl.org/qst

Radio Guide

P.O. Box 20975
Sedona, Arizona 86352
Phone: (928) 284-3700; *fax:* (866) 728-5764
Web site: http://www.radio-guide.com

Radio Magazine (formerly BE Radio)

9800 Metcalf
Overland Park, Kansas 66212
Web site: http://www.beradio.com

Radio World

IMAS Publishing (USA), Inc.
P.O. Box 1214
Falls Church, Virginia 22041
Phone: (703) 998-7600; *fax:* (703) 671-7409
Web site: http://www.radioworld.com

SMPTE Motion Imaging Journal

3 Barker Avenue
White Plains, New York 10601
Phone: (914) 761-1100; *fax:* (914) 761-3115
Web site: http://www.smpte.org/smpte_store/journals/

Television Broadcast

CMP Information, Inc.
460 Park Avenue South, 9th Floor
New York, New York 10016
Phone: (212) 378-0400; *fax:* (212) 378-2159
Web site: http://www.televisionbroadcast.com/

Television Week

Crain Communications, Inc
1155 Gratiot Avenue
Detroit, Michigan 48207
Phone: (888) 288-5900 (U.S. and Canada);
(313) 446-1665 (international);
fax: (313) 446-6777
Web site: http://www.tvweek.com

TV Technology

IMAS Publishing (USA), Inc.
P.O. Box 1214
Falls Church, Virginia 22041
Phone: (703) 998-7600; *fax:* (703) 671-7409
Web site: http://www.tvtechnology.com

TWICE (This Week in Consumer Electronics)

Reed Business Information
8878 S. Barrons Boulevard
Highlands Ranch, Colorado 80126
Phone: (800) 446-6551; *fax:* (303) 470-4280
Web site: http://www.twice.com

CHAPTER

1.3

FCC Organization and Administrative Practices

JOAN M.S. DOLLARHITE
National Association of Broadcasters
Washington, D.C.

STATUTORY AUTHORITY

The Federal Communications Commission (FCC) regulates television, radio, wire, satellite, and cable in all of the 50 states and U.S. territories. Congress, through adoption of the Communications Act of 1934, created the FCC as an independent regulatory agency directly responsible to Congress. Section I of the Act specifies that the FCC was created:

> For the purpose of regulating interstate and foreign commerce in communication by wire and radio so as to make available, so far as possible, to all the people of the United States, without discrimination on the basis of race, color, religion, national origin, or sex, a rapid, efficient, nationwide, and worldwide wire and radio communication service with adequate facilities at reasonable charges, for the purpose of the national defense, for the purpose of promoting the safety of life and property through the use of wire and radio communication, and for the purpose of securing a more effective execution of this policy by centralizing authority heretofore granted by law to several agencies and by granting additional authority with respect to interstate and foreign commerce in wire and radio communication. . . . (47 USC §151)[1]

THE COMMISSION

The FCC is directed by five Commissioners appointed by the President and confirmed by the Senate for staggered 5-year terms (47 CFR §0.1).[2] No more than three can be members of the same political party, and none can have a financial interest in any Commission-related business. The President designates one Commissioner as Chairman (47 CFR §0.3). The Commissioners make their decisions collectively by formal vote, although authority to act on routine matters is normally delegated to the staff.

FCC ORGANIZATION

The staff of the FCC performs the day-to-day functions of the agency, including processing of applications for licenses and other filings, analyzing complaints, conducting investigations, developing and implementing regulatory programs and rules, and participating in hearings, among other things (47 CFR §0.5). There are currently seven bureaus and 10 offices. Generally, the offices provide specialized support services. Bureaus and offices regularly join forces and share expertise in addressing FCC-related issues.

Since enactment of the Communications Act, the FCC has undergone numerous restructurings and reorganizations. In late 1999, the FCC announced the creation of two new bureaus, the Enforcement Bureau and the Consumer Information Bureau. In 2002, another reorganization was approved, and the Media Bureau, Wireline Competition Bureau, Consumer & Governmental Affairs Bureau, and International Bureau were established or realigned. Additionally, in 2006, the Public Safety & Homeland Security Bureau was formed. Normally, broadcasters deal with the Media Bureau (MB); however, actions by other elements of the agency may directly affect broadcasters.

[1]The U.S. Code (USC) is the codification by subject matter of the general and permanent laws of the United States. It can be accessed at http://www.gpoaccess.gov/uscode/index.html.

[2]The Code of Federal Regulations (CFR) is the codification of the general and permanent rules published in the *Federal Register* by the executive departments and agencies of the federal government. It can be accessed at http://www.gpoaccess.gov/cfr/index.html.

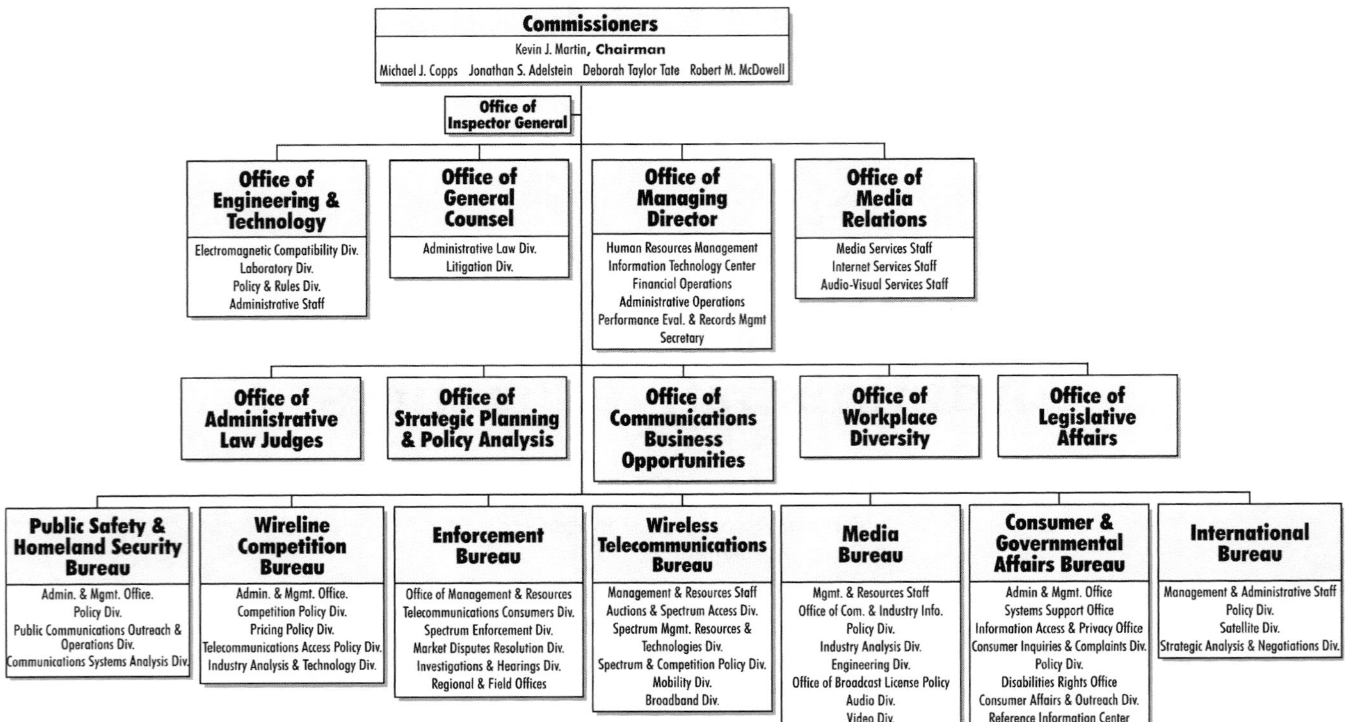

FIGURE 1.3-1 Operational/organizational chart for the Federal Communications Commission.

Each of the bureaus and offices is discussed in detail below, and the organizational units of the FCC are shown in Figure 1.3-1.

Office of the Managing Director

The Office of the Managing Director (OMD) (47 CFR §0.11) is responsible for activities involving the administration and management of the FCC. The Managing Director serves as the Chief Operations and Executive office, as supervised and directed by the Chairman. Under the direction of the Managing Director, the OMD:

- Develops and manages the agency's budget and financial programs
- Develops and oversees the agency's personnel management process and policy
- Designs and installs agency telecommunications and computer services
- Administers the fee program
- Develops and implements agency-wide management systems
- Oversees the agency's physical space and security, provides support services, and manages contracts and purchasing actions
- Through the Office of the Secretary, coordinates the FCC meeting schedule and manages the distribution and publication of official FCC documents

Office of Media Relations

The Office of Media Relations (OMR) (47 CFR §0.15) is responsible for the dissemination of information on FCC issues. The OMR is responsible for coordinating media requests for information and interviews on FCC proceedings and activities and for encouraging and facilitating media dissemination of Commission announcements, orders, and other information. The OMR manages the FCC *Daily Digest*, the FCC Web site, and the FCC's audio-visual center.

Office of Inspector General

The Office of Inspector General (OIG) (47 CFR §0.13) was established as an independent entity in 1989. The OIG provides independent and objective audits and investigations relating to agency programs and operations. The OIG also provides leadership and coordination and recommends policies to prevent and detect fraud and abuse in agency programs and operations. The Inspector General also provides a means for keeping the Chairman, Commissioners, and Congress fully informed about problems and deficiencies at the agency.

Office of Administrative Law Judges

The Office of Administrative Law Judges (OALJ) (47 CFR §0.151) is responsible for conducting the hearings ordered by the FCC. The hearing function includes

acting on interlocutory requests filed in the proceedings such as petitions to intervene, petitions to enlarge issues, and contested discovery requests. An Administrative Law Judge, appointed under the Administrative Procedures Act (APA; discussed below in the "The Rulemaking Process" section), presides at the hearing during which documents and sworn testimony are received in evidence, and witnesses are cross-examined. At the conclusion of the evidentiary phase of a proceeding, the presiding Administrative Law Judge writes and issues an Initial Decision, which may be appealed to the FCC.

Office of General Counsel

The Office of General Counsel (OGC) (47 CFR §0.41) serves as the chief legal advisor to the FCC and the various bureaus and offices. The General Counsel also represents the FCC before the federal courts, recommends decisions in adjudicatory matters before the Commission, assists the Commission in its decision-making capacity, performs a variety of legal functions regarding internal administrative matters, and advises the Commission on fostering competition and promoting deregulation in a competitive environment.

Office of Workplace Diversity

The Office of Workplace Diversity (OWD) (47 CFR §0.81) serves as the principal advisor to the Chairman and FCC on all aspects of workforce diversity, affirmative recruitment, equal employment opportunity, and civil rights within the Commission. The OWD develops, coordinates, evaluates, and recommends to the FCC internal policies, practices, and programs designed to foster a diverse workforce and to promote equal opportunity for all employees and applicants for employment.

Office of Communication Business Opportunities

The Office of Communication Business Opportunities (OCBO) (47 CFR §0.101) acts as the principal advisor to the Chairman and the Commissioners on issues, rulemakings, and policies affecting small, women-owned, and minority-owned communications businesses. The OCBO also represents the FCC in various matters coordinated with the U.S. Small Business Administration, including matters relating to the Regulatory Flexibility Act and the Small Business Act. In promoting telecommunications business opportunities for small, minority-owned, and women-owned businesses, the OCBO works with entrepreneurs, industry, public interest organizations, individuals, and others to provide information about FCC policies, increase ownership and employment opportunities, foster a diversity of voices and viewpoints over the airwaves, and encourage participation in FCC proceedings.

Office of Strategic Planning and Policy Analysis

The Office of Strategic Planning and Policy Analysis (OSP) (47 CFR §0.21) is responsible for working with the Chairman, Commissioners, bureaus, and offices to develop a strategic plan identifying short- and long-term policy objectives for the FCC. As part of this process, the OSP helps prepare the agency's annual budget, ensuring that budget proposals mesh with agency policy objectives and plans. The OSP also works with the Office of the Chairman and the Managing Director in developing a workforce strategy consistent with agency policy-related requirements. The OSP is responsible for monitoring the state of the communications industry to identify trends, issues, and overall industry health, and it produces staff working papers. The OSP acts as an expert consultant to the FCC in areas of economic, business, and market analysis and other subjects that cut across traditional lines such as the Internet. The OSP also reviews legal trends and developments not necessarily related to current FCC proceedings, such as intellectual property law, Internet, and e-commerce issues.

Office of Legislative Affairs

The Office of Legislative Affairs (OLA) (47 CFR §0.17) is the FCC's liaison to Congress. The OLA provides lawmakers with information regarding FCC regulatory decisions, answers to policy questions, and assistance with constituent concerns. The OLA also prepares FCC witnesses for congressional hearings and helps create FCC responses to legislative proposals and congressional inquiries. In addition, the OLA is a liaison to other federal agencies, as well as state and local governments.

Office of Engineering and Technology

The Office of Engineering and Technology (OET) (47 CFR §0.31) is responsible for developing overall policies, objectives, and priorities for OET programs and activities. The OET makes recommendations to the FCC on how the radio spectrum should be allocated and establishes the technical standards to be followed by users. In addition, the OET conducts engineering and technical studies to obtain theoretical and experimental data on new or improved techniques, including cooperative studies with other staff units and consultant and contract efforts as appropriate. The OET has four divisions:

- The *Policy and Rules Division* provides the FCC advice on technical and policy issues pertaining to spectrum allocation and use; prepares rulemaking items on spectrum allocations, radiofrequency devices, and industrial, scientific, and medical equipment; advises the Commission on technical, policy, and standards issues for the various radio services; and represents the Commission by participating in the work of national and international committees and other government agencies related to equipment authorization policy and rules and implementation of mutual recognition agreements.

- The *Electromagnetic Compatibility Division* plans and conducts studies on radiowave propagation and communications systems characteristics and develops analytical techniques and models that enable the FCC to improve spectrum use; performs mathematical and analytical studies for improved understanding of spectrum efficiency and other characteristics of telecommunications systems and devices; plans for future uses of the radio spectrum relative to developing technologies and services and domestic telecommunications requirements; supports implementation of new services and technologies; and provides expert advice to other parts of the Commission.
- The *Laboratory Division* acts on applications for certification of equipment or advance approval of subscription television technical systems; evaluates equipment subject to manufacturer documentation of compliance; identifies and evaluates new or novel products submitted for equipment authorization; designs test procedures for equipment subject to FCC regulation and maintains a laboratory facility used to conduct such tests; and provides customer service support in the equipment authorization program.
- The *Network Technology Division* develops strategic technology advice and directions on major issues and items in implementing the Communications Act; provides objective technical analysis and develops frameworks for assessing the competitive effect of market forces on new technologies; analyzes the effect of convergent architectures on the FCC's regulations; identifies and analyzes the impact of new network technologies that will necessitate significant changes in the Commission's regulations; and assists the bureaus and offices in identifying and resolving major technical issues.

Media Bureau

The Media Bureau (MB) (47 CFR §0.61) oversees the policy and licensing functions for electronic media, including cable television, broadcast television, and radio. The MB also handles post-licensing matters involving direct broadcast satellite and implements rules and policies to spur the transition to digital television and radio. The MB has five divisions and two offices:

- The *Audio Services Division* licenses commercial and noncommercial educational AM, FM, FM Translator, and FM booster radio services, as well as the noncommercial educational low-power FM radio service. This division provides legal analysis of broadcast, technical, and engineering radio filings and recommends appropriate dispositions of applications, requests for waivers, and other pleadings.
- The *Video Services Division* licenses commercial and noncommercial educational television, low-power television, Class A television, television translators, and television booster broadcast services. This division provides legal and technical analysis of applications and recommends appropriate dispositions of applications, requests for waivers, and other pleadings.
- The *Policy Division* conducts proceedings concerning broadcast, cable, and post-licensing direct broadcast satellite issues, including the Satellite Home Viewer Extension and Reauthorization Act (SHVERA), over-the-air-reception devices, digital transition, customer premises equipment, access to programming and distribution platforms, and other related matters. It also facilitates competition in the multichannel video programming marketplace by resolving carriage and other complaints involving access to facilities as well as petitions for findings of effective competition. In addition, this division administers the FCC's programs for political broadcasting and Equal Employment Opportunity (EEO) matters.
- The *Industry Analysis Division* conducts and participates in proceedings regarding media ownership and the economic aspects of existing and proposed rules and policies. This division reviews license transfers that implicate significant policy issues and collects, compiles, analyzes, and develops reports on relevant industry and market data and information, including preparation of the annual report to Congress on the status of competition in the market for the delivery of video programming.
- The *Engineering Division* conducts technical reviews of media-related matters, processes cable television relay service applications, special-relief and show-cause petitions involving technical matters, requests for ruling on technical matters, and requests for waivers of the rules.
- The *Office of Broadcast License Policy* develops, recommends, and administers policies and programs for the regulation of analog and digital broadcast services. It has direct responsibility for the work of the Audio and Video Services Divisions and maintains close working relationships with other divisions and offices of the Media Bureau.
- The *Office of Communications and Industry Information*, in coordination with the Office of Legislative Affairs, responds to inquiries from members of Congress and their staffs, prepares material for FCC personnel participating in Congressional hearings and meetings, and provides analysis of legislative proposals concerning specific matters within the jurisdiction of the Media Bureau. The Office also distributes official bureau decisions and reports and processes Freedom of Information Act (FOIA) requests.

Consumer & Governmental Affairs Bureau

The Consumer & Governmental Affairs Bureau (CGB) (47 CFR §0.141) interacts with consumers by responding directly to their inquiries and complaints and by conducting information and education campaigns. The CGB is responsible for consumer and governmental affairs policies that enhance the public's understanding of the FCC's work and improve the FCC's relationships with other governmental agencies, including reaching out to consumers to increase awareness about FCC rules, regulations, and policies. The CGB distributes information to enable consumers to make wise choices and find the best rates for telecommunications services and products and is the principal point of contact with the public seeking FCC records and documents. The CGB conducts consumer-related rulemakings and orders and interacts with public, federal, state, local, tribal, and other governmental agencies.

Enforcement Bureau

The Enforcement Bureau (EB) (47 CFR §0.111) is responsible for enforcing the Communications Act, as well as the FCC's rules, orders, and authorizations; promotes telephone service competition; protects consumers; and fosters efficient use of the spectrum while furthering public safety goals. The EB is also responsible for resolution of complaints against broadcast stations on matters such as indecency, broadcast of telephone calls, and contests. The EB has several regional and district field offices across the country (47 CFR §0.121). These offices conduct on-site investigations and inspections of possible FCC violations at broadcast stations and other operations regulated by the Commission.

International Bureau

The International Bureau (IB) (47 CFR §0.51) is responsible for rules, policies, and licensing concerning satellites and related spectrum issues, international telecommunications facilities and services, and compliance with international agreements. Its mission is to promote innovative, efficient, reasonably priced, widely available, reliable, timely, and high-quality domestic and global communications services. The IB advises and recommends to the FCC, or acts for the Commission under delegated authority, in the development and administration of international telecommunications policies and programs. In addition, the IB protects U.S. consumers and licensees from radio interference along U.S. borders and reviews foreign ownership of companies who provide service to U.S. consumers.

Wireless Telecommunications Bureau

The Wireless Telecommunications Bureau (WTB) (47 CFR §0.131) handles all FCC domestic wireless telecommunications programs and policies, except those involving satellite communications or broadcasting, including licensing, enforcement, and regulatory functions. Wireless communications services include cellular telephone, paging, personal communications services, public safety, and other commercial and private radio services. The WTB is also responsible for implementing the competitive bidding authority for spectrum auctions, given to the Commission by the 1993 Omnibus Budget Reconciliation Act. The WTB regulates the Broadcast Auxiliary Microwave Service.[3] Broadcast Auxiliary Microwave authorization is available to licensees of broadcast stations and to broadcast or cable network entities. Broadcast Auxiliary Microwave stations are used for relaying broadcast television signals. They can be used to relay signals from the studio to the transmitter or between two points, such as a main studio and an auxiliary studio. The Broadcast Auxiliary Microwave Services also include mobile television pickups, which relay signals from a remote location back to the studio.

Wireline Competition Bureau

The Wireline Competition Bureau (WCB) (47 CFR §0.91) develops and recommends policy goals, objectives, programs, and plans for the FCC on matters concerning wireline telecommunications. The WCB's overall objectives include ensuring choice, opportunity, and fairness in the development of wireline telecommunications services and markets; developing deregulatory initiatives; promoting economically efficient investment in a wireline telecommunications infrastructure; promoting the development and widespread availability of wireline telecommunications services; and fostering economic growth. The WCB is organized into four divisions and an Administrative and Management Office.

Public Safety & Homeland Security Bureau

The Public Safety & Homeland Security Bureau (47 CFR §0.191) is responsible for all FCC activities pertaining to public safety, homeland security, national security, emergency management and preparedness, disaster management, and other related issues. The Bureau is designed to provide an efficient, effective, and responsive organizational structure to address such matters and has three divisions to reach this goal: Policy Division, Public Communications Outreach & Operations Division, and Communications Systems Analysis Division. The Bureau is responsible for combined public safety-related functions that were previously dispersed among the other bureaus and offices and also serves as a clearinghouse for public safety communication information.

THE LICENSING PROCESS

Any qualified citizen, company, or group may apply to the FCC for authority to construct a standard (AM), frequency modulation (FM), or television (TV) broad-

[3]Additional information on licensing of the Broadcast Auxiliary Microwave Service may be found at http://wireless.fcc.gov/microwave/brdcstaux.html.

cast station. Licensing of these facilities is prescribed by the Communications Act of 1934, as amended, which sets up certain basic requirements (47 USC §308). In general, applicants must satisfy the FCC that they are legally, technically, and financially qualified, and that operation of the proposed station would be in the public interest. Full details of the licensing procedure and station operation are in Part 1 ("Practice and Procedure") and Part 73 ("Radio Broadcast Services") of the Commission's rules. Part 73 includes technical standards for AM, FM, and TV stations, and TV and FM channel (frequency) assignments by states and communities. Copies of the complete rules may be purchased from the Superintendent of Documents, Government Printing Office, Washington, D.C. 20402, (202) 512-1800, or can be viewed online at http://www.gpoaccess.gov/cfr/index.html.

Most applicants retain engineering consultants and legal counsel to perform frequency searches and help prepare the legal and technical portions of construction permit applications. The FCC does not maintain a list of or recommend any particular legal services or broadcast engineering consultants (see the "Additional Information" section at the end of this chapter for information on engineering consultants). Also, the FCC cannot tell applicants whether a frequency will be available in a particular location or help in the preparation of applications (except for addressing questions of a general nature).

The following is a summary of basic information required when applying for the authorization to build and operate a broadcast station:

- Any FCC rule can be retrieved at the Code of Federal Regulations Web site (http://www.gpoaccess.gov/cfr/index.html) or obtained in book form from the Government Printing Office by calling (202) 512-1800.
- Frequencies for radio and television services are always in heavy demand. Where broadcast frequencies remain available, competing applications are routinely received. In many areas of the country, no frequencies may be available on which a new station could commence operating without causing interference to existing stations, in violation of FCC rules. For that reason, the FCC recommends that applicants not purchase any equipment before receiving a construction permit.
- When conflicts occur between mutually exclusive commercial applicants (that is, where interference between the applicants would be created if all applications were granted), the conflict will be resolved by means of an auction (47 CFR §1.2103). The auction process was mandated by the Telecommunications Act of 1996. Fees and auction payments collected by the FCC are directed to the U.S. Treasury. Information about the auction process is available at http://wireless.fcc.gov/auctions/. The FCC cannot provide advance information as to when an auction or application filing window for a particular service might occur.
- Expansion of the AM or FM radio bands beyond the present 535 to 1705 kHz (for AM) and 88 to 108 MHz (for FM) is unlikely to occur. The FM band is constrained from expanding above 108 MHz by the presence of aeronautical operations from 108 to 136 MHz and is also prevented from expanding below 88 MHz by channel 6 television operations from 82 to 88 MHz. The AM band was recently expanded from 1605 to 1705 kHz after years of international negotiations; however, those frequencies are reserved for existing stations that were causing significant interference in the lower part of the band.
- Unlicensed operation of radio broadcast stations is prohibited, even at such low powers of 1 watt or less. The only unlicensed operation that is permitted in the AM and FM broadcast bands is covered under Part 15 of the FCC's rules and is limited to a coverage radius of approximately 200 feet. Unlicensed operation is also not permitted in the TV band. FCC Enforcement Bureau field offices routinely monitor for unlicensed transmissions and may shut down unauthorized operators, confiscate equipment, issue fines, and/or initiate criminal prosecution for the illegal operation of an unlicensed station.
- The FCC is rapidly moving toward electronic filing of its applications and the elimination of paper forms. This procedure has several advantages, including error checking of application entries before an application is accepted for filing, more rapid posting of data, and reduced processing time. At some future date applicants will not be able to file paper forms when applying for a broadcast station construction permit.
- For commercial AM, FM, and TV broadcast station applications, filing fees must be paid with the submission of any application. These fees are detailed in the Media Bureau's *Fee Filing Guide*, which may be retrieved at http://www.fcc.gov/fcc-bin/audio/appfees.html.

FCC Application Forms

FCC application forms and instructions may be obtained on the FCC Web site at http://www.fcc.gov/formpage.html. Many forms now require electronic filing, and the Media Bureau has set up the Consolidated Database System (CDBS) to administer this process. This system provides the public with the ability to fill out application forms and to file them electronically. The current system supports electronic filing for the following FCC Forms: 301, 301-CA, 302-CA, 302-DTV, 302-FM, 302-TV, 314, 315, 316, 317, 318, 319, 323, 323-E, 337, 340, 345, 346, 347, 349, 350, 381, 396, 396-A, and 396-C. As other forms are developed, they will be made available for public use. Forms not found on the Forms Menu screen must still be filed on paper using the standard form filing procedures. Also available are informal filings for Change of Address, Consummation, Engineering STA, Legal STA, and Silent STA.

The CDBS electronic filing system consists of an account registration function and a forms filing function. Before any forms can be filed electronically, the appropriate applicant account data must be entered into the account creation/maintenance screen. During account creation, the applicant's account ID number will be generated and the user-specified password will be saved. Account data can be updated at any time. In addition to applicant/licensee information, all forms require information about a contact representative. The account maintenance button must be used to fill in this information before beginning a form (as the data is copied to the form at that time).

When an account has been created, the filer chooses an online form to complete. After completion of the pre-form, which helps in the selection of the appropriate subsections of the form, each section of the form should be completed just as it would be on paper. If attachments or exhibits must be included with the form, each such question accommodates free-form text exhibits or saved versions of documents, spreadsheets, graphs, plots, etc. When a section is completed, it must be validated before moving on to the next section. When a filer must log out prior to validation of a partially completed section, the Save function will preserve any information entered in the section.

After all sections of the form have been validated, the form can be filed. The Application Reference Number (ARN) is automatically assigned when the form has been filed and has passed the filing edit checks. When the fee (if any) has been confirmed and other automatic processes occur, the application is given the status of "filed" and electronically moves on to the processing system (used by the FCC staff).

An applicant can correct erroneous data values entered in the electronic form before the application is filed. Erroneous data values on filed electronic applications can only be corrected by filing an electronic amendment.

For telephone assistance in filling out forms, contact:

- Radio Broadcast forms—(202) 418-2700
- Television Broadcast forms—(202) 418-1600
- Broadcast, cable, and multichannel video programming distribution (MVPD) EEO forms—(202) 418-1450
- Broadcast ownership reports—(202) 418-1625

AM Stations

An applicant must make a search for an AM frequency on which to operate without causing or receiving interference from existing stations and stations proposed in pending applications. AM stations operate within the frequency band from 535 to 1705 kHz. In the United States and in Region 2, the channels are spaced at 10 kHz intervals. Stations are designated channel by channel to serve various size areas and operate on clear, regional, and local channels. Dominant clear channel stations (Class A) operate at a maximum power of 50,000 watts; secondary clear channel stations (Class B) operate at a power between 250 and 50,000 watts. Class B stations operating on regional channels may now operate at power levels between 250 and 50,000 watts. Class C stations operate on local channels serving limited areas and operating with no more than 1000 watts day and night. Many stations operate as daytime-only stations. Applications for Class D (daytime-only) stations are no longer being accepted. Submission of an application for an AM station requires the payment of an application filing fee.

To be acceptable, an application for a new AM broadcast station must show that no interference will be caused to other U.S. and foreign AM stations on the same frequency or on the adjacent channels (out to 30 kHz above or below the desired frequency; see 47 CFR §73.37). Applications must also consider the second harmonic frequency and intermediate frequency relationships per 47 CFR §73.182(s) (*e.g.*, 2 × 800 kHz = 1600 kHz for the second harmonic relationship, or 800 kHz + 455 kHz if the frequency could affect reception on 1250 and 1260 kHz). In general, these complex engineering analyses require specialized knowledge and software and are best performed by broadcast engineering consultants.

Applications for new AM broadcast stations must be electronically filed on FCC Form 301 during specified application window periods. Noncommercial educational applicants should use FCC Form 340. Commercial applicants must include the new station application filing fee listed in the Media Bureau's *Fee Filing Guide* and include FCC Form 159 with the fee payment and application. Payments for commercial applications must be directed to the address listed in the *Fee Filing Guide*, not the FCC in Washington, D.C.

The Media Bureau periodically announces filing window periods during which new station applications and major change applications may be filed. Filing window announcements are made via Public Notices, and notices are also posted at several locations on the FCC's Web site.

FM Commercial Stations

FM commercial stations may be authorized on 92.1 to 107.9 MHz, corresponding to channels 221 through 300. Noncommercial educational FM stations may also be authorized in this band, but such applications must meet the spacing, city coverage, and other technical criteria applicable to commercial stations. An allotment must be created before applications can be accepted for commercial FM stations. The allotment is a set of reference coordinates that meet the spacing requirements for other U.S., Canadian, and Mexican stations. The allotment coordinates must also provide for a minimum signal strength of 70 dBuV/m over 100% of the area of the community of the license, using the maximum reference effective radiated power (ERP) and antenna height for the FM station class.

As FM commercial allotments are adopted by the FCC, they are added to the Table of Allotments (47 CFR §73.202). For each new allotment, an application filing window will be opened at some future date, during which applications on FCC Form 301 must be filed. Applicants may petition the FCC to add new FM commercial allotments. There is no form on which a petition for rulemaking must be filed. Each petition should contain:

- The community of license for which the channel is sought
- The frequency or channel
- A set of reference coordinates proposed for the allotment (latitude and longitude)
- An expression of interest in applying for the channel, if it is allocated

Only proposed allotments that meet the spacing and city coverage criteria will be put out for public comment. Other parties will then have an opportunity to file comments or counterproposals to the proposed allotment.

Petitions for rulemaking for new allotments should be directed (in triplicate) to the attention of the Audio Services Division (Media Bureau), c/o Office of the Secretary TW B204, Federal Communications Commission, 445 12th Street S.W., Washington, D.C. 20554. When the comment period is closed, the Commission may then make new allotments in keeping with its existing procedures. No finders' preference is made either at the allotment rulemaking or at the application stage. Applicants may find it helpful to examine FCC Form 301 to see what sort of information will be required with an application. Competing applications received during the application filing window will be set for auction, with the highest bidder receiving the construction permit for that allotment.

The Media Bureau periodically announces filing window periods during which new radio station applications and major change applications may be filed. Filing window announcements are made via Public Notices, and notices are also posted at several locations on the FCC's Web site.

FM Noncommercial Educational Stations

FM noncommercial educational stations may be authorized on 88.1 to 91.9 MHz, corresponding to channels 201 through 220. No commercial operation is permitted on these frequencies. FM noncommercial educational stations may also be authorized in the commercial FM band under the technical rules applying to that service. Contour protection is used to determine if interference with other stations will occur.

For noncommercial educational stations on channels 201 through 220, no allotment is established. Allocation is made via an on-demand system, with applicants receiving construction permits for facilities that will not cause interference with other stations. Interference calculations are made using specified signal strength contours, where protected service contours for one station generally cannot overlap the interfering contours for another station. Applicants must also protect pending applications that were filed before the announcement of the application filing window.

The Media Bureau periodically announces filing window periods during which new noncommercial educational station applications and major change applications may be filed. Filing window announcements are made via Public Notices, and notices are also posted at several locations on the FCC's Web site.

FCC Form 340 for noncommercial educational stations must be used to apply for this type of FM station. Although no application filing fee is involved, these applications must be filed (in triplicate) with the Office of the Secretary, Room TW B204, 445 12th Street S.W., Washington, D.C. 20554.

Low-Power FM Stations

Low-power FM (LPFM) stations operate with 1 to 100 watts of power and cover a radius of approximately 5.6 km (3.5 miles). Current information about the LPFM filing service is maintained on the LPFM page at http://www.fcc.gov/mb/audio/lpfm/index.html. The Audio Services Division has also assembled a program to help locate available FM channels for LPFM stations. Applications for new LPFM stations may only be filed during the dates specified for an application filing window. Applications received at other times will be returned without consideration. The FCC cannot provide advance information as to when the next application filing window may be, but when an announcement is made, it is posted on the Audio Services Division home page and on the LPFM main page on the FCC's Web site.

FM Translator Stations

FM translator stations rebroadcast existing FM stations to small areas. Noncommercial educational FM translators may be authorized on any frequency, while FM translators rebroadcasting commercial stations must stay on frequencies from 92.1 to 107.9 MHz (channels 221 to 300). Translator stations are prohibited from transmitting any programming not also transmitted on the originating or primary station at the same time.

FCC Form 345 for translator stations must be used to apply for this type of FM station. Commercial applicants must include the new station application filing fee listed in the Media Bureau's *Fee Filing Guide* and include FCC Form 159 with the fee payment and application. Note that commercial applications must be directed to the address listed in the *Fee Filing Guide*, not the FCC in Washington, D.C. Competing applications will be put up for auction, with the highest bidder receiving the construction permit for that allotment. Noncommercial applicants are not required to submit the application filing fee. Noncommercial FM translator applications for new stations must be filed (in triplicate) with the Office of the Secretary, Room TW B204, Federal Communications Commission, 445 12th Street S.W., Washington, D.C. 20554.

The Media Bureau periodically announces filing window periods during which new station applications and major change applications may be filed. Filing window announcements are made via Public Notices, and notices are also posted at several locations on the FCC's Web site. The FCC cannot provide advance information as to when the next application filing window might be.

Television Stations

Television in the United States is allocated through a Table of Allotments just as commercial FM allotments are made (47 CFR §73.606). At the time of publication of this handbook, television was in the midst of a conversion to digital transmissions. During this transition, each television station has been temporarily assigned a second TV channel on which to broadcast its digital signal, while the original channel continues broadcasting the analog signal. The date on which TV broadcasters must give up one of these two channels is February 17, 2009. Until the conversion to digital TV broadcasting is complete, the FCC is not accepting applications for new television stations. When applications for new television stations begin to be accepted by the FCC (to be announced in a Public Notice), these applications will be subject to broadcast auctions, with the possible exception of those few allotments specifically reserved for noncommercial educational television use.

Television Translators/Low-Power Television Stations

There are two types of secondary television stations—television translators and low-power television (LPTV). A television translator is a station that rebroadcasts signals from full-service television stations. LPTV stations operate with less than 150 kW of power and may rebroadcast a full-service station television signal in addition to originating their own programming. LPTV stations may also operate as a subscription service. These two secondary television services may operate on any available VHF or UHF channel, provided they do not cause interference with full-service stations or other authorized translator or LPTV stations.

How to Apply for an AM, FM, or Television Broadcast Station

To apply for a new AM station, a commercial FM station, or a commercial TV broadcast station for which a vacant allotment is available, the applicant must submit FCC Form 301 (Application for Construction Permit for Commercial Broadcast Station), with the appropriate application filing fee. Noncommercial educational television applicants and noncommercial educational FM station applicants filing for operation on channels 201 through 220 must submit FCC Form 340 (Application for Construction Permit for Noncommercial Educational Broadcast Station). No application filing fee is required for noncommercial educational applicants.

Other forms are used to make modifications to an existing facility. FM translator, TV translator, and LPTV applicants must submit FCC Form 346 (Application for Authority to Construct or Make Changes in an LPTV, TV Translator, or FM Translator Station). These forms require information about the citizenship, legal, and financial qualifications of the applicant, as well as engineering and technical specifications of the proposed or modified transmitter site.

Another way to obtain a broadcast station is to purchase an existing station that the owner is willing to sell. The FCC does not maintain a list of stations for sale and does not participate in the negotiations of the sales contract. Station brokers and communications attorneys can assist in identifying stations that are for sale. Potential applicants may also contact individual station owners directly to see if they are interested in selling their station. When a station for sale has been found and a contract signed to purchase the station, FCC Form 314 (Application for Consent to Assignment of Broadcast Construction Permit or License) must be submitted within 30 days accompanied by the appropriate filing fees. Applicants who apply to purchase a station may not take over operation until the FCC approves the application to purchase the station.

When the application is approved, the buyer must submit a letter of consummation within 90 days of the grant. FCC Form 323 (Ownership Report for Commercial Stations) or FCC Form 323-E (Ownership Report for Noncommercial Educational Stations) must also be submitted within 90 days of the grant. FCC Form 315 (Application for Consent to Transfer of Control of Corporation Holding Broadcast Station Construction Permit or License) must be submitted when a controlling block of shares of a broadcasting company is transferred to a new entity or an individual. FCC Form 316 (Application for Consent to Assignment or Transfer of Control) is used when a station is involuntarily transferred, such as to a trustee in bankruptcy. FCC Form 316 is also used for *pro forma* (changes in form, not substance) assignments and transfers, such as a sale from a person to a corporation controlled by that person.

Broadcast applications must be submitted in triplicate, with the appropriate application filing fee attached (see the Media Bureau's *Fee Filing Guide* for fee information). Applications for noncommercial educational stations do not require a filing fee.

Applicants Must Give Local Notice

All applicants for new broadcast stations (except LPFM) must give local notice in a newspaper of general circulation in the community in which the station is licensed or proposed to be licensed (47 CFR §73.3580). They must also afford an opportunity for the public to file comments on these applications with the FCC. Copies of the application must be maintained in the station's public files or at a location accessible to the public in the community where the station is proposed—for example, a public library or post office. Licensees who submit license renewal applications must give local public notice of the filing by broadcasting announcements over their stations.

What Happens When an Application Is Filed with the FCC?

Applicants for new and major modifications in existing facilities of commercial and noncommercial AM, FM, and TV broadcast stations are tendered for filing and then placed on a cutoff list when the application is accepted for filing. The cut-off Public Notice of acceptance triggers a 30-day filing period for any competing applicants interested in filing an opposition or petition to deny any of the applications listed on the Public Notice.

Applications generally are processed in the order in which they are filed. They are reviewed for engineering, legal, and financial data by the Media Bureau, which under delegated authority acts on routine applications. If an application is defective, the applicant will be afforded one opportunity to amend the application; thereafter, if all defects have not been corrected, the application will be returned. In most instances, all filing fees are retained by the FCC regardless of the final disposition of the application. If an application is complete and has no defects, petitions to deny, or competing applications, the application may be granted without a hearing and a construction permit issued. All dispositions of applications are announced by the FCC. Petitions for reconsideration of grants made without a hearing can be filed within 30 days of the release date of the Public Notice of such grant; however, when an objection is filed, the petitioner must show good cause why the objection was not raised before the grant.

Hearing Procedure

In instances where it appears that an application does not conform to FCC rules and regulations and petitions to deny or competing applications are filed, the Commission may designate the application for hearing (47 CFR §73.3593). The Commission will issue a Public Notice listing the applications that are designated for hearing. The hearing fee must be paid within 60 days of the release of that Public Notice. The hearing notice generally allows the applicant 20 days to file a Notice of Appearance. The Administrative Law Judge (ALJ) hearing the case is also assigned at this time. Amendments to the application may be filed as a matter of right within 30 days, with a showing of good cause, if the amendment relates to issues first raised in the Hearing Designation Order. After the hearing procedure and review of evidence and statements, the ALJ issues an Initial Decision. Initial Decisions may be contested within specified time limits. In all cases heard by an ALJ, the FCC or a review board it establishes may hear oral arguments and may adopt, modify, or reverse the ALJ's Initial Decision. Any action taken by the Commission may be appealed to the U.S. Court of Appeals for the District of Columbia Circuit (47 CFR §1.13).

Construction Permits

When an application is granted, the grantee is given a specified period of time for completing construction of the station. Construction permits for radio and television stations are issued for 3 years; 18 months for LPFM radio stations (47 CFR §73.3598). Grantees must request call sign assignment when the construction permit has been granted. If no request is received from the grantee for a specific call sign, the FCC will automatically assign one. If for whatever reason a grantee of a construction permit cannot complete construction of the station within the specified time, the grantee may request an extension of time on FCC Form 307 (Application for Extension of Broadcast Construction Permit or to Replace Expired Construction Permit) provided the grantee justifies the request by showing the amount of work completed and providing an estimate of the amount of time necessary to complete construction.

Upon completion of construction, the grantee may begin program tests. For a nondirectional AM or FM station or a nondirectional or directional TV or Class A TV station, a grantee may begin program testing after simply notifying the FCC; however, an application for a station license must be on file within 10 days of commencement of program tests. A grantee of an FM station with a directional antenna system must file an application for license requesting authority to commence program testing at least 10 days prior to the date on which full power operations are scheduled to commence (47 CFR §73.1620). If the grantee fails to complete construction within the allotted time or does not file a timely application for a license, the construction permit will automatically be forfeited on the expiration date.

Comparative Hearings

The 1996 Telecommunications Act substantially changed the FCC's processes for license renewal applications. Prior to the Act, stations filing renewal applications faced the possibility of other applicants filing for the same (or mutually exclusive) facilities. In those situations, the FCC was required to conduct a comparative hearing weighing the qualifications of the renewal application and the competitor. These hearing were always lengthy and expensive and, in rare instances, resulted in displacement of the incumbent licensee; however, in 2002, the FCC disposed of the last remaining comparative renewal proceeding.

Licenses and Ownership

All radio and television stations are licensed for 8 years (47 CFR §73.1020). With regard to ownership, in June 2003 the FCC released its decision reexamining its entire regulatory scheme governing media ownership. Generally, the FCC relaxed several of its ownership restrictions to permit increased consolidation;

however, in response to a petition from public interest groups, in September of that same year, the Third Circuit Court of Appeals issued a stay of the new ownership rules. In June 2004, the Third Circuit issued its opinion on review of the June 2003 FCC order. Overall, the court generally sided with the public interest groups' arguments against further media consolidation.

In June 2005, the Supreme Court declined to grant the requests of the National Association of Broadcasters and other media organizations to review the broadcast ownership case. This means that the 2004 decision of the Third Circuit stands. In June 2006, in response to the 2004 court decision, the FCC opened a new phase for a rulemaking proceeding concerning major broadcast ownership rules. The current ownership rules are given below (see 47 CFR §73.3555).

Local Radio Ownership

- In a radio market with 45 or more full-power commercial and noncommercial radio stations, a party may own, operate, or control not more than 8 commercial radio stations in total and not more than 5 commercial stations in the same service (AM or FM).
- In a radio market with 30 to 44 full-power commercial and noncommercial radio stations, a party may own, operate, or control not more than 7 commercial radio stations in total and not more than 4 commercial stations in the same service (AM or FM).
- In a radio market with 15 to 29 full-power commercial and noncommercial radio stations, a party may own, operate, or control not more than 6 commercial radio stations in total and not more than 4 commercial stations in the same service (AM or FM).
- In a radio market with 14 or fewer full-power commercial and noncommercial radio stations, a party may own, operate, or control not more than 5 commercial radio stations in total and not more than 3 commercial stations in the same service (AM or FM); however, no person or single entity (or entities under common control) may have a cognizable interest in more than 50% of the full-power commercial and noncommercial radio stations in such market unless the combination of stations is composed of not more than one AM and one FM station.

Local Television Ownership

A party may own, operate, or control more than one full-power commercial television broadcast station in the same Designated Market Area (DMA, as assigned by Nielsen Media Research) in accordance with the following conditions and limits:

- At the time the application to acquire or construct the station(s) is filed, no more than one of the stations that will be attributed to such party is ranked among the top four stations in the DMA.
- In a DMA with 17 or fewer full-power commercial and noncommercial television broadcast stations, a party may own, operate, or control no more than 2 commercial television broadcast stations; or, in a DMA with 18 or more full-power commercial and noncommercial television broadcast stations, a party may own, operate, or control no more than 3 commercial television broadcast stations.

Cross-Media Ownership

Cross-ownership of a daily newspaper and commercial broadcast stations or of commercial broadcast radio and television stations is permitted without limitation except as follows:

- In DMAs in which three or fewer full-power commercial and noncommercial educational television stations are assigned, no newspaper/broadcast or radio/television cross-ownership is permitted.
- In DMAs to which at least four but not more than eight full-power commercial and noncommercial educational television stations are assigned, a party that directly or indirectly owns, operates, or controls a daily newspaper may have a cognizable interest in one of the following:
  - —One, but not more than one, commercial television station in combination with radio stations up to 50% of the applicable local radio limit for the market
  - —Radio stations up to 100% of the applicable local radio limit if the party does not have a cognizable interest in a television station in the market

These limits on newspaper/broadcast cross-ownership do not apply to any new daily newspaper inaugurated by a broadcaster.

National Television Multiple Ownership

No license for a commercial television broadcast station will be granted, transferred, or assigned to any party (including all parties under common control) if it would result in such party having a cognizable interest in television stations that have an aggregate national audience reach exceeding 39%. For purposes of making this calculation, UHF television stations will be attributed with 50% of the television households in their DMA market.

Auctions for Certain Mutually Exclusive Initial License Applications

Mutually exclusive applications for new facilities and for major changes to existing facilities in the following broadcast services are subject to competitive bidding: AM, FM, FM translator, analog television, low-power television, television translator, instructional television fixed service (ITFS), and Class A television. Mutually exclusive applications for minor modifications of Class A television and television broadcast are also subject to competitive bidding. Mutually exclusive applications for broadcast channels in the reserved portion of the FM band (channels 200 to 220) or for television broadcast channels reserved for noncommercial educational use and initial licenses or construction permits for digital television services given to replace existing analog licenses are not subject to competitive bidding procedures. In addition, public safety radio services, including private internal radio services used by state and local governments and non-government entities and including emergency road services provided by nonprofit organizations, are not subject to competitive bidding (47 CFR §1.2102).

RULES AND REGULATIONS

The FCC Rules and Regulations can be purchased online at www.gpo.gov; by writing the Government Printing Office, Superintendent of Documents, Attn: New Orders, P.O. Box 371954, Pittsburgh, PA 15250-7954; or by calling (202) 512-1800. The rules on FCC Practice and Procedures are contained in the Code of Federal Regulations (CFR), Title 47, Parts 0–19. Broadcast rules are found in Title 47 of the CFR, Parts 70–79. An online version of the CFR that allows users to browse and search all titles and parts of the CFR is located at http://www.gpoaccess.gov/cfr/index.html.

The Rulemaking Process

The FCC, like other federal government agencies, enacts new rules and regulations through the terms of the Administrative Procedures Act (APA). The APA specifies how rules may be proposed, adopted, and appealed. The APA assures that the public has input into the rulemaking process. Part I of the FCC Rules and Regulations provides detailed information on the FCC's general rules of practice and procedure. The following is a general summary of common actions:

- *Petition for Rulemaking* (47 CFR §1.401)—Such petitions bring the desires of individuals or groups to the attention of the FCC, which will evaluate the petitions and either dismiss them or accept them for action (47 CFR §1.407). If accepted, a Public Notice will be released giving a brief description of the details of the petition. Other rulemakings may be initiated by direction of the Congress, President, or courts.
- *Notice of Inquiry (NOI)* (47 CFR §1.430)—The FCC releases an NOI to gather information about a broad subject or as a means of generating ideas on a specific issue. NOIs are initiated by the FCC following an internal study or an outside request.
- *Notice of Proposed Rulemaking (NPRM)* (47 CFR §1.412)—After reviewing comments from the public in response to an NOI or as the first step in the rulemaking process, the FCC may issue an NPRM. An NPRM contains proposed changes to the FCC's rules and seeks public comment on these proposals.
- *Further Notice of Proposed Rulemaking (FNPRM)* (47 CFR §1.421)—After reviewing comments to the NPRM, the FCC may choose to issue an FNPRM regarding specific issues raised in the comments. The FNPRM provides an opportunity for the public to comment further on a related or specific proposal.
- *Report and Order (R&O)*—After considering comments to an NPRM or FNPRM, the FCC issues a Report and Order, which may develop new rules, amend existing rules, or make a decision not to do so. Summaries of the R&O are published in the *Federal Register.* The *Federal Register* summary indicates when a rule change will become effective.
- *Petition for Reconsideration* (47 CFR §1.429)—If a party is not satisfied with the way an issue is resolved in the rulemaking R&O, they can file a Petition for Reconsideration within 30 days from the date the R&O appears in the *Federal Register.*
- *Memorandum Opinion and Order (MO&O)*—In response to a Petition for Reconsideration, the FCC issues a Memorandum Opinion and Order or an Order on Reconsideration amending the new rules or stating that the rules will not be changed.
- *Public Notice (PN)*—A PN is issued to notify the public of an action taken or an upcoming event. If comments are requested, a PN will generally have filing information within the notice, such as where to send comments and a closing date for comments.
- *En Banc*—An *en banc* is a meeting of the FCC to hear various presentations on specific topics, usually using panel groups. Specific witnesses are asked to present information at an *en banc* hearing, following issuance of a Public Notice announcing the hearing. The FCC questions the presenters. Comments and presentations can be used by the FCC when it makes rules or proposes rulemakings.

Ex Parte Communications (47 CFR §1.1200 et seq.)

An *ex parte* presentation in a rulemaking proceeding is any oral or written presentation (other than the party's formal comments) made to decision-making personnel after an NPRM is issued.

Ex parte rules ensure that all participants in an FCC proceeding are given fair opportunity to present information and evidence in support of their positions.

The rules governing these presentations play an important role in protecting the fairness of the FCC's proceedings by ensuring that FCC decisions are not influenced by impermissible off-the-record communications between decision makers and others. At the same time, the rules are designed to ensure that the FCC has sufficient flexibility to obtain the information that is necessary for it to make reasonable decisions. The *ex parte* rules apply to anyone who engages in the kind of communications covered by the rules, whether or not they are a party to the proceeding.

In some types of proceedings (restricted proceedings), the rules prohibit *ex parte* presentations to decision makers concerning issues in the proceedings. In other types of proceedings (permit-but-disclose proceedings), the rules require that summaries of such presentations be placed in the record. In still other types of proceedings (exempt proceedings), there are no restrictions on *ex parte* presentations. The rules describe which types of restrictions or requirements, if any, apply to *ex parte* presentations in the various types of Commission proceedings.

A presentation is a communication directed to the merits or outcome of a proceeding, including any procedural or other issues raised in the proceeding. There are some exceptions, however. Communications that are inadvertently or casually made are not presentations. Neither are routine inquiries about compliance with the FCC's procedural rules, such as when a pleading must be filed, as long as the question has not become the subject of dispute in the proceeding. Inquiries relating solely to the status of a proceeding are not presentations; however, a status inquiry is deemed a presentation if it states or implies a view as to the merits or outcome of the proceeding or a preference for a particular party, states why timing is important to a particular party (other than the need to avoid administrative delay), or indicates a view as to the date by which a proceeding should be resolved. An *ex parte* presentation is any presentation that, if written (including electronic mail), is not served on the parties to the proceeding, or, if oral, is made without advance notice to the parties and without opportunity for them to be present.

Decision-making personnel are those people at the FCC who are or who may reasonably be expected to be involved in formulating a decision, rule, or order in a proceeding. All FCC bureau or office staff are considered decision-making personnel with respect to decisions, rules, and orders in which their bureau or office participates unless they have been designated as part of a separate trial staff or otherwise formally excluded from the decisional process in the proceeding.

Persons making written *ex parte* presentations must, no later than the next business day after the presentation, submit notice of the presentation to the Commission's Secretary under separate cover for inclusion in the public record. The presentation (and cover letter) must clearly identify the proceeding to which it relates (including the docket number, if any) and be labeled as an *ex parte* presentation. If the presentation relates to more than one proceeding, notice must be filed for each proceeding.

Persons making oral *ex parte* presentations must disclose them if they present data or arguments not already reflected in that person's written comments, memoranda, or other filings in that proceeding. In that case, the person must, no later than the next business day after the presentation, submit to the Commission's Secretary, with copies to the Commissioners or Commission employees involved in the oral presentation, a memorandum that summarizes the new data or arguments. The subject matter of the presentation must be fully disclosed; a mere listing of the subjects discussed is not sufficient, and more than a one- or two-sentence description of the views and arguments presented is required.

The memorandum (and cover letter) must clearly identify the proceeding to which it relates (including the docket number, if any) and be labeled as an *ex parte* presentation. If the presentation relates to more than one proceeding, a copy of the memorandum must be filed for each proceeding. There is an exception to these requirements where, for example, presentations occur in the form of discussion at a widely attended meeting and preparation of a memorandum as specified in the rule might be cumbersome. Under these circumstances, the rule may be satisfied by submitting a transcript or tape recording of the discussion as an alternative to a memorandum.

When the FCC has issued a Public Notice that a matter will be considered at a Commission meeting (that is, the matter has been placed on the "Sunshine Agenda"[4]), restrictions are imposed on presentations to FCC decision makers in addition to the limitations otherwise applicable under the *ex parte* rules. While the sunshine period prohibition is in effect, all presentations to decision makers concerning matters listed on a Sunshine Agenda, whether *ex parte* or not, are prohibited unless they fall within certain exceptions. The sunshine period prohibition applies from the release of the Sunshine Agenda Public Notice until the FCC releases the text of a decision or order relating to the matter, issues a Public Notice stating that the matter has been deleted from the Sunshine Agenda, or issues a Public Notice stating that the matter has been returned to the staff for further consideration, whichever occurs first.

Filing Comments

When the FCC proposes new rules, a time period is established for the public to comment on these proposed rules (47 CFR §1.415). Each of the Commission's documents containing proposed rules clearly details the specific dates, deadlines, and locations for filing comments and reply comments. After initial comments are filed, there is an additional period for responding to the first set of comments. Reply comments are filed to support or disagree with what others have said in their initial comments. These two comment periods provide the commission with a "written debate" of the issues. The comments and reply comments received are reviewed and enter heavily into the Commission's final actions; however, a comment does not represent a vote for the proposed rules. The Commission must decide on each issue based on the public's interest, convenience, and necessity. Even if the majority of the comments opposes an item, the FCC can nevertheless adopt the proposal.

How to File Comments with the Commission

Comments may be filed with the FCC electronically or on paper. The FCC prefers and recommends that comments be filed electronically because of the ease and low cost of doing so. Documents can be filed with the FCC for all docketed and rulemaking proceedings through the Electronic Comment Filing System (ECFS)

[4]The FCC's rules regarding "Sunshine Agenda" may be found at 47 CFR §1.1203.

at http://www.fcc.gov/e-file/ecfs.html, with the exception of hearing cases and Tables of Allotments. The ECFS accepts documents 24 hours a day, with a midnight filing deadline. The official receipt for electronic filings will reflect Monday through Friday dates, except legal holidays.

If a commenter is unable to file comments electronically or prefers to file comments on paper, the following should be kept in mind:

- *Docket number*—Rulemaking proceedings at the FCC are assigned docket numbers. Each docket number lists a Bureau, a year, and a specific number assigned to the proceeding (e.g., MB 05-001 = 2005 Media Proceeding Number 1). If a document is being submitted that pertains to a docketed proceeding, the docket number must be included in the filing.
- *Copies*—Except as otherwise specifically provided in the Commission's Rules (See 47 CFR §1.51), the number of documents to be filed is as follows:
  - — Rulemaking proceedings:
    - Petitions, comments, reply comments required by the FCC (*original and 4 copies*); to include Commissioners (*original and 9 copies*); informal comments (*original and 1 copy*)
    - Table of Allotments (*original and 4 copies*)
    - *Ex parte* presentations (*original and 1 copy*) (*Note:* In *ex parte* filings, when referencing multiple docket numbers, file numbers, or rulemaking numbers in a single document, 2 copies of the document are required for each referenced docket number, file number, or rulemaking number.)
  - — Hearing proceedings:
    - Documents decided by Administrative Law Judges (*original and 6 copies*); full Commission (*original and 14 copies*)
    - Notices of Appearance (*original and 2 copies*)
    - Depositions (*original and 3 copies*)
    - Interrogatories (*original and 3 copies*)
  - — All other filings—pleadings, briefs, petitions, etc. (*original and 4 copies*)
- *Type size*—All filings must be in 10- or 12-point type or legibly written (47 CFR §1.49).
- *Contact name*—A contact name, address, and telephone number must be included on filed documents.
- *Signatures*—Filed documents require an original signature above the typed or clearly printed name.
- *Hand-delivered filings*—The person making the delivery should remove the filing package from its box or envelope before submission. The FCC will either sign for receipt of the filing or provide a stamped receipt copy, but not both. Hand-delivered documents are accepted Monday through Friday, except legal holidays, during the hours of 8:00 a.m. and 7:00 p.m. (Eastern Time). Questions may be directed to the Office of the Secretary by phone at (202) 418-0300 (voice) or (202) 418-2970 (TTY) or through the Secretary's Web site (http://www.fcc.gov/osec). Rulemaking comments may also be filed by e-mail.
- *Notifications*—If the document contains information to be withheld from public inspection, then "Confidential, Not for Public Inspection" must be written on the upper right-hand corner of each page. The documents should then be placed in an envelope also marked "Confidential, Not for Public Inspection." There are specific rules regarding requests for confidentiality in a rulemaking proceeding (47 CFR §0.459).
- *Filings sent by mail*—Filings in a rulemaking proceeding can be submitted by mail. Filers should include an extra copy of the first page of their filing and enclose a postage-stamped, self-addressed envelope to obtain a receipt from the FCC. The FCC will then stamp the page and return it. Filings sent by mail should be sent as follows:
  - — Hand-delivered or messenger-delivered paper filings for the FCC's secretary—236 Massachusetts Avenue N.E., Suite 110, Washington, D.C. 20002
  - — Other messenger-delivered documents, including documents sent by overnight mail (other than U.S. Postal Service Express Mail and Priority Mail)—9300 East Hampton Drive, Capitol Heights, MD 20743, between the hours of 8:00 a.m. and 5:30 p.m.
  - — U.S. Postal Service first-class mail, Express Mail, and Priority Mail—445 12th Street S.W., Washington, D.C. 20554

Appeal to the Courts

After the Commission has considered and reconsidered a matter, interested parties may appeal the decision to the federal courts (see 47 CFR §1.13). Ultimately, the Supreme Court could hear the case. Under current law, however, the Commission need not wait for the court decision before enacting the new rules and regulations. As long as the matter has been given full consideration under the Administrative Procedures Act, the FCC may place the new rule in effect until a court rules to the contrary or orders a "stay."

ADDITIONAL INFORMATION

A list of contract and consulting broadcast engineers is available from the Society of Broadcast Engineers (SBE) at www.sbe.org and the Association of Federal Consulting Engineers (AFCCE) at www.afcce.org. Additional information on communications law attorneys is available from the Federal Communications Bar Association (FCBA) at www.fcba.org/.

CHAPTER

1.4

Regulatory Compliance

GARRISON C. CAVELL
Cavell, Mertz & Davis, Inc.
Manassas, Virginia

Editor's note: This chapter is written from the perspective of an engineering consultant with first-hand experience in dealing with both station engineers and regulatory agency representatives. The intention of the discussion herein is to raise the awareness of broadcast engineers to a point where much of the initial contact with regulatory agencies can be handled by station representatives, saving the consultant's time and costs for more serious activities.

INTRODUCTION

Station engineers face increasingly expanded areas of professional responsibility beyond the technology aspects that attracted them to this business as entry-level employees. In addition to installing, operating, and maintaining station systems and equipment, today's engineers must become involved in adjunct technical matters that often require intellectual stretching beyond their comfort zones. The legal and regulatory environment has become far more complex as well. Inasmuch as a station's engineering team is often the first line of defense in regulatory compliance, it is important that upper-level technical staff become conversant and maintain fluency with the continually evolving regulations and policies imposed by the various federal, state, and local governing agencies.

Many regulatory entities have an interest in the broadcast industry or its activities. As such, broadcast engineers may eventually find themselves dealing with the Federal Aviation Administration (FAA), the Occupational Safety and Health Administration (OSHA), the Bureau of Land Management (BLM), the Environmental Protection Agency (EPA), and even the Army Corps of Engineers. Interaction, sometimes planned—sometimes not, can also occur with state and local regulatory entities when station construction, relocation, or renovation projects are underway. Although there are specialists and professionals that can be of assistance in addressing these matters—lawyers, electrical power engineers, electricians, civil engineers, structural engineers, heating-ventilation-air conditioning (HVAC) engineers, architects, space planners, system designers, and lighting consultants—the broadcast engineer will likely be the one "on-site" person who will personally have to deal "first hand" with building code compliance officers, electrical inspectors, insurance adjusters, and fire marshals.

A broadcast engineer's activities may even require active participation in supporting other professionals such as FCC consulting engineers, aeronautical consultants, FCC communications lawyers, and other law professionals. Many will even find themselves appearing before zoning boards and helping to negotiate with land use officials. It is therefore important that today's station engineers recognize that their responsibilities include more than maintaining and improving their technology skills. They must possess a well-rounded, diverse background, apply themselves continually to the learning process, become conversant with topics outside their traditional realm, and become familiar with information resources for an ever-broadening range of technical and regulatory topics.

Of course, the regulatory entity that plays the biggest role in the broadcast industry is the Federal Communications Commission (FCC). Accordingly, this chapter is principally designed to help radio and TV station engineers start the process of understanding the FCC, how it functions, and what should be known to help their stations achieve and maintain compliance with the FCC's Rules and Regulations. However, to facilitate the process of understanding regulatory compliance in general, "other" areas of technical regulation that may be encountered in station work will first be considered.

THE REGULATORY ARENA

As an engineer progresses from station to station, over the years, it is easy to lose focus of changes in FCC rules or become unaware of additions to regulatory demands. This is particularly true for areas that are not commonly regarded as being of concern to broadcast engineers, but that are increasingly creeping into these areas of responsibility. Further, new regulatory requirements continue to be imposed through the FCC's oversight process to address societal concerns such as environmental issues, public safety, historical preservation, and tower placement impact on native sites. Meanwhile, "other" regulatory entities become aware of the need for increased oversight and scrutiny of the broadcast industry. For station technical staff, these likely "other" regulatory parties may include the Occupational Safety and Health Administration (OSHA), the Environmental Protection Agency (EPA), the Federal Aviation Administration (FAA), and local authorities such as building inspectors, electrical code inspectors, and fire marshals. It is therefore helpful to consider the three important opportunities for regulatory concern and contact. They include

- Operation: The day-to-day running of a radio and television station,
- Facility Planning: Facility creation or changes—relocation—expansion, and
- Facility Construction: The actual execution of a successful plan.

Operation

Operation is the mode with which engineers are most familiar and comfortable, and may be the area where compliancy is the most "dangerous." Being unaware of a requirement is no excuse; plus such ignorance can create situations in which the station assets or personnel could be placed in jeopardy, even when one has the best of intentions. In point of fact, it is becoming common for an OSHA inspector or local fire marshal to visit a facility, either as a result of a complaint, or because of a targeted compliance or enforcement program. Their concerns typically involve "work place" safety, unsafe or exposed equipment, potential materials handling issues, and employee safety awareness. For example, open electrical panels, exposed rotating machinery, hot surfaces, falling objects, and tripping hazards can trigger a complaint or enforcement action. (Further information can be obtained by visiting www.OSHA.gov.)

Local fire marshals often inspect after building construction before an occupancy permit is issued, but they can also inspect an existing building at any time. Typically, local fire inspectors (or local fire companies) will look for

- clear escape routes,
- maintenance of these routes,
- clear marking on exits,
- storage of flammable materials (paper, boxes) stacked too close to a ceiling,
- sources of combustion,
- illegal use of extension cords,
- flammable materials storage practices,
- condition of fire alarms, and
- number, type, and placement of fire extinguishers.

The EPA (and local officials) may become concerned with an operating station if a potential environmental hazard is suspected or reported. Examples of likely issues include

- a suspected leaking generator fuel tank,
- a fuel spill,
- a PCB spill from a damaged or discarded capacitor or transformer,
- the presence of asbestos (in soundproofing, wall insulation, duct linings),
- the improper discharge or disposal of chemicals, such as cleaning fluids,
- discarded maintenance items containing suspected hazardous substances,
- old paint and oil supplies, and
- mercury.

More information can be found by visiting the EPA's Web site at www.EPA.gov.

The FAA is another federal entity broadcast engineers may encounter in day-to-day station operations. Most engineers are aware that the FAA reviews proposed new tower construction or the proposals for the alteration of existing structures. However, the FAA is also interested in an existing station's continuing maintenance of proper structure marking and lighting. Deviations are noted through reports from interested parties or by FCC inspectors, which can result in FCC enforcement action. An enforcement exception can be made when a station voluntarily reports a tower light outage to the nearest FAA Flight Service Station. The presumption is made that resumption of proper operation will be restored in a timely manner since aeronautical safety can be compromised by inaction. Proper tower painting, maintenance of the correct color and pattern of tower paint, proper tower lighting, and the maintenance of that lighting are also part of the continuing responsibility of a station licensee. (More information may be found at www.FAA.gov.)

Many engineers are unaware that the FAA also responds to pilot or aeronautical facility (air traffic control) reports of electromagnetic interference to navigation or communication systems. Remedial action is usually quickly sought through the FCC, and may require cessation of transmissions in the worst instance. While the risk for raising such a concern is most commonly encountered when a new facility is constructed or changes are made, "sudden" interference complaints can occur when a piece of equipment malfunctions or some other emitter commences operation in the vicinity, introducing "new" intermodulation interference that may be attributed to a station's operation. Being aware of a broadcast site's environment and having access to a spectrum analyzer can be important "first aid" items. It is also important to carefully document all communications and actions if a station is contacted about such a problem. Good records and apparent cooperation will go far in solving a mutual problem.

Facility Planning

Facility planning is the area where many engineers will actually become aware of the plethora of "other" regulations and regulatory interests. Often, engineers become acutely aware by what is learned and what should have been known. For example, when engineers are first tasked with building a new studio or transmitter plant, or more commonly, when facing relocating or expanding a studio or transmitter facility, it quickly becomes apparent that there is more involved than rack placement, ergonomics, and equipment selection. One must become familiar with the local building permit process and zoning requirements, and rapidly becomes immersed in the world of facility designers, architects, mechanical engineers, electrical engineers, structural engineers, environmental consultants, aeronautical consultants, consulting FCC engineers, civil engineers, HVAC engineers, lawyers, specialized tradesmen, and contractors. Regulatory entities rarely heard of may come into play such as the U.S. Bureau of Land Management (BLM) and the U.S. Fish and Wildlife Services.

While it is tempting to try to undertake construction planning without the benefit of professional assistance (architects, planners, and engineers), most projects will eventually involve securing building permits, which is difficult to accomplish when one is not intimately familiar with building codes, local permit processes, and inspection requirements.

For instance, the International Code Council (ICC) has developed guidelines, standards, and requirements ("codes") to ensure the safety and welfare of the general public in and around buildings. Previously known by some as the "BOCA Codes" (Building Officials and Code Administrators International, Inc.), the ICC codes are actually the result of a merger of the input of BOCA and two other nonprofit code-writing organizations. These codes are used by governing entities (states, counties, and local municipalities) as a basis for their building construction permitting and approval systems. Enforcement officials, contractors, architects, engineers, designers, and builders also reference and employ these codes in their work on broadcast facilities. (For further information, visit www.iccsafe.org.)

Recommendations and guidelines are published by such entities as the Institute of Electrical and Electronics Engineers (IEEE), in their "color book" series[1] (see www.IEEE.org), and the publications of the National Fire Protection Association, Inc. (NFPA), including their "Life Safety Code Handbook" and the "National Electrical Code" (NEC). (See www.NFPA.org for additional information.) While it may be possible to be somewhat conversant with portions of these topics, thorough knowledge of these requirements is not reasonably achieved by the busy, practicing station engineer. Hence, the assistance of experienced professionals such as architects, contractors, and specialized construction engineering firms is essential in any planning process.

Facility Construction

Facility construction typically brings station engineers into close contact with the people who monitor and enforce regulations governing how a plant is built and ensure safety for workers and the general public. The same entities that are involved in the regulation of building planning and permitting also play a role in ensuring compliance before occupancy is permitted. Engineers may encounter these professionals post-construction and on an ongoing basis, which can include

- zoning boards,
- electrical and building inspectors,
- OSHA inspectors,
- fire marshals, and
- other local authorities.

Of special interest to station engineers are the applicability of state, local, and national electrical codes (NEC); their interpretation by field inspectors; and the understanding of broadcasting by local electricians and electrical inspectors in particular. These codes are not limited to power systems; they also include lighting, RF systems, and low voltage wiring (such as audio, speakers, LAN, and control circuits). For instance, audio, data, speaker, microphone, and other low voltage cable runs must be done in accordance with the applicable standards. So if "non-plenum rated" cables are employed in a suspended ceiling that forms part of the air conditioning return air path, there may be difficulties getting approval, and these cables

[1]The IEEE "color book" series includes the "Green Book" covering electrical system grounding, the "Emerald Book" covering powering and grounding, the "Gold Book" covering Power Systems Reliability, the "Orange Book" covering Emergency and Standby Power, the "Buff Book" covering protection and coordination, the "Red Book" covering Electric Power Distribution, and the "Grey Book" covering Commercial Building Power Systems.

may have to be rerun in conduits or employ other, properly rated cables.

Additionally, many municipalities require that installed electrical equipment possess a special safety certification, which is usually a tag or sticker from the Underwriters Laboratories, Inc. (the "UL Label") or the Canadian Standards Association (the "CSA Label"). In order for a manufacturer to obtain the right to affix a UL or CSA label to their equipment, they must submit examples of each type of equipment to these laboratories for safety testing. This can be an expensive and lengthy process, which often involves the destruction of the equipment. However, if the equipment item or appliance meets the applicable fire and electrical safety criteria, the safety certification label can be displayed on the equipment. (See www.ul.com and www.csa.ca for more detailed information.) Unfortunately, a considerable number and types of broadcast equipment items have not been run through these programs such as RF phasing and coupling systems; many transmitters; and certain professional switchers, routers, and audio consoles. If the equipment does not bear an "appropriate" safety label, issues can be raised (by a building or electrical local inspector) during or after the installation process. This can lead to delays or even the denial of equipment use or occupancy permits.

There are often exceptions and provisions in the NEC and local codes that would nevertheless permit the use of professional, yet unlabeled, equipment; hence, it is wise to do a bit of advance research on local requirements and then employ the services of experienced architects, contractors, and electrical power engineers to ensure that your equipment items can be understood, and safely and successfully installed in compliance with local laws.

It is a major task to stay current with the broadcast rules, much less the requirements of day-to-day operations. However, there are professionals and resources available which will be of assistance when becoming conversant with likely issues. The resources include publications; Web-based information systems, professional and trade organizations (NAB, SBE, IEEE, AFCCE), subject matter experts (consultants), practice-specific legal experts (zoning, FCC, or environmental law practices), and subject matter experts such as consultants and engineers in the construction industry.

FCC RULES COMPLIANCE

Experienced engineers facing a troubleshooting problem recognize how important it can be to understand the overall system and its interrelated functions before diving in with a hot soldering iron. The FCC and rules compliance issues are best understood using a similar approach. To that end, the following sections will provide the basic FCC "system facts," cover internal organization "subsystems," describe how they interrelate, and finally provide troubleshooting hints for understanding the system and achieving FCC rules compliance.

About the FCC

The FCC makes all the decisions for the U.S. government on technical and nontechnical matters in broadcasting and telecommunications. It sets forth and makes decisions on domestic, interstate, and international rules and policies on television and radio broadcasting, as well as communications systems, whether by wire or "wireless." Its jurisdiction covers all U.S. possessions, the states, and the District of Columbia. It functions as an essentially independent federal agency, although it is responsible to the legislative branch of government (the Congress). It is composed of five "Commissioners," which actually form the "Commission." The president appoints commissioners, but they are confirmed by the Senate. The Commission holds regular meetings, which are open to the public.

Within the Commission itself, there exists supporting units called "Bureaus" and "Offices" that provide advice and administrative support for the Commissioners and their staffs. These seven Bureaus and ten Offices directly perform the "hands-on" regulatory functions of the organization.

The Bureaus are the entities that working broadcast engineers will most likely encounter. The Bureaus staff process applications for new stations or changes in existing stations, analyze complaints, conduct investigations, assist in administrative law hearings, and develop and implement regulations. The Offices generally provide support services to the Bureaus and the Commissioners. For additional information on the FCC organization, see Chapter 1.3, "FCC Organization and Administrative Practices."

A broadcast engineer will likely encounter the Media Bureau, the Enforcement Bureau, and the Wireless Telecommunications Bureau. On rare occasions, broadcast engineers (or their communications lawyer or consulting engineer) may have contact with the International Bureau and the Office of Engineering and Technology (OET). These broadcast-related bureaus and their functions are described in the following sections.

FCC BUREAUS MOST DIRECTLY RELATING TO BROADCASTERS

Media Bureau

The Media Bureau is the one entity that is directly involved with radio and television broadcasting industry on a daily basis. The people in this Bureau review and process applications for new stations and changes in existing stations, process applications for station licenses, review rule waiver requests, evaluate station sales and transfers, examine multiple ownership questions, grant call sign requests, and manage the processes that are essential to maintain a structure and order to our portion of the spectrum. If asked, the Media Bureau's Staff will willingly provide guidance to broadcasters on processes and procedures that need to be followed.

The Bureau structure includes the Office of the Bureau Chief, the Audio Division, Video Division, Policy Division, Industry Analysis and Engineering Divisions, along with the Office of Communication & Industry Information and necessary Management and Resources Staff.

Divisions within the Media Bureau

The radio broadcasting industry is principally handled by the Audio Division, which administers the licensing of both commercial and noncommercial AM, FM, Low Power FM, FM Translator, and FM Booster broadcast stations. Their staff includes lawyers, analysts, engineers, and support personnel who review the legal and technical aspects of broadcaster proposals for changes, and processes them for grants in accordance with prevailing FCC Policies and Rules. Additionally, they handle requests for exceptions to the rules and grant special temporary authority (STA) for operation of stations at variance to the rules, should unforeseen circumstances necessitate it.

The Video Division is structured similarly and administers the licensing of broadcast commercial and noncommercial TV, Low Power TV, Class A TV, TV Translators, and TV Booster stations. Like the Audio Division, the Video Division's technical and legal staff perform application review and processing functions, evaluate requests for exceptions to the rules, and grant special temporary authority to operate stations at variance to the rules.

The Media Bureau's Policy Division is one that the average engineer will not routinely directly encounter. It is involved with "policy and rules" matters, evaluating issues and conducting rulemaking proceedings on various broadcasting, CATV (cable television), and DBS (Direct Broadcast Satellite) issues. Particular areas of recent focus include the television digital transition, cable carriage complaints, the Satellite Home Viewer Improvement Act (SHVIA), political broadcasting issues, and Equal Employment Opportunity (EEO) matters.

The Engineering Division evaluates engineering matters that come before the Media Bureau. Broadcast engineers encounter them when faced with problems requiring a technical rule waiver request, need a rule interpretation, or to file a rulemaking petition. The Engineering Division also processes Cable Television Relay Service (CARS) applications, reviews cable television system signal leakage performance reports, and analyzes aeronautical frequency usage information.

Wireless Telecommunications Bureau

While it would seem otherwise, the Wireless Telecommunications Bureau (WTB) also has relevance to broadcasters. It is traditionally considered as being primarily focused on the wireless telecommunications industry including nonsatellite commercial, private, and "public safety" communications systems; two-way radios; certain microwave systems; paging systems; PCS; and cellular telephones. However, its role is much more expansive. For instance, the WTB is also responsible for administering spectrum auctions, the Universal Licensing System (ULS), the FRN (FCC Registration Number) program, the "Online Filing" and records system, the Antenna Structure Registration (ASR) program, and the licensing of "broadcast auxiliary" systems that include broadcast microwave STLs, RPUs, and ENG systems.

Its divisions also include the Auctions & Spectrum Access Division, Broadband Division, Mobility Division, Public Safety & Critical Infrastructure Division, Spectrum & Competition Policy Division, and the Spectrum Management Resources & Technologies Division.

Spectrum Auctions

This has to do with a process the FCC started in 1994, under the authority of the Omnibus Budget Reconciliation Act of 1993, to replace lotteries and the comparative hearing mechanism as a method for sorting out and awarding licenses to competing ("mutually exclusive" or "MX'ed") applicants for broadcast and non-broadcast spectrum. The previously employed "hearing" process was found to be expensive, inefficient, time consuming, and complicated since it involved the use of litigation before an FCC Administrative Law Judge. The involved judge was tasked with deciding who should be selected as a licensee when there was more than one interested applicant. A Review Board appeals process was in place, which could overturn a judge's decision. Absent some exceptions such as proposals for DTV licenses to replace analog licenses, or applications for noncommercial and public broadcast stations, auctions are now the primary method used to settle mutually exclusive proposals for new licenses.

Universal Licensing System

A broadcast engineer, when filing *any* application with the FCC, will run into something called the Universal Licensing System (ULS). The ULS, which is administered by the WTB, was established as an electronic, "on-line," Web-accessible, application filing and database system. It is intended to replace the old "paper" system that had previously been used for filing applications for new or modified stations. Some proposals, such as AM Radio Applications for License, still must be "paper filed." The ULS is also used to obtain operator's licenses or permits. Although still a "work in progress," it is a system that more readily allows public access to FCC information and the application process. It also facilitates FCC records research.

FCC Registration Number System

Any FCC filing activity, be it an application for a station or operator license or permit, now requires the applicant to obtain an FCC Registration Number (FRN). The FRN is used to uniquely identify the person or entity to the FCC when using the electronic filing system. Even if an applicant holds a "paper" license or permit that was obtained in the "pre-electronic filing" days, modifying or renewing that license will require obtaining an FRN. Fortunately, the FCC has made this process relatively painless; its Web site, www.fcc.gov, has easily found links to the required areas, the process is virtually self-explanatory, and responses to assistance requests are excellent.

Antenna Structure Registration

In late 1995, the FCC began requiring applicants proposing antenna structures that are near enough to an airport or tall enough (involving construction of more than 200 feet in height above ground) to potentially create an adverse aeronautical impact to have the structure registered in the Commission's Antenna Structure Registration (ASR) system. Under this program, any antenna structure, be it existing or newly proposed, that might (or actually does) require notification of the Federal Aviation Administration (FAA), must be registered with the FCC by its owner using FCC Form 854.

If the FCC determines that a structure registration is required, a construction permit will NOT be granted until one is obtained. The registration process potentially requires interaction with other regulatory entities (the FAA) before successful completion of the registration application. As such, unforeseen delays can be encountered in receiving a grant of an original station application proposal.

Station applicants, permittees, or licensees are not required to be the registering entity unless they are also (or proposing to be) the structure owner. The reason is that the FAA and the FCC are interested only in identifying a single "point of contact" for resolving tower/antenna-related problems, and particularly with respect to marking (painting) and lighting issues.

When it is unclear if a structure's proximity to an airport might cause it to require registration (or FAA notification), the FCC's WTB's Web page tool "TOWAIR" can be of assistance. This system allows the user to input the tower specifics to see if it would fail a proximity test known as a "slope test." If the proposed structure fails this test, then FAA notification is necessary, and FCC antenna structure registration is likely required as well. The Antenna Structure Registration rules are contained in Part 17 of the Commission's Rules.

It should also be understood that, in the Commission's view, an antenna structure could either be a tower (guyed or self-supporting) built either as a support or, in the case of an AM station, to act as an antenna, or it could be a structure mounted on some other human-made object. In these instances, the antenna structure (not the building, water tower, etc.) is the object that must be registered.

Other WTB Services: Tower Construction Notification

When considering the construction of a new antenna tower for a station, be aware of the numerous environmental requirements contained in Section 1.1305-1.1307 of the FCC's Rules.

Specifically, the FCC wants there to be some consideration of the possible environmental impact of any new construction. In particular, if facilities are proposed to be located in an officially designated wilderness area or wildlife preserve, or in an area that may impact listed threatened or endangered species or designated critical habitats, or in a flood plain, a formal "environmental assessment" (EA) may be required.

Additionally, an EA may be required if a proposal would involve a significant change in surface features (wetland fill, deforestation, or water diversion), or involve a supporting structure equipped with high-intensity white (strobe) lights in or near residential areas. If the proposed construction may be in locations that may affect Native American religious sites or be located near "districts, sites, buildings, structures, or objects significant in American history, architecture, archeology, engineering, or culture that are listed, or are eligible for listing, in the National Register of Historic Places," then an EA may be required. Most proposals are excluded from preparing and filing a full EA, but considerable effort must be expended to provide proper notification to potentially interested parties, and to ensure that the proposal would not adversely impact the environment.

To help the applicant cope with the above environmental requirement, the WTB's "Tower Construction Notification System" was developed to assist in providing the recommended separate formal "notifications of intended construction" (NIC) to potentially interested entities. Under this system, when an NIC is filed with the FCC, the FCC in turn forwards it to pertinent State Historic Preservation Officers (SHPOs), recognized Indian Tribes, and the Native Hawaiian Organization (NHO). While this process in and of itself may not satisfy all of the provisions of the last environmental requirement, it at least facilitates communication between these entities and tower proponents should there be concerns about the proximity of a proposed construction to a site of interest.

International Bureau

The International Bureau (IB) works behind the scenes to represent U.S. interests with other countries. Broadcasters become aware of this Bureau's existence if their proposal triggers, or potentially triggers, concerns from a foreign entity. Often, just being within a certain (seemingly large) distance from the U.S. international border will trigger a requirement for the IB to go to work on your behalf. Given the international reach of the satellite relay systems, the IB also handles satellite

uplink licensing. If a broadcaster has an uplink, or expects to build one, then IB staff will likely be involved during preparation, filing, and tracking of the necessary application.

Office of Engineering and Technology

The Office of Engineering and Technology (OET) provides guidance on complicated engineering matters to the FCC, its staff, and even the general public. This Bureau also administers the FCC Laboratory, which itself assists with, among other things, RF equipment technical standards acceptance and authorization. Broadcast engineers are probably most familiar with OET's work in providing guidance on human exposure to radiofrequency energy (RFR) through their bulletins and issued guidelines. OET also works extensively in the areas of signal propagation, coverage, and interference prediction.

Enforcement Bureau

The Enforcement Bureau serves on the front lines of rule-compliance monitoring and enforcement, and is the most likely "in person" contact a broadcaster will experience with the FCC. While the Enforcement Bureau conducts investigations and "on-site" inspections, responds to life safety issues, and investigates interference complaints, in recent years, its responsibilities have been broadened to include "public safety" and "homeland security" enforcement.

The Enforcement Bureau is composed of the Office of the Bureau Chief, the Investigations & Hearings Division, the Market Disputes Resolution Division, the Spectrum Enforcement Division, and the Telecommunications Consumers Division. This bureau also operates a network of Regional and Field Offices throughout the United States to facilitate its monitoring and compliance activities.

Most broadcasters are surprised that the FCC can be closer to their facilities than its Washington, D.C., headquarters. In fact, broadcasters may be visited by FCC representatives from one of the three Regional Offices, sixteen District Offices, or nine Resident Agent Offices; specific locations are shown in Table 1.4-1.

Broadcasters located near these field offices are specifically required to coordinate any new or modification proposals with these offices since excessive signal levels are not permitted toward these facilities.

Specific locations and contact information for these offices can be found in local telephone directories, the FCC's Web site, or within the FCC's Rules.

TABLE 1.4-1
FCC Regional, District, and Resident Agent Office Locations

Regional Offices	Chicago, IL (Northeast) Kansas City, MO (South-central)	San Francisco, CA (Western)
District Offices	Atlanta, GA Boston, MA Chicago, IL Columbia, MD Dallas, TX Denver, CO Detroit, MI Kansas City, MO	Los Angeles, CA New Orleans, LA New York, NY Philadelphia, PA San Diego, CA San Francisco, CA Seattle, WA Tampa, FL
Resident Agent Offices	Anchorage, AL Buffalo, NY Honolulu, HI Houston, TX Miami, FL	Norfolk, VA Portland, OR St. Paul, MN San Juan, PR
Field Offices (with monitoring facilities)	Allegan, MI Anchorage, AK Belfast, ME Canadaigua, NY Douglas, AZ Ferndale, WA Grand Island, NE	Kinsville, TX Laurel, MD Livermore, CA Powder Springs, GA Santa Isabel, PR Vero Beach, FL Waipahu, HI

Spectrum Enforcement Division

The Spectrum Enforcement Division of the Enforcement Bureau is responsible for handling complaints,[2] conducting enforcement actions in technical and public safety matters, and operating the FCC's direction-finding facilities. Examples of their activities include

- verifying tower registration,
- enforcing tower structure marking and lighting requirements,
- investigating suspected AM tower fencing violations,
- investigating unauthorized station construction and unlicensed station operation ("pirate stations"),
- investigating technical violations of operation at unauthorized location, power, mode, or frequency,
- monitoring compliance with the Emergency Alert System (EAS) rules, and
- investigating and resolving interference complaints.

Inspections

The FCC's Enforcement Bureau has the authority under the "Communications Act" to inspect most transmitting installations using the Bureau's Field Agents. On-site inspections can occur for many reasons. The focus of an inspection generally will be on

- suspected violations of antenna structure marking and lighting,

[2]Most complaints and public contact points are now served through the Consumer Inquiries and Complaints Division of the Consumer & Governmental Affairs Bureau.

- AM fencing requirements,
- EAS,
- public inspection file,
- operating mode and power levels.

Violations

Violations may also be investigated regarding the main studio rules and the authorized location of transmitting facilities.

When an inspection occurs, the agents will present their FCC identification card and badge, identify themselves by name and agency, state the purpose of the visit, and request permission to inspect the radio station. An agent's identity can be verified by calling the FCC's Communications and Crisis Management Center in Washington, D.C. Pertinent station records such as operator and facility licenses and authorizations must be provided upon request. Inspections can occur during any hours of station operation.

If a violation notice is issued, the terms "willfully" and "repeated" may seem alarming. Unfortunately, these terms have a slightly different meaning in the FCC's legal process than that provided by a dictionary. The term "willful" as used in Section 503(b) of the Communications Act means that the violation or omission was knowingly committed, irrespective of any actual intent to violate any rule. The term "repeated" means the commission or omission of such an act occurred more than once or for more than one day.

If a "notice" of "violation" or "inquiry" or "apparent liability" is received, the licensee will have a prescribed period within which to respond. It is usually at this point that an experienced communications lawyer, and perhaps a consulting engineer, should be called immediately since it may take time to help respond to such a letter.

Fines

Fines for violations can be substantial. Increases in the amount of a fine can be made for

- egregious misconduct,
  ability to pay,
- intentional violation,
- causing substantial harm,
- committing prior violations,
- achieving economic gain by virtue of the violation, and
- repeated or continuous violations.

Reductions in the amount of a fine can be awarded if

- a minor violation is involved,
- a good faith effort or voluntary disclosure was made,
- the station or individual has a history of overall compliance, or
- there is a clearly demonstrated inability to pay the fine (usually backed up by copies of income tax returns).

State Self-Inspection Programs

The FCC is primarily interested in achieving compliance through education and, if necessary, through unannounced on-site inspections, spectrum monitoring, and subsequent enforcement actions. The state broadcast association supported "Self-Inspection" or Alternative Inspection Program (AIP), which evolved through informal cooperation with the FCC's Enforcement Division, is another means of demonstrating compliance to the FCC.

This program, using published guidelines and checklist procedures, provides the means for participating broadcast entities to demonstrate rules compliance without undergoing a formal FCC inspection. Under this program, upon request, a person appointed by a state broadcaster association will inspect a station for a small fee. If a station is found to be operating *within* the rules, the FCC will be notified and a "Certificate of Compliance" will be provided to the station. If the AIP inspection finds compliance issues, the station will be advised of the issues. No certificate is issued and the FCC is not notified that the inspection took place. The station then has an opportunity to resolve the issues and can request a re-inspection to achieve a certificate.

If a station has received a Certificate of Compliance, under this program the FCC will agree not to conduct a surprise inspection of the station for a period defined under the agreement between the state association and the involved field office. However, an FCC inspection can still occur at a station that has received a certificate, if a targeted enforcement program is being conducted or a complaint is received about the station.

Information on the AIP and its availability in each state can be obtained through the FCC's Web site (http://www.fcc.gov/eb/bc-chklsts/) or by contacting the state broadcast association.

Other Compliance Resources

There are many resources available to help stations achieve and maintain compliance. These resources include

- the station's communications attorney, consulting engineer,
- books and information materials provided by the NAB,
- the training sessions provided during NAB, SBE, and IEEE conventions,
- the FCC's Rules and Regulations,
- the FCC's own Web site (www.fcc.gov),
- the FCC staff itself—a telephone and e-mail contact directory is provided on the FCC's Web site.

Commission staff are usually very informative and strive to help whenever possible.

For convenience, a summary of broadcast-related FCC Rules and Regulations is provided in Table 1.4-2.

TABLE 1.4-2
Broadcast-Related FCC Rules and Regulations

Part	Title
0	Commission organization
1	Practice and procedure
2	Frequency allocations and radio treaty matters; general rules and regulations
11	Emergency Alert System (EAS)
13	Commercial radio operators
15	Radio frequency devices
17	Construction, marking, and lighting of antenna structures
20	Commercial mobile radio services
21	Domestic public fixed radio services
25	Satellite communications
73	Radio broadcast services
74	Experimental radio, auxiliary, special broadcast, and other program distribution services
76	Multichannel video and cable television service
78	Cable television relay service
79	Closed captioning of video programming
101	Fixed microwave services

SUMMARY

As society becomes more complex, additional requirements are imposed through the FCC's oversight process (such as incorporating environmental and historical preservation issues). As discussed in the preceding sections, the FCC is a large and complicated organization whose mission has grown and expanded over the years as the broadcast industry has grown and changed. Nevertheless, FCC staff are genuinely interested in helping stations comply with the rules and work within the system. As other entities (such as OSHA, EPA, FAA) become aware of the further need for oversight of broadcasters, there will be a need to identify even more information resources.

Continuing self-education and awareness is important if broadcast engineers want to be a stronger part of the station team. This can be relatively easy to do. Keep an eye on industry publications and the FCC's own Web site. Obtain an up-to-date copy of, and periodically review, the Commission's rules. Monitor the resources provided by the NAB, other professional and trade organizations, and even private practice professionals, such as consultants and lawyers. Finally, find the time to regularly attend and participate in industry membership groups, societies, forums, and conventions. The information sharing at these functions is invaluable.

CHAPTER

1.5

Frequency Allocations for Broadcasting and the Broadcast Auxiliary Services

WILLIAM R. MEINTEL
Meintel, Sgrignoli & Wallace
Warrenton, Virginia

INTRODUCTION

This chapter provides an overview of terrestrial frequency allocations[1] and a listing of the frequencies available for AM, FM, and TV broadcast stations as well as for the "auxiliary" broadcast services that support broadcasting operations. Necessarily, such an overview must include a description of the decision-making process that is involved in allocating a frequency band for a specific purpose. This complex subject of allocations involves more than just the location of the service in the frequency spectrum; it includes decisions as to the number and width of channels and power and antenna limitations, as well as decisions concerning the technical standards that define how the spectrum will be shared with other users. Because each broadcast service presents unique service and interference objectives, the allocation process has been and will continue to be different for each of these services. Here it is possible to provide only a brief description of the allocation process before turning to the current situation for each of the services. Where appropriate, the discussion will touch on changes expected in the foreseeable future.

HOW SPECTRUM IS ALLOCATED

It is a fundamental characteristic of radiowave propagation that these waves follow the laws of physics and thus ignore political or geographic boundaries. As a result, decisions concerning radiofrequency allocations cannot be made solely at the local level but must take into account their anticipated impact outside the station's coverage area. Recognizing that coordination in the allocation and use of spectrum is essential, an international mechanism has been established to perform this function, and, in the U. S., a parallel coordination system has been established at the federal level.

INTERNATIONAL ALLOCATION PROCESS

At the international level, frequency allocation decisions are made by the International Telecommunication Union (ITU), a specialized agency of the United Nations, headquartered in Geneva, Switzerland. Like the United Nations, the ITU is a consortium of more than 180 governments, whose purpose is to propose, develop, revise, and administer worldwide frequency allocation plans. In the terminology of the ITU, participating governments are referred to as "states," "members," or "administrations." Such international cooperation serves to minimize interference and maximize use of the spectrum.

Although the ITU maintains a permanent staff, the power of the organization, as indicated by the structure shown in Figure 1.5-1, resides with the member nations who direct its activities through periodic meetings. At the highest level of this structure is the Plenipotentiary Conference, which adopts the fundamental policies of the organization and decides on the organization and activities of the Union through a treaty known as the International Telecommunication

[1]Technically the term "allocation" refers to the process by which a frequency band is made available for a specific purpose; however, it is used here in a broader sense that includes the *allotment* of frequencies within a band and the *assignment* of individual stations.

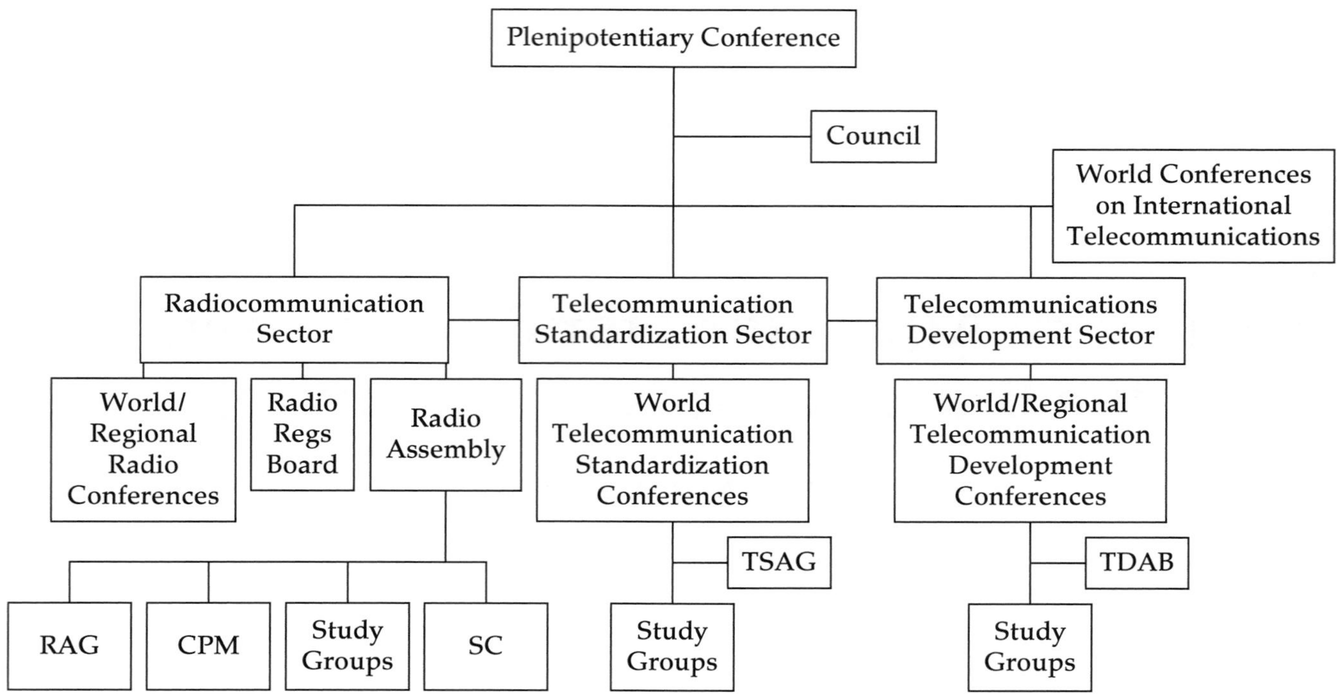

FIGURE 1.5-1 International Telecommunication Union (ITU) structure.

Constitution and Convention. Plenipotentiary Conferences are held approximately every 4 years and are open to ITU member countries, the United Nations and its specialized agencies, the International Atomic Energy Agency, regional telecommunications organizations, and intergovernmental satellites operators.

Proceeding down the structure, the ITU Council consists of a representative group of members elected by the Plenipotentiary Conference. The task of the council is to ensure that in the period between Plenipotentiary Conferences the ITU's policies and strategy are adequately responding to the constantly changing telecommunication environment. The Council is also responsible for ensuring efficient coordination of work, effective financial management, and assisting members in the implementation of the provisions and regulations of the Union. The day-to-day work of the ITU is performed in three Sectors, of which the Radiocommunication Sector is relevant to this discussion. The other sectors, as their names imply, deal with standardization and development of telecommunications.

Worldwide allocations for radio services are made through decisions made at international conferences of the Radiocommunication Sector called World Radio Conferences, or WRCs (pronounced "warcs"). These WRCs deal with frequency allocations and related matters of concern on a worldwide basis through review and revision of the international Radio Regulations. Through a process of give and take, decisions are made for the allocation of spectrum. WRCs are held every 2 to 3 years and deal with a specific agenda that is generally established 4 to 6 years in advance with a final agenda established by the ITU Council 2 years prior to the WRC. The early establishment of the agenda is intended to allow for proper preparation so the work of the conference can be concluded in a timely manner; however, due to the rapid changes occurring in radiocommunications, last minute requests for agenda changes are not uncommon.

Because of the geographic separation of certain parts of the world it was determined to be practical to divide the world into three regions (Figure 1.5-2). Region 1 consists of Europe and Africa; Region 2 consists of North and South America, Greenland, and the Caribbean; and Region 3 consists of Asia and Oceania. In view of this, the ITU sometimes convenes Regional Radio Conferences (RRCs) to consider questions that are unique to a specific ITU region. Often these RRCs consider implementation of decisions made at an earlier WRC.

Technical Submissions

Radiocommunication Assemblies are held in conjunction with and prior to the WRCs and provide the technical basis for the work of the WRC.[2] The Assemblies also approve and set priorities for the work of the study groups that operate under the Assemblies' supervision. In addition, the Assemblies also set up or

[2]In a reorganization of the ITU, the work of developing recommendations and providing reports dealing with technical issues formerly performed by the International Radio Consultative Committee (CCIR), with which some readers may be familiar, is now performed by the Radiocommunications Assemblies.

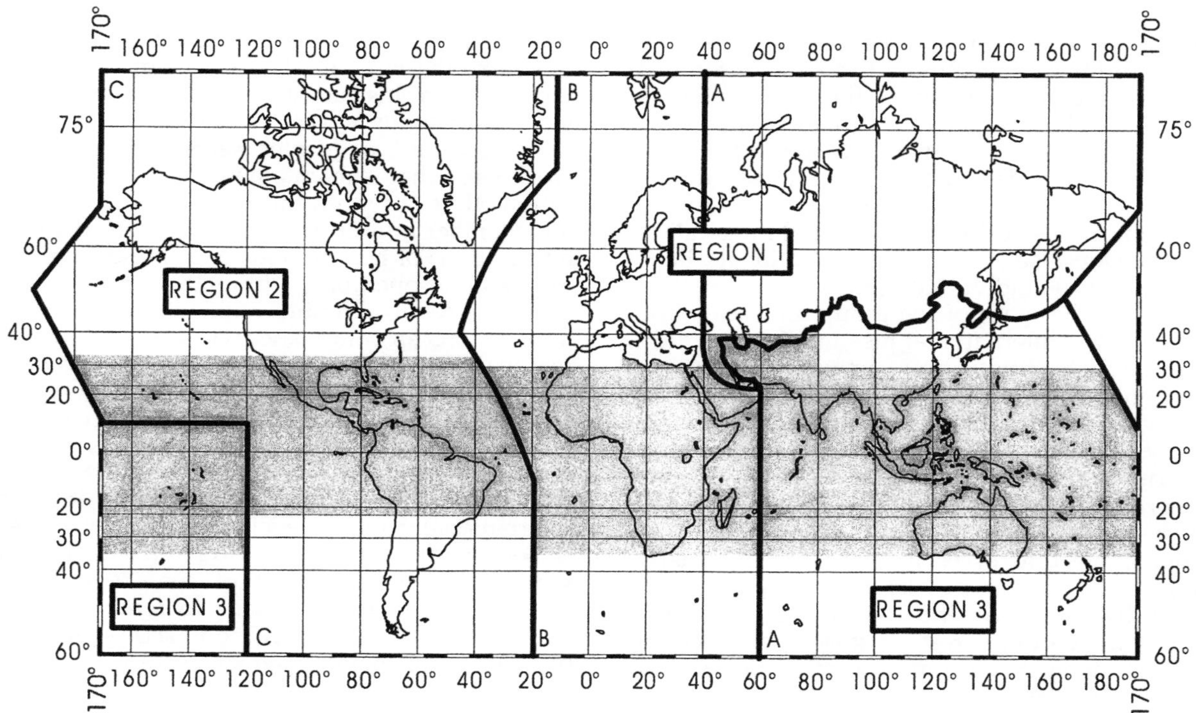

FIGURE 1.5-2 ITU regions. (From 47 CFR §2.104; the shaded area indicates the Tropical Zone as defined in 47 CFR §2.104(c)(4).)

abolish study groups as deemed appropriate. This work is currently carried out by seven Study Groups, listed in Table 1.5-1a, each of which is devoted to a particular radio communication service or specific technical issue. The work relevant to broadcasting is carried out by Study Group 6. This Study Group is further divided into subgroups or working parties that deal with the specific areas of interest listed in Table 1.5-1b. There also are joint Study Groups that deal with questions that are of relevance to more than one Study Group. Each Study Group has an international chairman and one or more vice-chairmen who are provided by interested participating administrations.

ITU Study Group work is carried out by the members of the ITU as well as recognized users and standard setting groups such as the European Broadcasting Union and the U.S. broadcasting networks. Within individual administrations, organized

TABLE 1.5-1a
ITU-R Study Groups*

Study Group	Subject
1	Spectrum management
3	Radiowave propagation
4	Fixed-satellite service
6	Broadcasting services
7	Science services
8	Mobile, radiodetermination, amateur, and related satellite services
9	Fixed service

*Currently, there are no Study Groups 2 or 5.

TABLE 1.5-1b
Subgroups of ITU Study Group 6

Subgroup	Subject
Working Party 6A	Program assembling and formatting
Working Party 6E	Terrestrial delivery
Working Party 6J	Program production, archiving, and international exchange
Working Party 6M	Interactive and multimedia broadcasting
Working Party 6Q	Performance assessment and quality control
Working Party 6S	Satellite delivery
Task Group 6/9	Large screen digital imagery
Joint Task Group 6–8–9	Use of the band 2500–2690 MHz by space services

structures often are created to provide input for the administration to submit to the Radio Assembly through its Study Groups. In the U. S., this activity is chartered by and operates under the Department of State (DOS). A U.S. National Chairman is appointed by the DOS, and there are two National Vice-Chairmen, one from the Federal Communications Commission (FCC) and one from the National Telecommunications and Information Administration (NTIA). The DOS also appoints a U.S. National Chairman for each of the individual Study Groups. Each is responsible for the work of the group and heads the U.S. delegation to international meetings of the Study Group. In the U. S., Study Group activities are open to participation by the public. Contributions to Study Group work in the U. S. come mainly from the private sector; elsewhere in the world, such work is performed primarily by government employees.

In general, the work of the Study Groups is timed to coincide with scheduled WRCs and RRCs, with an international meeting of each study group usually occurring about once a year. In the interim, work is carried out by correspondence. Administrations may suggest any matter of interest for study that is within the purview of the Radiocommunication Sector, but priority is given to issues that are relevant to a scheduled WRC or RRC.

The cumulative recommendations and reports of the various study groups are updated and published periodically in electronic format. Interested parties may obtain information on obtaining these as well as numerous other ITU publications, in both hard copy and electronic form, from the ITU Web site (http://www.itu.int). The Study Group reports and recommendations contain a wealth of information not only relating to spectrum allocation but also concerning measurement procedures and standards for audio and video recording equipment used for the exchange of broadcast programming.

The Allocation Process

The allocation of radiofrequency spectrum occurs as a result of a series of interrelated decisions. On the first, most basic level, blocks of frequencies are allocated on a worldwide basis by the ITU WRC process. Exactly which blocks of frequencies are allocated to particular services is determined by evaluating the many specific proposals submitted to the WRC for each frequency band. Technical input is obtained from propagation studies and other engineering analyses undertaken as part of the Study Group process and from the submissions of individual administrations. Frequently, blocks of spectrum are allocated for the same purpose on a worldwide basis but may also be allocated for different purposes on a regional basis.

Member nations theoretically retain the sovereign right to domestic use of the spectrum so long as such use is not in contravention of international radio regulations or the international agreements to which that administration is a party. However, as a practical matter, the flexibility of administrations to use the spectrum is limited by the worldwide allocation system and the need to avoid harmful interference.

Unlike the FCC, the ITU does not license users of the spectrum. Instead, it operates only as a coordinator, maintaining a Master International Frequency Register (MIFR, or Master Register) of radio stations worldwide that is maintained by the Radiocommunications Bureau within the Radiocommunications Sector. Member administrations have agreed to provide notifications of new stations or modifications in existing stations operating within their respective countries. The Radiocommunications Bureau studies these notifications for compliance with the existing world or regional agreements and provides the results of its studies to the member nations. Only those notifications that comply with the existing agreements are placed in the Master Register of stations. Once a station has been placed in the Master Register, the member nations are obliged to provide it with the internationally agreed level of interference protection.

Because the WRC and RRC agreement texts provide only a general framework, many specific matters are left to individual nations to resolve and implement. In the U. S., the FCC and NTIA share responsibility for implementing agreements to which the U. S. has assented, including the bilateral or multilateral agreements negotiated with our neighbors to deal with concerns that are unique to the countries involved.

There are differences in the treatment of the various broadcast services. Because lower frequency signals, such as those used by AM broadcasting, propagate over great distances, international decisions have a much greater impact on AM broadcasting than FM and television, for which VHF or UHF propagation is much more limited. This means that the restrictions on FM and TV allocations imposed by international agreements are usually applied only to areas near the borders; however, international agreement on technical transmission standards often is desired in order to foster the absence of interference and the worldwide free flow of communication.

DOMESTIC ALLOCATION PROCESS

Regulation of spectrum began with the U.S. Department of Commerce in the early 1920s, when the Secretary of Commerce granted the first AM broadcasting licenses. By 1927, the number of AM stations had increased to 733, and over 6 million radio receivers had been manufactured; however, because of an unfortunate court decision that precluded the Secretary of Commerce from dealing with the specific choice of location, power, and operating frequency, these matters were left largely to the discretion of the broadcaster. This led to chaotic use of the spectrum and widespread interference, a situation that led to creation of the Federal Radio Commission in 1927. Seven years later, the Federal Radio Commission was replaced by the Federal Communications Commission (FCC), formed pursuant to the Communications Act of 1934. Ever since, anyone desiring to operate a

broadcast station, or almost any kind of radio transmitting device, must apply to the FCC and be granted a license before commencing operation. Today, a broadcast license sets forth all essential technical parameters of station operation. The NTIA performs a similar function and coordinates the spectrum used by government agencies. The FCC works with the NTIA where there is a need for coordination between government and private uses of the spectrum.

Spectrum for use domestically must be allocated by FCC rule-making proceedings. Domestic allocation rule-making proceedings can be initiated by the FCC on its own motion or in response to requests from the public ("petitions"), but in so doing it must not act in contravention of international agreements to which the U. S. is a party. The FCC rule-making process is a complex subject in its own right, but for our purposes here only a brief description is required. Rule-making proceedings are based on a public record developed through responses to the issuance of a Notice of Proposed Rule Making (NPRM) that is filed by interested parties. In addition to filing comments on the FCC's proposal, the public may reply to the comments of other parties. After the Commission evaluates the responses to its Notice, it may decide to adopt the proposal as originally set forth, modify it based on the comments received, or possibly reject the proposal.[3]

Many FCC proceedings are controversial in nature. Allocation proceedings may be especially controversial, as a particular communications industry's livelihood may depend in part on how much spectrum is allocated. Thus, FCC allocation decision making is not simply a matter of technical evaluation but must be seen as part of the political process as well, as happens when entire industries may compete for a limited amount of spectrum. Where the number of users in a particular frequency band is expected to be relatively small or their use is sporadic, the FCC may propose sharing of this spectrum with other users. Such proposals may also be controversial, because sharing spectrum with a dissimilar service invites the possibility of interference and difficulties in coordinating the use of the frequencies. Because of these and other factors, FCC allocation proceedings consume a great deal of its time and energy and can impose burdens on the organizations that participate in them.

From time to time, alternative methods have been proposed for allocating spectrum. These ideas usually envision the removal of the FCC as the arbiter of mutually exclusive requests for spectrum and instead substituting marketplace forces. Under a market allocation system, frequencies would be used by entities that would pay for them; noneconomic, social, or public policy aspects would not be considered. In recent years, such a system has been employed for some frequency allocations in the form of spectrum auctions. In these cases, the Commission's role is reduced to that of a technical "traffic cop" of the airwaves.

At the time this chapter is being written, the FCC is considering final rules governing the implementation of digital radio broadcasting. In 2002, the FCC granted interim authorization to a system developed by iBiquity Digital Corporation.[4] This is an in-band/on-channel (IBOC) system, wherein a new digital signal is broadcast simultaneously in the same channel permitted to be occupied by the existing analog signal. While the FCC is considering these final rules, stations are permitted to operate using the iBiquity system upon notification to the FCC. Once the rules are finalized, they may have some impact on the allocation process discussed below.

U.S. AM Broadcasting Frequency Allocation

In the U. S., amplitude-modulated (AM) stations operate with carrier frequencies in the center of channels assigned every 10 kHz. For many years, the AM broadcast band in the U. S. included 107 channels in the band from 535 to 1605 kHz; however, the frequency band allocated to AM broadcasting now includes a total of 117 channels in the band from 535 to 1705 kHz. Currently, nearly 4800 commercial and noncommercial AM stations are operating in the United States. These stations operate with various power levels, up to a maximum of 50 kilowatts (kW). About half of these stations use multi-tower directional antennas to restrict radiation in certain directions for the purpose of controlling interference or maximizing radio service in particular directions.

Allocation decisions for the AM broadcast band are probably the most complex of the broadcast services. Because propagation varies with time of day, geographic latitude, soil conductivity, and frequency, the engineering analyses necessary to establish interference protection for other stations can be quite complicated. Engineers, the FCC, and the ITU have sophisticated computer programs that analyze the input of a new or modified AM station proposal. Before going into the details of the AM broadcast allotment system currently in place in the United States, it is useful to provide a brief history of AM broadcasting allocations.

Over the years, there have been many changes in the nature of AM broadcasting in the U. S. Initially, *clear channel* stations (high-powered, omnidirectional stations with a large coverage area) provided the only service available in many areas of the U. S., but with the end of World War II demand increased greatly and many AM stations were established in all areas of the country. Because of its early development of AM radio service, the U.S. experience has been used as a model for regional and bilateral agreements.

Early in the history of AM broadcasting, the countries in the North American area recognized the need to cooperate in the use of AM frequencies, and in 1937 they reached agreement on how to proceed. Soon, however, this agreement was found to be inadequate

[3]The nature of the Commission's rule-making process is described in greater detail in Chapter 1.3.

[4]See *First Report and Order: Digital Audio Broadcasting Systems and Their Impact on the Terrestrial Radio Broadcast Service*, 17 FCC Rcd 19990, 2002. A standard based on the iBiguity system was adopted by the National Radio Systems Committee (NRSC) in April 2005, called NRSC-5: IBOC Digital Radio Broadcasting Standard.

and negotiations began on a new agreement. Although the North American Regional Broadcasting Agreement (NARBA) was signed on November 15, 1950, it did not go into effect until 10 years later, on April 16, 1960. Signatories to NARBA include the U. S., Canada, Cuba, the Dominican Republic, and the United Kingdom on behalf of Jamaica and the Bahamas. Mexico, an earlier participant, removed itself from these negotiations, and a bilateral agreement between the U. S. and Mexico was reached in 1957. These international agreements became necessary principally because nighttime AM propagation has the potential for causing widespread interference to neighboring countries unless mutual allocation criteria and related technical standards could be agreed upon and implemented by the parties to the agreement. To this end, NARBA provided for a partitioning of AM broadcast channels into three basic classes.

Clear Channel Stations

The first of these classes includes the so-called *clear channels*, whose high-powered stations would have primary access to the frequency and other stations could use the channel subject to providing full protection to the dominant station(s). Clear channel dominant stations were designed to provide service over extensive areas by means of skywave as well as groundwave signals. NARBA set aside 60 of the 107 channels then available for clear channel use. Each NARBA country, except Jamaica, received a priority on one or more clear channels, with the U. S. receiving a major portion of available priorities. NARBA countries without a priority on a given clear channel could still assign stations on that channel, provided these stations protected the wide-area service of the dominant station in the country with the NARBA priority.

Regional Channels

The second class of channels, *regional channels*, occupies an additional 41 channels. Unlike clear channels, these channels were shared on an equal basis by all the NARBA countries. Stations operating on these channels were intended to provide service to a considerable area, but, unlike the clear channel stations that received protection for their skywave as well as groundwave service, only the groundwave service provided by the regional stations was protected.

Local Channels

The remaining six frequencies were the *local channels*, which provide an even more limited type of groundwave service. Only limited interference protection was provided to these relatively low-power operations.

1981 Rio Agreement

Recognizing the need for updating these agreements and for developing more efficient coordination throughout Region 2 (North and South America), the 1979 WRC called for a conference to be held in Region 2 to address AM broadcasting and sharing criteria. That conference was held in two sessions in 1980 and 1981 and resulted in the adoption of an agreement among most of the countries of the hemisphere. Included as part of that agreement, referred to as the 1981 Rio Agreement (for Rio de Janeiro, the location of the second conference session), was a list of all of the operating stations in the hemisphere along with information indicating whether or not the stations were receiving or causing harmful interference according to the technical criteria set forth in the agreement. Stations not causing interference were placed in the ITU Master Register and accorded protection from interference as defined by the agreement. In situations where interference already existed, the countries involved were asked to meet and work out mutually satisfactory solutions. Because the general framework of the Rio Agreement did not deal with the particular needs and desires of the U. S., Canada, and Mexico, separate new bilateral agreements have been negotiated that incorporate the required additional items concerning coordination and technical parameters.

Although the Rio Agreement applies throughout most of Region 2, as of this writing relations with the Bahamas and the Dominican Republic continue to be governed by NARBA, as neither country has taken the necessary steps to replace the NARBA provisions with the Rio 1981 Agreement. Relations with Cuba regarding the 535 to 1605 kHz band are governed solely by the international Radio Regulations rather than by agreement.

Daytime-Only Stations

As the AM broadcasting system in the U. S. continued to evolve, the demand for more stations also grew. One result of the demand for facilities was the increased use of directional antennas to provide required interference protection while enhancing coverage in other directions. With the spectrum becoming crowded, another development was a large increase in the number of daytime-only stations. These are stations that were authorized to operate only during daylight hours. Because propagation conditions during these daylight hours do not normally support significant skywave transmission, there are many locations where a station can be operated during the daytime without causing harmful interference to other stations. Based on this concept, the FCC over the years licensed approximately 2500 stations for daytime-only operation.

Recognizing that daytime-only stations (and even some full-time stations that operate with restrictive directional antenna patterns during the night) are unable to provide effective service during early morning hours, the FCC originally allowed these stations to operate during this early morning period so long as no interference complaint had been received. Ultimately, this proved to be unworkable, and a more formalized approach was adopted. The FCC began granting presunrise authorizations (PSRAs), permitting many of these stations to operate their daytime facilities with powers up to 500 W during the pre-sunrise period between 6:00 a.m. and local sunrise. Although some interference occurred, the FCC believed the interference was balanced by the public's need for local informational services during this important morning time period.

While pre-sunrise operation did provide some relief, it did not end the economic problems many stations faced in effectively competing with full-time stations. The FCC was pressed to provide relief in the form of post-sunset operation for daytime-only stations. With the removal of international impediments, the FCC did provide such relief. Once again, taking into consideration the need for more service of a local nature and recognizing the changing nature of propagation conditions in which full nighttime conditions do not exist until several hours after sunset, the FCC granted post-sunset authorizations (PSSA) for most of the daytime-only stations. These authorizations permit operation for periods of up to 2 hours past sunset with power reduced to prevent interference. The FCC also changed its rules concerning the minimum power at which a station is permitted to operate. This, in turn, led to a subsequent decision allowing many PSSA stations to operate throughout the night, albeit with reduced power.[5]

Class I-A Clear Channel Stations

Still another change that has occurred pertained to the use of the class I-A clear channels. At one time, only a single station was permitted to operate on these channels at night, but in two FCC decisions first some and now all of these channels have been broken down to permit the authorization of additional nighttime operations. Clear channel stations that at one time provided service for a major portion of the country during nighttime hours now are protected only out to a distance averaging 750 miles.

Use of the AM Expanded Band

At an ITU Regional Radio Conference that concluded in 1988, participating administrations reached agreement on the criteria for expanding the AM band in Region 2 by adding 10 new channels between 1605 and 1705 kHz. Although the regional agreement established specific technical criteria for the implementation of the new channels (including the granting of priority usage of certain channels), the U. S. still retains considerable latitude in its domestic implementation. In a large country like the U. S., the use of the channels is unrestricted except in the relatively few areas near the borders with its neighbors.

The basic criteria set forth in the regional agreement are as follows: Stations may operate with 1 kW of power with a nondirectional antenna height of 90°, or stations may operate with a power not in excess of 10 kW by employing a directional antenna to provide equivalent protection to stations in other countries. The channels allotted to the United States in the border areas vary from location to location, but, as noted, over a large portion of the United States all 10 of the channels may be used. Non-allotted channels are not precluded from use, but the allotted channels in the other countries must remain fully protected. After a lengthy proceeding, the FCC developed a system whereby the channels in the expanded band would be used to improve the current interference situation by giving selected stations the opportunity to move to the new band. In view of this, applications for new stations are not currently being accepted for operation in the 1605 to 1705 kHz band.

Persons seeking an authorization for a new AM broadcast station on frequencies below 1605 kHz may do so by filing an application with the FCC. Applications must provide documentation that the proposed operation will comply with all applicable FCC rules as well as the appropriate international regulations. The details and methodology for allocating AM radio stations along with basic design specifications for AM directional antennas can be found in the FCC Rules, §73.14–73.190.

AM Stations Reclassified

The 1981 Rio Agreement changed the NARBA station classifications. No longer are the channels themselves classified. Stations are now classified without regard to the channel on which they operate. Stations providing wide area service, both groundwave and skywave, are now designated class A stations, and stations providing the equivalent of regional and local services are designated as classes B and C, respectively. The 1981 Rio Agreement permits any class of station to operate on any channel so long as it provides protection to other stations based on their classification. Tables 1.5-2 through 1.5-4 show the relationships between these international classifications and those that existed in the U. S. prior to 1990. In 1990, the FCC proposed to align the U.S. domestic classification system for AM stations with the 1981 Rio Agreement system. In addition to the three international classes of stations, the FCC has added a fourth: class D. Class D stations are those stations that operate at nighttime with a power of less than 250 W and an equivalent RMS antenna field of less than 141 mV/m at 1 km. Such stations are usually former daytime-only stations that have been granted some limited amount of nighttime operation. The current status of AM station classification is contained in the FCC Rules, §73.21. In that the allocation system in the United States had been long established, the station classifications still are primarily related to frequency. The current scheme is shown in Tables 1.5-5 and 1.5-6.

FM Broadcasting Frequency Allocation

The 88 to 108 MHz frequency band is allocated for FM broadcasting in Region 2 and, with some exceptions, for Regions 1 and 3, as well; however, unlike AM, FM broadcast allotments are largely a domestic matter (especially in large countries such as the U. S.) due to

[5]The rules pertaining to PSSAs and PSRAs are contained in §73.99 of the FCC Rules and Regulations (47 CFR §73.99).

TABLE 1.5-2
International and Domestic Classifications of AM Stations and Channels Prior to 1990

International Classes of AM Stations	Corresponding U.S. Classes Prior to 1990	Classes of Channels Available in the U. S. for Each Class
Class A	I-A	Clear channels
	I-B	
	I-N	
Class B	II	Clear channels
	II-A	
	II-B	
	II-C	
	II-D	
	II-S	
	III	Regional channels
	III-S	
Class C	IV	Local channels

TABLE 1.5-3
Channel Utilization in the United States Prior to 1990

Frequency (kHz)	Class of Station*	Frequency (kHz)	Class of Station*	Frequency (kHz)	Class of Station*
540	II	900	II	1260	III, III-S
550	III, III-S	910	I-A, III, III-S	1270	III, III-S
560	III, III-S	920	I-A, III, III-S	1280	III, III-S
570	III, III-S	930	I-A, III, III-S	1290	III, III-S
580	III, III-S	940	II	1300	III, III-S
590	III, III-S	950	III, III-S	1310	III, III-S
600	III, III-S	960	III, III-S	1320	III, III-S
610	III, III-S	970	III, III-S	1330	III, III-S
620	III, III-S	980	III, III-S	1340	IV
630	III, III-S	990	II	1350	III, III-S
640	I-A	1000	I-B	1360	III, III-S
650	I-A	1010	II	1370	III, III-S
660	I-A	1020	I-A	1380	III, III-S
670	I-A	1030	I-A	1390	III, III-S
680	I-B	1040	I-A	1400	IV
690	II	1050	II	1410	III, III-S
700	I-A	1060	I-B	1420	III, III-S
710	I-B	1070	I-B	1430	III, III-S
720	I-A	1080	I-B	1440	III, III-S
730	II	1090	I-B	1450	IV

TABLE 1.5-3 *(continued)*
Channel Utilization in the United States Prior to 1990

Frequency (kHz)	Class of Station*	Frequency (kHz)	Class of Station*	Frequency (kHz)	Class of Station*
740	II	1100	I-A	1460	III, III-S
750	I-A	1110	I-B	1470	III, III-S
760	I-A	1120	I-A	1480	III, III-S
770	I-A	1130	I-B	1490	IV
780	I-A	1140	I-B	1500	1-B
790	III, III-S	1150	III, III-S	1510	I-B
800	II	1160	I-A	1520	I-B
810	I-B	1170	I-B	1530	I-B
820	I-A	1180	1-A	1540	I-B
830	I-A	1190	I-B	1550	II
840	I-A	1200	I-A	1560	I-B
850	I-B	1210	I-A	1570	II
860	II	1220	II	1580	II
870	I-A	1230	IV	1590	III, III-S
880	I-A	1240	IV	1600	III, III-S
890	I-A	1250	III, III-S		

*Refer to Table 1.5-2; in addition to the class I-A or class I-B stations that could be assigned to the above channels, various class II stations could also be assigned.

TABLE 1.5-4
Power Limitation by Class of Station

Class of Station	Power (kW)			
	Daytime		Nighttime	
	Minimum	Maximum	Minimum	Maximum
I-A	10.0	50.0	10.0	50.0
I-B	10.0	50.0	10.0	50.0
I-N	10.0	50.0	10.0	50.0
II	10.0	50.0	10.0	50.0
II-A	10.0	50.0	10.0	50.0
II-B	0.25	50.0	0.25	50.0
II-C	0.25	50.0	0.25	1.0
II-D	0.25	50.0	N/A	N/A
II-S	0.25	50.0	N/A	<0.25
III	0.25	5.0	0.25	5.0
III-S	0.25	5.0	N/A	<0.25
IV	0.10	1.0	0.10	1.0

TABLE 1.5-5
Current Channel Utilization in the United States

Frequency (kHz)	Class of Station	Frequency (kHz)	Class of Station	Frequency (kHz)	Class of Station
540	A, B, D	930	B, D	1320	B, D
550	B, D	940	A, B, D	1330	B, D
560	B, D	950	B, D	1340	C
570	B, D	960	B, D	1350	B, D
580	B, D	970	B, D	1360	B, D
590	B, D	980	B, D	1370	B, D
600	B, D	990	A, B, D	1380	B, D
610	B, D	1000	A, B, D	1390	B, D
620	B, D	1010	A, B, D	1400	C
630	B, D	1020	A, B, D*	1410	B, D
640	A, B, D*	1030	A, B, D*	1420	B, D
650	A, B, D*	1040	A, B, D*	1430	B, D
660	A, B, D*	1050	A, B, D	1440	B, D
670	A, B, D*	1060	A, B, D	1450	C
680	A, B, D	1070	A, B, D	1460	B, D
690	A, B, D	1080	A, B, D	1470	B, D
700	A, B, D*	1090	A, B, D	1480	B, D
710	A, B, D	1100	A, B, D*	1490	C
720	A, B, D*	1110	A, B, D	1500	A, B, D
730	A, B, D	1120	A, B, D*	1510	A, B, D
740	A, B, D	1130	A, B, D	1520	A, B, D
750	A, B, D*	1140	A, B, D	1530	A, B, D
760	A, B, D*	1150	B, D	1540	A, B, D
770	A, B, D*	1160	A, B, D*	1550	A, B, D
780	A, B, D*	1170	A, B, D	1560	A, B, D
790	B, D	1180	A, B, D*	1570	A, B, D
800	A, B, D	1190	A, B, D	1580	A, B, D
810	A, B, D	1200	A, B, D*	1590	B, D
820	A, B, D*	1210	A, B, D*	1600	B, D
830	A, B, D*	1220	A, B, D	1610	B, D
840	A, B, D*	1230	C	1620	B, D
850	A, B, D	1240	C	1630	B, D
860	A, B, D	1250	B, D	1640	B, D
870	A, B, D*	1260	B, D	1650	B, D
880	A, B, D*	1270	B, D	1660	B, D
890	A, B, D*	1280	B, D	1670	B, D
900	A, B, D	1290	B, D	1680	B, D
910	B, D	1300	B, D	1690	B, D
920	B, D	1310	B, D	1700	B, D

*Only one class A station will be assigned on these channels.

TABLE 1.5-6
Power Limitation by Class of Station

Class of Station	Power (kW)			
	Daytime		Nighttime	
	Minimum	Maximum	Minimum	Maximum
A	10.0	50.0	10.0	50.0
B	0.25	50.0	0.25	50.0 (below 1605 kHz)
B	0.25	10.0	0.25	10.0 (1605–1705 kHz)
C	0.25	1.0	0.25	1.0
D	0.25	50.0	N/A	<0.25

the limited nature of signal propagation at these frequencies. Although there are some international regulations regarding FM broadcasting, there is no region-wide FM agreement in Region 2. Instead, there are bilateral agreements between the U. S. and Canada and between the U. S. and Mexico. Both regulate the use of FM channels in the border areas and specify technical standards in order to ensure system compatibility. The FM broadcast band is divided into 100 channels, each 200 kHz wide. In the U. S., the lower 20 channels, located between 88 and 92 MHz, have been reserved for noncommercial broadcasting; however, such stations are not restricted solely to these channels. In addition, although it is part of television channel 6 (82–88 MHz), the frequency 87.9 MHz can be used for low-power noncommercial FM stations, but its use is severely restricted.

FM Station Classes and Broadcast Zones

As with AM broadcasting, different classes of FM stations are designed to provide different types of service. As the demand for more stations has increased so has the number of different classes of stations. In response to this demand, the FCC has significantly modified the criteria concerning the use of the frequencies. In June 1983, the FCC concluded a lengthy rule-making proceeding and modified the domestic allotment criteria for FM broadcasting. Prior to this action, there were three classes of stations in the 80-channel commercial band. Twenty of these channels were used for lower power class A stations having a maximum effective radiated power (ERP) of 3 kW and a maximum antenna height above average terrain (HAAT) of 91.4 m (300 ft). Class A stations had a 1 mV/m service radius of about 15 miles. Higher power class B or C stations operated on the remaining 60 channels. Whether a station was designated class B or C depended on where it was located (Figure 1.5-3).[6] Class B stations were located in Zone I or I-A. Zone I is the northeast United States, extending south to the Virginia–North Carolina border and west to the Mississippi River. Zone I-A is all but the northernmost portion of California, plus Puerto Rico and the Virgin Islands. Class C stations operate elsewhere in the country in what is referred to as Zone II.

Class B stations operate with a maximum ERP of 50 kW at 150 m (492 ft) HAAT and have a service radius of about 52 km (32 miles). Class C stations operate with a maximum power of 100 kW at 600 m (1968 ft) HAAT, for a service radius of approximately 92 km (57 miles). FM stations in each of these classes may elect to operate at a HAAT above the maximum, but in such cases they are required to make a compensatory reduction in ERP as noted in FCC Rules, §73.211.

This system was changed substantially by the above-noted 1983 proceeding, when the FCC:

- Permitted class A stations to operate on channels previously reserved for class B or C stations.
- Created three new classes of FM stations. Class B1 stations are permitted to operate in Zones I and I-A with a maximum ERP of 25 kW at 100 m (328 ft) HAAT; class C1 and class C2 stations are permitted to operate in Zone II. Class C1 stations are permitted a maximum ERP of 100 kW at 300 m (984 ft) HAAT and class C2 stations are permitted a maximum ERP of 50 kW at 150 m (492 ft) HAAT.
- Required stations that were previously licensed as class B or C and were not operating at the minimum level specified for their class under the new rules to upgrade their facilities within 3 years; otherwise, the under-minimum facilities would be reclassified to the appropriate lower class based on the facilities they used.
- Increased the maximum antenna HAAT for class A stations to 100 m (328 ft).

In 1989, the FCC further modified the rules to permit class A stations to operate with an ERP of 6 kW and a HAAT of 100 m and added an additional classification, C3, that permits operation in Zone II with a maximum of 25 kW at 100 m HAAT. In 2000, the FCC again modified the rules to create an additional station classification, C0. Stations in this class operate with a

[6]The concept of allowing different classes of stations in different areas or zones is based on the population density of the areas. The Commission has assumed that there is less need for wide areas of service in areas of dense population.

From: 47 CFR § 73.205

Federal Communications Commission
Office of Engineering and Technology
Michael R. Davis

FIGURE 1.5-3 FM broadcast Zones I and I-A.

maximum ERP of 100 kW with a HAAT of between 300 (980 ft) and 450 m (1476 ft).

In addition to the above, there also are class D stations that operate as noncommercial educational stations with power not in excess of 10 W; however, applications for this class of station are no longer being accepted (see 47 CFR §73.512(c)). A complete list of the station classes and a summary of FM allotment standards can be found in Tables 1.5-7 and 1.5-8.

FM Table of Allotments

Unlike AM broadcasting, where a new station may be applied for at any location where it can meet applicable criteria, the use of commercial FM channels (channels 221 to 300) is governed by the FM Table of Allotments found in the FCC Rules, §73.202. This table lists all FM channel allotments that have been made available for use. Most already are in use. If the table does not list a vacant channel in the desired community, the prospective applicant must file a rulemaking petition with the FCC seeking to add such a channel for the community. The rule-making petition proposing such addition must provide a showing that the proposal meets the separation requirements that are applicable to the class of station being proposed. A complete list of the spacing requirements, including

TABLE 1.5-7
Standards for FM Allotments in Puerto Rico and the Virgin Islands

Station Class	Maximum ERP kW	Maximum HAAT m (ft)	Expected Service Radius km (mi)
A	6	240 (787)	42 (26)
B1	25	150 (492)	46 (29)
B	50	472 (1549)	78 (49)

TABLE 1.5-8
Standards for FM Allotments for Locations Other Than Puerto Rico and the Virgin Islands

Station Class	Maximum ERP kW	Maximum HAAT m (ft)	Expected Service Radius km (mi)
A	6	100 (328)	28 (18)
B1	25	100 (328)	39 (24)
B	50	150 (492)	52 (33)
C3	25	100 (328)	39 (24)
C2	50	150 (492)	52 (33)
C1	100	299 (981)	72 (45)
C0	100	450 (1476)	83 (52)
C	100	600 (1968)	92 (58)

those that pertain to stations located near Canada and Mexico, is provided in Tables 1.5-9 through 1.5-12. Alternatively, a petitioner can propose to modify the table by deleting a vacant existing allotment or by changing the frequencies of an existing station and thereby achieve compliance with these spacing requirements.

Once a location has been added to the Table of Allotments, the FCC will announce a period of time (called a *window*) when it will accept applications for the location. If, as often is the case, there are multiple applicants, the winning applicant will be determined by an auction. It also should be noted, that in cases where it is not possible to locate a transmitter site that meets the mileage separation requirements, the FCC does permit the use of reduced power or antenna height, as well as directional antennas, in order to provide equivalent protection to other stations. However, this only applies to the filing of an application for a location that is already in the Table of Allotments. The FCC will not accept proposals to modify the table

TABLE 1.5-9
Minimum Distance Separation Requirements

Relation	Minimum Separation km (mi)			
	Co-Channel	200 kHz	400/600 kHz	10.6/10.8 MHz
A to A	115 (71)	72 (45)	31 (19)	10 (6)
A to B1	143 (89)	96 (60)	48 (30)	12 (7)
A to B	178 (111)	113 (70)	69 (43)	15 (9)
A to C3	14 (88)	89 (55)	42 (26)	12 (7)
A to C2	166 (103)	106 (66)	55 (34)	15 (9)
A to C1	200 (124)	133 (83)	75 (47)	22 (14)
A to C0	215 (134)	152 (94)	86 (53)	25 (16)
A to C	226 (140)	165 (103)	95 (59)	29 (18)
B1 to B1	175 (109)	114 (71)	50 (31)	14 (9)
B1 to B	211 (131)	145 (90)	71 (44)	17 (11)
B1 to C3	175 (109)	114 (71)	50 (31)	14 (9)
B1 to C2	200 (124)	134 (83)	56 (35)	17 (11)
B1 to C1	233 (145)	161 (100)	77 (48)	24 (15)
B1 to C0	248 (154)	180 (112)	87 (54)	27 (17)

Relation	Minimum Separation km (mi)			
	Co-Channel	200 kHz	400/600 kHz	10.6/10.8 MHz
C3 to C3	153 (95)	99 (62)	43 (27)	14 (9)
C3 to C2	177 (110)	117 (73)	56 (35)	17 (11)
C3 to C1	211 (131)	144 (90)	76 (47)	24 (15)
C3 to C0	226 (140)	163 (101)	87 (54)	27 (17)
C3 to C	237 (147)	176 (109)	96 (60)	31 (19)
C2 to C2	190 (118)	130 (81)	58 (36)	20 (12)
C2 to C1	224 (139)	158 (98)	79 (49)	27 (17)
C2 to C0	239 (148)	176 (109)	89 (55)	31 (19)
C2 to C	249 (155)	188 (117)	105 (65)	35 (22)
C1 to C1	245 (152)	177 (110)	82 (51)	34 (21)
C1 to C0	259 (161)	196 (122)	94 (58)	37 (23)
C1 to C	270 (168)	209 (130)	105 (65)	41 (25)
C0 to C0	270 (168)	207(129)	96 (50)	41 (25)
C0 to C	281 (175)	220 (137)	105 (65)	45 (28)

TABLE 1.5-9 *(continued)*
Minimum Distance Separation Requirements

Relation	Minimum Separation km (mi)			
	Co-Channel	200 kHz	400/600 kHz	10.6/10.8 MHz
B1 to C	259 (161)	193 (120)	105 (65)	31 (19)
B to B	241 (150)	169 (105)	74 (46)	20 (12)
B to C3	211 (131)	145 (90)	71 (44)	17 (11)
B to C2	241 (150)	169 (105)	74 (46)	20 (12)
B to C1	270 (168)	195 (121)	79 (49)	27 (17)
B to C0	272 (169)	214 (133)	89 (55)	31 (19)
B to C	274 (170)	217 (135)	105 (65)	35 (22)

Relation	Minimum Separation km (mi)			
	Co-Channel	200 kHz	400/600 kHz	10.6/10.8 MHz
C to C	290 (180)	241 (150)	105 (65)	48 (30)

TABLE 1.5-10
Minimum Distance Separation Requirements: Canadian Agreement

Relation	Minimum Separation km (mi)				
	Co-Channel	200 kHz	400 kHz	600 kHz	10.6/10.8 MHz
A1 to A1	78 (49)	45 (28)	24 (15)	20 (12)	4 (2)
A1 to A	131 (82)	78 (49)	44 (27)	40 (25)	7 (4)
A1 to B1	164 (102)	98 (61)	57 (35)	53 (33)	9 (6)
A1 to B	190 (118)	117 (73)	71 (44)	67 (42)	12 (7)
A1 to C1	223 (139)	148 (92)	92 (57)	88 (55)	19 (12)
A1 to C	227 (141)	162 (101)	103 (64)	99 (62)	26 (16)
A to A	151 (94)	98 (61)	51 (32)	42 (26)	10 (6)
A to B1	184 (115)	119 (74)	64 (40)	55 (34)	12 (7)
A to B	210 (131)	137 (85)	78 (49)	69 (43)	15 (9)
A to C1	243 (151)	168 (105)	99 (62)	90 (56)	22 (14)
A to C	247 (154)	182 (113)	110 (68)	101 (63)	29 (18)
B1 to B1	197 (123)	131 (82)	70 (44)	57 (35)	24 (15)
B1 to B	223 (139)	149 (93)	84 (52)	71 (44)	24 (15)
B1 to C1	256 (159)	181 (113)	108 (67)	92 (57)	40 (25)
B1 to C	259 (161)	195 (121)	116 (72)	103 (64)	40 (25)
B to B	237 (148)	164 (102)	94 (59)	74 (46)	24 (15)
B to C1	271 (169)	195 (121)	115 (72)	95 (59)	40 (25)
B to C	274 (171)	209 (130)	125 (78)	106 (66)	40 (25)
C1 to C1	292 (182)	217 (135)	134 (83)	101 (63)	48 (30)
C1 to C	302 (188)	230 (143)	144 (90)	111 (69)	48 (30)
C to C	306 (190)	241 (150)	153 (95)	113 (70)	48 (30)

Note: Class A1 stations are defined as having an ERP of 0.25 kW and an antenna HAAT of 100 m; U.S. class C2 stations are considered as class B. Class A1 stations are defined as having an ERP of 0.25 kW and an antenna HAAT of 100 m; U.S. class C2 stations are considered as class B.

TABLE 1.5-11
Minimum Distance Separation Requirements: Mexican Agreement

Relation	Minimum Separation km (mi)			
	Co-Channel	200 kHz	400/600 kHz	10.6/10.8 MHz
A to A	105 (65)	65 (40)	25 (15)	8 (5)
A to B	175 (110)	105 (65)	65 (40)	16 (10)
A to C	210 (130)	170 (105)	105 (65)	32 (20)
A to D	95 (60)	50 (30)	25 (15)	5 (5)
B to B	240 (150)	170 (105)	65 (40)	25 (15)
B to C	270 (170)	215 (135)	105 (65)	40 (25)
B to D	170 (105)	95 (60)	65 (40)	16 (10)
C to C	290 (180)	240 (150)	105 (65)	48 (30)
C to D	200 (125)	155 (95)	105 (65)	25 (15)
D to D	18 (11)	10 (6)	5 (3)	3 (2)

unless it is shown to meet applicable separation requirements.

Assignment of stations on the noncommercial educational channels (200–220) is accomplished more in the manner that is followed in AM, where an application includes a showing that interference will not be caused to other stations. In addition, for proposals to use channels 218, 219, and 220, compliance with applicable separation requirements to any allotments on higher, adjacent commercial channels is required.

Low-Power FM Stations

In addition to the provisions for noncommercial FM stations discussed above, the FCC also licenses low-power FM (LPFM) radio stations. This service was created by the Commission in January 2000 for noncommercial educational broadcasting only; stations may operate with a maximum ERP of 100 W with an antenna HAAT of 30 m (100 ft). The approximate service range of a 100 W LPFM station is a 5.6 km (3.5 mile) radius. LPFM stations are permitted a HAAT greater than 30 m with a reduction in ERP so as not to extend the service beyond that which would be produced at 30 m.

TABLE 1.5-12
Minimum Distance Separation Requirements to TV Channel 6 from FM Stations on Channel 253 (98.5 MHz)

FM Class	Minimum Separation km (mi)	
	TV Zone I	TV Zones II and III
A	17 (11)	22 (14)
B1	19 (12)	23 (14)
B	22 (14)	26 (16)
C3	19 (12)	23 (14)
C2	22 (14)	26 (16)
C1	29 (18)	33 (21)
C	36 (22)	41 (26)

The FCC permits two classes of LPFM stations: LP100 and LP10. LP100 stations are permitted the maximum facility discussed above with a minimum requirement of 50 W at 30 m or equivalent at a greater height. LP10 stations are authorized an ERP of 10 W at 30 m, with a minimum requirement of 1 W ERP. Like LP100 stations, HAATs greater than 30 m are permitted for LP10 stations with a commensurate reduction in ERP; however, LP10 stations will not be authorized power in excess of 1 W if the HAAT exceeds 100 m.

LPFM stations are not protected from interference that may be received from other classes of FM stations; however, LPFM stations must protect other authorized radio broadcast stations on the same channel (co-channel), as well as stations on first-, second-, or third-adjacent channels above or below the LPFM station. In addition, LP100 and LP10 stations must protect other LP100 stations, and LP10 stations must protect each other. The protection is accomplished through the use of minimum distance separation requirements that are detailed in the FCC Rules, §73.807.

LPFM stations are available to noncommercial educational entities and public safety and transportation organizations but are not available to individuals or for commercial operations. Also, current broadcast licensees with interests in other media (broadcast or newspapers) are not eligible to obtain LPFM stations.

TABLE 1.5-13
Minimum Distance Separation Requirements from LP100 to Other Stations

Relation	Minimum Separation							
	Co-Channel		200 kHz		400/600 kHz		10.6/10.8 MHz	
	km	mi	km	mi	km	mi	km	mi
LP100	24	15	14	9	0	0	0	0
D	24	15	13	8	6	4	3	2
A	67	42	56	35	29	18	6	4
B1	87	54	74	46	46	29	9	6
B	112	70	97	60	67	42	12	7
C3	78	49	67	42	40	25	9	6
C2	91	57	80	50	53	33	12	7
C1	111	69	100	62	73	45	20	12
C0	122	76	111	69	84	52	22	14
C	130	81	120	75	93	58	28	17

In addition to the above, LP100 and LP10 stations in Puerto Rico and the U.S. Virgin Islands must also meet the requirements listed in Tables 1.5-15 and 1.5-16.

LP100 and LP10 stations must also comply with the spacing requirements listed in Tables 1.5-17 through 1.5-20, with respect to stations in Canada and Mexico.

FM Translator and Booster Stations

In addition to the regular FM broadcast stations discussed above, two other types of stations are permitted to operate in the FM band on a secondary basis: *FM translator stations* and *FM booster stations*. An FM

TABLE 1.5-14
Minimum Distance Separation Requirements from LP10 to Other Stations

Relation	Minimum Separation							
	Co-Channel		200 kHz		400/600 kHz		10.6/10.8 MHz	
	km	mi	km	mi	km	mi	km	mi
LP100	16	10	10	6	0	0	0	0
LP10	13	8	8	5	0	0	0	0
D	16	10	10	6	6	4	2	1
A	59	37	53	33	29	18	5	3
B1	77	48	70	44	45	28	8	5
B	99	62	91	57	66	41	11	7
C3	69	43	64	40	39	24	8	5
C2	82	51	77	48	52	32	11	7
C1	103	64	97	60	73	45	18	11
C0	114	71	99	62	84	52	21	13
C	122	76	116	72	92	57	26	16

TABLE 1.5-15
Additional Minimum Distance Separation Requirements from LP100 to Other Stations in Puerto Rico and the U.S. Virgin Islands

	Minimum Separation							
	Co-Channel		200 kHz		400/600 kHz		10.6/10.8 MHz	
Relation	km	mi	km	mi	km	mi	km	mi
A	80	50	70	44	42	26	9	6
B1	95	59	82	51	53	33	11	7
B	138	86	123	77	92	57	19	12

TABLE 1.5-16
Additional Minimum Distance Separation Requirements from LP10 to Other Stations in Puerto Rico and the U.S. Virgin Islands

	Minimum Separation							
	Co-Channel		200 kHz		400/600 kHz		10.6/10.8 MHz	
Relation	km	mi	km	mi	km	mi	km	mi
A	72	45	66	41	42	26	8	5
B1	84	52	78	49	53	33	9	6
B	126	78	118	73	92	57	18	11

booster station retransmits the signal of a primary station on the primary station's channel in order to serve areas where the primary station's signal is inadequate. An FM translator station is similar to an FM booster station, except that the signal is not retransmitted on the same channel but instead is translated to a different channel. These are authorized in accordance with Part 74, Subpart L, of the FCC Rules which, among other things, requires that such stations provide protection from interference to all regular FM broadcast stations. For more information on FM translators and boosters, see Chapter 4.12.

TABLE 1.5-17
Minimum Distance Separation Requirements from LP100 to Other Stations in Canada

	Minimum Separation									
	Co-Channel		200 kHz		400 kHz		600 kHz		10.6/10.8 MHz	
Relation	km	mi	km	mi	km	mi	km	mi	km	mi
A1 and LP	45	28	30	19	21	13	20	12	4	2
A	66	41	50	31	41	26	40	25	7	4
B1	78	49	62	39	53	33	52	32	9	6
B	92	57	76	47	68	42	66	41	12	7
C1	113	70	98	61	89	55	88	55	19	12
C	124	77	108	67	99	62	98	61	28	17

TABLE 1.5-18
Minimum Distance Separation Requirements from LP10 to Other Stations in Canada

	Minimum Separation									
	Co-Channel		200 kHz		400 kHz		600 kHz		10.6/10.8 MHz	
Relation	km	mi	km	mi	km	mi	km	mi	km	mi
A1 & LP	33	21	5	3	20	12	19	12	3	2
A	53	33	45	28	40	25	39	24	5	3
B1	65	40	57	35	52	32	51	32	8	5
B	79	49	71	44	67	42	66	41	11	7
C1	101	63	93	58	88	55	87	54	18	11
C	111	69	103	64	98	61	97	60	26	16

TABLE 1.5-19
Minimum Distance Separation Requirements from LP100 to Other Stations in Mexico

	Minimum Separation									
	Co-Channel		200 kHz		400 kHz		600 kHz		10.6/10.8 MHz	
Relation	km	mi	km	mi	km	mi	km	mi	km	mi
Low Power	27	17	17	11	9	6	9	6	3	2
A	43	27	32	20	25	16	25	16	5	3
AA	47	29	36	22	29	18	29	18	6	4
B1	67	42	54	34	45	28	45	28	8	5
B	91	57	76	47	66	41	66	41	11	7
C1	91	57	80	50	73	45	73	45	19	12
C	110	68	100	62	92	57	92	57	27	17

TABLE 1.5-20
Minimum Distance Separation Requirements from LP10 to Other Stations in Mexico

	Minimum Separation									
	Co-Channel		200 kHz		400 kHz		600 kHz		10.6/10.8 MHz	
Relation	km	mi	km	mi	km	mi	km	mi	km	mi
Low Power	19	12	13	8	9	6	9	6	2	1
A	34	21	29	18	24	15	24	15	5	3
AA	39	24	33	21	29	18	29	18	5	3
B1	57	35	50	31	45	28	45	28	8	5
B	79	49	71	44	66	41	66	41	11	7
C1	83	52	77	48	73	45	73	45	18	11
C	102	63	96	60	92	57	92	57	26	16

TABLE 1.5-21
TV Channel Allotments*

Frequency Band (MHz)	TV Channels
54–72	2–4
76–88	5–6
174–216	7–13
470–806	14–69

*After February 17, 2009, all full-service television broadcasting will be restricted to channels 2 to 51. At some yet undefined date, all other television broadcasting will also likely be required to vacate channels above 51.

TELEVISION BROADCASTING: FREQUENCY ALLOCATION

Three different frequency bands are used for television broadcasting within the U. S. The plan, as shown in Table 1.5-21, includes the low-VHF band TV channels 2 to 4 (54–72 MHz) as well as channels 5 and 6 (76–88 MHz), the high-VHF band channels 7 through 13 (174–216 MHz), and the UHF band channels 14 through 69 (470–806 MHz). The greater portion of all of these bands is allocated for broadcasting throughout the world, but this allocation is not uniform and in many areas other uses such as land mobile are permitted on a secondary basis. This is also the case in the U. S., where certain UHF television channels are now used for land mobile in some major cities. In general, due to the limited extent of radiowave propagation in the television band, TV allocations, like FM, are basically a domestic matter, with few international regulations. Although no regional agreement exists, the U. S. does have agreements with both Canada and Mexico concerning television allocation.

It should be noted that, after the U. S. transitions to an all-digital television system, the UHF channels available for television broadcasting will be reduced to 14 through 51. The date for the final transition to digital for full service television stations in the U. S. is currently February 17, 2009. At this time, a date has not been set for low-power television stations, translators, or boosters.

Like FM broadcasting, TV allocations in the U. S. are governed by a Table of Frequency Allotments. The current FCC Rules (June 2006) contain two allotment tables—one for analog stations in §73.606 and one for digital stations in §73.622. These tables contain all commercial as well as noncommercial allotments (the latter are identified by an asterisk). Although the FCC is no longer accepting petitions to modify the analog table, a description of how it was constructed is given below for historical purposes.

The analog table is based on the separation criteria contained in §73.610 and §73.698 of the FCC Rules, provided in Tables 1.5-22 and 1.5-23, respectively. The tables also show there are different requirements for different zones or areas of the country which are shown in Figure 1.5-4.

Zone I is the northeast U. S. extending south to the Virginia–North Carolina border and west to the Mississippi River. Zone II consists of that portion of the U. S. that is not in Zone I or Zone III, including Alaska, Hawaii, Puerto Rico, and the Virgin Islands. Zone III is that portion of the southeast U. S. extending from the east coast of Georgia westward to the Mexican border. An exact description of the zones is contained in §73.609 of the FCC Rules. As with FM, the TV zones reflect the differing population densities in various parts of the country. In addition, differences in propagation conditions were also considered. The closer spacings in Zone I recognize the fact that, in the northeastern portion of the U. S., many large population centers require stations of their own and are close enough to one another to lessen the need for wide area service. Zone II is characterized by fewer population centers, usually smaller and farther apart; for them, wide area service is a necessity. Finally, because the area of the country along the Gulf Coast (Zone III) is susceptible to high levels of tropospheric propagation, stations in that area must be spaced farther apart to minimize interference. Although there are differences in the powers authorized for low-VHF, high-VHF, and UHF stations, unlike AM and FM broadcasting there are no class designations as such in television. The differing power limitations reflect the differences in signal propagation for low-VHF, high-VHF, and UHF.

In addition to the above, there are other constraints on UHF channels due to various types of problems

TABLE 1.5-22
Minimum Co-Channel TV Separation Distances

	Minimum Separation km (mi)			
	Co-Channel		Adjacent Channel	
Zone	Channels 2–13	Channels 14–69	Channels 2–13	Channels 14–69
I	272.7 (169.5)	248.6 (154.5)	95.7 (59.5)	87.7 (54.5)
II	304.9 (189.5)	280.8 (174.5)		
III	353.2 (219.5)	329.0 (204.5)		

that occur in television receivers that are caused by the mixing of signals of stations operating on different channels. A complete table of these restrictions, referred to as the "UHF Taboos," is contained in §73.698 of the FCC Rules.

Digital TV Allotments

The digital television (DTV) Table of Allotments was developed through a complex analysis that took into consideration actual computations of coverage and interference. The DTV Table contains an entry for each analog station that was in operation or for which an application had been filed prior to a cutoff date specified by the FCC. In addition, the table now contains some additional allotments that were added through the process discussed below. A petition for rule making may be filed with the FCC seeking to modify the table to include a new allotment. Such petitions must provide a showing that the proposal complies with the applicable separation criteria, shown in Table 1.5-24, that serve to prevent mutual interference. Furthermore, applications to construct a station on a new allotment must demonstrate that the proposed facility will provide applicable coverage to the city of license and that no new interference would be caused to existing stations either analog or digital.

TABLE 1.5-23
Minimum Distance Separation Requirements to TV Channel 6 from FM Stations on Channel 253 (98.5 MHz)

FM Class	Minimum Separation km (mi)	
	TV Zone I	TV Zones II and III
A	17 (11)	22 (14)
B1	19 (12)	23 (14)
B	22 (14)	26 (16)
C3	19 (12)	23 (14)
C2	22 (14)	26 (16)
C1	29 (18)	33 (21)
C	36 (23)	41 (26)

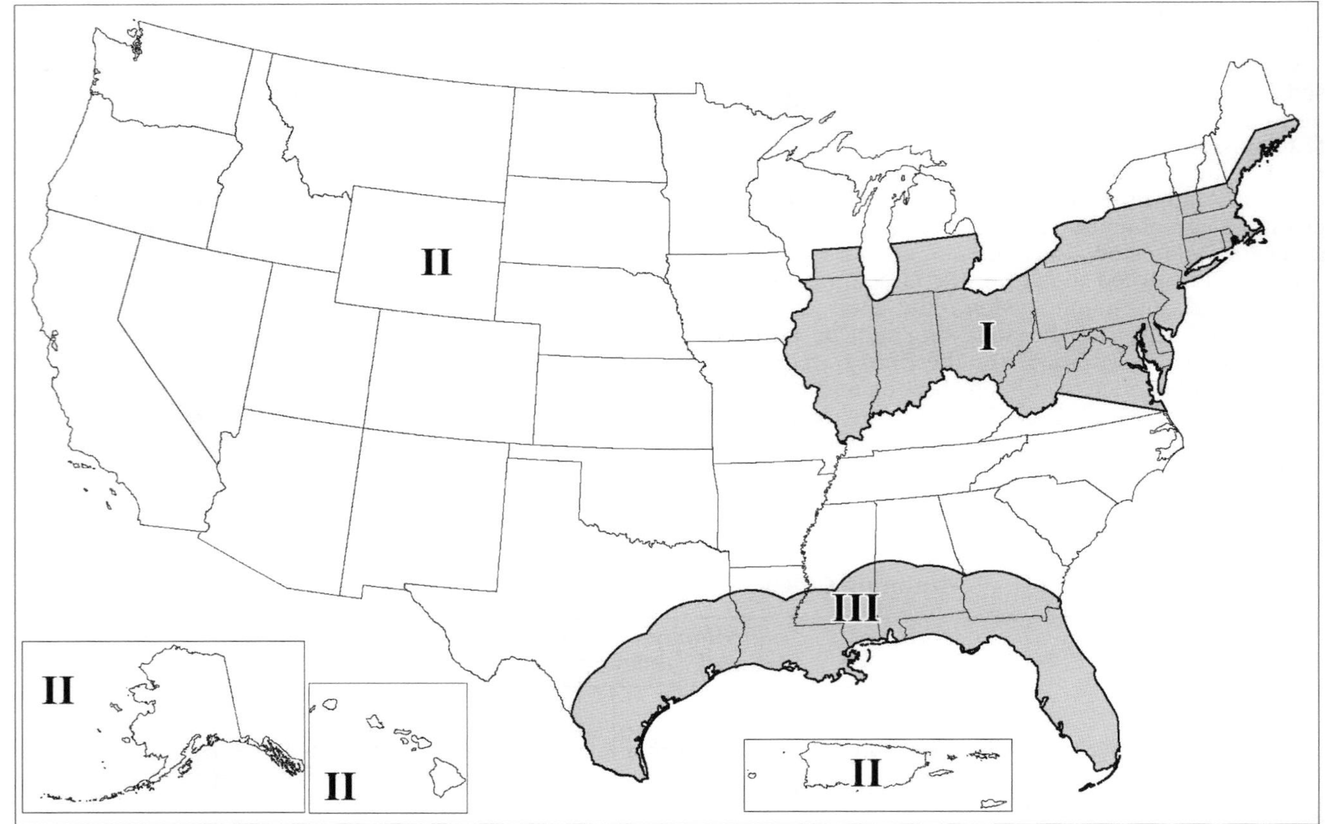

FIGURE 1.5-4 TV broadcast zones; Zone II is all of the United States not in Zones I or III.

TABLE 1.5-24
Required TV Separation Distances, DTV to Analog TV and DTV to DTV

Relation	DTV to DTV km (mi)	DTV to Analog TV km (mi)
	VHF Channels 2–13	
Co-channel, Zone I	244.6 (152.0)	244.6 (152.0)
Co-channel, Zone II and Zone III	273.6 (170.0)	273.6 (170.0)
Adjacent channel, Zone I	<20.0 (12.4) or >110.0 (68.4)	<9.0 (5.6) or >125.0 (77.6)
Adjacent channel, Zone II and Zone III	<23.0 (14.3) or >110.0 (68.4)	<11.0 (6.8) or >125.0 (77.6)
	UHF Channels	
Co-channel, Zone I	196.3 (122.0)	217.3 (135.0)
Co-channel, Zone II and Zone III	223.7 (139.0)	244.6 (152.0)
Adjacent channel, Zone I	<24.0 (14.9) or >110.0 (68.4)	<12.0 (7.5) or >106.0 (65.8)
Adjacent channel, Zone II and III	<24.0 (14.9) or >110.0 (68.4)	<12.0 (7.5) or >106.0 (65.8)
Taboo channels, Zone I*		<24.1 (15.0) or >80.5 (50.0)
Taboo channels, Zone II and Zone III*		<24.1 (15.0) or >96.6 (60.0)

*Values are only specified for DTV to analog TV and are for DTV channels ±2, ±3, ±4, ±5, ±7, ±8, +14, and +15 channels above the analog TV channel.

As mentioned above, the U. S. is transitioning to an all-digital television system with an overall reduction in the number of channels. As of this writing, the process to determine the final DTV channel for each full-service television station is almost completed, at which point a new post-DTV transition allocation table will be produced. Although it has not yet been stated, it is expected the process to modify the new (post-transition) table will be similar to the process that has been used to modify the original DTV allotment table.

Low-Power Television Stations, Translators, and Boosters

In addition to the regular full-service television broadcast stations, three other types of analog stations are permitted to operate in the television bands with low power on a secondary basis: *low-power television (LPTV) stations*, *TV translator stations*, and *TV booster stations*. A TV booster station retransmits the signal of a primary station on the primary station's channel in order to serve areas where the primary station's signal is inadequate. A TV translator station is similar to a TV booster station, except that the signal is not retransmitted on the same channel but instead is translated to a different channel. LPTV stations may retransmit the signals of another station or they may originate programming. All of these stations are authorized in accordance with Part 74, Subpart G, of the FCC Rules that requires full protection from interference to all regular television broadcast stations. These stations are not accorded any protection from the operation of regular television broadcast stations either analog or digital. As part of the transition to an all-digital television system, the FCC in September 2004 established rules to allow LPTV stations, TV translator stations, and TV booster stations to make the digital transition. These rules allow existing analog LPTV stations to file an application at any time to "flash cut" to digital. The rules also established interference protection criteria that will be used to assess these applications which are much like those used for regular DTV stations. In addition, these rules also set up criteria for allowing digital companion channels for LPTV stations similar to those established for full-service television stations. Applications for the companion channels were accepted during a filing window that closed on June 30, 2006. Like the current low-power stations, the low-power digital stations will be authorized on a secondary basis and will not be afforded protection from full-service stations.

Class A TV Stations

In March 2000, the FCC established a new class of television stations designated as class A stations. These are former LPTV stations that have been granted primary status based on meeting specific programming and operational requirements during a 90 day period ending November 28, 1999. To maintain their class A status, these stations must broadcast a minimum of 18 hours per day and broadcast an average of 3 hours per week of locally produced programming each quarter. Due to their primary status, these stations are afforded protection from increased interference from all other television broadcast stations. The rules that established class A stations also allow these stations to "flash cut" to digital at any time if they provide interference protection to all other primary and secondary stations.

AUXILIARY BROADCAST SERVICES: FREQUENCY ALLOCATION

Although all auxiliary broadcast services share a common role in support of AM, FM, or TV broadcast operations, there are important differences between them. The nature of the service they provide varies, as does the frequency band in which they operate. For these reasons the allotments available for each auxiliary service are discussed separately; however, before turning to a specific discussion of each service, some general comments are necessary. The steady growth in the number of AM, FM, and TV stations, including DTV, as well as their desire to use more advanced technology in every aspect of their operation, has greatly increased demand for spectrum in all auxiliary services. Because of the continuing demand for spectrum from a multitude of other broadcast and non-broadcast activities, it has been difficult for the FCC to allocate more spectrum to alleviate the congestion faced by the auxiliary services. At the same time, some relief has come through changes in the FCC rules to facilitate the use of newer, more spectrum-efficient, technology. When reading the material in this section, the reader should be aware that additional changes may take place and should check the current regulations governing that particular radio service. Recognizing that available spectrum is limited, it is necessary for broadcasters to use it efficiently. To do this, it is first necessary to understand what spectrum is allocated, how it may be used, and what advantages or disadvantages may be involved in the use of a particular band for each of the auxiliary services discussed below.

Remote Pickup Broadcast Stations

Remote pickup (RPU) stations are mobile or portable facilities used to transmit live on-the-air programming from a temporary remote location, such as a shopping center or football game, to the station's studio facilities. This material can be taped for later rebroadcast, or it can be incorporated into actual ongoing, live broadcasts. Radio stations typically have several RPUs that may be licensed to one or more frequencies. In November 1984, the FCC significantly revised its radio broadcast auxiliary frequency allocations to permit operational use of narrowband technologies in the Broadcast Remote Pickup Service. The goal was to foster spectrum efficiency in a flexible manner. Broadcasters and equipment manufacturers who wanted to operate narrowband equipment, the Commission believed, should not be precluded from doing so by rigid FCC rules; however, the FCC never formally implemented these changes due to delays in setting up the mechanism to handle the expected applications. In the meantime, the FCC accepted applications based on this revised allocation. Then, in November of 2002, the FCC once again revised the broadcast remote pickup rules, which are now in effect and are detailed below.

Frequency Allocations for Radio Broadcast Remote Pickup Stations

- *25.87–26.47 MHz*—There are 26 frequencies available in this band for use by remote pickup broadcast stations. Bandwidth is limited to 20 kHz except between 25.87 and 26.03 MHz, where 40 kHz is permitted. Note that the use of the frequencies between 25.87 and 26.09 MHz are subject to the condition that no harmful interference is caused to broadcast stations sharing this band. Note also that the frequencies between 26.100 and 26.175 MHz have been allocated on a worldwide basis to the Maritime Mobile Services and are shared with the Maritime Service in the U. S. Note further that the frequencies between 26.175 and 26.47 MHz are allocated on a worldwide basis for use by various types of fixed and mobile operations. Hence, when selecting a frequency in this band for remote pickup use, the implications of the above should be taken into consideration.
- *152.8625–153.3575 MHz*—There are 67 frequencies available in this band for use by remote pickup broadcast stations. Each channel is 7.5 kHz wide, and channels may be stacked to form a single channel with a maximum bandwidth of 30 kHz. This band is shared with the Private Land Mobile Radio Service, and operation of remote pickup stations is subject to the condition that no harmful interference is caused to these other services. Note that these frequencies are not available to network entities or for use on board aircraft.
- *160.860–161.400 MHz*—There are 73 frequencies available in this band for use by remote pickup broadcast stations in Puerto Rico or the Virgin Islands, where they are shared with the Public Safety and Industrial/Business Pool. Each channel is 7.5 kHz wide, and channels may be stacked to form a single channel with a maximum bandwidth of 30 kHz.
- *161.625–161.775 MHz*—There are 21 frequencies available in this band for use by remote pickup broadcast stations. Each channel is 7.5 kHz wide, and channels may be stacked to form a single channel with a maximum bandwidth of 30 kHz. These frequencies are not available to network entities and are not available for use in Puerto Rico or the Virgin Islands. Also, Public Safety and Industrial/Business Pool stations may continue to operate on these frequencies on a non-interference basis.
- *166.25 and 170.15 MHz*—These frequencies may be used by remote pickup stations with a maximum bandwidth of 12.5 kHz; however, the area in which they may be used is restricted. A description of the area in which they may be used is found in the FCC Rules, §74.402(e)(8).

- *450.00625, 450.0125, 450.01875, 450.025, 450.98125, 450.9875, 450.99375, 455.00625, 455.0125, 455.01875, 455.025, 455.98125, 455.9875, and 455.99375 MHz*—These frequencies may be used by remote pickup stations only for the transmission of operational communications, including tones for signaling and for remote control and automatic transmission system control and telemetry. They may be stacked for a maximum bandwidth of 12.5 kHz.

In addition to the noted restrictions, the above groups of VHF and UHF frequencies are also subject to the following restrictions; licensed stations or those with applications filed prior to April 16, 2003, must have complied with the above channel plan by March 17, 2006, or, alternatively, may continue to operate on a secondary, non-interference basis:

- *450.03125–455.61875 MHz*—There are 200 frequencies available in this band for use by remote pickup broadcast stations. Each channel is 6.5 kHz wide, and channels may be stacked to form a single channel with a maximum bandwidth of 50 kHz.
- *450.6375–455.8625 MHz*—There are 20 frequencies available in this band for use by remote pickup broadcast stations. Each channel is 25 kHz wide, and channels may be stacked to form a single channel with a maximum bandwidth of 50 kHz. Users committed to 50 kHz bandwidths to transmit program material will have primary use of these channels.
- *450.900, 450.950, 455.900, and 455.950 MHz*—These frequencies are available for use by remote pickup broadcast stations. Each channel is 50 kHz wide, and channels may be stacked to form a single channel with a maximum bandwidth of 100 kHz. Users committed to 100 kHz bandwidths to transmit program material will have primary use of these channels.

Aural Broadcast Auxiliary Stations

Aural broadcast auxiliary stations include studio to transmitter link (STL), intercity relay (ICR), and microwave booster stations used by radio broadcast stations. STL stations are fixed stations used for transmitting program material between the studio and the transmitter of a broadcasting station. ICR stations are fixed stations used for the transmission of program material between broadcasting stations, except international, for simultaneous or delayed broadcast. ICRs may also be used on a secondary basis to transmit program material from FM stations to co-owned FM translator or booster stations. Microwave booster stations are used to relay the signals of an STL or ICR station over a path that cannot be covered with a single station. They receive and transmit on the same frequency. One or more microwave booster stations may be authorized to licensees of STLs or ICRs. Note that stations in the aural broadcast auxiliary service may be authorized on a secondary non-interference basis to licensees of TV broadcast stations to transmit aural material. The following frequencies are available for assignment to STL, ICR, and microwave booster stations:

- *942.5, 943.0, 943.5, and 944 MHz*—These frequencies are available for use in Puerto Rico and the Virgin Islands. Also, stations licensed in other parts of the U. S. prior to November 21, 1984, may continue to operate on a co-equal primary basis in this band.
- *944–952 MHz*—There are 320 channels (25 kHz wide) available in this band. The channels may be stacked to form a single channel up to 300 kHz wide. Separately, stations also may be authorized additional 25 kHz wide channels up to a grand total of 20 channels. The use of these frequencies by ICR stations is subject to the condition that no harmful interference is caused to other classes of stations.
- *18,760–18,820 and 19,100–19,160 MHz*—There are 24 channels (5 MHz wide) available in these bands. These frequencies are shared on a co-primary basis with other fixed services, and their use is subject to the rigorous coordination requirements of the FCC Rules, §21.100(d). No new applications are being accepted for this band, and it is scheduled to be relocated by June 8, 2010. After that date, stations may continue to use this band but will lose their co-primary status, will not be protected from interference from other users with primary status, and must not cause harmful interference to services with primary status.

Television Broadcast Auxiliary Stations

The demand for the spectrum allocated for television auxiliary services is significantly greater than for spectrum allocated for radio broadcasting services. In addition to the extensive local demand, network remote units travel extensively and compete with local broadcasters for available frequencies, resulting in increased spectrum congestion. It is important to recognize that the variety of activities undertaken by television broadcasters usually requires more complex auxiliary systems than is the case in the radio industry. The following are the types of television broadcast auxiliary stations:

- *TV pickup stations* are land mobile stations used for the transmission of TV program material and related communications from the scenes of events to TV broadcast, class A TV, or LPTV stations.
- *TV STL stations* are fixed stations used for the transmission of program material and related communications from the studio to the transmitter of a TV broadcast, class A, or LPTV station.
- *TV relay stations* are fixed stations used for transmitting visual program material between TV broadcast, class A TV, or LPTV stations or for the relay of transmissions from a remote pickup station to a single TV stations.
- *TV translator relay stations* are fixed stations used for relaying programs and signals of TV broadcast stations or class A TV stations to class A, LPTV, or TV translator stations.

- *TV microwave booster stations* are fixed stations used to receive and amplify signals of TV pickup, TV STL, TV relay, or TV translator relay stations and retransmit them on the same frequency. These stations are used to transmit signals over a path that cannot be covered by a single transmitter.

The following bands are available for assignment to TV pickup stations:

- *2025–2110 MHz*—Seven 12 MHz wide channels are available in this band, as well as two 500 kHz wide channels to be used as data return links (DRLs). These frequencies are also available for assignment to all the other types of television broadcast auxiliary stations (see below). Note that the frequencies from 1990 to 2025 MHz had previously been available for TV pickup stations, and that the channel bandwidth in the band from 1990 to 2110 was wider but still allowed for the same seven channels.
- *2450–2483.5 MHz*—Two channels are available in this band, which is also available for assignment to all the other types of television broadcast auxiliary stations. This band is shared with industrial, scientific, and medical (ISM) devices and is not afforded any protection from these ISM devices.
- *6425–6525 MHz*—The channels available in this band are co-equally shared with mobile stations licensed under Parts 21, 78, and 94 of the FCC Rules. The available channel bandwidth varies from 1 to 25 MHz. Section 74.602 of the FCC Rules contains further explanation concerning the usage of this band.
- *6875–7125 MHz*—Ten channels, each 25 MHz wide, are available in this band, which is also available for assignment to all the other types of television broadcast auxiliary stations.
- *12700–13250 MHz*—This band contains 43 channels; however, the channels overlap. If use of this band is contemplated, then §74.602 of the FCC Rules should be consulted for a more complete understanding of their usage. This band is also available for assignment to all the other types of television broadcast auxiliary station.
- *38.6–40 GHz*—This band is available without channel bandwidth limitation on a secondary basis to fixed stations.

The following are available for assignment to TV STL, TV relay, or TV translator relay stations:

- *2025–2110 MHz*—Seven channels are available in this band. which is also available for assignment to TV pickup stations (see earlier discussion on TV pickup stations for information on recent changes in this band).
- *2450–2483.5 MHz*—Two channels are available in this band, which is also available for assignment to TV pickup stations.
- *6875–7125 MHz*—Ten channels, each 25 MHz wide, are available in this band, which is also available for assignment to TV pickup stations.
- *12,700–13,250 MHz*—This band contains 43 channels; however, the channels overlap. If use of this band is contemplated, then §74.602 of the FCC Rules should be consulted for a more complete understanding of their usage. This band is also available for assignment to TV pickup stations. In addition, the channels between 13,150 and 13,200 MHz are not available within 50 km of the top 100 markets.
- *17,700–19,700 MHz*—Frequencies in this band are shared on a co-equal basis with stations in the fixed service authorized by Parts 21, 78, and 94 of the FCC Rules. The available channel bandwidth varies from 2 to 80 MHz. Complete details concerning the use of this band are contained in §74.602 of the FCC Rules.

In addition to the above frequencies, TV STL and TV relay stations also may be authorized to use UHF-TV channels 14 to 69 on a secondary basis provided no interference is caused to TV and LPTV stations operating in this band. Furthermore, the aural portion of television broadcast program material may be transmitted over an aural broadcast STL or ICR station on a secondary, non-interference basis. Likewise, remote pickup stations may be used to transmit the aural portion of television program material.

SUMMARY

Frequency allocation is a complex matter and is subject to frequent changes. The reader is advised to consult a current version of the FCC Rules and Regulations for a complete description of the current allocation situation, including the procedures and policies being applied by the FCC concerning a particular band.

CHAPTER

1.6

Frequency Coordination for Auxiliary Services

DANE E. ERICKSEN, P.E., CSRTE, 8-VSB, CBNT
Hammett & Edison, Inc., Consulting Engineers
San Francisco, California

DAVID P. OTEY, CPBE
AZCAR Technologies
Centennial, Colorado

INTRODUCTION

The electromagnetic spectrum not only provides broadcasters the means for reaching listeners and viewers via the AM, FM, and TV bands, but under the FCC's Broadcast Auxiliary Service (BAS) Rules, the electromagnetic spectrum may be used by broadcasters for point-to-point links, remote-pickup links, voice channels for cueing and coordination, remote control of cameras and transmitters, wireless microphones, and other applications. Yet, as varied as these applications appear to be, their treatment in the FCC Rules is contained in only one of four Subparts in Part 74.[1] Further, these applications all have two things in common: All require some form of frequency coordination to minimize the likelihood of interference (since they operate in shared frequency bands; *i.e.*, they use frequencies which are also used by other services), and all are impacted by the ever-increasing demand for radio spectrum. This chapter provides an overview of the different ways frequency coordination is employed in an attempt to mitigate the effects of spectrum crowding on these essential auxiliary services.

The purpose of this chapter is not to provide an encyclopedic guide to all things encompassed by the phrase "frequency coordination." Rather, it is about frequency coordination services, whether those services are provided by a local frequency coordination committee, a volunteer appointed by a local chapter of the Society of Broadcast Engineers, Inc. (SBE), or a commercial firm in the business of providing frequency search services.

Frequency Coordination

A typical place to begin such a discussion would be with a definition of terms. Unfortunately, the phrase "frequency coordination" does not lend itself to a single, tidy definition. This simple term may mean different things in different radio services. Even within the range of BAS applications, frequency coordination will mean different things in different situations. It may mean working out how scarce TV pickup channels are to be shared on an hour-by-hour basis in the most crowded markets. Alternatively, it may mean knowing how to find an available channel for a fixed point-to-point link, and being able to demonstrate quantitatively why that channel can in fact be used on that link. These are quite different tasks requiring different skills.

Here, the FCC Rules provide some help. In the relevant Subparts of Part 74 are these statements: ". . . licensees shall endeavor to select frequencies or schedule operation in such manner as to avoid mutual interference . . ." (74.403); and ". . . they shall take such steps as may be necessary to avoid mutual interference, including consultation with the local coordination committee, if one exists . . ." (74.604); and ". . . licensees shall endeavor to select frequencies or schedule operation in such manner as to avoid mutual interference . . ." (74.803). The theme is clear: Licensees in the shared-frequency bands assume the obligation of choosing their frequencies and patterns of use so as

[1]Specifically, they are as follows: Subpart D—Remote Pickup Broadcast Stations; Subpart E—Aural Broadcast Auxiliary Stations; Subpart F—Television Broadcast Auxiliary Stations; and Subpart H—Low Power Auxiliary Stations.

to avoid interfering with one another. Frequency coordination in this context can be defined as the act of assisting licensees to meet that obligation, by means of knowledge of the uses being made by other affected licensees.

Local Coordination

Historically, such assistance has come through a local frequency coordination committee or from a local volunteer frequency coordinator. Either entity would most likely be equipped with some sort of database or directory of local BAS users, along with the locations of the various BAS facilities and the frequencies and other technical parameters of these BAS channels. The local licensees most likely know one another, share a common interest in avoiding interference, and achieve interference avoidance or mitigation primarily through the particular skill of knowing whom to call when interference is encountered or suspected. The process is informal, local, flexible, and generally effective.

While "local coordination" is not defined per se in the FCC's Rules, it continues to be the applicable requirement for most BAS services, other than for fixed links in the 950 MHz aural auxiliary band and fixed microwave links in the television auxiliary services.

Local coordination, however, is not the whole story. In recent years, FCC Rules have made a distinction between those services which may rely on informal, local coordination, and those for which one of two more formal protocols must be followed. As a result of a rule change which took effect in October 2003, broadcasters now must know which of three different frequency coordination protocols to observe, based on frequency range and whether operation is fixed or temporary/mobile. The rest of this chapter is devoted to describing and comparing the three different protocols, which break down as follows:

- For fixed and temporary operations in the band 6425–6525 MHz, and for fixed-link, point-to-point stations from the 950 MHz band upward, *except* the "2 GHz" band,[2] the Prior Coordination Notice (PCN) protocol of Part 101 applies.
- For fixed links in the 2 GHz band—now something of a rarity in the major TV markets owing to this band's widespread use for TV Pickup (colloquially referred to as Electronic News Gathering, or ENG)—the applicable standard is still local coordination, but with a supplemental procedure to provide quasi-PCN evidence of coordination.
- For all other BAS services, including all mobile or temporary fixed operations (except 6425–6525 MHz, as noted above), informal local coordination still applies.

[2]This band, most commonly used for television pickup (service code TP) but also for studio-transmitter links (TS) and intercity relays (TI), is at the time of this writing (September 2006) undergoing relocation from 1990–2110 MHz to a smaller portion of that band, 2025–2110 MHz.

FIXED-LINK COORDINATION

Frequency coordination of fixed links is fundamentally different from frequency coordination involving remote pickup (RPU) stations, Subpart H low-power auxiliary stations (e.g., wireless microphones), TV pickup, or ENG stations, where generally the available frequencies must be shared. For fixed links, frequency reuse is appropriate. For the purposes of this discussion, "fixed links" refers to point-to-point BAS microwave stations. Frequency reuse rather than frequency sharing is practical for these fixed-link stations because such stations use highly directive parabolic dish transmitting and receiving antennas, as opposed to often omnidirectional transmitting and receiving antennas used by RPU, wireless microphone, and some ENG systems.

Fixed-link frequency coordination involves first identifying all co-channel and adjacent-channel links sufficiently close to require consideration, and then selecting a frequency that will neither cause interference to these existing links nor receive interference from those links. While the concept is simple, the implementation can be complex. It is the intent of this portion of the chapter to give a sufficiently detailed overview of the fixed-link frequency coordination process to at least ensure that a station engineer knows the steps involved, and when the services of a commercial microwave frequency coordinator (CMFC) should be retained.[3]

Preparatory Steps

Before a fixed-link frequency coordination study is commenced, it should be established that the applicant is eligible to use the band in question and that an unobstructed microwave path exists. There is little point in going to all the work involved in coordinating a fixed link only to discover that the proposed use is not a permitted one, or that the proposed path is blocked by a nearby structure or by intervening terrain. The applicant should determine that a reasonable path length for the microwave band in question is involved:

- For 13 GHz, path lengths of more than about 20 miles (32 km) are generally to be avoided.
- For 18 GHz, path lengths of more than about 5 miles (8 km) are generally to be avoided.
- For 2, 2.5, and 7 GHz, path lengths 62 miles (100 km) long are entirely practical from a noise-limited standpoint. However, in most major metros, microwave paths become interference-limited long before they become noise-limited.

[3]See the WTB website, at http://wireless.fcc.gov/microwave/coordinators.html, for a list of CMFCs.

New BAS Fixed-Link Frequency Coordination Paradigm

As a result of the ET Docket 01-75 rulemaking, which undertook a general updating and harmonization of the Part 74 BAS Rules with other services, as of October 16, 2003, formal private operational fixed services (POFS) Section 101.105(d) frequency coordination protocols became applicable to 950 MHz aural BAS stations, and to 7 and 13 GHz TV BAS fixed-link stations (Part 101 frequency coordination protocols, often referred to as the "prior coordination notice," or PCN, process, already applied to 2.5 GHz TV BAS fixed-link stations, to 6.5 GHz TV BAS stations, and to 18 GHz aural and TV BAS stations). Frequency coordination for 2 GHz TV BAS fixed-link stations remains subject to "local" coordination, although a frequency coordination certification must be included with such applications; see Section 74.638(c) of the FCC Rules. This "certification" is a less formalized process than the Part 101 PCN process, but is still a significant change from the prior requirement, which did not require any frequency coordination showing at all; previously, a simple "yes" answer to Form 601, Schedule I, Question 7 ("Has frequency coordination been completed for this application?") was sufficient.

The PCN Process

Under the PCN process, one first identifies all existing co-channel and adjacent-channel links sufficiently close to the proposed link to require study. "Sufficiently close" is defined by Telecommunications Industry Association (TIA) Bulletin TSB-10F, "Interference Criteria for Microwave Systems."[4] As shown in Figure 1.6-1, this bulletin defines a coordination "keyhole" for determining whether a co-channel or adjacent-channel link is sufficiently close to require study. There are two "keyholes": one for stations below 15 GHz and a smaller keyhole for stations above 15 GHz. Further, Section 101.105(c) of the FCC Rules does not *require* use of TSB-10F methodologies: "Other procedures that follow generally accepted good engineering practice are acceptable to the Commission." But, the TSB-10F procedures can be thought of as a "safe harbor," in that if those procedures are followed, then the PCN would presumably not be second-guessed by the FCC.

Each of the stations identified as being sufficiently close to require study must then be checked for interference from the proposed newcomer link. Although not an FCC requirement, the PCN study should also check to ensure that each of these existing links will not cause interference *to* the proposed new link. Making these calculations requires detailed technical data for each of the studied stations, although in some cases conservative assumptions (conservative in that they would always over-predict the interference that the newcomer station would cause to an incumbent station), such as assuming only the use of Category B receiving antennas, parallel polarization, and a high receiving antenna height (*i.e.,* forgoing any terrain blockage that might exist). If, based on these conservative assumptions, the newcomer link is calculated not to be an interference threat to an existing link, one can probably forgo the time and effort to obtain often hard-to-find data for the to-be-protected path.[5] However, if the interference calculations are "close," or if moderate interference is predicted, then the actual facilities of the to-be-protected link must be ascertained.

If the first-cut interference calculations predict massive interference, an alternative frequency should be sought. The one exception would be if one suspects that an FCC Universal Licensing System (ULS) record is for a station that no longer exists, but the licensee has not informed the FCC of that fact (either due to oversight, or perhaps because the licensee thinks it might want the path at a future date). Such frequency "warehousing" is, of course, prohibited by the FCC Rules [Section 74.632(g) for TV BAS, Section 1.955(a)(3) for Wireless Telecommunications Services[6] in general], but this doesn't mean that the practice still doesn't exist. So, if only one ULS record is causing a preclusion, and if there are no other available channels, checking to see if the link is actually in use might prove fruitful.

Note that assumptions that are conservative as to the amount of interference, if any, a newcomer link would cause to an existing and must-be-protected link are generally also conservative insofar as predicting whether an existing link will cause interference to a proposed new link. For example, the assumption of

[4]TSB-10F was published in June 1994. Unfortunately, it is unlikely that a TSB-10G version will be released anytime soon. TSB-10F was written primarily for POFS and common carrier microwave links, although the standard does address FM video microwave links. However, TSB-10F is silent regarding interference criteria for 950 MHz Aural BAS microwave links.

[5]On April 4, 2003, the Society of Broadcast Engineers, Inc. (SBE) filed a Request for Stay of the PCN requirement for 950 MHz Aural BAS, and 2.5, 7, and 13 GHz TV BAS, fixed-link stations, scheduled to go into effect on April 16, 2003, asking for a one-year delay. The reason was the amount of missing or incomplete data for Part 74 fixed-link stations in the FCC's Universal Licensing System (ULS). SBE pointed out that an analysis of 21,033 fixed-link BAS records showed that 6163 of those records, or 29.3%, were either lacking receive-end geographic coordinates or had corrupted data for the receive-end coordinates (e.g., null coordinates, or coordinates that were identical to the transmit-end coordinates). In response, the FCC granted a sixth-month stay, until October 16, 2003, for the new PCN requirement.

A September 20, 2005, analysis of BAS records in the ULS by Karolj Lerinc of Micronet Communications, Inc., showed that missing and incomplete data was still an issue: Of 4,954 950 MHz Aural BAS records in the ULS, 740 were missing polarization information, 2047 (41%) had missing or corrupted receive-end coordinates, and 2 records has no transmit-end coordinates. Of 59 2.5-GHz TV BAS fixed-link records, 20 (34%) were missing receive-end coordinates. For 7 GHz TV BAS fixed-link records, 1235 of 2900 records, or 42%, had missing or corrupted receive-end coordinates, and 894 were missing polarization information. For 13 GHz TV BAS records, 405 out of 1015 records, or 40%, had missing or corrupted receive-end coordinates.

Therefore, conducting a valid PCN study often requires "detective work" by a CMFC, to obtain missing critical data for a to-be-protected, earlier-in-time, fixed-link station.

[6]Wireless telecommunications services (WTSs) include Parts 13, 20, 22, 24, 26, 27, 74, 80, 87, 90, 95, and 101 of the FCC Rules. Thus, all BAS stations are WTS stations and subject to the Part 1, Subpart F, WTS procedural provisions, in addition to any Part 74 provisions.

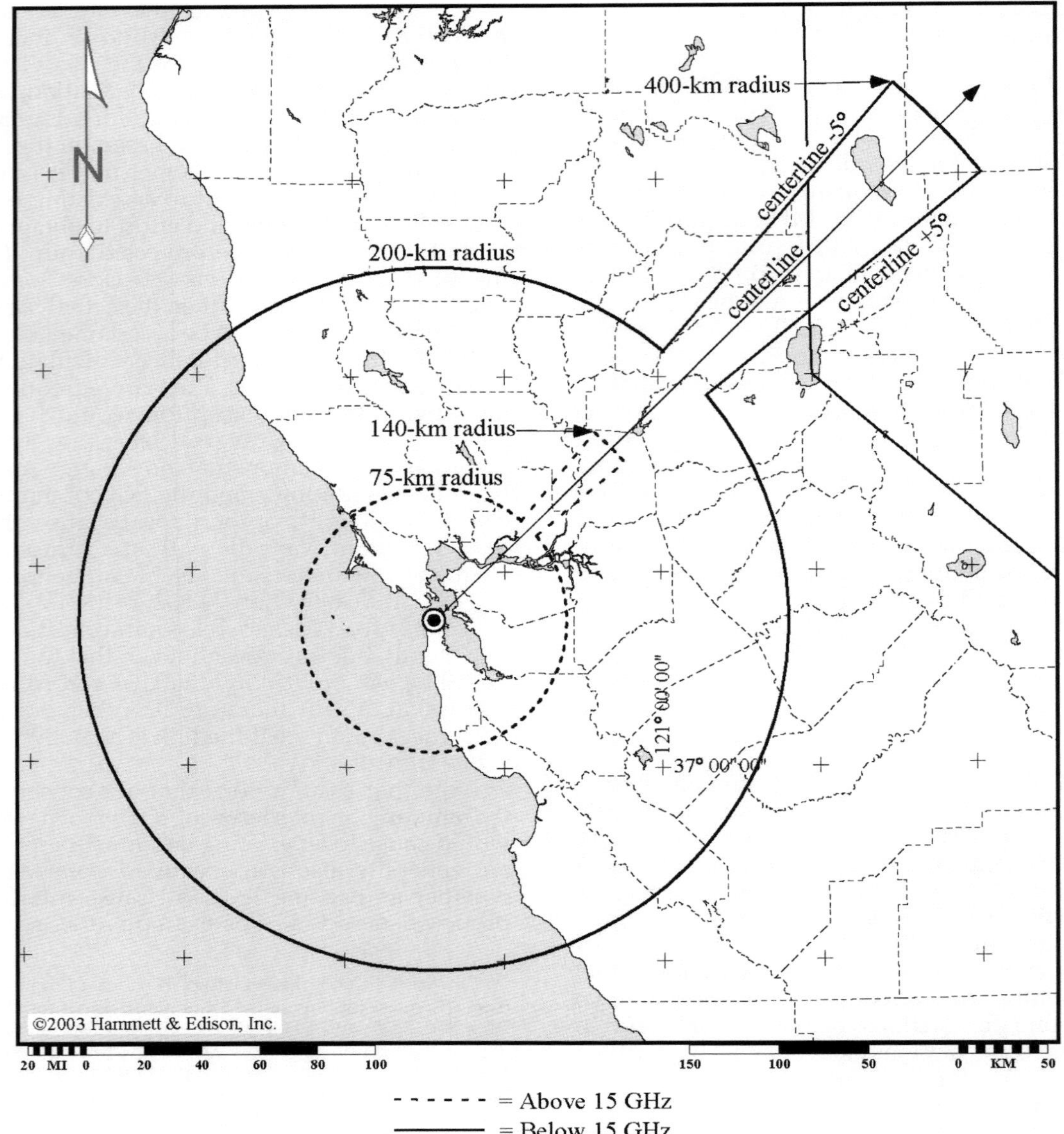

FIGURE 1.6-1 Example of TIA/EAI TSB10-F Coordination Keyholes (from Section 3.4.4 of TIA TSB-10 F standard. Albers equal area map projection; map data taken from Sectional Aeronautical Charts, published by the National Ocean Survey. Geographic coordinate marks are shown at 60-minute increments.)

only a Category B transmitting antenna at great height for an existing link would tend to overpredict any interference that existing link might cause to a proposed new link. Again, if based on such conservative assumptions no interference is predicted to the new link, it is not necessary to ferret out missing data for the existing, and potentially interfering, link.

Special Cases

When looking for potential to-be-interfered-with paths, beware of special cases, such as split-path links. Split-path links may use separate transmitting antennas, or a single transmitting antenna, splitting the difference in the azimuths and elevation (or depression) angles to two closely spaced receiving locations. Another situation to watch out for is an existing path with an unusually long receive-end waveguide run, which could cause the to-be-protected signal to be significantly (10 to 20 dB) weaker than one might otherwise predict.

Finally, some BAS microwave paths employ a passive reflector, meaning that the transmit-dish azimuth is not aimed toward the receive-end site, but rather toward the passive reflector that the receive-end dish can see (and, similarly, this means that the receive-end dish is aimed at the passive reflector, and not toward the microwave transmit site). While these situations are rare, the consequences of failing to properly study a split-path link or a link with a passive reflector could potentially be severe to the newcomer station, which could then be viewed as the at-fault party.

Interference Criteria

One of most difficult aspects of a PCN study, at least one involving BAS links, is deciding the criteria to use to define whether interference will exist. For analog FM video–into–analog FM video, interference criteria of 60 dB or better desired-to-undesired (D/U) signal ratio for co-channel and a 0 dB or better D/U ratio for adjacent-channel have been used for many years by this author with acceptable results. However, the interference criteria become more complex when the proposed or protected links are using digital modulation. Often a digital signal more fully occupies the channel bandwidth, resulting in greater signal strength near the channel edge than would exist for the FM video analog case; to allow for this, 10 dB adjacent-channel interference criteria, rather than just a 0 dB adjacent-channel interference criteria, is suggested.

When the protected link is one that uses digital modulation, the interference criteria become one of protecting the receiver's noise floor. TSB10-F allows up to a 1 dB degradation of an existing digital link's noise floor.[7] For example, if an existing digital link has a noise floor of –90 dBm, an undesired signal with the same bandwidth as the signal to be received at the protected receiver's input could be no stronger than –95.9 dBm, as –90 dBm plus –95.9 dBm = –89 dBm, or a 1 dB degradation of the noise floor.[8]

Another issue is what constitutes "co-channel" and "adjacent-channel" operation, especially when stations with different channel widths are involved. For example, consider a 25 MHz wide 13 GHz TV STL channel versus 6 MHz, 12.5 MHz, or 25 MHz wide Cable Television Relay Service (CARS[9]) microwave channels. One approach is to use the channel width of the protected station as the criteria: If any portion of the proposed channel overlaps any portion of the protected channel, no matter how small, the proposed station is treated as "co-channel." And if any portion of the proposed channel overlaps any portion of one channel bandwidth from the lower edge of the protected channel, or any portion of one channel bandwidth from the upper edge of the protected channel, then the proposed channel would be treated as "adjacent-channel."

[7]See TSB-10F, at Section 2.5.5.

[8]To add powers in dBm, each dBm value is first converted to mW, added, and then converted back to dBm.

[9]Formerly "community antenna relay service"; the "CARS" abbreviation was retained.

Under these criteria, a 6 MHz wide CARS channel would have adjacent-channel windows extending 6 MHz below the lower channel edge and 6 MHz above the upper channel edge, whereas a 25 MHz wide CARS channel would have 25 MHz adjacency windows. Finally, if a link is neither co-channel nor adjacent-channel, it need not be studied, on the assumption that the receiver's selectivity should be able to handle second and higher adjacencies.

Antenna Considerations

The FCC Rules require all fixed-link BAS stations to use directional antennas. Except for 950 MHz aural BAS stations, the FCC Rules specify the minimum acceptable performance of these antennas as either Category B or Category A. All fixed-link stations except 950 MHz aural BAS stations must normally use at least Category B transmitting antennas and in "frequency-congested areas" must normally use at least Category A transmitting antennas. However, in the most frequency-congested areas, even a Category A antenna often won't suffice, and "super" Category A antennas, with off-axis suppression ratios significantly better than those required to achieve Category A status, must be used.

There are presently no minimum antenna performance requirements for BAS fixed-link receiving antennas, except that the receiving antenna must be directional. However, Section 74.641(a)(3) of the FCC Rules provides a powerful incentive for also using highly directive receiving antennas, as follows:

> The choice of receiving antennas is left to the discretion of the licensee. However, licensees will not be protected from interference that results from the use of antennas with poorer performance than identified in the table of this section.

This means that an existing link in a non-frequency-congested area using a sub-Category B receiving antenna may be studied (and protected) as if the receiving antenna met Category B performance. Or, in a frequency-congested area, an existing link using just a Category B receiving antenna (or even a sub-Category B antenna) may be studied (and protected) as if the receiving antenna met Category A performance. Thus, it behooves one to always use at least Category B receiving antennas in non-frequency-congested areas, and to always use at least Category A receiving antennas in frequency-congested areas.

The FCC Rules define antenna performance using two of the following three parameters: the on-axis minimum gain or the maximum half power beam width (HPBW), and the minimum required suppression at larger and larger off-axis angles (i.e., up to ±180° from the main beam). This off-axis suppression is referred to as the antenna's Radiation Pattern Envelope (RPE). As an example, the RPE of an Andrew Model UHX10-59J 10-foot diameter, ultra high-performance (UHP), cross-polarized microwave antenna is shown in Figure 1.6-2. Note that the actual antenna performance may be even better than the suppression shown by the RPE, but this additional suppression is

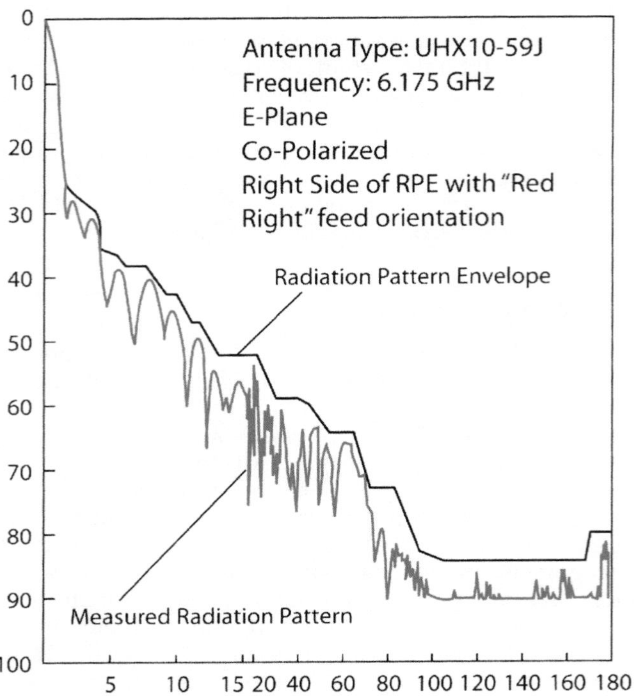

FIGURE 1.6-2 Radiation Pattern Envelope for an Andrew Model UHX10-59J antenna. (Courtesy of Andrew Corporation.)

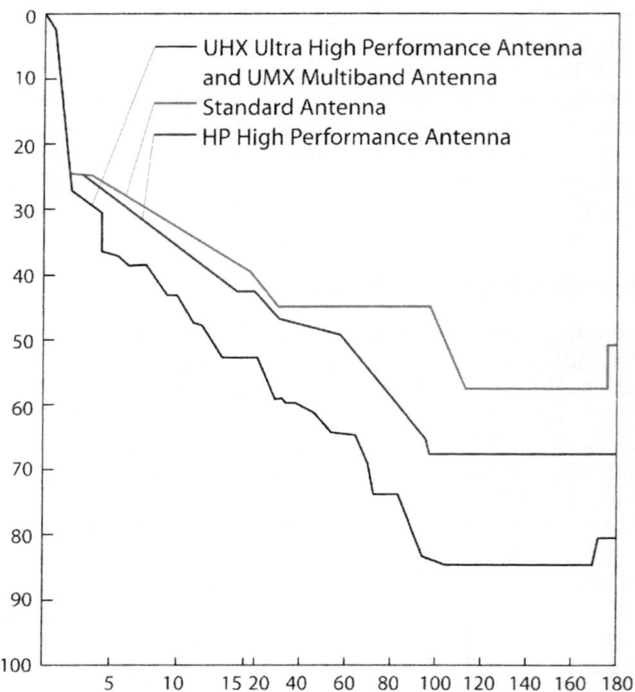

FIGURE 1.6-3 Comparison of the off-axis rejection capabilities of SP, HP, and UHP antennas having the same diameter. (Courtesy of Andrew Corporation.)

not guaranteed. Therefore, one should always base interference calculations on the guaranteed performance RPE.

It is important to note that for the same diameter dish, standard-performance (SP), high-performance (HP), and UHP antennas all have essentially the same gain and half-power beamwidth; it is only on the off-axis rejection that HP and UHP dishes differ from an SP dish. Because a UHP dish can cost up to four times more than an SP dish of the same diameter costs, and because HP and UHP dishes represent a greater structural load than an SP dish of the same size (diameter), HP or UHP antennas should be specified only when required for frequency coordination purposes. That is, use of HP or UHP dishes adds nothing to the path's link budget compared to SP dishes of the same size.

HP and UHP parabolic dish antennas are created by adding a "cake-pan" shroud around the periphery of the dish and then lining that shroud with microwave-absorbing material. Figure 1.6-3 shows a comparison of the off-axis rejection capabilities of SP, HP, and UHP antennas all having the same diameter. A UHP dish can have up to 40 dB better off-axis rejection than that of an SP dish of the same size, and so is a powerful, albeit expensive, frequency coordination tool. The benefits of an HP or a UHP antenna can apply at either the transmit end or receive end. If the problem is one of interference to an existing link, using an upgraded antenna at the transmit end of the new path can sometimes be a solution. If the problem is one of interference from an existing link to a newcomer link, using an upgraded antenna at the receive end of the new path can sometimes be a solution. In the most congested TV markets, using HP or UHP antennas for both ends of the new path is often the only way to make the new path frequency coordinate.

RH and LH Feeds

When a UHP antenna is used, especially a dual-polarized version with two feed horns, the RPE is generally no longer symmetrical around the main beam. This nonsymmetry can be used to the newcomer's advantage, by selecting a right-hand (RH) or left-hand (LH) feed orientation, so that the nonsymmetrical portion of the RPE with the greater suppression is aimed toward a problematic co-channel or adjacent-channel link. Thus, for UHP antennas, there are two degrees of freedom for minimizing interference: horizontal versus vertical polarization and RH versus LH feeds.

Frequency-Congested Areas

Section 74.641(b) of the FCC Rules defines "frequency-congested area" for the 13 GHz TV BAS band as one where a newcomer station shows that the use of a Category B transmitting antenna is precluding the new path, and that upgrading the existing station to a transmitting antenna meeting Category A criteria would eliminate the preclusion. However, this "frequency-congested" criteria applies only to the 13 GHz TV BAS

band. For 2, 2.5, 7, and 18 GHz band fixed links, the term "frequency-congested area" is undefined.

Section 74.641(c) provides exceptions to the use of Category A or B microwave antennas, where a showing can be made that structural, zoning, or other restrictions preclude use of a microwave antenna with a diameter sufficient to meet Category A or B specifications. However, the showing threshold is high, so in most cases the prudent choice will be to simply install the normally required Category A or B antenna.

Parallel Polarized and Cross-Polarized Signal Interference Calculations

Figure 1.6-4 shows how parallel polarized (PPOL) and cross-polarized (XPOL) signals between two fixed-link paths combine (for a given antenna, the RPEs for the PPOL and XPOL signal will always be different, as demonstrated in Figure 1.6-5). The drawings and tables in this figure reflect the fact that an HPOL or VPOL TX antenna will always have off-axis leakage for both polarizations, as will an RX antenna always be susceptible to off-axis interference for both polarizations. Thus, when the protected and interfering paths have different polarizations, both the PPOL portion of the interfering signal and the XPOL portion of the interfering signal must be taken into account. When the protected and interfering paths have the same polarizations, the PPOL leakage is always predominant.

Referring to Figure 1.6-4, there are four possible combinations between the desired and undesired signal paths. For the PPOL case, use (A + C) when calculating the level of the undesired signal. For the XPOL case, use either (A + D) or (B + C), whichever gives the smaller dB number (i.e., the worst-case off-axis rejection). If the (A + D) and (B + C) dB values are within 10 dB of each other, then the powers should be combined. For example, if (A + D) is 45 dB, and (B + C) is 41 dB, the "sum" is 39.5 dB of isolation. But if (A + D) is 45 dB and (B + C) is 55 dB, the "sum" is 44.6 dB. Thus, when (A + D) differs from (C + D) by more than about 10 dB, the smaller dB number is controlling, and the larger dB number can be ignored, with little loss in accuracy.

Design Example

Figure 1.6-6 shows a sample worksheet used for manually performing a microwave path interference calculation. The goal of this exercise is to determine the amount of interference created by a proposed, new co-channel path with respect to an existing path, WAA-1234, which is horizontally polarized. For this example, both paths are using Channel A07 (12,850–12,875 MHz) in the 13 GHz TV BAS band.

First, the unfaded received carrier level (RCL) of the existing signal, WAA-1234, is calculated in the right-hand column, by completing a "link budget," which consists of the following values:

- TX power output: Output power of the existing path transmitter (located at site 3)—for this example, equals 0.4 W (26 dBm);
- Line loss (TX): This is the loss (in dB) due to the waveguide between the transmitter and the antenna;
- TX antenna gain, in dBi (at site 3, on-axis): This is specific to the transmit antenna used;
- FSPL: free space path loss for 9.5 mile distance between site 3 and site 4 where

$$FSPL = 92.5 + 20\log(D_{miles}) + 20\log(F_{GHz})$$

- RX antenna gain in dBi: A function of the receive antenna used;
- Line loss (RX): Loss (in dB) due to waveguide between antenna and receiver;
- RX level: Sum of all of the above values—this is the unfaded signal level received at site 4, considered the "desired" signal level for the purposes of this exercise.

Next, the signal level of the new path (the "undesired" signal) at the input to the existing WAA-1234 receiver is calculated. For this part of the calculation, the RPEs of the newcomer TX dish (at site 1) and the to-be-protected RX dish (at site 4) must be known. This link calculation is performed in the left-hand column of Figure 1.6-6 as follows:

- TX power output: Output power of the new path transmitter (located at site 1)—for this example, equals 1.0 W (30 dBm);
- Line loss (TX): This is the loss (in dB) due to the waveguide between the transmitter and the antenna;
- Loss due to off-axis interference path (TX): Knowledge of the RPE of the TX antenna at site 1 is required here. First, the *off-axis angle* of the interference path is calculated—for this example, the baseline is the site 1 to site 2 path (since the TX antenna main beam is aligned with that path); in this example the angle is 15.3°. Next, both the HPOL and VPOL responses of the site 1 TX antenna at this angle must be determined since ultimately the potential interference for both cases needs to be determined (that is, case 1, where the new path is the same polarization as the existing path, and case 2 where the new path is cross-polarized to the existing path). For the antenna in this example, for 15.3° off-axis, the HPOL response is found to be 43 dB below the main beam gain, and the VPOL response is 54 dB below the main beam gain. Both values are written into the link budget, and both will be used later as discussed below;
- FSPL: Free space path loss for 12.3 mile distance between site 1 and site 4;
- RX antenna gain in dBi (on-axis, site 4): A function of the receive antenna used;
- Loss due to off-axis interference path (RX): Knowledge of the RPE of the RX antenna (site 4) is

required here. First, the *off-axis angle* of the interference path is calculated—for this calculation, the baseline is the site 3 to site 4 path (since the RX antenna main beam is aligned with that path); in this example the angle is 20.9°. Next, both the HPOL and VPOL responses of the site 4 RX antenna at this angle must be determined—for the antenna in this example, the HPOL response is found to be 45 dB below the main beam gain, and the VPOL response is 58 dB below the main beam gain. As before, both values are written into the link budget;

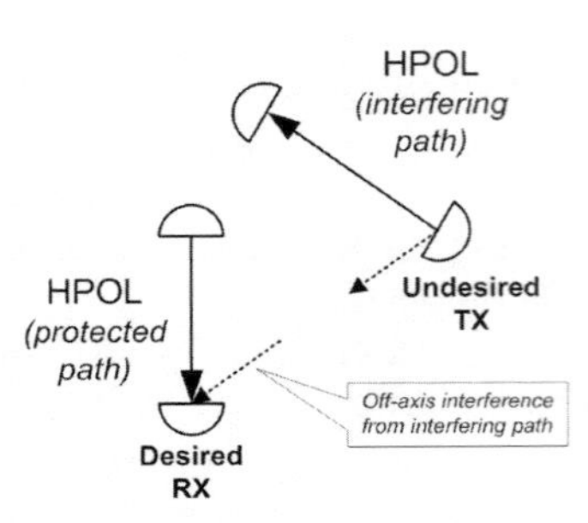

Off-axis attenuation	Undesired TX antenna	Desired RX antenna
HPOL	A	C
VPOL	B	D

Attenuation of off-axis interference at desired RX antenna:
(HPOL) = A + C (always predominant)
(VPOL) = B + D

CASE 1 - both paths HPOL

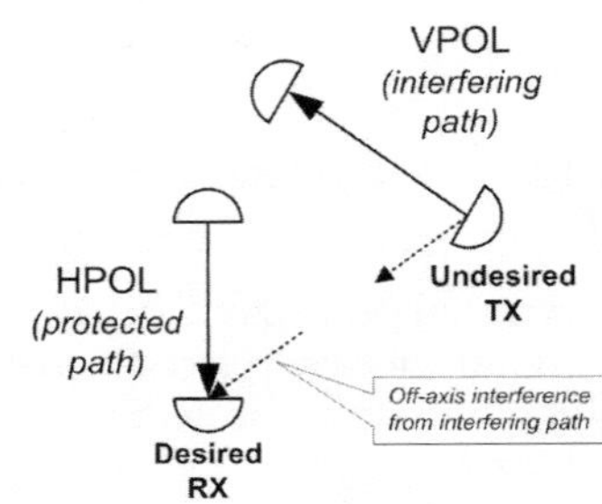

Off-axis attenuation	Undesired TX antenna	Desired RX antenna
HPOL	B	C
VPOL	A	D

Attenuation of off-axis interference at desired RX antenna:
(HPOL) = B + C
(VPOL) = A + D

CASE 2 - interfering path VPOL, protected path HPOL

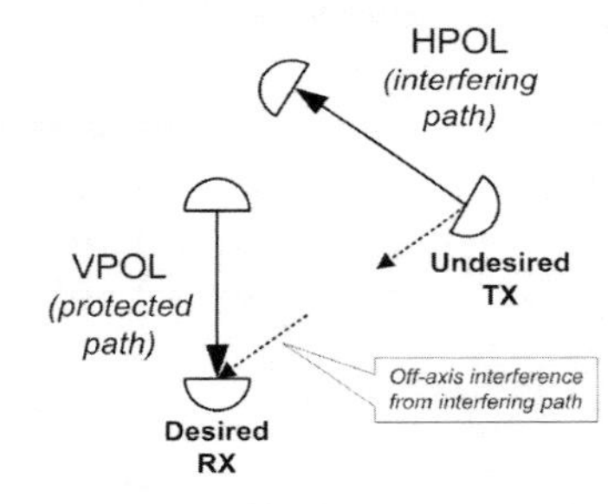

Off-axis attenuation	Undesired TX antenna	Desired RX antenna
HPOL	A	D
VPOL	B	C

Attenuation of off-axis interference at desired RX antenna:
(HPOL) = A + D
(VPOL) = B + C

CASE 3 - interfering path HPOL, protected path VPOL

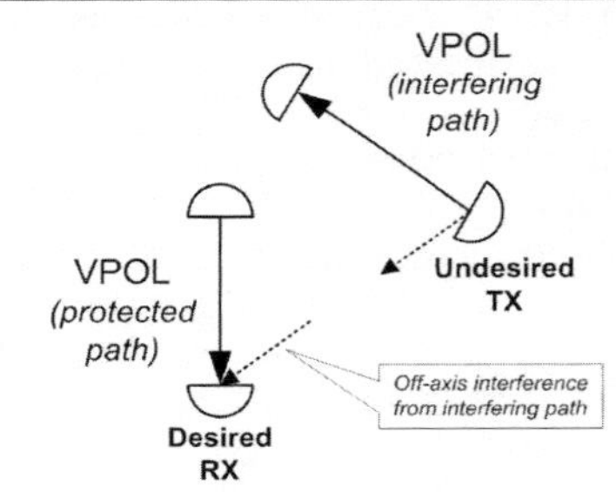

Off-axis attenuation	Undesired TX antenna	Desired RX antenna
HPOL	B	D
VPOL	A	C

Attenuation of off-axis interference at desired RX antenna:
(HPOL) = B + D
(VPOL) = A + C (always predominant)

CASE 4 - both paths VPOL

Notes:
1) For undesired TX antenna, A=PPOL off-axis suppression (dB), B=XPOL off-axis suppression (dB)
2) For desired RX antenna, C=PPOL off-axis suppression (dB), D=XPOL off-axis suppression (dB)
3) For PPOL case, (A+C) will always give lowest attenuation (worst interference)
4) For XPOL case, the smaller of (A+D) or (B+C) will give lowest attenuation (worst interference)

FIGURE 1.6-4 Examples of parallel polarized (PPOL) and cross-polarized (XPOL) interference.

- Line loss (RX): Loss (in dB) due to the waveguide between antenna and receiver;
- RX level, PPOL: If the newcomer link is HPOL (which means it is parallel-polarized to the existing path), then the HPOL RPE values are used for this calculation; that is, 30 dBm – 2.2 dB + 45.1 dBi – 43.0 dB (HPOL value for TX antenna) – 140.6 dB + 47.6 dBi – 45.0 dB (HPOL value for RX antenna) – 3.5 dB = –111.6 dBm (equals undesired signal level at site 4 RX antenna, PPOL).
- RX level, XPOL: If the newcomer link is VPOL, and thus cross-polarized to the existing protected path, then the smaller of the two XPOL sums (in dB) must be determined. The two possibilities are as follows: 43 dB (TX HPOL) + 58 dB (RX VPOL), or 54 dB (TX VPOL) + 45 dB (RX HPOL). Since the first combination totals 101 dB of isolation, and the second combination totals 99 dB of isolation, the smaller 99 dB figure is used. Note that the exact value is actually the power summation of these two

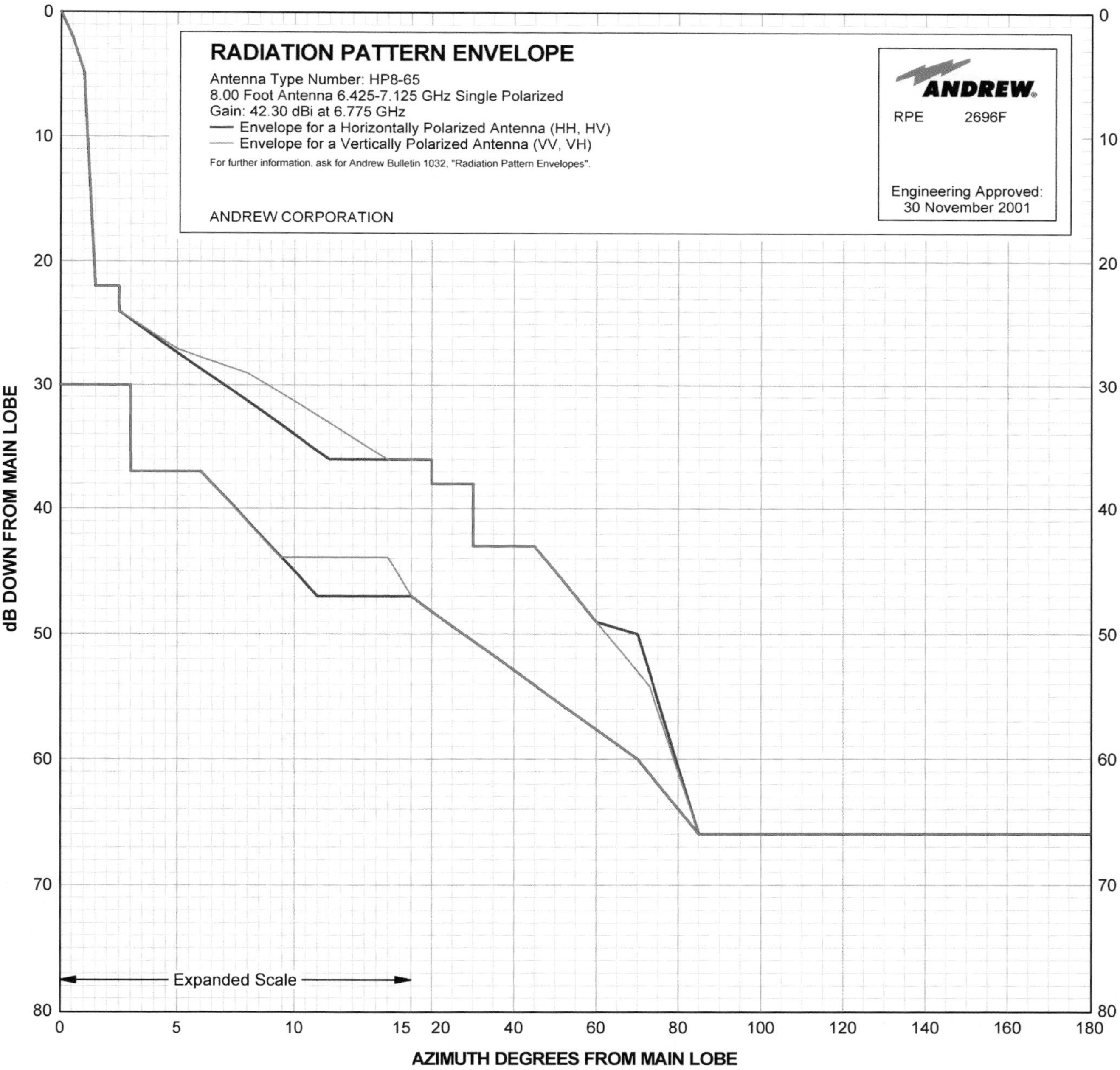

FIGURE 1.6-5 Illustration of differences in Radiation Pattern Envelope (RPE) as a function of polarization. (Courtesy of Andrew Corporation.)

MICROWAVE PATH CALCULATIONS

PROPOSED NEW PATH: Site 1 to Site 2, A07, polarization to be determined

EXISTING PATH TO BE CHECKED: WAA-1234, A07, Site 3 to Site 4

(√) Check for IX to existing path **( ) Check for IX from existing path**

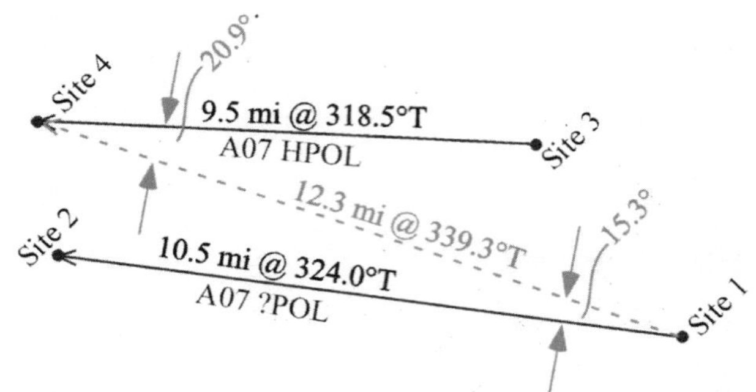

		WAA-1234 DIRECT PATH	
TX power output: 1W	30 dBm	**TX power output:** 0.4 w	26.0 dBm
Line loss:	-2.2 dB	**Line loss:**	-1.8 dB
TX antenna gain: Andrew P6-122F	45.1 dBi 72.9 dBm EIRP	**TX antenna gain:** 6' Prodelin 213-740	45.1 dBi 69.3 dBm EIRP
Loss 15.3° **off axis:**	-43.0 -54.0 dB		
FSPL, 12.3 **miles:**	-140.6 dB	**FSPL,** 9.6 **miles:**	-138.4 dB
RX antenna gain: 8' standard	47.6 dBi	**RX antenna gain:** 8' standard	47.6 dBi
Loss 20.9° **off axis:**	-45.0 -58.0 dB		
Line loss:	-3.5 dB	100' EW127 **Line loss:**	-3.5 dB
RX level, PPOL:	-111.6 dBm	**RX level:**	-25.0 dBm
RX level, XPOL:	-122.6 dBm		
D/U, PPOL:	84.4 dB	**D/U, Xpol:**	97.6 dB

FIGURE 1.6-6 Microwave path interference calculation worksheet: Example.

values; that is, (–99 dB) + (–101 dB) = (–96.9 dB). For manual calculations, though, this refinement can usually be skipped without significantly affecting the end result; *i.e.*, the worst-case error for this simplification is 3 dB.

Now the level of the undesired signal, as XPOL, is 30 dBm – 2.2 dB + 45.1 dBi – 54.0 dB (VPOL value for TX antenna) – 140.6 dB + 47.6 dBi – 45.0 dB (HPOL value for RX antenna) – 3.5 dB = –122.6 dBm (equals undesired signal level at site 4 RX antenna, XPOL).

Finally, the desired-to-undesired (D/U) ratios for the PPOL and XPOL cases can be calculated. For PPOL, the D/U ratio is –25.0 dBm (desired signal into RX antenna at site 4) – –111.6 dBm (undesired signal into RX antenna at site 4 in the PPOL case), which equals 84.4 dB. For XPOL, the D/U ratio is –25.0 dBm – –122.6 dBm (undesired signal into RX antenna at site 4 in XPOL case), which equals 97.6 dB. For co-channel analog links, a D/U ratio of 60 dB or better is generally required. In this case, that target ratio is met even if the newcomer link is PPOL to the existing link, so the newcomer would be free to choose either HPOL or VPOL. Additionally, in this case the undesired signal is calculated to be well below the –85 dBm typical 13 GHz receiver threshold, and so for this sample case the new link is no threat to the existing link.

Custom computer programs now typically do the interference calculations, using RPEs that have been tabulated from manufacturers' data so the program can "look up" the PPOL and XPOL suppression values at the pertinent off-axis angle for a specified microwave antenna. For such automated calculations, the software can routinely mathematically "add" the two suppression values by converting each dB suppression value to power, adding the two powers, and converting back to a combined dB suppression number, even when one XPOL dB value is 10 or more dB different from the lower, or controlling, XPOL dB value.

FIXED-LINK COORDINATION: 2 GHZ

As noted previously, for the band from 1990–2110 MHz (which soon will be limited to 2025–2110 MHz), no PCN process has been established on applications for new or major-modification fixed, point-to-point links. Rather, a procedure has been established which combines local coordination with quasi-PCN requirements. Here is the procedure for "Frequency coordination for all fixed stations in the band 1990–2110 MHz" as described in Section 74.638(c) of the FCC's rules:

- General requirements (from Section 74.638(c)(1) & (2)):
  - —Applicants are responsible for selecting the frequency assignments that are least likely to result in mutual interference with other licensees in the same area.
  - —Applicants may consult local frequency coordination committees, where they exist, for information on frequencies available in the area.
  - —Proposed frequency usage must be coordinated with existing licensees and applicants in the area whose facilities could affect or be affected by the new proposal in terms of frequency interference on active channels, applied-for channels, or channels coordinated for future growth.
  - —Coordination must be completed prior to filing an application for regular authorization, for major amendment to a pending application, or for major modification to a license.
  - —To be acceptable for filing, all applications for regular authorization, or major amendment to a pending application, or major modification to a license, must include a certification attesting that all co-channel and adjacent-channel licensees and applicants potentially affected by the proposed fixed use of the frequency(ies) have been notified and are in agreement that the proposed facilities can be installed without causing harmful interference to those other licensees and applicants.

LOCAL COORDINATION

The third paradigm—informal local coordination—is now addressed. As stated previously, coordination of shared frequencies is the act of facilitating frequency sharing, not performing path engineering to establish a basis for frequency reuse. This is the proper coordination paradigm for non-point-to-point services, notably remote pickup (service code RP, more commonly referred to as RPU), television remote pickup (TP, commonly referred to as TVPU or ENG), and low-power auxiliary devices (LP) such as wireless microphones and similar short-range devices typically used in a remote production environment.

The remainder of this chapter is intended as a primer on current issues affecting the practice of frequency sharing and coordination. Because the issues described here are in flux, the reader is encouraged to seek up-to-date information on the status of those matters of immediate interest. The two most useful resources for current information on spectrum management issues are the Federal Communications Commission Web site (www.fcc.gov) and the SBE publications and Web site (www.sbe.org).

Event-Specific Coordination

Local frequency coordination is seen in many locales as taking on ever-greater importance due to the prevalence of radio-frequency devices around major events. These may include sporting events, political conventions, or major news stories. The devices in use may comprise a broad range of RPU, ENG, and low-power auxiliary devices. The nature and location of the event often determines the mixture of local (or in-market) users and itinerant (or out-of-market) users. Quite a number of spectrum-sharing issues typically arise which require the attention (sometimes on-site and in real time) of a local frequency coordinator, for example:

- Itinerant users may travel with LP equipment that is incompatible with local TV "white spaces" (*i.e.*, unused TV channels) where most such devices operate;
- The sheer number of frequencies desired may overwhelm the sharing plan; this has been known to shift the coordination paradigm from one of "facilitating sharing" to the less-appropriate one of "gatekeeping";
- Irrespective of the fact that all the users competing for spectrum availability may be equally eligible, an unofficial priority-of-use paradigm sometimes insinuates itself, based on which entities have broadcast rights to a commercial sporting event. This can lead to conflict, underscoring the importance of the local coordinator as a disinterested third party.

To meet these challenges, there has been a growing acceptance of the role of a local, event-specific frequency coordinator. Perhaps the best known program of event coordination is the game day coordinator (GDC), a cooperative venture between the National Football League (NFL) and the SBE. The GDC program has been cited favorably by the FCC; in an October 2005 ruling, in response to the NFL's request for a waiver of station-identification rules on its wireless headset intercom system, the Commission explicitly noted the GDC's contribution to a more controlled RF environment:

> The NFL indicates that the game day frequency coordinators monitor the RF environment in real time during the game, and that they are careful to prevent any interference both inside and outside the stadium. In addition, the NFL indicates that by using prior coordination the potential for interference is greatly reduced if not eliminated.[10]

The specific case of NFL game coordination points to a more general trend that poses an ongoing challenge for local coordination. This is actually the convergence of two trends: the ever-greater use of wireless devices, especially in the LP class, and the shrinking availability of spectrum in which they can operate.

Section 74.802 of the FCC Rules states that the majority of the channels available for LP devices consists of television broadcast channels that are unused in a particular location (also known as "white spaces"). However, it is well known that the DTV conversion in the U.S. has the effect of greatly reducing the number of television channels available for BAS users. Upon completion of the conversion, broadcasters will lose the entire 700 MHz band, which has historically been a popular portion of spectrum for wireless microphones and similar LP devices. As of this writing (September 2006), no replacement spectrum has been identified. It remains to be seen what combination of procedural and technological changes may be able to mitigate this spectrum management issue.

One technological fix that some vendors are trying is an intelligent scanning receiver capable of communicating with the LP transmitter so as to set its operating frequency. On power-up, the receiver quickly scans the RF environment and programs the best choice of frequencies to the transmitter. While such a feature may be useful and even admirable in an isolated system, concerns have been raised about this type of automatic operation in a complex, but coordinated, event. Vendors, users, and coordinators are advised to proceed with caution when introducing such "smart" systems into an environment in which a human coordinator is tracking frequency usage and making recommendations to multiple users on site.

Regulatory Challenges

Other issues facing local event coordinators arise from decisions in the regulatory realm. For example, one effect of the ET Docket 01-75 rulemaking, mentioned previously, was to replace the existing RPU channel plan at 450 MHz and 455 MHz with a new one, compatible with similar Part 90 services and based on a narrower fundamental unit of bandwidth (6.25 kHz, down from 12.5 kHz).

This change has resulted in placing a mixture of voice-grade, program-quality, and telemetry channels in the same allocation. As of this writing, the effect of this rulemaking on the Part 74 RPU bands has not been fully realized. The SBE National Frequency Coordination Committee has been studying this and may have additional helpful information. The reader is encouraged to learn the latest status at either the FCC or SBE Web sites.

Yet another challenge arising from the regulatory environment is the trend toward admitting satellite earth stations into frequency bands once reserved for BAS, on the theory that terrestrial and satellite services are a compatible combination for frequency reuse. While this may be true for fixed terrestrial links, it may not be in the case of mobile operation. Again, the interested reader is referred to electronic sources for recent actions in these dockets.

SUMMARY

An understanding of the concepts presented in this chapter will help the broadcast engineer to know what type of frequency coordination services are required for projects making use of BAS spectrum. For events where heavy BAS usage is anticipated, especially involving use of LP auxiliary devices, the broadcaster is encouraged to seek out the coordinator responsible, either the local frequency coordinator or an event-specific coordinator such as a GDC. To ensure favorable outcomes, it is essential for both user and coordinator to be versed in the technological and regulatory challenges related to BAS bands.

With the advent of mandatory PCN protocols for 7 and 13 GHz TV BAS fixed-link stations and 950 MHz aural BAS stations, most stations planning new or modified facilities in these bands will likely find that it makes sense to retain the services of a CMFC. But, even when this task is so delegated, the user should be familiar with the methodology, so as to know what questions to ask when selecting a CMFC.

[10]See Order, DA 05-2870, in the matter of National Football League request for waiver, October 31, 2005.

CHAPTER

1.7

Distance and Bearing Calculations

DANE E. ERICKSEN, P.E., CSRTE, 8-VSB, CBNT

Hammett & Edison, Inc., Consulting Engineers
San Francisco, California

INTRODUCTION

The current Federal Communications Commission (FCC) Rules for distance calculations use two methods: *flat-earth* and *spherical-earth*. The flat-earth method assumes the distance between two points to be the hypotenuse of a right triangle whose sides are determined by the difference in latitude and longitude of the starting and ending points multiplied by the length per degree of latitude and longitude at the mid-latitude of the two points, as shown in Figure 1.7-1. The "flat-earth" term refers to the use of a right triangle to calculate the distance. Because the lengths of a degree of longitude and a degree of latitude used in the flat-earth method are derived from an ellipsoid rather than a spheroid model of the earth, the flat-earth method is actually more accurate than the spherical-earth method for short to moderate distances.

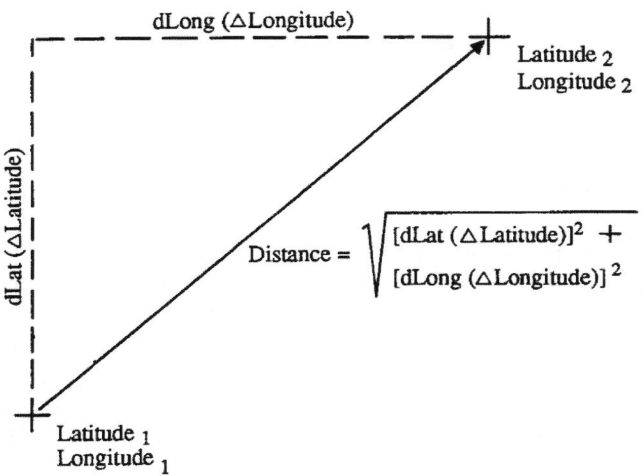

FIGURE 1.7-1 Flat-earth method.

The spherical-earth method uses conventional spherical trigonometry to determine the distance. Section 73.208 of the FCC Rules requires that the flat-earth method be used for distances up to and including 475 km. Distances greater than 475 km must be calculated using the spherical-earth method, which becomes more accurate than the flat-earth method for large distances. Section 73.208 is silent on how azimuths are to be calculated, nor does it specify the earth radius to be used for spherical-earth calculations.

FCC FLAT-EARTH METHOD

In FCC Docket 80-90 [1], formulas were substituted for the tables previously used for determining the length of a degree of latitude or of longitude as a function of latitude. However, the coefficients adopted in Docket 80-90 truncated to only two terms the trigonometric series used to generate the tables and adjusted the coefficients by a factor of (1.609/1.609347) because of the Docket 80-90 decision to define the conversion factor from U.S. statute miles to kilometers as 1.609 [2], rather than the value of (5280 ft/mile) × (1200/3937 m/ft) × (1/1000 km/m), or 1.609347219 km/mile (approximately) [3].

In the Second Report and Order to Docket 86-144 [4], the FCC corrected these problems by adopting the full-precision, nontruncated trigonometric series for the arc length formulas given in the 1966 edition of U.S. Naval Hydrographic Office Publication No. 9, also known as *H.O. 9*, the *American Practical Navigator*, or simply *Bowditch*, after Nathaniel Bowditch (1773–1838), its original author. These trigonometric series are based upon a binomial theorem expansion [5] of an ellipsoid model of the Earth corresponding to the Clarke spheroid of 1866, upon which topographic maps in the United States are currently based [6].

The trigonometric series defining the length of 1 degree of latitude and 1 degree of longitude for the Clarke spheroid of 1866 are:

$$dLat = 111.13209 - 0.56605\cos(2L) + 0.00120\cos(4L) \ldots$$

$$dLong = 11.41513\cos(L) - 0.09455\cos(3L) + 0.00012\cos(5L) \ldots$$

where dLat is the length in kilometers of 1 degree of latitude at latitude L and dLong is the length in kilometers of 1 degree of longitude, again at latitude L.

The latitude, L, is taken as the mid-latitude of the two points between which the distance is to be calculated, as follows:

$$L = (Latitude_1 + Latitude_2)/2$$

where $Latitude_1$ and $Latitude_2$ are the latitudes of the starting and ending points. Similarly, $Longitude_1$ and $Longitude_2$ are the longitudes of the starting and ending points. In all cases, north latitudes are treated as positive and south latitudes as negative, and west longitudes are treated as positive and east longitudes as negative.

The distance between two points is then given by the Pythagorean theorem:

$$D = \sqrt{\left[(dLat)(Lat_1 - Lat_2)\right]^2 + \left[(dLong)(Long_1 - Long_2)\right]^2}$$

Plots showing how the lengths of 1 degree of latitude and longitude vary with latitude are given in Figures 1.7-2 and 1.7-3. The FCC has a calculator that uses this method on the Media Bureau Web site (http://www.fcc.gov/mb/audio/bickel/distance.html). Hammett & Edison also has a bearing/distance calculator on their Web page (www.h-e.com).

Canadian Method

In August 1987 [7], the Canadian government adopted the truncated and adjusted arc-length formulas that had been implemented by the FCC in Docket 80-90; namely,

$$dLat = 111.108 - 0.566\cos(2L)$$

$$dLong = 111.391\cos(L) - 0.095\cos(3L)$$

At this point in time, the Canadian government has not yet adopted the corrected, more accurate formulas that were implemented by the FCC in Docket 86-144, nor was the flat-earth *versus* spherical-earth break point changed from 350 km to 475 km. This, then, is the source of current discrepancies between the U.S. and Canadian distance calculation methods. Although the differences between the two methods will usually not be significant when the calculated distance is rounded to the nearest kilometer (for FM) or to the nearest 1/10 km (for television), one should always check to see whether there is a difference between roundings. For calculations involving Canadian stations, the Canadian (Docket 80-90) version of the Clarke spheroid formulas is controlling.

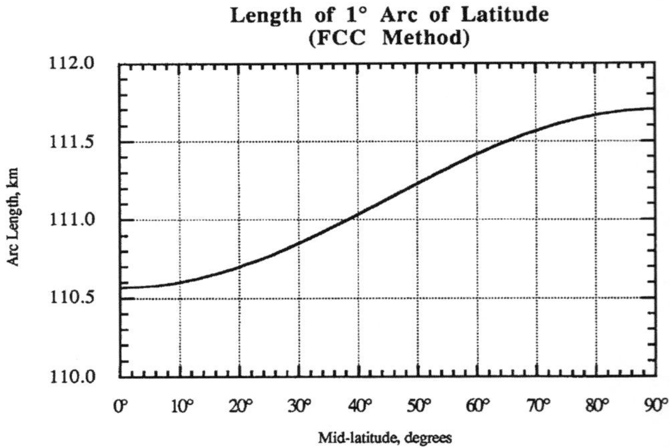

FIGURE 1.7-2 Length of 1° arc of latitude (FCC method).

Mexican Method

The August 11, 1992, U.S.–Mexican FM Agreement specifies the spherical-earth method exclusively; there is no provision for using the FCC flat-earth method for short to moderate distances. However, the Mexican

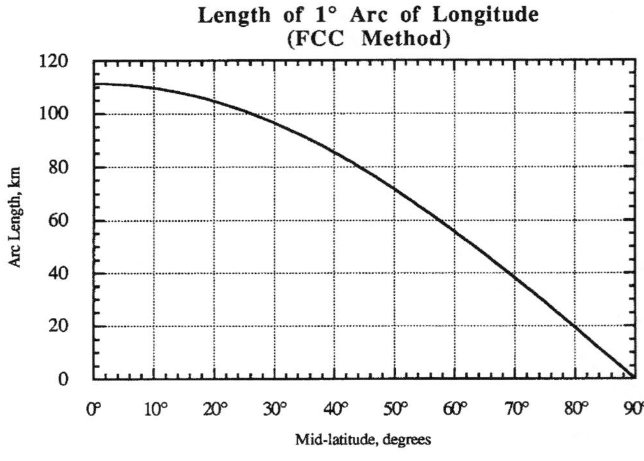

FIGURE 1.7-3 Length of 1° arc of longitude (FCC method).

Agreement specifies a spherical-earth arc of 111.18 km/degree, corresponding to an earth radius of 6370.14 km. The Mexican Agreement further specifies that azimuths are to be calculated on a spherical-earth basis. Finally, the Mexican Agreement specifies the rounding of distances to the nearest kilometer, so in that respect it matches the FCC rounding practice for FM station distances.

SPHERICAL-EARTH METHOD

The formula for the spherical-earth distance, or great-circle distance, is:

$$D = K\cos^{-1}[(\sin Lat_1)(\sin Lat_2) + (\cos Lat_1)(\cos Lat_2)\cos(Long_2 - Long_1)]$$

The constant K is in kilometers/degree and is determined by the radius of the sphere being modeled. The FCC has never defined the earth radius to be used for spherical-earth calculations. The example given in Section 73.185(d) of the FCC Rules suggests an earth radius of 6365 km (K = 111.090 km/degree). A 6373 km radius (K = 111.230 km/degree) is implied by the 5280 mile 4/3-earth radius given in Section 73.684(c)(1) of the FCC Rules. This 4/3-earth radius was also used in FCC Report No. R-6410 (*Elevation and Depression Angle Tables*; September 15, 1964). An earth radius of 6367 km (K = 111.125 km/degree) can be deduced from the 1852 m definition of a nautical mile [8]. Finally, an earth radius of 6371 km (K = 111.195 km/degree) corresponds to the mean radius of the Clarke spheroid of 1866. Until such time as the FCC so specifies, the author suggests using the mean Clarke spheroid value of 6371 km (K = 111.195 km/degree) [9].

AZIMUTH CALCULATIONS

Because the FCC Rules are silent on how azimuth, or bearing, calculations are to be performed, both the flat-earth and spherical-earth methods are commonly in use. The flat-earth method determines azimuth using the arctangent of the right triangle defined in the FCC method. The spherical-earth method uses standard spherical trigonometry. It is the author's recommendation that azimuth always be determined using the spherical-earth method, even when the FCC flat-earth method is used to determine distance. The flat-earth method will be in error by up to 2 degrees at distances approaching 500 km, whereas the spherical-earth azimuth will be correct within about 0.1 degree at such distances [10]. Figure 1.7-4 shows why this is so. Azimuths determined using the arctangent of the right triangle defined by the FCC flat-earth method assume that lines of longitude are parallel, whereas they are not. This is why the forward and back (reciprocal) azimuths using the spherical-earth method will generally not be exactly 180 degrees apart, whereas the forward and back azimuths using the flat-earth method are always 180 degrees apart.

Flat-Earth Model

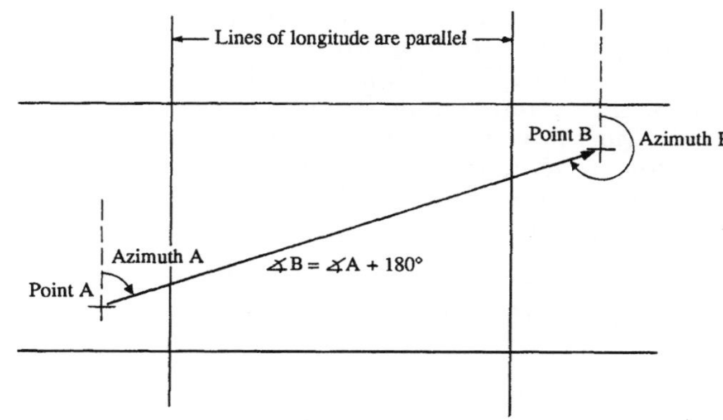

Spherical-Earth Model

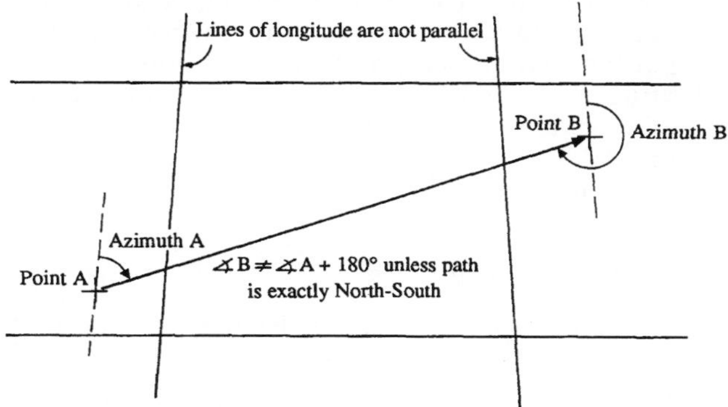

FIGURE 1.7-4 Azimuth calculations.

The formula for determining azimuth by the spherical-earth method is:

$$C = \cos^{-1}\frac{\sin Lat_2 - \sin Lat_1 \cos(D/K)}{\sin(D/K)\cos Lat_1}$$

If $\sin(Long_2 - Long_1) < 0$, azimuth = C.

If $\sin(Long_2 - Long_1) \geq 0$, azimuth = $360° - C$.

The ratio (D/K) is the great circle arc length in degrees and is obtained by re-arranging the formula for D given above, namely:

$$D/K = \cos^{-1}[(\sin Lat_1)(\sin Lat_2) + (\cos Lat_1)(\cos Lat_2)\cos(Long_2 - Long_1)]$$

It should be noted that the Industry Canada rules applying to FM broadcasting (BPR-3), the U.S.–Canada FM Agreement, and the U.S.–Mexico FM Agreement *do* specify that azimuth is to be calculated using the spherical-earth method and, further, that for Canadian azimuth calculations the bearing is to be rounded to the nearest degree [11]. It should also be noted that the radius of the sphere is irrelevant for azimuth calculations.

Clarke Spheroid versus *WGS Ellipsoid*

The original FCC distance tables were based on the Clarke spheroid (or ellipsoid) of 1866, with a major-axis radius of 6378.2064 km and ellipticity [12] of 1/294.98. The current edition [13] of the *American Practical Navigator* now bases Table 7, "Length of a Degree of Latitude and Longitude," on the World Geodetic System (WGS) ellipsoid of 1972. The coefficients for dLat and dLong for this ellipsoid are:

$$dLat = 111.13292 - 0.55982\cos(2L) + 0.001175\cos(4L) \ldots$$

$$dLong = 111.41282\cos(L) - 0.0935\cos(3L) + 0.000118\cos(5L) \ldots$$

The difference between the Clarke 1866 and WGS 1972 trigonometric series is inconsequential when coordinate databases are maintained only to the nearest second and distances are rounded to the nearest kilometer or 1/10 km. Over a 0 to 90 degree arc of latitude, at 1 degree increments, the RMS difference between the two arc length formulas is only 0.0041% for latitude and 0.0033% for longitude. For this reason, the FCC decided not to amend the formulas given in Section 73.208(c) of its Rules from the Clarke 1866 coefficients to WGS 1972 coefficients.

Rounding Practices

There continues to be an inconsistency in rounding practices between the FCC FM and TV rules. For FM distance calculations, Section 73.208(c)(8) specifies rounding to the nearest kilometer. For TV distance calculations, Section 73.611(d) specifies rounding to the nearest 1/10 km. Persons performing distance calculations must, therefore, be mindful to the different rounding criteria of the FM and TV rules. The AM rules and the Broadcast Auxiliary Service rules are silent on distance rounding practices. The Broadcast Auxiliary Service rules also do not prescribe how distances are to be calculated, even though portions of those rules include minimum distance requirements; for example, Subpart G (Low Power TV, TV Translator, and TV Booster Stations) specifies certain minimum distances between low-power television (LPTV)/TV translator stations and land mobile stations. Calculations regarding distances between TV translator and LPTV stations should, presumably, use the flat-earth method specified in the FCC's FM rules.

NAD27/NAD83 Datums

The Federal Geodetic Committee is in the process of converting all maps in the United States from the 1927 North American Datum (NAD27), which is based on the Clarke spheroid of 1866, to the 1983 North American Datum (NAD83). NAD83 differs from NAD27 in that it refers to the Earth's center of mass, making it fully compatible with satellite systems for position determination. The FCC has indicated it will eventually convert entirely to NAD83 to maintain accuracy and consistency with other government agencies. NAD83 is based on the Geodetic Reference System ellipsoid of 1980 (GRS 1980), with a major axis of 6378.135 km and an ellipticity of 1/298.26. These are the same parameters as for the WGS 1972 ellipsoid, and the trigonometric series for the 1-degree arc length formulas are identical to those for WGS 1972 [14].

The vast majority of 7.5-minute quadrangle maps available from the Geological Survey specify NAD27 as the reference datum, in the lower left corner of the map. At some point in the future, the Geological Survey will begin issuing maps based on the NAD83 datum. Initial attempts at printing dual-sided (NAD27/NAD83) maps proved impractical [15].

In order to prevent intermixing of numerical information using two different map datums, the FCC has stated [16] that the following procedures will be in effect until further notice:

1. Broadcast station applicants must continue to furnish coordinates based on NAD27.
2. All Form 601 Universal Licensing System (ULS) applications must use only NAD83 coordinates.
3. All Form 854 Antenna Structure Registration (ASR) filings must use only NAD83 coordinates.

Conversion between Datums

The National Geodetic Survey (NGS) has developed an algorithm known as the North American Datum Conversion program (NADCON). The program is bidirectional, meaning that it will convert NAD83 coordinates to NAD27 coordinates as well as NAD27 coordinates to NAD83 coordinates. It also applies in Alaska, Hawaii, and Puerto Rico. But, most importantly, it is the conversion algorithm recommended by statute. The *Federal Register* dated August 10, 1990 (Vol. 55, No. 155, p. 32681), stated:

> The intent of this notice is to standardize a horizontal datum transformation method when a mathematical transformation is desired. FGCC [Federal Geodetic Control Committee] selected the method incorporated in the software identified as NADCON. It is not the intent of the notice to declare when to use a datum transformation or by what method but only to declare that when a mathematical transformation is appropriate, NADCON is recommended.

Thus, although use of NADCON is not mandatory, use of conversion algorithms other than NADCON may result in a rounding error that will cause the results to be inconsistent with those obtained by the FCC or by the Federal Aviation Administration (FAA), both of which use NADCON.

NADCON utilizes a minimum-curvature approach to transforming between the two datums and comes with four reference files: a large file for the contiguous United States and smaller files for Alaska, Hawaii, and Puerto Rico. By having a separate set of polynomial coefficients for each 7.5-minute topographic map, NADCON achieves an improved conversion accuracy of approximately ±0.0003 seconds in latitude and ±0.0005 seconds in longitude. Because of the size of these data files, programmable calculators are unable to implement the program, but today's personal computers are certainly capable of loading the NADCON

data files. Information regarding the NADCON software is available at http://www.ngs.noaa.gov/TOOLS/Nadcon/Nadcon.html [17].

USE OF GPS TO DETERMINE COORDINATES

With the widespread availability of low-cost global positioning system (GPS) receivers and the termination of "selective availability," a technique formerly used to limit the full accuracy of GPS to non-U.S. government users, a typical consumer-grade GPS receiver that can simultaneously receive the signals from multiple GPS satellites will generally return geographic coordinates accurate to the nearest second of latitude and longitude, which is generally acceptable accuracy for FCC filings. A further potential source of error exists because most GPS receivers have user-driven menus allowing the display of coordinates in either the NAD27 or NAD83 datum. So, there is always the possibility that the user does not realize which datum the GPS receiver has been set to display. GPS-derived coordinates reported as NAD27 will be in error when in fact the GPS receiver is in its factory-default mode to display NAD83 coordinates. If NADCON is then used to convert the coordinates from NAD27 to NAD83, in the mistaken belief that the reported coordinates were NAD27, an erroneous set of coordinates, corresponding to neither datum, is then created. It is important, then, for persons using GPS receivers to determine coordinates to ensure that they are aware of which datum their receiver is displaying.

Notes

[1] Docket 80-90 Report and Order, May 26, 1983.

[2] *Ibid.*, p. 29, footnote 35.

[3] ANSI/IEEE Standard 268-1982, Metric Practice, p. 31, note 14.

[4] Docket 86-144 Second Report and Order, September 10, 1987.

[5] Personal correspondence between the author and Adam W. Mink, Chief, Hydrography and Navigation Department, Defense Mapping Agency, Washington, D.C., May 24, 1985.

[6] USDOI, *Maps for America*, 2nd ed., U.S. Department of the Interior, Geological Survey National Center, Reston, VA, 1981, p. 238.

[7] Broadcast Procedure No. 13 (BP-13), Issue 2, Broadcasting Regulation Branch, Department of Communications, Government of Canada (effective August 6, 1987); now Industry Canada BPR-3 (see footnote 11).

[8] U.S. Naval Hydrographic Office Publication No. 9, 1981 edition, Vol. 2, p. 862. H.O. No. 9 defines a nautical mile as 1 minute of any great circle of the Earth. In 1929, the International Hydrographic Bureau proposed a standard length of 1852 m (exactly), which is known as the *International Nautical Mile*. A nautical mile of 1852 m implies an earth radius of 6366.707 km ([1852 m/min × 60 min/degree × 360 degrees/circumference]/2π).

[9] U.S. Naval Hydrographic Office Publication No. 9, 1981 edition, Vol. 2, Appendix D, p. 648. The mean radius of an ellipsoid is defined as $(2a + b)/3$, where a is the major or equatorial radius and b is the minor or polar radius. The full precision value is 6370.9989 km (111.1949075 km/degree), but use of the 111.195 km/degree rounded value is suggested.

[10] Azimuth errors are referenced to values obtained from Andoyer–Lambert formulas. Andoyer–Lambert formulas model the Earth as a true ellipsoid and are used extensively in Loran computations. The complexity of Andoyer–Lambert formulas does not warrant their routine use for FCC calculations.

[11] See Industry Canada, *Broadcasting Procedures and Rules*, Part 3 (BPR-3), August 2002, Section C-4, for Canadian calculations; see Appendix 2, Item 2, of the U.S.–Mexico FM Agreement of 1992 for Mexican calculations.

[12] Ellipticity, or flattening (f) is defined as $f = (1 - b/a)$, where a is the equatorial radius and b is the polar radius.

[13] The *American Practical Navigator* is available online at http://www.irbs.com/bowditch/.

[14] The major axis and ellipticity for WGS1972 and GRS1980 are identical, according to the January 1987 NOAA Technical Memorandum NOS NGS-16, *Determination of North American Datum 1983 Coordinates of Map Corners (Second Prediction)*, by T. Vincenty, National Geodetic Survey. Because the binomial theorem expansion of an ellipsoid model starts with only two constants, the major axis dimension and the ellipsoid models of the Earth with the same major axis and ellipticity values must also be identical.

[15] USDOI, *Implementing North American Datum 1983 for the National Mapping Program (Ashaway Quadrangle)*, U.S. Department of the Interior, Geological Survey, Reston, VA (undated).

[16] FCC Public Notice, FCC Interim Procedure for the Specification of Geographic Coordinates, March 14, 1988.

[17] For a more detailed discussion of the NAD27/NAD83 datums, see "NAD83: What It Is and Why You Should Care," by the author and published in the proceedings of the 1994 SBE Engineering Conference. A copy of this paper is available on the Hammett & Edison Web site (www.h-e.com).

CHAPTER

1.8

Propagation Characteristics of Radio Waves

DOUGLAS VERNIER
V-Soft Communications
Cedar Falls, Iowa

INTRODUCTION

The broadcast industry, and virtually all communications services, is dependent on reaching people within a given area. The quality of coverage is a key factor. Knowing where the signal goes, through propagation prediction, is both science and art. It is based on the scientific modeling of the radio path as it travels from the transmitter to the receiver. A good model can accurately predict signal strength at various frequencies over distances as those signals are influenced by ground conductivity, atmospherics, and terrain. The propagation models currently in use were developed and tested over a period of time and have been adjusted to better account for observed variances. How information gleaned from propagation predictions is presented to the broadcaster is an art and an important part of making sense of what the predictions mean. In this chapter, the primary broadcast-oriented prediction models currently used in the United States are discussed, and information on how radio and television station engineers can use the models to assess the performance of their systems is provided.

The Purpose of Predicting Coverage

Predictions are used for all kinds of analyses of broadcast communications facilities. Coverage map predictions for estimating interference-free service area are required by the Federal Communications Commission (FCC) and every station must have a map available for inspection in the station's public file. Management uses coverage maps to provide the sales department with population information. Coverage maps are used universally in the process of estimating the population within the station's service area to evaluate upgrades and site relocations. Although field measurements can determine signal levels, they are difficult and time consuming to perform. Further, signal levels will often vary depending on the time of year, so predictions represent a quick and cost effective way to accurately estimate a station's coverage. The more accurate the coverage map, the more accurate the estimates of the size and location of the audience actually covered.

Radiowave propagation is the study of the transfer of energy at radio frequencies from one point (a transmitter) to another (a receiver). Radio waves are part of the broad electromagnetic spectrum that extends from the very low frequencies that are produced by electric power facilities up to the extremely high frequencies of cosmic rays. Between these two extremes are bands of frequencies that are found in everyday use: audio frequencies used in systems for the reproduction of audible sounds, radiofrequencies, infrared and ultraviolet light, and x-rays.

All electromagnetic waves propagate at the same velocity, regardless of the frequency. Light is an electromagnetic wave; thus, the propagation velocity is often referred to as the *speed of light* (c), which for a vacuum is approximately 3×10^8 m/sec. The velocity of any wave is dependent on the medium in which it is traveling, but for simplicity it is usually considered with respect to a vacuum. The frequency of a wave is defined in terms of the number of cycles per second or hertz (Hz) and is related to the wavelength (λ) by the expression $f = c/\lambda$. Figure 1.8-1a shows the ranges of various bands within the electromagnetic spectrum in terms of wavelength (λ) and frequency (F).

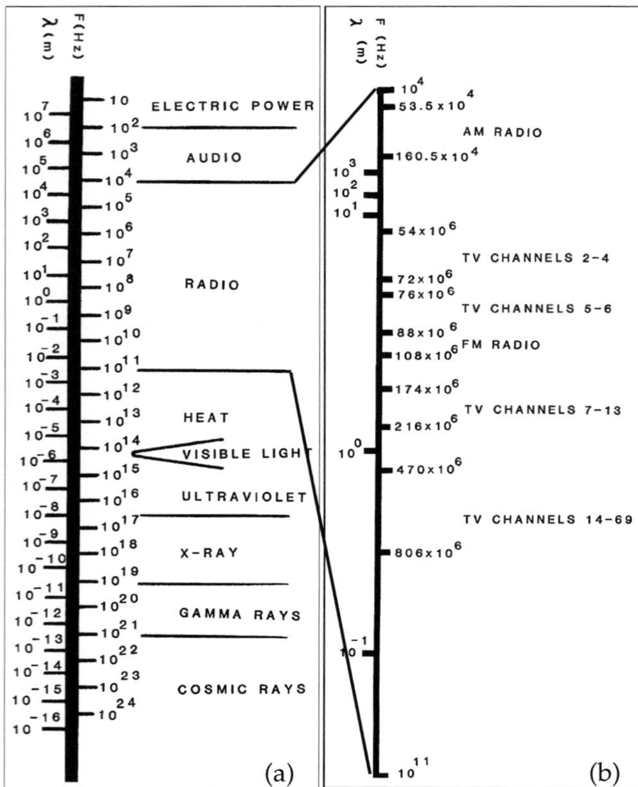

FIGURE 1.8-1 The spectrum.

Figure 1.8-1b is an expansion of the "radio" section of Figure 1.8-1a.

Free Space Propagation

To evaluate and compare radiowave propagation under various conditions, it is convenient to establish a reference standard. It is customary to consider as a standard the theoretically calculated loss for waves propagated in free space between two idealized antennas. The simplest case to investigate is the radiation emitted from an isotropic source: an ideal antenna that radiates energy with uniform intensity in all directions. The intensity of the energy varies proportionally to the inverse of the distance squared from the source—the *inverse square law.* The power of flux per unit area P_a (W/m^2) at a distance d (m) from a loss free isotropic antenna radiating power P_t (W) is given by:

$$P_a = P_t/4\pi d^2 \tag{1}$$

where $4\pi d^2$ is the surface area of a sphere at distance d (m) from the source. The power available from a loss free antenna (P_r) is the product of the power flux per unit area (P_a) and the effective aperture area of the receiving antenna (A_e). This area is related to the gain of the antenna by the expression:

$$A_e = (G\lambda^2)/4\pi \tag{2}$$

For a loss free isotropic antenna ($G = 1$), the basic free space transmission loss is defined as:

$$L_b = P_t/P_r = (4\pi d/\lambda)^2 \tag{3}$$

where d and λ have the same units. This equation can be rewritten in its more common form (expressing the loss in dB) as:

$$L_b = 32.44 + 20\log(F) + 20\log(d) \tag{4}$$

where F is the frequency in megahertz (MHz), and d is the distance between the antennas in kilometers. In the above equation, it should be remembered that ideal loss free isotropic antennas are being considered. In real world systems, antenna gain is a significant factor. The transmission loss (L) incorporates the antenna gains and is defined as:

$$L = L_b - (G_t + G_r + L_d) \tag{5}$$

where G_t and G_r are the free space antenna gains with respect to isotropic for the transmitting and receiving antenna, respectively. The term L_d is the aperture-to-medium coupling loss or polarization coupling loss between the antennas. The term L_d will have a value of 0 dB when the transmitting and receiving antenna have the same polarization.

In considering the potential service area coverage for a broadcast station, it is usual to express measurements in terms of field strength rather than transmission loss as previously presented. The root mean square (RMS) field strength, E (V/m), at a point where the power density of a plane wave is P_a (W/m^2), is given by:

$$E = \sqrt{120\pi P_a} \tag{6}$$

where the term 120π is the impedance of free space. The field strength is related to the power available from a loss free isotropic antenna by combining Equations (1), (3), and (6) as:

$$E = \sqrt{480\pi^2 P_r/\lambda^2} \tag{7}$$

A more useful form of free space field can be expressed in logarithmic terms above 1 microvolt per meter (dBμ) when F is in megahertz and P_r is power, expressed in decibels above 1 kW (dBk):

$$E(\mathrm{dB}\mu) = 107.2 + P_r + 20\log(F)\mathrm{dB}\mu \tag{8}$$

The electric field produced by a transmitter radiating a power P_t (W) at a distance d (m) in free space can be derived from Equations (1), (3), and (6) and is given by:

$$E = \sqrt{30P_t/d^2} \tag{9}$$

or, in logarithmic terms, where P_t is expressed in dBK, d is in kilometers, and a transmitting antenna has gain G_t in decibels above isotropic:

$$E(\mathrm{dB}\mu) = 105 + P_t + G_t - 20\log(d) \tag{10}$$

Using the same units, the field strength E(dBμ) for non-free space environments can be related to the basic transmission loss by:

$$L_b(\text{dB}) = 137 + 20\log(F) + P_t + G_t - E \quad (11)$$

These equations form the theoretical basis for characterizing propagation. They do not, however, take into account such real world factors as the presence of the earth, atmosphere, or obstructions.

Presence of Earth

When the transmitting and receiving antennas are placed over ground, the propagation of radio waves is modified from the free space models presented above. Radio waves that strike the earth are partially absorbed and partially reflected. Waves that are reflected by the earth experience changes in the phase of the wave, which affects the distribution of available energy. The extent to which the waves are reflected or absorbed is dependent on frequency and the ground constants: conductivity and permittivity.

Propagation over Plane Earth

Plane earth geometry is valid for antennas that are closely located so the curvature of the earth is not a factor, yet far enough apart from each other so the energy may be described as a plane wave and ray theory can be applied. The resultant received electric field can be represented as the sum of the direct and reflected rays:

$$E = E_d[1 + |R|e^{j(\phi_\Delta + \phi_r)}] \quad (12)$$

This equation is valid for small angles of θ and deserves some additional explanation. The term E_d is the free space electric field that is produced at distance d (m) by the direct ray. The terms $|R|$ and ϕ_r are the magnitude and phase of the complex reflection coefficient, respectively. This term is dependent on the nature of the surface (conductivity [σ] and permittivity [ε_r]), the angle between the surface and incident wave, the wavelength of the radio wave, the polarization of the wave, and the curvature of the earth. The magnitude of the reflection coefficient varies between –1 and +1. The term ϕ_Δ is the phase delay due to the longer path that must be taken by the reflected wave and has the form of:

$$\phi_\Delta = (4\pi h_1 h_2)/\lambda d \quad (13)$$

It is often sufficient to assume the ground approximates a large flat surface. In such a case, a sufficiently accurate expression is given by:

$$E = 2E_d \sin(2\pi h_1 h_2/\lambda d) \quad (14)$$

Some cases of special merit that can be derived from Equation (14) are:

Case I	$h_1h_2 = d\lambda/2$	$E = 0$
Case II	$h_1h_2 = d\lambda/4$	$E = 2E_d$
Case III	$h_1h_2 = d\lambda/12$	$E = E_d$

Therefore, depending on the antenna heights, distances, and wavelength, it is possible to cancel out the field at the receiver or magnify the wave to a field strength double that which could be achieved from a free space field. The variation of signal strength due to multipath effects can be minimized in point-to-point applications through the use of antennas with narrow beamwidths.

When considering the case of VHF antennas that are close to the ground, the effective antenna heights h_t (m) and h_r (m) will have to be substituted for h_1 and h_2, respectively, for Equation (14). The new antenna heights (h_t and h_r) allow for the effects caused by the relative permittivity (ε_r) and conductivity (σ) of the ground. The effective antenna heights are related to the physical antenna heights above ground level by:

$$h_t = \sqrt{h_1^2 + h_0^2} \quad (15.1)$$

$$h_r = \sqrt{h_2^2 + h_0^2} \quad (15.2)$$

where the term h_0 is dependent on the type of polarization being considered:

Vertical Polarization

$$h_0 = (\lambda/2\pi)[(\varepsilon_r + 1)^2 + (60\lambda\sigma^2)]^{1/4} \quad (16.1)$$

Horizontal Polarization

$$h_0 = (\lambda/2\pi)[(\varepsilon_r - 1)^2 + (60\lambda\sigma^2)]^{-1} \quad (16.2)$$

Medium Frequency Propagation

Medium frequency (MF) waves lie in the frequency range of 300 kHz to 3 MHz and are characterized by their long wavelengths (1000 to 100 m). The transmitting antenna is located right at the surface of the earth, and the receiving antenna is very close to the earth's surface with respect to a wavelength. In this case, the direct and ground reflected waves cancel, and the transmission is by means of the ground wave (also known as the surface wave) and the sky wave.

Ground Waves

Ground waves are guided along the earth's surface, similar to a transmission line. The field is attenuated in this propagation mode by losses in the ground; therefore, the composition of the soil ε_r, Relative Dielectric Constant, and σ, Conductivity, have a direct bearing on the amount of attenuation the wave will experience and subsequently how far reliable communications can be established. The attenuation is also dependent on the frequency and polarization type. The attenuation factor, A, is a measure of the amount of attenuation present. The term p is known as the numerical distance and b is the phase constant. Values for these terms can be calculated from the following equations:

$$\rho = (\pi d/\lambda x)\cos(b) \tag{17.1}$$

$$b = \text{Arctan}[(\varepsilon_r + 1)/x] \tag{17.2}$$

$$x = 18 \times 10^3 \sigma/F \tag{17.3}$$

To determine the electric field strength, the attenuation factor must be added to Equation (12):

$$E = E_d[1 + re^{j(\phi_\Delta + \phi_\Delta)} + (1 - R)Ae^{j(\phi_\Delta + \phi_\Delta)}] \tag{18}$$

It is interesting to note that the same earth which acts as a conductor at very low frequencies will act as a small-loss dielectric at very high frequencies. It is also noteworthy to observe that the losses for horizontally polarized waves are much greater than for vertically polarized waves. Thus, for practical applications only, vertically polarized waves should be considered. While the ground wave provides the major path for medium frequency propagation, the wave attenuates relatively quickly with distance and is reliable for distances of only a few hundred kilometers.

Sky Waves

Waves that propagate via the ionosphere are known as sky waves and can provide significant signal strength at distances up to a few thousand kilometers. The ionosphere is a constantly changing environment that begins approximately 65 km (40 miles) above the earth and extends to about 400 km (250 miles); see Figure 1.8-2. This region of the atmosphere is composed of three major sublayers: D, E, and F. These layers are not present at all times; for example, the D layer is present only during the day and is a major absorber of medium frequency waves. The E layer is a principal reflector of medium frequency waves; thus, during the day the majority of the medium frequency waves is absorbed by the D layer, but at night the D layer is not present, allowing the medium frequency waves to be reflected by the E layer.

- *D layer*—This layer exists at heights from about 50 to 90 km and is present only during daylight hours. The electron density is directly related to the elevation of the sun. This layer absorbs medium and high frequency waves.
- *E layer*—This layer exists at a height of about 110 km and is important in the nighttime propagation of medium frequency waves. The ionization of this layer is closely related to the elevation of the sun. At certain times, irregular cloud-like areas of high ionization may occur. These areas are known as sporadic E and occasionally prevent frequencies that normally penetrate the E layer from reaching higher layers. The sporadic E layer is prevalent during the summer and winter months. The sporadic E layer formed during the summer is the longest lasting from May to August, and the winter layer lasts about half as long beginning in December. During the mid-summer months when the electron density is at its greatest levels, TV signals in the lower VHF band can be transmitted over distances of hundred or thousands of kilometers [1].
- *F1 layer*—This layer exists at heights of about 175 to 200 km and is present only during the day. Waves that usually penetrate the E layer (3 to 30 MHz) will penetrate this layer and be reflected by the F2 layer. This layer introduces additional absorption of these waves.
- *F2 layer*—This layer exists at the upper boundaries of the atmosphere (250 to 400 km) and is present at all times, although the height and electron density will vary from day to night with the seasons and over sunspot cycles. During the night, the F1 layer merges with the F2 layer at about 300 km. This, in addition to the reduction of the D and E layers, causes nighttime field intensities and noise to be generally higher than during the day.

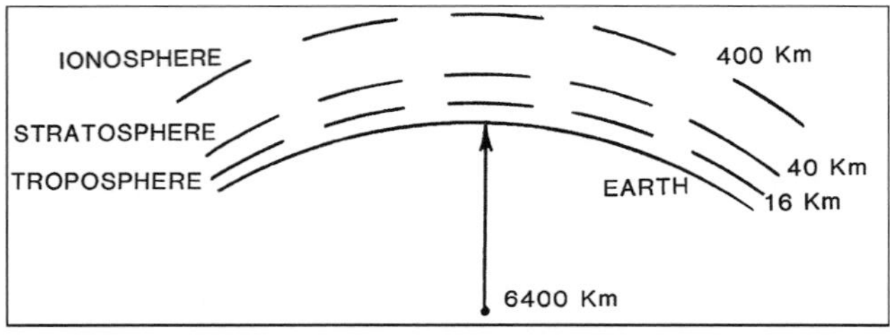

FIGURE 1.8-2 Atmospheric layers.

Interference between Ground Waves and Sky Waves

Interference to a receiver may occur from co-channel stations located many kilometers from the desired station. Because of the sky wave, sufficient signal strength may be received to interfere with the local station. This effect has been minimized by the FCC by limiting two factors in the operation of some AM stations: the operating power and time of operation.

Multipath interference occurs when the waves from a transmitting antenna reach a receiver from different paths in such a manner as to cancel or severely interfere with each other. This can happen at distances where both the ground wave and sky wave are sufficiently strong to interact. The direct ray will be a result of the ground wave, and the reflected wave will be from the ionosphere. At distances relatively close to the transmitter, the ionosphere will not reflect waves back to the earth, so the ground wave is predominant. At distances beyond a few hundred kilometers, the sky wave will dominate and the ground wave will be too weak to interfere. Multipath interference can also occur where the sky wave follows more than one path to the receiver.

Effects of Solar Activity

Interference to medium-frequency waves can also be caused by solar activity such as sunspots and flares, which manifest an increased or reduced emission of radiation from the sun. The changes in solar radiation levels can cause changes in the ionosphere layer that may result in unusual sky wave propagation condi-

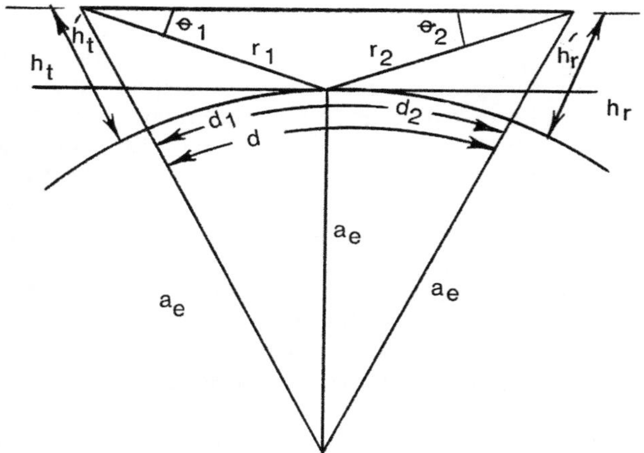

FIGURE 1.8-3 Reflection from smooth earth on line-of-sight path.

tions called *skip*, which, in turn, can cause inter-station interference. The effects of such solar activity will be strongest on propagation in the AM band during the first 5 to 10 days after the start of a storm. This has the effect of reducing sky wave field strengths. The effect has been observed to increase with frequency.

Propagation above 30 MHz

At frequencies above about 30 MHz the principal propagation mode is tropospheric. The surface wave is attenuated too severely to be of any practical long distance use and, though attenuated, the sky wave is usually passed through the ionosphere to space. For waves that propagate close to the earth's surface, the curvature of the earth will introduce additional effects that must be included in the plane earth model. First, the reflection coefficient (R) of the reflected wave has different characteristics than for a plane surface. Because the wave is reflected against a curved earth, the energy diverges more than is predicted by the inverse square law, and the reflection coefficient (R) in Equation (12) must be multiplied by the divergence factor (D), given by:

$$D = \sqrt{1 + 2d_1 d_2 / 2a_e(h'_t + h'_r)} \tag{19}$$

It should be noted that for smooth earth conditions, the transmitting and receiving antenna heights (h'_t and h'_r, respectively) above the plane tangent to the earth at the point of reflection are less than the antenna heights h_t and h_r above the surface of the earth.

Under normal propagation conditions, the refractive index of the atmosphere decreases with height so radio waves near the surface of the earth travel more slowly than at high altitudes. This variation in velocity as a function of height results in a bending of the radio waves. This may be represented as a modified earth radius commonly known as the effective earth radius (a_e), which allows the radio waves to be represented as straight lines. The ratio of the effective earth radius to true earth radius is commonly known as the k factor. Values of k can vary from 0.6 to 5.0, depending on the climate being considered. For temperate climates, the average value of k is 1.33, and most works refer to this as the 4/3 earth model when used in calculation.

Beyond Line of Sight Conditions

In order to determine when conditions exist where propagation is considered to be beyond line of sight, the respective distances from the transmitter and receiver to the radio horizon must be calculated. The radio horizon is the distance the horizon appears from an antenna, as defined by a plane from the antenna to the tangent of the earth's surface. The equation for the radio horizon in terms of d_{lt} (km) and h_t (m) and the k factor is of the form:

$$d_{lt} = 3.57\sqrt{h_t k} \tag{20}$$

When the sum of the distances to the radio horizon for the transmitter and receiver is less than the total distance of the path under consideration, a beyond line of sight condition exists. Diffraction makes it possible for radio waves to travel beyond that possible for line of sight transmission, though an additional loss term must be added to the free space loss. The amount of attenuation can be determined by diffraction methods.

Buildings

When planning for radio locations within built-up areas of cities or residential areas, buildings will have an effect on radio propagation. For radio relay stations, such as studio-to-transmitter links, it is the normal practice to select sites that will be clear of buildings; however, where this is not feasible and the path geometry is known (height and location of buildings) then the diffraction methods discussed for hills may be applied. In planning for broadcast systems, it is not practical to relate attenuation measurements made in built-up areas to the particular geometry of buildings. Therefore it is more conventional to treat the losses in a statistical manner, dividing the general classifications of building types into loss groups, so a loss can be derived for a particular type of building (multiple-story made of concrete and steel *versus* single story residence constructed of wood).

Within built-up areas there is much more back scatter than in open country. Additionally, due to the fact that buildings are more transparent to radio waves than the earth, there tends to be less shadow loss caused by buildings. However, the angles of diffraction due to buildings are usually much greater than in open country for natural terrain, thus the loss resulting from the presence of buildings tends to increase. Measurements indicate that at 100 MHz the median field strengths are 4 to 6 dB below that expected for a plane earth and drop off to about 10 dB for 200 MHz.[3] These measurements were made in areas containing some large buildings and open areas but mainly consisting of residential areas. Measurements conducted in the 850 MHz band indicate field strengths 20 to 34 dB below those expected for free space for path distances of 1 km to 25 km [2].

Vegetation

Among the many factors that have an effect on the determination of the losses present in a propagation path, vegetation is sometimes the most overlooked. Depending on the type of terrain in consideration (open or forest), the effect of vegetation can add a several decibel loss to the system. The amount of attenuation present is dependent on the frequency and polarization of the wave, as shown in Figure 1.8-4. The attenuation for a horizontally polarized wave for frequencies below about 1000 MHz is much less than for a vertically polarized wave. At around 1000 MHz, trees that are thick enough to block the field of vision can be modeled as an almost solid obstruction, and the attenuation over or around these obstructions can be predicted from knife-edge diffraction methods. The effect of vegetation on a radio path varies seasonally in the case of deciduous trees. During the winter months, the losses due to shadowing and absorption are less than those during the spring and summer. It is interesting to note that the greatest losses will occur during the spring because new growth has more sap and a greater moisture content which add to the absorption losses. When the antenna is raised above trees and other vegetation, the prediction of field strengths depends on the estimation of the height of the antenna above areas of reflection and the reflection coefficients. For areas of fairly uniform growth and for angles of incidence approaching grazing, the reflection coefficient will approach –1 at about 30 MHz. Even low growth that is uniform (*e.g.*, a wheat field) may yield a value of –0.3 for the reflection coefficient.

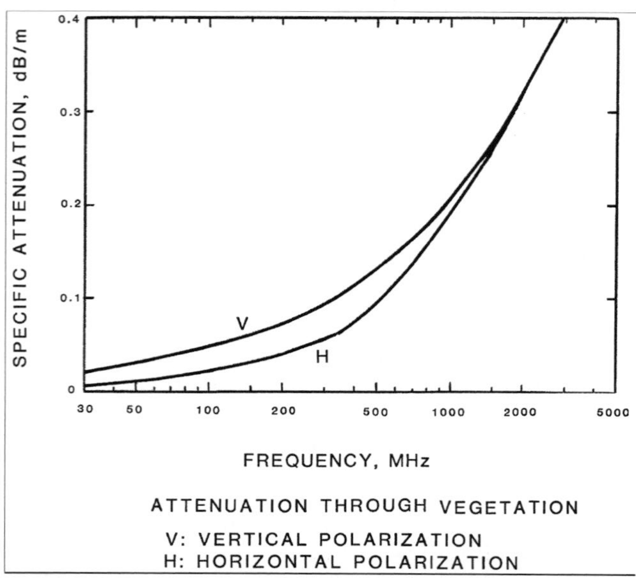

FIGURE 1.8-4 Vegetation land cover attenuation.

Atmosphere

The troposphere layer of the atmosphere is the major medium for propagation at VHF frequencies. The refractive index (n) of air has a value near unity (typically 1.00035). The index is dependent on the dielectric constant and can vary depending on the pressure and temperature of the air and on the amount of water vapor present; therefore, the refractive index changes with weather conditions and with the height above the earth. The velocity of radio waves is dependent on the refractive index of the atmosphere. As a general rule, the velocity of a wave is slower at the earth's surface than at higher altitudes, so a horizontally polarized wave will be refracted back toward the earth, although unusual atmospheric conditions may change this. Some simplifying assumptions are generally needed to obtain a solution under known meteorological conditions.

Ducting

Changes in the index of refraction of only a few parts per million can have dramatic effects on radio waves; therefore, it is usually more convenient to refer to the refractive index in terms of the refractivity N:

$$N = (n-1)x10^6 \tag{21}$$

Under meteorological conditions where the refractive index decreases rapidly with height over a large horizontal distance, radio waves can become trapped and experience low propagation loss over long distances. This phenomenon is known as *ducting*. Although ducting is frequent with some locations and meteorological conditions, due to its randomness and long-range unpredictability, it is not a reliable mode for communications. However, due to the strong fields over the horizon caused by ducting, inter-station interference can result. In addition, line of sight paths may be affected by severe fading.

In order for atmospheric ducts to occur, two conditions must exist. First, the refractive index gradient must be equal to or more negative than –157 N/km. The refractive index gradient is a measure of the change of the refractivity across a vertical height h (dN/dh). When this condition is present, the radio waves will remain close to the earth's surface beyond the normal horizon. Second, the refractive index gradient must be maintained over a height of many wavelengths. The duct may be thought of in the same manner as a transmission line waveguide; however, unlike metallic waveguides, natural ducts do not have sharp boundaries, although there is a wavelength cut-off above which waves will not propagate. Because the duct does not have sharp boundaries, the thickness (t) will not be rigid; therefore, the cut-off wavelength (λ) will not be fixed but an estimate can be obtained from Reference [1]:

$$\lambda = 2.5x10^{-3}t^{2/3}\sqrt{\frac{\delta N}{t} - 0.157} \tag{22}$$

where the wavelength and thickness are in meters. The term δN represents change in the refractive index across the duct. As an example, a duct near the ground that is 25 m thick and has a refractive index change of 10N (400N/km) will have a cut-off wavelength of 0.15 m (2 GHz). However, a duct with the same refractive index gradient will have to be about 87 m thick to propagate a wavelength of 1 m (300 MHz).

A duct spreads the energy within it in the horizontal direction but is contained in the vertical direction as the distance from the transmitter is increased. Thus, in principle it is possible for the field strength within a duct to be greater than the free space field at the same distance; however, a duct will leak or allow energy to escape at the boundary, thus adding to the transmission losses so that field strengths are seldom greater than free space values [1].

There are typically two types of ducts: *ground based* and *elevated*. A ground based duct forms close to the earth's surface. Energy is propagated in this duct by being refracted back to the earth, reflected off the earth, then refracted again, as shown in Figure 1.8-5a. An elevated duct forms above the earth's surface and is generally very short lived. Energy in an elevated duct is refracted back and forth between boundaries without coming in contact with the earth, as shown in Figure 1.8-5b, similar to the way coherent light propagates in a graded index optical fiber. Shadow regions are formed along the area outside of a duct where, due to the nature of the duct, radio waves are not present. Receiving antennas placed in such a region will experience a loss of signal. These regions cannot only form above the earth's surface from a ground based duct but can also form along the earth's surface in the case of an elevated duct; therefore, a shadow region that can result in loss of communications can form at a receiver that is located relatively close to the transmitter. Radio waves that leave the transmitting antenna at an angle greater than a certain angle, the critical angle, will not become trapped in a duct. These radio waves will propagate through the boundary of the duct, although they will experience some bending due to the change in the index of refraction at the duct's boundary.

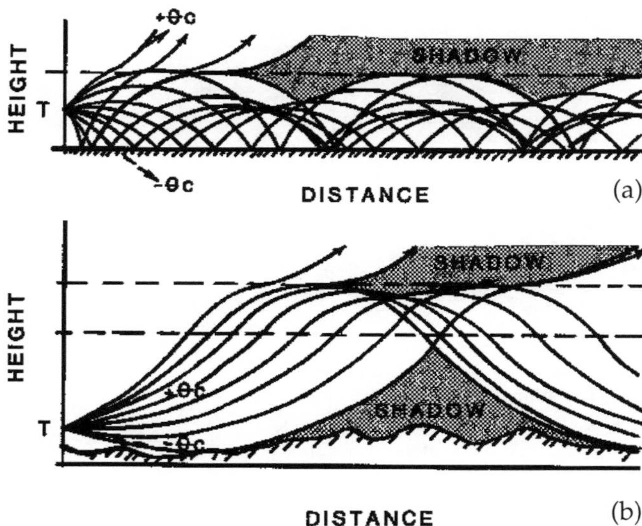

FIGURE 1.8-5 Ground based and elevated ducting.

FCC FM AND TV PREDICTION METHOD

The FCC method of determining FM and TV coverage involves calculating the average antenna height of a given transmission facility along a minimum of eight azimuths or radials (see Figure 1.8-6). For FM at least 50 terrain elevation points from 3 to 16 km from the transmitter must be evaluated on the 8 azimuth radials. TV uses the distance from 3.2 to 16.1 km.

Although topographic maps have been used to derive the elevation points, nearly all such work is performed by computers using either the National Geophysical Data Center's 30 arc-second digital terrain database or one of several available databases that offer higher resolution. Typically, when digital terrain elevation databases are in use the elevation points are spaced 0.1 km apart. The points are averaged to produce an average elevation for the radial. This figure is then subtracted from the antenna's center of radiation above mean sea level to determine the height above average terrain (HAAT) along the radial. For FM, the distance to a signal contour is calculated using the FCC's F(50,50) curves found under FCC Rules §73.333. The FCC has published two sets of curves: the F(50,50) for coverage and the F(50,10) for interference calculations. These curves are based on actual measurements with the receiving antenna at 9 m (30 ft) above ground. The F(50,50) curves show signal strength as a measure of distance under the statistical probability that the predicted signal will be at least 50% of the locations for at least 50% of the time. The interfering signal curves are based on a signal of certain strength for 50% of the locations for only 10% of the time; therefore, because the interfering signal

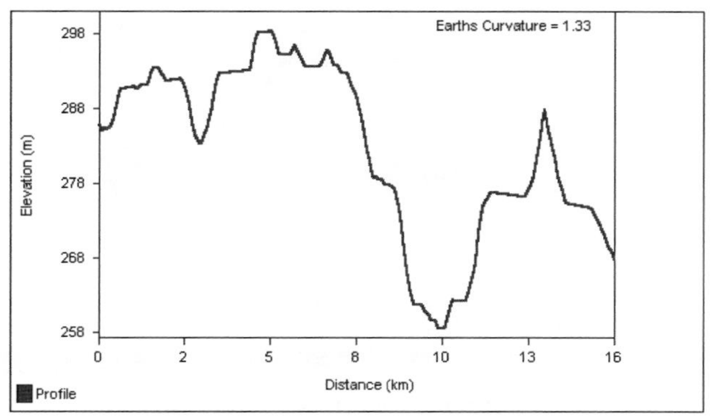

FIGURE 1.8-6 Terrain profile graph.

only has to be at a certain level for 10% of the time, it can also be said to be available at greater distances.

Six curves for analog TV can be found under FCC Rule §73.699.[1] These curves are divided into F(50,50) and F(50,10) curves for low VHF (channels 2–6), high VHF(channels 7–13), and UHF (channels 14–69). For digital TV, the coverage is predicted using F(50,90) calculations. In other words, the signal strength from a DTV station is present for at least 50% of the locations for at least 90% of the time. Section 73.625(b) describes how F(50,90) signal values can be calculated using the §73.699 F(50,50) and F(50,10) charts. Nearly all signal calculations are done with computers that use digitized versions of the Commission's curves.[2]

The FCC uses predictions in its various allocation schemes for radio and television broadcasting and for related auxiliary services. A protected station's signal strength at a given location is calculated and the interfering station's signal strength is also calculated at the same location. The undesired-to-desired (U/D) ratio is used to determine if the interfering signal and the protected signal meet the Commission's requirements.[3]

Although the U/D method has regular use for predicting the existence of interference at various points, for FM stations under short spacing, noncommercial educational (NCE) FM stations, FM translators, and analog low-power television (LPTV) and translators, the FCC typically uses a more rudimentary method of

[1]The FCC has recently placed new, easier to read FM and TV field strength curves on its Web site (http://www.fcc.gov/mb/audio/bickel/FM_TV_DTV_propagation_curves_graphs.html). For the first time, these curves include F(50,90) DTV curves that do not require use of the F(50,50) and F(50,10) curves to derive distance to contour values.

[2]The FCC offers a Web-enabled computer program that calculates FM and TV signals (http://www.fcc.gov/mb/audio/bickel/curves.html).

[3]The U/D ratios in use today for FM radio were originally established in 1947, when the FCC Laboratory Division conducted tests on FM radios and published reports on the characteristics of commercial FM broadcast receivers. Included in the project were tests concerning the interference rejection ratios on co-channel and adjacent channels. These measurements were the basis for the interference ratios used in the FM rules first adopted in 1951.

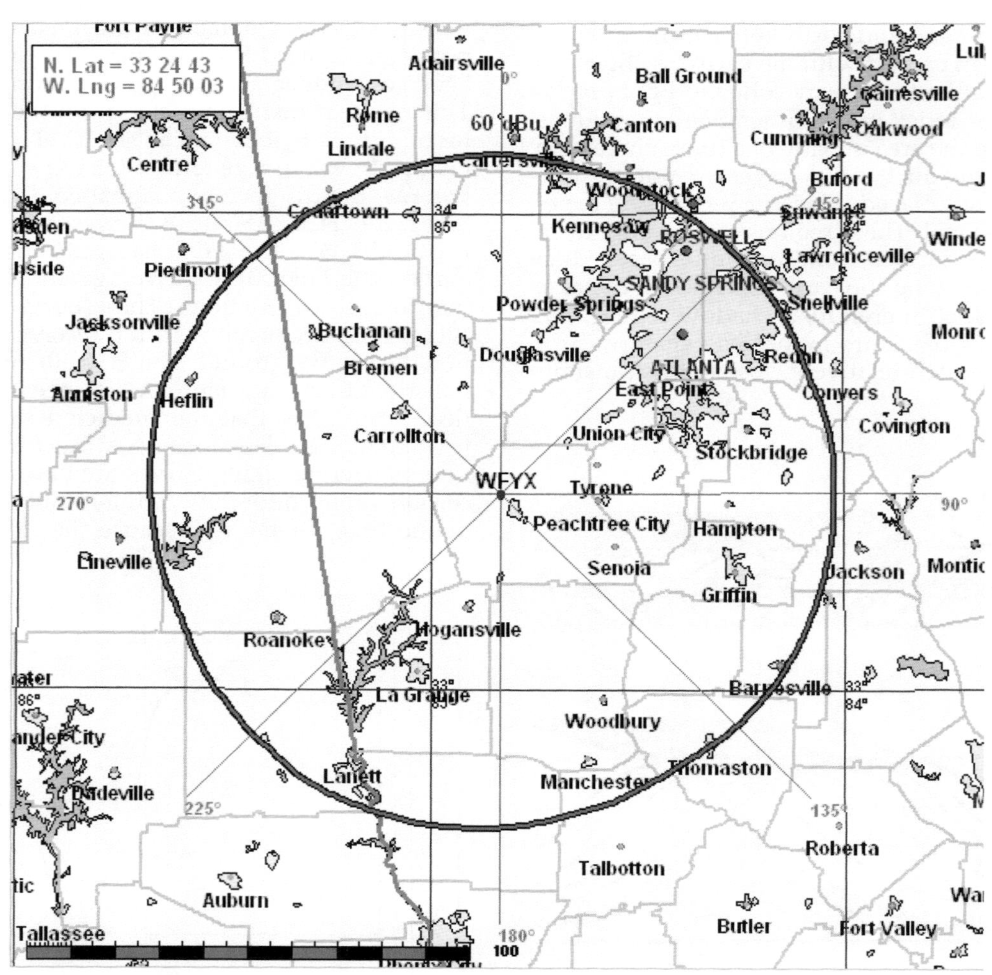

FIGURE 1.8-7 Typical 60 dBμV/m protected contour with cardinal radials. Refer to the CD for a color version of this figure.

overlapping contours. For FM class A, C3, C2, C1, C0, and C stations, the 60 dBμ F(50,50) signal contour is considered the protected contour (where dBμ is dB above 1 μV/m); see Figure 1.8-7. For class B1 it is 57 dBμ, and for class B it is 54 dBμ. In the case of co-channel stations, the applicable U/D ratio is –20 dB, so the 40 dBμ F(50,10) interference contour of an interfering station may not cross the F(50,50) 60 dBμ contour of a class A or one of the class C designated stations. The ratio for first adjacent stations is –6 dB, and for second- and third-adjacent stations the ratio is +40 dB.

For low-VHF, full service analog TV, the protected contour is the grade B contour at 47 dBμ; for high VHF, the 56 dBμ contour is protected; and, for UHF, the protected contour is 64 dBμ. LPTV or TV translators are protected to the 62, 68, and 74 dBμ contours for the three TV bands, respectively. DTV stations are protected within their noise-limited signal contours, which are 28 dBμ for low VHF, 36 dBμ for high VHF, and 41 dBμ for UHF.

PREDICTING AM COVERAGE

Although VHF and UHF frequency waves travel primarily via sky waves, standard band AM daytime propagation uses ground waves. The M3 map, shown in Figure 1.8-8, appears in §73.190 of the FCC's Rules. It shows the conductivity regions in various areas of the United States. A similar map for Region 2 defines the conductivity, in less detail, for the entire Western Hemisphere. Sections 73.183 and 73.184 refer to field strength curves that, when used with the M3 or R2 charts, will predict the distance to contours for AM stations during daylight hours.[4]

Nighttime propagation for standard band AM is calculated using a combination of the FCC's sky wave calculations and ground conductivity calculations. Class C stations use only ground wave for both day and night. Class A stations, sometimes called clear channels, are protected by U.S. stations on the co-channel to the 0.5 millivolt per meter (50 μV/m) ground wave contour. Stations operating at night also must protect the root sum square (RSS) limit, sometimes called *interference free*, signal contours of other stations. Calculation of nighttime interference-free service is accomplished by evaluating the signals on the co-channel and first-adjacent channel in order of decreasing magnitude by adding the squares of the values and extracting the square root of the sum, excluding those signals which are less than 50% of the RSS values of the higher signals already included.[5]

[4]Field strength curves for AM frequencies are available from the FCC's Web site (http://www.fcc.gov/mb/audio/73184/index.html).

[5]See FCC Rule §73.182(k).

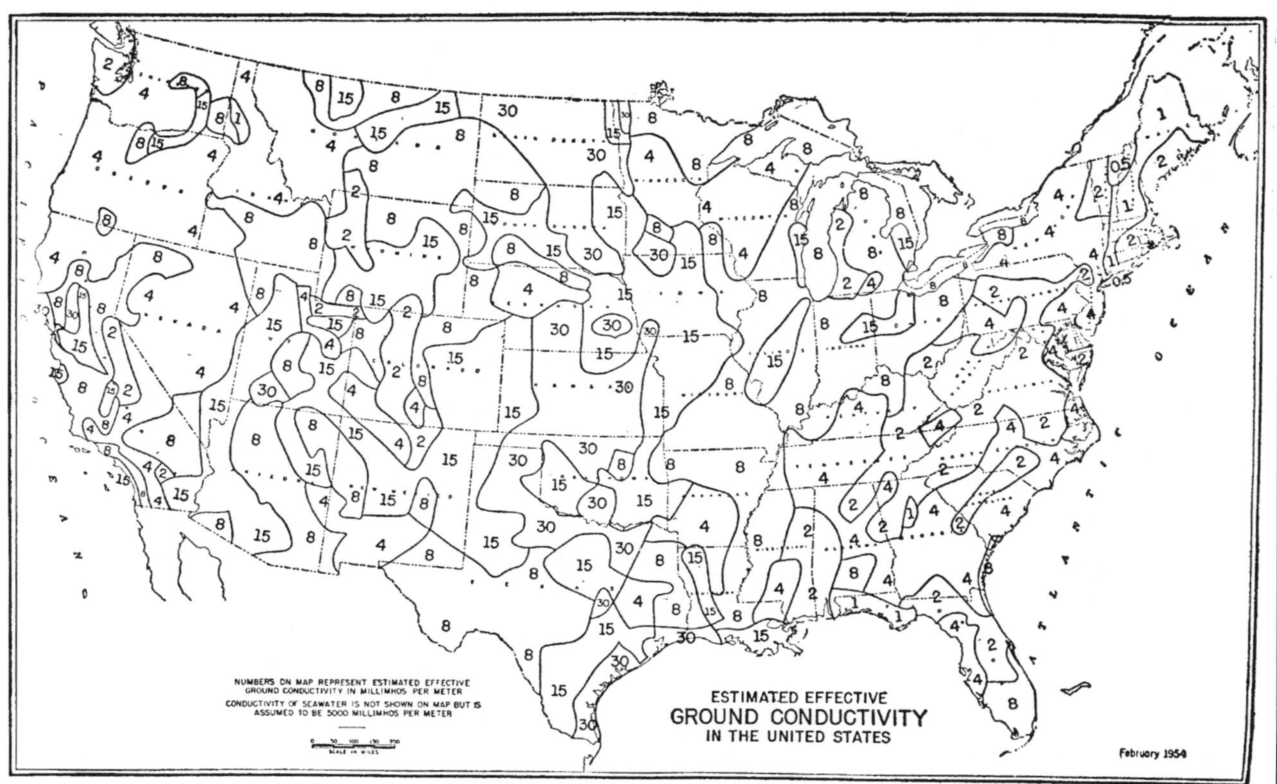

FIGURE 1.8-8 Ground conductivity map.

ALTERNATIVE PROPAGATION METHODS

Although the FCC's standard method has served the agency well over the years, it has its limitations. Because radial HAATs are evaluated from approximately 2 miles from the transmitter to 10 miles, the method does not account for mountains just beyond the 10 mile markers. Also, the FCC usually evaluates contour overlap using a 30 arc-second National Geophysical Data Center (NGDC) digital terrain set that was distilled from the original 3 arc-second database made by manually digitizing elevations from 1:250,000 topographic maps. Because these maps have minimal elevation contours when compared to 7-1/2 minute topographic maps, for example, errors such as actual elevation peaks are missing from the digitized product, resulting in overall lower average terrain elevations. A terrain resolution of 30 arc-seconds has an elevation point approximately every 3000 feet; consequently, because the average terrain is lower, this will result in an antenna that appears to be higher than actual. Thus, the FCC method, particularly when used with the low resolution 30 arc-second terrain elevation database, can often overpredict both coverage and interference distances. This may not be a bad thing, as the method's overpredictions will better protect stations from interference, which is the FCC's main goal.[6]

The Point-to-Point Propagation Method

In the 1998 Biennial Regulatory Review (Streamlining of Radio Technical Rules, in MM Docket No. 98-93, 98-117), the Commission proposed the point-to-point (PTP) method. Authored by Harry Wong of the FCC's Office and Engineering Technology, this method provided for an analysis of the entire path between the transmitter and receiver. The process is based on radio diffraction and attenuation of the free space path caused by irregular terrain entering the Fresnel zone. According to Wong, major determinants of this method include: "(1) the amount by which the direct ray clears terrain prominences or is blocked by them, (2) the position of terrain prominences along the path, (3) the strong influence of the degree of roundness of these terrain features, and (4) the apparent earth flattening due to atmospheric refraction." The original computer code for the PTP method used the 30 arc-second terrain elevation database and applied a static 5 dB of attenuation at points along the path to represent urban clutter. The Commission chose not to adopt this particular method but reported that it planned to do more work on the model, modifying it to use 3 arc-second terrain information and to provide for more flexible clutter calculations.[7]

[6]The FCC method can also underpredict certain paths where the base of an antenna is located on a hill or mountain that enters into the 3 to 16 km average elevation calculation.

[7]Second Report and Order: The 1998 Biennial Regulatory Review—Streamlining of Radio Technical Rules, in MM Docket No. 98-93, 98-117; parts 73 and 74 of the FCC's Rules.

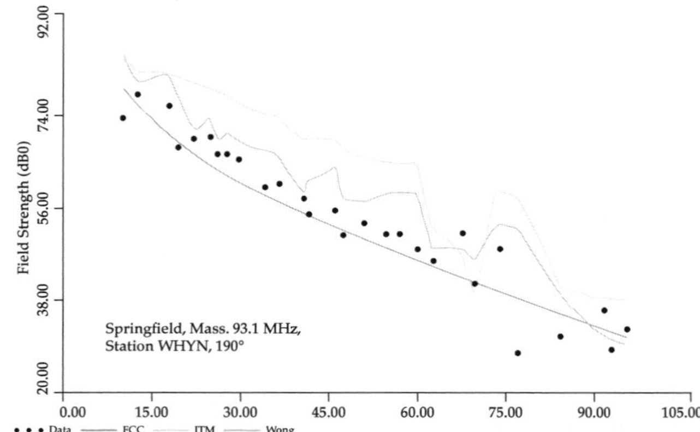

FIGURE 1.8-9 Various prediction models compared to measured data.

Wong updated his method in an abstract available from the FCC, dated November 1, 2002.[8] Here, he reports that: "Comparison with actual propagation measurements, and with the results of other prediction procedures, demonstrates that path loss values calculated by the PTP model are relatively accurate; moreover, the accuracy of the PTP model is as good or better than that achieved by alternative prediction procedures." A copy of the FORTRAN code developed for this method is available at the OET Web site (http://www.fcc.gov/oet/fm/ptp/). A graph from the Wong study is shown in Figure 1.8-9, which contrasts predictions using the FCC method, point-to-point, an irregular terrain model (ITM), and actual measurements (data points).

Longley–Rice Prediction Model

This propagation model is more commonly used to project coverage and interference relationships than the PTP method. In the mid-1960s, the National Bureau of Standards published Technical Note 101. P. L. Rice, A. G. Longley, A. Norton, and A. P. Barsis authored this two-volume propagation treatise in the course of their work at the Department of Commerce Institute for Telecommunications Sciences and Aeronomy in Boulder, CO. The concepts expressed in these documents were incorporated into a series of computer routines that came to be known as the "Longley–Rice model." This model was employed by the Commission to determine the digital television (DTV) allocation scheme. It has now become a standard alternative prediction method.

Going well beyond the FCC curves, the Longley–Rice method considers atmospheric absorption, including absorption by water vapor and oxygen as well as loss due to sky-noise temperature and attenuation caused by rain and clouds. It considers terrain rough-

[8]*Field Prediction in Irregular Terrain—The PTP Model*, by Harry Wong, FCC OET, November 1, 2002 (http://www.fcc.gov/oet/fm/ptp/report.pdf).

ness, knife-edge refraction (with and without ground reflections), loss due to isolated obstacles, diffraction, forward scatter, and long-term power fading. The following are the input parameters required of the user:

- Frequency (20–20,000 MHz)
- Transmitter antenna parameters (gain)
- Transmitter antenna height (above mean sea level; in meters)
- Transmitter antenna height (above ground; in meters)
- Transmitter power
- Transmitter antenna pattern.
- Receiver antenna height (above ground; in meters)
- System antenna polarization (vertical or horizontal)
- System ground conductivity (mhoS/m)
  - — 0.001 = Poor ground
  - — 0.005 = Average ground
  - — 0.020 = Good ground
  - — 5.000 = Sea water
  - — 0.010 = Fresh water
- System dielectric constant (permittivity)
  - — 4.0 = Poor ground
  - — 15.0 = Average ground
  - — 25.0 = Good ground
  - — 81.0 = Sea and fresh water
- System minimum monthly mean surface refractivity (adjusted to sea level):
  - — 200 to 450 (available from refractivity map; 301 N-units is default)
- Climate code:
  - — 1 = Equatorial
  - — 2 = Continental subtropical
  - — 3 = Maritime subtropical
  - — 4 = Desert
  - — 5 = Continental temperate (default for U.S. continent)
  - — 6 = Maritime temperate
  - — 7 = Maritime temperate overseas
- Probability factors:
  - — Q_t = Time variability—The percentage of time the actual path loss is equal to or less than the predicted path loss (standard broadcast coverage = 50%).
  - — Q_l = Location variability—The percentage of paths (all with similar characteristics) whose actual path loss is less than or equal to the predicted path loss (used with area mode only).
  - — Q_c = Prediction confidence or quality—The percentage of the measured data values the model is based on that are within the predicted path loss (standard broadcast = 50%; DTV = 90%).

Longley–Rice Computer Implementation

Because Longley–Rice evaluates the terrain along the entire path between the transmitter and receiver, a computer is essential to perform the large number of calculations required. The current Longley–Rice computer program used by the FCC is version 1.2.2.[9] In practice, Longley–Rice calculations are often used to evaluate signal strength within grid points. The grid point spacing can be set by the user. For display purposes, the signal strength value calculated for each grid point can then be coded either by color or by black and white line patterns. This procedure produces identifiable pools of coverage at certain signal values which makes it easy for a reader to understand how geography and terrain interact to affect signals. Further, population centroids with the grids can be interrogated to produce population totals and demographics within each selected signal value. An example of a coverage map using the Longley–Rice model with several signal level contours is shown in Figure 1.8-10.

OET Bulletin #69

The FCC's Office of Engineering Technology Bulletin #69 (OET 69) provides detailed information on using the Longley–Rice method to determine digital television coverage and interference.[10] The bulletin is divided into three parts: coverage or service calculation, interference calculations, and use of the Longley–Rice method. The Commission used OET 69 to analyze the service contours of existing analog TV stations in its allocation proceedings which resulted in the assignment of a second DTV channel having comparable coverage. Analog coverage was calculated within the grade B service area using the contour levels shown in Table 1.8-1. The Commission was able to approximately replicate analog coverage with digital coverage for the majority of the existing television stations. Table 1.8-2 defines the signal value at the noise-limited DTV coverage.

DTV coverage calculations were based on the DTV planning factors shown in Table 1.8-3. The planning factors are those assumed for home DTV receiving equipment including antenna systems. The values from Table 1.8-3 are calculated from the equation:

$$C/N = \text{Field} + K_d + K_a + G - L - N_t - N_s. \qquad (23)$$

Interference Calculations

The calculated service area is divided into square cells (typically 2 km on a side), and the Longley–Rice point-to-point propagation model is applied to a point in each cell to determine whether the predicted field

[9]Computer code for the Longley–Rice implementation can be found in an appendix of Report 82-100, A Guide to the Use of the I.T.S. Irregular Terrain Model in the Area Prediction Mode, by G. A. Hufford, A. G. Longley, and W. A. Kissack, U.S. Dept. of Commerce, April, 1982. A complete description of this model with downloadable FORTRAN computer code can be found at http://flattop.its.bldrdoc.gov/itm.html.

[10]Updated February 6, 2004; available at http://www.fcc.gov/Bureaus/Engineering_Technology/Documents/bulletins/oet69/oet69.pdf.

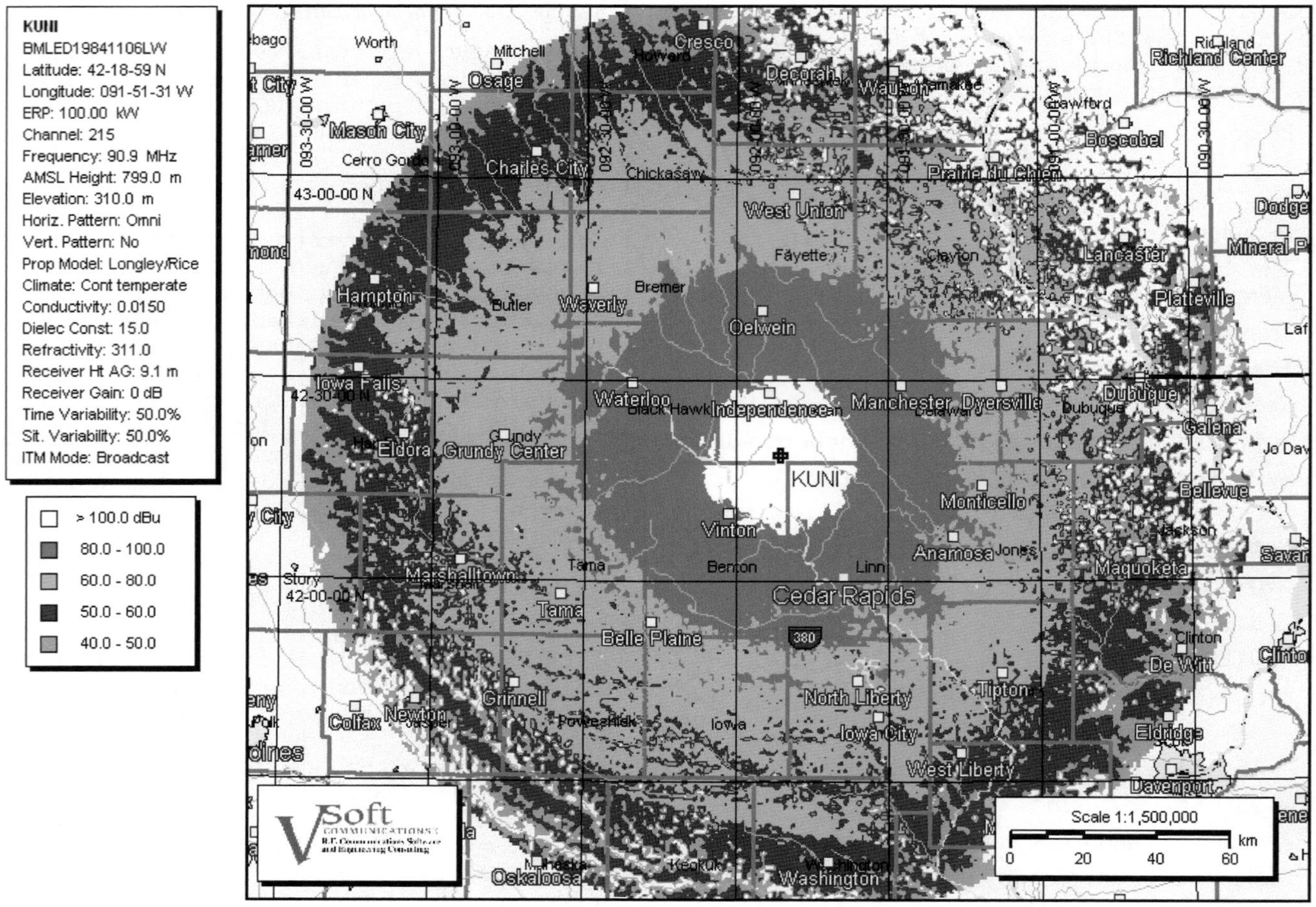

FIGURE 1.8-10 Longley-Rice coverage analysis. Refer to the CD for a color version of this figure.

TABLE 1.8-1
Field Strengths Defining the Area Subject to Calculation for Analog Stations

Channels	Defining Field Strength (dBμ/V) Predicted for 50% of Locations 50% of the Time
2–6	47
7–13	56
14–69	64–20log[615/(channel mid-frequency, in MHz)]

TABLE 1.8-2
Field Strengths Defining the Area Subject to Calculation for DTV Stations

Channels	Defining Field Strength (dBμ/V) Predicted for 50% of Locations 90% of the Time
2–6	28
7–13	36
14–69	41–20log[615/(channel mid-frequency, in MHz)]

strength is above the *threshold for reception,* which is the value shown in Table 1.8-1 or Table 1.8-2. For co-channel and adjacent-channel relationships, if the D/U ratio does not meet the minimum expressed in Table 1.8-4 the point is said to have interference. If the interference is masked at the point by another station's interference contour the interference is not counted. Tables for analog receiver intermediate frequency (IF) "taboo" protection and front-to-back antenna pattern discrimination are also found in the bulletin. Once the area of interference is determined, the population within the interference area is calculated using 1990 U.S. census population centroids.[11]

OET #69 Longley–Rice Implementation

The Commission has implemented the OET 69 method using FORTRAN code on its Sun Microsystem Enterprise 3500 and UltraSPARC computers. The FORTRAN code currently used by the Media Bureau is available for downloading on the FCC's Web site (http://www.fcc.gov/oet/dtv). The Commission

[11]The 1990 census continues to be used because it was the basis for the original service area calculations used by the Commission in the assignment of paired DTV channels.

TABLE 1.8-3
Planning Factors for DTV Reception

Planning Factor	Symbol	Low VHF	High VHF	UHF
Geometric mean frequency (MHz)	F	69	194	615
Dipole factor (dBm-dBµV/m)	K_d	–111.8	–120.8	–130.8
Dipole factor adjustment	K_a	None	None	See Table 1.8-2
Thermal noise (dBm)	N_t	–106.2	–106.2	–106.2
Antenna gain (dBd)	G	4	6	10
Downlead line loss (dB)	L	1	2	4
System noise figure (dB)	N_S	10	10	7
Required carrier-to-noise ratio (dB)	*C/N*	15	15	15

warns that: "The individual installing it should have computer programming skills and experience as a system administrator on the system on which it is being installed because linking the data files, which occupy 1.6 gigabytes of disk space, will be a site-specific task." What the Commission leaves unsaid is that to accurately replicate the program's answers requires an identical computer CPU. Because of rounding differences that occur in the processors, implementations on other systems, whether using FORTRAN or a substitute program language, may not deliver the required accuracy.

The FCC's Rules allow a DTV station to cause up to 2% interference to the population of a given station as long as this interference does not result in more than 10% interference from all interfering stations in total. LPTV stations may cause up to 0.5% to other LPTV stations, TV translators, and full service TV stations. DTV channel assignments that did not fully replicate the analog TV service area population are to be considered at their 10% maximums.

Individual Location Longley–Rice Model

In 1999, Congress enacted the Satellite Home Viewer Improvement Act (SHVIA). This legislation instructed the Commission to "develop and prescribe by rule a point-to-point predictive model for reliably and presumptively determining the ability of individual locations to receive signals in accordance to the signal intensity standard in effect under 119(d) (10) A of Title 17 (United States Code)" (47 USC §339(c)(3)). Section 339(c)(3) of the Communications Act provides that "[i]n prescribing such model, the Commission shall rely on the Individual Location Longley–Rice (ILLR) model set forth by the Federal Communications Commission in Docket 98-201 and ensure that such model takes into consideration terrain, building structures, and other land cover variations." The ILLR is used to determine whether a given viewer is within the qualifying signal strength of a local TV station. The presence of terrain impediments, man-made structures, and foliage in the radio path tends to reduce the strength of received signals. If the test determines that the viewer is not able to adequately receive a local station, the viewer is allowed to receive by satellite a more distant station having the same network.[12] The ILLR method uses Longley–Rice analysis in the individual point-to-point mode and then augments the results by considering land use and land cover (LULC) clutter losses.[13]

TABLE 1.8-4
Interference Criteria for Co- and Adjacent Channels

	D/U Ratio (dB)			
Channel Offset	Analog into Analog	DTV into Analog	Analog into DTV	DTV into DTV
–1 (lower adjacent)	–3	–14	–48	–28
0 (co-channel)	+28	+34	+2	+15
+1 (upper adjacent)	–13	–17	–49	–26

TIREM

The Terrain Integrated Rough Earth Model (TIREM) is licensed by Alion Science and Technology Corporation, Annapolis, MD. This model started with a Tech Note 101 base but has been modified over time to make up for believed inaccuracies in the Longley–Rice model. TIREM predicts median propagation loss from 1 MHz to 40 GHz. The techniques used to calculate these losses include:

- Free space spreading
- Reflection
- Diffraction
- Surface wave

[12]Refer to http://www.fcc.gov/oet/info/documents/bulletins/#72.

[13]The United States Geological Survey (USGS) maintains a database on land use and land cover, often called the LULC database: http://edc.usgs.gov/products/landcover/lulc.html.

- Tropospheric scatter
- Atmospheric absorption

In contrast to Longley–Rice, TIREM has built-in routines for evaluating radio paths over sea water. TIREM is used in numerous modeling and simulation tools at the Department of Defense. Because TIREM is a proprietary model, it is not possible to determine exactly what processing takes place, which makes the model less attractive to the FCC and other users.

TERRAIN DATABASES

While there are propagation models that calculate in the area mode, the computer models used for broadcasting depend on their links to digital terrain elevation databases. The accuracy and resolution of these databases are important to prediction accuracy. The digital terrain elevation databases discussed below are not meant to be exclusive with regard to propagation analysis systems. There are others in use; however, those listed below are the most popular implementations.

USGS 3 Arc-Second and 30 Arc-Second Databases

Most databases in use today have some degree of inaccuracy. The 30 arc-second database was derived from the original 3 arc-second U.S. Geological Survey (USGS) database, which was digitized from 1:250,000 scale maps. Because each second of latitude approximates 100 feet, the 30 arc-second terrain elevation database will have an elevation point every 3000 feet. The 3 arc-second database will have a point every 300 feet. Both databases are believed to have a number of errors such as mountain peaks being off as much as 15 seconds.

National Elevation Datum

In 2004, the USGS released the National Elevation Datum (NED) dataset.[14] This dataset was developed by merging the highest resolution, best quality elevation data available across the United States into a seamless raster format. NED is the result of maturation of the USGS effort to provide 1:24,000-scale Digital Elevation Model (DEM) data for the conterminous United States and 1:63,360-scale DEM data for Alaska. The dataset provides seamless coverage of the United States, including Hawaii, Alaska, Puerto Rico, and the Caribbean. NED has consistent projection (geographic), resolution (1 arc-second), and elevation (meter) units. The horizontal datum is NAD83, except for Alaska, which is NAD27. The vertical datum is NAVD88, except for Alaska, which is NAVD29. NED is a living dataset that is updated bimonthly to incorporate the best available DEM data. As more 1/3 arc-second (10 ft) data covers the United States, then this will also become a seamless dataset.

Shuttle Radar Topography Mission

The newest database is from the Shuttle Radar Topography Mission (SRTM).[15] In this mission, NASA obtained elevation data on a near-global scale to generate the most complete high-resolution digital topographic database of the earth. SRTM consisted of a specially modified radar system that flew onboard the Space Shuttle Endeavour during an 11-day mission in February of 2000. SRTM is an international project managed by the National Geospatial-Intelligence Agency (NGA) and the National Aeronautics and Space Administration (NASA). SRTM elevation data is available for the entire populated world. The database has unparalleled accuracy and contains buildings in its scans; however, there are some holes in the data caused by cloud cover at the time of the data capture. Use of this database requires interpolation of the measured points around some of the holes or a fallback to an existing database of lesser resolution at the few places where the holes were found to be large.

POPULATION DATABASES

All professional propagation software today has an associated population database. The database in common use is the U.S. Census Bureau's Summary File 1 (SF1).[16] This database contains 286 detailed tables focusing on population demographics such as age, race, sex, households, families, and housing units. SF1 presents data for the United States, the 50 states, and the District of Columbia in a hierarchical sequence down to the block level for many tabulations, but only to the census tract level for others. Typically, each block contains an associated latitude and longitude called a *centroid*. The usual method of counting these centroids is that, if a given signal contour includes the centroid, then the point is counted. The centroid is not counted if it is outside the signal contour, even though the population block may contain people on the edges.

PROPAGATION PREDICTION PROGRAMS

For personal computers, several excellent commercial programs are available that predict broadcast propagation and perform allocation studies. These programs seamlessly integrate geographic mapping with digital terrain and population databases. While price may be a significant obstacle for some users, others will find these programs valuable, if not essential, in the support of broadcast radio and television coverage analysis. Not all programs are the same in user conve-

[14]Additional information on the NED database can be found at http://ned.usgs.gov/.

[15]Refer to http://www2.jpl.nasa.gov/srtm/.

[16]Refer to http://www.census.gov/Press-Release/www/2001/sumfile1.html.

nience, accuracy, and presentation, so it is necessary to evaluate and take advantage of demonstrations offered by the software companies. The programs with the best price may not offer the best features, map clarity, or desired level of accuracy.[17]

Online Services

Online services offer the opportunity to use propagation or allocation systems on the pay-per-use method. If the cost of software is a concern, running studies on a cost basis may be an answer. However, note that, if numerous studies are to be run, online studies may end up costing more in the long run, particularly if the studies have to be repeated due to input errors. In general, online systems available today do not offer the higher resolution mapping available in some of the offline commercial propagation prediction PC programs that are available.[18]

SUMMARY

Propagation characteristics vary greatly with frequency. Broadcasters operate from the AM band through the UHF television band. The ability to predict where and how a station's signal arrives at the receiver is a key factor to delivering the broadcaster's product. Many advances in methodology and technology have brought tools that can predict real world coverage with great accuracy. Using these tools effectively to select an optimum transmitter site or directional antenna can ensure the best coverage and revenue generation.

ACKNOWLEDGMENT

Contributions were made by Martin H. Barringer, Frederick G. Griffin, Thomas Osenkowsky, and Kenneth D. Springer.

[17]Web addresses of some broadcast propagation computer program suppliers include www.radiosoft.com, www.rfsoftware.com/rfi/, www.softwright.com, and www.v-soft.com.

[18]Online systems include Dataworld® (http://dataworld.com/) and the Institute for Telecommunications Analysis, TA Services (http://tas.its.bldrdoc.gov/).

References

[1] Hall, M., *Effects of the Troposphere on Radio Communications*, P. Peregrinus, New York, 1979.

[2] Okumura, Y. et al., Field strength and its variability in VHF and UHF land-mobile radio service, *Rev. Elec. Comm. Lab.*, 16(9/10), 825–873, 1968.

Bibliography

Jordan, E. C., *Electromagnetic Waves and Radiating Systems*, Prentice Hall, Englewood Cliffs, NJ, 1950, pp. 608–688.

Norton, K. A., Ground wave intensity over a finitely conducting spherical earth, in *Proc. of IRE*, December, 1941, p. 623.

Bullington, K., Radio propagation variations at VHF and UHF, in *Proc. of IRE*, January, 1950, p. 27.

ARRL, Radio wave propagation, in *The ARRL Antenna Book*, 19th ed., American Radio Relay League, Newington, CT, 2000, chap. 23.

CHAPTER

1.9

NRSC Analog and Digital Radio Standards

STANLEY SALEK

Hammett & Edison, Inc. Consulting Engineers
San Francisco, CA

INTRODUCTION

The National Radio Systems Committee (NRSC)[1] is jointly sponsored by the National Association of Broadcasters (NAB)[2] and the Consumer Electronics Association (CEA).[3] Its purpose is to study and make recommendations for technical standards that relate to radio broadcasting and the reception of radio broadcast signals. The NRSC is a vehicle by which broadcasters and receiver manufacturers work together toward solutions to common issues and problems in radio broadcast systems.

Since beginning its active standards setting activities in the mid-1980s, the NRSC has produced five voluntary industry standards:

- Three are related to the transmission and reception of analog AM radio signals in the medium-wave broadcast band (535 to 1705 kHz).
- One documents the Radio Data System (RDS) used in the FM broadcast band (88 to 108 MHz).
- One covers in-band, on-channel (IBOC) digital radio broadcasting, used in both bands.

These standards are available at no charge on the NRSC Web site (www.nrscstandards.org/standards.asp).

[1]See http://www.nrscstandards.org for committee rules, membership information, a list of committees, and access to the standards.
[2]See http://www.nab.org.
[3]See http://www.ce.org.

AM RADIO BROADCASTING STANDARDS

The three NRSC AM radio broadcasting standards released between 1987 and 1990 cover recommended broadcast audio parameters, transmission systems, and receiver specifications, respectively. In the late 1980s, the Federal Communications Commission (FCC) incorporated portions of the first two standards into its Part 73 broadcast rules,[4] effectively making them mandatory for U.S. broadcasters.

NRSC-1

The NRSC AM Preemphasis/Deemphasis and Broadcast Audio Transmission Bandwidth Specifications (NRSC-1), adopted on January 10, 1987, specify the preemphasis audio characteristic of AM broadcasts, the complementary deemphasis characteristic contained in AM broadcast receivers, and a 10 kHz audio bandwidth limitation prior to modulation. The standard applies to both monophonic and the (L+R) component of AM stereophonic transmissions, as well as to both single- and multiple-bandwidth receivers. The purpose of the standard is to create a transmission and reception system that allows AM broadcast stations to know, with a degree of certainty, the audio response characteristics of receivers and receiver manufacturers to know, with a similar degree of certainty, the likely audio response characteristics of AM broadcasts.

[4]Title 47 of the Code of Federal Regulations, Part 73, §73.44.

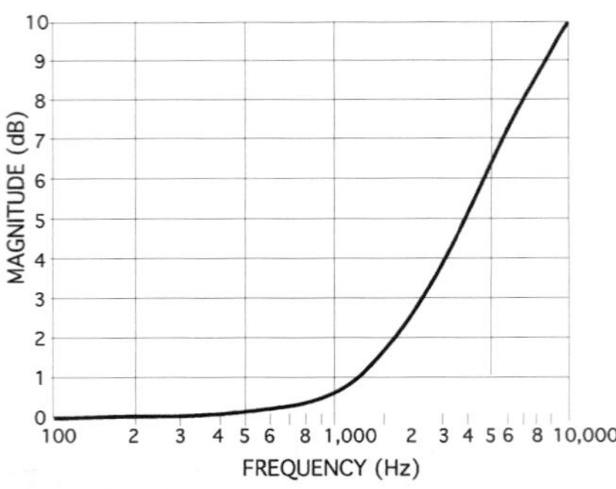

FIGURE 1.9-1 Modified 75 μsec AM standard preemphasis curve of NRSC-1.

Audio preemphasis is incorporated at the transmission end to boost high audio frequencies. The characteristic curve shown in Figure 1.9-1 is known as "modified 75 μsec" preemphasis, because its mathematical description (or transfer function) consists of a single zero at 2122 Hz, constituting 75 μsec preemphasis (as used in analog FM broadcasting), followed by a single pole at 8700 Hz, which limits the maximum boost to 10 dB at 10 kHz.

In the receiver, a complementary deemphasis curve is specified, employing a maximum of 10 dB attenuation at 10 kHz, theoretically resulting in a flat end-to-end response; however, most AM receivers recover much less than 10 kHz audio bandwidth, so the overall receiver response must be considered to determine the degree of deemphasis, if any, to employ that will best complement the preemphasis characteristic. For receivers that do have usable response out to 10 kHz, NRSC-1 recommends that notch filters be employed to attenuate the 10 kHz beat ("whistle") caused by first-adjacent-channel AM stations. Figure 1.9-2 illustrates the 10 kHz audio input spectrum limitation to the AM transmitter. The slopes and step in the response are included to accommodate the capabilities of practical analog low-pass filters and limiters available at the time of standard adoption.

NRSC-1 also includes sections on test signals and methods for determining compliance with the standard. Although normal program material can be used in conjunction with an audio spectrum analyzer or fast Fourier transform (FFT) analyzer operating in peak-hold mode, a standard test signal, known as pulsed-USASI noise, is also defined. Use of the standard test signal allows for easier comparison of compliant equipment from different manufacturers.

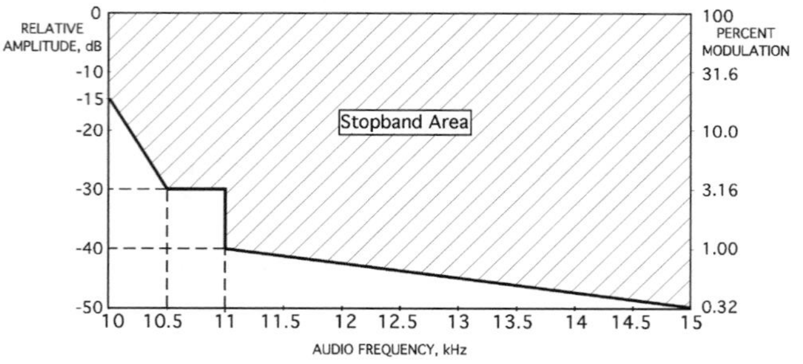

FIGURE 1.9-2 Audio stopband specification of NRSC-1.

NRSC-2

The NRSC Emission Limitation for AM Broadcast Transmission (NRSC-2) became effective on June 1, 1988. It builds upon the NRSC-1 transmitter audio input specifications, taking into account the AM transmission process, to provide for a maximum occupied bandwidth limit, as shown in Figure 1.9-3. The steps and slopes in the emission mask were designed to accommodate the capabilities of transmitters in existence at the time of standard adoption. The solid-line border represents the maximum limits of the standard, and the dashed-line border represents test limits, which are intended to serve as a guide for transmitter manufacturers using the NRSC-1 standard test signal. While the ultimate attenuation is 80 dB or greater below carrier reference level beyond ±75 kHz from the carrier frequency, a relaxation allowing lesser attenuation is incorporated for stations employing less than 5 kW carrier power.

Field measurement of mask compliance is made using a radiofrequency (RF) spectrum analyzer or other device that can effectively monitor emissions exceeding the RF mask. To accomplish this task, the RF spectrum analyzer should be set for 300 Hz resolution bandwidth in peak-hold mode for 10 minutes or more, referenced to the peak carrier level. Using normal program modulation, the system is in compliance

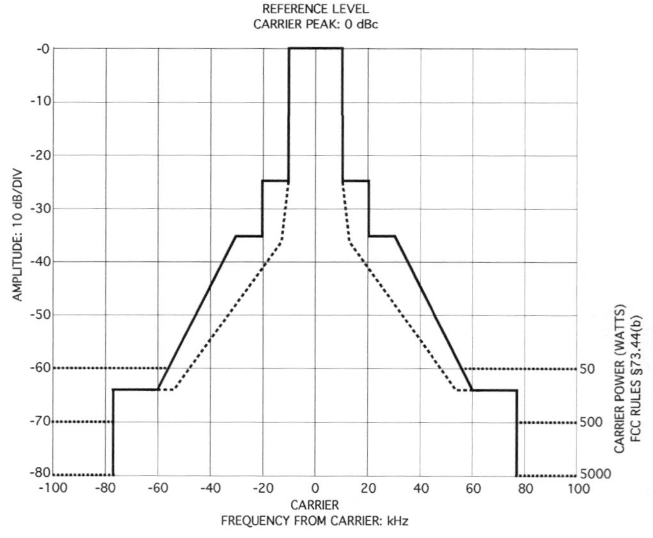

FIGURE 1.9-3 AM broadcast radiofrequency emission limits of NRSC-2.

if all stored spectrum products are contained within the mask area.

NRSC-3

The NRSC Audio Bandwidth and Distortion Recommendations for AM Broadcast Receivers (NRSC-3) became effective on October 15, 1990. The purpose of NRSC-3 is to provide specific performance goals for AM broadcast receivers, building upon the receiver elements of NRSC-1. Figure 1.9-4 illustrates the frequency response recommendation, which includes use of the NRSC-1 preemphasis characteristic at the test RF generator, resulting in a flat overall response. With modulation of an RF signal generator set to 15% at a 400 Hz reference frequency, the receiver under test must exhibit a frequency response within +1.5/–3.0 dB, from 50 Hz to 7.5 kHz. (NRSC-3 also includes receiver frequency response specifications when using an RF test signal generator that does not employ the NRSC-1 preemphasis characteristic.) The standard also recommends that attenuation at 10 kHz be at least 20 dB below the 400 Hz reference level for an RF test generator employing NRSC-1 preemphasis, or at least 30 dB below the reference level for use of a test RF generator without NRSC preemphasis.

The NRSC-3 standard specifies how RF energy is coupled into the receiver under test and how the audio output signal is to be obtained. Under the appropriate conditions for the specific receiver under test, total harmonic distortion plus noise (THD+N) must not exceed 2.0% at measurement frequencies between 50 Hz and 7.5 kHz for a modulation level of 80%, *without* the use of preemphasis at the test RF generator. NRSC-3 further specifies that all tests be conducted at carrier frequencies of 600, 1000, and 1400 kHz.

In 1992, NRSC sponsoring organizations NAB and CEA introduced AMAX, a certification program for AM radio receivers that met the technical specifications of NRSC-3 and that also exhibited other desirable characteristics, including adjustable reception bandwidth and the availability of an external antenna connection. Further, receivers meeting all of these conditions that also had stereo reception capability could use the designation "AMAX Stereo." Automotive receivers were granted limited relief to the bandwidth requirement, such that radios exhibiting at least a 6.5 kHz bandwidth could still receive certification.

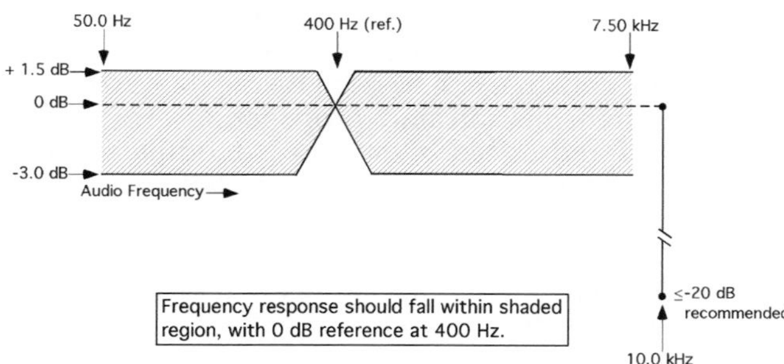

FIGURE 1.9-4 Receiver demodulator output response, including NRSC generator preemphasis, should fall within the range shown.

NRSC-4-A

The NRSC United States Radio Broadcast Data System (RBDS) Standard (NRSC-4) was released by the NRSC's RBDS Subcommittee in 1993. It was updated in 1998 and then again in 2005 when it was redesignated NRSC-4-A. The standard describes the Radio Data System (RDS) and is based on and essentially compatible with the European Broadcast Union (EBU)/Cenelec Standard EN50067:1998 (Specifications of the Radio Data System), with added features specifically intended for U.S. broadcasters.[5] The added features include North American program information and program type codes, a specification for an in-receiver database system, Emergency Alert System (EAS) integration, a specification for compatible radio paging, and a placeholder for a future AM broadcast-band RBDS specification. Note that although the U.S. version is the RBDS standard, all hardware implementations of this technology are referred to as RDS, and the logos that appear on consumer equipment are RDS logos. See Chapter 4.14 of this handbook for a more detailed description of RDS and NRSC-4-A.

The primary objective of the RDS system is to improve the functionality and user-friendliness of FM broadcast receivers through an added transmitted data stream. That data stream allows a compatible receiver to provide a visual display of the program service (station call sign or identifying name), potentially including a real-time updated display of the current program or song title being broadcast. RDS also includes a switching capability to allow a receiver to tune to the strongest signal offering the same program; this function is most commonly used in conjunction with nationalized networks outside the United States. A secondary objective of the NRSC-4-A standard is to provide for harmonization with similar program data streams in IBOC digital radio, such that digital–analog and analog–digital blends will maintain the appropriate receiver display and functionality. This aspect of the standard is covered in Appendix U of the standard, which was incorporated in 2005 with adoption of the "-A" version.

RDS information is transmitted on a 57 kHz subcarrier contained within the transmitted FM baseband that also carries stereophonic and possibly other subcarriers. The RDS subcarrier is locked to the 19 kHz stereo pilot signal and uses biphase suppressed-carrier modulation at 1187.5 bits per second. Figure 1.9-5 is a simplified block diagram of an RDS encoder, and Fig-

[5]In 2000, the Cenelec standard was retired and superceded by an International Electrotechnical Commission (IEC) standard, IEC 62106 (Specification of the Radio Data System [RDS] for VHF/FM Sound Broadcasting in the Frequency Range from 87.5 to 108.0 MHz), which is essentially an updated version (with only minor changes) of EN50067:1998.

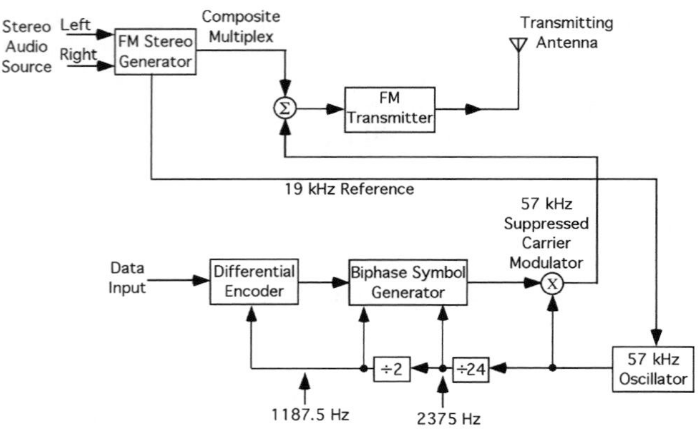

FIGURE 1.9-5 Simplified block diagram of RDS encoder subsystem.

ure 1.9-6 provides the complementary block diagram of an RDS decoder. Prior to modulation, the RDS data stream is formatted in accordance with a specified baseband coding structure. The largest element in the structure is called a *group*, which contains 104 bits of information. Each group is further broken down into *blocks* of 26 bits each, consisting of a 16 bit data word and a 10 bit *checkword*. Information such as program information (PI) and program type (PTY) codes, as well as real-time clock updates and potential paging, traffic, and emergency information, are formatted into the data stream. Encoders and receivers built to the NRSC-4-A standard may display the RDS logo, two variations of which are shown in Figure 1.9-7.

NRSC-5-A

The complex NRSC In-Band/On-Channel (IBOC) Digital Radio Broadcasting Standard (NRSC-5-A), released in 2005,[6] describes an IBOC digital radio broadcasting system developed by iBiquity Digital Corporation. The system is designed to permit a

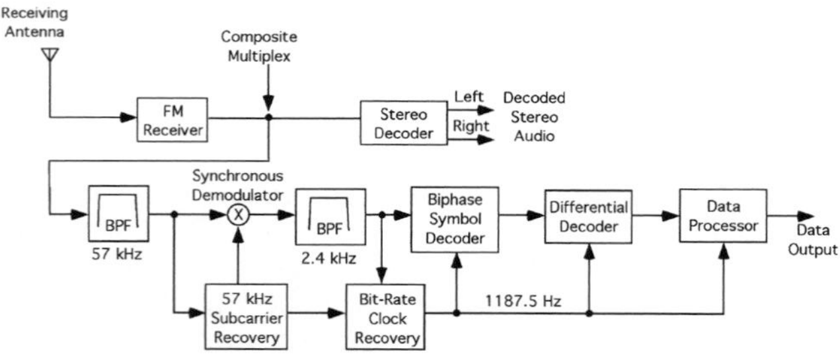

FIGURE 1.9-6 Simplified block diagram of RDS decoder subsystem.

FIGURE 1.9-7 RDS logos for use on RBDS-compatible equipment.

smooth evolution from current analog radio broadcasting to fully digital radio broadcasting, within the existing AM and FM terrestrial radiofrequency bands and utilizing the current FCC allocation system and channel assignments.[7] The system delivers digital audio and data services to mobile, portable, and fixed receivers from terrestrial transmitters on existing AM and FM broadcast channels. In the hybrid mode, broadcasters may continue to broadcast analog signals simultaneously with the IBOC digital signals, providing a means for listeners to make the transition from analog to digital radio. A more spectrally efficient all-digital mode, which is fully supported in hybrid-capable IBOC receivers, can be used in a future post-transition environment, after cessation of analog broadcasting. See Chapter 4.13 of this handbook for a more detailed description of NRSC-5-A and radio systems built to this standard.

The system accepts as input compressed digital audio bit streams and ancillary data information (which may or may not be related to the audio) and utilizes baseband signal processing techniques such as interleaving and forward error correction to increase the robustness of the signal in the transmission channel. These methods allow audio and ancillary data signals to be transmitted using power levels and spectral band segments selected to minimize interference to existing analog signals.

The block diagram of Figure 1.9-8 illustrates the three major subsystems specified by NRSC-5-A and how they relate to one another. The *RF/transmission subsystem* is an AM or FM broadcast facility that is compliant with references specified in the NRSC-5-A standard. This subsystem takes the multiplexed bit stream and applies coding and interleaving to the transmitted data that can be used by the IBOC receiver to reduce transmission and propagation errors. The multiplexed and coded bit stream is modulated onto orthogonal frequency-division multiplexed (OFDM) subcarriers and up-converted to the AM or FM broadcast channel.

The *transport and service multiplex subsystem* feeds the information to be transmitted to the RF/transmis-

[6]Two versions of the standard were released in 2005; the later NRSC-5-A standard added information relating to ADS that was not in the first version.

[7]IBiquity markets their technology under the trade name of "HD Radio." Where the term "HD Radio" is used within the NRSC-5-A standard, it is interpreted to reflect the generic term "IBOC" for an NRSC-5-A-compliant system.

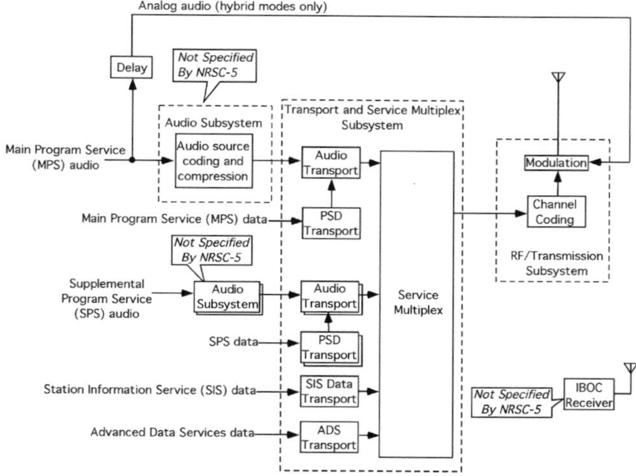

FIGURE 1.9-8 Overview of IBOC digital radio system.

sion subsystem. It takes the audio and data information, organizes it into packets, and multiplexes the packets into a single data stream. Each packet is uniquely identified as containing audio or data. Data packets containing program service data are added to the stream of packets carrying the associated audio information before they are fed into the multiplexer.

The *audio and data input subsystem* accepts coded and compressed digital audio, program service data (PSD), and advanced data services (ADS). Audio data consists of main program service (MPS) and supplemental program service (SPS) streams (also referred to as *multicast* streams), all of which have their own source coding, compression, and transport subsystems. NRSC-5-A does not include specifications for audio source coding and compression. MPS analog audio is delayed and modulated directly onto the AM or FM radiofrequency carrier portion of the IBOC signal to facilitate reception by analog-only receivers. The delay is set such that digital and analog information is decoded by IBOC receivers in a time-aligned fashion, enabling seamless transitions from digital to analog reception when the received signal quality is not sufficient for digital audio reception. This blend capability is also used for fast channel changes, allowing the IBOC receiver to demodulate the analog audio first and then blend into the digital audio stream. SPS streams do not have an analog backup.

Program service data is intended to describe or complement the audio program heard by the radio listener. Station information service (SIS) data provides more general information about the station's programming, as well as some technical information that is useful for non-program-related applications. ADS gives broadcasters the ability to transmit information that may be unrelated to the other audio or data streams and may be any content that can be expressed as a data file or data stream, including audio services.

The NRSC-5-A standard, along with ten incorporated reference documents, more fully describes each of these subsystems and features. Included are graphs and specifications for AM and FM spectral emission limits, which are fully compliant with current FCC requirements for AM and FM analog stations.

SUMMARY

The NRSC provides a forum for the broadcasting and receiver manufacturing industries to work together toward solutions to common issues and problems in radio broadcast systems. As the radio industry transitions to digital broadcasting, the NRSC stands ready to develop standards and recommended practices which will benefit the industry and the listening public.

CHAPTER

1.10

Worldwide Standards for Digital Radio

RICHARD MORRIS B.SC. (ENG), (HONS), M.I.E.T. (U.K.)

Commercial Radio Australia
Australia

INTRODUCTION

This chapter explores the major digital radio broadcast standards being developed and/or in use outside the U.S., and also provides a brief introduction to satellite radio. In the digital world, the division between television and radio services is becoming increasingly blurred. Systems primarily designed for television can in addition deliver audio services, and some systems originally intended for radio can deliver low–bit rate video services. The terrestrial systems discussed here are restricted to those with a spectrum bandwidth of below 1600 kHz that have been primarily designed to deliver audio services. Most of the systems discussed here are open standards, with the exception of the satellite systems. Only a brief introduction to the proprietary satellite systems is included as full details of their specifications are not publicly available.

COMMON ELEMENTS OF DIGITAL RADIO SYSTEMS

Audio Coding

All digital radio broadcast systems use powerful audio compression—the most common codecs in use are introduced here. For a more detailed description of audio coding methods refer to Chapter 3.7, Digital Audio Data Compression.

The MPEG-1 layer II codec, commonly known as "Musicam," is the main audio codec specified for the Eureka-147 Digital Audio Broadcast system (discussed in detail below). It is generally accepted that Musicam requires at least 224 kbps to have "near CD quality," 192 kbps can provide a quality equivalent to or better than analog FM stereo, and 128 kbps is considered the minimum bandwidth required for broadcasting stereo music. Musicam was developed in the early 1990s, and more efficient codecs are widely available today.

The Advanced Audio Coding (AAC) codec was developed by Fraunhoffer IIS in the 1990s, standardized as ISO/IEC 13818-7, and was added to the MPEG-2 family of standards in 1997. Various optional enhancements have been added to the original AAC specification. The main enhancements are spectral band replication (SBR) and parametric stereo (PS).

SBR systems apply a low pass filter to the source audio and encode this lower-frequency portion in the usual way. This band limited audio results in a lower–bit rate stream when compressed, but at the expense of higher-frequency components, which are not encoded. SBR systems counter this loss of high-frequency information by sending a very low–bit rate side channel that allows the receiver to synthesize higher frequencies when the audio is reconstituted. These low–bit rate channels work on the principle that the higher-frequency information can be roughly approximated by transmitting harmonics of the lower-frequency portion (e.g., in a piano note) or a noiselike signal (e.g., in clashing cymbals). The resultant audio generated at the receiver is not a true copy of the original, but sounds similar to a natural, full bandwidth sound.

PS is a system where only a mono audio signal is sent, along with a very small amount of side information, allowing the receiver to synthesize a stereo "ambience" at the end of the broadcast chain. Neither

SBR nor PS is intended to produce faithful, CD-quality representations of the original. The aim is to produce a low–bit rate audio signal with minimal audio artifacts that sounds pleasing to the ear. AAC with SBR is known as aacPlus V1, and AAC with PS and SBR is known as aacPlus V2. AAC with these enhancements has been added to the MPEG-4 audio codecs standard (ISO/IEC 14496-3:2005). Simple AAC is known as AAC LC (low-complexity AAC). aacPlus is described in MPEG-4 terms as AAC HE (high-efficiency AAC).

The codecs discussed so far are widely used in the broadcast industry to transmit speech and music programs. For low-quality, speech-only services, harmonic vector excitation (HVXC) and code-excited linear predictive (CELP) coders can be used. These are designed to carry speech only, and at very low bit rates. HVXC can reproduce electronic sounding but intelligible speech at 4 kbps, and CELP can reproduce better quality speech at 8 kbps. These coders are only designed to handle mono speech, so they are not suitable for transmitting music. The audio quality can be compared to a telephone service.

COFDM Modulation

Coded-orthogonal frequency division multiplexing (COFDM) modulation is an essential component of the terrestrial broadcasting systems described in this chapter. COFDM modulation theory is a very complex subject, and a full mathematical analysis is beyond the scope of this chapter. A more detailed treatment is available in a BBC document by P. Shelswell [1]. COFDM is not generally used for satellite broadcasting, as multicarrier modulation schemes have high peak-to-average ratios requiring extremely linear transmitters. Linear transmitters are not very power efficient and therefore unsuitable for satellite systems where efficiency is of paramount importance.

COFDM signals are comprised of a large number of closely spaced carriers (i.e., frequency division multiplexing, or FDM). These carriers (also referred to as "subcarriers," since hundreds or thousands of such carriers taken as a group, form a COFDM signal) have to be closely spaced for bandwidth efficiencies but must be prevented from interfering with each other, so it is important they are orthogonal (the "O" in COFDM). In typical multipath conditions a high proportion of these carriers may be "notched out" (i.e., severely attenuated due to destructive interference) resulting in transmission errors, so the transmitted signal must include powerful error correcting codes (the "C" in COFDM).

One of the simplest ways of transmitting a high–bit rate digital signal would be to modulate it onto a single quadrature phase shift keying (QPSK) carrier. QPSK is a means of encoding two bits of data into a "symbol" by altering the phase of a carrier (in 90° steps), which has constant amplitude and frequency, and is a commonly used digital communications technique. Such a simple system works well in point-to-point telecoms links, with fixed, directional antennas, but does not work well for broadcast systems aimed at portable receivers, such as in-car and in-building radios. Such receivers suffer from the detrimental effects caused by multipath reception.

The term *multipath* describes the situation when a receiver receives a wanted signal and also unwanted echoes from various directions as the signal bounces off nearby objects such as buildings and hills. These echoes will arrive later than the main signal, and will appear to the receivers as delayed versions of the original. Multipath degrades reception in three main ways: intersymbol interference, flat fading, and Doppler shift.

Intersymbol interference is caused when a multipath reflection is delayed such that the main signal is transmitting the "current" symbol, but the echo—which is also picked up by the receiver—is still modulated by the "previous" symbol. This distorts the wanted signal, and makes it harder for the receiver to correctly demodulate it.

Flat fading occurs when the path length difference between the wanted signal and the unwanted signal is such that one arrives in antiphase to the other, and the resulting destructive interference leaves very little or no resultant signal to be demodulated.

Doppler shift occurs when multipath reflections bounce off a moving object either moving toward or away from the receiver (or when the receiver itself is moving). In this case the reflections are shifted in frequency with respect to the main received signal. This again causes distortions that make it harder for the receiver to decode a signal.

COFDM modulation was specifically designed to reduce the problems caused by multipath. The process takes an input bit stream, divides it up into multiple, parallel, lower-rate bit streams, and modulates these onto a large number of individual carriers (an "ensemble" of carriers). The number of carriers can vary between a few hundred to several thousand, depending on the transmission specification and design parameters of the broadcast signal. These carriers are closely spaced in frequency and this would normally mean that adjacent carriers would interfere with each other.

The COFDM system avoids this self-interference by making use of the fact that if the carrier spacing (in hertz) is the reciprocal of the symbol length (in seconds) it is possible to demodulate each carrier individually and ignore the contribution of its neighbors. Under these conditions the carriers are said to be orthogonal to each other. The demodulation process (in practice) is typically accomplished using a fast Fourier transform (FFT). For this process to work, while fulfilling the orthogonality requirement, it is necessary that the receiver does not just sample the received signal at one instant, it has to sample it for a time period equal to the reciprocal of the carrier spacing. This time period is known as the *useful symbol period*.

In order to minimize the detrimental effect of delayed echoes, each symbol is transmitted for a length of time equal to the useful symbol period plus an additional guard interval. This means that if any echoes arrive after the wanted signal, but with a relative delay less than the guard interval, then the receiver can place its FFT decoding window over a period equal to the useful symbol length when both wanted and echo signals are transmitting the same symbol. In this way a COFDM demodulator can work effectively without being subject to intersymbol interference in both the frequency domain (due to adjacent carriers) and the time domain (due to delayed echoes).

There still remains the problem of frequency-selective fading. Any echo signal present at the receiver will have traveled a farther distance than the wanted signal. Because each carrier in the COFDM ensemble has a different frequency, this additional path length will equate to a different number of wavelengths for each carrier. Some of the echo carriers will be delayed by nearly a whole number of wavelengths, and arrive in phase, and others will arrive out of phase and destructive interference will occur.

Because the fading is frequency selective, it is unlikely to cancel out all the carriers in the ensemble—some carriers will be reduced in amplitude and some will constructively interfere and actually add up. The broader the bandwidth of the signal, the more likely it is that parts of the ensemble will always be decodable. This is the reason why schemes such as Eureka-147 DAB were designed to be a much wider bandwidth than was needed to carry a single audio signal. Carriers subjected to destructive interference will be reduced in amplitude and will therefore have lower carrier to noise ratios, making them more likely to be decoded erroneously by the receiver.

In addition to the information-carrying data bits, COFDM signals include a large number of error-correction bits. As a result of this a large number of COFDM carriers can be effectively "destroyed" (by multipath fading, etc.), but the overall signal still retains enough information to be correctly decoded. The actual rate of overhead of the convolutional decoding can be varied. A code rate of 1/2 refers to the fact that for every one bit of informational data, two bits of coded data are transmitted. A code rate of 2/3 refers to a less well-protected system where for every two bits of information, three bits of data are transmitted.

COFDM modulation was designed from the outset to be resistant to multipath effects caused by echoes. One fortunate result of this is that COFDM receivers can cope not only with delayed signals from the same transmitter, but also signals arriving from other transmitters. This is true because if the transmitters are close enough together, and if each transmitter transmits identical information on identical frequencies at the same time, then the receiver will see one signal as an echo of the other. COFDM works well with echoes provided the guard interval is longer than the delay spread of all the echoes. This fact allows *single frequency networks* (SFNs) to be built.

In an SFN, large geographical areas can be covered contiguously by a whole sequence of transmitters all synchronized together. The most complex COFDM SFNs yet built are probably those found in the United Kingdom using the Eureka-147 system (discussed below), where approximately 100 transmitters cover most of Great Britain, allowing an in-car receiver to be driven several hundred miles without retuning or losing its signal. The maximum transmitter spacing is governed by the guard interval; longer guard intervals allow transmitters to be spaced farther apart. The downside of increasing the guard interval is that a shorter proportion of time are spent transmitting usable data, resulting in a lower data throughput.

While SFNs allow a receiver to successfully decode the contributions of several transmitters, this does have an impact on the required carrier-to-noise ratio. SFN networks designed for mobile reception make it quite likely that receivers will be subjected to conditions where two or more received COFDM signals of equal amplitude will be present at the receiver location. In these conditions a higher amount of error protection is desirable, so convolutional code rates of 1/2 are commonly used to counter these effects.

The final mechanism of multipath interference to be discussed is Doppler shift. If a receiver is moving itself, and/or is in the presence of moving objects, then reflected signals will have a frequency shift imposed on them. This means that in effect the receiver sees multiple signals at its input with different frequencies. The amount of frequency shift that is acceptable is governed by the carrier spacing. Systems with closely placed carriers are more susceptible to problems caused by Doppler shift. This creates problems for mobile reception, which is a requirement for in-car receivers. The severity of the problem increases with the transmitted frequency. This results in a maximum speed that a receiver (i.e., motion of the receiver) will work at for a given modulation scheme and frequency.

Choosing a wider carrier spacing reduces the impact of Doppler effects, but this means that SFN transmitters must be more closely spaced. This wider carrier spacing means that symbol lengths have to be shorter in order to keep to the important rule that the symbol length has to be the reciprocal of the carrier spacing. This in turn means that guard intervals have to be shorter if data throughput is to be kept at a reasonable level, which in turn means that transmitters must be more closely spaced. For these reasons, Eureka-147 networks designed to work at VHF Band III frequencies (around 200 MHz) can be placed in an SFN up to 96 km (about 60 miles) apart, and still be received by vehicles traveling at up to 160 km/h (about 100 mph). Eureka-147 networks designed for mobile reception operating in the L Band at 1.5 GHz need more widely spaced carriers, and as a result transmitters have to be less than 24 km (about 15 miles) apart.

Time Interleaving

Time interleaving is a process used in a digital broadcast system to prevent bursts of errors encountered at the point of reception from causing the receiver to fail.

Not all COFDM modulation systems use time interleaving, but it is very common. In this process the bits of the data stream from the multiplexer are not transmitted through the RF transmission channel in their original order, but instead are "scrambled" (in time sequence) beforehand. This scrambling is performed in a buffer in the modulator, which is usually part of the transmitter. The bits are fed into this buffer sequentially, but fed out in a different but predefined sequence. At the receiver the reverse process is implemented to reconstitute the data stream in its original order. This means that adjacent bits in the bit stream are broadcast at widely spaced times in the COFDM signal.

Interleaving depths vary between a few hundred milliseconds to several seconds depending on the modulation scheme used. As a result of this process, the system is protected against short bursts of errors caused by low signal level or incoming noise. These errored bits may be received close together in time, but after they have been through the de-interleaver they appear as randomized bit errors distributed evenly over a longer period. The convolutional decoder can detect and correct large numbers of bit errors if individual errored bits are more widely spaced.

The longer the time interleaving period (i.e., the greater the interleaver depth), the better the signal can cope with disturbances in reception. The downside of this approach is that interleaving adds to both the total transmission delay before audio is played out from the receiver, and also can increase the radio channel changing time, when consumers tune to a different service. Long interleaving times also add to the memory requirements (and therefore the cost) of the receiver, especially for higher bit rate services, but this is becoming much less of an issue as memory prices continue to fall.

Reed-Solomon Coding

Reed-Solomon coding is a form of error protection that is very efficient at detecting and correcting sequential bursts of errors. In addition to the Reed-Solomon *block coding*, all COFDM transmission systems include *convolutional encoding* (this forms the "C" in COFDM). The convolutional encoding is very good at detecting and correcting large numbers of dispersed (nonconsecutive), errored bits. When the amount of errors in the transmission path increases above a certain threshold, bursts of consecutive (or closely spaced) errored bits will be fed to the convolutional decoder, and it will not be able to reliably detect and correct them. When this happens the result is bursts of uncorrectable errors at the output of the convolutional decoder. Therefore, it is desirable to have an additional error-correction tool that can correct bursts of errors. Reed-Solomon code is used for this purpose.

A typical scheme organizes data into 188 byte packets and transmits in addition 16 bytes of error protection known as *parity bytes*. This is referred to as a (204,188) Reed-Solomon scheme. A Reed-Solomon decoder using this scheme can detect and correct up to 8 errored bytes anywhere in the original 188-byte sequence. It doesn't matter if a single bit or all 8 bits in that byte are in error. These 8 bytes could be spread throughout the original packet or could consist of 8 consecutive errored bytes (a sequence of up to 36 errored bits). Reed-Solomon encoding is thus a powerful addition to a COFDM broadcast system.

EUREKA-147 DAB

Eureka-147 (E-147) digital audio broadcasting (DAB) was the first adopted digital sound-broadcasting standard. In fact, the acronym DAB is a trademark of the E-147 system. The original specification was developed in the early 1990s by a European consortium of broadcasters, research institutes, and equipment manufacturers, and adopted by the European Telecommunications Standards Institute (ETSI) in 1995 as standard EN 300 401. E-147 is an International Telecommunication Union–Radiocommunication Sector (ITU-R) recommended digital radio broadcasting system (ITU-R designation of "digital system A").

The first E-147 broadcast networks were rolled out in 1995, but it was another 7 years until the mass production of affordable receivers started. Since that date a number of countries around the world have implemented trial DAB networks (see Table 1.10-1). Early broadcasts and trials in each country tend to be simul-

TABLE 1.10-1
Countries Adopting E-147 DAB Standard

Country	Population Coverage	Country	Population Coverage
Australia	30%	Lithuania	20%
Austria	19%	The Netherlands	70%
Belgium	98%	Norway	70%
Canada	35%	Portugal	75%
China	2%	Singapore	100%
Croatia	30%	South Africa	18%
Czech Republic	40%	South Korea	21%
Denmark	99%	Spain	52%
Estonia	28%	Sweden	37%
Finland	40% (prior to switch-off)	Switzerland	60%
France	25%	Taiwan	90%
Germany	85%	Turkey	12%
Hungary	30%	United Kingdom	85%
India	1%		
Israel	85%		
Italy	65%		

Source: www.worlddab.org, July 2006.

casts of existing analog services, although it is usual for a mix of simulcast and new content to be offered once full services are launched. More and more countries are launching full DAB services. The United Kingdom and Denmark have achieved the highest market penetration rates of DAB receivers. Countries like Norway and Switzerland are actively rolling out and marketing new DAB services, and other countries, such as Australia and France, are currently in the process of implementing legislation that will allow the issuing of licenses to broadcasters. At least one country, Finland, has decided to switch off its DAB network, and is presently considering other alternatives to implementing digital radio broadcasting.

E-147 DAB was originally designed to broadcast high-quality sound with some data services to complement the audio. Recent developments have resulted in additions to the specification to carry video signals. These include T-DMB (terrestrial digital multimedia broadcasting) and DAB-IP. At the time of writing, WorldDMB (formerly WorldDAB, an international nongovernmental organization whose objective is to promote, harmonize, and coordinate the implementation of DAB digital radio services based on the Eureka-147 DAB system) is drafting a specification to include an updated audio delivery system to carry aacPlus encoded audio.

Audio Coding

Unlike traditional analog radio services that have a single main audio channel, each E-147 broadcast signal is designed to accommodate multiple audio channels, which taken together are referred to as an "ensemble." A typical E-147 broadcast will have between 5–10 audio signals in its ensemble.

The core DAB specification (ETSI EN 300 401) includes MPEG-1 layer II (Musicam) and MPEG-2 layer II audio, at a wide variety of bit rates. The majority of services being broadcast at present uses MPEG-1 layer II, which is sampled at 48 kHz. Lower–bit rate (but lower audio bandwidth) services are possible using MPEG-2 layer II audio, sampled at 24 kHz.

The WorldDMB community is investigating updating the standards using MPEG-4 AAC HE (also known as aacPlus) audio coding with the addition of Reed-Solomon error protection. The proposed system is more than twice as efficient as MPEG-1 layer II, and will allow many more audio services to be carried in a DAB ensemble at a similar quality. However, it will not be compatible with receivers built to the existing E-147 standard, hence it is unclear how a country with a large E-147 receiver population (such as the United Kingdom) would transition to this newer audio coding standard.

Multimedia Services

The DAB family of standards allows several methods of incorporating multimedia data. The original DAB standards included methods of embedding data services into the audio frames—this data is called program-associated data (PAD). Applications developed for transmission via this method include broadcast Web site (BWS) and broadcast slide show (BSS). These are simple applications designed to provide a visual and/or interactive component to radio services.

In recent years there has also been a substantial interest in adding to the E-147 system the capability to transmit video streams. The Electronic and Telecommunications Research Institute (ETRI, a nonprofit, government-funded research organization in Korea) developed the T-DMB system, which is designed to deliver MPEG-4 advanced video coding (AVC) streaming video with bit-slice arithmetic coding (BSAC) encoded audio to small mobile receivers. Services were launched in Seoul in December 2005, and are expected to roll out across the rest of the country in 2006 and 2007.

The Korean DMB broadcast ensembles typically allocate just under half their capacity to a single video service at about 500 kbps, and the remaining portion to audio services delivered in "traditional" MPEG-1 layer II DAB subchannels. The specifications for T-DMB were developed in Korea, but have since been incorporated into the ETSI DAB standard family. In Germany T-DMB services have been launched, using MPEG-4 AVC video, together with AAC HE audio.

BT in the United Kingdom has developed a system for delivering television over an E-147 subchannel called DAB–Internet Protocol (DAB-IP), and marketed under the brand name "Movio." This uses Windows Media Player 9–encoded video transmitted over an E-147-enhanced packet mode data channel. This system is designed to deliver live streaming video, but in addition it is intended that it will deliver additional programs in nonreal time that are stored in a cache on the receiver. The user can browse through these additional programs once they have been received in their entirety.

MULTIPLEX STRUCTURE

The transmitted DAB signal consists of three components transmitted in sequence (see Figure 1.10-1). The first (and smallest) component is the synchronization channel. This allows the DAB receiver to lock on to the DAB frames. The second component is the fast information channel (FIC), which contains multiplex configuration information (MCI) and service information (SI). The final component is the main service channel (MSC), which contains the payload audio, data, and video services to be delivered to the receiver. The MSC has a total bit rate of 2.3 Mbps, but typically half of this amount is used to carry error protection (convolution encoding), leading to a payload throughput of around 1.15 Mbps.

The synchronization channel consists of two portions. The first of these is the null symbol. For this period nearly all the DAB carriers are turned off. This acts as a coarse synchronization pulse to aid receivers. Following this null symbol is the phase reference symbol. This is an identical sequence transmitted every transmission frame to enable the receiver to carry out

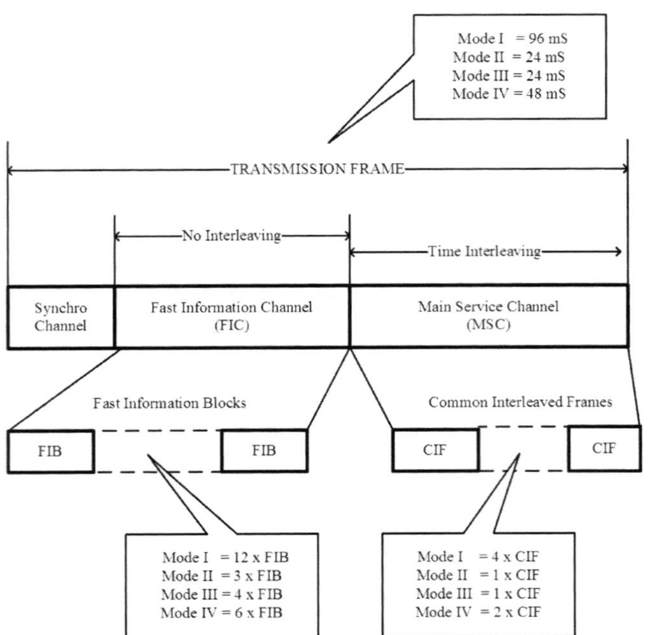

FIGURE 1.10-1 The E-147 DAB transmission frame.

radio frequency (RF) channel estimation (i.e., to assess the amount of impairment in the channel), and to act as a starting point for the demodulation of the symbols that follow.

The FIC follows the phase reference symbol. It largely consists of MCI, which describes the structure of the transmitted signal, and SI, which tells the receiver which services are carried in the ensemble, what their names are, what their bit rates are, what applications they may carry, what alternative frequencies they may be found on, along with other information useful to the receiver. This information is not time interleaved in order to make it quick and easy for a receiver to decode. It has no time interleaving or error-correction capabilities to protect it from transmission errors. Important information is sent in short messages (fast information groups, or FIGs) several times a second to ensure that the receiver can reliably build up its tables describing the services being transmitted and detailing where they can be found in the main service channel. These FIGs are grouped into fast information blocks (FIBs) for transmission in the FIC.

The MSC follows immediately after the FIC. This carries the payload of the DAB system. The payload for the audio, data, and video services is augmented with error-protection data and divided into 64 bit units, called *capacity units*; 864 of these capacity units form a common interleaved frame (CIF). The DAB system delivers one CIF (55,296 bits) every 24 milliseconds, in all modes. This provides a total MSC bit rate of 2.3 Mbps, with a typical payload throughput of 1.15 Mbps, with the balance being used to carry error protection. For example, with error protection of coding rate 1/2, there are as many error-protection bits as there are payload bits.

Channel Coding

E-147 DAB signals are composed of discrete subchannels each with a fixed bit rate. Each subchannel typically carries one audio service, one video service, or a selection of data services. For example, a Korean T-DMB broadcast will typically contain one subchannel carrying DMB MPEG-4 video, and two or more separate subchannels each carrying MPEG-1 layer II audio. Individual subchannels can have different levels of convolutional coding error protection as required.

All E-147 subchannels use convolutional encoding as their primary error protection. There are a number of convolutional code rates available specified for audio and data services. These vary from code rate 1/3 (best error protection—three bits transmitted for every information bit) to code rate 3/4 (four bits transmitted for every three information bits). Newer applications for DAB, such as the multimedia services mentioned above, add an additional Reed-Solomon code as well at the subchannel level.

Transmission Parameters

The E-147 DAB specification includes four modes for transmission. Each mode allows the same throughput of data, uses the same modulation scheme for each carrier (QPSK), and the same RF channel bandwidth. The different modes have varying spacing between carriers within the channel, as well as differing symbol lengths and guard intervals (Table 1.10-2). Modes that are designed for use at higher frequencies have fewer and more widely-spaced carriers, so they can better cope with the Doppler effects that become more severe as broadcast frequency increases. These modes have shorter guard intervals and symbol lengths in order to keep the channel capacity the same even though the number of carriers is now less. The shorter guard intervals mean the maximum allowable transmitter spacing is less for transmitters operating within an SFN. These parameters are also shown in Table 1.10-2.

Two maximum transmitter spacings are shown. The first is based on the European Broadcasting Union (EBU) recommendations that the length of the guard interval should determine absolutely the maximum transmitter spacing, and the second is the European Telecommunications Standards Institute (ETSI) recommendation, allowing for the fact that it is possible to design receivers that can cope with the situation when a delayed signal arrives just outside the guard interval.

Frequency and Bandwidth

The E-147 DAB specification allows for any channel center frequency between 30 MHz and 3 GHz that is divisible by 16 Hz. This gives potentially a vast possible range of center frequencies. In order to simplify receiver operation, the DAB implementation guidelines published by ETSI (TR_10149603) detail a list of recommended DAB frequencies in VHF band I (47–67

TABLE 1.10-2
DAB Mode Characteristics

Frequency	DAB Mode			
VHF band I, VHF band II, VHF band III	Mode I, Mode II, Mode IV			
VHF band IV, VHF band V, L-band (1,452–1,492 MHz)	Mode II and Mode IV			
Frequencies within 30–3000 MHz	Mode III			
Parameters	**DAB Mode**			
	I	**IV**	**II**	**III**
Carrier spacing (kHz)	1	2	4	8
Symbol length (ms)	1.0	0.5	0.35	0.125
Guard interval (μs)	246	123	62	31
Maximum frequency of operation (at 200 km/h) (MHz)	375	750	1,500	3,000
EBU—recommended maximum transmitter spacing (km/mi)	74/46	37/23	19/12	9/5.5
ETSI—recommended maximum transmitter spacing (km/mi)	96/60	48/30	24/15	12/7.5

MHz), VHF band III (175–240 MHz), and L-band (1,452–1,492 MHz). All countries that have implemented Eureka DAB networks have used frequencies in VHF band III or L-band. Consumer receivers that are available can usually tune to band III, or band III and L-band.

The DAB signal is 1.546 MHz wide, and requires a guard band of typically 176 kHz between adjacent DAB blocks. This signal bandwidth was chosen so as to be sufficiently wide enough to cope with frequency-selective fading in areas of strong multipath signals.

ISDB-TSB

The Integrated Services Digital Broadcasting–Terrestrial Sound Broadcasting (ISDB-Tsb) system was developed in Japan. The ISDB series of specifications was developed by the Association of Radio Industries and Businesses (ARIB) to produce a family of broadcast standards—for radio and television—that share common features. ISDB-S is intended for satellite broadcasting, ISDB-C is for cable networks, and ISDB-T is for digital terrestrial broadcasting.

The ISDB-T standards allow a wide range of bandwidths to be used as required. The basic element of bandwidth is called a *segment*, and is defined as 1/14 of the bandwidth of a video channel. Transmissions can be 1, 3, or 13 segments wide (Figure 1.10-2). The wider bandwidth version is suitable for transmitting high-definition television. One of the narrower bandwidth versions, known as ISDB-Tsb, is optimized for sound broadcasting. Another more recent narrow bandwidth variant developed in 2005 is known as 1Seg. This has been optimized for the transmission of video services to mobile devices.

Test transmissions of the ISDB system were carried out in Japan in 1999. The first radio services using IDSB-Tsb where launched in 2003 by a consortium of 32 broadcasters in Tokyo and Osaka. Currently, no country outside of Japan has adopted ISDB-Tsb for radio services. Brazil has, however, announced the adoption of ISDB-T for digital television services.

Audio Coding

ISDB-T uses AAC-HE for its audio services (also known as aacPlus). This same audio standard is used for the various variants of ISDB-T, whether for high-definition television, radio, or mobile TV services. High-quality radio services can be implemented using bit rates of up to 144 kbps to give near CD quality. 1Seg mobile TV services can use up to 64 kbps for their audio services.

Multimedia Services

ISDB-T systems were designed from the outset to carry both MPEG-4 AAC-LC audio, MPEG-4 AVC video, and data services. The recently developed 1Seg system allows MPEG-4 H.264 video and is aimed at broadcasting low-resolution television to handheld and mobile devices. The family of standards thus allows the broadcaster a wide choice of broadcasting options, depending on the channel capacity available.

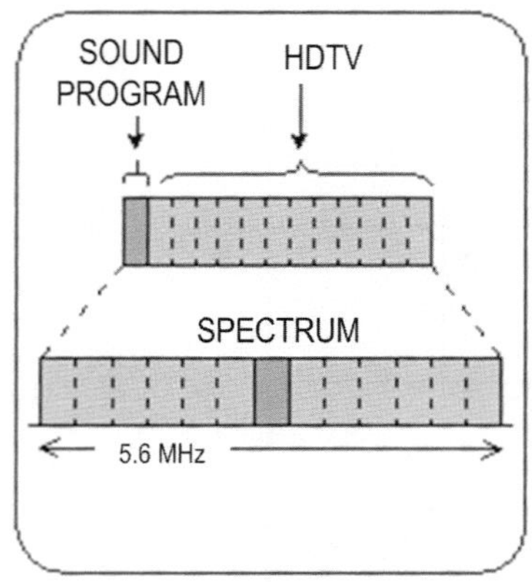

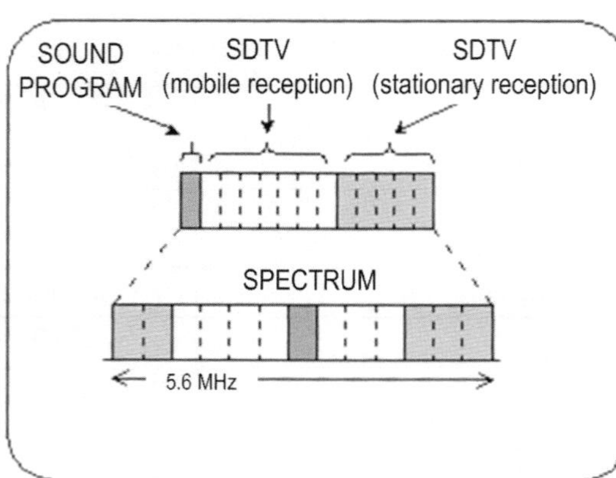

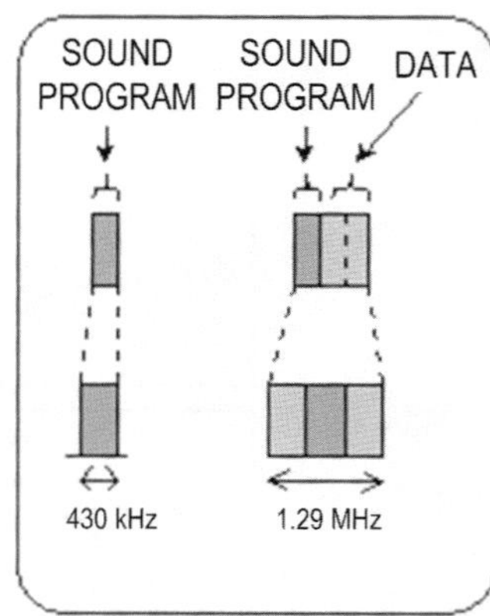

FIGURE 1.10-2 ISDB-T signals can be broadcast with a variety of bandwidths.

Multiplex Structure

ISDB-Tsb transmission is built around an MPEG-2 transport stream. This is a flexible arrangement that is capable of carrying a number of audio-compression algorithms, including MPEG-2 AAC. As an MPEG-2 transport stream is used, it is easy to insert low–bit rate video or data services if required. As many different services as are required can be sent in the same program stream.

Channel Coding

ISDB-T utilizes a robust combination of convolutional inner encoding supplemented with a Reed-Solomon (188,204) outer code. Transmissions may be implemented in different bandwidths, depending on the application. The building block of the signal is a "segment" (1/14 of the bandwidth of the TV channel being used—see Figure 1.10-2). This band segmentation approach is used to allow for simple receivers that will have longer battery life. The segments can be merged together with frequency interleaving across all segments to make the transmitted signal resistant to frequency-selective fading, or segments can be coded as "stand alone" transmissions, meaning that a receiver only has to decode a single segment, which is a narrow bandwidth, lower–bit rate transmission. This combination of narrow bandwidth and low data rate should allow the development of cheaper receivers with lower power consumption, and thus longer battery life.

The ISDB-Tsb sound broadcasting system uses channels that are either one segment wide or three segments wide. Three–segment wide transmissions are of a similar bandwidth to an E-147 DAB ensemble. If ISDB-T or ISDB-Tsb transmissions from separate program sources and multiplexers are closely synchronized and transmitted from the same site, they can be "butted up" against each other without leaving an unused guard band between them. This feature makes ISDB-Tsb very spectrally efficient.

Transmission Parameters

ISDB-T services have a choice of three transmission modes. These are detailed in Table 1.10-3. The modulation scheme is a band-segmented transmission–orthogonal frequency division multiplex (BST-OFDM). There are a wide variety of parameters that can be chosen:

- *Mode 1* has the widest carrier spacing at 108 carriers per segment and allows the fastest speeds for mobile reception, up to 800 km/h (at VHF band III frequencies).
- *Mode 2* has 216 carriers per segment and will work at speeds of up to 400 km/h.
- *Mode 3* has 432 carriers per segment, and will work at speeds of up to 200 km/h, again at VHF band III frequencies.

Four different carrier modulation schemes are allowed—differential QPSK (dQPSK), coherent QPSK, 16 quadrature AM (QAM), and 64 QAM. With a three-segment ISDB-Tsb signal, the outer two segments can have a different set of modulation parameters from the inner segment. The inner segment can have more robust error protection. Various guard intervals can also be selected and there are a number of options of code rate for the inner convolutional code. The stronger code rates are required if the intended target is mobile receivers.

Time and frequency interleaving is also applied to the transmitted data. Frequency interleaving can be applied across all segments in the transmitted signal provided that they all use the same modulation and channel-coding parameters. If different parameters are used for different segments, frequency interleaving cannot be carried out across these segments.

Frequency and Bandwidth

ISDB-T is designed to work in existing analog television broadcast spectrum. Analog TV in Japan uses 6 MHz–wide TV channels in VHF and UHF, but the

TABLE 1.10-3
Transmission Modes in ISDB-T

Parameter	Mode 1	Mode 2	Mode 3
Carrier spacing	3.968 kHz	1.984 kHz	0.99 kHz
Number of carriers per 429 kHz segment	108	216	432
Carrier modulation	4, 16, or 64 QAM dQPSK	4, 16, or 64 QAM dQPSK	4, 16, or 64 QAM dQPSK
Symbols per frame	204	204	204
Symbol duration	252 μs	504 μs	1,008 μs
Guard interval	$\frac{1}{4}, \frac{1}{8}, \frac{1}{16}$, or $\frac{1}{32}$	$\frac{1}{4}, \frac{1}{8}, \frac{1}{16}$, or $\frac{1}{32}$	$\frac{1}{4}, \frac{1}{8}, \frac{1}{16}$, or $\frac{1}{32}$
Inner code	Convolutional code $\frac{1}{2}, \frac{2}{3}, \frac{3}{4}, \frac{5}{6}, \frac{7}{8}$	Convolutional code $\frac{1}{2}, \frac{2}{3}, \frac{3}{4}, \frac{5}{6}, \frac{7}{8}$	Convolutional code $\frac{1}{2}, \frac{2}{3}, \frac{3}{4}, \frac{5}{6}, \frac{7}{8}$
Outer code	RS (204,188)	RS (204,188)	RS (204,188)
Time interleaving	0, 0.1, 0.2, 0.4, or 0.8 s	0, 0.1, 0.2, 0.4, or 0.8 s	0, 0.1, 0.2, 0.4, or 0.8 s
Bit rate per 429 kHz segment	280 kbps–1.8 Mbps	280 kbps–1.8 Mbps	280 kbps–1.8 Mbps

standards have been written to work with 7 MHz and 8 MHz channels as well. As previously mentioned, the system is based around the concept of a "segment," where a segment is defined as 1/14 of the bandwidth of a TV broadcast channel. In Japan, this means each segment has a bandwidth of 429 kHz. The sound broadcasting variant ISDB-Tsb can use 1 segment or 3 segments for a relatively narrowband signal intended primarily for audio. ISDB-T can use a channel up to 13 segments wide for broadcasting television services, both standard and high definition. ISDB single segment (or 1seg) is a recent development that uses 1 particular segment in a 13 segment–wide ISDB-T transmission to broadcast a low-resolution signal intended for mobile reception on handheld receivers.

ISDB-Tsb audio services are currently broadcast in VHF band III in Japan. The spectrum between 188 MHz and 192 MHz has been allocated for this purpose. It is anticipated that more spectrum will become available after analog television switches off in 2011. ISDB-T television services are broadcast in bands IV and V in the UHF band—between 430 MHz and 770 MHz.

DIGITAL RADIO MONDIALE

The Digital Radio Mondiale (DRM) consortium was founded by international broadcasters and equipment manufacturers in 1998 to develop a digital sound broadcasting system for use at frequencies below 30 MHz. These frequencies have been traditionally used for narrowband AM broadcasting. They are largely used for local, national, and international services in the medium wave (MW) and long wave (LW) bands, and for international broadcasting in the short wave (SW) bands. These bands have propagation characteristics that allow coverage of large areas and rugged terrain, but the primitive AM modulation has no protection against noise. The consortium's goal was to provide a nonproprietary digital standard that could be used on these frequencies.

The DRM standard was adopted by ETSI in 2001 as a technical specification. In 2003 it was granted the status of a full ETSI standard and published as ETSI ES 201980. Since then it has been adopted by the ITU-R as recommendation BS-1514, and the IEC as IEC 62272-1.

Audio Coding

DRM systems are designed to send audio services through narrow bandwidth channels between 4.5–20 kHz wide. The audio bit rate that can be carried by a DRM signal varies widely depending on the bandwidth and the configuration. A 5 kHz bandwidth signal configured for a short wave sky wave broadcast might only be able to carry 5 kbps of useful audio. A 20 kHz wide broadcast configured for a short-distance medium wave may be able to carry up to 70 kbps of useful audio. For this reason DRM has a range of audio codecs designed to work at different bit rates.

For very low bit rates, HVXC and CELP coders are used. These are designed to carry speech. HVXC can reproduce intelligible speech at 4 kbps, and CELP can reproduce good quality speech at 8 kbps. These coders are only designed to handle speech, and are not suitable for transmitting music.

At higher bit rates, MPEG AAC HE audio coding is used. This can provide AM quality audio (band limited to about 6 kHz) at about 16 kbps. With the addition of spectral band replication (SBR) and parametric stereo (PS) an acceptable quality stereolike signal is achieved at a bit rate of 20 kbps.

Whichever audio codec is used, an integer number of audio frames are formed together to make one 400 millisecond superframe. This superframe corresponds to the 400 millisecond RF transmission frame period.

DRM has the capability of carrying up to four audio services in one broadcast signal. Due to the very narrow bandwidths (and therefore bit rates) available in a DRM transmission, most broadcast services will not transmit this many programs, unless they are low–bit rate speech services such as, for example, news bulletins in four different languages.

Despite the low audio bit rates, DRM is expected to vastly improve the listener experience of many listeners to short wave and medium wave services. The analog services in these bands have traditionally used amplitude modulation (AM). This is a very basic modulation technique, with no protection possible against interference and, for long-range broadcasts, very susceptible to changes in signal amplitude (fading). The DRM signal is capable of providing a more constant signal and is much more immune to interference and fading.

Data Services

DRM has been designed to carry simple data services based on those already developed for DAB and, earlier for the FM subcarrier-based radio data system (RDS). Provision has been made for simple text messages and the broadcast of slide show and Web site data. Video is unlikely to be broadcast over DRM because of the very small bandwidths available. An important feature of DRM is the ability to send retuning information along with the signal. This feature will be attractive to short wave listeners in particular who are used to having to retune to programs whose frequency changes on a daily and perhaps seasonal basis due to atmospheric propagation variations.

International shortwave broadcasts often rely on reaching distant countries (several thousand miles away) by bouncing signals of the ionosphere layers in Earth's upper atmosphere. Different frequencies will bounce at different heights, and thus travel different distances. The height of the ionosphere is affected by the amount of radiation from the sun, and thus varies on a day/night and summer/winter basis. DRM has the possibility of making these frequency changes transparent to the user—the receiver will know which alternative frequencies may be available and will automatically check them to see if a better signal is there.

Multiplex Structure

The DRM transmitted signal consists of three distinct signal components:

- *Main service channel* (MSC) can carry up to four services. Each service can be an audio service or a data service. It employs the most complex modulation scheme.
- *Service descriptor channel* (SDC) contains information on the services in the MSC. It contains details of all the parameters necessary to decode each channel, and can also contain information such as a list of other frequencies and times of day when the program will be broadcast. This latter feature is very important for shortwave broadcasts, which use different frequencies throughout the day to cope with changing propagation conditions.
- *Fast access channel* (FAC) contains the information the receiver needs to demodulate the SDC. This includes information on the bandwidth of the signal, and the modulation parameters used for the SDC and the MSC. It also details the number of services present in the MSC, along with their names.

Channel Coding

DRM has a very complex modulation scheme. COFDM with convolutional coding and time interleaving is used with a wide variety of parameters available. Different parameters are used for different portions of the multiplex. As with E-147 DAB, it is possible to apply different levels of error protection to individual audio channels within the multiplex. Reed-Solomon coding is not used.

The FAC is broadcast over a fixed bandwidth of the signal (4.5 kHz for channels on a 9 kHz spacing basis and 5 kHz for channels on a 10 kHz spacing basis). For wider bandwidth signals this only occupies a portion of the channel, as shown in Figure 1.10-3. In this diagram, F_r is the reference frequency, that is, the frequency the receiver tunes to. When a receiver tunes to a frequency for the first time it will not know what

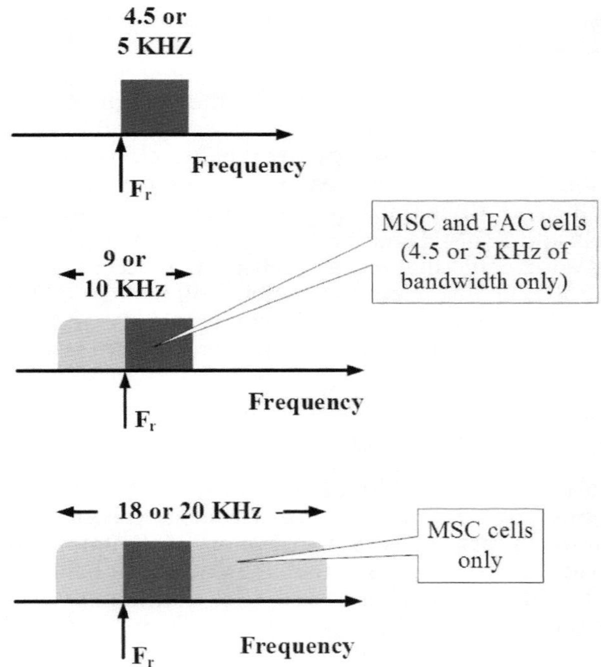

FIGURE 1.10-3 Relationship between FAC channel and overall bandwidth.

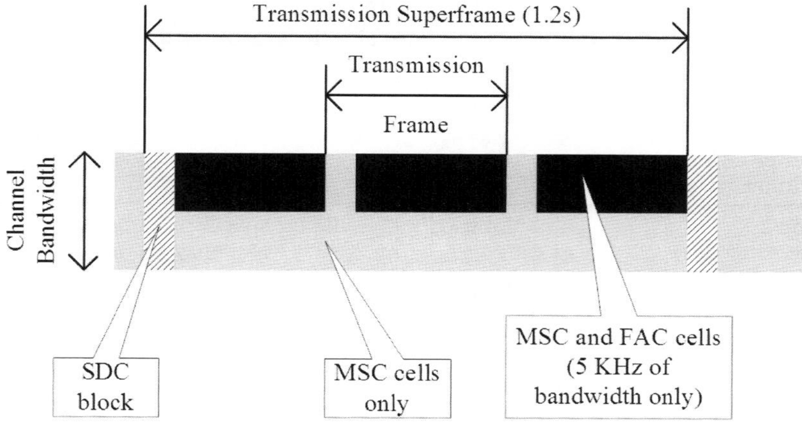

FIGURE 1.10-4 Time and frequency relationship of DRM components for a 10 kHz channel.

bandwidth or parameters the DRM multiplex is using. The receiver knows it only has to inspect the 4.5 or 5 kHz portion immediately above the reference frequency to decode the FAC. When it has decoded the FAC it can learn the full bandwidth and modulation parameters of the whole signal and start to decode first the SDC and then the MSC.

The SDC is transmitted in a burst at the start of each transmission superframe, as shown in Figure 1.10-4. While the SDC is being transmitted no MSC information is being broadcast. This means that if the MSC signal is weak, and alternative frequencies are available, the receiver can choose to ignore the SDC and use this time to "inspect" alternative frequencies to see if a good signal is available elsewhere. If another signal is present, and appears to be a stronger signal, then the receiver can choose to switch to the new frequency. This should be possible with very little audio disturbance to the listener.

Different portions of the multiplex use different modulation depths, all using coherent modulation. To enable demodulation of this, pilot cells are broadcast throughout the channel. These pilot cells have a known phase and amplitude, and help the receiver to decode each symbol. The pilot cells do not have a constant frequency. Different carriers in each symbol (i.e., in each time period) are used as pilot cells. They have a higher amplitude than normal carriers to ensure they have a good carrier-to-noise ratio. The FAC is designed to be quick and easy to decode, so it always uses QPSK (which is equivalent to 4 QAM). The FAC cells are interspersed among the MSC cells. Different carriers are used for the FAC during each symbol.

The MSC can use 64 QAM for maximum data throughput, or 16 QAM for maximum robustness. It uses time interleaving as well to protect itself during channel fades.

The SDC uses 16 QAM or 4 QAM, governed by the chosen modulation in the MSC. The SDC always uses a more robust modulation scheme, together with time interleaving, to ensure it is relatively easy to decode.

This modulation scheme is very complicated, as has been stated before. In certain transmission modes any individual carrier will spend part of its time modulated as 16 QAM as part of the SDC, followed by periods of 64 QAM as part of the MSC, and QPSK as part of the FAC. It will also from time to time be called on to perform the role of a pilot cell at increased amplitude.

Transmission Parameters

There are four DRM transmission modes, designed for different broadcasting scenarios (Table 1.10-4). Mode

TABLE 1.10-4
DRM Transmission Modes

Mode	A	B	C	D
Recommended modulation	64 QAM	64 or 16 QAM	64 or 16 QAM	16 QAM
Typical application	Ground wave (MW, LW)	Sky wave (MW, SW)	Vertical incidence (SW)	Sky wave (difficult propagation) (SW)
Interleave	Short	Long	Long	Long
Bit rate	High	Medium	Low	Medium/Low
Audio quality	High	Medium	Low	Medium/Low
Robustness	Medium	High	Very High	Very High

TABLE 1.10-5
Radio Channel Bandwidth versus ITU-R Region and RF Band

ITU-R Region	Long Wave	Medium Wave	Short Wave
Region 1 (Europe, Africa, Middle East, and Russia)	9 kHz	9 kHz	5 kHz
Region 2 (The Americas)	Not used	10 kHz spacing 20 kHz occupancy	5 kHz
Region 3 (Asia Pacific)	Not used	9 kHz spacing 18 kHz occupancy	5 kHz

A is used for ground wave broadcasting, where the service area is close to the transmitting tower. This is a typical configuration for "local" MW or LW broadcasts. Modes B, C, and D are intended for sky wave transmission.

Frequency and Bandwidth

The DRM standard covers terrestrial broadcasts below 30 MHz. The system was designed to be compatible with the existing ITU-R frequency channels in use in that band (Table 1.10-5). The DRM consortium is also currently developing a modification of the standard that will extend this frequency range up to 120 MHz (expected to be completed in 2009).

DRM provides one system that will work with all these bandwidths. Different DRM modes are available that occupy bandwidths of 4.5, 5, 9, 10, 18, or 20 kHz.

WORLDSPACE

WorldSpace, a proprietary, subscription-based digital radio service, was the first satellite digital audio radio service (SDARS) to be deployed, and presently has two satellites in geostationary orbit, AfriStar and AsiaStar, which serve Africa (and portions of Europe) and the Middle East and Asia, respectively (Figure 1.10-5). WorldSpace was also one of the founding shareholders of XM Satellite Radio in the United States and was responsible for some of the technology used in the XM system.

Audio Coding

Audio is coded using MPEG-2.5 layer III. This allows a mono signal to be encoded into a 16 kbps prime rate channel (PRC). Up to 96 of these channels can be added to one of the TDM carriers on each satellite. Alternatively, 192 lower-quality channels could be carried by each TDM carrier, or 24 channels of near CD quality, or any combination of these.

Multiplex Structure

The basic building block of the transmitted signal is a PRC of 16 kbps. One of these can carry a low-quality audio program, or two to eight PRCs can be joined together to carry higher-quality services. Ninety-six PRCs can be carried on each TDM carrier. Each TDM frame is divided into three fields: a frame synch word, a time slot control channel, and the data field containing the 96 PRCs. The channels are assembled into 255 byte blocks, consisting of 223 bytes of data and 32 bytes of Reed-Solomon coding (outer coding).

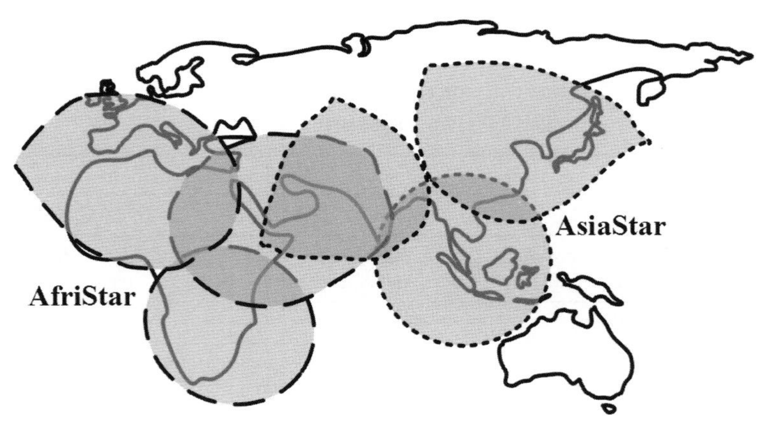

FIGURE 1.10-5 WorldSpace satellite coverage.

Channel Coding

The blocks are encoded by convolutional encoding (1/2 rate coding), and further protected by time interleaving. The data is modulated onto a QPSK carrier with a data rate of 3.68 Mbps.

Transmission Parameters

The WorldSpace system was designed with stationary reception in mind and was not originally envisioned as a service that would be used by mobile receivers (e.g., in moving automobiles). Consequently, this system had no time, space, or frequency diversity elements in its system design and the receiver had to rely on getting a good signal from the single satellite signal available to it. However, WorldSpace is in the process of adding a terrestrial repeater component to their service, which will allow for mobile reception. There are few technical details available regarding these repeaters but it is anticipated that the repeater signals will not be receivable by legacy WorldSpace receivers and that new receivers will be required.

Frequency and Bandwidth

WorldSpace currently has two operational geostationary satellites covering Africa and southeast Asia. A third satellite is due to be launched soon. Each satellite has three beams; each beam covers about 14 million square kilometers (see Figure 1.10-5). WorldSpace uses frequencies between 1,467–1,492 MHz in the L-band, and multiple satellite beams per satellite (using "spot beam" antennas). Each satellite beam has two TDM carriers, one with left-hand circular polarization, and one with right-hand circular polarization. Each carrier is QPSK modulated, with a bandwidth of 2.3 MHz and a raw bit rate of 3.68 MHz.

SIRIUS

Sirius is a subscription satellite radio service covering the United States. It is designed to provide around 100 channels of audio to mobile receivers. The network uses a proprietary broadcast system, so the exact details of transmission are not published, but the following is a brief guide of how the system works.

Sirius' original system configuration consisted of three satellites in a highly inclined elliptical orbit (HEO, Figure 1.10-6). They are equally spaced along the same orbit, which has a period of 24 hours. The three satellites follow each other, describing a complex path over the Earth's surface. Each in turn moves relatively slowly northward and then southward over the United States before heading off to carry out a faster loop over South America and the Pacific Ocean. This path ensures that two satellites are always over the United States at any one time, whilst the third completes the other part of the orbit. Thus, each satellite spends about 16 hours a day over the United States, and there is always one satellite at greater than 60 degrees above the horizon.

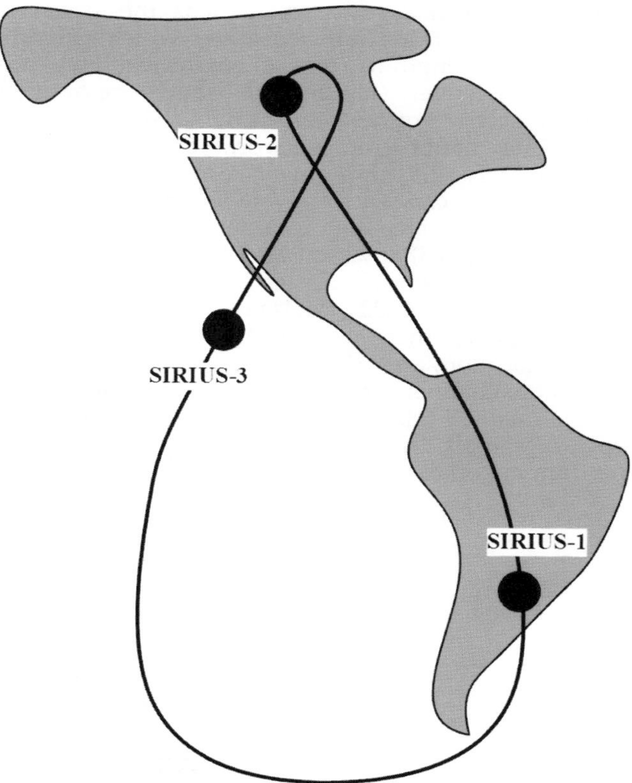

FIGURE 1.10-6 Sirius satellite orbit.

In 2006, Sirius announced that this satellite constellation was going to be augmented with a fourth satellite, launched into a geostationary orbit (expected launch date is 2008). This would create what Sirius calls a *hybrid constellation* and is expected to enhance redundancy, coverage, and performance.

The Sirius system has the following design features to ensure robust, reliable delivery to mobile receivers across the United States and parts of Canada:

Space diversity is provided by having two satellites transmitting from different regions of the sky. This makes it less likely that a receiver will be shadowed from the satellite by obstacles such as houses. Each satellite uses a different frequency band to avoid interference between the two signals.

Time diversity is provided by delaying the signal from one satellite by approximately 4 seconds with respect to other satellites. The receiver acquires both signals, and adds an appropriate delay to one channel so both are synchronized. It then chooses the best signal of the two to decode. A mobile receiver, therefore, has a good chance of decoding a signal, even if both satellites are blocked from view for a short period (i.e., if it is in a car traveling under a bridge). Providing both satellites are not shadowed for more than 4 seconds, the receiver can cope with an event like this without muting.

Terrestrial repeaters. In urban areas the signal is backed up by terrestrial repeaters, which provide good signal strength to areas where the satellites cannot reach effectively (Figure 1.10-7 shows the original planned locations of repeaters in the Washington, DC, area). The nationwide network design includes approximately 100 high-power terrestrial repeaters. Sirius receivers can seamlessly switch between terrestrial and satellite reception in order to take advantage of the best signal at any location.

The Sirius network occupies 12.5 MHz of space in the S band (2,320–2,332.5 MHz) that is split into three channels. Two of these channels, at the top and bottom of the allocation, are dedicated to the satellite downlink. These channels are 4.510 MHz wide. The middle channel, 4.012 MHz wide, is occupied by terrestrial repeater signals. All three channels carry the same program material.

The satellite downlink channel has a raw bit rate of 7.54 Mbps modulated onto a single time-division multiplexed (TDM) QPSK carrier. This modulation scheme is resistant to Doppler shifts, and receivers will work even if traveling "up to aircraft speeds."

The terrestrial "filler" repeaters are using COFDM at a data rate of 7.38 Mbps. These terrestrial transmissions provide good reception in urban areas subject to strong multipath signals, but are more susceptible to Doppler effects than the satellite downlinked signal. This is not a problem, however, as according to Sirius, receivers will work when traveling at speeds of up to 130 km/h (about 80 mph).

The Sirius system delivers 185 logical channels. Logical channels are bundled together to provide services at various data rates. Services may be audio, video, or data, but to date most broadcasts have been audio services. There are about 100 subscription audio channels on the Sirius system.

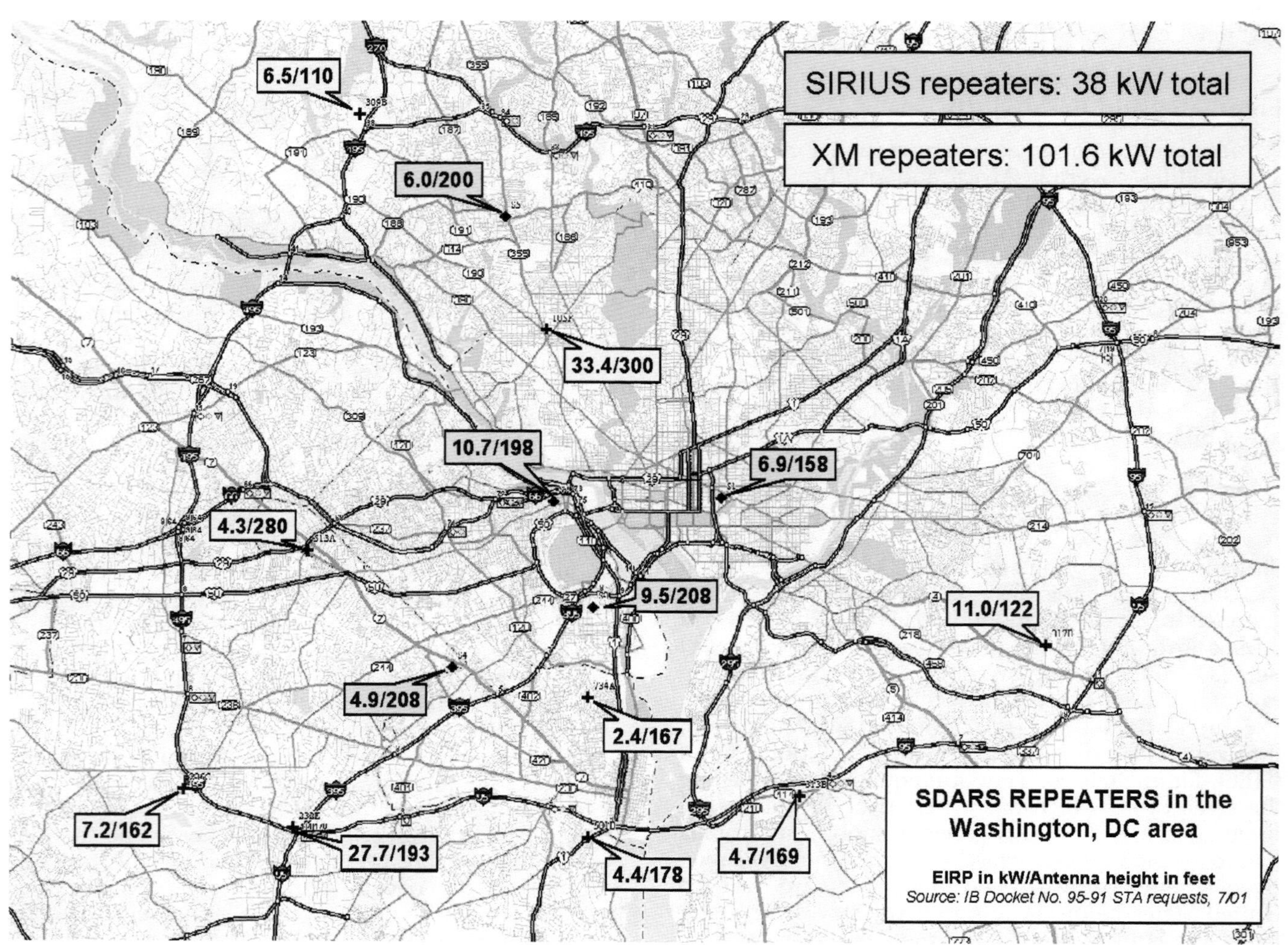

FIGURE 1.10-7 Initial (planned) SDARS repeater deployment in the Washington, DC, area. Refer to the CD for a color version of this image.

XM SATELLITE RADIO

XM Satellite Radio is another proprietary satellite system covering the U.S. Like Sirius, this is a proprietary system with few details published. It carries approximately 100 audio services.

The XM system relies on two geostationary satellites at 115 degrees west and 85 degrees west, situated over the equator (like all geostationary satellites). At launch, these satellites were, according to XM, the "highest power commercial satellites ever built." Each satellite has two 3 kW transponders that beam two 1.48 MHz bandwidth signals down to the coverage area. Each of the two signals contains over 100 audio channels. Both satellites broadcast identical signals, although one set of signals is delayed by about 4 seconds. Three satellites have been launched, XM-1 "Rock" and XM-2 "Roll" in 2001, and an XM-3 "Rhythm" launched in 2005. Rock and Roll are now positioned at the western station, and Rhythm has taken up position at the eastern station.

In order to make sure that the signal is received in any area, time and space diversities are used. Time diversity is achieved by delaying one of the satellite feeds by 4 seconds. Each receiver receives two streams, one of which is the delayed version. If the receiver passes under a bridge or similar obstruction, it can still receive a good signal by swapping between the two received signals. Space diversity is obtained by virtue of having two satellites, one over the eastern portion of United States, and one over the western portion.

The XM system utilizes over 1,000 terrestrial repeaters, with effective radiated powers (ERPs) of about 2 kW (on average), filling in urban areas where the satellites do not cover (Figure 1.10-7). As with Sirius, receivers can switch between any of the three feeds (two satellite, one terrestrial), with the goal of always using the best signal. XM needs this many terrestrial repeaters to provide good coverage throughout the service area as it relies on geostationary satellites. Such satellites are at lower elevation angles than those in the Sirius Satellite Radio HEO constellation, and, consequently, the XM signals are more susceptible to interruptions by shadowing from buildings and other terrain features than are the Sirius signals, and hence the larger number of terrestrial repeaters for XM.

The XM broadcast system uses the S band and occupies 12.5 MHz of bandwidth, between 2,332.5–2,345 MHz. This bandwidth is split into six channels. Channels 1 and 2 are satellite downlinks from one transmitter, each 1.48 MHz wide. Channels 3 and 4 are for the corresponding terrestrial fillers, each 2.53 MHz wide. Channels 5 and 6 are the downlinked signal from the second satellite.

Like Sirius, XM uses TDM QPSK modulation for the satellite signals and COFDM for its terrestrial repeaters. The audio channels are protected with convolutional encoding, and also by Reed-Solomon coding. XM uses the CT-aacPlus codec, otherwise known as MPEG-4 AAC HE.

SUMMARY

The dawn of the twenty-first century is a time of great change for radio. Around the world, many new systems are being developed to deliver audio content. At the same time boundaries between radio and television are becoming increasingly blurred, as low–bit rate video services start to appear on systems originally developed to deliver audio. In the future, a discussion on content delivery mechanisms is perhaps more likely to differentiate between fixed and mobile delivery technologies, rather than between radio and TV.

Although several different systems have been discussed here, these systems have many similarities. All the terrestrial systems use COFDM modulation. All the systems either use MPEG-4 audio variants or are in the process of being adapted to do so. The similarities between the systems are perhaps more striking than their differences.

Eureka-147 is the oldest system and has been adopted in the most countries. It is currently being revamped and modernized. DRM is the best system available for delivering international terrestrial broadcasts, and there is much interest developing in using it for local broadcasting as well. ISDB-Tsb is interesting because it is very bandwidth efficient, and allows a variety of different bandwidths to be used. The satellite systems are starting to become significant, and the biggest change they introduce is not the delivery mechanism but the business model of subscription radio.

Radio continues to evolve—in the not-too-distant future we will see more efficient modulation systems and the addition of even more powerful error-protection codes. Digital radio has a bright future.

References

[1] Shelswell, P., BBC RD 1996/8, The COFDM Modulation System. The Heart of Digital Audio Broadcasting. At www.bbc.co.uk/rd.

[2] MPEG-4 Audio, ISO/IEC 13818-7. See www.iso.ch.

[3] Eureka-147 DAB Core Standard ETSI EN 300 401. See www.etsi.org.

[4] DRM Core Standard ETSI ES 201 980. See www.etsi.org.

[5] ISDB-Tsb Core Standards, ARIB STD-B31, B32, and B24. See www.dibeg.org.

Further Information

Digital Radio Mondiale, at www.drm.org
Eureka-147 DAB, at www.worlddmb.org.
ISDB-Tsb, at www.dibeg.org.

CHAPTER

1.11

The ATSC Digital Television System

JERRY WHITAKER

Advanced Television Systems Committee
Washington, D.C.

INTRODUCTION

The Advanced Television Systems Committee (ATSC) digital television (DTV) standard has ushered in a new era in television broadcasting. The impact of DTV is more significant than simply moving from an analog system to a digital system; rather, DTV permits a level of flexibility wholly unattainable with analog broadcasting. An important element of this flexibility is the ability to expand system functions by building upon the technical foundations specified in ATSC standards such as the ATSC Digital Television Standard (A/53) [1] and the Digital Audio Compression (AC-3) Standard (A/52) [2].

With National Television System Committee (NTSC), and its Phase Alternation Line (PAL) and SECAM counterparts, the video, audio, and some limited data information are conveyed by modulating a radiofrequency (RF) carrier in such a way that a receiver of relatively simple design can decode and reassemble the various elements of the signal to produce a program consisting of video and audio and perhaps related data (*e.g.*, closed captioning). As such, a complete program is transmitted by the broadcaster that is essentially in finished form. In the DTV system, however, additional levels of processing are required after the receiver demodulates the RF signal. The receiver processes the digital bit stream extracted from the received signal to yield a collection of program elements (video, audio, or data) that match the services that the consumer has selected. This selection is made using system and service information that is also transmitted. Audio and video are delivered in digitally compressed form and must be decoded for presentation. Audio may be monophonic, stereo, or multi-channel. Data may supplement the main video/audio program (*e.g.*, closed captioning, descriptive text, or commentary), or it may be a stand-alone service (*e.g.*, a stock or news ticker).

The nature of the DTV system is such that it is possible to provide new features that build upon the infrastructure within the broadcast plant and the receiver. One of the major enabling developments of digital television, in fact, is the integration of significant processing power in the receiving device itself. Historically, in the design of any broadcast system—be it radio or television—the goal has always been to concentrate technical sophistication (when needed) at the transmission end and thereby facilitate simpler receivers. Because there are far more receivers than transmitters, this approach has obvious business advantages. While this trend continues to be true, for DTV the complexity of the transmitted bit stream and compression of the audio and video components require a significant amount of processing power in the receiver, which is practical because of the enormous advancements made in computing technology. Once a receiver reaches a certain level of sophistication (and market success), additional processing power is essentially free.

ATSC DTV System Overview

The ATSC DTV standard describes a system designed to transmit high-quality video and audio and ancillary data over a single 6 MHz channel. The system can deliver about 19 Mbps in a 6 MHz terrestrial broadcasting channel and about 38 Mbps in a 6 MHz cable television channel. This means that encoding high-definition (HD) video sources at 885[1] Mbps (highest rate progressive input) or 994[2] Mbps (highest rate interlaced picture input) requires a bit rate reduction by about a factor of 50 (when the overhead numbers are added, the rates become closer). To achieve this bit rate reduction, the system is designed to be efficient in utilizing available channel capacity by exploiting complex video and audio compression technologies. The compression scheme used for DTV service optimizes the throughput of the transmission channel by representing the video, audio, and data sources with as few bits as possible while preserving the level of quality required for the given application. The RF/transmission subsystems described in the DTV standard are designed specifically for terrestrial and cable applications. The structure is such that the video, audio, and service multiplex/transport subsystems are useful in other applications as well.

System Block Diagram

A basic block diagram representation of the ATSC DTV system is shown in Figure 1.11-1. According to this model, the digital television system consists of four major elements, three within the broadcast plant plus the receiver.

Video/Audio System

The video subsystem and audio subsystem, as shown in Figure 1.11-1, refer to the bit rate reduction methods appropriate for application to the video, audio, and ancillary digital data streams. The purpose of compression is to reduce the number of bits required to represent the audio and video information to a level that can be contained within the transmission channel capacity. ATSC employs the MPEG-2 video stream syntax (Main Profile at High Level) for the coding of video and the ATSC Standard Digital Audio Compression (AC-3) for the coding of audio. ATSC consumer receivers are designed to decode all high-definition TV (HDTV) and standard-definition TV (SDTV) streams to give program service providers maximum flexibility. The term "ancillary data" dates from the original drafting of A/53 and is a broad term that includes control data and supplementary data, including data associated with the program audio and video services. As standards were developed to define how to transport and process data, it became clear that different forms of data served very different purposes and different standards were needed for metadata and essence. Data delivered as a separate payload can provide independent services as well as data elements related to an audio- or video-based service.

[1] $720 \times 1280 \times 60 \times 8 \times 2 = 884.736$ Mbps (the 2 represents the 4:2:2 or color subsampling; an RGB with full bandwidth would be a 3).

[2] $1080 \times 1920 \times 29.97 \times 8 \times 2 = 994.333$ Mbps (the 2 represents the 4:2:2 or color subsampling; an RGB with full bandwidth would be a 3).

Service Multiplex and Transport

The service multiplex and transport system, as shown in Figure 1.11-1, refers to the means of dividing each bit stream into packets of information, the means of uniquely identifying each packet including packet type, and the appropriate methods of interleaving or multiplexing video bit-stream packets, audio bit-stream packets, and data bit-stream packets into a single transport mechanism. The structure and relationships of these essence bit streams are carried in service information bit streams, also multiplexed in the single transport mechanism. In developing the transport mechanism, interoperability among digital media—such as terrestrial broadcasting, cable distribution, satellite distribution, recording media, and computer interfaces—was of prime consideration. The ATSC system employs the MPEG-2 transport stream syntax for the packetization and multiplexing of video, audio, and data signals for digital broadcasting systems. The MPEG-2 transport stream syntax was developed for applications where channel bandwidth or recording media capacity is limited and the requirement for an efficient transport mechanism is paramount. It also provides the critical timing information for the receiver to perform video and audio synchronization.

RF Transmission System

The RF transmission system, as shown in Figure 1.11-1, refers to channel coding and modulation. The channel coder takes the packetized digital bit stream, reformats it, and adds additional information that assists the receiver in extracting the original data from the received signal which, due to transmission impairments, may contain errors. In order to protect against both burst and random errors, the packet data is interleaved before transmission and convolutional and Reed-Solomon forward error correction (FEC) codes are added. The modulation (or physical layer) uses the digital bit-stream information to modulate a carrier for the transmitted signal. The basic modulation system offers two modes: an eight-level vestigial sideband (8-VSB) mode for terrestrial broadcasting and a 16VSB mode intended for cable applications.[3] The 8-VSB mode was designed for spectral efficiency, maximizing the data throughput with a low receiver carrier-to-noise (C/N) threshold requirement, high immunity to both co-channel and adjacent channel interference, and a high robustness to transmission errors. The attributes of 8-VSB allow DTV channels to co-exist in a crowded spectrum environment that contains both NTSC analog and ATSC digital television signals. In addition, the lower power requirements (typically, 12 dB lower than analog NTSC) of 8-VSB allow ATSC DTV stations to operate on channels where analog stations cannot due to interference constraints. The spectral efficiency and power requirement characteristics of 8-VSB are essential to the conversion of terrestrial

[3] In the U.S., carriage of ATSC DTV signals is accomplished primarily (if not exclusively) using quadrature amplitude modulation (QAM).

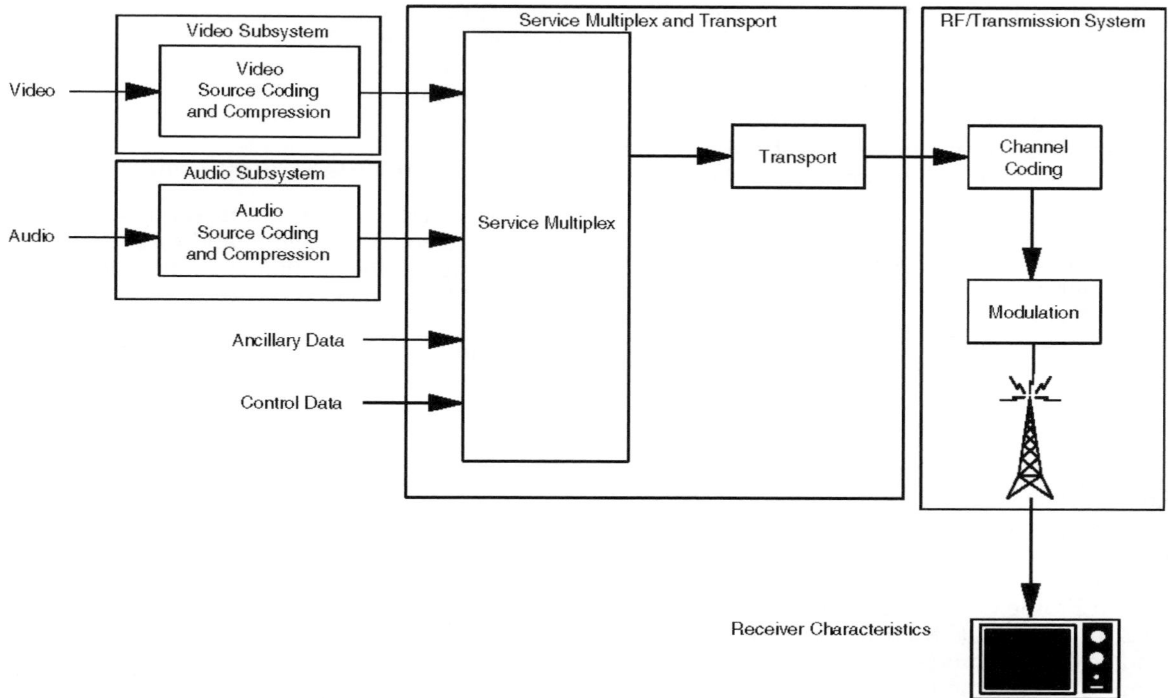

FIGURE 1.11-1 Block diagram of functionality in a transmitter/receiver pair.

broadcast transmission from analog to digital because new spectrum is not allotted during the transition phase.

The recently developed enhanced-VSB (E-VSB) mode involves the transmission of a backward-compatible signal within the standard 8-VSB symbol stream that can be received at a lower carrier-to-noise ratio than conventional 8-VSB. The E-VSB mode allows broadcasters to trade some of their data capacity for additional robustness. With an E-VSB transmission, some of the approximately 19.4 Mbps data is allocated to the robust mode and the remainder is allocated to the normal 8-VSB mode. However, the amount of delivered data (payload) is reduced for the robust mode because part of the payload is traded for additional FEC bits to correct bit errors that occur with reception under weaker signal conditions (resulting in up to a 6 dB improvement).

Receiver

An ATSC DTV receiver recovers the bits representing the original video, audio, and other data from the modulated signal. In particular, the receiver performs the following functions:

- Tunes the selected 6 MHz channel
- Rejects adjacent channels and other sources of interference
- Demodulates (equalizes as necessary) the received signal, applying error correction to produce a transport bit stream
- Identifies the elements of the bit stream using a transport layer processor
- Selects each desired element and sends it to its appropriate processor
- Decodes and synchronizes each element
- Performs product-specific video, audio, and data processing
- Presents the programming to the appropriate video or audio transducer

Special receiver circuits are designed to deal with elements of the terrestrial transmission path including noise, interference, and multipath fading. Innovations in equalization, automatic gain control, interference cancellation, and carrier and timing recovery improve signal reception and create product performance differentiation.

The decoding of transport elements that make up the programming is usually considered to be a straightforward implementation of the MPEG and AC-3 specifications, although significant opportunities for innovation in circuit efficiency or power usage exist. Innovations in video decoding offer opportunities for savings in memory and circuit speed and complexity. Product differentiation based on picture quality is also widespread, resulting from innovations in error concealment, format conversions, perceptual picture processing, and specific display-related processing. The user interface and new data-based services are other important areas for product differentiation.

The development of large-screen consumer displays has played an important role in the evolution of receivers. Whether intended for use in an integrated receiver or as a stand-alone display, the rapid deployment of new large, high-resolution flat-panel displays has substantially changed the video landscape; for example, one of the first concerns of technologists with regard to HDTV in the home was the size of the display device. In order to fully appreciate the image quality of HDTV at typical viewing distances, it is necessary to view the image on a large screen; however, with cathode ray tube (CRT) technology, a large screen also means a large, heavy enclosure. Flat-panel displays and projection systems with HDTV resolution have essentially eliminated this physical constraint.

VIDEO SYSTEM

The MPEG-2 specification is organized into a system of profiles and levels, so applications can ensure interoperability by using equipment and processing that adhere to a common set of coding tools and parameters [4]. The DTV standard is based on the MPEG-2 Main Profile (MP). The Main Profile includes three types of frames (*I*-frames, *P*-frames, and *B*-frames) and an organization of luminance and chrominance samples (designated 4:2:0) within the frame. The Main Profile does not include a scalable algorithm, where scalability implies that a subset of the compressed data can be decoded without decoding the entire data stream. The High Level (HL) includes formats with up to 1152 active lines and up to 1920 samples per active line, and for the Main Profile it is limited to a compressed data rate of no more than 80 Mbps. The parameters specified by the DTV standard represent specific choices within these constraints. The DTV video compression system does not include algorithmic elements that fall outside the specifications for the MPEG-2 Main Profile. Thus, video decoders that conform to the MPEG-2 MP@HL can be expected to decode bit streams produced in accordance with the DTV standard.

Overview of Video Compression

The ATSC DTV video compression system takes in an analog video source signal and outputs a compressed digital signal that contains information that can be decoded to produce an approximate version of the original image sequence. The goal is for the reconstructed approximation to be imperceptibly different from the original for most viewers, for most images, for most of the time. In order to approach such fidelity, the algorithms are flexible, allowing for frequent adaptive changes in the algorithm depending on scene content, history of the processing, estimates of image complexity, and perceptibility of distortions introduced by the compression. Figure 1.11-2 shows the flow of video signals in the ATSC DTV system. Analog signals, when presented to the system, are digitized and sent to the encoder for compression, and the compressed data then is transmitted over a communications channel. On being received, the compressed signal is decompressed in the decoder and reconstructed for display (following error-correction, as necessary).

Video Preprocessing

Video preprocessing converts the analog input signals to digital samples in the form needed for subsequent compression. The analog input signals are typically composite for standard-definition signals or components consisting of luminance (Y) and chrominance (Cb and Cr) signals.

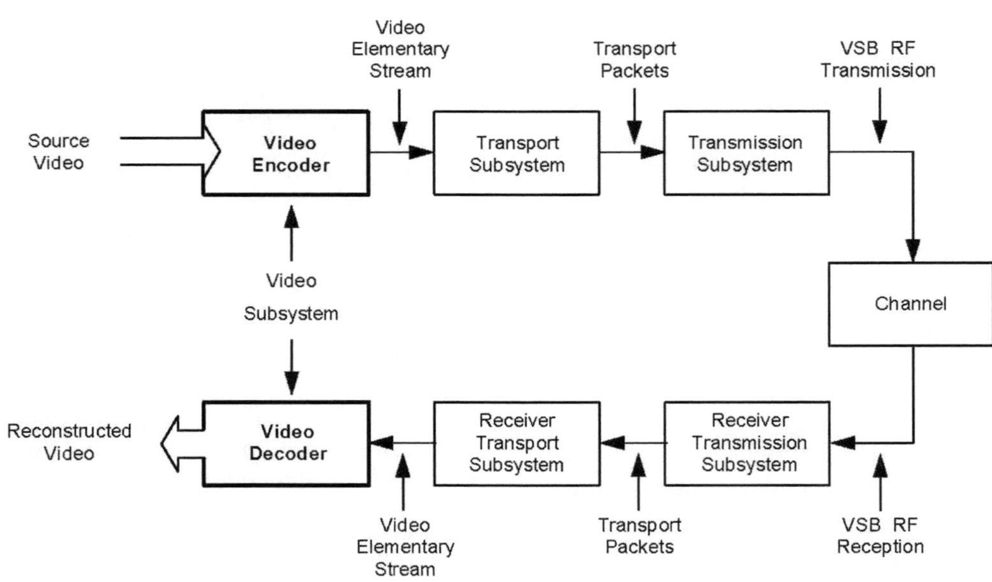

FIGURE 1.11-2 Video coding in relation to the ATSC DTV system.

TABLE 1.11-1
ATSC DTV Compression Formats

Vertical Lines	Pixels	Aspect Ratio	Picture Rate
1080	1920	16:9	60I, 30P, 24P
720	1280	16:9	60P, 30P, 24P
480	704	16:9 and 4:3	60P, 60I, 30P, 24P
480	640	4:3	60P, 60I, 30P, 24P

Video Compression Formats

Table 1.11-1 lists the compression formats allowed in the ATSC DTV standard. In Table 1.11-1, "Vertical Lines" refers to the number of active lines in the picture; "Pixels" refers to the number of pixels during the active line; "Aspect Ratio" refers to the picture aspect ratio; and "Picture Rate" refers to the number of frames or fields per second. In the values for picture rate, "P" refers to progressive scanning, and "I" refers to interlaced scanning. Note that both 60.00 Hz and 59.94 (60 × 1000/1001) Hz picture rates are allowed. Dual rates are allowed also at the picture rates of 30 Hz and 24 Hz. It should be noted that a greater range of video formats are allowed under SCTE 43 [5] and that consumers may expect receivers to decode and display these as well. One format likely to be frequently encountered is 720 pixels by 480 lines, matching the format of ITU-R BT 601-5 [6].

Possible Video Inputs

While not required by the ATSC DTV standard, there are certain digital television production standards, shown in Table 1.11-2, that define video formats that relate to compression formats specified by the standard. The compression formats may be derived from one or more appropriate video input formats. It may be anticipated that additional video production standards will be developed in the future that extend the number of possible input formats.

Sampling Rates

For the 1080 line format, with 1125 total lines per frame and 2200 total samples per line, the sampling frequency is 74.25 MHz for the 30.00 frames per second (fps) frame rate. For the 720 line format, with 750 total lines per frame and 1650 total samples per line, the sampling frequency is 74.25 MHz for the 60.00 fps frame rate. For the 480-line format using 704 pixels, with 525 total lines per frame and 858 total samples per line, the sampling frequency is 13.5 MHz for the 59.94 Hz field rate. Note that both 59.94 fps and 60.00 fps are acceptable as frame or field rates for the system.

For both the 1080 and 720 line formats, other frame rates, specifically 23.976, 24.00, 29.97, and 30.00 fps rates, are acceptable as input to the system. The sample frequency will be either 74.25 MHz (for 24.00 and 30.00 fps) or 74.25/1.001 MHz for the other rates. The number of total samples per line is the same for either of the paired picture rates. (See SMPTE 274M [7] and SMPTE 296M [8].) The six frame rates noted here are the only allowed frame rates for the ATSC DTV standard. In this discussion, any references to 24 fps include both 23.976 and 24.00 fps, references to 30 fps include both 29.97 and 30.00 fps, and references to 60 fps include both 59.94 and 60.00 fps.

For the 480 line format, there may be 704 or 640 pixels in the active line. The interlaced formats are based on ITU-R BT 601-5 [6]; the progressive formats are based on SMPTE 294M [9]. If the input is based on ITU-R BT 601-5 or SMPTE 294M, it will have 483 active lines with 720 pixels in the active line. Only 480 of the 483 active lines are used for encoding. Only 704 of the 720 pixels are used for encoding; the first eight and the last eight are dropped. The 480-line, 640 pixel picture format is not related to any current video production format. It does, however, correspond to the IBM VGA graphics format and may be used with ITU-R BT.601-5 sources by using appropriate resampling techniques.

Colorimetry

For the purposes of the ATSC DTV standard, "colorimetry" means the combination of color primaries, transfer characteristics, and matrix coefficients. Video inputs conforming to SMPTE 274M and SMPTE 296M have the same colorimetry. Video inputs corresponding to ITU-R BT 601-5 should have SMPTE 170M colorimetry [10].

Active Format Description (AFD)

The ATSC DTV standard makes provisions for conveying active format description (AFD) data [1]. The

TABLE 1.11-2
Standardized Video Input Formats

Video Standard	Active Lines	Active Samples/Line	Picture Rate
SMPTE 274M-1998	1080	1920	24P, 30P, 60I
SMPTE 296M-2001	720	1280	24P, 30P, 60P
SMPTE 293M-2003	483	720	60P
ITU-R BT.601-5	483	720	60I

term "active format" in this context refers to that portion of the coded video frame containing useful information; for example, when 16:9 aspect ratio material is coded in a 4:3 format (such as 480i) letterboxing may be used to avoid cropping the left and right edges of the widescreen image. The black horizontal bars at the top and bottom of the screen contain no useful information and in this case the AFD data would indicate 16:9 video carried inside the 4:3 rectangle. The AFD data solves a troublesome problem in the transition from conventional 4:3 display devices to widescreen 16:9 displays and also addresses the variety of aspect ratios that have been used over the years by the motion picture industry to produce feature films.

There are, of course, a number of different types of video displays in common usage—ranging from 4:3 CRTs to widescreen projection devices and flat-panel displays of various design. Each of these devices may have varying abilities to process incoming video. In terms of input interfaces, these displays may likewise support a range of input signal formats—from composite analog video to IEEE 1394. Possible video source devices include cable, satellite, or terrestrial broadcast set-top (or integrated receiver-decoder) boxes, media players (such as DVDs), analog or digital VHS tape players, and personal video recorders. Although choice is good, this wide range of consumer options presented two problems to be solved:

- No standard method had been agreed upon to communicate to the display device the active area of the video signal. Such a method would be able, for example, to signal that the 4:3 signal contains within it a letterboxed 16:9 video image.
- No standard method had been agreed upon to communicate to the display device, for all interface types, that a given image is intended for 16:9 display.

The AFD solves these problems and, in the process, provides the following benefits:

- Active-area signaling allows the display device to process the incoming signal to make the highest-resolution and most accurate picture possible. Furthermore, the display can take advantage of the knowledge that certain areas of video are currently unused and can implement algorithms that reduce the effects of uneven screen aging.
- Aspect ratio signaling allows the display device to produce the best image possible. In some scenarios, lack of a signaling method translates to restrictions in the ability of the source device to deliver certain otherwise desirable output formats.

Active Area Signaling

A consumer device such as a cable or satellite set-top box cannot reliably determine the active area of video on its own. Even though certain lines at the top and bottom of the screen may be black for periods of time, the situation could change without warning. The only sure way to know the active area is for the service provider to include such data at the time of video compression and to embed the data into the video stream.

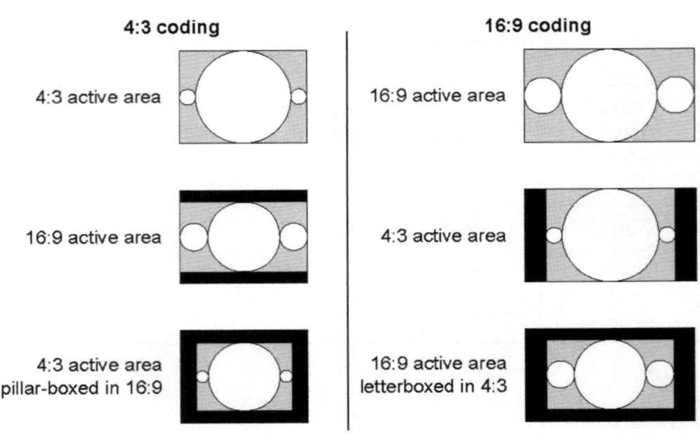

FIGURE 1.11-3 Video coding and active area.

Figure 1.11-3 shows 4:3 and 16:9 coded images with various possible active areas. The group on the left is coded explicitly in the MPEG-2 video syntax as 4:3 or the uncompressed signal provided in the NTSC timing and aspect ratio information (if present) indicates 4:3. The group on the right is coded explicitly in the MPEG-2 video syntax as 16:9; it is provided with the NTSC timing and aspect ratio signal indicating 16:9 or is provided uncompressed with 16:9 timing across the interface.

As shown in the figure, a pillar-boxed display results when a 4:3 active area is displayed within a 16:9 area and a letterboxed display results when a 16:9 active area is displayed within a 4:3 area. It is also apparent that double-boxing can also occur—for example, when 4:3 material is delivered within a 16:9 letterbox to a 4:3 display, or when 16:9 material is delivered within a 4:3 pillar-box to a 16:9 display.

For the straight letter- or pillar-box cases, if the display is aware of the active area it may take steps to mitigate the effects of uneven screen aging. Such steps could, for example, involve using gray instead of black bars. Some amount of linear or nonlinear stretching or zooming may be done as well using the knowledge that video outside the active area can safely be discarded. To address these situations, AFD (as defined in A/53 [1]) may be included in the video user data whenever the rectangular picture area containing useful information does not extend to the full height or width of the coded frame. (AFD data may also be included when the rectangular picture area does extend to the full height and width of the frame.)

AUDIO SYSTEM

As illustrated in Figure 1.11-4, the audio subsystem comprises the audio encoding/decoding function and resides between the audio inputs/outputs and the transport subsystem. An audio encoder is responsible for generating an audio elementary stream (there may

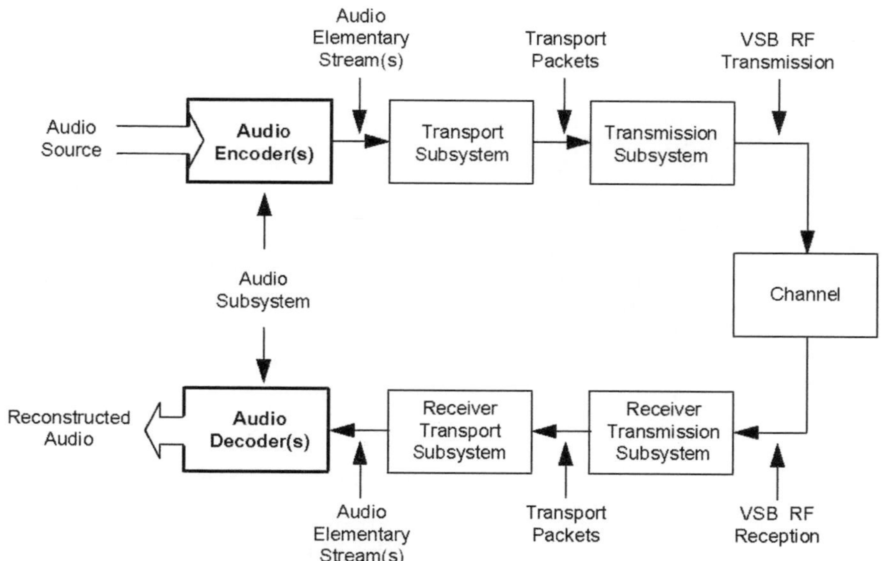

FIGURE 1.11-4 Audio subsystem within the ATSC DTV system.

be more than one stream), which is an encoded representation of the baseband audio input signals. The flexibility of the transport system allows multiple audio elementary streams to be delivered to the receiver. At the receiver, the transport subsystem is responsible for selecting which audio stream to deliver to the audio subsystem. The audio subsystem is responsible for decoding the audio elementary stream back into baseband audio.

An audio program source is encoded by a digital television audio encoder. The output of the audio encoder is a string of bits that represents the audio source, and is referred to as an *audio elementary stream*. The transport subsystem packetizes the audio data into packetized elementary stream (PES) packets, which are then further packetized into transport packets. The transmission subsystem converts the transport packets into a modulated RF signal for transmission to the receiver. At the receiver, the signal is demodulated by the receiver transmission subsystem. The receiver transport subsystem converts the received audio packets back into an audio elementary stream, which is decoded by the digital television audio decoder. The partitioning shown is conceptual, and practical implementations may differ. For example, the transport processing may be broken into two blocks: one to perform PES packetization and the second to perform transport packetization. Or, some of the transport functionality may be included in either the audio encoder or the transmission subsystem.

Audio Encoder Interface

The audio system accepts baseband audio inputs with up to six audio channels per audio program bit stream. The channelization is consistent with ITU-R Recommendation BS-775 (Multi-Channel Stereophonic Sound System With and Without Accompanying Picture). The six audio channels are left, center, right, left surround, right surround, and low-frequency enhancement (LFE). Multiple audio elementary bit streams may be conveyed by the transport system. The bandwidth of the LFE channel is limited to 120 Hz. The bandwidth of the other (main) channels is limited to 20 kHz. Low-frequency response may extend to DC but is more typically limited to approximately 3 Hz (–3 dB) by a DC-blocking, high-pass filter. Audio coding efficiency (and thus audio quality) is improved by removing DC offset from audio signals before they are encoded.

Input Source Signal Specification

Audio signals that are input to the audio system may be in analog or digital form. Audio signals should have any DC offset removed before being encoded. If the audio encoder does not include a DC blocking high-pass filter, the audio signals should be high-pass filtered before being applied to the audio encoder. For analog input signals, the input connector and signal level are not specified. Conventional broadcast practice may be followed. One commonly used input connector is the 3-pin XLR female (the incoming audio cable uses the male connector) with pin 1 ground, pin 2 hot or positive, and pin 3 neutral or negative. For digital input signals, the input connector and signal format are not specified. Commonly used formats such as the AES 3-1992 two-channel interface may be used [11]. When multiple two-channel inputs are used, the preferred channel assignment is:

- Pair 1—left, right
- Pair 2—center, LFE
- Pair 3—left surround, right surround

Sampling Frequency

The system conveys digital audio sampled at a frequency of 48 kHz, locked to the 27 MHz system clock. If analog signal inputs are employed, the A/D converters should sample at 48 kHz. If digital inputs are employed, the input sampling rate should be 48 kHz, or the audio encoder should contain sampling rate converters that convert the sampling rate to 48 kHz. The sampling rate at the input to the audio encoder must be locked to the video clock for proper operation of the audio subsystem. In general, input signals should be quantized to at least 16-bit resolution. The audio compression system can convey audio signals with up to 24-bit resolution.

Summary of Service Types

The following service types are defined in the Digital Audio Compression Standard (AC-3) [2] and in the ATSC Digital Television Standard [1]:

- *Complete main audio service (CM).* This is the normal mode of operation. All elements of a complete audio program are present. The audio program may be any number of channels from 1 to 5.1.[4]
- *Main audio service, music and effects (ME).* All elements of an audio program are present except for dialogue. This audio program may contain from 1 to 5.1 channels. Dialogue may be provided by a D associated service (that may be simultaneously decoded and added to form a complete program).
- *Associated service: visually impaired (VI).* This is typically a single-channel service, intended to convey a narrative description of the picture content for use by the visually impaired and intended to be decoded along with the main audio service. The VI service also may be provided as a complete mix of all program elements, in which case it may use any number of channels (up to 5.1).
  - —*Associated service: hearing impaired (HI).* This is typically a single-channel service intended to convey dialogue that has been processed for increased intelligibility for the hearing impaired and to be decoded along with the main audio service. The HI service also may be provided as a complete mix of all program elements, in which case it may use any number of channels (up to 5.1).
  - —*Associated service: dialogue (D).* This service conveys dialogue intended to be mixed into a main audio service (ME) that does not contain dialogue.
  - —*Associated service: commentary (C).* This service typically conveys a single-channel of commentary intended to be optionally decoded along with the main audio service. This commentary channel differs from a dialogue service, in that it contains optional instead of necessary program content. The C service also may be provided as a complete mix of all program elements, in which case it may use any number of channels (up to 5.1).
  - —*Associated service: emergency message (E).* This is a single-channel service that is given priority in reproduction. If this service type appears in the transport multiplex, it is routed to the audio decoder. If the audio decoder receives this service type, it will decode and reproduce the E channel while muting the main service.
  - —*Associated service: voice-over (VO).* This is a single-channel service intended to be decoded and added into the center loudspeaker channel.

[4]5.1 channel sound refers to a service providing the following signals: right, center, left, right surround, left surround, and low-frequency effects.

Multi-Lingual Services

Each audio bit stream may be in any language. In order to provide audio services in multiple languages a number of main audio services may be provided, each in a different language. This is the (artistically) preferred method, because it allows unrestricted placement of dialogue along with the dialogue reverberation. The disadvantage of this method is that as much as 384 kbps is needed to provide a full 5.1 channel service for each language. One way to reduce the required bit rate is to reduce the number of audio channels provided for languages with a limited audience. For instance, alternate language versions could be provided in two-channel stereo with a bit-rate of 128 kbps, or a mono version can be supplied at a bit rate of approximately 64 to 96 kbps.

Another way to offer service in multiple languages is to provide a main multi-channel audio service (ME) that does not contain dialogue. Multiple single-channel dialogue associated services (D) can then be provided, each at a bit rate of approximately 64 to 96 kbps. Formation of a complete audio program requires that the appropriate language D service be simultaneously decoded and mixed into the ME service. This method allows a large number of languages to be efficiently provided, but at the expense of artistic limitations. The single-channel of dialogue would be mixed into the center reproduction channel and could not be panned. Also, reverberation would be confined to the center channel, which is not optimum. Nevertheless, for some types of programming (sports, etc.), this method is attractive due to the savings in bit rate it offers. Some receivers may not have the capability to simultaneously decode an ME and a D service.

Stereo (two-channel) service without artistic limitation can be provided in multiple languages with added efficiency by transmitting a stereo ME main service along with stereo D services. The D and appropriate language ME services are combined in the receiver into a complete stereo program. Dialogue may be panned, and reverberation may be included in both channels. A stereo ME service can be sent with high quality at 192 kbps, while the stereo D services (voice only) can make use of lower bit rates, such as

128 or 96 kbps per language. Some receivers may not have the capability to simultaneously decode an ME and a D service. Note that, during those times when dialogue is not present, the D services can be momentarily removed and their data capacity used for other purposes.

Audio Bit Rates

The information in Table 1.11-3 provides a general guideline as to the audio bit rates that are expected to be most useful. For main services, the use of the LFE channel is optional and will not affect the indicated data rates. The audio decoder input buffer size (and thus part of the decoder cost) is determined by the maximum bit rate that must be decoded. The syntax of the AC-3 standard supports bit rates ranging from a minimum of 32 kbps up to a maximum of 640 kbps per individual elementary bit stream. The bit rate utilized in the ATSC DTV system is restricted in order to reduce the size of the input buffer in the audio decoder, and thus the receiver cost. Receivers can be expected to support the decoding of a main audio service, or an associated audio service that is a complete service (containing all necessary program elements), at a bit rate up to and including 448 kbps. Transmissions may contain main audio services, or associated audio services that are complete services (containing all necessary program elements), encoded at a bit rate up to and including 448 kbps. Transmissions may contain single-channel associated audio services intended to be simultaneously decoded along with a main service encoded at a bit rate up to and including 128 kbps. Transmissions may contain dual-channel dialogue associated services intended to be simultaneously decoded along with a main service encoded at a bit rate up to and including 192 kbps. Transmissions have a further limitation that the combined bit rate of a main and an associated service that are intended to be simultaneously reproduced is less than or equal to 576 kbps.

TABLE 1.11-3
Typical Audio Bit Rate

Type of Service	Number of Channels	Typical Bit Rates (kbps)
CM, ME, or associated audio service containing all necessary program elements	5	384–448
CM, ME, or associated audio service containing all necessary program elements	4	320–384
CM, ME, or associated audio service containing all necessary program elements	3	192–320
CM, ME, or associated audio service containing all necessary program elements	2	128–256
VI, narrative only	1	64–128
HI, narrative only	1	64–96
D	1	64–128
D	2	96–192
C, commentary only	1	64–128
E	1	64–128
VO	1	64–128

Enhanced AC-3

Enhanced AC-3 (E-AC-3) [2] is based at its core on AC-3 and is designed specifically to adapt to the changing demands of future audio delivery and storage systems while retaining backward compatibility with existing AC-3 decoders. Features of E-AC-3 include the following:

- It maintains quality at more efficient bit rates (<320 kbps for 5.1-channel audio) and supports data rates as high as 6 Mbps.
- Its higher coding efficiency complements new video compression technologies.
- It allows for fully discrete audio performances of 7.1 channels and beyond.
- It has interactive mixing and streaming capability.
- It allows multiple languages to be carried in a single bit stream.

All E-AC-3 decoders will also decode AC-3 bit streams. In addition, although E-AC-3 is not directly compatible with current AC-3 decoders, it is feasible to perform a modest-complexity conversion into a compliant AC-3 bit stream syntax, thus enabling backward compatibility to legacy decoders that have S/PDIF bit-stream inputs. Important technical capabilities of E-AC-3 that relate directly to ATSC broadcast applications include the following:

- *Expanded data rate flexibility.* E-AC-3 allows the number of blocks per sync frame and the number of compressed data bits per frame to be adjusted to achieve significantly more data rate flexibility than standard AC-3, including a greater maximum theoretical data rate and finer data rate granularity.
- *Spectral extension.* Enhanced AC-3 decoders support a new coding technique called *spectral extension*. Like channel coupling, spectral extension codes the highest frequency content of the signal more efficiently. Spectral extension recreates a signal's high-frequency spectrum from side data transmitted in the bit stream that characterizes the original signal, as well as from actual signal content from the lower frequency portion of the signal. Because it may be desirable, in some circumstances, to use channel coupling for a mid-range portion of the frequency spectrum and spectral extension for the higher-range portion of

the frequency spectrum, spectral extension is fully compatible with channel coupling. Both tools can be enabled at the same time, for different portions of the frequency spectrum.

- *Transient pre-noise processing.* This is an optional decoder tool that improves audible performance through the substitution of audio segments just before transients to reduce the duration of pre-noise distortions. This technique is referred to as *time-scaling synthesis*, where synthesized pulse-code modulation (PCM) audio segments are used to eliminate the transient pre-noise, thereby improving the perceived quality of low-bit rate audio coded transient material. To enable the decoder to efficiently perform transient pre-noise processing with no impact on decoding latency, transient location detection and time-scaling synthesis analysis are performed by the encoder and the information transmitted to the decoder. The encoder performs transient pre-noise processing for each full bandwidth audio channel and transmits "helper" information once per frame only when necessary (for example, when transients are present that will benefit from the technique).
- *Adaptive hybrid transform processing.* In 1995, the transform employed in A/52 AC-3—based on a modified discrete cosine transform (MDCT) of length 256 frequency samples—provided a reasonable trade-off between audio coding gain and decoder implementation cost. With continuing advances in silicon manufacturing processes over the years, the integrated circuit complexity that constitutes a reasonable level has now increased. This increase in chip performance provides an opportunity to improve the coding gain of AC-3 and hence perceptual audio quality at a given bitrate by increasing the length of the transform. This is accomplished through use of the adaptive hybrid transform (AHT), which adds a second transform in cascade in order to generate a single transform with 1536 frequency samples.
- *Enhanced coupling.* This is a new tool that improves the imaging properties of coupled signals by adding phase compensation to the amplitude-based processing of conventional coupling. Prior to down-mixing the coupled channels to a single composite signal, the encoder derives both amplitude and additionally interchannel phase information on a sub-band basis for each channel. The phase information includes a decorrelation scale factor as a measure of the variation of the phase within a frame. This side-chain information is transmitted to the decoder once per frame. The decoder uses the information to recover the multiple output channels from the composite signal using a combination of both amplitude scaling and phase rotation. The result is an improvement in soundstage imaging over conventional coupling. This improvement allows the technique to be used at lower frequencies than conventional coupling, thus improving coding efficiency.

Additional features of E-AC-3 of particular interest to applications outside of ATSC DTV include:

- *Channel and program extensions.* The E-AC-3 bitstream syntax allows for time-multiplexed substreams to be present in a single bit stream. With this capability, the E-AC-3 bit stream syntax enables a single program with greater than 5.1 channels, multiple programs of up to 5.1 channels, or a mixture of programs with up to 5.1 channels and programs with greater than 5.1 channels to be carried in a single bit stream. These extra channels do not affect a two-channel or 5.1-channel decoder in ATSC broadcast applications.
- *Sample rate processing.* Additional metadata is reserved for applications that involve source material sampled at 2× the nominal rate, such as 96 kHz and 88.2 kHz.
- *Mixing control processing.* Additional metadata are reserved for applications that involve the mixing of two program streams. These applications require control of the mixing process and resultant dynamic range control metadata; this feature reserves data capacity to accomplish this task.

ATSC DTV TRANSPORT

The ATSC DTV transport subsystem employs the fixed-length transport stream packetization approach defined in ISO/IEC13818-1 [12], which is usually referred to as the MPEG-2 Systems Standard. This approach is well suited to the needs of terrestrial broadcast and cable television transmission of digital television. The use of relatively short, fixed-length packets matches well with the needs and techniques for error protection in both terrestrial broadcast and cable television distribution environments.

The ATSC DTV transport may carry a number of television programs. The MPEG-2 term "program" corresponds to an individual digital TV channel or data service, where each program is composed of a number of MPEG-2 program elements (*i.e.*, related video, audio, and data streams). The MPEG-2 Systems Standard support for multiple channels or services within a single, multiplexed bit stream enables the deployment of practical, bandwidth-efficient digital broadcasting systems. It also provides great flexibility to accommodate the initial needs of the service to multiplex video, audio, and data while providing a well-defined path to add additional services in the future in a fully backward-compatible manner. By basing the transport subsystem on MPEG-2, maximum interoperability with other media and standards is maintained.

Referring to Figure 1.11-1, the transport subsystem resides between the application (*e.g.*, audio or video) encoding and decoding functions and the transmission subsystem. At its lowest layer, the encoder transport subsystem is responsible for formatting the encoded bits and multiplexing the different components of the program for transmission. At the receiver, it is responsible for recovering the bit streams for the

individual application decoders and for the corresponding error signaling. The transport subsystem also incorporates other higher-level functionality related to identification of applications and synchronization of the receiver.

Program and System Information Protocol (PSIP)

The Program and System Information Protocol (PSIP) addresses data transmitted along with a station's DTV signal that gives DTV receivers important information about the station and what is being broadcast. Described in ATSC Standard A/65 [3], the most important function of PSIP is to provide a method for ATSC DTV receivers to identify a DTV station and to determine how a receiver can tune to it. PSIP identifies both the DTV channel and the associated NTSC (analog) channel. It helps maintain the current channel branding because DTV receivers will electronically associate the two channels, making it easy for viewers to tune to the DTV station even if they do not know the RF channel number. In addition to identifying the channel number, PSIP tells the receiver whether multiple program channels are being broadcast and, if so, how to find them. It identifies, for example, whether or not the programs are closed captioned, it conveys V-chip information, and signals whether or not data are associated with the program.

PSIP Structure

PSIP is a collection of tables, each of which describes elements of typical digital television services [13]. Figure 1.11-5 shows the primary components and the notation used to describe them. The packets of the base tables are all labeled with a base *packet identifier* (PID). The base tables are:

- System Time Table (STT)
- Rating Region Table (RRT)
- Master Guide Table (MGT)
- Virtual Channel Table (VCT)

The Event Information Tables (EITs) are a second set of tables, whose packet identifiers are defined in the MGT. The Extended Text Tables (ETTs) are a third set of tables, and, similarly, their PIDs are defined in the MGT.

The STT is a small data structure that fits in one transport stream packet and serves as a reference for time-of-day functions. Receivers can use this table to manage various operations and scheduled events, as well as display the time of day. The RRT has been designed to transmit the rating system in use for each country using the ratings. In the United States, this is incorrectly but frequently referred to as the "V-chip" system; the proper title is Television Parental Guidelines (TVPG). Provisions have been made for multi-country systems. The MGT provides indexing information for the other tables that comprise the PSIP standard. It also defines table sizes necessary for memory allocation during decoding, defines version numbers to identify those tables that need to be updated, and generates the packet identifiers that label the tables. The VGT, also referred to as the Terrestrial Virtual Channel Table (TVCT), contains a list of all the channels that are or will be on-line, plus their attributes. Among the attributes given are the channel name and channel number.

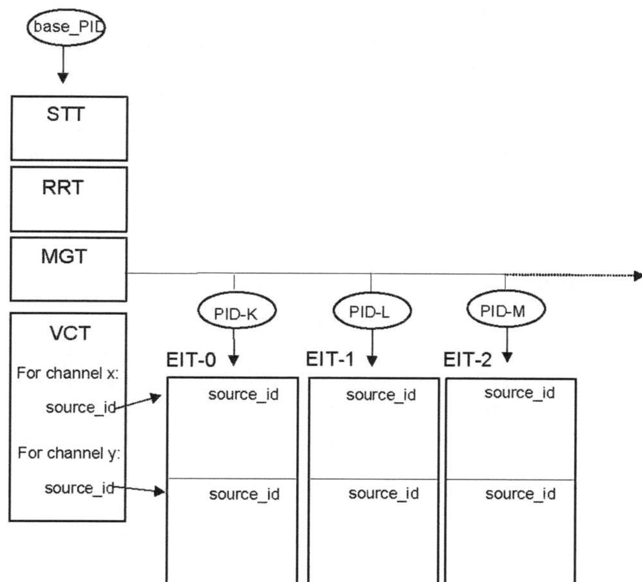

FIGURE 1.11-5 Overall structure of the PSIP tables.

There are up to 128 Event Information Tables (EIT-0 through EIT-127), each of which describes the events or television programs for a time interval of 3 hours. Because the maximum number of EITs is 128, up to 16 days of programming may be advertised in advance. At minimum, the first 4 EITs must always be present in every transport stream, and 24 are recommended. Each EIT-*k* may have multiple instances, one for each virtual channel in the VCT. As illustrated in Figure 1.11-6, there may be multiple ETTs, one or more channel ETT sections describing the virtual channels in the VCT, and an ETT-*k* for each EIT-*k*, describing the events in the EIT-*k*. These are all listed in the MGT. An ETT-*k* contains a table instance for each event in

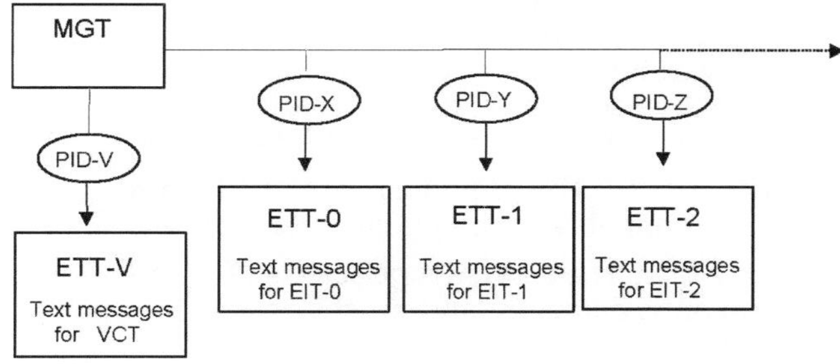

FIGURE 1.11-6 Extended text tables in the PSIP hierarchy.

the associated EIT-*k*. As the name implies, the purpose of the ETT is to carry text messages; for example, for channels in the VCT, the messages can describe channel information, cost, coming attractions, and other related data. Similarly, for an event such as a movie listed in the EIT, the typical message would be a short paragraph that describes the movie itself. ETTs are optional in the ATSC system.

PSIP Requirements for Broadcasters

The three main tables (VCT, EIT, STT) contain information to facilitate suitably equipped receivers to find the components needed to present a program (event). Although receivers are expected to use stored information to speed channel acquisition, sometimes parameters must change, and the VCT and EIT-0 are the tables that must be accurate each instant as they provide the actual connection path and critical information that can affect the display of the events. If nothing has changed since an EIT was sent for an event, then the anticipatory use of the data is expected to proceed, and when there is a change the new parts would be used. Additional tables provide TV parental advisory information and extended text messages about certain events. These relationships—and the tables that carry them—are designed to be kept with the DTV signal when it is carried by a cable system.

The Basics

There are certain "must have" items and "must do" rules of operation. If the PSIP elements are missing or wrong, there may be severe consequences, which vary depending on the design of the receiver. The following are key elements that must be set or checked by each station:

- *Transport stream identification (TSID).* The pre-assigned TSIDs must be set correctly in all three locations (PAT, VCT common information, and virtual channel-specific information).
- *System time table (STT).* The STT time should be checked daily and locked to house time. Ideally, the STT should be inserted into the TS within a frame before each seconds-count increment of the house time with the to-be-valid value.
- *Short channel name.* This is a seven-character name that can be set to any desired name indicating the virtual channel name—for example, a station's call letters followed by SD1, SD2, SD3, and SD4 to indicated various SDTV virtual channels or anything else to represent the station's identity (*e.g.*, WNABSD1, KNABSD2, WNAB-HD, KIDS).
- *Major channel.* The previously assigned, paired NTSC channel is the major channel number.
- *Service type.* The service type selects DTV, NTSC, audio only, data, etc., and must be set as the operating modes require.
- *Modulation mode.* This is a code for the RF modulation of the virtual channel.
- *Source ID.* The source ID is a number that associates virtual channels to events on those channels. It typically is automatically updated by PSIP equipment or updated from an outside vendor. Proper operation of this feature should be confirmed.
- *Service location descriptor (SLD).* The SLD contains the MPEG references to the contents of each component of the programs plus a language code for audio. The PID values for the components identified here and in the PMT must be the same for the elements of an event/program. Some deployed systems require separate manual setup, but PID values assigned to a VC should seldom change.

The maximum cycle time/repetition rate of the tables should be set or confirmed to conform with the suggested guidelines given in Table 1.11-4 for mandatory PSIP tables and Table 1.11-5 for optional PSIP tables.

It is recommended that broadcasters send populated EITs covering at least 3 days. The primary cycle time guidelines are illustrated in Figure 1.11-7. The recommended table cycle times given in this section result in a minimal demand on overall system bandwidth. Considering the importance of the information that these PSIP tables provide to the receiver, the bandwidth penalty is trivial.

Program Guide

Support for an electronic program guide (EPG) or interactive program guide (IPG) is another important function enabled by the PSIP standard. The concept is to provide a way for viewers to find out what is on directly from their television sets, similar to the *guide channels* that are typically available for cable and satellite broadcast services. In a terrestrial broadcast environment, there is no single authority that determines what programs are on all the channels, so each broadcaster needs to include this type of information within the broadcast stream. Receivers can scan all the available channels and create a program guide from the

TABLE 1.11-4
Mandatory PSIP Table Suggested Repetition Rates

PSIP Table	Transmission Cycle, Once Every:
Master Guide Table (MGT)	150 ms
Terrestrial Virtual Channel Table (TVCT)	400 ms
Event Information Table 0 (EIT-0)	0.5 s
EIT-1	3 s
EIT-2 and EIT-3	Minute
System Time Table (STT)	Second
Rating Region Table (RRT; not required in some areas, such as the United States)	Minute

TABLE 1.11-5
Suggested Repetition Rates for Optional PSIP Tables

PSIP Table		Transmission Cycle, Once Every:
DCCT*		
	DCC request in progress	150 ms
	2 seconds prior to DCC request	400 ms
	No DCC	N/A
DCCSCT†		Hour
Extended Text Table (ETT)		Minute
Event Information Table 4 (EIT-4) and higher		Minute
DET		A later version of this recommended practice will address data services.

*Directed Channel Change Table A/65 specifies these repetition rates for DCC per the specified conditions.
†Directed Channel Change Selection Code Table

consolidated information. The more viewers rely on these guides, the more critical the accuracy detail of the information they contain. Figure 1.11-8 shows a typical electronic program guide.

Building the On-Screen Display or EPG

The EIT has the dual functionality of announcing future programs and providing critical information about the current program. It contains program names and planned broadcast times as well as other information about an event. The data can be combined to build a receiver on-screen display such as that shown in Figure 1.11-9.

Programming and Metadata Communication Protocol (PMCP)

The Programming and Metadata Communication Protocol (ATSC A/76 [15]) provides the means to integrate the various information sources that are needed to compile the key PSIP tables. PMCP is designed to permit broadcasters, professional equipment manufacturers, and program service providers to interconnect and transfer data among systems that eventually

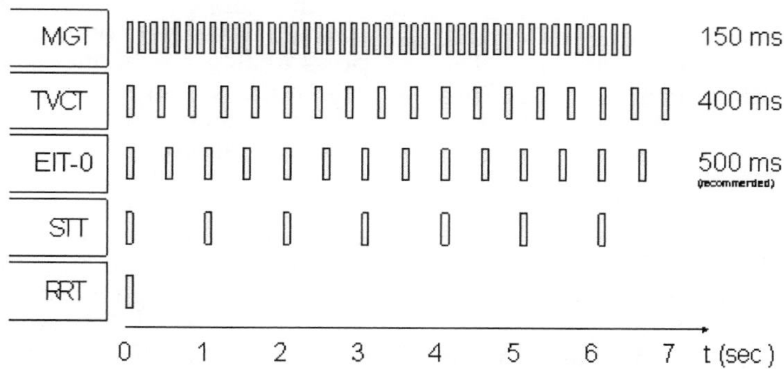

FIGURE 1.11-7 Recommended PSIP table cycle times.

Chan	Name	6:00 PM	6:30 PM	7:00 PM	7:30 PM	8:00 PM	8:30 PM
6-0	XYZ	City Scene		Travel Log		Movie: Speed II	
6-1	XYZ	City Scene		Travel Log		Movie: Speed II (HD)	
6-2	XYZ	Movie: Star Trek—The Voyage Home				Tune 6-1 for Movie: Speed II (HD)	
6-3	LNC	Local News		Airport Info		HD Program on 6-1	

FIGURE 1.11-8 Example electronic program guide.

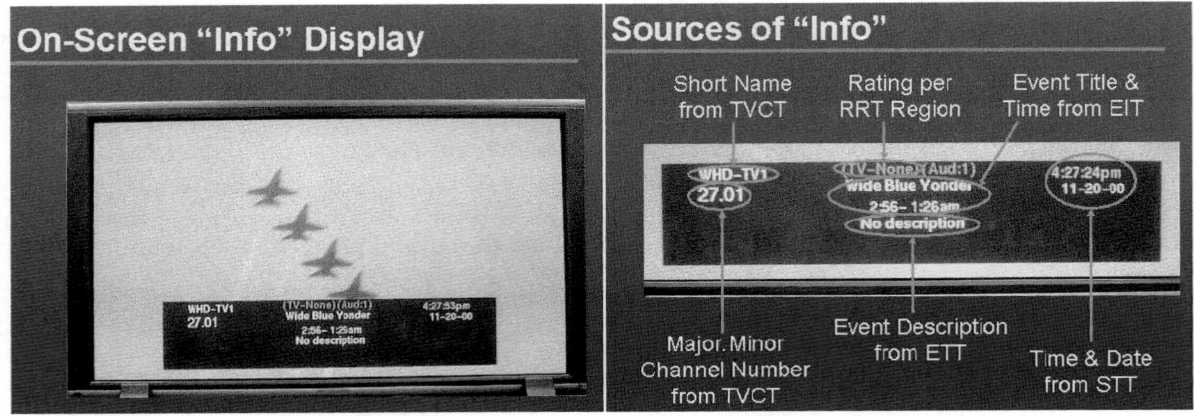

FIGURE 1.11-9 Illustration of how the various PSIP tables could combine to produce the on-screen display at the receiver.

must be communicated to the PSIP generator. These systems include:

- Traffic
- Program management
- Listing services
- Automation
- MPEG encoder

PMCP is based on a protocol utilizing eXtensible Markup Language (XML) message documents. The heart of the standard is an XML schema that defines the message structure, the elements allowed, their relationships, and attributes. XML is widely recognized as being flexible and usable for various system architectures. Equally important, it is capable of deployment without extensive development costs on the part of equipment vendors or service providers. XML is a World Wide Web Consortium (W3C) standard that allows structuring of information in a text document that is both human and machine readable. PMCP references and is complementary to other existing ATSC standards. It also supports the ISO V-ISAN standard for unique identification of program content.

RF TRANSMISSION

The ATSC transmission system offers two basic operational modes: a terrestrial broadcast mode (8-VSB) and a high-data-rate mode (16VSB) intended for cable applications. Both modes provide a pilot, segment syncs, and a training sequence (as part of data field sync) for acquisition and operation. The two system modes can use the same carrier recovery, demodulation, sync recovery, and clock recovery circuits. Adaptive equalization for the two modes can use the same equalizer structure with some differences in the decision feedback and adaptation of the filter coefficients. Furthermore, both modes use the same Reed–Solomon (RS) code and circuitry for *forward error correction* (FEC). The terrestrial broadcast mode is optimized for maximum service area and provides a data payload of approximately 19.4 Mbps in a 6 MHz channel. The high data rate mode, which provides twice the data rate at the cost of reduced robustness for channel degradations such as noise and multipath, provides a data payload of 38.8 Mbps in a single 6 MHz channel.

To maximize service area, the terrestrial broadcast mode incorporates trellis coding with added precoding that allows the data to be decoded after passing through a receiver comb filter, used selectively to suppress analog co-channel interference. The high-data-rate mode is designed to work in a cable environment, which is less severe than that of the terrestrial system. It is transmitted in the form of more data levels (bits/symbol). No trellis coding or precoding for an analog broadcast interference rejection (comb) filter is employed in this mode.

VSB transmission with a raised-cosine roll-off at both the lower edge (pilot carrier side) and upper edge (Nyquist slope at 5.38 MHz above carrier) permits equalizing just the in-phase (I) channel signal with a sampling rate as low as the symbol (baud) rate. The raised-cosine shape is obtained from concatenating a root-raised cosine in the transmitter with the same shape in the receiver. Although energy in the vestigial sideband and in the upper band edge extends beyond the Nyquist limit frequencies, the demodulation and sampling process aliases this energy into the baseband to suppress *intersymbol interference* (ISI) and thereby avoid distortion. With the carrier frequency located at the –6 dB point on the carrier-side raised-cosine roll-off, energy in the vestigial sideband folds around zero frequency during demodulation to make the baseband DTV signal exhibit a flat amplitude response at lower frequencies, thereby suppressing low-frequency ISI. Then, during digitization by synchronous sampling of the demodulated I signal, the Nyquist slope through 5.38 MHz suppresses the remnant higher frequency ISI. With ISI due to aliasing thus eliminated, equalization of linear distortions can be done using a single A/D converter sampling at the symbol rate of 10.76 million samples per second (Msamples/s) and a real-only (not complex) equalizer also operating at the symbol rate. In this simple case, equalization of the signal beyond the –6 dB points in the raised-cosine roll-offs at channel edges is dependent on the in-band equalization and cannot be set independently. A complex equalizer does not have this limitation, nor does a fractional equalizer sampling at a rate sufficiently above the symbol rate.

The 8-VSB signal is designed to minimize interference and RF channel allocation problems. The VSB signal is designed to minimize the peak-energy-to-average-energy ratio, thereby minimizing interference into other signals, especially adjacent and taboo channels. To counter the man-made noise that often accompanies over-the-air broadcast signals, the VSB system includes an interleaver that allows correction of an isolated single burst of noise up to 190 µs in length by the (207,187) RS FEC circuitry, which locates as well as corrects up to 10 byte errors per data segment. This was done to allow VHF channels, which are often substantially affected by man-made noise, to be used for DTV broadcasting. If soft-decision techniques are used in the trellis decoder preceding the RS circuitry, the location of errors can be flagged, and twice as many byte errors per data segment can be corrected, allowing correction of an isolated burst of up to 380 µs in length. The parameters for the two VSB transmission modes are shown in Table 1.11-6.

Bit Rate Delivered to a Transport Decoder by the Transmission Subsystem

As outlined previously, all data in the ATSC DTV system is transported in MPEG-2 transport packets. The useful data rate is the amount of MPEG-2 transport data carried end-to-end including MPEG-2 packet headers and sync bytes. The exact symbol rate of the transmission subsystem is given by:

$$\frac{4.5}{286} \times 684 = 10.76\ldots \text{Msymbols/s (megabaud)} \qquad (1)$$

The symbol rate must be locked in frequency to the transport rate.

The numbers in the formula for the ATSC symbol rate in 6 MHz systems are related to NTSC scanning and color frequencies. Because of this relationship, the symbol clock can be used as a basis for generating an NTSC color subcarrier for analog output from a set-top box. The repetition rates of data segments and data frames are deliberately chosen not to have an integer relationship to NTSC or PAL scanning rates to ensure that there will be no discernible pattern in the co-channel interference. The particular numbers used are:

- 4.5 MHz = the center frequency of the audio carrier offset in NTSC. This number was traditionally used in NTSC literature to derive the color subcarrier frequency and scanning rates. In modern equipment, this may start with a precision 10 MHz reference, which is then multiplied by 9/20.
- 4.5 MHz/286 = the horizontal scan rate of NTSC, 15,734.2657 + … Hz (note that the color subcarrier is 455/2 times this, or 3,579,545 + 5/11 Hz).
- 684 is a multiplier that gives a symbol rate for an efficient use of bandwidth in 6 MHz. It requires a filter with Nyquist roll-off that is a fairly sharp cut-off (11% excess bandwidth), which is still realizable with a reasonable surface acoustic wave (SAW) filter or digital filter.

In the terrestrial broadcast mode, channel symbols carry 3 bits/symbol of trellis-coded data. The trellis

TABLE 1.11-6
Parameters for VSB Transmission Modes

Parameter	Terrestrial Mode	High-Data-Rate Mode
Channel bandwidth	6 MHz	
Excess bandwidth	11.5%	
Symbol rate	10.76 million symbols per second (Msymbols/s)	
Bits per symbol	3	4
Trellis forward error correction (FEC)	2/3 rate	None
Reed-Solomon FEC	$T = 10$ (207,187)	
Segment length	832 symbols	
Segment sync	4 symbols per segment	
Frame sync	1 per 313 segments	
Payload data rate	19.39 Mbps	38.78 Mbps
Analog co-channel rejection	Analog rejection filter in receiver	N/A
Pilot power contribution	0.3 dB	
C/N threshold	14.9 dB	28.3 dB

code rate is 2/3, providing 2 bits/symbol of gross payload. Therefore, the gross payload is:

$$10.76 \times 2 = 21.52\ldots\text{megabits/second (Mbps)} \qquad (2)$$

To find the net payload delivered to a decoder it is necessary to adjust Eq. (2) for the overhead of the *data segment sync*, *data field sync*, and Reed-Solomon FEC.

To get the net bit rate for an MPEG-2 stream carried by the system (and supplied to an MPEG transport decoder), it is first noted that the MPEG sync bytes are removed from the data stream input to the 8-VSB transmitter, replaced with segment sync, and later reconstituted at the receiver. For throughput of MPEG packets (the only allowed transport mechanism), segment sync is simply equivalent to transmitting the MPEG sync byte and does not reduce the net data rate. The net bit rate of an MPEG-2 stream carried by the system and delivered to the transport decoder is accordingly reduced by the data field sync (one segment of every 313) and the Reed-Solomon coding (20 bytes of every 208):

$$21.52\ldots\text{Mbps} \times \frac{312}{313} \times \frac{188}{208} = 19.39\ldots\text{Mbps} \qquad (3)$$

The net bit rate supplied to the transport decoder for the high data rate mode is:

$$19.39\ldots\text{Mbps} \times 2 = 38.78\ldots\text{Mbps} \qquad (4)$$

Performance Characteristics of Terrestrial Broadcast Mode

The terrestrial 8-VSB system can operate in a signal-to-additive-white-Gaussian-noise (S/N) environment of 14.9 dB. The 8-VSB segment error probability curve including 4-state trellis decoding and (207,187) Reed-Solomon decoding shown in Figure 1.11-10 indicates a segment error probability of 1.93×10^{-4}. This is equivalent to 2.5 segment errors per second, which was established by measurement as the *threshold of visibility* (TOV) of errors in the prototype equipment [14]. Particular product designs may achieve somewhat better performance for subjective TOV by means of error masking.

The *cumulative distribution function* (CDF) of the peak-to-average power ratio, as measured on a low-power transmitted signal with no nonlinearities, is plotted in Figure 1.11-11. The plot shows that 99.9% of the time the transient peak power is within 6.3 dB of the average power.

Transmitter Signal Processing

A pre-equalizer filter is recommended for use in over-the-air broadcasts where the high-power transmitter may have significant in-band ripple or significant roll-off at band edges. Pre-equalization is typically required in order to compensate the high-order filter used to meet a stringent out-of-band emission mask, such as the Federal Communications Commission (FCC)-required mask [16]. This linear distortion can be measured by an equalizer in a reference demodulator ("ideal" receiver) employed at the transmitter site. A directional coupler, which is recommended to be located at the sending end of the antenna feed transmission line, supplies the reference demodulator a small sample of the antenna signal feed. The equalizer tap weights of the reference demodulator are transferred to the transmitter pre-equalizer for precorrection of transmitter linear distortion. This is a one-time procedure of measurement and transmitter pre-equalizer adjustment. Alternatively, the transmitter pre-

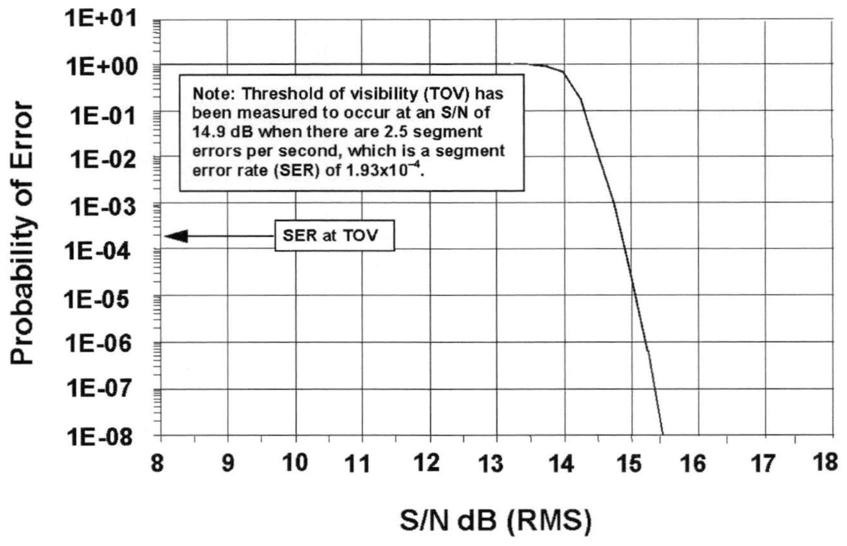

FIGURE 1.11-10 Segment error probability, 8-VSB with 4-state trellis decoding and (207,187) Reed-Solomon decoding. The plotted values shown are approximate.

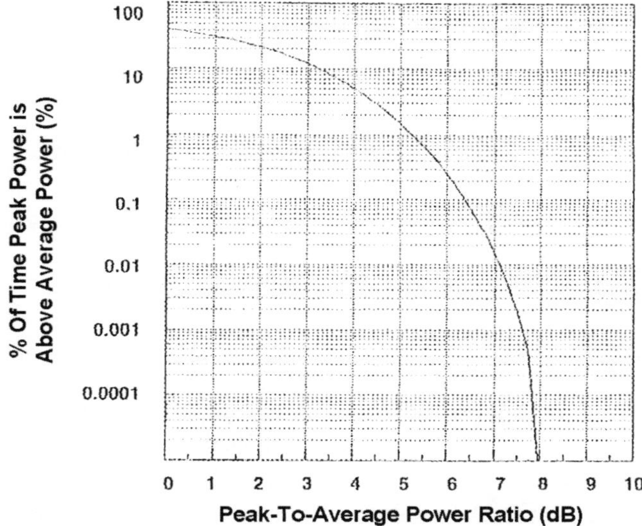

FIGURE 1.11-11 Cumulative distribution function of 8-VSB peak-to-average power ratio (in ideal linear system).

equalizer can be made continuously adaptive. In this arrangement, the reference demodulator is provided with a fixed-coefficient equalizer compensating for its own deficiencies in ideal response.

A pre-equalizer suitable for many applications is an 80 tap, feed-forward transversal filter. The taps are symbol spaced (93 ns), with the main tap being approximately at the center, giving approximately ±3.7 μs correction range. The pre-equalizer operates on the I channel data signal (there is no Q channel data signal in the transmitter) and shapes the frequency spectrum of the IF signal so that there is a flat in-band spectrum at the output of the high-power transmitter that feeds the antenna for transmission. There is no effect on the out-of-band spectrum of the transmitted signal. If desired, complex equalizers or fractional equalizers (with closer-spaced taps) can provide independent control of the outer portions of the spectrum (beyond the Nyquist slopes).

The transmitter vestigial sideband filtering is sometimes implemented by sideband cancellation, using the phasing method. In this method, the baseband data signal is supplied to digital filtering that generates in-phase and quadrature-phase digital modulation signals for application to respective D/A converters. This filtering process provides the root raised cosine Nyquist filtering and provides compensation for the $(\sin x)/x$ frequency responses of the D/A converters, as well. The baseband signals are converted to analog form. The in-phase signal modulates the amplitude of the intermediate frequency (IF) carrier at 0° phase, while the quadrature signal modulates a 90° shifted version of the carrier. The amplitude-modulated quadrature IF carriers are added to create the vestigial sideband IF signal, canceling the unwanted sideband and increasing the desired sideband by 6 dB.

Upconverter and RF Carrier Frequency Offsets

Modern analog TV transmitters use a two-step modulation process. The first step usually is modulation of the data onto an IF carrier, which is the same frequency for all channels, followed by translation to the desired RF channel. The digital 8-VSB transmitter applies this same two-step modulation process. The RF upconverter translates the filtered flat IF data signal spectrum to the desired RF channel. For the same approximate coverage as an analog transmitter (at the same frequency), the average power of the ATSC DTV signal is on the order of 12 dB less than the analog peak sync power (when operating on the same frequency). The nominal frequency of the RF upconverter oscillator in ATSC DTV terrestrial broadcasts will typically be the same as that used for analog transmitters (except for offsets required in particular situations). Note that all examples in this section relate to a 6 MHz ATSC DTV system. Values may be modified easily for other channel widths.

Nominal ATSC DTV Pilot Carrier Frequency

The nominal ATSC DTV pilot carrier frequency is determined by fitting the DTV spectrum symmetrically into the RF channel. This is obtained by taking the bandwidth of the DTV signal—5,381.1189 kHz (the Nyquist frequency difference or one-half the symbol clock frequency of 10,762.2378 kHz)—and centering it in the 6 MHz TV channel. Subtracting 5381.1189 kHz from 6000 kHz leaves 618.881119 kHz. Half of that is 309.440559 kHz, precisely the standard pilot offset above the lower channel edge. For example, on channel 45 (656–662 MHz), the nominal pilot frequency is 656.309440559 MHz.

Requirements for Offsets

There are two categories of requirements for pilot frequency offsets:

- Offsets to protect lower adjacent channel analog broadcasts, mandated by FCC rules in the United States and which override other offset considerations.
- Recommended offsets for other considerations such as co-channel interference between DTV stations or between DTV and analog stations.

Upper DTV Channel into Lower Analog Channel

This is the overriding case mandated by the FCC rules in the United States—precision offset with a lower adjacent analog station, full service, or low-power television (LPTV). The FCC Rules, Section 73.622(g)(1), state that:

> DTV stations operating on a channel allotment designated with a "c" in paragraph (b) of this section must maintain the pilot carrier frequency of the DTV signal 5.082138 MHz above the visual carrier frequency of any analog TV broadcast station that operates on the lower adjacent channel and is located within 88 kilometers. This frequency difference must be maintained within a tolerance of ±3 Hz.

This precise offset is necessary to reduce the color beat and high-frequency luminance beat created by the DTV pilot carrier in some receivers tuned to the lower adjacent analog channel. The tight tolerance ensures that the beat will be visually cancelled, as it will be out of phase on successive video frames.

Note that the frequency is expressed with respect to the lower adjacent analog video carrier, rather than the nominal channel edge. This is because the beat frequency depends on this relationship, and therefore the DTV pilot frequency must track any offsets in the analog video carrier frequency. The offset in the FCC rules is related to the particular horizontal scanning rate of NTSC and can easily be modified for PAL. Offset O_f was obtained from:

$$O_f = 455 \times \left(\frac{F_h}{2}\right) + 191 \times \left(\frac{F_h}{2}\right) - 29.97$$
$$= 5{,}082{,}138 \text{ Hz} \qquad (5)$$

where F_h = NTSC horizontal scanning frequency = 15,734.264 Hz.

The equation indicates that the offset with respect to the lower adjacent chroma is an odd multiple (191) of one-half the line rate to eliminate the color beat; however, this choice leaves the possibility of a luminance beat. The offset is additionally adjusted by one-half the analog field rate to eliminate the luminance beat. While satisfying the exact adjacent channel criteria, this offset is also as close as possible to optimal comb filtering of the analog co-channel in the digital receiver. Note additionally that offsets are to higher frequencies, rather than lower, to avoid any possibility of encroaching on the lower adjacent sound. (It also reduces the likelihood of the automatic fine tuning [AFT] in the analog receiver experiencing lock-out because the signal energy including the pilot is moved further from the analog receiver bandpass.)

Other Offset Cases

The FCC rules do not consider other interference cases where offsets help. The offset for protecting lower-adjacent analog signals takes precedence. If that offset is not required, other offsets can minimize interference to co-channel analog or DTV signals. In co-channel cases, DTV interference into analog TV appears noise-like. The pilot carrier is low on the Nyquist slope of the IF filter in the analog receiver, so no discernable beat is generated. In this case, offsets to protect the analog channel are not required. Offsets are useful, however, to reduce co-channel interference from analog TV into DTV. The performance of the analog rejection filter and clock recovery in the DTV receiver will be improved if the DTV carrier is 911.944 kHz below the NTSC visual carrier. In other words, in the case of a 6 MHz NTSC system, if the analog TV station is not offset, the DTV pilot carrier frequency will be 338.0556 kHz above the lower channel edge instead of the nominal 309.44056 kHz. As before, if the NTSC station is operating with a ±10 kHz offset, the DTV frequency will have to be adjusted in the same direction. The formula for calculating this offset is:

$$F_{pilot} = F_{vis(n)} - 70.5 \times F_{seg}$$
$$= 338.0556 \text{ kHz (for no NTSC analog offset)} \qquad (6)$$

where:

F_{pilot} = DTV pilot frequency above lower channel edge.

$F_{vis(n)}$ = NTSC visual carrier frequency above lower channel edge.
= 1250 kHz for no NTSC offset (as shown).
= 1240 kHz for minus offset.
= 1260 kHz for plus offset.

F_{seg} = ATSC data segment rate = symbol clock frequency/832 = 12,935.381971 Hz.

The factor of 70.5 is chosen to provide the best overall comb filtering of analog color TV co-channel interference. The use of a value equal to an integer +0.5 results in co-channel analog TV interference being out-of-phase on successive data segment syncs. Note that in this case the frequency tolerance is plus or minus 1 kHz. More precision is not required. Also note that a different data segment rate would be used for calculating offsets for 7 or 8 MHz systems.

Co-Channel DTV into DTV

If two DTV stations share the same channel, interference between the two stations can be reduced if the pilot is offset by one and a half times the data segment rate. This ensures that the frame and segment syncs of the interfering signal will each alternate polarity and be averaged out in the receiver tuned to the desired signal. The formula for this offset is:

$$F_{offset} = 1.5 \times F_{seg} = 19.4031 \text{ kHz} \qquad (7)$$

where F_{offset} is the offset to be added to one of the two DTV carriers, and F_{seg} = 12,935.381971 Hz (as defined previously). This results in a pilot carrier 328.84363 kHz above the lower band edge, provided neither DTV station has any other offset. Use of the factor 1.5 results in the best co-channel rejection, as determined experimentally with the prototype equipment. The use of an integer +0.5 results in co-channel interference alternating phase on successive segment syncs.

Summary: DTV Offsets

Table 1.11-7 summarizes the various pilot carrier offsets for different interference situations in a 6 MHz system (NTSC environment). Note that if more than two stations are involved the number of potential frequencies will increase. For example, if one DTV station operates at an offset because of a lower-adjacent-channel NTSC station, a co-channel DTV station may have to adjust its frequency to maintain a 19.403 kHz pilot offset. If the NTSC analog station operates at an offset of ±10 kHz, both DTV stations should compensate for that. Cooperation between stations will be essential in order to reduce interference.

TABLE 1.11-7
DTV Pilot Carrier Frequencies for Two Stations
(Normal Offset Above Lower Channel Edge: 309.440559 kHz)

	DTV Pilot Carrier Frequency Above Lower Channel Edge			
Channel Relationship	**NTSC Station, Zero Offset**	**NTSC Station, +10 kHz Offset**	**NTSC Station, –10 kHz Offset**	**DTV Station, No Offset**
DTV with lower adjacent NTSC	332.138 kHz ± 3 Hz	342.138 kHz ± 3 Hz	322.138 kHz ± 3 Hz	
DTV co-channel with NTSC	338.056 kHz ± 1 kHz	348.056 kHz ± 1 kHz	328.056 kHz ± 1 kHz	
DTV co-channel with DTV	+ 19.403 kHz above DTV	+ 19.403 kHz above DTV	+ 19.403 kHz above DTV	328.8436 kHz ± 10 Hz

Frequency Tolerances

The tightest specification is for an ATSC DTV station with a lower adjacent NTSC analog station. If both NTSC and DTV stations are at the same location, they may simply be locked to the same reference. The co-located DTV station carrier should be 5.082138 MHz above the NTSC visual carrier (22.697 kHz above the normal pilot frequency). The co-channel DTV station should set its carrier 19.403 kHz above the co-located DTV carrier. If there is interference with a co-channel DTV station, the analog station is expected to be stable within 10 Hz of its assigned frequency. While it is possible to lock the frequency of the DTV station to the relevant NTSC station, this may not be the best option if the two stations are not at the same location. It will likely be easier to maintain the frequency of each station within the necessary tolerances (through the use of a common external reference such as GPS or local high-precision reference frequency sources). Where co-channel interference is a problem, that will be the only option. In cases where no type of interference is expected, a pilot carrier-frequency tolerance of ±1 kHz is acceptable, but in all cases good practice is to use a tighter tolerance if practicable.

Enhanced VSB System

The enhanced-VSB (E-VSB) system [1] is fundamentally a method of adding further error protection coding to part of the 8-VSB signal. It is required to simultaneously provide a performance increase for the E-VSB coded portion, while not degrading the "normal" or "main" portion used by legacy ATSC receivers. Secondarily, but importantly, E-VSB applications require additions to the ATSC transport and PSIP standards to support functions such as synchronization of separate but related source material in the main and enhanced streams.

The basic technical advantage of the E-VSB stream is an improvement of at least 6 dB in SNR and interference thresholds. This is obtained in exchange for heavier FEC coding and therefore at the expense of payload data rate for the enhanced part of the transmission. Of course, designating a portion of the transmitted symbols as enhanced reduces the bandwidth of the main stream, just as in the case of multicasting, where each program uses only part of the 19.39 Mbps ATSC stream.

Applications envisioned for E-VSB include streams unrelated to the main stream, related streams, and synchronized related streams such as fallback audio and video. Unrelated streams are envisioned for use in carrying secondary channels or data that can be used by portable or PC-based devices with non-optimum antennas—for example, a subchannel carrying stock market information, news, and weather. Fallback audio is defined as a duplicate of the main audio that can be switched to in the receiver when the main signal is momentarily lost. The aim is to make this switch as seamless and unnoticeable as possible. This is the most demanding application envisioned, involving all primary and secondary aspects of E-VSB: the physical layer, synchronization of time stamps for the main and fallback, and enhanced PSIP that announces the availability of fallback to the enhanced-capable receiver.

Multiple Transmitter Networks

The conventional approach to covering a large television service area involves the placement of a single high-power transmitter at a central location. Under certain conditions, however, the conventional method may face economical and technical challenges that require careful considerations and engineering solutions. With the single-transmitter configuration, signal levels are not uniform throughout the service area. The radiated power of the transmitter is usually calculated so as to provide sufficient signal strength at the edges of the coverage area. In locations closer to the transmitter, the signal is stronger and may be considerably more than required for a satisfactory reception. The high cost of extending the coverage area of a transmitter by increasing its radiated power is another potential problem with the single-transmitter approach. Serving the last kilometer of coverage is far more expensive than the first kilometer. For example, for a UHF digital TV signal at about 80 km from a transmitter whose antenna is 300 m above average terrain, approximately a 3 dB (or 100%) increase in transmitted power would be required to increase coverage by 5 km. Thus, increasing coverage with raw transmitter power can be expensive to accomplish. Another issue is interference into neighboring service areas.

Based on calculations for the location and time availability of F(50,10) for interference and F(50,90) for coverage using FCC curves, it can be shown that co-channel interference from a digital UHF transmitter will extend on the order of three times the distance over which it can provide coverage. So, extending by 1 km the coverage area of a single transmitter by increasing its output power would add 3 more kilometers to its co-channel interference zone.

In situations such as those detailed above, one possible solution is to construct a multiple transmitter network and distribute the signal across the coverage area by using a number of lower power transmitters instead of a single central transmitter. Among the potential benefits of this approach are [16]:

- More uniform and higher average signal levels throughout the service area
- More reliable outdoor and indoor reception as a result of higher average signal levels
- Less overall effective radiated power (ERP) and/or antenna height requirements, resulting in less interference
- Stronger signals at the edges of the service area without increasing interference to neighboring stations

A multiple transmitter approach of sorts is used for analog TV systems in the form of translators. Such systems are mostly used to fill coverage gaps or to extend the coverage area. They are not usually intended to work with the main transmitter to uniformly distribute the signal across the service area. Instead, there is typically a master/slave relationship between the higher power central transmitter and the lower power translators. A primary limitation is the number of RF channels that must be used for an analog network. Usually for *N* transmitters, *N* channels are required.

Distributed Transmission Networks (DTxNs)

Distributed transmission (DTx) can be regarded as a way of covering a service area with a network of two or more transmitters, all synchronized and emitting exactly the same program and operating according to technical guidelines and standards specifically developed for this type of system. The number of channels used can be far less than the number of the transmitters that constitutes the network. Application of distributed transmission networks (DTxNs) is not limited to filling coverage gaps or extending the coverage area, which is usually the case in analog TV translators. It can also be used for creating a more uniform distribution of the transmitted signal in the main parts of the service area, as well as enhancing the signal in other parts by illuminating the area from different directions.

As a trade-off for their benefits, DTxNs may have to deal with certain operational restrictions under specific conditions. Such limiting conditions could exist if a single-frequency DTx network and a single transmitter operating on the adjacent channel co-exist in the same market area. Under these conditions, implementation of the single-frequency network (SFN) should be based on a careful and well-studied design to minimize interference to the single transmitter.

A generalized DTxN may include different combinations of distributed transmitters, distributed translators, and digital on-channel repeaters (DOCRs). Each of these structures can be implemented by using a number of methods that differ in their degree of complexity, mostly related to the achieving synchronization between different network transmitters. For example, the synchronized translators used in a DTx network can operate on the basis of RF–RF, or RF–IF–RF conversion, or the signal can be brought down to base-band and decoded, error corrected, and then re-encoded and up-converted to RF. Selection of each of these approaches is determined by the existing conditions. In the case when a number of studio-to-transmitter links (STLs) are used to bring the base-band signal to the transmitters, achieving frequency and time synchronization becomes more complex. Under these conditions, the synchronization process should be based on the methods specified in ATSC A/110 [17].

Transmitter Identification

As this book went to press, there were more than 1500 ATSC DTV transmitters in operation in the United States and Canada. As the number of DTV transmitters grows, there is an increasing need to identify the origin of each DTV signal received at different locations. ATSC A/110 [17] includes specifications for a spread spectrum sequence, embedded as an RF watermark, for transmitter identification (TxID) purposes. Transmitter identification techniques (or *transmitter fingerprinting*) can be used to detect, diagnose, and classify the operating status of radio transmitters. Transmitter identification also enables broadcast authorities and operators to identify the sources of interference, if any. More importantly, TxID can be used to tune various transmitters in a single-frequency network to minimize the effects of multipath interference caused by the destructive interference of several different transmissions or by the reflection of transmissions.

DATA BROADCAST AND INTERACTIVE CAPABILITY

The rollout of the digital television infrastructure has opened up new frontiers in communication, permitting powerful new applications that extend beyond regular television programming. Recognizing the interest in data broadcast applications, the ATSC developed a suite of data broadcast standards (documents A/90 to A/97) to enable a wide variety of data services that may be related to one or more video programs being broadcast or stand-alone services. Applications range from streaming audio, video, or text services to private data delivery services. Data broadcast receivers may include personal computers, televisions, set-top boxes, or other devices. Generally speaking, data broadcast applications targeted to consumers can be classified by the degree of *coupling* to the main video programming; specifically:

- *Tightly coupled* data is intended to enhance the TV programming in real time. The viewer tunes to the TV program and simultaneously receives the data enhancement along with it.
- *Loosely coupled* data are related to the program but are not closely synchronized with it in time; for example, an educational program might send supplementary reading materials or self-test quizzes within the broadcast stream.
- *Non-coupled* data are typically contained in separate "data-only" virtual channels. They may be data intended for real-time viewing, such as a 24-hour news headline or stock ticker service.

Advanced Common Application Platform (ACAP)

The Advanced Common Application Platform (ACAP) Standard (A/101) is a platform for interactive television services. ACAP was developed as the result of a landmark harmonization effort between the ATSC DTV Application Software Environment (DASE) Standard and Cable Television Laboratory's (CableLab's) Open Cable Application Platform (OCAP) specification. In essence, ACAP makes it appear to interactive programming content that it is running on a single platform, the so-called common receiver. This common receiver contains a well-defined architecture, execution model, syntax, and semantics. As a middleware specification for interactive applications, ACAP gives content and application authors assurance that their programs and data will be received and run uniformly on all brands and models of receivers.

The term *interactive television* (ITV) is broad and includes a vast array of applications including:

- Customized news, weather, and traffic
- Stock market data, including personal investment portfolio performance in real time
- Enhanced sports scores and statistics on a user-selective basis
- Games associated with program
- Online, real-time purchase of everything from groceries to software without leaving home
- Video on demand (VOD)

There is no shortage of reasons why ITV is viewed with considerable interest around the world. With the rapid adoption of digital video technology in the cable, satellite, and terrestrial broadcasting industries, the stage is set for the creation of an ITV segment that introduces to a mass consumer market a whole new range of possibilities. ACAP is intended to provide consumers with advanced interactive services while providing content providers, broadcasters, cable and satellite operators, and consumer electronics manufacturers with the technical details necessary to develop interoperable services and products.

References

[1] ATSC A/53E: ATSC Digital Television Standard (with Amendments No. 1 and No. 2), Advanced Television Systems Committee, Washington, D.C., December 27, 2005 (Amendment No. 1 dated 18 April 2006, Amendment No. 2 dated 13 September 2006).

[2] ATSC A/52B: Digital Audio Compression (AC-3, E-AC-3), Advanced Television Systems Committee, Washington, D.C., June 14, 2005.

[3] ATSC A/65C: Program and System Information Protocol for Terrestrial Broadcast and Cable (with Amendment No. 1), Advanced Television Systems Committee, Washington, D.C., January 2, 2006 (Amendment No. 1 dated 9 May 2006).

[4] ISO/IEC IS 13818-2:2000 (E), International Standard, Information Technology—Generic Coding of Moving Pictures and Associated Audio Information, video.

[5] ANSI/SCTE 43: Digital Video Systems Characteristics Standard for Cable Television, Society of Cable Telecommunications Engineers, Exton, PA, 2005.

[6] ITU-R BT.601-5: Encoding Parameters of Digital Television for Studios, 1994.

[7] SMPTE 274M: Standard for Television—1920 × 1080 Scanning and Analog and Parallel Digital Interfaces for Multiple Picture Rates, Society of Motion Picture and Television Engineers, White Plains, NY, 2005.

[8] SMPTE 296M: Standard for Television—1280 × 720 Progressive Image Sample Structure, Analog and Digital Representation and Analog Interface, Society of Motion Picture and Television Engineers, White Plains, NY, 2001.

[9] SMPTE 294M: 720 × 483 Active Line at 59.94 Hz Progressive Scan Production—Bit-Serial Interfaces, Society of Motion Picture and Television Engineers, White Plains, NY, 2001.

[10] SMPTE 170M: Standard for Television—Composite Analog Video Signal, NTSC for Studio Applications, Society of Motion Picture and Television Engineers, White Plains, NY, 2004.

[11] AES3-2003, AES Recommended Practice for Digital Audio Engineering—Serial Transmission Format for Two-Channel Linearly Represented Digital Audio Data (revision of AES3-1992, including subsequent amendments).

[12] ISO/IEC IS 13818-1:2000 (E), International Standard, Information Technology—Generic Coding of Moving Pictures and Associated Audio Information: Systems.

[13] ATSC Recommended Practice A/69: Program and System Information Protocol Implementation Guidelines for Broadcasters, Advanced Television Systems Committee, Washington, D.C., June 25, 2002.

[14] ATSC Recommended Practice A/54A: Guide to the Use of the Digital Television Standard, Advanced Television Systems Committee, Washington, D.C., December 4, 2003.

[15] ATSC A/76A: Programming Metadata Communication Protocol Standard, Advanced Television Systems Committee, Washington, D.C., September 18, 2006.

[16] FCC, *Memorandum Opinion and Order on Reconsideration of the Sixth Report and Order*, Federal Communications Commission, Washington, D.C., February 17, 1998.

[17] ATSC Recommended Practice A/111: Design of Synchronized Multiple Transmitter Networks, Advanced Television Systems Committee, Washington, D.C., September 3, 2004.

[18] ATSC A/110A: Synchronization Standard for Distributed Transmission, Advanced Television Systems Committee, Washington, D.C., July 14, 2004.

CHAPTER

1.12

NTSC Analog Television Standard

COMPILED BY NAB STAFF

INTRODUCTION

A variation of this chapter has been in every *NAB Engineering Handbook* since the 5th edition in 1960. By that time the NTSC color television standard was still in the early stages of development and implementation. However, understanding how color could be added to a black-and-white TV system in a compatible manner was a difficult concept for many television engineers (comparable to evolving from analog to digital systems). As a result, the 5th edition and subsequent editions of the *NAB Engineering Handbook* contained the following tutorial format of the fundamentals of the foundation technology used in color television broadcasting from camera to receiver. Although some of the material and terminology seems dated, it is still relevant to the NTSC color television system used today. While digital and high-definition systems use newer standards for video, NTSC still plays a major role in production, distribution, transmission, and reception and will for some time to come. The material in this chapter has been adapted from "Color Television, A Manual for Technical Training," RCA Corp. (with permission).

Nearly every branch of science, including chemistry and psychology, contributes in some way to the reality of color television. Through chemistry, improved phosphors are continually being found for use in color picture tubes.[1] Psychology enters into the selection of lighting arrangements and picture composition to obtain desirable interpretations by the viewer. But physics plays the leading role with intense application in optics and illumination as well as in the design of electronic circuitry and components for the complete television system.

Two specialized branches of physics, namely radio and television engineering, are responsible for the electronic techniques which make color television compatible with black-and-white (monochrome) television.

Compatibility

When introduced in 1953, the compatible color system offered tremendous economic advantages to the home viewer and the television broadcaster. Because of compatibility, color telecasts could be seen (in monochrome) on black-and-white television receivers without any changes or added devices. Also, color receivers could receive monochrome as well as color telecasts. Since compatible color is transmitted over the same channels as monochrome and within the same framework of standards, the television broadcaster could utilize the monochrome system as the transmitting nucleus when installing equipment to broadcast color.

Another important advantage of the compatible color system was the part it played in the conservation of the radio frequency spectrum. Compatible color required no additional space in the spectrum. However, it employed techniques which made much more efficient use of the standards originally set up for monochrome television.

[1]In this chapter the terms "CRT," "kinescope," and "picture tube" are used interchangeably.

A brief review of the fundamentals of monochrome television, particularly the areas wherein specialized color methods are employed, is presented in the next few paragraphs as an aid in describing the basic color concepts.

Television: A System of Communications

Basically, television is a system of communications consisting of the television station at one end and the television receiver at the other. The function of the television station is to divide and subdivide the optical image into over 200,000 picture elements, each of different light intensity; convert these light elements to electrical equivalents; and transmit them in orderly sequence over a radio-frequency carrier to the television receiver.

Reversing the process at the receiver, these electrical signals are each converted to light of corresponding brightness and reassembled to produce the transmitted image on the face of the picture tube.

Scanning

Picture elements to be transmitted in sequence are selected by a process of image scanning, which takes place in the television camera focused on the studio scene at the station. Within the camera, an electron beam in a pickup tube scans a sensitive surface containing an electrical image of the scene of action, as illustrated in Figure 1.12-1. The electron beam successively scans the image at great velocity, beginning at the upper-left corner and continuing left to right in a series of parallel lines to scan the image completely. Movement of the electron beam, which can be controlled magnetically by vertical- and horizontal-deflection coils surrounding the tube, is analogous to that of the eye in reading a printed page. The speed of movement is such, however, that 30 complete image frames of approximately 500 lines each are scanned every second. At the receiver an electron beam in the picture tube moves with the same speed and in synchronism with the camera tube beam so that the corresponding picture elements appear in the proper relative position on the television screen (see Figure 1.12-2). Modern cameras use charge coupled devices (CCD) in place of pickup tubes, and the scanning process occurs electronically. However, the tube-based explanation is the more straightforward approach.

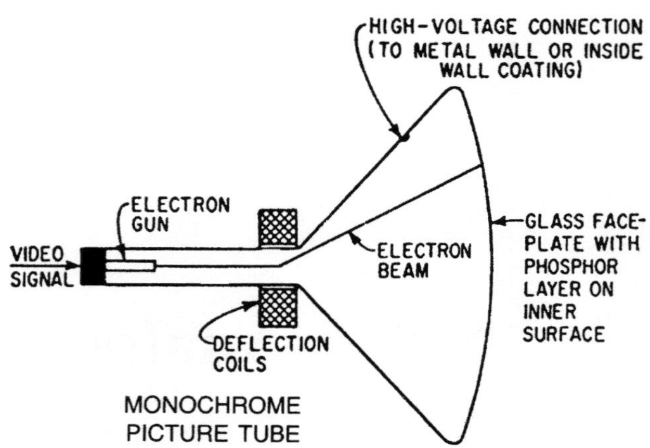

FIGURE 1.12-2 Diagram showing principal elements of the monochrome picture tube.

Owing to persistence of vision and the speed of scanning, these elements appear to be seen all at once as a complete image rather than individually. Thus, the impression is one of continuous illumination of the screen and direct vision.

Scanning standards have been established in this country to assure that all television receivers are capable of receiving programs broadcast by any television station within range. For the original monochrome television standard, the scanning pattern adhered to by manufacturers in the design of television receivers and broadcast equipment consists of 525 lines with odd-line interlaced scanning. Interlaced scanning, effective in eliminating perceptible flicker, is a method whereby the electron beam scans alternate rather than successive lines. For example, the beam begins by scanning odd-numbered lines (1, 3, 5, 7) until it

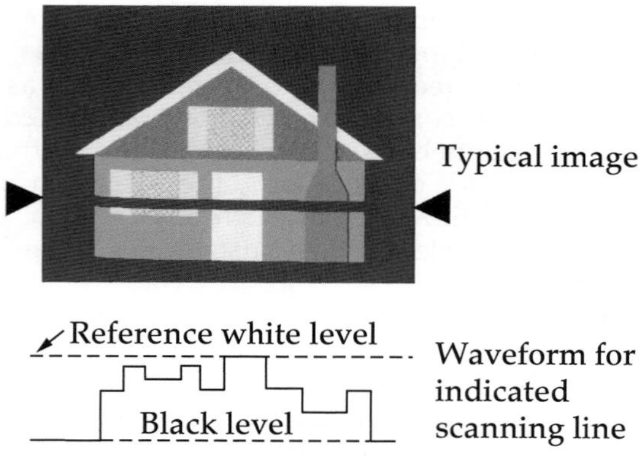

FIGURE 1.12-1 Typical image and camera output waveform produced by light and dark areas during one scan along line indicated by arrows.

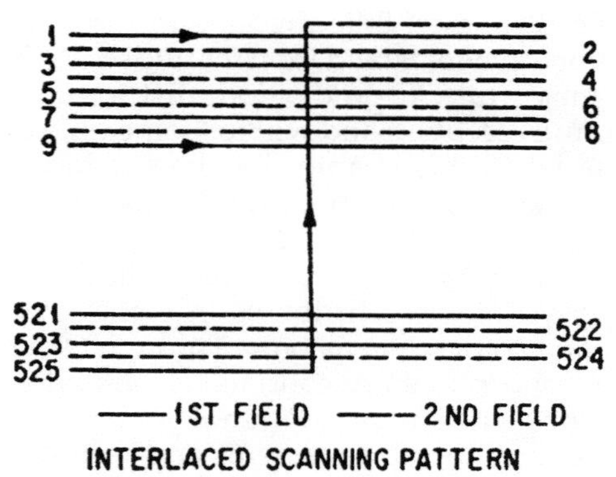

FIGURE 1.12-3 Diagram showing paths of the electron beam in both the pickup tube and CRT to produce the interfaced scanning pattern.

reaches the bottom of the image, whereupon it returns to the top of the image to scan the even-numbered lines (2, 4, 6, 8). Thus, each scan, or field, comprises only half of the total number of scanning lines, and two fields are required to produce the 525-line frame. Each field is completed in one-half the frame time. The vertical scanning frequency is 2 × 30 or 60 Hz, and horizontal scanning frequency is 30 × 525 or 15,750 Hz. This process is illustrated in Figure 1.12-3.

Resolution and Bandwidth

The degree of resolution or fine detail that can be seen in a televised image depends on the number of scanning lines used and the bandwidth of the transmitting and receiving system (see Figure 1.12-4).

The relationship between resolution and bandwidth can be seen by considering the number of picture elements that can be transmitted each second.

The standard 6 MHz broadcast channel provides a video bandwidth of approximately 4.1 MHz (the remaining bandwidth being required for a vestigial sideband plus the sound signal). Since each cycle of a sine wave is capable of conveying two picture elements (one black and one white), the maximum rate at which picture elements can be transmitted is 4,100,000 × 2, or 8,200,000 per second. Since 30 complete frames are transmitted per second, the number of picture elements per frame would be 8,200,000 ÷ 30, or 273,333 (if it were not for the retrace blanking problem, which requires interruption of the picture signal periodically by blanking pulses). Since the combination of horizontal and vertical blanking pulses requires nominally 25% of the total time, the maximum number of picture elements per frame is reduced in practice to 0.75 × 273,333, or approximately 205,000.

Synchronizing

In addition to the picture information, or video signals, blanking and synchronizing signals are transmitted by the television station to control the intensity and movement of the scanning beam in the CRT of the television receiver. Both of these signals are in the form of rectangular pulses. Moreover, their polarity and amplitude are such that they are received as black signals and therefore do not appear on the receiver screen.

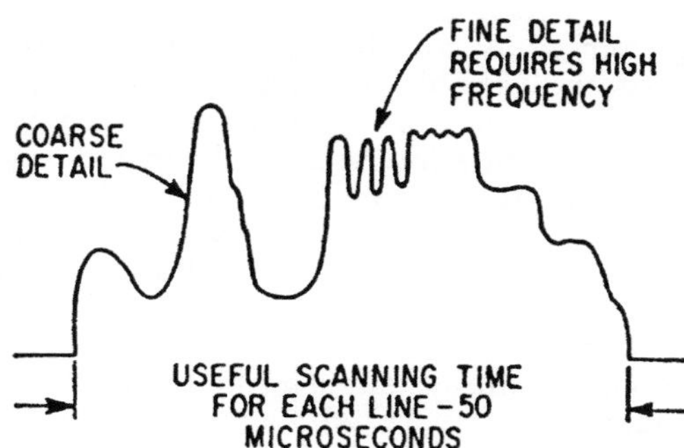

FIGURE 1.12-4 Diagram illustrating the relationship between picture detail and signal bandwidth.

Blanking pulses eliminate the *retrace* lines which would otherwise appear between scanning lines and at the end of each field from the bottom of the picture to the top. Horizontal blanking pulses, transmitted at the end of each line, or at intervals of 1/15,750 sec, blank the beam during retrace periods between lines. Vertical blanking pulses, transmitted at the end of each field, or at intervals of 1/60 sec, blank the beam during the time required for its return to the top of the picture. Because the vertical retrace is much slower than the horizontal, the vertical blanking periods are longer than the horizontal blanking periods. Vertical blanking pulses are about 20 lines' duration, whereas horizontal blanking pulses have a duration of only a small fraction of a line.

Synchronizing signals keep the scanning beam of the picture tube in step with that of the camera tube. These signals consist of horizontal and vertical pulses which are transmitted within the respective blanking periods. Although the sync pulses are of the same polarity as the blanking pulses, they are of greater amplitude (blacker than black) and thus easily separated in the receiver and fed to the deflection circuits of the picture tube.

Since the vertical sync pulses are quite long compared with the horizontal sync pulses and the two are of the same amplitude, separation at the receiver is accomplished through frequency discrimination. Serrations, or slots in the vertical pulses, prevent loss of horizontal sync during the vertical blanking period.

The Basic Television System

The major equipment in a typical television station consists of the aural and visual units illustrated in the block diagram of Figure 1.12-5. In the visual channel, the video signal leaving the camera is passed through processing equipment which inserts the blanking and synchronizing signals and performs other functions (such as aperture compensation and gamma correction). From the processing chain, the video signal is fed to a switching system which provides for selection from a number of video sources. The selected signal is then sent to the visual transmitter through coaxial cable or over a microwave relay link, depending on the distance between the television studio and transmitter. In the transmitter, the composite video signal amplitude modulates a carrier in the VHF or UHF range, which is radiated by the television antenna.

In the aural channel, the audio signal is fed from the sound sources through the switching system and to the aural transmitter. Frequency modulated output from the aural transmitter is combined with the visual output and radiated from the same antenna.

The Radiated Picture Signal

Amplitude relationships between the synchronizing pulses and the tonal gradations from white to black in

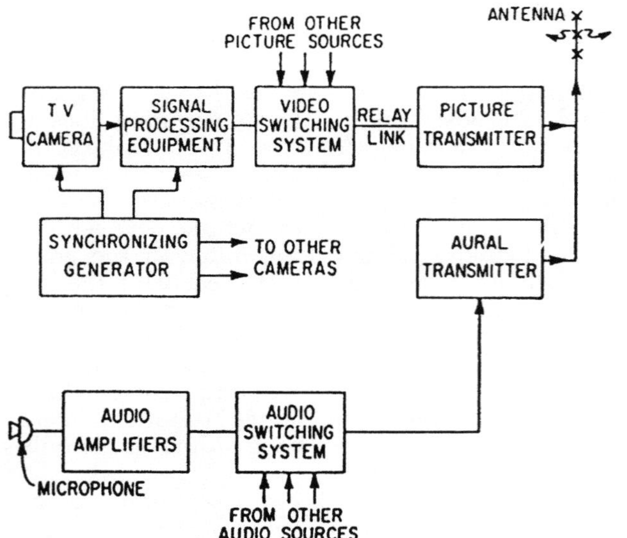

FIGURE 1.12-5 Simplified block diagram of the basic television station.

the picture are represented in the waveform of the radiated picture signal. From Figure 1.12-6, it can be seen that modulation takes place in such a way that an increase in the brightness of the picture causes a decrease in carrier output power. Note that the reference white line indicated on the sketch is relatively close to zero carrier level. Also, the synchronizing pulses are in the *blacker than black* region, representing maximum carrier power. The higher amplitude for the sync pulses makes it possible for home receivers to separate them by a simple clipping technique.

Receiver

The basic elements of the television receiving system are illustrated in Figure 1.12-7. The radiated television signal is picked up by an antenna and fed to a tuner, which selects the desired channel for viewing. Output from the tuner is passed through an intermediate-frequency amplifier, which provides the major selectivity and voltage gain for the receiver. A second detector

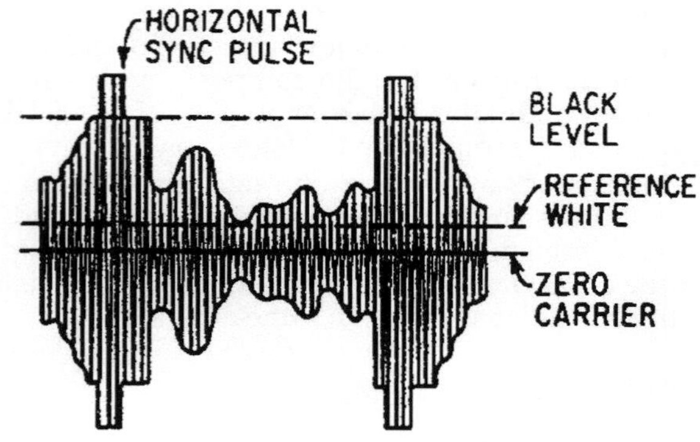

FIGURE 1.12-6 Waveform and radiated picture signal.

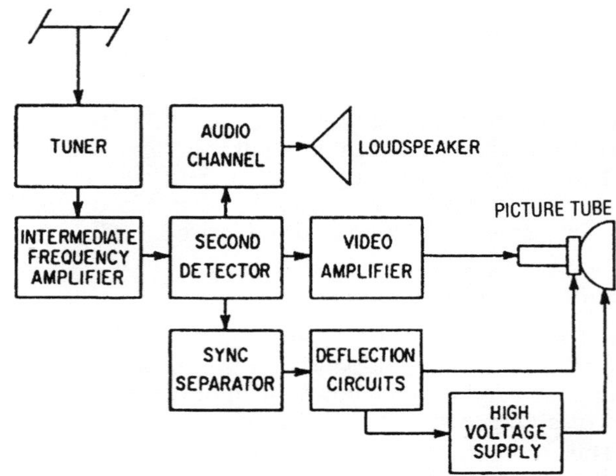

FIGURE 1.12-7 Block diagram of the basic television receiver.

then recovers a video signal which is essentially the same as that fed to the visual transmitter.

The sound carrier is usually taken off at the picture second detector in the form of a 4.5 MHz beat between the picture and sound carriers. The frequency modulated (FM) sound carrier is further amplified in an intermediate frequency (IF) stage, detected by a discriminator or ratio detector, and applied to the speaker through an audio amplifier.

Picture output from the second detector is fed to two independent channels. One of these is the video amplifier, which drives the electron beam in the picture tube, and the other is the sync separator, which separates the sync pulses from the picture information. The separated pulses are then used to control the timing of the horizontal and vertical deflection circuits. The high-voltage supply, which is closely associated with the horizontal deflection circuit, provides accelerating potential for the electron beam.

The Three Variables of Color

Color is the combination of those properties of light which control the visual sensations known as *brightness*, *hue*, and *saturation*. Brightness is that characteristic of a color which enables it to be placed in a scale ranging from black to white or from dark to light. Hue, the second variable of a color, is the characteristic which enables a color to be described as red, yellow, blue, or green. Saturation refers to the extent to which a color departs from white, or the neutral condition. Pale colors, or pastels, are low in saturation, whereas strong or vivid colors are high in saturation. For more information, see Chapter 5.2, "Principles of Light, Vision, and Photometry."

The monochrome system is limited to the transmission of images that vary with respect to brightness alone. Thus, brightness is the only attribute of a color which is transmitted over a monochrome television system. To produce a color image, therefore, provi-

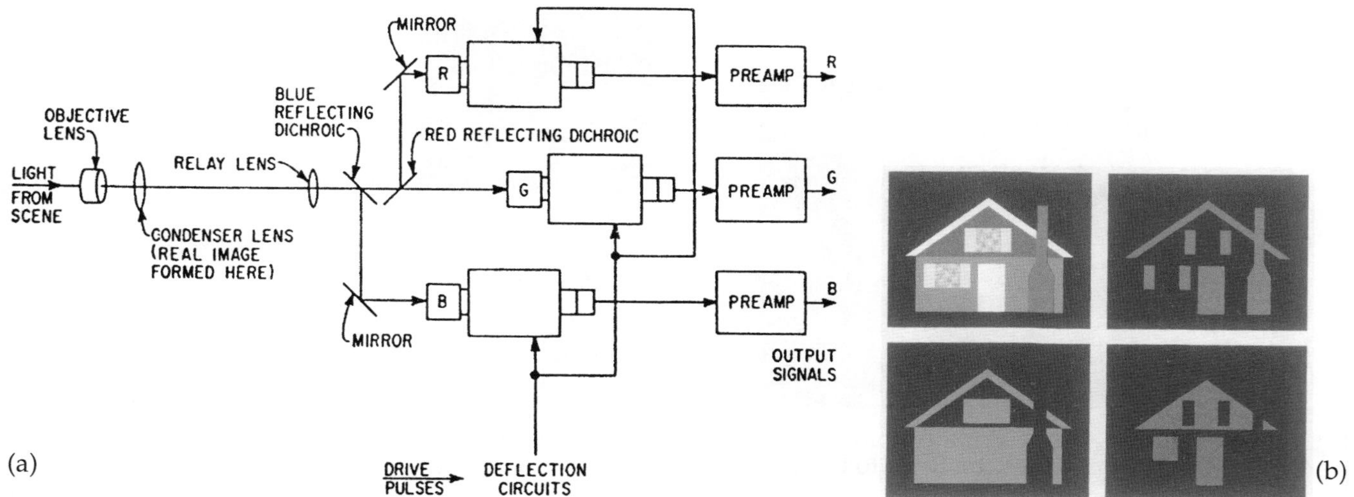

FIGURE 1.12-8 Simplified block diagram (a) of the optical and electrical components of the color camera, and (b) the outputs of each of the three color pickup devices. Refer to the CD for a color version of this figure.

sions must be made for the transmission of additional information pertaining to all three of the variables of color. However, since the primary color process can be employed, it is not necessary to transmit information in exactly the form expressed by the three variables.

Primary Colors in Television

Experiments have proved conclusively that virtually any color can be matched by the proper combination of no more than three primary colors. While other colors could be used as primaries, red, green, and blue have been selected as the most practical for color television use. Red combined with green produces yellow, red plus blue produces purple, and green plus blue produces cyan or blue-green. The proper combination of all three of the primary colors produces white, or neutral, as shown at the center of the illustration in Figure 1.12-9(a). By relatively simple optical means, it is possible to separate any color image into red, green, and blue, or RGB components.

Generating RGB Signals

The major components of a color television camera are shown in the simplified block diagram in Figure 1.12-8(a). While the monochrome camera contains only one pickup tube, or solid-state sensor, the color camera usually contains three separate pickup devices. An objective lens at the front of the camera forms a real image within a condenser lens, which is located where the pickup device is usually mounted in a monochrome camera. A relay lens transfers this real image to a system of dichroic (color separating) mirrors or prisms which shunt the red and blue light to the red and blue pickup devices and permit the green to pass straight through to the green tube or sensor. In this manner, the three pickup devices produce three separate images corresponding to the RGB components of the original scene, as shown in Figure 1.12-8(b). These images are scanned in the conventional manner by common deflection circuits.

A single scanning line through the simplified color image at a given point produces three separate waveforms, as shown in Figure 1.12-9(b). It is important to note the correlation between these waveforms and the image at the top. The yellow shutters in the image, for example, must be produced by a mixture of red and green, and the blue signal is not required. Thus, at this interval of scanning, the red and green signals are both at full value and the blue signal is at zero. The white door utilizes all three color signals. Of course, similar correlations can be seen for other parts of the image along the scanning line such as the red brick chimney and green painted house.

Displaying RGB Signals

RGB signals are displayed in color by the tricolor picture tube, the basic components of which are shown in Figure 1.12-10. Three electron guns produce three beams, which are independently controlled in intensity by the red, green, and blue signals. These three beams are all made to scan in unison by deflection coils around the neck of the tube. The three beams converge at the screen owing to the magnetic field produced by a convergence yoke.

The phosphor screen of the color picture tube consists of an array of very small primary color dots. Approximately 1/2 in. behind the phosphor screen is an aperture mask which has one very small opening for each group of red, green, and blue phosphors. Alignment of this aperture mask and screen is such that each beam is permitted to strike phosphor dots of only one color. For example, all the electrons emitted by the red gun must strike red phosphor dots on the aperture mask; they cannot strike either the green or

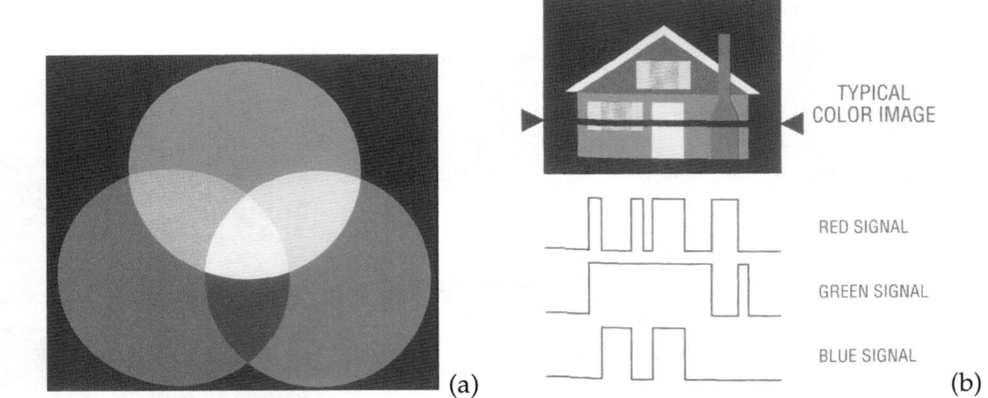

FIGURE 1.12-9 Primary colors in television (a) and simplified color image and related RGB waveforms (b). Refer to the CD for a color version of this figure.

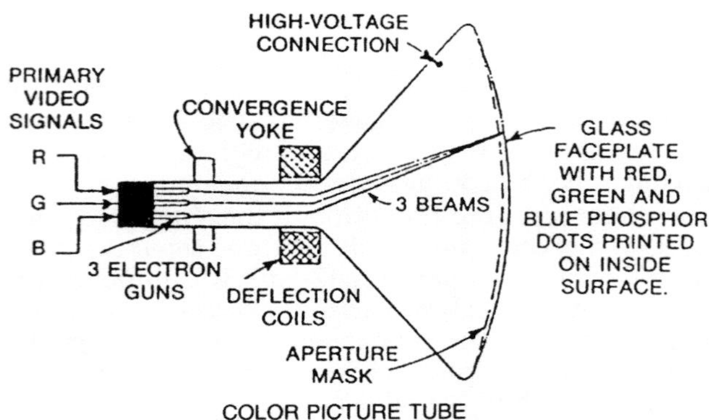

FIGURE 1.12-10 Diagram showing components of the three-gun tricolor picture tube.

blue dots because of the shadow effect of the mask. Likewise, the beams emanating from the other two guns strike only green or blue dots.

In this way, three separate primary color images are produced on the screen of the tricolor tube. But since these images are formed by closely intermingled dots too small to be resolved at the normal viewing distance, the observer sees a full-color image of the scene being televised.

ELECTRONIC ASPECTS OF COMPATIBLE COLOR TELEVISION

To achieve compatibility with monochrome television, color television signals must be processed so that they can be transmitted through most of the channels used for monochrome signals, and they must also be capable of producing monochrome pictures on monochrome receivers. Since color television involves three variables instead of the single variable (brightness) of monochrome television, an encoding process is required to permit all three to be transmitted over the one available channel. Likewise, a decoding process is required in the color receiver to recover the independent RGB signals for control of the electron guns in the color picture tube. Moreover, the process used must enable existing monochrome receivers to produce a monochrome picture from the color information (see Figure 1.12-11).

Encoding and decoding processes used in compatible color television are based on four electronic techniques known as *matrixing, band shaping, two-phase modulation*, and *frequency interlace*. It is these processes which make the color system compatible with monochrome and enable the color system to occupy the existing 6 MHz channel.

Matrixing

Matrixing is a process for repackaging the information contained in the red, green, and blue output signals from a color camera to permit more efficient use of the transmission channel. The matrix circuits which perform this function consist of simple linear cross-mixing circuits. They produce these signals, commonly

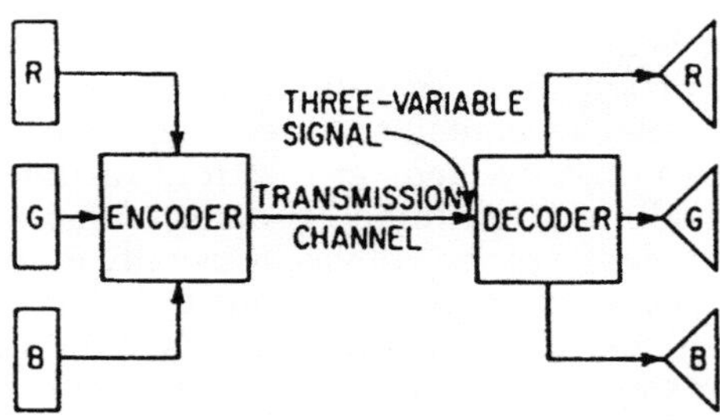

FIGURE 1.12-11 Encoding of the RGB signals provides a three-variable signal which can be transmitted over existing monochrome channels.

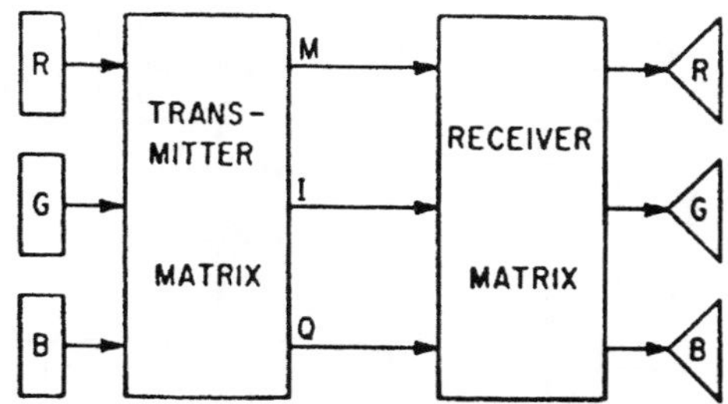

FIGURE 1.12-12 A part of the encoding process is the matrixing of R, G, and B signals to provide M, I, and Q signals.

designated M, I, and Q, each of which is a different linear combination of the original red, green, and blue signals (see Figure 1.12-12). Specific values for these signals have been established by FCC Rules section 73.682.

The M-signal component, or *luminance* signal, corresponds very closely to the signal produced by a monochrome camera, and is suitable for use on monochrome receivers. The M component is obtained by combining red, green, and blue signals in a simple resistor network (see Figure 1.12-13) designed to produce a signal consisting of 30% red, 59% green, and 11% blue.

The I and Q signals are *chrominance* signals which convey information as to how the colors in the scene differ from the monochrome, or neutral condition. The component I is defined as a signal consisting of 60% red, –28% green, and –32% blue. Minus values are easily achieved in the matrix circuits by use of phase inverters to reverse the signal polarity (see Figures 1.12-14 and 15). The Q signal is defined as 21% red, –52% green, and 31% blue.

Note that the quantities are related so that when red, green, and blue are equal, corresponding to the neutral condition, both I and Q go to zero. Thus, when the color camera is focused on an object having no color information, such as a monochrome test chart, the I-signal and Q-signal components are absent, leaving only the M component, or monochrome signal.

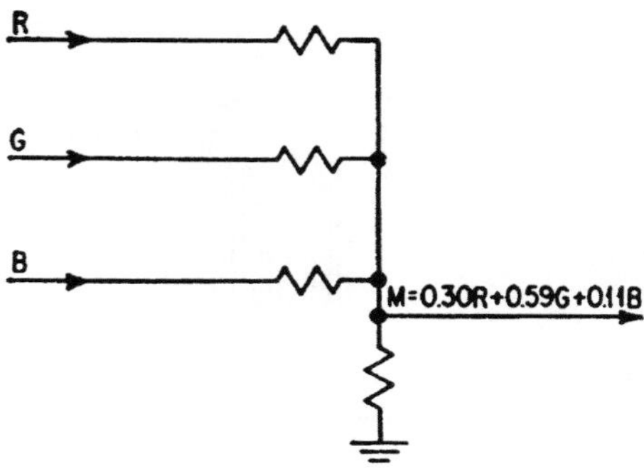

FIGURE 1.12-13 Diagram of resistance matrix circuit used to produce the M luminance signal.

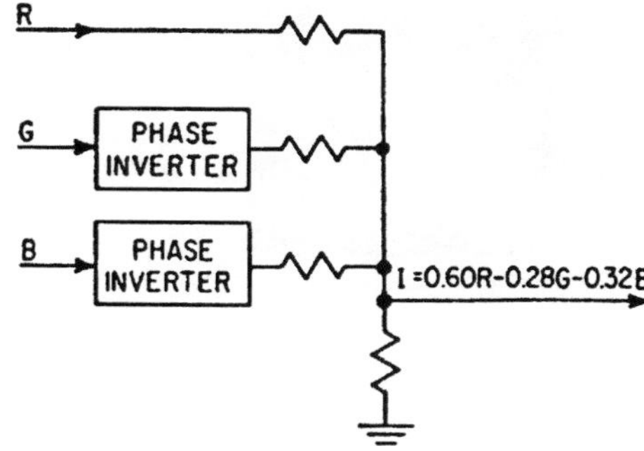

FIGURE 1.12-14 Diagram of I matrix showing phase inverters to produce minus green and blue quantities.

The matrix circuits, therefore, produce a new set of waveforms corresponding to the M, I, and Q components of the image. A comparison of the MIQ and RGB waveforms obtained from the image illustrates the correlation among the types of signals (see Figures 1.12-9 and 16). Note that the M signal remains in the region between black level and reference white. The I and Q signals, on the other hand, swing positive and negative around a zero axis and go to zero during the scan across the white portion (doorway) of the image (as indicated by the arrows in Figure 1.12-16).

Band Shaping

The eye has substantially less acuity in detecting variations in chrominance than it has for resolving differences in brightness. This important characteristic of human vision was considered when establishing the I and Q equations because it permitted a significant reduction in the bandwidth of these signals through use of low-pass filters. A bandwidth of approximately 1.5 MHz was found to be satisfactory for the I signal, which corresponds to color transitions in the range extending from orange to blue-green. For color transitions in the range from green to purple, as represented

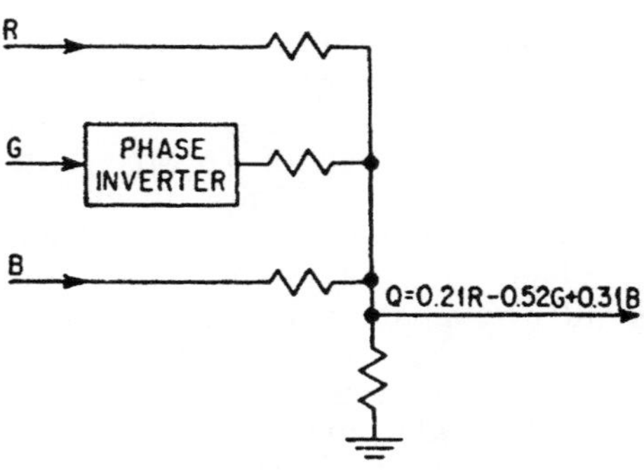

FIGURE 1.12-15 Diagram of the Q matrix showing phase inverter to produce required minus green signal.

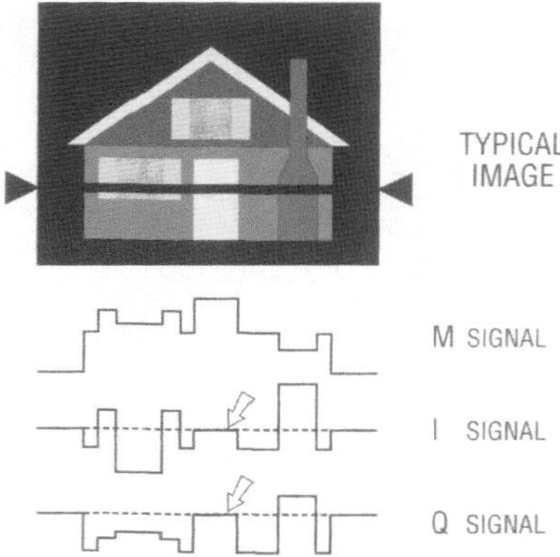

FIGURE 1.12-16 Sample color image and MIQ waveforms. Refer to the CD for a color version of this figure.

by the Q signal, the eye has still less acuity and the bandwidth was restricted to only 0.5 MHz. The M-signal component, which conveys the fine details, must be transmitted with the standard 4 MHz bandwidth.

Two-Phase Modulation: Generation of Color Subcarrier

Two-phase modulation is a technique by which the I and Q signals can be combined into a two-variable signal for transmission over a single channel. This is accomplished by adding the sidebands obtained through modulation of two 3.6 MHz carriers separated in phase by 90. The resultant waveform is the vector sum of the components. Elements of the transmitting and receiving system are shown in Figure 1.12-17. The two carriers, which are derived from the same oscillator, are suppressed by the balanced modulators. Thus, only the two amplitude modulated sidebands, 90 out-of-phase, are transmitted. At the receiving end of the system, the I and Q signals are recovered by heterodyning the two-phase signal against two locally generated carriers of the same frequency but with a 90 phase separation and applying the resultant signals through low-pass filters to the matrix circuits. Typical signal waveforms are illustrated in Figure 1.12-18.

The 3.6 MHz oscillator at the receiver must be accurately synchronized in frequency and in phase with the master oscillator at the transmitter. The synchronizing information consists of 3.6 MHz bursts of at least 8 Hz duration transmitted during the "back porch" interval following each horizontal sync pulse, as shown in Figure 1.12-19. The bursts are generated at the transmitter by a gating circuit which is turned on by burst keying pulses derived from the synchronizing generator. At the receiver, the two-phase modulated signal is applied to another gating circuit, known as a *burst separator,* which is keyed on by pulses derived from the horizontal deflection circuit. The separated bursts are compared in a phase detector with the output of the local 3.6 MHz oscillator. Any error

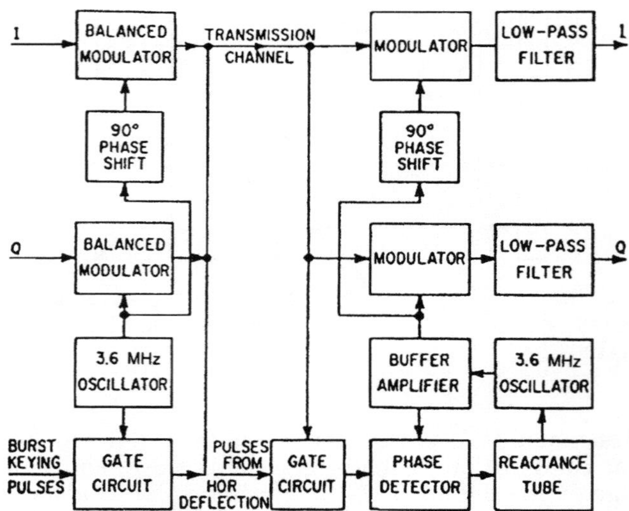

FIGURE 1.12-17 Simplified block diagram showing elements for transmitting and receiving the I, Q, and burst signals

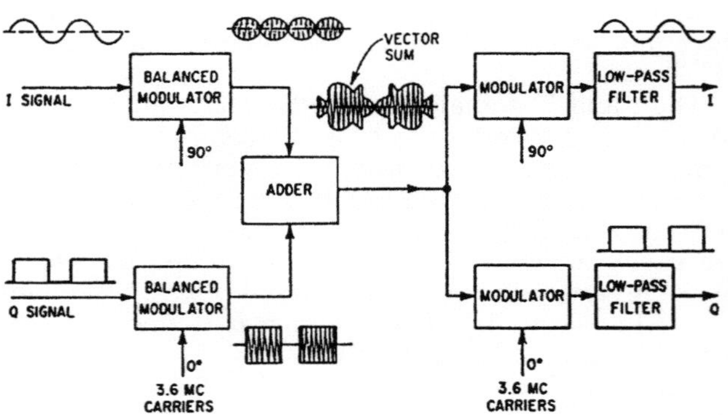

FIGURE 1.12-18 Representative waveforms of the separate I, Q signals and the vector sum of the suppressed carrier sidebands at the modulator output. Original I and Q signals are recovered by heterodyning in balanced modulators in the receiver.

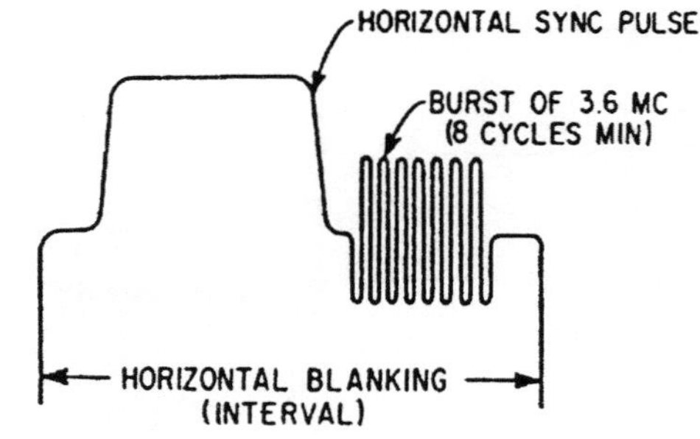

FIGURE 1.12-19 Diagram showing position of subcarrier burst during horizontal blanking interval.

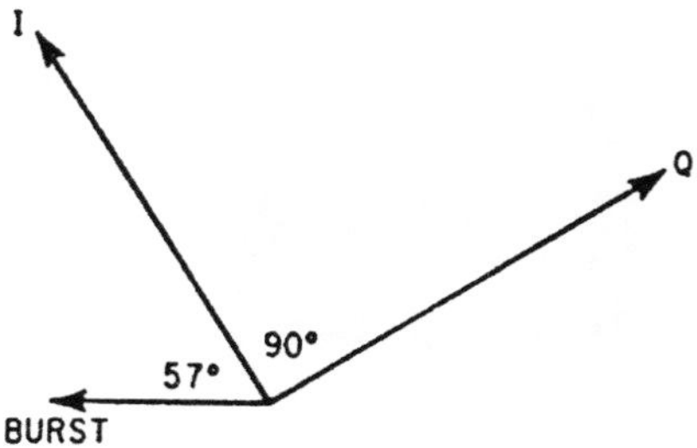

FIGURE 1.12-20 Diagram showing phase relationship of I, Q, and burst signals.

voltage developed is applied through a smoothing filter to a varactor which corrects the phase of the local oscillator.

FCC standard phase relationships between the I and Q signals and the color synchronizing burst are shown in the vector diagram of Figure 1.12-20. The I and Q signals are transmitted in phase quadrature, and the color burst is transmitted with an arbitrary 57 phase lead over the I signal.

Several interesting properties of the two-phase modulated signal are illustrated by the vector diagrams which represent the resultant signal under known transmission conditions. For example, when a pure red color of maximum amplitude is being transmitted, the green and blue components are at zero and the I and Q signals have levels of 60 and 21%, respectively. When modulated upon their respective carrier, these signals produce the resultant shown in Figure 1.12-21. The phase and amplitude shown are characteristic of pure red of maximum relative luminance. Figure 1.12-22 is a composite vector diagram showing the phase and amplitude characteristics of the three primaries and their complementary colors. This composite diagram indicates that there is a direct relationship between the *phase* of the resultant two-phase modulated signal and the *hue* of the color being transmitted. There is also a relationship (although indirect) between the *amplitude* of the resultant signal and the saturation of the color being transmitted. If the phase of the resultant subcarrier and the level of the mono-

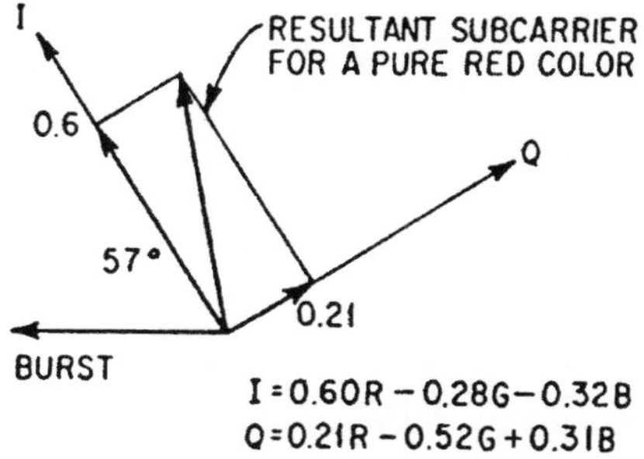

FIGURE 1.12-21 Vector diagram showing phase and amplitude of subcarrier for a pure red signal.

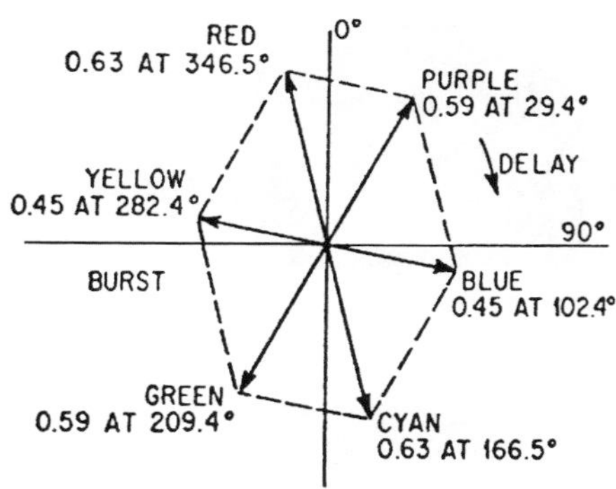

FIGURE 1.12-22 Composite vector diagram showing subcarrier phase and amplitude for each of six colors.

chrome signal both remain constant, then a reduction in the amplitude of the subcarrier indicates a decrease in color saturation. The composite vector diagram also shows an interesting symmetry between complementary colors (colors are complementary if they produce a neutral when added together); the resultants for any two complementary colors are equal in amplitude but opposite in phase.

Frequency Interlace

Since the 3.6 MHz carriers, consisting of the I and Q sidebands, fall within the video passband as shown in the diagram of the television channel (Figure 1.12-23), they become subcarriers and can be handled in many respects like unmodulated video signals. By use of frequency interlace, it is possible to add the

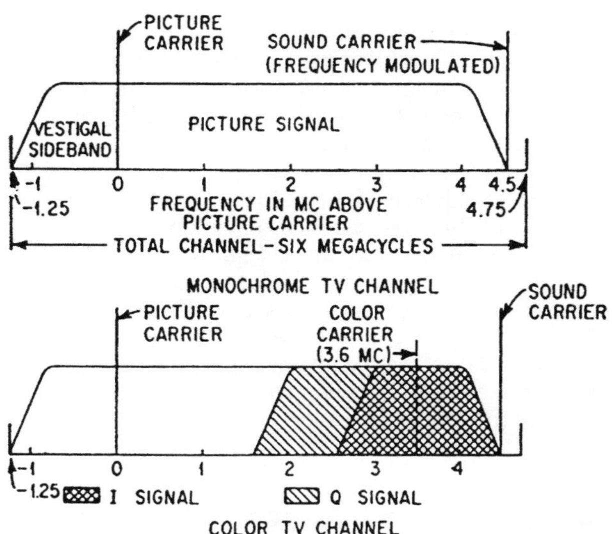

FIGURE 1.12-23 Diagram of television channel signal portions occupied by color and monochrome signal components.

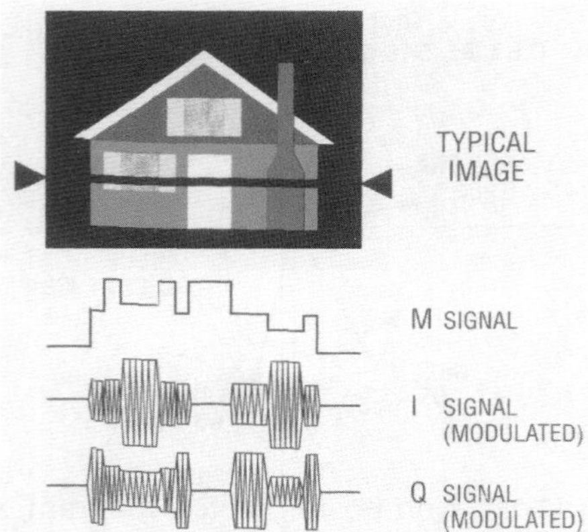

FIGURE 1.12-24 Sample color image and waveforms of the M signal and modulated I and Q signals. Refer to the CD for a color version of this figure.

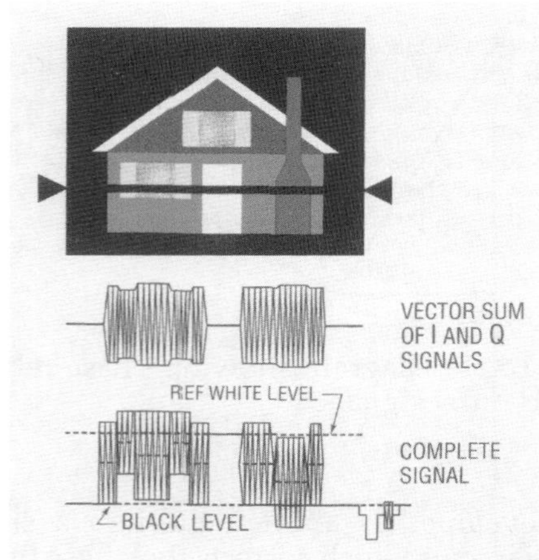

FIGURE 1.12-25 Sample color image and waveforms of combined I and Q signals and the complete signal containing the luminance and color signals. Refer to the CD for a color version of this figure.

several components of the chrominance and monochrome signals together without causing objectionable mutual interference.

The significance of the straightforward addition of signal components made possible by frequency interlace may be brought out by a study of waveforms derived from a simple color image. Figure 1.12-24 shows M, I, and Q signals after the latter two have been modulated upon 3.6 MHz subcarriers. Note that both the I- and Q-signal components are at zero during the scanning of the white door, a neutral area. Figure 1.12-25 shows the vector sum of the I and Q signals and also the complete compatible color signal formed by adding together all the components, including synchronizing pulses and color synchronizing bursts. The most significant fact about this signal is that it is still capable of providing good service to monochrome receivers, even though a modulated wave has been added to the monochrome signal component. Although the modulated wave is clearly a spurious signal with respect to the operation of the kinescope in a monochrome receiver, its interference effects are not objectionable because of the application of the frequency interlace principle.

Frequency interlace is a technique based on two factors: a precise choice of the color subcarrier frequency and the familiar persistence-of-vision effect. If the color subcarrier is made an *odd multiple of one-half the line frequency*, its apparent polarity can be made to reverse between successive scans of the same area in the picture. Since the eye responds to the average stimulation after two or more scans, the interference effect of the color subcarrier tends to be self-canceling, owing to the periodic polarity reversals (see Figure 1.12-26).

Color Frequency Standards

The relationships among the various frequencies used in a compatible color system are illustrated in the block diagram of Figure 1.12-27. The actual frequency of the color subcarrier, which has been referred to as 3.6 MHz, is specified by FCC Rules as 3.579545 MHz or exactly 455 multiplied by 1/2 the line frequency.

In broadcast practice, the frequency of the color subcarrier provides a frequency standard for operation of the entire system. A crystal oscillator at the specified frequency provides the basic control information for all other frequencies. Counter stages and multipliers derive the basic frequencies needed in the color studio. A frequency of nominally 31.5 kHz (actually 31.468.53 kHz) is required for the equalizing pulses, which precede and follow each vertical sync pulse and for the serrations in the vertical sync pulse. A divide-by-2 counter controlled by the 31.5 kHz sig-

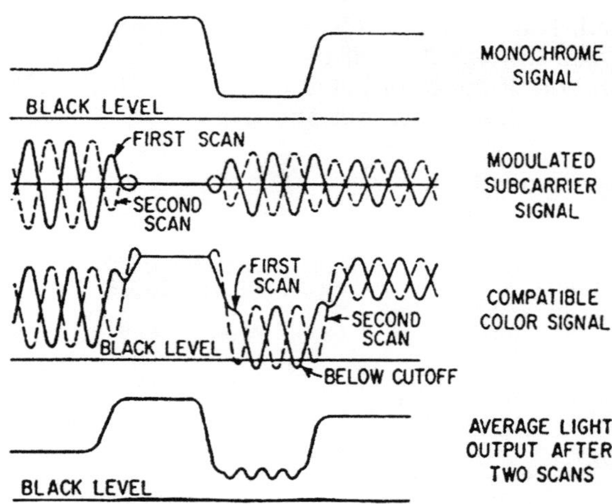

FIGURE 1.12-26 Waveforms showing superpositions of modulated subcarrier on scanning signals, compatible color signal, and effect of subcarrier on average light output.

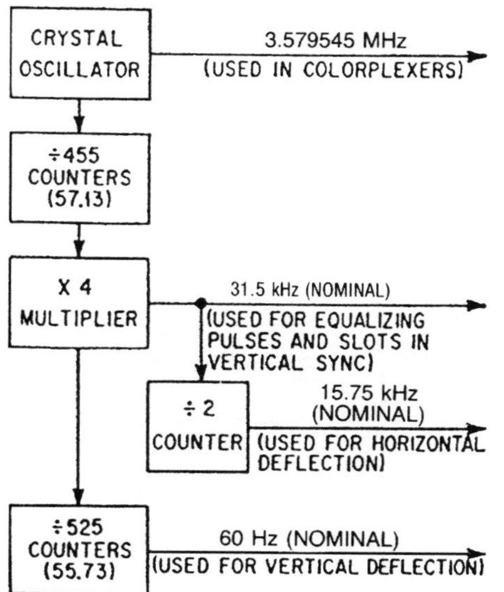

FIGURE 1.12-27 Block diagram showing relationship between various frequencies used in a color television station.

nal provides the line frequency pulses at nominally 15.75 kHz (actually 15.734.26 kHz) needed to control the horizontal blanking and synchronizing waveforms. Another counter chain provides the 60 Hz pulses (actually 59.97 Hz) needed for control of the vertical blanking and synchronizing circuits.

The Overall Color System

The major functions performed in transmitting and receiving color are shown in the overall block diagrams of the transmitting and receiving systems (see Figures 1.12-28 and 29).

At the transmitting end, camera output signals corresponding to the red, green, and blue components of the scene being televised are passed through nonlinear amplifiers (the gamma correctors), which compensate for the nonlinearity of the picture tube elements at the receiving end. Gamma-corrected signals are then matrixed to produce the luminance signal M and two chrominance signals I and Q. The filter section establishes the bandwidth of these signals. The 4.1 MHz filter for the luminance channel is shown in dotted lines because in practice this band shaping is usually achieved by the attenuation characteristics of the transmitter and the filter is not required.

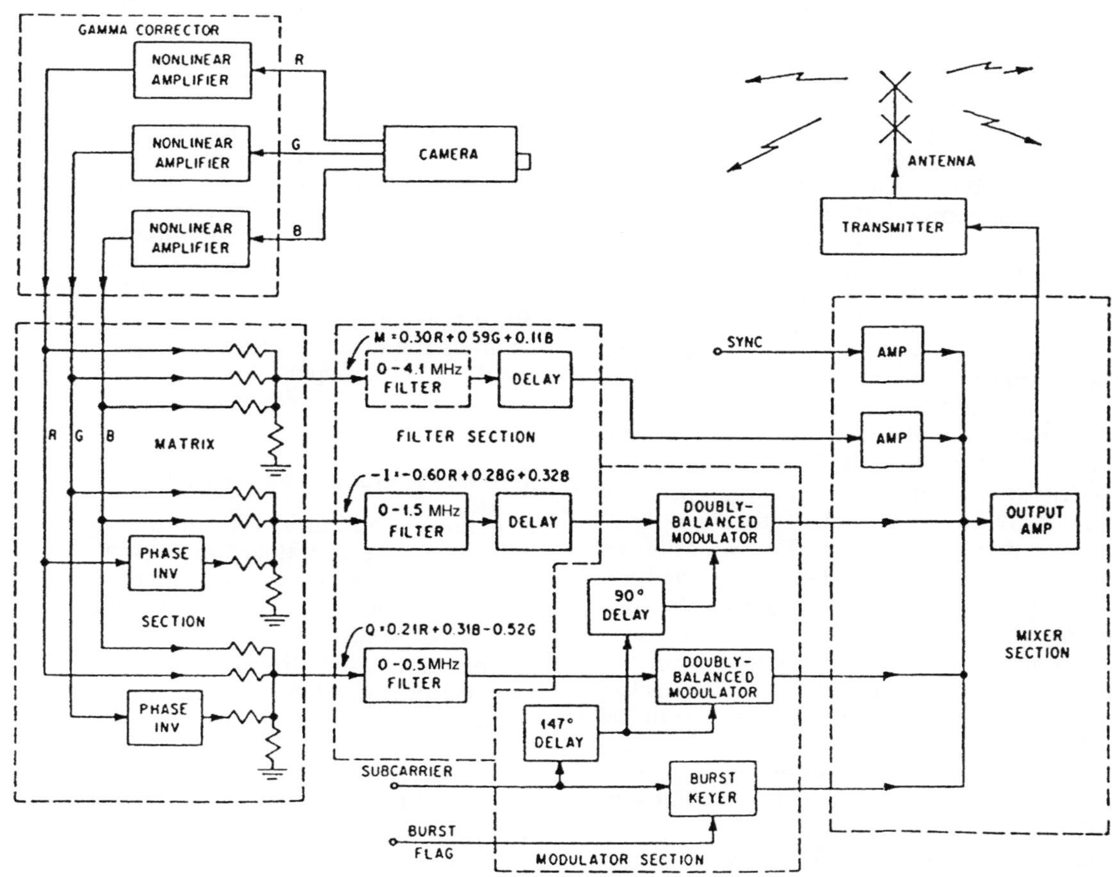

FIGURE 1.12-28 Block diagram showing major functions of color-transmitting system.

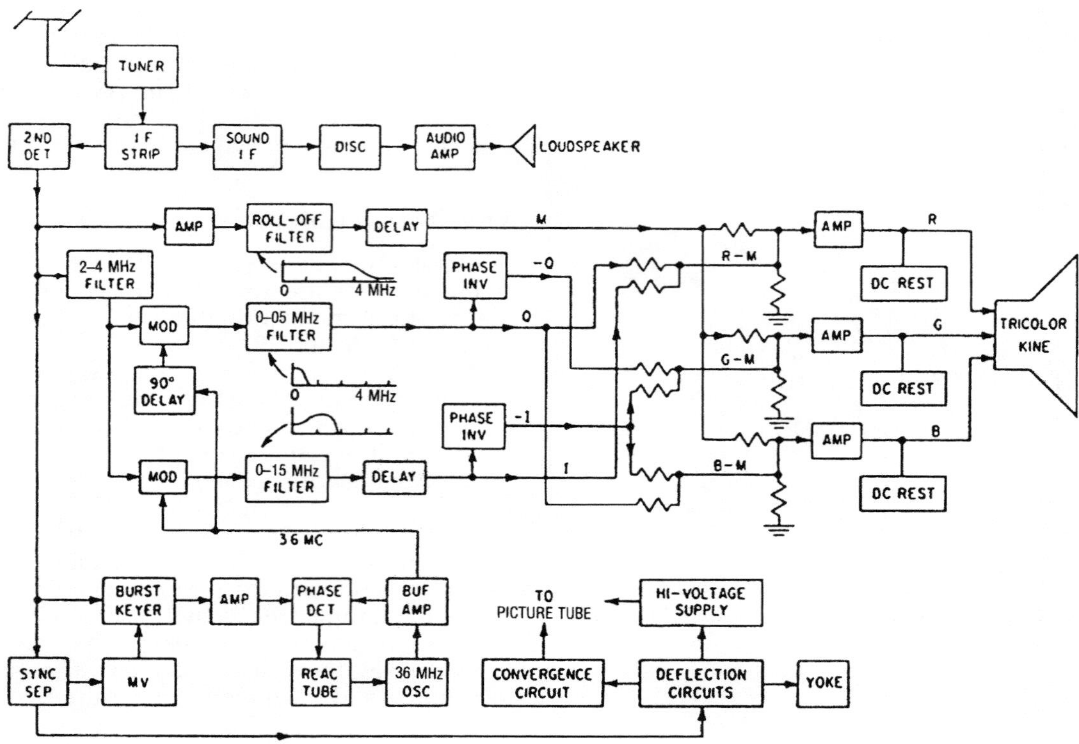

FIGURE 1.12-29 Block diagram showing major functions of color-receiving system.

The bandwidths of 1.5 and 0.5 MHz shown for the I and Q channels, respectively, are nominal only; the required frequency response characteristics are described in more detail in the complete FCC signal specifications. Delay compensation is needed in the filter section in order to permit all signal components to be transmitted in time coincidence. In general, the delay time for relatively simple filter circuits varies inversely with the bandwidth. The narrower the bandwidth, the greater the delay. Consequently, a delay network or a length of delay cable must be inserted in the I channel to provide the same delay introduced by the narrower band filter in the Q channel, and still more delay must be inserted in the M channel.

In the modulator section, the I and Q signals are modulated upon two subcarriers of the same frequency but 90° apart in phase. The modulators employed are the double balanced type, allowing both the carriers and the original I and Q signals to be suppressed, leaving only the sidebands. A keying circuit provides the color synchronizing bursts during the horizontal blanking intervals. To comply with the FCC signal specifications, the phase of the burst should be 57° ahead of the I component (which leads the Q component by 90°). This phase position was chosen mainly because it permits certain simplifications in receiver designs.

In the mixer section, the M signal, the two subcarriers modulated by I and Q chrominance signals, and the color synchronizing bursts are added together. Provision is also made for the addition of standard synchronizing pulses, so that the output of mixer section is a complete color television signal containing both picture and synchronizing information. This signal can then be put on-the-air by means of a standard television transmitter, which must be modified only to the extent necessary to assure performance within the reduced tolerance limits required by the color signal. Since the color signal places more information in the channel than a black-and-white signal, the requirements for frequency response, amplitude linearity, and uniformity of delay time are stricter.

The Color-Receiving System

In a compatible color receiver, the antenna, RF tuner, IF strip, and second detector serve the same functions as the corresponding components of a black-and-white receiver except that the tolerance limits on performance are somewhat tighter.

The signal from the second detector is utilized in four circuit branches. One circuit branch directs the complete signal toward the color picture tube, where it is used to control luminance by being applied to all kinescope guns in equal proportions. In the second circuit branch, a band-pass filter separates the high-frequency components of the signal (roughly 2.0 to 4.1 MHz) consisting mainly of the two-phase modulated subcarrier signal. This signal is applied to a pair of modulators which operate as synchronous detectors to recover the original I and Q signals. Note that the frequency components of the luminance signal falling between about 2 and 4.1 MHz are also applied to the modulators and are heterodyned down to lower frequencies. These frequency components do not cause objectionable interference, however, because they are frequency interlaced and tend to cancel out through persistence of vision.

The remaining two circuit branches at the output of the second detector make use of the timing or synchronizing information in the signal. A conventional sync separator is used to produce the pulses needed to control the horizontal- and vertical-deflection circuits, which are also conventional. The high-voltage supply for the kinescope can be obtained either from a flyback supply associated with the horizontal deflection circuit or from an independent RF power supply. Many color CRTs require convergence signals to enable the scanning beams to coincide at the screen in all parts of the picture area; the waveforms required for this purpose are readily derived from the deflection circuits.

The final branch at the output of the second detector is the burst gate, which is turned on only for a brief interval following each horizontal sync pulse by means of a keying pulse. This pulse may be derived from a multivibrator controlled by sync pulses, as illustrated in Figure 1.12-29, or it may be derived from the flyback pulse produced by the horizontal output stage. The separated bursts are amplified and compared with the output of a local oscillator in a phase detector. If there is a phase difference between the local signal and the bursts, an error voltage is developed by the phase detector. This error voltage restores the oscillator to the correct phase by means of a varactor connected in parallel with the tuned circuit of the oscillator. This automatic frequency control circuit keeps the receiver oscillator in synchronism with the master subcarrier oscillator at the transmitter. The output of the oscillator provides the reference carriers for the two synchronous detectors; a 90° phase shifter is necessary to delay the phase of the Q modulator by 90° relative to the I modulator.

The *filter section* in a color receiver is similar to the filter section of the transmitting equipment. The M, I, and Q signals must all be passed through filters in order to separate the desired signals from other frequency components which, if unimpeded, might cause spurious effects. The I and Q signals are passed through filters of nominally 1.5 and 0.5 MHz bandwidth, respectively, just as at the transmitting end. A step type characteristic is theoretically required for the I filter, as shown previously in Figure 1.12-23, to compensate for the loss of one sideband for all frequency components above about 0.5 MHz. This requirement is ignored in many practical receiver designs, resulting in only a slight loss in sharpness in the I channel. A roll-off filter is desirable in the M channel to attenuate the subcarrier signal before it reaches the CRT. The subcarrier would tend to dilute the colors on the screen if it were permitted to appear on the CRT grids at full amplitude. Delay networks are needed to compensate for the different inherent delays of the three filters, as explained previously.

Following the filter section in the receiver there is a matrix section in which the M, I, and Q signals are cross-mixed to re-create the original R, G, and B signals. The R, G, and B signals at the receiver are not identical with those at the transmitter because the higher frequency components are mixed and are common to all three channels. This mixing is justifiable because the eye cannot perceive the fine detail (conveyed by the high-frequency components) in color. There are many possible types of matrixing circuits. The resistance mixers shown provide one simple and reliable approach. For ease of analysis, the matrix operations at the receiver can be considered in two stages. The I and Q signals are first cross-mixed to produce R-M, G-M, and B-M signals (note that *negative* I and Q signals are required in some cases), which are, in turn, added to M to produce R, G, and B.

In the output section of the receiver, the signals are amplified to the level necessary to drive the kinescope, and the DC component is restored. The image which appears on the color CRT screen is a high-quality full-color image of the scene before the color camera.

The block diagram shown in Figure 1.12-29 is intended only to illustrate the principles used in color receivers and does not represent any specific model.

COLOR FIDELITY

Color fidelity, as used herein, is the property of a color television system to reproduce colors which are realistic and pleasing to the average viewer.

The following describes possible distortions in the color system and their effect on the picture and prescribes amounts or degrees of distortion that can be tolerated without adverse effects on picture quality.

Color System Analysis

Individual elements or areas of the complete color system are discussed in the following paragraphs with the aid of the diagrams shown in Figures 1.12-30 through 34.

Figure 1.12-30 is a theoretical color system in that it assumes linear camera pickup devices and a linear picture tube interconnected by a distortionless wire system. The only distortion that can result from this system is a flaw in colorimetry.

Figure 1.12-31 introduces linearity correctors to compensate for color errors produced by nonlinearities in the transducers (camera pickup device and picture tube).

Figures 1.12-32, 33, and 34 successively introduce the complexities of matrixing, band limiting, delay compensation, and the transmission system (shown as a dotted box in Figure 1.12-34). These diagrams, each representing a possible color system, introduce techniques used in compatible color television and permit the study of color distortions peculiar to each technique.

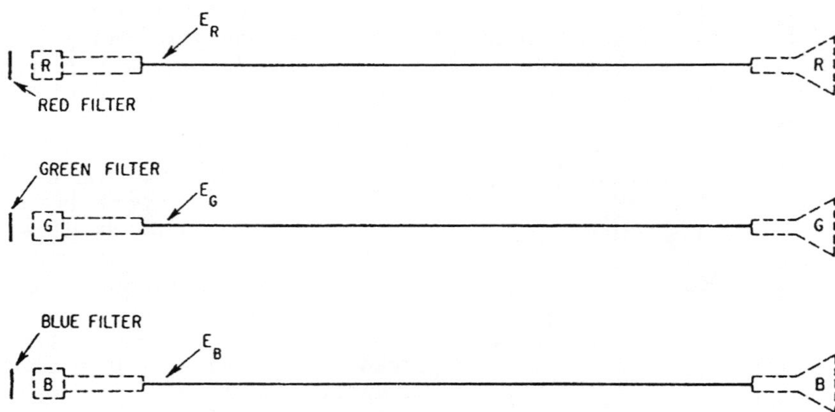

FIGURE 1.12-30 Diagram of theoretical color system showing linear RGB pickup elements and picture tubes interconnected by wire.

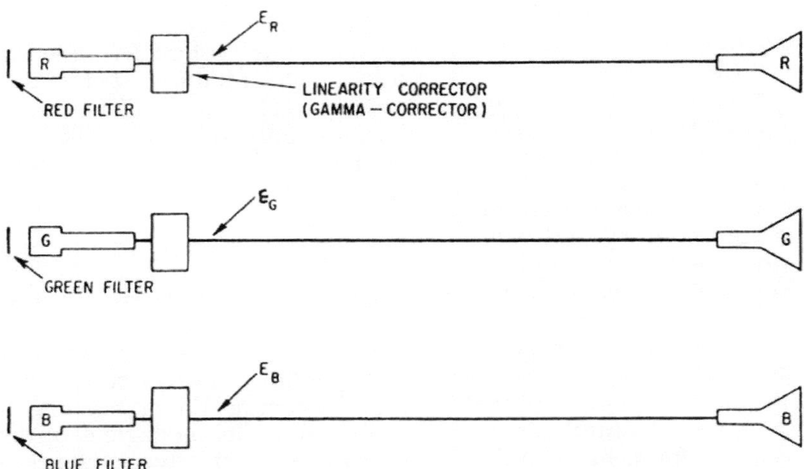

FIGURE 1.12-31 The basic color system with linear correctors to compensate for color errors introduced by the nonlinear transducers.

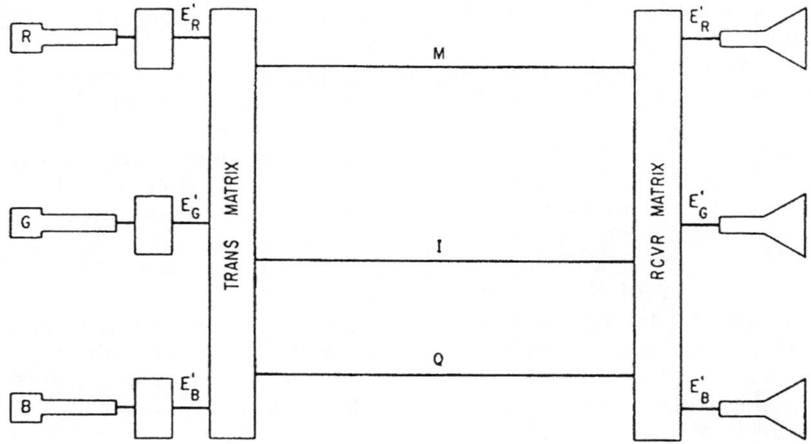

FIGURE 1.12-32 Diagram showing transmitter and receiver matrix functions in the color system.

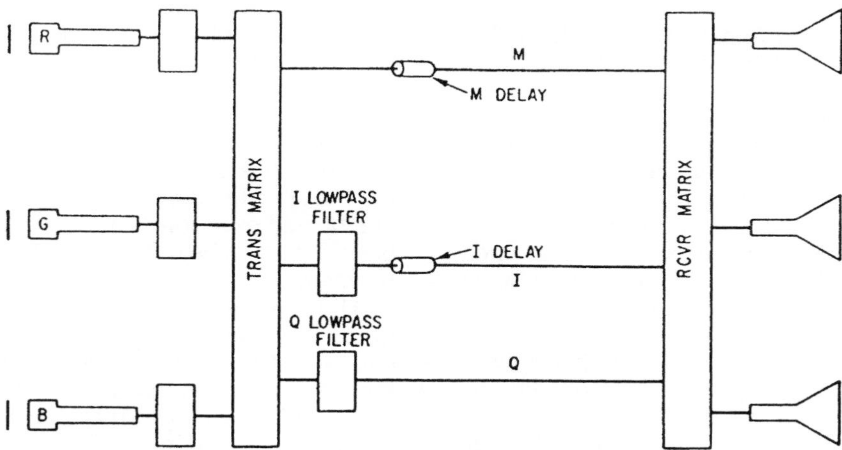

FIGURE 1.12-33 Basic color system with band limiting and delay compensation.

The systems diagrammed in Figures 1.12-30 and 31 are described under "Errors in Transducers," and those in Figures 1.12-32, 33, and 34 are described under "Encoding and Decoding Distortions." The system shown in Figure 1.12-34 is discussed under "Distortions in the Transmission System."

Characteristics of the Eye

To appreciate fully the significance of color fidelity, it is helpful to consider some of the characteristics of the eye that are associated with color perception and to analyze such terms as "color adaptation," "reference white," and "primary colors" and determine their relationship to a color television system.

Color Adaptation

One characteristic of the eye is the phenomenon known as *color adaptation*. It is this adaptation which enables one to describe accurately the color of an object under white light while viewing in nonwhite light. That is to say, recognition of color is surprisingly independent of the illumination under which an object is viewed. For example, if sunlight at high noon on a cloudless day is taken as white light, then, by comparison, the illumination from a typical 100 W incandescent bulb is very yellow light. Yet it is known that an object viewed under sunlight looks very little if any different when viewed under incandescent light. Moreover, it is obvious to the observer, after a very few minutes in a room illuminated with incandescent lights, that the light is not yellow at all; it is really white.

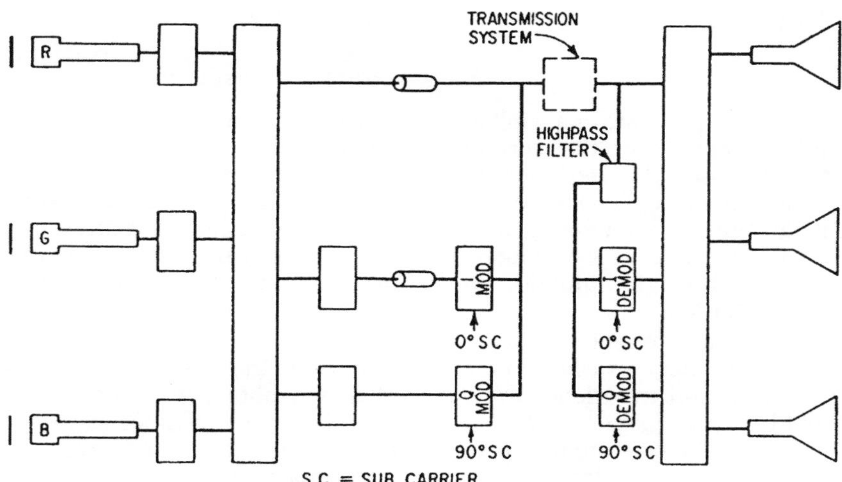

FIGURE 1.12-34 Basic color system showing all major elements, including the transmission system (dotted box).

It is apparent, then, that the color seen by an observer depends on the illumination to which that observer has been exposed for the past several minutes. This ambient illumination will have a marked effect on the choice of color to be called white.

This phenomenon can cause a loss of color fidelity under certain conditions. Consider, for example, a theoretically perfect color system with the camera viewing an outdoor scene under a midday sun while the reproduced picture is being viewed in a semidarkened room, with what little light is in the room also being derived from the midday sun. Under these conditions, the ambient illuminations at both camera and receiver are identical, so a person standing alongside the camera and one viewing the receiver would both see the same colors. Now, if a change in the weather at the camera location should cause a cloud to cover the sun, the ambient illumination at the camera location would shift toward a bluer color. This shift would not disturb the viewer standing alongside the camera, because the observer's eyes, bathed in the new ambient light, would rapidly adapt to the new viewing conditions and they would perceive the scene as being unchanged.

The person viewing the receiver would not be so fortunate. Assuming that the viewer is far enough away that this same cloud would not affect their ambient, the viewer would observe that everything on the screen had suddenly and inexplicably taken on a bluish cast, which would certainly be most disturbing.

Such errors in color fidelity can be corrected by making the camera imitate the human eye in adaptation. The eye adapts to changes in ambient illumination by changing its sensitivity to a certain color. For example, if a light source changes from white to blue-white (as in the example), the eye reduces its blue sensitivity until the light again appears to be white to the observer. Likewise, a camera operator can correct for the same situation by decreasing the gain of the blue channel of the camera or by attenuating the light reaching the blue camera tube. In this way, the camera is made to color adapt, and the reproduced picture on a receiver loses its bluish cast.

Reference White

Although color adaptation can generate a problem such as the one just described, it also simplifies certain requirements. Specifically, it eases the requirement that white be transmitted as a definite, absolute color, for there clearly can be no absolute white when almost any color can be made to appear subjectively white by making it the color of the ambient illumination to which an observer's eye has adapted.

In color television, this characteristic is taken advantage of in the following manner: A surface in the studio which is known by common experience to be white, for example, the EIA Gray-Scale Chart of a piece of Neutracor white paper, is selected to be reproduced as white on a home receiver. The relative sensitivities of the three-color channels of the camera are then adjusted so that the camera adapts to this white regardless of the studio illumination. The home receiver can then be adjusted to reproduce the surface as any white which the home viewer prefers, depending on his/her surroundings.

It has already been mentioned that the eye adapts readily to the illumination that surrounds conditions of an overcast day. This representative standard illumination has been adopted internationally as a base for the specification of the color of objects when they are viewed outdoors. This standard (Illuminant C) has been chosen to be the "standard-viewing-white" of the receiver.

The change in reference white between studio and home will inevitably produce errors in all reproduced colors, but the errors are small and, more important, tend to be subjectively self-correcting, so that any given object will produce the same color sensation whether viewed in relation to the studio reference white or the home reference white.

Consequently, a viewer may become familiar with an object such as a sponsor's packaged product and will recognize it on the television screen, under the fluorescent lighting of the supermarket, or under the incandescent lighting of the home and, furthermore, will note little difference in the colorimetric values of the package under the three conditions, even though the absolute colorimetric values would be appreciably different in the three situations.

Primary Colors

Of all the characteristics of the eye, there is perhaps none more fundamental to practical color television than that characteristic which allows viewers to choose certain colors called primary colors, and from these synthesize almost any other desired color by adding together the proper proportions of the primary colors. If it were not for this characteristic, each hue in a color system would have to be transmitted over a separate channel; such a system would be too awkward to be practical. Because of the eye's acceptance of synthesized colors, it is possible to provide excellent color rendition by transmitting only the three primary colors in their proper proportions.

Errors in Transducers

The block diagram of Figure 1.12-30 shows a fundamental color television system using red, green, and blue primaries and three independent transmission channels. The camera tubes and kinescopes are shown dotted to indicate that any inherent nonlinearities in these devices are to be disregarded, for the moment, in order to simplify the discussion of the colorimetry of the system.

The general plan is a system, such as Figure 1.12-30, to provide the three picture tubes with red, green, and blue phosphors, respectively, and to allow the corresponding camera tubes to view the scene through an appropriate set of red, green, and blue filters. If a phosphor and a filter have the same dominant wavelength—that is, if they appear to the eye to be the same color—it might be mistakenly supposed that they would be colorimetrically suited to be used as a filter and phosphor set for the channel handling that color. Actually, the basis for choosing filters and phosphors is much more complex and is based on the shape of the response curve of the filter, plotted against wavelength, and the shape of the light output curve of the phosphor, also plotted against wavelength. The following paragraphs will discuss briefly a technique which might be used to determine the required relationship between the phosphor curves and the filter curves.

The color characteristics of the phosphor are generally less easily changed than are filter characteristics. For this reason characteristics of phosphors are taken as the starting point, and characteristics of the filters are determined from them. A laboratory setup which could be used to determine these characteristics is shown in Figure 1.12-35.

In this figure, an observer (who must have normal vision) is simultaneously viewing two adjacent areas, one of which is illuminated by a source of single-wavelength light which can select any wavelength in the visible spectrum; the other of which is illuminated by a red picture tube, a green picture tube, and a blue picture tube. The phosphors of these CRTs are the phosphors which are to be used in the color system. Starting at the red end of the spectrum, a single-wavelength red is selected to illuminate the left-hand area, and the light from each of the three phosphors is varied until a color match is obtained between the left-hand and right-hand areas. The respective amounts of red, green, and blue lights needed to accomplish this match are recorded. Then another wavelength is chosen, the kinescope outputs varied to produce a match, and the new amounts of red, green, and blue needed for a match are recorded. Similarly, points are

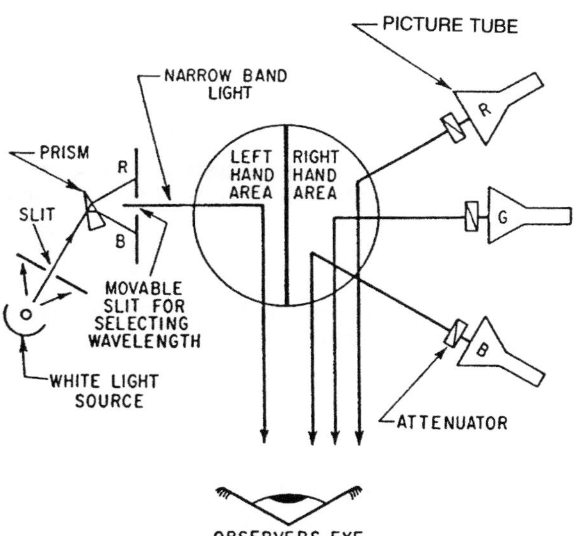

FIGURE 1.12-35 Diagram showing laboratory setup arranged to compare narrow-band light source and R, G, and B light produced by picture tubes to determine proper camera-filter color characteristics.

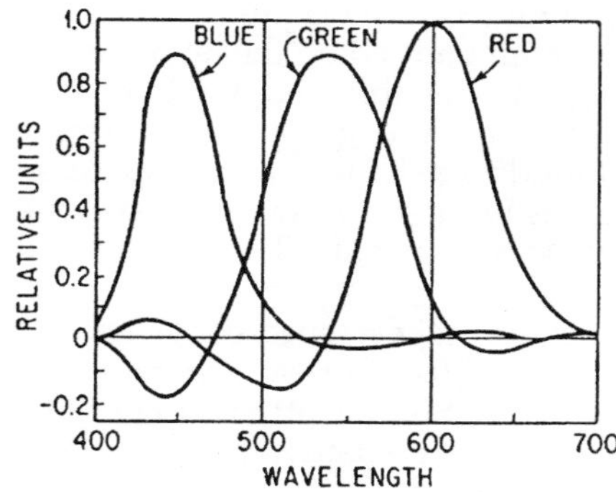

FIGURE 1.12-36 Curves showing relative quantities in camera output required to produce correct picture tube colors over the visible spectrum.

obtained throughout the entire spectrum, and a graph is plotted showing the various required outputs versus wavelength. The shapes of these three curves (one for red, one for green, and one for blue) are the required shapes for the three camera-filter response curves. The resulting curves would in general resemble Figure 1.12-36. To simplify the above discussion, it was assumed that the camera pickup devices responded equally well to all wavelengths. In practice, tubes show higher output at certain wavelengths than at others. The filter-response curves derived by the technique would have to be modified so that the combined response of filter and camera would be correct.

Certain practical difficulties could result in errors in the previous procedure. For example, if the observer had any deviations from normality in color vision characteristics (as most people do), these deviations would result in nonstandard matches and, hence, improper camera filter characteristics. Also, if the phosphors were contaminated in any way during their manufacturing process (as most phosphors are, at least to some small degree), the resulting phosphor characteristics would not be the proper ones and hence would give rise to improper camera filter characteristics. The observer errors can be normalized out by standard colorimetric procedures, but phosphor errors represent a basic error which may possibly be present not only in the previous experiment but also in varying degrees in a large number of receivers. Quality control of phosphor manufacture is sufficiently good, however, to make the net effect unnoticeable in home receivers.

A striking practical difficulty would also arise regardless of observer or phosphor errors. For most wavelengths, no combination of red, green, and blue picture tube outputs could be found which would produce a match. In order to obtain a match at these wavelengths, it would be necessary to move one or two of the CRTs over to the other side so that they could add their light to the single-wavelength light being matched. This procedure can be described mathematically, for graphing purposes, by saying that adding light to the left-hand area is the same as subtracting light from the right-hand area. Therefore, the amount of light added on the left would be considered as a negative quantity and would result in a point below the axis on the graph. Since this condition would be found to exist for several successive wavelengths, the resulting graph would show one or more minor lobes below the axis. These are called *negative lobes*.

These negative lobes represent a need for filters with negative light transmission characteristics at certain wavelengths. Simple attenuating filters cannot yield such a characteristic; much more elaborate means would be required.

It is theoretically possible to achieve these negative lobes with added camera complexity, but it has been shown that excellent color fidelity can be obtained by ignoring the negative lobes and using filters which yield the positive lobes only. Positive lobe processes such as color photography have gained wide acceptance for years. Masking techniques which employ electrical matrixing have been introduced which can modify the spectrum characteristics of a color camera. These techniques can be used to help compensate for deficiencies in the color fidelity such as the lack of negative lobes.

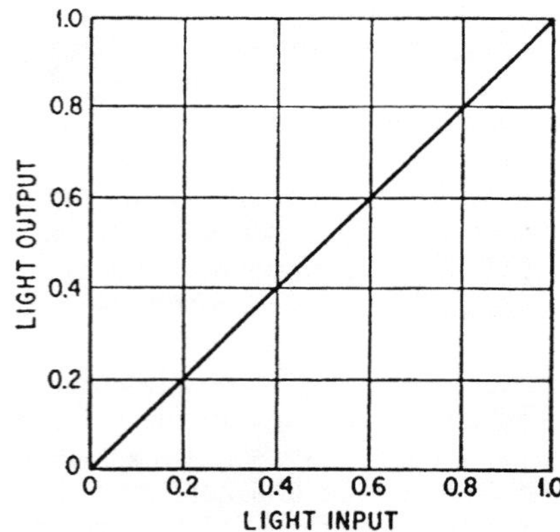

FIGURE 1.12-37 Curve showing light transfer characteristics of a perfectly transparent piece of window glass or a perfectly linear electronic system.

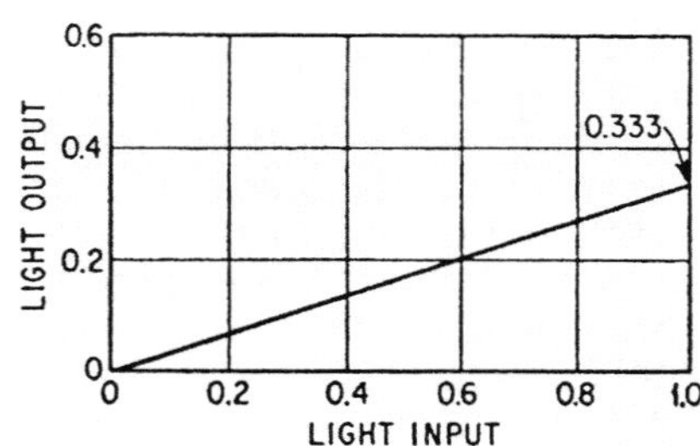

FIGURE 1.12-38 Curve showing transfer characteristic of a neutral density filter with 3-to-1 light attenuation.

Transfer Characteristics

A piece of window glass is perhaps the nearest approach to a perfect video system. For a piece of glass, the light output (to the viewer) is essentially identical with the light output (from the scene). This fact is shown graphically in Figure 1.12-37. This plot could be called the *transfer characteristic* of a piece of glass, since it describes the way that light is transferred through the system.

If the window glass is replaced by a neutral density filter, which attenuates light 3-to-1, the transfer characteristic will then be given by Figure 1.12-38. The difference between Figures 1.12-37 and 38 can be described by these simple relationships:

For the glass:
$$\text{Light output} = \text{light input}$$

For the neutral density filter:
$$\text{Light output} = k \text{ light input}$$

where $k = 1/3$ in this case.

Both systems are linear—doubling the light input of either will double its light output, tripling input will triple output, etc. A nonlinear system does not exhibit this simple proportionality. For example, consider a system described by

$$\text{Light output} = k\,(\text{light input})^2$$

Doubling the input to this system will quadruple its output, a threefold increase in input will result in a ninefold increase in output, etc. The transfer characteristic for this type of system is shown in Figure 1.12-39. Note that the characteristic is definitely nonlinear; that is, it is not a straight line, as were Figures 1.12-37 and 38.

In television and photography, nonlinearity is more common than linearity. For example, an ordinary picture tube is a nonlinear device, having a transfer characteristic which can be approximated by the expression:

$$\text{Light output} = k\,(\text{voltage input})^{2.2}$$

Camera tubes can be linear or nonlinear devices. For example, the characteristic of a vidicon (camera pickup tube) is approximately

$$\text{Current input} = k\,(\text{light input})^{0.65}$$

The general expression for a nonlinear transfer characteristic can be given approximately as

$$\text{Output} = k\,(\text{input})^{\gamma}$$

where the exponent is the Greek letter gamma.

Graphical Displays of Transfer Characteristics

The first reaction of any person asked to display two variables (like light input and light output) on a set of XY coordinates is to divide X and Y coordinates into equal increments and plot the variables in this manner. A typical result of such a plot has already been described (see Figures 1.12-37 and 38). Such a plot has

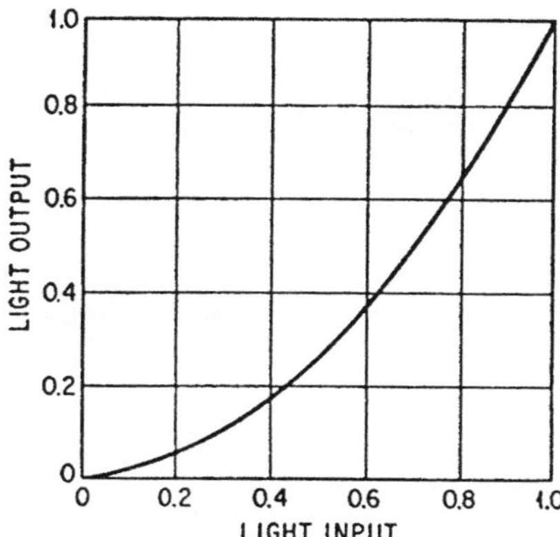

FIGURE 1.12-39 Curve showing a nonlinear transfer characteristic.

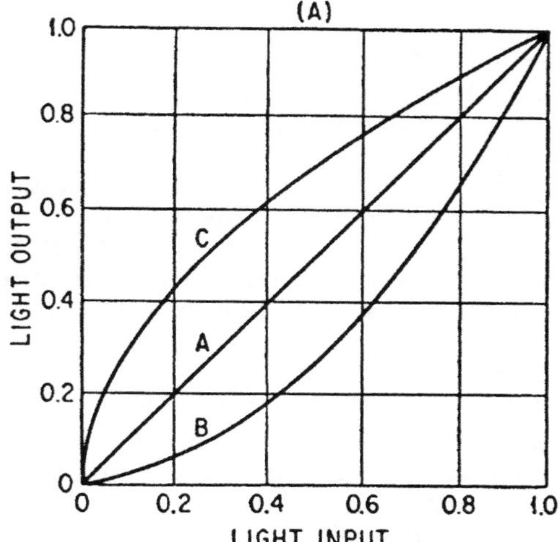

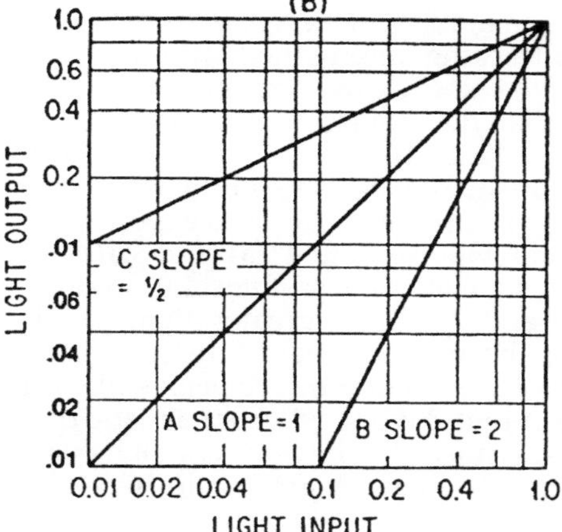

FIGURE 1.12-40 Graphs showing the curves obtained by plotting A, B, and C types of transfer characteristics on linear coordinates (a) and on log-log coordinates (b).

the advantage of showing at a glance the linearity of the device described by the variables. If the plot is a straight line, the device is linear; if curved, the device is nonlinear. Moreover, the slope of the line describes the attenuation (or gain) of the device. If the slope is unity (which occurs when the plot makes a 45° angle with the X axis), there is no attenuation. It is a very good piece of glass or electronic circuit. For the neutral density filter previously described, which has the equation (light output) = 1/3 (light input), the line has a slope of one-third (see Figure 1.12-38).

Such are the advantages of plotting transfer characteristics with equal increment divisions of the X and Y axes. However, other advantages (very important ones) can be obtained by dividing up the X and Y coordinates logarithmically. Such a plot is called a *log-log* plot.

Consider a system which has a transfer characteristic given by $L_0 = (L_{in})^{2.2}$. If this equation is plotted on axes which are divided logarithmically, the resulting plot is the same as though the logarithm of both sides of the equation were plotted on equal increment axes. Taking the logarithm of both sides, we obtain

$$\log L_0 = \log(L_{in})^{2.2}$$

Since $\log(L_{in})^{2.2}$ is the same as $2.2 \log(L_{in})$, then

$$\log L_0 = 2.2 \log L_{in}$$

Comparing the form of this equation with an earlier equation, light output = 1/3 light input, the attenuation, 1/3, was the slope of the earlier equation, so 2.2, the exponent, is the slope of the latter equation. The use of logarithmically divided coordinates yields a plot in which the exponent is given by the slope of the line. Therefore, this plot will show at a glance the magnitude of the exponent and will also show whether or not the exponent of the system is constant for all light levels. It also is advantageous in showing the effects of stray light.

Figures 1.12-40(a) and 40(b) compare the two types of plotting for three types of transfer characteristics.

The Effect of a Nonlinear Transfer Characteristic on Color Signals

In monochrome television, some degree of nonlinearity can be tolerated, but such is not the case for a color television system. It can be shown that a system exponent different from unity must inevitably cause a loss of color fidelity. For an example, consider a situation in which signals are being applied through linear amplifiers to the red and green guns of a perfectly linear (theoretical) picture tube. The green amplifier is receiving 1.0 V; the red amplifier, 0.5 V. If everything is perfectly linear, the proportions of the light output should be 1.0G + 0.5R = greenish yellow. However, if the CRT has an exponent of 2.0, the light output will be $(1.0)^2$G + $(0.5)^2$R = 1.0G + 0.25R = greenish yellow with an excess of green.

From the previous specific case, it may be correctly inferred that in general a system exponent greater than 1.0 will cause all hues made of the combination of two or more primaries to shift toward the larger or largest primary of the combination. Conversely, a system exponent less than 1.0 will shift all hues away from the largest primary of the combination.

In the previous example, an exponent of 0.5 would yield $(1.0)^{0.5}G + (0.5)^{0.5}R = 1.0G + 0.707R$ = a greenish yellow which is a shade off a pure yellow.

In addition, the reader can correctly conclude that white or gray areas, in which all the primaries are equal, will not be shifted in hue by a nonunity exponent.

Effect of differing exponents in each channel

The preceding discussion assumed that all three channels (in Figure 1.12-30) have the same exponent, whether in unity or not. In practical systems, however, there is always the possibility that the exponents of the channels may differ from one another. This situation will produce intolerable color errors if the differences become even moderately large. In general, the requirements for tracking among the light transfer characteristics of the individual channels are even more stringent than the requirement for unity exponent.

Figures 1.12-41(a), 41(b), 41(c), and 41(d) show the effects of unequal exponents in the three channels. In all four figures, the red and blue exponents are taken as unity; in Figures 1.12-41(a) and 41(b) the green exponent is taken as less than 1.0; and in Figures 1.12-41(c) and 41(d), as greater than 1.0. In Figure 1.12-41(a), the transfer characteristics are shown for the system adjusted to produce peak white properly. It can be seen that the bowed characteristic of the green channel will cause all whites of less than peak value to have too much green. A gray-scale step tablet before the camera would be reproduced properly only at peak white; the gray steps would all have a greenish tinge. Relative channel gains could be readjusted to reproduce one of the gray steps properly (Figure 1.12-40[b]), but then all highlight steps would be purplish while lowlight steps would still be greenish.

A green-channel exponent greater than unity would reverse those results (Figures 1.12-41[c] and 41[d]). With gains adjusted to reproduce peak white properly (Figure 1.12-41[c]), lowlights would be purplish. With gains readjusted to provide proper reproduction for one of the lower steps (Figure 1.12-41[d]), highlights would be green; and lowlights, purple.

The effect of stray light

If a picture tube is viewed in a lighted room, there will always be some illumination on the faceplate. Therefore, the eye will always receive some *light output* from the picture tube, regardless of the magnitude of the signal input voltage. Under this condition, a true black is impossible to obtain.

This condition is reflected in the transfer characteristic of the system. If, for example, the stray light were 5% of the peak highlight brightness of the picture, a linear plot of light output versus light input would have the entire transfer characteristic shifted upward by 5%. However, the most interesting change is found in the log-log plot, where, as seen in Figure 1.12-42, the stray light causes a change in the slope in the lowlight regions. Since the slope is equal to the exponent, this change shows that stray light causes an effective exponent error in the lowlight regions of the picture and hence will cause color fidelity errors, which will be most marked in lowlight regions.

These errors will be noted by an observer as improper hues and saturations, with the saturation errors (a washing out of the more saturated lowlight areas) being the more objectionable to a viewer.

Stray light is not the only cause of errors of this type. Similar effects will be noted whenever the CRT bias (brightness) is set too high, if the camera video output waveform black level (pedestal) is set too high, or if stray light enters the camera (whether through

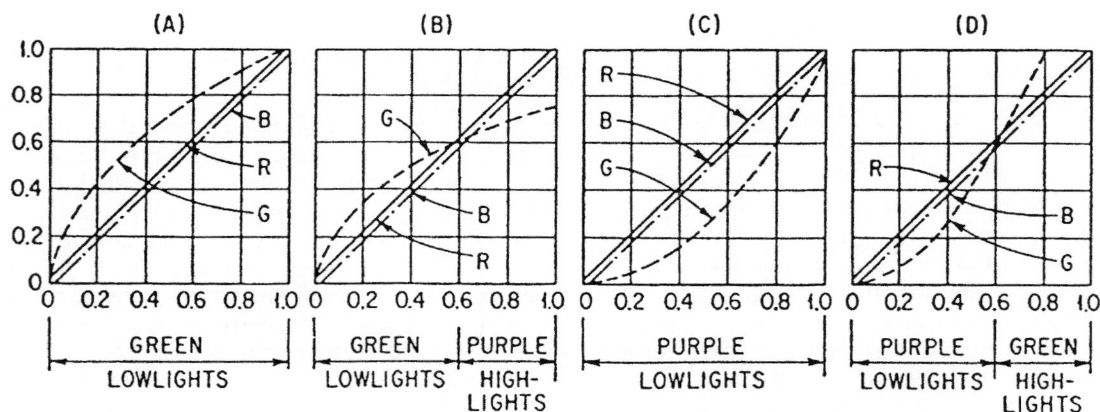

FIGURE 1.12-41 Linear plots showing graphically the effect of unequal exponents in the R, G, and B channels. In all four graphs the R and B exponents are taken as unity. In (a) and (b) the green exponent is taken as less than 1.0; and in (c) and (d), as greater than 1.0.

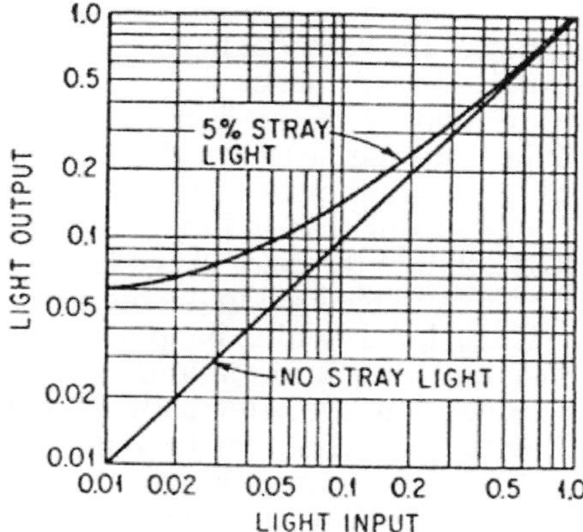

FIGURE 1.12-42 Log-log plot of system with stray light, illustrating change of slope in the low-light regions.

lens flare or any other source). In general, any condition which prevents the light output of the system from becoming zero when the light input is zero will cause errors similar to those caused by stray light.

Linearizing a system

It can be shown that a system using, for example, a vidicon camera pickup tube with an exponent of 0.675 to drive a CRT with an exponent of 2.2 will have an overall exponent given by the product 0.65 × 2.2 = 1.43, assuming that all devices in the system are linear. In general, the overall exponent of a system is the product of the exponents of the cascaded elements.

This knowledge provides an excellent tool for linearizing a system. For example, a system with an overall exponent of 1.43 could be linearized by inserting somewhere (in a video path) an amplifier having an exponent of 1/1.43 (= 0.7) so that the product becomes unity: 1.43 × 1/1.43 = 1.0.

In Figure 1.12-31, a nonlinear amplifier, or gamma corrector, is shown inserted in each of the three paths.

Encoding and Decoding Distortions

The second of the two systems discussed in the preceding section bordered on being a practical system but still required three independent 4 MHz channels. A fortunate characteristic of the human eye (the inability to see colored fine detail) allows us to modify this requirement to one 4 MHz channel for monochrome fine detail and two much narrower channels for color information. Before this modification can be made, the red, green, and blue signals must be combined to form three other signals, usually called M, I, and Q, such that the M signal alone requires a 4 MHz channel, and the I and Q channels, which contain the color information, are confined to narrower channels. This rearrangement of red, green, and blue to form M, I, and Q is called matrixing and was described in the previous section. A system which uses a matrix is block diagrammed in Figure 1.12-32. The illustration also shows that to recover the original red, green, and blue signals at the receiving end, a rearranging device is needed. This device is usually called the *receiver matrix*.

Matrixing alone offers no advantage unless steps are taken to limit the I-signal and Q-signal channels to the narrow bandwidths allowed. Figure 1.12-33 shows a system employing such band shaping. The band shaping filters themselves always introduce delay, which must be compensated for by placing delay lines in the wider band channels, as shown in the diagram.

To put both color and monochrome information in the spectrum space normally occupied by monochrome requires only that the color information overlap the monochrome. This overlap can be allowed for both I and Q signals, without incurring visible cross talk, if frequency interlace and two-phase modulation are employed. A system using these techniques, which were described in the section on "Electronic Aspects of Compatible Color Television," is block diagrammed in Figure 1.12-34.

Errors in the Matrixing Process

The entire matrixing process can be summed up in two sets of equations, the first describing how the transmitter matrix takes in red, green, and blue and turns out M, I, and Q:

$$M = 0.30R + 0.59G + 0.11B$$
$$I = 0.60R - 0.28G - 0.32B$$
$$Q = 0.21R - 0.52G + 0.31B$$

and the second describing how the receiver matrix takes in M, I, and Q and re-creates red, green, and blue:

$$R = 0.94I + 0.62Q + M$$
$$G = 0.27I + 0.67Q + M$$
$$B = 1.11I + 1.7Q + M$$

Both matrices continuously compute the desired output from the given input. The coefficients in the preceding six equations are usually determined by precision resistors or, in the case of negative numbers, by precision resistors and signal-inverting amplifiers. The basic error that can occur, therefore, is a change in a resistor value or an amplifier gain, resulting in a change in one or more coefficients. In general, the resulting picture error resembles cross talk among the primary colors.

More specifically, the transmitter matrix can have two distinct types of errors. The first involves the coefficients of the equation for M; the second, the coefficients for I and Q. An error in an M coefficient will brighten or darken certain areas. In a monochrome reproduction of a color signal, such an error, if small, would not be noticed; if large, it would still probably be tolerated by the average viewer. In a color reproduction, however, even a small error would be objectionable. For example, a reduction of the red coefficient from 0.3 to 0.2 would cause a human face to be reproduced with an unnatural ruddy complexion and dark red lips.

Note that the sum of the M coefficients is 1.0. An error in one coefficient would change this sum, so that peak white would no longer occur as 1 V. An operator could mistake this condition for a gain error and adjust either M gain or overall gain in an effort to obtain the correct peak white voltage. Changing M gain would cause errors to occur in all M coefficients; changing overall gain would put errors in all coefficients. Although such an error is rare in well-engineered equipment, it is a possible source of color error, which can be compounded by misdirected attempts at correction.

Note that the sums of the Q and I coefficients are each zero, which means that when R = G = B (the condition for white or gray), Q and I both equal zero. An error in a Q or I coefficient would cause color to appear in white or gray areas and, in addition, would cause general errors in colored areas resembling cross talk among the primaries. Controls are usually provided in the Q and I matrices, called *Q white balance* and *I white balance*, respectively, which allow the operator to adjust the sum of the Q or I coefficients by changing the value of one of the coefficients. If the coefficient controlled is the one in error, then adjusting white balance restores the condition that the sum of the coefficients is zero; that is, it removes the color from white and gray objects, but it does so by giving the controlled coefficient an error which just counteracts the error of a nonadjustable coefficient so that two coefficients are wrong instead of one. Again, such an error is rare in well-engineered equipment, for the adjustable coefficient is usually the one in error. However, the possibility of an error compounded by adjustment should be kept in mind.

A far more likely cause of white balance error is an error in input level, that is, a discrepancy between the peak white levels of input red, green, or blue. In such a case, an operator can still achieve white balance (Q and I = 0 for white input), but the entire system will be in error. The starting point for all investigations of the cause of white balance errors should be the levels of the red, green, and blue inputs.

In the receiver matrix, only one general type of error can occur instead of two as in the case of the transmitter matrix. This type of error, a *general coefficient error*, results in cross talk among the primary colors. For example, a change in the I coefficient for the red equation from 0.94 to 0.84 would yield about a 7% reduction in the peak red output available and would also result in unwanted red light output in green or blue areas at about 3.5% of the green or blue level.

Gain Stability of M, I, and Q Transmission Path

In the system of Figure 1.12-32, every gain device or attenuating device in the three transmission paths must maintain a constant ratio between its input and output in order to maintain the proper ratios among the levels of M, I, and Q at the input to the receiver matrix. A variation in the gain of one of these paths will result in a loss of color fidelity.

For example, a reduction in M gain must obviously cause a reduction in the viewer's sensation of brightness. Not quite so obvious are the effects of I and Q gain. Since these are color signals, their amplitude would be expected to influence the sensation of saturation, but the manner of this influence is not intuitively obvious until the factors which influenced the selection of I and Q compositions are recalled. It previously was pointed out that the eye has the greatest need for color detail in the color range from orange to blue-green (cyan) and the least in the range from green to purple. Hence I, the wider band signal, conveys mainly orange and cyan information, and Q, the narrower band signal, conveys principally the greens and purples. Therefore, a reduction in I gain could be expected to reduce the saturation sensation for colors in the orange and cyan gamut, leaving the greens and purples virtually unaffected. Conversely, Q gain will influence the greens and purples without causing much change in the appearance of orange and cyan objects.

Modulation and Demodulation

The system depicted in Figure 1.12-32, which introduced bandwidth limiting of the I and Q signals in accordance with the capabilities of the eye to see colored fine detail, is a fairly practical and economical system, except for the fact that three individual transmission channels are employed. To have a compatible system, however, these three channels must be reduced to one through some multiplexing technique. The technique used has already been described, and a system employing this technique is block diagrammed in Figure 1.12-34.

Errors in Modulation

Burst phase error

Perhaps the most fundamental error in the multiplexing process would be an error in the phase of the main timing reference, burst. Since the entire system is based on burst phase, an error in burst phase will appear as an opposite error in every phase except burst, because the circuits will insist that burst phase cannot be wrong. The general result will be an overall hue error in the reproduced picture. This effect can be better visualized by referring to Figure 1.12-43.

A phase error in burst produces the same result as holding burst phase stationary and allowing all other phases to slip around the circle an equal amount (but in a direction opposite to the burst phase error). Each color vector then represents a hue other than the one intended.

Burst amplitude error

In theory, the receiver circuits which extract timing information from the burst are insensitive to variations in burst amplitude as long as the burst is large enough to maintain a respectable signal-to-noise ratio and not so large that some type of clipping or rectification upsets the burst circuitry. But practical receivers always exhibit some degree of sensitivity depending mainly on the error in the subcarrier oscillator in the

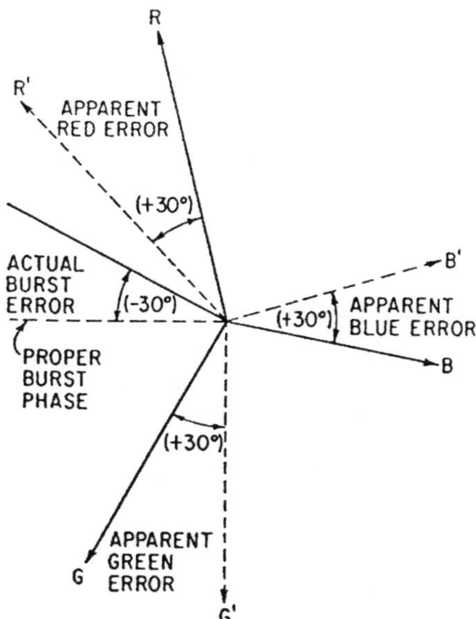

FIGURE 1.12-43 Vector diagram showing how error in subcarrier phase becomes an opposite error in all other phases.

receiver. If the free-running frequency of the receiver oscillator is very different from burst frequency—particularly if the difference is so great that the burst is in danger of losing control of the oscillator—then a fairly appreciable amplitude sensitivity will be noted. This sensitivity will take the form of a phase error, and the net result will be indistinguishable from a burst phase error, as previously discussed.

Most receivers have a circuit which automatically adjusts the gain of the color information channels so that the viewer always sees the proper saturations, regardless of errors, which might tend either to wash out or oversaturate the picture. Such a circuit, called an *automatic chroma control* (ACC), derives its control information from the amplitude of burst, which is presumed to bear a constant ratio to the amplitude of chroma. Transmission distortions, for example, might decrease the amplitude of both burst and chroma, but since the ratios of their amplitudes would be preserved, an ACC receiver could automatically modify its chroma channel gain to compensate for the decreased chroma amplitude. However, if a color encoder error should cause burst alone to decrease in amplitude, the ACC circuits would increase chroma gain just as in the previous case, with the result that a viewer would receive an oversaturated picture.

Two-phase modulation errors

The fidelity of color reproduction can be seriously affected if the phase separation of the Q and I subcarriers is not maintained at 90°. It can be shown that a slip in the angular position of the Q axis, for example, will result in cross talk of Q and I. The final result will be the same as cross talk among the primary colors.

Likewise, in a receiver, the phase relationship between the reference subcarriers must be maintained to avoid a similar error. Any deviation from the proper phase relationship will have a similar result, that is, cross talk of I into Q or Q into I, with the net picture result resembling cross talk among all the primary colors.

Carrier unbalance

In a properly operating doubly balanced modulator, the carrier component of the signal is suppressed in the modulator circuit. If some error in components or operation causes this suppression to be imperfect, the carrier will appear in the output. This condition is known as *carrier unbalance.*

The effect of carrier unbalance can be evaluated by considering the unwanted carrier as a vector of constant amplitude which adds itself vectorially to every vector present in the colorplexer output. In general, such a vector will shift all vectors and hence all hues seen in the picture toward one end or the other of the color axis represented by the unbalanced modulator. For example, a positive unbalance in the I modulator would shift all colors toward the color represented by the positive I axis, that is, toward orange. A negative I unbalance would shift all colors toward cyan.

To visualize this effect, refer to Figure 1.12-44, in which has been added to each color vector a small positive vector which is parallel to the I axis. This small vector represents the amount of carrier unbalance. The resultant vectors will all be rotated toward the positive I axis and changed in amplitude as well. Such changes represent errors in both hue and saturation.

Another error from carrier unbalance occurs in white and gray areas of the picture. In a normally operating colorplexer, a white (or gray) area in the scene causes the Q and I signals to become zero and

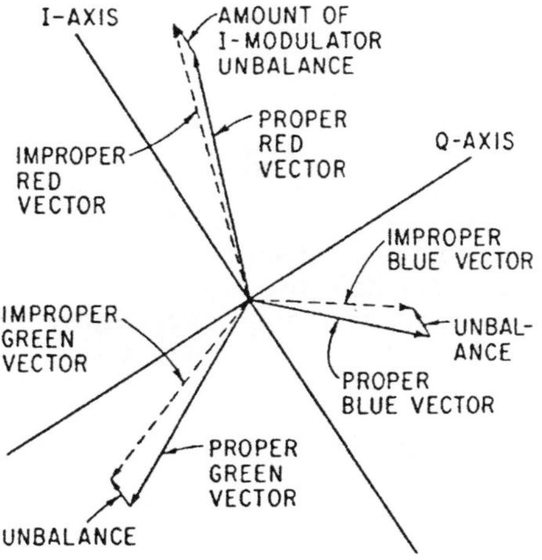

FIGURE 1.12-44 Vector diagrams of subcarrier phase and amplitude with positive vectors added to represent carrier unbalance in the I modulator.

thereby causes the modulator outputs to become zero. Hence, a white or gray area will normally appear in the signal as an interval of zero subcarrier amplitude. If one of the modulators begins to produce a carrier unbalance vector, however, a white or gray area will become colored because of the subcarrier which will be added in this interval. Moreover, certain areas, which are normally colored, may have their subcarrier canceled by the carrier unbalance vector and become white. Such white-to-color and color-to-white errors are very objectionable.

Video unbalance

A doubly balanced modulator derives its name from the fact that it balances out or suppresses both the carrier and the modulating video (Q or I). If, for any reason, the video suppression becomes less than perfect, the resulting condition is called *video unbalance.*

Video unbalance will cause unwanted Q or I video to appear in the modulator output, in addition to the desired sideband outputs. This unwanted video signal will be added to the luminance signal, thereby distorting the gray scale of the picture. For example, a slight positive unbalance in the Q modulator would slightly brighten reds and blues and slightly darken greens. A negative unbalance would have the opposite effect.

Subcarrier frequency error

The color subcarrier frequency is specified by the FCC to be 3.579545 MHz ±10 Hz. Deviations within this specified limit are of no consequence (provided they are slow deviations). Large deviations, however, can affect color fidelity. The effect does not usually become serious within the possible frequency range of a good crystal-controlled subcarrier source driving a properly designed receiver.

In receivers, the subcarrier timing information is extracted from the burst on the back porch (between sync and video) and used to control the frequency of a subcarrier-frequency oscillator in the receiver. As long as the unlocked frequencies of the burst and the receiver oscillator remain the same, the locked phase relationship between the two will remain the same. But if either the burst frequency or the receiver-oscillator frequency becomes different (and the difference between them is not so large that lockup is impossible), then the locked error, which obviously cannot be a frequency error, manifests itself as a phase error. This error can become as large as ±90° before the AFC circuit can no longer hold the receiver oscillator on frequency. The frequency range over which this phase shift occurs depends on the receiver design.

Distortions in the Transmission System

Preceding sections have described the processes involved in the generation and display of a color television signal. Errors in these processes are not the only possible source of distortion; when the signal is transmitted over great distances, the transmission system itself may contribute errors. This section discusses parameters which specify the behavior of a transmission system and describes the effects that errors in these parameters can have on the reproduced picture.

This section is divided into two parts. The first relates to the parameters of a perfectly linear transmission system, and the second part discusses the additional parameters required to describe the nonlinearities that are inevitable in any practical system.

The Perfectly Linear Transmission System

A perfectly linear and noise-free transmission system can be described by its gain and phase characteristics plotted against frequency as the independent variable.[2] Typical plots are shown in Figures 1.12-45 and 46, respectively. These two characteristics known, it is possible to predict accurately what effect the transmission system will have on a given signal.

Gain Characteristic

Figure 1.12-45 is usually known as the frequency response or gain characteristic of the system. Ideally, it should be perfectly flat from zero to infinite frequency, but this, of course, is impossible to attain. An amplifier has a definite gain bandwidth product, depending on the transconductance of its active elements (transistors, ICs, etc.), the distributed capacity shunting these elements, and the types of compensation (peaking) employed. The bandwidth of a given combination of transistors, stray capacitances, and peaking networks can be increased only by decreasing its gain, or conversely, its gain can be increased only by decreasing its bandwidth. There is a limitation, therefore, to the actual bandwidth that can be obtained. For a given scanning standard, the bandwidth required in a monochrome television system is determined by the desired ratio between the horizontal resolution and the vertical resolution. Although nominally a 4.0 MHz bandwidth is required for the monochrome standards, the requirement can be relaxed to the detriment of only the horizontal resolution. The subjective result is a softening of the picture in proportion to the narrowing of the bandwidth (neglecting the influence of the phase characteristic in the vicinity of the cutoff frequency). As pointed out in previous sections, the entire chrominance information of the color system is located in the upper 1.5 MHz of the prescribed 4.0 MHz channel; hence, any loss of response in this part of the spectrum can have a marked effect on the color fidelity of the reproduced picture.

One of the most serious forms of distortion inflicted on a color picture by bandwidth limiting is loss of saturation. Consider a case in which the bandwidth is so narrow as to result in no gain at the color subcarrier frequency. The output signal then contains no color subcarrier and hence reaches the color receiver as a monochrome signal, producing zero sat-

[2]If the filters in the system are of minimum-phase type, only one of the plots is needed, for either plot can be derived from the other for this type of filter. Almost all common interstage coupling networks are of the minimum-phase type.

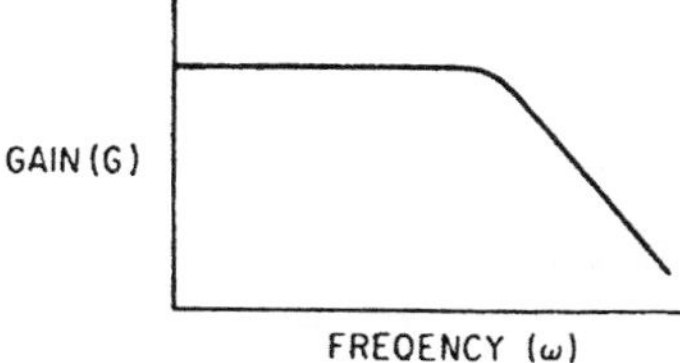

FIGURE 1.12-45 Typical curve showing a gain of a system plotted against frequency to determine its gain characteristic.

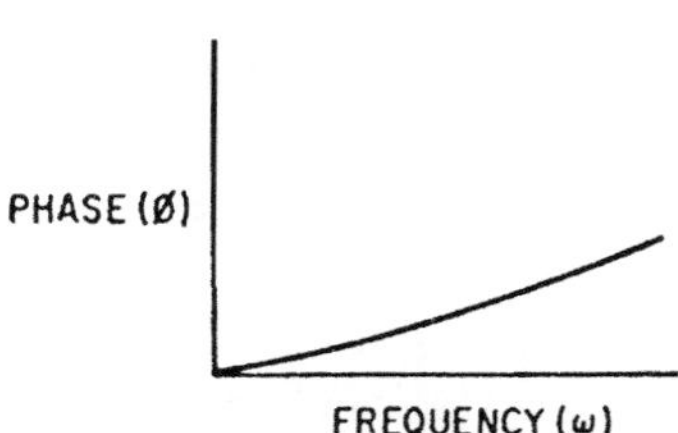

FIGURE 1.12-46 Curve showing phase characteristic of a system plotted versus frequency.

uration. Nearly as poor results can be expected from an amplifier with response such that the gain at 3.58 MHz is one-half the low-frequency gain. Since the saturation depends chiefly on the amplitude of the subcarrier, the saturation will be correspondingly reduced. The resultant color picture will have a washed-out look.

Loss of high-frequency response, which can be expected to contribute to loss of fidelity, is usually accompanied by phase disturbance, depending on the types of networks employed in the system. The intent in this section, however, is to treat each variable separately. Therefore, discussions are based on the effects of varying only one parameter of a system. It is suggested that the reader can determine the combined effect of two or more variables by comparing the results shown for the individual variables.

Phase Characteristic

An ideal system has a *linear* phase characteristic, as shown in Figure 1.12-47(a). Such a characteristic implies that all frequencies of a signal have exactly the same *time delay* in passing through this system, since the time delay is given by the phase angle divided by the (radian) frequency. It can be seen in Figure 1.12-47 that if three frequencies are chosen arbitrarily, then the corresponding phase angles must have values proportional to their corresponding frequencies (because of the geometric properties of a right triangle). To state it another way, if $\varphi_1/\omega_1 = 0.2$ µsec, then $\varphi_2/\omega_2 = 0.2$ µsec and φ_3/ω_3 also equal 0.2 µsec. Plotting these three values and drawing a straight line through them as in Figure 1.12-47(b) will show that the time delay for all frequencies is 0.2 µsec.

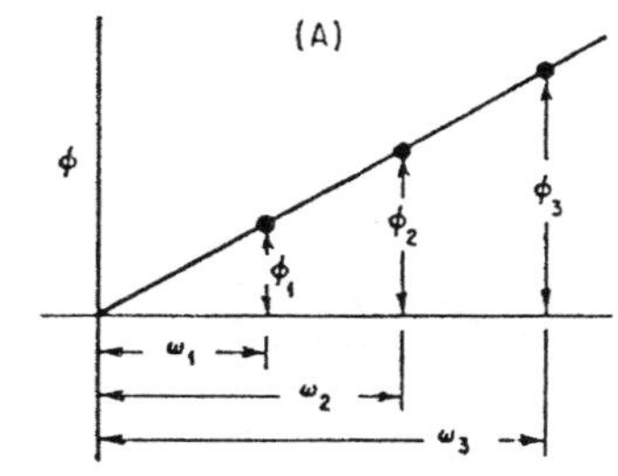

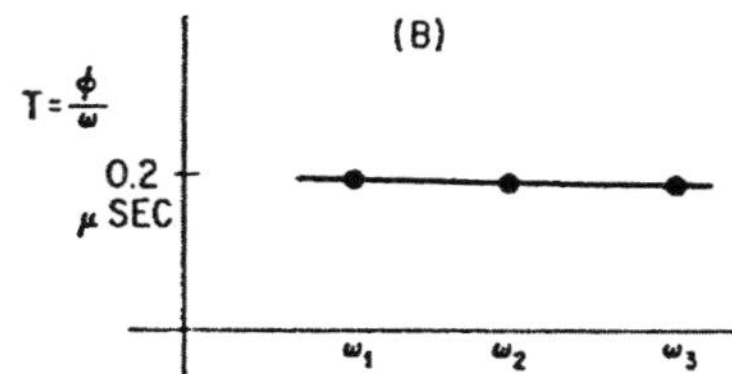

FIGURE 1.12-47 Curves illustrating a system with linear phase characteristics, which will give the same time delay for signals of all frequencies.

A signal is not distorted by delay as long as all parts of it are delayed by the same amount. However, when the phase characteristic is nonlinear (as in Figure 1.12-48[a]), the time delays for all parts of the signal are no longer equal (see Figure 1.12-48[b]). For example, if a complex waveform is made up of a 1 MHz sine wave and its third harmonic, these two components will suffer unequal delays in passing through a system having the characteristics of Figure 1.12-47. The resultant distortion can be seen by comparing Figures 1.12-49(a), 49(b), and 49(c).

Such distortion is detrimental to both the luminance and chrominance of a composite signal. The luminance signal will have its edges and other important details *scattered*, or *dispersed*, in the final image. Such a transmission system is said to introduce *dispersion*. (Conversely, if a system does not scatter the edges and other high-frequency information, it is said to be *dispersionless*.) The effect of phase distortion on the chrominance information is of a rather special nature and can best be explained by introducing the concept of *envelope delay*.

Envelope Delay

In the preceding discussion, the time delays φ_1/ω_1, φ_2/ω_2, and φ_3/ω_3 were always determined by measuring the frequencies and the phases from $\varphi = 0$ and $\omega = 0$. It might be said that the delay at zero frequency is commonly taken as the reference point for all other delays. This method is usually adequate for determining the performance of systems that do not carry any signals, which have been modulated onto a carrier. But a carrier, with its family of associated sidebands (Figure 1.12-50[b]), can be thought of as a method of transmitting signals in which the zero frequency reference is translated to a carrier frequency reference. This translation can be understood by referring to Figures

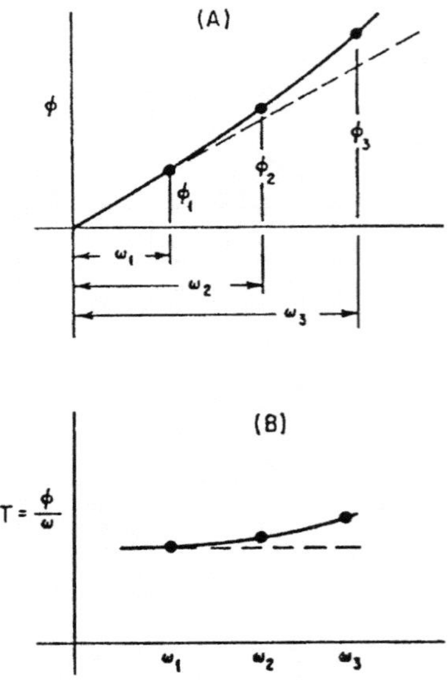

FIGURE 1.12-48 Curves showing the effect of nonlinear phase characteristic on time delay characteristic.

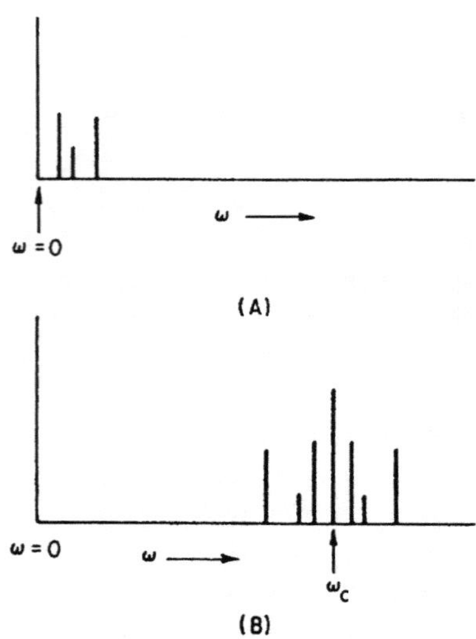

FIGURE 1.12-50 Illustration showing how a group of frequencies near $\omega = 0$ (a) can be translated by modulation onto a carrier to a group of sidebands near a carrier frequency ω_c (b).

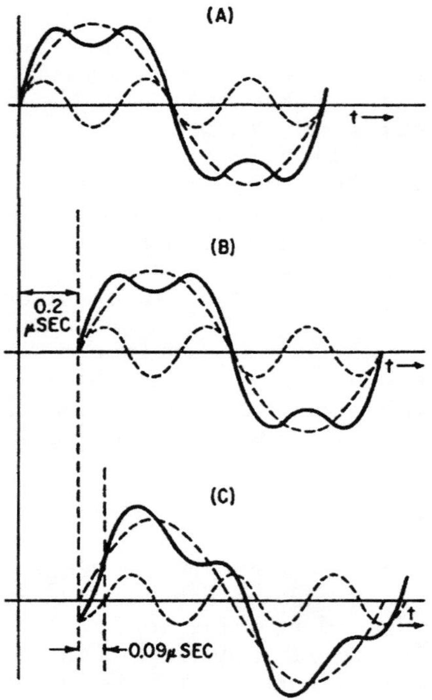

FIGURE 1.12-49 Curves showing that a complex wave (a) is not distorted by time delay (b) when both components (shown dotted) are delayed by the same amount. Unequal delays (c), however, cause distortion.

1.12-50(a) and 50(b). To calculate the delay of the carrier-borne signals *after* they have been demodulated, measurements of φ and ω must be referenced, not from zero frequency, but from *carrier* frequency.

In Figure 1.12-51(a), an impossible phase characteristic has been drawn to aid in further discussion of this subject. Such a characteristic, consisting of two perfectly straight lines, is never met in practice but makes a very simple system for developing the subject of envelope delay.

First, pass two frequencies ω_1 and ω_2 through this system. Let ω_1 be a carrier and ω_2 be a sideband which might be, for example, 1000 Hz higher. If ω_1 and ω_2 fall on the characteristic as shown in Figure 1.12-51(a), the delay which the 1000 Hz will show after demodulation can be found by putting new reference axes (shown dotted) with ω_1, the carrier, at zero on these new axes. Now, when ω_s and φ_s are measured as shown, the time delay after demodulation is φ_s/ω_s. In this case, the delay of the 1000 Hz after demodulation is the same as it would have been had it been passed through the system directly.

Second, pass two other frequencies ω_3 and ω_4 through this system as redrawn in Figure 1.12-51(b). This time, drawing in the new axes at ω_3, it can be seen that although ω_s is still 1000 Hz, φ'_s is larger than φ. Therefore, it can be concluded that the time delay φ'_s/ω_s for this second case is greater than for the first case. The 1000 Hz, when demodulated, will show a considerable error in timing.

Stressing the phrase *delay in a demodulated wave* should not be taken to mean that the demodulation process produces this delay or even makes it apparent where it was previously not detectable. Any delay that a demodulated wave shows was also present when the wave existed as a carrier having an envelope. In

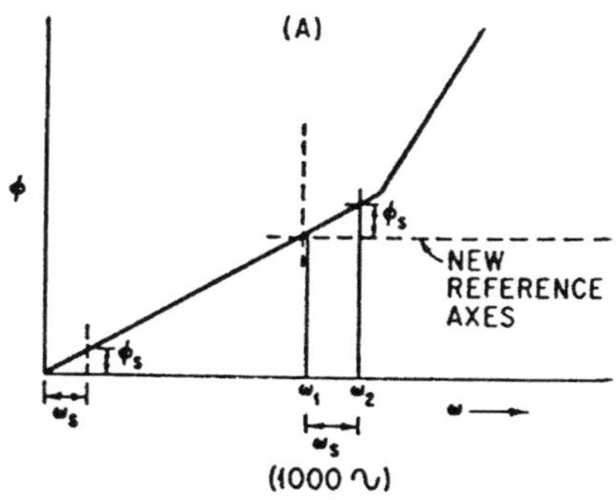

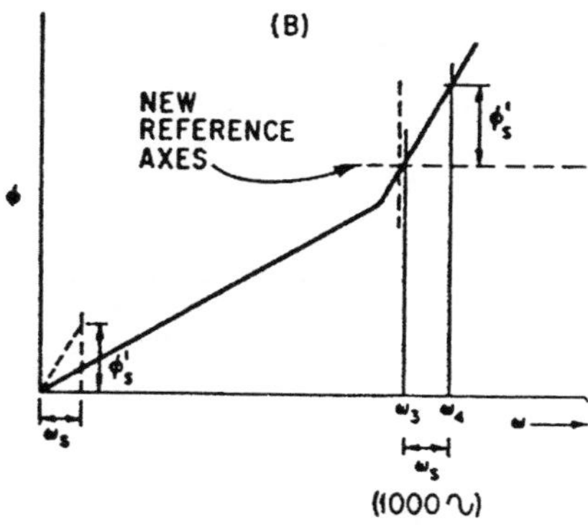

FIGURE 1.12-51 Idealized straight-line phase characteristics showing how carrier-borne 1 kHz signal can be delayed excessively when the carrier and sideband fall on a steeper portion of the phase characteristic.

short, the delay of the demodulated wave appears first as a delay of the envelope, hence the term *envelope delay*.

Envelope delay does not constitute a distortion. If a system such as the one shown in Figure 1.12-51(a) introduces a delay of 0.2 μsec to the 1000 Hz wave (measured after demodulation), then the envelope delay of the system is 0.2 μsec. However, it was shown that a 1000 Hz signal passed directly through the system (without first being modulated into a carrier) would also suffer a delay of 0.2 μsec. As long as the envelope delay φ_s/ω_s is the same as the time delay φ_1/ω_1 the envelope delay introduces no timing errors. But in the second system (Figure 1.12-51[b]) the demodulated 1000 Hz wave suffered a *larger* delay, say 0.29 μsec. A 1000 Hz signal passed directly through this system, however, would still be delayed only 0.2 μsec. Therefore, the second system has an envelope delay of 0.29 μsec and an *envelope delay distortion* of 0.09 μsec.

It is probably wise to point out that the time delay φ_3/ω_3 in Figure 1.12-51(b) is considerably less than the 0.29 μsec estimated for the value of envelope delay. Although φ_3/ω_3 would be greater than 0.2 μsec (say, for example, that φ_3/ω_3 is 0.22 μsec), the value would be optimistic about the amount of timing error that would be shown by the demodulated 1000 Hz signal. The need for knowledge of the envelope delay φ_s/ω_s of the system is therefore obvious.

Effect of envelope delay distortion on a color picture

A transmission system which exhibits envelope delay distortion will affect the time coincidence between the chrominance and luminance portions of the signal. This will result in misregistration between the color and luminance components of the reproduced picture. The following paragraph explains briefly how envelope delay distortion causes this error.

Any colored area in a reproduced picture is derived from two signals: a chrominance signal and a luminance signal. Since these two signals describe the same area in the scene, they should begin and end at the same time. The chrominance signal arrives at the receiver as a modulated subcarrier; the luminance signal does not. Therefore, as previously shown, the delay of the chrominance signal is determined principally by the envelope delay of the system, and the delay of the luminance signal is determined principally by the ordinary time delay φ/ω. If the two delays are not identical (that is, if there is envelope delay *distortion*), then the chrominance signal does not coincide with the luminance signal and the resultant picture suffers *color luminance misregistration* in a horizontal direction.

For example, in a system having the characteristic of Figure 1.12-51(b), the luminance signal is delayed by 0.2 μsec, but the chrominance signal is delayed by 0.29 μsec. The error in registration then amounts to 0.09 μsec, or about 0.2% of the horizontal dimension of the picture, which is about 0.3 in. on a 21-in. (diagonal) picture.

Although the subject of compatibility is outside the scope of this chapter, it is worth noting that envelope delay distortion adversely affects compatibility, since it causes wideband monochrome receivers to display a misregistered dot-crawl image in addition to the proper luminance image.

General method for envelope delay

The specific cases described (Figures 1.12-51[a] and 51[b]) made use of simple, idealized straight-line approximations to develop the concept of envelope delay. Practical circuits are not so simple. For example, a simple *RC* network has a φ versus ω plot, as in Figure 1.12-52. Finding the envelope delay of this curved-line plot will clarify what is meant by envelope delay.

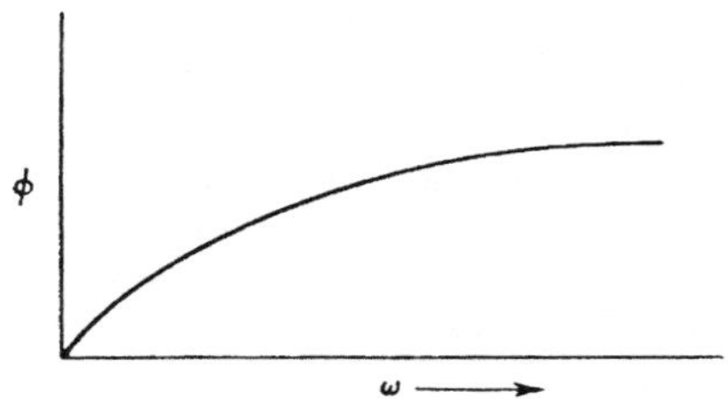

FIGURE 1.12-52 Phase characteristic of an RC network.

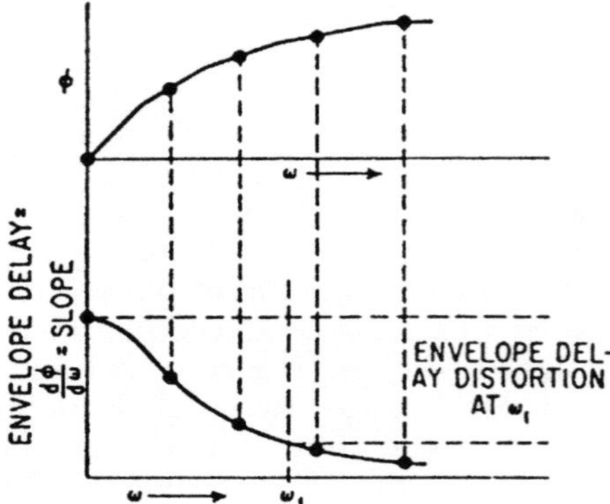

FIGURE 1.12-53 Graphs showing how a series of straight-line segments can be used to approximate the envelope delay characteristics (bottom).

Referring back to the plots of Figures 1.12-50(a) and 50(b), it can be seen that the characteristic of the plot that determines the value of envelope delay is its *slope*. The larger envelope delay, which was suffered by the ω_3-ω_4 pair (Figure 1.12-51[b]), was a result of their lying on the steeper slope. The envelope delay of *any* system is equal to the slope of the phase versus frequency characteristic. If this characteristic is a curved line (as for the *RC* network, Figure 1.12-52), then the slope is different at every frequency.

The slope of a curved line can be found by the methods of the differential calculus or to a good approximation by breaking up the line into a number of straight-line segments, as in Figure 1.12-53. If the slope of each of these straight lines is plotted against its corresponding frequency (that corresponding to the center of the line), the resulting curve will be approximately the envelope delay characteristic.

Nonlinearities of a Practical Transmission System

It is important to emphasize that the effect of nonlinearities in a color television system depends on whether these nonlinearities precede or follow the matrixing and modulation sections of the system. Nonlinearities in transfer characteristics detract from color fidelity; the same degree of nonlinearity after matrixing and modulation also affects color fidelity although in a different way. The purpose of the following paragraphs is to discuss how a nonlinear transmission system affects a *composite* color signal. It is assumed that all other nonlinearities in the entire system either are negligible or have been canceled by use of nonlinear amplifiers such as gamma correctors.

The major sources of nonlinearity in a transmission system are its amplifying devices. These devices have a limited dynamic range.[3] For example, if too much signal is supplied to them, an *overload* results. The transfer characteristic of such a system is illustrated in Figure 1.12-54(a).

[3]FM systems can have nonlinearity as a result of passive networks, but this is not considered here.

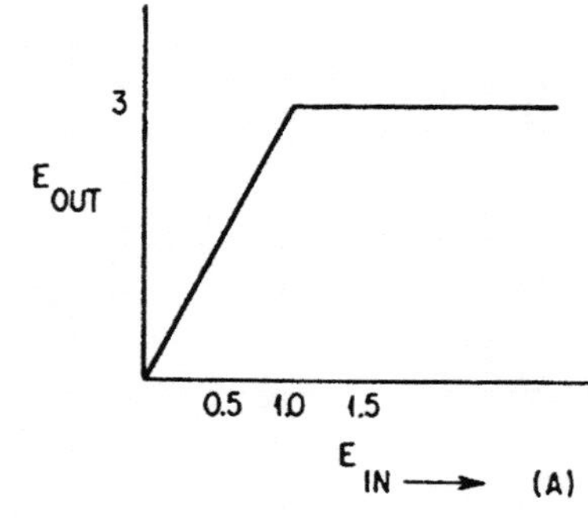

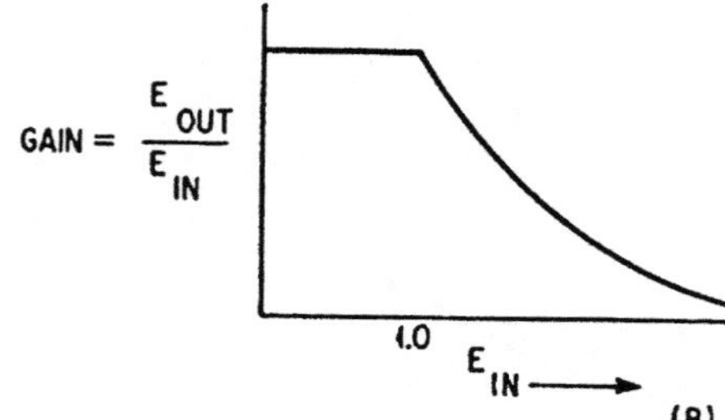

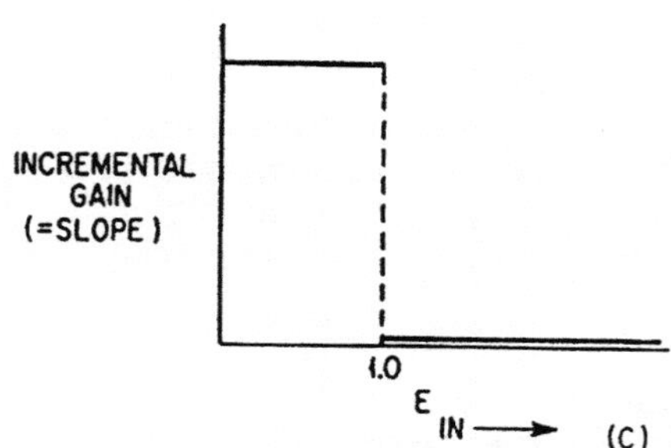

FIGURE 1.12-54 Idealized straight-line plots showing (a) output voltage of an amplifier versus input voltage, (b) gain of the amplifier versus input voltage, and (c) incremental gain of the amplifier versus input voltage. Curve (c) is the slope of curve (a).

Three types of nonlinearities commonly encountered in video transmission systems include

- Incremental gain distortion
- Differential gain
- Differential phase

The following paragraphs will show that differential gain is merely a special case of incremental gain distortion.

Incremental gain

The concept of the slope of a plot, developed in the discussion of envelope delay, will be useful here as well. Consider a plot as in Figure 1.12-54(a) which shows output voltage of an amplifier plotted against input voltage. Idealized straight-line plots are shown for simplicity. It can be seen that the amplifier has a

maximum output of 3 V for 1 V input. Larger input voltages result in no more output; the amplifier *clips* or *compresses* when inputs larger than 1 V are applied.

The gain of the amplifier is

$$Gain = \frac{E_0}{E_{in}} = \frac{3\ volts}{1\ volt} = 3$$

The gain is obviously constant below the clip point. For example, an input voltage of 0.5 V gives

$$Gain = \frac{1.5\ volts}{0.5\ volt} = 3$$

But at an input of 1.5 V, the output is still 3 V, so the gain is only 2. (The word "gain" is of doubtful use here because of the clipping involved.) The gain, defined as E_0/E_{in}, is plotted against E_{in} in Figure 1.12-54(b). It can be seen in this figure that the gain is constant only as long as the *slope* of Figure 1.12-54(a) is constant.

It is useful, then, to establish a new term, *incremental gain*, which will be defined as the *slope* of a plot such as Figure 1.12-54(a). For the particular plot of Figure 1.12-54(a), the slope is constant up to E_{in} = 1 V and then suddenly becomes zero. The corresponding plot of slope versus E_{in} is shown in Figure 1.12-54(c).

The importance of incremental gain in color television can be assessed by applying the input signal shown in Figure 1.12-55 to the distorting system of Figure 1.12-54(a). Before being applied to the distorting system, such a signal could be reproduced on a monochrome receiver as a vertical white bar and on a color receiver as a pastel-colored bar, say, for example, a pale green. After passing through the distorting system, the signal would still be reproduced as a white bar on the monochrome receiver with the only apparent error being a luminance distortion, that is, a slight reduction in brightness, which, for the magnitudes shown here, would probably pass unnoticed. The color receiver, however, would receive a signal completely devoid of any color information and would reproduce a white bar in place of the former pale green one.

A less extreme case is shown in Figure 1.12-56. For the system represented by this characteristic, the slope

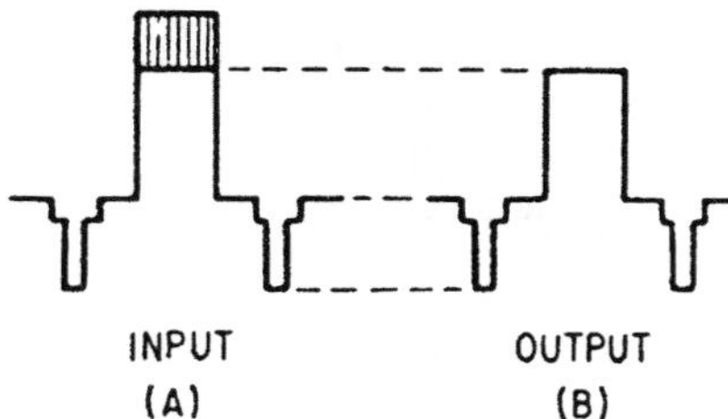

FIGURE 1.12-55 An extreme case of distortion is shown resulting from passing signal at left (a) through the amplifier represented by Figure 1.12-54. The output (b) has no color information remaining.

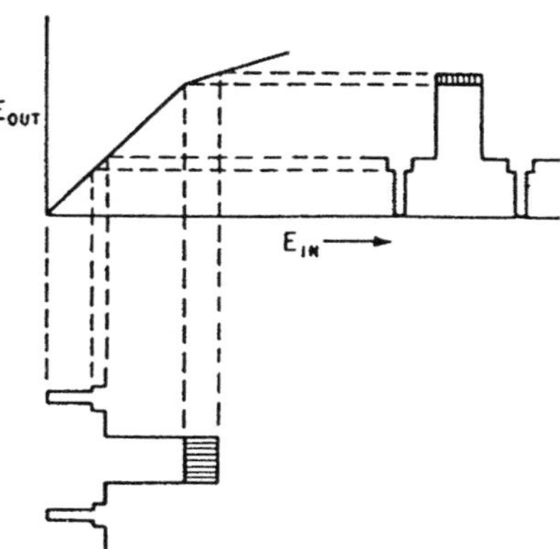

FIGURE 1.12-56 Diagram showing the effect of incremental gain distortion of reducing amplitude of the color portion of the signal.

(incremental gain) does not become zero for inputs above 1 V but instead falls to one-half of the excessive input value. The color signal of Figure 1.12-56 would not lose all color in passing through this system, but the amplitude of the subcarrier would become only one-half of its proper value. Since saturation is a function of subcarrier amplitude, the pale green of the undistorted reproduction would, in this case, become a paler green. The luminance distortion would also be less than in the extreme (clipping) case.

It can be seen, then, that unless the incremental gain of a system is constant, that system will introduce compression which will distort the saturation and brightness of reproduced colors. Usually, the error is in the direction of *decreased* luminance and saturation. For certain systems, however, exceptions can be found. For example, the effect that the system represented by Figure 1.12-56 will have on a signal depends on the polarity of the signal. For the signal as shown, the usual decrease in luminance and saturation is exhibited. For an inverted signal, however, the subcarrier amplitude would not be reduced, but the luminance signal would still be diminished. The subjective result of this distortion would be an *increase* in saturation. The unusual behavior of this particular system is attributable to its peculiar transfer characteristic, which was drawn with curvature at one end only to simplify the discussion. Most practical system transfer characteristics exhibit curvature at both ends and therefore have an effect on the signal, which is essentially independent of polarity.

Incremental gain can be measured in two ways, the first of which stems from its contribution to luminance distortion; and the second, from its contribution to chrominance distortion.

In the first method, an equal-step staircase waveform such as shown in Figure 1.12-57(a) is applied to the system to simulate a signal having equal luminance increments. If the system has constant incremental gain, the output will, of course, also have equal-step increments. But if the system does not have constant incremental gain, certain of the steps will be

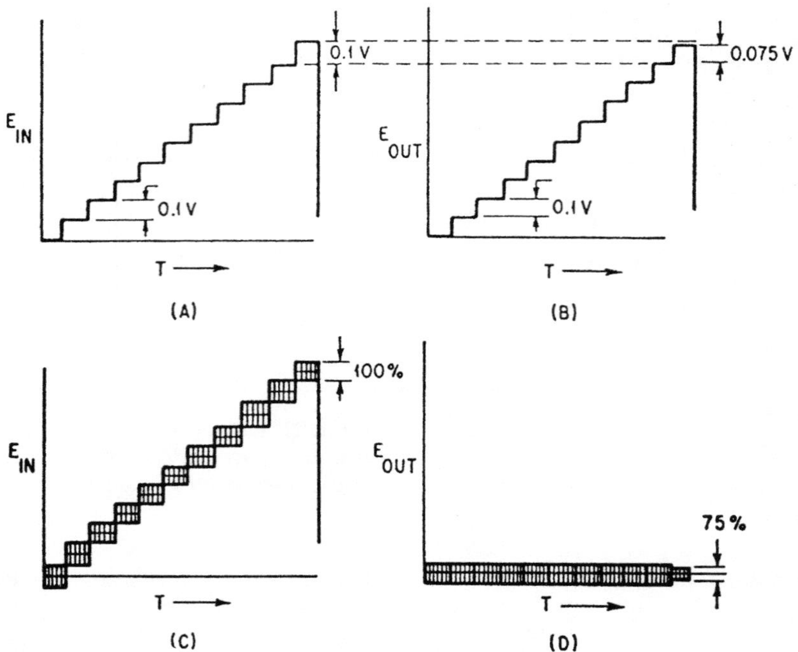

FIGURE 1.12-57 Diagrams showing two methods of measuring incremental gain distortion, namely, in (a) and (b) by its contribution to luminance distortion and in (c) and (d) chrominance distortion.

compressed, as in Figure 1.12-57(b). If the compression is as in the figure, the *incremental gain distortion* (IGD) is indicated by the distorted amplitude of the last step. Numerically, it can be stated as a percentage:

$$IGD = 1 - \frac{S_{distorted}}{D_{undistorted}} \times 100\%$$

where S is a step amplitude.

For example, if an undistorted step is 0.1 V and the distorted one is 0.075 V, then the incremental gain distortion would be 25%.

Using the chrominance distortion technique, an input signal consisting of the step wave plus a small, high-frequency sine wave, as shown in Figure 1.12-57(c), is applied to the system. After the signal has passed through the system, it is fed through a high-pass filter, which removes the low-frequency staircase. The incremental gain distortion then is indicated by the differences in the amplitude of the high-frequency sine waves (see Figure 1.12-57[d]). In this case, the high-frequency sine wave associated with the top step is shown as having 75% of the amplitude of the sine waves associated with the lower steps, which are assumed to be undistorted. Again, the incremental gain distortion is 25%.

A most important point must be made regarding the equivalence of these two techniques. Certain systems which show incremental gain distortion when tested by the luminance-step technique may or may not show the same distortion when tested by the high-frequency and high-pass filter technique. Moreover, a system which shows distortion by the second technique may or may not show distortion by the first. In other words, the incremental gain distortion may be different for different frequencies. Such differences are frequently found in staggered amplifiers, feedback amplifiers, or amplifiers having separate parallel paths for high and low frequencies, such as might be found in stabilizing amplifiers.

A thorough test of a system, therefore, should include a test of its incremental gain by both techniques. The staircase-plus-high-frequency waveform can be used to provide *both* tests by observing the system output (for this test waveform input) first through a low-pass filter and then through a high-pass filter. The first test will show low-frequency distortions; the second, high-frequency distortions.

Differential gain

On the basis of the previous discussion of incremental gain distortion, the important concept of *differential gain* can be presented merely as a simple definition. Differential gain is a special form of incremental gain distortion, which describes the IGD of a system for the superimposed high-frequency case only. In present color television practice, however, the ". . . high frequency sine wave . . ." is always a color subcarrier, and the low-frequency signal mentioned in the definition is a 15,750 Hz staircase, sine wave, or sawtooth.

Incremental phase and differential phase

The phase characteristic sketched in Figure 1.12-46 indicates that the system described by this plot will introduce a certain amount of phase shift for any

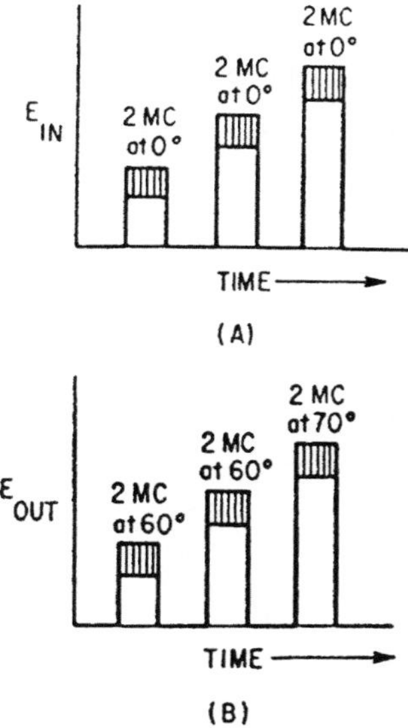

FIGURE 1.12-58 Graphs illustrating how a signal (a) may undergo different phase shifts (b) depending on where the zero axis at the sine wave falls on the system transfer characteristic. This distortion is called differential phase.

given frequency. For example, it might be found that a certain system would introduce a phase shift of 60° at 2 MHz. If the system in question were perfectly linear, this 60° phase shift would be produced regardless of how the 2 MHz signal might be applied to the system.

It can be shown, however, that some systems, when presented with a signal of the type shown in Figure 1.12-58, will introduce a delay different from 60°, depending on where the zero axis of the sine wave falls on the transfer characteristic of the system. For the case illustrated in the figure, a phase shift of 70° is drawn for the largest zero-axis displacement.

By analogy with the incremental gain and differential gain arguments, it is possible to define three quantities which pertain to this type of distortion. These quantities are *incremental phase, incremental phase distortion,* and *differential phase.* It can also be shown that of the three, differential phase is the most important quantity.

Incremental phase is the least exact analogue, since it is not very similar in form to incremental gain. Incremental *gain* is a *slope;* incremental phase is simply the absolute value of phase shift. In the previous system, the incremental phase was 60° or 70° (or somewhere in between), depending on the location of the zero axis.

Incremental phase distortion, like its analogue (*incremental gain distortion*), depends on the magnitude of the error. It should be zero for a perfect system. In the system of Figure 1.12-58 the 2 MHz signal with 70° incremental phase would be said to have 10° incremental phase distortion, so it is clear that the difference between two phases (one of which is assumed to be correct) gives the incremental phase distortion.

As previously stated, *differential gain* is identical with incremental gain distortion for the superimposed high-frequency case only. Similarly, differential phase is identical with incremental phase distortion, but there is no need to limit the definition to the superimposed high-frequency case, since there is no other case which is meaningful for phase distortion. Without the superimposed sine wave, no phase measurement is possible. Therefore, differential phase is identical with incremental phase distortion. In practical work, the first two terms are seldom used; for the last, differential phase, has been found completely adequate to describe this aspect of a system.

In summary, the differential phase of a system is the difference in phase shift of a small high-frequency sine wave signal when measured at different amplitude levels of a low-frequency signal on which the high-frequency signal is superimposed.

Effect of differential phase on color picture

The phase of a subcarrier in a composite signal carries information about the *hue* of the signal at that instant. If the signal passes through a system which introduces differential phase, the subcarrier phase (and hence, the hue) at the output will become dependent upon the amplitude of the luminance associated with the hue, since it is the luminance signal which determines the location of the zero axis of the subcarrier. For example, a system introducing 10° of differential phase might be adjusted to reproduce properly a low-luminance hue such as saturated blue or a high-luminance hue such as saturated yellow, but *not both.* One or the other would have to be in error.

THE COLOR ENCODER

The color encoder in the color television system performs the required encoding of the R, G, and B signals into a single color video signal conforming to FCC specifications. It is the heart of the modern color television system and represents a most ingenious application of many elements of communication circuit theory. Figure 1.12-59 shows a block schematic of a basic color television system indicating the functions and major components of the color encoder.

A more detailed block diagram of the color encoder showing the matrixing, bandwidth limiting, and quadrature modulation functions is shown in Figures 1.12-33 and 34.

Basic Functions

The principal operations and functions performed by the color encoder are

- Matrixing of R, G, and B video signals to produce luminance and chrominance signals
- Filtering of the chrominance signals to obtain the required bandwidth
- Delay compensation to correct for band-limiting time delay

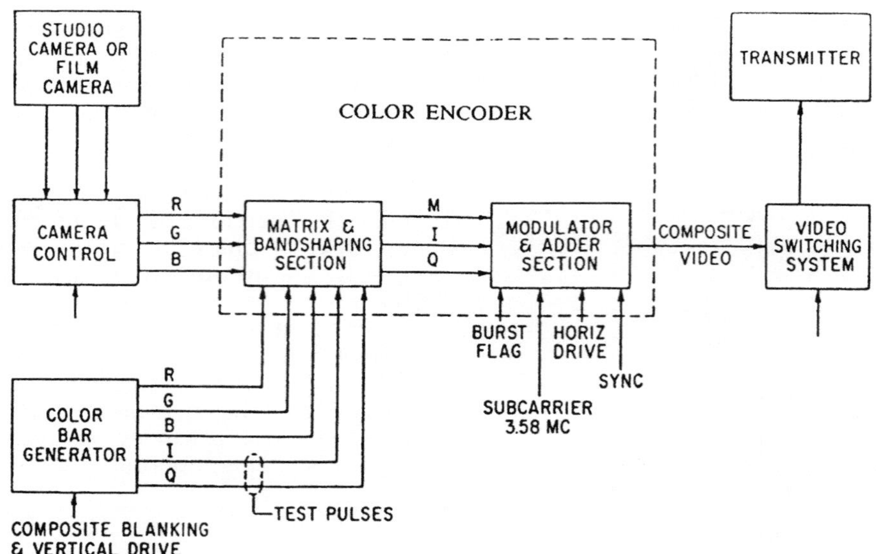

FIGURE 1.12-59 Basic color television system showing functions and major components of the color encoder.

- Modulation of 3.58 MHz carriers by chrominance signals
- Insertion of color sync burst
- Addition of luminance and chrominance signal to form a complete color signal
- Optional addition of sync

Color encoders of modern design are inherently stable and require only routine verification or adjustment. Most color encoders include a color bar generator to calibrate waveform monitors and color displays, as illustrated in Figure 1.12-60.

Colors at the top of this display pattern are arranged from left to right as white, yellow, cyan, green, magenta, red, and blue in their decreasing order of luminance. The lower portion of the pattern contains I, 100% white, Q, and black signal areas. The I and Q signals simplify subcarrier phase adjustments in the color encoder, and the white bar facilitates white balance adjustments. The specifications of the standard encoder color bar signal are given in EIA Standard RS-189-A.

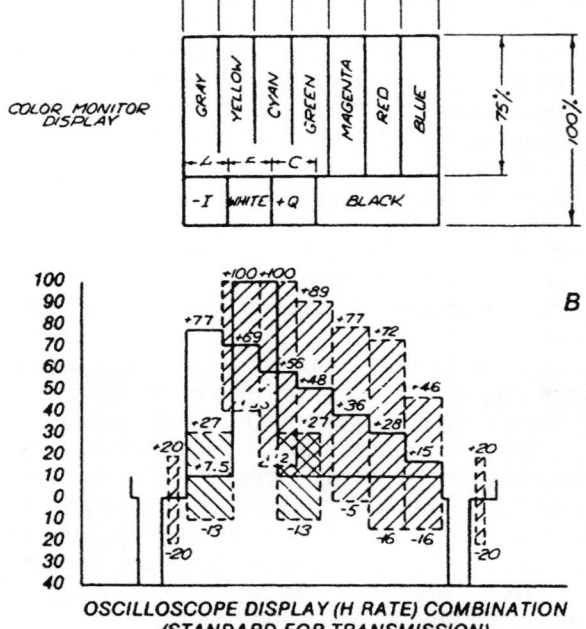

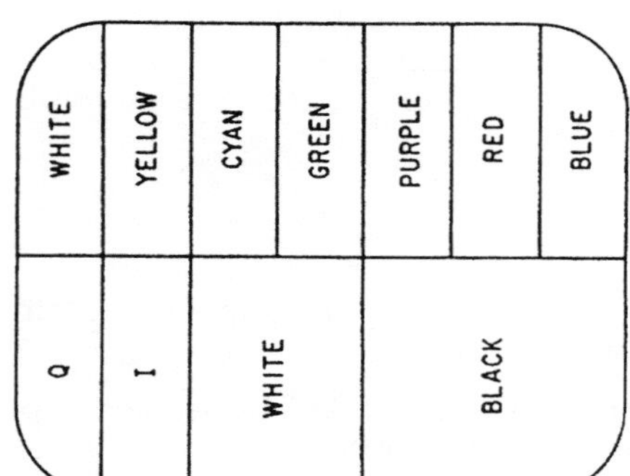

FIGURE 1.12-60 Diagram showing color monitor display of color and test bars electronically produced by a color bar generator.

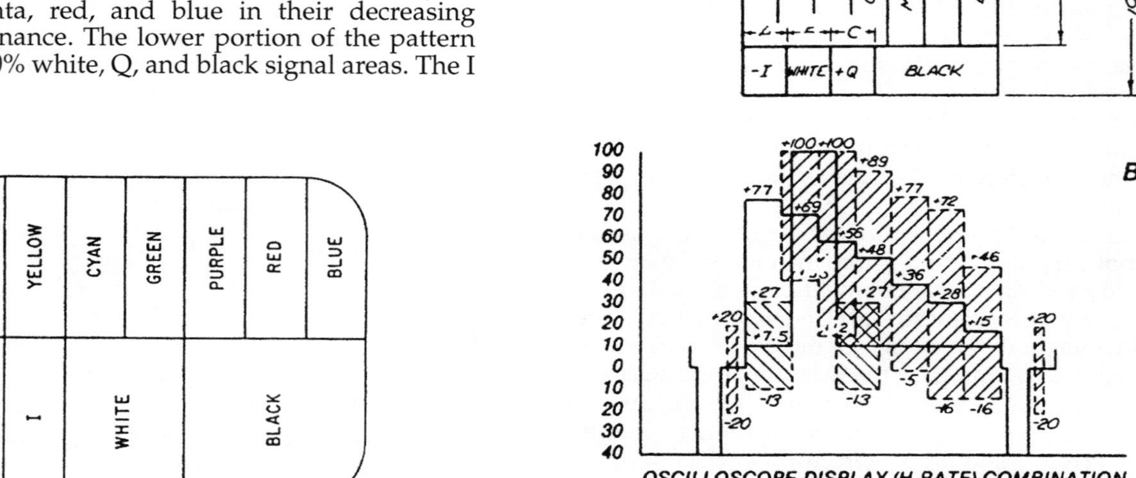

FIGURE 1.12-61 (a) Color monitor display and (b) oscilloscope display (H rate).

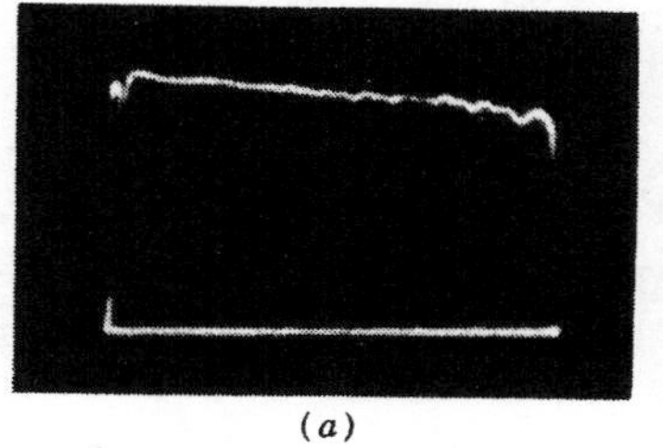
(a)

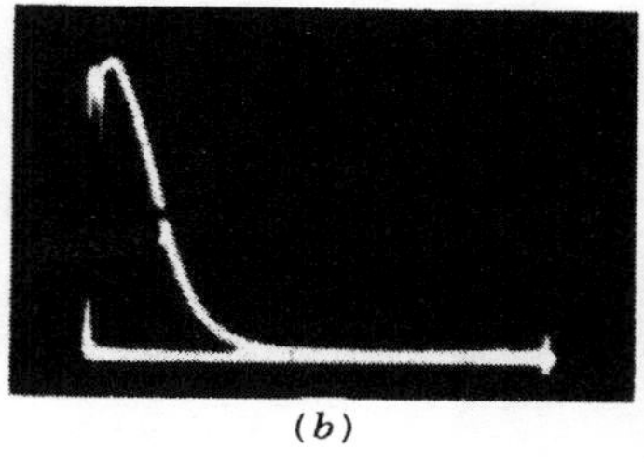
(b)

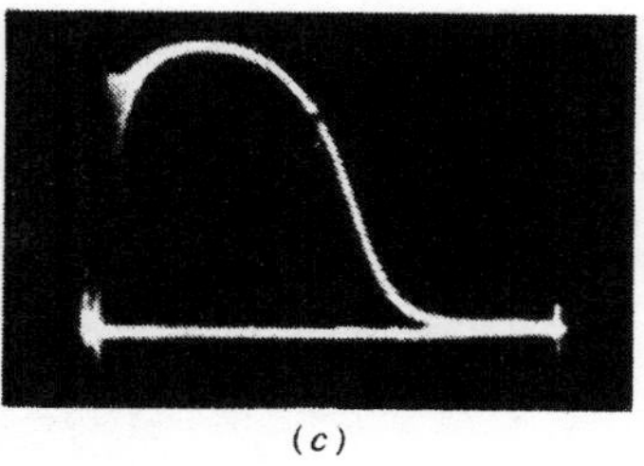
(c)

FIGURE 1.12-62 Waveforms showing response characteristics of monochrome, I, and Q channels. (a) Response of monochrome channel without aperture correction, marker at 8.0 MHz; (b) output of I filter, marker at 2.0 MHz; and (c) output of Q filter, marker at 500 kHz.

Waveforms

Figure 1.12-61 shows the oscilloscope waveforms at a horizontal sweep rate of the color bar signals displayed on the television raster. Note that this is a composite representation of waveforms of the top and bottom areas of the raster. The color sync precedes the color bar pulse information.

Figure 1.12-62 shows the various band-pass response characteristics of the luminance channel and of the I and Q channels of the color encoder.

A color encoder is set up and adjusted by using the calibrated color bars just described. The color encoder luminance gain is adjusted by using the white bar as a reference. When the luminance channel is switched off, the appropriate I and Q waveforms are available to set the proper peak amplitudes and the 90° phase separation. Either a wide-band oscilloscope or a vectorscope can be used for display in a variety of specialized setup procedures. The vector relationship of chrominance components is shown in Figure 1.12-63.

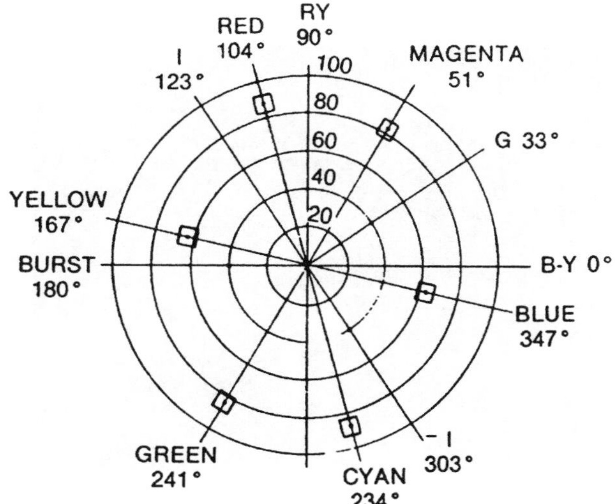

FIGURE 1.12-63 Vector relationship among chrominance components.

COLOR TEST EQUIPMENT

The color television broadcast station relies heavily on specialized test and monitoring facilities in order to maintain adequate standards of performance and to ensure compliance with FCC regulations. Test signals have become quite sophisticated and yield much useful information on the performance of the systems.

A stable high performance color monitor is an essential element of color test equipment. This, together with a waveform monitor, vectorscope, and a standard color bar generator for setup and calibration, serves as a means of evaluating performance and for rapid routine day-to-day check of the television system adjustments.

Additional test equipment needed for color TV performance evaluation falls into two categories: equipment to evaluate studio performance and equipment to evaluate microwave relay (STL) and transmitter performance.

The important electrical characteristics to be measured in either category are

- Frequency response and amplitude linearity
- Differential gain and differential phase performance
- Group delay characteristic
- Low frequency square-wave response

Full screen tests or special in-service test measurement methods, which include vertical interval test signals (VITSs), will give the required differential gain, phase, and group delay information.

Stair-Step Generator

A modulated stair-step generator waveform is shown in Figure 1.12-64. It consists of ten 30-IRE-unit steps with subcarrier modulation on each level. The amplitude-linearity or differential gain response of an amplifier can be determined directly from waveform monitor measurements. By the use of a high-pass filter, the differential gain characteristic can be displayed graphically (Figure 1.12-65, input; Figure 1.12-66, output which shows only a modest amount of distortion). Differential phase measurements can be obtained by

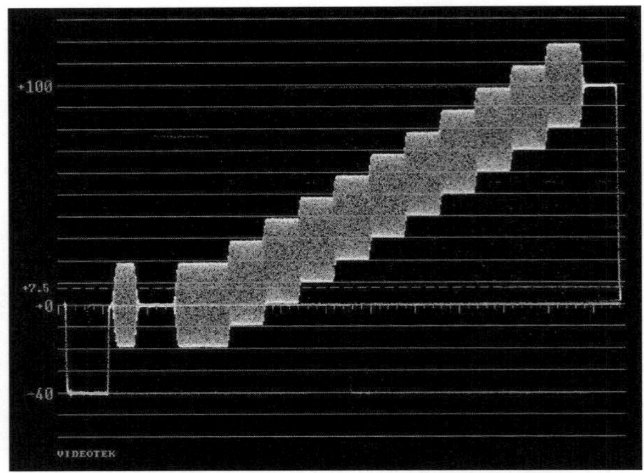

FIGURE 1.12-64 Modulated stair-step generator waveform. (Courtesy Videotek, a division of Leitch Technology.)

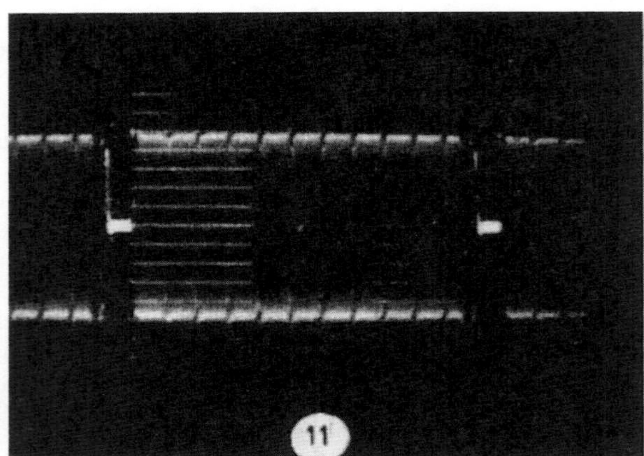

FIGURE 1.12-65 High-pass filter output with modulated stair-step waveform input. (Picture courtesy of Marconi Instruments.)

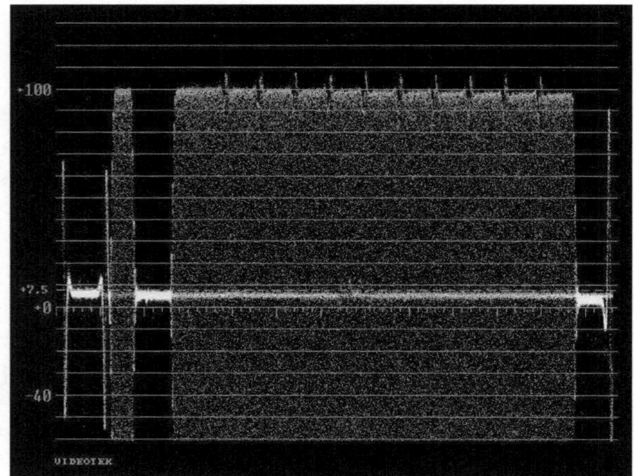

FIGURE 1.12-66 High-pass filter output of modulated stair-step waveform showing only a modest amount of differential gain error in amplifier under test. (Courtesy Videotek, a division of Leitch Technology.)

comparison of the subcarrier phase at each discrete level with phase of the color burst.

Sine-Squared Pulse and Bar

A second specialized waveform is the sine-squared pulse and bar with chrominance subcarrier modulation as shown in Figure 1.12-67. It evolved from the monochrome sine-squared pulse and bar shown in Figure 1.12-68. Use of this color test signal shows the presence of differential gain distortions, as in Figure 1.12-69, and delay distortions, as shown in Figure 1.12-70. Operationally, the method provides a direct display presentation where distortion limits may be checked by waveform monitor graticule overlay techniques.

Another frequently used waveform is the multiburst signal, as shown in Figures 1.12-71 and 72,

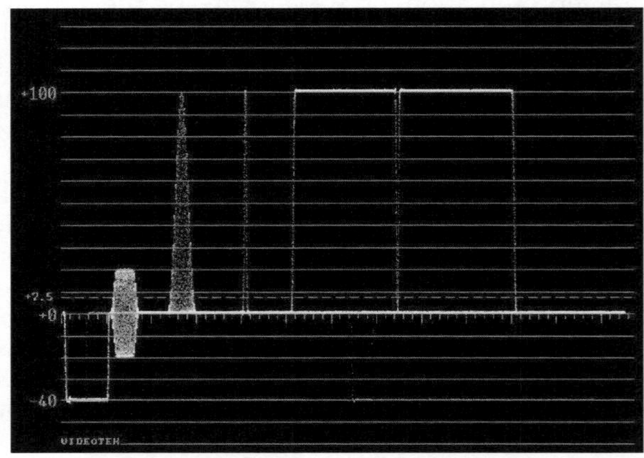

FIGURE 1.12-67 Combined luminance and chrominance sine-squared pulse and bar. (Courtesy Videotek, a division of Leitch Technology.)

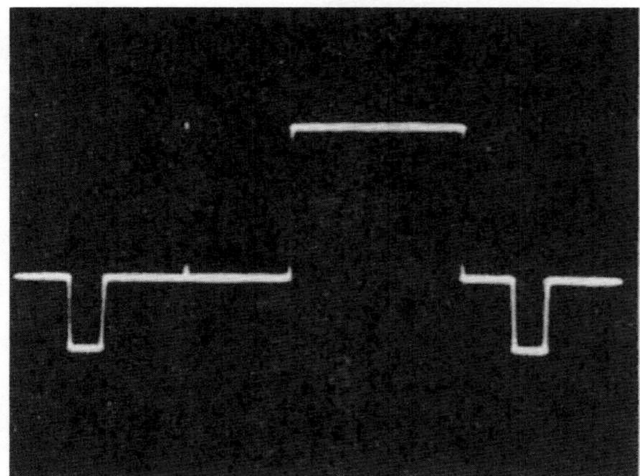

FIGURE 1.12-68 Monochrome sine-squared pulse and bar. (Picture courtesy of Marconi Instruments.)

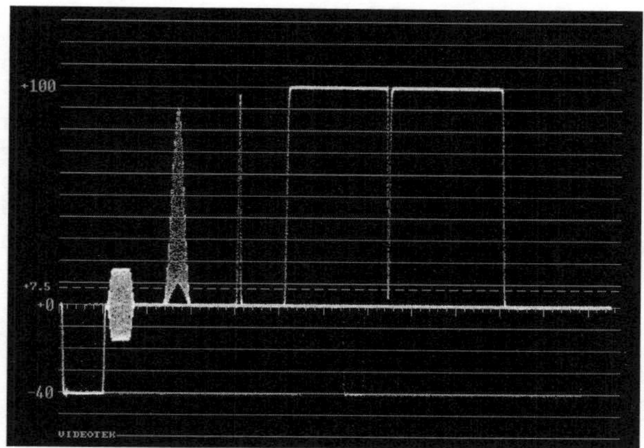

FIGURE 1.12-69 Gain inequality indicated by combined luminance and chrominance sine-squared pulse and bar. Compare with the waveforms of Figure 1.12-67. (Courtesy Videotek, a division of Leitch Technology.)

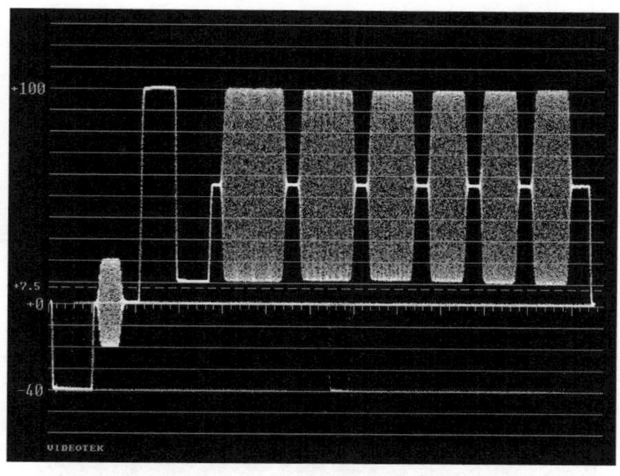

FIGURE 1.12-71 Multiburst test signal with white reference bar followed by sine wave bursts at 0.5, 1.0, 2.0, 3.0, 3.58, and 4.2 MHz. (Courtesy Videotek, a division of Leitch Technology.)

which provides a white bar for reference followed by a series of selected, constant-amplitude sine wave bursts of 0.5, 1.0, 2.0, 3.0, 3.58, and 4.2 MHz signals which provide sample frequencies in the video pass band that is convenient to use and to interpret in routine frequency response tests of broadcast equipment.

Vectorscope

The *vectorscope* is a measurement instrument developed especially for color TV system tests and monitoring. Its essential feature is the polar or vectorial display of chrominance information in which the radial deflection is proportional to saturation of a color, and the angular position is equal to the phase angle of that color subcarrier with respect to the color burst. The 360° polar coordinate display corresponds to a complete cycle of color subcarrier or 280 nsec in a time display. By convention, the color burst is normalized at 180°. If the color bar signal described in Figures 1.12-61 and 63 is applied to the input to the vectorscope and the burst is normalized at 180°, the display shown in Figure 1.12-73 is obtained on the graticule.

Note that for standard signal levels, each color vector in the color bar sequence falls within its appropriately marked box on the graticule. The outer boxes define the FCC maximum permissible errors of ±10° in phase and ±20% in amplitude. The inner boxes correspond to ±2.5° phase error and 2.5% amplitude error.

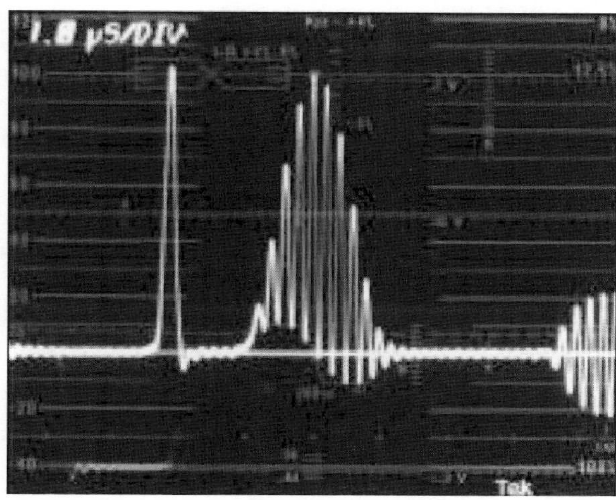

FIGURE 1.12-70 Luminance to chrominance delay inequality indicated by the distorted baseline of the modulated 20T pulse and measured in nanoseconds. (Picture from Tektronix "Television Measurements – NTSC Systems.")

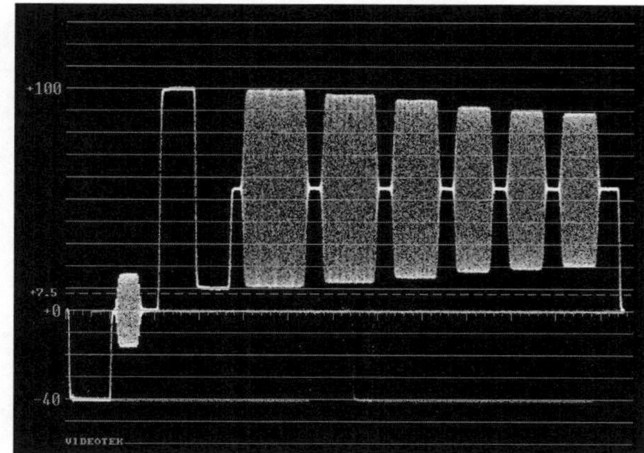

FIGURE 1.12-72 Multiburst output signal from an amplifier having distortion. Compare with Figure 1.12-71. (Courtesy Videotek, a division of Leitch Technology.)

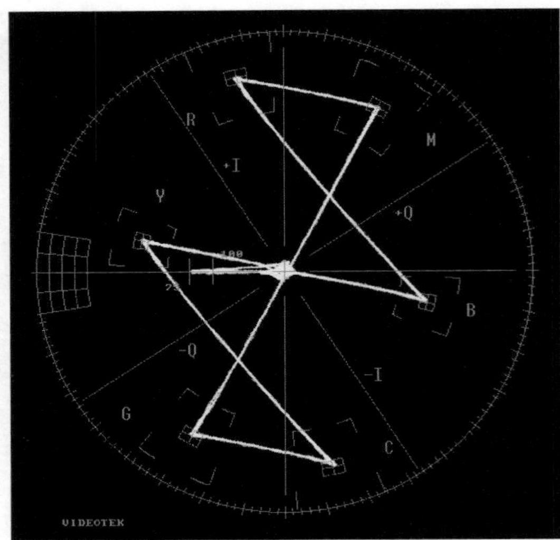

FIGURE 1.12-73 Vectorscope display. Split field color bars 75 % amplitude, 100% white reference, 10% setup. Conforms to EIA specification RS-189-A. (courtesy Videotek, a division of Leitch Technology.)

A feature of the vectorscope color bar technique is that it gives immediate reassurance on system performance with a color bar test signal display.

By alternating two signal sources at the input, one can obtain direct readings on differential phase and amplitude behavior of any selected picture sources.

Vertical Interval Test Signals

A *vertical interval test signal* (VITS) is used to verify transmission conditions using multiburst, sine-squared, or stair-step test signals. Such signals can be used for continuous in-service monitoring of TV system performance.

SUMMARY

This chapter is a tutorial of the NTSC color television system as employed for TV broadcasting in the United States. In spite of their apparent datedness, the theory and technology remain solidly in place. A thorough understanding of this material will provide a good foundation of how color television systems work in general and how certain characteristics of the human eye were used to craft the original color TV system.

CHAPTER

1.13

Worldwide Standards for Analog Television

D.H. PRITCHARD AND J.J. GIBSON

Updated for the 10th Edition by

ALDO CUGNINI

AGC Systems LLC
Long Valley, New Jersey

INTRODUCTION

The performance of a film-based motion picture system in one location in the world is generally the same as in any other location; thus, international exchange of film programming is relatively straightforward. This is not the case, however, with analog broadcast color television systems. The lack of compatibility has its origins in many factors, such as constraints in communications channel allocations and techniques, differences in local power source characteristics, network requirements, pickup and display technology, and political considerations relating to international telecommunications agreements.

The most outstanding effort—as well as the most controversial effort—of the Eleventh Plenary Assembly of the International Radio Consultative Committee (CCIR, now the ITU-R), held in Oslo in 1966, was an attempt at standardization of analog color television systems by the participating countries of the world. The discussions pertaining to the possibility of a universal analog system proved inconclusive; therefore, instead of issuing a unanimous recommendation for a single CCIR analog system, the CCIR was forced to issue only a report describing the characteristics and recommendations for a variety of proposed analog systems. It was left to the controlling organizations of the individual countries to make their own choice as to which standard to adopt.

This outcome was not totally surprising because one of the primary requirements for any analog color television system is compatibility with a coexisting monochrome system. In many cases, the monochrome standards already existed and were dictated by such factors as local powerline frequencies (relevant to field and frame rates) as well as radiofrequency channel allocations and pertinent telecommunications agreements. Thus, such technical factors as line number, field rate, video bandwidth, modulation technique, and sound carrier frequencies were predetermined and varied in many regions of the world. The ease with which the international exchange of program material may be accomplished is thereby hampered and has been accommodated over the years by means of standards conversion techniques, or *transcoders*, with varying degrees of loss in quality. While invariably introducing some compromises, standards conversion techniques now provide surprisingly good service with the use of satellite relays coupled with use of digital signal processing in both the video and audio domains.

MONOCHROME COMPATIBLE ANALOG COLOR TV SYSTEMS

In order to achieve success in the introduction of an analog color television system, it was viewed as essential that the color system be fully compatible with the existing black-and-white system. That is, monochrome receivers must be able to produce high-quality black-and-white images from a color broadcast, and color receivers must produce high-quality black-and-white images from monochrome broadcasts. The first such analog color television system to be placed into commercial broadcast service was developed in the United States. On December 17, 1953, the Federal Communications Commission (FCC) approved transmission standards and authorized broadcasters, as of January 23, 1954, to provide regular service to the public under these standards. This decision was the culmination of the work of the National Television System Committee (NTSC), upon whose recommendation the FCC action was based [1]; subsequently, this system, commonly referred to as the NTSC system, was adopted by Canada, Japan, Mexico, and others. That these standards are still providing, after more than 50 years, color television service of good quality testifies to the validity and applicability of the fundamental principles underlying the choice of specific techniques and numerical standards.

The previous existence of a monochrome television standard was two edged, in that it provided a foundation upon which to build the necessary innovative techniques while simultaneously imposing the requirement of compatibility. Within this framework, an underlying theme—that which the eye does not see does not need to be transmitted nor reproduced—set the stage for a variety of fascinating developments in what has been characterized as an "economy of representation" [1].

The countries of Europe delayed the adoption of an analog color television system, and in the years between 1953 and 1967 a number of alternative systems that were compatible with the 625 line, 50 field existing monochrome systems were devised. The development of these systems was to some extent influenced by the fact that the technology necessary to implement some of the NTSC requirements was still in its infancy; thus, many of the differences between NTSC and the other analog systems are the result of technological rather than fundamental theoretical considerations.

Most of the basic techniques of NTSC are incorporated into the other analog system approaches. For example, the use of wideband luminance and relatively narrowband chrominance, following the teachings of the principle of *mixed highs* (discussed in the next section of the chapter), can be found in all analog systems. Similarly, the concept of providing *horizontal interlace* to reduce the visibility of the color subcarriers is followed in all approaches. This feature is required to reduce the visibility of signals carrying color information that are contained within the same frequency range as the coexisting monochrome signal, thus maintaining a high order of compatibility.

An early analog system that received approval was one proposed by Henri de France of the Compagnie de Television of Paris. It was argued that if color could be relatively band limited in the horizontal direction, it could also be band limited in the vertical direction; thus, the two pieces of coloring information (*hue* and *saturation*) that must be added to the one piece of monochrome information (*brightness*) could be transmitted as subcarrier modulation that is sequentially transmitted on alternate lines—thereby avoiding the possibility of unwanted crosstalk between color signal components. At the receiver, a one-line memory, commonly referred to as a 1-H delay element, must be employed to store one line to then be concurrent with the following line, then a linear matrix of the red and blue signal components (R and B) is used to produce the third green component (G). Of course, this necessitates the addition of a line-switching identification technique. Such an approach, designated as Sequential Couleur Avec Memoire (SECAM; translates as "sequential color with memory"), was developed and officially adopted by France and the former U.S.S.R., and broadcast service began in France in 1967.

The implementation technique of a 1-H delay element led to the development, largely through the efforts of Walter Bruch of Telefunken Company, of the *Phase Alternation Line* (PAL) system. This approach was aimed at overcoming an implementation problem of NTSC that requires a high order of phase and amplitude integrity (skew symmetry) of the transmission path characteristics about the color subcarrier to prevent color quadrature distortion. The line-by-line alternation of the phase of one of the color signal components averages any colorimetric distortions to the observer's eye to that of the correct value. The system in its simplest form (simple PAL), however, results in line flicker (Hanover bars). The use of a 1-H delay device in the receiver greatly alleviates this problem (standard PAL). PAL systems also require a line identification technique. The standard PAL system was adopted by numerous countries in continental Europe, as well as in the United Kingdom. Public broadcasting began in 1967 in Germany and the United Kingdom using two slightly different variants of the PAL system.

NTSC, PAL, AND SECAM SYSTEMS OVERVIEW

To properly understand the similarities and differences of the conventional analog television systems, a familiarization with the basic principles of NTSC, PAL, and SECAM is required. As previously stated, because many basic techniques of NTSC are involved in PAL and SECAM, a thorough knowledge of NTSC is necessary in order to understand PAL and SECAM.

The same R, G, and B pickup devices and three primary color display devices are used in all systems. The basic camera function is to analyze the spectral distribution of the light from the scene in terms of its red, green, and blue components on a point-by-point basis as determined by the scanning rates. The three resulting electrical signals must then be transmitted over a band-limited communications channel to control the three-color display device to make the *perceived* color at the receiver appear essentially the same as the *perceived* color at the scene.

It is useful to define color as a psycho-physical property of light—specifically, as the combination of those characteristics of light that produce the sensations of brightness, hue, and saturation. *Brightness* refers to the relative intensity; *hue* refers to that attribute of color that allows separation into spectral groups perceived as red, green, yellow, and so on (in scientific terms, the *dominant wavelength*); and *saturation* is the degree to which a color deviates from a neutral gray of the same brightness—the degree to which it is "pure," "pastel," or "vivid." These three characteristics represent the total information necessary to define or recreate a specific color stimulus.

This concept is useful to communication engineers in the development of encoding and decoding techniques to efficiently compress the required information within a given channel bandwidth and to subsequently recombine the specific color signal values in the proper proportions at the reproducer. The NTSC color standards define the first commercially broadcast process for achieving this result.

A preferred signal arrangement was developed that resulted in reciprocal compatibility with monochrome pictures and was transmitted within the existing monochrome channel, as shown in Figure 1.13-1. One signal (luminance) is chosen in all approaches to occupy the wideband portion of the channel and to convey the brightness as well as the detail information content. A second signal (chrominance), representative of the chromatic attributes of hue and saturation, is assigned less channel width in accordance with the principle that, in human vision, full three-color reproduction is not required over the entire range of resolution (commonly referred to as the *mixed-highs principle*).

Another fundamental principle employed in all systems involves arranging the chrominance and luminance signals within the same frequency band without excessive mutual interference. Recognition that the scanning process, being equivalent to sampled-data techniques, produces signal components largely concentrated in uniformly spaced groups across the channel width led to introduction of the concept of *horizontal frequency interlace* (dot interlace). The color subcarrier frequency is so chosen as to be an odd multiple of one-half the line rate (in the case of NTSC) such that the phase of the subcarrier is exactly opposite on successive scanning lines. This substantially reduces the subjective visibility of the color signal dot pattern components.

The major differences among the three main analog systems of NTSC, PAL, and SECAM are in the specific modulating processes used for encoding and transmitting the chrominance information. The similarities and differences are briefly summarized here:

- All systems:
  - Three primary additive colorimetric principles
  - Similar camera pick-up and receiver display technology
  - Wideband luminance and narrowband color
- Compatibility with coexisting monochrome system:
  - Introduces first-order differences:
    - Line number
    - Field/frame rates
    - Bandwidth
    - Frequency allocation
- Major differences in color techniques:
  - NTSC—Phase and amplitude quadrature modulation of interlaced subcarrier
  - PAL—Similar to NTSC but with line alternation of the "V" component
  - SECAM—Frequency modulation of line sequential color subcarriers

NTSC Color System

The importance of the colorimetric concepts of brightness, hue, and saturation comprising the three pieces

- COMPATIBILITY WITH CO-EXISTING MONOCHROME SYSTEM.
- ENCODE WIDEBAND R, G, B COLOR PRIMARY SIGNALS.
  - WIDEBAND LUMINANCE (BRIGHTNESS)
  - NARROW-BAND MODULATION OF A COLOR SUBCARRIER (Hue and Saturation)

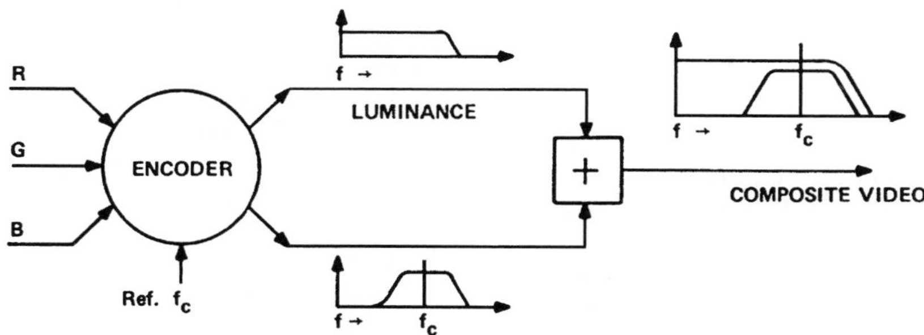

- SUBCARRIER FREQUENCY INTERLACE
  - ODD MULTIPLE OF ½ H TO REDUCE VISIBILITY OF CHROMINANCE INFORMATION SUBCARRIER.

FIGURE 1.13-1 Preferred approach to compatible analog color TV systems.

of information necessary to analyze or recreate a specific color value becomes evident in the formation of the composite color television NTSC format. The luminance, or monochrome, signal is formed by addition of specific proportions of the red, green, and blue signals and occupies the total available video bandwidth of 0 to 4.2 MHz. The NTSC, PAL, and SECAM systems all use the same luminance (Y) signal formation, differing only in the available bandwidths.

The Y signal components have relative voltage values representative of the brightness sensation in the human eye; therefore, the red, green, and blue voltage components are tailored in proportion to the standard luminosity curve at the particular values of the dominant wavelengths of the three color primaries chosen for color television. Thus, the luminance signal makeup for all systems, as normalized to white, is described by:

$$E'_Y = 0.299E'_R + 0.587E'_G + 0.114E'_B \quad (1)$$

The signal of Eq. 1 would be exactly equal to the output of a linear monochrome sensor with ideal spectral sensitivity if the red, green, and blue elements were also linear devices with theoretically correct spectral-sensitivity curves. In actual practice, the red, green, and primary signals are deliberately made nonlinear to accomplish *gamma correction* (adjustment of the slope of the input/output transfer characteristic). The prime mark (′) is used to denote a gamma-corrected signal. Table 1.13-1 gives the equations for the chrominance signal components.[1] Signals representative of the chromaticity information (hue and saturation) that relate to the differences between the luminance signal and the basic red, green, and blue signals are generated in a linear matrix.

These new signals are termed *color-difference* signals and are designated as R–Y, G–Y, and B–Y. These signals modulate a subcarrier that is combined with the luminance component and passed through a common communications channel. At the receiver, the color difference signals are detected, separated, and individually added to the luminance signal in three separate paths to recreate the original R, G, and B signals according to the equations:

$$E'_Y + E'_{(R-Y)} = E'_Y + E'_R - E'_Y = E'_R \quad (2a)$$

$$E'_Y + E'_{(G-Y)} = E'_Y + E'_G - E'_Y = E'_G \quad (2b)$$

$$E'_Y + E'_{(B-Y)} = E'_Y + E'_B - E'_Y = E'_B \quad (2c)$$

In the specific case of NTSC, two other color-difference signals, designated as *I* and *Q*, are formed at the encoder and are used to modulate the color subcarrier.

Another reason for the choice of signal values in the NTSC system is that the eye is more responsive to spatial and temporal variations in luminance than it is to variations in chrominance; therefore, the visibility of luminosity changes resulting from random noise and interference effects may be reduced by properly proportioning the relative chrominance gain and encoding angle values with respect to the luminance values. For this reason, the principle of *constant luminance* is incorporated into the system standard [1,2].

TABLE 1.13-1
Electronic Color Signal Values for NTSC, PAL, and SECAM

Luminance:
$E'_Y = 0.299\,E'_R + 0.587\,E'_G + 0.114\,E'_B$ (Common for all systems)
Chrominance:
NTSC
$E'_I = -0.274\,E'_G + 0.596\,E'_R - 0.114\,E'_B$ $E'_Q = -0.522\,E'_G + 0.211\,E'_R + 0.311\,E'_B$ $B\text{-}Y = 0.493\,(E'_B - E'_Y)$ $R\text{-}Y = 0.877\,(E'_R - E'_Y)$ $G\text{-}Y = 1.413\,(E'_G - E'_Y)$
PAL
$E'_U = 0.493\,(E'_B - E'_Y)$ $\pm E'_V = \pm 0.877\,(E'_R - E'_Y)$
SECAM
$D'_R = -1.9\,(E'_R - E'_Y)$ $D'_B = 1.5\,(E'_B - E'_Y)$

The voltage outputs from the three camera sensors are adjusted to be equal when a scene reference white or neutral gray object is being scanned for the color temperature of the scene ambient. Under this condition, the color subcarrier also automatically becomes zero. The colorimetric values have been formulated by assuming that the reproducer will be adjusted for *illuminant C*, representing the color of average daylight.

Figure 1.13-2 is a CIE chromaticity diagram indicating the primary color coordinates for NTSC, PAL, and SECAM. It is interesting to compare the available color gamut relative to that of all color paint, pigment, film, and dye processes.

In the NTSC color standard, the chrominance information is carried as simultaneous amplitude and phase modulation of a subcarrier chosen to be in the high-frequency portion of the 0 to 4.2 MHz video band and specifically related to the scanning rates as an odd multiple of one-half the horizontal line rate, as shown by the vector diagram in Figure 1.13-3. The hue information is assigned to the instantaneous phase of the subcarrier. Saturation is determined by the *ratio* of the instantaneous amplitude of the subcarrier to that of the corresponding luminance signal amplitude value.

[1]Apparently, there is an error in the original 1953 calculations of these reduction factors. A luminance matrix coefficient of 0.115 was used for blue instead of the correct 0.114. The reader is referred to SMPTE 170M [6], which also addresses calculation precision, handling of setup, revised chromaticity coordinates, and equiband color encoding.

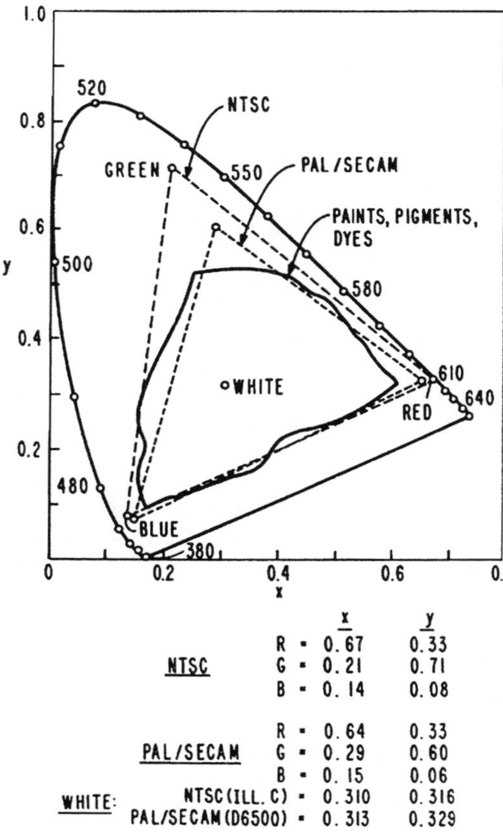

FIGURE 1.13-2 CIE chromaticity diagram.

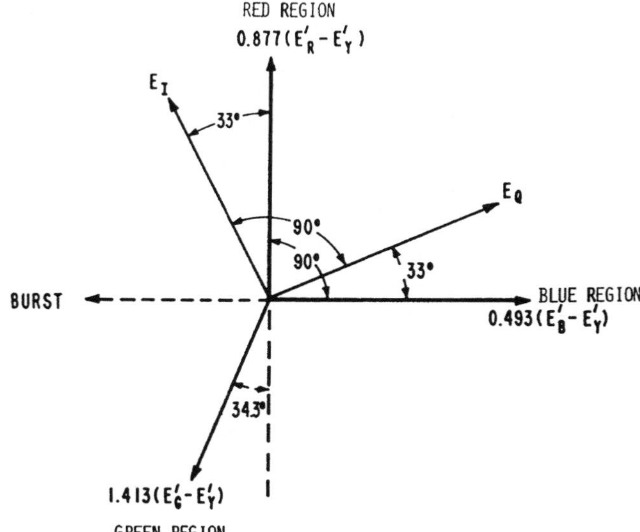

FIGURE 1.13-3 NTSC color modulation phase diagram.

The choice of the *I* and *Q* color modulation components relates to the variation of color acuity characteristics of human color vision as a function of the field of view and spatial dimensions of objects in the scene. The color acuity of the eye decreases as the size of the viewed object is decreased and thereby occupies a small part of the field of view. Small objects, represented by frequencies above about 1.5 to 2.0 MHz, produce no color sensation (*mixed highs*). Intermediate spatial dimensions (approximately in the 0.5 to 1.5 MHz range) are viewed satisfactorily if reproduced along a preferred orange–cyan axis. Large objects (0 to 0.5 MHz) require full three-color reproduction for subjectively pleasing results. The *I* and *Q* bandwidths are chosen accordingly, and the preferred colorimetric reproduction axis is obtained when only the *I* signal exists by rotating the subcarrier modulation vectors by 33°. In this way, the principles of mixed highs and *I, Q* color-acuity axis operation are exploited.

At the encoder, the *Q* signal component is band limited to about 0.6 MHz and is representative of the green–purple color-axis information. The *I* signal component has a bandwidth of about 1.5 MHz and contains the orange–cyan color axis information. These two signals are then used to individually modulate the color subcarrier in two balanced modulators operated in phase quadrature. The *sum products* are selected and added to form the composite chromaticity subcarrier. This signal—in turn—is added to the luminance signal along with the appropriate horizontal and vertical synchronizing and blanking waveforms to include the color-synchronization burst. The result is the total composite color video signal.

Quadrature synchronous detection is used at the receiver to identify the individual color signal components. When individually recombined with the luminance signal, the desired R, G, and B signals are recreated. The receiver designer is free to demodulate either at *I* or *Q* and matrix to form B–Y, R–Y, and G–Y, or, as in nearly all modern receivers, at B–Y and R–Y and maintain 500 kHz equiband color signals.

The chrominance information can be carried without loss of identity provided that the proper phase relationship is maintained between the encoding and decoding processes. This is accomplished by transmitting a reference burst signal consisting of eight or nine cycles of the subcarrier frequency at a specific phase [–(B–Y)] following each horizontal synchronizing pulse, as shown in Figure 1.13-4.

The specific choice of color subcarrier frequency in NTSC was dictated by at least two major factors. First, the necessity of providing horizontal interlace to reduce the visibility of the subcarrier requires that the frequency of the subcarrier be precisely an odd multiple of one-half the horizontal line rate. Figure 1.13-5 shows the energy spectrum of the composite NTSC signal for a typical stationary scene. This interlace provides line-to-line phase reversal of the color subcarrier, thereby reducing its visibility (and thus improving compatibility with monochrome reception). Second, it is advantageous to also provide interlace of the beat frequency (about 920 kHz) occurring between the color subcarrier and the average value of the sound carrier. For total compatibility reasons, the sound carrier was left unchanged at 4.5 MHz and the

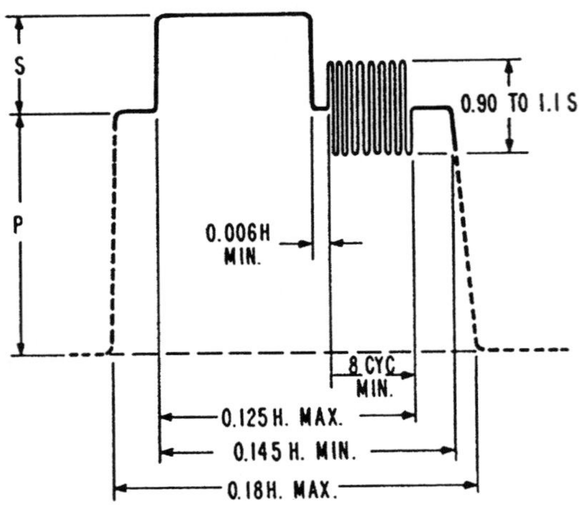

FIGURE 1.13-4 NTSC color-burst synchronizing signal.

line number remained at 525; thus, the resulting line scanning rate and field rate varied slightly from those for the monochrome values but stayed within the previously existing tolerances. The difference is exactly 1 part in 1000; specifically, the line rate is 15.734 kHz, the field rate is 59.94 Hz, and the color subcarrier is 3.578545 MHz.

PAL Color System

Except for some minor details, the color encoding principles for PAL are the same as those for NTSC; however, the phase of the color signal, $E_V = R - Y$, is reversed by 180° from line-to-line. This is done for the purpose of averaging, or canceling, certain color errors resulting from amplitude and phase distortion of the color modulation sidebands. Such distortions might occur as a result of equipment or transmission path problems.

The NTSC chroma signal expression within the frequency band common to both I and Q is given by:

$$C_{NTSC} = \frac{B-Y}{2.03}\sin\omega_{SC}t + \frac{R-Y}{1.14}\cos\omega_{SC}t \qquad (3)$$

The PAL chroma signal expression is given by:

$$C_{PAL} = \frac{U}{2.03}\sin\omega_{SC}t \pm \frac{V}{1.14}\cos\omega_{SC}t \qquad (4)$$

where U and ±V have been substituted for the B–Y and R–Y signal values, respectively.

The PAL format employs equal bandwidths for the U and V color-difference signal components that are about the same as the NTSC I signal bandwidth (1.3 MHz at 3 dB). There are slight differences in the U and V bandwidth in different PAL systems because of the differences in luminance bandwidth and sound carrier frequencies. (See the applicable ITU-R documents for specific details.)

The V component was chosen for the line-by-line reversal process because it has a lower gain factor than U and, therefore, is less susceptible to switching rate ($\frac{1}{2} f_H$) imbalance. Figure 1.13-6 provides a vector diagram for the PAL quadrature modulated and line-alternating color modulation approach.

The result of the switching of the V signal phase at the line rate is that any phase errors produce complementary errors from V into the U channel. In addition, a corresponding switch of the decoder V channel results in a constant V component with complementary errors from the U channel. Any line-to-line averaging process at the decoder, such as retentivity of the eye (simple PAL) or an electronic averaging technique such as the use of a 1-H delay element (standard PAL), produces cancellation of the phase (hue) error and provides the correct hue but with

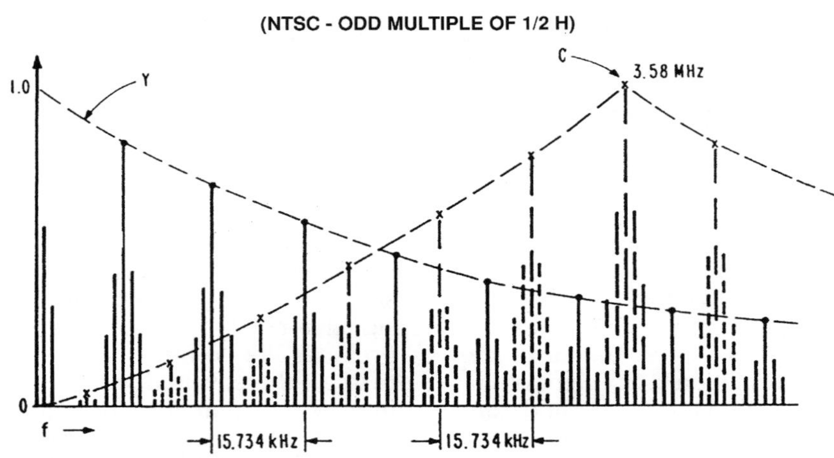

FIGURE 1.13-5 Luminance/chrominance horizontal frequency interlace principle (energy spectrum of the composite NTSC signal for a typical stationary scene).

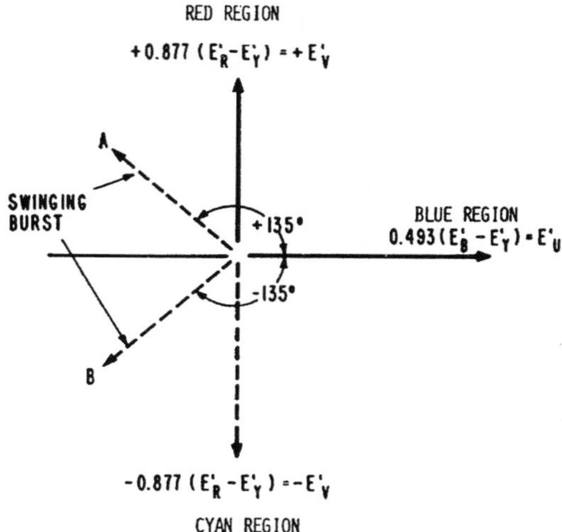

FIGURE 1.13-6 PAL color modulation phase diagram.

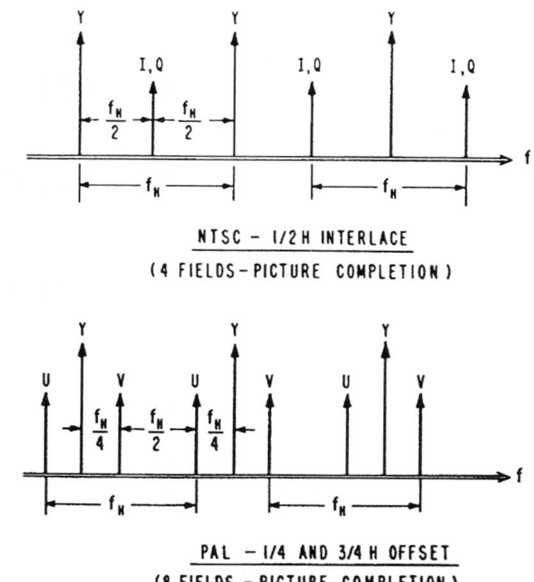

FIGURE 1.13-7 NTSC and PAL frequency interface relationships.

somewhat reduced saturation (this error being subjectively much less visible).

Obviously, the PAL receiver must be provided with some means by which the V signal switching sequence may be identified. The technique employed is known as *AB sync*, *PAL sync*, or *swinging burst* and consists of alternating the phase of the reference burst by ±45° at a line rate as shown in Figure 1.13-6. The burst is constituted from a fixed value of U phase and a switched value of V phase. Because the sign of the V burst component is the same sign as the V picture content, the necessary switching "sense" or identification information is available. At the same time, the fixed-U component is used for reference carrier synchronization.

Figure 1.13-7 explains the degree to which horizontal frequency (dot) interlace of the color subcarrier components with the luminance components is achieved in PAL, summarized as follows. In NTSC, the Y components are spaced at f_H intervals as a result of the horizontal sampling (blanking) process; thus, the choice of a color subcarrier whose harmonics are also separated from each other by f_H (as they are odd multiples of $\frac{1}{2}f_H$) provides a half-line offset and results in a perfect dot interlace pattern that moves upward. Four complete field scans are required to repeat a specific picture element dot position.

In PAL, the luminance components are also spaced at f_H intervals. Because the V components are switched symmetrically at half the line rate, only odd harmonics exist, with the result that the V components are spaced at intervals of f_H. They are spaced at half-line intervals from the U components which, in turn, have f_H spacing intervals due to blanking. If half-line offsets were used, the U components would be perfectly interlaced, but the V components would coincide with Y and, thus, not be interlaced, creating vertical stationary dot patterns. For this reason, in PAL a quarter-line offset for the subcarrier frequency is used, as shown in Figure 1.13-7. The expression for determining the PAL subcarrier specific frequency for 625 line/50 field systems is given by:

$$F_{SC} = \frac{1135}{4} f_H + \frac{1}{2} f_V \tag{5}$$

The additional factor $\frac{1}{2}f_V$ = 25 Hz is introduced to provide motion to the color dot pattern, thereby reducing its visibility. The degree to which interlace is achieved is, therefore, not perfect but is acceptable, and eight complete field scans must occur before a specific picture element dot position is repeated.

One additional function must be accomplished in relation to PAL color synchronization. In all systems, the burst signal is eliminated during the vertical synchronization pulse period. Because, in the case of PAL, the swinging burst phase is alternating line by line, some means must be provided for ensuring that the phase is the same for the first burst following vertical sync on a field-by-field basis; therefore, the burst reinsertion time is shifted by one line at the vertical field rate by a pulse referred to as the *meander gate*. The timing of this pulse relative to the A *versus* B burst phase is shown in Figure 1.13-8.

The transmitted signal specifications for PAL systems include the basic features discussed previously. Although a description of a great variety of receiver decoding techniques is outside the scope and intent of this chapter, it is appropriate to review—at least briefly—the following major features: Simple PAL relies on the eye to average the line-by-line color-switching process and can be plagued with line beats known as *Hanover bars* caused by the system nonlinearities introducing visible luminance changes at the line rate. Standard PAL employs a 1-H delay line element to separate U color signal components from V

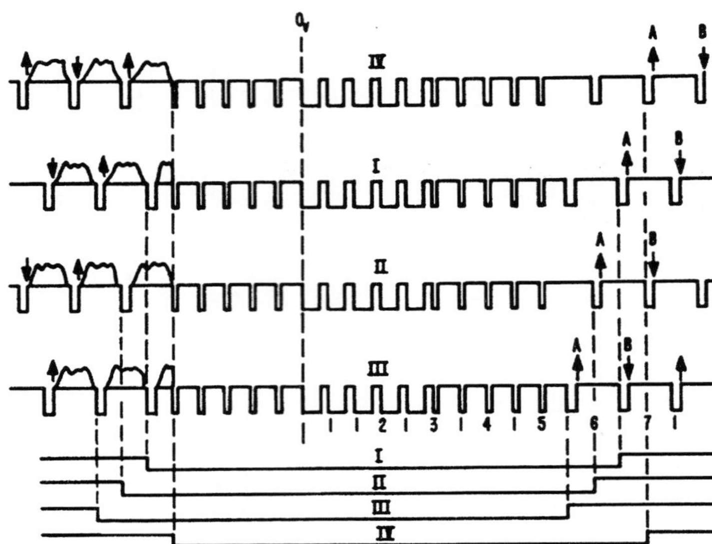

FIGURE 1.13-8 PAL meander burst blanking gate timing diagram for B, G, H, and I PAL.

color signal components in an averaging technique, coupled with summation and subtraction functions. Hanover bars can also occur in this approach if an imbalance of amplitude or phase occurs between the delayed and direct paths.

In a PAL system, vertical resolution in chrominance is reduced as a result of the line-averaging processes. The visibility of the reduced vertical color resolution as well as the vertical time coincidence of luminance and chrominance transitions differs depending upon whether the total system, transmitter through receiver, includes one or more averaging (comb filter) processes. PAL provides a system similar to NTSC and has gained favor in many areas of the world, particularly for 625 line/50 field systems.

SECAM Color System

The optimized SECAM system, known as SECAM III, is the system adopted by France and the former U.S.S.R. in 1967. The SECAM method has several features in common with NTSC, such as the same E'_Y signal and the same E'_B–E'_Y and E'_R–E'_Y color-difference signals; however, this approach differs considerably from NTSC and PAL in the manner in which the color information is modulated onto the subcarriers. First, the R–Y and B–Y color difference signals are transmitted alternately in time sequence from one successive line to the next, the luminance signal being common to every line. Because there is an odd number of lines, any given line carries R–Y information on one field and B–Y information on the next field. Second, the R–Y and B–Y color information is conveyed by frequency modulation of different subcarriers; thus, at the decoder, a 1-H delay element, switched in time synchronization with the line switching process at the encoder, is required to obtain the simultaneous existence of the B–Y and R–Y signals in a linear matrix to form the G–Y component.

The R–Y signal is designated as D'_R and the B–Y signal as D'_B. The undeviated frequency for the two subcarriers, respectively, is determined by:

$$F_{OB} = 272 f_H = 4.250000 \text{ MHz};$$
$$F_{OR} = 282 f_H = 4.406250 \text{ MHz} \tag{6}$$

These frequencies represent zero color difference information (zero output from the FM discriminator), or a neutral gray object in the televised scene.

As shown in Figure 1.13-9, the accepted convention for the direction of frequency change with respect to the polarity of the color difference signal is opposite for the D_{OB} and D_{OR} signals. A positive value of D_{OR} means a decrease in frequency, whereas a positive value of D_{OB} indicates an increase in frequency. This choice relates to the idea of keeping the frequencies representative of the most critical color away from the upper edge of the available bandwidth to minimize the instrumentation distortions.

The deviation for D'_R is 280 kHz; for D'_B, 230 kHz. The maximum allowable deviation, including preemphasis, for D'_R is –506 kHz and +350 kHz; the values for D'_B are –350 kHz and +506 kHz.

Two types of preemphasis are employed simultaneously in SECAM. First, as shown in Figure 1.13-10, a conventional type of preemphasis of the low-frequency color difference signals is introduced. The characteristic is specified to have a reference-level break point at 85 kHz (f_1) and a maximum emphasis of 2.56 dB. The expression for the characteristic is given as:

$$A = \frac{1 + j(f/f_1)}{1 + j(f/3f_1)} \tag{7}$$

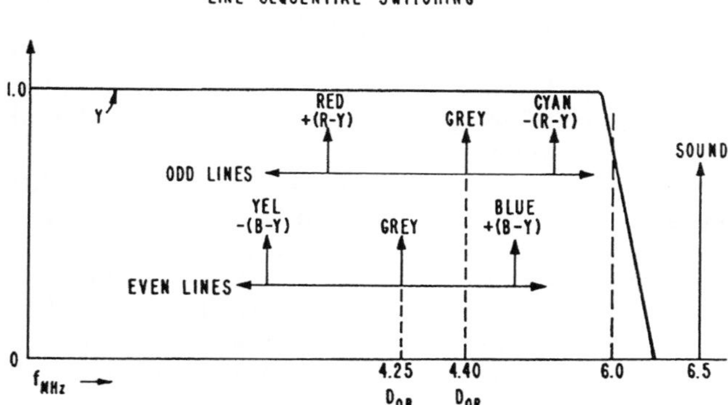

FIGURE 1.13-9 SECAM FM color modulation system.

A second form of preemphasis (Figure 1.13-10) is introduced at the subcarrier level where the amplitude of the subcarrier is changed as a function of the frequency deviation. The expression for this inverted bell shaped characteristic is given as:

$$G = M_0 \frac{1 + j16 \cdot \left(\frac{f}{f_c} - \frac{f_c}{f}\right)}{1 + j1.26 \cdot \left(\frac{f}{f_c} - \frac{f_c}{f}\right)} \tag{8}$$

where $f_c = 4.286$ MHz and $2M_0 = 23\%$ of the luminance amplitude (100 IRE).

This type of preemphasis is intended to further reduce the visibility of the frequency-modulated subcarriers in low-luminance-level color values and to improve the signal-to-noise ratio (SNR) in high-luminance and highly saturated colors; thus, monochrome compatibility is better for pastel average picture level objects but sacrificed somewhat in favor of SNR in saturated color areas.

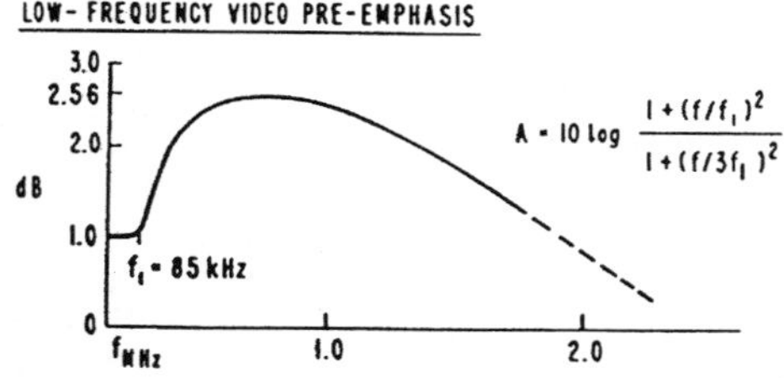

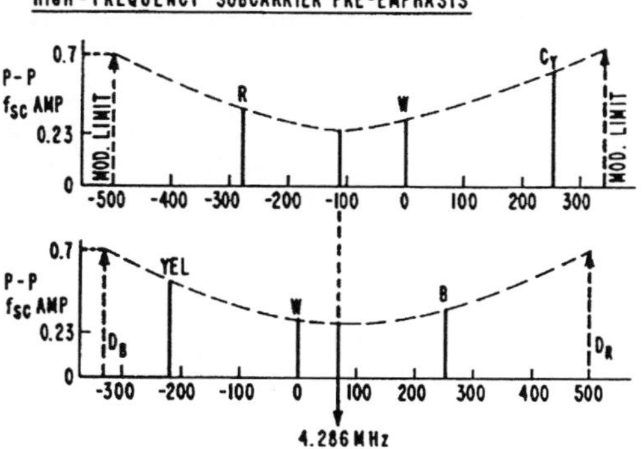

FIGURE 1.13-10 SECAM color signal preemphasis.

FIELD	LINE #	COLOR	SUBCARRIER φ
ODD (1)	n	f_{OR}	0°
EVEN (2)	n + 313	f_{OB}	180°
ODD (3)	n + 1	f_{OB}	0°
EVEN (4)	n + 314	f_{OR}	0°
ODD (5)	n + 2	f_{OR}	180°
EVEN (6)	n + 315	f_{OB}	180°
ODD (7)	n + 3	f_{OB}	0°
EVEN (8)	n + 316	f_{OR}	180°
ODD (9)	n + 4	f_{OR}	0°
EVEN (10)	n + 317	f_{OB}	0°
ODD (11)	n + 5	f_{OB}	180°
EVEN (12)	n + 318	f_{OR}	180°

Note: • 2 frames (4 fields) for picture completion.
• Subcarrier interlace is field-to-field and line-to-line of same color.

FIGURE 1.13-11 Color versus line and field timing relationships for SECAM.

Of course, precise interlace of frequency modulated subcarriers for all values of color modulation cannot occur; nevertheless, the visibility of the interference represented by the existence of the subcarriers may be reduced somewhat by the use of two separate carriers, as is done in SECAM. Figure 1.13-11 illustrates the line-switching sequence in that at the undeviated "resting" frequency situation, the two-to-one vertical interlace in relation to the continuous color difference line-switching sequence produces adjacent line pairs of f_{OB} and f_{OR} signals. To further reduce the subcarrier dot visibility, the phase of the subcarriers (phase carries no picture information in this case) is reversed 180° on every third line and between each field. This, coupled with the bell preemphasis, produces a degree of monochrome compatibility considered subjectively adequate.

As in PAL, the SECAM system must provide some means for identifying the line-switching sequence between the encoding and decoding processes. This is accomplished by introducing alternate D_R and D_B color identifying signals for nine lines during the vertical blanking interval following the equalizing pulses after vertical sync (see Figure 1.13-12). These bottle-shaped signals occupy a full line each and represent the frequency deviation in each time sequence of D_B and D_R at zero luminance value. These signals can be thought of as a fictitious green color that is used at the decoder to determine the line-switching sequence.

During horizontal blanking, the subcarriers are blanked and a burst of f_{OB}/f_{OR} is inserted and used as a gray-level reference for the FM discriminators to establish their proper operation at the beginning of each line; thus, the SECAM system is a line sequential color approach using frequency-modulated subcarriers. A special identification signal is provided to

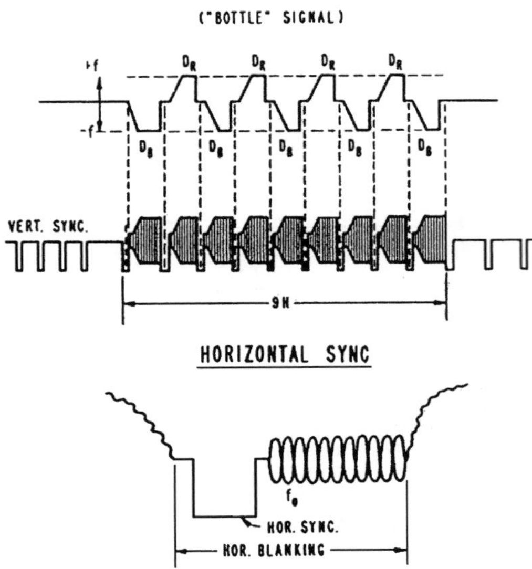

FIGURE 1.13-12 SECAM line identification signal.

identify the line-switch sequence and is especially adapted to the 625 line/50 field wideband systems available in France and the former U.S.S.R. It should be noted that SECAM, as practiced, employs amplitude modulation of the sound carrier as opposed to the FM sound modulation in other systems.

ADDITIONAL SYSTEMS OF HISTORICAL INTEREST

Of the numerous analog system variations proposed over the intervening years since development of the NTSC system, at least two others, in addition to PAL and SECAM, should be mentioned briefly. The first of these is additional reference transmission (ART), which involved the transmission of a continuous reference pilot carrier in conjunction with a conventional NTSC color subcarrier quadrature modulation signal. A modification of this scheme involved a multiburst approach that utilized three color bursts—one at black level, one at intermediate gray level, and one at white level—to be used for correcting differential phase distortion.

Another system, perhaps better known, was referred to as NIR or SECAM IV. Developed by the U.S.S.R. Nauchni Issledovatelskaia Rabota (NIR; translates as "scientific discriminating work"), this system consists of alternating lines of (1) an NTSC-like signal using an amplitude- and phase-modulated subcarrier and (2) a reference signal having U phase to demodulate the NTSC-like signal. In the linear version, the reference is unmodulated; in the nonlinear version, the amplitude of the reference signal is modulated with chrominance information.

Since the early 1990s, a number of enhanced-definition television (EDTV) systems were proposed for analog terrestrial service, offering a combination of widescreen picture and increased resolution. These systems include PALplus and Enhanced SECAM [7]. Although various EDTV systems based on NTSC were also proposed, that work eventually gave way to the development of analog-incompatible terrestrial digital television systems.

Of the various EDTV systems proposed, PALplus has seen the most widespread use. The system delivers a 16:9 picture with 574 active lines to a PALplus receiver. Appearing on a conventional television as a 16:9 letterboxed image with 430 active lines, the additional vertical luminance information is carried by a "helper" signal within the letterbox black bars. A transmission in this mode is indicated by the special Wide Screen Signaling (WSS) data carried on line 23. The specification for PALplus was standardized by the European Broadcasting Union (EBU) in 1997 [8], and the signal has been transmitted by various broadcasters since 1994.

SUMMARY AND COMPARISONS OF ANALOG SYSTEMS

History has shown that it is exceedingly difficult to obtain total international agreement on "universal" television broadcasting standards. Even with the first scheduled broadcasting of monochrome television in 1936 in England, the actual telecasting started using two different systems on alternate days from the same transmitter. The Baird system was 250 lines (noninterlaced) with a 50 Hz frame rate, while the Electric and Musical Industries (EMI) system was 405 lines (interlaced) with a 25 Hz frame rate.

These efforts were followed in 1939 in the United States with the broadcast of a 441 line interlaced system at 60 fields per second (the Radio Manufacturers Association [RMA] system). In 1941, the NTSC initiated the current basic monochrome standards in the United States of 525 lines (interlaced) at 60 fields per second, designated as system M by the CCIR. In those early days, the differences in powerline frequency were considered as important factors and were largely responsible for the proliferation of different line rates versus field rates, as well as the wide variety of video bandwidths. However, the existence and extensive use of monochrome standards over a period of years soon made it a top-priority matter to assume the reciprocal compatibility of any developing analog color system.

In 1998, the successor to the CCIR, the International Telecommunications Union, Radiocommunication Sector (ITU-R), formalized the definition of all worldwide analog television systems then in use in a Recommendation [9], which was subsequently amended [10]. The ITU-R documents define recommended standards for worldwide analog color television systems in terms of the three basic color approaches—NTSC, PAL, and SECAM [4]. The variations—at least 13 of them—are given alphabetical letter designations; some represent major differences, while others signify only very minor frequency-allocation differences in channel spacings or differences between the VHF and UHF bands. The key to understanding the CCIR/ITU designations lies in recognizing that the letters refer primarily to local monochrome standards for line and field rates, video channel bandwidth, and audio carrier relative frequency. Further classification in terms of the particular color system then adds to NTSC, PAL, or SECAM as appropriate.

As an example, the letter "M" designates a 525 line/60 field, 4.2 MHz bandwidth, 4.5 MHz sound carrier monochrome system; thus, M(NTSC) describes a color system employing the NTSC technique for introducing the chrominance information within the constraints of the basic monochrome signal values. Likewise, M(PAL) would indicate the same line and field rates and bandwidths but use the PAL color subcarrier modulation approach. In another example, the letters "I" and "G" relate to specific 625 line/50 field, 5.0 or 5.5 MHz bandwidth, 5.5 or 6.0 MHz sound carrier monochrome standards; thus, G(PAL) would describe a 625 line/50 field, 5.5 MHz bandwidth, color system utilizing the PAL color subcarrier modulation approach. The letter "L" refers to a 625 line/50 field, 6.0 MHz bandwidth system to which the SECAM color modulation method has been added (often referred to as SECAM III). System E is an 819 line/50 field, 10 MHz bandwidth, monochrome system. This channel was used in France for early SECAM tests and for system E transmissions.

Some general comparison statements can be made about the underlying monochrome systems and existing analog color standards:

- There are three different scanning standards: 525 lines/60 fields, 625 lines/50 fields, and 819 lines/50 fields.
- There are six different spacings of video-to-sound carriers: 3.5, 4.5, 5.5, 6.0, 6.5, and 11.15 MHz.
- Some systems use FM and others use AM for the sound modulation.
- Various schemes were developed in the 1980s to carry stereophonic audio and additional audio programs within an analog television transmission. The reader is referred to Chapter 6.3, "Multichannel Television Sound," of this handbook for further details.
- A number of countries using PAL transmission have approved the use of a two-sound-carrier FM system [11]. The second carrier is placed at a frequency $f_H \times 15.5 = 242.1875$ kHz above the first (traditional) carrier, at a power level 20 dB below peak visual power, and employing the same modulation as the first sound carrier. The second carrier can carry a second audio program or the right channel of a stereo program (wherein the first carrier carries the (L + R)/2 signal).
- A number of European countries using PAL or SECAM transmission have approved the use of an additional digital carrier for stereophonic or multichannel sound transmission [12]. Based on the NICAM-728 system introduced in 1987, a high-frequency subcarrier is digitally modulated with a 728 kb/s data stream consisting of 728-bit packets. Using a sampling rate of 32 kHz, the 14-bit samples are digitally companded to 10-bit words.
- Some systems use positive polarity (luminance proportional to voltage) modulation of the video carrier, but others, such as the U.S. (M)NTSC system, use negative modulation.
- There are differences in the techniques of color subcarrier encoding represented by NTSC, PAL, and SECAM, and in each case many differences can be found in the details of various pulse widths, timing, and tolerance standards.
- Various countries have developed their own schemes for using the vertical interval for some form of data. Although each analog standard can accommodate data in various ways, it is up to the individual permitting authority as to what data is allowed. The reader is referred to Chapter 5.24, "Closed Captioning Systems," and Chapter 5.25, "Data Broadcast Systems for Television," of this handbook for details on Teletext and other data systems in use over analog television systems.

It is evident that one must refer to the ITU documents for accurate information on the combined analog monochrome/color standards. Figure 1.13-13 presents a comparison of the relative bandwidths, color subcarrier frequencies, and sound carrier spacing for the major analog color systems used in the world today.

The signal in the M(NTSC) system occupies the least total channel width, which when the vestigial

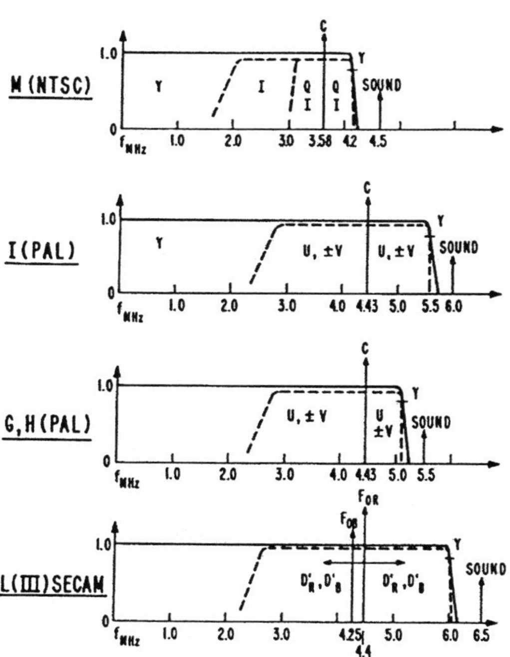

FIGURE 1.13-13 Bandwidth comparison between NTSC, PAL, and SECAM.

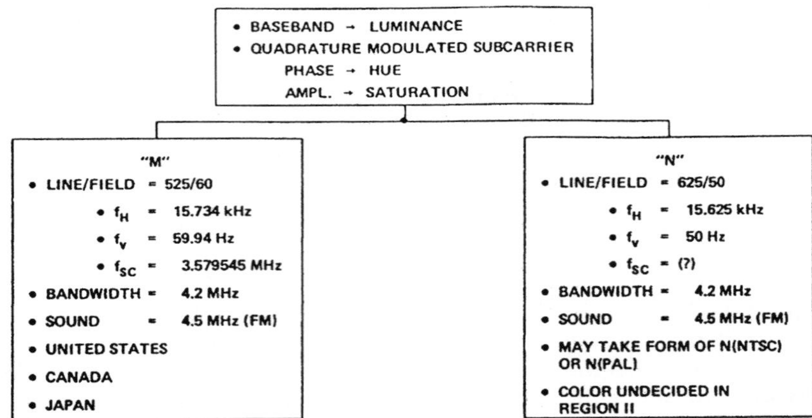

FIGURE 1.13-14 ITU-R designation for NTSC system (summary).

sideband plus guard bands are included, requires a minimum radiofrequency channel spacing of 6 MHz. The L(III) SECAM system signal occupies the greatest channel space with a full 6 MHz luminance bandwidth. Signals from the two versions of PAL lie in between and vary in vestigial sideband width as well as color and luminance bandwidths. NTSC is the only system to incorporate the *I*, *Q* color acuity bandwidth variation. PAL minimizes the color quadrature phase distortion effects by line-to-line averaging, and SECAM avoids this problem by only transmitting the color components sequentially at a line-by-line rate.

Figures 1.13-14 to 1.13-16 summarize, in organization chart form, the ITU-R designations for NTSC, PAL, and SECAM basic system identifications and characteristics. In Figure 1.13-14, M(NTSC) identifies the system used in the United States, Canada, Japan, Mexico, the Philippines, and several other Central American and Caribbean area countries. The N system may be implemented in color either in the NTSC or the PAL format [3]. An N version of PAL has been in use in Argentina. Figure 1.13-15 provides a summary of the PAL systems. PAL systems in one or another of the 625 line formats are predominately used in Continental Europe, the United Kingdom, some African countries, Australia, and various Asian countries, including India and China. An M (525 line) version of PAL has been in use in Brazil.

Figure 1.13-16 summarizes the SECAM III system, which is in use primarily in France and the former U.S.S.R. It should be noted that, as the SECAM system uses frequency modulation for the color subcarrier, the video signal cannot be directly switched or edited in this format; it typically originates in PAL and is transcoded prior to transmission. The proposed SECAM IV system almost gained favor in 1966 as a universal European approach but was never implemented [1]. The E system, mentioned in connection with early SECAM tests in France, was limited to monochrome broadcasts.

Table 1.13-2 provides a summary of the major analog color television system general characteristics. Table 1.13-3 characterizes the fundamental features relating to the differences between NTSC, PAL, and SECAM in the critical areas of color encoding techniques. Similarly, Table 1.13-4 illustrates the color

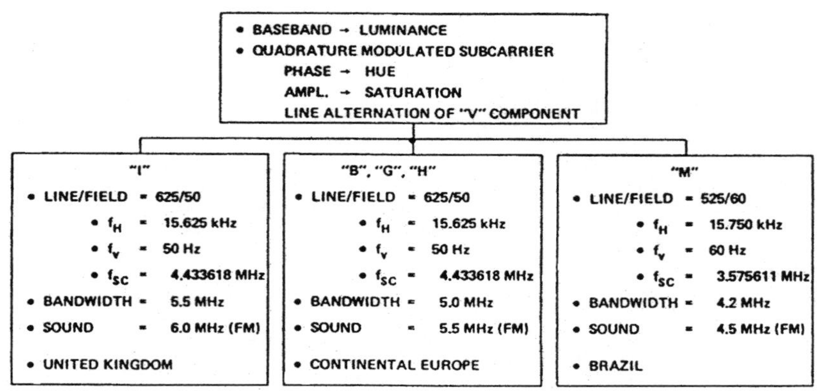

FIGURE 1.13-15 ITU-R designation for PAL system (summary).

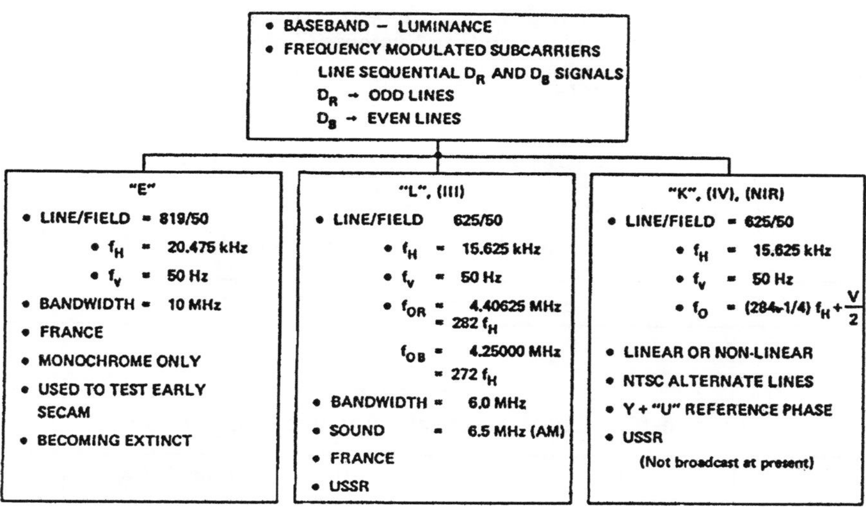

FIGURE 1.13-16 ITU-R designation for SECAM (summary).

TABLE 1.13-2
General Analog System Technical Summary

	NTSC	PAL	SECAM
TV system	M	B/G/H/I	D/K/L
Field rate (f_V) (Hz)	$60 \times \frac{1000}{1001} = 59.94$	50	50
TV lines	525	625	625
Line rate (f_H) (Hz)	15,734.265…	15,625	15,625
Luma BW (MHz)	4.2	5.0, 5.5	6.0
Sound subcarrier (MHz)	4.5 (F3)	5.5, 6.0 (F3)	6.5 (A3)
Gamma	2.2	2.8	2.8
White point	C	D_{65}	D_{65}

TABLE 1.13-3
Chrominance Encoding Systems Comparison

	NTSC	PAL	SECAM
Color subcarrier (Hz)	$5{,}000{,}000 \times \frac{63}{88} = 3{,}579{,}545.45\ldots$	4,433,618.75	$f_{OB} = 4{,}250{,}000$ $f_{OR} = 4{,}406{,}250$
Color subcarrier relationship to line rate	$f_{SC} = \frac{455}{2} f_H$	$f_{SC} = \left(\frac{1135}{4} + \frac{1}{625}\right) f_H$	$f_{OB} = 272\, f_H$ $f_{OR} = 282\, f_H$
Chroma encoding	Phase and amplitude quadrature modulation	Phase and amplitude quadrature modulation (line alternation)	Frequency modulation (line sequential)
Color difference signals	I, Q (1.3 MHz) (0.6 MHz)	U, ±V (1.3 MHz) (1.3 MHz)	D_R (f_{OR}) (>1.0 MHz) D_B (f_{OB}) (>1.0 MHz)

TABLE 1.13-3 *(continued)*
Chrominance Encoding Systems Comparison

	NTSC	PAL	SECAM
Color burst phase	–(B–Y)	U and ±V	f_{OR} and f_{OB} 180° phase switch every third line and every field
Color switch identification	Not required	Swinging burst ±45°	Nine lines of D_R and D_B during vertical interval
Additional signals	None	Meander gate $f_{H/2}$	$f_{H/2}, f_{H/4}, f_V, f_{V/2}$

TABLE 1.13-4
Line-to-Line Chroma Signal Sequence Comparison

		Line (N)	Line (N + 1)	Line (N+ 2)	Line (N+ 3)
NTSC	Chroma: Burst phase:	I, Q –(B-Y)	I, Q –(B-Y)	I, Q –(B-Y)	I, Q –(B-Y)
PAL	Chroma: Burst phase:	$U_1 + V$ $-U + V = +135°$	$U_1 - V$ $-U - V = +225°$	$U_1 + V$ $-U + V = +135°$	$U_1 - V$ $-U - V = +225°$
SECAM (FM)	Chroma:	D_R ± 280 kHz	D_B ± 230 kHz	D_R ± 280 kHz	D_B ± 230 kHz
	Burst phase:	D_R Deviation = +350 kHz –500 kHz		D_B Deviation = +500 kHz –350 kHz	
	Chroma Switch Ident. Lines During Vertical Interval				
	Line #:	7 8 9 10 11 12 13 14 15 320 321 322 323 324 325 326 327 328			
	Ident Signals:	D_R D_B D_R D_B D_R D_B D_R D_B D_R			
	Note: Phase reversed 180° every third line and every field.				

encoding line-by-line color sequence operation for the three systems. The information conveyed in these tables highlights the technical equalities and differences among the systems and attempts to show some kind of order as an aid to understanding the worldwide situation. It serves as well to point out the difficulties of achieving a "universal" analog system [5].

ACKNOWLEDGMENT

The original version of this article was first published in the *SMPTE Journal* and has been reprinted with permission of the Society of Motion Picture and Television Engineers.

References

[1] Herbstreit, J. W. and Pouliquen, J., International standards for color television, *IEEE Spectrum*, March 1967.

[2] Fink, D. G. (Ed.), *Color Television Standards*, McGraw-Hill, New York, 1955.

[3] Pritchard, D. H., U.S. color television fundamentals: A review, *SMPTE J.*, 86, 819–828, 1977.

[4] *CCIR Characteristics of Systems for Monochrome and Colour Television: Recommendations and Reports*, Recommendation 470-1 (1974–1978) of the Fourteenth Plenary Assembly of CCIR, Kyoto, Japan, 1978.

[5] Roizen, J., Universal color television: An electronic fantasia, *IEEE Spectrum*, March 1967.

[6] SMPTE Standard for Television, *Composite Analog Video Signal: NTSC for Studio Applications*, SMPTE 170M-1999, revision of ANSI/SMPTE 170M-1994, Society of Motion Picture and Television Engineers, White Plains, NY, 1999.

[7] Recommendation ITU-R BT.1117-2, *Studio Format Parameters for Enhanced 16:9 Aspect Ratio 625-Line Television Systems (D- and D2-MAC, PALplus, Enhanced SECAM)*, International Telecommunication Union, Geneva, Switzerland, 1994, 1995, 1997.

[8] European Telecommunication Standard ETS 300 731, *Television Systems; Enhanced 625-Line Phased Alternate Line (PAL) Television; PALplus*, European Telecommunications Standards Institute (ETSI), Sophia Antipolis, France, 1997.

[9] Recommendation ITU-R BT.470-6, *Conventional Television Systems*, International Telecommunication Union, Geneva, Switzerland, 1970, 1974, 1986, 1994, 1995, 1997, 1998.

[10] Recommendation ITU-R BT.1700, *Characteristics of Composite Video Signals for Conventional Analogue Television Systems*, International Telecommunication Union, Geneva, Switzerland, 2005.

[11] Recommendation ITU-R BS.707-5, *Transmission of Multi-Sound in Terrestrial Television Systems PAL B, B1, D1, G, H, and I, and SECAM D, K, K1, and L*, International Telecommunication Union, Geneva, Switzerland, 1990, 1994, 1995, 1998, 2005.

[12] Report ITU-R BT.2043, *Analogue television systems currently in use throughout the world*, International Telecommunication Union, Geneva, Switzerland, 2004.

CHAPTER

1.14

Worldwide Standards for Digital Television

ALDO CUGNINI
AGC Systems LLC
Long Valley, New Jersey

HIROSHI ASAMI
Ministry of Internal Affairs and Communications
Tokyo, Japan

DAVID WOOD
European Broadcasting Union
Geneva, Switzerland

INTRODUCTION

When the possibility of digital transmission of television signals became apparent in the late 1980s, various efforts were started to define a new, worldwide set of standards for digital image capture, manipulation, storage, and distribution. The most well known of these were assembled under the auspices of the Moving Picture Experts Group (MPEG), a working group of the International Organization for Standards (ISO) and the International Electrotechnical Commission (IEC). At about the same time, work was underway on the practical aspects of high-definition television (HDTV) production and distribution, leading to an effort to develop an analog HDTV system. In the early 1990s, ongoing work based on the encryption of satellite transmissions made it viable to encode and transmit HDTV using a 6 MHz broadcast channel. These interrelated developments spawned the first serious efforts to define digital television systems for satellite, cable, and terrestrial media around the world.

Earlier, in 1982, the Advanced Television Systems Committee (ATSC) was founded in the United States and was chartered with developing voluntary standards for the emerging technologies of advanced television systems—at first analog but including high-definition television. Ultimately, this led to the development of a digital terrestrial television standard, which became the U.S. digital television (DTV) standard in 1996. In Europe, meanwhile, work begun in 1991 by the European Launching Group (ELG) culminated in formation of the DVB Project, which led to a digital satellite standard in 1993 (DVB-S), followed later by cable and terrestrial standards. Parallel work in Japan also resulted in another variant of digital standards for satellite, cable, and terrestrial broadcasting for that country. The difference in initial transmission medium was probably as much responsible as any other factor for the divergence of the systems from a worldwide perspective.

The principle areas of deployment and the standards used in the fixed-service systems for terrestrial digital television broadcasting can be seen in Table 1.14-1. At the time of this writing, a new Chinese system was announced. At about the same time, it was reported that the governments of Argentina and Brazil will decide independently which digital TV standard each nation should choose, but they have agreed to work to implement a single standard for the Mercosur customs union in the future. In addition, not all regions have started full-time digital broadcasts. The table shows the decisions that have been made to date.

Comparison of Systems

The various characteristics of the different transmission systems are due to varying technologies, objectives, politics, industry structure, and channel usage at the time of the particular system development. To some extent, each region or grouping drew on what had been developed beforehand when it developed its own system. In addition, each implementation has planned on a degree of sophistication for the receiver, based on the ever-increasing power of semiconductor chips then available. Table 1.14-2 shows the main transmission characteristics of the systems.

TABLE 1.14-1
Worldwide Terrestrial Digital Television Systems

Region		Transmission	Transport	Video	Audio
Africa*		DVB-T	MPEG-2	MPEG-2	MPEG-1 & -2 layer II
Asia					
	China	TBD†	MPEG-2	MPEG-2, or AVS-China	MPEG-1 & -2 layer II or AVS-China
	India	DVB-T	MPEG-2	MPEG-2	MPEG-1 & -2 layer II
	Japan	ISDB-T	MPEG-2	MPEG-2	MPEG-2 AAC
	S. Korea	ATSC	MPEG-2	MPEG-2	AC-3
	Russia	DVB-T	MPEG-2	MPEG-2	MPEG-1 & -2 layer II
Australia		DVB-T	MPEG-2	MPEG-2	MPEG-1 & -2 layer II or AC-3
Europe		DVB-T	MPEG-2	MPEG-2, MPEG-4 AVC	MPEG-1 & -2 layer II, AC-3
North America					
	Canada	ATSC	MPEG-2	MPEG-2	AC-3
	Mexico	ATSC	MPEG-2	MPEG-2	AC-3
	United States	ATSC	MPEG-2	MPEG-2	AC-3
South America					
	Argentina	ATSC‡	MPEG-2	MPEG-2	AC-3
	Brazil	SBTVD-T**	MPEG-2	MPEG-4 AVC	MPEG-2 AAC

*The DVB Project reports that it has the support of all African nations for DVB-T.
†New Chinese standard announced; name to be determined.
‡Under discussion.
**Sistema Brasiliero de Televisão Digital Terrestre.

TABLE 1.14-2
Comparison of ATSC, DVB-T, ISDB-T, and Chinese Systems

Systems	ATSC	DVB-T	ISDB-T	Chinese System
Transmission	Single carrier	Multiple carrier (OFDM)		Single carrier (QAM)/ Multiple carrier (OFDM)
Bandwidth	6/7/8 MHz	6/7/8 MHz		8 MHz
Modulation	8-VSB	QPSK/16QAM/ 64QAM	DQPSK/QPSK/ 16QAM/64QAM	4QAM/16QAM/ 32QAM/64QAM
Error correction	Trellis code + RS	Convolutional code + RS		BCH outer code LDPC inner code
Characteristic	Distributed Transmission capability	SFN capability	SFN capability Segmented OFDM Time interleaving	Time interleaving Frequency interleaving
Proposing region	United States	Europe	Japan	China

Abbreviations: BCH, Bose-Chaudhuri-Hocquenghem; LDPC, low-density parity check; RS, Reed-Solomon; SFN, single-frequency network.

The first system to be developed in the early 1990s was the ATSC system. This uses eight-level vestigial sideband (8-VSB) modulation, with a fixed level of error correction and was designed to replicate the coverage of analog National Television System Committee (NTSC) terrestrial television using rooftop antennas for new digital HDTV broadcasting services. The history and technology of ATSC are covered elsewhere in this handbook.

When the DVB-T system was developed, it was possible to draw on the ATSC system and add new features. These included 2K and 8K coded orthogonal frequency-division multiplexing (COFDM), which addressed reception in difficult multipath environments, and an agile forward error correction (FEC) level for more flexibility. Several years later, when the ISDB-T system was developed in Japan, it was possible to draw on the DVB-T system and further add new features. These include a 4K OFDM mode and a multiple-segment structure to bring more flexibility in spectrum management.

Since ISDB-T was developed, the DVB-H system has emerged for broadcasting to handheld devices. Although DVB-H does not take advantage of the modular approach of ISDB-T and is designed only for the current analog TV channeling plans, it does include a 4K OFDM mode, and new features including "time slicing" help reduce receiver power consumption. It is also able to take advantage of technical progress and the evolution of compression tools in ISO/IEC JTC MPEG, and it allows the use of high-efficiency MPEG-4 AVC video compression.

At the time of this writing, the Chinese government had just introduced a new standard for terrestrial digital television. According to various sources, possible names for the new system included *Chinese Digital Multimedia Broadcast-Terrestrial* (CDMB-T) and *Digital Multimedia Broadcast-Terrestrial/Handheld* (DMB-T/H or DTMB), with the system designed for fixed and mobile reception, supporting both TVs and mobile devices.

DIGITAL VIDEO BROADCASTING

The Digital Video Broadcasting (DVB)[1] project is a European-originated worldwide alliance or consortium of about 250 to 300 companies. Its objective is to agree on specifications for digital media delivery systems, including broadcasting. It is an open, private-sector initiative with an annual membership fee, governed by a memorandum of understanding. Its formation in the early 1990s was the result of coinciding ideas about the need for such an initiative from European industry and the European Broadcasting Union (EBU), the collective organization of Europe's national broadcasters. The project is managed by the DVB Project Office, whose staff are employees of the EBU in Geneva, Switzerland, but who work exclusively in the interests of the members of the DVB Project. Over 100 million items of DVB equipment are now in the hands of the public throughout the world, many carrying the DVB logo. The members of the DVB Project develop and agree on specifications, which are then passed on to the European standards body for media systems—the EBU/CENELEC/ETSI Joint Technical Committee (JTC)—for approval. They appear as European Telecommunications Standards Institute (ETSI) standards or reports.

[1]The name "DVB" was suggested by Armin Silberhorn and others as a complement to the already developed Digital Audio Broadcasting (DAB). The scope of the DVB Project is now much wider than television broadcasting.

History

When the DVB Project began, the EBU was able to use its experience in organizing technical meetings and publications. Industry brought a key element—the belief that specifications are only worth developing if and when they can be translated to products that have a direct commercial value. The DVB specifications are market driven. The key elements that have contributed to the success of the project include the following:

- The committee structure of the project has made a critical contribution. The project is bicameral, with all the checks and balances this affords. A *commercial module* decides, without discussing how they should be achieved, what features or cost levels are needed to make a product a success. Then, a *technical module* is given the task of creating a technical specification that meets these needs. Finally, after the specification is prepared, the commercial module verifies that the technicians have done what was needed.
- Management of the project is almost entirely paperless; it has only one working language (English), and extensive use is made of the Internet. At the time of this writing, the DVB secretariat is led by Peter McAvock, and the chair of the key group, the technical module, is Professor Ulrich Reimers.
- The DVB Project has several Web sites [1,2], and copies of the DVB-developed specifications are available free of charge from the ETSI Web site [3] following a registration procedure. On request, the DVB Project Office also makes available a CD containing many of the key DVB documents.

The DVB Project began the first phase of its work in 1993. The project's philosophy was as follows:

- The initial task was to develop a complete suite of digital satellite, cable, and terrestrial broadcasting technologies in one pre-standardization body.
- Rather than having a one-to-one correspondence between a delivery channel and a program channel, the systems would be "containers" that carry any combination of image, audio, or multimedia. They would thus be open and ready for standard-definition television (SDTV), extended-definition television (EDTV), high-definition television (HDTV), surround sound, or any kind of new media that might arise over time.

- The work should result in ETSI standards for the physical layers, error correction, and transport for each delivery medium.
- It should also result in an ETSI report that outlines the baseband systems that are options for carriage.
- Wherever possible, there should be commonality across the different delivery platforms to lower costs for users and manufacturers. Only when there was no other choice would there be differences between different delivery media.
- The DVB Project should not re-invent anything, and should use existing open standards whenever they are available.

The DVB Project has used, and continues to draw extensively on, standards from the ISO/IEC JTC MPEG. The transport for all systems is the MPEG-2 transport stream [4]. The DVB specification for baseband systems [5] follows the systems developed in the JTC MPEG.

Overview

For convenience, the final documentation is arranged with sets of hyphenated initials that identify the area; for example, DVB-S is the specification for the first-generation version of the DVB digital satellite system. Other areas include DVB-S2 (the second-generation version of the digital satellite system), DVB-C (the digital cable system), and DVB-T (the digital terrestrial broadcasting system). More recently, DVB-H (the digital terrestrial broadcasting system serving handheld receivers), DVB-DATA (the data delivery system), DVB-SI (the service information system), and DVB-MHP (the multimedia delivery system or home platform) have been added. The various systems are summarized in Table 1.14-3. Although some of the features of DVB systems are outlined here, the territory is too large to cover fully here and readers are encouraged to consult the references, including the DVB and ETSI Web sites.

Satellite, Cable, and Terrestrial: DVB-S, DVB-C, and DVB-T

The DVB-S system [6] for digital satellite broadcasting was developed in 1993. It comprises a relatively straightforward transmission system using quadrature phase-shift keying (QPSK). The specification describes different tools for channel coding and error protection that are also used for other delivery media systems. The DVB-C system [7] for digital cable networks was developed in 1994. It is centered on the use of 64 quadrature amplitude modulation (QAM) and

TABLE 1.14-3
DVB Transmission Systems

	DVB-H	DVB-T	DVB-C	DVB-S	DVB-S2
Channel bandwidth	5/6/7/8 MHz	6/7/8 MHz	6/7/8 MHz	Any sat. transp. (*e.g.*, 26–72 MHz) and MCPC/SCPC	Any sat. transp. (*e.g.*, 26–72 MHz) and MCPC/SCPC
Modulation scheme	QPSK, 16QAM, 64QAM on COFDM	QPSK, 16QAM, 64QAM on COFDM	16–256QAM	QPSK	QPSK, 8PSK, 16APSK, 32APSK
Modulation modes*	2K, 4K, 8K	2K, 8K	CCM	CCM	ACM/CCM
FEC	RS (204,188), convolutional code, and MPE–FEC	RS (204,188) and convolutional code	RS (204,188) and byte interleaver (I = 12)	RS (204,188) and convolutional code	BCH outer code, LDPC inner code
Roll-off	N/A	N/A	0.15	0.35	0.35, 0.25, 0.20
C/N range	4.5–18.4 dB†	3.1–20.1 dB‡	23.1–25.1dB	+4 to +8 dB	–2.4 to +16 dB
Data rate in 8 MHz channel	4.98~15 Mb/s**	4.98–31.67 Mb/s	6.4 (16QAM)–48.1 (256QAM) Mb/s	28.1–45.4 Mb/s (in 36 MHz transp.)	30% improvement on DVB-S††

Abbreviations: ACM, adaptive coding and modulation; BCH, Bose-Chaudhuri-Hocquenghem; CCM, constant coding and modulation; LDPC, low-density parity check; MCPC, multiple channels per carrier; MPE, multiprotocol encapsulation; RS, Reed–Solomon; SCPC, single channel per carrier.

*Modulation mode is difficult to compare between single-carrier systems such as DVB-S, DVB-S2, DVB-C, and multicarrier COFDM systems such as DVB-T and DVB-H.
†The values are calculated using the theoretical C/N figures given in EN 300 744, plus an implementation margin of 2.5 dB, using the UK DTG D-book noise model with a receiver excess noise source value Px of –33 dBc and simulated MPE–FEC. An ideal transmitter is assumed. A quality criterion is Quasi Error Free IP-Stream (QEFIP), meaning an IP-packet error ratio 10^{-3} calculated over the entire multiplex.
‡C/N calculated for a Gaussian channel and a BER of 2×10^{-4} after Viterbi decoding, corresponding to Quasi Error Free operation. For DVB-T, the figures correspond to DVB-T in nonhierarchical mode, C/N figures are usually quoted for a Rayleigh channel and range from 5.4 to 27.9 dB.
**Data rate is same as for DVB-T (minus some overhead if optional MPE–FEC is used); however, the practical upper limit is approximately 15 Mb/s in an 8 MHz channel.
††Data rates for DVB-S2 are difficult to compare to other DVB systems given the wide range of options.

can, if needed for the European satellite and cable environment, convey a complete satellite channel multiplex on a cable channel. The DVB-CS [8] specification describes a version that can be used for satellite master antenna television installations. The digital terrestrial television system DVB-T [9] is more complex because it is intended to cope with a different noise and bandwidth environment and multipath. The system has several dimensions of receiver agility, where the receiver is required to adapt its decoding according to signaling. The key element is the use of orthogonal frequency-division multiplexing (OFDM). There are two modes: 2K carriers plus QAM and 8K carriers plus QAM. The 8K mode can allow more multipath protection, but the 2K mode can offer Doppler advantages where the receiver is moving. Guidelines for the applications are available [10]. There are two systems for multi-channel microwave distribution systems (MMDSs), one for systems that operate at radio frequencies below 10 GHz (DVB-MC [11]), which is like the DVB-C system, and one for systems that operate at radio frequencies above 10 GHz (DVB-MS [12]), which is like the DVB-S system. An MMDS system, like DVB-T (DVB-MT [13]) is also available.

DVB-S2 and DVB-H

A higher-efficiency digital satellite broadcasting system, DVB-S2 [14], has recently been developed which has both DVB-S backward-compatible and non-backward-compatible versions. The noncompatible version uses 8-phase-shift keying (PSK) and low-density parity check (LPDC) concatenated with BCH coding to achieve an efficiency increase of about 30% more data capacity for the same receiving dish size compared to DVB-S. DVB-S2 is likely to be used for all new European digital satellite multiplexes, and satellite receivers will be equipped to decode both DVB-S and DVB-S2. A more flexible and robust digital terrestrial system, DVB-H [15], has recently been developed. The system is intended to be receivable on handheld receivers and thus includes features that will reduce battery consumption (time slicing) and a 4K OFDM mode, together with other measures. DVB-H services will probably also use the more efficient video compression system MPEG-4 Advanced Video Coding (AVC). The DVB Project has also developed metadata and access systems that can be used with DVB-H.

DVB Conditional Access Tools

Broadcasters need a system to ensure that only authorized or paying viewers can view pay-TV or other pay services. The DVB Conditional Access system is a single private-key system. The simplified operation is as follows. At the broadcaster, the output from a complex pseudo-random binary sequence (PRBS) generator is added (modulo-2) to the digital stream to be transmitted. This process is termed *scrambling*. In the receiver is a matching PRBS generator that can descramble the signal. A signal or key is delivered over the broadcast path at intervals that resynchronize the two PRBS generators; this is the *control word* or scrambling key.

The version of the control word that is transmitted is itself scrambled, although the term *encryption* is used to avoid confusion with the scrambling of the program. The encrypted version is the *entitlement control message* (ECM). To decrypt the ECM requires another transmitted signal, the *entitlement management message* (EMM). Versions of this EMM are only transmitted to viewers who have paid their subscription or pay-per-view fee. They are encrypted with a key contained in the specific paid-up receiver itself. This is how overall control of the signal path is achieved.

The DVB Project has specified the PRBS generator [16] (the Common Scrambling Algorithm DVB-CSA); however, it is not available over the counter from ETSI, as applicants have to prove they have a genuine reason for needing it. Though currently secure, there are moves to agree on an even more sophisticated CSA. The ECM and EMM systems are not standardized, but are left to the individual broadcaster.

The DVB Project has suggested two strategies to make it possible to use the same receivers for several pay-TV services that have different ECM and EMM systems. The first strategy is termed *Multicrypt*. In this case, the receiver is equipped with a PC-Card (PCMCIA) slot. This is the DVB Common Interface (DVB-CI) [17]. The ECM and EMM systems can be included in secure PCMCIA cards, which can be changed depending on the pay services desired. This card interface interrupts the main signal path, so the slot can also have other uses [18]. The second strategy is termed *Simulcrypt*. In this case, the receiver is only able to cope with one of the ECM or EMM systems, but other pay-TV operators agree to send their signals in a form that can be understood by the native system of the receivers. Several sets of ECM and EMM keys are allowed to be transmitted at the same time. This requires quite complex rules [19] between the pay-TV operators.

Multimedia Broadcasting and Interactivity

Digital broadcasting has the capacity to deliver multimedia in addition to television programs. This can look like an electronic version of a magazine page or a Web page. It is either independent of the television program or allied to it in some way. It can be one-way multimedia that displays pictures and information on the screen (superimposed or separate), or it can be two-way multimedia that uses a return path system to the broadcaster to allow the viewer to interact directly with the broadcaster.

The information for the multimedia has to be delivered to the receiver in a way that can be predicted, and all the incoming information has to be coded in a language that is known to the receiver. The delivery of one-way material is usually arranged in a *carousel*. This means that information is available in a repeating cycle. The receiver grabs the information the viewer has requested (via his controls) as it goes by. There can be a finite waiting time for broadcast multimedia, whose length depends on chance and on how much overall multimedia is being offered by the channel. The DVB Project has developed a transport system for such data [20].

The language of the data for multimedia, the Application Programming Interface (API), was studied in the DVB Project for many years before agreement was reached on a specification. In fact, at the time agreement was reached, a number of different proprietary and open systems were in quite wide use. The DVB-developed API is thus an option rather than a mandatory part of the DVB family of systems. At the start of the discussions, there were already several proprietary APIs in use, with different capabilities. The project agreed that it could not take any specific one of these as a DVB system, but needed a new, outside, open system. The system developed was the Multimedia Home Platform (MHP) [21], which makes extensive use of JAVA™.

Multimedia content can have varying degrees of sophistication and can fall into two categories: *declarative content* and *procedural content*. In simple terms, declarative content simply gives a recipe for what should be on the screen at any time, as HTML does. Procedural content includes a list of instructions that are executed, on cue, in the receiver. The capability for procedural content is needed, for example, for sophisticated animated graphics. Different APIs offer the capability of procedural content only or both procedural and declarative content. MHP is designed to allow both.

To make it "future proof," MHP is arranged in a series of generations that will bring more tools to the disposal of the broadcaster when receiver complexity permits. The first-generation MHP 1.0 allows enhanced multimedia and interactivity. Small refinements have been made to MHP 1.0 based on experience, and the current version is MHP 1.0.3. The next generation of MHP (MHP 1.1) will offer more features, including seamless switching between broadcast multimedia and broadband- or narrowband-delivered Web pages. Work continues in the DVB Project on yet another next-generation MHP system that will work well with home personal video recorders (PVRs).

Great attention has been paid to mechanisms for checking whether implementations of MHP in receivers are able to fully and correctly decode MHP broadcasts, and an MHP test suite of software is available from ETSI to help receiver manufacturers in this respect. A compatible version of MHP is also available which can be used with non-DVB delivery systems—Globally Executable MHP (GEM) [22]—which at the time of this writing has been accepted in Japan and the United States.

Adoption and Use of the DVB Systems

The DVB-S system was adopted in 1994, and the first DVB broadcast services in Europe started in spring 1995, by pay-TV operator Canal Plus in France. The DVB-T system was adopted later, in 1997. The first DVB-T broadcasts began in Sweden and the United Kingdom in 1998. DVB-T services started in parts of Germany in 2002, and in 2003 Europe's first analog switch-off took place in Berlin. The DVB-S system is used across the world, although in some countries such as Japan and the United States other digital satellite systems are used as well as DVB-S. The DVB-C system is also widely used throughout the world. In the United States, the cable system is similar in some respects to DVB-C but was standardized by the Society of Cable and Telecommunications Engineers (SCTE). The DVB-T system is the least widely used, although the rollout of digital terrestrial television throughout the world has been slower than digital satellite and cable.

DVB in Australia

Because of some interesting regional variations, it is worth noting the implementation of DVB-T in Australia. The variations include, but are not limited to, the following:[2]

- 7 MHz channels are used both at VHF and UHF (*i.e.*, covering Australian channels 6–12 and 28–69). Transmissions may be in either 2 K or 8 K COFDM carrier modes.
- Transport streams include MPEG-encoded video at 25 frames, 50 fields, or 50 frames per second. The video formats broadcast are up to 1920 pixels by 1080 lines.
- Television services transmitted in SDTV will include at a minimum an MPEG-1 Layer II audio stream. This may be in mono, stereo, or stereo with surround components. In addition, any SDTV service may also contain a Dolby Digital™ (AC-3) audio stream with up to 5.1 discrete channels. The preferred audio stream for a television service transmitted in HDTV is a Dolby Digital™ (AC-3) audio stream with up to 5.1 discrete channels. Alternatively, an HDTV service may contain an MPEG-1 Layer II audio stream.
- Transport streams will include closed caption subtitles based on ETSI EN 300 472 or ETSI EN 300 743, or both.
- Transmissions may include data broadcasting as specified in ETSI EN 301 192.
- Transmission filter masks for Australia's 7 MHz channel spacing take into account the relative proximity of lower adjacent channel dual analog sound carriers.
- Australian broadcasters implement single-frequency networks in accordance with ETSI TS 101 191.
- Australian broadcasters may transmit DVB-T modes suitable for reception by mobile and portable receivers.
- Transmissions may include DVB-T hierarchical modes. The mode should be determined or con-

[2]From Australian Standard AS 4599.1–2005; reprinted by permission of SAI Global Ltd. (http://www.sai-global.com).

firmed from TPS signaling and identification of the high-priority and low-priority streams.

- Some Australian broadcasters' transmissions have a +125 kHz or –125 kHz frequency offset.
- Australian broadcasters have implemented a scheme for network identification coding through registration with the DVB Project office and the Australian Broadcasting Authority.

As specified by DVB, modulation is by means of COFDM, using QPSK, 16QAM, or 64QAM. When HDTV is broadcast, 64QAM is used with a 2/3 code rate and 19.4 MB/s net data rate. Also, the channel allocation plan has attempted to assign digital service on a channel adjacent to or within the same band as the analog service, and it is recommended that analog and digital transmitters are co-located.

The digital television conversion plan made by the Australian Broadcasting Authority required that digital and analog broadcasts be simulcast for at least 8 years from the start of digital transmission in a given area, that all programs be transmitted in digital SDTV mode, and that at least 1040 hours per year of programs be transmitted in HDTV mode. This requirement resulted in a so-called "triplecast" scenario when HDTV is transmitted.

Because of the different parameters in use, receivers must (dynamically) identify and receive all COFDM modes, including hierarchical modulation, and should allow for manual selection of modulation parameters. In order to deploy this as seamlessly as possible, Digital Broadcasting Australia (DBA) advises that receivers should check the DVB-T Transmission Parameter Signaling (TPS); in addition, the performance requirements of receivers are mandated, as defined in a digital television receiver standard [23].

Future of the DVB Project

Media development is clearly moving toward greater convergence between different delivery systems, broadcast and point-to-point transmission. The project has already agreed on a number of IP- and convergence-oriented systems—for example, DVB over IP networks [24]. The future lies in the seamless interoperation of digital telephony systems and digital broadcasting and the emerging in-home network environments.

INTEGRATED SERVICES DIGITAL BROADCASTING

Overview

The digital broadcasting system adopted in Japan is Integrated Services Digital Broadcasting (ISDB), which embodies the concept of expanding flexibility across the physical layer. A common format is used overall, while local modulation systems are defined according to transmission and reception characteristics. The history of ISDB goes back to the idea of a "digital broadcasting system for the 21st century," which was conceived at the NHK Science and Technical Research Laboratories in the 1980s. The ISDB of the time was aimed at a flexible system that would be open to new services, such as high-quality audio/video and data broadcasting, through the digitization of valuable broadcasting transmission channels and based on the following fundamental concepts:

- All information related to broadcasting, such as audio/video and multimedia data, is converted to a digital format in an integrated manner for processing.
- All digital signals are available, regardless of the type of transmission media, including satellite, terrestrial, and cable TV networks.
- By using the features of the respective transmission medium, users can easily view high-quality services, or obtain desired information either while at home or on the road.

ISDB has several features that are not available with either DVB or ATSC. One of the initial issues facing multiplexing technology for digital data was how to construct for various services effective, integrated handling methods for audio graphics, which combines HD still pictures and high-quality audio, as well as facsimile broadcasting, high-function teletext, closed captioning, and HDTV service. Thorough research was conducted in areas such as transmission packet length and the control functions needed for a multimedia broadcasting service. The Association of Radio Industries and Businesses (ARIB), the standardizing organization in Japan, examined and evaluated MPEG-2 systems based on these research results, resulting in the ISDB multiplexing scheme adoption.

ISDB is a broadcasting system that realizes HDTV-based multimedia services. It features a transmission scheme designed to make use of respective transmission media through the incorporation of MPEG-2 video/audio for its audio/video coding and MPEG-2 systems for its digital signal multiplexing scheme. The four variants of ISDB are summarized in Table 1.14-4.

Satellite: ISDB-S

Satellite broadcasting using frequencies in the 12 GHz band faces the problem of rain attenuation; precipitation is generated at a considerable hourly rate in Japan, as the country is located in a rainy region. The ISDB-S digital satellite television broadcasting system incorporates a hierarchical transmission scheme that modifies the transmission rate as a rain attenuation countermeasure. It defines a 204 byte slot in the bit stream, capable of allotting signals combined with BPSK/QPSK/trellis-coded 8PSK modulation and inner coding, by slot unit, based on programming data. This hierarchical transmission aims to provide continuous reception of comprehensible program at a low bit rate when severe rain attenuation prevents the transmission of the program in HDTV.

TABLE 1.14-4
Comparison of ISDB Systems

Systems		ISDB-T	ISDB-T$_{SB}$	ISDB-S	ISDB-C
Video		MPEG-2 video	—	MPEG-2 Video	
Audio		MPEG-2 audio AAC			
Multiplex		MPEG-2 systems			
Transmission		Multiple carrier (OFDM)		Single carrier	
Bandwidth		6 MHz	429 kHz	34.5 MHz	6 MHz
Modulation		DQPSK/QPSK/16QAM/64QAM (OFDM)		BPSK/QPSK/8PSK	64QAM
Error Correction	Inner	8PSK: Trellis Others: Convolutional code			—
	Outer	RS			

Terrestrial: ISDB-T

While satellite broadcasting provides nationwide programming, terrestrial broadcasting offers services tailored to each region. The transmission bandwidth for ISDB-T digital terrestrial television broadcasting was determined based on the required condition that HDTV is essential even in terrestrial service and on the requirement that it must coexist with current analog service. Using this transmission bandwidth, the band segmented transmission-OFDM (BST-OFDM) scheme was constructed with the purpose of providing services for mobile and portable reception, together with HDTV service. This scheme defines an OFDM segment by dividing a 6 MHz channel into 13 segments for signal transmission, as shown in Figure 1.14-1.

Data carriers and reference signal carriers are allotted in OFDM subcarrier segments at regular intervals, with the necessary transmission band consisting of a group of segments. Using this segment group unit, QPSK/16QAM/64QAM modulation is used, together with error correction inner coding and time interleaving. Each of these parameters can be changed according to the service purpose such as fixed reception or mobile reception. It is also feasible to realize mobile reception service that employs one segment that can be partially received (disregarding the others) simultaneously with HDTV broadcasting.

Terrestrial Sound: ISDB-T$_{SB}$

Digital terrestrial sound broadcasting employs OFDM segments with a structure identical to that of ISDB-T, sharing the same transmission scheme except for the number of segments. There are two types of transmission: a one-segment scheme and a three-segment

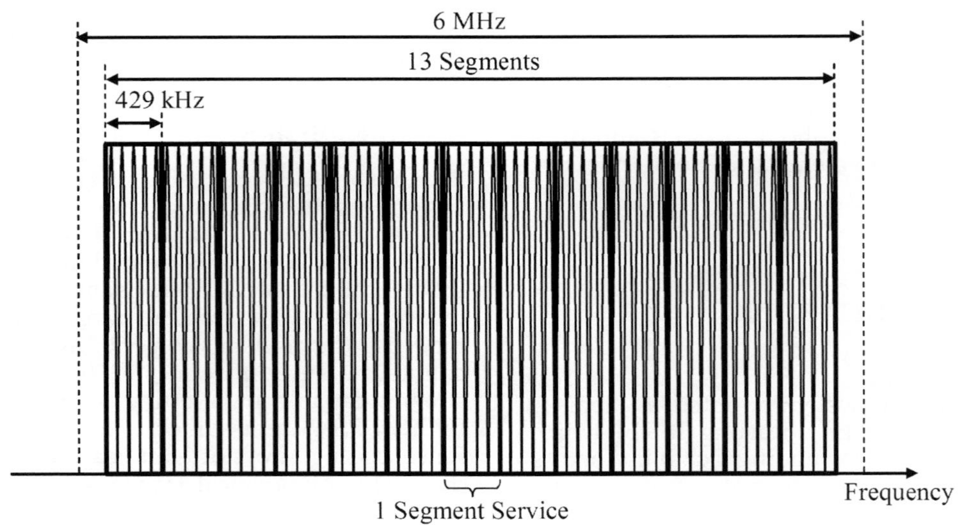

FIGURE 1.14-1 Band segmented structure of ISDB-T.

scheme. The central segment for the three-segment scheme can also be received partially, similar to ISDB-T. This allows a common one-segment receiver to be used with both ISDB-T and ISDB-T_{SB}. In the future, when ISDB-T_{SB} is established nationwide, integrated mobile terminals for the three media of cellular phone, ISDB-T, and ISDB-T_{SB} are expected to appear on the market.

Cable: ISDB-C

Full-scale cable-TV system digitization using ISDB-C started at the same time as digital satellite broadcasting. To transmit digital broadcasting, including HDTV programs, together with the re-transmission of conventional analog broadcasting, various re-transmission schemes were constructed. A *pass-through* scheme is utilized to transmit a digital program over a cable TV network without modulation scheme conversion, in order to employ direct-reception digital broadcasting receivers. The *transmodulation* scheme demodulates received signals first, then remodulates them to 64QAM. In this case, signal processing differs by the number of MPEG-2 transport streams (TSs) included in the received signals, categorized as either a single-TS scheme or a multiple-TS scheme. A *re-mux scheme*, suitable for independent broadcasting service, has also been standardized.

ARIB Standards

All of the ISDB systems have been standardized and published as ARIB standards, as shown in Table 1.14-5. ARIB standards are voluntary technical standards that supplement governmental regulations for telecommunications and broadcasting radio systems and are set for the purpose of guaranteeing compatibility of radio facilities and transmission quality as well as offering greater convenience to radio equipment manufacturers and users.

Service Features of ISDB-T for Digital Terrestrial Television

ISDB-T utilizes the OFDM transmission system, which is effective for single-frequency networks (SFNs), and is robust for multipath interference. Because of heavy usage of TV frequencies and ghost image interference by high-rise buildings, OFDM was chosen for Japan. ISDB-T separates the 6 MHz channel into 14 segments, using 13 segments for signal transmission and the remaining segment for a guard band between channels. Each segment occupies 429 kHz of bandwidth. By using 13 segments flexibly, multi-programming is possible. Various combinations are possible, including:

- 1 HDTV (12 segments) + mobile service (1 segment)
- 3 SDTV (3 × 4 segments) + mobile service (1 segment)

ISDB-T has some commonality with DVB-T such as the OFDM transmission system; however, the segment structure, time interleaving, and transmission and multiplexing configuration control (TMCC) are unique features.

Mobile Reception of ISDB-T

One-Segment Mobile Reception

In the ISDB-T system, mobile reception service for handheld terminals is also possible by using one segment located in the center of 13 segments. In this

TABLE 1.14-5
ARIB Standards

Systems	ISDB-T	ISDB-S	ISDB-T_{SB}
Transmission	STD-B31	STD-B20	STD-B29
Source coding and multiplex	STD-B32 Image encoding, sound encoding, and multiplexing		
Service information	STD-B10 Program lineup information		
Data broadcasting	STD-B24 (presentation engine; BML) STD-B23 (execution engine; GEM-based)		
Server-type broadcasting	STD-B38 System based on home servers		
Access control	STD-B25 Conditional access		
Receiver	STD-B21		STD-B30
Operational guideline	TR-B14	TR-B15	TR-B13

TABLE 1.14-6
One-Segment Parameters

Video	Bit rate	Max 384 kbps (operation example: 256 kbps)
	Encoding method	MPEG-4 AVC/ITU-T H.264
	Format	QVGA 4:3 or 16:9
	Frame rate	15 fps
Audio	Bit rate	Max 256 kbps (operation example: 48 kbps)
	Encoding method	MPEG-2 AAC
	Quantization	16 bit @ 24 kHz or 48 kHz sample rate
Still image	Encoding method	JPEG, GIF, aGIF
Caption	Number of characters/ number of lines	12 words × 4 lines (vertical display) 16 words × 3 lines (horizontal display)

manner, the service has the capacity for 412 kbps data transmission using QPSK modulation (see Table 1.14-6).

The strength of the ISDB-T system is that a 6 MHz channel can simultaneously carry an HDTV program using 12 segments and a one-segment video signal for cellular phone service. Stable TV reception has been demonstrated over a cellular phone by means of one-segment TV broadcasting using MPEG-4 AVC/H.264 video compression coding. A commercial service for simultaneous broadcasting of 12-segment HDTV and one-segment QVGA video programs was implemented in April 2006.

Cellular phones are a convenient tool for providing information, as they can display TV and data broadcasting within the same display. Recognizing that the data capacity for one-segment TV broadcasting is small and that all cellular phones in Japan have a wireless Internet connection facility, integrated services of data broadcasting and the wireless Internet are expected to be introduced to supplement data capacity and provide detailed information via the Internet. This kind of application is expected as a converged business model of broadcasting and communication.

HDTV Mobile Reception

Though one-segment TV broadcasting is appropriate in size for cellular phones and PDAs, it is not feasible for a large-screen display that requires higher resolution—and the demand for a large-screen display for mobile reception is expected to be high. Mobile reception of full HDTV pictures was considered difficult because the receiver direction changes rapidly and the signal frequency is often affected because of Doppler shift. To overcome these difficulties, an HDTV mobile reception system was developed using reception direction-control techniques and Doppler-shift compensation techniques, which enable the transmission parameters for fixed HDTV reception to be used for mobile reception (see Figure 1.14-2). The stability and feasibility of these techniques have been proved viable in a variety of applications. By utilizing prototype receivers installed in a large bus, public demonstrations were conducted in December 2004 showing stable HDTV reception at speeds of 100 km/hr. HDTV reception in sightseeing buses, trains, and ferryboats is expected to be realized together with car navigation systems and traffic information services within a few years.

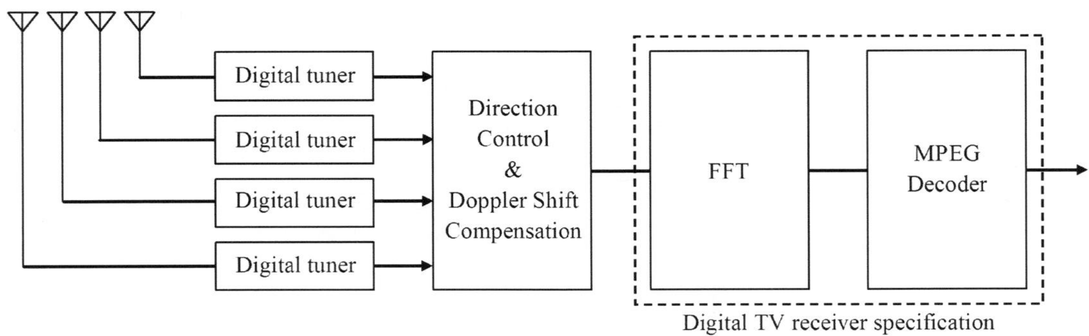

FIGURE 1.14-2 HDTV mobile reception system.

TABLE 1.14-7
Data Broadcasting in the World

	Presentation Engine (Declarative Content Format)	Execution Engine (Procedural Content Format)
Japan	BML	ARIB-J
Europe	DVB-HTML	MHP
United States, S. Korea	ACAP-X	ACAP-J

New Services of ISDB

Data Broadcasting

Data broadcasting services have started in Japan through both digital satellite and digital terrestrial television broadcasting. Some data programs are related to TV programs, providing supplemental information; others provide independent content, such as news, weather information, traffic reports, and other useful information. The viewer can obtain a variety of information by using the remote control. While digital satellite data broadcasting provides uniform information nationwide, regionally based digital terrestrial data broadcasting can convey data regarding the local community. The Broadcast Markup Language (BML) is used to describe these data programs in Japan. BML is based on the eXtensible Markup Language (XML), which has become the Internet standard and has growing utility in the PC environment. The different data broadcasting systems used in the world are shown in Table 1.14-7.

Through data broadcasting, interactive services are possible that allow the viewer to send information by using a communication line; for example, while watching television, a viewer can respond to a questionnaire, make a request, or participate in a program such as a quiz show. Using this function, the viewer not only views the television program, but also participates in it. These services are realized by a standard communication function in digital broadcasting receivers. Some programs are already adopting this style of viewer participation.

Server-Type Broadcasting

A new service called *server-type broadcasting* is now being developed. It provides services that take full advantage of the characteristics of digital broadcasting, communications, and digital large storage devices. In server-type broadcasting, the additional metadata information is sent out together with the broadcasting program and can be used by the receivers to browse and retrieve a desired program anytime; for example, searching for specific scenes of a program or extracting a digest version for digest viewing of a movie is possible. Moreover, the viewer can obtain detail information of the program or related program via the communication line (see Figure 1.14-3).

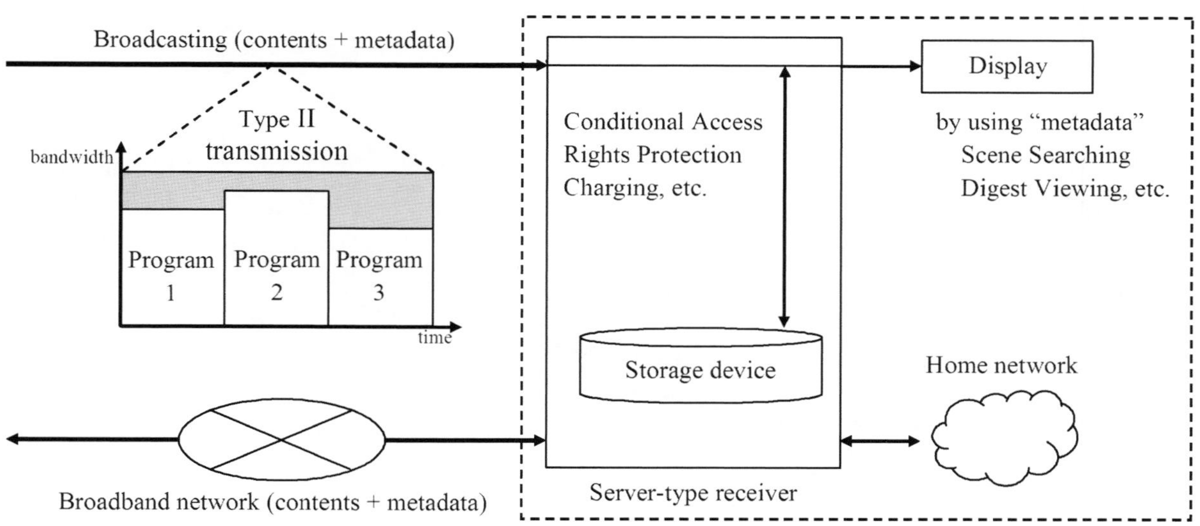

FIGURE 1.14-3 Server-type broadcasting.

The two types of content transmission systems in server-type broadcasting are Type I and Type II:

- Type I is a streaming service. Programs are sent in real time that can be viewed on a conventional television set in much the same fashion as with analog broadcasting; however, once they are stored in a storage device, with the use of metadata a variety of ways of viewing such as digest-viewing are possible.
- Type II is a file-type service. Data are first stored in a storage device and can then be viewed whenever desired. The time required for playback is different from the time required for receiving data. This transmission method is useful to maximize the efficiency of transmission capacity.

This type of sophisticated digital broadcasting service is planned to be launched in 2007. As the broadband service spreads widely, server-type broadcasting will further extend functionality and services by including not only the storage of received broadcasting contents but also new services provided through broadband linkage.

Status of Digital Terrestrial Television Service

Strategy to Promote Digital Terrestrial Television Broadcasting

Terrestrial television stations are part of the fundamental media in Japan. With 48 million households and 100 million TV sets, terrestrial broadcasters have established many relay stations to provide maximum coverage throughout the mountainous archipelago; there are more than 3000 transmitter sites. Due to the heavy usage of UHF channels by existing analog relay stations, it is impossible to assign digital channels without migrating analog stations, and many analog TV channels are being forced to shift to other UHF channels; therefore, digitization of terrestrial television broadcasting is a project of high importance to Japan. The government issued digital TV licenses to TV broadcasters in November 2003 with the following conditions and requirements:

- Assign 6 MHz channels for incumbent terrestrial broadcasters.
- Provide at least two-thirds digital/analog simultaneous broadcasting each day.
- Broadcast half of all programming each week in HDTV.
- Expand digital broadcasting coverage in the same areas as analog broadcasting, as soon as possible, until the analog service is discontinued.

Since the start of digital terrestrial television broadcasting in December 2003 in the three major metropolitan areas of Tokyo, Nagoya, and Osaka, service coverage has expanded; as of December 2004, 18 million households (38%) were covered. TV stations in eastern Japan began digital broadcasting in 2005, with 28 million households (60%) expected to be covered, and a nationwide service will be put in place in 2006, with coverage reaching 38 million households (82%). Smaller relay stations will be set up until 2011, when full digital migration will be completed. The switch-off of analog broadcasting is scheduled for 2011, which is mandated by the Radio Law. After that, VHF and the upper portion of UHF TV channels should be reallocated for mobile communication service.

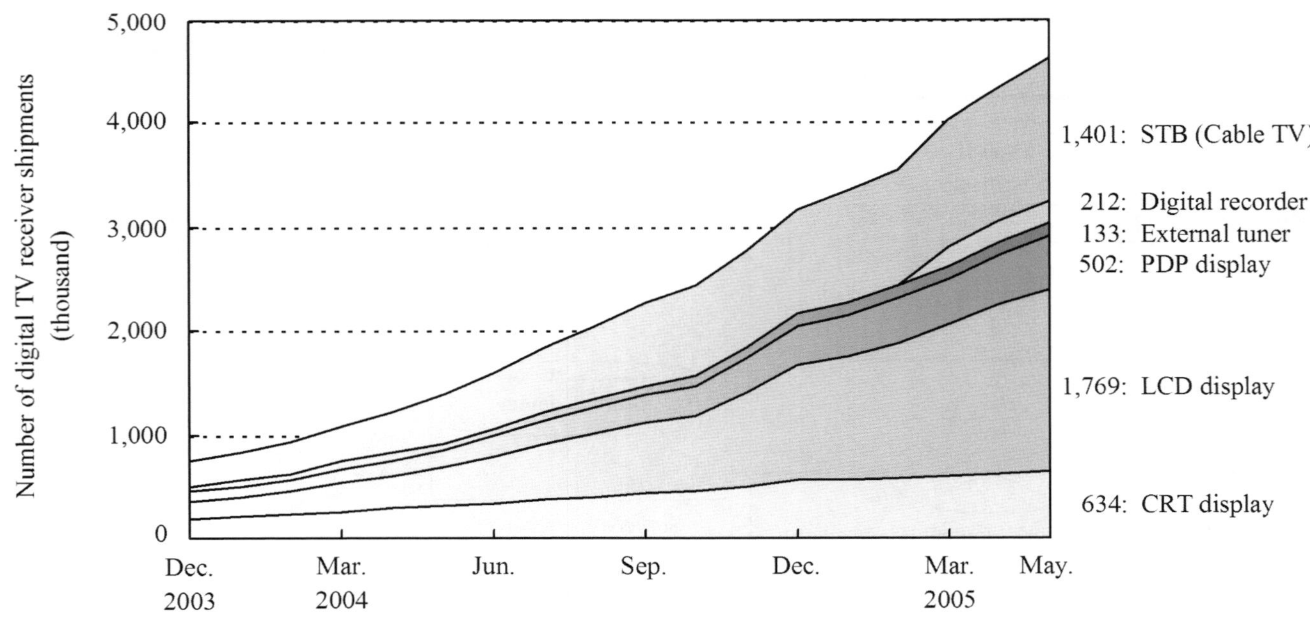

FIGURE 1.14-4 Growth of DTV receivers in Japan.

Penetration of Digital Terrestrial Television Broadcasting

HDTV enables high-quality video and high-fidelity sound with presence and is one of the key features to attract consumers to migrate from analog to digital. Broadcasters have already invested heavily in HDTV programming production facilities and enhancing the quality of programs. In 2004 and 2005, the number of pure HD programs produced in the 1080i format increased and covered 90% of all programs by the public broadcaster NHK and approximately 60% of all programs by the private broadcasters. Moreover, all broadcasters are providing data broadcasting carrying a variety of life information and supplemental program information.

From the start of digital terrestrial television broadcasting until mid-2006, about 15 million digital TV sets—including 8.5 million digital TV receivers with tuners and 3 million cable QAM HDTV tuners—have been shipped, as can be seen in Figure 1.14-4. In addition, various types of receivers, such as DVD/HDD recorders, PCs, and car navigation systems with digital tuners, are on the market.

As of 2005, more than 90% of digital TV receivers were HDTV-integrated receivers. All flat-panel display TVs including plasma display panel (PDP) and liquid crystal display (LCD) above 22 inches are HDTV-integrated receivers. The market trend is focused on integrated HDTV. All digital HDTV sets are equipped with the function of data broadcasting, and many sets have an Internet Web browsing function; people enjoy Web browsing without facing a complex PC function. In the near future, digital television is expected to evolve into a home gateway to the information and communications technology (ICT) society, converging broadcasting and communication functions.

CHINESE SYSTEM

After initially considering existing DTV systems, the Chinese government decided to develop their own system, due to a number of technological and economic factors. In 1999, a new terrestrial DTV transmission system was planned by the State Council of China. Five proposals were presented for evaluation, together with lab and field tests. From 2004 to 2005, a merging scheme was developed based on three proposals:

1. A single-carrier offset-QAM system, Advanced Digital Television Broadcast–Terrestrial (ADTB–T), developed by a group at Shanghai Jiao Tong University in Shanghai
2. A non-single-carrier system using time-domain synchronous-OFDM (TDS-OFDM), digital multimedia broadcast–terrestrial (DMB-T), developed by a group at Tsinghua University in Beijing
3. A multi-carrier COFDM system, terrestrial interactive multi-service infrastructure (TiMi), developed by a group at the Academy of Broadcasting Science (ABS) in Beijing

At the beginning of 2006, the draft standard was in its final stage, representing a combination of the above systems. Following comments and the results of lab and field testing, the draft standard was published in late 2006. Transmission of the signal allows several different options to be employed: a single-carrier mode, or a multiple-carrier mode with 3780 carriers with carrier spacing of 2 kHz. By some accounts, the latter mode is a hybrid of single-carrier and multiple-carrier modulation (*i.e.*, not a pure multiple-carrier system). As with other DTV systems, the provision of different options should in principle allow different modes to be used in different areas.

At the time of this writing, possible acronyms for the new transmission standard included CDMB-T, DTMB, and DMB-T/H, and the basic elements were as follows:

1. A data frame structure uses a PN sequence as the frame header for synchronization, channel estimation, and other purposes.
2. A layered frame structure features a calendar day frame, minute frame, super frame, and signal frame, from top to bottom. The minute frame synchronizes with real time, a function developed for power saving. The super-frames are composed of data frames and system information.
3. Two subcarrier options are:
   a. SC = 1 (direct pass) for single-carrier operation
   b. SC = 3780 (with IDFT module) for multiple-carrier operation
4. Modulation can be 4QAM, 16QAM, 32QAM, or 64QAM.
5. FEC uses a BCH outer code and LDPC inner code, with time and frequency interleaving options.
6. Occupied bandwidth of 7.56 MHz in an 8-MHz channel offers data throughput of up to 32 Mbps.

The system layer used is MPEG-2, with video coding similar to AVC (MPEG-4 Part-10) or VC-1, and audio coding similar to AAC (ISO/IEC 13818-7) or Dolby E-AC-3. For video and audio coding, MPEG-2 compression was being used provisionally, while a new standard (Audio-Video Standard, or AVS) was in preparation. The AVS video coding draft standard was published in March 2006, while the AVS audio coding draft standard was expected to be published soon afterward.

SUMMARY

There has been a cyclical pattern in the development of digital systems whereby each new system builds on previous systems and takes advantage of prevailing advances in potential receiver complexity. This is an understandable process, and this pattern may be expected to continue. Although the promise of a single worldwide digital television standard has been elusive, the systems developed have nonetheless succeeded in bringing about a vast change in the worldwide delivery of audio/visual entertainment.

ACKNOWLEDGMENTS

Thanks to Robert Graves of the ATSC Forum for guidance on terrestrial DTV deployment in South America and to Professor Xu Mengxia of Peking University and Professor Jian Song of Tsinghua University for information on the status of terrestrial DTV in China.

References

[1] Digital Video Broadcasting Project, http://www.dvb.org.
[2] Multimedia Home Platform, http://www.mhp.org.
[3] European Telecommunications Standards Institute, www.etsi.org.
[4] ISO/IEC Standard 13818-1, 2, and 3: Information Technology—Generic Coding of Moving Pictures and Associated Audio.
[5] ETSI TR 101 154: Digital Video Broadcasting (DVB)—Implementation Guidelines for the Use of MPEG-2 Systems, Video and Audio in Satellite, Cable and Terrestrial Broadcasting Applications.
[6] ETSI EN 300 421: Digital Video Broadcasting (DVB)—Framing Structure, Channel Coding and Modulation for 11/12 GHz Satellite Services.
[7] ETSI EN 300 429: Digital Video Broadcasting (DVB)—Framing Structure, Channel Coding and Modulation for Cable Systems.
[8] ETSI EN 300 473: Digital Video Broadcasting (DVB)—Satellite Master Antenna Television (SMATV) Distribution Systems.
[9] ETSI EN 300 744: Digital Video Broadcasting (DVB)—Framing Structure, Channel Coding and Modulation for Digital Terrestrial Television (DVB-T).
[10] ETSI TR 101 190: Implementation Guidelines for DVB-T.
[11] ETSI EN 300 749: Digital Video Broadcasting (DVB)—Microwave Multipoint Distribution Systems (MMDS) below 10 GHz.
[12] ETSI EN 300 748: Digital Video Broadcasting (DVB)—Multipoint Video Distribution Systems (MVDS) at 10 GHz and above.
[13] ETSI EN 301 701: Digital Video Broadcasting (DVB)—OFDM Modulation for Microwave Digital Terrestrial Television.
[14] ETSI EN 302 307: Digital Video Broadcasting (DVB)—Second Generation Framing Structure, Channel Coding and Modulation Systems for Broadcasting, Interactive Services, News Gathering and Other Broadband Satellite Applications.
[15] ETSI EN 302 304: Digital Video Broadcasting (DVB)—Transmission System for Handheld Terminals (DVB-H).
[16] ETSI ETR 289: DVB Common Scrambling Distribution Agreements.
[17] ETSI EN 50221: Common Interface Specification for Conditional Access and Other Digital Video Broadcasting Decoder Applications.
[18] ETSI TS 101 699: Digital Video Broadcasting (DVB)—Extensions to the Common Interface Specification.
[19] ETSI TS 101 197-1: Digital Video Broadcasting (DVB)—SimulCrypt, Part 1: Head-End Architecture and Synchronization.
[20] ETSI EN 301 192: Digital Video Broadcasting (DVB)—DVB Specification for Data Broadcasting.
[21] ETSI TS 101 812: Digital Video Broadcasting (DVB)—Multimedia Home Platform (MHP) Specification 1.0.2.
[22] ETSI TS 102 819: Digital Video Broadcasting (DVB)—Globally Executable MHP Version 1.0.2 (GEM 1.0.2).
[23] AS 4933.1–2005: Digital Television—Requirements for Receivers—VHF/UHF DVB-T Television Broadcasts.
[24] ETSI TS 102 033: Architectural Framework for DVB Services on IP.

Bibliography

Wood, D., Satellites, science, and success—the DVB story, *EBU Techn. Rev.*, Winter 1995.
Reimers, U., *DVB: The Family of International Standards for Digital Video Broadcasting*, Second ed., Springer, Berlin, 2005.
Morris, S. and Smith-Chaigneau, A., *Interactive TV Standards: A Guide to MHP, OCAP, and JavaTV*, Focal Press, Burlington, MA, 2005.
DVB Standards and Specifications, Version 7.0, CD issued by the DVB Project Office, August 2004.
Australian Standard AS 4599.1–2005: Digital Television—Terrestrial Broadcasting, Part 1: Characteristics of Digital Terrestrial Television Transmissions, April 29, 2005.
A Review into High Definition Television Quota Arrangements, Issues Paper, Australian Government, Department of Communications, Information Technology, and the Arts, May 2005.
Wan, E., DTV Standards Submitted to NDRC, *Pacific Epoch*, July 11, 2005.
Masaharu Tanaka, China embraces digital TV, *Nikkei Electronics Asia*, February 2004.
Jian Song, et al., *Technical Review of the Chinese Terrestrial Broadcasting Standard*, presented at the 56th Annual IEEE Broadcasting Symposium, Washington, DC, 2006.

C H A P T E R

1.15

Digital Audio Standards and Practices

CHIP MORGAN
CMBE
El Dorado Hills, California

RANDALL HOFFNER
ABC, Inc.
New York, New York

Updated for the 10th Edition by

TIM CARROLL
Linear Acoustic Inc.,
Lancaster, Pennsylvania

INTRODUCTION

Digital audio technology has supplanted analog audio technology in U.S. television and radio production and broadcast facilities. Like digital video, digital audio offers many advantages in production, editing, distribution, and routing. Digital audio is remarkably robust and far less susceptible to degradation from hum, noise, level anomalies, and stereo phase errors than analog audio. Each analog audio recording generation and processing step adds its own measure of noise to the signal, but in the digital domain audio is not subject to such noise buildup. However, perceptual coding artifacts can be a problem; see Chapter 3.7, "Digital Audio Data Compression," for additional information.

Digital audio may be stored on magnetic, optical, or magneto-optical discs and in solid state memories. When audio samples have been reduced to a series of numbers, processing and manipulation become largely mathematical operations, easily accommodated by microprocessors. Nonlinear editing is an example of a process that cannot be done in the analog domain. The technology of digital compression creates new economies in the storage and transport of digital audio and permits the broadcast of digital audio within a reasonable segment of spectral bandwidth.

The distribution and routing of audio in the digital domain present the broadcaster with new options as well, such as the capability to embed digital audio within a serial digital video signal, facilitating the carriage of video and multiple audio channels on a single coaxial cable. Although digital audio presents its own unique set of challenges, its advantages far outweigh its disadvantages.

Digital audio systems are inherently free of the hum and noise problems that can invade analog audio systems. The nature of the digital domain gives rise to a new set of considerations for the facility planner and designer. Digital audio signals operate in the multiple megahertz frequency domain that video engineers are well acquainted with, raising such considerations as signal reflections and impedance discontinuities. In digital audio system engineering, just as in analog audio system engineering, cognizance of the potential pitfalls to be avoided and the application of good engineering practices will result in facilities that function well.

Since the 9th edition of the *NAB Engineering Handbook* was published, digital audio standards and practices have not seen a dramatic change in their basic definitions, but their use has grown in popularity. Digital audio is no longer a format used only by the largest broadcasters and postproduction facilities; it has come to be relied upon as the only way to handle modern broadcast audio requirements. Digital audio has proven to be as robust and flexible as it was originally designed to be. Routing, distribution, storage, and signal processing have advanced to the point of making it difficult or impossible to find analog versions of these processes with the same features and capabilities. With multichannel and surround sound audio having gained remarkable popularity, the thought of handling these signals in the analog domain quickly becomes overwhelming and the efficiency and consistency of digital audio truly make it possible to process multichannel sound with the precision required.[1]

Some useful new formats and standards have emerged along with necessary revisions to existing standards. For example, new transport methods based on TCP/IP computer networks that, although not yet standardized, are gaining popularity, and the requirement for audio metadata in digital television is presenting new challenges. In addition, new audio data rate reduction (i.e., compression) schemes have found their place in the chain, and better standards have been developed to support them.

Proper synchronization of the digital audio plant still seems to be a challenge. It is difficult for audio-centric facilities and staff to realize that they now have timing requirements like their video counterparts, but AES-11 provides accurate guidance that, when followed, results in a properly timed plant and high-quality audio.

DIGITAL AUDIO STANDARDS

Following are standards that are relevant to digital audio in broadcast facilities. Many of them have been presented before and remain important parts of standardized digital audio systems. In recent years, there have been significant advances made by the Society of Motion Picture and Television Engineers (SMPTE), and it is necessary to consider these standards along with their Audio Engineering Society (AES) counterparts if they exist.

AES3-2003

AES3-2003 is the "AES Recommended Practice for Digital Audio Engineering—Serial Transmission for Two-Channel Linearly Represented Digital Audio Data." This is the baseline standard for digital audio developed by the AES and the European Broadcasting Union (EBU) and is commonly referred to as the AES/EBU standard.

AES3 defines a digital protocol and physical and electrical interfaces for the carriage of two discrete audio channels, accompanied by various housekeeping, status, and user information in a single serial digital bitstream. As its title indicates, AES3 was designed to carry linearly quantized (uncompressed PCM) digital audio. Compressed digital audio may be carried on the IEC 958 digital audio interface.[2] IEC 958 is identical to AES3 in protocol, but can have slightly different electrical characteristics for support of consumer electronics. It addresses a professional implementation (AES/EBU) and a consumer implementation (S/PDIF). The AES3 interface has the capacity to carry linearly sampled digital audio at bit depths from 16 to 24, data descriptive of such factors as channel status and sample validity, along with parity checking data and user data. Total bit count per sample, including audio and housekeeping, is 32 bits. An ancillary standard, AES5 (discussed later), recommends use of the professional audio sample rate of 48 kHz on AES3, while recognizing the use of sample rates of 44.1 kHz, 32 kHz, and 96 kHz. AES3 carries audio samples using time-division multiplexing, in which samples from each of the two represented audio channels alternate.

[1]It would be virtually impossible to handle multichannel audio to the precision required if the audio were handled via traditional analog means.

[2]Since 1997, the IEC document numbering system has added 60000 to the old IEC standard number, so the official number of this standard is now IEC 60958. Because the three-digit number is more widely known, this number will be used descriptively within the chapter.

Data Structure

The data carried on the AES/EBU interface is divided into blocks, frames, and subframes. An AES block is constructed of 192 frames, each frame being composed of two subframes, each subframe containing a single audio sample. A subframe begins with a preamble that provides sync information and describes what type of subframe it is, and ends with a validity bit, a user bit, a channel status bit, and a parity bit.

The subframe is divided into 32 *time slots*, each time slot being one sample bit in duration. The first four time slots are filled with a 4-bit preamble. The 24 time slots following the preamble may be filled in one of two ways. As shown in Figure 1.15-1(a),[3] an audio sample word of up to 24 bits may fill all the time slots. Figure 1.15-1(b)[4] illustrates that the first four time slots of the audio sample word space may be filled with auxiliary bits, which can represent user data or low-quality audio for informational or cueing purposes, for example. In all cases, the audio word is represented least significant bit (LSB) first, most significant bit (MSB) last. If digital audio words of bit depth less than the maximum are represented, the unused bits are set to logic 0. Time slots 28, 29, 30, and 31 are filled with a validity bit (V), a user bit (U), a channel status bit (C), and a parity bit (P), respectively. The subframes are assembled into frames and blocks as shown in Figure 1.15-2.

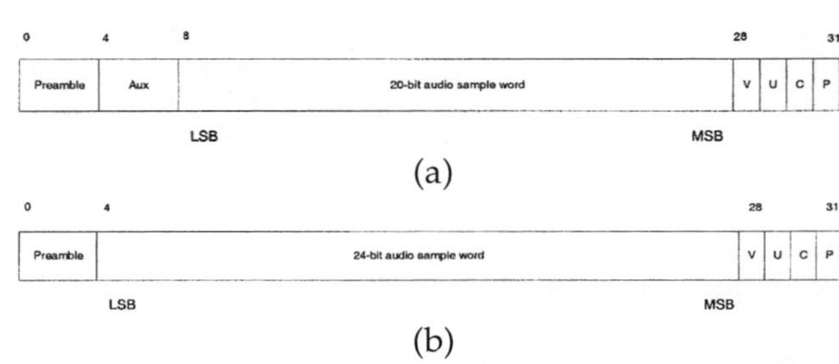

FIGURE 1.15-1 Subframe formats: (a) 16–20 bit audio word; (b) 16–24 bit audio word.

[3]AES3–1992 (r1997), AES Recommended Practice for Digital Audio Engineering—Serial Transmission Format for Two-Channel Linearly Represented Digital Audio Data, Figure 1.

[4]*Ibid.*

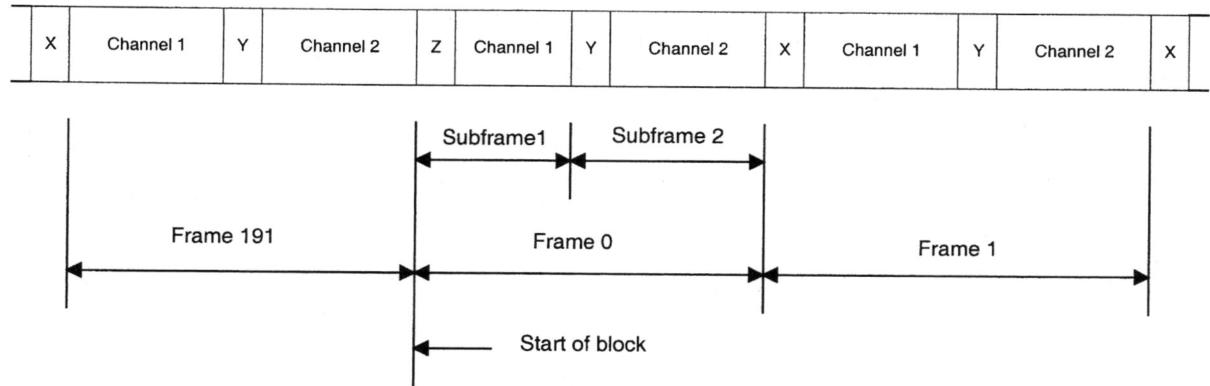

FIGURE 1.15-2 AES3 block and frame structure.

Each subframe begins with one of three preambles. The first subframe in the 192-frame block, a Channel 1 subframe, starts with Preamble Z. All other Channel 1 subframes in the block start with Preamble X. All Channel 2 subframes start with Preamble Y. Figure 1.15-2 represents the last frame of a block and the first two frames of the following block. Subframe 1 of Frame 0, the first subframe of the block, begins with Preamble Z, uniquely identifying the beginning of the block. After the first subframe, the successive subframes are marked by Preamble Y and Preamble X, to identify Channel 2 and Channel 1 subframes, respectively.

A frame, consisting of two 32-bit subframes, is made up of 64 bits, and the data rate of the interface signal may be readily calculated by multiplying the sampling rate times 64. In the case of the 48 kHz sample rate, the total data rate of the signal is 64 times 48,000 or 3.072 Mbps. As will be explained later, the interface employs an embedded clock signal that is twice the sample rate, making the actual frequency of this signal about 6.1 MHz.

Encoding

All time slots except the preambles are encoded using biphase-mark coding to prevent the transmission of long strings of logic 0's or logic 1's on the interface, and thereby minimize the dc component on the transmission line; facilitate rapid clock recovery from the serial data stream; and make the interface insensitive to the polarity of connections. The preambles intentionally violate the rules of biphase-mark coding by differing in at least two states from any valid biphase code to avoid the possibility of other data being mistaken for a preamble. Biphase-mark coding requires a clock that runs at twice the sample rate of the data being transmitted, and each bit that is transmitted is represented by a symbol that is composed of two binary states. Figure 1.15-3 illustrates these relationships.

The top sequence of Figure 1.15-3 illustrates the interface clock pulses, running at a speed twice the source coded sample rate. The middle sequence shows the source coding, which is the series of pulse code modulated (PCM) digital audio samples. The bottom sequence shows how the source coded data is represented in biphase-mark coding. In biphase-mark coding, each source coded bit is represented by a symbol that is composed of two consecutive binary states. The first binary state of a biphase-mark symbol is always different from the second state of the symbol preceding it. A logic 0 is represented in biphase-mark coding by a symbol containing two identical binary states. A logic 1 is represented in biphase-mark coding by a symbol containing two different binary states. This relationship may be seen by examining the first full source coding bit at the left in the figure, which is a logic 1. Note that the duration of this bit is two clock pulses. Because the symbol immediately before it ended with a logic 0, the biphase-mark symbol representing it begins with a logic 1. As the bit to be transmitted is a logic 1, the second state of the biphase-mark symbol representing it is different from the first, a logic 0. The second source coded bit to be transmitted is a logic 0. Its first biphase-mark binary state is a logic 1, because the immediately previous state was a logic 0, and the second state is also a logic 1. The fact that the first binary state of a biphase-mark signal is always different from the last binary state of the previous symbol ensures that the signal on the interface does not dwell at either logic 0 or logic 1 for a period longer than two clock pulses. Because biphase-mark coding does not depend on the absolute logic state of the symbols representing the source coded data, but rather on their relative states, the absolute polarity of a biphase-mark coded signal has no effect on the information transmitted, and the interface is insensitive to the polarity of connections.

Ancillary Data

The last four time slots in a subframe are occupied by various housekeeping and user data. The validity bit (V) indicates whether the audio sample word is suitable for conversion to an analog audio signal. The channel status bit (C) from each subframe is assembled into a sequence spanning the duration of an

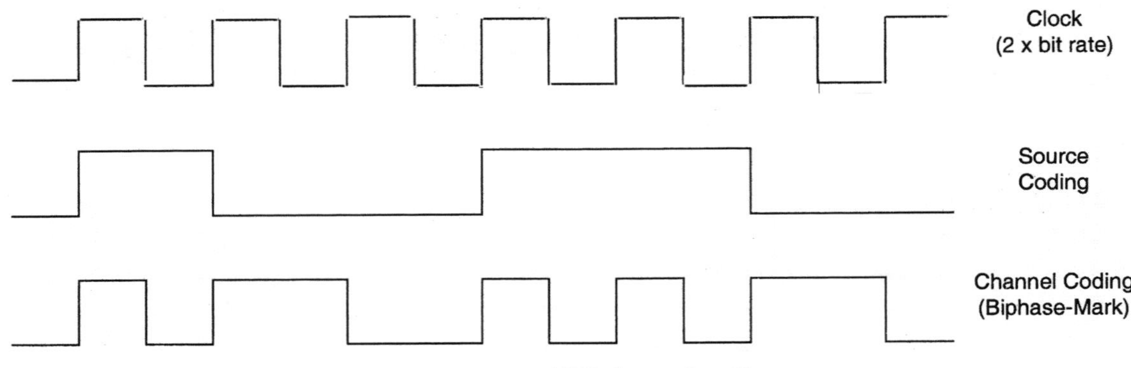

FIGURE 1.15-3 AES channel coding.

entire AES3 block, and these 192 bit blocks of channel status data describe a number of aspects of the signal. Examples of channel status data include

- the length of audio sample words,
- sampling frequency,
- sampling frequency scaling flag,
- number of audio channels in use,
- emphasis information,
- consumer or professional interface implemented,
- audio or data being transmitted on the interface,
- a variety of other possible information.

The 192-bit channel status bits (per block) are subdivided into 24-byte units. There is a separate channel status block for each audio channel, so channel status may be different for each of the audio channels. User data, or U-bits, may be used in any way desired. The parity bit (P) facilitates the detection of data errors in the subframe by applying even parity, ensuring that time slots 4–31 carry an even number of logic 1's and logic 0's.

Electrical Interface

The electrical interface specified by AES3 is a two-wire transformer balanced signal. The AES interface was devised by audio engineers, with the intent of creating a digital audio signal that could be carried on the same balanced, shielded, twisted pair cables and XLR-3 type connectors that are used for analog audio signals. The specified source impedance for AES3 line drivers and the specified input impedance for AES3 line receivers is 110 Ω, which is the approximate characteristic impedance of shielded twisted pair cable as used for analog audio. The permitted signal level on the interface ranges from 2–7 V peak-to-peak.

The balanced, twisted pair electrical interface can give rise to some problems in implementation. XLR type connectors and audio patch panels, for example, are not impedance matched devices. This is not critical when the highest frequency of interest is 20 kHz, but it can cause serious problems when a 6 MHz signal must be passed. These considerations, plus the familiarity of television engineers with unbalanced coaxial transmission of analog video, and the need for higher connector density for a given product size generated the requirement for standardization of an unbalanced, coaxial electrical interface for the AES3 signal. Such an electrical interface is standardized in SMPTE 276M, which describes carriage of the AES/EBU interface on standard 75 Ω video cable using BNC connectors, at a signal level of 1 V peak-to-peak. The fact that the 110 Ω balanced and 75 Ω unbalanced signal formats coexist in many systems frequently presents the requirement to translate between these two signals. Devices to perform such translations are readily available, and SMPTE 276M has an informative annex explaining how to build them. For density and compatibility issues, most modern multichannel audio equipment is being designed to support SMPTE 276M.

AES-2id–1996 (r2001) is an information document containing guidelines for the use of the AES3 interface. *AES-3id–2001* is an information document containing descriptive information about the unbalanced coaxial interface for AES3 audio.

AES5–2003

AES5–2003 is the "AES Recommended Practice for Professional Digital Audio—Preferred Sampling Frequencies for Applications Employing Pulse-Code Modulation." This companion document to AES3 contains the recommended digital audio sample rate for signals to be carried on the interface. The professional digital audio sample rate of 48 kHz is recommended, with recognition given to the use of the compact disc sample rate of 44.1 kHz, a low bandwidth sample rate of 32 kHz, and higher bandwidth sampling frequencies, also referred to as Double Rate (62–108 kHz) and Quadruple Rate (124–216 kHz) for applications requiring a higher bandwidth or more relaxed anti-alias filtering.

SMPTE EG 32, engineering guideline on AES/EBU audio emphasis and sample rates for use in television systems, also recommends that the 48 kHz sample rate be used. Variations on these sample rates are encountered. Varispeed operation requires the ability to adjust these sample rates by about +/− 12%, and of course, accommodation to 59.94 Hz video requires operation at 48 kHz/1.001.

AES10–2003

AES10–2003 is the "AES Recommended Practice for Digital Audio Engineering—Serial Multichannel Audio Digital Interface (MADI)." MADI is a multichannel digital audio interface that is based on AES3. It is designed for the carriage of a maximum of 64 audio channels (at 48 kHz sample rate) on a single coaxial cable or optical fiber. MADI preserves the AES3 subframe protocol except for the preamble. A MADI frame is composed of 64 channels, which are analogous to AES3 subframes. Each MADI channel contains 32 time slots, as does an AES3 subframe. The first four time slots contain synchronization data, channel activity status (channel on/off), and other such information. The following 28 time slots are filled in the same way as in an AES3 subframe—24 audio bits, followed by a V bit, a U bit, a P bit, and a C bit.

The MADI coaxial cable interface is based on the fiber distributed digital interface (FDDI) standardized in ISO 9314, for which chip sets are available. Data is transmitted using non-return-to-zero inverted (NRZI), polarity-free coding and a 4–5 bit encoding format, in which each channel's 32 bits are grouped into 8 words of 4 bits each, and each 4-bit word is then encoded into a 5-bit word. The data rate on the interface is a constant 125 Mbps, with the payload data rate running between approximately 50 and 100 Mbps, depending on the sample rate in use. Sample rates may vary from 32 to 96 kHz +/–12.5%. The specified coaxial cable length for the MADI signal is up to 50 m. A standard for carriage on optical fiber is under consideration.

MADI finds frequent use in multitrack audio facilities, for example, as an interface between multitrack audio recorders and consoles. It is conceivable that the MADI interface could be transmitted over very long distances, using, for example, a synchronous optical network (SONET) circuit.

AES-10id–2005 is an information document containing engineering guidelines for the implementation and use of the MADI interface.

AES11–2003

AES11–2003 is the "AES Recommended Practice for Digital Audio Engineering—Synchronization of Digital Audio Equipment in Studio Operations." This document describes a systematic approach to the synchronization of AES3 digital audio signals. Synchronism between two digital audio signals is defined as that state in which the signals have identical frame frequencies, and the timing difference between them is maintained within a recommended tolerance on a sample-by-sample basis.

AES11 recommends that each piece of digital audio equipment has an input connector that is dedicated to the reference signal. Four methods of synchronization are proposed: (a) the use of a master digital audio reference signal (DARS), ensuring that all input/output equipment sample clocks are locked to a single reference; (b) the use of the sample rate clock embedded within the digital audio program signal that is input to the equipment; (c) the use of video, from which a DARS signal is developed; and (d) the use of GPS to reference a DARS generator providing frequency and phase from one-second pulses, as well as time-of-day sample address codes in bytes 18–21 of channel status. Methods (a), (b), and (c) are preferred for normal studio practice, as method (b) may increase the timing error between pieces of equipment in a cascaded implementation.

The digital audio reference signal is to have the format and electrical configuration of the two-channel AES3 interface, but implementation of only the basic structure of the interface format, where only the preamble is active, is acceptable as a reference signal. A digital audio reference signal may be categorized in one of two grades. A grade 1 reference signal must maintain a long-term frequency accuracy within 61 ppm, whereas a grade 2 reference signal has a tolerance of less than 610 ppm.

AES17–1998 (r2004)

AES17–1998 (r2004) is the "AES Standard Method for Digital Audio Engineering—Measurement of Digital Audio Equipment." This standard defines a number of tests and test conditions for specifying digital audio equipment. Many of these tests are substantially the same as those used for testing analog audio equipment, but the unique nature of digital audio dictates that additional tests are necessary beyond those used for analog audio equipment.

AES18–1996 (r2002)

AES18–1996 (r2002) is the "AES Recommended Practice for Digital Audio Engineering—Format for the User Data Channel of the AES Digital Audio Interface." This standard describes a method of formatting the user data channels within the AES3 digital audio interface using a packet-based transmission format. This method has gained popularity in some broadcast facilities for carrying nonaudio ancillary data such as song titles and other information. It is critical to note, however, that user and other channel status bits are notoriously unreliable. In an effort to save data space, most storage equipment does not preserve this data and instead generates static values prior to output. If a facility design relies on using this data space, it is imperative to verify that all equipment in the chain supports it.

ATSC A/52B–2005 Digital Audio Compression (AC-3) Standard

Digital television broadcasting as described by the Advanced Television Systems Committee (ATSC) standard utilizes the AC-3 digital audio standard. Use of this standard will necessitate the carriage of AC-3 compressed digital audio streams between pieces of DTV equipment. An example of this is the interface between an AC-3 encoder and the program data stream multiplexer of a DTV transmission system. The former Annex B of the ATSC AC-3 Digital Audio Standard for digital television broadcast that describes the carriage of compressed AC-3 elementary streams on the IEC 958 digital audio interface has been replaced by IEC 61937.

IEC 60958 Digital Audio Interface

IEC 60958 (IEC 958) is logically identical to the AES3 digital audio interface. Electrically, it provides for both the 110 Ω balanced and the 75 Ω unbalanced interfaces. Two versions are described: a consumer version, the Sony/Philips Digital Interface (S/PDIF), in which bit 0 of the channel status word is set at logic 0; and a professional version, the AES/EBU interface, in which bit 0 of the channel status word is set at logic 1. Provision is made in the location of time slots 12–27, which are normally used to carry linear 16-bit PCM audio words, to permit some recording equipment to record and play back either linear 16-bit PCM audio or encoded data streams (compressed digital audio). The consumer implementation permits only the 32-bit mode, in which channel 1 and channel 2 subframes are simultaneously employed to carry 32-bit words. The professional implementation permits either the 32-bit mode or the 16-bit mode, in which each subframe carries a 16-bit digital audio word.

The consumer implementation may carry either two channels of linear PCM digital audio, or one or more compressed audio bitstreams accompanied by time stamps. The professional implementation may carry two channels of linear PCM digital audio, two sets of compressed audio bitstreams with time stamps, or one channel of linear PCM digital audio and one set of compressed audio bitstreams with time stamps. Note that the consumer implementation may also present output levels that are lower than the specified 1 V peak-to-peak of the professional version, and care is advised when connecting consumer and professional devices.

SMPTE STANDARDS AND RECOMMENDED PRACTICES CONCERNING THE USE OF AES DIGITAL AUDIO IN TELEVISION SYSTEMS

SMPTE 272M–2004

SMPTE 272M–2004 is "Television—Formatting AES/EBU Audio and Auxiliary Data into Digital Video Ancillary Data Space." This standard defines the embedding of AES/EBU digital audio into the standard definition serial digital interface specified in SMPTE 259M, *10-Bit 4:2:2 Component and 4fsc NTSC Composite Digital Signals—Serial Digital Interface*. With such embedding, up to 16 channels of digital audio in the AES3 format may be carried on the serial digital video interface signal that travels on a single coaxial cable.

SMPTE 276M–1995

SMPTE 276M–1995 is "Television—Transmission of AES/EBU Digital Audio Signals over Coaxial Cable." This SMPTE standard defines the unbalanced 75 Ω coaxial cable electrical interface for the AES3 bitstream.

SMPTE 299M–2004

SMPTE 299M–2004 is "Television—24-Bit Digital Audio Format for HDTV Bit-Serial Interface." This standard defines the embedding of AES/EBU digital audio data into the high-definition serial digital video interface specified in SMPTE 292M, *Bit Serial Digital Interface for High-Definition Television Systems*. This is the high-definition counterpart to SMPTE 272M.

SMPTE 302M–2002

SMPTE 302M–2002 is "Television—Mapping of AES3 Data into MPEG-2 Transport Stream." This SMPTE standard describes how the 20-bit audio payload of an AES/EBU signal is mapped into an MPEG-2 transport stream in a bit-for-bit accurate manner. This format can be found in most modern MPEG-2 encoders and integrated receiver/decoders (IRDs), and is a method used to carry uncompressed 20-bit PCM audio as well as mezzanine compressed audio such as Dolby E, high-density multiplexed AC-3, and Linear e-squared formats. Although it can be used to carry a single AC-3 stream, it is very inefficient and is incompatible with consumer equipment. IEC 13818 describes the proper manner for multiplexing an AC-3 stream into an MPEG-2 transport stream.

SMPTE 320M–1999

SMPTE 320M–1999 is "Television—Channel Assignments and Levels on Multichannel Audio Media." This often-overlooked standard defines proper channel ordering for multichannel audio soundtracks. The standard for television is as follows: 1 = Left front, 2 = Right front, 3 = Center, 4 = Low Frequency Effects (LFE or Subwoofer), 5 = Left surround, 6 = Right surround, 7 = Left or Lt ("left total," for matrix surround encoded systems), 8 = Right or Rt ("right total," for matrix surround encoded systems). It is possible for film format to differ slightly and the channel ordering is detailed in this specification, but for use within television facilities, film channel formatting should be corrected to match the order shown above.

SMPTE 337M through 341M

These standards are for "Formatting of Non-PCM Audio and Data in AES3 Serial Digital Audio Interface." They describe standardized methods for carrying compressed audio and other data types within AES3 signals, specifically:

- 338M–2000: Data types
- 339M–2000: Generic data types
- 340M–2000: ATSC A/52 (AC-3) data type
- 341M–2000: Captioning data type

They will become increasingly important as new professional equipment is developed to support compressed audio formats.

SMPTE RP 155–2004

SMPTE RP 155–2004 is "Reference Levels for Digital Audio Systems." This recommended practice describes a reference level lineup signal for use in digital audio recording on digital television tape recorders, and recommends the proper setting for the lineup signal on the recorder's digital audio level meters. The reference signal is the digital representation of a 1000 Hz sine wave, the level of which is 20 dB below the system maximum (full-scale digital). Meters are to be calibrated with this signal to indicate –20 dBFS (i.e., 20 dB below full-scale digital).

SMPTE EG 32–1996

SMPTE EG 32–1996 is "Emphasis of AES/EBU Audio in Television Systems and Preferred Audio Sampling Rate." This engineering guideline recommends that no emphasis be used on digital audio recordings for television applications and that the professional digital audio sample frequency of 48 kHz be used.

IMPLEMENTATION ISSUES

The key to realizing the benefits of digital audio on a systemwide scale is a thorough understanding of the principles underlying digital signal distribution, routing, and switching. There are, as explained, two electrical interfaces available for AES3 signals, and both require good engineering practices for successful implementation. Digital audio's data rate dictates that uncompressed digital audio signals occupy a bandwidth similar to that of analog video. Regardless of the electrical interface, a well-engineered interconnect requires proper match of source, destination, and characteristic cable impedances. Prior to the 1992 revision of AES3, any equipment manufactured to AES3–1985 violated this principle, as that standard specified a 250 Ω load impedance for receivers and a 110 Ω source impedance for transmitters. Beginning in 1992, AES3 specifies impedance matching among transmitter, receiver, and cable.

Choice of Cable

The use of the unbalanced coaxial cable interface for AES3 data transmission is often preferred by video engineers. SMPTE 276M and *AES3–id* provide guidance for using the 75 Ω unbalanced AES3 interface. Any high-quality video cable will be found quite acceptable for unbalanced AES3 signals. Those engineers designing facilities dealing only with audio may prefer the use of balanced, shielded, twisted pair cables with XLR-type connectors to carry AES3 signals, but should be aware of the cable length restrictions of this implementation and of the possibility that problems will arise from impedance mismatches at connectors and patch panels. For balanced transmission of AES3 signals, special low capacitance twisted pair cable intended especially for digital audio use is recommended over the standard twisted pair cables used for analog audio, as the higher capacitance of analog audio cable tends to distort square wave signals by rolling off the higher frequency components.

Digital Audio Distribution

The use of analog video distribution and routing equipment is generally not recommended for AES3 signals, as such equipment may distort AES signal shapes and rise times, adversely affecting the decoding of the signal at the receiving equipment. The spurious high frequency signal energy that may be generated by such distortions of signal shape can cause crosstalk-related bit errors that are difficult to detect and analyze. Distribution of the AES3 signal using high-quality digital audio distribution amplifiers will maintain the proper frequency and phase relationships, as well as signal shapes and rise times.

System Synchronization

When possible, all digital audio signals should be synchronous in order to avoid objectionable digital artifacts. In a large plant, it is necessary to provide a single master reference signal to which all interconnected systems are synchronized. The master reference, fed to all pieces of equipment, allows audio data to be retimed and synchronized within specified tolerances.

Large facilities, in particular, will benefit from the conversion of digital audio signals from sources without external sync capability to a standard, synchronized audio sample rate. Broadcast digital audio plants typically contain consumer and other nonsynchronizable equipment that requires sample rate conversion. Audio sample rate converters perform a function similar to video standard converters, in that a dynamic low pass filter continually adjusts the offending signal's phase at the output of the converter. In some cases, the output and input sample rates can be locked together via an integer relationship in a process known as *synchronous sample rate conversion*. For example, 48 kHz and 44.1 kHz are related by the integer ratio of 160 to 147. Modern sample rate conversion can be accomplished with full 24-bit resolution and THD+N below –140 dBm and as such has become an audibly lossless process. However, it is important to note that systems utilizing compressed audio, such as AC-3, Dolby E, and Linear esquared, bit-for-bit accuracy of the AES3 audio payload is imperative and will be corrupted by sample rate conversion—even when used in 1:1 modes for retiming (i.e., 48 kHz is reclocked to local 48 kHz reference).

MADI Synchronization

It is necessary for the equipment transmitting MADI data to include timing information that the receiving equipment can extract and use for synchronization. At least one sync code must be sent per frame; a sync code consists of two consecutive 5-bit words not used in the 4-bit to 5-bit encoding scheme. The total MADI interface data rate is higher than the payload data rate required, the difference between these two rates being sufficient to include sync codes within each frame. The fiber distributed digital interface (FDDI) chip set used for MADI implementation automatically handles the required synchronizing and coding operations.

AES3 Synchronization

AES3 is inherently synchronous, the clock signal being readily recovered from the AES3 bitstream. However, the use of a master digital audio reference ensures that all digital audio equipment in a system will be frequency and phase locked and free of cascaded timing errors, and is highly recommended by AES11. The master reference signal may come from the digital audio console in a facility on the scale of a single room, or from an external reference generator in larger facilities. The master sync signal should be sent to all equipment capable of accepting external sync signals.

Digital audio phase integrity must remain intact during the conversion of multiple audio channels between the digital and analog domains. Perfect phase synchronization requires use of an SDIF-2 word clock or an AES3 signal as the common master clock. Digital audio recording and processing equipment forces any AES3 input signal into a common AES3 frame phase. When such an AES3 frame alignment is performed, a phase error will result if there are any deviations in the frame phase of analog-to-digital (A/D) converters.

When digital audio signals are transferred to a piece of equipment that is not synchronized using a master sync signal, sample rate converters must be used at the inputs to the receiving equipment to prevent clicks and pops.

Word Clock Synchronization

SDIF-2 word clock, commonly referred to as simply *word clock*, is a square wave signal at the digital audio sample rate. Word clock is commonly used as a reference signal in small, audio-only facilities. In facilities that handle both video and audio, black burst is commonly used as the reference for both video and AES audio signal synchronization. Note that most professional audio equipment does not accept word clock as a reference signal, but instead relies on the AES11 standard whereby an AES/EBU signal with its embedded clock reference is used to derive proper synchronization. This eliminates the difficulties of distributing a high-frequency word clock square wave throughout a facility.

Signal Routing

Asynchronous routing is the simplest and most cost-efficient method of routing digital audio. It passes digital audio signals at any sample rate, a degree of flexibility that is ideal in situations in which a number of different audio sample rates are encountered. However, the lack of synchronization to a master reference makes it a poor choice for on-air applications or any other situation in which frame accurate switching or editing is required.

An asynchronous router may be thought of as an electronic patching system, functioning as though simple wires were used to connect inputs to outputs. In an asynchronous system, it is imperative that the destination equipment be capable of locking to the sample rate of the signal routed to it; otherwise, muting usually takes place.

The disadvantage of asynchronous routers is that their output signal is almost always corrupted when a switch is made between input signals. A switch typically results in one or more AES frames being damaged, and this may cause destination equipment to momentarily lose lock, causing muting or the generation of pops and clicks.

Synchronous routing ensures precise timing and no corruption of the data stream during switches. It is considerably more complex and costly than asynchronous routing, as it requires that a transition between two inputs be made at an AES frame boundary. All inputs to a synchronous router must be locked to a common digital audio reference. A digital audio console is essentially a synchronous router with many controls. Note that when routing compressed audio such as Dolby E, switching must occur not only at an AES frame boundary, but also at an AES frame boundary located near the video vertical interval switch point to prevent corruption of the compressed audio packets. Systems like the Linear Acoustic StreamStacker-HD require switching only on the AES frame boundary. Routing and switching AC-3 encoded signals are of greater difficulty, as the encoded packets from one stream to the next are not phase aligned.

Jitter

Jitter is short-term frequency variation in the input data stream to a digital audio device. It can result from a number of causes, including such things as the coupling of excessive noise into a transmission link. Some jitter buildup is inevitable in a system, as certain components of the system inherently generate some amount of jitter. For example, noise in the phase-locked loops that control clock frequencies in the components of the system unavoidably generates some jitter. The presence of out-of-specification jitter on a

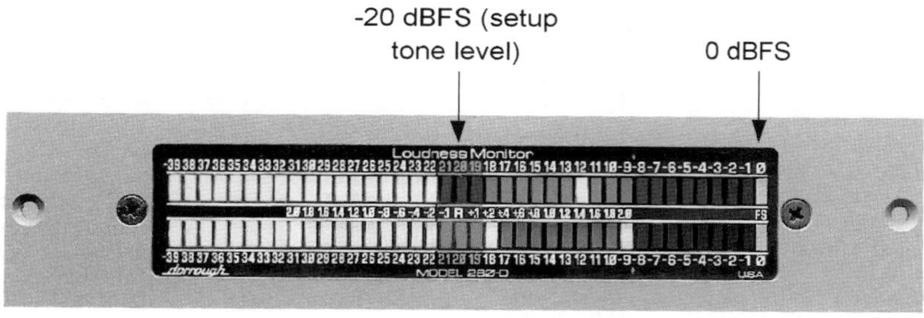

FIGURE 1.15-4 Representative digital audio level meter. (Courtesy Dorrough Electronics.)

digital audio signal or clock can result in bit errors that generate clicking and popping sounds. High levels of jitter may cause a receiving device to lose lock, while a relatively small amount may have no apparent negative effect unless present in devices performing analog-to-digital (A/D) or digital-to-analog (D/A) conversions. Excessive jitter is seldom a problem when only two pieces of equipment are involved, but typically builds up when larger numbers of equipment are interconnected. Jitter may be eliminated through the use of synchronizing digital-to-analog converters or a common synchronization signal. Jitter on the synchronization signal itself can cause degradation of all digital audio in devices locked to it.

Levels and Metering

When an analog audio signal is converted to digital, the greatest analog voltage level that may be represented digitally is called full-scale digital (FSD). When quantized, this voltage level causes all digital audio bits to be set to logic 1, and this level is called 0 dBFS (full scale). This is an inflexible limit, and any excursion of the analog signal above this level will be clipped off, as the digital audio word does not have the capacity to faithfully represent it. In practice, the FSD level is often set about 1 dB above the analog clip level in an effort to assure that digital clipping never occurs.

When signals are converted between the analog and digital domains, the analog reference levels of A/D and D/A converters may be set to any number of values. If the analog reference level is improperly calibrated in any of the converters in the path, A/D and D/A conversions may result in an increase or a decrease in the level of the recovered analog signal.

Consistency in the type of digital audio metering device used, good operator training, and the establishment of strict house standard reference levels and alignment practices are the best defenses when it comes to accurate audio level control.

There is no U.S. standard for a specific digital audio level meter. Digital audiometers are often of the instantaneous response type, with no integration time, permitting them to respond with full excursion to a peak as brief as a single digital audio sample. Contrast this with the standard volume indicator (VU meter), which is an average-responding device, and the typical peak-program meter, which does not respond with full excursion to peaks with durations less than 10 ms. Typically, digital audio metering devices display a maximum value of 0 dBFS, and reference level lineup tone is set to a designated point below 0 dBFS to accommodate peaks without digital clipping.

Figure 1.15-4 shows a representative digital audio meter, the display device of which is usually an array of light emitting diodes or other such devices. This representative meter displays a range of –40 dB to 0 dBFS, with lineup tone being calibrated at –20 dBFS.

For television applications, SMPTE RP 155 recommends adjusting the level of lineup tone to read –20 dBFS on digital audio meters used on digital videotape recorders. Other industry segments have variously used lineup tone levels of –15, –18, and –20 dBFS. These varying reference levels may cause inconsistent results when digital audio recordings are interchanged. It is therefore important to establish common digital audio reference and operating levels when exchanging digital audio recordings.

Loudness metering is best accomplished with meters designed to measure loudness. VU- and PPM-type meters are not truly appropriate for accurately judging loudness, as the results are often a mixture of meter readout and user interpretation and are thus unreliable for producing consistent results.

SUMMARY

Digital audio, with its many advantages, is inherently not susceptible to many of the problems that are encountered in analog audio systems. It does harbor some potential hazards of its own, however. With care and attention to good engineering practices in the design and maintenance of digital audio facilities and observance of the recommendations described in AES/EBU, IEC, and SMPTE standards, outstanding results will be realized.

Standards

[1] AES3–1992 (r2003) *AES Recommended Practice for Digital Audio Engineering—Serial Transmission Format for Two-Channel Linearly Represented Digital Audio Data*, New York, Audio Engineering Society, 2003.

[2] AES5–1998 (r2003) *AES Recommended Practice for Professional Digital Audio—Preferred Sampling Frequencies for Applications*

Employing Pulse-Code Modulation, New York, Audio Engineering Society, 2003.

[3] AES10–1991 (r2003) *AES Recommended Practice for Digital Audio Engineering—Serial Multichannel Audio Digital Interface (MADI),* New York, Audio Engineering Society, 2003.

[4] AES11–1997 (r2003) *AES Recommended Practice for Digital Audio Engineering—Synchronization of Digital Audio Equipment in Studio Operations,* New York, Audio Engineering Society, 2003.

[5] AES17–1998 *AES Standard Method for Audio Engineering—Measurement of Digital Audio Equipment,* New York, Audio Engineering Society, 1998.

[6] AES18–1996 *AES Recommended Practice for Digital Audio Engineering—Format for the User Data Channel of the AES Digital Audio Interface,* New York, Audio Engineering Society, 1996.

[7] ATSC A/52B–2005 *Digital Audio Compression (AC-3) Standard,* Washington, Advanced Television Systems Committee, 1995.

[8] IEC 60958 (1999) *Digital Audio Interface,* Geneva, International Electrotechnical Commission, 1999.

[9] SMPTE 259M–1993 *10-Bit 4:2:2 Component and 4fsc NTSC Composite Digital Signals—Serial Digital Interface,* White Plains, Society of Motion Picture and Television Engineers, 1993.

[10] SMPTE 272M–1994 (r2004) *Formatting AES/EBU Audio and Auxiliary Data into Digital Video Ancillary Data Space,* White Plains, Society of Motion Picture and Television Engineers, 2004.

[11] SMPTE 276M–1995 *Transmission of AES/EBU Digital Audio Signals over Coaxial Cable,* White Plains, Society of Motion Picture and Television Engineers, 1995.

[12] SMPTE 292M–1996 *Bit-Serial Digital Interface for High-Definition Television Systems,* White Plains, Society of Motion Picture and Television Engineers, 1996.

[13] SMPTE 299M–1997 (r2004) *24-Bit Digital Audio Format for HDTV Bit-Serial Interface,* White Plains, Society of Motion Picture and Television Engineers, 2004.

[14] SMPTE 302M–1998/2000 *Mapping of AES3 Data into MPEG-2 Transport Stream,* White Plains, Society of Motion Picture and Television Engineers, 2000.

[15] SMPTE 320M–1999 *Channel Assignments and Levels on Multichannel Audio Media,* White Plains, Society of Motion Picture and Television Engineers, 1999.

[16] SMPTE 337M through SMPTE 340M *Formatting of Non-PCM Audio and Data in AES3 Serial Digital Audio Interface,* White Plains, Society of Motion Picture and Television Engineers.

[17] SMPTE RP 155 *Audio Levels for Digital Audio Records on Digital Television Tape Recorders*, White Plains, Society of Motion Picture and Television Engineers, 2004.

[18] IEC 61937-1 *Digital Audio—Interface for Non-Linear PCM Encoded Bitstreams Applying IEC 60958, Part 1—General,* Geneva, International Electrotechnical Commission.

[19] IEC 61937-3 *Digital Audio—Interface for Non-Linear PCM Encoded Bitstreams Applying IEC 60958, Part 3—Non-Linear PCM Bitstreams According to the AC-3 Format,* Geneva, International Electrotechnical Commission.

Recommended Practices and Information Documents

AES2–id, 1996 *AES Information Document for Digital Audio Engineering—Guidelines for the Use of the AES Interface,* New York, Audio Engineering Society, 1996.

AES3–id, 1995 *AES Information Document for Digital Audio Engineering—Transmission of AES3 Formatted Data by Unbalanced Coaxial Cable,* New York, Audio Engineering Society, 1995.

AES10–id, 1995 *AES Information Document for Digital Audio Engineering—Engineering Guidelines for the Multichannel Audio Digital Interface (MADI) AES10,* New York, Audio Engineering Society, 1995.

SMPTE Recommended Practice RP 155–1997 *Audio Levels for Digital Audio Records on Digital Television Tape Recorders,* White Plains, Society of Motion Picture and Television Engineers, 1997.

SMPTE Engineering Guideline EG 32–1996 *Emphasis of AES/EBU Audio in Television Systems and Preferred Audio Sampling Rate,* White Plains, Society of Motion Picture and Television Engineers, 1996.

CHAPTER

1.16

Video Standards and Practices

JERRY C. WHITAKER

Advanced Television Systems Committee
Washington, D.C.

INTRODUCTION

Anyone concerned with the interchangeability of equipment or product should be concerned with standards. A prospective user hesitates to purchase equipment that does not conform to recognized interface standards for connectors, input/output levels, control, timing, and test specifications. A manufacturer may find a limited market for a good product if it is not compatible with other equipment in common use. Standards promote economies of scale that tend to produce more reliable products at a lower cost.

Standards ensure that the needs of the user are considered, and interconnection of equipment from different manufacturers is facilitated. The rollout of digital television (DTV) products at a record pace attests to the need for, and benefits of, standards. The progress made so far in the DTV era would have been wholly impossible without the considerable efforts of organizations such as the ATSC, SMPTE, SCTE, CEA, and NAB.

Rapid improvements in technology may tend to make some standards technically obsolete by the time they are adopted, but such is the nature of a rapidly expanding technology-based society. A standard provides a stable platform for manufacturers to market their product and assures the user of a degree of compatibility. Of the many standards-setting organizations in the professional video field, the most familiar are:

- Advanced Television Systems Committee (ATSC)
- Consumer Electronics Association (CEA)
- Institute of Electrical and Electronics Engineers (IEEE)
- International Standards Organization (ISO)
- Society of Cable and Telecommunications Engineers (SCTE)
- Society of Motion Picture and Television Engineers (SMPTE)

Standards, whether for a new television broadcast system or VTR connector pin assignments, are vital for the continued growth of the communications industry.

Web Resources

The data contained in this chapter draw heavily upon material made available by leading standards organizations. Web site addresses are given where applicable, and readers are encouraged to explore these valuable resources. Standards documents can be downloaded or ordered online from most of the sites. Because of the rapidly changing nature of digital audio and video implementation, readers are encouraged to check in regularly.

Another valuable resource is the SMPTE television standards on CD-ROM. This product, available for purchase from SMPTE, contains all existing and proposed standards, recommended practices, and engineering guidelines for television work. Similar products are available from other standards organizations, either on a per-document basis or as a complete package, as in the SMPTE offering. Subscription services are also available and should be considered for organizations that require having the latest standards, recommended practices, engineering guidelines, and related documents on hand.

VIDEO STANDARDS AND RELATED DOCUMENTS

It is impractical within the constraints of this chapter to list all standards, recommended practices, and related documents of interest to video engineers. The following sections, however, contain a representative sample of key documents relating to digital television.

Advanced Television Systems Committee

The Advanced Television Systems Committee is an international, non-profit organization developing voluntary standards for digital television. The ATSC has approximately 140 member organizations representing the broadcast, broadcast equipment, motion picture, consumer electronics, computer, cable, satellite, and semiconductor industries. The following is a partial list of ATSC standards and related technical documents.

ATSC A/52: Digital Audio Compression (AC-3, E-AC-3)

This document specifies coded representation of audio information and the decoding process, as well as information on the encoding process. The coded representation specified is suitable for use in digital audio transmission and storage applications and may convey from 1 to 5 (or more) full-bandwidth audio channels, along with a low-frequency-enhancement channel. A wide range of encoded bit rates are supported by this specification. Typical applications of digital audio compression are in satellite or terrestrial audio broadcasting, delivery of audio over metallic or optical cables, and storage of audio on magnetic, optical, semiconductor, or other storage media.

ATSC A/53: Digital Television Standard

The ATSC Digital Television Standard describes the system characteristics of the advanced television (ATV) system. The document and its normative annexes provide detailed specification of the parameters of the system including the video encoder input scanning formats and the preprocessing and compression parameters of the video encoder, the audio encoder input signal format and the preprocessing and compression parameters of the audio encoder, the service multiplex and transport layer characteristics and normative specifications, and the VSB RF/transmission subsystem. The system is modular in concept, and the specifications for each of the modules are provided in the appropriate annex.

ATSC A/54: Recommended Practice, Guide to the Use of the ATSC Digital Television Standard

This guide provides tutorial information and an overview of the digital television system defined by ATSC Standard A/53 (ATSC Digital Television Standard). In addition, recommendations are given for operating parameters for certain aspects of the DTV system.

ATSC A/64: Transmission Measurement and Compliance Standard for DTV

This document describes methods for testing, monitoring, and measurement of the transmission subsystem intended for use in the DTV system, including specifications for maximum out-of-band emissions, parameters affecting the quality of the inband signal, symbol error tolerance, phase noise and jitter, power, power measurement, frequency offset, and stability. In addition, it describes the condition of the radiofrequency (RF) symbol stream upon loss of MPEG packets.

ATSC A/65: Program and System Information Protocol for Terrestrial Broadcast and Cable

This document defines a standard for system information (SI) and program guide (PG) data compatible with digital multiplex bit streams constructed in accordance with ISO/IEC 13818-1 (MPEG-2 Systems). The document defines the standard protocol for transmission of the relevant data tables contained within packets carried in the transport stream multiplex. The protocol defined herein is referred to as the Program and System Information Protocol (PSIP).

ATSC A/69: Recommended Practice, PSIP Implementation Guidelines for Broadcasters

This document provides a set of guidelines for the use and implementation of the ATSC Program and System Information Protocol, as described in ATSC Standard A/65. The information contained in this document applies to broadcasters, network operators, infrastructure manufacturers, and receiver manufacturers.

ATSC A/75: Receiver Performance Guidelines

This recommended practice addresses the front-end portion of a receiver of digital terrestrial television broadcasts. The recommended performance guidelines enumerated in this document are intended to ensure that reliable reception will be achieved. Guidelines for interference rejection are based on the Federal Communications Commission (FCC) planning factors that were used to analyze coverage and interference for the initial DTV channel allotments. Guidelines for sensitivity and multipath handling reflect field experience accumulated by testing undertaken by the Advanced Television Technical Test Center (ATTC), Association for Maximum Service Television (MSTV), National Association of Broadcasters (NAB), and receiver manufacturers.

ATSC A/76: Programming Metadata Communication Protocol Standard

This standard defines a method for communicating metadata related to the Program and System Information Protocol, including duplicate data that must be entered in other locations in the transport stream. Communication is based on a protocol utilizing XML message documents generated in accordance with a

Programming Metadata Communication Protocol (PMCP) XML schema defined herein.

ATSC A/90: Data Broadcast

The ATSC Data Broadcast Standard defines protocols for data transmission compatible with digital multiplex bit streams constructed in accordance with ISO/IEC 13818-1 (MPEG-2 Systems). The standard supports data services that are both TV program related and non-program related. Applications may include enhanced television, Webcasting, and streaming video services. Data broadcasting receivers may include PCs, televisions, set-top boxes, or other devices. The standard provides mechanisms for download of data, delivery of datagrams, and streaming data.

ATSC A/91: Implementation Guidelines for the ATSC Data Broadcast Standard

This document provides a set of guidelines for the use and implementation of the ATSC Data Broadcast Standard (A/90). As such, it facilitates the efficient and reliable implementation of data broadcast services.

ATSC A/92: Delivery of IP Multicast Sessions over Data Broadcast Standard

This standard specifies the delivery of Internet Protocol (IP) multicast sessions, the delivery of data for describing the characteristics of a session, and usage of the ATSC A/90 Data Broadcast Standard for IP Multicast. This document defines a standard for the asynchronous transmission of Internet Protocols, specifically including multicast addressing compatible with the ATSC A/90 Data Broadcast Standard. This standard assumes the use of Session Description Protocol (SDP) as an integral part of the IP multicast-based data broadcast service.

ATSC A/93: Synchronized/Asynchronous Trigger Standard

This document defines a standard for the transmission of synchronized data elements and synchronized and asynchronous events. It specifically enables the synchronized delivery of data modules through decoupling of the timing from the delivery of the data element. It also enables the delivery of events to receivers, including application-defined events.

ATSC A/94: ATSC Data Application Reference Model

This standard defines an Application Reference Model (ARM) including a binding of application environment facilities onto the ATSC A/90 Data Broadcast Standard. This standard includes a systemwide resource naming scheme, a state model, data models, and constraints and extensions to A/53, A/65, and A/90 to implement application environments.

ATSC A/95: Transport Stream File System Standard

This document defines the ATSC Transport Stream File System (TSFS) standard for delivery of hierarchical name-spaces, directories, and files. This standard builds on the data service delivery mechanism defined in the ATSC A/90 Data Broadcast Standard.

ATSC A/96: ATSC Interaction Channel Protocols

This standard defines a core suite of protocols to enable remote interactivity in television environments. Remote interactivity requires the use of a two-way interaction channel that enables communications between the client device and remote servers. Examples of remote interactivity include E-commerce transactions during commercials, electronic banking, polling, e-mail services, or other services yet to be defined.

ATSC A/97: Software Download Data Service

This document specifies a data service that may be used to download software to a terminal device using an MPEG-2 transport stream via an appropriate physical layer. This service may be used to effect updates or upgrades of firmware, operating system software, device driver software, native application software, middleware, and other types of software that reside in a terminal device. This document specifies standard announcement, signaling, and encapsulation for the delivery of this download data service. The content and format of the software download data are not defined by this standard. The formats and interpretations of the software download payload are defined by each user of this standard.

ATSC A/101: Advanced Common Application Platform (ACAP)

This document defines the Advanced Common Application Platform (ACAP). This standard is intended to be used primarily by entities writing terminal specifications and/or standards based on ACAP. Second, it is intended for developers of applications that use the ACAP functionality and application programming interfaces (APIs). ACAP aims to ensure interoperability between ACAP applications and different implementations of platforms supporting ACAP applications. An ACAP application is a collection of information that is processed by an application environment in order to interact with an end-user or otherwise alter the state of the application environment.

ATSC A/110: Synchronization Standard for Distributed Transmission

This document defines a standard for synchronization of multiple transmitters emitting trellis-coded eight-level vestigial sideband modulation (8VSB) signals in accordance with ATSC A/53 Annex D (RF/Transmission Systems Characteristics). The emitted signals from transmitters operated according to this standard comply fully with the requirements of ATSC A/53. This document specifies mechanisms necessary to transmit synchronization signals to the several transmitters using a dedicated packet identifier (PID) value, including the formatting of packets associated with that PID and without altering the signal format emitted from the transmitters. It also provides for adjustment of transmitter timing and other characteristics through additional information carried in the specified packet structure.

ATSC A/111: Recommended Practice, Design of Synchronized Multiple Transmitter Networks

Many of the challenges of radiofrequency transmission are the same regardless of whether the information carried is in analog or digital form. Because of the signal processing applied when the information carried is digital, however, there are techniques to overcome some of those challenges that are more applicable to digital signals than to analog signals. Among such techniques is the use of multiple transmitters in single-frequency networks (SFNs) and multiple-frequency networks (MFNs). In the past, SFNs have been considered mostly for applications in multicarrier systems such as those using coded orthogonal frequency-division multiplexing (COFDM) modulation. This recommended practice applies SFNs to the single-carrier 8VSB system adopted by the ATSC and the FCC. SFNs can be implemented with digital on-channel repeaters (DOCRs), with distributed transmitters (DTxTs), with distributed translators (DTxRs), or with a combination of them. MFNs generally involve the use of translators. This recommended practice examines all three types of transmitters used in SFNs and MFNs and then concentrates on the design aspects of SFNs.

ATSC A/112: E-VSB Implementation Guidelines

The purpose of this document is to explain in detail the ATSC standards related to enhanced VSB (E-VSB) and provide guidelines to parameter selection and implementation scenarios where useful.

Consumer Electronics Association

The Consumer Electronics Association is a membership organization of about 2000 companies within the U.S. consumer technology industry. CEA produces standards and related documents relating to the consumer electronics industry. CEA also provides market research and educational programs and technical training to members. The following is a partial list of CEA standards and related technical documents.

CEA-CEB12-A: PSIP Recommended Practice

This document provides guidelines to receiver manufacturers regarding implementation of the Program and System Information Protocol used in the ATSC DTV Standard.

CEA-679-C: National Renewable Security Standard (NRSS)

The NRSS provides two physical designs. Part A defines a removable and renewable security element form factor that is an extension of the ISO-7816 standard. Part B defines a removable and renewable security element based on the Personal Computer Memory Card International Association (PCMCIA) ("C-Card") form factor. The common attributes allow either an NRSS-A or NRSS-B device to provide security for applications involving pay and subscription cable or satellite television services, telephone, and all forms of electronic commerce.

CEA-774-A: TV Receiving Antenna Performance Presentation and Measurement

This standard is intended to provide television receive antenna manufacturers with appropriate test and measurement procedures to examine antenna performance parameters necessary to comply with elements of the CEA TV Antenna Selector Map program. Essential elements include procedures to determine antenna gain, front-to-back ratio, directivity, and distortion performance of active antennas with integrated amplifiers.

CEA-775-B: DTV 1394 Interface Specification

This standard defines a specification for a baseband digital interface to a DTV using the IEEE-1394 bus and provides a level of functionality that is similar to the analog system. It is designed to enable interoperability between a DTV compliant with this standard and various types of consumer digital audio/video sources including digital set-top boxes (STBs) and analog/digital hard disk or videocassette recorders (VCRs).

CEA-796-A: NRSS Copy Protection Systems

The copy protection systems that have been included in CEA-796 are itemized for the purpose of identification. The systems outlined in CEA-796 all support the copy protection frameworks described in CEA-679-B, Parts A and B.

CEA-799-A: On-Screen Display Specification

This standard specifies syntax semantics for bitmapped graphics data typically used for on-screen display (OSD). The standard is applicable whenever it is necessary to specify a standard method for delivery of bitmapped graphics data. The pixel formats include optional alpha-blend and transparency attributes to support composition of graphics over analog or digitally decoded video within the display.

CEA-861-D: A DTV Profile for Uncompressed High-Speed Digital Interfaces

This standard defines video timing requirements, discovery structures, and a data transfer structure that is used for building uncompressed, baseband, digital interfaces on digital televisions or DTV monitors. A single physical interface is not specified, but any interface implemented must use the Video Electronics Standards Association Enhanced Extended Display Identification Data Standard (VESA E-EDID) for format discovery

CEA-2028: Color Codes for Outdoor TV Receiving Antennas

This standard defines color codes to be associated with minimum performance parameters of outdoor TV receiving antennas. When used in conjunction with the CEA TV antenna selector program at http://www.antennaweb.org, these color codes can help both consumers and professional installers select appropriate outdoor TV antennas for their particular reception environments.

CEA-2032: Indoor TV Receiving Antenna Performance Standard

This standard provides manufacturers of indoor television receive antennas with an appropriate standard for antenna characteristics and minimum performance requirements.

CEA-TVSB5r1: Multichannel TV Sound System BTSC System Recommended Practices

This recommended practice specifies the transmission of multichannel television sound (MTS) in accordance with the Broadcast Television Systems Committee (BTSC) system and the FCC rules governing its use. This document is intended for both manufacturers and broadcasters.

EIA-708-C: Digital Television Closed Captioning

This document is intended as a definition of DTV closed captioning (DTVCC) and provides specifications and guidelines for caption service providers, DTVCC decoder and encoder manufacturers, DTV receiver manufacturers, and DTV signal processing equipment manufacturers. This specification includes: (1) a description of the transport method of DTVCC data in the DTV signal, (2) a description of DTVCC-specific data packets and structures, (3) a specification of how DTVCC information is to be processed, (4) a list of minimum implementation recommendations for DTVCC receiver manufacturers, and (5) a set of recommended practices for DTV encoder and decoder manufacturers.

EIA-766-B: U.S. and Canadian Region Rating Tables (RRTs) and Content Advisory Descriptors for Transport of Content Advisory Information Using the ATSC A/65-A Program and System Information Protocol

This standard augments ATSC Standard A/65 and SCTE DVS-097 Rev. 7, both titled the Program and System Information Protocol for Terrestrial Broadcast and Cable. Along with the above two standards, this standard designates the RRT that provides the receiver with the definition of the rating system and the content advisory descriptor that provides the receiver with the specific program rating for each program. Specifically, this standard specifies the exact syntax to be used to define the U.S. and Canadian Rating Region Tables in accordance with ATSC A/65 Section 6.4, as well as the exact syntax to be used in the content advisory descriptors that convey the rating information for each program in accordance with ATSC A/65 Section 6.7.4. Thus, DTV receivers may block unwanted programs as determined by the user.

EIA/CEA-608-C: Line 21 Data Services

This standard serves as a technical guide for those providing encoding equipment and decoding equipment to produce material with encoded data embedded in line 21 of the vertical blanking interval of the NTSC video signal. It is also a usage guide for those who will produce material using such equipment. Revision B incorporates content advisory information.

EIA/CEA-770.3-C: High-Definition TV Analog Component Video Interface

This standard defines two raster-scanning systems for the representation of stationary or moving two-dimensional images sampled temporally at a constant frame rate. The first image format specified is 1920 × 1080 samples (pixels) inside a total raster of 1125 lines. The second image format specified is 1280 × 720 samples (pixels) inside a total raster of 750 lines. Both image formats must have an aspect ratio of 16:9.

EIA/CEA-818-E: Cable Compatibility Requirements

This standard defines the minimum requirements that must be met by digital cable TV systems and digital TV receivers such that the receivers may be connected directly to the RF output of the cable system to provide selected baseline services.

EIA/CEA-849-B: Application Profiles for EIA-775A-Compliant DTVs

This standard specifies profiles for various applications of the EIA-775-A standard. The application areas covered here include digital streams compliant with ATSC terrestrial broadcast, direct-broadcast satellite (DBS), OpenCable™, and standard-definition digital video (DV) camcorders.

EIA/CEA-909: TV Antenna Control Interface Standard

This standard enables receivers to automatically control antenna characteristics; it describes an antenna control physical interface and a control protocol. The standard also shows a number of antenna configurations that may be used and provides several example implementations of the antenna control signal processing circuitry. The standard allows for a second (optional) mode with many more control states and two-way communication.

Society of Cable and Telecommunications Engineers

The Society of Cable and Telecommunications Engineers is a professional association dedicated to advancing the careers of cable telecommunications professionals and serving the industry through excellence in professional development, information, and standards. Founded in 1969, SCTE has about 70 chapters. SCTE standards cover a wide range of industry needs from "F" connectors to protocols for high-speed data access over cable. SCTE is accredited by the American National Standards Institute (ANSI), recognized by the International Telecommunication Union (ITU), and works in cooperation with the European Telecommunications Standards Institute (ETSI). The following is a partial list of SCTE standards and related technical documents.

ANSI/SCTE 07 2006 (DVS 031): Digital Transmission Standard for Cable Television

This standard describes the framing structure, channel coding, and channel modulation for a digital multiservice television distribution system that is specific to a cable channel. The system can be used transparently with the distribution from a satellite channel, as many cable systems are fed directly from the satellite links. The specification covers both 64 and 256 quadrature amplitude modulation (QAM).

SCTE 18 (ANSI-J-STD-042-2002) (DVS 208): Emergency Alert Message for Cable

This standard defines an emergency alert (EA) signaling method for use by cable TV systems to signal emergencies to digital cable-ready devices. Use of the EA signaling method defined in this standard is designed for cable systems that support devices offered for retail sale and certified as cable-ready. Such devices include digital set-top boxes that are sold to consumers at retail, cable-ready digital TV receivers, and cable-ready digital VCRs. Cable terminals owned by cable operators may use this or other proprietary methods for EA signaling.

ANSI/SCTE 20:2004: Methods for Carriage of Closed Captions and Non-Real-Time Sampled Video

This document defines a standard for the carriage of vertical blanking interval (VBI) services in MPEG-2-compliant bit streams constructed in accordance with ISO/IEC 13818-2.

ANSI/SCTE 21:2001 R2006 (DVS 053): Standard for Carriage of NTSC VBI Data in Cable Digital Transport Streams

This document defines a standard for the carriage of VBI services in MPEG-2-compliant bit streams constructed in accordance with ISO/IEC 13818-2. The approach builds upon a data structure defined in the ATSC A/53 Digital Television Standard and is designed to be backward-compatible with the method.

ANSI/SCTE 43:2005: Digital Video Systems Characteristics Standard for Cable Television

This document describes the characteristics and normative specifications for the Video Subsystem Standard for Cable Television.

ANSI/SCTE 65:2002 (DVS 234): Service Information Delivered Out-of-Band for Digital Cable Television

This standard defines service information (SI) tables delivered via an out-of-band path on cable to support service selection and navigation by digital cable set-top boxes and other digital cable-ready devices. The SI tables defined in this standard are formatted in accordance with the Program Specific Information (PSI) data structures defined in MPEG-2 systems.

Society of Motion Picture and Television Engineers

The Society of Motion Picture and Television Engineers is the leading technical society for the motion imaging industry. SMPTE members are spread throughout 85 countries worldwide. SMPTE was founded in 1916 to advance theory and development in the motion imaging field. Today, SMPTE publishes ANSI-approved standards, recommended practices, and engineering guidelines, along with the highly regarded *SMPTE Journal* and its peer-reviewed technical papers. The following is a partial list of SMPTE standards and related technical documents.

SMPTE 12M: Time and Control Code

This standard specifies a digital time and control code for use in television, film, and accompanying audio systems operating at 30, 25, and 24 frames per second.

SMPTE 125M: Component Video Signal 4:2:2—Bit-Parallel Digital Interface

This standard defines an interface for system M (525/60) digital television equipment based on ITU-R BT.601. The standard has application in the television studio over distances up to 300 m (1000 ft).

SMPTE 170M: Composite Analog Video Signal—NTSC for Studio Applications

This standard describes the composite analog color video signal for studio applications: NTSC, 525 lines, 59.94 fields, 2:1 interlace with an aspect ratio of 4:3. This standard specifies the interface for analog interconnection and serves as the basis for the digital coding necessary for digital interconnection of NTSC equipment.

SMPTE 240M: 1125 Line High-Definition Production Systems—Signal Parameters

This standard defines the basic characteristics of the analog video signals associated with origination equipment operating in 1125 line high-definition television production systems. This standard defines systems operating at 60.00 Hz and 59.94 Hz field rates. The digital representation of the signals described in this standard may be found in SMPTE 260M. These two documents define between them both digital and analog implementations of 1125 line HDTV production systems.

SMPTE 259M: SDTV Digital Signal/Data—Serial Digital Interface

This standard describes a 10-bit serial digital interface operating at 143/270/360 Mbps. The serial interface may carry uncompressed SDTV signals, or data. This standard has application in the television studio over lengths of coaxial cable where the signal loss does not exceed an amount specified by the receiver manufacturer. Typical loss amounts may be in the range of 20 dB to 30 dB at one-half clock frequency with appropriate receiver equalization.

SMPTE 260M: Digital Representation and Bit-Parallel Interface—1125/60 High-Definition Production System

This standard specifies the digital representation of the signal parameters of the 1125/60 high-definition production system as given in their analog form by SMPTE 240M. This standard also specifies the signal format and the mechanical and electrical characteristics of the bit-parallel digital interface for the interconnection of digital television equipment operating in the 1125/60 high-definition production system.

SMPTE 266M: 4:2:2 Digital Component Systems—Digital Vertical Interval Time Code

This standard describes the signal format of a digital vertical interval time code suitable for use with the digital coding given in ANSI/SMPTE 125M (for 525 line, 59.94 Hz field rate, 4:2:2 component digital signals) or ITU-R BT.601 (for 625 line, 50 Hz field rate, 4:2:2 component digital signals).

SMPTE 272M: Formatting AES/EBU Audio and Auxiliary Data into Digital Video Ancillary Data Space

This standard defines the mapping of AES digital audio data, AES auxiliary data, and associated control information into the ancillary data space of serial digital video conforming to ANSI/SMPTE 259M. The audio data and auxiliary data are derived from ANSI S4.40, generally referred to as AES audio.

SMPTE 274M: 1920 × 1080 Image Sampling Structure, Digital Representation, and Digital Timing Reference Sequences for Multiple Picture Rates

This standard defines a family of raster-scanning systems for the representation of stationary or moving two-dimensional images sampled temporally at a constant frame rate and having an image format of 1920 × 1080 and an aspect ratio of 16:9.

SMPTE 276M: Transmission of AES/EBU Digital Audio Signals over Coaxial Cable

This standard describes a point-to-point coaxial cable interface for the transmission of AES/EBU digital audio signals throughout television production and broadcast facilities. The purpose of this standard is to ensure that a level of compatibility exists between signals generated to this standard and analog video equipment, such as nonclamping distribution amplifiers, switchers, cables, and connectors, as normally used in television applications.

SMPTE 291M: Ancillary Data Packet and Space Formatting

This standard specifies the basic formatting structure of the ancillary data space in the digital video data steam in the form of 10 bit words. Application of this standard includes 525 line, 625 line, component or composite, and high-definition digital television interfaces which provide 8 or 10 bit data ancillary data space. Space available for ancillary data packets is defined in the document specifying the connecting interface.

SMPTE 292M: Bit-Serial Digital Interface for High-Definition Television Systems

This standard defines a bit-serial digital coaxial and fiberoptic interface for HDTV component signals operating at data rates in the range of 1.3 Gbit/s to 1.5 Gbit/s. Bit-parallel data derived from a specified source format are multiplexed and serialized to form the serial data stream. A common data format and channel coding are used based on modifications, if necessary, to the source format parallel data for a given high-definition television system.

SMPTE 293M: 720 × 483 Active Line at 59.94 Hz Progressive Scan Production—Digital Representation

This standard defines the digital representation of stationary or moving two-dimensional images for television production. The representation is sampled linearly in the spatial domain and sampled temporally at a constant frame rate. The scanned image has an aspect ratio of 16:9.

SMPTE 294M: 720 × 483 Active Line at 59.94 Hz Progressive Scan Production—Bit-Serial Interfaces

This standard defines two alternatives for bit-serial interfaces for the 720 × 483 active line at 59.94 Hz progressive scan digital signal for production, as defined in ANSI/SMPTE 293M. Interfaces for coaxial cable are defined, each having a high degree of commonality with interfaces operating in accordance with ANSI/SMPTE 259M.

SMPTE 295M: 1920 × 1080 50 Hz—Scanning and Interfaces

This standard defines a family of raster scanning systems for the representation of stationary or moving two-dimensional images sampled temporally at a constant frame rate and having an image format of 1920 × 1080 and an aspect ratio of 16:9.

SMPTE 296M: 1280 × 720 Progressive Image Sample Structure—Analog and Digital Representation and Analog Interface

This standard defines a family of raster scanning systems for the representation of stationary or moving two-dimensional images sampled temporally at a constant frame rate and having an image format of 1280 × 720 and an aspect ratio of 16:9.

SMPTE 297M: Serial Digital Fiber Transmission System for ANSI/SMPTE 259M Signals

This standard defines an optical fiber system for transmitting bit-serial digital signals. It is specifically intended for transmitting ANSI/SMPTE 259M serial signals (143 through 360 Mbit/s). Its optical interface specifications and end-to-end system performance parameters are otherwise compatible with SMPTE 292M, which covers transmission rates of 1.3 through 1.5 Gbit/s.

SMPTE 298M: Universal Labels for Unique Identification of Digital Data

This standard defines universal labels, a universal labeling mechanism to be used in identifying the type and encoding of data within a general-purpose data stream. The labeling mechanism is intended to function across all types of digital communications protocols and message structures, allowing the intermixture of data of any sort.

SMPTE 299M: 24-Bit Digital Audio Format for SMPTE 292M Bit-Serial Interface

This standard defines the mapping of 24-bit AES digital audio data and associated control information into the ancillary data space of a serial digital video conforming to ANSI/SMPTE 292M. The audio data are derived from ANSI S4.40, better known as AES audio.

SMPTE 305M: Serial Data Transport Interface (SDTI)

This standard defines a data stream used to transport packetized data within a studio or production center environment. The data packets and synchronizing signals are compatible with ANSI/SMPTE 259M.

SMPTE 308M: MPEG-2 4:2:2 Profile at High Level

ISO/IEC 13818-2, commonly known as MPEG-2 video, includes specification of the MPEG-2 4:2:2 profile. Based on ISO/IEC 13818-2, this standard provides additional specification for the MPEG-2 4:2:2 profile at a high level. It is intended for use in high-definition television production, contribution, and distribution applications. As in ISO/IEC 13818-2, this standard defines bit streams, including their syntax and semantics, together with the requirements for a compliant decoder for a 4:2:2 profile at a high level but does not specify particular encoder operating parameters.

SMPTE 310M: Synchronous Serial Interface for MPEG-2 Digital Transport Streams

This standard describes the physical interface and modulation characteristics for a synchronous serial interface to carry MPEG-2 transport bit streams at rates up to 40 Mbit/s. It is a point-to-point interface intended for use in a low-noise environment.

SMPTE 325M: Opportunistic Data Broadcast Flow Control

This standard defines the flow control protocol to be used between an emission multiplexer and data server for opportunistic data broadcast. Opportunistic data broadcast inserts data packets into the output multiplex to fill any available free bandwidth. The emission mutliplexer maintains a buffer from which it draws data to be inserted.

SMPTE 330M: Unique Material Identifier (UMID)

This standard specifies the format of the unique material identifier (UMID). The UMID is a unique identifier for picture, audio, and data material which is locally created but globally unique. It differs from many unique identifiers in that the number does not depend wholly upon a registration process but can be generated automatically at the point of creation without reference to a central database.

SMPTE 332M: Encapsulation of Data Packet Streams over SDTI (SDTI-PF)

This standard specifies an open framework for encapsulating data packet streams and associated control metadata over the SDTI transport (SMPTE 305M). Encapsulating data packet streams on SDTI allows them to be routed through conventional SDI (SMPTE 259M) equipment.

SMPTE 333M: DTV Closed-Caption Server to Encoder Interface

This standard defines a standard for interoperation of DTVCC data server devices and video encoders. The caption data server devices provide partially formatted EIA 708 data to the video encoders using the request/response protocol and interface defined in this standard. The video encoder completes the formatting and includes the EIA 708 data in the video elementary stream picture-level user data field. This standard describes an interface for transmission of DTVCC data from a caption server to video encoder.

SMPTE 334M: Vertical Ancillary Data Mapping for Bit-Serial Interface

This standard defines a method of coding that allows data services to be carried in the vertical ancillary data space of a bit-serial component television signal conforming with SMPTE 292M or ANSI/SMPTE 259M.

SMPTE 335M: Metadata Dictionary Structure

The metadata dictionary structure defined in this standard covers the use of metadata for all types of essence (video, audio, and data in their various forms). Applications of individual dictionary entries will vary, but, when used, metadata must conform to the definitions and formats in this metadata dictionary structure standard and the associated metadata dictionary recommended practice (SMPTE RP 210).

SMPTE 344M: 540 Mbit/s Serial Digital Interface

This standard specifies a serial digital interface that operates at a nominal rate of 540 Mbit/s. This standard has application in the television studio over lengths of coaxial cable where the signal loss does not exceed an amount specified by the receiver manufacturer.

SMPTE 347M: 540 Mbit/s Serial Digital Interface—Source Image Format Mapping

This standard species the mapping of various source image formats onto the 540 Mbit/s serial digital interface. These formats include single-link 4:4:4:4 component digital signals (525i/59.94 and 625i/50), as well as progressive scan 4:2:2 component digital signals (525p/59.94 and 625p/50).

SMPTE 348M: High-Data-Rate Serial Data Transport Interface (HD-SDTI)

This standard provides the mechanisms necessary to facilitate the transport of packetized data over a synchronous data carrier. The HD-SDTI data packets and synchronizing signals provide a data transport interface that is compatible with SMPTE 292M (HD-SDI) such that it can be readily used by the infrastructure provided by this standard.

SMPTE 357M: Declarative Data Essence—Internet Protocol Multicast Encapsulation

This standard defines the encapsulation of declarative data essence using IP multicast. This is done in a transport-independent manner and relies solely on standard IP multicast techniques.

SMPTE 363.2M: Declarative Data Essence—Content Level 1

This standard defines a standard for the authoring of declarative data content intended to be combined primarily with video and/or audio services and distributed to data-capable television signal receivers. Declarative content is generally nonprocedural and most commonly in the form of HTML; however, procedural scripting is also defined.

SMPTE 364M: Declarative Data Essence—Unidirectional Hypertext Transport Protocol

This standard describes the Unidirectional Hypertext Transfer Protocol (UHTTP). UHTTP is a one-way data transfer protocol designed to deliver resource data in a one-way broadcast-only environment. This transfer protocol is appropriate for delivery of HTML and other content resources using IP multicast over television vertical blanking interval (IP/VBI) and other unidirectional transport systems.

SMPTE 366M: Declarative Data Essence—Document Object Model Level 0 (DOM-0) and Related Object Environment

This standard defines a document object model and an object environment for use in manipulating HTML documents using the ECMAScript environment of declarative data essence. The standard reflects the best current practice for continuing use in television and other applications.

SMPTE 372M: Dual-Link 292M Interface for 1920 × 1080 Picture Raster

This standard defines a means of interconnecting digital video equipment with a dual-link HD SDI (link A and link B), based on the SMPTE 292M data structure. The source formats for this dual-link interconnection are the picture raster formats, and digital interface representations as defined in SMPTE 274M.

ACQUIRING REFERENCE DOCUMENTS

Contact information is given below for a selection of standards developing organizations (SDOs) working in the digital video realm:

- *ATSC*—Advanced Television Systems Committee, 1750 K Street N.W., Suite 1200, Washington, D.C. 20006; phone 202-828-3130; fax 202-828-3131; http://www.atsc.org.
- *CEA*—Consumer Electronics Association, 2500 Wilson Blvd., Arlington, VA 22201-3834; phone 703-907-7600; fax 703-907-7675; http://www.ce.org.
- *IEEE*—Institute of Electrical and Electronics Engineers, Inc., 445 Hoes Lane, P.O. Box 1331, Piscataway, NJ 08855-1331; phone 800-678-IEEE (4333); outside United States and Canada, 732-981-9667; http://www.ieee.org.
- *ISO*—Global Engineering Documents, World Headquarters, 15 Inverness Way East, Englewood, CO 80112-5776; phone 800-854-7179; fax 303-397-2750; http://global.ihs.com.
- *ISO Central Secretariat*, 1, rue de Varembe, Case postale 56, CH-1211 Geneve 20, Switzerland; phone +41-22-749-0111; fax +41-22-733-34-30; http://www.iso.ch; e-mail central@iso.ch.
- *SCTE*—Society of Cable Telecommunications Engineers, Inc., 140 Philips Road, Exton, PA 19341; phone 610-363-6888; fax 610-363-5898; http://scte.org.
- *SMPTE*—Society of Motion Picture and Television Engineers, 595 W. Hartsdale Avenue, White Plains, NY 10607-1824; phone 914-761-1100; fax: 914-761-3115; http://www.smpte.org.

CHAPTER

1.17

Data Broadcasting Standards and Practices for Television

MICHAEL DOLAN
Television Broadcast Technology, Inc.
Del Mar, California

INTRODUCTION

Data broadcasting has, in the past, generally referred to the transmission of any information, carried along with the video or audio signal that is not the video or audio content itself. The classic example of "data" associated with radio is the radio data system (RDS), and with television, it is closed captioning. In analog television, *data* has also often been used to refer to anything encoded in the horizontal or vertical blanking interval (VBI) of the video signal, including what we today recognize more formally as metadata.

More commonly today, there is a distinction drawn between *metadata* (information about the video and audio content) and data *essence*, first coined in a Society of Motion Picture and Television Engineers (SMPTE) report [1]. Data essence refers to data items that convey their own information to the listener or viewer and not just information about the video or audio. For example, a digital program identifier encoded into the television VBI relates to the video and audio, has no intrinsic value on its own, and would rarely, if ever, be displayed to the viewer. It is thus considered metadata. In contrast, captioning is considered data essence since it has value on its own. Even though it is often a transcription of an audio track, it is not about the audio track. RDS and basic television electronic program guide (EPG) information may be considered to be metadata. However, more extensive and complex EPGs, especially containing future programming information, blur this distinction since they have intrinsic value to the viewer, allowing the viewer to schedule future viewing and recording, possibly including previewing programs and other advanced features.

The introduction of digital television (DTV) and digital radio systems enables a robust array of data broadcasting possibilities. This is primarily due to the fact that the broadcast is inherently digital, and (for television) that there is so much more bandwidth available. Being inherently digital makes the insertion and extraction of data easier, and provides a more robust (less error prone) delivery. In contrast, analog-based systems have limited bandwidth and are error prone in most distributions, especially when the signal is not strong at the receiving equipment. Reliable delivery in the analog environment often requires various elaborate redundancy techniques, such as forward error correction (FEC), which uses up more of the limited bandwidth.

This chapter introduces the principles and standards used for radio and television data broadcasting, in each case starting with the traditional analog systems, then describing the newer data systems associated with digital transmissions. For radio, this includes both data essence and metadata, while, for television, we are primarily concerned with data essence because the standards for DTV metadata (in particular ATSC PSIP) are covered elsewhere. Applications and implementation of radio and television data broadcasting are covered in other chapters in this book.

ANALOG TELEVISION SYSTEMS

Captioning and Text Services

In order to understand current developments in digital television data broadcasting, it is helpful to review analog data broadcasting, as it helps to explain the basic concepts and requirements of the digital systems. There are three main categories of analog data services in use today: captioning, text services, and Web services.

Captioning is a data service associated with the video and audio, and is considered a *synchronized* data service. That is, it provides a set of text and graphics that can be viewed simultaneously with its related video and audio. It is a *synchronized* service since the information is viewed along with specific video or audio segments, or even frames.

Captioning is also *streaming* since the encoding defines a stream of bytes that can be interpreted with very little context or structure. Closed captioning is encoded as text, can be turned on and off by the viewer, is normally in the same language as the primary audio, and is intended primarily as an aid to the hearing impaired. But, just as there may be multiple audio tracks, there can also be multiple caption channels. The Consumer Electronics Association (CEA) standard CEA 608 [2] defines how data services are encoded on line 21 of the VBI of the television signal. In addition to defining all the details of analog closed captioning, CEA 608 also defines the carriage of text services. Like captioning, text services are streaming but usually not related to the video and audio programming, nor are they synchronized. Text services have been used to deliver news wire applications, and often repeat, or "carousel," the feed. For more information on these closed captioning and text services, please refer to Chapter 5.24.

Teletext

There are a variety of authorized uses of VBI that exist but are sparingly deployed. Most of them are proprietary, but a few are standardized. One standard of note is North American Basic Teletext Specification (NABTS) [3]. Teletext in general is used widely in Europe, and NABTS is used in Canada, but teletext has never really been deployed in the United States. However, the design properties of a teletext service form an important foundation in digital data broadcasting.

Teletext is a stand-alone, asynchronous data service. That is, it provides a set of text and limited character-cell graphics symbols for viewing on the television display, usually unrelated to any video and audio programming. The information is organized into discrete pages, or blocks, and is not fundamentally a stream. The viewer may then navigate manually and asynchronously (relative to any video and audio) through these pages of information. Typical services that teletext provides are airplane and train schedules, weather, and sports scores.

Teletext pages are normally *carouseled* in the broadcast. A carousel is when the page is repeated at some time interval in order to facilitate its capture by the receiving equipment. This is needed for three main reasons. The first is when a viewer first tunes to the channel that is sending the service, the receiving equipment must be given a chance to acquire all the relevant pages. Since tuning can occur at any time, the pages must be repeated to allow capture of the pages regardless of when the viewer tunes. The second reason is to permit inexpensive receivers with virtually no memory to acquire the pages on demand from the viewer, thus not having to store them locally. Finally, a carousel provides some amount of error recovery. If a page or portion of a page is received with an error, it can be acquired again the next time it is sent. Carouseling algorithms often vary the frequency and order of the pages to try to improve the viewer experience. For example, the initial pages the viewer will see are sent faster than pages deeper in the organization of the pages. The general notion of carouseling and the algorithms for it remain very important in the digital domain.

Other Data Services

There are two important proprietary uses of VBI today worth noting: TVGuide Onscreen™ and Nielsen AMOL™. TVGuide Onscreen is a data service that provides program guide information to an application built in to many receivers. AMOL (automated measurement of lineups) is a metadata system that assists in the process of audience measurement (program ratings) and play-to-air commercial verification. While the details of both of these systems are proprietary, their use of analog VBI data is widespread today.

CEA 608, in addition to captioning and text services, also defines various items of metadata. Two important items being broadcast today in line 21, along with captions, include parental guidance ratings (V-CHIP) and copy protection, using the copy generation management system (CGMS).

There are numerous other standards and proprietary commercial ventures into analog television data broadcasting, including ATVEF, Cybercast [4], DDE [5], Digideck, Intercast, Moviebeam [6], Teleweb [7,8], WavePhore, WebTV [9], Wink, and Yes! Entertainment. However, many of these systems are no longer in use today.

TABLE 1.17-1
Summary of Features of Traditional Data Services

	Stand-Alone	Synchronized	Stream	Block/ Carousel
Captioning		X	X	
Text services	X		X	
Teletext	X			X

This review of analog television data broadcasting is important in providing the context for the development of the digital data models. From the above applications, we learn that we need both stand-alone and associated data services, both synchronized and asynchronous services, both streams of data and organized blocks of data, and the ability to carousel the blocks (pages) of information. This is summarized in Table 1.17-1.

These properties carry over to the infrastructure of the digital television data broadcast system.

DIGITAL TELEVISION SYSTEM DATA MODELS

The analog data services mentioned previously were introduced before any formal data models for such broadcasting were developed, and, in general, were each designed as individual special cases.

The introduction of digital television systems offered the opportunity to define a set of building blocks, or infrastructure, in which to carry some basic data models. While every digital transport has defined its own unique details, the basic data models that have been universally defined are Internet Protocol (IP) data packets, files (analogous to a computer file system), streams of bytes, and "triggers." These can be thought of in aggregate as a data broadcast infrastructure, which can then be used for the design and delivery of specific *data services* to the listener or viewer. Data services are an application of data broadcasting that result in some experience for the listener or viewer. Also, using this data infrastructure, interactive television (ITV) systems, such as the Advanced Television Systems Committee (ATSC) Advanced Common Application Platform (ACAP) [10], can add another layer of "tools" for the ultimate data service provider.

The International Standards Organization (ISO) and International Electrotechnical Commission (IEC) have defined many of the basic building block mechanisms used. Most come from the ISO/IEC 13818 standards series, part 6, and its amendments on Digital Storage Media—Command and Control (DSM-CC) [11]. The whole of the DSM-CC standard is complex and designed as an architecture and control mechanism for elaborate video-on-demand (VOD) systems. However, it includes some special MPEG-2 private sections (as defined in 13818-1 [12]) to carry a set of message formats that support the carriage of files and network IP packets. More information on the specific uses of the DSM-CC messages is provided in the following sections. First, the IP packets, files, streams, and triggers data models that apply generically to all of the transports are discussed, along with more details of their carriage using the DSM-CC mechanisms. Next, there is a brief overview of the ITV systems that use the data models; and finally the specific details for each of the digital facilities and the ATSC and other digital transports are discussed. It is assumed the reader is generally familiar with MPEG-2 and computer system concepts.

Internet Protocol Packets

IP packets are defined by a variety of Internet Engineering Task Force (IETF) [13] standards, which are beyond the scope of this chapter. However, because of the extensive amount of interoperable equipment available that can encode, decode, and route IP packets, their migration into broadcast television is not surprising. The television signal is effectively used as a single-hop, unidirectional subnetwork route. When used in the forward channel of broadcast systems, it is common to use IP multicast addressing, as it is generally not efficient to carry point-to-point (unicast) addressed packets in a broadcast medium. When used in MPEG-2 transports, IP packets are usually constructed according to the ISO MPEG-2 part 6 standard commonly known as DSM-CC (amendment no. 1). The encapsulation is a private table in MPEG-2 known as a DSM-CC addressable section. This encapsulation includes fields for physical layer addressing (media access control, or MAC, address), as well as the IP packet payload itself.

IP packets are typically asynchronous. A synchronized version of the encapsulation is also defined, but quite uncommon in practice. IP packets form the basis for many of the early new data services today found in several startup data broadcasting service companies' designs. In fact, IP packets are often used to carry proprietary file and stream formats, even though native file and stream mechanisms are defined.

Files

Files in data broadcasting are blocks of data just like files on a computer system and may be used to carry a variety of digitally-encoded information. When used in MPEG-2 transports, they are packaged and carried according to two kinds of mechanisms also defined in DSM-CC: the data carousel and the object carousel. The term *data carousel* is used interchangeably here with "download" to loosely describe all variant forms of the DSM-CC download protocols, which include the data carousel.

Data Carousel

The data carousel provides a basic packaging of a single file into a module, along with some signaling to identify and locate all the modules of the carousel. Each module is broken into one or more blocks. The module blocks are carried in a DSM-CC message called a Download Data Block (DDB). The DDBs are sequentially numbered and, when combined, form a complete module. Metadata for the modules, as well as additional signaling that assists in their acquisition, is communicated in a message called a Download Information Indication (DII). This message contains a list of modules, their size, and optionally a name. Some implementations support a two-layer hierarchy with the addition of the message, Download Server Initiate (DSI). This message references one or more DII messages. The organization of a simple two-layer data carousel, including the DSI message, the DII messages,

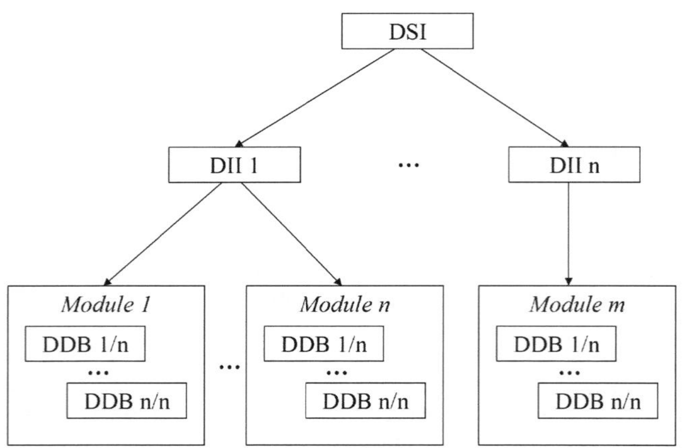

FIGURE 1.17-1 DSM-CC data carousel organization.

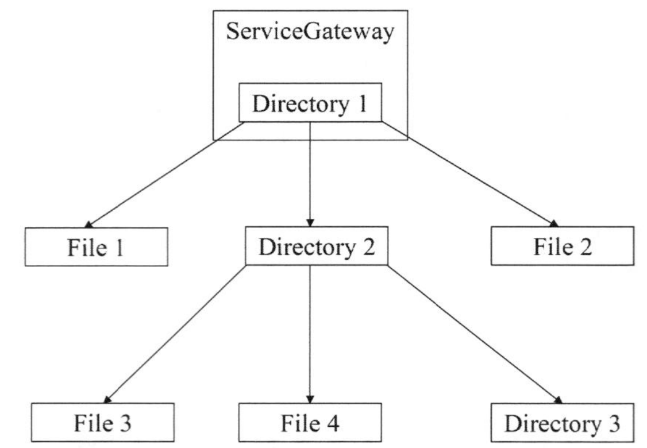

FIGURE 1.17-2 DSM-CC object carousel organization.

modules, and DDB messages, is shown in Figure 1.17-1.

The organization of the modules is, at most, a two-layer organization and is thus flat (no useful hierarchy), and, for small files, carrying them one per module can be relatively inefficient.

Object Carousel

The DSM-CC object carousel builds on the data carousel structure defined above and provides a more robust, although more complex, system. Relative to the data carousel, it adds a full hierarchical organization through the addition of directories (analogous to those found in computer file systems). This is done through the definition of another layer of structure generically called objects. The object carousel is an open-ended mechanism in which to deliver objects of any kind. Two important intrinsic objects defined by DSM-CC are File and Directory, which contain the obvious contents. Additionally, there is a support object called the ServiceGateway, which basically provides the entry point to the object carousel Directory hierarchy. The organization of these basic objects is shown in Figure 1.17-2.

The objects are mapped onto the elements of the data carousel as follows. The ServiceGateway object is signaled in the DSI. The ServiceGateway is a special kind of a Directory object and provides the "root" directory of the file system. All other objects (including Directory and File) are carried in data carousel modules. Objects may be packed more than one per module, thus making it more efficient for broadcasting many small (File) objects. This packing strategy provides another layer of opportunity to increase the acquisition efficiency. For example, one could aggregate all the Directory objects into a single module so that a receiver could more efficiently acquire Files more randomly.

Other DSM-CC intrinsic objects that are common to add to the object carousel in ITV systems are the Stream and StreamEvent. See the DSM-CC standard for more information on these. Some of the new ITV designs have chosen to expose only the object carousel objects rather than both that and the data carousel layer it builds on.

Streams

Streams of data bytes can be carried just like video and audio elementary streams. In the case of MPEG-2-based transports, they are most often packetized elementary stream (PES) encapsulated [12]. However, they can also be transmitted using a blocked format with the DSM-CC Download protocol. This form of download is not a carousel, but a single emission of data broken up into modules. This blocking offers some advantages over MPEG-2 PES. First, there is error detection through cyclic redundancy check (CRC) and block numbering. The latter allows for detection of entire missing blocks. Data cannot be reassembled out of order due to the block sequencing requirement. Finally, there is the advantage of a more well-formed buffer model for the *chunks* of the stream.

Streams can also be carried in what is known as *data piping*. Data piping is a formal name for carrying data in the raw MPEG-2 transport packet without any further standard structure (i.e., not PES or Download chunks). Streams are the least well-formalized carriage of all the models.

Triggers

Triggers are a special type of data model that are intended to link data to the video and audio elements, deliver an event to a receiver data application, and/or provide a timed sequence of stand-alone data essence. Triggers are inherently synchronized, although in practice they can be sent as a "best effort" and not linked to specific frames of video or audio. The trigger can also be used to get around some MPEG-2 program clock reference (PCR) and decoder model buffering issues when the data chunks are large and transmitted over a long period of time. Triggers have the most widely varying implementation across the transports, including uniform resource locators (URL) [14]

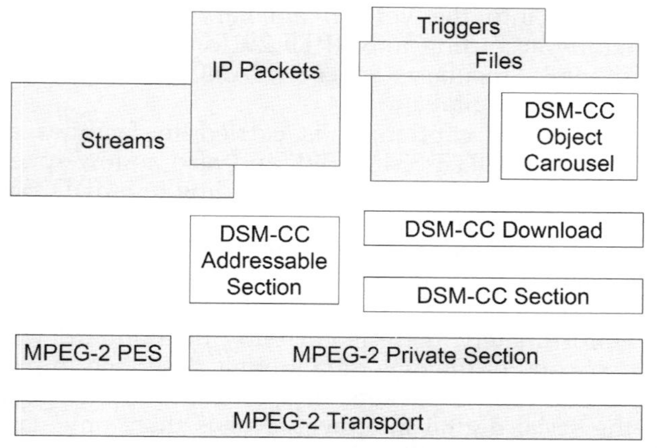

FIGURE 1.17-3 Summary of MPEG-2 data broadcast protocols.

strings, special DSM-CC Download payloads, and object carousel StreamEvent objects.

Summary of Data Models on MPEG-2 Transports

The general MPEG-2 encapsulations for all the above data models can be summarized in a stack of transport protocols defined by 13818-1 (MPEG-2 Systems), 13818-6 (DSM-CC), and as shown in Figure 1.17-3.

INTERACTIVE TELEVISION TECHNOLOGY

Interactive television means many different things today and generally applies to the collection of digital system viewer experiences ranging from electronic program guide (EPG) and video on demand (VOD) to the technology of various middleware systems that run on DTV receivers—virtually anything involving the remote control other than changing channels of volume. One use of the term considered here is the standard ITV environments defined by regional digital television standards organizations: ATSC [15], Society of Cable Telecommunications Engineers (SCTE) [16], and European Telecommunications Standards Institute (ETSI) [17]. The most recent ATSC ITV environment is ACAP [10], derived from OCAP and its predecessor design, the DTV Application Software Environment (DASE) [18]. The U.S. Cable ITV environment is the SCTE OpenCable Application Platform (OCAP) [19] developed by Cablelabs [20]. The ETSI ITV environment is the Digital Video Broadcasting (DVB) [21] Multimedia Home Platform (MHP) [22].

This section will provide a high-level overview of the technology common to these systems. The technology can be grouped into two main environments: the Java™ application environment and the Web application environment. There are also some special file formats supported that include JPEG and PNG for graphics and often the Bitstream™ font format.

Java Environment Technology

The Java environment includes a Java Virtual Machine (JVM) [23] and a set of standard *Java packages*. The JVM is a programming language (Java) interpreter that executes the Java *byte codes* in *class files* on the receiver's processor. The byte codes are the compiled version of the text Java programming language. The class files contain the compiled byte codes for the packages. The advantage to an interpretive language like Java is that it provides a hardware-independent environment for the ITV authors.

The Java packages (i.e., application programming interfaces) consist of several main components: Personal Java [24], Java Media Framework (JMF) [25], and Java TV [26] (for the latest commercial libraries that apply, see the product listings from the Java vendors such as Sun Microsystems). Personal Java is a profile of the more widely known Java Development Kit (JDK) available for most computer systems. It includes access for file data and Internet packets. JMF is a package used for streaming data access and control. The Java TV package is an abstraction of the MPEG-2 environment of tables and descriptors defined by DVB and ATSC for signaling and announcing the programs and services in the MPEG-2 transport. These packages have been aggregated recently by the DVB and given the name Globally Executable MHP (GEM) [27].

Web Environment Technology

The Web environment usually consists of a set of basic Web browser technology to process what amount to a set of Web pages not unlike today's desktop computer Web browser. The standards for this environment are derivations of those published by the World Wide Web Consortium (W3C) [28]. The Web technology is usually some form of W3C Markup Language [29], a corresponding W3C Style Sheet [30], and often includes a script interpreter called ECMAScript [31] (also known as JavaScript, but not to be confused with Java). The HTML derivation is more recently a profile (subset) of the newer Extensible Hypertext Markup Language (XHTML) [32], although there are deployed systems that use HTML versions 3.2, 4.0, and 4.1. A profile (subset) of cascading style sheets (CSS) has been defined by the W3C specifically for television, called CSS-TV [33]. CSS is a way to control the layout of the pages on the receiver's display (e.g., what fonts to use for each HTML element, etc.). ECMAScript is an interpretive programming language and is used quite often on Web pages on the Internet to provide animation, receiver-side form validation, and other simple programming tasks. Since it is a programming language, it too has a set of APIs that are somewhat loosely referred to as the document object model (DOM) [34], although strictly speaking the API goes beyond anything having to do with the HTML document itself.

DIGITAL CLOSED CAPTIONING AND TELETEXT

The data models for closed captioning and teletext in the digital domain mirror those of the analog domain. U.S. digital closed captioning, defined by CEA 708 [35], is a synchronized streaming text service. Its encapsulation does not use the ISO stream formats, but instead embeds the caption information in the video stream, due mostly to historical reasons. European digital teletext is defined by both ETSI [36] and the International Telecommunications Union (ITU) [37], which are different systems. Depending on the use (for subtitling or page-based services), teletext follows the general models for streams or files as appropriate. However, it is like ATSC in that the designs do not follow the DSM-CC encapsulations and use separate synchronized PES streams (although not in the video stream like ATSC). For more information on digital closed captioning, please refer to Chapter 5.24.

DIGITAL TELEVISION TRANSPORT SPECIFICS

Each of the data models and ITV environments described above has variations depending on the particular transport being used. This section discusses the variations for transport in the studio facility and in the U.S. terrestrial (ATSC) transport, the U.S. cable (SCTE) transport, and the European (ETSI/DVB) and Japanese (ARIB) transports.

Studio Facility

For the purposes of this section, it is assumed that the studio facility uses serial digital baseband video links defined by SMPTE standards SMPTE 259M and SMPTE 292M. Unlike the MPEG-2-based digital systems, the facility can carry data in multiple transports. Today's facilities generally route IP packets not on serial digital video links, but by using IP networks, with CAT-5 cable or similar. When data (of any kind) is carried on the serial digital systems, it is usually packaged into the vertical ancillary space (VANC) generally according to SMPTE 291M [38]. Use of the horizontal ancillary space (HANC) is typically reserved for digital audio.

U.S. digital captioning is carried in facilities as defined in SMPTE 334M [39], and also as low-speed serial data on dedicated links according to SMPTE 333 [40]. More information about the various configurations and interconnections can be found in a very informative guideline on facility captioning, SMPTE EG43 [41]; see also Chapter 5.24 of this book.

Unfortunately, there is currently no standard way to carry any of the new data models described in previous sections (IP packets, files, streams, and triggers) in the serial digital interfaces. Nor is there any standard way to carry the ITV system data, whether it is built on these data model layers or not. The assumption by the industry today is that there is a special device (a "data server") that formats these data models into the appropriate MPEG-2 transport format based on the data arriving at the server from unspecified means. This situation is less than desirable as it prevents any standard tight association of data on peer with its related video and audio, and it encourages proprietary solutions surrounding the "data server." More standardization work can be expected in this area.

U.S. Terrestrial (ATSC)

ATSC is an industry forum that publishes its own standards. The foundational work for data broadcasting in ATSC is A/90 [42] that adapts the data models and encapsulations of DSM-CC. All of the data models described here are more formally defined in A/94 [43], which may be helpful in understanding these not only for ATSC, but for the other transports as well. Guidelines for IP packet carriage for multicast sessions is defined in more detail in A/92 [44]. The ATSC object carousel that can be used for files (in addition to the data carousel) is defined in A/95 [45]. Note that this is intended to be compatible with the object carousel of ACAP. Triggers are defined in A/93 [46]. A more thor-

TABLE 1.17-2
Summary of ITV and Transport Standards

	ATSC	SCTE	ETSI	ARIB
Captioning	CEA 708	CEA 708	—	B-36
Teletext	—	—	300-706, ITU BT.653-2	—
IP packets	A/90, A/92	42	301-192	—
Files	A/90, A/95	90-1	301-192, 102-812	B-24
Streams	A/90	—	301-192	B-24
Triggers	A/93, A/100, A/101	90-1	102-812	B-24
Java	A/100, A/101	90-1	102-812	—
HTML	A/100, A/101	—	102-812	B-24

ough discussion of the ATSC data broadcast specification can be found in the Chernock-Crinon-Dolan-Mick text [47].

The A/9x series of standards provides a general framework for constructing specific data broadcasting systems, but does not define any application. One such application of A/9x is software download, defined in A/97 [48]. This service provides a standard mechanism for manufacturers to update their decoder devices via the broadcast.

U.S. Cable (SCTE/Cablelabs)

U.S. digital cable standards are developed jointly between SCTE and a private organization, Cablelabs. U.S. cable standards are, in practice, a mixture of ATSC and DVB standards, and the carriage of data is no exception. Cable defined the carriage of IP packets over MPEG-2 following the DVB encapsulation and documented it in SCTE 42 [49]. However, due to the existence of bidirectional communication inherent in cable, IP packets are normally carried in the out-of-band channel and not in the forward MPEG-2 transport. Although cable has made reference to the A/9x series of ATSC, it has not deployed them as an infrastructure. Software download continues to use a proprietary mechanism. The ITV standard for U.S. cable is a derivation of the DVB MHP work, known as OCAP.

Europe (ETSI/DVB)

DVB has addressed the digital data broadcast needs through several standards ultimately published by ETSI (see Chapter 1.11). The foundation data broadcasting specification for DVB is ETSI 301-192 [50]. It defines the basic packaging for IP packets, files, and streams. Files may be carried in either the data carousel or the object carousel.

A software download service in the United Kingdom is defined and deployed by the forum, the Digital TV Group (DTG) [51], and is very similar to the ATSC download service. European ITV is defined by both a lightweight system, known as MHEG [52], as well as the ETSI/DVB MHP/GEM mentioned before.

Japan (ARIB)

Subtitling in Japan is carried much like it is in Europe, and is defined in standards from the Association of Radio Industries and Businesses (ARIB) [53], ARIB B-36 [54]. Its carriage in the facility is covered by B-37 [55], analogous to SMPTE 334 for U.S. captions. The ITV system used in Japan is the Broadcast Markup Language (BML) [56].

Summary of Digital MPEG-2 Data Transport Standards

A summary of the standards that cover the ITV and transport encapsulations is shown in Table 1.17-2. IP encapsulation and the object carousels vary only slightly between ATSC, SCTE, and DVB. Triggers are entirely different, mainly since the ATSC trigger is not dependent on the object carousel as is the SCTE/ETSI trigger.

SUMMARY

The capability to carry data along with audio and video signals has been available to broadcasters for some time and (particularly with digital television) provides some powerful tools for services from broadcasters. The opportunities enabled by these tools have not as yet been fully exploited (at least in the United States) but data broadcasting, in its various forms, is likely to play an increasingly important role in the age of digital broadcasting. This chapter has introduced the main principles and standards that form the basis for the technology. Much more information on particular topics is available in the listed references and more examples of applications for data broadcasting for both television and radio can be found in Chapter 5.25 of this book.

ACKNOWLEDGMENTS

Major portions of this chapter have been adapted from the *Broadcast Engineer's Reference Book*, chapter 11, Data Broadcast, by Michael Dolan, published by Focal Press, and used with permission of the publisher.

Minor portions of this chapter draw on material from the ninth edition of the *NAB Engineering Handbook*.

References

[1] SMPTE/EBU, Joint Task Force for Harmonized Standards for the Exchange of Program Material as Bit Streams, Final Report: Analyses and Results, July 1998.
[2] CEA 608-C, line 21, Data Services.
[3] CEA 516, North American Basic Teletext Specification (NABTS).
[4] EN Technology Cybercast, at http://www.entechnology.com.
[5] SMPTE, SMPTE 363M Standard for Television: Declarative Data Essence, Content Level 1.
[6] Disney Moviebeam, at http://www.moviebeam.com.
[7] IEC 62297, Triggering Messages for Broadcast Applications.
[8] IEC 62298, Teleweb Application.
[9] Microsoft WebTV, or MSN-TV, at http://www.webtv.com.
[10] ATSC A/101, Advanced Common Application Platform (ACAP).
[11] ISO/IEC 13818-6, Information Technology: Generic Coding of Moving Pictures and Associated Audio Information, Part 6: Extensions for DSM-CC.
[12] ISO/IEC 13818-1, Information Technology: Generic Coding of Moving Pictures and Associated Audio Information, Part 1: Systems.
[13] IETF Internet Engineering Task Force, at http://www.ietf.org.
[14] IETF RFC 3986, Uniform Resource Identifier (URI) Syntax.
[15] Advanced Television Systems Committee (ATSC), at http://www.atsc.org.
[16] Society of Cable Telecommunications Engineers (SCTE), at http://www.scte.org.
[17] European Telecommunications Standards Institute (ETSI), at http://www.etsi.org.
[18] ATSC A/100, DTV Application Software Environment.

[19] SCTE 90-1, SCTE Application Platform Standard, Part 1: OCAP 1.0 Profile.
[20] Cablelabs Cable Television Laboratories, Inc., at http://www.cablelabs.org.
[21] Digital Video Broadcast (DVB), at http://www.dvb.org.
[22] ETSI 102-812, Digital Video Broadcasting (DVB); Multimedia Home Platform (MHP) Specification 1.1.
[23] Lindholm, T., and Yellin, F. *The Java™ Virtual Machine Specification*, 2nd edition. Boston: Addison-Wesley, 1996.
[24] Sun Microsystems Personal Java Product, at http://java.su.
[25] Sun Microsystems JMF Product, at http://java.sun.com/products/java-media/jmf.
[26] Sun Microsystems JavaTV Product, at http://java.sun.com/products/javatv.
[27] ETSI TS 102-819, Digital Video Broadcasting (DVB); Globally Executable MHP (GEM) Specification.
[28] World Wide Web Consortium (W3C), at http://www.w3.org.
[29] W3C Markup Languages, at http://www.w3.org/MarkUp/.
[30] W3C Cascading Style Sheets, at http://www.w3.org/Style/CSS/.
[31] ISO, ISO Standard 16262, Information Technology: ECMAScript Language Specification.
[32] W3C XHTML 1.0: The Extensible HyperText Markup Language, at http://www.w3.org/TR/2000/REC-xhtml1-20000126.
[33] W3C CSS TV Profile 1.0, at http://www.w3.org/TR/css-tv.
[34] W3C Document Object Model (DOM), at http://www.w3.org/DOM/.
[35] CEA 708-C, Digital Television (DTV) Closed Captioning.
[36] ETSI 300-706, Digital Video Broadcasting (DVB); Enhanced Teletext Specification.
[37] TU-R BT.653-2, Teletext Systems.
[38] SMPTE 291M, Ancillary Data Packet and Space Formatting.
[39] SMPTE 334M, Vertical Ancillary Data Mapping for Bit-Serial Interface.
[40] SMPTE 333M, DTV Closed-Caption Server to Encoder Interface.
[41] SMPTE EG43, System Implementation of CEA-708-B and CEA-608-B Closed Captioning.
[42] ATSC A/90, Data Broadcast Standard.
[43] ATSC A/94, Data Application Reference Model.
[44] ATSC A/92, Delivery of IP Multicast Sessions over Data Broadcast Standard.
[45] ATSC A/95, Transport Stream File System Standard.
[46] ATSC A/93, Synchronized/Asynchronous Trigger Standard.
[47] Chernock, R., Crinon, R., and Dolan, M., et al. *Data Broadcasting: Understanding the ATSC Data Broadcast Standard*. New York: McGraw-Hill, 2001.
[48] ATSC A/97, ATSC Standard: Software Download Data Service.
[49] SCTE 42, IP Multicast for Digital MPEG Networks.
[50] ETSI 301-192, Digital Video Broadcasting (DVB); DVB Specification for Data Broadcasting.
[51] The Digital TV Group (DTG), at http://www.dtg.org.uk.
[52] ISO 13522-5, Information Technology: Coding of Multimedia and Hypermedia Information, Part 5: Support for Base-Level Interactive Applications.
[53] Association of Radio Industries and Businesses (ARIB), at http://www.arib.or.jp.
[54] ARIB B-36, Exchange Format of the Digital Closed Caption File for Digital Television Broadcasting System.
[55] ARIB B-37, Structure and Operation of Closed-Caption Data Conveyed by Ancillary Data Packets.
[56] ARIB B-24, Data Coding and Transmission Specification for Data Broadcasting.

CHAPTER

1.18

Emergency Alert System

DAVID E. WILSON
Consumer Electronics Association
Arlington, Virginia

OVERVIEW

The purpose of the Emergency Alert System (EAS) is to enable the President of the United States to speak directly to the public over broadcast and cable channels during a national emergency. A secondary purpose is to enable state and local government leaders to speak directly to the public during state and local emergencies. The EAS requirements are described in Part 11 of the Federal Communications Commission's (FCC's) Rules [1]. The specific details outlined in this chapter are current as of the November 10, 2005, release of the FCC Report and Order in EB Docket No. 04-296 ("In the Matter of Review of the Emergency Alert System").[1] This chapter analyzes the EAS from five different perspectives:

- EAS responsibilities of individual broadcasters
- Flow of messages within the EAS network
- Methods used to generate the EAS signal
- Coding structure of the digital message
- EAS transmission via the radio data system (RDS)

BROADCASTER EAS RESPONSIBILITIES

Participation in the EAS by broadcasters, cable systems, and satellite systems is voluntary, although entities that opt not to participate must receive prior approval from the FCC. Broadcasters, cable systems, and satellite systems that opt not to participate are required to cease transmissions during times of national emergency when the EAS is activated. These stations are called *nonparticipating national* (NN) stations.

The FCC strictly enforces its EAS rules. In 2005, it averaged about one citation per month related to EAS. Broadcasters were typically cited for failing to have EAS equipment installed, properly maintained, or in automatic mode at unattended transmitter sites or for failing to conduct required weekly and monthly tests. Under Section 503(b) of the U.S. Code, the FCC has the authority to fine stations up to $25,000 for each violation or each day of a continuing violation, up to a maximum of $250,000 for any single violation. The fines issued for EAS violations in 2005 were typically around $10,000.

Decoders

With only a few exceptions, all analog and digital broadcast stations, cable operators, and satellite radio and television systems, including nonparticipating ones, are required to have EAS decoders at their facilities so they can detect national alerts. Broadcast stations that operate as satellites or repeaters rebroadcasting 100% of the programming of another station and low-power TV stations that operate as translators for full-power stations are not required to have any of their own EAS equipment. Also, the home satellite dish systems that have been around since the days before DIRECTV® and Dish™ Network are not required to have any of their own EAS equipment.

[1]At the time of publishing, the FCC was considering comments and reply comments to a "Further Notice of Proposed Rulemaking" in this docket.

Encoders

Full-power AM, FM, and TV stations (both analog and digital) and Class A TV stations (both analog and digital) are required to have EAS encoders unless they operate as satellites or repeaters rebroadcasting 100% of the programming of another station or they have received permission from the FCC to be nonparticipating stations. Stations that operate as satellites or repeaters are required to rebroadcast the EAS messages from the hub station or common control point. Analog and digital cable systems, including wireless cable systems (Broadband Radio Service [BRS] or Educational Broadband Service [EBS] stations), are required to have EAS encoders. Satellite digital audio radio service (SDARS) systems and direct broadcast satellite (DBS) systems are also required to have EAS encoders, although EAS requirements for DBS systems do not become effective until May 31, 2007. Class D FM stations, low-power TV (LPTV) stations, and low-power FM (LPFM) stations are not required to have EAS encoders.

Individual Station Operating Requirements

Individual stations have six basic EAS responsibilities:

1. Own, install, and keep in good working order an EAS encoder/decoder, as required by the FCC Rules.
2. Have a current copy of the EAS handbook.
3. Send a test message once every week, and log the fact that it was sent. The weekly test must be sent on random days within the week and at random times throughout the day or night. A video message is not required to be part of the weekly test. No weekly test is required during any week that a monthly test occurs nor is a weekly test required during any week when a state or local emergency causes the station to transmit an EAS message. If, however, the station receives one or more EAS messages concerning state or local emergencies during a week and the station elects not to retransmit any of these messages, then the station must still broadcast the weekly EAS test message sometime during the week. Weekly EAS tests by digital radio stations must be sent over all audio streams, and weekly EAS tests by digital TV stations must be sent over all program streams. Weekly EAS tests by analog and digital cable systems with 5000 or more subscribers per headend and by wireless cable systems with 5000 or more subscribers must be sent over all programmed channels. Weekly tests by analog and digital cable systems with fewer than 5000 subscribers per headend and by wireless cable systems with fewer than 5000 subscribers must be sent over at least one programmed channel. Weekly EAS tests by SDARS providers must be sent over all channels. DBS providers, analog and digital class D noncommercial educational FM stations, and analog and digital LPTV stations are not required to transmit the weekly test but must log receipt of weekly tests.
4. Log the date and time that weekly test messages and those state or local alert messages that have been selected by the broadcaster are received from assigned monitoring sources. If a test is not received from any assigned source, then the reason for the failure to receive the test must be determined. This can be done by checking the station's EAS equipment and by calling each monitoring source from which no EAS messages were received during the week. The station log must reflect the findings of any investigation into why EAS messages were not received. Also, any necessary corrective action must be taken to ensure that EAS tests and EAS alerts will be received properly as soon as possible.
5. Transmit the monthly test message within 60 minutes of receiving it every month, and log both the receipt and transmission of this message. The monthly test must be conducted between 8:30 a.m. and sunset, local time, in odd-numbered months, and between sunset and 8:30 a.m., local time, in even-numbered months. During the week that the monthly test occurs, no weekly test is required. Monthly test messages must be rebroadcast in their entirety, except for the transmitting station identifier (ID) code. Each station that rebroadcasts a monthly test message must insert its own transmitting station ID code. State primary (SP) and local primary (LP) sources originate these monthly tests. (SP and LP sources are described in the "Message Flow within EAS" section of this chapter.) If, during a particular month, a station transmits an actual EAS emergency message that is of the same format as the monthly test, then the station does not have to transmit the monthly test. Analog and digital Class D noncommercial educational FM stations and analog and digital LPTV stations are required to transmit only the test script during the monthly tests. Effective May 31, 2007, DBS providers must monitor SP or LP sources and retransmit their monthly test messages on at least 10% of their channels each month (excluding local-into-local channels for which the monthly transmission tests are passed through by the DBS provider). The DBS channels tested must vary from month to month so that over the course of a given year all system channels are tested.
6. Immediately broadcast any national level alerts that are received. Although broadcasters can choose to air, or not to air, any state or local alert, they have no discretion when a national level alert is received. If the audio for a national alert is available from an alternative source of higher audio quality, this alternative source may be rebroadcast in lieu of the audio portion of the EAS message; however, retransmission of a national alert may not be delayed to wait for a higher quality audio source. Both automatically operated and manually operated stations must interrupt their programming immediately when a national level alert is received. Digital radio and TV stations, digital

cable systems, SDARS systems, and DBS systems must transmit national alerts on all audio and video program streams, except that analog and digital cable systems with fewer than 5000 subscribers per headend and wireless cable systems with fewer than 5000 subscribers need only transmit EAS messages on one channel, but they still must provide a video interruption and an audio alert message on all channels, and the audio alert message must state which channel is carrying the EAS video and audio message. Stations that are participating in the EAS must put the national alert on the air. Stations that have opted not to participate in the EAS (by providing written notification to the FCC's EAS office) must immediately rebroadcast the digital header and two-tone attention signal for national level alerts, followed by an announcement indicating that a national level alert is about to be broadcast by other stations, and that this station is about to go off the air. The NN station must then go off the air for the duration of the national level alert and wait for the Emergency Action Termination message.

Although not required by the FCC rules, many stations choose to voluntarily rebroadcast certain EAS messages regarding state and local emergencies. If an SDARS or DBS system is not capable of receiving and transmitting state and local EAS messages on all channels, it must inform its subscribers on its Web site and in writing on an annual basis and indicate which channels are and are not capable of supplying state and local messages.

Broadcasters may operate their EAS equipment in either an automatic or a manual mode. In the automatic mode, the broadcaster programs the EAS unit to automatically relay certain selected EAS messages that are received by its decoder. Stations that are operating without a transmitter duty operator will typically set their EAS equipment in automatic mode. When EAS equipment is operated in manual mode, the broadcaster must assign a person to monitor the EAS equipment and make decisions about which alerts to relay. Stations that operate their EAS equipment in manual mode are allowed to control their EAS equipment from a remote location, provided their remote control system enables them to put national level alerts on the air immediately and the required monthly tests on the air within 60 minutes.

Every national EAS message and every test message that is received or transmitted by a broadcaster must be logged. State and local emergency messages must be logged if they are retransmitted by the broadcaster.

MESSAGE FLOW WITHIN EAS

The EAS network is a blending of daisy-chain and Web-like signal distribution. It is like a daisy chain in the sense that there is a well-defined hierarchy of broadcasters and cable systems in which national EAS alerts enter the system at primary entry point (PEP) stations and flow from there to state primary (SP) stations, state relay (SR) stations, local primary (LP) stations, and finally to the other local stations and cable systems in the local area. It is Web-like in the sense that each station is required to monitor at least two sources for EAS alerts, thus ensuring that a failure at a single station does not prevent a message from getting through to all stations below it. State-level alerts can enter the system at SP stations, and local alerts can enter the system at LP stations.

In the EAS, all broadcast stations fall into one of six basic categories:

- National primary (NP) source
- State primary (SP) source
- State relay (SR) source
- Local primary (LP) source
- Participating national (PN) source
- Nonparticipating national (NN) source

The decisions regarding what type of source an individual station will be and what stations it must monitor are made by its State Emergency Communications Committee (SECC). Some localities may also have a Local Emergency Communications Committee (LECC) that specifies procedures for local emergencies. SECCs and LECCs are made up mostly of broadcasters that serve on a voluntary basis. These committees develop the state and local emergency communications plans for their regions. Their plans must first be approved by the FCC before they can go into effect. Once approved by the FCC a state plan, or a state plan amendment, is incorporated into the FCC's EAS mapbook for that state. The FCC EAS mapbook for a state lists all the broadcast stations in the state organized by their EAS local area and any special designations needed to identify each station's role in the EAS. Broadcasters may obtain a copy of the FCC EAS mapbook from the FCC's EAS office. An example of how messages can flow within the EAS is illustrated in Figure 1.18-1.

There are three possible entry points for an EAS alert:

1. The federal government can initiate a message by feeding it to an NP source.
2. State governments can initiate EAS messages by feeding them into SP sources. Typically, an SP source will be located in the state capital, although it can be located anywhere within the state. Most states are large enough that a single broadcaster's signal cannot cover the entire state. In these situations, it is necessary to have relay stations pick up state-level alerts and relay them out to distant counties. These relay stations are referred to as SR sources.
3. Local governments, local emergency management offices, and the National Weather Service can initiate EAS messages by feeding them into LP sources. LP sources are assigned numbers (LP-1, LP-2, etc.) to indicate the sequence in which they are to be monitored by other broadcast stations in the local area.

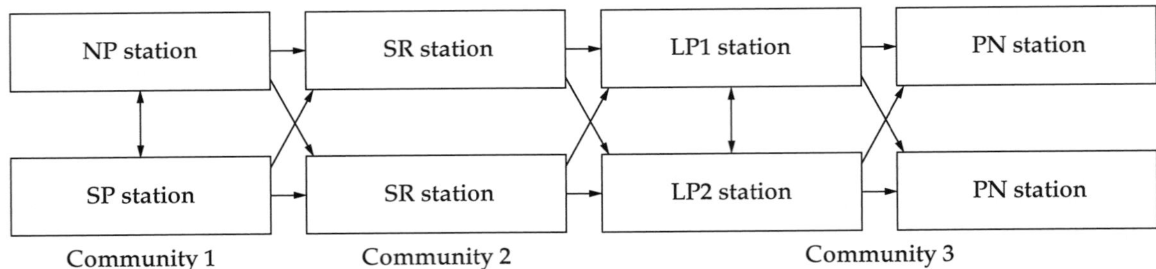

FIGURE 1.18-1 Example of message flow within the EAS.

EAS MESSAGE PROTOCOL

An EAS message consists of four elements, in the following order: digital header code (repeated three times); two-tone attention signal; audio, video, or text message describing the actual alert; and digital end-of-message code (repeated three times). EAS message elements are illustrated in Figure 1.18-2. Four different frequency tones are used to create an EAS message: 853, 960, 1562.5, and 2083.3 Hz. The first two tones are the same as those used to produce the well-known two-tone attention signal in the old Emergency Broadcast System (EBS). They continue to be used to produce the two-tone signal in the EAS, although their duration is not as long as it was under the EBS, and they are used less frequently. The second two tones are used to produce the digital header and end-of-message codes, which are the principal improvements of the new EAS over the old EBS.

In the digital header and end-of-message codes, the 2083.3 Hz tone represents a mark (or "1") and the 1562.5 Hz tone represents a space (or "0"). When a digital bit is transmitted, its associated tone is transmitted for precisely 1.92 msec, which is the amount of time it takes to transmit exactly three cycles of the 1562.5 Hz tone and four cycles of the 2083.3 Hz tone. Figure 1.18-3 illustrates, in the time domain, how these tones can be merged together to create an 8 bit digital byte—or one ASCII character. The character depicted in Figure 1.18-3, for example, is the capital letter "U," as defined in ANSI X3.4-1977, which is the ASCII character set standard referenced in the FCC's EAS rules. In this form of ASCII text transmission, the least significant bit is transmitted first, and the most significant (seventh) bit is transmitted last, followed by an eighth null bit (either 0 or 1) to create an 8 bit byte. In Figure 1.18-3, the character being transmitted is "X1010101," which corresponds to the decimal ASCII character value of 85 and the capital letter "U." The two-tone attention signal is produced by simultaneously transmitting an 853 Hz tone and a 960 Hz tone. Figure 1.18-4 is an illustration of the composite two-tone signal.

When the digital components of an EAS message are transmitted, FCC rules require that they modulate the transmitter at no less than 50% of full channel modulation. When the two-tone attention signal is transmitted, FCC rules require that each individual tone modulate the transmitter at no less than 40% of full channel modulation—and that the modulation levels of the two tones be within 1 dB of one another. When the audio portion of an EAS message is transmitted, it is subject to the same modulation requirements as the station's normal audio programming. This means that, for AM, FM, and TV stations, the modulation level of the audio portion of an EAS message should be no less than 85% on peaks of frequent recurrence.

None of the components of an EAS message should exceed the maximum modulation limits specified for the relevant service. For AM stations, the maximum modulation level is 100% on negative peaks of frequent recurrence and 125% on positive peaks of frequent recurrence. For FM and TV stations the maximum modulation level is 100% on peaks of frequent recurrence, except that FM stations may increase their total modulation 0.5% for every 1% of subcarrier injection, up to a maximum level of 110% total modulation.

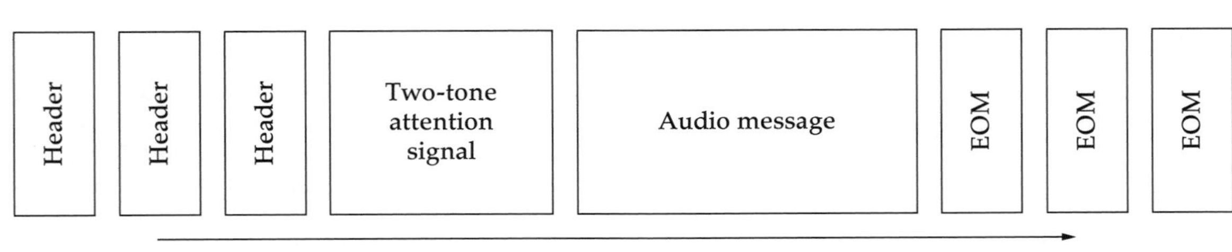

FIGURE 1.18-2 EAS message elements.

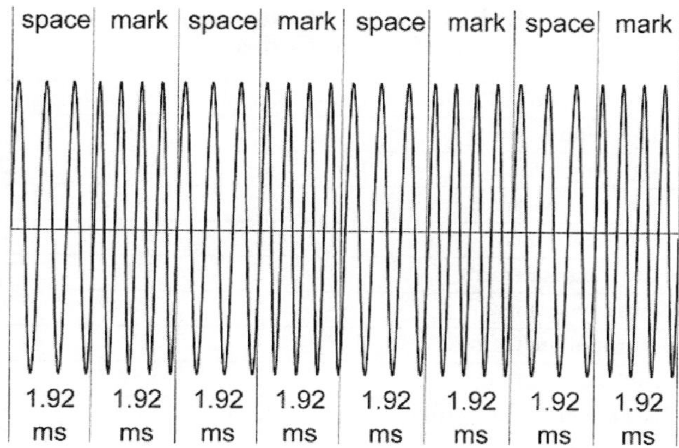

FIGURE 1.18-3 Time-domain illustration of an EAS digital byte.

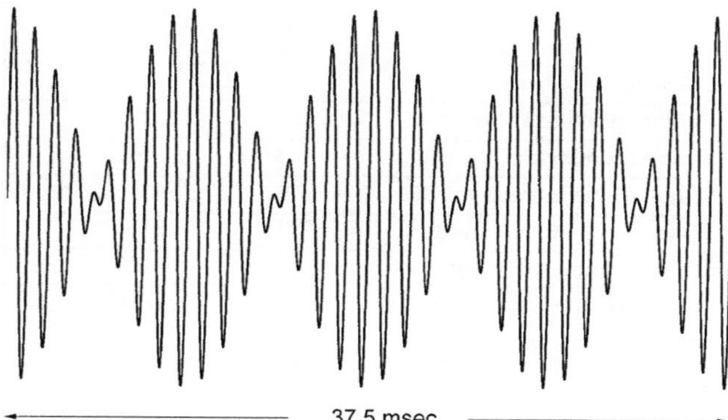

FIGURE 1.18-4 Time-domain illustration of two-tone EAS attention signal.

EAS CODING METHOD

The header codes and the end-of-message codes in an EAS message are composed of a series of digital bytes and are transmitted using the 1562.5 Hz and 2083.3 Hz tones. The header and end-of-message codes are each repeated three times to ensure that they are received correctly. For all EAS transmissions except the weekly test, television stations must transmit the header code data both aurally and visually. If they use a video crawl to do this, the crawl must be displayed at the top of the television screen or at another location where it will not interfere with other visual messages such as closed captioning. The two-tone attention signal is created by simultaneously transmitting the 853 Hz and 960 Hz tones, the same two-tone signal that was used in the old EBS. The actual message is simply an audio, video, or text signal.

Except for the actual audio message itself, the header is the portion of the EAS message that contains the most information. The two-tone attention signal is not intended to serve any purpose other than to audibly alert the audience that an EAS message is about to be broadcast, and the end-of-message code is simply a series a four ASCII "N" characters used to indicate that the alert is over and that the EAS equipment should reset itself to its normal, non-alert state. Both the header and end-of-message codes are each preceded by a preamble that is a string of 16 bytes (128 bits) consisting of ASCII character 171 ("½") repeated 16 times. The purpose of this preamble is to prepare EAS receivers to accept data by setting their automatic gain control (AGC) circuits and asynchronous clocking cycles.

The header code contains 42 bytes of information. Its components are illustrated in Figure 1.18-5. The first code segment in the header code consists of the characters "ZCZC," which indicates that an ASCII code sequence is about to begin. The next code segment contains the originator code, which indicates who initiated activation of the EAS for the particular alert in question. A list of acceptable originator codes is provided in Table 1.18-1.

The event code, which follows the originator code, indicates the type of emergency that is the subject of the alert. The event codes used in the EAS are a subset of those used by the National Oceanic and Atmospheric Administration (NOAA) Weather Radio, which themselves are a subset of those defined in the Public Alert™ receiver standard developed by the Consumer Electronics Association (CEA-2009-A). Event codes are listed in Table 1.18-2. CEA-2009-A defines requirements for receivers that are meant to work with weather radio broadcasts in both the United States and Canada, which is why it specifies a set of alert codes that is slightly larger than the sets used by the FCC and NOAA Weather Radio.

The location code follows the event code and indicates the state (or U.S. territory), county, and county subdivision (northwest, south, etc.) affected by an EAS alert. Fifteen special location codes have also been assigned for specific offshore marine areas. The valid values for the three segments of the location code are described in Tables 1.18-3, 1.18-4, and 1.18-5, which, because of their length, are provided at the end of the

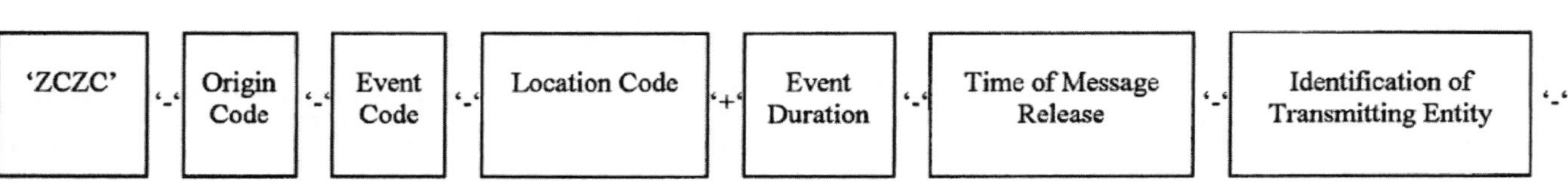

FIGURE 1.18-5 EAS header code.

TABLE 1.18-1
EAS Originator Codes

Originator	Originator Code
Broadcast station or cable system	EAS
Civil authorities	CIV
National Weather Service	WXR
Primary entry point system	PEP

TABLE 1.18-2
EAS Event Codes

Nature of Activation	Event Code
National codes (required):	
Emergency action notification (national only)	EAN
Emergency action termination (national only)	EAT
National Information Center	NIC
National periodic test	NPT
Required monthly test	RMT
Required weekly test	RWT
State and local codes (optional):	
Administrative message	ADR
Avalanche warning	AVW*
Avalanche watch	AVA*
Biological hazard warning	BHW†
Blizzard warning	BZW
Boil water warning	BWW†
Chemical hazard warning	CHW†
Child abduction emergency	CAE*
Civil danger warning	CDW*
Civil emergency message	CEM
Coastal flood warning	CFW*
Coastal flood watch	CFA*
Contagious disease warning	DEW†
Contaminated water warning	CWW†
Dam watch	DBA†
Dam break warning	DBW†
Dust storm warning	DSW*
Earthquake warning	EQW*
Evacuation immediate	EVI
Evacuation watch	EVA†
Fire warning	FRW*
Flash flood warning	FFW
Flash flood watch	FFA
Flash flood statement	FFS
Flash freeze warning	FSW†
Flood warning	FLW
Flood watch	FLA
Flood statement	FLS
Food contamination warning	FCW†
Freeze warning	FZW†
Hazardous materials warning	HMW*

TABLE 1.18-2 *(continued)*
EAS Event Codes

Nature of Activation	Event Code
High wind warning	HWW
High wind watch	HWA
Hurricane warning	HUW
Hurricane watch	HUA
Hurricane statement	HLS
Iceberg warning	IBW†
Industrial fire warning	IFW†
Landslide warning	LSW†
Law enforcement warning	LEW*
Local area emergency	LAE*
National audible test	NAT†
National silent test	NST†
Network message notification	NMN*
911 telephone outage emergency	TOE*
Nuclear power plant warning	NUW*
Power outage advisory	POS†
Practice/demo warning	DMO
Radiological hazard warning	RHW*
Severe thunderstorm warning	SVR
Severe thunderstorm watch	SVA
Severe weather statement	SVS
Shelter in place warning	SPW*
Special marine warning	SMW*
Special weather statement	SPS
Tornado warning	TOR
Tornado watch	TOA
Transmitter backup on	TXB†
Transmitter carrier off	TXF†
Transmitter carrier on	TXO†
Transmitter primary on	TXP†
Tropical storm warning	TRW*
Tropical storm watch	TRA*
Tsunami warning	TSW
Tsunami watch	TSA
Volcano warning	VOW*
Wildfire warning	WFW†
Wildfire watch	WFA†
Winter storm warning	WSW
Winter storm watch	WSA

*Effective since 2002, broadcast stations, cable systems, and wireless cable systems may upgrade their existing EAS equipment to add these event codes on a voluntary basis until the equipment is replaced. All models of EAS equipment manufactured after August 1, 2003, must be capable of receiving and transmitting these event codes. Broadcast stations, cable systems, and wireless cable systems replacing their EAS equipment after February 1, 2004, must install equipment that is capable of receiving and transmitting these event codes.

†These event codes are included in CEA-2009-A (Performance Specification for Public Alert Receivers) but are not included in the FCC's EAS rules. Public Alert™ receivers are designed to automatically activate upon receipt of an alert from an NOAA Weather Radio transmitter.

chapter. The offshore marine areas do not have any "county-like" subdivisions, so their location codes will typically end with "0000" (*e.g.*, 730000, 750000). The FCC expects most alerts to have a location code that shows the alert to be effective over an entire county. It only expects the county subdivision feature to indicate something other than the entire county in cases involving oddly shaped or unusually large counties.

The valid time period of the alert, in HHMM format, follows the location code. It indicates the length of time after the origination of the alert during which it is effective. The valid time period of an EAS alert must be in 15 minute increments for periods that are no longer than 1 hour and in 30 minute increments for periods that are longer than an hour.

The origination time of the alert is the time the message is released. It is in JJJHHMM format, where JJJ represents the day of the year in Julian calendar days (*e.g.*, July 4 would be 185 in a non-leap year), and HHMM is in 24 hour Universal Coordinated Time (UTC).

Following the origination time is the transmitting station identifier. This is the identification of the station *transmitting* the alert. It is not the station that originated the alert but instead the station that is actually transmitting or retransmitting it. When the EAS is activated and an alert is making its way through the network of stations, the transmitting station identifier will be the only part of the header that gets changed as the message is relayed from one station to the next. All other components of the header will remain unchanged.

ASCII character 45, the dash/hyphen character ("-"), is used at the end of all but one of the code segments within the header code as a separator. The location code ends with ASCII character 43, the plus character ("+").

The EAS header code contains enough information to enable the receiving broadcast station to generate its own audible emergency message. Many stations choose to do this simply because they prefer to have their own announcers provide the emergency information to their audiences. During national level alerts and the required monthly test, however, stations are required to rebroadcast the audio that they receive without modification, though alternative audio sources of higher audio quality may be used during national alerts provided they do not cause any delay in the transmission of the alert. One of the advantages of using the received audio for state and local emergencies is that, in some cases, the audio portion of the EAS message received by a station may contain more information than the header alone. For example, the header might indicate that there is a tornado watch for a particular county, but the audio portion of the received alert might indicate that a tornado has actually touched down at a particular location in the county.

EAS TRANSMISSIONS VIA RDS

EAS information may be transmitted via an FM station's radio data system (RDS) subcarrier, although transmitting EAS information via RDS does not relieve a station of its responsibility to transmit EAS information in its main audio channel. One of the advantages of using the RDS subcarrier for EAS transmissions is that it enables silent emergency warning networks to be developed that can provide very locally specific alerts to audiences that have RDS EAS decoding equipment. For example, by using RDS a broadcaster could transmit an alert about a chemical spill at a factory that would only be heard by people who are living and working in the immediate vicinity of the factory. This would be done by coding the RDS EAS message with an access code that would prevent receivers that are not programmed with the same access code from decoding the alert message.

The radio data system uses a 1187.5 baud data stream that is transmitted on an FM station's 57 kHz subcarrier. The format of RDS transmissions in the United States is specified in NRSC-4-A, the U.S. Radio Broadcast Data System (RBDS) standard, which was adopted by the National Radio Systems Committee. The following is a description of a typical sequence of events that occurs when an EAS message is transmitted via RDS:

1. The *warning activation* bit in Group 3A of the RDS signal is set to 1 to tell sleeping battery-operated receivers to wake up.
2. The *program type* code (rock, country, jazz, etc.) is set to "ALERT!" to tell consumer RDS equipment to interrupt current activity (CD/cassette playback, radio off state, etc.) and switch to FM reception.
3. The actual audio alert message is transmitted in the FM station's main audio channel, and a text version of the alert message is transmitted to radio-text-equipped RDS receivers using Group 2A or 2B.
4. The *Specific Area Message Encoding* (SAME) codes are transmitted using Group 9A. These codes may be useful to specialized RDS receivers equipped to decode SAME data; however, consumer RDS radios generally do not make use of this data because of the additional signal processing circuitry that would have to be included to decode the information. Consumer radios rely on the actual audio alert message and any radio text information that is transmitted.
5. At the end of the EAS alert, the *program type* code is set back to its appropriate state (rock, country, jazz, etc.), and the *warning activation* bit is set back to 0 to tell battery-operated receivers that they may return to their sleep state.

SUMMARY

For most broadcasters, the EAS is an improvement over the EBS because it enables them to automate the emergency warning function at their stations, thus making it easier for them to operate unattended transmitters and thereby reduce their operating expenses. Broadcasters that make use of RDS technology to enhance their EAS alerts can provide added services to their audiences by enabling RDS-equipped radios to do such things as automatically turn on in the middle of the night to warn of approaching danger.

Reference

[1] U.S. Government Printing Office, Code of Federal Regulations, Title 47, Part 11; October 1, 2005 (http://www.access.gpo.gov/nara/cfr/waisidx_05/47cfr11_05.html).

TABLE 1.18-3
EAS Location Codes: State, Territory, and Offshore Marine Area Codes

State/Territory/ Marine Area	Code	State/Territory/ Marine Area	Code	State/Territory/ Marine Area	Code
Alabama	01	Alaska	02	Arizona	04
Arkansas	05	California	06	Colorado	08
Connecticut	09	Delaware	10	District of Columbia	11
Florida	12	Georgia	13	Hawaii	15
Idaho	16	Illinois	17	Indiana	18
Iowa	19	Kansas	20	Kentucky	21
Louisiana	22	Maine	23	Maryland	24
Massachusetts	25	Michigan	26	Minnesota	27
Mississippi	28	Missouri	29	Montana	30
Nebraska	31	Nevada	32	New Hampshire	33
New Jersey	34	New Mexico	35	New York	36
North Carolina	37	North Dakota	38	Ohio	39
Oklahoma	40	Oregon	41	Pennsylvania	42
Rhode Island	44	South Carolina	45	South Dakota	46
Tennessee	47	Texas	48	Utah	49
Vermont	50	Virginia	51	Washington	53
West Virginia	54	Wisconsin	55	Wyoming	56
American Samoa	60	Federated States of Micronesia	64	Guam	66
Marshall Islands	68	Northern Mariana Islands	69	Palau	70
Puerto Rico	72	U.S. Minor Outlying Islands	74	U.S. Virgin Islands	78

State/Territory/ Marine Area	Code	State/Territory/ Marine Area	Code
Eastern North Pacific Ocean and along U.S. west coast from the Canadian border to the Mexican border	57*	Lake Superior	91*
North Pacific Ocean near Alaska and along Alaska coastline, including the Bering Sea and the Gulf of Alaska	58*	Lake Michigan	92*
Central Pacific Ocean, including Hawaiian waters	59*	Lake Huron	93*
South Central Pacific Ocean, including American Samoa waters	61*	Lake St. Clair	94*
Western Pacific Ocean, including Mariana Island waters	65*	Lake Erie	96*
Western North Atlantic Ocean and along U.S. east coast from the Canadian border south to Currituck Beach Light, NC	73*	Lake Ontario	97*
Western North Atlantic Ocean and along U.S. east coast, south of Currituck Beach Light, NC, following the coastline into Gulf of Mexico to Bonita Beach, FL, including the Caribbean	75*	St. Lawrence River above St. Regis	98*
Gulf of Mexico and along the U.S. Gulf Coast from the Mexican border to Bonita Beach, FL	77*		

*Effective May 16, 2002, broadcast stations, cable systems, and wireless cable systems may upgrade their existing EAS equipment to add these marine area location codes on a voluntary basis until the equipment is replaced. All models of EAS equipment manufactured after August 1, 2003, must be capable of receiving and transmitting these marine area location codes. Broadcast stations, cable systems, and wireless cable systems replacing their EAS equipment after February 1, 2004, must install equipment that is capable of receiving and transmitting these location codes.

TABLE 1.18-4
EAS Location Codes: County/Area FIPS Codes

State	County/Area	FIPS Code
AK	Aleutians East Borough	013
AK	Aleutians West census area	016
AK	Anchorage Borough	020
AK	Bethel census area	050
AK	Bristol Bay Borough	060
AK	Denali Borough	068
AK	Dillingham census area	070
AK	Fairbanks North Star Borough	090
AK	Haines Borough	100
AK	Juneau Borough	110
AK	Kenai Peninsula Borough	122
AK	Ketchikan Gateway Borough	130
AK	Kodiak Island Borough	150
AK	Lake and Peninsula Borough	164
AK	Matanuska–Susitna Borough	170
AK	Nome census area	180
AK	North Slope Borough	185
AK	Northwest Arctic	188
AK	Prince of Wales–Outer Ketchikan census area	201
AK	Sitka Borough	220
AK	Skagway–Yakutat–Angoon census area	231
AK	Southeast Fairbanks census area	240
AK	Valdez–Gordova census area	261
AK	Wade Hampton census area	270
AK	Wrangell–Petersburg census area	280
AK	Yukon–Koyukuk census area	290
AL	Autauga County	001
AL	Baldwin County	003
AL	Barbour County	005
AL	Bibb County	007
AL	Blount County	009
AL	Bullock County	011
AL	Butler County	013
AL	Calhoun County	015
AL	Chambers County	017
AL	Cherokee County	019
AL	Chilton County	021
AL	Choctaw County	023
AL	Clarke County	025
AL	Clay County	027
AL	Cleburne County	029
AL	Coffee County	031

TABLE 1.18-4 *(continued)*
EAS Location Codes: County/Area FIPS Codes

State	County/Area	FIPS Code
AL	Colbert County	033
AL	Conecuh County	035
AL	Coosa County	037
AL	Covington County	039
AL	Crenshaw County	041
AL	Cullman County	043
AL	Dale County	045
AL	Dallas County	047
AL	DeKalb County	049
AL	Elmore County	051
AL	Escambia County	053
AL	Etowah County	055
AL	Fayette County	057
AL	Franklin County	059
AL	Geneva County	061
AL	Greene County	063
AL	Hale County	065
AL	Henry County	067
AL	Houston County	069
AL	Jackson County	071
AL	Jefferson County	073
AL	Lamar County	075
AL	Lauderdale County	077
AL	Lawrence County	079
AL	Lee County	081
AL	Limestone County	083
AL	Lowndes County	085
AL	Macon County	087
AL	Madison County	089
AL	Marengo County	091
AL	Marion County	093
AL	Marshall County	095
AL	Mobile County	097
AL	Monroe County	099
AL	Montgomery County	101
AL	Morgan County	103
AL	Perry County	105
AL	Pickens County	107
AL	Pike County	109
AL	Randolph County	111
AL	Russell County	113
AL	St. Clair County	115
AL	Shelby County	117
AL	Sumter County	119
AL	Talladega County	121
AL	Tallapoosa County	123
AL	Tuscaloosa County	125
AL	Walker County	127
AL	Washington County	129
AL	Wilcox County	131

TABLE 1.18-4 *(continued)*
EAS Location Codes: County/Area FIPS Codes

State	County/Area	FIPS Code
AL	Winston County	133
AR	Arkansas County	001
AR	Ashley County	003
AR	Baxter County	005
AR	Benton County	007
AR	Boone County	009
AR	Bradley County	011
AR	Calhoun County	013
AR	Carroll County	015
AR	Chicot County	017
AR	Clark County	019
AR	Clay County	021
AR	Cleburne County	023
AR	Cleveland County	025
AR	Columbia County	027
AR	Conway County	029
AR	Craighead County	031
AR	Crawford County	033
AR	Crittenden County	035
AR	Cross County	037
AR	Dallas County	039
AR	Desha County	041
AR	Drew County	043
AR	Faulkner County	045
AR	Franklin County	047
AR	Fulton County	049
AR	Garland County	051
AR	Grant County	053
AR	Greene County	055
AR	Hempstead County	057
AR	Hot Spring County	059
AR	Howard County	061
AR	Independence County	063
AR	Izard County	065
AR	Jackson County	067
AR	Jefferson County	069
AR	Johnson County	071
AR	Lafayette County	073
AR	Lawrence County	075
AR	Lee County	077
AR	Lincoln County	079
AR	Little River County	081
AR	Logan County	083
AR	Lonoke County	085
AR	Madison County	087
AR	Marion County	089
AR	Miller County	091
AR	Mississippi County	093
AR	Monroe County	095
AR	Montgomery County	097

TABLE 1.18-4 *(continued)*
EAS Location Codes: County/Area FIPS Codes

State	County/Area	FIPS Code
AR	Nevada County	099
AR	Newton County	101
AR	Ouachita County	103
AR	Perry County	105
AR	Phillips County	107
AR	Pike County	109
AR	Poinsett County	111
AR	Polk County	113
AR	Pope County	115
AR	Prairie County	117
AR	Pulaski County	119
AR	Randolph County	121
AR	St. Francis County	123
AR	Saline County	125
AR	Scott County	127
AR	Searcy County	129
AR	Sebastian County	131
AR	Sevier County	133
AR	Sharp County	135
AR	Stone County	137
AR	Union County	139
AR	Van Buren County	141
AR	Washington County	143
AR	White County	145
AR	Woodruff County	147
AR	Yell County	149
AZ	Apache County	001
AZ	Cochise County	003
AZ	Coconino County	005
AZ	Gila County	007
AZ	Graham County	009
AZ	Greenlee County	011
AZ	La Paz County	012
AZ	Maricopa County	013
AZ	Mohave County	015
AZ	Navajo County	017
AZ	Pima County	019
AZ	Pinal County	021
AZ	Santa Cruz County	023
AZ	Yavapai County	025
AZ	Yuma County	027
CA	Alameda County	001
CA	Alpine County	003
CA	Amador County	005
CA	Butte County	007
CA	Calaveras County	009
CA	Colusa County	011
CA	Contra Costa County	013
CA	Del Norte County	015
CA	El Dorado County	017

TABLE 1.18-4 *(continued)*
EAS Location Codes: County/Area FIPS Codes

State	County/Area	FIPS Code
CA	Fresno County	019
CA	Glenn County	021
CA	Humboldt County	023
CA	Imperial County	025
CA	Inyo County	027
CA	Kern County	029
CA	Kings County	031
CA	Lake County	033
CA	Lassen County	035
CA	Los Angeles County	037
CA	Madera County	039
CA	Marin County	041
CA	Mariposa County	043
CA	Mendocino County	045
CA	Merced County	047
CA	Modoc County	049
CA	Mono County	051
CA	Monterey County	053
CA	Napa County	055
CA	Nevada County	057
CA	Orange County	059
CA	Placer County	061
CA	Plumas County	063
CA	Riverside County	065
CA	Sacramento County	067
CA	San Benito County	069
CA	San Bernardino County	071
CA	San Diego County	073
CA	San Francisco County	075
CA	San Joaquin County	077
CA	San Luis Obispo County	079
CA	San Mateo County	081
CA	Santa Barbara County	083
CA	Santa Clara County	085
CA	Santa Cruz County	087
CA	Shasta County	089
CA	Sierra County	091
CA	Siskiyou County	093
CA	Solano County	095
CA	Sonoma County	097
CA	Stanislaus County	099
CA	Sutter County	101
CA	Tehama County	103
CA	Trinity County	105
CA	Tulare County	107
CA	Tuolumne County	109
CA	Ventura County	111
CA	Yolo County	113
CA	Yuba County	115
CO	Adams County	001

TABLE 1.18-4 *(continued)*
EAS Location Codes: County/Area FIPS Codes

State	County/Area	FIPS Code
CO	Alamosa County	003
CO	Arapahoe County	005
CO	Archuleta County	007
CO	Baca County	009
CO	Bent County	011
CO	Boulder County	013
CO	Chaffee County	015
CO	Cheyenne County	017
CO	Clear Creek County	019
CO	Conejos County	021
CO	Costilla County	023
CO	Crowley County	025
CO	Custer County	027
CO	Delta County	029
CO	Denver County	031
CO	Dolores County	033
CO	Douglas County	035
CO	Eagle County	037
CO	Elbert County	039
CO	El Paso County	041
CO	Fremont County	043
CO	Garfield County	045
CO	Gilpin County	047
CO	Grand County	049
CO	Gunnison County	051
CO	Hinsdale County	053
CO	Huerfano County	055
CO	Jackson County	057
CO	Jefferson County	059
CO	Kiowa County	061
CO	Kit Carson County	063
CO	Lake County	065
CO	La Plata County	067
CO	Larimer County	069
CO	Las Animas County	071
CO	Lincoln County	073
CO	Logan County	075
CO	Mesa County	077
CO	Mineral County	079
CO	Moffat County	081
CO	Montezuma County	083
CO	Montrose County	085
CO	Morgan County	087
CO	Otero County	089
CO	Ouray County	091
CO	Park County	093
CO	Phillips County	095
CO	Pitkin County	097
CO	Prowers County	099
CO	Pueblo County	101

TABLE 1.18-4 *(continued)*
EAS Location Codes: County/Area FIPS Codes

State	County/Area	FIPS Code
CO	Rio Blanco County	103
CO	Rio Grande County	105
CO	Routt County	107
CO	Saguache County	109
CO	San Juan County	111
CO	San Miguel County	113
CO	Sedgwick County	115
CO	Summit County	117
CO	Teller County	119
CO	Washington County	121
CO	Weld County	123
CO	Yuma County	125
CT	Fairfield County	001
CT	Hartford County	003
CT	Litchfield County	005
CT	Middlesex County	007
CT	New Haven County	009
CT	New London County	011
CT	Tolland County	013
CT	Windham County	015
DC	District of Columbia	001
DE	Kent County	001
DE	New Castle County	003
DE	Sussex County	005
FL	Alachua County	001
FL	Baker County	003
FL	Bay County	005
FL	Bradford County	007
FL	Brevard County	009
FL	Broward County	011
FL	Calhoun County	013
FL	Charlotte County	015
FL	Citrus County	017
FL	Clay County	019
FL	Collier County	021
FL	Columbia County	023
FL	Dade County	025
FL	DeSoto County	027
FL	Dixie County	029
FL	Duval County	031
FL	Escambia County	033
FL	Flagler County	035
FL	Franklin County	037
FL	Gadsden County	039
FL	Gilchrist County	041
FL	Glades County	043
FL	Gulf County	045
FL	Hamilton County	047
FL	Hardee County	049
FL	Hendry County	051

TABLE 1.18-4 *(continued)*
EAS Location Codes: County/Area FIPS Codes

State	County/Area	FIPS Code
FL	Hernando County	053
FL	Highlands County	055
FL	Hillsborough County	057
FL	Holmes County	059
FL	Indian River County	061
FL	Jackson County	063
FL	Jefferson County	065
FL	Lafayette County	067
FL	Lake County	069
FL	Lee County	071
FL	Leon County	073
FL	Levy County	075
FL	Liberty County	077
FL	Madison County	079
FL	Manatee County	081
FL	Marion County	083
FL	Martin County	085
FL	Monroe County	087
FL	Nassau County	089
FL	Okaloosa County	091
FL	Okeechobee County	093
FL	Orange County	095
FL	Osceola County	097
FL	Palm Beach County	099
FL	Pasco County	101
FL	Pinellas County	103
FL	Polk County	105
FL	Putnam County	107
FL	St. Johns County	109
FL	St. Lucie County	111
FL	Santa Rosa County	113
FL	Sarasota County	115
FL	Seminole County	117
FL	Sumter County	119
FL	Suwannee County	121
FL	Taylor County	123
FL	Union County	125
FL	Volusia County	127
FL	Wakulla County	129
FL	Walton County	131
FL	Washington County	133
GA	Appling County	001
GA	Atkinson County	003
GA	Bacon County	005
GA	Baker County	007
GA	Baldwin County	009
GA	Banks County	011
GA	Barrow County	013
GA	Bartow County	015
GA	Ben Hill County	017

TABLE 1.18-4 *(continued)*
EAS Location Codes: County/Area FIPS Codes

State	County/Area	FIPS Code
GA	Berrien County	019
GA	Bibb County	021
GA	Bleckley County	023
GA	Brantley County	025
GA	Brooks County	027
GA	Bryan County	029
GA	Bulloch County	031
GA	Burke County	033
GA	Butts County	035
GA	Calhoun County	037
GA	Camden County	039
GA	Candler County	043
GA	Carroll County	045
GA	Catoosa County	047
GA	Charlton County	049
GA	Chatham County	051
GA	Chattahoochee County	053
GA	Chattooga County	055
GA	Cherokee County	057
GA	Clarke County	059
GA	Clay County	061
GA	Clayton County	063
GA	Clinch County	065
GA	Cobb County	067
GA	Coffee County	069
GA	Colquitt County	071
GA	Columbia County	073
GA	Cook County	075
GA	Coweta County	077
GA	Crawford County	079
GA	Crisp County	081
GA	Dade County	083
GA	Dawson County	085
GA	Decatur County	087
GA	DeKalb County	089
GA	Dodge County	091
GA	Dooly County	093
GA	Dougherty County	095
GA	Douglas County	097
GA	Early County	099
GA	Echols County	101
GA	Effingham County	103
GA	Elbert County	105
GA	Emanuel County	107
GA	Evans County	109
GA	Fannin County	111
GA	Fayette County	113
GA	Floyd County	115
GA	Forsyth County	117
GA	Franklin County	119

TABLE 1.18-4 *(continued)* EAS Location Codes: County/Area FIPS Codes

State	County/Area	FIPS Code
GA	Fulton County	121
GA	Gilmer County	123
GA	Glascock County	125
GA	Glynn County	127
GA	Gordon County	129
GA	Grady County	131
GA	Greene County	133
GA	Gwinnett County	135
GA	Habersham County	137
GA	Hall County	139
GA	Hancock County	141
GA	Haralson County	143
GA	Harris County	145
GA	Hart County	147
GA	Heard County	149
GA	Henry County	151
GA	Houston County	153
GA	Irwin County	155
GA	Jackson County	157
GA	Jasper County	159
GA	Jeff Davis County	161
GA	Jefferson County	163
GA	Jenkins County	165
GA	Johnson County	167
GA	Jones County	169
GA	Lamar County	171
GA	Lanier County	173
GA	Laurens County	175
GA	Lee County	177
GA	Liberty County	179
GA	Lincoln County	181
GA	Long County	183
GA	Lowndes County	185
GA	Lumpkin County	187
GA	McDuffie County	189
GA	Mclntosh County	191
GA	Macon County	193
GA	Madison County	195
GA	Marion County	197
GA	Meriwether County	199
GA	Miller County	201
GA	Mitchell County	205
GA	Monroe County	207
GA	Montgomery County	209
GA	Morgan County	211
GA	Murray County	213
GA	Muscogee County	215
GA	Newton County	217
GA	Oconee County	219
GA	Oglethorpe County	221

TABLE 1.18-4 *(continued)* EAS Location Codes: County/Area FIPS Codes

State	County/Area	FIPS Code
GA	Paulding County	223
GA	Peach County	225
GA	Pickens County	227
GA	Pierce County	229
GA	Pike County	231
GA	Polk County	233
GA	Pulaski County	235
GA	Putnam County	237
GA	Quitman County	239
GA	Rabun County	241
GA	Randolph County	243
GA	Richmond County	245
GA	Rockdale County	247
GA	Schley County	249
GA	Screven County	251
GA	Seminole County	253
GA	Spalding County	255
GA	Stephens County	257
GA	Stewart County	259
GA	Sumter County	261
GA	Talbot County	263
GA	Taliaferro County	265
GA	Tattnall County	267
GA	Taylor County	269
GA	Telfair County	271
GA	Terrell County	273
GA	Thomas County	275
GA	Tift County	277
GA	Toombs County	279
GA	Towns County	281
GA	Treutlen County	283
GA	Troup County	285
GA	Turner County	287
GA	Twiggs County	289
GA	Union County	291
GA	Upson County	293
GA	Walker County	295
GA	Walton County	297
GA	Ware County	299
GA	Warren County	301
GA	Washington County	303
GA	Wayne County	305
GA	Webster County	307
GA	Wheeler County	309
GA	White County	311
GA	Whitfield County	313
GA	Wilcox County	315
GA	Wilkes County	317
GA	Wilkinson County	319
GA	Worth County	321

TABLE 1.18-4 *(continued)* EAS Location Codes: County/Area FIPS Codes

State	County/Area	FIPS Code
HI	Hawaii County	001
HI	Honolulu County	003
HI	Kalawao County	005
HI	Kauai County	007
HI	Maui County	009
IA	Adair County	001
IA	Adams County	003
IA	Allamakee County	005
IA	Appanoose County	007
IA	Audubon County	009
IA	Benton County	011
IA	Black Hawk County	013
IA	Boone County	015
IA	Bremer County	017
IA	Buchanan County	019
IA	Buena Vista County	021
IA	Butler County	023
IA	Calhoun County	025
IA	Carroll County	027
IA	Cass County	029
IA	Cedar County	031
IA	Cerro Gordo County	033
IA	Cherokee County	035
IA	Chickasaw County	037
IA	Clarke County	039
IA	Clay County	041
IA	Clayton County	043
IA	Clinton County	045
IA	Crawford County	047
IA	Dallas County	049
IA	Davis County	051
IA	Decatur County	053
IA	Delaware County	055
IA	Des Moines County	057
IA	Dickinson County	059
IA	Dubuque County	061
IA	Emmet County	063
IA	Fayette County	065
IA	Floyd County	067
IA	Franklin County	069
IA	Fremont County	071
IA	Greene County	073
IA	Grundy County	075
IA	Guthrie County	077
IA	Hamilton County	079
IA	Hancock County	081
IA	Hardin County	083
IA	Harrison County	085
IA	Henry County	087
IA	Howard County	089

TABLE 1.18-4 *(continued)*
EAS Location Codes: County/Area FIPS Codes

State	County/Area	FIPS Code
IA	Humboldt County	091
IA	Ida County	093
IA	Iowa County	095
IA	Jackson County	097
IA	Jasper County	099
IA	Jefferson County	101
IA	Johnson County	103
IA	Jones County	105
IA	Keokuk County	107
IA	Kossuth County	109
IA	Lee County	111
IA	Linn County	113
IA	Louisa County	115
IA	Lucas County	117
IA	Lyon County	119
IA	Madison County	121
IA	Mahaska County	123
IA	Marion County	125
IA	Marshall County	127
IA	Mills County	129
IA	Mitchell County	131
IA	Monona County	133
IA	Monroe County	135
IA	Montgomery County	137
IA	Muscatine County	139
IA	O'Brien County	141
IA	Osceola County	143
IA	Page County	145
IA	Palo Alto County	147
IA	Plymouth County	149
IA	Pocahontas County	151
IA	Polk County	153
IA	Pottawattamie County	155
IA	Poweshiek County	157
IA	Ringgold County	159
IA	Sac County	161
IA	Scott County	163
IA	Shelby County	165
IA	Sioux County	167
IA	Story County	169
IA	Tama County	171
IA	Taylor County	173
IA	Union County	175
IA	Van Buren County	177
IA	Wapello County	179
IA	Warren County	181
IA	Washington County	183
IA	Wayne County	185
IA	Webster County	187
IA	Winnebago County	189

TABLE 1.18-4 *(continued)*
EAS Location Codes: County/Area FIPS Codes

State	County/Area	FIPS Code
IA	Winneshiek County	191
IA	Woodbury County	193
IA	Worth County	195
IA	Wright County	197
ID	Ada County	001
ID	Adams County	003
ID	Bannock County	005
ID	Bear Lake County	007
ID	Benewah County	009
ID	Bingham County	011
ID	Blaine County	013
ID	Boise County	015
ID	Bonner County	017
ID	Bonneville County	019
ID	Boundary County	021
ID	Butte County	023
ID	Camas County	025
ID	Canyon County	027
ID	Caribou County	029
ID	Cassia County	031
ID	Clark County	033
ID	Clearwater County	035
ID	Custer County	037
ID	Elmore County	039
ID	Franklin County	041
ID	Fremont County	043
ID	Gem County	045
ID	Gooding County	047
ID	Idaho County	049
ID	Jefferson County	051
ID	Jerome County	053
ID	Kootenai County	055
ID	Latah County	057
ID	Lemhi County	059
ID	Lewis County	061
ID	Lincoln County	063
ID	Madison County	065
ID	Minidoka County	067
ID	Nez Perce County	069
ID	Oneida County	071
ID	Owyhee County	073
ID	Payette County	075
ID	Power County	077
ID	Shoshone County	079
ID	Teton County	081
ID	Twin Falls County	083
ID	Valley County	085
ID	Washington County	087
IL	Adams County	001
IL	Alexander County	003

TABLE 1.18-4 *(continued)*
EAS Location Codes: County/Area FIPS Codes

State	County/Area	FIPS Code
IL	Bond County	005
IL	Boone County	007
IL	Brown County	009
IL	Bureau County	011
IL	Calhoun County	013
IL	Carroll County	015
IL	Cass County	017
IL	Champaign County	019
IL	Christian County	021
IL	Clark County	023
IL	Clay County	025
IL	Clinton County	027
IL	Coles County	029
IL	Cook County	031
IL	Crawford County	033
IL	Cumberland County	035
IL	DeKalb County	037
IL	De Witt County	039
IL	Douglas County	041
IL	DuPage County	043
IL	Edgar County	045
IL	Edwards County	047
IL	Effingham County	049
IL	Fayette County	051
IL	Ford County	053
IL	Franklin County	055
IL	Fulton County	057
IL	Gallatin County	059
IL	Greene County	061
IL	Grundy County	063
IL	Hamilton County	065
IL	Hancock County	067
IL	Hardin County	069
IL	Henderson County	071
IL	Henry County	073
IL	Iroquois County	075
IL	Jackson County	077
IL	Jasper County	079
IL	Jefferson County	081
IL	Jersey County	083
IL	Jo Daviess County	085
IL	Johnson County	087
IL	Kane County	089
IL	Kankakee County	091
IL	Kendall County	093
IL	Knox County	095
IL	Lake County	097
IL	La Salle County	099
IL	Lawrence County	101
IL	Lee County	103

TABLE 1.18-4 *(continued)*
EAS Location Codes: County/Area FIPS Codes

State	County/Area	FIPS Code
IL	Livingston County	105
IL	Logan County	107
IL	McDonough County	109
IL	McHenry County	111
IL	McLean County	113
IL	Macon County	115
IL	Macoupin County	117
IL	Madison County	119
IL	Marion County	121
IL	Marshall County	123
IL	Mason County	125
IL	Massac County	127
IL	Menard County	129
IL	Mercer County	131
IL	Monroe County	133
IL	Montgomery County	135
IL	Morgan County	137
IL	Moultrie County	139
IL	Ogle County	141
IL	Peoria County	143
IL	Perry County	145
IL	Piatt County	147
IL	Pike County	149
IL	Pope County	151
IL	Pulaski County	153
IL	Putnam County	155
IL	Randolph County	157
IL	Richland County	159
IL	Rock Island County	161
IL	St. Clair County	163
IL	Saline County	165
IL	Sangamon County	167
IL	Schuyler County	169
IL	Scott County	171
IL	Shelby County	173
IL	Stark County	175
IL	Stephenson County	177
IL	Tazewell County	179
IL	Union County	181
IL	Vermilion County	183
IL	Wabash County	185
IL	Warren County	187
IL	Washington County	189
IL	Wayne County	191
IL	White County	193
IL	Whiteside County	195
IL	Will County	197
IL	Williamson County	199
IL	Winnebago County	201
IL	Woodford County	203

TABLE 1.18-4 *(continued)*
EAS Location Codes: County/Area FIPS Codes

State	County/Area	FIPS Code
IN	Adams County	001
IN	Allen County	003
IN	Bartholomew County	005
IN	Benton County	007
IN	Blackford County	009
IN	Boone County	011
IN	Brown County	013
IN	Carroll County	015
IN	Cass County	017
IN	Clark County	019
IN	Clay County	021
IN	Clinton County	023
IN	Crawford County	025
IN	Daviess County	027
IN	Dearborn County	029
IN	Decatur County	031
IN	De Kalb County	033
IN	Delaware County	035
IN	Dubois County	037
IN	Elkhart County	039
IN	Fayette County	041
IN	Floyd County	043
IN	Fountain County	045
IN	Franklin County	047
IN	Fulton County	049
IN	Gibson County	051
IN	Grant County	053
IN	Greene County	055
IN	Hamilton County	057
IN	Hancock County	059
IN	Harrison County	061
IN	Hendricks County	063
IN	Henry County	065
IN	Howard County	067
IN	Huntington County	069
IN	Jackson County	071
IN	Jasper County	073
IN	Jay County	075
IN	Jefferson County	077
IN	Jennings County	079
IN	Johnson County	081
IN	Knox County	083
IN	Kosciusko County	085
IN	Lagrange County	087
IN	Lake County	089
IN	La Porte County	091
IN	Lawrence County	093
IN	Madison County	095
IN	Marion County	097
IN	Marshall County	099

TABLE 1.18-4 *(continued)*
EAS Location Codes: County/Area FIPS Codes

State	County/Area	FIPS Code
IN	Martin County	101
IN	Miami County	103
IN	Monroe County	105
IN	Montgomery County	107
IN	Morgan County	109
IN	Newton County	111
IN	Noble County	113
IN	Ohio County	115
IN	Orange County	117
IN	Owen County	119
IN	Parke County	121
IN	Perry County	123
IN	Pike County	125
IN	Porter County	127
IN	Posey County	129
IN	Pulaski County	131
IN	Putnam County	133
IN	Randolph County	135
IN	Ripley County	137
IN	Rush County	139
IN	St. Joseph County	141
IN	Scott County	143
IN	Shelby County	145
IN	Spencer County	147
IN	Starke County	149
IN	Steuben County	151
IN	Sullivan County	153
IN	Switzerland County	155
IN	Tippecanoe County	157
IN	Tipton County	159
IN	Union County	161
IN	Vanderburgh County	163
IN	Vermillion County	165
IN	Vigo County	167
IN	Wabash County	169
IN	Warren County	171
IN	Warrick County	173
IN	Washington County	175
IN	Wayne County	177
IN	Wells County	179
IN	White County	181
IN	Whitley County	183
KS	Allen County	001
KS	Anderson County	003
KS	Atchison County	005
KS	Barber County	007
KS	Barton County	009
KS	Bourbon County	011
KS	Brown County	013
KS	Butler County	015

TABLE 1.18-4 *(continued)*
EAS Location Codes: County/Area FIPS Codes

State	County/Area	FIPS Code
KS	Chase County	017
KS	Chautauqua County	019
KS	Cherokee County	021
KS	Cheyenne County	023
KS	Clark County	025
KS	Clay County	027
KS	Cloud County	029
KS	Coffey County	031
KS	Comanche County	033
KS	Cowley County	035
KS	Crawford County	037
KS	Decatur County	039
KS	Dickinson County	041
KS	Doniphan County	043
KS	Douglas County	045
KS	Edwards County	047
KS	Elk County	049
KS	Ellis County	051
KS	Ellsworth County	053
KS	Finney County	055
KS	Ford County	057
KS	Franklin County	059
KS	Geary County	061
KS	Gove County	063
KS	Graham County	065
KS	Grant County	067
KS	Gray County	069
KS	Greeley County	071
KS	Greenwood County	073
KS	Hamilton County	075
KS	Harper County	077
KS	Harvey County	079
KS	Haskell County	081
KS	Hodgeman County	083
KS	Jackson County	085
KS	Jefferson County	087
KS	Jewell County	089
KS	Johnson County	091
KS	Kearny County	093
KS	Kingman County	095
KS	Kiowa County	097
KS	Labette County	099
KS	Lane County	101
KS	Leavenworth County	103
KS	Lincoln County	105
KS	Linn County	107
KS	Logan County	109
KS	Lyon County	111
KS	McPherson County	113
KS	Marion County	115

TABLE 1.18-4 *(continued)*
EAS Location Codes: County/Area FIPS Codes

State	County/Area	FIPS Code
KS	Marshall County	117
KS	Meade County	119
KS	Miami County	121
KS	Mitchell County	123
KS	Montgomery County	125
KS	Morris County	127
KS	Morton County	129
KS	Nemaha County	131
KS	Neosho County	133
KS	Ness County	135
KS	Norton County	137
KS	Osage County	139
KS	Osborne County	141
KS	Ottawa County	143
KS	Pawnee County	145
KS	Phillips County	147
KS	Pottawatomie County	149
KS	Pratt County	151
KS	Rawlins County	153
KS	Reno County	155
KS	Republic County	157
KS	Rice County	159
KS	Riley County	161
KS	Rooks County	163
KS	Rush County	165
KS	Russell County	167
KS	Saline County	169
KS	Scott County	171
KS	Sedgwick County	173
KS	Seward County	175
KS	Shawnee County	177
KS	Sheridan County	179
KS	Sherman County	181
KS	Smith County	183
KS	Stafford County	185
KS	Stanton County	187
KS	Stevens County	189
KS	Sumner County	191
KS	Thomas County	193
KS	Trego County	195
KS	Wabaunsee County	197
KS	Wallace County	199
KS	Washington County	201
KS	Wichita County	203
KS	Wilson County	205
KS	Woodson County	207
KS	Wyandotte County	209
KY	Adair County	001
KY	Allen County	003
KY	Anderson County	005

TABLE 1.18-4 *(continued)*
EAS Location Codes: County/Area FIPS Codes

State	County/Area	FIPS Code
KY	Ballard County	007
KY	Barren County	009
KY	Bath County	011
KY	Bell County	013
KY	Boone County	015
KY	Bourbon County	017
KY	Boyd County	019
KY	Boyle County	021
KY	Bracken County	023
KY	Breathitt County	025
KY	Breckinridge County	027
KY	Bullitt County	029
KY	Butler County	031
KY	Caldwell County	033
KY	Calloway County	035
KY	Campbell County	037
KY	Carlisle County	039
KY	Carroll County	041
KY	Carter County	043
KY	Casey County	045
KY	Christian County	047
KY	Clark County	049
KY	Clay County	051
KY	Clinton County	053
KY	Crittenden County	055
KY	Cumberland County	057
KY	Daviess County	059
KY	Edmonson County	061
KY	Elliott County	063
KY	Estill County	065
KY	Fayette County	067
KY	Fleming County	069
KY	Floyd County	071
KY	Franklin County	073
KY	Fulton County	075
KY	Gallatin County	077
KY	Garrard County	079
KY	Grant County	081
KY	Graves County	083
KY	Grayson County	085
KY	Green County	087
KY	Greenup County	089
KY	Hancock County	091
KY	Hardin County	093
KY	Harlan County	095
KY	Harrison County	097
KY	Hart County	099
KY	Henderson County	101
KY	Henry County	103
KY	Hickman County	105

TABLE 1.18-4 *(continued)*
EAS Location Codes: County/Area FIPS Codes

State	County/Area	FIPS Code
KY	Hopkins County	107
KY	Jackson County	109
KY	Jefferson County	111
KY	Jessamine County	113
KY	Johnson County	115
KY	Kenton County	117
KY	Knott County	119
KY	Knox County	121
KY	Larue County	123
KY	Laurel County	125
KY	Lawrence County	127
KY	Lee County	129
KY	Leslie County	131
KY	Letcher County	133
KY	Lewis County	135
KY	Lincoln County	137
KY	Livingston County	139
KY	Logan County	141
KY	Lyon County	143
KY	McCracken County	145
KY	McCreary County	147
KY	McLean County	149
KY	Madison County	151
KY	Magoffin County	153
KY	Marion County	155
KY	Marshall County	157
KY	Martin County	159
KY	Mason County	161
KY	Meade County	163
KY	Menifee County	165
KY	Mercer County	167
KY	Metcalfe County	169
KY	Monroe County	171
KY	Montgomery County	173
KY	Morgan County	175
KY	Muhlenberg County	177
KY	Nelson County	179
KY	Nicholas County	181
KY	Ohio County	183
KY	Oldham County	185
KY	Owen County	187
KY	Owsley County	189
KY	Pendleton County	191
KY	Perry County	193
KY	Pike County	195
KY	Powell County	197
KY	Pulaski County	199
KY	Robertson County	201
KY	Rockcastle County	203
KY	Rowan County	205

TABLE 1.18-4 *(continued)*
EAS Location Codes: County/Area FIPS Codes

State	County/Area	FIPS Code
KY	Russell County	207
KY	Scott County	209
KY	Shelby County	211
KY	Simpson County	213
KY	Spencer County	215
KY	Taylor County	217
KY	Todd County	219
KY	Trigg County	221
KY	Trimble County	223
KY	Union County	225
KY	Warren County	227
KY	Washington County	229
KY	Wayne County	231
KY	Webster County	233
KY	Whitley County	235
KY	Wolfe County	237
KY	Woodford County	239
LA	Acadia Parish	001
LA	Allen Parish	003
LA	Ascension Parish	005
LA	Assumption Parish	007
LA	Avoyelles Parish	009
LA	Beauregard Parish	011
LA	Bienville Parish	013
LA	Bossier Parish	015
LA	Caddo Parish	017
LA	Calcasieu Parish	019
LA	Caldwell Parish	021
LA	Cameron Parish	023
LA	Catahoula Parish	025
LA	Claiborne Parish	027
LA	Concordia Parish	029
LA	De Soto Parish	031
LA	E. Baton Rouge Parish	033
LA	E. Carroll Parish	035
LA	E. Feliciana Parish	037
LA	Evangeline Parish	039
LA	Franklin Parish	041
LA	Grant Parish	043
LA	Iberia Parish	045
LA	Iberville Parish	047
LA	Jackson Parish	049
LA	Jefferson Parish	051
LA	Jefferson Davis Parish	053
LA	Lafayette Parish	055
LA	Lafourche Parish	057
LA	La Salle Parish	059
LA	Lincoln Parish	061
LA	Livingston Parish	063
LA	Madison Parish	065

TABLE 1.18-4 *(continued)*
EAS Location Codes: County/Area FIPS Codes

State	County/Area	FIPS Code
LA	Morehouse Parish	067
LA	Natchitoches Parish	069
LA	Orleans Parish	071
LA	Ouachita Parish	073
LA	Plaquemines Parish	075
LA	Pointe Coupee Parish	077
LA	Rapides Parish	079
LA	Red River Parish	081
LA	Richland Parish	083
LA	Sabine Parish	085
LA	St. Bernard Parish	087
LA	St. Charles Parish	089
LA	St. Helena Parish	091
LA	St. James Parish	093
LA	St. John the Baptist Parish	095
LA	St. Landry Parish	097
LA	St. Martin Parish	099
LA	St. Mary Parish	101
LA	St. Tammany Parish	103
LA	Tangipahoa Parish	105
LA	Tensas Parish	107
LA	Terrebonne Parish	109
LA	Union Parish	111
LA	Vermilion Parish	113
LA	Vernon Parish	115
LA	Washington Parish	117
LA	Webster Parish	119
LA	W. Baton Rouge Parish	121
LA	W. Carroll Parish	123
LA	W. Feliciana Parish	125
LA	Winn Parish	127
MA	Barnstable County	001
MA	Berkshire County	003
MA	Bristol County	005
MA	Dukes County	007
MA	Essex County	009
MA	Franklin County	011
MA	Hampden County	013
MA	Hampshire County	015
MA	Middlesex County	017
MA	Nantucket County	019
MA	Norfolk County	021
MA	Plymouth County	023
MA	Suffolk County	025
MA	Worcester County	027
MD	Allegany County	001
MD	Anne Arundel County	003
MD	Baltimore City	510
MD	Baltimore County	005
MD	Calvert County	009

TABLE 1.18-4 *(continued)*
EAS Location Codes: County/Area FIPS Codes

State	County/Area	FIPS Code
MD	Caroline County	011
MD	Carroll County	013
MD	Cecil County	015
MD	Charles County	017
MD	Dorchester County	019
MD	Frederickv	021
MD	Garrett County	023
MD	Harford County	025
MD	Howard County	027
MD	Kent County	029
MD	Montgomery County	031
MD	Prince George's County	033
MD	Queen Anne's County	035
MD	St. Mary's County	037
MD	Somerset County	039
MD	Talbot County	041
MD	Washington County	043
MD	Wicomico County	045
MD	Worcester County	047
ME	Androscoggin County	001
ME	Aroostook County	003
ME	Cumberland County	005
ME	Franklin County	007
ME	Hancock County	009
ME	Kennebec County	011
ME	Knox County	013
ME	Lincoln County	015
ME	Oxford County	017
ME	Penobscot County	019
ME	Piscataquis County	021
ME	Sagadahoc County	023
ME	Somerset County	025
ME	Waldo County	027
ME	Washington County	029
ME	York County	031
MI	Alcona County	001
MI	Alger County	003
MI	Allegan County	005
MI	Alpena County	007
MI	Antrim County	009
MI	Arenac County	011
MI	Baraga County	013
MI	Barry County	015
MI	Bay County	017
MI	Benzie County	019
MI	Berrien County	021
MI	Branch County	023
MI	Calhoun County	025
MI	Cass County	027
MI	Charlevoix County	029

TABLE 1.18-4 *(continued)*
EAS Location Codes: County/Area FIPS Codes

State	County/Area	FIPS Code
MI	Cheboygan County	031
MI	Chippewa County	033
MI	Clare County	035
MI	Clinton County	037
MI	Crawford County	039
MI	Delta County	041
MI	Dickinson County	043
MI	Eaton County	045
MI	Emmet County	047
MI	Genesee County	049
MI	Gladwin County	051
MI	Gogebic County	053
MI	Grand Traverse County	055
MI	Gratiot County	057
MI	Hillsdale County	059
MI	Houghton County	061
MI	Huron County	063
MI	Ingham County	065
MI	Ionia County	067
MI	Iosco County	069
MI	Iron County	071
MI	Isabella County	073
MI	Jackson County	075
MI	Kalamazoo County	077
MI	Kalkaska County	079
MI	Kent County	081
MI	Keweenaw County	083
MI	Lake County	085
MI	Lapeer County	087
MI	Leelanau County	089
MI	Lenawee County	091
MI	Livingston County	093
MI	Luce County	095
MI	Mackinac County	097
MI	Macomb County	099
MI	Manistee County	101
MI	Marquette County	103
MI	Mason County	105
MI	Mecosta County	107
MI	Menominee County	109
MI	Midland County	111
MI	Missaukee County	113
MI	Monroe County	115
MI	Montcalm County	117
MI	Montmorency County	119
MI	Muskegon County	121
MI	Newaygo County	123
MI	Oakland County	125
MI	Oceana County	127
MI	Ogemaw County	129

TABLE 1.18-4 *(continued)*
EAS Location Codes: County/Area FIPS Codes

State	County/Area	FIPS Code
MI	Ontonagon County	131
MI	Osceola County	133
MI	Oscoda County	135
MI	Otsego County	137
MI	Ottawa County	139
MI	Presque Isle County	141
MI	Roscommon County	143
MI	Saginaw County	145
MI	St. Clair County	147
MI	St. Joseph County	149
MI	Sanilac County	151
MI	Schoolcraft County	153
MI	Shiawassee County	155
MI	Tuscola County	157
MI	Van Buren County	159
MI	Washtenaw County	161
MI	Wayne County	163
MI	Wexford County	165
MN	Aitkin County	001
MN	Anoka County	003
MN	Becker County	005
MN	Beltrami County	007
MN	Benton County	009
MN	Big Stone County	011
MN	Blue Earth County	013
MN	Brown County	015
MN	Carlton County	017
MN	Carver County	019
MN	Cass County	021
MN	Chippewa County	023
MN	Chisago County	025
MN	Clay County	027
MN	Clearwater County	029
MN	Cook County	031
MN	Cottonwood County	033
MN	Crow Wing County	035
MN	Dakota County	037
MN	Dodge County	039
MN	Douglas County	041
MN	Faribault County	043
MN	Fillmore County	045
MN	Freeborn County	047
MN	Goodhue County	049
MN	Grant County	051
MN	Hennepin County	053
MN	Houston County	055
MN	Hubbard County	057
MN	Isanti County	059
MN	Itasca County	061
MN	Jackson County	063

TABLE 1.18-4 *(continued)*
EAS Location Codes: County/Area FIPS Codes

State	County/Area	FIPS Code
MN	Kanabec County	065
MN	Kandiyohi County	067
MN	Kittson County	069
MN	Koochiching County	071
MN	Lac qui Parle County	073
MN	Lake County	075
MN	Lake of the Woods County	077
MN	Le Sueur County	079
MN	Lincoln County	081
MN	Lyon County	083
MN	McLeod County	085
MN	Mahnomen County	087
MN	Marshall County	089
MN	Martin County	091
MN	Meeker County	093
MN	Mille Lacs County	095
MN	Morrison County	097
MN	Mower County	099
MN	Murray County	101
MN	Nicollet County	103
MN	Nobles County	105
MN	Norman County	107
MN	Olmsted County	109
MN	Otter Tail County	111
MN	Pennington County	113
MN	Pine County	115
MN	Pipestone County	117
MN	Polk County	119
MN	Pope County	121
MN	Ramsey County	123
MN	Red Lake County	125
MN	Redwood County	127
MN	Renville County	129
MN	Rice County	131
MN	Rock County	133
MN	Roseau County	135
MN	St. Louis County	137
MN	Scott County	139
MN	Sherburne County	141
MN	Sibley County	143
MN	Stearns County	145
MN	Steele County	147
MN	Stevens County	149
MN	Swift County	151
MN	Todd County	153
MN	Traverse County	155
MN	Wabasha County	157
MN	Wadena County	159
MN	Waseca County	161
MN	Washington County	163

TABLE 1.18-4 *(continued)*
EAS Location Codes: County/Area FIPS Codes

State	County/Area	FIPS Code
MN	Watonwan County	165
MN	Wilkin County	167
MN	Winona County	169
MN	Wright County	171
MN	Yellow Medicine County	173
MO	Adair County	001
MO	Andrew County	003
MO	Atchison County	005
MO	Audrain County	007
MO	Barry County	009
MO	Barton County	011
MO	Bates County	013
MO	Benton County	015
MO	Bollinger County	017
MO	Boone County	019
MO	Buchanan County	021
MO	Butler County	023
MO	Caldwell County	025
MO	Callaway County	027
MO	Camden County	029
MO	Cape Girardeau County	031
MO	Carroll County	033
MO	Carter County	035
MO	Cass County	037
MO	Cedar County	039
MO	Chariton County	041
MO	Christian County	043
MO	Clark County	045
MO	Clay County	047
MO	Clinton County	049
MO	Cole County	051
MO	Cooper County	053
MO	Crawford County	055
MO	Dade County	057
MO	Dallas County	059
MO	Daviess County	061
MO	DeKalb County	063
MO	Dent County	065
MO	Douglas County	067
MO	Dunklin County	069
MO	Franklin County	071
MO	Gasconade County	073
MO	Gentry County	075
MO	Greene County	077
MO	Grundy County	079
MO	Harrison County	081
MO	Henry County	083
MO	Hickory County	085
MO	Holt County	087
MO	Howard County	089

TABLE 1.18-4 *(continued)*
EAS Location Codes: County/Area FIPS Codes

State	County/Area	FIPS Code
MO	Howell County	091
MO	Iron County	093
MO	Jackson County	095
MO	Jasper County	097
MO	Jefferson County	099
MO	Johnson County	101
MO	Knox County	103
MO	Laclede County	105
MO	Lafayette County	107
MO	Lawrence County	109
MO	Lewis County	111
MO	Lincoln County	113
MO	Linn County	115
MO	Livingston County	117
MO	McDonald County	119
MO	Macon County	121
MO	Madison County	123
MO	Maries County	125
MO	Marion County	127
MO	Mercer County	129
MO	Miller County	131
MO	Mississippi County	133
MO	Moniteau County	135
MO	Monroe County	137
MO	Montgomery County	139
MO	Morgan County	141
MO	New Madrid County	143
MO	Newton County	145
MO	Nodaway County	147
MO	Oregon County	149
MO	Osage County	151
MO	Ozark County	153
MO	Pemiscot County	155
MO	Perry County	157
MO	Pettis County	159
MO	Phelps County	161
MO	Pike County	163
MO	Platte County	165
MO	Polk County	167
MO	Pulaski County	169
MO	Putnam County	171
MO	Ralls County	173
MO	Randolph County	175
MO	Ray County	177
MO	Reynolds County	179
MO	RipleyV	181
MO	St. Charles County	183
MO	St. Clair County	185
MO	Ste. Genevieve County	186
MO	St. Francois County	187

TABLE 1.18-4 *(continued)*
EAS Location Codes: County/Area FIPS Codes

State	County/Area	FIPS Code
MO	St. Louis County	189
MO	St. Louis City	510
MO	Saline County	195
MO	Schuyler County	197
MO	Scotland County	199
MO	Scott County	201
MO	Shannon County	203
MO	Shelby County	205
MO	Stoddard County	207
MO	Stone County	209
MO	Sullivan County	211
MO	Taney County	213
MO	Texas County	215
MO	Vernon County	217
MO	Warren County	219
MO	Washington County	221
MO	Wayne County	223
MO	Webster County	225
MO	Worth County	227
MO	Wright County	229
MS	Adams County	001
MS	Alcorn County	003
MS	Amite County	005
MS	Attala County	007
MS	Benton County	009
MS	Bolivar County	011
MS	Calhoun County	013
MS	Carroll County	015
MS	Chickasaw County	017
MS	Choctaw County	019
MS	Claiborne County	021
MS	Clarke County	023
MS	Clay County	025
MS	Coahoma County	027
MS	Copiah County	029
MS	Covington County	031
MS	DeSoto County	033
MS	Forrest County	035
MS	Franklin County	037
MS	George County	039
MS	Greene County	041
MS	Grenada County	043
MS	Hancock County	045
MS	Harrison County	047
MS	Hinds County	049
MS	Holmes County	051
MS	Humphreys County	053
MS	Issaquena County	055
MS	Itawamba County	057
MS	Jackson County	059

TABLE 1.18-4 *(continued)*
EAS Location Codes: County/Area FIPS Codes

State	County/Area	FIPS Code
MS	Jasper County	061
MS	Jefferson County	063
MS	Jefferson Davis County	065
MS	Jones County	067
MS	Kemper County	069
MS	Lafayette County	071
MS	Lamar County	073
MS	Lauderdale County	075
MS	Lawrence County	077
MS	Leake County	079
MS	Lee County	081
MS	Leflore County	083
MS	Lincoln County	085
MS	Lowndes County	087
MS	Madison County	089
MS	Marion County	091
MS	Marshall County	093
MS	Monroe County	095
MS	Montgomery County	097
MS	Neshoba County	099
MS	Newton County	101
MS	Noxubee County	103
MS	Oktibbeha County	105
MS	Panola County	107
MS	Pearl River County	109
MS	Perry County	111
MS	Pike County	113
MS	Pontotoc County	115
MS	Prentiss County	117
MS	Quitman County	119
MS	Rankin County	121
MS	Scott County	123
MS	Sharkey County	125
MS	Simpson County	127
MS	Smith County	129
MS	Stone County	131
MS	Sunflower County	133
MS	Tallahatchie County	135
MS	Tate County	137
MS	Tippah County	139
MS	Tishomingo County	141
MS	Tunica County	143
MS	Union County	145
MS	Walthall County	147
MS	Warren County	149
MS	Washington County	151
MS	Wayne County	153
MS	Webster County	155
MS	Wilkinson County	157
MS	Winston County	159

TABLE 1.18-4 *(continued)*
EAS Location Codes: County/Area FIPS Codes

State	County/Area	FIPS Code
MS	Yalobusha County	161
MS	Yazoo County	163
MT	Beaverhead County	001
MT	Big Horn County	003
MT	Blaine County	005
MT	Broadwater County	007
MT	Carbon County	009
MT	Carter County	011
MT	Cascade County	013
MT	Chouteau County	015
MT	Custer County	017
MT	Daniels County	019
MT	Dawson County	021
MT	Deer Lodge County	023
MT	Fallon County	025
MT	Fergus County	027
MT	Flathead County	029
MT	Gallatin County	031
MT	Garfield County	033
MT	Glacier County	035
MT	Golden Valley County	037
MT	Granite County	039
MT	Hill County	041
MT	Jefferson County	043
MT	Judith Basin County	045
MT	Lake County	047
MT	Lewis & Clark County	049
MT	Liberty County	051
MT	Lincoln County	053
MT	McCone County	055
MT	Madison County	057
MT	Meagher County	059
MT	Mineral County	061
MT	Missoula County	063
MT	Musselshell County	065
MT	Park County	067
MT	Petroleum County	069
MT	Phillips County	071
MT	Pondera County	073
MT	Powder River County	075
MT	Powell County	077
MT	Prairie County	079
MT	Ravalli County	081
MT	Richland County	083
MT	Roosevelt County	085
MT	Rosebud County	087
MT	Sanders County	089
MT	Sheridan County	091
MT	Silver Bow County	093
MT	Stillwater County	095

TABLE 1.18-4 *(continued)*
EAS Location Codes: County/Area FIPS Codes

State	County/Area	FIPS Code
MT	Sweet Grass County	097
MT	Teton County	099
MT	Toole County	101
MT	Treasure County	103
MT	Valley County	105
MT	Wheatland County	107
MT	Wibaux County	109
MT	Yellowstone County	111
MT	Yellowstone National Park	113
NC	Alamance County	001
NC	Alexander County	003
NC	Alleghany County	005
NC	Anson County	007
NC	Ashe County	009
NC	Avery County	011
NC	Beaufort County	013
NC	Bertie County	015
NC	Bladen County	017
NC	Brunswick County	019
NC	Buncombe County	021
NC	Burke County	023
NC	Cabarrus County	025
NC	Caldwell County	027
NC	Camden County	029
NC	Carteret County	031
NC	Caswell County	033
NC	Catawba County	035
NC	Chatham County	037
NC	Cherokee County	039
NC	Chowan County	041
NC	Clay County	043
NC	Cleveland County	045
NC	Columbus County	047
NC	Craven County	049
NC	Cumberland County	051
NC	Currituck County	053
NC	Dare County	055
NC	Davidson County	057
NC	Davie County	059
NC	Duplin County	061
NC	Durham County	063
NC	Edgecombe County	065
NC	Forsyth County	067
NC	Franklin County	069
NC	Gaston County	071
NC	Gates County	073
NC	Graham County	075
NC	Granville County	077
NC	Greene County	079
NC	Guilford County	081

TABLE 1.18-4 *(continued)*
EAS Location Codes: County/Area FIPS Codes

State	County/Area	FIPS Code
NC	Halifax County	083
NC	Harnett County	085
NC	Haywood County	087
NC	Henderson County	089
NC	Hertford County	091
NC	Hoke County	093
NC	Hyde County	095
NC	Iredell County	097
NC	Jackson County	099
NC	Johnston County	101
NC	Jones County	103
NC	Lee County	105
NC	Lenoir County	107
NC	Lincoln County	109
NC	McDowell County	111
NC	Macon County	113
NC	Madison County	115
NC	Martin County	117
NC	Mecklenburg County	119
NC	Mitchell County	121
NC	Montgomery County	123
NC	Moore County	125
NC	Nash County	127
NC	New Hanover County	129
NC	Northampton County	131
NC	Onslow County	133
NC	Orange County	135
NC	Pamlico County	137
NC	Pasquotank County	139
NC	Pender County	141
NC	Perquimans County	143
NC	Person County	145
NC	Pitt County	147
NC	Polk County	149
NC	Randolph County	151
NC	Richmond County	153
NC	Robeson County	155
NC	Rockingham County	157
NC	Rowan County	159
NC	Rutherford County	161
NC	Sampson County	163
NC	Scotland County	165
NC	Stanly County	167
NC	Stokes County	169
NC	Surry County	171
NC	Swain County	173
NC	Transylvania County	175
NC	Tyrrell County	177
NC	Union County	179
NC	Vance County	181

TABLE 1.18-4 *(continued)*
EAS Location Codes: County/Area FIPS Codes

State	County/Area	FIPS Code
NC	Wake County	183
NC	Warren County	185
NC	Washington County	187
NC	Watauga County	189
NC	Wayne County	191
NC	Wilkes County	193
NC	Wilson County	195
NC	Yadkin County	197
NC	Yancey County	199
ND	Adams County	001
ND	Barnes County	003
ND	Benson County	005
ND	Billings County	007
ND	Bottineau County	009
ND	Bowman County	011
ND	Burke County	013
ND	Burleigh County	015
ND	Cass County	017
ND	Cavalier County	019
ND	Dickey County	021
ND	Divide County	023
ND	Dunn County	025
ND	Eddy County	027
ND	Emmons County	029
ND	Foster County	031
ND	Golden Valley County	033
ND	Grand Forks County	035
ND	Grant County	037
ND	Griggs County	039
ND	Hettinger County	041
ND	Kidder County	043
ND	LaMoure County	045
ND	Logan County	047
ND	McHenry County	049
ND	McIntosh County	051
ND	McKenzie County	053
ND	McLean County	055
ND	Mercer County	057
ND	Morton County	059
ND	Mountrail County	061
ND	Nelson County	063
ND	Oliver County	065
ND	Pembina County	067
ND	Pierce County	069
ND	Ramsey County	071
ND	Ransom County	073
ND	Renville County	075
ND	Richland County	077
ND	Rolette County	079
ND	Sargent County	081

TABLE 1.18-4 *(continued)*
EAS Location Codes: County/Area FIPS Codes

State	County/Area	FIPS Code
ND	Sheridan County	083
ND	Sioux County	085
ND	Slope County	087
ND	Stark County	089
ND	Steele County	091
ND	Stutsman County	093
ND	Towner County	095
ND	Traill County	097
ND	Walsh County	099
ND	Ward County	101
ND	Wells County	103
ND	Williams County	105
NE	Adams County	001
NE	Antelope County	003
NE	Arthur County	005
NE	Banner County	007
NE	Blaine County	009
NE	Boone County	011
NE	Box Butte County	013
NE	Boyd County	015
NE	Brown County	017
NE	Buffalo County	019
NE	Burt County	021
NE	Butler County	023
NE	Cass County	025
NE	Cedar County	027
NE	Chase County	029
NE	Cherry County	031
NE	Cheyenne County	033
NE	Clay County	035
NE	Colfax County	037
NE	Cuming County	039
NE	Custer County	041
NE	Dakota County	043
NE	Dawes County	045
NE	Dawson County	047
NE	Deuel County	049
NE	Dixon County	051
NE	Dodge County	053
NE	Douglas County	055
NE	Dundy County	057
NE	Fillmore County	059
NE	Franklin County	061
NE	Frontier County	063
NE	Furnas County	065
NE	Gage County	067
NE	Garden County	069
NE	Garfield County	071
NE	Gosper County	073
NE	Grant County	075

TABLE 1.18-4 *(continued)*
EAS Location Codes: County/Area FIPS Codes

State	County/Area	FIPS Code
NE	Greeley County	077
NE	Hall County	079
NE	Hamilton County	081
NE	Harlan County	083
NE	Hayes County	085
NE	Hitchcock County	087
NE	Holt County	089
NE	Hooker County	091
NE	Howard County	093
NE	Jefferson County	095
NE	Johnson County	097
NE	Kearney County	099
NE	Keith County	101
NE	Keya Paha County	103
NE	Kimball County	105
NE	Knox County	107
NE	Lancaster County	109
NE	Lincoln County	111
NE	Logan County	113
NE	Loup County	115
NE	McPherson County	117
NE	Madison County	119
NE	Merrick County	121
NE	Morrill County	123
NE	Nance County	125
NE	Nemaha County	127
NE	Nuckolls County	129
NE	Otoe County	131
NE	Pawnee County	133
NE	Perkins County	135
NE	Phelps County	137
NE	Pierce County	139
NE	Platte County	141
NE	Polk County	143
NE	Red Willow County	145
NE	Richardson County	147
NE	Rock County	149
NE	Saline County	151
NE	Sarpy County	153
NE	Saunders County	155
NE	Scotts Bluff County	157
NE	Seward County	159
NE	Sheridan County	161
NE	Sherman County	163
NE	Sioux County	165
NE	Stanton County	167
NE	Thayer County	169
NE	Thomas County	171
NE	Thurston County	173
NE	Valley County	175

TABLE 1.18-4 *(continued)*
EAS Location Codes: County/Area FIPS Codes

State	County/Area	FIPS Code
NE	Washington County	177
NE	Wayne County	179
NE	Webster County	181
NE	Wheeler County	183
NE	York County	185
NH	Belknap County	001
NH	Carroll County	003
NH	Cheshire County	005
NH	Coos County	007
NH	Grafton County	009
NH	Hillsborough County	011
NH	Merrimack County	013
NH	Rockingham County	015
NH	Strafford County	017
NH	Sullivan County	019
NJ	Atlantic County	001
NJ	Bergen County	003
NJ	Burlington County	005
NJ	Camden County	007
NJ	Cape May County	009
NJ	Cumberland County	011
NJ	Essex County	013
NJ	Gloucester County	015
NJ	Hudson County	017
NJ	Hunterdon County	019
NJ	Mercer County	021
NJ	Middlesex County	023
NJ	Monmouth County	025
NJ	Morris County	027
NJ	Ocean County	029
NJ	Passaic County	031
NJ	Salem County	033
NJ	Somerset County	035
NJ	Sussex County	037
NJ	Union County	039
NJ	Warren County	041
NM	Bernalillo County	001
NM	Catron County	003
NM	Chaves County	005
NM	Cibola County	006
NM	Colfax County	007
NM	Curry County	009
NM	De Baca County	011
NM	Dona Ana County	013
NM	Eddy County	015
NM	Grant County	017
NM	Guadalupe County	019
NM	Harding County	021
NM	Hidalgo County	023
NM	Lea County	025

TABLE 1.18-4 *(continued)*
EAS Location Codes: County/Area FIPS Codes

State	County/Area	FIPS Code
NM	Lincoln County	027
NM	Los Alamos County	028
NM	Luna County	029
NM	McKinley County	031
NM	Mora County	033
NM	Otero County	035
NM	Quay County	037
NM	Rio Arriba County	039
NM	Roosevelt County	041
NM	Sandoval County	043
NM	San Juan County	045
NM	San Miguel County	047
NM	Santa Fe County	049
NM	Sierra County	051
NM	Socorro County	053
NM	Taos County	055
NM	Torrance County	057
NM	Union County	059
NM	Valencia	061
NV	Carson City	510
NV	Churchill County	001
NV	Clark County	003
NV	Douglas County	005
NV	Elko County	007
NV	Esmeralda County	009
NV	Eureka County	011
NV	Humboldt County	013
NV	Lander County	015
NV	Lincoln County	017
NV	Lyon County	019
NV	Mineral County	021
NV	Nye County	023
NV	Pershing County	027
NV	Storey County	029
NV	Washoe County	031
NV	White Pine County	033
NY	Albany County	001
NY	Allegany County	003
NY	Bronx County	005
NY	Broome County	007
NY	Cattaraugus County	009
NY	Cayuga County	011
NY	Chautauqua County	013
NY	Chemung County	015
NY	Chenango County	017
NY	Clinton County	019
NY	Columbia County	021
NY	Cortland County	023
NY	Delaware County	025
NY	Dutchess County	027

TABLE 1.18-4 *(continued)*
EAS Location Codes: County/Area FIPS Codes

State	County/Area	FIPS Code
NY	Erie County	029
NY	Essex County	031
NY	Franklin County	033
NY	Fulton County	035
NY	Genesee County	037
NY	Greene County	039
NY	Hamilton County	041
NY	Herkimer County	043
NY	Jefferson County	045
NY	Kings County	047
NY	Lewis County	049
NY	Livingston County	051
NY	Madison County	053
NY	Monroe County	055
NY	Montgomery County	057
NY	Nassau County	059
NY	New York County	061
NY	Niagara County	063
NY	Oneida County	065
NY	Onondaga County	067
NY	Ontario County	069
NY	Orange County	071
NY	Orleans County	073
NY	Oswego County	075
NY	Otsego County	077
NY	Putnam County	079
NY	Queens County	081
NY	Rensselaer County	083
NY	Richmond County	085
NY	Rockland County	087
NY	St. Lawrence County	089
NY	Saratoga County	091
NY	Schenectady County	093
NY	Schoharie County	095
NY	Schuyler County	097
NY	Seneca County	099
NY	Steuben County	101
NY	Suffolk County	103
NY	Sullivan County	105
NY	Tioga County	107
NY	Tompkins County	109
NY	Ulster County	111
NY	Warren County	113
NY	Washington County	115
NY	Wayne County	117
NY	Westchester County	119
NY	Wyoming County	121
NY	Yates County	123
OH	Adams County	001
OH	Allen County	003

TABLE 1.18-4 *(continued)*
EAS Location Codes: County/Area FIPS Codes

State	County/Area	FIPS Code
OH	Ashland County	005
OH	Ashtabula County	007
OH	Athens County	009
OH	Auglaize County	011
OH	Belmont County	013
OH	Brown County	015
OH	Butler County	017
OH	Carroll County	019
OH	Champaign County	021
OH	Clark County	023
OH	Clermont County	025
OH	Clinton County	027
OH	Columbiana County	029
OH	Coshocton County	031
OH	Crawford County	033
OH	Cuyahoga County	035
OH	Darke County	037
OH	Defiance County	039
OH	Delaware County	041
OH	Erie County	043
OH	Fairfield County	045
OH	Fayette County	O47
OH	Franklin County	049
OH	Fulton County	051
OH	Gallia County	053
OH	Geauga County	055
OH	Greene County	057
OH	Guernsey County	059
OH	Hamilton County	061
OH	Hancock County	063
OH	Hardin County	065
OH	Harrison County	067
OH	Henry County	069
OH	Highland County	071
OH	Hocking County	073
OH	Holmes County	075
OH	Huron County	077
OH	Jackson County	079
OH	Jefferson County	081
OH	Knox County	083
OH	Lake County	085
OH	Lawrence County	087
OH	Licking County	089
OH	Logan County	091
OH	Lorain County	093
OH	Lucas County	095
OH	Madison County	097
OH	Mahoning County	099
OH	Marion County	101
OH	Medina County	103

TABLE 1.18-4 *(continued)*
EAS Location Codes: County/Area FIPS Codes

State	County/Area	FIPS Code
OH	Meigs County	105
OH	Mercer County	107
OH	Miami County	109
OH	Monroe County	111
OH	Montgomery County	113
OH	Morgan County	115
OH	Morrow County	117
OH	Muskingum County	119
OH	Noble County	121
OH	Ottawa County	123
OH	Paulding County	125
OH	Perry County	127
OH	Pickaway County	129
OH	Pike County	131
OH	Portage County	133
OH	Preble County	135
OH	Putnam County	137
OH	Richland County	139
OH	Ross County	141
OH	Sandusky County	143
OH	Scioto County	145
OH	Seneca County	147
OH	Shelby County	149
OH	Stark County	151
OH	Summit County	153
OH	Trumbull County	155
OH	Tuscarawas County	157
OH	Union County	159
OH	Van Wert County	161
OH	Vinton County	163
OH	Warren County	165
OH	Washington County	167
OH	Wayne County	169
OH	Williams County	171
OH	Wood County	173
OH	Wyandot	175
OK	Adair County	001
OK	Alfalfa County	003
OK	Atoka County	005
OK	Beaver County	007
OK	Beckham County	009
OK	Blaine County	011
OK	Bryan County	013
OK	Caddo County	015
OK	Canadian County	017
OK	Carter County	019
OK	Cherokee County	021
OK	Choctaw County	023
OK	Cimarron County	025
OK	Cleveland County	027

TABLE 1.18-4 *(continued)*
EAS Location Codes: County/Area FIPS Codes

State	County/Area	FIPS Code
OK	Coal County	029
OK	Comanche County	031
OK	Cotton County	033
OK	Craig County	035
OK	Creek County	037
OK	Custer County	039
OK	Delaware County	041
OK	Dewey County	043
OK	Ellis County	045
OK	Garfield County	047
OK	Garvin County	049
OK	Grady County	051
OK	Grant County	053
OK	Greer County	055
OK	Harmon County	057
OK	Harper County	059
OK	Haskell County	061
OK	Hughes County	063
OK	Jackson County	065
OK	Jefferson County	067
OK	Johnston County	069
OK	Kay County	071
OK	Kingfisher County	073
OK	Kiowa County	075
OK	Latimer County	077
OK	Le Flore County	079
OK	Lincoln County	081
OK	Logan County	083
OK	Love County	085
OK	McClain County	087
OK	McCurtain County	089
OK	McIntosh County	091
OK	Major County	093
OK	Marshall County	095
OK	Mayes County	097
OK	Murray County	099
OK	Muskogee County	101
OK	Noble County	103
OK	Nowata County	105
OK	Okfuskee County	107
OK	Oklahoma County	109
OK	Okmulgee County	111
OK	Osage County	113
OK	Ottawa County	115
OK	Pawnee County	117
OK	Payne County	119
OK	Pittsburg County	121
OK	Pontotoc County	123
OK	Pottawatomie County	125
OK	Pushmataha County	127

TABLE 1.18-4 *(continued)*
EAS Location Codes: County/Area FIPS Codes

State	County/Area	FIPS Code
OK	Roger Mills County	129
OK	Rogers County	131
OK	Seminole County	133
OK	Sequoyah County	135
OK	Stephens County	137
OK	Texas County	139
OK	Tillman County	141
OK	Tulsa County	143
OK	Wagoner County	145
OK	Washington County	147
OK	Washita County	149
OK	Woods County	151
OK	Woodward	153
OR	Baker County	001
OR	Benton County	003
OR	Clackamas County	005
OR	Clatsop County	007
OR	Columbia County	009
OR	Coos County	011
OR	Crook County	013
OR	Curry County	015
OR	Deschutes County	017
OR	Douglas County	019
OR	Gilliam County	021
OR	Grant County	023
OR	Harney County	025
OR	Hood River County	027
OR	Jackson County	029
OR	Jefferson County	031
OR	Josephine County	033
OR	Klamath County	035
OR	Lake County	037
OR	Lane County	039
OR	Lincoln County	041
OR	Linn County	043
OR	Malheur County	045
OR	Marion County	047
OR	Morrow County	049
OR	Multnomah County	051
OR	Polk County	053
OR	Sherman County	055
OR	Tillamook County	057
OR	Umatilla County	059
OR	Union County	061
OR	Wallowa County	063
OR	Wasco County	065
OR	Washington County	067
OR	Wheeler County	069
OR	Yamhill County	071
PA	Adams County	001

TABLE 1.18-4 *(continued)*
EAS Location Codes: County/Area FIPS Codes

State	County/Area	FIPS Code
PA	Allegheny County	003
PA	Armstrong County	005
PA	Beaver County	007
PA	Bedford County	009
PA	Berks County	011
PA	Blair County	013
PA	Bradford County	015
PA	Bucks County	017
PA	Butler County	019
PA	Cambria County	021
PA	Cameron County	023
PA	Carbon County	025
PA	Centre County	027
PA	Chester County	029
PA	Clarion County	031
PA	Clearfield County	033
PA	Clinton County	035
PA	Columbia County	037
PA	Crawford County	039
PA	Cumberland County	041
PA	Dauphin County	043
PA	Delaware County	045
PA	Elk County	047
PA	Erie County	049
PA	Fayette County	051
PA	Forest County	053
PA	Franklin County	055
PA	Fulton County	057
PA	Greene County	059
PA	Huntingdon County	061
PA	Indiana County	063
PA	Jefferson County	065
PA	Juniata County	067
PA	Lackawanna County	069
PA	Lancaster County	071
PA	Lawrence County	073
PA	Lebanon County	075
PA	Lehigh County	077
PA	Luzerne County	079
PA	Lycoming County	081
PA	McKean County	083
PA	Mercer County	085
PA	Mifflin County	087
PA	Monroe County	089
PA	Montgomery County	091
PA	Montour County	093
PA	Northampton Count	095
PA	Northumberland County	097
PA	Perry County	099
PA	Philadelphia County	101

TABLE 1.18-4 *(continued)*
EAS Location Codes: County/Area FIPS Codes

State	County/Area	FIPS Code
PA	Pike County	103
PA	Potter County	105
PA	Schuylkill County	107
PA	Snyder County	109
PA	Somerset County	111
PA	Sullivan County	113
PA	Susquehanna County	115
PA	Tioga County	117
PA	Union County	119
PA	Venango County	121
PA	Warren County	123
PA	Washington County	125
PA	Wayne County	127
PA	Westmoreland County	129
PA	Wyoming County	131
PA	York County	133
RI	Bristol County	001
RI	Kent County	003
RI	Newport County	005
RI	Providence County	007
RI	Washington	009
SC	Abbeville County	001
SC	Aiken County	003
SC	Allendale County	005
SC	Anderson County	007
SC	Bamberg County	009
SC	Barnwell County	011
SC	Beaufort County	013
SC	Berkeley County	015
SC	Calhoun County	017
SC	Charleston County	019
SC	Cherokee County	021
SC	Chester County	023
SC	Chesterfield County	025
SC	Clarendon County	027
SC	Colleton County	029
SC	Darlington County	031
SC	Dillon County	033
SC	Dorchester County	035
SC	Edgefield County	037
SC	Fairfield County	039
SC	Florence County	041
SC	Georgetown County	043
SC	Greenville County	045
SC	Greenwood County	047
SC	Hampton County	049
SC	Horry County	051
SC	Jasper County	053
SC	Kershaw County	055
SC	Lancaster County	057

TABLE 1.18-4 *(continued)*
EAS Location Codes: County/Area FIPS Codes

State	County/Area	FIPS Code
SC	Laurens County	059
SC	Lee County	061
SC	Lexington County	063
SC	McCormick County	065
SC	Marion County	067
SC	Marlboro County	069
SC	Newberry County	071
SC	Oconee County	073
SC	Orangeburg County	075
SC	Pickens County	077
SC	Richland County	079
SC	Saluda County	081
SC	Spartanburg County	083
SC	Sumter County	085
SC	Union County	087
SC	Williamsburg County	089
SC	York County	091
SD	Aurora County	003
SD	Beadle County	005
SD	Bennett County	007
SD	Bon Homme County	009
SD	Brookings County	011
SD	Brown County	013
SD	Brule County	015
SD	Buffalo County	017
SD	Butte County	019
SD	Campbell County	021
SD	Charles Mix County	023
SD	Clark County	025
SD	Clay County	027
SD	Codington County	029
SD	Corson County	031
SD	Custer County	033
SD	Davison County	035
SD	Day County	037
SD	Deuel County	039
SD	Dewey County	041
SD	Douglas County	043
SD	Edmunds County	045
SD	Fall River County	047
SD	Faulk County	049
SD	Grant County	051
SD	Gregory County	053
SD	Haakon County	055
SD	Hamlin County	057
SD	Hand County	059
SD	Hanson County	061
SD	Harding County	063
SD	Hughes County	065
SD	Hutchinson County	067

TABLE 1.18-4 *(continued)*
EAS Location Codes:
County/Area FIPS Codes

State	County/Area	FIPS Code
SD	Hyde County	069
SD	Jackson County	071
SD	Jerauld County	073
SD	Jones County	075
SD	Kingsbury County	077
SD	Lake County	079
SD	Lawrence County	081
SD	Lincoln County	083
SD	Lyman County	085
SD	McCook County	087
SD	McPherson County	089
SD	Marshall County	091
SD	Meade County	093
SD	Mellette County	095
SD	Miner County	097
SD	Minnehaha County	099
SD	Moody County	101
SD	Pennington County	103
SD	Perkins County	105
SD	Potter County	107
SD	Roberts County	109
SD	Sanborn County	111
SD	Shannon County	113
SD	Spink County	115
SD	Stanley County	117
SD	Sully County	119
SD	Todd County	121
SD	Tripp County	123
SD	Turner County	125
SD	Union County	127
SD	Walworth County	129
SD	Yankton County	135
SD	Ziebach County	137
TN	Anderson County	001
TN	Bedford County	003
TN	Benton County	005
TN	Bledsoe County	007
TN	Blount County	009
TN	Bradley County	011
TN	Campbell County	013
TN	Cannon County	015
TN	Carroll County	017
TN	Carter County	019
TN	Cheatham County	021
TN	Chester County	023
TN	Claiborne County	025
TN	Clay County	027
TN	Cocke County	029
TN	Coffee County	031
TN	Crockett County	033

TABLE 1.18-4 *(continued)*
EAS Location Codes:
County/Area FIPS Codes

State	County/Area	FIPS Code
TN	Cumberland County	035
TN	Davidson County	037
TN	Decatur County	039
TN	DeKalb County	041
TN	Dickson County	043
TN	Dyer County	045
TN	Fayette County	047
TN	Fentress County	049
TN	Franklin County	051
TN	Gibson County	053
TN	Giles County	055
TN	Grainger County	057
TN	Greene County	059
TN	Grundy County	061
TN	Hamblen County	063
TN	Hamilton County	065
TN	Hancock County	067
TN	Hardeman County	069
TN	Hardin County	071
TN	Hawkins County	073
TN	Haywood County	075
TN	Henderson County	077
TN	Henry County	079
TN	Hickman County	081
TN	Houston County	083
TN	Humphreys County	085
TN	Jackson County	087
TN	Jefferson County	089
TN	Johnson County	091
TN	Knox County	093
TN	Lake County	095
TN	Lauderdale County	097
TN	Lawrence County	099
TN	Lewis County	101
TN	Lincoln County	103
TN	Loudon County	105
TN	McMinn County	107
TN	McNairy County	109
TN	Macon County	111
TN	Madison County	113
TN	Marion County	115
TN	Marshall County	117
TN	Maury County	119
TN	Meigs County	121
TN	Monroe County	123
TN	Montgomery County	125
TN	Moore County	127
TN	Morgan County	129
TN	Obion County	131
TN	Overton County	133

TABLE 1.18-4 *(continued)*
EAS Location Codes:
County/Area FIPS Codes

State	County/Area	FIPS Code
TN	Perry County	135
TN	Pickett County	137
TN	Polk County	139
TN	Putnam County	141
TN	Rhea County	143
TN	Roane County	145
TN	Robertson County	147
TN	Rutherford County	149
TN	Scott County	151
TN	Sequatchie County	153
TN	Sevier County	155
TN	Shelby County	157
TN	Smith County	159
TN	Stewart County	161
TN	Sullivan County	163
TN	Sumner County	165
TN	Tipton County	167
TN	Trousdale County	169
TN	Unicoi County	171
TN	Union County	173
TN	Van Buren County	175
TN	Warren County	177
TN	Washington County	179
TN	Wayne County	181
TN	Weakley County	183
TN	White County	185
TN	Williamson County	187
TN	Wilson County	189
TX	Anderson County	001
TX	Andrews County	003
TX	Angelina County	005
TX	Aransas County	007
TX	Archer County	009
TX	Armstrong County	011
TX	Atascosa County	013
TX	Austin County	015
TX	Bailey County	017
TX	Bandera County	019
TX	Bastrop County	021
TX	Baylor County	023
TX	Bee County	025
TX	Bell County	027
TX	Bexar County	029
TX	Blanco County	031
TX	Borden County	033
TX	Bosque County	035
TX	Bowie County	037
TX	Brazoria County	039
TX	Brazos County	041
TX	Brewster County	043

TABLE 1.18-4 *(continued)*
EAS Location Codes: County/Area FIPS Codes

State	County/Area	FIPS Code
TX	Briscoe County	045
TX	Brooks County	047
TX	Brown County	049
TX	Burleson County	051
TX	Burnet County	053
TX	Caldwell County	055
TX	Calhoun County	057
TX	Callahan County	059
TX	Cameron County	061
TX	Camp County	063
TX	Carson County	065
TX	Cass County	067
TX	Castro County	069
TX	Chambers County	071
TX	Cherokee County	073
TX	Childress County	075
TX	Clay County	077
TX	Cochran County	079
TX	Coke County	081
TX	Coleman County	083
TX	Collin County	085
TX	Collingsworth County	087
TX	Colorado County	089
TX	Comal County	091
TX	Comanche County	093
TX	Concho County	095
TX	Cooke County	097
TX	Coryell County	099
TX	Cottle County	101
TX	Crane County	103
TX	Crockett County	105
TX	Crosby County	107
TX	Culberson County	109
TX	Dallam County	111
TX	Dallas County	113
TX	Dawson County	115
TX	Deaf Smith County	117
TX	Delta County	119
TX	Denton County	121
TX	DeWitt County	123
TX	Dickens County	125
TX	Dimmit County	127
TX	Donley County	129
TX	Duval County	131
TX	Eastland County	133
TX	Ector County	135
TX	Edwards County	137
TX	Ellis County	139
TX	El Paso County	141
TX	Erath County	143

TABLE 1.18-4 *(continued)*
EAS Location Codes: County/Area FIPS Codes

State	County/Area	FIPS Code
TX	Falls County	145
TX	Fannin County	147
TX	Fayette County	149
TX	Fisher County	151
TX	Floyd County	153
TX	Foard County	155
TX	Fort Bend County	157
TX	Franklin County	159
TX	Freestone County	161
TX	Frio County	163
TX	Gaines County	165
TX	Galveston County	167
TX	Garza County	169
TX	Gillespie County	171
TX	Glasscock County	173
TX	Goliad County	175
TX	Gonzales County	177
TX	Gray County	179
TX	Grayson County	181
TX	Gregg County	183
TX	Grimes County	185
TX	Guadalupe County	187
TX	Hale County	189
TX	Hall County	191
TX	Hamilton County	193
TX	Hansford County	195
TX	Hardeman County	197
TX	Hardin County	199
TX	Harris County	201
TX	Harrison County	203
TX	Hartley County	205
TX	Haskell County	207
TX	Hays County	209
TX	Hemphill County	211
TX	Henderson County	213
TX	Hidalgo County	215
TX	Hill County	217
TX	Hockley County	219
TX	Hood County	221
TX	Hopkins County	223
TX	Houston County	225
TX	Howard County	227
TX	Hudspeth County	229
TX	Hunt County	231
TX	Hutchinson County	233
TX	Irion County	235
TX	Jack County	237
TX	Jackson County	239
TX	Jasper County	241
TX	Jeff Davis County	243

TABLE 1.18-4 *(continued)*
EAS Location Codes: County/Area FIPS Codes

State	County/Area	FIPS Code
TX	Jefferson County	245
TX	Jim Hogg County	247
TX	Jim Wells County	249
TX	Johnson County	251
TX	Jones County	253
TX	Karnes County	255
TX	Kaufman County	257
TX	Kendall County	259
TX	Kenedy County	261
TX	Kent County	263
TX	Kerr County	265
TX	Kimble County	267
TX	King County	269
TX	Kinney County	271
TX	Kleberg County	273
TX	Knox County	275
TX	Lamar County	277
TX	Lamb County	279
TX	Lampasas County	281
TX	La Salle County	283
TX	Lavaca County	285
TX	Lee County	287
TX	Leon County	289
TX	Liberty County	291
TX	Limestone County	293
TX	Lipscomb County	295
TX	Live Oak County	297
TX	Llano County	299
TX	Loving County	301
TX	Lubbock County	303
TX	Lynn County	305
TX	McCulloch County	307
TX	McLennan County	309
TX	McMullen County	311
TX	Madison County	313
TX	Marion County	315
TX	Martin County	317
TX	Mason County	319
TX	Matagorda County	321
TX	Maverick County	323
TX	Medina County	325
TX	Menard County	327
TX	Midland County	329
TX	Milam County	331
TX	Mills County	333
TX	Mitchell County	335
TX	Montague County	337
TX	Montgomery County	339
TX	Moore County	341
TX	Morris County	343

TABLE 1.18-4 *(continued)*
EAS Location Codes: County/Area FIPS Codes

State	County/Area	FIPS Code
TX	Motley County	345
TX	Nacogdoches County	347
TX	Navarro County	349
TX	Newton County	351
TX	Nolan County	353
TX	Nueces County	355
TX	Ochiltree County	357
TX	Oldham County	359
TX	Orange County	361
TX	Palo Pinto County	363
TX	Panola County	365
TX	Parker County	367
TX	Parmer County	369
TX	Pecos County	371
TX	Polk County	373
TX	Potter County	375
TX	Presidio County	377
TX	Rains County	379
TX	Randall County	381
TX	Reagan County	383
TX	Real County	385
TX	Red River County	387
TX	Reeves County	389
TX	Refugio County	391
TX	Roberts County	393
TX	Robertson County	395
TX	Rockwall County	397
TX	Runnels County	399
TX	Rusk County	401
TX	Sabine County	403
TX	San Augustine County	405
TX	San Jacinto County	407
TX	San Patricio County	409
TX	San Saba County	411
TX	Schleicher County	413
TX	Scurry County	415
TX	Shackelford County	417
TX	Shelby County	419
TX	Sherman County	421
TX	Smith County	423
TX	Somervell County	425
TX	Starr County	427
TX	Stephens County	429
TX	Sterling County	431
TX	Stonewall County	433
TX	Sutton County	435
TX	Swisher County	437
TX	Tarrant County	439
TX	Taylor County	441
TX	Terrell County	443

TABLE 1.18-4 *(continued)*
EAS Location Codes: County/Area FIPS Codes

State	County/Area	FIPS Code
TX	Terry County	445
TX	Throckmorton County	447
TX	Titus County	449
TX	Tom Green County	451
TX	Travis County	453
TX	Trinity County	455
TX	Tyler County	457
TX	Upshur County	459
TX	Upton County	461
TX	Uvalde County	463
TX	Val Verde County	465
TX	Van Zandt County	467
TX	Victoria County	469
TX	Walker County	471
TX	Waller County	473
TX	Ward County	475
TX	Washington County	477
TX	Webb County	479
TX	Wharton County	481
TX	Wheeler County	483
TX	Wichita County	485
TX	Wilbarger County	487
TX	Willacy County	489
TX	Williamson County	491
TX	Wilson County	493
TX	Winkler County	495
TX	Wise County	497
TX	Wood County	499
TX	Yoakum County	501
TX	Young County	503
TX	Zapata County	505
TX	Zavala County	507
UT	Beaver County	001
UT	Box Elder County	003
UT	Cache County	005
UT	Carbon County	007
UT	Daggett County	009
UT	Davis County	011
UT	Duchesne County	013
UT	Emery County	015
UT	Garfield County	017
UT	Grand County	019
UT	Iron County	021
UT	Juab County	023
UT	Kane County	025
UT	Millard County	027
UT	Morgan County	029
UT	Piute County	031
UT	Rich County	033
UT	Salt Lake County	035

TABLE 1.18-4 *(continued)*
EAS Location Codes: County/Area FIPS Codes

State	County/Area	FIPS Code
UT	San Juan County	037
UT	Sanpete County	039
UT	Sevier County	041
UT	Summit County	043
UT	Tooele County	045
UT	Uintah County	047
UT	Utah County	049
UT	Wasatch County	051
UT	Washington County	053
UT	Wayne County	055
UT	Weber County	057
VA	Accomack County	001
VA	Albemarle County	003
VA	Alexandria City	510
VA	Alleghany County	005
VA	Amelia County	007
VA	Amherst County	009
VA	Appomattox County	011
VA	Arlington County	013
VA	Augusta County	015
VA	Bath County	017
VA	Bedford City	515
VA	Bedford County	019
VA	Bland County	021
VA	Botetourt County	023
VA	Bristol City	520
VA	Brunswick County	025
VA	Buchanan County	027
VA	Buckingham County	029
VA	Buena Vista City	530
VA	Campbell County	031
VA	Caroline County	033
VA	Carroll County	035
VA	Charles City County	036
VA	Charlotte County	037
VA	Charlottesville City	540
VA	Chesapeake City	550
VA	Chesterfield County	041
VA	Clarke County	043
VA	Clifton Forge City	560
VA	Colonial Heights	570
VA	Covington City	580
VA	Craig County	045
VA	Culpeper County	047
VA	Cumberland County	049
VA	Danville City	590
VA	Dickenson County	051
VA	Dinwiddie County	053
VA	Emporia City	595
VA	Essex County	057

TABLE 1.18-4 *(continued)*
EAS Location Codes: County/Area FIPS Codes

State	County/Area	FIPS Code
VA	Fairfax City	600
VA	Fairfax County	059
VA	Falls Church City	610
VA	Fauquier County	061
VA	Floyd County	063
VA	Fluvanna County	065
VA	Franklin City	620
VA	Franklin County	067
VA	Frederick County	069
VA	Fredericksburg City	630
VA	Galax City	640
VA	Giles County	071
VA	Gloucester County	073
VA	Goochland County	075
VA	Grayson County	077
VA	Greene County	079
VA	Greensville County	081
VA	Halifax County	083
VA	Hampton City	650
VA	Hanover County	085
VA	Harrisonburg City	660
VA	Henrico County	087
VA	Henry County	089
VA	Highland County	091
VA	Hopewell City	670
VA	Isle of Wight County	093
VA	James City County	095
VA	King and Queen County	097
VA	King George County	099
VA	King William County	101
VA	Lancaster County	103
VA	Lee County	105
VA	Lexington City	678
VA	Loudoun County	107
VA	Louisa County	109
VA	Lunenburg County	111
VA	Lynchburg City	680
VA	Madison County	113
VA	Manassas City	683
VA	Manassas Park City	685
VA	Martinsville City	690
VA	Mathews County	115
VA	Mecklenburg County	117
VA	Middlesex County	119
VA	Montgomery County	121
VA	Nelson County	125
VA	New Kent County	127
VA	Newport News City	700
VA	Norfolk City	710
VA	Northampton County	131

TABLE 1.18-4 *(continued)*
EAS Location Codes: County/Area FIPS Codes

State	County/Area	FIPS Code
VA	Northumberland County	133
VA	Norton City	720
VA	Nottoway County	135
VA	Orange County	137
VA	Page County	139
VA	Patrick County	141
VA	Petersburg City	730
VA	Pittsylvania County	143
VA	Poquoson City	735
VA	Portsmouth City	740
VA	Powhatan County	145
VA	Prince Edward County	147
VA	Prince George County	149
VA	Prince William County	153
VA	Pulaski County	155
VA	Radford City	750
VA	Rappahannock County	157
VA	Richmond City	760
VA	Richmond County	159
VA	Roanoke City	770
VA	Roanoke County	161
VA	Rockbridge County	163
VA	Rockingham County	165
VA	Russell County	167
VA	Salem City	775
VA	Scott County	169
VA	Shenandoah County	171
VA	Smyth County	173
VA	South Boston City	780
VA	Southampton County	175
VA	Spotsylvania County	177
VA	Stafford County	179
VA	Staunton City	790
VA	Suffolk City	800
VA	Surry County	181
VA	Sussex County	183
VA	Tazewell County	185
VA	Virginia Beach City	810
VA	Warren County	187
VA	Washington County	191
VA	Waynesboro City	820
VA	Westmoreland County	193
VA	Williamsburg City	830
VA	Winchester City	840
VA	Wise County	195
VA	Wythe County	197
VA	York County	199
VT	Addison County	001
VT	Bennington County	003
VT	Caledonia County	005

TABLE 1.18-4 *(continued)*
EAS Location Codes: County/Area FIPS Codes

State	County/Area	FIPS Code
VT	Chittenden County	007
VT	Essex County	009
VT	Franklin County	011
VT	Grand Isle County	013
VT	Lamoille County	015
VT	Orange County	017
VT	Orleans County	019
VT	Rutland County	021
VT	Washington County	023
VT	Windham County	025
VT	Windsor County	027
WA	Adams County	001
WA	Asotin County	003
WA	Benton County	005
WA	Chelan County	007
WA	Clallam County	009
WA	Clark County	011
WA	Columbia County	013
WA	Cowlitz County	015
WA	Douglas County	017
WA	Ferry County	019
WA	Franklin County	021
WA	Garfield County	023
WA	Grant County	025
WA	Grays Harbor County	027
WA	Island County	029
WA	Jefferson County	031
WA	King County	033
WA	Kitsap County	035
WA	Kittitas County	037
WA	Klickitat County	039
WA	Lewis County	041
WA	Lincoln County	043
WA	Mason County	045
WA	Okanogan County	047
WA	Pacific County	049
WA	Pend Oreille County	051
WA	Pierce County	053
WA	San Juan County	055
WA	Skagit County	057
WA	Skamania County	059
WA	Snohomish County	061
WA	Spokane County	063
WA	Stevens County	065
WA	Thurston County	067
WA	Wahkiakum County	069
WA	Walla Walla County	071
WA	Whatcom County	073
WA	Whitman County	075
WA	Yakima County	077

TABLE 1.18-4 *(continued)*
EAS Location Codes: County/Area FIPS Codes

State	County/Area	FIPS Code
WI	Adams County	001
WI	Ashland County	003
WI	Barron County	005
WI	Bayfield County	007
WI	Brown County	009
WI	Buffalo County	011
WI	Burnett County	013
WI	Calumet County	015
WI	Chippewa County	017
WI	Clark County	019
WI	Columbia County	021
WI	Crawford County	023
WI	Dane County	025
WI	Dodge County	027
WI	Door County	029
WI	Douglas County	031
WI	Dunn County	033
WI	Eau Claire County	035
WI	Florence County	037
WI	Fond du Lac County	039
WI	Forest County	041
WI	Grant County	043
WI	Green County	045
WI	Green Lake County	047
WI	Iowa County	049
WI	Iron County	051
WI	Jackson County	053
WI	Jefferson County	055
WI	Juneau County	057
WI	Kenosha County	059
WI	Kewaunee County	061
WI	La Crosse County	063
WI	Lafayette County	065
WI	Langlade County	067
WI	Lincoln County	069
WI	Manitowoc County	071
WI	Marathon County	073
WI	Marinette County	075
WI	Marquette County	077
WI	Menominee County	078
WI	Milwaukee County	079
WI	Monroe County	081
WI	Oconto County	083
WI	Oneida County	085
WI	Outagamie County	087
WI	Ozaukee County	089
WI	Pepin County	091
WI	Pierce County	093
WI	Polk County	095
WI	Portage County	097

TABLE 1.18-4 *(continued)*
EAS Location Codes: County/Area FIPS Codes

State	County/Area	FIPS Code
WI	Price County	099
WI	Racine County	101
WI	Richland County	103
WI	Rock County	105
WI	Rusk County	107
WI	St. Croix County	109
WI	Sauk County	111
WI	Sawyer County	113
WI	Shawano County	115
WI	Sheboygan County	117
WI	Taylor County	119
WI	Trempealeau County	121
WI	Vernon County	123
WI	Vilas County	125
WI	Walworth County	127
WI	Washburn County	129
WI	Washington County	131
WI	Waukesha County	133
WI	Waupaca County	135
WI	Waushara County	137
WI	Winnebago County	139
WI	Wood County	141
WV	Barbour County	001
WV	Berkeley County	003
WV	Boone County	005
WV	Braxton County	007
WV	Brooke County	009
WV	Cabell County	011
WV	Calhoun County	013
WV	Clay County	015
WV	Doddridge County	017
WV	Fayette County	019
WV	Gilmer County	021
WV	Grant County	023
WV	Greenbrier County	025
WV	Hampshire County	027
WV	Hancock County	029
WV	Hardy County	031
WV	Harrison County	033
WV	Jackson County	035
WV	Jefferson County	037
WV	Kanawha County	039
WV	Lewis County	041
WV	Lincoln County	043
WV	Logan County	045
WV	McDowell County	047
WV	Marion County	049
WV	Marshall County	051
WV	Mason County	053
WV	Mercer County	055

TABLE 1.18-4 *(continued)*
EAS Location Codes: County/Area FIPS Codes

State	County/Area	FIPS Code
WV	Mineral County	057
WV	Mingo County	059
WV	Monongalia County	061
WV	Monroe County	063
WV	Morgan County	065
WV	Nicholas County	067
WV	Ohio County	069
WV	Pendleton County	071
WV	Pleasants County	073
WV	Pocahontas County	075
WV	Preston County	077
WV	Putnam County	079
WV	Raleigh County	081
WV	Randolph County	083
WV	Ritchie County	085
WV	Roane County	087
WV	Summers County	089
WV	Taylor County	091
WV	Tucker County	093
WV	Tyler County	095
WV	Upshur County	097
WV	Wayne County	099
WV	Webster County	101
WV	Wetzel County	103
WV	Wirt County	105
WV	Wood County	107
WV	Wyoming County	109
WY	Albany County	001
WY	Big Horn County	003
WY	Campbell County	005
WY	Carbon County	007
WY	Converse County	009
WY	Crook County	011
WY	Fremont County	013
WY	Goshen County	015
WY	Hot Springs County	017
WY	Johnson County	019
WY	Laramie County	021
WY	Lincoln County	023
WY	Natrona County	025
WY	Niobrara County	027
WY	Park County	029
WY	Platte County	031
WY	Sheridan County	033
WY	Sublette County	035
WY	Sweetwater County	037
WY	Teton County	039
WY	Uinta County	041
WY	Washakie County	043
WY	Weston	045

TABLE 1.18-4 *(continued)*
EAS Location Codes: County/Area FIPS Codes

Territory	Area	FIPS Code
American Samoa	Eastern District	010
American Samoa	Manu'a District	020
American Samoa	Rose Island	030
American Samoa	Swains Island	040
American Samoa	Western District	050
Federated States of Micronesia	Chuuk	002
Federated States of Micronesia	Kosrae	005
Federated States of Micronesia	Pohnpeit	040
Federated States of Micronesia	Yap	060
Guam	Guam	010
Marshall Islands	Ailinginae	007
Marshall Islands	Ailinglaplap	010
Marshall Islands	Ailuk	030
Marshall Islands	Arno	040
Marshall Islands	Aur	050
Marshall Islands	Bikar	060
Marshall Islands	Bikini	070
Marshall Islands	Bokak	073
Marshall Islands	Ebon	080
Marshall Islands	Enewetak	090
Marshall Islands	Erikub	100
Marshall Islands	Jabat	110
Marshall Islands	Jaluit	120
Marshall Islands	Jemo	130
Marshall Islands	Kili	140
Marshall Islands	Kwajalein	150
Marshall Islands	Lae	160
Marshall Islands	Lib	170
Marshall Islands	Likiep	180
Marshall Islands	Majuro	190
Marshall Islands	Maloelap	300
Marshall Islands	Mejit	310
Marshall Islands	Mili	320
Marshall Islands	Namorik	330
Marshall Islands	Namu	340
Marshall Islands	Rongelap	350
Marshall Islands	Rongrik	360
Marshall Islands	Toke	385
Marshall Islands	Ujae	390
Marshall Islands	Ujelang	400
Marshall Islands	Utrik	410
Marshall Islands	Wotho	420
Marshall Islands	Wotle	430
Northern Mariana Islands	Northern Islands	085
Northern Mariana Islands	Rota	100
Northern Mariana Islands	Saipan	110
Northern Mariana Islands	Tinian	120
Palau	Aimeliik	002

TABLE 1.18-4 *(continued)*
EAS Location Codes: County/Area FIPS Codes

Territory	Area	FIPS Code
Palau	Airai	004
Palau	Angaur	010
Palau	Hatoboheit	050
Palau	Kayangel	100
Palau	Koror	150
Palau	Melekeok	212
Palau	Ngaraard	214
Palau	Ngarchelong	218
Palau	Ngardmau	222
Palau	Ngatpang	224
Palau	Ngchesar	226
Palau	Ngernmlengui	227
Palau	Ngiwal	228
Palau	Peleliu	350
Palau	Sonsorol	370
Puerto Rico	Adjuntas	001
Puerto Rico	Aguada	003
Puerto Rico	Aguadilla	005
Puerto Rico	Aguas Buenas	007
Puerto Rico	Aibonito	009
Puerto Rico	Anasco	011
Puerto Rico	Arecibo	013
Puerto Rico	Arroyo	015
Puerto Rico	Barceloneta	017
Puerto Rico	Barranquitas	019
Puerto Rico	Bayamo'n	021
Puerto Rico	Cabo Rojo	023
Puerto Rico	Caguas	025
Puerto Rico	Camuy	027
Puerto Rico	Canovanas	029
Puerto Rico	Carolina	031
Puerto Rico	Catano	033
Puerto Rico	Cayey	035
Puerto Rico	Ceiba	037
Puerto Rico	Ciales	039
Puerto Rico	Cidra	041
Puerto Rico	Coamo	043
Puerto Rico	Comerio	045
Puerto Rico	Corozal	047
Puerto Rico	Culebra	049
Puerto Rico	Dorado	051
Puerto Rico	Fajardo	053
Puerto Rico	Florida	054
Puerto Rico	Guanica	055
Puerto Rico	Guayama	057
Puerto Rico	Guayanilla	059
Puerto Rico	Guaynabo	061
Puerto Rico	Gurabo	063

TABLE 1.18-4 *(continued)*
EAS Location Codes:
County/Area FIPS Codes

Territory	Area	FIPS Code
Puerto Rico	Hatillo	065
Puerto Rico	Hormigueros	067
Puerto Rico	Humacao	069
Puerto Rico	Isabela	071
Puerto Rico	Jayuya	073
Puerto Rico	Juana Diaz	075
Puerto Rico	Juncos	077
Puerto Rico	Lajas	079
Puerto Rico	Lares	081
Puerto Rico	Las Marias	083
Puerto Rico	Las Piedras	085
Puerto Rico	Loiza	087
Puerto Rico	Luquillo	089
Puerto Rico	Manati	091
Puerto Rico	Maricao	093
Puerto Rico	Maunabo	095
Puerto Rico	Mayaguez	097
Puerto Rico	Moca	099
Puerto Rico	Morovis	101
Puerto Rico	Naguabo	103
Puerto Rico	Naranjito	105
Puerto Rico	Orocovis	107
Puerto Rico	Patillas	109
Puerto Rico	Penuelas	111
Puerto Rico	Ponce	113
Puerto Rico	Quebradillas	115
Puerto Rico	Rincon	117
Puerto Rico	Rio Grande	119
Puerto Rico	Sabana Grande	121
Puerto Rico	Salinas	123
Puerto Rico	San German	125
Puerto Rico	San Juan	127
Puerto Rico	San Lorenzo	129
Puerto Rico	San Sebastian	131
Puerto Rico	Santa Isabel	133
Puerto Rico	Toa Alta	135
Puerto Rico	Toa Baja	137
Puerto Rico	Trujillo Alto	139
Puerto Rico	Utuado	141
Puerto Rico	Vega Alta	143
Puerto Rico	Vega Baja	145
Puerto Rico	Vieques	147
Puerto Rico	Villalba	149
Puerto Rico	Yabucoa	151
Puerto Rico	Yauco	153
U.S. minor outlying islands	Baker Island	050
U.S. minor outlying islands	Howland Island	100
U.S. minor outlying islands	Jarvis Island	150

TABLE 1.18-4 *(continued)*
EAS Location Codes:
County/Area FIPS Codes

Territory	Area	FIPS Code
U.S. minor outlying islands	Johnston Island	200
U.S. minor outlying islands	Kingman Reef	250
U.S. minor outlying islands	Midway Islands	300
U.S. minor outlying islands	Navassa Island	350
U.S. minor outlying islands	Palmyra Atoll	400
U.S. minor outlying islands	Wake Island	450
U.S. Virgin Islands	St. Croix	010
U.S. Virgin Islands	St. John	020
U.S. Virgin Islands	St. Thomas	030

TABLE 1.18-5
EAS Location Codes:
County Subdivision Codes

Subdivision	Code
All	0
Northwest	1
North	2
Northeast	3
West	4
Central	5
East	6
Southwest	7
South	8
Southeast	9

SECTION

2

TECHNICAL MANAGEMENT AND SAFETY

CHAPTER

2.1

Engineering Management

DR. SAM SWAN
School of Journalism and Electronic Media
University of Tennessee
Knoxville, Tennessee

INTRODUCTION

Most broadcast station engineering staff plan a career in which they will work in studio or remote operations, maintenance, or transmitters. Few think about or plan for a career in management. What happens in many stations is that engineers are promoted from within to become managers, in positions such as chief engineer or director of engineering. However, while engineers may be excellent at the technical aspects of their work, they may not be at all prepared to handle the challenges of managing an engineering department. Very few broadcast station engineers have had specialized courses or training in management. The result is that many chief engineers feel unprepared dealing with personnel and financial aspects of their new positions. Some fail, and either resign or are replaced.

This need not be the case if engineers who aspire to lead engineering teams fully understand the many challenges they will face as managers. This chapter covers the concepts of engineering management for radio and television broadcast stations, and discusses the relationship of the broadcast engineer with other members of the station management team.

The term *chief engineer* (sometimes known simply as "the chief") is used throughout the chapter as a general term for the head of a station's engineering department. Different organizations may have different names for specific positions. The term *engineer* is also used here as a general term; staff working in an engineering department may include persons who are not actually qualified engineers, such as technicians and others. Use of the word "he" applies to both male and female staff.

THE BUSINESS OF BROADCASTING

Mission

One of the first principles chief engineers must understand is the *mission* of the broadcast station. All stations in the United States are licensed by the Federal Communications Commission (FCC). The FCC requires that stations operate in the "public interest, convenience, and necessity." This is often referred to as the *public service* mission of broadcasters. All stations are required to provide programming that serves the public.

A second mission for all commercial stations, though unwritten in FCC code, is that stations must *earn a profit*. As a result, station general managers seek ways to serve their audience and make money. A chief engineer, and all engineers for that matter, must understand how they fit into these two missions. The station cannot achieve either mission without a quality product to put on the air. Engineers must provide the technical expertise to make that happen. They must also do it in a manner that takes the station's profit margin into consideration.

Businesses

Engineers need to understand that commercial broadcast stations are unlike most other businesses, in that they operate in at least five different business areas.

The first business is that of *news*. Engineers must assist the news department in providing equipment and resources to cover breaking news. The public depends on broadcast stations to provide the latest news and weather information.

The second business relates to *information*. Stations produce a wide variety of programs designed to inform their audiences. Engineering assists with the production of these programs to ensure quality.

A third business is in *education*. This is especially true for public stations but also applies to commercial stations. Programming for children, documentaries, and other programs are designed to educate the audience. Engineers must provide the technical expertise to develop these educational programs.

The fourth business is that of *entertainment*. Many people listen to radio for the music and many people watch television for the entertainment programs. Engineers must provide the best signal quality possible so that the audience will want to continue to listen or watch the station.

The last business broadcast stations are in is *advertising*. Advertisers pay the bills. Stations produce and broadcast commercials in exchange for revenue. Advertisers are more interested in placing their spots in programming from technically superior facilities that reach the largest possible audience.

The result is that engineers must work on a daily basis with *all* departments of a broadcast station. Chief engineers work alongside the general manager, program director, news director, promotions manager, and sales manager to meet the missions of the station.

Customers

Broadcast stations serve four distinct constituent groups, or "customers." Engineers need to understand who these customers are and how they influence decisions that are made.

The first group of customers is the *audience*. All decisions made by the station management team should be made with the audience in mind. The simple fact is that broadcast stations strive to attract and keep the largest possible audience. The station with the "biggest stick" or largest transmitter usually has the potential to reach the largest audience. Quite apart from any programming decisions, a station needs to have a strong, quality signal to make it possible to compete for audience ratings.

The second group of customers the broadcaster serves is *advertisers*. Advertisers want to associate with the "best" station, as they perceive it. They can see or hear the station and they will notice obvious technical problems.

The third customer group is *employees*. Employees of the station are a vital resource. Employees who are proud of their station will usually perform at a higher level than those who are indifferent.

The fourth customer group is the *investors* or *stockholders*. They are primarily interested in making money.

These four groups largely determine the priorities for the station. The management team should consider the impact of major decisions on the audience, advertisers, employees, and ownership.

MANAGERS AND LEADERS

There is a difference between management and leadership and many engineers tend to be better managers than leaders. A manager may manage things—money, equipment, schedules, buildings, vehicles, and so forth. A manager can move things around as he sees fit. In contrast, a leader leads people. Without special knowledge and abilities in leadership, it is not always possible to get people to do what you want them to do.

It is possible for a person to be a manager without being a leader. For instance, a single engineer for a small radio station will most likely function as a manager. He manages the equipment of the station but may have no personnel to lead. On the other hand, a chief engineer for a larger radio station or for a television facility must function as both a manager and leader because he will have equipment to manage and people to lead. As a result, chief engineers must develop an appropriate leadership style.

Leadership Styles

There are four major leadership styles often found in broadcast stations, depending on the department and the situation. Considerable research has been done to examine the effectiveness and appropriateness of each style to each position.

Autocratic Style

The first style some chief engineers may adopt is what social scientists call the X or autocratic style of leadership. This style of leadership is often compared with the military style of leadership. The leader is in command of his troops. He tells them what to do and how to do it. The X leader directs his employees to perform their tasks according to his wishes. Getting the job done is the most important goal for the X manager. He is socially distant from his employees. He barks out commands and demands performance. If he does not get it, he fires the employees.

This style of leadership may be found in broadcast stations. Directors in the control room are clearly X-style leaders—they bark out commands to all crew members in the studio. But, is this style effective? The research has examined that question on three measures—employee morale, job turnover, and job productivity. The results are that morale is generally low, job turnover is generally high, but productivity is also high. The job gets done because the manager demands that it is done. The X style seems to be most effective for employees who are not self-directed. Engineering units may have technicians who need to have managers tell them what to do and how to do it. In that case, the X style may be most appropriate.

Democratic Style

The second style of leadership some chief engineers may adopt is the opposite of the X style. It is often called the Y style or democratic style of leadership. Y leaders are people-oriented. The style is follower-driven. Managers will likely have many meetings and take votes on every course of action. The overarching philosophy of the Y manager is that "happy workers will be productive workers." The Y manager is socially close to his employees. He sees all of his employees as members of his extended family and treats them that way.

In broadcast stations, Y managers may include sales managers, creative services directors, and perhaps some others. The research shows that employee morale under a Y manager is generally high and job turnover is very low. Employees like their managers and they enjoy going to work everyday, but productivity is low. The reason seems to be that employees have no fear and spend much of their time socializing with coworkers. The kinds of employees who would thrive with Y managers are creative and self-directed employees. That is why creative services, programming, or sales managers are Y managers. Very few chief engineers find this style appropriate for engineers.

Goal-Oriented Style

The third style of leadership found in broadcast stations is the Z style. Z leaders are goal-oriented. They understand the importance of people to the organization and the need to get the job done. Managers participate in management by objectives. They ask all employees to set measurable and realistic goals for themselves. From individual goals, Z managers develop plans for their units.

Many general managers and sales managers function as Z managers. They are used to setting revenue goals for the station. The findings for this style of leadership are that employee morale is relatively high, job turnover is relatively low, and productivity is relatively high. The main problem with this style is that it does not work in all departments and with all people. Some people simply cannot set goals for themselves. Some units cannot set goals. Very few chief engineers seem to function exclusively in this style.

Contingency Style

The fourth style of leadership found in broadcast stations is the most common and is a combination of the other three styles. It is often called the contingency or C style of leadership. This style can change according to different situations and different people. For instance, in times of crisis, a type-X management style would be best. When weather emergencies occur, chief engineers must go into a type-X mode. When a general manager wants the chief engineer to create plans and budgets for the coming year, the "chief" must function with type-Z style in communicating with others on the management team. On other days, the chief engineer may use the type-Y style to get the most from some of his employees.

Which Style to Use

The research indicates that the contingency type C style of leadership is most commonly found among news directors and may be a good choice for chief engineers as well. There are some engineers who are professional and self-directed individuals. A type-Y approach may be the best for them. Other staff may need more of a type-X style to motivate them. What is important for the chief engineer to understand is that the management team may be comprised of very different personalities and leadership styles. For instance, the general manager and sales manager are most likely Z or goal-oriented leaders. That is why they seem to see eye-to-eye on most issues. Program directors, production managers, and creative services or promotions directors may tend to be more Y or people-oriented personalities. The news director and chief engineer may be contingency-style leaders and be more likely to see eye-to-eye on many issues.

FUNCTIONS, DUTIES, AND QUALITIES

All chief engineers must perform four main functions as part of their responsibilities. This section discusses the functions of planning, leading, organizing, and controlling, the duties that go with them, and some of the qualities that lead to success.

Planning

Broadcast stations engage in three kinds of planning—short-term, long-term, and strategic.

Short-term planning is done each year by most stations, principally in preparing the annual budget. From three to six months ahead of the fiscal year, stations begin the planning and budgeting process. Each department will be asked to submit plans for the coming year along with a budget. The chief engineer must submit an operating budget including personnel, operating, and equipment expenditures. The short-term plan may also call for special equipment purchases needed in the coming year.

The long-term plan is typically a five-year plan. The chief engineer will be asked to put together a plan related to equipment maintenance and replacement over a five-year period so that the station's general manager can set aside money to meet the needs. This plan will normally include major system upgrades and capital purchases.

Strategic planning is also done periodically by many stations. Chief engineers may be asked to put together what are called "what if" plans. For example: "What if the FCC changes the ownership rules to allow television stations to own and operate more than one station in a market?" How will the station respond from an engineering perspective?

Many chief engineers will spend 25% or more of their time engaged in planning. The process requires time away from the day-to-day operation of the station.

Leading

Leadership of the engineering department involves all personnel decisions. Many engineers are more prepared to deal with the management of equipment and other resources but may not be adequately prepared to handle the myriad problems leading people can bring. The larger the engineering unit, the more personnel problems the manager will usually face. Many of them can be avoided with careful hiring, training, motivating, and evaluating strategies. These will be addressed later in this chapter. In addition, communicating with and coordinating all employees are other functions of a leader. Many managers will spend more than 25% of their time on personnel issues.

Organizing

The chief engineer will be expected to create systems to ensure the proper functioning of the unit. These systems include an organizational structure for the unit and ways to monitor productivity. Time and resource management is an important organizational skill.

Controls

The chief engineer must establish quality-control procedures and systems for his department and for the station. This includes regular maintenance checks on all equipment. It would also include monitoring the quality of the signal, the sound, and/or look of the station, and monitoring the quality of studio productions from a technical standpoint. These tasks may be delegated to other staff or to some extent be automated, but the responsibility rests with the chief engineer.

Another responsibility is cost control. The goal is to be both efficient and effective; this is covered later on.

Duties

In addition to these four basic functions that all chief engineers must perform, several other management duties should be mentioned. Many chief engineers will say that their most important duty is to protect the license. Broadcast stations are granted licenses by the FCC for a period of time. If the station operates according to FCC rules and regulations, the station will most likely be relicensed. Many of the FCC rules and regulations that result in fines are technical in nature. Station general managers rely on their chief engineers to operate the station on the right frequency or channel and at the licensed power. If the station loses its license, it is out of business.

Another key duty is financial management. This includes the budgeting process (mentioned previously), and monitoring the budget and expenditure. This area of management is important and more attention will be given to it later in this chapter.

Another duty is to manage the station's inventory. This usually includes all technical equipment—its vehicles, buildings, electrical, and mechanical systems. In other words, most of the tangible assets of the station.

The chief engineer must also manage the time and resources of his unit in relation to all other departments of the station. Engineers are usually required to help with all aspects of the station's technical operation. Allocating engineering resources for this is a duty of the chief engineer.

Other duties include liaison to all departments in the broadcast station. The chief engineer is part of the management team of the station and must work together with programming, production, news, and sales departments. He may also be asked to work with engineering unions and perhaps be involved in community relations.

Qualities of the Successful Engineering Manager

It should be clear by now that broadcast engineering managers play a very important role in broadcast stations. Not everyone can do this well but there are some qualities that seem to be characteristic of those who are successful.

First, effective chief engineers must be excellent communicators. This is a problem for some engineers who find themselves in leadership roles, but it is very important for chief engineers to communicate to the general manager, sales manager, and program director about the needs and issues facing the engineering unit.

It is equally important for the chief engineer to be able to communicate from the management team to engineering employees. Chief engineers must become excellent listeners and team players if they are going to be successful. They must also be able to motivate employees to work together as a team to achieve the goals of the station.

Finally, they must be good decision makers. Many decisions regarding equipment are made each year and the general manager depends on the chief engineer to make good decisions about all equipment purchases. It is vital that the chief engineer should stay abreast of all the technological changes occurring in broadcasting.

ORGANIZATION

An engineering department at a broadcast station can range in size from one person at a small radio station to perhaps 10–20 engineers and other staff at a medium-market television station, to a much larger staff for a large market or network operation. One of the most important functions of a chief engineer is to help the station general manager determine the right number of employees needed to operate and maintain

the station properly. Too many staff results in waste and inefficiency. Too few results in burnout and excessive overtime wages. The structure and organization of the unit should be periodically evaluated to determine if each area of the unit has the right number of people. Needs have changed over the years as stations have moved from analog to digital operations, to an increasing use of automation, and to information technology (IT)–based systems. The station may perhaps need fewer transmitter and studio engineers and more IT specialists, or perhaps staff with a mix of different skills.

Where there are lower-level managers below the chief engineer, they are generally responsible for scheduling employees and day-to-day management in their respective areas. It is important for structures to be functional. If the organizational structure is too tall with too many levels between the workers of the station and the top management, there could be problems of coordination and communication. Generally, the flatter the organizational structure, the better the coordination and communication within the unit and across the organization. Stations should strive to minimize the number of managers while ensuring that the responsibilities and workload do not result in overload for any particular position.

Organization Structure

A well-defined organization structure is important for units to function effectively. The purpose of an organizational chart is to indicate chains of command and relationships among staff members and the various areas of the station or unit. However, in small- and medium-sized markets the positions perhaps tend to be filled by ability and willingness rather than going strictly by organizational flowcharts.

The organizational structure of the engineering and operations departments may vary widely from station to station. Three typical examples follow. Other structures may, of course, be selected for particular requirements.

Figure 2.1-1 represents the simplest station engineering department structure, which may apply to either radio or television. In this example, the chief engineer heads the department that is responsible for equipment installation, maintenance, and FCC technical compliance. He is usually responsible for preparing engineering operating and capital budgets and is the designated chief operator for the station. Directly under his supervision is the assistant chief engineer, who is more responsible for the day-to-day maintenance and scheduling maintenance personnel. The number of technical maintenance personnel will vary depending on how much equipment the station has. The hardware includes that in the studio, master control, news department, transmitters, translators, low-power/satellite stations, and sometimes all IT computer systems.

In this arrangement, staff working in technical operations for studios, news, and master control would report through a separate chain of command to

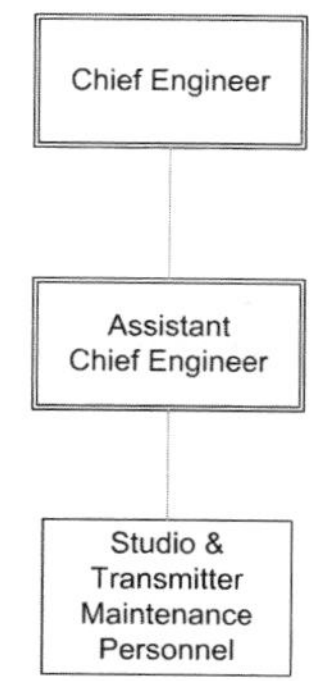

FIGURE 2.1-1 Simple engineering department organization.

the program director or news director for the station. Responsibility for buildings and grounds may come under the chief engineer or with staff reporting separately to the general manager.

Figure 2.1-2 outlines a very common structure for a television station in which the chief engineer and assistant chief have responsibilities for the technical aspects of the "on-air" station operations as well as the maintenance responsibilities as outlined above. This includes master control, audio, video, and tape operators, as well as IT. Again, sometimes buildings and grounds are added. A radio station equivalent structure would be similar.

In this arrangement for television, staff in "creative" technical positions, such as camera and studio floor crew, would normally report to the production manager and not to engineering.

Figure 2.1-3 illustrates a staffing structure where engineering and operations are combined into a single department with four (or more) subdepartments. Each subdepartment is responsible for a different aspect of station operations and maintenance and all report to a broadcast operations and engineering (BO&E) manager. This arrangement may improve coordination of the operational and engineering functions of the station and is often the most efficient and cost-effective

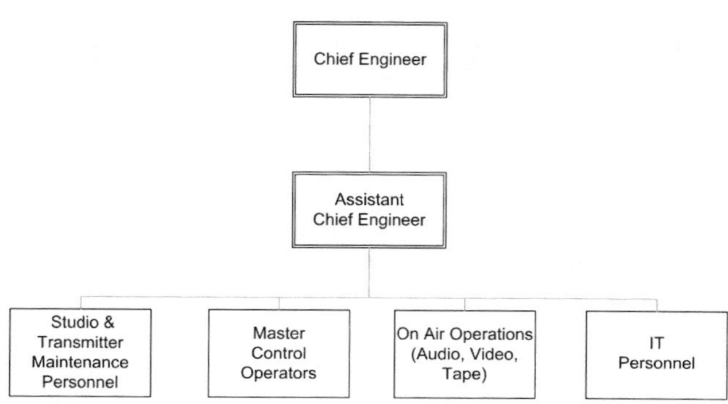

FIGURE 2.1-2 Typical television station engineering organization.

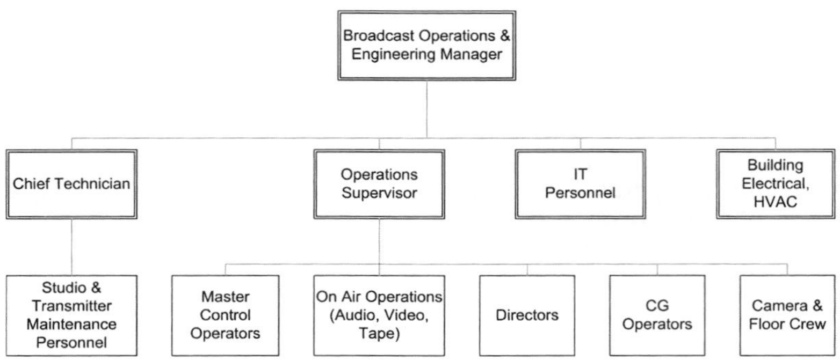

FIGURE 2.1-3 Joint operations and engineering organization.

arrangement. Clearly, the BO&E manager should have a good understanding of the technical aspects of all the areas and how they tie together.

Broadcast facilities today consist of many sophisticated systems, including heating and air conditioning, telephone, and security, as well as the broadcast systems for newsroom, newscast automation, master control, remote control, transmitters, traffic, and many more. In modern plants most such systems rely on computer technology and IT networks and many of them have to communicate with each other. A structure that combines responsibility for all these systems under one department has many advantages.

Job Descriptions

Whatever the particular organizational structure, it is important to prepare and maintain detailed job descriptions for each position. A job description should include the job title, specific duties, and desired qualifications. This is useful both for staff recruitment and as a guide for the employee on what is expected from him.

MANAGING PEOPLE

Recruitment

One of the biggest problems facing general managers and chief engineers is finding and hiring qualified engineers for vacant positions. Few technical or vocational schools today are adequately preparing people for broadcast engineering positions. Nevertheless, vacant positions must be filled. What is important is that chief engineers approach the process of filling positions in a positive manner, not just by finding someone who is readily available. The process should be approached as an opportunity to improve the overall quality of the unit and the station. There are three main goals in filling key positions in the engineering department:

- Try to find someone who is better at the specific job than the person who is leaving. If you replace the person with someone who is not as good, your department's performance will likely decline.
- Aim to fill the vacant position with someone who is better than other employees on your staff performing similar duties. By bringing in someone from the outside that is better, you will cause others to work harder and, therefore, improve the quality of the unit and the station.
- If possible, hire someone better than the chief engineer is at the specific task needing to be done. Recognize that the chief engineer is a manager and cannot be an expert at every kind of engineering, particularly for the more specialized technologies in a modern broadcast facility.

The first place to look for qualified applicants should be at other stations within the same market. Broadcast engineering is a relatively small fraternity and most chief engineers will likely know who the good engineers are at local radio and television stations. By hiring someone away from your competition, you are helping your station and hurting your competition. A second strategy is to look for engineers who are employed at radio or television stations in smaller stations in adjacent markets. Most stations in larger cities can pay larger salaries than stations in smaller markets. It may be possible to recruit engineers from smaller cities to larger cities. The third strategy is to seek engineers from related fields. There may be engineers working for other companies who might want to learn broadcast engineering.

Job advertisements can be placed in the trade press or on the Internet, including the station's website. NAB [1] maintains a "Career Center" web page that has job postings for NAB member stations and also lists a large number of other broadcasting "Job Bank" websites that carry broadcast job advertisements. The most appropriate places to advertise should be selected based on the type of job and geographical area. Another possibility, particularly for larger organizations, is participation in job fairs organized by state broadcasting associations and at the annual NAB convention in Las Vegas, NV.

Training

Once a person is hired, many managers believe the task of filling the job position is complete—but it is

actually often just beginning. Chief engineers must take responsibility for training to ensure the new hire has the knowledge and skills to do the job properly.

The chief engineer should be the most knowledgeable and experienced engineer on the staff. When new employees need to be trained, the chief should either do it himself or, where appropriate, delegate the responsibility to others with appropriate knowledge and experience. There are three main levels of training required: orientation, on-the-job, and continuing training.

Orientation

For a new hire, orientation should take place over the first week of employment. Employees should not be just left to their own devices and "thrown to the wolves." They should be introduced to all employees of the engineering unit and taken on a tour of the station. The new engineer should meet key members of staff and managers, especially the general manager. Orientation should also include a detailed explanation of the specific job responsibilities and a review of the station policies and producers manual or station handbook.

On-the-Job Training

Many chief engineers will assign on-the-job training to an experienced engineer in the specific area in which the new employee will be working. This person will serve as a mentor for the new employee and show him how to do his job properly. Depending on the work and skills involved, this kind of training can take from one to two weeks to many months.

Continuing Education

The third type of training is continuing education. Technology is constantly changing and nearly all engineering staff can benefit from ongoing training. Stations should take advantage of specialized training offered by equipment manufacturers. Sometimes companies will send trainers to the station to do on-site training. Alternatively, engineers may be sent to the company for specialized training at the factory, with courses typically from one to three weeks in duration.

Chief engineers should encourage staff to attend relevant meetings, seminars, and training events arranged by professional organizations such as the NAB, the numerous state associations of broadcasters, the Society of Broadcast Engineers (SBE) [2], and the Society of Motion Picture and Television Engineers (SMPTE) [3]. Some of these take place at the local level in larger cities where SBE and SMPTE have local sections or chapters. Other events take place at a regional or national level and require a greater commitment of time and money for staff to attend but may be justified for key personnel.

Engineering staff should be encouraged to become certified broadcast engineers. Full details of qualification levels, training materials, and examinations may be found on the SBE website.

Chief engineers should make a commitment to ongoing education for themselves to maintain their awareness of new technologies.

Motivation

Hiring and training are extremely important, but motivating engineering staff is equally important. Motivation can be accomplished essentially in two ways—intrinsic and extrinsic. Intrinsic motivation involves recognition and praise. Many managers fail to adequately recognize the contributions of their employees, but sometimes simple words of praise can go a long way in keeping people motivated. Managers should praise publicly and criticize privately.

Staff can also be motivated with extrinsic means, money in particular. Stations can use merit salary increases tied directly to performance evaluations as one effective way to motivate individuals. Those who exceed expectations should receive larger salary increases than those who only meet expectations. Those who do not meet expectations should not receive a salary increase.

For professional development, it is highly desirable for engineering staff to join and take part in the activities of organizations, such as the SBE and SMPTE, the Audio Engineering Society [4], and the IEEE Broadcast Technology Society [5]. Additional motivation for staff can be provided if stations agree to pay the annual subscriptions and cost of attending special events for one or more of these organizations.

Evaluation

Evaluating employees is the fourth personnel responsibility of managers. All employees should be evaluated regularly. Most stations do this at least once a year. The evaluation should be based on the employee's job description or goals statement. The evaluation should be a personal one-to-one meeting, and the manager should meet with each member of staff individually to go over the employee's performance. A summary should be sent following the meeting and placed in the employee's file.

COMMUNICATIONS

Communicating with employees is a major responsibility for chief engineers. There are three kinds of employee communication: downward, upward, and across the organization.

Downward Communication

Chief engineers are members of the station management team and must communicate policies and procedures from company management to employees in clear and understandable ways. There can be many misunderstandings if downward communication is not handled properly. Some stations have tried to communicate policies and procedures by having meetings or sending out memos or emails. What happens all too often is that some employees fail to understand what they are told or do not read what they are sent. What has worked in many organizations is a more formalized way of communicating to engineers from the management. The technique is the signed memorandum. Any policy or procedure change is sent individually to all employees asking for a signature from the employee that he has read the communication.

Upward Communication

Another form of communication within the unit is upward communication. Engineering employees may want to share their opinions with the station's top management. The chief engineer's job is to take their concerns and communicate them accurately to the management. Some stations will place suggestion boxes for employees but few employees ever submit suggestions. Another approach sometimes used is the so-called open-door policy. Employees are encouraged to take their concerns directly to their manager. The problem with this strategy is that, again, few employees take advantage of the opportunity because they do not want to be perceived as whistle-blowers by their colleagues. An effective technique is known as MBWA or "management by walking around." The chief engineer should spend some time each day in each area of the operation just to see how things are going and to make himself available to employees who may want to communicate directly with him. The chief should develop effective listening skills and develop the ability to serve as an advocate for his staff.

Interdepartmental Communication

The third form of communication is communication across the organization. What happens in many broadcast stations is that each department functions as a separate entity. Programming does its own thing. Sales departments operate independently, and so forth. One of the most effective ways to break down walls separating departments is to institute a system of project teams. Project teams are made up of employees of all relevant departments involved in a specific project. For instance, the station may decide to do a live broadcast from an advertising client's place of business, and station management will appoint a team to see the project through to completion. The sales person who works with the client will be on the team, a promotions person will also be added, a programming/on-air person will be needed, a production person, and an engineering person. The project team will work together to plan and execute the event. The result will be that engineers will feel a more integral part of the overall station effort. The chief engineer does not need to be part of every special project team.

STATION OPERATIONS AND MAINTENANCE

The broadcast station relies on technical equipment, for which the engineering department is always responsible for installation, alignment, and maintenance. Responsibility for technical operations is less clear-cut and operation of equipment may be a responsibility split between engineering and program departments. For critical technical systems, staff operating and monitoring the equipment usually report to the engineering department. However, creative technical operators, such as video editors, camera operators, lighting directors, and others, may report to the program department. As discussed in the Organization Structure section, the structure with a BO&E manager avoids this split.

Operations

Many broadcast engineers see their role as a narrow one of installing and maintaining equipment. However, all engineering staff should be encouraged to see their role in the overall operation of the station. Chief engineers serve on the management team with a program director, news director, and sales manager, and other engineering staff may serve on project teams with employees from other station units. It is extremely important for the chief engineer to communicate the goals of the station clearly to his staff. Similarly, the chief must communicate the concerns of engineers to the management team.

Conflicts can occur when communication is not clear. One of the areas of most frequent conflict is between engineering and programming. Program departments may come up with ideas for productions or live remote broadcasts. It often seems to programming personnel that engineers most often throw up red flags or reasons why the station cannot or should not do something programming wants to do. However, the chief engineer and his staff should view their role as a service unit for the rest of the station. They must be perceived as problem-solvers rather than obstructionists to progressive ideas presented by other units in the station. The engineering department should offer help and guidance to help the station achieve its goals.

Maintenance

Maintenance is one of the most important responsibilities for the chief engineer and his staff. Some stations rotate engineers among studio, transmitter, and maintenance work. This approach may keep staff members from getting bored. However, other stations find that assigning engineers to a specific area is more productive in the long run, as maintenance skills become increasingly specialized. Assignments should generally be made by the chief engineer.

Techniques for maintaining broadcast equipment have changed as equipment has become more sophisticated, but there are still two basic categories: scheduled maintenance and troubleshooting.

Scheduled Maintenance

Scheduled maintenance can be predicted and planned for. It includes tasks as varied as cleaning filters for cooling systems, lubrication of mechanical assemblies, cleaning and adjusting tape transports, and alignment of signal levels. With computer-based systems, it also now includes functions such as regular data backup. Much of the regular maintenance has to be performed in-house, but in some cases it may be more efficient to send equipment back to the manufacturer for routine maintenance. Getting the maximum life out of existing equipment should be the goal of good engineering departments and the chief engineer should establish schedules and routines for maintenance and ensure that they are carried out. Discrepancy reports should be filed when problems are discovered.

Troubleshooting

Modern broadcast equipment has greatly improved reliability compared to older technologies. While this reduces the chances of failure, it means that, when faults do occur, it is very likely that the maintenance engineer will not have seen the fault before. This potentially increases the time for faultfinding and repair. Good system design, with fault-tolerant systems, built-in redundancy, and automatic fail-over systems, can greatly reduce the likelihood that equipment failures will take the station off the air.

By their nature, most equipment failures are unpredictable. The chief engineer should analyze potential failure scenarios, establish routines to be used in each circumstance, and make sure that maintenance staff know what is expected of them. Traditionally, maintenance staff often did faultfinding to component level. That is usually not possible with modern equipment with high levels of circuit integration and with multilayer printed circuit boards. These days, the best approach is often to replace modules or complete items of equipment with a spare, allowing the defective unit to be returned to the manufacturer for service.

Techniques for troubleshooting IT-based systems are fundamentally different from traditional equipment. This highlights the need for specialist staff with training in computer and network technologies and software-based systems.

Spares

Comprehensive sets of equipment spares are vital to avoid disruption to station operations from equipment failures and to allow maintenance staff to get systems back to full functionality as soon as possible. However, spare parts, held in inventory, that are not used are a drain on station capital. The chief engineer should establish spares levels needed for equipment and systems based on the quantities of equipment used, manufacturers' recommendations, and his own experience.

Test Equipment and Tools

Considerable quantities of expensive test equipment and specialized tools are required for modern equipment and systems maintenance and alignment. The chief engineer should establish what is essential and what is desirable, again balancing the need to maintain and align equipment to the highest standards with the need to control costs.

Service Contracts

An increasing amount of broadcast equipment is sold with the expectation that the user will purchase a service contract. Service may be provided without such contracts, but typically with slower response and with greater cost for parts and labor. Particularly for software-based systems, service contracts can be a significant proportion of the original purchase price, and may effectively be a mandatory purchase if prompt support and upgrades for critical systems are required. It may be possible to obtain more favorable terms for long-term support contracts and the time to do this is usually when the initial equipment purchase contracts are being negotiated.

Support Systems

Software-based systems may be used to provide support for many aspects of asset management, including equipment management and maintenance. These may range from simple customized spreadsheets and databases using standard office software to more sophisticated purpose-built systems. An example of a system suitable for large organizations is the Maximo® Enterprise Suite from MRO Software (now part of IBM) [6] and, for mid-sized operations, the Datastream 7i® from Infor [7]. These provide a multitude of record-keeping and automated support functions for tasks including:

- Equipment inventory
- Parts inventory and ordering management
- Maintenance schedules
- Trouble tickets
- Work orders
- Repair and maintenance logs
- Purchase orders
- Financial records

Such systems may include capabilities for staff scheduling and similar functions.

A smaller-scale integrated support package that was designed especially for broadcast stations was known as *WinBCAM* based on a relational database system called *MainTrack* from Integrated Tracking Systems. This is still used by many stations but is no longer a supported product.

INVENTORY MANAGEMENT

Part of the chief engineer's duties is to keep track of all broadcast equipment inventory at the station. As well as being useful records for the engineering department, most stations require there to be an annual inventory of all equipment for the annual financial report.

TABLE 2.1-1
Sample Equipment Inventory Form

Inventory No.	Asset Type	Manufacturer/ Model	Serial Number	Location	Building/ Room	Date of Purchase	Cost

The inventory should identify the equipment, indicate where it is located, when it was purchased, and its acquisition cost. A simple inventory form might include categories as shown in Table 2.1-1.

Preparing and maintaining this inventory can be a major task and the software-based support systems mentioned in the previous section all include modules to help chief engineers keep track of their equipment. Such systems require staff to enter data into the system when equipment arrives and when changes are made. They may use techniques such as barcode technology to track all equipment including parts.

An alternative approach used by some stations is to outsource the inventory process to an inventory company that comes into the station and for a fee does a complete inventory of all equipment.

SAFETY AND SECURITY

One of the most important duties of chief engineers involves safety and security. The safety of all employees at a broadcast station is of paramount importance and it is the responsibility of the chief engineer to ensure the physical plant is free of safety hazards for employees and visitors. Broadcast stations must comply with Occupational Safety and Health Administration (OSHA) standards for workplace safety and with FCC rules. Because the nature of the broadcasting business can easily result in dangerous situations arising from electrical and radio frequency (RF) sources, safety procedures should be clearly communicated, and training should be provided as required so that accidents can be prevented. Tower safety procedures are vital and only trained engineers should be allowed in the transmitter and tower area. Drivers of electronic newsgathering (ENG) microwave or satellite newsgathering (SNG) vehicles should be thoroughly trained in safety procedures to prevent electrocutions and other accidents involving microwave masts and satellite antennas.

The security of the physical plant is equally important. Efforts must be made to provide secure facilities for all employees and for station operations. Many stations institute procedures allowing only employees and registered guests of the station into secure areas including the studio, master control, and other critical areas. The transmitter and tower area must be secured so that no one can interfere with the broadcasts of the station.

For more on managing workplace and environmental hazards, RF hazards, electrical shock, and facility security and disaster planning, see Chapters 2.4, 2.5, 2.6, and 2.7.

LIAISON WITH THE FCC

The chief engineer should be the authority on all FCC matters pertaining to technical matters. He should be knowledgeable about all FCC rules and regulations affecting the technical operation of the station and make sure the station abides by the rules. He may utilize consultants to assist in the interpretation of rules and regulations. The chief engineer must keep the general manager and the station management team apprised of any possible violations or potential issues and should bring all new policy changes to their attention.

WORKING WITH UNIONS

Many broadcast stations have unionized engineering departments. Chief engineers are usually asked to represent the station in dealing with the engineering union. Chief engineers will be required to work with the shop steward and local union officials in negotiating the union contract with the station. For many chief engineers this aspect of their job may be the most difficult. They may have been members of the union before becoming part of management. The advantage of having a union is that virtually all policies and procedures can be negotiated with the union rather than having to be agreed with individual employees. The disadvantage is that all engineers are often treated the same—not allowing for individual motivation and evaluation. The key is to maintain constant communication between the station and the union.

FINANCIAL MANAGEMENT

A key area of responsibility for the chief engineers is financial management. This involves planning, budgeting, and controlling costs.

Budgets

As mentioned earlier, planning and budgeting work hand-in-hand. Budgeting for broadcast stations may be done in a variety of ways. Some stations follow a prior-year basis. What this means is that stations look at expenses and revenues for the previous year in making projections for the coming year. Changes may be based on the rate of inflation or an expectation of increased expenses in one category or another. One drawback with budgeting in this manner is that the process does not take into account unexpected expenses or expansion plans.

Another option for budgeting is the zero-based budget. This is based on the premise that all departments within broadcast stations begin the budgeting process with zero allocated for expenses. Each department head is expected to propose a plan for expenditures for the coming year based in part on goals and objectives as well as prior experience. Many zero-based budgets identify fixed- and variable-cost projections in their budgets. Fixed costs refer to recurring expenses including salaries, operating expenses, supplies, and other expenses. Variable expenses include possible nonrecurring expenses or one-time expenses to buy new equipment or to allow for a contingency budget in anticipation of unforeseen expenses.

The budget process begins approximately six months before the beginning of the fiscal year. Each unit is asked to set goals for the coming year. Programming and production will submit goals for both purchasing programs and producing local programs. Engineering will be asked to determine approximate costs for supporting these plans and will also develop its own goals for replacing equipment.

Once a station budget is developed, the management team will be asked to monitor its execution to make sure the station is operating profitably. It is, therefore, vital for the chief engineer to establish and maintain good financial records. The software-based support systems mentioned previously include modules to help keep track of engineering department expenditure.

Financial Statements

While chief engineers are not expected to become experts in financial analysis, it is important for them to understand the two basic documents that relate to accounting: the income statement and balance sheet. They should know how operations within the engineering department affect the line items on these documents. They can then contribute to and understand the significance of financial decisions being discussed in management team meetings.

Typical examples of these documents are included in Tables 2.1-2 and 2.1-3.

Income Statement

The income statement (also known as the profit-and-loss statement or statement of operations; see Table 2.1-2) is a summary of a company's profit or loss during any one given period of time, such as a month or a year. It lists all revenues during this given period, as well as the operating expenses.

The statement enables the management team to monitor the operating performance of the business to be determined over a period of time, compared with the previous year and goals for the current year. It should be able to show what areas of the business are overbudget or underbudget and identify specific items that are causing unexpected expenditures.

Balance Sheet

The balance sheet (see Table 2.1-3) is a snapshot of the business' financial condition at a specific moment in time, usually at the close of an accounting period. It is often used by stations when borrowing money or buying new equipment. It comprises a simple list of assets, liabilities, and owners' or stockholders' equity. At any given time, assets must equal liabilities plus owners' equity. An asset is anything the business owns that has monetary value. Liabilities are the claims of creditors against the assets of the business. The difference between assets and liabilities is known as net worth.

For input to this statement, the general manager will ask the chief engineer to make a list of all tangible assets, including equipment, vehicles, and buildings, to help determine the station's value. This inventory, and software-based systems that make it easier to prepare, is mentioned in a previous section.

Controlling Costs

General managers are constantly looking for ways to control all costs for operating the station. This is because they are usually evaluated on their ability to return a profit to the owners of the station. One approach to this strategy is to control costs so that the total costs for operating the station do not increase more than 4–5% per year. Stations also set goals to increase revenues from 8–10% per year. The goal using this strategy is to increase the station's annual profit margin while building the value of the station over the long term.

Chief engineers will be asked to help with controlling costs. There are four main areas of the annual budget that they need to monitor carefully—personnel, operating expenses, equipment, and program production.

Personnel

Personnel costs often amount to as much as 80% of the total budget for the station. It is extremely important, therefore, to have the right number of engineers on staff. It is equally important for those engineers to be fully occupied without being overworked. One personnel cost that can sometimes lead to conflicts with the general manager is overtime. Chief engineers must manage their units carefully to reduce or eliminate overtime wherever possible.

Another arrangement sometimes used for managing personnel costs—more often in radio than in television—is employment of freelance staff for some activities. By employing freelance engineering staff on an "as-needed" basis rather than full-time permanent members of staff, the station may save some of the costs of a fully staffed engineering department.

TABLE 2.1-2
Profit-and-Loss Statement Form

INCOME STATEMENT						
As of ________ (Date)						
MONTH: ________				YEAR TO DATE:		
Actual	Budget	Last Year		Actual	Budget	Last Year
			OPERATING REVENUE			
			Local			
			National			
			Network			
			Other Broadcast			
____	____	____	Misc. Revenue	____	____	____
			Gross Revenues			
____	____	____	Less: Commissions	____	____	____
____	____	____	Net Revenue	____	____	____
			OPERATING EXPENSES			
			Technical			
			Programming			
			News			
			Sales and Traffic			
			Research			
			Advertising			
			General and Admin.			
____	____	____	Depreciation and Amortization	____	____	____
____	____	____	Total Operating Expenses	____	____	____
____	____	____	Operating Profit (Loss) before Taxes	____	____	____
____	____	____	Provision for Taxes	____	____	____
____	____	____	Net Income	____	____	____
Reprinted courtesy of the Broadcast Cable Financial Management Association.						

Operating Expenses

There are many operating expenses under the control of the chief engineer, such as replacement parts and supplies. Even utility costs incurred by the station may be included in the engineering budget. Chief engineers must monitor all procedures to make sure that resources are used efficiently and waste does not occur.

Equipment

This major area of expense is the most obvious. Equipment replacement may be necessary. The general manager may want to spend the least possible for replacement equipment, but the chief engineer should consider the life of the equipment before making a recommendation. It may be more cost effective to buy better equipment designed to last three to five years or more rather than to buy something that will likely wear out in one year. If it is at all possible to repair equipment to keep it operating, engineers should strive to do so.

Chief engineers need to become familiar with the concept of depreciation for equipment and plant. Accounting rules for tax purposes allow the cost of capital purchases to be written off as expenses against income over different numbers of years, depending on the type of asset. Periods may range from as short as 3 years, to 5 years, to 10 years, or more.

One strategy some stations employ as a means of controlling equipment costs is through leasing equipment rather than purchasing it. This technique avoids large upfront expenditures while allowing the station

TABLE 2.1-3
Balance Sheet Form

BALANCE SHEET	
As of ___________ (Date)	
ASSETS	
Current Assets	
Cash	
Marketable Securities	
Accounts Receivable	
Program Rights—Current	
Prepaid Expenses and Deferred Charges	
Other Current Assets	
Total Current Assets	_____
Property, Plant, and Equipment at Cost	
Less: Accumulated Depreciation	
Program Rights—Long Term	
Other Noncurrent Assets	
Intangible Assets	
Total Assets	_____
LIABILITIES AND STOCKHOLDERS' EQUITY	
Current Liabilities	
Accounts Payable	
Notes Payable	
Accrued Expenses	
Income Taxes Payable	
Program Rights Payable	
Other Current Liabilities	
Total Current Liabilities	_____
Long-term Debt	
Deferred Income Taxes	
Program Rights—Payable Long Term	
Other Noncurrent Liabilities	
Total Liabilities	_____
Stockholders' Equity	
Capital Stock	
Paid-in Capital	
Retained Earnings	
Treasury Stock	
Total Stockholders' Equity	_____
Total Liabilities and Stockholders' Equity	_____

Reprinted courtesy of the Broadcast Cable Financial Management Association.

to account for actual equipment costs as recurrent expenses each year.

Program Production

The last main area of operating expense is in local production and news. Programming and news departments often want to produce live programs, newscasts, or cover major events. Each program or event must be budgeted within the budgeting guidelines of the station. If specific projects are going to be very costly from an engineering perspective, the management team must be made aware of it. It is often necessary for stations to cover hurricanes, tornadoes, or other breaking news. It should be made clear whether these costs are news or engineering expenses.

CAPITAL PROJECTS

The chief engineer will regularly be asked to present proposals to the management team for new equipment and other capital projects. Sometimes other departments request new equipment, such as camcorders, editing systems, and weather graphics for news. The engineering department will be asked for expert advice on the selection and purchasing of this equipment. On other occasions, engineers will request major system upgrades and items of equipment.

It may be difficult for engineers to explain the rationale for a new transmitter, tower, amplifiers, and other technical equipment that nonengineers do not understand. Therefore, it is important for engineers to learn the "language" of the other managers when presenting proposals to purchase new equipment. Engineers should avoid using jargon in presentations. Simple, easy-to-understand language should be used for all nonengineers, without talking down to them. Present proposals with a rationale detailing long-term cost savings due to greater efficiency or effectiveness. Describe the proposal in terms of making the programming more appealing to more people. Program directors and sales managers understand the benefits of improving signal quality as a way to attract and keep more viewers or listeners.

Once authorization is given to buy new equipment, agreed procedures should be followed for procurement. Specifications for the desired new equipment should be drawn up and made available to all potential vendors. Bid requests may include quotations for installation, on-site and/or factory training for station engineers, and sometimes the cost of spare parts. Bids should be sought from three to five companies so that cost comparisons can be made. Bids should be carefully reviewed to ensure that all specifications are being met and, in general, chief engineers should go with the lowest compliant bid.

MANAGING THE MANAGER

Working with station general managers (GMs) can be a daunting task for many chief engineers, especially if they do not fully understand the missions and business of the station and its owners or the motivations of the GM. The more the chief engineer knows about the GM and what "drives" him, the more successful he is likely to be. GMs are interested in meeting the two missions of all licensed broadcast stations—to serve the public and to make a profit. The GM entrusts the service component to the programming, production, and news departments, but an important part of being able to serve the public is the engineering support provided to these other units.

Ratings

The GM is interested in achieving the highest ratings possible for all programs. Whatever you can do as the chief engineer to assist him in achieving higher ratings will be received in a positive light. Ways to assist in achieving higher ratings is to make sure that all equipment in the news department, including microwave and satellite equipment, is functioning at the highest level. Local television stations know that as the news department goes, so goes the station. The news director will need your help in providing him with the latest information and guidance for camcorders, editing systems, and ENG/SNG systems. The weather radar and graphics systems should also be maintained at the highest possible level.

Revenue and Costs

The second major goal of general managers is to return a profit to owners and investors. Chief engineers should do everything they can to assist the GM to control unnecessary costs. Engineers also need to assist sales departments in generating revenues for the station. Sales departments often want to produce live remote broadcasts for advertising clients or produce quality commercials. In both cases station engineers need to be as cooperative as possible in assisting sales personnel in meeting the needs of advertising clients. There are various technologies that stations can use to increase revenue, in which engineers may provide leadership. Whether it is services using SCA channels for FM radio stations or opportunities such as data broadcasting for television, engineers can help with the technical requirements.

When chief engineers present budgets or proposals for capital expenditures, they should present their proposals within the framework of achieving higher ratings and generating greater revenues. Engineers need to speak in the language that general managers understand—ratings and revenues. A new transmitter or tower will be seen by management as a major expense unless it can see the benefits as measured in increased ratings and revenues. New and improved technology should improve the quality of the station's signal or extend the reach. In both cases, the station should increase its chances of attracting a larger audience. The more people who listen or watch, the higher the ratings; the higher the ratings, the greater the revenues.

SUMMARY

Chief engineers have a challenging position in broadcast stations, requiring careful preparation and training. Leading a unit of engineering staff requires skills above and beyond solving engineering problems. The chief engineer must see himself as a leader responsible for leading a staff and managing the resources of the station. How well he is able to balance the leadership and management functions of his position will determine not only his success but also the success of the station. Chief engineers should seek continuing education in management and leadership, available through professional organizations such as the NAB, the state broadcasting associations, the SBE, and elsewhere.

ACKNOWLEDGMENTS

Thanks are extended to Bob Williams, chief engineer at WATE-TV, Knoxville, TN, for providing the three organization charts and associated text on alternative engineering staffing structures. Thanks to Clyde Smith of Turner Broadcasting System, Atlanta, GA, and Ron Peters of WUSA-TV, Washington, DC, for information on software-based asset management support systems.

References

[1] National Association of Broadcasters, see information at http://www.nab.org/.
[2] Society of Broadcast Engineers, see information at http://www.sbe.org/.
[3] Society of Motion Picture and Television Engineers, see information at http://www.smpte.org/.
[4] Audio Engineering Society, see information at http://www.aes.org/.
[5] IEEE BTS, see information at http://www.ieee.org/organizations/society/bt/index.html.
[6] IBM MRO Software, see information at http://www.mro.com/.
[7] Infor Datastream, see information at http://www.datastream.net/English/Default.aspx.

CHAPTER

2.2

Broadcast Engineering Documentation Management

DAVID M. BADEN

Chief Technology Officer
Radio Free Asia
Washington, D.C.

INTRODUCTION

Documentation is a statement of pride and ownership. Documentation is a tangible public declaration of competence and ability. This chapter covers various document creation, maintenance, and information management processes. It will outline basic "best practices" and hopefully inspire a more collaborative and comprehensive approach toward the broadcast facility's technical information management.

Even small broadcast facilities are a compilation of multiple complex components and systems. By utilizing the wide variety of design, computer-aided design (CAD) and documentation tools available, and by following best practices as outlined in this chapter and instituting a proactive document management process, all broadcast engineers should be capable of maintaining comprehensive facility documentation.

The necessity for documentation has increased with the complexity of broadcast facilities. With new generations of digital components and systems, not only do hardware updates need to be constantly monitored, but also firmware and software. Furthermore, there is an ever-expanding interconnectivity between systems and components. This interconnectivity is not only through the normal audio, video, and control interfaces but also through local area networks (LAN) and wide area networks (WAN) such as the Internet. These multiple connections are dynamic in nature and require constant upgrading. No longer are a few technical schematics of static systems adequate to document the complex and dynamic systems used in the modern broadcast facility.

The proper management for the creation and maintenance of critical information in technical documentation is essential for preventing downtime or degradation of service in a broadcast facility. Regardless of the complexity of a broadcast facility, proper documentation will help to ensure a successful and efficient daily operation.

The technical documentation process starts in the design phase of broadcast facility construction (or upgrade) and is part of a continuous process that never ends. From construction to normal online operations to the daily routine maintenance of any facility, a comprehensive approach toward documentation is beneficial. The relevance of the information conveyed in a facility's technical documentation can only be maintained if it is:

- *Comprehensive.* In modern broadcast facilities, independent standalone systems have rapidly become a thing of the past. Even in the simplest radio facility with one air studio and a single production studio, all components are in some way a modular part of larger systems as a whole. This interconnectivity is needed to support the transfer of digital content over LAN-to-network file servers and to a common storage platform to be controlled by a shared software application, such as a traffic control system. In an environment where any small change can impact the facility as a whole, all information pertaining to these changes must be tracked and incorporated into the facility's documentation.
- *Accurate.* With the wide variety of components, computers, software, and systems that comprise the

modern broadcast facility, the accuracy of all information is critical. Consider, for example, a networked broadcast component that is only compatible with a limited number of manufacturer-certified network switches. Failure to accurately note this compatibility could cause a system failure at startup or in the future during what should be a routine network upgrade. Inaccurate information can cause facility downtime; hours of troubleshooting can be avoided with proper documentation.

- *Dynamic.* In a broadcast facility there will never be a point in time where the technology will be static. Computer and network equipment in particular is advancing at a rapid pace. With a short component life cycle comes a never-ending need to upgrade and patch operating systems, firmware, and application software. With rapid obsolescence of equipment and systems there is an associated rapid obsolescence of documentation—documentation needs constant upgrading and revision if it is to be useful.
- *Accessible.* The most concise and well-structured documentation is worthless if it is not readily available to the engineers who need it, when they need it. Technical documentation should be online, searchable, and available to staff needing the information.

CAD, word processor, spreadsheet, and database software provide all the tools needed to enable broadcast engineers to effectively create and manage technical information. The larger challenge facing today's broadcast engineer is the long-term management of this wide variety of electronic information and resources.

DOCUMENT CREATION

The creation and maintenance of technical documentation is a cyclical process. The depth and complexity of the documentation will mature with the broadcast facility. Technical documentation mirrors the same life-cycle phases of the facility—conceptual, build, and as-built (Figure 2.2-1). While the documentation process overall is simpler if it starts and grows with a broadcast facility, the process can be started in any phase.

Conceptual Phase

The conceptual phase is defined as the point where a new broadcast facility, or a facility-wide upgrade, is a proposed project. This is the "what we want to do" stage of documentation. The conceptual phase is where an overall vision for the facility is conceived and work processes are defined. The facility vision should include how the facility will function as a whole and a working idea of how the various systems will be integrated. Most importantly, it is also the point of the project where the operational parameters of the new facility are stated.

Documentation tools used for conceptual design can be as simple as a basic CAD program, a spreadsheet, or a word processor (Figure 2.2-2). Documentation at this phase generally consists of workflow diagrams, basic system block diagrams, manufacturers' supplied product data, and discussion papers geared toward sharing the vision of the new facility with the user community. It is important that flexible documentation tools are used as the documentation created during the conceptual phase will serve as the starting platform for all the documentation that will follow.

Workflow diagrams are needed when designing an upgrade as it is important to document how a facility currently operates and how any facility changes will impact these workflow processes. Only by documenting and comparing workflow processes can possible negative impacts on normal daily operation be discovered and designed out. For any change to be implemented, it must have an overall beneficial impact, because if not, then the need for change is negated.

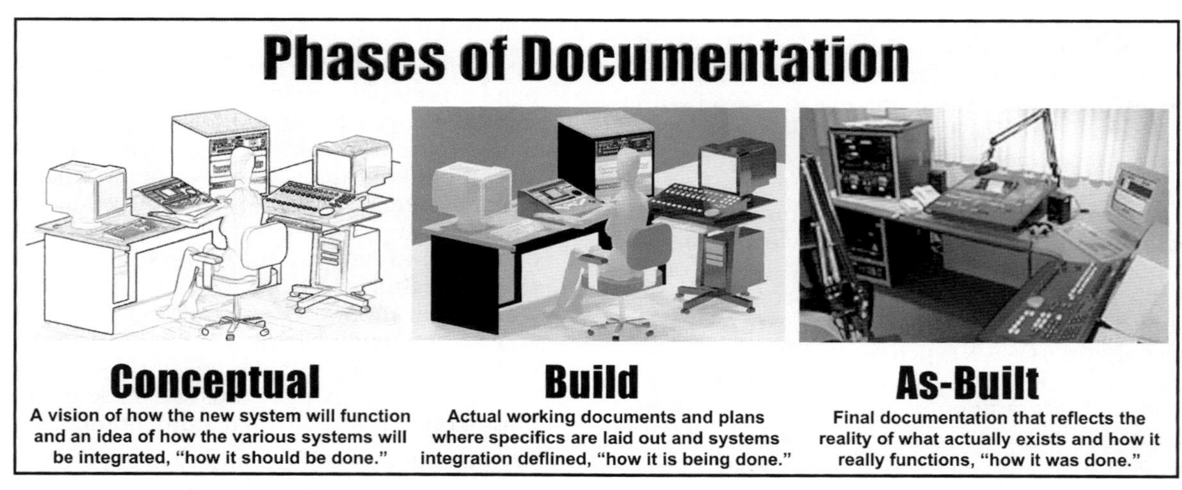

FIGURE 2.2-1 The three phases of documentation—conceptual, build, and as-built.

FIGURE 2.2-2 Virtually all technical documents are now created using CAD and other computer-based documentation tools.

Basic, initial system block diagrams are useful for primary overall systems planning. These can show what new systems are to be used, what existing systems and resources will be incorporated, and how these will all interface. In creating these initial block diagrams, design tools should be used that allow for systems to be easily interchanged as the designs are fleshed out and more suitable alternatives are discovered.

Not to be overlooked are the user "white papers," which can be invaluable. These are high-level descriptions of how users will ultimately work in the facility, for example, where each task will be performed, the tools for accomplishing these tasks, etc. The measure of success for any new system or upgrade is ultimately based on the end user's acceptance. This is usually subjective and based on perceptions of the benefits of any new system versus the old. Bringing the users in at the early stages of any project creates realistic expectations and promotes total ownership within an organization.

Whether the facility is a new physical build or a rebuild of an existing facility, it is during the conceptual phase that initial space use is decided. Where will the studios, satellite support room, master control, electric closets, and related facilities be located? During this phase the broadcast engineers will find themselves working closely with architects or general contractors, or will become the ad-hoc design architect. This close working relationship increases the likelihood that these technical rooms are in the most optimum location. If it can be avoided, technical space usage decisions should never be left solely to an architect who may want to favor aesthetics over practicality.

Regardless, if ultimately the facility will be installed by an outside integrator or by the in-house engineering staff, as much of the conceptual design should be done in-house as possible since it is the in-house engineering staff that will ultimately have to live with and maintain all new systems. Additionally, internal control of the design process will ensure control over future expenditures.

Build Phase

During the build phase actual working documents and plans are created—this is the "how it is going to be done" stage of documentation. It is at this stage that all specifics are laid out and documented (Figure 2.2-3). Specifics include:

- What equipment will be used
- Where it will reside
- How it will be interfaced
- Where it will be purchased
- What its role is in the system as a whole
- Numbers and types of cables needed to connect the systems and cable routes

Regardless of whether the facility installation or upgrade is being done by in-house engineering staff or is outsourced, this documentation is critical—it must exist or it would be impossible to order the equipment, order various systems and all the basic hardware needed to complete the job, and accomplish the installation in an efficient manner.

The document tools that are used for the build phase should be as varied as necessary for the required project documentation. The use of databases, either standalone or embedded in a CAD program, should be considered or at minimum a master spreadsheet used in order to track multiple items, connections, cable numbers, and other facility specifications. Databases that are integrated with or linked to CAD software whenever available are recommended. The use of such capabilities within a CAD program eliminates a redundancy of effort. With many CAD pro-

FIGURE 2.2-3 Actual construction of a broadcast facility is the culmination of work started during the conceptual phase of documentation.

grams the drag-and-dropping in of a graphical representation of a piece of equipment will also update the associated database regarding how many extra connectors need to be ordered, how much additional rack space is required, and other relevant information.

Facility Standard

Early in the build phase of a broadcast facility, a facility standard should be determined, finalized, and documented. The facility standard is a statement of what standards and conventions, on multiple levels, will be used within the broadcast facility. This is a listing of the multiple choice "givens" in a facility that need to be decided on and, once decided on, will not change. It is important to clearly define, document, and adhere to these standards before any actual work starts in order to ensure facility-wide compatibility. Facility standards should include (but are not limited to):

- Grounding architecture
- Cable conventions (shielded cable, plenum cable, manufacturer, etc.)
- Wiring conventions (shields grounded, hot pin, etc.)
- Wiring details (terminations, cross connects, location of, raceways, etc.)
- Facility signal level standards
- Digital media format (sampling rates, bit rate, clocking, etc.)
- Electrical power (backup, location of breaker boxes, breaker types, etc.)
- Network type and topology
- Shutdown procedure
- Evacuation procedure
- Location of exits, alarms, and fire extinguishers
- Location and use of lockouts
- Firewall rules
- Virus protection rules
- Intrusion protection rules
- Allowed system access levels

Having a facility standard, especially at this stage of the documentation process, will help to ensure that any and all work (in progress or future) will be accomplished in a consistent manner. Further, understanding and having this information readily available will make routine maintenance and troubleshooting more effective and efficient.

As-Built

As-built documentation is the updated and final build documentation that reflects the reality of what actually happened during the construction of a facility and how it now really works. This is the "how it was done" stage of documentation. As-built documentation should consist of three main sections: physical, operations, and maintenance.

The as-built documentation is the dynamic documentation that will be maintained and updated throughout the life of the broadcast facility. As such, when using only one shared set of documentation, special attention should be paid to setting up a documentation access system so as to preserve the integrity of the information. The level of access for the all sections of the as-built documentation should be set to "write" privileges for a small trusted number of the engineering staff members. Read-only privileges should be set for the wider user population requiring access.

It is recommended that two sets of as-built documentation be maintained. One set should be the "working set" and kept in access-controlled directories in the original file formats that it was created in. This is the set of documentation that is routinely upgraded and revised by the engineering staff that is responsible for its maintenance.

A second, more publicly accessible set should be published to a more widely-accessible online directory in a more common format. This is the "user online" set of the facility documentation. The user online set of documentation should be published from the working set into either HTML or PDF files. This would allow for the creation of a document set that is in a read-only format and one that can be placed in a read-only location. The user online documentation, regardless of the format used, should be fully indexed and searchable.

If this two-set model is used, it is important for the staff that is updating the working set of documentation to make sure that they routinely publish these changes to the user online set. This can be accomplished as simply as performing a "save as" or "print" function (e.g., in the case of PDF file creation) after the "save" function when updating files.

Physical

Often there is a wide discrepancy between what was planned and what was actually built. What actually made it from the documents of the build phase to the real world needs to be reflected in the as-built documents. The physical as-built is the documentation of how all systems, components, and hardware were actually installed and work. As far as documenting these changes, the same tools that were used to create the build plans should be used to revise and update the as-built documentation.

Facility Standard Revisited

An update of the facility standard should be included in the physical as-built section of the documentation. The revised facility standard should be updated to include all previously unknown critical information. This facility standard update should include such information as:

- How to disconnect utility power to service-line voltage wiring within equipment racks

- Where breakers are located, and how they are marked
- What equipment is on UPS power, generator, or utility power
- A list of telephone numbers for important contacts (e.g., power company, building systems maintenance, etc.)
- Where keys are located and how they are marked
- Telecommunications circuit designations, associated vendor contact information, and "D-mark" locations
- A facility shutdown plan
- An occupant emergency plan
- A facility recovery plan
- An organizational continuance of operation plan

All additional essential information should be updated into the facility standard. This information will greatly assist the engineering staff with routine operations, emergency situation response, basic maintenance, and facility troubleshooting. Having this information in one location that is easily accessible, preferably online, will accelerate troubleshooting and problem solving in the future and keep the technical staff all on the "same page."

Operations

Not all documentation is created for the engineering staff. Documentation for the facility users should also be included as a section of the as-built facility documentation. This documentation includes operational instructions on how to use all equipment, rooms, systems, and software.

Operational documentation designed for the actual users should also include a tutorial on how the facility is operated as a whole and the proper workflow processes that the facility was designed to work under. This should be the final "as-built" updated version of the user white papers that were created during the conceptual phase of the documentation process.

User manuals are generally already available from the manufacturers in their product manuals. These manufacturers' manuals should be incorporated into the facility's overall operations documentation whenever possible. Additionally, the engineering staff should create a corresponding quick reference guide for equipment and systems if it is not available from the manufacturer. A quick reference guide is usually a one-page sheet that is a brief step A to step B single-line narrative that describes minimal operations.

If the manufacturers' manuals are only available in hard copy then they should be electronically scanned (as shown in Figure 2.2-4) or else filed in a common centrally accessible area in logical fashion, either alphabetically or by work areas. Most manufacturers provide electronic copies of their product manuals. Electronic manuals from the manufacturers should be stored online as part of the complete facility's documentation package. As electronic documents, depending on the format that they were originally provided in, they can be not only stored in a common physical location but they can also be actively linked to a central index.

FIGURE 2.2-4 The flatbed scanner (shown at right in this photo) is an invaluable technical documentation tool.

It is important to keep the operations section of the as-built documentation as current as possible. A large volume of it will be dedicated to software systems. Software, by nature, is upgraded often. With software upgrades, application functions, tools, and interfaces often are changed. As a result the operations section of a facility's documentation can rapidly become useless if it does not change along with these software version upgrades.

Maintenance

The most dynamic section of the as-built documentation will be the maintenance section. Each piece of equipment within a broadcast facility can potentially be modified, reconfigured, or removed from service. This is especially true for the digital broadcast facility where equipment routine maintenance regularly includes software and firmware upgrades. Maintenance records should be kept on four hierarchical levels:

- *System wide.* These records are especially relevant in a digital broadcast facility where systems are often based on personal computers as a core platform. There are several overall common denominators in these systems that need upgrading or maintenance as a whole. Examples include what version of operating system (OS) all servers and workstations are running on and what patch level for the OS has been installed. System-wide records ensure overall compatibility and uniformity between the common pieces of individual equipment and components that comprise any given system as a whole.
- *Location.* When problematic locations are identified it is easier for possible causes to be identified (e.g., unstable power, improper grounding, a problem user) and corrected. When was the last time a par-

Master Equipment List

	A	B	C	D	E	F	
4171	Barcode	Description	SerialNumber	Purchase Date	Code	Location	Room
4172	110535	Sony MZB50 MiniDisc Recorder	700308	9/29/2000	TC	DC	Help Desk
4173	110536	Sony MZB50 MiniDisc Recorder	700123	9/29/2000	TC	DC	Help Desk
4174	110538	Sony MZB50 MiniDisc Recorder	700307	9/29/2000	TC	DC	Help Desk
4175	110539	Sony MZB50 MiniDisc Recorder	700121	9/29/2000	TC	DC	Help Desk
4176	110540	PC	Case	2002	CE	DC	252-06
4177	110544	PC	Case	2002	CE	DC	252-02
4178	110549	PC	Case	2002	CE	DC	252-10
4179	110550	PC	Case	2002	CE	DC	247-22
4180	110551	PC	14501550	2002	CE	DC	260-05
4181	110553	HP LaserJet 3200 Printer/FAX	USDH116038	6/4/2001	CE	DC	4th Floor General
4182	110554	HP LaserJet 3200 Printer/FAX	USDH119326	6/4/2001	CE	DC	4th Floor General
4183	110555	8 Port Millennia KVM Switch Pt#MILL-OSD-8	1ML2-0622-38	6/26/2001	TS	DC	MC
4184	110556	Supermicro SC750-A Full Server ATX case 300 watt power supply		6/25/2001	TS	DC	MC
4185	110558	Supermicro SC750-A Full Server ATX case 300 watt power supply		6/25/2001	TS	DC	MC
4186	110559	Supermicro SC750-A Full Server ATX case 300 watt power supply		6/25/2001	TS	DC	MC
4187	110561	Antek IPC 35809B dual redundant 400 W rack Mount server case		6/26/2001	TC	DC	MC
4188	110562	Antek IPC 35809B dual redundant 400 W rack Mount server case		6/26/2001	TC	DC	MC
4189	110563	VAIO FX170 Notebook (windows Me)	2.83E+14	7/2/2001	CE	DC	Help Desk
4190	110564	VAIO FX170 Notebook (windows Me)	2.83E+14	7/2/2001	CE	DC	Help Desk
4191	110565	VAIO FX170 Notebook (windows Me)	2.83E+14	7/2/2001	CE	DC	Help Desk
4192	110566	PC	Case	2002	TC	DC	MINI 13 Rm278
4193	110567	PC	Case	2002	TC	DC	MC
4194	110568	VAIO FX170 Notebook (windows Me)	2.83E+14	7/2/2001	CE	DC	Help Desk
4195	110569	VAIO FX170 Notebook (windows Me)	2.83E+14	7/2/2001	CE	DC	Help Desk
4196	110571	Paladin PC Cable Check model 1570		6/29/2001	TS	DC	Help Desk
4197	110578	JKAudio-TWAT-1	USCF028491		SE	DC	401-B
4198	110579	Viewsonic E70F 17 Perfect Flat Display	2.18E+11	7/19/2001	CE	DC	MC
4199	110587	AOC: 7ELRA 17inches 1280x1024 60hz Monitors with speakers	7TCN16C065798	7/24/2001	CE	DC	B-1
4200	110606	Desktop Director		9/15/1999	TC	DC	Studio 6
4201	110609	Telephone	MB011062		TC	DC	MC
4202	110611	Telephone			TC	DC	MC
4203	110612	Telephone	D120B0050024		TC	DC	MC

FIGURE 2.2-5 Equipment inventory spreadsheet.

ticular studio received a complete preventative maintenance service? How often does a particular editing suite break down? Downtime can be reduced with proper routine maintenance when problematic areas can be identified.

- *Equipment by model*. Maintenance records by model can be used to track manufacturer-driven hardware, firmware, and software upgrades. These records are also useful for tracking the performance regarding overall reliability of a given type of equipment in the facility.
- *Equipment by piece*. With proper tracking, a single piece of equipment that is uniquely problematic can be identified and replaced. An excerpt of a master equipment inventory list developed using a computer spreadsheet is shown in Figure 2.2-5. While a database program would offer the optimal search capacity, a spreadsheet is a quick starting point from which data can be later easily exported. One feature of spreadsheet programs that comes in handy here is the "sort" feature, which allows for sorting of the inventory list by any of the included columns. The list in the figure is arranged by "barcode" (a number assigned each item by the owner) but could just as easily be arranged by description, room, etc. When the equipment is software based, on a computer workstation, it is highly unlikely that all like equipment, on the individual level, is in fact alike. Different workstations running the same application can have differences in motherboards, bios, memory type, CPU, and hard drives that all create a unique maintenance profile.

PERFORMANCE-MONITORING DOCUMENTATION

Performance-monitoring documentation (standalone or as part of an overall facility quality assurance program) should not be overlooked as an essential part of a broadcast facility's documentation. With performance monitoring, small problems within a facility can be identified in systems or equipment before they become major problems. Major problems generally tend to be noticeable as they can be system stoppers or adversely affect other systems (and become perhaps harder to trace down). If performance can be shown to have measurably degraded, there must be a reason; it is difficult to establish performance degradation if past performance is not documented.

There are several performance measurement tools to help document baseline technical systems and equipment. Where applicable, these tools should be dynamic monitors that establish a documented baseline and then constantly monitor the systems or

equipment. Dynamic monitoring is the real-time comparison of current performance against initial baselines. Real-time alert notifications can then be received when a system's performance starts to degrade. As a result these problem systems can be identified and corrective measures taken with near real-time reaction.

Documented baselines can also be used for planning replacements and upgrades since these baselines can be compared to the performance levels of new systems and equipment when they become available. For example, a parameter can be established such that when an online system performs at 50% below the newest available systems version, then at that time the old system will be upgraded or replaced.

Finally, proper and objective performance documentation can help to eliminate subjective user's complaints in regards to subjective perceived performance issues. For example, a common complaint for a digital editing station could be that "it is running slower." This complaint can be objectively verified or dismissed if past performance measurements for that workstation have been documented.

THE USE OF DATABASES

Several CAD programs offer application-specific internal databases as well as the ability to link to external multi-user database engines. The use of either internal or external databases provides expanded search capability for documents and information as well as facilitating multiple document linking and updating. Figure 2.2-6 shows an example of such a linked data-

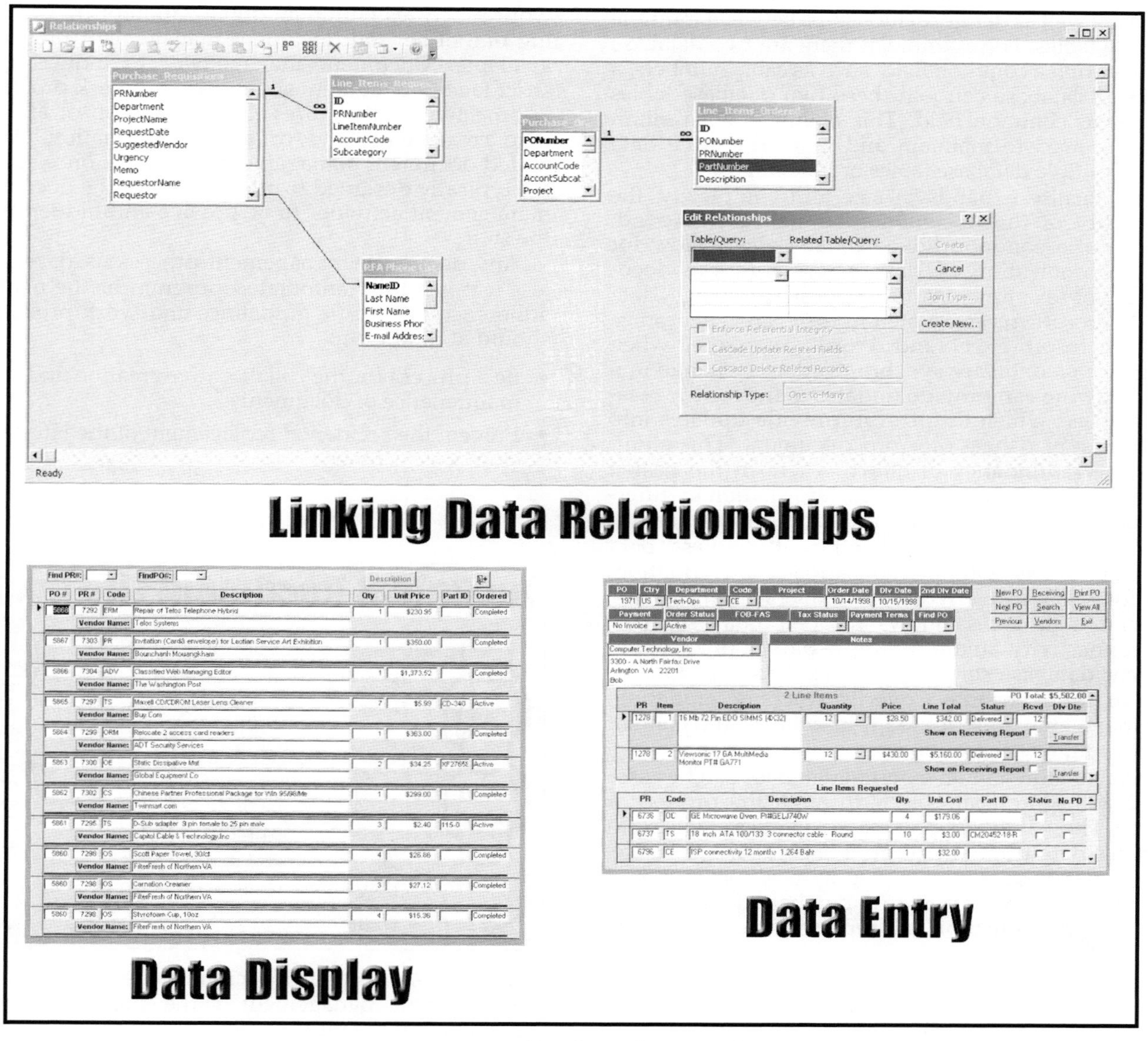

Linking Data Relationships

Data Entry

Data Display

FIGURE 2.2-6 Database functionality.

base as it appears on a computer screen (top) along with the data display (bottom left) and data entry (bottom right) screens. A data record, such as for a piece of equipment, can include several separate items of information about the equipment stored in data fields. Records can then be linked either by all or selected fields and then linked into tables for data display. For example, a table could be defined that shows all equipment housed in a rack in order to show the total power consumption for the rack or the total cost of the equipment. Data entry can also be allowed in individual records by individual fields or updated from defined tables that would simultaneously update several records.

A database is a collection of logically related information, usually presented in a table format similar to a spreadsheet. Databases are extremely flexible and can be modified by adding new fields or removing existing ones from the data tables. Likewise, one can add or delete records from database tables and edit existing records as well as define relationships between multiple database tables. For example, if there are two database tables sharing a common field, such as cable numbers, a relational database can join these two database tables using their common field. This ability is what eliminates the need to maintain and input duplicate data across multiple database tables or linked files.

As a facility expands in size and complexity, the complexity of the documentation will also expand. The use of a common linked database can allow for the synchronization of data across multiple documents. For example, AutoCAD can exchange data with Microsoft Access, which in turn can synchronize with Microsoft Excel and Word. Therefore, when using Access database synchronization, a change in a CAD drawing can provide updates to the facility master database, which in turn can provide updates into an Excel spreadsheet or Word document. This eliminates the redundancy of effort, where in the past it was required to update the same information in different document files using separate applications.

Allowing applications to efficiently access data where it resides without replication, transformation, or conversion is referred to as universal data access. Universal data access allows an open database interface connectivity among multiple data sources. Independent database services, such as Access or Oracle, provide for distributed queries, caching, updating, data marshaling, distributed transactions, and content indexing among these sources.

Data contained in an external database can be associated with graphical objects through the process of linking. These links are pointers to database tables that reference data from one or more records in that table. Linked database records to graphical objects in a drawing can provide powerful database queries to filter and sort information and represent the new results graphically.

The use of databases is especially important in facility documentation for tracking facility-common systems and interfaces, such as cabling origination points, destination points, and cable numbering or LAN network topography, which also includes cable numbering and routes. Database use can also be expanded to include equipment maintenance records such as firmware and software updates.

DOCUMENT MANAGEMENT

A well-implemented document management process is important to realizing the true potential of a broadcast facility's documentation in regards to long-term information access and retention (Figure 2.2-7). Document management entails more than organizing and archiving large numbers of documents. Document management facilitates collaboration, ensures standard compliance, and improves work processes overall. Access to critical information when needed empowers all staff members in a broadcast facility.

Document management can be implemented either by enforcing a well-defined document standard and work processes within an organization or by employing one of several document management programs available. Document management by adhering to strict facility document standards and work processes is proactive and requires the cooperation of all staff who create edits and/or access information. Automated document management programs, once set up, are typically designed to allow much of the document management activities to be passive and hidden from the staff.

Any document management process used, regardless if it is a fully automated program or the manual adherence to specific guidelines and work processes, should at minimum:

- Be a process for the collaborative creation and maintenance of documents
- Prevent the accidental replacement of the latest version of a document

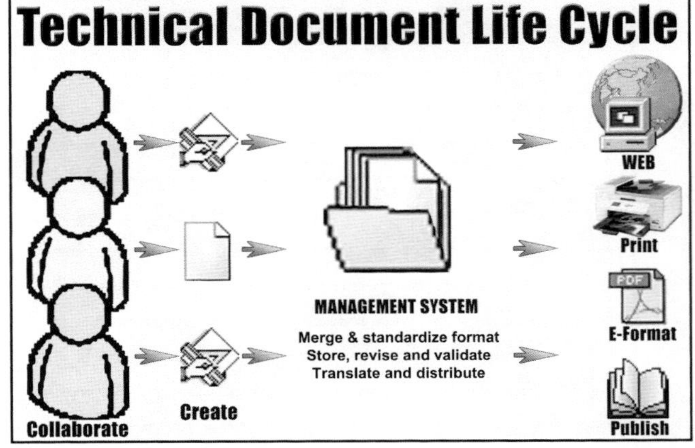

FIGURE 2.2-7 Technical document life cycle requires the intelligent management of the information from the creation to the multiple-format distribution of documentation.

- Prevent the accidental (or purposeful) deletion of the latest version of a document by casual users
- Provide a method to track changes to documents by date, time, and by who did what
- Provide a logical structure in which documents can easily be located

Document management, regardless if the process is automated or manual, is simply the adherence to and enforcement of a set of rules and guidelines in regards to how documents are created, what style they follow, what format they are saved in, where they are stored, and who has access rights. The failure of any document management system is usually not tied to a specific system methodology but instead is due to a negative productivity impact that in turn leads to a failure of acceptance by the end users. Successful document management must create a positive improvement in process in order to have long-term viability.

Document Standard

The document standard itself is the first step required for any document management process as it is what is required to be defined in a setting management program. The document standard defines how documents are to be created, stored, and accessed, as well as what level of access is granted to what user groups. A well-defined document standard that is adhered to by all users is in itself a document management process. Before a single file is created, a document standard should be established for a broadcast facility. A well-defined document standard, at minimum, should include the following:

- *File type*. Most software document and CAD tools have the capability of saving files in multiple file types or formats (e.g., DWG, DXF, DOC, HTML, RTF, etc.). File types should be selected for each type of document being created based on the maximum compatibility of all documentation tools and CAD programs being used. For example, the standard for all CAD drawings could be to save as DXF (drawing exchange format) files. This would ensure that multiple CAD programs could access the CAD file. Proprietary file types that can only be opened by a single program should be avoided whenever possible.
- *File-naming conventions.* The purpose of a file should be reflected in the file title. Therefore, a naming system needs to be standardized. A file name such as "drawing110.dxf" conveys nothing to a user browsing a list of files. In this case the file would need to be opened in order to determine exactly what it was. On the other hand, a file name "STU1_RCK1.dxf," even not knowing the defined naming schema, could easily be read as a drawing of Rack 1 in Studio 1. In this naming schema the first four characters in the file name are reserved to denote location, the fifth character is a "_" used as a space separator, and the next four characters designate the subject of the drawing. File-naming schemas can be as simple or complicated as needed but should be structured in a way that most users can, at a glance, be able to determine what information the file contains.
- *Layer-naming conventions*. CAD programs allow multiple layers in a single drawing; layers are provided to separate information within a CAD drawing. Layers can then be controlled individually (e.g., turned on or off, plot or not, locked). Therefore, a schema for layer naming should also be established. The layer-naming schema like the file-naming schema should be structured to allow the average users to identify what information is contained in the layer at a glance.
- *Document properties.* Document properties are details about a file that help to identify it; typically they are metadata or information about a file that can be accessed without opening the file. These properties include a descriptive title, author, subject, and keywords to identify important information in the file. Often document properties are overlooked and left blank. The document standard should define what is required to be entered into the document property information box. This information should include the original author as well as later authors who edited the file and a notation of what these edits were.
- *Document format and styles*. Format and styles for a document are internal definitions that describe the presentation of a document. These include the font type, font size, page size, page layout, margins, tabs, outline numbering conventions, paragraph justification, spacing, line weights (in CAD), line types, and other parameters. While this may seem trivial, the need for standardization becomes apparent when merging or linking documents into a consolidated documentation package. Without format and style standardization, multiple merged documents will appear to be disjointed rather than one cohesive document. In most programs, styles and format can be defined in templates that can be used by all users when creating a document.
- *File storage organization.* Where document files are to be stored and how they are organized. File storage organization is critical as files that become lost when stored in the wrong place might have well have never been created. Directories and folders on file servers are today's filing cabinets. Just as with a filing cabinet, a filing system needs to be established. A well-defined file organization structure will alleviate any confusion on where to store and subsequently retrieve files. File storage organization should also include how the original creation documents are organized and archived. Since multiple files are often used to create one single final published file it is important to have ready access to all files that went into the creation processes to ensure that the document can be properly revised as needed.
- *Access rights*. Who has rights to create, edit, and view a document. Access needs to be defined in order to minimize potential data corruption or acci-

dental deletion of information. Access rights can be set in various ways either at the file level or at the server level. For the best practices, access rights should be set at all levels possible.

- *Common master index.* This is an overall index of a facility's total documentation and indicates where each file resides in hierarchical importance. A master index can be used when publishing the documentation as it can be converted to a common-use file, such as HTML, and then linked to the various relevant associated files.
- *Database links.* Whether external or internal databases are used to synchronize data, a definition of these links and their relationship to each other need to be defined. Defining these links early helps to establish the authoritative source database that all other database applications and linked applications use as a master and synchronize to.
- *Retention policy.* How long documents are to be kept. Will documents be deleted when no longer relevant or are they retained as historical reference? How are transient documents that were used to create final documents treated? A retention policy defines how long documents are kept, how a document is determined to no longer be of use, and a deletion policy.

Trustworthiness

A well-defined document standard that is adhered to, regardless if it is an enforced process or a document management program, will provide trustworthiness for the information it manages. Information trustworthiness refers to the assumed reliability of the capture, maintenance, preservation, and presentation processes for information. Trustworthiness is important to ensure management accountability, operational continuity, and historical legacy for information.

It is also important, as with a facility standard, to make sure that the document standard is widely published, understood, and trusted. The user community should be informed that technical information within the facility is to be well managed and the maintenance of data integrity is foremost.

When multiple-linked databases are used, the defined authoritative database is the most critical point of accurate data entry. As this master database is used to synchronize and update multiple files, the entire facility's documentation will only be as valid and trustworthy as this primary data source.

Collaboration

Special consideration for collaboration should also be factored into the document management processes that a facility establishes. Collaboration improves communication, shortens work cycles, and increases overall ownership and accountability.

If a document management program is not being used, then collaboration is controlled by access rights and the file storage structure defined by the document standard. Various work folders can be created, and access controlled, by allowing write privileges to any particular work folder to only a limited number of people. On a more microscale, write access in a folder can be defined on a file-by-file basis.

The trustworthiness of the file that was created by a joint collaboration effort can be maintained by defining who has final authority to approve a file or file set as complete and accurate. Once a file has gone through the approval process, it can then be moved into the public folder and be accessible with read-only privileges.

In the cases where collaborative work involves engineers from multiple organizations, online editing access via the Internet should be considered. Online collaboration allows for people not physically located in the same facility to create, edit, or provide input to facility documentation. There are several document tools and services available that can provide online collaboration.

For online access, and also included in many CAD programs, document tools, and document management programs, markup and/or commenting is provided as a form of collaboration input. Comments and markup provide a vehicle that allows for a change request and comments to be made on documents in progress while not altering the information within a document. Comments and markups can then either be accepted or rejected by the controlling authority for the document.

Version Control

In a sense, the document standard when properly adhered to acts as a version control system for documentation. The document standard defines who is allowed write access, where files are stored, and what change notations are to be made (i.e., in the property box). A version-control program also tracks similar changes to files (Figure 2.2-8). Version-control programs typically offer advanced automated features, such as allowing for file rollback to a previous version, the automated processing of file and directory compare, and document merge. Version-control programs provide not only a recorded history of a file but the ability to restore a file back to a point in that history.

Additionally, version-control programs assist collaboration within groups working on the same files by preventing the accidental overwriting of document changes. This is accomplished by preventing files from being modified by multiple users at the same time. Version control usually prevents overwrite conflicts by insulating the different editors from each other by requiring that individual editors work in a privately owned directory. The version-control program then coordinates the merging of all work back into the common repository directory.

Version control is not a substitute for a proper review and approval process. Version control ensures that the most currently available file version is foremost in the depository. Version control cannot assume responsibility for data integrity regarding accuracy or

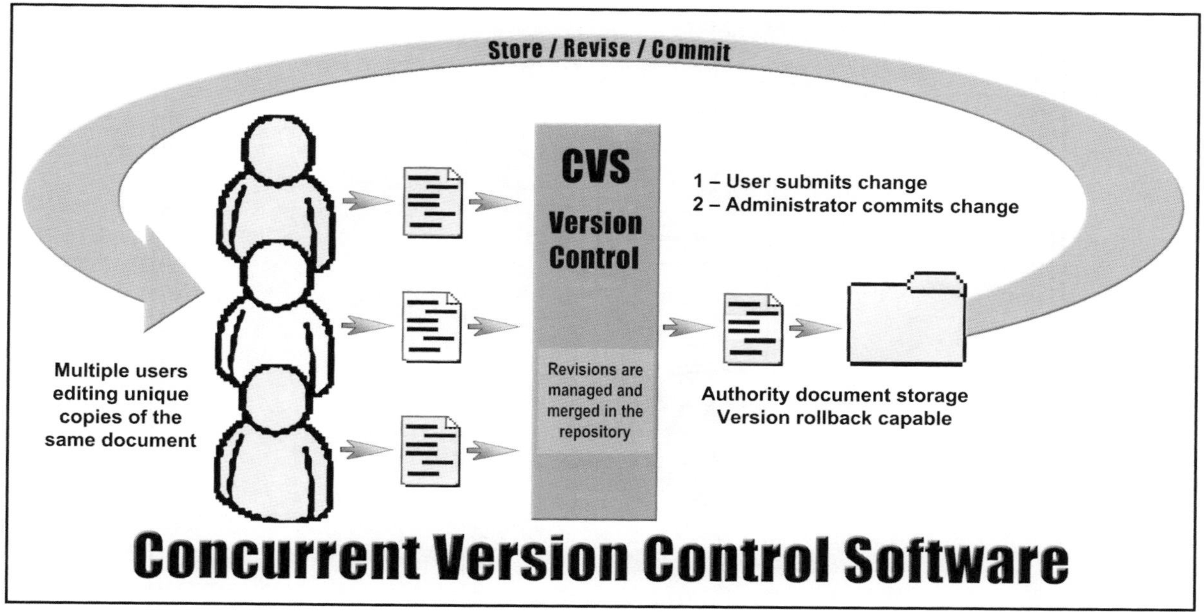

FIGURE 2.2-8 Version-control software. Version control provides control for documents that are under constant revision by multiple users. Multiple users submit changes and an administrator accepts and commits the change to the authoritative document. Version control also allows for version rollback in the event a previous version of a document proves to be preferred.

correctness. When groups of individuals are responsible for making multiple changes in documents, the areas of responsibilities for these changes still need to be communicated. Work, when completed, still needs to be reviewed whenever possible and files still need final approval before they can be moved to the next level of access, such as publishing for online access.

FILE TYPES

While file types are established in a broadcast facility's document standard it is important to understand the various types that are available. In some cases a document will be required to be saved in multiple formats depending on the final functions the information is to fulfill.

For example, a tutorial operator's manual designed for users in training might be created using the Microsoft Word program. For editing, the file would remain in the Word DOC format, but, for placing the document into an online repository for users to access, it might also be converted into an HTML- or PDF-format file. Likewise, the illustrations used in the tutorial document may have been created by several different programs or devices (e.g., CAD, screen capture, a scanner, a digital camera). The illustrations might all be edited in a common graphics program, such as Adobe PhotoShop, in order to maintain a consistent insertion format (size and style) within the tutorial. Therefore, to create one tutorial document several files and file types would be required.

File types are also an important consideration regarding portability. Not all document and CAD programs with like functions can open or import like files.

CAD File Types

The following file types are typical of those the broadcast engineer often encounters when working with software documentation programs.

DWG

DWG is the native binary file type used for saving CAD vector graphics and is the default file format for the Autodesk AutoCAD program (Figure 2.2-9). Since Version 1.0, Release 1, in December of 1982, AutoCAD has been the leading CAD software for design documentation. Shown is the twentieth release, AutoCAD 2006, released on March 2005. Autodesk has controlled and updated the DWG format since its founding and has made programming libraries available for other applications to access the DWG format since 1994.

The DWG format can be exported or imported into most CAD programs. Further interoperability between programs for the DWG format is being pursued in the CAD industry by such organizations as the Open Design Alliance. The Open Design Alliance is committed to the OpenDWG format, a distinct open standard that is based on Autodesk's DWG format. More information on the Open Design Alliance and OpenDWG can be found at: http://www.opendwg.org/.

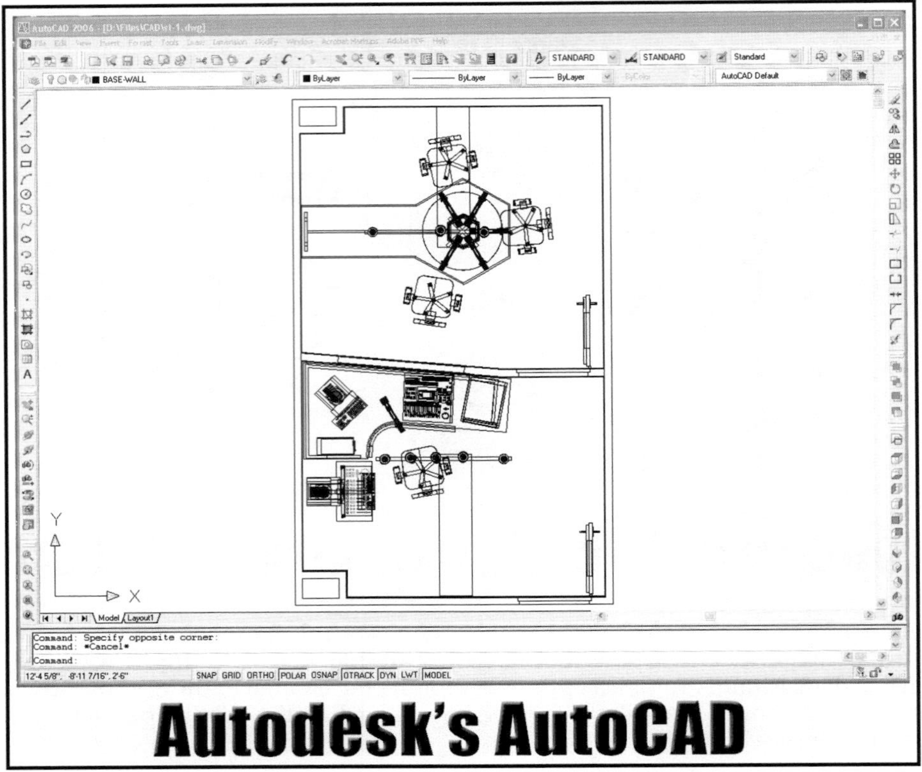

FIGURE 2.2-9 AutoCAD drawing of a radio broadcast studio (upper room) and control room (lower room). (See also Figure 2.2-11 for additional two-dimensional (2D) and three-dimensional (3D) renderings of this studio.)

DXF

DXF, the drawing interchange format, is an ASCII or binary file format for CAD vector graphics. The DXF format is used for exporting and importing CAD drawings between CAD applications. ASCII-format DXF files can be read with a text editor and are compatible with a wide range of CAD programs and applications. Binary-format DXF files contain all of the information of an ASCII DXF file but are more compact. DXF files, while more portable between applications, are typically 2.5 to 3 times larger than DWG files. The DXF format was created by Autodesk as a human-readable format that could be used by programs that needed to access CAD data.

DWF

Design Web Format (DWF) is an Autodesk format that provides a highly compressed file created from a DWG file. DWF is designed to be viewed on the Internet. DWF files can be displayed using Microsoft Internet Explorer 5.01 or later. DWF files support real-time panning and zooming and the display of layers and named views. Additionally, DWF files can be used for comment and markup distribution by using DWF-specific tools.

ANSI-Standard CAD Formats

ANSI-standard format for digital representation and exchange of information between CAD/CAM systems includes the STEP and IGES formats. STEP (standard for the exchange of product data) is an ISO standard data representation and exchange format. The STEP file structure is a "class" modular structure that is easy to adapt for specific CAD uses. IGES (initial graphics exchange specification) is the most popular neutral file format for the transfer of 3D files. A neutral file format is nonproprietary and can usually allow for basic data to be imported by a wide variety of applications (i.e., the TXT text format). A disadvantage of neutral format translation is that it lacks interoperability. Native CAD files include a wide range of data types, not just CAD vector data. STEP is more extensible and as such has the ability to deal with more varied data types and applications than IGES. However, the ever-expanding scope of data availability in native CAD formats requires a high cost for these formats to be continually developed and supported.

Graphic File Types

Graphic file types can be separated into two basic types: vector or raster. Understanding the difference between vector and raster images and when it is best

to use one or the other can greatly enhance the appearance, portability, and ultimately the readability of facility documentation. A raster image inserted improperly into a document file can at times be illegible if resized to fit using the wrong resolution.

All CAD programs utilize vector graphics. Vector graphics store information about a drawing as a data set of described shapes that make up an image, including their attributes and their coordinates in a drawing. Vector drawings do not have a set resolution and will display and/or print at the highest resolution the output device supports.

Raster images, also referred to as bitmap images, have a fixed resolution as they are a stored collection of pixels. A computer screen generally displays 72 pixels per inch; printers operate on average anywhere from 300–2,400 pixels per inch. Raster images describe a resolution grid and how the grid is laid out.

Vector-based images, being resolution independent, have an advantage when being inserted into a document—the resolution can be resized to fit without the loss of image quality. The disadvantage of vector images is that most photographs cannot be described in vector format. Continuous-tone images, such as photographs, are more efficiently described as raster images. Further, for Web use there is limited support for vector images.

Drawing programs, such as Adobe Illustrator and Freehand, produce vector-based images. Adobe Photoshop and PaintShop produce and edit raster-based images. Raster image file types include tif, jpg, gif, bmp, and png. The difference between raster file types is usually in the amount of compression that they are capable of supporting. Compression is used to make the file size smaller (trading off image quality for file size) for use in limited bandwidth applications such as displaying multiple images on the Internet. bmp and tif files use very little compression and are typically large files compared to compressed bitmap formats such a jpg, gif, or png. Some differences in quality as a function of file type are shown in Figure 2.2-10. Note in particular the definition and sharpness of edged surfaces and the smoothness of color and shading transitions.

Publishing Formats

When facility documentation needs to be made available to a wide audience, the overall documentation package needs to be converted to a format that can be commonly displayed. All users cannot be expected to have at their disposal the numerous database, document, CAD, and graphics programs required to display the multiple file types that encompass the entirety of a facility documentation package. Common publishing formats are PDF (portable document format), HTML (hypertext markup language), and XML (extensible markup language). All programs used to create any type of document file should be capable of exporting to one or all of these formats.

Publishing formats provide two significant advantages for a broadcast facility's technical documentation—navigation and format independence. Using the master index that should be part of the facility document standard, all files converted to either HTML, PDF, or a combination of both can be indexed and linked. These common formats will provide a quick load onto a user's workstation with easy navigation provided by an actively linked master index and page-to-page links. The use of common formats, such as HTML and PDF, also minimizes the licensing requirements for specialized programs that are usually required to open CAD and other project files.

PDF

Portable document format (PDF) is a file format that preserves the fonts, images, and layout of source doc-

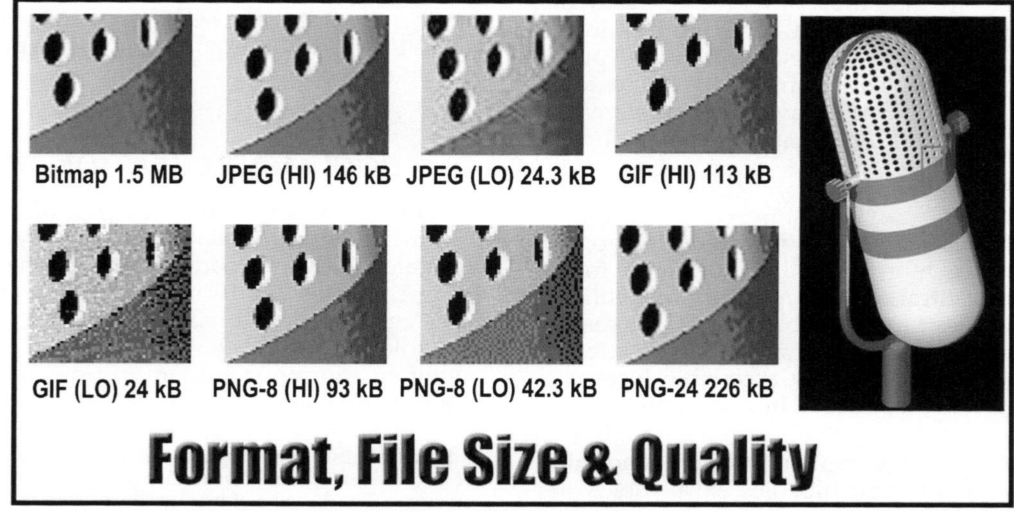

FIGURE 2.2-10 Graphics file type compression examples.

uments regardless of which application or platform they were created on. PDF files are relatively compact documents that are viewable with a wide range of free reader applications including web browser plug-ins. These combined makes the PDF format ideal for electronic document distribution.

For the broadcast engineer the PDF is one of the most common formats used by manufacturers to supply systems and equipment documentation. Fortunately there are tools available, such as Adobe Acrobat Professional, which will allow the broadcast engineer to easily manipulate these manuals in order to incorporate them into a facility's documentation package. Multiple source PDF files can be consolidated and re-indexed into one common PDF file. Or, multiple manufacturers' source PDF files can have their indexes exported into a common new index file that is actively linked to a collection of document files.

No matter how PDF documents are consolidated, the re-indexing of multiple manuals will enhance searching speed and efficiency. These electronic manuals can be organized in the same manner that a file cabinet would be (e.g., alphabetically, one PDF file: manuals A through E; or by location, one PDF file: Studio #1).

HTML

Web browsers use a variety of methods for addressing and communicating with Internet servers, called protocols. The most common protocol is hypertext transfer protocol (HTTP). HTTP was originally created to publish and view linked text documents but has been extended to display and run a variety of multimedia content. Text displayed in HTTP protocol is commonly accomplished with HTML (hypertext markup language). HTML, originally a simple system for publishing documents on the Web, has rapidly evolved to include features that can be used to create more sophisticated, interactive applications. In HTML, navigation to other documents or within a document is accomplished by the use of hyperlinks.

Most document creation tools provide some form of export to HTML documents. In some applications, such as AutoCAD, this function is found in the "publish to Web" command. This command will create a web page of the current CAD drawing complete with HTML-indexed links and DWF drawing conversions.

Help Files

As many broadcast systems are software-based applications, manuals are, more often than not, in the form of "help files." A central online repository of a collection of these help files can be established. This is easily accomplished by copying individual application's help files into a central publicly accessible location. Help files can usually be found in an installed program's local directory.

Help files are easily identified by chm and hlp extensions. If both chm and hlp files exist, they will both need to be moved and copied into the common index. If a program is using HTML help files, then the associated files can usually be found in a "help" or "docs" directory. These multiple HTML files are accessible from a common index file usually titled "index.html" or "help.html." The easiest way to determine which HTML help file is used as the master index is to click on one and see if it opens the help file master index. Another method is to check in the applications preference menu. Sometimes there is a help file preference designation with the name and directory location of the HTML index file.

Once these help files are copied into a central publicly accessible location, an HTML index page that provides an active link to the collection of help files in the central repository can easily be written.

XML

XML was designed to fill a deficiency identified in other Internet and markup standards by supporting expanded data definition and information-processing requirements. XML is a self-describing, domain-specific markup language syntax.

XML is not in itself a language but a common syntax for expressing structure in data. XML allows for the creation of a document-unique language that describes a class of data objects that are contained within the document. The XML syntax also provides descriptive instructions for the behavior of XML processors or parsers. These parsers are software modules that work on behalf of an application (e.g., Explorer, Word) to interpret XML documents. Parsers provide both data content and data structure access when translating XML documents.

XML separates data structure and content from the presentation of that data. The ability to separate data from presentation allows for the potential of XML to become the standardized mechanism for data exchange as well as a universal document translation platform.

The use of XML in a broadcast facility's documentation should be considered by the broadcast engineer. XML is positioned as a future translation platform and the fact that it uses self-described structured data makes it capable of providing multiple publishing and presentation options.

DOCUMENT CREATION TOOLS[1]

Document creation tools in the past often consisted of a drafting table, t-square, triangle, typewriter, and templates. Today's broadcast engineer has a wide variety of computer platforms and documentation, database, graphic, and CAD software applications available for use. These modern tools are capable of saving an enormous amount of time and effort when creating and maintaining technical documentation.

[1]Specific software applications mentioned in this section should not be considered as either an independent review or a product endorsement. For the sake of brevity, all products available could not be covered. Likewise, the omission of any products should not be misconstrued to represent a nonendorsement.

Computer-Aided Design Tools

The benefit of computer-aided design (CAD) relies on its ability to organize, share, and reuse information. Most CAD programs offer connectivity to databases and are capable of creating robust database applications.

In the past, the selection of CAD options was somewhat limited. CAD software applications were both platform limited and costly. This situation has since changed with the increase in commonly available high-powered processor workstations. There are now dozens of CAD programs on the market, available for all workstation platforms and in all price ranges, including several free open-source solutions.

2D or 3D

When a facility is documented using CAD, consideration should be given to the use of 3D. While the initial drawing process will take longer using 3D, on the backend more accurate and flexible drawings will be realized.

For example, to draw a studio desk table in 2D, multiple views have to be drawn—top, front, and isometric. With a 3D CAD drawing the object only has to be drawn once. Once drawn the 3D object can be represented from any angle and rendered into a photo-realistic image. 3D CAD drawings also allow for ergonomic studies and a virtual walkthrough of facilities prior to actual construction. The 2D drawing (Figure 2.2-11(a)) resembles any standard architectural plan and shows two small offices at left and four studio/control room combinations to the right of the offices. The studios can be recognized by the hexagonal-shaped interview tables, and the control rooms by the workstation tables with computers located in the corners. The same multistudio facility drawing begins to take on depth and perspective in Figure 2.2-11(b) as an isometric wireframe. Again, obvious in this drawing is the hexagonal-shaped table (with boom microphones now more apparent) and the workstation areas in the control rooms. Finally, a photo-realistic 3D rendering of the same drawing is shown in Figure 2.2-11(c) completing the transition from flat plan conceptualization to virtual realization.

The most time-consuming task in migrating documentation to a 3D platform is creating the numerous drawings for objects that make up a facility. These drawings are the individual objects (e.g., broadcast equipment items and furnishings) that are inserted into multiple final drawings. The availability of preexisting 3D object drawings can save a broadcast engineer's time when migrating to a 3D platform. Therefore, broadcast equipment manufacturers should be asked if 3D CAD drawings of their products exist.

AutoCAD

AutoCAD has had the longest tenure in the CAD software market and is the most widely used CAD program. Most other CAD programs can either convert or import directly AutoCAD DWG format drawings. AutoCAD has internal and external database support as well as several additional tools to manage data objects and multiple drawing sets, such as sheet manager, block manager, and reference manager.

AutoCAD also provides compatibility to several programming scripts that allow third-party vendors and users to create custom programs and macroroutines inside of AutoCAD. Scripting languages supported include Microsoft .NET, Visual Basic, Visual C++, Delphi, Java, and Visual LISP.

Included since the AutoCAD 2002 release are drawing standards checking tools. These tools ensure that established CAD drawing standards are being applied during the creation of a CAD file. Additional batch standard checking tools allow for the comparing of the established standard against the standard used in existing documents. This function runs externally to AutoCAD and allows multiple files to be checked against a standards file and logs any discrepancies into an XML-based report. More information on AutoCAD and all Autodesk products can be found at the Autodesk website at http://www.autodesk.com.

IntelliCAD

While not specifically a CAD program, IntelliCAD is a CAD engine that is used in various commercially available CAD programs. IntelliCAD's native file format is DWG and, therefore, when opening an AutoCAD-created DWG file, IntelliCAD does not perform a file conversion. The IntelliCAD engine is compatible with the AutoCAD command set, as well as programming APIs such as LISP. http://www.intellicad.org is the website for the IntelliCAD Technology Consortium (ITC). ITC is a nonprofit organization committed to cooperatively developing the CAD system IntelliCAD.

Data Visualization Tools

Data visualization programs are capable of synchronizing directly with data sources to provide up-to-date diagrams usually in the form of flowcharts or network diagrams. While appearing on the face to be 2D stencil-based programs, they often contain powerful database interfaces that are capable of updating, or in some cases, generating complex drawings from information contained in the database. 2D stencils generally represent various manufacturers' equipment and specifications on the product are integrated into a database.

Visio

While primarily thought of as a 2D stencil-based drag-and-drop program, Visio is a diagramming program capable of organizing complex technical diagrams. Visio can automate data visualization by synchronizing directly with data sources to provide up-to-date diagrams. As a Microsoft product, Visio is part of the Office Suite and can integrate business processes and systems by extracting data from Visio diagrams and importing or exporting them into Microsoft Office

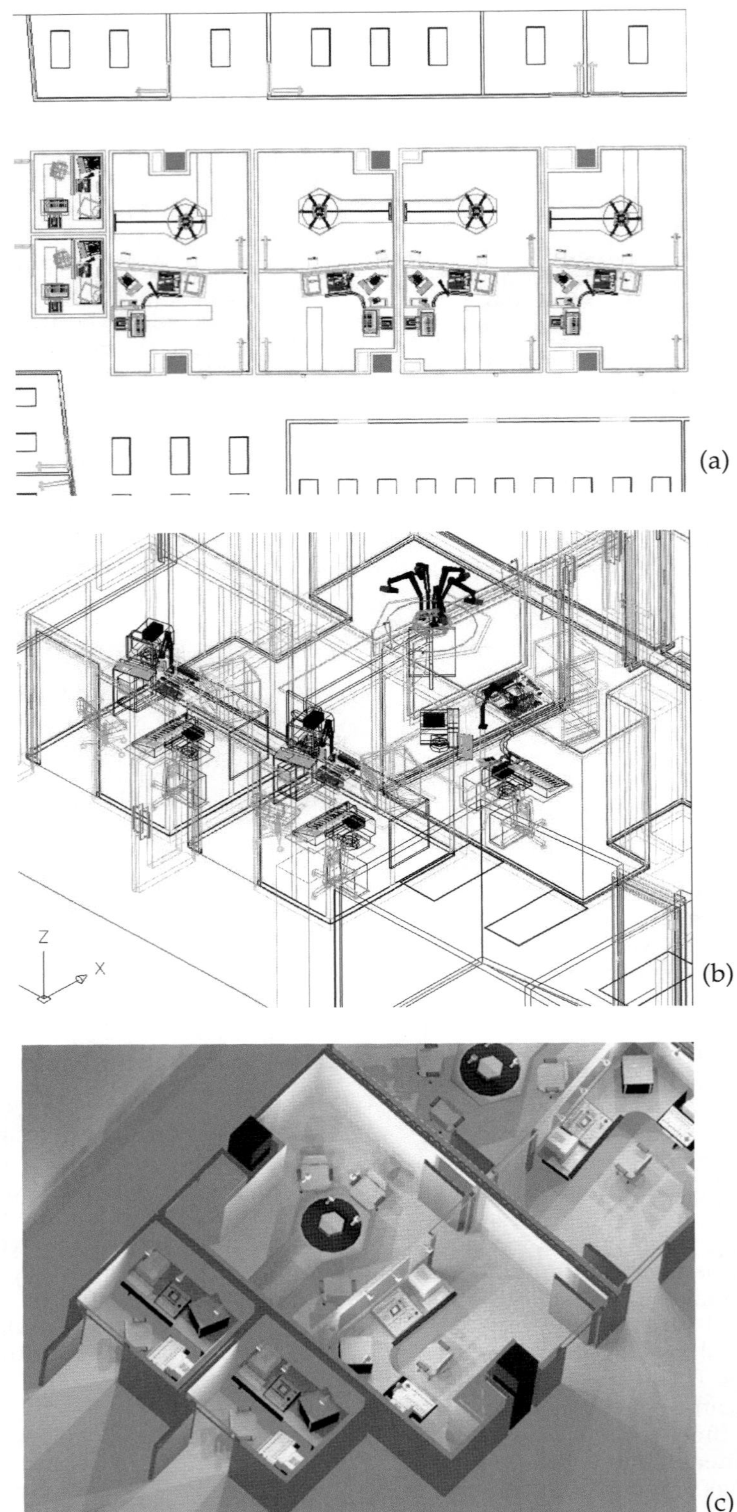

FIGURE 2.2-11 (a) A studio shown in a 2D rendering, (b) a studio drawing as 3D wire-frame, and (c) a studio drawing 3D photo-realistic.

Access, Excel, and Word, and Microsoft SQL Server. Visio is also capable of opening and exporting to native AutoCAD DWG files. The Visio home page is www.microsoft.com/office/visio.

NetViz

NetViz is a widely-used program within the information technology (IT) and telecommunications industries. With modern broadcast facilities becoming more and more IT-based, NetViz is a capable tool and its use should be considered by broadcast engineers. NetViz is also a drag-and-drop stencil-based 2D program that can automate data visualization by synchronizing directly with data sources. NetViz's website is http://www.netviz.com.

Project Design Applications

Project design applications are software tools that interface with one or more programs, such as AutoCAD, and provide common database access control to manage document file sets that together comprise a project.

VidCAD

VidCAD is a broadcast-specific tool that is an extension and augmentation of AutoCAD. VidCAD has an integrated equipment product database with the capability to add new equipment. The VidCAD program includes sophisticated diagram generation, report generation, and wire-routing capabilities. Product line options in VidCAD, such as global editing, allow all drawings and data files to be synchronous with automatic updates. The enterprise version, VidCAD ESP, allows collaboration with multiple facilities, remote engineers, and integrators at remote locations around the world. More information on VidCAD's documentation programs, secure infrastructure documentation, and management solutions for documentation can be found at http://www.vidcad.com.

System Integrator 4.5

System Integrator, another extension and augmentation tool, is a design solution that is capable of interfacing with both AutoCAD and Visio. D-Tools System Integrator ties together the processes of design, engineering, estimating, documentation, and installation into a single portable project file. Using both Microsoft Visio and AutoCAD for the graphic interface allows for the use of precise drawing tools to automatically create elevation, flow diagrams, and line views from database information.

SUMMARY

In the modern broadcast facility even small facilities are a compilation of multiple complex components and systems. By utilizing the wide variety of design, CAD, and documentation tools available, and by following best practices as outlined in this chapter and instituting a proactive document management process, all broadcast engineers should be capable of maintaining comprehensive facility documentation.

CHAPTER

2.3

System Integration and Project Management

JOHN LUFF

Media Technology Consultant
Sewickley, Pennsylvania

INTRODUCTION

The design and installation of systems in a broadcast plant are a discipline that requires a set of skills that is not easily learned on the job as an engineer in a broadcast station. The skills required are both technical and business related. Fundamentally, the most important skills are technical system design analysis and implementation, skilled wiring, and project management. Understandably, one of the mistaken technical approaches often used is to replicate a design or layout that one has seen in other installations. This leads to systems that are not optimized for local needs and conditions and that will not often achieve goals.

The difficulty of learning project-related skills inside a station is greatly exacerbated by the fact the stations do not undergo substantial change regularly. Change is often incremental, perhaps the installation of an additional edit bay or a new audio console. However, large-scale projects do not come up often enough that personnel become skilled at the process. Between projects, team members may change jobs or move to other departments in a station and take responsibilities that do not leave them free to participate in the completion of large projects. As a result, skills may need to be learned, relearned, or perhaps outside resources brought in to assist. In any event it is important that those in the station understand the process of planning and executing a project even if the intent is to use skilled outsiders to facilitate or complete much of the work.

This chapter explores the process of system design and implementation with the goal of establishing a methodology that will lead to success. The process is undertaken by individuals organized in a manner similar to how the departments in a station are organized. Each step in the process requires a manager to ensure that the goals of that step are monitored and achieved, and that team members are contributing their unique talents to the success. While the text and examples that follow focus on TV facilities, the principles and concepts are applicable to radio facilities as well.

THE PROJECT MANAGER

The overall project should be headed by a project manager. Of the disciplines involved in developing and completing projects, the one skill most foreign to most broadcast engineers is project management, which is often not a skill learned in professional or academic training by broadcast engineers.

The project manager acts as the traffic cop, assuring all that goals are met and timelines maintained, as well as monitoring budget performance against the budget that was established at the onset. Formal training in project management is useful for this person, but the tasks are understandable by anyone with reasonable skills of analysis and communication. It requires discipline and attention to detail to manage a project successfully, and the skills of the project manager may determine the overall outcome, functionally and financially. As such it is important that this person is chosen carefully and that he or she understand his or her role in the process.

One of the prime responsibilities of the project manager is to maintain communication with the entire

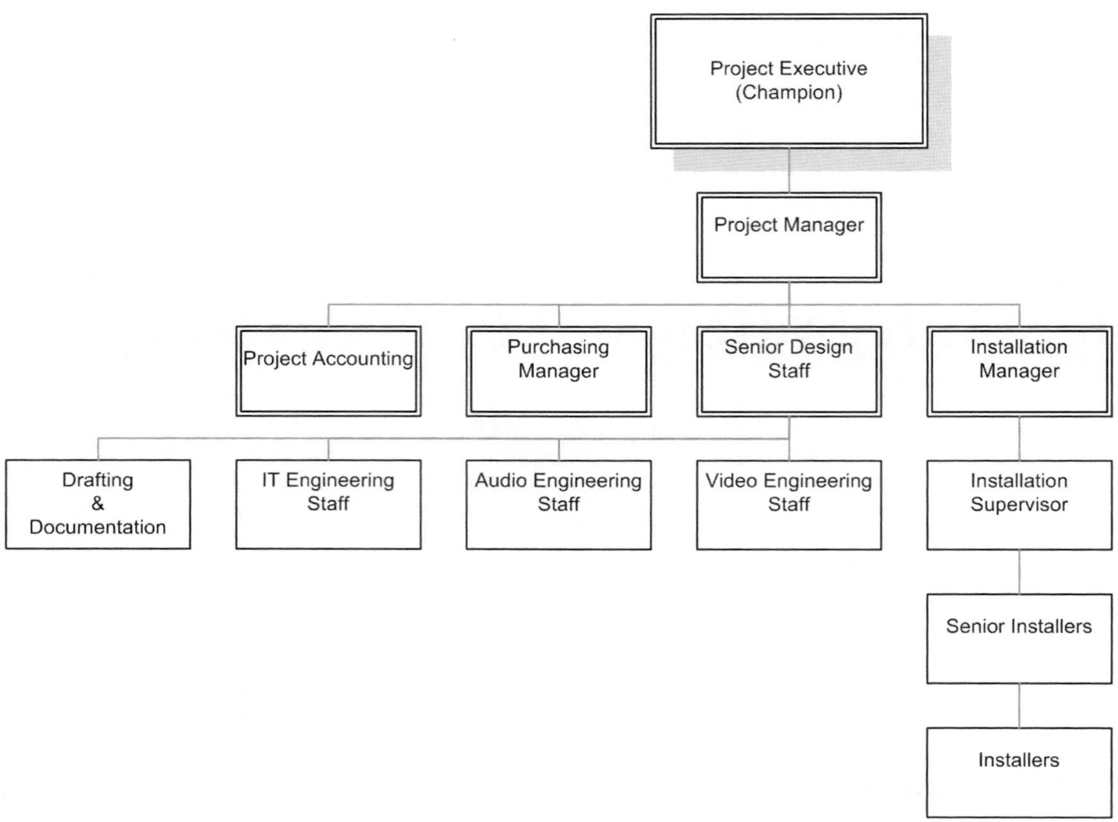

FIGURE 2.3-1 Example of an organization chart for management of complex technical project.

team as well as the management of the station. Communication should be formalized and regular, with set intervals for delivery of a project report for all to review. The project report could be considered chapters in a book (e.g., "The History of the Project"), and constitutes the written representation of the flow of the project and timeline. The project report should summarize project status, schedule for upcoming activities, and assignments for tasks, including a description of the task and the date it is expected to be finished. Regular distribution of the project report to the team helps to focus the team on their individual goals, and ensure that everyone is aware of progress and barriers to success.

Classic time-management training says that individual goals should be short term and easily achievable. Making the goal for an individual the completion of the whole design and a timeline of 6 months will assure that procrastination is practiced. Reducing the task to one that can be executed in 2 weeks or less, perhaps the layout of the floor plan for a rack room, gives one an achievable and manageable task that can be monitored. A series of short-term goals can be strung together into a large overall task, and the project manager can keep track of progress and help individuals meet their goals and tasks, which helps other members of the team complete their tasks, which often depend on others' tasks.

PROJECT "CHAMPION"

It might be assumed that the project manager "owns" the project, but in reality the project manager is only the conduit, a manager who keeps everyone else focused on short-term and long-term goals of the project and the successful completion of the work plan. The "champion," who might be considered the originator of the project, takes ultimate responsibility for the project, and might well be considered the "client" to the rest of the project team. For example, if the project is to implement a new nonlinear workflow for the news department, the champion may be the news director. If the project is a new building and technical facility, the champion might be the general manager or, perhaps, the corporate director of engineering. It is the champion who signs off on the goals of the project and it is the project manager who organizes the rest of the team and the work plan to achieve the goals identified.

DESIGN PERSONNEL

In addition to the project manager, in a broadcast technology project it is normal to have a "lead designer" who takes responsibility for the details of the equipment selection and implementation plans. This person may also produce the documentation of the design, or

may have additional team members who execute the details. Draftsmen, design engineers, IT and network specialists, commissioning staff from vendors and station personnel, and others may well be involved. In small projects, it may only be necessary to have the lead engineer, but as the size and complexity of the project grows, it will be necessary to grow the team to ensure timelines are met and details are adequately covered.

After designs are complete, other members of the team are required to implement the design. Though many broadcast engineers are skilled at wiring, the complexity of modern systems with heavy emphasis on IT systems may well require specialists. For example, someone certified to install and terminate CAT 6 cabling or fiber-optic termination may be critical to the team even if the engineers who do the design are also tasked with completing the wiring. In large systems, the management of large cable bundles is a skill that will require installation specialists who have experience with planning and executing the wiring of the plant. A lead installer or installation supervisor should be identified or hired to complete both the planning and execution of the detailed installation. He or she may well be a key participant during the design phase as well, assuring the wiring paths are thought through carefully so that the installation has the space needed and room for growth as well.

PROJECT PLANNING

A key tool of the project manager, indeed the entire team, is a breakdown of the project into tasks and assignments. It is quite easy to accomplish this using project planning software (PS) readily available off the shelf from a number of vendors. Microsoft Project and other similar products are tools that are easy to learn, and local resources for training in their use are available in most major cities. Such software allows graphical representations of the flow of the project, like *Gantt charts* (see Figure 2.3-2) and *PERT charts* (see Figure 2.3-3), and lists of resource assignments. It permits the project manager to show in easily understood ways the progress of each task and where current missed deadlines will affect other work.

Entering all the project details into the PS package allows project budgets to be tracked without resorting to a report from the accounting department, which may not be available quickly enough to allow the project manager to make useful judgments of success of the project team. Though costs for labor are important, as is the cost of capital hardware and construction renovation as required, recording the hours spent on each task provides a quick reference to the ability of the team to complete their work according to the original plan.

Before PS software can be loaded with the details of the project and the schedule built, the work must be broken down into tasks and organized so that each task can follow in lock-step with those that must precede and follow. For example, planning a routing switcher installation must include deciding on the number of inputs and outputs, assigning sources and destinations, and detailing the connections to patch panels and equipment. These tasks become meaningless out of order, and when one task stalls, the rest cannot be finished. These steps are entered into the software and "linked" to each other, noting their mutual dependency. The PS system when properly implemented and kept up-to-date, will allow critical items to be identified and personnel resource overloads to be identified. Each time the weekly report is issued a copy of the PS system Gantt chart should be issued as well. It is easy to send a copy to all concerned electronically or in print. Electronic delivery may allow the program manager to note when project staff members have read the document by requesting "read receipts" when email is sent. If any team members do not read the document, the program manager would thus know and be able to nudge the recalcitrant team members to uphold their part of the project.

Some software also allows required equipment purchases and other costs to be tracked as well. If not, or if training is not available on the full capabilities of the PS software chosen, a spreadsheet can easily be created to track the status of the required equipment, as shown in Figure 2.3-4. The spreadsheet should identify the following:

- Manufacturer
- Model
- Quantity
- Purchase source (direct or reseller)
- List price
- Budgeted cost
- Discounted price
- Variance against budget (calculated field)
- PO number
- Date ordered
- Expected delivery
- Status (order open or delivered)

Each large-scale task in a project has a number of steps under it. A complete breakdown must be done showing the person doing each task, when it is to be completed, and giving team members clear guidance for what is expected at each step along the way. Management trainers often say that no individual task should be so large that it cannot be completed within 2 calendar weeks. That allows individuals to make progress and measure their success. As individuals get used to fitting their work into the overall interlocking project plan, they will gain confidence in their ability to contribute and push the project along.

Near the end of a project, when wiring is nearly finished, the emphasis switches to the process of turning on equipment and commissioning of the systems. At this point station personnel may well be at the highest risk of lacking the necessary skills, as it is likely the equipment is new to them. IT-based systems will require attention by networking specialists, perhaps with knowledge of detailed configuration of

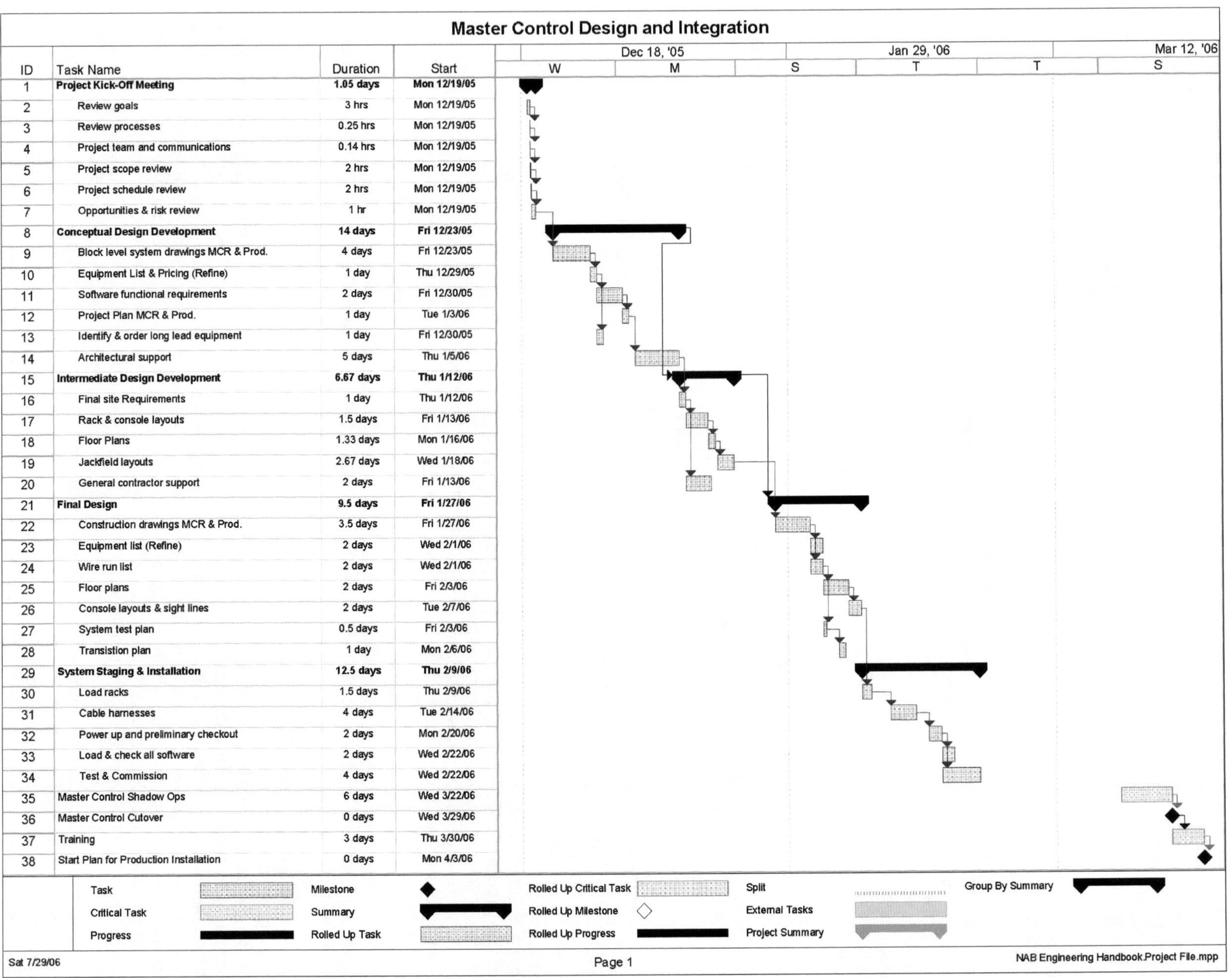

Master Control Design and Integration

ID	Task Name	Duration	Start
1	**Project Kick-Off Meeting**	**1.05 days**	**Mon 12/19/05**
2	Review goals	3 hrs	Mon 12/19/05
3	Review processes	0.25 hrs	Mon 12/19/05
4	Project team and communications	0.14 hrs	Mon 12/19/05
5	Project scope review	2 hrs	Mon 12/19/05
6	Project schedule review	2 hrs	Mon 12/19/05
7	Opportunities & risk review	1 hr	Mon 12/19/05
8	**Conceptual Design Development**	**14 days**	**Fri 12/23/05**
9	Block level system drawings MCR & Prod.	4 days	Fri 12/23/05
10	Equipment List & Pricing (Refine)	1 day	Thu 12/29/05
11	Software functional requirements	2 days	Fri 12/30/05
12	Project Plan MCR & Prod.	1 day	Tue 1/3/06
13	Identify & order long lead equipment	1 day	Fri 12/30/05
14	Architectural support	5 days	Thu 1/5/06
15	**Intermediate Design Development**	**6.67 days**	**Thu 1/12/06**
16	Final site Requirements	1 day	Thu 1/12/06
17	Rack & console layouts	1.5 days	Fri 1/13/06
18	Floor Plans	1.33 days	Mon 1/16/06
19	Jackfield layouts	2.67 days	Wed 1/18/06
20	General contractor support	2 days	Fri 1/13/06
21	**Final Design**	**9.5 days**	**Fri 1/27/06**
22	Construction drawings MCR & Prod.	3.5 days	Fri 1/27/06
23	Equipment list (Refine)	2 days	Wed 2/1/06
24	Wire run list	2 days	Wed 2/1/06
25	Floor plans	2 days	Fri 2/3/06
26	Console layouts & sight lines	2 days	Tue 2/7/06
27	System test plan	0.5 days	Fri 2/3/06
28	Transistion plan	1 day	Mon 2/6/06
29	**System Staging & Installation**	**12.5 days**	**Thu 2/9/06**
30	Load racks	1.5 days	Thu 2/9/06
31	Cable harnesses	4 days	Tue 2/14/06
32	Power up and preliminary checkout	2 days	Mon 2/20/06
33	Load & check all software	2 days	Wed 2/22/06
34	Test & Commission	4 days	Wed 2/22/06
35	Master Control Shadow Ops	6 days	Wed 3/22/06
36	Master Control Cutover	0 days	Wed 3/29/06
37	Training	3 days	Thu 3/30/06
38	Start Plan for Production Installation	0 days	Mon 4/3/06

FIGURE 2.3-2 Gantt charts are a valuable tool for helping a project manager plan and track the progress of a project.

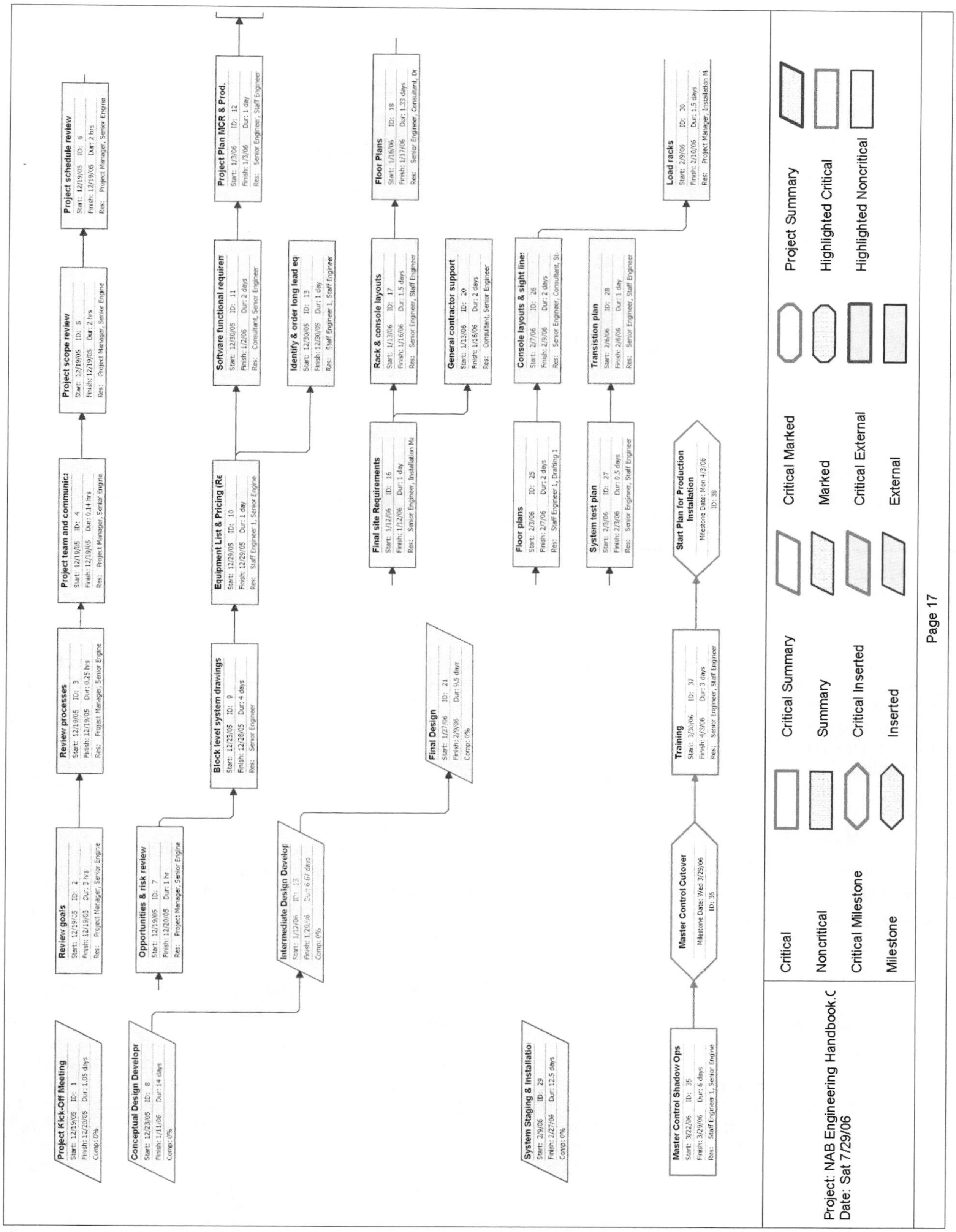

Project Kick-Off Meeting
Start: 12/19/05 ID: 1
Finish: 12/20/05 Dur: 1.05 days
Comp: 0%

Review goals
Start: 12/19/05 ID: 2
Finish: 12/19/05 Dur: 3 hrs
Res: Project Manager, Senior Engine

Review processes
Start: 12/19/05 ID: 3
Finish: 12/19/05 Dur: 0.25 hrs
Res: Project Manager, Senior Engine

Project team and communic
Start: 12/19/05 ID: 4
Finish: 12/19/05 Dur: 0.14 hrs
Res: Project Manager, Senior Engine

Project scope review
Start: 12/19/05 ID: 5
Finish: 12/19/05 Dur: 2 hrs
Res: Project Manager, Senior Engine

Project schedule review
Start: 12/19/05 ID: 6
Finish: 12/19/05 Dur: 2 hrs
Res: Project Manager, Senior Engine

Conceptual Design Developr
Start: 12/23/05 ID: 8
Finish: 1/11/06 Dur: 14 days
Comp: 0%

Opportunities & risk review
Start: 12/19/05 ID: 7
Finish: 12/20/05 Dur: 1 hr
Res: Project Manager, Senior Engine

Block level system drawings
Start: 12/23/05 ID: 9
Finish: 12/28/05 Dur: 4 days
Res: Senior Engineer

Equipment List & Pricing (Re
Start: 12/29/05 ID: 10
Finish: 12/29/05 Dur: 1 day
Res: Staff Engineer 1, Senior Engine

Software functional requiren
Start: 12/30/05 ID: 11
Finish: 1/2/06 Dur: 2 days
Res: Consultant, Senior Engineer

Project Plan MCR & Prod.
Start: 1/3/06 ID: 12
Finish: 1/3/06 Dur: 1 day
Res: Senior Engineer, Staff Engineer

Intermediate Design Develop
Start: 1/12/06 ID: 15
Finish: 1/20/06 Dur: 6.67 days
Comp: 0%

Identify & order long lead eq
Start: 12/30/05 ID: 13
Finish: 12/30/05 Dur: 1 day
Res: Staff Engineer 1, Staff Engineer

Final site Requirements
Start: 1/12/06 ID: 16
Finish: 1/12/06 Dur: 1 day
Res: Senior Engineer, Installation Ma

Rack & console layouts
Start: 1/13/06 ID: 17
Finish: 1/16/06 Dur: 1.5 days
Res: Senior Engineer, Staff Engineer

Floor Plans
Start: 1/16/06 ID: 18
Finish: 1/17/06 Dur: 1.33 days
Res: Senior Engineer, Consultant, Dr

Final Design
Start: 1/27/06 ID: 21
Finish: 2/9/06 Dur: 9.5 days
Comp: 0%

General contractor support
Start: 1/13/06 ID: 20
Finish: 1/16/06 Dur: 2 days
Res: Consultant, Senior Engineer

System Staging & Installatio
Start: 2/9/06 ID: 29
Finish: 2/27/06 Dur: 12.5 days
Comp: 0%

Floor plans
Start: 2/3/06 ID: 25
Finish: 2/7/06 Dur: 2 days
Res: Staff Engineer 1, Drafting 1

Console layouts & sight line
Start: 2/7/06 ID: 26
Finish: 2/9/06 Dur: 2 days
Res: Senior Engineer, Consultant, St

System test plan
Start: 2/3/06 ID: 27
Finish: 2/3/06 Dur: 0.5 days
Res: Senior Engineer, Staff Engineer

Transistion plan
Start: 2/6/06 ID: 28
Finish: 2/6/06 Dur: 1 day
Res: Senior Engineer, Staff Engineer

Master Control Shadow Ops
Start: 3/22/06 ID: 35
Finish: 3/29/06 Dur: 6 days
Res: Staff Engineer 1, Senior Engine

Master Control Cutover
Milestone Date: Wed 3/29/06
ID: 36

Training
Start: 3/30/06 ID: 37
Finish: 4/3/06 Dur: 3 days
Res: Senior Engineer, Staff Engineer

Start Plan for Production Installation
Milestone Date: Mon 4/3/06
ID: 38

Load racks
Start: 2/9/06 ID: 30
Finish: 2/10/06 Dur: 1.5 days
Res: Project Manager, Installation M.

Project: NAB Engineering Handbook.C
Date: Sat 7/29/06

Critical | Critical Summary | Critical Marked | Project Summary
Noncritical | Summary | Marked | Highlighted Critical
Critical Milestone | Critical Inserted | Critical External | Highlighted Noncritical
Milestone | Inserted | External

Page 17

FIGURE 2.3-3 PERT charts offer an alternate representation of the project based on the same database used to create a Gantt chart.

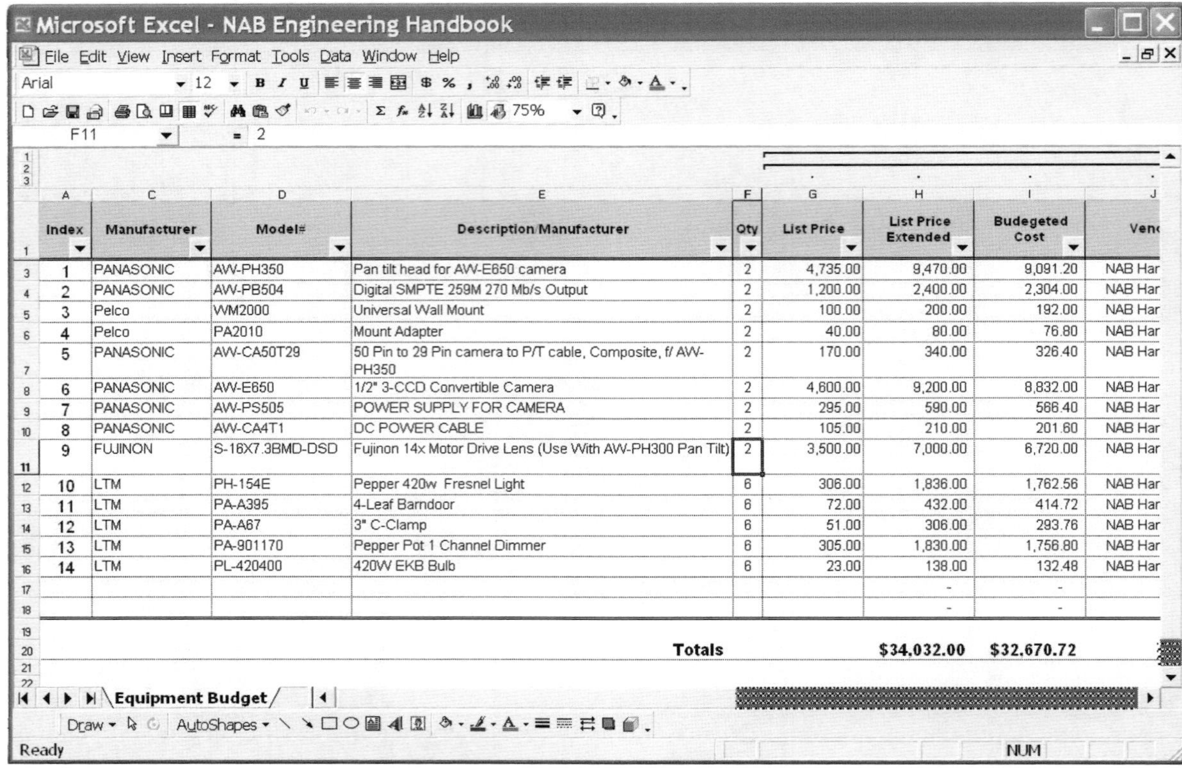

Index	Manufacturer	Model#	Description/Manufacturer	Qty	List Price	List Price Extended	Budegeted Cost	Ven
1	PANASONIC	AW-PH350	Pan tilt head for AW-E650 camera	2	4,735.00	9,470.00	9,091.20	NAB Har
2	PANASONIC	AW-PB504	Digital SMPTE 259M 270 Mb/s Output	2	1,200.00	2,400.00	2,304.00	NAB Har
3	Pelco	WM2000	Universal Wall Mount	2	100.00	200.00	192.00	NAB Har
4	Pelco	PA2010	Mount Adapter	2	40.00	80.00	76.80	NAB Har
5	PANASONIC	AW-CA50T29	50 Pin to 29 Pin camera to P/T cable, Composite, f/ AW-PH350	2	170.00	340.00	326.40	NAB Har
6	PANASONIC	AW-E650	1/2" 3-CCD Convertible Camera	2	4,600.00	9,200.00	8,832.00	NAB Har
7	PANASONIC	AW-PS505	POWER SUPPLY FOR CAMERA	2	295.00	590.00	566.40	NAB Har
8	PANASONIC	AW-CA4T1	DC POWER CABLE	2	105.00	210.00	201.60	NAB Har
9	FUJINON	S-16X7.3BMD-DSD	Fujinon 14x Motor Drive Lens (Use With AW-PH300 Pan Tilt)	2	3,500.00	7,000.00	6,720.00	NAB Har
10	LTM	PH-154E	Pepper 420w Fresnel Light	6	306.00	1,836.00	1,762.56	NAB Har
11	LTM	PA-A395	4-Leaf Barndoor	6	72.00	432.00	414.72	NAB Har
12	LTM	PA-A67	3" C-Clamp	6	51.00	306.00	293.76	NAB Har
13	LTM	PA-901170	Pepper Pot 1 Channel Dimmer	6	305.00	1,830.00	1,756.80	NAB Har
14	LTM	PL-420400	420W EKB Bulb	6	23.00	138.00	132.48	NAB Har
						-	-	
						-	-	
			Totals			$34,032.00	$32,670.72	

FIGURE 2.3-4 A spreadsheet program can be used to track the status of equipment purchases and other costs incurred by the project.

routers and firewalls. Even thoughtful and skilled broadcast engineers with some degree of network training may not be able to get the most out of systems without proper training. Hired resources are often a critical part of the completion of the testing and commissioning.

PROJECT PHASES

The phases of a project can be broken down to the following (see also Figure 2.3-5):

- Business analysis and high-level planning
- Conceptual design
- Detailed design and documentation
- Construction
- Configuration and testing
- Training and support

In some cases the scale of a project may make it appropriate to combine some phases, though it is much more likely that these would be expanded on in order to add detail inside each one to allow further breakdown of the work and facilitate assignments to individual team members.

The completed project schedule could well extend into hundreds of lines. In addition, if new buildings or renovations are involved, the construction project will likely have its own project manager, architect, and engineers dealing with construction disciplines. Their construction schedule will contain key dates and milestones that must be loaded into the technical project plan so that all tasks that are dependent can be tracked together. For example, if a new computer floor is to be installed, it will be necessary to coordinate the completion of that work before the laying of cabling or the installation of racks.

Those wishing to get formal training, including certification as a project manager, are directed to the Project Management Institute (www.PMI.org) or local academic organizations that may offer in-depth training on project management topics.

Business Analysis and High-Level Planning

In each phase of a project unique skills are required. In modern broadcast organizations it is often assumed by managers that the chief engineer or director of engineering is a design specialist and can plan and execute complete plans. As a first step, however, it is best to analyze the problem, and do a skills analysis that leads to a decision to either use internal or external resources for each phase of the project. Knowing the borders of the problem, and setting the goals out in complete and descriptive detail, is critically important to the success of a project. One might say the goal is to install a new production control room, but that needs to be broken

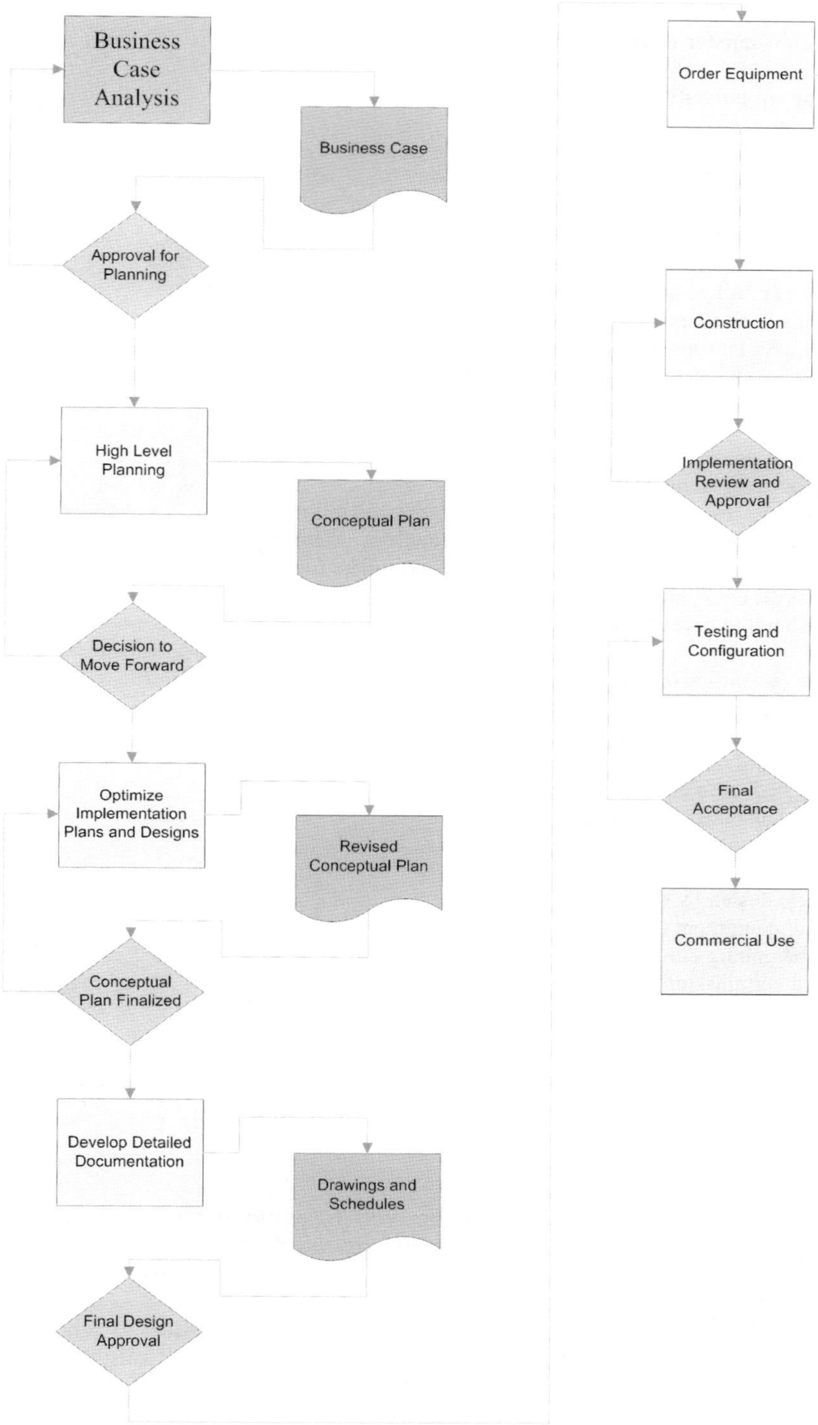

FIGURE 2.3-5 Project phases.

down into a set of goals against which all of the decisions can be compared. Figure 2.3-6 offers an example of how a project can be broken down into a set of goals. In this example the goals are organized according to scope (functions to be supported), cost (cost-containment goals), and schedule (schedule goals).

This business-planning phase of a project is also often used to get those involved "on board" with the project's goals and responsibilities, and serves as the foundation for management as the project moves forward.

Goal: Build a new master control room

Functions to be supported:

- Ingest from tape
- Transfer from commercial delivery services directly to the video server
- Ingest from satellite
- Automated selection of satellite as directed by dub list in traffic software
- Automated unattended assignment of record destinations
- Conflict resolution and assignment of ports on server and VTRs as a backup
- Control of routing switcher and routing through converters as necessary
- Reporting back to traffic and automation that content has been recorded
- Control over HD and SD paths to DTV transmission
- Simultaneously switch both paths, accounting for path delays
- Switch to upconverted local commercials when HD spots do not exist

Cost containment goals:

- Re-use existing technology when possible
- Use in-house resources where possible
- Leverage vendors to research and propose solutions where possible (may require writing requirements documents for them to assimilate for proposals)
- Check skills bank internally for persons with experience needed
- Establish not to exceed budget with station management

Schedule goals (assume work on project begins on 1/1):

- Meet with internal departments to gather information by 2/15
- Project definition completed by 3/1
- Gather proposals from vendors and collate by 3/15
- Propose budget to management for review by 4/1
- Final project budget review by 4/15
- Project goals and schedule established by 5/1
- Kickoff meeting 5/15
- Complete design by 6/15
- Order all equipment by 7/15
- Install beginning 8/30
- Test and commissioning complete by 9/30
- Training and shadowing complete by 10/30
- On air date 11/1

FIGURE 2.3-6 Breaking down a project into a set of goals.

Conceptual Design

Conceptual design is sometimes confused with sketchy information about what will be done. Ideally it is the disciplined, technology-specific approach to the overall project plan developed during the business analysis and high-level planning phase. A conceptual design is not intended to be make-and-model specific, indeed it is quite the opposite. By doing a design that is not tied to a single product, one forces the discipline of understanding the technical goals in the context of workflow and signal flow without being constrained by a single manufacturer's solution.

As an example, review the goals list in Figure 2.3-6. This list does not speak about specific options for implementation but rather more general goals (e.g., ingest from tape). The conceptual design would define how that might be done; how many boxes need to be assembled to achieve that goal, how they interplay with existing systems like DTV transmission, network reception, traffic, and automation; and other functions.

Once a technical option is chosen, a high-level block diagram is created. It might show nothing more than network reception, local production, master control processing, and emission, a total of four major blocks. How these interplay must first be understood, and as each successive view of the project is developed it is useful to think of this as "zooming in" on the details. Sometimes this is accomplished by looking "back into a wire," in other words, defining an output signal and then looking back into the system to see what it takes to develop the needed signal. By doing this one gains a greater understanding of exactly what must happen in the signal path to achieve the desired output signal.

As the conceptual design develops, the project team can create a list of hardware solutions that must be selected and integrated together. For example, in an MPEG master control it might be:

- File translation software and hardware.
- Splicers to integrate local content into network delivered signals.
- Encoders as necessary for signals that are baseband.
- Keyers and MPEG switching hardware to concatenate signals.
- Servers for record and playback.

As the process moves to the next level of detail, manufacturer-specific solutions are mapped onto the plan, which in some cases will identify interfaces needed, or perhaps combine multiple features into one monolithic solution from a manufacturer, which in itself might include more than one physical device, but on the conceptual plan might be labeled simply "MPEG switcher/splicer."

Price and delivery quotes are then gathered and a spreadsheet is created that shows how the equipment budget is shaping up. If the total exceeds the target budget established in the business plan, steps must be taken to control the "creep" (slow expansion in cost and/or scope) of the project. This generally takes the form of a reiterative value engineering process in which the numbers drive the technological decision to find other alternative approaches. At times the budget would be too high, and at times the features supported too low, but at the end of the conceptual design process a "happy medium" should be reached in which the functionality is maximized and the budget maintained.

For example, the concept may have been to build a cutting-edge MPEG master control, but the conceptual design process is unable to achieve that result at the target budget with the necessary functions supported. This "bifurcates" the problem and may force retreat back to a higher level of conceptual design thinking and the use of a different approach. An analogy would be a project that started out with the goal of obtaining a house boat, but after conceptual planning the specification was for the Queen Mary. In this situation what is required is replanning based on what can be afforded instead of what was wanted.

It is important to note that the conceptual plan is never developed to the point that every detail is identified and documented. One might represent the inputs to a master control switcher, such as eight lines from a routing switcher represented by a single line on paper. Control connections would not identify the protocols and IP connections in detail, but rather provide a broad-stroke definition sufficient to know that electrically and from a software standpoint they are *conceptually* correct. This distinction is important, for the detail necessary to validate each and every connection and software interface would constitute the completion of detailed design, which is the next step. Everyone in the project must understand that some questions and detail will not be completely vetted during conceptual planning, and as a result the budget will only be approximate at the conclusion of conceptual design.

Once the conceptual plan has been validated and is within budget, it is time for a formal review. All parties are provided the complete document set, including drawings, specifications, budget analysis, manufacturers' proposals for major systems, and other pertinent information. A formal review should be conducted in which the project manager leads everyone through the project plan and each of the interest groups (engineering, operations, traffic, promotions, sales, management) is asked to comment on the outcome. Using a predefined approach a decision is then taken to either reenter conceptual design to correct flaws in the plan, or to advance to the next stage, accepting compromises to budget or technical results.

It is important to stress that by being organized about the plan at the beginning of conceptual design, and defining goals that must be met and timelines and budgets to be achieved, the designers are given clear instruction in what the outcome of their work must support. Consultants and system integrators are used to working in this manner, but internal team members must learn the discipline and methods. It is a different mind-set than people are used to having in their everyday job, and everyone will need to have patience.

Detailed Design and Documentation

With a solid set of conceptual documents as the guide, the next step needed is to turn conceptual drawings into single-line flow diagrams and plans for software interconnection and integration as well. Software plans would show the complete workflow and must define the interface between applications. For example, in a newsroom it might be sufficient to identify that the newsroom computer system provides an MOS-compliant device control interface, and the teleprompter and character generator are MOS compliant as well. But one must also carefully look to be sure that there are no firewalls that restrict communication that must happen between applications, and that the IP addressing structures that the IT department must apply are compatible with the vendor's software and hardware. These are not details that happen in conceptual design, but rather are vetted and documented in the detailed design process.

The document output from this part of the process is just short of what is necessary for construction. For example, the documents might show details of the fronts of equipment racks, identifying where all equipment will be mounted. However, during the construction phase the details of how wiring will be laced within a rack are documented to be sure installation proceeds along repeatable lines. During the detailed design, the space sufficient for cabling would be determined, such as making sure the rack depth can accommodate the wiring, but the "dress plan" for the cable bundles would take place later.

The detailed design documents should identify all electrical connections. During the process of taking conceptual plans to the next level, some interfaces will be found that were not identified in conceptual plans. This is not a fault in the project plan; it is in fact the essence of the project plan. The precise and detailed documentation of the complete system would burden conceptual design with the need to be 100% correct, slowing the process and limiting the creativity with which conceptual planning is done.

During detailed design, vendor quotes are obtained for all hardware, complete with negotiations on final pricing. It is also part of detailed design to work out a commercial purchasing plan that might include aggregating purchases with a smaller number of vendors in order to leverage quantity pricing that might be available. That effort should be collaborative between the purchasing department and the engineering staff defining the technology. By working together to achieve the goal of lowering the capital cost, the ability to include the maximum amount of technology is achieved.

As part of the detailed design one must consider the workflow impact of each and every major decision. Certainly this does not apply to, for example, the selection of patch panels, but it might apply to a decision about whether patching is desirable at all. One of the goals established in the definition of the project could have been to reduce the cost to the largest extent possible. The maintenance department may have indicated that it needs patches throughout the system. In contrast, the operations department may have indicated its need to minimize the possibility that patches are not cleared before the night shift begins, when less-skilled personnel are on duty, and thus may want to use routing switcher ports to achieve the same goals. The final vetting of the interplay of all of these competing project inputs happens as the final technical model is detailed and each member of the team sees the effect of all of the perhaps conflicting inputs from the conceptual design process.

Stage by stage the documents become more detailed and complete. Team members dealing with software interface issues have analyzed vendor responses, and in some cases acted as moderators for conversations between vendors who need to solve interface issues that are outside of the control of the project team. An example of this is the connection between traffic and automation software. The interface is not complex, but each traffic implementation is a bit different, and either the traffic or automation vendor needs to write interface code for each and every implementation.

By now, the document set has grown to include a range of documents with substantial detail. Drawings have expanded, with floor plans done in conceptual design, perhaps expanding to 3D renderings of, for example, the new master control room with a clear indication of the sight lines and appearance of the final room. Then, the time has come for another comprehensive review.

At the conclusion of the detailed design phase another meeting is held, where each and every document is reviewed and scrutinized to be sure it is accurate and complete. Every detail must be verified, which is often best done by individuals a bit removed from the detail being reviewed. This allows proofreading in a way that individuals will spot the mistakes or incongruities produced by others. The marked-up drawings are then corrected and readied for the final push to implementation.

At this point the budget should be complete and accurate to a high degree. Vendor quotes and negotiations are entered into the project budget and cash flow projections are made so that funding can be in place when invoices are received later in implementation.

During the detailed design process implementation planning begins. Quotes on labor for installation are assembled, and the final schedule is fed into the project plan with all known data identified that might affect the overall schedule. As the date of implementation begins to approach crew schedules are blocked to assure the labor is prepared and ready when the go-ahead is given.

Late in the detailed design process the details for construction are added to the drawing set. Wire numbers, cable run schedules, wiring dress plans, and all other details are documented in preparation for handing off the design to the implementation team. The project manager queries the implementation team and leads a second review of the design documentation to verify that all of the detail they need for installation of the systems has been entered on the drawings.

If software system integration is involved at this stage, all parties are engaged in a conversation about their readiness to deliver the goals of the project. Proof-of-concept work may be required to establish that the user displays and software interfaces will be sufficient and workable.

At the conclusion of detailed design the final decision to complete the project is made. Until this point the project is on paper only, and the only dollars expended are those necessary to complete the design. At this point, "hard" dollars are about to be spent and the team should gather for one last meeting at which all parties agree to move forward with implementation.

Construction

Now the exciting part begins for others on the team who have waited while the design team handled the details and planned for the implementation of the project. Like design, implementation requires specific skills. The installation supervisor (IS) or lead installer takes on the role of managing a myriad of details at the most fundamental level about the physical installation of the technology. It is the responsibility of the IS to ensure that installation materials (wire, connectors, cable ties, cable labels, rack screws, blank panels, and other "expendables") are on-site at the right time. That responsibility includes making sure tools, ladders, temporary desks or work benches, and any other tools of the installation trade are accounted for in the planning.

As the date of the installation draws near, the IS reviews the drawings and wire run lists, gleaning details about the work and planning the detailed actions the crew needs to take to get the project under way and maintain the schedule. The IS works with the project manager to make sure the work breakdown in the project plan is complete and workable. Once installation begins the IS may take charge of the crews as the on-site supervisor, moving labor to the most important part of the project on an hourly basis. A good IS can accelerate the project by managing and motivating the crew, or can slow things down by taking a less aggressive approach.

Often construction begins by prefabricating cables before the facility is ready for the final installation push. This can accelerate a project, much as "fast tracking" is done in the construction industry. If sufficiently accurate detail on cable assemblies is available, both ends of the cable may be completed without having any racks or hardware available. For instance, when wiring a routing switcher to a patch panel the length of all cables can be specified as a standard length for timing purposes, therefore, they can be premade. If space is available for an off-site preassembly of the system, it is possible to complete a great deal of work without having the actual architectural facility available. This can include full checkout and software interface verification at one end of the spectrum, or just partial confirmation of basic functionality at the other end.

In modern facilities with high bandwidth cabling it is important that the cable manufacturer's specifications for installation are strictly observed. Many thoroughly competent engineers are surprised to hear that neat and extremely compact cable bundles may actually represent installation problems. When cables are compressed, the impedance can be affected. The most obvious case of this is when someone steps on a cable, partially crushing it. But more insidious is the routine practice of using cable ties in a regular pattern to bundle cables. While a bundle of cables, with a cable tie every 6 inches and cables "combed" straight as a stick and pulled into close contact with the side of a rack, look impressive, doing this can produce standing wave patterns that degrade return loss and affect the ability of cables to pass signals without impairment. Irregularly spaced ties applied with just enough force to keep bundles neat are a more appropriate strategy. With both fiber and coax cables it is important to observe minimum bending radius specifications as well. With twisted pair networking cables, it's a good idea to have at least one person on the installation team attend certification classes on installation technique.

It is also valuable to plan the installation so that subsystems can be finished in an orderly progression. This will allow the beginning of testing and commissioning earlier, shortening the length of time the project takes. It will also allow more time for corrective action if technical problems are discovered with the design when commissioning commences.

At the conclusion of installation the final, and often the most difficult, phase of installation occurs. Cutover to new systems in an existing facility requires careful orchestration of the move and connection of existing equipment as well as the decommissioning of old systems and turn on of new. It is at this time that issues often arise if planning has been incomplete.

Part of the final review of design issues must be to get full agreement on any requirements for plenum-rated cable. Plenum should be avoided where possible due to cost, and additional installation time required due to the difficulty of working with cable that is much stiffer and difficult to handle. However, don't make the mistake of ignoring plenum requirements. It will come back later as a poor choice when the building inspection fails due to the use of nonrated cable where plenum rated is necessary.

Configuration and Testing

At some point near the end of the installation phase of the project, the process of turn-on and checkout of the technology can begin. There are three distinctly different activities that must occur—testing, configuration, and commissioning:

- *Testing.* Hardware is turned on and basic functionality is confirmed. This verifies that cables are in the right place and continuity has been achieved. Signal quality, for instance SMPTE 259 format verification and error rate, testing is performed. Thorough records of testing should be kept. This makes it possible to demonstrate to manufacturers that the wiring is not faulty, and can serve as a benchmark for later facility monitoring over long periods of time.
- *Configuration.* After basic testing is done it is time to configure hardware and software to accomplish the tasks that each item was purchased for. In isolation each device is set up to the correct modes of operation, software is loaded and current versions verified. Basic configurations are recorded to establish a baseline of where things started. This is really useful when software interfaces must be checked with other devices later.
- *Commissioning.* In this final phase the entire facility is checked end-to-end and all signals are proven out and software interfaces between devices are exercised and tested. In modern facilities this may well prove to be a much more lengthy process than anticipated. At the conclusion of this phase, the system is "accepted," and hardware is ready for use on-air, but station staff must now be trained to operate what may be a radically different operation than they are used to.

Having the best test instruments available for turn-on and checkout is important. With analog video systems a waveform monitor and vectorscope would suffice. With compressed digital signals and high bandwidth data circuits more sophisticated test equipment is required, which may include:

- Digital waveform monitor (SD and HD capability)

- MPEG- and ATSC-format test instruments (with PSIP capability)
- CAT 5/6 and fiber test instruments
- SMPTE 259 and 292 test instruments with error and format checking
- AES, Dolby AC3, and Dolby E audio test instruments
- Acoustic test instruments

Training and Support

With the increase in hardware complexity introduced by new systems it is important that training be part of the program starting at the beginning of the project and continuing to the end. For example, a new master control installation may replace a manual operation with VTRs and automated cart machines, and replace decades-old switchers with grafted-on boxes for push back, graphics, weather and school closing crawls, and EAS. The new operation will likely be based largely on computers and automation. The mind-set is quite different and the operations department will need considerable help to get up to speed on the new systems.

This phase should not be left to manufacturers to train on only their part of the system. The detailed engineering phase likely has considerable flexibility and the ability to work around likely failure modes embedded in the completed facility. That kind of local design training cannot be done by anyone except the design and installation team. This phase should be treated every bit as formally as the previous steps, including providing time in the project plan for completion of training on all aspects of the facility.

It is important to coordinate the training in a sequence that allows the full staff to be part of the process. If timing will not allow extensive training for all, then a "train the trainer" approach might be chosen. In any event, formal processes should be used and thoughtful time put into writing down the aspects of the project that project staff must know, which will result in less confusion in the long run. Take time to prepare graphics for presentation, which might be no more complicated than hanging drawings on the wall, or perhaps creating electronic slide presentations to clarify a sequence in advance of delivery. Ask manufacturers to provide a syllabus for training purchases, and ask for multiple copies of all handouts.

Create a binder with training materials kept in an organized fashion. There will be new employees in the future who can benefit from the same information, so storing it in a way that it will be useful months or years later is appropriate.

After the training is done, or perhaps during the training period, shadow operations should be considered. This is particularly important in the case of new automation and control systems in master control. Shadowing can include recording programming on VCRs or servers or streaming media recordings of output signals. A split-screen arrangement helps to keep the time code superimposed over the signals to allow for linking video and audio to the printed log or script. When problems are noted this will be a useful tool for figuring out what happened and how to take corrective action.

DISASTER PLANNING

Lastly, especially in the case of a new master control room project, create a disaster plan and train personnel to use it before first commercial use of the new facility. Document the disaster plan for training new hires. Regularly exercise this plan and be sure everyone knows what it contains and what it accomplishes.

CHOOSING IN-HOUSE OR OUT-OF-HOUSE RESOURCES

So far this chapter has concentrated on internal planning without looking at how outside resources might augment a project and allow things to run smoother or faster. It should not be considered a sign of a weak internal staff that outside resources are brought in to help complete a complex project. To the contrary, a strong and effective project team will complement the needed skills with outside resources and make recommendations as to what gaps they have in their ability to complete the complex aspects of the project.

Key outsourcing skills include project management, design consulting, documentation, installation supervision, and general installation labor. Outsourcing all or only a portion are options. As discussed earlier, the most important single person on the project team is the project manager. If there is no person on staff who has both the time and the skills to function as the project manager, then look outside the organization. There are professionals who work as freelance project managers. It is important that they are not making the technical decisions, and so their credentials should be looked at in light of the particular skills needed in an effective project manager. Those skills include an ability to communicate and understand the complex interrelationships within the project, experience as a project manager (hopefully in a similar project type), strong leadership ability, and analytical skills to control the finances of the project effectively. Often the chemistry internally is such that an outside project manager can be more effective than a similarly skilled person internally. This arises from the lack of competition between full-time managers for control of the agenda. An outsider comes in with no preconceived notions of how a particular project should be executed.

Outside technology consultants and designers are also available on a contract basis. Some projects are done by using a consultant to help define the goals and explore the technical options. Then, after this preliminary work is complete, their role may switch to a strictly advisory role and a request for proposal (RFP) is created for delivery to a system integration company, which then carries the work on to completion. The consultant may remain involved with either an outside system integration company or with internal staff to the extent that his or her skills are useful to the project completion. On large projects that span a period of many months, a team of temporary employees may be assembled who then act as full-time staff, but coordinate the work of an outside system integration company.

It is important to recognize how temporary staff fit into the organization as well as the project. They are no less important than full-time permanent staff, and in some ways their unique abilities will allow the rest of the staff on the project to grow (professionally) over the length of the project.

Working with a Consultant

An outside consultant can add much to a complex technology project. One must be careful to check credentials, define the scope of work to be accomplished, and negotiate a fixed price for the entire period of time the consultant is to be involved.

There are specialists in many areas whose skills are narrow, though important to the outcome of a project. For instance, a studio and lighting consultant might be able to help specify both lighting and rigging in the studio, as well as light a news set in the new studio. But expertise in studios is not all-encompassing and may not, for example, include acoustics. A good noise-control specialist can explain how to prevent noise from traversing ductwork and how to build acoustic barrier walls, but might not be of much help in designing the interior acoustics in a control room or studio. By the same token a studio design acoustician might be suitable for major sound recording stages but might not be able to deal with isolating the noise of a large uninterruptible power supply (UPS).

It is common to think of consultants in the broadcast business in the context of technology consultants but experience in designing major network origination centers does not necessarily transfer to small-market television transmitter installations. One should find a consultant whose unique skills are an appropriate match to the specific project. He or she may be able to bring along additional expertise as a subcontract to fill in the unusual gaps in knowledge that relate to, for example, the studio acoustics and lighting systems. It is, in fact, quite common for nationally recognized experts to have associations with other subject-matter specialists.

Do not be afraid to engage a consultant in a minor advisory role. Bringing a consultant into the project on an infrequent basis may well be enough to fill in gaps and point to approaches not identified by a less-skilled design team, internal or external.

Working with an Integrator

If the internal team is too busy running day-to-day business then it might be advisable to bring in a design and integration company as a "turnkey" (i.e., "one-stop shop") supplier. These firms have existed for over 25 years, and some have grown quite large. They can offer unique advantages and may well reduce the cost and time to complete the project to such an extent that one should consider giving them broad latitude to run the project.

It is often incorrectly thought that an integration firm that sells hardware is both a conflict of interest and not the most cost-effective way to purchase hardware. There are cases where integration firms have "pushed" one solution over another due to their ability to make a greater profit by suggesting one manufacturer's hardware. However, for every case of that inappropriate behavior one finds, there seem to be many more cases where the breadth of experience and knowledge of many product lines allows a good firm with strong engineering skills to select the most appropriate solution regardless of commercial factors.

Integration firms are resellers in one sense. Resellers have an important advantage over end users: they buy hardware in large volume from many suppliers annually. This gives them considerable leverage to assist in negotiating the best possible deal. It is not inappropriate for them to make a profit on that hardware, and indeed they may be able to sell hardware for several percentage points less compared to buying the hardware elsewhere, while themselves making a profit.

Selecting an integration firm is a complex mix of reputation, cost, geographic location, specific technical skills, and interpersonal chemistry. Due-diligence reviews can take considerable time to conduct. One approach to the selection process is to write a request for information (RFI), which asks for general information about the firm, skills, experience, references, and a description of approach to project planning and execution, such as those listed in Figure 2.3-7. After reviewing the responses of perhaps half a dozen firms, pick one or two to interview. Select one that seems to be a good fit and negotiate a contract that achieves the project's budget targets. This can be done by opening up and telling them how much money is available to get the project done and then using a value-engineering approach similar to what was described early in this chapter to achieve maximized results.

Another approach often used is the request for proposal (RFP). This allows the customer to look at completely fleshed out proposals with technology choices from several potential vendors. However, writing an effective RFP is a difficult process. It may require a significant portion of the conceptual planning be completed to allow a comprehensive RFP to be written. Responses are only as good as the RFP on which they are based. To the extent that the RFP is complete and exhaustive it will be possible to get easily compared responses. But if the RFP is more general and asks firms to "use creativity and best practice" to recommend a solution, the wide range of responses received may lead to a decision to hire a firm based on price when the least-expensive solution is perhaps the one that will produce an inferior result and an undue number of change orders during the project.

A factor not usually considered by end users is the cost to the system integration firm for responding to

Outline for Integration RFI
Introduction and description of project

Name of Firm:
Address:
Years in Business:
Number of Employees:
Number of Design Engineers/Consultants:
Number of Project Managers:

Number of Current Projects:
Approximate Dollar Value of Average Project:
Average Project Time Scale:

Identify Project Specialties:

Provide a narrative on how you would approach the design and implementation of the project.

Provide an estimate of project costs based on the above intended approach, as well as a suggested time line necessary for successful implementation.
Describe your approach to project management and controls.

FIGURE 2.3-7 Outline for an RFI that would be useful when seeking an outside integration firm.

RFPs, which can become a substantial overhead cost for that firm. Inevitably bids to RFPs will be lost, and as a result firms must recover costs of preparing these unsuccessful bids by raising the cost of subsequent proposals. By selecting a firm with whom to negotiate a contract, their cost of sales is lowered, which allows them to minimize the cost to you. Multiple rounds of negotiation or successive rounds of bidding based on an RFP may reduce the price, but at a cost that may well come back in less flexibility to deal with changes in the project without affecting the final price through change orders.

When selecting an integration firm, be sure to treat it just like a hiring decision. Insist on meeting and interviewing the project manager, lead engineer(s), and other key personnel who will be part of the project. Just trusting the sales person will not be enough, for the skills and approach of the professionals who will be working on the project on a daily basis are important to the success of the project. Ask them to bring sample reports from other projects for your review, sample drawings of a similar project, and the names of other customers who can act as references. Ask them for their worst experience, and insist on knowing how they salvaged a good outcome from adversity.

In the end a good integration firm can become a partner for a long time, through many projects. The effect proceeding in this fashion is to reduce the internal labor force and leverage their considerable skills to the project's benefit. The more they know about an organization's business over a long time period the better they can understand the proposed project and produce superior outcomes at minimized cost.

Interfacing with Contractors and Architects

The interface between the technical project team and others is an aspect that often is overlooked at the onset of a project. While everyone wants to put their two cents into architectural planning, a well-run project may be better off by letting the architect take the lead on issues related to construction.

During installation there will be considerable detail to be worked out between the project manager, installation supervisor, lead engineer, and a contractor renovating or building new space. For this to run smoothly it is important that planning be detailed and thorough and that all dependencies in the project schedule that are affected by the work of others are documented. By placing the appropriate milestones in the technical project plan it is possible to see the interplay of any delays by either construction or installation on the overall completion of the project.

The key to avoiding unnecessary and unforeseen delays is communication. This is where the thorough communication plan, put into place at the beginning of the project (and maintained by the project manager), shows its true value. By documenting issues, who is responsible for resolution, and the expected date of the completion of that resolution, the project manager makes it possible for all parties to clearly understand their individual responsibilities and the effect the resolution of these issues will have on the total project. Communication is particularly important when parties who are not managed by a single entity are involved, for in this case there will likely be more than one manager.

Architects are highly trained professionals and many have experience in broadcasting. However, the peculiarities of broadcasting are not in the normal trade of many competent firms or practitioners. In such a case it is particularly important for the technology team to work in close harmony with the architect. It becomes a subplot in the project to explain the particular details that relate to the design of studios, control rooms, electrical grounding, space-planning needs, and the impact of design on workflow. It is more difficult to permit an architect to do his or her craft when he or she has little experience and the technology team members dominate the conversation, which is all too often the case. It takes particular vigilance on the part of the project manager to make sure the entire team focuses on listening and providing guidance that enhances the ability of the architect to produce a superior design in such circumstances.

Equal vigilance is needed regarding the details of construction. Many construction companies have little experience in the development of broadcast facilities. Often the small details, such as the construction of acoustic isolation walls, with details seldom used in

other types of construction are not well understood. The project manager and the rest of the design team need to spend additional time monitoring the smallest details to be sure the completed facility is precisely what they want and what was documented by the architect. Don't be shy about asking the architect to step in and manage the minutiae that if not executed properly will hurt the completed facility. This level of involvement is normally an architect's job as part of the standard form contract (American Institute of Architects, AIA, issued standard contract). In any event the architect often is a neutral party who can help effect solutions to the problems.

There are firms with specialization in broadcast architecture. Often they provide coordination at the schematic design and programming phases of a project, with a local firm completing the detailed drawings for construction. In such a case the local architect would normally participate in all meetings and site reviews and would act as the go-between when issues arise between the owner and the contractor. The consulting architect becomes an advisor to the project who is available for specialized problem solving without paying the expense for flying them into town for regular meetings that can be handled by the local firm.

SUMMARY

Good project management and thorough planning are vital in the execution of design and installation of systems in a broadcast plant. This chapter has provided an overview of the various aspects of project management, planning, and execution.

Bibliography

General References to Project Management Techniques

Bruce, Andy, and Langdon, Ken. *Project Management*. New York: Dorling Kindersley, 2000.

Dinsmore, Paul C. *The AMA Handbook of Project Management*. Bladwells Book Services.

Kerzner, Harold. *Project Management: A Systems Approach to Planning, Scheduling, and Controlling*. New York: Wiley, 1998.

Williams, Paul B. *Getting a Project Done on Time: Managing People, Time, and Results*. New York: Amacom, 1996.

Wysocki, Robert K., Beck, Jr., Robert, and Crane, David B. *Effective Project Management: How to Plan, Manage, and Deliver Projects on Time and within Budget*. New York: Wiley, 1995.

Project Management Software

BaseCamp (Web-based tool) (http://www.basecamphq.com): A good tool for simple projects. Very affordable and requires no software to be loaded on the desktop since it is Web based. Monthly and license fees for full features (shareware, $150/month, unlimited users). Few graphics tools.

EasyProjects.net (web-based tool) (http://www.easyprojects.net/): Somewhat more full featured than BaseCamp, but still less complex than MS Project or Primavera. Supports simple Gantt charts.

Microsoft Project (standard, professional, and enterprise versions) (http://office.microsoft.com/project): MS Project is a full-featured product with Gantt charts and other representations of the project data. It manages schedules, resources and conflicts, timesheets (enterprise edition), and will interface to various accounting packages.

Primavera Project Planner (P3) (http://www.Primavera.com): Like MS Project, this is a full system intended to support complex and large projects and many projects at once. By reputation it requires considerable training to use effectively. It is sold by VARs in most circumstances and may not be suitable for a single project unless it is quite extensive. A contractor version is available as a shrink-wrapped package.

Software Training Text

Stover, Teresa. *Microsoft Project Version 2002 Inside Out*. Microsoft Press.

CHAPTER

2.4

Human Exposure to Radio Frequency Energy

WILLIAM F. HAMMETT AND ROBERT D. WELLER

Hammett & Edison, Inc.
Consulting Engineers
San Francisco, California

INTRODUCTION

The use of electromagnetic energy for communications has increased dramatically since the last edition of the *NAB Engineering Handbook* was published. This chapter updates previous versions, including new and revised standards in force, emphasizing the requirements on broadcasters, and describing methods for measuring radio frequency (RF) energy. Three issues important to broadcasters are still in development:

- The International Committee on Electromagnetic Safety (ICES) has revised IEEE Standard C95.1, further harmonizing it with the standard published by the International Commission on Non-Ionizing Radiation Protection (ICNIRP). IEEE Standard C95.7, specifying recommended practices for RF safety programs, was also recently published. Although these standards are now "in force," regulatory agencies have not yet taken any position on their adoption or use.
- In addition, at the time of this writing, the FCC is considering the comments it received on a 2003 Notice of Proposed Rulemaking, ET Docket 03-137, to update its regulations concerning human exposure to RF energy.
- The federal Occupational Safety and Health Administration (OSHA) has apparently abandoned an internal proposal to revise administratively long out-of-date regulations concerning RF exposure in the workplace, but consideration is being given to rule changes under a public process.

These three ongoing activities mean that the current regulatory state is in flux, so the reader is urged to check, perhaps using the web resources given at the end of this chapter, for the latest information.

STANDARDS

The first standards limiting human exposure to RF energy were promulgated in 1953 by Bell Telephone Laboratories to protect against reduction in visual acuity [1]. At that time, this adverse effect was observed at a power density of 100 mW/cm^2 (1,000 W/m^2), and the exposure limit was set 30 dB below that level. A succession of private and public standards appeared between 1954 and 1966, when a uniform exposure limit of 10 mW/cm^2 (100 W/m^2) was established as a standard by the United States of America Standards Institute. Exposure limits were uniform with frequency until 1982, when the standard was changed to reflect research demonstrating that the typical human body exhibits a strong resonance at frequencies near 70 MHz owing to its physical dimensions. In 1985, the FCC adopted ANSI Standard C95.1-1982, one of the first standards that exhibited frequency dependence, to be used for evaluating human exposure to RF energy.

Since 1986, there has been a trend toward two-tiered safety standards limiting human exposure to RF energy. Today, nearly all of the prevailing standards are two-tiered, typically setting more restrictive limits on public exposure than on worker exposure. The reasons for adopting different limits are largely nonscientific. That is, the general public should not be regarded as being more susceptible to injury from RF exposure than workers. Rather, workers are allowed higher

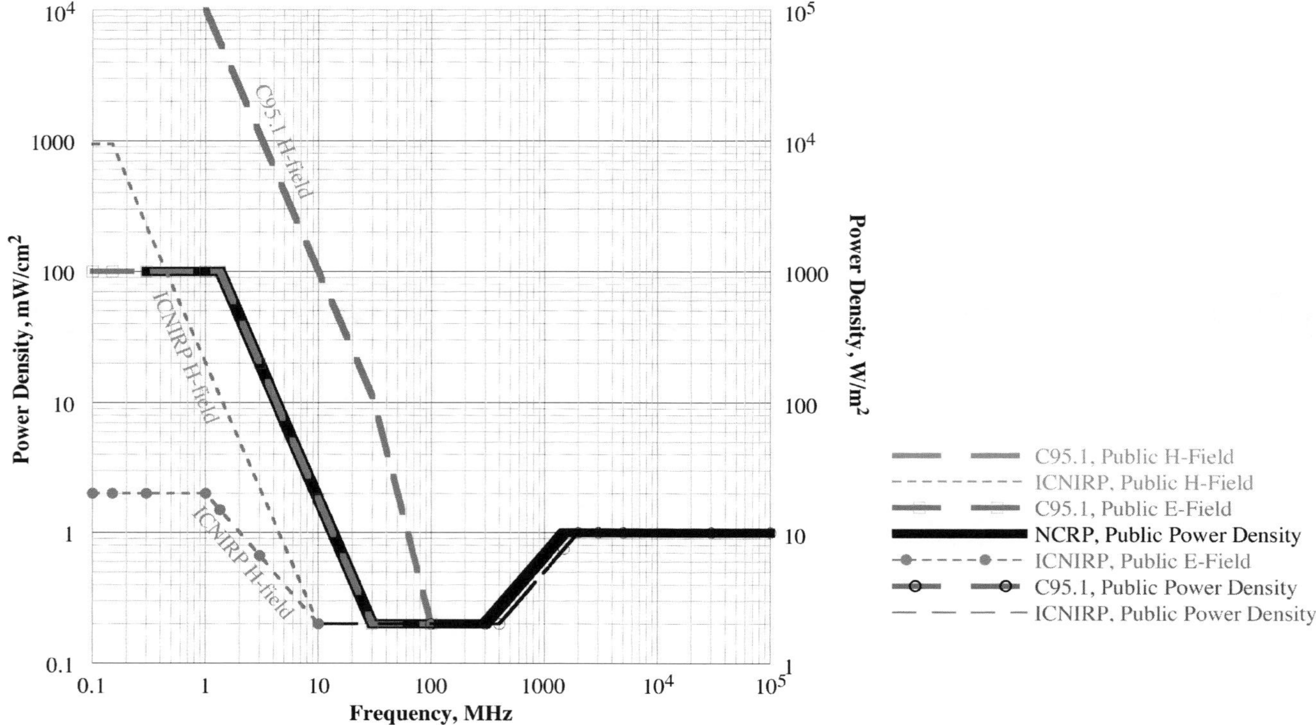

FIGURE 2.4-1 Public exposure limits.

levels of exposure, because they are aware of their exposure, are educated about its risks, and are able to control the level and extent of their exposure. The more restrictive public limits do afford some measure of protection against nonbiological hazards, such as interference with implanted pacemakers.

Summaries of the NCRP-86 power density guidelines, the C95.1 exposure limits, and the ICNIRP limits are shown in Figures 2.4-1 and 2.4-2 for public and occupational exposures, respectively. Contact and induced body current limits are given in Table 2.4-3 at the end of the chapter.

NCRP Report No. 86

In 1996, the FCC adopted the recommended power-density limits specified in a 1986 report published by the congressionally chartered National Council on Radiation Protection and Measurements (NCRP) [2]. The NCRP report covers 0.3 MHz to 100 GHz, which includes all of the broadcast and broadcast-auxiliary bands, except for long wave (below 300 kHz). Although other standards, such as C95.1, were more recent, the Environmental Protection Agency (EPA) and other U.S. health agencies endorsed the scientific basis of the NCRP guidelines as being superior to those of other standards. Rather than updating its 1986 guidelines, the NCRP anticipates publishing by 2007 a report that will provide a critical evaluation of RF radiation exposure guidelines promulgated by other national and international organizations. An important component of this report will be the development of a risk-based framework for future RF exposure guidance in occupational and public settings.

ANSI/IEEE Standard C95.1

The C95.1 standard is part of a family of standards (including techniques of measurement, terminology, RF safety programs, and extremely low frequency (ELF) and static field exposure safety levels) developed by the International Committee on Electromagnetic Safety (ICES), under sponsorship of the Institute of Electrical and Electronics Engineers (IEEE) [3]. Standard C95.1 covers 3 kHz to 300 GHz. Although the current edition was published in 2005, the federal OSHA applies the 1991 edition (including revisions dated 1999) of the IEEE standard (also recognized by the American National Standards Institute, or ANSI) to workplace situations, including radio stations and tower sites. Under IEEE policy, standards must be reaffirmed, rewritten, or cancelled every 6 years, so the 2005 edition will likely be current until at least 2011.

Safety Code 6

In Canada, Health Canada (formerly the Health and Welfare Ministry) [4] adopted Safety Code 6 (SC-6, currently published as the 1999 edition). SC-6 is a standard very similar to C95.1, but is largely practical in scope, providing exposure limits, safety procedures,

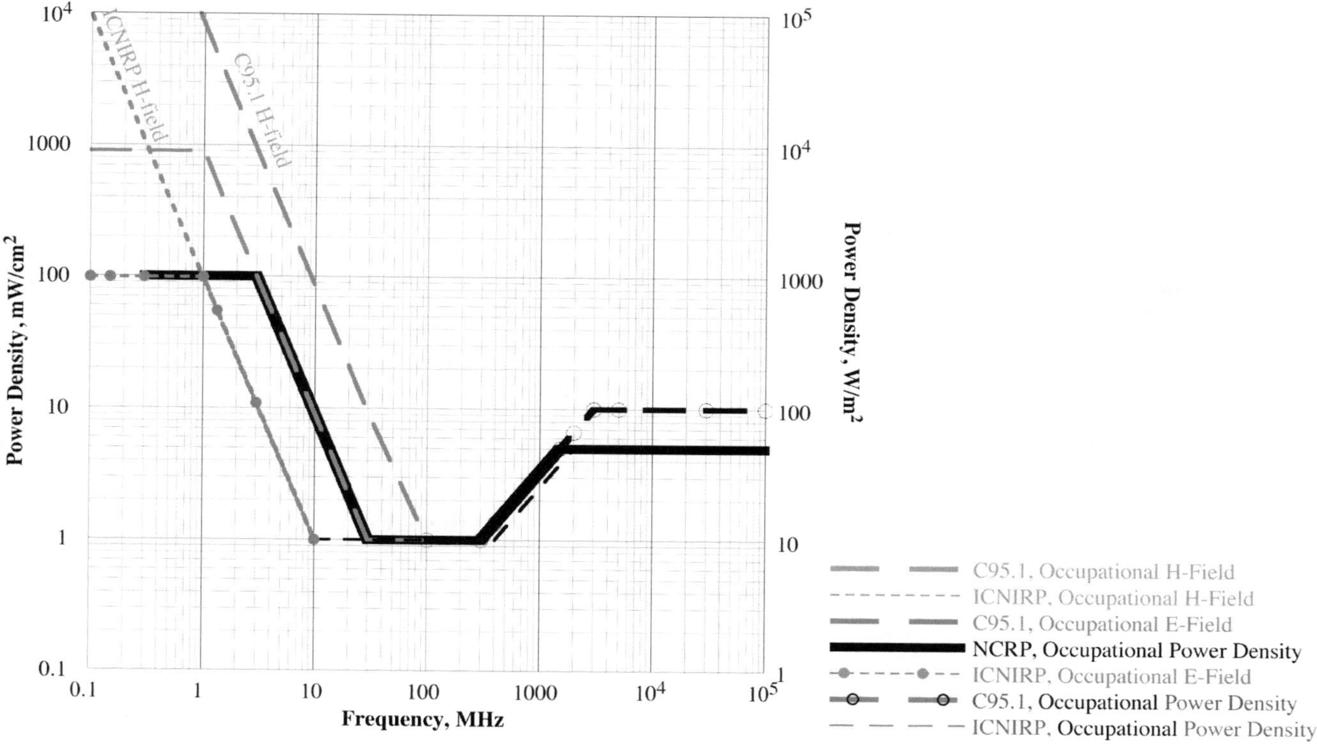

FIGURE 2.4-2 Occupational exposure limits.

and specifications for signage, while leaving lengthy literature reviews to other publications.

ICNIRP

In most countries other than the United States and Canada, the 1998 recommendations of the ICNIRP [5] are used to assess human exposure to RF energy.

Unit Conversion

The NCRP power-density exposure limits are specified in terms of equivalent plane-wave power densities, while the other standards express the limits in terms of both E- and H-fields (typically below about 300 MHz), and power density (typically above 300 MHz). Since most practical equipment measures either the E- or H-field only, it is helpful to define the relationship between these various exposure metrics. The conventional relationship is limited *only* to free space locations (where the intrinsic impedance is approximately 120π ohms), which are in the far field of the antenna (so that the E- and H-field vectors are orthogonal). In SI units, the relationship is:

$$\left|\vec{S}\right| = \frac{\left|\vec{E}^2\right|}{120\pi} = 120\pi\left|\vec{H}^2\right| \quad (1)$$

where $\vec{S}$ is the power density (Poynting) vector in W/m^2, $\vec{E}$ is the electric field vector in V/m, and $\vec{H}$ is the magnetic field vector in A/m. In traditional units, the relationship is

$$\left|\vec{S}\right| = \frac{\left|\vec{E}^2\right|}{1200\pi} = 12\pi\left|\vec{H}^2\right| \quad (2)$$

where $\vec{S}$ is in units of mW/cm^2.

FCC Guidelines

As discussed above, the FCC presently uses the public and occupational power density guidelines recommended by the NCRP. At typical broadcast sites, compliance with the power density limits is determined on a spatial average basis. It is important to note that, although most standards include protection against shock and burn through established limits on induced body current and contact current, the FCC does not *presently* enforce these current limits. Enforcement of those limits is one topic of ET Docket 03-137, so this policy may be subject to change. Additionally, OSHA *does* enforce those limits in the workplace. Thus, it is important to ensure that significant RF potentials and currents are not created on guy wires, fall-arrest cables, crane cables, metal fences, or other locations that may be subject to human contact.

OSHA Guidelines

The OSHA requires employers to "furnish to each employee a place of employment which is free from recognized hazards that are causing or are likely to cause death or serious physical harm to its employees" [6]. Additionally, some 26 states have OSHA-approved plans and have adopted their own standards and enforcement policies. For the most part, these states have adopted standards that are identical to federal OSHA. However, some states have adopted different standards applicable to this industry or may have different enforcement policies, so review of state OSHA regulations is recommended.

Some of the federal OSHA regulations are considered to be out-of-date, and the exposure limit specified in its regulations (10 mW/cm^2 or 100 W/sq. m) is not enforced. Additionally, the regulation specifying the design of an RF warning sign is also out of date. Instead of providing enforceable, up-to-date regulations, OSHA applies a complex patchwork of policies and interpretations that are available from its website [6].

Lockout/Tagout

In addition to exposure guidelines, OSHA applies a "lockout/tagout" requirement in many situations involving maintenance and servicing of high-power RF transmitting equipment. For example, climbing a TV transmitting antenna to replace a top-mounted aviation warning beacon would require that the associated transmitter be shut down. To prevent unexpected activation of the transmitter while the climber is aloft, the circuit breaker feeding the transmitter should be locked (using a padlock) into the off position, and a warning tag placed to indicate that the transmitter may not be operated until the lock and tag are removed by the person who installed them [7].

ESTABLISHING A CONTROLLED ENVIRONMENT

The first step in considering whether or not an area complies with FCC or other guidelines is to determine whether it is a controlled or uncontrolled environment. Controlled environments are those where persons within are exposed under known conditions and have knowledge of and control over their exposure to RF energy. However, the less-restrictive occupational exposure limits cannot generally be applied *unless* access to the area in question is controlled by fencing, locked passage, or other measures. An educational program or worker training may also be necessary. In limited circumstances, the mere posting of warning signs may be sufficient, but most commonly, access to areas where the public limits may be exceeded must be restricted by a physical barrier and all persons working within must understand the levels of potential RF exposure and the nature of the environment.

Fencing

Fencing is *required* under FCC rules around series-fed AM towers to prevent people from being injured through direct contact with energized components. For liability reasons, it is preferred that *all* tower bases be fenced to prevent climbing into areas having high RF fields and to protect unauthorized climbers from falling. The fence should provide an effective barrier and be difficult to climb. For example, a few strands of barbed wire that can be easily climbed through probably are not adequate. A 7 ft chain-link fence with a locked gate almost certainly is adequate.

Near AM towers and some other towers, use of nonconductive fencing materials may be preferred to avoid potential shock and contact current hazards, and to reduce the possibility of intermodulation distortion created by "rusty joints" as the fence ages. However, nonmetallic fences are often less effective as physical barriers and may be subject to rapid degradation by sunlight (e.g., ultraviolet rays), so there is often no choice but to construct a fence of metal. A metal fence should be adequately grounded and electrically continuous. This may mean that jumper cables be used to bond the fence to the gates and posts.

Signs

RF exposure warning signs should include at least the following components:

- International RFR symbol
- Appropriate signal word (Danger, Warning, Caution, or Notice)
- An explanation of the hazard (exposure to RF energy)
- Instructions to avoid the hazard (e.g., "Do not climb tower while antennas are energized")
- Contact information

Commercial warning signs are available for a variety of circumstances [8] but often custom signs will be required. Signs should be posted at locations where they can be read without entering a high-RF area, and there should be a sufficient number that they will not be missed because the area is being accessed by other than the main entrance. If non-English speaking persons (including subcontractors and trespassers) are likely to be in the area, signs should be posted in other languages.

The choice of the appropriate sign depends on the severity of the hazard. ANSI Standard Z535.2-2002 specifies four "signal words" that designate an appropriate degree of alerting, as follows:

- *DANGER.* Indicates an imminently hazardous situation that, if not avoided, *will* result in death or serious injury.
- *WARNING.* Indicates a potentially hazardous situation that, if not avoided, *could* result in death or serious injury.
- *CAUTION.* Indicates a potentially hazardous situation that, if not avoided, *may* result in minor or moderate injury.
- *NOTICE.* Indicates a statement of policy related to the safety of personnel or protection of property.

TABLE 2.4-1
Recommended Categories for RF Safety Programs and Example Actions Required

Category	Characteristic	Example Actions Required
1	No areas above *public* limit	None
2	Public limit exceeded, but no *accessible* areas above *occupational* limit	Signs plus access control
3	Accessible areas above occupational limit, if mitigating controls not applied	Above item 2 plus training program
4	Accessible areas above occupational limit	Above item 3 plus lockout/tagout, physical barriers, output must be reduced to achieve category 1, 2, or 3

IEEE Standard C95.7 (2005 edition) offers recommendations concerning RF safety programs, including signage. This new standard divides RF environments into four categories, as described in Table 2.4-1.

The minimum letter height recommended in ANSI Z535.2 is 1 unit for every 150 units of viewing distance for the signal word, and one-half that height for the rest of the sign text. When visibility conditions may not be favorable (e.g., poor lighting), the letter height should be at least doubled. Examples of some suitable signs are shown in Figure 2.4-3.

Other Factors

In 1986, the FCC published guidance to broadcasters in the form of a list of situations that are likely to arise where the public may be exposed above the applicable (public) limit, and an explanation of the mitigation required. Those situations are paraphrased as:

- On tower exposure (above ground). If the tower is marked by appropriate warning signs, it may be assumed that there is adequate protection of the general public.
- Ground-level exposure in a remote area. If the area of concern is remote and/or there are natural barriers to access *and* marked by appropriate warning signs, it may be assumed that there is adequate protection of the general public. It is recommended that fences also be used where feasible.
- Ground-level exposure in an accessible area (not remote). Fencing and signs are required.
- Exposure in an accessible area (not remote), where access cannot be restricted. An environmental assessment must be filed with the FCC to determine what actions are required.

Significant Contributors

The FCC recognizes all stations contributing more than 5% of their exposure limit in an area that exceeds the standard as being equally responsible for bringing that area into compliance. Calculations demonstrating that a station should not contribute more than 5% of its exposure limit do not provide a "safe harbor," if field measurements demonstrate a contribution above 5%. There is also some day-to-day variation in relative contribution levels. It should *not* be assumed that other stations will take care of problem areas. All stations may need to be involved in ensuring compliance.

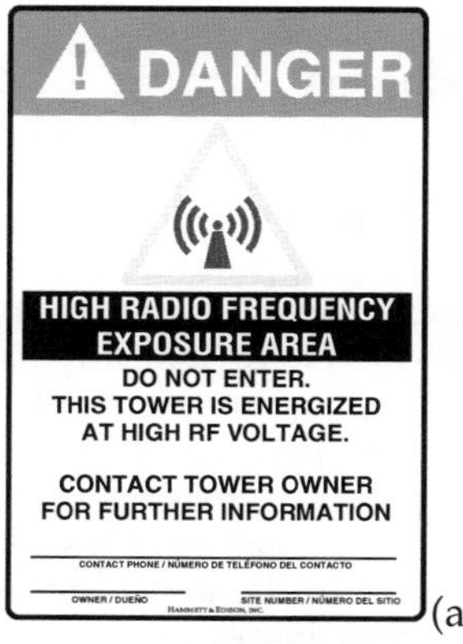

(a)

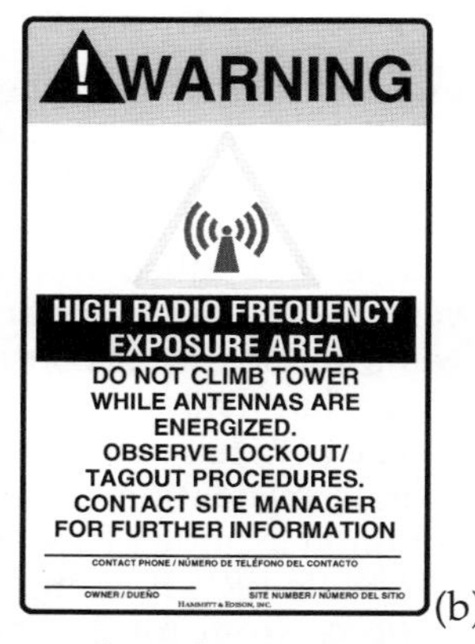

(b)

(c)

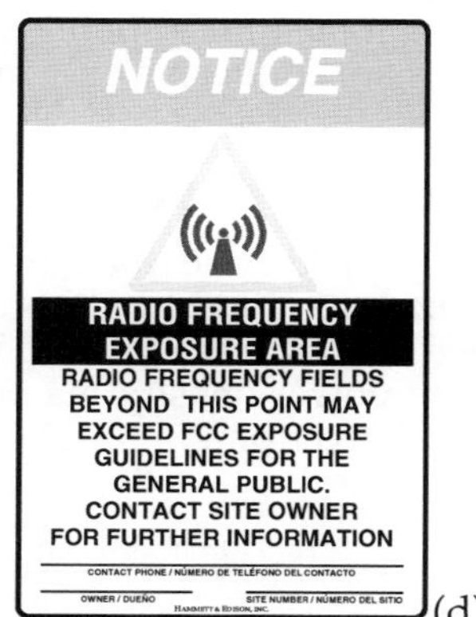

(d)

FIGURE 2.4-3 Examples of suitable RF hazard warning signs: (a) Danger, (b) Warning, (c) Caution, and (d) Notice.

CALCULATING COMPLIANCE

In many practical situations, a determination of presumptive compliance with the prevailing standards can be done by calculation. Basic to the use of far-field calculation methods is an assumption that the source can be modeled as a point source with known azimuth and elevation plane patterns. This assumption is reasonable in the case of VHF and UHF antennas installed on tall towers. Far-field calculations may not be a good choice near the antennas, however, as in the case of rooftop installations, on-tower situations, or in the vicinity of AM towers.

Near-Field and Far-Field Conditions

The exposure from a transmitting antenna's radiated field is characterized by the power density (power radiated per unit area), which is a complex quantity called the Poynting vector, usually abbreviated $\bar{S}$. Mathematically, $\bar{S}$ is defined as the vector cross product of the electric and magnetic field vectors (times a constant). Close to the antenna (called the "near-field zone"), $\bar{S}$ is imaginary, meaning that all of the power is reactive, while further away (called the "far-field zone"), $\bar{S}$ is real, meaning that all of the power radiates. A transition occurs between the reactive and radiating regions, where the power is composed of both reactive and radiating components.

A determination of whether the area in question is in the near-field, transition, or far-field region of the antenna is critical to determining whether calculations are appropriate and what measurements must be taken. As a rule-of-thumb, one is in the far field of the antenna at a distance, *r*, when:

$$r > \frac{2D^2}{\lambda} \text{ and/or } r > 5D \tag{3}$$

where *D* is the largest dimension of the antenna and λ is the wavelength of operation; both must be in the same units. The former test is generally applicable to VHF frequencies and above, while the latter test is most applicable at LF (LW) and MF (AM) frequencies. The manufacturer's published directional antenna patterns are applicable only in the far field.

Far-Field Calculations

In the far-field zone, the antenna directional patterns, both azimuth and elevation, are assumed to be fully formed. For typical VHF and UHF transmitting antennas, the power-density magnitude at distance, *r*, can be calculated from:

$$|\bar{S}| = \frac{2.56 \times 1.64 \times 100 \times RFF_{AZ}^2 \times RFF_{EL}^2 \times ERP}{4 \times \pi \times r^2} \tag{4}$$

where the factor 2.56 accounts for typical ground reflection, the factor 1.64 is the gain of half-wave dipole over an isotropic radiator, the factor of 100 converts to units of mW/cm^2 (use 1000 to convert to W/m^2), the *RFF* factors are the relative field factors for the azimuth and elevation antenna patterns, *ERP* is effective radiated power in kilowatts, and *r* is in meters. This is the formula recommended by the FCC Office of Engineering and Technology (OET) Bulletin No. 65 (1997).

Some caveats must be applied to the application of this equation; namely, the azimuth pattern is typically defined only in or near the plane of the main beam of the antenna. At steep elevation or depression angles (above or below the antenna), the azimuth pattern degrades and should not be included in the calculation (that is, assume a relative field of 1 for RFF_{AZ}).

Near-Field Calculations

Because many VHF and UHF broadcast antennas consist of a vertical array of elements, they can be treated to a first-order approximation as a line source having the input power distributed uniformly over the antenna aperture. Within the near-field boundary of the antenna, the power-density magnitude at distance, *r*, can be calculated from:

$$|\bar{S}| = \frac{180 \times 0.1 \times P_{IN}}{\theta_{HPBW} \times \pi \times D \times r} \tag{5}$$

where P_{IN} is the average input power to the antenna, θ_{HPBW} is the half-power azimuth beam width of the antenna (an omnidirectional antenna has an HPBW of 360°), and *D* is the antenna aperture. The units of power density in this equation are mW/cm^2; eliminate the 0.1 factor to convert to W/m^2. This near-field calculation is applicable *only* in the aperture of the antenna and may very well overstate levels at locations above and below the antenna. For broadcast antennas, there are several ways to refine these calculations, although none is specified in OET Bulletin No. 65.

Intrinsic Compliance

In some situations, the radiating elements of the antennas are insulated from direct contact and the power levels are so low that the public limit cannot be exceeded, even at the point of closest approach. Such sources are called intrinsically compliant. In earlier versions of the same standards, power levels of, say, 6 watts and less were considered to be intrinsically compliant, but that exemption is no longer recognized.

The FCC has identified certain circumstances where routine evaluation is not required. Two of the more common include LPTV and FM translators of less than 100 watts ERP. It is important to understand that this *exclusion* from routine evaluation is not an *exemption* from compliance, and such stations may well exceed the exposure limits in the vicinity of the antennas requiring further, specific consideration.

TABLE 2.4-2
Recommended Distances (in Meters) from Tower Base to Fence for One-Quarter Wavelength AM Towers at Various Power Levels

Frequency (kHz)	Transmitter Power (kW)			
	50	10	5	1
535–1540	4 m	2 m	2 m	1 m
1540–1705	5 m	2 m	2 m	1 m

From FCC/OET Bulletin No. 65.

ANTENNA FARMS

At sites with several antennas on several towers, power reductions can often be used to enable work on a particular tower without the need for all stations to shut down. Computer modeling or, less practically, measurements can be performed to determine for any section of any tower which antennas have the largest impact on localized power density levels, and which of those sources should be cut back, and by how much, in order to bring total power-density levels within the occupational limit. The results of these calculations typically are organized as tables and can form the basis for the cooperative agreements anticipated by the FCC: "The licensee also certifies that it, in coordination with other users of the site, will reduce power or cease operation as necessary to protect persons having access to the site, tower, or antenna from radiofrequency electromagnetic exposure in excess of FCC guidelines" [9].

AM Sites

Most types of AM towers present a shock-and-burn hazard at or near the base, and the FCC requires that the area surrounding the base be fenced, as described in Table 2.4-2. Using a variant of the Numerical Electromagnetic Code (NEC), the FCC developed tables of the predicted distances necessary for compliance with its limits for public exposure. These limits are functions of frequency, tower height, and power, so only the most common case—that of one-quarter wavelength tall towers—is presented here. Additional guidance is found in Supplement A to FCC/OET Bulletin No. 65 (August 1997 edition).

MEASURING COMPLIANCE

Broadband Measurements

Surveys of broadcast transmitting sites, especially multi-user sites, are generally taken using broadband instruments. Manufacturers of calibrated survey instruments include:

- Narda Safety Test Solutions [10]
- ETS-Lindgren [11]
- Instruments for Industry [12]
- Herley Farmingdale [13]

Broadband instruments typically include an integrated readout and probe, which cover a substantial frequency range, such as 0.3–3,000 MHz. While broadband measurements can suffer from reduced accuracy due to interaction of several frequencies within the instrument [14], an awareness of its limitations and proper operating techniques can result in the rapid survey of a large area.

There are several manufacturers of handheld EMF meters that include RF capabilities. Note that most of these meters do not have traceable calibration and may have wildly different responses with frequency, modulation, or orientation. While these units are priced at a few hundred dollars, you should expect to pay at least $3,000 for a survey-quality instrument and substantially more if special features are included.

Narrowband Measurements

Narrowband measurements, typically of a single station in isolation, are sometimes useful in assessing the contribution of that station without taking other stations out of service. Typically, narrowband measurements are conducted using a field strength meter, such as the Potomac Instruments FIM-71, FIM-72, or equivalent, or using a spectrum analyzer with a calibrated antenna. Care must be taken to account for all planes of polarization, and this can become burdensome at lower frequencies. Note that some narrowband instruments, such as the FIM-41, use a magnetic field antenna, but report results in equivalent electric field units. This situation makes the determination of compliance complex.

Narda has recently introduced a combination broadband/narrowband instrument, the SRM, which is small enough to be used for surveys. Essentially, it is a spectrum analyzer with near real-time digital signal-processing capabilities. The digital signal processor (DSP) engine allows the same meter to display in different units, accounts for frequency dependence of various safety standards, integrates power over a particular frequency range, and has other useful features. This instrument can measure the contribution of individual stations directly as a percentage of any standard. Its main drawback is that the available triaxial probe does not cover frequencies below 75 MHz, making a comprehensive survey difficult if there are nearby stations operating at or below TV Channel 4. In addition, it is important that the resolution bandwidth (RBW) and amplitude range are appropriate to the emission(s) being measured. Because they have different bandwidths, simultaneous measurement of, for example, FM and DTV emissions, may not be possible using a single RBW setting. [15]

E-Field Measurements

Measurements of the electric field are most commonly conducted with electrically short dipole or monopole elements. Care must be taken to account for all planes of polarization, and many manufacturers make probes that are designed to provide an isotropic response, so that repetitive measurements in each plane are not necessary. At typical installations above 30 MHz, only measurements of the electric field are necessary at ground level, since the observer will be in the far field. Below 30 MHz, and especially at AM frequencies and below, both E- and H-field measurements are required, because the observer is typically not in the far field. E-field measurements are typically reported in units of volts per meter (V/m), although some instruments report in units of $(V/m)^2$.

So-called "hot spots" may be encountered near conductive objects, such as tower bracing. These hot spots are generally created by reradiation and hence are often a near-field phenomenon. Because much of the energy in the near field is reactive, there is little real power present that can result in a hazardous condition. To account for this, the C95.3 standard recommends that measurements near reradiating objects be conducted at a distance of at least 20 cm (about 8 inches).

H-Field Measurements

Except in the far field, measurements of both the electric and magnetic fields are recommended. As a practical matter, there are few calibrated magnetic field antennas available for use above 300 MHz, so magnetic field measurements at higher frequencies may not be possible using conventional instruments. H-field measurements are typically most important below 30 MHz, where conventional inductors are often used for impedance matching, and can generate substantial magnetic fields. High magnetic fields are often encountered near AM antenna–tuning units, tower bases, and phasors, and it is often the magnetic field that predominates.

Some H-field probes have out-of-band resonances, so care should be used when surveying a site having sources in several frequency bands.

Shaped Probes

Recently, some manufacturers began to offer electric field probes that have a frequency dependence matching the standard, so that exposure at multi-user sites could be assessed directly. These probes offer the ability for rapid assessment, but typically are less accurate than their "flat response" counterparts, because the abrupt frequency transitions in the standards cannot be precisely realized. The cost of a shaped probe is often twice that of a flat one, *and* it becomes obsolete if the standard changes.

Spatial Averaging

Although most standards contain both spatial and time-averaging provisions, reliance on these factors at marginally compliant sites is not recommended. There is simply too much uncertainty associated with RF exposure assessment to state unambiguously that a location having a peak power density of, say 110% of the standard, but an average power density of just 90%, will always be in compliance. The FCC uses spatial averaging in its measurements, and this can work to a station's advantage, since spatially averaged measurements will always be lower than peak measurements. Typically, spatially-averaged results at FM/TV sites are about 60% of peak readings.

For several reasons, the use of spatial averaging is generally not recommended as a means to achieve compliance.

- There is no single approved method of obtaining a spatial average, so different observers may obtain different results.
- There is a need to protect against localized levels at the eyes and testes.
- The use of spatial averaging is subject to administrative review and could be eliminated at any time.

If spatial averaging is to be used, a common method (though not the most accurate) is to take measurements over a vertical line, from 0.2–1.8 m above ground with the probe oriented in each of the four compass directions. The results of the four linear averages are themselves averaged together. A plastic pipe can be used to ensure that each measurement is taken at the same location and over the same vertical segment.

Time Averaging

Like spatial averaging, the use of time averaging to achieve compliance in marginal situations is not recommended. Certainly, in most uncontrolled environments, it is not possible to control the actions of the public, including, for example, how long they might remain at some location admiring the view. In controlled environments, it may be possible to control how long a worker remains at a particular location, but such restrictions are often inconvenient or difficult to enforce. The use of time averaging in situations typically encountered by broadcasters should be limited to situations where the source is highly variable, such as a weather radar system, which employs a low-duty cycle, pulsed emission and typically rotates. In that situation, the worker is unlikely to be exposed to the peak power of the radar for any length of time, and time averaging of the emission may be appropriate.

Uncertainties and Variability

The power density at broadcast sites is subject to variation over time, due to changes in atmospheric conditions, soil conductivity, foliage, and physical changes to the site environment. The day-to-day variation at FM/TV sites can be on the order of 20–30%, so a site having measured power densities exceeding 70–80% of the applicable limit may actually be out of compliance at times. The FCC has imposed fines on stations that were found to exceed the 5% threshold of "significant contribution" in areas exceeding the public limit at one time but were measured to be under that threshold at another time [16].

The worst-case uncertainty of broadband measurement equipment comes about if all the possible sources of error were at their extreme values and in such directions as to add together constructively, and are on the order of 3–6 dB. Most of the various contributing errors are not systematic, so random errors of perhaps 1 dB can be expected in a well-conducted survey [15].

Detectors using metal barrier or Schottky diodes are perhaps the most common type used in broadband instruments. Diodes have a *major limitation* when used at typical broadcast sites, often, since they do not respond properly in a multiple-station environment [14]. Multiple source and frequency (MSF) errors can result in errors up to 3 dB, with an average error of about 1.2 dB. Meters that include circuitry to compensate for this effect appear to be less susceptible to MSF errors.

Contact and Induced Currents

Contact (or touch) currents flow when the human body comes into contact with a metallic object at a different electric potential. Typically, this occurs when a person touches a tower or wire at RF potential, but can also occur when fields are coupled to implanted medical devices. The exposure limits *are not* intended specifically to protect against such situations. Induced currents are those induced in the human body immersed in an RF field, which flow to ground through a foot or both feet.

Although the FCC does not presently enforce contact or induced current limits, they *are* specified in both the C95.1 and ICNIRP standards, and are *enforced* by OSHA. Table 2.4-3 summarizes the limits over the range 0.1–110 MHz (there are no limits above 110 MHz).

Contact and induced currents can be measured using a toroid current probe, which surrounds the limb being measured. Induced currents can also be measured using a parallel-plate current meter.

Except in the case of direct contact with an energized radiator, the contact and induced currents are a direct function of the electric field strength in which the object or body is immersed. The new C95.1 standard contains charts that show the fields below which contact or induced currents do not require measurement.

Contact Voltage

In addition to the limits specified for contact and induced currents, the C95.1 standard also suggests a limit for the open-circuit voltage that exists on objects in the range 0.1–100 MHz with which an individual may come into contact. The limit is generally 140 volts RMS between any two points of contact with the body.

TABLE 2.4-3
Contact and Induced Body Current Limits from C95.1-2005 and ICNIRP

	Current Limit (mA)
C95.1-2005—100 kHz to 110 MHz	
Public	
Induced through one foot	45
Induced through both feet	90
Contact through one finger	16.7
Occupational	
Induced through one foot	100
Induced through both feet	200
Contact through one finger	50
ICNIRP - 10 to 100 MHz (field limits only apply below 10 MHz)	
Public	
Induced through any limb,	45
Contact through one finger	20
Occupational	
Induced through any limb	100
Contact through one finger	40

INTERACTION OF RADIO FREQUENCY ENERGY WITH ORGANIC MATTER

The prevailing standards in the United States are intended to protect against established adverse health effects identified in the reviewed studies. An important distinction is that not all reported effects are established or adverse. The standards generally define *basic restrictions*, which are based directly on the established effect. Depending on the frequency of exposure, the physical quantities of the basic restrictions are current density, specific energy absorption rate (SAR), and power density. Only power density outside the body can be readily measured for exposure of individuals, so the maximum permissible exposure limits established in the standards are derived from the basic restrictions (plus a safety factor) but are specified in indirect, equivalent units of electric or magnetic field strength, power density, and current flow.

Below about 5 MHz, the standards minimize effects associated with electro-stimulation to prevent adverse effects on nervous system functions. In the frequency range 100 kHz to about 5 GHz, the standards protect against adverse health effects associated with whole-body heating and excessive localized heating. Above about 5 GHz, the standards protect against excessive heating at or near the body surface.

Although the standards protect primarily against the effects of tissue heating, the biological bases for limiting exposure considered a wide variety of studies on potential or reported effects, including epidemiological studies of cancer risk, reproductive outcome (teratology), cellular effects, behavioral effects, and central nervous system (CNS) effects. For exposure to radio frequency fields (as opposed to contact and induced currents) the lowest threshold of adverse effect was found to be behavioral change in animals due to thermal stress.

ANSWERS TO COMMONLY ASKED QUESTIONS

I work at a radio station. Why am I allowed to be exposed to greater RF levels than a member of the public? The basic restrictions on exposure are the same for both workers and the public. It is the underlying philosophy of the safety factor that differs. Since workers are aware of their exposure, the safety factor is smaller, but there is no credible evidence that anyone has been harmed by RF exposure at or below the occupational maximum permissible exposure (MPE) level.

Can a local jurisdiction establish its own arbitrary exposure limits? In the case of "Personal Wireless Facilities," the U.S. Congress prohibits local government entities from applying exposure limits more restrictive than the FCC limits. However, neither Congress nor the FCC has established a regulatory preemption for RF exposure from broadcast facilities.

What is the ambient exposure of a typical person in the United States? The EPA conducted several surveys and found that exposure levels varied considerably depending on whether a location was urban or rural. Typical exposures in urban areas today are 1 $\mu W/cm^2$ or less.

Do the new digital modes (ATSC television and IBOC radio) present different and unknown biological effects? Some research has suggested that RF signals containing substantial modulation at extremely low frequencies (below about 200 Hz) may lead to so-called nonthermal effects. However, none of the broadcast modes contain significant components at those frequencies [17].

I've heard that some countries (e.g., Italy, Russia, and China) have exposure limits many times more restrictive than in the United States. Is that true? Yes. In the case of Italy, there is a more restrictive limit that applies near hospitals and schools, but it has no scientific basis. In the case of Russia and China, the more restrictive limits are largely historical (dating from the 1960s), and are due to a different philosophical basis for standard setting. Specifically, the more restrictive standards are believed to have been set to preclude all known effects, regardless of whether the effects were adverse or even whether they occurred in whole beings (as opposed to cell cultures).

I worked on a tower on which one of the antennas that was supposed to be turned off was in fact energized. What adverse effects should I expect? Likely, none. The safety standards are not set at the boundary of adverse health effects. Rather, they include a significant safety margin below that threshold. So, even at exposure at levels of five times the standard, there would be little likelihood of health problems. If the exposure was sufficiently intense to cause health problems, it is likely that a heating or tingling sensation would have been felt while on the tower. See a doctor if any physical symptoms, such as blurred vision or headaches, occur.

What are the cumulative effects of all of these broadcast towers along with all of the cell sites and other RF equipment that is around? Cumulative usually means that the effects accumulate with time, which is not the case with exposure to RF energy at levels below the standards. A better term might be additive, because the exposure levels from all of the sources mentioned add together. A broadband survey would include the contributions of all of these sources.

I read about some research that suggests new adverse effects. What do you know about that? Largely because of the widespread use of personal communications devices, the public has become sensitized to the issue of RF exposure and has demanded further assurances that exposure is safe. Research continues, and a positive result from one study is often offset by negative results from other studies. Information on many specific studies and how those results might be interpreted can be found on the Internet [18]. Findings of particular effects usually must be replicated by other researchers before they are considered *established* effects.

Key Terms

Contact current: The electric current between a metallic object and a biological medium as a result of contact with a source of RF current. Normally, finger contact is assumed, although other limits may apply for so-called "grasping" contact by trained individuals. Typically, milliamperes (mA).

Electric field: The effect of an electric force exerted on an electrically charged object, divided by the charge. Typically, volts per meter (V/m).

Induced body current: The electric current in a biological medium that occurs when the medium is in an electric or magnetic field, but not in direct contact with a source of RF current. Typically, milliamperes (mA).

Magnetic field: The effect of moving electric charges that exert a force on other moving charges. Typically, amperes per meter (A/m).

Maximum permissible exposure: Those limits on exposure allowed for persons in particular environments, defined as either controlled (occupational) or uncontrolled (public).

Power density: The magnitude of the cross product of the electric and magnetic field vectors. Also, equivalent far-field power density, derived from individual electric or magnetic field. Conventionally, microwatts or milliwatts per square centimeter ($\mu W/cm^2$ or mW/cm^2), but now in MKS units of watts per square meter (W/m^2).

Radiation: Streams of photons or waves from a source. Radio frequency radiation is nonionizing, that is, of insufficient energy to

break atomic bonds. Not to be confused with radioactivity. Typically, for radio frequencies, expressed in terms of power density.

Radio frequency: That portion of the electromagnetic spectrum in which waves or photons can be generated by an alternating current fed to an antenna. Typically taken as the frequency range from 3 kHz to 300 GHz.

Spatial average: At frequencies above 100 kHz, the average of the power densities, computed over the planar area of a human. Typically, expressed in power-density units.

Specific absorption rate: The measure of energy deposition. Typically, watts per kilogram (W/kg). The SAR forms the basic restrictions from which the MPE limits are derived.

References

[1] Hammett, W. F. *Radio Frequency Radiation*. New York: McGraw Hill, 1997, p. 45

[2] See http://www.ncrponline.org/.

[3] See http://grouper.ieee.org/groups/scc28/.

[4] See http://www.hc-sc.gc.ca/ewh-semt/pubs/radiation/99ehd-dhm237/index_e.html.

[5] See http://www.icnirp.de/.

[6] See http://www.osha.gov/SLTC/radiofrequencyradiation/index.html.

[7] OSHA Regulations, Title 29 Code of Federal Regulations §1910.147.

[8] See http://www.nabstore.com, http://www.radhaz.com, and http://www.rfsigns.com.

[9] Environmental Effects section of FCC licensing application forms. See http://www.fcc.gov.

[10] http://www.narda-sts.com/. Narda Safety Test Solutions, 435 Moreland Road, Hauppauge, NY 11788; Tel.: 631-231-1700.

[11] http://www.holadayinc.com/. 1301 Arrow Point Drive, Cedar Park, TX 78613; Tel.: 512-531-6400.

[12] http://www.ifi.com/. 903 South Second Street, Ronkonkoma, NY 11779; Tel.: 631-467-8400.

[13] http://www.herley.com/index.cfm?act=prodsforfamily&fam=20. Herley Farmingdale, 425 Smith Street, Farmingdale, NY 11735; Tel.: 631-630-2000.

[14] Randa J., and Kanda, Motohisa. "Multiple-Source, Multiple-Frequency Error of an Electric Field Meter," *IEEE Trans. on Antennas and Propagation*, AP-33(1), Jan. 1985, pp. 2–9.

[15] Weller, R. D., and Salek, S. "FCC RFR Guidelines: Is your facility really in compliance?" Proc. NAB Engineering Conference, 1999.

[16] *Forfeiture Order*, FCC 04-281, released Dec. 10, 2004.

[17] "Biological Effects of Modulated Radio-frequency Fields," *NCRP Commentary* No. 18, 2003.

[18] Medical College of Wisconsin, see http://www.mcw.edu/gcrc/cop/cell-phone-health-FAQ/toc.html.

C H A P T E R

2.5

Managing Workplace and Environmental Hazards

DAVID MAXSON
Broadcast Signal Lab
Medfield, Massachusetts

INTRODUCTION

Broadcast engineers are perhaps most familiar with workplace safety issues relating to human exposure to radio frequency (RF) energy, which can be found in Chapter 2.4 of this handbook. However, as places of employment, broadcast facilities also must comply with a broad range of workplace safety requirements. In addition, *materials* are present at broadcast facilities that could be hazardous to humans or the environment that must be handled in accordance with environmental requirements.

This chapter provides a review of key aspects of federal regulations regarding workplace safety and environmental protection. When preparing safety programs or addressing environmental protection issues, broadcasters are encouraged to read further and seek the advice of specialists and attorneys in these fields.

Two federal agencies administer workplace safety and environmental protection—the Occupational Safety and Health Administration (OSHA) and the Environmental Protection Agency (EPA). In addition to these, it is important to be aware of any local or state laws or regulations that may differ from the federal ones.

This chapter covers four main areas:

- OSHA background and the General Duty Clause
- OSHA standards of particular interest to the broadcasting workplace
- Toxic and hazardous substances
- Environmental impacts of new facilities

OSHA AND THE GENERAL DUTY CLAUSE

OSHA was formed by the U.S. Department of Labor under the 1970 Occupational Safety and Health Act (OSH Act). Its mission is to "assure so far as possible every working man and woman in the Nation safe and healthful working conditions" [1]. The act contains a section commonly called the "General Duty Clause" [2], which is the umbrella that covers all workplace safety.

> Section 5. Duties
>
> (a) Each employer—
>
> (1) shall furnish to each of its employees employment and a place of employment which are free from recognized hazards that are causing or are likely to cause death or serious physical harm to his employees;
>
> (2) shall comply with occupational safety and health standards promulgated under this act.
>
> (b) Each employee shall comply with occupational safety and health standards and all rules, regulations, and orders issued pursuant to this act which are applicable to his own actions and conduct.

The General Duty Clause makes it clear that both the employer and the employee have duties to operate safely. At one extreme, if the employer has a sound record of informing, training, and enforcing compliance among its employees, it has reduced its legal exposure if an employee is injured because he or she failed to comply. Here the employee has failed to meet the General Duty Clause. At the other extreme, in the absence of the employer providing suitable workplace safety practices, the employee has no guidance on what compliance is. Here, the employer has failed to meet the General Duty Clause.

Of course, the causes of workplace accidents and injuries are often not so clear-cut. Much workplace litigation revolves around who was at fault for an accident. If not from the obvious legal and ethical responsibilities, then at least from the perspective of self-preservation, broadcasters should implement a thorough and consistent workplace safety program to prevent injury and provide a strong defense in the event of an injury.

The employer's responsibility under the General Duty Clause is two-pronged. While the second part, (a)(2), requires employer compliance with OSHA regulations (called standards) on certain workplace practices, standards compliance is not enough. It is the first part of the General Duty Clause, (a)(1), that unambiguously makes a safe workplace the responsibility of the employer. If an OSHA standard does not exist on a topic, the employer must rely on other expert sources for guidance on safe practices. If an OSHA standard does exist, and there is a more restrictive national consensus standard, OSHA, via the General Duty Clause, may rely on the more restrictive national consensus standard [3].

Under the General Duty Clause, four conditions must be met to result in a violation [4]:

1. The employer failed to keep the workplace free of a hazard to which employees of that employer were exposed.
2. The hazard was recognized.
3. The hazard was causing or was likely to cause death or serious physical harm.
4. There was a feasible and useful method to correct the hazard.

Section 6 of the OSH Act addresses the creation of OSHA Occupational Safety and Health Standards. New OSHA standards must pass a test comparing them against national consensus standards. If a new OSHA standard "differs substantially from a national consensus standard" [5] OSHA must publish the reasons why OSHA's "will better effectuate the purposes of this act than the national consensus standard."

Why is the consensus standard principle important? OSHA cannot be expected to maintain up-to-date standards for all working environments, including broadcasting. Industry standards are regularly reviewed and revised as needed. Therefore, it is incumbent on broadcasters to identify or create national consensus standards that apply to their unique workplace practices, and to rely on other standards that may be applicable in broadcasting. For example, national consensus electrical safety standards and RF safety standards certainly apply to broadcast facilities.

In addition to OSHA requirements, states and municipalities may have different requirements or more restrictive ones that apply to broadcasting facilities. Broadcast engineering managers must identify such requirements as they prepare workplace safety programs.

Broadcast facilities also must be compliant with state workplace standards if they are more restrictive than federal standards. Streamlining workplace safety regulation in 26 states, OSHA has approved these states' OSHA programs. OSHA monitors and provides funding to support these state programs. So long as they implement standards that are at least as restrictive as OSHA's, these states take on OSHA responsibilities for monitoring workplace safety, providing safety management consultation, and enforcing standards.

States operating OSHA-approved plans:

Alaska	*New Jersey**
Arizona	*New York**
California	North Carolina
*Connecticut**	Oregon
Hawaii	Puerto Rico
Indiana	South Carolina
Iowa	Tennessee
Kentucky	Utah
Maryland	Vermont
Michigan	*Virgin Islands**
Minnesota	Virginia
Nevada	Washington
New Mexico	Wyoming

*These four jurisdictions have implemented state plans that apply only to public sector employees' workplaces. State-owned broadcast facilities in these places may be subject to these plans, but private workplaces are not.

Specific OSHA standards that may be relevant to the broadcasting workplace are discussed below. While this chapter provides an overview of several OSHA requirements, it is not a substitute for a professional evaluation of the broadcasting workplace.

OSHA STANDARDS OF PARTICULAR INTEREST TO THE BROADCASTING WORKPLACE

Reviewing the OSHA regulations in §1910 of part 29 of the Code of Federal Regulations, several topics are applicable to broadcasters. What follows is a brief treatment of some of the more germane regulations.

§1910.151, First Aid

OSHA requires that:

- First aid kits are required at the workplace. Consider kits that are compliant with ANSI Z308.1-2003, "Minimum Requirements for Workplace First Aid Kits."
- If medical help is not available "in proximity" to the workplace, first aid trained employees should be available.
- Provide suitable personal protective equipment (PPE) to protect employees from blood or other potentially infectious materials (§1910.1030(d)(3)). At a minimum, latex gloves and pocket CPR masks should be in first aid kits.

- Add suitable gear to the kits for industry-specific hazards.

In addition to these OSHA requirements, it is a good idea to maintain an oxygen bottle for first response to certain emergencies. Be sure first responders are familiar with the reasons for administering oxygen and seek advice from emergency medical services in its application. As well, even if not OSHA-required, employees benefit by employer-sponsored first aid training programs. To keep first aid kits up to date, consider subscribing to a service that periodically checks and updates first aid kits. These companies can also provide advice on regional or site-specific first aid supply needs.

§1910.132-148, Personal Protective Equipment

Personal protective equipment (PPE) is anything that is used to provide employees (or guests) protection from specific hazards. These sections of the regulations address requirements for protecting eyes and face, head, hands, feet, and lungs from workplace hazards. PPE for electrical hazards is also addressed here. In general, these regulations are nonspecific and reference various ANSI standards for protective footwear, eyewear, head protection, and the like.

Eye protection should be readily available in broadcast operations that involve power tools or flying or blowing debris. Employees involved in construction work or operations that require moving or lifting materials or equipment should have protective footwear and handwear to prevent crush or pinch accidents (e.g., steel-toe shoes, leather gloves). In addition to requiring employees who work with power tools to wear eye protection, operations managers might look at other potential eye hazards that may warrant requiring protection. For example, an operations manager might consider questions like:

- Should field crews and reporters be required to wear eye protection outdoors in very windy conditions or at sites where particles might be airborne?
- Does the facility perform certain construction or cleanup work that may generate airborne particles in addition to those that involve power tools?
- Do any employees handle materials, such as glass lamps or lenses, that may shatter accidentally?
- Do any lamps have removable UV filters that are necessary for the protection of individuals under the lights?

In addition to the OSHA PPE standards cited here, there are many kinds of PPE addressing many hazards. It is important to ensure employees know when and how to use their PPE. Also, PPE wears out, or is subject to retirement after use. Gloves or shoes may wear out, while a safety harness may either wear out or need to be replaced after experiencing "impact loading" when its wearer falls. Personnel should have the ability to determine whether their PPE is in good order and have the authority to cease hazardous work until PPE issues are resolved.

Fall Protection

Broadcast operations may include work that involves climbing or working near high openings. Roof hatches, rooftops, towers, stages, staging, sets, stage lighting, microphone placement, scaffolding, and moveable lifts are among the special conditions in which broadcast facilities may place workers at risk of falling. OSHA regulates fall hazards in myriad standards. Subpart D of the Occupational Safety and Health Standards (29 CFR 1910.21-30) contains OSHA standards for general industrial fall risks. These sections address industrial stairs, fixed and portable ladders and scaffolding, and floor and wall openings. Subpart F, Powered Platforms, Manlifts, and Vehicle-Mounted Work (29 CFR 1910.66-68), addresses these mechanized systems for elevating workers. Generally, these sections address design characteristics and appropriate uses of the various safety measures. Railings, toe-kicks, gates, and the like are required as applicable. Fall arrest systems are required in some circumstances, such as on lifts and high ladders.

OSHA references several American National Standards Institute (ANSI) standards, which can be obtained from www.ansi.org:

- American National Safety Standard for Manlifts, ANSI A90.1-1969
- Safety Code for Mechanical Power Transmission Apparatus, ANSI B15.1-1953 (R 1958)
- Safety Code for Fixed Ladders, ANSI A14.3-1956
- Safety Requirements for Floor and Wall Openings, Railings and Toeboards, ANSI A12.1-1967
- Safety Requirements for Personal Fall Arrest Systems, ANSI Z359.1-1992

The OSHA construction regulations provide more detailed standards for fall protection. 29 CFR §1926.1053 addresses ladders used in construction work. Ladders that are part of a rise of more than 24 ft must be equipped with a ladder safety device (such as a climbing lifeline with rope-grab attachment), a self-retracting lifeline with rest platforms at regular intervals, or a cage or well wrapped around the ladder climbing space. Other locations with exposed heights that are not satisfactorily fall-protected by railing, fencing, or netting are also subject to fall-arrest requirements.

Workers involved in activities that require fall protection must be trained periodically in the safe setup and use of fall protection systems. 29 CFR §1926.502, Fall Protection Systems Criteria and Practices, addresses the various fall protection methods, including railings, fencing, safety nets, fall arrest systems, and positioning devices.

Full-body harnesses are necessary for fall arrest systems, with the attachment point being at the center of the upper back of the worker. A deceleration device, such as a tearing or deforming lanyard or a retractable lifeline, is employed to absorb the shock of a fall. Any equipment that has borne the impact of a fall must be retired from service. Climbing belts are not suitable as fall arrest equipment, but may be used with a proper lanyard as a positioning device, if the worker so positioned can fall no further than 2 ft.

Generally, it is easily assumed that work on a broadcast tower will require fall arrest and fall positioning devices. Tower ladders should be equipped with fall arrest lines or rails that should be used by all climbers. It is also important, however, to apply fall protection and fall arrest criteria to other locations frequented by broadcast personnel. Work on rooftops, whether it is to install or orient antennas, to examine rooftop mechanical systems, or to shoot pictures, requires fall protection. If suitable railings and such are not available, a fall arrest system may be required. When exiting a roof hatch, it may be necessary to be wearing a harness and attach to a fall-arrest point.

Overhead work, for instance, in theatrical environments placing microphones, cameras, or lights, may require the use of fall-arrest systems, particularly if one leaves the protected confines of a properly railed or fenced catwalk.

§1910.101, Compressed Gases (General Requirements)

OSHA references Compressed Gas Association standards and publications for dealing with compressed gas cylinders. Broadcasters who have air compressors with storage tanks, who feed dry nitrogen from tanks to charge their transmission lines, or who maintain propane or natural gas storage tanks should review their cylinder-handling practices.

In general, compressed gas cylinders should be kept at least 20 ft away from highly flammable materials and secured from falling. When moving a cylinder, a cover or shield should be in place to protect the valve assembly, and should only be moved with a suitable wheeled hand truck. Cylinders held for long periods of time may require periodic inspection and pressure testing. Check with a compressed gas supplier for advice on how to determine what a cylinder's maintenance requirements are.

Compressed gases, other than air, have the potential to displace oxygen in a workspace. Depending on the space, the gas, and the normal ventilation of the place where the cylinder is used or stored, precautions should be taken to inform employees about on-site gas storage hazards and, if warranted, set up sensors to trigger an alarm in the event of a gas leak or displaced room air. Compressed air for cleaning should have an outlet pressure of less than 30 PSI (§1910.242).

§1910.242, Hand and Portable Powered Tools and Equipment (General)

While an employer may permit employees to supply their own tools or equipment, the employer is responsible for assuring that all tools used on the job are in safe condition.

§1910.147, The Control of Hazardous Energy (Lockout/Tagout)

Broadcasting facilities contain equipment that is permanently connected to a power source and that requires maintenance and repair work with the equipment turned off. Lockout/tagout regulations require each employee to have positive control over the power source(s) to the equipment upon which he or she is working.

The most obvious equipment that fits this category is the broadcast transmitter. When powered, it is capable of generating lethal or injurious voltages. In addition to electrical or RF energy hazards, kinetic and potential energy hazards are a concern as well. The transmitter is likely to contain moving parts in its air-handling system that must also be made safe prior to and during maintenance.

Examples of other equipment at broadcast facilities that may require energy control are:

- Chillers, heaters, and air handlers
- Robotic studio equipment (lighting and camera supports)
- Cranes, booms, lifts, and the like
- Elevators (building or tower)
- Steerable antennas (satellite, radar, ENG, etc.)
- Motor and solenoid actuated devices (e.g., coaxial switches, AM RF relays, ventilation doors, and louvers)
- Motorized vehicles (e.g., remote trucks)
- Transmitters and other equipment hard-wired to power sources

Equipment that can simply be unplugged, with the plug remaining under the control of the affected employee(s), does not have to be locked out.

To manage these hazards, OSHA §1910.147 states:

> (c) General—(1) Energy control program. The employer shall establish a program consisting of energy control procedures, employee training, and periodic inspections to ensure that before any employee performs any servicing or maintenance on a machine or equipment where the unexpected energizing, startup, or release of stored energy could occur and cause injury, the machine or equipment shall be isolated from the energy source and rendered inoperative.

To accomplish this important task, OSHA requires that equipment be "locked out." Each employee working on a device must be able to physically secure the power source(s) in the deenergized position while working on it. Broadcasters should design their facilities with lockout capabilities on their circuit breaker and shutoff switch systems. Lockout fixtures are available from electrical and safety supply houses that permit multiple employees to each apply their own locks to the selected shutoff switch.

Tagging out a power source is an alternative that is permitted if locking out is not possible. Tagging out is also permitted if it demonstrably provides the same degree of safety as locking out. Tagging out a power source involves the placement of a clearly understood tag on the control to the power source with a nonreusable fastener that requires at least 50 lbs of force to break, to ensure that inadvertent loss of the tag does not occur. Fortunately, the common cable tie is explicitly mentioned by OSHA as a suitable tagout device. The tag must warn of the hazard and give a clear instruction, such as Equipment Being Serviced—Do Not Energize. Locks and tags must contain the employee name.

In addition to the actual locking/tagging requirements, OSHA requires that personnel be trained and their use of the procedures be inspected at least annually. The inspector must individually review each affected employee's knowledge of the procedures and understanding of the hazards. Lockout/tagout programs have employer recordkeeping requirements as well.

The employer or an authorized employee must inform affected employees of the application of and removal of lockout/tagout devices. This is to ensure there is no confusion among a crew about systems going off- or on-line. Locks and tags may only be applied and removed individually by each employee working on the system. An exception can be made for a single group lockout instead of individual lockouts, with a number of additional procedural requirements to ensure worker safety.

In preparing a written lockout/tagout program, the whole of §1910.147 should be reviewed by the program's author to absorb the details of program creation and management. Section 1910.147 also contains a model lockout/tagout procedure document.

Subpart S, OSHA Electrical Standards

For electrical systems, Subpart S of the OSHA regulations apply additional criteria to the lockout/tagout procedures. Once a system is deenergized and locked/tagged out, it is not considered safe to work upon until it has been tested. Section 1910.333(b)(2)(iv)(B) states:

> A qualified person shall use test equipment to test the circuit elements and electrical parts of equipment to which employees will be exposed and shall verify that the circuit elements and equipment parts are deenergized. The test shall also determine if any energized condition exists as a result of inadvertently induced voltage or unrelated voltage backfeed even though specific parts of the circuit have been deenergized and presumed to be safe. If the circuit to be tested is over 600 volts, nominal, the test equipment shall be checked for proper operation immediately after this test.

High-capacity capacitors should be discharged, shorted, and grounded if they might endanger personnel. Interlocks, selector switches, and pushbutton switches are not sufficient meet lockout/tagout requirements. OSHA does anticipate that under some circumstances it may be necessary for a qualified person to temporarily defeat an interlock to service a device. When a qualified person is working near energized components, he or she must maintain a safe distance from energized components or employ appropriate PPE. While OSHA provides Table S-5 (shown here in Table 2.5-1) for qualified persons to maintain safe distances from overhead power lines,

TABLE 2.5-1
29 CFR 1910.333(c)(3)(ii), Table S-5, Approach Distances for Qualified Employees—Alternating Current

Voltage Range (Phase to Phase)	Minimum Approach Distance
300 V and Less	Avoid Contact
Over 300 V, not over 750 V	1 ft. 0 in. (30.5 cm)
Over 750 V, not over 2 kV	1 ft. 6 in. (46 cm)
Over 2 kV, not over 15 kV	2 ft. 0 in. (61 cm)
Over 15 kV, not over 37 kV	3 ft. 0 in. (91 cm)
Over 37 kV, not over 87.5 kV	3 ft. 6 in. (107 cm)
Over 87.5 kV, not over 121 kV	4 ft. 0 in. (122 cm)
Over 121 kV, not over 140 kV	4 ft. 6 in. (137 cm)

the table may be informative for broadcast engineers working near high-voltage power supplies.

The "qualified person" or "qualified employee" in Subpart S is one who "has training in avoiding the electrical hazards of working on or near exposed energized parts." Broadcast engineers familiar with the hazards of high-voltage power supplies may be likely to employ safe work practices with electrical systems, however, it is the responsibility of the employer to provide the necessary safety program and training to ensure that qualified persons are truly qualified and have the procedures and equipment necessary to work safely with electrical power.

Antennas Near Power Lines

There are instances where broadcast operations may come near overhead power lines, such as remote trucks with extendable antenna masts or the installation of an antenna and mast on a roof or an antenna and tower on the ground. OSHA requires a minimum distance of 10 ft from any part of the antenna or mast to the power line. The distance is increased for certain high-voltage lines. In §1910.333(c)(3)(iii)(A), OSHA states:

> Any vehicle or mechanical equipment capable of having parts of its structure elevated near energized overhead lines shall be operated so that a clearance of 10 ft (305 cm) is maintained. If the voltage is higher than 50 kV, the clearance shall be increased 4 in. (10 cm) for every 10 kV over that voltage.

This regulation concerning operations near energized overhead power lines is contained within Subpart S, and is subject to the subpart's personnel protection requirements. Personnel must be trained to follow corporate safety procedures, including identifying and observing safety distances from overhead power lines when erecting antennas and masts. It may be prudent to require greater distances than the OSHA minimums. Safety is improved with the use of alarm/interlock systems that detect the proximity of high-voltage lines, but such systems cannot be a substitute for human observation. Companies such as Sigalarm and Will-Burt offer proximity warning devices for mobile mast systems.

Title 8 of the California Code of Regulations, Chapter 4, Article 40 contains ENG antenna safety requirements for operators in California that operators in other states may find instructive. In addition to antenna safety, operators of ENG news trucks have numerous other safety management concerns. Website www.engsafety.com, provided by Mark Bell, is a useful resource for all ENG safety matters. ENG operators must be aware of vehicle maintenance, safe driving practices, weather-related safety, generator power and carbon monoxide safety, protection of/from bystanders, among others.

29 CFR §1910.95, Occupational Noise Exposure

Occupational noise exposure is limited by a combination of duration and intensity. Noise exposure is evaluated by considering the amount of time in the workday an employee is exposed to certain levels of noise. Noise level is expressed in dBA ("A-weighted" sound pressure level) as measured by a "slow response" instrument that integrates sound pressure level in one-second long samples.

First, employees may not be exposed to peak sound levels in excess of 140 dBA. Exposure to lesser levels are regulated by duration.

As shown in OSHA's Table G-16 (shown here in Table 2.5-2), it is permissible to be exposed to sound pressure levels of 90 dBA for an 8 hour shift. However, when the average daily exposure is greater than 85 dBA, a "hearing conservation program" must be implemented. OSHA describes the components of such a program in §1910.95(c). Whenever practicable, it would be less cumbersome simply to ensure some workers not exceed the 85 dBA daily average.

In an environment where producers, talent, and technical people have enclosed spaces with speakers (studios and production rooms) and have headphones available, it may be difficult to be certain that average daily exposures are below the 85 dBA threshold. For comparison, normal level conversations tend to fall in the 60–65 dBA range. A pair of headphones turned up to a level that blocks out conversational level and ambient sounds could be in the 75–85 dBA range. It is not uncommon to find some users running headphones at higher levels.

TABLE 2.5-2
29 CFR §1910.95, Table G-16, Permissible Noise Exposures

Duration Hr/Day	Sound Level dBA, Slow Response
8	90
6	92
4	95
3	97
2	100
1.5	102
1	105
0.5	110
0.25 or less	115

The assessment of daily average noise exposure is complicated by the level-dependent time limits. For instance, Table 2.5-2 shows that there is a 2 hour daily limit on exposure to 100 dBA noise. If an exposure occurred for 1 hour at 100 dBA, that would be an exposure of one-half of the daily limit. Meanwhile, the exposure limit to 90 dBA is 8 hours. If an exposure of 4 hours at 90 dBA were to occur that day, that would also be half of the daily limit. Thus, the employee who is exposed to 100 dBA for 1 hour and 90 dBA for 4 hours has reached the daily limit, and the remaining 3 hours of the 8 hour work day must be below 85 dBA.

Total daily exposure is presented as a sum of the weighted ratios of the various levels of exposure during the day. The calculation is performed with the variable C_n representing an exposure duration to a certain noise level, n, and T_n representing the allowed time limit of exposure to that level, n. The duration of exposure to each sound level is divided by the exposure limit for that sound level, and the ratios are summed:

$$C_1/T_1 + C_2/T_2 + \ldots + C_n/T_n = \text{daily noise exposure} \quad (1)$$

The daily noise exposure is not permitted to exceed the value of 1, which represents 100% of the exposure limit.

Facilities managers should determine whether workers are persistently exposed to sound levels in excess of 85 dBA. If so, then a more detailed evaluation of occupational exposure in the facility should be undertaken. It may be necessary to establish a hearing conservation program, provide training, monitoring and control of exposure levels, and perform a medical surveillance program in which affected employee hearing is routinely tested.

Radio Frequency Energy Exposure (§1910.97, Nonionizing Radiation)

The OSHA standard on RF energy exposure maintains several antiquated elements, including the use of the term "nonionizing radiation," which has generally fallen out of favor; the use of a fixed power density level of 10 mW/cm^2 from 10 MHz to 100 GHz; and an old radiation sign design that has been superseded by several ANSI standards.

Consequently, in radio-transmitting facilities OSHA relies on the exposure standard adopted by the FCC in 47 CFR §1.1310. OSHA applies the General Duty Clause not only in the application of the FCC standard, but also in the expectation that employers will develop a published safety program and implement employee training.

Broadcast facilities should have an RF energy exposure program developed under the same principles that OSHA applies to other workplace safety programs. The following elements are excerpted from OSHA's *Elements of a Comprehensive RF Protection Program: Role of RF Measurements* [6]:

> Element 1: Utilization of RF source equipment that meets applicable RF and other safety standards when new and during the time of use, including after any modifications.
>
> Element 2: RF hazard identification and periodic surveillance by a competent person who can effectively assess RF exposures.
>
> Element 3: Identification and control of RF hazard areas.
>
> Element 4: Implementation of controls to reduce RF exposures to levels in compliance with applicable guidelines (e.g., ANSI, ICNIRP), including the establishment of safe work practice procedures.
>
> Element 5: RF safety and health training to ensure that all employees understand the RF hazards to which they may be exposed and the means by which the hazards are controlled.
>
> Element 6: Employee involvement in the structure and operation of the program and in decisions that affect their safety and health, to make full use of their insight and to encourage their understanding and commitment to the safe work practices established.
>
> Element 7: Implementation of an appropriate medical surveillance program.
>
> Element 8: Periodic (e.g., annual) reviews of the effectiveness of the program so that deficiencies can be identified and resolved.
>
> Element 9: Assignment of responsibilities, including the necessary authority and resources to implement and enforce all aspects of the RF protection program.

For additional information on human exposure to radio frequency fields, see Chapter 2.4.

Perhaps the only element that requires elaboration in this context is medical surveillance (element 7). Medical surveillance involves collecting and acting upon medical information relevant to maintaining workplace safety. OSHA explains that events, such as RF burns or heating sensations, should be recorded in a log to enhance evaluation and improvement of protection programs. Maintaining exposure records collected from personal RF exposure monitors is a form of medical surveillance. As a means of accident or injury prevention, there may be circumstances where it is appropriate to have employees report, for instance, any medical implant devices that might be affected by or cause injury from exposure to high RF energy levels. An employer may have a policy that requires a medical examination in the event of an injury or overexposure. Whatever the elements of the medical surveillance program are, they should be part of a written RF safety program. In summary, RF energy exposure should be managed under an appropriate workplace safety program that should

- Be written and available.
- Be constructed with worker input.
- Be reviewed periodically.
- Include periodic training.
- Provide for reporting and recordkeeping appropriate to the conditions.

TOXIC AND HAZARDOUS SUBSTANCES

This section addresses toxic and hazardous substances. Since this is a regulatory matter of equal interest in protecting workers and the environment, OSHA regulations and EPA regulations are discussed in tandem. The topics discussed below include:

- Asbestos
- Hazard communication
- Material safety data sheets
- Hazardous substance release into the environment
- Underground storage tanks
- PCBs (polychlorinated biphynels)
- Environmental impacts of new broadcast facilities

29 CFR Subpart Z (§§1910.1000–1450) contain OSHA standards for managing toxic and hazardous substances in the workplace. More than 30 specific substances have their own standards in these regulations. In addition, scores of substances are listed under the classification of air contaminants. Broadcast professionals who have responsibilities for administering worker safety are encouraged to review the OSHA substances and air contaminants to identify any that might be in the workplace and to implement the OSHA standards relating to them.

Asbestos

One substance, asbestos, which is also a potential air contaminant, might be encountered by broadcasters in their facilities. Its OSHA standard is found at §1910.1001. Asbestos is a fire-resistant fiber that is used to reinforce materials that may be exposed to high heat. When airborne, the fibers can be breathed into the lungs where they can provoke disease. Asbestos-containing materials (ACM) that are "friable" are considered hazardous. When a material can be readily crushed or broken in a fashion that releases fibers into the atmosphere, it is friable.

Asbestos may be found in an older broadcasting facility in a variety of places. Boilers and heating pipes could be wrapped or coated in an asbestos material. Some vinyl-composition tile may contain asbestos. Older buildings may have asbestos-containing shingles. Poured-in insulation, such as that consisting of the mineral vermiculite, may be contaminated with asbestos. Roofing felt may be an ACM. Some types of mineral board used as a fireproof backing or liner to fire-prone spaces (boiler rooms, workshops, and the like) may be ACMs.

So long as the ACM remains undisturbed it may be considered safe. Rather than removing such materials, it may be sufficient to *encapsulate* or *envelop* them to prevent them from being disturbed. Encapsulation is the act of sealing the ACM in a material that keeps it stable and prevents the release of fibers into the air.

Enveloping an ACM is the act of containing the ACM within a confined space to prevent the disturbance of the material and the release of the material outside the envelope.

CAUTION: If any removal or repair work involves asbestos-containing materials it should be evaluated by a licensed asbestos removal professional.

Nonfriable materials, such as vinyl-asbestos floor tile, mineral board, or roof felt, should not be cut, broken, or sanded. Before removing such materials, consult a professional to be certain the removal process will not create friable conditions.

The EPA publishes several guides on dealing with asbestos in buildings, including the *Green Book* on managing asbestos in place, the *Blue Book* on controlling ACM in buildings, and the *Custodial Brochure,* a guide for maintenance personnel, and guidelines on maintaining asbestos flooring. Visit www.epa.gov/asbestos/buildings.html for more information. The National Institute of Building Sciences also publishes a book on the subject, available at www.nibs.org.

§1910.1200, Hazard Communication Standard

In addition to the substances that have specific OSHA standards, all potentially hazardous substances in the workplace are subject to the OSHA Hazard Communication Standard (HCS), with limited explicit exceptions. From §1910.1200(b)(1):

> This section requires . . . all employers to provide information to their employees about the hazardous chemicals to which they are exposed, by means of a hazard communication program, labels and other forms of warning, material safety data sheets, and information and training.

Potentially hazardous materials should be evaluated and include "any chemical which is known to be present in the workplace in such a manner that employees may be exposed under normal conditions of use or in a foreseeable emergency." The most applicable exception to this is "where the employer can show that it is used in the workplace for the purpose intended by the chemical manufacturer or importer of the product, and the use results in a duration and frequency of exposure that is not greater than the range of exposures that could reasonably be experienced by consumers when used for the purpose intended." In short, if a chemical or mixture of chemicals is used in the workplace to a degree greater than a consumer would use it and if it presents a potential hazard in regular use or in a foreseeable emergency (fire, spill, etc.), then it is subject to the HCS.

Nonionizing radiation is not considered a covered "substance" in this context. OSHA has separate requirements for it.

Material Safety Data Sheet

The focal point of the HCS is the material safety data sheet (MSDS). This document contains specific information about the chemical or material, including its name, composition, hazard mechanisms, safe exposure levels, target organs, safety procedures, and the like. All producers and distributors of chemicals and substances are required to make an MSDS available to the purchaser. A corporate hazard communication program will review the MSDS of all chemicals at the workplace and determine which ones, if any, should be included in the program.

Once hazardous materials are identified, several steps must be taken:

- A written program should be prepared.
- An individual should be assigned responsibility for maintaining the program and executing periodic training. A procedure for periodically reviewing the program's effectiveness should also be implemented.
- A list of hazardous materials and their respective MSDS must be posted in a location where employees have access to it. A binder containing the written program and MSDS is suitable. Access via computer is also acceptable.
- Employees should receive regular training on the use, handling, and emergency response to each hazardous material with which they work.
- Safety equipment, and if necessary, medical monitoring, should be provided as appropriate.
- All containers should be labeled in a manner that conforms to the standard, including the identity of the material as given on the MSDS and a description of the hazard(s).

Broadcast facilities may contain, for example, cleaners, solvents, paints, compressed gases, fuels, industrial fire extinguishing material, and solders that could be used in a manner that causes greater than consumer-level exposure. Some materials may be in solid, granular, or particulate form instead of liquid or gas, such as wood cuttings, sawdust, dust from cleaning operations, grinding dust, theatrical smoke, real smoke, or breakable lamps containing heavy metals. These materials may present *physical hazards* and/or *health hazards.* For example, paint may present the risk of fire or explosion (physical hazard) and the risk of toxicity or disease (health hazard). Hazard communication programs are supplemented with the provision of suitable PPE, such as respirators, glasses, gloves and such, and the provision of suitable protection systems, such as ventilated and filtered work booths, spill-containing curbs, fire-suppression systems, and the like.

Hazardous Substance Release

Once such hazardous materials are identified, catalogued, and labeled at the broadcast facility, another aspect of these materials must be considered—environmental effects in use or accidental release. This is the purview of the EPA. The enactment of the Comprehensive Environmental Response, Compensation, and Liability Act of 1980 resulted in certain EPA regu-

lations relating to environmental release of hazardous substances. Those substances are listed in 40 CFR §302.4 along with their "reportable quantities." If an accidental release exceeds the reportable quantity of the substance, the EPA must be notified.

Underground Storage Tanks

CERCLA also controls the use of underground storage tanks (USTs) that may contain petroleum products or other hazardous substances. EPA regulations in 40 CFR §280 apply to the care and use of underground storage tanks. A tank is considered an underground storage tank if at least 10% of its capacity is below grade. Tanks storing heating oil for on-premises use are not considered USTs and are exempt.

Many broadcast facilities have standby power systems fueled by petroleum products (e.g., diesel or gasoline). USTs of 110 gallons or less capacity are exempt from these regulations. Larger USTs are subject to certain design requirements intended to prevent accidental release (leaks), such as corrosion protection, double-wall construction, corrosion-free materials, and/or overfill protection. They are also subject to complex leak-monitoring requirements in 40 CFR §280 Subpart D. Fortunately, USTs for standby power systems are exempt from the leak-monitoring requirements.

Old USTs must meet current requirements through upgrade or replacement if necessary.

State and local regulations may be more stringent than federal regulation of USTs, so it is important to be familiar with local requirements. EPA has permitted 26 states and jurisdictions to administer the EPA UST program.

List of States with EPA-Approved UST Programs:

Alabama	North Carolina
Arkansas	North Dakota
Connecticut	Oklahoma
Georgia	Rhode Island
Iowa	South Dakota
Kansas	Tennessee
Louisiana	Texas
Maine	Utah
Massachusetts	Vermont
Mississippi	Virginia
Nevada	West Virginia
New Hampshire	Puerto Rico
New Mexico	

Polychlorinated Biphenyls: Occupational Exposure

Except in its standards for workplace air contaminants, OSHA does not specifically address polychlorinated biphenyls (PCBs) in the workplace. PCBs are a persistent environmental pollutant. Their manufacture in the United States ended in 1977. PCBs were used as a heat-stable material in various applications.

With respect to broadcast facilities, PCBs were often used in capacitors and transformers associated with high-power equipment such as transmitters and power supplies. Because of the manner in which they are used at broadcast facilities, PCBs are not likely to become airborne. However, because they are not soluble in water, PCBs were often dissolved in solvents to perform their function. Thus, the carrier fluid that contains PCBs may also contain other hazardous materials such as benzenes [7].

A PCB MSDS provides the necessary information about PCBs in the event of "accidental release" (a leak or spill) of the chemical, and should be kept in the company MSDS file if there are PCB devices in the facility. Generally stable at room temperature, hot PCB fluids may produce irritation in eyes and lungs. PCBs can be absorbed through the skin by direct contact with the fluid. According to the GE Industrial MSDS the evidence of possible PCB connection to chronic human disease is not strong, but a significant exposure could be injurious to the subject.

PCBs: Environmental Exposure

The EPA regulates the use, storage, and disposal of PCBs as potential environmental pollutants in 40 CFR §761. Table 2.5-3 contains a matrix of the different classes of electrical equipment that may contain PCBs, the degree of contamination in parts per million of PCBs in the fluid, and the disposal requirements.

PCB items that are not in use and are not stored for a clearly identified reuse are considered PCB waste. If a broadcaster has equipment removed from service that contains PCB items, or has PCB items, such as capacitors, sitting separately on a shelf, it is advisable to determine whether each item can be reused or is intended to be reused. If not reusable, the items should be properly disposed of within certain time limits. Generally, there is a 1 year grace period from removal from service to disposal (40 CFR §761.65) with some exceptions.

Disposal of PCB items must conform to EPA requirements. Only those PCB items listed in Table 2.5-3 as being eligible for disposal in municipal waste require no special treatment. All other PCB items require licensed services to handle, store, transport, and dispose of them. When disposing of such items, contact a qualified waste handler or environmental services provider. For the remainder of its existence, the PCB item that your waste handler picks up from your facility is tracked with transportation manifests, storage records, and ultimately, a disposal document.

If a PCB item has leaked at the broadcast facility, an environmental cleanup may also be necessary. Tools, flooring, and such may need to be decontaminated according to EPA criteria. The cleanup materials will also have to be disposed by the waste carrier.

PCB items must be marked with the EPA PCB mark (40 CFR §761.50; see Figure 2.5-1). The equipment or container within which the PCB item resides should also be marked.

TABLE 2.5-3
PCB Equipment Disposal Matrix

Equipment	Disposal	<50 ppm (Non-PCB)	<500 ppm (PCB Contaminated)	≥500 ppm (PCB Transformer or PCB Capacitor)
Transformers with <3 lbs fluid	Municipal solid waste	Assumed		
Circuit breakers, reclosers, oil-filled cable, rectifiers	Municipal solid waste	Assumed		
Mineral oil-filled equipment manufactured before July 2, 1979	EPA incineration		Assumed	
Pole and pad-mount transformers with >3 lb fluid manufactured before July 2, 1979, or if date is uncertain	1) EPA incineration, or 2) clean and dispose in chemical waste landfill, then dispose of fluid separately via EPA incineration		Assumed	
Capacitor stamped at time of manufacture "No PCBs"	Normal disposal	Assumed		
Capacitor or electrical equipment made after July 2, 1979	Normal disposal	Assumed		
Capacitor made before July 2, 1979, or with uncertain date of manufacture	See large and small capacitors below			Assumed
Small capacitor: Less than 3 lbs fluid, or assume less than 3 lbs fluid if less than 100 cubic inches overall size, or less than 200 cubic inches overall size and less than 9 lbs	Municipal solid waste			Assumed
Large high voltage capacitor 2000-volt or greater rating and greater than 3 lbs fluid	EPA incineration			Assumed
Large low-voltage capacitor: <2,000 volt rating and greater than 3 lbs fluid	EPA incineration			Assumed
In lieu of assumptions indicated above, PCB concentration may be determined by fluid test, permanent label at time of manufacture, or by full documentation of fluid servicing history. Documented communication with manufacturer to obtain concentration is also permissible.				
Source: From 40 CFR §§761.2, 761.60.				

FIGURE 2.5-1 The EPA PCB mark.

ENVIRONMENTAL IMPACTS OF NEW FACILITIES

The construction of broadcast facilities is subject to the National Environmental Policy Act (NEPA) when they involve FCC-licensed facilities such as transmitters. In addition to any local regulations for permitting and constructing such facilities, 47 CFR §1.1306 et seq., implements NEPA for FCC licensees.

Of all the environmental impacts with which broadcasters could be concerned, the most familiar may be RF exposure of the general public. Radio-frequency emissions of broadcast facilities are among the criteria to be assessed under the FCC NEPA regulations. Also covered by these regulations are more common national environmental issues, listed in 47 CFR §1.1307 and summarized below.

Environmental Assessment

If there is a potential environmental impact from a new or changed facility, such as a broadcast tower, it is subject to "environmental processing." Environmental processing begins with the preparation of an environmental assessment (EA). If the EA finds that there is an environmental impact, the FCC must respond by preparing an environmental impact statement (EIS).

Some facilities are assumed to have no environmental impact under NEPA and are "categorically excluded" from making an EA. According to 47 CFR §1.1306 and §1.1307, if a proposed licensed facility does not involve any of the following, it is categorically excluded from environmental processing, and need not prepare an EA:

1. Officially designated wilderness area.
2. Officially designated wildlife preserve.
3. Existing or proposed endangered species and/or their habitats.
4. Historical, cultural, archeological, architectural, or engineering places or structures that are listed, or are eligible for listing, in the National Register of Historic Places.
5. Indian religious sites.
6. Flood plain.
7. Significant change in surface features (filling wetlands, diverting rivers, clearcutting, etc.).
8. High-intensity white FAA lights in residential neighborhoods.
9. Human exposure to radio-frequency energy in excess of the applicable safety standards.

With a little legwork, a civil engineering or real estate development firm employing a common NEPA checklist can address the first seven points. It will need the broadcaster's input on the last two—tower lighting and RFE exposure.

If one or more of these conditions apply, the broadcaster must prepare and file an EA with the FCC. The FCC must then determine if the EA demonstrates there will be no significant environmental impact, in which case the construction of the new or modified facility may proceed. However, if there is a significant environmental impact, the FCC must prepare an EIS. The EIS process is cumbersome, time-consuming, and costly. It may result in an unfavorable decision. Consequently, licensees work intently first to avoid the nine conditions above by carefully designing their proposed facilities. Only if necessary, a broadcaster may have to propose a facility that requires the preparation of an EA. With a well-crafted facility and a well-crafted EA, it may be possible to demonstrate that there is no significant environmental impact, despite an impact on one or more of the items in the checklist. However, local opposition may organize and challenge any favorable conclusions of the EA.

The National Historic Preservation Act

In addition to the specialized nature of items 8 and 9 above, tower lights and RF energy exposure, items 4 and 5 are subject to special treatment at the federal level. The National Historic Preservation Act (NHPA) outlines requirements that the federal government must observe relating to impacts on historical sites and native religious sites. To streamline its handling of its responsibilities, the FCC has entered into a "nationwide programmatic agreement for review of effects on historic properties for certain undertakings approved by the Federal Communications Commission" (NPA) with historic preservation authorities.

The FCC explains that the NPA is a "win-win" agreement because it succeeds in [8]:

- Refining the process for identifying "eligible properties," by requiring the use of records in the SHPO (state historic preservation offices) or in THPO (tribal historic preservation offices).
- Excluding certain categories of undertakings from review that, as determined by the working group, are not likely to adversely affect historic properties. These undertakings include:
  - —Enhancements to towers,
  - —Temporary towers,
  - —Replacement towers,
  - —Certain towers constructed in industrial and commercial areas,
  - —Certain towers constructed in utility corridor rights-of-way, and
  - —Towers constructed in SHPO/THPO designated areas.
- Establishing specific procedures for contacting SHPOs, including a provision authorizing tower constructors to proceed with construction if a SHPO does not respond within thirty days.
- Creating standard forms, known as the submission packet, establishing a uniform, nationwide standard for filings to SHPOs. On January 18, 2005, OMB approved FCC forms 620 and 621 (see OMB No. 3060-1039).
- Establishing guidelines for consulting with federally recognized Native American tribes and NPOs (national preservation offices).

SUMMARY

This chapter has provided a review of key aspects of federal regulations regarding workplace safety and environmental protection. When preparing safety programs or addressing environmental protection issues, broadcasters are encouraged to read further and seek the advice of specialists and attorneys in these fields.

References

[1] OSH Act of 1970, 29 USC 651 (b).

[2] OSHA, 29 USC 654.

[3] Curtis, Robert, OSHA.

[4] OSHA Standard Interpretation, Elements Necessary for a Violation of the General Duty Clause, Dec. 18, 2003.

[5] OSHA, 29 USC 655 (b)(8).

[6] Curtis, Robert A., director, U.S. DOL/OSHA Health Response Team, Presentation at the National Association of Broadcasters Broadcast Engineering Conference, Las Vegas, NV, April 12, 1995.

[7] GE Industrial, PCB MSDS.

[8] FCC, Learning Interactive Unit: Nationwide Programmatic Agreement, see http://wireless.fcc.gov/siting/npa/intro.html.

CHAPTER

2.6

Electric Shock—What to Do When It Occurs

ANNE DEVINE
Medtronic Emergency Repsonse Systems
Redmond, Washington

CLIFFORD D. FERRIS
University of Wyoming
Laramie, Wyoming

INTRODUCTION

This chapter deals with electrical shock—an explanation of how it occurs in the body and some of the newer techniques (e.g., automated external defibrillators [AEDs]) to revive a victim of electrical shock. It is important, must-read material for all broadcast engineers and anyone who may encounter an electrical shock victim. In this chapter, Anne Devine discusses the immediate treatment using an AED, then Clifford D. Ferris provides general information about electrical shock.

POST-SHOCK CARE FOR AN UNCONSCIOUS VICTIM

In the case of an unconscious, unresponsive victim, the rescuer should immediately summon emergency medical services (EMS) personnel, check to see if the victim is breathing and has signs of circulation, and initiate cardiopulmonary resuscitation (CPR) as appropriate. Rescuers should be familiar with current CPR guidelines published by the American Heart Association (AHA), which change periodically.

If the victim is not breathing, cardiopulmonary resuscitation should be initiated immediately in order to help maintain blood flow to the heart and brain. When the oxygen supply to the brain is cut off, brain death starts to occur within 4–6 minutes without treatment [1].

DEFIBRILLATION AND CARDIAC ARREST

It is ironic that the treatment for cardiac arrest caused by electric shock (with ensuing ventricular fibrillation or VF) is a shock from an electronic device called a defibrillator. Sudden cardiac arrest caused by ventricular fibrillation may come on suddenly and without warning, or as a result of precipitating conditions, such as an electric shock. A defibrillation shock to the heart stills the random fluttering produced by VF and gives the heart the opportunity to start beating in a normal rhythm. Manual external defibrillators are available for use only by emergency medical personnel and other trained medical professionals.

AUTOMATED EXTERNAL DEFIBRILLATORS FOR LAYPERSON USE

AEDs are a different matter, however. An AED is a small computerized device that enables anyone with minimal training to deliver a potentially lifesaving defibrillation shock to a heart in cardiac arrest—before paramedics arrive. Designed especially for the first person at the scene, an AED can be used by anyone with CPR and AED training. It is easy and intuitive to use, removing the burden of decision making from the rescuer by explaining what to do, step by step. Although rescuers should be trained in CPR and AED use, one study has shown that untrained school children could demonstrate the skills for effective and fast use of an AED [2]. An AED can be purchased for about the same cost as a computer and software, and is available through a number of manufacturers.

The rescuer should follow the CPR/AED protocols as learned in his or her nationally recognized CPR/AED course. After turning the defibrillator on, the rescuer applies adhesive electrode pads to the chest as shown in the instructions. Voice commands give instructions for each step. The AED's internal computer automatically analyzes the heart's rhythm and then advises a shock *only* if it detects a lethal rhythm such as VF. Some AEDs will prompt the rescuer to push a shock button while others will deliver the shock automatically.

The AED uses voice prompts to lead the rescuer through CPR and defibrillation until the EMS team arrives to take over and provide advanced emergency medical care. Some sudden cardiac arrest victims may require more than one shock for the heart to resume a normal rhythm. The rescuer should continue to provide CPR and defibrillation shocks as instructed by the AED.

When providing defibrillation to a victim of cardiac arrest, it is crucial that the rescuer makes sure that no one (including the rescuer) is touching the victim when the shock is delivered. Touching the victim when the shock is delivered could cause serious injury to the rescuer.

When the victim is successfully resuscitated and stable, turn him/her sideways to help maintain a clear airway. Stay with the person until the emergency medical services team arrives.

THE CASE FOR AEDs

About 95% of the people who suffer sudden cardiac arrest die before reaching the hospital because early CPR and defibrillation are not available soon enough [3]. The American Heart Association recommends defibrillation within 5 minutes of arrest for sudden cardiac arrests occurring outside the hospital [4]. It can take emergency medical services 6–12 minutes to arrive [5]—longer if they are delayed by traffic, remote locations, elevators, and multiple calls. Chances of surviving sudden cardiac arrest drop 7–10% for every minute without defibrillation [6] and brain damage can begin within 4–6 minutes [1] without treatment. One study showed that survival rates for sudden cardiac arrest brought on by VF (in the absence of electric shock) were as high as 74% with defibrillation given within 3 minutes of collapse [7].

ELECTRIC SHOCK

The possibility of electric shock must be taken very seriously in the operation of electrical, electronic, and broadcast systems because it can produce permanent injury and death. The best approach to electric shock is prevention. Human tissues individually, and the human body as a whole are electrical conductors. An electric shock is produced when a potential difference is applied across human tissues such that a response current results. The severity of the resulting shock is a function of skin quality, contact pressure, nature of the voltage source (AC or DC), and frequency when alternating current sources are involved.

General Definitions

Human body tissues, both solid and fluid, are classified as *electrolytes*. Thus, they are electrically conductive and may be characterized at DC and low frequencies as ohmic resistances. As frequency increases, human tissues also manifest capacitive properties. This latter attribute is of considerable importance with respect to broadcast and communications equipment.

Two general categories of electric shock are recognized: *microshock* and *macroshock*. The former describes an inadvertent electric shock that occurs within the body, and associated electric current levels range from 10–100 μA. Such a shock may result from improper electrical grounding techniques associated with various surgical and diagnostic medical procedures (such as cardiac catheterization) in which electrically operated sensors are placed internally in the body.

Macroshock (electric shock) is the result of simultaneous physical contact between the external body surface and two or more electrical conductors at different potentials. Electric shock can be induced by body contact between energized conductors, such as two of the phase conductors in a three-phase system, or by contacting a hot wire and neutral or ground.

CURRENT LEVELS

The electric current levels and associated physiological responses are somewhat different between AC and DC. Table 2.6-1 presents typical data for alternating current shocks at 60 Hz (50 Hz data are similar).

Alternating Current Shock

Generally speaking, women react to electric shock at current levels that are about two-thirds of those necessary to elicit the same response in men. For example, current level versus frequency studies published by C. F. Daziel indicate that the mean *let-go threshold* at 60 Hz for women is approximately 10 mA as opposed to 16 mA for men.

By a strange quirk of nature, the electrical impedance of the human body is at a minimum between approximately 30–80 Hz. This situation, along with other factors to be addressed subsequently, renders a sustained electrical shock at 60 Hz and 120 VAC rms (root mean square) potentially extremely lethal. The threshold levels indicated in the table increase markedly on either side of the low-impedance frequency window.

Direct Current Shock

The average value for the let-go current at DC is on the order of 75 mA. This value is greater than the corresponding AC peak-to-peak value at the 20 mA rms

TABLE 2.6-1
60 Hz Electric Shock Current rms Levels and Physiological Responses

Current Level	Physiological Response
1 mA	Lower limit for sensation of shock
5 mA	Upper limit for what is normally considered a harmless shock
10–20 mA	Let-go threshold: contractions of *flexor muscles* are stronger than extensions of extensor muscles; victim cannot let go of conductor and begins to perspire
30–40 mA	Tetany with sustained contraction and cramping of muscles
50–70 mA	Intense pain with physical exhaustion, fainting, permanent nerve damage, possible respiratory arrest and asphyxiation; potential for ventricular fibrillation (heart)
100 mA	Death by cardiac arrest if the current passes through the body trunk
>100 mA	Ventricular fibrillation; if shock is survived: severe burns, amnesia, severe electrolysis at body contact sites
>5 A	Little chance for survival

level. The DC threshold of sensation level is increased to 5 mA. On the high-frequency side of the window, the let-go threshold increases to an average value of approximately 60 mA at 5 kHz. A momentary direct shock current path through the heart of 60 mA rms at 60 Hz can induce ventricular fibrillation, while 300–500 mA is required for DC.

PRIMARY AND SECONDARY EFFECTS

There are both primary and secondary affects associated with electric shock. The primary or direct effect is the passage of an electric current through body tissues. Secondary effects include immediate physiological damage, psychological shock, and involuntary muscle contractions. In some cases, the latter are strong enough to cause the victim to be thrown some distance from the source of the shock, which can then result in broken bones and/or death if massive objects are impacted.

Various factors control the severity of an electric shock. The intensity of the electric current that is driven through the body as a consequence of the externally applied potential difference is the cause of the reversible and irreversible effects produced by an electric shock. The electrical resistance of the body volume involved and the contact pressure and surface area at the electrical contact sites have a major influence on the severity of a shock. The potential seriousness of an electric shock is influenced by the paths that the shock current takes through the body. Current paths that do not involve the body trunk or head are normally not fatal, although permanent injury may result in the form of partial or complete paralysis of a limb, neuromuscular damage, loss of limbs, and scars. A severe electric shock to the head may produce loss of memory, loss of motor function (voluntary muscle motion), or death. Shock current paths that pass through the body trunk in the region of either the heart or lungs can cause death. A current path through the heart may induce cardiac arrest, while a path through the lungs can paralyze the nerves that control the breathing diaphragm. The two most dangerous current paths through the chest are from hand to hand and hand to leg or foot, especially the path from the left hand to the left leg/foot (see Figure 2.6-1).

Human response to an electric shock is both objective and subjective. The condition of the skin surface that comes into contact with the shock source and the contact pressure are major factors, since they determine the contact interface resistance. If an adult tightly grasps the probes of an ohmmeter between the thumb and forefinger of each hand, the resistance that is subsequently measured will vary widely across individuals. Individuals whose hands are calloused from physical activity may generate resistance readings on the order of 1 or 2 Megohms with dry hands, and on the order of 100 kilohms when the hands are wet. Individuals with soft and pliant skin may generate readings of 5–10 kilohms with dry hands, and as low as a few hundred ohms when the hands are wet or damp with perspiration. The resulting current changes by a factor of 1,000 from a 1 Megohm contact resistance to a 10 kilohm contact resistance.

Individuals who experience the same low-level electric shock may experience very different sensations. Among the sensations reported are tingling (pins and needles), burning, buzzing, and jolts. To a large extent, the body location of the shock and the emotional state of the individual at the time may strongly influence the response. Response to an electric shock tends to be amplified in individuals who are already tense. Shocks that occur close to major nerves or nerve endings that lie close to the skin surface also tend to elicit amplified reactions. An analogous example is the reaction when one bumps an elbow and triggers the "funny-bone" response.

There are several secondary effects of electric shock that may produce serious injury even when the shock itself is not life threatening. Most commonly the startle reflex is triggered, which may cause the victim to flail

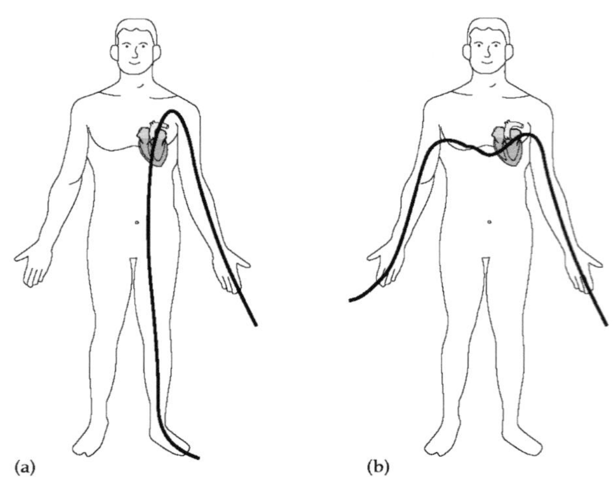

FIGURE 2.6-1 Most lethal electric shock paths: (a) left hand to left foot, and (b) hand to hand.

limbs, fall down, drop objects, or otherwise move in a manner that causes injury. High voltages, including those produced by capacitor discharges, can produce generalized severe muscle contractions of sufficient intensity to throw a person across a room. Because of the violence of such reactions, the victim may sustain serious bone fractures, internal injuries, or even death from impact with nearby objects. The violence of the muscle contractions alone may be sufficient to fracture spinal vertebrae, produce shoulder dislocations, and other injuries.

DIRECT PHYSIOLOGICAL EFFECTS

Table 2.6-1 shows that the medical consequences of an electric shock range in severity from no effect at all to minor burns, muscle and nerve damage, and, in the extreme case, death. Factors that affect the physiological consequences of electric shock are:

- The effective electrical resistance between the shock contact points on the body.
- The portion of the body traversed by the shock current path(s).
- Type of electrical source that produced the shock.
- Victim's body weight.

The first two items have been addressed to some extent in the introductory section. As contact resistance at the shock points decreases, the resulting shock current increases along with the potential for severe physiological damage. Current paths through the head and the body trunk are potentially the most lethal. The electrical properties of the human body are frequency dependent such that minimum electrical impedance and consequently maximum shock currents occur in the frequency range from 30–80 Hz. Shock threshold parameters are lower for women than men by approximately 30%, and may result from differences in skin condition and associated electrical resistance.

Shocks that exceed the let-go threshold are particularly dangerous because the victim begins to perspire. The resulting increase in perspiration moisture on the skin surface reduces contact resistance at the shock sites, thus increasing the shock current and the effects thereof.

Alternating current shocks at 50–60 Hz are potentially more serious than DC shocks because the body impedance has a minimum value. Radio frequency (RF) shocks have some additional attributes that will be discussed subsequently.

Soft Tissue and Skeletal Damage from Electric Shock

Human tissues are electrolytes and they can be grossly characterized electrically by the simplistic parallel resistance capacitance model illustrated in Figure 2.6-2 (VS represents the applied electric shock voltage). Both the resistance and capacitive components of tissue are functions of frequency, which is why the model elements are shown as R(f) and C(f). Only the resistive element is significant at DC. As frequency increases into the RF range, both the resistance and capacitance values decrease. The relative change in these parameters results in the minimum value of body impedance in the 30–80 Hz range. At radio frequencies, the capacitive component of the current is significant.

Electrolysis is a primary effect of the passage of an electric current through tissue. Ionic dissociation of the electrolyte (tissue) occurs with the production of heat. When current level is high, steam is generated from the tissue water component, with the consequent rupture of cell membranes and cell destruction. Sustained passage of current causes coagulation of proteins and tissue death along the current path. Medical effects include surface burns at the shock contact sites, penetrating burns along the current path (especially when RF currents are involved), and the potential for permanent damage to both nerve and muscle tissues.

Traditionally, thermal burn severity has been classified as first, second, or third degree, with a designation of fourth degree for certain burns resulting from electric shock. Thermal burns resulting from localized contact with a hot object differ from those produced by electric shock in that only one circumscribed portion of the body surface is affected. With electric shock, burns may occur at all contact sites. Burns may be categorized as follows:

- *First degree.* Superficial with local discoloration or reddening of the skin in the burn area, mild swelling, and associated pain. Healing is normally rapid and without medical complications unless the victim has some underlying medical condition such as diabetes.
- *Second degree.* Burn damage penetrates more deeply into the skin than in the first-degree case. The skin at the burn site develops a red or mottled aspect

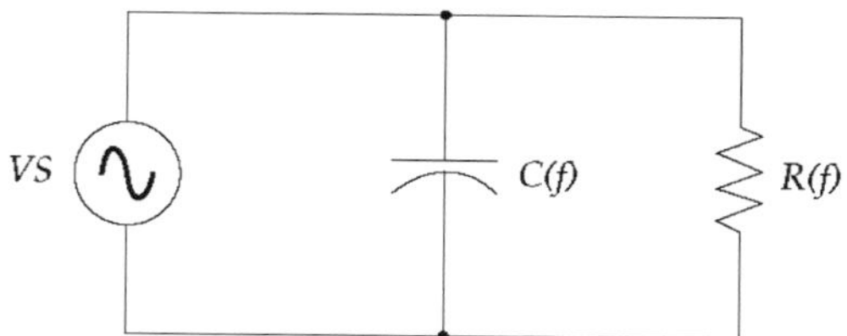

FIGURE 2.6-2 Elementary electrical model for tissue impedance.

with blister formation and substantial pain. The area may swell considerably over the course of several days. The skin surface becomes wet from loss of plasma from the damaged skin layers. Medical attention should be sought to prevent secondary infection, and especially in cases of underlying medical disorders.

- *Third degree.* Deep penetration into the skin occurs, often into the underlying tissue, with coagulation of the skin tissues, destruction of red blood cells, and charring of the skin. The skin surface may initially appear white or charred, or it may resemble a second-degree burn. Full healing of the skin may occur only at the edges of the burned area, with scar tissue replacing the normal tissue in the central portion of the area injured. Risk of infection at the burn site is high and medical attention is important.
- *Fourth degree (electrical burns).* Overlying tissues are charred such that underlying bone is exposed. This type of injury can be sustained from contact with high-voltage transmission lines or the high-power output section of an RF transmitter. If the incident does not result in death, permanent physical disability can be anticipated.
- *High frequency.* Deep penetration of the tissues at the contact sites. Intense electrolysis and searing of the tissue. Fourth-degree burns may result. The shock current may pass over the skin surface (because of the capacitive component of the skin tissue) as well as through a portion of the body. Multiple damage sites may occur, especially when the current passes over the skin surface and then arcs to grounded objects in the near vicinity. Because high-frequency burns are deep with severe electrolysis, they are very slow to heal and very painful. Immediate medical attention is important and necessary.

Burns produced by high-frequency energy have beneficial medical uses although they may have grave consequences in the workplace. *Electrosurgery apparatus* (ESA), also called an *electric scalpel*, is used to provide the surgical functions of cutting, cauterizing, or simultaneous cutting and cauterizing. In effect, a controlled high-frequency shock current is applied at the surgical site (shock entrance site). A large contact-area return path (exit) electrode is located elsewhere on the body. The current density at the return electrode is normally very low so that skin injury does not occur. ESA apparatus typically operates in the 300–400 kHz range. Appropriate modulation of the waveform produces the surgical functions required. Even when ESA is used under highly controlled conditions, extraneous burns may occur to the patient if the surface conduction component of the current finds an alternative path to ground (such as an area of the skin surface that has accidentally contacted some grounded metal object that is present in the operating room).

Electric arcs are another source of electrical burns, which may or may not be associated with electric shocks. Burns can be produced by heat and flying hot debris from electric arcs even if the victim is not in electrical contact with the associated circuit. The temperatures within an electric arc can be on the order of 5,000° C and above. Fourth-degree burns may be expected if an arc occurs in conjunction with an electric shock.

Skeletal damage may be a secondary effect produced by an electric shock. High voltages, either from momentary contact with a continuous source of electrical energy or a charged capacitor, generally cause violent muscle contractions. Such muscle contractions may be of sufficient intensity to break bones or cause joint dislocations. In some cases, the victim is propelled away from the shock source. Broken necks and other serious skeletal injuries have been reported when the victim has been thrown against some object (such as a table or workbench).

Shock Paths—Limbs and Extremities

In most cases, electric current paths restricted to the limbs and extremities (excluding the head) are not fatal, although permanent damage and/or disability may result. Loss of a limb may occur, and permanent skin scars, partial or total limb paralysis, and physical impairment are not uncommon.

Shock Paths—Head

High-intensity shocks to the head may cause permanent injury to the neck and spine (as a result of body spasms), loss of memory and/or motor function, and permanent brain damage. Various bones in the body may be fractured as a result of severe and generalized muscle contractions. At one time electroshock therapy, in which high-intensity shock currents were passed through the temples, was used medically to treat patients with certain mental disorders. There are records in the medical literature of the many side effects of such treatments.

Shock Paths—Body Trunk

When the shock current passes through the heart or lungs in the body trunk, grave consequences may occur. The human body is a complex electrical machine. Sensory information from our environment is translated into electrical pulses that are transmitted along nerve pathways to sites in the brain for central processing. Our senses of sight, smell, hearing, temperature, and touch depend on electrical pulses. Our voluntary actions associated with muscle movement and locomotion, as well as involuntary actions such as digestion, heartbeat, and respiration, depend on electrical pulses. Muscle motion is activated when pulses are transmitted along nerves to *myoneural junctions* where they then initiate muscle extension or contraction.

The heart mechanically is a simple four-chambered pump that is composed of muscle tissue. Its pumping action is controlled by electrical signals generated by a physiological multivibrator, called the *pacemaker* or *sinoatrial (SA) node*. The SA node is a small region of specialized tissue located at the top of the heart. A stylized diagram of the heart is presented in Figure 2.6-3.

The right and left atria that form the two upper chambers of the heart are reservoirs that store the blood that is returned to the heart from the lungs and body organs. Through appropriate valves, these atrial chambers supply blood to the ventricles, which are the two lower and larger pump chambers. When the ventricular muscles contract to produce a heartbeat, blood from the right ventricle is pumped through the pulmonary artery to the lungs (the pulmonary circulation); blood from the left ventricle is pumped through the aorta for distribution to the body organs (the systemic circulation). Additional valves in the ventricles prevent blood backflow. The beating of the heart is controlled by the regular electrical pulses that are generated by the pacemaker, which cause the ventricles to contract in unison (one pulse per heartbeat). The repetition frequency of the pacemaker in a normal heart automatically adjusts to the body's demand for oxygen through a complex physiological feedback system.

Contraction and relaxation of the skeletal muscles also generate electrical pulses. Contraction of the heart is initiated by electrical pulses generated by the pacemaker, but at the same time, the contraction and relaxation of the heart muscle tissue produces electrical pulses. This electrical activity caused by the beating of the heart can be detected on the external body surface because the human body is an electrical volume conductor. The electrical impulses produced can be detected and recorded electronically to produce an *electrocardiogram* (ECG). There are several protocols and body locations for placing electrodes on the surface of the body to record an ECG. When a reference electrode is placed on the inside of the right ankle and two recording (active) electrodes are placed respectively on the insides of the left ankle and left wrist, a signal of the form illustrated in Figure 2.6-4 can be obtained.

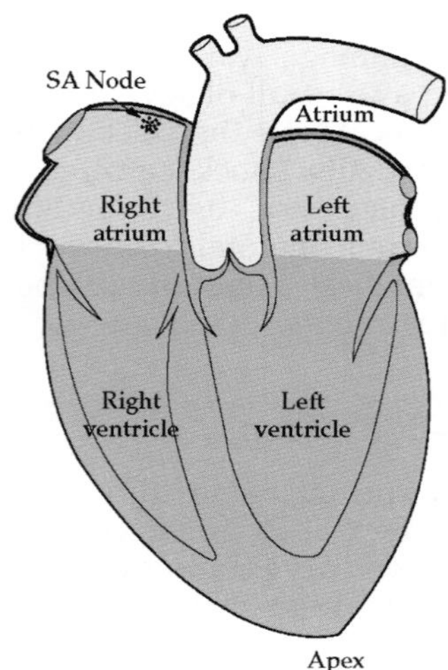

FIGURE 2.6-3 Stylized anatomical cross section of the heart.

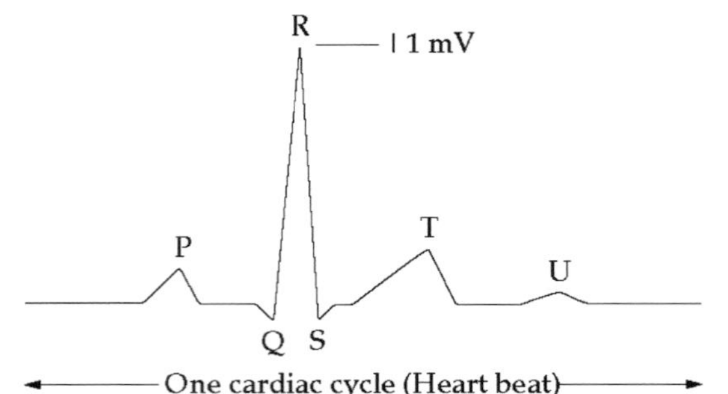

FIGURE 2.6-4 Stylized ECG waveform for one cardiac cycle.

The figure is a representation of the lead II signal produced by one of the standard methods for recording the electrical activity from the heart. The presentation has been linearized and stylized for purposes of discussion. Actual ECG records are somewhat less linear and the peaks of some of the pulses are rounded. The illustration represents one cardiac cycle or heartbeat. In the figure, the time axis has been normalized to a length of one heartbeat. The major events in the record (waves) are designated as P, Q, R, S, T, and U. The peak-voltage amplitude produced by the R wave on the body surface typically ranges from approximately 1–1.5 mV. These waves or events relate to the beating of the heart as follows:

- The P wave represents the electrical activity produced when the atria contract to force blood into the ventricles.
- The R component of the QRS complex represents the contraction of the ventricles and the pumping of blood out of the heart and into the aorta and pulmonary artery.
- The T wave represents the electrical activity of the relaxation (called repolarization) of the ventricular muscles following their contraction.
- The electrical signal produced by the relaxation (repolarization) of the two atria is masked by the high-amplitude R wave.
- The origin of the U wave is somewhat unclear, but is thought to be produced by the repolarization of the aorta after the blood volume produced by the left ventricle has passed through its length.

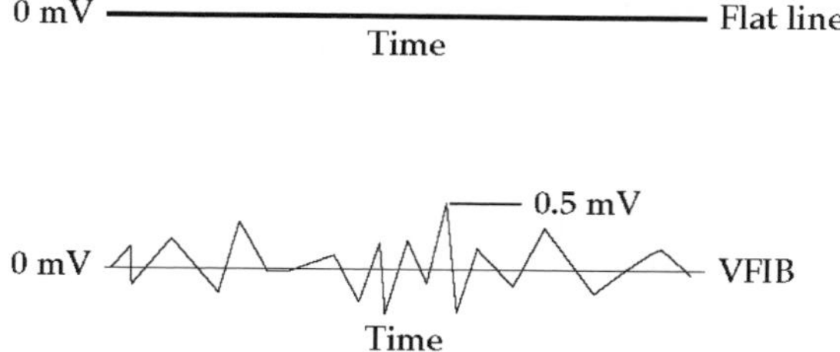

FIGURE 2.6-5 Cardiac arrest ECG waveform.

When an intense electrical shock current passes through the chest cavity, it may cause the entire heart muscle to contract and cease beating during the time duration of the current. If this shock current is of short duration, normal heart rhythm usually returns following the passage of the current and there is no irreversible cardiac tissue damage. Less intense shock current levels can momentarily disrupt the regularity of the normal heart rhythm. The heart is most sensitive to external electrical stimulation during the occurrence of the T wave. A shock current amplitude that does not ordinarily disrupt cardiac rhythm during other portions of the cardiac cycle may produce cardiac arrest if it occurs during the T wave event.

Either of two conditions is described under the designation cardiac arrest: *cardiac standstill* and *ventricular fibrillation (VF)*. As the name implies, the heart does not beat in the condition of standstill, and the ECG is a flat line. Ventricular fibrillation describes a condition in which individual muscle groups of the ventricles contract randomly and out of unison. No blood is pumped and the blood pressure drops to zero, but a low-level random ECG signal is produced, as is shown in Figure 2.6-5. To differentiate between the conditions of flat line and VF, an ECG recorder is required.

Involuntary contraction of the respiratory muscles is often a consequence of the passage of an electric shock current through the chest region. Since respiration ceases, the victim can die from asphyxiation if the current is not interrupted. Cessation of breathing (respiratory paralysis) can occur if the electric shock current that passes through the body trunk damages the nerve that controls the diaphragm. In such a case, the victim typically inhales deeply; the chest then expands and becomes rigid and remains so after the shock current has ceased. Initially, the heart may or may not be affected. In the latter case, the heart continues to beat, but the beats become irregular and erratic with time as the blood carbon dioxide level increases (acidosis). Acidosis produces irritability of the heart muscle tissue and makes restoration of a normal heartbeat difficult even if respiration can be restored.

DANGERS OF 120 VAC RMS 60 HZ ELECTRIC SHOCKS

The electrical impedance of the human body is at a minimum value for 60 Hz. A voltage level of 120 VAC rms is generally not sufficiently high to cause muscle contractions that throw the victim away from the source of the shock current. It is also low enough that a strong startle reflex may not occur. Consequently, the subject may not be able to let go of the source if the conductor has been firmly grasped. If the effective impedance presented by the body to the shock source is 500 Ω, then the shock current will be 240 mA rms, which is more than enough to produce ventricular fibrillation. Even if a higher impedance was presented initially, the continuation of the shock produces perspiration buildup on the skin that lowers the effective resistance. Shock charges on the order of 100 mA seconds (rms current values) are sufficient to produce VF. The cardiac cycle at a heart rate of 70 beats per minute is 857 ms or 0.857 seconds. At a current level of 240 mA, the shock energy generated during one cardiac cycle is 206 mA seconds. Thus, VF can be anticipated as one of the consequences of a sustained 120 V 60 Hz shock.

ELECTRICAL SHOCK PREVENTION

Because death or permanent disability may be a consequence, electric shock must not be taken lightly. Shock prevention is far superior to medical remediation after the fact. Some general guidelines are:

- *Workplace safety regulations.* Observe all company and local regulations regarding electric shock prevention. In many industrial settings, the regulations promulgated by the Occupational Safety and Health Administration (OSHA) apply as well. See also Chapter 2.5, Managing Workplace and Environmental Hazards.
- *Clothing and jewelry.* When working in close proximity to exposed energized electrical circuits, loose clothing (remove or secure neckties and scarves) and jewelry (rings, wristwatches, bracelets, necklaces) should not be worn. Exposed skin should be minimized (do not wear shorts, sandals, or open-toed shoes, etc.).
- *Protective clothing.* Where applicable, observe all company regulations regarding the use of safety clothing and accessories. Safety goggles, nonmetal hard hats, electrically insulating safety shoes, and various safety garments (e.g., electrically insulating gloves) may be required. OSHA regulations generally require the use of goggles, hard hats, and safety shoes in many industrial venues.
- *Equipment servicing.* Power switches should always be turned off and electrical/electronic equipment should be disconnected from the power mains before opening the equipment cabinet for servicing or calibration of internal circuits. Never rely exclusively on switches to disconnect power. Sometimes they are defective, and sometimes a switch may be miswired or mislabeled.
- *High-voltage capacitors.* High-voltage capacitors associated with power supplies, cathode ray tubes, radio transmitters, and other high-voltage electronic and RF equipment must be discharged prior

to equipment repair or servicing. Use appropriate discharge cables designed for this application. Always discharge the external surfaces of cathode ray tubes and other devices on which high-static charges can develop.

- *Live circuits.* When diagnostic tests or calibration procedures must be conducted on energized equipment, use all appropriate electrical safety procedures. A large body of modern electronic equipment utilizes switching/regulating power supplies that are potentially lethal because portions of the internal printed circuit boards are hot at the AC power line voltage. For servicing, an isolation transformer must be used between the power mains and the equipment to reduce electric shock hazard. The hazard results from a bridge rectifier that is connected directly across the AC power mains when the equipment is energized. This condition causes the circuit board ground to be at a different potential from the AC power line neutral or earth ground. Without the use of an isolation transformer, accidental contact between earth ground and the circuit board ground may result in a lethal electric shock. As general practice, an individual should not work alone on energized circuits. There should always be at least one noninvolved observer who can disconnect electric power and obtain help as required in the event of an accident.
- *Earth ground connections.* Never disable or remove the earth ground connections on power cords and within equipment.
- *Equipment design and modification.* All appropriate codes and guidelines should be observed, such as those promulgated by the National Electrical Code (NEC) of the National Fire Protection Association (NFPA); Medical Devices Safety Act of 1971, 1973, and subsequent amendments; American National Standards Institute (ANSI); Institute of Electrical and Electronics Engineers (IEEE); American Society of Mechanical Engineers (ASME); Underwriters Laboratories (UL); Canadian Safety Association (CSA); industry standards for equipment design and manufacture; and other regulations and guidelines as appropriate
- *Antenna installation.* Erection of antenna towers and antenna installation requires skilled workers. Mechanical assembly should follow good engineering practices and be in compliance with all local and federal codes. Ensure that overhead power lines and other electrical distribution systems are not accidentally touched by personnel, equipment, or structural components. Where appropriate, structures must be correctly grounded.
- *Transmitters.* The high-voltage power supplies and high-voltage RF-tuned circuits associated with transmitters present strong potential for electric shock. Personnel who work in close proximity to such installations must exercise caution. Corona discharge to electrically conducting objects (including human flesh) that come in close proximity to energized circuits is an ever present danger. Areas of high energy concentration, such as low radius bends in conductors, connections to high-voltage capacitors and inductors, and the like, are potential sources for corona discharge. The danger for both electric shock and severe RF burns cannot be ignored.

High-Frequency Hazards

There are other electrical hazards in addition to direct electric shock that are associated with high-frequency equipment. Chapter 2.4 covers the subject of Human Exposure to Radio Frequency Energy.

Ungrounded conducting surfaces located in high-flux electromagnetic fields can develop large induced circulating currents. When high-power radar and communications transmitters are positioned in close proximity, such as on some ships, an electric discharge hazard can develop. This situation is the basis for the requirement that aircraft be electrically grounded before fueling, or arming in the case of military aircraft, and that radio transmitters be turned off in the vicinity of blasting operations. These induced circulating currents may be of sufficient intensity to activate ordinance if a discharge path is generated, and to produce serious electrical shocks in personnel. Caution should be observed when working close to energized transmitting antennas, regardless of the frequency. Nearby ungrounded conductors may be at dangerous potentials.

FIRST AID

Prevention is the best first aid for electric shock. When a person has sustained or is sustaining an electric shock, the following actions may serve to save a person's life or minimize injury.

Shock in Progress

In the situation when an individual is sustaining an electric shock and cannot let go of the electrical source, the safest action is to disconnect the circuit. This could be accomplished by tripping a circuit breaker in the power mains supply, or by pulling the power line plug that feeds the equipment involved if this can be accomplished safely. Shout for help! Under no condition should the rescuer touch the victim, since the rescuer's body also may then be placed in the dangerous current path. In the event that neither the circuit breaker nor the power plug can be located, then an attempt should be made to separate the victim from the electrical source by physical means using an insulating object. The implement employed should not contain metal or other electrically conducting materials. The optimum device is a *hot stick* (with the use of insulating lineman's gloves) to pull or push the victim away from the hazard. Pulling on a hot stick normally provides more

motion control than pushing. Lacking a hot stick, a wooden broom handle or wooden chair/stool may be used in an attempt to free the victim. The rescuer must be careful not to touch either the victim or the shock source, and thus also become a victim. Once the victim has been completely separated from the shock source (either by disconnecting the power or by physical removal), first-aid measures should be initiated immediately.

Post-Shock Care with Conscious Victim

In situations when the victim is conscious and moving about, convince the victim to sit or lie down. Often there is a delayed action to electric shock that causes the individual to faint or collapse. Shout for help! Summon the appropriate in-plant paramedical personnel immediately, or call 911. If the arrival of emergency medical personnel will be delayed, check for electrical burns and any additional immediately obvious injuries. In instances of severe electric shock, there will be burns at both the current entrance and current exit sites. These burns should be covered with (preferably sterile) dry dressings. Examine the victim for possible bone fractures and dislocations if the individual sustained spasms associated with the shock or was violently thrown from the source (whether or not there was impact with nearby objects). As required, splints should be applied if the rescuer has appropriate training and suitable materials are available. Use a coat or a blanket to cover the victim when the environmental temperature is low, or if the individual complains of feeling cold.

Post-Shock Care with Unconscious Victim

In the case of an unconscious victim with no movement or response, call 911 immediately or ask someone else to do it. Open the airway and check for breathing. If the victim is not breathing normally give two rescue breaths over one second each. Begin chest compressions in the ratio of 30 compressions to two breaths (100 compressions per minute). When the AED arrives, turn it on, bare the victim's chest and apply the electrode pads. Then follow the visual and voice prompts given by the AED.

Defibrillator Use

See the first part of this chapter. The massive countershock from a defibrillator causes a massive contraction of the heart and stills the random fluttering produced by ventricular fibrillation when it exists. When conditions are ideal, a normal heartbeat starts spontaneously following the countershock. However, if a defibrillating shock is indiscriminately applied to an unconscious person whose heart is beating, cardiac arrest can be induced, just the condition that the instrument was designed to correct.

Key Terms

Atrial: Pertaining to the atrium or two reservoir chambers at the top of the heart.

Automatic external defibrillator (AED): Portable defibrillator designed for layman use.

Cardiac: Pertains to the heart.

Coagulation: The formation or production of a blood clot.

Defibrillator: An electronic instrument used in an attempt to shock the heart into beating normally.

Electrocardiogram (ECG or EKG): A visual record of the heart's electrical activity as recorded from electrodes placed on the body surface.

Electrolyte: A substance that chemically decomposes by the passage of electric current through it.

Extensor muscle: A muscle that extends a joint.

Fibrillation: Spontaneous and random contraction of individual muscle groups.

Flexor muscle: A muscle that flexes a joint.

Pacemaker: A small area of specialized tissue at the top of the heart that produces the electrical pulses that initiate heartbeats.

Tetany: The state of a muscle when it is in sustained contraction.

Ventricular: Pertaining to the main pump chambers of the heart.

References

[1] American Heart Association. Sudden Deaths from Cardiac Arrest—Statistics, 2004; see at http://www.americanheart.org.

[2] Gundry, J. W., et al. Comparison of Naïve Sixth-grade Children with Trained Professionals in the Use of an Automated External Defibrillator, *Circulation*, 100, 1999, pp. 1703–1707.

[3] American Heart Association, CPR Facts and Statistics, March 2006; see at http://www.americanheart.org.

[4] American Heart Association 2005 Guidelines for Cardiopulmonary Resuscitation and Emergency Cardiovascular Care, *Circulation*, 112, 2005, pp. IV-204–IV-205.

[5] Medtronic Emergency Response Systems. Review of Published Literature.

[6] Cummins, R. O. "From Concept to Standard of Care: Review of the Clinical Experience with Automated External Defibrillators," *Annals of Emergency Medicine*, 18, 1989, pp. 1269–1275.

[7] Valenzuela, T. D., et al. Outcomes of rapid defibrillation by security officers after cardiac arrest in casinos. *New England Journal of Medicine*, 2000, 343-120-1209.

Bibliography

American Academy of Orthopaedic Surgeons. *Emergency Care and Transportation of the Sick and Injured*, 3rd. ed. Menashi, WI: George Banta, 1981.

American Red Cross. *First Aid: Responding to Emergencies*. St. Louis, MO: Mosby Life Line, 1991.

Berkow, R., ed. *The Merck Manual of Diagnosis and Therapy*, 16th ed. Rahway, NJ: Merck, 1992.

Dalziel, C. F. Electric Shock, in *Advances in Biomedical Engineering*, vol. 3, eds. J. H. U. Brown and J. F. Dickson. New York: Academic Press, 1973, pp. 223–248.

Ferris, C. D. *Introduction to Bioinstrumentation*. Clifton, NJ: Humana Press, 1978.

Lee, R. C., et al., eds. *Electrical Injury: A Multidisciplinary Approach to Therapy, Prevention, and Rehabilitation*, Annals of the New York Academy of Sciences, vol. 720. New York: The New York Academy of Sciences, 1994.

Valenzuela, T. D., et al. "Outcomes of Rapid Defibrillation by Security Officers after Cardiac Arrest in Casinos," *New England Journal of Medicine*, 343, 2000, pp. 1206–1209.

Additional Sources of Information

In addition to the publications listed under the Bibliography, readers can find detailed information on human physiology, cardiac function, electric shock, and electrical properties of tissue in the following sources:

Ackerman, E. *Biophysical Science*. Englewood Cliffs, NJ: Prentice-Hall, 1962.

Schwan, H. P. Electrical Properties of Tissues and Cell Suspensions, in *Advances in Biological and Medical Physics*, vol. V, eds. J. H. Lawrence and C. A. Tobias. New York: Academic Press, 1957, pp. 147–209.

Vander, A. J., et al. *Human Physiology*, 6th ed. New York: McGraw Hill, 1994.

Webster, J. G. ed. *Medical Instrumentation, Application, and Design*, 2nd ed. Boston: Houghton Mifflin, 1992.

CHAPTER

2.7

Broadcast Facility Security, Safety, Disaster Planning, and Recovery

RICHARD RUDMAN
Remote Possibilities
Studio City, California

THOMAS G. OSENKOWSKY
Radio Engineering Consultant
Brookfield, Connecticut

INTRODUCTION

Risk assessment and disaster preparedness and recovery plans may look good on paper, but the ultimate test is always the next real emergency. This chapter is for designers and managers of critical communications facilities that may be at risk from natural and man-made disasters and terrorist attacks. The premises are the same here as in any other area of disaster and contingency planning:

- Disaster readiness should be an integral part of the design process.
- It is not possible to protect against every possible risk.
- Periodic testing is the only way to know if a plan has a chance of working, short of an actual event.

Adapting to major emergencies that take a station from normal on-the-air operations into the uncertain world of survival in a split second can be daunting. Offering specialized emergency training courses to key personnel can be beneficial to maintaining continuity in the face of emergencies. Personnel in a broadcast organization can receive training so that they know how to shift their work style to emergency-response mode at a moment's notice.

LEGAL OBLIGATIONS AND ISSUES

Before embarking on a disaster planning process, a station should carefully research all legal obligations for preparedness under federal, state, and local obligations as they apply to location, ownership, and investment makeup. This is best accomplished by working in concert with legal counsel.

If outside workers set foot in the facility, some local and state statutes require preparation of a special safety manual, which should include installation standards, compliance with lockout/tagout[1] procedures, and emergency contact names and phone numbers. Ensure outside contractors carry proper insurance and are qualified, licensed, or certified to do the contracted work. If audiences are hosted at the facility, there may be an obligation to plan for their well-being for some specified period of time should a disaster occur while they are on the premises.

WORKPLACE SAFETY BEFORE THE EMERGENCY

Employers are expected to assure safety in the workplace at all times. Some states, such as California, have passed legislation that mandates that most employers identify hazards and protect workers from them. At the federal level, rules issued by the Occupational Health and Safety Agency (OSHA) are relevant to broadcasters. A strong foundation of built-in day-to-day safety practices can lessen the impact of major

[1]*Lockout/tagout* is one of a set of standard safety procedures that assures that energy is removed from equipment during installation and maintenance. It assures that every member of a work detail is clear before power is reapplied.

emergencies. Some safety procedures are not obvious, for instance:

- Plate glass in doors should have a high safety rating to avoid an accidental workplace injury.
- An earthquake, hurricane, or a tornado can hurl heavy objects through the air at lethal velocities.
- Equipment racks and storage shelves are often not properly secured to floors or load-bearing walls that would prevent them from tipping over during an earthquake.
- Equipment racks should be tethered, rather than firmly bolted.
- While securing heavy objects is mostly common sense, consult experts for special cases.
- Ensure that safety chains and wires are attached to heavy objects in the studio, such as TV studio lights, video monitors, or large speakers.
- Special anchors and fasteners should be installed that can restrain heavy desktop peripherals such as computer monitors if the work surface should fall over.
- For mission-critical areas, consider bolting workstations to the floor and securing heavy equipment to wall studs with properly rated fasteners.
- Cables can snap, monitors can implode, and delicate electronics can be smashed.

While some regions have had no recent seismic activity, note that some seismically quiet parts of the earth have been given a long overdue rating by respected seismologists.

Maintaining safety standards is difficult in any size organization. A regularly updated written safety manual with specific practices and procedures for normal workplace hazards as well as emergency-related identified hazards may be required under law, and could help lower insurance rates.

EMERGENCY PLANNING PROCESS

Emergency planning is integral to a functional and reliable facility. The designer, consultant, or person responsible for maintenance of a facility that will be able to survive and operate under emergency conditions, must first obtain operational and financial commitment from management. The planning process must support the main mission as well as the people who must implement the plan under very challenging circumstances.

The technical support group of a critical communications facility may have responsibility over the total physical structure. This group must work closely with top management and legal counsel. Without competent oversight, critical electronic systems may be subject to damage or destruction from environmental hazards. Internal emergencies can be triggered by failures in air-supply systems, roof leaks, or uncoordinated telephone, computer, or AC wiring changes. While most of these emergencies are foreseeable and preventable, it requires the right people to have oversight and access to management in order to plan, test, and refine emergency response.

Successful practitioners of emergency systems' design and support must realize the potential for daily internal emergencies as well as major external emergencies in the overall planning process. Seemingly innocent acts such as plugging an electric space heater into the wrong AC outlet to which an automation or environmental computer controller is connected, for example, can take down an entire facility for hours.

FACILITIES OPERATIONS AND MANAGEMENT CONSIDERATIONS

Government and business often attempt to shift to a different management mode when disaster strikes. Doing this requires the staff to shift gears at the worst time possible—in the middle of chaos and disaster. Since the bookshelf plan is rarely exercised, these shifts often add to confusion and delayed response and recovery. The most successful facilities response and recovery plans have their roots in day-to-day operations.

It may be appropriate to review on a regular basis, such as quarterly or annually, the operation's mission, goals, and objectives to determine if there could be a benefit from a change in day-to-day operations and management. The day-to-day challenges and minor crises (e.g., sewer backup, or temporary loss of water, power, and telephone due to local construction) of the typical broadcast facility lend themselves to lessons learned from emergency management practice. Compounding the problem would be a cluster of multiple studio facilities under one roof.

Incident Commander

Programming, editorial, or engineering management personnel at single or multiple studio locations may be already doing many of the functions of an emergency incident commander.[2] A person holding such a job should have good judgment and the ability to stay cool and calm—traits that are vital to incident management during real emergency conditions.

RISK ASSESSMENT FOR COMMUNICATIONS FACILITIES

Stations should perform a realistic assessment of the risks and all the steps needed to avert or mitigate them. Consider that the ultimate expert threat-risk advice might be to move the facility before a disaster strikes. In addition to larger scene events, such as severe weather and earthquakes, do not overlook the obvious but often smaller risks. If computers, transmitters, or

[2]For more on disaster management organization, please refer to the FEMA Web site at http://training.fema.gov/EMIWeb/IS/IS100CM/ICS0103summary.htm.

telephone equipment depend on cool air, it is unlikely they can continue to operate during a heat wave when the air conditioner has failed. If a water main or wastewater pipe runs through the central power and telecommunications room, a leak in that pipe could disable nearby equipment.

Reliability calculation is an educated prediction based on a number of factors. Emergencies can be counted on to introduce variables into the reliability equation. A system with a calculated reliability of 99.999% seems as if it should be quite reliable. However, even this high percentage may represent several minutes of outage over a 1 year period that may occur in multiple short instances or one prolonged period. Strive to design beyond that level of reliability to have a greater chance to stay on-line during major emergencies. Double, triple, and even quadruple redundancy might be realistic options when 100% uptime is mandated by facility management. However, facility designers must balance realistic redundancy with realistic uptime expectations.

Today's facility designer must have a deep professional commitment to ensure continued operation during and after a major catastrophe. Disaster planning, for example, for a force 3 hurricane, may simply entail boarding up the windows, installing a satellite receiver to provide programming continuity, and leaving town for the duration. Some facilities plan for uninterrupted operation, and even staffing in the face of severe weather. In some communities the communication facility's mission may be an essential tool of local government emergency management.

Realistic Risk Assessment

A *realistic risk* list must contain specific hazards based on local conditions, such as:

- Finding the high water marks for the 100 and 150 year storms, and other flood potentials and whether they will affect the facility.
- Assessing social, political, and governmental conditions.
- Determining commercial electrical power reliability.
- Tracking weather conditions, including icing conditions, lightning, and tornado potential.
- Noting the geography and the related risks from adjacent structures or trees.
- Researching the local geology for earth movement.
- Assessing the potential for terrorism and security risks.

Assess the specific local hazards (e.g., towers, power systems) that could be triggered by deliberate vandalism by present and former employees or external parties.

Other factors that can make the facility an easy target include

- Nearby man-made hazards or risks (e.g., dams).
- Special on-site hazards (e.g., towers or antennas).
- Disgruntled neighbors.
- Construction of the facility with respect to earth movement or other local geologic conditions.[3]
- Hazardous materials on the premises (e.g., propane tanks).
- Communications links to the outside world (e.g., satellite antennas and internet e-mail.
- Electrical power (e.g., transformers).
- Other utilities (e.g., exposed gas meters).
- Buried pipelines (e.g., gas).
- Computer virus and security breaches.

Examine the effect specific threats to critical points in the facility's infrastructure would have on the ability to maintain service and provide safety to personnel:

- Damage to towers, Antenna Tuning Units (ATUs), transmission lines, buildings, generators, and fuel tanks.
- Fire.
- Component failure due to neglect or wear.
- Tower failure due to overloading.
- Equipment failure due to vehicular contact.
- Power surges, improper mains voltage, or phase loss.
- Flooding due to pipe breakage.
- Prolonged outage due to lack of spares.
- Tower failure due to improper tensioning, out of plumb, or galvanic action eroding guy anchors.

FIRST RESPONDER COMMUNICATIONS BACKUP SYSTEM

Broadcasting operations vary by market from single-story licensee-owned buildings, to rentals in a high-rise office building, to locations in office parks. Transmitter sites in some of the larger cities are often located in high-rise buildings, such as the Empire State Building, 4 Times Square, Sears Tower, or Prudential Center. All need to have plans to cover varying types of incidents whether natural or man-made and the involvement of the first responders whose task it is to safely deal with the occupants of the property.

Disaster planning is no longer a back-burner issue, but an everyday business continuity necessity. Included in that planning are issues concerning first responders and how they will deal with the incident at hand and how they will safely evacuate station personnel.

[3]Information stemming from the Northridge earthquake in 1994 has influenced seismic building codes in California and beyond for many types of structures. The Northridge earthquake showed that some high-rise structures thought to be quake safe really are not. Designers should also be aware that seismic building codes usually only allow for safe evacuation. They do not embody design criteria that prevent major structural damage. Quake safe may not be quake proof just as water safe may not be waterproof.

Starting with the simple stand-alone station, first responders, whether police, firefighters, EMS, or any other municipal agency, have to assess the incident, establish a working perimeter, and be able to respond to the needs of the occupants and deliver them to safety. The occupants, hundreds or even thousands, may be on additional floors, above and below the incident.

A backup communications system for these first responders within the structure is as important as their own communications outside the building and, in some cases, more so. Most radio communications used by municipal law enforcement, fire departments, and emergency services work well in an open environment (outside), but may fail quickly inside steel and concrete structures. Emergency stairwells, elevator lobbies, below-grade parking, building support equipment, and storage areas are of primary concern and these are the areas most lacking in communications reliability.

Whether the structure is owned by the station or the station is a tenant/licensee in a multi-user building, as part of a station's disaster planning, there needs to be a resilient and redundant communications system that covers the entire structure must be in place for use by the building's emergency team as well as first responders. Both primary and secondary communications infrastructures are necessary so that a failure of one system, whether due to electronic, natural, or man-made events, does not negate the communications for the first responders and even the station's communications system within the building/facility.

By using different building risers and having the primary and secondary systems in upper and lower building areas, the chance of having at least one system operable during an incident is substantially increased. External high-power repeaters can also be fed into the system should there be an internal catastrophic system failure disabling both primary and secondary systems. With such a system a wireless interoperability system can be installed in the buildings, thus enabling the fire department, police, EMS, OEM, or any other agency to communicate when outside systems fail or are inadequate. The incident commander could, for example, issue an evacuation order to all first responders instantly, without having to worry about whether or not the information was relayed between services.

Voice Over Internet Protocol (VOIP) control of the communications systems can also be provided that will allow a remote site (e.g., fire headquarters) to communicate directly to the firefighters at the incident, query the equipment remotely to check operating parameters, and send data from the incident to the surrounding hospitals about incoming patients' vital signs, electrocardiograms, and other information.

Data networks, fiber-building backbones, distributed antenna systems, and other innovative techniques are among the present-day technologies being discussed and utilized. When preparing disaster plans, make certain to remember the first responder's communications. Lives may depend on it and stations can help implement this kind of planning.

INSURANCE

An insurance policy can cover a broadcaster for a variety of emergency conditions. In many cases, law or corporate mandate may require insurance coverage. Some special areas of coverage worth considering are:

- Flood insurance to cover rising water damage.
- Loss of income or business interruption coverage.
- General liability coverage.
- Inland property coverage.
- Marine policy to cover items specific to broadcasters.

Federal flood insurance coverage is dependent on geographic eligibility. Mountaintop transmitter sites and studios located in high-rise office buildings, for example, are not good candidates for flood insurance since it is unlikely these facilities have or will fall victim to rising waters. Flood insurance is determined by location in or near a 100 year flood zone and nearest body of water.

Flooding caused by ruptured pipes, leaking roofs, or accidental spills, for example, is covered by a property coverage policy and not by flood insurance. Loss of income or business interruption insurance provides compensation when it is no longer possible for a facility to broadcast normally. When a station is unable to broadcast, listeners or viewers will select another station. Insurance cannot predict the amount of loss the station will suffer due to the listeners who may never return to the station once they have switched away. In addition, it is not possible to accurately gauge the degree of lost ratings points and subsequent revenue loss during the outage period.

Business interruption insurance may also compensate the insured for expenses incurred for temporary measures taken to resume broadcasting. For example, the rental of a box truck used to house an AM and FM transmitter for a period while a new transmitter building is installed following an arson fire that destroyed the transmitter building. The coverage could also provide payments for security guards on duty 24 hours a day during the construction project.

General liability insurance is important for broadcasters whether they be locally or remotely programmed. It is possible that a listener, advertiser, company, or individual may sue the station for comments aired during a broadcast. Of specific concern are talk-formatted stations. Liability insurance may also cover the station for the words and actions of its employees while on duty.

Inland property coverage will insure the broadcaster for general items in the facility such as furniture, fixtures, office equipment, and computers.

So-called "marine" policies generally underwrite equipment that is specific to broadcasters. Examples include towers, ATU's, antennas, transmission lines, isocouplers, power generators, ground systems, transmitters, STL's, audio processors, remote controls, and non-stationary equipment such as that employed for remote broadcasts.

Most insurance policies provide for "replacement cost" or "new for old without depreciation." This coverage will provide for replacement of an item with its closest available match. For example, an aged tube transmitter may be replaced with its modern-day solid-state equivalent.

Most insurance policies will compensate for the loss of an item and the labor to properly install and test it. Not all policies will cover services 100%. For example, a fallen tower may be replaced with one of the exact same height and physical construction. However, if a road must be constructed to allow access to the tower site, insurance may not cover this cost. The coverage may not compensate for soil boring tests or other such activities even though required by local or state regulation. Insurance may not cover other regulatory items such as removal of underground storage tanks, etc.

Underwriters will often send inspectors to a station to verify that good engineering practices have been followed and that the insured has taken reasonable care of assets. Frequently, the station engineer will be asked specific questions pertaining to tower structures. The engineering data on towers should be kept readily available for this purpose. Towers with elevators may be subject to periodic city, county, or state inspection. The inspector may also check the main electrical service panel and other utilities.

Consider seeking help from experienced emergency communications contingency planning professionals to help devise a well-written and comprehensive emergency plan. This is especially useful if personnel tasked with day-to-day work to keep the facility operating do not have time to devote to the planning. Professionals can also help with research on factors such as geology and dealing with hazardous materials.

COMMONSENSE PRECAUTIONS

There are commonsense precautions a broadcaster can take to ensure reliable and safe operation. For facilities in locations prone to hurricanes or high winds, the tower should be evaluated to ensure it is able to withstand expected wind forces given its present loads (antennas, dishes, etc.). Many tower companies now offer such evaluation services, as do independent structural consultants. Tornados produce extreme wind forces. While few towers or building structures can survive direct hits, recent revisions to tower standards (TIA-222-G) have addressed tower design and structural integrity in high wind–prone areas of the country. See Chapter 7.2 for additional information.

Lightning protection for towers and equipment buildings is an important aspect of maintaining continuity of transmission. See Chapter 7.3 for a detailed discussion.

Falling ice is responsible for a great deal of damage to broadcast equipment and buildings. An ice bridge can protect horizontal runs of transmission line between tower and building. Small antennas on the tower, such as those used by STL or RPU systems, can be protected by an ice shield above them. Transmitter buildings and carports should have icebreakers installed above them. Guy wire ends should have icebreakers installed to prevent damage to anchor plates from ice sliding down the cables. Hard hats should be provided at ice-prone transmitter sites. Safety procedures should be established including prudent measures to protect personnel from slip-and-fall injuries on icy catwalks or access ladders.

Flooding caused by rising waters can be anticipated. Transmitter buildings in low-lying areas should be raised above the level of the worst expected water level. In some instances transmitters and racks can be mounted on 2 × 4 pressure-treated lumber so that floor-mounted components are reasonably protected in case of either rising or flooding water.

Earth movement may be defined as earthquakes, mudslides, avalanches, or unstable earth. Facilities located in earthquake-prone areas can be designed to withstand these forces. Equipment inside these buildings can and should be anchored to prevent damage from shaking or horizontal movement. Unstable earth can be buttressed with suitable retaining walls. A licensed structural engineer specializing in geologic structural mitigation should be consulted on these matters.

Some earth movements may be attributed to man-made causes such as nearby blasting. Blasting can be dangerous in high radio frequency (RF) environments due to the possibility of blasting cap detonation by radio waves. Nearby blasting may loosen earth and cause guy wire anchor movement and possible tower failure.

Large nearby trees also pose a potential hazard. High winds, either alone or combined with ice accumulation, can cause otherwise healthy trees to topple. Heavy rain can saturate earth and destroy the ability to adequately hold tree roots in place, causing what appears to be a healthy mature tree to fall onto guy wires and buildings. Trees or limbs that could fall on buildings, guy wires, generators, or equipment should be kept properly trimmed by a licensed tree expert.

FACILITY SECURITY

FCC Rule 73.127 and others state, "the licensee or permittee must retain control over all material transmitted in a broadcast mode via the station's facilities with the right to reject any material that it deems inappropriate or undesirable."

While this rule covers programming on subcarrier services on AM, FM, TV, LPFM, LPTV, as well as all digital broadcast services, it could be extrapolated to mean transmitting inappropriate or undesirable programming on the main broadcast services as well as preventing unauthorized people from entering a broadcast facility to go on camera or on mike.

Aside from protection of employees while in the workplace and protection of assets critical to staying on the air, every broadcast facility has that legal duty under its license to prevent unauthorized access to the airwaves. Stations that exercise due diligence to assess risks to life and property should extend security measures aimed at controlling on-the-air programming.

No one in a broadcast facility is better trained and equipped than the technical staff to carry out the safety and security assessment process, or act as key advisors for outside experts brought in to help. The technical staff will not only know all the facility's vulnerable points, but are in the best position to advise on and oversee implementation of recommendations of the experts.

While design of a broadcast facility reception area is beyond the scope of this chapter, security concerns require a review of new and existing designs to limit entrance of unauthorized persons. Some facilities are placing receptionists behind bullet-resistant glass with intercom communication to visitors. This approach may not be practical in high-rise buildings with multiple tenants or at stations with limited space, but simply adding better locks, security cameras, and peepholes can reduce the potential for casual walk-ins. Further security measures may be needed at the lobby and garage levels as well as parking lots.

Hiring trained security personnel to guard facilities on a 24 hour basis may represent a significant ongoing expense, but it may be prudent in clustered facilities or in cases when physical security cannot be improved. Before proceeding, station management should consult its legal department or outside counsel. Making insurance underwriters aware of risks and obtaining their guidance is advisable as well.

High-quality video surveillance systems with motion-sensing digital recording capability and off-site internet monitoring can be a reasonable and wise investment.

Reports of transmitter site break-ins, thefts, and vandalism have increased since remote control came into practice. While the best option for transmitter site security is live guards, few stations have taken this step. A transmitter site security package that includes video surveillance is now an important and necessary requirement for many facilities. Many sophisticated security features are now economical choices for broadcasters. Most modern electronic security systems can easily be set up for motion sensing. Alarm closures built into the security system can be wired to existing remote control systems even if there is no means for internet visual supervision. A studio operator monitoring the remote control can make a call to the police, or the remote control can dial a series of numbers including security, police, and the fire department. Many systems can be programmed to do this in addition to calling for transmitter technical failures.

Other types of detectors are readily available that can be wired in to remote control channels that can give a measure of advanced warning to many conditions:

- Heat
- Smoke
- Water
- Motion
- Outdoor and indoor temperatures
- Door and window position switches

Perimeter Security

Perimeter security starts with assessing the condition of all doors, windows, and access hatches for a structure. A careful inspection of door frames, window frames, windows, hinges and existing locks, deadbolts, and hasps and other fasteners for signs of rot or termites, may reveal a general lack of integrity. The assessment may suggest installing metal doors in place of wooden doors, and solid doors in place of hollow-core doors.

Eliminating some windows or installing iron bars or sturdy metal grilles over vulnerable windows may reduce security risk. Roof hatches along with all doors and windows that can be opened should have security switches installed that are tied in to a master alarm panel. At-risk windows could be equipped with glass shatter sensors tied in to the alarm system. Interior motion sensors can be installed in areas secured for nights and weekends. These need to be set so they do not falsely trigger on rodents and other wildlife that frequent many remote transmitter sites. A reputable and experienced security company can assess risks and do installation, but it usually needs help to cover all the site's vulnerabilities known to the technical staff.

There is a wide choice of secure keys, swipe cards, and biosensors for entrances to studio and technical facilities. The best key systems have features that make it almost impossible for anyone, including employees, to make duplicates.

Card-swipe and biosensor systems not only make it easy to give everyone unique access, but many systems also have the capability of printing out entry and exit logs, and allow selective lockout of employees in off hours. Coupled with a video security system, they make for a powerful security combination. Readily available in the biosensor category are reasonably priced fingerprint recognition sensor pads and even retinal scanners.

High-intensity outdoor lighting should be considered for parking lots and transmitter sites. Some utility companies offer a special fixed rate for such lighting systems. Consider installing current monitors on outdoor lighting circuits. These monitors, similar to the same ones used to monitor tower lights, can be tied in to remote control systems to warn if exterior lights are not working, or to expose tampering.

A functional alarm system is essential and must report to live humans to be truly effective. The alarm should monitor for intrusion, heat, smoke, and fire. It is wise to have a wireless alarm in the event of intentional telephone line failure. Under some circumstances the transmitter building can be completely surrounded by a fence at least 12 ft. high with razor wire installed inside the fence perimeter at the upper level. Each tower should have its own fence. Another protection approach would be to install an anticlimbing gate on the tower itself. In some cases, it may be necessary to locate external video cameras to monitor the most vulnerable points within the transmitter site.

Locks

Padlock quality is sometimes overlooked in security reviews. Some locks in common use can be opened with keys purchased with new locks, or a medium-sized bolt cutter. If padlocks must be used, remove the key codes from the lock bodies.

For critical locking situations, consult a competent and trustworthy locksmith. There are locks used by contractors, railroads, and the motor freight industry that are close to tamper proof. These padlocks are often referred to as contractor's locks. Taking an example from the storage container industry, it may be possible to add a welded cage around the hasp, making it practically impossible to cut off the padlock. Replace inexpensive padlocks with heavy-duty barn door–type hardware. Door locks may have to be a specific type depending on code.

Assess how facility gates are constructed and hung, and make sure they cannot be unbolted, disconnected, or easily climbed over or under to bypass locks and chains. Trip-wire sensors woven into chain-link fences or buried on or near ground level give a facility added warning if there is a breach.

Special Security Precautions

It is a fact of modern life that man-caused disasters must now enter into the planning and risk-assessment process. The FCC's Media Security and Reliability Council (MSRC) published a comprehensive list of recommendations to harden facilities to survive.[4] Often overlooked, basic security precautions have led to serious incidents at a number of places throughout the country. Here are the basics:

- Approve visits from former employees through their former supervisors.
- Always have nonemployees escorted in critical areas.
- Assure that outside doors are never propped open.
- Secure roof hatches from the inside, and have alarm contacts on the hatch.
- Use photo ID badges when employees might not know each other by sight.
- Check for laws that require a written safety and security plan.
- Use video security and card key systems where warranted.
- Install entry alarms at unattended sites. Test weekly.
- Maintain fences, especially at unmanned sites.
- Consider remote video surveillance using the Internet for monitoring.
- Redesign to limit places bombs could be planted.
- Redesign to prevent unauthorized entry.
- Redesign to limit danger from outside windows.
- Plan for fire, bomb threats, hostage situations, and terrorist takeovers.
- Plan a safe way to shut the facility down in case of invasion.
- Plan guard patrol schedules to be random, not predictable.
- Plan for off-site relocation and restoration of services on short notice.

[4]See http://www.fcc.gov for more information on the MSRC, or http://www.mediasecurity.org.

FIRE PREVENTION

Fires can start through intentional or unintentional means. Local codes will likely require fire extinguishers at certain locations throughout the facility. Be certain that the extinguishers are the appropriate type for a particular area or equipment. Local ordinances and good practice require monthly checks to make sure the pins in fire extinguishers are in place, they are holding pressure, they are mounted at the correct height, and they are not obstructed. Yearly checks by a licensed extinguisher inspection company is not just a good idea, it is usually the law. Most jurisdictions and insurance underwriters also mandate entries on the tag on each extinguisher to be initialed and dated by the person making the check each month. Automatic sprinkler systems may be a requirement depending on state and local regulations.

OTHER DISASTER PREVENTION MEASURES

Component failure can occur due to neglect, such as overlooking cleaning air filters, lubricating blower motors, and control systems. Preparing and adhering to a maintenance schedule can prevent failures before they occur. A site-specific inspection schedule form and maintenance shifts that are long enough to allow personnel time to do really careful inspections can, in many cases, catch small problems in time to prevent much larger ones. For example, a transmission line dehydrator that is not working at rated capacity can be identified easily as needing maintenance with simple periodic inspections that include log entries for status and readings. Not catching such a fault can lead to the inability of the dehydrator to keep up with a transmission line leak. Excessive moisture in the transmission line can cause arcing that can take a station off the air with spectacular ad expensive results.

Towers sometimes fail due to improper or excessive loading in an attempt to obtain extra rental income. Prior to the addition of any appurtenance to a tower, a structural analysis should be performed. Such an analysis can be contracted to many tower firms or independent structural consultants. A few extra income dollars may not be worth the risk of a total tower loss.

Vehicular contact with tower guy anchor points, power generators, or propane storage tanks can occur if these critical components are located near roadways or parking lots or side roads. Most local regulations require concrete or metal pilings to protect natural gas tanks. The same precaution can be taken for guy anchors and generators. Where guy wires cross low over a parking lot or driveway the lowest guy wire should be decorated with red fluorescent flagging to help prevent contact with ground vehicles. Aircraft contact with towers falls under FAA jurisdiction. While this is less common, it does occur so it is important that the tower lights are fully functional at all times.

Power surges may be caused by other users on the facility's utility mains feed as well as new equipment installed on an old service at the site. Surge suppressors should be installed on all incoming AC power mains. Uninterruptible Power Supply (UPS) devices should be installed on each computer. Occasionally a utility will accidentally feed improper mains voltage or phasing. See Chapter 9.2 for a discussion on AC power-conditioning issues.

Strategic location of water feed pipes can prevent accidental flooding. Runs along ceilings in general should be avoided but avoid routing pipes near or over electrical equipment. Overhead pipe leaks and bursts can cause thousands of dollars in equipment and time loss. If overhead pipes cannot be avoided then a water shield above sensitive equipment may be worth consideration.

Many prolonged outages are caused by not having spare parts on hand. While managers consider spare parts to be an interruption in cash flow, to the engineer they are insurance. Adequate spares should be maintained for each critical piece of equipment. The station manager and chief engineer may differ on which equipment is critical. Both could list mission-critical equipment. The chief engineer can then compile a list of reasonable spare parts that might include fuses, IC chips, tubes, pots, switches, power supplies, and functional modules. The component budget can then be used to determine the reasonable cost for what amounts to an insurance policy against prolonged outage or failure.

OUTSIDE PLANT LINKS

Interfacility links such as wire, microwave, or fiber communications circuits are single-point failure opportunities that could cause significant outages. Outside plant links discussed below presuppose proper installation by people and organizations not under station control. Despite this lack of control, a visual inspection of the local utility and telecommunication infrastructure could be revealing. For wire and fiber, this means an inspection to check for adequate service loops so quake and wind stresses will not snap lines. It means that telephone and telecommunications service providers have installed terminal equipment so it will not be vandalized, fall over in a quake, or easily be flooded out. No matter what the results of the survey, a range of backup options are available and should be considered.

Outside Plant Wire

Local telephone companies still use wire, even for T1 and DSL connections. If the facility is served only by wire on telephone poles, or underground in flood-prone areas, investigate alternate telecommunications routing. Alternate routing from the facility to another central office (CO) may be very costly since the next nearest CO is rarely close. But, if an alternate route is only to the next block, or duplicating existing telephone pole or underground risk, the advantage gained may be minimal.

Many telephone companies can designate as an essential service a limited block of telephone numbers at a given location for lifeline communications. Lines so designated are usually found at critical facilities like hospitals and public safety headquarters. Contact the local phone company to ascertain if the broadcast facility can apply for such service. Applicants with close ties to local government emergency management and who are vital links to the public during emergencies (such as broadcasters) may easily qualify.

Microwave Links

Wind, explosions, and seismic activity likely will cause microwave dishes to go out of alignment. Quake-resistant towers and mounts can help prevent alignment failure, even in the face of high wind or explosions. Redundant systems should be considered part of the solution, but an exact duplicate microwave system may lead to a false sense of security. Consider a non-microwave backup such as a fiber link.

Fiber Optics Links

If a broadcast facility is not a fiber customer today, it probably will be tomorrow. The facility may be fortunate enough to be served by separate fiber vendors with separate fiber systems and routing to enhance reliability and uptime. Special installation techniques are essential to make sure fiber links will not be bothered by earth movement, subject to vandalism, or other single-point failure causes.

Fiber should be installed underground in a sturdy plastic sheath with a warning ribbon above the cable. The sheath offers protection from sharp rocks or other forces that might cause a nick or break in the armor of the cable, or actually sever one or more of the bundled fibers. The ribbons are usually colored bright orange to make them stand out in trenches, manholes, and other places where careless digging and prodding could spell disaster. From a risk-assessment standpoint, their color easily identifies what they hold as targets to technically aware terrorists or vandals.

Cable systems that use aerial rights-of-way on utility poles for their fiber may not prove as reliable in some areas as fiber installed underground. Terminal equipment for fiber should be installed in quake-secure (where applicable) equipment racks and away from flood hazards. Fiber terminal electronics should have at least two DC power supplies in parallel that are, in turn, on line with rechargeable battery backup.

Satellite Links

Ku or C Band satellite is a costly but effective way to link critical communications elements. C Band has an added advantage over Ku during heavy rain or snow conditions. A significant liability of satellite transmission for ultrareliable facilities is the possibility that a storm could cause a deep fade, even for C Band links. Another liability is short but deep semiannual periods of "sun outage" when a link is lost while the sun is focused directly into a receive dish. While these periods are predictable and last for short periods, service will be lost unless there is alternate service on another satellite with a different sun outage time, or terrestrial backup.

SINGLE-POINT FAILURES

Single-point failure can occur in any system without backups or bypasses. Single-point failure analysis and prevention are based on a simple concept: A chain is as strong as its weakest link, but two chains, equally strong, should never share the same weak link. The lesson may be that each chain should be constructed with totally independent links, or, even better, where the risk and the consequences are both high, use three chains.

STANDBY POWER SYSTEMS

Chapter 9.4 discusses the variety of standby power systems that are available in the event of commercial power interruption. Chapter 9.2 addresses AC power conditioning. Clean, reliable, and stable AC power is essential in modern broadcast facilities due to the use of computer- and microprocessor-based technology currently employed in nearly all stations.

AIR-HANDLING SYSTEMS

People and equipment both do not operate well when their environment becomes too hot. Adequate, clean, cool, dry, and pollutant-free air from the facility's heating, ventilating, and air conditioning (HVAC) system is important for modern communications equipment and operating staff. If the facility occupies leased space in a high-rise building, total control of the facility air system may not be possible. Many large buildings have no backup systems and do not provide night and weekend supervision or emergency service. The best protection is to establish precise conditions for air heating and cooling in the lease. Adding a separate backup system in such cases is prudent but may be costly. A short-term emergency option might be portable industrial-strength air-conditioning systems that are available from many rental companies. A contract for emergency equipment that can be obtained with a phone call could save hours or days of downtime. Consider buying a portable emergency HVAC unit if the facility and its operations provide essential services, or negotiate a contract for emergency rapid delivery of a portable HVAC with a vendor providing such a service.

HVAC Full Recirculation

Wherever cooling air comes from, there may be occasions when there is a need to set the system to recirculate air within the building. This temporarily makes it a closed system. Smoke or toxic fumes from a fire in the neighborhood can enter an open system and incapacitate the staff, and even cause some equipment to malfunction. Also consider the effect of smoke detectors accidentally setting off a sprinkler system from outside events. With advanced warning, having the ability to configure the air-handling system to full recirculation might avoid or forestall personnel discomfort and increase their safety. It could provide enough time to arrange an orderly evacuation and transition to an alternate site, or time for outside toxic conditions to dissipate.

WATER HAZARDS

Water in the wrong place at the wrong time can be part of a larger emergency or be its own emergency. A simple mistake like locating a water heater where it can flood out electrical equipment can cause fatal electrical problems when it finally wears out and begins to leak. Unsecured water heaters can tear away from gas lines, possibly causing an explosion or fire in an earthquake. At best, the water in that heater could be lost, depriving employees of a source of emergency drinking water.

ELECTROMAGNETIC PULSE PROTECTION

The electromagnetic pulse (EMP) protection phenomenon associated with nuclear explosions can disable almost any electronic component in a communications system. EMP energy can enter any component or system coupled to a wire or metal surface directly, capacitively, or inductively. Some chemical weapons can produce EMP, albeit on a smaller scale.[5] The Federal Emergency Management Agency (FEMA), now part of the Department of Homeland Security, has been involved in EMP protection since 1970 and is charged

[5]FEMA publishes a three-volume set of documents on EMP referenced in the Bibliography. They cover the theoretical basis for EMP protection, protection applications, and protection installation.

at the federal level with the overall direction of the EMP program. FEMA provides detailed guidance and, in some cases, direct assistance on EMP protection to critical communications facilities in the private sector. People responsible for a site that needs EMP protection should discuss the issue with a knowledgeable consultant before installing such protection devices for RF circuitry.

PREPAREDNESS TIPS BASED ON EXPERIENCE

Experience is often the best teacher when considering security and preparation of a site to reduce safety and potential risks. Examples include:

- Evidence of previous high-water marks on walls and in cable and transformer vaults means it could happen again. Such evidence could present itself as calcium or mineral salt deposits that stand out against normal wall or soil color.
- Evidence of roof and wall leaks that may have happened years ago could cause new problems when major rain or wind storms strike.
- Review old maintenance logs looking for disasters (not just water related) that occurred in the past and could happen again if no remedial measures are taken.
- In shared facilities, such as in high-rise buildings, get to know the neighbors. Find out if others near the station are using hazardous chemicals or otherwise conducting their business in a way that could affect station operations.

ALTERNATE SITES

Prudent disaster planning should include the potential of abandoning the facility for a substantial time period. Government emergency planners usually arrange for an alternate site for their emergency operations centers (EOCs). Communications facilities can sign mutual aid agreements with each other. This may be a way to access telephone lines, satellite uplink equipment, microwave, or fiber on short notice.

PRESERVING VITAL SITE-SPECIFIC INFORMATION

Management should have an information access document containing:

- The supervisor username password for the network.
- Telephone system control and voicemail usernames and passwords.
- Engineering area keys and lock combinations.
- Transmitter site alarm codes, keys, and lock combinations.
- Passwords to digital audio processors.
- Names and telephone numbers for tower, telephone system repair, and computer service personnel.
- Other contractors, such as snow plowing contractors, consulting engineers, electricians, plumbers, HVAC contractors, and others.
- Circuit numbers for telephone, ISDN, and communications systems for use when contacting repair services.

A security and risk-assessment checklist should be compiled and updated quarterly or annually. Check the facility maintenance calendar annually for critical items that apply to each predicted risk or after some serious event. For example, guy points and tower plumb need to be checked periodically as well as after every seismic event.

SELF-DISPATCHING PLANNING FOR EMERGENCIES

A key part of each station's disaster planning should be a section that clearly outlines where and when key personnel should self-dispatch when emergency conditions exist. This is especially important when major disasters take out phone service, including cell phones. Management should know in advance where engineering resources will be when their expertise is required.

SUMMARY

Preparing for natural and man-made disasters, ensuring personnel safety, and securing critical facilities are all key elements in keeping broadcast stations on the air when they are needed most. Assessing risks within all aspects of station operations, studio, transmitter, and towers, and planning how to deal with disruptions, can reduce downtime, improve the safety and welfare of station personnel, and instill confidence that the facility can continue to operate when disasters of all kinds occur.

ACKNOWLEDGMENTS

The authors thank John Lyons, assistant vice-president and director of broadcast communications for the Durst Organization in New York City, for material for the First Responder Communications Backup System section.

Key Terms

Business impact analysis (BIA): A formal study of the impact of a risk or multiple risks on a specific business. A properly conducted BIA becomes critical to the business recovery plan.

Business recovery plan (BRP): A blueprint to maintain or resume core business activities following a disaster. Three major goals of a BRP are the resumption of products and services, customer service and cash flow, and recovery of the business to normal operations.

Central office (CO): The building where local switching is accomplished.

Emergency operations centers (EOCs): Locations where emergency managers receive damage assessments, allocate resources, and begin recovery.

Electromagnetic pulse (EMP): A high burst of energy associated most commonly with nuclear explosions. EMP can instantly destroy the functionality of many electronic systems and components.

HVAC: Architectural acronym for heating, ventilation, and air conditioning.

Incident commander (IC): The title of the person in charge at an emergency scene from a street corner incident to an emergency involving an entire state or region.

Incident command system (ICS): An effective emergency management model invented by firefighters in California.

USAR: Emergency management acronym for urban search and rescue.

Bibliography

Baylus, Ellen. *Disaster Recovery Handbook.* Chantico Publishing Company.

Federal Emergency Management Agency (FEMA), *Electromagnetic Pulse Protection Guidance*, vols. 1–3, Washington, DC.

Fletcher, Robert. *Federal Response Plan*, Federal Emergency Management Agency, Washington, DC.

Hadden, Josh. "Disaster Preparedness and Business Continuity: Ideas, Financing, Implementation," *Broadcast Engineering Conference Proceedings*, National Association of Broadcasters, 2006.

Handmer, John, and Parker, Dennis. *Hazard Management and Emergency Planning*. James and James Science Publishers.

Janitschek, Andrew. "Common Sense Rules to Uncommon Events," *Broadcast Engineering Conference Proceedings*, National Association of Broadcasters, 2006.

Osenkowsky, Thomas G. "Disaster Planning . . . Before and After the Fact," *Broadcast Engineering Conference Proceedings*, National Association of Broadcasters, 1997.

Osenkowsky, Thomas G. "Emergency Planning for Radio and Television," *Broadcast Engineering Conference Proceedings*, National Association of Broadcasters, 2000.

Rothstein Associates. *The Rothstein Catalog on Disaster Recovery and Business Resumption Planning*. Rothstein Associates.

Further Information

The firms listed below are shown as representative only and are not specifically endorsed by the authors.

Associations/Groups

Association of Contingency Planners (http://www.acp-international.com).

Business and Industry Council for Emergency Planning and Preparedness (BICEPP), P.O. Box 1020, Northridge, CA 91328.

Disaster Recovery Institute (DRI), 1810 Craig Road, Suite 125, St. Louis, MO 63146. DRI holds national conferences and publishes the *Disaster Recovery Journal*.

Earthquake Engineering Research Institute (EERI), 6431 Fairmont Avenue, Suite 7, El Cerritos, CA 94530.

National American Red Cross, 2025 E Street, N.W., Washington, DC 20006.

National Center for Earthquake Engineering Research, State University of New York at Buffalo, Science and Engineering Library-304, Capen Hall, Buffalo, NY 14260.

National Coordination Council on Emergency Management (NCCEM), 7297 Lee Highway, Falls Church, VA 22042.

National Hazards Research & Applications Information Center, Campus Box 482, University of Colorado, Boulder, CO 80309.

Business Recovery Planning

Harris Devlin Associates, 1285 Drummers Lane, Wayne, PA 19087.

Industrial Risk Insurers (IRI), 85 Woodland Street, Hartford, CT 06102.

MLC & Associates, Mary Carrido, President, 15398 Eiffel Circle, Irvine, CA 92714.

Price Waterhouse, Dispute Analysis and Corporate Recovery Dept., 555 California Street, Suite 3130, San Francisco, CA 94104.

Resource Referral Service, P.O. Box 2208, Arlington, VA 22202.

The Workman Group, Janet Gorman, President, P.O. Box 94236, Pasadena, CA 91109.

Life Safety/Disaster Response

Caroline Pratt & Associates, 24104 Village #14, Camarillo, CA 93013.

Industry Training Associates, 3363 Wrightwood Drive, Suite 100, Studio City, CA 91604.

Emergency Supplies

BEST Power Technology, P.O. Box 280, Necedah, WI 54646.

Extend-A-Life, Inc., 1010 South Arroyo Parkway, #7, Pasadena, CA 91105.

Exide Electronics Group, Inc., 8521 Six Forks Road, Raleigh, NC 27615.

Velcro® USA, P.O. Box 2422, Capistrano Beach, CA 92624.

Worksafe Technologies, 25133 Avenue Tibbets, Building F, Valencia, CA 91355.

Construction/Design/Seismic Bracing

American Institute of Architects, 1735 New York Avenue, N.W., Washington, DC 20006.

DATA Clean Corporation (800-328-2256), Geotechnical/Environmental Consultants.

H. J. Degenkolb Associates, Engineers, 350 Sansome Street, San Francisco, CA 94104.

Leighton and Associates, Inc., 17781 Cowan, Irvine, CA 92714.

Miscellaneous

Data Processing Security, Inc., 200 East Loop 820, Forth Worth, TX 76112.

EDP Security, 7 Beaver Brook Road, Littleton, MA 01460.

ENDUR-ALL Glass Coatings, Inc., 23018 Ventura Blvd., Suite 101, Woodland Hills, CA 91464.

Mobile Home Safety Products, 28165 B Front Street, Suite 121, Temecula, CA 92390.

SECTION

3

AUDIO PRODUCTION AND STUDIO TECHNOLOGY

CHAPTER

3.1

Planning an Audio Production Facility

STEVE CAMPBELL
Aurora Media LLC
Las Vegas, Nevada

ROY W. RISING
ABC-TV
Hollywood, California

INTRODUCTION

This chapter is a composite of material from previous editions of the *NAB Engineering Handbook* combined with updated information from professionals in the field of audio production facility design and implementation. The first part of the chapter draws on the authors' long experiences both as designers and end users of radio and television audio facilities, and the subjects presented are those that should not be overlooked. The latter part of the chapter addresses the implementation of radio facilities on a limited budget. It also provides some sample facility floor plans for small, medium, and larger radio stations.

The objective of planning for new or upgraded facilities is to consider every possible detail and develop an integrated scheme that leads to the best final result. There will be tradeoffs; no single element is so important that it cannot yield to another's requirements. It helps to begin by giving equal weight to every aspect. In the end, the larger and smaller issues will usually find their own places.

The following sections provide guidance for the broadcast engineer. It is assumed there will be support by an architect with a good understanding of acoustical considerations, and specialist consultants for other aspects of the design.

INITIAL PLANNING CONSIDERATIONS

Design is the art and science of intelligent compromise. All projects have limitations—some due to space, some due to time, and some due to budget—and all have conflicting needs and priorities. It is up to the design team to determine the best compromises to achieve the design specifications and meet the end users' requirements.

Design Team

Depending on the scope of the project, the makeup of the design team will vary from a few people to many. The team may include architects, acoustical experts, cabinet and enclosure designers, structural engineers, mechanical and HVAC engineers, electrical engineers, radio and television engineers, equipment manufacturers and vendors, and system integrators.

The first step in designing any successful project is to appoint a project manager with responsibility to coordinate the members of the design team and assure that communications paths are open and flowing as the design progresses. The constant guidance of a single person helps keep focus on the goal and coordinates all the diverse talent being applied to the project.

In most general commercial construction, the architect has the role of project manager. However, in specialized construction like audio production/broadcast facilities, the project manager may be an audio/broadcast engineer for initial planning and possibly for the duration of the project, or with responsibilities shared with the architect. There are architectural firms that specialize in audio production and broadcast facilities and their use is particularly appropriate for large projects.

The choice of the architect is sometimes beyond the control of the project manager, due to building requirements, the owner's preference, or the scope of the project; and if the architect is not experienced with this type of facility, the role of the in-house project manager to guide the specialized portions of the design may be increased.

An architect's primary role in every project is to utilize space efficiently and to make the environment friendly, attractive, safe, code compliant, and productive for its occupants and for the intended purpose. By taking account of the special factors and design considerations discussed in the following sections, the architect will be better able to design a good facility and provide comprehensive design specifications. Depending on the experience of the architect and the other consultants with the specialized requirements of broadcast facilities, the in-house project manager must pay close attention to the specific details of acoustical and technical design and confirm space requirements and technical aspects as the design progresses. Architects usually welcome the opportunity to explain their design to station engineers and users and, through this process, the design goals are best met.

Architects typically retain several building trade engineers as their consultants. These include the structural, mechanical/HVAC, electrical, and fire protection engineers, and other specialist consultants. Many architects who have implemented audio production/broadcast facilities before have engineering contacts with the necessary experience. More information on project management and the role of the design team may be found in Chapter 2.3.

Goals

The owner will usually have a major input in establishing what the goals are for the facility, the location, and the scope of construction desired, and in setting an initial budget for the project. The planning process leading to the design specifications should start by determining in detail what the goals are. A significant amount of prognostication is needed for this phase to take account of future program needs and changes in technology. The fundamental points to establish are:

- The purpose of the facility.
- The amount and type of production.
- The on-air transmission requirements.
- The scale of the programming and business operations.

The audio/broadcasting industry is constantly changing and the best information on the direction of the subject facility will help increase the productive life of the facility. This is at best a guess, but the more educated the guess, the better the facility.

Most audio production facility projects are part of larger facilities improvement or upgrade projects. The staffing levels now and (as far as can be predicted) for the future, and the appropriate work groupings for these personnel, should be detailed along with any special requirements the staff members may have. This information is necessary for the architect to start the design work.

Armed with the owner's goal, the project manager and design team must further refine the design and specifications of the project. The following sections identify some of the many items that must be addressed.

Location

The normal considerations for locating a facility include easy access for staff, talent, and business clients; availability of parking; neighborhood security; and zoning restrictions. In addition, if the transmitter is not colocated, many broadcast facilities require, or are greatly simplified, by the availability of a clear microwave path from the studio facility to the transmitter site or sites. This requirement, including zoning implications, is often overlooked when selecting a site. The availability of a clear path, or the ramifications of a lack of a path, must be determined early in the planning process.

Many audio facilities, particularly those for radio and television broadcast, use a variety of satellite programming services. A suitable location for the satellite antennas with a clear view of the southern sky may be critical to smooth operation of the facility. Large satellite antennas preferably should be on the ground, as roof mounting can subject the system to interference problems and introduce significant structural concerns. Local zoning codes can also create difficulty in the placement of satellite antennas. The effect on the project of satellite location difficulties or zoning problems should be determined early in the design phase.

Transmitter

Considerations of the intended service area of the stations and of radio frequency (RF) propagation determine the best location for the transmitter and antenna. The transmitter is frequently separate from the studio facility due the need for maximum elevation for FM antennas and for large plots of land for AM antennas, particularly directional AM systems. However, if the location of the transmitter site is appropriate for the business operations of the station, meeting the concerns previously addressed, colocating the studios and office complex and the transmitter site can be convenient and cost effective. There will be some additional installation costs and engineering time associated with colocating a facility, but the potential cost savings of colocation in the long run may be worth the extra initial costs. For example, colocation may provide the opportunity to reclaim waste heat from the transmitters to assist in heating the studio buildings.

User Input

An important part of the planning process is a continuing dialog with the end users of the facility. As plans for various elements begin to take shape, circulate sketches and descriptions among the users. Establish an open return channel for constructive criticism and follow up for comments even when not volunteered.

Maintain a log of memos and conversations. This may be helpful in resolving later disputes.

End users are quite often justifiably nearsighted. Their focus is on those things that matter the most to them personally. Part of the planner's task is to moderate the areas where different users' needs overlap. First, develop a consensus among users who share the same areas. Next, design a proposed zone of transition between adjacent areas. Then circulate the proposal and make adjustments based on the responses.

LAYOUT AND DESIGN

Few production facilities have the good fortune to be built new from the ground up. More often an existing structure is used. By using an existing production center, the design can benefit from the good aspects of the design and where possible avoid its previous problems. Basic types of audio production facilities used in broadcasting include:

- Private project studios
- Radio studios
- TV studios
- Video postproduction

This chapter concentrates on small studios and control rooms for radio stations. The audio control rooms and ancillary areas for different applications have different layouts, but many of the requirements are similar and follow common patterns and design considerations. These apply both for new facilities and reuse of an existing building.

Radio Control Room/Studios

Most present-day radio control rooms also double as studios and are called *combos* for combination control and studio. This is where most of the station's announcing is performed, and in a small station may be the only studio area. The production studio is almost always a combo room. Combos in a small station may contain most of the audio equipment of the station.

In many instances, the newsroom will have live microphones for late-breaking news events. Combo rooms and active news studios make the problem of acoustics more complex and, because radio sells with sound, good acoustic design is essential for a successful operation.

Designs for combo control rooms are a compromise of several design factors:

- Location of the control room within the studio building.
- Isolation from unwanted internal and external sound and noise.
- Construction—special walls, ceilings, and floors may be required.
- Reverberation control—size of rooms and special wall, ceiling, and floor coverings to eliminate or reduce reflections and resonances that degrade the studio sound.
- Ventilation and air conditioning—technical equipment generates heat and studio guests add to the heat load that must be accommodated.
- Size and arrangement of equipment—plan some flexibility for future changes for new requirements.

Master Control Room

When the number of studios in an audio production center exceeds two, a *master control,* sometimes called *central switching, rack room,* or *tech center,* is a desirable addition. This is a central place for access to the circuits that tie the rooms together. It offers a location for the intercom matrix, central digital storage systems, and any other equipment that is shared between studios. This area also manages the incoming and outgoing broadcast and communications feeds and their associated processors, receivers, and interfaces, lessening the demand on the audio production room's space. The electronics for STL and satellite systems are usually located in the master control room with cable runs to the antennas, although alternatively such electronics may be in a shelter nearer to the antenna, depending on the facility design.

Acoustical Considerations

The first consideration for layout is sound isolation. Rooms with high-level loudspeaker monitoring should not be adjacent to each other or to rooms with live microphones. The obvious exception is where a performance studio must be connected to the control room. Here, extra care must be taken for acoustic isolation.

In all but manned radio air studios, a good approach to layout places the sonically sensitive rooms in a central core surrounded by offices and other support areas. These other areas serve as acoustic buffers against noise from outside the building. Production rooms may be thought of as islands in the core separated by corridors, equipment rooms, and storage areas. It is helpful to provide access to control rooms from both sides of the core but not to the exclusion of a corridor across the core.

Appropriate interroom acoustical isolation design and intraroom reverberation control are important to an audio production facility and represent expensive line items in the construction budget. Many of the factors that affect the costs of acoustic construction are under the designer's control. The application also affects the acoustical requirements. For example, a "top 40" radio station whose talent works only a few inches from a single microphone, relies less on room acoustics for its live "sound" than does an all-talk format station requiring several open microphones to pick up guests at a variety of distances. In the latter situation, to maintain broadcast quality, studio acoustics must be carefully controlled.

The more audio signal compression and limiting a station employs, the more evident will be the annoyance of poor studio acoustic characteristics. Conversely, the compression on stations with better studio acoustics will be less noticeable during live broadcasts and recordings.

Sound isolation problems are very difficult to correct after the facility is built. For project success, the following points must have high priority:

- Determination of the required attenuation between adjacent rooms.
- Proper specification of isolation requirements.
- Adequate design of the isolation to achieve these requirements.
- Appropriate testing to confirm the isolation requirements have been met before acceptance of the construction.

Acoustic treatment to adjust the reverberation characteristics and sound of the audio production room after construction, while a significant issue, is easier than improving the isolation. Even so, reverberation control within the audio production room should receive appropriate attention during the design phase.

The special construction of the walls to obtain the desired sound isolation may make the wall unique and its specification may not be consistent with the tested wall assemblies required for fire rating within the facility. This point should be addressed early in the design phase. Likewise, the materials required for acoustic treatment may present concerns to the local fire officials. A determination of *flame spread* and *smoke production* requirements of all interior surfaces as specified by local code and the fire officials should be obtained. All required certifications for the materials to be used should be obtained and cataloged for reference and compliance demonstration.

Chapter 3.2 of this book provides additional information on techniques for acoustical isolation and reverberation control within audio production facilities.

Traffic Patterns

Give careful attention to traffic patterns when determining where surrounding offices are located; the most used paths should be the shortest. Back-to-back control rooms may have doors to the perimeter corridors but this can lead to staff taking shortcuts instead of using the crossover corridor between them. If the control room doors are positioned on the crossover, they should not be opposite each other and should be offset by at least the width of one doorway.

A curious aspect of human nature is to congregate in the middle of high-traffic areas. Two people will stand just outside a control room door or at the intersection of two corridors and talk, with apparent disregard for those who would enter or leave the room or traverse the corridor until someone tries to get by. Small groups will often congregate at corridor intersections, perhaps because the space seems large enough. Attempts to provide convenient alcoves for this purpose only partially mitigate this behavior. The best solution is to make the corridors wider. Two or three people occupy a 4 ft diameter circle suggesting 6 ft 6 in. as a minimum corridor width for a very busy facility.

Transmitter Location

When the transmitter facility is separate from the studio complex its design can be considered independently. However, if it is colocated, various factors must be taken into account. It is generally recommended that the site should be laid out so that the studio and office complex are beyond the edge of the ground system of an AM antenna system, and to provide a minimum of a 50 ft work area around the base of an FM or TV tower. Building codes may require greater distances. When colocating near tall towers, appropriate design consideration must be given to lightning protection, grounding, and to protecting the studio and office building as well as personnel from falling ice in environments where tower icing may occur.

In the past, constant monitoring of transmission equipment was required by the FCC. It was common practice in colocated facilities to locate the transmission equipment within or adjacent to the main control room. Improved stability and the ease of remote control and monitoring for modern broadcast equipment along with relaxed FCC rules now allow transmitters to be located elsewhere. However, if the transmitter is located at the studio facility, additional space will be needed within the main building or in an auxiliary building for the transmitting and support equipment.

The transmitter room should be provided with AC power conditioning and sufficient independent cooling. It may be possible to reclaim the waste heat from the transmitter to assist heating the office facility.

Colocation of studio and transmission facilities does create some difficulties and requires careful audio installation techniques and extra time to resolve audio problems caused by the high RF levels at the site. Special attention to grounding will be required and it is recommended to run all data and telephone interconnections, as well as that for audio, with shielded cable or in metal conduit.

HUMAN FACTORS

Ergonomics is the science of designing and arranging things people use for safe and efficient interaction. Most devices in an audio system have been designed with some ergonomic consideration. When these elements are brought together to form a production facility it is easy to lose the efficiency each one may have on its own. Careful design will preserve the efficiencies of each piece of equipment as well as of the completed system. Many good references are available on ergonomic design, but one of the most useful tools is the human dimension data presented in the Architectural Graphics Standards [1].

Audio Mixer

The *audio console* or *mixing desk* is the heart of most audio production/broadcast facilities and its ergonomics deserve some consideration in the facility initial design. Products from various manufacturers may look alike at first glance but they can be quite different in detail. Look for good layout of the most-used controls—with a larger mixer, the operator may have to stand and lean to reach some of them.

Ancillary Equipment

In addition to the mixer, other equipment will be incorporated in the initial system design. Various players and recorders, processors, and effects devices fill out the system for flexibility among projects. Some installations place these devices below and to the side of the mixer where the drawers of an office desk would be. This causes continual leaning and bending during normal operation, which is ergonomically undesirable, and also places the equipment at risk of damage from rolling chairs.

Ancillary equipment is better placed in a rack or console to the side of the operator, installed from elbow height to just above eye level. A pullout shelf at script apron height provides a place to put materials before and after use. The lower part of the rack may contain patch bays, distribution amplifiers, and other equipment where quick access is not usually required. Computer or video monitors should be centered at or just below eye level. Continuous viewing of an object more than 15 degrees above eye level can lead to nerve and muscle damage in the neck and shoulders.

Systems built with metal racks and cabinets are highly functional but somewhat sterile and mechanical in their appearance. Custom cabinetry using wood accents and plastic laminates is attractive but can be inflexible. In either case, when designing the system seek to use a modular approach with separate cabinets that fit together to form an integrated appearance. This will simplify changes of major elements as the installation evolves with production requirements and technology. When designing or selecting the cabinet system, pay particular attention to adequate provisions for cooling and appropriate wire ways for neat installation.

Communications

Good visual and verbal communication may be taken for granted in a production facility but it does not happen automatically. It is important to ensure open sight lines between the participants. A nod, wink, or gesture might be critical to a project's success. Where windows are used between adjoining rooms, they must be wider and taller than normal. This is because the parties might be seated or standing and groups may spread out for comfort and ease of interaction. The basic assumption is that almost every place in each room should be visible to the other.

Voice communications range from the simple talk back built into most mixers to sophisticated multistation intercoms tailored to larger facilities. Almost every production center benefits from some form of modern matrix intercom with built-in *interrupted fold back* (IFB) and telephone interface. The advantage of a central matrix is that all users can be interconnected as production activities require. An important feature is a separate volume control on the master stations for each incoming channel. This allows the user to adjust each source as needed to avoid interference or distraction from program sound.

Additional information on audio production equipment is contained in the other chapters in Section 3 of this book.

AUDIO INFRASTRUCTURE

Cable Ways

Several miles of cable may be hidden in a studio center. Installations built with *computer floors* in corridors and technical rooms benefit from easy access to subfloor cable raceways. Computer floors are a system of removable modular floor panels supported on pedestals, providing an underfloor space for installation of cabling and other services. However, a computer floor in an audio control room can be a disadvantage because the floor is resonant, which affects the room acoustics, and the underfloor void makes sound isolation from adjacent rooms more difficult. Building corridors with computer floors is one solution but few facilities have the luxury of combining solid floors in some areas with computer floors in others due to the necessity for different concrete slab levels.

Cableways above suspended ceilings are a good alternative. A network of overhead raceways or cable trays can follow the corridors, dropping down in closet spaces for passage into the rooms. Multiple 2 to 4 inch diameter thin-wall electrical metallic tubing (EMT) conduits should be provided, not only for isolation of different cable categories and the various audio levels, but to ensure room for expansion. Some newer audio and video network systems and associated equipment rely on fiber-optic cable for connectivity. Fiber-optic cables are fragile and difficult to splice if damaged. Fiber-optic cables should be placed in a special orange hard plastic inner duct that is clearly marked "Fiber-optic Cable."

The traditional way to maintain acoustic integrity where conduits pass between rooms is to stuff the ends of the conduit with cloth. Remember to leave a nylon pull rope in each conduit when it is installed.

Due to the diverse signal levels, the following types of circuits should be placed in separate conduits or bundled apart on the overhead raceways:

- Microphone level audio
- Line level audio

- Digital audio, timecode, and computer network cables
- Loudspeaker level audio
- AC power

Inside audio control rooms, a raised floor at the sides and behind the equipment provides a tidy way to route the cabling from the wall to the cabinet if no portion of the cabinet is mounted directly against a wall. Carpeted sections of ¾ inch plywood resting on 4 inch × 4 inch supports make it easy to access the wiring when changes are needed. When old cables are permanently abandoned, remove them to keep the area open and clean.

When designing cableways, be aware of the requirements imposed by the National Electrical Code and other regulatory agencies on cables used in plenums. The use of computer floors and above ceiling cable trays down corridors may mandate the use of plenum-rated (type P) cable, which is more difficult to obtain and more expensive. Fully enclosed metal raceways may be preferable to open cable trays in these installations.

Interconnection Cabling

Audio interconnect cabling is provided as part of the technical system, but it is appropriate to consider the requirements in the facility design process because of the need to provide appropriate cable routes, raceways, and conduits.

When a master control area is provided, all inter-room interconnections, perhaps with the exception of microphone-level circuits from a dedicated studio to its control room, should be routed through this area. The benefits of a systematic approach and maintainability for room interconnections will greatly outweigh the additional installation cost.

During the audio system design, it will become apparent how many interconnect lines of each signal type will be required between each control room and master control. These include permanent circuits used for the normal operation, such as studio outputs, incoming program lines, and sources from a central routing switcher and possibly a central digital storage vault. Other circuits include off-air monitoring, intercom connections and other fixed functions, interconnections for temporary circuits for particular program configurations, and spare interconnects.

In planning for the future, it is wise to install significantly more of each type of circuit, and while there is no fixed guide, more is always better. Practice has shown that eventually some operation will arise that exceeds the number of tie lines available. Amortized over the life of the system, the cost of providing, say, 24 rather than 12 spare tie line trunks from every control room to master control is minimal. With the addition of 5.1 and 7.1 surround-sound production, it may be necessary to increase the undedicated tie line count even further.

Patching and Routing

The flexibility of audio interconnections should be determined by the function of the facility and by the capability of its operating personnel. In facilities that are subject to significant varied demands, like television and video postproduction where facilities are generally manned by well-trained engineers, every circuit should be accessible for reconfiguration. Traditional installations have used patch bays to achieve this level of reconfigurability. Newer technologies with router-based systems are making inroads for system configuration, but patch bays will have their place in audio facilities for the foreseeable future.

Within the control room where reconfigurability is important, there should be patch bay or alternate router access to every possible circuit, including:

- Incoming microphone circuits
- Line level sources
- Mixer inputs, inserts, and outputs
- Machine inputs and outputs
- DA inputs and multiple outputs
- In-room trunks to panels on the side or rear walls
- IFB program inputs
- IFB outputs
- Power amplifier inputs and outputs
- Loudspeaker inputs

In smaller facilities like project studios and radio control rooms, reconfigurability of circuits is much less important and patch bays are less necessary. In addition, the expertise of the typical operator is such that multiple patch bays may prove to be confusing. In these facilities, the use of patch bays should be limited to accessing circuits that are necessary for quick restoration of operations due to equipment failure, and are generally limited to patching studio outputs to final program chain inputs. Often these patch bays are located only in master control.

Proper installation and "normaling" of patch bays are critical to their efficient use and is discussed in Chapter 3.5.

FACILITY INFRASTRUCTURE

Beyond the audio system infrastructure, many other systems including electrical, HVAC, fire, telecom, and the data infrastructure all require special attention.

Electrical Systems Design

Master control and audio control rooms are completely dependent on the electrical system supplying them with power. The importance of the intended purpose for these rooms will dictate the precautions necessary in the electrical system design for the station. The electrical system issues to be addressed by the project manager and/or the electrical engineer include the following.

Capacity

The first thing is to determine the total electrical load to be supplied now and over the projected life of the facility. An inventory of all equipment to be installed in the facility with its electrical consumption should be made. When addressing a remodeling project with limited equipment replacement, measurement of the current load with all equipment in its most consumptive state is a good starting place. Planning for the future requires some estimation, but designing the electrical system for 100% increase in electrical load (increase by a factor of two) is not unreasonable. In addition to the equipment load, an estimate of the power required for lighting and other electrical devices is needed, with an allowance to accommodate future additions. As lighting technologies are advancing to more efficient systems, a smaller design factor for the future is acceptable.

Protection

The electrical supply for the technical equipment should be "clean" (without interference), conditioned, and well protected. The electrical engineer will be familiar with the requirements and arrangements for overcurrent protection, but perhaps less so with the need for surge protection caused by lightning strikes and other power line disturbances. The cost of technical equipment is high and protecting it with multiple levels of surge protection at several stages along the power feeders to the studios is a smart investment that should not be minimized. In general, staged protection has worked best, with a protection device located at the main entrance point to the building, a protection device located at each distribution panel, and protection installed at the point of connection to the equipment.

There are two broad classes of surge protectors: series/shunt and shunt-only protectors. In high lightning areas, the first protector should be a series/shunt unit. All other protectors in the system will then be sufficient if they are appropriately installed (well grounded with short installation leads that are mostly straight) shunt suppressors. Often the downstream protectors are built into other equipment like uninterruptible power supply (UPS) systems and plug strips. Plug-in shunt surge suppressors are available for AC power distribution panels that can provide protection directly on the supply bus of the panel, providing significant intermediate protection to the system.

Uninterruptible Power Supplies

Depending on the importance of the audio production facility and operation (on-air, recording, master control), UPS systems and generators should be considered. It is certainly unwise to operate any computer system without a UPS system and an automatic safe shutdown mechanism. This will protect unsaved work that is in progress within a computer system and also protect the file system from corruption due to loss of power. With the large number of computer systems that constitute major parts of modern studio equipment, this is particularly important. In many facilities, reliance on small desktop UPS systems for this protection is common. This is acceptable if the small systems are monitored and checked often (at least monthly) to assure they are functioning correctly and their batteries are serviceable. However, in most facilities, the maintenance staff does not have time to do this, so power failures may be accompanied by a computer problem due to a UPS failure. It is better to install one or more large UPS systems supplying all the critical loads of the facility. Maintenance on a few large high-quality UPS systems is much more reliable than maintenance on dozens of small units hidden in odd places. For a full discussion, see Chapter 9.2, AC Power Conditioning.

With all but the largest UPS system, access at all locations to both UPS-conditioned and non-UPS power is desirable. Non-UPS power may be required for large maintenance and utility devices. Each outlet should be marked clearly to make its intended use known as well as identifying its source to aid in troubleshooting. It is a great embarrassment to be the engineer that plugs in a large vacuum cleaner to a fully loaded UPS system and takes the on-air studio off the air!

If continuous audio output from the facility is mission-critical to the enterprise, a high-quality emergency generator set is essential. Electrical systems are usually reliable, but the electric supply from the utility company is outside the control of the station. Sooner or later every electrical supply will fail for a period that is longer than the runtime of the batteries installed in the UPS system. An on-site generator system will be a show saver, provided it is well maintained, frequently tested, and has an abundant fuel source.

One difficulty that has caught more than a few facility managers during emergency operations is that many UPS systems do not recognize the generator electrical supply as normal power. Variations in generator frequency and voltage with changes in load may cause a UPS to fail to transfer to the generator power after a main's power failure, or the UPS may determine that another power failure is imminent when the generator is running, and transfer back to battery. The end result may be that the generator is running, but the UPS reaches the end of its battery life and shuts down. This defeats the objective of installing both the generator and the UPS system. Many UPS systems are adjustable to compensate for this problem but some are not. Tests and adjustments to the system to assure generator/UPS compatibility should be performed *before* the loss of power occurs.

UPS systems that are designed for the information technology (IT) industry can generate noise on the power lines, which may cause an increase in the noise floor in some audio equipment. This is a particular problem with some online UPS systems where this increase in the noise floor is always present. Consultation with UPS vendors about using their products in audio systems, and confirmation of previous successful use in audio systems, is recommended during the selection process.

Technical Infrastructure

The electrical contractor will most likely install the infrastructure raceways for the audio and technical systems. The requirements for the raceways should be established by the broadcast engineer working with the architect. The electrical engineer then provides the details. It is easy to forget the small things when planning these systems, like the location of recording/on-air warning lights and their controls, clock outlets, tally and annunciator panel signals and other alarms, and emergency lighting.

If backup power is not supplied to the entire facility, it is easy to overlook providing adequate lighting in critical operational areas. In that case, critical lighting needs to be connected to the conditioned power circuits that are supplied with backup power.

Grounding System

The importance of a well-designed grounding system in a studio center cannot be overemphasized, and this is particularly the case if the transmitter is located on the same site. The grounding system is necessary for four distinct purposes that are sometimes at odds with one another. It must:

- Provide a common reference for all metallic objects within a facility to protect personnel from shock hazards.
- Provide a common ground reference for all signals within a facility.
- Provide for safe conduction of fault currents to the power source in the event of a power fault.
- Provide a low-impedance surge path for unexpected discharges into the electrical system (static discharge and lightning).

Well-designed ground systems achieve these four functions in harmony, but a poorly designed system will be a constant source of noise and unexpected potential differences within the installation.

A single-point ground system, as illustrated in Figure 3.1-1, is the most effective for technical facilities. The ground system within each room is a single-point or star ground system, often connected to a ground bus bar directly attached or adjacent to the main console within the room. Each rack and major piece of equipment is grounded to this bus bar. Each bus bar is then connected to a single-point ground, or star ground, in the master control with a conductor that is routed along the same path as the signal conductors connecting to the master control. The single-point ground in the master control is tied to the technical ground electrode system.

The ground electrode system may comprise the building steel structure, ground rods, ground mats and rings, metal underground water pipes, or other appropriate grounds available in the facility as the technical system ground connection. If the electrode system for this technical ground is separate from the building's electrical system ground, it should be confirmed that this separate technical ground system is bonded to the electrical ground system as required by code. Avoiding multiple connections from all equipment to external ground points or electrodes will prevent ground loops and their unwanted currents. This arrangement is called a *star on a star* configuration.

To aid in minimizing ground loops between the power ground and the technical ground, the technical power should be installed to each studio along the same general path as the studio ground and the signal conductors. By installing the technical equipment's dedicated electrical distribution panel(s) for master control and all the control rooms in or very near master control, the branch circuits can be routed along the same path as the technical ground to each studio. The dedicated electrical distribution panel should be grounded to the technical ground (center of the main star) in master control. In the best of installations, the supply to this panel will be from a separately derived source, such as a shielded dry-type transformer or a large UPS system (National Fire Protection Association's NFPA 70 National Electrical Code, Articles 250.20 and 250.30), and this separately derived source's neutral can be bonded to the technical ground within the panel. As the path and destination of the technical electrical system ground are the same as the technical ground, they serve as a parallel ground conductor and are less likely to create a ground loop that must be isolated, often by lifting the ground conductor in the equipment power cord. It is bad practice to fail to utilize the electrical system ground by using ground lift adapters on power plugs or cutting off the third pin of a power cord.

This design arrangement provides a single, reasonably low-impedance path to a common system ground. Depending on the local situation, the size of the technical ground conductor to each studio can be small since the conductor is not attempting to swamp out multiple different points of grounding. A #6 AWG ground cable from master control to the control rooms has proven to be adequate in many installations without ground loops; however, in the presence of high-power transmitters, a 2 or 4 inch ground strap may be more appropriate. Again, a common path from master control to the audio control rooms for all signals, power, and the ground strap is preferred.

When the transmitter and antenna are colocated with the studios, appropriate grounding must be installed at the antenna, where the antenna transmission lines enter the building, and at the master control star ground point described above.

Further discussion of facility grounding practices is available in Chapter 9.3 of this book and in the *NEC 2005 Handbook* [2].

HVAC Design

Heating, ventilating, and air-conditioning (HVAC) systems in a studio complex are important to a successful operation and this area of design is usually

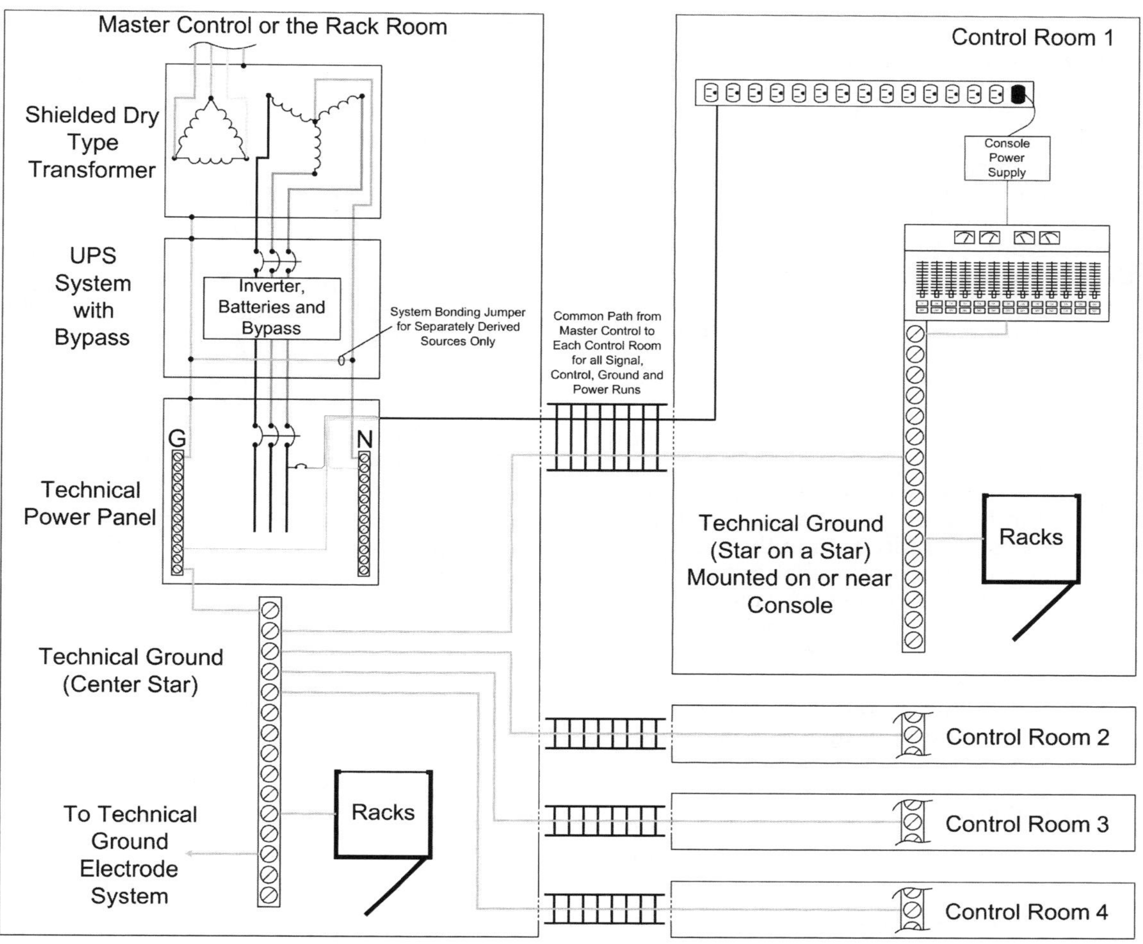

FIGURE 3.1-1 Star on a star grounding system.

best left to a specialist. However, even an experienced HVAC engineer may not have experience with the specific requirements of an audio production facility, therefore, it would be good to bring some of the issues mentioned below to his or her attention at the onset of a project, and check in design review how the engineer addresses these requirements.

The key issue in designing an HVAC system for audio facility use is to maintain the temperature and humidity within the studio or control room in the comfort zone (typically considered 72–78°F with a 20–60% relative humidity) while also maintaining acceptably low noise levels for the space. The system must not contribute appreciable noise to the program audio being produced and must not interfere in any way with the quality control monitoring of that audio. There is a belief that the performance of performers in studio environments is enhanced by temperatures that are below the normal comfort zone normally used by HVAC engineers and this should be taken into account. It is also necessary to maintain the code-required amount of fresh air in audio control rooms and studios.

To start the design of any HVAC system, the HVAC engineer will need to know many things. First and most obvious is the size of the space, its exposure to uncontrolled temperature areas (the outside in general), and the amount of heat load within the studio. The engineer will also need to know the sound levels that the system will need to achieve while satisfying the maximum cooling or heating demand of the system.

In most studios, because of the constant equipment heat generation and the super-insulated nature of the construction to achieve the sound attenuation from outside noise sources, heating is the secondary concern. Often the system may need to provide cooling when the outside ambient temperature is very low, and the HVAC engineer must consider this because special precautions may be required for the refrigeration system to operate with its condenser units at unusually low ambient temperatures. This condition also provides for significant energy savings advantageous in the use of economizers, which use fresh outside air for cooling when the outside ambient temperature is low.

Heat Sources

The physical size, insulation, and heat load from external sources are items that the HVAC engineer can easily determine from the architectural drawings of the studio complex. The internally generated heat load of the technical equipment for the studio, however, is often underestimated due to incomplete data being provided by the broadcast engineer. Virtually all the technical power input into the studio complex will be turned into heat. Keep in mind that although a solid-state device runs cooler than its vacuum tube ancestor, the equipment density is higher and heat loads may be larger than expected. Additionally, the majority of the power supplied to the lighting within the studio complex becomes heat.

To effectively design the cooling system for a studio, the HVAC engineer must know the amount of power dissipated in these two electrical systems and the *demand factor* for these loads now and in the future. The demand factor is the ratio of time these facilities will be in operation over a period that is short compared to the thermal change in the room (usually 100% for most broadcast facilities). The future load considerations should be coordinated with the reserve capacity design in the electrical system as discussed above. The continuous nature of these loads can make their impact significant on the cooling demands of the audio production facility.

The relationship of electrical power to cooling requirements is as follows:

- 1 kWh of electrical energy dissipated in the studio = 3412 BTU
- 12,000 BTU = 1 ton of refrigeration
- 3.5 kWh of electrical energy dissipated in the studio = 1 ton of refrigeration required

Another heat load in the studio is from the occupants and their level of physical activity. That factor should be evaluated and the highest number of occupants and their most strenuous level of activity provided to the HVAC engineer.

With the HVAC loads determined above, the HVAC engineer can determine the amount of cooling and heating that will be required to keep the studio in the modified comfort zone. The amount of cooling will dictate the quantity of air that must be moved through the studio. This is where the design for studio use departs from more traditional designs due to the need for especially quiet systems.

Acoustical Considerations

The broadcast engineer in conjunction with the acoustical consultant will determine the required *noise criteria* required for each area (for information on noise criteria, see Chapter 3.2). With the desired levels established, the HVAC engineer must carefully review the design to assure his or her selection of materials and installation techniques are adequate and well enough documented to meet the noise criteria while providing adequate cooling. Some of the more important noise control techniques include:

- Locating the air handlers a significant distance from the studio.
- Use of inside lining of fiberglass or other more advanced materials on supply and return ducts.
- Using oversized duct work to reduce the air velocity yet retain desired volume.
- Use of at least three 90 degree turns with fiberglass turning vanes in the lined duct work on both the supply and return ducts between the equipment or tap point and the studio or control room. This requirement is also true for any transfer ducts between areas that may be required.
- Use of oversized supply and return registers selected on their noise ratings in the important rooms, and providing more supply and return registers than would normally be used for the cooling design.
- Installation of the required adjustment dampers for all supply registers at the air-handler tap of the duct, not at the register.
- Independent runs from the air handler to each room rather than tapping larger supply and return ducts.
- Installation of vibration-isolating collars at the appropriate places in the design, which include at the air handler and whenever the duct passes through the studio sound walls.
- Use of commercial muffler assemblies where needed.

These and other techniques are used to maintain the noise levels below the designed noise criteria rating level while providing the necessary amount of cooling. Many HVAC engineers are experienced with quiet installations but some are not, so it is necessary for the project manager to confirm that the appropriate design is implemented.

HVAC Control Systems

With the large heat loads presented by some audio production facilities in small areas, simple cooling systems that cycle large amounts of cooling on and off can produce temperature variations with a noticeable and undesirable large range in a short time. This can be mitigated by using more advanced systems with variable cooling capability, such as variable air volume systems with bypass dampers and hot gas bypass systems, but these have cost impacts on the project. Determining the appropriate control for the cooling system is important for optimum comfort.

A constant source of difficulty in any building occupied by more than one person is disagreement about the correct temperature of any room. This is complicated in audio production facilities because of large differences in heat loads between rooms. In general, each room should have its own temperature control. When this is not practical, the thermostat for the rooms should be located in the most critical room. The thermostats should be locked down to eliminate undesired large changes in settings, particularly temperature, but

also system selection. Modern thermostats will allow a limited time temperature set point change to satisfy a temporary comfort need, but can revert to standard operating set points after a few hours. It is important that no normally accessible controls be permitted to shut down the cooling system. An inadvertent system mode change that kills the cooling system can have undesirable effects, particularly if the audio production room is unmanned for long periods of time.

Fresh Air

The primary cause of the so called "sick building syndrome" is a lack of fresh air being continuously supplied to occupied areas. This part of the design cannot be overstressed, and the amount, as well as the "tempering," of the fresh air to all occupied areas should be reviewed. Air-to-air heat exchangers and heat pipes can be used effectively to aid in economically providing this tempered fresh air, and their use should be considered.

Redundancy

The last concern for the HVAC system is redundancy. All mechanical systems will eventually fail or require to be taken out of service for maintenance. Depending on the application and station design, the audio production facility being serviced may not be able to be removed from service during these periods and, if that is the case, the design should include a contingency plan for emergency cooling. While it might be possible to provide sufficient cooling for equipment with fans set in the door to the corridor, it is doubtful that a national talk show host will be happy doing his or her show under these conditions. Determining the application and planning for unknown events during the design is important.

Fire Protection

The fire protection engineer is primarily responsible for the design of the fire suppression system, and the fire alarm and annunciator system. Both of these systems can have a drastic impact on the operation of an audio production/broadcast facility.

The majority of fire suppression systems are water charged and use water as the suppression agent. However, water can severely damage the electronic components of an audio production facility or master control room. This is particularly true with the discharge of a sprinkler system because the water in the sprinkler system becomes black and heavily contaminated with scale and iron byproducts from the piping system. With the approval of the local fire authorities, there are alternative agents that can be used to suppress fire in sensitive areas. These agents require alteration of not only the fire suppression system, but also the fire suppression system control and the ventilation system within the protected area. Use of these alternate agents may also require protection of the personnel who may occupy these areas as oxygen replacement is part of the method they use to control a fire. Alternatively, delayed release or preaction systems may be employed that require the heat activation of a sprinkler head as in a normal system to be confirmed with an additional signal from the fire control system. These confirming systems include smoke detectors, manual activation, or a timeout of the system without an abort command to allow manual confirmation of the existence of a fire. Consultation with the fire protection engineer and the local fire authorities should be sought on the implementation of a suppression system to be installed in any of the equipment-intensive areas of the facility.

While false alarms and periodic testing of the fire alarm and fire annunciator system will cause no damage to facilities, it can be very disruptive to operations. With the cooperation of the fire protection engineer and the local fire authorities, delayed alarm systems may be permitted to be installed. Such systems sound an alarm in a staffed area and the responsible person in that area has a preset time to confirm the alarm or to stop the system from going into general alarm. Such systems might be particularly desirable in audio production complexes where live programming is under way.

During the initial design or remodeling an audio production facility, a generous allocation of fire strobes/horn/speaker annunciators should be included throughout. With the soundproof nature of studios and control rooms and the possibility of occupants spending long periods listening to audio on headphones or at high levels with monitor loudspeakers, the failure to place a horn/speaker-strobe unit within sight of these users could have life threatening results during an emergency. Do not rely on alarm annunciators from outside the audio production rooms.

Data and Telecommunications

With the widespread use of computer-based systems and networks for the storage and routing of program audio, the data infrastructure of an audio production facility deserves the best design possible. Current trends show that the needs for network connections and dedicated subnets within a facility are increasing more than any other connection pathways. Overbuilding the data infrastructure is a wise investment. Utilizing the latest approved category cable and devices in the data infrastructure will pay dividends in prolonging the life of the installation. While an argument can be made to physically separate the business and general-use data networks from those used by the audio production technical infrastructure, technology convergence and maintainability concerns point to a common data infrastructure with separate subnets, physical segments, and/or VLANs (virtual local area networks). Furthermore, if the routers, switches, and other equipment for the single data infrastructure are located in or near master control, they can take advantage of the continuous conditioned power, cooling, grounding, and technical oversight present in that location.

The telephone industry has moved to IT-based systems with the introduction of voice-over-IP phone systems. These can eliminate an entire cable structure from the office and audio production facility, and should be considered as an alternative to more conventional designs. Whatever type of phone system is chosen, installing the system's common equipment in or near the master control area is also appropriate for the same reasons as stated above for the data system. See Chapter 3.10 for telco interface information.

The audio production rooms and master control area will have special telephone, carrier circuits, and other needs, and these will be easier to implement if all of the telecommunications operations are located in one area. It is, however, preferable to have the primary entrance and surge protection for the telephone company (*telco*) demarcation (*demark*) interface away from master control, and to run sufficient cable and fiber from the telco demark point to master control where secondary surge protection can be applied to all metallic circuits and the ground can be clamped to the center of the star ground.

PROJECT PLANNING AND IMPLEMENTATION

The design process is an interactive process between all the participants. Eventually, a complete set of plans and specifications will be developed and all parties to the planning process will agree that the design is satisfactory and will meet the intended goals.

Plan Checking and Contracting

It is important to review the final design set carefully with the owner and managers as well as with the primary users. Errors in the plans prior to construction are much less expensive to correct than the change orders needed to correct these errors after contracts have been signed. While there will always be change orders to a project because there are always unexpected conditions that alter the plan, it is important to minimize the need for them. Once the final plans are finished, bidding or contract negotiations begin.

Timeline

In addition to preparing a quotation for the cost of a project, the firm doing the construction will accept the construction timeline developed by the architect or the project manager, or will provide a proposed timeline for the project. These timelines will have many logical milestones and should be designed with the intent of tracking the project as it progresses. If the timeline must be modified or if a time constraint is placed on the project, this should be done as part of the final construction contract. Time constraints and timeline estimates do have a cost impact, but are important to keep a project from extending beyond expectations. With the contractor's timeline and construction contract, the project manager can create a master project schedule using project management software, such as Microsoft's Project®, and can track expenditures and change orders with a spreadsheet program or one of many construction tracking programs. The use of these tools will aid in the management of the project in significant ways.

Construction Phase

Once construction starts, the project manager should use all information available to track the costs of the project and to ensure that the project is running on the timeline determined during the contracting phase. Deviations from either timeline or costs should be discussed and justified, and should be brought to the owners attention promptly. Weekly project meetings, headed by the project manager, which include all active participants should be held at the site so that issues can be reviewed and resolved promptly. A record of these meetings should be kept as part of the construction project documents. The meetings should review the progress to date, the anticipated progress over the next week, and any obstacles or issues that threaten to slow the project or change the cost of the project. Attention to these details will help keep the project on budget and on schedule.

The project manager should pay particular attention to the construction details. An audio production facility has many facets of construction that will be new to some contractors on the job. It is important that the details of construction be followed precisely to obtain the desired results. As an example, an electrician unfamiliar with sound wall construction may bridge the two walls in a split wall system with conduit. A contractor may allow a duct to rest on a floating ceiling when it should be hung from the concrete structure, or may miss an important vibration isolation collar. Leaving out some inexpensive acoustical caulk can drastically reduce the attenuation of an expensive sound wall.

Daily or more often inspections of the construction are important during the construction of critical phases of the project. The project manager must understand these construction details and bring discrepancies in their implementation to the construction supervisor immediately upon their discovery. Where appropriate and issues of design are involved, the architect or other consultants should also be informed. Documentation of these discrepancies is important and should be part of the construction meeting record.

More guidance on project management and system integration is in Chapter 2.3.

Final Documentation

Upon completion of the project, a final construction file set should be established. If possible, this set should be in electronic as well as hardcopy form and should include the *as-built* drawings for the facility provided by the contractors. These should include all the technical drawings and diagrams in a maintainable electronic

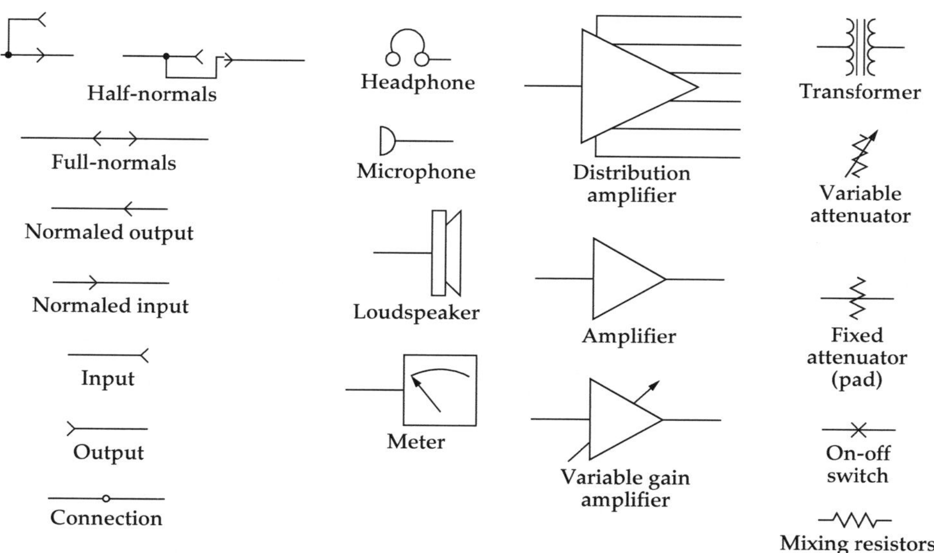

FIGURE 3.1-2 Common symbols for audio equipment.

form for the audio production facility. Much of this data can be maintained in tabular form that lends itself to spreadsheets, and the rest in drawing form using programs such as AutoDesk's AutoCAD®, Microsoft's Visio®, or CommSYS Design's VidCAD®. These drawing sets will aid greatly in troubleshooting the installation, maintaining the facilities, and will provide a starting point for a future remodeling project.

The documentation file should include instruction manuals for all equipment installed in the facility. While most of this is under the control of the broadcast engineers or system integrator, manuals should also be requested from all the contractors building the different systems (electrical, HVAC, fire, security, etc.) within the facility.

Single-Line Diagram

When the facility has been completed, there will be many pages of detailed installation drawings for the audio and associated technical systems. These will reside in the electronic and hardcopy files of the technical maintenance shop. Each production room should also be provided with a simplified diagram to help the operators find their way through the system. The audio single-line diagram minimizes the details in favor of clarity. Rather than every wire and connection, only one line is used to represent a path.

A typical input channel along with the console output and monitoring section should be shown. Ancillary production equipment in the system and all utility circuits and devices, along with their patch bay locations, should also be included. Figure 3.1-2 shows the symbols commonly used to abbreviate the components. Figure 3.1-3 is an example of a very simple path from an input to the mixer's output. It also shows the gains, losses, and nominal levels.

For a larger system it may take some time to understand the many paths and capabilities. For this reason it is helpful to find a convenient wall space for the single-line diagram. Note that if the drawing has a hard

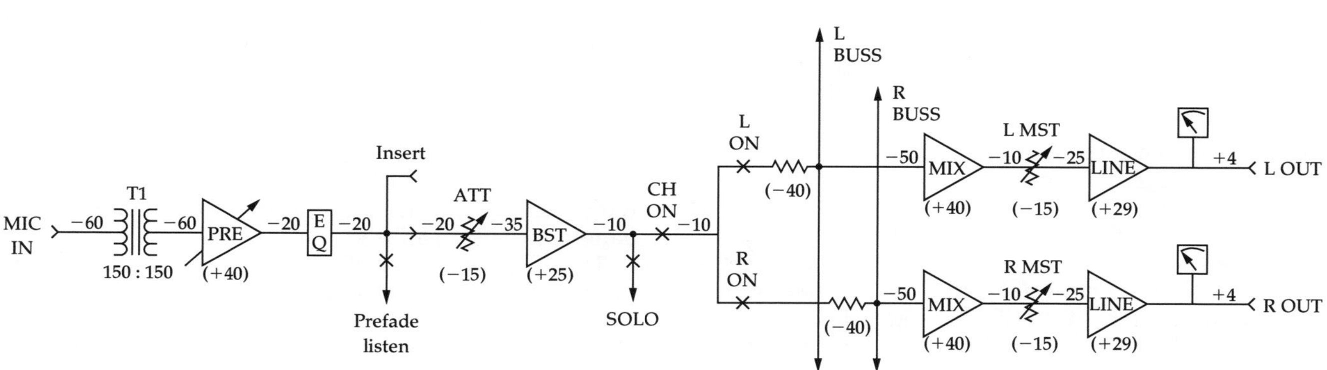

FIGURE 3.1-3 Single-line diagram of a typical analog channel strip.

surface it may affect the room's acoustics. Often there is a wall behind the equipment racks where such a reflector will do little harm.

More guidance on engineering documentation can be found in Chapter 2.2.

Maintenance Schedules

In addition to the construction and facilities documentation set, a checklist and maintenance schedule for all equipment installed in the facility should be developed. Formalized, periodic maintenance is more likely to occur if a written schedule is established.

As well as technical equipment for the station, the checklist and schedule should include such items as the air conditioning and heating system, UPS, backup generator, security systems, tower, antenna, lights, and the general condition of the facility and surrounding environment. The HVAC and electrical contractors can be of great assistance in establishing the initial maintenance schedule for the equipment and systems that have been installed.

RADIO STUDIOS ON A LIMITED BUDGET

Different sized stations require different construction criteria. A 50,000 watt AM clear channel or Class C FM station in a major market and a 250 watt AM daytime or Class A FM station in a rural area are going to differ dramatically. In some cases, the selection of a studio site in a small market is dictated by space costs. It is not unusual for space to be traded out in part or in full for advertising.

Basic Considerations

For the construction or renovation of a smaller facility, the chief engineer, contract engineer, or engineering consultant may be presented with an existing suite of offices, a storefront, a mall space, or even an old house to be converted into offices and studios/control rooms. After determining the nature of the broadcast programming, the facility can be designed to meet the requirements. Most stations use combo control rooms and studios, a simple production facility (that can be used as an emergency control room), and a newsroom with recording and editing facilities capable of on-air use.

If needed, a larger studio can be fashioned from the lobby area, the general office area, or the conference room. With some advance planning, these areas can be acoustically treated and wired for use as acceptable studios for those occasional talent shows, church programs, talk shows, or special group events that do not occur often enough to justify having a full-featured studio available. Plan initially for what might be needed in two or three years even if implementation must be delayed.

Physical Layout

While the total amount of space may be limited, note that one or more people will occupy combo control rooms for long periods and the room should be considered as their office. The room should be large enough to house the equipment, move around, and have visual access to other areas (including the outside if possible), and be comfortable and well lit.

When laying out the broadcast facility, draw a sketch of the available floor space and existing walls. Make a number of copies of this floor plan and start drawing in the studios, offices, newsroom, reception areas, and engineering. Make three or four versions of the plan. If the existing layout will allow the installation of new walls, avoid cubic rooms. A room that is 8 ft × 8 ft × 8 ft is going to have a very definite and unpleasant resonance. Strive for rooms with a 3:4:5 proportion and add some extra wall thickness for the studios and control rooms.

Consider laying out the facility using scale cutouts of typical office furniture, studio equipment, and consoles, plus storage cabinets and countertops, or invest in a limited CAD system to do the scaled layout electronically. Be sure to add some closets and undefined storage spaces for the future. Keep the main studio and control room from beneath heavily traveled stairways and corridors and air conditioners, and use caution when considering front windows opening on busy streets. Hallways take up space. Every square foot of hallway that can be eliminated yields an equal amount of space that can be used for something productive.

Sound locks, while desirable, are not essential, but the further from high-traffic corridors to the studio doors, the better. In commercial construction, doors are generally 3 ft wide and all doors within the facility should be this width if possible. The narrowest hallways should be no less than 3 ft 8 inches in width. All designs should be checked for compliance with the Americans with Disabilities Act as well as local codes.

Where the location of the facility allows, windows to the outside should be considered. In some instances control rooms are built in the interior of a building and exterior windows are not possible. However, when a radio studio or combo facility is built on an outside wall, a small exterior window will allow occupants to know a little more about the weather and have a connection with the outside world (and know whether it is day or night). Of course, if the facility is located next to an airport or major highway, the window may result in more background noise than is acceptable.

If the transmitter is at the studio site, locate it and the main control room so that meters on the transmitter and its associated monitoring equipment can easily be seen and accessed from the operating position (via windows), or install the appropriate remote control panel for the transmitter.

The reception area of any business should be the most presentable, spacious, and elegant room in the building. Above all, it should be neat and clean. Designing the reception area to have a view into the air studio may add a great deal of interest, but the on-air talent may be distracted by this design. It is said that every building should have at least three "Wow!" areas (a look that attracts attention for a good reason). The building front should be one of the Wow! areas and the reception area should be another. Pick the third Wow! area for additional impact—perhaps a control room or transmitter.

Construction Details

The control room floor should be very solid, the most desirable material being concrete. A wooden floor can be used if it is reinforced to prevent bouncing and resonance. If space is available, locate the console in the middle of the room as a desk might be in an office, but useable space within the room may be improved by installing the system against one or more walls.

Doors to studios and control rooms should have solid core doors with weather stripping (felt, not rubber that will squeak) and a good quality door closer with hold-open capability. Do not use a latch on the door that will click each time the door is opened and closed.

Plan on plenty of lights but place the ballasts of fluorescent lights out of the production areas or use modern electronic ballasts to eliminate acoustic noise and interference.

Equipment

In a low-budget situation it is important that the cost of equipment be considered against the actual need for it. For example, buying a $500 directional microphone to do a one-time remote broadcast may be poor business practice. Use this logic when equipping the station. Create a list of the functions to be performed and spend only what is needed to get on the air and achieve those functions. However, plan for the future if at all possible.

Design around a console with enough inputs to accommodate current and unanticipated needs. A four-fader board is likely not enough for any on-air operations and may be quite limiting for simple productions. Even the smallest stations should have provision for eight to twelve channels. While it is rare to have more than three inputs active at one time, easy access to all the inputs that make up a show is important.

Invest in a good-quality telephone interface system as virtually all stations now make phone calls an important part of their programming.

Pay close attention to the installation of a digital audio storage and automation system because, for many facilities, this is practically the entire radio station.

Consider used equipment with great care. While older equipment was built for long life, obtaining spare parts may be a problem and the general quality level may not meet today's needs.

Do not forget the patch panel in the facility. There is a growing tendency among broadcast engineers to eliminate this important routing capability. All major equipment inputs and outputs should appear here. The patch bay should include main processor inputs and outputs, studio to transmitter link (STL) inputs, console outputs, automation outputs, and may contain recorder inputs plus a few console input channels. Open positions on the patch bay for the unplanned functions will be helpful. Be sure to make the layout as logical and as intuitive as possible and use the appropriate normaling techniques as discussed in Chapter 3.5 of this book.

Consider every piece of equipment in the audio chain as a potential point of failure and plan on how to bypass it while it is being repaired. That is the purpose of the patch panel or alternative routing system. This also provides a method of connecting either the mixing console or the automation system direct to the input of the processor, program chain input, or to an alternate STL or a telco dialup circuit. During an emergency, poor-quality audio is better than dead air. Full reconfigurability is perhaps not needed in most radio control rooms, but the ability to bypass major component failures with ease is important.

Consider providing a backup power generator. For a small station a modest-sized generator can power all the essential loads of the facility.

RADIO STATION FLOOR PLANS

A major requirement for a successful broadcast station is the careful layout of studio, production, and administrative areas to achieve maximum effectiveness of space and personnel. No two stations and facilities have exactly the same requirements and constraints, so no two designs will be the same. Sample floor plans are provided here for some typical radio station layouts and serve to help develop other designs per the various requirements as discussed in this chapter.

Figure 3.1-4 shows the layout of an approximately 2500 square ft single broadcast station designed to occupy a retail space in a commercial shopping mall. This space provides retail space to market concert tickets and radio station promotional material. The main air studio is viewable from the mall, but the operator position is such that the operator will not be distracted while on the air.

Figure 3.1-5 is a plan for a freestanding preengineered steel building that is part of a retail development on the outskirts of a town. The studio and office facility is approximately 3000 square ft. This design incorporates an approximately 700 square ft garage to protect the promotional van and provide storage for the promotional materials, and has a 1250 square ft covered patio on the back of the facility for a weekly "picking party" featuring local musicians.

Figure 3.1-6 is an approximately 6600 square ft complex to support two active FM stations. This design features space for a large sales staff with several sales managers. It is an adaptation of an existing multitenant office building.

These designs assume that the main transmitter plant is not colocated with the studio complex. If the transmitter is to be colocated with the station, additional space must be allotted somewhere in the

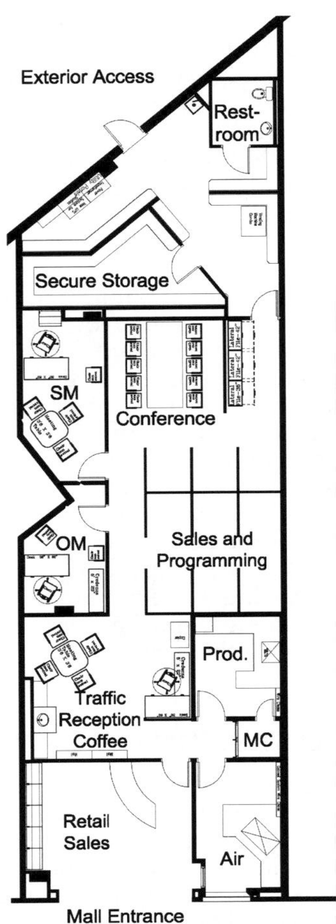

FIGURE 3.1-4 Example 2500 sq. ft single radio station facility.

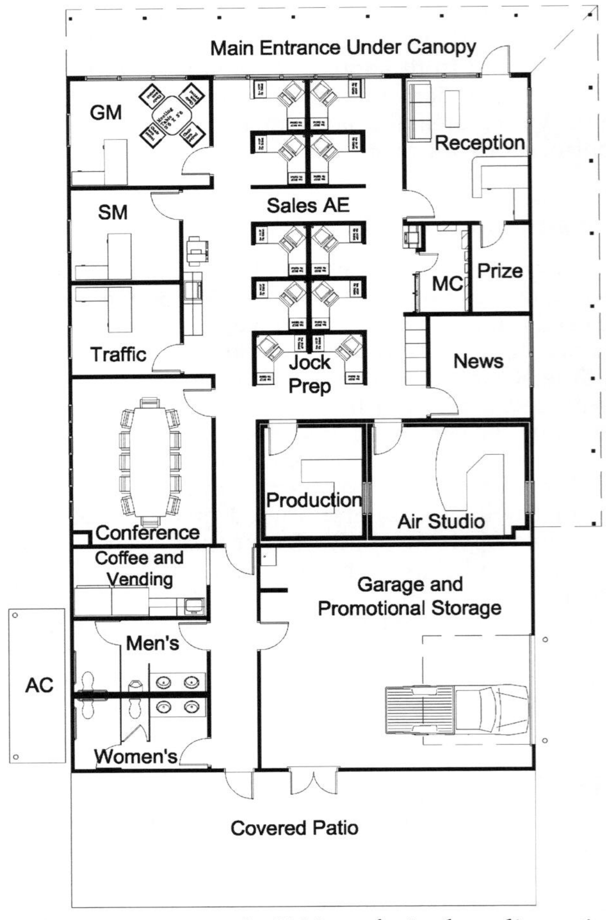

FIGURE 3.1-5 Example 3000 sq. ft single radio station with garage and patio for local producti3ons.

building for the transmitter room and related services (HVAC, AC power). The tower and antenna of course require further consideration. Locations for the STL and satellite antennas must be determined by the orientation of the site and local conditions.

SUMMARY

The planning and implementation of an audio production center can be a difficult task but every project reaches a point where the deadline is met and the budget must be closed. However, when everything is up and running, the job may still not be done. It is not unusual to see only 85% completion at this point. The owner may ask, "We're in production, why go any further?"

The exhilaration of a new facility will give the production and operations people the energy to work around the uncompleted parts of the project, but as time goes by, the end users will tire of these workarounds and the product invariably suffers. It is important to keep the original goal and design intentions in mind, along with the fine tuning that comes with any new facility, and achieve 100% completion of the project, including 100% documentation of what has been built. The construction group will generally have moved on prior to the 100% completion point so the local technical staff will need to bring the facility to completion in addition to carrying out routine activities. Budget some overtime for finishing up, and work with everyone involved to bring the project to full completion. The process may take time, perhaps three to six months, so keep communications open with the users so they will be patient as they see progress occurring.

With good design, a look toward the future, and full completion of the plan, every audio production facility can be attractive, professional, and productive. At the end of the project, the engineer should be able to step back and be proud of the resulting audio production facility.

ACKNOWLEDGMENTS

Parts of the material for this chapter were contributed for the ninth edition of the *NAB Engineering Handbook*

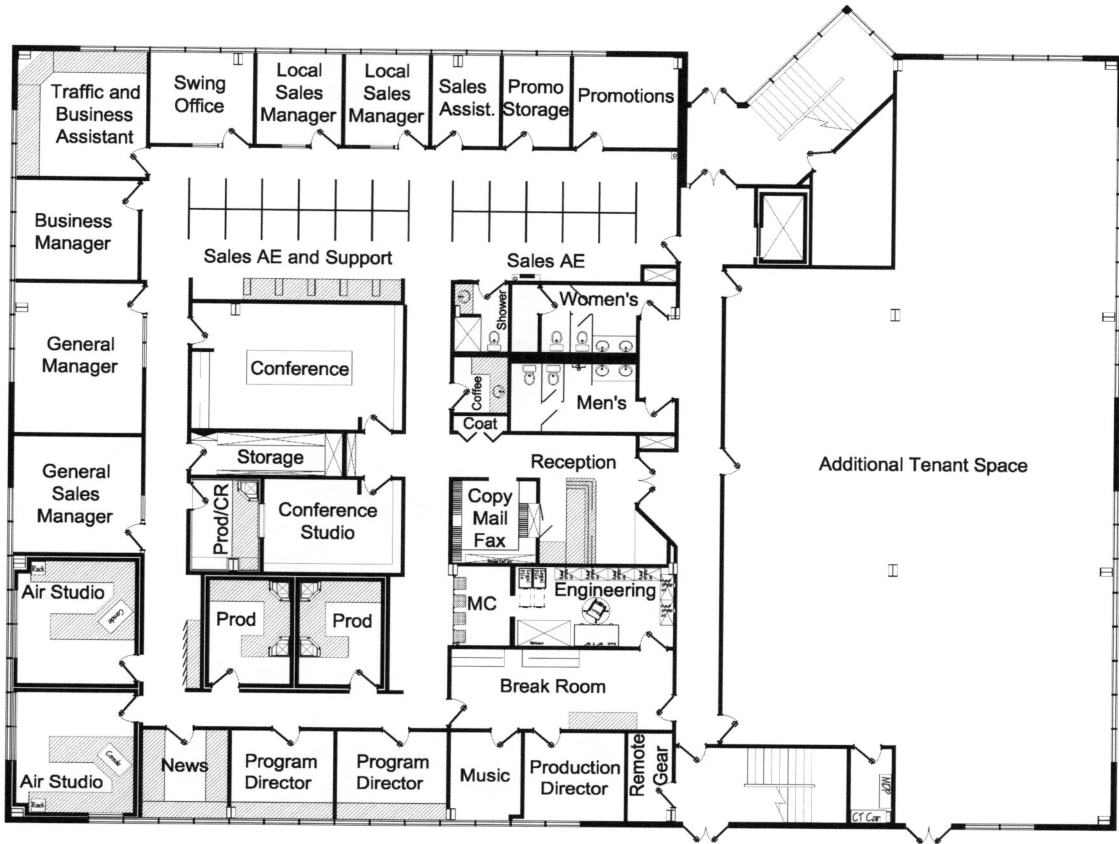

FIGURE 3.1-6 Example of a 6600 sq. ft facility for a two-radio station complex that is part of an office building.

by Malcolm M. Burleson of Alexandria, VA, and David Carr of Dallas, TX. Thanks are extended to Martin Stabbert of Citidel Communications Corporation, Reno, NV, and Matt Lightner of Lightner Electronics, Claysburg, PA, for their helpful reviews and suggestions in updating the chapter for this edition.

References

[1] Ramsey, C.G., and Sleeper H.R., *Architectural Graphic Standards*. Sponsored by the American Institute of Architects. New York: John Wiley & Sons, 2000.

[2] Early, M.W., Sargent, J.S., Sheehan, J.V., and Caloggero, J.M., *NEC 2005 Handbook*. Quincy, MA: National Fire Protection Association, Inc., 2005.

CHAPTER

3.2

Principles of Acoustics and Noise Control for Broadcast Applications

HOWARD K. PELTON, TED N. CARNES, AND GARY L. MCAULIFFE*
Pelton, Marsh, Kinsella
Dallas, Texas

INTRODUCTION

Acoustics is both the science and the art of understanding sound and the interaction of sound with the rest of the physical world. The science encompasses the study and analysis of vibrations at any frequency (subsonic to audible to ultrasonic) through the principal states of matter (solid, liquid, and gaseous). The art of acoustics applies the principles learned through science to practical engineering and design matters. It includes the development and use of empirically derived formulae, data, and observations into usable experiential guides to designing spaces for acoustic and amplified speech and music, recording, monitoring, storage and transmission, and, most importantly, listening. The art of acoustics also requires an understanding of human auditory perception and of the acoustical properties and application of modern construction materials. It requires a recognition of the realities of construction methods that are practical within the confines of schedules, budgets, and level of construction trade skills that are available and appropriate to the project at hand.

This chapter provides a general briefing on both the science and the art of acoustics, with an emphasis on practical acoustical criteria for broadcast spaces. It includes advice on how to plan radio and television facilities to meet those criteria, and provides some background information on the physics and measurement of sound. It is intended to allow the broadcast industry professional to effectively communicate facility acoustical performance requirements to the architects and engineers planning renovated or new radio and television studio facilities.

In many cases the broadcast professional, architect, or mechanical/electrical/plumbing engineer(s) will require the assistance of an acoustician. In all cases, with or without an acoustician, an acoustical program and a quality-control mechanism, to ensure that what is intended in design is actually constructed, is recommended so that the acoustical results can be quantified and evaluated.

Further sources of information can be obtained from several related professional and technical organizations as listed at the end of the chapter.

PRACTICAL ACOUSTICS

The branches of practical acoustics applicable to radio and television facilities are architectural acoustics, MEP noise control, environmental noise, and (in the case of control room monitoring and recorded listening playback environments) electroacoustics.

Architectural Acoustics

Architectural acoustics concerns the audible optimization of an interior space to suit a user-defined purpose by providing acoustical isolation from adjacent spaces; setting room dimensions and geometries to avoid

*This chapter draws extensively on material first published in the 9th edition of the *NAB Engineering Handbook*, Chapter 3.2, "Principles of Acoustics for Broadcast Applications," by Lynn Claudy.

modes, resonances, and standing waves; and by controlling echoes and reverberation.

Isolation

Building shell and room partition construction techniques are the foundation upon which all other critical space acoustical parameters rest. An otherwise acoustically well-behaved space, free of resonances, reflections, and reverberation, is useless for recording or critical listening if the envelope cannot exclude external noises from outside the building envelope and from adjacent spaces, to produce a commensurate and complementary low noise floor.

The basic tools for creating high-acoustical isolation between spaces are the mass of the partitions and "dead" air, in the sense that air cannot move freely between spaces through door jambs, duct penetrations, and electrical device wall penetrations. Decoupling of partition faces and the sealing of connections to other structural elements are also important factors in the overall rating of the completed partition system.

Modes and Resonances

Room modes, resonances, and standing waves are best avoided by careful planning of room dimensions, the ratios of length to width and height, and good control of overall geometry by having at least one wall not parallel to its opposite. In unavoidable room dimension or geometry conditions, acoustical traps based on the Helmholtz principle can provide some correction.

Reverberation

Optimizing reverberation for the intended use of the space involves the controlling of reflected sound, by either directing, eliminating, or diffusing specular acoustical reflections in the space. Reflections can be desired or undesired depending on the use of the space. Desired reflections generally fall within very short arrival times, as in the case of an orchestra shell in a concert hall, where it is necessary for musicians to hear one another without the perception of echo. In most broadcast facility spaces, it is necessary to control and suppress acoustical reflections.

Control of unwanted reflections is accomplished by directing the unwanted sound energy away from the listener or microphone, eliminating reflected energy by absorption. Absorption, in essence, changes sound power to small amounts of heat by slowing propagation and disrupting the incident wavefront in a porous material, or by diffusing the incident wavefront by scattering the energy across a large number of reflected angles with much lower energy at each angle.

Methodology

Obtaining the desired acoustical results in each type of space requires setting appropriate acoustical performance criteria and selection of the appropriate building and room construction methods and materials. It also requires selection of the appropriate surface-applied or built-in acoustical treatments to reflect, absorb, or diffuse sound energy. Further considerations are the arrangement of the broadcast electronic equipment and control surfaces so as to not create undesirable acoustical reflection zones or heat gradients and, finally, monitoring of the construction process to enforce the settled criteria. In critical recording and monitoring environments, the process also requires measuring and documenting the acoustical results of the completed building construction and electronic equipment installation effort.

MEP Noise Control

Mechanical, electrical, and plumbing (MEP) noise control acoustics concerns the proper sizing, location, isolation, and noise control treatment of MEP equipment to prevent unwanted noise intrusions and the elimination of flanking paths for noise transmission between sensitive adjacent spaces. The noise floor metric for the room is usually quantified in NC or RC terms, as explained later in this chapter. In most small room spaces, the largest determinant of acoustical quality is the noise floor of the heating, ventiliation, and air conditioning (HVAC) system. No amount of after-the-fact room treatment can correct for noisy HVAC systems. Local building codes may affect duct design by restricting common acoustical duct lining treatments and, as in all building code matters, competent design professional advice is recommended. Quiet HVAC systems must maintain laminar flow with minimal turbulence along duct paths and avoid noise regeneration at flow regulating and terminating devices.

In control room or studio spaces with significant equipment or lighting heat loads, it is imperative that the anticipated electrical and heat loads be accurately estimated to allow the MEP engineer to provide appropriate cooling capacity with air-flow velocities consistent with acoustical criteria. Also, large electrical transformers and lighting dimming systems equipment can produce significant structure-borne noise transmission and require special isolation considerations.

Vibration

Vibrations affecting studio acoustical performance may be a result of internal and external building mechanical equipment, or externalities including subway, rail, or vehicular traffic, nearby industrial plant noise, or heavy construction activity. The MEP engineer and acoustician will, in the course of the professional design, address building mechanical equipment vibration issues. This involves the use of appropriate equipment suspension, structural breaks, isolators, and mass inertia bases. When the proposed or existing building location cannot avoid vibration from external transportation sources, special structural engineering considerations, including examination of the natural frequency of the building structural system, are

needed before possible foundation isolation systems can be considered.

Environmental Noise

Environmental noise conditions and community noise regulations play a part in broadcast facility site selection and building design and are therefore part of practical acoustics. Environmental noise assessment should include an evaluation of local aircraft traffic conditions, as a typical commercial buildings' lightweight roof construction is an often overlooked source of noise ingress. Local weather conditions should also be considered as a potential source of studio noise, including rain and hail roof drumming. In the case of large television stations with helicopter-equipped news operations, there will be additional building shell isolation and community noise concerns that should be professionally addressed and are beyond the scope of this chapter.

Selection of the type, location, and the physical isolation of large noise-producing mechanical equipment and emergency generators must be considered in the light of local noise ordinances as well as the building shell construction.

Psychoacoustics

Psychoacoustics is the study of the human perception of sound and vibration. The ability of the ear and brain combination to selectively filter, isolate, and understand conversations in crowded and noisy conditions is well known. While most radio, and much of television, broadcast production operations use close microphone techniques, which, to a great extent, reduce background noise and reflections and improve intelligibility, attention to acoustics is still necessary. The broadcasting system, by definition, removes the listener from the original acoustical environment and eliminates the psychoacoustic cues necessary to enhance intelligibility. It also may remove the original directional and positional information that allows the brain to select information from noise.

Electroacoustics come into play in considering the performance of electronic transducer devices in the acoustic environment, in particular microphones and monitor loudspeakers. This is a specialized subject in its own right and is not considered further here.

Planning Criteria

Broadcast space planning should begin with collecting design criteria data for architectural, MEP, and acoustical requirements for each noise-sensitive room and presenting the information in a summary fashion. Particular attention should be paid to electrical and lighting loads. This data serves to communicate needs and expectations to architects and engineers during the design phase and allows the engineer to correctly size mechanical systems and the acoustician to evaluate and suggest corrections to the design. It also serves as the basis of an acoustical quality-control program. Table 3.2-1 summarizes some typical criteria and acoustical data that need to be gathered.

TABLE 3.2-1
Sample Room Data and Criteria Sheet

Architectural Design Criteria	**Value/Type**
Wall finish	See acoustical data
Millwork—base building provided	
Floor finish	Carpet
Floor type/construction	Concrete
Ceiling finish	Acoustical
Ceiling type	Lay-in fiberglass
Glazing	Double
Room occupancy count	6
MEP Design Criteria	
Equipment HVAC load	
Technical power requirements	
Utility power requirements	
Work/maintenance lighting	
Intensity (FC)	
Task lighting	
Television lighting requirements	
Lighting HVAC load	
Oversize neutrals	
Separate ground every circuit	
Dedicated ground bus system	
Acoustical Design Requirements	
HVAC noise (NC)	25–30
Wall STC	60
Acoustical panel type and coverage	Fabric-wrapped, 65% coverage
Door type and STC	Acoustical, 55
Door seals/type	Magnetic
Door hinge type	Cam lift
Other door hardware	See plan
Ceiling NRC	.95 min.
Ceiling (TL) STC	60, with concrete structure
Glazing STC	60–65
Floor rating (IIC)	60–65
Floor above rating (IIC)	60–65, use carpets above

Acoustic Treatment

For acoustical purposes, radio and television studio spaces are categorized by function, area, and room volume. Functional requirements include recording, control, editing, monitoring, and final critical listening playback.

Radio Studios

In most radio broadcast environments, reverberation is not a significant factor because room dimensions and volumes, except in the large studio sound stages, are relatively small, and true reverberation cannot develop. The general rule of thumb is that any room with a maximum dimension (length, width, or height) less than 30 ft is considered small in acoustical terms.

In smaller spaces, control of specular reflections is the goal. Most radio facilities have acoustical ceilings and carpeted floors that provide good absorption, while television studio floors are usually concrete and the ceilings are open to the structure above. Assuming a well-isolated shell, quiet HVAC, good room geometry, and treated floors and ceilings, it is still necessary to treat the walls to prevent specular reflections. Acoustical open cell foams or fabric-wrapped semi-rigid fiberglass covering 60–80% of the wall space is adequate in most applications. Fabric should be an open weave or a perforated vinyl where durability is a concern. In cases where wall space is restricted, panel absorption can be improved by increasing the thickness of the panel. Generally speaking, it is better to distribute the treatment rather than to concentrate it on fewer walls, but it should be located above counter height and around the path between the talent and the microphones. Whatever treatment material is selected, it is important to specify it as class A flame spread rated.

Television Studios

Television studios usually have larger volumes than radio studios, with dimensions in excess of 30 ft, frequently with high ceilings and, by nature, more hard surfaces including floors, cycloramas, sets, and flats that cannot be acoustically treated and which therefore contribute to the room reverberation. This generally requires that all available wall surfaces be treated with a high-performance absorption treatment, which also needs to be durable. Various combinations of fiberglass and cementitious fiberboard are suggested later in this chapter. The ceiling presents the largest treatable area that is generally free of reflection causing hard obstructions. Suggested acoustical treatments include upgraded lay in ceilings and spray-on materials. The latter are preferable in larger studios for cost reasons. Open-to-structure studio ceilings also have the advantage of greater room volume to dissipate lighting heat in to a greater volume of air.

Television studio doors also tend to be larger than radio studio doors for easy delivery of large set pieces, and in the case of audience participation programs, entry and egress of the general public. Whenever possible when floor space, Americans with Disabilities Act (ADA) building code egress requirements, and traffic-flow permits, it is desirable to create door airlock vestibules. Two door sets of a lower sound transmission class (STC, see later section) rating and cost will outperform a single more expensive higher STC-rated door.

Windows

Radio studios tend to have more window requirements than television facilities and these require special attention. STC-rated glazing assemblies can be premanufactured or built in the field. Typical measures to improve glazing STC include using acrylics, laminated glass, increasing glass thickness, employing dual and triple panes with dead-air spaces, using different thickness glass on each pane, and avoiding parallel panes in the assembly to control resonance and decouple the panes.

Television "fishbowl" studios with views onto public streets have become increasingly popular in the last decade. The glazing systems for these often include all of the above acoustical measures and often have requirements for ballistic rating (UL 752) and outdoor light color correction as well.

Typical studio sizes, functions, recommended criteria, and basic acoustical treatments are shown in Table 3.2-2.

Methodology

Common sense is often the most important ingredient in noise control and acoustical room design and also often the first forgotten when practical matters arise that introduce conflicts. In optimizing noise-control design for broadcast applications the following checklist identifies the highest priority items that should be considered, listed in approximate order of importance:

1. Select site for least noisy area.
2. Perform noise survey to determine required acoustical isolation.
3. Select position of room within building.
4. Select sound-isolation techniques.
5. Control noise within building including structure-borne and airborne noise.
6. Design room geometry for good diffusion and room mode spacing.
7. Select and place sound-absorbing material for optimum room acoustic response.

Numbers 1–5 refer to the control of noise and intentionally are given a higher priority for broadcast applications. While room shape and optimum reverberation time are certainly important, they have a lower priority for the discussion in the section. A quiet room is the most important design goal and also potentially the most expensive and difficult to achieve.

TABLE 3.2-2
Studios and Control Rooms for Broadcasting—
Characterizations and Suggested Baseline Acoustic Criteria

Studio Type	Typical Floor Area (Sq. ft)	Typical Room Volume (Cu. ft)	HVAC Noise Criteria (NC/RC(N)*)	Shell Wall Rating (STC)	Door Rating (STC)	Max. Rt60 (Secs)	Ceiling (NRC)	Wall Panel (NRC)
Radio								
Announce booth	64	610	20–25	65+	60+	0.5	0.90	1.0
News studio	120	1140	20–25	60	55	0.5	0.90	1.0
On-air combo operator studio	120	1140	20–25	60	55	0.5	0.90	1.0
Panel discussion studio	260	2470	20–25	60	55	0.5	1.0	1.0
Dedicated control room	168	1600	25–30	55	50	1.0	1.0	1.0
DAW/edit bay	100	950	25–30	55	50	1.0	1.0	1.0
Client listening/conference room	240	2300	25–30	55	50	0.5	0.90	1.0
Television								
Announce booth	100	950	20–25	65+	60+	0.5	0.90	1.0
Working newsroom set	—	—	25–30	60	55	1.0	0.90	1.0
Small insert studio	1200	28,800	20–25	60	55	0.5	1.0	1.05
Midsize production studio	2400	57,600	20–25	60	55	0.5	1.05	1.05
Audience studio with live music	4800	115,200	25–30	55	50	1.8	1.05	—
Large production studio	4800	115,200	20–25	60	55	1.0	1.05	1.10
Large soundstage	12,000+	408,000	20–25	60	55		1.05	1.15
Video production control room	750	7875	25–30	55	50	1.0	0.90	1.0
Audio control room	200	1900	20–25	60	55	1.0	1.0	1.0
NLE edit bays	120	1140	25–30	55	50	1.0	1.0	1.0
Master control room	—	—	25–30	55	50	1.0	0.90	1.0

*RC(N) is the preferred criteria when the quality of the space dictates that background sound should be neutral and unobtrusive. Intrusive noise sources must be at least 7 dB below the noise HVAC noise levels in each octave band of the audible spectrum.

PLANNING FOR QUIET

Broadcast studios require extremely low background-noise levels. Good noise-control planning can help control, and perhaps reduce, the construction costs for the studio buildings. If all studios could be sited in the quiet of the countryside, costs would certainly be much less; however, that is not usually the case. Generally, studio complexes are required to be located close to the center of cities, which are inherently noisy environments. Since extraneous noise will generally be picked up by the studio microphones, the planning and structure of the building must be carefully designed to provide maximum attenuation of noise into the studio areas from the exterior, as well as from noisy areas that may be inside the building.

The steps that must be taken in the initial planning phase involve all or some of the following:

- Review site location for environmental noise sources, such as traffic, trains, subways, helicopters, civilian, and military aircraft.
- Carry out a detailed noise survey at the proposed site, including one-third octave band analysis of air-borne and ground-borne vibration levels.
- If necessary, carry out long-term environmental noise measurements to ascertain maximum levels from identified sources.
- Review the preliminary building layout to ensure that noise-sensitive spaces, such as studios and control rooms, are placed far from internal noise-

producing equipment, such as HVAC equipment and, insofar as possible, away from environmental noise sources.

Building Location and Environmental Noise Sources

From an acoustical perspective, it is always preferable to locate a studio complex away from major transportation centers, such as freeways, railroads, and airports, since there will always be more cost associated with the planning, design, and construction of the building if it is located adjacent to such sources of noise. If that is not possible, then more care has to be taken with the ambient noise survey and assessment of the noise environment, and special consideration will be needed for the building shell design and construction.

A detailed noise assessment is required in a noisy urban environment. Surveys resulting in a single number of values, such as dBA, are not suitable for this type of evaluation. Normally, a one-third octave band analysis is required using environmental modes on a real-time spectrum analyzer. This will provide the low-frequency noise levels that help in selection of building products for the exterior envelope of the building.

Construction Noise

When adding new facilities to an existing studio complex, the effect of construction noise on normal studio operations has to be considered and special construction techniques planned where necessary to reduce noise from intruding into the existing building.

The disturbance caused by construction noise and from general building operations can be divided into three basic categories:

- Noise from construction that affects adjacent third-party buildings and, in this case, may be governed by environmental noise ordinances of the local entity in which the building is being constructed.
- Noise from construction that affects the occupants of the existing studio building and its operation. Both airborne and structure-borne noise may need to be carefully controlled since normal broadcast operations typically cannot be changed and/or eliminated during the construction phase.
- Noise, such as that from demolition operations, which could be done at different times but still must be thought through in the planning stage.

The noise levels from construction sites can generally be reduced by one or more of the following means:

- Increasing the distance from the noise source to the existing building. This is the most effective means because each doubling of distance between the noise source and receiver may provide a noise reduction of 6 dB depending on the terrain, adjacent buildings, and weather conditions. For example, placement of air compressors, generators, or other heavy equipment far away from the existing building as possible will be very beneficial.
- Reducing noise at the source and along the pathway. For example, most noise-generating equipment can be provided with enclosures, barriers, hoods, and other noise-control methods that can reduce the noise. This requires investigations and construction specifications for equipment noise levels ahead of time.
- Using quiet building techniques. For example, pile driving and compaction operations generate extremely high noise levels in both airborne and ground-borne vibration, which will be carried into adjacent buildings. In order to reduce noise, piles can be bored rather than driven and many times this can be done at a time when the studios or other noise-sensitive spaces are not in use. It is important to coordinate with the contractors well ahead of time to ensure that "quiet" methods can be used.
- Finally, the use of a specification for the building site that will alert the contractor that special precautions are required. This specification can be very detailed. In special cases, noise and vibration monitoring systems can be installed that will alert the contractor, owner, and consultants when a certain limit has been reached and that project construction activity should cease until changes in procedure can be made. This is a common means of keeping the noise levels to a minimum when construction must be carried out in parallel with standard facility operations.

Studio Complex Layout

As the initial planning is carried out, it is important that the design team think through all the possibilities related to ingress and transmission of noise, whether the studio complex is a single ground floor level design or a multifloor studio complex. This includes environmental noise from outside and MEP and other noise from within the building. Figure 3.2-1 illustrates a typical layout for a small radio studio complex, with the studios located away from support areas and mechanical equipment. Figure 3.2-2 illustrates the pathways by which noise and vibration can enter the structure and pass between internal areas.

For planning purposes, a good starting point would be the room characteristics and baseline acoustic criteria shown in Table 3.2-2.

Mechanical Systems Locations

Intrusive noise sources within the studio complex, such as emergency generators, refrigeration machines, boilers, and other heavy equipment, should be sited as far away as possible from noise-sensitive studios. In addition, all such equipment and plant must be completely isolated from the studio building structure using vibration-isolation materials, expansion joints, and other means to prevent structure-borne noise and

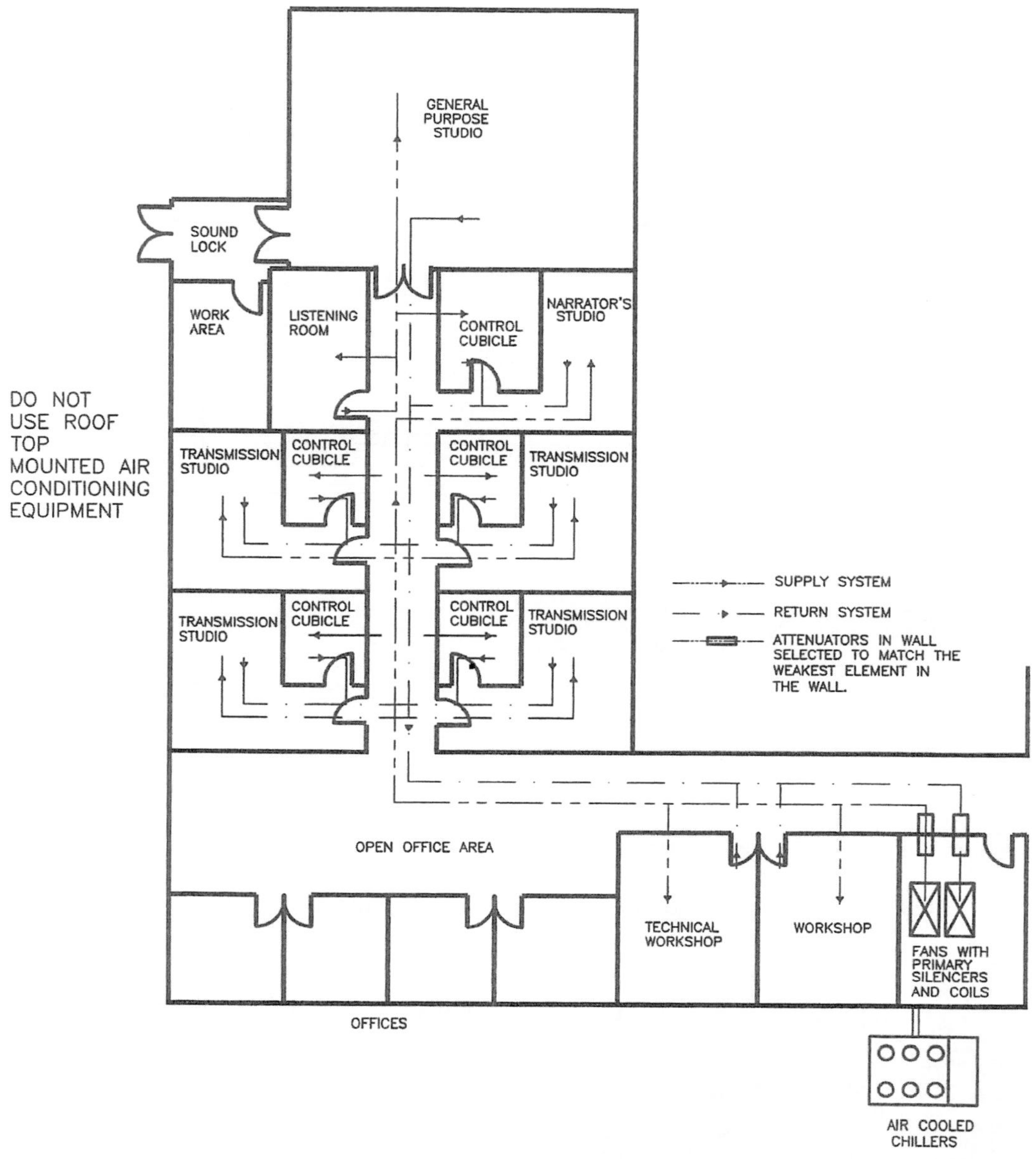

FIGURE 3.2-1 Small radio studio complex. Plan for quiet—locate noisy systems as remote as possible from noise-sensitive areas.

vibration transmission into noise-sensitive areas. Every possible means should be taken to reduce structure-borne vibration.

Air-handling units, while required by ductwork limitations to be reasonably close to the noise-sensitive areas that they serve, will need to be isolated both from structure-borne and airborne noise pathways. For example, in a large studio complex, the air-handling equipment may be located outside on an adjacent mechanical structure, with ductwork penetrating the upper walls of the studios. For a smaller studio complex, suitable interior locations for air-handling units can usually be found but may require the use of a *floating floor.*

Methods for airborne and structure-borne noise and vibration isolation are discussed in the section on sound-isolation techniques and materials later in the chapter.

Studio and Control Room Layout and Design

In development of the studio and control rooms layout and designs and "planning for quiet," the starting point should be the development of a room data sheet for each noise-sensitive area, as discussed previously under planning criteria and shown in Table 3.2-1. This will allow constructive discussion and thinking about the individual spaces and their adjacencies.

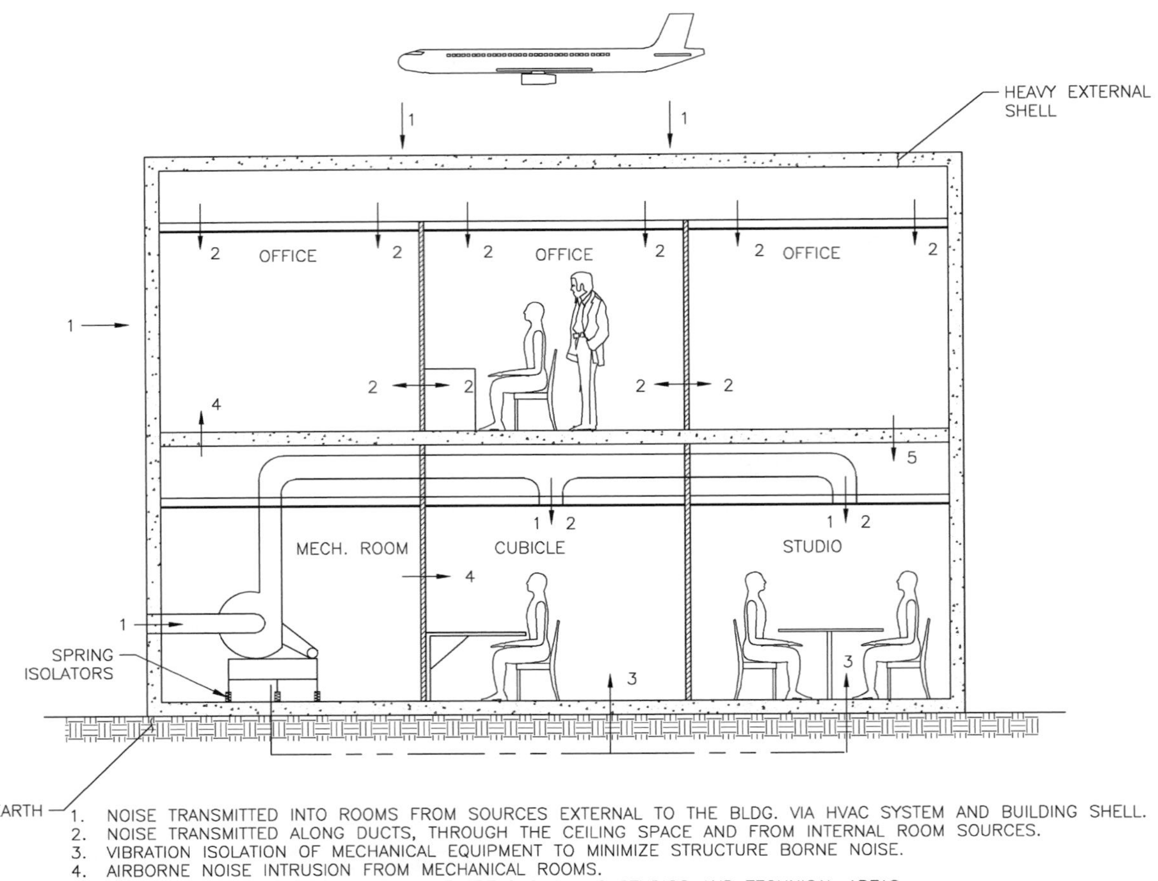
HEAVY EXTERNAL SHELL
OFFICE
OFFICE
OFFICE
MECH. ROOM
CUBICLE
STUDIO
SPRING ISOLATORS
EARTH
1. NOISE TRANSMITTED INTO ROOMS FROM SOURCES EXTERNAL TO THE BLDG. VIA HVAC SYSTEM AND BUILDING SHELL.
2. NOISE TRANSMITTED ALONG DUCTS, THROUGH THE CEILING SPACE AND FROM INTERNAL ROOM SOURCES.
3. VIBRATION ISOLATION OF MECHANICAL EQUIPMENT TO MINIMIZE STRUCTURE BORNE NOISE.
4. AIRBORNE NOISE INTRUSION FROM MECHANICAL ROOMS.
5. POTENTIAL IMPACT NOISE TO SENSITIVE AREAS SUCH AS STUDIOS AND TECHNICAL AREAS.

FIGURE 3.2-2 Noise and vibration paths into and inside a studio building.

FIGURE 3.2-3 Radio studio complex.

Part of the planning process is determining the type and use of each studio. As listed in Table 3.2-2, radio and television studios have somewhat different criteria. Radio and small television studios tend to need highly absorptive acoustic treatment, although this is dependent on the type of activities in the studio. Larger television studios are typically less absorptive, while dedicated studios for live music performances need to be more reverberant ("live") and have more reflections. Control rooms often have a combination of absorption and diffusion on the interior surfaces to optimize the environment for critical listening using loudspeaker monitors.

A control room with a view of the studio (see Figure 3.2-3) will require specialized acoustical windows, doors, and partitions (see details later in the chapter), and an acoustic design optimized for high-quality listening. If on the other hand this is a combination radio studio, then the acoustic design will be more like a highly absorbent small radio studio. The Bibliography lists a BBC paper on detailed design considerations for critical listening in studios and control rooms.

ACOUSTIC CRITERIA

Criteria are needed for the design, commissioning, and possible diagnostics or troubleshooting of the room acoustics and background sound in the facility.

The reverberation time (RT_{60}), as discussed later, is one descriptor of the room acoustics. Background sound levels, also discussed below, are another. Various descriptors are also used to quantify speech intelligibility aspects in studies.

Statistical Descriptors of Speech Intelligibility

Several statistical descriptors are used to classify the intelligibility of speech both in rooms and over audio networks. These include descriptors, such as articulation index, speech transmission index, rapid speech transmission index, and speech and preferred speech interference levels.

Articulation index (AI) is the most often used of these metrics and can be measured using a simple sound source and sound level meter. It is the weighted proportion of speech that is usable to convey information and is calculated from the scores of a group of experienced listeners with normal hearing, who write sentences, words, or syllables read to them from specially selected lists.

Speech transmission index (STI) is similar to AI but uses a modulation transfer function in its evaluation, and as such requires more sophisticated instrumentation. The STI procedure is often used for expressing the loss in articulation of consonants (%Alcons) produced by room reverberation and background sound.

The rapid speech transmission index (RASTI) is similar to STI, but uses fewer modulation frequencies and incorporates only speech and background sound levels in only two octave bands—500 Hz and 2000 Hz. Measurement of RASTI also requires sophisticated instrumentation.

As two final descriptors, speech interference level (SIL) is the arithmetic average of the sound pressure level in four octave bands whose center frequencies are 500, 1000, 2000, and 4000 Hz. Also the preferred speech interference level (PSIL) is the arithmetic average of the sound pressure level in three octave bands whose center frequencies are 500, 1000, and 2000 Hz. See the discussion below on the use of the PSIL as used with the room criteria (RC) sound descriptor.

Noise and Room Criteria

The *noise criteria* (NC) curves were developed through experimentation and knowledge of the sensitivities of the human hearing system to provide a single number figure of merit for maximum permissible background noise level for a given activity. The NC number is given approximately by the value of the NC curve in the 1000–2000 Hz frequency band range. Figure 3.2-4 shows the NC family of curves.

The newer *room criteria* (RC) curves, which better discipline the speech and quality of sound, are shown in Figure 3.2-5.

For assigning an NC rating to an arbitrary room, the following procedure is used:

- Sound pressure level (SPL) readings are taken in the 8 octave bands from 63 to 8000 Hz.

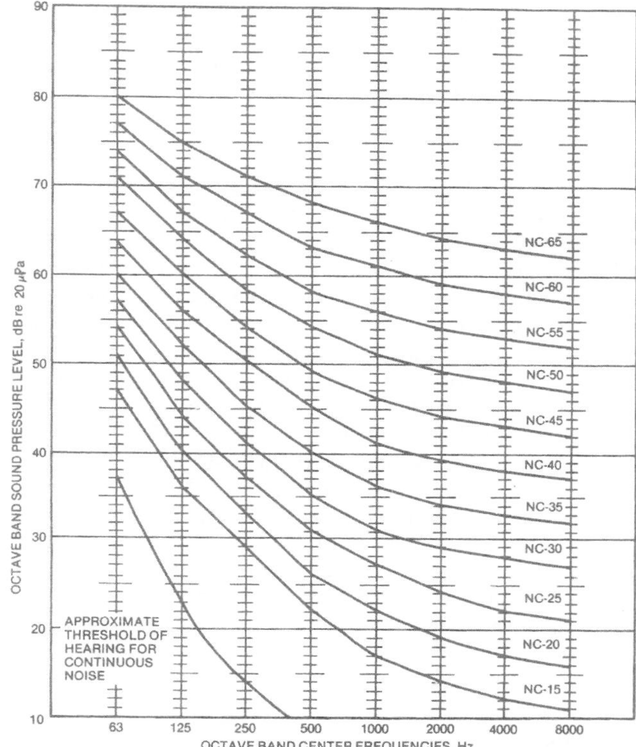

FIGURE 3.2-4 Noise criteria curves.

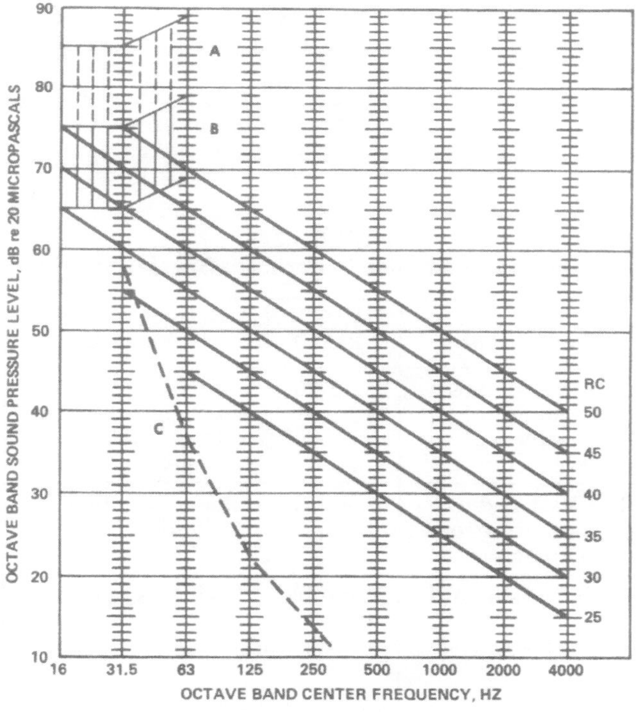

FIGURE 3.2-5 Room criteria curves.

- The NC rating is defined as the lowest value NC curve that lies wholly above the measured data. This is referred to as the Noise Criteria Tangency Method and the point of tangency setting the value should be stated.

For radio and small television broadcast studios, a rating of NC-20 to NC-25 or less is desirable, larger television studios can have somewhat higher levels, as indicated in Table 3.2-2. Table 3.2-3 provides more stringent guidelines for specialized sound studios.

The nature of the background noise in a space is important to consider, especially where speech is concerned. Thus, the use of RC is suggested for those situations where not only the level but also the sound quality need to be considered. The magnitude of RC is simply the PSIL as mentioned above. Then a *sound quality* descriptor is added to indicate possible variation of the sound spectrum from the desirable *neutral* "N" balanced condition. The RC room criterion curves are straight lines with a slope of –5 dB/octave. While the RC curves are officially defined only in the RC of 25 to 50 range, the methodology can be used to assess both higher and lower values.

Note that the RC curves are defined over a wider frequency range than NC with the 16 and 31.5 Hz octave bands considered. These two very low-frequency octave bands rarely can be assessed in the design phase but often must be considered for diagnostic purposes.

The two most common sound-quality descriptors used to indicate an imbalance in the sound spectrum are those for low-frequency rumble ("R") and high-frequency hiss ("H"). The PSIL value defines the RC curve at 1000 Hz. If the sound levels exceed the particular curve by more than 5 dB below 1000 Hz, then an "R" descriptor is added; if levels exceed the curves by more than 3 dB at and above 1000 Hz, then an "H" descriptor is added; so we have, for instance, RC-35(R), RC-35(H), or even RC-35(RH). Again, a balanced neutral background sound is always desired regardless of the overall levels, and the "N" descriptor is used, for example, RC-35(N). If R and H sound-quality descriptors are found, the level above the particular criterion curve should be reported.

In general, the background sound levels in studios need to be as low as possible although, in reality, this is frequently not practical. The majority of steady-state background noise in most studios comes from the air conditioning and ventilation system, but the cost of HVAC equipment and air-distribution systems can become absurdly high to meet extremely low NC and RC criteria. As a conservative approach, it is well to design for perhaps 3 to 5 dB below the set NC and RC criteria but a greater amount will surely impose greater expense. Be conservative, but be reasonable.

It is best to discuss what is really required and expected in a studio with the owner and users, based on program type, intended audience, and the cost/performance tradeoffs, and then set the background sound criteria limits as deemed appropriate using input from all parties.

For other spaces in a facility, consult guidance from references such as Table 34 in the *ASHRAE Applications Handbook* (see Bibliography).

TABLE 3.2-3
Recommended Noise Criteria for Special Studios

Room Type	HVAC NC/ RC(N)*	Remarks
Broadcast studios (distant microphone pickup used)	15–20	As the size and volume increase, the low values are desired
Broadcast studios (close microphone pickup only)	20–25	Budget will probably set value
Recording studios	"Audible" threshold (see criterion curves)	As low as practical

*RC(N) is the preferred criteria when the quality of the space dictates that background sound should be neutral and unobtrusive. Intrusive noise sources must be at least 7 dB below the noise HVAC noise levels in each octave band of the audible spectrum.

REVERBERATION

Sound Decay in a Room

When a sound source stops in a room, the SPL at a given location will not decrease to zero instantaneously as in the free field case. Rather, the sound energy in the room will decay over a period of time. This is due to reflected sound energy gradually dissipating as a result of the absorptive qualities of the room surfaces. This sound decay is called *reverberation*. The reverberation time of a room (RT_{60}) is defined to be the amount of time required for sound to decay by 60 dB:

$$RT_{60} = 0.49V / S\bar{\alpha} \qquad (\text{when } \bar{\alpha} < 0.1)$$

where:

V is room volume (ft^3)

S is surface area of room (ft^2)

$\bar{\alpha}$ is average absorption coefficient

This is the classic *Sabine* formula for RT_{60}.

For rooms in which the average absorption coefficient for all surfaces ($\bar{\alpha}$) is greater than 0.1 or where the absorption of various surfaces is significantly different, more complex equations govern. The reader is directed to *Acoustic Design and Noise Control* by Michael Rettinger or other advanced texts on architectural acoustics (see Bibliography).

In very large rooms or at high frequencies, the effect of excess absorption due to humidity and other effects must be taken into account:

$$RT_{60} = 0.49V / (S\bar{\alpha} + 4mV)$$

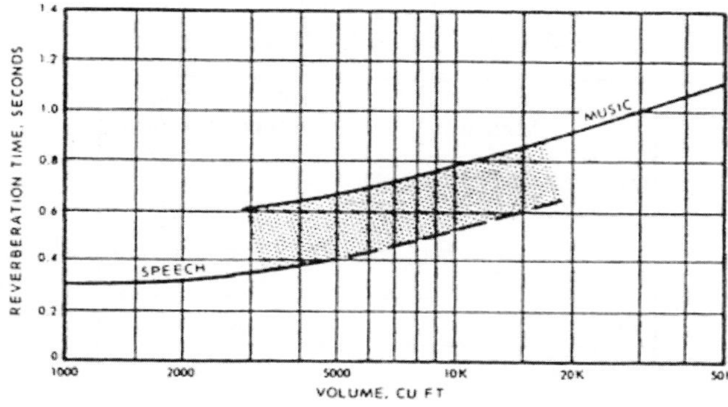

FIGURE 3.2-6 Optimum reverberation time (at mid-frequencies). (Reprinted with permission from book #1696 *Acoustic, Techniques for Home and Studio,* 2nd ed. By F. Alton Everest. © 1984 by TAB Books, A Division of McGraw-Hill, Inc., Blue Ridge Summit, PA.)

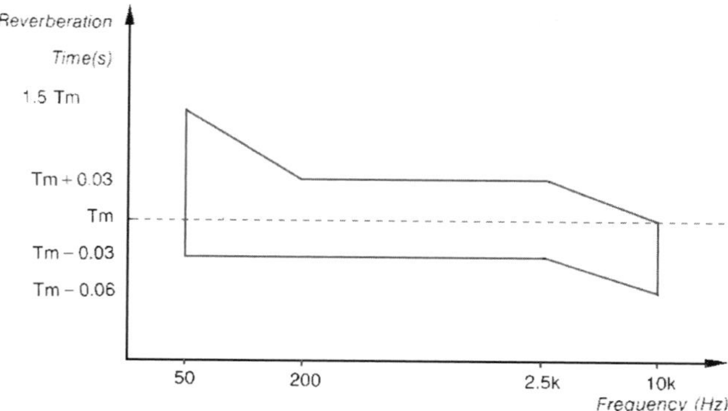

FIGURE 3.2-7 EBU recommendation on optimum reverberation time as a function of frequency. (Reprinted from EBU Technical Recommendation R22-1985: Acoustical Properties of Control Rooms and Listening Rooms for the Assessment of Broadcast Programs.)

where *m* is the excess sound attenuation in dB/100 ft (see later section on the effect of humidity).

Generally accepted optimum reverberation times for speech and music studios of different volumes are shown in Figure 3.2-6.

Limitations of RT_{60}

A basic premise of the RT_{60} formula is that the room exhibits a uniform rate of decay of sound. This in turn would require that the sound field is completely diffuse, an assumption generally more close to being true in large "live" rooms than small "dead" rooms. In small, very absorptive rooms, where all significant sound energy dies away in a few reflections, the validity of a statistically based tool like reverberation time becomes questionable. In these cases, the reflection profile itself must be considered on a more specific basis. In critical cases, in addition to addressing the statistical decay of the room versus frequency as represented by reverberation time, it is important to consider the strategic placement of absorptive, reflective, and/or diffusive materials to provide control of first- or second-order reflections. For example, strategic reflection control may be important in a studio control room on the side walls and ceiling between the loudspeakers and listening position, as well as the front wall between the loudspeakers and the rear wall, which is often intentionally made diffusive according to several popular studio design philosophies.

Optimum reverberation time has received a lot of attention over the years. Being a subjective figure of merit, numbers that can be universally agreed upon for all circumstances will probably never exist. Certainly, the optimum RT_{60} varies with the size of the room and the intended application. For control rooms and listening rooms the European Broadcasting Union (EBU) recommends an RT_{60} of 0.3 seconds ±0.1 seconds at mid-frequencies (250–2000 Hz) as shown in Figure 3.2-7.

In general terms, most studio environments should have as little reverberation as possible. This is due to the simple fact that reverberation can be created electronically, but to date, it is virtually impossible to remove excess reverberation from an electronic signal.

RT_{60} varies as a function of frequency. When not specified, mid-frequencies around 500 Hz are usually the assumed frequency range. For critical listening applications, the optimum reverberation characteristic is sometimes deemed to be flat as a function of frequency. However, for a natural-sounding environment with "warmth," it is often desirable to have higher RT_{60} at low frequencies compared to mid- and high frequencies. Figure 3.2-7 shows the recommendations of the EBU in this regard.

THE PHYSICS AND THEORY OF SOUND

This section will discuss the science of sound, which provides a basis for understanding the concepts in this field. There are many detailed texts on sound acoustics and noise listed in the Bibliography that provide derivations of many equations related to this field.

Sound may be described as a disturbance propagating as vibrations, or a pressure wave, through a physical medium, such as air, water, wood, steel, or other materials. The vibration of molecules in the medium of air creates minor pressure changes that are sensed by the ear. This is termed *sound pressure* and it can be measured by a transducer (microphone). The amplitude is designated *sound pressure level* (SPL) and noted as L_p.

Sound Wave Characteristics

The wave phenomenon of sound propagation can be illustrated in two ways. The graph in Figure 3.2-8 shows the variation of pressure with time at a particular point as a sound wave passes.

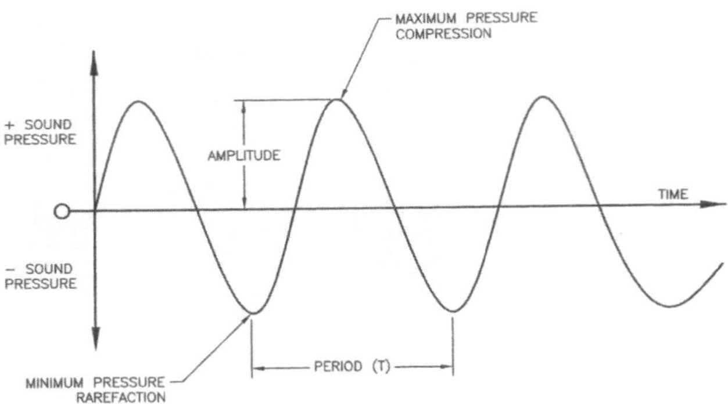

FIGURE 3.2-8 Graph of a sound wave.

In the second illustration, the air molecules in the wave move back and forth in the same direction as the wave motion. All molecular motion occurs with respect to any motion of the air; there is no net motion of the air particles within the air. Figure 3.2-9 shows a sound wave traveling down a tube at two points in time: the distribution of air particles and regions of high and low pressure in the tube are also shown. As is evident, the air is alternately compressed and rarefied. In the second part of the figure, at an instant of time later, it is apparent that the compression/rarefaction cycle is moving to the right.

Speed of Sound

The speed of sound depends principally on the nature of the propagation medium in which it is traveling. Table 3.2-4 lists the speed of sound in various materials.

Looking at Table 3.2-4, it is easy to understand situations such as when banging on a pipe in the basement of a building results in the perception of a double bang in an upper floor room. This is due to the arrival of sound energy transmitted via the pipe itself

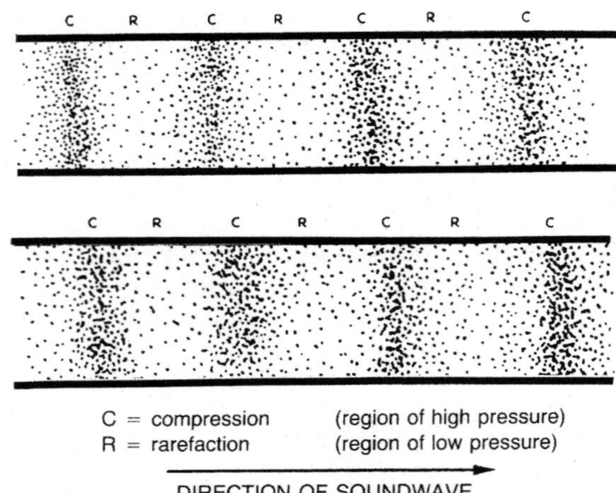

FIGURE 3.2-9 Illustration of sound wave propagation in a tube.

TABLE 3.2-4
Speed of Sound in Different Media

Medium	Approximate Speed of Sound
Air (room temperature)	1,130 ft/sec
Soft wood	11,000 ft/sec
Gypsum board	22,300 ft/sec
Steel	16,600 ft/sec
Water	4,900 ft/sec
Glass	17,000 ft/sec
Concrete	11,200 ft/sec
Hydrogen	4,100 ft/sec
Helium	3,000 ft/sec

reradiating in the room versus the slower airborne path through the room partitions and oor/ceiling structures. Also, it is worth keeping in mind that loudspeakers closely coupled to a wall structure can in some cases transmit sound via the structure to be reradiated from another point in the structure. Thus, since sound travels much faster in solid structures than in the air, sound energy can arrive at a listener or microphone before the direct sound path arrives, causing a weak but annoying pre-echo.

The speed of sound in air at normal room temperature is approximately 1130 ft/sec. However, the speed of sound is also dependent on temperature. At 32°F and 760 mmHg, the speed of sound has been experimentally verified to be approximately 1087 ft/sec. A reasonably accurate simplified formula for the speed of sound is to assume a 1.1 ft/sec increase or decrease for each degree F above or below 32°:

$$\textit{Speed of sound} = 1087 + 1.1(T - 32)\ \text{ft/sec}$$

where T is temperature in degrees Fahrenheit.

It is worth noting that the ambient air pressure is *not* a factor in the speed of sound.

Relative humidity changes the density of air and has a small effect on the speed of sound. At typical room temperatures, the percentage change in velocity between 0 and 100% RH is less than 0.5%. This factor is usually ignored in practical situations.

Sound Level Definitions

A decibel (dB) is a mathematically convenient way for expressing the ratio of two powerlike quantities:

$$\text{dB} = 10\log\left(\frac{P_1}{P_2}\right)$$

where P_1 and P_2 are power quantities.

Units expressed in decibels are designated as *levels* and are defined as follows:

$$\textit{Intensity level} = \mathrm{IL} = 10\log\left(\frac{I}{I_{\mathrm{ref}}}\right)$$

where:

I is sound intensity (watts/m^2)

$I_{\mathrm{ref}} = 10^{-12}$ watts/m^2

$$\textit{Acoustic power level} = \mathrm{PWL} = 10\left(\frac{W}{W_{\mathrm{ref}}}\right)$$

where:

W is acoustic power (watts)

W_{ref} is 10^{-12} watts

$$\textit{Sound pressure level} = \mathrm{SPL} = 20\log\left(\frac{p}{P_{\mathrm{ref}}}\right)$$

where:

p is sound pressure (N/m^2)

P_{ref} is 2×10^{-5} N/m^2 or 0.0002 microbar

In a free field,

$$I = \frac{P^2}{4\rho_o C}$$

where:

ρ_o is air density

C is speed of sound

SPL $\approx$ IL in a free field

In a diffuse field,

$$I = \frac{P^2}{4\rho_o C}$$

$$SPL = PWL - 10\log\left(4\pi r^2\right) + .5$$

where r is the distance from the sound source in meters, and the source is omnidirectional.

When adding decibel quantities, it must be remembered that power levels themselves can be added directly but decibels cannot. To add decibel quantities, the decibel notation must be rearranged so that the associated power levels are added and then converted back to decibel form:

$$\mathrm{dB}_{\mathrm{total}} = 10\log\left(10\log^{\mathrm{dB}_1/10} + 10^{\mathrm{dB}_2/10} + 10^{\mathrm{dB}_n/10}\right)$$

For an approximate rule of thumb in adding decibels, refer to Table 3.2-5.

TABLE 3.2-5
"Rule of Thumb" for Adding Decibels

When adding two dB values that differ by:	Add to the higher value to obtain the total:
0 or 1 dB	3 dB
2 or 3 dB	2 dB
4 to 8 dB	1 dB
9 or more dB	0 dB

Sound Pressure Level

SPL Attenuation as a Function of Distance

In a free field, sound intensity varies inversely with the square of the distance from the source. Hence, like light, sound follows so-called inverse square law. The difference in dB between the SPL at two different distances, d_1 and d_2, is then:

$$\textit{Difference in } \mathrm{dB} = 20\log\left(\frac{d_2}{d_1}\right)$$

This is equivalent to a 6 dB loss per doubling of distance or a 20 dB loss per tenfold increase of distance.

Directivity of Sound Sources

In the practical case, sound sources are not omnidirectional and have an axis of main radiation. This can be quantified in the concept of directivity, or Q. Directivity (Q) is the ratio of intensity along a given axis to the intensity that would be measured at the same distance if the same quantity of total acoustic power were being radiated omnidirectionally. The designated axis is usually taken to be the axis of maximum radiation, so $Q \geq 1$. This can be expressed as:

$$Q = \frac{(\text{On-axis pressure})^2 \text{ at some distance}}{(\text{Mean sound pressure})^2 \text{ at same distance}}.$$
averaged over all directions

See Figure 3.2-10 for examples of directivity.

It can be shown that for a sound source having theoretical horizontal and vertical coverage angles of a and b,

$$Q = \frac{180}{\sin^{-1}\left(\sin\frac{a}{2}\sin\frac{b}{2}\right)}$$

Directivity is often expressed in decibel notation:

$$\textit{Directivity factor} = \mathrm{DF} = 10\log Q$$

The attenuation of SPL as a function of distance given previously can be modified to include the factor of source directionality:

$$\mathrm{SPL} = \mathrm{PWL} + \mathrm{DF} - 10\log(4\pi r^2)(42),$$

where r is the distance from the source in meters.

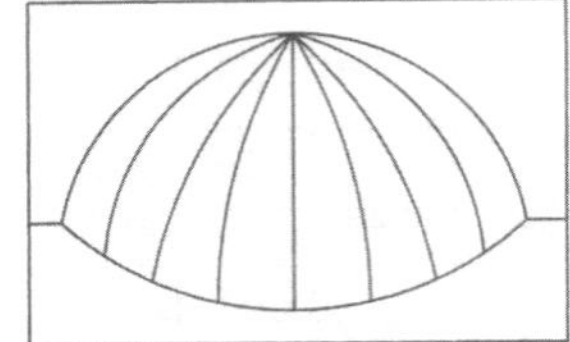

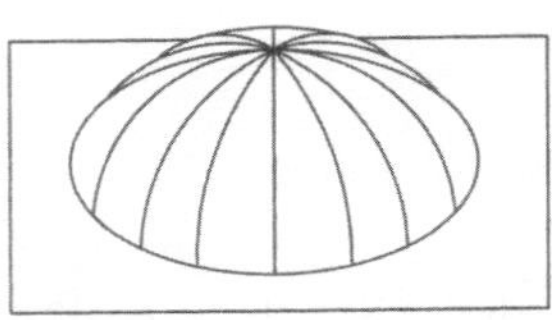

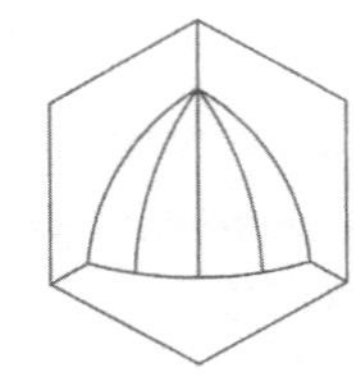

FIGURE 3.2-10 Illustration of directivity: Q = 1: For free-field radiation and there are no reflecting surfaces. Q = 2: Where the sound source is on the floor in the center of a large room or on ground level outdoors. Q = 4: When the source is near the intersection of the floor and a wall. Q = 8: When the source is near the intersection of the floor and two walls.

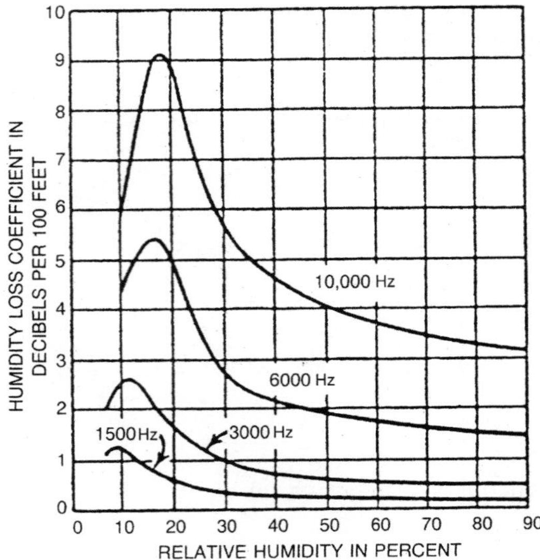

FIGURE 3.2-11 The effect of humidity on sound transmission (dB loss per 100 ft).

Effect of Humidity on Sound Level

Humidity in the air will increase sound level loss in excess of that predicted by inverse square law alone. Figure 3.2-11 shows a family of curves at different frequencies showing dB loss per 100 ft as a function of RH. In general, the attenuation increases with frequency and absorption is greatest between approximately 10 and 40% RH, decreasing above and below these levels for all frequencies.

The curves show why it is difficult to acoustically transmit high-frequency energy long distances. For example, the SPL at a distance of 4 ft required to produce 80 dB SPL at 1000 ft for 10 kHz at 20% RH is:

$$\underbrace{80 + 20\log\left(\frac{1000}{4}\right)}_{\text{inverse square law term}} + \underbrace{9\,\text{dB} \times 10}_{\text{humidity term}} = 218\,\text{dB SPL!}$$

Sound Level as a Function of Distance in an Enclosed Room

In an enclosed room, the situation is more complicated than the case of a free field in an environment with no reflected sound energy. The total SPL at any point in the room will be the result of contributions directly from the source (referred to as the direct sound field) and sound energy associated with multiple reflections from the room surfaces (referred to as the diffuse or reverberant sound field). The direct sound field varies inversely with the square of the distance from the source: the diffuse sound field is, by definition of being diffuse, equal at all points in the room. It can be theoretically shown that, for an enclosed room, at a distance r (in feet) from a sound source:

$$\text{SPL} = \text{PWL} + 10\log\left[\frac{Q}{4\pi r^2} + \frac{4}{R}\right] + 10.5$$

where

$$R = \frac{S\bar{a}}{1-\bar{\alpha}}$$

Q is the directivity of source

$\bar{\alpha}$ is the average absorption coefficient

S is total room surface area

R is sometimes called the room constant

The graphical form of this equation is shown in Figure 3.2-12.

At distances remote from the sound source, the first fraction ($Q/4\pi r^2$) in the equation above becomes small as r^2 becomes large and the reverberant sound pressure level may be approximated as:

$$\text{SPL} = \text{PWL} - 10\log R + 16.4$$

Airborne Noise Reduction

For sound energy striking a partition, one can define a transmission coefficient for the partition:

$$\tau = \frac{\text{Energy transmitted through the partition}}{\text{Energy incident on the partition}}$$

Transmission loss (TL) is then defined as:

$$\text{TL} = 10\log\left(\frac{1}{\tau}\right)$$

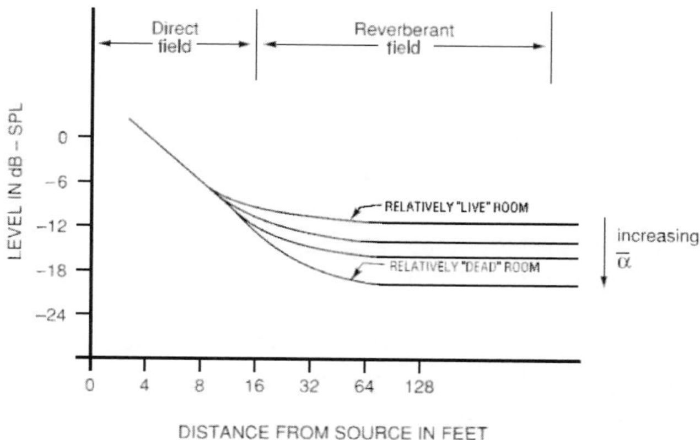

FIGURE 3.2-12 Sound attenuation as a function of distance and the amount of total room absorption.

where TL = 10log and represents the decibel reduction of sound energy through the partition. The overall TL through a partition composed of several areas (e.g., a wall with a door and window) can be calculated as follows:

$$Composite\ TL = 10\log\left[\frac{S_r}{\tau_1 S_1 = \tau_2 s_2 + \cdots \tau_n S_n}\right]$$

where

S_T is total surface area of the partition.

τ_n and S_n are the transmission coefficient and surface area of the nth element of the composite partition.

The disastrous effect of cracks and air gaps on achieving high-TL partitions can easily be illustrated with this formula. For example, using $\tau = 1$ for a crack, consider the effect of an 1/8-inch crack under a door having a TL of 30 dB. The composite TL is then approximately 26 dB. Applying this same situation to a door with TL = 50 dB, yields a composite TL of approximately 28 dB, a 22 dB loss in isolation due to the air gap!

In partition design, the need for avoidance of cracks and gaps and the importance of gasketing and sealing cannot be overstated. High sound isolation simply cannot be achieved if these factors are ignored or compromised.

Absorption of Sound

In general, when sound waves strike a surface:

- A portion of the energy is transmitted through the surface.
- A portion of the energy is reflected back into the room.
- A portion of the energy is absorbed.

The metric for the absorptive quality of a material is called the absorption coefficient, α:

$$\alpha = \frac{\text{Energy absorbed by the surface}}{\text{Energy incident on the surface}}.$$

The values of α range between 0 and 1, with 0 being a perfect reflector and 1 being a perfect absorber. In practical terms, absorption coefficients of surfaces are used in room design to aid in controlling reflections. For room design, an open window is considered a perfect absorber since no energy that strikes the opening is reflected back into the room.

The total absorption of a surface is defined as the product of its absorption coefficient and the surface area. The unit of absorption is the *sabin*, named for the early twentieth-century acoustician, Wallace Clement Sabine.

In many applications, it is convenient to define the average absorption coefficient of the room: $\overline{\alpha}$. This is defined as the total absorption in the room (in sabins) divided by the total surface area, as in the following equation:

$$\overline{\alpha} = \frac{\alpha_1 s_1 + \alpha_2 s_2 + \cdots \alpha_n \alpha_n}{s_1 + s_2 + \cdots s_n} = \frac{\sum_{i=1}^{n} \alpha_i s_i}{s_\tau}$$

where α_n and S_n are the absorption coefficient and surface area of a portion of the room, and s_τ is the total surface area of the room.

The average absorption coefficient is accurate only for the frequencies at which the absorption coefficient values are valid. Absorption coefficients vary as a function of frequency and are usually listed in tables at 125, 250, 500, 1000, 2000, and 4000 Hz.

Often materials are specified by their *noise reduction coefficient* (NRC) value. NRC was intended to be a convenient single-number index of average sound-absorbing efficiency. It is dened as the arithmetic average of absorption coefficients at 250, 500, 1,000, and 2,000 Hz, rounded off to the closest multiple of .05.

$$\text{NRC} = \frac{\alpha_{250} + \alpha_{500} + \alpha_{1,000} + \alpha_{2,000}}{4}.$$

While this has some merit, NRC has the considerable disadvantage that low- (and high-) frequency absorption is not considered. These are important criteria to consider in critical broadcasting applications. Also, the variation of $\overline{\alpha}$ with frequency is an important consideration in critical applications where optimizing the total absorption in a room at all frequencies is important. Since NRC is an unweighted average, two materials with the same NRC could have drastically different absorption versus frequency characteristics. Wherever possible, absorption in a room should be analyzed separately in the various frequency ranges of interest.

SOUND IN ENCLOSED SPACES

Room Modes

Resonant Frequencies

It is well known that a guitar or violin string resonates at only selected discrete frequencies, determined by the length of the string and the tension on the string. It can be easily shown that these resonant frequencies include higher-order harmonics, by placing a finger on the exact center of the string, damping out the fundamental frequency tone, but not the even harmonics, which continue to resonate.

Similarly, the resonant tone produced by an organ pipe is dependent on the length of the tube, atmospheric pressure, and the end conditions—whether the pipe is open or closed. For a pipe closed at each end, the resonances are those of the standing waves in a closed tube, as shown in Figure 3.3-13. Once again, the resonant frequencies include the fundamental, for which the length of the pipe is one-half the wavelength (i.e., $\lambda/2$) and other harmonics for which the length of the pipe is some multiple of one-half of the wavelength (λ, $3\lambda/2$, $\lambda 2$, etc.)

Modes

Like closed tubes, rooms exhibit analogous characteristics of resonance—those frequencies at which sound is sustained or reinforced by the geometry and boundary conditions of the space. For example, it has been said that the room is the most important stop in a pipe organ. That is to say, room acoustics have everything to do with the results of the sound that is heard by the listener; whether it is over the air in radio or television studios, or in a performance hall. However, the situation is now three-dimensional instead of one-dimensional. Three types of resonance or room *modes* can be described:

- *Axial modes:* Resonant condition involving two parallel surfaces of room.
- *Tangential modes:* Resonant condition involving four surfaces and parallel to two surfaces of room.
- *Oblique modes:* Resonant condition involving all six surfaces of room.

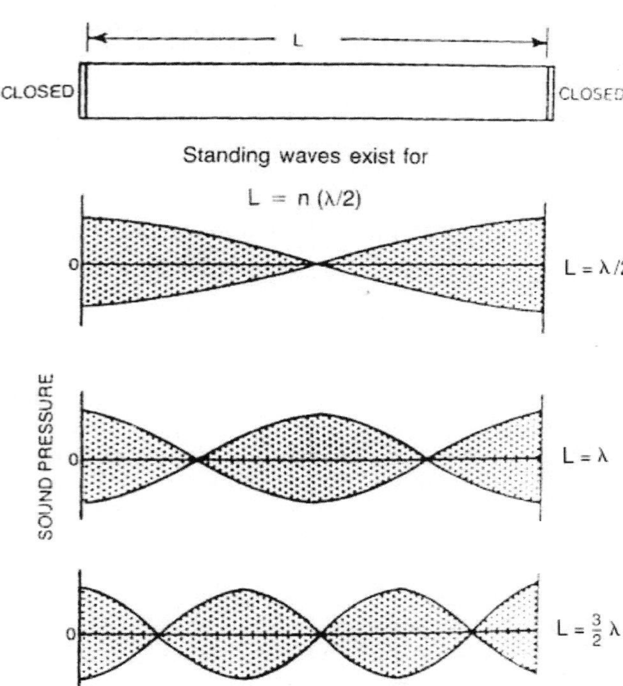

FIGURE 3.2-13 Standing waves in a closed tube.

Because of the resonance problems of small rooms, minimum sizes have been investigated over the years. The EBU recommends a preferred volume of 80 cubic meters (2825 ft^3) for control rooms and 100 cubic meters (3530 ft^3) for listening rooms. Less than 40 cubic meters (1410 ft^3) is not recommended for use due to mode spacing problems.

The axial, tangential, and oblique room modes for a rectangular room can be calculated as follows:

$$F = \frac{C}{2}\left[\left(\frac{p}{L}\right)^2 + \left(\frac{q}{W}\right)^2 + \left(\frac{r}{H}\right)^2\right]^{1/2}$$

where:

F is the mode frequency

L, W, H are the length, width, and height of room

C is the speed of sound

p, q, r are the integers 0, 1, 2, 3, ...

A set of p, q, and r represents one mode of vibration.

For the axial modes, only one of p, q, or r will be nonzero. For tangential modes, two of p, q, and r will be nonzero. Oblique modes have p, q, and r all nonzero.

The above formula can be simplified for the axial modes:

$$\text{Axial mode frequencies} = \frac{Cp}{2L}, \frac{Cq}{2W}, \text{ and } \frac{Cr}{2H}.$$

For p, q, r = 1, 2, 3, … .

Room Dimensions

At lower frequencies, modal frequencies are spaced farther apart. As frequency is increased, the modes become spaced closer together. At low frequencies, the distribution of modes with respect to frequency determines the coloration of room response. In general, room dimensions that lead to an even and uniform spacing of room modes result in the most natural-sounding environment. Room dimensions that lead to common modal frequencies should be avoided since increased response at these coincident frequencies may cause irregular *boominess*. Similarly, a situation with large spacings (approximately 20 Hz or larger)

TABLE 3.2-6
Recommended Room Dimension Ratios for Optimum Mode Spacing

	Height	Width	Length
Set 1	1.00	1.41	1.73
Set 2	1.00	1.26	1.59
Set 3	1.00	1.52	3.18

between adjacent modes will also result in an unnatural responses.

Rooms where ratios of length, width, and height are related by small whole numbers should be avoided since this leads to coincident modes and consequent boominess at these frequencies. A room in the shape of a cube would be the worst case since axial modes associated with all three dimensions of the room would overlap. Beyond these simple guidelines, more concrete design criteria become complex. The search for optimum room ratios has been going on since the 1940s. Recent trends suggest the use of a geometric series that close upon an even number when continued past three dimensions. See Table 3.2-6, in which the first set is created from the square roots of successive numbers,

$$\sqrt{1}, \sqrt{2}, \sqrt{3}$$

while the second set is the successive powers of the cube root of two,

$$\sqrt[3]{2^0}, \sqrt[3]{2^1}, \sqrt[3]{2^2}$$

For rooms in which one or more dimensions must exceed twice the smallest dimension, even multiples of any of the preferred ratios can be used. Set 3 utilizes the ratios of set 2, but with the length and width each multiplied by 2.

Room mode analysis and optimization are particularly important in small rooms, such as studios, control rooms, announce booths, and so forth. A look at the example in Figure 3.2-14 shows why. Small rooms have few modes at the lowest audio frequencies and the spacing between them may be excessively large. The lowest mode in a room is equal to $565/L$, where L is the longest dimension of the room. This is $\lambda/2$ for a sound speed of 1130 ft/sec. Audio energy below this frequency will not be supported at all by resonance, and room response at these frequencies will be attenuated.

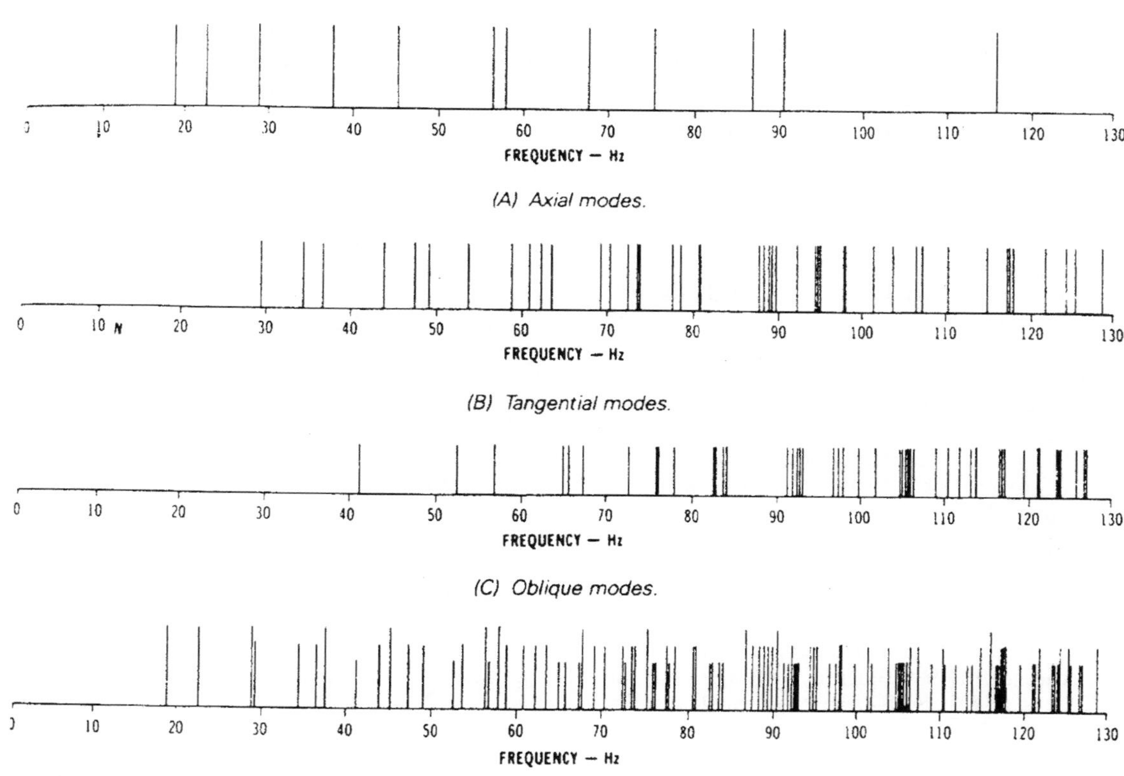

FIGURE 3.2-14 Modal frequencies for a room 30 ft × 25 ft × 19.5 ft. (From Glen Ballou, *Handbook for Sound Engineers: The New Audio Cyclopedia*. H. Sams, 1987.)

As stated previously, three types of room modes exist. While all three are significant, a reasonable first-order design and/or analysis can be made using only the axial modes. The tangential and oblique modes involve four or six room surfaces, respectively, and are likely to suffer more attenuation due to the absorption of the surfaces and greater likelihood of meeting physical obstructions. In general, an optimized axial mode pattern will lead to an overall optimized modal pattern since the tangential and oblique modes will tend to fill in the gaps between the axial modes.

SOUND ISOLATION TECHNIQUES AND MATERIALS

Insulation and Absorption

Airborne sound insulation, sound absorption, and thermal insulation are often confused by using the terminology "insulation" or "absorption," and have been mistaken to have similar properties and solutions whereas they are really quite different.

Airborne *sound insulation* requires mass, discontinuity, and/or resilience to mitigate sound energy from transferring from one area to another by a building structural path. Sound *absorption*, however, relies on the materials within an enclosed space that will absorb the sound energy over the audible frequency range. The most efficient materials for this purpose are generally lightweight in nature and are applied directly to a solid substrate in the room.

Thermal insulation requires sealed cavities or composite construction to retain or impede the flow of heat energy using lightweight materials, such as glass fiber, polystyrene, urethane, and mineral wool, which can be sandwiched into a structure or wrapped around various elements to impede the flow of thermal energy.

Fiberglass and mineral wool provide sound absorption, but any improvement gained to the overall sound insulation between two rooms by their use is provided solely by the reduction of reverberant sound energy within a room. These materials are also used within composite construction to absorb sound in partition cavities, but do not materially add to the mass of the structure that impedes the flow of sound energy.

A common mistake that is usually made with sound-absorbing materials is that they will dramatically improve the sound "insulation value" of a structure, but as noted above, insulation is only a part of a composite construction and it does not materially add to the mass of the structure. In order to achieve a 5 to 6 dB increase in the sound "insulation," the mass must be doubled. That is to say, an 8-inch block wall would have to be increased to 16 inches to significantly increase the sound transmission loss. It is often not very practical to do this, and many lightweight composite construction techniques have been developed to provide the amount of sound transmission loss that is required without significant thickness increase in building elements.

Noise Reduction between Rooms

The formulae for calculating transmission loss of a partition were discussed in the Physics and Theory of Sound section, under Airborne Noise Reduction.

Noise reduction (NR) is the actual difference in SPL measured in a room containing an offending noise source and the room under test. NR is determined by the area of the dividing partition (*S*), the total absorption in the receiving room (*a*), and the TL of the partition:

$$\mathrm{NR} = \mathrm{TL} + 10\log\left(\frac{a}{S}\right)$$

Thus, it can be seen that the actual isolation between rooms can vary both above and below the TL of the partition, although in practice, NR will typically be within 6 dB of the TL.

STC Rating

The *sound transmission class* (STC) rating represents a single number figure of merit for overall acoustic isolation of a material. TL is plotted on one-third octave bands and compared with a standard contour curve to determine the STC rating. The higher the STC, the better the material for reducing sound transmission. While convenient as a figure of merit, in critical applications, such as broadcasting, it is always preferable to consider transmission loss at different frequencies to meet specific design objectives. There is no shortcut that can overcome the fact that transmission loss varies as a function of frequency.

The British Broadcasting Corporation (BBC) has developed a *Guide to Acoustic Practice* that provides a great deal of detailed information (see Bibliography). Of particular interest in the section on STC is tabulation showing STC values versus various adjacencies.

Transmission Loss of Solid Materials

The TL of a solid wall tends to have three dened frequency regions as shown in Figure 3.2-15: a mass-controlled region at low frequencies, a plateau region (also known as a coincidence dip), and another mass-controlled region above the plateau region. For rigid materials, especially at low frequencies, TL increases about 5 dB for each doubling of surface weight. In general, heavier materials have greater sound-isolation capability.

Examples of masonry partitions are shown in Figure 3.2-16. Adding gypsum board material to masonry can further improve transmission loss performance (see Figure 3.2-17). Typical standard drywall construction for comparison to masonry are illustrated in Figures 3.2-18 and 3.2-19 and used in office area locations. Lighter weight construction might be required, where weight limitations are a concern on upper floors of existing structures, thus a discontinuous construction is required to meet the stringent noise-control requirements, for instance between studios and control rooms.

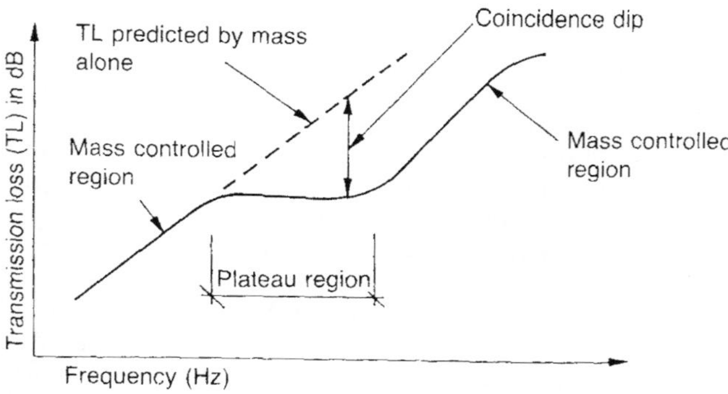

FIGURE 3.2-15 General transmission loss profile of solid materials.

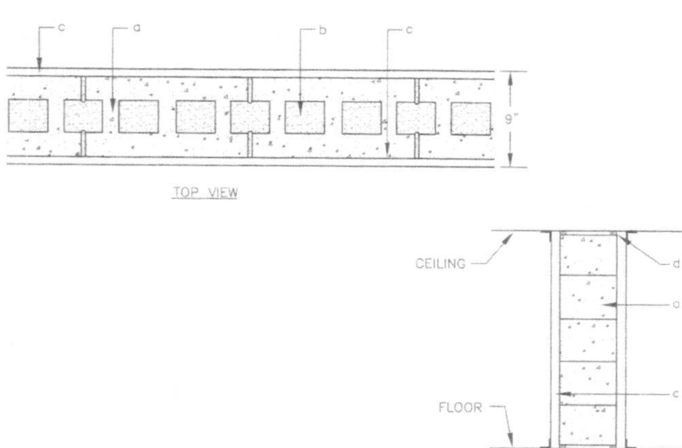

FIGURE 3.2-16 Example of single layer CMU wall: STC-58. (a) 8 × 8 ×18 inch three-cell lightweight concrete masonry units (33 lbs/block). (b) Grout in cells (sand may be used depending on structural requirements). (c) Two coats of heavy latex paint on both sides. (d) Acoustical caulk (expansion joint required depending on structural conditions).

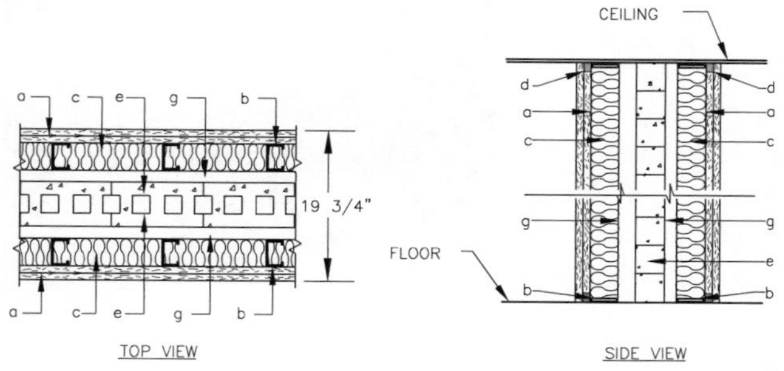

FIGURE 3.2-17 Example of composite concrete block/gypboard wall: STC-65: (a) 5/8-inch gypboard, (b) 3-5/8-inch metal stud, (c) 3-1/2-inch batt insulation, (d) acoustical caulk, (e) 8 × 8 × 16-inch CMU (average dry weight 35 lbs/ft), (f) dry sand fill (g) 1-inch airspace.

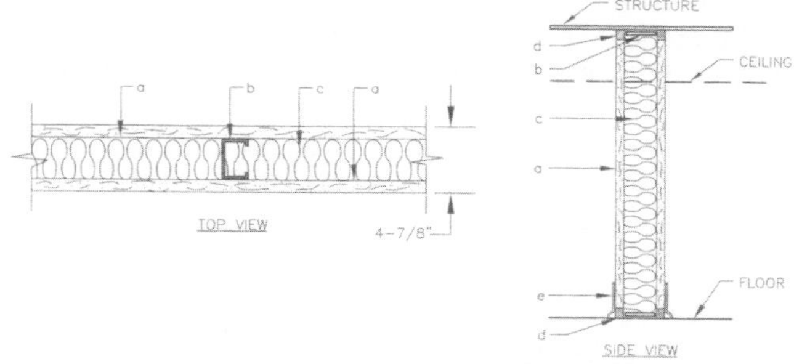

FIGURE 3.2-18 Example of single-layer gypboard wall metal stud construction: STC-47: (a) 5/8-inch gypboard, (b) 3-5/8-inch metal stud, 24-inch O.C., (c) 3-1/2-inch batt insulation, (d) acoustical sealant, (e) cove base.

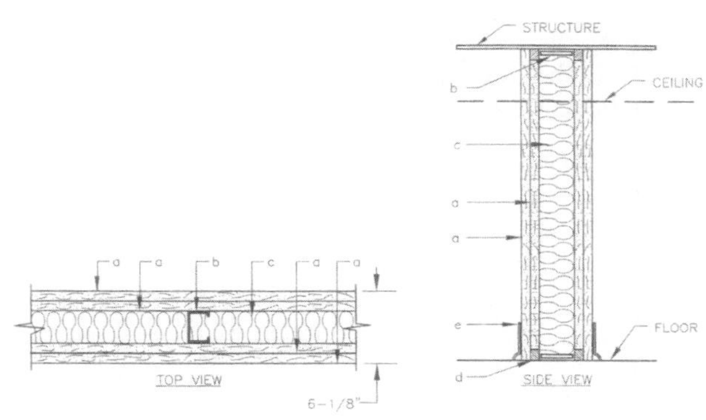

FIGURE 3.2-19 Example of double layer gypboard wall metal stud construction: STC-47: (a) 5/8-inch gypboard, (b) 3-5/8-inch metal stud, 24-inch O.C., (c) 3-1/2-inch batt insulation, (d) acoustical sealant, (e) cove base.

Discontinuous Construction

Increasing the mass of a wall is an effective method of reducing the transmitted sound energy; however, each doubling of mass will accomplish only about a 5 dB reduction in additional sound transmission loss. A more effective approach is the isolation of the two sides of the wall with an airspace and some form of resilient connection, often called *discontinuous construction*. It is important that the two sides of the wall not be rigidly tied together, as with wood studs, which will transmit sound energy as structure-borne sound.

Staggered or double-stud partitions with multiple layers of drywall are but two examples of discontinuous construction. These are shown in Figures 3.2-20 and 3.2-21. A resilient channel can be added to improve the effectiveness by reducing structure-borne sound. Within normal limits, the additional sound attenuation provided by an isolated airspace will also increase as the width of the airspace increases. Placing porous insulation (glass fiber or mineral wool) within the airspace can also increase the STC-rating by 3 to 6 dB, due to damping of the resonant coupling of the wall panels. If space is at a premium, a single stud can be used with multiple layers of drywall and resilient connections. This is shown in Figure 3.2-22.

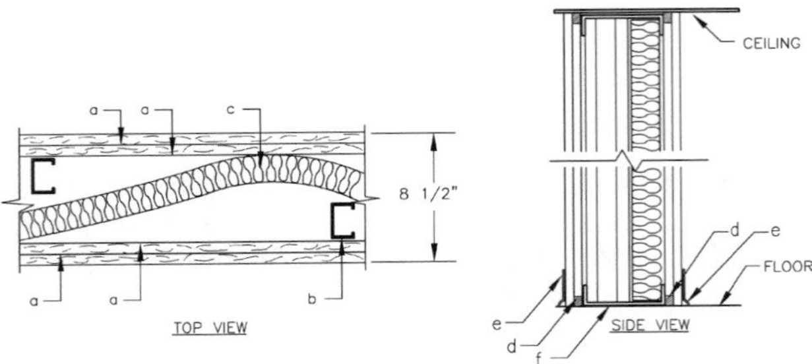

FIGURE 3.2-20 Example of double-layer gypboard wall staggered-stud construction: STC-58: (a) 5/8-inch gypboard, (b) 3-5/8-inch metal stud, 24-inch O.C., (c) 3-1/2-inch batt insulation, (d) acoustical caulk, (e) cove base, (f) 7-1/4-inch head and track.

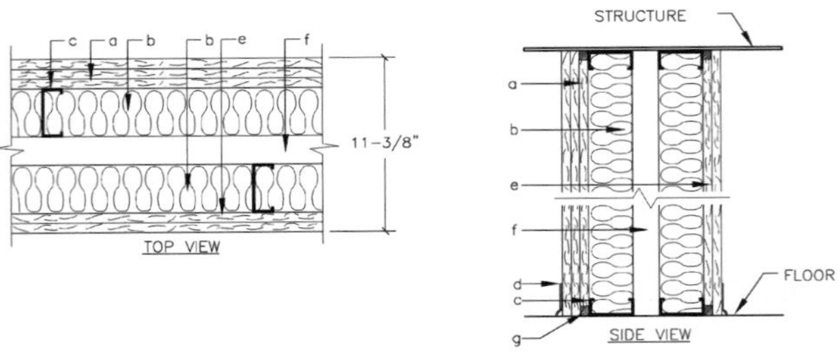

FIGURE 3.2-21 Example of double-stud gypboard wall: STC-76: (a) three layers of 5/8-inch gypboard, staggered, (b) 3-inch glass fiber or mineral wool batt sound blanket (×2), (c) 3-5/8-inch metal stud (×2), (d) cove base, (e) two layers of 5/8-inch gypboard, staggered, and butted joints, tape all layers, (f) 1-inch air gap, and (g) acoustical sealant.

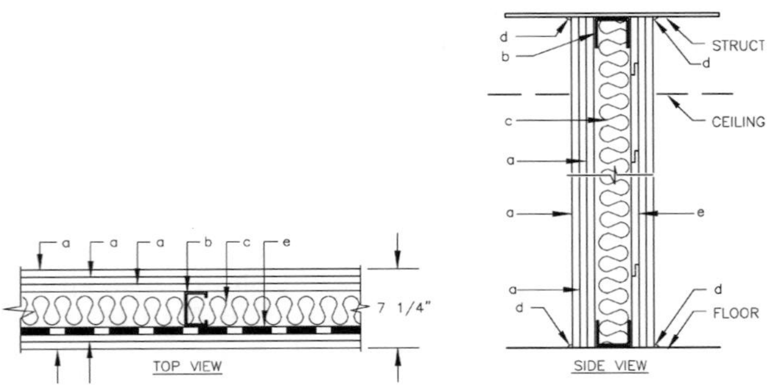

FIGURE 3.2-22 Example of multiple layer gypboard wall metal-stud construction: STC-62: (a) 5/8-inch gypboard, (b) 3-5/8-inch metal stud, 24-inch O.C., (c) 3-1/2-inch batt insulation, (d) acoustical caulk, (e) resilient 1/2-inch channel, 24-inch O.C.

Windows

Achieving high acoustic isolation with windows requires a somewhat complex structure. A double pane with a 6-inch air gap between the panes is desirable. See Figure 3.2-23 for a typical window installed in a radio station. The glass must be resiliently mounted within the frame without cracks or air gaps. For best performance, the two panes should be different thicknesses. The perimeter of the air gap between the panes should be covered with sound-absorbing material. In factory-made acoustical windows, the absorptive liner is often a perforated metal with acoustical foam beneath. When this type of construction is found, often one pane is angled with respect to the other, a characteristic that has more advantages in visual glare reduction than affecting the TL significantly. This is shown in Figure 3.2-24 and is field constructed. Note the important discontinuous assembly of the window frame, using foam as a spacer. Acoustical foam could be placed all around the perimeter between the window panes to improve the sound transmission loss although this is not shown.

Doors

High-isolation doors are difficult to build and maintain since the gasketing and sealing of the door must be extremely precise and not degrade with time or wear and tear. Figure 3.2-25 shows standard acoustical door frame and gaskets. Figure 3.2-26 shows an installed door. The acoustical door itself must be of high mass low-resonance solid construction. Figure 3.2-27 shows increased isolation achieved with a double door arrangement built into a combined masonry and drywall partition.

FIGURE 3.2-23 Factory-fabricated double-pane acoustical window unit.

Because of the difficulties of constructing and maintaining doors with high isolation, it is often more economical and reliable to create high isolation entry ways through use of a *sound lock*, that is, an outer door leading to a vestibule with an inner door, as shown on the studio layout drawing in Figure 3.2-46, later in the chapter. The overall transmission loss from the outside to inside of the room via the vestibule can be very high even with only moderately stringent construction techniques many times without acoustical doors. Also,

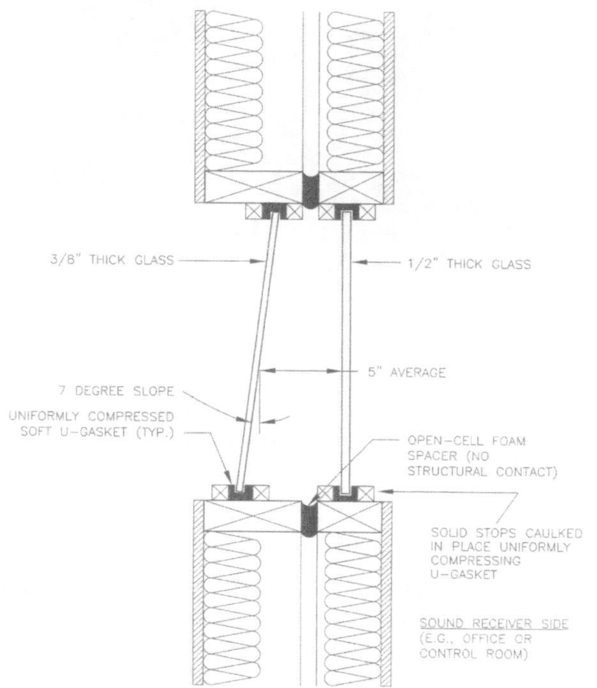

FIGURE 3.2-24 Typical double-pane, sound-isolating window details.

FIGURE 3.2-26 Installed acoustical door.

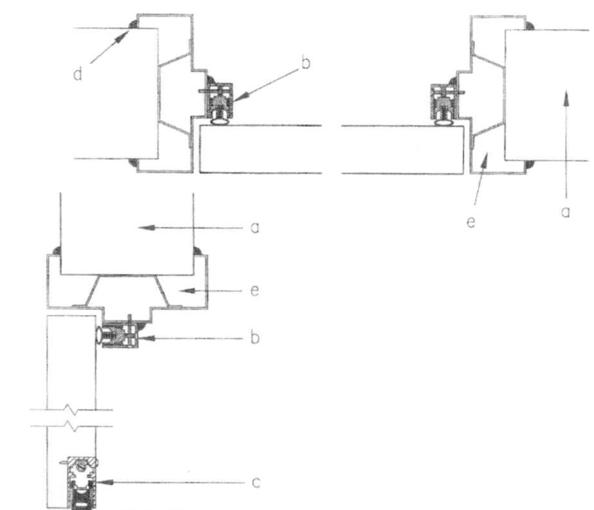

FIGURE 3.2-25 Typical acoustical door detail: (a) designated partition, (b) adjustable acoustical seal, (c) adjustable semi-mortised seal, (d) continuous bead of acoustical caulk, (e) grout-fill.

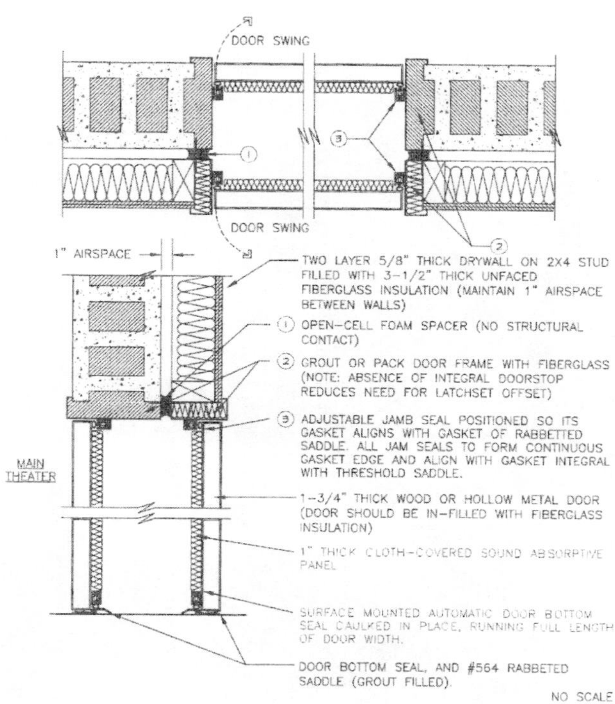

FIGURE 3.2-27 Double door arrangement in studio wall.

by sequentially entering or exiting a sequence of doors, the room is never exposed to the total loss of isolation inherent in using a single door partition.

Exterior Curtain Wall and Roofs

The use of lightweight building materials for the building exterior and roof is always discouraged because of noise intrusion, especially low frequency. For a new building, the use of exterior concrete built-up walls will provide the required mass to mitigate the flow of low-frequency sound energy into the studio area. Concrete roofs are also encouraged and can be built-up composite construction. Rain noise might be a consideration and any type of metal roofing is highly discouraged even though it may have a better architectural look. The function of the roof for sound control, thermal energy control, and rain noise control is more functional than decorative.

A rubber membrane roof has the best potential for reducing rain noise. The rubber membrane roof over some minimal fiberglass material and a highly efficient thermal insulation on a concrete roof will provide the most efficient rain noise reduction without providing a floating ceiling in the space below. This is an important issue that should be considered in the "quiet planning" phase.

Many radio studios are built into existing buildings with exterior curtain walls, often with large areas of glass that provide minimal noise reduction from adjacent freeways, railroads, or aircraft noise. Additional isolation is then required and one example of an interior curtain wall noise-control method is shown in Figures 3.2-28 and 3.2-29. An acoustical laminated glass panel running from floor to ceiling is added to the interior of the curtain wall, with an airspace to provide noise reduction.

FIGURE 3.2-28 Added interior laminated acoustical glass curtain wall in an existing building.

FIGURE 3.2-29 Close view of added laminated glass panel.

Floating Construction

Floating construction is a combination of discontinuous construction and resilient mounting techniques. Floating floors are solid slab floors that are completely isolated from the structural floor by a resilient underlayment or resilient isolators. Walls may be built attached to the floating floor and a ceiling may be resiliently hung from the structural ceiling, resulting in an actual room-within-a-room with very high isolation possibilities, as shown in Figure 3.2-30.

This type of construction can be extremely expensive but is sometimes the only avenue to achieving high levels of sound isolation, especially at low frequencies, in high noise environments or when very low ambient noise levels are required. There are many variants of floating construction that can be used depending on the actual need. One example is a prefabricated studio using a room-within-a-room concept designed and constructed of factory-fabricated metal acoustical panels. These have built-in acoustical doors and quiet ventilation ducts.

Vibration Isolation

Vibrations and sound energy produced by mechanical equipment can be transmitted throughout a building via vibration of the structure and reradiated as sound energy in a particular room. In general, vibrating equipment can be effectively isolated from the building structure by mounting the equipment on resilient mounts. The mass of the equipment and the compliance of the resilient mounting form a resonant system. Vibrations of the equipment at frequencies much higher than this resonant frequency will effectively be

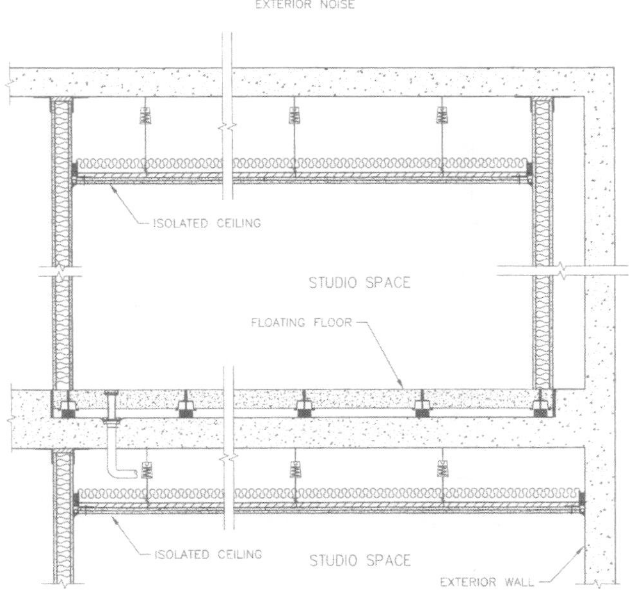

FIGURE 3.2-30 Floating construction concept for studios.

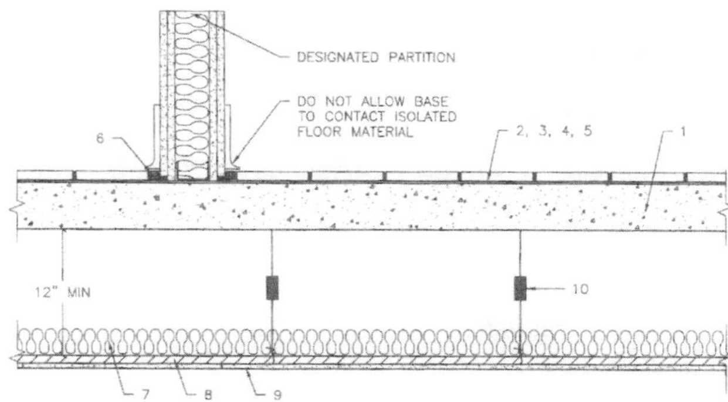

FIGURE 3.2-31 Example of impact isolation control under hard-surfaced floor: 1. substrate structure; 2. isolation sheet bonding agent; 3. sound-isolation sheet properly sealed; 4. tile or stone bonding agent; 5. tile, stone, marble, wood, or other hard-surface material; 6. acoustical sealant between hard-surfaced flooring and perimeter partition; 7. R19 sound attenuation batt; 8. metal framing system sufficient to carry load; 9. 1-5/8-inch sheet rock; 10. neoprene isolator equal to Mason Model WHR.

prevented from being transmitted into the building structure. As a rule of thumb, the resonant frequency should be one-third or less than the lowest desired frequency of effective isolation. The lower the resonant frequency, the lower the level of transmitted vibrations for a given vibration frequency. The ASHRAE 2003 Application Guide provides details for vibration isolation in Chapter 43.

Impact Noise Reduction

Impact noise, as the name implies, refers to such mechanisms as footsteps, objects dropped on floors, slamming doors, and so forth. The use of hard-surfaced flooring in corridors that are adjacent to studios should be discouraged from the design standpoint early on as part of the "quiet planning" for the interior studio design. Carpeted floor surfaces in corridors and support areas, where appropriate, will help alleviate the problem. Reduction of impact noise may, in many cases, be effectively helped by the obvious, such as a rug on the floor or castered chairs. To obtain improvements where hard-surface floors have to be used, techniques to improve TL, such as discontinuous construction, will be needed. An example of such construction with a hard-surfaced floor over a rubber-based resilient sound isolation sheet, with a sound-isolated ceiling below, is shown in Figure 3.2-31.

Similar to the concept of STC, impact isolation class (IIC) is a single-number rating system to assess a barrier's effectiveness at arresting transmission of impact noise. The IIC method is based on the use of a standard *tapping machine* that supplies a known impact noise profile. SPL readings are then taken in one-third octave bands in the receiving room and compared with a standard contour to determine the IIC rating.

SOUND ABSORPTION AND DIFFUSION TECHNIQUES AND MATERIALS

Sound absorption and diffusion are carried out within the room and change its acoustic characteristics, that is, how the room *sounds.* Two terms that are often used in relation to the sound are *live* (a lot of reverberation) and *acoustically dead* (no reverberation). The characteristics of reverberation and echoes are typically controlled fixing materials or devices to the inside of the walls, ceiling, and floor of the room to absorb and/or diffuse the sound. The human body also absorbs sound and the presence of a large number of people in a room typically has a significant effect on reverberation.

The theory of absorption of sound was covered earlier in the Physics and Theory of Sound section, where the concepts of *absorption coefficient* and *noise reduction coefficient* (NRC) were explained. Different materials have very different sound-absorption properties and coefficients. The table in Appendix A at the end of this chapter lists the absorption coefficients of various commercial sound absorbers and numerous general building materials.

Sound-Absorbing Material Applications

Echo Control

Sound-absorbing material can be used very effectively in specific trouble areas to stop echoes or flutter echoes (delayed sound reflections of sufficient intensity to be heard discretely above the general reverberant sound level).

Noise Reduction

Sound-absorbing material can be used to control noise within a room by lessening the amount of reverberant (reflected) energy present.

$$\textit{Noise reduction}\ (\text{in dB}) = 10\log\left(\frac{\alpha_{\text{after}}}{\alpha_{\text{before}}}\right),$$

where is the total absorption in the room in sabins before and after room treatment.

Note that the total absorption must be doubled to lower the noise 3 dB and doubled again to achieve a total 6 dB reduction. A practical limit is quickly reached in attempting to achieve more than 6 to 10 dB of noise reduction using this technique.

Reverberation Control

The *liveness* or *deadness* of a room can be controlled by the introduction of sound-absorbing material, which is perhaps its most common use. It is desirable to maintain uniformity of reverberation across the frequency band and, while it is comparatively easy to obtain sound absorption at high frequencies, it is often one of the challenges of acoustic design to introduce sufficient low-frequency absorption.

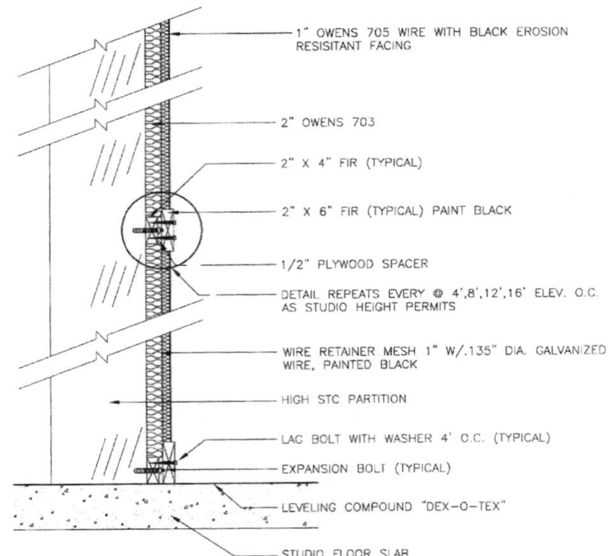

FIGURE 3.2-32 Example of acoustical absorbing wall treatment for a TV studio.

Types of Absorbers

There are three basic types of absorber: porous, panel, and cavity resonator:

- *Porous absorber:* Characterized by a material with deep pores and cavities. Sound energy entering the pores is dissipated by frictional and viscous resistance and/or vibrations of fibers of the material. Glass fiber, mineral wool, heavy drapes, and carpeting are all porous absorbers.
- *Panel absorber:* Sound energy forces a panel into vibration. The vibrational activity converts the sound energy into heat.
- *Cavity (Helmholtz) resonator:* Analogous to blowing air across the mouth of a jug. At the resonant frequency, air in the jug neck vibrates back and forth as a single air mass. Sound energy is dissipated by frictional resistance in and around the neck.

Porous Absorbers

Acoustical Blankets and Batts

Acoustical blankets, such as glass fiber, are common porous absorbers. Such materials are also available in semi-rigid batts, which are more suitable for some types of installation. The absorption of a given material depends on its thickness, density, and relative porosity. In general, increasing thickness increases absorption (mainly at low frequencies) and absorption increases as the density of the material increases.

A blanket must be composed of interconnected open pores (for example, closed cell foam is a poor absorber). As a guideline, if a blanket will pass smoke under moderate pressure, it will probably be a good absorber.

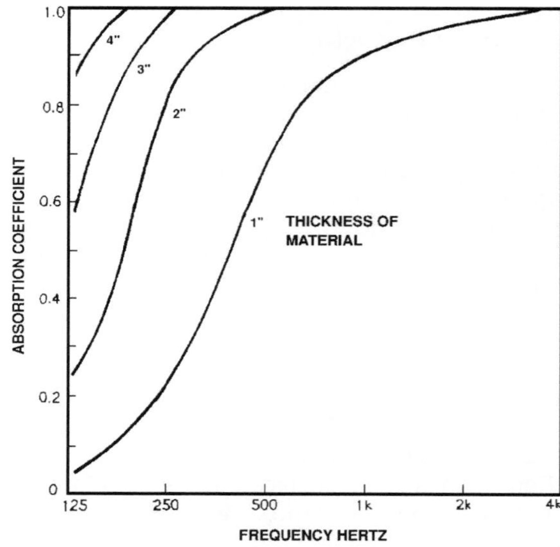

FIGURE 3.2-33 Effect of thickness on absorption characteristics of glass fiber (3 lbs/cu. ft density) mounted directly on a hard surface. (Reprinted with permission from #3096, *The Master Handbook of Acoustics*, 2nd ed., by F. Alton Everest, © 1981, 1989 by TAB Books, a division of McGraw-Hill, Inc., Blue Ridge Summit, PA.)

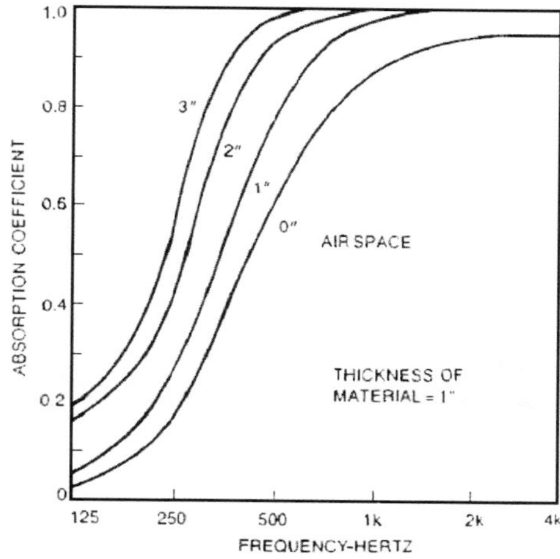

FIGURE 3.2-34 Effect of airspace on absorption characteristics of glass fiber (3 lbs/cu. ft density). (Reprinted with permission from #3096, *The Master Handbook of Acoustics*, 2nd ed., by F. Alton Everest, © 1981, 1989 by TAB Books, a division of McGraw-Hill, Inc., Blue Ridge Summit, PA.)

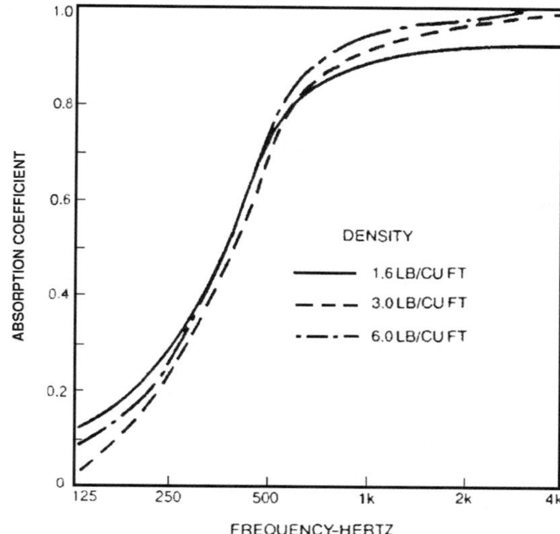

FIGURE 3.2-35 Effect of packing density on absorption characteristics of glass fiber (1 inch thickness). (Reprinted with permission from #3096, *The Master Handbook of Acoustics*, 2nd ed., by F. Alton Everest, © 1981, 1989 by TAB books, a division of McGraw-Hill, Inc., Blue Ridge Summit, PA.)

Glass fiber is certainly one of the workhorse materials for sound-absorption applications. A typical example of a television studio wall treatment using fiberglass absorption is shown in Figure 3.2-32. The thickness and mounting arrangement of the blanket or batt affects the absorptive qualities of this material. High-frequency absorption is more a function of the surface texture and is relatively independent of thickness. Figure 3.2-33 shows that increased thickness mainly affects the low-frequency performance, and Figure 3.2-34 shows the effect of increased low-frequency absorption as a function of an airspace between the glass fiber and the wall. Figure 3.2-35 shows that the effect of packing density on the absorption performance is rather small.

Cellular Foam Absorbers

Commercially available cellular foam absorbers are commonly used in many studios and an example is shown in Figure 3.2-36.

Although irregularly sculpted foam is popular for its appearance, the user should be aware that the geometric pattern of wedges and cavities does not increase the sound absorption of cellular foam except at very high frequencies, where the depth of the wedges is equal to $\lambda/2$, typically above 10,000 Hz. Foam of uniform thickness equal to the average thickness will be equally effective at most frequencies. Installing cellular foam over an airspace will be far more effective at increasing sound-absorption coefficients than varying the surface of the foam.

Only open-cell (soft) foams are acoustically absorptive. Rigid, closed-cell foam, although a good thermal insulator, does not provide significant acoustical absorption.

Drapes and Curtains

Drapes and curtains are often used to attenuate sound energy. However they tend to have high absorption at high frequencies but low absorption at low frequencies. Listed below are some guidelines to increase the low-frequency absorption:

Use heavy base material with a lining.

Use 100 to 200% gathering of the drape.

Hang at least 6 inches from the wall.

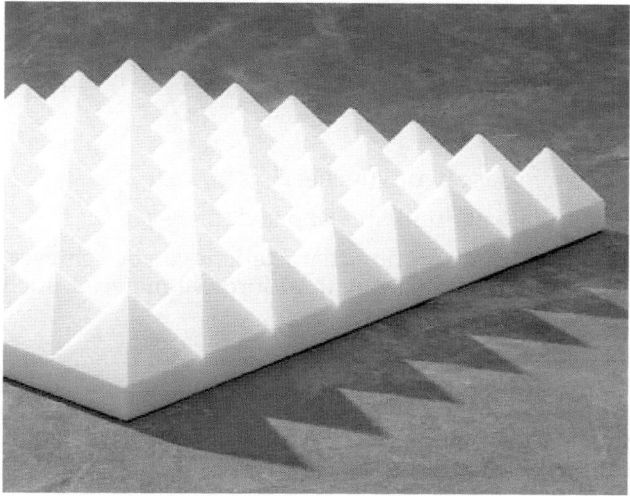

FIGURE 3.2-36 An open cell foam acoustical material. SONEX (Courtesy of Illbruck, Inc., SONEX Acoustical Products Division.)

Figures for the absorption of different types of drapes are listed in Appendix A.

Carpeting

Carpeting is often used as an absorber in radio studios and control rooms. However, its absorption at low frequencies is relatively poor. In general, a foam rubber or hair felt underlayment can improve low-frequency absorption significantly. Typical absorption figures (with concrete as a reference) are listed in Appendix A.

Effect of Mounting Method of Material on Absorption

Different methods of mounting a given material may give different absorption coefficient results. Tables of absorption coefficients must indicate the mounting method used for the figures measured to be useful.

In general terms, materials mounted against a hard backing will exhibit the lowest sound-absorption coefficients. Sound absorption, particularly at low frequencies, is increased by the presence of an airspace behind the absorptive material.

Figure 3.2-37 shows some of the various mounting designations used for standard measurements.

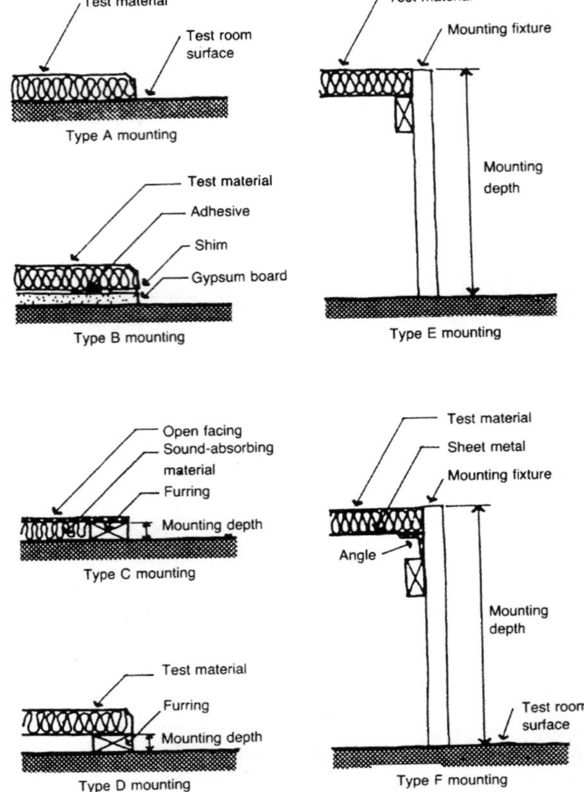

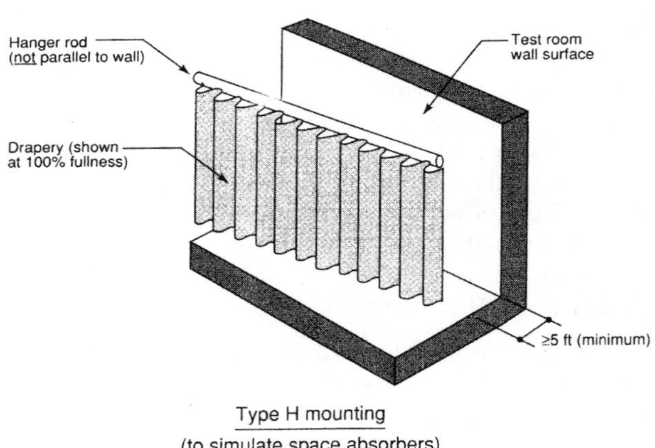

FIGURE 3.2-37 Standard mounting methods for measuring the absorption coefficient of a material, ASTM C 423.

Acoustic Absorption Characteristics of People

Absorption by people is specified in absorption tables as either the number of sabins per person or as an absorption coefficient for people and surrounding objects provided by an audience based on normal seat and aisle spacings. Typical values are listed in Appendix A.

Panel Absorbers

A panel with an enclosed air space behind it forms a resonant system with the air mass behaving as a spring and the panel as a mass. If the panel is thin, the resonant frequency, and hence the frequency of maximum absorption coefficient, can be shown to be:

$$f_{\text{resonant}} = \frac{170}{\sqrt{md}},$$

where

d is air space depth (inches)

m is mass per unit area of panel (lbs/sq. ft)

A broader absorption characteristic can be achieved by filling the air space with absorptive material, such as fiber glass.

Cavity Resonators

There are three basic types of cavity resonators:

Individual units

Perforated panels

Slit resonators

Individual Units

An example of an individual prefabricated cavity resonator is shown in Figure 3.2-38, which is essentially a slotted concrete block. Another type of commercially available individual unit is shown in Figure 3.2-39.

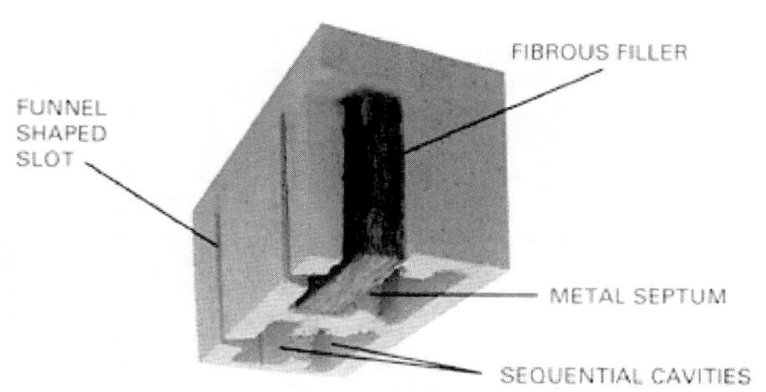

FIGURE 3.2-38 A concrete cavity resonator used in construction. (Type RSC SOUNDBLOX® photo courtesy of the Proudfoot Co., Inc.)

Perforated Panels

A perforated panel of significant thickness (greater than approximately 1/8 inch) spaced away from a rigid backing exhibits the absorptive behavior of a cavity resonator. The resonator has an approximate resonance frequency of:

$$f_{\text{resonant}} = 200\sqrt{\frac{P}{dt}},$$

where:

P is the percentage of open area of panel

d is airspace depth (inches)

t is panel thickness + 0.8 × hole diameter (inches)

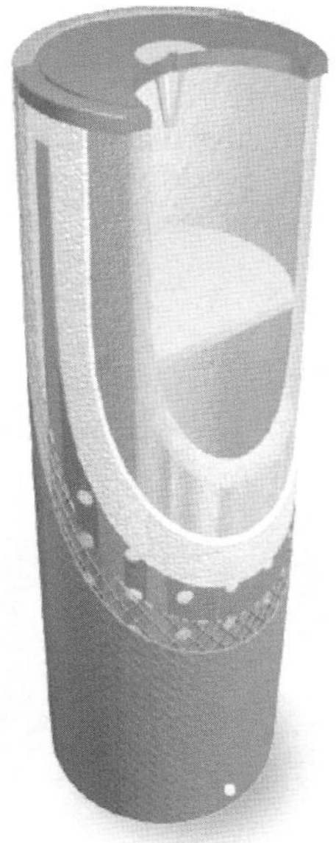

FIGURE 3.2-39 Cutaway view of a portable bass trap cavity resonator. (TubeTrap™ photo courtesy of Acoustic Sciences Corp.)

Slot Resonators

A slot resonator consists of a number of slats spaced away from a rigid backing with airspaces (slots) in between the slats as shown in Figure 3.2-40.

The resonant frequency can be calculated as follows:

$$f_{\text{resonant}} = 2160\sqrt{\frac{s}{dD(w+s)}}$$

where:

s is width of slot (inches)

d is thickness of slat (inches)

D is depth of airspace (inches)

w is width of slat (inches)

In the example shown in Figure 3.2-40, a resonant frequency of about 250 Hz results. Broader absorption characteristics can be achieved by using slats and slots of varying widths or nonparallel airspaces. Such resonant cavity absorbers are particularly useful for increasing low-frequency absorption in a studio.

Diffusion

Diffusion of sound in a room refers to the extent to which sound energy is uniformly distributed throughout a room. Making an analogy with baking a cake, diffusion would represent an indicated measure of how well the batter was mixed and individual ingredients of the recipe were dispersed throughout the mixture. In pursuing good room acoustics, maximum

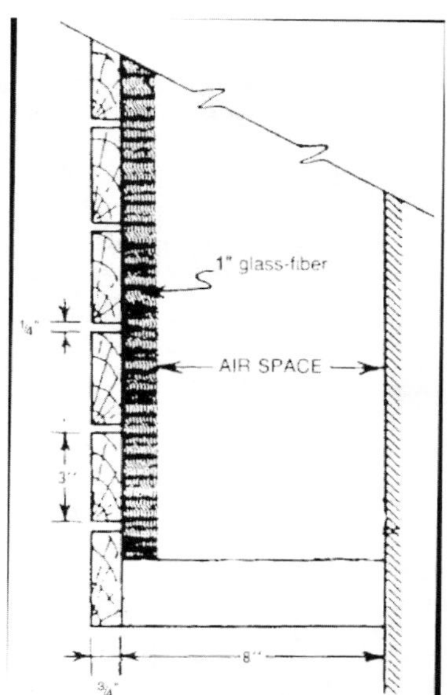

FIGURE 3.2-40 Cross-section of a slot resonator.

diffusion is desirable, all other things being equal. In a purely diffuse sound field:

- At a given location, sound waves are equally likely to be traveling in any direction.
- The sound pressure will be equal at all locations throughout the room.

These criteria are determined by the pattern of reflections within the room. Diffusion in a room can be maximized in several ways. The introduction of oddly shaped protrusions aids in increasing diffusion. The success of many famous concert halls built in the nineteenth century can be largely attributed to the florid architectural features that offer diverse reflective and scattering properties for incident sound. Also, the intentionally irregular distribution of absorptive material in patches will increase diffusion and increasing the randomness of the location of the patches results in the greatest diffusion.

The possible effects of surface treatment on incident sound energy are shown in Figure 3.2-41. As shown in (b), a hard, flat surface will reflect sound in a specular manner just as light is reflected from a mirror, following the relationship that the angle of incidence (relative to a line perpendicular to the surface) equals the angle of reflection and is equal in magnitude. At a location where both direct and reflected waves arrive, the reflected wave is a delayed replica of the incident wave. Similarly, sound-absorptive material applied to a surface yields an attenuated and delayed replica of the incident wave, as shown in (a). Finally, a diffusive surface, as shown in (c), reflects the incident energy equally over a wide angular range and also corresponds to a widening of the energy received as a function of time.

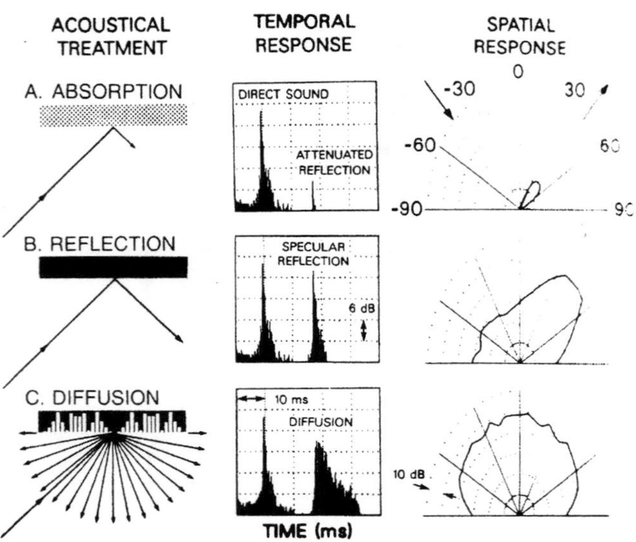

FIGURE 3.2-41 Absorption, reflection, and diffusion from acoustic surface treatments. (Courtesy of RPG Diffusor Systems.)

FIGURE 3.2-42 Example of commercially available diffusor. (Photo courtesy of RPG Diffusor Systems.)

Diffusing elements designed specifically for this purpose are available and the design of these units is quite refined. While traditional surface relief ornamentation, as referenced above, is useful as a diffusion element, it generally does not provide broad-bandwidth, wide-angle diffusion. An ideal diffusor would provide sound diffusion that is not a function of frequency, angle of incidence, or observation angle. An example of a commercially available product designed with these goals in mind is shown in Figure 3.2-42.

Based on the construction of a series of wells of different depths derived mathematically, the diffusor shown achieves a uniform angular distribution of reflected energy from a wide range of incident angles for mid and high frequencies.

FIGURE 3.2-43 Flutter-free type of diffusion panel.

FIGURE 3.2-44 Studio wall with diffusers and acoustic materials shown.

Figure 3.2-43 illustrates diffuser panels below a control room window. Cloth is used to cover the diffusion providing the desired finished appearance.

Examples of studio wall and ceiling treatment are shown in Figures 3.2-44 and 3.2-45. The ceiling uses RPG BAD® panels with RPG Quatratic® panels on the rear wall. Two inch fiberglass is shown on other portions of the wall. A decorative, acoustically transparent cloth is stretched on the ceiling and walls to provide the desired finished appearance.

Use of Nonparallel Walls

When the room is not rectangular, room mode problems do not vanish, they just become difficult to calculate. The modal distribution is certainly affected by room shape changes, but room resonances are not eliminated since the basic presence of room modes is more associated with the volume of the room than with the shape alone. Nonparallel walls are, however, used successfully to promote good diffusion at higher frequencies and prevent flutter echoes. Flutter echoes are produced by repeated reflections from two flat parallel surfaces sufficiently distant from each other to produce a distinct echo.

Usually an angle of 5 degrees or a splay of 1 ft in 12 running feet is enough to destroy flutter echo problems. However, other more cost-effective techniques exist to solve these problems, typically using acoustic treatment of the surface.

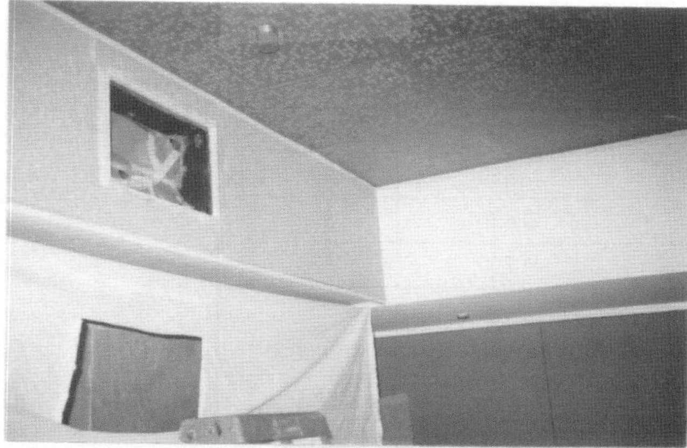

FIGURE 3.2-45 Studio wall and ceiling shown with diffusion on ceiling and absorption on walls prior to stretching acoustically transparent cloth.

MECHANICAL EQUIPMENT NOISE CONTROL

The HVAC system in a studio or control room is often the primary source of noise introduced into the room. Also, the HVAC ductwork configuration may determine the level of acoustic leakage between rooms. The design of the air-distribution systems in studios is critical if these two parameters are to be maintained at acceptable levels, because relatively high volumes of conditioned air and large ducts are required, especially for television studios where high thermal loads are imposed by the lighting system. To accommodate the high airflows and maintain suitable sound levels, the air-distribution systems must be carefully designed from the fan systems through the supply air ductwork to the studio and back to the fan systems through the return air systems.

Use of control dampers and registers must be minimized in all studio designs as they tend to add turbulence and noise. Proper air balance in the design and commissioning phase is critical.

Duct Arrangements

Figure 3.2-46 illustrates a typical television studio layout plan for a large television studio and a small studio, showing the supply ductwork, silencers, and outlet grille locations.

Air Velocity

The amount of noise from a duct system is strongly dependent on the air velocity in the duct. Table 3.2-7 lists the maximum air velocities needed for various NC values. Low-noise design requires low air velocities that, for the same amount of total airflow, lead to the use of multiple ducts or larger duct cross-sections, both of which imply higher costs. Diffusers designed to minimize turbulence-induced noise at the entry point to the room are also important.

Overhead Outlets

With studio lighting thermal load, air buoyancy forces the warm air upward. However, in many studios the cold air supply is usually introduced at a high elevation with return air inlets at a low elevation. In larger studios the use of overhead distributed systems can provide good air coverage, away from any curtains, sets, and scenery that may obstruct the airflow. With

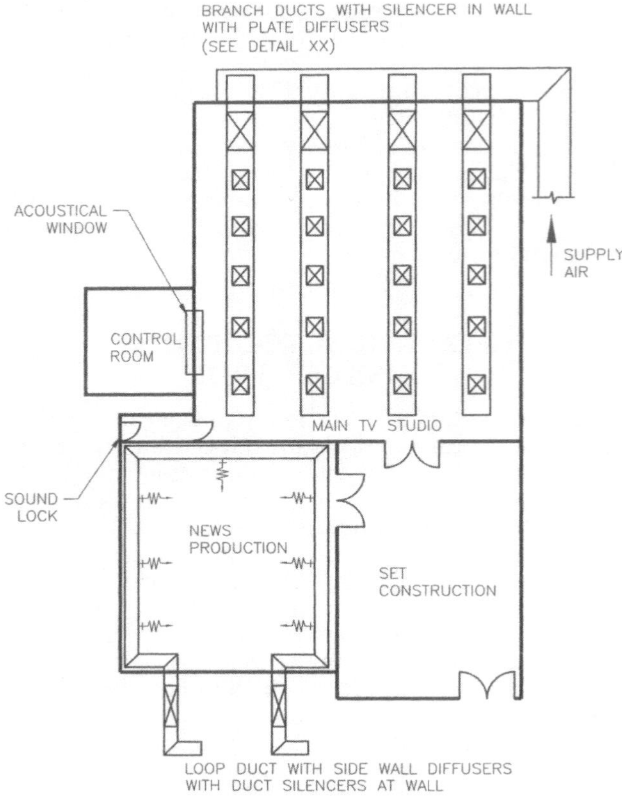

FIGURE 3.2-46 Example of layout of two TV studios showing HVAC duct design and location of duct silencers and outlets.

outlets at higher elevations, there tends to be space for larger ducts at closer spacing and low discharge velocities and flow can be utilized. Furthermore, the high inlet location places these sound sources farther away from microphones. This type of system becomes even more important as the volume increases and/or background noise criteria are lower. Figure 3.2-46 shows an example of overhead ductwork layout in the large studio. Figures 3.2-50 and 3.2-51 shown later show two types of overhead plaque diffusers.

Sidewall Outlets

On smaller studios and those with less demanding airflow quantities and sound level limits, the use of sidewall air grilles and diffusers is quite acceptable. With proper selections of grilles and diffusers, adequate airflow and low sound levels can be obtained. Figure 3.2-46 shows an example sidewall ductwork layout in the small studio.

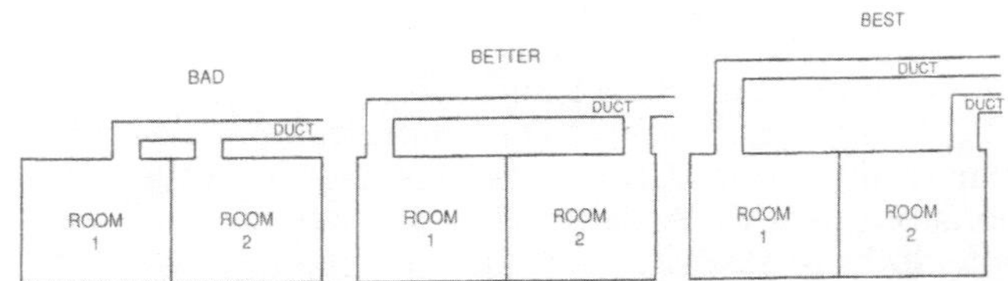

FIGURE 3.2-47 Effect of duct arrangement on sound isolation.

TABLE 3.2-7
Criteria for Air-Distribution Systems Serving Noise-sensitive Areas

	Recommended Maximum Air Velocities for Lined Duct, in ft/sec*			
	Slot Speed at Terminal	Initial 10 ft of Duct before Terminal	10–20 ft	20–30 ft
NC 15 supply	250	300	350	425
NC 15 return	300	350	350	500
NC 20 supply	300	350	425	550
NC 20 return	350	425	500	650
NC 25 supply	350	425	550	700
NC 25 return	425	500	650	800
NC 30 supply	425	500	700	850
NC 30 return	500	600	800	950
NC 35 supply	500	600	800	1000
NC 35 return	600	700	900	1150
NC 40 supply	575	675	875	1200
NC 40 return	675	775	975	1255

*If duct is unlined, velocities must be reduced by 20%.

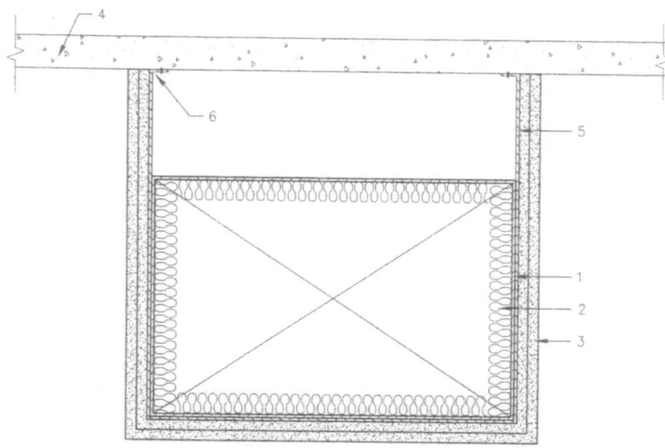

FIGURE 3.2-48 Supply duct with internal acoustical duct liner and external sound insulation: 1. steel duct; 2. acoustical duct liner, 1-inch or 2-inches as required; 3. Double-layer 5/8-inch sheetrock, off-set and caulk all joints; 4. structure; 5. hanger; 6. seal to deck and caulk all joints.

For both overhead and sidewall outlets, the return air paths must also be considered as a potential noise source just as the supply air paths and duct insulation and silencers may be required.

- The layout of supply and return ducts can be a hidden source of poor acoustic isolation between rooms. As shown in Figure 3.2-47, higher isolation is achieved when the duct path length between two rooms is increased on supply or return ducts that feed both rooms. The optimum situation is to have completely separate ducts back to the fan source, however, this may not always be practical.

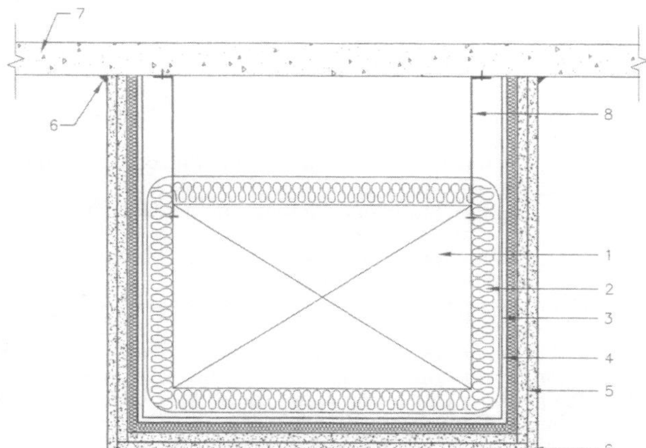

FIGURE 3.2-49 Supply duct with external lagging and external sound insulation: 1. steel duct; 2. 2-inch 3 to 4 lbs/ft fiberglass insulation; 3. and 4. Steel-stud frame, do not allow to touch duct; 5. Double-layer 5/8-inch sheet rock, off-set joints; 6. acoustical caulking all joints; 7. structure; 8. hanger.

Duct Insulation and Silencers

A major contributor to HVAC noise is the fan noise itself that propagates down the supply ducts and subsequently enters the room (see Figure 3.2-2, earlier in the chapter). Ventilation ducts can be lined with absorbing material such as 1-inch glass fiber to reduce the noise at the end of the duct and brought into the room. Instead of, or in addition to, duct liners, prefabricated duct silencers may be used, which are placed in line with the duct and offer significant sound attenuation characteristics. Structure-borne transmission of fan noise into the duct itself can be stopped by coupling the duct surrounding the fan motor assembly to the distribution duct system via a canvas or rubber coupling to break the vibration path.

While internally lined ductwork is always preferred to attenuate the duct-borne sound, the use of unlined and externally insulated ductwork is quite common and can be beneficial in some instances when coupled with duct silencers. However, more care must be exercised in the design of unlined systems. More severe limits must be placed on airflow velocities and more care taken to minimize possible sources of air turbulence. It is recommended to followguidelines from the Sheet Metal and Air Conditioning Contractor's National Association (SMACNA) ductwork systems.

Even with lined ducts, noise from the ducts may radiate into a very quiet studio area. This may require further duct sound insulation for isolation, as shown in Figures 3.2-48 and 3.2-49. Figures 3.2-50 and 3.2-51 show two arrangements for noise-reducing outlet diffusers. If low-frequency noise levels are high enough, a further step could be required: adding an isolated ceiling, as shown previously in Figure 3.2-31 and in Figure 3.2-51. In all cases, the ASHRAE guidelines for good HVAC noise control design practice should be followed.

Wall Penetrations

Wherever ducts or pipes pass through walls of noise-sensitive areas, care must be taken to seal the opening to prevent sound ingress, and also to isolate the duct

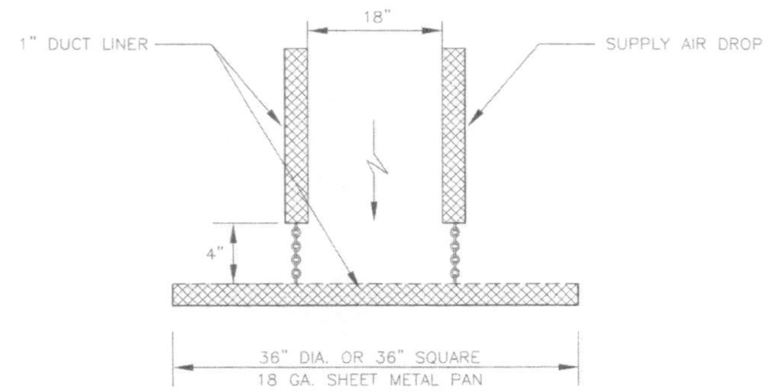

FIGURE 3.2-50 HVAC lined plaque diffuser on drop from main trunk duct.

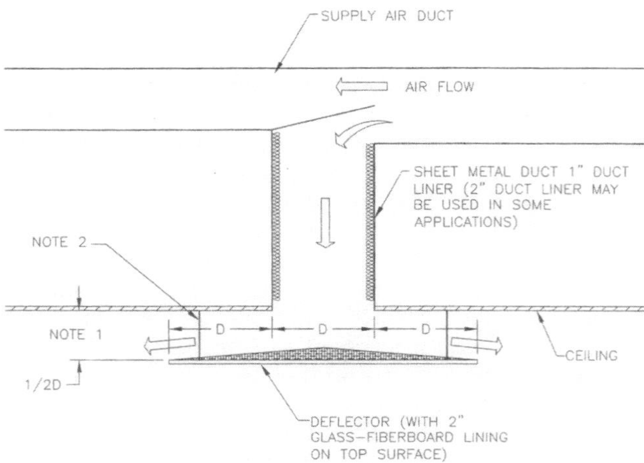

FIGURE 3.2-51 Example of studio ceiling deflector diffuser. This square deflector, is mounted on threaded rods and can be adjusted for airflow.

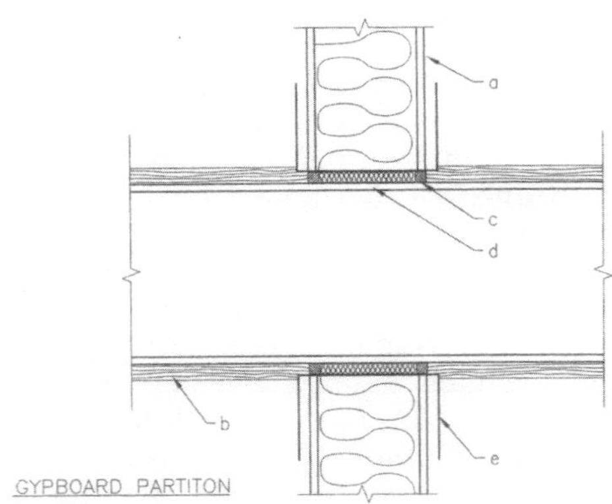

FIGURE 3.2-52 Example of external-insulated duct penetration construction detail: (a) gypboard wall construction, (b) exterior duct insulation, (c) acoustical sealant, (d) pack with fiberglass insulation (3# to 6# density), and (e) metal framing.

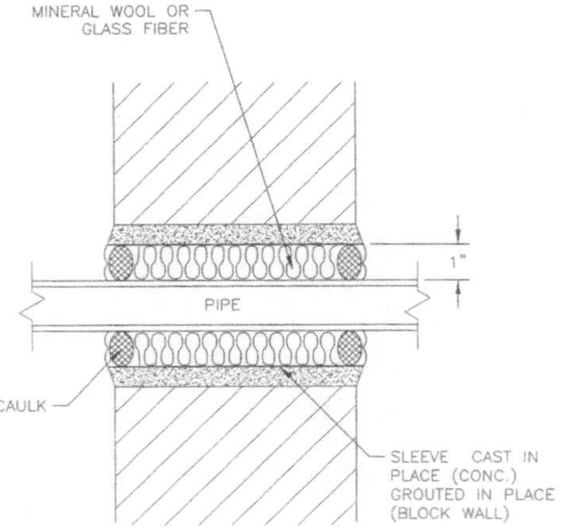

FIGURE 3.2-53 Wall pipe penetration detail.

or pipe from the structure of the wall to prevent transfer of vibration.

Figures 3.6-52 and 3.2-53 show typical duct and pipe penetrations through partitions. In high-STC partitions, this is critical to achieve desired results and an important part of the overall noise-control system.

SUMMARY

This chapter introduces something of both the science and the art of acoustics for the broadcast environment. It presents guidelines for the design and implementation of studio facilities that will lead to good acoustical environments for program making. This is a broad subject and, although the science is mature and many techniques are well established, new methods of solving problems are still being developed. The reader is referred to the Bibliography and other sources of information listed below for more comprehensive data.

ACKNOWLEDGMENTS

Thanks are extended to Lynn Claudy of NAB, who wrote the original chapter in 9th edition of the Handbook and provided many of the figures.

Bibliography

ASHRAE Handbook, HVAC Applications, American Society of Heating, Refrigerating, and Air Conditioning Engineers, Inc., 2003, Chapter 47: Sound and Vibration Control, and Chapter 37: Testing, Adjusting, and Balancing.

ASHRAE Handbook, Fundamentals, American Society of Heating, Refrigerating, and Air Conditioning Engineers, Inc., 2005, Chapter 7: Sound and Vibration.

Backus, J. *The Acoustical Foundations of Music*. New York: W. Norton, 1977.

Ballou, G. *Handbook for Sound Engineers: The New Audio Cyclopedia*, 2nd ed. Boston: Focal Press, 1991.

BBC Engineering. *Guide to Acoustic Practice*, 2nd ed. London: British Broadcasting Corporation, 1990. See http://www.bbc.co.uk/rd/pubs/archive/pdffiles/architectural-acoustics/bbc_guideacousticpractice.pdf.

BBC Research and Development Report, The British Broadcasting Corporation:.

Baird, M. A Wideband Absorber for Television Studios, BBC RD 1994/12.

Burd, A., and Sproson, W. Acoustic Scaling: Subjective Appraisal and Guides to Acoustic Quality, BBC RD 1974/28.

Fletcher, J. A Practical Study of a Vibration Isolated Room, BBC RD 1990/8.

Mathers, C. Some Properties of Antivibration Mounts Used in Building Isolation, BBC RD 189/3.

Plumb, G. Lightweight Partitions Having Improved Low-frequency Sound Insulations, BBC RD 1995/6.

Plumb, G. The Sound Insulations of Metal-framed Partitions Having Structural Loadbearing Properties, BBC RD 1994/8.

Walker, R. Optimum Dimension Ratios for Studios, Control Rooms, and Listening Rooms, BBC RD 1993/8.

Walker, R. Acoustic Noise Criteria for Listening Rooms and Control Rooms, BBC RD 1994/6.

Walker, R. Acoustic Criteria and Specification, R&D White Paper, WHP 021, January 2002.

Walker, R. The Control of the Audible Effects of Ground Vibrations in Building Structures, BBC RD 1989/2.

Beranek, L. L. *Acoustics*. Acoustical Society of America, 1986.

Beranek, L. L. *Acoustic Measurements*. Acoustical Soc. of America, 1988.

Beranek, L. L *Music, Acoustics and Architecture.* New York: John Wiley & Sons, 1962 (out of print).
Beranek, L. L., and Ve'r, I. L. *Noise and Vibration Control Engineering.* New York: John Wiley, 1992.
Cooper, J. *Building a Recording Studio.* Recording Institute of America, 1978.
Crocker, M. J. *Handbook of Acoustics.* New York: John Wiley & Sons, 1998.
Davis, D. *Acoustical Tests and Measurements.* H. Sams, 1965 (out of print).
Davis, D. *Sound System Engineering,* 2nd ed. H. Sams, 1987 (out of print).
Doelle, L. *Environmental Acoustics.* New York: McGraw-Hill, 1972 (out of print).
Egan, M. D. *Architectural Acoustics.* New York: McGraw-Hill, 1988.
Everest, F. A. *Acoustic Techniques for Home and Studio,* 2nd ed. Tab, 1984 (out of print).
Everest, F. A. *How to Build a Small Budget Recording Studio from Scratch ... With 12 Tested Designs,* 2nd ed. Tab, 1988 (out of print).
Everest, F. A. The Master Handbook of Acoustics, 3rd ed. Tab, 1994.
Everest, F. A. *Sound Studio Construction on a Budget.* New York: McGraw-Hill, 1996.
Hall, D. E. *Basic Acoustics.* Krieger, 1992.
Knudsen, V., and Harris, C. *Acoustical Designing in Architecture.* Acoustical Society of America, 1978.
Mankovsky, V. *Acoustics of Studios and Auditoria.* Hastings House, 1971 (out of print).
Rettinger, M. *Acoustical Design and Noise Control,* vols. 1 and 2. Chemical Publ. Co., 1977 (out of print).
Rettinger, M. *Handbook of Architectural Acoustics and Noise Control.* Tab, 1998.
Rettinger, M. *Studio Acoustics.* Chemical Pub. Co., 1981 (out of print).
Yerges, L. F. *Sound, Noise, and Vibration Control.* Krieger, 1978.

Other Sources of Information

The National Council of Acoustical Consultants (NCAC, see http://www.ncac.com/), which can provide information and guidance on working with qualified acousticians.

The American Society of Heating, Refrigerating, and Air Conditioning Engineers (ASHRAE, see http://www.ashrae.org/) publishes the indispensable *ASHRAE Handbook.*

The Acoustical Society of America (ASA, see http://asa.aip.org/) has a comprehensive collection of useful information and research papers.

The Sheet Metal and Air Conditioning Contractors' National Association (SMACNA, see http://www.smacna.org/xs) publishes guidelines on ductwork design.

APPENDIX A:
ABSORPTION COEFFICIENTS OF VARIOUS MATERIALS

Reprinted with permission from *Architectural Acoustics*, by M. D. Egan, copyright 1988 by McGraw-Hill.

Material	Sound Absorption Coefficient						NRC*
	125 Hz	250 Hz	500 Hz	1000 Hz	2000 Hz	4000 Hz	
Walls(1–3, 9, 12)							
Sound-Reflecting:							
1. Brick, unglazed	0.02	0.02	0.02	0.04	0.05	0.07	0.05
2. Brick, unglazed and painted	0.01	0.01	0.02	0.02	0.02	0.03	0.00
3. Concrete, rough	0.01	0.02	0.04	0.06	0.08	0.10	0.05
4. Concrete block, painted	0.10	0.05	0.06	0.07	0.09	0.09	0.05
5. Glass, heavy (large panes)	0.18	0.06	0.04	0.03	0.02	0.02	0.05
6. Glass, ordinary window	0.35	0.25	0.18	0.12	0.07	0.04	0.15
7. Gypsum board, 1/2-inch thick (nailed to 2 × 4s, 16-inch oc)	0.29	0.10	.05	0.04	0.07	0.09	0.05
8. Gypsum board, 1 layer, 5/8-inch thick (screwed to 1 × 3s, 16-inch oc with airspace filled with fibrous insulation)	0.55	0.14	0.08	0.04	0.12	0.11	0.10
9. Construction no. 8 with two layers of 5/8-inch thick gypsum board	0.01	0.01	0.01	0.01	0.02	0.02	0.00
10. Marble or glazed tile	0.01	0.02	0.02	0.03	0.04	0.05	0.05
11. Plaster on brick	0.01	0.02	0.02	0.03	0.04	0.05	0.05
12. Plaster on concrete block (or 1-inch thick on lath)	0.12	0.09	0.07	0.05	0.05	0.04	0.05
13. Plaster on lath	0.14	0.10	0.06	0.05	0.04	0.03	0.05
14. Plywood, 3/8-inch paneling	0.28	0.22	0.17	0.09	0.10	0.11	0.15
15. Steel	0.05	0.10	0.10	0.10	0.07	0.02	0.10
16. Venetian blinds, metal	0.06	0.05	0.07	0.15	0.13	0.17	0.10
17. Wood, 1/4-inch paneling, with airspace behind	0.42	0.21	0.10	0.08	0.06	0.06	0.10
18. Wood, 1-inch paneling with airspace behind	0.19	0.14	0.09	0.06	0.06	0.05	0.10
Sound-Absorbing:							
19. Concrete block, coarse	0.36	0.44	0.31	0.29	0.39	0.25	0.35
20. Lightweight drapery, 10 oz/yd^2, draped to half area (*Note:* The deeper the airspace behind the drapery (up to 12 inches), the greater the low-frequency absorption)	0.03	0.04	0.11	0.17	0.24	0.35	0.15
21. Medium weight drapery, 1/4 oz/yd^2, draped to half area (i.e., 2 ft of drapery to 1 ft of wall)	0.07	0.31	0.49	0.75	0.70	0.60	0.55
22. Heavy-weight drapery, 18 oz/yd^{2,} draped to half area	0.14	0.35	0.55	0.72	0.70	0.65	0.60
23. Fiberglass fabric curtain, 8-1/2 oz/yd^{2,} draped to half area	0.09	0.32	0.68	0.83	0.39	0.76	0.55
24. Shredded-wood fiberboard, 2 inch thick on concrete (mtg. A)	0.15	0.26	0.62	0.94	0.64	0.92	0.60
25. Thick, fibrous material behind open facing	0.60	0.75	0.82	0.80	0.60	0.38	0.75
26. Carpet, heavy, on 5/8-inch perforated mineral fiberboard with airspace behind	0.37	0.41	0.63	0.85	0.96	0.92	0.70

Material	Sound Absorption Coefficient						NRC*
	125 Hz	250 Hz	500 Hz	1000 Hz	2000 Hz	4000 Hz	
Walls[1–3, 9, 12] ***(continued)***							
Sound-Absorbing:							
27. Wood, 1/2-inch paneling, perforated 3/16-inch diameter holes, 11% open area, with 2 1/2-inch glass fiber in airspace behind	0.40	0.90	0.80	0.50	0.40	0.30	0.65
Floors[9, 11]							
Sound Reflecting:							
28. Concrete or terrazzo	0.01	0.01	0.02	0.02	0.02	0.02	0.00
29. Linoleum, rubber, or asphalt tile on concrete	0.02	0.03	0.03	0.03	0.03	0.02	0.05
30. Marble or glazed tile	0.01	0.01	0.01	0.01	0.02	0.02	0.00
31. Wood	0.15	0.11	0.10	0.07	0.06	0.07	0.10
32. Wood parquet on concrete	0.04	0.04	0.07	0.06	0.06	0.07	0.05
Sound Absorbing:							
33. Carpet, heavy, on concrete	0.02	0.06	0.14	0.37	0.60	0.65	0.30
34. Carpet, heavy, on foam rubber	0.08	0.24	0.57	0.69	0.71	0.73	0.55
35. Carpet, heavy, with impermeable latex backing on foam rubber	0.08	0.27	0.39	0.34	0.48	0.63	0.35
36. Indoor-outdoor carpet	0.01	0.05	0.10	0.20	0.45	0.65	0.20
Ceilings[6, 8–10]†							
Sound Reflecting:							
37. Concrete	0.01	0.01	0.02	0.02	0.02	0.02	0.00
38. Gypsum board, 1/2-inch thick	0.29	0.10	0.05	0.04	0.07	0.09	0.05
39. Gypsum board, 1/2-inch thick, in suspension system	0.15	0.10	0.05	0.04	0.07	0.09	0.05
40. Plaster on lath	0.14	0.10	0.06	0.05	0.04	0.03	0.05
41. Plywood, 3/8-inch thick	0.28	0.22	0.17	0.09	0.10	0.11	0.15
42. Acoustical board, 3/4-inch thick, in suspension system (mtg. E)	0.76	0.93	0.83	0.99	0.99	0.94	0.95
43. Shredded-wood fiberboard, 2-inch thick on lay-in grid (mtg. E)	0.59	0.51	0.53	0.73	0.88	0.74	0.65
44. Thin, porous sound-absorbing material, 3/4-inch thick (mtg B.)	0.10	0.60	0.80	0.82	0.78	0.70	0.75
45. Thick, porous sound-absorbing material, 2-inch thick (mtg. B), or thin material with airspace behind (mtg. D)	0.38	0.60	0.78	0.80	0.78	0.70	0.75
46. Sprayed cellulose fibers, 1-inch thick on concrete (mtg. A)	0.08	0.29	0.75	0.98	0.93	0.76	0.75
47. Glass-fiber roof fabric, 12 oz/yd^2	0.65	0.71	0.82	0.86	0.76	0.62	0.80
48. Glass-fiber roof fabric, 37 1/2 oz/yd^2 (*Note:* Sound-reflecting at most frequencies)	0.38	0.23	0.17	0.15	0.09	0.06	0.15
49. Polyurethane foam, 1-inch thick, open cell, reticulated	0.07	0.11	0.20	0.32	0.60	0.85	0.30
50. Parallel glass-fiberboard panels, 1-inch thick by 18-inches deep, spaced 18-inches apart, suspended 12-inches below ceiling	0.07	0.20	0.40	0.52	0.60	0.67	0.45

Material	Sound Absorption Coefficient						NRC*
	125 Hz	250 Hz	500 Hz	1000 Hz	2000 Hz	4000 Hz	
Ceilings(6, 8–10) ***(continued)***							
Sound Reflecting:							
51. Parallel glass-fiberboard panels, 1-inch thick by 18-inches deep, spaced 6 1/2-inches apart, suspended 12-inches below ceiling	0.10	0.29	0.62	1.12	1.33	1.38	0.85
Seats and Audience(1, 5, 7, 9)‡							
52. Fabric well-upholstered seats, with perforated seat pans, unoccupied	0.19	0.37	0.56	0.67	0.61	0.59	
53. Leather-covered upholstered seats, unoccupied**	0.44	0.54	0.60	0.62	0.58	0.50	
54. Audience, seated in upholstered seats§	0.39	0.57	0.80	0.94	0.92	0.87	
55. Congregation, seated in wooden pews	0.57	0.61	0.75	0.86	0.91	0.86	
56. Chair, metal or wood seat, unoccupied	0.15	0.19	0.22	0.39	0.38	0.30	
57. Students informally dressed, seated in tablet-arm chairs	0.30	0.41	0.49	0.84	0.87	0.84	
58. Persons adult (total number of sabins)	2.5	3.5	4.2	4.6	5.0	5.0	
Miscellaneous(3, 9, 11)							
59. Gravel, loose and moist, 4-inches thick	0.25	0.60	0.65	0.70	0.75	0.80	0.70
60. Grass, marion bluegrass, 2-inches high	0.11	0.26	0.60	0.69	0.92	0.99	0.60
61. Snow, freshly fallen, 4-inches high	0.45	0.75	0.90	0.95	0.95	0.95	0.90
62. Soil, rough	0.15	0.25	0.40	0.55	0.60	0.60	0.45
63. Trees, balsam firs, 20 ft^2 ground area per tree, 8-ft high	0.03	0.06	0.11	0.17	0.27	0.31	0.15
64. Water surface (swimming pool)	0.01	0.01	0.01	0.02	0.02	0.03	0.00

*NRC (noise reduction coefficient) is a single number rating of the sound absorption coefficients of a material. It is an average that only includes the coefficients in the 250–2000 Hz frequency range and therefore should be used with caution.

†Refer to manufacturers' catalogs for absorption data, which should be from up-to-date tests by independent acoustical laboratories according to current ASTM procedures.

‡Coefficients are per square foot of seating floor area or per unit. Where the audience is randomly spaced (courtroom or cafeteria for instance), mid-frequency absorption can be estimated at about 5 sabins per person. To be precise, coefficients per person must be stated in relation to the spacing pattern.

**The floor area occupied by the audience must be calculated to include an edge effect at aisles. For an aisle bounded on both sides by audience, include a strip 3 ft wide; for an aisle bounded on only one side by audience, including a strip 1½ ft wide. No edge effect is used when the seating butts walls or balcony fronts (because the edge is shielded). The coefficients are also valid for orchestra and choral areas at 5 to 8 ft^2 per person. Orchestra areas include people, instruments, music racks, and so on. No edge effects are used around musicians.

Test Reference

Standard Test Method for Sound Absorption Coefficients by the Reverberation Room Method, ASTM C 423. Available from American Society for Testing and Materials (ASTM), 1916 Race Street, Philadelphia, PA 19103.

Sources

1. Beranek, L.L. "Audience and Chair Absorption in Large Halls," *Journal of the Acoustical Society of America*, January 1969.
2. Burd, A.N., et al. "Data for the Acoustic Design of Studios," British Broadcasting Corporation, BBC Engineering Monograph no. 64, November 1966.
3. Evans, E.J., and Bazley, E.N. "Sound Absorbing Materials," H. M. Stationery Office, London, 1964.
4. Hedeen, R.A. *Compendium of Materials for Noise Control*, National Institute for Occupational Safety and Health (NIOSH), Publication no. 80-116, Cincinnati, Ohio, May 1980. (Contains sound absorption data on hundreds of commercially available materials.)
5. Kingsbury, H.F., and Wallace, W. J. "Acoustic Absorption Characteristics of People," *Sound and Vibration*, December 1968.
6. Mariner, T. "Control of Noise by Sound-Absorbent Materials," *Noise Control*, July 1957.
7. Moore, J.E., and West, R. "In Search of an Instant Audience," *Journal of the Acoustical Society of America*, December 1970.
8. Moulder, R., and Merrill, J. "Acoustical Properties of Glass Fiber Roof Fabrics," *Sound and Vibration*, October 1983.
9. "Performance Data, Architectural Acoustical Materials," Acoustical and Insulating Materials Association (AIMA). (This bulletin was published annually from 1941 to 1974.)
10. Purcell, W.E. "Materials for Noise and Vibration Control," *Sound and Vibration*, July 1982.
11. Siekman, W. "Outdoor Acoustical Treatment: Grass and Trees," *Journal of the Acoustical Society of America*, October 1969.
12. "Sound Conditioning with Carpet," The Carpet and Rug Institute, Dalton, Ga., 1970.
13. Knudsen, V., and Harris, C. "Acoustical Designing in Architecture," American Institute of Physics, 1978.

CHAPTER

3.3

Microphones for Broadcast Applications

TY FORD
Ty Ford Audio and Video
Baltimore, Maryland

GREG SILSBY
Fellowship of Technical Ministries
Snohomish, Washington

INTRODUCTION

Improving a broadcast facility's sound by choosing the right microphone is the direct result of knowing enough about microphones to make the right decisions. The combination of studio acoustics, type of microphone, directional characteristics, and preamplifier plays a role in the selection of the microphone that sounds best for any given situation. A microphone is an electro-acoustic transducer which, when activated by acoustic energy from a sound source, converts or transduces that energy to another form—an electric current. In the microphone, acoustical energy (sound waves impinging on the diaphragm) is converted to a varying voltage that is the electrical analog of the sound. This chapter describes the basic types of microphones, their construction, and how their different characteristics can be used in various applications for best results.

ACOUSTIC ENVIRONMENT

The successful use of any microphone depends on the acoustic environment in which it is used and the particular voice or instrument on which the microphone is used. The problems of poor acoustic environments cannot always be overcome by using more expensive microphones. The end result might be technically excellent audio and a great performance in what can be heard as an obviously bad acoustical environment. Too many large glass windows and other hard reflective surfaces create a difficult acoustical environment in which no mics perform well. In addition, glass windows are often mounted with the glass presenting a downward angle to sound. While this works in large music studios, it doesn't work well in small rooms because the sound usually bounces back down into the rear or side of a microphone. Minimizing the amount of glass, using window treatments such as curtains, and angling the glass to reflect the sound upward and away from the microphone are much better strategies.

In addition to acoustic issues, individual voices vary greatly. Historically, part of the reason some men and women were chosen for on-air work was for the quality of their voices. This is less frequently the case now. There is only so much compensation a mic and preamp can do to improve speech quality. Great sounding audio, then, is the result of a great source, a great environment, and the right microphone and microphone preamplifier.

MICROPHONE CHOICES

Of the many microphones available, broadcasters usually choose handheld, boom-mounted, and lavaliere styles. The handheld microphone is often used for on-camera and electronic newsgathering (ENG). Boom-mounted microphones are usually found in the broadcast studio, and some handheld microphones also are boom mounted. The lavaliere microphone is usually used in TV news, where a handheld, desktop, or boom-mounted microphone would be obtrusive. A miniature microphone can also be mounted on a headset device, with or without headphones. Lavaliere microphones may be useful in some radio talk show programming where the guest may be unfamiliar with good microphone technique or intimidated by a large boom-mounted microphone.

MICROPHONE PREAMPLIFIER

Because of the low signal level from a microphone, preamplifiers are often used to boost the level to that needed for the mixing bus in an audio console. The preamplifier may be located in the console or externally in an equipment rack. From a technical perspective, the input impedance of the preamplifier will affect how well a microphone will match with a preamplifier. However, there is no hard data as to the best impedance, or combination of resistance, capacitive reactance, and inductive reactance. Some microphones are relatively immune to these variables, whereas others are not. Quality and price are not necessarily factors. High-quality microphones and preamps can substantially improve the sound of a station. The same high-quality microphone can be compromised by connecting it to a low-quality microphone preamplifier.

One solution, albeit a costly one, has been to design preamplifiers with tunable capacitive and inductive front ends, as in some phonograph preamps. These preamps allow the user to vary the input impedance of the preamplifier and, thus, the sound of the microphone/amplifier combination. On a more practical level, when one is considering both budget and application, there are usually a number of suitable choices.

MICROPHONE ATTRIBUTES

The four attributes of a microphone most broadcasters need to consider are *self-noise*, *directionality*, *sensitivity*, and *ruggedness*.

Self-Noise

Self-noise is normally perceived as a hiss. It is the electronic noise generated by the passive or active circuitry of the microphone itself. Condenser microphones use a vacuum tube or field effect transistor (FET) to convert the high impedance of the condenser capsule to a lower impedance, more suited for the input of a microphone preamp. Vacuum tube condenser microphones are generally noisier than condenser microphones that use FETs. However, careful selection of the vacuum tube can result in a tube microphone being quieter than its FET counterpart.

Self-noise is an important factor in studio recordings, where every effort is made to eliminate extraneous noise. Self-noise is less important in applications in which the ambient noise level is considerably higher. The ambient noise level in a broadcast air studio may be quite low, equaling that of a well-designed music studio. On the other hand there may be noise from HVAC systems, tape machines, computer hard drives and fans, and sound transferred through the structure from the surrounding environment. If the combined noise from any or all of these sources cannot be reduced, spending extra money for a microphone with exceptionally low self-noise is not the best use of that money. Finding the best-sounding directional microphone and positioning it to reject as many of the ambient noises as possible may be a better approach.

The self-noise of dynamic or ribbon microphones is usually much less because these microphones do not use active electronics. However, the output of a dynamic or ribbon microphone is usually lower than that of a condenser. Because of this, more gain is required of the microphone preamplifier to bring the dynamic and ribbon microphones up to the appropriate operating level. Less expensive or poorly designed microphone preamps also generate noise, also usually perceived as hiss.

Directionality

Omnidirectional and *directional* microphones each have their place in broadcasting. Omnidirectional microphones pick up most sounds from all directions equally. However, they are usually directional at high frequencies. Omnidirectional microphones are often used in news gathering interviews in which one or more voices from different directions are to be picked up. Their wide pattern is useful because the position of the microphone does not have to be moved as accurately from person to person. Most omnidirectional microphones are also less susceptible to popping and wind noise than more directional microphones. However, some small-diaphragm omnidirectional condenser microphones are quite sensitive to popping. If there is high ambient noise, or if only one voice is needed, a directional microphone may be more effective.

Directional microphones with *cardioid* (heart-shaped), *hypercardioid*, and *supercardioid* patterns are used in broadcasting to reduce ambient noises. *Shotgun* microphones, sometimes referred to as supercardioids, offer high directionality at upper frequencies, but are much more omnidirectional at middle and lower frequencies. While their "reach," or ability to hear sounds at a distance, is greater than hypercardioid patterns, they are best used on a large well-damped soundstage or in quiet exterior spaces.

Hypercardioid patterns perform much better than shotgun mics in tight studio environments. Because they are more directional than a shotgun, they don't pick up as much sound reflected from the ceiling, walls, floor, and other hard, flat surfaces.

Sensitivity

Output level, or the sensitivity of a microphone, is important because all microphone preamps add a certain amount of noise, especially at the upper end of their operating range. The higher the output of the microphone, the less amplification will be required from the microphone preamp. Condenser microphones with their internal amplifier are more sensitive than dynamic and ribbon microphones.

Ruggedness

Ruggedness is more important in ENG (electronic news gathering) and EFP (electronic field production) recording than in controlled studio situations. Microphones for these applications must withstand rough handling and harsh elements (cold and hot weather and rain and very dry conditions). Ribbon microphones should not be used in the field because of the fragile ribbon element. Condenser microphones designed for recording studio use are better suited for more controlled environments because they are sensitive to humidity, temperature changes, and wind. However, some new condenser microphones are almost as rugged as dynamic microphones.

High-Frequency Response

Condenser microphones are typically more sensitive to high frequencies than dynamic microphones. Special care must be taken in choosing condenser microphones if the intended environment is small, noisy, and has many reflective surfaces. Their sensitivity to high frequencies in that environment will result in an increase in unwanted room sound.

Hot Spot

The *hot spot* (sometimes called a *sweet spot*) of a microphone is that area within its pickup pattern in which a particular sound source sounds best. In a quiet and well-damped environment, the hot spot may be larger. However, if the sound source is thin and would benefit from more bass response, the hot spot for a cardioid microphone might be within the range of its "proximity effect," usually two to four inches from the microphone.

If the sound source, an announcer for example, moves around too much, a hypercardioid pattern may provide too narrow a hot spot, resulting in noticeably uneven levels and sense of presence. If the announcer cannot be trained to stay within the hot spot, a broader cardioid pattern may be a better choice.

MICROPHONE TYPES

Microphones can be typed according to their electric generating element (transducer):

- Ribbon
- Dynamic
- Condenser

Ribbon Microphone

The *ribbon*, or *velocity*, microphone utilizes a very thin, corrugated metallic foil ribbon suspended within the flux field of a strong permanent magnet. While the ends of the ribbon are held in place, the rest of it is allowed to move freely back and forth in a sympathetically induced mechanical recreation of the amplitude and frequency of the sound presented to it.

As the metallic ribbon is moved across the magnetic flux lines, it induces a small AC current through the ribbon. Wires from the ends of the ribbon connect it to a step-up transformer which converts the low impedance of the ribbon (approximately 1 ohm) to a value between 50 and 500 ohms more suitable for matching the input to the preamplifier. Ribbon microphones are available in a variety of fixed or variable patterns and are known for delivering a very warm sound, due to the ribbon's sensitivity to low frequencies. Blowing into a ribbon microphone or using it in a windy environment can destroy it. Even rapid panning on a studio boom has caused ribbon failure. Newer designs, however, have provided considerable improvements in durability and a lower failure rate.

Some ribbon microphones should not be connected to a microphone preamp with active phantom or A/B powering because the voltage from these supplies can damage the ribbon.

Dynamic Moving Coil Microphones

Although the ribbon microphone is a type of dynamic microphone, in common usage, the term *dynamic* microphone usually refers to a microphone with a moving coil. The dynamic microphone has a diaphragm attached to a voice coil, as illustrated in Figure 3.3-1. This lightweight coil of wire is suspended in a magnetic field supplied by a permanent magnet structure. The ends of the voice coil are brought out to stronger leads, which connect either to a transformer or the microphone's output connector.

Sound waves reaching the diaphragm cause it to move back and forth. The attached voice coil cuts the lines of flux in the magnetic field, causing a small AC current. This signal closely emulates the sound waves in frequency and amplitude. The diaphragm must be

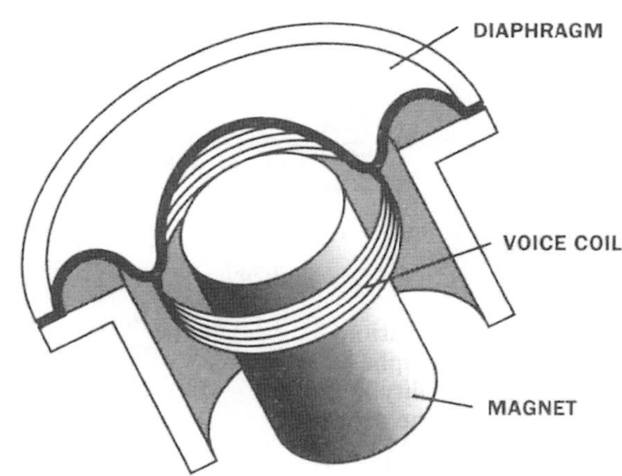

FIGURE 3.3-1 Dynamic moving coil element. (Courtesy of Shure, Inc.)

highly compliant to allow effortless excursion at all frequencies of interest. In addition, this movement must be accomplished with maximum linearity and a minimum of break-up modes. Break-up modes occur when a portion of the diaphragm resonates independently of the rest of the surface. Phase cancellation results and, with it, response anomalies occur.

The design and construction of a high-quality dynamic microphone suitable for broadcasting use blend science and art. As is true in other areas of engineering, design trade-offs are numerous and the laws of physics tend to win in the end.

Size plays an important role in the performance of the dynamic microphone. Small dynamic mechanisms tend to have low acoustic sensitivity and high mechanical sensitivity. The result may be a poor system signal-to-noise ratio (SNR) and excessive handling noise or noise transmitted through the microphone stand. Internal shock-mount systems may be used to reduce the mechanical excitation, but the design goal of small size may then be defeated. Small size usually means sacrificed low-frequency response in dynamic microphones. This is not to say that large dynamic microphones will always have an extended low-frequency response.

Another physical characteristic of the dynamic microphone that affects its performance is the mass of the diaphragm/voice coil assembly. The greater the mass, the more limited will be the high-frequency response. Common design practice includes the use of Helmholtz resonators[1] immediately in front of the diaphragm, as illustrated in Figure 3.3-2, that creates

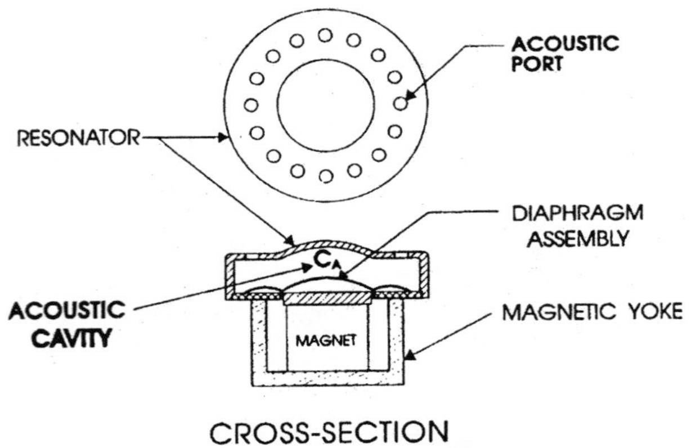

FIGURE 3.3-2 Cutaway drawing of a dynamic microphone element showing Helmholtz resonators. (Courtesy of Electro-Voice, Inc.)

[1]Named for Hermann von Helmholtz, German physicist who explained certain acoustic and psychoacoustic principles. For more information, see http://www.phys.unsw.edu.au/~jw/Helmholtz.html and http://physics.kenyon.edu/EarlyApparatus/Rudolf_Koenig_Apparatus/Helmholtz_Resonator/Helmholtz_Resonator.html.

peaks and effectively extend high-frequency response beyond the normal limits of the system. Advances in metallurgy, specifically the use of stronger neodymium magnets, have resulted in dynamic microphones with higher outputs.

Most broadcast-quality dynamic low impedance (Z) microphones exhibit an impedance that is a function of the number of turns and gauge of the voice coil wire. Some older, more public-address-oriented microphones employ a transformer within the housing to correct for design trade-offs in the voice coil. The transformer adds to the microphone's cost and may also restrict performance, if it is not a high-quality unit, by limiting the frequency response and possibly increasing distortion. Properly designed dynamic microphones can be the most rugged of the high-quality transducer types. Some have truly become legendary for their capability to provide high-quality broadcast audio with virtual bullet-proof construction.

Condenser Microphone

In the *condenser* microphone, a capacitor forms the generating element, as illustrated in Figure 3.3-3. One side of the capacitor is the diaphragm; the other is the fixed backplate. Air between these two plates acts as a dielectric. The capacitor, of course, must possess a positive electrical charge on one plate and a negative charge on the other. The conventional or discrete condenser receives this polarizing or bias voltage from an external DC power supply. *Phantom* power is used in most conventional condenser systems to deliver the required DC voltage to the microphone over the same cable used to carry the audio signal.

Upon activation of the power supply, a voltage is quickly built up on the surface of the diaphragm or backplate of the capsule. This causes an electrical current to flow through the resistor until the surface of the backplate finally receives an opposite charge of equal value. As sound waves (air pressure changes) strike the diaphragm, causing it to move back and forth, the distance between the two plates rapidly increases and decreases. This causes proportional changes in the capacity of the condenser. The result is an AC current flow in the resistor and a voltage across

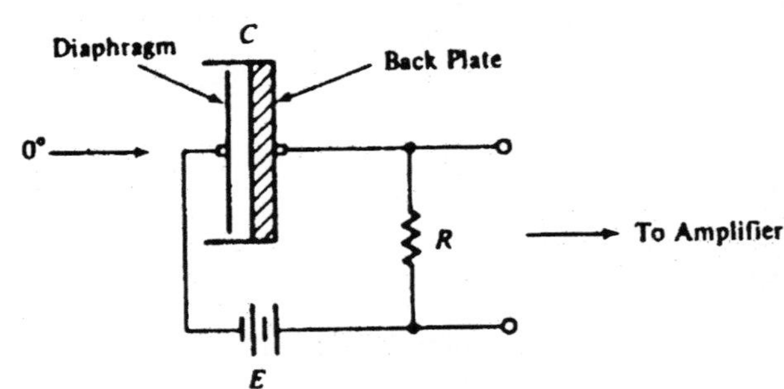

FIGURE 3.3-3 Conventional capacitor microphone system.

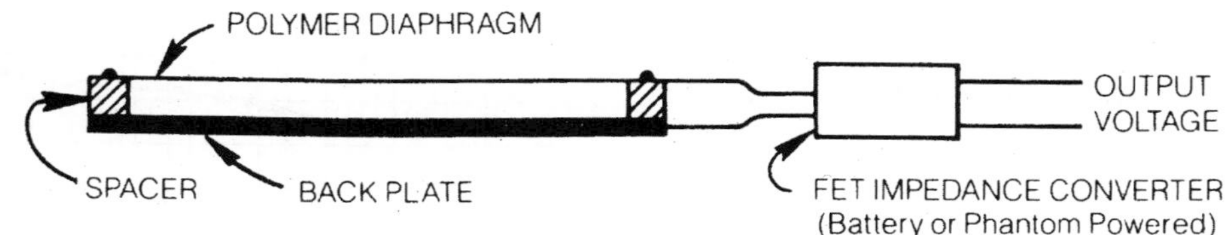

FIGURE 3.3-4 Electret condenser element. (Courtesy of Audio-Technica U.S., Inc.)

the resistor that corresponds to the excursion of the diaphragm.

While this voltage effectively represents the output voltage of the microphone, the source impedance is too high to be carried for any distance over microphone cable. This output signal, then, is presented to an impedance converter circuit, usually a vacuum tube or FET inside the microphone. Power for the impedance converter is derived from the same source that provides the polarizing voltage for the element. The impedance converter delivers a low impedance output that can be fed through long microphone cables with minimal loss.

Electret Condenser

The *electret* condenser microphone utilizes a material which has the capability to hold a charge applied during the manufacturing process. Most high-quality electrets apply this material to the fixed backplate of the capacitor, as shown in Figure 3.3-4. Some designs employ a charged diaphragm instead. Lowering the weight of the diaphragm, by moving the electret material to the backplate, results in lower handling noise, extended frequency response, and improved transient response.

Although the electret functions much like the discrete condenser, but produces its output voltage without the need for an external high-voltage DC supply, an impedance converter is still required. The low voltage needed to power it may be derived from internal or external batteries or an external AC-powered supply.

PHANTOM POWER

Phantom power, or *simplex power,* provides one means for remotely powering condenser microphones and may range from 9 to 52 VDC from an external supply, usually the audio console or mixer. While many electrets will operate over a wide range of voltage, most modern discrete condenser designs require 48 volts. The phantom supply voltage for nonelectret condenser microphones is often stepped up by an internal circuit to provide a sufficient capacitor-polarizing charge for good signal-to-noise figures.

The amount of current delivered by the power supply is also a factor. While some condenser microphones can operate with less than 1 mA, others require up to 2.5 mA or more. If insufficient current is provided, the level of the microphone is reduced and the signal is distorted.

In a phantom power circuit, the positive side of the DC supply is applied equally to both of the signal-conducting leads of a balanced microphone line and the negative to the shield. This is accomplished by means of either a build-out of matched precision resistors or via a center-tapped transformer. In each case, the return path is the shield. In the microphone, the DC may be similarly tapped via the resistor or center-tapped transformer method to provide the power it needs. The DC is prevented from appearing at the impedance converter output by DC-blocking capacitors or the internal center-tapped transformer.

If a balanced-output dynamic microphone is connected to a line with phantom power present, performance should not be altered, nor should damage occur to the dynamic element. The voice coil or output transformer winding connects across the two signal leads and should see no potential difference between them. Because there is no connection between either lead and the shield (the DC return path), there is no current flow. If an unbalanced dynamic microphone is connected to a phantom supply, the DC current will pass through the voice coil and may destroy it. A less common powering system called *A/B* or *T* powering is not compatible with dynamic or phantom-powered microphones. A/B power puts the positive side of the DC on one signal lead and the negative on the other. This will damage even a balanced output dynamic microphone.

MICROPHONE CABLES

Regardless of the style or pattern of microphones chosen, the best results require the use of a high-quality, well-shielded microphone cable. Quality cable is more important here than anywhere in the audio chain because of the high amount of gain applied by the microphone preamplifier. Poorly shielded or low-quality cable may not reject electromagnetic (from nearby AC power lines) or radio frequency (from nearby transmitters) induced noise. As a result, that low-level noise is amplified by the microphone preamplifier and can become audible.

MICROPHONE SPECIFICATIONS

Some microphones come with individual graphs for frequency response and polar pattern. While specifications tell part of the story, there is nothing better than trying a microphone in a specific application. Note that the same microphone may sound different when amplified through different preamps.

Frequency Response

One of the first specifications considered on a microphone data sheet is the frequency response range. Often, more attention is given to the response limits than to how the microphone actually sounds in its intended application. Nonlinearities of the response often contribute more to the listener's subjective impression of sound transmitted by the microphone than the response range. A specification that reads "frequency response 40 to 18,000 Hz," by itself, says little about a microphone's actual sound in use. Adding to that some amplitude limits, such as ± 3.0 dB, improves the specification but, depending on other characteristics of the microphone, the result may still be unsatisfactory.

The shape of the frequency response and polar pattern characteristics contribute to the character or personality of the microphone's sound. Response nonlinearities can create acoustic feedback in sound reinforcement, a nasality sound, poor intelligibility, excessive sibilance, muffled sound, or any of a variety of other acoustic problems. On the other hand, a microphone's response may be deliberately tailored by the design engineer to solve problems rather than create them. A rolled-off low-frequency response and a rising high-frequency response may be employed in a microphone that is intended to reduce the effects of unwanted low-frequency information, such as traffic noise or the rumble of air-handling systems.

Some microphones that are intended to be worn on the body exhibit a response that compensates for the chest cavity resonance which they tend to pick up in a lapel or clip-on mounting position. A rolled-off low-end response may help considerably in attenuating handling or stand-borne noise as well as wind noise or the breath blasts of plosives in speech.

Published specifications can serve as only one guide in understanding a microphone. Even curves from well-respected manufacturers may be difficult to compare due to the variety of test procedures and standards. For example, the frequency response chart is an X-Y graph that compares output to frequency. If that chart is compressed vertically or stretched horizontally, the response curve may appear to be more linear. Such variables as the recording speed, damping, or even the direction of the tone sweep (low to high or high to low) may result in substantially different curves.

Directional microphones present a phenomenon known as the *proximity effect*, which results in a bass-boosted output when used close to a small sound source. Proximity effect is neither good nor bad; its value depends on the intended application. Various designs will differ in the amount of proximity effect that is possible to attain. The response curve of a particular directional microphone, tested at a specific (or perhaps even unknown) distance, may be of little value to someone who wishes to use the microphone at another distance. Ideally, directional microphone data should include close and distant curves, as shown in Figure 3.3-5.

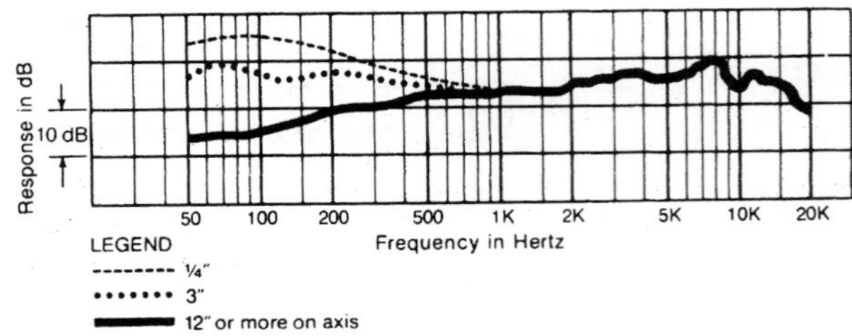

FIGURE 3.3-5 Influence of proximity effect on a directional microphone response. (Courtesy of Audio-Technica U.S., Inc.)

Polar Pattern

As difficult as it might seem at times to pick up a desired audio signal, the real problem often lies in eliminating unwanted sounds. Microphones with various directional patterns may be often used to improve the ratio of desired signal-to-ambient noise or other unwanted sounds. Ambient noise, leakage from other instruments in a band or orchestra, room reverberation, and feedback potential from public address systems or monitor speakers are some of the reasons why it is important to know the off-axis response of a microphone. The best view of the microphone's off-axis response is obtained by examining several different polar plots. These should be drawn at low, mid-band, and high frequencies. Overlaid, these plots should reveal how well the microphone maintains its directionality at each frequency, as illustrated in Figure 3.3-6.

There are several broad categories of polar patterns to which most microphones' directional characteristics conform to some extent or another:

- Omnidirectional
- Bidirectional
- Cardioid
- Supercardioid
- Hypercardioid

Figure 3.3-7 shows the directional patterns obtained from mathematical models representing the perfect polar characteristic for each example. Actual microphone patterns may vary from near perfection to close resemblance. In practice, the microphone design engineer must go beyond math equations to accomplish a desired axial response and sensitivity while maintaining polar uniformity. The chart in Figure 3.3-7 shows how the various patterns should relate to the reference omni in their ability to reject unwanted energy arriving from various points off-axis.

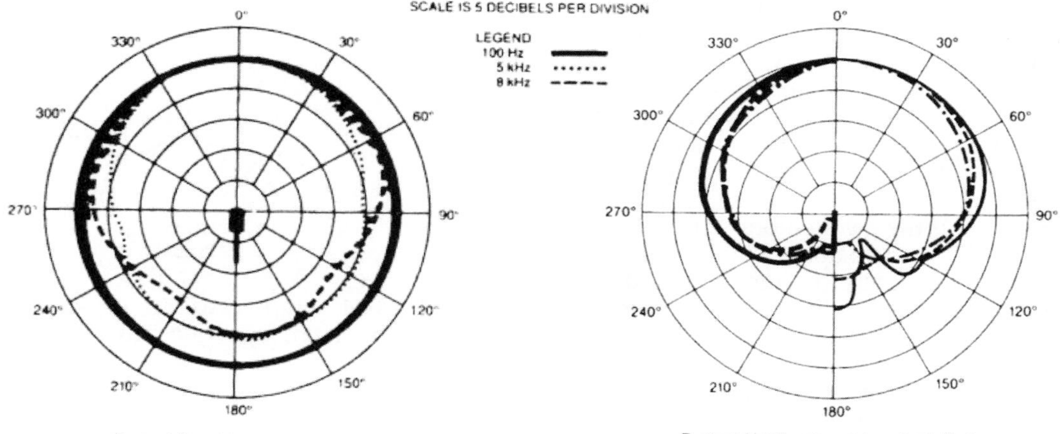

FIGURE 3.3-6 Polar patterns drawn at several frequencies. (Courtesy of Audio-Technica U.S., Inc.)

Polar Scaling and Range

It is important when reading polar patterns to observe the several variables in the way that they are represented, as illustrated in Figure 3.3-8.

First, determine whether the scale is logarithmic or linear. A log scale (the most commonly used) will show a fairly modest inward curve of the cardioid pattern at 90°, indicating a 6 dB drop in level. The linear

CHARACTERISTIC	OMNI-DIRECTIONAL	CARDIOID	SUPER-CARDIOID	HYPER-CARDIOID	BIDIRECTIONAL
Polar response pattern	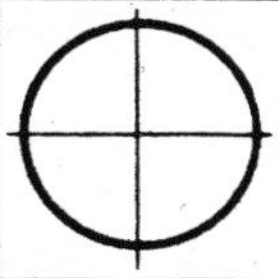	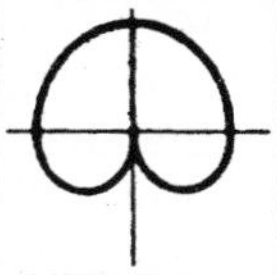	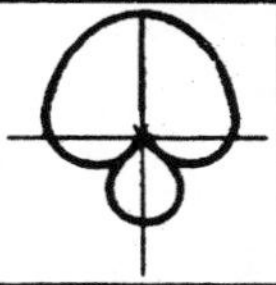	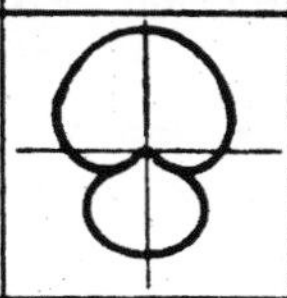	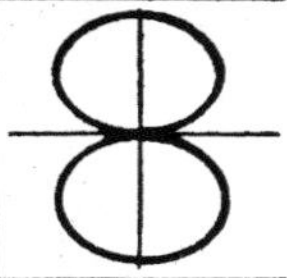
Polar equation	1	.5 + .5 cos θ	.375 + .625 cos θ	.25 + .75 cos θ	cos θ
Pickup ARC 3 dB down (1)	—	131°	115°	105°	90°
Pickup ARC 6dB down	—	180°	156°	141°	120°
Relative output at 90° dB	0	−6	−8.6	−12	−∞
Relative output at 180° dB	0	−∞	−11.7	−6	0
Angle at which output = 0	—	180°	126°	110°	90°
Random energy efficiency (REE)	1 0dB	.333 −4.8dB	.268 −5.7dB (2)	.250 −6.0dB (3)	.333 −4.8dB
Distance factor (DF)	1	1.7	1.9	2	1.7

NOTE:
1 = Drawn shaded on polar pattern
2 = Maximum front-to-total random energy efficiency for a first order cardioid
3 = Minimum random energy efficiency for a first order cardioid

FIGURE 3.3-7 Microphone polar patterns and characteristics.

scale for the same microphone will show a polar pattern that appears much more directional. The outside circle of the linear polar represents 100% output, whereas the center of the circle equals zero output. Because a 6 dB loss is equal to a 50% drop in voltage, on a linear scale the polar curve at 90° sweeps to half the distance between the outside of the circle and its center.

Second, determine if the graduations between concentric circles are 5 dB or 10 dB apart. Finally, take note of the range of the polar pattern. This may be determined by counting graduation lines inward from the point where the polar crosses 0°, in 5 or 10 dB steps (as marked) to the smallest inner circle. Patterns may be found in most any range, with 25, 30, and 40 dB all being common. These differences will also alter the shape of a polar pattern.

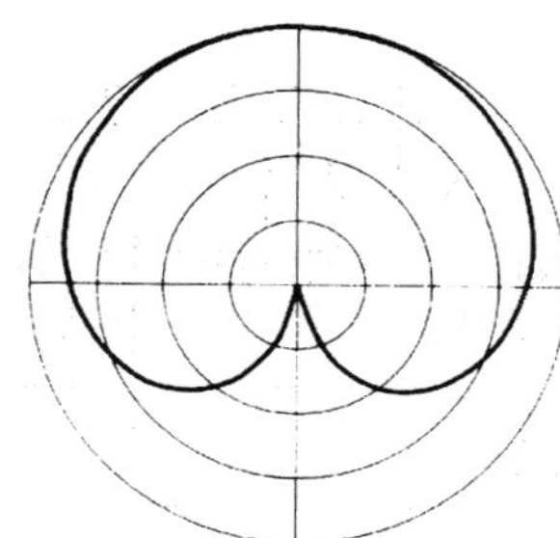

FIGURE 3.3-8a Cardioid log scale polar. Scale is 10 dB per division.

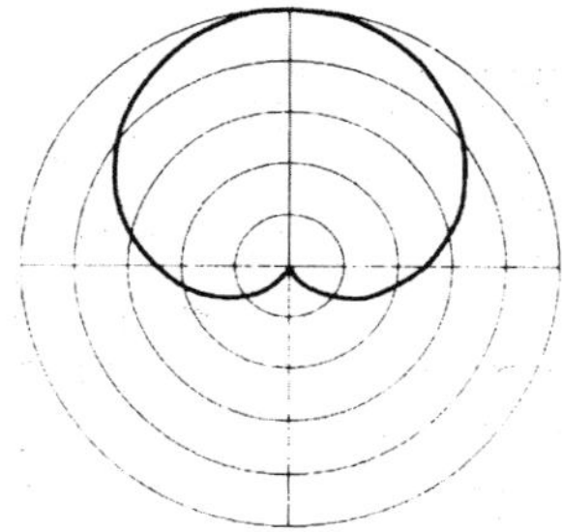

FIGURE 3.3-8b Cardioid linear scale polar.

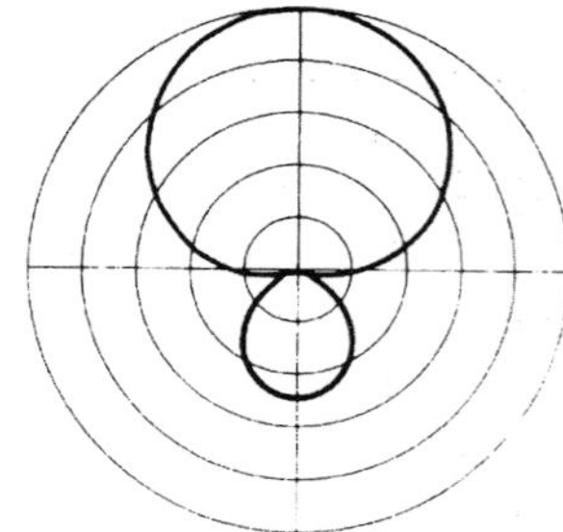

FIGURE 3.3-8c Hypercardioid linear scale polar. Dynamic range is 50 dB.

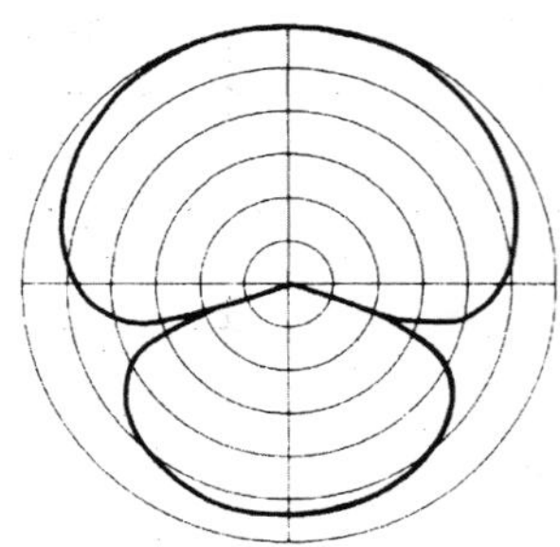

FIGURE 3.3-8d Hypercardioid log scale polar. Scale is 10 dB per division. Dynamic range is 40 dB.

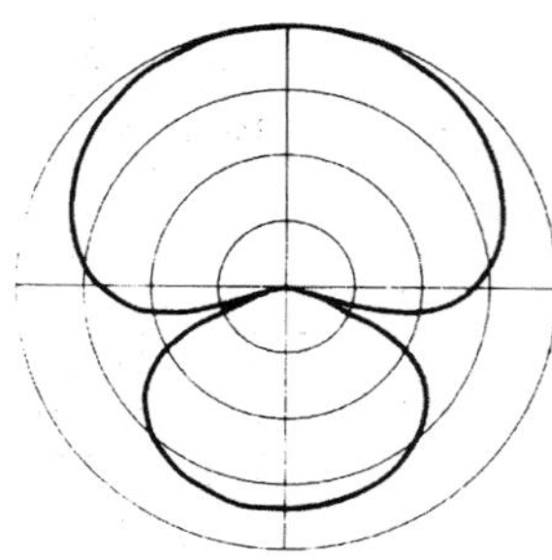

FIGURE 3.3-8e Hypercardioid log scale polar. Scale is 10 dB per division. Dynamic range is 60 dB.

Note that the polar pattern represents a cross-sectional, two-dimensional diagram of a three-dimensional function. The 131° pickup arc, for example, in the chart in Figure 3.3-7, that is described by the 3 dB down points on either side of the axis of the cardioid microphone, can really best be thought of as a conical area within which the microphone is virtually uniformly sensitive. This area is often referred to as the microphone's *angle of acceptance* or *included angle*.

Omnidirectional

The omnidirectional microphone consists of a diaphragm and generating element backed by a totally sealed case. When placed in a sound pressure field, the perfect omni disregards the direction of the sound's origin. A positive pressure (air expanding) at the diaphragm, for example, causes the diaphragm to move inward regardless of the sound's point of origin, as illustrated in Figure 3.3-9. Such a microphone may also be referred to as a *pressure microphone*.

Most omnidirectional microphones, however, are not truly omnidirectional. The case of the microphone represents a barrier to higher frequencies arriving from off-axis. Because of this case effect, most omni mics are increasingly directional at higher frequencies. The smaller the omni mic, the more truly omnidirectional it may be. In addition to the case effect, energy arriving at the diaphragm from on-axis is reinforced at those frequencies to which the size of this *baffle area* is significant. The baffle effect causes a rise in the micro-

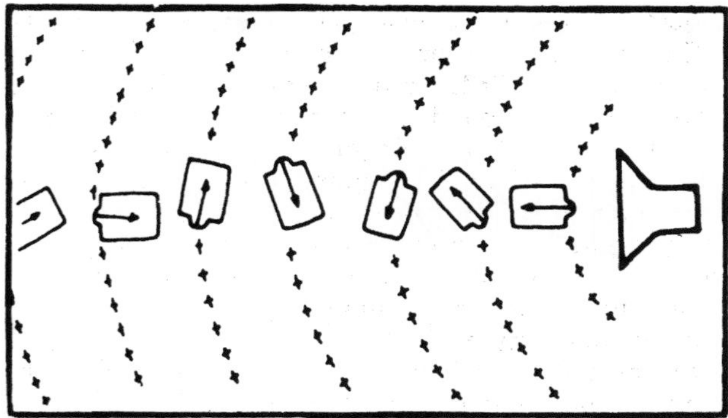

FIGURE 3.3-9 Omnidirectional microphone principle.

phone's high-frequency output, but only with respect to energy arriving on-axis.

Cardioid

Directional microphones employ a damped porting system in their element design that allows sound waves to act upon the rear of the diaphragm as well as the front. The design introduces varying amounts of phase shift for sound arriving from off-axis, resulting in cancellation. The rear entry ports of most directional microphones are spaced at a single distance or "D" from the diaphragm, illustrated in Figure 3.3-10. Multiple port systems are also available and are designed to reduce the proximity effect.

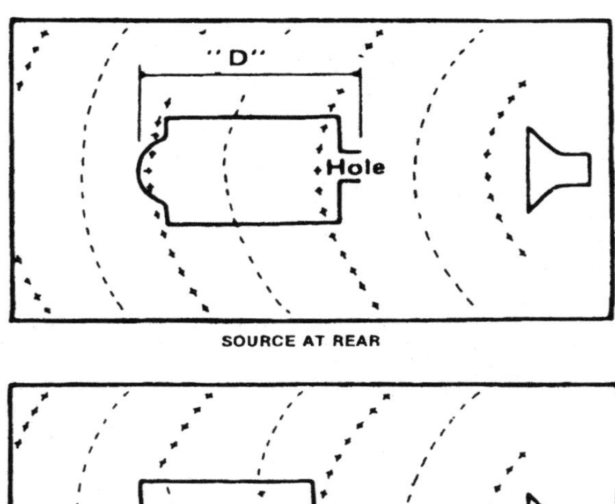

FIGURE 3.3-10 Single-D cardioid microphone operating principle.

When using any directional microphone, avoid obstructing the ports with the microphone clip, hand, clothing, gaffer's tape, or logo flags. Covering any of the ports results in serious degradation of the microphone's directional characteristics and overall sound quality.

A sound source that delivers, for example, 60 dB sound pressure level (SPL) to a cardioid microphone on-axis from one foot away will drop 6 dB when the microphone is rotated to position the sound source at 90° off-axis. Here, a properly designed cardioid emulates well its mathematical model. At 180° off-axis, however, the cardioid cannot live up to the equation. The chart indicates zero output. In reality, well-designed cardioids are capable of something only on the order of a 20 dB differential. That is equivalent to moving the sound source to 10 times its actual distance from the microphone.

The 180° response curves (back curve) of many cardioid microphones show their tendency to more closely resemble an omni mic at both the low and high frequencies. The much more impressive cancellation in the midrange sometimes causes a manufacturer to release a data sheet that shows only one polar pattern and for an unknown frequency. One of the most beneficial performance advantages to look for in a well-designed microphone is off-axis linearity, as illustrated in Figure 3.3-11. Note the comparatively smooth back response of the higher quality condenser microphone in (b), campared to that of the dynamic microphone in (a).

Omni versus Cardioid

While the omni picks up sound from all angles, the cardioid reduces the pickup of ambient noise and reverberant energy from behind the microphone. The comparison chart (Figure 3.3-7) shows that the *random energy efficiency* (REE) of the cardioid is 0.333 compared to an REE of 1 for the omni.

The random energy efficiency is a measurement that compares a microphone's sensitivity to random (or reverberant) energy to its on-axis sensitivity. While this shows the cardioid to be one-third as sensitive to random ambient noise as the omni, note that discrete sound sources positioned at the null of the cardioid will be attenuated to a much greater extent. This will prove true in outdoor performances where sounds arrive at the microphone directly with minimal reflections. Indoors, the advantage of the cardioid's deep rear null is appreciated only when the microphone is situated within *critical distance* of the offending sound source. Within the critical distance, the direct sound is greater in intensity than the reflected energy. After that point, the two remain approximately equal. Other characteristics of the cardioid mic include

- It is more susceptible to the problems of "pop" (the blast of plosives from words that begin with "B," "P," and "T" in speech), wind noise, and handling or mechanical noise.

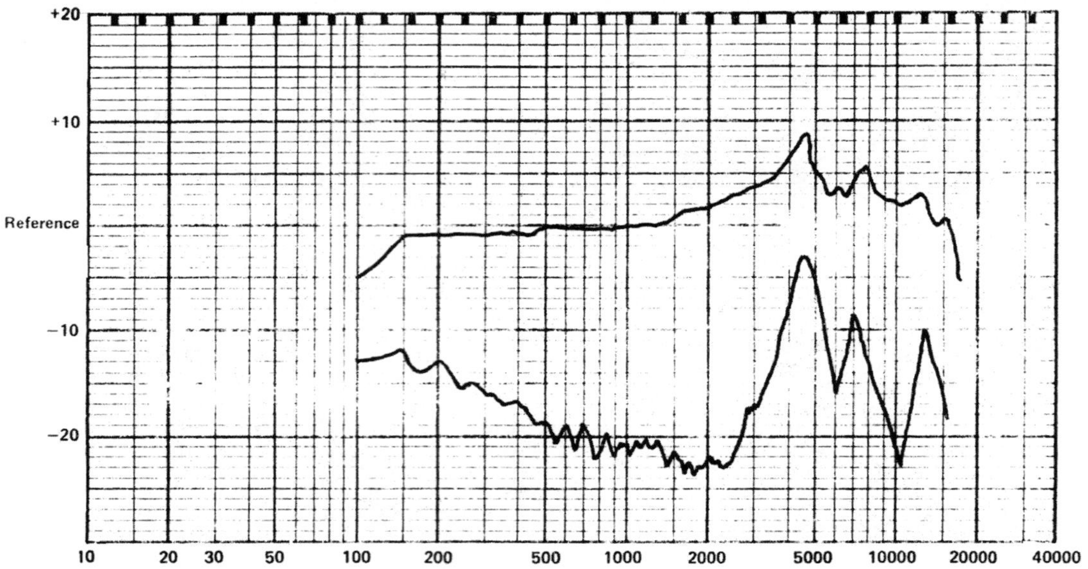

FIGURE 3.3-11a Front and back curves of a typical cardioid dynamic handheld vocal microphone. Vertical scale is in dB; horizontal scale is in Hz.

- It is more complex to design and construct and generally costs more than an omni of apparent equal audio quality.
- More complex construction makes it generally less rugged than a omni mic.
- It has greater resistance to feedback in most sound reinforcement applications due to its lower REE and is further aided by the proximity effect.
- It increases the effective working distance. From Figure 3.3-7 note the *distance factor* (DF) of 1.7 for the cardioid mic, meaning that it has a working distance advantage over the omni of 1.7:1. This factor is calculated on the assumption of a perfect cardioid, in a totally diffuse noise field. An ideal cardioid, then, could be used at a distance of 1.7 times that of the omni for a given ratio of desired, on-axis signal to ambient noise.

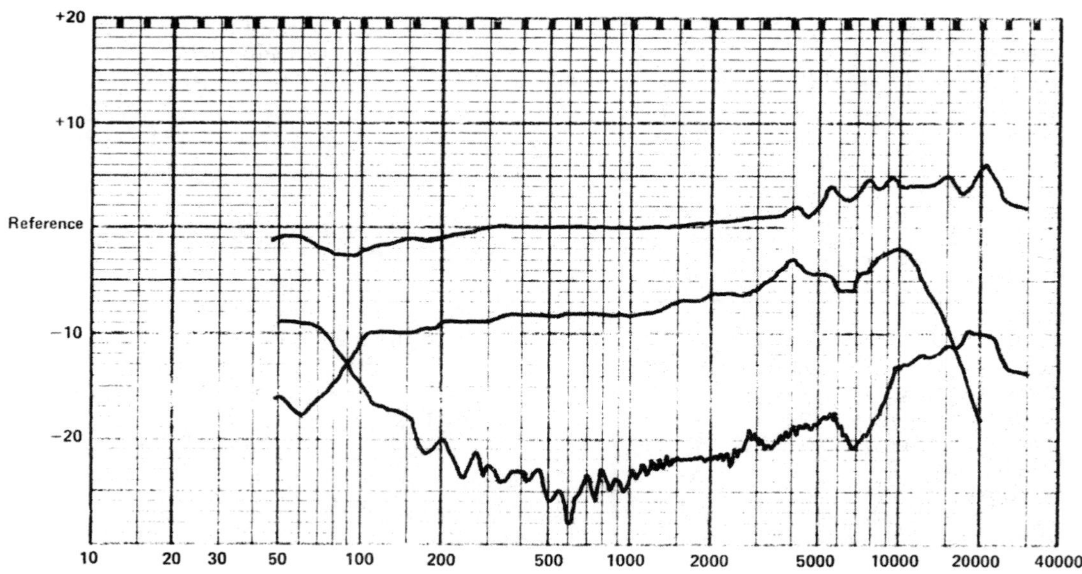

FIGURE 3.3-11b Front, side, and back curves of a high-quality cardioid condenser microphone. Vertical scale is in dB; horizontal scale is in Hz.

- The nonlinear polar response and the inability of cardioid microphones to achieve total cancellation at their null would seem to reduce the cardioid's working distance advantage. However, increasing the working distance often has more to do with attenuating a single, offending noise source than with overcoming a diffuse noise field. Directing the deep null of a good cardioid microphone at an offending noise source in the field or studio may offer more than a 1.7:1 working distance advantage over an omni.
- The cardioid exhibits proximity effect. While some designs are quite low in proximity effect, all exhibit some bass-boost effect when used close to the source. Although this may be considered an enhancement in many close-mic applications, it is important to avoid preamp input overload or loss of intelligibility that may result from excessive proximity effect.

Other Patterns

Again referring to Figure 3.3-7, note how the three other polar patterns compare to the omni and the cardioid. The hypercardioid, for example, combines a tight acceptance angle with superior side rejection and offers the lowest REE. The bidirectional pattern offers the best side rejection, but with no advantage over the cardioid in REE. Bidirectional (also called *figure of eight*) microphones are typically ribbons or dual-diaphragm condensers. They are useful in eliminating unwanted sounds from the side and for picking up two sound sources (such as two people opposite each other) while incurring no phase problems.

Output Impedance

The impedance (Z) of a microphone is a measurement of its AC resistance looking back into the transducer. Broadcast microphones should be low impedance, ranging typically from 50 to 600 ohms. Dynamic moving-coil microphones achieve their low impedance by either a low-Z voice coil winding or a transformer. Condenser microphones use an impedance converter circuit to step down the capacitor's high-Z output.

Low impedance offers the advantages of low susceptibility to hum and electrical noise pickup, and the capability to use relatively long cables with a minimal loss of level or high frequencies. Unlike matching power amplifier impedances to speaker systems, which may be desired for best power transfer, microphones need load impedances on the order of 10 times their internal impedance. This assures maximum voltage transfer. A microphone that looks like a resistive source of 150 ohms, looking into a load resistor of 150 ohms, for example, will suffer a 6 dB voltage drop compared to an open-circuit connection.

Dynamic Range

The difference between a microphone's own self-noise and the maximum sound pressure level it can handle before distortion is the *dynamic range.* In many field applications, ambient noise provides sufficient masking to make the self-noise specification of minor interest. The importance of this specification increases as greater working distances are demanded or ambient noise levels are lowered.

The impedance converter of condenser microphones, like any active circuit, will create some noise which will vary from one design to another. The impedance converter design also determines the headroom or maximum SPL that the microphone can handle. A maximum SPL of as high as 141 dB is achieved in several high-quality condenser.

Dynamic microphones contribute virtually no self-noise. When they are greatly amplified, only the noise of the thermal agitation of air molecules is detected. While this is very low in level, the dynamic microphone does not automatically rank as the first choice in a low-noise system. Because the output level of the dynamic is often lower than that of a condenser system, the user may end up working into the noise floor at the upper extremes of the preamplifier in order to provide sufficient system gain.

Some new dynamic microphones employ powerful rare-earth magnets to increase the efficiency of their motor mechanisms. Their higher output, while still not as high as many condenser microphones, can provide a considerable S/N advantage over earlier designs.

WORKING DISTANCE

Sometimes it is not sufficient to reduce ambient noise merely by using a polar pattern that offers the lowest REE. In very noisy environments (for example, in an aircraft, or a factory, or at a sporting event), it may be desirable to differentiate between close sound (an announcer, for example) and distant sound. Microphones that offer considerable proximity effect may be used to advantage in these situations by having the announcer work the microphone very close and roll off the low end as needed to flatten the response.

Noise-Canceling Microphone

In extreme situations, a noise-canceling (differential) microphone may be required. Because of the special design (rear ports and back damping systems) of the differential microphone, sound arriving from a distance strikes both sides of the diaphragm with equal intensity and in phase. A positive pressure on the front, for example, would encounter a positive pressure on the rear of the diaphragm, causing the signal to be canceled. A combination of inverse square law and port damping causes sound that originates very close to the front of the noise-canceling microphone to be lower in intensity and to exhibit some phase error by the time it arrives at the rear of the diaphragm. The noise-canceling microphone is able, therefore, to differentiate between close and distant sound sources.

The audio quality of such systems normally limits them to voice communication applications.

Inverse Square Law

The easiest, and certainly the least expensive, way to limit the apparent working distance of a microphone is by positioning the microphone very close to the sound source. Inverse square law shows that decreasing the distance between the microphone and the sound source by one-half (for example, from 8 to 4 ft) increases sound intensity at the microphone by a factor of four, or 6 dB, as shown in Figure 3.3-12. As the input sensitivity control of the mixer or recorder is lowered to compensate for the additional 6 dB available from the now-closer sound source, the microphone, in effect, becomes less sensitive to distant sounds.

Working at a Distance

In applications in which the sound source is at a significant distance, the effective maximum working distance may be determined by the electronic signal-to-noise ratio of both the microphone and subsequent amplifiers. For example, the selection of an ideal boom microphone for picking up dialogue in a quiet environment, with no reverberation problems, may have little to do with polar patterns. Instead, the desirable microphone would have a high output and low self-noise. Most often electronic signal-to-noise and signal-to-ambient noise ratios are the main concern.

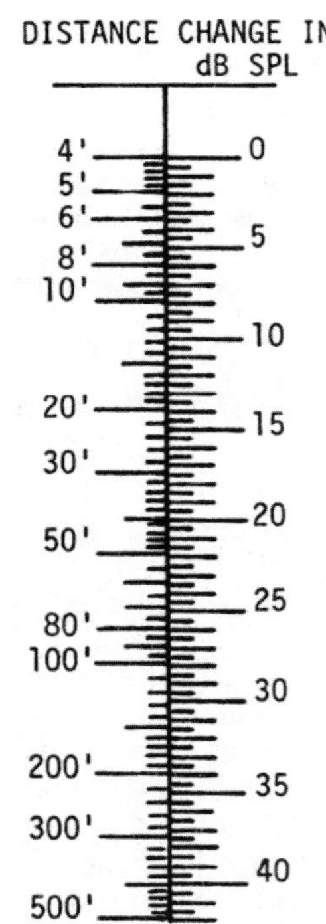

FIGURE 3.3-12 Inverse square law.

HEADSET MICROPHONES

Headset microphones provide benefits gained from always being a fixed distance from the speaker's mouth. Background noise is reduced (because of the inverse square law) and levels remain consistent. Omni, cardioid, and differential elements are available in headset systems. Cardioids offer the best combination of ambient noise suppression and acceptable broadcast quality.

SHOTGUN MICROPHONES

Effective working distances beyond those afforded by cardioid, supercardioid, or hypercardioid systems may be realized through the use of a shotgun microphone, which uses a long, slotted interference tube ahead of the element to provide a high degree of cancellation at the sides. Sound waves arriving on-axis are essentially unaffected by the tube. Sound arriving from slightly off-axis, however, is forced to turn and travel down the tube to the element. This results in numerous out-of-phase conditions being set up in the tube, with cancellation increasing as the microphone is rotated to 90°.

Newer shotgun microphone designs from several manufacturers provide superior pattern control using shorter interference tubes than those required by older standards. The new generation of shorter, lighter products are much easier to handle in boom applications.

Some shotgun microphones are much less uniform in off-axis response than conventional hypercardioids, as shown in Figure 3.3-13. Even with their multilobed

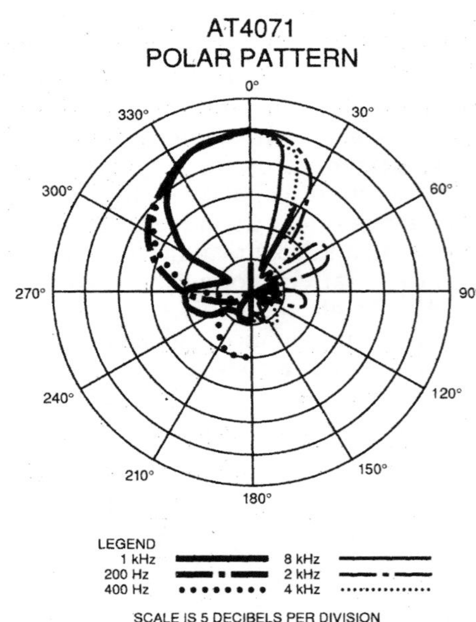

FIGURE 3.3-13 Shotgun polar patterns. (Courtesy of Audio-Technica U.S., Inc.)

polar patterns, however, their increased reach over hypercardioids' reach may make them a logical choice in the right acoustical environment.

Shotguns work best outdoors and in controlled acoustic environments such as well-designed studios. Distant micing down a hallway will not be assisted greatly by the use of a shotgun microphone. A shotgun microphone pointed upward toward an actor wearing a wide-brimmed hat may work better than one boomed from above, unless there is also an HVAC duct in the path above the actor.

ACOUSTIC GAIN DEVICES

While shotgun microphones increase working distance by rejecting off-axis sound energy, thereby narrowing the acceptance angle, some devices increase working distance by providing *acoustical gain*. The most commonly used acoustic gain device is the *parabola* or *parabolic reflector*, as illustrated in Figure 3.3-14. The parabolic reflector is shaped so that sound is reflected onto a focal point a short distance in front of the centerpoint of the dish. An omnidirectional microphone placed at this point receives multiple reflected sound waves in phase, which add to produce significant gain. The response of such systems is ragged and limited. Low-frequency response is extended as the dish diameter is increased. While unacceptable for most broadcast applications, the audio quality achieved with the parabolic microphone is often deemed adequate for sound effects pickup such as at sporting events.

A second type of acoustic gain device is the *horn*. Low cost re-entrant horns are often used for talk-back in paging systems and are quite directional and sensitive. Installed on the side of a building, the small horn is virtually as inconspicuous as a light fixture and is seldom thought of as a microphone. Some horns are built with 45 ohm voice coils, providing higher output signals to microphone inputs.

Acoustic gain is also realized by using a microphone in the very close vicinity of a large, hard, reflective surface. Omnidirectional microphones may be flush mounted into the barrier, facing out. In this position the microphone is in a half-space environment, or looking into only half the world. The output for sound arriving on-axis is increased by 6 dB. As the sound source is rotated off-axis, however, the microphone output drops. At 90° off-axis, the output is down 6 dB, or equal to the omni in free space. The resulting polar pattern resembles a cross-section of a cardioid cut through the microphone at 90°.

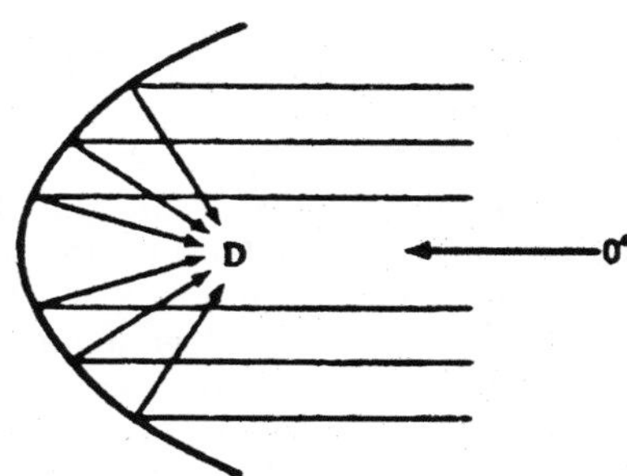

FIGURE 3.3-14 Parabolic microphone principle.

Frequency Response and Distant Micing

Distant micing may result in noticeable or even dramatic changes in the spectrum of the sound being recorded. High frequencies, attenuated by the air, may require boosting to restore a normal sound. Similarly, high-pass (low-cut) filters may prove helpful in reducing low-frequency room reverberation or background noise, thereby extending the useful working distance.

SENSITIVITY RATINGS

Microphone sensitivity is rated in several ways including open circuit output voltage or power level into a given impedance with a specific sound pressure level. Sensitivity is specified according to a test procedure in which a 1 kHz tone at 94 dB SPL applied at the input of the microphone element produces an open circuit voltage that is measured or the power into a specified impedance is measured. A 94 dB SPL is also described as the pressure of 1.0 Pa (pascal) or 10 dynes/sq cm, and the output voltage is measured in terms of mV/Pa (millivolts per pascal) according to standards IEC 60268-4 and EIA SE-105.[2]

Open Circuit Output Voltage

Microphones are often specified as having a particular output voltage when looking into an open circuit. In most modern equipment, microphone preamplifier inputs are at least 10 times the measured impedance of the microphone and may be regarded as an open circuit. Specifications may be given as an actual output voltage or as decibels below one volt at a sound pressure level of 74 dB (1 dyne/cm^2 or 0.1 Pa) or 20 dB below the standard procedure. These ratings are referred to as *the open circuit output voltage* rating or *open circuit sensitivity.*

The open circuit sensitivity may be expressed in dB by means of the following formula:

$$V_{OC} = 20 \log E_O - SPL + 74$$

where,

V_{OC} = Open circuit voltage in dB (referenced to 1 V/0.1 Pa)

E_O = Microphone output in volts

SPL = Actual SPL at the microphone

[2]IEC is International Electrotechnical Commission and EIA is Electronic Industries Alliance.

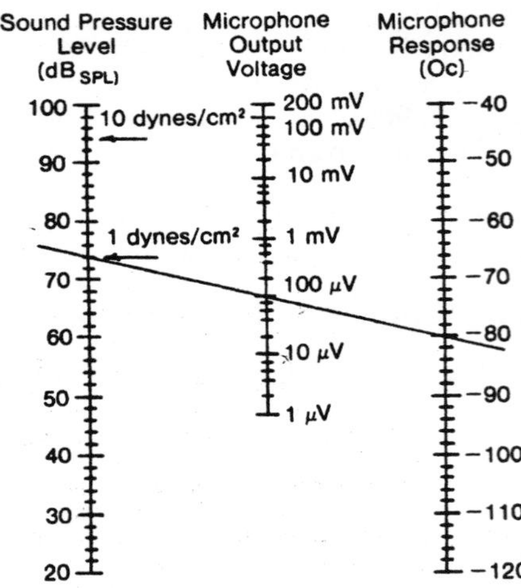

FIGURE 3.3-15 Nomograph: Open circuit voltage rating.

The nomograph shown in Figure 3.3-15 provides an easy method of calculating the open circuit voltage.

Power Level

Microphone sensitivity may also be specified in terms of its output power level. This equivalent power-level rating takes into consideration the open circuit rating and either the actual impedance of the microphone or the rated impedance. Specifications given would be in dBm (or just dB) referenced to 0 dB = 1 mW/10 dynes/cm^2 or 0 dB = 1 mW/Pa.

The following formula is used for calculating the power-level rating:

$$P_E = V_{OC} - 10\log_{10} Z + 44 \quad \text{(dB)}$$

where,

P_E = Equivalent power level

Z = Impedance of the microphone

A nomograph for determining the equivalent power rating is shown in Figure 3.3-16.

EIA Sensitivity Rating

The EIA sensitivity rating is sometimes specified but rarely used. The formula for determining EIA sensitivity is

$$ESR = V_{OC} - 10\log_{10} R_{MR} - 50 \quad \text{(dB)}$$

where,

ESR = EIA sensitivity rating

V_{OC} = Open circuit voltage in dB

R_{MR} = Center value of the nominal impedance range

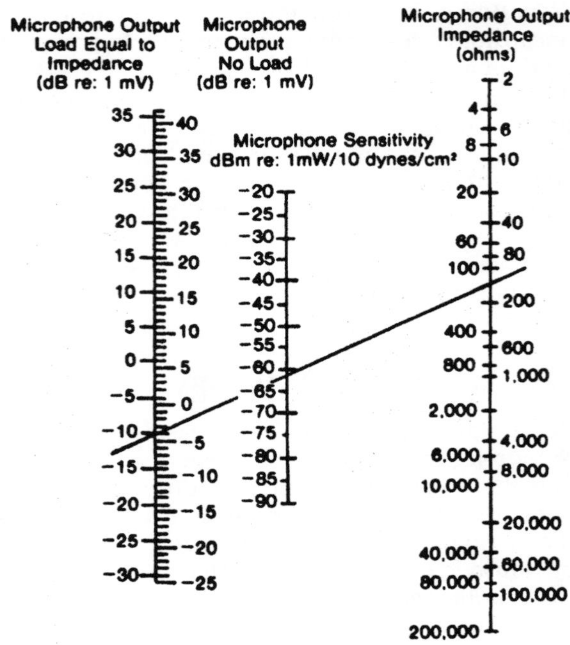

FIGURE 3.3-16 Nomograph: Equivalent power rating.

Table 3.3-1 may be used for determining R_{MR}. An EIA sensitivity rating nomograph is given in Figure 3.3-17.

LINE-LEVEL MICROPHONES

For situations involving long mic cables, using a portable preamp/limiter combination will reduce loss and hum pickup. Line-level microphones designed for remote use may incorporate a microphone, preamplifier, limiter, and power supply in one handheld, or a separate unit with belt clip, battery-operated package.

HIGH ACOUSTIC LEVEL APPLICATIONS

Properly designed dynamic microphones are difficult to drive to audible distortion. The distortion heard when a dynamic microphone is subjected to the lips-touching proximity of a very loud rock and roll vocalist is usually caused by the peak clipping in the electronics following the microphone. Outputs of 1 volt or more may actually be delivered in such applications as rock and roll music. Preamp or amplifier clipping may be avoided by attenuating the microphone output or adjusting the trim (gain adjustment) of the specific input on the mixer. Note that many mixers offer adjustment only after a gain stage or transformer, either of which may distort before any control is possible. In-line attenuators, or pads, are commercially available that allow the selection of 10, 20, or 30 dB of attenuation and plug directly into the

TABLE 3.3-1
Center Value for Nominal Impedance Ranges

Range (ohms)	Center Value (ohms)
20–80	38
80–300	150
300–1250	600
1250–4500	2400
4500–20,000	9600
20,000–70,000	40,000

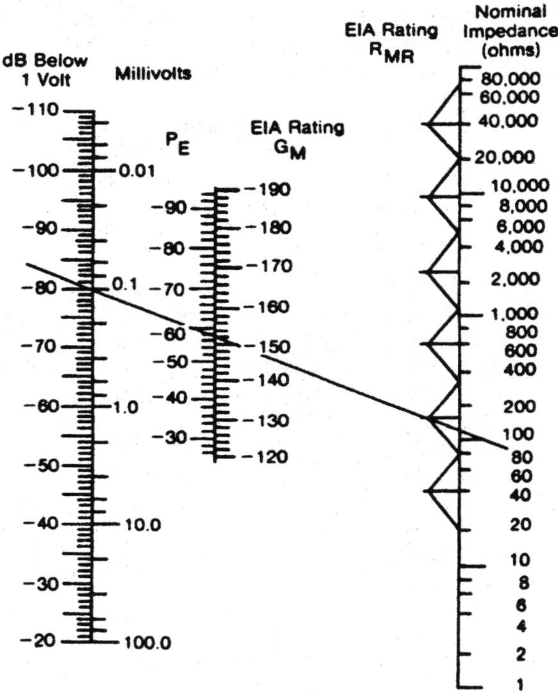

FIGURE 3.3-17 Nomograph: EIA sensitivity rating.

microphone line at the input of the mixer. Before using any in-line device, verify that it is compatible with the powering system being used for condenser microphones. A typical in-line attenuator is shown in Figure 3.3-18.

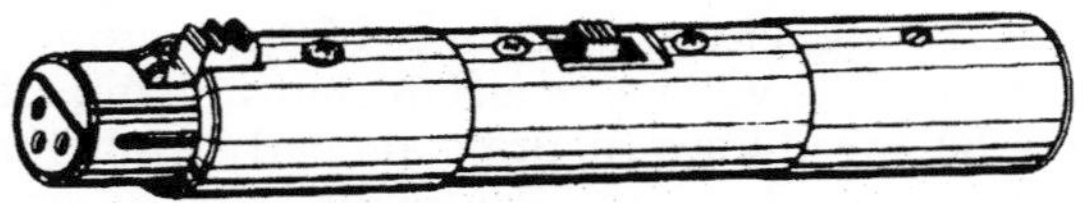

FIGURE 3.3-18 Microphone in-line attenuator. (Courtesy of Audio-Technica U.S., Inc.)

AVOIDING NOISE PROBLEMS

Unwanted signals include wind noise, "P-pop," mechanical or handling noise, AC hum, and radio frequency interference (RFI). The reduction or elimination of each of these can be handled both through microphone design and user technique.

RFI problems can usually be traced to a point in the low-level circuitry at which the signal leads are unbalanced, high-Z, or both. Condenser microphones, for example, may sometimes be sensitive to RFI at or around their impedance converter. If RFI is a problem, the manufacturer should be consulted for low-pass modifications or information.

P-pops due to the plosives from words beginning with "B," "D," and "P" in speech may be reduced in several ways:

- **Use an omni microphone.** Directional microphones are much more prone to P-pop problems than are omnidirectional ones.
- **Position the microphone out of the area of the breath blast.** In an announce application, speak across the microphone with the microphone 45° left or right of the person speaking. Stand-ups and handheld interview micing should be done with the microphone capsule below the axis of the mouth.
- **Use a pop filter.** This is often the same as the manufacturer's windscreen. Test the combination carefully for frequency response and directional characteristics before putting it into service. Windscreen/pop-filter foam is specially designed reticulated foam that comes in a variety of densities. Even very acceptable open-cell foam may be too thick for use on some microphones. Nonreticulated foam (such as "Nerf balls") rolls off high-frequency response and alters the polar patterns of directional microphones.
- **Fashion a pop filter.** For radio and other off-camera micing, a piece of silk may be suspended a short distance in front of the microphone diaphragm. A number of systems are commercially available that use a frame made of wood or plastic, and some allow the insertion of multiple screens. Other systems use perforated metal discs. Some of these pop filters install directly on the microphone suspension mount. Others clamp onto the mic boom.
- **Use a high-pass filter.** Most of the disturbing plosive energy of a P-pop is very low in frequency. Try using a very abrupt high-pass (low-cut) filter in the microphone line. Rolling this energy off before it gets to the mixing board or recorder input will further reduce distortion in the audible range.

Wind Noise

While wind noise may be dealt with similarly to P-pop, superior results with shotgun microphones can be attained by using a well-engineered fabric/mesh cylindrical screen which provides an air space between the material and the microphone, as shown in

FIGURE 3.3-19 Zeppelin-type windscreens for shotgun microphones. (Courtesy of Audio-Technica U.S., Inc.)

Figure 3.3-19. To handle severe cases (gale-force winds), special fur-like socks are available to wrap around the tubular windscreen. While using these materials will result in some performance trade-offs, recordings made under such conditions are typically not intended for critical listening. Windscreens must cover all openings to the element—front and rear.

Use a High-Pass Filter

A microphone with a limited low-frequency response will help minimize wind noise. Extended-response condenser microphones can produce very high outputs of infrasonic energy when moved or when air around them is moved by wind or air-handling systems. The result may be preamp overload or undesirable compressor or limiter action. Again, windscreens and/or high-pass filters may solve these problems.

Handling or Mechanical Noise

The problem of mechanical, nonacoustic noise is one that is encountered by the user whether the microphone is hand held, body worn, or hardware mounted. The reduction of a transducer's sensitivity to such noise, or the improvement of its acoustic-to-mechanical noise sensitivity ratio, starts with the basic element design.

Microphone elements are often internally shock mounted by the manufacturer to avoid the transfer of noise from the case to the element. Lowest noise is achieved through the combining of omni or omni condenser systems with internal shock mounts, as illustrated in Figure 3.3-20.

External shock mounts are often employed in stand or boom-mounted microphone applications. Properly designed shock mounts allow excursion on-axis, or perpendicular to the diaphragm plane. The combination of mechanical isolation and an internally shock-mounted microphone provides the best results.

Another method of reducing mechanical noise is to raise the resonant frequency of the mechanical drive system. An example would be that of bracing wooden platforms, tables, or lecterns to eliminate the very audible, drum-like sound produced when they are struck. The use of high-density materials for microphone support systems will result in a higher resonant frequency. A microphone stand set onto concrete or into sand also takes advantage of this density.

Mechanical Cable Noise

Mechanical noise transfer to the diaphragm may also be reduced through decoupling the diaphragm from tensile forces, converting them to lateral forces. This may be demonstrated by selecting a microphone that has some noticeable handling noise problems and plugging it into a talk-out system, raising the gain until the handling noise is evident. Now hold the microphone face up (diaphragm horizontal), with the cable hanging straight down, and tap on the cable. This should produce a low thump. Next, rotate the microphone 90° so that the cable is hanging at a right angle to the microphone axis. Tap the cable again, and the thump should be all but gone. This effect may be applied to custom hardware designs and the dressing of cables on body mics or as they enter stand or boom-mounted microphones. A loop of cable or a small coiled cord lowers mechanical noise transfer by this method.

AC Hum Rejection

Microphones may also be sensitive to noise induced in the element and cables by electromagnetic or electrostatic radiation, which may be the result of proximity to power transformers, fluorescent ballasts, high-voltage AC lines, and SCR dimmers, as examples.

- **Ensure that lines are balanced and low impedance.** The higher the impedance of the microphone, the greater the voltage of electrostatically induced hum. Balanced lines ensure that nearly equal hum will be induced on each conductor. Little differential is seen at the amplifier input, resulting in common mode rejection of the hum.

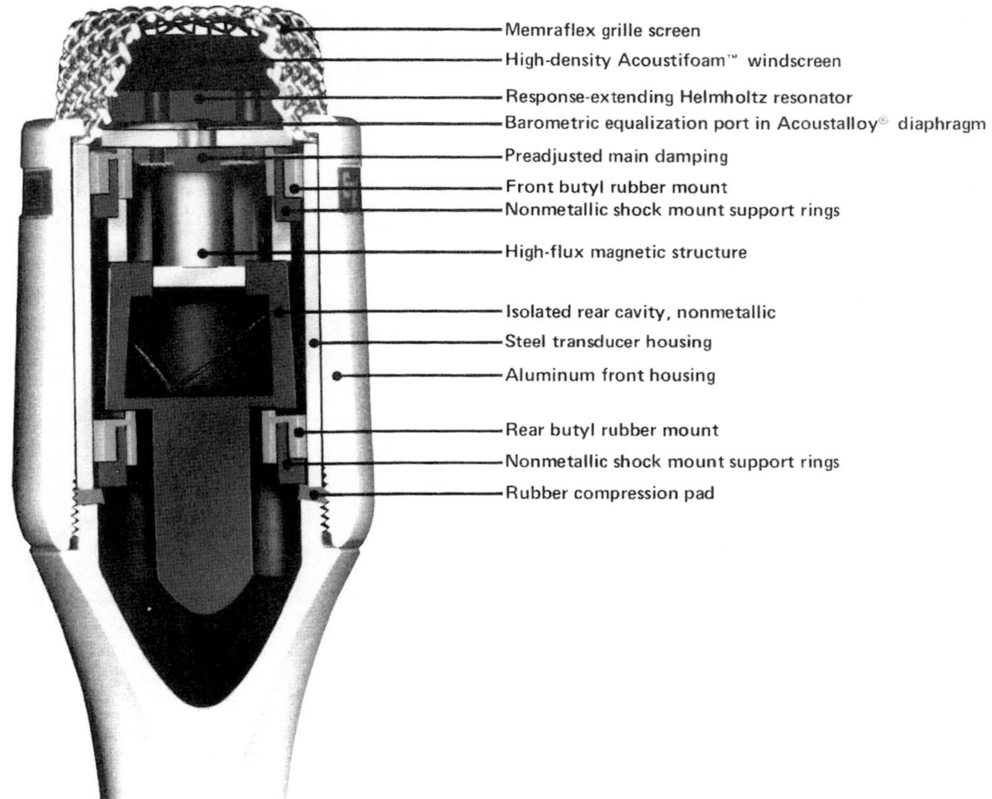

FIGURE 3.3-20 Cutaway view of shock-mounted omni. (Courtesy of Electro-Voice, Inc.)

- **Route cables with caution.** Avoid running low-level signal cables long distances near AC power cables. When such cables must cross, they should do so at right angles. If more than one AC line must cross microphone lines, separate the AC lines so that they cross at different points.
- **Use twisted pair cable.** Leads should be twisted inside the microphone and out. The virtually identical positioning this provides for the two conductors within the hum field and the fact that they are out of phase with each other will further reduce induced hum.
- **Use well-shielded cable.** Installed cable may use a foil shield, as flexibility and low-flex memory are not factors. Good stage and field cable normally has a braided shield. Cables which offer the best combinations of flexibility and good shielding use conductive cloth or conductive vinyl under the braided shield.
- **Follow good grounding practices to avoid ground loops.**

In general, dynamic microphones are much more sensitive to induced hum than are condenser microphones. The voice coil is a very effective inductor. Hum-buck coils are employed in some designs that lower electromagnetic hum sensitivity by about 20 dB. The hum-buck coil is wrapped around the outside of the motor mechanism and wired in series with the voice coil but out of phase. When both coils are placed into an electro-magnetic field, equal energy is induced onto each. Because they are out of phase with each other, the offending signal is canceled.

Transformers located within microphones should be avoided if electromagnetic hum is a possible problem. Some transformers are constructed with hum-bucking characteristics that reduce hum induction.

In severe problem situations, operation at line level rather than microphone level may be required.

ACOUSTIC PHASE INTERFERENCE

Another micing problem relates not to sounds that are added to the output, such as popping or hum, but to portions of the spectrum that are greatly attenuated. This change in the microphone's apparent frequency response is the result of acoustic phase cancellation that may occur when two or more microphones are mixed, or even when a single microphone is subjected to an overdose of reflected sound. The response charts in Figure 3.3-21(b) through (e) show the severe phase cancellation problems that can result from incorrect use of multiple microphones laid out as in Figure 3.3-21a.

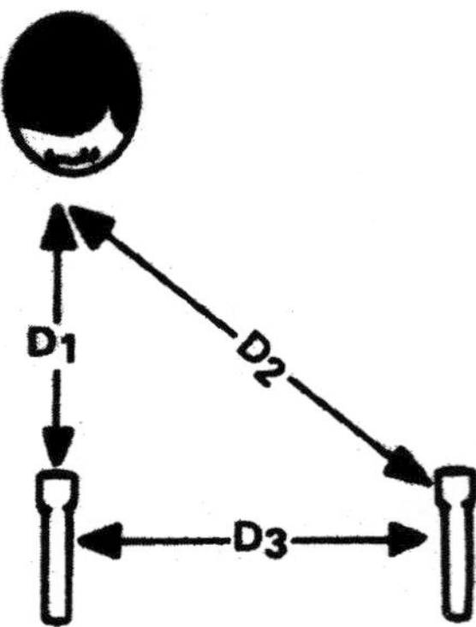

FIGURE 3.3-21a Phase cancellation with multiple microphones.

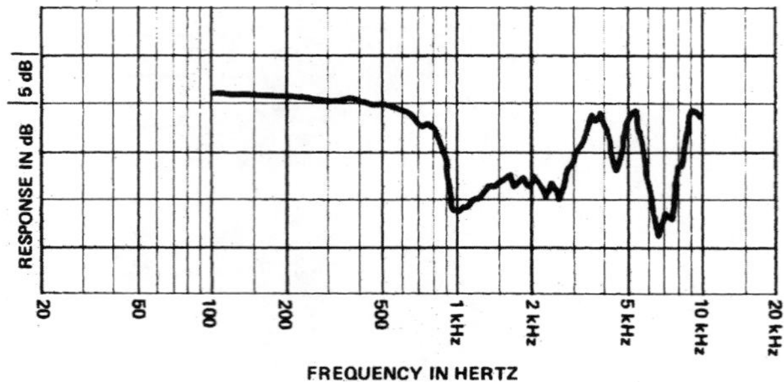

FIGURE 3.3-21b $D_1 = 12''$, $D_2 = 21.6''$, $D_3 = 18''$.

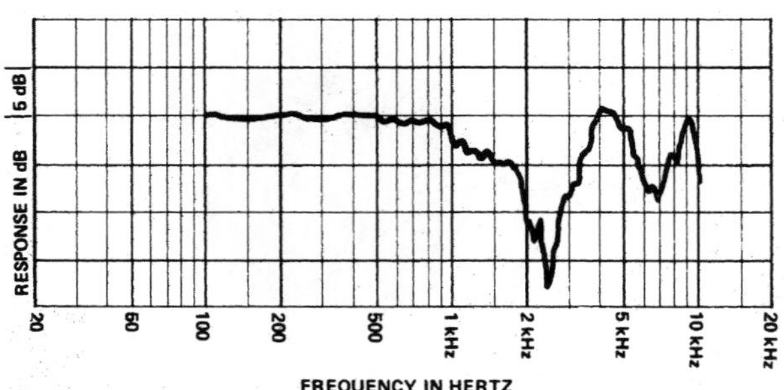

FIGURE 3.3-21c $D_1 = 18''$, $D_2 = 21.6''$, $D_3 = 12''$.

Although sound arriving at each microphone is identical, originating at the same source, it arrives at the two microphones by paths of varying lengths. This causes a difference in the arrival times and results in phase cancellation at certain frequencies. The curves given for each of the examples are Fast Fourier Transform (FFT) derived displays of the actual frequency response of the two microphones combined, with respect to a sound source positioned as shown. The FFT analyzer and its companion microprocessor were used to compare the combined output of two matched, calibrated microphones to the output of one of the two microphones by itself. If no phase cancellation occurred, no trace variations would appear on the X/Y plot, and the plot would be a straight line.

The response charts show that no matter which way the microphones and sound source are oriented, the summed response of the two microphones is poor. These experiments reveal graphically what the ear often perceives as a comb-filter or notch-filter effect that sweeps up and down in frequency (and even changes Q) as the variables D_1, D_2, and D_3 change with the movement of the microphones or sound source. In more subjective terms, the resultant sound may be described as hollow, as if the sound is being forced through an empty cardboard tube.

Situations that cause acoustic phase cancellation arise quite frequently. One classic example occurs with a pair of microphones on a podium, spaced apart to provide on-microphone coverage as the speaker turns his or her head to address all of the audience in front. The curves shown in Figures 3.3-21b and 21d are typical of the problems caused by this approach. If the output of these two microphones is summed and fed simultaneously to a house sound system, feedback problems may occur as well.

The simplest solution to the problems caused by this spaced-pair podium micing technique is to use one microphone only, placing it in front of the person speaking and toward the center of the podium. When redundancy micing is desired for critical applications,

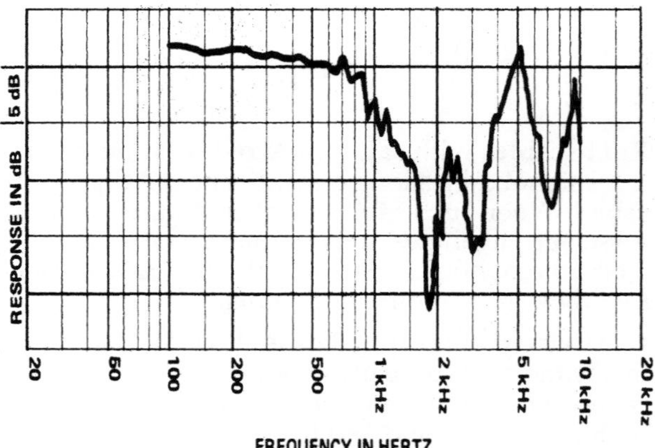

FIGURE 3.3-21d $D_1 = 24''$, $D_2 = 30''$, $D_3 = 18''$.

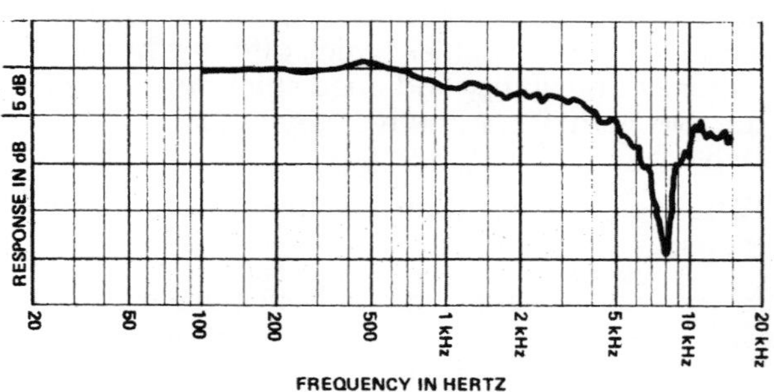

FIGURE 3.3-21e $D_1 = 5.6''$, $D_2 = 6''$, $D_3 = 2''$.

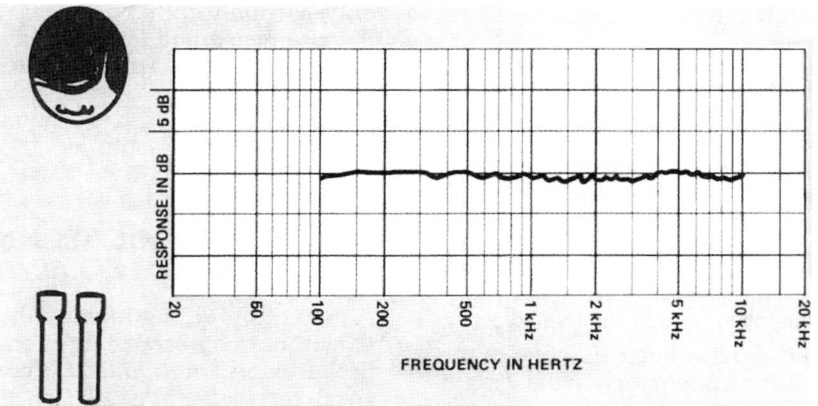

FIGURE 3.3-22 Response of two microphones when colocated for redundancy. (Courtesy of Shure Inc.)

the two microphones should be placed immediately adjacent to each other, preferably one above the other. Figure 3.3-22 shows the response of two adjacent microphones. Normally, only one of these microphones would be open at a time; the second is strictly a backup. Sometimes multiple microphones are used to feed separate systems, such as for house PA, government agencies, and broadcast. Each may still be used as a backup.

Adjacent pairs of cardioid microphones may at times be angled inwardly with their axes crossed and their diaphragms closely spaced. This arrangement will broaden the acceptance angle of the two microphones while still maintaining some cancellation at the rear. The close proximity allows their diaphragms to occupy nearly the same point in space, thus reducing sonic time path differences. This ensures that negligible phase cancellation will occur should their output be summed. The same formation is often used as a two-microphone stereo pickup technique and has the added benefit of good mono compatibility.

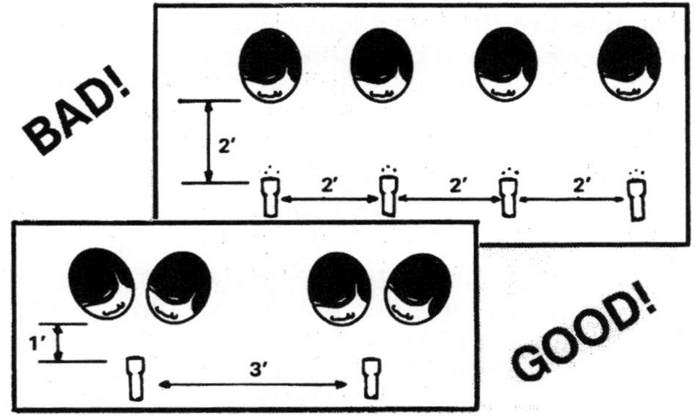

FIGURE 3.3-23 Obeying and violating the 3:1 ratio rule.

There are occasions when the outputs of two or more open microphones must be mixed and, in this case, phase cancellation must be avoided. The problems shown earlier in Figure 3.3-21 can be substantially reduced by employing the 3:1 ratio rule, where D_3 must always be at least three times D_1. Figure 3.3-23 shows examples of compliance with and violation of the 3:1 ratio rule. Subjective tests have shown that an amplitude difference of at least 9 dB between the two signals will reduce phase cancellation to an inaudible level. The 3:1 ratio rule is a means by which this 9 dB minimum difference may be quickly approximated in most multiple-microphone setups.

The amplitude variance desired may also be obtained through the mixer's gain or fader controls. In general, only microphones in actual use should be opened to their normal operating levels; others should be lowered in level or preferably off.

Acoustic Phase Cancellation with a Single Microphone

Acoustic phase cancellation can also occur in a single microphone system when reflected energy from a nearby barrier such as a music stand, podium, table, or floor is introduced at the microphone's diaphragm at a sound pressure level within 9 dB of the direct sound. Such problems may be avoided in several ways:

- Increase the reflected path length.
- Shorten the direct sound path length.
- Reduce the reflectivity of the barrier.

It may be possible to cover the barrier with an acoustically absorptive material or construct it out of an acoustically transparent material. For example, use an acoustically transparent, visually opaque screen in chroma keying to eliminate reflections into a weather person's lavaliere mic.

MICROPHONE POLARITY REVERSAL

Phase cancellation will also occur if the outputs of two microphones, positioned in the same sound field, are combined with their polarities reversed (which could occur if pins 2 and 3 are reversed at one end of one cable). The sound energy from the two microphones will cancel, and the degree to which cancellation will occur depends on how far apart the microphones are spaced, how closely matched their frequency responses are, and the relative levels of the two mixed signals. Note that the terms *phase* and *polarity* do not mean the same thing. Phase refers to a difference in the relative timing of two signals. Polarity refers to the wiring of a microphone or connectors in its circuit and, when reversed, results in a shift of 180° in the phase of the signal. Having noted this distinction, in common usage, the terms *in-phase* and *out-of-phase* are often used to refer to matters of polarity. Most microphones will be wired to what is sometimes called the *XLR pin count,* which is

Pin #1 shield

Pin #2 high (+)

Pin #3 low (–)

This conforms to IEC standards 268-12 and 268-4. Refer also to EIA-221 (paragraph 3.3) in which the in-phase terminal shall be the red (or other than black) conductor and that the out-of-phase terminal shall be the black conductor. The terms *in-phase terminal* or *high* (pin #2) indicate the terminal that has a positive voltage present when a positive pressure is applied to the microphone diaphragm.

Checking Polarity

While there are commercially available devices that use a pulse generator to check for polarity reversal, microphones and their cables may also be checked by simply bringing them together and summing their output while speaking into them from a foot or so away. Two microphones which are "in-phase" will deliver a higher output under such a test; if one is reversed in polarity, the output should drop noticeably.

Polarity Reversal as a Tool

While inadvertent polarity reversal in a microphone line can result in some very bad audio, deliberate polarity reversal is sometimes employed as a problem solver.

Reducing Background Noise

A pair of microphones may be reversed in polarity to reduce the pickup of ambient noise. This technique is sometimes employed with two microphones in fixed locations, such as in a press box at a sporting event. If these microphones are brought together, a noise-canceling or differential microphone is created. The speaker must now talk into one of the microphones only, virtually in a lip-touching position. Because of the inverse square law, the amplitude of the voice at that microphone will be much greater than at the other, resulting in a reasonable output level. Distant sound will be picked up equally well, however, by both elements and canceled.

STEREO MIC TECHNIQUES AND MONO COMPATIBILITY

A goal of providing high-quality stereo audio is to do so without compromising monophonic audio quality. Broadcasters must maintain compatibility with monaural receivers (or stereo receivers operating in the monaural mode and FM receivers that blend to mono as the received signal drops off, etc.).

Spaced-Pair Microphones

The need for mono compatibility normally excludes the use of spaced-pair microphone techniques involving omnidirectional or cardioid microphones. Spaced microphones depend on a combination of amplitude and timing (phase) differences to provide stereo separation. They do not sum well for mono, as the very phase differences that aid in separation result in multiple comb filter effects in the mono mix.

Coincident Microphone Techniques

Coincident microphone techniques utilize two microphones whose diaphragms are placed as near to the same point in space as possible. They offer the potential for good stereo without adversely affecting the mono signal with the phase anomalies introduced by spaced microphones. Coincident microphones depend only on amplitude differences for stereo separation and imaging and provide excellent mono compatibility. There are several coincident microphone schemes including X-Y, M-S, and Blumlein.

X-Y Microphone Technique

The simplest of the coincident techniques is called *X-Y,* which crosses two directional microphones so that their patterns meet at their 3 dB down points, as shown in Figure 3.3-24a. The two microphones should be positioned so that one capsule is directly above the other and on the same vertical axis as shown in Figure 3.3-24b. This minimizes any reflection or shadowing of high frequencies that each might contribute to sound arriving on the horizontal plane.

An ideal cardioid microphone would have an acceptance angle of 131° and so would be 3 dB down at 65.5° off-axis. If the angle is too great, sound sources at the center of the stereo image are placed farther off-axis of each microphone and are thereby attenuated, making them sound as if they are farther away. Similarly, too narrow an angle results in near-center sources sounding louder, or appearing to be closer.

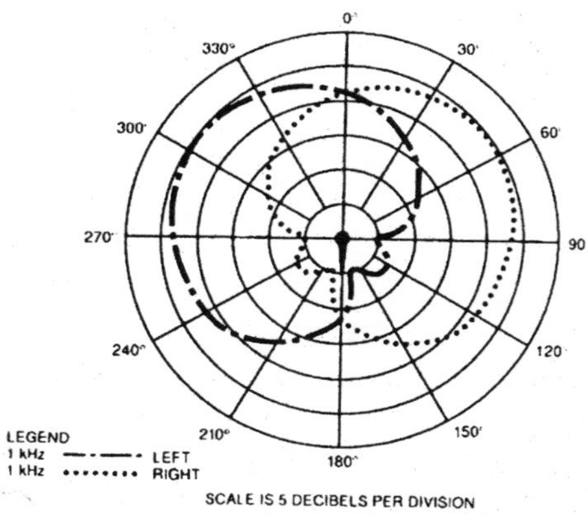

FIGURE 3.3-24a X-Y pattern orientation. (Courtesy of Audio-Technica U.S., Inc.)

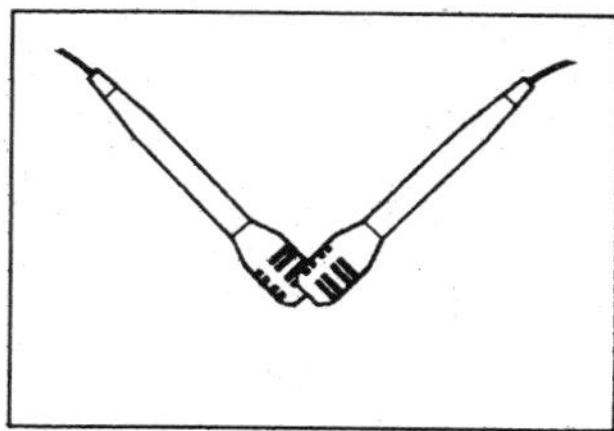

FIGURE 3.3-24b X-Y capsule orientation. (Courtesy of Audio-Technica U.S., Inc.)

Crossing the patterns to overlap at their 3 dB down points ensures that sound arriving from the center of the stereo stage will be summed such as to provide uniform sensitivity from left through center to right.

While a 131° angle would be correct for ideal cardioid capsules, the optimum angle for real microphones will likely be somewhat less. It is important for the microphones used in the X-Y technique to have uniform polar patterns because the patterns of most cardioid microphones tend to collapse at higher frequencies, suggesting that a better X-Y positioning would be 90°. This narrow spacing, however, often results in too much overlap of the patterns. Stereo separation suffers and center-channel information tends to be brought forward of where it should lie in the stereo image.

The optimum angle for many cardioid elements will be approximately 120°. Experimentation and a thorough knowledge of the patterns of the microphones chosen to use will obtain the best X-Y results. Even highly directional shotgun microphones may be used successfully in X-Y, particularly if some of the newer models are selected that have greatly improved polar pattern uniformity. Cross the microphones at the elements, not the ends of the microphones.

Several X-Y stereo microphones are available that integrate two directional elements into one housing that greatly simplifies microphone placement. It may seem obvious that in most cases the axes of the left and right microphones should be near horizontal. Some X-Y microphones, however, hide their capsules in round housings or windscreens that do not permit a quick visual indication of just what is horizontal. It is not good practice to use an X-Y microphone for a close-up announcer or reporter application. Even a slight side-to-side head movement can cause the voice to shift dramatically from one channel to the other.

Testing Mics Outdoors

A quick check outdoors can be made of the angle adjustment of an X-Y pair using the following procedure. Sum the outputs of the two microphones into a mono audio monitor and provide equal gain for each. Feed pink noise into a small powered speaker about 5 to 6 ft in front of the microphones.

Rotate the X-Y pair horizontally at their capsules and monitor the pink noise from far left channel to far right. Pay particular attention to the amplitude at center. There should be a smooth transition from left to right channel. If there appears to be a hole in the middle, the angle is too great. If the noise seems suddenly closer at the center, try increasing the angle.

Mid-Side Coincident Microphones

The most versatile of the coincident microphone types for stereo broadcasting is the M-S, or "mid-side," microphone, whose polar pattern is shown in Figure 3.3-25. The M-S microphone is a combination of a *middle microphone*, typically a cardioid or hypercardioid, and a bidirectional *side microphone*. The capsules of the two are placed as close together as possible. An M-S pair may be constructed using mid and side microphones plus a matrixing system. A matrixing network combines their outputs and decodes them as left and right channel information, as illustrated in Figure 3.3-26. The information derived from the matrix is nearly identical to that delivered by an X-Y pair, but with some important control advantages.

Sound originating from directly on-axis of the M-S microphone will be picked up by the mid element and delivered equally to left and right channels through the matrix. The side microphone, with one lobe facing left, and the other lobe facing right, is insensitive to sound arriving from the center of the stereo stage, as the sound is arriving at 90° off-axis, where the null is deepest. It is, of course, sensitive to sound arriving from each side—a figure of eight pattern.

This is part of the process by which the M-S microphone derives directional cues. Sounds arriving from the left are picked up by the mid and side elements and, because they are in-phase, are summed and sent to the left channel. Because the rear of the side element is out of phase with the mid microphone, their sum cannot be used to produce right channel information. Instead, an inverted-polarity version of the side microphone output is mixed with the mid microphone and delivered to the right channel. This

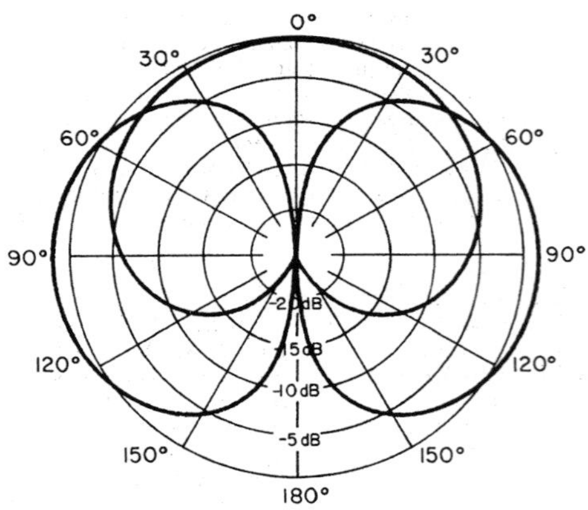

FIGURE 3.3-25 M-S microphone pattern orientation. (Courtesy of Shure Inc.)

processing happens in the sum-and-difference matrix according to the following equations:

$$\text{Left} = \text{mid} + \text{side}$$

$$\text{Right} = \text{mid} - \text{side}$$

Commercially available M-S systems offer well-matched capsules, easy operation, and considerable control flexibility such as a choice of outputs: mid and side or stereo. A mixer may also be used for deriving L/R information from the M-S pair, as shown in Figure 3.3-27.

The M-S technique offers several control capabilities that may be exercised in either production or postproduction. Adjusting the relative levels of the M and the S signals will narrow or broaden the perceived stereo image. This may be done in the field using the M-S microphone's matrix system or by recording the outputs of the M and S capsules on separate tracks, saving the matrixing of them for postproduction. Matrixing in post will allow the audio perspective to be adjusted to make sense with the video image.

Panning the M signal off-center may be done to deliberately shift the stereo image. For example,

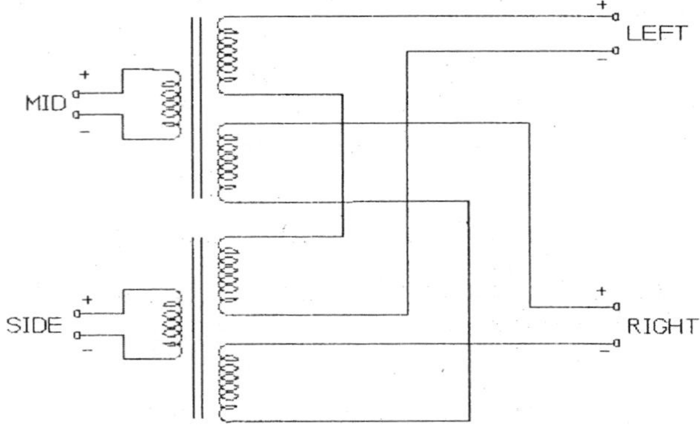

FIGURE 3.3-26 A passive M-S matrix.

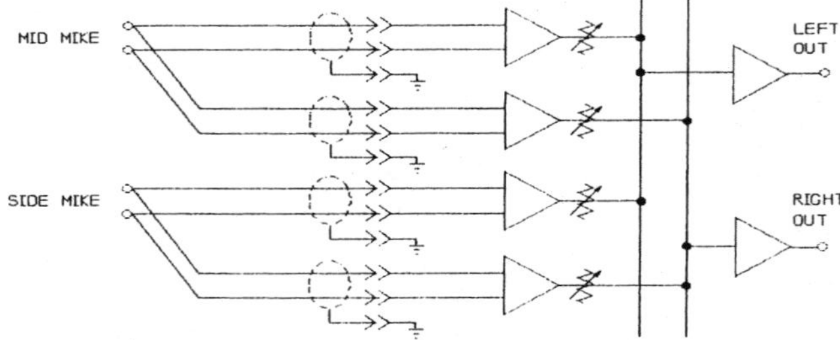

FIGURE 3.3-27 Mixer used as an M-S matrix.

crowd noise at a sporting event may be shifted to appear more closely balanced left and right of a microphone position without moving either the microphone or several thousand fans.

Substituting various patterns, from omni to hypercardioid, for the mid microphone will affect the apparent microphone-to-sound source distance as well as the signal-to-ambient or reverberant noise ratio.

Blumlein Technique

The Blumlein micing technique[3] employs coincident crossed bidirectional elements and, like the M-S, responds to amplitude differences to achieve stereo separation. The sound source is placed between the 90° arc of the front capsules of the two bidirectional elements. The stereo sound achieved by this approach can be quite natural, and mono integrity is well maintained. The Blumlein technique is more sensitive to ambient noise and reverberation than the M-S, and placement is critical.

MAINTENANCE

- Microphones require a certain amount of care in their handling and storage. Misuse, or even some attempts to service or clean the microphone, could affect some manufacturers' warranties. When in doubt, ask or return microphones to the manufacturer's recommended service organization for maintenance.
- Use a windscreen or pop filter to protect the microphone if it is to be subjected to airborne contaminants such as dust or smoke.
- A foam windscreen will also protect a microphone from exposure to rain or snow. Over time, the cells will fill with water, resulting in high-frequency loss and level drop. The foam may be squeezed to

[3]The Blumlein difference technique employs two microphones aimed forward and separated by the width of the human head with an acoustic absorption material in between. It was invented by British sound engineer Alan Blumlein in 1931. For more information on the technique, see http://homepage.ntlworld.com/chris.burmajster/Blumlein.htm. Blumlein is also credited with the invention of stereo sound and many other audio and recording techniques. See http://www.doramusic.com/blumlein.htm.

reduce the moisture content or a dry screen substituted as required.

- Foam windscreens will accumulate deposits of dust and other contaminants. The result will be a deterioration of frequency response and, perhaps, even altered polar response. Foam may be cleaned with soap and water. Rinse well to remove all residue. Nondetergent soaps work well.
- Many microphones may be carefully opened to remove a foam pop filter and sometimes a cloth insert. Do so only in a very clean environment. These filters should be cleaned as detailed previously.
- Avoid placing microphones on workbenches or other areas where metal particles or metallic dust may be attracted to their internal magnet structures. Very small metal particles can work their way onto the diaphragm and alter the response greatly. In some cases, the dynamic microphone can be opened to reveal the diaphragm for examination. Metallic particles may be very carefully removed onto the magnetized tip of a screwdriver. The screwdriver shaft should be steadied on the edge of the microphone case and the tip very carefully lowered to attract particles which would likely be held immediately above the voice coil gap.
- Avoid subjecting electret condenser microphones to high temperatures such as in the trunk or glove compartment of a car left in the sun on a hot day or on a boom very close to hot lights. The result may be a loss of charge on the capacitor element and a drop in level.
- Avoid moisture with all microphones but especially with condenser microphones.
- If given a choice between using mercury or alkaline batteries to power a microphone, note that mercury cells die much more suddenly than alkalines. The gradual drop in level with an alkaline battery may save a production. Mercury batteries also drop in output level in cold weather and may give off a gas that can corrode the contacts.
- Avoid unnecessary mechanical shocks. Store in clean, padded enclosures.
- Moving a condenser microphone from a cold environment to a warm one may cause noise problems from condensation.
- Avoid moisture in cables and connectors, particularly where phantom power is being used.

Bibliography

Abbagnaro, Louis. "Microphones," AES Anthology Series, Audio Engineering Society, New York.

Alexander, Robert Charles. *The Inventor of Stereo, the Life and Works, of Alan Dower Blumlein*, Focal Press, 1999.

Burroughs, Lou. *Microphones: Design and Application*, Sagamore Publishing Company, Inc., 1974.

Davis, Don, and Carolyn Davis. *Sound System Engineering*, Second edition, Focal Press, 1997.

Eargle, John. *Microphone Book*, Focal Press, 2004.

Eargle, John. *Sound Recording*, Van Nostrand Reinhold Co., New York, 1976.

Ford, Ty. *Advanced Audio Production Techniques*, Focal Press, 1993.

Ford, Ty. *Audio Bootcamp Field Guide*, Second edition, http://www.tyford.com.

Long, James. "Layman's Guide to Microphone Specifications," *Audio Magazine*, August, 1969.

Long, James. *The Microphone Handbook*, John Eargle, ed., Elar Publishing Company, Inc.

"Microphones: An Anthology of Articles on Microphones." *Journal of the Audio Engineering Society*, AES, New York.

Olson, Harry. *Modern Sound Reproduction*, Van Nostrand Reinhold Co., New York.

Sank, Jon R. "Microphones," *Journal of the Audio Engineering Society*, AES, New York, July/August 1985, vol. 33, no. 7/8.

Tremaine, Howard M. *Audio Cyclopedia*, Howard W. Sams & Co., Indianapolis, IN, 1974.

Woram, John M. *The Recording Studio Handbook*, Elar Publishing Company, Inc., 1982.

CHAPTER

3.4

Audio Recording

MICHAEL STARLING
National Public Radio
Washington, D.C.

JOHN EMMETT
Broadcast Project Research Ltd.
Teddington, United Kingdom

INTRODUCTION

Anyone who has been away from the field of audio recording for a number of years may be surprised by the range of technologies that are discussed in this chapter. That is not to say that the process of sound recording has changed fundamentally, but the specific recording equipment has changed and, in many cases, even the physical media might appear to have vanished. In large part, audio has followed text into becoming a generic Information Technology (IT) data application.

For much of the 20th century, analog magnetic tape was the broadcaster's principal medium for audio recording. In the 1980s and 1990s, the industry focused on the standardization and implementation of digital recording techniques using magnetic and optical media. Recent years have seen the near-completion of the migration to digital systems, including solid-state storage with no moving parts and the continuing refinement of hard-disk-based storage technologies.

This chapter covers a wide range of recording technologies and devices both past and present. It includes material originally published in previous editions of the *NAB Engineering Handbook*[1] on both analog and digital recording, followed by new sections that discuss the migration to IT-based recording.[2] In-depth coverage of computer-based audio recording systems may be found in Chapter 3.6, "Radio Station Automation, Networks, and Storage."

[1]Mike Starling, "Audio Recording Systems," *NAB Engineering Handbook*, 9th edition, National Association of Broadcasters, 1999, pp. 321–340.

[2]John Emmett, "Sound Recording," *Broadcast Engineers Reference Book*, Focal Press, 2004, pp. 599–607. Used with permission.

HISTORY

Until about 130 years ago, all sound was *live* sound. Despite the developments of mechanical recording, it was only with the harnessing and commercial use of electricity that amplification and broadcasting could be developed, and that subsequently led to a wider need to store and reproduce sound waves. It also offered the tools that were needed to develop ever more sophisticated recording systems.

Mechanical Recording

Audio recording preceded and helped fuel the introduction of broadcasting. The earliest recorded audio was Thomas Edison's 1877 mechanical cylindrical *phonograph* employing a constant velocity vertical recording groove. The phonograph's cylindrical media mandated that each recording be a master and stymied mass production.

In 1887, Emile Berliner patented a successful system of sound recording on flat discs. The first discs were made of glass, later zinc, and eventually plastic. An acoustic horn was used to collect the sound, which was converted into vibrations recorded in a spiral groove that was etched into the disc. For replay, the disc rotated on a turntable while a pickup arm held a needle that read the grooves in the record. The vibrations in the needle produced the sound that was transmitted mechanically to the horn speaker, the whole machine being known as a *gramophone*.

The flat disc recordings could be duplicated by creating molds from the master recordings, allowing copies to be mass-produced. Berliner's discs, which became known as *records*, dominated recorded audio for the next half-century.

In the 1920s, the technology for disc recording improved greatly with the introduction of microphones and electronic amplifiers to drive an electromagnetic cutting head on a *lathe* to produce the master disc, instead of relying on the acoustic horn.

Early broadcast use of recorded media exploded in the late 1920s. This development coincided with the rapid proliferation of AM broadcasting. Among the first actions of the Federal Radio Commission in 1928 was the deletion of several stations due to their heavy reliance on airing commercial records, which the FRC cited as "provision of a service which the public can readily enjoy without the service." The new FRC favored original programming, and this stimulated the use of disc recording lathes by the burgeoning population of radio stations. While many, if not most, early broadcast facilities acquired recording lathes for production of recorded audio, the widespread introduction of the more forgiving and affordable magnetic tape recording did not begin until after World War II.

Magnetic Recording

Danish telephone engineer Valdemar Poulsen demonstrated a magnetic wire recorder as early as 1898, as shown in Figure 3.4-1. It was 30 years later that German researchers pioneered magnetically coated, paper-based tape for good-quality recorders/reproducers.

By 1936, German scientists had advanced magnetic-based recording using cellulose-based tape and achieved remarkably good sound quality.

After the war, the AKG Magnetophone (see Figure 3.4-2) was copied and commercially exploited worldwide. A host of benefits compared to disc-based recording, including portability, immediacy of playback, ease of storage, wide dynamic range, low distortion, and freedom from ticks and pops, propelled magnetic recording to the forefront in broadcasting.

Principles of Magnetic Recording

Magnetic audio tape recording uses a tape composed of a plastic base material, with a thin coating or emulsion of *ferric oxide* powder. This is a ferromagnetic material, meaning that if exposed to a magnetic field, it becomes magnetized by the field.

The tape recording device uses a *record head* that applies a *magnetic flux* to the oxide on the tape as it is moved past the record head at a constant speed. The oxide then responds to the flux as it passes. The record head is a small, circular electromagnet with a small *gap* in it. During recording, the audio signal is sent to the record head to create a magnetic field in the core. At the gap, magnetic flux forms a fringe pattern to bridge the gap, and this flux is what magnetizes the oxide on the tape. During playback, the motion of the tape pulls a varying magnetic field across the gap in the *playback head* (which may be the same head used for recording, or a different head). This creates a varying magnetic field in the core and therefore a signal in the coil of the electromagnet. This signal is amplified to drive the speakers.

FIGURE 3.4-1 Valdemar Poulsen's Telegraphone won a Grand Prix Award at the 1900 Paris World's Fair. The device used pole pieces located on each side of the wire. (From Finn Jorgensen, *The Complete Handbook of Magnetic Recording*, 4th edition, 1995, McGraw-Hill. Reproduced with permission of the McGraw-Hill Companies.)

FIGURE 3.4-2 Early German Magnetophone brought to the USA after capture by officers in the U.S. Army Signal Corps. (From Finn Jorgensen, *The Complete Handbook of Magnetic Recording*, 4th edition, 1995, McGraw-Hill. Reproduced with permission of the McGraw-Hill Companies.)

From the earliest days, various techniques were used to improve the quality of the recorded analog signal. They include adding a high-frequency *bias* signal to the signal applied to the record head; this improves the linearity of the magnetic recording process and greatly reduces *distortion* levels. Another technique is to provide record and playback audio *equalization* that improves frequency response and signal-to-noise ratio.

Further Developments

Numerous variants of the analog magnetic tape recorder were developed for different applications, all based on the same basic recording principles. These variants included different arrangement of the reels or cassettes, as shown later in the chapter. A range of magnetic emulsion formulations were developed with features specific to the application. In some cases, chromium dioxide or metal particles were used rather than ferric oxide. Machines with different tape speeds, tape widths, track widths, bias, equalization, and level standards were all developed for specific purposes. There were decades of incremental refinements in frequency response, signal-to-noise, print-through, emulsion composition, backing media, lubrication, and adhesive composition, all of which interact with one another and require trade-offs depending on the intended purpose of the recording device. From the 1960s onward, various techniques for *noise reduction* were introduced, including the well-known systems from Dolby Laboratories.

SOUND RECORDING FORMATS

Because of the long history of sound recording, broadcast sound operators may be presented for many years to come with source recordings in many and varied forms, so the most significant are discussed in the following sections.

In order to recognize the majority of sound recording formats for what they are, and especially with all the languages in European broadcasting, the European Broadcasting Union (EBU) has worked on the International Broadcast Tape Number (IBTN) scheme, and an associated bar-code label specification given in the EBU document Tech 3279.

The IBTN scheme can be applied to any broadcast tape and related items and enables them to be uniquely identified, from the earliest stages of the production process. The bar-code representation of the IBTN allows broadcast tapes to be scanned as they move from production facilities to broadcasting outlets and during transfers between broadcasters.

For convenience, recording formats can be divided into those carrying analog or digital signals and then subdivided by the type of mechanical carrier. Table 3.4-1 provides an extract of the most common IBTN sound recording format codes, from which these subdivisions can be made.

The formats listed in Table 3.4-1 currently form the bulk of broadcast industry sound archives.

TABLE 3.4-1
The Most Common International Broadcast Tape Number (IBTN) Scheme Sound Recording Format Codes

Code	Material
16T	16 mm sepmag analog audio film
17T	17.5 mm sepmag analog audio film
33L	33 rpm LP phonogram analog audio disc
35T	35 mm sepmag analog audio film
45D	45 rpm phonogram analog audio disc
78D	78 rpm phonogram analog audio disc
A01	6.3 mm (¼") analog audio tape, full track
A02	6.3 mm (¼") analog audio tape, 2-channel
A04	6.3 mm (¼") track half-width analog audio tape, stereo
A08	12.5 mm (¼") analog audio tape, 8-channel
A16	25.4 mm (1") analog audio tape, 16-channel
A32	25.4 mm (1") analog audio tape, 32-channel
AI1	AIT (Advanced Intelligent Tape) digital data tape, 25 GB capacity
AI2	AIT (Advanced Intelligent Tape) digital data tape, 50 GB capacity
AI3	AIT (Advanced Intelligent Tape) digital data tape, 36 GB capacity
AIX	AIT (Advanced Intelligent Tape) digital data tape, extended length, 35 GB capacity
AS2	6.3 mm (¼") analog audio tape, 2-channel stereo
AT2	6.3 mm (¼") analog audio tape, 2-channel stereo and TC
CCA	Compact Cassette format analog audio tape, cassette
CDA	Compact Disc Audio digital audio disc
CDD	CD-ROM digital data disc
CDR	Recordable CD digital data disc
D24	25.4 mm (1") DASH format digital audio tape, 24-track
D32	25.4 mm (1") PD format digital audio tape, 32-channel
D48	25.4 mm (1") DASH format digital audio tape, 48-track
DA2	DAT format digital audio tape, 2-channel
DAT	DAT format digital audio tape, stereo
DCC	DCC format digital audio tape
DD2	6.3 mm (¼") DASH format digital audio tape, 2-channel
DP2	6.3 mm (¼") PD format digital audio tape, 2-channel

TABLE 3.4-1 *(continued)*
The Most Common International Broadcast Tape Number (IBTN) Scheme Sound Recording Format Codes

Code	Material
H8A	Hi-8 format 8-channel digital audio tape, cassette
LAQ	Lacquer phonograph analog audio disc
MDA	MD (MiniDisc) digital audio disc
NAB	NAB cartridge analog audio tape
SVA	A-DAT 8-channel digital audio tape
WAX	Wax cylinder phonogram analog audio disc

ANALOG RECORDING FORMATS

Quarter-Inch and Cassette Tape Formats

The media for most common professional analog audio tape formats are shown in Figure 3.4-3. On the left is a typical analog radio station mainstay, the 10-1/2-inch diameter "NAB" center 1/4-inch tape reel providing one hour of recording time at 7-1/2 ips. Alongside is a full 7-inch reel of Ampex 600 tape providing 1/2 hour of recording time. Run time is reduced by half for 15 ips speed. Portable 1/4-inch tape recorders for field reporting could also record at 3-3/4 ips, giving longer record times with smaller tape reels. Specialized recorders for station logging recording could run at speeds as low as 1-7/8 ips or even 15/16 ips to give extra long record times at reduced quality.

The NAB standardized endless loop cartridge, or *cart*, on the right in Figure 3.4-3, was the staple contribution format for radio station commercials and inserts from the 1960s until the 1990s. This format used 1/4-inch tape with a lubricated backing, usually running at 7-1/2 ips. The center cue track carried several automation tones, which could be used to cue and trigger cart players in order to automate commercial breaks. Distributed tapes were usually recorded on open-reel recorders, or duplicators, and then physically transferred to the cartridge for distribution. For archival or restoration purposes, the tape can be extracted for replay on an open reel recorder, and this is to be preferred for transfer, as the tape path control in cart players is necessarily poor. The format is still in use by some broadcasters, but for quality and reliability reasons, there has been a steady migration to digital storage such as the MiniDisc and, later, hard disk and solid-state storage.

FIGURE 3.4-3 The main analog tape media.

In the background is a *compact cassette*, which typically recorded 45 minutes per side stereo at 1-7/8 ps. These cassettes were initially developed by Sony and Philips and introduced for consumer recording in 1962, but from the late 1960s, and with the use of noise-reduction techniques, they became increasingly popular as rugged interview recorders for professional users. More recently, they were superseded by MiniDisc and solid-state recorders.

For many years, the workhorse of broadcast sound recording was the 1/4-inch open-reel recorder, of which an example is shown in Figure 3.4-4.

In dealing with legacy analog tape program material that has to be played back, one first needs to establish the tape speed and track layout of the original recording so a suitable playback machine can be found. The record characteristics for the tape should also be determined if possible, although this may be more difficult to establish. High-frequency and (sometimes) low-frequency equalization is applied during analog tape recording to compensate for nonuniform response in the head-tape system. Equalization has to be applied during playback to achieve a flat overall frequency response. The characteristic was standardized in the United States according to NAB-published criteria and by CCIR and IEC in Europe.

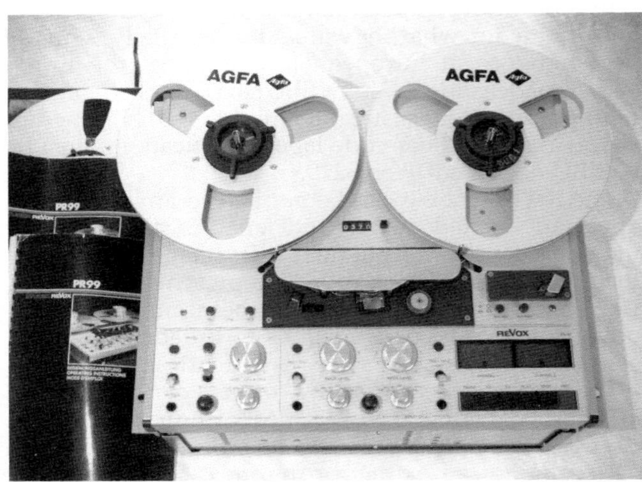

FIGURE 3.4-4 Analog 1/4-inch tape recorder. Mounting a recorder that would take 10.5-inch spools into a 19-inch rack was always a challenge, and the Revox PR99 was one machine with this capability.

TABLE 3.4-2
The IEC and NAB Record Equalization Characteristics for 1/4-Inch and Cassette Tape

	CCIR/IEC Time Constants		NAB Time Constants	
Tape Speed Use	**Bass Boost**	**High Roll-Off**	**Bass Boost**	**High Roll-Off**
15 ips Studio	none	35 µs	3180 µs	50 µs
7-1/2 ips Studio	none	70 µs	3180 µs	50 µs
7-1/2 ips Studio carts	none	50 µs	none	50 µs
3-3/4 ips Reporter & home	3180 µs	90 µs	3180 µs	90 µs
1-7/8 ips Logging & home	3180 µs	120 µs	3180 µs	90 µs
1-7/8 ips Fe-cassette	3180 µs	120 µs		
1-7/8 ips Cr-cassette	3180 µs	70 µs		

Table 3.4-2 lists the most common recording equalization characteristics for 1/4-inch tapes. The break frequencies of the filters are described by the time constants of simple RC low-pass networks. If a replay machine with playback settings matching the tape cannot be found, the correction equalization can be quite easily applied, post replay, inside a digital audio workstation.

The 1/4-inch analog tapes recorded with center time code or sync tracks for video synchronization require a special replay machine that may be increasingly difficult to find.

Multitrack Tape and Sepmag Film Recording Formats

Multitrack analog audio tapes, especially in the 2-inch wide 24-track form, still form a major interchange format within recording studios for remixing archive music sessions, but in broadcast use these tapes are rare.

Sprocketed analog magnetic film, known as separate magnetic (*sepmag*) recordings, was for many years associated with the huge quantities of 16 mm broadcast film material used for television. There are few variations in playback standards, although any sepmag transfer will require a specialized playback deck.

MAINTENANCE, CARE, AND STORAGE OF MAGNETIC TAPE RECORDINGS

Although many analog recordings have held up in good condition for decades, they are quite sensitive to permanent physical damage from improper handling, machine malfunction, and environmental hazards. Winding tapes *tails out* immediately after complete playback is the most important safeguard in preventing physical edge damage to audiotapes. Environmental damage is most directly safeguarded by cleanliness and carefully controlling temperature and humidity

Tape wind or pack must be even to prevent protrusion scatter between layers that will crease and permanently damage tape edges during subsequent playback. Scatter-wound tape is susceptible to edge damage from the pressure exerted on flanges during careless handling. For this reason, reels should be handled by their hubs rather than by the flanges. Similarly cinching of layers with actual foldover is possible during rapid acceleration/deceleration from jerky transport operation.

Many professional recorders have a *library wind* mode that operates at a higher than normal operating speed but with constant tension to assure a smooth pack. Tape libraries invariably have professional tape winding equipment that is optimized for gentle handling during higher speed precision winding. At professional libraries, preventive maintenance includes periodic rewinding to minimize print-through and depletion of lubrication, and to interrupt *stiction* buildup from adhesive action. The recommended period between rewindings varies greatly with storage conditions. Tapes stored at 20°C should be rewound every 3000 hours. Tapes stored at 30°C should be rewound roughly every 300 hours.

Minute particles can cause serious system degradation. Static buildup, scraping, scratching of the tape surface, and separation in pack and head contact can cause dropouts and permanent damage to tape and equipment. Thus, frequent cleaning of heads, guides, capstan, and pinch roller, typically after each recording or playback session, is imperative. Careful demagnetization of heads is also required for best performance, at intervals depending on the number of hours of operation. Oils and salts from fingerprints will attract foreign particles and can themselves interfere with reliable head-to-tape interface.

Hydrolysis is a chemical reaction with water that affects polyester-based recording tapes. High temperature and high humidity will accelerate hydrolysis reactions in any polyester-based tape stock. However, from roughly 1977–1983 industrywide polyester binder phenomena, referred to as *sticky-shed* syndrome, exacerbated the rate of hydrolysis reactivity.

Tapes from the sticky-shed era typically exhibit slip-stick phenomena as carboxylic acid and alcohol are sloughed from the binder as debris products. Tapes of this vintage are frequently unusable due to residue buildup that causes transports to squeal and bind. Fortunately, this phenomenon has been extensively documented and can be reversed temporarily with no apparent damage to the tape recording. The reversal process consists of warming (or *baking*) the tapes in a convection oven at 120°F for 24 hours. The tapes will then be usable upon cooling for several weeks before hydrolysis again sheds sufficient

TABLE 3.4-3
Recommended Magnetic Tape Storage Conditions

Storage Temperature	Maximum Relative Humidity
50°C	39%
40°C	47%
30°C	60%
20°C	79%
10°C	100%

amounts of debris to interfere with transport functioning. Recommended humidity and temperature conditions for storage are shown in Table 3.4.3.

DIGITAL AUDIO PRINCIPLES[3]

Before discussing the methods of recording digital audio, it is instructive to review some of the principles of digital audio itself.

Using the principles of discrete time sampling and quantization, a sampled signal can be processed, transmitted, or stored and, through conversion, can reconstruct an accurate representation of the original analog signal. This is how a *pulse-code modulation* (PCM) digital representation is created, but it is of interest to consider that there are many other ways of digitally representing audio. Certainly, Fourier Transforms, Walsh functions, and a number of other representations have been tried in the past, and are capable of equally high quality performance as PCM. The reason that PCM is so universal today is probably down to the ease of processing, along with the relative simplicity and low *latency* (delay) of the analog-to-digital (A/D) converters.

Discrete Time Sampling

An analog waveform such as an acoustic pressure function in air exists continuously in time over a continuously variable amplitude range. Such an analog function may be discrete time sampled; moreover, the sample points can be used to reconstruct the original analog waveform. This digitization of audio forms the basis for the encoding and decoding of the audio signals in any digital audio format.

The Nyquist theorem states that given correct, band-limited conditions, sampling can be a lossless process. However, the relationship between sampling frequency and audio frequencies must be observed. The Nyquist theorem defines the relationship: if the sampling frequency is at least twice the highest audio frequency, complete waveform reconstruction can be accomplished.

The choice of sampling frequency determines the frequency response of the digitization system; S samples per second are needed to represent a waveform with a bandwidth of S/2 Hz. As the sampled frequencies become higher, given a constant sampling rate, there will be fewer samples per period. At the theoretical limiting case of critical sampling, at an audio frequency of half the sampling frequency, there will be two samples per period. A low-pass filter is placed at the output of every audio digitization system to remove all frequencies above the half-sampling frequency. This is required because sampling, through modulation, generates new frequencies above the audio band. The output filter removes all spectra above the half-sampling frequency. This is summarized in Figure 3.4-5.

By definition, audio samples contain all the information needed to provide complete reconstruction. However, band-limiting criteria must be strictly observed; a too high frequency would not be properly encoded and would create a kind of distortion called *aliasing*. An input frequency higher than the half-sampling frequency would cause the digitization system to alias. If S is the sampling frequency and F is a frequency higher than half the sampling rate, then new frequencies are also created at S ± F, 2S ± F, 3S ± F, and so on. An input low-pass filter will prevent aliasing if its cutoff frequency equals the half-sampling frequency. To achieve a maximum audio bandwidth for a given sampling rate, filters with a very sharp cutoff characteristic, *brick wall filters,* are employed in either the analog or digital domain.

Amplitude Quantizing

The amplitude of each sample yields a number that represents the analog value at that instant. By definition, an analog waveform has an infinite number of amplitude values; however, quantization selects from a finite number of digital values. Thus, after sampling, the analog staircase signal is rounded to a numerical value closest to the analog value. The difference between the original values of the signal and values after quantization appears as an error.

The number of quantization steps available is determined by the length of the data word in bits; the number of bits in a quantizer determines resolution. Sixteen bits yields 2^{16} = 65,536 increments. Every added bit doubles the number of increments; hence, the magnitude of the error is smaller. The accuracy of a quantizing system provides an important performance specification. In the worst case, there will be an error of one-half the least significant bit of the quantization word. The ratio of maximum expressible amplitude to error determines the signal-to-error (S/E) ratio of the digitization system. The S/E relationship can be expressed in terms of word length as S/E (dB) = $6.02n + 1.76$ where n is the number of bits.

Although a 16-bit system would yield a theoretical S/E ratio of 98 dB, as the signal amplitude decreases, the relative error increases. Consider the example of a

[3]This section is adapted from a contribution by Ken C. Pohlmann, *NAB Engineering Handbook,* 8th edition, National Association of Broadcasters, 1992, pp. 863–875.

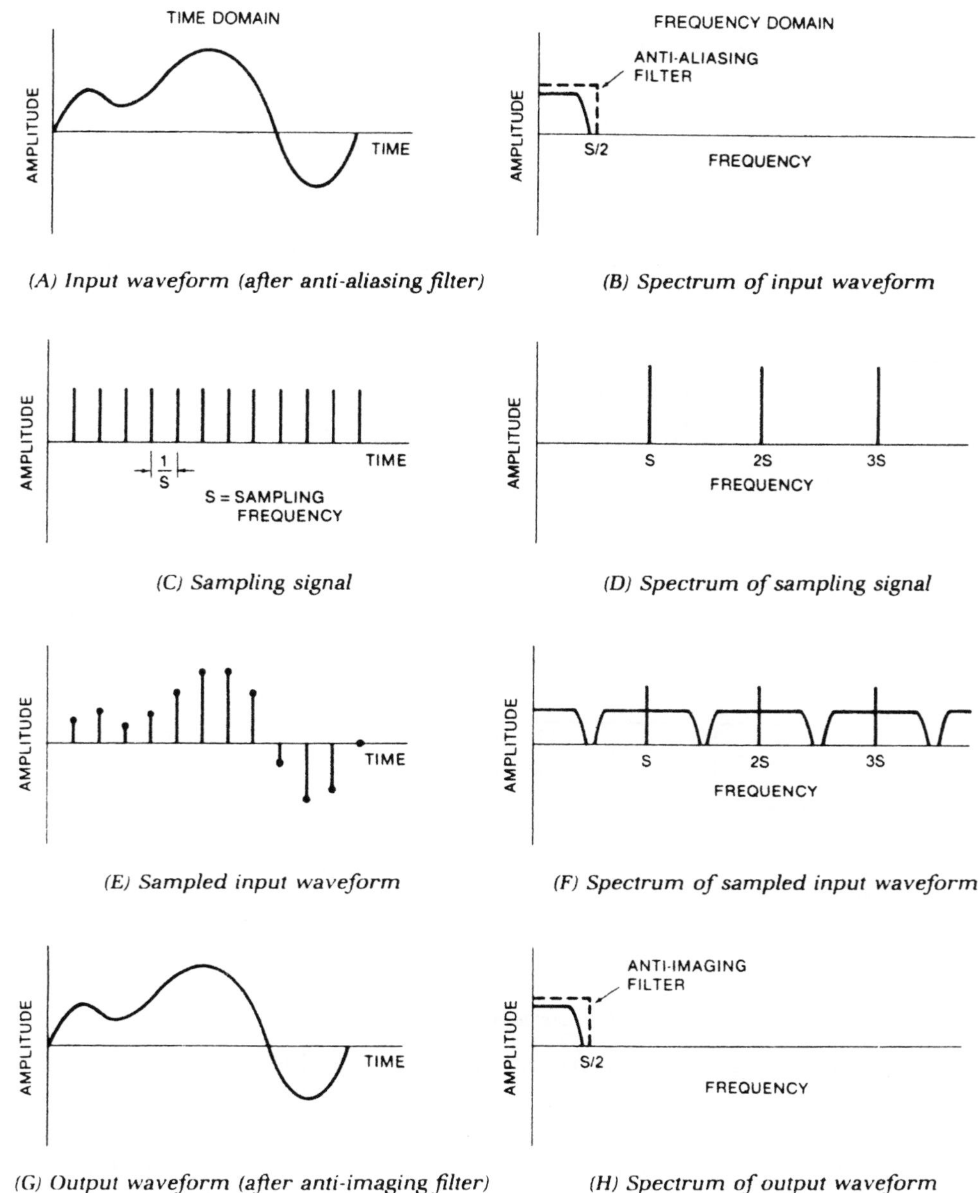

FIGURE 3.4-5 Summary of discrete-time sampling, shown in the time and frequency domains. (From Ken C. Pohlmann, *Principles of Digital Audio.*)

signal with amplitude on the order of one quantization step. The signal value crosses back and forth across the threshold, resulting in a square wave signal from the quantizer. *Dither* suppresses such quantization error. Dither is a low-amplitude analog noise added to the input analog signal (similarly, digital dither must be employed in the context of digital computation when rounding occurs).

When dither is added to a signal with amplitude on the order of a quantization step, the result is duty-cycle modulation that preserves the information of the original signal. The average value of the quantized signal can move continuously between two steps; thus, the incremental effect of quantization has been alleviated. Audibly, the result is the original waveform, with added noise. That is more desirable than the clipped quantization waveform. With dither, the resolution of a digitization system is below the least significant bit.

The recording section of a pulse-code modulation (PCM) system, shown in Figure 3.4-6(a), consists of input amplifiers, a dither generator, input (antialiasing) low-pass filters, sample-and-hold circuits, analog-to-digital converters, a multiplexer, digital processing circuits for error correction and modulation, and a storage medium such as digital tape. The reproduction

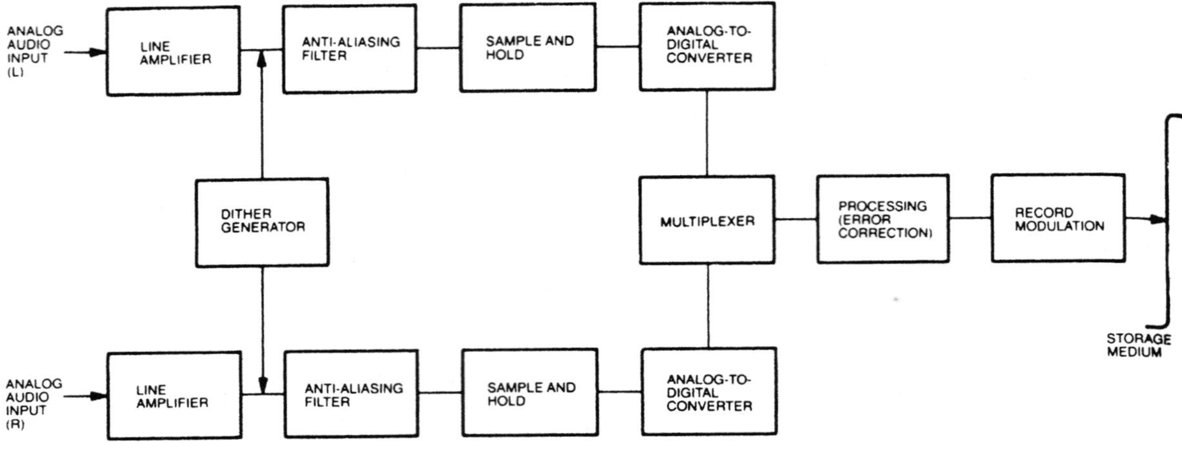

(A) *Recording section*

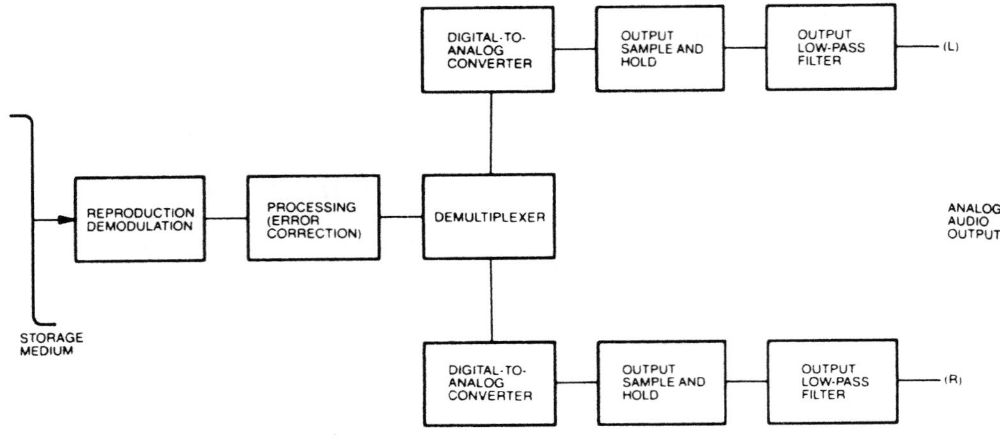

(B) *Reproduction section*

FIGURE 3.4-6 Block diagram of the recording (a) and reproduction (b) sections of a linear PCM system. (From Pohlmann, *Principles of Digital Audio*.)

section, shown in Figure 3.4-6(b), contains processing circuits for demodulation and error correction, a demultiplexer, digital-to-analog converters, output sample-and-hold circuits, output (anti-imaging) low-pass filters, and output amplifiers. In most contemporary designs, digital filters are used in both the input and output stages. The output section forms the basis for a compact disc player.

DIGITAL AUDIO RECORDING SYSTEMS[4]

Removable Media for Digital Recordings

Digital audio can be recorded using a wide variety of optical, magnetic, magneto-optical, and solid-state media. The media for most common professional audio tape formats are shown in Figure 3.4-7.

In the 1980s, rotary head developments from analog video recording and the low importance of print-through for digital recordings led to a big jump forward in sound recording density. The stereo R-DAT tape on the left in Figure 3.4-7 is essentially a miniaturized video tape cassette using 4 mm wide tape, while for the multitrack DA-88, introduced in 1992, the tape was 8 mm wide and actually used the same cassette as the Hi-8 consumer analog video recorder. The compact disc (CD)-sized, 120 mm diameter, optical disc, first introduced as the audio CD in 1983, can carry any one of a number of recorded formats. The optical discs had an immediate advantage of quick access, rather than lengthy tape spooling, but that meant that a discipline of recording *metadata* was needed. The 80 mm disc in the caddy is uniquely a MiniDisc for recording stereo audio, introduced in 1992.

[4]This section is adapted from a contribution by Ken C. Pohlmann. *NAB Engineering Handbook*, 8th edition, National Association of Broadcasters, 1992, pp. 863-875, with new additions.

FIGURE 3.4-7 The most-used digital recording media, left to right: R-DAT tape, compact disc, MiniDisc.

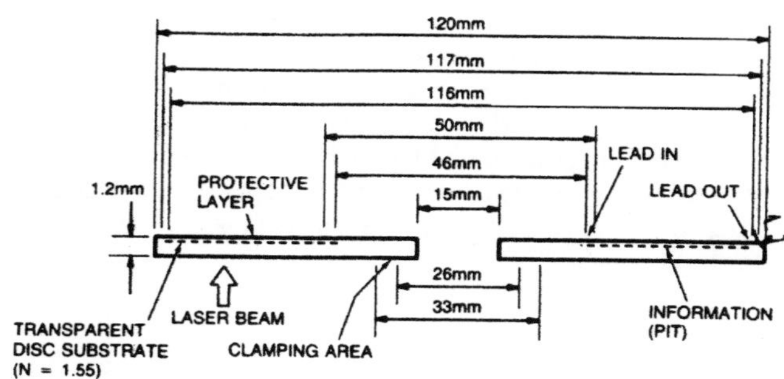

FIGURE 3.4-8 Compact disc physical specifications. (From Ken C. Pohlmann, *The Compact Disc.*)

The various devices and technologies that utilized these media are discussed next.

Compact Disc

The compact disc was developed to store up to 74 minutes of stereo digital audio program material of 16-bit PCM data sampled at 44.1 kHz. The total user capacity is over 650 MB. In addition, for successful storage on a nonperfect medium, error correction, synchronization, modulation, and subcoding were required. The CD was originally conceived as a distribution medium to replace vinyl records, but its high quality of uncompressed digital audio firmly established it as the medium of choice for playback of prerecorded music in the professional domain. Later variants were introduced for recording and playing back both data and audio.

Although the audio CD has now to some extent been superseded by more recent optical developments such as the DVD family which have come out of it and, for professional purposes, the streaming format used for recording has been largely replaced by file-based audio, a study of the design process is common to all optical storage systems and provides a good insight into data storage in general.

Compact Disc Physical Design

The diameter of a compact disc is 120 mm, its center hole diameter is 15 mm, and its thickness is 1.2 mm. Data are recorded in an area 35.5 mm wide. It is bounded by a lead-in area and a lead-out area, which contain nonaudio subcode data used to control the player's operation. The disc is constructed with a transparent polycarbonate substrate. Data are represented by pits that are impressed on the top of the substrate. The pit surface is covered with a thin metal (typically aluminum) layer 50–100 nm thick, and a plastic layer 10–30 μm thick. A label 5 μm thick is printed on top. Disc physical characteristics are shown in Figure 3.4-8.

Pits are configured in a continuous spiral from the inner circumference to the outer. The pit construction of the disc is diffraction-limited; the dimensions are as small as permitted by the wave nature of light at the wavelength of the readout laser. A pit is about 0.5 μm wide. The track pitch is 1.6 μm. There is a maximum of 20,188 revolutions across the disc's data area.

The disc rotates with a constant linear velocity (CLV) in which a uniform relative velocity is maintained between the disc and the pickup. To accomplish this, the rotation speed of the disc varies depending on the radial position of the pickup. The disc rotates at a speed of about 8 rev/s when the pickup is reading the inner circumference, and as the pickup moves outward, the rotational speed gradually decreases to about 3.5 rev/s. The player reads frame synchronization words from the data and adjusts the speed to maintain a constant data rate.

The CD standard permits a maximum of 74 minutes, 33 seconds of audio playing time on a disc. However, when encoding specifications such as track pitch and linear velocity are modified, it is possible to manufacture discs with over 80 minutes of music. Although the linear velocity of the pit track on a given disc is constant, it can vary from 1.2–1.4 m/s, depending on disc playing time. All audio compact discs and players must be manufactured according to the *Red Book*, the CD standards document authored by Philips and Sony.

Compact Disc Encoding

CD encoding is the process of placing audio and other data in a frame format suitable for storage on the disc. The information contained in a CD frame prior to modulation consists of a 27-bit sync word, 8-bit subcode, 192 data bits, and 64 parity bits. The input audio bit rate is 1.41×10^6 bps. Following encoding, the channel bit rate is 4.3218×10^6 bps. Premastered digital audio data are typically stored on a 3/4 in. U-Matic video transport via a digital audio processor with a 44.1 kHz sampling rate and 16-bit linear quantization.

A frame is encoded with six 32-bit PCM audio sampling periods, alternating left and right channel 16-bit samples. Each 32-bit sampling period is divided to yield four 8-bit audio symbols. The CD system employs two error correction techniques: interleaving to distribute errors and parity to correct them. The standardized error correction algorithm used is the *Cross Interleave Reed-Solomon Code* (CIRC), developed specifically for the compact disc system. It uses two correction codes and three interleaving stages. With error correction, over 200 errors per second can be completely corrected, and indeed, on such a storage medium, possible error rates of this size are to be expected.

Subcode Data

Following CIRC encoding, an 8-bit subcode symbol is added to each frame. The 8 subcode bits (designated as P, Q, R, S, T, U, V, and W) are used as 8 independent channels. Only the P or Q bits are required in the audio format; the other 6 bits are available for video or other information as defined by the CD + G/M (Graphics/MIDI) format. The CD player collects subcode symbols from 98 consecutive frames to form a subcode block with eight 98-bit words; blocks are output at a 75 Hz rate. A subcode block contains its own synchronization word, instruction and data, commands and parity. An example of P and Q data is shown in Figure 3.4-9.

The P channel contains a flag bit that can be used to identify disc data areas. Most players use information in the more comprehensive Q channel. The Q channel contains four types of information: control, address, data, and *cyclic redundancy check code* (CRCC) for subcode error detection. The control bits specify several playback conditions: the number of audio channels (two/four); pre-emphasis (on/off); and digital copy prohibited (yes/no). The address information consists of 4 bits designating three modes for the Q data bits. Mode 1 data are contained in the table of contents (TOC) that is read during disc initialization. The TOC stores data indicating the number of music selections as a track number and the starting points of the tracks in disc running time. In the program and lead-out areas, Mode 1 contains track numbers, indices within a track, track time, and disc time. The optional Mode 2 contains the catalog number of the disc. The optional Mode 3 contains a country code, the owner code, the year of the recording, and a serial number.

EFM Encoding and Frame Assembly

The audio, parity, and subcode data are modulated using eight-to-fourteen modulation (EFM) in which symbols of 8 data bits are assigned an arbitrary word of 14 channel bits. When 14-bit words with a low number and known rate of transitions are chosen, greater data density can be achieved. Each 14-bit word is linked by three merging bits. The 8-bit input symbols require 256 different 14-bit code patterns. To achieve pits of controlled length, only those patterns are used in which more than two but less than ten 0's appear continuously. Two other patterns are used for subcode synchronization words. The selection of EFM bit patterns defines the physical relationship of the pit dimensions. The channel stream comprises a collection of 9 pits and 9 lands that range from 3–11T in length, where T is one period. A 3T pit ranges in length from 0.833–0.972 μm, and an 11T pit ranges in

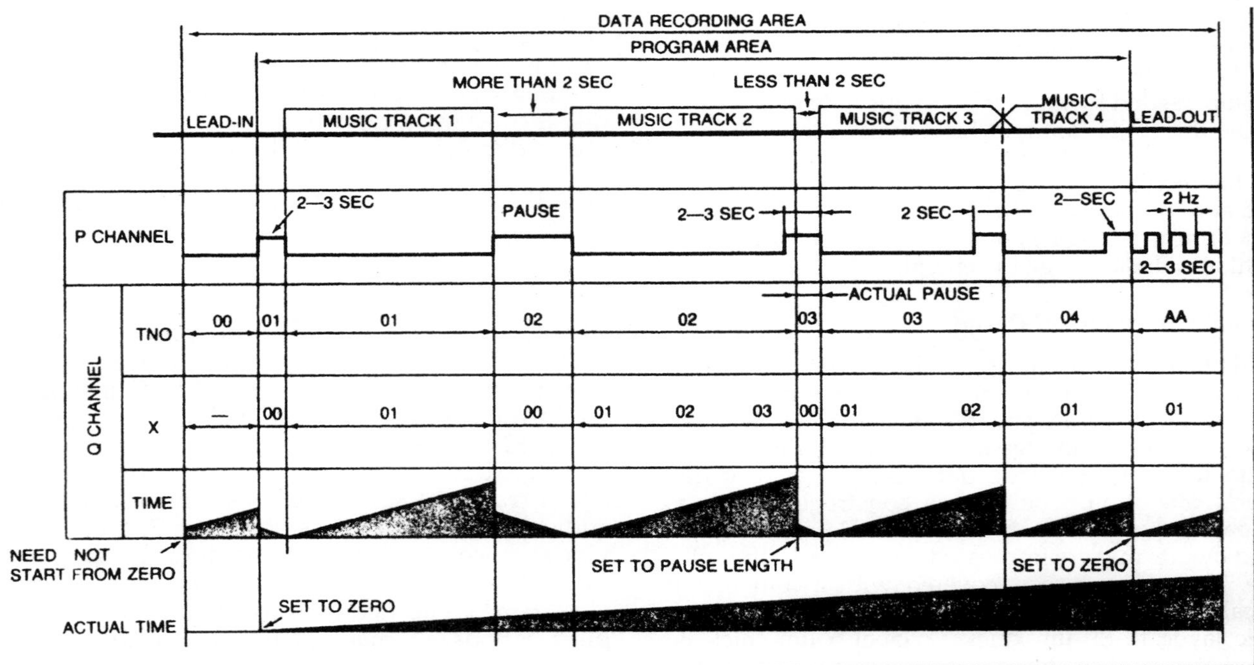

FIGURE 3.4-9 Typical subcode contents of the P and Q channels. (From Pohlmann, *The Compact Disc*.)

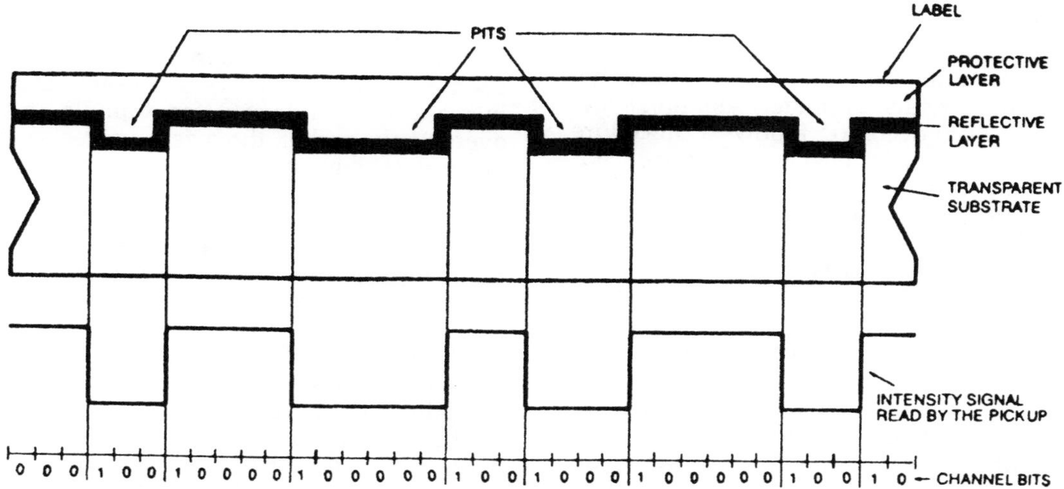

FIGURE 3.4-10 Channel bits as represented by the pit structure. (From Pohlmann, *Principles of Digital Audio.*)

length from 3.054–3.560 µm, depending on pit track linear velocity. Each pit edge, whether leading or trailing, is a "1" and all increments in between, whether inside or outside a pit, are 0's, as shown in Figure 3.4-10.

The start of a frame is marked with a 24-bit synchronization pattern, plus three merging bits. The total number of channel bits per frame after encoding is 588, composed of 24 synchronization bits, 336 (12 × 2 × 14) data bits, 112 (4 × 2 × 14) error correction bits, 14 subcode bits, and 102 (34 × 3) merging bits.

Data Readout

CD pickups use an aluminum gallium arsenide (AlGaAs) semiconductor laser generating laser light with a 780 nm wavelength, which was the most economical type to be developed during the late 1970s. Developments of the CD use shorter wavelength lasers in order to record smaller pits and hence denser data. The beam passes though the substrate, is focused on the metalized pit surface, and is reflected back. Because the disc data surface is physically separated from the reading side of the substrate, dust and surface damage on the substrate do not lie in the focal plane of the reading laser beam; hence, their effect is minimized. The polycarbonate substrate has a refractive index of 1.55; because of the bending of the beam from the change in refractive index, thickness of the substrate, and the numerical aperture (0.45) of the laser pickup's lens, the size of the laser spot is reduced from approximately 0.8 mm on the disc surface to approximately 1.7 µm at the pit surface. The laser spot on the data surface is an Airy function with a bright central spot and successively darker rings, and spot dimensions are quoted as half-power levels.

When viewed from the laser's perspective, the pits appear as bumps with height between 0.11–0.13 mm. This dimension is slightly less than the laser beam's wavelength in polycarbonate of 500 nm. The height of the bumps is thus approximately one-fourth of the laser's wavelength in the substrate. The reflective flat surface of a CD is called *land*. Light striking land travels a distance one-half wavelength longer than light striking a bump, as shown in Figure 3.4-11. This creates an out-of-phase condition between the part of the beam reflected from the bump and the part reflected from the surrounding land. The beam thus undergoes destructive interference, resulting in cancellation. Optically, if the CD pit surface is considered as a two-dimensional reflective grating, the focused laser beam diffracts into higher orders, resulting in interference. The disc surface data thus modulate the intensity of the reflected light beam. In this way, the data physically encoded on the disc are recovered by the laser.

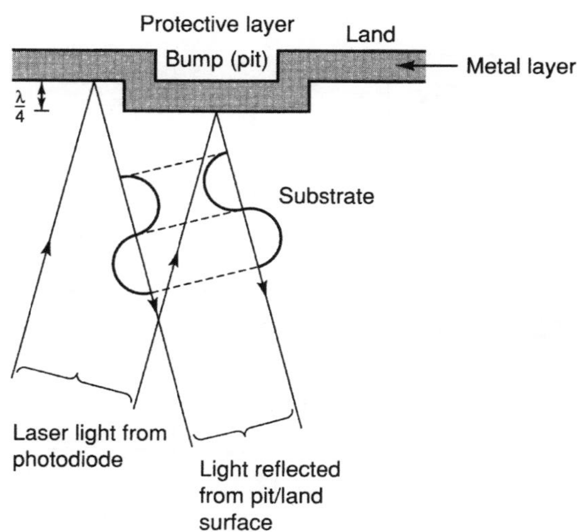

FIGURE 3.4-11 A pit causes cancellation through destructive interference.

Data Decoding

A CD player's data path, shown in Figure 3.4-12, directs the modulated light from the pickup through a series of processing circuits, ultimately yielding a stereo analog signal. Data decoding follows a procedure that essentially duplicates, in reverse order, the encoding process. The pickup's photodiode array and its processing circuits output EFM data as a high-frequency signal. The first data to be extracted from the signal are synchronization words. This information is used to synchronize the 33 symbols of channel information in each frame, and a synchronization pulse is generated to aid in locating the zero crossing of the EFM signal.

The EFM signal is demodulated so that 17-bit EFM words again become 8 bits. A memory is used to buffer the effect of disc rotational wow and flutter. Following EFM demodulation, data are sent to a CIRC decoder for de-interleaving, and error detection and correction. The CIRC decoder accepts one frame of 32 8-bit symbols: 24 audio symbols and 8 parity symbols. One frame of 24 8-bit symbols is output. Parity from two Reed-Solomon decoders is utilized. The first error correction decoder corrects random errors and detects burst errors and flags them. The second decoder primarily corrects burst errors, as well as random errors that the first decoder was unable to correct. Error concealment algorithms employing interpolation and muting circuits follow CIRC decoding.

In most cases, the digital audio data are converted to a stereo analog signal. This reconstruction process requires one or two D/A converters, and low-pass filters to suppress high-frequency image components. Rather than use an analog brick wall filter after the signal has been converted to analog form, the digitized signal is processed before D/A conversion using an oversampling digital filter. An oversampling filter uses samples from the disc as input and then computes interpolation samples, digitally implementing the response of an analog filter.

A *finite impulse response* (FIR) transversal filter is used in most CD players. Resampling is used to increase the sample rate; for example, in a four-times oversampling filter, three zero values are inserted for every data value output from the disc. This increases the data rate from 44.1 kHz to 176.4 kHz. Interpolation is used to generate the values of intermediate sample points—for example, three intermediate samples for each original sample. These samples are computed using coefficients derived from a low-pass filter response.

The spectrum of the oversampled output waveform contains image spectra placed at multiples of the oversampling rate; for example, in a four-times oversampled signal, the first image is centered at 176.4 kHz. Because the audio baseband and sidebands are separated, a low-order analog filter can be used to remove the images, without causing phase shift or other artifacts common to high-order analog brick wall filters.

Traditionally, D/A conversion is performed with a multibit PCM converter. In theory, a 16-bit converter could perfectly process the 16-bit signal from the disc. However, because of inaccuracies in converters, 18-bit D/A converters are often used because they can more accurately represent the signal. Alternatively, low-bit (sometimes called 1-bit) D/A converters can be used. They minimize many problems inherent in multibit converters such as low-level nonlinearity and zero-cross distortion. Low-bit systems employ very high oversampling rates, noise shaping, and low-bit conversion.

Also present in the audio output stage of every CD player is an audio de-emphasis circuit. Some CDs were encoded with an audio pre-emphasis characteristic with time constants of 15 and 50 μsec. Upon playback, de-emphasis is automatically carried out, resulting in an improvement in S/N.

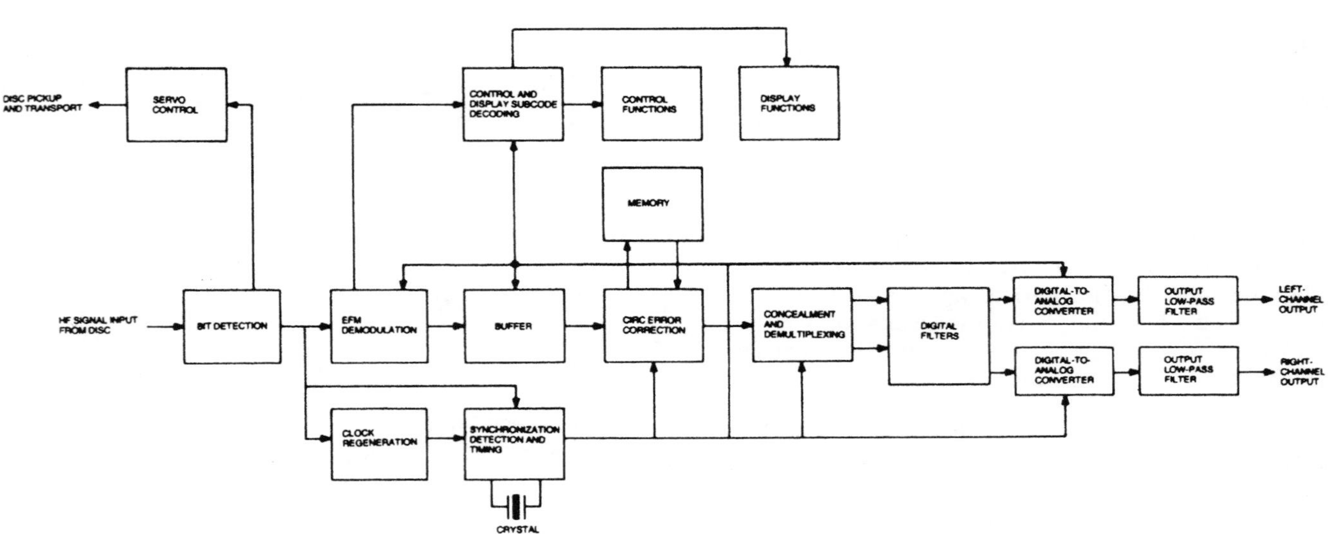

FIGURE 3.4-12 Block diagram of a CD player with digital filtering. (From Pohlmann, *The Compact Disc*.)

Recordable CD-R

With a CD-R (or CD-WO) write-once optical disc recorder, the user may record data until the disc capacity is filled. Recorded CD-R discs are playable on conventional CD players. A block diagram of a CD-R recorder is shown in Figure 3.4-13. An encoder circuit accepts an input PCM signal and performs CIRC error correction encoding, EFM modulating, and other coding and directs the data stream to the recorder. The recorder accepts audio data and records up to 74 minutes in real time. In addition to audio data, a complete subcode table is written in the disc TOC, and appropriate flags are placed across the playing surface.

Write-once media are manufactured similarly to conventional playback-only discs. As with regular CDs, they employ a substrate, reflective layer, and protective top layer. Sandwiched between the substrate and reflective layer, however, is a recording layer composed of an organic dye. Together with the reflective layer, it provides a typical in-groove reflectivity of 70% or more. Unlike playback-only CDs, a pregrooved spiral track is used to guide the recording laser along the spiral track; this greatly simplifies recorder hardware design and ensures disc compatibility. Shelf life of the media is said to be 10 years or more at 25°C and 65% relative humidity. However, the dye used in these discs is vulnerable to sunlight; thus, discs should not be exposed to bright sun over a long period.

The CD-R format is defined in the *Orange Book Standard* authored by Philips and Sony. In CD recorders adhering to the *Orange Book I Standard*, a disc must be recorded in one pass—start-stop recording is not permitted. In recorders adhering to the *Orange Book II Standard*, recording may be stopped and started. In many players, tracks may be recorded at different times and replayed, but because the disc lacks the final

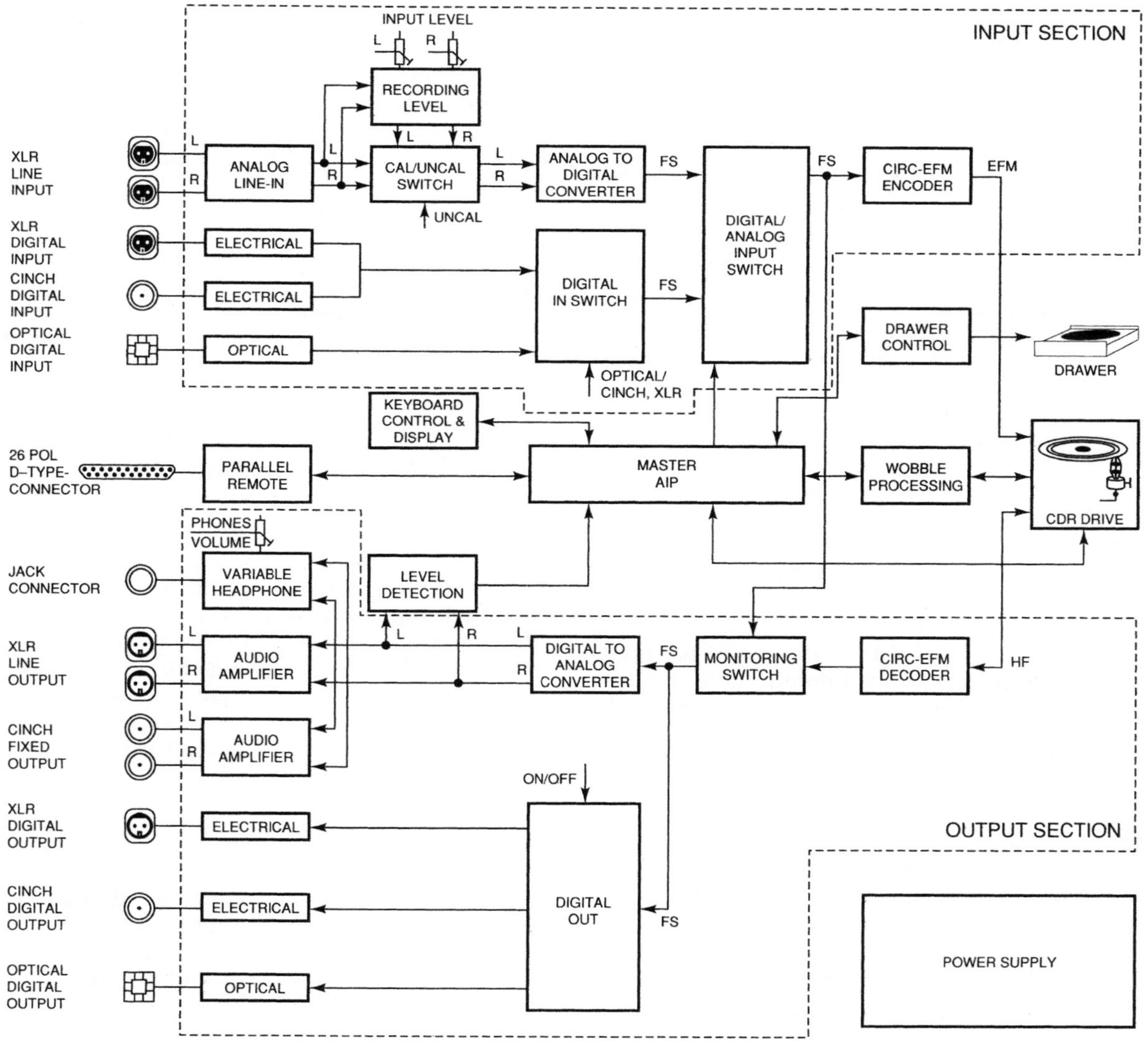

FIGURE 3.4-13 Block diagram of a CD-R recorder. (From Pohlmann, *The Compact Disc*.)

TOC, it can be played only on a CD-R recorder. When the entire disc is recorded, the interim TOC data are transferred to a final TOC, and the disc may be played in any CD audio player. The program memory area (PMA) located at the inner portion of the disc contains the interim TOC record of the recorded tracks. In addition, discs contain a power calibration area (PCA); this allows recorders to automatically make test recordings to determine optimum laser power for recording. Some recorders exceed the *Orange Book II Standard;* they generate an interim TOC that allows partially recorded discs to be played on playback-only CD players.

CD-R recorders are useful because they eliminate the need to create an edited master tape prior to CD recording. If a passage is not wanted, it can be marked prior to writing the final TOC so that the recorder will not play it back. For example, dead air during a live performance can be marked so it is deleted whenever the disc is played back. The data physically continue to exist on the disc, however.

Magnetic Digital Recording Design Considerations

A great advantage of digital magnetic recordings is that system performance is no longer limited by performance of the storage medium. Since *transitions* are the fundamental language of digital recording systems, rather than perfect waveforms, neither AC bias nor particularly high signal-to-noise ratio is required. In fact, distorted waveforms are the norm. However, since a massive amount of transition density must be stored for high fidelity audio, higher bandwidth and more precision magnetic emulsions are needed. Linear density, or kilobits per inch, is the critical factor. Several techniques are employed to maximize density capabilities, as well as to minimize density requirements.

The need for higher storage densities for digital audio accelerated research and development in tape composition and magnetic head design. At higher recording densities, error vulnerability requires ever smoother recording media and revolutionary designs of recording and playback heads.

Due to decreased signal-to-noise ratio requirements, print-through effects are operationally nonexistent, and much thinner tape base thicknesses and oxide layers are commonly employed. However, coercivity is much higher on digital magnetic media and typically ranges from 800–1500 Oe versus the more typical 300–400 Oe in analog recordings. Thus, digital recordings are deep and robust.

Acicular magnetic particles are cigar-shaped fragments employed in most magnetic digital recording media. Because transitions are the basis of digital recording, *saturation recording* is employed and is typically of the traditional longitudinal format. However, for greater storage density, the acicular particles can be oriented perpendicularly to the direction of the recording medium's travel. A balance is required between too low a density, which requires excessive tape consumption, and too high a density, which requires additional error correction to combat dropouts and intersymbol interference.

Isotropic recording utilizes longitudinal and vertical modes simultaneously. In isotropic recordings, the vertical field erases the longitudinal fields near the tape's surface. Thus, the tape is recorded to saturation with longitudinal fields and is multiplexed with vertical fields near the surface. The longitudinal field is structured for dominance at low frequencies, and the vertical field carries the higher frequencies. Because the head gaps in isotropic recordings are so minute, there is essentially no intersymbol interference because only a small area at the trailing edge of the gap is recorded.

Thin film heads used for digital recording are of a substantially different design from analog heads. These heads are manufactured using photolithography to achieve a minute, precise shape. Multiturn thin-film inductive record heads (IRH) are used for recording but do not have good playback characteristics at slow speeds. However, magneto-resistive (MR) heads are useful due to the output being independent of tape speed. With MR heads, the head never touches the tape, and thus both head and media life are prolonged. Both crosstalk and signal-to-noise characteristics are excellent in such systems.

In order to minimize damage and errors due to head-to-media contact in systems with high media velocities, a load-carrying air film is formed at the interface between record head and magnetic media. Physical contact should only occur as the media starts and stops its motion. The air film must be thick enough to conceal any near-contact surface irregularities and thin enough to provide a reliable record and playback signal. Head-to-medium separation ranges from about 50 nm to 0.3 μm, and the roughness of the head and medium surfaces ranges from 1.5 to 10 nm rms.[5]

Rotary-Head Digital Audio Tape

The rotary-head digital audio tape (R-DAT or DAT) format was originally designed as a consumer medium to replace the analog cassette. However, the format has found wider application as a low-cost professional digital recording system, and although now obsolescent, it represented the state of the art in rotary head recording for many years, and a study of the specification will be of great help in understanding many such tape formats. An example of a portable R-DAT recorder is shown in Figure 3.4-14.

Format Specifications

The DAT format is based on a tape only 3.81 mm wide, using rotating heads to achieve the head-to-tape speed necessary for digital recording. It supports four record/playback modes and two playback-only modes. The standard record/playback and both playback-only modes, *wide* and *normal*, are implemented on every DAT recorder. The standard mode offers 16-

[5]See Bharat Bhushan, "Tribology of the Head-Medium Interface," *Magnetic Recording Technology*, McGraw-Hill, 1996, Chapter 7.

FIGURE 3.4-14 Sony PCM 2000 portable R-DAT recorder. This recorder continued some of the rugged engineering of its analog predecessors, weighing some eight pounds, while the cassette weighs merely an ounce. Recording duration on battery power was always an issue with portable R-DAT recorders.

bit linear quantization and 48 kHz sampling rate. Both playback-only modes use a 44.1 kHz sampling rate, for user- and prerecorded tapes. Three other record/playback modes, called *Options 1, 2,* and *3,* all use 32 kHz sampling rates. Option 1 provides two-hour recording time with 16-bit linear quantization. Option 2 provides four hours of recording time with 12-bit nonlinear quantization. Option 3 provides 4-channel recording and playback, also using 12-bit nonlinear quantization. These specifications are summarized in Figure 3.4-15.

The user can write and erase nonaudio information into the subcode area: start ID indicating the beginning of a selection, skip ID to skip over a selection, and program number indicating selection order. These subcode data permit rapid search and other functions. Although subcode data are recorded onto the tape in the helical scan track along with the audio signal, they are treated independently and can be rewritten without altering the audio program, and entered either during recording or playback. With the ID codes entered into the subcode area, desired points on the tape such as the beginning of selections can be searched for at high speed by detecting each ID code. During playback, if the skip ID is marked, playback is skipped to the point at which the next start ID is marked, and playback begins again.

In the DAT format, the recorded area is distinguished from a blank section of tape with no recorded signal, even if the recorded area does not contain an audio signal. Unlike blank areas, the track format is always encoded on the tape even if no signal is present. If these sections are mixed on a tape, search operations may be slowed. Hence, blank sections should be avoided. A consumer DAT deck with an interface meeting the specifications of the Sony/Philips digital interface format (SPDIF) will identify when data have been recorded with a copy inhibit Serial Copy Management System (SCMS) flag in the subcode (ID6 in the main ID in the main data area) and will not digitally copy that recording. In other words, SCMS permits first generation digital copying, but not second generation copying. Analog copying is not inhibited.

ITEM \ MODE	DAT (REC/PB MODE)				PRERECORDED TAPE(PB ONLY)	
	STANDARD	OPTION 1	OPTION 2	OPTION 3	NORMAL TRACK	WIDE TRACK
CHANNEL NUMBER [CH]	2	2	2	42	2	2
SAMPLING FREQUENCY [kHz]	48	32	32	32	44.1	
QUANTIZATION BIT NUMBER [BIT]	16 (LINEAR)	16 (LINEAR)	12 (NONLINEAR)	12 (NONLINEAR)	16 (LINEAR)	16 (LINEAR)
LINEAR RECORDING DENSITY [kBPI]	61.0	61.0			61.0	61.1
SURFACE RECORDING DENSITY [MBPI²]	114	114			114	76
TRANSMISSION RATE [MBPS]	2.46	2.46	1.23	2.46	2.46	
SUBCODE CAPACITY [KBPS]	273.1	273.1	136.5	273.1	273.1	
MODULATION SYSTEM	8–10 MODULATION					
CORRECTION SYSTEM	DOUBLE REED-SOLOMON CODE					
TRACKING SYSTEM	AREA SHARING ATF					
CASSETTE SIZE [mm]	73 × 54 × 10.5					
RECORDING TIME [MIN]	120	120	240	120	120	80
TAPE WIDTH [mm]	3.81					
TAPE TYPE	METAL POWER					OXIDE TAPE
TAPE THICKNESS [μm]	13 ± 1 μ					
TAPE SPEED [mm/s]	8.15	8.15	4.075	8.15	8.15	12.225
TRACK PITCH [μm]	13.591				13.591	20.41
TRACK ANGLE	6°22'59.5"					6°23'29.4"
STANDARD DRUM SPECIFICATIONS	ø30 90° WRAP					
DRUM ROTATIONS [rpm]	2000		1000	2000	2000	
RELATIVE SPEED [m/s]	3.133		1.567	3.133	3.133	3.129
HEAD AZIMUTH ANGLE	±20°					

FIGURE 3.4-15 DAT standard specifications. (From Pohlmann, *Principles of Digital Audio.*)

DAT Recorder Design

From a hardware point of view, a DAT recorder utilizes many of the same elements as a CD-R recorder: A/D and D/A converters, modulators and demodulators, error correction encoding and decoding. Audio input is received in digital form, or is converted to digital by an A/D converter. Error correction code is added and interleaving is performed. As with any helical scan system, time compression must be used to separate the continuous input analog signal into segments prior to recording and then rejoin them upon playback with time expansion to form a continuous audio output signal. Subcode information is added to the bitstream, and it undergoes eight-to-ten (8/10) modulation. This signal is recorded via a recording amplifier and rotary transformer.

In the playback process, the rotary head generates the record waveform. Track-finding signals are derived from the tape and used to automatically adjust tracking. Eight-to-ten demodulation takes place, and subcode data are separated and used for operator and servo control. A memory permits de-interleaving as well as time expansion and elimination of wow and flutter. Error correction is accomplished in the context of de-interleaving. Finally, the audio signal is output as a digital signal, or through D/A converters as an analog signal.

The DAT rotary head permits slow linear tape speed while achieving high bandwidth. Each track is discontinuously recorded as the tape runs past the tilted head drum spinning rapidly in the same direction as tape travel. The result is diagonal tracks at an angle of slightly more than 6° from the tape edge, as shown in Figure 3.4-16. Despite the slow linear tape speed of 8.15 mm per second (5/16 in. per second), a high relative tape-to-head speed of about 3 m per second (120 in. per second) is obtained. A DAT rotating drum (typically 30 mm in diameter) rotates at 2000 rpm, typically has two heads placed 180° apart, and has a tape wrap of only 90°. Four head designs provide direct read after write, so the recorded signal can be monitored.

Azimuth recording (*or guard-bandless recording*) is used in which the drum's two heads are angled differently with respect to the tape; this creates two track types, sometimes referred to as A and B, with differing azimuth angles between successively recorded tracks. This ±20° azimuth angle means that the A head will

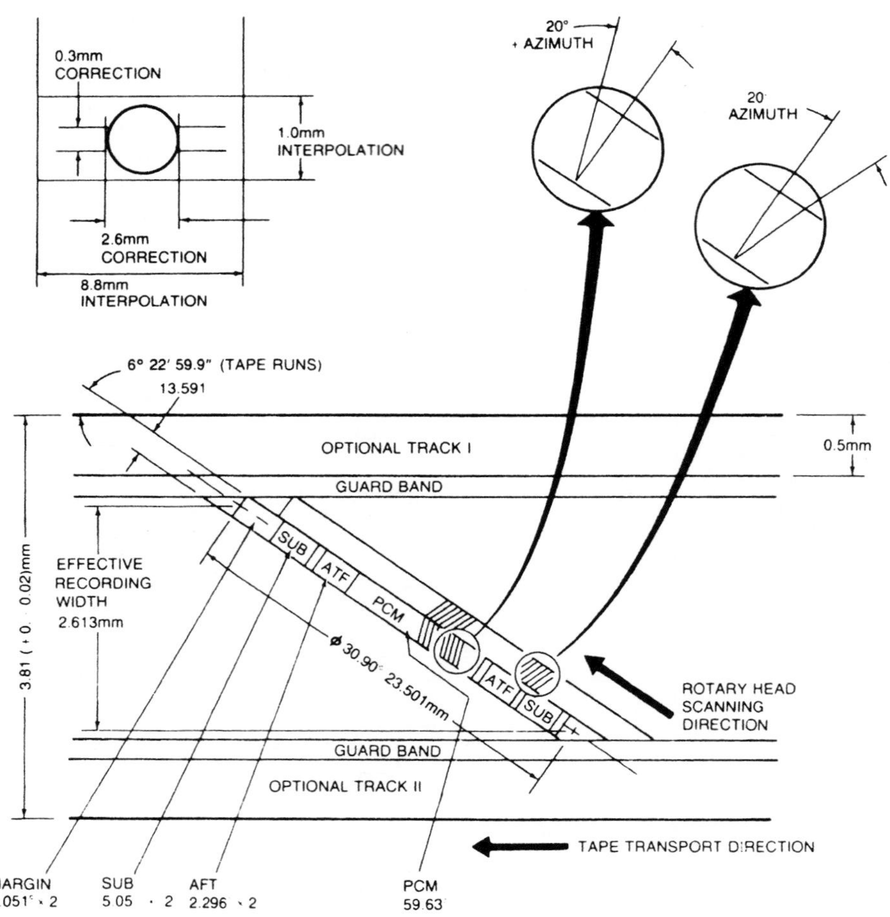

FIGURE 3.4-16 DAT track configuration. (From Pohlmann, *Principles of Digital Audio*.)

read an adjacent B track at an attenuated level due to phase cancellation. This reduces crosstalk between adjacent tracks, eliminates the need for a guardband between tracks, and promotes high-density recording. Erasure is accomplished by overwriting new data to tape such that successive tracks partially write over previous tracks. Thus, the head gaps (20.4 microns) are approximately 50% wider than the tracks (13.59 microns) recorded to tape.

The length of each track is 23.501 mm. Each bit of data occupies 0.67 microns, with an overall recording data density of 114 Mb per square inch. With a sampling rate of 48 kHz and 16-bit quantization, the audio data rate for two channels is 1.536 Mbps. However, error correction encoding adds extra information amounting to about 60% of the original, increasing the data rate to about 2.46 Mbps. Subcode raises the overall data rate to 2.77 Mbps.

The primary types of data recorded on each track are PCM audio, subcode, and automatic track finding (ATF) patterns. Each data (or sync) block contains a sync byte, ID code byte, block address code byte, parity byte, and 32 data bytes. In total, there are 288 bits per data block; following 8/10 modulation, this is increased to 360 channel bits. Four 8-bit bytes are used for sync and addressing. The ID code contains information on pre-emphasis, sampling frequency, quantization level, tape speed, copy-inhibit flag, channel number, and so on. Subcode data are used primarily for program timing and selection numbering. The subcode capacity is 273.1 kbps. The parity byte is the exclusive or sum of the ID and block address bytes, and is used to error correct them.

Since the tape is always in contact with the rotating heads during record, playback, and search modes, tape wear necessitates sophisticated error correction. DAT is thus designed to correct random and burst errors. Random errors are caused by crosstalk from an adjacent track, traces of an imperfectly erased signal, or mechanical instability. Burst errors occur from dropouts caused by dust, scratches on the tape, or by head clogging with dirt.

To facilitate error correction, each data track is split into halves, between left and right channels. In addition, data for each channel are interleaved into even and odd data blocks, one for each head; half of each channel's samples are recorded by each head. All of the data are encoded with a doubly encoded Reed-Solomon error correction code. The error correction system can correct any dropout error up to 2.6 mm in diameter, or a stripe 0.3 mm high. Dropouts up to 8.8 mm long and 1.0 mm high can be concealed with interpolation.

Other Rotary Head Digital Tape Formats

ADAT and DA88 (Multitrack)

Once the R-DAT format was established and adopted by professional users, it was a small step to produce multitrack versions using S-VHS cassettes (ADAT) or 8 mm video cassettes (DA88 type). Professional users tended toward the generic DA88 series, which had track bounce and integral time code features (see Figure 3.4-17). Originally carrying 16-bit audio tracks (and integral time code), a newer series of these machines will record 20-bit PCM tracks. Although these newer machines will play back 16-bit recordings, the earlier machines will not play back 20-bit recordings.

FIGURE 3.4-17 Tascam DA-88 digital multitrack recorder.

Interconnection of these multitrack recorders in the digital domain is via proprietary interfaces, a parallel electrical TDIF interface in the case of the DA88s, and a serial optical "lightpipe" for the ADAT.

"1610"-Type Videotape Recording (Stereo)

The vital need in the mid-1970s for compact disc source recordings in an editable digital form created one of the most bizarre audio recording formats ever, and it has left us the curious legacy of a sample rate of 44.1 kHz samples per second, as used for the compact disc. As an idea, it started with the then-emerging helical scan video recorders, which were capable of recording (via a special adapter) three 16-bit (stereo) samples of digital audio on each television line. The most practical video recording format in those days was the U-Matic cassette with 3/4-inch tape. These tapes could be assemble edited using multiple machines and an edit controller. Because analog recorders have no storage, the time when the rotary heads switched over at the tape edges needed to be avoided for recording, and the system therefore used only 588 lines of the 625-line 25-frame system for recording the audio. The 588 lines times 3 samples times 25 per second gave 44,100 samples per second. The equivalent U.S. system used 490 lines out of 525, but the slightly lower frame rate of 29.94 gave 44,056 samples per second. The 0.1% difference in these sample rates was soon forgotten, and 44.1 kHz lives on.

Fixed-Head Digital Tape Formats

DASH (Stereo and Multitrack)

During the 1980s several *digital audio stationary head* (DASH) recording formats were developed, often based on the platform of existing analog tape decks.

The attraction at that time was the illusory economic advantages of razor blade editing, combined with digital audio quality. The most enduring of these DASH formats, especially for storage and exchange in the music recording industry, is probably the Sony multitrack version using 2-inch wide tape, and capable of recording up to 24 audio channels. In broadcasting circles, multitrack machines such as this were used only for serious music recording backup, although even today, the obvious successor in terms of bulk multitrack storage has yet to be found.

DCC Cassettes (Stereo)

The advent of the compact disc in 1982 exposed the inadequacies of the compact cassette in home recording applications and, as a result, there was a three-sided development of digital recording formats for the consumer market. Ultimately, two of these—the R-DAT and the MiniDisc—were to find a ready acceptance in professional radio production applications. The third format started out life as the *stationary head digital audio tape* (S-DAT) format, using a compact cassette-sized tape, but with uncompressed 16-bit PCM stereo recording. It became the *digital compact cassette* (DCC) when MPEG layer I bit rate reduction was applied. A few examples of these cassettes may be found in libraries; however, the recorders were discontinued around 1996.

Optical and Magneto-Optical Disc Formats

CD and DVD Audio

CD-A (the commercial record format described in the "Compact Disc" section earlier), whether in CD-R, RW, or glass-mastered form, is not a particularly good format to use for broadcast recording, although it may have attractions for easily played *samplers*. It possesses limited error correction, a table of contents must be written at the start of the disc, and the encoding is limited to 16-bit PCM. As a physical *carrier*, however, the optical medium of the CD (or equally the similar DVD) has a lot to commend it, not least the fact that the IT industry has greatly reduced the recorder and blank media prices by adopting them in such large numbers.

CD/DVD-R and RW

The computer industry, and particularly Kodak with its Photo-CD application, pushed forward the CD-R (and the related RW) formats into affordable computer components, and the derivatives of them now form such economical carriers for fast audio file exchange. CD-R media can be "closed" to ISO 9660 format, which was the basic universal CD-ROM format. However, CD-RW discs cannot be closed to the ISO 9660 format. More recent formats use a derivative such as UDF that was basic to the recordable DVD; these formats have various benefits including allowing longer names to be entered.

FIGURE 3.4-18 Sony pocket MiniDisc recorder of the early 1990s.

MiniDisc

The Sony MiniDisc recorder (see Figure 3.4-18) uses a 2.5 in. *magneto-optical* disc housed in a *caddy* for protection. It uses the proprietary adaptive transform acoustic coding (ATRAC) audio data rate reduction system, based on block frequency transforms. Magneto-optical recording technology combines magnetic recording and laser optics, utilizing the record/erase benefits of magnetic materials with the high density and contactless pickup of optical materials.

With magneto-optics, a magnetic field is used to record data, but the applied magnetic field is much weaker than conventional recording fields. It is not strong enough to orient the magnetic particles. However, the coercivity of the particles sharply decreases as they are heated to their Curie temperature. A laser beam focused through an objective lens heats a spot of magnetic material and only the particles in that spot are affected by the magnetic field from the recording coil, as shown in Figure 3.4-19(a). After the laser pulse is withdrawn, the temperature decreases and the orientation of the magnetic layer records the data. In this way, the laser beam creates a small recorded spot, thus increasing recording density.

The Kerr effect may be used to read data; it describes the slight rotation of the plane of polarization of polarized light as it reflects from a magnetized material. The rotation of the plane of polarization of light reflected from the reverse-oriented regions differs from that reflected from unreversed regions, as shown in Figure 3.4-19(b). To read the disc, a low-powered laser is focused on the data surface, and the angle of rotation of reflected light is monitored, thus recovering data from the laser light.

Available as both studio models and affordable portable recorders, the MiniDisc was a popular format for reporters as well as for radio playback applications, where it formed a direct replacement for NAB tape cartridge players.

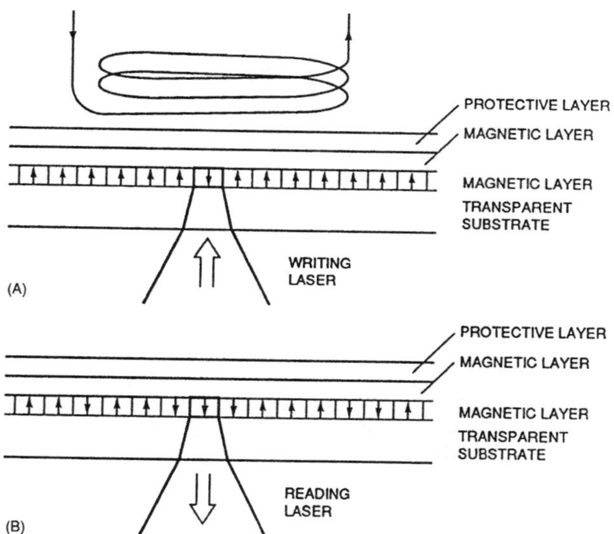

FIGURE 3.4-19 Magneto-optical recording (a) and playback (b).

The MiniDisc system uses a TOC, very much like that on a CD-A, and when one is recording, this TOC is written automatically *after* the stop button is pressed. This detail was sometimes forgotten by the journalist users, and physical vibration could sometimes impair recordings (including that all-important TOC), but large buffer memories helped to minimize this, and the low bit rate makes playback access to any track particularly fast, a detail that did make it popular with any previous users of compact cassette recorders.

Solid-State and Hard Disk Recorders

Solid-State Recorders

An obvious solution to possible mechanical vulnerability of portable digital recorders is to record directly to a solid-state memory. The memory itself may be removable or fixed, and in order to keep the power consumption down and the recording time up, bit rate reduction may be used. Examples of solid-state recorders are shown in Figures 3.5-15 and 3.5-16 in Chapter 3.5.

Probably the most important consideration is how quickly and conveniently the recording can be dumped to a computer or sent back to the home station. This is an important function so that the recorder can be quickly reused.

Another consideration if bit rate reduction is used is to what extent post processing (which might be as simple as equalization) will compromise the audio quality. The most efficient bit rate reduction systems such as MP3 are efficient only because the artifacts are evenly distributed across the threshold of our hearing. A simple equalization change can quite easily uncover those artifacts in a particular and audible area of the audio spectrum.

Metadata may seem an expensive luxury for a simple interview recording, but in the case of solid-state recorders, the transfer of the material will leave no space for any label, and some form of explanation or traceability becomes vital. Fortunately, an initiative between industry, the AES, EBU, and a UK-based broadcast craft group called the Institute of Broadcast Sound has realized the problems of entering (or not entering) consistent manual metadata. They have instituted an XML-based automatically recorded data chunk, *iXML*, which allows the originating machine and take time to be traced uniquely and passed along through the sound workflow.[6]

NAGRA and Similar Portable Recorders

The NAGRA series of 1/4-inch analog recorders made by the Swiss Kudelski company were highly considered for location recording and especially for feature film applications. Several digital versions of this old favorite have appeared using solid-state memory, as well as disk packs or tape. In particular, the NAGRA-D offered four 24-bit PCM channels on tape, a highly attractive top-quality capture format in its time. The only thing to consider with all tape-based storage is the need for real time to dump these recordings to computer files, but equally, removable disk or solid-state packs can make for an expensive alternative to a tape library.

Digital Dubbers

The name *digital dubber* is a film industry misnomer for solid-state or disk-based temporary stores for (typically) eight tracks of up to 24-bit PCM audio, used during audio editing and dubbing, hence the name. These recorders are rarely used for the actual interchange of recordings and can be viewed as a sort of local server.

Computer Disks

Hard-disk-based storage now forms the heart of most modern broadcast facilities with digital audio workstations (DAW) and audio servers. Some of the implications of this revolution are discussed later in this chapter, and the systems and equipment are covered in depth in Chapter 3.6.

DIGITAL SIGNAL PROCESSING

Digital signal processing (DSP) has improved the performance of many existing audio functions such as equalization and dynamic range compression, and permits new functions such as ambience processing, dynamic noise cancellation, and time alignment. DSP is a technology used to analyze, manipulate, or generate signals in the digital domain. It uses the same principles as any digitization system; however, instead of a

[6]See http://www.ixml.info/.

storage medium such as CD or DAT, it is a processing method.

DSP Applications and Design

DSP employs technology similar to that used in computers and microprocessor systems; however, there is an important distinction. A regular computer processes data, whereas a DSP system processes signals. It is accurate to say that an audio DSP system is in reality a computer dedicated to the processing of audio signals.

Some audio functions that DSP can perform include error correction, multiplexing, sample rate conversion, speech and music synthesis, data compression, filtering, adaptive equalization, dynamic range compression and expansion, crossovers, reverberation, ambience processing, time alignment, acoustic noise cancellation, mixing and editing, and acoustic analysis. Some DSP functions are embedded within other applications; for example, the error correction systems and oversampling filters found in CD players are examples of DSP. In other applications the user has control over the DSP functions.

Digital processing is more precise, repeatable, and can perform operations that are impossible with analog techniques. Noise and distortion can be much lower with DSP; thus, audio fidelity is much higher. In addition, whereas analog circuits age, lose calibration, and are susceptible to damage in harsh environments, DSP circuits do not age, cannot lose calibration, and are much more robust. However, DSP technology is an expensive technology to develop. Hardware engineers must design the circuit or employ a DSP chip, and software engineers must write appropriate programs. Special concerns must be addressed when writing the code needed to process the signal. For example, if a number is simply truncated without regard to its value, a significant error could occur, and the error would be compounded as many calculations take place, each using truncated results. The resulting numerical error would be manifested as distortion in the output signal. Thus, all computations on the audio signal must be highly accurate. This requires long word lengths; DSP chips employ digital words that are 32 bits in length or longer.

In addition, even simple DSP operations may require several intermediate calculations, and complex operations may require hundreds of operations. To accomplish this, the hardware must execute the steps very quickly. Because all computation must be accomplished in real time—that is, within the span of one sample period—the processing speed of the system is crucial. A DSP chip must often process 50–100 million instructions per second. This allows it to run complete software programs on every audio sample as it passes through the chip.

DSP products are more complicated than similar analog circuits, but DSP possesses an inherent advantage over analog technology: it is programmable. Through the use of software, many complicated functions can be performed entirely with coded instructions. Figure 3.4-20(a) shows a band-pass filter using conventional analog components. Figure 3.4-20(b) shows the same filter, represented as a DSP circuit. It employs the three basic DSP operators of delay, addition, and multiplication. However, this DSP circuit may be realized in software terms. Figure 3.4-20(c) shows an example of the computer code (Motorola DSP56001) needed to perform band-pass filtering with a DSP chip. There are many advantages to this software implementation. Whereas hardware circuits would require new hardware components and new circuit design to change their processing tasks, the software implementation could be changed by altering parameters in the code. Moreover, the program could be written so different parameters could be employed based on user control.

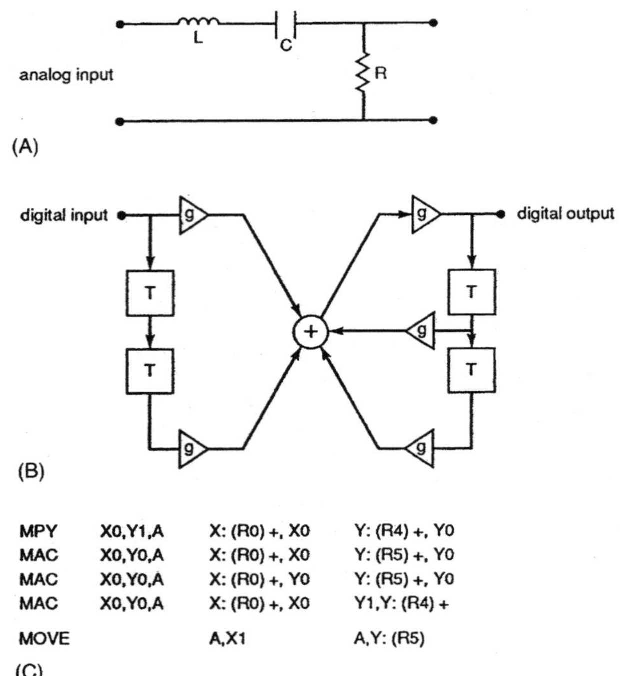

FIGURE 3.4-20 (a) A band-pass filter represented by an analog circuit, (b) digital signal processing circuit, and (c) digital signal processing instructions.

As noted, DSP can be used in lieu of most conventional analog processing circuits. The advantages of DSP are particularly apparent when various applications such as recording, mixing, equalization, and editing are combined in a workstation. For example, a personal computer, combined with a DSP hardware card, hard disk drive, appropriate software, and a DAT or CD recorder forms a complete postproduction system. Such a system allows comprehensive signal manipulation including the capability to cut, paste, copy, replace, reverse, trim, invert, fade in, fade out, smooth, loop, mix, change gain and pitch, crossfade, and equalize. The integrated nature of such a workstation, its low cost, and high-processing fidelity make it clearly superior to analog techniques.

SOUND RECORDING AS A PROCESS

Recording delays an audio signal, and it also enables the material to be shared. For a radio interview the delay element might involve only a few seconds, just enough for a "top and tail" edit. On the other hand a transcription recording of a live concert could well be recorded and then lie dormant for many years.

Nowadays, economic pressures in the broadcast industry demand a recording process that is far more complex than these two examples represent. For instance, if a recording can be made available to multiple operators soon after the start of that recording, the editing processes for different distribution paths and different programs could take place in parallel.

Workflow

Therefore, we now need to think of broadcast sound recording as a part of general *workflow*, and in this context the audio could usefully accompany other *essence* such as text and pictures. These essence items are all linked together by the *metadata*, which takes the place of the information once carried on the label and package of a disc or tape recording, although in the digital world the metadata by itself possesses a much greater potential power than a label ever did.

For the overall workflow process to be a success, several presumptions are made about the sound recording. The highest source quality needs to be preserved as far along the workflow chain as possible in order to preserve the possibilities for later processing or for future (and possibly lucrative) applications of the recorded material. The technical quality of the recording must be adequate to survive the numerous signal processing and editing procedures that may be applied at any stage along the workflow route. This does not mean that current sound recording practices need to be perfect, but it does mean that other items in the signal chain must produce a greater level of impairments than the recording. This has implications for any bit rate reduction that is being considered.

In fact, the days of needing bit-rate-reduced recordings for the convenience or economic use of IT applications are long over, although low bit rate contribution of interviews and similar recordings over telephone circuits is likely to remain a prime use of bit rate reduction for many years to come, especially in areas without widespread Internet access.

Advantages of Digital Recording

The "transparent" quality of digital systems should not give the impression that sterility has crept into sound recording. In reality, the creative opportunities available in the most basic computer recording equipment well exceed those existing in the most complex analog studio facilities of 10 years ago. What has gone are the subtle perceptual sound effects that were inherent in analog recording or noise-reduction processes. However much of a cult has built up around some of those effects over the years, there can be no doubt about the technical efficiency of a digital recording process. Digital recording has enabled big and helpful changes in audio production, mainly as a result of the transparency factor. For instance, the recording of two-channel stereo in the form of "M" and "S" (mid-side or mono-stereo) channels was not considered using analog equipment, because any recording artifacts such as noise on the M channel would appear as a coherent center image in the reproduced sound field. "A and B" (left and right channel) analog recording resulted in a much more diffusely reproduced field of any recording noise; therefore, this method became the standard for two-channel stereo recording. M and S recording can prove quite useful as the apparent width of the stereo image can be varied just by adjusting the S channel level.

Perhaps this is a good point to step back from the actual details of the recording technology and ask whether we have actually experienced some kind of digital audio revolution.

A Digital Audio Revolution

While there have undoubtedly been great changes, what has taken place in the field of sound recording has been perhaps more an *evolution* than a revolution and, what is more, it has taken place in two distinct steps.

First, digital encoding of the recorded signal circumvented the imperfections inherent in analog recording systems. This led to the development of dedicated digital audio recording formats, mostly using magnetic tape or optical disc as the storage media.

The second stage came when the encoded digital audio signals from stage one became available for computer-based editing. This led to an obvious use of generic IT storage for sound recordings. The main disadvantage of any tape-based system lies in the inevitable slow access to any given part on the tape, although the archive life of tapes may not be exceptional either. On the other hand, IT-based storage media are often not so easy to exchange, sometimes for physical reasons and sometimes because of format incompatibilities.

Returning to the "audio" part of the digital audio evolution, the key stage was the initial encoding of the audio in a digital form. The later digital recording and manipulation technologies were both largely developed by the generic IT industry. Audio coding in the broadcast industry in the form of pulse-code modulation (PCM) was first put into practical use during the 1970s for the accurate long-distance transmission for the two channels of stereo radio programs. Prior to use of this system, there was a limit of a hundred miles or so for the distance over which a sufficiently good match of two analog landline paths, as necessary for stereo operation, could be maintained. At this time, a similar form of weakness was also beginning to show up in the recording studio, as the number of generations or layers of recording that could be employed before the signal quality was compromised was beginning to limit multitrack recording techniques, especially in some popular overdub formats. When accurate digital audio coding arrived (and accuracy in audio conversion did take some years to develop), it quickly bypassed the need for the expensive mechanical precision in professional analog recorders. This need for precision had passed to the electronics, and that development was paid for, not by the small audio market, but by the millions of computers, mobile phones, and countless other electronic items in the mass consumer market.

The huge market for IT-based products rapidly created an economic jump that reduced the cost of processing, mixing, and editing of the audio. These processes have now become integrated in the form of what has become known as an audio *workstation*.

Meanwhile, the advances made in distributing the audio material, in finished or unfinished form, have been crucial factors in creating even further jumps in recording economics. These advances are nowhere more visible than in the often "invisible" contributions from radio reporters in the field. These contributions are now sent via e-mail on telephone lines or satellites, and have led to fast audio file transfer between workstations anywhere in the world. In this way the audio rushes from Hollywood can be sent during the evening to London, England, for editing during the day, and the finished material can then be sent back ready for the following morning on the West Coast.

In parallel with these operational advances, a previously mentioned but often overlooked (and slow) technical advance has taken place in the quality of the basic audio digital coding and decoding. If anyone doubts this, compare the D-to-A performance of any early CD player with the much lower cost equivalent of today, where what is actually heard approaches the theoretical coding quality promised in 1982.

Revolution or not, one solid piece of advice when facing any new regime is to have a long-term strategy established. All too often, the economic advantages of one technical advance have been reversed when the next advance came along. Planning any sound recording process for broadcasting requires much thought, especially as the sound is now often linked with other workflow patterns such as video production, and these other patterns are still in the process of changing inside their own digital revolutions.

Inherent complexities in IP-based topologies dictate that contingencies for on-air product reliability should be built into all design and operating initiatives using digital audio networked systems. All systems will occasionally fail and the mission-critical dependence on such systems is a potential single point of failure not typical of the analog systems being replaced.

STREAMING AND FILE FORMATS

In discussing the migration to file-based audio, it is useful to consider the distinction between real-time audio streams and audio data files.

Principles

A microphone channel produces *streaming* audio, and a real-time output *data stream* will be required in many stages along the signal route to feed loudspeakers or headphones. Any streaming format can be thought of as one with the capability for audio data delivery in real time with a low and controlled latency. Streaming audio cannot be slowed down or speeded up, and it cannot usually be interrupted without undesirable consequences. It can start up immediately when the signal is available, and it can go on streaming indefinitely. As it exists only in real time, it effectively carries its own timing information with it. While this could apply to any analog audio signal, when adapted to digital forms, there arises a need to declare the original sampling frequency at the very least in the associated metadata.

A file, on the other hand, never existed in the analog world, although a finished physical recording was not a dissimilar concept. A file has to wait until a definable portion of the streaming program is available to be packaged. The size of any file is limited and needs to be declared, as do all the conditions necessary to rebuild a streaming output (that is, replay it). Once this metadata information has been gathered and stored in a header *chunk*, tightly coupled with the audio essence (bare data samples), the file is fully formed and only then can it be handled in the same way as any data file.

In a typical recording production chain, it might at first seem easy to see where both streaming and files sit. This is not necessarily true, as for instance microphone signals might stream into a recorder via a mixer of some kind. The recorder, however, may then record the signal as a series of tiny files, and even on a digital broadcasting system the signal will be formed of *packets*, which may be viewed as files of a tiny size. These packets must be delivered at the output as a streaming format. There is therefore some crossover between streaming and files, and the fundamental penalty for any use of intermediate files is signal latency (delay) between the input and the output streams.

File formats, on the other hand, are fairly well-defined and pre-agreed arrangements for storing and exchanging data of any kind at unspecified speeds. For audio, any file format must, at the very minimum, contain some form of header containing the information necessary to accurately rebuild the audio *stream* from which the files were initially built. Summing up, therefore, file formats and streaming formats are inextricably linked, and in some cases such as in those data formats used for packet transmission, the division between streaming and file may be blurred.

Streaming Audio Formats

The AES3 digital audio format is the most important streaming format at the heart of modern broadcast audio systems. It was originally designed around the existing cable and routing infrastructure in the analog studios of the early 1980s, so that the connectors first specified were balanced XLR, allowing analog patch cables to be used. Unbalanced signals on 75 ohm video coax using BNC connectors can also be found, and the consumer version of this interface (SPDIF) in the IEC[7] standard was similarly (but not identically) electrically based on unbalanced signals on RCA phono connectors. Both in the PCM IEC60958 form and the packet-carrying version 61937, which is used for reduced bit rate multichannel carriage, an optical fiber link is often used on consumer equipment. Between these professional and consumer PCM interfaces, the audio essence is identical, and the electrical interface differences are in reality no more problematic than those found in analog practice. There are, however, different metadata formats (called Channel Status) in these two standards.

Audio File Formats

There are many types of possible audio file exchange, and it is a relatively trivial task to convert between formats. However, there are always some penalties in conversion, and the EBU decided in 1996 that the demands of the professional broadcaster would be best met with an extensible generic format. The EBU adopted an interesting approach, in the form of the Broadcast Wave File (BWF). The BWF is a development of the existing WAV format, used on many digital audio workstations and computers. A *Wave* file is an audio file that is one type of the more general Resource Interchange File Format (RIFF) file. RIFF was developed by the IBM and Microsoft corporations.

However, no computer is concerned with what a file is for or what it actually does. The computer can be asked to look at the file-name extension when given a file, and if a suitable application is available—either inside or connected to the computer—it will offer that file to the application. WAV audio sample files can be recognized by many different types of audio applications, and these applications will look at the *chunks* of data in order to see if they recognize any data relevant to that application. It will leave all other chunks alone, a fact that enables broadcast-specific information to be inserted into generic Wave files without disabling any low-cost existing applications such as simple players. If the mandatory file chunk known as "fmt-ck" contains parameters such as sample rate and data information which suit that application, then the application will be satisfied that it can play the data. It will then send the data chunk itself (and no other content) to a buffer, from which it will play the file through a pre-arranged port. The beauty of the Wave procedure is that not only can we expand files by adding other chunks of broadcast-specific information, but we can also specify different audio formats for the data chunk itself. Some of the first practical applications of the BWF used MPEG layer II coding, for example.

The full BWF file format is defined in the EBU Tech document 3285,[8] and this basic recommendation is further extended into the complexities of a *native* file format for use in computer editing systems, within the AES Standard AES-31. The beauty of the whole procedure is that it lends itself to being extended even further, allowing industry users to incorporate their own enhancements to the basic arrangement. For example, when the BWF file goes to form part of a multimedia package, the BWF file can easily become a component of an exchange format such as MXF, or it could be incorporated in an assembly of files and metadata such as in the AAF structure.

[7]See http://www.iec.ch.
[8]Available at http://www.ebu.ch/

Piggy-Back Audio Formats

"Piggy-back" audio networks are a relatively new approach, although we are effectively riding an IT "wave" with all audio recording applications. These networks, in the broadest sense, use existing generic standards to carry smaller specialized audio formats within, or on top of, them. They have arisen out of the simple economics of using generic IT or telecom-based distribution methods, or even just cabling, for carrying multiple streams of digital audio. However, strictly audio formats, such as IEC61937, actually use the consumer version of the AES/EBU streaming format in order to carry packet audio information for multichannel compressed audio streams, so in that case it is maybe more of a "Trojan Horse" than a "piggy-back" application.

In the case of the current professional proposals such as that used in AES-47, the actual audio is carried in multiple AES/EBU streams as the signal source, although the multichannel audio digital interface (MADI) might be seen as an earlier stripped-down version of a piggy-back network. An important point to re-emphasize is that many IT-packet-based applications will carry streaming audio only up to a fairly low level of occupancy, maybe less than 10%, so the dream of streaming a Gigabit per second of audio reliably down a Gigabit network remains just that, a dream.

PROGRAM RECORDING LEVELS

The header on any recording needs to follow a fixed structure so that exchange is easily possible between broadcasters who may not speak the same language or follow the same production procedures. In the case of a file format such as BWF, this header will be contained in the header data "chunks" which serve the same purpose. In Europe, EBU recommendations are normally followed for these headers, such as the multichannel audio line-up described in Tech 3304, whereas the SMPTE is the usual reference agency for North America. For direct delivery to a broadcaster, a number of technical and program delivery requirements will need to be followed, and they can vary from broadcaster to broadcaster, as well as between world regions.

Listed among these recommendations are working practices, such as the requirement for broadcasting not to exceed the ITU-R 645 defined "Permitted maximum level",[9] how these levels should be metered, and any line-up tone requirements,[10] as well as the preferred track allocations on multitrack formats.

Sound recordings that accompany pictures, be it for television or film, require *time code* to be recorded at the point of capture and the audio sample rate locked to the video. The time code used relates to the *picture* format (identifying the video frames), not the audio, and where there has been picture format conversion in postproduction, this time code may need to be changed, even if the audio essence is unaffected. Fortunately, much of the audio recording equipment and software used for postproduction[11] works happily with a number of time code types, translating between them and using the locked audio sample rate as the timing reference.

Once the recording is delivered to the broadcaster or postproduction stage, it may pass through a quality control check, and at this stage, some kind of an audit trail of the recording history[12] can be of enormous benefit in solving any problems.

SUMMARY

Now that ever more economical IT system components have caught upwith and exceed most audio requirements, it does seem at last that some sort of "golden age" of audio recording has arrived. Much as the look, ergonomics, and even the smell of a classic audio tape recorder may be missed by some audio engineers, there is little doubt that the recording quality, flexibility, accessibility, and economics of IT-based audio recording are all vastly superior.

The technology is now in place for brilliant audio recordings to be made and we need to explore the opportunities and possibilities that this allows. The transparency of the modern digital recording actually makes it more important than ever to get the microphone and positioning just right, with equal care being taken in composing and focusing on the content.

Glossary

AL: Alignment Level. A reference level, usually recorded as a header recording, and used to check the alignment of recorded signals, and signal paths.

DAW: Digital audio workstation

dB: Decibel. A logarithmic scale for power, but for audio, it is usually understood in signal voltage terms. One decibel is quoted as being the smallest gain change the human ear can distinguish under good conditions.

dBA: "A" weighted sound pressure level

dBC: "C" weighted sound pressure level

dBm: The level of an analog signal referred to 1 mW across a 600 ohm load. (dBm refers to power, this relates to a voltage of 0.775V.)

[9]Typically, not to exceed a level 9 dB below digital FSD level.

[10]EBU line-up tone at a level 18 dB below digital FSD level can be audibly recognized from the left channel being interrupted every three seconds or so. SMPTE line-up tone for U.S.markets will be at 20 dB below digital FSD.

[11]Such as the DA88 family of recorders.

[12]Such as may be carried in the "Coding History" chunk in the BWF.

dBu: The level of an analog signal referred to 0.775 Volt RMS. This voltage level produces 1 milliwatt of power into a load of 600 ohms.

dBv: The level of an analog signal referred to 1.0 Volt RMS.

Headroom: The maximum signal level beyond MPL before unacceptable distortion is produced, or sometimes it is quoted as the same level above AL.

LFE: Low Frequency Effects

Mean AC Voltage: The average rectified value, as read on a moving coil meter. (See also *RMS.*)

ML: Measurement Level. A level of analog signal chosen to be low enough not to cause problems with crosstalk, etc., but to be high enough to be useful for continuous tone technical measurements.

MPL: Maximum Permitted Level. The point at which analog transmitter signal limiters are set to operate. Normally 8 or 9 dB above AL.

Pink Noise: Random or quasi-random noise with equal energy per octave of bandwidth. (White noise has equal energy per linear bandwidth.)

RMS: Root Mean Square volts. The value of a DC voltage which would produce the same average power as the AC signal quoted. (Not the same as the mean voltage value, except for sine waves.)

SPL: Sound Pressure Level

Resources

The organization that sets the legally recognized standards for radiocommunications is the ITU-R (formerly known as the CCIR), based in Geneva. (http://www.itu.int)

Standards for the electrotechnical equipment used in broadcasting, including program meters and interconnection, are the responsibility of the IEC. However, in this case, the actual standards produced are often published in parallel by national standards organizations. (http://www.iec.ch)

These two international organizations draw their expertise and technical inputs from a number of specific industry groups which themselves produce their own standards and recommendations documents.

The Audio Engineering Society covers many key areas of audio standards and practices (http://www.aes.org)

The Society of Motion Picture and Television Engineers covers a wide area of interest including the motion picture industry. (http://www.smpte.org)

Bibliography

Alten, Stanley R., *Audio in Media*, Fourth Edition, Wadsworth Publishing Company, Belmont, California,1994.

Audio Engineering Society, "AES Recommended Practice for Digital Audio Engineering-Serial Transmission Format for Linearly Represented Digital Audio Data," *Journal of the Audio Engineering Society*, vol. 33, no. 12, December, 1985.

Carasso, M.G., Peck, J.B.H., Sinjou, J.P., "The Compact Disc Digital Audio System," *Philips Technical Review*, vol. 40, no. 6, 1982.

EBU (European Broadcasting Union). "Specification of the Digital Audio Interface." EBU Doc. Tech. 3250.

IEC (International Electrotechnical Commission). "Draft Standard for a Digital Audio Interface." IEC Report TC 84/WG11, November, 1986.

Jorgensen, Finn, *The Complete Handbook of Magnetic Recording*, 4th Edition, McGraw-Hill, 1996.

Mee, Dennis C. & Daniel, Eric D., editors, *Magnetic Recording Technology*, Second Edition, McGraw-Hill, 1996.

Pohlmann, Ken C., *Principles of Digital Audio*, Third Edition, McGraw-Hill, 1995.

Pohlmann, Ken C. *The Compact Disc*, Second Edition. AR Editions, Madison, WI, 1992.

Pohlmann, Ken C., *Principles of Digital Audio*, Second Edition. Howard W. Sams and Co., Carmel, IN, 1989.

Pohlmann, Ken C., editor, *Advanced Digital Audio*. Howard W. Sams and Co., Carmel, IN, 1991.

Tremaine, Howard M., *Audio Cyclopedia*, Howard W. Sams, Inc., now out of print but still available via specialty bookstores.

Watkinson, John, *RDAT*, Focal Press, Oxford, 1991.

CHAPTER

3.5

Studio Audio Equipment and Systems

STEVE CAMPBELL
Aurora Media, LLC
Las Vegas, Nevada

Additional material provided by

MARTIN STABBERT
Citadel Broadcasting Company
Reno, Nevada

MATHEW LIGHTNER
Lightner Electronics
Claysburg, Pennsylvania

JAY TYLER
Wheatstone Corporation
New Bern, North Carolina

INTRODUCTION

A successful radio station is created by building a format that is in demand by a large number of listeners and packing it with a fresh and exciting presentation to make it distinguishable from the competition, be it terrestrial, satellite, or portable systems. However, even a popular format, packaged by the best programming team, will suffer if the end product, *sound*, is produced by inadequate equipment.

When coupled with excellent programming, state-of-the-art audio equipment producing excellent sound for the station will equate to ratings and sales, and that will translate to revenue for the enterprise. State-of-the-art audio equipment and systems can also make the production of audio significantly less labor intensive and less expensive, and can greatly help the cost side of the ledger.

Over the past decade, technology convergence has been a hot topic. In the audio field, convergence is synonymous with computers and digital signal processing. Almost every facet of audio production, storage, and presentation has become a digital process. Even turntables to play vinyl records now have digital outputs. Complete broadcast facilities can be built where the audio is converted to a digital format at the earliest possible stage after the microphone and remain in the digital format right up to the transmitter (and, for HD Radio, including that also). Faithful reproduction of the input signal, whether it is good or bad, is maintained by these systems.

Most broadcast facilities, however, are a combination of new technology equipment and legacy equipment retained due to the need to replay legacy-recorded materials, traditional mindsets of the programming personnel, and also cost concerns. This melding of differing technologies brings with it interfacing and control issues that have to be addressed in studio equipment design and utilization.

This chapter describes most of the equipment usually found in a broadcast studio, followed by a discussion regarding the layout of studio equipment. Common installation and interconnection practices are then discussed, with an explanation of the terms and techniques used when interfacing equipment. Finally, there is information about the nature of the digital audio signal, the factors affecting its quality, and the standards for digital audio interconnections.

PLANNING FOR EQUIPMENT PROCUREMENT

Know What Is Available

Knowledge of the different types of modern audio equipment is crucial. The years of experience and successfully perfected practices of personnel responsible for producing the audio have caused much of the new digital equipment to emulate traditional analog equipment, and it is often designed using the same terms that were applied to traditional systems. Mixing desks or consoles may now be analog, digitally controlled analog, digital/analog hybrids, or digital workstations, but all may be designed to look much like a version of their analog predecessors to the user. For example, consoles still typically use 120 mm sliding faders, the on/off buttons are big and familiar colors, and CD players have *scrub wheels* to allow traditional record scratching. This appearance gives great comfort and proven performance to the users while taking advantage of improved technologies.

However, knowledge about studio equipment that goes much deeper than its appearance aids greatly in making good choices. The choice of equipment is vast and ever-increasing, and the market is dynamic. Each year there are innovations and new offerings in almost every type of equipment that makes up a studio. This is particularly true in the recording studio and live sound production portion of the industry that are driven by larger sales volumes and greater competition, and equipment from this side of the industry is increasingly finding its way into broadcast studios.

Sources of Information

One excellent source of information is the many trade publications targeting the broadcast and recording industries. In addition, most manufacturers maintain websites with large amounts of information on their products, including equipment manuals and, often, white papers on system design and technical issues. Much detailed information about the equipment's capabilities can be determined prior to purchase.

Most manufacturers in broadcast, and to a lesser degree in the recording industry, have technical support staffs that can assist greatly with any questions, both pre- and post-purchase. Attendance at the NAB trade shows is a wonderful opportunity to gain hands-on experience with most equipment that will be installed in a studio, and it is an opportunity to visit with knowledgeable representatives of the manufacturer.

Equipment Vendors and Systems Integrators

Before beginning to build a new studio, one needs to establish a good working relationship with one of the many excellent equipment vendors and/or an established systems integrator. The audio studio equipment market is highly competitive, and the cost differences between vendors for a given volume of purchases are small. A good working relationship with a trusted sales representative can be of great assistance in recommendations and studio purchases, and in keeping current with the ever-changing equipment landscape.

Based on the scope of the project and the engineering staff's current workload, hiring a systems integrator might be appropriate. With increased demands being placed on fewer station/studio engineers, it can be difficult to focus on new projects. The rate of change in technology makes it a constant battle to keep up with the changes, and an integrator or consultant can bring specialist knowledge that is not otherwise available. If the installation is planned to utilize a control surface-based router system and the vendor has been selected, ask the vendor for recommendations on studio integrators familiar with, or factory-certified to install, the equipment. Many equipment manufacturers have a systems integrator training and certification program to assure the systems integrator's staff is properly qualified to install the product.

It is important to talk to the integrator in the planning stages of the project. Since systems integrators build studios and systems regularly, they are familiar with multiple manufacturers' equipment and how it interacts to become the complete system. The key to a good systems integrator is organization and documentation. An integrator must have a good project manager who closely monitors the whole project, from design through construction. The system will require support after the systems integrator leaves, so the systems integrator must provide detailed system documentation that can be updated easily.

Today's systems integrator is more than someone who can terminate wires. With technology convergence, most integrators have specialists in the programming of control-surface-based router systems. During initial system implementation, it is important to work with the integrator's programming specialist to learn about the programming of the system software and how to make future changes.

The systems integrator's staff needs to be trained in computer networking since newer audio network systems use very similar technologies and components as generic data networks. In fact, in modern studio systems, a large part of the project can be specialized computer and information technology (IT) work. A good integrator will certainly have the tools and specialized test equipment to test, verify, and document the Category 5 (CAT5), Category 6 (CAT6), Category 7 (CAT7), or fiber optic cable systems that are a major part of today's facilities.

AUDIO CONSOLES

The heart of most studios is the *audio console* or *mixing desk*. This device is the central focus of all the audio originating from the room (although in radio today it can be argued that the automation system frequently has a similar indispensable status). There are many consoles on the market, all targeted at a variation of the same function. The function is to mix many audio inputs to one or more audio outputs and to provide monitoring of the individual inputs and each of the outputs, both visually by metering and aurally. The function of the studio room will determine the features necessary and, perhaps nearly as important, the features not wanted on the console. For radio broadcast use, the vast majority of mixers currently in use are designed for two-channel stereo operations. Consoles available at the time of writing (2006) include those based on both analog and digital technologies, so we have analog consoles, electronically or digitally controlled analog consoles, digital consoles, and also digital workstations.

Analog Consoles

Analog consoles are the traditional type of audio mixer and have been used since the earliest days of broadcasting.

In its simplest form, each *input* of an analog mixer is buffered by a *transformer* or an *amplifier*, may be processed by *equalization* or *dynamics control*, and then is applied to an *audio fader* (rotary or linear). The output of the fader is applied to *bus selector* switches and then fed to a common *mixing bus*, which is amplified, buffered, perhaps processed in some manner, and sent to the mixer outputs. Associated with this process, there will be a variety of inputs and outputs, monitoring, and various other controls and ancillary functions.

Analog Console Features

Taps are taken from the input module, pre fader, that can be supplied to a *cue* or *pre fade listen* (PFL) bus for monitoring, and each output bus can be selected for monitoring. *Metering* is provided to monitor the primary output, with other metering that can be selected to monitor other buses and/or the cue/PFL bus. The console may have multiple monitoring sections to permit a loudspeaker monitor in the control room as well as monitor outputs for multiple studios that may need to monitor a different source than the control room, and may be equipped with a *talkback* microphone and circuitry for communications with the studio.

When intended for the broadcast market, *line inputs* will usually be *stereo* (with *left* and *right* channels) and microphone inputs will often be *mono* (single channel, which can be routed to left, right, or both channels to place the sound where desired on the sound stage). The console will have circuitry to *mute* the monitors and cue/PFL speakers within the control room and/or studio when a microphone associated with that room is turned on. It may also have *count-up timers* that are reset when selected channels are turned on, and may contain a *real-time clock* and perhaps a *thermometer.* It will have *machine control* outputs to permit control of *audio source* equipment from the console and *channel control* inputs to permit some functions of the console to be controlled remotely from elsewhere.

All the controls associated with an input channel are usually mounted in line on a *channel strip,* whereas other controls for buses, outputs, and ancillary functions are mounted on other parts of the control panel or on separate modules.

If a console is originally designed for the recording studio or sound reinforcement market, some of the features just listed may need to be added to make the console functional in the broadcast environment, and the broadcast accessory makers build add-ons for this purpose.[1]

Input channels may be designed for either microphone or line-level inputs or may be configurable to be operated as either with configuration jumpers or switches. A switch to select alternate "A" and "B" inputs is common. Specialty input modules may also be available for the console to provide more complex input and the *mix-minus* configurations utilized for telephone calls (see later section).

[1]Broadcast Tools, Inc., is just one example; see http://www.broadcasttools.com/.

Configuration

Some consoles are built in a modular form, placing each channel strip, monitor, output amplifier, and specialty control card on a separate module that can be removed from the front or top of the console, often while the console is in operation. Modular consoles greatly simplify troubleshooting and repair but add additional electrical connections that can become problematic. Other consoles may be a monolithic design, constructed on a single or a few printed circuit boards for all channels. Monolithic designs may reduce the cost of the console but can make repairs and troubleshooting more difficult. Figure 3.5-1 shows a popular monolithic analog console used in radio broadcast.

The major cause of failure for analog consoles is spilled beverages, which can adversely affect either method of construction.

Electronically or Digitally Controlled Analog Consoles

Traditional analog consoles share the common trait that the front panel switches and audio faders directly control all audio. As these devices become dirty and worn, they inject noise into the program signal or may cause a channel to drop out. Therefore, many newer analog console designs are implemented with digitally controlled solid-state analog switches and voltage controlled amplifier (VCA) or digital controlled variable-gain amplifiers. The operator interfaces may still be front-panel controls with the same layout as a traditional analog console, but they provide DC control voltages to electronically switch and fade the analog signal. This improvement has made the consoles more reliable and, in some cases, less expensive to construct. The analog console shown in Figure 3.5-1 is of this type.

The use of electronic controls also allows another configuration in which the analog audio circuitry is

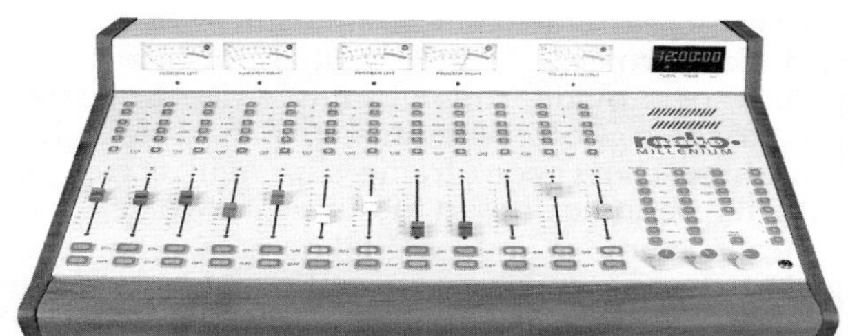

FIGURE 3.5-1 A Radio Systems Millennium Series monolithic electronically controlled analog audio console. (Courtesy of and copyrighted by Radio Systems, Inc. 2006.)

mounted in one unit, typically rack mounted, while the control surface is located remotely in the location for the operator to use. The interconnection between the two carries the control signals. This allows more flexible arrangements for system design and implementation.

The use of electronic controls makes the adverse effects of dirty and worn controls much less objectionable. With solid-state electronic switching, both stereo channels are either present or they are missing. With a remotely controlled amplifier, when an audio fader becomes intermittent, instead of hearing audio crackles and stereo tracking errors, the channel gain just becomes unsmooth and not proportionate to the control position.

The features of electronically controlled consoles intended for broadcast use are typically a slightly enhanced version of those described for traditional analog consoles. With this arrangement, control features are more easily implemented, and some console designs have micro-controllers on each channel to minimize component count while further increasing flexibility.

FIGURE 3.5-2 Digital console primarily targeted for recording studios and sound reinforcement that can be effectively used in radio and television production. (Courtesy of Yamaha Corporation of America.)

Digital Consoles

The most recent development in console design is the fully digital console. The first digital consoles were introduced in the mid 1980s for high-end recording use, but it was not until the mid 1990s that reasonably priced digital consoles showed up on the market, intended for the sound reinforcement and recording studio market. These later migrated into the broadcast market; a popular version of such a console is shown in Figure 3.5-2. Digital consoles designed especially for broadcast applications have since been introduced by numerous manufacturers.

Digital Console Features

Digital consoles either convert analog audio inputs (typically from microphones) to the digital domain upon entering the console, or they accept the input already in digital form, typically from an AES/EBU serial digital interface. With digital inputs, the console may have *sample rate converters* to resample and lock the incoming audio to the sample rate used by the console. Without internal sample rate converters, it is necessary to either lock all audio sources to the console, lock the console and the audio sources to a master clock, or provide external sample rate converters locked to the console. Without source/console lock or sample rate conversion, frame slips will occur that sound like periodic pops in the audio, or worse.

Once the audio is in the digital domain, all processing, gain changes, mixing, and routing take place with mathematical processes within one or more digital signal processors (DSPs). Features for digital consoles typically include at least all those available for analog systems. However, in the digital domain, many specialized mixes can be derived without the addition of additional mixing buses of traditional consoles. For example, it is possible to easily construct a console that can provide a mix-minus output for each of its inputs. Many digital console manufacturers also package digital effects processors into their consoles, which previously had been available only with outboard devices. The limits on what can be done to the signals depend only on the power of the DSP engine(s) and the forethought of the software designer.

Digital broadcast consoles were first introduced principally so as to be able to handle digital audio inputs and outputs and to realize the economies of technology convergence. As the use of such systems increased, it was quickly realized that the paradigm of a one-to-one correspondence between the channel strip on the console and the console audio input (or the two inputs where A/B switching was provided), as used in traditional analog designs, served little purpose in the digital domain. While a tight electrical connection between an input and its control channel strip was necessary for analog systems, for digital systems, an input signal could be accessed by any channel strip and routed with just a few lines of computer code. Similarly, any function that was made available to one channel is now available for all channels.

Control Surface/Router Concepts

With the addition of reliable high-speed serial data communication capable of carrying many channels of audio between two points, the paradigm further changed by moving the audio inputs/outputs and the *mixing engine* away from the mixer control panel, which now became known as a *control surface*. It became practical to distribute the inputs and outputs between multiple chassis. Two technologies in broadcast—the router technology for switching signals and the console technology for audio mixing—converged, and the newest high-end consoles for broadcast are, in

FIGURE 3.5-3 A console that operates both as a stand-alone console and as part of a control-surface-based router system. (Courtesy of Harris Corporation.)

whole or in part, a control-surface-based multilocation input/output configuration where entire facilities become a control-surface-based system using a single router.

Some consoles realize this integration for only a portion of the console's inputs and outputs, and the remainder of the console operates as a conventional digital or analog console, as is the case for the digital console shown in Figure 3.5-3. Other consoles are implemented utilizing a total control surface/router design, as is the case for the work surface shown in Figure 3.5-4. The installation and operational benefits of this convergence are enormous; however, care must be taken with the system architecture to avoid the possibility of a single-point failure affecting multiple studios.

Integration with Automation and Other Systems

Integration of control-surface-based/router audio mixing systems with other equipment, such as automation systems and video switchers, is easier than with their analog counterparts. The digital network backbone of the control-surface-based routers may be a proprietary protocol, but some systems are converging to conventional IP-based Ethernet networks. One system that uses IP-based Ethernet is shown in Figure 3.5-5. In addition to routing audio between multiple *nodes* on the system, each node can generally support general-purpose input/output (GPIO) control for other equipment within the system. The access to the GPIO controlling a machine can follow the machine's audio and be controlled by multiple surfaces. Additional developments of inter-system IP-based control connectivity between the source equipment and control-surface-based router systems allow automation systems to control the router system, and the router systems to control the automation systems, without the labor-intensive implementation of controls by GPIO.

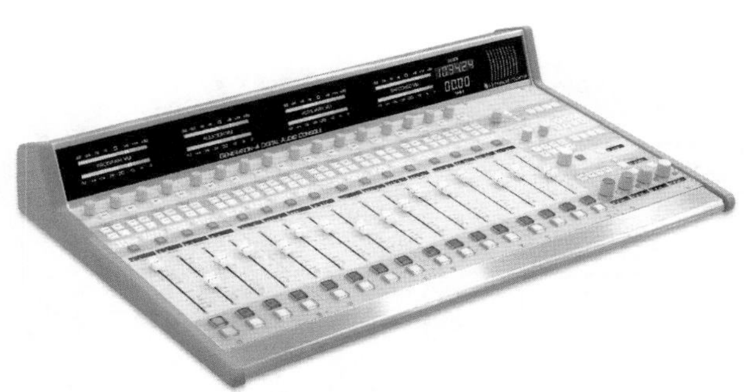

FIGURE 3.5-4 A console that is implemented completely as a work surface in a control-surface-based router system. (Courtesy of Wheatstone Corporation.)

FIGURE 3.5-5 A control-surface-based router system work-surface-based on IP Ethernet connectivity for both control and audio. (Courtesy of Telos Systems.)

A frame that holds all the input, output, mixing, and GPIO is shown is Figure 3.5-6. As the scope and complexity of the control-surface-based router systems grow, the need for consultation with the manufacturer during the system design phase also grows. Most manufacturers offer expert guidance on their products.

Digital Workstations as a Console

As the use of personal computers (PCs) for digital audio workstations grows, so do the choices for audio interfaces and processing systems to transport audio into and out of the computer. From the original poor-sounding audio cards, audio quality has grown to rival the best audio consoles. The number of channels that can be handled has also grown due to the inclusion of powerful DSPs in the audio interface. It is now possible to control 96 channels or more of high-quality audio with a PC.

The input/output module for one of these systems is shown in Figure 3.5-7. The operator interface and all of the functions of a traditional console have been replaced by a graphical representation of an audio mixing desk, as illustrated in Figure 3.5-8. In some applications where quick control and traditional tactile operator feedback are not required, the computer and its audio interfaces can become the

FIGURE 3.5-6 Front and back view of the rack-mounted chassis for a control-surface-based router system, which contains all the audio and control connections to the studio. (Courtesy of Wheatstone Corporation.)

audio console for a studio. Additionally, digital audio workstation controllers are available to provide the desired tactile feel of a traditional console to the workstation. Some of the audio interfaces are able to operate as a stand-alone mixer with a compact operator interface without using the computer except for its computer-based functions. While currently not a good solution for active mixing of a dynamic program, this convergence of technologies can be useful in off-line production facilities where the console is used more as a source selector than a mixer. These audio interfaces may also serve well for remote and portable operations where size is of concern, as sub mixers or auxiliary mixers in a larger studio, and other lower-demand applications.

Continued progress in operator interfaces and the comfort level of younger operators with computer interfaces will no doubt see workstation-based consoles making inroads into more traditional console applications.

Console Selection Considerations

The audio console is the single most important equipment selection of any studio. When this selection is made from the large number of products available, targeted at many different markets, the following information should be compiled as a guide.

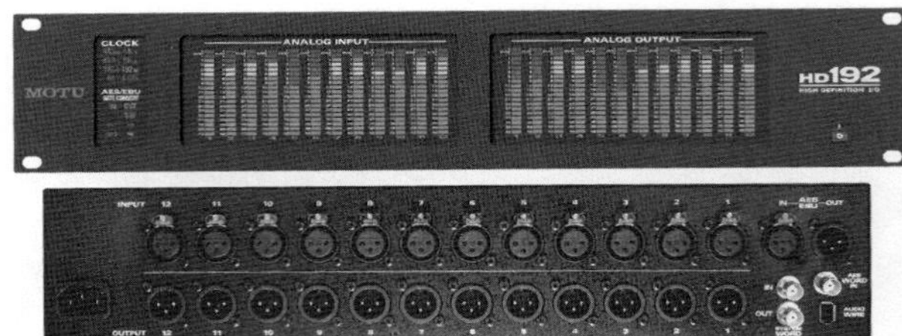

FIGURE 3.5-7 Many channels of audio can be controlled by computers and may eliminate the need for traditional audio consoles in some installations. The I/O module shown is the interface between the audio equipment and the computer. (Courtesy of MOTU, Inc.)

Number and Types of Inputs

The number of *sources* and their names should be compiled for each studio. This is best done in a spreadsheet and should include the following details for each source:

- Is the source mono or stereo?
- Is the source digital (format AES/EBU, S/PDIF, or other) or analog (mic, consumer level, or balanced professional level)?
- Does the source require machine control and machine status?
- Will the channel require remote control and remote indicators (remote on/off, cough switch, tally lights)?
- What other studios must have access to this source?

When this list of inputs is compiled, it is easy to overlook seldom-used but important sources such as the studio television receiver, talent's laptop computers, the other studios within the facility that might need to be added to the air, and remote programs such as Traffic. The list should contain a reasonable number of uncommitted source inputs for the future.

For a system design with traditional individual consoles, the list of inputs should be compiled for each studio, with cross lists for sources shared between stu-

FIGURE 3.5-8 A virtual audio console represented on a computer screen (Courtesy of MOTU, Inc.)

dios. For a system that includes a central router, the shared sources should be planned as inputs to the router and several router channels planned as inputs to each studio's console.

When this list is compiled for a facility-wide control-surface-based router system, the list of inputs should be compiled for the entire complex, eliminating all duplicate sources but including a sixth item of information:

- Where will the source be located?

In a control-surface-based router system, it is also necessary to include each air monitor or other monitored source as an input to the system.

It should be noted that, except in highly complex live production, a console that is too large can be a disadvantage. It is possible to have too many input channels. A simple rule suggests if you cannot reach it, do not buy it. Many mixer channels are on 1.5 in. centers. Thirty-six inputs cover 54 in.—a reasonable limit to the average person's reach.

Number of Outputs Needed

A list of the *outputs* needed should also be compiled for each studio; again, a spreadsheet is recommended. Traditional consoles have little flexibility in the number of outputs. There is an output for each bus, for the control room monitor speakers, and sometimes for the headphone section and the cue section. To this are added outputs for any optional studio room monitors, outputs for the mix-minus in any specialized telephone modules, and often an unbalanced or balanced output for an insert point within the channels. Integrated digital/console systems are more flexible, but each output must be explicitly specified; the systems normally do not have any standard outputs.

An output channel must be specified for each stereo and mono mix bus, the headphones, studio monitors, control room monitors, and all mix-minus outputs. Other less obvious output needs may be for each restroom, the background speaker system, and the input to the telephone music-on-hold. They should all be detailed in the spreadsheet and should include

- If the destination is mono, stereo, or multichannel;
- If the destination is analog or digital and, if digital, its format;
- The physical location of the output since it need not reside within the studio.

Number of Inputs Needed during Any Session

The number of installed channel strips on a console or work surface should be sufficient to permit the normal operation of a session or show to be executed without re-assigning faders, selecting "B" inputs to the module, or selecting audio from a multiselector switch. "B" inputs, multiselectors, or dynamic re-assignment of console inputs in a digital console should be used only for *specialty* demands. In general, something that happens within every hour of a session should be considered as normal. Sources needed less often are specialty and are candidates for "B" inputs or multiselectors in traditional consoles. When "B" inputs or multiselectors are used, care is needed not to place two sources that are needed at the same time on the same input module or multiselector.

The number of inputs needed during a session plus the number of inputs needed for the multiselector inputs will determine the number of channel strips needed in the console. This number should be increased by a channel strip or two to allow for the unexpected. Many modular consoles are available in multiple frame sizes, and it is common practice to order a frame size larger than the number of channel strips required. The extra slots allow for future expansion, and spaces can be used to logically group channel strips for the ease of the operator. Any other differentiations between channel strips, including clear labeling and fader knob color-coding, will make the system easier to use.

Qualification of the Operator Using the Equipment

The best equipment will produce only mediocre results if not used correctly. Some studios are primarily used by personnel only marginally proficient in broadcast technology and engineering. Selecting a console with too many options and controls may only cause intimidation and confusion, whereas the installation of a simple broadcast console without equalization or dynamic control may suffice. In contrast, a proficient user working on a complicated mix will be severely limited by such a selection. Matching the console to the users is critical. An engineer making the console choice can quite easily overlook that what he considers intuitive may be far less than intuitive to the users. The end user should be kept in mind at all times.

Criticality and Function of the Studio

As discussed previously, consoles can be constructed in a modular form or as a single circuit board. If the studio is mission critical for the enterprise, modular consoles offer significant advantage for repair and troubleshooting. Additionally, redundant power supplies are available for some consoles and can add to the reliability of the installation. Conversely, an uncomplicated production room or a voice track studio may be well served by just an input selector or a simple mixer. Some production rooms may be very well served by sound reinforcement or recording studio consoles, particularly those that control stereo inputs on single channel strips.

Television Studios

Audio for television studios can be challenging because of the need for many complex audio mixes and, often, for many live audio sources. Television studio audio systems require at minimum a stereo program feed, on-camera talent monitor mixes, usually with interruptible fold-back (IFB), possibly a studio audience PA mix, mix-minus for telephone feeds and remote shots, and audio monitoring for the preview channels. Add to this the complexity of second language information for the secondary audio channel (SAP), and the possibility of generating 5.1 and 7.1 mixes, and it is clear that the capabilities of the audio console for television must be carefully considered.

Consoles for television must be designed specifically for these mixing tasks and may also need to integrate with the video switcher to help seamlessly handle the workload. Consoles of this complexity should include full equalizers and dynamic control sections to allow sweetening of the audio. The equalization and dynamic control section on digital consoles can be a sub-work surface shared by all of the channels, which can greatly simplify the routine operation of the console.

Scope, Future, and Budget

Console selection should take into account the integration needed with the rest of the facilities. If the facility is currently structured with a control-surface-based router system, the selection should also be integrated. When the plans for the future include a control-surface-based router system, the current selection should also integrate with future purchases, and this may require selecting the system architecture well before the major installation. If a control-surface-based router system is not in the future plans, purchasing one for a single studio may not be the appropriate use of resources. When the cost of a console system is evaluated, the installation costs should be taken into account. While the equipment costs of a control-surface-based router system are high, the installation requires significantly less interconnection, fewer wires and connectors, fewer distribution amplifiers, fewer switchers and routers, and much less installation time. When these factors are considered, the additional hardware cost of a control-surface-based router system may be easier to justify.

Conversely, small studios may be well served by a rack-mounted digital audio work surface audio input module and full on-screen control of the workstation. Console choices are vast, and each is the perfect fit for some application. The matching of the console to the project is probably the single most important choice in the studio.

Mix-Minus Issues

In many circumstances, it is common to derive a separate mix for monitoring of a live program. Television and sound reinforcement have been using *monitor mixes* for years, and any large concert has two mixers with two separate engineers: one on stage creating a unique mix for each member of the band and a front-of-house mixer creating the product for the audience. Until recent times, radio broadcast has made less use of separate mixes for monitoring. However, the increased use of telephone on air, where the caller must hear the entire program, and the increase in the use of remote broadcast systems with significant delay have made the use of separate mixes for monitoring more commonplace in radio broadcasting.

One special-case mix arrangement used primarily for monitoring is called a *mix-minus*. This mix contains the entire program channel minus the audio from the channel that will be monitoring this mix. Because phone systems are typically a two-wire system, carrying the send and caller audio on the same circuit, sending a copy of the caller audio back down the phone line where it will become part of the caller audio causes several problems, including frequency selective cancellation, echo, and feedback. The use of a mix-minus feed to the phone system eliminates the closed return loop. Feeding monitor audio without mix-minus to remote lines with long delays will cause the talent at the end of the line to hear himself delayed, which will cause most talent difficulty and create cadence problems with speaking. The use of a mix-minus feed to the talent eliminates this delay problem while maintaining the capability to monitor the main studio program. Mix-minuses are often coupled with an IFB circuit such that the program being fed to the remote talent can be interrupted for operational communications. This IFB can be done with a button on a console selecting the main studio microphone or may be done as part of a more elaborate intercom system.

Mix-minuses can be derived in several ways. Many consoles provide custom modules that cascade two mixing buses, such that the sum from the first bus of everything but the target channel is available for feeding to the target channel's monitor circuit, summing the target channel with previously summed channels to create the final program. A simplified block diagram of a console telephone module is shown in Figure 3.5-9. Specialized telephone modules are available for most broadcast consoles, which create one or two mix-minuses, and often the buses involved in the mix-minus for each phone channel can change based on the channel on/off status or other settings.

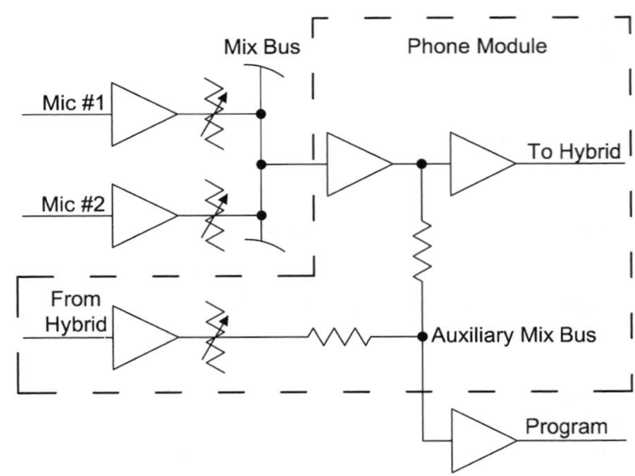

FIGURE 3.5-9 Simplified block diagram of an analog console telephone module.

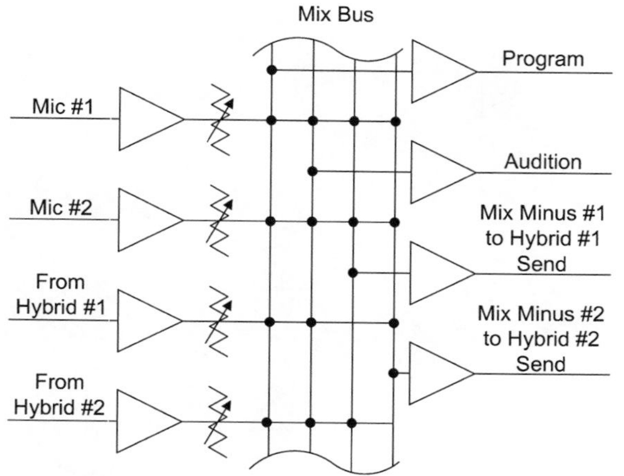

FIGURE 3.5-10 Cross-fed mix-minus outputs created with console auxiliary buses.

One of the most common methods is to create the mix-minus feed on a separate bus by summing all the desired channels into both the main bus and an auxiliary bus, omitting the target channel from the auxiliary bus, as shown in Figure 3.5-10. On a multibus console, several different cross-feed mix-minuses can be created in this manner with the advantage that each feed can be simplified and customized for each receivers' need.

In analog consoles, a mix-minus requires additional circuitry or requires additional console buses, but digital techniques have greatly simplified the process. In digital consoles, a mix-minus is simply an additional software buffer, additional code to do multiple audio sums, additional DSP utilization, and a destination digital-to-analog output channel. In some cases, the mix-minus can be handed off in digital form to an interface that will encapsulate it with the data being sent to the remote location. There are digital consoles that have mix-minuses available for every input channel.

MONITOR LOUDSPEAKERS

The ability to hear the product is critical to producing a good product. In a production studio, the quality of the audio produced will be colored by the accuracy of the monitoring system. Often audio will be adjusted to compensate for the characteristics of the monitors, and the product will not sound right except on those monitors. The air signal in a radio broadcast control room is generally not sweetened to the extent that it is in a production room, but the ability to spot problems with the product of the radio station is equally important.

When a monitor loudspeaker system, often known simply as *monitors* or *speakers*, is chosen, the room size largely dictates the power-handling capacity required and in part the cabinet size. A large studio should invest in monitors with a 12-inch woofer. Smaller studios may be limited to cabinets that contain a 5- to 6-inch woofer. As technology has advanced, improvements in low-frequency response for smaller speakers, at reasonable audio levels, are remarkable.

Theoretically, the best loudspeakers are coaxial in construction. Next are those with drivers aligned in a vertical array. There should be only one driver for each frequency range; two-way systems tend to be more accurate than those with three drivers. This is due to inevitable distortions in the crossover range. Systems with drivers horizontally offset create acoustical phase distortion. This is due to different arrival times, at the ears, of sounds in the crossover band. Loudspeakers built in mirror image pairs are the first to delete from the list.

"Specsmanship" is a game that manufacturers play with all equipment, and speakers are susceptible to difficult-to-interpret claims, but specifications are important. When purchasing a speaker, one should look for low distortion numbers and a flat response with a pattern that is compatible with the installation. However, there is no substitute for listening.

Evaluation of Loudspeakers[2]

A/X/Y testing of loudspeakers maintains objectivity and overcomes the variables of other methods. A is a known loudspeaker; X and Y are unknowns to be compared to A and to each other. The testing may be done in a large room, parking lot, or open field. If an existing control room is available, listening should be done there. Otherwise, a small mixer, headphones, and a few power amplifiers may be used.

The process consists of placing two candidate systems together and laying a microphone on the floor or other hard surface about 15 ft away. An omnidirectional dynamic microphone with wide, smooth frequency response is recommended.

Using program material containing music and speech, set up the mixer to compare the direct program with that returning from the microphone. Switch repeatedly between the direct and each of the test loudspeakers. The more accurate one will reveal itself easily; replace the other with the next test unit. To ease any doubts about the control room monitor, use a similar monitor as a test system. Any inaccuracies will be magnified, not masked. Finally, rerun the test using the initial winner as a control room reference to check the top two or three candidates. The results should not vary significantly.

While not as systematic as the preceding method, a trip to a large professional sound dealer with a good demonstration room will reveal much about the many monitor speakers available on the market today.

Speaker Location and Mounting

The monitor speaker is only part of the equation, however; the *room acoustics* are also part of the monitoring

[2]This section from the *NAB Engineering Handbook*, 9th edition, Chapter 3.1, p. 268, by Roy W. Rising.

system. If the project has an acoustical consultant, his advice should be followed carefully regarding the speaker placement and mounting methods. Without a consultant, follow the manufacturers' recommendations as closely as possible and strive for symmetrical location about the primary operator's listening position.

Care should be given to the mounting system to minimize mechanical coupling to the wall and to another studio. As the weight of the monitor speakers increase, the attention to the support capability of the mount must also increase. Be sure that the monitor speakers are securely mounted and, if in doubt, install a safety cable to prevent personnel injury should the mount fail.

Near-Field Speakers

The popular use of small *near-field* monitors lessens the effect of the room acoustics from the monitoring system. The closer the listener is to the monitor, the less the reverberant field colors the audio. Using near-field monitors as recommended by their manufacturer is proving to be the most effective truthful monitor for many broadcast applications.

FIGURE 3.5-11 A near-field powered monitor loudspeaker. (Courtesy of Tannoy Ltd. North America.)

Power Amplifiers

The *power amplifier* is a vital component of the monitoring system. Matching the size of the power amplifier to the monitor speakers selected is important. Many speaker manufacturers will recommend an amplifier power rating for their speakers, and this advice is appropriate. A power amplifier that is too small may be driven into clipping too easily, and one that is too large may melt voice coils too often. However, if in question, err on the high power side for the amplifier. If damage to speakers is a problem, consider installing fuses in the speaker feeds or a limiter in the input circuit to the amplifier. The amplifier selected should have a balanced input and, if it is to be mounted in the studio, should be convection cooled. Fans in any equipment within a quiet studio should be avoided if possible. The speaker wire is important and #14 or #12 AWG should be considered a minimum for all but the shortest runs. It is wise to attempt to keep all speaker runs the same length.

Many monitors are available in a self-powered version, and one popular unit is shown in Figure 3.5-11. In earlier offerings of monitor speakers, and in the offerings of the current lower-priced self-powered monitors, the performance is not good. The newer offerings of the professional powered monitor speakers are now coupling well-designed amplifiers to their speakers, and the performance is excellent. Self-powered monitors are worth considering. In the studio design, be sure to specify appropriate power at the monitor locations if a self-powered speaker is to be used.

In general, buy the best loudspeaker monitor the budget will allow and install it per the manufacturer's instructions. Except in extraordinary conditions, do not install equalizers to compensate for the monitor speakers, and if they are used, lock them down. It is critical that the monitor system be consistent. The ear is very adaptable, and minor inaccuracies will be integrated out of the listeners' perception with time. If the monitor system is inconsistent by changes in placement, balance, drive, equalization, or nearby environment, the adaptation to the room's sound will be greatly hampered, and the mixes will suffer.

Often, critical monitoring of the final mix using multiple alternative sets of monitors is desired. Engineers have installed home stereo speakers, car stereo speakers, or speakers designed with characteristics similar to conventional radios and televisions in order to assure that the end product will deliver the expected results in a variety of environments. It is not uncommon for a recording engineer to take his final mix for a ride in his car as a final check.

Cue and Intercom Speakers

The *cue* and any *intercom* speaker systems also deserve attention, but in the opposite way. A cue or intercom speaker system is not intended for critically analyzing audio quality. Its purpose is to get a quick communication from another location or to locate the beginning of audio in a source. It is important there is no confusion

between the cue and the normal monitors within the studio, and the operator must distinguish between the monitor and the cue/intercom source without thought. Often, cue speakers are better with limited bandwidth, opting for intelligibility rather than full fidelity. On-air errors have occurred when the cue system sounded as good as the monitors and an operator believed the source was playing on air when it was in cue.

AUDIO SOURCE EQUIPMENT

Digital Audio Storage and Automation Systems

The digital audio storage/automation system has all but replaced every audio source for music and commercial content playback and archival in most broadcast stations. It is an important area in broadcast, and Chapter 3.6 is devoted to automation and digital audio storage systems.

In the past, hardware expense and size limitations made the use of digital audio compression techniques critical to the implementation of digital audio storage/automation systems. With the clock rates of modern workstations now exceeding 3 GHz, Gbit LANs, the availability of terabyte hard drives, and inexpensive RAID controllers, the need for digital audio compression in digital audio storage has largely disappeared, and its use should be avoided wherever possible.

Most audio *compression algorithms* that yield significant compression ratio are *lossy;* the restored data is not an exact copy of the original data but rather only that which the algorithm designers believe necessary for the algorithm to sound appropriately like the original. These algorithms generally perform well, but they can start to accentuate the discarded parts of the audio when they are cascaded with multiple passes through the same or different algorithms.

HD Radio uses digital audio compression for transmission, and it is sometimes necessary to use it for limited-channel bandwidth remote audio and limited-bandwidth studio-to-transmitter links, so the elimination of digital audio compression in studios, and at any stage that it can reasonably be eliminated, is good engineering practice. Digital audio data compression is covered in detail in Chapter 3.7.

Digital Audio Workstation

A *digital audio workstation* (DAW) is a reasonably powerful computer with a high-quality multichannel sound card and specialized audio editing software. The editing screen image of a popular DAW software package is shown in Figure 3.5-12. The computer may also include a custom user interface, such as the one shown in Figure 3.5-13, called a DAW controller, which will allow audio-console-like control of the channels within the audio editing software. It provides tape-machine-like buttons to control the digital recorder within the audio editing software and a scrub wheel to emulate the rocking of reels back and forth to locate an exact segment of audio. These devices have all but eliminated the use of analog tape machines in audio production because they bring many advantages and capabilities that can be accomplished in an analog tape environment only by extensive overdubbing, mixing, and editing, all processes which degrade the audio and build up the noise floor.

FIGURE 3.5-12 A popular digital audio editor program. (Courtesy of Adobe Systems, Incorporated.)

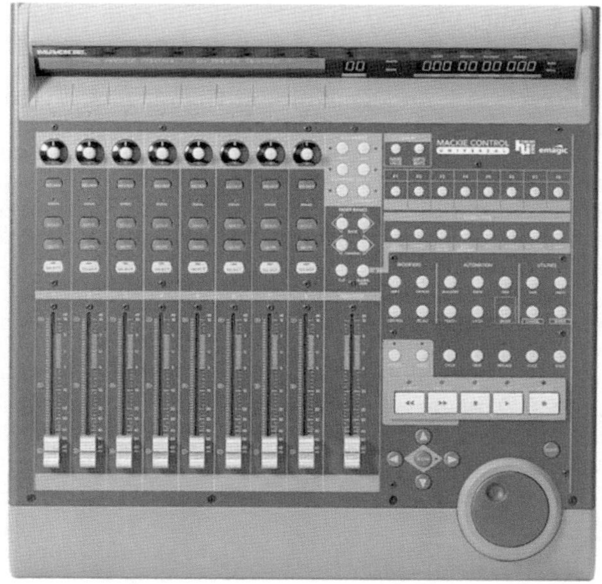

FIGURE 3.5-13 A production digital audio workstation control surface used to emulate the functions of traditional equipment. (Courtesy of Loud Technologies, Inc.)

DAW Editing

Digital editing software allows nonlinear editing, which is the random access to audio in a manner that is not continuous from start to finish. Additionally, multiple audio tracks can be slipped in time with respect to one another, which cannot be done with tape without overdubbing. A single piece of audio can be replicated many times within the production and placed with individual sample time accuracy, without resorting to creating a copy and overdubbing. In short, all audio manipulation can be done by recording the indexes to segments of audio and playing the audio back at the appropriate time. These changes can be nondestructive such that the entire content of all the elements making up a project remains intact if the production needs to be updated or redone.

DAWs can be integrated into the digital audio storage network so the finished production can be copied at file-transfer speeds for use in the rest of the facility. As digital audio storage and automation systems have replaced cart machines in on-air studios, digital audio workstations have replaced nearly all tape equipment in a modern production room.

The cost of this equipment is within reach of most audio professionals, so project studios are often built at home, allowing production work to be done off-premises and allowing staff the ability to earn extra income doing independent projects.

CD Players and Recorders

Next to computerized digital audio storage systems, the *CD player* is the most common music source within a radio station. The choice in high-quality commercial CD players is large. Rack-mount CD players such as the one shown in Figure 3.5-14 are very functional in broadcast studios. All professional CD players provide the capability for remote control, which allows console control of CD player starts. The technology of the CD player is discussed in detail in Chapter 3.4 of this handbook.

Consumer-grade CD players can be used in a studio, but it is likely that a semipro to pro converter will be needed to modify the audio level, and remote controllability is likely to be limited. Cueing consumer quality CD players accurately is difficult and, with the wide selection of reasonably priced professional grade units on the market, consumer grade machines should be avoided.

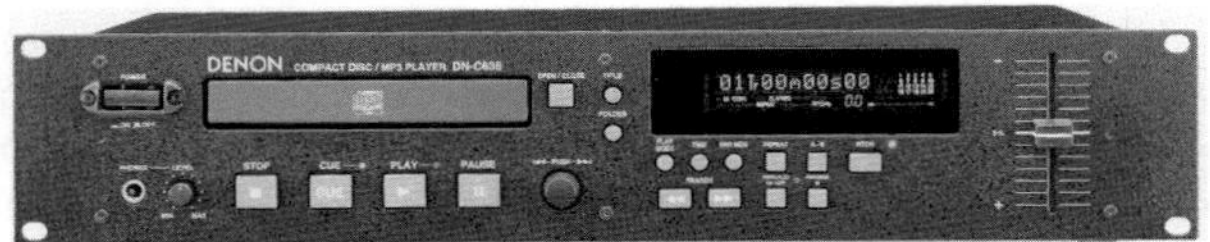

FIGURE 3.5-14 A professional rack-mount CD player appropriate for use in a broadcast environment. (Courtesy of D&M North America.)

CD recorders have moved out of the computer and into stand-alone rack-mounted equipment. There is a wide offering of CD recorders capable of recording CD audio in standard format. Some of these recorders are also capable of recording data CDs containing a large number of MP3 tracks.

Turntables

For many years, *vinyl records* were the mainstay of recorded music for broadcast stations. Many stations have record libraries dating back to the beginning of the station, and some of this material is available only on their archived vinyl disks. *Turntables* to play these disks are still available, but the newer units are not constructed as robustly as the turntables of old with their 1/8 horsepower rim-drive motors. Their quality, however, has greatly improved. The newer turntables are generally direct-drive units; the rotor is a permanent magnet attached to the platter, and the stator is electronically driven by a closed loop speed controller. Most modern turntables have an integral base with vibration isolators that virtually eliminate the need for custom mounting techniques. Most have a *tone arm*, with a *pickup cartridge*, designed for the turntable and factory mounted. Newer entries offer integrated preamps, some with digital outputs.

Most broadcast operations should have at least one turntable with an appropriate preamp installed in a production room. The use of turntables on air is less reliable than CDs or digital audio storage systems, and the dubbing of the needed vinyl record into the digital audio system is preferred. The exception is for specialty shows that might do *beat mixing* or scrubbing, as is often done in some contemporary music stations. A complete setup, such as a club disk jockey (DJ) might use, is often installed as a source to the main console so that a DJ can perform his show just as he would at the club. This setup generally includes two turntables and a club-type audio mixer, and might include one or more club CD players that have scrub capability.

MiniDisc Player

Announced in 1991 by Sony, the *MiniDisc* (MD) was designed to be the replacement for the consumer cassette machine and as a digital data storage device, but it has not enjoyed wide acceptance in either role. The MD did, however, establish itself as a reliable, compact, and inexpensive medium for professional use. The small recordable disc, enclosed in a protective case, or *caddy*, uses a magneto-optical recording system and the Sony proprietary ATRAC compression system.

The MD is capable of recording between 74 and 80 minutes of compressed audio per disc, has indexing capabilities, and can cue quickly. Sony rates the rerecord capability of the medium at 1 million passes. While sensitive to physical shock during record and index writing, the player is immune to physical shock and skipping at playback due to the large buffers and high read rates of the player. The formats of MD have

advanced since its introduction, and in 2000, Sony announced MiniDisc Long Play (MDLP), which updated the compression algorithm and permitted three recording-quality standards: SP, L2, and L4, which have data rates of 292, 132, and 66 Kbps, respectively, providing record times of 80, 160, and 320 minutes on an 80-minute MD. Subsequently, Sony added Hi-MD, another improvement to its compression algorithm and a significant increase in the recording density of the medium. The original MD had approximately 140 MB of raw data storage capacity, whereas the Hi-MD is approximately 1 GB of raw storage. The addition of compression modes and of 16-bit uncompressed to the format lineup has improved the quality of the audio. The later releases of MD machines allow uploading and downloading of audio to a computer via USB connections utilizing proprietary Sony software.

Products to use the MD as a studio cart machine replacement were offered, but by the time they were available, the trend to computer-based digital audio storage was apparent and the products were short lived. However, the MD has found several niches in broadcast, most notably as a device to capture news and promotional actualities at remote locations, as air-check machines, to "skim" studio microphones for show analyzing, and for some archive use. Rack-mounted MD systems are available.

The transition of audio systems to computer technologies is making inroads into all the niches for MD, and the format is losing popularity in studio broadcast applications, but the platform is still a viable field-audio recording device. The technology of the MD is discussed further in Chapter 3.4.

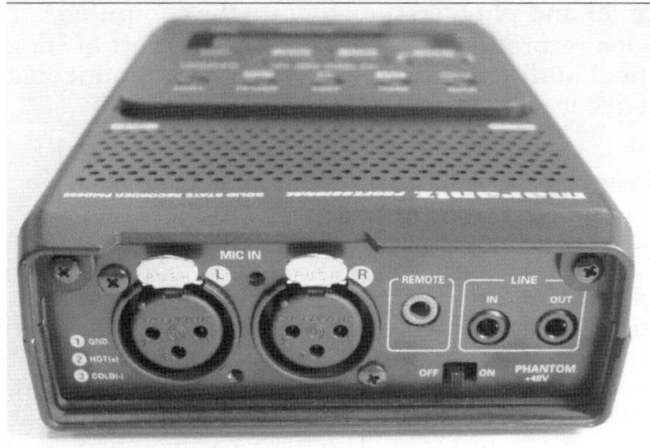

FIGURE 3.5-15 Portable solid-state audio recorders. (Courtesy of D&M North America.)

Compact Cassette Machines

Cassette tape machines have always been somewhat troublesome for use within broadcast facilities but, because of their widespread consumer acceptance and convenience, they have found niches in studios. Their principal use in broadcast studios was in support roles, primarily to remotely record news and promotional events, to use as air-check machines, and to deliver spec spots to clients. Many of these functions have been replaced by computer-based technologies, and cassettes are disappearing from broadcast stations as they are for consumer use. A *spec spot* delivered on a recordable CD now has a much higher possibility of being played than one on a cassette.

Solid-State and Compact Hard Disk Recorders

Solid-state and compact *hard disk* recorders are the miniaturization of their big brother DAWs used in studios. Some of the devices, such as the one shown in Figure 3.5-15, are equipped with professional balanced mic inputs; others combine a small audio mixer with a digital recorder into a battery-powered device that can be held in one hand. Solid-state recorders are also available that are built into a hand-held microphone, as shown in Figure 3.5-16. Others utilize unbalanced mic connectors and have limited control features, whereas still others have built-in stereo microphones.

Some consumer devices such as cell phones and personal digital assistants (PDAs) can be used as solid-state recorders and the well-known iPod[3] is a hand-held hard disk recorder.

Some of these devices utilize audio compression, but others are capable of stereo linear 16-bit recording at 48 kHz sample rate or better. The audio quality of solid-state recorders ranges from acceptable for taking notes to first-class high-quality music recording. These devices utilize either fixed or removable solid-state flash memory as the storage medium or have low-power hard drives to hold the recorded data. Due to their lack of moving parts, most are very rugged.

One of the main advantages of these recorders is the quick download of recorded audio information into the studio digital editor. Many of the devices are accessible to the computer system as an external drive

[3]Trademark of Apple Computer, Inc.

FIGURE 3.5-16 Solid-state recorder built into a microphone. (Courtesy of HHB Communications Limited.)

through USB or FireWire interfaces. With others, the flash memory card can be unplugged from the recorder and plugged directly into the computer.

Such recorders are becoming the recorders of choice for field audio recording work, and rack-mount solid-state recorders are available for studio use.

DAT Tape Machines

Originally designed for the consumer market by Sony and Philips in the mid 1980s, *digital audio tape* (DAT) machines became widely accepted in the broadcast industry for long-form program storage and archiving of audio when open-reel analog tape started to be phased out. DAT machines utilize small tape cassettes and rotating-head helical scan technology similar to a video cassette recorder. They record two-channel 16-bit uncompressed audio with sample rates of 32, 44.1, or 48 kHz, although some machines are capable of four-channel recording, other bit depths, and sample rates.

DAT machines offer fast and automatic cueing and instant starts, and the professional units provide for remote control and time-code compatibility. Potentially, this made the machines a practical digital replacement for open-reel tape. Their complexity, however, made the machine reliability somewhat problematic, and their difficulty of service made them less popular than other technologies. The improvements in computer-based digital audio storage systems, the decreasing cost of hard drive space, and the advent of recordable CD devices reduced the demand for DAT machines, and the technology is being phased out of the broadcast marketplace.

DAT technology migrated to the computer industry, where it was adapted as a digital data backup format called *digital data storage* (DDS) and is still in production for this purpose. A detailed description of DAT technology is covered in Chapter 3.4 of this handbook.

Other DAT-like technologies that record on larger format video cassette tape systems have been built and have found acceptance in recording studios and concert venues for recording both digital stereo masters and large numbers of tracks for later album production. The machines, which are also discussed in Chapter 3.4, are typically stereo or eight-track machines but can be locked with other machines to build a recorder capable of recording as many tracks as necessary. The medium is well suited for archiving of large audio sessions and events. Again, computer-based digital audio storage systems are making significant inroads into this market.

Audio Cart Machines

For most of the past 45 years, analog *tape cartridge,* or *cart,* machines have been the true workhorses in the broadcast industry. While cart machines have been mostly replaced by digital audio storage/automation systems and other stand-alone digital playback units, these devices are still used in some radio stations. However, few manufacturers are now producing cart machines, and parts are becoming scarce.

The audio cart machine was first shown at the NAB trade show in 1959. It quickly became the preferred method for playing commercials and music in broadcasting until the widespread adoption of digital automation systems. The cartridge houses a continuous loop of ¼" magnetic tape and was available in three different sizes. The most common AA size could hold 10 minutes of audio at 7.5 ips, the standard speed. The system recorded on one or two tracks on the 1/4" tape for mono or stereo, respectively. An additional track, called the cue track, carried signaling in the form of three mono frequency tones, and later provided confirmation logging by the addition of an FSK data stream.

The three tones recorded in the cue track were called *primary* (1 kHz), *secondary* (150 Hz), and *tertiary* (8 kHz). The primary tone activated the stop function and caused the machine to stop just prior to the beginning of the next audio cut on the cart. The secondary tone was most often used to start the next audio source, so facilities could be installed to sequence multiple cart machines, providing for early operator-assist mode or automation. The tertiary tone was most often used as a warning indicator to inform the operator that the end of the cut was approaching.

Maintenance of cart machines was labor intensive, requiring daily cleaning and periodic alignment to maintain desired audio performance. Correct preparation of a cart for recording and correct recording techniques for tight presentations were almost an art. Computer systems often utilize the name "cart" and the names of the tones used in cart technology to describe the modern computer counterparts.

Digital Cart Machines

As it became apparent that the analog cart machine was reaching the end of its product life, several versions of *digital cart machines* appeared on the market. Most relied on media imported from the computer industry, including high-density floppy drives, removable hard drives, MDs, and other storage systems to

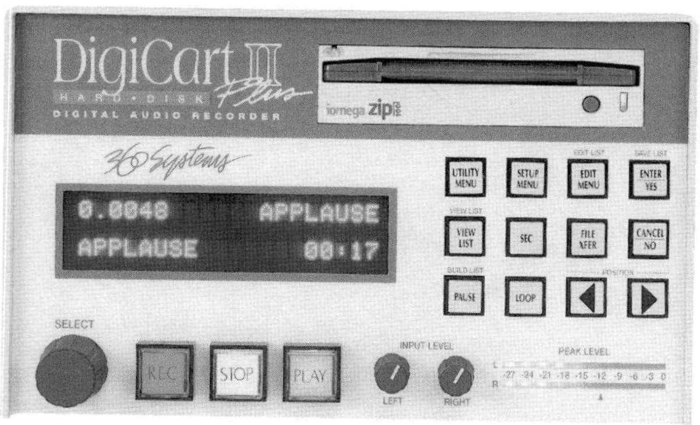

FIGURE 3.5-17 The 360 System DigiCart II. (Courtesy of 360 Systems.)

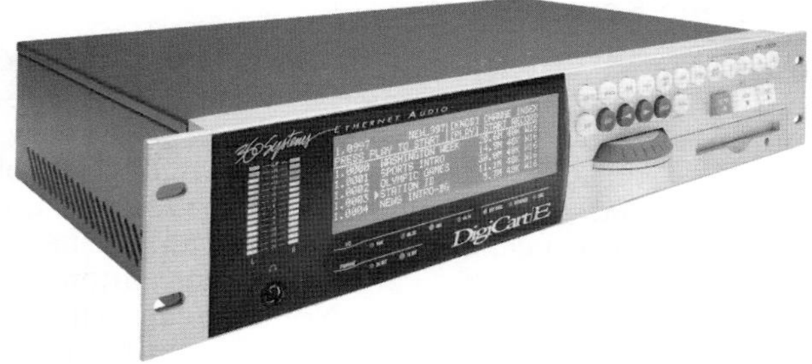

FIGURE 3.5-18 The 360 Systems DigiCart/E. (Courtesy of 360 Systems.)

emulate a cart machine. 360 Systems,[4] produced one of the first successful digital cart machines, which has now become one of the workhorses of the television industry. The DigiCart and the DigiCart II, shown in Figure 3.5-17, moved away from removable media, except for archiving and audio copying, and provided machines in a form factor similar to analog cart machines but with self-contained hard drive digital audio storage systems. With front panel cut selection and operations that were patterned after well-accepted analog cart machines, the multiple cart machines in a control room and an entire wall of storage for the carts were replaced by one or more small boxes. The DigiCart allows networking and sharing of audio between machines, it has both analog and digital inputs and outputs, and it provides a full complement of remote control options, including serial control, to allow integration into many systems.

The more advanced DigiCart E machine, shown in Figure 3.5-18, allows connection to a local area network, enabling interfacing to other systems and making sharing of audio with equipment much easier. In radio broadcast, computer audio storage systems and automation systems handle most of the audio internally, but in applications where large amounts of audio must be easily accessible, the DigiCart family has become the industry standard.

Reel-to-Reel Tape Equipment

Open-reel analog tape recording equipment is still in use in some broadcast stations, but with the cost of computer hardware and hard drives spiraling down, open-reel tape is heading in the direction of the cart machine. The demise of the tape machines was confirmed when both Ampex and 3M announced their exit from the audio magnetic tape business in 1995.

Open-reel tape machines were available in many configurations. Most often-used machines were 1/4" one- and two-track machines capable of holding a 10-1/2" reel of tape. Multitrack machines were often used in production studios to provide artistic flexibility, allowing multiple tracks to be laid down and later mixed into the final product. In broadcast facilities one could find 1/4" 4-track through 2" 24-track machines. These machines found use in delaying programming, recording requests, and contest winners, and extensively in production. Dubs of commercial content were circulated between radio stations on smaller reels of 1/4" tape. All of these functions have been replaced by computers and e-mail in most radio stations.

Other open-reel machines often found in a broadcast station were play-only machines that would hold either 10-1/2" or 15" reels to play back prerecorded music in automation systems. These systems used 25 Hz and 35 Hz tones much like a cart machine to control an automation system. All of this has been replaced by computers.

Telephone Interface Equipment

Telephone calls are an important part of programming for most broadcast radio stations. The public's most convenient method of communicating with the spoken word is the telephone, and broadcasters use this to keep in touch with their listeners through requests, contests, topical conversation, and talk shows. The ability to carry on a conversation with a caller and either record the conversation or air the conversation simultaneously, or delayed, is not as simple as it might appear.

With the conventional public telephone network interface, the sending audio and the receiving audio are on a single balanced audio telephone circuit. A device called a *hybrid*, which is an analog two-wire to four-wire converter, is used to break the signal into its two parts—the send audio and the receive audio—so it can be interfaced to the studio audio system. Most stations have many phone lines for use on air and often need to select from multiple callers and/or conference two or more callers together. Small custom telephone switches are used to route many telephone lines to one or more hybrids.

Chapter 3.10 provides detailed information on telephone network interfacing.

[4]See http://360systems.com/.

Telephone Recorders

Closely related to telephone systems is the need to avoid airing undesirable material. With the requirement imposed by the FCC to be in control of everything aired on a broadcast station, and the lack of control of what a caller might say, broadcasters are wise not to allow a caller directly on the air. This may be done by recording the call in a digital editor and trimming the call before airing. Not only can inappropriate or offensive speech be removed from the playback, but the call can be edited to flow more dynamically, creating a better presentation. Several editors are available that are specifically designed for this purpose; the 360 Systems Shortcut[5] shown in Figure 3.5-19 and VoxPro[6] are two of the most popular phone editors on the market. It is customary to utilize these stereo editors as two-track recorders, recording the send audio on one channel and the caller audio on the second channel. This allows postrecording balance between the caller and the host levels, and can create a better caller signal for recycles and promos. Often an excited host will talk over a perfect response from a caller, and the caller-only audio may make the response usable for promotional recycling. For playback on the air, the two channels are summed to mono either in the console, the recorder, or through an external network. Often the stereo feed is available on the console's B input for the recorder channel, and the mono sum is available on the A input.

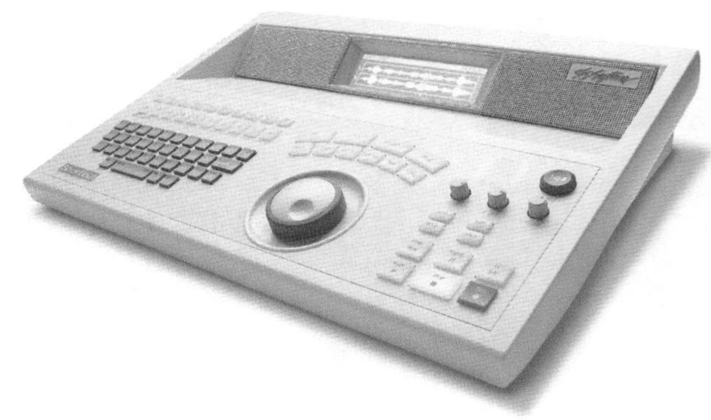

FIGURE 3.5-19 A stand-alone editor often utilized for telephone call editing in radio applications. (Courtesy of 360 Systems.)

Profanity Delay

Live talk radio cannot be done by recording and editing comments. Profanity *delay systems* are used when "almost" live callers are needed. The original delay systems were tape based. A three-head cart machine was placed in the program line from the console. The head arrangement was playback-erase-record, and the cart length was generally 7 to 10 seconds long. The machine would continuously record the current audio and, about 7 to 10 seconds later, that audio would be played, the tape erased and rerecorded with new audio. If something inappropriate was said, the call would be terminated and the delay system bypassed, eliminating the last 10 seconds before it could be broadcast. This action is called *dump*.

Digital delay systems such as the one shown in Figure 3.5-20 have eliminated the tape delays. The audio is recorded in a high-quality RAM-based ring buffer. A ring buffer is a memory system whose address space closes upon itself. Data is read from a cell in the buffer and sent to the audio output stream, and the new data is written into the cell from the audio input stream. The address is advanced in a continual process that loops once each delay time when the buffer is full. The typical digital delay line is a stereo 16- to 24-bit sample width, 48 kHz sample rate, with ring buffers as long as 80 seconds, though 7 to 10 seconds is typical. Newer profanity delays with longer buffers permit a partial dumping of the buffer such that the program is still delayed after a dump, continuing to provide program content protection should a second dump be quickly needed.

Both the original tape delays and the digital delays create a few problems. First, the buffer must be filled with the program before anything is played on the air. This means that the program may need to start up to 80 seconds, though typically 7 to 10 seconds, before the actual airtime, and each time the program is dumped, there must be a delay time to be made up. Often this is done by playing a prerecorded fill liner to allow the delay buffer to be filled with audio. Newer digital profanity delays have a feature that will time shift the incoming programming over a longer period of time to gradually fill the delay buffer, so the program can continue as normal as heard by both the talent and the listener after a dump while the delay buffer is built up.

Another issue while utilizing a profanity delay is that the talent must monitor the input to the profanity delay rather than the station output. The actual program will lag the input by between 0 and 80 seconds, making it impossible to listen to the air signal while producing the program. Care must be used to assure that the aired signal is correct and on the air, a function that is normally done by off-air monitoring, at least until the advent of HD broadcasting with its delayed audio. Lastly, the caller must be cautioned not to attempt to listen to the radio while on a call. It is almost impossible to carry on a normal tempo conversation when hearing one's own voice offset in time by more than a few tens of milliseconds. Listening to the 7-second delay confuses most callers, and the "Turn

FIGURE 3.5-20 Digital broadcast profanity delay unit. (Courtesy of Symetrix, Incorporated.)

[5]Trademark of 360 Systems.
[6]Trademark of Audion Laboratories, Inc.

your radio down" phrase has become one of the most common in talk radio.

PROCESSING EQUIPMENT

Equipment to "sweeten" audio is available in almost unlimited types and functions. The most basic types of processing are *dynamics control* equipment, *equalizers*, and *effects processors*, each intended for a different purpose. Dynamics control equipment includes *compressors*, *limiters*, and *expanders*, each designed to automatically change the signal levels based on the input signal. Equalizers are used to change the *relative frequency response* of the audio. Effects processors are used to add *special effects* to the audio. These devices, which are available as dedicated hardware, all have counterparts available as software plug-ins for digital audio workstations. If there is a desired alteration of sound, there is a processor or a plug-in to achieve it.

Primary Broadcast Processors

In broadcast, the main audio chain is always passed through a *processor* similar to the one shown in Figure 3.5-21 before it is sent to the transmitter. This unit is the focus point for creating the *sound* and the *loudness* desired by the program director for the station; Chapter 3.8 covers this subject in detail.

The main station processor primarily performs *dynamics control* and *equalization*. The broadcast material is generally pre-emphasized, its frequency response purposely altered prior to transmission to be similar to the transmission noise floor response so that, when de-emphasized on reception, the noise floor is reduced. The equalization and the dynamics control must be tailored to the selected pre-emphasis in order to function appropriately. Newer processors designed for the HD Radio market may also include delays to match analog and digital audio. Additionally, digital processors introduce delays because of their algorithms.

The location of this processor depends on the arrangements for the studio transmitter link (STL). The processor's best location is at the transmitter, but it may be necessary to protect the STL with either the main processor or an additional processor coordinated with the main processor. Caution should be used when audio compression is employed in the STL, as heavily processed input to the compression algorithm can cause its artifacts to be exaggerated.

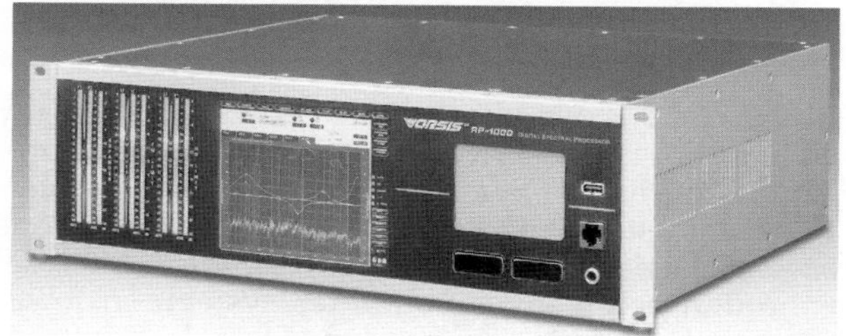

FIGURE 3.5-21 Audio processor. (Courtesy of Wheatstone Corporation.)

Compressors, AGC, and Limiters

There are many uses for compressors and limiters in studios. Often dynamics control is necessary to help the operator ride gain on signals, and it can help achieve good level control when the operator is distracted. An example of the use of automatic gain control (AGC) devices or compressors would be on the record inputs of a recorder to capture both sides of a telephone conversation on a phone-in line. Assistance riding the gain of both the caller and the local host will help in an effective playback if the host is busy during the call and his attention cannot be focused on adjusting audio levels. Compressors can also be arranged to *duck* the caller to assure the host will win all shouting contests.

Compression

Dynamics control can take three general forms. The first form of dynamics control is *compression*. Compression is primarily used to increase the *density* of the audio being processed. Audio has a dynamic range that is loosely defined as the level of soft passages of the music compared to the level of the loud passages. Often, more consistent audio levels are desired, an effect achieved by evening out the average level. With the addition of gain after the compressor, the perceived loudness and density of the audio are increased.

A compressor is an amplifier whose gain can be reduced by a control signal. The control is generally derived from the level of the audio being fed to the input of the device. As the level of the audio reaches a preset threshold, the gain of the controlled amplifier is reduced to keep the output level closer to the threshold level. Several parameters may be adjusted to tailor the response of the gain control signal. The *threshold* control is set to the incoming signal level above which compression should start, and this control is generally calibrated in dBm. The *ratio* control determines how aggressively the compressor will reduce the amplifier gain and is generally adjustable with a range of 1:1 to 20:1 or greater. The higher the ratio, the more aggressive the compression effect on the audio. A ratio of 1:1 will do nothing, and a ratio of 5:1 would mean that an increase of the input of 5 dB above the threshold would yield an increase in the output of the compressor of 1 dB. Some compressors offer a *soft knee* setting, which causes the compression ratio to change from 1:1 to the setting on the ratio control gradually, as the input signal passes above the threshold. The gain function of the soft knee takes on a curved transition at the threshold, and the effects on the audio are less severe. Other compressors offer two threshold and ratio settings to allow a two-stage transition of the gain. The *attack* and *release* parameters are often adjustable so that the length of time that a signal is above the threshold before the gain reduction occurs and the amount of time that the signal is below the threshold before the gain reduction is removed can be set to preserve some quick dynamic transients and to prevent audio-level *pumping* of the signal. The combination of all the controls on a compressor affects the audio from no effect, through a subtle increase in density and average level, to a *brick wall* of sound. Beyond a certain point, the more aggressively a compressor is set, the more fatiguing the audio will be to listen to, and the shorter the time that the listener will enjoy staying with the program before changing to another station.

Limiting/Clipping

The second form of dynamics control is *limiting* and the more aggressive *clipping*. A limiter behaves much like a compressor with the attack and release time set to minimum and the ratio control set to infinity. A limiter will instantaneously reduce the level of the signal and will not let signals above the threshold pass without attenuation. The most severe dynamics controller is a clipper, which effectively shears off excursions above the clipper threshold, a process that creates distortion.

In broadcast applications, limiters must be designed to be aware of the pre-emphasis applied to the audio by the local broadcast standard, which adds a frequency-sensitive component to the threshold. Limiters and clippers are used primarily in situations in which an absolute limit in the audio level is required, most generally to prevent overmodulation of broadcast transmitters. The limiters and clippers should be as close to the input to the transmitter they protect as possible because group delays and other transitions in the signal between the limiter/clipper can cause less than absolute modulation control.

Automatic Gain Controls

Automatic gain controls are similar to compressors with slow attack times, very slow release times, and low compression ratios. These devices are set up to act as a slow "hand on the pot" to keep the audio level in the "sweet spot" for other, more aggressive processing. In addition to slowly reducing the gain of the AGC amplifier for signals above the threshold, an AGC amplifier may also increase the gain of the AGC when the signal is below the threshold, possibly with a limit on the amount of gain increase to prevent bringing up the noise floor during long periods of silence.

Most of these dynamic control systems are stereo or multichannel devices. The control signals of multiple channels can generally be tied together or linked so the stereo image is not affected by the processing. Many devices, particularly compressors, have the capability for introduction of a *side chain* signal. This signal replaces the audio feed to the control circuitry, which is normally the incoming audio to be processed, with another signal. The main signal is compressed based on the level of another signal, and this can be used for frequency-conscience compression, automatic duckers, and key gating.

Equalizers

Equalizers are amplifiers whose gain varies with frequency, and this variance can be adjusted. There are two broad types of equalizers: *graphic* and *parametric*. Graphic equalizers break the audio spectrum into equal segments, typically 1 or 1/3 octave segments, creating between 7 and 30 bands of control, with the controls displayed as boost/cut sliders, appearing like a graph. Graphic equalizers are the quickest and easiest to adjust. Parametric equalizers generally have fewer channels, typically 3 to 5 channels, but each channel has multiple parameters that can be adjusted for more precise control (for instance, to take out a narrow range of frequencies from some particular interference). In addition to the boost/cut control of a graphic equalizer, each channel of a parametric equalizer also has a frequency adjustment to set the center frequency of the channel and a "Q" adjustment to set the bandwidth of the channel. The channels of a parametric equalizer are band-pass filters, but two can typically be set to be shelving channels, making them high-pass or low-pass filters.

Microphone Processors

A *microphone processor* is a multifunction processor that combines a high-quality preamplifier, a compressor, a downward expander, de-esser, a phase rotator, and an equalizer. All mic processors accept the low-level analog signal from the microphone, and most have both analog and digital outputs. Some are digitally controlled or digitally implemented to allow multiple presets and advanced processing. Some models include a tube stage to obtain the perceived warmth of microphone systems of old. Examples of four mic processors are shown in Figure 3.5-22.

FIGURE 3.5-22 Four microphone processors typically used in radio. The first two are analog microphone processors. (Courtesy of Symetrix, Inc. and Harmon International Industries.) The third is a digital microphone processor. Digital microphone processors are often set up through access by a computer system, so fewer front panel controls are needed. Digital processors allow multiple presets to be recalled as needed. (Courtesy of Symetrix, Inc. and Wheatstone Corporation.)

Most commercial music played on a broadcast station is processed, sometimes quite heavily, during its production. The purpose of the mic processor, beyond being a preamp, is to increase the density of the announcer's voice through compression, essentially elevating lower-level sounds to match the higher-level sounds, to help match the density of the local voice to the density of the processed music, thereby helping the voice cut through the mix.

Some voices, often accentuated by the addition of compression, have a large amount of energy in the brilliance area of the audio spectrum, particularly in the pronunciation of "S's," which can cause a splattering in processed and limited audio. A *de-esser* is a circuit included in most microphone processors that quickly limits the brilliance portion of the audio spectrum (6–15 kHz) if it becomes overpowering. Lastly, an equalizer is provided to alter the warmth and clarity of a voice and obtain a more dynamic and pleasing sound. Some program directors also like a small amount of *reverb* added to microphones, and a dedicated reverb system is added after the microphone processor. Since the signal from the microphone processor is further processed by the main station processor, the sound of the station should be finalized with the main station processor and with the majority of the station's program content before a large amount of time is spent obtaining the desired sound from the microphone system.

Effects Processors

Effects processors are the fun devices used in production. The majority of these units is multifunction digital devices or software plug-ins for digital audio workstations. Effects processors alter the audio by frequency shifting, delaying, modulating, panning, and various combinations, often time varying, of these functions. Effects processors generally have equalizers and dynamic controls used in conjunction with the other effects to produce *reverbs*, *chorus* effects, *flanging*, *pitch changes*, *time compression* and *expansion*, and many other effects. Most effects processors allow combining of all their capabilities to create new effects tailored to the project at hand and to allow the storage of these effects for quick recall. Most processors also allow MIDI control to punch the effects in and out or to key effects under instrument control.

EAS Equipment

At the time of writing in late 2006, the *Emergency Alert System* (EAS) is in the process of being changed or supplemented. The current system required by the FCC requires the station to monitor two or more remote audio sources, looking for a frequency shift keying (FSK) data stream that identifies the audio to follow as an emergency-related message. The EAS equipment acts as an immediate interrupter or a store-and-forward audio source, depending on the way the system is set up. The equipment is installed as an interrupter for the primary program channel and has the capability to monitor the receivers feeding the system and to review the stored audio. Depending on the class of the station, it may also be necessary to be able to input program audio to the EAS for encapsulation within an EAS message.

The importance of emergency broadcast and the EAS is so high that it requires the primary attention of a broadcaster and is a requirement of the broadcaster's license. The EAS is covered in detail in Chapter 1.18.

AUXILIARY STUDIO EQUIPMENT

Semipro to Pro Converters

Pro to semipro adapters such as the one shown in Figure 3.5-23 are devices that allow a consumer-grade analog device to be interfaced to professional equipment with balanced audio inputs and outputs. Often devices with unbalanced RCA phono inputs and outputs must be integrated into a studio. These adapters provide the level, impedance, and balancing necessary to achieve proper installation. A pro to semipro converter also allows a personal computer's native sound card to be interfaced for special functions.

Along the same lines, adapters are available to convert S/PDIF, the consumer grade of digital output from many audio devices, to AES/EBU. While some consoles' AES/EBU inputs will accept S/PDIF directly by using a cable adapter only, there are differences in electrical levels and in the format of the channel status data that may make an adapter necessary to interconnect between the two standards.

Distribution Amplifiers

When it is necessary to send a single balanced analog audio source to multiple locations, this can usually be done with modern equipment by bridging multiple balanced bridging inputs across a single low impedance balanced output. However, bridging has several inherent problems that should be considered. First, there is no individual level control for each of the destinations; they will all receive the same signal level. Second, a fault in the wiring of any of the bridged feeds, or any noise picked up by any connection, will appear on every connection and, third, if the lines are long, significant capacitive loading can affect the high-frequency response of every connection. The more appropriate method of feeding multiple destinations

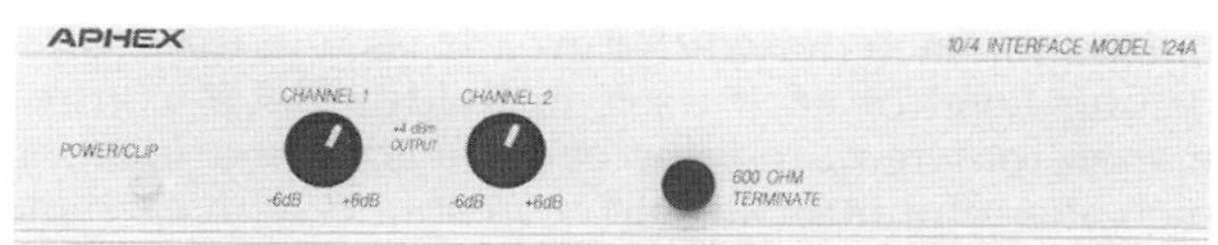

FIGURE 3.5-23 Pro to semipro audio-level interface amplifier. (Courtesy of Aphex Systems.)

FIGURE 3.5-24 A typical single rack unit height distribution amplifier (DA). (Courtesy of Symetrix, Inc.)

with a single source is the use of a *distribution amplifier* (DA), such as the unit shown in Figure 3.5-24.

An analog DA is a device that accepts a single analog balanced input per section and provides usually between four and eight balanced audio outputs that are individually amplified and isolated from one another. Generally, each output has an individual gain control, and the DA can have many different options including metering, input level control, loss of signal alarms, and redundant power supplies. With modern electronics, the DA is essentially sonically transparent, and it can solve many problems in an installation.

Digital DAs are also available to distribute a single AES/EBU signal to multiple AES/EBU devices.

Interface Panel

In any studio, there are sometimes special events that require an unexpected piece of equipment to be connected into the console or where an unexpected visitor needs to be fed an audio feed of the current program. For example, a newsworthy event involving the station may bring a television camera crew into the studio, needing an audio feed of the on-air signal, or a band may bring its own small mixer setup into the studio and want to provide the station a feed. All studios should be set up to handle the unexpected, and should have an available *jack field*. Commercially constructed jack fields have been built to provide just this function. One example is the Harris World Feed Panel[7] shown in Figure 3.5-25, which provides buffered, level-adjusted inputs and outputs in a variety of connector formats.

Headphone Amplifiers and Intercom Systems

Headphones are a necessary tool for all professional announcers to execute their craft. Most consoles provide a headphone output as part of their monitoring system, which can be switched to any of the output buses of the console or to one of several external inputs such as generally used to monitor an off-air signal. Most consoles do not provide more than one controllable headphone output, so headphone amplifiers are necessary to distribute the signals to multiple members of the broadcast team. Headphones vary widely in impedance and efficiency, so the drive required for each may differ greatly, as will the desired listening level of each user. Individual level controls are therefore required for each headphone location, which at a minimum should correspond with each mic location within a studio.

[7]Trademark of Harris Corporation.

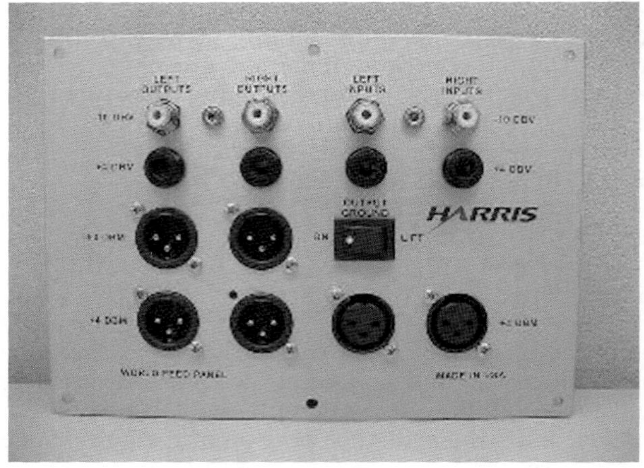

FIGURE 3.5-25 Harris *World Feed Panel* used as a convenient interface to temporary equipment in a radio studio. (Courtesy of Harris Corporation.)

Most commercially available *headphone amplifiers* provide individual level controls on the amplifier, but if the amplifier cannot be conveniently located with easy access to all users, an alternate control is needed, located near the user. Often this is addressed by the installation of L-pads, a high-power resistive attenuator used for speaker output control, or by mounting individual headphone amplifiers or jacks in the mic control station near the talent, as shown in the installation in Figure 3.5-26. Other manufacturers offer distributed headphone amplifier solutions that gang the multiple headphone stations over CAT5 cable, and there are also headphone solutions connectable over CAT5 that allow a tailored mix to each headphone. One solution places all electronics in a rack-mount unit and provides individual control at the remote location by means of a voltage controlled amplifier (VCA) control, remoting only the two-wire level control along with the headphone amplifier output to the mic position.

Often it is convenient for the talent to receive alternate communications, particularly talkback from the producer or cues from the console, in one or both ears of their headphones. This may or may not be appropriate for each team member within the studio. Some headphone systems permit selection of an alternate feed to each headphone independently.

Closely related to the headphone system is the *intercom system* (ICM) with *interruptible fold-back* (IFB). The proper design capability for the intercom/IFB system is to allow any user with an intercom terminal, such as in a live studio, newsroom, or other work area, to communicate audibly and quickly with any other user, without leaving her position and without using a telephone.

Intercom systems may also be extended to off-site remote broadcast positions via analog dial-up, ISDN, leased circuits, or the Internet, to allow remote air talent or reporters to communicate with a studio engineer, director, or newsroom. An IFB feature of the

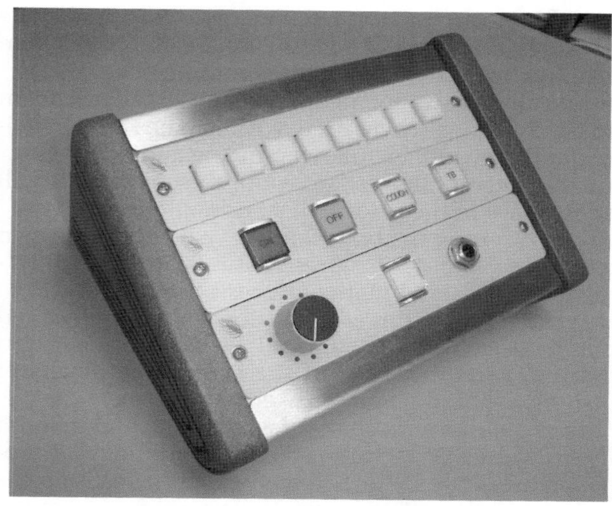

FIGURE 3.5-26 A mic control station including headphone jack and level control. Additional controls are shown for mic operation and other utility functions. (Courtesy of Wheatstone Corporation.)

intercom allows the remote talent to hear a specialized mix of the live program, called *background audio*. The studio director or news editor may press an intercom key to talk to the remote talent, which interrupts the background program feed and allows the editor or director to be heard in the remote talent's headphones.

The configuration of the ICM and IFB is as important as the overall design. The best ICM and IFB systems are designed for easy configuration—such as assigning an IFB to an ICM port. In addition, incorporating add-on wireless features allows the remote talent to move around while on the air.

Digital processing and delivery systems for intercom and IFB add a fixed delay to the signal passing through them due to D/A and A/D conversions, packetizing, DSP processing algorithm delays, error correction, and other factors. When the talent speaks into a microphone and also hears the system output through headphones, or hears the return from an active loudspeaker, there is a threshold of around 6 to 15 milliseconds of delay in which the return audio becomes distracting to the talent and adversely affects the talent's cadence in speaking. This delay is referred to as *latency* or *bulk delay*. The practice of monitoring the off-air signal is common in most broadcast operations, in order to allow the operator to adjust the mix as it will be heard by the audience and to confirm that the program is on the air. When the delay introduced by digital chain becomes large enough to be distracting, an alternate source must be derived from a lower-latency feed to supply the live monitor. This signal should be processed in a similar manner to the main signal in order to allow the mix to be appropriately created.

Some main station processors provide an output that has a lower delay than the program output, which can be used for this purpose. Another solution is to use the backup analog processor to feed the headphone system. In either case, it is still desirable to monitor the off-air signal on the studio monitors to assure the station is on air and the program is being appropriately delivered. This may require switching between different inputs to the monitoring system when studio microphones are in use. The advent of HD Radio broadcasting has increased the transmission delay to approximately 8 seconds, which leads to more extensive monitoring system changes and the need for *silence-sensors* or *off-air alarms* to compensate for the lack of live monitoring of the off-air received signal.

Annunciator Panels

Studios often operate with open microphones. This precludes getting the attention of the operator by using aural devices when an event demands his immediate attention. Providing this information to the operator visually is the purpose of an annunciator system. As part of the design of a studio, the designer should determine what information must be presented to the operator while he or others may be on air with a microphone within the studio.

In a broadcast studio, this information usually includes hot-line phone ringing, contest lines ringing, transmitter faults requiring attention, EAS alerts, primary audio loss or dead air, and perhaps the station doorbell. The different solutions for an annunciator system include light bars with labels, commercial annunciator panels such as the system shown in Figure 3.5-27, or LED message display panels. They should all be installed within the line of vision of the operator at the operator's position and may include remote display panels where needed. Often it is necessary to include a strobe to call attention to the panel if there is a status change that requires prompt attention. An additional annunciator from the fire alarm system should also be included in the same general area as the broadcast annunciator panel. The fire alarm annunciator device should not be part of the broadcast annunciator panel and must be installed in accordance with local fire codes.

Keyboard, Video, Mouse Switches and Extenders

Keyboard, video, mouse (KVM) switches and extenders do not carry any audio within the facility but are becoming an important part of every large studio complex. With the need for many computers in each studio, the logistics of locating and mounting the computer's central processing units (CPUs) is becoming difficult. CPUs tend to be power-hungry, heat-generating, and often noisy devices, and are better installed in a controlled technical environment such as a rack room or master control area rather than a studio. Remoting the necessary operator interfaces to the studios as needed is the role of KVM switches and extenders. These active extender units permit all operator interfaces to be controlled remotely over a single

FIGURE 3.5-27 Visual annunciator panel to provide information to talent within the studio. (Courtesy of Enberg Electronics.)

CAT5 cable, and a properly designed switch and patch bay installation will allow remote administration and troubleshooting of all systems with minimal interruption of the studio. This will allow the easy transfer of control from one set of interface equipment to another should it become necessary to move to another studio after a failure. Caution must be used when designing a KVM switch/extender system to assure that no single point of failure is created. While common control over all operator interfaces is of great help within a facility, having all operator interfaces within a facility fail at the same time is clearly not.

STUDIO EQUIPMENT LAYOUT

Once the console and other equipment selections have been made, designing the studio layout and the cabinetry to house the equipment will often be the next item of concern. Studio layout and furniture design entail, like all design, the art of intelligent compromise. In a studio, the major items of equipment, the talent, the guests, and the visual attractions of the studio are all vying to be front and center of the operator. The best approach is to prioritize everything in terms of necessity to complete the prime task and start by laying out the equipment in that order.

In general, the mixing desk or console will be the center of the operator's workspace, as it is the focus of the studio operations, although this is becoming less true in off-line production rooms where the console is starting to become more of a switcher that is preset and left. In the latter situations, the computer monitors and the computer input devices (keyboard, mouse, trackball, and/or digital editor work surface) might be more the center of operations for the studio.

Moving down the prioritized attention list, one should place each item as close to the operator's center of attention as possible and start making the intelligent compromises of good design. Generally, after the console, the next most important item in a studio is the monitoring system, and placing the speakers in the correct location relative to the operator should be given high attention. A mix will not be any better than the audio heard by the person doing the mix.

Typically, the other people in the broadcast team and the locations where they work will be the next priority, and they will be followed by the audio computer systems. Eventually, all the rest of the equipment will find its place in the priority list and will find a home.

The layout can be planned with hand-drawn sketches, scale cutout representations, or other traditional techniques, but the process is well suited to being done with three-dimensional computer-aided design. Once one or several designs have been achieved, careful review of the ergonomic details of the design should be undertaken.

Many studios with similar tasks tend to end up with similar layouts. Most radio station air studios and production areas use U- or J-shaped installations with the primary focus point, the console, being at the bottom of the U or the J. An example of a well-designed and efficient radio studio is shown in Figure 3.5-28.

Ergonomics

Ergonomics is the scientific discipline concerned with the understanding of interactions among humans and other elements of a system, and the profession that applies theory, principles, data, and methods to design in order to optimize human well-being and overall system performance.[8] Many professionals practice to improve the ergonomics of almost everything being built today. The results of their work have likely been applied in the design of each piece of equipment being installed in the facility. Utilizing the equipment in accordance with its design principles and recommended installation arrangements, often specified in the instruction manual, will help ensure that the entire system design is ergonomically sound.

Hiring experts in ergonomics to consult on the equipment layout is beyond most projects' resources, but certainly considering the ergonomics of the design is important.

Working Height

One of the first factors that influence the ergonomic design is the basic height of the work surface. Many program directors in broadcasting believe a better delivery is obtained from the talent while standing, so many air studios are built to stand-up height. The selected height of these studios must be high enough to permit the average person to comfortably reach the work surface without stooping, but not too high to be unusable by persons on the short end of standards. This is typically in the 36" to 40" range. Accommodation of disabled persons must also be considered in all parts of the design, as dictated by the Americans with Disabilities Act (ADA), which places additional needs for accommodation in the ergonomic design of the installation. One solution to the varying height

[8]The definition adopted by the International Ergonomics Association in 2000.

FIGURE 3.5-28 Modern radio studio. Note the limited source equipment necessary with the use of computer-based automation systems. (Courtesy of Lightner Electronics.)

demands is adjustable-height work surfaces for studios, and these are now being used in many high-end control rooms. The height of sit-down facilities should also be considered, and it is generally desirable to have the work surface slightly higher than a normal 29" desk to better present the details of the control to the operator. It is also good practice to provide slide-out keyboard trays that can fit in the area between the operator edge of the work surface and the console if the operator must do extensive typing. The use of ergonomically designed keyboards may be of help to the operator.

Personnel and Equipment Positioning

After the work surface height is determined, the normal position of the operator in the room relative to the equipment should be reviewed. The operator should be able to see everything most often needed without the need to turn greater than 45° horizontally from his normal position, with the optimal viewing area restricted to 30° off center. The vertical viewing area for objects that must be in the normal visible area should be restricted to 30° upward and 45° below level sight from the operator's position, with the viewing angle to often-used objects restricted to 15° upward to 38° downward. Objects outside these limits will not be seen without conscious effort. The lower visual limit is about 70° below level sight. The operator's reach should also be of concern, more in sit-down operations than when standing, as lateral motion is more difficult.

Men's easy reach is an approximate 31" arc from normal position; women's, approximately 28". Therefore, an attempt to have the most often-used controls within this region is appropriate.[9]

Other ergonomic concerns should be the seating arrangements for the work area. Chairs should be adjustable such that the operator can place his feet squarely on the floor with a comfortable forward knee angle when seated at the workstation. If the work surface is high enough that a stool must be used, providing a stool with an appropriate footrest is important to prevent harm to lower body circulation while at work. It is important that sufficient leg space be provided under the operator locations; between 18" and 24" is recommended. The thickness of the work surface and any keyboard drawers must not prevent comfortable posture while seated.

When the location of the console or the computer workstation is planned, it is appropriate to leave sufficient room to place an 11" piece of paper between the equipment and the operator's edge of the countertop as a free and clear workspace. This space helps avoid scripts, logs, and other papers spreading over the audio console controls and also provides a natural armrest while using the console.

[9]Information from Charles George Ramsey and Harold Reeve Sleeper, *Architectural Graphic Standards*, John Wiley and Sons, Inc., 1994.

Lighting and Air Conditioning

The correct lighting of the work area is an important ergonomic consideration. The lighting should be sufficient to read the necessary materials and control legends with ease without causing glare or washout of the monitors being observed within the studio. This generally dictates the use of directional lighting. Separately switched bright general area lighting, often fluorescent, should also be provided for maintenance operations.

The environment must also be temperature and humidity regulated within the comfort zone, with ample fresh makeup air from the HVAC system.

Cabinetry

The studio cabinetry must enable the correct location of the equipment as required for the operator to use it efficiently and also provide an appropriate environment for the equipment. Equipment must be securely supported with mounting arrangements as approved by the manufacturer, generally within 19" equipment racks, mounted on the countertop, or in cutouts in the countertop. Often, operational equipment is installed in angled short racks mounted above the countertop, with less-used support equipment located in closed cabinets or racks under the countertop that also provide the countertop supports, electrical and communications wireways, and ventilation.

Where equipment is mounted at low levels, the cabinetry should protect the equipment from physical damage such as might be caused by sliding a chair into it.

Radio talk studios contain much less equipment than the control room, but often have the need for many computer displays for the host and possibly the guests. The operator interfaces for the talk studio are usually mounted in small tabletop turrets containing a few microphone remote controls and headphone control. These studios are generally conference-room-like in appearance with the electronics installed under the tabletop.

Cable Management

The cabinetry must provide a systematic method of routing all the cabling needed between equipment and should contain a terminal block area for outside connections to the cabinet. The cable routing capabilities should include the capability to separate the cable routes for diverse signal-level cabling, keeping mic level, line level, speaker level, and power apart as much as possible and crossing one another where necessary at right angles.

All cabling between cabinets or between the wall and the cabinet system must be installed in a raceway that will not be a trip hazard.

Equipment Ventilation

All equipment must be supplied with sufficient cooling air and, if possible, this is best provided by natural convection, as the addition of forced cooling creates noise that can be problematic in a studio. It is often desirable to install the cabinetry on adjustable legs and to leave the typically 4" base area below the cabinetry open to promote airflow within the room. Cabinetry built flush to the floor can create dead spots in the room for air circulation, which can cause uncomfortable areas within the room and create cooling problems for some equipment locations.

Electrical Supply and Grounding

Ample electrical supply to the equipment must be provided as well as appropriate equipment grounding installed in harmony with the electrical system, the subject of parts of Chapters 3.1 and 9.3. Access to any live electrical circuits must be protected by the cabinetry from persons working outside the cabinetry and should protect persons working inside the cabinetry. Under no circumstances should terminals whose voltages are greater that 25 volts to ground or between each other, or whose currents could be great enough to cause heating if shorted, be openly exposed either on the outside or on the inside of the cabinetry. These terminals must have protective covers that are not easily removed.

Cabinet Construction

Many manufacturers provide modular systems that allow the assembly of predesigned modules to make up the needed basics of the studio, and some offer predesigned countertops to match the base modules. A stock system for a small radio station is shown in Figure 3.5-29. Some of these systems are quite utilitarian metal structures from rack manufacturers, whereas others are beautifully constructed fine furniture with plastic laminate, solid-surface, or granite countertops. While the application and budget will dictate the studio cabinetry, one should remember that radio and television are show business, and the stage should look good.

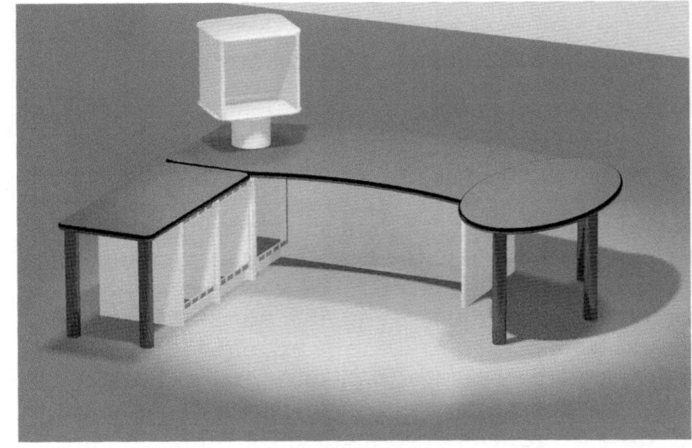

FIGURE 3.5-29 A stock modular studio cabinet system capable of handling a small radio station's needs. (Courtesy of Wheatstone Corporation.)

Many manufacturers will design and construct custom cabinetry arrangements or hybrids of custom and standard modulars to fit special circumstances. Local cabinet talent can be successfully hired to construct studio cabinet systems, but the local cabinetmaker may not have experience in ergonomic design of a radio studio and use of a local cabinet shop places most of the design responsibility on the station engineer or the systems integrator. Many station engineers and all systems integrators are capable of this level of design, but utilizing the expertise of specialist broadcast cabinet suppliers can simplify a project greatly.

AUDIO INTERCONNECTION AND ROUTING

Considerations

When the facility contains more than one studio, interconnections between areas are invariably required. Audio routing is necessary to allow audio resources to be shared by multiple users. After equipment selection and ergonomic layout of a studio, audio routing is one of the most important parts of audio equipment utilization. Having the audio routed to where it is needed and when it is needed is critical to a good operation.

An analysis is required of what sources are needed in which locations and what is likely to change. For instance, a studio's dedicated CD player is not likely to be required as a source on a console in another room, while sources such as off-air monitoring will be needed in all rooms but are not likely to be changed in normal operation. The component parts of a studio's digital audio storage/automation system are not likely to be needed in any other room, but the mix of all of the components may well be. The output of a studio to the input to the processor/transmitter chain should not be easily disrupted, but it must be able to be changed quickly for recovery during abnormal operations.

In television and video postproduction, the flexibility required of the audio facility may be more demanding than in radio. In some facilities, total reconfiguration of the audio system might be necessary for a given event. All of these are issues that must be considered in assessing the requirements for audio interconnection and routing.

Routing Philosophies

Two schools of thought exist on audio routing. The first is to provide quick access, usually by patch bays, to every input and output of a system. Complete access to every input and output provides complete flexibility that may be needed in some facilities, but is expensive and can be confusing to lesser-experienced personnel.

The second school of thought is to design the routing system based purely on the need. Applying the traditional 80/20 rule, 80% of the reconfiguration needs can usually be achieved by 20% of the reconfigurable routing.

The function of the facility and the level of expertise of the operating personnel should guide the selection of the method used.

Technology convergence in audio console systems have, in fact, tended to blend the two schools of thought. Newer digital control-surface-based router systems such as the *Wheatstone Bridge*, *Telos Axia*, *Logitek Mosaic*, *Klotz*, *Harris*, *Sierra Automated Systems*, and others systems marry traditional console control with router technologies. When installed as a complete digital system, each audio source is connected to an input that is then available under software control to any mix within the facility. With the exception of multiplexing seldom-used sources to save on relatively expensive audio channels, and providing bypass capabilities during catastrophic failures that will occur with all equipment, routing beyond the framework of the digital audio system is unnecessary. These systems are capable of remoting a great number of inputs a significant distance from the target mixer over a pair of fiber-optic cables or a CAT5 or CAT6 network cable. The simplification of the installation and expansion of the utility of a large facility by these systems is enormous.

The entire audio interconnect backbone of a large multistation broadcast facility is shown in Figure 3.5-30. Here, all audio sources are tied to the router frame within the room of the source, and every audio source is available in every room. Note the redundant paths for audio, eliminating single points of failure that could affect more than one control room. Most of the backbone connection is copper Ethernet, but fiber-optic cable is used to isolate the chassis tied to the satellite systems, to improve lighting immunity.

The method for interconnection and routing chosen for a given installation is varied and part of the art of studio design. In broad terms, the methods for interconnect and routing are discussed in following sections in ascending levels of complexity and flexibility.

Point-to-Point Installation

In simple one-studio installation, remote studio applications, and temporary studio or studio supplement setups, *point-to-point* connections are often best. Point-to-point installations are constructed by positioning the equipment in the appropriate location and plugging the equipment together with premanufactured or custom-constructed source cables. This is the method used for almost all project studios and can be appropriate for small and uncomplicated broadcast stations. Done neatly, with cable ends labeled, and dressed in some form of cable management system, this method can look professional and can be maintainable. Point-to-point installation can be the quickest and least expensive method of installation, particularly if all the source cables are specified, measured, and constructed prior to the installation. There are many reasonable-cost wire labeling solutions that print on heat shrink or on self-laminating wrap on labels. This method,

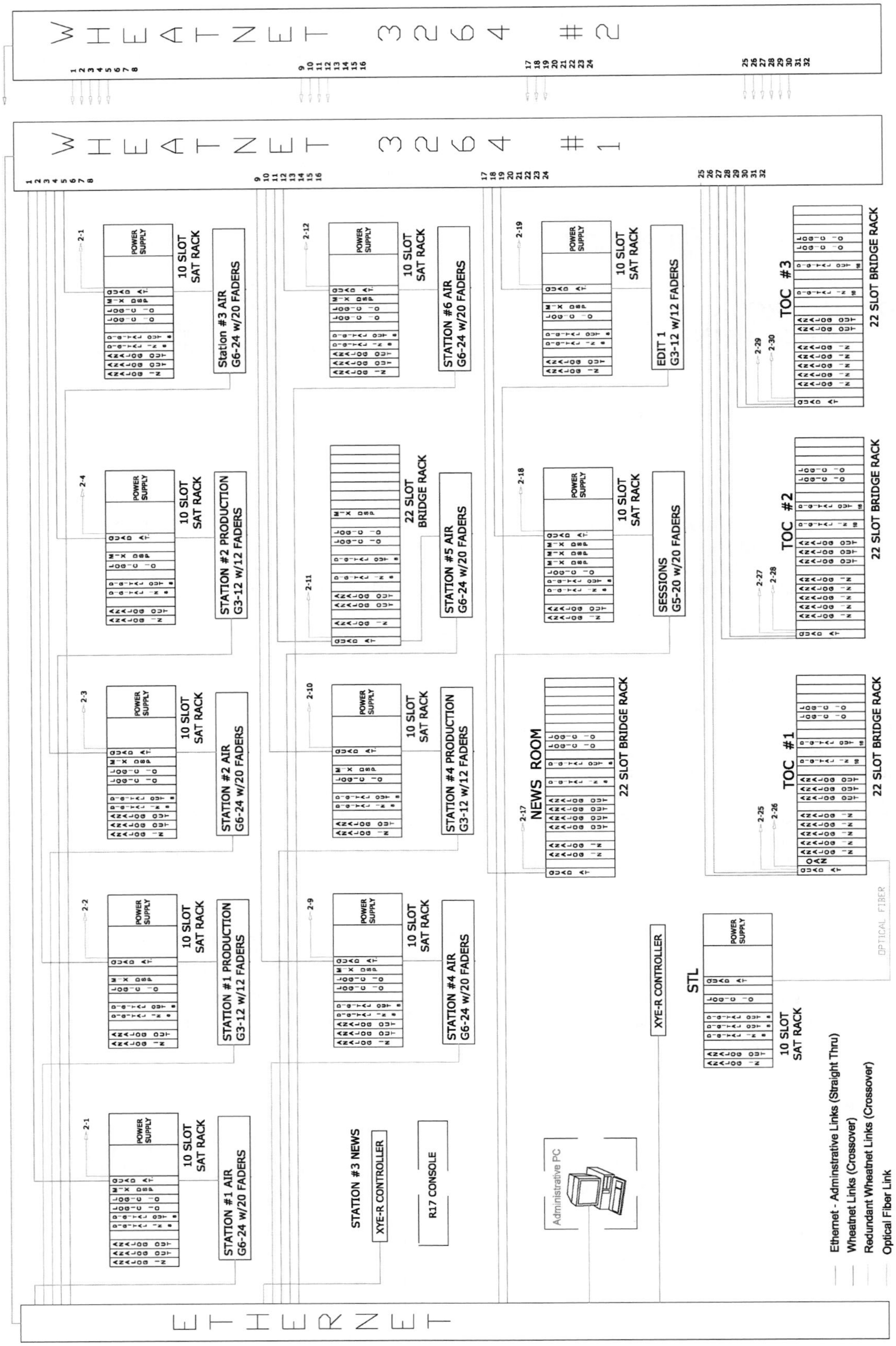

FIGURE 3.5-30 The entire audio backbone for a large multistation broadcast facility. (Courtesy of Wheatstone Corporation.)

however, should be limited to small installations. As the number of inputs and source equipment variety increase, this method can become unruly and nonsystematic, making it difficult to maintain and troubleshoot.

Reconfigurability of point-to-point installations is carried out by reconnecting the source cables. This is not recommended on the fly for a live studio, as disconnecting the incorrect cable can easily cause major disruption to the program. Point-to-point wiring is also made more difficult by the trend of many console manufacturers to combine multiple channel input signals, or multiple channel control lines used for machine controls, to a single connector. While this reduces connector count, it becomes much more difficult to prebuild source cables, as the source cables may have multiple audio or control locations to interface.

Console Demarc and Interfacing

To address the difficulty imposed by console manufacturers that combine multiple audio inputs and/or multiple selectors' controls on a single connector, and to borrow on well-proven wire management techniques used by telephone systems, *demarcing,* or breaking out, all connections to the console in a *cross-connect* area is a common practice. With this method of installation, a convenient location, but not prime studio real estate, is designated as the studio cross-connect area. All signals to and from the console are systematically connected to *termination blocks,* and either all source equipment is connected to another set of terminal blocks and *jumpered* to the console blocks, or the source equipment is connected to additional terminals that may be available on the console termination blocks.

Terminal Blocks

There are many choices for these terminal blocks and, due to their ready availability, the use of telecommunication products is common. Personal preference will determine the choice of the many terminal blocks systems available, but a common choice is the S-66M1-50[10] (66 block) -style block, as shown in Figure 3.5-31 mounted on an S89B, a telephone company standard. This 66 block has four terminals across, separated in the center to form two pairs of two terminals, and the use of bridging clips to join the four terminals helps in temporary changes to the facility and is an aid in troubleshooting the installation. All the variants of the 66 block were designed to use solid #22 to #26 AWG insulated solid wire but have been successfully used with #22 AWG stranded audio wire such as Belden 8451 and 9451. The introduction of higher bandwidth controlled impedance wire used in AES/EBU circuits, with its thicker insulation, has presented a problem to standard 66-type blocks, and the wire/block compatibility should be carefully checked before the installation.

[10] Siemon Company Part Number; the block is available from numerous manufacturers.

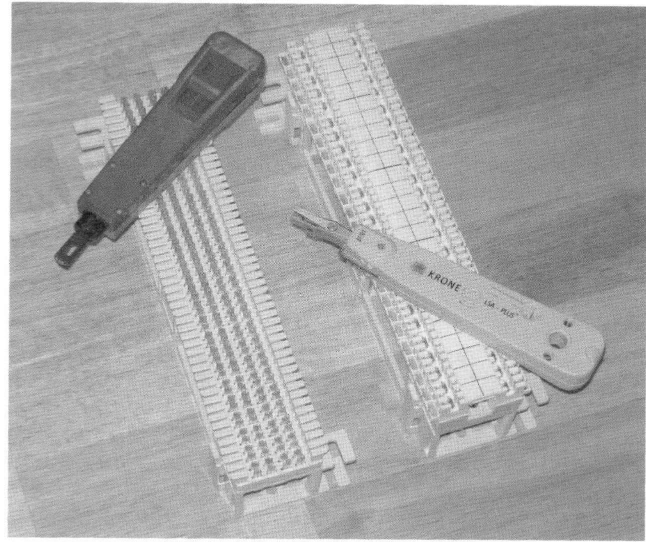

FIGURE 3.5-31 A standard "66"-type punch block and a Krone punch block. (Courtesy of Siemon and ADC Telecommunications, Inc.)

A newer telecommunications block of the same form factor of the S-66M1-50 block is the Krone Series 2 K110 50 pair block, also shown in Figure 3.5-31. The advantages of K110 blocks are that they are rated for #22–#26 AGW insulated solid or stranded wire, and they are rated to accept up to two wires of the same gauge per contact. These blocks are also CAT5 rated, making them candidates in impedance-controlled installations. The K110 blocks have twice the density of the standard 66 block and have integrated bridging clips. With the aid of testing cables, available from the block manufacturer, it is possible to bridge or to break into the circuit on the block and "look" both ways. This feature is helpful for troubleshooting and can be used to patch special signals into the system for short-time reconfigurations.

It is now possible to produce a complete studio wiring and *demarc* system based around shielded CAT5 cable.[11] Systems are available that contain connector blocks and adapter cables to simplify the studio install. Various companies have been producing different connector block systems for years that have found their use in broadcast studios.

Whichever termination block is chosen, it should be used throughout the facility. A common block for each studio and for the technical center, which is the central hub of all studio interconnections, will simplify maintenance and help to keep order in the system. The arrangement of wire pairs on the block should also be consistent throughout the facility and should be in harmony with the features of the block. One important purpose of utilizing a demarced installation is to provide a systematic order to the installation. The documentation of all signals on the blocks is easier, and the

[11] See "Studio Hub" from Radio Systems, Inc., at http://www.studiohub.com/.

effort required to document the blocks and keep the documentation cannot be overstressed.

Patch Bays

In all facilities to one extent or another, it is important to be able to quickly reroute audio for special configurations or for emergencies. The *patch bay* has been the industry standard since the early days of broadcasting and is still the simplest and most reliable way to accomplish reconfiguration. A patch bay used in broadcast generally consists of two rows of *jacks* (sockets) that are interconnected. These *jackfields* can be of several connector types depending on the density of the installation and the types of signals to be transmitted. Patch bays are available for balanced audio, AES/EBU signals, video, four-pair Ethernet, and any other common signal. The utility of a patch bay above a simple jackfield is its layout and *normaling* capability.

Temporary cross connections are made with a double-ended *patch cord,* comprising a length of flexible cable with a jack plug at each end.

Normaling

In a two-row patch bay, each column is typically for one channel, with the source jack on the top row and the destination jack on the bottom row. Each jack has a *normal* connection for each signal conductor that returns the signal to an alternate connection in the absence of a plug being inserted in the jack. By appropriate connection of the normal signals, the patch bay will have a normal path, a completed signal path established between two jacks in the absence of any patch cord with jack plugs. The term *normaled* is used to describe this path, as the name implies.

There are three distinct types of normaling in general use in patch bays. The first is *full normaling,* shown in Figure 3.5-32, which has its roots in the telephone industry. With full normaling, the insertion of the plug in either jack breaks the circuit; any signal on the plug is injected into the pair connected to the jack, or the signal on the jack is routed to the plug. Full normaling is used primarily where impedance concerns do not permit bridging or where bidirectional signal routing is necessary on either circuit.

In modern audio installations, driving source impedances are low, load impedances are generally high, and controlling the impedance of the circuit is

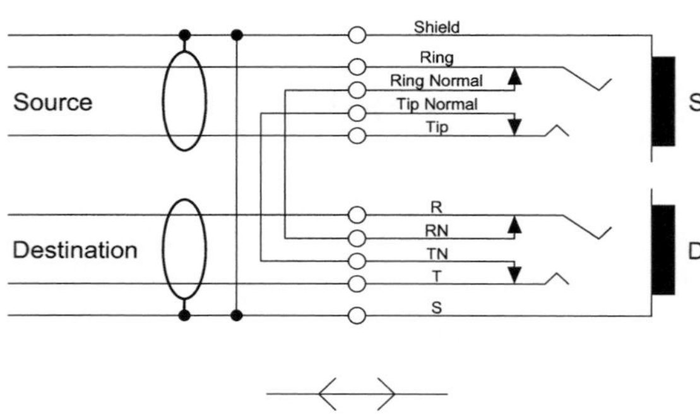

FIGURE 3.5-32 Full normal patch bay channel.

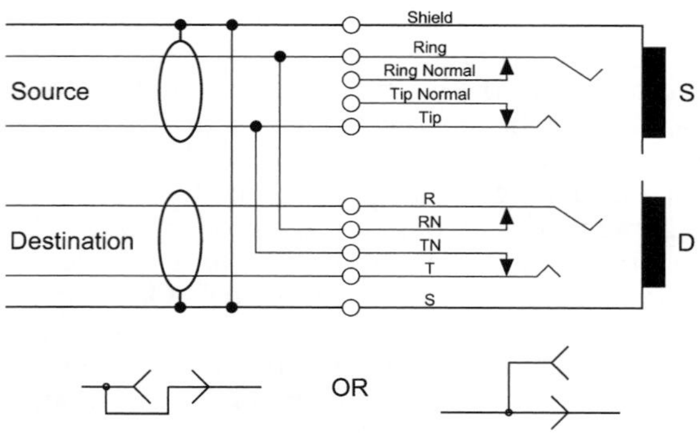

FIGURE 3.5-33 Half normal patch bay channel.

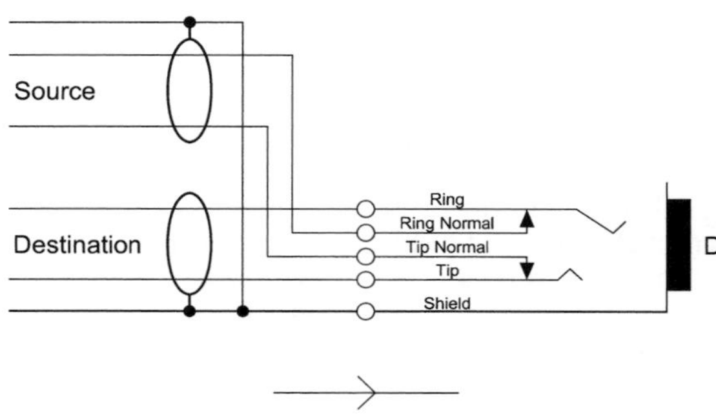

FIGURE 3.5-34 Blind normal patch bay channel.

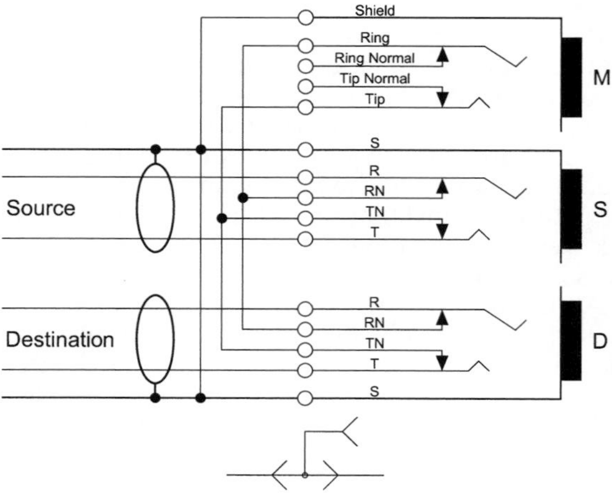

FIGURE 3.5-35 Full normal patch bay channel with monitor.

not as critical. Additionally, the need to send a signal up the source cable may not be a concern. Under these conditions, *half normaling*, shown in Figure 3.5-33, is often used. Half normaling connects the source to the source jack as in full normaling, but also parallels it to the return contact of the destination jack. Plugging into the source jack does not interrupt the output to the destination jack or its circuit, and the signal obtained on the patch cord can be used for monitoring the circuit or for bridging the circuit to another destination. Plugging into the destination jack will interrupt the normal source and will inject the signal on the patch cord in its place. Care must be used to avoid patching between two source jacks, as this will parallel two sources and generally not produce a desired result. Likewise, any signal on the patch cord will be paralleled with the source, and shorting the patch cord will short the source it is plugged into.

The final normal type that is sometimes used in broadcast is the *blind normal*. A blind normal is a half normal with no source jack, as shown in Figure 3.5-34. It is useful if monitoring of the normal source or bridging the normal source to another destination is not required. A blind normal requires only one jack per channel and eliminates the possibilities of disturbing the source through mispatching the source jacks.

Another available style of patch bay utilizes three jacks per channel circuit, allowing for full normal of the source and destination and a separate monitor jack above, but these are most often found in the telecommunications industry and are not normally used in broadcast. Figure 3.5-35 shows this configuration. Most modern patch bays are constructed for ease of configurability of the normaling method used in each jack.

Patch Bay Layout

Patch bay layout requires careful thought and, above all, clear labeling. When patch bays are laid out, like signals should be grouped together. Mic patch bays should be segmented from analog line level patch bays, which should be segmented from AES/EBU digital patch bays. Generally, patch bays are more easily understood if within a segment, the channels are grouped by destination and not by source. Clear labeling is critical, and some color-coding is helpful, but not at the expense of legibility. White printing on bold colors is best because the eye adjusts to the white information rather than the darker background.

In addition to console inputs, outputs, and studio tie-lines, access to specialty interface items on the patch bay can aid in last-minute signal accommodation or to correct a difficulty. Some of these specialty interface items are

- External device circuits connected to panels of XLR connectors near where equipment is likely to rest for temporary use;
- At least six male and female XLR connectors strategically located within the studio for analog utilities;
- Three males and three females for AES/EBU digital service;
- Several jacks connected in *parallel.* to allow bridging of multiple signals.

In many installations, the patch bay will be directly wired to the console, and the nonconsole side sources and destinations of the patch bay will be terminated on terminal blocks in the same way as a demarced console, discussed previously, connected directly to the source equipment or connected to the main XLR jackfield. Reconfigurability of the patch bay sources and destinations will dictate the correct choice.

Routers

Electronic crosspoint routers have made significant inroads for studio interconnections. These devices can be used to connect any of their sources to any or all of their destinations. The majority of router designs prior to the last 10 years was analog crosspoint routers, where one or more large X-Y multiple channel matrix analog switches selectively connected a single output of an input buffer amplifier to one or more inputs of line driver, all under computer control. This technology is still used for small crosspoint switchers, generally under 32 × 32 channels.

Newer router technologies utilize time division multiplexing (TDM) to share signals, similar to large telecommunications switches. These devices start by digitizing the incoming analog signal or accepting a digital input and gating it on the TDM bus during the appropriate time slot. After the appropriate propagation delay, the receivers scheduled to receive the audio all latch the data from the TDM bus and output the signal in digital form or convert it into an analog signal as required. Further refinements allow the spanning of the TDM bus between multiple frames and add digital signal processors (DSPs) to allow mixing and other processing to be performed on the signals that pass through the router. These natural evolutions led to the new breed of control-surface-based router systems discussed previously.

Router Control

All but the earliest routers are digitally controlled, and the range of operator interface control stations available to control the routers is vast. They range from a simple thumbwheel with a take button, a digital display with an encoder knob and a take button, a wall plate with multiple push buttons and a volume control for bathroom speakers, a full computer interface with time macros to do automated reconfiguration, to full broadcast console work surfaces—virtually anything is possible.

Sizing

How much router to purchase is a matter of careful design, and then upsizing with a significant safety margin for the unexpected. Each studio output and each air monitor receiver for the facility should be on the router so that they are available in every studio. Feeding the processor/transmitter system through the router should be considered carefully, balancing convenience with additional elements and failure points in the primary audio chain. Using the router is handy if it becomes necessary to switch programming to another studio in an emergency, but a patch bay's reliable passive nature might be a better choice for this critical circuit. Each network and emergency response feed should be a source on the router. Every studio should have several available channels from the router as inputs to the console, and every recording device should be a destination on the router. The remaining configuration is up to the individual design, but planning for the unexpected by increasing the size of the router is important.

Contingency

One final consideration is that, like all electronic equipment, sooner or later the router will almost certainly fail so contingency planning for failure modes is important, with bypass and manual patching available for critical connections.

Extreme caution should be taken when running wired outputs from a router over long distances. Lightning induced on a cable run to the roof of a high-rise building can have disastrous effects on a router. Avoid feeding the output of an audio router directly into a metallic pair (telco line or private buried audio cable) outside the building unless proper isolation or line conditioning or surge suppressors are employed. Fiber-optic interfaces provide a good solution to keep lightning from ruining expensive equipment and causing disruptions in programming.

CABLING AND INSTALLATION

Cable Selection

The selection of the type of cable to be used for studio interconnections is important. In environments where the cable is subject to mechanical motion, as in vehicle-mounted equipment or as the final jumper to equipment that is not secured in place, the conductors should be amply secured and should be of stranded wire. In fixed building installations, the securing is still important, but either stranded or solid wire may be used.

The choice of cable for interconnect is broad. Many traditional installations have utilized multipair individually shielded stranded twisted pair cable for all interconnections between rooms. Others have used multiple one- and two-pair individually shielded twisted pair cables such as Belden 8451 or 8723 or have used two-pair overall shield CAT3 telephone cable for the interconnects. The use of individual one- and two-pair cables is slightly more labor intensive during the installation but may be easier to install in a conduit and is less expensive than the multipair cable. Installations have been successfully done utilizing 25-pair unshielded telephone cable, providing that all signals are comparable in level and all equipment inputs and outputs are well balanced, but crosstalk issues can become problematic if a circuit becomes unbalanced by a connection problem or incorrect installation techniques. Most professional installations refrain from the use of unshielded telephone cable for any use other than telephones and logic control signaling and then, when available, these cables are obtained with an overall shield.

With the transition to digital audio equipment interconnection (AES/EBU), consideration should be given to installing the entire audio interconnect facility with impedance-controlled AES/EBU-compatible cable. AES/EBU cable is well suited for carrying conventional audio, but the converse is not true. Many successful installations have been performed utilizing shielded CAT5 cable, which can carry both analog and AES/EBU signals, and this has proven to be a cost-effective overall cable.

A newer network cable that may be suitable for audio system interconnections is the yet-to-be-adopted CAT7 cable. This is a four-pair cable featuring individually shielded pairs with an overall shield. When coupled with the Tera®[12] connector proposed as a replacement for the 8-pin modular connector, this standard could become very effective. The individual shielding will allow freer use of the pairs, and the new connector allows the use of one, two, or four pairs of each cable in separate jacks. The cable will perform well on data beyond 350 MHz; will carry balanced audio, AES/EBU, and conventional telephone; and may be able to carry video with the use of baluns.

Conduits

When the costs of an interconnect system are compared, the comparisons are best done by calculating the cost per pair installed. This cost should include the cost of the conduit required between rooms, as the cable and the outer jacket thickness greatly affect the number of conductors that can be installed in a conduit. The National Electrical Code (NEC) specifies that a 40% area fill of a conduit with multiple wires or cables is the maximum. While this NEC section is not specifically targeted at low-voltage audio wiring, it is based on experience pulling cables. Filling a conduit beyond 40% is difficult and stressful on the installed wire, and many cable manufacturers recommend a more conservative fill ratio.

While it is good practice to leave a nylon pull string in any empty or partially filled conduit, care must be exercised that the pull string does not become wrapped around the installed conductors. Great care should be used when pulling an additional cable into a partially filled conduit, as the existing cables are easily damaged during this operation. Except for short conduits, good practice is to fill the conduit to capacity with the needed and spare cables during the first installation and then not disturb the conduit afterward. Extra conduits should be provided for future needs.

[12]Registered trademark of The Siemon Company, Watertown, CT.

Cable Separation

Balanced circuits and shielding go a long way to protecting an installation from interference and crosstalk, but there are limits to their protection from cross-cable coupling. To assure a "quiet" installation, spatial separation between signals of significantly different signal levels is highly recommended. Even a few inches of separation can make a large difference in coupling. Grouping cables in separate conduits by signal type is an important practice. The following are the broad classes of signal levels and types that should be separately grouped:

- Low level (microphones and other pickup devices);
- Analog-balanced line level;
- Digital-balanced line level (AES/EBU audio, RS-422/485, and Ethernet cables);
- Analog telephone and control circuits;
- High-level audio (speakers);
- Power.

The analog line level, digital line level, and analog telephone are comparable level signals. They can generally be run together, but in ideal conditions, they will be separated.

Shield Grounding

Professional installations are usually carried out with shielded or screened cable for all signal connections, and this is good recommended practice. In most environments, connection of the shield at one end only is appropriate, as connection at both ends can lead to currents flowing in the shield if there is a voltage difference between the equipment ground terminals at each end, and this current can induce noise into the shielded conductors. In high RF environments, however, grounding the shield at both ends may provide better attenuation of interference to the signal conductors. If grounding at both ends produces objectionable low-frequency shield currents in a high-frequency environment, directly connecting the shield at one end and connecting the other end to ground through a small ceramic or Mylar capacitor prove to be most effective. When studio interconnect cable is installed at terminal blocks, connecting the shield on all terminal blocks along with the signal wires allows easy adjusting of shield grounding locations by changes in bridging clips or connections if necessary. Shields should not be allowed to float at both ends.

AUDIO INTERFACING THEORY

No single piece of equipment will make an audio system for broadcast. Audio systems are many pieces of equipment interconnected to perform their individual functions, which as a composite arrangement create a vehicle for programming. Incorrect interconnection of these multiple pieces of equipment can cause the quality of the signal to degrade significantly or fail altogether. Understanding the various standards, techniques, and terms used for interconnection is critical to success. The following is a description of terms and techniques used with equipment at its interfaces.

There are two broad classes of interconnections to be considered, analog and digital, and any modern facility will typically contain both. The one thing that they have in common is they carry power with electrical energy, which means that there is a voltage inducing a current in load impedance. This concept will be discussed first, followed by consideration of the decibel, and then the use of balanced circuits for signal interconnections. The subsequent section on digital audio signals discusses the nature of digital audio and its carriage on a serial digital interface.

Voltage, Current, and Impedance

After the input transducers (microphones and other pickup devices) and before the output transducers (speakers), the audio is carried as time-varying electrical signals on wires or in modulated light in optical fibers, the vast majority on the former. Time-varying signals induce a *voltage*, which is carried to its destination over a wire or transmission line to a load, and causes a time-varying *current* to be created in that load and to deliver power to the load. The *impedance* is the ratio of the voltage at the load to the current imposed on the load. This term, expressed in ohms is often specified in data sheets for the equipment.

Characteristic Impedance

A cable or a transmission line has *characteristic impedance*, which is its capacitance per unit length divided by its inductance per unit length. An infinitely long piece of this wire, when driven by a time-varying voltage, will present a load of its characteristic impedance to the source. There are no infinitely long lines, but for maximum energy transfer, it is important for the source impedance, the transmission line characteristic impedance, and the load impedance to be the same, a condition called *matched*. If they are not, energy will be reflected from the load back to the source, and there will a standing wave on the line.

Matching is critical in radio frequency circuits because of the desire to maximize the transmission of power to the load, and standing waves on the line affect the performance of the system. Generally, in analog audio, with the advent of modern electronics where amplifier voltage and power gain is cheap and easy, and where the length of the transmission lines is short compared to a wavelength of the signal to be carried, impedance matching is not critical. In modern audio equipment interfacing, the desire is to drive the transmission line and the load with the lowest impedance practical to swamp out the effects of the line and load reactance, and to receive the signal in a load of as high impedance, within limits, as possible to lower demands on driving circuits and reduce power requirements.

Matching Impedances

Sometimes, however, matching of the transmission line to the source and the load is important:

1. Matching is important when the line is long compared to a wavelength of the signal to be passed. This occurs when the frequency of the signal is high, as is the case in digital signals and video. In these cases, the energy being reflected back from the load will appear as a ghost signal to video, and can cause the edges and levels critical for decoding a digital signal to become corrupt and errors added to the signal. In general, this impedance matching is provided within the equipment being utilized, but it is necessary to assure that the transmission line (wire) being used is acceptably near the design impedance of the input and output circuits. Difficulty occurs when attempting to bridge multiple devices to a single source, and this must be handled appropriately with distribution amplifiers or removing terminations of bridged devices to keep the system matched.
2. Long audio pairs or loops fall into the classification of lines that are long compared to the wavelength of the signal to be passed. The reactive nature of these circuits is designed to be driven by and be loaded by specified impedance. It is often necessary to match the source and load impedances to the design value of the line, and this is often done by the use of resistive pads, build-out resistors, resistive loads, and transformers. Often these long lines require equalization to overcome the loss of high-frequency response due to their reactive nature. Errors in setting these equalizers can occur if the driving source impedance, which is usually connected to the line or loop by a transformer or repeat coil, changes from the test set used during equalization to the device driving the line or loop in operation. Active drivers and receivers that remain on line for both test and operation are recommended.
3. Transformers are passive devices that change the impedance of a circuit between their input and their output and provide metallic isolation between the two circuits. The signal, energy, and power are preserved and transmitted between the two circuits by the magnetic coupling within the transformer. Many times transformers are a wonderful solution to signal coupling, but their use must be in accordance with their design. These devices are highly reactive; they have a large amount of inductance and stray capacitance due to their construction. If not driven by and loaded by their design impedances, the highly reactive nature of the transformers will adversely affect the frequency response of the system, can cause loss of transient response to the system, and can cause ringing to be added to the signal. Matching is often done by building out the driving source impedance and by loading the transformer with resistance or a resistive pad.

Several impedances are often encountered in audio work. The most common is 600 Ω, and is the impedance that the audio industry inherited from early telephone work. The characteristic impedance of #6 wire strung on telegraph poles at 12" centers, typical in the early days of telephone, was 600 Ω. This standard impedance has remained with us to this day. Early audio equipment was built with transformer inputs and outputs to produce a balanced 600 Ω signal during a time when voltage and power gain within the circuits (tube circuits) was expensive to create and matching of input and output impedances of equipment transmitted maximum energy. With the advent of transistors and integrated circuits, voltage and power gain have become inexpensive and easy. All professional equipment is capable of driving 600 Ω loads, but most professional equipment input impedances are significantly higher than 600 Ω, typically around 10 kΩ, requiring much less current from the source than lower impedance loads. These inputs are often referred to as *bridging inputs*. While for control and fault reasons the practice of paralleling or bridging many devices' inputs on one output can be problematic, it is acceptable.

Microphone impedances are most often standardized at 150 Ω, which is close to the characteristic impedance of microphone cable. This impedance was chosen to maximize energy transfer. Typical cables used in installations within a studio have impedances in the 90 to 150 range, and the AES/EBU standard for digital audio transmission specifies 110 Ω as the design impedance for the circuits.

To drive speaker systems with the power needed at reasonably low voltages and improve damping factors—the ability of the source to control the current in a speaker, which is both a load and a generator—speaker systems utilize impedances significantly lower than the rest of the audio system. The impedances are in the 4 Ω to 16 Ω range. Underloading power amplifiers of modern design is not a problem, but overloading them is. The combined impedance of the speakers connected to each amplifier output should not be lower than the load rating of the amplifier employed.

The Decibel or dB

The range of voltages found in a broadcast facility is large, with the smallest voltage being in the order of 1 μV (0.000001 volt; a quiet passage from a microphone) and the largest being 100 volts or greater (power connections and very high power amplifiers). This represents power levels of 6 fW (0.000000000000006 W) to kWs (>1,000 watts). To simplify working with such a large range of numbers, the decibel, or dB = 0.1 bel (B; first used by Bell Telephone Laboratories and named in honor of its founder, Alexander Graham Bell), is extensively used and misused within the sound, audio, broadcasting, and electronic industry.

The decibel is defined as 10 times the log (base 10) of the ratio of subject powers to the reference power, and is a unitless quantity. Stating that something is 22.5 dB is incomplete. When the term *dB* is appended to a value, either the reference power is stated or

implied by the context of the discussion; the value is a change in some other known signal understood from the context, like the gain in an amplifier; or the term is being misused.

The dB is often used as a unit of measure, implying that it has a unit. This is done in many contexts in many different industries. To make the unitless quantity of a dB into a quantity that has a specific meaning, a reference power is defined and the reference power becomes the denominator in the definition of the dB. The most common references in the broadcast industry are

dBm or dBmW	Power referenced to one milliwatt
dBW	Power referenced to 1 watt
dBkW	Power referenced to 1 kilowatt

In a constant impedance system, the dB is equally a measure of the ratio of two voltages as it is the ratio of two powers. Most often, measuring the voltage of a signal is much easier than measuring its power. As the power delivered to a load is proportional to the square of the voltage applied to the load, the unit of a dB is 20 times the log (base 10) of the subject voltage to the reference voltage. Common references used in the broadcast industry are

dBV	Voltage referenced to 1 volt RMS
dBu	Signal voltage referenced to an unterminated circuit that would provide one mW to a standard 600 Ω load or 0.775 volts RMS
dBspl	Typically the sound pressure of 20 μP with varying weighting filters specified

Other less direct measurements are also expressed in dB. The dB is often used as a relative measurement. When it is used as a relative measurement, the context is often specified as part of the symbol as in the case of

dBa, dBc	A modification to a sound pressure measurement in dBspl specifying a weighting factor to adjust the spectrum to mimic the inverse response of the human ear's equal loudness response at low levels and high levels, respectively.
dBd, dBi	Gain of an antenna above a reference dipole or isotropic antenna, respectively.
dBc	Power of the noise floor or sideband power referenced to the carrier power.
dBfs	Amplitude referenced to the maximum possible signal, or full scale, of the circuit. In an analog signal, this is the clipping point. In a digital system, it is the maximum value that can be represented by the digital word sampling the signal. Unlike most other applications of the dB, this is referenced to a peak value, not a power or RMS value.
dBr	Intensity referenced to some arbitrary reference.
dBvu	Intensity referenced to the established reference level of the equipment or facility.

Balanced and Unbalanced Circuits

Any electrical circuit that moves information requires a current. Current must flow in completed circuits: The current flows from the source, through the lines and loads, and returns to the source. With most internal analog audio, this signal is returned or referenced to the analog ground within the equipment. In consumer grade equipment interconnections, the signal is presented on a single conductor and returned on the shield wire to the analog ground. This works well as long as the analog ground references of all equipment are the same. If they are not, and a voltage difference exists between the grounds of two pieces of equipment, this voltage difference is inserted in series with the desired signal voltage, and noise is introduced to the equipment connection. In installations containing more than just a few pieces of equipment, it is difficult or impossible to assure that no voltage differences exist between the analog grounds of all connected equipment.

To eliminate the noise introduction by differing references for the analog signals, *balanced* or *differential inputs* and *outputs* are used on virtually all professional equipment. Rather than referencing the signal to the analog ground of the equipment, two balanced wires are used to convey the signal. The input of the equipment responds to the difference in voltage between these two conductors, not to the voltage with respect to ground. The outputs are likewise driven differentially, with one terminal going positive, while the other is going equally negative with respect to ground. This implementation is done with transformers, at which time there may not be a reference to ground, or in modern equipment, with active differential line drivers and receivers.

Common Mode Rejection Ratio

One important parameter of a differential input is the Common Mode Rejection Ratio (CMRR), expressed in dB, which is the ratio of the output when driven by a signal driving the input normally to the output when the same signal is driving both input terminals tied together, referenced to ground. When equipment with voltage differences between the two grounds is interconnected, the differences in ground voltages are imposed equally on both of the input terminals. As the desired signal is the difference between these two terminals, the difference in ground voltage between the two analog grounds is canceled. This cancellation is not perfect, but rather defined by the CMRR.

Noise Rejection

In addition to noise potentially introduced by analog ground voltage differences between two connected pieces of equipment, noise can be imposed on the cable between the two pieces of equipment by magnetic or capacitive coupling. By use of balanced circuits and good cable, this noise is induced equally and in phase on both conductors. Again, since the receiving end is sensitive to the difference between its two input terminals, the signal that is the same on the two input terminals is canceled or reduced to the level defined by the CMRR. To ensure that the induced signal is the same in both signal conductors, the driving impedance of both terminals of the transmitting side should be the same, as should the load impedance of both sides of the receiving side. Additionally, the connecting cable should be symmetrical about the two signal wires, and these two wires should be twisted around one another such that the induced signals are similarly imposed on each wire. Shielding or screening each twisted pair also helps reduce the imposed signal. Likewise, the two currents of a balanced circuit on twisted pair cable cancel and radiate less energy. Balanced circuits cause less interference to other circuits.

Maintaining Phase Relationship

It is important that the instantaneous phase relationship of all channels that compose a complete audio signal maintains the same relative phase with respect to one another. With unbalanced circuits, it is difficult to alter the phase relationship between channels. With balanced circuits, it is easy to alter the phase relationship between channels by reversing the connection of any pair of signal wires, so a systematic method of wiring and testing should be implemented to maintain relative phase of the systems at all times. While this is particularly critical when dealing with the multiple channels of a given source, standards recommend that relative phase be considered for entire plants. By standards (AES26-2001), a positive pressure on a microphone will create a positive voltage in the unbalanced circuit, which will produce a positive pressure in the speaker. In a balanced circuit, one conductor must be defined as positive; this does not reference a polarity but rather a positive pressure/positive voltage relationship, and this is generally the nonblack wire. Standards specify that this positive wire should be connected to the terminal that is marked with some variant of +, should be connected to the tip of a tip-ring-sleeve connector, and should be connected to pin #2 of an "XLR"-type connector. It is helpful to follow industry standards, but whatever standard is used by the studio facility in question, the standard should be documented, posted, and followed for all work within the studio to simplify maintenance and to reduce the number of errors during installation.

DIGITAL AUDIO SIGNALS AND INTERFACES

An analog audio signal is a time-varying voltage whose amplitude is in proportion to the air pressure applied to the microphone that is altered, amplified, processed, stored, recalled, distributed, and otherwise handled until it is applied to a loudspeaker to create air pressure in proportion to its amplitude. The digital audio process inserts at least two other steps in the chain and changes most of the steps defined above. At some point, preferably early in the chain, the analog voltage is coded into a string of digital words, and near the end of the chain, these digital words are decoded into an analog voltage that drives the speaker. In the middle, the signal is altered, amplified, processed, stored, recalled, distributed, and otherwise handled, but this is done with computers and digital signal processors rather than precision analog electronic equipment. In some systems, the audio is moved from the analog domain to the digital domain several times before finally being returned to the analog domain for consumption.

There are many advantages of handling audio in the digital domain. Some are:

- More readily maintained noise floors;
- Faithful reproduction of the information encoded;
- Less degradation of stored signals with time;
- Reduction in cost of equipment needed for elaborate manipulation of the audio;
- Faster access to the audio material stored in nonlinear storage devices.

The penalty for using digital audio is generally more complexity in studio implementation and interfacing.

Quality Issues

Analog-to-digital (A/D) and *digital-to-analog* (D/A) converters are the boundaries of the two domains. An analog signal is supplied to the A/D converter and converted to a digital word that is an instantaneous representation of the voltage being sampled.

Bit Depth

The accuracy of the digital sample, which is the smallest change in the signal voltage that can be quantified, is dictated by the number of bits in the sample. The most common sampling system is linear *pulse-code modulation* (PCM), which applies equal weight to digital step of sampling and represents each sample value as a 2's complement integer. A sample that is 8 bits wide can hold 256 discrete levels of quantification, and the maximum signal level it can contain is a peak-to-peak signal of 256 steps. The minimum signal level it can represent above no change is a peak-to-peak signal 1 step. The ratio of the maximum to minimum signal is called the *dynamic range,* and with 8 bits, it is 256 or 48 dB (20 log 256). This is poor telephone quality. Each additional bit dded to the sample doubles the number of quantification levels or increases the dynamic range by about 6 dB, since the amplitude of the signal is being sampled, not its power.

In professional audio, the minimum normal linear PCM signal encountered is 16 bits, or 65,536 quantification levels, which can represent a dynamic range of about 96 dB (20 log 65,536). A technique to add a small

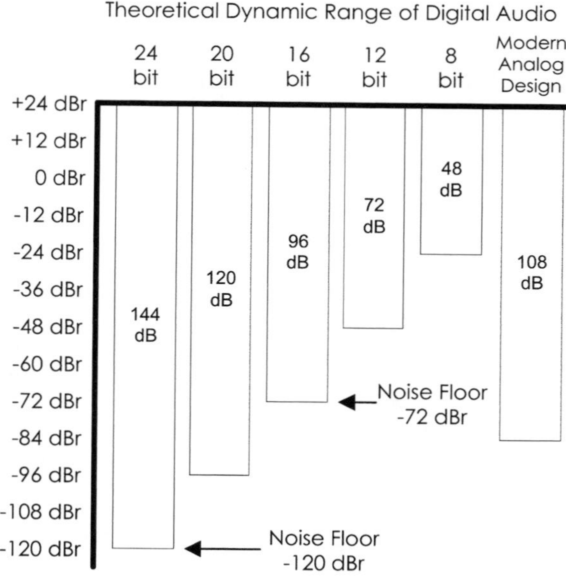

FIGURE 3.5-36 Dynamic range comparisons.

amount of random noise, with amplitude about that of the smallest quantification step, or the least significant bit, is called *dithering* and can be applied to the digital signal in an attempt to mask and extend the quantification noise.

Standard word lengths used in professional audio systems are 16, 20, and 24 bits, providing theoretical dynamic ranges of approximately 96, 120, and 144 dB, though the analog electronics involved do not generally have the capability to realize the higher dynamic ranges.

When digital and analog equipment specifications are compared, care must be taken with the way the figures are presented. Specifications for analog equipment often give the noise floor below the operating level, which is about 20 dB lower than the clipping level of the equipment. The dynamic range of a digital system is the ratio of the noise floor to the digital clipping point. One needs to be sure to compare like values when comparing specifications. The theoretical dynamic ranges are shown in Figure 3.5-36.

Sampling Frequency

The maximum frequency of the signal being sampled determines the required sample rate for the digital system. The Nyquist-Shannon sampling theorem states that with a band-limited signal, a complete representation of the signal is obtained if the bandwidth is less than half of the sampling rate. To characterize a 20 kHz signal, the sampling frequency must be at least 40 kHz. The Nyquist frequency is the maximum signal frequency that can be sampled and is one-half of the sampling frequency. Attempting to sample spectrum above the Nyquist frequency will create aliasing, which is a particularly unpleasant nonharmonically related distortion added to the recovered signal. It is therefore necessary to sharply filter the incoming signal to frequencies below the Nyquist frequency, and these *brick wall* filters can present unpleasantness in the decoded audio. For this reason, the standard sampling frequencies are slightly higher than defined by the Nyquist-Shannon sampling theorem to ease the requirements on the anti-aliasing filters. There are other techniques using *oversampling* and implementation of *digital filters* to lessen the effects of severe filtering. Filtering of D/A converters is less critical than A/D converters, but is necessary to smooth the step transitions created in the process. Again, techniques of oversampling and digital filtering are used to improve the audio performance of these converters. The filters are the key to the aural performance of all converters.

Serial Interfaces

Standards

The Audio Engineering Society (AES) in concert with the European Broadcasting Union (EBU) established a standard that is widely utilized for the transport of two-channel digital audio as a serial bitstream between two pieces of equipment. This standard is known as AES3-2003 or AES/EBU. A close cousin for consumer equipment has been standardized by the International Electrotechnical Commission (IEC) and is commonly referred to as SPDIF (Sony/Phillips Digital Interface Format). It is specified in the IEC Report TC 84/WG11 and IEC Publication 958. Details of these standards may be found in Chapter 1.15.

These standards have permitted diverse manufacturers' digital audio equipment to be almost assured of interoperability.

AES-3 recommends a balanced data circuit for noise immunity and to help prevent crosstalk into other circuits, with cable runs up to 100 meters. It specifies three pin "XLR"-type connectors. While relative phase should be maintained in all cables, the standard states that the audio signal is cable phase insensitive. The signal levels at 3–10 Vpp are essentially the same as line level analog audio. Implementations of AES-3 have also been standardized under AES-3id-2001 utilizing unbalanced 75 Ω coax cable for longer runs without repeaters and more easy adoption into a television infrastructure. More advanced implementations of multichannel AES-3 have been specified on 50-pin D-Subminiature connectors.

S/PDIF is a close but not fully compatible cousin to AES-3. The fundamental encoding between the two is identical, but the channel status data and the user data channel (subcode information) are not the same. The consumer format also contains provisions for the Serial Copy Management System (SCMS) in the channel status bits. In particular, bit 2 is used to indicate whether copying is permitted and bit 15 distinguishes between original and copied material. In the IEC or SPDIF format, video coaxial cable with phono plugs can be used to convey data with an unbalanced 0.5 to 1V signal. There is also a fiber-optic version of the interface. Converters are available to transform between S/PDIF and AES-3, and often the two will work together without converters.

While the AES-3 format can operate over short distances on many different cables, because of the high data rates, the cable chosen is important. The AES-3 standard (as well as the S/PDIF format) specifies two 32-bit subframes per sample period. As the standard uses *biphase mark coding*, which creates two binary states for each bit transmitted, the binary state transmission rate for the connection is 128 times the sample rate. With a sample rate of 96 kHz, the data transmission rate is 12 Mbps. This data rate requires that any cable of useful length must have low capacitance and the appropriate impedance to match the transmitter and receiver.

Cables

All major cable manufacturers offer cable specifically designed for use with the AES/EBU standard, and its use is necessary for successful implementation of digital audio transmission. The special dielectrics and the thickness of the insulation of AES/EBU-rated cable can present difficulty to some insulation displacement connectors such as standard 66 blocks. Confirmation of compatibility between the cable and the connector chosen is important. Where the transmission distance required exceeds 100 meters, implementation on 75 Ω coaxial cable should be considered.

Clock Reference and Sample Rate Converters

When multiple independent sources of digital audio are mixed, it is imperative that the samples be clocked at the exact same rate. This can be accomplished in two ways. One approach is to lock all equipment to a master clock or time base. The details of this are covered in the AES11-2003 Standard, which outlines a digital audio reference signal (DARS). Assuming that all the sources are the same standard sample rate, proper synchronization and mixing will occur; however, there can be several different sample rates used commonly within a facility.

Most high-quality installations have standardized on a sample rate of 48 kHz with some at 96 kHz. CDs use 44.1 kHz sampling. Often the studio transmitter links are limited by available bandwidth to 32 kHz sampling. Locking these divergent sample rates to a master clock can be difficult and may be impractical. Sample rate converters are available, both stand-alone and as part of the digital inputs to many consoles and other equipment, that will lock to the incoming signal and resample and reclock the signal to the internal standard of the device. Sample rate converters do an excellent job, but they are not perfectly sonically transparent. It is desirable to use sample rate converters judiciously.

SUMMARY

This chapter has covered most of the types of studio equipment, but the specifics of any of the equipment mentioned are likely to change between the time of writing and the release of this handbook, certainly over the next few years. What can be assured is that the convergence of audio and IT technologies will continue. Digital signal processors and computers, coupled with the established global libraries of code to drive them, are increasing in power and decreasing in cost more rapidly than the requirements of the application are changing. This combination assures intense competition and innovation for the manufacturer to maintain market share. The result will be more advanced equipment at a lower cost.

Studio systems will continue to be less about specialized equipment and more about specialized software. The studio equipment of the future may be nothing but a large touch screen, an audio input/output module, and a single high-power computer connected by a single fiber-optic cable to the world. Embracing and marrying these new products to the existing systems will be a challenge. The future is bright for the broadcast engineer or systems integrator who can keep current on the changes, respond accordingly, and implement systems with the features that users require, in a systematic, maintainable, and well-documented manner.

Standards

AES3-2003, *Serial Transmission Format for Two Channel Linearly Represented Digital Audio Data*, Audio Engineering Society, Inc., 2003

AES-3id-2001, *Transmission of AES3 Formatted Data by Unbalanced Coaxial Cable*, Audio Engineering Society, Inc., 2001

AES-26-2001, *Conservation of the Polarity of Audio Signals*, Audio Engineering Society, Inc., 2001

Audio Engineering Society, General References, http://www.aes.org/publications/standards/

Bibliography

Ramsey, C. G., and Sleeper, H. R. *Architectural Graphic Standards*, 10th edition. John Wiley & Sons, Inc., 2000.

Ballou, G. *Handbook for Sound Engineers*, 3rd edition. Focal Press, 2005.

CHAPTER

3.6

Radio Station Automation, Networks, and Audio Storage

DAVID T. TURNER

ENCO Systems, Inc.
Southfield, Michigan

INTRODUCTION

The term *station automation* is a very broad concept. It generally refers to the use of devices, processes, and system interconnections designed to make a broadcast station (radio or TV) run a series of scheduled events automatically (i.e., without operator intervention). This includes every scheme ever devised from massive reel-to-reel tapes and interconnected cart machines to all-encompassing control systems that steer the directional array, turn on the transmitter, play the audio, control the console cross-fades, and generate transmitter logs and billing information.

While the individual pieces of hardware involved in broadcasting have evolved to include powerful automatic functions and interconnect capabilities, nothing has advanced the "state of the art" of station automation more than the personal computer. The purpose of this chapter is to bring the reader up to date with what the state of *this* art currently is. To do this, we will explore some background on how personal computers and their powerful data manipulation and communications capabilities have provided the necessary "glue" to bond all of a station's subsystems together. And how these strengths, coupled with powerful and intuitive user-interface features, have created a complete automation solution. In the process, we will examine the evolution of practical network architectures and cabling standards and the current "favorites." We will discuss network bandwidth requirements, storage capacity and speed requirements, multiuser performance issues, and data redundancy options. We will finish with a look at some current storage architectures that fulfill these necessities, and point out a few popular methods of protecting these systems from catastrophic failure.

The majority of the details discussed in this chapter comes from practical radio station automation systems; however, most of the concepts are also applicable to television station automation.

THE PC HAS REVOLUTIONIZED STATION AUTOMATION

It wasn't too long ago that radio station automation referred to a 15-inch reel of tape playing at 1 7/8 inches per second on an autoreversing tape deck. This automation arrangement could provide most of a day's programming with minimal operator intervention, but the audio quality was mediocre due to the slow tape speed, and it required production personnel to invest a great deal of time assembling the contents of the tapes. Once the tapes were assembled, it was very difficult to make any changes, especially if the tape was playing on-air.

A more flexible arrangement used a stack of tape cartridge (cart) machines with the secondary cue output of one wired to the start input of the next, and so on with the last machine connected back to the first. This allowed the operator to load several songs and commercials to play automatically in sequence, freeing up short amounts of time for him or her to do other important operations (like running to the restroom). This crude form of automation was followed by numerous variations of mechanical beasts designed to hold and cycle dozens and sometimes hundreds of carts through multiple playback decks.

These electromechanical marvels were capable of automatically sequencing several days' worth of programming.

These machines were incredible achievements in station automation and have faithfully served the broadcast industry for many years. However, their dependency on extensive mechanical transports made them high-maintenance devices and limited their flexibility. Even the audio quality was hard to maintain due to the number of capstans and tape heads involved. Broadcasters needed newer, higher quality, easier to maintain, more versatile automation systems.

As compact disc (CD) players and digital audio tape (DAT) machines became commonplace in the home and in the studio, the audience demand for CD-quality broadcast audio also grew. Along with the improved audio, this new generation of equipment brought improved control features that allowed broadcasters to elevate automation to a new level. In addition to the standard Start and Stop functions, a much more complete set of instructions including shuttling and indexing were now available, usually through a serial data protocol. This enabled a single electronic controller to "talk" to multiple devices, directing them to cue up then play individual tracks from CDs and tapes containing multiple tracks. With CD jukebox devices available that could hold and play tracks from over 300 CDs and DAT tape machines that could cue and play hours worth of programming, a station could now automate for days at a time, if the controller was programmed properly.

While several manufacturers developed automation systems for their own equipment, these were mostly proprietary hardware and not able to communicate with devices from other manufacturers. These units often employed special keypads for data entry and ran ROM-based programs that were difficult to upgrade or modify. This is where the personal computer (PC) made its biggest mark on station automation. PCs were already being used in business and at home to run database applications, perfect for storing a *log* of scheduled events. PCs had excellent serial (and parallel) communications capability and could be configured/programmed to "speak" any protocol required. PCs used a standard user interface (keyboard and monitor). PC-based programs were stored on disk and ran from RAM and were, therefore, relatively simple to modify and reload. The PC was an excellent platform for station automation controllers.

A number of manufacturers created custom software packages using standard PC hardware to automate CD jukeboxes, DAT players, and fire relays to play standard cart machines. These were quite effective and could provide a good level of *walk-away* automation, but some of the stations' most important material (e.g., commercials, IDs, and promos) were left to the weakest link—the cart machine. It became increasingly obvious that a more sophisticated CD-quality version of the cart machine was needed to play spots, IDs, jingles, and other material.

As digital cart machines began appearing as direct hardware replacements for the existing analog units, some clever computer people were realizing that everything needed for a digital cart machine was already in a PC, except the analog-to-digital and digital-to-analog conversion electronics, and this could easily be added by building a custom signal-processing circuit board designed to plug into one of the existing PC expansion slots. Since PCs were already driving the development of the hard disk storage technology, they had access to the capacity required to store many hours of digital audio. This could be divided into any number of any size pieces, allowing an extensive inventory; and unlike tape or floppy disk–based systems, PC-based systems could provide *random* access to any and all of the material they contained.

Adding these powerful audio capabilities to the automation strengths of the PC has created the perfect platform for the continued evolution of station automation.

KEY FEATURES OF AN AUTOMATION SYSTEM

The personal computer is capable of providing all of the key functions required for a powerful station automation system in a single box:

- Random access high-quality audio
- Database operations
- Communications capabilities
- Programmability
- User-friendly operator interface

But to be an effective automation system, the hardware and software must be designed to provide certain key features.

CD-Quality Audio and Sound Cards

Let's start with high-quality audio. When the compact disc came out, it set a new standard of audio quality that became a widely used benchmark when discussing audio performance. The term *CD quality* implies the following basic specifications:

- Frequency response: 20 Hz–20 kHz (± 0.5 dB)
- Dynamic range: >90 dB
- Signal to noise: >90 dB
- THD + N: <0.05%
- Phase error between L-R channels: <1 degree @ 15 kHz

These specifications are very close to what signal-processing theory would predict for a digital system using 16-bit quantization. Current multi-gigahertz PC platforms are quite capable of manipulating this (and considerably higher resolution) data while maintaining its purity. Therefore, a determining factor in the overall audio quality of a PC-based digital audio system is the quality of the electronics used to convert between the analog and digital worlds. In a PC-based

system, this process occurs in what is generally referred to as the *sound card*.

These devices include the analog input stage, the input anti-aliasing filter, the analog-to-digital (A/D) and digital-to-analog (D/A) converter devices, the output reconstruction filter, and the analog output stage for each audio channel. It is very important that the analog stages have very low noise, wide dynamic range, and be very linear. The input low-pass filter must have a steep enough response to preserve the maximum high-frequency information for a given sample rate, should have minimal passband ripple, minimal ringing, and have a linear phase response. The A/D converter must have very accurate quantization levels (high linearity) and must complete each conversion very quickly (short conversion time). The D/A converter and reconstruction filter should use higher resolution (more bits) than the A/D to keep from raising the noise floor and limiting the headroom of the output signal.

It is often challenging to obtain these specific details without reviewing extensive technical documentation. To complicate things further, many manufacturers will measure these parameters and report the resulting specifications using differing terminology and references. But fortunately for the user, these components have advanced to the point that most every PC-based sound system available today will provide CD-quality audio when configured and connected correctly. The type of connections required will generally determine the type of sound card required.

Sound cards exist in several forms. Typically, they are plug-in cards designed to fit into a free expansion slot inside the computer. Many contain the A/D and D/A electronics on the card itself and this has proven to be a successful format for many years. Some manufacturers have questioned the engineering wisdom of putting highly sensitive analog electronics inside a computer's "noisy" digital environment. They have chosen to design their computer cards to contain only the digital electronics while moving all analog functions to an external chassis. While there is merit to the audio purity concern of this concept, one of the best features is the flexibility this design creates by allowing the use of different audio chassis designed for different standards (e.g., analog/digital, balanced/unbalanced).

Similar to this is the relatively new concept of the *remote audio card*. Since audio signals inside a PC are really just streams of data, this design uses a software driver to convert these audio streams to packetized data and transmit them to the audio hardware via an Ethernet-based network. Some designs, such as CobraNet [1], use a dedicated audio network architecture while others, such as Axia–Livewire [2], use standard Internet protocol (IP) on standard network interface cards (NIC) over a conventional switched Ethernet network. These audio networks not only allow the audio electronics to be in a properly conditioned environment, but the inputs and outputs can exist in different physical locations throughout a facility. This concept not only handles traditional sound card functions, but can serve as the primary audio routing system and provide traditional audio console functions as well.

In addition to these "add-on" solutions, many current PCs include audio hardware as a part of the main board (motherboard). This hardware usually falls into the "consumer grade" or "multimedia" category. While this hardware can provide good audio performance, consumer units typically have more high-frequency roll-off for a given sample rate, more passband ripple, more interchannel phase error, and higher distortion due to less expensive A/D and D/A converters. Generally speaking, products intended for the professional market (radio, TV, recording studio) use very high-quality components and provide measurably better audio quality than consumer-grade multimedia units. Pro units also usually provide balanced analog and digital audio connections where consumer units do not. And most pro units provide onboard digital signal processing (DSP) "horsepower" to perform real-time data compression and audio effects where consumer units do not. But with PC processing power growing in leaps and bounds, many of these functions are being provided in software (plug-ins) and offloaded to the main central processing unit (CPU). This architecture can provide very cost-effective audio features; therefore, these consumer devices have their place.

So it appears we have a number of ways to provide high-quality audio from a PC-based system. Thus, a key feature of a good automation system is to be flexible enough to make the best use of the different audio resources available to it.

Databases Keep It Organized

With the ability to record and play high-quality audio established, the next most important feature an automation system must possess is a way to organize its inventory. This is where the PC's database capacities are utilized. The system should be able to hold numerous fields of information for each piece of audio recorded into the system. Things like the title, length, out cue, record date, start date, and kill date should be stored along with a unique cut identifier (numeric or alpha-numeric) for every recording. The system should be able to sort the database by any of these fields, and should offer search functions that are able to find a string from any of the fields. It should be easy to change, update, or remove data as well as to generate electronic and/or hardcopy reports of the contents. More advanced systems might also include features like storing segue points, auxiliary cue points, back timer points, number of plays counters, and the date and time of each cut's last use. Most importantly, the information and tools to manipulate it must be presented clearly and logically to the user. This is a function of the user interface that will be discussed later in the chapter.

From this overall database of the complete system inventory, an automation system must be able to create (or import from an external source) sub-databases that include the elements that will be used (played) in succession. These *playlists* might contain elements for a single stop set or might contain an entire day's programming and thus represent the station's daily program schedule or log. In addition to the actual audio elements, the playlists need to include automation information that instructs the system how to sequence from one event to the next automatically. The ability to program lists to cover multiple days or the ability for one list to sequence to another list is also very important.

Since these playlists might typically hold hundreds or possibly thousands of elements, scheduling every event and programming the automation manually would be very tedious. Again, the power of the PC comes to the rescue. Many PC-based programs are available that use database functions combined with user-defined sets of rules to manage and schedule commercials/promos (traffic) and music. Each category is handled somewhat differently. Programs that schedule commercials generally take sales department information (orders) and apply rules based on airtime rates, time of day, time of year, and adjacent programming to generate a daily schedule or log. These programs are usually also tied to the billing department and are often able to accept *as-played* logs from the automation system and reconcile them with the original schedule to generate the appropriate billing invoices. Music scheduling programs, on the other hand, must choose cuts based on the station's format, time of day, music category, and artist, and may apply rules, such as prohibiting two artists from the same music category from playing back-to-back or limiting specific cuts from playing more than once per hour or once per day.

While some automation systems provide these scheduling features as part of an overall comprehensive automation package, it is not necessarily a great advantage to obtain everything from a single vendor. With all the excellent scheduling software that is already available from a multitude of vendors, often the most flexible approach is to choose each package separately based on its individual merits (possibly already familiar or in use), and then select an automation system that can interface with this system to import and merge the traffic and music schedules and provide seamless integration of scheduling and playback. The automation system then exports an as-played log representing what actually aired to the traffic system, which uses the data to expedite the billing process as described above. In short, to be a good automation system, it is not necessary to possess scheduling features, but a good system must be able to use the schedules generated by other applications and then report back to them the on-air as-played results.

Sequencing, Cross-Fades, and "Play while Record"

A good automation system will include several options for controlling when and how the system automates from event to event. Features to preset and/or execute transitions at a specific time of day are important for cueing up *stopsets* and starting programs. Transitions that tie the end of one event to the beginning of the next are useful for connecting multiple spots in a single stopset. Advanced systems often provide the ability to overlap events to provide segues between audio elements and to add voiceover announcements over the transitions. The ability to perform these overlaps with dynamic level control (e.g., cross-fades, ducking) all within a single stereo playback channel is highly desirable.

Another desirable automation system feature is the ability to record into the device while simultaneously playing audio. This allows production personnel to create new inventory without having to wait for specific breaks in the normal on-air programming. Most systems provide this function by offering multiple record and/or play audio channels. Some systems are scalable, allowing the channels to be added as needed by inserting additional audio boards.

An important extension of this function is the ability to play a file while it is still being recorded. One excellent use of this feature is to perform short interval time shifts. Once a program is being recorded, playback of that recording may start at any time, even before the program (and recording) is completed. This allows long-form programs to be delayed by as little as a few minutes or even seconds. Another powerful use is the ability to edit and play excerpts of a recording while the recording is still in progress. News feeds provide an excellent example. Sound bites can be trimmed and used on air before the originating news feed recording is complete. While most automation systems require a recording be in progress for a specified amount of time before the cut may be accessed for edit or playback, this interval is typically only a few seconds, making this feature very handy.

Peripheral Interfaces

A PC-based automation system by itself rarely contains everything necessary for complete walk-away operation. For it to be truly effective, a system needs to be able to interface with various other broadcast devices, like audio switchers, audio consoles, satellite receivers, CD players, DAT players, station clocks, and more. These interfaces are created in two parts: the hardware interface, which provides proper electronic interconnection, and the software interface, which communicates specific commands to and from the peripheral.

Systems based on PCs can take advantage of a multitude of off-the-shelf hardware plug-in input/output (I/O) boards that can provide any mix of control interfaces using RS-232, RS-422, RS-485, and large numbers of opto-isolated contact closure inputs and dry contact outputs. Systems that are able to address these generic I/O products are more flexible than those that offer their own custom I/O hardware that may offer a mix of these features.

In addition to these traditional communication interfaces, many new consoles and routing switchers feature network interface ports and are able to communicate via IP over standard Ethernet networks. This is a perfect match for the network hardware of a PC platform and can greatly simplify the physical interconnections for systems that are capable of communicating through this channel. This can allow PC-based automation systems to control input routing, fader labeling, bus assignment, and even automate mixes through a simple network connection.

With the hardware connections becoming more standardized and commonplace, the real strength of an automation system is in the software portion of this interface—how it uses the hardware to communicate with the various devices external to the PC. A wide variation of command syntax and order exists among the various peripheral devices an automation system might connect to. Command strings, string lengths, and the use of carriage return and/or line feed vary from device to device. Even simple contact closure-controlled devices can often require specific closure sequences to perform a given function. Therefore, it is very important that the automation system software support the creation and transmission of custom serial protocols, be capable of storing and running control macros that execute specific command sequences, and provide these features through a user-programmable mechanism so that each station's unique complement of equipment can be administered and maintained.

Many system manufacturers provide software modules (drivers) that perform the data translations necessary to allow the automation system to communicate with other popular devices. In many cases, that is a simple and adequate way to provide interoperability, as long as the configuration never changes. But things will change. An automation system that is able to utilize industry standard methods of communication and provides a mechanism to "tweak" communication details is more flexible and able to adapt to future developments.

Data communication abilities of automation systems are now extending beyond peripheral control and are now being used to create a better user experience through ancillary data services. An automation system can send database information about its audio (often referred to as metadata) to external data channels that can provide relevant "now playing" information to the end user. Examples of this include Radio Data System (RDS) encoders feeding conventional FM transmitters to display title, artist, and station ID text strings on user's receivers and roadside billboards. IBOC HD Radio includes an enhanced version of this feature through its Program Service Data (PSD) channel. In addition, metadata over IP is now allowing broadcasters to tightly integrate their on-air product with their station websites and streaming services. By and large, these services are employing standardized information exchange formats like eXtensible Mark-up Language (XML), which can provide a very powerful and flexible framework for future development.

Automation-Production System Integration

This communication flexibility and standardization are also providing a path toward tighter integration of automation and production systems. Many facilities utilize multitrack production systems to create content because of advanced production tools available in those environments. But completed projects then need to be moved to the automation system to benefit from their more advanced play-to-air sequencing and control features. Oftentimes these systems run on different hardware platforms and employ different operating systems to run their application software. In the past, this created a significant roadblock to data sharing and required the transfer to be accomplished in real time by dubbing from one system to another. This also required additional time and effort to manually re-enter the metadata into the destination system's database.

Today, networking technology allows us to connect even cross-platform devices, allowing them to share data at the file system level. Most systems have adopted standardized audio file formats, such as WAV and AIFF, allowing direct file transfers from one system to another. Even if the standards are different, many production and automation applications have file conversion applets that allow non-native files to be reformatted during transfer. And if not, there are a number of generic format converter applications available that can harmonize each side of a connection.

In addition to the audio, we need to be able to share the metadata. This has been somewhat problematic in the past due to the many types of databases in use and the specific fields of information stored and manipulated by differing systems. However, an increasing number of system vendors are offering application programming interfaces (APIs) with their systems that allow an outside application to extract and inject relevant information (metadata) to and from their native database structure. This allows the vendor of one system to create an interface module to another vendor's application that allows it to access, manipulate, and return updated information to the host application. When properly implemented, this can create integration so tight that the user does not realize he or she is changing environments. An example of this is clicking an Edit button inside an automation system and having the audio file with its associated metadata open inside a different vendor's multitrack editor. Upon completing the edit session, the user clicks Save in the multitrack editor and returns to the automation system's screen with the edited audio and metadata instantly available.

Even if specific integration modules are not available, many systems are starting to support generic ways of moving metadata. A simple approach to this is to place key information into the file name of the exported audio file using a special character as the delimiter. At the destination, an application like ENCO System's DropBox can be configured to parse this information from the file name and use it to populate the associated metadata fields in its system's database.

Some systems go a step further and import and export metadata in XML files that are linked to the audio file. XML is well suited to this purpose since data tags are defined within the data itself making the structure flexible enough to handle any type and size of metadata.

Metadata can also be contained within the audio file itself. Many system vendors are supporting extensions to standard file formats such as the Broadcast Wave Format (BWF) and Cart Chunk format. These are both nonproprietary standards that allow metadata to be contained within a "chunk" that is embedded into a standard WAV file such that only "aware" applications see it and non-aware applications ignore it. Systems supporting MP3 files are making use of their ID3 tags, which serve a similar metadata function but are only applicable to MP3 audio files. Other file exchange formats that combine media and metadata, such as Open Media Framework Interchange (OMFI), Media eXchange Format (MXF), and Advanced Authoring Format (AAF) are also becoming more widespread.

Graphical User Interface

Even if an automation system possesses all of the key features described thus far, it must also have a clear, well-organized, intuitive operator interface, or much of its power will go unused. Many of the early automation systems were quite powerful but displayed busy, text-based screens and required multiple-key keyboard commands to operate. Some operators were afraid of these units, making them useful to only the more computer literate station personnel. Graphical user interfaces (GUIs), like Mac O/S, IBM O/S2, Unix X-Windows, ENCO System's ENCOWare, and various versions of Microsoft's Windows, have made it possible to display screens that provide more visual clues as to function and use. The necessary text-based data remains, but it is surrounded with buttons and scroll boxes that allow the operator to navigate the screen and operate the system with a simple pointing device like a mouse or trackball (see Figure 3.6-1).

Some of the more advanced automation systems have taken this a step further by allowing the pointing device to be the operator's finger. This touch-screen technology coordinates a touch sensitive x-y matrix placed over the display screen with the graphics displayed through it by the program. The operator can press a button-like area on the screen with the illusion that a physical button is being pressed due to visual clues like color change and shadow reversal. This type of interface has engendered PC-based automation systems into common use by even staff members with little computer skills, and has enhanced the role of the automation system as an on-air live assist device. Screens that present lists of cuts to be played sequentially along with their hit times, transition information, a countdown clock, a time of day clock, and audition functions work well for tightly formatted live-assist periods (see Figure 3.6-2). Screens that present panels of programmable playback buttons provide the instantaneous playback capabilities needed for spontaneous live shows (see Figure 3.6-3). Finally, script-based programs like newscasts can benefit from systems that allow the integration of text and audio to provide sound bite playback from buttons that are embedded in and scroll with the script (see Figure 3.6-4). Examples such as these illustrate how a well-designed human-machine interface can broaden an automation system's usability and enhance its effectiveness.

NETWORKED AUTOMATION SYSTEMS

Of all of the contributions the personal computer has made to advance the state of the art of station automation, perhaps none has had as pronounced an effect as networking. A single, stand-alone PC with the features described thus far can do an excellent job providing the basic production, live-assist, and automation features required for a single station. If multiple sta-

FIGURE 3.6-1 PC user interface combining audio and database functions.

tions need to be automated, there are two choices: try to *squeeze* the additional resources from a single workstation or add additional workstations to handle the additional stations. Using a single workstation simplifies inventory management and allows the production and scheduling to be shared; however, as you try to do more and more within a single box, the unit's capabilities will be spread thinner and thinner until the overall performance is no longer satisfactory. If you add more independent units, you maintain each system's performance level but add the burden of managing multiple inventories, multiple production facilities, and multiple scheduling systems. To achieve the best from these two scenarios, the workstations need to be interconnected so they can share audio inventory, production elements, and scheduling resources while they each provide high-performance live-assist and/or automation features to their individual stations. This kind of interconnection is exactly what PC networks are designed to provide.

Networking the individual workstations together enables them to move audio, automation schedules,

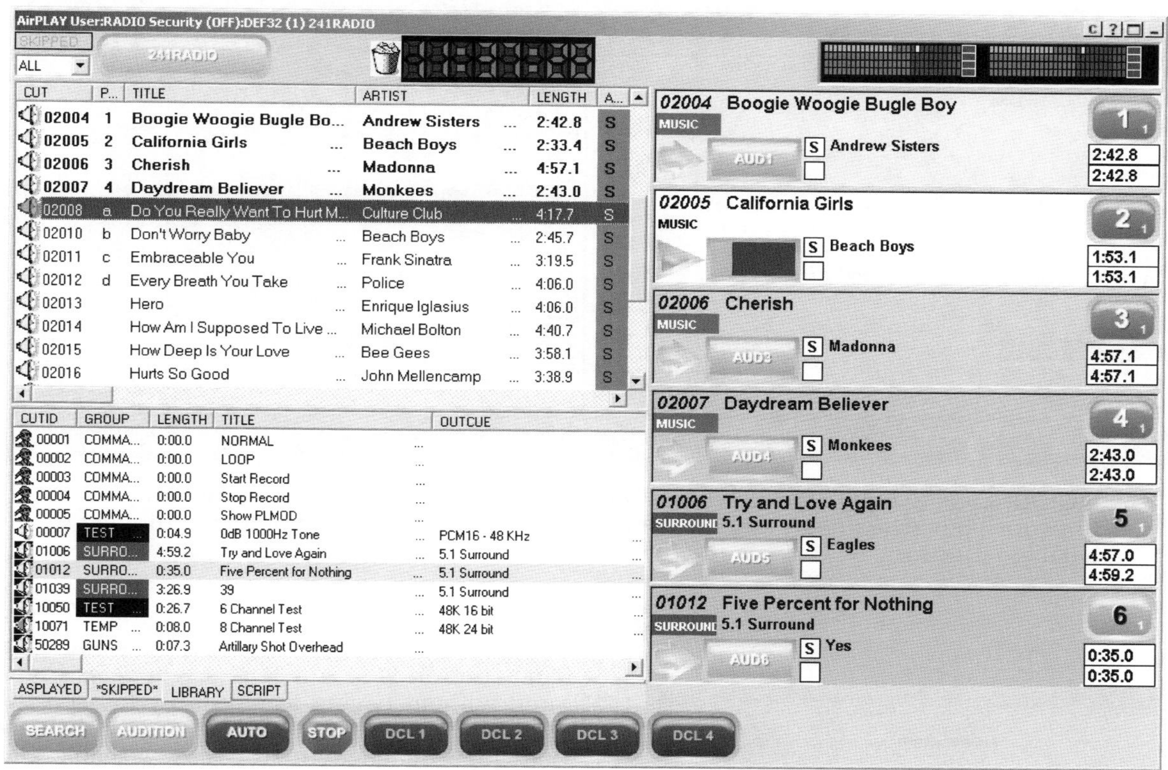

FIGURE 3.6-2 Samples of touch-screen controlled live-assist graphic user interface.

FIGURE 3.6-3 Sample of touch-screen controlled graphic user interface providing features for both scheduled events and random access playback, combining audio and database functions.

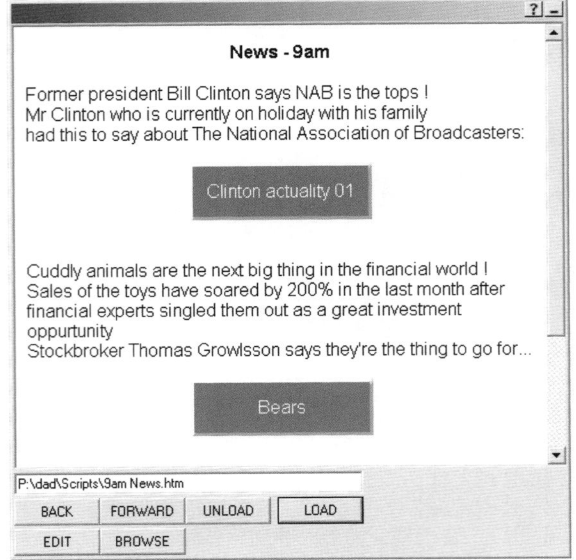

FIGURE 3.6-4 Touch-screen controlled playback buttons integrated with text.

as-play logs, scripts and wire copy text, configuration and control data, and software updates from unit to unit without having to physically transport the data via removable media (such as floppy disk, Bernoulli cartridge, magneto-optical disk, writable CD/DVD, flash memory drive, and so on). Eliminating the time and effort associated with such a sneaker net is usually by itself enough to justify the expense of the network hardware and software. But, in addition to providing a conduit to expedite the exchange of information between one unit and another, networking can actually increase the overall efficiency of a group of workstations by eliminating unnecessary replication of resources and by allowing individual workstations to specialize in a particular operation. For instance, it is usually more efficient in both equipment and manpower, to equip a single workstation with the production facilities to create spots for *all* of the stations than it is to perform production at each of the on-air workstations. This also makes the on-air workstations more efficient by allowing them to devote their resources to their primary on-air functions. The same is true of other groupwide functions, such as wire service integration, news feed collection, commercial/music scheduling, and overall inventory management. There are further benefits to be realized from specific network interconnections, and therefore some in-depth discussion of computer network architecture is in order.

COMPUTER NETWORKS

Network Architectures

A number of computer network architectures are presently being utilized for broadcast applications. Each has certain advantages and disadvantages. The three most common configurations are the *peer-to-peer* network, the *audio (video) server*-based network, and the *dedicated file server*-based network.

Peer-to-Peer

A peer-to-peer network is a simple and inexpensive way to link two or more workstations at the file system level, permitting file transfer between them. This arrangement has achieved reasonable popularity since it only requires the addition of network interface cards (NIC) to each workstation and the loading of some relatively simple network operating system (NOS) software. The basic concept of a two-workstation network is illustrated in Figure 3.6-5.

Peer-to-peer networks allow each workstation to access the contents of another workstation's drive. In early generations of PC hardware, a workstation's capabilities could easily be taxed by the overhead required to service a network client trying to access its hard drive. This would often affect the host workstation's on-air performance. For this reason, audio was usually not shared directly, but was copied at a slow rate across the network to the client's local hard drive before it could be used. This added a considerable delay between production and playback, especially if the cut to be transferred was quite large. This method of file sharing also usually dictated that each peer run from its own separate database, which required considerable file maintenance to keep the multiple separate inventories organized.

Current PC workstations with 3 GHz (or faster) processors, UDMA or SATA hard drives, and 100 Mbps LAN hardware are much more capable of "serving" while maintaining high-performance local operations. This has made effective peer-to-peer configurations very practical and commonplace. However, a workstation is not a server and these configurations should be limited to just a few workstations. This limitation is also reinforced by most operating systems, which usually limit client (peer) connections to 10 or less.

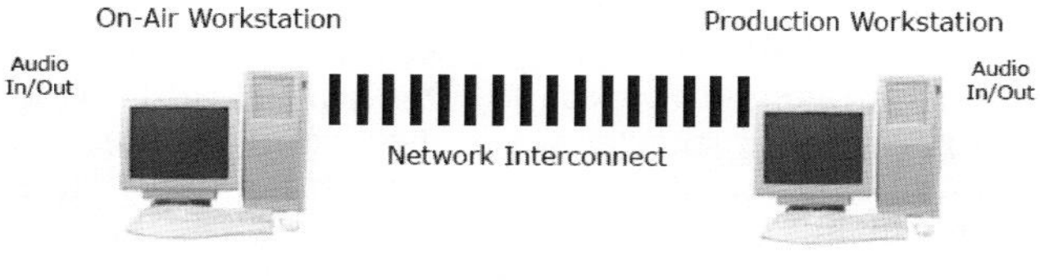

On-Air Workstation
Production Workstation
Audio In/Out
Audio In/Out
Network Interconnect

FIGURE 3.6-5 Peer-to-peer network.

Audio Server

An audio server-based network usually consists of one or more workstations communicating with a dedicated and often proprietary computer, which houses both the audio processing hardware (DSP boards) and audio file hard disk storage. The workstations act as control interfaces that command the central audio server to perform all record and play functions. A nice feature of the arrangement is that all the audio data is stored centrally making it instantly accessible by any workstation. Audio recorded by a production workstation can be immediately played by an on-air workstation without requiring any file copying to the local hard drive. A serious disadvantage to this network configuration is that all audio processing is performed by the audio server, which has a physical limit to the number of audio channels it can contain. This limits the number of workstations that can be supported by the system and makes it difficult and expensive to expand. Also, since all audio inputs and outputs occur at the audio server, audio interconnect wiring is required to and from each studio adding to the installation complexity and potential points of failure. And since the workstations contain no audio hardware themselves, they have no functionality in the event of an audio server or network failure. An audio server is illustrated in Figure 3.6-6.

File Server

A dedicated file server-based network combines the good points of these first to architectures. The network generally consists of one or more workstations linked to a central high-performance, nonproprietary computer that runs a standard high-performance NOS like Novell NetWare, Unix NFS, or Windows Server 2003. The server presents itself to the rest of the system as a large shared disk drive and manages the digital storage of all audio and system data. As with the audio server concept, the audio stored centrally is therefore instantly accessible by any of the workstations. The major advantage this configuration has over the audio

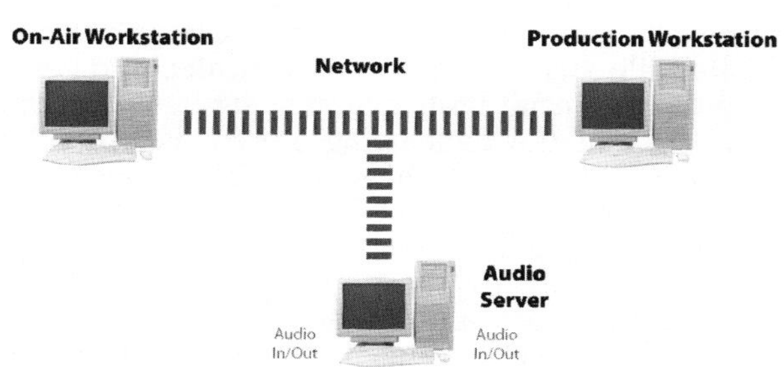

FIGURE 3.6-6 Audio server network.

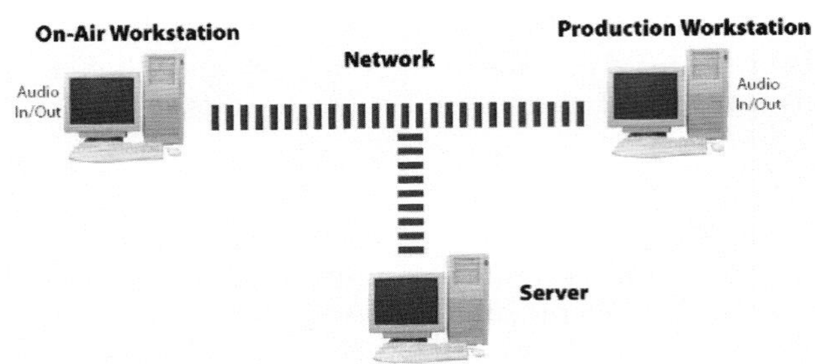

FIGURE 3.6-7 Dedicated file server network.

server configuration is that the audio processing hardware (sound card) is located in each of the individual workstations as illustrated in Figure 3.6-7. This places the audio inputs and outputs at the workstation's location with the network used to transfer packets of digitized audio data between the workstation and server. Not only does this allow all workstations to share the centrally-stored audio, but it also provides audio distribution via the data network. An audio source local to a workstation at one end of a facility can be recorded and then immediately played by a workstation at the other end of the facility. A practical example of this is a workstation installed in an equipment room with the station's satellite receivers and programmed it to automatically record multiple news feeds throughout the day. Workstations anywhere within the facility can access, edit, and play portions of these feeds without having any direct audio feed from the receivers themselves. The interesting thing is that the complete interconnect for distributing the normal system data and audio to and from a workstation is a single piece of inexpensive cable!

The fact that the file server contains no audio hardware is also very significant to server functionality, because this means it is not bound by any physical number of channels. The server is designed to service large numbers of data streams bound only by performance and capacity of its disk I/O subsystem and the speed of the network. With proper server design, any number of workstations/audio channels can be supported, making this type of network very scalable.

Another benefit of the dedicated file server-based network is the various data redundancy (backup) options available. With a number of workstations relying on the file server to store all of their work, it is very important that this storage remain online at all times and be protected in the event of catastrophic hardware failure such as a hard disk failure. NOS software for dedicated file servers is quite sophisticated and generally contains special fault-tolerant features to protect the integrity of the data being stored. Features like disk mirroring, server mirroring (clustering), and RAID (described later in this chapter) are available, along with support for a considerable range of industry-standard hardware components designed to provide performance and redundancy. Also, the dedicated file server architecture is able to provide a level of redundancy via the workstation. The primary mode of operation is for each workstation to record and play directly to and from the server; however, since each workstation contains its own audio processing hardware, local hard disk storage can be included for emergency use in the event of file server or network maintenance or failure.

Each of these network architectures provides useful connectivity between workstations and has its place in station automation. But, all things considered, the dedicated file server-based network offers the best collection of features from which to build a multistation automation system.

Now let us examine how these network architectures are physically accomplished.

Network Topology

The way cables are arranged to provide interconnection between devices in a network is known as their *physical topology* or shape. The way the signals are actually routed through the network is known as their *logical topology*. The physical topology and logical topology may not necessarily be the same, as we will see shortly.

There are two building blocks for all physical topologies: the *bus* link and the *point-to-point* link.

The bus topology uses a single cable that each workstation and server attaches to and shares. When any node (connected devices are often referred to as nodes) transmits, it broadcasts over this transmission line with every node hearing every transmission. Communications would become hopelessly confused if multiple workstations tried to transmit on this shared line simultaneously; therefore, bus networks must employ a control mechanism to make sure only one transmitter is active at a time. This media access control mechanism will generally lower the network's efficiency well below its theoretical value. The bus topology can be easily expanded by making the bus cable longer and attaching more workstations up to the electrical limitations of the cable. It is also possible to expand beyond this limit through the use of repeater electronics and multiple cable segments. Figure 3.6-8(a) represents the basic concept of the bus topology.

Point-to-point networks, on the other hand, do not broadcast on a shared cable, but as the name implies, transmit to exactly one receiver and receive from exactly one transmitter. As Figure 3.6-8(b) illustrates, the transmitter of one workstation is connected directly to the receiver of another with the transmitter of that workstation connected back to the receiver of the first. Since transmission and reception occur over separate wires, this topology has the advantage of supporting duplex operation (simultaneous transmit and receive), which doubles communication speed. However, point-to point networks are somewhat more complicated to expand. One way to add workstations is to provide a point-to-point connection between every pair of workstations in the network as shown in Figure 3.6-8(c). Known as a *mesh*, this can provide very efficient point-to-point communications

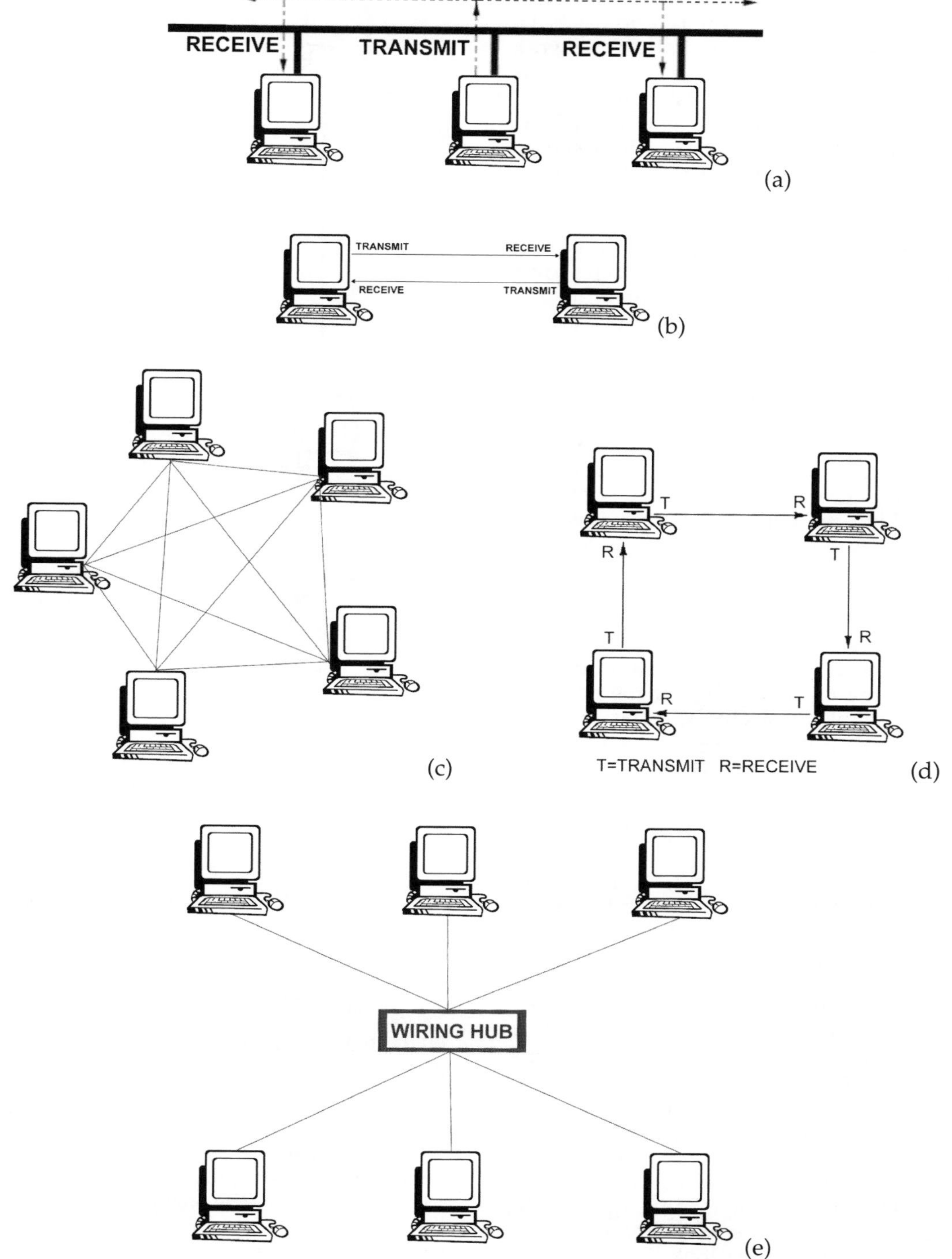

FIGURE 3.6-8 Physical network topologies: (a) bus, (b) point-to-point, (c) mesh, (d) ring, and (e) star.

but it is obvious that these connections would become overly complex for even a small group of computers.

Another way to expand a point-to-point link network is to connect the workstations in a *ring* arrangement as shown in Figure 3.6-8(d). The transmitter of one workstation is connected to the receive port of the next workstation in the ring. Data, transmitted in individually addressed units called *packets*, travels from point to point to point around the ring until it reaches its destination address. With full duplex communications, this can be an effective arrangement; however, since packets are transferred around the ring in a single direction, there can be situations where a packet must travel the entire circumference of the ring to reach the workstation immediately adjacent to it. If the network is very large, this can add a considerable delay to the exchange of data, resulting in a sluggish network.

By far, the most popular implementation of the point-to-point link network is the *star* topology shown in Figure 3.6-8(e). In this arrangement, all workstations and servers have a point-to-point connection with a centrally located device (often known as a *hub*) that provides the signal routing between the nodes. A star network offers some distinct advantages over other topologies. Point-to-point connections can provide more efficient full duplex (FDX) communications. Centralized connections provide for easy expansion and reorganization of the network. Connection ports in the hub typically include signal conditioning electronics to clean and reshape waveforms, which improve the performance, reliability, and range of the network. In addition, the conditioning electronics allow the hub to ignore a disconnected or defective port or cable making the network less sensitive to a single point failure. And most hubs also include traffic monitoring and management functions, which allow system administrators to analyze and fine tune their networks.

How the central device (internally) connects the nodes determines how the network communications take place; that is the *logical topology* of the network. While a network might physically be a star topology using point-to-point connections, the central hub might logically connect the workstations as a bus or ring as shown in Figure 3.6-9(a) and (b).

Switched Ethernet

These logical topologies have been useful in the past; however, the current state of the art in local area networking (LAN) is *switched Ethernet*. In this type of star-connected network, the central connection device, called a *switch*, provides high-speed port-to-port packet switching functions that create a mesh logical topology. As described earlier, a mesh network provides very efficient point-to-point communications; exactly what we need for multi-workstation-to-server and workstation-to-workstation transfers.

This arrangement requires a sophisticated switching device. A switch is "smart" in that it decodes the addressing of each packet it receives and routes it to the specific destination without loading up parts of the network that are not involved in the exchange. This effectively makes each port look like a separate segment, which isolates the traffic and provides full Ethernet bandwidth on each port.

The switch has electronics to do the decoding (processor), packet buffers (RAM) to store packets until forwarded, memory (more RAM) to store the address table of destinations, and a fast matrix (backplane) to move the packet to the desired port. There are also several different algorithms that can be used to decode, check, and process the packets before they are sent out.

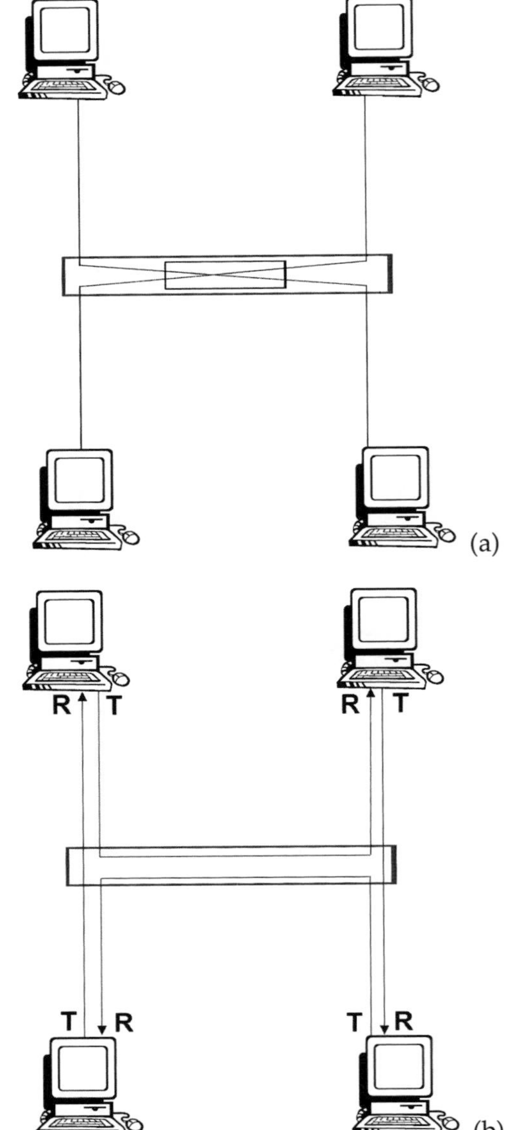

FIGURE 3.6-9 Logical topologies: (a) star as bus, and (b) star as ring.

Each feature described above is important to the overall performance of a switch. The faster the processor and the higher the backplane speed, the higher the throughput. The greater the onboard memory, the larger the packet buffer size and the larger the address table storage.

Frequently, these individual specifications are not provided by the manufacturer. Fortunately, all of them combine to determine the real benchmark of interest, which is *latency*. This is a measure of how long it takes for a packet to get through the switch and on to its destination. An automation system requires its packets arrive in a very timely manner to keep its audio and control features from being interrupted. High latency can kill a workstation's ability to keep its audio buffers full and can make the user interface too sluggish to use effectively.

Therefore, an important part of creating and maintaining a successful station automation system is to build it upon a backbone with very low latency. The good news is even "economy" switches these days generally have latency less than 100 microseconds and will work quite well in small- to moderate-size systems. High-quality switches have latency figures in the low microsecond range and should be used in large systems and/or if high data rates (like PCM audio) are required.

It is important to note is that even high-quality switches with great throughput and low latency specifications can still degrade system performance if not properly configured. A *managed switch* allows a user to control and program its features through a serial or IP connection, using a telnet session or Web-style interface. These switches often contain various additional software algorithms that can be valuable in certain situations but that should be disabled in some specific situations. Features to limit broadcast traffic, spanning tree protocol (STP) to find and block redundant links, port trunking/teaming/link aggregation to provide higher bandwidth to critical devices, and the ability to manage *multicast* traffic are all powerful and useful features but they can sometimes cause severe interactions with features of an automation system. Typically the factory default settings for a switch will provide proper performance, but be aware of the potential complications and contact your system vendor if in doubt.

Network Cabling and Hardware

So what does all this mean to the station engineer who needs to install and connect the station automation system? To get to the real-world nuts and bolts of networking, we must discuss cabling standards.

Ethernet was the first and is by far the most favored network standard in use. It has a number of different physical implementations that have existed over the years including 10Base5, 10Base2, and 10BaseT, which are 10 Mbps technologies; 100BaseT4, 100Base-TX, and 100BaseFX, which are 100 Mbps technologies; and 1000BaseT, 1000BaseLX, and 1000BaseSX, which are 1000 Mbps (Gigabit) technologies.

Ethernet uses *baseband* signaling, which is indicated by the inclusion of "base" in each of the specific designations. This means that only a single Ethernet signal occupies a transmit or receive path at a time.

10Base5 and 10Base2 were coaxial cable-based networks that have largely disappeared from the networking landscape. 10Base5 used 0.5 inch diameter RG-8 style backbone and was dubbed *Thicknet* due to the wire size. The trunk had a maximum length of 500 meters, hence "5" in the IEEE designation. 10Base2 used a 0.2 inch RG-58 style trunk and was called *Thinnet* due to the thinner cable. The trunk could extend approximately 200 meters, thus the "2." These coaxial bus networks were popular in their day, but the fact that a single point failure could kill the entire network caused most installations to move to twisted pair infrastructures.

10BaseT, as you can probably guess, stands for 10 Mbps over twisted pair wiring. This was the popular standard until 100BaseT, dubbed *Fast Ethernet*, emerged with two different cabling standards: 100BaseTX and 100BaseT4. More recently, 1000BaseT, known as *Gigabit*, has become widely available. All of these employ a star physical topology and utilize unshielded twisted pair (UTP) cabling to make point-to-point connections between each workstation or server and a central hub or switch. Twisted pair networks use affordable, easy-to-apply RJ-45 connectors. Individual cable runs can be up to 100 meters (328 ft). Each connection port is isolated, which reduces the sensitivity to single point failure and provides convenient patch points that can ease expansion and load balancing. The low cost of cabling and the ease of installing, connecting, and maintaining these networks have made the twisted pair configuration the most popular cabling arrangement for almost all modern LANs.

The quality of UTP cable in use is vital to system performance and must meet certain specifications based on the Ethernet standard in use. UTP cable includes four separate twisted pairs in a single sheath, and, although seemingly simple in structure, many factors influence its electrical quality. Things like the insulating material, the spacing of the wires in a pair, and whether the wires in a pair are bonded together all have an effect. Each pair is twisted at a different rate per length to minimize crosstalk between the pairs, but this can also lead to a skew in timing between the pairs. These and many other parameters determine the frequencies and data rates a cable can reliably support.

To help standardize cabling practices, the Telecommunications Industry Association (TIA), which represents the communications sector of the Electronic Industries Alliance (EIA), has developed a series of three standards for designing and implementing cabling systems. These are formally know as ANSI/TIA/EIA-568-B.1-2001, -B.2-2001, and -B.3-2001, but are often collectively referred to as just TIA/EIA-568-B. Within this standard, several *Categories* (abbreviated CAT) of performance are defined to qualify twisted pair cabling. Ratings commonly found in the field include CAT3, CAT5, CAT5e, and CAT6. CAT6a and CAT7 are also starting to be utilized in high-end installations.

10BaseT connections require only two of the pairs (one transmit, one receive) and require UTP cable-rated Category 3 or higher. 100BaseTX also only uses two of the pairs but requires Category 5 or higher. 100BaseTX links are limited to 100 meters (328 ft) with an overall segment length limit of 250 meters (820 ft). 100BaseT4 uses all four twisted pairs and can use Category 3, 4, or 5. 100BaseT4 was originally developed to allow a simple upgrade path to 100 megabits for existing 10BaseT installations that utilized Category 3 wiring. But this 100 Mbps derivative never really caught on and this hardware is not widely available. So for

TABLE 3.6-1
Summary of Common Ethernet Networks

Network Media	Ethernet Specification	Speed (Mbps)	Cable	Description
Coaxial cable	10Base2	10	RG-58	a.k.a. "Thinnet," max length 607 ft (185 m)
	10Base5	10	RG-8	a.k.a. "Thicknet," max length 1640 ft (500 m)
Twisted pair	10BaseT	10	CAT3	Basic twisted pair Ethernet, max length 328 ft (100 m)
	100BaseT4	100	CAT3	Uses four pair to get 100 Mb on CAT3, max length 328 ft
	100BaseTX	100	CAT5	a.k.a. "Fast Ethernet," uses two pair, max length 328 ft
	100BaseVG-AnyLAN	100	CAT3 CAT5	A cross between Fast Ethernet and Token Ring. Uses four pair; max length 328 ft on CAT3, 492 ft on CAT5
	1000BaseT	1000	CAT5e CAT6	a.k.a. "Gigabit" or "GigE," uses four pair, max length 328 ft
	10GBaseT	10,000	CAT6a CAT7	a.k.a. "XGbE" or "10GbE," should be adopted sometime in 2006. Max length 328 ft on CAT7
Fiber optics	100BaseFX	100	Multimode, single mode	Fiber-based Fast Ethernet, 1.2 mile (2 km) on multimode, 6 mile (10 km) on single mode fiber
	1000BaseLX	1,000	Multimode, single mode	Long-wave laser, 1800 ft (550 m) on multimode, 3 mile (5 km) on single mode
	1000BaseSX	1,000	Multimode	Short-wave laser, 1800 ft (550 m) on multimode fiber

CAT3-based infrastructures, rewiring is the only realistic way to upgrade.

Currently, 100BaseTX using Category 5 or 5e wiring is the most common network infrastructure, but 1000BaseT use is growing steadily. This Gigabit Ethernet uses echo cancellation and a five-level pulse amplitude modulation (PAM-5) scheme to put 250 MHz of data bandwidth on each of the four pairs and thus requires a higher grade of cable. CAT5e is the "enhanced" specification capable of handling 1000BaseT, but any new facility would be wise to use CAT6 or the highest standard available when all new wiring is to be installed. Table 3.6-1 presents a comparison of these common Ethernet configurations.

Ethernet can also utilize optical fiber. 100BaseFX was an early 100 Mbps technology that has given way to two Gigabit specifications: 1000BaseSX and 1000BaseLX. All use a star physical topology with each connection using two strands of fiber-optic cable, one for transmit and one for receive. 1000BaseSX uses short-wave (0.85 micrometer) laser light over multimode fiber to provide link distances up to 550 meters. 1000BaseLX uses long-wave laser light, which can provide the same 550 meter link when transmitted over 50 micron multimode fiber, but can also provide long distance links of up to 5000 meters when sent over 9 micron single-mode optical fiber.

It is also interesting to note at this point that Gigabit Ethernet is no longer the fastest Ethernet standard. A supplementary standard ratified in 2002, IEEE 802.3ae, defines 10 Gigabit Ethernet (XGbE or 10GbE). While both copper and fiber interfaces have been defined, most implementations so far have been based on fiber with distances up to 80 km being supported! A copper implementation using special cable and connectors called *Infiniband* has been successful at providing short (<15 meters) connections. However, the IEEE is working on 10GBaseT, which will provide 10 Gigabit Ethernet over CAT6a unshielded or CAT7 individually shielded twisted pair cable. This should be approved sometime in 2006.

STORAGE REQUIREMENTS FOR DIGITAL AUDIO

Now that we have a better idea of what a station automation system is and how individual units can be networked to enhance efficiency, we should turn our attention to the storage requirements of a networked digital audio system. The first thing we need to know is how much data is there?

Data Size

Audio is encoded for digital storage through a modulation technique known as *sampling*. This is rather like taking a snapshot of the audio waveform at regular intervals. Sampling a continuous waveform produces a stream of pulses, and is thus a form of pulse modulation. The amplitude of each pulse is then converted into a digital number that can be stored as data, which is known as *coding*. Therefore, the overall process of sampling an analog signal into a stream of digital data is known as *pulse coded modulation* (PCM) and is often referred to as *linear PCM* to emphasize that no data compression or reduction schemes have been applied. The amount of data generated by this process depends on how fast we sample, how many bits are devoted to each sample, and how many signals we sample simultaneously (mono, stereo, or multichannel).

In order to completely describe the incoming audio waveform, the *Nyquist sampling theory* states that the sampling must occur at a rate at least twice as fast as the highest frequency in the incoming audio. Due to the limitations of real-world filtering required to keep samples from being contaminated (known as *anti-aliasing filters*), the sample rate must actually be slightly higher than twice the highest input frequency expected. 32 kHz sampling has a reasonable frequency response up to 15 kHz, 44.1 kHz sampling is good to about 20 kHz, and 48 kHz sampling can produce frequencies up to 22 kHz or so. In recent years, much higher sample rates like 96 kHz and even 192 kHz have been added by the audio hardware manufacturers. These high sample rates provide more accurate high-frequency reproduction with less coloration by allowing the anti-aliasing filters to roll off smoothly, farther up the spectrum from the audio frequencies of concern.

The number of bits used to describe each sample is known as the level of quantization and determines the resolution or accuracy of the sample. The number of bits also determines the quantization (digital) noise floor that relates directly to the overall dynamic range and signal-to-noise ratio of the digitized signal. Sixteen-bit quantization provides a theoretical dynamic range of 98 dB, which is very near the limit of human perception. While this is adequate for most broadcast purposes, 24-bit quantization with its 146 dB of dynamic range is becoming more commonplace with some applications even using 32-bit quantization.

As for channels, this number is pretty simple—stereo produces twice as much data as mono. And there are now multichannel formats to be considered. Six- and eight-channel files are being used to store multilingual programs and surround-sound material. These files contain six and eight times as much data as mono or three and four times as much as stereo audio files.

Putting these numbers together, we arrive at the first equation in Table 3.6-2. This shows us the rate at which data is created as a function of the number of channels, the sample rate, and the quantization level. Applying some basic conversions and rearranging this equation, we obtain two more useful relationships for analyzing storage requirements. The first gives the number of megabytes required to store a minute of audio while the second yields the number of hours of audio that can be stored per gigabyte of storage capacity. Evaluating these equations for common 44.1 kHz sampled 16-bit audio provides us with approximately 10.1 MB/min and 1.69 hour/GB, respectively.

Let's do a sample calculation. Suppose a station has a basic inventory of 1000 songs with an average length of 3 minutes. It also maintains 100 commercials averaging 60 seconds and 100 station promotions averaging 30 seconds. It also carries five separate 1 hour–long syndicated shows that are recorded and played back weekly. All audio is stored as 16-bit linear PCM sampled at 44.1 kHz. This adds up to:

$$(1000 \times 3) + (100 \times 1) + (100 \times 0.5) + (5 \times 60) = 3450 \text{ minutes of stereo audio.}$$

At a sample rate of 44.1 kHz, the second equation in Table 3.6-2 shows we require 10.1 MB/min for a total of 34,823 MB, which is about 35 GB of total storage. By itself, this is not very large compared to current storage standards. But combine several stations together, as is most common today, and provide both production and archive space, and it quickly starts to add up.

Data Rate

Knowledge of the *data rate*, as defined in the first equation in Table 3.6-2, is important to determine: 1) how many simultaneous streams of audio the network can handle, and 2) how fast must the storage system be to keep up with the reads and writes associated with this amount of systemwide I/O. Again, using common numbers: a stereo signal sampled 44,100 times per second and quantized to 16 bits generates a data stream of 1,411,200 bits per second or about 1.41 Mbps. Dividing by 8 this represents 176,400 bytes per second or about 172 KB/s.[1]

In the not-so-distant past, when network speeds topped out at 10 Mbps and SCSI buses could only handle 5 MB/s, these data rates posed a significant challenge. A single network segment would only support about seven streams with no other traffic and a single disk drive channel would not even support 30 streams. Therefore, in order to build a useful, networked digital audio system, these data rates and storage needs had to be reduced.

Compression

This brings us to the issue of *data compression*. This term generally refers to a mechanism, usually a digital algorithm, which is used to reduce the amount of data needed to accurately represent an object like a data file or a digitized signal. The amount of reduction is usually expressed as a ratio (know as the *compression ratio*) and is a factor that can be applied to the formulas of Table 3.6-2 to obtain the effective data rate or required storage capacity. Table 3.6-3 compares Linear PCM, Dolby AC-2, and two MPEG compression formats in terms of effective compression ratios and required storage capacity for the sample rates and data rates commonly used by radio station automation systems, assuming 16-bit quantization.

The details of specific data compression algorithms are quite complex and beyond the scope of this

[1]KB/s refers to 1024 bytes of data per second.

TABLE 3.6-2
Digital Audio Data Rates and Storage Requirements

$$\text{Data rate (bytes/sec)} = \text{channels} \times \frac{\text{samples}}{\text{second}} \times \frac{\text{bits}}{\text{second}} \times \frac{\text{1 byte}}{\text{8 bits}}$$

For a stereo, 44.1 kHz sampled, 16-bit audio stream: **Data rate = 176,400 bytes/second**

Adding some conversions we get a more useful number for determining capacity requirements:

$$\text{Data rate (MB/min)} = \text{channels} \times \frac{\text{samples}}{\text{second}} \times \frac{\text{bits}}{\text{sample}} \times \frac{\text{1 byte}}{\text{8 bits}} \times \frac{\text{1 megabyte}}{1024^2\ \text{bytes}} \times \frac{\text{60 sec}}{\text{1 min}}$$

For a stereo, 44.1 kHz sampled, 16-bit audio stream this becomes: **Data rate = 10.1 MB/min**

Another useful way to arrange this equation yields the amount of audio that can be stored per unit of storage:

$$\text{Audio storage (hrs/GB)} = \frac{1}{\text{channels}} \times \frac{\text{seconds}}{\text{sample}} \times \frac{\text{samples}}{\text{bits}} \times \frac{\text{bits}}{\text{bytes}} \times \frac{1024^3\ \text{bytes}}{\text{1 gigabyte}} \times \frac{\text{1 hour}}{\text{3600 sec}}$$

For a stereo, 44.1 kHz sampled, 16-bit audio stream this becomes: **Audio storage = 1.69 hours/GB**

discussion; however, a few observations are in order (see Chapter 3.7 for more information). In the data world, compression algorithms (e.g., WinZip) work by eliminating *redundant* information. This information is "extra" and can be removed without affecting the inverse algorithm's ability to completely restore each and every bit of the original source. This is called *lossless compression*. Unfortunately, sampled audio has very little redundant information and does not compress with these utilities. For audio, a different class of algorithms has been developed that removes the *irrelevant* information—meaning information not important to how a human perceives sound. Since these algorithms throw away data that cannot be brought back by an inverse operation, they are referred to as *lossy compression*.

This does not mean these processes are flawed. Through a single encode/decode cycle, lossy algorithms, which include Dolby AC2 and the whole MPEG family, are quite transparent and imperceptible to all but the highly trained ear. At a time when network bandwidth and hard disk capacity were very costly, these compression schemes made great economic sense and most systems employed them.

But some information *is* sacrificed in the process, and unfortunately the automation system is not the only place compression exists. Satellite downlinks, studio-transmitter links (STL), and even satellite (XM and Sirius) and terrestrial (HD Radio) delivery mechanisms all employ some form of data compression. There has always been concern that audio quality might not hold up through multiple encode/decode cycles or through a cascade of different lossy compression schemes. As we move forward in the digital age, this is becoming the primary concern, and many automation system users are migrating back to full bandwidth PCM to keep the source material as pure as possible.

The good news is that in the last decade, network bandwidth and hard drive capacity have both increased by two orders of magnitude (100×) while the basic unit cost of each has dropped roughly in half. In the same timeframe, disk drive transfer rates have doubled six times. Compression is no longer a necessary evil.

Bandwidth Considerations

Let us look at some numbers in use in 2006. A dedicated 100TX network link in a full duplex switched environment can carry 100 Mbps in both directions simultaneously. This represents 100 Mbps/1.41 Mbps = 70 simultaneous stereo streams both in and out. A Gigabit link should be able to handle 10 times this number or about 700 streams in both directions. Current-day workstations often include four or more physical outputs and each may layer up to three streams at a time during overlap sequences. Even 10 channels, all playing 3 streams simultaneously, only represent half of the available bandwidth of a 100TX link, leaving the rest for database transactions and other miscellaneous I/O. Clearly, a 100TX link is adequate for connecting even a full-featured audio workstation to the network.

But a single 100TX connection to a central storage device can become a potential bottleneck, particularly if all workstations are reading and writing to and from this same device. Just three full-featured workstations as described above could overload a single 100TX connection to a server. Fortunately, most modern managed network switches and server NICs support link aggregation, which allows multiple physical links to be combined to effectively multiply the bandwidth (and additionally provide fault tolerance) between the two devices. Obviously, Gigabit connections are very beneficial for server connections, and when combined with link aggregation, dozens, even hundreds, of workstations can be supported.

TABLE 3.6-3
Disk Storage-Capacity Requirements (16-Bit Quantization)

	Sampling Rate, kHz	Data Rate, kbps/ch	Compression Ratio	MB per Stereo Minute, Divide by 2 for Mono	Stereo Hours per GB, Multiply by 2 for Mono
Linear PCM	32.0	512	N/A	7.32	2.33
	44.1	705.6	N/A	10.09	1.69
	48.0	768	N/A	10.99	1.55
Dolby AC-2	32.0	N/A	6.0	1.22	13.98
	44.1	N/A	6.0	1.68	10.15
	48.0	N/A	6.0	1.83	9.32
ISO/MPEG Layer 1	32.0	32	16.0	0.46	37.28
	32.0	64	8.0	0.92	18.64
	32.0	96	5.3	1.38	12.35
	44.1	32	22.1	0.46	37.37
	44.1	64	11.0	0.92	18.60
	44.1	96	7.4	1.36	12.51
	44.1	128	5.5	1.84	9.30
	48.0	32	24.0	0.46	37.28
	48.0	64	12.0	0.92	18.64
	48.0	96	8.0	1.37	12.43
	48.0	128	6.0	1.83	9.32
	48.0	160	4.8	2.29	7.46
ISO/MPEG Layer 2	32.0	32	16.0	0.46	37.28
	32.0	48	10.7	0.68	24.93
	32.0	56	9.1	0.80	21.20
	32.0	64	8.0	0.92	18.64
	32.0	80	6.4	1.14	14.91
	32.0	96	5.3	1.38	12.35
	44.1	32	22.1	0.46	37.37
	44.1	48	14.7	0.69	24.86
	44.1	56	12.6	0.80	21.30
	44.1	64	11.0	0.92	18.60
	44.1	80	8.8	1.15	14.88
	44.1	96	7.4	1.36	12.51
	44.1	112	6.3	1.60	10.65
	44.1	128	5.5	1.84	9.30
	48.0	32	24.0	0.46	37.28
	48.0	48	16	0.69	24.86
	48.0	56	13.7	0.80	21.28
	48.0	64	12.0	0.92	18.64
	48.0	80	9.6	1.14	14.91
	48.0	96	8.0	1.37	12.43
	48.0	112	6.9	1.59	10.72
	48.0	128	6.0	1.83	9.32
	48.0	160	4.8	2.29	7.46

Disk Drive Systems

So it appears we have no problem moving full bandwidth PCM audio to and from our network. Now, how and where do we store this sizeable volume of data? As described earlier under the Network Architectures section, we can choose to provide storage in each workstation, in a centralized server, or in both. But no matter where we choose to store the data, the current state of the art dictates that it will end up on hard disk drives. This technology has made huge advances in capacity, speed, and reliability that have helped propel digital computing systems into the mainstream of nearly all broadcast operations. But due to their core mechanical properties, hard disk drives remain the weakest link of all automation systems and therefore deserve some special understanding.

The greatest point of failure in any computer-based system is the hard disk drive. The issue is not *if* a disk drive will fail, but *when* it will fail. Current premium quality, high-performance drives are quoting figures of 1 million hours mean time between failure (MTBF), which is over 100 years! However, a large percentage of the high-capacity drives used for multimedia systems fail far short of this mark—many failing within the first 6 months of operation. There appear to be several reasons for this. A few stem from manufacturing difficulties while most are related to environmental conditions such as heat, shock, and vibration.

Although the drive manufacturers are continuously improving the reliability of disk drives by reducing the internal parts count and using higher levels of integration in the onboard electronics, the ultimate storage mechanism is mechanical, and therefore subject to mechanical failure.

In the very early releases of "high-capacity" drives, primarily the 4 and 9 GB sizes, several manufacturers had problems with the bearings used to support the spindle and its associated media platters. The bearings would wear prematurely causing the spindle and platters to shift and wobble, thereby degrading track alignment and causing loss of data. It was not uncommon to see drives wear so badly that they would rumble loudly then come to a horrible grinding halt.

While this sort of problem has largely disappeared through improved design and manufacturing processes, mechanical problems associated with handling remain.

Disk drives are extremely sensitive to shock and vibration. A short fall to a workbench or floor is often enough to destroy a drive. The manufacturers have addressed these kinds of problems through better head parking designs and improved packaging and handling procedures, but what shipping companies and stock room personnel do with the drive before it gets to you is another story. The best advice is to only deal with reputable system vendors and distributors.

Once the drive has made it to the final destination and is up and running, the biggest environmental factor is *heat*. Disk drive cabinets must have adequate cooling fans and the fan filters must be kept clean. Air flow through the enclosure should be free flowing and unimpeded by doors or covers. Some cabinets offer temperature-sensitive fans that will increase flow as the temperature rises. Others provide temperature-monitoring devices and alarms to alert maintenance personnel of a problem. Even though most drives are rated to operate in temperatures up to 50°C (122°F), the cooler they are, the longer they will run.

Unfortunately, as CPU clock speeds go up and as drives spin faster, they consume more power and generate more heat. And as more and more equipment is packed into technical rooms, it usually becomes necessary to employ additional air-conditioning equipment to remove the heat from the room and keep the computers and their disk drives cool. So it is necessary to determine how much cooling is needed.

A good estimate of the heat load produced by a system can be determined by estimating the total amount of AC electrical power it will consume and then assume that all of this power is turned into heat. Obviously the units will draw less than the maximum (probably closer to half the maximum), but designing for the maximum leaves headroom to cover all situations.

To estimate power consumption, count the total number of PC power supplies that will be operating within the space being considered. Multiply the number of supplies by the power rating of the supply (e.g., 350 watts). Then add the total number of picture monitors involved multiplied by an average of, say, 200 watts per monitor. Add the rating of any other major power consuming devices to get the total power in watts. Then use the conversion:

$$1.0 \text{ kWH} = 3400 \text{ BTU},$$

which yields the following relationship:

$$\text{Heat load (BTU/hour)} = \text{total power (watts)} \times 3.4 \text{ (BTU/WH)}.$$

An HVAC professional can take this heat load and determine how many tons of air conditioning are required and whether the current HVAC system has the needed capacity or if additional air-conditioning equipment will be needed.

Disks Fail—Plan for It

In an expected lifetime of 100 years, failures within 6 months to a year could be considered infant mortality—certainly within the hardware warranty. But a radio or TV station can become very dependent on such a system in that timeframe and will trust it with a considerable amount of irreplaceable data. It is, therefore, imperative that the data storage system be able to handle disk failures while continuing to function. This is achieved through *disk redundancy.*

The basic concept of disk redundancy is to store extra (redundant) data along with the target data, and arrange it among multiple hard drives such that this extra data can be used to replace the original data should one of the disk drives fail. This can be achieved in several ways. The most basic method is called *disk mirroring*. In this configuration, each disk drive is paired with an identical disk drive. The data on the first drive is automatically replicated onto the second drive, either by the operating system or via a special disk controller (as in RAID technology to be explained later). If either disk should fail, the operation will be unaffected because the system will just use the other disk for all reads and writes. It is also possible to mirror more than two disks for additional redundancy, however, this is usually not very cost effective.

Disk mirroring can also enhance system performance. Since two (or more) copies of the data are available on independent drives, the controller can pull from both, using whichever drive is better positioned to service a particular request. This effectively increases the number of heads available to gather data.

To increase disk performance even further, a disk configuration called *spanning* or *striping* is often employed. The idea behind this scheme is to subdivide the data between two or more drives. The operating system or a specialized disk controller takes the data stream and divides it into smaller units then writes one unit to each drive. Since each drive is able to write (or read) simultaneously with the other drives, the disk I/O throughput is increased by a factor equal to the number of drives. This also means the total capacity is equal to the sum of all the drives.

However, this configuration offers no redundancy whatsoever and actually puts a facility at greater risk of overall failure. If any of the drives in the span fail, all of the data on *all* of the drives are lost. Since more drives increase the total parts count, the probability of failure due to one of the components increases. For this reason it is never recommended to use spanning or striping by itself.

A good solution is to combine mirroring with spanning/striping. In this configuration, a set of drives is spanned, each of which has its own associated mirror drive. This provides the increased performance and the additive capacity of spanning/striping as well as the redundancy and performance advantages of mirroring. In fact, this arrangement performs so well that it is possible for it to experience multiple drive failures without significantly affecting system performance. This makes it very important to diligently monitor the status of the drive subsystem and correct any failures as soon as possible. This configuration offers great scalability with both performance and redundancy.

The obvious downside of disk mirroring is that for a given capacity, the disk drive requirements are doubled. For systems requiring large capacities, this can be very costly, not to mention the added complexity, space, power, and heat concerns. These issues, along with advances in disk controller technology, have caused the simple but bulky mirrored disk arrangements to be replaced by a more complex, but more efficient redundant disk configuration know as *RAID*.

RAID

Although RAID is often perceived to be a relatively new innovation, this technology was first described in 1987 by Berkeley authors David Paterson, Garth Gibson, and Randy Katz. RAID is an acronym that stands for *redundant array of inexpensive disks*, but when you price a unit worthy of multimedia storage requirements, you might wonder where they got the "inexpensive" part. Because of this misnomer, manufacturers have adopted a slightly modified definition of the acronym: redundant array of *independent* disks. Either way, the important point is that the technology has matured, become very dependable and offers great value.

These disk subsystems are usually comprised of a specialized multichannel disk controller combined with three or more disk drives. The reason for their popularity is that they are capable of providing high capacity, excellent performance, and data redundancy with fewer disks than any other configurations.

RAID units can function in several modes referred to as RAID levels. The various levels offer different combinations of capacity, performance, and redundancy. RAID level 0 is also known as disk striping and provides scalable capacity and high performance but offers no redundancy, as described earlier. RAID level 1 is also known as disk mirroring and provides redundancy but doubles the number of disks required for a given capacity. RAID controllers can also provide a combination of RAID 0 + RAID 1 (often referred to as RAID 10), which provides scalable capacity, excellent performance, and redundancy, but still gives away one-half of the capacity to accomplish the redundancy. RAID level 2 uses multiple check disks to allow error correction without having to completely duplicate all data. This can provide fully redundant storage with fewer than double the number of disks, but since higher RAID levels are even more efficient, level 2 is rarely used. RAID level 3 provides redundancy by dividing and striping each data block across a set of synchronized drives, with stripe level parity calculated and written to a single dedicated parity drive. Level 3 can provide excellent performance, but storing parity at the byte level requires a large number of synchronized drives, which add considerable expense to the system.

The real efficiency comes with RAID levels 4 and 5. In both of these modes, the data stream is subdivided into smaller blocks (which can be many bytes) with each block being written to a separate disk, just as in disk striping (level 0). But in addition, the blocks written to each drive are compared by an algorithm that generates parity data that is written to another drive. If any drive in the stripe fails, this parity information is combined with the other good drives to recreate the data from the missing drive. When the failed drive is replaced, a similar process is used to rebuild the new disk to the original data structure.

The main difference between the two levels is that level 4 always stores the parity information for every stripe on a single drive, while level 5 stores parity information for each stripe on a different drive. RAID level 4 can only access one stripe at a time limiting its effective I/O for random transfers; however, its parity calculations are somewhat simpler, making it effective for long sustained sequential operations as in single-

TABLE 3.6-4
RAID Features

RAID Level	Name	Description	Minimum Disks	Performance	Fault Tolerance
0	Striping	Data stream is divided into blocks that are distributed equally between all disks increasing the overall I/O transfer speed above that of the individual disks. No data redundancy; loss of a single component can lead to unrecoverable data loss.	2	Very high read and write I/O with virtually no overhead.	None
1	Disk Mirroring	Each disk is paired with a duplicate disk resulting in twice the cost per byte. All data is 100% duplicated on the mirrored disk.	2	Writes equal to single disk, reads two times better.	Excellent Highest of all levels.
0+1	Striping and Mirroring (Mirrored stripes)	Combination of RAID levels 0 and 1. Data is striped across several physical disks. Each stripe has a mirror stripe generated and written across the same set of disks. This level provides performance through striping and redundancy through mirroring.	3	High read and write data rates.	Good Same as RAID 5.
2	Disk array	Hamming code error checks across the disks. Data can be recovered without complete duplication of all data, although several check disks are needed. All disks in a group must be accessed, even for transfers, and the slowest must finish before the transfer is complete.	3	High data rates possible with more check disks. Transaction rate equal to single disk.	Very good, Between RAID 5 and RAID 1.
3	Disk array	Data block is divided and striped across data disks. Redundancy is accomplished through stripe parity data that is generated and written to a single parity disk. Disk spindles are typically synchronized. All disks must be accessed and the slowest must complete the process before a transfer is complete.	3	Very high read and write data rates but transaction rate equal to single disk.	Good Handles single drive failure.
4	Disk array (Striping with fixed parity)	Full data block written to each data disk with block level parity stored on a single parity disk. This results in faster individual disk reads for small transfers but the parity check disk becomes a throughput bottleneck on writes.	3	Very high read transaction and transfer rates. Very low write transaction and transfer rates.	Good Handles single drive failure.
5	Disk array (Striping with floating parity)	Data blocks and parity data generated as in RAID 4, but parity information is spiraled across all data disks. This eliminates the parity disk throughput bottleneck problem of level 4. Distributed parity increases write performance but introduces high overhead to track the location of parity addresses.	3	High read transaction rate, medium write transaction rate. Good I/O data transfer rates.	Good Handles single drive failure.
10	Striping mirrored disks	Combination of RAID levels 1 and 0. Each physical disk is paired with a mirror disk. Data is striped across the mirrored pairs. This level provides redundancy through mirroring and increased performance through striping.	4	Very high read and write data rates with highest redundancy.	Excellent Same as RAID 1.
50	Striping multiple disk arrays	Combination of RAID levels 5 and 0. Multiple RAID 5 arrays are striped together to combine their capacities. Striping can also improve overall performance above that of the individual RAID 5 arrays.	6	High read/write transfer rates and high I/O data rates for small requests.	Very good Greater than RAID 5 unless two failures in one array
JBOD	Just a bunch of disks	Each disk is operated independently like a normal disk controller, or multiple disks can be spanned and seen as a single large disk. No data redundancy.	1	Just equal to individual disk performance.	None

user multimedia applications. RAID level 5 is able to process multiple stripes simultaneously, making it better suited for random scattered disk requests as in multiuser centralized storage environments. The parity structure of RAID level 5 is a bit more complicated than other levels, but given the proper horsepower and caching to offset this overhead, it is the best choice for a wide variety of file I/O operations.

The complexity of RAID 5 operations can often impose a limitation on the maximum size of a single RAID 5 array. To overcome this limitation, most controllers will allow multiple RAID 5 arrays to be combined under a RAID 0 stripe to create an array with the aggregate capacity. This is commonly referred to as RAID 50.

Table 3.6-4 summarizes the features of the various RAID levels.

STORAGE ARCHITECTURES

Disk arrays can be built with the capacity and performance necessary to handle just about any size automation system, but the array cannot be accessed directly from the network. Important functions like rights management and file-locking need to be provided by a layer in between the actual storage and the network users. These functions are traditionally performed by a server.

File Server

Currently, the most common storage architecture for automation systems is to use a centralized file server. This is basically just another networked PC with a lot of storage capacity. However, server hardware is usually more powerful with multiple high-speed CPUs, multiple gigabytes of RAM, multiple Gigabit network interface connections, and a large RAID disk array. The operating system is also usually a specialized "server" version that optimizes and prioritizes file I/O functions over user-interface features and is designed to handle many simultaneous client connections to its resources. Network operating systems (NOS) like Windows Server 2003 and Novell Netware are the most common, but various forms of UNIX and Linux have their share of applications as well.

The most essential component of any NOS is to provide access control and rights management. The important data you will be entrusting to your storage system needs to be protected from nonauthorized users. Clients must log in and authenticate to the server NOS to gain access to the files they need. Even authorized clients must be controlled as to which files they can read, which they can execute, which they can modify, and which they can write or delete. And it even gets trickier when one client needs to read from one part of a file while another clients needs to write to the same file. Sophisticated file locking at the byte level coupled with caching and transaction tracking mechanisms are required for this level of interaction to be successful. These are the functions accomplished by a good NOS. Without it there would be data anarchy and chaos.

Servers can also provide more that just file services. Servers can be used to centralize and manage printing operations for an entire network. It usually makes much better economic sense to provide networkwide access to a few, high-end printers as opposed to providing many locally connected basic printers. Servers can also serve applications to the clients. This means the application software resides on the server and a client pulls the components necessary to run the application across the network and loads them into memory for each session. This reduces the local storage requirements and makes networkwide software updates much easier to manage since the applications reside in a single centralized place.

However, all of these features tend to make servers somewhat complex to set up and maintain. Highly trained administrators are usually required to make changes and keep system configurations up to date. This may not be an issue if your facility has its own information technology (IT) department to provide this needed expertise. For the moderate- to large-size facilities where computing systems are involved in almost every general business application, in-house IT makes good business sense. But for small installations, this is usually not the case and server administration can be intimidating and costly. Unfortunately, this often leads to naivety, complacency, and the server being ignored. This can be very dangerous for the reliability of the broadcast system.

Many automation systems don't need anything more than basic file services. If the I/O needs are moderate and the automation system runs in an isolated environment where complicated access control is not necessary, there is a somewhat simpler solution.

NAS

NAS stands for network attached storage. What it basically comprises is a "black-box" connected (attached) to the network, which acts like a big disk drive that anyone can use for storage. The structure of a NAS is similar to that of a server but on a simpler, more appliance-like level. Hardware-wise it is a specialized PC with good network connectivity and a large disk array. But the NAS concept is all about simplicity: connect it to the network, plug it in, and turn it on. NAS is designed to work with very little interaction and generally doesn't even have a keyboard or monitor attached. Configuration is typically done through a simple web page accessed via a standard web browser on any PC on the network. Since little user interface is desired, NAS devices generally utilize compact efficient operating systems like Linux that can maximize file system performance by running through a simple command line interface that lends itself to telnet or web server control.

A NAS device can generally be used in conjunction with an existing server. Setup is simple with a configuration option that allows it to follow the access control list and rights that are set up on the existing server or domain. This is the beauty of NAS: a simple way to add more drive space and locate it where the network needs it.

But a NAS can also be used as the primary (and only) storage device. In this role, the main difference between a file server and a NAS is in the granularity of control. NAS devices are simple to configure because they have fewer options, but this also means there are fewer aspects you can control. Access control is simpler, but that can make the file system less secure. Rights management is less tiered but fewer choices make it more of an "all or nothing" sort of assignment.

Simpler configuration can impact performance as well. Many NAS units only allow RAID 1 (mirrored) configurations. Even those that provide RAID 5 do not provide much flexibility in block sizes, partitioning, or hot spare configurations. Just pick the RAID level and the file system, and say go. This certainly makes things simple and can be completely acceptable with the high-performance drives available today. But this is another area in which NAS generally provides a compromise.

While most servers use SCSI disk drives, most NAS units employ arrays of less-expensive IDE disk drives. This makes the units more affordable but can impact the level of performance. Many systems now use serial ATA (SATA) drives that provide many performance and economic benefits over traditional parallel (PATA) IDE devices and these have provided respectable results. However, even SATA drives currently have lower rotational velocities and lower bus transfer rates than the SCSI drives typically used in servers. This means that more drives are needed in the array to achieve the same level of performance. But, more drives also means more potential for drive failure. Also, SATA drives offer lower MTBF figures and are generally regarded as being less robust than SCSI. Furthermore, the currently available multichannel SATA RAID controllers are not as mature or as full-featured as the SCSI offerings. Fortunately, many NAS systems retain the drive failure-alerting mechanisms that we have come to expect with servers.

Overall the future of NAS looks bright. There are currently high-end SATA-based NAS devices that easily handle 50 linear PCM streams in and out. This indicates that NAS and SATA are up to the challenge. How these units will hold up to the rigors of 24 hours a day, 7 days a week, 365 days a year of digital audio abuse remains to be seen, but we all know that SCSI drives fail too. The bottom line is the performance/price value that SATA-based NAS is able to deliver is impressive and hard to beat. For many applications, it can adequately handle the job right now and it will only get better. For these reasons it is likely the application of SATA-based NAS will grow in the future, but there will also be more application of SATA-based storage with file servers as well.

SAN

Another acronym that should be explained at this point is SAN. It often gets confused with NAS since it is the same three letters, but SAN stands for *storage area network* and is something quite different.

A storage area network is a specialized high-speed network that lives between a server and its storage disks. Clients cannot access this network directly; they access the server as usual via a standard Ethernet network, and then the server uses the SAN to communicate with its disk drives. High-speed, high-bandwidth connections like Fibre Channel, SSA, and SAS are utilized to provide similar data rates as if the disks were directly attached, but being a network, SAN allows multiple servers to connect to and share a pool of storage disks. This improves system administration by allowing a facility to manage all of its disk drives as a single resource instead of having to manage multiple isolated arrays distributed among multiple independent servers.

SANs are very scalable and provide efficient, redundant storage. SAN management software allows disk drives to be added or removed at any time and merged in and out of redundant RAID configurations. The aggregate storage of the combined disks can be partitioned and assigned to individual servers as needed. With all disk drives managed as a single entity, disk maintenance and backup operations are much easier to schedule and control. And routine backups take far less time to complete because backup devices can connect directly to the SAN and move data at SAN speed without interfering with client/server operations.

SAN does not necessarily offer any performance advantage to a server over directly attached disk drives (often referred to by yet another acronym, DAS, for direct attached storage). And the hardware and management software needed to build and maintain a SAN can be very expensive. The main advantage of a SAN is the ability to consolidate all of the storage needs for every in-house data system (accounting, automation, email, graphics, news, sales, traffic, etc.), and supply the overall required storage from a single array of standardized drives managed by easy-to-reconfigure software. SAN does not make much sense for small facilities due to its cost, but in large facilities with many systems to share the price tag, SAN can be an excellent and economic solution.

BACKUP FOR A DIGITAL STORAGE SYSTEM

Although all of these redundant systems provide excellent fault tolerance, they usually require user intervention after a failure to return the system to full capability. It is important that the station personnel responsible for these systems be aware of potential failures and be prepared to take corrective action as soon as possible to guarantee data safety.

But what if the unthinkable happens and all your data is destroyed? This is a remote but real possibility.

For instance, in a RAID 5–based storage system, all it takes is for two disk drives to fail at the same time to lose 100% of the data. These catastrophic events can happen and must be planned for.

Tape

There are several ways to protect data from catastrophic failure. The most basic approach is to use a tape backup device and to keep complete data backups in a secure location. Tapes are available in a number of different formats: QIC, Travan, DAT, 8 mm, Mammoth, AIT, DLT, Super DLT, ADR, LTO, and VXA. The "right" format for a particular installation depends on many factors including capacity per tape, transfer rate, and whether a single tape drive or an autoloader is required to do a complete backup. All of these factors affect the overall price and must be considered carefully. It is beyond the scope of this section to lay out a complete and comprehensive backup plan, but a brief discussion of some key components and backup architectures is in order.

Tape systems use backup software to stream the hard disk data onto the tape, which can then be removed and stored safely in another location until needed. The data stored on tape is a snapshot of the files at the time of the backup, so it is only as current as the last backup. While many backup software packages allow you to schedule periodic backup sessions, these will always lag behind the rapid changes made daily in a broadcast automation system.

Once the software has been configured, this process can usually be unattended unless it takes multiple tapes to contain the full backup. In this case, an operator must switch tapes when instructed by the backup application, or an expensive tape changer or autoloader must be employed to perform these changes automatically. Currently, the second generation LTO cartridges have the greatest capacity at 200 GB each, but with many systems moving into the terabyte range, multiple tapes are a practical necessity.

But the biggest problem with tape devices is that the data transfer rate is very slow compared to a disk drive, so backup sessions can take long periods of time. Again, the second generation LTO tape drive is one of the fastest with a 30 MB/s transfer rate. This is incredible for a tape-based system. But even at this rate, a single 200 GB cartridge will take almost 2 hours to fill.

The same is true of restoration. If the primary data is destroyed, the system is generally useless until all the data can be restored from tape. At 2 hours per tape, this can lead to considerable downtime. Even if only a partial restoration is needed, the data on the tape must be accessed linearly and sequentially, so restoring even a single file can be very time consuming.

With these facts in mind, it should be obvious that tape-based backups cannot be considered an "online" resource. It takes good planning, diligent execution, and lots of time to collect and safely store tape backups. However, they are worth their weight in gold if they are ever needed.

Disk-to-Disk Backups

A strategy that has recently become popular is to use disk drives as the backup media. With large 300–400 GB IDE (SATA) drives now available, it is becoming very affordable to keep data backups on disk drives. Portable disk enclosures that connect via "plug-and-play" USB 2.0 or IEEE 1394 FireWire interfaces make it a simple task to plug in and backup entire workstations. Server storage can be divided and written to several disks with the backup drives easily disconnected and stored in a secure location. Transfer rates are at very high disk-to-disk speeds, which greatly reduces backup, and restore times.

Another useful form of this is to include large disk drives in each of the workstations and use them to backup the relevant portions of the server content. This could be done either manually or by using an application designed to identify and copy the correct elements by following a user-defined set of rules. An example of such an application is shown in Figure 3.6-10. In a system with a server holding data for several separate stations, each on-air workstation could backup the content necessary for its own operation. This way, each station would remain operational should the server or network go down. By combining the data from all the workstations there would be a complete copy of all the data should the server ever need to be restored.

A secondary or backup server is another popular disk-to-disk architecture. In this arrangement, another server with capacity greater than or equal to the

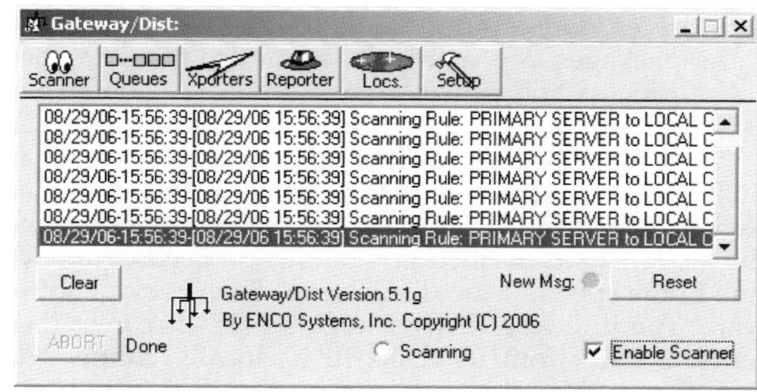

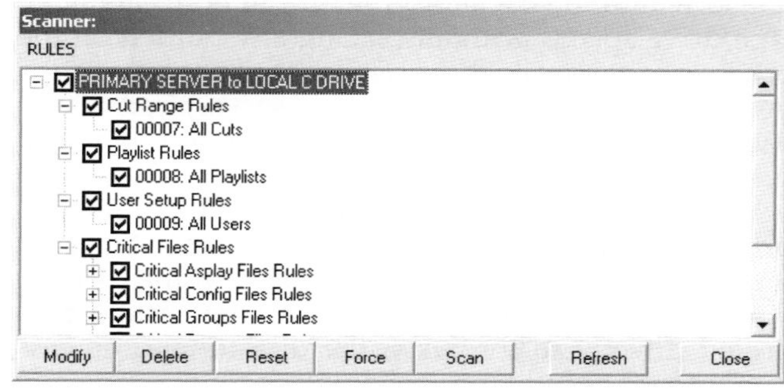

FIGURE 3.6-10 Example of a rule-based file synchronizer application used for disk-to-disk backups.

primary server holds the backup copy of the data. Some systems utilize an intelligent application running on the backup server that is constantly scanning for changes and actively synchronizes the file systems of the two servers. Other systems make use of server clustering algorithms that are implemented at the operating system level. Generally the client workstations can access data from either server making this a truly "online" solution.

There is also an extremely interesting version of this mechanism available in some SAN-based systems. Using disk imaging algorithms, a "snapshot" of the contents of a disk array can be written to another array with the same SAN. This can be completed in a fraction of the time it takes to actually copy the data. Hundreds of gigabytes can be captured in minutes. This is the epitome of disk-to-disk backup schemes, but is very expensive.

Disk-to-Disk-to-Tape Backups

One more extension of this architecture that deserves mention is disk-to-disk–to-tape backup. One of the most time-consuming attributes of tape is that it can only read or write a single data stream at a time. If several sources need to be copied, they must be done sequentially adding considerable time to the procedure. By adding a disk-to-disk layer ahead of the tape process, the best attributes of disks—speed and random access—can be combined with the best attributes of tape—portability, long shelf life, and zero power off-site storage. With the cost of large disk drives dropping (see NAS above), this arrangement is often more cost effective than upgrading to newer, faster tape drives, and certainly adds to the flexibility. Software packages are available that can manage the entire process and utilize almost any arrangement of existing hardware.

As a basic example of this process, let's examine a dual server system with a tape drive installed in the backup server. The disk-to-disk copy occurs between the servers as described above and proceeds quickly due to the speed of the disk arrays. Then backup software running on the backup server streams the server's data from its disks to its locally connected tape drive. Since this part of the process if fully contained within the backup server, it can occur at native tape drive speeds without placing any burden on the primary server or its disks. The overall process results in both an online backup (backup server) and a tape backup that can be moved off site and stored in a secure place in case of catastrophe.

This principle can also be applied within a SAN to provide extremely efficient backups. The snapshot mechanism explained above can perform the disk-to-disk image in minutes, and with a tape system connected directly to the SAN, the disk-to-tape process can proceed at full tape speed without burdening any servers or clients.

SUMMARY

This chapter has covered a broad range of station automation features and explored the wide variety of options available to network, store, share, and protect the digital audio assets of a station. These general discussions should provide a framework with which to evaluate your own specific station automation needs.

To summarize, here is my "top 10" (actually 17) list of important features for a state-of-the-art networked digital audio automation system (in no particular order):

1. CD-quality audio with balanced analog and AES/EBU digital I/O. High sample rates, 24-bit quantization, and multichannel audio a plus.
2. Comprehensive database of entire network inventory with easy to use search, sort, edit, modify, and report-generating functions.
3. Import daily logs from scheduling system of choice.
4. Modify logs while on-air, right up to air time.
5. Report as-played log back to billing system of choice.
6. Track voiceovers and program segues for live sound during automation.
7. Automatically record news feeds independent of on-air playback.
8. Accept closures from satellite receiver to play stopsets, IDs, and jingle/liner rotations.
9. Control routing switcher, console on/off keys, telephone systems.
10. Send "now playing" data to HD transmitter, RDS encoder, and/or website via IP.
11. Easy to learn and use operator interface with touch-screen operation for live assist.
12. Integration of third-party multitrack production equipment.
13. Integration of wire services, text, and audio.
14. Dedicated file server-based network architecture.
15. Switched Ethernet 100TX/1000T network using Category 5e or better UTP.
16. Multiprocessor, multi-NIC file server, or NAS with integrated disk array or SAN.
17. High-speed (15,000 rpm), high-capacity (300 GB), high transfer rate (>300 MB/s) disk drives configured as a RAID level 5 array.

Station automation is a continuously evolving entity. With the personal computer at the heart of it, this evolution will go on due to the rapid advances being made in computing platforms, networks, and the software applications that run on them. One thing the first decade of these systems has shown us is that advancements occur so quickly that what is leading-edge technology today will surely be obsolete long before it wears out. Therefore, it is wise when building or upgrading a system to use the most advanced, stable, and reliable technology available at that time.

I hope this chapter has given the reader some insight as to the current state of this technology and provided enough history to help the user prepare for what emerging technology may bring in the future.

Key Terms

1000BaseLX: 1000 Mbps fiber-optic Ethernet using long-wave laser.
1000BaseSX: 1000 Mbps fiber-optic Ethernet using short-wave laser.
1000BaseT: 1000 Mbps twisted-pair Ethernet (four pair of CAT5E or CAT6).
100BaseFX: 100 Mbps fiber-optic Ethernet.
100BaseT4: 100 Mbps twisted-pair Ethernet (four pair of CAT3, CAT4, or CAT5).
100BaseTX: 100 Mbps twisted-pair Ethernet (two pair of CAT5).
100BaseVG-AnyLAN: 100 Mbps star-wired network using DPP media access control mechanism over four twisted pairs of CAT3 or higher.
10Base2: 10 Mbps RG-58-style coax bus-cabled Ethernet.
10Base5: 10 Mbps RG-8-style coax bus-cabled Ethernet.
10BaseT: 10 Mbps twisted-pair Ethernet (two pair of CAT3).
10GBaseT: 10,000 Mbps twisted-pair Ethernet (CAT6a or CAT7).
10GbE: 10 Gbps Ethernet.
AAF: Advanced Authoring Format; an interchange file format developed by the AAF Association, Pro-MPEG Forum, and SMPTE primarily for postproduction interchange of media and metadata.
A/D: Analog-to-digital conversion.
AES: Audio Engineering Society.
AES/EBU: Audio Engineering Society/European Broadcast Union; used to refer to the standardized AES3 professional format for serial digital audio.
Anti-aliasing filter: A low-pass filter used to limit input frequencies to an A/D converter to one-half the sampling rate to satisfy Nyquist criteria and prevent sample contamination.
As-played log: A log file containing date, audio file ID, and start and stop time information for every element aired.
ATA: AT attachment specification that defines the IDE interface; AT refers to the IBM PC/AT, which was the first computer to use the IDE interface.
Audio server: A file server that contains both the audio file hard disk storage and the audio processing hardware (DSP boards) to enable it to record and play audio directly.
Baseband: Communication in which only a single signal occupies a transmit or receive path at a time.
Bus link: A communications link where every node shares a single communication medium.
BWF: Broadcast Wave Format; an extension of the WAV file format developed by the EBU that embeds metadata into a broadcast EXTension (BEXT) "chunk" contained within the audio file; described in EBU Document Tech. 3285 and its supplements.
Cart Chunk: Cart Chunk format; an extension of the WAV file format developed by D. Pierce and G. Steadman with input from the radio broadcast community that embeds metadata into a "chunk" contained within the audio file; a public domain, license-free interchange standard ratified by the AES as AES46-2002.
Cart machine: Traditional analog tape machine that uses an endless loop cartridge.
CAT3: Category 3 voice-grade unshielded twisted pair cable (UTP); frequencies up to 16 MHz; used for telephone and for data up to 10 Mbps.
CAT5: Category 5 data grade UTP cable; frequencies up to 100 MHz; used for up to 100 Mbps.
CAT5e: Category 5 enhanced data grade UTP cable; frequencies up to 125 MHz; used for up to 1000 Mbps.
CAT6: Category 6 data grade UTP cable; frequencies up to 250 MHz; used for up to 1000 Mbps.
CAT6a: Category 6a data grade UTP cable; frequencies up to 500 MHz; used for up to 10 Gbps.
CAT7: More formally known as ISO/IEC 11801 Class F cable; has four individually shielded twisted pairs (SCTP) inside an outer shield; frequencies up to 600 MHz; used for 10 Gbps.
CD jukebox: A CD player that houses hundreds of compact discs and uses robotics to move the CDs from storage to an active CD transport for playback.
CD-quality: Frequency response flat from 20 Hz to 20 kHz, dynamic range >90 dB, signal-to-noise ratio >90 dB.
Cheapernet: Slang name for 10Base2; comes from low cable cost.
Chunk: The basic data building block of a file based on the Microsoft Resource Interchange File Format (RIFF), such as a WAVE file.
Coding: The process of converting signal amplitude to a digital word.
Collision: When two or more transmitters access a shared bus simultaneously and their data packets interfere with each other.
Control macros: Sets of control functions executed in sequence as a single command.
CPU: Central processing unit.
CSMA/CD: Carrier sense multiple access with collision detection; contention-based network control mechanism used by Ethernet networks.
D/A: Digital-to-analog conversion.
DAS: Direct attached storage; refers to disk drives connected directly to a server or workstation as opposed to network-based architectures like SAN and NAS.
Data compression: The process of reducing the amount of data needed to accurately represent an object.
Data rate: The speed at which data is produced or consumed.
Data redundancy: Storing extra data such that original data can be recovered in the event of a storage component failure.
Disk mirroring: Data written to two (or more) identical disk drives for redundancy.
Dolby AC-2: Digital audio data reduction algorithm developed by Dolby Labs.
DPP: Demand priority protocol; network access mechanism used by 100BaseVG.
DSP: Digital signal processing; refers to digital processing hardware used to manipulate audio data and execute data reduction algorithms.
Duplex communications: The ability to transmit and receive simultaneously.
EBU: European Broadcast Union.
Ethernet: The first and still most popular LAN cabling standard that provides shared access through the CSMA/CD media access control mechanism; includes 10Base2, 10Base5, 10BaseT, 100BaseFX, 100BaseT4, 100BaseTX, 1000BaseLX, 1000Base SX, and 1000BaseT specifications.
Fibre Channel: A high-speed transport mechanism used primarily to connect devices in a SAN; most often implemented with fiber-optic connections although three types of electrical (copper) connections exist.
File server: A computer with an integrated network operating system that presents itself to the rest of the system as a large shared disk drive and manages the storage of all audio and system data.
FDX: Full duplex; communication that provides simultaneous transmit and receive.
GUI: Graphical user interface; computer screens designed for point-and-click operation as opposed to keyboard-only operation.
Hit times: Time of day that a program element is scheduled to air.
Hub: A central wiring cabinet or device used in star-wired networks.
HVAC: Heating, ventilation, and air conditioning.
IEEE: Institute of Electrical and Electronic Engineers.
ID3: IDentify an MP3; a metadata standard for MP3 files developed in 1996 by Eric Kemp; current version 2 is known as ID3v2.
IP: Internet protocol.
IT: Information technology; generally refers to all computer and network equipment and the departments and personnel that manage and maintain them.
LAN: Local area network; computer network existing in one geographical location.
Latency: Time delay between the request for data and the receipt of data.
Linear PCM: Uncompressed PCM; sampled, quantized, full-bandwidth digital audio.
Link aggregation: The process of combining individual network connections (links) into a single logical link that offers higher bandwidth and/or link redundancy.
Logical topology: How a network routes communication.

Lossless compression: Algorithms that eliminate redundant information and can completely restore each bit of the original data through an inverse algorithm because no unique data is discarded.

Lossy compression: Algorithms that remove irrelevant information but cannot completely restore each bit of the original data through an inverse algorithm because some minimally useful data has been discarded.

Macros: A set of functions programmed to execute in sequence for singular effect.

Managed switch: An Ethernet switch that provides a user interface (usually serial, telnet, or web based) to a menu of internal configuration parameters.

Mbps: Megabits per second; millions of data bits per second.

Media access control: How a network limits transmission to a single node at a time.

Mesh topology: A network where each node has a point-to-point connection with every other node.

MPEG: Moving Pictures Experts Group; used to refer to the digital audio data reduction algorithms developed by the group.

MTBF: Mean time between failures; statistic relating to longevity of a device.

Multicast: A part of the IP protocol suite that allows a single source packet to be delivered to many destination addresses.

MXF: Material eXchange Format; an interchange file format developed by the AAF Association, Pro-MPEG Forum, and SMPTE primarily for the interchange of essence and metadata between storage, broadcast, and play-out systems.

NAS: Network attached storage.

NIC: Network interface card; a plug-in card that attaches a computer to the network wiring; also used to refer to integrated main board electronics that serve the same purpose.

Node: Any device that communicates on a network.

NOS: Network operating system; software that manages a shared file system.

Nyquist: Mathematician who developed the sampling theory that states a signal must be sampled at more than twice its highest frequency.

Off-the-shelf: Nonproprietary computer hardware that is available from common computer suppliers.

OMFI: Open Media Framework Interchange; an open interchange standard pioneered in 1992 by Avid Technology, Inc. along with industry partners to provide a platform-independent method of exchanging media and metadata.

Packet: The basic unit of information exchanged between nodes on a network; consists of data packaged with address information.

PATA: Parallel ATA; the original IDE interface introduced in 1986.

PCM: Pulse-coded modulation; sampled, quantized, full-bandwidth digital audio.

Peer-to-peer network: A basic network link between two or more workstations at the file system level that permits file sharing and transfer between the workstations.

Physical topology: How a network is connected; also known as its shape.

Playlist: A database representing elements to be played in sequence.

Point-to-point link: A communications link where exactly one transmitter communicates with exactly one receiver.

Port trunking: Another name for *link aggregation* often used in Ethernet switches.

Quantization: The process of breaking continuous analog levels into discrete digital levels or steps.

RAID: Redundant array of inexpensive (or independent) disks; a disk technology that provides data redundancy through additional error-correction disks.

RAID Level 0: Also known as *disk striping.*

RAID Level 1: Also known as *disk mirroring.*

RAID Level 5: A disk array where redundancy is provided by a single additional disk.

RAM: Random access memory; fast rewritable memory.

Reconstruction filter: A filter used to smooth the pulsed output of a D/A converter back to a continuous analog waveform.

Ring topology: An arrangement of point-to-point links that connect in a circle or ring.

RJ-45: The eight-position modular connector used in twisted-pair Ethernet wiring.

ROM: Read-only memory; slow, nonvolatile memory.

Sampling: the process of periodically measuring a signal's value.

Sample rate: The frequency at which a signal is sampled.

SAN: Storage area network.

SAS: Serial attached SCSI; designed as a serial communication-based replacement for parallel (traditional) SCSI, it offers simpler interconnects and higher speed data transfers; uses standard SCSI command set easing integration and is also compatible with SATA technology.

SATA: Serial version of the ATA (originally defined as parallel) IDE interface.

SCSI: Small computer systems interface; refers to a standard interconnection and command set used to transfer data between computers and peripheral devices; most commonly used for connecting workstations and servers to disk and tape drives, but is designed for device independence and can be used with other devices such as printers, scanners, and CD and DVD burners; SCSI has evolved through several generations using 8- and 16-bit wide parallel data buses to provide data transfer rates from 5 MBps to 320 MBps.

Sneakernet: Slang term for transferring data from one computer to another by running (in sneakers) a floppy disk or other medium between the two.

Sound card: A plug-in card that contains the electronics necessary to allow a PC to receive as input and send as output audio signals; also used to refer to integrated main board electronics that serve the same purpose.

Spanning: Combining multiple disk drives to act as a single drive with the sum capacity of all the individual disks.

SSA: Serial Storage Architecture; a serial transport protocol developed by IBM to provide redundant data paths to up to 192 disk drives in direct attached storage configurations; utilized SCSI command set and was adopted as an ANSI X3T10.1 standard; promoted by the SSA Industry Association as an open standard but lost favor to the more widely adopted Fibre Channel protocol.

Stand-alone: Refers to a single computer that functions by itself.

Star topology: A network where each node has a point-to-point connection with a central hub.

Station automation: The use of devices, processes, and interconnections to make a broadcast facility function automatically.

Stopset: A set of audio elements that play in succession after a program segment has completed (stopped); usually refers to a group of commercials or other announcements placed between program segments.

STP: Shielded twisted-pair cable.

Striping: Dividing and writing a data stream to multiple drives to improve performance.

Switched Ethernet: An Ethernet network in which the central connection device, called a *switch*, provides high-speed port-to-port packet switching functions that create a logical mesh topology.

Teaming: Another name for *link aggregation* often used by NIC manufacturers.

Thicknet: Another name for *10Base5*; comes from thick RG-8-style cable.

Thinnet: Another name for *10Base2*; comes from thin RG-58-style cable.

Token: A special data frame used as a network access control mechanism.

Token Ring: 4/16 Mbps star-wired, twisted-pair network that operates as a ring with token-based access control.

UTP: Unshielded twisted-pair cable.

Walk-away: Station automation so complete, the operators can leave the equipment unattended.

XML: eXtensible Mark-up Language; a standardized, self-describing information exchange format that allows very flexible organization and description of the data payload; this simplifies communicating complex data between applications.

XGbE: 10 Gigabit Ethernet.

References

[1] CobraNet website, at http://www.cobranet.info/en/support/cobranet/.

[2] Axia–Livewire website, at http://www.axiaaudio.com/livewire/default.htm.

Bibliography

AAF and MXF: A Complementary Pair (n.d.), Sept. 15, 2006, from the AAF Association website, at http://www.aafassociation.org/.

Bird, Robert. Hard Disk Recording for Broadcast Use, *Broadcast Engineering*, Aug. 1991, p. 60.

Cabot, Richard. Performance Aspects of Digital Oversampling, *Broadcast Engineering*, July 1990, pp. 26–93.

Carlson, Bruce. *Communication Systems*, 2nd ed. New York: McGraw-Hill, Inc., 1975, pp. 294–326.

Carter, R. Scott, and Stevenson, Stephanae Ann. Strike Up the Bandwidth, *NetWare Connection*, July/Aug. 1994, pp. 17–27.

Category 6 Cabling Overview, FAQs, and Whitepapers (n.d.), Sept. 14, 2006, from the Telecommunications Industry Association website, at http://www.tiaonline.org/standards/technology/cat6/faq.cfm.

Category 7 Cable (n.d.), Sept. 13, 2006, from Wikipedia, The Free Encyclopedia website, at http://en.wikipedia.org/wiki/Category_7_cable.

Chalmers, R. The Broadcast Wave Format: An Introduction, *EBU Technical Review*, 274, Winter 1997, pp. 1–6.

Definition of Gigabit Ethernet (n.d.), Dec. 1, 2005, from the PC Magazine.com Encyclopedia website, at http://www.pcmag.com/encyclopedia_term/0,2542,t=Gigabit+Ethernet&i=43779,00.asp.

Gigabit Ethernet (n.d.), Dec. 1, 2005, from the TechWeb TechEncyclopedia website, at http://www.techweb.com/encyclopedia/defineterm.jhtml?term=Gigabit+Ethernet.

Conover, Joel. ATM Fast, Fast Ethernet, Fibre Channel Fast, *Network Computing*, Nov. 1995, pp. 46–60.

Direct Attached Storage (n.d.), Sept. 13, 2006, from Wikipedia, The Free Encyclopedia website, at http://en.wikipedia.org/wiki/Direct_Attached_Storage.

EBU Document Tech. 3285 (July 2001), Specification of the Broadcast Wave Format—Version 1, Sept. 14, 2006, from the EBU website, at http://www.ebu.ch/CMSimages/en/tec_doc_t3285_tcm6-10544.pdf.

ENCO Systems, Inc. *DAD_{PRO32} Digital Audio Delivery System Reference Manual*. Southfield, MI: ENCO Systems, Inc., 2001, pp. 12-5–12-17.

Heywood, Drew, Dulaney, E., Homer, B., Niedermiller-Chaffins, D., Orr, S., Stevens, S., and Stone, H. *Inside NetWare 3.12*, 5th ed. Indianapolis: New Riders Publishing, 1995.

Introduction (n.d.), Sept. 14, 2006, from the Cart Chunk Organization website, at http://cartchunk.org:8080/introduc.htm.

Katron Technologies, Inc. *Installation Guide: Fast Ethernet PCI Adapter* [Manual]. Miami: Katron Technologies, Inc., 1996, pp. 1–19.

Marks, Howard. Review: Disk-to-Disk-to-Tape Software, Escape the Tape, *Network Computing*, Sept. 14, 2005, pp. 67–72.

Meggyesi, Zoltan (Aug. 15, 1994). Fibre Channel Overview, Jan. 11, 2006, from the CERN High-Speed Interconnect Pages website, at http://hsi.web.cern.ch/HSI/fcs/spec/overview.htm.

PC Technology Guide (Nov. 15, 2003). Tape Storage, Dec. 21, 2005, from the PCTechGuide website, at http://www.pctechguide.com/15tape.htm.

Pizzi, Skip. Digital Audio Workstations Diversify, *Broadcast Engineering*, Aug. 1991, pp. 56–66.

Pohlmann, Ken. *Principles of Digital Audio*, 2nd ed. Carmel, IN: SAMS, 1989, pp. 41–98.

RAID Tutorial (2005), Sept. 13, 2006, from the Advanced Computer and Network Corporation website, at http://www.acnc.com/04_01_00.html.

Rodgers, Adam. Understanding RAID Technology, *Inside NetWare*, 1(9), Sept. 1992.

Serial Attached SCSI (n.d.), Sept. 13, 2006, from Wikipedia, The Free Encyclopedia website, at http://en.wikipedia.org/wiki/Serial_Attached_SCSI.

Smyth, Stephen. Digital Audio Data Compression, *Broadcast Engineering*, Feb. 1992, pp. 52–60.

Storage Computer Corporation. *RAID Aid: A Taxonomic Extension of the Berkeley Disk Array Schema*. Nashua, NH: Storage Computer Corporation, 1991, pp. 1–4.

Tagging Introduction (n.d.), Sept. 14, 2006, from the ID3 Organization website, at http://www.id3.org/intro.html.

Whitmann, Art. Hey, Hewlett-Packard! Let's Stop the Insanity! *Network Computing*, May 1996, pp. 103–105.

CHAPTER

3.7

Digital Audio Compression Technologies

SCHUYLER QUACKENBUSH
Audio Research Labs
Scotch Plains, New Jersey

FRED WYLIE
Audio Processing Technology, Ltd.
Belfast, Northern Ireland

INTRODUCTION

Virtually all applications of digital audio deal with enormous amounts of data. Even though modern computer networks have a capacity that may eliminate the need for audio signal compression, there remain many channels, including nearly all wireless communications channels, for which signal compression is essential for commercial viability of applications. Because of this, compression is an integral part of all digital radio and digital television systems, including

- The Advanced Television Systems Committee (ATSC) digital television broadcast system;
- The iBiquity Digital Corporation's HD Radio System (standardized by the National Radio Systems Committee);
- Commercial satellite radio broadcast systems (such as XM and Sirius Satellite Radio);
- Other terrestrial DAB systems (such as Digital Radio Mondiale, Eureka-147, and ISDB-T) and nascent cellular streaming audio systems.

Audio compression involves taking the "full" digital representation of an audio signal (usually obtained by digitally sampling an analog audio signal from a microphone or other audio source) and sending it through an *encoder* which removes information from this full representation in a prescribed manner. This encoded signal is sent through some medium (for example, a broadcast channel), and then on the receiving end is sent through a *decoder* which either restores the audio signal to its original quality or in some cases, by design, to a signal of lesser quality as a result of some limitation of the system (for example, severely restricted transmission bandwidth).

Many existing and proposed audio compression systems employ a variety of processing techniques. Any scheme that becomes widely adopted can enjoy economies of scale and reduced market confusion. Timing, however, is critical to market acceptance of any scheme. If a scheme is selected well ahead of market demand, more cost-effective or higher-performance approaches may become available before the market takes off. On the other hand, any particular scheme may be merely academic if it is established after alternative schemes already have become well entrenched in a specific marketplace.

These forces are shaping audio technology of the future. Numerous scenarios have been postulated for the hardware and software systems that will drive the digital audio production and transmission facility in the 21st century. However, one thing is certain: It will revolve around audio compression.

The professional audio industry continues to demand ever more complex equipment for the capture, storage, postproduction, exchange, distribution, and transmission of high-quality audio, whether it is mono, stereo, or multichannel surround sound. This demand is being driven by end users, broadcasters, filmmakers, and the recording industry, all of which are moving rapidly toward an all-digital environment. Over the last two decades, there have been continuing advances in digital signal processing (DSP) technology, such that it is possible to create low-cost, real-time hardware for digital audio signal compression that implements complex and powerful signal processing algorithms. Such systems can signicantly lower the bandwidth and storage requirements for the transmission, distribution, and exchange of high-quality audio.

The introduction in 1982 of the compact disc (CD) digital audio format set a quality benchmark that the manufacturers of professional audio equipment had to meet or exceed. The introduction in 1998 of the DVD and later DVD-Audio and Super Audio CD (SACD) raised this quality benchmark by providing 5.1 channels of high-resolution audio (higher sampling frequency and greater sample word length than those used in the CD). Some consumers expect the same quality from radio and television receivers, which represents a potential challenge for the broadcaster.

The infrastructure required to support the shift from analog-to-digital audio can be expensive. It is a rather complex technical exercise to fully implement a digital pulse code modulation (PCM) infrastructure. To demonstrate the advantages of distributing compressed digital audio over wired or wireless systems and networks, consider again the CD format as a reference. The CD uses a 16-bit linear PCM representation, which has excellent sonic qualities, but with the handicap of the amount of bandwidth the digital signal occupies in a transmission system. A stereo CD transfers data at 1.411 Mbps, which would require a bandwidth of approximately 700 kHz to avoid distortion of the digital signal. In practice, additional bits are added to the signal for channel coding, synchronization, and error correction, which increases the bandwidth further. The commonly quoted bandwidth requirement is 1.5 MHz such that a circuit is capable of carrying a CD or similarly coded linear PCM digital stereo signal. This can be compared with the 20 kHz needed for each of two circuits to distribute the same stereo audio in the analog format, a 75-fold increase in bandwidth requirements.

For the most part, this chapter focuses on two-channel stereo signals. However, there is a growing trend toward the delivery of multichannel audio programs to the consumer. This is evident in audio for high-definition television, but there is interest on how audio-only broadcast channels might be extended to deliver multichannel sound. In the case of radio, there is the issue of making a transition to stereo digital radio and then a further transition to multichannel digital radio. Obviously, the amount of audio data in a multichannel broadcast is significantly larger than in a stereo broadcast, so it seems certain that new audio compression technology will play an important role in making multichannel digital radio a practical reality.

AUDIO BIT RATE REDUCTION

Lossless and Lossy Systems

Audio compression systems can be either lossless or lossy. A lossless system is able to reconstruct an output signal that is bit-identical to its input signal. Therefore, a lossless system that compresses a signal in CD format can truly be said to have CD-quality. By comparison, a lossy system reconstructs an output signal that is not identical to its input, but, if the system does not use overly aggressive compression, is still *perceived* by listeners to be identical to its input. Lossless compression systems typically remove 50% of the input data, while lossy compression systems typically remove 90% (or more) of the input data.

Redundancy and Irrelevancy

An audio signal is very complex and contains a great deal of information. Some aspects of the information are highly predictable and are therefore redundant. Such signal components can be removed in the encoder and replaced in the decoder, a process often referred to as *statistical signal compression*. Other aspects of the information are not perceived by the listener and are therefore *irrelevant*. Such signal components, once removed in the encoder, don't need to be replaced in the decoder (since they are irrelevant). They are irretrievably lost in the compression process but are not *perceived* to be missing, a process referred to as *perceptual signal compression*. Coders that use this process are perceptual coders and, if their compression is not too aggressive, can result in a perceptually lossless compression system. Hence, lossless compression systems remove only the redundancy in a signal, whereas lossy compression systems remove both redundancy and irrelevancy.

Some signals, such as pure tones, are high in redundancy and low in irrelevancy. They can be compressed quite effectively, almost totally as a statistical compression process. Conversely, other signals, such as complex audio or noisy signals, are low in redundancy and high in irrelevancy. They also can be compressed quite effectively, but almost entirely as a perceptual compression process.

The Human Auditory System

A variety of phenomena exhibited by the human auditory system have been studied and form the basis of what is called the *psychoacoustic model*. The sensitivity of the human ear is greatest at the lower end of the audible frequency spectrum, around 3 kHz. At 20 Hz, the bottom end of the spectrum, and 17 kHz, the top end, the sensitivity of the ear is reduced by approximately 50 dB relative to its sensitivity at 3 kHz. The curve shown in Figure 3.7-1 represents the typical limit of human hearing, such that sounds having a loudness below the threshold are not audible.

Another aspect of the hearing process is that a loud sound (or tone) at a given frequency will mask a quieter sound at a nearby frequency. This effect is called *simultaneous masking*. For a given separation in frequency between loud and quieter tones, it is more pronounced when the loud tone has a frequency that is lower than the quieter tone. Furthermore, it is more pronounced as the loud tone increases in frequency. For example, as illustrated in Figure 3.7-2, with a 2 kHz tone at a level of 70 dB SPL, tones at 1.5 kHz or 3.4 kHz would require levels greater than 40 dB SPL to be heard.

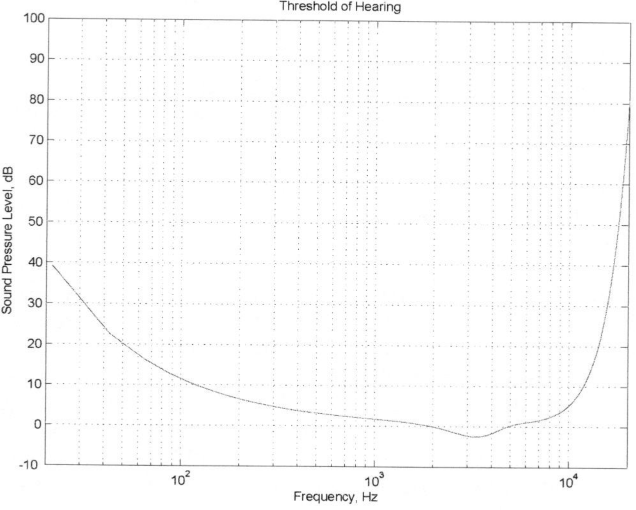

FIGURE 3.7-1 Typical threshold of hearing curve.

Finally, the ear is also subject to *temporal masking*, where a strong sound will mask a weaker sound that precedes or follows the stronger sound. This effect is strongest when the weaker sound follows the stronger one.

By exploiting these three aspects of the human auditory system—threshold of audibility, simultaneous masking, and temporal masking—an audio compression system can identify the irrelevant components of the input signal that the ear is unable to hear, not code or transmit those components, and thereby achieve data compression. The maximum tolerable level of these irrelevant components which can be removed, across the range of frequencies and for a given segment of the signal, is called the *noise masking threshold*.

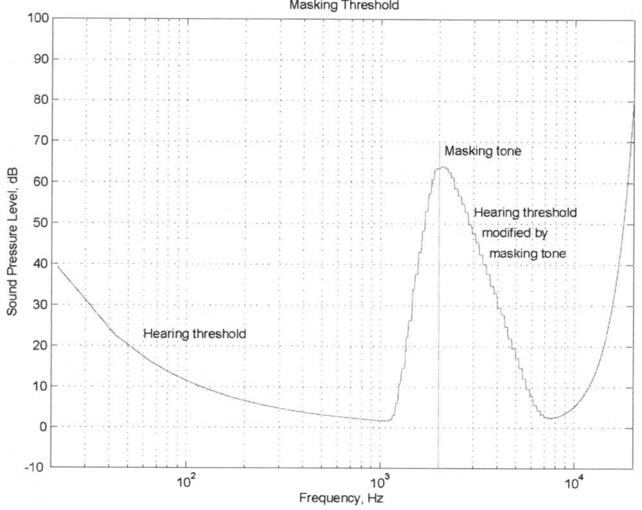

FIGURE 3.7-2 Masking effect of a loud tone at 2 kHz (shown as vertical line in figure).

Sampling and Quantization

The analog signals that comprise the audio and visual stimuli that we perceive in the world around us are continuous-time and continuous-valued functions, which is to say that they take on a continuously varying range of values over a continuously varying range of time. Sampling is the process of converting a continuous-time signal into a discrete-time signal, which is done by sampling the continuous function at a sequence of discrete, typically uniform, instants in time. Quantization is the process of converting a continuous-valued function into a discrete-valued function, which is done by mapping a given interval of values to a single value (typically the interval's midpoint). Pulse Code Modulation (PCM) is one of the most common forms of quantization, in which a continuous value is represented by an N-bit digital word.

An analog-to-digital (A/D) converter performs simultaneous sampling and quantization so that it can convert an analog signal to its representative digital format. For example, the CD represents an analog signal that has been converted into a sequence of 16-bit PCM values uniformly sampled at 44.1 kHz. The level of each audio sample, therefore, is one of 65,536 discrete levels or steps. A complementary process, that of digital-to-analog (D/A) conversion, converts a digital audio signal back into analog form suitable for human listening.

The process of sampling and quantization typically entails some loss of information. While a continuous-time variable can have a theoretically infinite range of frequency content, an important rule of information theory known as Nyquist's theorem states that the highest frequency component in a signal that is to be sampled must be no greater than one-half of the sampling frequency. The reason is that the process of sampling creates copies, or *aliases*, of the original signal's spectrum at integer multiples of the sampling frequency. If the lower sideband of the first alias (which has its DC or zero frequency component located at 1 times the sampling frequency) overlaps with the upper region of the baseband audio, audible and objectionable aliasing effects occur. Hence signals that are to be sampled must first be filtered to ensure that they do not contain frequencies greater than half the sampling rate. In practice, the sampling rate is set to slightly above the highest desired frequency, making such "anti-aliasing" filters much less expensive to produce. This is the reason that the CD sampling rate is 44.1 kHz; so that all audible frequencies (20 to 20 kHz) are preserved while permitting some margin for a practical anti-aliasing filter to transition from the passband at 20 kHz to the stopband at 22.05 kHz.

Quantization is a nonlinear process in which a continuous-valued input is mapped to a discrete-valued output, as is shown in Figure 3.7-3, where vertical lines indicate the quantizer decision boundaries, and the "X" indicates the set of discrete output values. For

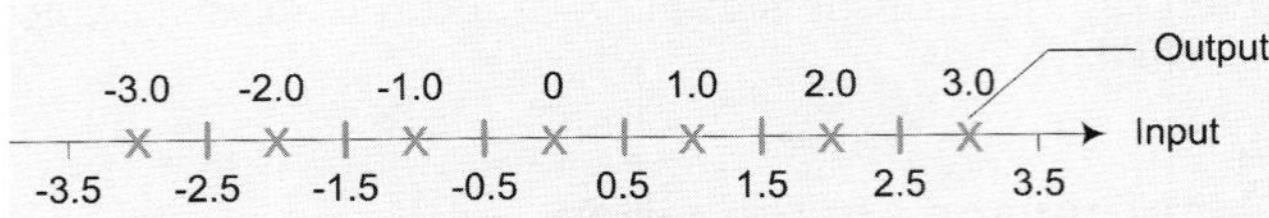

FIGURE 3.7-3 Seven-level uniform quantizer function for converting a continuous input value to a discrete output value. Horizontal axis is input range, vertical strokes show input decision boundaries, and "X" indicates quantized values.

example, an input value anywhere in the range of 1.5 to 2.5 is mapped to the output value 2.0. As in the case of the seven-level quantizer illustrated in the figure, the quantization process results in an error in the quantized output value relative to the unquantized input value. The error is referred to as quantization noise, and quantization can be modeled as a process which adds quantization noise to the signal to be quantized.

It is good operating practice to limit the number of A/D and D/A conversions in the audio processing chain, and similarly it is good practice to limit the number of compression stages in the audio chain. Typical lossless or lossy compression systems entail more than one quantizer, each of which injects quantization noise. Such noise can be minimized by operating the compression systems at as high a bit rate (i.e., as low a compression ratio) as is practical. Otherwise, after a number of A/D and D/A conversions and passes of compression coding, the accumulation of quantization noise and other unpredictable signal degradations inevitably will exceed the noise-masking threshold and be perceived by the listener.

Bit Rate and Compression Ratio

The bit rate, B, of a digital signal is defined as

$$B = \text{sampling frequency} \times \text{word length} \times \text{number of audio channels}$$

Hence, the bit rate of a CD is

$$B = 44{,}100 \text{ samples/sec} \times 16 \text{ bits/sample} \times 2 \text{ audio channels} = 1.4112 \text{ Mbps}$$

for a two-channel stereo signal. The compression ratio, R, is defined as the ratio of the bit rate of the input signal to the bit rate of the output signal of the compression system:

$$R = B_{in}/B_{out}$$

For example, if a CD signal is compressed to 128 kbps, the resulting compression ratio is

$$R = 1.4112 \text{ Mbps}/128 \text{ kbps} = 11.025$$

The ITU has recommended (in ITU-R BS.1115-1) the following bit rates when incorporating MPEG-1 Layer 2 data compression in an audio chain:

- 128 kbps per mono channel (256 kbps for stereo) as the minimum bit rate for any stage if further compression is anticipated or required;
- 180 kbps per mono channel (360 kbps for stereo) as the minimum bit rate for the first stage of compression in a complex audio chain.

These recommendations indicate that a 4:1 compression ratio is quite safe, even when extensive additional processing is envisioned. However, more aggressive compression ratios, up to 30:1, are possible. Hence, high levels of compression early in the processing chain can lead to problems if any subsequent stages of compression are required or anticipated.

With successive stages of compression, the noise floor and the audio bandwidth will be set by the stage operating at the lowest bit rate. Therefore, it is worth noting that when processing chains include such a low bit rate stage, the result cannot be subsequently improved by a stage operating at a higher bit rate.

A stage of compression may well be followed in the audio chain by another digital stage, either of compression or other processing, but which operates at a different sampling frequency. If tandem D/A and A/D conversion is to be avoided, a digital sample rate converter must be used. This can be a stand-alone unit, or it may already be installed as a module in existing equipment. If the stages have the same sampling frequencies, a direct PCM or AES/EBU digital link can be made, thus avoiding the need for sample rate or D/A conversion.[1]

Processing Delay and Computational Complexity

A signal is processed by a system in which there is an input and an output. The processing delay, or throughput *latency*, is the time between the arrival of a signal component at the input of the system and its reproduction at the output of the system. For the purpose of throughput latency, an audio compression system can be viewed as a single module consisting of encoder, transmission channel, and decoder, since the signal, as a waveform, exists only at the input of the encoder and the output of the decoder. Although one can assign a time to a frame of compressed data produced by the encoder, it is primarily the processing delay of the encoder-channel-decoder system that is of practical interest.

The processing delay of an audio compression system is composed of two elements: algorithmic delay and implementation delay. The former depends on the compression algorithm used and is greater for coders which process long intervals of the input signal all at once (i.e., using long block sizes), and such delay is unavoidable given the compression algorithm. The latter depends on how the compression algorithm is

[1]AES/EBU link refers to a standard digital audio interface commonly used in broadcast audio equipment. For additional information, see "AES-3id-2001: AES information document for digital audio engineering—Transmission of AES3 formatted data by unbalanced coaxial cable," Audio Engineering Society, www.aes.org.

implemented, such as the size of input or output signal buffers, or the extent of pipelining in the calculations of the algorithm. A more powerful DSP chip may reduce the implementation delay; hence, the computational complexity of a compression algorithm may be a consideration, not only due to the cost of such chips, but also due to its potential effect on processing delay. The processing delay of an audio compression system can range from a few milliseconds (ms) to tens and even hundreds of ms, and must take into account the additional delay imposed by the communications channel that connects encoder to decoder.

Most audio compression systems operate in real time, which is to say that signals can be processed "on the fly." However, due to algorithmic and implementation delay, all systems have some processing delay that introduces a measurable delay into the audio chain. The amount of delay will be important if the equipment is to be used in an interactive or two-way application. As a rule of thumb, more than 40 ms of round-trip delay in a two-way audio exchange is problematic, and in some cases the delay due to transmission of the compressed data over the channel may be non-negligible. While a two-way hookup over a 1000 km, full-duplex, digital telecommunications link has a propagation delay of only 3 ms in each direction, the propagation delay in satellite and long terrestrial circuits is considerably longer. Hence, the processing delay of a codec and transmission delay of the associated channel must be taken into consideration when designing or implementing a two-way interactive audio system.

Editing Compressed Data

The minimum temporal resolution of a compressed data format may or may not be adequate to allow direct editing in the compressed domain (i.e., direct editing of the compressed audio signal waveform). The minimum set of audio samples that can be manipulated via editing of the coded signal is determined by the size of the time block associated with a frame in the compressed domain and directly corresponds to the number of samples in a block in the case of block-processing compression algorithms (see the discussion of transform coders in the next section). The larger the time block, the more temporal granularity occurs in the editing process.

FILTERBANK TIME-FREQUENCY ANALYSIS

The Time-Frequency Domain

Most audio coders that have a compression ratio of 4:1 or more are perceptual coders, which achieve compression by the removal of both redundancy and irrelevancy. Since irrelevancy is primarily due to masking, and masking principles are best applied in the frequency domain, most perceptual coders involve a transformation of the signal from the time domain to the frequency domain. This typically involves using a filterbank that divides the audio signal spectrum into from 4 to 1024 (or more) subbands.

Historically, audio coders employing such filterbanks were divided into two classes: subband coders and transform coders. Both employ filterbanks, and from a mathematical point of view they are not different. *Subband coders* typically have fewer subbands (e.g., 32) with sharper rejection of frequency components in adjacent subbands due to their relatively long filter lengths. *Transform coders* typically have more subbands (e.g., 1024) with less rejection of frequency components in adjacent subbands due to their relatively short filter lengths. Most subband coders have filter lengths that are several times longer than the number of subbands (e.g., 4 times), while transform coders have filter lengths that are at most a few times longer than the number of subbands (e.g., 2 times).

Examples of subband coders are G.722, an ITU-T standard introduced in the mid 1970s, and MPEG-1 Layers 1 and 2, standardized in 1993. Examples of transform coders are Dolby AC-3, a proprietary algorithm introduced in the early 1990s, and MPEG-2 Advanced Audio Coding (AAC), standardized in 1997. In the remainder of this chapter the term "subband coder" will refer to both subband and transform coders.

Subband Filtering

Redundancy Reduction

The algorithms mentioned in the preceding section process the PCM signal by splitting it into a number of frequency subbands, in one case as few as 2 (G.722) or as many as 1024 (AAC). MPEG-1 Layer 1, with 4:1 compression, has 32 frequency subbands.

Subband filtering enables the frequency domain redundancies within the audio signals to be exploited, such that the coded bit rate can be reduced relative to PCM while maintaining the same signal fidelity. Spectral redundancies are present whenever the signal energies in the various frequency bands are unequal at any instant in time. When the bit allocation for each subband is altered, typically by dynamically adapting it according to the energy of the signal in each subband, the quantization noise can be reduced across all bands.

On its own, subband coding, incorporating PCM in each band, is capable of providing compression, or *subband gain*, compared to that of full-band PCM coding. This is due to the reduction in redundancy provided by the subband coding gain. Subband coding gain is defined as the improvement in the signal-to-noise ratio (SNR) when using subband coding as compared to full-band PCM coding, given that the bit rate in each domain is constant. This can be quantitatively expressed as the ratio of the variance of the quantization errors when coding in the full-band PCM domain and the variance of the quantization errors when coding in the subband domain, again with the constraint that the bit rate in each domain is constant. Subband coding gain increases as the number of subbands increases, although the complexity of the compression algorithm also increases.

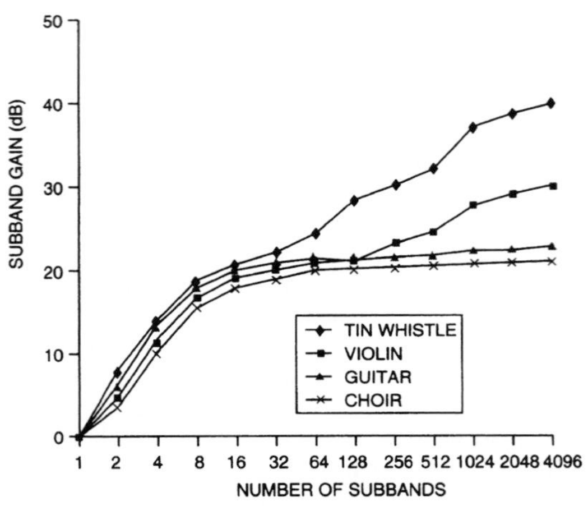

FIGURE 3.7-4 Subband gain as a function of the number of subbands.

Figure 3.7-4 charts subband gain as a function of the number of subbands for four essentially stationary,[2] but differing, complex audio signals. This figure suggests that subband coders will deliver greater compression as the number of bands is increased, even to 4096 bands or more. However, the fallacy in this is that the subband gain shown is possible only for stationary signals. A large number of subbands typically require long filter lengths, and if, during a time interval comparable to the filter length, the character of a signal varies in a way that is not predictable, the redundancy in the signal is sharply decreased. Hence, the subband coding gain is sharply reduced.

Irrelevancy Reduction

The most important contribution to compression in perceptual coders comes from irrelevancy reduction. As stated previously, this comes from exploiting the three types of auditory masking: threshold of hearing, simultaneous masking, and temporal masking. The human auditory system itself can be modeled as a subband filterbank, but one with nonuniform bandwidths that are narrow at low frequency and wide at high frequency. This is equivalent to having low temporal resolution at low frequency and high temporal resolution at high frequency. Simultaneous masking is best exploited with a high-frequency-resolution filterbank, whereas temporal masking is best exploited with a high-time-resolution filterbank. Although a nonuniform bandwidth subband filterbank would seem to be the obvious solution, this type of filterbank is typically not used, as it can result in reduced signal compression and increased computational complexity.

In the end, the number of subbands in a coding system reflects the trade-off between higher frequency resolution that permits greater redundancy reduction and higher temporal resolution that prevents errors in irrelevancy reduction. The AC-3, MPEG-1 Layer 3 and MPEG-4 AAC coders address this trade-off by using filterbanks that can seamlessly switch between a high-frequency-resolution mode and a high-time-resolution mode.

[2]In this context, "stationary" refers to a random process (in this case, an audio signal) where its statistical properties do not vary with time.

Subband Quantization and Coding

Exploiting the irrelevancy properties of the auditory system requires computing an estimate of the short-time spectrum of the signal, since masking models are best manipulated in the frequency domain. A common method of doing this is to use a Fast Fourier Transform (FFT) that is aligned in time and frequency with the subband filterbank. This means that the FFT frequency bins are aligned with the band edges of the subband filterbank so that the power in the FFT is an estimate of the short-time power spectrum of the subband signals flowing out of the encoder filterbank.

The short-time power spectrum is converted from Hz to *Bark* (at any given frequency, unit intervals on the *Bark scale* correspond to a width of one auditory critical band). The advantage of the Bark scale is that simultaneous masking curves are, to first order, identical at every frequency. Based on the power spectrum, along with notions as to whether a frequency region is tone-like or noise-like, a masking threshold is computed. This threshold, which specifies the "just noticeable" quantization noise at each frequency, can be used to directly set the quantizer step sizes within each subband. These step sizes are typically transmitted as side information. Note, however, that maintaining a level of "coding margin," which keeps the actual quantization noise some amount below the estimated masking threshold, is important if further compression is contemplated as part of any additional postproduction or transmission steps.

Different coding algorithms may use different strategies to code the quantized subband values. For example, subband values can be coded in vectors, and entropy (i.e., lossless) coding can be used to further compress the codes assigned to these vectors.

Finally, some systems exploit the significant redundancy between the two channels (left and right) in a two-channel stereo (or more for multichannel) signal by using some form of joint channel coding. For stereo signals, this might result in coding regions of the stereo spectrum as a single amplitude-panned spectrum; or computing left, right, sum, and difference spectra; and then, for each region of the spectrum, selecting the domain (left/right or sum/difference) that provides the greatest coding gain.

DIGITAL AUDIO COMPRESSION ALGORITHMS

This section presents the technical details of several audio compression systems that are widely adopted in

the marketplace or are anticipated to have a significant impact on the marketplace.

Algorithms and Compression

- *Dolby AC-3 (also known as Dolby Digital):* Standardized in 1996, this audio compression scheme is used in the ATSC DTV system and is also included in the DVB digital TV standard. It can encode five surround audio channels plus a low-frequency effects channel at a (typical) total bit rate of 384 kbps. This configuration is referred to as 5.1 channel surround sound.
- *ISO/MPEG-1 Layer 2:* Standardized in 1993, this audio compression scheme is widely used as a storage format in audio production, as it permits signals to undergo many stages of compression without suffering perceivable degradation. It is also the audio compression format used in the Eureka 147 DAB system.
- *ISO/MPEG-1 Layer 3 (MP3):* Standardized in 1993, this audio compression format is used in the majority of portable music players worldwide and has significant presence in Internet download and streaming music applications and services.
- *ISO/MPEG-4 Advanced Audio Coding (AAC):* Standardized in 1997, this audio compression format is used in portable music players and in music download services. It is also the audio compression format used by the Integrated Services Digital Broadcasting (ISDB) service in Japan.
- *ISO/MPEG-4 High-Efficiency Advanced Audio Coding (HE-AAC):* Standardized in 2003, this audio compression scheme is used in satellite-delivered DAB in the United States, and in Digital Radio Mondiale. The core technology in HE-AAC (Spectral Band Replication) is used in HD Radio, the audio compression scheme used by the in-band/on-channel (IBOC) Digital Audio Broadcast system for AM and FM in the United States.
- *HD Radio Audio Compression*: The "HD Codec" (HDC) was developed by iBiquity Digital Corporation and is used in its HD Radio in-band/on-channel (IBOC) AM and FM digital radio systems.[3]
- *Dolby E-AC-3 (Enhanced AC-3, also known as Dolby Digital Plus):* This audio compression technology was incorporated into ATSC Standard A/52B in June 2005. It provides additional compression relative to AC-3 by incorporating a high-frequency-resolution transform, spectral extension, enhanced coupling channel, and "pre-echo" noise suppression tools, while retaining compatibility with AC-3 at the level of data framing and metadata.
- *ISO/MPEG Surround:* Standardized in 2006, this is a backward-compatible multichannel audio compression scheme, in that it can encode a 5.1 channel audio program as a compressed stereo program plus a small bit rate of side information. Legacy receivers can decode this as stereo, and advanced receivers can decode it as 5.1 channel surround.

Table 3.7-1 shows the typical operating parameters used by each of these audio coding algorithms when coding a stereo signal.

The following sections will discuss each of these audio compression algorithms.

[3]The HD Radio system was standardized in 2005 by the National Radio Systems Committee as NRSC-5-A, "In-Band/On-Channel Digital Radio Broadcasting Standard"; however, the audio codec is not specified in this standard.

TABLE 3.7-1
Typical Audio Compression Algorithm Operational Parameters from Compressing a Two-Channel Signal

Coding System	Typical Compression Ratio	Subbands	Bit Rate, kbps (stereo)*	Typical Implementation Delay, ms†
MPEG-1 Layer 1	4:1	32	384	19
MPEG-1 Layer 2	6:1	32	256	40
AC-3	7:1	256	192	50
MPEG-1 Layer 3	9:1	576	160	120
MPEG-4 AAC	11:1	1024	128	130
E-AC-3	15:1	256/1536	96	50–150
HDC	15:1	(not disclosed)	96	340
MPEG-4 HE-AAC	29:1	64/1024	48	340
MPEG Surround	32:1	64/1024	140	340

*Typical values shown; each system is capable of operating at various bit rates.
†The total one-way delay (encoder input to decoder output), neglecting any additional channel delay.

AC-3

The AC-3 audio compression algorithm specified in the ATSC DTV system can encode from 1 to 5.1 channels of PCM audio into a serial bitstream at data rates ranging from 32 kbps to 640 kbps.[4]

A typical application of this compression algorithm entails the following steps. A 5.1 channel audio program is converted from a PCM representation requiring more than 4 Mbps (6 channels × 48 kHz × 16 bits = 4.608 Mbps) into a 384 kbps serial bitstream by the AC-3 encoder. This may be multiplexed with other data (e.g., compressed video) and have timing, synchronization, and transport information added to the stream. Radio frequency (RF) transmission equipment converts this stream into a modulated waveform that is applied to a terrestrial broadcast signal. The amount of bandwidth thus required for the transmission of the audio information has been reduced by a factor of 12 by the AC-3 digital compression system. The received signal is demodulated and demultiplexed to recover the 384 kbps serial bitstream, which is decoded by the AC-3 decoder. The result is the reconstructed 5.1 channel audio program.

Encoding

The AC-3 encoder accepts PCM audio and produces the encoded audio bitstream. The AC-3 algorithm achieves high coding gain by coarsely quantizing a frequency domain representation of the audio signal. A block diagram of this process is shown in Figure 3.7-5. The first step in the encoding chain is to transform the representation of audio from a sequence of PCM time samples into a sequence of blocks of frequency coefficients, which is done by the analysis filterbank. Overlapping blocks of 512 time samples are multiplied by a time window and transformed into the frequency domain. These PCM blocks have 50% overlap, so that each new block contains only 256 new samples. However, because the filterbank is *critically sampled*, the 512 PCM samples processed by the transform result in only 256 frequency coefficients, so there is no growth in data rate (i.e., 256 new PCM samples per block result in 256 frequency coefficients). Because of the overlapping blocks, each PCM input sample is represented in two adjacent transformed blocks.

The individual frequency coefficients are represented in block floating point, with a common exponent for a set of coefficients and a mantissa for each coefficient in the set. The exponents are encoded into a coarse representation of the signal spectrum, referred to as the spectral envelope. This spectral envelope is used by the core bit allocation routine, which determines how many bits should be used to encode each individual mantissa. The spectral envelope and the coarsely quantized mantissas for six audio blocks (1536 audio samples) are formatted into an AC-3 frame. The AC-3 bitstream is a sequence of AC-3 frames.

[4]See ATSC Standard A/52B.

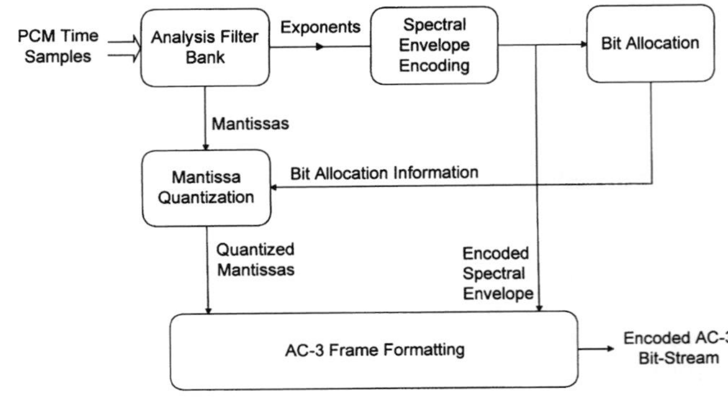

FIGURE 3.7-5 Block diagram of AC-3 encoder.

The actual AC-3 encoder is more complex than shown in the simplified system of Figure 3.7-5. Most notably, the following functions also are supported in the encoder:

- An AC-3 frame has a header that contains a synchronization word and other information (bit rate, sample rate, number of encoded channels, and additional data) such that the decoder can synchronize to and decode the bitstream.
- Error-detection codes are inserted to allow the decoder to verify that a received frame of data is error-free.
- The analysis filterbank spectral resolution may be dynamically altered to better match the time/frequency characteristic of each audio block.
- The spectral envelope may be encoded with variable time/frequency resolution.
- A more complex bit allocation may be performed, with parameters of the core bit allocation routine modified to produce a more optimum bit allocation.
- The channels may be coupled at high frequencies to achieve higher coding gain for operation at lower bit rates.

In the two-channel mode, a rematrixing process may be selectively performed to provide additional coding gain, and to allow improved results to be obtained in the event that the two-channel signal is decoded with a matrix surround decoder.

Decoding

The decoding process is essentially the inverse of the encoding process. The basic decoder, shown in Figure 3.7-6, must synchronize to the encoded bitstream, check for errors, and deformat the various types of data (the encoded spectral envelope and the quantized mantissas). The bit allocation routine is run, and the results are used to unpack and dequantize the mantissas. The spectral envelope is decoded to produce the exponents. The exponents and mantissas are transformed back into the time domain to produce the

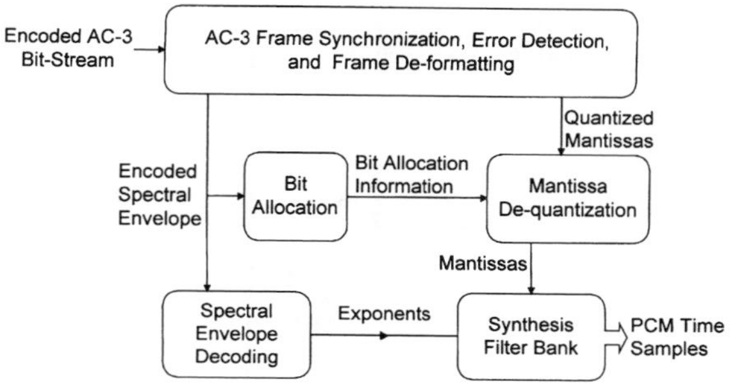

FIGURE 3.7-6 Block diagram of AC-3 decoder.

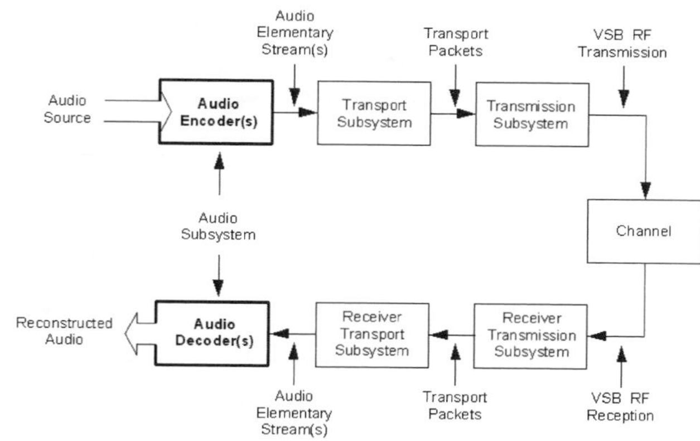

FIGURE 3.7-7 Block diagram of ATSC audio subsystem.

decoded PCM time samples. Additional steps in the audio decoding process include

- Error concealment or muting may be applied in the event a data error is detected.
- Channels with high-frequency content coupled must be decoupled.
- Dematrixing must be applied (in the two-channel mode) whenever the channels have been rematrixed.
- The synthesis filterbank resolution must be dynamically altered in the same manner as the encoder analysis filterbank was altered during the encoding process.

AC-3 in the ATSC Audio System

A simplified block diagram of the audio subsystem of the ATSC DTV standard is shown in Figure 3.7-7. It comprises the audio encoding and decoding functions that connect the audio inputs and outputs to the transport subsystem. The audio encoder is responsible for generating the audio elementary stream, which is an encoded representation of the baseband audio input signals. (Note that more than one audio encoder may be used in a system.) The transport subsystem formats the audio elementary stream into packetized elementary stream (PES) packets, which are then further formatted into transport packets. The transmission subsystem converts the transport packets into a modulated RF signal for transmission to the receiver. At the receiver, the signal is demodulated by the receiver transmission subsystem and the receiver transport subsystem converts the received audio packets back into an audio elementary stream. The flexibility of the transport system allows multiple audio elementary streams to be delivered to the receiver. At the receiver, the transport subsystem is responsible for selecting which audio streams to deliver to the audio decoder subsystem. The appropriate audio decoder is then responsible for decoding the audio elementary stream back into baseband audio.

The partitioning shown in Figure 3.7-7 is conceptual, and practical implementations may differ. For example, the transport processing may be broken into two blocks: the first would perform PES packetization, and the second would perform transport packetization. Or, some of the transport functionality may be included in either the audio coder or the transmission subsystem.

System Timing Issues

The AC-3 system conveys digital audio sampled at a frequency of 48 kHz, and this sampling clock must be phase locked to the 27 MHz system clock referenced in the ATSC DTV Standard. If analog signal inputs are employed, the A/D converters should sample at 48 kHz. If digital inputs are employed, the input sampling rate should be 48 kHz, or the audio encoder should contain sample rate converters that translate the sampling rate to 48 kHz. Obviously, the sampling rate at both the input to the audio encoder and the output of the audio decoder is locked to the 27 MHz system clock, but they also must be locked to the video clock to ensure proper synchronization of the audio and video subsystems.

In general, input signals should be quantized to at least 16-bit resolution, although the audio compression system can convey audio signals having up to 24-bit resolution.

ISO/MPEG-1 Layer 2

A block diagram of the ISO/MPEG-1 Layer 2 encoding and decoding systems is shown in Figure 3.7-8.

Encoder

In the encoder, the incoming linear PCM signal is formed into blocks of 1152 samples and filtered by a polyphase analysis filterbank to divide the input signal into 32 equal-bandwidth subband signals of 36 samples per subband. *Polyphase* refers to a specific implementation of a filterbank that permits significant computational efficiency. At a 48 kHz sampling rate, the duration of each input block is 24 ms, and the bandwidth of each subband is 750 Hz. The filterbank, which displays moderate delay and minimal complexity, has sufficiently high frequency resolution to

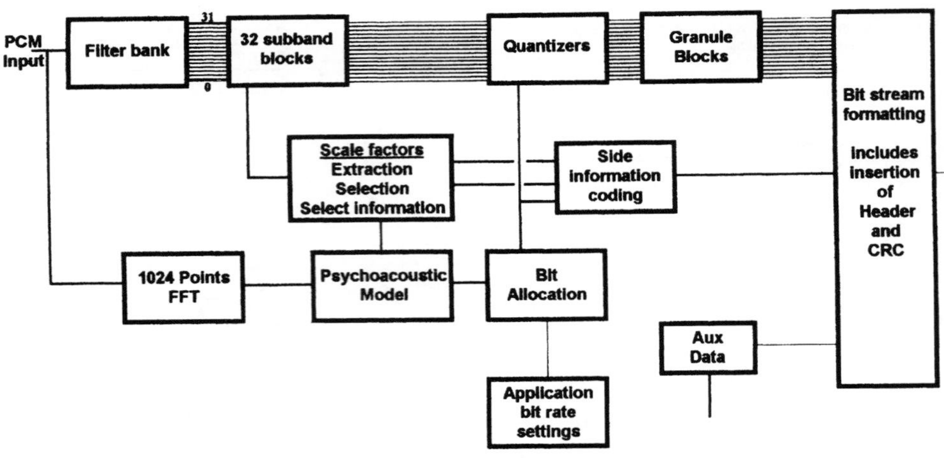

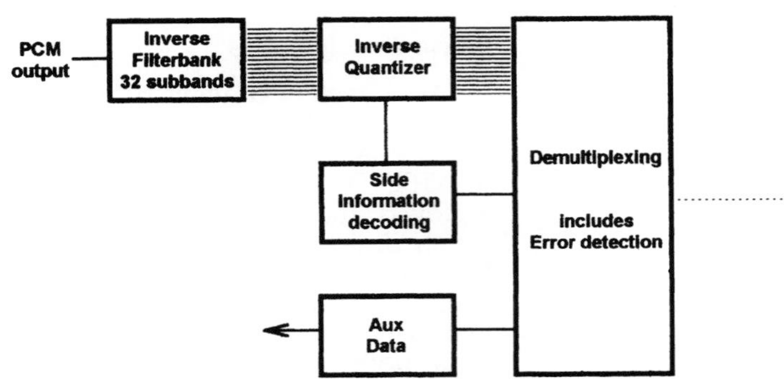

FIGURE 3.7-8 Block diagram of MPEG-1 Layer 2 encoder (top) and decoder (bottom).

effectively exploit simultaneous masking and sufficiently high time resolution to assure that temporal masking is modeled correctly. Multiple quantizers can be used for the subband samples, each of which has an adaptive step size. Furthermore, the encoder has an adaptive bit allocation among the set of subbands. In the decoder, after demultiplexing and decoding of the audio data and side information in a frame, the synthesis filterbank reconstructs the output block of 1152 PCM samples.

The PCM signal is fed both to the filterbank and to an FFT, which computes the short-time power spectrum of the input signal. This information is used by the psychoacoustic model to determine the masking threshold for the set of subband data. This, in turn, determines the quantizer step sizes such that the quantization noise is below the masking threshold in each of the subbands. More precisely, it determines a signal-to-mask ratio (SMR) threshold, which is the difference between the signal power (in dB) and masking threshold (in dB), computed in each subband. The actual number of levels for each quantizer is determined by the bit allocation.

MPEG standards specify only the format of a bitstream of compressed audio data and the process for decoding the data. An MPEG encoder is not specified other than to be a process that produces a standard-conforming bitstream. Since the psychoacoustic model is found only in the coder, it can evolve over time to permit better estimates of the masking threshold and hence better signal compression. For example, in a broadcasting system that utilizes this method of audio coding, as advances in coder technology are made, only the encoder portion of the system need be upgraded. The decoder portion (in receivers) does not need to be changed to experience this improvement. This is an important feature of the technology since typically there are few encoders but many receivers in use.

Scale Factors and Side Information

A scale factor is determined for each set of 12 samples in each subband. The maximum of the absolute values of these 12 samples determines a scale factor word consisting of 6 bits. With a quantization step size of 2

dB, this results in a dynamic range of more than 120 dB. Because each frame of audio data in Layer 2 corresponds to 36 samples in each of the 32 subbands, this process will generate three scale factors per subband per frame. However, the transmitted data rate for these scale factors can be reduced by exploiting redundancy in the data. Three successive subband scale factors are analyzed and a pattern is determined. This pattern, which is obviously related to the nature of the audio signal, will decide whether one, two, or all three scale factors are required. The four possibilities of the two additional scale factors being present or absent are signaled by the scale factor select information (SCFSI) data word of 2 bits.

In the case of a fairly stationary tonal-type sound, there will be very little change in the scale factors, and only the largest one of the three is transmitted, so the corresponding data rate will be (2+6) or 8 scale factor bits per subband. However, in a complex sound with rapid changes in content, the transmission of two or even three scale factors may be required, producing a maximum scale factor bit rate demand of (2+6+6+6) or 20 bits per subband.

The number of data bits allocated to a frame's bit pool is determined by key operating parameters, those being sampling frequency, compression ratio, and, where applicable, limitations imposed by the transmission medium. In the case of 20 kHz stereo being transmitted over ISDN, if the maximum data rate is 384 kbps (e.g., the aggregate of 6 B channels, at 64 kbps per B channel), and the signal is sampled at 48 kHz, then this necessitates a compression ratio of 4:1. A B channel is a 64 kbps telecom channel that is a "building block" for ISDN systems. Forty-eight thousand samples/sec × 16 bits/samp/chan × 2 chan = 1.536 Mbps/384 kbps = 4, hence 4:1 compression.

After the number of side information bits required for scale factors, bit allocation codes, cyclic redundancy check (CRC), and other functions have been determined, the remaining bits left in the pool are used for coding of the audio subband samples. The allocation of bits for the audio is determined by calculating the SMR, via the FFT, for each of the 12 subband sample blocks. The bit allocation algorithm then selects from 15 available quantizers, such that the overall bit rate limitations are met and the quantization noise is masked, or at least masked to the greatest extent possible. Note, that if the signal is identically zero, no bits are allocated so that no quantizer needs to be signaled. If the composition of the audio signal is such that there are not enough bits in the pool to adequately code the subband samples, then the quantizer step sizes are increased so as to obtain a best-fit solution that should minimize the impact of the quantization noise on the subjective quality of the decoded audio signal.

If the signal block being processed lies in the lower one-third of the 32 subbands, a 4-bit code word is generated to signal the selected low-subband quantizers, with this word carried as side information in the main data frame. Similarly, a 3-bit word signals the selected mid-subband quantizer, and a 2-bit word signals the high-subband quantizer. This allows for at least 15, 7, and 3 different sets of quantizer ranges, respectively, in each of the three subband groupings.

As with the scale factor data, some further redundancy can be exploited in the coding of quantization levels. For the lowest quantizer ranges (3, 5, and 9 levels), three successive subband sample blocks (3 groups of 12 subband samples) are grouped into a *granule* and a single quantizer range can be specified for this entire set of samples using a single code word. This is particularly effective in the higher frequency subbands where the quantizer ranges are invariably set to the lower end of the scale.

Framing

Error detection information can be relayed to the decoder by inserting a 16-bit cyclic redundancy check (CRC) word in each data frame. This parity check word allows for the detection of up to three single-bit errors or a burst of errors of up to 16 bits in length. A codec detecting an uncorrectable set of errors can mute the signal, or if it incorporates an error concealment strategy, it might replace the impaired data with a previous, error-free, data frame or some other suitable estimate of the damaged signal block. The typical data frame structure for MPEG-1 Layer 2 audio is given in Figure 3.7-9.

MPEG-1 Layer 3 (MP3)

MPEG-1 Layer 3, referred to more commonly as MP3 (a technique very popular for use in portable digital music players), offers significantly greater compression than that provided by Layer 2. It is typically used to compress a stereo signal sampled at 44.1 kHz to a data rate of 128 kbps to 160 kbps. It can compress to a fixed bit rate so that it can operate over fixed-rate channels or can operate in variable rate mode. In the latter case, the encoder attempts to maintain a constant signal-to-mask ratio for each signal block and permits the length of the encoded frame of data to vary frame by fame, subject only to that constraint.

A block diagram of the Layer 3 encoder algorithm is shown in Figure 3.7-10.

MPEG-1 Layer 3 Encoder

The filterbank used in MPEG Layer 3 is a hybrid filterbank which consists of a polyphase filterbank and a Modified Discrete Cosine Transform (MDCT). This hybrid form provides a common framework with Layer 1 and Layer 2. The further subdivision of the 32-band polyphase filterbank by the 18-band MDCT filterbank produces a division into 576 subbands and provides for a greater opportunity to exploit signal redundancy as a means to gain compression. It also permits a finer tracking of the masking threshold at lower frequencies, thus permitting the coder to better exploit signal irrelevancy. In the case that the normal "high-frequency-resolution" mode of the MDCT filterbank would have insufficient time resolution to avoid poor coding performance due to temporal masking

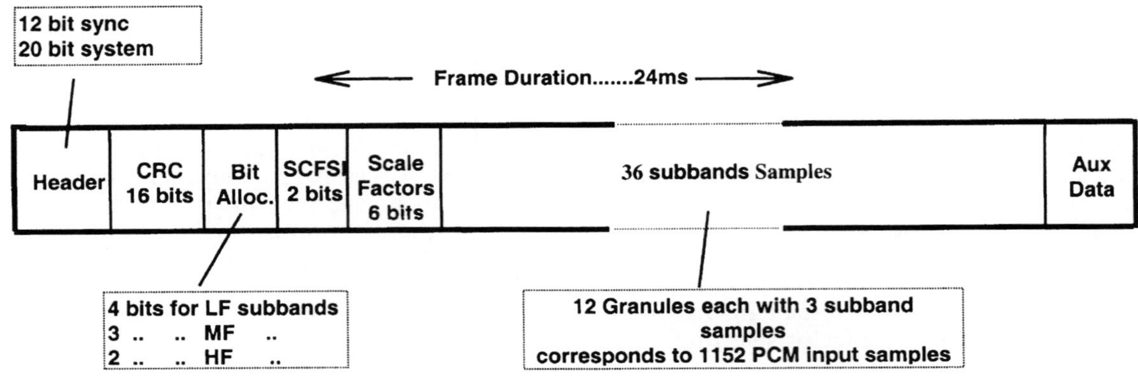

FIGURE 3.7-9 MPEG-1 Layer 2 bitstream frame structure.

issues (the so-called "pre-echo" problem), the MDC filterbank can switch to a "high-time-resolution" mode, with a time resolution of 4 ms by limiting the division of the signal into only 192 subbands.

The perceptual model is the principal component that determines the quality of a given Layer 3 encoder implementation. It typically uses an FFT to compute the short-time energy of the signal, which is then used to calculate the masking threshold. It is important that this energy estimate is aligned in time and frequency with the subband signal that is being coded. The output of the perceptual model consists of a signal-to-mask ratio, or the allowed quantization noise in each subband. If the bit rate is sufficient to keep the quantization noise below the masking threshold, then the reconstructed signal should be perceptually indistinguishable from the original.

Joint stereo coding takes advantage of the fact that left and right channels of a stereo channel pair often contain nearly the same information. These stereophonic irrelevancies and redundancies are exploited to reduce the total bit rate. Joint stereo coding is used in cases in which both high compression and stereo output are desired.

Quantization and coding to a fixed bit rate typically involve setting quantization step sizes and determining the size of the coded frame, including bits allocated from the bit reservoir. A system of two nested loops is the common solution for quantization and coding in a Layer 3 encoder. The quantization process effectively allocates noise to each subband based on the signal-to-mask ratio. Since some signal blocks are harder to code than others (based on the instantaneous statistics of the signal), they will inherently require more bits to meet the requirements posed by the ratio. Rather than suffer inconsistent audio quality under the constraint of constant bit rate, Layer 3 solves this problem with the concept of a bit reservoir. With this structure, the number of bits comprising a bitstream frame is allowed to vary based on the requirements imposed by the masking threshold. When a frame requires greater than the nominal number of

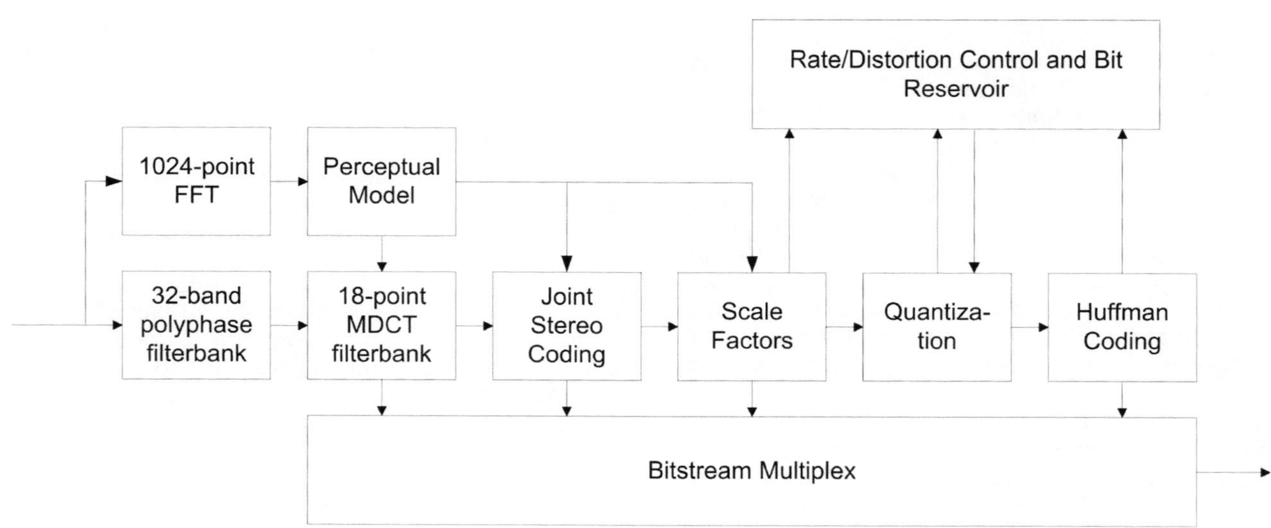

FIGURE 3.7-10 Block diagram of the MPEG-1 Layer 3 (MP3) encoder.

bits, they are supplied by the bit reservoir; when it requires fewer, the surplus is returned to the reservoir. While this has the advantage of raising the audio quality of the most difficult-to-code signal blocks, it has the trade-off of imposing an additional throughput delay equal to the size of the bit reservoir.

Quantization is done via a nonuniform power-law quantizer. In this way, larger values are automatically coded with less accuracy so that a mild degree of noise shaping is built into the quantization process. The quantized values are entropy coded using Huffman codes; this is a lossless process. Since this process does not inject any noise into the signal, it is also referred to as noiseless coding.

Finding the optimum gain and scale factors for a given signal block, bit rate, and signal-to-mask ratio threshold is usually done using two nested iteration loops in an analysis-by-synthesis manner:

- *Inner iteration loop (rate loop):* The Huffman code tables assign shorter code words to more frequently occurring (which are typically smaller) quantized values. If the number of bits resulting from the coding operation exceeds the number of bits available to code a given block of data, this can be corrected by adjusting the global gain, resulting in larger quantization step sizes, leading to smaller quantized values. This operation is repeated until the resulting bit demand from Huffman coding fits within the current frame's bit budget. This loop is called the rate loop because it modifies the overall coder bit rate until it is within the desired range.
- *Outer iteration loop (noise control/distortion loop):* To shape the quantization noise according to the masking threshold, scale factors are used to normalize the subband signal within each scale factor band. Initially, the scale factors have a default factor of 1.0 for each band. If the quantization noise in a given band is found to exceed the masking threshold in a given band, the scale factor for that band is adjusted to reduce the quantization noise. Since achieving a smaller quantization noise requires a larger number of quantization steps and thus a higher bit rate, the rate adjustment loop has to be repeated every time there are new scale factors. In other words, the rate loop is nested within the noise control loop. The outer (noise control) loop is executed until the actual noise (computed from the difference between the original spectral values and the quantized spectral values) is below the masking threshold for every scale factor band.

Bitstream Framing

In the context of transmission channels, Layer 3 can operate over a constant-rate isochronous link, and has constant-rate headers (as do Layers 1 and 2). However, Layer 3 is an instantaneously variable rate coder, which adapts to the constant-rate channel by using a "bit buffer" and "back pointers." Each header signals the start of another block of audio signal. However, due to the Layer 3 syntax, the frame of data associated with that next block of audio signal may be in a prior segment of the bitstream, pointed to by the back pointer. See Figure 3.7-11 and specifically the curved arrows pointing to main_data_begin.

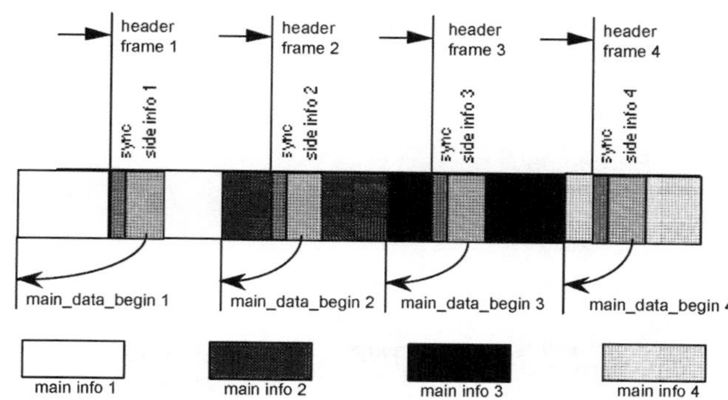

FIGURE 3.7-11 MPEG-1 Layer 3 frame structure during active use of bit buffer.

MPEG-4 Advanced Audio Coding

MPEG-4 Advanced Audio Coding (AAC) is a sophisticated perceptual coder that builds upon the principles of MPEG-1 Layer 3 in a number of ways to increase coding efficiency. First, it has a filterbank that provides higher frequency resolution (1024 subbands compared to 576 in Layer 3) so as to better exploit signal redundancy and also signal irrelevancy at low frequencies. Second, it has improved joint stereo coding, in that coding decisions can be made on a group-of-subbands basis rather than for the entire block of subbands (as in Layer 3). Third, it has improved Huffman coding, in that it uses more codebooks and they better represent a range of signal statistics. Finally, it incorporates new coding tools: Temporal Noise Shaping (TNS) and Perceptual Noise Substitution (PNS). With its coding efficiency, an AAC system operating at 96 kbps produces the same sound quality as ISO/MPEG-1 Layer 2 operating at 192 kbps—a 2:1 reduction in bit rate. A block diagram of the AAC system is shown in Figure 3.7-12.

AAC was first standardized in 1997 as part of MPEG-2. MPEG-4 AAC is equivalent to MPEG-2 Low Complexity AAC (the most widely used version of MPEG-2 AAC), with the addition of the PNS tool and the adoption of the MPEG-4 system layer and file format.

MPEG-2 AAC provides the capability of up to 48 main audio channels, 16 low frequency effects channels, 16 overdub/multilingual channels, and 10 data streams. By comparison, ISO/MPEG-1 Layer 1 provides 2 channels, and MPEG-2 Layer 2 provides 5.1 channels (maximum). AAC is not backward compatible with the Layer 1 and Layer 2 codecs.

AAC Encoder

Each of the blocks in the encoder is described here.

- *Filterbank:* AAC uses a resolution-switching filterbank which can switch between a high-frequency-

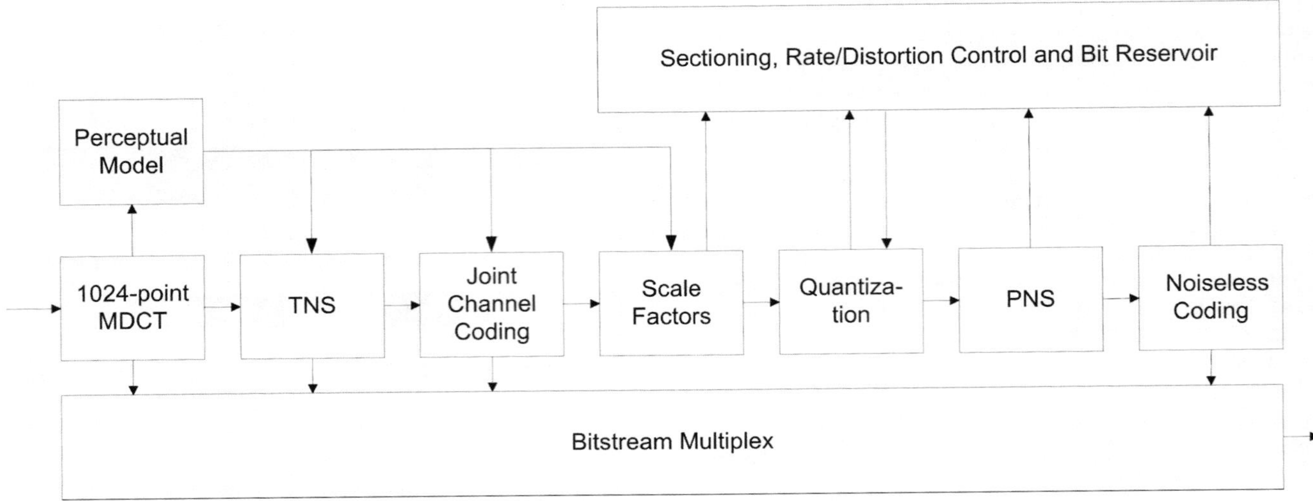

FIGURE 3.7-12 Block diagram of the MPEG-4 AAC encoder.

resolution mode of 1024 bands (for maximum statistical gain during intervals of signal stationarity) and a high-time-resolution mode of 128 bands (for maximum time-domain coding error control during intervals of signal nonstationarity). The Modified Discrete Cosine Transform (MDCT) is used to implement the filterbank, which has the properties of perfect reconstruction of the output (when no quantization is present), critical sampling (so that the number of new PCM samples processed by the transform exactly equals the number of time/frequency coefficients produced), and 50% overlapping transform windows (which provide signal smoothing on reconstruction).

- *Temporal Noise Shaping (TNS):* The TNS tool modifies the filterbank characteristics so that the combination of the two tools is better able to adapt to the time/frequency characteristics of the input signal. It shapes the quantization noise in the time domain by doing an open loop linear prediction in the frequency domain so that it permits the coder to exercise control over the temporal structure of the quantization noise within a filterbank window. TNS is a new technique which has proved very effective for improving the quality of signals such as speech when coded at low bit rates.
- *Perceptual model:* The perceptual model estimates the masking threshold, which is the level of noise that is subjectively just noticeable given the current input signal. Because models of auditory masking are primarily based on frequency domain measurements, these calculations typically are based on the short-term power spectrum of the input signal. Threshold values are adapted to the time/frequency resolution of the filterbank outputs. The threshold of masking is calculated relative to each frequency coefficient for each audio channel for each frame of input signal so that it is signal-dependent in both time and frequency.
- *Joint channel coding:* This block actually comprises three tools—intensity coding, mid/side (M/S) stereo coding (also known as "Sum/Difference coding"), and coupling channel coding—all of which seek to protect the stereo or multichannel signal from noise imaging, while achieving coding gain based on exploiting the correlation between two or more channels of the input signal. M/S stereo coding, intensity stereo coding, and L/R (independent) coding can be combined by selectively applying them to different frequency regions, and by using these tools, it is possible to avoid expensive overcoding when using Binaural Masking Level Depression to correctly account for noise imaging and very frequently to achieve a significant savings in bit rate.
- *Perceptual Noise Substitution (PNS):* This tool identifies segments of spectral coefficients that appear to be noise-like and codes them as random noise. It is extremely efficient in that, for the segment, all that needs be transmitted is a flag indicating that PNS is used and a value indicating the average power of the noise. The decoder reconstructs an estimate of the coefficients using a pseudo-random noise generator weighted by the signaled power value.
- *Scale factors, quantization, coding, and rate/distortion control:* The spectral coefficients are coded using one quantizer per scale factor band, which is a division of the spectrum roughly equal to one-third Bark. The psychoacoustic model specifies the quantizer step size (inverse of the scale factor) per scale factor band. As with MPEG-1 Layer 3, AAC is an instantaneously variable rate coder that similarly uses a bit reservoir. If the coded audio is to be transmitted over a constant rate channel, then the rate/distortion module adjusts the step sizes and number of quantization levels so that a constant rate is achieved.

The quantization and coding processes work together. The first quantizes the spectral components, and the second applies Huffman coding to vectors of quantized coefficients in order to extract additional redundancy from the nonuniform probability of the quantizer output levels. In any perceptual encoder, it is very difficult to control the noise level accurately, while at the same time achieving an "optimum quantizer" (in the minimum mean square error sense). It is, however, quite efficient to allow the quantizer to operate unconstrained and to then remove the redundancy in the quantizer outputs through the use of entropy coding.

The noiseless coding segments the set of 1024 quantized spectral coefficients into sections, such that a single Huffman codebook is used to code each section. For reasons of coding efficiency, section boundaries can only be at scale factor band boundaries so that for each section of the spectrum, one must transmit the length of the section, in scale factor bands, and the Huffman codebook number used for the section. Sectioning is dynamic and typically varies from block to block, such that the number of bits needed to represent the full set of quantized spectral coefficients is minimized.

- The rate/distortion tool adjusts the scale factors such that more (or less) noise is permitted in the quantized representation of the signal which, in turn, requires fewer (or more) bits. Using this mechanism, the rate/distortion control tool can adjust the number of bits used to code each audio frame and hence adjust the overall bit rate of the coder.
- *Bitstream multiplexer:* The multiplexer (MUX) assembles the various tokens to form a coded frame, or access unit. An access unit contains all data necessary to reconstruct the corresponding time-domain signal block. The MPEG-4 system layer specifies how to carry the sequence of access units over a channel or store them in a file (using the MPEG-4 file format).

AAC has a flexible bitstream syntax for a coded frame that permits up to 48 main channels and up to 16 LFE channels to be carried in an access unit, but in a manner that does not incur any overhead for the additional channels. In this respect, it is as efficient for mono, stereo, and 5.1 channel representations.

MPEG-4 High-Efficiency AAC

The MPEG-4 High-Efficiency AAC (HE-AAC) audio compression algorithm consists of an MPEG-4 AAC *core coder* augmented with the MPEG-4 Spectral Band Replication (SBR) tool. The encoder SBR tool is a preprocessor for the core encoder, and the decoder SBR tool is a postprocessor for the core decoder, as shown in Figure 3.7-13. The SBR tool essentially converts a signal at a given sampling rate and bandwidth into a signal at half the sampling rate and bandwidth, passes the low-bandwidth signal to the core codec, and codes the high-bandwidth signal using a compact parametric representation. The lowband signal is coded by the core coder, and the highband parametric data are transmitted over the channel. The core decoder reconstructs the lowband signal, and the SBR decoder uses the parametric data to reconstruct the highband data, thus recovering the full-bandwidth signal. This combination provides a significant improvement in performance relative to that of the core coder by itself, which can be used to lower the bit rate or improve the audio quality.

SBR Principle

A perceptual audio coder, such as MPEG-4 AAC, provides coding gain by shaping the quantization noise such that it is always below the masking threshold. However, if the bit rate is not sufficiently high, the masking threshold will be violated, permitting coding artifacts to become audible. The usual solution adopted by perceptual coders in this case is to reduce the bandwidth of the coded signal, thus effectively increasing the available bits per sample to be coded. The result will be a cleaner sound, but duller due to the absence of high-frequency components.

The SBR tool gives perceptual coders an additional coding strategy (other than bandwidth reduction) when faced with severe bit rate restrictions. It exploits the human auditory system's reduced acuity to high-frequency spectral detail to permit it to parametrically code the high-frequency region of the signal. When the SBR tool is used, the lower-frequency components of the signal (typically from 0 to between 5 kHz and 13 kHz) are coded using the core codec. Since the signal bandwidth is reduced, the core coder will be able to code this signal without violating the masking threshold. The high-frequency components of the signal are reconstructed as a transposition of the low-frequency components followed by an adjustment of the spectral envelope. In this way, a significant bit rate reduction is achieved while maintaining the same audio quality, or alternatively an improved audio quality is achieved while maintaining the same bit rate.

SBR Technology

Using SBR, the missing high-frequency region of a low-pass filtered signal can be recovered based on the existing low-pass signal and a small amount of side information, or control data. The required control data is estimated in the encoder based on the original wide-band signal. The combination of SBR with a core coder (in this case MPEG-4 AAC) is a dual-rate system, where the underlying AAC encoder/decoder is operated at half the sampling rate of the SBR encoder/decoder. A block diagram of the HE-AAC compression system, consisting of SBR encoder and its submodules, AAC core encoder/decoder, and an SBR decoder and its submodules, is shown in Figure 3.7-13.

A major module in the SBR tool is a 64-band pseudo-quadrature mirror analysis/synthesis filterbank (QMF). Each block of 2048 PCM input samples processed by the analysis filterbank results in 32 subband samples in

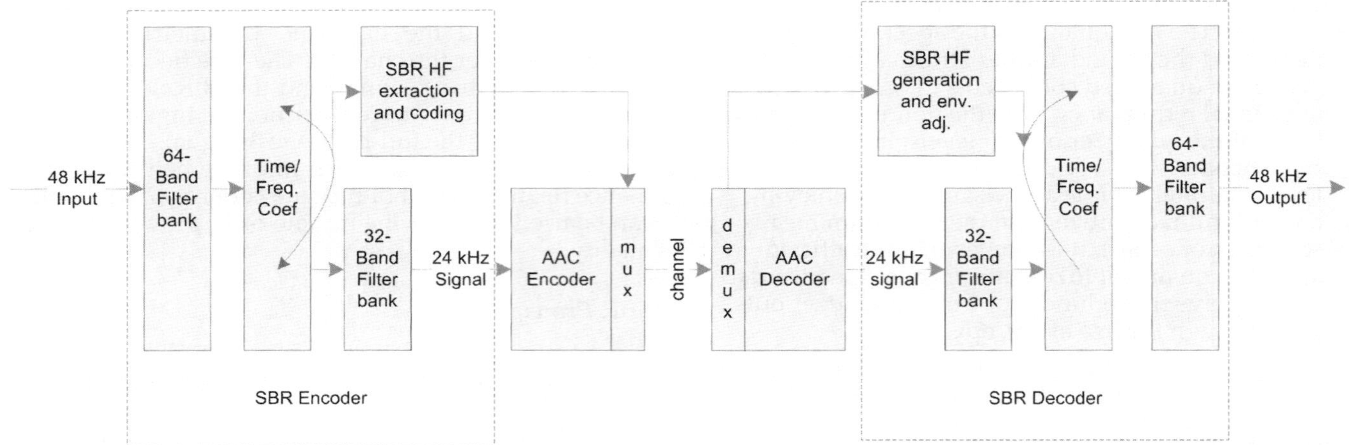

FIGURE 3.7-13 Block diagram of HE-AAC encoder and decoder.

each of 64 equal-width subbands. The SBR encoder contains a 32-band synthesis filterbank whose inputs are the lower 32 bands of the 64 subbands and whose output is simply a band-limited (to one-half the input bandwidth) and half-sampling rate version of the input signal. Actual implementations may use more efficient means to accomplish this, but the illustrated means provides a clearer framework for understanding how SBR works.

The key aspect of the SBR technology is that the SBR encoder searches for the best match between the signal in the lower subbands and those in the higher subbands (indicated by the curved arrow in the Time/Freq. Coef. box in the figure), such that the high subbands can be reconstructed by transposing the low-subband signals up to the high subbands. This transposition mapping is coded as SBR control data and sent over the channel. Additional control parameters are estimated in order to ensure that the high-frequency reconstruction results in a highband that is as perceptually similar as possible to the original highband. The majority of the control data is used for a spectral envelope representation. The spectral envelope information has varying time and frequency resolution such that it can control the SBR process in a perceptually relevant manner while using as small a side information rate as possible. Additionally, information on whether additional components such as noise and sinusoids are needed as part of the highband reconstruction is coded as side information. This side information is multiplexed into the AAC bitstream (in a backward-compatible way).

In the HE-AAC decoder, the bitstream is demultiplexed, the SBR side information is routed to the SBR decoder, and the AAC information is decoded by the AAC decoder to obtain the half-sampling rate signal. This signal is filtered by a 32-band QMF analysis filterbank to obtain the low 32 subbands of the desired time/frequency coefficients. The SBR decoder then decodes the SBR side information and maps the low-band signal up to the high subbands, adjusts its envelope, and adds additional noise and sinusoids if needed. The final step of the decoder is to reconstruct the output block of 2048 time-domain samples using a 64-band QMF synthesis filterbank whose inputs are the 32 low subbands resulting from processing the AAC decoder output and 32 high subbands resulting from the SBR reconstruction. This results in an up-sampling by a factor of two.

Target Applications

The technology is suited for any application where the full audio bandwidth cannot be sufficiently well coded by a waveform coder. This makes it an excellent tool for applications such as

- Digital radio, such as Digital Radio Mondiale;
- Digital TV transmission, such as DVB; and
- Mobile music services such as streaming and music download services.

HDC

The HD Codec (HDC) is a subsystem of the iBiquity Digital Corporation's HD Radio in-band/on-channel (IBOC) digital radio system. As such, all the bit rates described here refer to the bit rates made available to the codec and its associated transport and do not include other overheads associated with other parts of the HD Radio system.

HDC incorporates many techniques to maximize audio quality at lower bit rates. Three of the more important techniques are filterbank optimization, bit allocation and quantization, and Spectral Band Replication (SBR). Filterbank optimization is used to improve encoding of transients in the audio signal (e.g., sharp attacks in castanets, triangles, drums, etc.). Distortion of attacks (abrupt transitions in level and response) creates noticeable artifacts at lower bit rates. In HDC, a signal-adaptive switched filterbank, which switches between two Modified Discrete Cosine

Transforms (MDCT), is employed for analysis and synthesis. This technique substantially reduces the artifacts normally associated with audio attacks and allows for higher frequency resolution and transient control. Sophisticated bit allocation and quantization strategies which rely on nonuniform quantization, analysis-by-synthesis, and entropy coding are introduced to allow reduced bit rates and improved quality. Perhaps the single most important technique is the use of SBR, a method for highly efficient coding of high frequencies that can be used with nearly all audio codecs. By efficient encoding of the high-frequency information, the underlying perceptual codec is responsible only for accurately reproducing the lower frequencies, which results in the significant reduction of noticeable artifacts, especially at the lower bit rates.

An important feature of HDC is that it has been optimized for use in the HD Radio digital radio system. It incorporates powerful error-concealment techniques for mitigating the effects of channel errors. It also offers the ability to split the bitstream into a core bitstream, which is independently decodable, and an enhanced bitstream, providing the optimal audio quality for the prevailing channel conditions and/or interference scenarios. Moreover, HDC supports all bit rates (in 8 bps increments) from 96 kbps down to 11.5 kbps.

Perceptual Coding and HDC

HDC employs source coding techniques to remove signal redundancy and perceptual coding techniques to eliminate signal irrelevancy. In addition, it uses the SBR tool to parametrically code the high frequency portion of the signal spectrum. The basic principle of SBR relies on the fact that the higher one or two octaves are psychoacoustically less relevant but still require a significant part of the overall bit consumption when coded with a perceptual codec. SBR is an efficient method to code the high-frequency part of the spectrum by extracting guidance information of the original high-frequency characteristic, whereas only the low frequency information will be coded by the perceptual coder. In the decoder the high-frequency content will be reconstructed through guided transposition of the low-frequency portion to match the original frequency response. The amount of SBR guidance information is reduced when compared with conventional perceptual coding. With SBR an increase in efficiency can be achieved while maintaining the original sonic quality.

Combined, these methods yield a high compression ratio while ensuring maximal quality in the decoded signal. The result is a high-quality, high-compression-rate coding algorithm for audio signals.

HDC Structure

Figure 3.7-14 shows a more detailed block diagram of the HDC encoding algorithm and illustrates the flow of data between the algorithmic blocks.

Enhanced AC-3

Enhanced AC-3 (E-AC-3) extends and enhances the existing ATSC AC-3 standard. E-AC-3 provides superior compression as compared to AC-3, is backward compatible such that all E-AC-3 decoders can also decode AC-3 bitstreams, and was designed so that E-AC-3 bitstreams can be easily converted into AC-3 bitstreams. This last feature provides a low-complexity means for enhanced services that use the E-AC-3 compression format to support the installed base of nearly 20 million A/V receivers containing 5.1-channel Dolby Digital (AC-3) decoders.

Conversion of bitstreams from E-AC-3 format to AC-3 format utilizes a special form of transcoder that minimizes quality degradations resulting from tandem coding. This is possible due to E-AC-3 and AC-3 sharing the same input signal framing, Modified Discrete Cosine Transform (MDCT), and bit allocation strategies. This permits the use of a frame-synchronous partial E-AC-3 decoder and partial AC-3 encoder that are designed to minimize changes in the original quantization of mantissas, and thus provides significantly less distortion as compared to an asynchronous tandem coding in which the signal is decoded to baseband PCM as an interim step.

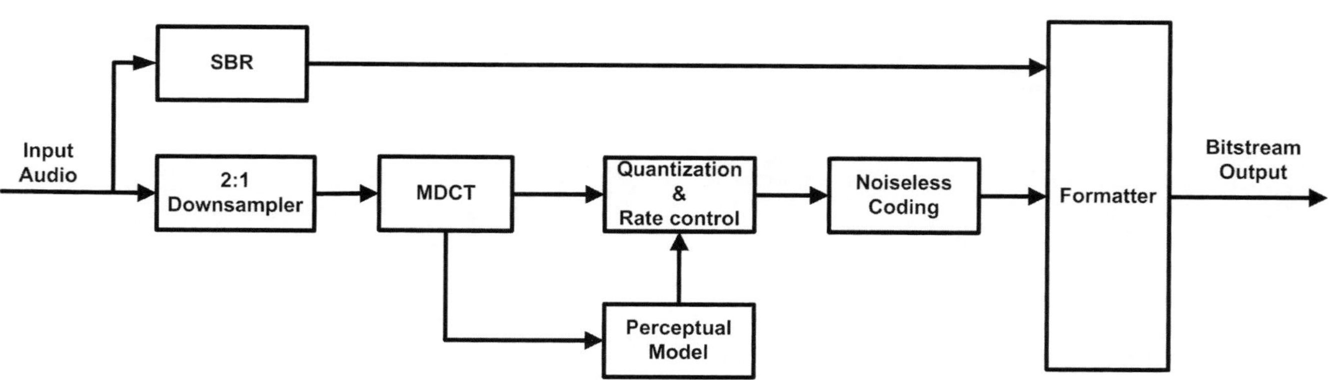

FIGURE 3.7-14 Block diagram of the basic SBR-enhanced HDC audio encoding structure.

A sample application of this special form of tandem coding, or *transcoding*, is "fallback" audio for terrestrial broadcasting. In this application it is desired to have a more robust simulcast of the audio program than provided for in the main channel, for example, by including a lower bit rate signal in conjunction with a more robust but lower throughput modem signal constellation, along with the main channel signal constellation. Whenever transmission channel conditions deteriorate such that the main signal is received with an unacceptable error rate, the receiver "falls back" to the robust, low-rate simulcast signal. If AC-3 is used for the regular audio program, and E-AC-3 is used for the fallback program, then the receiver can synchronously switch from AC-3 to E-AC-3 (since their signal framing is identical). Also, if the receiver audio is nominally being routed (as a coded AC-3 bitstream) to an A/V receiver, then during the fallback mode the E-AC-3 decoder can transcode the robust E-AC-3 signal to AC-3 and hence the A/V receiver need not have an E-AC-3 decoder to accommodate the fallback mode of transmission. This is illustrated in Figure 3.7-15.

New Technology in E-AC-3

E-AC-3 incorporates a number of new coding tools that give greater coding efficiency and increased flexibility. It preserves the framing, metadata, and filter-bank structure found in AC-3, and adds the following:

- *Adaptive hybrid transform (AHT):* AC-3 uses a Modified Discrete Cosine Transform (MDCT) which produces 256 time/frequency samples. When standardized in 1996, this provided a reasonable trade-off between audio coding gain and decoder implementation cost, in terms of computation and memory requirements. However, it is well known that a higher spectral resolution can provide increased compression for stationary signals. Therefore, E-AC-3 incorporates a second transform: a Type II discrete cosine transform (DCT). The two transforms in cascade form the AHT, which takes six sets of 256 MDCT coefficients and transforms them into a single set of 1536 time/frequency coefficients. The bitstream provides signaling such that the AHT can be adaptively enabled or disabled in response to signal statistics.
- *Spectral extension:* This coding tool synthesizes high-frequency transform coefficients without the need to actually code and transmit them. This is done by copying regions, or bands, of low-frequency coefficients to high-frequency regions. The energy envelopes of the transposed coefficients are adjusted to match that of the original spectrum. A shaped noise spectrum can optionally be added, since in audio signals high frequencies tend to be more "noise-like" than low frequencies. The spectral extension technique is schematically illustrated in Figure 3.7-16. The frequency region labeled "A" is the copy region, which is repeatedly transposed, or copied, to the synthesized high-frequency region (indicated by the arrows). Next, the high-frequency envelope is adjusted (indicated by the decreasing envelope with increasing frequency for the copied regions). Finally, a noise spectrum is added (indicated by the hatched regions).
- *Enhanced coupling:* This tool improves the imaging properties of coupled channels by adding phase compensation and decorrelation to the amplitude-envelope processing of conventional coupling. Prior to downmixing the coupled channels to a single composite signal, the encoder derives both amplitude and phase information for each coupled channel. Phase information is obtained by using a Modified Discrete Sine Transform (MDST) in conjunction with the existing MDCT. The decoder uses the information to recover the coupled signals from the composite signal using a combination of amplitude scaling, phase rotation, and decorrelation. The result is an improvement in soundstage imaging as compared to conventional coupling, which allows the technique to be used at lower frequencies than conventional coupling, thus improving coding efficiency.

  Because it may be desirable in some circumstances to use channel coupling for a mid-range portion of the frequency spectrum and spectral extension for the higher-range portion of the frequency spectrum, spectral extension is fully compatible with channel

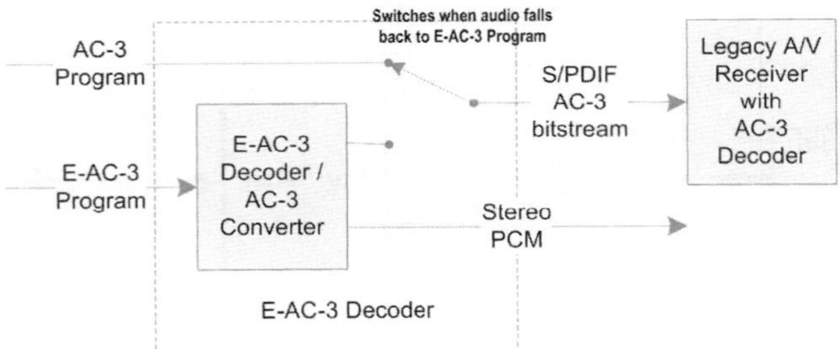

FIGURE 3.7-15 Broadcast receiver that can "fall back" from AC-3 "normal" transmission to E-AC-3 "robust" transmission.

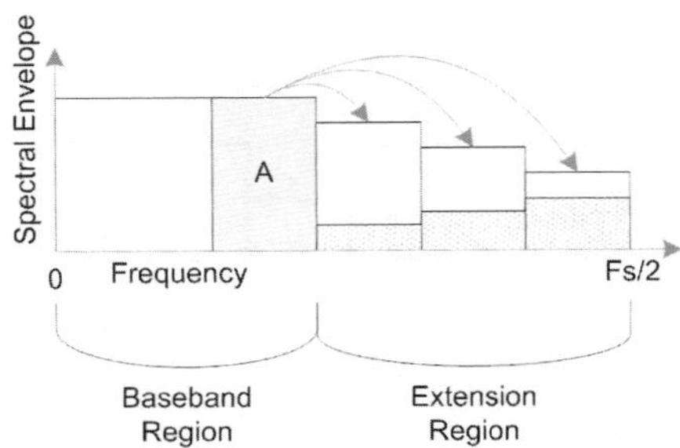

FIGURE 3.7-16 Illustration of spectral extension in E-AC-3.

coupling, with both tools being enabled at the same time, but for different portions of the frequency spectrum.

- *Transient prenoise processing:* Perceptual coders must balance the requirements of high-frequency resolution for maximum redundancy removal during stationary signal intervals against high-time resolution for minimum errors in irrelevancy reduction during nonstationary signal intervals. E-AC-3 can disable the AHT and use the 256-length window "short-block" mode of the AC-3 transform during signal nonstationarity. However, at low bit rates even this level of time resolution may not be sufficient to guard against transient "pre-echo" coding artifacts.

  The transient prenoise tool detects this condition and, for each audio channel, replaces the noise with a portion of the signal from the prior transform block. This replacement is done in the time domain by extending the signal from the prior transform block into the block with the transient up to the point of the onset of the transient. The extension is done using aspects of time-scale modification, and the exact channels and signal intervals that receive the prenoise processing are determined using aspects of auditory scene analysis.
- *Increased number of audio channels and increased bit rate range:* Encoding of up to 13.1 channel audio programs is supported, which easily encompasses the 6.1 or 7.1 channel audio programs that are envisioned today. The bitstream can range in rate from 32 kbps to 6.144 Mbps. The higher rates support not only the larger number of audio channels, but also a higher rate per channel, as is beneficial in applications that serve legacy equipment by transcoding from E-AC-3 to AC-3.

MPEG Surround

The MPEG Surround algorithm achieves compression by encoding an N-channel (e.g., 5.1 channels) audio signal as a two-channel stereo (or mono) audio signal plus side information in the form of a secondary bitstream which contains "steering" information. The stereo (or mono) signal is encoded and the compressed representation sent over the transmission channel along with the side information, such that a spatial audio decoder can synthesize a high-quality multichannel audio output signal in the receiver. Although the audio compression technology is quite flexible, such that it can take N input audio channels and compress those to one or two transmitted audio channels plus side information, discussed here is only a single sample configuration, which is the case of coding a 5.1 channel signal as a compressed stereo signal plus some side information. This configuration is illustrated in Figure 3.7-17.

Referring to the figure, the MPEG Surround encoder receives a multichannel audio signal, x_1 to $x_{5.1}$, where x is a 5.1 channel (left, center, right, left surround, right surround, LFE) audio signal. Critical in the encoding process is that a "downmix" signal is derived from the multichannel input signal, and that this downmix signal is compressed and sent over the transmission channel rather than the multichannel signal itself. A well-optimized encoder is able create a downmix that is, by itself, a faithful two-channel stereo equivalent of the multichannel signal, and which also permits the MPEG Surround decoder to create a perceptual equivalent of the multichannel original. The downmix signal, x_{t1} and x_{t2} (two-channel stereo), is compressed and sent over the transmission channel. The MPEG Surround encoding process is agnostic to the audio compression algorithm used. It could be any of a number of high-performance compression algorithms such as MPEG-1 Layer 3, MPEG-4 AAC or

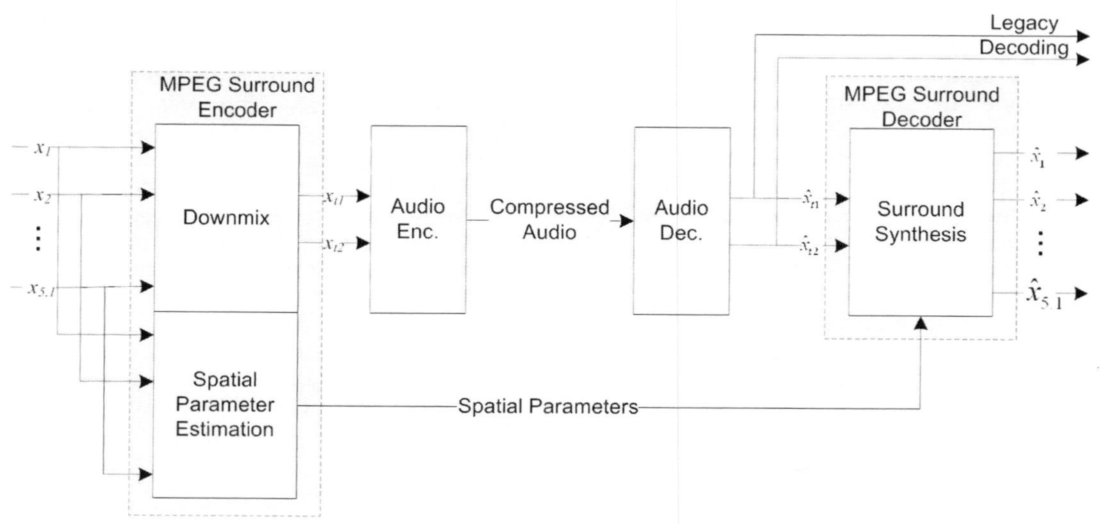

FIGURE 3.7-17 Principle of MPEG Surround.

MPEG-4 High-Efficiency AAC, HDC, or even PCM. The audio decoder reconstructs the downmix as $\hat{x}_{t1}$ and $\hat{x}_{t2}$. Legacy systems would stop at this point; otherwise, this decoded signal plus the spatial parameter side information are sent to the MPEG Surround decoder, which reconstructs the multichannel signal $\hat{x}_1$ to $\hat{x}_{5.1}$.

The heart of the encoding process is the extraction of spatial cues from the multichannel input signal that captures the salient perceptual aspects of the multichannel sound image (this is also referred to as "steering information" in that it indicates how to "steer" sounds to the various loudspeakers). Since the input to the MPEG Surround encoder is a 5.1-channel audio signal that is mixed for presentation via loudspeaker, sounds located at arbitrary points in space are perceived as phantom sound sources located between loudspeaker positions. Because of this, MPEG Surround computes parameters relating to the differences between each of the input audio channels. These cues are composed of

- Channel Level Differences (CLD), representing the level differences between pairs of audio signals;
- Inter-Channel Correlations (ICC), representing the coherence between pairs of audio signals;
- Channel Prediction Coefficients (CPC), able to predict an audio signal from others; and
- Prediction error (or residual) signals, representing the error in the parametric modeling process relative to the original waveform.

A key feature of the MPEG Surround technique is that the transmitted downmix is an excellent stereo presentation of the multichannel signal. Hence, legacy stereo decoders do not produce a "compromised" version of the signal relative to MPEG Surround decoders, but rather the very best version that stereo can render. This is vital, since stereo presentation will remain pervasive due to the number of applications for which listening is primarily via headphones (e.g., portable music players).

In the decoder, spatial cues are used to upmix the stereo transmitted signal to a 5.1 channel signal. This operation is done in the time/frequency (T/F) domain, as is shown in Figure 3.7-18. Here, an analysis filterbank converts the input signal into two channels of T/F representation, where the upmix occurs (schematically illustrated in Figure 3.7-19), after which the synthesis filterbank converts the six channels of T/F data into a 5.1 channel audio signal. The filterbank must have frequency resolution comparable to that of the human auditory system and must be oversampled so that processing in the time/frequency domain does

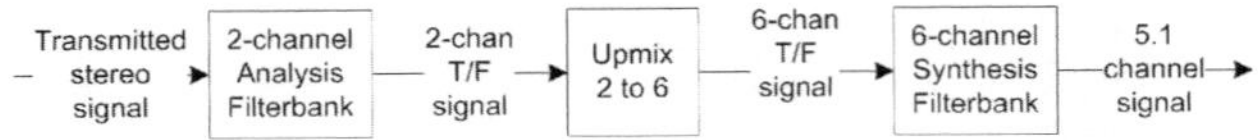

FIGURE 3.7-18 Block diagram of MPEG Surround spatial synthesis.

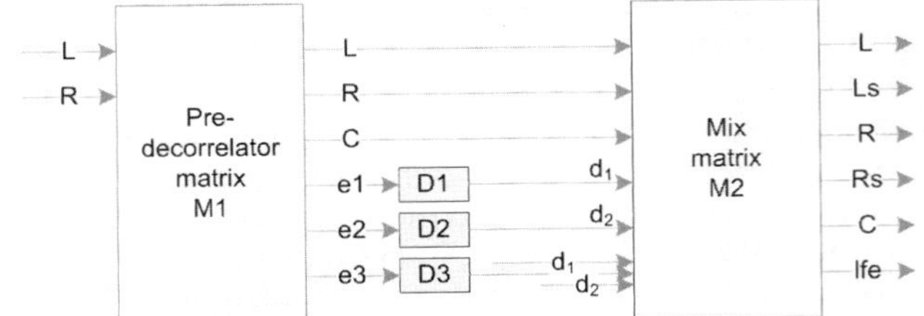

FIGURE 3.7-19 Block diagram of MPEG Surround upmix.

not introduce aliasing distortion. MPEG Surround uses the same filterbank as is used in HE-AAC, but with division of the lowest frequency bands into additional subbands using an MDCT.

The upmix process applies mixing and decorrelation operations to regions of the stereo T/F signal to form the appropriate regions of the 5.1 channel T/F signals. This can be modeled as two matrix operations (M1 and M2) plus a set of decorrelation filters (D_1–D_3), all of which are time-varying. Note that in the figure the audio signals (L, R, C, Ls, Rs, LFE) are in the T/F domain.

In addition to encoding and decoding multichannel material via the use of side information, as just described, the MPEG Surround standard also includes operating modes that are similar to conventional matrix surround systems. It can encode a multichannel signal to a matrixed stereo signal, which it can decode back to the multichannel signal. This mode is MPEG Surround with "zero side information," and is fully interoperable with conventional matrixed surround systems. However, MPEG Surround has an additional feature in this mode: It can produce a matrix-compatible stereo signal and, if there is a side information channel, it can transmit information permitting the MPEG Surround decoder to "un-do" the matrixing, apply the normal MPEG Surround multichannel decoding, and present a multichannel output that is superior to that of a matrix decoder upmix.

MPEG Surround has a unique architecture that, by its nature, can be a bridge between the distribution of stereo material and the distribution of multichannel material. The vast majority of audio decoding and playback systems is stereo, and MPEG Surround maintains compatibility with that legacy equipment. Furthermore, since MPEG Surround transmits an encoded stereo (or mono) signal plus a small amount of side information, it is compatible with most transmission channels that are currently designed to carry compressed stereo (or mono) signals. For multichannel applications requiring the lowest possible bit rate, MPEG Surround based on a single transmitted channel can result in a bit rate savings of more than 80% as compared to a discrete 5.1 multichannel transmission.

Target Applications

Perhaps the greatest potential for multichannel digital audio broadcast is playback in the automobile, since the majority of radio listening occurs in cars. The automobile environment is quite suitable for enjoying multichannel music, in that many automobiles already have five or more loudspeakers and a subwoofer and, in contrast to a home environment, the automotive listener is in a fixed position relative to the loudspeakers. Although automobiles are a difficult environment to achieve proper spatial imaging due to a mismatched speaker and listener position, surround sound greatly improves this situation in that it provides a larger optimum listening area (i.e., "sweet spot").

This suggests that surround sound may be an excellent match to terrestrial digital radio broadcasting in that it provides, in addition to the existing benefits of delivering a digitally coded stereo signal to legacy receivers, an additional benefit to new receivers: highly discrete multichannel sound. MPEG Surround requires almost no additional resources to transmit the multichannel signal as compared to a stereo signal, and the low bit rate spatial side information could be transmitted using existing data fields in the transmission multiplex.

QUALITY MEASUREMENTS

Perceptual audio coding has revolutionized the processing and distribution of digital audio signals. However, one aspect of this technology that is rarely discussed is how difficult it is to determine the quality of perceptually coded signals. Audio professionals could benefit from a simple yet accurate method for signal characterization, in that it would provide a reliable means to check the audio quality within a given facility.

Most often quality assessment involves *subjective evaluations* of audio quality, in which groups of listeners compare reference audio material to coded audio material and then judge the level of impairment caused by the coding process. It is widely acknowledged that whenever an audio processing system involves perceptual audio coding, subjective evaluation is the most reliable means of assessing audio quality. However, such a procedure is both time consuming and costly. A relatively new alternative is to use fully objective assessment methods. Although these are not considered as accurate as subjective methods (when perceptual coding is involved), they are fast and inexpensive, and can be performed on a personal computer in real time. Both subjective and objective assessment methods will be reviewed in the following sections.

Subjective Quality Measurements

In subjective quality measurement, human judgment determines the audio quality rating. Typically, a set of relevant audio test items (informally referred to as *audio clips* or *audio cuts*) are identified, the systems to be evaluated process the test items, and a set of subjects listen to each processed item and judge its audio quality. For a given audio processing system, the average score of all listeners is the score given to that system. This is intuitively satisfying, in that humans, after all, are the ultimate consumers of audio programs. However, humans demonstrate both "noise" in their judgments and also personal differences in taste, which lead to random and systematic differences in listener judgments, respectively. Accurate assessment is obtained only by using a relatively large number of subjects, such that these differences are reduced by averaging over the set of subject responses.

A procedure for the subjective assessment of very high-quality systems has been standardized in ITU-R Recommendation BS.1116-1, "Methods for the Subjective Assessment of Small Impairments in Audio Systems Including Multichannel Sound." This standard specifies the test environment, the test procedure, and the analysis process. Concerning environment, it is paramount that the tests be conducted in an acoustically conditioned listening room or sound booth. The listening space must have little to no background noise or reverberation so that potential audio processing distortions are not masked by the listening environment itself. It also specifies the configuration of loudspeakers and the position of the listener.

The procedure requires that the listener hear the unprocessed audio ("reference") followed by two additional audio samples ("A" and "B"), one of which is the reference and one of which is processed. For this reason it is also called the "triple-stimulus hidden reference" method. Using this method of presentation, both the processed and reference items are in the listener's short-term aural memory, permitting the assessment of very small differences in the processed item relative to the reference item. The listener grades both A and B, and one must receive a score of 5.0 indicating that it is judged to be the hidden reference. In order to eliminate any systematic errors in assessment, the location of the hidden reference (A or B) and the order of presentation of test items are randomized, preferably with different randomizations for each listener. Any difference between the reference and the processed audio is to be regarded as a distortion or impairment, which listeners rate using the ITU-R five-grade impairment scale shown in Figure 3.7-20.

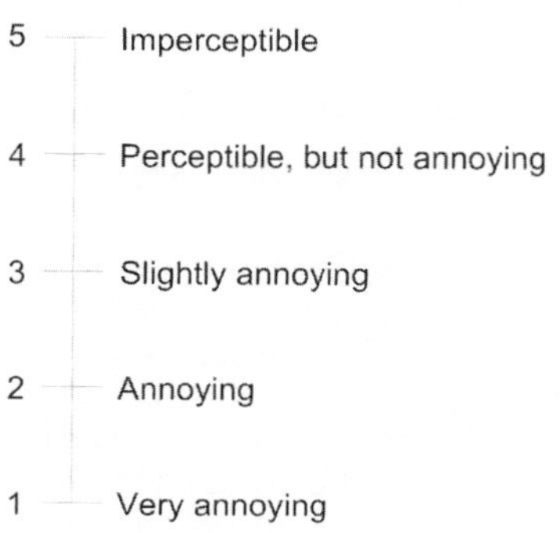

FIGURE 3.7-20 ITU-R five-grade impairment scale.

The results are analyzed by computing the sample mean and 95% confidence interval of the mean score associated with a given processing system as taken, for example, over all listeners or all listeners and all test items. When analyzing the results of the assessment, *diffgrades* are used, which are equal to the grade given to the coding system minus the grade given to the hidden reference. Systems with scores that approach zero are of very high quality. Figure 3.7-21 shows an example of a subjective test result using the BS.1116 methodology. The 95% confidence interval is indicated by the vertical stroke with end caps, and the sample mean is indicated by the marker graphic.

In many broadcast satellite or terrestrial digital radio applications, the target bit rate for delivery of audio programs is such that near-transparent audio quality is not possible. A procedure for the subjective assessment of intermediate quality systems has been standardized in ITU-R Recommendation BS.1534-1 (3/2001), "Method for the Subjective Assessment of Intermediate Audio Quality." It is also referred to as MUlti-Stimulus with Hidden Reference and Anchor (MUSHRA). This relies on BS-1116 for the specification of acoustic environment and other aspects of the testing procedure. As with BS-1116, any difference between the reference and the processed audio is regarded as a distortion or degradation, which listeners rate using the 100-point scale shown in Figure 3.7-22.

The procedure requires that, for each test item, the subject be able to listen to the unprocessed test item

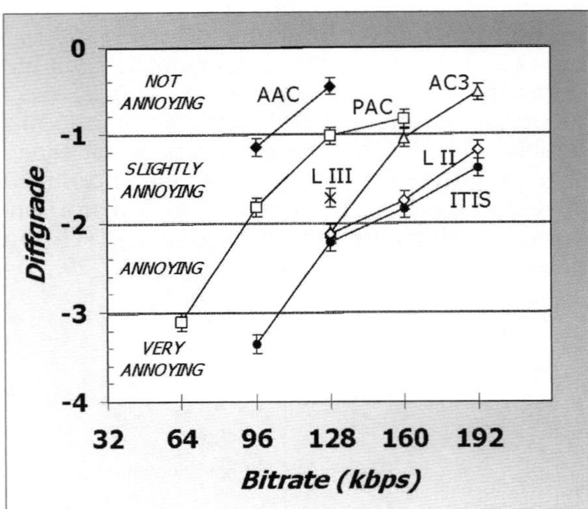

FIGURE 3.7-21 Example of BS.1116 Subjective Assessment (data taken from March 1998 *Journal of the Audio Engineering Society*). The processed audio material is stereo. In the graph AAC is MPEG-2 Advanced Audio Coding; PAC is Lucent Perceptual Audio Coding; AC3 is Dolby AC-3; L II and L III are software implementations of MPEG-1 Layer 2 and 3, respectively; and ITIS is a hardware implementation of MPEG-1 Layer 2.

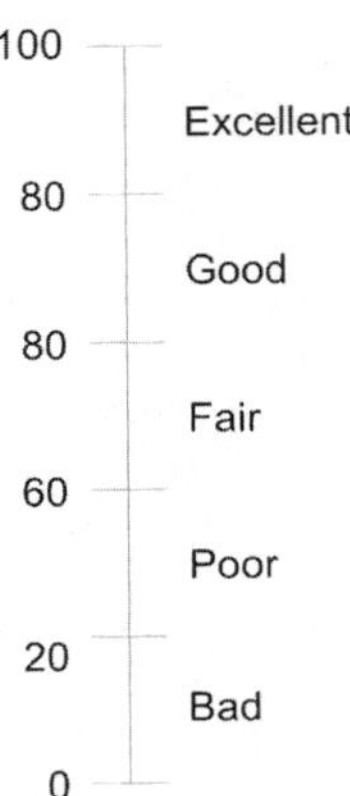

FIGURE 3.7-22 The 100-point scale used in MUSRA testing.

(the "open reference") or any of the processed test items, and to grade all processed items before proceeding to the next item. The listener can listen to the reference and processed items in any order and as often as desired in the course of making the evaluations. This is possible only because of an innovation in the MUSHRA methodology, which recommends that computer-controlled presentation and scoring be used. An example of a MUSHRA graphical user interface is shown in Figure 3.7-23 (from the Audio Research Labs subjective quality assessment software tool STEP).

The systems to be evaluated using the MUSHRA procedure are often a considerable distance from the open reference, in terms of points on the grading scale. While the open reference represents the high end of the grading scale (i.e., transparency), a second, lower-

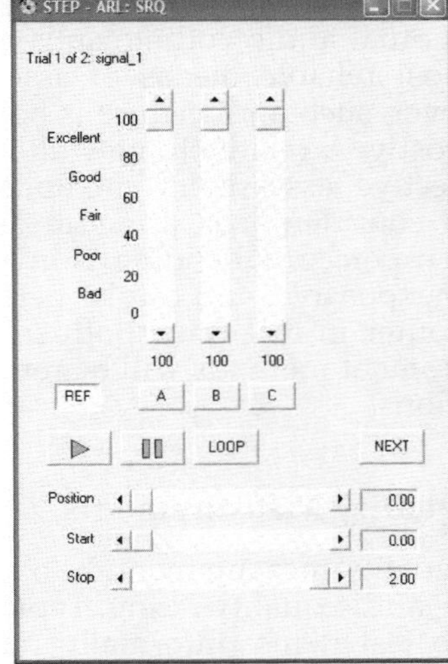

FIGURE 3.7-23 Example of BS-1534 MUSHRA graphical user interface (from Audio Research Labs' software subjective evaluation tool STEP).

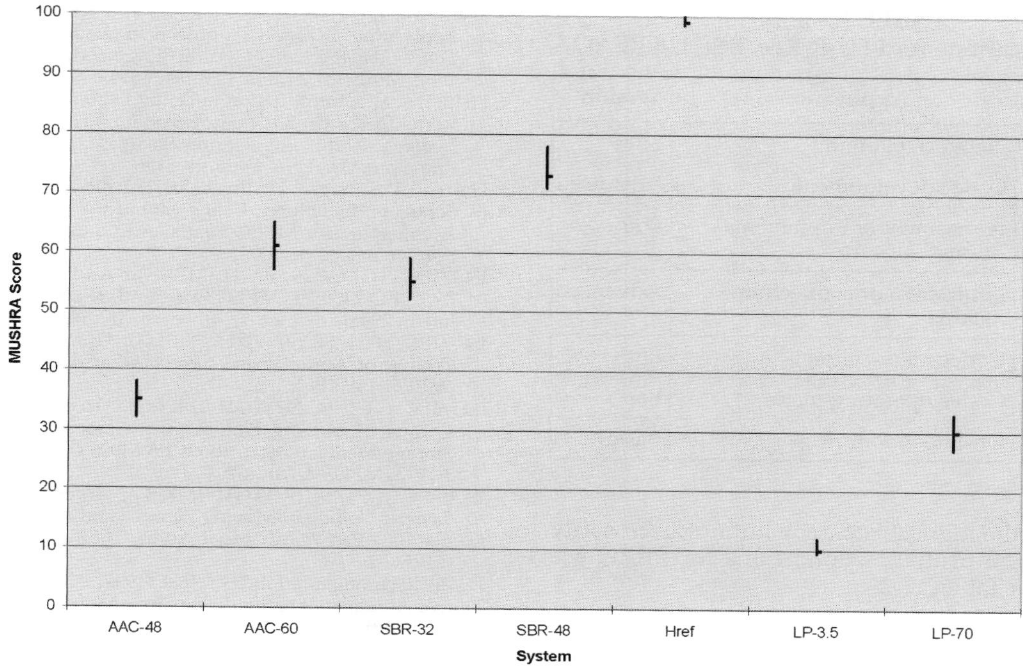

FIGURE 3.7-24 Sample MUSHRA subjective assessment (data taken from MPEG document N6009, October 2003, Brisbane, Australia). The processed audio material is stereo and in the graph AAC-48 is MPEG-4 AAC at 48 kbps; AAC-60 is MPEG-4 AAC at 60 kbps; SBR-32 is MPEG-4 HE-AAC at 32 kbps; SBR-48 is MPEG-4 HE-AAC at 48 kbps; Href is the hidden reference; and LP-35 and LP-70 are 3.5 kHz and 7.0 kHz low-pass filtered versions of the reference, respectively.

quality reference called the *low anchor* is mandated in the procedure in order to give a point of reference at the low end of the scale. Specifically, a 3.5 kHz and optionally a 7.0 kHz low-pass filtered version of the reference are used as low anchors. Additionally, a hidden reference is included, which is often used as a means to test listener reliability.

For each test item, the listener must rate all of the processed items, which are denoted by letters (A, B, C, etc.) underneath the vertical scroll bars, while the reference is indicated by the button REF to the left of the letter buttons. The scroll bars are used to designate the score for each item. For the example shown in Figure 3.7-23, clicking on any of the buttons instantly switches the audio to that processed item. In this way, the listener can repeatedly compare and contrast the processed items as a means to converge on a final set of scores. When the listener is satisfied with the scoring, the NEXT button is pressed to proceed to the next trial.

Figure 3.7-24 shows an example of a subjective test result using the MUSHRA methodology. The graph in this figure shows results aggregated over the listener population which participated in this test. The 95% confidence interval (a statistical measure of reliability) is indicated by the vertical stroke, and the sample mean (i.e., the average value of the results over the listener population) is indicated by the horizontal tick mark.

Quality measurements made with properly executed subjective evaluations are widely accepted and have been used for a variety of purposes, from determining which of a group of perceptual coders performs best, to assessing the overall performance of an audio broadcasting system.

Objective Quality Measurements

Traditional objective measures of audio performance, such as signal-to-noise ratio or total harmonic distortion, are not reliable measures of the audio quality delivered by a perceptual audio encoder. To remedy this situation, the ITU-R developed Recommendation BS.1387-1, "Method for Objective Measurement of Perceived Audio Quality (PEAQ)." The method requires feeding the original audio signal and the processed audio signal into the PEAQ measurement device. For each block of the signals, PEAQ calculates the degree to which noise in the processed signal is above a masking threshold derived from the original signal and accumulates parameters based on that impairment over the duration of the signal. From the impairment parameters, an estimated total signal quality is computed.

PEAQ can be realized in software or hardware. It offers a basic version, which is designed to support real-time implementations, and an advanced version

TABLE 3.7-2
Target Applications for ITU-R Rec. BS.1116 PEAQ

Category	Application	Version
Development	Network planning	Both
	Codec development	Advanced
Diagnostic	Assessment of implementations	Both
	Equipment or connection status	Advanced
	Codec identication	Both
Operational	Codec operating mode	Basic
	On-line monitoring	Basic

optimized for the highest accuracy but not necessarily in real time. The primary applications for PEAQ are summarized in Table 3.7-2.

SUMMARY

In the design of an audio processing chain, a balance must be struck between the degree of compression selected and the level of distortion that can be tolerated. This is true whether the processing chain is the result of a single coding pass or the result of a number of passes, such as in a complex audio production and delivery system. There have been many outstanding successes for digital audio data compression in communications and storage, and as long as the limitations of the various compression systems are fully understood, the number of successful implementations will continue to grow in number.

Looking to the future, it is clear that the processing power of DSP chips will continue to increase, even as their price continues to drop. This will permit ever more complex audio compression algorithms to be considered for practical implementations. Therefore, the level of audio compression available for use in commercial audio production and broadcast applications can be expected to continue to increase.

Bibliography

ATSC, *Digital Audio Compression Standard (AC-3), Revision A,* Advanced Television Systems Committee, Washington, D.C., Doc. A/52A, August 20, 2001.

ATSC, *Digital Television Standard (A/53), Revision D, Including Amendment No. 1,* Advanced Television Systems Committee, Washington, D.C., Doc. A/53D, July 19, 2005.

Bosi, M., Brandenberg, K., Quackenbush, S., Fielder, L., Akagiri, H., Fuchs, H., Dietz, M., Herre, J., Davidson, G., and Oikawa, Y., "ISO/IEC MPEG-2 Advanced Audio Coding," *J. Audio Eng. Soc.,* vol. 45, no. 10, October 1997, pp. 789–814.

Brandenburg, K., "Introduction to Perceptual Coding," in *Collected Papers on Digital Audio Bit-Rate Reduction,* N. Gilchrist and C. Grewin, Eds., 1996, pp. 23–30.

Brandenburg, K., and Stoll, G. "ISO-MPEG-1 Audio: A Generic Standard for Coding of High Quality Digital Audio," *92nd AES Convention Proceedings,* Audio Engineering Society, New York, NY, 1992, revised 1994.

Dietz, M., Liljeryd, L., Kjörling, K., and Kunz, O., "Spectral Band Replication, a Novel Approach in Audio Coding," *112th AES Convention Proceedings,* Audio Engineering Society, May 10–13, 2002, Munich, Germany. Preprint 5553.

Fielder, et al., "Introduction to Dolby Digital Plus, an Enhancement to the Dolby Digital Coding System," *117th AES Convention Proceedings,* Audio Engineering Society, October 28–31, 2004, San Francisco, CA, USA. Preprint 6196.

Herre, J., Purnhagen, H., Breebaart, J., Faller, J. Disch, J., Kjörling, K., Schuijers, E., Hilpert, J., and Myburg, F., "The Reference Model Architecture for MPEG Spatial Audio Coding," *118th AES Convention,* Barcelona 2005.

ISO/IEC JTC1/SC29/WG11 (MPEG), "Audio Coding Technology," ISO/IEC 23003-1, MPEG Surround. (Expected to be published late in 2006).

ISO/IEC JTC1/SC29/WG11 (MPEG), "Information Technology—Coding of Audio-Visual Objects—Part 3: Audio," 14496-3:2006 (MPEG-4 Audio).

ISO/IEC JTC1/SC29/WG11 (MPEG), "Information Technology—Coding of Moving Pictures and Associated Audio for Digital Storage Media at up to about 1.5 Mbit/s—Part 3: Audio," 11172-3:1993 (MPEG-1 Audio).

ISO/IEC JTC1/SC29/WG11 (MPEG), "Information Technology—Generic Coding of Moving Pictures and Associated Audio Information—Part 7: Advanced Audio Coding (AAC)," 13818-7:2006 (MPEG-2 AAC).

ITU-R Recommendation BS.1116-1 (1997), "Methods for the Subjective Assessment of Small Impairments in Audio Systems including Multichannel Sound."

ITU-R Recommendation BS.1387-1 (2001, November), "Method for Objective Measurement of Perceived Audio Quality."

ITU-R Recommendation BS.1534-1 (2001, March), "Method for the Subjective Assessment of Intermediate Audio Quality."

Jayant, N. S., and Noll, P., *Digital Coding of Waveforms.* Englewood Cliffs, NJ: Prentice Hall, 1984.

Moore, B. C. J., "Masking in the Human Auditory System," in *Collected papers on Digital Audio Bit-Rate Reduction,* N. Gilchrist and C. Grewin, Eds., 1996, pp. 9–19.

National Radio Systems Committee, *NRSC-5-A In-Band/On-Channel Digital Radio Broadcasting Standard.* September 2005.

Painter, T., and Spanias, A, "Perceptual Coding of Digital Audio," *Proc. of the IEEE,* vol. 88, no. 4, April 2000, pp. 451–513.

Recommendation ITI-R BS.1115-1 "Low Bit-Rate Audio Coding."

Todd, C., et. al., "AC-3: Flexible Perceptual Coding for Audio Transmission and Storage," *96th AES Convention Proceedings,* Audio Engineering Society, New York, February 1994. Preprint 3796.

Zwicker, E., and Fastl, H., *Psychoacoustics Facts and Models.* Berlin, Germany: Springer-Verlag, 1990.

CHAPTER

3.8

Transmission Audio Processing

ROBERT ORBAN
Orban/CRL
Tempe, Arizona

FRANK FOTI
Omnia Audio, A Telos Systems Company
Cleveland, Ohio

INTRODUCTION

Transmission audio processing is both an engineering and artistic discipline. The engineering goal is to make the most efficient use of the signal-to-noise ratio and audio bandwidth available from the transmission channel while preventing its overmodulation. The artistic goal is set by the organization using audio processing. It may be to avoid audibly modifying the original program material at all, or it may be to create a distinct *sonic signature* for a broadcast by radically changing the sound of the original. Most broadcasters operate somewhere in between these two extremes.

If the transmitted signal meets regulatory requirements for modulation control and radio frequency (RF) bandwidth, there is no well-defined right or wrong way to process audio. Like most areas requiring subjective, artistic judgment, processing is highly controversial and likely to provoke exceedingly opinionated arguments among its practitioners. Ultimately, the success of the audio processing of a broadcast must be judged by its results—if the broadcast gets the desired audience, then the processing must be deemed satisfactory regardless of the opinions of audiophiles, purists, or others who consider processing an unnecessary evil.

One mark of the professionalism of broadcast engineers is their mastery of the techniques of audio processing. The canny practitioner has a bag of tricks that can be used to achieve the processing goal specified by the station's management, whether it is purist or "squashed against the wall."

FUNDAMENTALS OF AUDIO PROCESSING

Compression reduces the dynamic range of program material by reducing the gain of material whose average or root mean square (RMS) level exceeds the *threshold of compression*. The amount by which the gain is reduced is called the *gain reduction*. Above threshold, the slope of the input/output curve is the *compression ratio*. Low ratios provide loose control over levels but generally sound more natural than high ratios, which provide tight control. The *knee* of the input/output level graph can show an abrupt transition (*hard-knee*) into compression or a gradual transition (*soft-knee*), in which the ratio becomes progressively larger as the amount of gain reduction increases.

The *attack time* is, generally, the time that it takes the compressor to settle to a new gain following a step increase in level. There is no generally agreed upon precise definition for how to measure attack time. Some measure it as the *time constant*—the time necessary for the gain to achieve 67% of its new value. Others measure it as the time for the gain to reach 90% of its new value for a given amplitude step (often 10 dB). The *release time* is the time necessary for the gain to recover to within a certain percent of its final value after the level of the input signal to the compressor has been reduced below the compression threshold. It is sometimes convenient to specify the release time in dB per second if the shape of the release time is a straight line on a dB-*versus*-time graph; however, this shape often is not linear. *Multiple time constant* (sometimes called *automatic*) release time circuits change the release rate (in dB/sec) according to the history of the program and according to how much gain reduction is

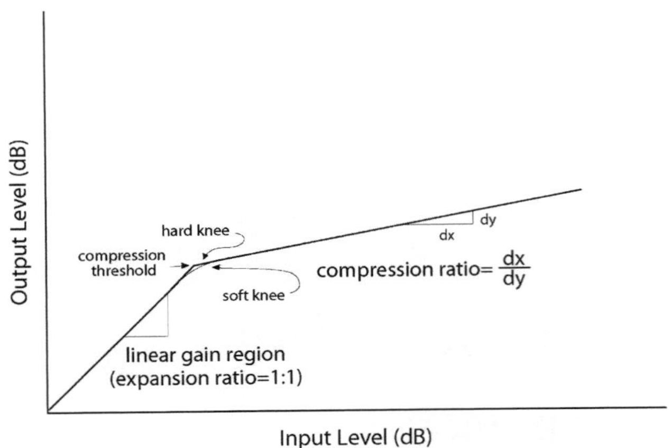

FIGURE 3.8-1 Input *versus* output levels for compressors.

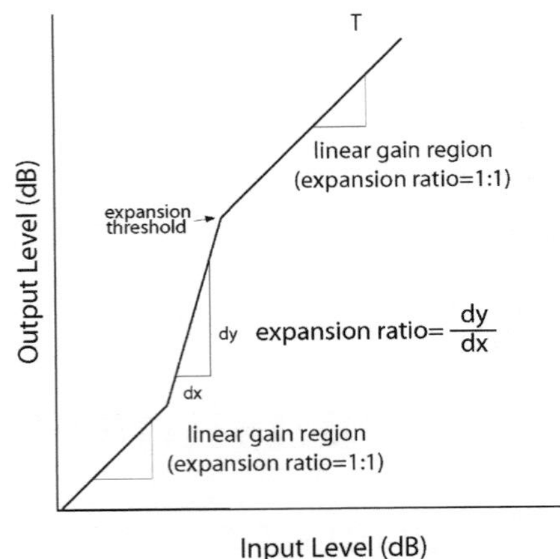

FIGURE 3.8-2 Input *versus* output levels for expanders.

in use. For example, the release time will temporarily speed up after an abrupt transient to prevent a hole from being punched in the program by the gain reduction. The release time may slow down as 0 dB gain reduction is approached to make compression of wide-dynamic-range program material less obvious to the ear.

Delayed release holds the gain constant for a short time (typically less than 20 msec) after gain reduction has occurred. This prevents fast release times from causing modulation of individual cycles in the program waveform, thus reducing the tendency of the compressor to introduce harmonic or intermodulation distortion when operated with fast attack and release. The foregoing compression parameters are illustrated in Figure 3.8-1.

Expansion

Expansion increases the dynamic range of program material by reducing gain when the program level is lower than the *threshold of expansion* (see Figure 3.8-2). The primary purpose of expansion is to reduce noise, either electronic or acoustic. Expanders are often coupled to compressors so low-level program material is not amplified, thus reducing the noise that would otherwise be exaggerated by the compression. Expanders have attack times, release times, and expansion ratios that are analogous to those for compressors.

Peak Limiting and Clipping

Peak limiting is an extreme form of compression characterized by a very high compression ratio, fast attack time (typically less than 2 μsec), and fast release time (typically less than 200 μsec). In modern audio processing, a peak limiter, by itself, usually limits the peaks of the envelope of the waveform, as opposed to individual instantaneous peaks in the waveform. These are controlled by *clipping*. As a matter of good engineering practice, peak limiters are usually adjusted to produce no more than 6 dB of gain reduction to prevent offensive audible side effects. The main purpose of limiting is to protect a subsequent channel from overload, as opposed to compression, the main purpose of which is to reduce the dynamic range of the program.

Peak clipping is a process that instantaneously chops off any part of the waveform that exceeds the *threshold of clipping*. This threshold can be either symmetrical or asymmetrical around 0 volts. Although peak clipping can be very effective, it causes audible distortion when overused. It also increases the bandwidth of the signal by introducing both harmonic and intermodulation distortion into its output signal. Manufacturers of modern audio processors have therefore developed various forms of *overshoot compensation*, which is essentially peak clipping that does not introduce significant out-of-band spectral energy into its output.

Radio frequency clipping (RF clipping) is peak clipping applied to a single-sideband RF carrier signal (a typical RF carrier frequency is 1 MHz). All clipping-induced harmonics fall around harmonics of the carrier (*e.g.*, for a 1 MHz RF carrier, the first harmonic is at 2 MHz). Upon demodulation, these harmonics remain at high frequencies and are removed by a low-pass filter; thus, RF clipping produces only intermodulation distortion and no harmonic distortion. Ordinary or *audio frequency clipping* (AF clipping) produces both. RF clipping is substantially more effective than AF clipping on voice because intermodulation distortion is considered less objectionable than harmonic distortion in this application. On the other hand, RF clipping is considered much more objectionable than AF clipping on music. The *Hilbert Transform Clipper* combines the features of RF and AF clippers [1]. It acts as an RF clipper below 4 kHz (the region in which most voice energy is located) and as an AF clipper above 4 kHz to prevent excessive intermodulation distortion with music.

Unless a limiter has an attack time of less than about 10 μsec, it will exhibit *overshoots* at its output. If the goal of the processing is to precisely constrain the instantaneous values of the waveform to a given threshold, it usually sounds best to control these overshoots by a limiter with a 2 msec attack time followed by a clipper. Attempting to provide all peak control with the limiter does not sound as good, because the clipper affects only the offending overshoot and does not apply gain reduction to the surrounding signal.

When used in this way, clippers can cause audible distortion on certain program material; however, fast-attack limiters will cause audible clipping of the first half-cycle of certain program material, such as solo piano, harp, and nylon-string acoustic guitar. Such distortion can be eliminated by a *delay line limiter,* sometimes called a *look-ahead limiter* [2]. In its basic form, this device consists of two audio paths. The audio is applied to a pilot limiter, which has a very fast attack time. The gain control signal generated by the pilot limiter is applied to a low-pass filter to smooth its sharp edges and then to the gain control port of the variable gain amplifier that passes the actual program signal. To compensate for the group delay in the control voltage low-pass filter (which delays application of gain control to the audio), the audio is delayed equally by a delay line prior to the input of the variable gain amplifier.

Gating

The two fundamental types of gates are the *compressor gate* and the *noise gate*. The *compressor gate* prevents any change in background noise during pauses or low-level program material by freezing the compressor gain when the input level drops below the *threshold of gating*. Because it produces natural sound, it is very popular in broadcasting. Instead of freezing, many compressor gates will cause the gain to move very slowly to a nominal value (typically 10 dB of gain reduction) if the gating period is long enough. This prevents the compressor from getting stuck with an unusually high or low gain.

The *noise gate* is an expander with a high expansion ratio. Its purpose is to reduce noise. Because it causes gain reduction when the input level drops below a given threshold, the ear is likely to hear the accompanying gain reduction as a fluctuation in the noise level, sometimes called *breathing*. This can sound unnatural; therefore, the noise gate is most useful when applied to a single microphone in a multi-microphone recording. Usually, the other microphones will mask any breathing, yet the noise reduction provided by the noise gate will still be appreciated during quiet program material.

Multiband Compression and Frequency-Selective Limiting

These techniques divide the audio spectrum into several frequency bands and compress or limit each band separately (although some interband coupling may be used to prevent excessive disparity between the gains of adjacent bands). This is the most powerful and popular contemporary audio processing technique, because, when done correctly, it eliminates spectral gain intermodulation. This occurs in a wideband compressor or limiter when a voice or instrument in one frequency range dominates the spectral energy, thus determining the amount of gain reduction. If other, weaker elements are also present, their loudness may be audibly and disturbingly modulated by the dominant element. Particularly unpleasant effects may occur if the dominant energy is in the bass region; the ear is relatively insensitive to bass energy, so the dominant bass energy pushes down the loudness of the midrange, seemingly inexplicably.

Another type of frequency-selective limiting uses a program-controlled filter. The filter's cutoff frequency, its depth of shelving (explained in the next section), or a combination of these parameters are varied to dynamically change the frequency response of the transmission channel. Such program-controlled filters are most often used as *high-frequency limiters* to control potential overload due to preemphasis in preemphasized systems such as FM radio and television audio (National Television System Committee [NTSC] and Phase Alternation Line [PAL]), as well as in frequency modulated transmission channels, such as microwave links and satellite circuits.

Equalization

Equalization is changing the spectral balance of an audio signal and is achieved by use of an *equalizer*. In broadest terms, an equalizer is any frequency-selective network (filter) placed in the signal path. In audio processing, an equalizer is usually a device that can apply a *shelving* or *peaking* curve to the audio. A shelving curve starts off at a certain gain. As frequency changes, the gain increases (boost) or decreases (cut) asymptotically. Finally, the gain shelves off and does not change with further changes in frequency. A peaking curve is bell-shaped on the frequency axis. As opposed to a shelving curve, it has a well-defined peak frequency. The shape of the curve can be uniquely defined by three parameters: *amount of equalization* (in dB), *frequency of maximum equalization* (in Hz), and Q, which is a dimensionless number that describes whether the curve is broad or sharp.

A *parametric equalizer* provides several peaking equalizers in which the user has control of all three parameters. This type of equalizer is generally considered to be the most flexible and musical-sounding equalizer. Some parametric equalizers can also be used as notch filters. A *graphic equalizer* provides a number of peaking equalizers (usually 8 to 31) distributed on frequency centers spaced by octave or fractions of an octave (1/4 or 1/3) throughout the audible range. The controls for the amount of equalization are linear-throw faders, which are arranged on the panel in order of frequency. The positions of the controls, when considered together, thus provide a very rough graphic display of the amount of equalization provided by the entire equalizer. The advantage of a graphic equalizer is that it is easy to understand and quick to adjust. Its primary disadvantage is lack of flexibility. Usually, only the amount of equalization is adjustable, as the *Q* and center frequency are fixed; however, a few manufacturers make parametric equalizers with graphic-style controls which provide the advantages of both types.

Low-pass and *high-pass filters* remove spectrum at the top and bottom of the audible range, respectively. They are usually used to remove unwanted high- or low-frequency noise and can also produce special effects (such as telephone simulation). These filters usually come with their rate of cutoff fixed in multiples of 6 dB/octave, although 12 dB/octave and 18 dB/octave are also popular. In addition, the shape of the region around the cutoff frequency has a considerable effect on the listening quality of such filters. Bessel (constant-delay) filters have a gentle transition into cutoff and sound pleasant and musically neutral. Butterworth (maximally flat magnitude) filters have a sharper transition into cutoff; they are more effective at removing noise than Bessel filters but have a more colored listening quality.

Equalizers are sometimes used online in transmission to create a certain sonic signature for a broadcast. Any of the types above may be used. Commercial audio processors may include equalizers for program coloration or for correcting the frequency response of previous or subsequent transmission links. Sometimes the various bands of a multiband compressor or limiter are used as an equalizer by adjusting the gains of the various bands to achieve the desired equalized frequency response.

Loudness

One of the main uses of audio processing is to increase perceived loudness within the peak modulation constraints of a transmission channel. Assessing the effectiveness of audio processing thus requires a means of measuring loudness. Loudness is subjective; it is the intensity of sound as perceived by the ear/brain system. No simple meter, whether peak program meter (PPM) or volume unit (VU) meter, provides a reading that correlates well to perceived loudness. A meter that purports to measure loudness must agree with a panel of human listeners.

Three important factors correlate to subjective loudness:

- The spectral distribution of the sound energy—The sensitivity of the ear depends strongly on frequency; the ear is most sensitive to frequencies between 2 and 8 kHz, and sensitivity falls off fastest below 200 Hz.
- Whether the sound energy is concentrated in a wide or narrow bandwidth—For a given total sound power, the sound becomes louder as the power is spread over a larger number of *critical bands* (about 1/3 octave). This is referred to as *loudness summation*.
- The duration of the sound—A given amount of sound power appears progressively louder until its duration exceeds about 200 msec, at which point no further loudness increase will occur.

Torick and Jones have published a paper describing a meter for measuring the loudness of broadcast signals [3]. The Federal Communications Commission (FCC) did an informal validation of the results of this meter and concluded that it was effective in assessing whether commercials in television were noticeably louder than the surrounding entertainment programming [4]. Additionally, the independently developed loudness measuring methods of Stevens [5] and of Zwicker [6] have both become international standards.

Recently, the Audio Engineering Society (AES) has published a considerable amount of work on the technology of loudness measurement. This work, done in Europe, was also summarized in the *AES Journal* [7]. These summaries provide references to the original papers. Unfortunately, none of the authors of the new work compared the accuracy of their algorithms with that of the Jones and Torick algorithm.

Dolby Laboratories manufactures a loudness meter [8] that has been demonstrated to be accurate in indicating the loudness of speech but which is inappropriate for measuring other program material. Accordingly, this meter is more relevant to television than to radio. The Dolby meter uses the Leq(A) algorithm [9], which is basically a wideband meter with frequency weighting and appropriately selected attack and release time constants. Unlike the CBS algorithm, it does not model loudness summation. Dorrough Electronics manufactures several meters that it calls "Loudness Monitors." These combine peak and quasi-average metering on one scale [10].

According to the Network Performance, Reliability, and Quality of Service Committee of the Alliance for Telecommunications Industry Solutions (ATIS), a *psophometer* is an instrument that provides a visual indication of the audible effects of disturbing signals of various frequencies [11]. A psophometer usually incorporates a weighting network. The characteristics of the weighting network depend on the type of circuit under investigation, such as whether the circuit is used for high-fidelity music or for normal speech. The standard for a psophometer intended to measure undesired signals on a wideband program line is the International Telecommunication Union, Radiocommunication Sector (ITU-R) Recommendation 468-2. For telephone lines, it is the ITU, Telecommunication Standardization Sector (ITU-T) Recommendation Vol.V.P.53.

GENERAL PERFORMANCE REQUIREMENTS FOR TRANSMISSION AUDIO PROCESSORS

The audio processor must control the peak modulation of the RF carrier to the standards required by the governing authority, such as the FCC in the United States. For AM broadcasting, this usually means that negative carrier pinch-off must not occur at any time because this would cause splatter interference into adjacent channels. In FM and analog television (NTSC and PAL), the peak deviation of the carrier must be controlled so the modulation monitor specified by the governing authority does not indicate overmodulation

[12]. Because the rules often permit the modulation monitor to ignore very brief overshoots, the instantaneous peak deviation might exceed the peak modulation as indicated on the modulation monitor.

The requirements for peak control and spectrum control tend to conflict, which is why sophisticated nonlinear filters are required to achieve highest performance. Applying a peak-controlled signal to a linear filter almost always causes the filter to overshoot and ring because of two mechanisms: *spectral truncation* and *time dispersion*. One can build a square wave by summing its Fourier components together with correct amplitude and phase. Analysis shows that the fundamental of the square wave is approximately 2.1 dB higher than the amplitude of the square wave itself. As each harmonic is added in turn to the fundamental, the phase of a given harmonic is such that the peak amplitude of the resulting waveform *decreases* by the largest possible amount. Simultaneously, the RMS value *increases* because of the addition of the power in each harmonic. This is the fundamental theoretical reason why simple clipping is such a powerful tool for improving the peak-to-average ratio of broadcast audio: Clipping adds to the audio waveform spectral components whose phase and amplitude are precisely correct to minimize the peak level of the waveform while simultaneously increasing the power in the waveform.

If a square wave (or clipped waveform) is applied to a low-pass filter with constant time delay at all frequencies, the higher harmonics that reduce the peak level will be removed, increasing the peak level and with it the peak-to-average ratio. This is a manifestation of the "Gibbs phenomenon"[15], which Figure 3.8-3 illustrates.

For squarewaves with a frequency that is very low compared to the cutoff frequency of the system's low-pass filter, the overshoot is approximately 0.74 dB. The worst case is a squarewave with a frequency that is higher than one-third of the cutoff frequency of the system lowpass filter. In this case, the filter removes all of the squarewave's harmonics and only the fundamental sinewave remains. The peak value of this sinewave is approximately 2.1 dB higher than the peak of the original squarewave. Thus, even a perfectly phase-linear, low-pass filter will cause overshoots as large as 2.1 dB [13].

If the system lowpass filter group delay characteristic is minimum-phase instead of constant-delay, the filter will exhibit a sharp peak in group delay around its cutoff frequency [14]. Because the filter does not produce the same delay at all frequencies, it will remove the higher harmonics required to minimize peak levels and will change the time relationship between the lower harmonics and the fundamental. These lower harmonics become delayed by different amounts of time, causing the shape of the waveform to change. This time dispersion further increases the peak level; more than 70% overshoot is possible with the most difficult program material. For this reason, minimum-phase filters are less desirable than constant-delay filters when used as bandwidth-controlling elements in an audio processor.

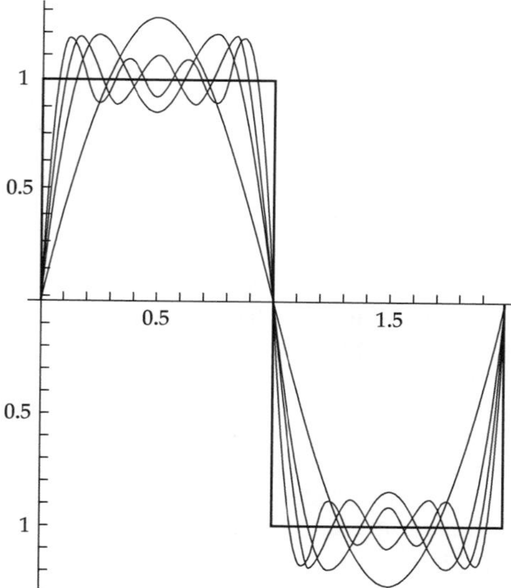

FIGURE 3.8-3 The Gibbs phenomenon, showing overshoot.

Because even constant-delay linear filters produce overshoots, overshoot-free spectral control to FCC or ITU-R standards requires nonlinear processing. Broadcast audio processors typically use proprietary techniques that combine clippers and low-pass filters so overshoots are eliminated and the bandwidth is controlled. The three functions of *clipping*, *low-pass filtering*, and managing *overshoots* are equally challenging when processing in either analog or digital.

Application Considerations

Except as required to achieve very specific artistic goals (most notably in some major-market, high-energy, hit-music formats), the processed audio should be free from unnatural subjective side effects, such as:

- *Pumping*, a sense that the gain is constantly and unnaturally changing and a characteristic side effect of wideband compressors and limiters when driven heavily
- *Breathing*, an audible pulling up of background noise, cured by a compressor gate
- *Hole-punching*, a sudden drop in loudness after a program transient, caused by the transient's inducing a large amount of gain reduction which then does not decay quickly and is cured by multiple time-constant release time circuitry

The audio processor itself should conform to the following general requirements:

- It must be packaged so it is easy to operate and maintain and can work in high RF fields without compromise.

- It should have setup controls with enough versatility to enable the subjective effect to be readily tuned to the requirements of the broadcasting authority operating it. For mixed-format applications, the processor may have several presets, selectable by remote control, that permit the operator to set the amount of compression, limiting, clipping, and other parameters to complement the program material being transmitted.
- It should ordinarily be equipped with sufficient remote control facilities to enable it to be interfaced efficiently with modern, automated plants. Most of the required facilities are specific to the application: for example, AM (MW), shortwave (HF), FM (VHF), or television. Modern processors are capable of communicating via network (TCP/IP), modem, and serial communications. Firmware upgrades are possible on some units via the remote application.
- It should have sufficient metering to permit it to be easily set up with tones or program material. The metering should also provide operations and diagnostic capabilities. Metering usually includes input level, output level, and gain reduction (the amount of limiting or compression) occurring in each variable-gain stage.

Processing for Stereo

Processing for two-channel stereophonic transmission is similar to processing for monophonic transmission, except that two audio processing chains are used. To preserve stereo imaging, the gains of the left and right automatic gain control and compression circuitry must be identical. Conversely, experience has shown that fast peak limiting and high-frequency limiting circuits sound best when operated independently (without stereo coupling), because the ear does not perceive channel-imbalance-induced spatial shifts with these fast time constants; however, the ear can perceive the loudness of one channel's being modulated unnaturally by a dominant element in the other channel when the channels are coupled.

The gain of the coupled elements is determined by the requirements of the transmission service. In FM, the channel requiring the greatest amount of limiting determines the gain of both channels. The processor operates by sensing the higher of the left and right channels and determining the gain of both channels such that the higher channel does not exceed a given level at the processor's output. In AM, the gain of both channels is controlled by sensing and controlling the level of their sum (L+R), because the envelope modulation represents the sum of the channels.

Another popular method is matrix, or sum and difference, processing. In this configuration, the left and right channels are matrixed into the L+R and L–R format. These are processed independently of each other using multiband automatic gain control (AGC)/compressor sections. This method creates a pseudo stereo-enhancement/sound-field-management effect as the gain of the L–R quotient is maintained within a targeted level. Care must be observed with this method, as the sound field can become significantly altered from the original intent. Additionally, if the L–R is compressed too much, the effects of multipath can become exaggerated due to the increased RMS level of the L–R modulation.

SYSTEM CONSIDERATIONS

Reducing the peak-to-average ratio of the audio increases loudness. If peaks are reduced, the average level can be increased within the permitted modulation limits. The level with which this can be accomplished without introducing objectionable side effects (such as clipping distortion) is the single best measure of audio processing effectiveness.

Density is the extent to which the amplitudes of audio signal peaks are made uniform (at the expense of dynamic range). Programs with large amounts of short-term dynamic range have low density, and highly compressed programs have high density. *Compression* reduces the difference in level between the soft and loud sounds to make more efficient use of permitted peak-level limits, resulting in a subjective increase in the loudness of soft sounds. It cannot make loud sounds seem louder. Compression reduces dynamic range relatively slowly in a manner similar to "riding the gain."

Limiting and clipping, on the other hand, reduce the short-term peak-to-average ratio of the audio. *Limiting* increases audio density. Increasing density can make loud sounds seem louder but can also result in an unattractive, busier, flatter, and denser sound. It is important to be aware of the many negative subjective side effects of excessive density when setting controls that affect the density of the processed sound. *Clipping* sharp peaks does not produce any audible side effects when done moderately. Excessive clipping will be perceived as audible distortion.

Building a System

Combining several audio processors into a good-sounding system is tricky because of headroom and time constant considerations. The device driving a given processor must be able to drive that processor into full compression or limiting. If the driving device (for example, distribution amplifier) runs out of headroom before full limiting occurs in the driven device, then that device cannot achieve its full capability. This consideration is particularly critical when setting up the input analog-to-digital (A/D) converter of a digital audio processor. The analog drive level to the A/D converter must be set so the A/D converter does not clip when receiving levels sufficient to cause full gain reduction in the following audio processor.

Beware of interactions between the attack times and release times when cascading several processors. It is wise to start the system with the slowest device. This is usually a compressor or AGC with slow attack and release times and a compressor gate to prevent noise breathing. Such a processor does not significantly increase the *density* of the audio; it simply does gentle gain riding to ensure that following stages are driven at the correct level.

A multiband compressor with moderate attack and release times often follows the slow AGC. Correctly designed multiband processors have these time constants optimized for each frequency band. The low frequency bands have slower time constants than the high-frequency bands. This multiband compressor usually does most of the work in increasing program density.

The amount of *gain reduction* determines how much the loudness of soft passages will be increased (and, therefore, how consistent overall loudness will be). Our hypothetical system reduces gain with the broadband AGC and the multiband compressor. The broadband AGC is designed to control average levels and to compensate for a reasonable amount of operator error. It is *not* designed to substantially increase the short-term program density; the multiband compressor and peak limiters handle that function.

Modern audio processing systems usually add other elements to the basic system described here; for example, it is common to incorporate an equalizer to color the audio for artistic effect. The equalizer may be any of the types described earlier and is usually found between the slow AGC and the multiband compressor. The multiband compressor itself can also be used as an equalizer by adjusting the gains of its various bands.

Various low-pass filters are often included in the system to limit the bandwidth of the output signal to 15 kHz (for FM), 10 kHz (AM in NRSC countries), 4.5 kHz (AM in European Broadcast Union countries, and shortwave worldwide), or other bandwidths as required by the local regulatory authority. The final low-pass filter in the system is almost always overshoot compensated to prevent introducing spurious modulation peaks into the output waveform. High-pass filters may be incorporated to protect the transmitter. This is particularly important in high-power AM and shortwave installations exceeding 100 kW carrier power. A transmitter equalizer that corrects the pulse response of the transmitter is found on high-end AM processors.

In Europe, some countries are required to control their multiplex power according to ITU-R Recommendation BS412. The integrated power of the composite multiplex signal (including the stereo pilot tone) in any arbitrary 60 sec window must be less than or equal to the integrated power of a sine wave that modulates the FM carrier ±19 kHz. Many modern FM processors include an automatic multiplex power controller. This usually works by feedback. The controller measures the multiplex power and reduces the drive into the audio processor's peak-limiting system to ensure that the standard is obeyed. Figure 3.8-4 shows a simplified block diagram of a modern audio processing system.

Location of System Components

The best location for the processing system is as close as possible to the transmitter, so the output of the processing system can be connected to the transmitter through a circuit path that introduces the least possible change in the shape of the carefully peak-limited waveform at the processing system output. One possible configuration is shown in Figure 3.8-5, in which a studio-to-transmitter link (STL) is utilized, and the processor is located between the STL receiver and the broadcast transmitter. The STL might be telephone or post lines, analog microwave radio, or various types of digital paths. Sometimes, it is impractical to locate the processing system at the transmitter, and it must instead be located on the studio side of the link connecting the audio plant to the transmitter. This situation is not ideal because artifacts that cannot be controlled by the audio processor can be introduced in the STL to the transmitter or by additional peak limiters placed at the transmitter. Such additional peak limiters are common in countries where the transmitter is operated by a different authority than the one providing the broadcast program.

In this case, the audio output of the processing system should be fed directly to the transmitter through a link that is as flat and phase linear as possible. Deviation from flatness and phase linearity will cause spurious modulation peaks because the shape of the peak-limited waveform is changed. Such peaks add nothing to average modulation; thus, the average modulation must be lowered to accommodate those peaks within the carrier deviation limits dictated by FCC modulation rules. This implies that if the transmitter has built-in high-pass or low-pass filters (as some do), these filters *must* be bypassed to achieve accurate waveform

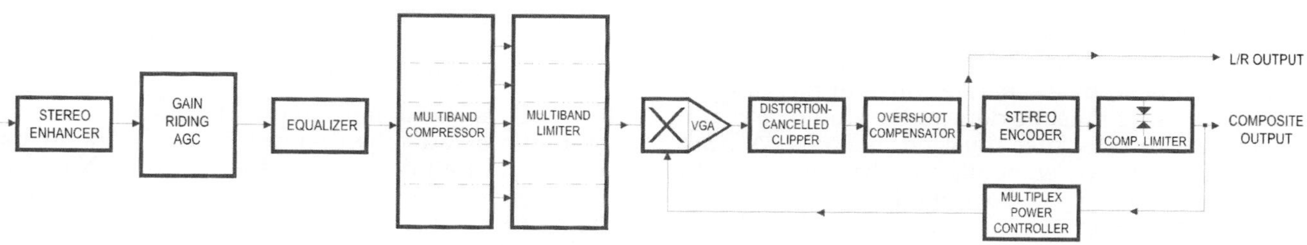

FIGURE 3.8-4 Modern audio processing system.

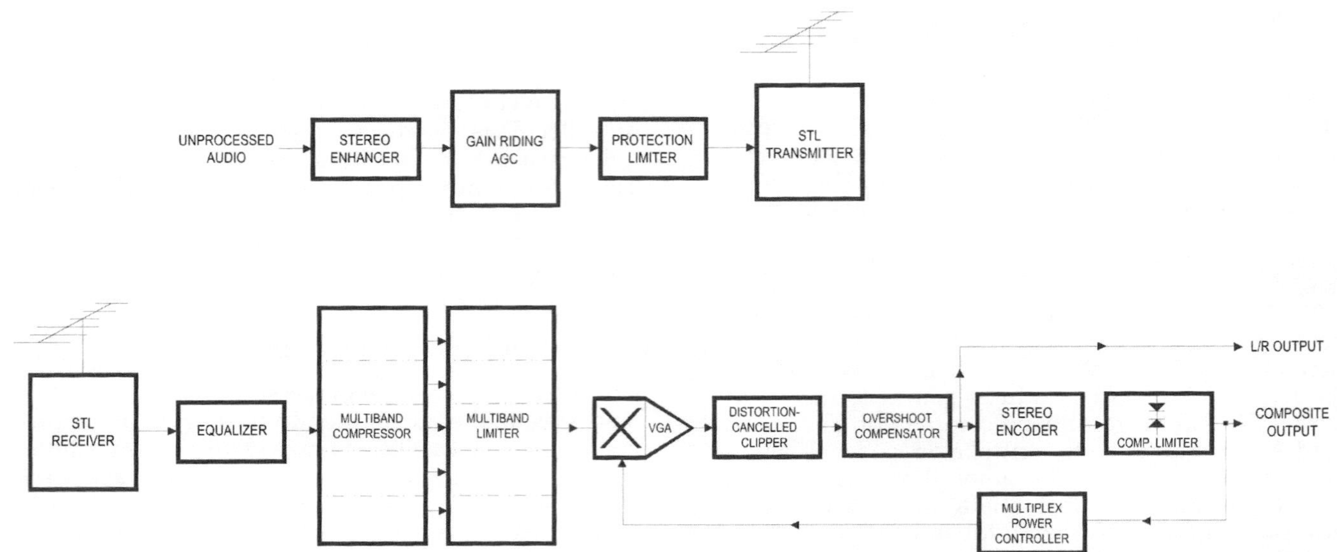

FIGURE 3.8-5 Equipment configuration with processing after the STL.

fidelity. Modern processing systems contain filters that are fully able to protect the transmitter but which are located in the processing system where they do not degrade control of peak modulation.

Where Access to the Transmitter Is Available

The audio received at the transmitter site should be of as good quality as possible. Because the audio processor controls peaks, it is not important that the audio link (STL) feeding the input terminals of the processing system be phase linear; however, the link should have low noise, flat possible frequency response from 30 to 15,000 Hz, and low nonlinear distortion. If the audio link between the studio and the transmitter is noisy, performing the compression function at the studio site can minimize the audibility of this noise. Compression applied before the audio link improves the signal-to-noise ratio because the average level on the link will be greater. If the STL has limited dynamic range, it may be desirable to compress the signal at the *studio* end of the STL. To apply such compression, split the processing system, placing the AGC and multiband compressor sections at the studio and the peak limiter at the transmitter.

Where Access to the Transmitter Plant Is Not Available

In some situations, the organization originating the program does not have access to the transmitter, which is operated by a separate entity. In this case, all audio processing must be done at the studio, and any damage that occurs later must be tolerated. A peak limiter would, however, be used at the transmitter to provide protection against overmodulation.

If it is possible to obtain a broadband phase-linear link to the transmitter, the processing system at the studio location can feed the STL. The output of the STL receiver is then fed directly into the transmitter with no intervening processing. A *composite* STL (ordinarily used for FM stereo baseband) has the requisite characteristics and can be used to carry the output of the processing system to the transmitter; however, the output of a typical composite STL receiver is often at the wrong level and impedance to directly drive a typical transmitter (most of which require +10 dBm into 600 Ω). In this case, the transmitter may have to be modified to make it compatible with the composite STL, or it might be necessary to employ an intervening drive amplifier of sufficient output capacity. Use of a composite STL has many ramifications, and the installation of the processing system at the transmitter may be the less complicated approach.

Where only an audio link is available, feed the audio output of the processing system directly into the link. If possible, transmitter protection limiters should be adjusted for minimum possible action as the processing system does most of that work. Transmitter protection limiters should respond only to signals caused by faults or by spurious peaks introduced by imperfections in the link. Where maximum quality is desired, it is important that all equipment in the signal path after the studio be carefully aligned and qualified to meet the appropriate standards for bandwidth, distortion, group delay, and gain stability and that such equipment is requalified at reasonable intervals.

Requirements for STLs

If the STL is ahead of the audio processor, the STL signal-to-noise ratio (SNR) must be sufficient to pass unprocessed audio. This means that the SNR of the link must be better than the sum of the desired SNR of the transmitted signal plus the maximum gain of the audio processor plus about 6 dB (a useful rule of thumb). If the STL follows the audio processor, its

SNR should be 6 dB better than the desired SNR of the transmitted signal. To ensure that the STL does not distort the shape of the audio waveform, the frequency response must be flat (±0.1 dB) throughout the operating frequency range, typically 20 to 15,000 Hz. To prevent the introduction of overshoot into peak-limited waveforms that are applied to the STL input, the group delay must be essentially constant throughout this range. Deviation from linear phase should be less than ±10°. Phase correction can be applied to meet the requirement at high frequencies.

At low frequencies, by far the best way to meet the audio specification is to extend the –3 dB frequency of the STL to 0.15 Hz or lower and to eliminate any peaking in the infrasonic frequency response prior to the roll-off frequency. It is common for a microwave STL frequency to bounce because of a large infrasonic peak in its frequency response caused by an underdamped automatic frequency control (AFC) phase-locked loop. This bounce can increase the peak carrier deviation of the STL by as much as 2 dB which can result in a reduction in the average modulation. Many commercial STLs have this problem, but it can be corrected by (1) modifying the AFC loop, or (2) applying equalization prior to the STL transmitter that is complementary to the existing low-frequency roll-off, such that the overall system frequency response rolls off smoothly at 0.15 Hz or below. This solution is far better than clipping the tilt-induced overshoots after the STL receiver because clipping will introduce nonlinear distortion, while the equalizer is distortion free. For highest quality, the nonlinear distortion of the STL system should be less than 0.1% total harmonic distortion (THD) throughout the operating frequency range.

Digital links may pass audio as straightforward pulse-code modulation (PCM) encoding, or they may apply data-rate-reduction processing to the signal to reduce the number of bits per second (bps) required for transmission through the digital link. Such processing will almost invariably distort peak levels; therefore, such links must be carefully qualified before being used to carry the peak-controlled output of the audio processor to the transmitter or stereo encoder. For example, the MPEG-1 Layer 2 algorithm can increase peak levels up to 4 dB at 160 kB/sec by adding large amounts of quantization noise to the signal. Although the desired program material may psychoacoustically mask this noise, it is nevertheless large enough to substantially affect peak levels. For any lossy compression system, the higher the data rate, the less the peak levels will be corrupted by added noise, so use the highest data rate practical in the system. Even with the maximum available data rate, overshoot will probably be large enough to require use of an overshoot compensator at the STL receiver (or in the equipment being driven by it). Some modern FM exciters and stand-alone stereo encoders are now equipped with such compensators.

Other links may use straightforward PCM without lossy data rate reduction. These can be very transparent and can exhibit accurate pulse response provided that their input anti-aliasing filters and output reconstruction filters are rigorously designed to achieve constant group delay over the frequency range that contains significant program energy. This is not particularly difficult to do with modern oversampled converter technology. Near Instantaneous Companded Audio Multiplex (NICAM) is essentially a hybrid of PCM and data-rate-reduced systems. It uses a block-companded floating-point representation of the signal with J.17 preemphasis [16]. NICAM links can exhibit low overshoot when designed for good low-frequency response and are equipped with phase-linear anti-aliasing and reconstruction filters [17]. Because the output spectrum of most modern audio processing systems is already tightly bandlimited, any anti-aliasing filters in digital links driven by such systems may be bypassed. This ensures the most accurate possible transient response.

Using Lossy Data Reduction in the Studio

Many stations are now using lossy data reduction systems to increase the storage capacity of digital playback media. In addition, source material is often supplied through a lossy data reduction system, whether from satellite or over landlines. Sometimes, several encode/decode cycles will be cascaded before the material is finally presented to the input of the audio processor.

All such algorithms operate by increasing the quantization noise in discrete frequency bands. If not psychoacoustically masked by the program material, this noise may be perceived as distortion, *gurgling*, comb filtering, or other interference. Psychoacoustic calculations are used to ensure that the added noise is masked by the program material and cannot be heard. In addition, at least two other mechanisms in broadcasting can cause the noise to become audible at the receiver. First, a modern multiband transmission audio processor performs an *automatic equalization* function that can radically change the frequency balance of the program. This can cause noise that would otherwise be masked to become unmasked because the psychoacoustic masking conditions under which the masking thresholds were originally computed have changed. Second, the frequency response of the radio receiver (particularly in AM) can remove frequencies that were used to make the psychoacoustic masking calculations and that would otherwise have masked the added quantization noise. Accordingly, if lossy data reduction is used in the studio, then the highest data rate possible should be selected. This maximizes the headroom between the added noise and the threshold where it will be heard. Also, the number of encode/decode cycles should be minimized, because each cycle moves the added noise closer to the threshold where it will be heard.

Transmission Levels and Metering

Engineers at the transmitter and the studio consider transmission levels and their measurements differently. Transmission engineers need to know the peak level of a transmission commonly measured by an oscilloscope. Studio engineers need to know the line-up (or reference) level of a transmission commonly measured by a VU meter (as the approximate RMS level) or by a peak program meter (as the PPM level) [18].

Metering

The VU meter is an average-responding meter (measuring the approximate RMS level) with a 300 msec rise time and decay time; the VU indication usually lags the true peak level by 8 to 14 dB. PPM indicates a level between RMS and the actual peak. The PPM reading has an attack time of 10 msec, slow enough to cause the meter to ignore narrow peaks and lag the true peak level by 5 dB or more.

Transmission Levels

The transmission engineer is primarily concerned with the peak overload level of a transmission to prevent overloading. This peak overload level is defined differently system to system. In tape, it is defined as the level producing the amount of harmonic distortion considered *tolerable*—often 3% THD at 400 Hz. In FM, microwave, or satellite links, it is the maximum-permitted RF carrier deviation. In AM, it is negative carrier pinch-off. In analog telephone transmission, it is the level above which serious crosstalk into other channels occurs or the level at which the amplifiers in the channel overload. In digital, it is the largest possible digital word.

Studio Levels

The studio engineer is primarily concerned with what is commonly called the *reference level*, *operating level*, or *line-up level*. This line-up level aids studio engineers in providing adequate headroom between the line-up level and the overload level of equipment to allow for the peaks that the meter does not indicate. In facilities that use VU meters, the line-up level is usually at 0 VU, which corresponds to the studio standard level, typically +4 or +8 dBm. In systems that use PPM, the line-up level may be at PPM 4 (for the BBC standard) or at the studio standard level (often +6 dBm). In studios using digital links such as AES3, there are two commonly used line-up levels: –20 dB relative to digital full-scale (dBfs) and –18 dBfs. The –20 dBfs level is most commonly used and is preferable because it allows a generous amount of headroom (20 dB).

Transmission-Link Limiting

Transmission-link-limiting devices are sometimes used ahead of transmission links (such as STLs, satellite uplinks, interstudio digital links) to protect them from overload. These devices are usually used below threshold (that is, with no gain reduction unless the threshold is reached) as protection limiters to control peak levels. They only produce gain reduction when abnormally high levels are applied to their input due to operator error or unforeseen level variations at the source. This is useful to transmission engineers concerned with overload and to studio engineers concerned with headroom. For the needs of both engineers, the output of such a limiter must be adjusted to be at or slightly below the peak overload level of the transmission channel.

To properly match the studio line-up level to the transmission protection limiter, the desired headroom must be known. For example, assume that the transmission protection limiter produces 0 dBm at its output at 100% modulation of the transmission link. Further, assume that the line-up level in a production facility is designed to allow 8 dB of headroom. The input attenuator of the transmission protection limiter would then be adjusted so the studio line-up tone produces –8 dBm at the output of the transmission protection limiter.

This assumes that the amplifier or other link between the studio and the input of the transmission protection limiter has enough headroom to drive the transmission protection limiter into gain reduction without clipping this link. The transmission protection limiter only protects a link connected to its output. In the previous example, if the transmission protection limiter provides 15 dB of maximum protection, the system prior to the transmission protection limiter requires 8 + 15 = 23 dB of headroom above the studio line-up level. If the link is simply an amplifier, this should be achievable without difficulty if the absolute level of the studio line-up tone is chosen carefully. In our example, if the amplifiers in the system clip at +21 dBm, the level of the studio line-up tone can be no greater than –2 dBm (*i.e.*, 23 dB below +21 dBm).

AUDIO PROCESSING REQUIREMENTS FOR MW AND HF BROADCAST STATIONS

In amplitude-modulated transmission services, reception is usually compromised by noise and interference and may be further compromised by acoustic noise in the listening environment (such as an automobile); thus, the processor must compensate for noise (electrical and acoustic) and interference by reducing dynamic range. This is most readily done by multiband compression and limiting to achieve the lowest peak-to-average ratio without significant processing-induced side effects.

The processor must provide absolute negative peak control to prevent AM carrier pinch-off, which would otherwise cause out-of-band emissions. Additionally, the processor must incorporate overshoot-free filtering to control the audio input spectrum to the transmitter, thus preventing out-of-band emissions and interference. National (FCC) or international (most notably ITU-R) broadcast authorities usually specify the permissible occupied bandwidth to make most efficient use of available radio frequency spectrum. The processor may also be equipped with a receiver equalizer that compensates for the limited frequency response of the typical mediumwave or shortwave radio due to narrowband RF and intermediate frequency (IF) stages.

Transmitter Equalization

The processor may provide a transmitter equalizer to eliminate tilt, overshoot, and ringing in the transmitter and antenna. Accurate reproduction of the shape of the processed waveform requires that the transfer function between the audio input and the modulated RF envelope represents a constant delay (which may be any positive number or 0) at all frequencies contained within the audio input signal. Failure to meet this criterion can result in tilt, overshoot, and *ringing* in the modulated RF envelope. The causes of overshoot and ringing as spectrum truncation and time dispersion at the high-frequency end of the system bandpass were discussed earlier. Tilt, on the other hand, is caused by problems at low frequencies.

Figure 3.8-6 shows the response of a 10 kW plate-modulated transmitter to a 50 Hz square wave. The transmitter causes the waveform to tilt, which increases peak modulation in both positive and negative directions. The magnitude of the frequency response of the transmitter is essentially flat to 50 Hz; the problem is caused by infrasonic roll-off. This roll-off is equivalent to that of a high-pass filter and is minimum phase, which introduces time dispersion, causing the shape of the waveform to change and further increasing the peak level. This is an example of a transmitter with an inadequate low-frequency response that requires transmitter equalization to avoid introducing tilt and overshoot. The rule of thumb is that the equalized transmission system must have its –3 dB frequency at 0.15 Hz or lower in order to avoid significant tilt-induced peak modulation overshoot. It does not matter whether the transmitter is tube-type or solid state—the basic requirements for equalized system frequency response are identical; however, modern solid-state transmitters are likely to have been designed with the required 0.15 Hz cutoff frequency so they do not require external equalization.

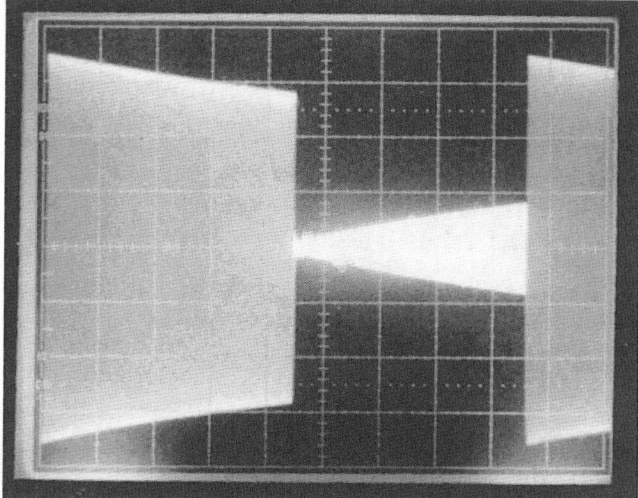

FIGURE 3.8-6 Waveform of tilt in a plate modulated transmitter.

Some transmitters contain high-pass filters at their audio inputs to protect high-power stages. *This location is absolutely inappropriate*; these filters can easily increase the peak-to-average ratio of the input audio by 3 to 4 dB. The correct location for a protection high-pass filter is in the audio processor, where measures can be taken to prevent the high-pass filter from increasing the peak-to-average ratio at the output of the audio processor.

Bounce

Predistorting the waveform in the audio processor (*i.e.*, adjusting the waveform in anticipation of known errors in the transmission path) can equalize linear errors; however, one major nonlinear error, commonly referred to as *power supply bounce*, is caused by resonances in the inductance–capacitance (LC) filter elements of the high-voltage power supply of the transmitter. These resonances superimpose a subaudible modulation onto the power supply voltage, resulting in a form of very fast carrier shift that is too quick to be seen on a conventional carrier shift meter. The net result is to compromise the control of modulation peaks, particularly on strong bass transients which cause momentarily large current demands on the power supply and which excite the resonance.

In some older transmitters, bounce has been known to compromise achievable modulation by up to 3 dB. Because bounce is not linearly related to the modulation, small-signal equalization cannot cure it. The most successful cure has been the use of a 12-phase power supply in the transmitter. The AC ripple from such a supply is down about 40 dB without filtering; a simple filter capacitor is all that is necessary to achieve adequate smoothing. Because there are no chokes in the power supply filter, resonance cannot occur. In all cases, bounce can be minimized by preventing excessive bass energy from being applied to the transmitter.

Slew Rate Limiting (Transient Intermodulation Distortion)

Transmitters using pulse-duration modulation (PDM) schemes are prone to problems with slew rate limiting. Because the PDM low-pass filter is located within the audio feedback loop of the transmitter and because this filter is typically a multi-pole elliptic function filter with a cutoff frequency below 70 kHz, it will introduce substantial delay into the feedback loop. This has two consequences: Stability requires the amount of feedback applied around the transmitter to be limited, and stability also requires that the open-loop gain of the modulator be rolled off at a very low frequency. The first issue makes it difficult to design PDM transmitters with THD below 1 to 2% at midrange frequencies, and the second renders *transient intermodulation distortion* (TIM; nonlinear behavior of the amplification stage prior to the frequency compensation stage) probable [19]. To minimize the probability that TIM will be bothersome, any amplification stage before the frequency compensation stage should be designed to be very linear to its clipping point and to have sufficiently high headroom to accommodate the maximum rate of change to be expected at the transmitter's audio input [20].

A transmitter can be qualified for TIM by one of the various difference-frequency intermodulation distortion tests. If the tests indicate that the transmitter has a low slew rate, it will not respond well to preemphasized audio, and preemphasis will have to be reduced until the first derivative of the processed audio waveform seldom, if ever, exceeds the slew rate limit of the transmitter. Because of the benefits of preemphasis at the receiver, it is desirable to modify such transmitters to increase their slew rate, even if this means somewhat compromising harmonic distortion performance at low frequencies.

NRSC-1 Audio Standard

As the North American AM band became more crowded, interference from first and second adjacent stations became more of a problem. Receiver manufacturers responded by producing receivers with decreased audio bandwidth, so the encroachment of an adjacent station's modulation extremes would not be audible as interference. This truncating of the bandwidth had the effect of diminishing the high-frequency response of the receiver, but it was decided that lower fidelity would be less annoying than interference. To address these problems, the National Radio Systems Committee (NRSC) in 1987 formalized NRSC-1, a standard for preemphasis and low-pass filtering for AM broadcast to provide brighter sound at the receiver while minimizing interference [21]. (See Chapter 1.9, "NRSC Analog and Digital Radio Standards," for more information on NRSC work.)

AM Stereo Introduces a Preemphasis Dilemma

Certain AM receivers manufactured since 1984 for sale in North America, particularly those designed for domestic AM stereo reception, have a frequency response that is substantially wider than that of the typical mono AM receiver. The frequency response was widened largely to enhance the sales potential of AM stereo by presenting a dramatic, audible improvement in fidelity in the showroom. Were these new receivers to become more prevalent, broadcasters would have to choose whether the station's preemphasis would be optimized for the new AM stereo receivers or for the existing conventional receivers that form the vast majority of the market. If the choice was for conventional receivers (which implies a relatively extreme preemphasis), the newer receivers might sound strident or exceptionally bright. If the choice favored the newer receivers (less preemphasis and probably less processing), the majority of receivers would be deprived of much high-end energy and would sound duller and have less loudness.

NRSC Standard Preemphasis and Low-Pass Filtering

In response to this dilemma, the NRSC undertook the difficult task of defining a voluntary recommended preemphasis curve for AM radio that would be acceptable to broadcasters (who want the highest quality sound on the majority of their listeners' radios) and to receiver manufacturers (who are primarily concerned with interference from first- and second-adjacent stations). A modified 75 μsec preemphasis/deemphasis standard (NRSC-1) was adopted (see Figure 3.8-7) that provides a moderate amount of improvement for existing narrowband radios while optimizing the sound of wideband radios. Most importantly, it generates substantially less first-adjacent interference than do steeper preemphasis curves.

The second part of the NRSC-1 standard calls for a sharp upper limit cutoff of 10 kHz for the audio presented to the transmitter (see Figure 3.8-8). This essentially eliminates interference to second and higher adjacencies. Although some broadcasters believe that this is inadequate and that 15 kHz audio should be permitted, it is not likely that interference-free 15 kHz audio could be achieved except by a reallocation of the AM band. The practical effect of widespread implementation of the 10 kHz standard is that 10 kHz radios are feasible, and the bandwidth perceived by the average consumer (now typically limited by the receiver to 3 kHz) can be dramatically improved. The difference between AM and FM reception will then become less pronounced.

The NRSC and the Consumer Electronics Association (CEA; formerly the Electronic Industries Association [EIA]), worked together to define a standard for wideband, NRSC-compliant AM stereo radios which is referred to as NRSC-3 and trademarked as AMAX. Several manufacturers (most notably Delco) introduced AMAX radios into the marketplace. Although the NRSC standards calls for a 10 kHz audio bandwidth at all times, one can seriously argue that any North American AM station whose programming is

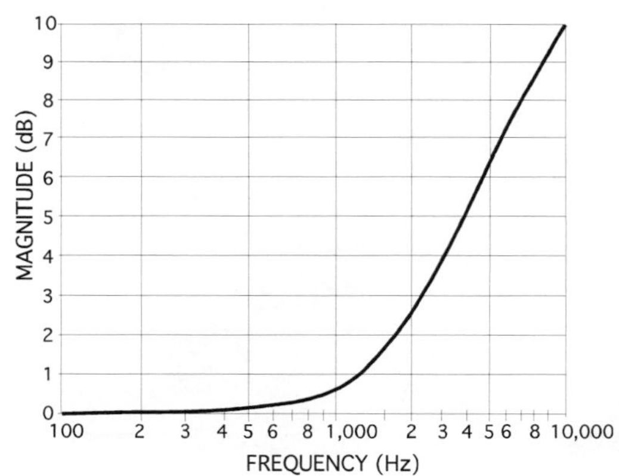

FIGURE 3.8-7 NRSC-1 preemphasis curve.

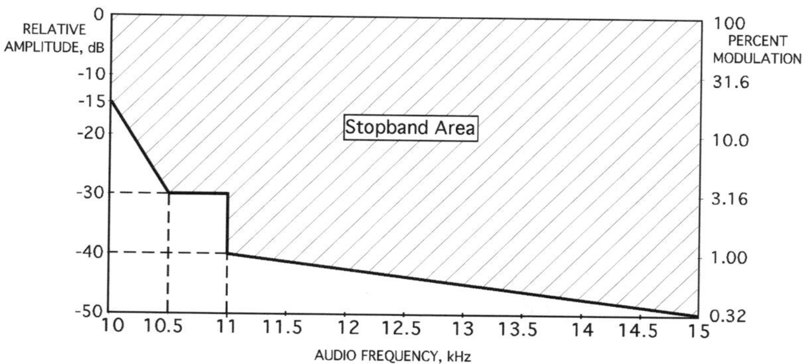

FIGURE 3.8-8 NRSC-1 low-pass filter curve.

primarily talk should consider voluntarily reducing its audio bandwidth to 5 kHz at night. This will cause little or no audible change in frequency balance as heard on the average AM radio (for which the audio bandwidth at –3 dB continues to be approximately 3 kHz), but it will completely prevent any interference from being applied to first-adjacent channels. (At night, skywave propagation can convey such interference to stations hundreds of miles from the interferer.) In addition, the station is likely to find that it can process its audio harder for a given amount of processor-induced distortion, increasing on-air loudness and coverage. This is particularly important at night when interference increases.

NRSC-2 Audio Standard

In 1989, the FCC released a Report and Order that amended Section 73.44 of the FCC Rules by requiring all U.S. AM stations to comply with the occupied bandwidth specifications of the NRSC-2 standard by June 30, 1990. The NRSC-2 standard is an *RF mask* that was derived from the NRSC-1 audio standard by the NRSC. The purpose of the NRSC-2 RF mask is to provide a transmitted RF occupied bandwidth standard that any station with a properly operating transmitter will meet, provided that NRSC-1 audio processing is used prior to the transmitter and provided that the station is not overmodulating.

Audio Processing for AM Stereo

In all AM stereo systems, the envelope modulation is forced to a close approximation of the sum of the left and right channels to ensure compatibility with mono radios equipped with envelope detectors. To ensure minimum loudness loss compared to monophonic transmission, it is necessary to process stereo audio in the *sum and difference* format. This means that the left and right channels are passed through matrix circuits to create L+R (sum) and L–R (difference) signals. These signals are then passed separately through those parts of the processing that control modulation. To prevent clipping and distortion in the C-Quam decoder (C-Quam is the AM stereo method adopted by the FCC in 1993) in the receiver, the negative-going modulation in the left and right channels must be no greater than 75% modulation (where 100% modulation is carrier pinch-off). Thus, an audio processor for C-Quam must have both a sum and difference processor (which does the main processing) and a safety limiter to protect the left and right channels. This safety limiter is usually inactive and typically only comes into play when the input program has sections that are momentarily single channel, such as ping-pong stereo.

AUDIO PROCESSING REQUIREMENTS FOR FM (VHF) BROADCAST TRANSMISSION

The audio processor for FM broadcasting should provide a comfortably listenable dynamic range in domestic and automotive listening environments by applying subtle compression to the signal. Unless the program director requests otherwise for competitive reasons, such compression should be undetectable to the ear unless the original source is available for comparison. The processor must provide high-frequency limiting to complement the preemphasis employed (50 μsec or 75 μsec, depending on the region in which the transmission occurs).

The processor must provide accurate peak control (as measured by a modulation monitor meeting the standards of the governing authority) in both the positive and negative directions. To ensure that absolute peak control will be retained at the system output, any system elements following the processor must have a flat frequency response (±0.1 dB) and constant group delay (deviation from linear phase less than ±10°). Because the preemphasis networks and low-pass filters ordinarily found in stereo encoders may not meet these requirements, it may be desirable that they be bypassed (check stereo encoder specifications first); thus, the processor should provide preemphasis and bandlimiting for the transmission system. Its output must contain negligible energy above the bandwidth limit of the transmission system. In FM stereo broadcasting by the world-standard pilot-tone method, this bandwidth is limited to less than 19 kHz to prevent aliasing from the stereo subchannel into the main channel, and *vice versa*.

To protect the pilot tone itself (ensuring correct operation of the phase-locked loop subcarrier regeneration circuitry in the receiver's stereo decoder) the bandwidth must be further limited to no greater than 17 kHz. In practice, it is customary to begin the high frequency roll-off at slightly above 15 kHz to minimize group delay distortion in the low-pass filters used to affect the bandwidth limit. Nonlinear low-pass filters are usually used to prevent overshoot, enabling the processor to control peak deviation absolutely.

The processing system must be readily adjustable to achieve the subjective effect desired by the broadcasting authority operating it. To achieve a competitive sound in markets where many stations compete for listeners, it may be necessary to add additional multi-bandlimiting to the basic audio processing system (which usually consists of compressor, high frequency limiter, and peak limiter/clipper). Adding additional multi-bandlimiting can create greater program density than the basic processing system alone without introducing spectral gain intermodulation.

AUDIO PROCESSING REQUIREMENTS FOR DIGITAL RADIO TRANSMISSION

The United States has adopted the HD Radio system, an in-band-on-channel (IBOC) digital transmission system developed by iBiquity Digital Corporation [22]. Other popular digital radio transmission systems in use around the world include the Eureka-147 and the Digital Radio Mondiale (DRM) systems [23]. To digitally encode the audio, all digital radio systems use lossy data reduction with no preemphasis. The specific system of lossy data reduction that is used depends on the particular digital radio system. DRM uses a codec based on aacPlus (an extension of MPEG-2 Advanced Audio Coding developed by Coding Technologies), and the original Eureka-147 standard specified the MPEG-1 Layer 2 algorithm. HD Radio uses the HDC codec, a proprietary technique developed by iBiquity Digital Corporation.

Differences between HD Radio Digital Audio (FM Band) and FM Analog Audio

There are several differences between the audio broadcast over an FM analog channel and that broadcast using an HD Radio digital channel in the FM band. FM analog has a theoretical audio bandwidth limit of slightly less than 19 kHz but is most commonly limited to 15 kHz or slightly above. HD Radio digital audio for the FM band has a maximum audio bandwidth of 20 kHz; for the AM band, it has a maximum audio bandwidth of 15 kHz. Another significant difference is that HD Radio, Eureka-147, and DRM do not have any form of preemphasis specified for use in the audio path.

One of the unique features of the HD Radio digital radio system is that the main channel digital audio signal is broadcast along with the main channel analog audio signal in a simulcast fashion. To prevent audio muting, HD Radio receivers automatically crossfade between the digital and analog audio whenever the receiver cannot demodulate the digital signal (which is at a much lower power level). To achieve an unobtrusive crossfade, the two signals should have nearly the same loudness [24]. By design, the gain in the digital path of the HD Radio receiver is 5 dB higher than the gain in the analog FM path so, to a first approximation, the digital path has 5 dB more peak headroom than the analog path. Because the preemphasis of analog FM increases the peak-to-average ratio, holding this channel's peak level constant (which is required to prevent overmodulation) pushes its average modulation down and increases the headroom advantage of the digital channel even more. This advantage is clearly audible; the digital audio sounds noticeably punchier than the analog FM audio. The difference is particularly audible on drums and percussion.

Processing for HD Radio, Eureka-147, and DRM

Digital audio channels challenge an audio processor differently than do analog audio channels. In the analog system, the processor must provide precision peak control to guard against overmodulation, must manage the preemphasis curve to avoid audible distortion generated by the processing, and must include a highly selective low-pass filter to protect the 19-kHz pilot signal of the multiplex stereo system. Analog FM's preemphasis indirectly produces what some industry insiders call the "sounds like radio" effect. Preemphasis boosts high-frequency energy. The preemphasized audio usually feeds a final limiter employing high-frequency limiting and distortion-controlled clipping. This produces the "radio-like" sound that many listeners are used to.

A processor for digital audio broadcast has a different set of requirements because it must work with lossy codecs. There is no high-frequency boost, so the audio processor does not need to do high-frequency limiting or clipping. Accordingly, the high-frequency content in the digital channel will sound clearer and more open. Particularly at the lower bit rates used in digital radio systems, these codecs can sometimes produce unique artifacts not encountered in analog transmission. Just as a processor for FM analog must manage the preemphasis curve, a digital broadcast processor must manage the audio spectrum to minimize codec artifacts. Given knowledge of the codec, it is possible to predict what spectral conditions will stress it. Dynamic algorithms in the processor can ameliorate these conditions and prevent unwanted artifacts, especially at higher bit rates such as 96 kbps. In essence, the audio processor can improve the efficiency of the encoder.

Peak Limiting Technology for Digital Radio

Probably the most common method of achieving precision peak control is the hard limiter (sometimes called a peak clipper), which controls peaks by truncating any part of the input waveform that exceeds the threshold of clipping. Most audio processors designed over the last twenty years employ some form of distortion masking or cancellation to suppress the nonlinear distortion that the clipper generates. This makes it possible to clip harder, increasing loudness.

In a digital transmission system, it is possible to use a clipper as a peak limiter; however, this can cause problems with the codec. Clipping adds spectral energy to its input signal even if distortion masking is used. In digital radio, data rates are low and efficient coding is crucial. When a codec wastes bits encoding clipping distortion products, it has fewer bits left to encode the remaining, undistorted spectral components. This lowers perceptual quality, particularly in the high-frequency range. Another form of peak limiter that is the perfect companion for the digital radio application is the look-ahead limiter. The reason why it suits this application so well is that, although it provides excellent peak control, it does so while adding very little, if any, added spectral content.

One can model any peak limiter as a multiplier that multiplies its input signal by a gain control signal. This is a form of amplitude modulation. Amplitude modulation produces sidebands around the "carrier" signal. In a peak limiter, each Fourier component of the input signal is a separate "carrier," and the peak limiting process produces modulation sidebands around each Fourier component. Considered this way, a hard clipper has a wideband gain control signal and thus introduces sidebands that are far removed in frequency from their associated Fourier "carriers"; hence, the "carriers" have little ability to mask the resulting sidebands psychoacoustically. Conversely, the gain control signal of a look-ahead limiter has a much lower bandwidth and produces modulation sidebands that are less likely to be audible. Accordingly, the psychoacoustic model in the codec rejects these sidebands as inaudible, and the codec does not attempt to encode them.

Following is a quick review of how a basic look-ahead processor operates. In essence, the processor has the ability to calculate the peak level of a signal over a specified period of time. While that is occurring, the audio is delayed by a like amount. Then, as the control signal is applied to the audio gain function, the audio peak is reduced at the precise time that the control signal reaches the maximum control level, and the crest of the peak is reduced without truncation. This is how clipping is avoided. The diagrams in Figure 3.8-9 show a simple view of how a look-ahead processor operates.

Processing Configurations for the HD Radio System

The HD Radio transmission system uses time diversity to improve reception quality in moving vehicles.

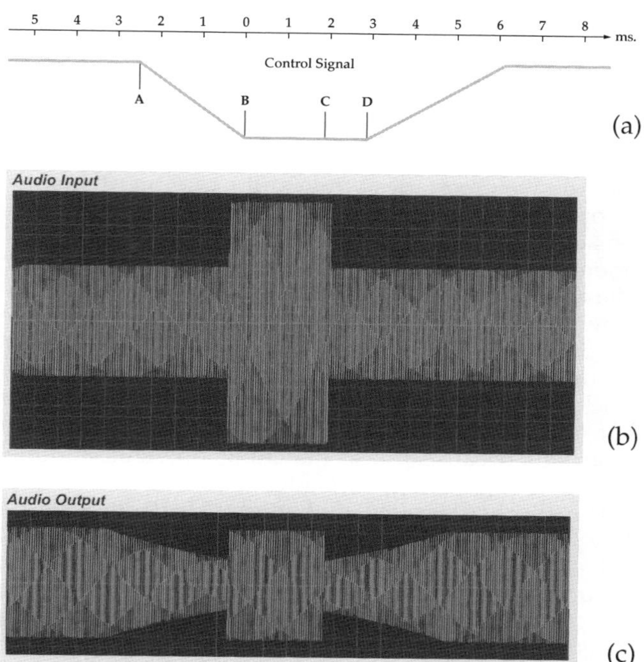

FIGURE 3.8-9 (a) A. Look-ahead limiter concept; A large amplitude peak enters the look-ahead window. By means of interpolation, the control signal starts aiming at the lower gain value that is needed 2 ms ahead. B. The final gain value is reached exactly as the amplitude peak arrives. C. The amplitude decreases. The gain value stays low another 1 ms to prevent the gain from riding on close consecutive peaks. D. The normal release function takes over. (b) Tone burst input to a look-ahead limiter. (c) Controlled output from a look-ahead limiter

To accomplish this, at the transmitter modulation of the analog FM signal is delayed by several seconds compared to modulation of the digital signal. The receiver then delays the digital signal to reestablish time synchronization between the two signals at the crossfade mixer of the receiver. This increases the likelihood that, when an RF dropout occurs, either the analog or the digital signal will be intact at the crossfade mixer because a given RF dropout will affect a different time segment of each signal's content.

There are two basic audio processing configurations for an HD Radio system. The first uses two separate audio processors and is mainly appropriate for broadcasters who want their analog and digital channels to have a noticeably different texture at the expense of having analog/digital crossfades sound obvious. A broadcaster taking this route usually believes that to sell the advantages of HD Radio to the public, there must be an obvious, audible difference between the analog FM and digital channels; in other words, the digital audio must have a "wow!" factor that the analog does not. It is usually possible to achieve this style of processing while avoiding crossfades that are so annoying that they are likely to cause tune-outs. This is the trickiest style of processing to set up and administer effectively.

The second configuration uses a processor that simultaneously provides dedicated processing for both the IBOC and FM analog signals. This configuration is ideal for the broadcaster who wants to ensure unobtrusive crossfades while still exploiting the ability of the digital channel to sound punchier, for reasons discussed earlier in this chapter. Moreover, dual-mode processors often have enough independent adjustability in the analog FM and digital channels to allow them to sound noticeably different if that is the station's goal. Dual-mode processors make it easy to ensure that the analog and digital signals remain time synchronized at the crossfade mixer of the radio and that the two chains are phase matched, which is important because phase matching prevents audible comb filtering during crossfades.

The block diagrams in Figures 3.8-10 and 3.8-11 show the internal functions of stock processors for FM and AM broadcasting. The key functions required of an audio processor are illustrated. Audio processors in this configuration must be inserted into the overall system (Figure 3.8-12) for use with HD Radio. The heavy line represents the audio path for the analog transmission.

All HD Radio exciters offer a diversity delay line for the analog signal. This accepts and emits AES3 digital signals in left/right form. (In the AM system, only one channel is used.) If the audio processor does not contain a diversity delay, the analog-processed output of the processor must be delayed by routing it through the HD Radio exciter. This creates an additional point of failure within the facility, as any failure of the HD Radio exciter will shut down the conventional audio path, forcing both the digital and analog systems off the air. Of course, it is possible to install an external bypass switch to allow for a remedy to this situation (either automatically or by remote control) in the event it should occur.

For HD-FM, a further advantage of including the diversity delay in the processor is that the processor's internal stereo coder and composite limiter (if any) can be used, which eliminates the need to use an external stereo encoder or the exciter's built-in stereo encoder (see Figure 3.8-13). For HD-AM, the primary advantage of building diversity delay into the processor is potentially higher reliability, assuming that the diversity delay in the processor is more reliable than the diversity delay in the exciter. It does not improve system performance, however, because the analog AM output of the processor is already correctly peak limited and is ready to drive the transmitter without further processing. The HD exciter's diversity delay is transparent to this signal and does not compromise peak control.

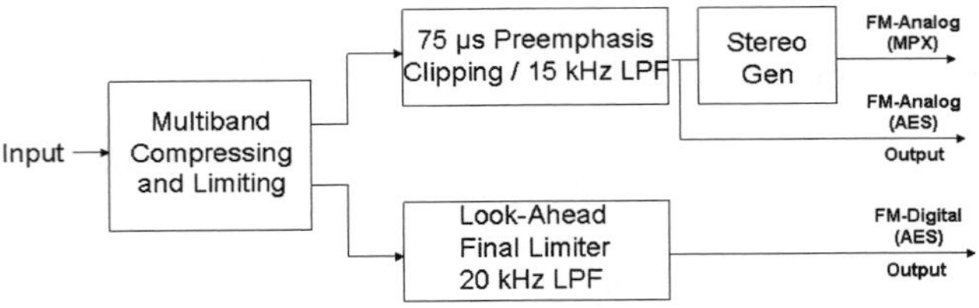

FIGURE 3.8-10 Block diagram of combined FM audio processor.

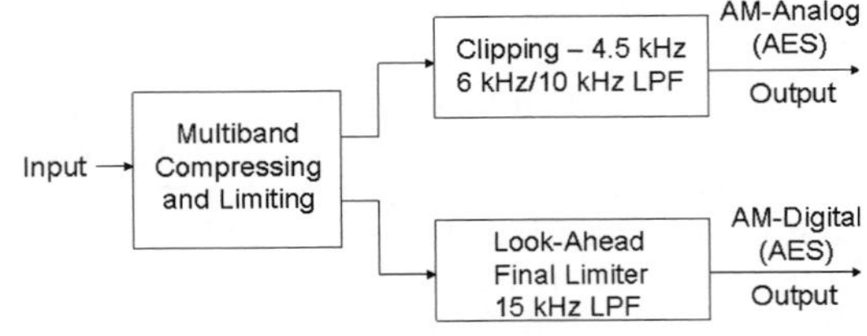

FIGURE 3.8-11 Block diagram of combined AM audio processor.

Audio Processing for HD Radio Multicasting

The HD Radio system can be used to support *multicasting*. This enables multiple audio channels to be multiplexed into the HD Radio data stream (nominally 96 kbps for the FM hybrid mode; other modes, in particular the extended hybrid mode, will support higher bit rates). Multicast audio channels are not simulcast on the analog audio signal and do not benefit from the analog backup feature of the system as does the main channel audio signal. Broadcasters can choose how they wish to parse the payload and initiate additional audio channels. Processing designed for streaming audio or netcasts is useful here. Each audio stream requires a separate processor.

Because of the potentially low bit rate of a multicast channel, one important function of the processor should be to condition the audio to minimize codec artifacts; however, many broadcasters will also want to process the signal to achieve source-to-source consistency, both spectrally and dynamically. Failure to do this could irritate listeners who have come to expect this kind of smooth, professional presentation in a broadcast. To achieve consistency while minimizing codec artifacts, the audio processor must not significantly increase the density of the signal, must not aggressively peak-limit the signal, and must prevent the excessive build-up of high-frequency energy. It is possible to use multiband processing to achieve these goals, but the processor must be carefully set up. It is wise to use little or no stereo enhancement, slow release times, a well-chosen compressor-gating threshold, and very little look-ahead limiter gain reduction. Moreover, the thresholds of the highest frequency bands of the processing must be set low enough to tame excessive high-frequency energy.

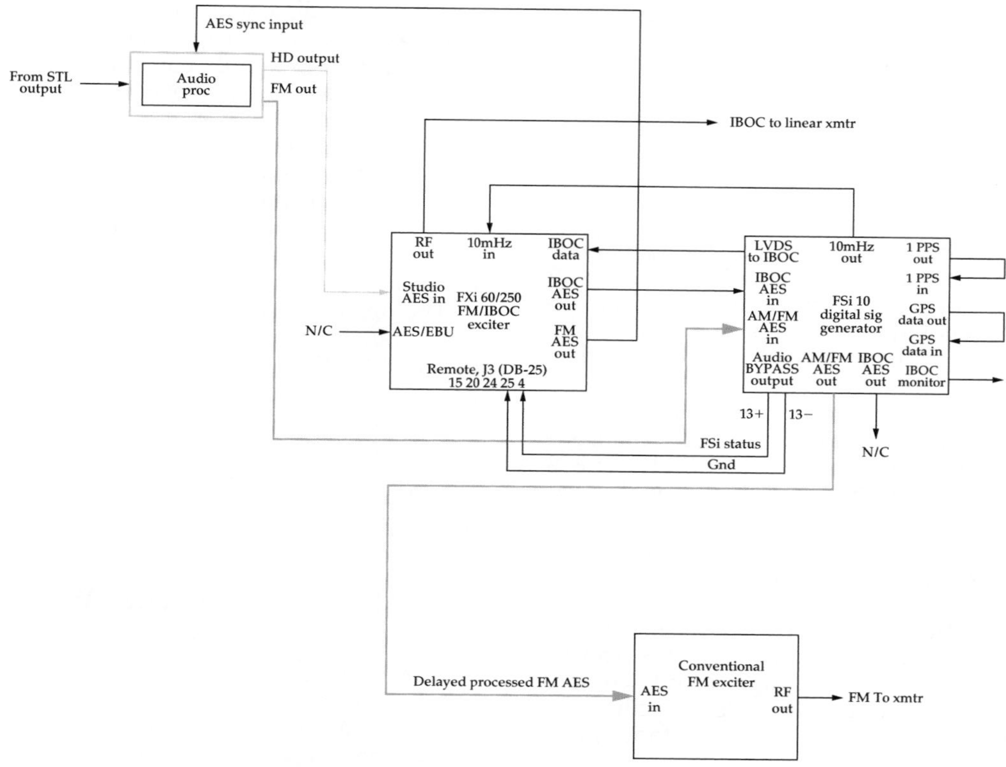

FIGURE 3.8-12 HD Radio stock setup.

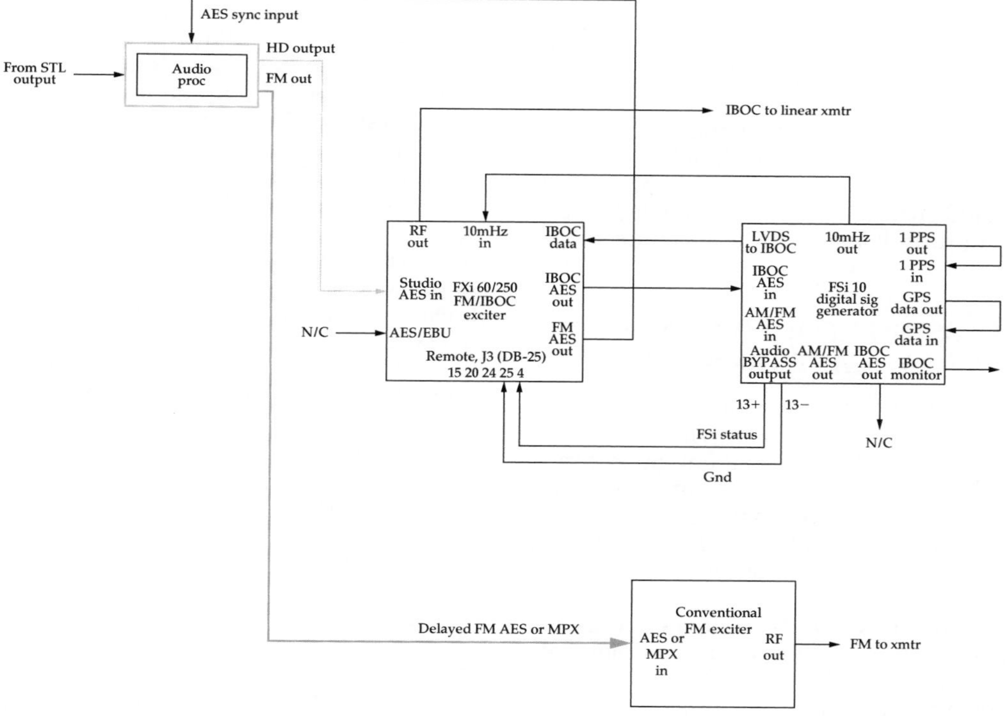

FIGURE 3.8-13 HD Radio setup illustrating how analog and digital portions of an IBOC signal can be isolated.

Audio Processing for FM Surround Sound

As of this writing, a number of systems (specifically, Neural Audio®, Dolby® Pro Logic® II, and SRS Circle Surround®) have been developed to offer 5.1 surround sound via the HD Radio system. Although each method employs a different surround mechanism, they all make use of a two-channel signal transport; thus, conventional two-channel processing will work for surround. The only caveat is that the gain functions must remain strapped or coupled together or else the surround field will become disjointed or distorted. Any independent left/right or matrix sum/difference processor will not work correctly when processing for surround.

All of the above-mentioned systems are forms of so-called matrix encoding, which typically increases the level of the stereo difference channel (L–R). This is known to increase multipath distortion in analog FM transmission (whether the L-R increase is due to surround processing or some other factor). It is important to test such a system carefully before broadcasting audio encoded with it on the FM analog channel. Although it is possible to broadcast a normal stereo signal on the analog channel and a matrix surround signal on the digital channel, this significantly complicates almost every aspect of the broadcast facility and will also cause comb filtering during crossfades in an HD-FM receiver.

Another surround technology is MPEG Surround, which is still under development as of this writing [25]. This technology allows a normal two-channel stereo signal to be broadcast on both the analog FM and HD-FM channels. MPEG Surround makes use of a *spatial encoder*, which derives a low-bit-rate digital steering signal from the original 5.1-channel source. This is transmitted as a sidechain data signal on the HD-FM data stream.

It is practical to use a two-channel audio processor to process the spatial encoding audio signal. The steering signal must pass around the audio processor through a dedicated channel that applies a delay equal to the input and output delays of the audio processor.

AUDIO PROCESSING REQUIREMENTS FOR TELEVISION BROADCAST TRANSMISSION

The processor should provide a comfortably listenable dynamic range in domestic listening environments by applying subtle compression to the signal. Such compression should be undetectable to the ear unless the original source is available for comparison. Usually an available gain reduction range of 25 dB is adequate to handle the level variations encountered in typical operations.

In analog TV using an FM aural carrier, the processor must provide high-frequency limiting to complement the preemphasis employed (25 μsec or 50 μsec, depending on the region in which the transmission occurs). The processor should provide accurate peak control (as measured by a modulation monitor meeting the standards of the governing authority) in both the positive and negative directions. In general, the comments on FM (given previously) apply here as well.

The processor should control subjective loudness to prevent unpleasant inconsistencies when transitions occur between various program elements. This is most accurately achieved using technology similar to that developed for loudness measurement (discussed earlier). In essence, the processor uses a loudness meter in a servo loop to control loudness and ensure consistency of loudness between one program source and the next.

The processor should handle voice cleanly. The Hilbert Transform Clipper and delay-line limiter are effective for this, because neither creates audible clipping distortion on voice, even when the source is narrowband (such as optical film or telephone). Such narrowband sources are extremely difficult for a conventional audio-frequency clipper to process without introducing some audible harmonic distortion on voice.

Audio Processing for Stereo Television

The general requirements for stereo television processing are not very different from the general requirements enumerated previously. As discussed in the "Processing for Stereo" section earlier in this chapter, the processing elements with slow-release time constants must be coupled to preserve the stability of the stereo image. In the North American Broadcast Television Systems Committee (BTSC) system, the peak modulation criteria are complex; however, it can be shown that FM-stereo-style processing will always prevent overmodulation in BTSC stereo, although it will not necessarily allow the greatest L–R modulation theoretically possible in this system. This style of processing is also appropriate for the other international stereo systems [26], as it will always prevent overmodulation.

Because of the close proximity between the edge of the audio passband (approximately 15 kHz) and the stereo pilot tone (15.734 kHz), the BTSC system requires sharp low-pass filters to prevent aliasing. It is impractical at the current state of the art to apply nonlinear overshoot compensation to these filters. Such overshoots do not cause interference or problems in television receivers; thus, they must be accepted as inherent to the BTSC system and must be ignored by modulation monitors designed as a reference for setting modulation levels. If these overshoots are not ignored, average modulation will be set too low, and the viewer will experience annoying increases in loudness when switching from stereophonic to monophonic channels.

Audio Processing for Digital Television (DTV)

The digital television system specified by the Advanced Television Systems Committee (ATSC) and approved by the FCC for use in the United States specifies 5.1-channel audio. The channels are left, center, right, left surround, right surround, and a limited-bandwidth subwoofer channel (the 0.1) for effects. The audio is digitally compressed using the Dolby AC-3 system. This system specifies two auxiliary data channels. The first, dialog normalization (DIALNORM), provides information to the receiver about the nominal level of the dialog so the receiver can hold this constant. This lets the broadcaster trade off headroom (for loud sound effects, for example) against noise floor without changing the loudness of dialog at the receiver. The second auxiliary data channel is a dynamic range control (DYNRNG) channel that provides a wideband gain reduction signal that can be used by the receiver, under the viewer's control, to selectively compress the dynamic range of the broadcast.

At the time of this writing, it is unclear how these signals will be operationally implemented in a typical television broadcast environment, particularly with regard to older material mixed without these signals and live news. Each U.S. television network has its own technique for accomplishing this and for routing signals around the facility. Most local stations either do nothing with DIALNORM and DYNRNG or in some cases pass through the network processing signals.

TECHNICAL EVALUATION OF AUDIO PROCESSING

Common swept frequency response, harmonic distortion, and intermodulation distortion tests are often used to evaluate audio processors; therefore, it is useful to discuss why these tests may at times produce misleading results.

Definition of Linearity

A system can be tested for linearity as follows. Apply an input signal A to the system and measure its output; let X be the output signal caused by input A. Then, remove A from the input and apply another signal B; let Y be the output signal caused by input B. The system is linear if the following things happen:

1. If the input waveform is multiplied by a factor k to scale it, the output waveform also becomes scaled by a factor of k, but its shape is not distorted by the process of scaling.
2. If inputs A and B are applied to the system simultaneously, the output of the system is X + Y (*superposition*).

It is clear that expanders, compressors, and limiters are strongly nonlinear systems. The output of such devices is not scaled proportionally to their input; they are expanded or compressed. Similarly, when two signals are applied to such devices, their output is not the same as the sum of their response to either signal individually; superposition does not hold. Clippers are similarly nonlinear.

Sine-Wave Measurements and Nonlinearity

When predicting the response of a system to program material by measuring its response to individual sine waves, certain assumptions are made. The first assumption is that program material can be adequately represented as a sum of sine waves (Fourier analysis). The second assumption is that superposition holds, so the response of the system to single sine waves also applies when several sine waves are added together at the system's input. Thus, the sine-wave results can be extrapolated to program material.

Because dynamic audio processing (compression, limiting, clipping, expansion, gating) is strongly nonlinear, the usual assumptions of superposition and scaling, which permit sine-wave measurements to be extrapolated to complex program material through Fourier analysis, do not hold. Conventional harmonic and intermodulation distortion measurements, historically designed to measure slight departures from linearity in weakly nonlinear systems, are of very limited usefulness. Swept or spot frequency-response measurements are not useful.

When making distortion measurements with tones, their relevance must be assessed psychoacoustically. Does the system output *sound* distorted when listening to the tones? For example, when measuring harmonic distortion using fundamentals in the 50 to 1000 Hz region, the higher harmonics are more significant than the lower harmonics because the higher harmonics are less readily masked by the desired fundamental. However, as the fundamental frequency is increased, the harmonics become less troublesome because the ear becomes less and less sensitive to them. Eventually, their frequency exceeds the passband of the system, and they become irrelevant.

Similarly, the Society of Motion Picture and Television Engineers (SMPTE) intermodulation (IM) distortion methods measure the level of 50 or 60 Hz sidebands around a high-frequency tone induced by system nonlinearity [27]. Because these sidebands are within a single critical band (approximately 1/3 octave) of the high-frequency tone, they are maximally masked by it; therefore, rather high amounts of measured SMPTE IM distortion are not necessarily cause for concern. On the other hand, ITU-R difference–frequency intermodulation distortion measurements measure the low-frequency difference tone caused by two high-frequency tones. Because the difference tone is far removed in frequency from the desired tones, it is not well masked by them, and high amounts of ITU-R IM are of some concern. (See Chapter 8.1, "Audio Signal Analysis," for more information on distortion and measurements.) In all cases, it is not appropriate to attempt to extrapolate the results of tone tests to program material, because superposition does not hold.

Subjective Listening Tests

Few, if any, measurement techniques can adequately predict whether the subjective effect of an audio processor will be satisfactory. The only effective way to evaluate nonlinear broadcast audio processing is by *subjective listening tests*. These must be done over a long time period, using many different types of program material, because a processor that sounds good on a certain type of program material may sound unsatisfactory on other program material having markedly dissimilar spectral balance or dynamics. Usually, the subjective goal of broadcast processing is to have its action undetectable to the audience. In the case of processing in highly competitive major market stations, some degradation of the program (as perceived on a high-quality monitor) is often accepted for the sake of maximizing punch and loudness. Moderate quality compromises are usually masked on smaller and lower cost radios and are noticeable only on higher-quality radios by critical listeners.

Clarifying Audio Processing Objectives

Success in installing and getting the most out of a processing product is directly related to how well the objectives for the system have been developed. With a clear set of objectives, processing tasks will be more clearly defined. Whether the goals are better overall quality or specific spectral improvements, goals and objectives should be articulated and written down. Is the goal a little more loudness or "presence" on the dial? Are there certain characteristics of the sound of other stations in the local radio market that it would be nice to emulate? Are there any to be avoided? If multiple products are being auditioned, it is equally important to have a list of specific factors to be used in making a comparison with the present audio processing system. Engineering, programming, and management should all participate in the development of the objectives. After all, processing can have a direct effect on the bottom line of the station. Agreement before installation can avoid disagreement afterwards.

Available Time

It takes time, a good deal of serious time, to process station audio. Audio processors are expensive and multifaceted. Avoid installing or modifying processing settings in between other major projects or when staff members who help set audio processing objectives are not available for consultation. Working with a system for at least a week is a good starting point, and this should be done during a week when a few hours every day can be spent on the project.

Installation and Adjustment Considerations

Installing or adjusting processing equipment requires more than mounting it in the rack, connecting some cables, and then putting it on the air. Among the factors to be considered are:

1. *A good monitor location* with which the system can be monitored along with a good tuner, with good reception, feeding a set of studio monitor speakers should be sufficient. While car radio and other typical listener situation settings are important, they should not be the main reference points.
2. *Good, clean source material* is essential. When poor source material or poorly performing playback equipment is used, maximum sonic benefit cannot be expected. Anomalies that are perceived to be processing problems may actually be source problems exaggerated by the processing.
3. *Microphone processing* may appear trivial, but the perceived sound of "live" voices over the air can change dramatically with different processing systems. Whatever the effect, on-air microphone sound will probably change when processing is changed. If separate mic processing is utilized, it may have to be adjusted to suit the operation of the new processing system. Most announcers develop a comfort zone with respect to the sound of their voices over the air. When that comfort zone is changed or modified, the common response is that something is wrong. Mic processing personalized to a particular person can be a very important part of the overall station sound.
4. *Operating level* is another simple area where trouble can develop. Make sure the input and output levels of the system are operating at the proper level within the system (operating a processor with insufficient levels into an STL system may cause loss of modulation and loudness). When comparing different processors, be sure they are both operating at the same levels. The modulation monitor is an important tool in any processing comparison.
5. *Pick a starting point for reference*. Proceed from a processing level similar to that currently used by the station and then, if desired, become more aggressive. This is less likely to draw hasty and negative opinions.
6. *Listen for awhile, then adjust*. Try to avoid the temptation to fiddle with adjustments moments after making initial changes. Processing changes should be evaluated over time, rather than moment to moment. When making adjustments, create worksheets that can assist in establishing improved settings of operation.
7. *Avoid making hasty or radical changes or making many different adjustments at once*. If too many parameters are changed at one time, it is difficult to determine which change made the difference. It can be frustrating trying to figure out which and whether the change made the sound better or worse!
8. *Use the "sleep on it" method*. Spend time adjusting and then listening, and when the system gets to a point where it sounds good stop for the day. When making changes, there does come a time when the ears become less and less sensitive to adjustments performed. That is why spreading the adjustment period over a number of days is recommended. If it still sounds good after having slept on it, quit adjusting. If it does not, continue with this method

until you are satisfied. If the procedure is working, it is likely that each day the differences will be smaller and fewer adjustments will have to be made.

9. *When the sound is what is wanted, STOP!*

SUMMARY

Audio processing offers broadcasters an opportunity to create a signature sound and to help compensate for problems in the audio and broadcast chains, but it must be used with care and with the knowledge of what it can and cannot do. The material in this chapter provides broadcasting engineers with information on how processors work, some pitfalls to avoid, and how to put them to best use.

Notes

[1] BBC, *The Dynamic Characteristics of Limiters for Sound Programme Circuits*, Research Report No. EL-5, British Broadcasting Corporation, Engineering Division, London, 1967.

[2] Haller, R. A., *An Update on the Technology of Loud Commercial Control*, OST Technical Memorandum FCC/OST TM83-1, 1983.

[3] Jones, B. L. and Torick, E. L., A new loudness indicator for use in broadcasting, *J. SMPTE*, Sept., 772, 1981.

[4] Orban, R., Increasing coverage of international shortwave broadcast through improved audio processing techniques, *J. Audio Eng. Soc.*, June, 419, 1990.

[5] Stevens, S. S., The measurement of loudness, *J. Acoust. Soc. Am.*, 27, 815–829, 1955.

[6] ISO 532-1975(E), Method B.

[7] Anon., How loud is my broadcast?, *J. AES*, 52(6), 662–666, 2004; Anon., Program loudness revisited, *J. AES*, 53(10), 945–948, 2005.

[8] Dolby Laboratories Model LM-100.

[9] Reidmiller, J., Robinson, C. Q., Seefeldt, A., and Vinton, M., *Practical Program Loudness Measurement for Effective Loudness Control*, AES Preprint 6348, paper presented at the 118th AES Convention, Barcelona, Spain, May 28–31, 2005.

[10] http://www.dorrough.com/dorrough/about_us/about_vision.html.

[11] ATIS is a U.S.-based industry standards setting organization; its Network Performance, Reliability and Quality of Service Committee (formerly designated T1A1) develops and recommends standards, requirements, and technical reports related to the performance, reliability, and associated security aspects of communications networks, as well as the processing of voice, audio, data, image, and video signals, and their multimedia integration.

[12] SECAM customarily uses AM sound, with the usual requirements for preventing carrier pinch-off.

[13] A phase-linear filter has constant delay with frequency.

[14] A minimum phase filter has no zeros in the right half of the *s*-plane. As its name implies, there is no filter with the same magnitude response that can have less phase shift. Given the magnitude response of a minimum-phase filter, its phase shift can be computed (with the Hilbert transform). This means that, if a minimum-phase filter has constant group delay in its passband, this is associated with a certain type of magnitude response that rolls off gently around the filter's cutoff frequency. A minimum-phase filter with constant group delay in the passband cannot simultaneously have a highly selective magnitude response. Many textbooks provide the well-known mathematical details. See, for example, Blinchikoff, H. J. and Zverev, A. I., *Filtering in the Time and Frequency Domains*, Wiley & Sons, New York, 1976, pp. 89–94.

[15] Gottlieb, D. and Shu, C.-W., On the Gibbs phenomenon and its resolution, *SIAM Rev.*, 39(4), 644–668, 1997.

[16] J.17 is a first-order shelving preemphasis with a zero at 477 Hz and a pole at 4134 Hz.

[17] For additional information on NICAM, see Bower, A. J., NICAM 728: digital two-channel sound for terrestrial television, *IEEE Trans. Consumer Electron.*, Aug., 286, 1987.

[18] A good discussion of the theory and operation of PPM may be found on the Internet at http://en.wikipedia.org/wiki/Peak_programme_meter; for additional information on the VU meter, see Section 21.2 in *Handbook for Sound Engineers: The New Audio Cyclopedia*, Ballou, G. (Ed.), Howard Sams & Co., Indianapolis, IN, 1987.

[19] Simply stated, almost all feedback systems contain a filter that forces the open loop characteristic to be either low pass (all-pole, *lag compensation*) or low-pass shelving (poles and zeros; *leadlag compensation*). Feedback forces the amplifier before this filter to present a preemphasized signal to the filter's input such that the total response of the system is flat. If the filter rolls off at 6 dB/octave starting at 15 Hz (a typical situation in an op amp like the TL072 or the LF353), this preemphasis *rises* at 6 dB/octave starting at 15 Hz. High frequencies applied to this system will obviously challenge the headroom of the amplifier prior to the filter; for example, 20 kHz will be up 62.4 dB! If high frequencies drive the amplifier prior to the filter into clipping or substantially nonlinear operation, transient intermodulation distortion occurs. Because a filter with a low-pass characteristic follows the clipping process, harmonics generated by clipping will be deemphasized, so difference–frequency IM tests are more sensitive than THD tests to this mechanism.

[20] For a maximum audio bandwidth *f*, the required slew rate in percent modulation per microsecond is 0.0002π% per μsec. For 4.5 kHz, this is 2.827% per μsec.

[21] NRSC standards are available on the Internet at www.nrscstandards.org.

[22] See NRSC-5-A, Digital Radio Broadcasting Standard, for additional information on the iBiquity HD Radio™ system (as standardized by the National Radio Systems Committee); available at www.nrscstandards.org.

[23] See European Telecommunications Standards Institute (ETSI) standard ETSI EN 300 401 V1.3.3 (2001-05), Radio Broadcasting Systems; Digital Audio Broadcasting (DAB) to Mobile, Portable, and Fixed Receivers, for additional information on the Eureka-147 system, and ETSI TS 201 980 v2.1.1 (2003-12), DRM System Specification, for additional information on the DRM system (www.etsi.org).

[24] As discussed earlier, this is not true for loudness measurement with arbitrary program material; however, here we are comparing two signals that represent the same program material. They differ only in their peak-to-average ratios and possibly in their high-frequency spectra if substantial high-frequency limiting has been applied to the analog FM channel.

[25] Herre, J., Purnhagen, H., Breebaart, J., Faller, C., Disch, S. *et al.*, *The Reference Model Architecture for MPEG Spatial Audio Coding*, Preprint 6447, paper presented at the 118th AES Convention, Barcelona, Spain, May 28–31, 2005.

[26] West German (dual-carrier); Japanese (FM subcarrier); English NICAM (block-companded digital).

[27] See RP 120-1994, SMPTE Recommended Practice, Measurement of Intermodulation Distortion in Audio Systems.

CHAPTER

3.9

Radio Remote News and Production

SKIP PIZZI
Microsoft Corporation
Redmond, Washington

JERRY WHITAKER
Advanced Television Systems Committee
Washington, D.C.

Updated for the 10th Edition by

JIM BOSTON
DTV Engineering
Auburn, California

INTRODUCTION

Radio stations use remote location broadcasts to bring the listener an added sense of realism and excitement of an eyewitness account of events. Although the concept of the *remote*, as it is better known, has not changed substantially over the years, the means to accomplish the task have made significant improvements in terms of performance, ease of operation, and reliability. The object is still to provide the necessary production equipment on location to capture the required action and to transmit the sound back to the radio station. This chapter addresses the two major technical areas in remote broadcasting:

- The portable production systems used to create the program (field production); and
- The method of backhaul (the means of delivering the signal from the remote site to the broadcast studio location).

The latter, often referred to as Radio Electronic News Gathering (RENG), utilizes either wired technology (i.e., telephone lines) or wireless RF links in the broadcast auxiliary or other spectrum. Technology for delivery over "plain old telephone service" (POTS) and Integrated Services Digital Network (ISDN) lines has advanced dramatically in recent years. As with any other area of telecommunications, the key to a successful RENG system is thoughtful planning.

PLANNING THE RENG

The importance of careful planning of a RENG system cannot be overemphasized. The network should be configured based on the specific needs of the broadcast and location. Ideally, everyone involved in the use of the system should be consulted to determine what will be needed to accomplish the requirements of the RENG system in concert with the station's technical capabilities. Whether a broadcaster's format is news, talk, or music, staff in the news, production, and engineering departments should work together to define the requirements of the assignment. At such gatherings, engineers should be open to any unique requirements that the assignment might require.

Most RENG systems have been built on a piecemeal basis, as needs and economics allowed. The lack of a unified plan has often led to RENG systems that are cumbersome to operate and, in the long run, more expensive to build than necessary for a given level of performance. The size and layout of the broadcaster's market will have a substantial effect on how the RENG network is designed. A system intended to cover a sprawling urban area of 10,000 square miles will be configured much differently than a tightly clustered urban center covering 2,000 square miles. The number of broadcasters in the market that are involved in RENG activity may also affect how a system is designed and what types of equipment are used. Broadcasters in major metropolitan areas that are expanding their remote operations may find that few, if any, frequencies are available for RENG activity. Program material can be returned from the field to the studio through either of two common routes: wired lines or wireless transmission systems of various types. The route back to the studio will depend on a number of factors, including the location of the event, availability and type of telephone service in the area, amount of setup time provided, and duration of the broadcast. Wireless cellular technology as applied to phones and computer networks is allowing backhaul connectivity not contemplated by broadcasters a generation ago.

Wired versus Wireless

Until the 1960s, the word "remote" was rarely spoken without reference in the same sentence to the telephone company. Wired systems, either using the dial-up POTS network or leased broadcast loops, provided most interconnections from remote broadcast sites to a station's studio facilities. Since that time, however, RF systems have assumed an important role in remote activities because they offer greater flexibility, offer higher-quality audio, and are more cost effective than leased program circuits from the telephone company.

Radio systems are ideally suited for broadcasts from several different locations during a short span of time. Longer lasting or regularly occurring remotes from sports arenas or press rooms and other regularly scheduled events may be best handled by a wired arrangement. The amount of frequency congestion in the origination area will also have an effect on which method a station will choose for the greatest reliability. Urban areas in which secure remote pickup unit (RPU) channels or line of sight (LOS) transmission paths are difficult to find are also good candidates for wired links.

The amount of lead time provided before a remote is scheduled to occur will also have a significant effect on the backhaul method chosen. Remotes scheduled weeks in advance are well suited for use of a telephone company (telco) service. Breaking news dictates a rapid remote method provided by an RF system.

The cost of telco facilities must also be considered. Unless the service is to be left in place for a long period of time, installation costs can become prohibitive, especially if equalized loops or ISDN is needed. Many stations are able to justify the cost of a RENG RF system based solely on the telco savings that can be anticipated.

Wired Connectivity

The best approach may be to use both wired links and a wireless system. Large-scale systems are often built using both interconnection methods, either as various links in the chain or as redundant services for recovery from a partial system failure. While earlier dial-up (switched) and dedicated (leased-line) telco services were exclusively analog, newer digital systems are commonplace in most areas. For dedicated paths, the T-1 service (also referred to as DS1) may be appropriate for broadcast use, particularly when multiple and/or bidirectional circuits are required between remote site and station. The T-1 format is a bidirectional digital service with a data transmission rate of 1.544 Megabits per second (Mbps) in each direction. It is primarily used by telcos to carry 24 voice-grade channels of 3.5 kHz bandwidth each, but can be easily reconfigured to carry a smaller number of wider-bandwidth signals by means of changing plug-in cards in terminal racks at each end of a dedicated T-1 path.

Program audio can be carried on T-1 in uncompressed or compressed digital form. A T-1 circuit can carry one uncompressed CD-quality stereo audio signal in each direction (occupying most of the 1.5 Mbps bandwidth), while it may handle six or more such signals in compressed form. Alternatively, a mix-and-match configuration can be established with a variety of different bandwidths depending on the quality of service needed and whether the service is needed on a temporary or permanent basis (such as a news bureau or sports arena). For example, a T-1 link could be configured to carry several 15 kHz stereo program circuits bidirectionally (for transmission from, and monitoring returns back to, the remote site), plus bidirectional 5 kHz communication paths, bidirectional fax lines, voice circuits, and data control signals.

In some cases, a half-DS1 circuit or some other partial-carrier arrangement may be obtained, when capacity needs are not so great. While not available in all areas, this service is called Fractional T-1. In most cases, the T-1 digital service will be significantly less costly than its analog program circuit equivalents (particularly when considering that the latter are not inherently bidirectional).

ISDN

Another digital telco service is the Integrated Services Digital Network (ISDN). ISDN is a dial-up digital service that was originally intended to provide a transport vehicle that delivers digital data services (voice and data) to the end user. Other methods for delivering those services at higher bit rates have evolved, but ISDN has found some niche usage in the broadcast market. It uses a telco's existing network of copper lines. The most common form of ISDN service, basic rate interface (BRI), provides two 64 kbps channels for program audio; this is the form of ISDN of most interest to broadcasters and audio professionals. Using a single pair of ordinary phone wires, BRI offers two "bearer" channels at a 64 kbps transmission rate and one "data" channel at 16 kbps. This configuration is often referred to as 2B+D. ISDN is full duplex, and calls are dialed and routed just like analog calls. The two "B" channels can be used for bidirectional audio (transmitted as digital data), ancillary RS-232 data, and inter-unit signaling. The "D" channel is reserved exclusively for telephone network signaling. The ISDN terminal gear acts much like a modem does for data on an analog network.

There is also ISDN Primary Rate Interface (PRI). In North America PRI offers 23 "B" channels and one "D" channel. While most broadcast equipment does not support PRI directly, special equipment or a PBX switch can break a PRI into multiple BRIs.

Using a technique called inverse multiplexing (IMUX), the two bearer channels can be combined to provide a 128 kbps path, which allows good stereo or excellent mono audio fidelity in a compressed digital form. ISDN therefore puts real-time, high-quality audio backhaul only a (digital) phone call away. Like POTS, either local or long-distance ISDN paths can be routed via dialing. ISDN fee schedules are also similar to or slightly higher than POTS costs.

The program or source audio is converted by a codec to data, typically via MPEG Audio Layer III (15 kHz mono over a single ISDN B channel) or G.722 (7.5 kHz mono, low delay) over a single ISDN B channel. Encoding delays should be taken into account. G.722 has a low encoding delay of only 6 ms. MPEG Audio Layer III is typically much longer—in the several hundred millisecond range.

Note that broadcasters using ISDN services for high-quality audio mono or stereo transport must purchase or lease specialized terminal equipment (including codecs, when compressed audio is used). An initial investment of several thousand dollars for a portable codec and a rack-mounted companion (decoder) may be amortized over the number of remotes scheduled or anticipated.

DSL versus ISDN

Another technology available from telcos is DSL (Digital Subscriber Line). This is a technology that is being offered by telcos and costs less to implement than ISDN. Digital Subscriber Line (DSL) technology extends the capacity of standard twisted pair circuits to T-1 data rates and higher. DSL is being implemented in a number of different forms. The most common is the Asymmetrical Digital Subscriber Line (ADSL), which provides high-speed service in one direction (typically from the central office to the subscriber) and lower-speed service in the other, plus simultaneous POTS—all on standard copper telco loops (two twisted pairs). Figure 3.9-1 shows a typical ADSL spectrum.

ADSL can provide speeds ranging from 64 kbps to 1 Mbps upstream and from 1.5 Mbps to 8 Mbps downstream plus POTS. ADSL is limited to 18,000 ft from the central office (CO). High-bit-rate DSL (HDSL) is a symmetrical service, offering from 2 Mbps to 6 Mbps in both directions. Very high bit-rate DSL (VDSL) offers extremely high data rates over shorter distances from a CO such as 13 Mbps up to one mile, or up to 51 Mbps at 1,000 ft. DSL (particularly its HDSL variant) could have significant value for wired remote audio backhaul without data compression where local conditions (distances to CO) permit.

There is confusion by users and even telcos themselves as to when or where ISDN makes more sense than DSL. ISDN is not the best choice for packet connections to the Internet. Packet-data-using protocols like TCP/IP allow for data to be lost and then resent. Packet data does not support "real-time" applications

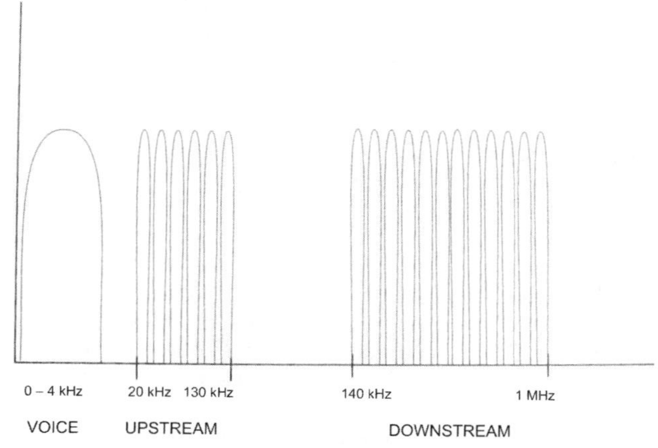

FIGURE 3.9-1 ADSL spectrum: The lowest 20 kHz is reserved for voice; carriers spaced 4.3125 kHz between 20 kHz and 130 kHz are used for uploading, and carriers between 140 kHz and 1 MHz are used for downloading. The methodology allows for up to 640 kbps upload speed and up to 8 Mbps download speed.

very well. Systems designed for Internet use have large buffers which allow time for packet retransmission. These large buffers result in potentially large delays in audio applications.

ISDN is still the better choice where circuit-switched connections are needed. Just as with a POTS connection, once a connection is established, the path has a full-time connection. Many telephone companies that offer DSL use proprietary technology, which means it is necessary to use their "Data Communications Equipment" (DCE, or modem). In addition, a DSL connection might go through several networks to get to the local ISP, unlike a T1 connection, which goes directly to the local ISP.

POTS Systems

For most remote backhaul needs, the conventional analog dial-up telephone network combined with cellular telephone technology provides a broadcaster with the greatest degree of flexibility in terms of when and where it is used, including bidirectional capability. Today, telephone connectivity is a mix of wired and wireless technology. POTS/cell audio quality leaves a great deal to be desired, as it has a bandwidth of about 4 kHz. For this reason, a number of encode-decode algorithms have been developed that improve the fidelity of audio signals that pass through a POTS line.

One improvement to POTS audio transmission combines the technologies of computer modems with the perceptual coding (or data compression algorithms) used with digital audio. This class of devices, generically called *POTS codecs*, provides approximately 7 kHz audio transmission in real time on a single POTS line. The POTS codec turns a monaural analog audio input into a digital signal, applies a relatively heavy (around 30:1) data compression process that results in a data stream of about 24 kbps, and feeds it to the POTS line via a standard 28.8 kbps modem. At the receive end, the process is reversed for a real-time audio output (actually, the term "real-time" in this context is inexact, since the throughput delay on a POTS codec path can be 500 ms).

Some POTS codecs can adapt to variable phone line conditions and reduce their transmitted data rates to accommodate degraded line conditions, with corresponding reductions in audio bandwidth. For example, a connect rate of 9.6 kbps yields 4.7 kHz audio bandwidth, and a connect speed of 33.6 kbps can carry up to 14 kHz audio. In some cases (particularly for international calls), line quality may be so poor that POTS codecs cannot operate reliably or at all. Some POTS codecs also include capacity to handle auxiliary control data or contact closures to remotely trigger an event back at the studio.

Modern remote codecs are compact devices. Most are equipped to handle full-duplex audio, with up to 15 kHz audio bandwidth, over standard dial-up POTS. The system is designed so that audio quality scales automatically to compensate for poor phone lines. Program audio to the station and interruptible foldback (IFB) mixed with real-time program audio returning from the station can be sent over the same phone line. This is advantageous when off-air monitoring is delayed from real time, such as when using HD Radio™ systems or when using a profanity delay so that the on-air talent does not experience a long delay in hearing returning program. IFB is often sent via a "mix-minus," which is a separate mix or feed out of the audio mixer that has all audio components of the program with the exception of the talent's own voice. On-air talent, when hearing their own voices returning from the station that is over a few hundred milliseconds delayed, will find it distracting and will generally pull the IFB ear piece out of their ear and then put it back in when they stop talking.

FIGURE 3.9-2 Remote equipment can be as simple as a recorder and a codec, as shown in the photograph. The device between the headset and keyboard is a two-channel audio mixer that can interface with POTS, ISDN, and GSM services. The monitor is for an on-board PC that contains audio-editing software. This setup allows a driver/reporter to go solo to, and report back from, a breaking news event.

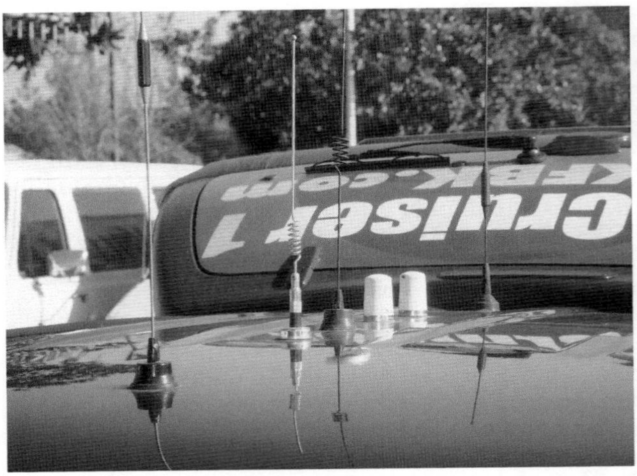

FIGURE 3.9-3 A RENG vehicle usually has an array of antennas on its roof, allowing communications in the traditional 455.9 MHz RPU and two-way radio bands, along with GSM, and even 802.11 WiFi services.

Full-duplex FM-quality (15 kHz) audio bandwidth is accomplished at connect rates of 24 kbps and above. Full-duplex audio scales from 5 kHz to 12 kHz at connect rates of 9.6 kbps to 21.6 kbps. Most devices allow for an ancillary data channel and extra forward error correction. Phone line quality is monitored continuously, and the modem renegotiates if the line degrades during the call. Renegotiation may be disabled to prevent dropouts at inconvenient times.

One very compelling reason broadcasters are adopting the new POTS codec equipment is the simplicity of the equipment (see Figures 3.9-2 and 3.9-3). This allows remote planners to pull operators from a greater pool of non-technical employees such as the actual on-location talent or promotion assistants. The need to budget expensive engineers can be drastically reduced.

Setting Up a Dial-Up Remote

A program transmitted back to the studio via a standard POTS line—without any bandwidth extension—will usually be brief in duration, if for no other reason than the poor audio quality typical of such an arrangement. Spot news reports are common examples of this method of program return. Small, battery powered mic-to-line amplifiers are available to drive dial-up telephones through direct connection to the tip and ring circuits of the phone company cable or through clip leads at the handset microphone pins. The direct connection method of coupling is preferred over the handset connection because the former bypasses the telephone hybrid coil assembly with its associated level loss and possible distortion. For this direct connection method to work, the device feeding the phone line—known as a coupler—must have the capability to seize the line (meaning to hold it open, in an off-hook condition). The advantage of the seizing coupler is twofold:

- The handset microphone will not contribute its output to the phone feed, so the signal transmitted will be only what comes from the amplifier or mixer plugged into the coupler's input; and,
- The transmitted level of the audio signal will be higher without the instrument off-hook, maximizing the signal-to-noise ratio of the feed.

The output level of the device feeding a phone line (whether by coupler or handset) should be carefully controlled, since the dynamic range of the dial-up network is somewhat limited. Transmit level should not exceed +8 dBm for program material, or 0 dBm for test tones. Be aware that the received end noise floor may be only 25–30 dB below this on local calls, and considerably less on some international long-distance calls.

The other option for dial-up remotes is ISDN, which is used for many permanent remote sites (concert halls, convention centers, sports venues, etc.) already wired with ISDN circuits ordered by broadcasters. In some areas the lead time for ordering new ISDN service is substantially longer than for POTS, and installation and service fees are also somewhat higher for ISDN.

Setting Up Leased-Line Remotes

The most important aspect in using dedicated, leased-line telco service for remote backhaul is getting the order with the telco placed properly and with sufficient advance notice. T1 service is often used for leased-line service for broadcast backhaul, since it provides a higher quality of service than the analog equalized broadcast loop or program circuit, or for that matter ISDN service. T1 can be used in applications that previously required analog broadcast loops, particularly for longer backhauls and local events of short duration or infrequent recurrence. T1 interface equipment are multiplexers that take multiple services, such as audio and data, and create a single T1 data stream. These multiplexers can also act as LAN bridges between locations.

Changing tariffs continue to alter the comparative costs of these services (analog loops, ISDN, and T1). Another advantage of T1 and ISDN is that traditional program loops are one-way paths, whereas T1 and ISDN are bidirectional. ISDN is pay-by-the-minute service, whereas analog loops and T1 are charged on a flat fee per month basis.

WIRELESS

While a wired service has traditionally been the better choice for long-distance remotes, wireless is generally a more convenient route for local remotes. Today's competitive marketplace requires sound from the field to be clean and quiet with good frequency response and low distortion. Radio stations can create powerful remote gathering and acquisition systems without building the traditional remote pickup unit (RPU) system. However, because most radio stations still have RPU systems, they are discussed in the following section.

RPU Systems

Radio channels used for traditional RENG work are typically shared by multiple broadcasters within a market, and so receipt of a license has been and continues to be no guarantee of unlimited interference-free operation. The frequency coordination process is a complicated procedure that requires careful thought and planning, and generally a great deal of lead time. The need for coordination is important to all persons involved in RENG activity in the region. The main driving force behind coordination efforts has been the Society of Broadcast Engineers (SBE), which encourages and supports local coordination efforts, and provides whatever support might be needed in this regard.

Licensing Procedures

RENG activity takes place on several bands of frequencies set aside by the FCC for RPU operation. A number of frequency groups are allocated near 150 MHz and 450 MHz. Some assignments are also made on frequencies near 26 MHz. A particular broadcast station is not restricted to a maximum number of RPU systems that it may put into operation.

Most RENG activity is currently centered in the 150 MHz and 450 MHz bands. In these slices of spectrum, three major license classifications exist: Automatic Relay Station (ARS), base station, and remote pickup mobile station:

- *ARS systems* are designed to receive program material on one frequency and retransmit on another. With multiple relays, the average area of the RENG system can be extended considerably.
- *Base stations* are, as might be expected, fixed-position transmitters used for communication between the central point and one or more remote points. Base stations may, in the event of emergency conditions, be used as a program relay channel for Emergency Alert System information.
- *Remote pickup mobile stations* consist of vehicle-mounted and portable (hand-carried) transmitters. They are usually licensed as a system in conjunction with a principal base station or stations. Remote pickup mobile station licenses generally specify a minimum and maximum number of mobile transmitters allowed in the RPU system. Standard divisions include from 1 to 4 stations, 4 to 12 stations, 10 to 20 stations, and 20 to 50 stations.

The Commission's Rules require that the transmitter power for an RPU station be limited to a level necessary for satisfactory coverage of the service area. In any event, not more than 100 W of transmitter power output will be licensed. RPU transmitting equipment operating onboard an aircraft is normally limited to a maximum transmitter power of 15 W. A mobile station consisting of a hand-carried or pack-carried transmitter is restricted to not more than 2.5 W power output.

All RPU transmitting equipment must be type-accepted by the Commission and must be checked each year (for units with more than 3 W output) for frequency accuracy, deviation, and RF power output. FCC Rules also require that RPU transmitters rated for 3 W or greater must be equipped with a circuit that will automatically prevent modulation in excess of the authorized limits. Typical audio bandwidth options are 7.5 kHz, 5 kHz, and 20 kHz. RPU transmitter deviation is +/– 20 kHz. A good choice of RPU transmitter is one with a mic/line option mixer built in. This eliminates a separate mixer (and the power to run it) and a possible failure point. There are virtually no operator requirements for the use of a unit in the RPU service. Any person designated by and under the control of the licensee of the station may operate the equipment.

Efficient Use of RPU Systems

The first rule of spectrum efficiency is to use only the effective radiated power (ERP) necessary to do the job. There is no justification for putting 15 W into the air when 5 W will provide the desired (or acceptable) signal-to-noise ratio to the receiver. Ideally, all transmitters in a RENG system would be equipped with continuously variable power output stages. The operator at the remote site would then run the transmitter with only enough power output to reach the required S/N figure at the receive (studio) point. With some types of units, this method of operation is possible, but in most cases, continuously variable power output transmitters are not available. User modification of existing equipment is not an acceptable solution, since such work would most likely invalidate the transmitter's FCC type acceptance.

A more practical solution, therefore, is to purchase RENG transmitters of several different power levels operating on the same frequency (or frequencies). All of the popular RENG broadcast equipment manufacturers offer units with different power output levels. With some equipment, a low-power transmitter is used and an optional power amplifier module is added between the transmitter and the antenna to give the needed RF output.

Directional receive and transmit antennas are appropriate from both an efficiency and coordination standpoint. The use of a pair of high gain antennas makes it possible to achieve a much greater ERP for the same transmitter power. Of equal benefit in a crowded urban area is the elimination of any nonessential radiation. Through the use of directional transmit and receive antennas, stations can establish more secure channels by placing the radiated energy where it will do the most good (from the transmit end) and rejecting unwanted signals from other directions (at the receive end).

A simple and sometimes effective coordination tool is *cross polarization*. Two stations on adjacent frequencies may achieve as much as 25 dB RF isolation through the use of orthogonal polarizations (horizontal and vertical, or left-hand circular and right-hand circular) of transmit antennas, matched by like polarization at their respective receive antennas. Cross polarization results in varying degrees of success, depending on the frequency of operation and the surrounding terrain. Line-of-sight paths usually will provide good cross-pol isolation results, but urban centers with their highly reflective buildings generally cause polarity shifts in the transmitted signal that may significantly reduce the benefits of cross polarization.

Mobile RPU transmitters can be configured to be operated by relatively unskilled operators. The entire remote process can be reduced to the following steps:

- Position the remote van and antenna for best LOS (line of sight) to receive site;
- Raise mast (avoiding overhead power lines and buildings) and point Yagi antenna array in the correct direction (at the receive site);
- Connect mics to RPU audio inputs; and,
- Transmit (radiate).

RPU serves news reporting very well with a vehicle-mounted transmitter and omni directional antenna. This type of operation depends on multiple receive sites, especially in dense metro areas, and, consequently, RPU planners should consider multiple receive sites with some type of dedicated backhaul circuit such as dedicated loops or remote-accessed autodial ISDN.

Cell Phones

Cell phones have become an integral tool used in newsgathering. Cell providers are generally assigned 832 frequencies in each market that they serve. The duplex requirement of the service requires two frequencies per call. Thus, there are 395 voice channels, plus 42 control channels per carrier. These voice channels are distributed across various cell sites so that the same channels are not found in adjacent cell sites. Each cell has 56 voice channels. Each cell averages 10 square miles of coverage. Low power phones and many cell sites make this concept possible.

Individual cell phones are tracked, as they travel from cell to cell, by a unique set of numbers. Each phone has a unique 32-bit Electronic SN (ESN) and a 10-digit phone number known as the Mobile ID # (MIN), plus the carrier's 5-digit System ID (SID).

As in other areas of technology, digital services have rapidly replaced analog approaches. Two primary approaches to implementing digital cellular service are in current use. The first and currently most prevalent is known as Code Division Multiple Access (CDMA), and it is used in second-generation (2G) and third-generation (3G) wireless systems in the 800 MHz and 1.9 GHz bands. CDMA does not assign a specific frequency to a user; instead, it uses the full available spectrum in a spread spectrum approach. The frequency use during a call varies according to a defined pattern or code. The receiver or cell site uses the same code to follow the transmission. The opposite takes place between cell and end user. GSM (Global System for Mobile Communications) service, the longtime global standard, is spreading throughout North America. Many modern codecs are now capable of making GSM phone calls. Most of these devices work as standard mobile phones when not using codec features, and they accept a standard SIM card from GSM providers.

GSM uses TDMA on a band that is 30 kHz wide and 6.7 milliseconds long. TDMA allows each band to be split time-wise into three time slots. Each connection gets the radio for one-third of the time. Thus, GSM allows three times the capacity of analog technology. This approach is possible because voice data that has been converted to digital information is compressed so that it takes up significantly less transmission space. GSM calls are either based on data or voice. Voice calls use audio codecs called half-rate, full-rate, and enhanced full-rate. Data calls can turn the cell phone into a modem operating at 9.6 kbps. An extended GSM feature is high-speed circuit-switched data, allowing the phone to transmit up to around 40 kbps. GSM operates in the 900 MHz band (890–960 MHz) in Europe and Asia and in the 1900 MHz band in the United States. It is used in digital cellular and PCS-based systems. GSM is also the basis for Integrated Digital Enhanced Network (iDEN), a popular system introduced by Motorola and used by Nextel.

The GSM network can be divided into three broad parts. The mobile station is the end user's mobile device. The mobile device communicates with the base station subsystem, which controls the radio link with the mobile station. The network subsystem, the main part of which is the Mobile Switching Center (MSC), performs the switching of calls between the mobile users, and between mobile and fixed network users. The MSC also handles the mobility management operations as the user moves from cell to cell. General Packet Radio Service (GPRS) is an extended service of GSM Network, which adds the capability to surf the Internet on a GSM system at 4 to 5 kilobytes per second. Of particular interest to the remote planner is that high-quality audio systems have evolved for remote broadcasting using GSM.

GSM codecs may pass up to 7.5 kHz audio depending on cell service to the user. This type of GSM modem is ideal for news reporting or long-form shows. GSM codecs are being successfully used by radio remote people from countries supporting GSM technology. These codecs support return IFB and multiple audio inputs for mics and recorded interviews.

Lastly, a GSM codec is used much like a desktop phone dialer, so it does not require any technical skills. These new units can be battery powered for up to three hours per battery, or by a vehicle power outlet.

Path Engineering for Fixed Stations

Base station and ARS systems are fixed-position installations that cannot always be located in the best possible geographic locations because of space availability problems, excessive construction or site rental costs, or local or federal licensing difficulties. Careful path engineering should be performed prior to any licensing work to determine if the proposed locations of base station and ARS installations will be able to achieve the desired results without using excessive amounts of transmitter power.

Planning for any RENG system should begin with an accurate, detailed U.S. Geological Survey (USGS) map covering the proposed path. Note should be made of any natural obstructions or Fresnel clearance obstructions (such as mountains, hills, or vegetation) or man-made obstructions (such as buildings, water tanks, or transmitting towers) in the proposed path. The transmitting and receiving antennas should be plotted so that a minimum of 0.6 Fresnel Zone clearance is obtained over 4/3 earth radius.

When one is planning a RENG path, a profile drawing of the transmitting and receiving antenna sites, the terrain and any obstructions in between, should be made on graph paper or with software set to 4/3 earth radius. This will compensate for the curvature of the earth and the normal refraction of VHF and UHF frequency signals when determining Fresnel Zone and obstruction clearance. Simple height above sea level is insufficient to determine whether a natural or man-made obstruction will interfere with the RENG signal on a long-distance path. Once a proposed path has been drawn, a visual inspection must be made of the area for any problems that could degrade the performance of the system. Particular attention should be paid to items not documented on the USGS maps, such as buildings and towers.

The terrain from the transmitting antenna to the receiving antenna must be examined not only for obstructions but for reflection possibilities as well. A large body of water will usually cause problems for a RENG system operating in the UHF frequencies. If the water is an even number of Fresnel Zones from the direct path, signal attenuation will likely occur at the receiver. Temperature changes and tidal conditions will also have an effect. Likewise, thick vegetation or forested areas can be reflective to RF signals when wet. Generally, the solution to reflection problems is to change either the transmitting or receiving antenna height or to employ a diversity reception system.

The site selection process for repeaters and receivers should also take into consideration the RF environment in which the equipment will be working. Multiuser locations can be very good transmit sites but are often terrible receive sites. For such situations, a remotely located receiver (especially in the case of a repeater system) should be considered. The two sections of the ARS station would then be tied together via analog or digital techniques. Digital telco services can be helpful because of their higher audio fidelity and freedom from cross talk and their ease in bidirectional, multiline applications.

Determining the Fade Margin

A gain and loss balance sheet should be computed to determine the fade margin of the proposed system. An adequate fade margin is vital to reliable performance of the system because a link that is operating on the edge of the minimum-acceptable receiver quieting will encounter problems later down the road. The RENG system fade margin can be computed by using the following equations:

$$G_s = G_t + G_{ta} + G_{ra}$$

where:

G_s = total system gain (dB)

G_t = transmitter power output (dBm)

G_{ta} = transmit antenna gain (dBi)

G_{ra} = receive antenna gain (dBi)

The values for G_{ta} and G_{ra} are gathered from the antenna manufacturer's literature.

The value for G_t is given by the following formula:

$$G_t = 30 + 10 \log P_o$$

where:

G_t = transmitter power output in dBm

P_o = transmitter power output in watts

Next, the system losses are computed:

$$L_s = L_p + L_t + L_c + L_m$$

where:

L_s = total system losses (dB)

L_p = path loss (dB)

L_t = transmission line loss (dB)

L_c = connector losses (dB)

L_m = miscellaneous losses (dB)

The values for L_t and L_c can be determined from the manufacturer's literature. Figure 3.9-4 shows typical loss values for 1/2-in. foam-filled transmission line. A reasonable value for connector loss with components normally used in 1/2-in. transmission line installations is 0.5 dB.

The value for L_p can be found by using the following formula:

$$L_p = 36.6 + 20 \log F + 20 \log D$$

where

L_p = free space attenuation loss between two isotropic radiators (dB)

F = frequency of operation in MHz

D = the distance between the antennas in statute miles

Now, the fade margin can be calculated:

$$\text{Fade Margin (dB)} = G_s - L_s - R_m$$

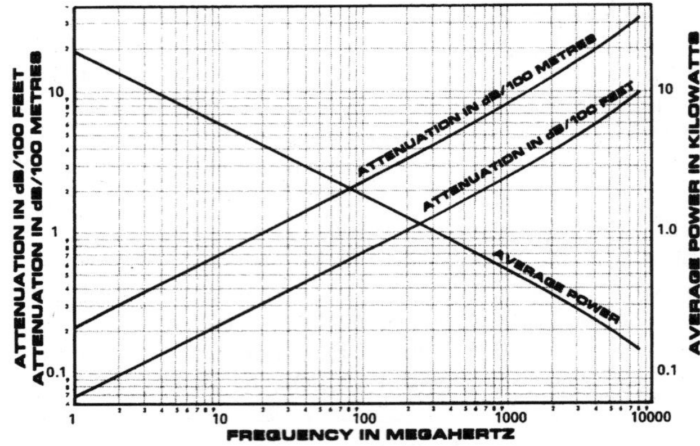

FIGURE 3.9-4 The attenuation and power handling ratings for ½ in. foam dielectric coax. (Courtesy of Andrew Corp.)

where:

G_s = total system gain (dB)

L_s = total system losses (dB)

R_m = receiver sensitivity – minimum signal strength required for target S/N (dBm, usually a negative number)

G_s and L_s are determined by the equations previously shown. R_m (receiver sensitivity) is determined from the receiver manufacturer's specifications. If the manufacturer gives a receiver sensitivity figure in microvolts, the following formula can be used to convert to dBm:

$$R_m = 20 \log \times (V_r \times 10^{-6}/0.7746)$$

where:

R_m = minimum required signal strength (dBm)

V_r = receiver sensitivity (microvolts)

R_m can also be measured by using a communications service monitor, available at most commercial and industrial communications radio service shops.

In order to predict accurately the performance of a RENG radio link, the value of R_m must be determined carefully. Many receiver manufacturers specify V_r for 20 dB of receiver quieting. This level is a convenient reference point; however, it should not be used for fade margin calculations. For maximum system performance and reliability, the fade margin determination should be made based on the signal level required to provide the minimum acceptable receiver signal-to-noise performance.

The recommended fade margin for a 150 MHz band RPU system is at least 10 dB plus 2 dB for each 10 miles of line-of-sight path distance greater than 10 miles. At 450 MHz, the fade margin should be increased to a minimum of 15 dB plus 3 dB for each 10 miles of path distance greater than 10 miles. These fade margins are designed to limit periods of performance degradation of the radio link to 1% or less during worst-case environmental conditions. The fade margin assumes transmit and receive antenna clearance above the ground and all obstructions of 50 to 100 ft.

While it is important to provide an adequate fade margin, needlessly high fade margins should be avoided because of the spectrum congestion problems that may result.

Other Planning Considerations

Path engineering for remote location broadcasts is seldom done for RENG activities because of the transient nature of such events, but general coverage surveys should be conducted to understand what areas are likely problem remote sites. The best location for a RENG system is not always the highest building in town. Placing a receive antenna at a high elevation in a metropolitan area can result in poor performance of the system in the downtown area, since the gain of many omni directional vertically polarized antennas decreases as the antenna is raised above the transmitting point. Tall buildings are excellent for point-to-point relay transmissions but are generally unsatisfactory for wide area coverage in a metropolitan region.

An inexpensive installation option is available to AM broadcast stations that do not want to erect a separate RENG transmitting tower at the main transmitting site. An isocoupler can be installed at the AM tower base that will pass the RPU transmitter signal with good efficiency (90% is typical), while at the same time presenting a high impedance to the AM band energy. Isocouplers are available in various frequency and power ranges. Installation of these devices may change the base impedance of the AM tower slightly; thus, an engineering consultant should be contacted before installation work begins.

Every effort should be made to locate the receiving antennas of a RENG system as far away from high-power transmitting antennas as possible. This should be attempted regardless of the frequency separation between the receive unit and the suspect high-power transmitting antenna. Failure to achieve adequate physical separation may require the installation of filters of various types on the receiver front-end.

Additional receiver interference isolation from adjacent transmitters can be obtained by installing cavity resonators. Cavity resonators, in series or parallel, may be inserted between the antenna and receiver to minimize intermodulation (IM) from paging transmitters and other RPU transmitters. The cavities introduce some insertion attenuation, but they are a real problem solver if a receiver, or antenna, is located in an RF-rich environment such as a leased tower. "Desensing" interference can be decreased by up to 50 dB. A low noise, in-line RF amplifier may be inserted after the cavity to boost the signal. Again, careful planning and a little "cut and try" can be beneficial.

In order to keep system losses to a minimum, a low-loss transmission line should be used, such as the 1/2-in. foam-filled coax or hard line. The transmission line and connectors must be made watertight if exposed to the elements. Each connector should be sealed with a silicone dielectric compound and then wrapped with good quality tape. Unless this is done, rain may eventually work its way into the connector and cause signal loss or VSWR problems. The line should be grounded (using a recommended grounding kit) at the point where it leaves (or enters) the equipment building and where it starts its climb up the tower (unless the vertical distance to the antenna is less than 10 ft). This will prevent any high-voltage transients caused by lightning from entering the equipment building, and thus the RENG equipment. The advantages of using a low-loss line are illustrated in Table 3.9-1.

In the table, note the poor performance figures for RG-58/U compared to hard line. LDF4-50 is 1/2-in. foam-filled line, LDF5-50 is 7/8-in. line, and LDF7-50 is 1 5/8-in. line.

The two ends of the transmission line (at the receiver and transmitter) are probably the easiest parts of hard line in which loss can be introduced, so care should be taken to install the lines and connectors according to good engineering practice. A short length of flexible coax is generally used on each end of the two transmission lines for connection to the equipment and antennas (when 1/2 in. or larger coax is used). This pigtail is normally no more than 18 in. long (see Figure 3.9-5).

TABLE 3.9-1
Typical Signal Loss for Popular Types of Transmission Line*

	150 MHz		450 MHz		950 MHz	
Type	**Loss (dB)**	**Efficiency (%)**	**Loss (dB)**	**Efficiency (%)**	**Loss (dB)**	**Efficiency (%)**
RG-58/U	6.0	25	12.0	6.3	20.0	1
RG-8/U	2.5	55	5.0	31	9.0	13
LDF4-50A	0.85	83	1.7	67	2.5	55
LDF5-50A	0.48	90	0.9	85	1.55	71
LDF7-50	0.28	94	0.56	88	0.88	84

*Loss = Loss in dB per 100 feet; Efficiency = approximate power transmission efficiency of 100-ft length. (Table data courtesy of Scala Electronics.)

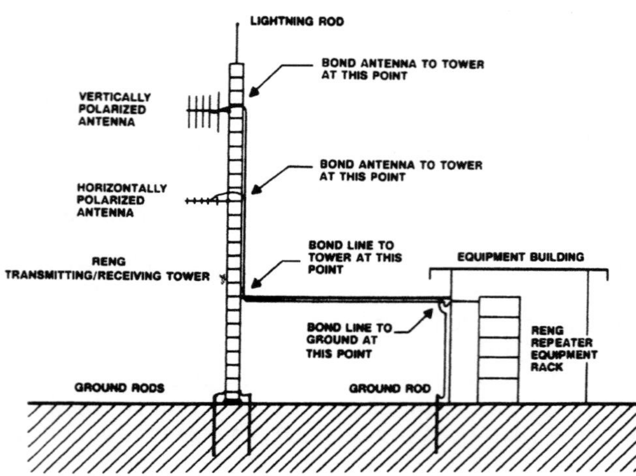

FIGURE 3.9-5 The recommended installation practices for RENG antennas and transmission lines.

FIGURE 3.9-6 Typical mast-mounted vertically polarized directional RENG antenna array.

Antenna Considerations

The selection of an antenna for use in a RENG system is an important decision because of the effect the antenna has on system performance and spectrum usage. The traditional RENG antenna has been the omni directional vertical whip with a small amount of gain. Now system planners are forced by interference concerns to use directional antennas with moderate amounts of gain. The omni directional base station antennas commonly used in the 150 MHz and 450 MHz bands are vertically polarized units with 4 to 6 dB gain. Electrical beam tilt is sometimes available. Depending on the manufacturer, up to 20° downtilt can be provided on 150 MHz antennas, and up to 11° is common for 450 MHz omni directional units. Large amounts of beam tilt are normally used when the antenna is to be mounted on a structure that is substantially above the surrounding terrain, thereby improving the antenna's close-in coverage.

The typical directional RENG antenna would be a medium gain Yagi that provides about 9 to 10 dB gain over a reference dipole, with a front-to-back ratio of approximately 14 to 18 dB. These antennas weigh only a few pounds and thus are small and light enough to be used on remote broadcasts. They are also suitable for permanent installations using either horizontal or vertical polarization. These antennas may be stacked in two- and four-bay arrays (with suitable phasing harnesses) for additional gain and directivity, as shown in Figure 3.9-6. Most Yagi antennas are made to match the specific frequency requirements of the equipment being installed. Multiple frequency operation using a single antenna is possible with reasonable VSWR numbers as long as the operating frequencies are not removed from the cut center frequency by more than 1 to 2%. Figure 3.9-7 shows the radiation pattern for a commonly used five-element 150 MHz Yagi.

Another addition to the RENG user's bag of electronic tricks is the broadband log periodic antenna, which can be used on any channel within a wide band of frequencies. Such antennas provide a smooth pattern with minimal side lobe radiation and a high front-to-back ratio (typically 25 dB in the 150 MHz band). Nominal gain for 150 MHz operation is 7 dB. Units can also be stacked to provide additional gain and directivity. Such antennas are usually larger and heavier than the familiar Yagi; however, they allow use of the antenna for virtually any frequency within the specified band at low VSWR levels (a maximum of 1.5-to-1 is typical). Horizontal or vertical polarization is available. Figure 3.9-8 shows the radiation pattern of a log periodic antenna designed for use in the 450 MHz band.

Just as a TV or FM broadcast antenna must be protected against icing problems, so should antennas used in RENG applications. Although antenna deicers are not used in RENG installations, a radome is often available for an antenna to protect it from damage or degradation in performance due to snow, ice, or salt spray.

Transmitter-Receiver Considerations

Receiving equipment usually presents the greatest challenge to a system designer, as it is most often subjected to conditions that may make good performance difficult. A receiver that has sufficient dynamic range and headroom should be selected to allow the system to deal with strong adjacent-channel signals, as well as very weak desired and very strong co-channel signals from transmitters in the network. A receiver with inadequate headroom will clip and cause distortion. Wide dynamic range active devices should be used in

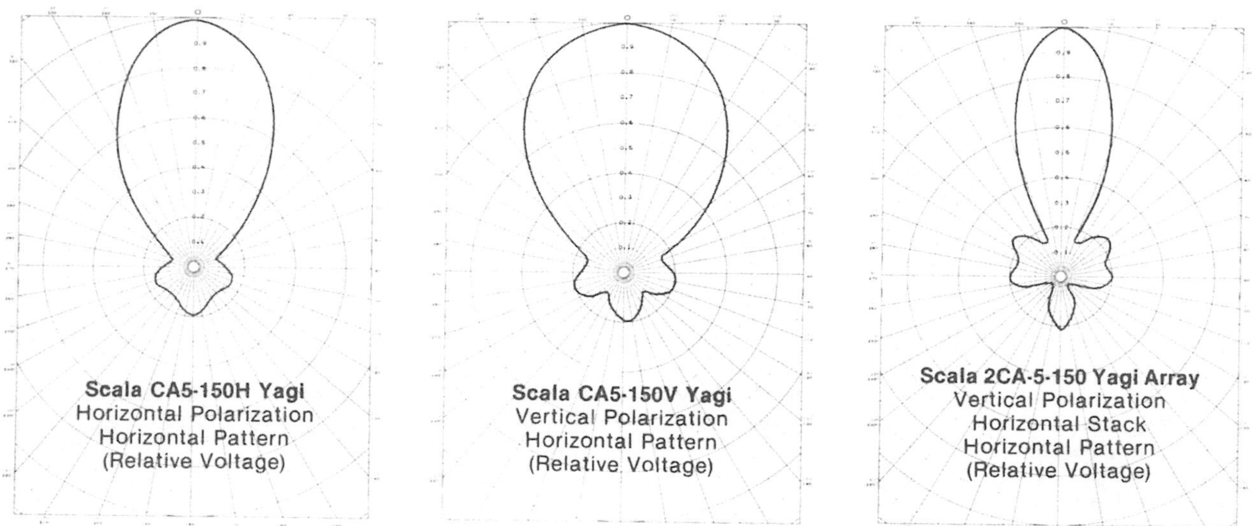

FIGURE 3.9-7 Radiation patterns for the CA5-150 five-element Yagi antenna made by Scala Electronics for use in the 150 MHz frequency band.

the receiver front-end, such as gallium arsenide field effect transistors (GAsFETs).

The need for a preamplifier or cavity preselector network ahead of the receiver's first RF stage should also be considered. RF preamplifiers can add sensitivity, but they can also cause overload conditions in the presence of medium-level co-channel signals. Preselectors are often necessary at mountaintop or antenna farm locations because of the high-level RF signals present at such sites. It is not uncommon to have a 1 kW land mobile paging transmitter operating in the 454–455 MHz range located near an RPU band receiver that is working in the 455–456 MHz

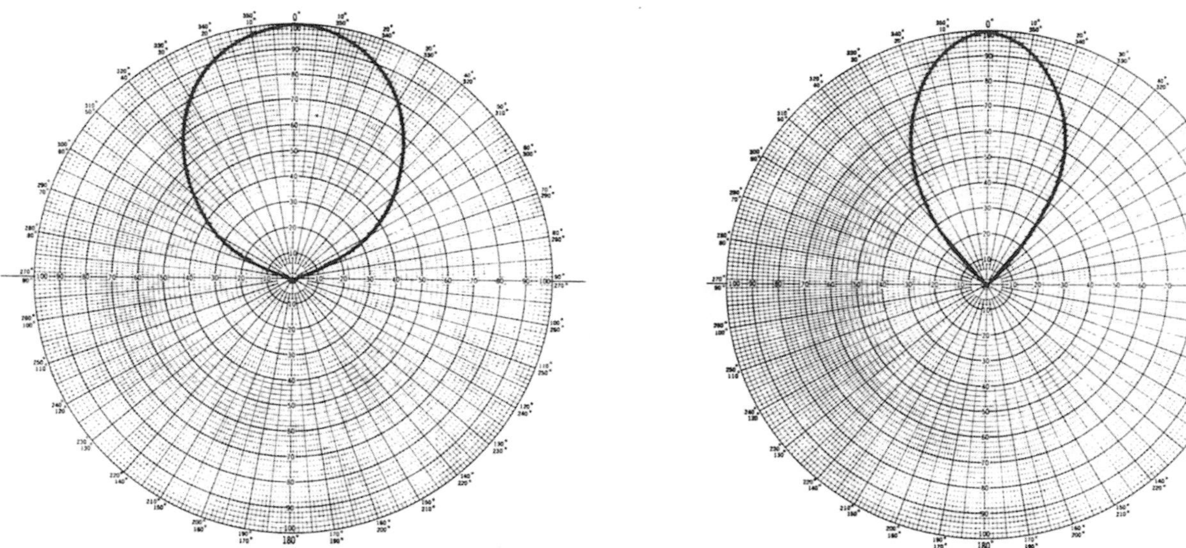

FIGURE 3.9-8 The radiation patterns for the Scala CL-400 broadband log periodic antenna, designed for use in the 450 MHz RPU frequency band. (Courtesy of Scala Electronics.)

frequencies. High-power FM or TV transmitters can also cause desensitizing of the receiver front end, unless adequate bandpass filtering has been included in the receiver design.

The locations commonly used for relay sites are seldom ideal from an environmental standpoint. For these reasons, rugged equipment should be selected. Temperature extremes can also cause problems for frequency determining elements, as well as accessories such as cavity filters, preselectors, and preamplifiers. Since relay sites are often difficult to reach, equipment should be designed for easy maintenance, preferably through module replacement. A spare stock of modules should be kept at the site so that the system can be quickly returned to operation and the defective module can then be serviced away from the relay site.

If trouble is experienced with a piece of receiving equipment, the possibility of interference from other services should not be overlooked. A spectrum analyzer is essential for such work.

System Configuration

When a RENG system is configured, several standard system configurations can be modified to fit the requirements of most users. They range from the simple point-to-point program relay system to complicated multipoint relay installations with automatic signal-quality voting circuits.

Simple systems can use only omni directional antennas for simplicity, but such an arrangement is not practical in an increasing number of urban areas because of spectrum congestion problems and the need to cover large geographical areas. Receive sites can vary based on market geography and facility topology. An RPU antenna can be side mounted on the station's broadcast tower, and in that case the RPU receiver system can be housed in the shack or transmitter building along with the station's broadcast transmission equipment. If the antenna is located remotely from the studio, RPU audio can be routed from the receiver output back to the studio via a transmitter-to-studio link (TSL). Simple control of the RPU receiver can use spare relays or contact closures from the broadcast transmitter's remote control system.

The range of a RENG system can be greatly extended through the use of ARSs (Figure 3.9-9). Such systems also make it possible to use lower power transmitters in the field, since the transmitter at the program origination point need only be powerful

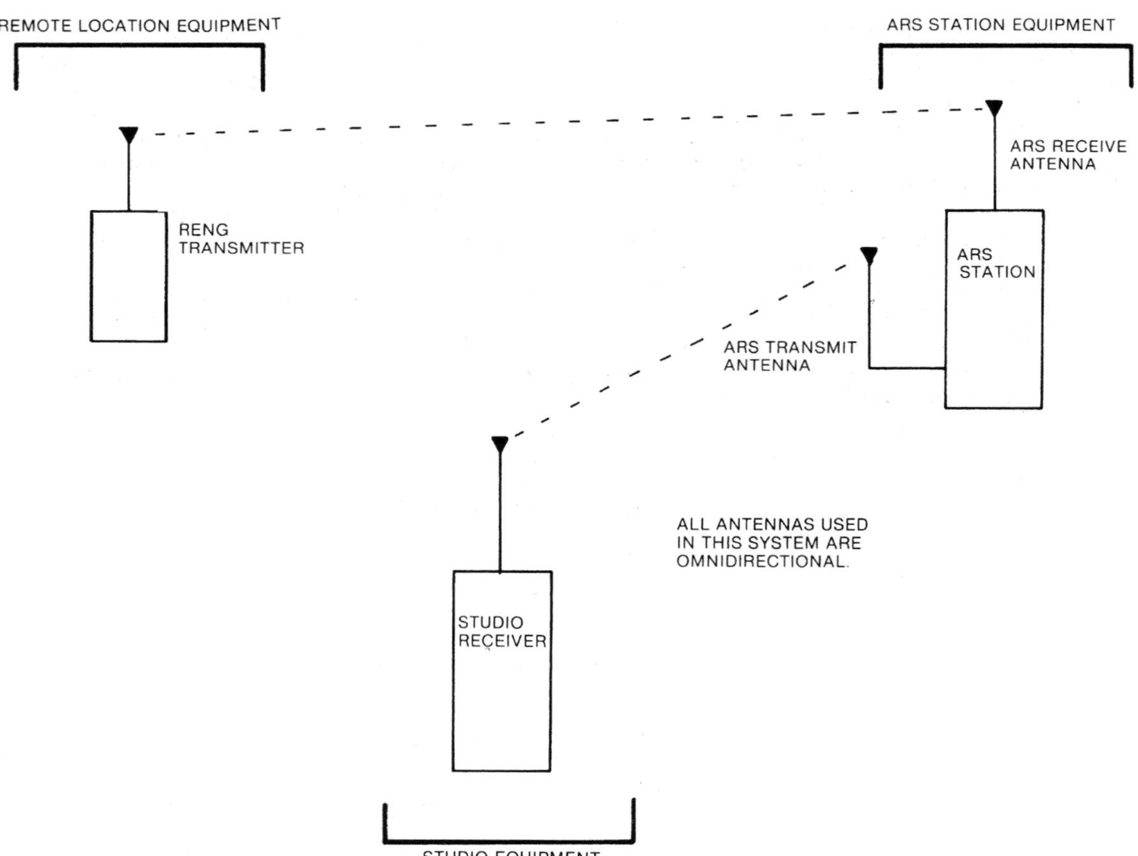

FIGURE 3.9-9 The basic RENG program relay configuration using an ARS station between the remote location and the studio.

enough to reach the nearest ARS site. This often allows the use of smaller and lighter remote transmitters, usually hand-carried or pack-carried units.

Figure 3.9-10 shows a high performance two-point RENG system designed for operation in spectrum-congested areas. At the remote site, two transmitters and two antennas are used. Usually, a communications transceiver is used in conjunction with RENG transmitter/receiver for conveying setup information between technical personnel at either end, and to relay cues and coordinating information. At the studio site, a communications transceiver, feeding an omni directional antenna, is used for the setup information, cues, and coordination work.

When remote broadcasts are not occurring and when beginning the initial setup procedure for a new remote, the omni directional antenna is patched into the broadcast-quality RPU band receiver at the studio through the coaxial antenna relay. Once contact has been established with the remote crew, one of the directional antennas, which are mounted on a common mast driven by a remote-controlled antenna rotor, is switched into the studio receiver. The polarization of the transmission from the remote site is planned before the remote crew leaves the studio. Selection of either horizontal or vertical polarization is made during the frequency coordination process or at the discretion of the user. Engineers may find that a particular polarization may yield better results from certain geographical areas, and in such cases, that polarization would be chosen.

Once the proper antenna has been selected, the antenna rotor is adjusted for maximum received signal strength. The studio operator then talks the remote crew into the best position for its Yagi transmit antenna. At this point, the antennas are locked down and the link is ready for the remote broadcast. If a variable power transmitter is used at the remote site, or transmitters of various power levels are available, the transmitter power would next be adjusted to the point necessary to achieve the required S/N performance at the receiver, typically 20 dB or better. After power output adjustment, the antennas on both the receive and transmit ends should be checked again for correct positioning. This will assure a high-quality, secure RF link from the field to the studio and will also result in a minimum of unwanted radiation to other RPU band users.

If an ARS is used in the system, the antenna switching and positioning work is done by remote control. The link for this remote control system can be a subcarrier on the main station broadcast signal, a separate dedicated radio link, a dial-up telephone patch, or a leased telco data or voice loop. A standard broadcast transmitter remote control system is used, with the

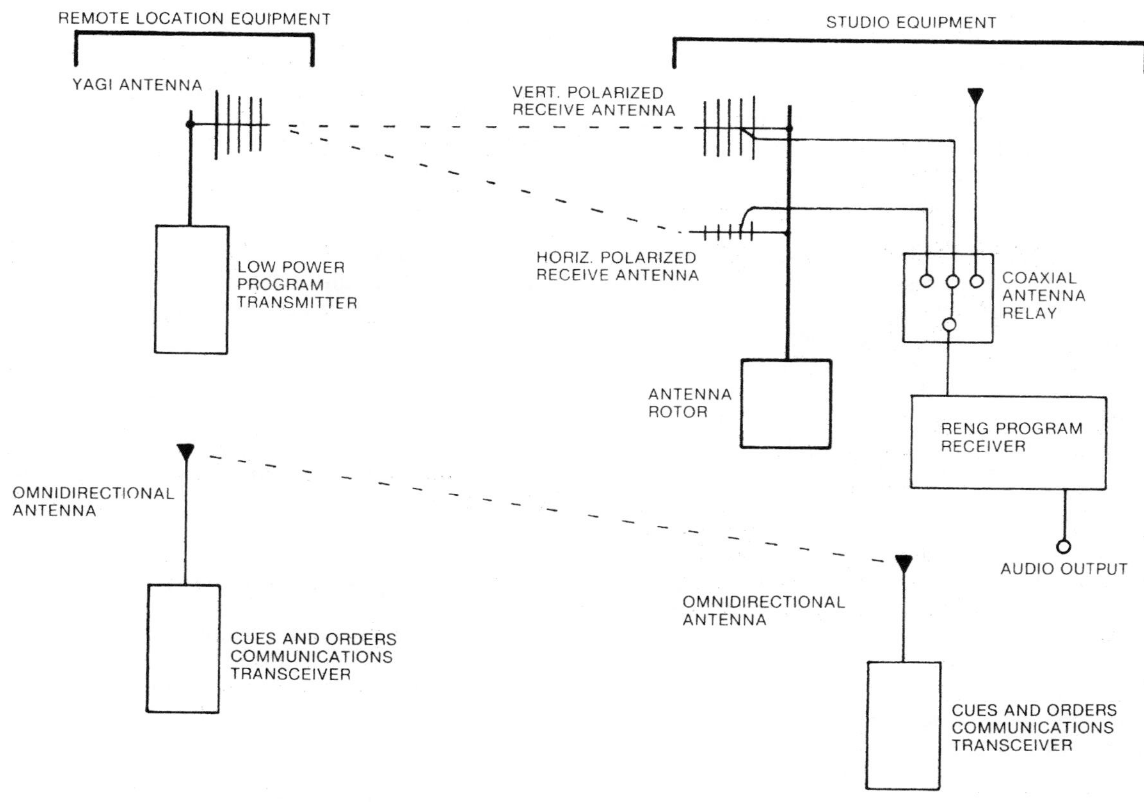

FIGURE 3.9-10 A high performance two-point RPU system designed for operation in frequency-congested areas.

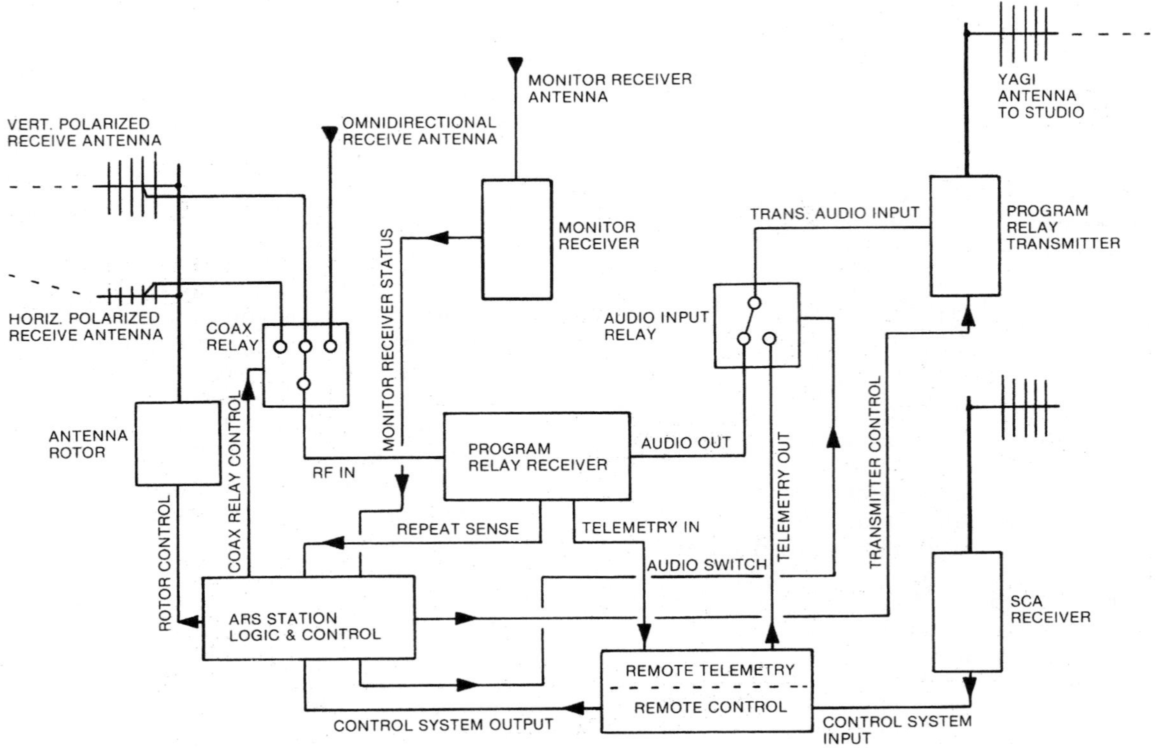

FIGURE 3.9-11 A high-performance secure channel ARS station with remote control of system functions.

common channel on-off/up-down functions performing the control at the remote site.

For stations with multiple site capability on their main transmitter remote control system, the ARS remote points can be simply treated as other transmitter sites and controlled as such from the master unit. A monitor receiver is included at each ARS installation to inhibit activation of the ARS transmitter if a transmission is already in progress on that frequency. As shown in Figure 3.9-11, control commands are commonly received over a subcarrier receiver from the main station transmitter. The relay station logic interfaces the remote control unit with the receive antenna coaxial switch and the antenna rotor control box. During setup, the telemetry section of the remote control unit provides an audio FSK signal that is sent back to the studio control unit via a telco line or the ARS transmitter.

In a system where two or more of the ARS stations are used, the studio remote control unit is used to determine which of the ARS stations is allowed to repeat the program traffic. Those stations that will not be used to repeat the program material would be instructed by the studio operator to remain inactive. For multiple-site ARS operation, individual directional antennas or a single directional receive antenna mounted on an antenna rotor may be used to receive the ARS traffic at the studio.

A mobile ARS relay transmitter can be used to support roving talent or reporters on site. As mentioned previously, the communications transceiver is used for IFB (see the following section), cues, and orders from the studio location. The program channel signal can consist of a hand- or pack-carried transmitter, which directly feeds the studio receiver or one or more ARS systems on site. This arrangement works when the event requires added range and a high-power transmitter is required to reach the studio or to provide the talent extra flexibility because a small hand-carried transmitter can be used there, rather than a larger unit with antenna and power cables attached.

There is a limit, of course, to the number of times a signal can be repeated and still maintain good audio quality. Moreover, each added hop in the path between the remote site and the studio increases the chances of a spurious signal interrupting the remote feed.

Remote Cues and Orders (IFB)

Communications with a remote crew from the studio can be accomplished in one of several ways. The simplest method is an over-the-air cue in which the talent simply listens to the station's air signal and takes cues from the studio announcer or a prerecorded introduction. This method will become unavailable as radio stations transition to in-band/on-channel (IBOC) digital radio technology due to the several second delay inherent in the IBOC signal processing. Other methods

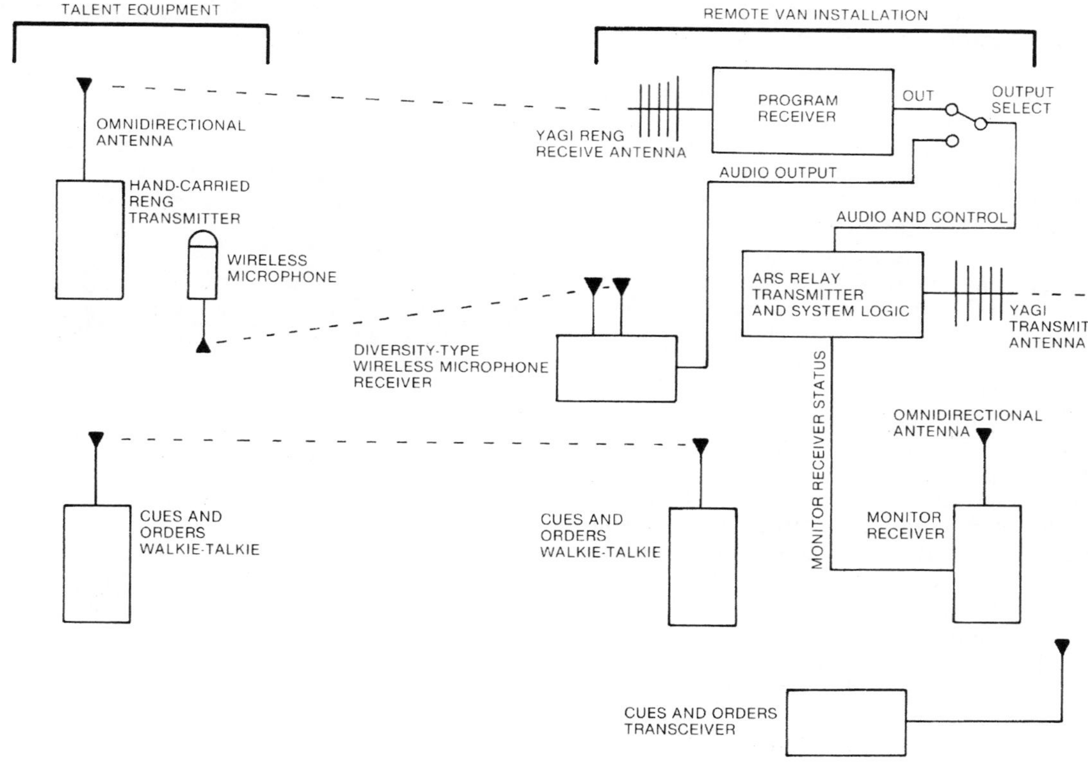

FIGURE 3.9-12 The use of an ARS system at the event site for added range and talent flexibility.

include use of an FM station's subcarrier signal for cueing information or a separate, dedicated, radio link specifically used for cueing instructions, either from the remote truck (as shown in Figure 3.9-12) or from the main studio. If a station needs a more sophisticated intercommunication system, a trunked 800 MHz radio system can be considered. A 5- or 10-channel trunked repeater acts like a small telephone exchange in which the number of users (telephones) exceeds the number of channels (trunk lines). Telephone system theory is used to predict the busy level that can be expected during periods of heavy radio traffic. Three-minute time-out timers are usually included in mobile transmitters to enforce time limits.

These trunked systems can tie into the regular telephone system at hilltop repeater sites or at trunked base stations. Broadcasters interested in 800 MHz trunked radio should contact their local area land mobile operator to see if such a system is available. In certain situations, a station may be able to design and license a UHF business radio system for dispatch and coordination of RENG crews. These systems offer the user the luxury of not encountering a busy signal, as may occasionally happen in a trunked system. As with the trunked network, no programming is allowed on a UHF business radio system.

A more recent solution available in most cases is the use of the cellular telephone system. A portable or transportable cellular phone has become a common part of most RENG systems today. When more than just simple voice communication is required, such as interruptible foldback (IFB) or other audio signals to talent headphone or monitor speakers, an audio interface to and from the cellular phone may be used, on both ends of the link. Unlike the standard dial-up system, the cellular phone system is 4-wire end-to-end (it uses a pair of RF channels, one for transmit and one for receive), so that no hybrid or gating circuitry is required. The same type of hardware may be used for feeding RENG program audio from the field when no other method is possible. Audio performance will typically be somewhat degraded from what is expected from the standard dial-up phone feed.

REMOTE PRODUCTION EQUIPMENT

Most multiple audio source productions are centered around a small audio mixer which feeds either a wired or wireless telco service or via an RPU. Careful attention should be given to the connection of the mixer output to a POTS line. A phone coupler should be used between the mixer and the telephone unless the mixer is specifically designed to work directly into a hot dial-up line (one with DC voltage across it). This caution applies to a connection made either to the phone line tip-and-ring wire or to the telephone set through the handset terminals.

The complexity of a remote event can require that multiple mixers cascade into each other such as when a local stage mixer's output is fed to a broadcast mixer. When multiple mixers (such as the PA and separate radio truck just mentioned) each require access to the same microphones, a number of options are available. Instead of placing multiple microphones for each mixer, a shared arrangement is best, using proper microphone splitting techniques. Good transformer isolated microphone splitters are recommended for this. One mixer remains directly connected to the microphones, while the others are fed by the secondary of a bridging input transformer. The microphones see only one load, and the consoles remain electrically isolated from one another. Typical microphone splitters also provide separate ground lift switches for each channel (or in some cases, a single, ganged switch), by which ground loops between consoles can also be eliminated. If condenser microphones are used, their phantom power can come only from the console receiving the direct feed, since DC supply voltage from the transformer-fed consoles will not pass back across the transformers.

Sound reinforcement (PA) feeds of musical acts are not always inclusive of all the elements required, or are improperly balanced, for the radio mix. For this reason, a separate radio mix is often required if proper balance and control of the stage event's mix is important.

A recorder which might utilize cassette, digital audio tape (DAT), "mini disc" (see Figure 3.9-13), hard disk, or flash memory technology is often useful at a remote broadcast because it gives added flexibility to the remote crew. The recorder input signal can be taken from an auxiliary output on the audio mixer, allowing interviews or material from the PA system to be mixed and recorded for later use on the air. In cases in which a separate stage mixer or truck is used for the stage premix, it is this "dry premix" that is usually recorded on site. These production recordings therefore include only the mix of the actual event and do not contain any continuity or other local production elements added at the broadcast mixer. The full broadcast mix can be recorded for archival purposes either

FIGURE 3.9-13 Example of portable mini disc recorder and disc media.

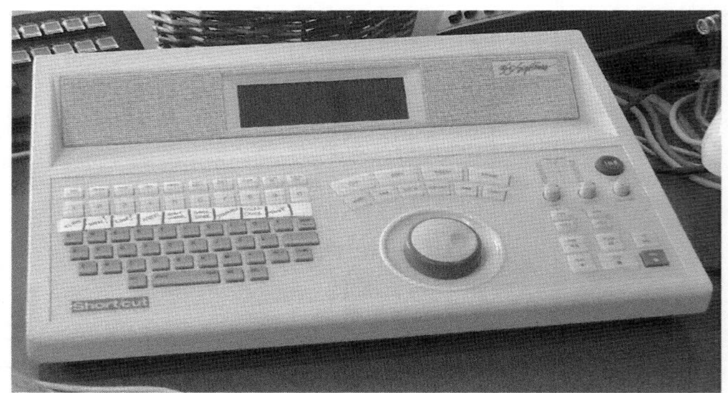

FIGURE 3.9-14 Nonlinear editing can be used for on-site or station editing, playback of news stories, and sound bites. It is also often used to prerecord phone calls for on-air use.

at the remote site or at the station. Disk- and flash-based systems allow nonlinear editing on site or back at the station and controlled playback through newsroom editing systems. Figure 3.9-14 shows an example of a nonlinear editing console.

If multistation monitoring and communication are necessary at the remote site, an IFB system may be required. An IFB system combines the function of intercom and monitor, such that a director, either on site or elsewhere, can communicate with and cue various personnel at the remote (talent, stage managers, floor directors, engineers) via headset or loudspeaker, using a multistation intercom. The director's control panel for the IFB system is typically equipped with push-to-talk individual station and all-call (talk to all stations) buttons. What differentiates the IFB from a standard intercom is that when the director is not talking to an IFB station, that station receives program audio. The director's cues temporarily interrupt the program audio, which returns after the director's message has ended, hence the term "IFB." Unlike a traditional intercom, the IFB system is usually not bidirectional, but rather feeds one way from the director out to the various receive stations. Figure 3.9-15 shows a typical IFB layout in a complex remote situation.

Most IFB systems allow the use of two or more different program audio sources to be selectively routed to different stations, so one station could be fed the dry stage mix, for example, while another station hears the whole remote transmission with continuity included. Both would hear the director when their channels were designated for communications, regardless of which program audio channel was selected. Off-site IFB (in which the director is not at the remote location but back at the studio) is often referred to as a private line (PL); dial-up or leased telco lines are used for PLs, along with wireless return links or cellular phones, in many ENG cases. Some IFBs are stereo capable. Besides allowing the system to provide a stereo program feed to all stations, these systems are usually set up to place the

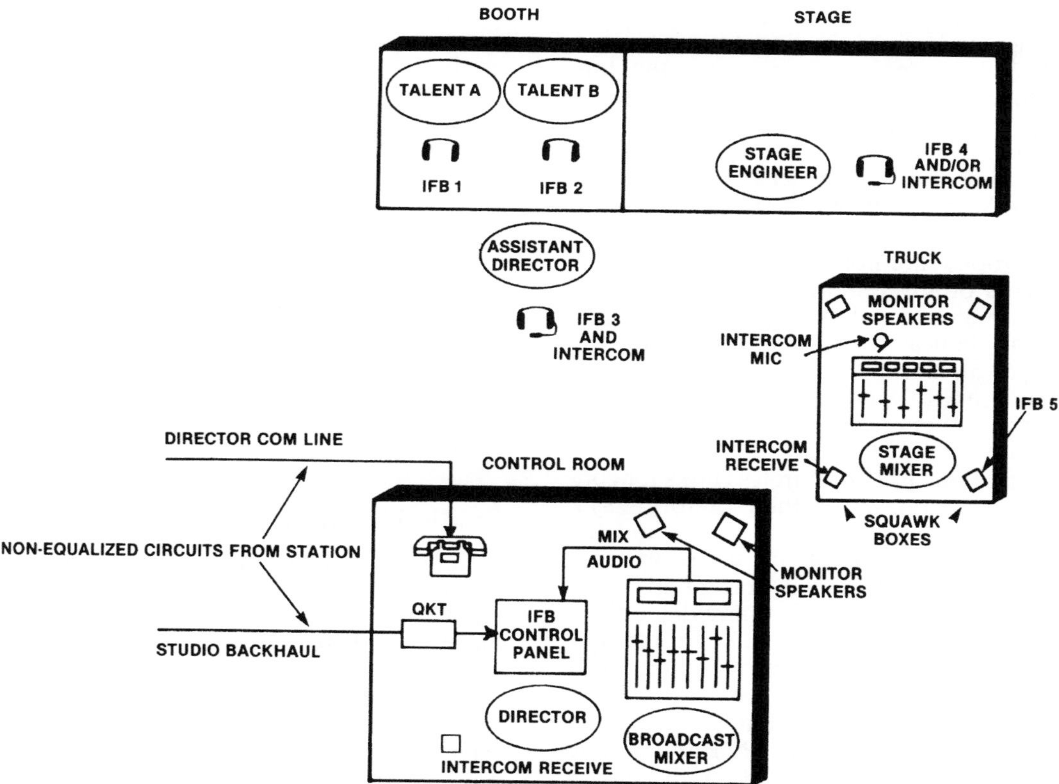

FIGURE 3.9-15 Communications layout for a complex live remote broadcast. The stage mixer can use a simple intercom (with no program audio) for communications to stage and other personnel, while the main control room uses an interruptible foldback (IFB) system to coordinate the entire broadcast.

director's communications into one channel only. This makes it easier for talent (listening on stereo headphones) to distinguish between program and communications audio. It can also allow the communications to interrupt only one channel, keeping program audio continuous (or only slightly attenuated, if desired) in the other channel. Most radio stations that do news have an ingest center like the one shown in Figure 3.9-16 for setting up remote paths along with monitoring other local media and various contribution services.

FIGURE 3.9-16 Radio station ingest center.

Setup and Testing Procedures

A tone oscillator is useful in setting up the remote broadcast. Most mixers designed for remote applications include an oscillator that can be switched on to the program channel. If the oscillator has multiple frequencies, a rough order of path bandpass response can be determined. Path tests for frequency response, S/N, headroom, and relative polarity for stereo broadcasts should be performed. Phase response and distortion tests are also useful, if time and test equipment permit. If any noise reduction or other enhancement devices are being used across the line, testing should be performed with them bypassed. After the line proves satisfactory, engage these devices and recheck. Multifrequency checks are less critical with digital backhaul systems, because these are typically quite flat within the passband. Nevertheless, simple level and polarity checks are still worth performing with ISDN and T1 services.

Having the proper monitoring facilities is important to the success of any remote broadcast. A loudspeaker (or well-matched pair for stereo broadcasts) and a set of headphones should be provided for the remote crew. Not all portable mixers can support a loudspeaker and multiple headphone outputs, so a separate power amplifier and headphone booster may be needed.

After checking phone lines or wireless links for continuity back to the station, set up the monitoring system first and check it for clean audio. Then add the other elements or subsystems of the setup one at a time, checking the monitors for continued clean response after each. In this way, when any deleterious effects are heard (hum, buzz, hiss, etc.), it will be fairly simple and quick to track down the offending hardware or interface method.

An off-air receiver is a requirement for nearly all live remote broadcasts. The receiver gives the remote crew a way of checking the total link and (if the station is analog and not IBOC) allows easy cueing of talent at the event. A separate dedicated telephone set is suggested for complicated remote broadcasts or broadcasts on IBOC stations. The phone provides an easy means of communicating with the studio. It can also serve as a backup line for program audio in case the RPU system or telco program service should fail. And in cases in which the remote broadcast originates from outside the coverage area of the station, it is essential for monitoring as well as communicating. If digital telco service is employed, it is generally not much more expensive to have this return (or backfeed) line be of the same fidelity as the transmit circuit.

Wireless Microphones

The use of wireless microphones for the talent at a remote broadcast is gaining popularity with stations involved in RENG activity. The advantages to the talent include complete freedom of movement and nothing to carry around but a microphone and air monitor receiver. There are no controls or meters for talent to worry about. The range of a wireless mic is somewhat limited, but a properly designed system for remotes that are more or less stationary can provide simple setup and coverage of an event.

The receiver used in conjunction with the wireless microphone may use either diversity or nondiversity reception techniques. A nondiversity receiver is used where multipath cancellation is not a problem, such as in open areas or when conducting fixed-position interviews. If, on the other hand, the wireless mic is to be used in several places and the possibility of multipath cancellation exists due to nearby reflective objects, a diversity receiver is recommended. The diversity receiver uses two antennas that are physically separated by a distance varying between several inches to several dozen feet. In some diversity receivers, two complete RF sections are used. The diversity receiver automatically selects one of the two signals based on its criteria for better reception. (Some receivers combine the two signals through a phase-shifting network.) The switching of RF sources occurs silently without any squelch-type noise bursts.

Many wireless microphone systems include audio companding circuits to extend the dynamic range and lower the apparent noise floor. Both VHF and UHF frequencies are used. An often-overlooked FCC regulation requires licensing of most wireless microphone systems.

Studio Planning for a Remote

Setting up a remote into the studio requires careful planning by a qualified engineer. The remote signal has to go to the studio so it can be managed and blended with spot breaks, callers, in-studio guests, or breaks for news.

The number one consideration is not to return the remote audio back down the IFB/program line to the talent, as an annoying echo will throw off the talent's performance. This type of circuit is called a "mix-minus." The mix-minus is all sources the talent needs to hear minus the voice of the talent.

Radio engineers set up the studio audio board using multiple audio buses sometimes called "utility" and "audition." Some very capable audio boards have additional "send" buses. The IFB/program return circuit is connected to one of these buses. Simply select the program sources the talent needs to hear, music, callers, liners, live announcers, for example, and "lift," or isolate, the audio module the remote talent is using. This is sometimes called "splitting the board."

Planning and designing mix-minus circuits can be complex, and can be the reason some remotes are successful or not. The remote planner must consult with an experienced engineer for this phase of the operation.

EMERGING TECHNOLOGIES

As always, new technical developments will find their way into the radio remote kit. A number of these systems are emerging at this writing that will likely affect both backhaul and remote production systems.

Backhaul Systems

Under Part 15 of its Rules, the FCC has established two bands for unlicensed use of spread-spectrum transmission (47 CFR §15.247): one in the UHF band (902–928 MHz) and one in the S-band (2,400–2,485 MHz). Some digital audio transmission equipment has become available for these bands, particularly the

latter, which allows RPU-like operation without licensing or frequency coordination. Many of these new systems are being developed for wireless Internet and WAN systems. With Yagi-type antennas, these low-power (100 mW) systems can be used on line-of-sight paths of approximately 20 miles. Their use is not yet widespread in broadcasting, but other industries (primarily wireless computing) are beginning to use these bands. While spread-spectrum transmission is fairly resistant to interference, there is some concern that as the bands become more popular, the reliability of such audio links will not remain adequate for broadcast use. Nevertheless, in most cases, the audio quality and robustness of S-band spread-spectrum RPU have proven more than adequate for broadcasters' requirements, among the relatively few using the systems to date.

Remote Production Equipment

A new item for on-location audio hardware is the fiber-optic snake. Here, a single, robust, yet lightweight cable carries several dozen separate microphone or line-level channels from one part of the remote site (typically the stage area) to another (typically the mix position), replacing the traditional heavy copper multipair cable. Also unlike its predecessor, the fiber-optic system is impervious to the pickup of hum or EMI, nor does it generate any. Microphone-level audio is preamplified at the head or stage end, then digitally encoded and multiplexed, converted to the optical domain, and sent down the fiber. The reverse process takes place at the console end, terminating in line-level analog-balanced outputs, to be plugged into a conventional mixer's line inputs.

Splitting can also be accomplished without difficulty or degradation in the digital (electrical, not optical) domain with these systems. A helpful feature is the addition of some remote control ability for microphone preamp gain from the mix position, since the preamps are located at the head end of the snake. Because this is an active rather than passive system, some provision must be built in for return signals running in reverse from the console back to the stage. A typical configuration is 56 send and 8 return paths on a single cable, about the size of a standard mic cable.

SUMMARY

A RENG network should be planned and constructed with long-term service and frequency coordination requirements in mind. Areas that currently do not experience spectrum congestion problems may encounter them in the future. It pays, therefore, to design a system that is spectrum efficient and relatively immune to interfering signals. New digital services, both telco supplied and of an RF variety, should also be considered, and the progress in these areas carefully monitored for application on future links. It is always easier, and cheaper, to do the job right the first time. No matter how good the backhaul link, if the remote audio program is not properly created at the site, it will not make a successful broadcast. Well-equipped, simple, and reliable systems are essential for the remote crew to be able to set up quickly and make the instant creative decisions often required on location. The well-engineered remote can make radio programming exciting and unique—the kind of compelling material that keeps listeners coming back for more.

Bibliography

Bates, Regis, and Gregory, Donald. *Voice & Data Communications Handbook,* 3rd ed., McGraw-Hill, 2000.

"Common Carrier Audio Program Services." *NAB Engineering Handbook,* 10th ed., Elsevier, 2007.

Parker, Rich. "Audio TX STL-IP." *BE Radio,* May 2005.

Pizzi, Skip. "Beyond ISDN." *BE Radio,* January 1998.

Scherer, Chriss. "The Alphabet Soup of Codecs." *BE Radio,* April 2004.

Scherer, Chriss. "Reliable Connections." *BE Radio,* January 2005.

"Telephone Network Interfacing," *NAB Engineering Handbook,* 10th ed., Elsevier, 2007.

Internet Resources

The Society of Broadcast Engineers: www.sbe.org

Federal Communications Commission, Audio Division: www.fcc.gov/mb/audio

CHAPTER

3.10

Telephone Network Interfacing

STEVE CHURCH AND ROLF TAYLOR

Telos Systems
Cleveland, Ohio

INTRODUCTION

Without connection to the outside world via the telephone network, radio and television station programming would be decidedly less interesting.

Because of their ubiquity and ease of use, the telephone and Internet Protocol networks that enmesh the globe are often taken for granted. But they are, indeed, a remarkable tool for broadcasters. An ordinary telephone and the dial-up network allow almost anybody from anywhere to be immediately live on the air. Digital ISDN lines and modern audio compression techniques have made instant full-fidelity remotes a commonplace. As the Internet proliferates, VoIP (Voice over Internet Protocol) is beginning to offer yet another opportunity to broadcasters to connect with the world outside their studios.

Despite being accomplished everyday, interfacing telephone and IP networks to broadcast studio equipment can be a challenge. It's not so difficult to attach a commercial phone hybrid between a POTS line and the studio mixing console. But what if the line is noisy or the level is too low or varies too much from caller to caller? Then, knowing how the network works internally will be an advantage. For advanced operations such as required for call-in shows, or for sharing lines among studios, broadcast vendors offer sophisticated and specialized telephone-to-broadcast interface equipment. Knowing how this gear works is essential to getting the most from it. ISDN, IP, and mobile phones present opportunities, but knowledge is again the key to successful application of these new technologies. Today, voice calls may be delivered by analog lines, via ISDN, or with VoIP. Which is best? High-fidelity remotes are possible with all of these technologies, as well. But what are the tradeoffs? The intention here is to inform and assist the broadcast engineer who needs to successfully navigate through the many choices, make it all work, and find the cause of the problem when it doesn't.

THE TRADITIONAL TELEPHONE NETWORK

As broadcast facilities make the transition to digital systems, it is noteworthy that the Public Switched Telephone Network (PSTN) has been almost entirely digitized for many years. The watershed event was Illinois Bell's 1962 installation of a T-carrier system—the first widespread commercial application of digital audio. Telephone engineers appreciate digital technology for the same reason broadcasters do: reduced susceptibility to noise and other disturbances and improved ability to switch, monitor, and maintain the circuits.

Although the worldwide dial-up telephone network is a significant achievement, it is mostly made from a simple ubiquitous element: digital circuit-switched channels of 64 kbps each. *Circuit-switched* means that the channel is connected end-to-end with the entire capacity available for the duration of the call. This is in contrast to *packet-switched* systems, such as the Internet, where capacity is shared among users and there is often no guaranteed bandwidth or time of arrival.

Most of the network is digital, but the last-mile copper connections from the central office to the

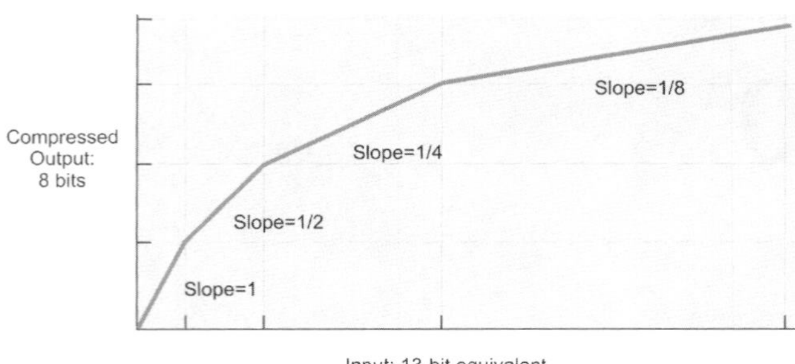

FIGURE 3.10-1 G.711 μ-law companding curve.

customer site generally are not. The vast majority of users connect to the network via an analog technology that is little different from that employed since the beginning of telephone service. This has begun to change with the introduction of digital last-mile technologies such as Integrated Services Digital Network (ISDN), T1, Asynchronous Digital Subscriber Line (ADSL), and, in some cases, optical fiber to the premises.

Speech Coding

A digital channel operating at 64 kbps supports phone-grade speech audio encoded with a modified pulse-code modulation (PCM) technique. When a call is made, speech is sampled at an 8 kHz rate and encoded into a digital word 8 bits long. Telco engineers call this 64 kbps bit rate a *digital signal level 0* (DS-0) channel. The word length is what determines dynamic range. An 8 bit word would only permit 48 dB were it used in standard PCM linear fashion. A primitive kind of compression is used to stretch the dynamic range: mu-law (μ-law) in North America and much of Asia (see Figure 3.10-1), and A-law in Europe. This is a scheme that equalizes the step size in decibel terms across the dynamic range. A smaller step size on low-level signals reduces quantization noise and improves effective dynamic range to the equivalent of about 13 bits; thus, the quantization noise (and distortion) is approximately a fixed percentage of the signal amplitude, regardless of its level. The process of conversion and companding (combination of compression at the sending end and expansion at the receiving end) is done in specialized analog-to-digital (A/D) and digital-to-analog (D/A) integrated circuits called *voiceband codecs* (COder/DECoders). The method is standardized by the International Telecommunication Union (ITU) as G.711 [1].

Analog lines (twisted pair copper) connect most subscribers to Telco central offices. Both speech directions are mixed together on these lines, but that is not the way signals are handled within the telephone transmission and switching network. Digital switching operates on independent send and receive signals. Non-copper transmission media, such as microwave radio, satellite, and fiberoptic cables, are one way only, so the two speech directions are independent, as illustrated in Figure 3.10-2. Even on copper, long-distance links keep the send and receive separated so amplifiers can be inserted. The usual local subscriber analog circuit is called *two-wire,* because it uses a two-wire pair. The network is referred to internally as *four-wire,* so named because in the past a four-wire circuit required a pair for each of the send and receive transmission directions (*i.e.,* four wires altogether).

The POTS Analog Line

The plain old telephone service (POTS) lines provided by Telcos are known officially as subscriber loops, trunks, or simply CO (central office) lines. Trunks are usually lines destined for private branch exchange (PBX) systems and sometimes include special signaling as well. Because these are two-wire circuits, the CO uses a two-to-four-wire converter (also called a *hybrid*) to interface the analog lines to its internal four-wire system, as shown in Figure 3.10-2. This process takes place on the line card, which is also responsible for digitization, talk battery insertion, off-hook detection, and ring generation.

Talk Battery and Ringing

The *talk battery* DC voltage and the conversation audio appear together on the phone pair. The talk battery leaves the exchange at –48V and is limited to 20 to 100 mA by a series resistor and the loop resistance; the series resistance value is engineered with the

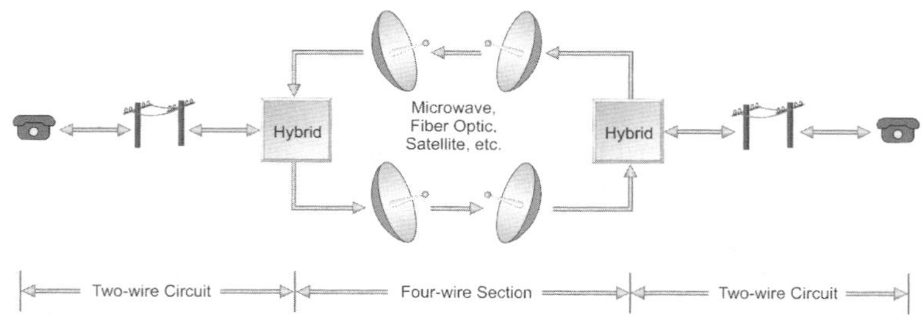

FIGURE 3.10-2 Two-wire circuits have both directions on a single pair of wires, which are separated with hybrids for switching and long-distance transmission into four-wire signals.

TABLE 3.10-1
U.S. Telephone Line Characteristics

Parameter	Typical Values	Operating Limits
Talk battery voltage	–48 VDC	–47 to –105 VDC
Loop current	23 to 80 mA	20 to 120 mA
Loop resistance	0 to 1300 Ω	0 to 3600 Ω
Loop loss	8 dB	17 dB, depending on type
Distortion	–50 dB	Not applicable
Ringing voltage	90 V RMS at 20 Hz	40–130 V RMS at 16–60 Hz
Noise (objective)	–69 dBm0, 0 to 180 mi –50 dBm0, 180 to 3000 mi (–16 dBm0 talk level) (C-message weighting)	—

resistance of the loop in mind. The DC resistance of the loop itself varies from less than 100 to 1300 Ω, depending on length. Because of this series resistance, when a line is off-hook the voltage at the customer equipment drops to around –12 V, but this value varies widely. For ringing, an AC voltage of 90 V RMS at 20 Hz is superimposed on the line. Talk battery is maintained during ringing, so the resulting signal has a sinusoidal shape shifted 48 V to negative. Talk signals are AC coupled with a nominal impedance of 600 Ω; however, some CO equipment uses complex impedance coupling, and the nature of the telephone network usually results in the actual impedance as presented to the user rarely being the specified 600 Ω. This turns out to be an important issue for interfacing with broadcast facilities. The basic parameters are summarized in Table 3.10-1.

Frequency Response

Audio bandwidth for POTS calls is strictly limited to a 3.4 kHz bandwidth by the sharp low-pass filters required for proper digitization. The phone network's 8 kHz sampling rate permits a theoretical Nyquist upper frequency limit of 4 kHz, but a 600 Hz transition band is necessary for anti-aliasing and reconstruction filtering (see Figure 3.10-3).

Noise and Level

A 1971 Bell System survey of the phone network nationwide determined that average conversation has a level of –16 dBm. Because of variations in line length and the arrangement of systems at the CO, the actual level arriving at the broadcaster interface may have a range of from –40 to –4 dBm, as illustrated in Figure 3.10-4.

Audio sent into the telephone line must be limited to the average –9 dBm as specified in Part 68.308 of the Federal Communications Commission (FCC) Rules. Audio loss on any given local loop is limited by tariff to 8 dB or less. This loss limit, however, applies only to the loop from the CO to the subscriber and does not include the rest of the signal path. Also, the 8 dB loss may occur at each end of a conversation path, once at the calling party end and again at the called party end, for a total loss of up to 16 dB [2].

Telephone engineers measure noise upside-down, defining a reference noise floor and then measuring *up* from there. The reference noise level is 1 picowatt (pW), which corresponds to –90 dBm; thus, a noise level of –60 dB relative to 0 dBm would be reported as 30 dBrn noise (dBrn = dB above reference noise). In contrast to conventional audio measurements, the higher this number, the worse the noise.

Note that *idle channel noise* used in telephone engineering is not the same as the familiar signal-to-noise ratio used in professional audio because the presence of a signal causes noise in a digital audio system to increase. This effect is called *modulation* or *quantization* noise and is primarily dependent upon the number of bits per sample.

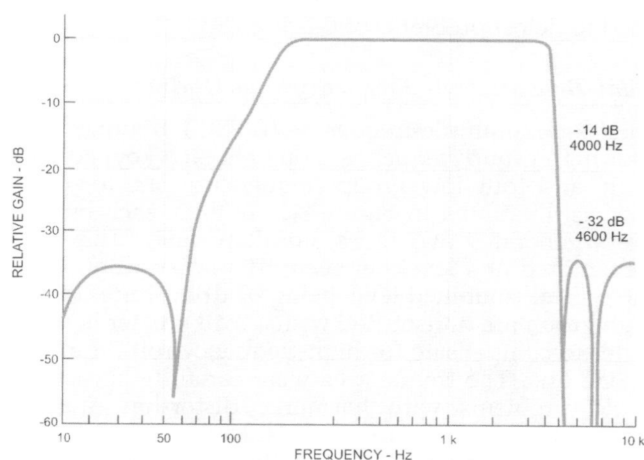

FIGURE 3.10-3 Frequency response of POTS system to permit 8 kbps sampling and remain within the 4 kHz Nyquist frequency limit.

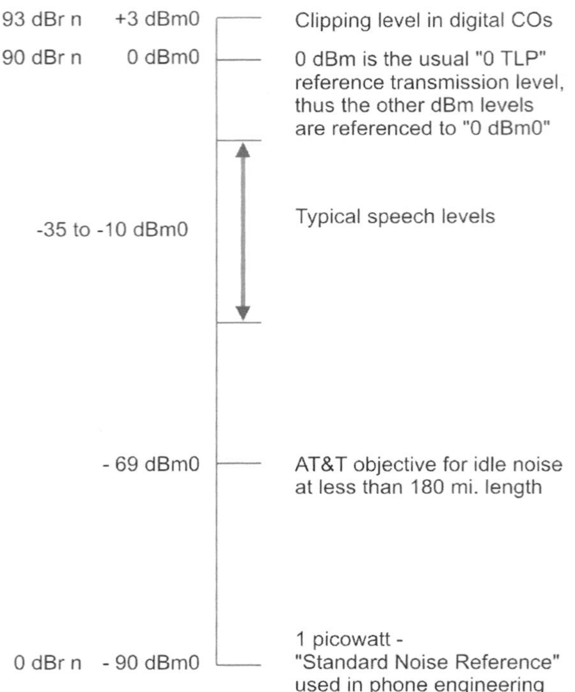

FIGURE 3.10-4 Signal and noise level references used in telephone engineering.

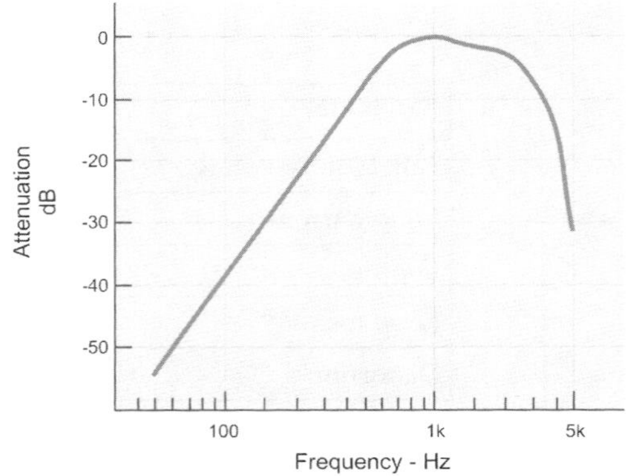

FIGURE 3.10-5 The C-message noise weighting curve used when measuring noise on telephone company channels.

A *C-message weighting* filter is employed when measuring the phone line signal-to-noise ratio (S/N). The C-message curve as shown in Figure 3.10-5 was developed to simulate the frequency response of an old-style telephone earpiece and, accordingly, it has considerable low-frequency roll-off. This means that a telephone line can have significant hum and other low-frequency noise and still meet the noise specifications. Since broadcast telephone interfaces often have good low-frequency response, such noise can be heard on the air. The noise meters used by telephone company technicians do have a "flat" setting, which can be used to detect problems of this kind.

Dual-Tone Multiple-Frequency Tone Dialing

Dual-tone multiple-frequency (DTMF) dialing uses four high group frequencies, one for each keypad column, and four low group frequencies, one for each row, as illustrated in Figure 3.10-6; thus, each button press generates two tones simultaneously. They are transmitted at a level between –10 and –6 dBm, with the typical combined level being –2 dBm. Tones in the high group are transmitted with a 2 dB greater level in order to compensate for high-frequency roll-off in the phone line. The frequencies were carefully chosen to avoid problems with harmonic distortion causing incorrect detection. Frequency tolerance is ±1.5% at the transmit side and ±2% for the digit receiver. The digit receiver is designed to avoid "talk-off," which is the tone detector accidentally misinterpreting speech or noise as a key-press tone. The time required to recognize any digit tone is 50 ms, with a minimum interdigit interval of 50 ms. If a valid tone pair is accompanied by significant additional audio energy, a well-engineered DTMF receiver assumes noise is present and suppresses detection.

Loop Start and Ground Start

Analog lines come in *loop start* and *ground start* variants. Loop start is the most common. In this circuit, the CO provides talk battery to the line at all times and detects that an off-hook condition is occurring when the terminal equipment allows current to flow between the tip and ring conductors. The terms *tip* and *ring* originated with the description of the circuits being on the tip and ring of the patch cords that were used by telephone operators. With ground start circuits, the CO waits for a connection from the ring wire to ground before connecting the talk battery, at which time the terminal equipment removes the ground con-

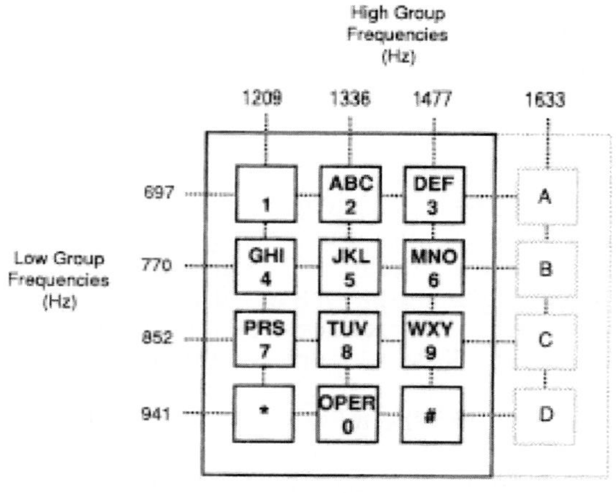

FIGURE 3.10-6 DTMF tone keypad frequency assignments. The four tone pairs in the last column (ABCD) are for special applications.

nection to establish a balanced talk path. When the calling party hangs up, a ground start circuit removes the talk battery. A loop start circuit may or may not provide a momentary interruption or reversal of the talk battery when the calling party terminates. One quirk of ground start lines is that for incoming calls they will generally work with loop start equipment. Some clever broadcast engineers have exploited this quirk as a simple way to easily get incoming-only lines. Many PBXs are designed to work with the ground start circuits because the possibility of collision (called *glare* in telephony circles) is reduced. Collision occurs when the phone system tries to seize a line for an outgoing call just as that line is ringing in.

Disconnection: Calling Party Control

Loop current momentary interruption occurs on most POTS lines when the calling party hangs up. This is sometimes referred to as *calling party control* (CPC), because the calling party controls local equipment when the party hangs up. The CPC may turn off an answering machine, for example, or extinguish the winking light on a held line on a key phone. Perhaps a by-product of early mechanically switched, relay-controlled exchanges, the CPC interruption was probably never intentional; thus, some phone lines do not provide this function or provide it unreliably. With the proliferation of answering machines that rely upon CPC, however, most central office equipment now has this capability designed in, although the duration of the interruption varies from about 100 ms up to 1 s. In some cases, it is necessary to specifically request this feature from the phone company on a per-line basis.

Loop current reversal, on the other hand, has long been a phone company signaling method. First used within networks between central offices, loop reversal was later employed to communicate with some large PBX systems. Lines that are set up for PBX use or that originate at central offices with large concentrations of business customers sometimes use this method; however, the preferred and more modern situation for PBX control is to use ground start or digital lines.

Every Telco CO in the United States eventually returns dial tone to POTS lines after the calling party hangs up; thus, the presence of dial tone can be used as a back-up when the loop current detection methods fail. As with DTMF detection, an important consideration with broadcast interface equipment is to prevent false disconnection from noise, applause, or other spectrally rich audio. A software-based statistical detection method will ensure that the dial tone is really present before terminating the connection. Most PBXs do not generate CPC, so the dial tone detection method must be used for equipment connected to these lines.

Caller ID

Caller ID (CID) allows the called party to know the phone number of the calling party. This capability is useful for call-in shows, where it might be desirable to deny access to problem callers. The technology is simple. Between the first and second ring, the information is sent in a packet using a 1200 baud modem. This is the same modulation scheme used in computer modems operating at this rate. Customer equipment normally suppresses the first ring, so the answering user does not take the call before the CID information is fully transmitted.

Loading Coils

A typical 24 gauge phone pair has 2.5 dB attenuation per mile at 3 kHz due to capacitive effects. On an 8 mile (12.9 km) long line, high-frequency attenuation would be about 20 dB, a significant amplitude reduction. Telephone engineers invented loading coils to flatten frequency response. These are inductors in the phone lines that counter the effects of the phone pair's natural capacitance. Although the coils are effective at correcting the response within the voice band, the roll-off above 3.5 kHz is rapid, as shown in Figure 3.10-7. This is not much of a problem because the low-pass filters required for digitization cut off at 3.4 kHz anyway, but the loading coils also introduce impedance irregularities, which make a studio hybrid interface's work more difficult. Also, they must be removed for a line to be able to transport digital signals such as ISDN and Digital Subscriber Line (DSL), because these depend on the line being able to carry signals with spectrum extending well past the speech range.

Physically, loading coil banks are long cylinders, with the individual donut-like coils stacked one on top of the other inside. They are typically placed at 3000 ft (0.9 km), 4500 ft (1.4 km), or 6000 ft (1.8 km) intervals along the phone cables. Generally, loading coils are found only on cables greater than 3 miles (4.8 km) in length.

FCC Regulations

In the U.S., FCC requirements for connecting equipment to POTS phone lines are standardized by ANSI and described in ANSI TIA 968-A-2002 – Telephone Terminal Equipment: Technical Requirements for Connection of Terminal Equipment to the Telephone Network. [2]

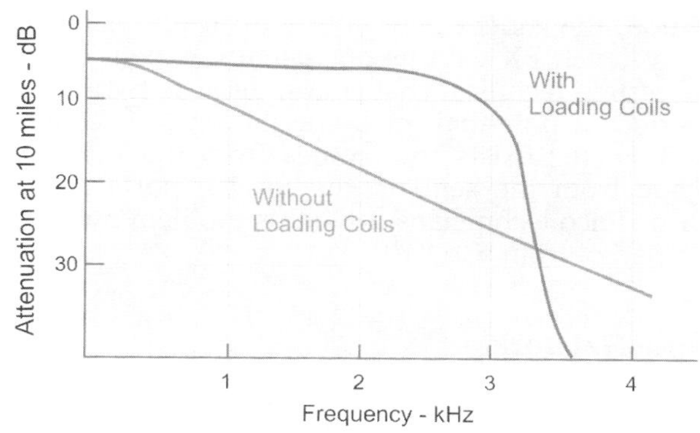

FIGURE 3.10-7 Frequency response of a 10 mile loop with and without loading coils.

Foreign Exchange (FX) Loops

Foreign exchange (FX) provides local telephone service from a central office that is foreign to the subscriber's exchange area. If a station is located in the suburbs and the choke network CO is downtown, FX loops will be necessary to connect the lines to the station. Dial tone comes not from the local suburban CO, but from the downtown office. FX service is also sometimes used to extend the service area into another city, so people can call the station without paying a toll charge and calls can be made within that city without incurring toll charges. For example, if the studio is in Cleveland and the goal is to serve listeners in Akron as if they were local, FX service could be the answer.

At one time, FX loops were based on various analog four-wire or carrier technologies; however, today they are usually fed over channels of a T1 circuit. Regardless of the technology employed, the requirements at the two ends are different. At the CO side the interface from the FX equipment is referred to as FXO (Foreign eXchange Office). The FXO termination of an FX line appears to the CO as if it were a telephone set. It goes off-hook (a low impedance state) or on-hook (a high impedance state) to indicate its status, and it detects ring current from the CO to detect the presence of incoming calls. At the customer end of the FX circuit, the interface to the customer is the FXS (Foreign eXchange Station). This interface is capable of generating loop current and responding to the flow or interruption of this current by the customer's going off- or on-hook, as well as generating ring current to alert the customer's equipment of incoming calls.

The FXS and FXO terminology is frequently seen in contexts outside of the provisioning of FX loops—for example, T1 channel banks use it to describe their ports. Note that in addition to FXS or FXO one must know if a given port or line is loop start or ground start, and whether dial pulse (DP) or DTMF signaling is supported.

Because FX loops add an extra layer of hardware to the phone line audio, they can be a source of problems for on-air interfacing, particularly if using analog technology. They usually are engineered to have a few decibels loss and add to the impedance complexity of the line. Another problem is that the ringing current supplied by FXS interfaces may be nonstandard. In addition, the codecs may not be up to the usual CO quality level. FX circuits are usually expensive and pose certain technical challenges. Because hybrids are imperfect, a potential for a special kind of feedback called *singing* exists that comes from the inevitable leakage from the send to the receive ports at each hybrid. Telco technicians solve this problem by inserting a pad of from 5 to 8 dB.

Choke Networks

Most stations are required to use special high-volume exchanges for their contest and request lines. These lines are referred to as Public Response Calling Service (PRCS) or High-Volume Call-In (HVCI) network lines.

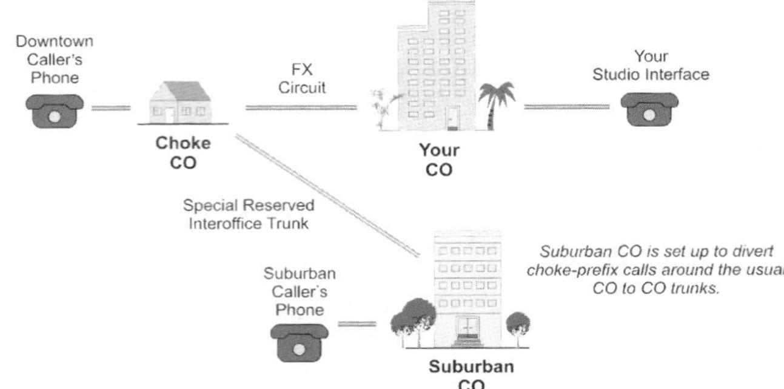

FIGURE 3.10-8 Choke exchange setup used for radio station high-volume applications.

This requirement probably results from the days when aggressive program directors desired the publicity that overloading a CO would generate. The choke network works by diverting calls beginning with the unique choke prefix around the local serving CO's usual interoffice trunking and instead sending them directly to the choke switching exchange, usually located downtown, via dedicated trunks as illustrated in Figure 3.10-8. The usual switching and routing process is bypassed. The phone company dedicates very few trunks to the task of connecting the caller's serving CO choke ports to the choke exchange; for example, in the densely populated Los Angeles area, it seems only three connections exist from most central offices to the choke exchange. In addition, the poorest facilities are often given over to the high-volume service. The net result is that if one station is having a contest, other stations on the choke exchange will be unable to receive calls.

Unless the station is near the choke central office, FX circuits will probably be employed to connect the choke CO to the station's serving CO. This is one of the reasons why choke circuits often have a lower audio level than standard lines. Because of their higher complexity, choke lines also usually have bumpier impedance curves, making good hybrid performance difficult to achieve. In some areas, FX circuits are being replaced by virtual call forwarding. This means that a published number (which has no actual line associated with it) is software forwarded to a real number originating from the local serving CO. The main advantage to this approach is lower cost, as there is no premium to pay for the FX circuit; however, there usually is a call-forwarding charge.

Centrex

Centrex originally stood for central exchange service, and it goes by various names, but the principle is that the Telco's equipment replaces customer-owned PBXs, with each phone set having a direct connection to the CO. The idea is to eliminate customer up-front costs and transfer maintenance responsibility to the Telco. After divestiture of the Bell System in 1984, this was the only way for the traditional LECs to offer

PBX-like services, and it was marketed heavily at that time. More recently, deregulation has allowed local Telcos to sell PBX and key equipment. Traditional analog Centrex is declining in popularity. The latest twist is that several Telcos now offer *hosted IP telephony* services under the name *IP Centrex*. There are also ISDN versions.

T1 Digital Service

T1 is a subscriber loop carrier (SLC) technique that multiplexes multiple lines onto one or two copper pairs. T1 is possible because an ordinary copper phone pair can carry a much wider bandwidth signal than the 3.4 kHz required for a single voice conversation. Indeed, a metallic path of appropriate length is easily capable of passing frequencies in excess of 100 kHz; thus, digitization and multiplexing can be used to carry a number of voice channels over a single pair of wires.

In residential areas, the T1 station-side equipment is usually located in an underground vault or in a cabinet above ground. From there, POTS copper pairs connect to houses and small businesses. For larger businesses, the T1 line is brought directly into the subscriber premises. Broadcast stations that require a large number of lines will often have their phone service via a T1 link, with an on-site box performing the conversion.

To create the T1 bit stream, 24 64 kbps DS-0 channels are assembled serially, and the equivalent of another 8 kbps channel is added for synchronization. Thus, the ultimate data rate becomes 1.544 Mbps, a rate called *DS-1*. The signal is then converted into a digital bipolar bit stream in a special format called *binary 8 zeroes suppression* (B8ZS). The voltage is modulated between –3 V and +3 V. Early T1 systems used robbed-bit signaling, taking the lower audio bit for call control, and leaving 56 kbps for the available capacity. Modern systems use Extended Super Frame (ESF), which uses some of the 8 kbps framing channel for signaling and restores the speech channels to clear 64 kbps. In this case, the technique for conveying the signaling bits is called Channel Associated Signaling (CAS). With CAS, there are four signaling bits in each direction, rather than two, although in most cases the actual signaling uses only the first two bits and is identical to the robbed-bit method.

Telephone networks in Europe and other parts of the world use E1 rather than T1. The technology is similar, but there are 32 channels rather than 24, and the total bit rate is 2.048 Mbps. One E1 channel is used for framing and another for signaling, so 30 are available for speech.

Most long-distance carriers in the United States offer service on T1 connected directly to their point of presence (POP). Because the long-distance carrier does not have to pay the usual fee to the local Telco for routing over its CO and lines, the user cost can be lower.

T1 arrives at the subscriber site as one or two conventional copper pairs. If it is a classic T1, there must be a repeater within 3000 ft of each end and every 5000 ft in between. The physical connector is the common RJ-48C, an eight-position modular plug. The usual components of a terminal system for a T1 circuit include:

- CSU and DSU—The T1 line is first connected to a piece of equipment called the *channel service unit* (CSU). The CSU used to be considered part of the network but is now almost always customer provided and may also be merely included as an adjunct section in a complete T1 interface solution. The CSU contains the last signal regenerator as well as a number of testing and maintenance features such as provision for loopback testing by the central office. It may also include a system to collect and report error statistics. The *data service unit* (DSU) handles the remaining digital housekeeping functions and data conversion from the bipolar T1 format to standard serial data and is used for data T1s only.
- Multiplexer and channel cards—The multiplexer, sometimes called a *channel bank,* is where the multiple voice (or data) channels are combined into the single bit stream required for T1 transmission. Each voice channel is converted to and from digital using codecs. In order to simulate typical Telco lines, a talk battery is added, ringing voltage is generated, and loop current is detected (this is typically referred to as an FXS port). Classic multiplexers are constructed using a modular circuit card approach so the available digital bandwidth may be configured as desired, but many modern ones are small and tightly integrated.

In theory, T1-derived POTS lines offer better quality audio than direct analog lines, particularly when the alternative is long runs and loading coils; however, some low-grade channel banks have poor-quality codecs, making modem and hybrid performance poor as well. Some also provide nonstandard ringing current. Most are locally powered and typically offer less battery back-up time than a CO if local power goes off.

Many modern PBX systems, and at least one broadcast on-air system, are able to accept T1 (or ISDN Primary Rate Interface [PRI]) lines directly. This is a near ideal approach, with a low-cost direct digital connection into the Telco network. In this case, no multiplexer and channel cards are necessary because the connection is made directly to the CSU. Some PBX equipment incorporates the DSU. Unfortunately, in many cases, voice T1s are constructed by feeding analog POTS lines into a Telco CO-located channel bank, so an extra pair of A/D and D/A conversions is made. ISDN PRI is a better choice, since it is likely to be a true digital connection directly into the network.

Common Channel Interoffice Signaling

Prior to 1974, signaling among COs and between COs and the long-distance network was via DC voltage on separate pairs, reverse battery signaling on the talk pair, or *in-band signaling*, using tones over the talk channels themselves. A major problem with this approach was that if a lengthy connection were established by switching together various network elements, and the far end was then found to be busy, these network elements would be used to return a busy tone without generating any revenue. However, the system was vulnerable to hackers (so-called *phone phreaks*) because once a busy or toll-free connection had been established, tones could be used to clear the call and establish a new one without being billed. It was also vulnerable to the talk-off phenomenon where certain female voices would cause trunks to be released.

To solve these problems, in 1974 Telco engineers installed the first *Common Channel Interoffice Signaling* (CCIS) facilities into the toll network between a crossbar switch in Madison, Wisconsin, and an electronic switch in Chicago. Common channel signaling, as the name implies, uses a common packet-data channel (originally at 4800 bps) for all signaling between two COs. Because this channel is only connected to Telco equipment, vulnerability to hacking was eliminated. Another feature of CCIS is the fact that a network path can be prenegotiated in an orderly fashion; if no path is possible or the far end is busy, the dialing CO can return an *all trunks busy* or a *busy* tone, and the reserved channels are immediately released, rather than expensive long-distance trunks being used to convey these tones to the user. Because near-instantaneous audio cut-through is required upon answer, audible ringing is still generated at the far end and returned to the dialing party in-band. CCIS also made possible a number of innovative network features such as caller ID and the ability for multiple carriers to provide toll-free 800 services.

Several versions of CCIS were deployed, with *Signaling System 7* (SS7) finally being standardized internationally by the CCITT in 1980. The ability of SS7 to perform intelligent decision making, by consulting network databases, has led to many innovations in telephony service. ISDN, first standardized in 1987, builds on SS7 and brings common channel signaling to the customer's premises. Indeed, a PBX on ISDN PRI functions very much as a peer to the network switches; the PBX provides a busy tone to local users in case the far-end line is in use and generates the ringing tone sent (in-band) to far-end callers.

INTEGRATED SERVICES DIGITAL NETWORK (ISDN)

By the mid-1980s, much of the telephone network was using digital voice channels, and common channel signaling had made network operations more flexible and secure. In addition, the amount of data traffic being sent over the switched voice network was growing; data exceeded voice traffic in the late 1990s, and by 2003 data traffic exceeded voice by 10 times. Translating data into modem tones and using 100-year-old transmission and signaling techniques were necessary only because customer access was via analog connections. This approach meant that 64 kbps network channels were being used to convey only about 33 kbps of user data, and reliability was poor. A much better way would be to provide direct access to the digital network, which was done when the Integrated Services Digital Network was developed. It uses the digital switched telephone backbone for either voice or data service. Because ISDN provides connection directly in digital form to the telephone network, it often improves voice quality on normal calls. A characteristic of ISDN important to broadcasters is that the B channels are true four-wire full duplex, with no crosstalk between the send and receive signal paths. More importantly, users may bypass the normal POTS speech-coding methods and supply their own algorithms, such as those standardized by Moving Pictures Expert Group (MPEG). There are two ISDN services: the *Basic Rate Interface* (BRI), which consists of two bearer channels of 64 kbps each, and the *Primary Rate Interface* (PRI), which has 23 bearer channels.

ISDN Basic Rate Interface

With a BRI circuit, two 64 kbps bearer channels are provided, each of which can be used for voice or data, as well as one 16 kbps D (data) channel multiplexed together. The D channel is used for call setup and status communication and is usually not available for use by the user.

The BRI S and U Interfaces

The line from the CO is a single copper pair physically identical to a POTS line. It arrives at the subscriber at the *U interface*. The U interface converts to an S/T interface with a small box called an *NT1*. In the United States, NT1 functionality is usually included in the terminal equipment, as shown in Figure 3.10-9. In Europe, the telephone company provides the NT1. Only one NT1 may be connected to a U interface, but as many as eight terminals may be paralleled onto an S bus.

Service Profile Identification Numbers (SPIDs)

Service profile identification numbers are required when using several of the ISDN protocols offered in the United States. These are only used with BRI. This number is given to the user by the phone company and must be entered into the terminal equipment in order for the connection to function. SPIDs usually consist of the phone number plus a few prefix or suffix

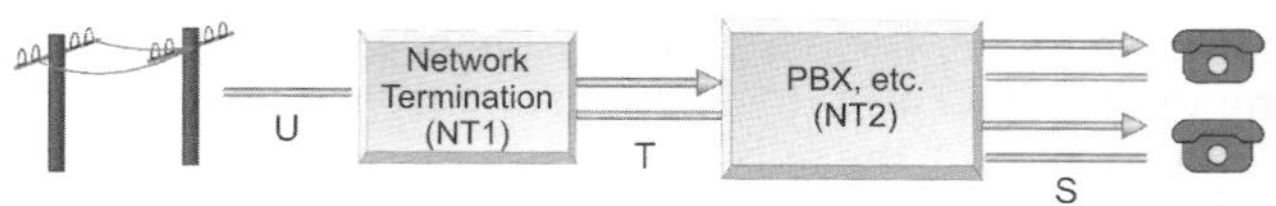

FIGURE 3.10-9 ISDN termination technology.

digits. The intention of the SPID was to allow the Telco equipment to automatically adapt to various user requirements by sensing different SPIDs from each of several types of user terminal on a shared S interface; for example, custom phones could have certain features that an ISDN data terminal would not need. Typically, there are only two SPIDs per line, so the ability to support multiple terminals is in fact quite limited on U.S. ISDN BRI circuits. None of this matters for most broadcasters, but the SPIDs must be entered nevertheless. If a station is using the National I-1, National I-2, DMS Custom, or AT&T 5ESS Custom multipoint protocol, the Telco service representative must provide one or two SPID numbers for each line ordered. Upon power-up, connection of the ISDN line, or boot, the TA and the Telco equipment go through an initialization/identification routine. The TA sends the SPID and, if it is correct, the network signals this fact. Thereafter, the SPID is not sent again to the switch. The SPID number must be provided and it must be correct, or the system will not work. Do not let an installer depart without leaving the SPID numbers.

CSD and CSV

Recall that each ISDN BRI has two possible B channels. It is possible to order a line with one or both of the B channels enabled, and each can be enabled for voice or data use. Phone terminology for this class of service is *circuit-switched voice* (CSV) and *circuit-switched data* (CSD). CSV is for standard voice phone service and allows ISDN to work with analog phone lines and phones using normal G.711 voice codecs. CSD is required for MPEG codec connections. Even though only voice is sent, the codec bit stream output looks like computer data to the phone network. For MPEG codec applications, a station may want POTS speech capability, because some support this feature; therefore, order CSV as well as CSD on one or both B channels.

Directory and DID Numbers

Directory numbers (DNs) are the telephone numbers assigned to an ISDN line. A station may be assigned one or two, depending on the line configuration. If two ISDN B channels are active, there will usually be two DNs; however, the physical channels are independent from the logical numbers. A call coming in on the second number will be assigned the first physical B channel, if it is not already occupied; therefore, there must be some way for the TA to sort out which call goes to which channel or line. The DN is used for this function. When a call rings in, it contains setup information, which includes the DN that was dialed by the originating caller. The last seven digits are matched with the DNs programmed into the TA and the proper assignment is made. On Euro ISDN BRI lines, the DNs are called *multiple subscriber numbers* (MSNs). Euro ISDN actually supports four or more ISDN devices on the shared S interface, so the MSNs can optionally be used to cause only certain terminals to respond to calls for a specific MSN. If no MSN is entered, a terminal will respond to all calls. Typically, three MSNs are provided by default. On a PRI, the DNs are called *direct inward dial* (DID) numbers and work in the same way. One difference: The number of digits may not be seven, so it is necessary to confirm the length with the Telco for programming local equipment. With PRI, it is quite common to have more DIDs than there are channels. This feature is very popular for business users because every telephone in an office can be directly dialed.

Protocols

Ideally, all ISDN terminal equipment would work with all ISDN lines without regard for such variations as 5ESS, DMS100, CSV/CSD, SPIDs, etc. In fact, there are a few different standards in use for the communications protocol. This is the language the Telco switch and the ISDN user equipment use to talk with each other on the D channel for setting up calls and the like. In the United States, the Bellcore standards *National ISDN-1* (NI-1) and *National ISDN-2* (NI-2) are by far the most common; however, both AT&T and Nortel CO switches have custom protocol versions that predate the national standard versions. These have been mostly phased out but are sometimes still encountered. In Europe, the usual protocol is Euro ISDN, following the ETS300 standards. Euro ISDN has also been adopted in a number of places outside of Europe, including Australia and much of South America. One strange variant that sometimes pops up for PRI is a custom protocol from the 4ESS switch, which is normally used only for the long-distance core network. This protocol is only supported by a few larger PBXs and may not be supported by your equipment. Do not believe the Telco if they tell you this is the same as the custom 5ESS protocol—it isn't. When ordering an ISDN line, be sure to learn which protocol will be used so the terminal equipment can be configured to correspond.

Digital Long-Distance

Circuit-switched data long-distance connectivity is required for MPEG codec calls. There are a few long-distance companies that offer this service, but most do not. The "dial 1+" default carrier may be chosen at the time the line is ordered, just as with traditional voice lines. Just as with voice lines, a carrier can be chosen on a per call basis by prefixing the number with a 101XXXX carrier selection code. Be sure to confirm with the carrier that it supports CSD connections. If the ISDN is used only for voice, any long-distance carrier will do, although the bargain-basement brands may not have as good quality. Note that costs can be reduced by arranging a service plan directly with the long-distance carrier. When using 1+ dialing without contacting the carrier, a standard rate plan that has the highest cost of any of the pricing tiers may be activated. Recall that some long-distance connections are limited to 56 kbps/channel due to robbed-bit signaling used within the network infrastructure. If the connection is limited to 56 kbps, the equipment will have to be configured to work at this rate.

ISDN PRI

U.S. PRI ISDN has a data rate equivalent to T1 circuits, providing 23 64 kbps bearer channels and a 64 kbps D channel for control (23B+D). European PRIs have 30B+1D+1 framing channels. PRI ISDN is delivered over the T1 or E1 physical interfaces. Because it offers superior signaling, it is slowly replacing the earlier systems. Although the terminology is not used very often, it also has U and S interfaces. The U interface is the T1 DS1 signal using Extended Super Frame (ESF) framing and the B8ZS line coding. The line is terminated with a CSU, and the S interface is the DSX1 signal. This is nearly identical to the U interface but has a limited maximum distance and there is no embedded operation channel on the DSX side. Some equipment can share the D channel for one PRI with a second PRI; in that case, the second PRI would have 24 bearer channels.

Trunk Groups for PRI

Broadcast telephone systems are likely to be flooded with large numbers of incoming calls when running a contest or a talk show. Because most PRIs are not served by choke networks, this can fill all capacity on the PRI, partially blocking calls to other numbers on this PRI and making outbound calls impossible. *Trunk groups* prevent this problem by reserving certain groups of channels for use with a number (or numbers). Incoming calls hunt across channels in their assigned group and, if the group is full, the CO returns busy. The channels in other groups remain unaffected.

By default, the Telco switch has no way of knowing that additional "call appearances" are not available for a particular number. If all call appearances are in use, the user equipment rejects the call, and the channel is released within a few hundreds of milliseconds; however, because each incoming call is assigned temporarily to a bearer channel until it is rejected, this can fill up the PRI. The result is that a large number of calls into the system could prevent access for outbound calls or block access to other DID numbers on the system, even though the number involved may have no call appearances left. For these reasons, a way to restrict the number of call setup messages coming from the Telco is needed. In the United States, a way to do this is to create multiple trunk groups (AT&T 5ESS) or to use group sizing (Nortel DMS-100) on the PRI to restrict the number of inbound calls to specified numbers. Not all telephone companies use these same terms, so there may be a need to explain. In the United Kingdom, a similar function is called *virtual facilities grouping*.

VOICE OVER INTERNET PROTOCOL NETWORKS

Until the emergence of the Internet, users were forcing data to travel over the circuit-switched network that was designed for voice. The opposite, voice as one application among many over a data-centric network offers significant advantages, but also poses challenges. For over a century, telephones have been much the same, consisting of a handset and a numeric "dial." Important innovations over this span of time have been push buttons in place of rotary wheels and caller ID. Contrast these to the Internet and the World Wide Web, where, in little over a decade, major innovations have occurred. The key is the nature of the Internet Protocol (IP) packets, which are an open and universal transport used to communicate nearly anything. As Internet applications demonstrate, voice, text, pictures, and video may efficiently co-exist. The flexibility of IP encourages new applications and user interface approaches.

Despite being supported by a digital infrastructure, traditional telephones can hardly be called "high-fidelity." The PSTN gives us only one choice: the G.711 codec, which is limited to 3.4 kHz audio bandwidth and 3% distortion, a quality level that was chosen 40 years ago by Telco engineers and has been used since. ISDN permits the first step away from this one-codec-fits-all restriction, since users can choose their own alternatives, such as those from MPEG. IP takes this further, removing the fixed 64 kbps channel constraint and allowing bandwidth to be scaled to any desired value.

VoIP offers a wide variety of endpoint styles. From basic telephones to phones with lots of buttons and built-in Web browsers to PCs with rich user interfaces to sophisticated conferencing and on-air systems, there is something to fit every application. Figure 3.10-10 shows an example of an IP "telephone" and

FIGURE 3.10-10 A VoIP phone with large color HTML display.

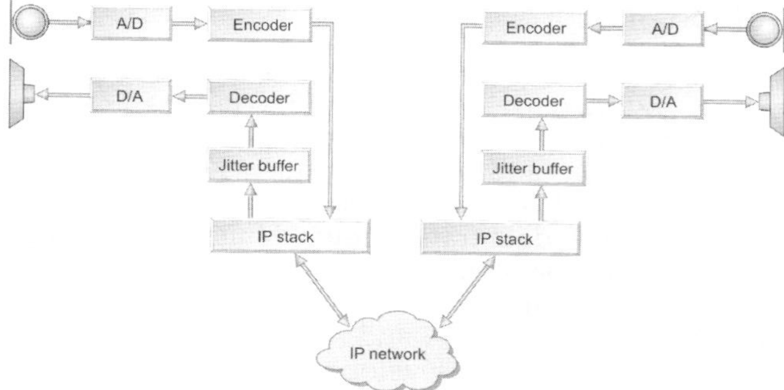

FIGURE 3.10-11 End-to-end VoIP call path.

Figure 3.10-11 shows a simplified block diagram of how a VoIP system works.

Packet data networks are more efficient than circuit-switched networks with respect to their ability to multiplex multiple users onto a particular digital pipe. In the traditional network, when a call is connected, a constant 64 kbps is consumed and cannot be shared. The telephone company cannot use this bandwidth for anything else and therefore bills it to you by the minute. The logic of the multiplexing used in packet data networks is one of the features of VoIP. Being able to use a single network for voice and data offers both simplification and cost savings. Within many broadcast facilities are an Ethernet network for computer data, a telephone PBX system, an audio routing system, various kinds of STL links, etc. These may all be combined into a single network.

Ensuring Quality

The flexibility of VoIP does not always lead to higher quality. Poor-quality, low-bit-rate codecs are as common as higher-quality ones. Low network quality-of-service levels (too much packet loss or jitter) can cause drop-outs. Codecs and packet buffers can make for significant delay. Delay can cause echo. These are all correctable problems, and VoIP can be made to be as reliable as the circuit-switched network, but they are new issues and need to be understood and addressed.

Packet Loss

IP networks can drop packets when there is more demand for bandwidth than capacity. This is a completely normal condition and is an inherent part of the design. A computer file transfer would take all of a network link's bandwidth if it could, but there is a speed governor built into the network interface of attached computers: the Transmission Control Protocol part of TCP/IP, which adjusts a computer's transmission speed downward when it senses packet loss. It also recovers any lost packets by requesting a resend from the source.

When voice and data share the same packet network, voice must have priority. Modern Ethernet switches and IP routers include a function that detects a packet's priority level and processes it accordingly. Voice packets are tagged with a higher priority than general data so the switching equipment knows to put them first. TCP works in concert with prioritization by regulating the rate of non-voice general data so it does not flood the network. This is a simple but effective way to ensure that speech quality is not impaired by packet loss. With prioritization, packet loss is controlled and speech flows smoothly.

Ethernet switches use the priority mechanisms specified by the IEEE in its standards for Ethernet equipment. IP routers have used DiffServ for the same purpose. MultiProtocol Label Switching (MPLS) is quickly catching on as the state-of-the-art architecture for large-scale mixed voice and data networks. In an MPLS-enabled network, only the first router in the path examines the packets for priority level. This router attaches a label to the packet, which is used by each router along the way to send the packet via a pre-provisioned path to its destination. A "heartbeat" is also sent along the path before the packet so a backup path can be used in the event of a network outage. MPLS is being promoted by a number of Internet Service Providers (ISPs) and Telcos as the way they will be providing voice service.

Codecs used for VoIP are designed to conceal packet loss up to 2 to 3%. Quality drops off quickly with loss over 5%, and networks used for voice must be engineered to prevent this from happening. Figure 3.10-12 illustrates how both IP and ISDN networks are combined for voice-quality transmission.

Delay

Packet-based networks inherently cause speech delays. The transmitting equipment needs to accumulate enough audio samples to fill up the packet before sending it. A buffer is provided in the receiving equipment to compensate for jitter in the arrival time of the

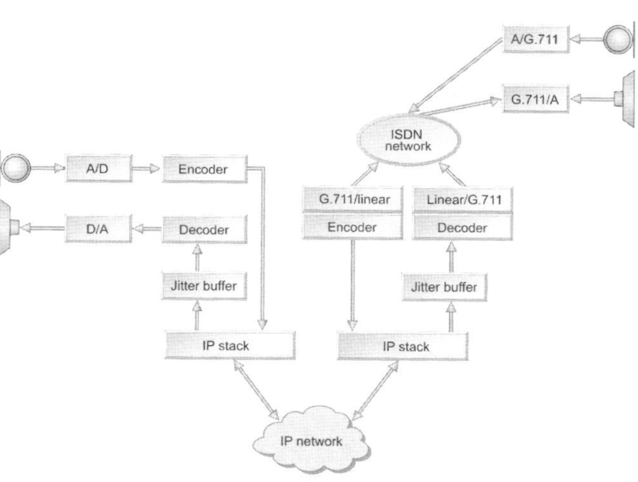

FIGURE 3.10-12 VoIP system using both IP and ISDN networks.

packets. The various routers along the way each impose a delay as they inspect packets before switching and sending them on. Most codecs cause delays because they need to work on a block of samples for their analysis and conversion. Delay and packet loss are significant problems on uncontrolled networks such as the public Internet, but, with properly engineered local-area networks (LANs) or wide-area networks (WANs) specifically designed to support speech, the negative effects of these problems can be eliminated. On switched LANs, for example, there is no packet loss and delays can be kept below a couple of milliseconds.

Echo

Echo is caused by any leakage of the send signal into the receive signal at the other end of the VoIP packetization and transmission process. The delay inherent in VoIP is what turns crosstalk into echo. Indeed, echo that is inaudible in the circuit-switched network may become noticeable with packet transmission because of the increased delay, as illustrated in Figure 3.10-13. Interconnections between packet networks and circuit-switched networks are especially susceptible to echo impairment, mainly from reflections caused by the hybrids used to interface to analog lines. This is why the digital hybrid function in broadcast interfaces is even more important than in the usual analog line case. Even when a link is end-to-end digital, there are sources of send/receive crosstalk such as acoustic coupling in the far-end terminal. The echo audibility depends on both the delay and the level of the echo. As delay increases, echo must be kept lower because there is not so much masking from the natural mouth-to-ear acoustic transmission. The ITU-T recommends that total delay be not more than 150 ms when the echo level is reasonably low. More than this and delay starts to be noticeable and conversation becomes inconvenient. VoIP networks are designed to satisfy this goal, taking into account all of the causes of delay.

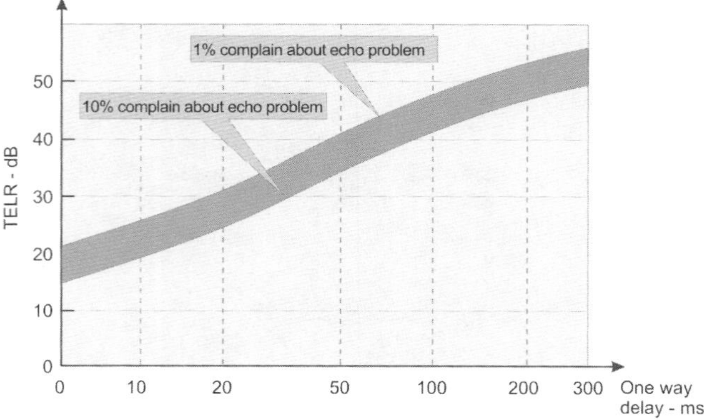

FIGURE 3.10-13 ITU Recommendation G.131 describes the relationship between delay, talker echo loudness rating (TELR), and listener perception [3].

VoIP Signaling

Signaling within the old analog network is primitive—drawing current from the 48 V talk battery signals a lifted receiver, 90 V AC rings a bell to signal a call, and various audible tone frequencies and lengths indicate if a call is ringing through or busy. With the full power of IP data transmission available, signaling for VoIP can be considerably more sophisticated. There are two standards for VoIP call setup and control. One, based on ISDN and standardized by ITU-T, is H.323. The second, developed by people working exclusively with Internet telephony, is the Session Initiation Protocol (SIP).

Many within the VoIP development community regard H.323 as being overly complicated and prefer the simpler and more Internet-friendly style of SIP. Certainly, SIP is now growing in popularity. Microsoft® Windows® XP, for example, includes SIP as part of its Windows® Messenger™ application. Some VoIP equipment will work in either H.323 or SIP mode and will automatically adjust to whichever is needed by the endpoint. Both allow normal telephone numbers as destination addresses to call people who have traditional phones rather than VoIP endpoints. H.323 and SIP calls can also use text names in the same form as e-mails (*e.g.*, user@domain.com or sip:user@domain.com). This is based on the same Universal Resource Locator (URL) and Internet Domain Name Service (DNS) system that e-mail and the Web use to look up text names and translate them to the physical IP address assigned to endpoint equipment. Call addresses can also be specified directly in dotted-decimal IP number form, such as 234.56.267.32. Proxy servers or registration servers are needed when users move around and change their IP addresses, when they use automatic IP assignment, or when they are behind routers that hide their addresses from the outside. In this case, a database lookup finds the physical address of the called party. The main practical consequence of having the two signaling methods is be sure when specifying gear that all the pieces are compatible with each other; they are then configured to communicate with each other.

Network Transport

VoIP audio and control are carried over IP networks. In theory, IP packets can be transported over a variety of local network technologies. In practice, Ethernet is the only LAN being currently used to interconnect phones, gateways, servers, etc. Within the Telco network, another networking technology called *Asynchronous Transfer Mode* (ATM) is sometimes being used in the backbone, but this is transparent from the user perspective. Control for call setup and other functions is usually via the same TCP/IP that e-mail and the Web use, but the audio uses User Datagram Protocol (UDP)/IP. The difference is that UDP has no flow control or lost-packet recovery mechanisms. For that reason, it is considered an unreliable transport. Using the more reliable TCP retransmission would take too

much time. When a lost packet finally shows up, it would likely be too late to use it. A very long buffer would be required to wait for these packets to arrive, but then there would be too much speech delay. Another protocol on top of UDP, the *Real-Time Transport Protocol* (RTP), extends UDP's very basic capabilities to support sequence numbering and time stamps, which are necessary for packet audio.

VoIP Codecs

The right codec makes all the difference. Compare the sound of a typical mobile phone to a broadcast ISDN codec using MPEG AAC. When VoIP sounds bad, the codec is probably to blame. Conversely, when it sounds good, the codec deserves the praise. Several codec standards are shown in Table 3.10-2. Within a facility, the telephone standard G.711 codec is likely to be used to connect phone sets to PBXs, but, when connected to the outside world, the codec supported by the service provider will be used (and perhaps the equipment at the other end of the call). Phones, gateways, and PBXs that are intended to connect to outside networks will generally support a variety of codecs. The call setup messages automatically negotiate and establish which codec will be used. Often PBXs will convert from a codec used for offsite transport to G.711 for internal use. This means that phone sets do not require the complexity of supporting all the codecs that could be found on outside connections.

As with all perceptual codecs, the usual objective quality measurements such as total harmonic distortion (THD) and S/N do not tell the whole story. A subjective measurement widely used within the telephone industry is the mean opinion score (MOS) [4]. Listeners rate codecs and apply a scale of 1 (bad) to 5 (excellent). The results for the various codecs are given in Table 3.10-2. The G.722.2/AMR-WB codec is the first codec to be standardized for both mobile and fixed-line use. It is expected to be deployed in third-generation Global System for Mobile Communications (GSM) phones. Because it is a relatively high-quality wideband codec, it offers the surprising possibility that calls from mobile phones could sound better on-air than those from landlines. For this to happen, the studio gear would have to be connected to the mobile network directly or via an IP path, though. If the call were to pass through the switched PSTN, the quality would be lost. Advanced Audio Coding–Low Delay (AAC-LD) is not routinely used for VoIP systems, but it is widely used for IP teleconferencing systems.

HDSL and ADSL

High-Data-rate Subscriber Lines (HDSLs) are at the T1 rate or higher (DS1 or 1.544 Mbps) and are symmetrical. These are frequently used, in disguise, by Telcos to deliver T1 service. Whereas classic T1 requires two copper pairs and repeaters every 5000 feet, HDSL technology can deliver this rate up to 12,000 feet on two pairs using 2B1Q line coding. HDSL2 technology uses trellis-coded pulse amplitude modulation (TC PAM) coding and can provide T1 rates to 12,000 feet on a single pair. In the majority of cases, the Telco's *smart jack* is used to convert the HDSL line back to a T1 signal at the network interface, although in theory HDSL or HDSL2 equipment could be used to terminate such a line directly. HDSL/HDSL2 offer some interesting possibilities for homebrew T1 in cases where customer-owned copper cable is in place, as no repeaters are needed, and easy-to-use modules can be used to create what looks to user-equipment to be a T1.

Asymmetric Digital Subscriber Lines (ADSLs) promise connections at speeds of up to 3 Mbps in the direction from the CO to the user. The upstream speed is limited to some much smaller value, which explains the asymmetric part of the name. An important advantage is cost. Because this service is aimed at the residential consumer market, it is typically priced at around the same level as ISDN BRI. There are numerous offerings, and paying more often provides a higher degree of tech support as well as higher speeds.

Initially, this technology was viewed by the Telco industry as a way to compete with cable TV for the

TABLE 3.10-2
VoIP Codecs

Codec	Bit Rate (kbps)	Delay (msec)	Quality (MOS)	Notes
G.711	64	0.0125	4.2	μ-Law PCM, usual phone codec
G.723	6.4/5.3	37.5	3.9/3.7	Popular, low-bit-rate VoIP codec
G.726	40/32/24	20/20/15	4.2/4.0/3.2	ADPCM, not often used
G.729	8	15	4.0	Good quality at low bit rate, widely used
G.722.2/AMR-WB	6.6–23.85	20	—	50 Hz to 7 kHz audio, new GSM codec
AAC-LD	64	50	(Hi-Fi)	Same rate as G.711, but full fidelity

Abbreviations: AAC-LD, Advanced Audio Coding–Low Delay; ADPCM, adaptive differential pulse-code modulation; AMR-WB, Adaptive Multi-Rate–Wideband; GSM, Global System for Mobile Communications; MOS, mean opinion score; PCM, pulse-code modulation; VoIP, Voice over Internet Protocol.

delivery of video services. Combined with an MPEG video/audio encoder, the bit rate offered by ADSL would permit full-quality National Television Systems Committee (NTSC) video. These projects now appear to be stalled, and current efforts are focused on high-speed Internet connectivity. As a result, the majority of ADSL lines connected directly to an ISP do not offer a direct connection to the PSTN nor direct end-to-end dedicated service. Because the Internet is a packet-based system with no bandwidth guarantee, broadcasters that want to use DSL for audio transmission face the risk of dropouts and longer outages. Nevertheless, given the ubiquity and low cost of this service, some stations could find utility in it for noncritical applications. The challenges with packet loss, delay, and jitter outlined above with respect to VoIP systems apply. As providers gear up to provide high-grade VoIP services, the same services could be used to support broadcast IP-enabled codecs.

The Cellular Mobile Telephone Network

With regard to mobile phones, broadcasters are often concerned about the low fidelity of calls from the public, and want to know what can be done to achieve an improvement. As well, they are excited by the opportunity to use mobile data services for high-fidelity remotes. Mobile services use a number of different technologies and their utility for broadcaster applications depend upon which is being considered:

- First-generation (1G) mobile phones used analog narrow-band FM technology on the 800 MHz band.
- Second-generation (2G) networks took different paths in the United States and most of the rest of the world. In the United States, time-division multiple access (TDMA) was the original technology, but it is now being phased out because it does not use bandwidth very efficiently and does not support data transmission. Code-division multiple access (CDMA) is widely used in the United States today. The Global System for Mobile Communications (*Groupe Speciale Mobile*, or GSM) is the standard for most of the world and is growing more popular in the United States, where 850 MHz is the usual frequency band. In Europe, 900 and 1800 MHz are used.

For 1G and 2G networks, the simplest way to improve quality is to connect an external professional microphone to a standard mobile phone. Most phones have a jack for this purpose. An automatic gain control (AGC) in the interface could also improve audio quality.

To achieve a significant improvement, the standard codec must be replaced by a better one. GSM phones offer a so-called *high-speed circuit-switched data* (HSCSD) capability. Perhaps the designation "high-speed" is a bit optimistic—with data rates of only 9.6 or 14.4 kbps, there is not much capacity for dramatic improvement. It might be possible to combine two channels to double the bandwidth, but the problem in the uplink direction is that only one channel is guaranteed. A connection might start with both active, but when the network becomes busy one could be taken away mid-call.

Packet data services over second-generation networks such as General Packet Data Service (GPRS) or Enhanced Data Rates for GSM Evolution (EDGE)—so-called 2.5G services—offer faster burst data rates, but with no guarantees. These seem to be replacing the HSCSD service. One could imagine using an IP-enabled codec over these channels for transmission of high-fidelity audio, but the lack of guaranteed bandwidth makes the proposition risky. Because this service is for IP data intended for e-mail, Web browsing, and the like, the connection to studio equipment would be via an Internet link, not the PSTN, and the quality of this link will affect the overall results, as well.

- Third-generation (3G) networks have more bandwidth, so they can offer higher bit rate services. A 3G network can provide rates from 384 kbps to 2 Mbps in a stationary environment. These could well turn out be useful for high-fidelity broadcast remotes with appropriate codecs. Again, these will be IP-based services, but because vendors are talking about live full-motion video as one of the attractions, they presumably are engineering the networks to have real-time media transmission capability. As mentioned in the context of VoIP, the new G.722.2 Adaptive Multi-Rate–Wideband (AMR-WB) codec has been standardized as routine for 3G phones. With its 7 kHz audio bandwidth, there is the chance that even off-the-shelf mobile phones will offer quite good fidelity. Delivered to studio equipment over IP networks that can reliably support the wide bandwidth, mobile calls from both the public and remote announcers could sound much better.

The new 3G CDMA service called EV-DO (Evolution Data Optimized) Rev. A now being offered by some U.S. carriers is perhaps the first to be useful for practical high-fidelity remotes. It offers a 3.1 Mbps rate in the downlink direction. More important to broadcast application is that it is the first service to offer a fast uplink rate—1.8 Mbps—and quality of service guarantees for latency and packet loss. The interface on the field side is via PC Cards or a mobile phone USB connection. The connection at the broadcast studio site is via an IP service such as the Internet. Since this is an IP data service, suitable codec equipment would need to support this protocol. It would also need to deal with packet loss, jitter, and probably varying bandwidth.

TELEPHONE TO BROADCAST INTERFACING

This section discusses connecting the telephone network in one form or another to studio equipment. It may connect to analog POTS lines, to ISDN, or to VoIP networks. For this, specialized gear will be needed that interfaces the two systems and adapts the audio in a way that achieves maximum quality. When airing programming with significant use of telephone segments, operators need a comfortable way to organize and select calls from multiple lines for air. Equipment for broadcast interface ranges from very basic couplers to multi-studio systems with sophisticated features for call control.

One-Way Interfacing Using Couplers

It is often necessary to take or send audio from a phone line in only one direction at a time. Newsroom "phoners" are a common application. If there is no requirement for a two-way conversation, a simple coupler interface will do. A simple one could consist of nothing more than a transformer, a capacitor, and a couple of Zener diodes, as shown in Figure 3.10-14. The capacitor provides DC blocking so the transformer does not become saturated from talk battery current. Because it will not draw loop current to hold the line, this is intended to be used in parallel with a phone set. If the coupler is to hold the line, the capacitor is eliminated and a transformer that can withstand the loop current without distortion is used (such as the SPT117 from Prem Magnetics). When sending audio into the phone line, it is important that the average level be limited to –9 dBm, which is the function of the Zener diodes. Commercial couplers are available that are a little fancier than the simple device described here, some offering auto-answer and disconnect capability, AGC, and remote control.

When using a coupler, it is most convenient to have the telephone instrument online and equipped with a push-to-talk switch on its receiver. This is because the phone's receiver has to be off-hook while a feed is coming in; the switch turns off the receiver's mouthpiece microphone when it is not depressed, thus ensuring that noise from the studio side will not be included in the recording. Because this coupler works in both directions, it can be used to send audio down the phone as well—useful in the production studio for letting clients hear their commercials before they go into the control room.

Two-Way Interfacing Using Hybrids

The simple coupler's limitations become apparent when it is necessary for the caller to hear the announcer and the audience to hear the caller simultaneously. A more sophisticated method is needed because independent send and receive connections are required to go to the studio equipment. These must be isolated. If too much of the send audio leaks into the receive connection, unacceptable problems will arise:

- There will be no control of the relative levels of the local announcer audio and the caller. The mixer fader for the telephone will have both signals, with the announcer at a level much higher than the caller.
- There will be distortion in the announcer's voice. The telephone line will change the phase of the send audio before it returns, with varying shifts at different frequencies. The announcer audio will suffer degradation as the original and leakage audios are mixed at the console and combine in and out of phase at the various frequencies. When this occurs, the announcer sounds either "hollow" or "tinny" as the phase cancellation affects some frequencies more than others.
- Feedback can result from the acoustic coupling created when callers must be heard on an open loudspeaker.

The simplest way to isolate the send and receive is to switch off the receive audio whenever there is send audio (Figure 3.10-15), as is done in most speakerphones. Electronic switches or voltage-controlled amplifiers connect either the send or the receive path, but never both simultaneously. When the announcer speaks, the caller disappears annoyingly, and noises in the studio can sometimes cause a caller to disappear momentarily, especially on weak calls. That is why commercial broadcast interface products no longer use this scheme.

Hybrids were invented to separate the send and receive signals on phone lines. They were necessary because long analog lines require amplifiers, and hybrids are necessary to attach amplifiers, as they are one-way devices. Building on this principle, hybrids are used to extract the send and receive signals from analog lines in order to convert the "to" and "from" signals to the isolated digital channels that make up the PSTN. Early hybrids were made from transformers with multiple windings. Today, most hybrids are

FIGURE 3.10-14 Simple coupler interface. The capacitor is for DC isolation and is not required when a transformer is used that can sink loop current. The Zener diodes are chosen to properly limit transmission levels to the required –9 dBm.

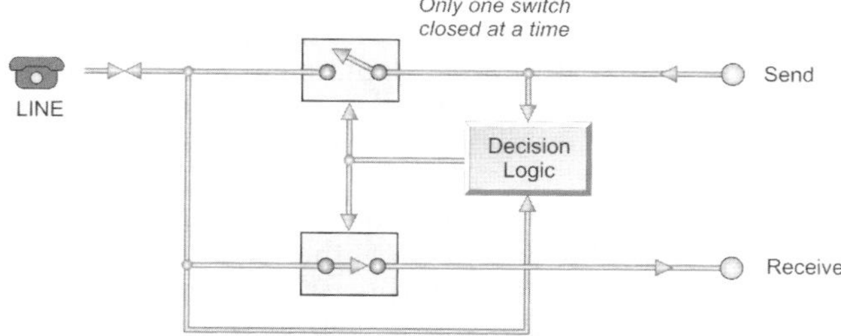

FIGURE 3.10-15 The switching interface allows two-way conversation but only one way at a time.

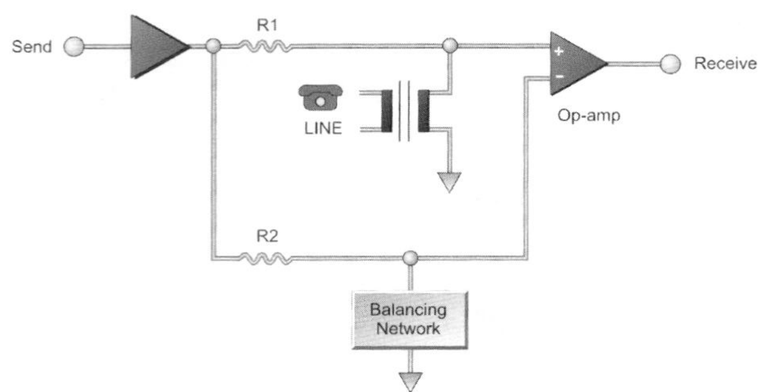

FIGURE 3.10-16 Op-amp hybrid. The second op-amp is used as a differential amplifier to perform the required subtraction for nulling the send audio in the receive output signal.

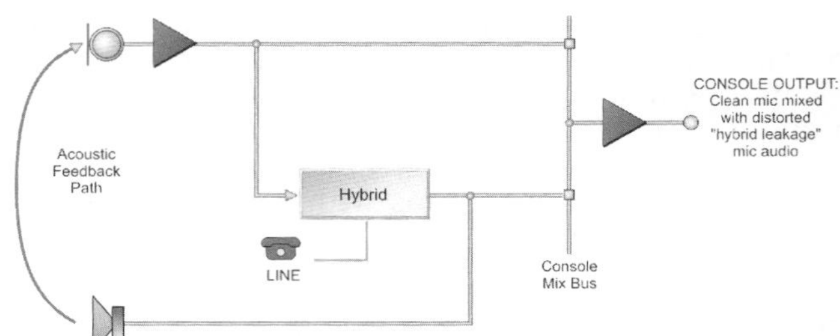

FIGURE 3.10-17 Block diagram of the typical studio arrangement with a telephone hybrid. Announcer audio is combined with the hybrid output, potentially causing problems with announcer audio distortion. The acoustic path is a possible source of audible feedback.

made with operational amplifiers (op-amps) and are known as *active hybrids*. Both circuit types use the same principle and achieve the same effect. In Figure 3.10-16, the first op-amp is simply a buffer, and the second is used as a differential amplifier; the two inputs are added out of phase (subtracted). If the phone lines and the *balancing network* have identical characteristics, then the send signals at the second differential amp will be identical, and no send audio will appear at the output.

The balancing network is a circuit consisting of capacitance, resistance, and sometimes inductance, forming an impedance network. Depending on the hybrid's application, this circuit can be very simple or it can be comprised of a large number of components and have a very complex impedance characteristic. R1 and the phone line form a voltage divider, as does R2 and the balancing network. If the phone line and balancing network are pure resistances, then, clearly, the phone line and the balancing network must have the same value in order for the signals at the differential amplifier to have the same amplitude and for complete cancellation to occur. The phone line, however, is not purely resistive but rather has a complex impedance, causing both the amplitude and phase to vary as the send signal frequency varies. Two-to-four wire converters, transformers, repeaters, T-carrier systems, and other Telco systems cause significant impedance bumps. Loading coils also have a deleterious effect on the performance of hybrid interfaces because the coils can create resonant peaks and phase anomalies in the phone line impedance curve, which are difficult to null out. Only when the impedance of the balancing network is the same as the phone line and the signals at the differential amplifier are matched in both amplitude and phase will full cancellation of the send signal be achieved; otherwise, leakage results—the main problem with hybrids.

In broadcast applications, the studio mixing console combines the output of the hybrid and the announcer's microphone audio, as illustrated in Figure 3.10-17. As discussed, the hybrid output consists of both the desired caller audio and the undesired leakage—the announcer audio is phase shifted because of the phone line's reactance. If trans-hybrid loss is not kept sufficiently high, there will be problems with announcer voice distortion. When lines are conferenced, there could be another problem: When the gain around the loop of the multiple hybrids is greater than unity, feedback singing will be audible. So, a capable broadcast hybrid must ensure that leakage is kept acceptably and consistently low.

The plots of phone line impedance *versus* frequency and phase shift shown in Figure 3.10-18 are the result of measurements performed on phone lines at a radio station in Indianapolis, Indiana. They indicate the wide variation seen on typical Telco lines as provided to broadcasters. The lines with smooth curves have impedance characteristics that could be emulated with a simple compromise-valued resistor–capacitor combination. These lines would work fairly well with a simple hybrid, because an RC balance network would match the impedance characteristic closely enough to make the cancellation of send audio at the hybrid output good enough to prevent coloration of the announcer audio.

Although it would theoretically be possible to construct a balance network to match the difficult lines, practical considerations usually keep this approach from being used. The impedance characteristic required is too difficult to produce using resistors and capacitors. If the hybrid is to be switched among a number of lines, the line characteristic would have to be consistent from call to call and have nearly the same impedance curve.

End-to-end ISDN and VoIP are fully send/receive isolated networks that theoretically have no need for echo-canceling hybrids; however, calls originating on ISDN or VoIP usually terminate to an analog line at the other end, where a significant source of echo would occur from both the Telco hybrid on the CO line card and the phone itself. Acoustic coupling, where the microphone picks up the output of the receiver, is another potential source of echo. Electrical pickup between analog circuits (crosstalk) is yet another. Even low echo levels can be audible when there is a long delay, such as with mobile or VoIP connections; thus, good hybrids are important, even on

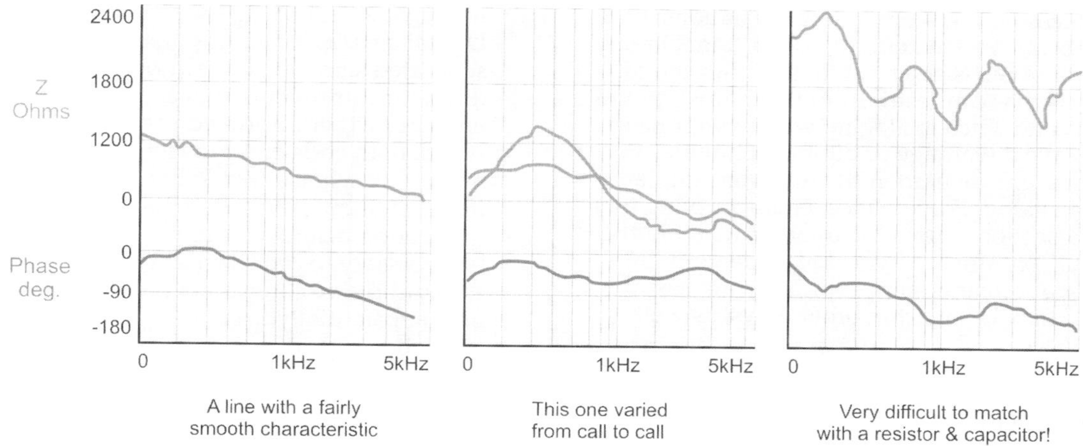

FIGURE 3.10-18 Impedance *versus* frequency curves for several typical phone lines.

digital lines. Actual practice confirms this necessity. In one instance, to troubleshoot a T1 circuit the incoming audio on various channels of a T1 were monitored. On some calls, only half of the conversation could be heard; however, on other calls it was easy to hear the outbound audio even when listening only to the receive side. A hybrid would be needed to cancel this leakage.

Digital Signal Processing Hybrids

Digital signal processing (DSP) offers a very powerful and effective technology to improve hybrids. DSP is the process of operating on analog signals that have been converted into the digital domain. Because the signals are numbers, mathematical operations can be performed to manipulate them before they are returned to analog. Complex processing functions either impractical or impossible to perform with analog circuit elements are achievable with DSP. With the DSP hybrid, natural simultaneous conversation is possible without distortion of the announcer audio. To

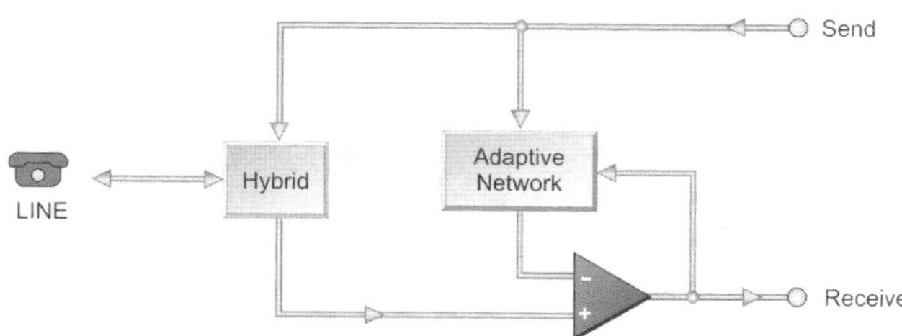

FIGURE 3.10-19 In the DSP hybrid, the digital balancing network continuously adjusts to the phone line impedance characteristics. When the adaptive network transfer function is identical to the phone line, perfect cancellation is achieved. Because the adaptive network is a digital filter, it can create almost any required transfer curve, so system performance is superior to the analog hybrid alone.

accomplish this, the announcer and caller audio signals are digitized and processed in a system that makes use of a specialized DSP microprocessor. The digital hybrid incorporates software programmed to perform the hybrid cancellation function. The technique, *convolutional least mean square adaptive filtering*, is capable of very accurate synthesis of the required balancing transfer function for maximum nulling. Unlike resistor/capacitor analog schemes, the adaptive filter can create the complex multiple break-point impedance *versus* frequency curves required by difficult-to-match phone lines. The send and receive signals are constantly compared in a feedback loop, with the leakage becoming an error control signal which drives adjustment of the adaptive digital balancing network, as shown in Figure 3.10-19.

The performance advantage of the digital hybrid technology is significant. On a typical phone line with a fairly smooth impedance curve, an analog hybrid might attain 15 to 20 dB trans-hybrid loss. A digital hybrid will likely produce 40 dB or better. On lines with difficult impedance curves, the analog hybrid's performance will usually be so poor as to prevent its use, while a digital hybrid would perform acceptably. On ISDN lines better than 65 dB is typical.

When a call is initially established, a brief mute/adaptation period provides an opportunity for the broadcaster interface system to adjust to the phone line prior to the call going on air. The caller hears a noisy tone, but none of this tone is heard on the air because the output is muted. This has the incidental benefit of removing the line switching transient. Adaption continues as the conversation proceeds, using voice as the reference signal.

While in the digital domain, other operations in addition to the hybrid adaptive balancing can be performed:

- *Automatic gain control* (AGC)—This function should be provided on both the send and receive audio paths. On the send side, it is necessary to smooth the wide level variations that arise from usual studio practices. On-air talent is accustomed to having audio processing take care of level variations and generally is not very careful at riding gain. On the receive side, AGC is essential to deal with the different levels that can result from the many types of phone sets and Telco analog network components. AGC can take advantage of digital techniques to significantly improve upon the functions implemented in analog. For instance, cross coupling to the hybrid section is possible in order to avoid the output AGC, confusing hybrid leakage with low-level caller audio and inappropriately increasing gain. AGC may be smart in other ways, as well.

- *Adaptive floating expansion threshold*—Improves noise-gating quality.
- *Acoustic echo cancellation* (AEC)—There is often an acoustic path between the received caller audio and the send audio signal. This results from having a loudspeaker in the studio that produces sound that couples into the microphones. When the on-air talent uses headphones for monitoring callers, this is not a problem, but sometimes it is not practical to convince guests to wear headphones, and television stations generally do not want talk show talent to wear earplugs. In these cases, a combination of adaptive cancellation and dynamic gain reduction will reduce the coupling electronically.
- *Dynamic equalization*—With phone sets having a very wide variety of microphone characteristics, a multiband automatic equalizer helps callers have a reasonable spectral consistency.
- *Caller "ducking" or "override"*—A form of send/receive switching can be used to improve effective trans-hybrid loss. This function reduces receive gain when send audio is present; thus, leakage also is minimized. Because the level from the phone is also reduced when the announcer is speaking, there is a sacrifice of full-duplex operation which is often a desirable trade-off. A user adjustment permits variation of the amount of receive ducking, allowing full duplex operation when the hybrid alone produces sufficient rejection or switching to speakerphone-like operation. This can also be used for purely aesthetic purposes to give announcers automatic control over callers who continue to talk.

ISDN Hybrids

ISDN can provide a direct digital connection into the PSTN analog network, so it can be used to enhance the quality of on-air calls. A call set-up message sent from the hybrid to the network tells the CO switch to go into POTS inter-working mode. In most areas, the cost of ISDN service is not a barrier. An ISDN BRI, with two channels, costs about twice as much as a POTS line. Pricing varies depending on the Telco but ranges from a 20% discount to a 30% premium while the average is probably around a 10% premium. For studio applications, ISDN offers a number of advantages over analog POTS lines:

- ISDN lines are inherently four-wire—As discussed, analog lines use a single pair of wires for both signal directions, mixing the send and receive audio. Digital circuits inherently offer independent and separated signal paths. Although DSP hybrids are a dramatic improvement over analog hybrids, ISDN enables further improved performance. In the case where both ends of a connection are digital, there is no mixing whatsoever. When the far end is a POTS line and phone, the audio paths will not be fully independent, and a digital hybrid function will still be necessary to cancel residual leakage, but moving the studio side connection away from POTS can help substantially because it gives the hybrid a much better starting point.
- Better digital-analog conversion quality—The codecs used in telephone COs are not as good as the converters commonly used in regular audio equipment. In a professional interface for studio applications, converters can be designed with better performance than available in a telephone company's equipment. Noise-shaping functions permit a larger word-length converter to provide significantly better distortion and S/N performance. In all-digital studios, an ISDN hybrid can preserve the digital path all the way into the studio mixer.
- Lower noise—As digital circuits, ISDN lines are not susceptible to induced noise. Analog lines are exposed to a wide variety of noise and impulses and hum as they are strung across town on poles and through a building. Hum is the main problem, given the line's proximity to transformers and AC power lines, but there are also sources of impulse noise from motors, switches, and other sources. Digital lines convey the bits precisely and accurately from the network to studio equipment without any perturbation, so the audio remains clean.
- Better call setup and supervision—The sophisticated transactions on the ISDN D channel are able to keep both ends of a call accurately informed about what is happening. ISDN call setup times are often a few tens of milliseconds, enhancing production of a fast-paced show. Perhaps more importantly, when a caller disconnects while waiting on hold, the ISDN channel communicates this status change instantly. This contrasts with the usual 11-second delay for CPC on most analog lines. One of the most common complaints of talk show hosts is that when they go to a line where they expect a caller to be waiting they are met instead with a dial tone. The chance of this happening with an ISDN line is reduced to near zero. Another common error is when on-air talent punches up a line that looks free but is actually just about to begin ringing and connects to a surprised caller. This condition results from the delay in the ring signaling, which is due to the nature of the analog line's ringing cadence. This is less likely to occur with ISDN because the ambiguous status period is eliminated.
- Levels—ISDN does not have the FCC required –9 dBm send level limit. Audio may be adjusted to fill up the digital word, resulting in higher send signal

volume. Of course, an excessive level will cause undesirable clipping in a digital circuit.

- Reduced feedback during multi-line conferencing—When conferencing is required on two-wire circuits, high-performance hybrids are needed to separate the two audio paths to permit adding gain in each of the several directions. When the gain around the loop exceeds unity, feedback singing is possible. Because the conference path usually includes four AGC functions, the hybrid must be sufficiently good to cover the additional gain that may be dynamically inserted. Because of the four-wire nature of ISDN, the hybrid function is more effective and more reliable across a variety of calls. That means more gain can be inserted between calls before feedback becomes a problem.
- Line monitoring—Because there is a full-time connection between the central office and the terminal on the D channel, it is possible to detect when a line is a not working. On an analog line, a problem is discovered only from a failed attempt to use the line.

Most of the functions performed by an ISDN interface are similar to those of an analog DSP hybrid, but there are some differences, both in the required functions and in the implementation of the common features:

- *Send/receive separation*—Despite the fact that ISDN lines have independent send and receive, the DSP hybrid function is still needed.
- *Sampling-rate conversion*—When the studio connection is via digital AES3 or IP, no analog/digital conversion is required, but it will be necessary to adapt the sampling rate of the telephone network to the studio rate. ISDN sampling rate is 8 kHz and studio equipment will usually operate at 44.1 or 48 kHz. A process is required to perform the required up-and-down conversion, while suppressing aliasing and reconstruction of audio components.
- *Caller ID*—ISDN naturally conveys caller ID information. ISDN interface equipment can detect and deliver this information to users.

VoIP Hybrids

A hybrid designed for VoIP application will have many of the same audio processing functions as hybrids intended for analog and ISDN connection. A greater demand will be placed on the canceling function, however. An echo canceling hybrid attached to a VoIP line might need to accommodate a time span extending to 350 ms. VoIP's delay makes any crosstalk much more audible than it would be in an analog or ISDN hybrid. The network interface will be via Ethernet to IP, rather than to ISDN, of course. It will have to address the delay, jitter, echo, and dropped-packet issues that are unique to VoIP. The network interface should support all the codecs that are expected to be encountered. A unit intended to be used only on the station side of a PBX would probably require only a G.711 codec, but one connecting to the outside would have to keep a number of codecs on-hand to accommodate the various incoming call formats. These would be automatically selected and switched-in to the audio path during the call setup negotiation.

Evaluating Hybrid Performance

Broadcast hybrids are required to cancel the leakage resulting from the local analog line connection. The amount of cancellation—the trans-hybrid loss—directly affects the on-air audio quality and is the most critical measure of hybrid performance. The test of hybrid performance is determined by measuring the amount of rejection across the entire audio frequency range, preferably with pink noise as a test signal at the send input. A hybrid with an adjustable R and C balance network can produce high rejection at a single frequency, as both phase and amplitude at one frequency can be adjusted for good cancellation, but this will not translate to good actual performance. Although the two are related, the trans-hybrid loss is not the same as the observed difference between the caller level and the leakage at the hybrid's output. Assume a phone call is arriving with a –20 dBm level. The send level is –10 dBm to the line. That means that the hybrid has to use 10 dB of its trans-hybrid loss just to break even. Other important performance characteristics include S/N ratio, distortion, and number of bits in the audio path. The operations of the dynamic functions—AGC, noise gate, and override ducking—also make a significant contribution to a hybrid's effective performance.

Interfacing to PBX Systems

Modern PBXs use a digital link similar to ISDN BRI to connect the common equipment to phone sets. It is usually possible to interface to PBX phones for on-air use using a handset attachment or a tap-off at the internal speaker terminals; however, this is best reserved for casual phone use such as for the occasional request or contest winner call. For applications where phone calls are a significant programming element, consider the specialized on-air systems from broadcast-oriented manufacturers:

- *POTS lines from PBXs*—Because fax machines and modems require connections that look like CO lines, many systems provide ports for this use. They may be connected to broadcast interfacing gear as if they were CO lines. Sophisticated PBXs have programming features that allow these ports to be configured in various and potentially useful ways; for example, they may be set up for private line ringing so that when a given incoming CO line rings the call may be directly sent to the selected port. Unfortunately, with most PBX systems, awkward operation may result, because the only way to move a call from a phone set to the port may be to transfer it using multiple button punches rather than the usual simple place-on-hold-and-pick-up-elsewhere operation. Taking calls in sequence on-air may be difficult or impossible.

- *ISDN PBX ports*—Many ISDN-capable PBXs can emulate a network ISDN BRI or PRI line and so can be connected to ISDN hybrids. These offer some interesting advantages, such as the sharing of trunks and the ability for PBX users to transfer callers to the on-air telephone system, as well as a fully digital connection. Transfers from the on-air system to the PBX are problematic and typically use the *trombone* method, where two B channels are used (also sometimes called a *hairpin* transfer). Note that the ISDN protocol delivered by some PBXs is not necessarily compliant with the relevant standards. In most cases, it can be made to work but will probably require patience. An ISDN protocol analyzer will be helpful in this situation.
- *Speakerphone tap-off*—One way to get low-cost interfacing is to take advantage of the switching-type interface that many phone set internal speakerphones provide. The procedure is to tap-off the speaker with a transformer and pad to match the console's required input level. An external send audio source can be a substitute for the phone's internal microphone. Again, a pad and probably a transformer will be needed. The input feed must be set so appropriate switching action and proper send levels are obtained.
- *Handset adapters*—Adapters are available that plug into the phone set's handset modular jack and convert the microphone and earpiece signals into a signal that emulates a standard CO line, which can be connected to a hybrid. While useful in some applications, this approach is likely to offer a lower quality feed because the phone set's network remains in the signal path, causing impedance bumps and other problems. When a hybrid is connected to the handset jack, it cannot determine when a new call is selected, so it cannot adjust its null to the new line before the conversation starts. Because the hybrid can null on voice during conversation, null will be achieved in perhaps 4 seconds. This is acceptable if only a portion of the call is to be aired, as is common with on-air requests, contest winner calls, and the like. The line switching transient will not be muted, although this is not a problem when calls are not aired sequentially live. Poorly designed PBXs may have more than the usual noise and distortion.
- *Intercepting the serial data stream*—It would be desirable to emulate an electronic phone set by generating and decoding the phone system's serial data; however, PBX vendors insist on keeping their data protocols secret. That means that broadcast manufacturers are unable to design direct emulation equipment. Of course, even if the protocols were available, there is the problem of accommodating the dozens of communication methods employed by PBX designers.

Multi-Line Studio Telephone Interface Systems

With phones a key part in the programming for many stations, interface systems that enable convenient, high-quality, on-air integration of phone conversations are essential. These integrate hybrids and line selection controllers to provide a solution to adapting the telephone network to the broadcast studio environment. Although many business phone systems offer similar functions—line selection, status indication, conferencing—they are generally awkward to operate in an on-air environment and may have other limitations such as poor audio quality.

Broadcast phone systems are designed with studio requirements in mind; for example, a broadcast phone system output should be free of inappropriate switching sounds, and on-air talent should be able to access and manipulate lines live without any pops or transients being audible to listeners. Good audio quality must be extracted from the limited fidelity phone network. Line selection and other functions must be performed intuitively and with a minimum of hassle. Unlike a telephone set, broadcast line selection panels should have large buttons and easily visible status indicators, as shown in the examples in Figure 3.10-20. To avoid operator confusion, features are limited to those necessary for on-air application. Control panels that drop into an open position in the studio mixing console can locate the line selection buttons conveniently near the faders assigned to phones.

(a)

(b)

FIGURE 3.10-20 The console on the top (a) is an example of a multi-line on-air call controller console designed for easy operation by on-air talent; this particular Telcos model is expandable to 24 lines with an add-on module and 4 hybrids. The console on the bottom (b) is another example of a phone line controller; this particular model is expandable to 12 lines with an add-on module.

These systems may use analog, T1/E1, ISDN BRI, ISDN PRI, or VoIP Telco line connections. A system using ISDN PRI, T1/E1, or VoIP may be able to share lines among a number of studios, with connections to both hybrids and codecs. Features offered by multiline systems may include:

- *Conferencing*—Most broadcast systems allow any number of lines to be switched to air, even if only via a single hybrid, but unless excellent phone lines are used additional hybrids will be needed with a multiple mix-minus arrangement to provide amplification between callers. Without multiple hybrids, callers might have difficulty hearing each other with only the Telco-delivered line level. ISDN and VoIP make conferencing less problematic because of their four-wire connection and more consistent levels.
- *Busy/unbusy*—To prepare for a contest, all lines may be busied-out and then returned to readiness after the contest has been announced.
- *Automatic next line selection*—Pressing a "Next" button picks up the line that has been holding the longest. If no line has been holding, the longest ringing-in line is selected.
- *Call length timer*—The call duration time is displayed.
- *Held caller timer*—This tells which line has been holding longest and for how long.
- *Talk show production/call screening software*—This PC software lets a talk show screener or producer communicate to the on-air talent which caller is on the line waiting to talk. An example of the screen for this kind of software is shown in Figure 3.10-21. It replaces the paper-notes-on-the-window system employed for years at many talk stations. The better packages offer features such as direct control of the on-air telephone system using the computer, display of liner messages and other information, and storage of caller data for demographic analysis. Several modern systems use caller ID and a caller database to automatically bring up information about frequent callers and also have the ability to keep a list of unwanted callers to automatically command the telephone system to drop them.

An Ethernet or serial port can connect the PC to the telephone system. Via an Ethernet/IP connection, the software may be used to extend full control capability and status display to a remote site over the Internet. One system implements the call screener function as a Web page for access without a custom client.

Integration of Multi-Line On-Air Systems with PBXs

To interconnect the on-air system with PBXs, there are a number of possibilities:

- *Segregate the studio and office phone lines*—Ports from the PBX that are configured to look like CO lines feed an input or two on the studio system so calls taken by the receptionist can be put on the air.

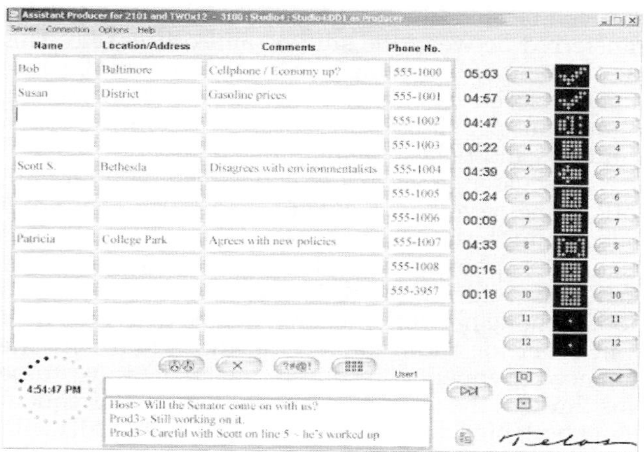

FIGURE 3.10-21 Monitor display of talk show production/call screening software.

- *Route all lines through the PBX*—The studio lines are programmed in the PBX to be forwarded to the ports that feed the studio system. Some audio degradation may result if analog ports are used. Audio quality is not an issue if an ISDN connection from the PBX is used, but in this case it is essential that the PBX implements the ISDN protocol according to the relevant standards or compatibility may be an issue. Note that on-air and business trunks are to be configured in such a way that heavy on-air traffic does not block business traffic.
- *Parallel the two systems*—With no cross coupling of line status information, there could be trouble if a line is inadvertently picked up on one system while the other is being used.
- *Route the on-air lines through the broadcast system*—This is possible if the broadcast system brings out a loop-through connection. This scheme prevents PBX phones from picking up active on-air lines.
- *Use a VoIP PBX with an IP-enabled on-air system*—VoIP systems can achieve tight integration of the PBX and on-air systems. Because VoIP uses open standards, it is possible to share lines and pass calls between the systems, in addition to other features.

Using VoIP in a Broadcast Facility

Whether all telephone service in a broadcast facility is via VoIP or not, the technology provides a valuable way to interconnect equipment and route calls within studio facilities. An important benefit is the possibility to integrate all of a facility's networks into one infrastructure. Most stations now have an Ethernet data network, a phone PBX system, and AES3 and analog audio connections. A pro-audio over IP system such as Axia's LiveWire allows audio and data to share one network [5]. VoIP is the final piece that consolidates all the different pieces into one unified system. Another benefit is the possibility to have tighter integration of the office PBX system and the on-air phone system, so lines can be shared among studios, calls can be easily transferred between the studios and offices, etc. This has always been a problem with traditional PBXs due to their closed design and the limitations of analog connection signaling. VoIP solves this with its open standards and rich signaling protocol.

As discussed, a variety of codecs can be used with VoIP. Some have better quality than traditional voice codecs, and these have the potential to enhance on-air quality for phone segments. In particular, the new AMR-WB wide-band codec standardized for 3G mobile phone networks will have to be delivered over IP in order for audio to arrive at the studio with the enhanced frequency response intact. Passing these calls through the usual PSTN would result in frequencies at both high and low ends being cut off. Calls with IP at both ends could use a high-quality codec to improve caller fidelity. Current Internet telephony applications such as Windows® Messenger™ and Skype™ demonstrate the potential.

An important distinction between VoIP on local networks and over WANs that carry calls offsite is that LANs have plenty of bandwidth and are completely under local control, so high quality is achieved with a bit of care. Transmitting a call over the public Internet is quite another matter, indeed. There is no guarantee of any kind with the Internet, so good quality may occur on one day and annoying drop-outs the next. The MPLS networks that Telcos are now proposing for voice services are intended to be engineered to support high-quality voice service, but whether it will work is yet unknown. It is for this reason that VoIP could well make sense within a studio, while traditional circuit-switched technology is used for outside connectivity. Today's VoIP PBX systems can be used in this way, and they provide gateways to ISDN or even to analog Telco lines.

Broadcast on-air systems may use VoIP in a number of ways. Simple hybrids could interface to VoIP PBXs. Elaborate multi-studio setups could use VoIP as their internal interconnect, and IP could be used as a transport for offsite connections to remote announcers and intercoms.

The elements needed to assemble a VoIP system are shown in the block diagram in Figure 3.10-22 and summarized below:

- *Gateway*—The element that interfaces the PSTN to the IP network. It converts the audio from either analog or ISDN to the IP packet format. It also translates the signaling from one to the other. The gateway is only needed during the transition period when phone calls are coming to the studio via the traditional circuit-switched network. At some future time, when all calls are conveyed over IP, this equipment may not be needed.
- *Softswitch*—The controller for the call processing functions; for example, it listens to the keystrokes from IP telephone sets and initiates calls by sending commands to the gateway. Even though it is called a "switch," no audio passes through it. The actual interconnection of audio is done in the Ethernet switch. This offers an important benefit—the "soft" in the name emphasizes that its characteristics can be much more easily changed than in the traditional system with the control and switching tightly bound in a single device. Some vendors refer to their softswitch products as *call servers* or *communications servers*.

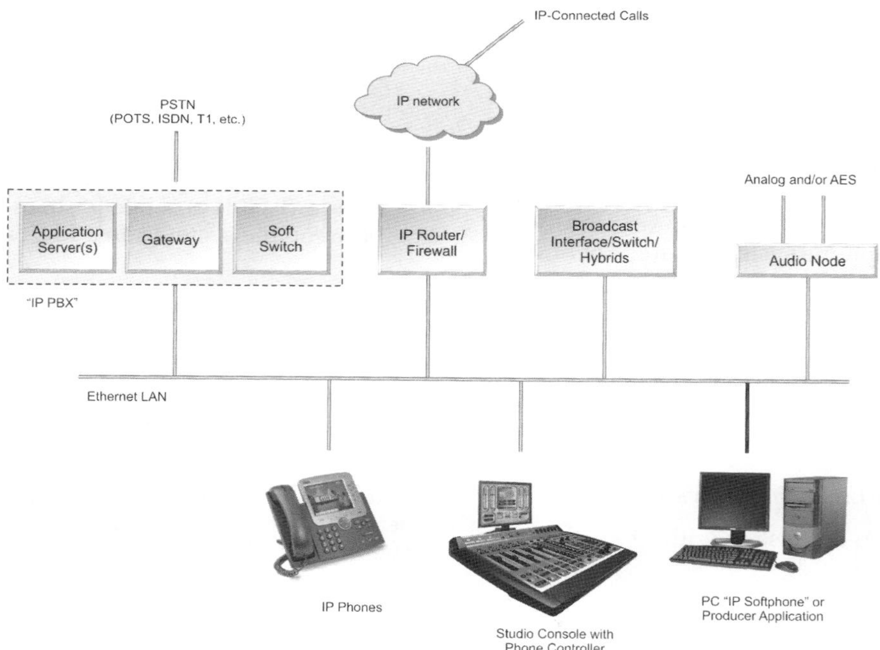

FIGURE 3.10-22 A VoIP system as it might be installed in a radio studio.

- *Application server*—Usually a PC-based device that provides functions beyond basic switching. Voice-mail is the most common, but other sophisticated tasks such as might be required by a high-volume call center would be handled here. Any number of these may be attached to the network to perform a range of tasks. The gateway, softswitch, and application servers could be independent physical boxes, each individually connected to the Ethernet, but most products for small businesses have these functions integrated into a single package. Products in this category are provided by many of the well-known PBX manufacturers such as Nortel, Siemens, Avaya, and Mitel. Vendors from the computer networking world such as Cisco and 3Com also have popular products.
- *IP router/firewall*—The device typically used for connecting Ethernet to the Internet or private company WANs. This device examines the packets on the network to determine which should be sent outside or kept in the local network. It performs the same function in the other direction, usually with a firewall service that prevents unwanted traffic from entering the broadcaster's network. Phone calls that arrive at the studio via IP will pass into the network via this box rather than the gateway. Sometimes the router and firewall are included as part of the gateway and softswitch.
- *Broadcast interface*—A specialized device made by a broadcast equipment vendor to support studio telephone operations. It converts the audio from the phone format to the professional audio format necessary to connect to studio mixers, including sample rate conversion. It provides the hybrid function to cancel echo as well as the broadcast-specific switching and conferencing.
- *Audio nodes*—Required to interface to legacy analog and Audio Engineering Society (AES) digital studio equipment. When studio equipment based on IP over Ethernet is being used, there is no need for this interface because the packets can flow directly to the audio routing and mixing elements.
- *IP phones*—Connect to the Ethernet, but otherwise work like usual business telephone sets. They range from simple single-line sets to ones that have multiple features and line buttons as well as a color liquid crystal display (LCD) screen that can display HTML (Web) graphics.
- *Studio console*—May be equipped with a call controller module that talks to the broadcast interface to select lines for on-air transmission. This generally will be a control-only device, with no audio passing through it.
- *IP softphones*—Run on PCs, providing similar functions as the dedicated hardware phone sets. A headset/microphone plugged into the soundcard or USB jack is used in place of the usual handset.
- *Producer/screener applications*—These are similar to the producer/screener packages described earlier, but because VoIP systems put the phone audio on the network, the PC can serve as a softphone as well as a controller, or it can work in conjunction with an associated VoIP telephone set.

One of the advantages of VoIP is that it can readily scale from serving a single studio to accommodating the demands of sophisticated multiple studio facilities. Ethernet switch ports are numerous and low cost. Plugging in more phones, controllers, and PCs for expansion is straightforward.

Improving Phone Audio Quality

Whether extracted from analog or digital lines, audio that originates from a traditional telephone set is hardly studio quality. Until everyone has a high-fidelity VoIP set, techniques to improve inherent phone audio need to be considered. One important quality limitation results from the anti-aliasing and reconstruction filters in PSTN G.711 codecs. These filters usually have an ultimate roll-off of around 35 dB. Audio above the 4 kHz Nyquist frequency will alias and appear in the 300 Hz to 3.4 kHz band as distortion; thus, typical codecs have distortion of 2 to 3% caused by aliasing. The raspy noise that seems correlated with the speech sometimes heard on a telephone circuit is a result of the effects of this kind of distortion combined with audible quantization errors. Older codec filters used switched-capacitor technology, which can be noisy.

An ISDN or VoIP connection solves half of the problem, because at least one of the Telco's codecs is bypassed. The other end, and the majority of broadcast connections, will remain analog for some time to come. Fortunately, there are some remediation possibilities. Filtering, equalization, gating, and dynamic compression are the primary tools. Most commercial interfaces have some or all of these processes built-in:

- *Filtering*—On a dial-up phone line, there is very little audio above 3.4 kHz, but there are noise and digital reconstruction distortions. Thus, a filter with a steep roll-off above the telephone passband will reduce phone line noise significantly without affecting conversation audio. Low-frequency hum is often a problem, as well. This is usually 60 Hz mixed with its second harmonic, 120 Hz. It is often a good idea, then, to have a sharp roll-off starting at about 200 Hz.
- *Equalization*—An equalizer (EQ) used to shape the frequency response of the phone line within its audio bandwidth can result in marked improvements in perceived quality. A typical phone line has an excess of energy at around 400 Hz and considerable roll-off at both the top and bottom ends of its passband, so the idea is to compensate by adding gain at both. Boosts at 2.5 kHz and 250 Hz and attenuation at 400 to 500 Hz with a parametric equalizer will help achieve better sound.

When it is not possible or practical to make custom adjustments, an adaptive multiband EQ can be an effective tool. Some advanced hybrids include this function. The principle is much the same as implemented in broadcast transmission processors. Audio is filtered into multiple bands, and an automatic gain adjustment is performed on each spectral segment. Given the limited frequency range of telephone calls, three bands are sufficient.

- *Noise gating*—Another effective processing device is the expander or noise gate. These devices may be used to reduce gain between the words of a conversation, thus making phone line noise less objectionable. On extremely noisy lines, however, the gating action can make noise more distracting by causing it to come and go with the words. In such cases, it might sound better to leave the gate off and let the noise remain present at a constant level.
- *Dynamic compression*—Levels on phone calls vary widely. A compressor helps to smooth this out. An AGC that maintains a constant compression ratio regardless of average gain reduction produces more consistency. Freeze gating is also important so gain does not increase during caller speech pauses.

Recording Phone Calls

At some stations, calls that DJs take off-air are routinely recorded for later playback. One technique is to have the mix-minus go to one track of a stereo recording device, while the other channel gets the hybrid output with the caller audio. The result is a two-track recording with the announcer and caller separated. For playback, the console's input mode is set to mono and the relative balance can be adjusted as needed. The production department can use this two-channel file to facilitate extraction of contest winner squeals and the like.

BROADCAST CODECS: HIGH-FIDELITY REMOTES OVER ISDN, POTS, AND IP

ISDN makes high-quality remotes possible with dial-up convenience. Convenient, reasonably priced, studio-quality audio from almost anywhere in the world is now routinely available. POTS codecs use modems over analog lines to digitally convey coded audio. Using this technology, audio frequency response may be extended to as much as 20 Hz to 15 kHz, albeit with audible coding artifacts. Nevertheless, codecs have improved over recent years, and the resulting fidelity has become useful for remotes that do not require ISDN-grade quality.

IP is the new frontier. Audio quality on IP network-connected codecs ranges from unusable to ear-grating fuzzy to full studio quality, depending upon the codec used and the quality of the IP connection. Because IP remote equipment uses the same underlying technology as VoIP, the same network performance issues apply. VoIP telephony and IP remotes may well eventually converge to use the same equipment.

IP connectivity can be provided by wire lines such as T1 and DSL, by radio links such as WiMax, and by satellite and mobile phone networks. The public Internet is attractive due to its ubiquity and low cost, but it offers no performance guarantees; however, it may be useful for noncritical applications. New IP-based telephone services are appearing that are said to provide service levels similar to ISDN. These are intended for VoIP application but could certainly be used for broadcast remotes.

Broadcast codecs are usually single-box solutions that include an ISDN TA and/or IP interface and a number of selectable coding algorithms. Most are full duplex, with provision for transmitting and receiving simultaneously. Some offer a feature to allow connecting to POTS phones for voice communications or to POTS codecs for higher-than-phone fidelity over analog lines. End-to-end, parallel contact closures offered by many codecs may be used to control recorders and other devices. Some portable units include a mixer for multiple audio inputs and outputs. Many include a receive-side mixer to combine the mix-minus signal from the studio with the local audio.

Audio Coding

Audio coding in the context of VoIP has been discussed. A new class of codecs has been designed for better fidelity than those made for VoIP application. High-fidelity audio coders are able to reduce the bit rate of a digital audio signal to as little as 5% of the original size, while preserving fidelity very close to the original. There are a number of these audio coding methods, each with advantages and drawbacks.

ADPCM Coding

Adaptive differential pulse-code modulation (ADPCM) predates MPEG perceptual coding and is much simpler than the perceptual methods but has much poorer performance. It achieves data reduction by transmitting only the difference between successive samples. G.722 dates from the late 1970s. It is a sub-band ADPCM codec that has a frequency response extending to 7 kHz at 56 or 64 kbps bit rate. Unless there is no alternative, it should be used only for voice feeds, as music transmitted via G.722 has a distinct fuzzy quality. It is good also for cueing and intercom channels. Only 2 bits are allocated per sample for audio frequencies above 4 kHz—sufficient for conveying the sibilance in voice signals but not adequate for intricate musical sounds. Also, the predictor model used to determine the step size in the adaptive function is designed only for speech. G.722 has the lowest delay of all popular coding methods, about 20 ms. For this reason, it is often used as a return channel so round-trip delay is reduced, even when a higher fidelity method is used for the on-air feed. G.722 uses a procedure called *statistical recovery timing* (SRT) or *statistical framing* to lock the decoder to the data stream. The process usually happens instantaneously but can take up to 30 seconds. The locking can be sensitive to audio present on the G.722 path, as it relies upon the properties of the audio bit stream itself. Some audio material

and tones can prevent lock altogether. Low-level noise is the most reliable signal for locking, and undistorted voice is usually acceptable. There can be problems with sine-wave tones and distorted voice or music signals, in which case removing or lowering the audio by about 12 dB or so for a few seconds will generally cause lock to occur. In rare cases, it may be necessary to disconnect and redial. Tones and noises may be present before locking occurs, and some continuous audio tones may cause momentary unlocking.

Perceptual Coding: MPEG

By far, the most popular perceptual coders rely upon techniques developed under the MPEG umbrella. The MPEG process is open and competitive. A committee of industry representatives and researchers meets to determine goals for target bit rate, quality levels, application areas, and other parameters. Interested organizations that have something to contribute are invited to submit their best work. A careful, double-blind listening test series is then conducted to determine which technology delivers the highest performance. The subjective listening evaluations are done at various volunteer organizations around the world that have access to both experienced and inexperienced test subjects. Finally, results are tabulated, a report is drafted, and a standard is issued [6]. In 1992, this process resulted in the selection of three popular related audio coding methods, each targeted to different bit rates and applications. In 1997, another algorithm, Advanced Audio Coding (AAC) was added to the MPEG suite [7–9].

All of the MPEG codecs rely upon the *acoustic masking principle*—a unique property of the human aural perception system. When a tone is presented at a particular frequency, we are unable to perceive audio at nearby frequencies that are sufficiently low in volume, as illustrated in Figure 3.10-23. As a result, it is not necessary to use precious bits to encode these inaudible, masked frequencies. In perceptual coders, a filter bank divides the audio into multiple bands. When audio in a particular band falls below the masking threshold, few or no bits are devoted to encoding that signal, resulting in a conservation of digital bandwidth that can then be used for the bands where it is needed. Masking also occurs in the time domain, with low-level signals occurring shortly before or after a masking event being inaudible.

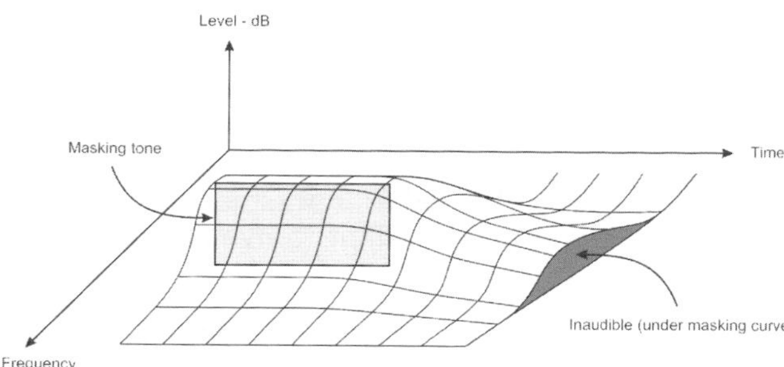

FIGURE 3.10-23 Masking occurs in both the frequency and time domains. MPEG perceptual codecs use both masking principles to achieve high compression ratios.

Predominant MPEG coding technologies include:

- *MPEG Layer 2*, which is widely used on satellite links and high-capacity terrestrial paths such as Primary ISDN or T1 channels. Layer 2 is the standard for European Eureka 147 terrestrial digital broadcasting.
- *MPEG Layer 3* (MP3), which is perfectly matched to the bit rates available on ISDN BRI lines, permitting full FM broadcast quality. Full-fidelity 15 kHz mono is possible on a single ISDN B channel and near-CD-quality 20 kHz stereo is achievable using both ISDN B channels. It is widely supported in broadcast codec equipment from a number of manufacturers.
- *MPEG AAC* is a very powerful audio coding method. According to careful tests, it achieves quality indistinguishable from the original at 64 kbps per mono channel and has approximately 100% more coding power than Layer 2 and 30% more power than the former MPEG performance leader, Layer 3.
- *MPEG High-Efficiency AAC*, also called *aacPlus*, which uses spectral band replication (SBR) to add yet more coding power to AAC [10]. It has reasonably good fidelity down to 24 kbps mono and 48 kbps stereo. The codec used for the U.S. HD Radio™ system also uses SBR and is similar to this one.
- *MPEG AAC-LD*, which is a low-delay variant of AAC. It is not quite as powerful as AAC but has about a third of the delay.

Choosing the Appropriate Coding Method

Table 3.10-3 compares some of the important characteristics of codecs used for broadcast ISDN and IP. Note the trade-off between delay and audio performance. Generally, longer delay equates to better fidelity. AAC's excellent audio performance requires a significant delay, because some of its power comes from the ability to analyze the audio over a long period. An important principle is that a given codec will have better fidelity and more cascading margin at higher bit rates. The most flexible broadcast codecs permit the coding mode for the send and receive paths to be independently chosen, so the choice may be optimized for the specific requirement of each direction.

Cascading Codecs

When multiple codec stages are used in sequence, there is the chance that audio could be impaired by the additive effects of the combined coding processes. Discovering which combinations are satisfactory and which create audible problems is often a matter of trial

TABLE 3.10-3
High-Fidelity Codecs

Codec	Bit Rate Mono (kbps)	Delay Mono/64 kbps (msec)	Delay Stereo/128 kbps (msec)	Frequency Response Stereo/128 kbps	Coding Power	Notes
AAC	64–384	190	240	20 kHz	16–18:1	Today's standard
AAC-LD	64–256	56	68	14 kHz	12:1	Low delay
aacPlus	14–64	600	—	20 kHz*	21–23:1	Best low-bit-rate codec
Layer 3	64–256	316	326	16 kHz	12:1	Internet standard
Layer 2	128–384	208	220	9.8 kHz	6–8:1	Basic hi-fi codec
G.722	64	14†	—	7 kHz†	4:1	Ubiquitous, very low delay

Notes: Delay will vary depending on ISDN network delay and other factors. Typical figures for encode-to-decode delay are given (network delay is not included). Frequency response is for swept sine-wave test. Response with program material will vary due to the dynamic nature of the coding process. "Coding Power" is the ratio of input to output bit rate. Delay and frequency response is for a 48 kHz sampling rate.

*aacPlus frequency response is for mono 19 kbps at a 48 kHz sample rate (AAC half-sampled at 24 kHz).
†G.722 values are for mono at 56 kbps at a 16 kHz sample rate.

and error. As a general rule, the goal is to get as much coding headroom as possible at each stage. This is achieved by using the highest possible bit rate at each stage of the signal chain or using the most powerful coding method of those available at each stage.

ISDN Codecs

ISDN is the premium high-fidelity transport service. It offers fully guaranteed bandwidth, very low delay, and sufficient bit rate for full studio-grade audio when combined with an audio codec such as MPEG Layer 3 or AAC. These codecs are a near-perfect fit to ISDN's capacity: One B channel is enough for mono, while both are combined for stereo operation. ISDN is well established, and most radio stations have at least one ISDN codec on hand. The downside is that charges for circuit-switched long-distance are per minute, and the time-based costs can add up. ISDN is not widely used for either general telephony or Internet access, so it is not nearly as ubiquitous as POTS or some kind of IP connection. It usually has to be special ordered in advance of need. Broadcast codec equipment is used much like a telephone. In the United States, the ISDN Basic Rate Interface Telco line is connected via the U interface. In Europe, the Telco provides the NT1 device and connection is via the S interface. Most codecs accommodate both. Calls are dialed using familiar telephone numbers on a standard keypad. Users select from among the various provided codecs, with care that both ends are operating with the same algorithm. Often the receive-side codec can automatically detect and adapt to the incoming signal.

IP High-Fidelity Codecs

IP connectivity can be provided by wire lines such as T1 and DSL, by radio links such as WiMax, and via satellite and mobile phone networks. As noted earlier, the public Internet is attractive due to its ubiquity and low cost, but it offers no performance guarantees; however, it may be useful for noncritical applications, especially if adaptive technology is used to adjust the codec optimally to network conditions. New IP-based telephone services are appearing, such as MPLS, which are said to provide service levels similar to ISDN. These are intended for VoIP application, but could certainly be used for broadcast remotes.

The core technology is the same as used for VoIP telephone systems, but better codecs enable higher fidelity. Not all high-quality codecs are useful for IP applications, though. A suitable codec must have effective mechanisms for concealing the audible effects of dropped packets that occasionally occur on almost all IP links. Delay is another concern. Because IP networks will have more delay than ISDN or POTS, codecs with low delay are preferred for two-way interactive applications to keep the total delay within acceptable range.

When IP codecs are to be used on non-guaranteed networks such as the public Internet, adaptive buffers are an important part of the receiver system. These detect the jitter in the network and dynamically adjust for optimum performance. When jitter is low, the buffers can be small, so delay can be minimized. When jitter is high, the buffer is extended so drop-outs are reduced. Of course, this comes at the expense of longer delay. A time-stretching/contracting algorithm allows the buffer to adjust inaudibly while the program is ongoing. A further refinement makes the codec bit rate variable to adapt to available network bandwidth.

IP codecs may sometimes have to be used behind firewalls or Network Address Translation (NAT) devices [11]. A *transversal server* placed outside the firewall or NAT allows remote codecs to call and connect to ones inside the firewalled network. It works on the principle that firewalls and NATS will usually open an incoming path in response to an outgoing one

being initiated. When it is powered up, the inside codec makes an outgoing connection to the transversal server, which then opens a return path and keeps it open with occasional pings. The remote codec can then ask the transversal server to mediate a connection request to the inside codec. Name servers are often integrated with the transversal server to allow text names, rather than IP numbers, to be used for dialing.

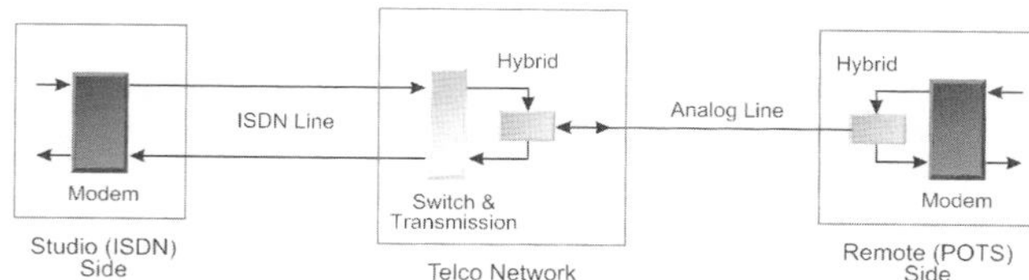

FIGURE 3.10-24 The POTS to ISDN solution takes maximum advantage of the modern telephone network.

POTS Codecs

It is not always possible to get an ISDN line at a remote site. Sometimes they are not available from the Telco, or they are not practical because of the cost or the delay and trouble of getting one installed. Low-cost POTS dial-up lines are everywhere, so it makes sense to find ways to use them for program remotes. The problem is that the 300 Hz to 3.4 kHz frequency response and limited dynamic range that the PSTN provides are not generally adequate for modern broadcast needs. POTS codecs combine a high-power codec with a fast modem. "Fast" is relative here. Recall that ISDN supplies 64 kbps, while the fastest modem is limited to 33.3 kbps upstream speed and very rarely achieves it, usually settling in at around 24 kbps. Because the goal is to achieve something approaching broadcast quality, this is a very challenging bit rate for audio coding technology. The aacPlus codec is a good fit to these low rates. Because aacPlus has a quite high delay, it is typically used together with a lower delay algorithm for the return (studio to remote) direction. Proprietary coding algorithms optimized for low bit rates are widely used for POTS codecs, as well. Modem speed is highly variable and depends on a number of factors along the connection path. When distortion levels are high (usually caused by too many analog/digital/analog conversions) or there is too much noise, the modem is not able to work optimally and will negotiate to a low transmission rate. The codec will adjust to this rate, but quality will be poor at the lowest bit rates. Some POTS codecs will switch to a non-coded coupler mode when the rate falls too low for good audio fidelity. Sometimes a line's characteristics will change during a call and the modem will re-train, disturbing the audio flow.

POTS to ISDN Codecs

A better solution to the POTS codec challenge takes advantage of having an ISDN line at the studio end. Because ISDN interworks with analog lines, a remote POTS codec can call into an ISDN-connected studio unit, as shown in Figure 3.10-24. By eliminating the four-wire to two-wire conversion required at the studio end, along with removing the noise and other impairments present on that POTS line, modem performance is more reliable. This is a well-established technique, often used by Internet service providers during the pre-DSL era, to provide customers with 56 kbps dial-up modem connections. Because only one studio-side unit is used for both ISDN and POTS, savings on Telco line and equipment costs, as well as console/router inputs, are realized.

MIX-MINUS

Both broadcast codecs and phone hybrids require *mix-minus* to operate properly within typical studio setups. An example of a mix-minus setup is shown in Figure 3.10-25. All perceptual codecs have too much delay for on-air talent on remote to hear themselves via a round-trip loop; therefore, a mix-minus arrangement is required so the remote on-air talent's voice is not heard via the studio return. Instead, that microphone is mixed locally with a studio feed that has a mix of everything minus the remote audio—thus the *mix-minus* designation. European broadcasters refer to this as *M-1*. A non-delayed version of the announcer's voice is fed to the headphones mixed with a slightly delayed version of all of the studio-originated audio.

A second problem is caused by hybrid leakage. The mix-minus sent to the remote site will contain residual leakage of the announcer audio being sent to the caller. Because the codec causes this to be delayed, the announcer experiences this as a low-level echo in the headphone feed. Because the longer the delay, the more noticeable this echo, choosing coding algorithms with reduced delay is part of the solution. A digital hybrid with maximum trans-hybrid loss is required. If it has variable override (caller ducking), the amount can be increased when remotes are in progress to enhance the effective trans-hybrid loss. Because it has the best performance, use an ISDN hybrid when possible.

With regard to delay, it's the round-trip that counts—the sum of both send and return path delays. This can be reduced in a typical remote broadcast by using a low-delay algorithm such as G.722 for the return cueing path. Recall the delay *versus* level echo graph discussed in the VoIP section. The same principle applies here, with the degree of annoyance being a function of both delay and level. Reducing either of these factors decreases irritation from echo.

Even without codecs in the picture, hybrids require mix-minus feeds. Without them, hybrids will generate feedback when forced to "chase their tails." A multiple mix-minus setup is also essential for conferencing multiple lines.

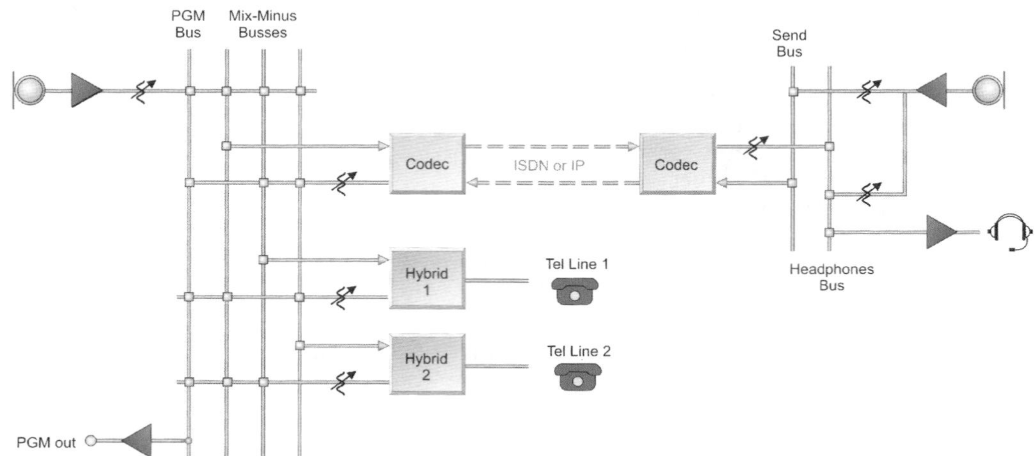

FIGURE 3.10-25 Simplified block diagram shows the mix-minus required for hybrids and codecs.

Most modern consoles have built-in mix-minus, often with provision for selective switching of sources into the feed. With codecs and hybrids playing such an important role in modern broadcasting, the trend is for consoles to have a mix-minus return capability associated with each source and fader. Indeed, it may be wise to think of audio connections as having audio flowing in both directions. In TV applications, intercoms could be merged with other audio in the plant and be thought of as just another audio source. PC-based automation systems have both outputs and inputs. People speaking into microphones usually use headphones or IFB earplugs.

Mix-minus busses may fed be pre- or post-fader. Post-fader is preferred, so callers and remote sites hear the on-air mix; however, phone hybrids are often used as fancy speakerphones for off-air conversations or recording. In this case, microphones need to feed the phone even when their mix channels are off and faders down. Some consoles are able to automatically switch between the two modes depending on whether or not the phone channel is active to air. If the console does not have a dedicated mix-minus, aux sends can be configured to create mix-minus busses. This is common in TV audio, where hybrids, codecs, and satellite feeds are routine. Studios with older consoles that do not have either mix-minus or aux busses will need an outboard mixer or a homemade summing system of some kind. As an alternative, for an older analog console a device made by Henry Engineering could be used that creates mix-minus by subtracting a fader's audio from the console program output with a differential amp scheme [12]. In simple installations, a feed taken from the main announce microphone may be all that is necessary. On the codec remote end, it is necessary to create the local headphone mix one way or the other, perhaps with a small mixer that has aux sends. Codecs intended for remote application often include internal mixers for the local headphone feed.

INTERFACING PRODUCTION INTERCOM SYSTEMS

To aid communication with the field crew during remote broadcast projects, connecting the production intercom system to dial-up telephone lines is often required. Smooth integration of live news remote feeds, for example, requires that production personnel at all locations be able to communicate with each other in a simple, trouble-free fashion. This is especially true when multiple remote sites are involved, such as for election coverage, major sporting events, and telethons. Ideally, crews at each location would use the intercom system without regard for the distances involved. Most often, access to the dial-up phone network is available by wire or cellular, so an interconnection of the intercom system to the telephone network may be the solution.

Four-Wire Intercom Systems

Four-wire systems are those in which the two speech directions are kept separated in the switching and distribution process. Although it would be possible to use special four-wire Telco circuits (or two standard loops) to maintain independent signal paths to remote sites, it is more economical and convenient to be able to use a single phone line. To accomplish this, an effective conversion between the two-wire phone line and a port on the four-wire intercom matrix is needed. It will be necessary to separate the send and receive speech signals on the phone line with a hybrid. One approach is illustrated in Figure 3.10-26.

Trans-hybrid loss performance will be important when intercom stations with open loudspeakers and mics are to be used and when conferencing of multiple telephone lines is desired. In the first case, the acoustic coupling between the speaker and mic completes a feedback path which includes the hybrid. Clearly, the better the hybrid's isolation, the higher the feedback margin will be. In the second case, a feedback path

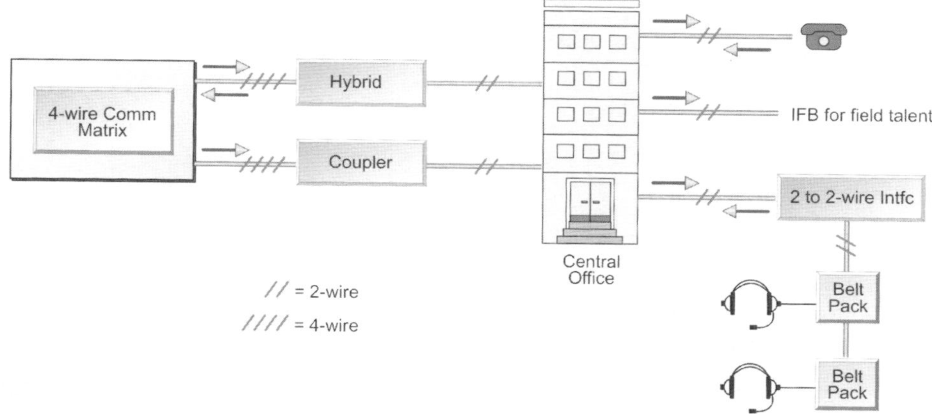

FIGURE 3.10-26 An arrangement that integrates a four-wire intercom switching matrix with Telco lines, an interruptible fold-back (IFB), and a two-wire party-line intercom system for field production work.

exists from each active hybrid through all of the others that are conferenced to it. When the total gain exceeds unity, feedback results. The goal is to have the best possible trans-hybrid loss so the maximum line-to-line gain may be achieved.

An auto-answer and disconnect function may be required for unattended operation. This circuit responds to a phone line ringing signal by activating the hybrid and deactivates the hybrid when the calling party hangs up. As discussed in the section on calling party control (CPC), a dial-tone detector may be necessary to ensure reliable operation. The tone detectors are connected so as to respond to signals on the hybrid's separated Telco receive audio signal. Were this not the case, and the detector was merely connected across the phone line, there would be a major problem when multiple lines are used together in a conference. This is because the tones would be conveyed to each line in use (through the intercom switching matrix) from every other line, causing all of the detectors to respond to the tones from all of the other lines as well as their own. When one line's interface receives a "disconnect," all of the others would turn off as well. Therefore, there is a requirement in this setup that trans-hybrid loss must be sufficient to be certain that any cross coupling is below the threshold of the tone detectors. The same situation applies with any DTMF detection that is used on a per-line basis.

Two-Wire Systems

Two-wire systems are the popular party-line systems. Here, the interface requires two hybrids (Figure 3.10-27). The hybrids are connected back-to-back so the intercom hybrid receive output is fed to the phone line hybrid send input and *vice versa*. Appropriate gain and processing stages are inserted in the four-wire path. This system is what telephone engineers refer to as a *two-wire-to-two-wire repeater*. High-quality hybrids are required to prevent feedback. The signals can feed around the loop and feedback could build up. This happens when the combined trans-hybrid loss of the hybrids is not at least as great as the gain in the two amplifiers. As telephone circuits have widely varying and unpredictable end-to-end transmission characteristics, interfacing intercom systems to phone lines without gain and AGC is not likely to work very well.

ISDN and VoIP for Intercoms

Because digital ISDN and VoIP connections are inherently four-wire, they are perfect for the intercom application. Used with a four-wire intercom system, speech paths may be kept separated end-to-end. Applied to a two-wire intercom system, the problem of maintaining sufficient hybrid balance is eased. ISDN and VoIP lines are less expensive and easier to get than the special analog four-wire lines sometimes used in the past for intercom interconnection. With ISDN, two channels are available so that production and talent busses may be kept separate. Codecs that combine good fidelity and low delay such as G.722, AAC-LD, or AMR-WB are good choices for the intercom application.

TROUBLESHOOTING

When it involves the phone network, troubleshooting can often be quite challenging. Engineers that are comfortable fixing 50 kW transmitters may be lost when dealing with the Telcos. Perhaps the main source of frustration is that, unlike the rest of the facility, telephone services are not under the engineer's full control. The authors are in a unique position to understand, having been involved for many years with assisting station engineers and phone companies to effectively connect. The first round of troubleshooting should involve simple substitution such as trying a different line or trying different equipment on the line in question. If that does not quickly resolve the

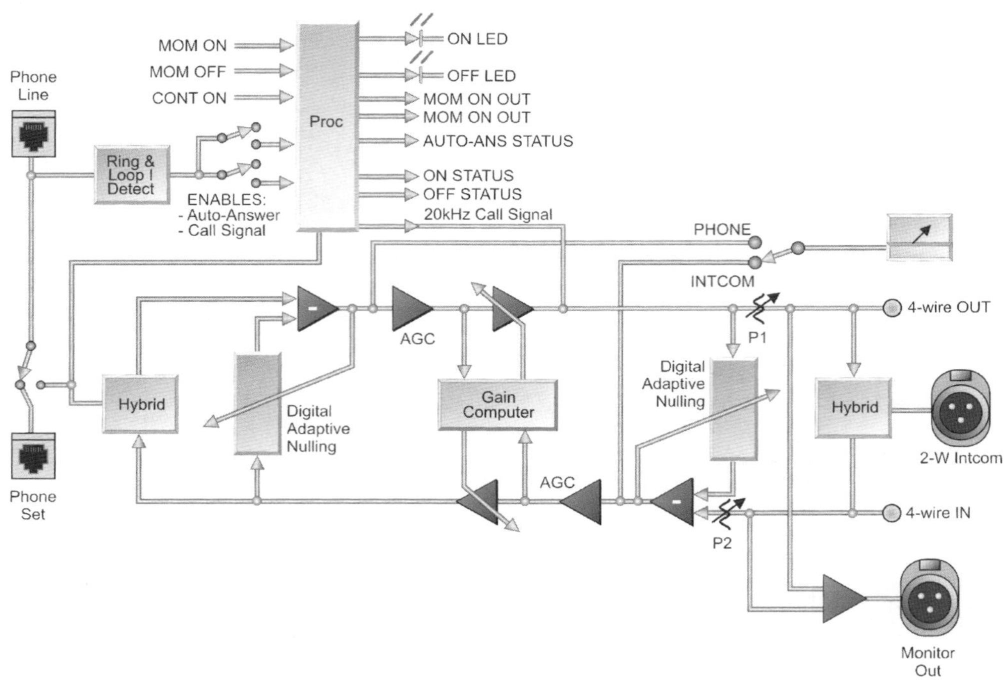

FIGURE 3.10-27 Telcos Link, an example of a two-wire intercom-to-telephone line interface.

problem, a more sophisticated approach may be necessary.

About Problem Solving

The principles here apply whether solving problems at the system level or at the component level. At the different levels, the actions required can be very different, but the process is the same. The basic philosophy of troubleshooting any system is to follow these steps:

- Observe the behavior to find the apparent problem.
- Observe collateral behavior to gain as much information as possible about the problem.
- Look at previously troublesome equipment.
- Generate a hypothesis about the problem.
- Generate an experiment to test the hypothesis.
- Fix the problem.
- Repeat, if necessary, to attack additional problems.

The step-by-step troubleshooting sequence in detail is:

- Step 1—Observe the behavior to find the apparent bug. In other words, determine the bug's symptoms. Remember that many problems are subtle and reveal themselves via a confusing set of symptoms.
- Step 2—Observe collateral behavior to gain as much information as possible about the problem. Does the problem only occur with a specific phone line or piece of gear? Does the audio problem correlate to an alarm LED or odd behavior in some other portion of the system? Try to avoid studying a problem in isolation, but also be wary of trying to fix too many symptoms at the same time. Make the assumption that there is only one problem and work to solve it. Once it is solved, tackle the remaining symptoms, if any.
- Step 3—Look at previously troublesome equipment. At the system level, always suspect the configuration settings, the cables, the Telco's line setup, the punch-blocks, etc.
- Step 4—Generate a hypothesis. Before changing things, formulate a hypothesis about the cause of the problem. To generate the hypothesis, more data on the problem may be needed. Sometimes you will have no clue as to what the problem might be. Start with the basics of looking for a bad connection or cable, loose connector, or other seemingly unrelated cause.
- Step 5—Generate an experiment to test the hypothesis. Change the ISDN connection to a known good line; call a known good phone or hybrid at the other end. If long-distance does not work, try a local call. Plan the tests to eliminate 50% of the possible problems in one test, if possible. Keep track of what is being done to determine what has been eliminated. Keeping notes will also make it easier to run the scenario by a fellow engineer or to discuss the problem with the equipment manufacturer's tech support staff to get a second opinion and reality check.
- Step 6—Fix the problem.

Constantly apply sanity checks; for example, just because the Telco line was checked last night and it

was fine that does not mean that it is okay now. Remember that the system worked well at one time and likely will work well again.

Tools and Test Equipment

Given the importance of various telephone and networking to our industry, it is important to have the appropriate tools and test equipment on hand to quickly solve problems. The importance of the various types of lines to the broadcast facility will determine how extensive the collection of telephone-related test equipment should be. In some cases, test equipment must be on hand for immediate use; in other cases, a particular test item may only be needed on rare occasions and could be shared among a number of stations. Online auction sites have made it much easier to purchase test equipment at reasonable prices, in anticipation of future needs. Typical useful test equipment for telephone technology troubleshooting may include:

- For POTS—A *subscriber loop analyzer* is a simple line tester that allows measurement of loop current, ring voltage, on- and off-hook loop voltage, as well as resistance and capacitance. A *transmission test set* is basically a variable tone generator and a dBm meter in one device. Some units only cover the voice band, while the better units are intended for testing program channels and therefore can operate at a wider bandwidth. Typically, one is placed on each end of a circuit to be tested, with one sending tones and the other receiving. When in the receiving mode, the meter can be used to measure noise and includes the appropriate bandpass filters to give the same measurements used by the Telcos. When connected to a milliwatt-tone number at the local CO, it can determine circuit loss. Most transmission test sets do not include the functions of a loop analyzer, although some loop analyzers do include selected functions of a transmission test set. It is important that the correct termination method be used to get accurate level and noise measurements. If the user equipment is left connected to the line (and off-hook), then the transmission test set should be set to *bridge* the line, whereas if the test set replaces the user equipment it must be set to *terminate* the line and the appropriate termination selected (usually 600 ohms).
- For digital circuits—Different digital technologies offer differing low level signals, so different test equipment will be required for different types of lines (some offer interchangeable modules to cover more than one type of interface). For ISDN BRI, a tester that can handle both the four-wire (AMI) S interface as well as the two-wire (2B1Q) U interface is desirable. It is quite useful if the unit can emulate an NT1. The ability to act as a telephone is typical and can give a quick non-quantitative test of basic function. An important function that most ISDN BRI testers offer is the ability to do a *bit error rate test* (BERT). Most can do so by dialing from B1 to B2 (thereby testing both B channels simultaneously); however, some require dialing into a piece of equipment that is able to loop back the bit stream. BERT function sends a test pattern and then examines the bits coming back to determine if errors have been introduced.
- For T1—For T1 (and low-level ISDN PRI) testing, a similar test set is used. Typically, these can be placed across the incoming or outgoing half of the T1 and can be used to passively monitor the framing bits for improper patterns. They can also determine the level of the incoming signal from the Telco (or CSU if placed after it). If framing errors or bipolar violations are detected, this is indicative of a problem. Most also permit monitoring audio and the robbed-bit signaling bits. A more complete test is to put the T1 tester on the line in place of the user's equipment. In this case, end-to-end bit error rate testing can be done by placing another tester at the far end or by placing the far end into loopback mode. Most T1 test sets include digital transmission test set functions that work just like an analog transmission test set but over a specified channel of the T1. Some T1 testers can emulate a CSU for substitution testing.
- For ISDN PRI—ISDN PRI test sets include some, or all, of the functionality of a T1 tester, but they also have the ability to emulate a piece of PRI user equipment. In this mode, it can make or place calls just as the PBX or telephone system would. Some units can emulate the network, allowing connection of user equipment in the absence of a working PRI. Basic Layer 3 protocol monitoring is often included.

  Protocol analyzers are available for both ISDN BRI and PRI (some units handle both). These typically do not include other test functions but do allow viewing the D channel protocol transaction in detail. These are only rarely needed on Telco-provided ISDN but are quite useful for troubleshooting compatibility problems with PBX-generated ISDN circuits. User equipment such as broadcast codecs may have built-in protocol analyzers, and some basic ISDN test sets offer limited protocol monitoring.

Troubleshooting POTS

For problems with calls hanging up while in use, check the loop current by opening up one side of the line and inserting an ammeter in series with it, or by using a loop analyzer. The loop current should be at least 23 mA, typically 30 mA or greater. Problems dialing could be due to loss. Problems where calls fail to clear when the far end disconnects may be due to lack of CPC or may possibly be related to the user equipment's CPC detection threshold. A differential oscilloscope (or battery-operated unit without a ground reference) will allow for easy measurement of the duration of interruption in loop current present when the far end hangs up, if this occurs.

Noise problems can be difficult to isolate. Note that because a typical telephone instrument has considerable loss at the high and low ends of the spectrum, often the line in question will sound okay on a phone. Testing with a transmission test set (see above) will yield useful information. If the difference in noise (measured across the line) is more than a few decibels different between the 3 kHz and the C-message filters, then the problem may be circuit balance. This is particularly likely if the complaint is 60 Hz hum. Even if the C-message noise is barely within the Telco tariffs (usually 20 dBrnC), the Telco may assist if it can be proved that the *longitudinal balance* is not up to specs. To do so, measure the noise to ground (sometimes called *power influence*) which generally should be 80 dBrnC or less. Subtracting the tip-to-ring number (called *metallic noise*) from the noise-to-ground number will give the figure for balance, which should be 60 dB or less. (Note that this calculation will be incorrect unless the figure for noise to ground is greater than 70 dBrnC.)

Troubleshooting ISDN BRI

There are two basic classes of troubleshooting for BRI circuits. The first relates to the physical layer (the S or U interface as appropriate to the region). Equipment designed for use on the U interface has a built in NT1, which normally has an LED that shows NT1 status. If the U interface of the NT1 remains in the initializing state (or will not stay initialized for very long), then this usually means the line is bad or marginal. However, it is possible that the NT1 is malfunctioning. It could be that the jack is mislabeled. Try an analog telephone connected to the jack; if white noise or clicking is heard, then it is probably ISDN, whereas if a dial tone is heard it is not. The trouble may be a bad cord, jack, or inside wire. Try connecting the ISDN device directly to the network interface.

If the S interface is used, chances are the NT1 will have an LED indicating the status of the S interface. The S interface requires two pairs and is usually in the form of an eight-pin/eight-position miniature modular jack (*e.g.*, RJ-45 style). There are several different allowable wiring configurations, and these have different termination resistor requirements. A direct connection of a single device to an NT1 using a short 1 m cord should always work, however, so that makes a good test. The NT1 documentation should include more on the allowable configurations.

If the U and S (if present) interfaces are both synchronized and remain so for 60 s or more, then the equipment should be able to establish Layers 2 and 3 and perform basic dialing functions. If the equipment in question supports multiple calls, then calling from one B channel to the other is a good basic test. If the equipment supports both data and voice calls try both types of calls, as appropriate.

If the physical layer comes up and seems stable but will not dial or receive incoming calls, then the problem may be a Layer 2/3 problem. In this case, first check the equipment configuration to determine that the ISDN protocol setting is matched to the line and, if using one of the U.S. protocols that require SPIDs, ensure that they have been entered correctly. Check that the directory numbers (DNs) are correct and entered in the same order as the SPIDs. It may be necessary to reboot the equipment after changing any of these parameters. If non-U.S. ISDN is being used, multiple subscriber numbers (MSNs) are optional. If these are incorrect they will prevent receiving incoming calls, so remove them while troubleshooting.

If you can only place voice or data calls, but not both, then the line may have been provisioned for just one type of service. The correct provisioning to include both is called *alternate voice data* service. If in doubt, send the Telco the equipment's specific requirements. In many cases, problems are due to advanced key-system emulation features being enabled when they are not needed; most broadcast equipment does not need anything but the most basic features.

If the problem is only with long-distance calls, there may not be a long-distance carrier (IXC) or the chosen IXC may not handle calls. This can be determined by attempting both types of calls. Also, check that each type can be received, as an IXC is not required for incoming calls.

Troubleshooting T1

The first place to check when troubleshooting a T1 circuit is the user device and the CSU. Typically, these will have a *loss-of-signal* alarm that will illuminate if the incoming signal can no longer be detected. The CSU will generally have such an LED for both the user side as well as the network side. CSUs have the ability to maintain an error log that shows the number of various errors for each 15 minute period for at least 24 hours. If an ESF T1 is being used, the Telco can poll the CSU for this data. If the CSU has a serial port or LCD, it may be possible to access this important information on the instrument. The smart jack or HDSL/HDSL2 transceiver may have error LEDs or a serial port that, if the cabinet is unlocked, can yield useful information (try 9600 bps). Don't forget to configure the CSU and user equipment to accommodate the amount of cable between them. Once all of the above sources of data have been checked, the next step would be to bridge a T1 tester into the incoming T1 signal before the CSU. Check for level and for a lack of CRC errors (ESF only) and bipolar violations (primarily for SF). When this group of tests has been completed, move the tester to other points on both sides of the CSU (four points total). If all else fails, for a dedicated T1 place one end into loopback (a simple adaptor can be made, or the equipment may have this function built in), and do a bit error rate test. Be sure to test for at least 15 minutes with each of the test patterns supported by the tester.

Troubleshooting ISDN PRI

First eliminate low-level problems by following the steps for T1 testing, above. Most Layer 2/3 problems

are due to the wrong protocol settings. In one case, a public switch was unable to support one of the usual protocols and was not compatible. In another case, a periodic maintenance process by the CO was not supported by the user equipment and caused an interruption of calls at the same time every evening.

One common source of problems is where two (or more) high-volume direct inward dial numbers are pointed at the same PRI. In this case, one of the numbers experiencing a flood of calls can block incoming calls for the other number, as well as outbound calls. This is because by default all B channels are configured as a single trunk group for both numbers. The solution is to carefully consider the specific needs and request that the Telco configure two or more trunk groups as required. Incoming calls for a specific number will then return busy when the associated trunk group is full, rather than temporarily using a B channel for each call attempt until the user-equipment rejects each excess call. For example, ask the Telco to create a trunk group of 10 B channels for the call-in number for one station, with a second trunk group of 10 B channels for the call-in number for the second station. A third trunk group would include other, non-call-in numbers such as the hot and warm lines.

When a PBX is used to emulate a network ISDN line, a number of issues must be considered, such as whether the PBX can emulate the network or only user equipment. Not all PBXs can do so. If it can, determine what protocols it can emulate. Not all PBX implementations are standards compliant. It can also take a great deal of time for a PBX vendor to figure out the correct configuration settings. Often, monitoring the ISDN protocol can be helpful to determine where the problem is occurring. When the ISDN protocol has been captured, the equipment vendor should be able to help interpret it.

Troubleshooting the PSTN

When the line and equipment have been eliminated as the source of problems, troubleshooting PSTN problems can be difficult. Only through an understanding of the network is it possible to locate the source of the problem. Moreover, because the Telco is bound to be skeptical, it is best to be sure of the problem before going to it. Usually the network can be ignored when considering telephone problems. The simplistic view of the network "cloud" is generally sufficient to allow finding local problems—after all, everything in the "cloud" is the Telco's responsibility, and most network problems will frequently resolve of their own accord. However, sometimes there is no time to wait for someone else to discover and fix the problem.

The key to understanding network behavior is understanding trunk groups and how they function. Consider a local call between telephone line A and telephone line B. Line A is served by one Central Office ("CO1") and Line B is served by a second Central Office ("CO2"). When A dials B's telephone number, CO1 consults a routing table to determine a path or route to CO2. In this example, there is a six channel trunk group directly between these two COs. The switch's next step is to choose an available channel on that trunk group for this call. This process is called *hunting*, a term that dates back to the days when an operator or relays scanned or hunted through the trunks to find a vacant channel. The usual hunt method is to begin at the lowest numbered channel and then sequentially check each channel to see if it is in use. When a vacant channel is located, the call can proceed. This is called *bottom-up hunting*. *Top-down hunting* works identically, except that the highest numbered trunk is tested first, and the switch hunts downward.

Trunk Group Behavior When a Trunk Is Bad

Trunk hunting is important to understanding the patterns of behavior typically experienced when the network is the cause of the problems. Using the same example, note what would happen if different members of the trunk group were having problems. Recall that this is a bottom-up trunk group. Assume that the first member of this group has the problem that it has audio in one direction only. Trunks are nearly always four-wire circuits, so problems that occur in only one direction are the usual case. Line A dials B and gets no response due to the problem. Line A then hangs up and dials again. Because this is the first member of the group, this will happen repeatedly until, by chance, another call is on member 1 while A is dialing, causing A's call to hunt up to member 2 or higher. In this case, the symptoms will occur quite often and will occur frequently regardless of how busy the network is. Now, let's assume that member 6 has the problem instead of member 1. In this case, during slow times of the day, the problem will not occur until 5 other calls are up; the symptoms will then be similar to the first case. The average incidence will be lower than the previous example, even during the busy hour, because whenever a party on members 1 through 5 hangs up, the call will go through rather than hunting all the way to member 6.

Troubleshooting Methods

The key to troubleshooting network problems is persistence. If enough calls are made, and one does not fail, this reveals several things:

- The problem is not local. Terminal equipment (such as a telephone or codec) should not care how many calls are made; it should act similarly in each case.
- The problem is an "acting like a network" (*e.g.*, a trunk) problem in that it is non-absolute; rather, it is probabilistic.

Generally, 15 calls are made, carefully keeping track of the number of calls where the problem occurs. Next, reverse the direction of the call, and place 15 more calls. If the success rates are markedly different, it is likely that this is a network problem. The logic for this conclusion is as follows: On each call, the same customer equipment and same CO switches will be used; however, as discussed, trunk selection is dynamic. Another clue that the problem may be network related is if the success rate varies substantially depending on the time of day. Note that sometimes the success of circuit-switched data (CSD) calls at 56 kbps may differ *versus* CSD calls at 64 kbps, and both may act differently *versus* voice calls. It is important to do multiple tests. If dialing from A to B fails and dialing from B to A works, very little is proved; however, if 7 of 15 calls dialed from A to B fail while 15 out of 15 calls from B to A succeed, then an important clue has been found.

Eliminate the Easy Stuff First

Test the line and equipment. If ISDN is used, there is the ability to dial from one channel to the other. Do so on each end, using a mode that requires both B channels. It is fairly easy to eliminate both the codec and the line. Then do end-to-end tests using the same mode used for the local test (the local codec might have a problem specific to a certain mode).

If the problem is on a POTS line, see if the same problem occurs on another line from the same CO switch. Also, determine if the problem occurs when dialing between two lines on the same switch (if so, it is not a network problem). If the problem only occurs when dialing in one of the two directions, it is likely to be a network problem. If the problem is occurring only with long distance calls, try some local calls or *vice versa*. If the problem is limited to only one of these types of calls, it is probably a network problem. If the problem happens only on long distance, try placing the problem calls with several 101XXXX access codes. If the problem occurs only with a certain carrier, contact them, explain the problem, and work with them to solve it. Not all carriers can handle CSD calls, so be sure to keep this in mind.

Working with the Phone Company

If the suggestions in this section have been followed, then the local line and equipment and all sources of problems other than the network have been eliminated. If many test calls have been made, one or all of the following attributes may indicate a network problem:

- Problem is limited to local or long distance, but not both.
- Probability of the problem occurring varies significantly between incoming *versus* outgoing calls.
- Probability of the problem occurring varies significantly between CSD calls *versus* voice calls.
- Probability of the problem occurring varies significantly between CSD calls at 56 kbps *versus* 64 kbps.
- Probability of the problem occurring varies significantly depending on the time of day.

With this information, contact the phone company—that is, the local dial-tone provider or long-distance company as indicated by the symptoms.

Working with Long-Distance Carriers

Here the task is reasonably straightforward. Nearly all long-distance carrier problems are network related, so they will not doubt the claims of problems as local Telcos often do. Generally, all that is needed is to explain that "the problem only occurs when using their network and disappears when dialing with another 101XXXX code" and they will start investigating. Be prepared to place test calls until the problem occurs. If the problem only occurs on inbound calls, be sure to have someone standing by elsewhere to assist. The process is that the tester sets up a "trace" to capture information about calls from the originating number. It will be necessary to keep placing calls until the problem occurs. The network will then examine the routing information for that call and then tell you to hang up and dial again. After three to six "bad" calls, the problem is likely to be found. At that point, the next step is usually to "busy out" the trunk group in question to prevent calls from using the affected trunks, and there should be no additional failures. Make sure to test this based on the previous investigation, but note that if only 1 of 10 calls failed in the tests, then make 20 calls just to be sure. On the other hand, if 12 of 15 calls were failing, then it is only necessary to make 4 or 5 successful calls to know the problem is solved. When the problem trunk group is found, it will be left busied out and fixed later.

Working with Local Dial-Tone Providers

Generally speaking, with proper homework, there is no reason for the Telco to dispatch a technician to the station. In fact, this is likely to slow getting the problem resolved. However, explaining in detail the troubleshooting that has been done (and even the nature of the problem) is likely to be futile when talking to the business office. It is better to give a very brief description of the problem and politely insist that they have this ticket referred to the "trunking group" because it is a "network problem" and to have someone call back. When someone does call back, be sure to inquire if that person is with the trunking group, then explain in detail what the problem is, the testing that has been done, the results thereof, and the conclusion that a network problem seems likely.

In some cases, local procedure will insist that a technician be dispatched. Make a friendly attempt to explain that this is wasting money and time, but do not make too big an issue of it; however, when the technician arrives, stay around and be friendly and helpful as they work. Explain that you are sorry that this is probably a waste of their time, as you believe it is a network problem. Make this person an ally and be sure to find out how to get back in touch later. The goal is to understand that the problem is having an effect on the listeners and to work together to solve it.

In the meantime, pay attention to what the technician is doing. If the problem really is a network problem, the technician may be unable to find any problems with the line without your assistance. Do not let the technician depart at this point, but explain and demonstrate how to create the problem. This may require having the far end call into the technician's test equipment or dialing long distance, as necessary. It may be possible to demonstrate the problem using the station equipment. Again, the goal is to convince the technician that there is a problem and to call to the trunking group to get it fixed.

Once someone from the trunking group is involved, describe the tests conducted and the results. Some negotiations may be involved, but a Telco technician at the station could verify the results of the tests and explain the problem in Telco terms. At this point, the procedure is identical to what was described above for working with long-distance carriers. The person from the trunking group will set up a trace and the problem can be demonstrated to determine where the problem is originating.

Troubleshooting IP Networks

With IP networks, there is a chance for trouble either within the station's LAN or with the external IP link. While there is some overlap, techniques for troubleshooting the two parts of the system are quite different. Problems with LANs are often solved by examining the maintenance and statistics Web pages served by Ethernet switches and IP routers. Ethernet "sniffers," such as the free PC software package Ethereal [13], can be attached to the LAN to display the packet traffic. Most network equipment has a logging function that can help guide the engineer to a solution. WANs are out of the station engineer's direct control, and troubleshooting will usually require the help of the network vendor. The Internet is comprised of segments from many vendors connected together, so it may be difficult to find someone who will take responsibility. If the problem can be isolated to a particular vendor's segment, there is a chance for resolution, particularly if the segment is the last-mile link that connects the station's facility to the network. In the case of a quality of service (QoS)-guaranteed link that is provided and maintained by a single vendor, the situation is much better. When the Internet is being used, if both ends of the link are from the same vendor, then the chance is very good that the backbone segment will also be provided by the same vendor, and the probability of being able to find a solution is high. As with PSTN troubleshooting, developing a picture of the nature of the problem and its probable cause is the key to obtaining effective attention from the ISP technicians.

IP Debugging

The PC applications *ping* and *tracert* (or traceroute) are the first-line tools for any IP debugging. Ping is used to detect if a device is connected, if it has basic functionality, and if a working path exists to it. Traceroute goes further to show the path the packet is taking through the network. These applications are included in both Windows and Linux operating systems. Under Windows, start the Command Prompt program, and then enter `ping` or `tracert`, followed by the target address. Under Linux, these are accessible at the command prompt. Both accept either an IP address or text domain name as input. Here is an example of `ping` in action:

```
F:\Documents and Settings\steve>ping 80.232.227.100
Pinging 80.232.227.100 with 32 bytes of data:
Reply from 80.232.227.100: bytes=32 time=11ms TTL=123
Reply from 80.232.227.100: bytes=32 time=10ms TTL=123
Reply from 80.232.227.100: bytes=32 time=10ms TTL=123
Reply from 80.232.227.100: bytes=32 time=10ms TTL=123
Ping statistics for 80.232.227.100:
Packets: Sent = 4, Received = 4, Lost = 0 (0% loss),
Approximate round trip times in milli-seconds:
Minimum = 10ms, Maximum = 11ms, Average = 10ms
```

The ping application connected four times with the destination device in order to develop the average statistics reported at the end. Here is `tracert`, with a text domain name as input:

```
F:\Documents and Settings\steve>tracert www.google.com
Tracing route to www.1.google.com [66.249.85.99] over
a maximum of 30 hops:
1 2 ms 1 ms 1 ms 81.198.53.54
2 10 ms 9 ms 10 ms 81.198.232.1
3 10 m 10 ms 10 ms 80.232.232.65
4 11 ms 11 ms 11 ms core1.telecom.com [195.13.173.21]
5 18 ms 18 ms 19 ms 195.250.170.49
6 18 ms 19 ms 18 ms noe.estpak.com [194.126.97.194]
7 45 ms * 44 ms ffm.estpak.com [194.126.123.14]
8 45 ms 45 ms 45 ms main.google.com [80.81.192.108]
Trace complete.
```

The traceroute application tries to connect with each router three times and reports the delay in milliseconds for each attempt. When the delay is too long, an asterisk is printed in place of the number. The rightmost column is the IP number and name, if available, of each router hop in the path. When there is a problem in the network, such as a broken link or congestion, it will usually be visible in the traceroute output.

SUMMARY

Interfacing with telephone companies and Internet service providers can be a complex and frustrating experience, but, with knowledge of how the system works and appropriate test equipment and training, local station engineering personnel will be better able to produce the quality expected of the interface facility and determine how to fix problems when they occur.

References

[1] International Telecommunication Union, Recommendation G.711: Pulse Code Modulation (PCM) of Voice Frequencies, November 1988, www.itu.int/rec/T-REC-G.711/en.

[2] ANSI/TIA Standard TIA-968-A-2002: Technical Requirements for Connection of Terminal Equipment to the Telephone Network, www.part68.org/SecureDocuments/TIA-968-A-Final.pdf.

[3] International Telecommunication Union, Recommendation G.131: Talker Echo and Its Control, November 2003, www.itu.int/rec/T-REC-G.131/en.

[4] International Telecommunication Union, Recommendation P.800.1: Mean Opinion Score (MOS) Terminology, March 2003, www.itu.int/rec/T-REC-P.800.1-200303-I/en.

[5] Axia Audio, www.axiaaudio.com (pro studio-grade IP audio over LANs).

[6] MPEG, www.chiariglione.org/mpeg.
[7] ISO/IEC JTC1/SC29/WG11 MPEG, International Standard ISO/IEC 11172-3: Coding of Moving Pictures and Associated Audio for Digital Storage Media At Up to About 1.5 Mbit/s, 1992, http://isotc.iso.org/livelink/livelink/fetch/2000/2489/Ittf_Home/ITTF.htm.
[8] ISO/IEC JTC1/SC29/WG11 MPEG, International Standard ISO/IEC 13818-7: Generic Coding of Moving Pictures and Associated Audio: Advanced Audio Coding, 1997, http://isotc.iso.org/livelink/livelink/fetch/2000/2489/Ittf_Home/ITTF.htm.
[9] ISO/IEC JTC1/SC29/WG11 MPEG, International Standard ISO/IEC 14496: MPEG-4, http://isotc.iso.org/livelink/livelink/fetch/2000/2489/Ittf_Home/ITTF.htm.
[10] Dietz, M. and Meltzer, S., aacPlus: a state of the art audio coding scheme, *EBU Techn. Rev.*, July 2002, www.ebu.ch/trev_291-dietz.pdf.
[11] Srisuresh, P. and Egevang, K., Traditional IP Network Address Translator (Traditional NAT), RFC 3022, January 2001, www.rfc-editor.org/rfc/rfc3022.txt.
[12] Henry Engineering MixMinus Plus, www.henryengineering.com/hemmp.html.
[13] Ethereal, www.ethereal.com (free, open-source, network analyzer).

Additional Information

MPEG-4 Industry Forum, www.m4if.org.

Bosi, M., Brandenburg, K., Quackenbush, S., Fielder, L., Akagiri, K. et al., ISO/IEC MPEG-2 Advanced Audio Coding, *J. AES*, 45(10), 789–814, 1997.

Stoll, G. and Kozamernik, F., EBU Listening Test on Internet Audio Codecs, *EBU Techn. Rev.*, June 2000, www.ebu.ch/trev_283-kozamernik.pdf.

AES Technical Committee of Coding of Audio Signals, *Perceptual Audio Coders: What To Listen For*, www.aes.org (CD-ROM with tutorial information and audio examples).

Fraunhofer IIS, www.iis.fraunhofer.de/amm (MPEG AAC and MP3 information from the inventors).

Brandenburg, K., MP3 and AAC explained, in *Proc. of AES 17th Int. Conf. on High Quality Audio Coding*, www.telos-systems.com/techtalk.

Coding Technologies, www.codingtechnologies.com (aacPlus information).

Church, S., *On Beer and Audio Coding: Why Something Called AAC Is Cooler Than a Pilsner, and How It Got To Be That Way*, www.telcos-systems.com/techtalk.

Internet Engineering Taskforce, www.ietf.org; www.rfc-editor.org (repository for Internet standards, or Requests for Comments [RFCs], including those pertaining to VoIP SIP call control and RTP transport).

Kurose, J.F. and Ross, K.W., *Computer Networking: A Top-Down Approach Featuring the Internet*, Third ed., Pearson Education/Addison-Wesley, 2005, www.aw-bc.com/kurose-ross.

Comrex Corp., www.comrex.com (broadcast telephone interface and codec resources and product information).

Telos Systems, www.telos-systems.com (broadcast telephone interface and codec resources and product information).

CHAPTER

3.11

Audio Contribution and Distribution Channels

SKIP PIZZI

Microsoft Corporation
Redmond, Washington

INTRODUCTION

This chapter deals with the transmission of high-quality audio by broadcasters prior to, or "upstream" of, actual broadcast—in other words, any point-to-point transmission of audio signals involved in the production of broadcast content.

This chapter does *not* cover transmission via voice-grade telephone service; for that, refer to Chapter 3.10. It also does not cover the specialized, licensed RF transmission systems owned and operated by broadcasters for local remote backhaul (Remote Pick-Ups [RPUs]), or studio to transmission-site links (STLs), which are covered in Chapters 3.9 and 9.1, respectively—although some of the same audio and data transmission technologies covered here may be applied on such links.

Note also that although most new connectivity in this space uses some form of digital interconnection, there still may be some analog audio circuits in use by broadcasters in any given locale. Thus, this chapter considers both the analog legacy and the current-day digital services in use by broadcasters, and concludes with a look at the ongoing emergence of next-generation technologies for these purposes. Over time, however, much of the legacy information contained here will become outdated for a growing number of locations across the U.S., and remain useful only for historical context.

Terminology

In recent years, providers and users of audio transmission circuits have developed a nomenclature that establishes a hierarchy of signal pathway types, namely *Contribution*, *Distribution*, and *Emission* channels:

- Contribution channels are those used typically for transmission from program origination sites to broadcast assembly centers. They are often referred to as "backhaul" circuits, and common examples are the signal paths used from a remote site to a broadcast station or from a sports venue to a broadcast network operations center.
- Distribution channels are typically those used by broadcast networks to send programming content from operation centers to broadcast stations or transmission centers. Examples are the satellite distribution paths used by television and radio networks to send content to affiliate stations.
- Emission channels are those used in the "last mile" path from stations' transmitters to consumer receivers.

This is a true hierarchy, in that the quality metrics of Contribution channels are higher than those of Distribution channels, which are in turn higher than those of Emission channels. Such an arrangement allows each part of the chain to be optimized to the audio content it is likely to encounter (e.g., the audio levels of live events are more unpredictable than those of a finished program, so a Contribution channel needs wider dynamic range capacity than Distribution or Emission channels).

Note also that Contribution and Distribution channels are generally point-to-point in nature, whereas Emission channels, as the name suggests, are point-to-multipoint broadcast transmissions.

This chapter will consider the Contribution and Distribution channels used for remote backhaul and network interconnections of audio content. These circuits require special attention because they are usually not owned, controlled, or operated by broadcasters who use them, but by third parties—typically telephone companies or other telecommunications carriers ("telcos").

For clarity, the term "circuit" will be used to designate a telco audio path, supplanting the other commonly heard (and potentially confusing) synonyms of "loop," "private line," "leased line," "program circuit," and the like. Likewise, "telco" will be used to refer generically to all third-party audio and data carriers, both wired and wireless. "LDS" (long-distance service) will be used to refer generically to all common carriers providing service between local telcos.

At this writing, the telecommunications industry has nearly completed its quiet revolution from analog to digital distribution, so greater emphasis will be placed on the latter.

Dealing with Telco

Broadcasters and telcos have been allied since the earliest days of both industries and will likely remain so. This chapter considers the opportunities for audio signal interconnections afforded to broadcasters by telcos and examines how broadcasters can implement these hookups.

To minimize problems with a telco, the station engineer should establish a good working relationship with the appropriate personnel and to understand as much as possible about the company's service and operation. If the station's staff comes across as friendly and knowledgeable, but also professionally firm and businesslike, things should go well. If possible, keep the station's liaison to the telco limited to one staff member, and try to always deal with the same person at telco as well.

Of course, the latter may be difficult, as telcos continue to merge and morph into different forms over time. In the process, the distinction between local and long-distance services and carriers is becoming increasingly blurred, so the references to these differences in this chapter may be (or become) moot. At this writing, the celebrated breakup of "the Bell System" (AT&T) in the 1980s is gradually being reversed by the ongoing re-merging of the "Baby Bells" (RBOCs). Thus, maintaining stable business relationships with telco operators has become a challenge for broadcasters, although striving for as much of it as possible remains worthwhile.

For ordering digital circuits, another option that may be available is the data-line brokerage service. Here, the station gives its time, place, and quality service requirements to a third party, who books the line for the station at no charge. (Like travel agents, these services receive commissions from the telcos whose circuits they book.) This service can be especially helpful in long-distance applications, where two local telcos (one foreign to the station) and an LDS are involved. Following the travel-agent model, data line brokers may also be able to book the service at a lower cost. In some cases, any extra customer-provided equipment (CPE) required for the circuit may be rented from these brokers as well.

More recently, as telcos have instituted services as Internet Service Providers (ISPs), a department of the telco itself can act as a data-interconnection consultancy for stations. In many cases, digital interconnection services for both digital audio interconnection and computer data-networking needs can be bundled into a single agreement with a local service provider, for increased cost-effectiveness and stable operations.

Note also that this transition has moved much of what broadcasters need from telco for audio transmission into the typical consumer domain, whereas in earlier times it had been available only to commercial broadcast business clients. While this may considerably decrease costs for broadcasters, it may also reduce the level of telco customer service from what broadcasters formerly enjoyed to that of the typical consumer. Bear this in mind when dealing with telco and arranging any bundled service packages.

An Abundance of Options

There have never been more choices of technologies and pathways available for broadcasters' transport of audio programming. Many readers will recall when the only choice broadcasters had was the bandwidth of an analog circuit (3.5 kHz, 5 kHz, 8 kHz, or 15 kHz), from a single service provider. Today the options are myriad, including the last vestiges of such dedicated analog paths, to dedicated digital paths (e.g., T-1), to switched digital circuits (e.g., ISDN), to packet-based networks (telco IP and "bandwidth on demand" services), to broadband access to the packet-based "cloud" of the Internet (e.g., DSL, DOCSIS), to newly emerging wireless broadband Internet access (e.g., WiFi, 3G cellular, WiMAX, BGAN, etc.).

This chapter will address all of the above, in the order of their emergence. Naturally, since most of these technologies are implemented as local terrestrial services, new systems are introduced gradually, region by region. Not all options will be available in all areas at any given time, although most portions of the U.S. already have at least some choice among multiple service providers today. Over time, currently emerging technologies will become broadly available in all but the most remote areas. Consult the local telecommunications services providers for currently available options. (Readers who have no access to or interest in legacy analog audio interconnection are advised to skip ahead to the "Digital Audio Circuits" section.)

ANALOG CIRCUITS

Analog audio circuits are still offered by some telcos under the schedule shown in Table 3.11-1, but in major metropolitan areas very few if any new lines are being installed. Telcos are instead replacing the existing analog circuits that remain with digital equivalents. Meanwhile, LDSs have already phased out such analog services for their long-haul paths.

Where they are still offered, costs for both service and installation of analog program circuits continue to increase. (The exception is in some rural telephone service areas, where alternative digital service is not yet a viable option.) Installation of an analog program circuit generally requires several hours of an experienced technician's work to equalize the line, as opposed to digital services, which take much less time and trouble to pass spec at installation.

Past practice with local analog circuits often provided wider bandwidth than what was ordered, whereas digital services typically cut off exactly as specified. Increasingly, though, analog services that have to pass between telco central offices (COs) or switching centers will make the trip bundled on an interoffice digital carrier anyway, whereupon any excess bandwidth will likely be removed.

Obtaining and Testing Analog Telco Circuits

As Table 3.11-1 explains, analog circuits are available in a variety of bandwidths and under temporary or permanent status. Check with the local telco to determine which, if any, are available and at what cost. Installation charges for permanent lines may be much higher than those for temporary service, because the telco may want to actually install new wiring rather than permanently occupy any pairs on its existing network cables. In an increasing number of areas, analog audio circuits will not be offered at all, in which case the reader may skip the remainder of this section. For those legacy users of analog audio program circuits, however, the following is provided.

Types 6008 and 6009 (15 kHz circuits) service may be ordered as a stereo pair, incurring a one-time installation surcharge for *stereo conditioning*. (Some telcos also offer it for 6007/8 [8 kHz] service.) This ensures that both lines are routed together throughout their runs, so that interchannel phase differences will be minimized. A third line can be ordered for backup, and this too should be included in the stereo conditioning. Although billed routing (as the crow flies) may be a short distance, actual routing of the circuits may be much longer and indirect, providing ample opportunity for phase differences to occur. (Approximately 5µsec time difference occurs for every mile of path length difference.)

TABLE 3.11-1
Classes of Service for Analog Telco Audio Circuits

Class of Service	Approximate Bandwidth	Full- or Part-Time
Type 6002	200–3500Hz	PT
Type 6003		FT
Type 6004	100–5000Hz	PT
Type 6005		FT
Type 6006	50–8000Hz	PT
Type 6007		FT
Type 6008	50–15,000Hz	PT
Type 6009		FT

Lines should be ordered well in advance of the need for them. Check with the telco for its preferred lead time. Always specify a start date at least one business day earlier than the actual requirement, to allow time for station tests on the lines to be performed. Check frequency response, signal-to-noise ratio (S/N), distortion, and headroom. For stereo pairs, check relative phase response and relative polarity.

Frequency response should be at least within ±3 dB of what was ordered. Be sure to check outside the passband, because response may not roll off but instead rise beyond the cutoff frequency. On occasion, in order to get a line to meet specifications, telco equalizers may be used to boost the extreme frequencies, and if done improperly, the response may indeed be flat to the cutoff frequency but then increase for another octave or more before finally rolling off. This will result in audible consequences from the altered response and reduced headroom, particularly if a noise reduction system that preemphasizes high frequencies is used on the circuit.

Use caution with any complementary analog noise reduction (NR) system used on a telco circuit. As a general rule, be sure that the audio fed to the NR encoder is prefiltered to match the line's response so that the audio passband seen by the encoder is relatively similar to that seen be the decoder after the audio has passed through the circuit. If the decoder sees a substantially narrower bandwidth than the encoder, it will not decode the noise reduction process in a complementary fashion.

The proper procedure for calibrating levels on analog program circuits employs sinewave test tones of 400 Hz or lower, fed at the telco program operating level (POL) of +8 dBm. Frequencies above 400 Hz must be sent at the telco test level of 0 dBm to minimize cross talk into other circuits via capacitive coupling of higher frequencies. To keep things simple, run all frequency response tests at 0 dBm.

For measuring noise, telcos use a slightly unorthodox approach. They consider a noise level of –90 dBm to be *absolute quiet* or noise free, and measure noise from that reference point. The unit used is dBRN (RN for reference noise). Therefore, a –50 dBm noise level would be called 40 dBRN by telco. If the telco's specified audio reference level of +8 dBm is used by the customer on this circuit, a 58 dB SNR is achieved in this case.

The greater the distance an analog circuit travels, the noisier it becomes. The wider a circuit's bandwidth, the quieter it needs to be. Although specs vary between telcos, noise specifications generally reflect those observations. A typical noise level for local 15 kHz circuits is 33 dBRN or lower, providing 65 dB or better SNR. Long-distance analog circuit noise levels generally hit 40 dBRN (58 dB SNR) for 15 kHz circuits. Noise levels for a 5 kHz service are around 46 dBRN (52 dB SNR) for local and 56 dBRN (42 dB SNR) for long-distance service. Again, check with the local telco for circuit specifications, and when connected always verify that those specifications are met.

Some telcos offer a lossless or zero loss option, in which the circuit acts as a unity gain device. This is contrasted to a standard circuit, which may exhibit up to 30 dB of loss. Audio signals in copper cable drop at about 1 dB/mi from broadband resistive losses. Frequency selective reactive losses occur at greater rates, but these are remedied by receive end equalization. The additional telco amplifiers required by lossless lines may have a detrimental effect on the distortion and headroom performance, and may not do much to reduce the overall static noise floor, but they are useful in densely trafficked urban areas where cross talk and impulse noise are prevalent.

Static or random noise is far less objectionable than those coherent noises sometimes found on telco circuits. Such noises can be caused by capacitive coupling between adjacent pairs in multipair cable, dial pulses and other switching, inadequate common mode rejection, and carrier beating (causing high-pitched sings or tones). Because of this, circuits should be carefully auditioned at the receive-end for a while after installation, without any audio on the line. Check circuits again prior to each on-air use. Report any cross talk or impulse noise problems to telco at the first sign of trouble.

Although there are usually no published specs for distortion or phase response on telco circuits, total harmonic distortion (THD) should be <0.25% on 15 kHz lines. For stereo pairs, relative phase response should be within 30° across the passband. Widely divergent frequency response between the two circuits in a stereo pair is a tip-off to check phase response carefully. When a spare third circuit is ordered, the two closest to each other in frequency and phase response should be designated as the main pair, assuming distortion and noise are equivalent across the three.

The maximum level guaranteed on telco circuits is +18 dBm, which implies that only 10 dB of headroom will exist above the +8 dBm reference level. A sensible alternative is to use +4 dBm as a reference level (most professional audio hardware uses this level anyway), thus allowing a more realistic 14 dB of headroom, at the expense of 4 dB less SNR—generally a worthwhile trade-off.

Interfacing Procedures

Figures 3.11-1 through 3.11-5 illustrate some do's and don't's of audio interfacing to analog telco circuits. This is one area of today's audio where 600-ohm impedance matching is still important. Telco equalizes its circuits for flat response under the conditions of a 600-ohm source impedance and 600-ohm termination. Because of the reactive components in long wired paths (their transmission line behavior implies complex impedances), varying these impedances will affect the frequency response of audio carried on the circuit. Most contemporary audio equipment expects a bridging interface arrangement, so output source impedances are typically much lower than 600 ohms, and input impedances are much higher. Yet the telco does not guarantee flat response without 600-ohm matching conditions.

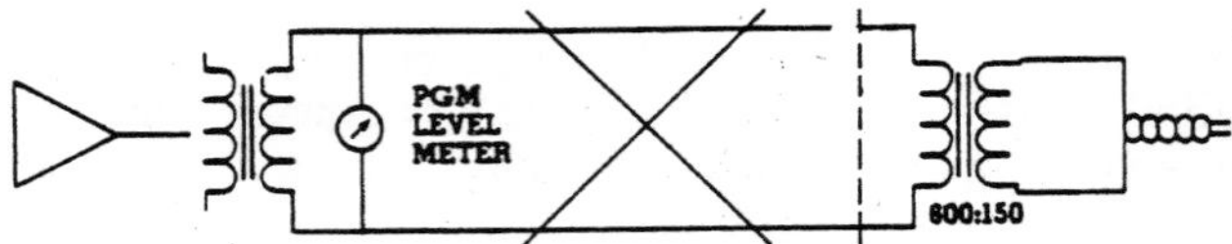

FIGURE 3.11-1 DON'T meter the signal across the input to the repeat coil (transformer at right), and DON'T feed the repeat coil directly from another transformer, if possible.

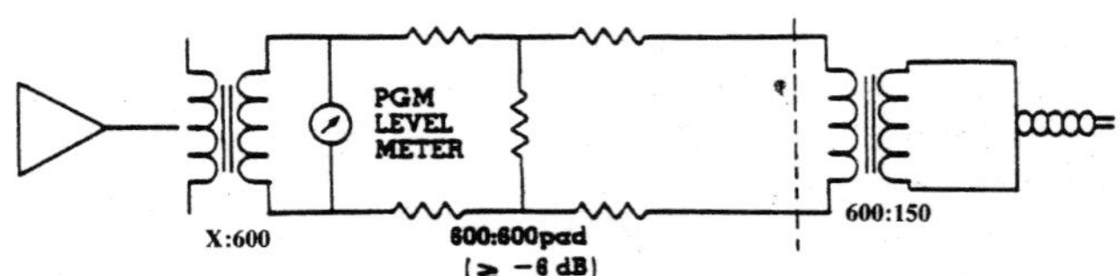

FIGURE 3.11-2 DO place a pad between the output device's transformer and the repeat coil, and DO place a level meter before the pad, calibrated for the voltage across the pad's output when terminated with a 600 Ω resistor.

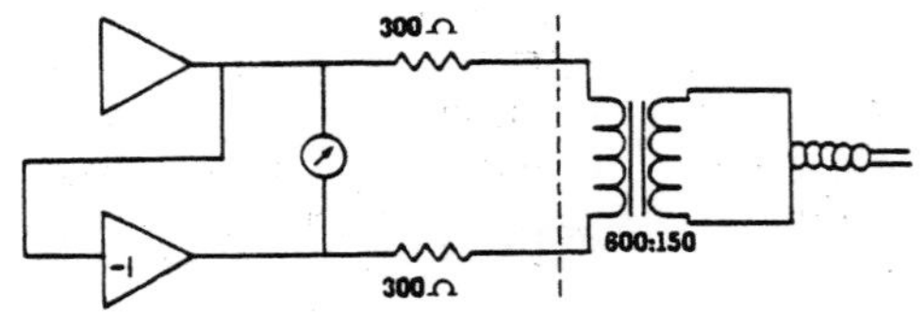

FIGURE 3.11-3 DO feed low-impedance sources (e.g., op-amps) through a 600 Ω differential balanced pad, and DO place the level meter before the pad, calibrated as in Figure 3.11-2.

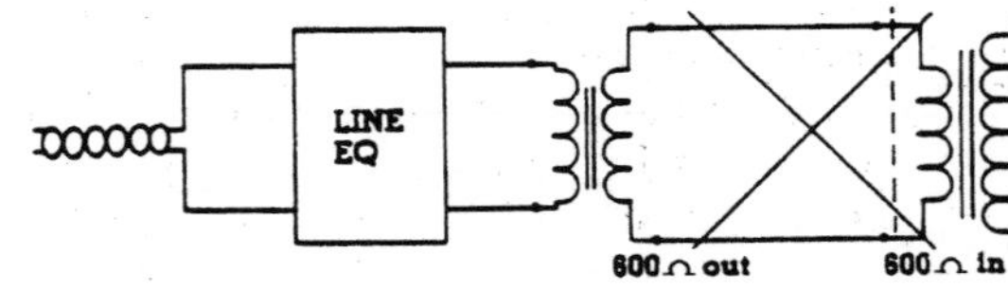

FIGURE 3.11-4 DON'T terminate the receive-end repeat coil (transformer at right) with another transformer, if possible. Its loading may vary with frequency from the true 600 Ω resistive termination used in line-up, causing level and frequency response variations.

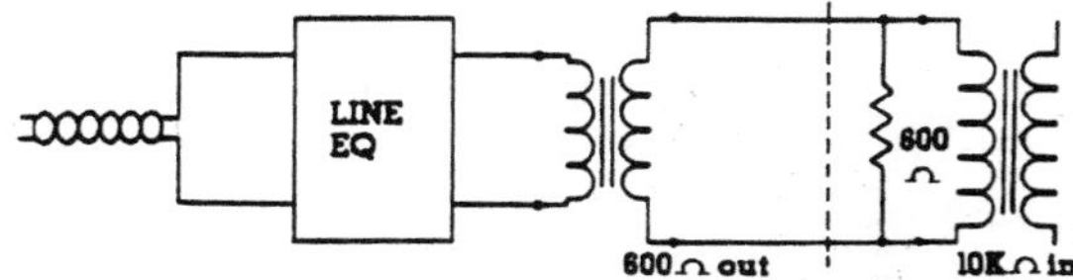

FIGURE 3.11-5 DO terminate the receive-end of a program circuit with a 600 Ω resistive load and bridge the load with a high-impedance transformer or active balanced input. *Note:* Common mode rejection of the transformer or active input must be considered if the distance from the repeat coil is great or is near other lines which induce cross talk.

Ideally, the secondary of the repeat coil should be resistively terminated. An active, balanced input circuit does this nicely, provided any RF is bypassed before the first stage of amplification. An alternative is the use of an input transformer with a high-impedance, bridging input. This allows the 600 Ω resistor, as shown in Figure 3.11-5, to match the repeat coil. Another version, often seen on the input of broadcast line amps and modulation limiters, is a 600 Ω H-pad, effectively the reverse of Figure 3.11-2. Following the steps shown in Figures 3.11-1 through 3.11-5 usually ensures that transmission loss and frequency response closely match those the telco set up.

If a typical low impedance mixing console output is used to feed a telco circuit directly, it will generally cause a rising high-frequency response to be received. This is not telco's responsibility, but rather the customer's. It is therefore essential that the device directly feeding the line have a 600-ohm source impedance, and be capable of driving such a balanced load to at least +18 dBm across the passband. Many mixers—even expensive ones—are not designed to do this, so an appropriate outboard line amp or distribution amplifier (DA) fed by the mixer should be used.

A DA will also provide isolation between the mixer and any other inputs, and more importantly, between all these devices and the telco circuit. Drive the telco circuit with its own DA output; no other inputs (even bridging ones) should be paralleled to it. DC voltages may appear on telco circuits, so this isolation is essential. For the same reason, a transformer-coupled DA is the better choice over a transformerless design in this application. High-quality transformers can be quite helpful on remotes, especially when interfacing with a telco. On most 15 kHz circuits (and some others) telcos provide repeat coils on each end in order to provide some degree of isolation. In this case, actively balanced outputs can be used safely, but flat frequency response is guaranteed only if they are interfaced to the repeat coil in the manner shown in Figure 3.11-3. A caveat here: Remember that the active DA still has to drive a 600 Ω balanced load to +18 dBm across the passband, and many popular operational amplifiers do not have the current-drive capability to do so. To avoid cross talk to other circuits, do not conduct lengthy tests at +18 dBm.

The "H" pad in Figure 3.11-2 does not provide a 600 Ω source to the line by itself. If the secondary of the output transformer on the left of the diagram were not 600 Ω, the pad would not set things right. The pad is there to provide isolation of the meter or a test oscillator from the repeat coil.

A simple test to determine the actual source impedance of a device's output is as follows. Place a bridging input voltmeter across the device's output while feeding 1 kHz at reference level. (No other inputs should be connected.) Terminate the output with various resistances, and watch the meter. The device's source impedance is the value of the resistor that drops the level by 6.02 dB. Then verify that this level drop is consistent at other frequencies.

Note also that the meter on the transmit end shown in Figures 3.11-2 and 3.11-3 should be used only for initial absolute-level calibration, and not for relative levels in frequency response tests. For verification that consistent level is being transmitted at each frequency during response tests, and for realistic end-to-end results, put the oscillator farther upstream, and use a more isolated meter for reference. This typically is done by feeding the oscillator into a properly telco-interfaced mixer or DA input (or using the mixer's inboard multifrequency oscillator, if it has one) and recalibrating for oscillator drift using the meter on the oscillator, mixer, or DA, not the downstream meter shown in the figures.

Failure to follow the procedures outlined in the preceding paragraphs will result in poor frequency response, distortion, calls from the telco warning of too high a level, too much noise, and general unhappiness with the service.

DIGITAL AUDIO CIRCUITS

As with much of the technological progress in our industry, the digital audio revolution that has dramatically affected the broadcast world began at the telco. The transmission of data is nothing new to telco, but the high data rates required for digital audio transmissions had previously rendered the availability and cost of such service out of the practical range. Data compression (or bit-rate reduction) systems have made possible broadcast applications of data transmission paths that previously were useful only for computer interconnection. Reductions from earlier data rates for digital audio transmissions of 4:1 or higher have become commonplace, and compression ratios of 12:1 or more without significant audible penalty are now frequently used.

This implies that audio program circuits are ordered from telcos not as audio services per se, but simply as digital circuits with appropriate bandwidth (i.e., data rate) for the transmission task at hand. Unlike dedicated audio circuits, the actual signal(s) passing through the line are of little consequence; a data channel on a digital circuit cares not whether the content it carries is a symphony or a spreadsheet.

Data Compression

Although the data compression algorithms used to enable such audio transmissions are viewed today as major breakthroughs, history will likely look upon them as natural evolutions and consider the earlier linear pulse-code modulation (PCM) systems as dinosaurs. While the straightforward nature of linear PCM may have been helpful in making the transition from analog systems, especially where bandwidth was cheap and available, it is an inefficient method for encoding digital audio. The resolutions of today's linear PCM systems are often excessive in terms of the actual needs of most listeners, and significant reductions in transmitted data can be achieved by applying data compression algorithms to the datastreams that linear PCM conversion produces. At present, linear PCM of as high a resolution as economically feasible is still a good idea for the original conversion of analog signals to the digital domain, and for any digital production or signal processing. But for signal delivery systems (and in some cases, for long-term storage), data compression is an appropriate tool. In fact, it has become so commonplace that the term "uncompressed" is now typically used to describe a linear-coded digital audio signal.

Earlier compression systems (again pioneered by telcos) used a purely statistical or numerical analysis of the datastream's coding redundancies. These so-called "lossless" systems could be used to reduce the data rate of any kind of data transmission and were designed in such a way that the datastream after decoding was an exact, bit-for-bit representation of the original bitstream. More recent "lossy" data-compression systems are designed for exclusive use on audio signals. Their algorithms take into account the constraints of a human listener's aural perception abilities and exploit them to achieve much higher data-compression ratios than lossless systems. Thus, they are generally referred to as "perceptual coders." They are therefore based on psychoacoustic models and owe their coding efficiencies to an appreciation of the audience's tolerances.

Unlike the lossless coder's statistical analysis of the bitstream itself, perceptual coders analyze the audio waveform and allow a recoding of the linear digital audio signal in such a way that far fewer bits are required to represent it, without substantially noticeable aural impairment in the process. Also, unlike lossless systems, this reduction is permanent, in that the original linear signal is never regenerated exactly as it was. But a good perceptual coder will arrange the signal on this lesser number of bits in such a way that the listener is not aware of the difference. These systems' impact is sizeable, allowing as much as 90% reduction in the data rate required from that of the original linear PCM signal. Thus, they are key enablers for affordable, high-quality digital audio transmission and are heavily employed by broadcasters in this environment. (For more on perceptual coding, see Chapter 3.7.)

Data Rates

Like their analog counterparts, digital circuits come in various bandwidths. But rather than specifying cutoff frequencies of the audio passband, telcos specify digital circuits in terms of their data rates. Broadcasters now need to consider bandwidth requirements in two dimensions: that of the audio signal itself and the data-transmission path. The former is specified directly (in Hz), whereas the latter is specified as a data rate (in bps). The resultant audio quality through a digital transmission path is a function of the bandwidth of the digital circuit and the performance of the coding algorithm at that data rate. For example, linear PCM will provide only POTS-like results at 64 kbps, but ISO/MPEG-1 Layer III Audio Coding can provide near CD-quality mono audio on the same 64 kbps circuit.

Therefore, broadcasters must develop acuity for the appraisal of various data rates' capabilities, both with and without data compression. In the linear PCM (uncompressed) mode, the data rate requirement of a given audio signal is determined by simply multiplying its sampling frequency (in Hz) by its resolution (in bits). For example, CD-quality audio uses 44.1 kHz sampling at 16 bits per sample resolution, requiring a 705.6 kbps data rate, per channel (stereo requires doubling that data rate to 1.411 Mbps), before adding any error correction overhead. A digital audio compression algorithm capable of 4:1 data rate reduction can reduce that signal's resolution to an average of 4 bits/sample (while leaving its sampling frequency alone), therefore providing a 176 kbps data rate for mono audio. Table 3.11-2 shows some other data compression ratios for audio and their resultant data rates at several common sampling frequencies.

Because the sampling rate is not changed by a data compression system, frequency response remains the same as it is in linear PCM conversion. Throughput delay is introduced by these codecs, however, and it is generally in direct proportion to the amount of data compression applied.

Telephone company installations and tariffs provide a variety of services, with more new services continually being deployed in many cities. Table 3.11-3 shows some common wired, point-to-point data services and their capabilities.

One of the most important differences to the broadcaster between analog and digital circuits is that like dial-up, but unlike analog program circuits, digital services are almost always provided bidirectionally. This fact should not go unnoticed when making cost comparisons to analog circuits. Although interfacing hardware for return path channels must still be provided, their circuits require no separate costs or orders, as they likely would have with analog paths.

Note also that telcos have always had their own insider vocabulary and set of acronyms, but since the introduction of digital services, this has expanded dramatically. Many of these are explained in the "Glossary of Telco Terms" at the end of this chapter.

For overseas links, rough equivalents to each of the domestic services shown in Table 3.11-3 do exist outside the U.S., but their actual data rates differ. Format

TABLE 3.11-2
Data Compression Table*

Resolution (avg. bits/sample)	Compression Ratio	Output Data Rates (kbps, per audio channel)		
		f_s=48 kHz	f_s=44.1 kHz	f_s=32 kHz
16	1:1	768	706	512
4	4:1	192	176	128
3	5.3:1	144	132	96
2.67	6:1	128	118	85
2	8:1	96	88	64
1.45	11:1	70	64	46
1.33	12:1	64	59	43

*Data compression table showing a range of compression ratios and their resultant output data rates at a variety of sampling rates (f_s). Audio bandwidth is approximately one-half of f_s. Data rates shown are for a single audio channel (mono).

conversions are therefore required for international transmissions, but most LDSs can handle this for the broadcaster. For example, European telcos offer an E-1 rate of 2.0 Mbps, in lieu of the North American T-1 rate of 1.5 Mbps.

DS1 or T-1 Service

Digital audio transmission on DS1 (or T-1) lines has become widely available and is often cheaper than standard analog circuits in both service and installation charges. (See the "Glossary of Telco Terms" for the distinction between DS1 and T-1 nomenclature.) DS1 is a bidirectional 1.544 Mbps serial data link. The previous data rate calculations show how DS1 can carry a single, linear PCM stereo audio signal or several such compressed channels.

DS1 service is extremely reliable. Its bit error rate (BER) of 10^{-9} (the probability of error reflected by the specification of no more than one erroneous bit in a billion carried by the circuit) is among the lowest available. By way of reference, IEEE and ITU-T have both established 10^{-6} as the BER required for data customers' satisfaction.

The data carried on a DS1 circuit is actually a multiplex of 24 data channels, or slots, of 64 kbps each. (An additional 8 kbps is reserved for sync data.) These individual 64 kbps slots are called DS0 channels. For standard telco T carrier use, each DS0 carries a digital voice grade circuit, using the so-called μ Law (nonlinear 8 bit) algorithm on 8 kHz sampled audio. When a customer leases a DS1 circuit, it can be configured to carry any bandwidth channel that DS1 hardware is available for (3.5 kHz, 5 kHz, 7.5 kHz, 15 kHz) in any combination, up to the customer's payload data limit. When a customer leases a full DS1, the telco may use one DS0 slot for framing and other overhead, in addition to the 8 kbps synchronization slot, leaving around 1.4 Mbps for customer data. Check with the telco if it is important to know the exact rate.

A rack of coding and multiplexing hardware appears on each end of the DS1 line, usually as customer-provided equipment (CPE), and the circuit can be reconfigured simply by changing the appropriate cards in the proper slots in the racks at both ends. The customer can perform these reconfigurations at any time, without telco involvement or notification.

Unlike the labor intensive installation and equalization of an analog circuit, putting in a T-1 circuit has become nearly as routine as a standard dial-up telephone service installation. This and the excess capacity in some areas continue to lower costs for DS1 service, although other newer services—also cheaper than DS1 in most cases—are also being aggressively

TABLE 3.11-3
Some Current U.S. Digital Telco Data Services and Their Characteristics*

Service	Data Rate (bits/sec)	Mode/Carrier
Switched 56	56 k	Switched/copper
ISDN-BRI	128 k	
ISDN-PRI	1.472 M	
DS 0	64 k	Dedicated/copper
DS1 (T-1)	1.544 M	
DS2	6.312 M	
DS3	44.736 M	
OC1	51.84 M	Dedicated/optical
OC3	155 M	
OC12	622 M	
OC48	2.5 G	
OC192	9.6 G	

*The ISDN data rates shown are for B-channels only.

marketed in some regions. Customers' use of digital compression systems on these channels can also increase this economy. Whereas a 15 kHz (mono) audio channel had originally required six DS0 slots, today the use of perceptual coding can reduce this to two (or even a single) DS0 slots.

Fractional T-1

In some areas, Fractional T-1 service is available, generally for intraLATA (local) applications only (intraLATA refers to telephone communications wihin a "local access and transport area," or LATA). This service allows a customer to lease only the number of DS0 slots on a DS1 circuit that are needed for a particular application. Although installation charges will be about the same as a full DS1, service charges may be substantially reduced for many remote audio applications.

Switched 56

Another telco digital offering that has been used by broadcasters is the Switched 56 service. This facility had been available for a number of years for local service in many metropolitan areas in the United States, and from some long-distance carriers, but it is now largely obsolete. It provided a bidirectional 56 kbps data path for use with dial-up terminals. Like POTS, a monthly service/network access fee was charged (which often included at least some free local calling), with long-distance calls billed by the minute, at rates similar to regular dial-up long-distance service. In some cases, local calls were also billed for connect time. An installation fee was also typically charged for initiation of service.

A switched channel service unit (CSU)—the equivalent of a telephone instrument and data interface—was provided by the telco, or could be purchased by the customer. It allowed voice or data interconnection and dial-up routing of the data path to any other similarly equipped destination on the network. For broadcast use, external codecs were required, as shown in Figure 3.11-6, to feed wideband audio. The typical codec used for Switched 56 service was ITU-T G.722, which provides 7.5 kHz mono audio (see Figure 3.11-6).

The appeal of this service was strong for the broadcasters, and many employed it heavily when it became widely available in the 1980s. An obvious savings was possible with such a switched approach, since dedicated circuits need not be installed between a radio station and all of its remote sites. The station and each site only had to be wired for the switched service (with each line going to the telco CO only, and not end-to-end, as with dedicated circuits), and the station could then dial up any remote site as needed. Some stations installed two or more lines at their studio locations for accessing multiple remote sites simultaneously or in quick succession. (However, it could take several minutes to sync up the codecs and pass audio bidirectionally after establishing a connection on this service.)

Switched 56 CSUs were available in two-wire or four-wire versions, with two-wire types costing less. The choice between two- and four-wire versions was not up to the customer, but to the local telco serving the area.

There are few if any new installations of Switched 56 service, and maintenance of previously existing services is becoming rare, with most telcos (and users) having moved to Integrated Services Digital Network (ISDN) or other more modern services.

Some telcos may still offer a similar unswitched service, in which a single DS0 channel can be leased on a monthly basis. For heavy point-to-point users, this may be more economical than a switched approach. As single DS0s, these circuits operate at 64 kbps, and their terminal hardware is less expensive because it need not accommodate switch signaling.

ISDN Service

During the 1990s, ISDN became widely available in most U.S. urban centers and eventually became almost universally deployed. The most common variety of ISDN service is the Basic Rate Interface (BRI), which provides two 64 kbps paths (bearer or "B"-channels) and one 16 kbps circuit (data, delta, or "D"-channel). As a result, this service is also referred to as 2 B+D.

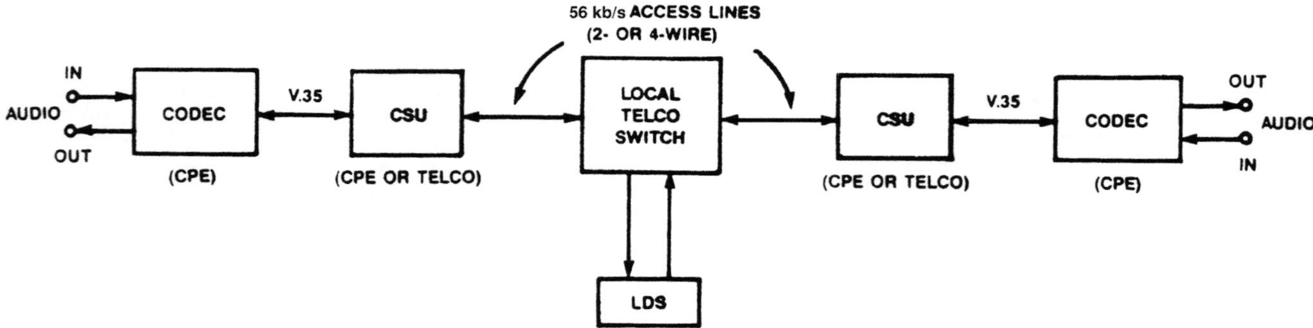

FIGURE 3.11-6 Block diagram of Switched 56 circuit path. (See "Glossary of Telco Terms" for acronym definitions.)

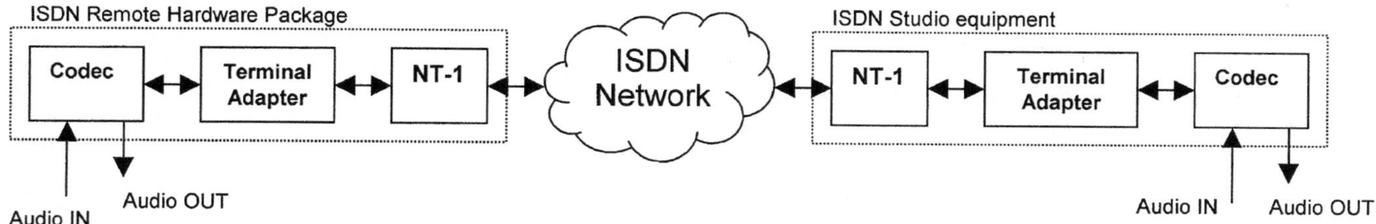

FIGURE 3.11-7 Block diagram of typical ISDN-BRI service used for a radio remote.

Bearer channels carry customer data, whereas data channels carry signaling and call routing data. This feature is a significant departure from POTS and Switched 56 service, in which the call signaling data is routed via the same data path as the program audio or data. (This is why DTMF tones and call-waiting beeps or pulses are audible on a POTS call, for example.) Such interruptions can cause problems to data communications, which benefit from continuous, bidirectional connections. ISDN's use of such *out-of-band signaling* provides significant improvement in connection speed, throughput, and robustness for switched data communications.

ISDN-BRI is intended as a standard residential service, while business service is provided by ISDN's *Primary Rate Interface* (PRI). This service provides twenty-three 64 kbps B channels and one 64 kbps D-channel (23 B+D). (The D-channel in ISDN-PRI operates at a higher rate than the BRI service's D-channel because it has many more B-channels to manage.) The data payload capacity of the ISDN PRI service (1.472 Mbps) is roughly equivalent to the data payload capacity of a T-1 circuit (1.536 Mbps).

While the preceding data rates are common for ISDN, it should be noted that some ISDN systems in the United States offer users only 56 kbps of payload data on their B-channels.

ISDN is a bidirectional, customer-switched service, operating as a dial-up, billed minute data network, allowing both circuit-switched and packet-switched operations. Its multichannel nature allows simultaneous voice and data to separate destinations (like multiline POTS service) or other applications in which multiple B-channels are combined to provide a single, higher-bandwidth call. This is done via a process called *inverse multiplexing* (IMUX). Each line is still billed separately in such applications, however. Today this is often called "bonding" (a name that originated from an actual acronym for an early IMUX technique, which has since been replaced by more flexible and dynamic multiplexing algorithms, but the generic terminology remains in use for all such methods).

The ISDN equivalent of the POTS telephone instrument is called the *terminal adapter* (TA), and it is available in several varieties, for the control of one or more BRIs or PRIs. Each TA is connected to the ISDN network via a standard *network termination* (NT), which varies with the type of terminal hardware and software installed by the local telco. In North America, all telcos have agreed to use a single standard interface, which requires the use of the NT-1 termination. While originally supplied as a separate unit for easier transport of a TA between service areas, the standardization encouraged manufacturers to include the NT-1 internally in most terminal equipment sold in the United States.

By the late 1990s, ISDN-BRI had become the delivery method of choice for most radio remotes, offering mono or stereo 15 kHz broadcast audio plus communications to/from a remote site. Often a single B-channel is used with one of several audio codecs capable of providing high-quality mono audio transport at a 56 or 64 kbps rate. In other cases, two B-channels are IMUXed for high-quality stereo at 112 or 128 kbps. Numerous devices are available for such applications, and most include an integrated ISDN terminal adapter, NT-1 and codec, plus machine-control closures (see Figure 3.11-7). Most of these devices also include multiple codecs, each with variable data rates. Appropriate coding algorithms include ISO/MPEG-1 Audio Layers II and III, APT x-100, Microsoft Windows Media Audio Professional (WMA Pro), Apple QuickTime, or MPEG-4 AAC. (The latter is available in several forms, one optimized for low delay, and another for extended frequency response, called High-Efficiency AAC [HE-AAC, also known as AAC+].)

For more information on ISDN, including its use in voice transmission applications, see Chapter 3.10, "Telephone Network Interfacing." This chapter also contains information on the use of POTS codecs, which can provide moderate-quality mono audio on standard dial-up phone lines.

EMERGING SYSTEMS

As this handbook goes to press, the audio interconnection industry's transition from analog to digital is nearly complete, but a second-generation digital transition has begun. This new phase moves away from systems that packetize and directly modulate carriers using traditional telco data transmission formats to a method that encapsulates digital audio packets into the Internet Protocol (IP) for networked transmission.

IP Audio

This so-called IP Audio approach offers additional transmission efficiencies, and leverages the cost-effectiveness of the IP networking architecture that has been so widely deployed as the result of the Internet's penetration.

The fundamentals of IP networking are beyond the scope of this chapter, but are explained in many other widely available general texts on the subject. More details on the use of IP for media transmission are also available in Chapter 6.13 of this handbook. Some specifics of IP Audio interfacing are worth noting here, however.

By way of some simple background, the classical data networking architecture is based on a layered approach formulated by ISO, called the Open Systems Interconnect (OSI) model. OSI used a seven-level stack consisting of Physical, Data Link, Network, Transport, Session, Presentation, and Application layers.

More recently, the development of the Internet has led to a variant known as Transmission Control Protocol/Internet Protocol or, much more commonly, "TCP/IP" networking. TCP/IP simplifies and combines the seven layers of the OSI model into four layers:

- Link: Defines the network hardware and device drivers, such as Ethernet or WiFi; uses Media Access Control (MAC) "hard" addressing of individual devices, in the form of six colon-separated pairs of hex digits (e.g., 0F:A7:00:B4:92:FF).
- Network: For addressing, routing, and other basic communication, via IP, the Internet Control Message Protocol (ICMP), and the Internet Group Management Protocol (IGMP); uses IP addresses for "soft" addressing of individual devices currently on the network, in the form of four dot-separated decimal numbers between 0 and 255 (each represented by one byte of address data; e.g., 168.21.422.7).
- Transport: For communication between programs, via TCP or the UDP.
- Application: For user functionalities over the network, such as File Transfer Protocol (FTP), Real-time Protocol (RTP), Simple Mail Transfer Protocol (SMTP), Hypertext Transfer Protocol (HTTP).

While the Network layer (IP) is ultimately responsible for delivering data packets from device to device via packet-address management, the Transport layer (TCP or UDP) is charged with protecting the integrity of the data thus delivered. This protection is done on a "best-efforts" basis, meaning that accurate delivery is not guaranteed in all cases.

IP networking allows both point-to-point ("unicast") and point-to-multipoint ("multicast") transmission styles. When the latter approach is used, the UDP (rather than TCP) transport is often engaged. UDP/IP offers less robust error recovery, but it provides a simpler and more direct way to send and receive data packets.

Audio-Specific Techniques

Digital networking is primarily designed for the transfer of data files, which is inherently not a real-time process. Part of basic network functionality is the mitigation of data collisions for optimum use of the network by many simultaneous users. In the interest of minimizing file-transfer time and the possibility of collisions, increased networking speeds are constantly sought after. Today's state-of-the-art for LAN applications of IP is Gigabit Ethernet architecture, providing approximately 1 Gbps of network bandwidth on CAT-5 wiring.

These networking speeds can manage faster-than-real-time transfer of very large audio files, but with proper care, they can also provide reliable real-time, multichannel audio streaming on LANs. Various network applications have been developed to provide this kind of network management, such as CobraNet. Originally developed by Peak Audio, CobraNet is now implemented by several manufacturers for Ethernet-based digital audio routing.

Protocols such as CobraNet and others address the specific needs of real-time audio transmission over a digital network, in particular the reduction of TCP- and UDP-related latencies associated with high-bandwidth (uncompressed audio) streams and improved quality of service (QoS) management. (See the "Latency" section in this chapter for more information on this topic.)

Thus, the IP Audio approach formats digital audio signals into a structure that can easily pass over any IP-enabled network, whether it be an intra-facility LAN, a dedicated inter-facility network, or the public Internet.

Of course, the use of the Internet for a critical broadcast signal path presents potential problems, due to the uncertainties of network conditions over time. The Internet's "best efforts" service implies that no particular quality of service (QoS) is specified or guaranteed. This stands in stark contrast to more traditional digital telco transmission services, where robustness is high and a minimum BER is typically specified.

This does not imply that IP Audio is inherently unreliable, however. It is important to maintain a clear distinction between IP and the Internet. The latter is simply one of the physical networks ("link" layer) that utilize IP as a networking protocol. Many applications of IP Audio do not use the Internet, but travel on more robust broadcaster-operated networks, which may even have dedicated bandwidth assigned for a critical audio signal path.

For the applications considered in this chapter, however, it is implicit that transmission across non-broadcaster-operated networks (i.e., telco networks) will often be required. IP Audio is the format of choice for an increasing number of these data transmission offerings.

IP Audio Applications

There are three basic methods available to broadcasters today for interfacility transmission of their audio content via IP. The first is the use of traditional dedicated digital transmission services such as T-1, which can be repurposed for IP transmission with the appropriate terminal hardware at each end.

The second approach is to simply purchase IP bandwidth from telco. The actual signal-transport architecture used will be determined by telco. This may be a more cost-effective approach if the requirements on a given path are only occasional, or particularly if bandwidth needs vary widely over time, in which case a "bandwidth on demand" option may be provided by telco.

The third method is the use of broadband Internet service, with each involved facility connected via broadband Internet access of some form, such as a telco Digital Subscriber Line (DSL) or a cable modem (DOCSIS), both described later.

Of particular interest to broadcasters for remotes is an emerging variant on this option, which offers *wireless* broadband Internet access, via a wide range of technologies, each offered by different wireless service providers (see "Wireless Broadband" later in this chapter).

Using the Public Internet for Broadcast Audio

The first two methods described in the preceding section require lead time for ordering and dedicated installation, plus relatively high cost for CPE hardware and/or service, but they provide high reliability and guaranteed QoS.

The third method—in all its variants—offers the option of easy access with no dedicated-line installation, and if Internet access is already available on site, no lead time or service orders at all. Connection hardware is also generally inexpensive and off-the-shelf. This flexibility is highly attractive to broadcasters, but connections via the Internet do not offer guaranteed availability or quality of service. As a result, broadcasters typically have avoided this option for live-to-air or other critical audio transport applications. Yet some other emerging technologies have reduced the risks involved with use of the public Internet, as have overall—and continuing—improvements in robustness, availability, and bandwidth of the Internet and its access paths.

These devices include generic data as well as audio-specific devices. Among the former are LAN-contention and prioritization devices (sometimes called Broadband Boosters or Accelerators) that can provide improved robustness for a particular connection requiring continuous, real-time connection within the upstream data on a local area network. These devices are targeted at the growing Voice-over-IP (VoIP) telephony and on-line gaming markets, but can be applied for any real-time signal usage over a shared Internet connection. These devices work by monitoring all the traffic on a LAN and determining which applications within it require real-time priority. Of course, if an Internet connection path is used wholly for a broadcast audio application, without traveling through a LAN sharing the Internet access, these devices are of no value.

At this writing, audio-specific devices for this application are best represented by a system developed by Comrex Corporation called *Broadcast Reliable Internet Codec* (BRIC), which allows adjustable and adaptive redundant packetization of audio, such that a stable audio connection can be maintained across an Internet connection with minimum latency (typically 100 to 250 ms). Such processing is likely to substantially improve results on any type of broadband Internet connection, although guaranteed service over the Internet can still never be given.

Other basic networking techniques can also be used to improve robustness over the Internet, such as the use of fixed IP addresses and, in some cases, connection via Virtual Private Networking (VPN).

Another issue to consider is the asymmetry of most broadband Internet connections. Upstream data rates are generally far lower than downstream rates. This implies that the upstream speed of the broadband service at the originating source location of an audio signal path will be the limiting factor in any Internet-based Contribution channel.

Regardless of the technology or access method employed, using the Internet for real-time audio transport should always be considered as a second choice to dedicated point-to-point service due to the inherently lower reliability of Internet connections. But the differential between dedicated service and broadband Internet service is continually shrinking, and when considered with cost/benefit ratios and availability issues, the use of a broadband Internet connection for broadcast audio contribution or distribution may be an appropriate choice in an increasing number of cases.

Good recommended practice for using the public Internet therefore includes the following:

- Choose the most appropriate, efficient, and fault-resistant codec;
- Set the codec's output to the lowest tolerable bit rate for the application;
- Fortify the stream and set up the network appropriately at the sending end (as described above);
- Perform adequate testing;
- For live broadcasts, have a hot-standby (typically POTS or POTS-codec) backup.

Finally, many radio remotes may require Internet connectivity for other reasons, such as reporters' need for access to news sites, etc. For this reason, the use of a broadband Internet connection can serve double duty for both audio backhaul and a staff research resource. On the other hand, remember that the connection's bandwidth is shared by all users, so any critical audio signal paths traveling over the broadband Internet link should take priority, with adequate bandwidth reserved for their use whenever required.

DSL Service

One of two popular methods being deployed for broadband Internet connectivity (although it can be used for other purposes, such as IPTV), and which can play a role in broadcast audio transport, is the *Digital Subscriber Line* (DSL) service. It provides multimegabit connectivity over the telcos' existing twisted-pair infrastructures, which is accomplished by implementing adaptive DSP at both ends of the signal path, thereby compensating for the distortion inherent in long copper lines. Path lengths for DSL are limited, but the most widely deployed systems are designed to accommodate the typical nonrural telco customer-to-CO runs of 12,000 to 18,000 ft. DSL is provided by telcos as a data plus voice service on the same pair of wires, with a standard switched POTS service combined with dedicated broadband data service. The two services are split at either end of the line. In many areas, standalone or dry DSL service is also offered (often called "Naked DSL"), which provides the broadband data only without POTS.

DSL paths are dedicated, point-to-point services, and intermediate repeater amplification is not used. Two different transmission technologies were originally employed, *Discrete Multitone* (DMT, developed by Amati Communications), and *Carrierless Amplitude/Phase modulation* (CAP, from AT&T), but most deployments today use DMT, which is essentially similar to OFDM. Importantly, DSL's data service is not switched (so it is not a direct substitute for ISDN), and it generally is installed between a customer's service location and a telco central office, not from service location to service location (so it is not a direct substitute for T-1). Nevertheless, it may provide an economical alternative to either service for transmission of audio files via the Internet, or in some cases, for real-time audio backhaul (see the "IP Audio" section earlier).

DSL service comes in a range of data rates (typically from 128 kbps to 24 Mbps), and is offered in both symmetrical (SDSL) and asymmetrical (ADSL) forms. The latter offers a higher data rate from the telco CO to the customer's service location ("downstream") than in the opposite direction (from customer to CO, or "upstream"). Actual speeds of any service are affected by physical wiring path lengths between the service location and termination at the telco CO, but it is not uncommon for ADSL service to be limited to an upstream data rate of 64 kbps. Because the upstream path is what a broadcaster would use when feeding audio from a DSL-connected remote site, it is the more important data rate for such applications, although telco typically identifies and markets the service by its higher downstream rate. So take care when ordering the service to specify the minimum *upstream* data rate from the service location. If this service is shared with other networked users at the originating site, audio quality will be further constrained, so it is recommended that any such circuit be exclusively dedicated to audio backhaul when used for broadcast applications.

More appropriate to broadcasters may be the *high bit rate (or very high bit rate) Digital Subscriber Line* (HDSL or VDSL), which can be configured to provide symmetrical high-speed capacity, although over shorter distances than noted previously. For example, a typical VDSL can provide around 10 Mbps in each direction across a one-mile path. Lower speeds may be possible over longer path lengths. In some cases, multiple physical DSL lines can be bonded to achieve higher aggregate bandwidths.

If voice telephony service is otherwise provided at the site, Naked DSL may be an appropriate choice for broadcasters who wish to use it simply for broadband Internet access at a given location.

DOCSIS Service

The other primary wired method of broadband Internet access to consumers is over cable television systems' coaxial cable, typically via the *Data Over Cable Service Interface Specification* (DOCSIS). At this writing, the DOCSIS 2.0 standard is being deployed, which is better suited for broadcast applications (again, given the previous caveats regarding use of the public Internet), due to the 2.0 spec's higher upstream speeds and improved QoS capabilities over the predecessor DOCSIS 1.0 and 1.1 standards.

DOCSIS uses QAM modulation (from 16-QAM to 256-QAM are specified), with a maximum upstream throughput of 30 Mbps per channel in DOCSIS 2.0 (DOCSIS 1.x was limited to 10 Mbps), and a maximum downstream capacity for all versions of 38 Mbps. The "channel" in this case is a 6 MHz-wide TV channel on the cable system infrastructure, and each of these is typically shared by multiple (often hundreds of) cable-modem customers. Cable operators therefore set limits ("caps") on each customer's upload and download rates; a typical example at present is 384 kbps up and 6 Mbps down (again, for broadcast audio transport, the upstream value is more critical). Some higher-end services offer up to 2 Mbps upstream at this writing, however.

These caps on DOCSIS service speed are adjustable by the cable operator on an individual customer basis, so it may be possible for a broadcaster to arrange for higher-speed service from a particular location if necessary. A future standard currently under development, DOCSIS 3.0, is expected to allow higher maximum speeds in both directions, through the use of channel bonding.

WIRELESS BROADBAND

"Wireless broadband" has two distinct meanings today. One refers to the IEEE 802.11x ("WiFi") family of standards and its future relatives (802.16x, 802.20x), whereas the other refers to mobile telco data deployments. Another way to distinguish these is "unlicensed" versus "licensed," respectively, although some applications of 802.16x ("WiMAX") and beyond may be licensed services.

First, all these services are simply first/last-mile access routes to/from the Internet, so they are subject to the same caveats noted previously on use of the public Internet for broadcast audio transport applications.

TABLE 3.11-4
Selections from the IEEE 802.11x Family of Wireless LAN Standards and Their Basic Characteristics

Format	Release Date	Op. Freq.	Data Rate (Typical)	Data Rate (Max)	Range	Comments
802.11	1997	2.4 GHz	1.0 Mbps	2.0 Mbps	?	Legacy
802.11a	1999	5 GHz	25 Mbps	54 Mbps	30m	Less interference but poorer propagation than 2.4 GHz
802.11b	1999	2.4 GHz	6.5 Mbps	11 Mbps	30m	Original "WiFi"
802.11g	2003	2.4 GHz	25 Mbps	54 Mbps	30m	Backward compatible to 802.11b
802.11n	2007	2.4 GHz	200 Mbps	540 Mbps	50m	Adds MIMO antennas

Next, consider that while the 802.11x family has adequate bandwidth for even uncompressed broadcast audio application, it is generally provided as a private, short-range service, in the unlicensed spread spectrum bands of 2.4 or 5 GHz. (See Table 3.11-4 for details.)

Note that the maximum data rates specified are achieved only under optimal signal conditions; when signal strength or quality fades, data rates are reduced in steps until the link fails. For example, the 802.11b standard provides 11 Mbps bidirectional connectivity under optimal conditions, shifting down to 5.5 Mbps, then to 2.0 Mbps, and then to 1.0 Mbps, as signal strength or quality decreases. All signal ranges quoted assume indoor transmission and reception, using omni directional antennas. Somewhat longer ranges can be expected outdoors or with directional antennas. (MIMO = Multiple Input/Multiple Output antennas; see "Glossary of Telco Terms.")

Note also that the data overhead of 802.11x limits the available throughput for streaming content to about 60% of available bandwidth for TCP and about 70% for UDP. Thus, on a fully available, best-case 802.11b link, for example, maximum throughput for audio streaming is never more than 7 Mbps or less, and often lower. Nevertheless, this is still adequate for a single, uncompressed stereo audio feed.

More importantly, however, the availability and reliability of 802.11x service are variable and, on the whole, lower than wireless services provided by actual wireless data service providers (telcos), due to their generally private and highly localized ("hotspot") deployments. Note also that even if an 802.11 service is available, it may be blocked for usage by the access point owner. So "WiFi" is useful for broadcast audio transmission only in certain very specific, predetermined, short-haul cases.

At this writing, it is expected that 802.16x (WiMAX) systems may be of somewhat greater utility to broadcasters, given their anticipated greater coverage. While 802.11x is considered a wireless local area network (WLAN) standard, 802.16x is a wireless metropolitan area network (WMAN) standard, which is anticipated to operate in a "mesh" fashion, covering a large area from multiple antenna sites, with customer devices that have an operating range up to several miles. It will operate at various spectral locations between 2 GHz and 66 GHz (typically 2–11 GHz), offering 40 Mbps or greater bidirectional customer connectivity. The 802.16e variant will offer mobile reception capability (dynamic handoffs between nodes, like cellular telephony), but initial deployments are expected to be for fixed usage only. If licensed services are deployed by communications services operators, these may provide adequate reliability for broadcasters to consider using for Internet access, particularly for quick availability from random remote sites around their metro areas.

Meanwhile, most mobile telephone companies are developing so-called 3G (third-generation) wireless data services, as extensions to their existing voice and low-speed data networks. These are almost all packet-based, asymmetrical formats designed for Internet page browsing on handheld terminals with small, low-resolution screens, so they are not particularly applicable to real-time broadcast backhaul use—particularly given their low upstream capacity. One proposed service might prove to be an exception, however. It is called EV-DO Rev A, for Revision A to the *Evolution-Data Optimized* standard (also called 1xEV-DO Rev A), which will offer upstream capacity of up to 1.8 Mbps, along with a Wideband Audio Element optimized for real-time, low-latency streaming.

Broadband Global Area Network (BGAN)

While the wireless broadband systems discussed in the preceding sections are all terrestrial services, the international satellite telecom service provider Inmarsat has developed its own form of broadband connectivity called BGAN. This service provides voice plus data, with adequate upstream bandwidth (about 256 kbps) for real-time audio uplinking via the Internet from nearly anywhere on earth. BGAN may be useful to broadcasters for "extreme" remotes where no other form of connectivity is available.

Latency

All digital transmission links suffer from some degree of throughput delay, or "latency." These delays—if longer than a few hundred milliseconds—can cause complications when the circuit is used for real-time broadcast applications, since it will require mix-minus to be used for monitor return feeds from the studio to the remote site (see "Communication Lines" next). It may also result in pauses or interruptions when studio talent engages in conversation with remote talent.

When the Internet is involved, latency can become extreme—extending to several seconds—as well as quickly variable and unpredictable. Wireless broadband links add yet another layer of high and variable latency. Thus, broadcasters should be prepared to face the reality that even if reliable, high-fidelity connectivity can be achieved over a digital circuit that involves wired or wireless Internet connections, real-time two-way communication suitable for live broadcast may not be possible.

COMMUNICATION LINES

For remote broadcasts, circuits for transmission of program audio often must be complemented with separate communication, interruptible foldback (IFB) or talkback lines. When analog program circuits are used, communications are usually carried on standard POTS lines. When digital services are used for program backhaul, their bidirectional nature usually allows monitoring and communications to be carried on the return side of the same lines. Nevertheless, it is still a good idea to have at least one POTS line on hand as a backup return line in such cases.

Communications lines may be used purely for talkback if off-air monitoring is possible on site—although this will no longer be the case if transmission delays from long transmitter-link paths (especially those including a satellite segment) or if HD Radio "diversity delay" is implemented in the station's signal (and the analog signal is synchronized).

In any case, the codecs used for digital audio transport can add significant delay—on the order of 100 ms in each direction in some cases—in the contribution (backhaul) path, as can studio-to-transmitter links and other components of the air chain. The cumulative delay from these devices can cause monitoring problems for talent listening to their own voices or in conversation with others. For this reason, dedicated analog communications lines may be useful even when air monitoring or digital return is possible.

When POTS communications lines are used, one or more standard dial-up telephone circuits can be ordered for program audio monitoring and communications. Alternatively, an RF communication system (handheld radio or cellular phone) could be used. If wired telephone service is chosen, the station should provide its own instrument, which may be equipped with visual ring/bell cutoff and headset attachment.

There are also some variations to consider over a standard dial-up service. A communications phone line (comline) can be set up as an *off premises extension* (OPX) of the station's *private branch exchange* (PBX) system, or the line can be a dedicated *private line* (PL) that runs only between the station and remote site, in an unswitched form. In the latter case, such a PL is normally equipped with *automatic ringdown* (PLAR), in which special hotline-type phones are installed by the telco, having no keypad or dial; the phone at one end rings whenever the phone at the other end is picked up. Services of this type (and their specific nomenclature) may vary among telcos. These services are also typically provided by a different department than the ones that handle the ordering of program audio circuits or ISDN—if those even still exist—so coordinate carefully. Typically, this part of the telco is the same one that handles all consumer dial-up installation and service calls from the general public, so lead-time requirements and service-call response time can be lengthy.

T-1 Applications

Using the return path on T-1 circuits for communications and backfeeding requires appropriate terminal hardware on each end. Channel configuration need not be the same as the circuit's other direction, and because narrowband (lower data rate) channels are usually all that are required, the number of communication channels on the return path can exceed the number of program audio channels on the transmit side, if necessary. Another possibility is the luxury of wideband backfeed and communications, such that the receive-audio quality at the remote site is as good as the transmit quality.

For permanent interfacility hookups, a T-1 circuit could carry audio transmissions between sites plus telephone service, such that both locations can have their phones connected to the same PBX. If the locations are far apart, a *foreign exchange* (FX) arrangement can be established, whereby one facility can place calls to the other facility's telco service area without incurring toll charges. The PBX system is programmed to recognize area codes of the two cities involved (or special internal access codes), and it directs appropriate calls to FX lines on the T-1 instead of placing them as regular long-distance calls.

COST OF SERVICES

For local service, there is normally only one provider of telco circuits, analog or digital, which is, of course, the local telephone operating company. So there is little competitive choice in the matter for intraLATA service pricing. Check with the local telco frequently to determine which services are more economical, for either permanent or temporary applications. Analog services continue to increase in price in most areas (if they are even available), whereas the cost of digital services declines at varying rates around the country—although some telcos are also moving away from point-to-point services entirely in favor of their broadband Internet businesses. Therefore, it is worthwhile to keep a close eye on these changing rates and service offerings.

For long-distance paths, there is significant competition for interLATA digital service, so prices are kept low and continue to drop (interLATA refers to telephone communications between "local access and transpot areas," or LATA; there is little or no inter-LATA analog service available anymore). Shop around for the best deal on interLATA digital service, or use a dataline broker. Secondary services may also be offered, such as switching and monitoring.

Backup is also an important issue, particularly for live broadcasts. Many broadcasters establish primary and secondary backhaul and communications paths, using different services for each. (For example, ISDN primary and POTS secondary services.) This adds a level of redundancy because the services are typically routed differently. Costs of each service are an important factor, particularly when broadcasters are paying for something (the secondary service) that they hope they never have to use. The use of a lower-cost service for backup therefore makes both technical and economic sense, and it is a highly recommended practice.

For those broadcasters who still book a significant number of traditional point-to-point telco program circuits (or for historical reference), it also may be helpful to understand how telco billing works for these services. Figure 3.11-8 shows the traditional route-billing concepts followed by local and long-distance common carriers for dedicated (nonswitched, point-to-point) lines.

On the other hand, costs for today's broadband Internet connectivity from telcos are generally quite straightforward, with flat monthly rates for a given bandwidth service being the norm. Remember that many services are asymmetrical, so *two* rates—upstream and downstream—are being quoted. Telcos are used to stressing the downstream rate to their general customers, but as noted previously, broadcast customers using Internet connectivity for remote backhaul are more concerned with the upstream rate. It is also important to understand that the actual access bandwidth may vary over time. While this is particularly true for DOCSIS and wireless services, even a DSL connection's bandwidth can vary with overall network load conditions. Here again, *upstream* rate consistency is the more critical consideration. Note that the quoted bandwidths refer only to first/last-mile access to/from the ISP; throughput bandwidth and latency over the Internet backbone are "best effort" only and never guaranteed.

Numerous Internet sites and applications are available to test or verify the bandwidth of a given connection. They should be used by broadcasters whenever new service is established or when any change in service or recovery from an outage takes place. Note that many of these services test only the downstream rate, so be sure to use a test that considers upstream performance as well. Bandwidth test results should reach at least 90% of the specified rate (in both directions). Contact the service provider if results are significantly and consistently below this value.

The importance to radio of interconnection, immediacy, and fidelity underscores the importance of a long-term relationship between broadcasters and telcos. Good awareness of and rapport with communications service providers are essential to the daily work of the broadcaster. In today's context, that also means keeping abreast of the changes in telecommunications that affect broadcasting as the digital revolution continues to roll along.

LEGEND

CM — Channel Mileage (per air-mile billing basis)
CO — Central Office (LEC switching center; also called SWC—Serving Wire Center)
CT — Channel Termination (flat rate billing)
IOC — Interoffice connection on LDS network (distance-sensitive)
LATA — Local Access and Transport Area (telco service zone)
LDS — Long-Distance Service (common carrier)
LEC — Local Exchange Company (local telco)
POP — Point of Presence [LDS's office in each LEC; also called SO (Serving Office)]

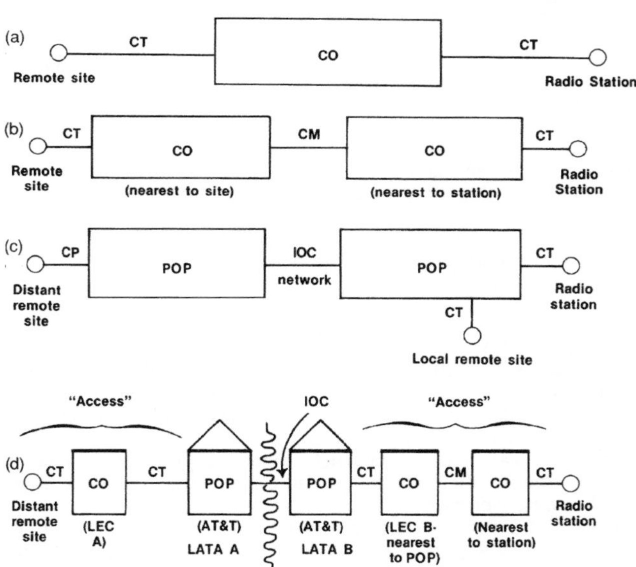

FIGURE 3.11-8 Traditional billing methodology for dedicated telco circuits. In (a), the two ends of a local circuit are served by the same central office, or "rate center." In (b), two different central offices are involved. In (c) and (d), long-distance service is depicted. LEC connections to LDS are referred to as "access." Access shown in LATA B will be more expensive because the radio station is in a different rate center of LEC B than the long-distance carrier's POP.

Glossary of Telco Terms

1xEV-DO: See *EV-DO*.

3G: General name applied to third-generation wireless telephony, which includes significant broadband data capability.

802.11x: IEEE standards for wireless LAN (local area network), often called "WiFi." Uses unlicensed spread spectrum in 2.4 and 5 GHz bands, offering nominal data rates of 11 Mbps to 540 Mbps (minimum 1 Mbps under impaired signal conditions), with nominal range of 30m (100 ft). Includes numerous variants (802.11a, 802.11b, etc.); entire family referred to as 802.11x, not to be confused with 802.1X, which is a remote authentication protocol for wired or wireless Ethernet networking.

802.16x: Developing IEEE standards for wireless MAN (metropolitan area network), often called "WiMAX." Expected to offer 40 Mbps or higher data rates over cellular-type network of metro-area connectivity, from fixed or mobile terminals, in either unlicensed or licensed forms, in various bands of spectrum located between 2 GHz and 66 GHz.

Access point: The name used for an IEEE 802.11x ("WiFi") transceiver, which connects the wireless service area to a wired network.

ADPCM: Adaptive Differential Pulse Code Modulation. A form of digital coding more efficient than linear PCM because it codes only the difference between one sample and the next, instead of assigning a fully discrete value to each sample. It also adapts its coding to the signal values currently under process. Considered a form of statistical data compression.

ADSL: Asymmetric Digital Subscriber Line. A high-speed digital service running on telcos' existing copper (twisted-pair) infrastructure, which provides data at a higher rate from the telco CO to the customer's premises ("downstream") than in the opposite direction ("upstream"). For example, service may run at 512 kbps from the telco CO to the customer, but only at 64 kbps from customer to telco. Typically used for broadband Internet access or IPTV service by telcos.

AMI: Alternate Mark Inversion. The binary modulation code used by the telephone company for data and digital voice transmission. It uses RZ coding in an alternate bipolar scheme, with logical 0's corresponding to zero volts, and logical 1's alternating between +3V and –3V. (The first logical 1 produces a +3V output, the next 1 produces –3V, the next +3V, and so on.) Self-synchronization is possible with this approach, but the number of continuous zeros must be limited.

Baud: Symbols per second.

B-channel: In ISDN service, a channel designated for customer data transmission, uninterrupted by any signaling data.

Bellcore: Bell Communications Research. The R&D firm that feeds technology and standards to the RBOCs, and is funded by them. Formerly Bell Labs.

Bonding: The generic term for any of several protocols (such as Multilink-PPP) that allows a data signal to be inverse multiplexed (IMUXed) over several physical circuits, to increase the effective bandwidth of a signal path. (Originally, BONDING was an acronym for a specific IMUX protocol, but this protocol is no longer widely used, and the term has become generically applied to all IMUX applications.)

BRIC: Broadcast-Reliable Internet Codec. A proprietary system from Comrex that fortifies an audio streaming signal for more reliable transport over the public Internet.

Carrier: In telco parlance, a multiplexed digital interoffice signal, containing many individual calls or signals in a single cable or fiber.

CCITT: Consultative Committee of International Telephone and Telegraphy. The international standards setting organization for telephone systems established by the UN.

CO: Central Office. The generic name given to a telco's switching and service center, where all the telco circuits in a given physical area are terminated.

Codec: Coder/decoder. Any device that includes digital transmission/encoding and reception/decoding circuitry in the same chassis.

Contribution channel: A digital audio channel used between an outside origination site and the first permanent broadcaster facility the signal reaches; in other words, the circuit used from a remote site to a broadcast operations center or radio station's studio.

CPE: Customer-provided equipment. Any network interface hardware not provided by telco.

CSU: Channel Service Unit. Terminal hardware for a telco data line, either CPE or telco provided. Also referred to as CSU/DSU (DSU=Data Service Unit) in T-1 applications. Interfaces unipolar NRZ computer style datastreams to the RZ bipolar (AMI) telco data format. A switched CSU includes a keypad for call direction and other switch control.

D-channel: In ISDN service, a channel designated for signaling data only.

DDS: Dataphone Digital Service. The first telco data service in the United States, originated in the mid-70s by AT&T.

Distribution channel: A digital audio channel used to deliver finished programs from a broadcast operations facility to transmission facilities that will deliver it to listeners, such as a satellite feed from a broadcast network to its affiliates, or an STL.

DOCSIS: Data Over Cable Service Interface Specification (v 1.0, 1.1, 2.0, 3.0), an asymmetrical, high-speed data service deployed by Cable TV operators to offer broadband Internet access over their coaxial cable networks.

DS0: Digital Service 0. A 64 kbps data channel.

DS1: Digital Service 1. A 1.544 Mbps data service usually configured as 24 DS0 channels plus an 8 kbps sync channel.

DS2: Digital Service 2. Four DS1 channels multiplexed together for transmission. Generally reserved for telco interoffice transmission, and not offered to customers directly.

DS3: Digital Service 3. Twenty-eight DS1 channels multiplexed together with additional control data, providing a data rate of 44.736 Mbps (generally quoted as 45 Mbps). Used for compressed NTSC and high-definition video distribution.

DSL: Digital Subscriber Line. See also *xDSL*.

Emission channel: A digital audio channel used to deliver a broadcast signal to end users such as IBOC, DAB, or SDARS services.

EV-DO: Evolution-Data Optimized (or Evolution-Data-Only). A format for broadband wireless data service via PCS. (Also known as 1xEV-DO.) The Revision A (EV-DO Rev A) of this format includes broadband upstream capacity and added robustness for high-quality streaming media delivery.

First-mile: The signal path between a program's origination site and its entry point to a common carrier's network or a private satellite uplink. Usually a terrestrial RF link or a local telco circuit.

G.722: A CCITT standard for audio data compression. It uses two sub-band ADPCM coding to put 7.5 kHz audio into 64 kbps.

HDSL: High bit rate Digital Subscriber Line. A digital service that uses telcos' existing copper (twisted-pair) infrastructure to provide high-speed bidirectional connectivity over limited distance. Adaptive terminal hardware allows higher speed than previous equipment was capable of on the same physical paths. Bidirectional rates of 6 Mbps or higher are possible. The smart hardware allows quick installation without extensive testing and tweaking of lines. HDSL is replacing T-1 service in some areas.

IMUX: Inverse multiplexing. See also *Bonding*.

InterLATA: Telco service or rates between LATAs, or long-distance service.

IntraLATA: Telco service or rates within a LATA, or local service.

Inverse multiplexing: See *Bonding*.

IP: Internet Protocol. The networking layer used in the Internet; serves as an intermediate layer between the Data Transport layer (such as TCP or UDP) and the Network Link (physical) layer (such as Ethernet or WiFi).

IP Audio: Streaming audio packets encapsulated into an Internet Protocol (IP) datastream.

ISDN: Integrated Services Digital Network. A new telco service designed to eventually replace POTS with flexible digital service. It will be offered in basic rate (2 B+D) service, intended for home use, and primary rate (23 B+D) service for business customers.

ISP: Internet Service Provider.

J.41: A CCITT standard for digital audio encoding. Using 1411 PCM encoding (14 bits for lower-level signals, 11 bits for higher-level signals), it places 15 kHz audio on 384 kbps.

Last-mile: The short haul signal path between a long-distance network terminal point (or private satellite downlink) and the customer's receive point. Usually a local telco circuit.

LATA: Local Access and Transport Area. The service area of a local exchange company (LEC).

LDS: Long-Distance Service. A carrier of long-distance (interLATA) telecommunications, such as AT&T, MCI, Sprint, and others.

LEC: Local Exchange Company. A local telco. Each RBOC contains one or more LECs. Also refers to independent, non-Bell local telcos.

Mark: The telco term for high-level data pulse, usually corresponding to logical 1. (See also *Space*).

MIMO: Multiple Input/Multiple Output. Antenna technology allowing increased data throughput and extended range on wireless links via use of multiple antennas for spatial diversity and spatial multiplexing.

Naked DSL: A DSL line provided with broadband data only, and without the POTS service included in telco's standard DSL offerings. Also called dry or stand-alone DSL.

NRZ: Non-return zero. The most basic form of binary modulation coding, in which logical 1's and 0's are directly represented by high and low levels, respectively. Because no level transition occurs between continuous strings of like logical values, this form of modulation is not self-synchronizing and requires an external bit clock output for synchronous operation.

OC: Optical Carrier. Specifies the speed of fiber-optic networks in the SONET standard. Base rate (OC-1) is 51.84 Mbps. Common multiples used by telcos internationally include OC3, -12, -48, and -192.

Packet switching: A sort of data partyline, in which data is transmitted in addressed bursts or packets, occupying the transmission channel only for the duration of the packet, after which the channel is free for other packets to or from the same or other users. Many users can be interconnected to the same line, but data can be sent discretely to each destination.

PDN: Public Data Network. Telco data services, including both switched and leased lines.

POTS: Plain Old Telephone Service. Refers to the Public Switched Telephone Network (PSTN).

PSTN: Public Switched Telephone Network. The standard dial-up phone system.

QoS: Quality of Service. The ability of a network to guarantee a specified level of accuracy and/or timeliness in its signal delivery.

RBOC (or BOC): (Regional) Bell Operating Company. The seven *Baby Bells* created when AT&T divested itself of its local telephone operations.

RZ: Return zero. A form of digital modulation coding in which logical 1's and 0's are directly represented by high and low levels, respectively, but where coding output returns to low level following each high pulse.

SDSL: Symmetrical Digital Subscriber Line. A form of DSL in which upstream and downstream paths operate at the same data rate (see also *ADSL*).

Slot (or Time-slot): Generally refers to a DS0 channel within a DS1 signal.

SONET: Synchronous Optical Network. An international standard for connecting fiber-optic transmission systems. SONET establishes Optical Carrier (OC) levels from 51.8 Mbps (OC-1) to 9.95 Gbps (OC-192).

Space: The telco term for low-level data pulse, usually corresponding to logical 0. (See also *Mark.*)

Switch: Generic name for any telco call routing and connection hardware.

Switched 56: A switched digital service offering 56 kbps data service on a dial-up network, generally no longer available in most areas, having been largely replaced by ISDN.

T-carrier: See *Carrier.*

T-1: The copper network and hardware used to carry DS1 service.

TA: Terminal Adapter. The terminal equipment ("telephone") used at the customer's premises in ISDN service.

Tariff: A schedule of services and their prices that a telco will provide to a given service area, subject to approval by the appropriate regulatory agency.

TCP: Transmission Control Protocol. One of the two most commonly used Transport layer protocols in the Internet. Serves as an Intermediate layer between the IP Networking layer and the Application layer, such as HTTP. (See also *UDP.*)

UDP: Universal Datagram Protocol. One of the two most commonly used Transport layer protocols in the Internet. Serves as an intermediate layer between the IP Networking layer and the Application layer, such as RTP. UDP uses less overhead than the other common IP Transport layer, TCP. Therefore, UDP provides less reliability than TCP, but faster operation for time-critical applications, so UDP is preferred for streaming media transport.

VDSL: Very high bit rate Digital Subscriber Line. A digital service that can be configured either symmetrically or asymmetrically; in the latter form it can provide a maximum of 52 Mbps downstream and 12 Mbps upstream over a single twisted pair of wires. (See also *HDSL.*)

VSAT: Very Small Aperture Terminal. Refers to Ku band satellite earth stations for fixed or portable use with dish diameters on the order of 1.5 m or less.

V.35: An older CCITT telco standard for low-speed data I/O to a CSU, with a unique multipin connector.

WiFi: See *802.11x.*

WiMAX: See *802.16x.*

xDSL: Generic terminology for any of the variety of Digital Subscriber Line services offered by telcos, which provide high-speed connectivity across limited distances on the existing copper (twisted-pair) infrastructure. (See also *ADSL, HDSL, SDSL, VDSL.*)

SECTION

4

RADIO TRANSMISSION

CHAPTER

4.1

Planning Radio Transmitter Facilities

W.C. ALEXANDER
Crawford Broadcasting Company
Denver, Colorado

JOHN M. LYONS
The Durst Organization
New York, New York

INTRODUCTION

The ultimate success or failure of any broadcast transmitter site is often determined during the planning process. Choices made in the planning stages have considerable leverage over the long-term viability and operation of the site. Factors such as size; layout; electrical power; heating, ventilation, and air conditioning (HVAC); radio frequency (RF) screening, and many others must be considered early. Regulatory matters such as local tower regulation, zoning, and building codes must also be researched before committing to a particular locale. In this chapter, each of these areas and more will be examined and discussed at length with the ultimate goal of producing a transmitter site that works well both now and in the future.

DEFINING THE APPLICATION

The initial planning of any transmitter site must begin with a clear set of objectives. Consider the following questions:

- Is the site for AM or FM, or both?
- Is it a stand-alone or shared facility?
- Will an AM antenna be directional or nondirectional?
- Is the area within two or so miles already home to other AM antenna sites?
- Are any potential reradiators such as powerline towers, cell towers, or elevated water storage tanks nearby?
- For FM sites, will there be a main and auxiliary transmitter?
- Will there be main and auxiliary antennas?
- Will they be on the same tower or building or at different locations?
- What about studio-to-transmitter links (STLs) and intercity relay links (ICRs)?
- Is HD Radio™ operation planned?
- Will common amplification, high-level combined or separate, dual-input antennas be employed?
- Will tower space be rented to other broadcasters or nonbroadcasters?
- Will tower space be rented from a tower site owner or another broadcaster?
- Will a master antenna/combiner be used?
- Will the new facility be an upgrade or a replacement?

Once the objectives have been defined, and the above questions answered, other variables can be considered in light of those objectives.

FACILITY DESIGN ISSUES

The complexity of the installation defines the intricacies of the design. Issues that cause major delays in implementing the project in a multi-user site (Figure 4.1-1) or in an urban area may not be issues in a stand-alone or suburban installation (Figure 4.1-2). Some items are concerns of both types of installations and are addressed here.

FIGURE 4.1-1 Consolidated single owner multi-station space. Left to right: WWPR, WLTW, WAXQ, WHTZ, and WKTU (Clear Channel, New York). (Photograph courtesy of John M. Lyons.)

FIGURE 4.1-2 Single FM station equipment space (WNYC-FM, New York). (Photograph courtesy of John M. Lyons.)

Often overlooked as a major construction factor at the start of a project is the weather. The effect of adverse weather on the schedule and costs can be considerable. Once the project starts, all the trades (crafts) are on the clock, and trades that are affected by delays will probably seek compensation for the downtime.

Hiring a tower crew is another area for advance consideration. Good tower crews are usually booked many months in advance. The crew of choice must be locked in for the project, taking into account the anticipated on-air date. Written contracts should be obtained from the tower crew and antenna company early in the project planning (Figure 4.1-3).

Tower crews are usually busy. Building in northern climates must take into account the short construction season and antenna construction schedules. Building in southern areas raises other construction concerns, such as rainy periods and hurricanes. The hurricane season stretches over 6 months, with the peak occurring between August 15 and September 15.

Complying with the local zoning process and obtaining building department approvals and permits are areas often overlooked with respect to the amount of time required to complete the procedures. Also, construction moratoriums, community or citizen objections, and other unforeseen or unanticipated events can affect the construction schedule and have an impact on the construction budget.

Construction that takes place in a preexisting, multi-user site has the added factor of the lease or license agreement. Permitted hours of work and elevator usage (in large buildings) are two of a myriad of building rules and regulations that must be followed in this environment.

Permitting, along with zoning, is a time-consuming process that can involve any of the following agencies or issues:

- Federal Aviation Agency (FAA)
- Environmental issues dealing with use of wetlands
- Federal Communications Commission (FCC)
- Previous usage of the location
- Presence of hazardous materials

In addition, it is important to research the property fully for liens or covenants on the deed that could affect the usage of the site for the intended transmitter facility. Because all of these issues must be considered regardless of the location, it is often wise to commission a Phase I environmental study of the site very

FIGURE 4.1-3 Antenna mount/TX line installation with joint iron worker/electrician crew (4 Times Square, New York). (Photograph courtesy of John M. Lyons.)

early in the project. Such a study will often uncover these and other issues.

Certain long-lead items, such as the transmitter, tower, antenna, and mechanical and electrical systems, may have to be purchased well in advance, even before the transmitter facility design is completed, in order to have the equipment arrive on site at the proper time. These items generally have a turn-around time of 4 to 6 months from the time of order and down payment to on-site delivery. An important note is that the date of arrival on the job site is the determining date for the order, not the date that the specific items ship from the manufacturer. Transcontinental or overseas shipments may take longer than expected and could throw off a tight schedule if they are delayed en route. For a more detailed look at many of these issues, refer to Chapter 6.1, "Planning TV Transmitter Facilities."

ASSEMBLING THE DESIGN TEAM

The design team may consist of very few people for a single user, low power transmitter facility design, or it may include architects, consultants, engineers, and managers in a large or multi-user design.

Choosing a Builder

The low bidder may not be the best choice when it comes to constructing a transmitter facility. Discuss the project with the engineers in the market to find a builder with whom they have had a good experience. Check with consultants on this project for recommendations on builders with whom they have had previous experience.

Avoid the use of a residential contractor, as this type of builder usually does not have experience with the structural, mechanical, and electrical requirements of a broadcast transmitter site. In interviews with potential builders, discuss the specifics of the project, especially mechanical and electrical. Determine what subcontractors the builder uses in these critical trades, and ask if they specialize in commercial work. Avoid an electrical contractor that is not familiar with three-phase power, if that is required on the project.

Solicit at least three, preferably five, bids for the project. Allow potential bidders to pick up drawings at least a month before the bid deadline. Schedule a pre-bid meeting and site walk-through 2 weeks before the deadline. Host that meeting with the architect or engineer to answer the bidders' questions. General contractors often will bring their key subcontractors to the pre-bid meeting. The better the information that can be provided at the pre-bid meeting, the more accurate will be the bid and the fewer change orders (and accompanying cost overruns) that will occur during the course of the project.

For new construction, some local regulatory bodies require soil tests prior to issuance of a building permit, and this geotechnical report should be included in the bid instructions. This will give the excavator and foundation contractors an idea of what they are dealing with. For example, if the water table is close to the surface, it may be necessary to pump the excavations during foundation installation. If the soil is very sandy, it may be necessary to install temporary reinforcements to keep the excavations from caving in. In any case, a more accurate bid can be made if all the parties know what they are dealing with.

Plan for a staging area for materials storage (see Figure 4.1-4), as well as a truck entrance and exit (Figure 4.1-5). It is rare that a paved road leads up to a remote building site, and wet weather may make it impossible for trucks and construction equipment to get to the building location. One of the first steps

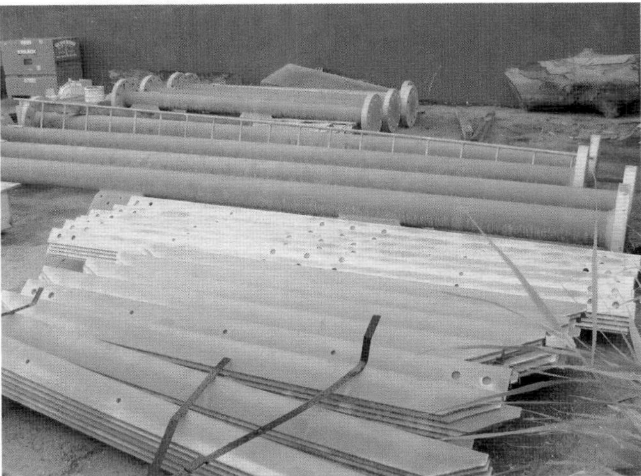

FIGURE 4.1-4 Staging yard for delivery of tower sections to the job site. (Photograph courtesy of John M. Lyons.)

FIGURE 4.1-5 Load-in considerations ensure that the delivery vehicle will fit the loading area. (Photograph courtesy of John M. Lyons.)

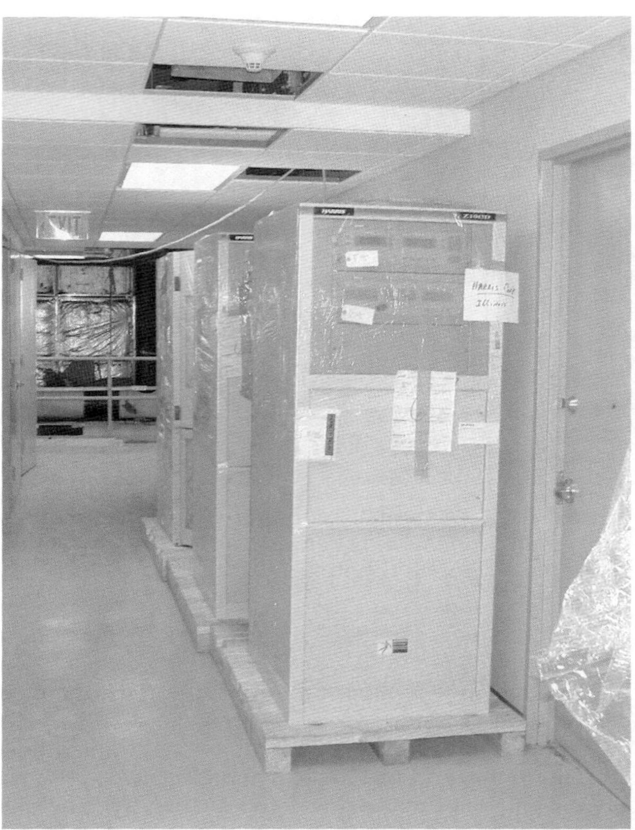

FIGURE 4.1-7 Staging of transmitters during construction. (Photograph courtesy of John M. Lyons.)

should be to prepare site egress by grading and installing gravel and road base or other material to keep the way passable (Figure 4.1-6).

If the building will be at a new or existing multiuser inner-city site, elevators must be scheduled and measured to make certain the equipment will fit and can be maneuvered into and out of them (Figure 4.1-7). Transmission line routes must be planned from the equipment room spaces to the combiner room and other locations in between. Requiring equipment to be broken down into smaller parcels and reassembled on the job site will affect the budget and schedule. This kind of information must be included in the bid instructions.

FIGURE 4.1-6 WOR-AM transmitter site in New York. (Photograph courtesy of Tom Ray.)

BUILDING SIZE AND TYPE

Engineers have a wide range of choices for transmitter building configuration, size, and construction. Lower power, single-station operations may require only a simple frame shelter; however, the greater the transmitter power, the more space that is required. Also, the greater the weight of the equipment, the greater the demands on the structure that must support that equipment (Figure 4.1-8).

In the early days of radio, transmitter buildings were quite often built around the transmitters they housed. This often included elaborate floor troughs, subfloor and overhead air ducts, plenum and air mixing rooms, high-voltage vaults, and other features. With today's modern solid-state and hybrid transmitters, such features may no longer be necessary. Still, careful thought must be given as to the requirements of the equipment as well as external environmental

FIGURE 4.1-8 Placing considerable weight in a space (5 tons across 20 linear feet). (Photograph courtesy of John M. Lyons.)

factors, such as climate, proximity to the tower, and drainage. Modern transmitter buildings generally fall into one of four categories: frame, masonry, prefabricated concrete, or steel.

Frame Structures

A simple frame structure is almost always the easiest, least expensive shelter. Such a shelter may consist of a prefabricated building such as Tuff Shed® or similar product, or it may be assembled from scratch on site. Disadvantages of the frame shelter are poor security, poor fire safety, suboptimum insulating properties, and, if a wood (plywood or oriented strand board [OSB]) floor deck is employed, limited weight-bearing capability. Frame structures are also prone to damage from the weather, such as high winds, falling ice, and heavy snow loads. In some cases, though, a simple frame shelter may be the best choice (particularly for simple, low-power, nondirectional AM or FM installations).

Masonry Block

A popular transmitter configuration is the masonry block building with a slab-on-grade foundation. Such a building can be constructed in virtually any size. It is relatively inexpensive and provides fire safety, security, and a good amount of unobstructed interior space. If desired, wire troughs can be set into the poured slab. The main variables in the construction of such a building are size and roof composition. The same size factors that apply to all transmitter buildings also apply to this type of structure. Roof composition can be flat with steel trusses and a corrugated steel deck, flat with wood trusses and wood deck (plywood or OSB), or pitched with wood rafters and deck. In temperate climates, the flat roof (which should be slightly pitched with scuppers for drainage) works well. Where snow load is a factor, a pitched roof built according to local standards may be a better choice. A third option, which can be used in temperate or more extreme climates, is a flat concrete roof. Concrete roofs work well when the building is situated near the base of the tower, where ice shedding may otherwise puncture a wood or steel deck roof.

Prefabricated Concrete

Another popular option is the prefabricated concrete building. This type of building was popularized by the cellular telephone industry, where rapid development and deployment of sites with a minimum of on-site construction were standard practice. Prefab concrete buildings are, as the name implies, fabricated at a factory, prewired, factory fitted with a self-contained HVAC system, and then shipped to the location by flatbed truck. At the site, the building is unloaded by crane and placed on a bed of gravel or rock or, alternatively, on a concrete foundation. Because the roof is also concrete, placing such a building at the base of a tower prone to shed ice is not generally a problem.

The advantage of this type of building is that it considerably shortens set-up time for the transmitter site. The engineer employing such a building will not have to arrange for and coordinate the various trades (electrical, mechanical, etc.). Most of that work will have been done at the factory. Such a building can be ordered with electrical drops in place for transmitters, racks, and other equipment so the engineer can simply roll the transmitter to its installed location and connect the power. Similarly, mechanical provisions can also be ordered factory-installed so transmitter intake and exhaust can easily and quickly be ducted to precut louvered openings. Halo-type grounding comes standard in many such buildings, as does waveguide egress with a grounding block (Figure 4.1-9). The disadvantage of this type of building is that there are some limitations on size and configuration. Larger buildings can be shipped in separate halves or parts and assembled on site, but the size of the sections is limited by what can be shipped over the public highway system by truck. Overall, however, the prefabricated concrete building is an excellent value and a good choice for a radio transmitter shelter.

FIGURE 4.1-9 Six-port transmission line wall sleeve. (Photograph courtesy of John M. Lyons.)

Steel

In years past, steel offered an economical choice for large transmitter buildings. Such structures generally feature a steel I-beam frame with corrugated steel roof and siding. Because of the structural strength of the frame, large spans are possible in such a building, creating roomy spaces unhindered by support columns. This configuration is ideal for large transmitters, antenna phasors, and combining equipment. With steel prices increasing, however, the price of such construction is less attractive than it once was. Transmitter equipment is also much smaller than in the past. Long the "big rigs" of the broadcast transmitter world, 50 kW AM transmitters are now not much larger than a good-sized refrigerator. It is thus rare that the large uninterrupted spaces afforded by steel buildings are needed. Although steel buildings by nature provide some measure of RF shielding, the overlapping panels are seldom bonded electrically and, over a period of time, the resulting nonlinearities in the joints are prone to produce rectification of substantial RF currents flowing therein. In short, the steel structure, with its metal siding and fascia, can often cause more RF-related problems than it prevents.

Starting from Scratch and Determining Building Size

The required size of a transmitter building is quite often a deciding factor in the type of building. Beginning with a blank page or an empty piece of land, the engineer has the advantage of designing a building to fit the application instead of fitting the transmitter installation into an existing building. As such, the best course of action in the initial design process is to start with the ideal equipment layout, drafted without regard to the eventual size or type of building. Computer-aided design (CAD) programs are particularly well suited to this type of work (Figure 4.1-10).

Start with the transmitters, and add the ancillary equipment at the most advantageous position, then work through the transmission lines, combiner, phasor, and other equipment—all the way to the transmission line egress and tower. Think through required operator and engineer stations, and position equipment for the most efficient human interaction with the equipment; for example, in directional AM installations, it is necessary to adjust the phasor controls while observing the antenna monitor. Situate the equipment rack containing the antenna monitor close to and at a right angle to the phasor so the monitor is easily visible when the operator's hands are on the phasor controls.

One aspect often overlooked in the early design phase is the area over the working space where transmission lines, conduits, ductwork, and utilities are most often located. It is important to think through these factors early in the design process, not only for planning the elevation but also for dealing with equipment location and spacing. Manufacturers of transmitters, phasing and coupling equipment, and other gear can provide the height of the equipment as well as ducting requirements, electrical connection specifications (location and size), transmission line connections, harmonic filter mechanical specifications and other essential information. A large external harmonic filter, for example, may preclude locating a transmitter immediately adjacent to a wall or another transmitter. The same may be true because of ductwork and transmission line connections.

In the case of phasing and coupling equipment, virtually all of which is custom manufactured to the user's specifications, the design engineer has the option of specifying the transmission line and control cable egress location. The two most common locations are the top and the underside. This should be considered carefully in the design process. Unless the building has a crawl space or basement, it is often considered convenient to specify an underside egress, which keeps the overhead clear for other items; however, this egress location will make later replacement of the phasor difficult. A better option is often a top egress, which requires considerable planning to deal with transmission line routings but makes replacement of the equipment much easier. Either way, the phasor transmission line egress has considerable bearing on the design of the building.

There are many ways to partition the interior of a transmitter building. One way, popular in times past, was to put the transmitters and other equipment in a line and construct a partition around the equipment, placing the front (operating position) of the equipment in one room and the back (service area) in another (Figure 4.1-11). With such an arrangement, it is possible to install a drop ceiling with lighting and HVAC supply and return registers in the front room, providing a well-lighted, relatively quiet operating position. The back room would then have a high (or unfinished) ceiling allowing for transmission line, electrical, and mechanical overhead runs, with the partition wall confining the bulk of the mechanical noises to the nor-

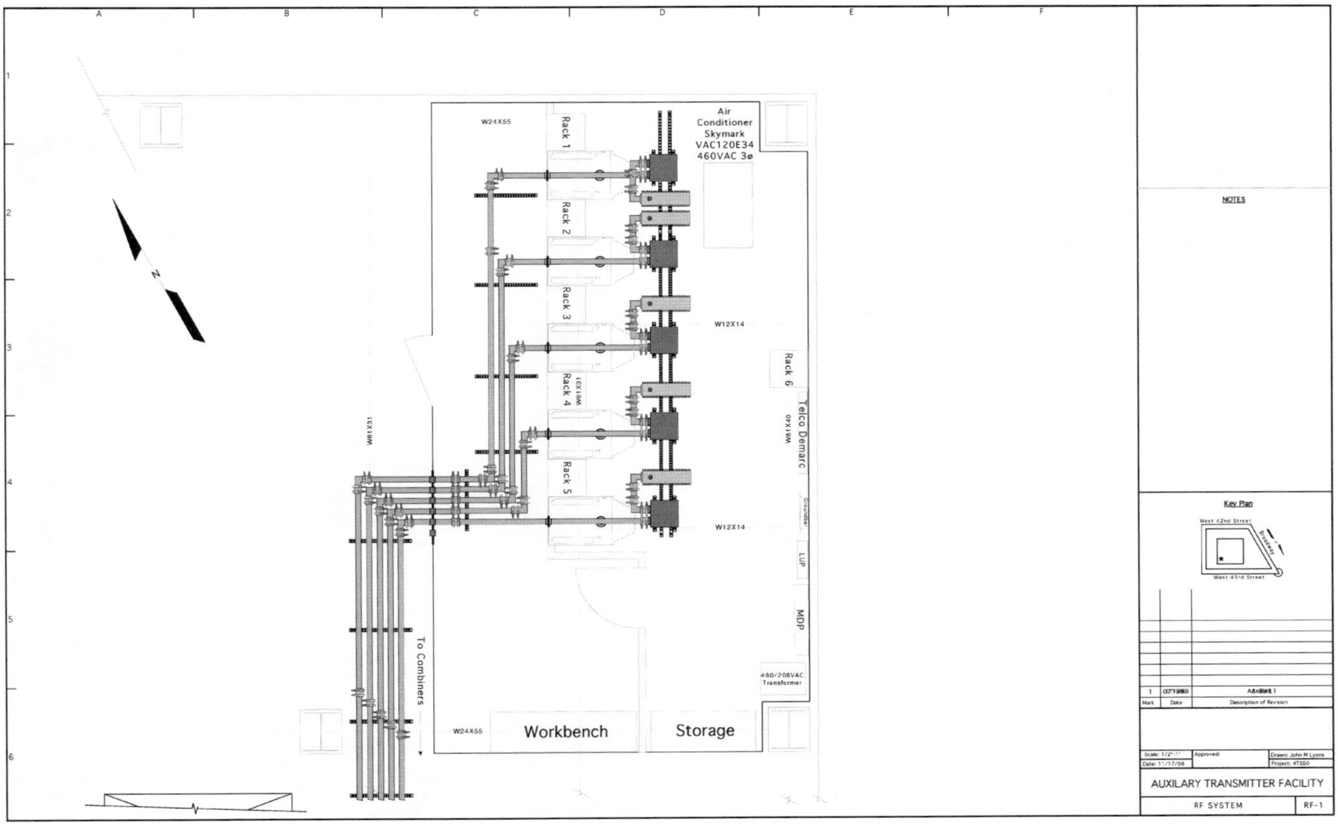

FIGURE 4.1-10 Computer-aided design (CAD) drawing.

mally unoccupied back room (see Figure 4.1-12). This method is still a popular construction method in transmitter sites where large, noisy transmitters are used. A disadvantage is that in installations with a long line of equipment, it can be a long walk from the front of the equipment to the back. Another is that some demolition and reconstruction are necessary to replace a transmitter, phasor, or rack.

FIGURE 4.1-11 Framing common space for multiple stations. (Photograph courtesy of John M. Lyons.)

Another scheme uses no partitions, and all of the equipment is located in the center of a single large room. The room or building is sized to provide 36 to 48 inch aisles all the way around, providing good access to every part of the equipment. The HVAC is simply ducted into the room, often from a single point of entry containing both supply and return and with no HVAC ductwork within the room or building. Hanging or ceiling-mounted fixtures both in front of and behind the equipment provide lighting. As transmitters have become increasingly solid-state with lower required air volumes and reduced operating mechanical noise, this method has become increasingly popular. It is the easiest to construct and generally has the lowest cost. Because of the increased noise in this kind of installation, it is generally more suited to remote controlled facilities.

Whatever partition configuration is decided upon, it is essential to provide for storage for spare parts, test equipment, and other items that must be securely kept at the site but not necessarily in the open. An area must be provided for electrical panels, generator transfer switch, and other electrical systems. The National Electrical Code requires a 3 foot clearance from the front surface of such equipment and 30

FIGURE 4.1-12 Placing coaxial switches and dummy loads above transmitters for better space utilization. (Photograph courtesy of John M. Lyons.)

FIGURE 4.1-13 Planning code requirements for electrical equipment. (Photograph courtesy of John M. Lyons.)

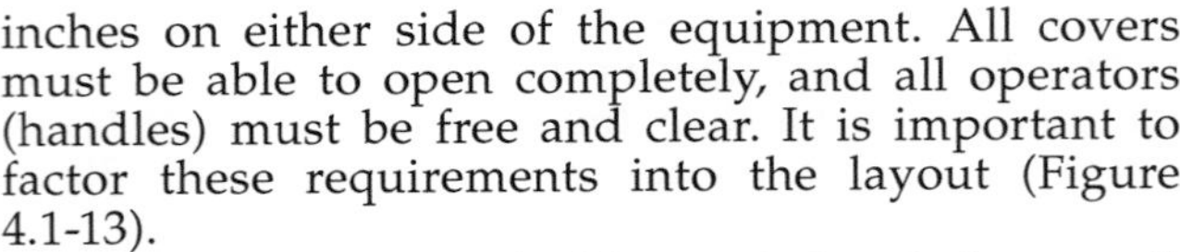

inches on either side of the equipment. All covers must be able to open completely, and all operators (handles) must be free and clear. It is important to factor these requirements into the layout (Figure 4.1-13).

With a satisfactory ideal layout in hand, the overall size of the building can be determined. This is an excellent time to employ an architect or design/building firm to help make carefully considered choices in building size and configuration. For example, the ideal layout that has been designed may come just a little short of fitting into a particular size of structure, thus requiring mid-span supports for roof beams. By making small changes to the layout, a smaller unsupported span might be possible, which can reduce costs and simplify the layout. A design professional can help with such decisions and avoid costly mistakes, optimizing the overall design.

MECHANICAL PLANNING

Important factors to consider in planning the mechanical systems in a radio transmitter facility include the heat loads produced by the equipment (primarily the transmitters, but also the ancillary equipment) and the ambient heat load of the building from solar heating and other non-equipment-related sources. The cooling system must be sized to maintain a safe temperature inside the building on the hottest day with transmitters operating at full power. Keep in mind that the safe temperature is a few degrees below the maximum rated operating temperature of the most sensitive piece of equipment in the facility. With many CPU-equipped devices now common in transmitter equipment, periods of 120°F temperatures are no longer tolerated.

HVAC Systems

If the facility involves a tube-type power amplifier (PA) with a significant amount of air flow, a decision must be made whether to operate the transmitter in a closed-loop system or to dump the hot PA exhaust directly to the outside. The closed-loop option has the advantage of providing for a very clean installation, as it minimizes the introduction of outside contaminants such as pollen, dirt, and insects. The building can effectively be sealed or slightly positively pressurized. The disadvantage is that much more cooling power is required to overcome several tons of heat load from

the PA. This requires a much larger air conditioning system and will result in significantly higher operating costs.

Dumping the hot PA exhaust outside has the advantage of requiring much less cooling (only the radiation or convective heat load of the transmitter must be overcome as opposed to that plus the PA heat load). It has the disadvantage of requiring either venting make-up air into the room, thus introducing a significant volume of air at outside ambient temperature into the space, or ducting an outside air supply into the PA intake. Generally speaking, ducting outside air into the PA intake is a better option, minimizing cooling load on the rest of the building. Either way, the outside air source must be well filtered and the filtration checked regularly.

Such a system must also be made fail-safe, by providing a barometric damper on the intake and exhaust ducts inside the building to maintain airflow should the outside intake or exhaust vents become blocked. It is also necessary for wintertime venting that part or all of the PA exhaust goes into the transmitter room for "free" building heat.

Solid-state transmitters, both AM and FM, generally utilize low-velocity, high-volume cooling air, generally provided by many small (8 inch or smaller) fans strategically located within the transmitter. Warm exhaust air is generally vented out the top or front of the cabinet. These transmitters are designed to be operated in a cooled room or building with plenty of air flow within the room. One critical factor in such installations is that the HVAC system must be fail-safe. This can be achieved either by dual units, both of which have the capacity to adequately cool the facility, or by an emergency ventilation system that is operated by a mechanical thermostat and circulates a large volume of outside air through the building when the inside temperature exceeds a preset value. Either system may require a transmitter power reduction to reduce heat load which can be accomplished manually by an operator upon observation of a temperature alarm or automatically. Some transmitters have built-in thermal protection that reduces power when a high internal temperature is detected.

A wide variety of choices of air conditioners are available for transmitter sites. Some installations may call for a pad-mounted condenser unit and ceiling-mounted evaporator. Others may call for all-in-one, roof-mounted units that require a single roof penetration. One option popular with cellular facilities is the self-contained end-mounted unit that is easy to install and requires just a single wall penetration for supply and return. Because of their relatively small size, two units can easily be mounted side by side and operated in tandem in a lead–lag configuration.

Advice from a mechanical engineer can be of great help in designing a trouble-free HVAC system—one that does a good job of keeping the building and equipment at a safe temperature at all times while minimizing construction and operating costs.

ELECTRICAL SYSTEM PLANNING

There is often much more to planning the electrical system in a transmitter building than for any other similarly sized commercial installation. Employing an electrical engineer to assist in this part of the design (and some venues may require it) may be useful, but as is the case with the space planning, the initial design is up to the station engineer. Begin with the overall power requirements. First, for high-power transmitters, the use of 480 volts will yield considerable savings on conductor and switch-gear size, but a step-down transformer will be required for the 240/208 and 120 volt loads.

Overall peak power consumption of the site must also be calculated. Examples include:

- The peak power consumption of each transmitter at full power with modulation
- The peak power consumption of the HVAC system
- The peak power of ancillary electrical loads, such as equipment racks and lighting
- The peak power of tower lighting (a tower with two beacon levels and two obstruction levels could have a peak power draw of close to 4 kW)

It is generally accepted that a "wye" power system is superior to a "delta" with respect to lightning and surge immunity and radiofrequency interference (RFI) immunity, because each of the three legs has a fixed reference to ground. For this reason, a 480 or 208 volt wye system is recommended whenever possible in radio transmitter facilities. If a standby generator (see Figure 4.1-14) or uninterruptible power supply (UPS) will be employed, either initially or in the future, allowances should be made in the design, even if the generator or UPS is years in the future.

FIGURE 4.1-14 Standby power generator. (Photograph courtesy of John M. Lyons.)

Grounding Systems

The heart of any effective lightning protection scheme is a central ground system. Some call this a "star" grounding scheme because of the way all the ground conductors return to a central point or reference ground. If the transmitter building is located near the tower, this ground can be the same as that for the tower itself. In most cases, however, there will be some distance between the tower and transmitter building, and in those instances another array of ground rods should be provided. While such an array can be added after the fact, the installation will be simpler and less costly if allowances for grounding are made at the design stage. (See Chapter 9.3, "Facility Grounding Practice," for more information on grounding systems.)

The best place for the array of grounding rods is on the tower side of the building. At least four or more rods, long enough to reach below the deepest frost line into the water table, should be employed. The rods should be placed a distance of at least two to three times their length away from one another and should be joined together with bare copper wire at least 1/0 in size. Cad welding is the preferred method of connection, as mechanical clamps do not provide a joint with low-enough resistance. A cad-welded joint will not oxidize or corrode as a mechanical clamp junction is prone to do.

All conductors operating at ground potential that enter or leave the transmitter building should be bonded to this ground rod array. That includes the outer conductors of all transmission lines. A conductor from the ground rod array should be brought into the transmitter building via the shortest and straightest route possible. The point where it enters the building becomes the center of the "star," or the point to which everything in the building is grounded. This is referred to as the *station reference ground*. All grounds in the building, including the safety ground of the electrical system (service entrance ground) and the ground conductors from all the equipment and outlets are connected to this point.

Important note: It is wise to leave some room for expansion wherever a ground rod or conductor penetrates a concrete slab or wall. During a lightning strike, such a conductor may well carry thousands of amps for a few milliseconds. Considerable heating and expansion will occur during the transient, and if there is no room for the conductor to expand (up to twice its nominal size) the surrounding concrete may crack, crumble, or explode. A good practice is to pass the ground wire through the slab or wall in an oversized conduit or through a piece of PVC conduit. Expansive foam can be used to seal around the rod or conductor.

SHARED USE FACILITIES

Collocation is a word often used with regard to communications towers and facilities. There is often considerable community resistance to the construction of new towers, and a solution often sought is to collocate facilities. This means that broadcast facilities, even AM stations, sometimes must consider sharing a site with another broadcaster. It also means that broadcast towers are often sought as homes for non-broadcast uses (PCS, paging, two-way, and other communications services; see Figures 4.1-15 and 4.1-16). Such site sharing offers considerable revenue potential. If there is even a remote possibility that site sharing may someday be considered, making allowances for such use at the design stage will pay substantial dividends later.

Designing for the collocation of services is highly dependent on what the anticipated shared uses will be. If, at a high-power FM site, it is anticipated that another high-power FM station may someday move in, all of the above-discussed layout, space, electrical, HVAC, and other considerations will have to be multiplied by as much as, if not more than, two. If each station is to have its own private space, room must be allowed for a dividing wall to establish boundaries for each facility. If both stations are to share one room, some allowances should be made for each station's private storage (such as spare parts and test equipment). Be sure to allow room for additional transmission line egress to the tower (Figure 4.1-17).

FIGURE 4.1-15 Multi-user, multiple-antenna installation. (Photograph courtesy of John M. Lyons.)

FIGURE 4.1-16 Shared-use facility main tower (shown under construction) for FM/TV broadcasters; ancillary tower with four levels for STL, electronic news gathering (ENG), microwave, etc. (Photograph courtesy of John M. Lyons.)

FIGURE 4.1-17 Spacing transmission lines in a multiuser site for ease of removal or change out. (Photograph courtesy of John M. Lyons.)

Other issues include:

- Will a common electrical feed be used, or will each station be metered separately?
- Will one HVAC system be used for both stations, or will each station use its own system?
- Will a combiner be used with a shared antenna?
- How much space is required for the combiner, and how much heat will it produce?

If there is the possibility that nonbroadcast users will be added at the site, consider a separate "communications suite" with its own entrance, HVAC system, and transmission line egress/grounding bar. This will keep the PCS, paging, and two-way technicians out of the broadcast transmitter space. Using a code lock on the communications suite entry door makes it relatively easy to manage tenant access. The floor of the suite can be marked off in numbered 4 square-foot bays with a 2 foot buffer between each bay. Tenant space is assigned by bay number, and tenants with larger space needs can be assigned (and charged for) additional bays.

Keep in mind that poor grounding practices by nonbroadcast users can allow lightning to enter their equipment and damage broadcast equipment. Make it easy for nonbroadcast users to properly ground their transmission lines and equipment to the site's central grounding system by:

- Providing a grounding bar at the transmission line egress
- Providing a ground bus around the suite with several connection points at each bay
- Requiring users to properly ground transmission line and equipment cabinets
- Requiring toroid cores to be placed on all transmission lines just inside the building on the building side of the grounding bar (Figure 4.1-18)

FIGURE 4.1-18 Grounding bar, showing connections. (Photograph courtesy of John M. Lyons.)

FIGURE 4.1-19 Urban installation of tower atop office building; hoisting of transmission lines 700 feet above the street. (Photograph courtesy of John M. Lyons.)

FIGURE 4.1-20 Combiner system for multiple FM users showing 5 combiner modules in place in a system that will handle 20 stations. (Photograph courtesy of John M. Lyons.)

Building at an Existing Multi-User Site

Existing multi-user sites present challenges that are not present in single- or two-user sites. Most of the FM installations in major markets are atop tall office buildings (Figure 4.1-19), such as the Empire State Building and 4 Times Square in New York, the Sears and Hancock buildings in Chicago, the Prudential in Boston, and others. These sites were generally designed for the seamless installation of additional stations. The combining systems are designed for expansion, and space has been planned to accommodate the transmission equipment required by the additional broadcasters (Figure 4.1-20). Some systems allow the incoming user to simply become the final station in the combiner chain, while others require insertion in a frequency-specific order, thus raising the need for additional transmission lines and downtime for all users during the installation and testing phases. These factors must be considered in the budgeting and scheduling of the on-air date for the facility.

Leases and License Agreements

Most leases and license agreements for multi-user sites stipulate that the newest user is responsible for any interference caused by that user's addition to the system, and the agreements generally specify all frequencies in use at the facility before the additional station arrived. In this way, the ongoing operations of stations previously operating at the site are protected, and the full responsibility for interference lies with the newcomer. Remedies include a shutdown of that new operation or a reduction in power to a level that eliminates the interference until the cause can be ascertained and corrected. Multi-user AM sites differ in that they are generally spread over a considerable tract of land, sometimes with interleaved guy wires. The issues here include interference with the operational surrounding stations and a host of other issues.

SUMMARY

Careful advance planning cannot be overemphasized in developing and building a successful radio transmitter facility. Decisions made early during the design phases have great leverage over the project as a whole, with the effects extending 30 or more years into the future. Carefully consider all the aspects of the project, present and future. Carefully choose a design team

and contractors, and plan how the various trades will be employed. Pay close attention to local regulatory requirements, leases, license agreements, and site rules and regulations. Through careful and thoughtful planning, most surprises can be avoided.

ACKNOWLEDGMENTS

John Lyons would like to acknowledge the instruction of Richard Lambeck, P.E., Professor in the Graduate Construction Management Program at New York University.

CHAPTER

4.2

AM Transmitters and AM Stereo

DANIEL L. DICKEY
Continental Electronics Corporation
Dallas, Texas

EDWARD J. ANTHONY
Quincy, Illinois

INTRODUCTION

This chapter combines the former "AM Transmitters" chapter, written by Daniel L. Dickey, with the "AM Stereo" chapter from the 9th edition, written by Edward J. Anthony and revised by Thomas Osenkowsky, which has been retained and placed at the end of this chapter.

The earliest form of wireless audio broadcasting uses the technique known as *amplitude modulation* (AM). As other forms of broadcasting compete with AM in the marketplace, AM broadcasting has evolved by adopting stereo capability, more aggressive modulation techniques, and more recently digital modulation formats. These digital formats are better able to exploit many of the advantages of the frequency bands traditionally assigned to AM broadcasting.

The advent of digital radio broadcasting is blurring the once distinct lines that existed between the different forms of radio. In the past, it was easy to segment broadcasting by the modulation method employed. The medium and low frequencies, below 30 MHz, are natural candidates for AM because AM is spectrally efficient, requiring a bandwidth only twice the highest audio frequency transmitted. Digital modulation creates a dilemma in classification because no longer is the modulation method purely amplitude or phase. For the purposes at hand, this chapter continues to refer to AM transmitters as those employed in the bands below 30 MHz, even though the transmission may be AM plus digital or entirely digital.

Digital radio is arguably one of the most disruptive events since the introduction of AM; therefore, it is as important as ever that detailed technical information be made available to the broadcast engineer and the many other stakeholders in AM radio. The genesis of AM broadcasting is found not in the field of transmitter technology but rather in receivers. Beginning in the late 1800s and into the early 20th century, the only receiver technology that allowed detection of information via radio waves was AM. Because it was simple to convert the amplitude of a received signal into audio, the first broadcasts to the public were necessarily AM. Then, as now, receiver complexity and therefore cost were the primary drivers behind selection of the modulation technique. The same will be true for any new digital modulation method for AM broadcasting bands.

In the mid-1920s a second primary modulation technique known as angle modulation was proposed [1]. Today all modulation formats, including digital, can be described mathematically through their instantaneous amplitude and phase (or angle) of the assigned carrier frequency. Because nearly all modern AM transmitters can be both amplitude and angle modulated, they often can be adapted to newer digital formats.

AMPLITUDE MODULATION THEORY

The early AM transmitters for long distance communication involved on/off keying of a radio carrier wave. The pattern or *code* of the on/off keying process determined the content of the transmitted information. In the case of audio broadcasting via AM, sound waves are converted to analog electrical waves. These audio frequency waves are used to modulate the amplitude of the carrier wave which the receiver converts back into a reasonably close replica of the original sound waves. The radio carrier wave signal onto which the analog amplitude variations are to be impressed can be expressed as:

$$e(t) = E\cos(\omega_c t) \qquad (1)$$

where:

$e(t)$ = instantaneous amplitude of carrier wave as a function of time (t).

E = amplitude of the instantaneous envelope of the carrier wave.

ω_c = angular frequency of carrier wave (radians/second).

This equation is easily recognized as being an ordinary sinusoidal wave. If E is constant, then the peak amplitude or power of the carrier wave is constant and no amplitude modulation exists. In order to impart audio information on the carrier, the amplitude component is varied or modulated. AM broadcasting uses a particular type of AM known as double sideband full carrier (DSBFC). Thus, the amplitude component (E) becomes nonconstant and can be expressed mathematically as:

$$E = E_c(1 + m(t)) \qquad (2)$$

where:

E_c = amplitude of carrier at times when no modulation is present; coupled with other usually fixed parameters, this value determines the carrier power of the transmitted signal.

$m(t)$ = instantaneous value of the audio signal, which varies between –1 and 1 or slightly higher in the case of positive overmodulation.

Often the peak amplitude of the audio signal is expressed as a percentage relative to the peak values of $m(t)$. When the value of $m(t)$ is exactly 1, this is considered +100% modulation. When the value of $m(t)$ is exactly –1, this is considered –100% modulation. It is common for the peak amplitude of the audio signal to reach as high as 1.25, or 125%. But, modulation beyond –100% is not practical, as it causes significant distortion in a receiver's envelope detector and interference to adjacent channels and thus is not permitted under regulatory rules for AM broadcast transmitters.

Periodic modulation of the carrier wave exists if the envelope (E) is caused to vary with respect to time as, for instance, a sinusoidal wave. In this case, the equation for the modulating signal $m(t)$ might be:

$$m(t) = E_m\cos(\omega_m t) \qquad (3)$$

where:

E_m = peak amplitude of the modulating signal.

ω_m = angular frequency (radians per second) of the modulating signal.

Substituting all of these equations back into Eq. (1) gives us:

$$e(t) = E_c(1 + E_m\cos(\omega_m t))\cos(\omega_c t) \qquad (4)$$

This is the wellknown basic equation for periodic (sinusoidal) modulation of an AM transmitter, and when all multiplications and a simple trigonometric identity are performed the result is:

$$e(t) = E_c\begin{bmatrix}\cos(\omega_c t) + \\ \frac{E_m}{2}\cos((\omega_c + \omega_m)t) + \\ \frac{E_m}{2}\cos((\omega_c - \omega_m)t)\end{bmatrix} \qquad (5)$$

Equation (5) can be represented graphically in at least two practical ways: in the envelope time domain representation, as shown in Figure 4.2-1, and in the frequency domain, as shown in Figure 4.2-2 ($\omega = 2\pi f$). The graphical representations shown are for a single tone with modulation index (M) of 0.7; the peak modulating voltage is 70% of the peak carrier wave voltage (Em/Ec = 0.7). Figure 4.2-2 also shows the occupied bandwidth of an AM signal with single tone modulation. From this figure and its defining equation, Eq. (5), it is clear that the bandwidth of an AM signal is equal to twice the highest modulating frequency when no system distortion is present.

High quality music reproductions include frequency components as high as 15 kHz or higher; therefore, the required theoretical bandwidth of a DSBFC AM signal capable of high quality music reproduction would be at least 30 kHz. Most AM broadcast systems deployed intentionally limit the highest frequency of the modulation component to improve allocation efficiency and reduce adjacent

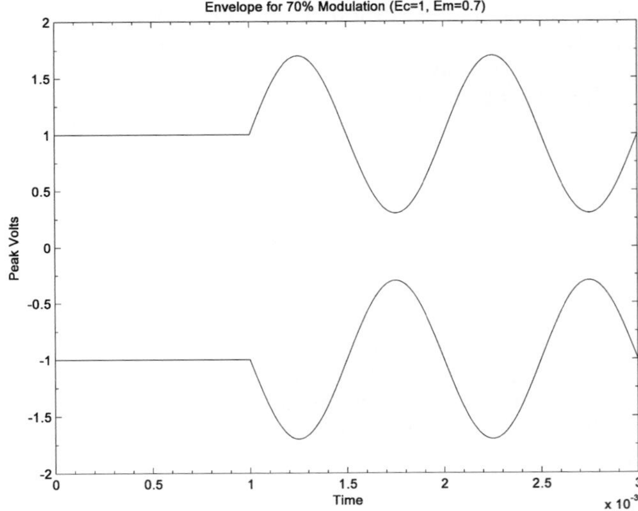

FIGURE 4.2-1 Time domain representation.

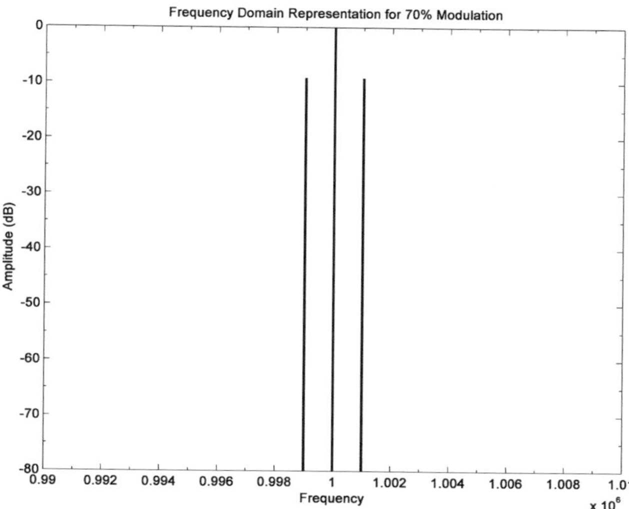

FIGURE 4.2-2 Frequency domain representation.

channel interference. In the United States, the National Radio Systems Committee (NRSC) has published recommendations on the bandwidth limitation that should be employed by broadcast AM stations. In much of the rest of the world, the bandwidth of the modulating signal is limited to 4.5 kHz because the channel allocations are on a 9 kHz spacing; however, there are variations from these practices in many regions of the world. In a later section of this chapter, it will be shown that the amplitude modulation bandwidth is significantly higher for newer digital modulation formats that may require 50 kHz or greater bandwidth to adequately accommodate certain digital formats.

Harmonic and intermodulation distortions have the effect of widening the effective occupied bandwidth of the system; however, modern transmitters have sufficiently low distortion characteristics such that bandwidth broadening due to system nonlinear distortions is not a significant problem for analog transmission. For digital systems, this assumption cannot be made. The ability to meet certain spectral occupancy masks for digital transmitters requires strict limits on distortion components. The occupied bandwidth characteristics of analog AM broadcast transmissions are discussed in more detail in the sections on factory tests and audio processing and preemphasis.

AM TRANSMITTER BASICS

All AM transmitters have a radiofrequency (RF) power amplifier (PA). In most cases this amplifier has an efficient means for controlling (modulating) its output level. This is generally how the amplitude modulation signal is imparted onto the RF, thus creating AM modulation. RF power amplifiers also require an input that contains the carrier frequency component. This signal is usually derived from a stable oscillator operating at the carrier frequency. From these two signals, the RF PA can produce AM. A basic block diagram of an AM transmitter is shown in Figure 4.2-3.

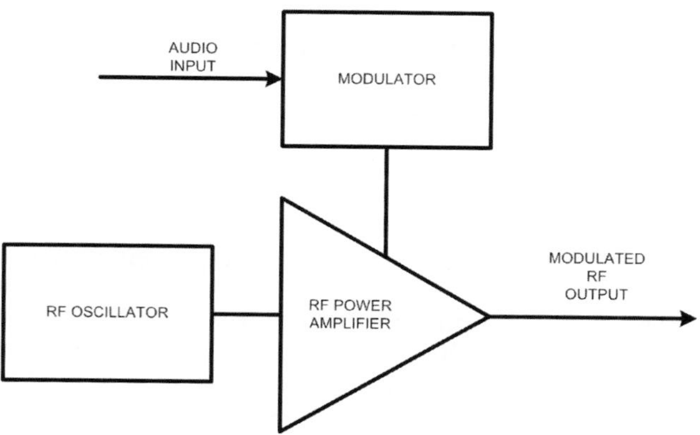

FIGURE 4.2-3 Basic block diagram of an AM transmitter.

RF Power Amplifiers

There are numerous different designs of RF power amplifiers. The main requirements for any particular design are the desired power output and highest possible efficiency. Efficiency is important because any input power that is not converted to RF is wasted as heat and must be removed from the amplifier by some means of cooling. The cooling system's power requirements are generally proportional to the amount of heat that must be removed, so the higher the efficiency of the RF amplifier the less power that must be used to remove the excess heat. Another important requirement is the method of imparting the AM modulation to the RF carrier. The RF power amplifier must be able to produce widely varying levels of RF output without significant change to its efficiency while also maintaining a low level of amplitude distortion. As the details of RF power amplifier design for AM transmitters are examined, these seemingly simple requirements can become difficult to achieve. As AM broadcasting transitions from analog modulation techniques to digital methods additional constraints are placed on the RF power amplifier. Many RF amplifiers employed in AM transmitters within the past decade are well suited to digital modulation systems. Digital modulation systems make use of both amplitude and phase variations to create complex signals. The conversion from a relatively simple AM signal to a complex one brings significant challenges to both the transmitter design as well as the overall system design of a complete AM broadcast facility.

CLASS D RF POWER AMPLIFIER

A popular RF amplifier at least in terms of number of units in the field is the *class D* switching design. This amplifier is highly efficient with theoretical efficiency approaching 100%. It is also very easy to modulate with a separate audio amplifier. At RF frequencies up to about 2 MHz, this amplifier type is used almost exclusively in all modern AM transmitters. Because it is easy to combine the outputs of multiple class D solid-state amplifiers, almost any power level is achievable without the use of tubes. There are several examples of solid-state AM medium wave band transmitters capable of producing carrier powers of 2000 kW.

The class D amplifier is distinguished by its simple two-state design. The amplifier is producing a voltage output of either one state or another state. It switches between these states as quickly as possible. The period between state changes is one-half the period of the transmitter carrier frequency. The reason this type of amplifier is very efficient is because the devices used to produce the two output states have very low voltage or power loss in either state. Because the power loss is proportional to the voltage loss, the lower this value the higher the efficiency. If a device could be designed with zero voltage loss, then the resulting efficiency would be 100%. Although it is impossible to achieve 100% efficiency in practice, it is possible to achieve efficiencies in excess of 90%.

A simplified diagram of a class D switching amplifier is shown in Figure 4.2-4. In the state depicted, the output voltage across the load resistor is equal to the negative supply voltage minus the small voltage across the switching device. After one-half of the RF frequency's period, the amplifier is switched instantly from this state to the state shown in Figure 4.2-5. In this state, the output voltage across the load resistor is equal to the positive supply voltage minus the voltage across the switching device. This switching cycle repeats indefinitely at a rate equal to the transmitter's RF frequency. The voltage across the load resistor (R1) will be as shown in Figure 4.2-6.

Several interesting observations can be made about the waveform in Figure 4.2-6. The voltage is essentially a square wave. It has significant energy at odd harmonic frequencies but quite low levels of even harmonic frequencies. Usually this harmonic energy is undesirable in a transmitter assigned to a single frequency, and there must be a means of removing these harmonics. Another undesirable feature is that the current in the switches changes throughout the switching transition. This additional wasted energy should be avoided. The solution is to put a filter between the switches and the resistive load. This filter is referred to as an *output network.*

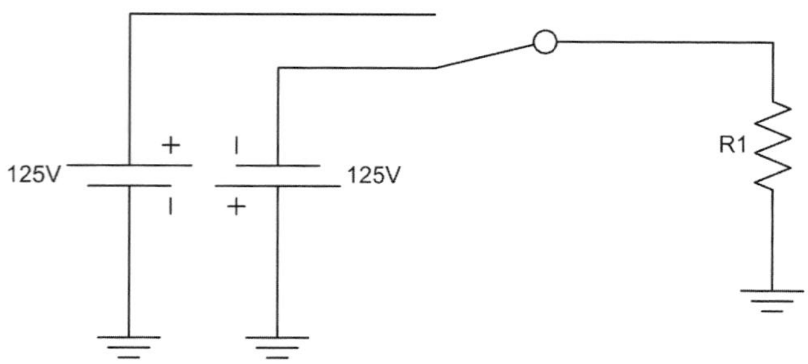

FIGURE 4.2-4 Simplified diagram of a class D switching amplifier, negative state.

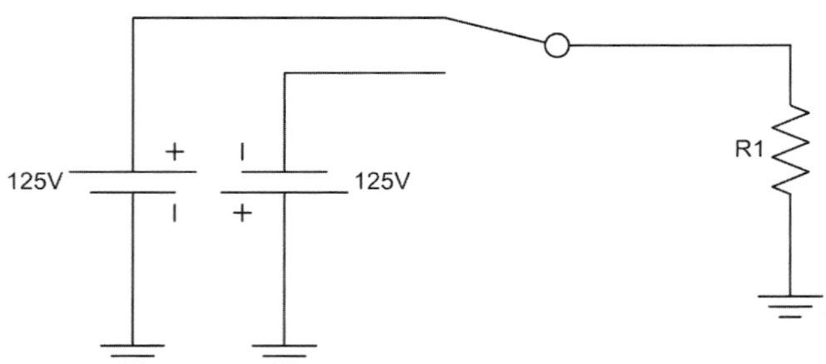

FIGURE 4.2-5 Simplified diagram of a class D switching amplifier, positive state.

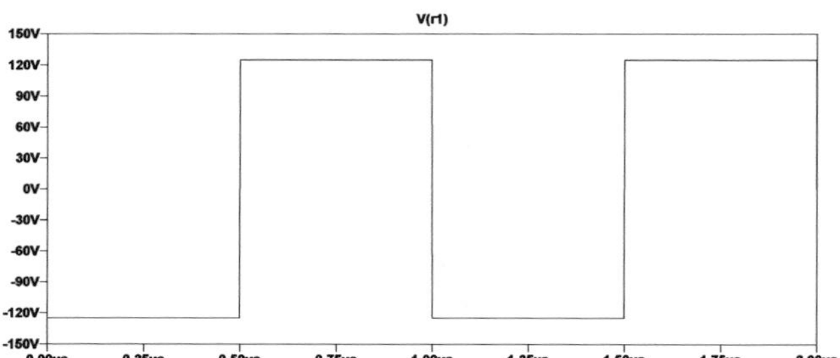

FIGURE 4.2-6 Class D amplifier switching cycle.

Class D Output Network

There are specific requirements for an output network or filter for a class D amplifier. The filter must remove the odd harmonic energy inherent in the voltage waveform and present a high impedance to all odd harmonic currents. The reason for this requirement is that the class D amplifier is most efficient when the current through the switching devices is as near to zero as possible at the instant of the switch transition. If this condition is met, then during the switching transition the product of voltage loss and current will be nearly zero and thus very little power is dissipated during each transition. A final important requirement is that the bandwidth of the filter be sufficient for the broadcast signal of interest. For a digital or high quality AM signal, the bandwidth will be about 30 kHz. Filters that have this type of frequency response and impedance characteristic generally fall into two categories. The first is a bandpass topology with the first elements being series tuned. Another equally useful topology is a low pass filter with an input series inductor that presents a high impedance to frequencies above the fundamental carrier. The simplest filter having the required characteristics is a series tuned inductance–capacitance (LC) filter in series with the

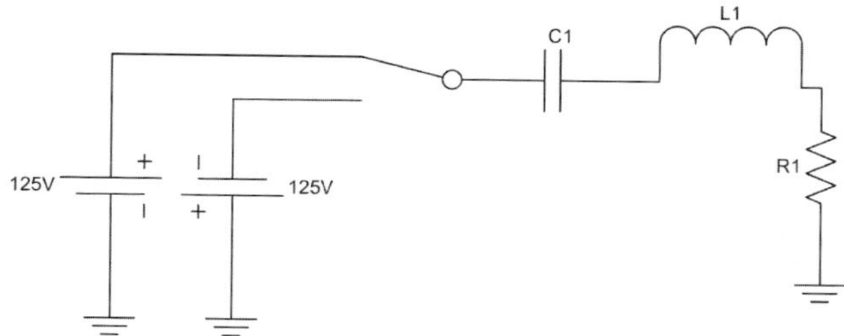

FIGURE 4.2-7 Class D amplifier with series-tuned band-pass filter.

output load resistor, as shown in Figure 4.2-7. The voltage waveform at the input to the series tuned network is still a square wave at the carrier frequency, but the voltage waveform at the load resistor will be transformed into a sine wave that is very low in harmonic energy, as shown in Figure 4.2-8.

The power lost in the switching device is very low because the current is nearly zero at each switching instant and the voltage loss across the switching device is very low at all other times when the current is nonzero. The output sinusoidal voltage at the load resistor has a higher peak voltage than the switching voltage waveform. This is the result of the filter removing the odd harmonic energy. A Fourier analysis of the square wave reveals that the amplitude of the fundamental frequency component is $4/\pi$ or about 1.27 times the amplitude of the square wave. The output power of the class D amplifier is 62% higher than if it were a sine wave output of the same peak amplitude. Normally, the simple bandpass filter shown here is not sufficient for removing all of the odd harmonic energy so most transmitters employ bandpass filters with three or more tuned sections. Another type of filter that can meet the requirements of the class D amplifier is the inductive input low pass filter, also with three or more sections. These two filter classes make up the vast majority all solid-state AM transmitter output networks.

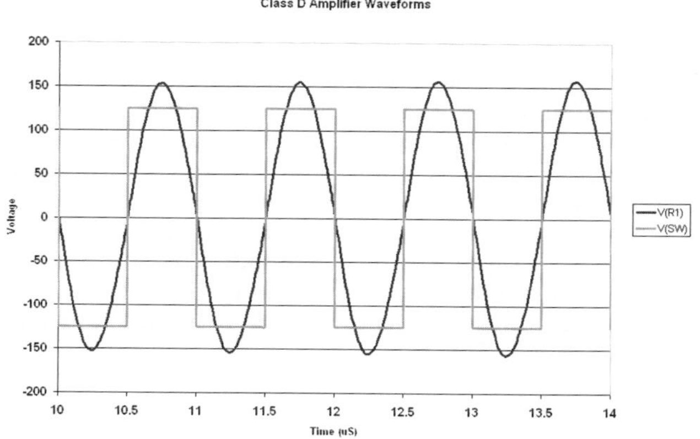

FIGURE 4.2-8 Class D amplifier before and after filter.

The class D RF power amplifier is distinguished by another characteristic that may seem obvious but is important to understand when it is the RF amplifier of an overall AM broadcast system. The output of the switching amplifier switches between two voltages. In some cases, these voltages are fixed, and in other cases these voltages are varied in proportion to the AM modulation. No matter how the modulation is accomplished, the class D amplifier always switches between two voltage states. Such an amplifier can be considered as a nearly ideal voltage source. This observation allows analysis of the amplifier in the context of an overall system so optimal performance can be obtained. By observing that the class D switching amplifier acts as an ideal voltage source, it can be understood how the transmitter will perform in nearly all types of antenna systems. Unless the amplifier is terminated in a pure resistance, there will be power generated that is both real and reactive. The reactive power plays no part in the signal that radiates from the antenna. Only the so-called "real power" can be converted by the antenna into radiated energy, and some of that power will be lost to heat depending on the efficiency of the antenna system. Because most antenna systems present the transmitter with an impedance that varies with frequency, it is necessary to know what type of output terminal impedance variation is optimal for a given AM transmitter.

Optimizing Impedance for Best Performance

The optimal impedance presented to an AM transmitter is the one that causes the antenna radiated power gain to be constant across the RF channel at all locations in the station's coverage area. For analog AM broadcasting, the channel bandwidth can vary from 9 to 30 kHz, depending on regulatory requirements and assumed receiver characteristics. For digital broadcasting, the channel bandwidth varies across a similar range depending on the type of digital system that is employed. For in-band, on-channel (IBOC) systems, the channel bandwidth is 30 kHz. In the case of the Digital Radio Mondiale (DRM) system, the broadcaster can select from several available options that require channel bandwidth between a minimum of 9 kHz and a maximum of 20 kHz.

In this discussion, the primary interest is in the transmitter itself so a simplifying assumption will be made about the antenna. If the amplifier delivers real power to its load such that the gain is constant over the required channel bandwidth, then the assumption will be that the antenna system converts this real power to radiated energy with equal efficiency across the channel. Although this assumption is usually met in practice, it should be verified whenever practical. The most common case where this assumption will not be valid is in the null regions of directional antenna arrays. In this case, the antenna design must be rigorously investigated if performance in the null regions is critical to the success of the station. Chapter 4.4 addresses the design of AM antenna and coupling systems.

The optimal impedance characteristic for an AM transmitter may be determined by considering that the power amplifier acts as a voltage source that only delivers constant power when the real part of the complex load admittance is also constant. The simplest example is a parallel RLC circuit. The real part of admittance is 1/R and is constant. Because only the resistor (R) dissipates power, it will be constant as long as the voltage is constant. This is true regardless of frequency. At some frequency sufficiently removed from the resonance condition, the current in the inductor (L) or capacitor (C) will be too high and the amplifier will overload, but the real power delivered will not change.

The simplest circuit that meets this requirement is a fixed resistance in parallel with any reactance. In practice class D amplifiers have some maximum current output rating so this places a lower limit on the reactive impedance of the parallel circuit. The ideal resistance value will be determined by the amplifier designer. For example, in the case of a 50 ohm design resistance, the optimal impedance for the class D amplifier is one that has a constant Y equal to 0.02 mhos. This constant real admittance condition can be met by a parallel resonant network because the real part of admittance is a constant even though the reactive part changes with frequency.

The class D amplifier output filter must present a high impedance at all harmonic frequencies, and either a series tuned bandpass or inductive input lowpass filter will meet this requirement. The concept of constant real admittance and a series tuned resonant load seems to contradict itself; therefore, it is important to note that the condition of constant real admittance need only be met across the desired RF channel. Typically, the channel is less than one octave, which presents minimal difficulty in meeting this criteria. To determine the optimal sideband impedance characteristic requires knowledge of certain characteristics of the transmitter's output network. Output networks are designed to perform two functions: Satisfy the class D amplifier requirement of presenting a sufficiently high impedance at the harmonics so the amplifier current remains sinusoidal while the voltage waveform is rectangular, and transform the output terminal characteristic impedance (*e.g.*, 50 ohms) to the optimal characteristic impedance of the RF switching amplifier. Knowing the output network phase shift allows the system engineer to determine the optimal impedance variation at the output terminals of the transmitter. Knowing an accurate output network phase shift value makes it a simple matter to determine if a particular measured terminating impedance is optimal and if not what must be done to make it so.

The necessary data points are usually available from an impedance sweep at the transmitter output terminals along with the output network phase shift. There are two main methods of making the necessary computations. The first involves a graphical method by plotting the impedance at the transmitter output terminal on a Smith chart. The transmitter output network phase shift is then applied, and a new Smith chart plot is obtained. These data represent the impedance or voltage standing wave ratio (VSWR) curve at the terminals of the actual RF amplifying devices. When this plot falls as near as possible to a constant admittance line passing through the center of the Smith chart, then the transmitter terminal impedance is optimized. If the measured data when rotated by the output network phase shift do not fall near the constant admittance line, additional phase shift may be obtained as necessary to achieve the optimal condition. This additional phase shift can be physically implemented through a lumped element network (*e.g.*, Tee network). If only a few degrees of phase shift are required, an appropriate length of transmission line may be employed.

The second method of solving the optimization problem is accomplished using a computer. Fortunately, it is possible to use a basic spreadsheet program such as Microsoft® Excel® to perform the required computations and the optimization step. For the case of the class D RF power amplifier, a spreadsheet named "Class D Optimizer.xls" is provided on the companion CD. This spreadsheet is applicable for nearly all high efficiency solid-state transmitters. The input data include the measured impedance at the transmitter output terminals along with the manufacturer specified output network phase shift. The spreadsheet computes the additional phase shift that is required to optimize the system.

It is not necessary to have a detailed schematic of the transmitter output network with exact element values. To make an exact calculation of the RF switching device impedance it may be necessary to have this detail. However, unless the bandwidth of the output network is very narrow compared to the bandwidth of interest (30 kHz), the impedance values obtained by an element-by-element detailed analysis should not vary by as much as the uncertainty in the swept impedance measurements. All AM transmitters designed for modern modulation levels have adequate bandwidth such that the two optimization methods outlined above should prove to be adequate.

This discussion has been centered on optimization of class-D-based solid-state transmitters because nearly all production AM transmitters are of this type. A significant number of vacuum tube based AM transmitters are in active service. Vacuum tube AM transmitters in the medium wave band often can benefit from optimizing the load impedance. The mathematics and techniques are the same for these transmitters as for the solid-state case. The primary difference is in the type of RF power amplifier. Most vacuum tube based transmitters are of the class C type. Class C amplifiers operate on the principle of a current source instead of a voltage source. Instead of operating the RF power amplifier on a load line of constant real admittance, it must instead operate on a constant real impedance line. This is equivalent to a 180° rotation on the Smith chart or 90° in electrical phase. The direction of the rotation (clockwise or counterclockwise) is determined by the practical abilities of the resulting phase shift circuit. The spreadsheet described above may be used for class C optimization by adding 90° to the output network phase shift specified by the transmitter manufacturer. In this case, the spreadsheet will compute the optimal network phase shift required to achieve a constant real part of impedance at the plate of the final amplifier.

CLASS C RF POWER AMPLIFIER

The class C RF power amplifier has been in widespread use for many decades. It is practical to implement a class C amplifier using either solid-state or vacuum tube devices; however, because class C is somewhat less efficient than class D, the latter is found almost exclusively in solid-state designs. In the modern high power AM transmitter, the class C amplifier will usually be implemented using a single vacuum tube. Another consideration when selecting an amplifier type is the frequency and mode of operation. Because vacuum tubes are able to handle much greater power levels than transistors, it is convenient to use them in shortwave transmitters that must change frequency on a regular basis, to compensate for changing propagation conditions, often without direct manipulation by an operator. Shortwave transmitters are often required to generate carrier power levels from 100 up to 500 kW, and this is practical using a single vacuum tube.

The basic tuned anode vacuum tube amplifier is shown in graphical form in Figure 4.2-9. The tube can be a triode, tetrode, or pentode, with tetrodes being the most common type found in modern high power vacuum tube designs. The RF excitation voltage is supplied to the grid of the power amplifier tube and the ratio of DC grid bias to voltage peak RF excitation voltage, shown sinusoidal in Figure 4.2-9, determines the conduction angle (θ) of anode current, given as:

$$\theta = 2 \times \arccos\left(\frac{Ecc}{Ecc + Eg}\right) \tag{6}$$

where the exciting grid signal (Eg) is sinusoidal, as shown in Figure 4.2-9a.

The shape of the anode current pulse is determined by the device transfer characteristics and input wave shape. The pulse of current thus generated, Figure 4.2-9c, is supplied by the DC power supply (EBB) and passed through the resonant tank circuit shown in Figure 4.2-9d. The resonant tank circuit is assumed to have sufficient operating Q to force anode voltage (ep) to be essentially sinusoidal and of the same periodic frequency as the RF excitation voltage and resultant anode current pulse. The instantaneous anode dissipation, shown in Figure 4.2-9e, is the product of instantaneous tube anode voltage drop and anode current. The tube transfer characteristic is a variable that is dependent upon many tube factors as well as the max-

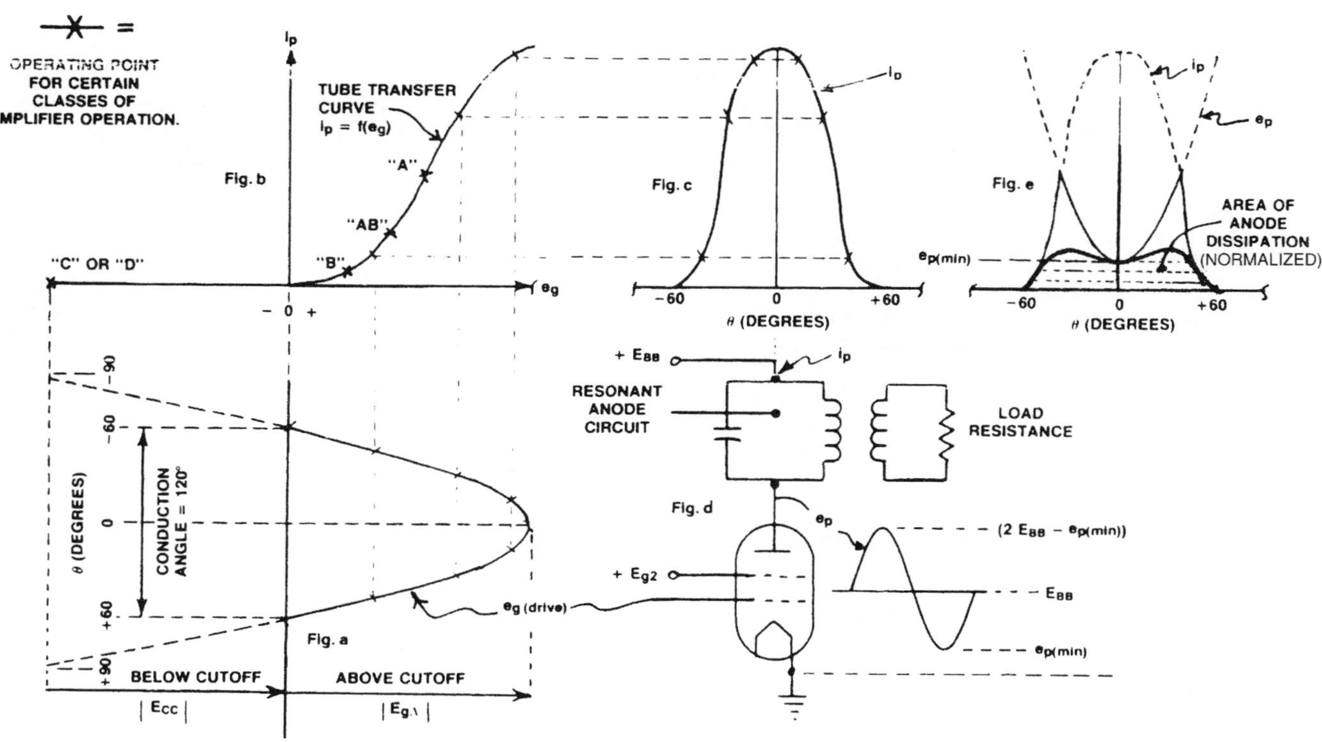

FIGURE 4.2-9 Classical vacuum tube class C amplifier with sinusoidal grid drive, 120° anode current conduction, and resonant anode load.

TABLE 4.2-1
Comparison of Tuned RF Amplifier Efficiencies

Amplifier Class	Conduction Angle (degrees)	Anode Efficiency (%)	Defined Conditions of Operation
A	360	30	Eb(min) = 0.10*EBB
A–B	240	60	Eb(min) = 0.10*EBB
B	200	67	Eb(min) = 0.10*EBB
C	120	84	Eb(min) = 0.05*EBB
C–D	120	90	Eb(min) = 0.05*EBB (odd harmonic resonators)
D	180	95	Eb(min) = 0.05*EBB (square-wave voltage shape)

imum drive signal (Eg). The exact shape and magnitude of the current waveform are normally obtained from an operating line plot on constant current characteristic tube curves supplied by the tube manufacturer. The resonant anode load impedance is chosen and adjusted to allow ep(min) to be as low as possible without causing excessive screen grid (in the tetrode case) or control grid dissipation. Some manufacturers increase anode efficiency beyond the limits for typical class C amplifiers by using a circuit employing a third harmonic resonator between the output anode connection and the fundamental resonant circuit. This has the effect of squaring up the anode voltage waveform (ep), thus causing the integral of the (ep ip) product, or anode dissipation, to be smaller and resulting in lower anode dissipation for a given RF power output. An amplifier employing the third harmonic anode trap is commonly referred to as *class C–D*, suggesting an efficiency rating somewhere between conventional class C operation (nominal 120° conduction angle) and true class D operation with rectangular anode voltage waveforms. Anode efficiencies can be increased to about 90% for transmitters up to approximately 10 kW carrier power and about 85% for transmitters higher than 10 kW carrier power by using the third harmonic trap technique. Table 4.2-1 shows a comparison of anode efficiency for six classes of high power tuned RF amplifiers under certain defined conditions.

MODULATION METHODS

All AM transmitters employ some method of imparting the amplitude of the low-level audio input onto the RF output. More correctly, this is referred to as *envelope modulation*. It is adequate for the purposes at hand to consider amplitude and envelope to be the same. It was previously noted that the most common RF power amplifier is the solid-state class D high efficiency type. Although one might conclude that there must be an equally common method of amplitude modulation, this is not the case. There are several different forms of amplitude modulation even for solid-state transmitters.

Pulse Width Modulation

Pulse width (or pulse duration) modulation (PWM or PDM) is the most common form of amplitude modulation in widespread use today. The introduction of PWM into AM transmitters in the 1960s revolutionized the AM broadcast industry due to the increase in efficiency over then existing modulation systems. PWM also made high power solid-state transmitters practical.

PWM amplification is a switching scheme in which the pulse width of each switching interval is governed by the amplifier's input signal amplitude. There are many different methods of determining the exact switching instants to maintain the proportionality between the pulse width and the input signal level. In the earliest methods, an analog sawtooth or triangle waveform was compared to the analog input signal. In recent years, digital signal processors precisely control the pulse width. No matter how the pulse width is determined, it is amplified using a very efficient switching amplifier. Because the amplifier spends nearly all the time in either the on or the off state, there is very little wasted power that must be removed as heat. Passing the pulses through an averaging filter removes the harmonics of the switching rate, leaving only the average of the pulse width. Because the pulse width is exactly proportional to the input signal, a very accurate reproduction of the input signal is obtained. A simplified schematic diagram of a PWM amplifier is shown in Figure 4.2-10. The resistor is where the audio energy is delivered with efficiency approaching 98%. When this resistor is replaced with a class C or class D RF amplifier, it will be modulated with a replica of the audio input.

An important variation of the original PWM concept was patented by Harris Corporation in the late 1970s [2]. This version recognized that multiple PWM amplifiers could be combined in series either by connecting the actual switching amplifiers in series or by combining the associated RF amplifier outputs in series. Then, using a fixed phase relationship between the different PWM amplifiers, the effective switching frequency can be increased without actually changing

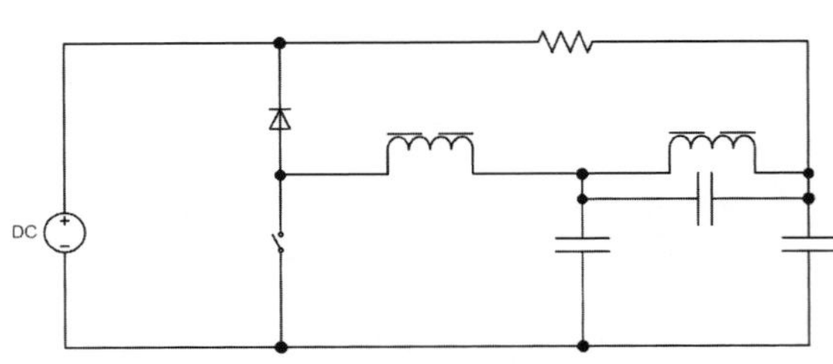

FIGURE 4.2-10 Simplified schematic diagram of a pulse-width modulation (PWM) amplifier.

the switching frequency of an individual PWM amplifier. This innovation was termed *polyphase PDM*, and it greatly reduced the complexity of the averaging filter and improved the audio bandwidth.

PWM also made it practical to use solid-state devices to modulate high voltage vacuum tube transmitters. These modulators are known by various names, such as *pulse-step modulation* or simply *solid-state modulators*. By connecting a sufficient number of switching amplifiers in series, it is practical to obtain audio power levels of several megawatts. These modulators are widely used in high power short-wave transmitters.

Typically, the PWM amplifier controls the anode voltage of a class C amplifier. In nearly all modern shortwave transmitters, this amplifier uses a high power tetrode. A tetrode is a voltage-controlled current device. If the screen voltage is held constant, the anode current is a function of grid voltage. Variations in the anode voltage have only a minimal effect on the anode current. The previous description of the class C amplifier revealed that the RF output level is determined by the amplitude of the anode current pulse; therefore, it would seem counterintuitive to apply high level modulation to the anode if little variation in RF output level would result.

The real amplitude control of a tetrode class C amplifier with a high level PWM anode modulator is accomplished by varying the screen voltage proportionally to the modulation input, then the variation in the anode serves to maintain the high efficiency of the class C amplifier at all levels of amplitude. Conveniently, the tetrode amplifier can be screen modulated using a technique known as *screen impedance modulation*. This technique uses a fixed DC power supply for the screen voltage, with a large inductor (typically 10 Hy) connecting this power supply to the tetrode screen terminal, as shown in Figure 4.2-11.

This circuit works by applying the correct modulation to the screen in sympathy with the anode modulation even though there is no active device controlling the screen modulation. The DC power supply is only adjusted to the correct fixed value for a given carrier power. When the anode voltage rises in response to an increasing modulation input signal, the screen current decreases (in other words, its impedance increases). This decrease in screen current is resisted by the large inductor. Because the inductor seeks to maintain constant current through the screen, the output voltage of the inductor rises sufficiently to achieve this equilibrium point. Similarly, when the anode voltage falls, the screen current will increase (in other words its impedance decreases). The inductor resists this increase in current, and the voltage output of the inductor falls. These attempts to change the screen current by modulating the anode are said to modulate the screen impedance.

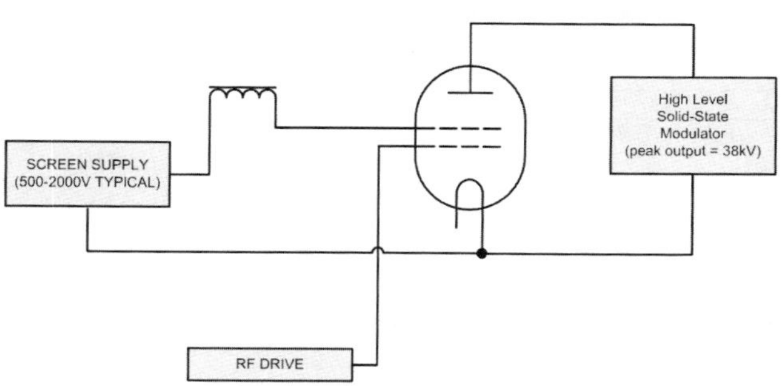

FIGURE 4.2-11 Schematic of tetrode amplifier screen impedance modulator.

MODULATION WITHOUT A MODULATOR

There are techniques for imparting amplitude modulation that use only the RF amplifier. These techniques are generally restricted to frequencies and power levels that lend themselves to solid-state RF amplifiers. The class D RF amplifier is already a switching design. By controlling when and for how long each amplifier is switched on or off, the modulation can be controlled very accurately. There are two basic systems in use today. The first uses a coding technique known as pulse-code modulation (PCM) and the second uses a technique that is sometimes called *RF PWM*.

Harris DX Series Pulse-Code Modulation

The PCM technique was first introduced in the Harris Broadcast DX series of medium wave AM broadcast transmitters employing direct digital techniques to generate the amplitude modulated wave. The technique has no modulator as such, and the complete AM wave, carrier, and modulation sidebands are produced by direct digital pulse-code control of several low power solid-state RF amplifier modules. The system has many inherent advantages over conventional AM systems. It is solid state and employs no separate modulation of the final RF amplifiers (the amplitude-modulated wave is directly achieved by digitally selecting series-connected small RF output modules). It is capable of high quality modulated wave characteristics such as lower modulation noise, distortion, and transient modulation characteristics. The Harris DX system block diagram is shown in Figure 4.2-12.

The incoming analog signal is converted to a digital code by conventional sample-and-hold and pulse-code modulation techniques (A/D) and circuitry. The top 6 (most significant) bits of the 12 bit A/D PCM system are converted into 64 equal steps. The top 22 steps are not used, leaving 42 equal steps to control 42 equal-output series RF amplifier modules. This is equivalent to keeping the lower 5.4 of the 6 top bits of the PCM digital code ($2^{5.4} = 42$). The bottom 6 (least significant) bits are then used to control 6 half-step RF amplifiers (amplifiers with outputs of 1/2, 1/4, 1/8, 1/16, 1/32, and 1/64 of the major 42 equal-step amplifiers). Figure 4.2-13 shows how this process creates a digital reconstruction of the analog input waveform of 2709 discrete steps ((42 + 1)*63 = 2709). This is roughly

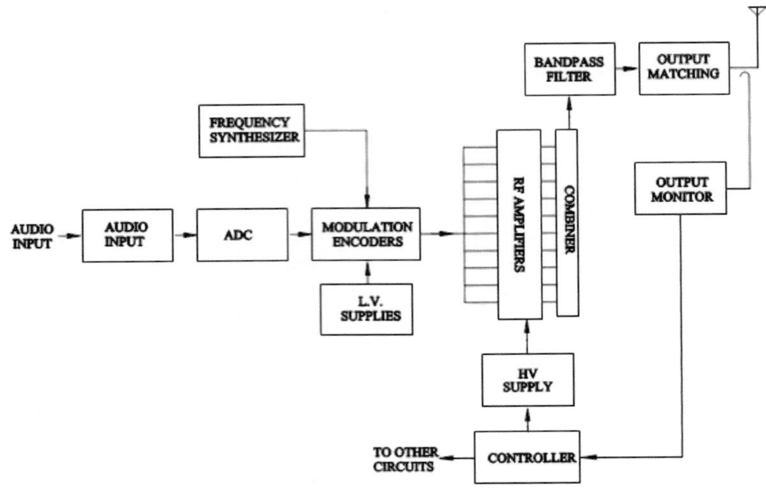

FIGURE 4.2-12 Harris DX series AM transmitter direct digital modulation block diagram.

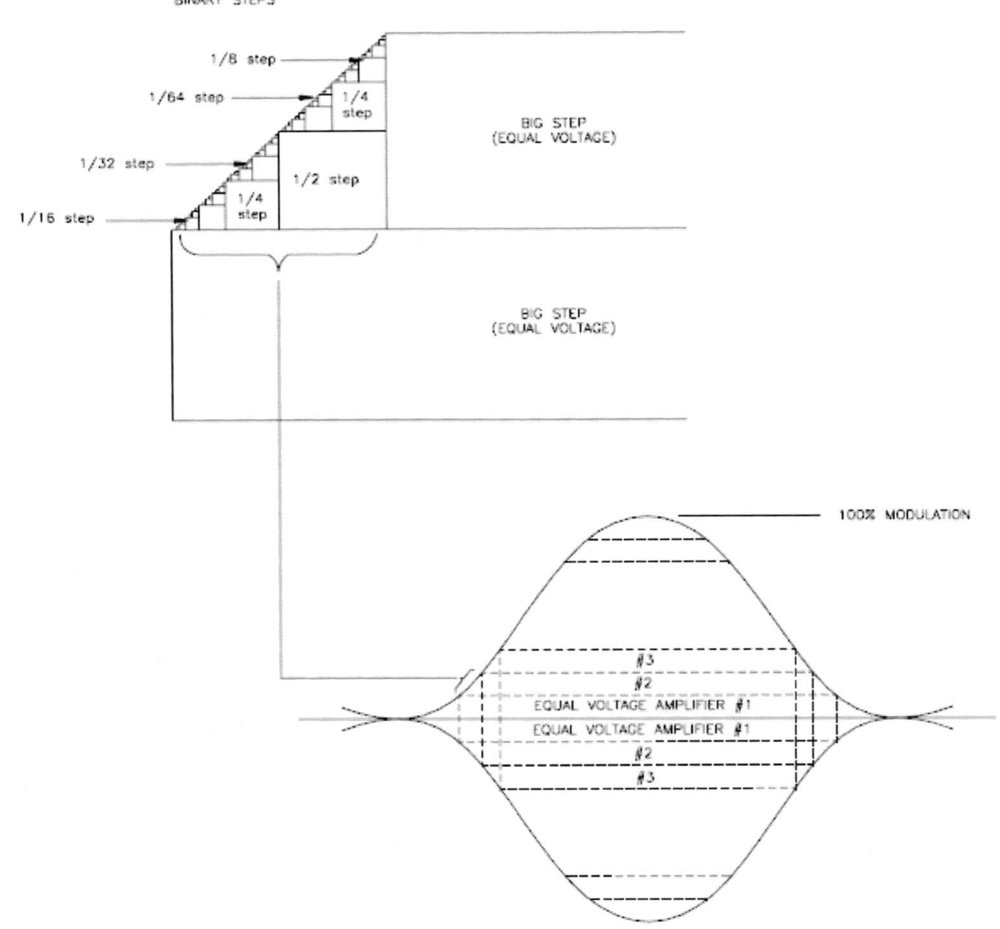

FIGURE 4.2-13 Harris DX transmitter digital modulator showing big-step and binary amplifiers.

equivalent to a PCM system with a total resolution of 11.4 bits ($2^{11.4} = 2702$). The theoretical capabilities of such a PCM system are well known, yielding a theoretical signal-to-noise ratio of approximately 70 dB. If the total harmonic distortion plus noise (THD+N) were calculated, it would have a value of approximately 0.03% referred to 100% modulation. Typical measured values of THD+N are approximately 10 to 30 times this theoretical figure due to the practical difficulty in achieving balanced output contribution from all of the series output modules; however, even these levels of harmonic distortion (less than 1% maximum) are well below accepted norms for modern high-power AM broadcast transmitters. Typical values of measured audio noise are close to the theoretical maximum of –70 dB referred to 100% tone modulation at full carrier power.

Each high power amplifier module is a solid-state amplifier operating in the class D mode and having approximately 95% power efficiency, including output circuit losses. Because there are no additional modulator losses (the modulation process takes place in the RF circuitry itself), the auxiliary circuits of cooling, control, excitation, and drive represent the only other input power requirements. The total auxiliary input power of a 50 kW transmitter is approximately 5 kW. This means that the total input power of a 50 kW DX transmitter with 95% RF efficiency at carrier power (no modulation) is approximately: $50/0.95 + 5 = 57.63$ kW, yielding an overall efficiency of approximately $(50/57.63)*100\% = 86.76\%$. Recently, Harris updated the design of the DX to 3DX to signify the third generation design. The modulation mathematics are basically the same as the original DX, but the method of distributing the RF drive and PCM control pulses has been significantly revised, resulting in fewer components and easier maintenance. All RF and digital modulation signals are distributed to each module at a low level using high speed balanced cables. This method can be called *direct digital drive*, or 3D.

Broadcast Electronics 4MX Transmitters

Another technique that requires no separate modulator is a method called 4MX. Broadcast Electronics makes a 25 kW and a 50 kW transmitter based on such a system. The 4MX represents innovations for which patents are pending. Because the system relies on modulating the first term of the Fourier series of the RF waveform, the marketing term 4MX is used. A block diagram of the 4MX is shown in Figure 4.2-14.

The fundamental component of the class D square wave is $4/\pi$ times the peak switching voltage. If the pulse width of the class D waveform is varied such that it is no longer a square wave, then the fundamental component varies as the sine of the pulse width. Therefore, by comparing the modulation input waveform to a sine wave and varying the RF pulse width accordingly, the fundamental component of the switching frequency will vary in exact proportion to the modulation input. A typical RF switching waveform is shown in Figure 4.2-15. Note that the waveform is not a square wave as in the typical class D amplifier. This waveform changes its on time continuously in response to variations in the modulation input. If the output network removes all harmonics, then a perfectly modulated AM signal will result.

Because there is no need to vary the supply voltage of the class D amplifier, the power supply can be simplified with the same benefits as the DX system. Broadcast Electronics has elected to use a number of power factor corrected DC supplies and thus save significant weight over conventional 60 Hz rectifier transformers.

The 4MX output network uses low loss capacitors, as well as Litz wire in several critical inductors to minimize heating losses. These steps result in an overall AC to RF efficiency of greater than 87%. The 4MX has the additional ability to act as an RF sweep generator. By using the built-in reflectometer, the transmitter's control computer is able to measure the RF impedance presented by the antenna system at the transmitter

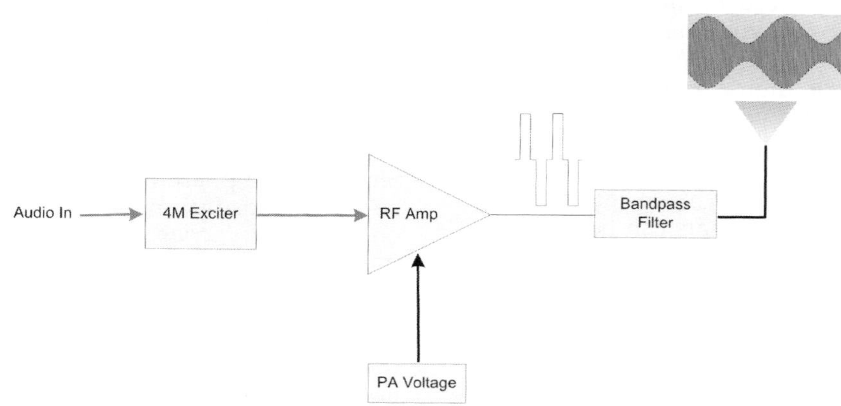

FIGURE 4.2-14 Block diagram of Broadcast Electronics' 4MX modulation technique. (Reproduced with permission.)

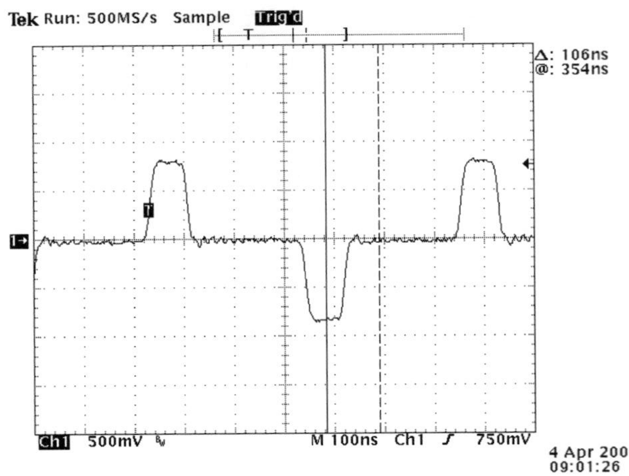

FIGURE 4.2-15 Broadcast Electronics' 4MX modulator switching waveform.

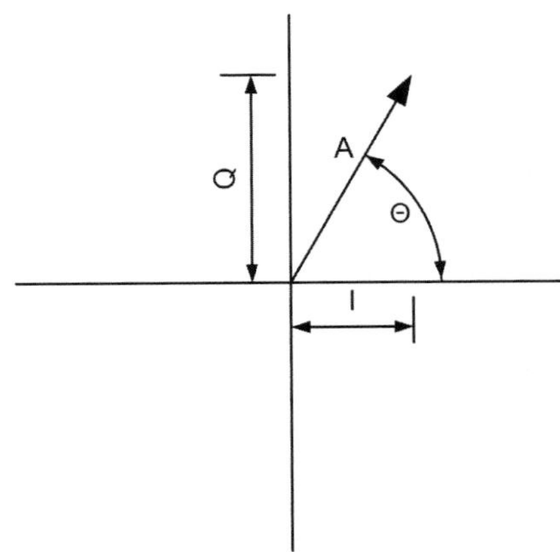

FIGURE 4.2-16 Relationship of amplitude (A) and phase (θ) to I and Q.

output terminals. Displays that are intuitive for presenting such data include numerical VSWR as well as graphical Smith chart representations. The 4MX uses a digital signal processor to generate the RF pulse width modulation and is capable of producing arbitrary modulation formats limited only by the ingenuity of the DSP program writer. The processor can be programmed to generate IBOC as well as DRM modulation formats.

DIGITAL MODULATION METHODS

Digital modulation formats have been standardized and need to be understood in order to be exploited to their maximum potential. Fortunately, the knowledge already attained by the AM transmitter engineer is not obsolete in the digital world. In fact, digital modulation formats are inherently feasible in nearly all modern AM transmitters. Unlike other bands, such as VHF FM, where new transmitter technologies are being introduced to implement digital, the AM transmitter is proving its flexibility in a growing digital world.

An understanding of digital modulation can be found in Chapter 4.13, "AM and FM IBOC Equipment and Systems," but an in-depth understanding of the modulation formats is not necessary to understanding how an AM transmitter can be made into a digital transmitter. All modulation formats can be described in one of two ways. The first is an amplitude and phase representation. The second is the complex representation of I and Q. Figure 4.2-16 shows the relationship between these representations.

The mathematical relationships between these two formats are shown in Eq. (7). The quantities r and θ represent amplitude and phase, respectively. This is the familiar phasor representation that is often used in making power system calculations. The quantities I and Q represent the two vectors that when summed together result in the same vector as described by r and θ:

$$
\begin{aligned}
I &= r \times \cos(\theta) \\
Q &= r \times \sin(\theta) \\
r &= \sqrt{I^2 + Q^2} \\
\theta &= \arctan 2\left(\frac{Q}{I}\right)
\end{aligned}
\tag{7}
$$

The last equation uses the function arctan2() which may be unfamiliar to some readers. Arctan2() is the same as arctan() except that it returns the correct angle depending on the signs of the two input arguments (I and Q). The fact that there are exact and fairly simple formulas for converting between complex (I and Q) and amplitude/phase (r and θ) means that AM transmitters can also be converted for use by complex modulation techniques employed in digital systems. Nearly all complex modulation systems such as IBOC and DRM are defined in terms of their real-time representation of I and Q. Either r and θ or I and Q define everything about the signal that should be transmitted except for the carrier frequency. All modern AM transmitters are by their nature capable of transmitting arbitrary amplitude and phase. For traditional AM, the amplitude is modulated but the phase is not. By computing the amplitude of digital I and Q signals, the digital amplitude modulation is straightforward. Similarly, the phase of the digital I and Q signals can be computed. By impressing this phase modulation onto the RF carrier determining device of the AM transmitter, it is possible to combine the two into the desired modulation. The realization that AM transmitters can be converted to modulation formats other than AM by these simple formulas is credited to Leonard Kahn [3]. The method is called the *Kahn method* or *envelope elimination and restoration* (EER).

It would be advantageous if the actual implementation of digital modulation through AM transmitters were as simple as the mathematics involved; however, as is the case with many engineering problems, the details sometimes obliterate the simplicity of the result. Perhaps the biggest problem with the amplitude/phase approach is the unspoken assumption that the transmitter correctly performs the computations in Eq. (7). This is seldom the case. The AM transmitter has two distinct paths. One path contains only constant-amplitude RF waves (possibly phase modulated), and the other path contains the amplitude modulation component. Slight time delay variations between these paths will mean that the amplitude and phase signals arrive at the final amplifier at different times. Because the phase and amplitude of the signal are not correct, there will be errors in the transmitted vector. This quantity can be measured as an error between the actual transmitted vector and the desired vector. This is called *error vector magnitude* (EVM) or *modulation error ratio* (MER). A receiver perceives errors in the transmitted vector as unwanted noise (MER). These error measurement concepts allow a single measurement value to describe the transmitted signal quality. Each digital modulation standard will have a prescribed MER or EVM that can be considered acceptable for the particular digital format. Time delay differences between the RF and amplitude paths of a transmitter are easily compensated by the circuits or programming that converts the I and Q data into amplitude and phase representations. A simple delay of the RF path is commonly employed because it is usually the amplitude path delay that is the longer of the two.

Another source of errors is the frequency response of the amplitude path. An examination of Eq. (7) shows that amplitude r is the nonlinear result of two linear components (I and Q) because the square and square root function are not linear. This implies that the bandwidth of the amplitude path will likely be greater than the bandwidth of the transmitted signal. This is usually the case in digital formats. The exact amount of bandwidth expansion is dependent on the particular digital format, but it is quite common for the amplitude bandwidth to be four or five times the bandwidth of the transmitted signal. This is much larger than an AM-only signal. Recall that the bandwidth of the amplitude path of an AM-only transmitter is half that of the transmitted signal. For digital broadcasting, the amplitude bandwidth path may have to be an order of magnitude higher than for a traditional AM transmitter. It is quite common for digitally modulated AM transmitters to require amplitude path bandwidths greater than 50 kHz. Another consequence of the amplitude/phase approach has to do with the low frequency response of the amplitude path. It is usually the case that there is energy in the computed amplitude signal down to and including DC; therefore, all amplitude modulators intended for most digital operating modes should be direct coupled with no high pass filtering.

RF path bandwidth is also a concern because the arctan2() function is also nonlinear and a similar increase in bandwidth occurs in the phase signal computed from I and Q. Most AM transmitters have sufficient bandwidth in the RF path to accommodate this increase, but it should be verified with the manufacturer before assuming that the transmitter can be upgraded to a particular digital format. A rule of thumb is that the RF path bandwidth is approximately twice the required bandwidth of the amplitude path.

HISTORICALLY SIGNIFICANT MODULATION METHODS

Several modulation methods are important from a historical perspective even though they are not in active production.

Heising High-Level Anode Modulation

The first practical method of generating the AM signal was Heising constant current, a method of applying audio modulation to the anode supply voltage of a class C RF amplifier [4,5]. This general type of modulation has since been referred to as high level anode (or plate, or collector) modulation, as shown in Figure 4.2-17. The Heising modulator was used as early as 1920 to modulate a low power RF amplifier or master oscillator stage that was followed by several linear amplifier stages until the desired power level was attained. In some cases, the Heising modulator was used to modulate the final RF amplifier stage of lower power transmitters. The Heising shunt modulator operated in class A mode and therefore was low in operating efficiency. The early linear amplifiers were tuned class B amplifiers operating with carrier level anode efficiencies of about 30%. The Heising and similar systems of audio amplification were also used to modulate the grid bias level of RF amplifier stages in

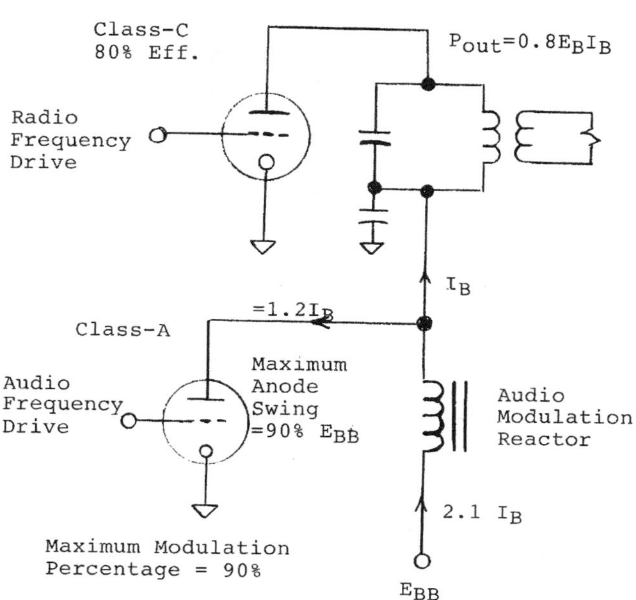

FIGURE 4.2-17 Schematic diagram of Heising amplitude modulation system.

order to obtain the AM signal to be used for further linear amplification. Heising constant current anode modulation was also popular in military and aviation radio sets used during World War II.

Class "B" High Level Anode Modulation

Until the 1970s, the most popular method of applying the audio modulating voltage to the anode circuit of a class C RF power amplifier was by a high power push–pull class B audio amplifier. This type of modulation was first used to improve the operating efficiency and to increase the output power of AM broadcast transmitters. Class B push–pull audio amplification was first used to improve distortion and output power of telephone transmission amplifiers. The invention was soon recognized by broadcast engineers and applied to high level anode modulation. With the final RF power amplifier operating at approximately 80% anode efficiency and the class B audio modulator total static currents approximately one-tenth that of an equivalent Heising modulator, total anode efficiencies at carrier level rose to approximately 72% compared to 37% for the Heising system and 30% for conventional linear amplification. A simplified drawing of a typical high level class B anode modulation system is shown in Figure 4.2-18. The vacuum tubes shown may be triodes, tetrodes, or pentodes. The output circuit of the class B modulator shows the output modulation transformer (MT), an audio coupling capacitor (C), and a DC shunt feed inductor (L).

This arrangement was used in all high level class B high power broadcast transmitters until about 1960 because of a transformer design constraint that would not economically allow unbalanced direct currents to magnetize the transformer core material without poor low frequency distortion. Advanced technology core materials and careful magnetic transformer design allowed elimination of the coupling capacitor and the DC feed shunt inductor, initially in some 100 kW European transmitter designs in the early 1960s, and in an American shortwave transmitter design in the later 1960s. Many of the more advanced modern AM broadcast transmitter designs still using high level class B anode modulation have eliminated the extra C and L components from the modulator output circuitry, and the DC current to the modulated RF amplifier anode flows directly through the secondary of the output modulation transformer. Elimination of these C and L components is necessary for optimum operation of modern AM stations. With the extra C and L components, the modulator output is effectively a three-pole high pass filter that causes low frequency transient distortion to be generated when driven with the complex waveforms that are produced by many modern and popular program audio processors. Eliminating the C and second L component causes the output modulator circuit to become a single-pole high-pass filter, greatly reducing low frequency transient distortion.

Another problem with transformer coupled high-level class B anode modulation is with high frequency audio transient distortion. Stray internal winding capacitances and leakage inductances form multi-pole low pass filtering at the high frequency end of the

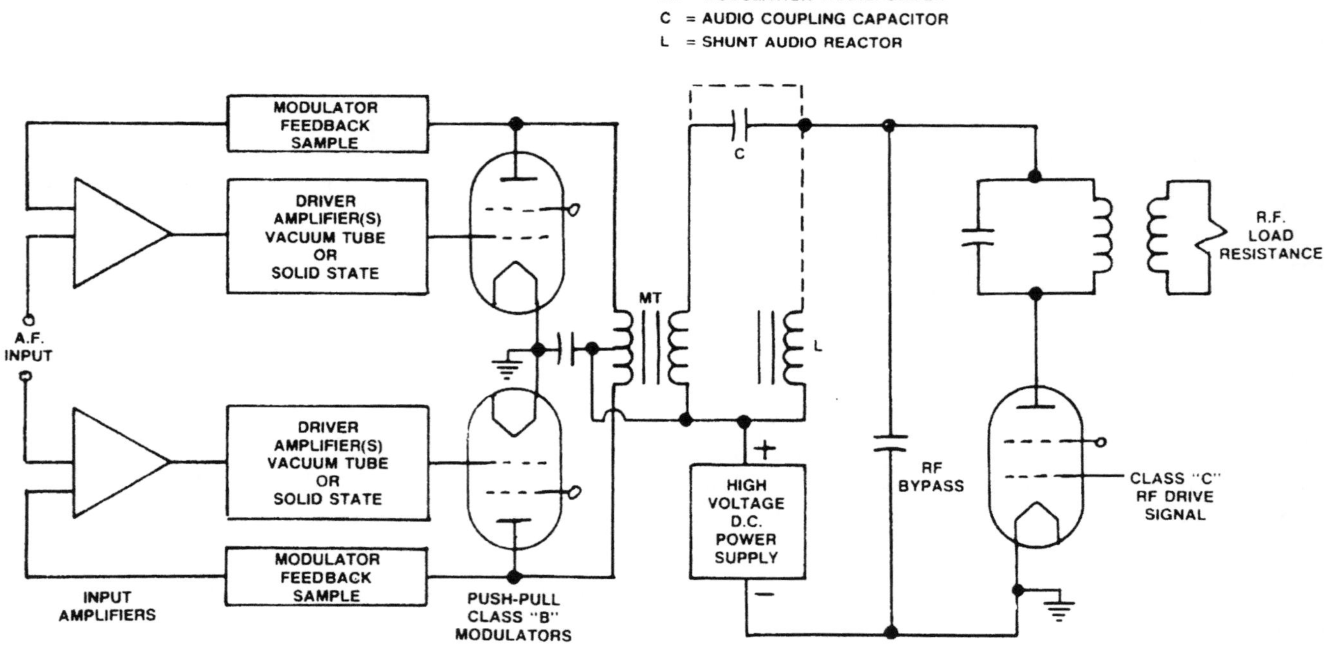

FIGURE 4.2-18 Schematic diagram of class B high-level anode amplitude modulation system.

audio spectrum. This equivalent multi-pole low pass filter generates transient overshoot distortion when driven by the same type of processed complex program waveforms previously mentioned. Transient overshoot up to 12% is typical for square wave modulator input signals and results in a required modulation level reduction of the same 12% in order to maintain peak modulation levels within FCC allowed limits. This high frequency transient overshoot distortion can be effectively minimized by filtering the audio input to the transmitter with linear phase filters, resulting in somewhat lower high frequency audio response or by careful control of the modulation transformer equivalent circuit, yielding more linear audio phase characteristics for the entire modulator circuitry. Significantly, the poor low frequency response renders the class B modulator unusable in modern digital systems because they require the frequency response of the amplitude modulator to include all low frequencies including DC. Balanced modulator negative feedback is used to reduce modulator nonlinear distortion and noise. Negative feedback, however, usually worsens high audio frequency transient distortion characteristics.

Doherty High-Efficiency Linear Amplifier

The Doherty High-Efficiency Linear Amplifier was first described in the technical literature in 1936 by its inventor, W.H. Doherty [4]. So contrary were the terms *linear* and *high efficiency* in the context of amplitude-modulated waves that many engineers in broadcasting refused to accept the concept as workable, similar to the reaction Armstrong received when he proposed that frequency modulation was a practical mode of radio transmission. Nevertheless, the Doherty system was proven to work by 1938. It has been used at power levels up to 1000 kW carrier power in both the original and in the patented Weldon modified form in many installations throughout the world on the medium-wave broadcast and international shortwave broadcast bands, as well as the long wave broadcast band in Europe [5,6]. Its implementation was the result of true inventive genius, using one or more known basic scientific principles to create a totally new and needed product. The Doherty Linear Amplifier is shown graphically in Figure 4.2-19. As with conventional linear amplifiers, the AM signal is generated at low levels and applied to the input of the final amplifier stage. The Doherty system employs two output amplifier stages, one defined as the carrier amplifier and the second as the peak amplifier.

The outputs of the two stages are combined in phase at the anode of the peak amplifier tube. At carrier level, the carrier tube is operated as a nearly saturated class B amplifier and thus delivers almost all of the carrier power at class B efficiencies (approximately 70% anode efficiency). The peak tube at carrier condition is biased and driven just above cut-off and therefore supplies a small amount of carrier power (approximately 2 to 6%). The anodes of the two tubes are connected together through a 90° impedance inverting RF network. As the modulated signal increases in the positive direction to both peak and carrier tubes, the current supplied to the output load by the peak tube increases. The saturated voltage drop at the anode of the carrier tube remains constant over the entire positive modulation half-cycle, thus causing the current at the output of the inter-anode 90° network also to be constant during the same positive modulation half-cycle. The rising current from the peak tube anode has the effect of raising the impedance presented to the inter-anode network. Because the current from the network is constant, the net effect is an increase in output power from the carrier tube ($I^2 \times R_L$ increases because R_L increases). At the 100% positive modulation crest, both tubes are producing exactly twice carrier power to the load, satisfying the requirement that peak envelope power (PEP) equals $4 \times P_{carrier}$. During the negative half-cycle of modulation, the peak tube is cut off and the carrier tube behaves as a normal linear amplifier, allowing the envelope power output to drop linearly to zero at the 100% negative modulation trough. The anode efficiency of the Doherty High-Efficiency Linear Amplifier at carrier level is more than twice the efficiency of conventional AM class B linear amplifiers.

The Doherty system also has two other important advantages for high power broadcast transmitters. First, and most important, the peak anode voltage at either tube is only about one-half that required for an equivalent carrier power high level PWM or class B anode modulated transmitter, thus increasing reliability and usable tube life significantly. Second, no large modulation transformer or special filtering components are needed in the final amplifier stages that would cause transient overshoot distortion, as previously discussed for class B anode modulation or pulse width anode modulation.

The main problems with the Doherty Linear Amplifier are nonlinear distortion and an increase in the complexity of tuning. The major sources of nonlinear distortion are the nonlinearity of the carrier tube at or near the 100% negative modulation crest and the nonlinearity of the peak tube at or near carrier level when it is just beginning to conduct. Both sources of distortion are effectively reduced by use of moderate amounts of overall envelope feedback. The tuning complexity problem is usually overcome by simplified tuning procedures aided by built-in test equipment.

TRANSMITTER PERFORMANCE MEASUREMENTS

Certain basic AM transmitter performance parameters should be measured periodically in order to ensure that certain minimum broadcast quality standards are provided to the listening public. Some are specifically required by the FCC, but others are a matter of good engineering practice:

1. Spurious emissions (FCC Rules Section 73.44); this measurement must be made annually, with no more than 14 months between readings (Section 73.1590(a)(6))

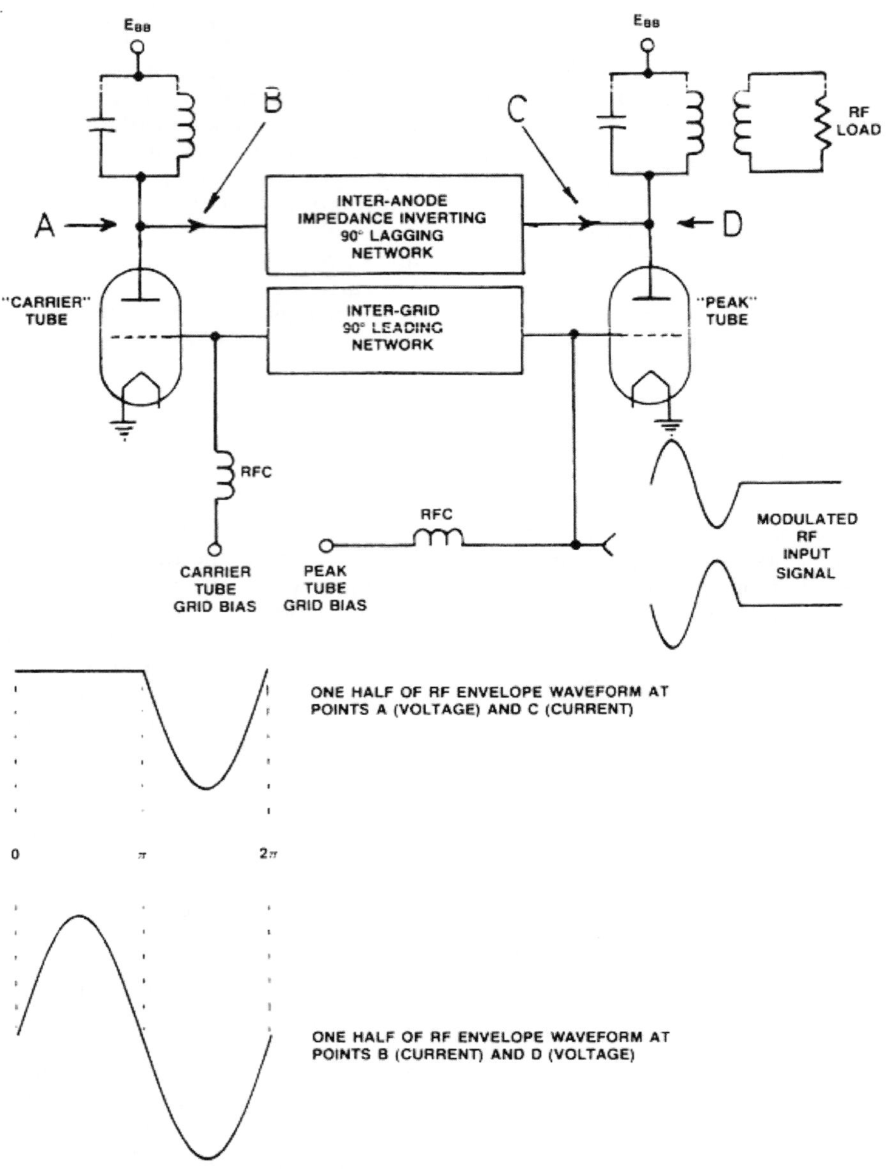

FIGURE 4.2-19 Doherty High-Efficiency Linear Amplifier AM modulator.

2. Operating power (Section 73.51)
3. Carrier output power delivered to the antenna system (Section 73.54)
4. Modulation capability
5. Total audio frequency harmonic distortion
6. System frequency response
7. Carrier-amplitude regulation (carrier shift)
8. Hum and noise output level
9. Carrier frequency tolerance (Section 73.1545(a))

All audio measurements should be made from a demodulated voltage or current sample at the antenna system common point.

Regarding item (3), the measurement of transmitter power output by the direct method described in FCC Rules Sections 73.51 and 73.1215 is potentially subject to more than 13% error if allowed FCC measurement inaccuracies are taken to the limit. For example, a common point impedance of 50 ohms resistive can be measured within 2% accuracy with a radiofrequency impedance bridge, a realistic tolerance. Further, a direct-reading RF ammeter having a full-scale reading of 100 amp and an FCC allowed tolerance of ±2% of full scale is used for this measurement and indicates 33.33 amp of common point current (just meeting the minimum FCC accuracy and indication requirements in FCC Rules Sections 73.1215(b)(2) and (3)) and producing 55.554 kW. Under these conditions, the actual

power delivered to the antenna is between the limits of $I^2R = 35.33 \times 35.33 \times 51 = 63.65$ kW, as a maximum, and $I^2R = 31.33 \times 31.33 \times 49 = 48.10$ kW, as a minimum, yielding a total measurement error of approximately +14/–13%. Using an RF ammeter with a 50 amp full-scale reading, the same FCC allowed inaccuracies, and the same meter indications as above would result in a power output measurement error of approximately ±8%.

FACTORY TESTS

When looking for an AM broadcast transmitter to buy for replacement or new equipment, it is advisable to ask manufacturers for specific and detailed test and performance data at the start of the investigation. Before a final decision is made, specific and detailed tests should be made at the manufacturer's factory under strict control of an experienced engineer or engineering consultant. Following are some suggestions on the kinds of tests and a discussion on the details of each test.

Audio Frequency Response

At one time, the term *broadcast quality* meant a standard to which all other system equipment, by comparison, was considered inferior. Today the audio quality of consumer high fidelity and stereo equipment often surpasses, with one exception, the signal quality that can be broadcast by any AM transmitter manufactured today or likely to be made in the future. The one exception is audio frequency response. Practically all mass production AM receivers made in any country of the world have intermediate frequency (IF) and audio amplifier bandpass characteristics that limit receiver –3 dB high end audio frequency response to between approximately 2500 and 5000 Hz, with 2500 Hz being the more common of the two figures. This limited high frequency response of the receiver is not a limitation of technical capability but rather a conscious decision on the part of receiver manufacturers to limit high end response in order to minimize perceived nighttime skywave co-channel and adjacent channel interference on the medium-frequency AM band. Typical low end –3 dB frequency response of consumer AM radios is between 100 and 300 Hz, with approximately 200 Hz being a common value. It is important to consider whether digital operation will be required for a particular transmitter. If digital operation will be required in the future, the audio frequency response will likely extend down to DC and up to about 50 kHz for the known digital systems.

Audio Harmonic and Intermodulation Distortion

Current production AM broadcast transmitters typically produce less than 2% total harmonic distortion (THD) at up to 90% modulation at any frequency of modulation between 50 and 10,000 Hz for monophonic transmission. Most of the modern solid-state digital systems produce typical audio distortion of less than 1% THD at most conditions of modulation. Intermodulation distortion (IMD) is considered to be a more disturbing kind of distortion than harmonic distortion, though both are important. The International Telecommunications Union, Radiocommunication Sector (ITU-R) method of IMD measurement is the preferred method for radio broadcast transmitters. With this method, two equal level audio tones separated by approximately 170 Hz are fed into the transmitter and the peak modulation level adjusted to between 85 and 95% modulation. The levels of odd and even order products are measured using an audio wave or spectrum analyzer connected to the test output terminals of a high quality modulation monitor. Two IMD measurements should be taken, one with the two tones near the middle audio band (*e.g.*, 400 and 570 Hz) and one with the two tones near the upper audio end (*e.g.*, 7000 and 7170 Hz). High quality broadcast transmitters should produce odd order IM distortion products more than 30 dB below the level of either of the two modulating tones. The root sum square (RSS) values of all IM products, relative to the level of either modulating tone, should also be less than 5% at 90% peak modulation levels.

Residual AM Hum and Noise

AM noise should be about 60 dB below the 400 Hz/100% modulation level and can be achieved by most current production AM broadcast transmitters. The bandwidth of noise measuring equipment should be at least 20 kHz. Typical modulation monitor demodulated audio bandwidth is approximately 25 kHz to accommodate these measurement requirements.

Residual PM Hum and Noise

Residual phase modulation (PM) noise is normally not a problem for modern AM broadcast transmitter designs for monophonic broadcasting. The quartz crystal oscillator circuits and even moderately careful RF component mechanical designs produce quite acceptable phase noise characteristics. However, for AM stereo applications, where a frequency synthesizer may be used for the RF signal generation, it is wise to perform a test to determine the purity of the synthesizer circuitry. At some level PM noise can convert to AM noise over certain nighttime skywave propagation paths and appear to distant listeners as objectionable AM hum and noise. An acceptable value of PM noise is –25 dB RMS relative to 1 radian peak, measured in a 15 kHz bandwidth, for monophonic medium wave band AM transmission. Transmitters with quartz crystal oscillator exciters typically exceed this recommendation by 25 dB or more. Transmitters used for international shortwave broadcasting (4 to 26 MHz) require PM noise levels of approximately –45 dB relative to a 1 radian peak because of more severe skywave PM to AM conversion at higher frequencies.

Incidental Phase Modulation

Like residual PM, incidental phase modulation (IPM) is more important in stations using or planning to use digital modulation formats or AM stereo. IPM is defined as the peak phase deviation of the carrier frequency (in radians) resulting from the process of AM. IPM values of several radians were common in the early days of AM broadcasting. Typical values of IPM for modern transmitters that have not been specifically designed or adjusted to minimize IPM range from about 0.1 to 0.5 radians peak (approximately 6 to 30°). A maximum acceptable value of IPM required for AM stereo operation is generally considered to be 0.05 peak radians with a desired value of less than 0.02 peak radians. State-of-the-art modulation meters such as the Hewlett-Packard® Model HP-8901B and other similar instruments are preferred for accurate PM and IPM measurements.

Carrier Amplitude Regulation (Carrier Shift)

The amount of carrier level shift in a given transmitter has more importance to overall transmitter performance than many broadcast engineers realize. Large values of negative carrier shift can have as much of an effect on effective transmitted sideband power as poor transient overshoot distortion. The term *carrier shift* may be somewhat confusing, especially to newcomers in radio broadcasting whom often equate the terminology with frequency shift instead of level or amplitude shift, hence the new terminology of *carrier amplitude regulation*. The ITU-R refers to the same characteristic as *carrier level shift*, which appeals to many engineers because of its closer adherence to the original terminology, but without the ambiguities. Carrier level shift is the effective shift in apparent carrier level due to the AM process. Carrier level shift can be caused by either poor power supply regulation or modulation even-order harmonic distortion, which generates a DC offset component in the modulated RF envelope, or both. Carrier level shift can be either positive or negative, although is usually negative because power supply regulation is most often the major source of carrier level shift and power supply regulation is most generally negative in sign (lower voltage output at higher current loads). Poor power supply regulation is not always caused by the transmitter power circuitry. It can also be caused by poor supply-line voltage regulation or more often a combination of the two.

Another common misconception regarding carrier level shift is that the defined level shift is a direct shift in carrier power. Actually, carrier level shift is defined as the shift in effective carrier *voltage or current* due to the process of modulation. This means that a carrier level shift of –5% is equivalent to a carrier power shift of approximately –10%:

$$P_{carrier(mod)} = P_{carrier} \times (1 - 0.05)^2 = 0.9025 \times P_{carrier}$$

Transmitters having no carrier level shift produce an average output power of 1.5 times the carrier power level with 100% sinusoidal tone modulation. Transmitters that exhibit –5% carrier level shift produce an average output power only 1.35 times the carrier level at the same conditions of symmetrical sinusoidal tone modulation. A broadcast station engineer or engineering consultant should have a complete understanding of these and other equally important transmitter characteristics in order to be able to make an intelligent buying decision based upon measurable and proven technical merit.

Audio Phase Linearity

Proper attention is given to phase linearity by most station engineers and engineering consultants. Station managers and engineers are concerned about the sound of their stations, and loudness, or perceived loudness, is a common criteria of quality in many stations with diverse programming formats. However, it is not uncommon for station engineers and program directors to spend more time researching the various objective and subjective merits of program processing equipment than the one characteristic in their transmitter that could partially neutralize the potential advantage from a new or different program processor. That one characteristic is audio phase linearity.

Audio phase nonlinearity and its major detrimental results, transient overshoot and low frequency tilt, have been discussed in previous sections of this chapter. Before the popularity of modern broadcast audio processing and some modern broadcast programming formats, audio phase linearity of a transmitter or any component in the program audio chain was not the concern that it is today. The FCC authorized 125% positive program modulation, allowing AM transmissions to accommodate certain naturally occurring asymmetry in voice and music, thus achieving a gain of about 2 dB of real loudness or actual program sideband power (20 log[1.25/1.00] = 1.94 dB). As stated earlier, some AM broadcast transmitters exhibit as much as 12% overshoot of a square wave input, due to phase nonlinearity, which has the effect of taking away one of those 2 dB (20 log[1/1.12] = –0.98 dB).

A simple way to determine the effects of phase nonlinearity is to require the prospective transmitter manufacturer to demonstrate rectangular wave modulation characteristics of the transmitter under investigation. When such a test is performed, the overshoot produced by the transmitter will be directly visible and measurable on the oscillographic display of the transmitter output envelope.

Occupied Bandwidth

Occupied bandwidth of an AM broadcast transmitter can best be measured with the use of a band-limited colored Gaussian noise source, similar to pink noise used in certain acoustical tests, to provide a continuous wideband modulating signal to the transmitter. Pink noise has an equal energy bandwidth ratio (equal energy per octave, per third octave, per tenth octave, etc.). White noise has equal

energy per bandwidth (equal energy per Hz). Both noise signals can have a Gaussian or pseudo-Gaussian probability density function, a probability density function that closely resembles that of typical voice and music. One method of specifying and measuring occupied bandwidth is by use of the NRSC-1 [7] standard published by the National Radio Systems Committee. This standard offers a recommended practice for audio preemphasis as well as measuring the resulting spectrum of the transmitted envelope. In addition to envelope spectral shaping, NRSC-1 recommends a dynamic measurement that uses a pulsed noise source to simulate program material. Use of a noise source with specified spectral shape allows repeatable measurements under controlled conditions. CCIR Recommendation 559-1 describes another method of occupied bandwidth measurement. Like the pulsed noise test of NRSC-1, this measurement of occupied bandwidth is a dynamic measurement that effectively summarizes nonlinear distortion effects on several important transmitter parameters such as IMD, THD, and IPM.

Harmonic and Spurious Output

Harmonic and spurious output of an AM broadcast transmitter or transmission system can be measured in two ways. One way is to use a sample of the transmitter RF output signal when operating into a dummy load with a known or accurately measurable impedance characteristic out to approximately the tenth RF harmonic. Then, with a calibrated measuring system, measurements and calculations are made to compute the output power of each harmonic according to the formula:

$$P_N = V_N{}^2 \times R_{PN} \tag{8}$$

where:

P_N = the power at the Nth harmonic.

V_N = the corrected measured voltage of the Nth harmonic at the calibrated impedance point PN.

R_{PN} = the parallel resistive component of the load impedance at the Nth harmonic.

A second way to measure the harmonic and spurious output of an AM broadcast transmitter is to measure the actual power radiated from the antenna system at each harmonic frequency or suspected spurious output frequency by standard field intensity measurement techniques. The field intensity method is the most meaningful of the two techniques because it allows the measurement to be made under actual conditions of operation with all systems interconnected. It is normally the joint responsibility of the new transmitter manufacturer and users to correct any actual interference problems to other broadcast or non-broadcast communication services even when the particular interfering signal meets the standard FCC requirements (FCC Rules Section 73.44.(c)).

Carrier Output Power

The most accurate method of measuring the RF output power of a transmitter is by the calorimetric method, a method that uses the very accurately known and measurable physical and thermal characteristics of water or other similarly well defined liquids. This measurement is usually only done in a transmitter manufacturer's factory because the capital investment required to purchase and maintain calibration of this kind of infrequently used equipment is usually not justified for AM broadcasting operations. Water has a thermal capacity very close to 4.186 Joules per °C per gram weight at a mean temperature of 60°C. A Joule is equal to 1 watt-second. Therefore, the capacity of water to absorb power is 69.8 watts per °C per liter of water flow per minute, or:

$$Power(kw) = Flow(lpm) \times \Delta T° \times 0.0698 \tag{9}$$

for water ow measured in liters per minute (Lpm) or,

$$Power(kw) = Flow(gpm) \times \Delta T° \times 0.2461 \tag{10}$$

for water flow measured in U.S. gallons per minute (gpm).

In both cases, the temperature differential must be measured in degrees Centigrade. The flow of water can be measured with an accuracy of approximately ±1%, by even the most common methods. Differential temperature measurement accuracy of approximately ±0.1°C is practical, which, for temperature differentials of 20°C, is equivalent to ±0.5% accuracy. Therefore, using calorimetric measurement techniques, the output power of AM transmitters can be measured with total accuracies better than ±2%. The RF output amplifier efficiency factor (F), referred to in Section 73.51(e), can then be determined for future operating and proof-of-performance reference; however, even with this method of determining factor F, an accuracy of less than ±2% cannot be maintained over the life of the equipment. The required FCC transmitter voltage and current meters, employed during indirect method power calculation, have a ±2% accuracy, in addition to the multiplied power accuracy of approximately ±4%. When accuracy levels are combined, this yields a total uncertainty of ±6% for efficiency factor F. Still, this is considerably better than the accuracies obtainable by the direct measurement technique.

Operating Efficiency and Input Power

The measurement of transmitter input power can be made under actual or simulated operating conditions. This requires program or simulated program signals for the transmitter modulating source during the period of power input measurement. The preferred method of measuring AC input power is by use of a standard rotating disk watt-hour meter. The measurement accuracy of these familiar watt-hour meters is typically better than 0.5%, better than four times the accuracy of any other conventional direct AC power measurement technique, and can be obtained with accuracies better than 0.1%. The watt-hour metering system should be connected in the main power feed line to the transmitter under test. Sinusoidal test signals, useful for other tests of transmitter performance, are not recommended for power consumption tests because of the distinctly opposite statistical characteristics of sinusoidal signals and typical voice and music program material and the effect that this difference has on actual operating power consumption measurements. This difference between periodic sinusoidal signals and mathematically random types of signals such as human voice or music is well known by manufacturers. The effect this difference produces in AM broadcast transmitter power consumption and operating efficiency measurements, however, was first documented in 1980, by investigators in Europe and was later verified and further explained by investigators in the United States and other countries [8,9].

The critical difference between sinusoidal signals and typical program types of signals is explained in Figure 4.2-20 which shows the amplitude density characteristics of a sinusoidal waveform (U-shaped curve) and an amplitude density characteristic of typical broadcast program modulation (pyramid curve). The different shapes of the two curves cause the measured transmitter efficiency at identical root mean square (RMS) modulation levels also to be different. The pyramid shaped curve was generated in 1984 from measurements taken off the air of five differently formatted FM radio stations in Dallas, TX. Data taken continuously for 24 hours for each station and averaged for presentation are also shown in Figure 4.2-20. FM stations were used in the collection of the data because of their consistent day/night signal levels, symmetrical modulation characteristics for easier comparison to the sine wave, and omnidirectional emission pattern. Some transmitters, or more correctly modulation systems, are more efficient with program modulation than with equivalent RMS sinusoidal test signals, while others have poorer efficiency with program modulation. It has been shown that modulation systems or techniques that yield higher carrier efficiencies also yield higher program efficiencies. This results in lower program power consumption than with an equivalent RMS sinusoidal modulation test signal [6]. The program input power consumption test

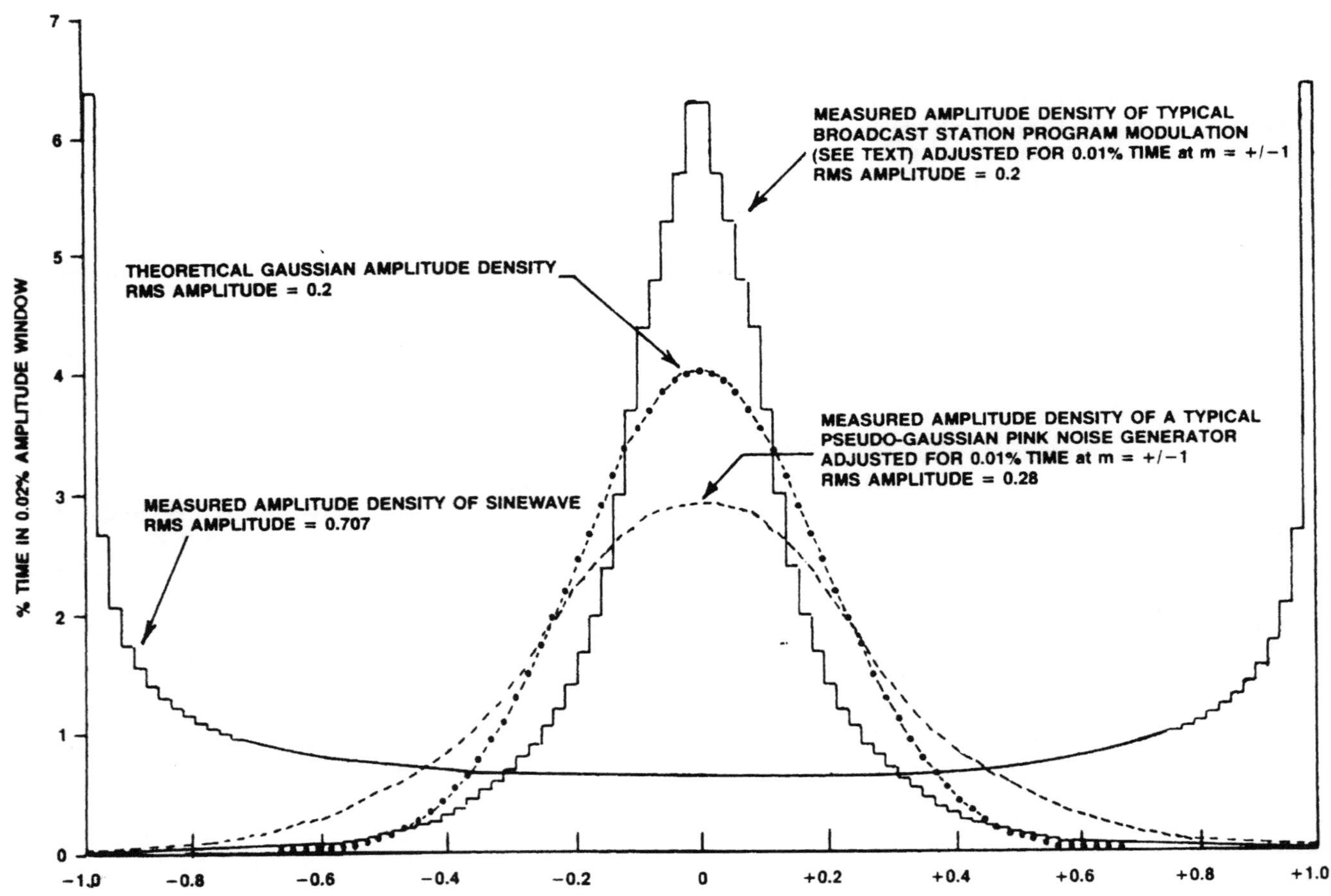

FIGURE 4.2-20 Graph showing differences between measured amplitude densities of sinusoidal signals and typical program types of signals for determining transmitter power consumption.

is the single most meaningful test for broadcasters to understand because it represents actual broadcast programs, rather than sinusoidal tones. In this discussion, the terms *program efficiency* and *program input power consumption* have generally been given equal weighted value. It is correct to equate the weighted value of these two parameters for transmitter-to-transmitter comparisons if it is assumed or defined that average transmitter-to-transmitter output power is constant for a defined program input. Such an assumption is correct except that transmitters with excessive transient overshoot will have correspondingly less average modulated RF power output for given peak levels of processed program modulation. Experimental methods have been devised to measure average long-term transmitter output power (RF kilowatt-hour output) [10]. With such equipment it would be convenient to accurately compute actual program efficiency knowing both kilowatt-hours input and kilowatt-hours output. Until such equipment becomes commercially available, however, only the transmitter power input can be accurately measured with single-phase and multi-phase kilowatt-hour instruments.

Figure 4.2-20 also shows a third curve (dashed curve) which is the amplitude density function of a popular pink noise generator. To conclude, either a pink noise signal, with amplitude density characteristics similar to the curve in Figure 4.2-19, or a recording of actual program modulation should be used as a program source for factory input power consumption tests. Program processing equipment, if used, and peak modulation levels should be adjusted to satisfy normal station operating procedures and transmitter input power measurements then taken during at least a 30 minute segment of the program source material. (See FCC Rule Sections 73.1570(a) and (b) for modulation setup procedures.) The average input power determined by the method described will be very close to the transmitter power consumption during its operating life and therefore can be used to accurately predict actual operating energy costs. This is the only method of transmitter power consumption tests that will provide accurate energy consumption forecasting data.

SHORTWAVE TRANSMITTERS

Shortwave broadcast transmitters are similar in many respects to medium wave transmitters and are very different in others. The similarities are in methods of modulation, control, and monitoring. The differences are, generally, that shortwave transmitters are higher power, more complex in tuning, and more difficult to operate and maintain than medium wave standard broadcast transmitters. Although there are numerous exceptions, the general rule is that the minimum usable carrier power level is 50 kW and the maximum economical carrier power level from a single transmitter is 250 to 500 kW. It is not unusual for a shortwave transmitter to operate on five to ten separate frequencies every broadcast day. Modern broadcasting schedules of the prestigious broadcasting organizations are very tight, thus necessitating built-in automatic frequency changing circuitry, which allows the transmitter to tune to several programmed frequencies in approximately 10 to 30 seconds, with little or no operator intervention. The trend in shortwave transmitter operation is toward unattended or minimum attended sites, with program and frequency changes done by remote or computer control.

ONLINE RESOURCES

A number of online resources may prove helpful to the engineer interested in AM transmitters:

- http://www.fcc.gov (online regulations for broadcasting in the United States)
- http://www.ibiquity.com (information on IBOC digital broadcasting)
- http://www.nrscstandards.org (information on industry standards in the United States)
- http://www.drm.org (information on DRM digital broadcasting)
- http://www.etsi.org (information on industry standards worldwide)

AM STEREO BROADCASTING

The proper operation of stereo transmission for AM is relatively more difficult than its FM counterpart. This section discusses the more practical aspects of the FCC accepted Motorola System stereo preparation, installation, and maintenance to aid the AM broadcaster in achieving a high quality stereo AM transmission.

Station Preparation for AM Stereo

In addition to preparing the transmitter for AM stereo, considerable changes may be necessary at the studio and the interconnecting link (studio-to-transmitter link, or STL).

The Audio Chain

Unless the current AM studio is outfitted with stereo equipment, it will be necessary to install new stereo sources such as CD players, consoles, and other distribution facilities. One area of concern is to ensure proper audio phase and amplitude matching of left and right channels throughout the facility. Without adequate tracking between both channels, monophonic frequency response and stereo imaging will be degraded.

Another consequence of poor channel matching is a rapid degradation in L+R (monophonic) to L–R (stereo) and LR to L+R crosstalk. This parameter typically degrades with frequency, reducing high frequency monophonic coverage and altering the high frequency stereo image. For example, a 0.1 dB amplitude error combined with 1° of phase mismatch will limit L+R to

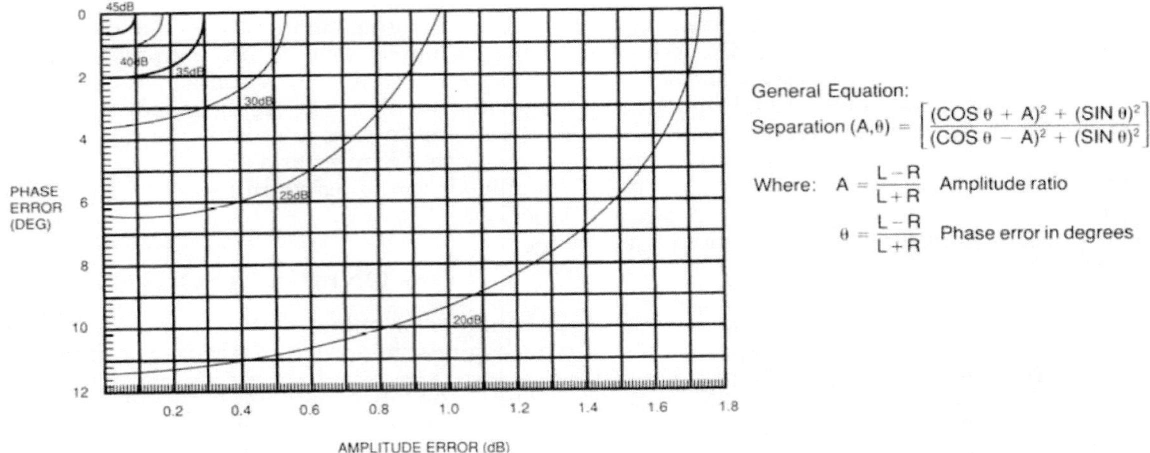

FIGURE 4.2-21 Crosstalk (separation) versus amplitude and phase errors in a matrix system.

L–R and L–R to L+R crosstalk to about 40 dB. This relationship holds true for stereo separation during transmitter equalization. Figure 4.2-21 shows the correlation of phase error and amplitude and the resulting change in stereo separation.

AM Stereo Processing

Due to the nature of AM stereo transmission, the audio paths are different than its FM multiplex counterpart. Rather than operating in a discrete left and right channel mode, AM stereo transmission requires the use of matrix mode for transmission, because the L+R (monophonic) audio comes from amplitude modulation of the RF carrier and the L–R (stereo) information comes from phase modulation of the RF carrier. Unlike composite FM stereo, these paths are independent of each other and may also be processed independently. In fact, because the two audio paths are combined in matrix form before the actual modulation occurs, processing for AM stereo is most effective if it is also done in matrix form.

Figure 4.2-22 is a representation of a common and very effective tool used in AM stereo installation and maintenance: the XY or Lissajous pattern of a two-channel oscilloscope. Standard operation is with the X input driven by the left channel, and the Y input driven by the right. This will produce a straight line display of 45° in the first and third quadrants for total L+R modulation and a straight line display of 45° in the second and fourth quadrants for total L–R information.

If conventional left and right channel processing is used, the limiting must be done symmetrically and at a level equivalent to 50% L+R and L–R modulation if only one channel were being applied. In this manner, 100% limiting will occur when L = R for the envelope (amplitude) modulation, and when L = –R for the phase modulation. However, under single channel conditions, there is a 6 dB loss in both monaural loudness and coverage; this undesired effect can be improved if the processing is done in matrix form.

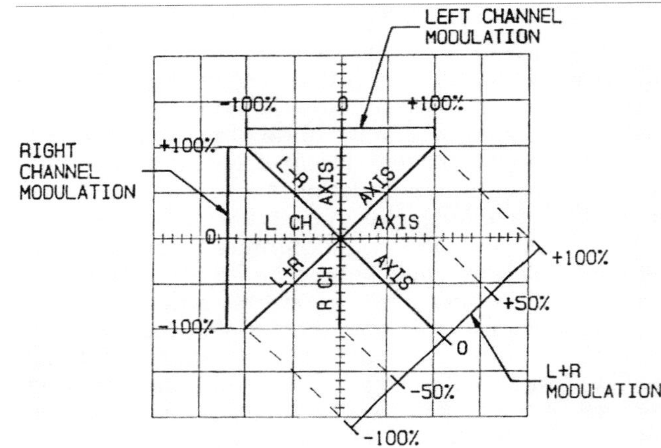

FIGURE 4.2-22 Conventional X–Y oscilloscope display.

Figure 4.2-23 shows the classic diamond shape produced by full matrix stereo processing. With this method, left and right channels are increased under single channel conditions and under heavy stereo conditions to maintain full envelope modulation. This mode of operation is called *full monaural support matrix stereo limiting.*

Unfortunately, instead of the 6 dB loss in monaural loudness associated with the conventional processing method, the full matrix method results in a 6 dB *increase* in stereo single channel loudness; however, this has been found to be less objectionable than the alternative and, combined with the infrequent nature of single channel conditions, is a more desirable side effect for AM stereo.

There is still one aspect of full monaural support matrix stereo limiting that makes it unacceptable. The decoder for the C-QUAM® AM stereo system (Appendix A) requires that the left and right channels be limited to –75%, where –100% is equivalent to full envelope modulation caused by a single channel

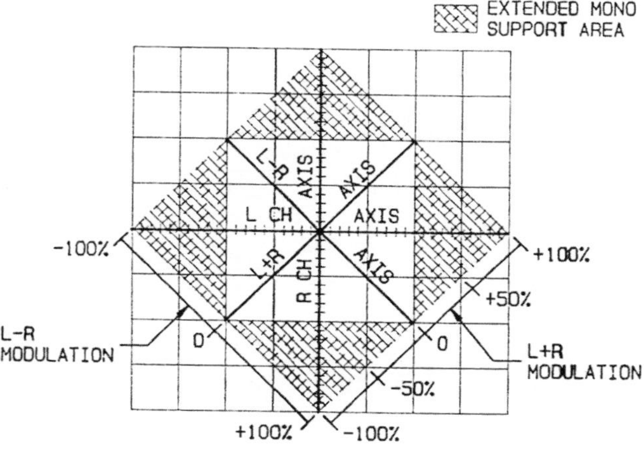

FIGURE 4.2-23 Matrix stereo processing.

input. All professional AM stereo matrix processors include a single channel limiter to prevent these problems. Proper alignment and operation of any matrix AM stereo processor require a thorough understanding of these departures from conventional FM stereo processing, as well as the practical limitations imposed by AM stereo hardware designs. Without a proper grasp on these concepts, the processing can do more harm than good and has often been the source of great frustration during installation and maintenance.

Studio to Transmitter Link

Many AM stations operate studios at locations other than that of the transmitter; therefore, there must be some form of studio-to-transmitter link to carry the audio information to the transmission facility. Traditionally, this has taken one of two forms: equalized phone lines or a microwave radio link. Stereo transmission requires a second link to be installed that, if not done carefully, can be a serious limitation to proper stereo operation.

Equalized Telephone Lines

Installing a second phone line is often the first choice for AM stations that are currently using one. Two of the more common complaints with this approach are the increase in monthly fees associated with the second line and the troublesome problem of maintaining the amplitude and phase matching between the two lines. Equalized phone lines typically have complex filters and frequency shaping equalizers. This makes them difficult to match and also prone to drift with time and temperature. The complexity of the circuits can also increase the total possible phase shift, making complete monophonic cancellation, or *combing*, a problem under extreme conditions. This same phenomenon will also cause mono to stereo image rotation, which will cause unexpected stereo imaging and excessive phase modulation. Unfortunately, these paths are not under the direct control of the station and are dependent on the local telephone company for support.

Radio Links

Installing a new radio link is considered a more favorable alternative providing the frequency allocation is available. It also relieves the monthly financial burden associated with rented phone lines. However, discrete left and right STL transmission can suffer some of the same problems with amplitude and phase matching, but usually not to the same degree as phone lines. The STL transmitters and receivers will have audio, RF, and IF filters, and, depending on the design, construction and alignment, which must be well matched.

Transmitter Preparation

Depending on the age and model of the current transmitter, preparing it to accept AM stereo can range from minor modifications to the nearly impossible. Unfortunately, there are no well defined procedures for making a transmitter stereo ready. In general, the work that will be required can be broken down into a few basic categories.

General Maintenance

The first step to high quality AM stereo is to have high quality AM monaural. Many AM transmitters have been neglected for years. A monaural proof-of-performance will show if the system is up to specifications. If not, it must be fixed so that it produces a decent proof. This is money well spent, as it will improve on-air sound quality for all listeners, both monaural and stereo. It will also simplify stereo installation because there will be enough other problems to compensate for without the added burden of poor monaural performance.

Factory Modification Kits

Most manufacturers of AM transmitters will provide support to help upgrade their transmitters for AM stereo because they know where the stereo problems will be for any given model. In addition, some manufacturers have standard modification kits available, with instructions, to prepare the transmitter for stereo.

Incidental Phase Modulation Reduction

Incidental phase modulation (IPM) is broadly defined as any undesired angular phase shift of the RF carrier and falls into two subcategories: IPM caused by power supply ripple (or induced magnetic fields) and IPM caused by envelope modulation. The first form of IPM most often results from insufficient power supply bypassing. Prior to AM stereo, transmitter manufacturers were not concerned about IPM, because it was not decoded by conventional envelope detectors and generally did not affect the AM signal. It was simply not worth the additional cost to provide the extra bypassing. The effect of IPM is to limit the decoded L–R signal-to-noise ratio (S/N), and it is often the dominant stereo noise component. Any improvement in this area will improve the decoded stereo S/N. It should be possible to reduce this form of IPM to 45 to 55 dB below 100% L–R modulation.

The second form of IPM is much more troublesome to reduce and more detrimental to quality AM stereo. This mode of IPM is created by amplitude modulation of the carrier and can be caused by several mechanisms. For tube type transmitters, it is often the result of poor neutralization of the final PA stage and can result in equivalent LR modulation levels in excess of 25%. The interelectrode capacitances found in vacuum tubes can cause instabilities. In particular, the plate-to-grid feedback capacitance provides a path for positive feedback. Figure 4.2-24 shows the static interelectrode capacitances found in a tetrode. The C_{gc} and C_{gp} are the two capacitances of importance for proper neutralization.

The intent of neutralization is to cancel the grid-to-plate capacitance (C_{gp}) to improve stability and reduce IPM. The most popular method of neutralization is the *capacitance bridge*. Figure 4.2-25 shows a simplified tetrode configuration employing classic bridge neutralization. C_{gp} and C_{gc} are the grid-to-plate and grid-to-cathode interelectrode capacitances, respectively. C is comprised of any input capacitance, including stray capacitance, and is required for complete neutralization of C_{gc}. C_n is the neutralization capacitance required to balance C_{gp}. For proper balance (proper neutralization), the capacitor ratios must satisfy the following relationship:

$$\frac{C_n}{C} = \frac{C_{gp}}{C_{gc}}$$

With proper neutralization, IPM levels should be able to be reduced, typically to 35 to 45 dB below 100% L–R modulation.

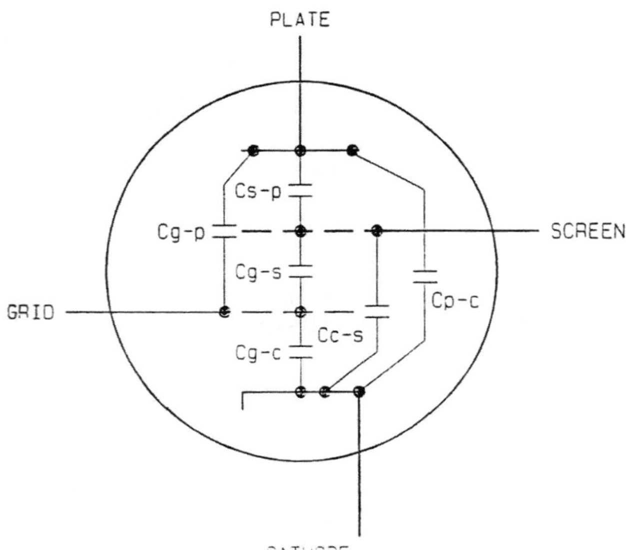

FIGURE 4.2-24 Static interelectrode capacitances of a tetrode tube.

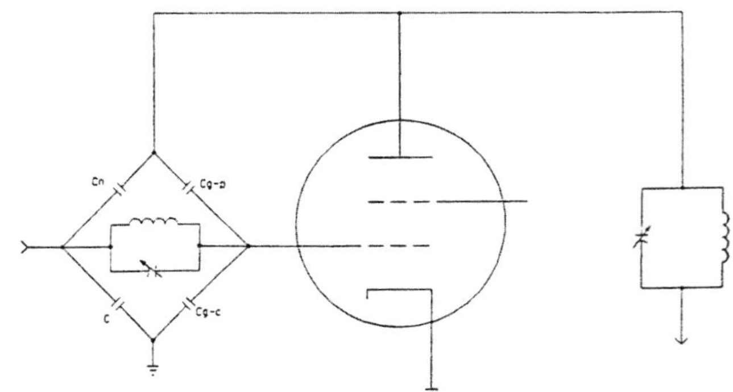

FIGURE 4.2-25 Bridge neutralization.

For solid-state transmitters, IPM is most often caused by the nonlinear output capacitance of the solid-state device. The value of this capacitance is a function of the applied collector or drain voltage. Because most solid-state transmitters use high level modulation, this voltage varies at the audio rate that will affect the RF phase angle. Fortunately, most solid-state transmitters are relatively recent designs, and many new transmitters are shipped stereo ready, requiring no additional effort to reduce IPM.

Tuned Circuits

Some older transmitters used narrow bandwidth-tuned RF circuits as input, interstage, and output coupling networks. Proper stereo operation requires that the amplitude and phase relationship of the L+R (envelope) information match that of the L–R (phase modulated) information. These narrowband-tuned networks have three undesired effects. First, they introduce a nonuniform time delay to the RF carrier, resulting in complex phase equalization problems. Second, they cause L–R high frequency response problems, further aggravating the equalization requirements. Third, the narrowband nature of the network alters the sideband structure of the phase modulation, resulting in increased distortion. It is a fairly common practice to increase the bandwidths of these circuits to improve the stereo signal; however, reducing the Q of a tuned stage will result in a decrease in voltage at the output of the network which may result in inadequate signal to drive the following stage. Consult the factory or a knowledgeable consultant before attempting this modification.

Antenna and Phasor Alignment

All AM stereo transmissions are sensitive to amplitude and phase nonlinearities to one degree or another. For best performance, the antenna system should be as broadband and as symmetrical as possible. The C-QUAM® system especially requires good transmission bandwidth and phase linearity. Chapter 4.4 discusses the phasing and coupling networks employed in AM nondirectional and directional operations, and Chapter 4.5 addresses modern computer analysis and modeling of AM antenna systems.

Monaural Proof-of-Performance

One final step should be carried out before the actual stereo conversion begins. A complete monaural audio proof-of-performance should be conducted to ensure the system is ready to accept stereo and as a record to compare final monaural performance after installation.

STEREO EXCITER INSTALLATION

With a properly operating monaural transmission chain in place, the actual transmitter conversion to AM stereo may begin. Mount the AM stereo exciter close enough to the transmitter so the interconnecting cable is no more than 30 ft long.

RF Interfacing

Because AM stereo transmission uses some form of phase modulation of the RF carrier, the AM stereo exciter will have a phase modulated RF output that will replace the transmitter's internal RF oscillator. The rated output power varies from exciter to exciter, from a few watts up to 10 W. It will be necessary to pick a suitable RF insertion point to break the internal RF chain and replace it with the RF from the exciter. Stereo-ready transmitters will be outfitted with a BNC connector, usually requiring a transistor–transistor logic (TTL) level (5 V, peak-to-peak) signal for proper operation. Other transmitters require modification to provide a path into the RF sections. If the particular type of transmitter uses narrow tuned RF stages, select the farthest point in the transmitter that the exciter can provide an adequate amount RF drive.

Providing the Proper Termination

The RF output of an AM stereo exciter is designed to operate in a 50 Ω system, so 50 Ω coax must be used along with a 50 Ω termination to prevent reflections on the line. This is important if the RF output is a square wave, which it is most likely to be. Reflections on the line can cause excessive ringing which for TTL inputs results in improper logic operation and may damage the input stage due to excessive voltages. For TTL-compatible input levels, a 1/4 W resistor is adequate, provided that the input to the transmitter is AC coupled, then DC restored. If higher power levels are required, select the proper resistor power rating based on the following formulas:

$$P_{diss} = \frac{(V_{rms})^2}{50} \text{ (for sine wave input)}$$

$$P_{diss} = \frac{(V_{peak})^2}{50} \text{ (for square wave input)}$$

Internal RF Oscillator

Because the transmitter's original RF oscillator has been replaced by the AM stereo exciter some problems may occur. Coupling between the original oscillator and the new RF signal may result in a beat frequency equal to the difference in frequency of the two oscillators. To cure the problem, remove the original crystal from the transmitter or disable the power supply to the oscillator. This minor modification will still allow quick replacement should the original oscillator be needed in an emergency.

Audio Interfacing

The left and right audio channels must be connected to the exciter and the levels set according to the manufacturer's recommendations. The existing monaural input to the transmitter should be replaced by the monaural output from the AM stereo exciter. All audio connections should be made using a high quality shielded cable to prevent RF pickup on the cable. To prevent ground loops, ground the shield at only one end.

AM STEREO EQUALIZATION

Because half of the stereo information is contained in the monaural (L+R) connection to the transmitter and half is in the form of phase modulated RF (L–R), both must arrive at the modulator portion of the transmitter at the same time. This assures that left and right channels may be properly recovered in the receiver after dematrixing. The amount of delay and amplitude equalization required for one type of transmitter is different from any other. In addition, the circuits in the transmitter that cause the problems are often complex, so the best that any AM stereo exciter equalizer can do is to approximate the response shape. The effectiveness of the equalizer is dependent on the type of transmitter and the complexity of the antenna pattern at any given station. Each installation is unique. Also, equalization may be different for different modes of antenna operation (day, night, critical hours). As a result, equalization is often a trial and error routine, especially at the higher audio frequencies where the response shape is most complex; however, AM stereo exciters designed specifically for one brand of transmitter equalize the two paths almost perfectly.

Types of Equalization

There are three different equalization sections in an AM stereo exciter. Each is intended to fix a particular frequency range. Some installations require all three, others only two, and, very rarely, only one will be needed.

Group Delay

The first equalization section which is almost universally required is the group delay equalizer. Its purpose is to provide a constant time delay for the L+R and L–R paths. It is primarily used to equalize the low to middle frequencies (up to a few kilohertz). If the response shape of the offending path in the transmitter is low pass (or bandpass for the RF path), the time delay for the lower frequencies will be constant so the alternate path can include this equalizer to compensate adequately. A common example would be any AM transmitter employing pulse-width modulation; in this case, the L+R path would include a PWM low-pass filter in the modulator just ahead of the RF amplifier. The L+R audio will have significantly more delay than the L–R (RF phase modulation) path. The group delay equalizer would be added to the LR path to slow it down so it arrives at the RF amplifier at the same time as the L+R signal does. Some transmitters will have more low/middle frequency delay in the LR path due to several RF interstage bandpass filters. In this instance, the group delay equalizer would be added to the L+R path to compensate.

High Frequency Equalization

The second audio equalizer is designed to approximately complement the complex high frequency response shape of most AM transmitters. The amount and shape of high frequency correction vary from transmitter to transmitter, which makes this section the most difficult adjustment of the installation process. Because this section is only meant to approximate the high frequency amplitude and delay characteristics, it will not correct at all frequencies. It is not uncommon to be caught in a seemingly endless loop of adjusting for one frequency, only to find another frequency becomes worse. The key to a successful installation is the ability to recognize a good overall compromise and stop. This is one area where an engineering consultant with AM stereo installation practice will be helpful.

Low Frequency Equalization

The third type of equalizer found in the exciter is the least often used section. It is only necessary if the low-frequency response of the L+R path is nonuniform. This is found primarily in older plate modulated transmitters where the reactance of the plate transformer causes phase and, to a lesser degree, amplitude nonlinearities. Some transmitters will also include an active audio high pass filter in the monaural input that can be effectively corrected by the low frequency equalizer.

Equalization Path Selection

The proper alignment of the three equalization sections is essential for high quality AM stereo. Before they can be adjusted, however, they must be inserted into either the L+R path or the L–R path, depending on where they are needed. What results is an iterative method. It may be necessary to change paths during the installation procedure as other equalizer sections are included. This could happen, for example, when placing the high frequency equalizer in series with the group delay equalizer forces a constant delay to be added to the other path to compensate for low- and mid-frequency delay found in the high frequency equalization.

Equalization Procedure

Figure 4.2-26 shows a typical test setup for AM stereo equalization. Test equipment required can be as simple as a low-distortion audio oscillator/analyzer, an AM stereo modulation monitor, and a dual-trace oscilloscope with XY capability. A spectrum analyzer capable of resolving audio sidebands at AM band frequencies can be useful, but is not necessary.

The XY Oscilloscope

The main problem in AM stereo equalization is determining the type and path of equalization required. In addition, it is necessary to determine if the system requires phase or amplitude equalization. A dual-trace oscilloscope operating in the XY or Lissajous pattern mode is capable of resolving between amplitude and phase errors (or a combination thereof), as well as showing a multitude of other information such as harmonic distortion, IPM, negative limiting, and modulation compression. It is one of the most useful tools in AM stereo installation without which it would be almost impossible to equalize the system.

AM Stereo Modulation Monitor

The final performance results of a stereo installation will only be as good as the equipment used to measure them. An important piece of test equipment for proper installation and routine maintenance is the AM stereo modulation monitor.

An Equalization Example

AM stereo exciters will have similar equalization adjustments but may be labeled differently. The *day*

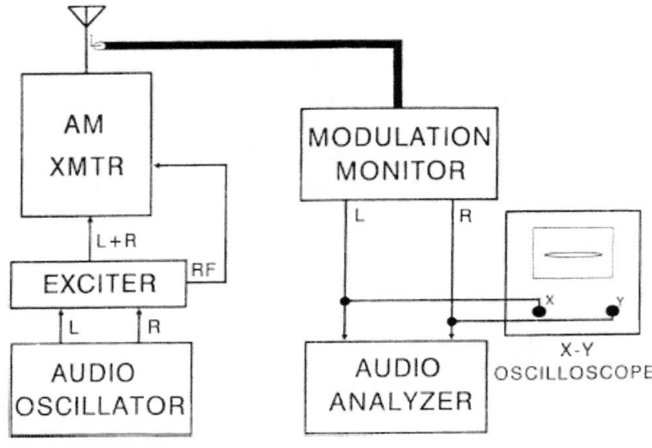

FIGURE 4.2-26 Stereo equalization test setup.

adjustments are used in this example; however, the *night* procedure is identical. The first adjustment to be made is to set the exciter's L+R, or envelope, level to get the correct amplitude modulation for a corresponding L=R input. This is, at this stage, only a coarse adjustment and can be done at any convenient modulation percentage. For example, input 1 kHz L=R to the exciter such that the L+R meter on the exciter reads 80%, then adjust the envelope output level of the exciter until the transmitter is modulating 80% as measured on the modulation monitor.

Next, apply a 1 kHz tone to the input of the left channel only, at a level equivalent to 50% amplitude modulation. The oscilloscope should show a horizontal line (assuming the X input is the left channel). Most likely, it will be neither exactly on the X axis nor a perfectly straight line. If it is not possible to obtain satisfactory results, the polarity of the L+R connection to the transmitter may be wrong. This can easily be fixed by reversing the plus and minus terminals at the transmitter input.

Next, the fine adjustments to maximize separation can be made. While monitoring both the residual right channel separation and the oscilloscope, insert the left channel group delay in either the L+R or L–R path and adjust the coarse group delay. If the display opens up rather than closes and the separation worsens, it has been inserted in the wrong path. Reverse the location, and adjust the coarse and fine group delays until the display closes, then fine-tune the envelope level until the display lies exactly on the X axis and the left to right channel separation nulls.

Apply a 1 kHz tone to the input of the left channel only, at a level equivalent to 50% amplitude modulation. The oscilloscope should show a vertical display with only phase error. Insert the right channel group delay in the same path as the left channel and adjust until the display closes. The residual separation in the left channel will also null. Theoretically, the envelope level adjustment should not have to be adjusted for the right channel, but often the right to left separation can be improved slightly by fine-tuning the level at the expense of left to right separation. It is best to compromise on the two to obtain the same amount of separation left to right as right to left. Some exciters are equipped with a balance control to help this problem. Refer to the individual service manual for the correct adjustment procedure.

The next step is to check the low frequency separation. A good frequency to use is 100 Hz. If the separation has degraded significantly, the low frequency equalizer can be used to correct the problem. Again, this is usually only necessary in plate modulated transmitters or ones that employ lowlevel audio input high pass filters. The procedure is very similar to the group delay adjustment. If low frequency equalization is used, recheck separation at 1 kHz and adjust the group delay, if necessary, to compensate.

At this point, the transmitter should be equalized for good separation (30 to 40 dB) from 50 Hz to a few kilohertz. A quick check of separation at 5 to 10 kHz will most likely show a rapid degradation in separation performance. With the left channel high frequency (HF) equalization controls set at minimum, insert the HF equalizer in one of the paths. Input 7.5 kHz to the left channel at a level sufficient to produce 50% envelope modulation. Observe the oscilloscope display while adjusting the HF cutoff control. If the phase degrades (display opens), then switch the paths. If it improves, adjust until it closes. Input 1 kHz and readjust the group delay to compensate for the added mid-frequency delay, and repeat until 1 kHz and 7.5 kHz are equalized for phase.

If the display at 7.5 kHz is not on the X axis, adjust the HF peaking and then the HF cutoff until the display lies on the X axis with minimum phase error. Again, adjust the group delay control at 1 kHz to compensate, and repeat the procedure until maximum separation is achieved at both 1 kHz and 7.5 kHz. Repeat the procedure for the right channel. When this is done, spot check the separation performance from 1 to 10 kHz. Using good judgment, it may be necessary to compromise the performance at 7.5 kHz to improve it elsewhere. The procedure is identical with the exception of X kHz in place of 7.5 kHz. In practice, it should be possible to achieve greater than 20 dB separation out to 10 kHz.

The Stereo Proof

When satisfactory separation performance is obtained, the final step in the AM stereo installation is to run a full proof-of-performance, both monaural and stereo. In particular, pay attention to high frequency distortion. It may be necessary to fine-tune the equalization to reduce distortion at the expense of a few decibels of separation. An acceptable proof should show less than 3% THD and greater than 20 dB separation at all frequencies. This proof is a valuable tool during routine maintenance to gauge performance and to assess whether the equalization requires adjusting.

ROUTINE MAINTENANCE

Any change in the transmission system amplitude or phase response will cause a degradation in stereo performance. These changes will occur for several reasons, including seasonal environment changes, routine transmitter tuning, and vacuum tube variations over time. It is a good practice to include a spot check of stereo performance as a routine maintenance item every 6 months. Left unattended, normal system variations can cause a serious degradation in audio quality, but a small investment in time for routine maintenance will ensure continued high quality AM stereo performance.

SUMMARY

Properly installed, AM stereo can offer good-quality AM audio with good stereo separation. Proper planning, execution, and follow-up to the installation are essential to avoiding the creation of new problems.

References

[1] Carson, J.R., Notes on modulation theory, *Proc. IRE*, 10(1), 57, 1925.
[2] Swanson, H.I., Polyphase PDM Amplifier, U.S. Patent No. 4,164,174, August 1979.
[3] Kahn, L., Single-sided transmission by envelope elimination and restoration, *Proc. IRE*, 40, 803–806, 1952.
[4] Doherty, W.H., A new high efficiency power amplifier for modulated waves, *Proc. IRE*, 24(9), 1163, 1936.
[5] Weldon, J.O., Amplifiers, U.S. Patent No. 2,836,665, May 1958.
[6] Heizing, R.A., Transmission System, U.S. Patent No. 1,655,543, January 1928.
[7] National Radio Systems Committee, NRSC-1 NRSC AM Pre-emphasis/Deemphasis and Broadcast Audio Transmission Bandwidth Specifications, July 1998.
[8] Sempert, M. and Tschol, W., Efficiency of high power broadcasting transmitters in regular programme service, *BBC Rev.*, July 1980.
[9] Woodard, G.W., Efficiency comparison of AM broadcast transmitters, *IREE J. (Australia)*, 2(2), 1982.
[10] Woodard, G.W., Simulating typical program modulation measurements of operating efficiency and modulation capability of AM broadcast transmitters, *Radio Electr. Eng. (U.K.)*, 53(9), 325.

APPENDIX A: MOTOROLA C-QUAM® AM STEREO SYSTEM

Greg Buchwald

Motorola, Inc., Schaumburg, IL

C-QUAM System Equations

Any broadcast signal can be broken into three major components: amplitude, frequency, and phase. The equation for a monophonic transmission can be described by:

$$E_R = (1 + L + R)\cos(\omega_c t + \varphi) \quad \text{(A1)}$$

where 1 represents the carrier, $L + R$ represents the monophonic (or left and right) information to be sent, $\omega_c t$ represents the carrier frequency, and φ represents phase modulation information that is, ideally, zero.

In fact, if one uses ±100% modulation of $L + R$ as an example and substitutes ±1 into the equation, it is obvious that at $L + R = +1$ the carrier level is instantaneously twice as high, and at $L + R = -1$ the carrier level is instantaneously 0.0, or cutoff has occurred. From this equation, it is also obvious why negative modulation is limited to 100% while positive modulation can exceed 100%. In the United States, the positive limit is 125%, or 1.25 in the equation.

To insert stereophonic information, one could alter the amplitude, frequency, or phase of the transmitted signal. Altering the amplitude is to be avoided because the envelope would no longer represent $1 + L + R$, but instead, a distorted component containing $L + R$. Substantial alteration of the frequency is also to be avoided. This leaves only the phase component available for modulation.

One method of adding a second channel of information to an existing carrier and utilizing the same spectral assignment is to make use of quadrature modulation (QUAM) techniques. Linear QUAM, not unlike that used to convey the chroma information in NTSC color transmission, can be used to convey the second channel of information. Advantages of QUAM include:

- No increase in occupied bandwidth
- Little S/N degradation
- No loss of existing coverage
- The potential for synchronous reception techniques

The major disadvantage, however, is that the envelope term is not $1 + L + R$, but instead is:

$$\sqrt{(I+L+R)^2+(L-R)^2} \quad \text{(A2)}$$

which leads to high levels of distortion in current monaural, envelope detector receivers and some difficulty in the conversion of existing broadcast transmitters due to additional requirements placed on the modulator stage.

The C-QUAM system was derived from QUAM; therefore, it retains, to a large extent, the advantages of

QUAM. In fact, C-QUAM can also use synchronous detection techniques particularly when conditions warrant. One difference between generation of the C-QUAM signal *versus* the QUAM signal is found in the envelope audio term. By substituting the distorted term required by QUAM with the simple summation of the left and right channels, the envelope is made compatible with existing receivers. Mathematically, the system has been designed as follows.

The equation for the QUAM signal is:

$$E_R = \sqrt{(I+L+R)^2+(L-R)^2}\cos(\omega_c t+\phi)$$

where:

$$\phi = \tan^{-1}\left[\frac{L-R}{1+L+R}\right]$$

The desired envelope component is $E = (1 + L + R)$. The cosine of the instantaneous phase modulation is:

$$\cos\phi = \frac{I+L+R}{\sqrt{(I+L+R)^2+(L-R)^2}}$$

If QUAM is multiplied by $\cos\varphi$:

$$\frac{I+L+R}{\sqrt{(I+L+R)^2+(L-R)^2}} x\sqrt{(I+L+R)^2+(L-R)^2}\cos(\omega_c t+\phi)$$

Then the resultant is:

$$E_R = (I+L+R)\cos(\omega_c t+\phi)$$

where:

$$\phi = \tan^{-1}\left[\frac{L-R}{1+L+R}\right]$$

In the process, the envelope term is made compatible by sending $1 + L + R$ while the in-phase (I) and quadrature (Q) components are multiplied by $\cos\varphi$; therefore, the broadcast C-QUAM signal has the following characteristics:

Envelope (E) = $1 + L + R$

In-phase (I) = $(1 + L + R)\cos\varphi$

Quadrature (Q) = $(L - R)\cos\varphi$

It can be seen that the mono $(1 + L + R)$ signal may be directly derived from the envelope detector output, while a quadratine detector combined with division by $\cos\varphi$ may be used to demodulate the $L - R$.

The C-QUAM Encoding System

There are several forms by which the C-QUAM encoder may be implemented. In the first form, a series of linear, balanced multipliers are employed to generate a quadrature modulated signal. The signal is then amplified to the point of limiting in order to remove the QUAM envelope term. The summation of the left and right channels provides a distortion-free monophonic audio term that modulates the transmitter. The resultant phase modulated signal is used to replace the crystal oscillator stage of the transmitter. The existing broadcast transmitter conveys the L–R information in the form of complex phase modulation, and the envelope conveys the L+R information to both the existing monophonic as well as newer stereophonic receivers. Other modulation techniques include the matrix switching method and the time-division multiplex method. The latter eliminates the need for audio matrices in the stereo modulator path. To help visualize the generation of the signal, refer to Figure 4.2-A1, the block diagram for the exciter. Because the linear, balanced modulator technique is the easiest to understand, it will be described herein. Interested readers are encouraged to study literature produced by other C-QUAM broadcast equipment manufacturers to obtain a full understanding of the various modulation techniques.

Beginning with the RF stages, a crystal at four times the carrier frequency ($4F_c$) is fed to a digital divider circuit that results in an output that is an on-frequency carrier signal at 0°, 90°, 180°, and 270°. These RF signals are utilized in pairs as differential references for three balanced modulators. The purpose of each balanced modulator is to suppress the carrier feed-through and produce sideband information only when audio is present. It can be seen that the first two modulators are fed with the 0° and 180° reference signals. These modulators, therefore, produce an output referenced in-phase or synchronous to the carrier frequency. The third modulator is, conversely, fed with the 90° and 270° outputs from the RF divider and therefore forms a quadrature modulator stage. The first modulator is fed with a DC voltage that causes a precise offset from the null of the 0° carrier reference, thereby producing a precise DC carrier term at its output. The second modulator is fed with L+R and therefore produces in-phase L+R modulation sidebands. The final modulator is fed with L–R audio information and produces quadrature sidebands relating to the stereophonic information. It is important to note that the carrier-producing modulator is used in preference to simply unbalancing the in-phase (L+R) modulator for lower distortion.

The three balanced modulator outputs are summed and bandpass filtered to remove odd-order harmonics of the desired carrier frequency. Removal of odd-order harmonics prior to limiting is important because odd-order products represent a DC term that can unbalance the subsequent limiting stage, thereby introducing distortion. After bandpass filtering, the RF signal is amplitude limited by multiple stages that exhibit approximately 50 dB of limiting gain. The envelope component is therefore removed, leaving only the phase modulated carrier component. This constant amplitude RF signal is then further amplified to a level adjustable from 0 to 30 V, peak-to-peak. It is this amplified, constant amplitude carrier signal that is interfaced to the transmitter to create the composite stereo transmission.

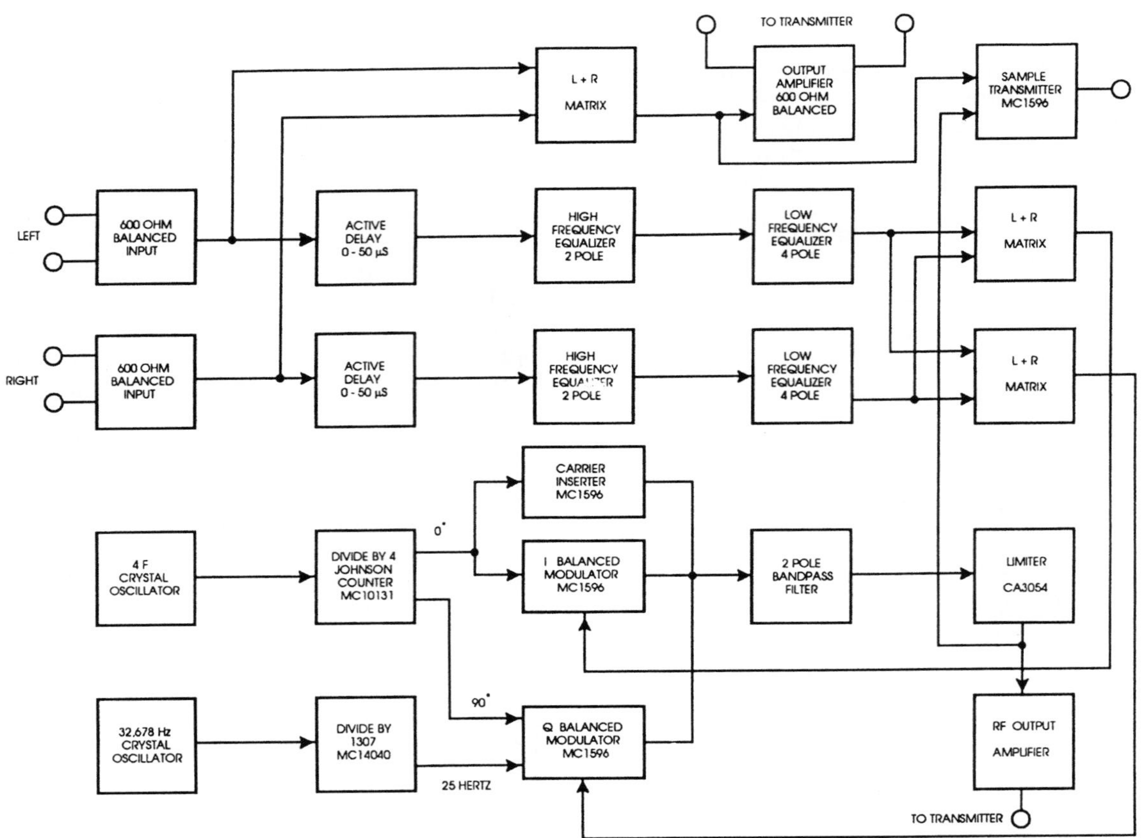

FIGURE 4.2-A1 Block diagram of C-QUAM® stereo AM exciter.

Unlike FM stereo, where a composite signal is fed into a wideband modulator input, the AM stereo signal is constructed at the power amplifier stage of the converted broadcast transmitter. A portion of the signal from the exciter enters the transmitter through an RF interface and follows one path to the RF power amplifier, while the remaining element, L+R, enters through the audio input terminals and follows a different path to the modulator, where it is finally combined with the RF phase modulated information. Because two different circuit paths are used, the time delay along each path may be different, often by 40 µsec or greater. This level of delay can be understood when one considers the phase shift through the low-pass filter in a conventional PDM transmitter that may exhibit a fifth or seventh order elliptical function. If the delay between the paths is not matched, both reduced separation and increased distortion will result. Loss of separation is easily understood because proper dematrixing of the audio signals in the receiver will occur only if the phase and amplitude of L+R and L–R are closely matched. The increase in distortion is a concept that can only be clearly understood when one examines the phase modulated component of the signal and finds that an L+R term is found in the PM component. If this L+R term does not match the envelope term, an increase in distortion would result; therefore, the audio equalizer is a very important section of the exciter.

In the simplest form of equalization, a delay circuit in the audio path to the quadrature modulator section of the exciter would be utilized, thereby introducing time delay into the L+R and L–R audio fed into the QUAM generator so it matches the L+R audio delay through the broadcast transmitter. Although adequate results will occur with this approach, a modification of this approach allows for additional correction of the signal to compensate for bandwidth or sideband symmetry problems commonly associated with broadcast antenna systems, particularly directional arrays.

A once common form of single sideband generation, particularly popular in the 1950s, utilized a scheme known as the *phasing method*. The process was simple because one could simultaneously amplitude and phase modulate a signal, essentially generating a QUAM signal. If the process is taken one step further so both the RF terms and the audio terms fed to each modulator are shifted 90°, a single sideband transmission results without the use of expensive, sharp cutoff RF filters. The same holds true for AM stereo signal generation. If a small amount of phase shift is introduced into the audio driving the phase modulated path of the transmitter with reference to the envelope audio path, the signal will take on an unsymmetrical sideband structure. If the antenna system exhibits an overall tilt toward the upper sidebands, for example, one would only need to predistort the signal so the lower sidebands were favored in the transmitter, thereby restoring symmetry in the radiated signal.

The exciter can perform this task by inserting the delay sections into the left and right channels prior to matrixing as shown in the block diagram. The audio is delayed independently in the left and right channels. It is then matrixed to produce L+R and L–R that feed the QUAM generator. An uncompensated summation of L+R is used to feed the external broadcast transmitter. Such equalization, known as *differential equalization*, can be quite powerful in correcting situations where asymmetrical antenna sideband radiation would otherwise limit stereo performance.

The audio equalizer is broken into several components. The first is the *constant time delay circuitry* that is used to compensate for the bulk of differential transit delays through the transmitter. The second section is the *high frequency equalizer* that is used to compensate for additional phase shift (nonlinear group delay) characteristics in the broadcast transmitter audio path. These nonlinear delay characteristics are introduced by roll-off in the modulator stages, phase shift in PDM filters, and reflected antenna impedances. The final section is the *low frequency equalizer*, which, not unlike tilt correction commonly found in audio processors and limiters, anticipates the phase shifting and amplitude roll-off action of modulation transformers, reactors, and coupling capacitors in older, plate-modulated transmitters. By introducing a similar roll-off and phase shift into the phase modulated audio path, separation may be maintained at frequencies below 50 Hz.

The last section of the exciter to be discussed is the *pilot generator*. The *pilot tone*, a 25 Hz sine-wave audio component, is injected into the L–R modulator at a 5% modulation level in the L–R channel. The purpose of the pilot tone is to indicate the presence of stereophonic information. This is different from the pilot tone signal in the FM stereo system where the 19 kHz signal is used as a synchronizing term in the demodulator. Indeed, AM stereo receivers could be built without the pilot tone detector; however, the consumer has grown accustomed to seeing a stereo indicator on the radio, hence it is incorporated into the system.

C-QUAM Decoding/Receiving Techniques

There are over a dozen ways to decode a C-QUAM transmission. The most common approach is the *feedback decoder* technique shown in Figure 4.2-A2. In this figure, there are three detectors: envelope, in-phase, and quadrature. The envelope detector demodulates the monophonic, L+R information. It may be a simple diode detector; however, most stereo demodulator integrated circuits utilize a limiter/multiplier approach that offers superior performance. Distortion measurements in the 0.1 to 0.3% region are commonly found at 99% negative modulation when this technique is used.

The in-phase and quadrature demodulators are identical in action to the balanced modulators found in the exciter. Each has a reference signal, either 0° for the I detector or 90° for the Q detector, and is phase locked to the incoming, received signal. However, referring back to the system equation, the actual received I signal is not $1 + L + R$, but instead $(1 + L + R)\cos\varphi$, and the received quadrature signal is $(L - R)\cos\varphi$ rather than $L - R$ as desired. In a proper C-QUAM decoder, the incoming RF/IF signal must be divided by the term

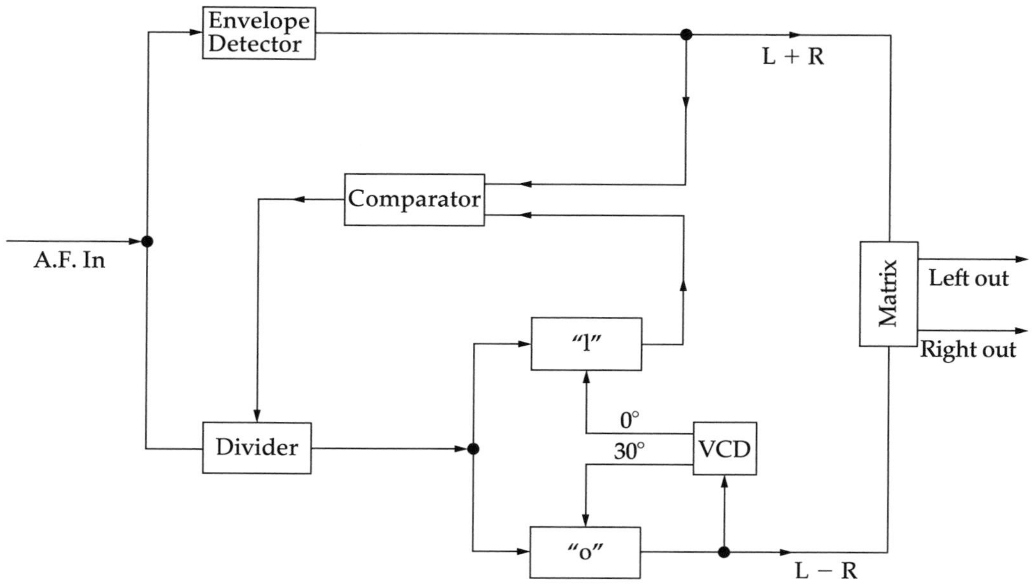

FIGURE 4.2-A2 Block diagram of feedback decoder.

cosφ, generally prior to quadrature demodulation. In the feedback decoder, the divider stage accomplishes this task. The derivation of the cosφ term conveyed in the phase modulated component is, however, an interesting process.

As indicated earlier, because it is both phase and amplitude sensitive, the I detector output would be $(1+L+R)\cos\varphi$, assuming no divider action. If the cosφ component is eliminated from the I detector output, the I detector would produce a signal identical in theory to the envelope detector. Assuming further that the I and Q detectors are fed from the same IF signal path, as cosφ is removed from the I detector output it is also removed from the Q detector output, leaving $L-R$ as the result. Because it is known that the IF signal must be divided by cosφ, a divider stage is placed in the IF path feeding the I and Q detectors. If the output of the I detector is then analyzed against the envelope detector output in a high-gain comparator, the resultant is cosφ. By connecting the output of the comparator to the control port of the divider, a feedback loop results that will cause the cosφ signal normally found at the output of the I detector to be effectively transferred to the input of the divider. Because the IF path is therefore divided by cosφ, the Q detector performs the task of directly demodulating $L-R$. The remaining circuitry in the decoder detects the pilot tone and dematrixes the $L+R$ and $L-R$ audio terms into the original left and right components.

Performance Considerations

The C-QUAM system is capable of on-air performance figures in excess of 40 dB separation and under 1% distortion from 100 to 5000 Hz, particularly on newer solid state broadcast transmitters. It is not uncommon to obtain separation figures in excess of 30 dB from 50 Hz to the 10 kHz NRSC limit. RF spectrum occupancy is also well controlled within limits of occupied bandwidth specified in FCC Rules Section 73.44 and the NRSC standard. The modulation constraints of the system are as follows:

- L+R (monaural modulation): +125, –100% (or FCC limits)
- L–R (stereo difference channel only): ±100% (± 45° phase modulation)
- L, R only (single channel, ref. to L+R): +125, –75% (limit of 71.56° phase modulation)
- *Pilot tone* (injected into L–R channel): 25 Hz ± 0.1 Hz, 5% +1/–0% injection level, sine wave

In addition, audio preemphasis may be used. Motorola recommends use of the NRSC modified 75 μsec standard. Although audio bandwidths up to 15 kHz produce good results, the FCC Rules limit audio bandwidth and subsequent RF emissions to 10 kHz as defined by the NRSC standard.

CHAPTER

4.3

AM Broadcast Antenna Systems

RONALD D. RACKLEY*

du Treil, Lundin & Rackley, Inc.
Sarasota, Florida

INTRODUCTION

Medium-wave broadcast (AM) antenna systems can reach a high degree of sophistication, much of which is based on advanced mathematics. The purpose of this chapter is to provide the station engineer with an understanding of some of the basic concepts of antenna design and an appreciation for the complexities of this specialty. An extensive bibliography is included at the end of the chapter for further study of the subject.

The chief purpose of a broadcast antenna system is to radiate efficiently the power supplied to it by the transmitter. A simple antenna can do this job quite well. This is often a single vertical tower that radiates its signal equally in all directions along the ground in a so-called nondirectional or omnidirectional pattern. A second purpose of an AM antenna system is often to concentrate the power in desired directions to cover populated areas and to suppress it in other directions to protect the coverage of other stations sharing the same or closely adjacent channels. This directionality may require a complicated directional antenna system with several towers if the requirements are stringent.

The antenna is the last point in the system under the control of the broadcaster. The signals radiated from the antenna are propagated through space to each receiving antenna. The factors affecting the strength of the received signal include the strength of the signal radiated by the broadcasting station in a particular direction, the distance to the receiving site, losses incurred by the less-than-perfect conductivity of the ground along the propagation path, terrain obstructions (large hills cast shadows even at AM frequencies), and, in the case of skywave transmission, the ionospheric conditions that determine how much of the radiated signal will be reflected back to each distant receiving location. Signal strength in a particular direction can also be affected by the presence of structures such as buildings or towers near the transmitting antenna.

The polarization of the transmitted waves is also a factor; for medium-wave broadcast stations, vertical polarization is used because of its superior groundwave propagation. The Federal Communications Commission (FCC) has established maximum transmitter power limits for each of the four classes of AM channels (A, B, C, and D), so the only variables available to the design engineer attempting to maximize the coverage of a radio station involve the antenna location, the pattern design, and a limited choice of power levels. These factors go hand in hand when designing a directional antenna system. Severe constraints are usually imposed on transmitter site selection because of aeronautical, zoning, environmental, and coverage requirements. The constraints encountered in the pattern design relate to the size and shape of the transmitter site, the extent to which the necessary signal suppression can be achieved at the desired transmitter power level, and the cost to design, construct, adjust, and maintain the multi-tower system. The pattern design can also seriously affect the stability, efficiency, and bandwidth of the completed system.

*With appreciation for material adapted from the previous edition by Carl E. Smith (1906–1998).

TRADITIONAL AND MODERN ANTENNA ANALYSIS METHODS

For simplicity of analysis, AM antennas have been traditionally deemed to function as though the currents carried by their elements were purely sinusoidal in nature, such as would theoretically be produced in a standing-wave pattern by two sinusoidal traveling waves of identical magnitude "passing" in opposite directions. This may be visualized as the forward wave that is imposed by the current that is conducted into the base of a tower "passing" the wave that is reflected at the tower top on its way back down. The sum of the two form a sinusoidal standing wave that may be readily integrated using classical mathematics, which is very convenient for relating an antenna's current to its far-field radiation characteristics.

The problem with the sinusoidal current distribution assumption is that it simply cannot be true if radiation is taking place, as the forward and reflected waves on the tower are subject to continuous attenuation as they travel its length due to the energy that leaves the antenna to form the far-field radiation pattern. In other words, the current magnitude of the forward wave grows smaller from the bottom to the top of the tower, and the current magnitude of the reflected wave grows smaller still as it travels back down from the top of the tower to the base. Summing the two waves, therefore, does not form a perfectly sinusoidal current distribution on the tower.

It has always been recognized that antennas cannot actually have sinusoidal current distribution, but integrating true current distribution would have been exceedingly difficult in the early days of AM antenna design. Studies were done in the 1920s, and it was determined that, for determining far-field radiation, sinusoidal current distribution assumptions gave results without sufficient error to warrant more complicated analysis. It was thereafter generally agreed that antenna theory based on sinusoidal current distribution would be used for antenna analysis. The methods for calculating far-field radiation performance in textbooks and the FCC Rules have been based on sinusoidal current distribution ever since.

For directional antennas, an additional layer of error is associated with the sinusoidal current distribution assumption due to the effects of mutual coupling between the towers of an array. This may be understood by visualizing the current distribution of a tower fed at its base and operating in the nondirectional transmitting mode and again with the tower base connected to ground through an impedance while it receives the energy that induces current in it from an external field operating in the receiving mode. The transmitting mode current distribution will approximate a sinusoidal shape; the receiving mode current distribution will depend on the terminating impedance at the tower base and might not resemble the transmitting mode current distribution at all. A common method for detuning a quarter-wave tower, for instance, is to place a reactance at its base to produce a sharp minimum in tower current at approximately one-third of its height—where the current is near maximum in the transmitting mode. As each tower in a directional antenna functions in both the receiving mode and the transmitting mode at the same time, its current distribution will be determined by both. In general, the current distributions of the towers of a directional antenna will differ from one another. Nonetheless, sinusoidal current distribution has traditionally been assumed for AM directional antenna analysis because the resulting far-field radiation errors are not great when compared with other known errors in the processes used to analyze interference between stations, such as are inherent in the statistically derived nighttime skywave signal propagation models that are used for allocation studies.

Although sinusoidal current distribution is a very convenient fictional assumption for AM antenna far-field radiation analysis—where its errors have historically been held to be acceptable—much is left to be desired when it is used to deal with practical matters related to feeding energy into antennas. There can be very large errors in base impedance, particularly for towers near one-half wavelength in height, and in the ratios and phases of array element base currents necessary to produce the far-field radiation parameters of a desired directional antenna pattern. Because of these errors, the FCC has always required that a proof-of-performance based on field strength measurements be made after the operating tower currents have been adjusted to produce the desired directional antenna radiation pattern, generally through experimentation. The field adjustment process has historically bridged the gap between the ideal design world of sinusoidal current distribution and the real world of the field strength proof-of-performance. The unknowns involved have always dictated that phasing and coupling systems be designed to match a range of expected tower base impedances based on experience as well as calculation.

Great advances in antenna analysis along with the development of modern computers facilitate the numerical analysis of mathematical problems that defy solution using classical methods. It is now possible to model an AM directional antenna (through what is known as *moment method modeling*) as a large number of small conductor segments and to take into account the contributions of current that are both conducted from adjacent segments and induced through mutual coupling from all of the other segments. This makes it possible to calculate tower base impedances and drive currents using close approximations to their real-world current distributions, which facilitates the design of performance-optimized phasing and coupling equipment and minimizes the amount of experimentation involved in tuning new systems. When careful checks of an antenna monitoring system are made, it is not unusual, when the system has been set up, to find that no further parameter adjustments are necessary to the tower currents predicted using moment method modeling.

It is known that moment method modeling can also be used to refine the process of AM antenna far-field analysis, but the traditional methods based on sinusoidal current distribution remain standard for that as they form the basis for both the FCC Rules and international regulations and treaties. For that reason, and because traditional methods are more familiar than moment method modeling for antenna analysis today, this section will employ them while acknowledging the significant advantages of moment method modeling for certain tasks.

TRADITIONAL AND MODERN DIRECTIONAL ANTENNA DESIGN METHODS

The process of determining what array geometry and field parameters are necessary to produce a desired directional antenna radiation pattern has become more automated with the advent of modern digital computers that can perform far-field calculations very rapidly. It is common today for a new pattern design that is filed with the FCC to be the last of a series of hundreds or even thousands of designs that were evaluated automatically in a search for the optimum solution to a specified set of radiation characteristics. Most directional antenna patterns in use today were designed long before such computations were possible, however, using straightforward mathematical techniques. Inline patterns having more than two towers and parallelogram patterns were generally developed from two-tower "building blocks" using the pair multiplication process.

Directional antenna patterns are explained in this section using the traditional mathematical approach, as it is applicable for understanding most existing patterns and is good for establishing the starting point for further iterative computer optimization in new pattern design. It should be remembered, however, that every directional antenna pattern is determined by how the individual tower vector contributions add in various directions and at various vertical angles, no matter what process was used to design it. Pair multiplication is just a shortcut to finding the parameters that will produce a multi-tower pattern having nulls in the desired directions, for example, instead of a method for designing patterns that are distinctly different from patterns designed by other methods.

RADIATION AND FIELD STRENGTH

Two independent factors determine the signal strength at any given point within a station's service area. First is the strength of the signal radiated in that direction; second is the path attenuation between the transmitting and receiving antennas. Attenuation is determined by distance and the conductivity and dielectric constant of the ground along the propagation path. The dielectric constant influences the effective depth to which the currents associated with the vertically polarized electromagnetic wave flow beneath the surface, and the conductivity determines the loss that occurs within that effective depth. For calculating AM station signal strengths, the FCC has published families of curves of field strength *versus* distance for a number of frequency spans covering the AM broadcast band. See Figure 4.3-1. Because all of the curves were plotted with an assumed dielectric constant of 15, which was chosen to represent the average value for soil in the United States, it is customary to simply refer to the conductivity value of the curve that best fits measured data along a radial as the *conductivity* of that radial.

Field strength measurements are often graphically analyzed with conductivities that differ from the actual soil conductivity, when the dielectric constant differs significantly from 15, because it is more convenient to use the existing graphs published by the FCC than to develop alternative curves based on dielectric constants other than the assumed 15. Analyzed ground conductivity values that cascade downward in segments with increasing distance, for instance, sometimes indicate that the dielectric constant is significantly lower than 15 and that the data would plot along a single conductivity curve calculated for the actual dielectric constant. It is customary, however, to avoid consideration of differing dielectric constants by using the published FCC curves except in rare cases of extremely poor soil where it is necessary to custom develop curves based on dielectric constants lower than 15 to be able to fit the field strength data to curves. Conductivity is normally the only term that is mentioned to describe the characteristics of earth for groundwave propagation analysis.

It is customary to express the radiation in units of millivolts per meter at 1 km, unattenuated. This is the field that would exist at 1 km over perfectly conducting earth. In this case, the field strength would follow a straight line depicting values inversely proportional to the distance from the transmitting antenna; hence, the radiation is also described as the *inverse distance field*. It is convenient to present antenna radiation pattern information in terms of unattenuated field strength because it standardizes antenna analysis without regard to the soil conductivity in the vicinity of the antenna. The unattenuated radiation cannot be measured directly but can be inferred with great accuracy if sufficient field strength measurements are made to determine the ground conductivity.

The standard distance for specifying unattenuated radiation in North America was changed from 1 mile to 1 km in the early 1980s. To convert field at 1 km to field at 1 mile, the multiplication factor is 0.62137.

Field strength coverage is always dependent on radiation, distance, and ground conductivity. It is important to realize that specification of antenna radiation is an entirely different process than specification of field strength coverage.

The term *efficiency* is sometimes used to refer to unattenuated radiation. As applied for decades in FCC practice to define radiation, the word is utilized in an unconventional sense. It does not express an output/input relationship of an antenna in percent as it is used to define amplifier efficiency. It expresses the unattenuated field strength in the horizontal plane of

FIGURE 4.3-1 Field strength versus distance for family of conductivity curves.

a nondirectional antenna or the *root mean square* (RMS) of the horizontal plane radiation pattern of a directional antenna pattern with a reference input power level of 1.0 kilowatt.

THE SINGLE TOWER NONDIRECTIONAL ANTENNA

The majority of single-tower antennas is neither top loaded nor sectionalized, and most of them are insulated from ground. For such simple towers, the current is deemed to be sinusoidal and to reach a maximum 90 electrical degrees down from the top (or at the base if the tower is shorter than 90° in height). The distance along the height of a tower, measured in electrical degrees, differs from the physical distance slightly because the velocity of propagation along the tower structure is slower than the velocity of propagation in free space. This is primarily because the cumulative effect of the small currents that flow onto cross members from the legs of a tower, due to the cross members' individual capacitances to ground, makes the tower structure function in a sense as a delay line. The amount of delay depends on the cross section of the tower and the size and number of its cross members. Rather than attempting to model the intricacy of a tower's cross-member structure, it is customary to consider it to be a simple wire extended in length to account for velocity-of-propagation effects when calculating its base impedance. Past experience with towers having similar structural characteristics is often considered in selecting the velocity factor for analysis. A typical guyed tower that is 90° high physically is known to be about 95° to 98° high electrically when analyzed as a wire conductor, for instance.

Although it is beneficial to consider velocity-of-propagation effects when calculating tower impedances, antenna radiation characteristics have traditionally been calculated assuming thin wires equal in height to the towers they represent. The FCC Rules are based on these assumptions. The approximate shape of the current distribution on a thin wire of uniform cross section is given by:

$$i_a = I_a \sin(G - y)$$

where:

i_a = current (in amperes) at height y

I_a = maximum current (in amperes)

G = tower height (in degrees)

y = height (in degrees) of the current element i_a

As an example, the general shape of the current and voltage distribution on a thin tower 210 electrical degrees high is shown in Figure 4.3-2. For shorter towers, the distribution would approximate that shown but with the lower portions cut off; there always being a current node and a voltage maximum at the top of any such tower that does not employ top loading. It is important to visualize the shape of the voltage distribution along the tower because of the need for good

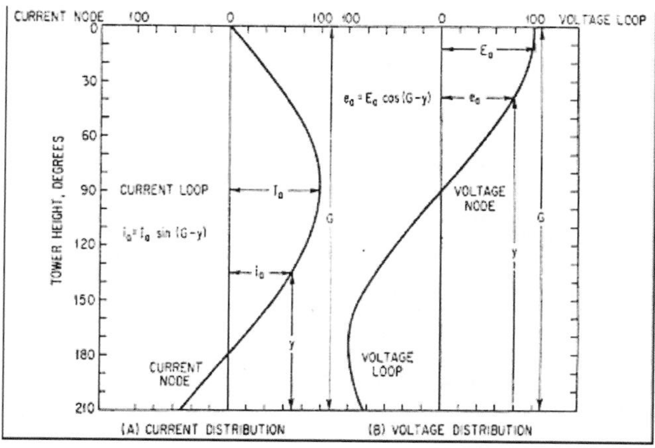

FIGURE 4.3-2 Theoretical current and voltage distribution on a vertical radiator.

insulators at the high-voltage points; otherwise, corona or arc-overs may result and disrupt broadcasting service.

The real-world tower current and voltage are not zero at the nodes shown along the tower; rather, they reach minimum values and shift rapidly approximately 180° in phase when traversing the node region. When towers considerably taller than 180° in height are considered, the current near the base is in the opposite direction from that in the upper portion of the tower. Under these conditions, when viewed in the horizontal plane, the radiation from the lowest part of the tower is canceling a portion of the radiation from the part above the current minimum. Any increase in tower height above the optimum would actually reduce horizontal plane radiation.

VERTICAL RADIATION CHARACTERISTICS

Maximum groundwave radiation occurs for a tower 225 electrical degrees high (5/8 wavelength). The variation in tower current distribution with increasing tower height defines the shape of the radiation characteristic in the vertical plane. Figure 4.3-3 shows the size and shape of the vertical plane radiation patterns for a single tower of various heights atop a perfect ground system fed with 1 kW of power. The high-angle lobe of the vertical pattern of a 225° tower shows why that height is generally avoided by stations that are concerned about self-interference due to skywave signal within their groundwave coverage areas at night. Most Class A clear-channel stations that provide wide-area skywave coverage at night use towers no higher than approximately 190° for this reason.

INSULATED TOWER BASE IMPEDANCE

The base impedance of a single nondirectional tower is determined principally by its electrical height, its cross section, the extent of the ground system, and the

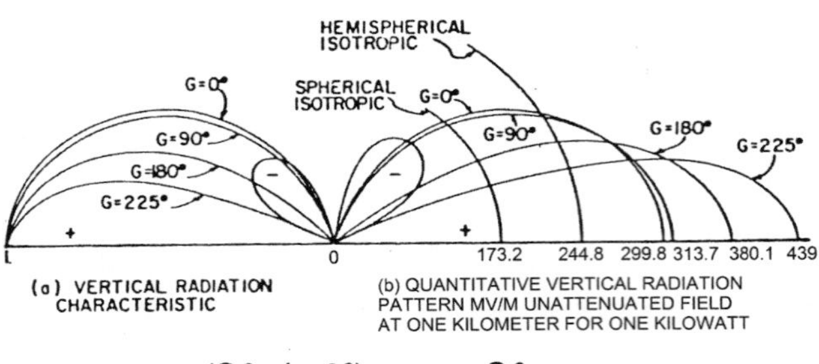

$$f(\theta°) = \frac{\cos(G° \sin\theta°) - \cos G°}{(1 - \cos G°)\cos\theta°}$$

FIGURE 4.3-3 Radiation characteristics in a vertical plane.

elevation of the feed point above ground. For typical guyed towers of uniform cross section that are base insulated and fed 4 or 5 feet above ground level, the resistive and reactive components of the base impedance approximate the values shown in Figure 4.3-4. The base impedance of self-supporting towers departs radically from the values shown, not only because of their large and tapering cross section, but also because of the high capacitance about the bases of such towers.

Electrically short towers are inefficient radiators, not only because of the shape of their vertical radiation characteristics, as shown in Figure 4.3-3, but also because of proportionately higher ground losses. For example, a tower 48° high with a base resistance of only 9 Ω will waste approximately 18% of the available power in the ground system resistance losses (typically 2 Ω with a full ground system).

It must be remembered that the base input impedance of a tower, when measured at the output terminals of the antenna tuning unit that is used to feed it, includes the shunt effects of stray capacitance and any conductive circuits that are connected across its base insulator, as well as the series inductance of the conductor used to make the connection to the tower base. The shunt effects include those of isocouplers, isolation coils, lightning chokes, and the like. The series hookup inductance can add as much as several tens of ohms of reactance to the feedpoint impedance when a conductor having a "lightning retard coil" of one or more turns is employed as the feedline. A matching network must be designed to match the load impedance presented by the calculated tower base impedance with these other factors considered to the extent that they can be known before construction. To the extent that they cannot be known before construction, the design must allow for a sufficiently large range of adjustment.

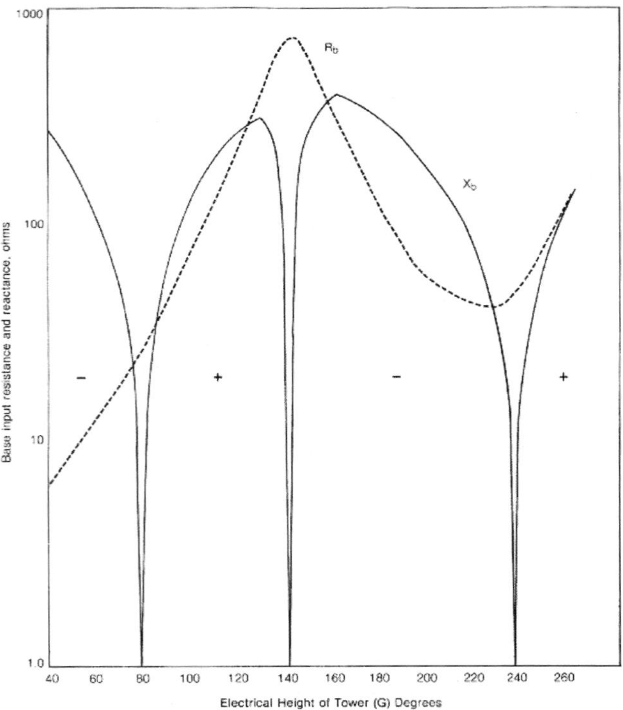

FIGURE 4.3-4 Typical base input resistance and reactance of a uniform-cross-section, base-insulated, guyed tower.

GROUNDED TOWERS AND SHUNT-FED AND FOLDED MONOPOLES

Occasionally, towers without insulated bases must be utilized as AM radiators. Such structures include towers that are also used for land-mobile communication and FM and TV stations. Although the impedance at the base of such a tower is necessarily essentially zero, the impedance rises with increasing height of the feed point. It is a simple matter to determine experimentally the height at which a shunt-fed tower must be driven to provide a desirable input impedance. A common technique is a *slant-wire*[1] feed, where a wire is attached to the tower at a selected height above ground and brought down to near ground level at an angle approximating 45° to serve as the antenna input terminal. A slant-wire feed can distort the otherwise omnidirectional pattern of a single tower and tend to suppress radiation over the sector on the side where the slant wire is attached. This effect is much greater when towers on the order of one-half wavelength are shunt fed rather than with towers on the order of one-quarter wavelength, but it can be avoided for all height towers if, instead of the slant wire, the feed conductors are insulated at the base, brought up outside of the tower, and bonded to the tower 90° above ground to form a folded monopole. The conductors, in this concentric arrangement, in effect form the outer conductor of a coaxial transmission line with a short to the tower at the 90° point and an open at the base insulators. This quarter-wave, open-circuit transmission line in effect puts an insulator at the tower base. The currents that flow up on the outer conductors and down on the tower essentially cancel as far as radiation is concerned. An additional component of current

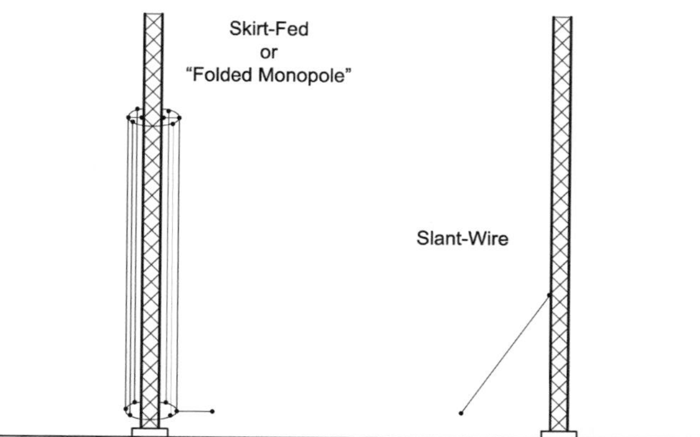

FIGURE 4.3-5 Shunt-fed grounded towers.

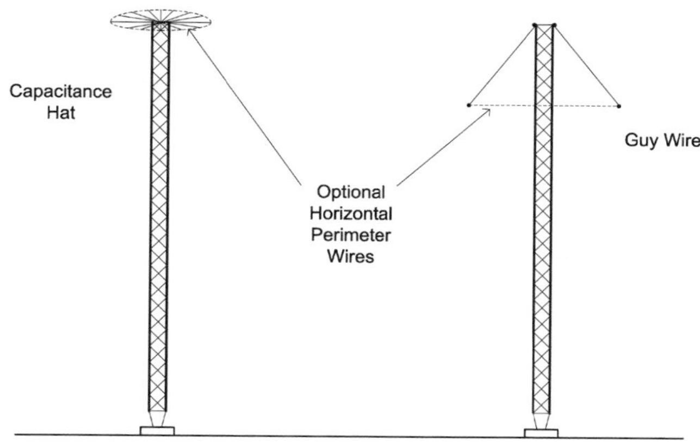

FIGURE 4.3-6 Top-loading methods.

flows up the skirt and onto the tower, if the tower is taller than the skirt, and goes to zero at the top to produce the radiation, so the tower with this insulated skirt performs essentially as a base insulated tower. The concentric arrangement of conductors, usually six, are tied together above the ground level conductor insulators and fed like a base insulated tower.[1]

Although the traditional method for matching both slant-wire and folded monopole antennas has involved experimentation with regard to the physical connection points of the feed wires, moment method modeling is sometimes used today to design optimized feed arrangements. Examples of both slant-wire and folded monopole shunt feeding are shown in Figure 4.3-5.

TOP LOADING

The performance of an electrically short tower (significantly less than 90°) can be improved, both as to radiation efficiency and bandwidth, by means of top loading. Top loading is also sometimes used to provide vertical radiation characteristics that would otherwise require construction of taller towers where tower heights are sufficient such that radiation efficiency and bandwidth are not the major concerns. Top loading is accomplished by increasing the capacitance to ground from the top of the tower. The physical realization can take the form of either a flat, more or less circular horizontal disk (usually consisting of a number of conductors fanning out from a central point) attached to the top of the tower or sections of guy wires bonded to the top of the tower and extending down a useful distance before encountering the first of the guy wire insulators (see Figure 4.3-6). The former arrangement is commonly called a *top hat*.

Moment method modeling can be used to design the physical arrangement of top-loading conductors for new towers with quite good accuracy before they are built. Most existing top-loaded towers were designed using approximate mathematical formulas based on the capacitance added by the top-loading conductors and the calculated average characteristic impedance of the tower viewed as a single wire transmission line. Because of the approximate nature of the traditional methods for top-loading design, the FCC requires that current distribution measurements be made on top-loaded towers before the stations using them are issued licenses.

[1]Slant-wire feeds are no longer accepted by the FCC for critical hours and night operation.

Many variations of top loading are possible. Most recent installations use sections of the three upper guy wires for top loading, although some have used 6 or even 12 nonstructural wires for top loading. By interconnecting the lower ends of the top-loading wires, the capacitive loading is increased for a given guy wire length. As the current flowing downward in each guy wire produces a far field that opposes that produced by the tower itself, which should be minimized for optimum performance, interconnecting the lower ends is recommended where relatively large amounts of top loading are required. Top loading is less desirable than increased tower height but is useful where towers must be electrically short due to either extremely low carrier frequencies or to aeronautical limitations. Top loading increases the base resistance and lowers the capacitive base reactance, thus reducing the Q and improving the bandwidth on towers less than 90° high. When the tower height is of the order of 130°, top loading can be used to increase the electrical height of the tower to improve groundwave radiation and minimize skywave radiation at critical elevation angles.

SECTIONALIZED TOWERS

A utopian vertical radiator would have a constant current of unchanging phase throughout its height, but in real life the current must ultimately reduce to zero at the tower top or at the end of the top-loading cables. The current can be made to diminish less rapidly by inserting an inductance in series with the tower at a point partway up its height. This is the same technique as the familiar loading coil near the

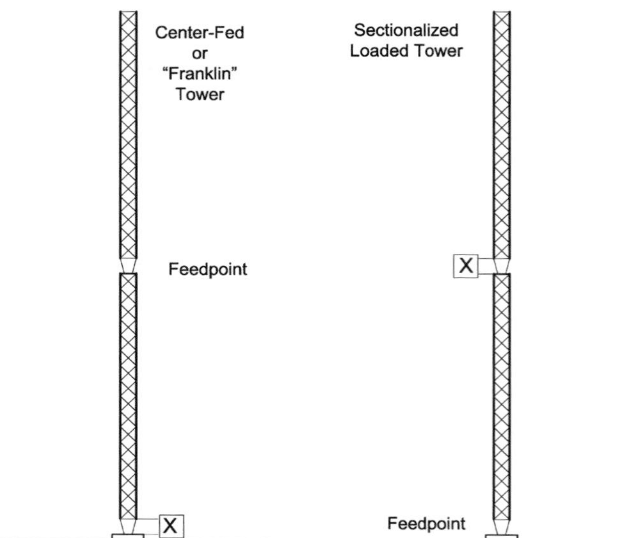

FIGURE 4.3-7 Sectionalized towers.

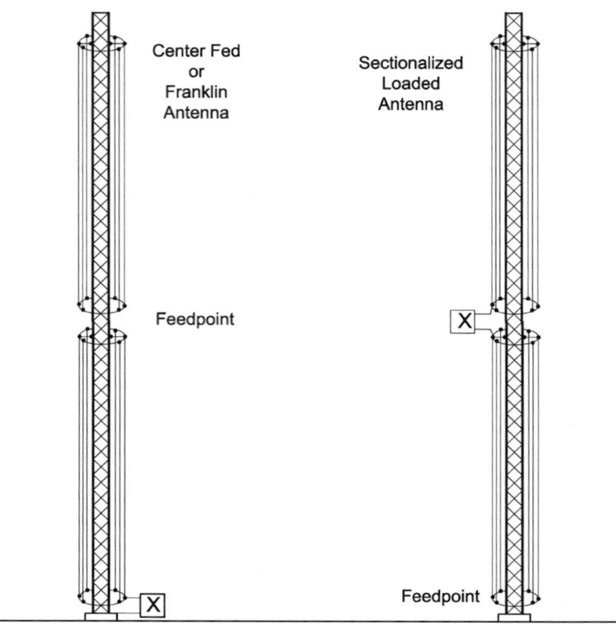

FIGURE 4.3-8 Sectionalizing with skirts.

center of the vertical whips often used for mobile radio systems. Towers approaching one wavelength in height can be employed to provide increased horizontal plane radiation and greater suppression of high-angle radiation when they are fed at approximately half of their physical height. Such center-fed towers are commonly known as *Franklin* antennas. Examples of how both Franklin and sectionalized, loaded antennas can be constructed by placing insulators within tower structures at appropriate places are shown in Figure 4.3-7.

It is also possible to use the technique of skirt-wire feeding for sectionalizing towers where it is not feasible to use insulators, as shown in Figure 4.3-8. Although not shown because the towers are not drawn to scale, the skirt wires would typically be shorted to the tower at approximately 90° from the open ends of the skirts. It is also possible to eliminate the need for tuning boxes when skirts are used by adjusting the points at which the skirt wires are bonded to the tower to produce the required net reactances across the open skirt ends.

The FCC Rules contain formulas for calculating the vertical radiation characteristics of sectionalized and Franklin antennas and specify how the parameters that describe their physical characteristics must be specified in applications. Because most existing sectionalized and Franklin antennas were licensed before the current Rules were enacted (when there were no standard methods for their analysis) many of them are grandfathered and require custom analysis to determine their vertical radiation characteristics. The information on their physical characteristics contained in the FCC's engineering database does not conform to the standards outlined in the Rules, and different mathematical formulas are used in the software employed for allocation studies on a case-by-case basis.

TOP-LOADED SECTIONALIZED TOWER

For a simple vertical radiator, the radiation characteristic can be improved by increasing the tower height up to 225° for maximum groundwave signal where skywave self-interference from the high-angle lobe that is present for tower heights greater than 180° during nighttime or transition hours is not a concern. This in effect raises the position of the current loop with respect to the ground. This principle can also be applied to the top section of a sectionalized tower. The purpose of top loading a sectionalized tower is to provide a means of further controlling the current distribution on the tower. Considering efficiency and stability, it is often possible to achieve a more favorable radiation characteristic of the whole tower by employing top loading and sectionalization together (see Figure 4.3-9). In the case of tall towers used to support FM or TV antennas, it may not be practical to employ top loading.

GROUND SYSTEMS

The current on a tower does not simply disappear. It returns to earth through the capacitance between the earth and each incremental element of the tower and the top-loading conductors, if used. For towers not exceeding 90° in height, the tower current is greatest at the base. For such towers, the radial ground current is greatest near the tower and decreases with increasing distance from the tower. For single towers, the ground currents are radial from the tower base. The ground losses are greatly reduced if the tower has a radial copper ground system, so the ground current will be in the low-loss copper ground system rather than in the

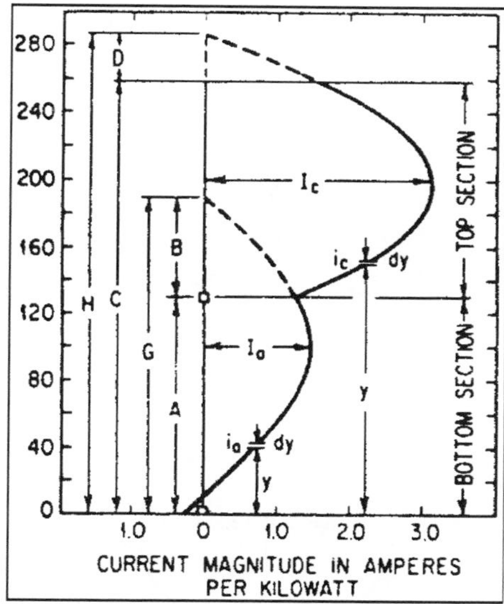

FIGURE 4.3-9 Theoretical current distribution on top-loaded sectionalized tower.

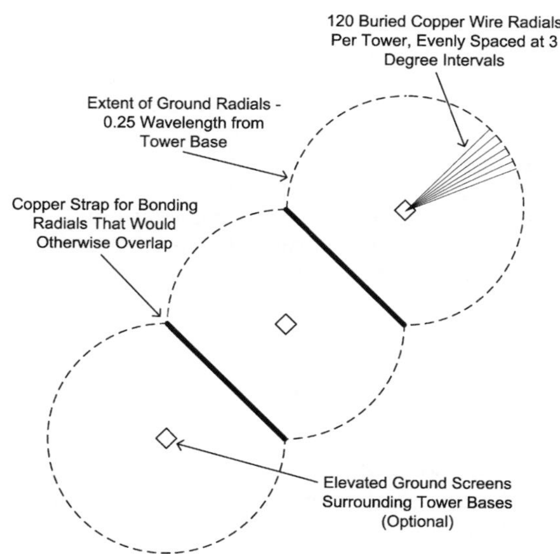

FIGURE 4.3-11 Directional antenna ground system.

earth, which has a much higher resistance. A solid copper sheet of infinite radius would be the ultimate ground system, but experiments and experience have defined the dimensions of an adequate ground system. A system of 120 radial ground wires, each 90° long and equally spaced out from the tower base, constitutes a standard ground system (see Figure 4.3-10).

This arrangement is sometimes augmented with an additional 120 interspersed buried short radials (often 50 feet long) or an expanded copper-mesh ground screen, 25 to 50 foot square and centered at the tower. A superior ground screen material is the copperweld, mesh ground mat often utilized by power companies for lightning protection under electrical substations. Where wet/dry or seasonal soil variations might impact antenna stability, elevated ground screens consisting of mesh ground mats above beds of crushed rock or elevated short radials are sometimes employed.

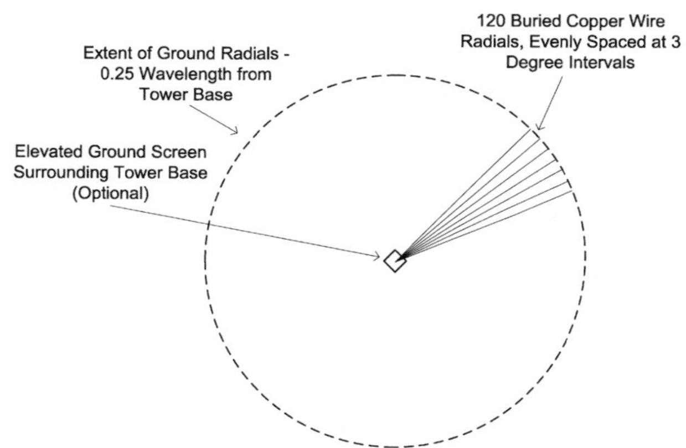

FIGURE 4.3-10 Nondirectional antenna ground system.

There is no magic in a standard ground system for nondirectional towers; such a system simply represents a reasonable balance between cost and radiation efficiency. The antenna system loss including the tower and ground system is normally assumed to be 2 Ω and is added to the base resistance of the tower for simplified analysis. Most ground systems under directional antenna arrays consist of the usual 120 radials per tower truncated and bonded to traverse copper straps where the radials from the towers would otherwise intersect, as shown in Figure 4.3-11.

Ground system losses are minimized if the radial wires are placed above ground; thus, the E-field voltage from the tower and top-loading cables (if any) terminate on these radial conductors so the H-field current can return to the tower base without penetrating the lossy earth. Some ground systems have been installed with elevated ground radials, and it has not been found necessary to have the full 120 radials in such cases. Ground radials, however, are usually buried 6 to 8 inches underground for mechanical protection. Burial up to 24 inches is feasible where necessary to permit deep plowing for agricultural crops; however, the ground system should be very near the earth surface in the immediate vicinity of the tower. The earth losses are greater for the buried ground system. Changes in weather conditions can change the dielectric constant and conductivity of any unshielded earth to the detriment of base current stability, but the effects are minimized when 120 buried radials are used.

TWO-TOWER DIRECTIONAL ANTENNA

When a nondirectional antenna, with a given power, does not radiate enough field strength to serve the community of interest or fails to protect other radio stations, then it is logical to resort to a directional antenna system to achieve these objectives. The FCC Rules spell out the protection requirements to be provided to the various classes of stations, both daytime and nighttime, on the same and adjacent channels. These limits, which must be met in the directional antenna design, tend to define the shape and size of the required antenna pattern. Because the distances and directions to the other stations requiring protection are rarely the same, most directional antenna patterns are tailored to meet the specific requirements in various directions.

A directional antenna functions by carefully controlling the amplitude and phase of the radiofrequency currents fed to each tower. The resulting field in any direction is the vector sum of the individual tower radiation components. To visualize the resulting pattern in the horizontal plane, one must consider the individual tower radiation components when viewed from distant points in different directions. The relative amplitudes from the individual towers remain unchanged, but the relative phases shift with azimuth because the signal from the closest tower arrives first. In a directional antenna system, one tower is defined as the reference tower, and the amplitude and phase of each other tower are measured relative to this reference. The ratio of the field from each other tower relative to the reference tower field is a fractional number sometimes expressed as a percent of the reference tower field.

The phase of the field, radiated by each tower relative to the reference tower, has two components when viewed from any distant point of observation. The phase resulting from the phase of the current fed to the tower is one component and is adjustable. The second component is the phase that appears to lead or lag the reference tower by virtue of being more distant or closer than the reference tower to the point of observation. This is termed the *space phase component* and varies continuously for each tower in a sinusoidal manner as the observation point is moved in azimuth along a distant circle around the array.

Figure 4.3-12 shows three simple directional antennas and their resulting patterns, which are easy to visualize. Figure 4.3-12a shows two towers arranged along a north–south line separated by 180° and fed with equal currents in phase. When viewed from the east or west, the fields from the two towers are in phase and the maximum field strength results. When viewed from the north or south, the field from the more distant tower is delayed by the 180° of additional distance, thus canceling the field of the closer tower so as to result in a minimum or null. The deepest minimum or null occurs only when the fields are exactly equal in amplitude and opposite in phase. Figure 4.3-12a is termed a *broadside array* because the maximum radiation is broadside to a line through the towers. Figure 4.3-12b shows a similar arrangement, but with the phase of the current in the north tower shifted by 180°. In this case, the fields from the two towers cancel each other when viewed from the east or west but would produce maximum radiation from north or south. This arrangement would be termed an *end-fire array*, because maximum radiation coincides with a line through the ends of the array. Figure 4.3-12c alters the spacing to 90° and phasing to 90° so as to produce a cardioid pattern. Other combinations of tower spacing and phasing can produce a great variety of pattern shapes.

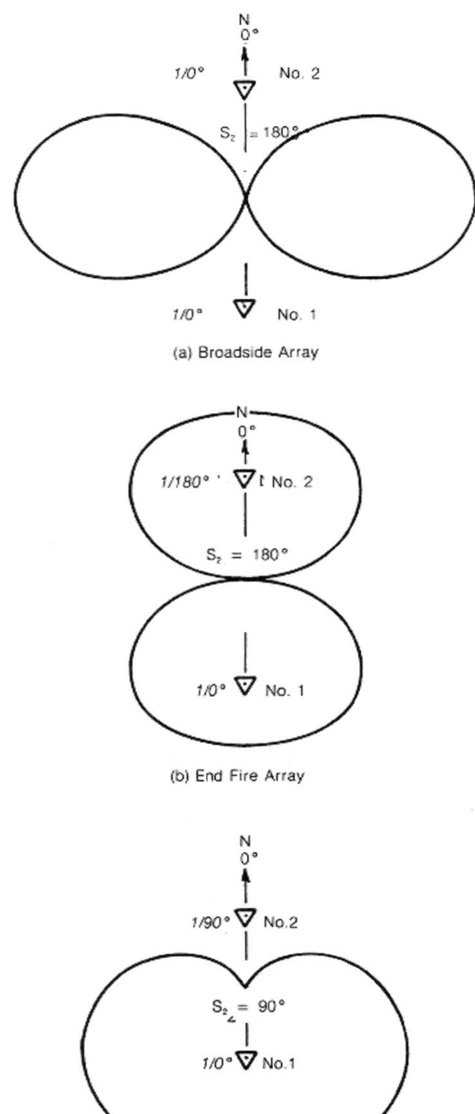

FIGURE 4.3-12 Three simple directional antenna patterns.

MULTIPLICATION OF TWO TOWER PATTERNS

Perhaps the most widely used method of controlling pattern shape involves the multiplication of two-

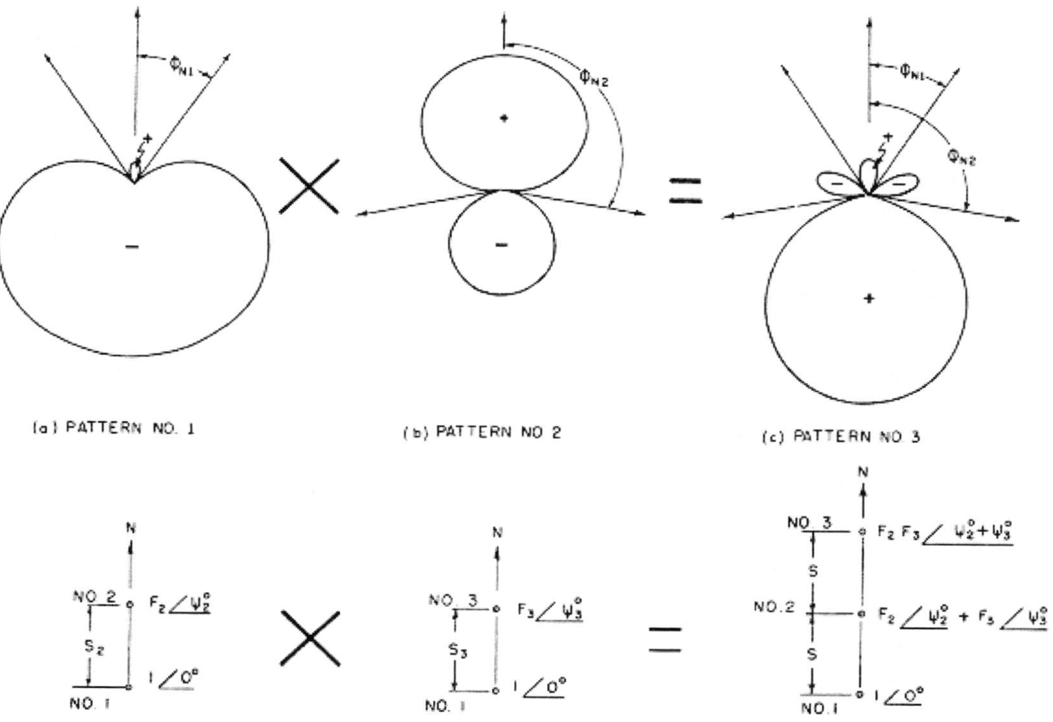

FIGURE 4.3-13 Multiplications of patterns to produce a three-tower inline array.

tower patterns and is known as *pair multiplication*. This is illustrated in Figure 4.3-13. When a two-tower pattern such as pattern 1 with nulls at $\pm\phi_{n1}$ is multiplied by pattern 2 with nulls at $\pm\phi_{n2}$, the result is pattern 3 in a three-tower array. The directions of the two-tower pair nulls are maintained in the three-tower array. This is a very powerful design technique for protecting other stations and still serving a desired service area. In this special case, the spacings S_2 and S_3 are equal, resulting in an inline array with the fields of towers 2 and 3 being added in the center tower, and the end tower of the three-tower array is the multiplication of these fields, as shown in pattern 3 of Figure 4.3-13.

In the event that the protection directions are not symmetrically located, the two tower pairs can be placed on different azimuth angles and have different spacings to produce a four-tower parallelogram array (as shown in Figure 4.3-14). The nulls of the pair 1 and pair 2 patterns are maintained in the resulting four-tower parallelogram array. This approach can be expanded for a larger number of design pairs by using a larger number of towers. Figure 4.3-15 shows how the parameters of a six-tower parallelogram array can be developed from three design pairs. The design includes two pairs having the same spacing, S_1, and the same orientation multiplied by an additional pair with a different spacing, S_2, and a different orientation. With this approach, a wide variety of asymmetrical patterns are possible with relatively simple design calculations.

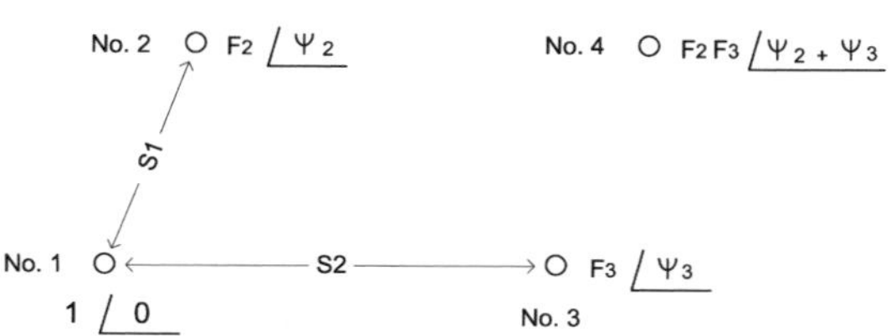

FIGURE 4.3-14 Four-tower parallelogram.

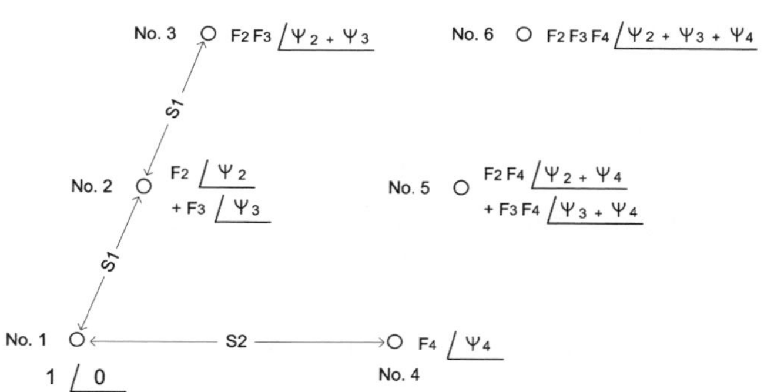

FIGURE 4.3-15 Six-tower parallelogram.

PATTERN INVERSION (OR MODING)

Directional antenna pattern designs using towers of identical height that do not have zero-field nulls often offer a choice of base impedances and power division.

This occurs when a pattern has a nonunity field ratio or one or more embedded design pairs have nonunity field ratios and the same degree of null fill can be achieved by field ratio inversion. A two-tower pattern that can be produced with a field ratio of 0.8, for example, can also be produced with the same phase and a field ratio of 1.25—and each option will have a distinct set of operating base impedances and power division. The embedded pairs in a pattern can be inverted in any combination. For example, a three tower pattern derived from two design pairs has four possible sets of parameters: no inversion, pair 1 inverted, pair 2 inverted, and pairs 1 and 2 both inverted. A pattern without any embedded design pairs can be inverted at least once by rotating all of the towers about the center point (adding 180° to each azimuth) and changing the sign of each phase angle.

It is often beneficial from the standpoint of bandwidth performance to consider all possible parameter inversions before a pattern design is considered final. In general, better performance is achieved by selecting the design with the most nearly equal operating resistances and, where there are towers having negative power flow, with the minimum total negative power. A parameter inversion study is recommended whenever a new pattern is designed and when a replacement phasing and coupling system is designed for an existing system having a geometry that makes it possible without moving towers. Software is available for reverse-engineering patterns larger than two towers to determine their embedded design pairs, if any.

PATTERN DESIGN USING MODERN COMPUTER METHODS

Most existing patterns were originally designed using the pairs multiplication process. Those designed prior to about 1970 may have had their parameters developed with the design engineer's slide rule and had their radiation pattern calculations done on paper with the assistance of tables of trigonometric functions and mechanical calculators. In those days, complicated pattern shapes were developed by specifying where pattern nulls would be produced, and, if the radiation in other directions was found be satisfactory after the overall pattern calculations were completed, they were considered final. Pattern shapes were often biased toward meeting interference protection requirements without, for instance, optimization of null fill on the less critical protection azimuths.

It is now possible to use computers to develop designs to fit specified radiation pattern shapes iteratively, with the major benefit being the ability to optimize radiation over all important spans of azimuth rather than simply meeting the radiation limit requirements at specific azimuths chosen to have nulls when pairs are multiplied. Software is available that varies directional antenna tower field parameters, locations, and heights while comparing calculated radiation with the specified requirements in an attempt to converge on the optimum design. In addition to radiation pattern optimization, it is sometimes possible to design patterns with fewer towers than would be necessary in a parallelogram designed using the pair multiplication technique. For a given set of pattern requirements, however, approximately the same size property is required. The pair multiplication technique remains useful for quickly estimating the property requirements for a new directional antenna system and for giving the computer software used for design optimization a good starting point to minimize the amount of random effort required of it.

RADIATION PATTERN SIZE

The pattern size is usually determined by integrating the energy flow outward through an imaginary hemispherical surface surrounding the directional antenna array. This method does not give information regarding the distribution of power radiated from the various towers of the directional antenna array; however, it is very useful for making comparisons of pattern size. This computation method is available in digital computer programs and is used by the FCC. There are other methods of determining pattern size, such as the *mutual resistance method*, which employs Bessel functions, and the *driving point impedance method*, which uses mesh circuit equations with self- and mutual impedance information.

The FCC Rules require that directional antenna radiation patterns submitted in proposals for new and changed facilities be calculated based on the loss assumption of 1.0 Ω per tower. It is known that the best efficiency possible with a full ground system about each tower corresponds more closely to what would be calculated for 2.0 Ω loss, but the 1.0 Ω loss assumption is required for calculations related to new directional antenna patterns so the standard patterns that are used for interference calculations in station-to-station allocation studies might be based on worst-case (for interference) assumptions.

Moment method computer software, which is described elsewhere in this chapter, has in recent years become available for determining current distribution on towers and top-loading cables, base driving point impedances, and the patterns of directional antenna arrays. It has found common use for predicting the drive characteristics of array elements to use in phasing and coupling system design and predicting tower base current ratios and phases to produce desired far-field pattern shapes for adjustment purposes. The calculation of pattern shapes and sizes for regulatory and international notification purposes, however, is not done using moment method software because of the requirements of regulations and international treaties based on older methods in use when they were originally enacted.

RSS-TO-RMS RATIO

Each directional antenna pattern calculated to modern standards has specified for it both an RMS, which is

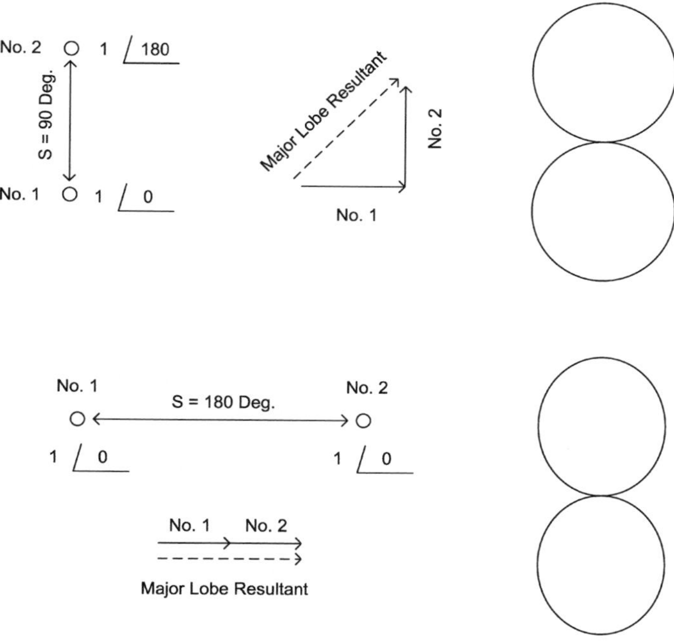

FIGURE 4.3-16 High and low RSS design approaches.

the root mean squared of its horizontal plane radiation, and an RSS, which is the root sum squared of the individual field values radiated by the various towers to produce the pattern. The RMS corresponds to the area inside a directional pattern that is plotted to scale in millivolts per meter. It is a measure of how much radiation leaves the antenna system. The RSS, on the other hand, is a measure of how much field is required from the towers in aggregate to produce the far-field pattern. For a nondirectional antenna, the RMS and RSS are both equal to the radiated field that is the same in all directions in the horizontal plane, and the RSS-to-RMS ratio is 1.0.

The RSS-to-RMS ratio for a given directional antenna pattern is the closest thing available to a quality factor for judging its characteristics relative to other patterns. A high ratio means that the combination of array geometry and pattern shape forces the individual tower fields to be high for the amount of power that is radiated into the far field. As an indicator of relatively high stored energy within a system, a high RSS-to-RMS ratio indicates that a pattern can be expected to be more difficult to adjust and maintain in adjustment because the individual tower contributions to the vector summations at null azimuths are larger than would be the case for a low RSS pattern. It is also an indicator of relatively poor expected bandwidth, as high tower fields translates to high tower currents which, in turn, requires lower base resistances to satisfy the requirement for conservation of energy. Because of their higher element currents, directional antenna patterns having high RSS values tend to be more sensitive to loss resistances associated with towers and their ground systems. In general, high RSS patterns have lower RMS efficiency than do low RSS patterns.

An RSS-to-RMS ratio value as close as possible to 1.0 is to be desired. The selection of array geometry is the biggest factor in determining what the ratio will be. Figure 4.3-16 shows two possible solutions for producing a two-tower figure-eight directional pattern with nulls to the east and west. It is obvious from the lengths of the tower field vectors relative to the major lobe resultants—where they fall most nearly in phase—that the broadside array with the towers spaced 180° apart has a significantly lower RSS than the end-fire array with towers 90° apart. In general, it is desirable to employ one or more broadside tower pairs where it is necessary to squeeze a pattern side to side. End-fire tower pairs are best used for cardioid patterns where nulls are used to reduce radiation on one side only. Parallelogram arrays often use a combination of widerspaced broadside and closer spaced end-fire pairs to make the best use of both in producing an overall required pattern shape. In order to achieve the best pattern design with a computer program that uses iterative techniques, it is a good idea to first design a pattern having low RSS characteristics to use as the starting point.

NEAR-FIELD VERSUS FAR-FIELD CONDITIONS

Theoretically, a directional antenna pattern is not fully formed except at an infinite distance, where the separate towers can be considered as one point source. Figure 4.3-17 shows the space phasing between two

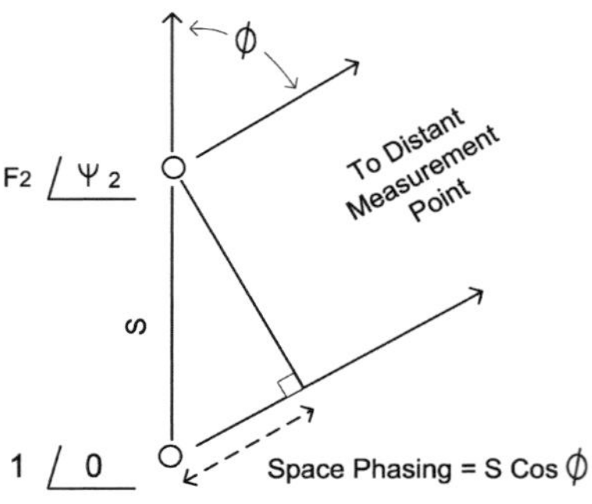

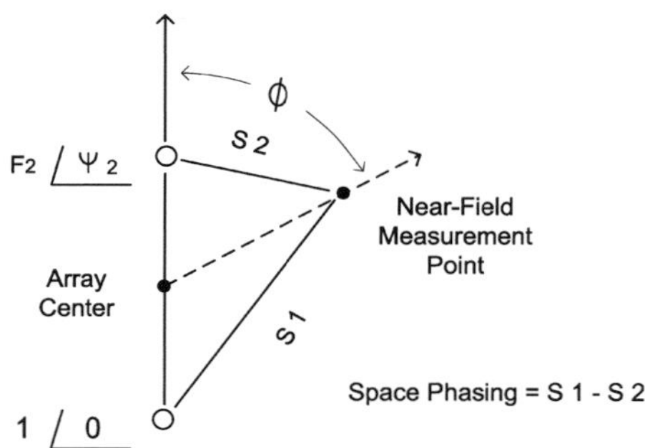

FIGURE 4.3-17 Space phasing and near-field errors.

towers of an array for far-field conditions, where the rays pointing from the two towers to a very distant observation point are virtually in parallel, and for an observation point in the near field, where the right-triangle assumption for calculating the space phasing term does not apply.

Measurements made at points in the near field, therefore, will not fall on the unattenuated field strength *versus* distance line calculated for far-field conditions. As a practical matter, near-field effects can persist as far as 32 km (20 miles) from an antenna before far-field conditions prevail. This is especially true in the deep minima of wide-spaced arrays; however, misleading measurement results can often occur under apparently innocent circumstances. Near-field calculations involve consideration of the actual inverse distance attenuation and the actual phase delay from each antenna element to the observation points along a radial.

Figure 4.3-18 shows the results of such calculations on a minimum radial and the resulting analysis of field strength measurements. Line A is the inverse distance line for the theoretical unattenuated radiation at 1 km. Curve B is the result of the near-field calculations assuming only inverse distance attenuation (that is, no soil losses). It converges with the inverse distance line with increasing distance. Curve C represents a soil conductivity of 10 mmhos/m as drawn in the conventional manner from analysis of nondirectional measurements on the radial. Curve D is a composite of curves B and C. It includes the near-field calculations and is attenuated with distance in accordance with the soil conductivity previously established. This composite line converges with the near-field calculations at short distances where soil attenuation is negligible and converges with the soil conductivity line at great distances where near-field effects disappear. Because curve D accounts for both near-field effects and soil losses, it is the proper curve against which the directional field strength measurement data should be fitted. Note the good fit to the measurement data, both close to the array and at distant points, even though the first 19 measurement points fall considerably above the inverse distance line A.

When using near-field analysis on directional antenna field strength data, it must be remembered that a statistical analysis of field strength data with "corrections" applied to the individual point ratios can be very misleading if the pattern is not adjusted to the parameters for which the near-field values were calculated in the first place. When properly analyzed, the measurement data should generally conform to the shape of the calculated near-field curve as shown in Figure 4.3-18.

SEASONAL VARIATION OF FIELD STRENGTH

Field strength measurements on a previously licensed directional antenna may appear to indicate a change in pattern shape or size when the change was in fact due to changes in soil conductivity. Such changes affect distant measurements more than close-in measurements. In some areas of the United States, the conductivity is typically higher during winter and spring months when the soil is more moist than in summer and fall months, with the conductivity being the highest when the ground is frozen. Seasonal conductivity variations are not observable in some portions of the country, yet are extreme in other areas. One well-documented case showed a seasonal doubling of signal strength at 32 km (20 miles) in the main lobe of a correctly adjusted system operating on 1380 kHz. To avoid the misleading effects of seasonal conductivity changes that might appear to distort measured directional antenna patterns in size or shape, the FCC requires that all the field strength measurements in a directional antenna proof-of-performance be made under "similar environmental conditions."

STANDARD PATTERNS

Theoretical (also called calculated) patterns can have nulls wherein the radiation at specific azimuths goes completely to zero. In practice, it is not possible to prove by field strength measurements that a null exists. Reradiation and scatter from objects external to the array limit the depth to which a pattern minimum can be proven. Additionally, operational variations in phase and ratio parameters will increase radiation in any direction where the deepest possible minimum has been previously established. To accommodate

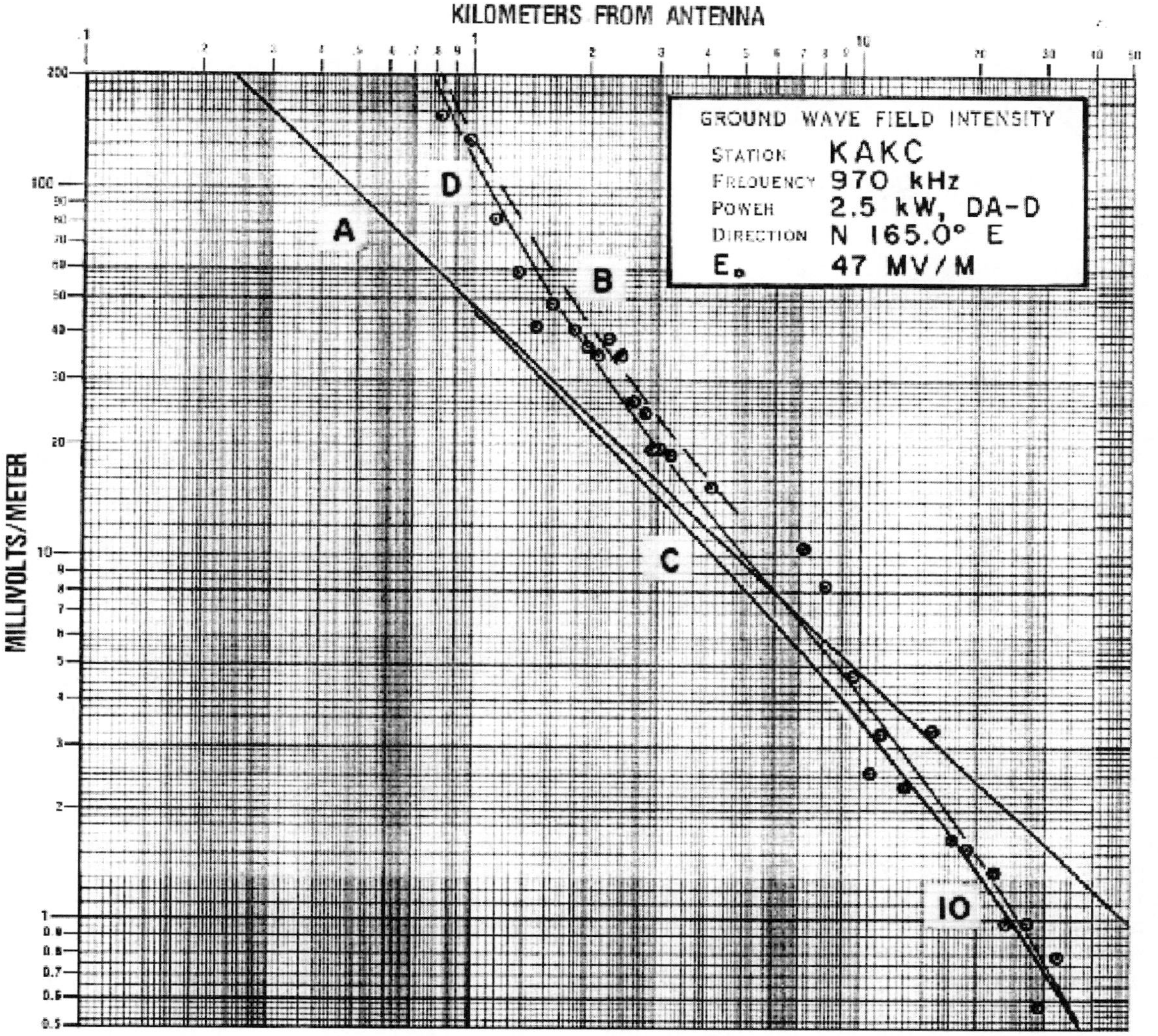

FIGURE 4.3-18 Near-field effects.

these limitations, the FCC authorizes a standard pattern for each directional antenna station. Standard patterns exceed the theoretical pattern at all azimuths by specified and easily calculated amounts. It is required that the radiation from a directional station not exceed its standard pattern. All U.S. stations employing directional antennas have FCC-specified standard patterns. These supersede all earlier patterns based on theoretical calculations or on field strength measurements. The standard pattern radiation values are now used exclusively in all calculations of coverage and interference.

AUGMENTED PATTERNS

Augmentation is applied to the standard pattern when the measured field strength is exceeded in discrete directions but does not cause interference to other stations. When augmentation is required, it is achieved by applying Eq. B-9 in Appendix B. Figure 4.3-19 shows the theoretical and augmented standard patterns of a simple two-tower example directional antenna.

DRIVING POINT IMPEDANCE

The input impedance of each tower in an array (called the *driving point impedance*) is not what it would be if the tower were used as a nondirectional antenna. This is because of the effects of mutual coupling with the other towers of the array. The base current in one tower in a two-tower array, depicted as I1 in Figure 4.3-20, induces a voltage across the base of the other tower, E2, and vice versa.

The ratio of E1 to I1 without the second tower present would be its self-impedance. The ratio of E1 to I2 is the mutual impedance between the two towers. When both towers are radiating to produce a

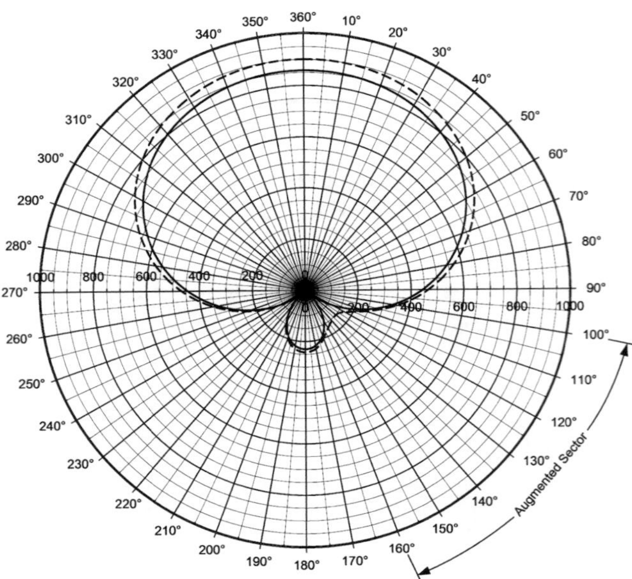

FIGURE 4.3-19 Theoretical and augmented standard directional antenna radiation patterns.

directional antenna pattern, each tower base voltage is the sum of the value produced by the base current flowing into its self-impedance and the value produced by the current flowing into the other tower through the mutual impedance. The operating, or driving point, impedance of each tower in an array is the ratio of its total base voltage to its base current while functioning in the array. In other words, the driving point impedance contains the self-impedance plus the mutual impedances multiplied by the current ratios that exist with other towers in the array as driven to produce the desired pattern. The driving point impedance will depend on the array parameters and can even have a negative resistance component so the tower draws power from the other towers and delivers it back to the phasing system.

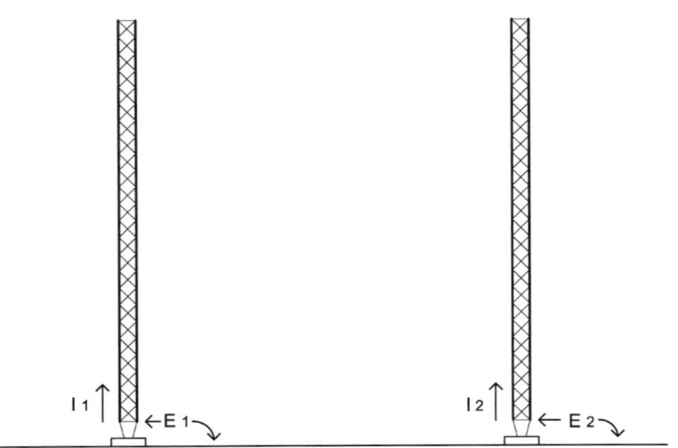

FIGURE 4.3-20 Mutual coupling.

Calculated driving point impedances are used in the design of new phasing and coupling equipment that is designed before the towers are erected, but it is sometimes desirable to measure the operating impedances of towers in an existing directional antenna system. Because the operating impedance is affected by the currents in the other towers, it can only be measured by an operating impedance bridge inserted at the tower feed point while the other towers are operating with their correct current magnitudes and phases. For precise measurement, it is necessary to readjust the system with the bridge in place to produce the correct parameters, because the series inductance and stray capacitance that it adds can significantly impact tuning.

BASE CURRENTS VERSUS RADIATED FIELDS

In a directional array, the tower base current ratios will usually depart substantially from the calculated radiated field ratios when the pattern is correctly adjusted. This is caused by the nonsinusoidal current distribution of each tower as described elsewhere in this section. Thus, it has been standard practice for directional antenna patterns to be initially proven to have their correct shapes by means of a series of field strength measurements in significant radial directions from the station rather than by assuming that measurement of tower currents and phases can establish the correct pattern. It is possible, with moment method modeling, to determine base current ratios and phases that correspond very well with those necessary to produce a required directional antenna pattern shape. Using these methods, the amount of experimental adjustment work necessary before acceptable proof-of-performance measurements can be made is greatly reduced and, in many cases, eliminated.

DETUNING STRUCTURES NEAR AM ANTENNAS

It is sometimes necessary to detune a tower on a directional antenna system's property, such as a tower that is used in one pattern but not another. For tower heights below one-half wavelength, this can usually be done by placing a reactance from the base feedpoint to ground to cause the current distribution on the tower to have the general shape shown in Figure 4.3-21. The far-field radiation in the horizontal plane is reduced to virtually zero because the contributions from the currents that flow on the tower above and below the minimum, at approximately 1/3 of the tower height, are equal and 180° out of phase. The height of the minimum for optimal detuning ranges from approximately 33% of the total tower height for towers in the range of 90° up to approximately 38% for towers approaching 180° in height. It, and the reactance value necessary at the tower base, may be computed using moment method modeling.

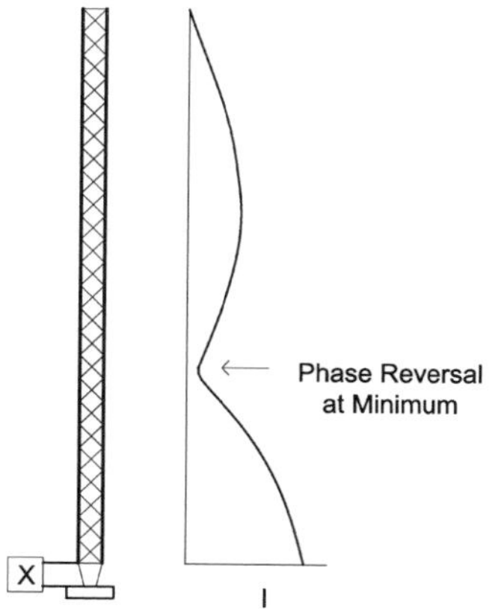

FIGURE 4.3-21 Detuning with base termination.

Where the degree of detuning is not critical, such as with communication towers and other structures that are located well away from the towers of the directional antenna system, but close enough to scatter enough field to fill out pattern nulls if they are untreated, it is usually possible to eliminate reradiation problems by breaking up current on them as shown in Figure 4.3-22. The tower on the left shows the method applied to a tower on the order of 90°, while the tower on the right shows a tower on the order of 180°. For taller towers, it may be necessary to break the current up into more than two pieces. The general principle is to limit reradiation by breaking the structure into current conducting pieces on the order of 90° in length, or shorter. This can be done by placing insulators within the structures, as shown in Figure 4.3-22, or by surrounding the structures with wire skirts, as discussed for feeding towers elsewhere in this chapter.

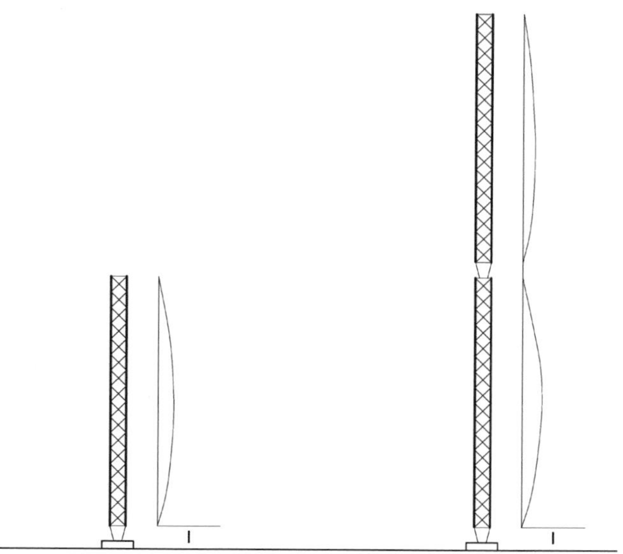

FIGURE 4.3-22 Detuning by breaking up current.

DETUNING POWER LINES NEAR AM ANTENNAS

Power lines near AM antennas are more complicated to analyze than simple vertical structures, as their support poles or towers are generally interconnected at their tops by one or more horizontal ground conductors. Reradiation can take place at levels far in excess of what would be expected for the vertical height involved when the circuit path between adjacent support structures, including the ground between them and the top ground wires, is near resonance at the AM station's frequency. This phenomenon defies simple theoretical analysis because of the influence of the physical characteristics of the support structures and the electrical characteristics of the soil between them. The most expedient remedy when powerline reradiation is a problem is to insulate, or RF isolate, the horizontal ground wires from the tops of the support structures that are closest to the AM antenna. This can be accomplished by installing insulators with lightning gaps and/or RF chokes across them at the tops of the vertical structures. In many cases, no further measures (such as detuning the vertical structures with wire skirts) will be required.

MONITORING DIRECTIONAL ANTENNA OPERATING PARAMETERS

Antenna monitoring systems are used by AM stations that employ directional antennas to monitor the ratios and phases of the currents flowing in the array elements so they can be maintained at the values that are known to produce the required pattern shapes, as shown in Figure 4.3-23. Antenna monitors are designed to meet the FCC requirements for accurate monitoring of the ratios and phases of the current samples that are fed into them, through the sampling lines, from the current sampling devices. The sampling lines

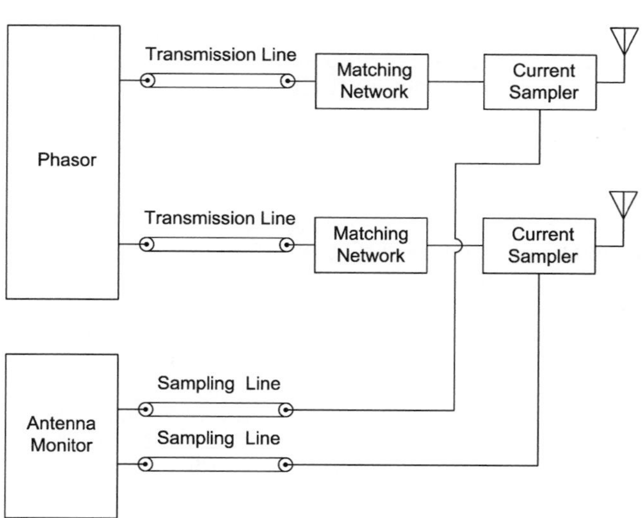

FIGURE 4.3-23 Antenna monitoring system.

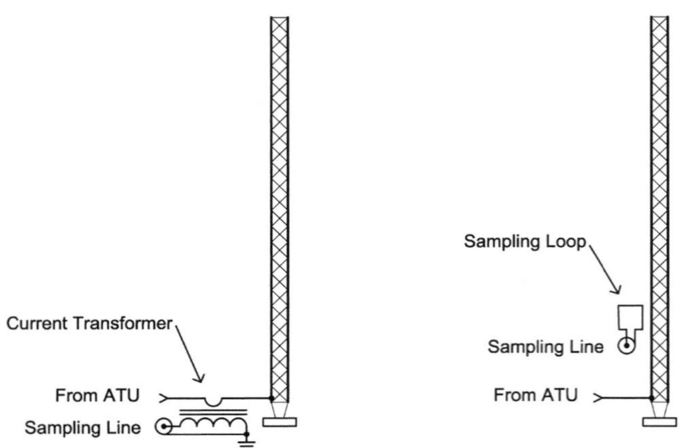

FIGURE 4.3-24 Base and loop sampling.

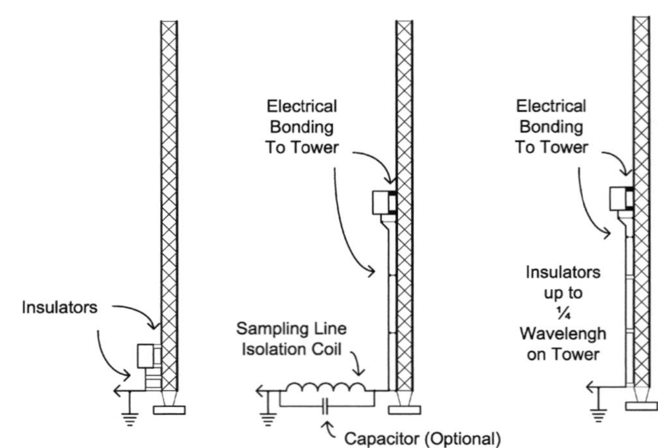

FIGURE 4.3-25 Insulated and uninsulated sampling loops.

are typically 3/8 or 1/2 inch foam dielectric transmission lines and are semiflexible with solid outer conductors. The current sampling devices are either current transformers through which the tower base currents pass or tower-mounted inductive pickup loops, as shown in Figure 4.3-24. In either case, the sampling devices must be rated to produce voltages within the acceptable range of the antenna monitor with full power into the antenna system.

Base current transformers differ from tower-mounted sampling loops in a very important way: The current that is sampled is the sum of the actual tower current and other currents flowing to ground after the transformer due to stray capacitance in the base region and any circuits—such as lightning chokes, ring transformers, and isocouplers—that are across the base insulator. For this reason, it has been customary to use loop instead of current transformer sampling for towers much over 90° in height, where base impedances are high. For towers of any height, the relationship of the base sampled current to the actual tower current can change with changes to the circuits across the base. It should be understood that it might be necessary to operate with a different antenna monitor ratio and phase to maintain the same radiation pattern if changes are made across the base of a tower employing current transformer sampling (such as changing an isocoupler or adding a new one) and that a partial proof-of-performance might be necessary to have the FCC license modified to reflect the new parameters.

Three methods for coupling the sampling line from a tower-mounted loop to ground are shown in Figure 4.3-25. For sampling loops that essentially monitor the base current, which are typically mounted 10 feet above the tower base, it is possible to insulate them from the tower and have them operate at ground potential. The added capacitance of the loop and sampling line on the lower portion of the tower can have a significant effect on the base impedance, but that can be compensated for at the time of the initial adjustment. Most antenna systems that employ tower-mounted sampling loops use isolation coils wound of sampling line across their tower bases. Sampling line isolation coils typically have inductances in the range of 100 to 150 μH and have minimal, and sometimes beneficial, effects on the tower base impedances. Where it is desired to compensate for the effects of the parallel inductance of an isolation coil across a tower base, a capacitor can be employed in parallel with it for that purpose. Some systems with towers on the order of one-half wavelength use quarter-wave isolation (with the sampling lines insulated from the tower for the bottom 90° of height) instead of isolation coils.

Figure 4.3-26 shows the heights at which sampling loops are typically mounted on towers. Historically, sampling loops have been mounted where the current

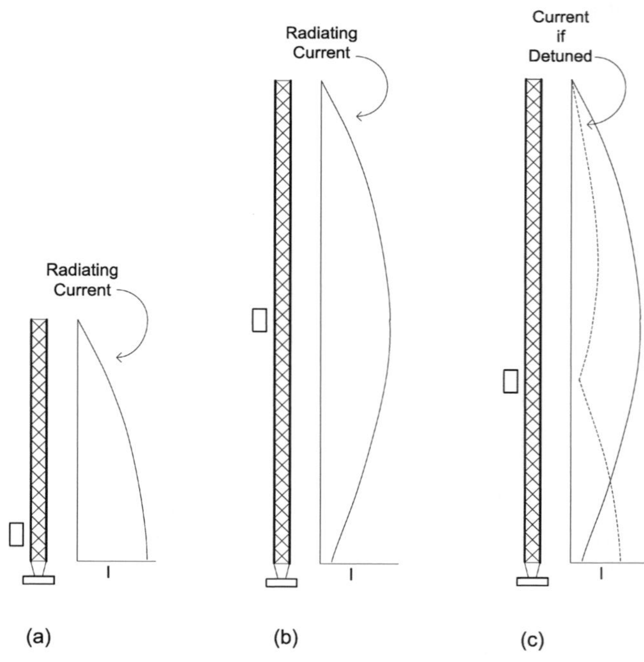

FIGURE 4.3-26 Sampling loop tower placement.

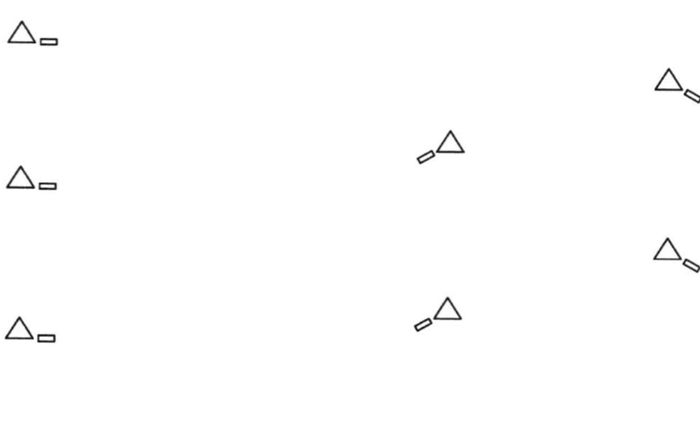

FIGURE 4.3-27 Sampling loop orientation on array elements.

is theoretically maximum, assuming sinusoidal current distribution. Most existing antenna systems employing sampling loops have them mounted as shown in Figure 4.3-26a (but no lower than about 10 feet above ground level to avoid stray effects from conductors near ground level) on towers 90° or less in height and as shown in Figure 4.3-26b for towers greater than 90° in height. Figure 4.3-26c shows where the sampling loops may be mounted to give indications closest to the theoretical field parameters, at the heights where the current minimum occurs for the detuned condition. This height may be determined by moment method modeling and can give measured tower currents bearing a direct 1:1 relationship to field parameters where the towers are identical. Where the towers are not identical, there is still an advantage in that the incremental changes in ratio and phase will agree between the sampled currents and tower fields even when their absolute values are not in agreement.

When placing sampling loops on the towers of an array, care should be taken to minimize pickup from other towers. The sampling loops should be mounted with their planes including the structures of the towers on which they are located while excluding the closest nearby towers as much as possible, as shown on Figure 4.3-27.

ELECTRICALLY SHORT ANTENNAS

There is considerable interest in AM transmitting antennas that are much shorter than the typical quarterwave tower. Such antennas are useful in situations where conventional towers cannot be constructed because of environmental or aeronautical concerns, or for emergency backup antennas at stations that have conventional towers. The difficulties with such antennas center on radiation efficiency and bandwidth issues, as they typically have low base resistances and, therefore, high base currents for the power that is fed into them.

The FCC is primarily interested in the efficiency and radiation properties of antennas. Before a station can be licensed with an antenna that is shorter than would be necessary for a conventional tower antenna, the FCC must be satisfied that its efficiency meets the minimum requirements specified in the Rules and that its vertical radiation pattern is known. If an application for a construction permit is filed specifying an antenna of insufficient height under the Rules, the FCC may grant it with the added requirement that a proof-of-performance be run to demonstrate that its radiation efficiency meets its minimum requirement before a license will be issued. If there is any question about the current distribution on the antenna as it might impact the vertical radiation pattern, such as arise when top loading and center loading are employed, current distribution measurements will also be required.

Antenna designs have been proposed in recent years that claim to overcome the disadvantages of short antennas by employing controversial principles that are not acknowledged by the larger peer group of antenna design engineers, such as synthesis of the electric and magnetic field components of a propagating wave separately in a small space using methods that are outside the bounds of analysis using Maxwell's equations. Despite claims of exceptional performance during informal tests and considerable discussion of a theoretical nature, several examples of such antennas have been the subject of much speculation without evidence ever being provided that they meet the FCC's minimum efficiency requirements based on objective, scientific tests.

There are well-known methods for obtaining better efficiency than is expected for short conventional tower antennas, however. Several principles for optimizing the efficiency of very short antennas, as shown in Figure 4.3-28, have been recognized for decades. Figure 4.3-28a shows how top loading can be used to maximize the value of the integral of antenna current over the vertical length of a short conductor, which is

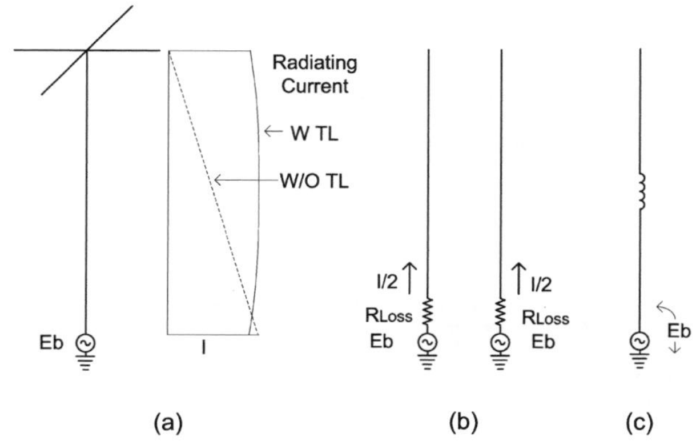

FIGURE 4.3-28 General principles for short antenna efficiency improvement.

the condition for maximum radiation. Figure 4.3-28b shows how the radiating current can be divided between multiple closely coupled antenna conductors to reduce ground losses. Rather than intensifying directly under one conductor, the ground return currents are divided to flow under the multiple conductors. Figure 4.3-28c shows how what is called *center loading* (although it does not necessarily have to be placed at the exact center of the antenna to be effective) can be used to reduce the voltage across the base of a short antenna that would otherwise have a high capacitive reactance component and, thus, reduce stored energy in the region of the feedpoint. These principles may be used together to improve the performance of electrically short antennas.

The electrically short antennas that have been licensed by the FCC have generally used one or more of the principles illustrated by Figure 4.3-28. Two short antennas available from manufacturers have been approved for nondirectional use by stations in the United States. One consists of four top-loaded vertical wires supported by wooden poles over a quarter-wave radial ground system and has been preapproved for both efficiency and vertical radiation so it can be employed without the need to make proof measurements following construction. It must be constructed with its height, top-loading dimensions, and ground system scaled for frequency. Another consists of a fiberglass pole with both top and center loading, which requires a proof to establish its efficiency following construction, as it is a standard height rather than being scaled for frequency, but does not require current distribution measurements.

Bibliography

Directional Antenna Patterns, Smith Electronics, Inc., Cleveland, OH.

Theory and Design of Directional Antennas, Smith Electronics, Inc., Cleveland, OH.

Standard Broadcast Antenna Systems, Smith Electronics, Inc., Cleveland, OH.

Design and Operation of Directional Antennas, Smith Electronics, Inc., Cleveland, OH.

Directional Antenna Pattern Shapes, Smith Electronics, Inc., Cleveland, OH.

Radiation Characteristics of Transmitting Antennae: An Introduction to Directional Antenna Pattern Design, Smith Electronics, Inc., Cleveland, OH.

Directional Antenna Design Example, Smith Electronics, Inc., Cleveland, OH.

Parasitic Reradiation, Smith Electronics, Inc., Cleveland, OH.

Introduction to Directional Antenna Systems, Smith Electronics, Inc., Cleveland, OH.

Instructions for Installation of Radio Broadcast Stations Ground Systems, Smith Electronics, Inc., Cleveland, OH.

Log Periodic Antenna Design Handbook, Smith Electronics, Inc., Cleveland, OH.

Radio Broadcast Ground Systems, Smith Electronics, Inc., Cleveland, OH.

Ballantine, S., On the optimum transmitting wavelength for a vertical antenna over perfect earth, *Proc. IRE*, 12, 833–839, 1924.

Brown, G. H., A critical study of the characteristics of broadcast antennas as affected by current distribution, *Proc. IRE*, 24, 48–81, 1936.

Harmon, R. N., Some comments on broadcast antennas, *Proc. IRE*, 24, 36–47, 1936.

Smith, C. E., A critical study of two broadcast antennas, *Proc. IRE*, 24, 1329–1341, 1936.

Brown, G. H., Directional antennas, *Proc. IRE*, 25(1), 79–145, 1937.

Brown, G. H., Lewis, R. F., and Epstein, J., Ground systems as a factor in antenna efficiency, *Proc. IRE*, 25(6), 753–787, 1937.

Morrison, J. F. and Smith, P. E., The shunt excited tower, *Proc. IRE*, 25, 673–696, 1937.

Smith, C. E. and Johnson, E. M., Performance of short antennas, *Proc. IRE*, 35, 1026–1038, 1947.

Jeffers, C. L., An antenna for controlling the nonfading range of broadcasting stations, *Proc. IRE*, 36, 1426–1431, 1948.

Smith, C. E., Hall, J. R., and Weldon, J. O., Very high-power long-wave broadcasting antennas, *Proc. IRE*, 42(8), 1222–1235, 1954.

Smith, C. E., A Critical Study of Several Antennas Designed to Increase the Primary Coverage of a Radio Broadcasting Transmitter, thesis, Ohio State University, Columbus.

Smith, C. E., Hutton, D. B., and Hutton, W. G., Performance of sectionalized broadcasting towers, *IRE Trans.*, Dec., 22–34, 1955.

Anderson, H. and Pinion, D., Short low loss antennas, *IEEE Trans. Broadcasting*, BC-32(3), 1986.

Dawson, B. F., Sharing AM transmitter sites by diplexing antenna systems, in *Proc. of the 41st Annual National Association of Broadcasters Broadcast Engineering Conf.*, March, 1987, pp. 31–36.

Hatfield, J. B. and Leonard, P. W., A comparison of the fields of a medium wave directional antenna as calculated by the FCC method and the numerical electromagnetic code, in *Proc. of the Applied Computational Electromagnetics Society Conf.*, Monterey, CA, March, 1987.

Trueman, C. W., Kubina, S. J., and Baltassis, C., Ground loss effects in power line reradiation at standard broadcast frequencies, *IEEE Trans. Broadcasting*, 34(1), 24–38, 1988.

Trueman, C. W., Modelling a standard broadcast directional array with the numerical electromagnetics code, *IEEE Trans. Broadcasting*, 34(1), 39–49, 1988.

Christman, A., Radcliff, R., Adler, D., Breakall, J., and Resnick, A., AM broadcast antennas with elevated radial ground systems, *IEEE Trans. Broadcasting*, 34(1), 75–77, 1988.

Hatfield, J. B., Analysis of AM directional arrays using method of moments, in *Proc. of the 42nd Annual National Association of Broadcasters Broadcast Engineering Conf.*, April, 1988, pp. 84–87.

Hatfield, J. B., Verifying the relationships between AM broadcast fields and tower currents, in *Proc. of the Applied Computational Electromagnetics Society Conf.*, Monterey, CA, March, 1989.

Hatfield, J. B., Relative tower currents and fields in an AM directional array, *IEEE Trans. Broadcasting*, 35(2), 176–184, 1989.

Dawson, B. F., Analysis of a sectionalized tower as an element in a medium wave phased array using the method of moments, *IEEE Trans. Broadcasting*, 35(2), 185–189, 1989.

Chiodini, T., Moment method predicted impedances compared to actual measured impedances of directional arrays, *IEEE Trans. Broadcasting*, 35(2), 1909–1192, 1989.

Bruner, P. and Waniewski, B., Directional MF antennas using self-supporting towers with driven wire cages, *IEEE Trans. Broadcasting*, 35(2), 193–199, 1989.

Westberg, J. M., Matrix method for relating base current ratios to field ratios of AM directional stations, *IEEE Trans. Broadcasting*, 35(2), 172–175, 1989.

Christman, A. and Radcliff, R., Impedance stability and bandwidth considerations for elevated-radial antenna systems, *IEEE Trans. Broadcasting*, 35(2), 167–171, 1989.

Smith, C. E., Short low loss antennas, *IEEE Trans. Broadcasting*, 35(2), 237–249, 1989.

Trueman, C. W., Roobroeck, T. M., and Kubina, S. J., Stub detuners for free standing towers, *IEEE Trans. Broadcasting*, 35(4), 325–338, 1989.

Trainotti, V., The nocturnal service area of MF AM broadcast stations using vertical arrays, *IEEE Trans. Broadcasting*, 36(1), 74–81, 1990.

Trainotti, V., Height radius effect on MF AM transmitting monopole antenna, *IEEE Trans. Broadcasting*, 36(1), 82–88, 1990.

Trueman, C. W. and Kubina, S. J., Power line tower models above 1000 kHz in the standard broadcast band, *IEEE Trans. Broadcasting*, 36(3), 207–218, 1990.

Dawson, B. F., Modern analysis methods for medium wave antenna design, in *Proc. of the International Broadcasting Convention of the IEE*, Brighton, U.K., September, 1990.

Rackley, R. D., Modern methods in mediumwave directional antenna feeder system design, in *Proc. of the 45th Annual National*

Association of Broadcasters Broadcast Engineering Conf., April, 1991, pp. 43–54.

Christman, A. and Radcliff, R., Using elevated radials with ground mounted towers, *IEEE Trans. Broadcasting*, 37(3), 77–82, 1991.

Frese, G. M., Paran antenna for use where horizontal and vertical space is limited, *IEEE Trans. Broadcasting*, 38(3), 163–165, 1992.

Lundin, J. A. and Rackley, R. D., Medium frequency broadcast antennas, in *Antenna Engineering Handbook*, Johnson, R. C. (Ed.), McGraw-Hill, New York, 1993, pp. 25-1–25-28.

Trueman, C. W. and Kubina, S. J., Scattering from power lines with the skywire insulated from the towers, *IEEE Trans. Broadcasting*, 40(2), 53–62, 1994.

Christman, A. and Radcliff, R., AM broadcast antennas with elevated radials: varying the number of radials and their height above ground, *IEEE Trans. Broadcasting*, 42(1), 10–13, March, 1996.

Hatfield, J. B., Computer analysis of antenna systems, in *The Electronics Handbook*, Whitaker, J. C. (Ed.), CRC Press, Boca Raton, FL, 1996, pp. 1353–1367.

Smith, J. L., A method for modeling array elements when using NEC and MININEC, *IEEE Trans. Broadcasting*, 44(2), 186–193, 1998.

Rackley, R. D. and Folkert, M., Antennas for medium-frequency broadcasting, in *Wiley Encyclopedia of Electrical and Electronics Engineering*, Webster, J. G. (Ed.), John Wiley & Sons, New York, 1999, pp. 578–594.

Hatfield, J. B., Computer simulation of AM radio antenna systems, in *NAB Engineering Handbook*, National Association of Broadcasters, Washington, D.C., 1999, chap. 4.10.

Smith, J. L., A method to determine the detuning reactance for unused elements in directional arrays, *IEEE Trans. Broadcasting*, 47(3), 259–262, 2001.

Trainotti, V., Short medium frequency AM antennas, *IEEE Trans. Broadcasting*, 47(3), 263–284, 2001.

Trainotti, V., MF AM folded monopole characteristics, *IEEE Trans. Broadcasting*, 48(4), 324–330, 2002.

Rackley, R. D., Engineering exhibit star-H experimental antenna, in *Proc. of the 57th Annual National Association of Broadcasters Broadcast Engineering Conf.*, April, 2003, pp. 401–439.

Breakall, J. K., Jacobs, M. W., King, T. F., and Resnick, A. E., Testing and results of a new, efficient low-profile AM medium frequency antenna system, in *Proc. of the 57th Annual National Association of Broadcasters Broadcast Engineering Conf.*, April, 2003, pp. 235–243.

Beverage, C. M., Compact medium wave transmitting antennas, *IEEE Trans. Broadcasting*, 50(2), 142–147, 2004.

Leonard, P. W., Computer models of Kabul, Afghanistan, medium-wave antennas, *IEEE Antenna Propagation Soc. Mag.*, 48(2), 38–41, 2006.

APPENDIX A: DIRECTIONAL ANTENNAS FOR PATTERN SHAPE

Space Configuration

The plan configuration of the kth tower in an array is shown in Figure 4.3-A1. A space view of the kth tower and observation point P is shown in Figure 4.3-A2.

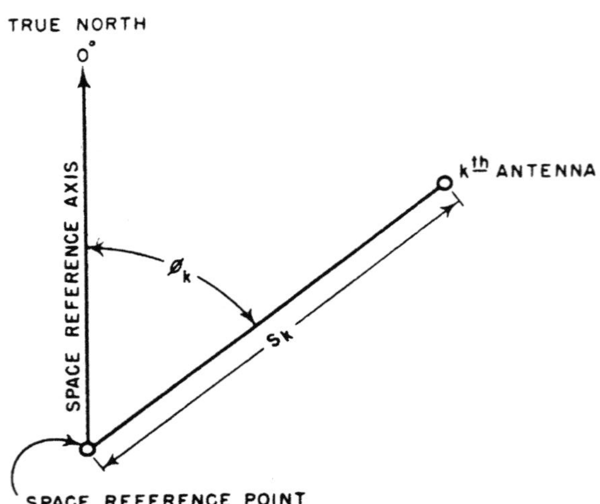

FIGURE 4.3-A.1 Plan view of space configuration of kth antenna.

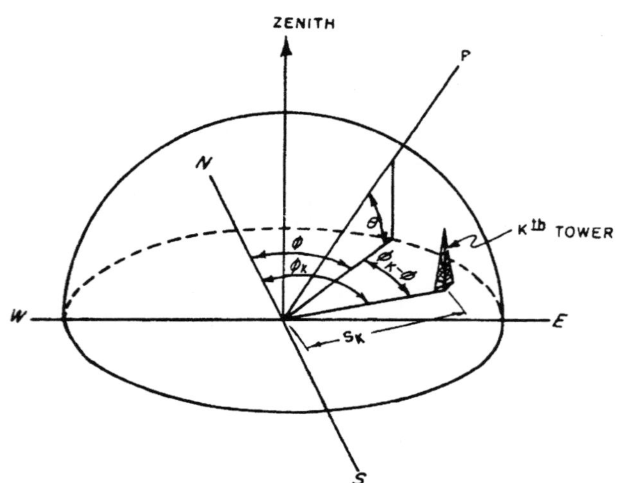

FIGURE 4.3-A.2 Space view of observation point P and the kth tower.

Vector Diagram

The field strength at point P in space for the kth tower is shown in Figure 4.3-A3. The space phasing in the horizontal plane is shown in Figure 4.3-A4, and in the elevation plane the space phasing is reduced further, as shown in Figure 4.3-A5.

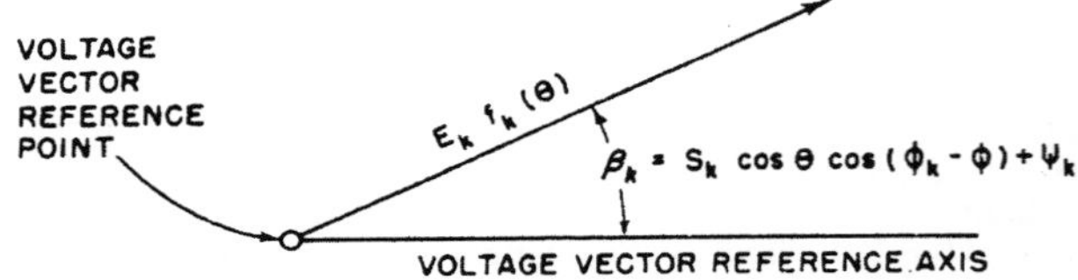

FIGURE 4.3-A.3 Voltage vector diagram for the kth antenna.

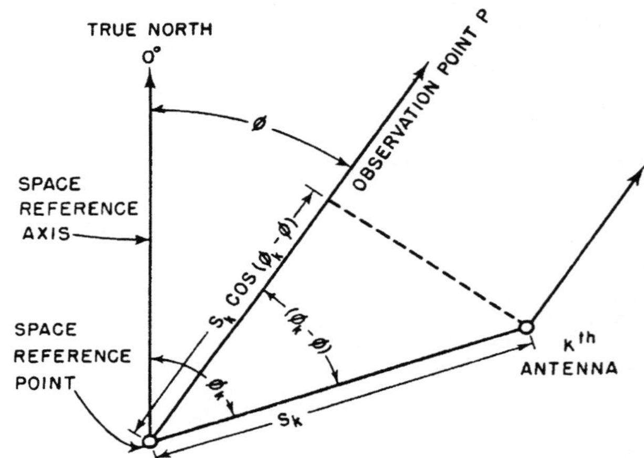

FIGURE 4.3-A.4 Plan view of kth antenna showing space phasing in the horizontal plane.

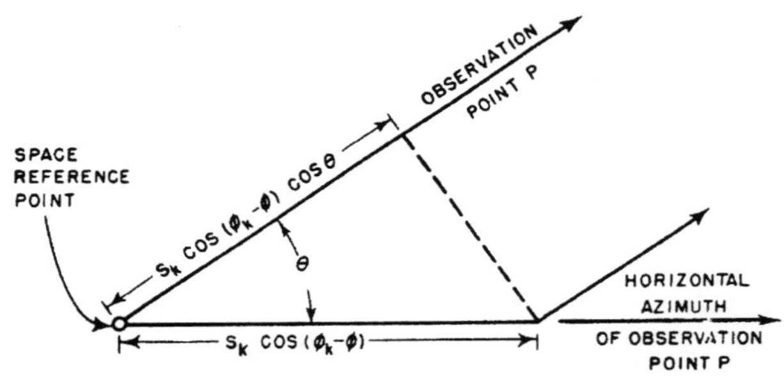

FIGURE 4.3-A.5 Elevation angle u shortens the spacing S_k to the value of $S_k$cosu.

Generalized Equation

The vector equation to express the vectors in Figure 4.3-A6 is the generalized equation that can be used to express the pattern shape for a directional antenna array of n towers. The equation in condensed form is:

$$E = \sum_{k=1}^{k=n} E_k f_k(\Theta)\beta_k \tag{A1}$$

where:

E = the total effective field strength vector at unit distance (P) for the antenna array with respect to the voltage vector reference axis. This vector makes the angle β with respect to this axis, as shown in Figure 4.3-A6.

k = the kth tower in the directional antenna system.

n = the total number of towers in the directional antenna array.

E_k = the magnitude of the field strength at unit distance in the horizontal plane produced by the kth tower acting alone.

$f_k(\Theta)$ = vertical radiation characteristic of the kth antenna as given in Eq. A3.

Θ = elevation angle of the observation point P measured up from the horizon in degrees.

$$\beta_k = S_k \cos\Theta \cos(\phi_k - \phi) + \psi_k \tag{A2}$$

is the phase relation of the field strength at observation point P for the kth tower taken with respect to the voltage vector reference axis.

$$S_k \cos(\phi_k - \phi)\cos\Theta \tag{A3}$$

is the space phasing portion of β_k due to the location of the kth tower, and ψ_k is the phasing portion of β_k.

S_k = electrical length of spacing of the kth tower in the horizontal plane from the space reference point.

ϕ_k = true horizontal azimuth orientation of the kth tower with respect to the space reference axis.

ϕ = true horizontal azimuth angle of the direction to the observation point P (measured clockwise from true north).

ψ_k = time phasing portion of β_k due to the electrical phase angle of the voltage (or current) in the kth tower taken with respect to the voltage vector reference axis.

The shape of any directional antenna pattern can be computed by applying the above equations; however, many directional antenna arrays can be designed by simplified versions of this equation. For a vertical antenna having a sinusoidal current distribution with

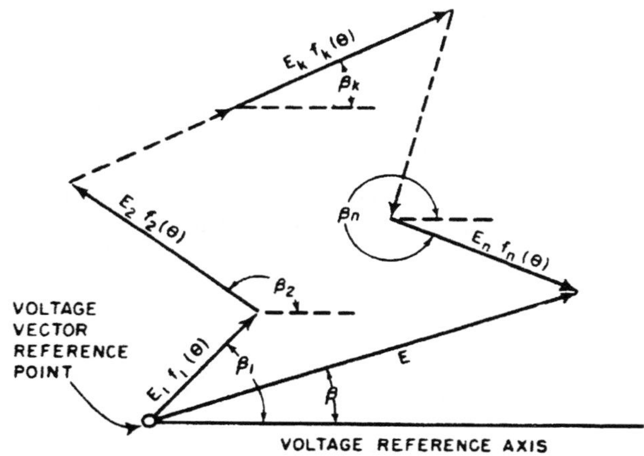

FIGURE 4.3-A.6 Summation of field strength vectors for n antennas in the directional antenna array.

a current node at the top, the vertical radiation characteristic takes on the form:

$$f(\Theta) = \frac{\cos(G\sin\Theta) - \cos G}{(1-\cos G)\cos\Theta}$$

where:

$f(\Theta)$ = vertical radiation characteristic.

G = electrical height of the antenna in electrical degrees.

Θ = elevation angle of the observation point measured up from the horizon in degrees.

The vertical radiation characteristics in Eq. A3 are graphed in Figure 4.3-A7.

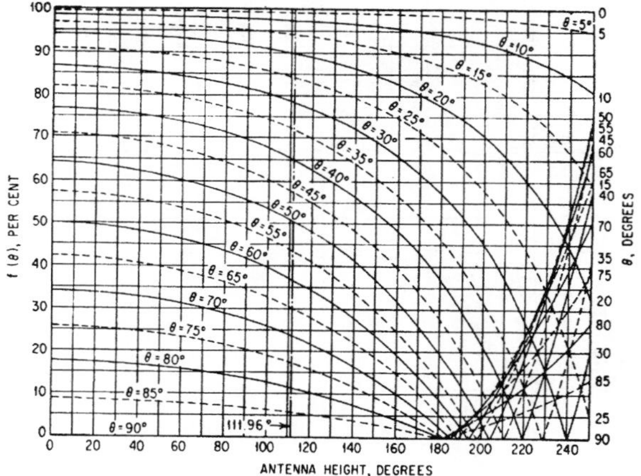

FIGURE 4.3-A.7 Vertical-radiation characteristics as a function of electrical tower height for various values of elevation angle.

For a top-loaded tower the formula is:

$$f(\Theta) = \frac{\cos B \cos(A \sin\Theta) - \cos G - \sin B \sin\Theta \sin(A \sin\Theta)}{\cos\Theta(\cos B - \cos G)}$$

This is the vertical radiation characteristic for a top-loaded tower of height A and top loaded to a height of $G = A + B$. For a two-section top-loaded tower as shown in Figure 4.3-A5, the formula is:

$$f(\Theta) = \frac{\cos B \cos(A \sin\Theta) - \cos G + \dfrac{\sin B \cos(H - C)\cos(C \sin\Theta)}{\sin(H - A)} - \dfrac{\sin B \sin\Theta \sin(H - C)\sin(C \sin\Theta)}{\sin(H - A)} - \dfrac{\sin B \cos(H - A)\cos(A - \Theta)}{\sin(H - A)}}{\cos\{\cos B - \cos G + [\sin B / \sin(H - A)](\cos H - C \cos H - A)\}}$$

This is the vertical radiation characteristic equation for a two-section sectionalized tower. The same procedure can be applied if more than two sections are involved (see Figure 4.3-A6).

Theoretical Self-Loop and Base-Radiation Resistance

It is useful to know the theoretical loop and base resistance of a vertical radiator. This information is presented graphically in Figure 4.3-A8, along with the theoretical inverse field strength at 1 km.

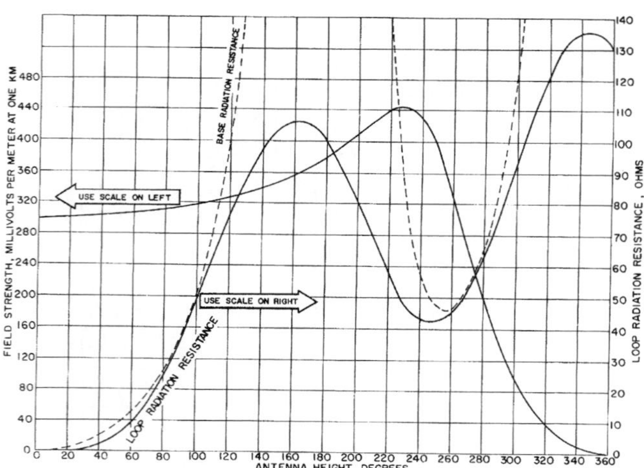

FIGURE 4.3-A.8 Inverse field strength at 1 km for 1-kW loop and base radiation resistance as a function of tower height over a perfectly conducting earth.

Mutual Impedance Curves

The value of mutual impedance for most tower heights and spacing is given in Figure 4.3-A9. The loop mutual impedance between quarter-wave towers is shown in Figure 4.3-A10.

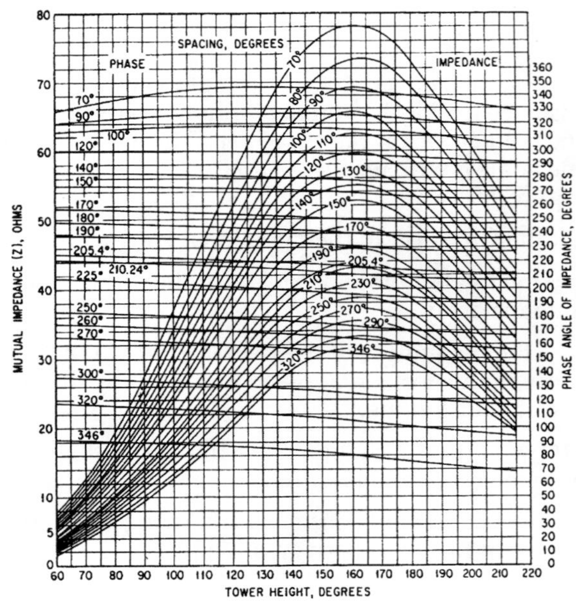

FIGURE 4.3-A.9 Loop mutual impedance and phase angle between two towers of equal height.

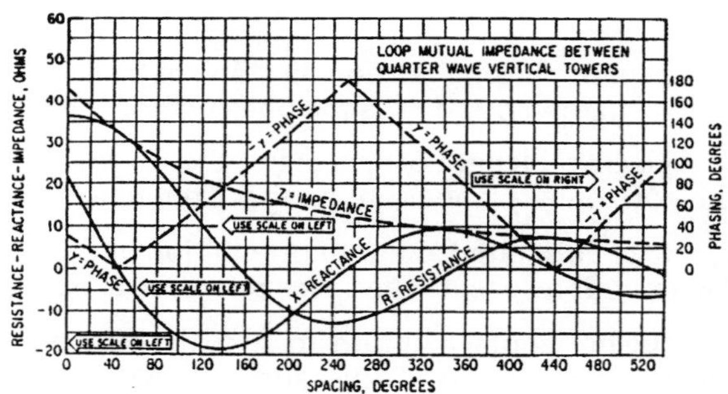

FIGURE 4.3-A.10 Loop mutual impedance between quarter-wave vertical towers.

Horizontal RMS Field Strength

The field strength gain or loss of a two-tower array for various values of phasing and spacing is shown in Figure 4.3-A11.

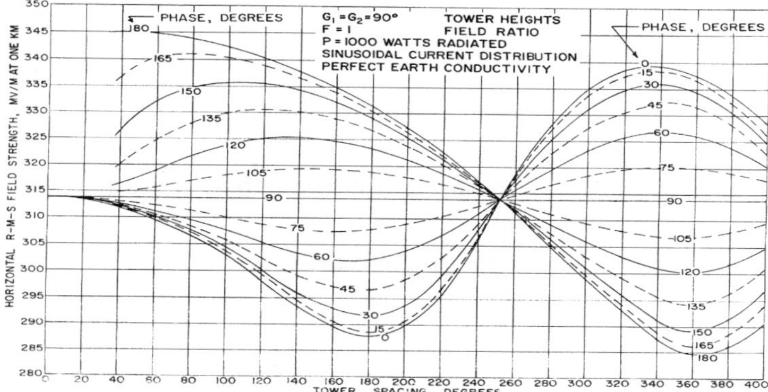

FIGURE 4.3-A.11 Horizontal RMS field strength of two-tower directional antenna.

APPENDIX B: PATTERN DEVELOPMENT OF DIRECTIONAL ANTENNAS

Theoretical Pattern Equation

The theoretical pattern equation of Appendix A can be written as follows by changing the *k*th tower to the *i*th tower to conform with the FCC practice, thus:

$$E(\phi,\Theta)th = \left| k\sum_{i=1}^{n} F_i(\Theta) \Big/ S_i \cos\Theta\cos(\phi_i - \phi) + \psi_i \right| \quad \text{(B1)}$$

where *k* is a multiplying constant that determines pattern size.

Standard Pattern Equation

The standard pattern equation is obtained from Eq. B1 by adding the quadrature term *Q* to fill minimums and increase the size by 5%, thus:

$$E(\phi,\Theta)_{std} = 1.05\sqrt{\left[E(\phi,\Theta)th\right]^2 + Q^2} \quad \text{(B2)}$$

where *Q* is the greater of the quantities:

$$0.025g(\Theta)E_{rss} \quad \text{(B3)}$$

or

$$10.0g(\Theta)\sqrt{P_{kw}} \quad \text{(B4)}$$

where $g(\Theta)$ is the vertical plane distribution factor, $f(\Theta)$, for the shortest element in the array (see Eq. B2 above; also see FCC Rules Section 73.190, Figure 5). If the shortest element has an electrical height in excess of 0.5 wavelength, $g(\Theta)$ is computed as follows:

$$g(\Theta) = \frac{\sqrt{\{f(\Theta)\} + 0.0625}}{1.030776} \quad \text{(B5)}$$

$$E_{rss} = \sqrt{\sum_{i=1}^{n} E_t^2} \quad \text{(B6)}$$

As an example, consider a two-tower array. The theoretical pattern equation (Eq. B1) becomes:

$$E = E_1 f_1(\Theta) \; \underline{/0^\circ} + E^2 f_2(\Theta) \; \underline{/S_2\cos\Theta\cos(\phi_2 - \phi) + \psi_2} \quad \text{(B7)}$$

Now, for 5-kW, 90° towers and the following parameters:

Tower Number	Height (G^0)	Field Ratio	Spacing (S^0)	True Bearing (ϕ^0)	Phase (ψ)
1	90	1.0	0	0	0
2	90	1.0	90	0	–90

E_{rss} (theoretical pattern) = 691.92 mV/m.

Q (quadrature term) = 21.60 mV/m.

E_{rms} (standard pattern) = 726.87 mV/m.

A plot of the theoretical and standard patterns are shown in Figure 4.3-B1.

The minimum horizontal field strength (at 1 km) when the theoretical field strength goes to zero is given by Eq. B2 for a standard pattern along the ground using Eq. B4 with $g(\Theta) = 1.0$. For 1 kW and under, Q is 6 according to FCC Rules. For various FCC licensed values of power, the minimum field strength values are as follows:

P_{kw}	Q	E_{min}
0.25	6.0	10.14
0.50	6.0	10.14
1.00	6.0	10.14
2.50	9.49	16.03
5.00	13.42	22.67
10.00	18.97	32.06
25.00	30.00	50.69
50.00	42.43	71.69

The minimum field strength (at 1 km) for any elevation of a standard pattern, by Eq. B5, is:

$$g(\Theta) = \frac{\sqrt{0 + 0.0625}}{1.030776} = 0.2425 \qquad \text{(B8)}$$

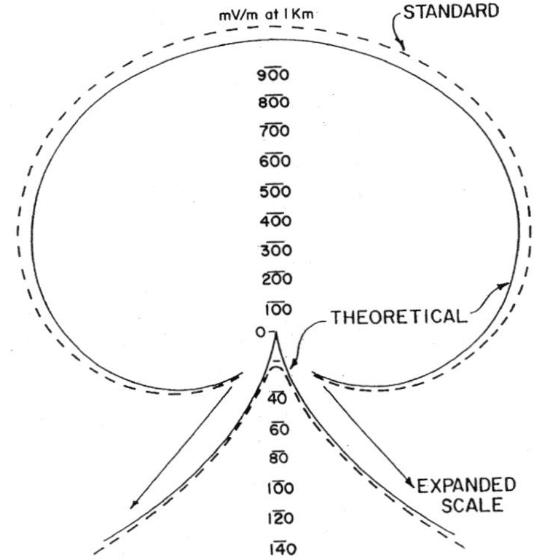

FIGURE 4.3-B.1 Theoretical and standard pattern.

Augmented Pattern Equation

The augmented pattern equation is obtained by adding an augmentation quadrature term to the standard pattern, as given here:

$$E_{(\phi,\Theta)aug} = \sqrt{\left\{E_{(\phi,\Theta)std}\right\}^2 + \left\{g(\Theta)\cos\left(180\frac{D_A}{S}\right)\right\}} \qquad \text{(B9)}$$

where:

$E_{(\phi,\Theta)aug}$ = augmented pattern radiation value at azimuth, elevation

$E_{(\phi,\Theta)aug}$ = standard pattern radiation value at azimuth, elevation

A = augmentation constant

$$g(\Theta) = \frac{\sqrt{\{f(\Theta)\}^2 + 0.0625}}{1.030776}$$

D_A = angular distance from center of span.

S = span of augmentation in degrees

The principle of augmentation is illustrated in the cardioid pattern of Figure 4.3-B2.

The FCC has converted all augmented directional patterns to a table for each station, as shown in the example of Figure 4.3-B3. In this case, there were six augmentations, as tabulated in Figure 4.3-B4 and shown on the polar chart of Figure 4.3-B5. It should be noted that, where the spans overlap, Eq. B9 is applied repeatedly, once for each augmentation, proceeding clockwise from true north.

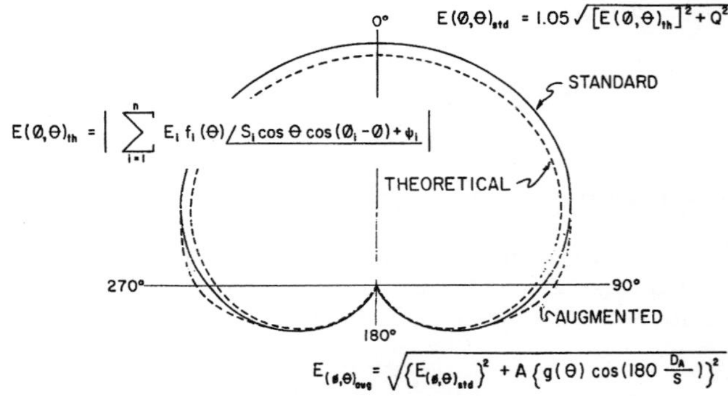

FIGURE 4.3-B.2 Theoretical, standard, and augmented patterns.

Technical Parameters Resulting From Conversion of AM Broadcast Stations To Standard Patterns

STANDARD PATTERN CONVERSION NO.: 1280–22

FREQ. KHZ	CALL LETTER	CITY	STATE	PATTERN HRS.	PATTERN STATUS	CLASS
1280	WHVR	HANOVER	PA	N	LIC.	B

POWER KW	LATITUDE	LONGITUDE	PAT-MULT MV/M	TH-RMS MV/M	STD/AUG RMS-MV/M	PAT-RSS MV/M	Q-FACTOR
.500	39-49-11	77-00-25	131.27	143.00	150.54	185.65	6.0000

TOWER NO.	PHYS-HT (A)-DEG	TL-HT (B)-DEG	TOT-HT (C)-DEG	TL-HT (D)-DEG	FIELD RATIO	PHASE DEG.	SPACING DEG.	ORIENT DEG-TR	REF FLG
1	91.0	.0	.0	.0	1.000	149.5	.0	.0	
2	91.0	.0	.0	.0	1.000	.0	90.0	178.0	

AUGMENTATION DATA

CENTRAL AZIM. DEGREES TRUE	SPAN DEGREES	FIELD AT AZIM. MV/M
64.0	12.0	17.0
260.5	55.0	103.0
288.0	14.0	7.5
288.0	10.0	21.2
295.0	14.0	30.0
295.0	10.0	43.3

HORIZONTAL PLANE STANDARD/AUGMENTED RADIATION VALUES

AZ. DEG	FIELD MV/M	AZ. DEG	FIELD MV/M	AZ. DEG	FIELD MV/M	AZ. DEG	FIELD MV/M	AZ. DEG.	FIELD MV/M	AZ. DEG	FIELD MV/M
0	136.8	60	29.1	120	174.0	180	239.4	240	163.9	300	42.2
10	132.8	70	9.0	130	196.3	190	237.0	250	136.2	310	71.0
20	123.1	80	43.5	140	213.5	200	231.5	260	104.7	320	95.4
30	107.5	90	80.0	150	225.9	210	221.4	270	69.3	330	114.4
40	86.3	100	115.0	160	234.0	220	207.2	280	31.3	340	127.7
50	59.8	110	146.7	170	238.4	230	188.0	290	20.4	350	135.1

CONSTRUCTION PERMIT LIMITS

AZIMUTH DEG. TRUE	PRESENT MV/M	NEW MV/M
64.0	17.0	17.0
231.0	179.0	185.8
288.0	33.0	21.2
352.0	131.0	135.9

PATTERN MINIMA

AZIMUTH DEG. TRUE	FIELD MV/M
68.5	6.8
284.5	17.3
290.0	20.4
299.0	40.9

PATTERN MAXIMA

AZIMUTH DEG. TRUE	FIELD MV/M
178.0	239.4
288.0	21.2
296.0	44.2
358.0	136.9

FIGURE 4.3-B.3 FCC method of specifying augmentation.

CENTER AZIMUTH OF AUGMENTATION	SPAN DEGREES	EXTENT OF SPAN	FIELD AT CENTER SPAN AT 1 KM
64	12 (°6)	(58 -70)	27.4
260.5	55 (°27.5)	(233 - 288)	165.8
288	14 (°7)	281 -295)	12.1
288	10 (°5)	(283 - 293)	34.1
295	14 (°7)	(288 - 302)	48.3
295	10 (°5)	(290 -300)	69.7

FIGURE 4.3-B.4 Table of augmentation data.

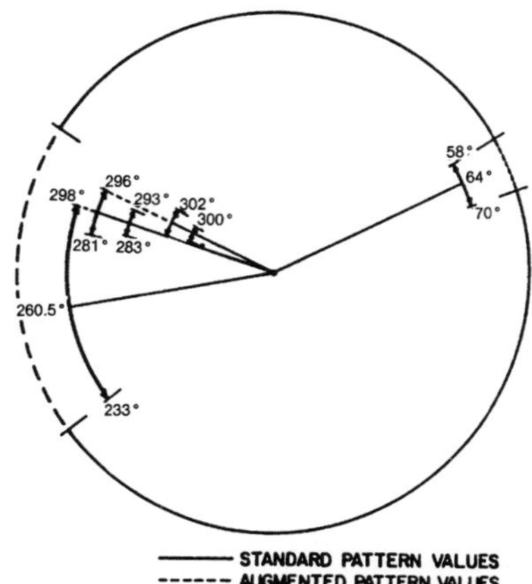

FIGURE 4.3-B.5 Augmented pattern flowchart showing overlapping spans.

C H A P T E R

4.4

AM Antenna Coupling and Phasing Systems

JERRY WESTBERG

Broadcast Electronics
Quincy, Illinois

INTRODUCTION

It is the purpose of this chapter to provide reference material to those who may desire further understanding of phasing and matching circuitry for AM antenna systems. Design equations are given for the most common phasing, matching, and diplexing circuits, procedures are given for the design of both phasing and diplex systems, and some rules of thumb for the successful design and implementation of these circuits are also provided.

THE NOT SO SIMPLE L NETWORK

An L network is the simplest way to transform one resistance to another. It uses the fewest components of any of the standard circuits, requiring only two reactive elements. The elements are arranged in an inverted "L" configuration, which is why the circuit is so named. The series element will always be on the low-resistance side, and the shunt element will be on the high-resistance side. The magnitude of the phase shift is set by the transformation ratio. A leading (positive phase shift) or a lagging (negative phase shift) network may be chosen and will work equally well when the transformation is purely resistive. The phase shift across an L network is defined as the voltage on the larger resistance compared to the current in the smaller resistance in the direction of power flow. In general, the L network will have the best bandwidth of any of the common networks (tee or pi).

Equations

It seems best to calculate the magnitude of the reactive components using the covariant Q. This method gives additional information about the circuit Q, and the equations are easily remembered.

Let R_H be the higher of the resistive values and R_L be the lower resistance. Then, the Q of an L network that will transform one impedance to the other will be:

$$Q = \text{sqrt}(R_H/R_L - 1)$$

The series element will have a reactive magnitude of:

$$X_{\text{Series}} = R_L \times Q$$

The shunt element will have a reactive magnitude of:

$$X_{\text{Shunt}} = R_H/Q$$

If a lagging network (negative phase shift) is desired, the shunt element will have a capacitive reactance and the series element will have an inductive reactance; the opposite would be true for a leading network. The magnitude of the phase shift of the network is:

$$|\text{Phase shift}| = \arctan(X_{\text{Series}}/R_L) = \arctan(Q)$$

Consider the following example. An L network should match a 230 ohm open-wire transmission impedance to a 50 ohm coax impedance using a lagging network:

$$Q = \text{sqrt}(R_H/R_L - 1) = \text{sqrt}(230/50 - 1) = 1.9$$

$$X_{\text{Series}} = R_L * Q = 50 \times 1.9 = 94.9 \text{ ohms}$$

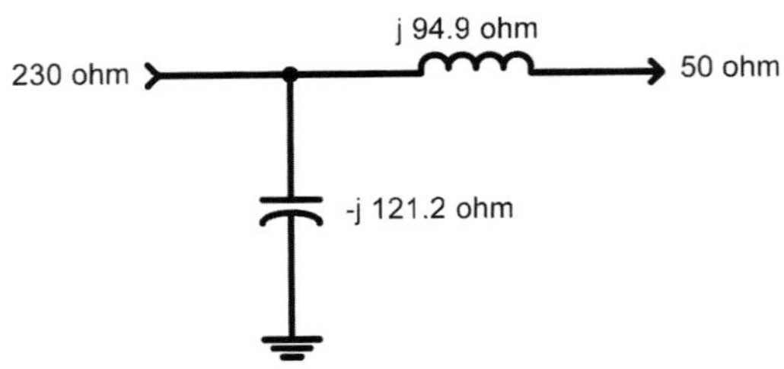

FIGURE 4.4-1 Typical L network.

The shunt element will have a reactive magnitude of:

$$X_{Shunt} = R_H/Q = 230/1.9 = 121.2 \text{ ohms}$$

$$|\text{Phase shift}| = \arctan(Q) = \arctan(1.9) = 62.2°$$

The final design is shown in Figure 4.4-1.

So far the L network seems simple. First, calculate the Q of the network; from that covariant, calculate the magnitude of the series and shunt elements and then the phase shift. But, much more is involved when the impedance being matched has a reactive component.

L Network Example

Match an impedance of 15 + j35 ohms to a 50 ohm transmission line. Four possible L networks will match this impedance, as described in the following sections.

Case 1. Negative Phase Shift: Series Element Next to the Load

Because the equations for an L network were defined using only purely resistive impedances, the reactance must be resonated out. One way to remove the 35 ohms of reactance is to place a series element of –j35 ohms. This reactance can then be included in the series calculation. The equation for the series element has been modified below:

$$Q = \text{sqrt}(R_H/R_L - 1) = \text{sqrt}(50/15 - 1) = 1.53$$

For a negative phase shift, the series element will have a reactance of:

$$\begin{aligned} X_{Series} &= +(R_L \times Q) - \text{load reactance} \\ &= (15 \times 1.53) - 35 = -12.1 \text{ ohms} \end{aligned}$$

Note that the "+" sign was added because a lagging network was chosen.

The shunt element will have a reactance of:

$$X_{Shunt} = -R_H/Q = 50/1.53 = -32.7 \text{ ohms}$$

Note that the "–" sign was added because a lagging network was chosen.

The phase shift of the network is:

$$\text{Phase shift} = -\arctan(Q) = -\arctan(1.53) = -56.8°$$

Note that the "–" sign was added because a lagging network was chosen.

Case 2. Positive Phase Shift: Series Element Next to the Load

The positive phase shift network is done similarly to the negative phase shift:

$$Q = \text{sqrt}(R_H/R_L - 1) = \text{sqrt}(50/15 - 1) = 1.53$$

For a positive phase shift, the series element will have a reactance of:

$$\begin{aligned} X_{Series} &= -R_L \times Q - \text{load reactance} \\ &= 15 \times 1.53 - 35 = -57.9 \text{ ohms} \end{aligned}$$

Note that the "–" sign was added because a leading network was chosen.

The shunt element will have a reactance of:

$$X_{Shunt} = +R_H/Q = 50/1.53 = +32.7 \text{ ohms}$$

Note that the "+" sign was added because a leading network was chosen.

The phase shift of the network is:

$$\text{Phase shift} = +\arctan(Q) = +\arctan(1.53) = +56.8°$$

Note that the "+" sign was added because a leading network was chosen.

Case 3. Negative Phase Shift: Shunt Element Next to the Load

The load, 15 + j35 ohms, is normally considered as a series combination of 15 ohms of resistance and 35 ohms of reactance. It may also be considered as a parallel combination. The equations for converting a series to a parallel combination are as follows:

$$R_P = \left(R_S^2 + X_S^2\right)/R_S$$

$$X_P = \left(R_S^2 + X_S^2\right)/X_S$$

So, for this problem:

$$R_P = \left(R_S^2 + X_S^2\right)/R_S = \left(15^2 + 35^2\right)/15 = 96.7 \text{ ohms}$$

Note: Because R_P is larger than 50 ohms, an L network can be designed where the shunt element is next to the load. If the parallel resistance is not larger than the 50 ohms, this case could not be designed. For example, let us say that the reactance was j20 ohms instead of j35 ohms. Then, the parallel resistance would be:

$$R_P = \left(R_S^2 + X_S^2\right)/R_S = \left(15^2 + 20^2\right)/15 = 41.7 \text{ ohms}$$

Because this resistance is smaller than the 50 ohms, the case where the shunt element is toward the load could not be designed. The shunt element always goes toward the higher resistance.

The parallel reactance is given by:

$$X_P = \left(R_S^2 + X_S^2\right)/X_S = \left(15^2 + 35^2\right)/35 = 41.4 \text{ ohms}$$

Because the equations for an L network were defined for only purely resistive loads, the parallel reactance, X_P, must be resonated out of the design. The easiest way to remove the +j41.4 ohms of reactance is to place a parallel element of –j41.4 ohms. This reactance can then be included in the shunt element calculation. The equation for the shunt element has been modified below:

$$Q = \text{sqrt}(R_H/R_L - 1) = \text{sqrt}(96.7/50 - 1) = .97$$

For a negative phase shift, the series element will have a reactance of:

$$X_{Series} = +R_L * Q = 50 \times .97 = + 48.3 \text{ ohms}$$

The shunt element will have a reactance of:

$$\begin{aligned} X_{Shunt} &= -(R_H \times X_P)/(Q \times X_P + R_H) \\ &= -(96.7 \times 41.4)/(.97 \times 41.4 + 96.7) \\ &= -29.3 \text{ ohms} \end{aligned}$$

The phase shift of the network is:

$$\text{Phase shift} = -\arctan(Q) = -\arctan(.97) = -44.0°$$

Case 4. Positive Phase Shift: Shunt Element Next to the Load

Again because the equations for an L network were defined for only purely resistive loads, the parallel reactance, X_P, must be taken into account. This reactance will be combined with the shunt element. The equation for the shunt element has been modified below:

$$Q = \text{sqrt}(R_H/R_L - 1) = \text{sqrt}(96.7/50 - 1) = .97$$

For a positive phase shift, the series element will have a reactance of:

$$X_{Series} = -R_L \times Q = -50 \times .97 = -48.3 \text{ ohms}$$

The shunt element will have a reactance of:

$$\begin{aligned} X_{Shunt} &= (R_H \times X_P)/(Q \times X_P - R_H) \\ &= (96.7 \times 41.4)/(.97 \times 41.4 - 96.7) \\ &= -70.7 \text{ ohms} \end{aligned}$$

The phase shift of the network is:

$$\text{Phase shift} = +\arctan(Q) = -\arctan(.97) = +44.0°$$

Choosing the Best Network

The best choice will always be the network with the lowest Q. This means that cases 3 and 4 are better than 1 and 2. But, which phase shift should be chosen, leading or lagging? For cases 1 and 2, where the reactive load is on the series element side, choose the phase shift that would be opposite in sign of the reactance sign; thus, a lagging network would be better than a leading network. Case 1 would be better than case 2. When the reactive load is on the shunt element side, choose the phase shift that would be the same in sign of the reactance sign; thus, a leading network would be better than a lagging network. Overall, case 4 would be the design of choice. "Best choice" means that the bandwidth of the network is the optimum with the least stress on component elements.

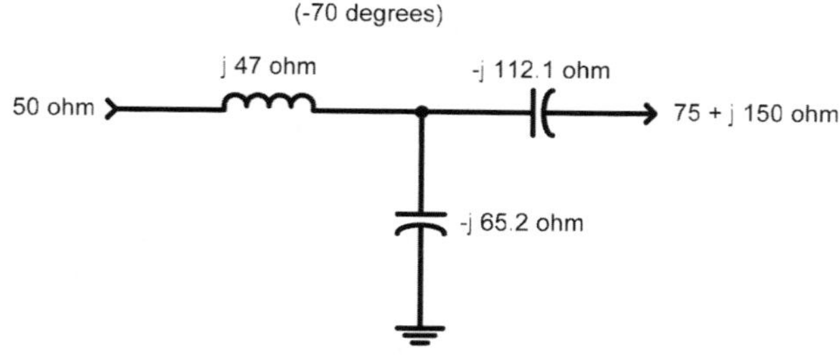

FIGURE 4.4-2 Tee network.

TEE NETWORKS

Tee networks are used extensively in phasing and matching systems because not only do they match an impedance to a resistive load but also a desired phase shift can be chosen. A tee network has three component parts, as shown in Figure 4.4-2. The parts include two series elements and one shunt element. The tee network has one element in series with the load and one element in series with the matched resistance. The shunt element is in between the two series elements, forming a letter "T." The phase shift across a tee network is defined as the *in voltage* to the load current in the direction of power flow.

Equations

The following value definitions will be used in the equations for a tee network:

R_L = Load resistance.

X_L = Load reactance.

R_O = Matched resistance.

θ = Phase shift of network.

X_{Shunt} = Shunt reactance.

X_{In} = In leg series reactance (toward R_O).

X_{Out} = Out leg series reactance (toward load).

If the reactance of the shunt element is calculated first and this value is used in the other element calculations, the equations for a tee network can be easily memorized:

$$\begin{aligned} X_{Shunt} &= \text{sqrt}(R_O \times R_L)/\sin\theta \\ X_{In} &= -X_{Shunt} + R_O/\tan\theta \\ X_{Out} &= -X_{Shunt} + R_L/\tan\theta - X_L \end{aligned}$$

Consider the following example:

R_L = 75 ohms

X_L = 150 ohms

R_O = 50 ohms

θ = –70°

$$\begin{aligned} X_{Shunt} &= \text{sqrt}(R_O \times R_L)/\sin\theta \\ &= \text{sqrt}(50 \times 75)/\sin(-70°) = -65.2 \end{aligned}$$

$$X_{In} = -X_{Shunt} + R_O/\tan\theta$$
$$= 65.2 + 50/\tan(-70°) = 47.0$$

$$X_{Out} = -X_{Shunt} + R_L/\tan\theta - X_L$$
$$= 65.2 + 75/\tan(-70°) - 150 = -112.1$$

Is a 90° phase shift best for a tee network? Although a 90° tee network does have some properties that are desirable in certain applications, it may not be the phase shift of choice for all applications. In many respects, choosing the phase shift is like choosing the length of a transmission line. It is almost inconsequential, but there is one important difference that puts constraints on the choice of a phase shift for a tee network that is in the bandwidth of the circuit. The best bandwidth for a tee network is when the phase shift is the same as the L network that would make the same transformation. It is also true that the bandwidth does not change much if the phase shift is varied from the L network phase shift. Consider matching 10 ohms to 50 ohms. The phase shift network with the best bandwidth would be the L network:

$$Q = \text{sqrt}(R_H/R_L - 1) = \text{sqrt}(50/10 - 1) = 2.0$$

$$|\text{Phase shift}| = \arctan(Q) = \arctan(2.0) = 63.4°$$

Moving away from this phase shift, the Q will increase, but a 90° phase has a Q of only 2.7. A 40° phase has a Q of only 2.8. Both of these would be an acceptable bandwidth for a 5:1 impedance transformation in most, if not all, applications.

Compare these Qs to a 120° phase shift network. The 120° phase shift network has a Q of 4.25. This is more than double the Q required to make the transformation. Note that the current in the shunt leg of this 120° network will be 40% higher than for the L network.

A rule of thumb would be to keep the phase shift less than 100° or 110°; less than 90° is desirable. These phase shifts use circuitry that will be tunable and give good performance. A higher phase will have larger currents than necessary in the shunt leg of the tee network. This means larger components with higher stresses and poorer bandwidth.

When choosing the sign of the phase shift it is best to pick a sign that is opposite the sign of the load reactance. In the example above, where the load reactance was a positive 150 ohms, a –70° network would be preferred over a +70°. This is because the load reactance will be used as part or all of the reactance needed in the output leg of the tee network.

Tee with a Filter

Occasionally it is convenient to add a filter in the shunt leg of a tee network. If the frequency being trapped is higher than the carrier frequency, a lagging network is used. If the frequency being trapped is lower than the carrier frequency, a leading network is required.

Lagging Network Equations

$$C = 1/(W_L X) - W_L/(W_H^2 X)$$

where X is the reactance of the shunt leg of the tee network, and

$$L = 1/(W_H^2 C)$$

Leading Network Equations

$$L = XW_H/(W_H^2 - W_H^2)$$

$$C = 1/(W_L^2 L)$$

PI NETWORK

The pi network, although rarely used, does have its place in the design of phasing and matching networks. A pi network has three components, as shown in Figure 4.4-3. The parts include two shunt elements and one series element. The pi network has one element shunt with the load and one element in shunt with the matched resistance. The series element is in between the two shunt elements, forming a Greek letter Π. The phase shift across a pi network is defined as the voltage in to the voltage out in the direction of power flow. This is different from the tee network, where the phase shift is defined as the voltage in to current out in the direction of power flow. This means that the networks are only direct replacements in a phasing system if the load impedance is purely resistive. Two sets of equations are given for the pi network calculations. The pi network equations are defined for admittance, not impedance. If the admittance parameters are used, the equations are similar to the tee network. These equations are easily memorized, but, because most people are not as familiar with admittance as impedance and because many will simply program them into a computer or calculator, the impedance equivalent equations are also presented here.

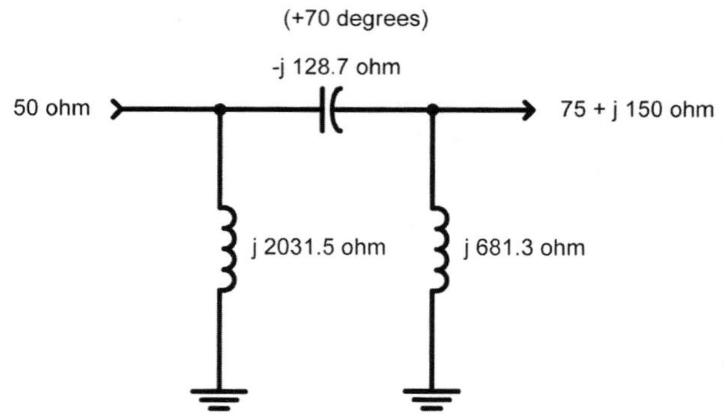

FIGURE 4.4-3 Typical pi network.

Equations

The following value definitions will be used in the equations for a pi network:

R_L = Load resistance

X_L = Load reactance

R_O = Matched resistance

θ = Phase shift of network

X_{Series} = Series reactance = $-1/B_{Series}$

X_{In} = In shunt leg series = $-1/B_{In}$

X_{Out} = Out shunt leg reactance = $-1/B_{Out}$

The following equations convert impedances into admittances:

G_O = Matched conductance = $1/R_O$

G_L = Load conductance = $R_L/(R_L^2 + X_L^2)$

B_L = Load susceptance = $X_L/(R_L^2 + X_L^2)$

The following will be calculated using the admittance equations:

$B_{Series} = \text{sqrt}(G_O \times G_L)/\sin\theta$

$B_{In} = -B_{Series} + G_O/\tan\theta$

$B_{Out} = -B_{Series} + G_L/\tan\theta - B_L$

These equations are simple, but it is necessary to convert from impedance to admittance. The following equations use the impedances only:

$X_{Series} = -\sin\theta/\text{sqrt}[R_L/(R_O \times R_L^2 + R_O \times X_L^2)]$

$X_{In} = -R_O \times X_{Series} \times \tan\theta/(R_O \times \tan\theta + X_{Series})$

$$X_{Out} = \frac{-X_{Series} \times \tan\theta \times (R_L^2 + X_L^2)}{\left[\tan\theta \times (R_L^2 + X_L^2) + X_{Series} \times R_L + X_{Series} \times X_L \times \tan\theta\right]}$$

Consider the following example:

R_L = 75 ohms

X_L = 150 ohms

R_O = 50 ohms

$\theta = +70°$

$X_{Series} = -\sin\theta/\text{sqrt}[R_L/(R_O \times R_L^2 + R_O \times X_L^2)]$
$= -\sin 70°/\text{sqrt}[75/(50 \times 75^2 + 50 \times 150^2)]$
$= -128.7$

$X_{In} = -R_O \times X_{Series} \times \tan\theta/(R_O \times \tan\theta + X_{Series})$
$= -50 \times (-128.7) \times \tan 70°/(50 \times \tan 70° + (-128.7))$
$= 2031.5$

$X_{Out} = -X_{Series} \times \tan\theta \times (R_L^2 + X_L^2)/[\tan\theta \times (R_L^2 + X_L^2) + X_{Series} \times R_L + X_{Series} \times X_L \times \tan\theta]$
$= 128.7 \times \tan 70° \times (75^2 + 150^2)/[\tan 70° \times (75^2 + 150^2) + (-128.7) \times 75 + (-128.7) \times 150 \times \tan 70°$
$= 681.3$

When choosing the sign of the phase shift, it is best to pick the sign that is the same as the sign of the load reactance. In the example above, where the load reactance was a positive 150 ohms, a +70° network would be preferred over a –70°. This is because the load reactance will be used as part or all of the reactance required in the output leg of the pi network.

ZERO DEGREE PHASE SHIFTER

When designing a phasing system, many times it is necessary to use a circuit that has some adjustment but is close to zero degrees of phase shift. Such a network can be designed by using a resonant coil and capacitor in series. The reactance of each component is usually around 50 ohms. If the coil is made variable, the phase shift through the network can be adjusted by changing the series combination from its resonant point.

DESIGNING A PHASING SYSTEM

To design a phasing system for a multiple tower array, the drive impedances, drive currents, and power for each tower have to be calculated. This is done with computer programs that accept information about the antenna array. It is also necessary that the transmission line lengths and types are known. The phasing system accepts energy from the transmitter and distributes it to each tower so the power distribution and current for each tower matches the calculated values. The impedances of each tower must be matched to the transmission line impedance, and the input to the system must be matched to the transmitter impedance. It will also be necessary to give each network enough range to adjust for differences between the final tuned phasing system and the calculated values. Figure 4.4-4 shows a block diagram of the parts of the phasing system.

Input Matcher

The input matcher, which is usually a tee network, transforms the bus point impedance to the transmitter output impedance. The phase of this network is arbitrary. Usually the network is lagging and less than or equal to 90°.

Power Divider

A power divider takes the energy at the bus point of the phasing system and divides it among the towers. Each power divider will have some phase shift associated with it that must be accounted for in the final design. Equations are shown for the shunt power divider; this is the divider most commonly used in modern designs. Other common dividers are also discussed.

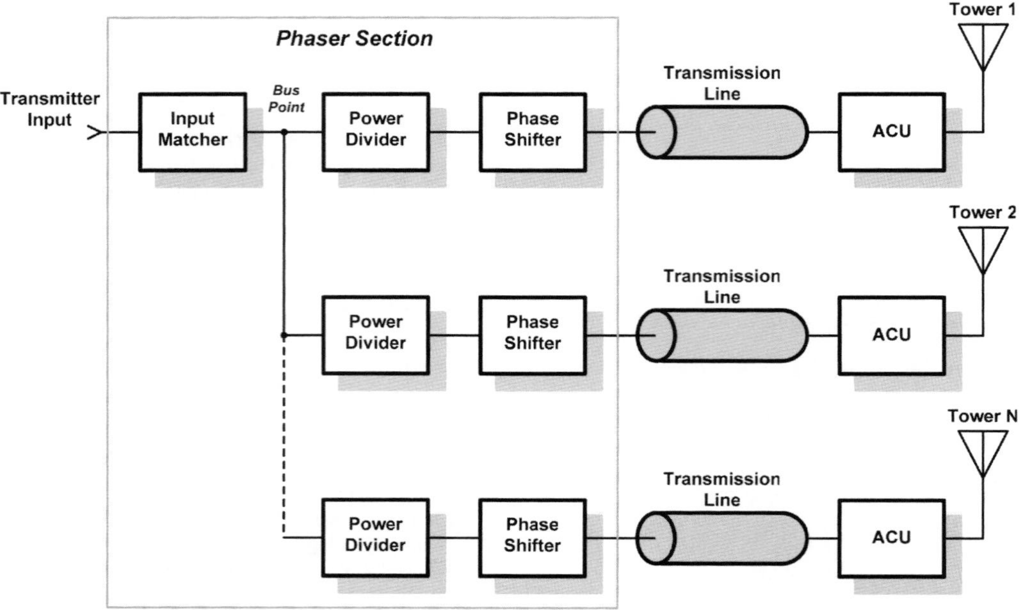

FIGURE 4.4-4 Typical multi-tower phasing system.

Phase Shifter

The phase shifter is used to provide the additional phase shift required to ensure that each tower receives power at the proper phase. The phase shifter is usually a tee network with the input leg variable and sometimes both input and output leg variables.

Transmission Line

The transmission line transfers energy from the phasing section to the antenna coupling units (ACUs) of the phasing system. The transmission line will have a phase shift associated with it that must be taken into account in the design.

Antenna Coupling Units

The ACUs match the tower impedance to the characteristic impedance of the transmission lines. These units are usually tee networks. The phase of the tee network will be determined when doing the total phasing of the system.

Phasing the System

These general guidelines apply to the design of a phasing system. First, choose the phase shift for the ACU of the high-power tower. The high power tower is the most important tower in determining the bandwidth of the system. It is important to choose a phase shift close to ideal for this tee network. Next, start by choosing a zero degree phase shifter for the high power tower. When calculating the phase shifts for the system, it may be necessary to change this network to a phase shift different than 0°. In general, choose the fewest components or this tower, as it has the most power.

Starting at the tower, calculate the bus point phase by subtracting the phase of the ACU, transmission line, and phase shifter from the current phase feeding this tower. For example, let the system phase shifts chosen for the high-power tower be:

- Phase shifter, 0°
- Transmission line, –100°
- ACU, –70°
- Tower current, –75°

The bus point phase shift would be –75 – (–70) – (–100) – 0 = 95°.

If the number is above 180° or below –180°, 360° can be added or subtracted to the calculated bus phase to yield an equivalent value. If the bus point phase calculation gives 276°, an equivalent phase shift would be 276 – 360 = –84°.

A power divider must be designed for each additional tower in the system. The design equations for a shunt power divider are given later in this chapter. A power divider is not typically used for the high-power tower because it is the power reference for the system. The phase shift across the power divider must be calculated and taken into account when choosing phase shifts for the other networks.

The phase of the phase shifter and ACU must be chosen so the system will yield the correct current phase going into each tower. To calculate the additional phase, subtract the bus phase, power divider phase, and the transmission line phase from the tower's feed current phase. For example, if the bus phase, power divider phase, transmission line phase, and feed current phase are 95°, –31°, –100°, and 25°, respectively, then the phase shift of the ACU and phase shifter must add up to 61° [25 – 95 – (–31) – (–100) = 61]. Again, 360° can be added or subtracted to this number and still have an equivalent value.

The options that will satisfy the design are infinite; however, it is desirable to use a zero degree phase shifter wherever possible. As an example, one solution for this tower would be to use a zero degree phase shifter and 61° for the ACU. Another solution would be to use –90° for the ACU and two –104.5° networks for the phase shifter. This would be more parts but would work. The total phase shift of this solution would be 61 – 360 = –299°. After the phase shifts for each circuit are chosen, each network should be designed and the components chosen. Equations and examples are given for the most common networks.

Software for Phasor Designs

Most designers have at least some part of the phasing system designed by a computer rather than manually [1].

LOW-RESISTANCE TOWERS

Low-resistance tower analysis has produced a myriad of solutions to a variety of problems. Transformer designs and multiple-stage networks are typical examples. A slight change in impedance (just a few ohms of inductance or resistance) will produce a large change in the match on the other end of the network. This is because a change in impedance must be compared to the resistive value to determine the voltage standing wave ratio (VSWR).

Consider a tower with a 50 ohm resistance. If the resistance or reactance changes or the network is misadjusted by 1 ohm, a VSWR of 1.02:1 would be obtained. If this same change were to be applied to a tower of 1 ohm of resistance, the VSWR would be 2:1. So, even if the network is matched perfectly, it would be easily mismatched due to weather-induced changes in the system. This would then be reflected back through the system, causing changes in phase and amplitude to that tower.

There are two ways to reduce the effect of these phase changes. The first may be considered unusual, but it is the best way to stabilize a system with a low-resistance tower: Place a small resistance, 5 to 10 ohms, in series with that tower. The drawbacks are obvious. Power will be lost in the resistor that must be accounted for, and a transmitter of a higher power may be necessary. The advantage is that the system will be more tolerant of changes in impedance, will be more stable, and will exhibit better bandwidth.

If resistors are not an option, there is a way to minimize the effect this tower will have on the system. Because matching a tower impedance of 1 ohm is not practical, it should not be attempted. Instead, place a circuit in series with the tower to resonate out the reactance, then make the phase shift from the bus point to the tower an odd multiple of 90°. This is referred to as *quadrature phasing*. This approach fixes the magnitude and phase of the current going into the tower. Consider the example shown in Figure 4.4-5.

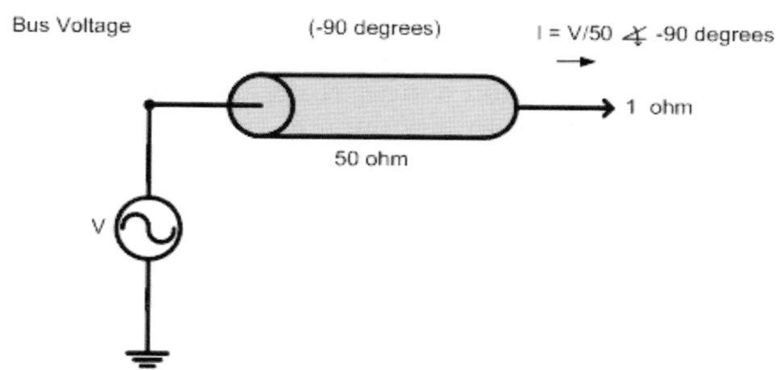

FIGURE 4.4-5 Low resistance tower feed concept.

The current going into the tower will be V/50 at an angle of –90°. If the resistance changes from 1 to 2 ohms, the current going into the tower will still be V/50 at an angle of –90°. By using the principle of a 90° circuit, the current into the tower is fixed. The rest of the system must then be designed around the phase to this tower. This design technique works even if the resistance of the tower becomes negative.

HIGH-RESISTANCE TOWERS

High-resistance towers are not to be avoided; in fact, they can be beneficial, and a proper circuit on a high-resistance tower can be no more complicated to match than any other tower impedance. For a single tower, high-resistance towers fall in the area of 140°. It should

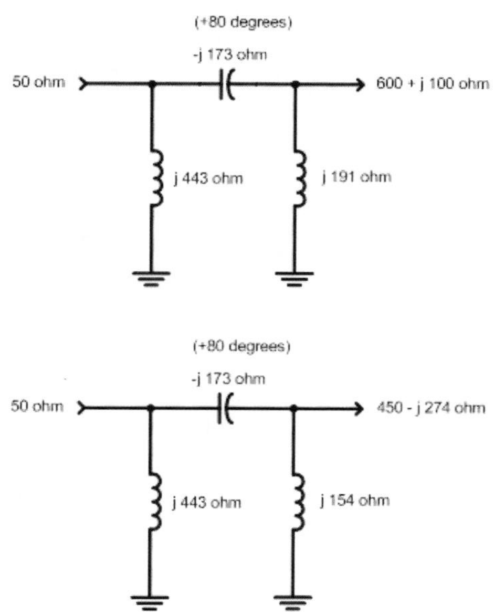

FIGURE 4.4-6 Impedance matching using pi network.

be noted that this height tower has the most bandwidth of all the tower heights. For this reason, it should not be avoided simply for matching purposes. The high-resistance tower tends to be voltage driven, which means that the voltage on the tower will be the same with changes in weather even though the impedance may seem to change. This is an excellent application for a pi network. Another option producing equally good results would be to use a tee network followed by an inductor shunting the tower.

Consider a tower impedance of 600 + j100 ohms. If this tower had a shunt change in capacitance of 200 pF, due to rain or snow, the impedance would change (assuming 1 MHz) to 450 – j274 ohms. Although a tee network can be designed to match these changes in impedance, the pi network also offers a solution, as shown in Figure 4.4-6. It can be seen that only slight changes in this network (and only the output leg) would be necessary to adjust the system for a wide variation in impedance. This would not be the case for a tee network.

NEGATIVE POWER TOWER

Negative power towers have been a source of anxiety for both field and design engineers. As long as the tower is definitely negative (*i.e.*, significant power relative to the other towers in the array), these towers are not any more difficult to design for or tune than any other towers. Two important characteristics are (1) the system or tower currents are defined going into the tower, and (2) all networks, including transmission line phase shifts, are defined in the direction of power flow. This means that phase shift calculations should start at the tower and work backward toward the bus point.

It is necessary to add 180° when figuring the phase shift of a negative power tower. The best way to look at the system is to start at the tower. The current going into the tower is defined by the array parameters; however, the power is going in the other direction (*i.e.*, from the tower toward the bus point), so it is necessary to determine the current going from the tower into the ACU because the network phase shifts are defined in that direction. The current going into the tower will be 180° from the current going into the ACU. Consider the example shown in Figure 4.4-7.

The current going into the tower is at a phase of 131°; therefore, the current going into the ACU from the tower will be 131° + 180°, or 311°. Calculating backward:

$$311 + X - 80 + 0 - 31 = 115°$$

$$X = -85°$$

SHUNT POWER DIVIDER

A widely used power divider is the shunt divider. It consists of a single variable coil connected from the bus point to ground. The rotor connection is tied to the load. To adjust the power of the tower, the rotor is turned toward the bus point for higher power or toward ground for lower power. The actual design equations are not quite as simple as the adjustment concept. There is coupling between the top part and the bottom part of the coil. The inductor can be modeled as a coupled inductor, as shown in Figure 4.4-8.

Where L_1 is the inductance of the top part of the coil and L_2 is the inductance of the bottom part of the coil, M can be determined by iteration of one of the following equations:

$$P/R = \frac{V^2\left\{L/2 + \text{sqrt}\left[(2M-L)^2/4 - M^2/K^2\right]\right\}^2}{R^2L^2 + W^2M^4\left(1-K^2\right)/K^4}$$

$$P/R = \frac{V^2\left\{L/2 - \text{sqrt}\left[(2M-L)^2/4 - M^2/K^2\right]\right\}^2}{R^2L^2 + W^2M^4\left(1-K^2\right)/K^4}$$

where:

V is the bus voltage and must be larger than sqrt(PR).

P is the power into the load.

R is the load resistance.

L is the total inductance used.

M is the mutual inductance.

K is the coefficient of coupling between the top part and bottom part of the inductor ($K = 0.25$ is typical).

W is $\omega = 2\text{pi}F$, where F is the operating frequency.

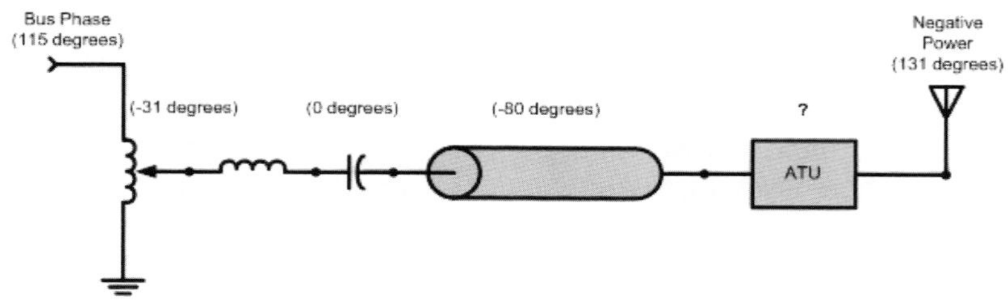

FIGURE 4.4-7 Matching network for negative power tower.

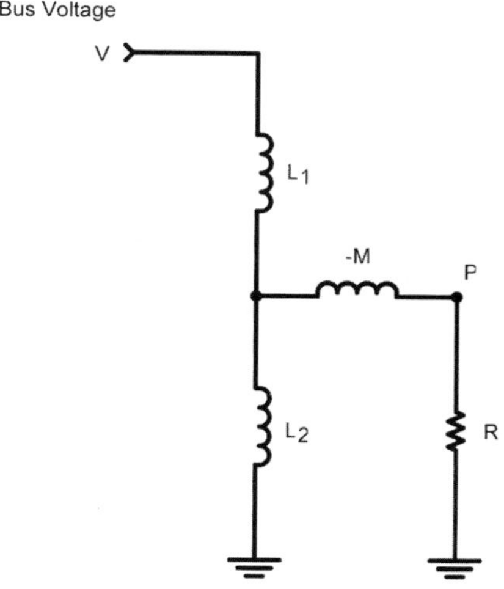

FIGURE 4.4-8 Shunt-type power divider network.

To assist in the calculation of M, the limits of M are:

$$0 < M < LK/(2 + 2K)$$

$$L_2 = L/2 + \text{sqrt}[(2M - L)^2/4 - M^2/K^2]$$

or

$$L_2 = L/2 - \text{sqrt}[(2M - L)^2/4 - M^2/K^2]$$

depending on the equation used to solve M. The signs should match in both equations.

$$L_1 = L - L_2$$

$$\theta = -\arctan\,[W(1 - K^2)M^2/(RLK^2)]$$

To determine which equation should be used for iterating for M, place the maximum value of M into the equation and solve for P. If the calculated P is lower than the actual value of power, then the first equation with a "+" sign should be used. If the actual power is lower than the calculated P, then the second equation with a "–" sign should be used. For example, if V = 200 volts, P = 75 watts, R = 50 ohms, L = 25 μH, K = .25, and F = 1 MHz, then the coil division, mutual coupling between the top and bottom part of the coil, and the phase shift of a series divider can be calculated.

First, calculate the maximum value of M:

$$M_{\text{Max}} = LK/(2 + 2K) = 25 \times .25/(2 + 2 \times 25)$$
$$= 2.5$$

This means that M must be between 0 and 2.5. These values can be used as limits of an iterative program in solving for M. Also, a value in between 0 and 2.5 may be used to "seed" a solver computer program on a calculator.

To determine which equation should be used to iterate to find M, choose M to be equal to 2.5 and solve for P in either equation. It should be noted that the portion of the equation in the square root will be equal to zero in this case.

$$P/R = \frac{V^2\left\{L/2 + \text{sqrt}\left[(2M - L)^2/4 - M^2/K^2\right]\right\}^2}{R^2L^2 + W^2M^4\left(1 - K^2\right)/K^4}$$

$$P/R =$$

$$\frac{200^2\left\{\left(25\times10^{-6}\right)/2\right\}^2}{50^2\left(25\times10^{-6}\right)^2 + \left(6.283\times10^6\right)^2\left(2.5\times10^{-6}\right)^4\left(1 - .25^2\right)^2/.25^4}$$

$$P = 163.657 \text{ watts}$$

Because the desired power of 75 watts is smaller than the calculated P of 163.657 watts, the equation with the "–" sign will be used to iterate for M.

Using the following equation and a solver program, find M:

$$P/R = \frac{V^2\left\{L/2 - \text{sqrt}\left[(2M - L)^2/4 - M^2/K^2\right]\right\}^2}{R^2L^2 + W^2M^2\left(1 - K^2\right)^2/K^4}$$

$$M = 2.315\ \mu\text{H}$$

Solving for L_2, the following is obtained:

$$L_2 = L/2 - \text{sqrt}[(2M - L)^2/4 - M^2/K^2]$$
$$= (25 \times 10^{-6})/2 - \text{sqrt}\,[(2(2.315 \times 10^{-6}) - (25 \times 10^{-6}))^2/4 - (2.315 \times 10^{-6})^2/.25^2] = 8.26\ \mu\text{H}$$

$$L_1 = L - L_2 = 25 \times 10^{-6} - 8.26 \times 10^{-6}$$
$$= 14.43\ \mu\text{H}$$

$$\theta = -\arctan[W(1 - K^2)M^2/(RLK^2)]$$
$$= -\arctan[6.283 \times 10^6(1 - .25^2)(2.315 \times 10^{-6})^2/(50(25 \times 10^{-6}).25^2)]$$
$$= -22^\circ$$

OTHER POWER DIVIDERS

Many other power dividers can be used. Among the most popular is the series divider. A series divider consists of a large coil (on the order of 30 μH) that has a pair of taps going to small variable coils (less than 10 μH). The rotor of the small variable coil feeds a tower. The advantage of such a power divider scheme is its versatility. A single design can tune any phasing system, even if the power to each tower is different from the predicted values. The difficulty with a series divider design is that it becomes difficult to tune. If a four-tower array uses a series divider, it would have eight taps and four small variable coils. When one tower is adjusted, there is a mutual coupling through the main coil to all the other towers. This would be true even if the system is terminated with resistors.

TEE NETWORK DIVIDER

Another popular divider is a tee network divider. The concept is to use a tee network to match the transmission line resistance to a resistance that will divide the power to each tower properly. The tee network divider is versatile in that the phase can be changed as well as the matched impedance.

EQUAL AND OPPOSITE REACTANCE DIVIDER

A specific application of a tee network divider is the equal and opposite reactance (EOR) divider. For this divider, the tee network is fixed at ±90°. The EOR divider consists of a variable inductor and a capacitor. The variable inductor is tapped to resonate with the capacitor at the carrier frequency. The EOR provides a –90° phase shift from the bus voltage to the output current regardless of the tuning on the circuit or the load provided to the divider. This circuit can be useful if a particular tower is being quadrature phased, as mentioned in the low-resistance tower section.

COMPONENT STRESSES

Selecting capacitors and inductors that are rated correctly can be confusing in that components are rated in different ways and the limiting factor may be current or voltage. For the broadcast frequency range, for example, vacuum capacitors will reach their voltage limit well before the current limit. For a mica capacitor, the opposite is true.

Vacuum Capacitors

A vacuum capacitor is rated for a DC voltage. A manufacturer's derating for RF may be 60% of the DC rated value. The peak voltage can be calculated by multiplying the root mean square (RMS) value by the square root of 2 (1.414). To calculate the voltage at peak modulation levels, multiply by 1 + the modulation percentage. For 125% modulation peaks, multiply by 2.25. For 150% modulation, multiply by 2.5. The last factor is for safety. In general, a 25 to 50% safety factor is sufficient.

$$\text{Derating} \times \text{RMS to peak} \times \text{modulation factor} \times \text{safety}$$

$$1.6 \times 1.414 \times 2.25 \times 1.25 = 7.07$$

If the unmodulated RMS stress on a capacitor is 1.4 kV, the capacitor would be 7 × 1.4, or 9.8 kV. A good choice would then be a vacuum capacitor rated at 10 kV.

Mica Capacitors

Mica capacitors reach their current rating before their voltage rating except when they are used in multiple-frequency applications. Also, the current is rated at different frequencies. In general, a mica capacitor is rated for higher current at higher frequencies. For frequencies in between the published frequencies, it is safe to use a linear interpolation. The stress for the current is the maximum with a 100% tone modulation. To calculate this stress, take the unmodulated RMS current and multiply by the square root of 1.5 (1.225). Phasing system components often encounter different impedances than predicted. As an example, with a 2:1 VSWR, the increase in current could be as much as the square root of 2 (1.414) times. A safety factor for mica capacitors would be the product or 1.73 times the unmodulated RMS value.

Two factors should be examined. Although the factor of 1.73 is safe for almost any possible design, it will be high for most designs. Generally, a lower rating will be sufficient. The manufacturers' ratings are conservative, so capacitors can be run right up to the published rating. Also, rarely is a 100% tone used (except during transmitter testing). Furthermore, a VSWR of 2:1 may be extremely high for the design application. So, although a factor of 1.73 is safe for all applications, a lower factor may be acceptable.

DIPLEX SYSTEMS

Diplexing (combining with another AM station) is becoming popular with AM radio because the cost of constructing a new site is higher than the cost of diplexing several stations together. In the past, it was assumed that the filters used in the diplex system would degrade the bandwidth and thus the sound quality of the stations. Today, with the aid of the microcomputers and design software, antenna systems can be designed to minimize this effect as discussed in this section.

A diplexer is a set of filters that allow the operation of two frequencies on the same antenna. Figure 4.4-9 shows a schematic representation of the parts of a diplexer. Although all parts are not necessary in every case, the basic diplexer has five parts:

- Prematcher
- Main filter
- Antenna resonator
- Auxiliary filter
- Antenna coupling unit

The *prematcher*, usually consisting of a coil or capacitor in series with the tower, is extremely important to the bandwidth of the system and also will improve the isolation between the two transmitters. The prematcher adjusts the impedance at the diplex point at both frequencies. The diplex point is where the two signals first come together. The impedances at this point are used to calculate the stresses and Qs for the filter circuits.

The *main filter* is used to attenuate the reject frequency. This filter provides a high impedance at the reject frequency (on the order of 10 K ohms) and a low impedance at the pass frequency (on the order of 1 ohm). There are two main filters in a diplexer, one for each frequency.

The *antenna resonator* is used to force the auxiliary filter to improve the bandwidth of the system. It will be shown that if the impedance at the point where the auxiliary filter is placed is close to resonant then this filter will improve the bandwidth of the system. A coil

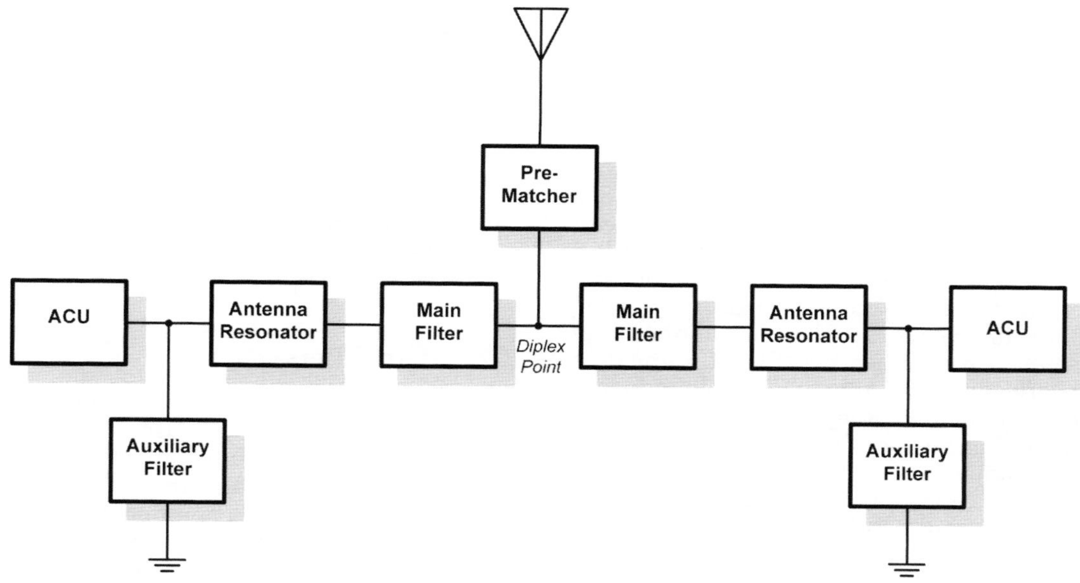

FIGURE 4.4-9 Typical network block diagram for diplexing two stations into one tower.

is used for an antenna resonator if the impedance at that point is capacitive, and a capacitor is used if the impedance is inductive. This technique is known as Q-matching because the same Q is used for the main and auxiliary filters.

The *auxiliary filter* is used to further attenuate the reject frequency. It provides a high impedance to ground at the pass frequency and a low impedance to ground at the reject frequency.

The *antenna coupling unit* (ACU) is used to match the impedance of the diplex system to the Z_O of the transmission line.

The Prematcher

The prematcher can be one of the most powerful circuits in improving the port-to-port isolation, bandwidth, and component stresses. For this reason, it is worth the effort to design a circuit that will yield desirable impedances for both frequencies. It is desirable to have the diplex point impedance at each frequency to be close to resonance. Also, it is helpful for the parallel resistance to be somewhere between 50 and 200 ohms. This of course is a general guideline but has sound reasoning behind it. If the diplex point has a reactance component, the voltage at that point will be higher than if it were purely resistive. A higher voltage at the diplex point will mean that the port-to-port attenuation will be lower or the filters will have to be of a higher Q or a combination of both. Both conditions will adversely affect the bandwidth of the system. If the resistance is too low, the series filter will have a higher Q and current than for an equivalent filter with a higher impedance. If the resistance is too high, the voltage will be high.

Each main filter stores energy at the pass and the reject frequencies. Using the 50 to 200 ohm criterion will provide enough series resistance so the Q of the main filter will not be too high at the pass frequency. Also, a parallel resistance of 50 to 200 ohms is not so high that the Q of the main filter will be too high at the reject frequency. To demonstrate the use of a prematcher, consider the following example (each station is assumed to operate at 1000 watts):

Frequency	Impedance	R_P	Voltage	Minimum Main Filter Q
800 kHz	90 + j150	339	582	6.6
1000 kHz	381 + j310	633	796	2.3

R_P is the parallel resistance of the impedance calculated by:

$$R_P = (R^2 + X^2)/R$$

These are typical numbers for a diplex system. The voltage at the diplex point is important because this voltage will be attenuated by the filter design. Ideally, the voltage will be zero at the transmitter. This, of course, is impossible, but it is good to have the voltage at the diplex point as low as practical to aid in the design effort. The minimum Q values represent the stored energy in the main filters and are calculated by choosing components for the filter that will affect high and low frequencies the same (*i.e.*, the filter has the same Q for the high and low frequencies). The higher the Qs, the more narrow the bandwidth of the system. The total minimum Q for this system is 8.9. If a .001 μF capacitor is placed in series with the antenna, all the numbers improve:

Frequency	Impedance	R_P	Voltage	Minimum Main Filter Q
800 kHz	90 – j49	117	342	5.5
1000 kHz	381 + j151	441	664	1.4

Now the filter system on the low-frequency (800 kHz) side will only have to attenuate 342 volts instead of 582 volts. This is a 4.6 dB improvement. The filter system on the high-frequency (1000 kHz) side will only have to attenuate 664 volts instead of 796 volts. This is a 1.6 dB improvement. The total minimum Q for this system is 6.9. Other benefits of using this capacitor in series with the antenna include reducing the stress on the main filter components.

In the above example, both the low frequency and high-frequency antenna impedances are inductive, and it is clear that a capacitor should be used for a prematcher. If both the high and low frequency impedances are capacitive, an inductor would be used for a prematcher circuit. A more complicated circuit would have to be designed if the impedance at one frequency is inductive and the impedance at the other frequency is capacitive. If a capacitor is used for the prematcher, the frequency with the inductive impedance diplex point impedance will improve, but the diplex point of the other frequency will change for the worse. If an inductor is used, the frequency with the capacitive impedance will improve, but the frequency with the inductive impedance will worsen.

Using both a coil and a capacitor is not a good way to adjust the diplex point impedances. Although it successfully adjusts both impedances, it will detract from the bandwidth of the system. Consider the following example:

Frequency	Impedance	R_P	Voltage	Minimum Main Filter Q
800 kHz	130 + j200	437	662	3.5
1200 kHz	300 – j370	756	869	1.8

If a 120.2 μH coil is placed in series with a 247 pF capacitor, the diplex point impedance for both frequencies will be resonant. This gives the following data:

Frequency	Impedance	R_P	Voltage	Minimum Main Filter Q
800 kHz	130 + j0	130	361	2.2
1200 kHz	300 + j0	300	548	.96

Although the diplex point improved considerably, the 120.2 μH inductor at the low frequency increased the Q to 4.6. The 247 pF capacitor at the high frequency has added an additional Q of 1.8 to its system. The following data should be compared to the unmodified impedances, which include the negative effects of the series L–C circuit:

Frequency	Impedance	R_P	Voltage	Minimum Main Filter Q
800 kHz	130 + j0	130	361	6.8
1200 kHz	300 + j0	300	548	2.8

Although there may be advantages to the filter design, the bandwidth of the overall system is worse than the system without any prematching circuit. In this case, the best prematch may be none at all.

The Main Filters

Choosing the main filter components is probably one of the most important aspects of designing a diplexer. Each main filter affects the bandwidth at both frequencies and the attenuation of the unwanted signal. The components in the main filter will have the highest stress of any components in the system because they handle the stress of both the pass and reject frequencies.

The two types of main filters used for a diplex system are the *parallel* main filter and the *series* main filter. The series main filter, shown in Figure 4.4-10, is most often used. The parallel main filter, shown in Figure 4.4-11, is equal in overall performance to the diplex system. Each type of filter will have one component type change (capacitor to inductor or inductor to capacitor), depending on whether it is located on the high or low frequency side of the system. The equations for the design of each type of filter are given below (the equations assume you have chosen a value for C_1):

$W_L = 2 \times \text{pi} \times F_L$, where F_L is the low frequency.

$W_H = 2 \times \text{pi} \times F_H$, where F_H is the high frequency.

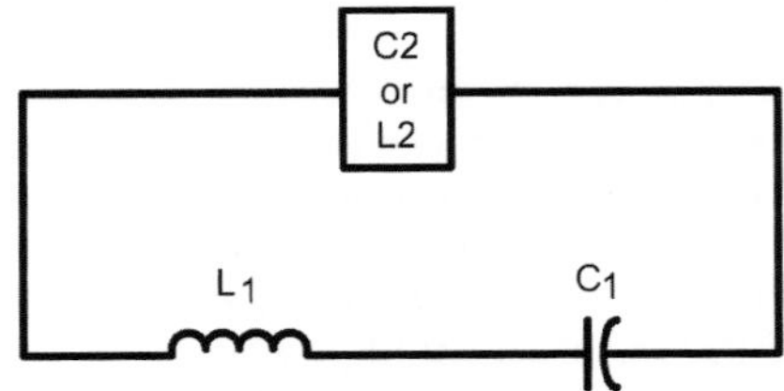

FIGURE 4.4-10 Series main filter network.

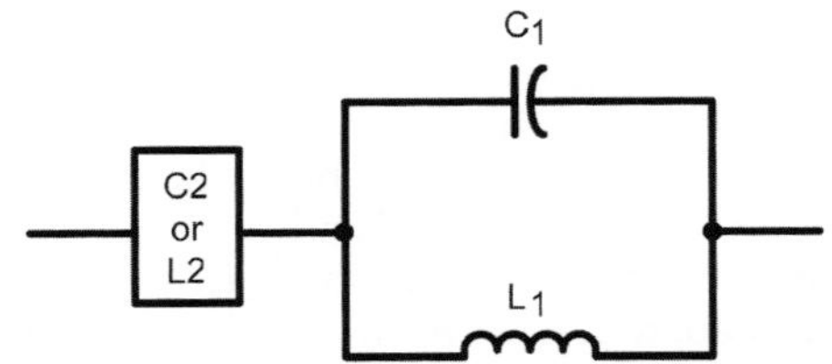

FIGURE 4.4-11 Parallel main filter network.

$F = F_L/F_H$; F is the frequency ratio.

$Q_M = 1/(1 - F^2)$; Q_M is the Q multiplier.

$R_L + jX_L$ = Diplex point impedance at the low frequency.

$R_{PL} = (R_L^2 + X_L^2)/R_L$; R_{PL} is the parallel resistance of the diplex point at the low frequency.

$R_H + jX_H$ = Diplex point impedance at the high frequency.

$R_{PH} = (R_H^2 + X_H^2)/R_H$; R_{PH} is the parallel resistance of the diplex point at the high frequency.

Series Main Filter, Low-Frequency Side

$L_1 = 1/(W_L^2 \times C_1)$.

$C_2 = (C_1 \times F^2)/(1 - F^2)$.

$Q_L = 1/(W_L \times C1 \times R_L)$; Q_L is the Q added to the system due to this filter at the low frequency.

$Q_H = Q_M \times R_{PH}/(W_H \times L_1)$; Q_H is the Q added to the system due to this filter at the high frequency.

Series Main Filter, High-Frequency Side

$L_1 = 1/(W_H^2 \times C_1)$.

$L_2 = (1 - F_2)/(WL_2 \times C_1)$.

$Q_L = Q_M \times R_{PL}/(L_2 \times WL)$.

$Q_H = 1/(W_H \times C_1 \times R_H)$.

Parallel Main Filter, Low-Frequency Side

$L_1 = 1/(W_H^2 \times C_1)$.

$C_2 = C_1 \times (1 - F^2)/F^2$.

$Q_L = Q_M/(W_L \times C_2 \times R_L)$.

$Q_H = R_{PH} \times W_H \times C_1$.

Parallel Main Filter, High-Frequency Side

$L_1 = 1/(W_L^2 \times C_1)$

$L_2 = 1/[(W_H^2 - W_L^2) \times C_1]$

$Q_L = R_{PL} \times W_L \times C_1$

$Q_H = (W_H \times L_2 \times Q_M)/R_H$

After examining the equations for each of the filter types, there are two factors to be addressed when designing the main filter. The first is choosing C_1 for any particular filter type. The answer to this problem is twofold. If auxiliary filters are not used, C_1 should be chosen so that Q_L and Q_H are equal. This will give the minimum total Q added to the system. Choosing Q_L and Q_H to be equal in value tends to give the lowest stresses on the components of the filter.

If auxiliary filters are going to be used, the Q value on the side the filter is placed can be chosen to be higher than the other Q value; for example, if auxiliary filters are to be used when designing the main filter on the low frequency side, Q_L can be chosen to be higher than Q_H. The additional stored energy on the low-frequency side can be cancelled by the stored energy in the auxiliary filter. This technique, known as Q-matching, involves choosing the Q of the auxiliary filter to be the same as the Q for the main filter and will be discussed briefly later in the next section. The second factor in determining the main filter design is selecting which type of filter to use: series or parallel. As noted earlier, both filter types will give the same bandwidth and attenuation performance; therefore, when designing for performance, it does not make any difference.

Component stresses should be the determining factor in choosing which type of filter to use. Components in the series main filter for a particular design may be easier to find or less expensive than the parallel main filter equivalent. Because the series main filter is the more popular design, the parallel main filter is often overlooked. Both filter types should be considered before determining the type that is actually used. Only stresses and the availability of parts should be the determining factors when choosing the type of main filter.

Stresses on components for each frequency can be calculated separately. Current stress is calculated by adding the current for each frequency in quadrature; for example, if the currents in a particular component are 8 amp and 15 amp for the two frequencies, the current stress can be calculated as follows:

$$\text{Current stress} = \text{sqrt}(8^2 + 15^2) = 17 \text{ amps}$$

Voltage stresses for each frequency must be summed; for example, if the voltage on a particular part is 7 kV and 12 kV for the two frequencies, then the voltage stress on that part would be 19 kV or the sum of the two voltages.

Using the above factors, the main filter for the low-frequency side can be designed. If a .001 μF capacitor is used in series with the antenna, the following values apply:

Frequency	Impedance	R_P	Voltage	Minimum Main Filter Q
800 kHz	90 – j49	117	342	5.5
1000 kHz	381 + j151	441	664	1.4

To find the value of C_1, it is best to have a computer program with all the equations programmed into it and have it iterate through different possibilities for C_1 and calculate the other parameters. If an auxiliary filter is not going to be used but a series main filter will be used, a C_1 of 400 pF will give Q_L and Q_H both equal to 5.5. This would be a total system Q of 11 (Q_L + Q_H). L_1 would then be 99 μH and C_2 would be 711 pF.

If an auxiliary filter is used to compensate for the stored energy in the main filter (Q-matching technique), C_1 could be as small as 200 pF. This gives a Q_L of 11.1 and a Q_H of 2.7. The auxiliary filter should be chosen to also have a Q of 11.1. This will compensate for most of the stored energy, giving an equivalent Q_L of 3.0. This new system will have a total Q of 5.7 (2.7 + 3.0). The Q-matched design will give the optimum bandwidth performance of the system. It should be noted that the component stresses are usually higher in the Q-matched design.

Auxiliary Filters

In a diplex system for AM radio transmission, it is sometimes necessary to further attenuate the reject frequency beyond what can be achieved by the main filters. This can be accomplished by the auxiliary filter which can also be used to improve the bandwidth of the system. Often a high Q circuit is used to improve the bandwidth of any antenna system. This stored energy counteracts the stored energy in the system over a small frequency passband (the passband for AM is typically 10 kHz, or 50–80 kHz for digital radio).

These circuits can be parallel or series resonant. Because an auxiliary filter is a parallel resonant circuit, it can be used for this purpose. For the auxiliary filter to be used to improve the bandwidth of the system, the impedances at the sideband frequencies must be properly oriented. If not, the stored energy in the circuit can add instead of subtract from the total stored energy of the system. To accomplish this, it is necessary for the impedances of the sideband frequencies (*e.g.*, 10 kHz) to have the same parallel resistance. An antenna resonator is used for this purpose. The equation to calculate the parallel resistance from an impedance is $R_P = (R + X)/R$.

The antenna resonator is placed between the main and auxiliary filters to adjust the impedance at the sideband frequencies so their parallel resistances are of equal value. It turns out that the amount of reactance required will generally bring the impedance close to purely resistive, thus the name *antenna resonator.* The antenna resonator consists of an inductor or a capacitor.

When this is done, the auxiliary filter can be designed. Like the main filter, the auxiliary filter has two types, series and parallel, as shown in Figures 4.3-12 and 4.3-13. Both filter types can be used and have the same bandwidth characteristics. For this reason, the type of filter should be chosen based on the stresses and availability of the components. Below are the design equations for the auxiliary filter types; it is assumed that the value of C_1 is chosen:

$W_L = 2 \times PI \times F_L$, where F_L is the low frequency

$W_H = 2 \times PI \times F_H$, where F_H is the high frequency

$F = F_L/F_H$, where F is the frequency ratio

$Q_M = 1/(1 - F^2)$; Q_M is the Q multiplier

$R_L + jX_L$ = Impedance at the auxiliary filter on the low-frequency side

$R_{PL} = (R_L^2 + X_L^2)/R_L$; R_{PL} is the parallel resistance at the low frequency

$R_H + jX_H$ = Impedance at the auxiliary filter on the high-frequency side

$R_{PH} = (R_H^2 + X_H^2)/R_H$; R_{PH} is the parallel resistance at the high frequency

Series Auxiliary Filter, Low-Frequency Side

$L_1 = 1/(W_H^2 \times C_1)$

$L_2 = (1 - F^2)/(W_L^2 \times C_1)$

$Q_L = (R_{PL} \times Q_M)/(W_L \times L_2)$; Q_L is the Q added to the system by this filter at the low frequency

$Q_H = 0$; Q_H is the Q added to the system by this filter at the high frequency

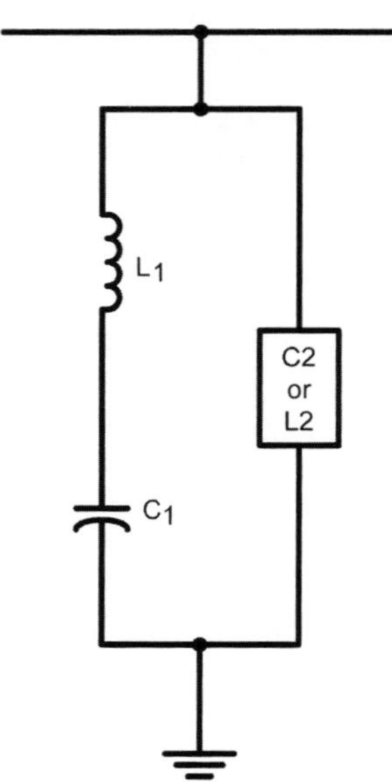

FIGURE 4.4-12 Series auxillary filter network.

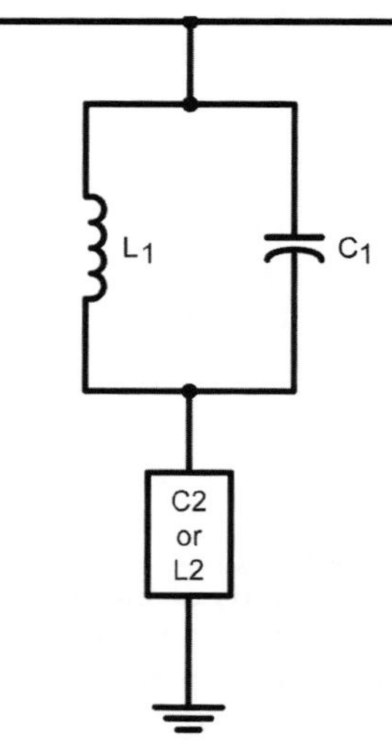

FIGURE 4.4-13 Parallel auxillary filter network.

Series Auxiliary Filter, High-Frequency Side

$L_1 = 1/(W_L^2 \times C_1)$

$C_2 = C_1 \times F \times Q_M$

$Q_L = 0$

$Q_H = R_{PH} \times W_H \times C_2 \times Q_M$

Parallel Auxiliary Filter, Low-Frequency Side

$L_1 = 1/(W_L^2 \times C_1)$

$L_2 = 1/[(W_H^2 - W_L^2) \times C_1]$

$Q_L = R_{PL} \times W_L \times C_1$

$Q_H = 0$

Parallel Auxiliary Filter, High-Frequency Side

$L_1 = 1/(W_H^2 \times C_1)$

$C_2 = C_1/(F^2 \times Q_M)$

$Q_L = 0$

$Q_H = W_H \times R_{PH} \times C_1 \times Q_M$

Two concerns must be addressed. The first is the need for an auxiliary filter. This depends on the individual antenna system. In general, the diplex system requires an auxiliary filter if the frequencies are within 10%. Also, an auxiliary filter may not be needed if the two frequencies are more than 25% apart. The real test is the attenuation from one port to the other. A good rule of thumb is 60 dB attenuation. If the attenuation is only 40 dB with a main filter alone, an auxiliary filter must provide another 20 dB. The second concern assumes that an auxiliary filter will be placed in the design and addresses how to choose C_1 for an optimum system. Equations were given above to calculate the Qs of a main filter at both frequencies. For optimum bandwidth, the value of C_1 should be chosen so the Q of the auxiliary filter will be equal to the Q of the main filter at that frequency. This may not always be possible because the stresses on the components go up as the Q of the circuit is increased. So it may become necessary to use an auxiliary filter with a lower Q.

As an example, an antenna resonator was chosen to give the following impedances at the point where the auxiliary filter is to be attached. This filter is on the low-frequency side when the upper frequency is 1000 kHz.

Frequency	Impedance	R_P	Voltage	Minimum Main Filter Q
800 kHz	90 – j49	117	342	5.5
1000 kHz	381 + j151	441	664	1.4

An antenna resonator was chosen (an inductor) to make the parallel resistance at the 10 kHz sideband frequencies the same:

Frequency	Impedance	R_P
790 kHz	87 – j38	103.6
800 kHz	90 – j6	—
810 kHz	96 + j27	103.6

Notice two things about these data. First, the impedance at the carrier frequency (800 kHz) is not quite resonant but is close. Second, the parallel resistances at the sideband frequencies are of equal value. This will allow the auxiliary filter to improve the bandwidth of the system. The main filter for this design has a Q of about 8. A series auxiliary filter with C_1 = 2000 pF produces a Q of 7.1, which is close to the required Q of 8. L_1 will then be 12.7 μH, and L_2 is 7.1 μH.

Software for Diplexer Designs

Like the design of phasing systems, diplexers are usually designed with the aid of a computer [1].

PHYSICAL CONSTRAINTS AND CONSIDERATIONS

Coil Layout

When doing a mechanical design for any phasing, matching, or filter system it is important that the inductors in the circuitry not couple with each other. Coupled circuitry will tend to be difficult to tune and, in the case of diplex filters, provide much less isolation than the paper design would predict. To implement a good paper design, it is necessary to consider the layout of the inductors.

There are two competing design criteria for good circuitry layout. The first is to bring the circuitry close together so stray inductances will be small and the total circuitry package will be minimal. The other is to space inductors as far apart as practical to reduce coupling between inductive components. A good layout will be minimum coupling and yet fit into a small package. There is a misconception that as long as the inductors are aligned 90° with respect to each other they will not couple. This is only true if one inductor is pointing directly to the middle of the active portion of the other inductor, as shown in Figure 4.4-14.

Only the second layout will provide a low coupled design, as one inductor points directly at the center of the active portion of the other inductor. This is because the flux lines linking the two sides of the top inductor produce currents that oppose (thus canceling) each other. The flux lines linking the two inductors of the first layout produce currents in the same direction; therefore, the interaction between the two inductors will be greater. Another method for placing inductors in close proximity to one another is by shielding, as shown in Figure 4.4-15. Proper shielding for inductors is obtained by providing ground on five sides of the inductor. With the inductors surrounded by ground

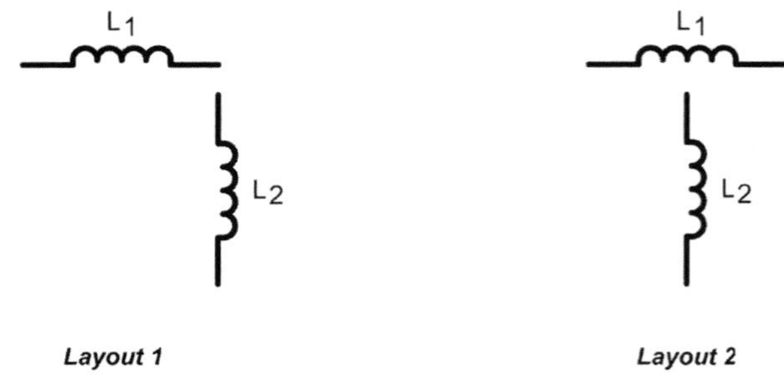

FIGURE 4.4-14 Inductor layouts.

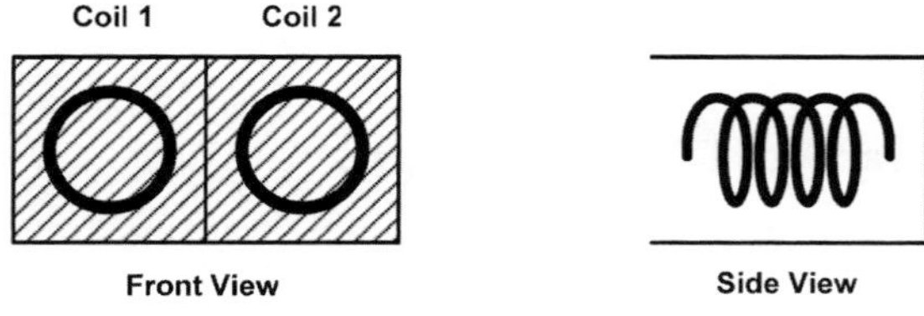

FIGURE 4.4-15 Inductor shielding.

walls as shown and with the back side covered, the inductor is shielded. It is not necessary to place a ground plane on the front side.

Voltage Considerations

It is important with any high power design to calculate the voltage to ground at every point in the circuitry. This will not only help with choosing insulators but may also provide information on improving the design layout. On the paper design, the order of placing a series coil and capacitor will make no difference in the design, but it may have a substantial effect on the voltage to ground between the two components. Consider the following example:

- 10 kW into an antenna that is 200 –j150 ohms
- Design calls for a capacitor and coil of –j150 ohms and +j150 ohms in series

Consider putting the capacitor toward the tower:

Voltage Inductor	Voltage Capacitor	Voltage Antenna
1777	2550	1777

Now consider the same circuitry, but change the order of the inductor and capacitor:

Voltage Capacitor	Voltage Inductor	Voltage Antenna
1777	1414	1777

Both designs provide equivalent circuitry, yet the voltage to ground of the first case is nearly twice the voltage of the second. The point is to keep the reactance as close to zero as possible. Placing the capacitor close to the antenna will increase the reactance between the two components. Placing the inductor close to the antenna will decrease the reactance in between the two components, thus giving a lower voltage to ground. This principle can be used if a series of capacitors and inductors are needed. Consider the following example:

> A 10 kW system, 50 ohms followed by two capacitors, 100 ohms each, and one inductor of 200 ohms

If the two capacitors are placed together the voltages to ground will be as follows:

Voltage Capacitor #1	Voltage Capacitor #2	Voltage Inductor	Voltage Load
707	1581	2915	707

Now consider the voltages to ground if the inductor is placed between the two capacitors:

Voltage Capacitor	Voltage Inductor	Voltage Capacitor	Voltage Load
707	1581	1581	707

The highest voltage to ground using the second arrangement is nearly half that of the first component arrangement.

SUMMARY

Several networks have been shown that may be used to optimize an AM transmission system. Proper selection of matching and phasing system components is an important aspect for obtaining optimum bandwidth and stability. The latter are important not only for traditional AM broadcasting, but also for in-band on-channel (IBOC), for which the specifications are more critical. Diplexing has gained popularity as tower-site real estate values have increased, zoning processes for new tower construction and alterations have become more stringent, and the ability to share a common tower is economically appealing.

References and Bibliography

[1] Software for phasor designs, http://www.westbergconsulting.com.

[2] Ball, W. G., Analytic tools for broadbanding AM antennas for high fidelity sound, in *Broadcaster Engineering Conference Proceedings*, National Association of Broadcasters, 1986, pp. 28–37.

[3] Westberg, J. M., Sideband analysis of medium wave broadcast antenna systems, in *Broadcaster Engineering Conference Proceedings*, National Association of Broadcasters, 1986, pp. 68–77.

[4] Parnau, A. W., Broadbanding AM antennas for higher fidelity and stereo, in *Broadcaster Engineering Conference Proceedings*, National Association of Broadcasters, 1987, pp. 9–16.

[5] Osenkowsky, T. G., Analytical methods to improve directional antenna efficiency, in *Broadcaster Engineering Conference Proceedings*, National Association of Broadcasters, 1987, pp. 63–70.

[6] Westberg, J. M., Diplexer Design: *Q*-Matching Techniques, in *Broadcaster Engineering Conference Proceedings*, National Association of Broadcasters, 1989, pp. 103–110.

[7] Westberg, J. M., Design considerations for AM directional phasing equipment, in *Broadcaster Engineering Conference Proceedings*, National Association of Broadcasters, 1990, pp. 384–387.

[8] Rackley, R. D., Modern methods in mediumwave directional antenna feeder system design, in *Broadcaster Engineering Conference Proceedings*, National Association of Broadcasters, 1991, pp. 43–54.

[9] Bingeman, G. W., Optimizing impedance and pattern bandwidths of a phased array, *BM/E*, Feb., 84–94, 1980.

[10] Morton, L. L., Increasing the high frequency response of AM stations, *Broadcast Engineering*, Sept., 44–52, 1978.

[11] Bingeman, G. W., Solving AM bandwidth problems, *BM/E*, June, 65–79, 1978.

[12] Bingeman, G. W., Negative towers, *BM/E*, Nov., 73–76, 1980.

[13] Bingeman, G. W., Designing AM coupling networks, *Broadcast Eng.*, July, 64–78, 1985 (correction printed in *Broadcast Engineering*, Nov., 135, 1985).

[14] Bingeman, G. W., Use circuit to match Xmtr *RF* impedance, *Radio World*, March 15, 1984, p. 10.

[15] Bingeman, G. W., How mutual inductance affects coil performance, *BM/E*, Oct., 88–92, 1980.

[16] Jubera, D., Toward improved control of medium wave directional antennas, *IEEE Transactions on Broadcasting*, vol. BC-27 no. 4, December 1981.

[17] Bingeman, G. W., Broadband your antenna with an external network, *BM/E*, April, 263–272, 1984.

CHAPTER

4.5

Computer Simulation of Broadcast Antenna Systems

JAMES B. HATFIELD

Hatfield and Dawson, Consulting Engineers, Inc.
Seattle, Washington

INTRODUCTION

Antenna systems have been analyzed and designed using computer simulation for many years by academics and those designing antennas for the military. By comparison to electromagnetic simulation of aircraft and ship structures, modeling medium wave antenna systems is an almost trivial exercise. It has been the experience of this writer and other consultants that a relatively short time spent modeling an AM array can save a great deal of trial and error time tuning up a directional array (DA). FM and television antennas can also be modeled using moment method programs for analysis of the effects of nearby antennas and other reflecting structures on antenna pattern distortion and human radiofrequency (RF) exposure.

AM DIRECTIONAL ARRAY FCC EQUATIONS AND MOMENT METHODS

Equations set forth in the Federal Communications Commission (FCC) Rules, Section 73.150(b)(1)(i), have been the basis of AM directional array design, with few modifications, since the 1940s. Additional discussion may be found in Chapter 4.3, "AM Antenna Systems." The term *moment method* refers to the fact that electromagnetic fields from antennas are proportional to the area under the tower current distribution curve. The units of this area are (amperes) × (length) which is of the same form as the mechanical turning moment used in civil engineering. Moment method programs find the area under the current distribution curve by a process of numerical integration, where the incremental area is the product of the tower current at a point and an incremental distance.

Most moment method programs compute far-field horizontal plane pattern shape by using the same mathematical expressions as the aforementioned FCC equations, although with different notation. The major difference between moment method computations and the FCC equations is that the FCC formulas for vertical plane field and pattern size computations assume sinusoidal tower current distributions, but all moment method field and pattern size computations are derived from Maxwell's equations in integral form and are scaled to the specified input power. The moment method equation for the horizontal plane far fields from a tower can be reduced to:

$$E = \frac{\pi}{3}\sum i \cdot \Delta l$$

This equation states that the inverse distance field (in mV/m) at 1 km from a tower is 1.0472 times the sum of the incremental tower current moments (in amp-degrees). This summation of incremental current moments over the length of the tower is usually referred to as the *tower moment*.

Because the field from each tower in an array is 1.0472 times the tower moment of that tower, the field ratio of each tower is the ratio of the tower moment of that tower to the tower moment of the reference tower. The tower moment is a polar number, with a magnitude and angle, so the phase angle of the field ratio is found by subtracting the angle of the reference tower moment from the tower moment angle of the tower in question.

PRODUCING THE CORRECT PATTERN WITH MOMENT METHOD PROGRAMS

The FCC equations use variables called *field ratios* and *current phases*. These field parameters are specified in FCC 301 AM application forms and on licenses as *theoretical parameters*. The current phase, in practice, refers to the relative phase angle of the *antenna loop* current. The loop refers to the maximum antenna current. The FCC equations are based on the assumption that the loop current phase angle and the phase angle of the contribution of the tower to the far field are the same. It is important to note that the FCC equation uses the field ratio as the relative contribution of the tower in question to the far field. The phase angle used in the equation is the phase angle of that far field contribution. In fact, the ratio and phase angle of the field parameters are determined by the behavior of the tower as a whole. The ratio and phase angle of the tower loop current are seldom, if ever, the same as the actual field ratio and phase angle. This difference is the reason why antenna monitor readings do not usually match the theoretical parameters. Calculating the antenna monitor indicated ratios and phase angles avoids a great deal of trial and error during the array tune-up process.

The antenna system can be thought of as a black box where the inputs are voltages and the outputs are fields. The antenna system is linear, so the inputs and outputs can be related by a series of constants even if the towers are not all of the same height. If one includes mutual impedance, the relationship between the voltage drives and fields for a two-tower array is of the following form:

$$E_1 = V_{11}T_{11} + V_{12}T_{12}$$
$$E_2 = V_{21}T_{21} + V_{22}T_{22}$$

where the E values are the FCC field parameters (ratios and phases); V_{11} and V_{22} are the tower base drive voltages; V_{12} and V_{21} are the tower base voltages induced by the other towers; and the Ts are the constants that relate them. The T constants are found by shorting the respective towers to set the pertinent base voltages to zero.

Moment method computer programs devoted to broadcast antennas solve these equations for the complex base voltages necessary to produce the proper pattern. Base drive parameters for the correct FCC pattern can be computed by some method moment programs for top-loaded or self-supporting towers. Antenna monitor ratios and phase angles can be found by calculating the ratios and relative phase angles of the computed tower currents at the location on the tower where the antenna monitor samples the current. Some programs include a provision for current drives. If the antenna sample system is well characterized (*i.e.*, sample line lengths and sample transformer sensitivities are known), the correct base current ratios and phases can be determined and used as base drives for computation of the pattern as adjusted.

AM DIRECTIONAL ARRAY NEAR-FIELDS AND PROXIMITY EFFECT

The FCC theoretical equation calculates pattern shape using the assumption that lines are parallel between the towers and the location where the field is being calculated. This assumption is generally true only at some distance from the array. Closer than this to the towers the pattern is not properly formed. The measured fields close to the antenna on null radials will not be the same as the FCC equations would predict, even though the pattern is properly adjusted. This has been called the *proximity effect* and may or may not be an important factor depending upon array geometry. The proximity effect occurs in an area called the *array near field*. Accurate computations of the fields in the array near field region can be made with moment method programs. These near fields can then be compared to theoretically calculated fields computed at the same locations for analysis. Field-strength meters use shielded magnetic field-sensing loops and convert the magnetic field component of the radiated signal to equivalent electric field units. The magnetic field is multiplied by 377 to convert amperes per meter to volts per meter. The conversion factor of 377 only applies to plane-wave fields and, in the presence of reradiating objects, may result in an improper indication of the equivalent electric field. Magnetic fields and electric fields are computed separately by the near-field computational portion of the Numerical Electromagnetic Code (NEC) and MININEC family of programs. As a result, measurement anomalies resulting from non-plane-wave conditions can be determined.

IMPEDANCE

Tower self-impedances computed by method of moment program are benchmarked to values computed in King's *Theory of Linear Antennas* [1]. The King values for monopole antennas are based on measurements made at high frequencies using the center conductor of a coaxial cable projecting through a copper sheet ground plane to which the outer conductor is connected. This produces resistance and reactance values that differ significantly from the resistance and reactance of a base-insulated guyed tower over a radial wire ground system. The shunt capacitance of the tower base insulator can have a significant transforming effect on high tower base impedances. Tower feed connections also add a significant inductive reactance to the tower reactance, as was demonstrated by Bloomer [2]. These usually are in the form of a single or double turn lightning retard loop and are connected from the antenna tuning network (ATU) to the tower. These defects in the computation of tower impedance can be corrected by the addition of reactance to the tower model or by increasing the tower height by about 6.7%. Note, however, that the fields and tower radiation efficiencies computed by moment method programs for actual tower heights are close to measured values and also to those shown in Figure 8 in the FCC Rules, Section 73.190. Operating impedances cal-

culated by the moment method have been far more accurate than previous algorithms. An additional advantage is the ability to predict current distribution along a tower, allowing optimum placement of current sampling loops. The current distribution is dependent on the array parameters and can vary significantly between modes of operation.

DETUNING TOWERS FOR AM DIRECTIONAL ARRAYS

When taking field-strength readings it is sometimes necessary to detune nearby re-radiating objects which cause localized perturbations in the measured fields. The moment method programs can be used to determine whether the re-radiating object affects the stations pattern or only the measurements near the re-radiating object. An effective method for detuning towers is to treat the tower as a part of the array. The field parameters for the detuned tower are set to zero as a program data input. When the computed drive voltage (and phase) is applied to the tower to be detuned, along with the proper drives to the array, and the computation is performed, the impedance of the tower that it is desired to detune is noted. If the reactive component is approximately ten times the resistive component for the detuned tower, it can be detuned by loading it with the conjugate of the computed impedance. If the computed reactance is –j450 ohms, the tower would be detuned by +j450 ohms across its base to ground.

If the resistive part of the detuned tower impedance is large and negative, it may, in addition, have to be loaded by a resistance in series with the detuning reactance. If the resistive component of the detuned tower impedance is large and positive it will have to be driven with a small amount of power.

The effects of the shunt capacitive reactance of the tower base insulator must be taken into account when detuning a tower. The best procedure for being sure that a tower is properly detuned is to send someone a third of the way up the tower with a current-sensing device. (If the tower is over about 130° tall, the height will be greater.) The detuning reactance is adjusted until a tower current minimum is observed at the detuning point. A field set with the loop shield shorted out through use of a sheet metal screw can provide a sensitive indication of when the tower current minimum is reached.

This procedure is based on the fact that towers are detuned in the horizontal plane by having the area under two halves of the tower current distribution curve being equal and opposite in phase. If the current on a 90° tower is nearly sinusoidal, the area under the curve is given by the cosine function. On a per-unit basis, the cosine is 0.5 at 30° and one-half of the area is one-third of the way up the tower. Although detuning drastically lowers the fields in the horizontal plane, it can cause large increases in vertical plane fields. For a daytime directional, this effect may not be important, but it could have a significant effect on radiation minima at night at high pertinent vertical angles.

MATCHING ANTENNA MONITOR AND THEORETICAL FIELD PARAMETERS

The location on a detuned tower where the antenna current goes through a minimum and a reversal of phase angle is at the center of the area of the current distribution curve. The current at this location is proportional to the tower moment and hence the radiated field from the tower. If a sample loop is placed at this location and the sample system is the same for all towers of equal height, then the antenna monitor parameters will be the same as the theoretical field parameters. For towers below 110°, the sample loop should be placed one-third of the way up the tower. It is necessary, of course, for the rest of the sample system to be identical, with equal length and same type of sample lines up to the loops and identical sample loops on all towers, with all loops oriented in the same direction. When the towers are of different heights, the monitor ratios can be determined by taking the ratios of the computed currents one-third of the way up the towers. The monitor phase angle relative to the reference tower will be the same as the relative theoretical field phase angle.

The ratio of the current at a point on a tower to the tower moment is nearly constant for varying drive conditions. For different height towers, the monitor ratio one-third of the way up the tower for tower two of a two-tower array is:

$$(\mathbf{M}_1/\mathbf{I}_1)\times(\mathbf{I}_2/\mathbf{M}_2)\times\mathbf{F}_2$$

where **M** refers to the tower moment for the respective numbered towers, **I** is the current on the tower one-third of the way up the tower (slightly higher for towers over 130°), and **F** is the field ratio for the non-reference tower; the monitor phase angle for the non-reference tower is the same as the theoretical field phase angle. The tower moment and corresponding sample current can be calculated for each tower separately, independently of drive conditions, and the correction factor for the antenna monitor current ratio of the non-reference tower will be accurate for all field parameters.

When the antenna monitor sample loops are located at the correct height on the tower, the antenna monitor sample current ratios for towers whose heights are different from the reference tower can be found using the tower moment to tower current ratios according to the procedure outlined above.

Because the fields differ from **M** by a constant, one could, in theory, determine the correction to the monitor ratio for unequal height towers by measuring the tower currents at the sample location and the fields at the same distance from each tower (or use Figure 8 in the FCC Rules, Section 73.190) under identical tower drive conditions and compute the correction to the antenna monitor current ratio from the above relationship. This is not always a practical procedure, however, due to the limitations of measurement accuracy and the difficulty of isolating the tower to be measured.

ARRAYS NEAR RE-RADIATING OBJECTS

A typical problem arises when an AM array is near a power transmission line. Several effects occur simultaneously:

- The current flowing through the transmission line towers and skywires causes large changes in the ratio of the electric and magnetic components of the radiated field of the array.
- The power line may affect only the accuracy of nearby field strength readings and not the inverse distance pattern of the station.
- The power line may affect both the field strength readings and the station pattern.

These effects can be separated out by modeling the power line and the AM array and then computing the near electric and magnetic fields along with the far-field pattern as influenced by the power line. A 1 ohm loss pattern can be computed by subtracting the square of the loop currents (maximum currents on the printout) from the station input power. Corrections to the field strength meter readings can be calculated from the magnetic near-field data. In one instance, three tall guyed communications towers were within a mile of an AM array. The array and the guyed towers were modeled, and the station's pattern was scalloped. The computed pattern did not exceed the standard pattern so no attempt was made to detune the communications towers.

TOP LOADING AM TOWERS

To determine the effects of modest amounts of guy wire top loading, a 76° non-directional tower was modeled as top loaded to 8° and 14°. The top loading, in electrical degrees, was simply the length of uninsulated guy wire connected to the top of the tower. Horizontal skirts were also added to the ends of the guy sections to determine their effects. The 8° of top loading increased the resistive component of the base impedance by 45%, while the reactance was reduced to 12% of the non-top-loaded base reactance and went from capacitive to inductive. Adding horizontal skirts caused a further 16% increase in base resistance, while the reactance became more inductive by a factor of five. The 14° of top loading produces results that are, practically speaking, indistinguishable from skirted 8° of top loading with regard to resistance and reactance. Adding skirts to 14° of top loading tripled the inductive reactance and increased the resistance by 28%.

The conclusions from the above are that top loading makes short antennas more inductive, and the Q of the antenna is worse with skirted top loading and optimum with modest (8°) unskirted top loading. For all top loading from 0° to 14°, skirted or not, it was found that the inverse field at 1 km only varied from 310 to 314 mV/m, so one can conclude that 8° to 14° of top loading has only a minor effect on tower radiation efficiency.

COMPUTING HUMAN EXPOSURE TO RF FIELDS FROM AM ARRAYS

Human exposure to magnetic and electric fields from AM towers can be computed using moment method programs. Licensed AM radio facilities in this country are restricted to a maximum power of 50 kW so the FCC exposure limits are reached at a distance from each tower that is much less than the distance between adjacent towers. For this reason, each tower can be treated as a separate case if no other sources of exposure are present at the site. Tables 1 through 4 of distances to fences around AM towers shown in Supplement A of *OET Bulletin* 65 [3] were based on moment method computations. The Commission has accepted such computations for RF guidance level determination for many years. The measured and computed electric and magnetic fields agree quite closely with the computed values. One must model the tower carefully if correct results are to be achieved.

FM AND TELEVISION ANTENNA ENVIRONMENTS

Moment method modeling can be used to analyze the effects of nearby towers, appurtenances, and antennas on the operation of FM and TV antennas. In one situation, careful measurements showed that the reflected fields from an adjacent tower were responsible for the human RF exposure fields at ground level exceeding the FCC maximum permissible exposure (MPE) levels for the general public. A model was constructed of the FM antenna and the adjacent tower. The computed ground-level fields agreed with the results of the measurements. Analysis of the effects of nearby scattering object upon the patterns of television antennas can be made using moment method modeling. The antenna can be represented by a vertical array of properly phased orthogonal horizontal dipoles. Slot or traveling wave antennas can be treated as cylinders of appropriate cross-section. Enough tower detail must be shown, at a reasonable fraction of a wavelength, for significant reflection geometry to be included.

SUMMARY

Modern moment method computer analysis of AM, FM, and TV antennas can yield accurate field and impedance predictions, resulting in efficient and broadband antenna system operation. Tune-up time is reduced, and stable and reliable long-term performance is frequently realized. With increasing use of digital broadcasting technology, antenna system performance must be broadband and stable for optimal reception quality. The use of moment method analysis

can ensure accurate prediction of real world performance.

Terms

Antenna monitor parameters: The relative antenna current magnitudes and phase angles at a specified location on the towers.
Current loop: The maximum current location on a tower; this is at the base of the tower for tower heights of a quarter wave or less.
Electrical degrees: A unit of distance proportional to the free space wavelength at the frequency of interest. One wavelength is 360° and a quarter-wave tower has a height of 90°.
Field parameters: The relative magnitude and phase of the contribution to the far electric field of each tower in an AM array.
Inverse field: The electric far field of an AM antenna at 1 km that is not attenuated by earth losses and varies with distance **R** proportional to 1/**R**.
Method of moments: In the method of moments, the integral equations relating antenna currents to radiated electric and magnetic fields are approximated by a set of linear equations. In these equations, the differential distance is expanded to finite size, called a *segment*, and the integral is replaced by a summation of a fixed number of current-segment products. When the matrix formed by these linear equations is inverted, the unknown variables can be expressed in terms of known input variables and numerically evaluated.
Sinusoidal current distribution: Antenna current that varies with antenna height proportional to the sine or cosine of the antenna height in electrical degrees.

References

[1] King, R.W.P., *The Theory of Linear Antennas*, Harvard University Press, Cambridge, MA, 1956.
[2] Bloomer, T.M., Antennas, in *Industrial Electronics Reference Book*, Westinghouse Electric/John Wiley & Sons, New York, 1948.
[3] Evaluating compliance with FCC guidelines for human exposure to radiofrequency electromagnetic field, *OET Bull.*, 65 (Suppl. A), Edition 97-01, 1997.

Bibliography

47 CFR: Broadcast Radio Services, Part 73, Subpart A, FCC §73.190, Figure 8.
Lahm, K., Excitation sampling in MW directional arrays, in *Proc. of the IEEE Broadcast Technology Society 44th Annual Symp.*, 1994.
Li, S.T., Logan, J.C., Rockway, J.W., and Tam, D.W., MININEC, in *Microcomputer Tools for Communications Engineering*, Artech House Publishing, Dedham, MA, 1984.
Westberg, J.M., Matrix method for relating base current ratios to field ratios of AM directional stations, *IEEE Trans. Broadcast.*, 35(2), 172–175, 1989.
Whitaker, J.C., Computer analysis of antenna systems, in *The Electronics Handbook*, CRC Press/IEEE Press, Boca Raton, FL, 1996, pp. 1353–1357.

Additional Sources of Information

Applied Computational Electromagnetics Society (ACES), http://aces.ee.olemiss.edu/.
Expert MININEC Broadcast Professional computer program source, http://www.emsci.com.
Kildal, P.-S. and Yang, J., Lecture 5, *Moment Method, Monopoles and Dipoles, Loop Antennas*, http://www.elmagn.chalmers.se/elmagn/antenna/courses/Lecture5.pdf.
Phasor Professional computer program source, http://www.westbergconsulting.com.

C H A P T E R

4.6

AM Antenna Maintenance

JOHN F. WARNER
Clear Channel Radio

INTRODUCTION

It is important to understand that antenna systems do not operate without considerable supervision and maintenance. Difficulties can arise either within the antenna system itself or in the environment in which it operates. Maintenance of AM antenna systems can be broadly divided into issues of:

- Electrical changes or failures
- Mechanical changes or failures

Problems typical of AM antenna system can be generally defined as:

- Catastrophic, such as caused by lightning
- Gradual degradation or "drift"

PREVENTATIVE MAINTENANCE

A schedule of routine preventative maintenance is essential to reliable operation and should be established and rigidly adhered to by appropriate technical staff. In the past, many transmitter sites were staffed and rigid preventative maintenance plans were in place. This is not to say that such plans cannot be continued under different circumstances. While deregulation and advances in equipment reliability have removed operators from most transmitter sites, preventative maintenance remains key to reliable and Federal Communications Commission (FCC)-compliant operation. Recordkeeping is an important part of any maintenance program. Although computerized databases are normally thought of for maintenance scheduling, a paper logbook or notebook is satisfactory. Historical documentation of previous problems and their resolution is essential to solving problems. Records dating back to the original tune up, including the consultant's field notes, are often found to be useful. Copies should be made of all documents and stored at a location other than the transmitter site. The loss of a transmitter building to fire or natural disaster not only destroys the facility but also likely destroys the information needed to replicate it as well. Appropriate lock-/tag-out practices should be in force at all transmitter sites. Working alone in a transmission facility can be dangerous. If it is necessary to do so, the employee should inform a colleague or superior of the planned activities and expected return time.

MECHANICAL CONSIDERATIONS

Reliability of mechanical components such as radio frequency (RF) contactors is dependant on cleanliness, good mechanical condition, and proper lubrication. Tuning unit enclosures should be sealed against the elements and the entry of insects, vermin, and snakes. If the enclosure is well sealed and in good repair, routine cleaning of the area every few months should be sufficient. A good "spring cleaning" should be scheduled to remove anything that has taken up residence over the winter. The use of a shop type vacuum cleaner is preferred to blowing dirt around with an air compressor. Standard cleaning techniques can be used on the building or enclosure. Avoid using soap based cleaning products on electrical components as they may leave a residue. Denatured alcohol or ammonia is often used for this purpose, but the area should be well ventilated, and proper personal protective equipment such as gloves should be used. Follow label cautions as to mixing various cleaning agents. Lubricants that are oil based are not appropriate in areas where dust can collect. Pivoting parts of contactors should be cleaned as necessary with a solvent and lubricated with graphite. Solvent label instructions and warnings should be followed. If possible, the contactor should be removed to a well ventilated area. Many stations stock a spare for each type of contactor that allows for removal of a unit from service for maintenance.

Repeated use of abrasives on silver plated surfaces is not recommended. Spring finger contacts should be replaced if they have become burned or distorted. Although overheated contacts can often be bent back into shape, their spring properties will be lost. Spring contacts from various vendors may look the same, but it is better to buy the correct ones. Close examination of contact surfaces is needed to differentiate between contacts that have burned black and the natural oxidation process. Silver oxide is black and shiny and should not be disturbed as it is still a good conductor.

Improper operation of micro switches is often a source of intermittent problems with directional antenna systems. Micro switch mounting hardware should be tight, and the position of the switch in relation to the moving arm of the contactor should be as specified by the manufacturer. In general, when the movable arm of the contactor is fully seated, the associated micro switch should be fully depressed and a slight amount of pressure exerted on the movable actuator of the switch. A switch that is working properly will usually emit an audible "click" when its actuator is fully depressed. An ohmmeter can be used to check the electrical functioning of a micro switch once the associated control circuit has been de-energized. Most contactors have two pairs of micro switches. One pair is a tally of contactor position for the related pattern change logic and the other is an interlock to ensure that only one contactor solenoid can be energized at a time, in event of logic failure. In the case of traditional pattern control logic systems, the failure of one of the micro switches in series with the solenoid coils will prevent that coil from being energized when a pattern change command is applied. The failure of the tally micro switch will make the controller think the contactor has not moved and will hold the transmitter off. Many approaches have been taken to pattern change logic design. They often contain a large number of traditional relays that can be sources of failure or intermittent operation.

Systems that use highly reliable programmable logic controllers (PLCs) are now available. PLCs are used by the millions in industrial applications, and they perform well in systems that are exposed to voltage transients and other anomalies. The timing and sequencing functions are programmed using a ladder logic program and then loaded into the PLC. External relays are sometimes used to buffer the controller from high AC switching voltages used on older systems. Modern systems use low voltage DC for control and tally indicators. Figure 4.6-1 and Figure 4.6-2 show a well-designed and laid out control and tally panel for a multi-tower dual AM transmitter facility. Note that the control buttons and tally indicators are clearly marked. A key switch disables remote control functions to eliminate accidental changes while conducting local maintenance. All control and status indicators can be sent to a remote control and metering point.

FIGURE 4.6-1 Control panel of PLC transmitter and antenna system controller. (Photograph compliments of Tunwall Radio.)

Mechanical drive mechanisms and turn counters in phasors should be checked for freedom of movement and setscrews should be checked for tightness each time the system is shut down for inspection and maintenance. As with tuning units, the phasor should be cleaned on a regular basis. Knife switches used for meter switching should be kept clean, and their contact pressure should be firm enough to ensure proper operation.

FIGURE 4.6-2 PLC controller of current design. Logic is contained within the PLC and can be reprogrammed to change timing functions as needed. (Photograph compliments of Tunwall Radio.)

Evidence of heating or discoloration at connections points is often a sign of loose hardware. Bolted connections should be checked for tightness when the system is de-energized. When the system is locked-/tagged-out, carefully touch the components and connections in the system to determine that excessive heating is not occurring.

A record should be kept of all turn counter positions used in the system. If line or base current meters are present, their indications should be noted before the system is shut down. Taking time to provide thorough documentation of the antenna system will pay off in the long run.

ELECTRICAL MAINTENANCE

The amount of electrical maintenance needed by an antenna system is directly related to its age, complexity, and maintenance history. Two major problems in phasing and coupling systems are the deterioration and failure of mica capacitors and intermittent operation of variable or tapped inductors. A mica capacitor is essentially a stack of metallic foil sheets interleaved with layers of mica insulating material. In effect, these layers can be viewed as a group of capacitors in series. As depicted in Figure 4.6-3, the total capacitance of the device follows the formula for capacitors in series. Should one section of the capacitor short, the total capacitance of the device will increase, with the potential of causing a domino effect. As the capacitance increases, the capacitive reactance decreases. A decrease in reactance allows more current to flow, increasing the heating of the capacitor. Additional heating can cause an additional section or sections to fail and the process repeats itself, often to the point of a short circuit of the device. Heating can be detected by touching the component when the system has been de-energized, or through the use of an infrared temperature sensor. Infrared sensors are important tools for the broadcast engineer. The cost of these sensors has dropped to the point where they are affordable by the smallest station or contract engineer. Another sign of impending failure of a capacitor or inductor or transformer is the appearance of potting material oozing from the component.

The gradual failure of a mica capacitor is usually accompanied by the gradual drift of a parameter or parameters than cannot be attributed to environmental effects and can reach the point where corrective steps are necessary. To compensate for the apparent

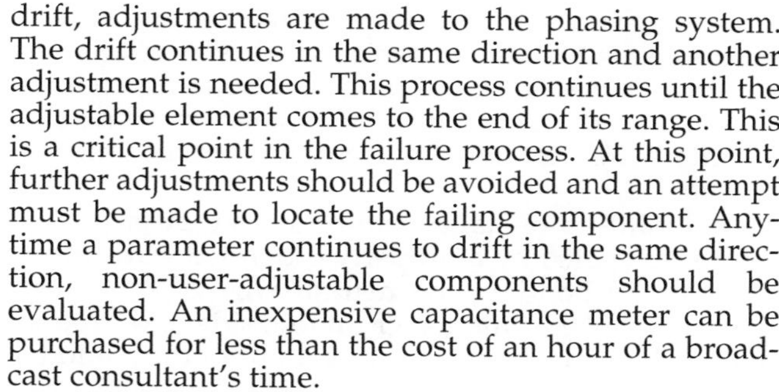

drift, adjustments are made to the phasing system. The drift continues in the same direction and another adjustment is needed. This process continues until the adjustable element comes to the end of its range. This is a critical point in the failure process. At this point, further adjustments should be avoided and an attempt must be made to locate the failing component. Anytime a parameter continues to drift in the same direction, non-user-adjustable components should be evaluated. An inexpensive capacitance meter can be purchased for less than the cost of an hour of a broadcast consultant's time.

Unlike mica capacitors, vacuum capacitors are not known to change in value. They either short or are physically destroyed by lightning. Inductors with a roller contact are often sources of trouble if they have to be moved after years of sitting in the same position. The protective circuits in modern transmitters are so fast and so sensitive that visible arcing usually does not develop. Bad contacts in roller coils can be heard to be "microphonic," in concert with modulation. Coils can be cleaned by using silver polish sparingly, or they can be removed, disassembled, and cleaned ultrasonically.

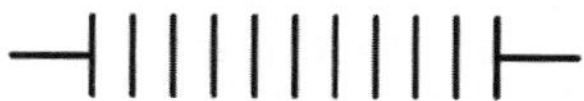

$$1/Ct = 1/C1 + 1/C2 + 1/C3 + \ldots$$

FIGURE 4.6-3 Schematic diagram of mica capacitor showing multiple capacitive elements in series.

LIGHTNING AND SUDDEN COMPONENT FAILURE

Sudden component failures are often related to lightning activity or the inadvertent application of power exceeding design values. Most lightning discharges can be effectively dealt with on insulated towers by the proper installation and adjustment of spark gaps. Gaps should always be installed across tower base insulators and can also be installed across vacuum caps and at the ends of transmission lines. The feed from an antenna tuning unit to a tower feed point can include a few turns of conductor wound about 12 inches in diameter to act as a high impedance at the rise time of a lightning strike. This high reactance is there to force the lightning stroke current toward the spark gap as opposed to through the tuning unit. A preliminary spark gap setting can be calculated based on the voltage present using Ohm's law referenced to the complex base impedance (Z_b) and the base current (I_b):

$$E = I_b Z_b$$

The voltage at carrier multiplied by a factor of four (×4) will yield an approximation of the voltage to be expected with full modulation. Under dry conditions and with large conducting surfaces, the breakdown voltage of air is considered to be 30 kV/inch. This spacing can be used to calculate the initial gap setting when ball gaps are used, and they are large in relation to the gap dimension. As a practical matter, begin with this spacing and decrease the gap until either arcing or transmitter protective reaction is noted and then open it slightly till the arcing stops.

If lightning damage is a chronic problem, the grounding and bonding around the tower base should be investigated. Ground straps at the tower base can be supplemented with ground rods driven adjacent to the tower base. All grounding and bonding connections should be made with silver solder or by brazing. For more information on lightning protection, see Chapter 7.3.

The build-up of static charges on the tower proper can be dissipated by a properly grounded static drain choke or by bonding the neutral wire of the tower lightning choke to the tower on one side and ground on the other side. Local interpretations of the National Electrical Code may forbid this practice in some jurisdictions so a local electrical contactor or the regulating body should be consulted.

Tall towers at the low end of the AM band are often plagued with static discharges across guy insulators. The electromagnetic fields generated by these discharges cause the operation of the protective circuits of modern transmitters and nuisance carrier interruptions. Inductive chokes or high-value resistors across guy insulators may be helpful in these situations, but the assistance of an experienced consulting engineer will probably be necessary.

FIGURE 4.6-4 Multiple ground straps connect the antenna tuning unit to the ground radial system. The straps have been bolted to the tuning unit legs for mechanical stability and to reduce vibration due to wind.

GROUND SYSTEMS, BONDING, AND LIGHTNING PROTECTION

Grounding and bonding of antenna tuning units and other hardware in the proximity of tower bases are essential to maintain stable antenna system operation. All ground radials should terminate at a common ring of copper tubing surrounding the antenna base pier. Ground straps that cross under the base insulator should be bonded to the ground ring as well. These straps provide a point of attachment for the tower spark gap. The ground ring should be supplemented with a minimum of four driven ground rods for lightning protection purposes. A minimum of one 4 inch ground strap should bond the antenna tuning unit ground surface to the tower ground system described above. The outer conductor of the transmission line as well as the sample line should be bonded to the antenna tuning unit ground surface. The use of running multiple ground straps to the antenna tuning unit is encouraged as they provide a low impedance connection between the tuning unit and ground. An example of base grounding is shown in Figure 4.6-4. Multiple ground straps from the antenna tuning unit cabinet are shown as well as the ground straps that cross the tower base, under the base insulator. All straps have been brazed to each other and to the copper ring that terminates the radials at the tower. To keep vegetation under control, the area surrounding the tower base has been leveled and covered with several layers of visquene sheeting. The sheeting was then covered with washed gravel to hold the sheeting in place and to keep the ultraviolet rays of the sun from causing deterioration. This method prevents weed seeds from reaching the soil beneath the visquene and sprouting.

When mechanical connections are not silver soldered or brazed, be careful not to mix dissimilar metals. Discussions of the connection of dissimilar metals and the galvanic series can be found in engineering texts. In general, copper should not be mechanically connected to galvanized steel enclosures with bolts or screws alone. If antenna tuning units are to be housed in metal cabinets or buildings, the seams of the cabinets should be bonded and the structure then bonded to the antenna ground system. The same procedure should be applied to metal doors on frame or masonry buildings, especially when dealing with high power levels.

If a new transmitter building is being built or an existing building refurbished, it is important that attention be given to grounding and bonding. In a new installation, a ground strap buried outside the perimeter will serve to keep lightning and other transients out of the building. The strap should be buried below grade and supplemented with driven ground rods at intervals equal to the length of the rods (*e.g.*, 8 ft rods at 8 ft intervals). All equipment that is installed in the building should be bonded to ground straps that are attached to the floor. These straps should leave the building at grade and attach to the buried perimeter strap. The electrical service grounding conductor as well as telephone company grounds should be bonded at a common point outside the building. If a prefabricated transmitter building is to be used, slots can be provided at floor level for the passage of ground straps to the outside. Slots can be cut into existing structures as well, as shown in Figure 4.6-5. The outer conductors of the transmission lines have been bonded to this same common strap where they enter the building using grounding kits supplied by the cable manufacturer. The purpose of this important

FIGURE 4.6-5 Interior grounding straps through slot in side of building bonded to building perimeter strap and electrical service grounding conductor.

grounding practice is to have all potentials rise and fall together. A lightning strike that enters a transmitter building on a power line and exits via the RF transmission line can cause considerable damage. For more information on facility grounding see Chapter 9.3, "Facility Grounding Practices."

GROUND SYSTEMS AND RF PERFORMANCE

Array instability and a gradual deterioration of coverage are indicative of ground system defects or deterioration. The definitive paper on this subject, authored by G.H. Brown, R.F. Lewis, and J. Epstein, is entitled "Ground Systems as a Factor in Antenna Efficiency" (*Proceedings of the IRE*, 25(6), June 1937). The antenna should be considered as a series circuit consisting of the radiation resistance of the antenna, the reactance of the antenna, and the ground resistance. In the case of antenna efficiency, the reactive power dissipates no power and can be ignored. The amount of energy radiated is controlled by the ratio of the radiation resistance divided by the sum of the ground and radiation resistances and is expressed as $R_r/(R_r + R_g)$, where R_r is the radiation resistance and R_g is the ground or loss resistance. Note that as the ground resistance increases, the efficiency of the system decreases. In determining efficiency, the FCC assumes a ground loss resistance of 1 ohm. A standard AM ground system consists of 120 radials at least one-quarter wavelength long. Radials around towers higher than 90° tall are often as long as the tower is tall. In addition, a ground screen or mesh is also installed in the area around the base of the tower. The size of the screen varies from installation to installation but usually extends 30 to 40 ft from the tower base. Radial wires should be bonded to the perimeter of the screen as well as to the tower base ring. Instead of using ground screens, the practice of installing an additional 120 shorter radials between the full length radials is a popular option. These shorter radials are usually 50 ft in length.

In the case of the non-directional antenna, the base resistance can be measured and compared to historical measurements of base resistance as an indicator of ground system condition. If the ground system has deteriorated, the measured resistance will rise as the sum of the radiation resistance and the ground resistance increases. In the case of the directional antenna, deterioration of the ground system is often indicated by a decrease of all the tower base currents compared to historical values.

A physical inspection of radials in the area of the tower base can be performed by careful excavation of the area with a shovel. Unfortunately, soft or lead solder was occasionally used in the past, and radials have often become disconnected from the rest of the system. Radials can be evaluated by using a shielded loop connected to the external input of a field strength meter or by using a locator such as those used by utility companies. The latter method is used when the antenna system is de-energized. If adjacent radials can be found at their outer ends, an ohmmeter can be used to determine if they are continuous to the tower base and back. The resistance of the entire circuit to and back from the tower base should be comparable to that calculated from wire resistance values found in textbooks.

Another, although time consuming, method of evaluating ground systems is to make field strength measurements to determine the unattenuated field strength at 1 km. The unattenuated field strength is determined by analysis of field strength measurements made between 0.5 and 3.0 km. Measurements should be made at intervals as close as possible to every 0.2 km. The data are plotted on log–log paper and compared to a family of curves found in the FCC Rules. Field strength data may also be provided to a qualified consulting engineer for analysis and recommendations.

It is important to pay particular attention to that portion of a ground system that is installed in a marshy area or in areas of clay, peat, or other expansive soils. These areas move more during freeze and thaw cycles, and the periodic movement of the conductors can cause work hardening and breakage. Stations in the northeast part of the United States appear to suffer more ground system deterioration than other areas due to acidic soil caused in part by acid rain. Acid soil conditions can be readily identified by soil testing and the presence of acid loving plants. A local agricultural extension agent can be consulted for help in remedying acid coils conditions.

TRANSMISSION AND SAMPLE LINES

Transmission lines are often buried and therefore "out of sight, out of mind." Little can be done to inspect buried transmission lines visually. The polyethylene-like material used for modern coax jackets are suited for direct burial if not damaged during the burial process. Cables should be buried below the frost line in soil that is free of sharp rocks or rubble. If stone layers such as shale are present, the trench should first be backfilled with a 6 inch layer of clean sand, followed by the cables and another cushioning layer of sand. Sharp stones should be removed from the backfill material to the extent possible. Where the budget allows, installation of large diameter conduits is preferred. The additional material expense will likely be a small percentage of the cost of the excavation.

Cable Fault Location

A broken outer conductor of a transmission line is very difficult to locate. If this is suspected, simple DC continuity measurements may be of limited value in locating the fault due to the existence of numerous parallel ground paths. A time domain reflector (TDR) can be used for fault location, but the operator should be experienced in its use. If an incorrect velocity factor is used with a TDR, the distance to the fault (in feet or meters) will be wrong. It is important to measure the distance to the fault from both ends of the cable and use that information to confirm the fault location proportionally. If the cable is determined to be open or shorted by an ohmmeter, an RF bridge can be used to locate the fault as described below. Current leakage through the dielectric of a cable can be quantified with a high voltage leakage detector, known as a "Megger." This device applies several hundred volts across an open line and measures the current. A good cable will have leakage in the range of a few microamperes or tens of microamperes.

The reactance values presented by a resonant transmission line are important in antenna array analysis, design, and troubleshooting. These principles are explained in depth in various books on antenna and transmission line theory as well as in the *ARRL Radio Amateurs Handbook*, and some are explained below.

If a transmission line is terminated in a short, the short will repeat itself every half wavelength. When using an impedance bridge for these measurements, the reactance at resonance will be zero but there may be a few ohms of resistance due to the resistance of the copper conductors. A short length of copper strap should be used to short the line at the far end. A clip lead is not suitable for this application. Look for the lowest frequency at which the impedance bridge indicates zero reactance. The length of the line under test, in degrees, can be calculated as a simple ratio:

Null frequency (kHz)/180 = Operating frequency (kHz)/Line length in degrees

Likewise, an open line that is one-quarter wavelength long will appear as a short at the far end. The ratio for this calculation is:

Null frequency (kHz)/90 = Operating frequency (kHz)/Line length in degrees

Pressurized lines will help to keep moisture out and provide more reliable operation. While foam dielectric lines are generally less troublesome they can create problems if the foam becomes waterlogged. In addition to a decrease in insulation resistance, water in the foam will change the dielectric constant and therefore the electrical length of the line. Since dielectric constant can not always be reliably determined with a TDR, impedance bridge measurements as described above are usually more accurate. A bridge can be used to determine the actual characteristic impedance of a line as well. The line is measured both shorted and open at any convenient frequency. The frequency selected should not be near the resonant frequency of the cable under test. The open and shorted resistance and reactance values are first converted to polar form. An inexpensive hand calculator can be used for the conversion.

Cable Test Example

A length of old RG-11 cable was measured both open and shorted at 1.5 MHz. The reactance readings were normalized for frequency with the bridge in use and the results were as follows:

- Open resistance, 4.7 ohms
- Reactance, –j105 ohms
- Shorted resistance, 2.1 ohms
- Reactance, +j56 ohms

Using the handheld calculator, the rectangular values above were converted to polar values as follows:

- Open, 105 at –87°
- Shorted, 56 at +87°

The polar values are multiplied using standard polar convention by multiplying the magnitudes (105 × 56 = 5880) and adding the angles (–87 + 87 = 0). Because the angles cancel, they can be ignored and the characteristic impedance of the cable is found by taking the square root of the magnitude (5880), which yields a characteristic impedance of 76.68 ohms. This is close to the characteristic impedance (75 ohms) published for this cable by the manufacturer. When making this or any other RF bridge measurement, it is important to use a very short conductor when short-circuiting the cable. A small piece of copper strap or braid is preferred over a clip lead.

Sample System Issues

As often as not, perceived array problems may actually be sampling system problems. If the monitor point values are within limits and the common point is presenting the proper load to the transmitter, chances are the improper antenna monitor values are due to a failure of some part of the sample system. Antenna monitor values are meant to be an indication of the fields in the individual elements of an array. Samples from each tower are extracted either by loops mounted on the tower or by toroids that sample the current at the base of the tower. Toroids, although popular due to their simplicity, are inappropriate in

the case of electrically tall towers and towers that support other systems such as FM antennas. This is because the toroid cannot distinguish between the current flowing in the tower and the current flowing in an isocoupler or other circuit element that is shunted across the tower base.

Samples from the towers are fed to the antenna monitor via lengths of small diameter coaxial cable. Specifications for antenna sample systems can be found in the FCC Rules as well as FCC policy statements on good engineering practice. Sample lines should be of equal length to preserve the phase relationships of the towers with those indicated on the monitor. In an array where the sample system has been properly designed, installed, and operated, there will be close agreement between the monitor values and the theoretical array parameters.

Difficulties with sample systems employing loops are usually mechanical in nature. Broken leads on the sample loops or poor connections to the loops are often the cause of incorrect or intermittent measurements. The hardware used to mount the coax connector to the loop often comes loose due to vibration caused by wind. Circular breaks of the outer conductor may also occur. These breaks are often hidden under the polyethylene jacket of the cable. An open sample line outer conductor in the vicinity of the loop will cause a higher than usual voltage to be produced because the effective area of the loop is increased. Because unterminated sample lines can develop voltages high enough to cause painful burns, be careful when handling them.

If the reference tower sample is missing or very low, the antenna monitor indications of the other towers may be random and variable. If the reference sample is high due to a poor cable outer connection, the higher voltage present will make *all* the other ratio indications low in amplitude, but the phase indications will be correct or nearly so. Because the antenna monitor is essentially an RF voltmeter with phase measuring capabilities, a dual-trace oscilloscope can be substituted for troubleshooting. The voltages indicated on the scope can be compared to find the voltage ratio values, and the time difference between zero crossings can be used to mathematically determine phase.

A length of flexible coax and an RF oscillator can also be used to determine proper antenna monitor operation. The oscillator is fed to the flexible cable and then to the reference input of the monitor through a tee connector. The other end of the coax is attached to the other monitor inputs in sequence. The loop ratio indication of the reference and tower under test should be the same within a few percent. The tower under test will have a negative phase indication that is related to the length of the flexible cable. A 100 ft length of polyethylene dielectric coax (velocity factor = 0.66) is about 55° long at 1 MHz or 83° long at 1.5 MHz. When this method is used, the monitor channel being tested will lag the reference by these values. In the case of those antenna monitors that are made to operate on only one frequency, the numbers given above can be scaled to find lag at the frequency of interest.

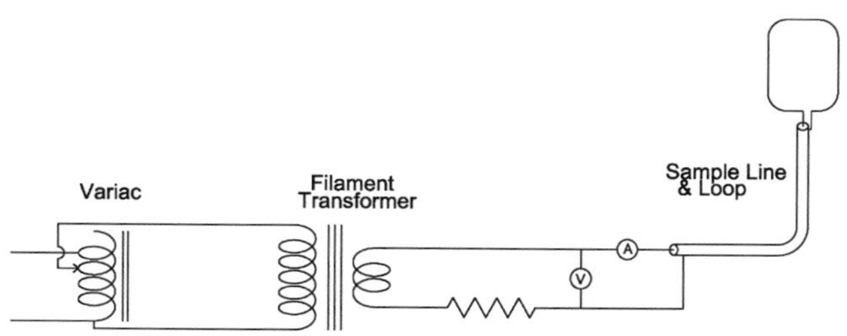

FIGURE 4.6-6 Sample line variable voltage and current test system.

Aging of power supply filter capacitors and the failure of mercury wetted relays are common problems in antenna monitors, along with lightning damage. Although power supply problems can usually be repaired locally, returning the monitor to the manufacturer for repair and calibration may be the best choice.

Although a simple continuity check can be used to test the integrity of a sample loop and line, a traditional ohmmeter is often affected by RF energy in the vicinity. A simple alternative involves the use of a transformer with a low voltage secondary, such as a filament transformer, an AC ammeter, or a Variac, as shown in Figure 4.6-6. The primary of the transformer is connected to the AC line through the Variac and the secondary is connected across the sample line at the transmitter building. A current limiting resistor of one or two ohms can be connected in series with the secondary as well. The Variac is adjusted to produce 2 or 3 amps of current flow in the secondary. The secondary voltage of the transformer is then read and the resistance of the loop and line calculated using Ohm's law.

The resistance of the loop and line should be in the low single ohms range. More importantly, if all sample lines are the same length, as is usually the case, the resistance readings should be very close to the same value. A significantly higher resistance on one line indicates a problem with a connector, a broken outer conductor, or a poor mechanical connection on the loop or connector.

Sample lines used along with toroids can be checked for continuity by shorting the far end with a good low resistance conductor and checking resistance of the line. Toroids can be checked by moving them around in the system to determine if an abnormal reading follows one particular toroid. Most toroids used for sampling employ an internal termination resistor that can be damaged by lightning. Toroids are not field repairable and should be sent back to the manufacturer for repair and calibration. When removing and replacing toroids, note the arrow stamped on top of the case. Although reversing the toroid will not make a difference in the sample amplitude, the phase will be reversed by changing the current sense relationship between its winding and the conductor passing through it.

BASE CURRENT METERS

The FCC Rules no longer require base current meters for AM directional stations, but they are still an effective troubleshooting tool. Provision should be made for measuring base currents in newly designed systems. Meter jacks should be present at the output of each antenna tuning unit, and appropriate meters should be purchased and kept safely in the transmitter building. Meters stored this way will be safe from lightning, and comparisons can be made with currents measured when the array is built and tuned. If the budget allows, toroidal meters can be installed in each antenna tuning unit.

MONITOR POINTS

Monitor points are selected on each protected radial in a directional array when its initial proof of performance is conducted. Points are selected to be indicative of the signal strength on the radial in question. Points that were selected many years ago may no longer be suitable due to changes in the electromagnetic environment. If the field strength at a monitor point is above the licensed limit, further measurements should be made before any adjustments are made. Eight to ten additional points should be measured on the radial and the measurements compared to previous measurements. If the entire radial is found to be above its construction permit limit, measurements should be made on the other monitor point radials to establish a baseline for readjustment. Although this may seem time consuming it is important to note that parameter adjustments affect radiation in all directions, not just the direction with the abnormal measurement. The concept can be visualized by putting a rubber band on a table. If you take your forefinger and thumb and squeeze on the rubber band, it will get smaller where the pressure is applied, but bulge out in other directions.

In the past, the "talk-down" method of pattern adjustment was used by many broadcast engineers. With this method, observers with field meters and two-way radios were placed on each of the monitor points and the parameters were varied via the phasor controls in an attempt to bring all monitor points below their prescribed limits. Often these phasor adjustments were random, and the intentional adjustment resulted in unintentional changes to other parameters that went undetected and countered the desired result. An important rule in any kind of adjustment technique is to change only one parameter at a time and evaluate the results of that change. If any ratio parameter has moved more than 5% from its licensed value or any phase moved more than 3°, a partial proof of performance will have to be conducted in order to license the new parameters. The partial proof of performance process is discussed later in this chapter, as is the vector analysis method of adjustment.

RE-RADIATORS AND THEIR TREATMENT

Before any attempt is made at readjustment, the area around the array should be scouted for new tower or power-line construction. These re-radiators are usually close to the array if they have enough influence on the array to raise the level of a whole radial. Re-radiators are usually found in the main lobe of the array, where they are illuminated with a high signal level as opposed to in the null area, where levels are much lower. If the re-radiator is within a few wavelengths of the antenna, it is possible but not probable that the antenna monitor values have changed. Re-radiators that are located in pattern minima may affect the readings obtained at a monitor point but usually will not move an entire radial above its limit.

The increase in the number of cell and other communications towers built in the last 20 years has had significant impact on AM directional arrays and the engineers who maintain them. FCC Rules require owners of these structures to notify broadcasters when they are built within 1 km of non-directional antennas and 3 km of directional facilities. Measurements must be made both before and after construction of the new tower to ensure that the antenna pattern of the station has not been affected. The FCC policy is to require a partial proof of performance to be conducted on directional stations. In the case of a non-directional station, 8 points should be measured on 6 equally spaced radials between 3 km and 16 km. Measurement programs consisting of fewer measurements are often proposed to broadcasters, but these should be evaluated on a case-by-case basis, and the counsel of a qualified consulting engineer may be in order to ensure that all relevant factors are taken into account. The directional station will have to cooperate in any measurement process by operating with its night pattern during daytime hours. Before and after measurements should be conducted within as short a time span as possible so changes in environmental conditions do not affect the results. The process can be expedited by supplying the person making the measurements with copies of relevant measurement locations found in proofs as well as copies of maps used previously.

Identifying Re-Radiators

AM antennas are often located in places where extensive development may be taking place. Their environment is made up of the ground on which the antenna is built, the surrounding terrain, the ground radial system that is installed, and other nearby manmade structures that may be new or modified. The likelihood of significant re-radiation is related to the height and effective radius of any new metallic structure and its proximity to the array. A method of determining if an object is re-radiating significantly is the use of a field strength meter in the vicinity of the suspected structure. This method takes advantage of the highly directional characteristics of the loop antenna of the instrument. A point is located a distance away from the suspected structure approximately equal to the

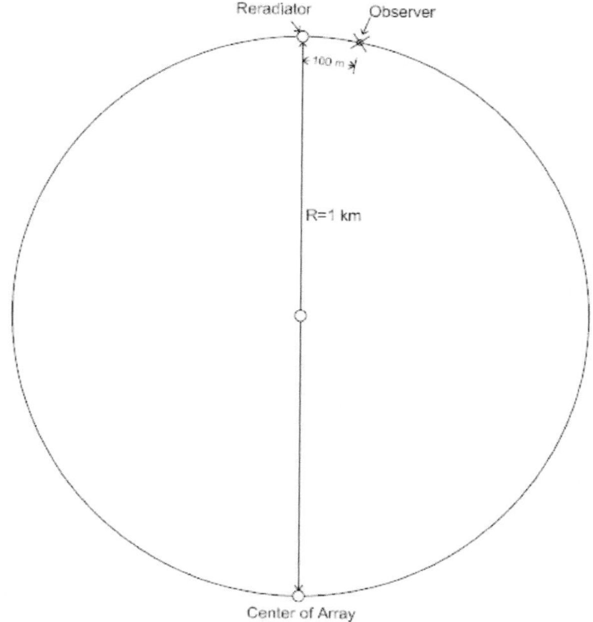

FIGURE 4.6-7 Establishing a measurement location for a re-radiating structure.

height of the structure. The point should be located such that when the field strength meter antenna is pointed directly at the array, it is oriented orthogonally, or 90°, off-axis of the re-radiating object, as shown in Figure 4.6-7.

If the observer and the re-radiating object are on the circumference of a circle with its origin on a line through the center of the array and the re-radiator, and, with the radius equal to one-half the distance between the two, the angle between the re-radiator and the array at the observation point will always be 90°. This places the minima of the loop antenna of the field intensity meter at 90° to the undesired signal. It is assumed the distance from the observation point is approximately equal to the height of the re-radiator and that this distance is on the order of 5% or less of the distance to the center of the directive array.

For this example, assume that the suspected re-radiator is 100 meters tall. The observation point is located on the circumference of a circle whose radius is 1 km, 100 m from the re-radiator, and 2 km from the center of the array. Because radiated fields in antenna systems are normalized to 1 km for comparison purposes, the measurements in this example will be normalized to the unattenuated field at 1 km. The meter is pointed at the center of the array, and a field strength of 100 mV/m is measured. When normalized to 1 km, the unattenuated field strength on this azimuth would be 100 mV/m × 2.0 = 200 mV/m. The meter is then turned toward the re-radiator (100 m from the monitor point) and 30 mV/m is measured. Normalized to 1 km, this represents an unattenuated field of 30 mV/m × 100 m/1000 m = 3.0 mV/m.

Inexpensive versions of the MININEC electromagnetic modeling program are now available. Versions available to the amateur radio community are based on the same numerical code as professional versions and can be helpful to the station engineer in assessing the effects of re-radiating structures. A model of the array can be constructed using the theoretical array parameters found on the station license, and the pattern can be calculated and presented graphically. An additional element driven at zero field can be added to the model, representing the suspected re-radiator and the resulting pattern examined for distortion.

Responsibility for the installation and continued maintenance of detuning hardware lies with the owner of the re-radiator. When changes are made on the structure, the measurement process should be repeated. Even though the physical height of the structure may not be increased, additional platforms and antennas are often added. These additions can be considered as top loading, which can increase the electrical height of the towers. It is a good practice to drive the area within 3 km of the antenna system on a regular basis in order to be aware of any construction that may alter the electromagnetic environment.

PROOF OF PERFORMANCE PROCESS

Arrays that have performed properly for many years often reach the point of requiring readjustment due to cumulative changes in the electromagnetic environment that cannot be controlled or corrected. These are usually related to development in the area of the array. What was once an area surrounded by farms and open fields may now be a commercial or residential neighborhood. Often the topography of the area is changed by moving hills and rerouting streets and power lines. The first step in the process is to gather all available proofs and construction notes for study. If these documents have been lost in the transfer of stations from owner to owner, copies can be obtained from the FCC files through the consulting engineer or a researcher familiar with the process. The end result of the proof process is to bring the radiation on the FCC designated monitored radials below the limits prescribed in the construction permit. The most recent full proof of performance conducted on the station will include these limits. This information can also be extracted from the FCC databases by a qualified consulting engineer.

When the radials to be measured have been identified, 10 to 12 points between 3 km and 16 km should be measured on each radial. Points used in the last full proof should be used if at all possible to simplify the measurement and analysis process. Descriptions of measured points in old proofs may no longer be valid. Locations such as "by the marked tree" or "across from the yellow house" are ambiguous and may no longer be valid. While it may appear simple to use a handheld GPS to find distance and bearing, this method should not be followed without study. Maps that were prepared years ago may be inaccurate compared to today's standards. A radial line drawn diagonally from map to map, out to a distance of 20 or 30 miles may be off by several degrees. Likewise, the distance to points on the maps further out may contain cumulative errors amounting to several tenths of kilometers. The original maps are best if they are available, or refer to the reduced scale maps found in the original proof. The most important consideration is to be where you believe the original measurement was made, even if that point is of a lesser degree of accuracy than is possible today. Once locations have been determined, a notation of that location via global positioning system (GPS) coordinates is helpful in relocating the point in the future. The written description should be updated as well as using descriptions that include house numbers, utility pole numbers, or references to objects not likely to move, such as road intersections, culverts, and bridges.

There are several mathematical ways to compare the data gathered to the existing proof data. Directional measurements taken can be "ratioed" to the directional measurements in the original proof for each point measured. These ratios can be further analyzed by either of two methods. The mathematical average of the ratios can be found, or the log of each ratio can be found and the mean of the logs determined. The antilog of this value is then found. The average ratio or the antilog is multiplied by the analyzed directional field strength found in the reference proof. The resulting field strength found should be below the limit for that radial, also found in the proof. Alternatively, the directional field strengths found can also be "ratioed" to the non-directional field strengths measured in the original proof and the analyzed ratio multiplied by the unattenuated, non-directional field strength at 1 km found in the original proof. This value is then compared to the directional limit found in the proof.

Proof of Performance Rules

The FCC Rules on proofs have been simplified in the past few years, and new measurement options are now available. In cases where original measurement points cannot be positively identified or have become unsuitable due to local environmental changes (such as the installation of power lines), new measurement points can be added. The added points must be measured both non-directionally and directionally. Because only one method of analysis in a partial proof may be used, the addition of points means the non-directional-to-directional analysis must be used. In the case of points previously measured, the option of using the originally measured non-directional value or repeating the non-directional measurements can be chosen. The FCC Rules require that a minimum of eight directional points be measured and analyzed on each monitored radial. The designated monitor point must be one of the points measured. A minimum of four radials must be measured. In cases where there are less than four monitored radials, radials adjacent to the monitored radials have to be measured to meet the requirement of four measured radials. The points measured should be between 3 and 16 km from the array. In the past, the practice was to compare partial proof data to previous partial proofs. That approach is no longer allowed, and reference must be made to the most recent full proof. In cases where features of the land have changed significantly, along with the electromagnetic environment, a new full proof may be indicated. Although an experienced station engineer can make the measurements required, analysis is best left to a qualified consulting engineer.

Full Proof of Performance

The first step in a full proof of performance is to make new non-directional measurements, including near-field measurements, also known as "walk-ins" because they are measured by walking rather than driving to each closely spaced point. Non-directional measurements are made, beginning at a distance equal to 10 times the distance equal to the spacing of the farthermost towers in the array; for example, if the farthermost towers are 100 m apart, the first measurement would be made at a distance of 1000 m. From this point, a measurement is made every 0.2 km out to 3 km. This would require 11 measurements. Due to access problems encountered today, the Rules require that a minimum of 7 measurements be made in this region. Efforts should be made to measure the maximum number of points, spaced as closely to every 0.2 km as possible (a minimum of 7 points). From 3 to 5 km, a measurement should be made every kilometer (a minimum of 3 points). From 5 to 15 km, measurements should be made every 2 km (a minimum of 5 points) for a total of 15 non-directional points. A concerted effort should be made to measure as many points as possible in each range so sufficient data are available to conduct a partial proof if access to some of the original measurement points is lost.

A full proof must include a minimum of 6 measured radials for simple arrays; a maximum of 12 radials can be required for complex arrays. One radial measured must be in the main lobe of the pattern. Additional radials are designated in the construction permit as monitored radials. If the requirements of the construction permit are met, along with measurements in the main lobe, additional radials must be measured to meet the minimum six radial requirement. No two radials can be more than 90° apart. In the case of complex patterns, a maximum of 12 radials will be required, and the concept of pattern symmetry is used to ensure proper operation of symmetrical patterns. The FCC Rules have been simplified and are quite clear. An additional reference that explains the rationale behind the Rules is FCC Report and Order FCC 01-60, MM Docket No 93-177. This document can be found at http://www.fcc.gov/Bureaus/Mass_Media/Orders/2001/fcc01060.doc.

FIELD INTENSITY MEASUREMENTS

STATION:		AZIM:				ENG:	
FIM:		KHZ:					
SERIAL:						TRIAL:	

POINT #	DIST km	DATE NDA	TIME NDA	mv/m NDA	DATE DA	TIME DA	mv/m DA	POINT DESCRIPTION
4 [MP]	1.8	9/2/64	1353	45.0	11/3/04	1216	8.4	40 01 44.9N 76 21 12.6W On Rock, Country Club entrance.
5	2.75	9/2/64	1341	30.0	11/3/04	1228	5.8	40 01 02.8N 76 21 40.8W Private Rd @ end of left curve, North side of road.
6	3.44	9/2/64	1326	18.8	11/3/04	1234	4.6	40 00 25.8N 76 21 59.7W In front of Toymaker farm house.
7	3.89	9/2/64	1318	14.0	11/3/04	1237	3.1	40 00 05.8N 76 22 13.1W Old Farmhouse Road, between 2nd & 3rd power pole
8	4.57	9/2/64	1310	9.7	11/3/04	1243	2.3	39 59 33.5N 76 22 31.8W Entrance to Windhaven farm...238 Letort Rd.
9	5.29	9/2/64	1300	6.5	11/3/04	1249	1.5	39 58 59.1N 76 22 52.6W MB 328 Owls Bridge Road, Middle of the road.
10	6.39	9/2/64	1254	4.7	11/3/04	1254	1.6	39 58 10.1N 76 23 21.4W Indian Run Road, crest of the hill, West side.
11	7.30	9/2/64	1244	3.4	11/3/04	1259	1.3	39 57 24.6N 76 23 51.2W Safe Harbor Road, West side, across from farmhouse curve.
12	8.16	9/2/64	1235	2.0	11/3/04	1304	0.7	39 56 45.0N 76 24 12.6W Pittsburg Valley Road, North side, behind grotto.
13	8.83	9/2/64	1225	1.4	11/3/04	1319	0.7	39 56 12.6N 76 24 30.3W Oak Road, 500' west of Observation Road @ gate with posts, South side of road.

FIGURE 4.6-8 Typical field-measurement sheet.

DOCUMENTATION

The need for accurate and complete recordkeeping is as important as the measurements made in the field. Many forms and formats have been devised over the years, but they all most include the same information. The form shown in Figure 4.6-8 was created in Microsoft® Excel® and contains all the needed information in a form that is convenient for analysis. This electronic form allows data to be easily transmitted to a consultant or anyone doing the analysis and directing the proof effort. By using the Analysis Pak add-in of Excel, the sheet can be made to automatically ratio the data and report the results.

ARRAY ADJUSTMENT

Much progress has been made in the pattern adjustment process in recent years. Moment method modeling using the Numerical Electromagnetics Code (NEC) or the Mini Numerical Electromagnetics Code (MINI-NEC) can now be used to place sample elements correctly, resulting in sample phases and ratios much closer to indicating the true fields in array elements. Through the use of moment method analysis, operating parameters are determined and sample loop placement is based on the actual current distribution in the individual elements of the array as opposed to the sinusoidal current assumption that was made in the past. Many new arrays are being built today, adjusted to the modeled parameters, and found to be in adjustment.

Instead of placing the sampling loops one-quarter wavelength down from the top of the tower, as was done in the past, most loops are now mounted approximately one-third up the tower from the base. At this location, the sampling system indications are very close to the theoretical parameters of the array. Sampled parameters in arrays that use toroid sampling generally vary from the theoretical and are determined by tower height. NEC computer modeling of the array will detail the current distribution of each element. The current distribution is a function of operating parameters and may not be the same for each mode of operation.

Chapter 4.3, "AM Antenna Systems," presents the pattern shape formula. This formula will calculate the theoretical field strength in any direction from the array. It makes use of vectors that represent the fields from the individual array elements. The combination of these vectors represents the far-field radiation in the direction of interest.

PATTERN CALCULATOR

Antenna Parameters							
Tower	Field	Phase	Spac.	Orien.	G/A	B	TRS
1	1.000	0.0	0.0	0.0	75.1	5.4	0
2	0.885	-132.2	79.1	154.0	75.1	5.4	0
3	0.789	-122.6	229.6	212.7	75.1	5.4	0
4	0.792	2.4	192.9	232.4	75.1	5.4	0
5	0.000	0.0	0.0	0.0	0.0	0.0	0
6	0.000	0.0	0	0.0	0.0	0.0	0
7						0.0	0
8						0.0	0
9						0.0	0
10						0.0	0
11						0.0	0
12						0.0	0

Power:	5	kW
Vertical:	0	deg
K:	532	mV/m
THEOrms:	661	mV/m
THEOrss:	926	mV/m
STDrms:	695	mV/m
MODrms:	695	mV/m
RSS/RMS:	1.40	
Qfactor:	23.2	mV/m
Rloss:	1.0	ohms

Single Point		Theo	Std
Azim	Elev	Field	Field
0.0	0.0	241	254

Augmentations			
No.	Bear	Span	Field
1			
2			
3			
4			
5			
6			
7			
8			
9			
10			
11			
12			
13			
14			
15			
16			
17			
18			
19			
20			
21			
22			
23			
24			
25			
26			
27			
28			

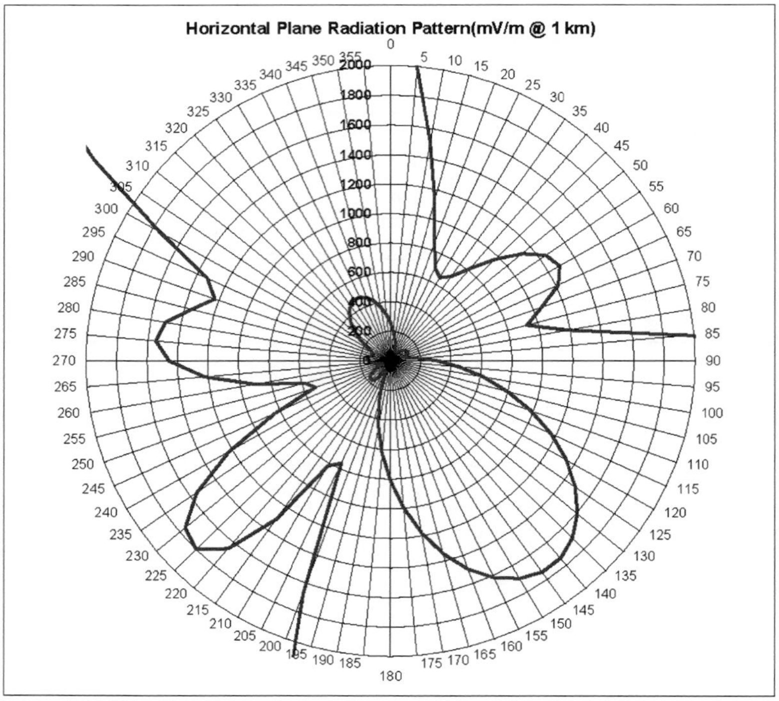

Horizontal Plane Radiation Pattern(mV/m @ 1 km)

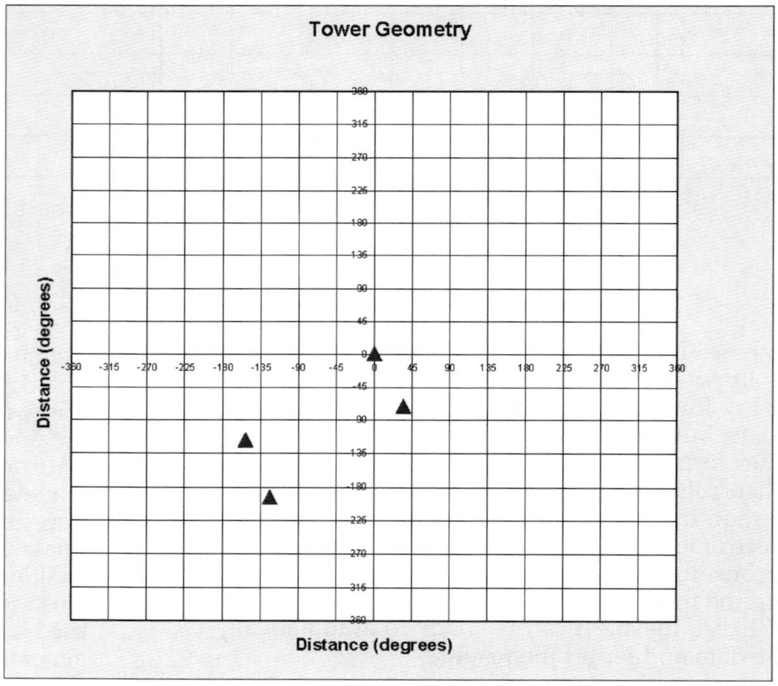

Tower Geometry

Distance (degrees)

FIGURE 4.6-9 Example of calculated pattern documentation.

Each array is defined by a contour of the radiation in all directions. Most of the energy is radiated in the main lobe which usually covers the city of license. More important in the adjustment of an array is the limiting of radiation in the pattern minima where co-channel and adjacent channel stations are located. The FCC construction permit for each station defines these protected directions and the maximum radiation in each defined direction. An example of a calculated pattern is shown in Figure 4.6-9. The pattern maxima and minima can be readily seen. This array has protected radials at 29.5°, 75°, 207°, 252°, and 291°. The field vectors on these azimuths can be calculated, and two are drawn on polar diagrams as shown in Figure 4.6-10.

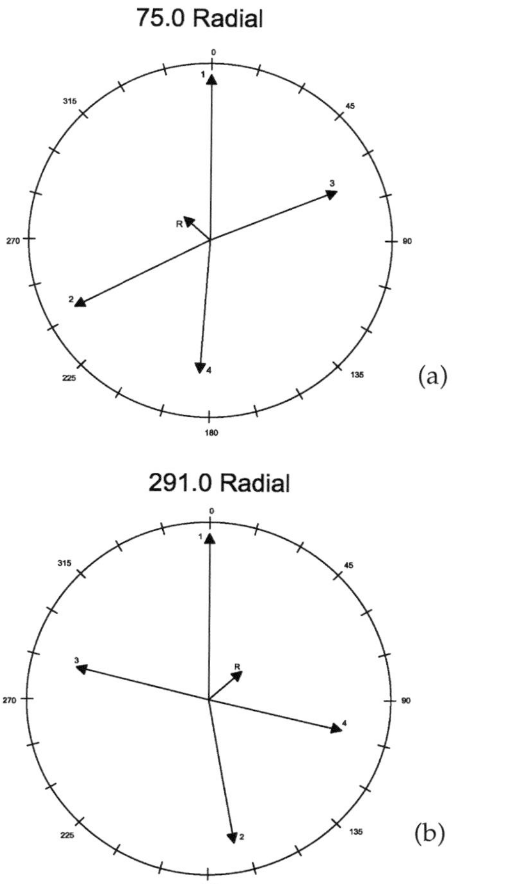

FIGURE 4.6-10 Field vectors for (a) 75° and (b) 291° radials.

For clarity, the 0° azimuth is pointed toward the top of the graph as a compass would be depicted, as opposed to the pure mathematical representation, which would put 0° to the right side of the graph. Note that these graphs depict the theoretical field vectors when the array is in correct adjustment. If the array is not in adjustment, the actual vector relationships are somewhat different, and it is necessary to reverse engineer the array to represent what has been measured in the field. Note also, that the vectors depicted here are theoretical and do not take the array environment, sample element location, and sample line length differences into account.

In the past, the practice has been to measure and compare the monitor point values, noting which go up and which go down. In today's complex electromagnetic environment, this practice can be misleading due to re-radiation and other anomalies in the vicinity of the monitor point. Five or six points should be measured and analyzed in each trial to determine the effect of the adjustment. After three or four parameter changes and analysis of the resulting fields, a familiarity develops for what parameters must be manipulated to achieve proper adjustment. Make small changes initially, on the order of 5° and 5%, to see which direction values are going. This may seem to be a tedious process, but it is a more reasoned approach given today's environment as compared to random phasor cranking with people positioned on monitor points.

One caution should be noted relating to the comparison of new to old measurements. In certain areas of the country, the Northeast in particular, the ground conductivity can change by a factor of two or more from winter to summer. This means that measurements made in the winter can be significantly higher than those made in summer. If a winter readjustment is contemplated, it would be wise to repeat the non-directional measurements as well as the directional. The ratio of these measurements would then be multiplied by the non-directional inverse distance field measured in the original proof.

Test Equipment

Measurements of AM radio frequency energy require the use of specialized test equipment. The most commonly used devices are the RF impedance bridge and an RF generator/detector. All bridges operate on the principle of a Wheatstone bridge in which unknown quantities of resistance and reactance are compared to closely controlled and known values of resistance and reactance in opposing arms of the bridge. Each type requires a signal source and a means to detect the balanced condition of the bridge.

The two common types of bridges are described by the means by which they operate. Bridges such as the General Radio Model 916 or Model 1606 are known as *cold bridges* because they are used with the circuit under test de-energized. The cold bridge is calibrated against standards of resistance and reactance before use. The unknown is then connected to the bridge, along with an RF generator and detector, and the resistance and reactance controls are adjusted until a minimum signal is indicated at the detector output. The dial indications are then read and corrected for the operating frequency in use by dividing the dial numbers by the operating frequency (in megahertz). Many of these fine old bridges can be found around transmitter sites but may be missing their operating instructions. Copies of these manuals can sometimes be found on the Internet. The other type of bridge commonly found is the Delta operating impedance bridge. This bridge is known as a *hot bridge* because it can be used in an energized circuit as long as the power applied is below the rating of the bridge. This mode of operation is shown in Figure 4.6-11.

In this example, the bridge is measuring the impedance at the input of an antenna tuning unit. Note that the J plug has been removed to allow the bridge to be inserted into the circuit. The input on this type of bridge is on the right side of the instrument, and the output is on the left. Most phasing system manufacturers strive to design their products so these bridges can be inserted without crossing the leads. For a circuit with negative power flow, the leads are connected in the opposite direction. In that case, the tuning unit

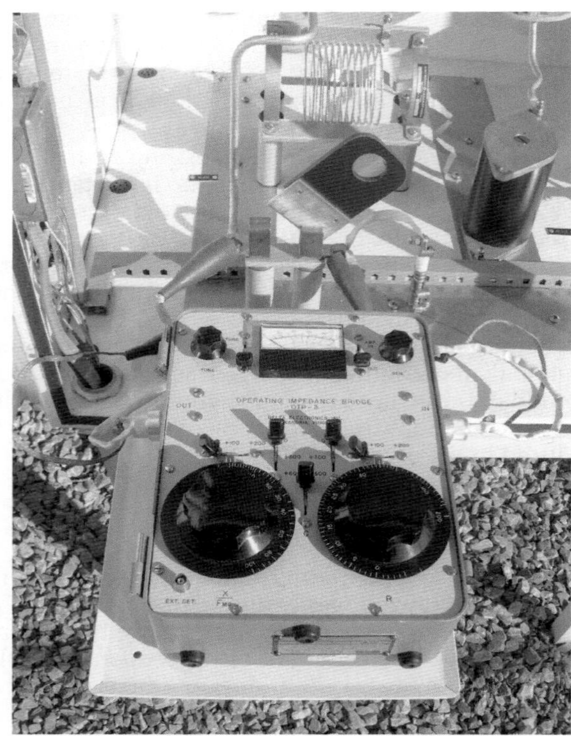

FIGURE 4.6-11 Operating impedance bridge at input to antenna tuning unit.

becomes the input and the transmission line becomes the output or load. The dial readings on this type of bridge are multiplied by the operating frequency (in megahertz) to obtain the correct values. This is the opposite of the General Radio bridge.

SUMMARY

The maintenance of an AM antenna system includes not only the transmitter and tower but all components related to the system as a whole, such as antenna tuning units, sample lines, and monitoring. The work must be meticulous and well documented. Testing takes patience and the proper test equipment. The result of conducting AM antenna system maintenance in a methodical and careful manner will be reliable and stable operation for the long run.

ACKNOWLEDGMENTS

The author wishes to thank to Ron Rackley, P.E., and Jack Sellmeyer, P.E., for their support and counsel in this effort. Thanks also to Jeff Frey of Clear Channel Radio for help in preparing the graphics.

CHAPTER

4.7

FM and Digital Radio Broadcast Transmitters

GEOFFREY N. MENDENHALL*

Harris Broadcast Communications Division
Mason, Ohio

INTRODUCTION

Although the mathematical principles explaining frequency modulation (FM) have been known for many years, the advantages and practical application to radio broadcasting were not realized until the 1930s, when Major Edwin H. Armstrong conducted extensive developmental work proving that FM radio transmissions were possible and practical. Many theoreticians claimed to have proof that Armstrong's experiments were impossible based on mathematical models claiming that an infinite transmission bandwidth would be required. He never received proper credit for his many contributions to the radio communications industry during his lifetime [1]. The advantages of FM include freedom from static, wide audio bandwidth, and the ability of an FM receiver to capture the stronger of two signals transmitted on the same carrier frequency. More recently, an in-band, on-channel (IBOC) digital radio system, also known as "HD Radio™," has been added to many FM broadcast stations [26,33,34]. This chapter will cover conventional analog FM technology, as well as digital, HD Radio transmission technology.

FCC TRANSMISSION STANDARDS

The Federal Communications Commission (FCC) regulates and enforces the technical standards that apply to radio broadcasting in the United States. In theory, this ensures that the public is provided with a consistently high standard of transmission quality from station to station. The rules and regulations covering radio broadcast services, including those for FM broadcast transmitters, are set forth in Part 73 of Title 47 of the Code of Federal Regulations (CFR), available from the U.S. Government Printing Office in Washington, D.C. The rules and regulations are changed from time to time to keep pace with new technology and changes within the broadcast industry. Every broadcast engineer should have access to a current copy of these rules and regulations so the station's technical performance is maintained within the prescribed limits.

FREQUENCY MODULATION THEORY

Angular Modulation

Frequency modulation (FM) and phase modulation (PM) are both special cases of angular modulation. In any angular modulation system, both the frequency and phase of the carrier vary with time as a function of the modulating signal. The relationship between the frequency deviation of the carrier, the phase deviation of the carrier, and the sinusoidal modulating frequency is defined as the modulation index (m), where:

$$m = \frac{\text{frequency deviation(peak-to-peak Hertz)}}{\text{modulating frequency(Hertz)}}$$

*With contributions by George Cabrera, Randall Restle, Anders Mattsson, and Dmitri Borodulin.

Because FM and PM are both subsets of angular modulation, they are virtually indistinguishable from one another except in the modulator characteristics.

In a PM system, the modulating signal causes the phase of the carrier wave to vary according to the instantaneous amplitude of the modulating signal. A phase modulator generates a constant amount of phase deviation of the carrier with a constant amplitude modulating signal, independent of the frequency of the modulating signal. The frequency deviation of the carrier produced by a phase modulator does increase as the modulating frequency is increased even though the level of the modulating voltage is held constant. The net effect is that the phase modulator behaves as if it were a frequency modulator with a 6 dB/octave rising slope on the modulating signal input.

An FM modulator generates a constant frequency deviation of the carrier with a constant amplitude modulating signal, independent of the frequency of the modulating signal. The phase deviation of the carrier produced by a frequency modulator decreases as the modulating frequency is increased even though the level of the modulating voltage is held constant. The net effect is that the frequency modulator behaves as if it were a phase modulator with a 6 dB/octave falling slope on the modulating signal input. In FM broadcasting, the RF carrier should have frequency deviation that is proportional to the amplitude of the modulating signal but independent of the frequency of the modulating waveform.

The instantaneous frequency (rate of change of phase) of the RF output wave differs from the carrier frequency by an amount proportional to the instantaneous amplitude of the modulating waveform. For example, consider a 100 MHz carrier wave frequency modulated by a 1000 Hz audio tone and assume that a 1 volt input to the modulator causes ±20 kHz of frequency deviation on the positive and negative peaks of this tone. If the audio input amplitude is increased to 2 volts, the peak deviation will become ±40 kHz varying in sine-wave fashion from one peak of deviation to the other and back again at the 1000 Hz rate. In FM broadcasting, 100% modulation results in a peak frequency deviation of ±75 kHz of the RF carrier.

When preemphasis is used ahead of the frequency modulator, the system becomes a phase modulator at audio frequencies above the turnover point of the preemphasis network. This is because the frequency response of the preemphasis network rises at the rate of 6 dB/octave above this point. FM broadcasting with preemphasis really becomes a mixture of FM at low modulating frequencies and PM at high modulating frequencies.

PRE-EMPHASIS

The standards adopted for FM broadcasting in the U.S. allow the use of preemphasis. The standard preemphasis curve is defined in FCC Rules Section 73.317(e) as an ideal resistance capacitance network with a time constant equal to 75 microseconds. The 3 dB point for 75 microsecond preemphasis is at a frequency of:

$$f = \frac{1}{2\pi(RC)} = \frac{1}{2\pi(75 \times 10^{-6})} = 2{,}122 \text{ Hz}$$

The 75 microsecond curve and the tolerance allowed by the FCC are shown in Figure 4.7-1. The frequency response characteristics in decibels for several popular preemphasis time constants are given in Table 4.7-1.

The noise voltage in a narrow bandwidth (for example, 1 Hz) increases directly with frequency; therefore, the power spectral density increases as the square of frequency. When deemphasis is used in the receiver, the noise voltage is attenuated above 2.1 kHz so it remains constant with frequency. The power spectral density is also constant above 2.1 kHz. The high-frequency noise at the receiver would be much greater without deemphasis.

Preemphasis is practical because program content energy tends to peak at several kilohertz and then falls off rapidly at the higher frequencies. For this reason, the higher frequencies may be boosted in amplitude without causing an excessive increase in modulation level. Modern audio processing equipment takes the preemphasis curve into account when controlling peak modulation levels.

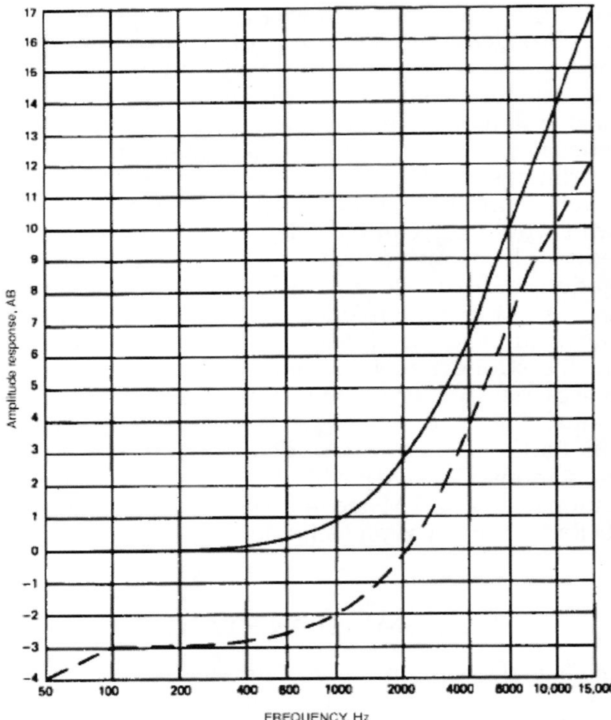

FIGURE 4.7-1 Typical 75 μs preemphasis curve (solid line) and tolerance limits between solid and dashed lines.

TABLE 4.7-1
Precise Amplitude Response Values for Various Amounts of Preemphasis

Freq (Hz)	Preemphasis Amplitude Response *Versus* Time Constant			
	25 μs (dB)	50 μs (dB)	75 μs (dB)	150 μs (dB)
50	0.000	0.001	0.002	0.010
100	0.001	0.004	0.010	0.038
400	0.017	0.068	0.152	0.577
1,000	0.106	0.409	0.871	2.761
2,000	0.409	1.445	2.761	6.583
3,000	0.871	2.761	4.769	9.540
4,000	1.445	4.115	6.583	11.822
5,000	2.087	5.400	8.164	13.656
6,000	2.761	6.583	9.540	15.182
7,000	3.442	7.661	10.749	16.486
8,000	4.115	8.643	11.822	17.623
9,000	4.769	9.540	12.785	18.630
10,000	5.400	10.362	13.656	19.534
11,000	6.005	11.120	14.451	20.353
12,000	6.583	11.822	15.182	21.103
13,000	7.135	12.475	15.858	21.793
14,000	7.661	13.084	16.486	22.433
15,000	8.164	13.656	17.073	23.029

Notes: Values in shaded area are subject to roll-off by audio low-pass filter section of subcarrier generator. Values shown in decibels are relative to $f_m = 0$ Hz.

The location of the preemphasis network in the system depends on the operating mode. Stereo transmission requires that the FM modulator have a flat response to the composite baseband signal from the stereo generator, so the individual preemphasis networks for the left and right channels are located in the stereo generator before the left and right audio channels are multiplexed into the composite baseband signal. In the case of a digital stereo generator, the Audio Engineering Society (AES) [22] serial audio data contains both the left and right channel information in alternating data frames. The digital stereo generator or the upstream digital audio processor applies the preemphasis required and 15 kHz audio low-pass filtering using digital signal processing (DSP) techniques.

FM Sideband Structure

The frequency modulated RF output spectrum contains many sideband frequency components, theoretically an infinite number. Consider, as an example, a radio frequency (RF) carrier of frequency *fc* at a frequency much greater than the modulated sinusoidal signal with frequency *fm* (*fm* << *fc*). The spectrum consists of pairs of sideband components spaced from the carrier frequency by multiples of the modulating frequency. When the modulation index is small ($m = 0.5$), the amplitude of the second and higher order sidebands is small so the output consists mainly of the carrier and the pair of first-order sidebands, as illustrated in Figure 4.7-2(A). The total transmitter RF output power remains constant for a given modulation waveform, but the distribution of that power into the sidebands varies with the modulation index such that power at the carrier frequency (and lower order sidebands) is reduced by the amount of power added to the higher order sidebands.

As the modulation index is increased (as in wide deviation FM broadcasting), the higher order sidebands become more prominent. The amplitude and phase of the carrier as well as the sidebands can be expressed mathematically by making the modulation index (m) the argument of a simplified Bessel function, as shown in Figure 4.7-3.

In a monophonic FM broadcast transmitter, the modulation index can become very high at low modulating frequencies. With a 50 Hz audio input signal of sufficient amplitude to produce 75 kHz deviation (100% modulation), the modulation index is:

$$m = \frac{75{,}000}{50} = 1{,}500$$

With a 15,000 Hz input at the same deviation (also 100% modulation), the modulation index is only:

$$m = \frac{75{,}000}{15{,}000} = 5$$

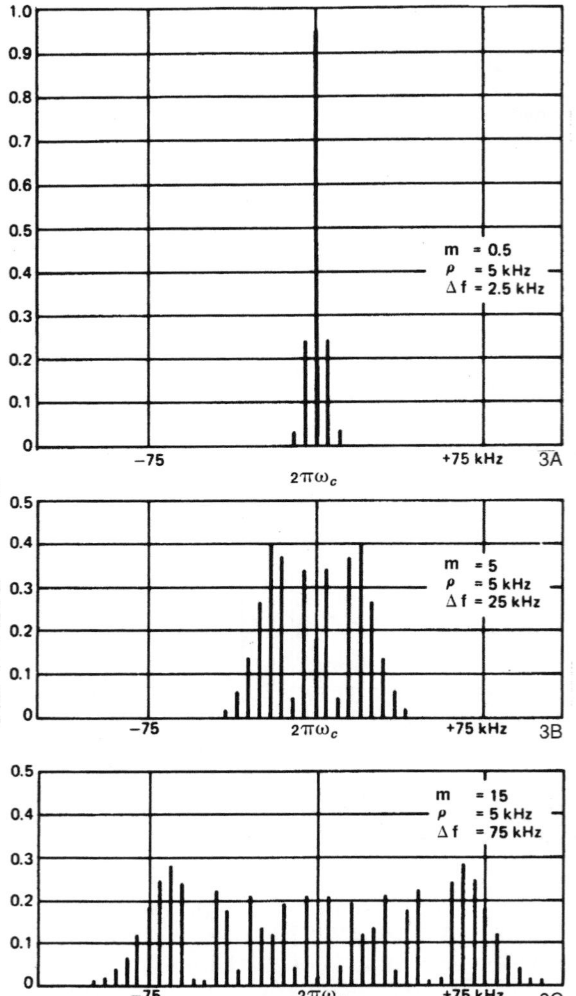

FIGURE 4.7-2 RF spectrum with modulation indexes of (a) 0.5, (b), 5.0, and (c) 15.

Figure 4.7-2 illustrates the frequency components present for modulation indices of 0.5, 5, and 15. Note that the number of significant sideband components becomes very large with a high modulation index. For a given modulation index, the total bandwidth occupied can extend beyond ±75 kHz from the carrier depending on the modulating frequency. This single tone modulating frequency analysis is useful in understanding the general nature of FM and for making tests and measurements. When audio program modulation is applied, there are many more sideband components present. They vary so much that sideband energy becomes distributed over the entire occupied bandwidth rather than appearing at discrete frequencies.

Bessel Nulls

At certain modulation indices, the amplitude of the carrier component of the signal goes to zero, with all the transmitted power being distributed at frequencies other than the carrier frequency [2]. This carrier null phenomenon is useful as an extremely accurate method for measuring the frequency deviation and to check the calibration of modulation monitors. Referring again to Figure 4.7-3, note that the carrier amplitude goes to zero and reverses sign at several values of modulation index, including 2.405, 5.520, and 8.654. Figure 4.7-4 is a photograph taken from an RF spectrum analyzer showing the first Bessel null ($M = 2.405$) of a carrier at a frequency of 100 MHz.

To determine the audio input level required to achieve 75 kHz deviation, apply an audio tone of exactly 8667 Hz (75,000 divided by 8.654). Starting from zero amplitude, increase the audio level until the carrier disappears for the third time (as 8.654 corresponds to the third Bessel null, as shown in Table 4.7-2). At this audio level, the deviation is exactly

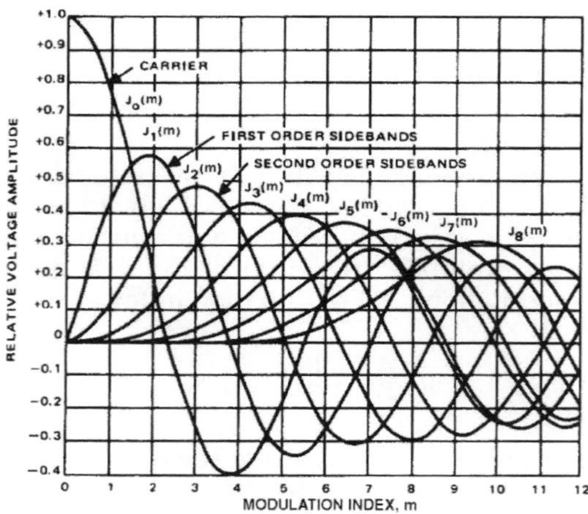

FIGURE 4.7-3 Relationship of carrier and sideband amplitudes to modulation index.

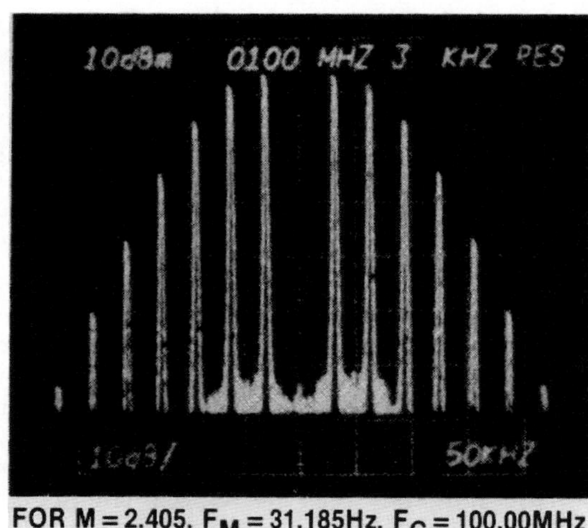

FIGURE 4.7-4 Photgraph taken from an RF spectrum analyzer showing the first Bessel null (M = 2.405) of a carrier at a frequency of 100 MHz.

TABLE 4.7-2
Sideband Nulls as a Function of Modulation Index and Modulating Frequency

Null	Modulation Index (M)		*Fm* for 75 kHz Deviation	
	Carrier	First Sidebands	Carrier	First Sidebands
1st	2.405	3.832	31,187	19,574
2nd	5.520	7.016	13,587*	10,690
3rd	8.654	10.174	8667	7372
4th	11.792	13.324	6361	5629
5th	14.943	16.471	5023	4554
6th	18.071	19.616	4150	3823
7th	21.212	22.759	3536	3295

* This tone is recorded on track 38 of the NAB Broadcast Audio System Test CD.

75 kHz. The carrier amplitude (null) detector must have sufficient selectivity to separate the carrier from the sidebands and could be a spectrum analyzer or a receiver with a narrow IF bandwidth. The FM signal can be heterodyned down to a convenient frequency for measurement. Heterodyning does not alter the modulation index; however, when a frequency (or phase) modulated wave is multiplied or divided, this also multiplies or divides the frequency deviation and the modulation index by the same amount. A listing of useful carrier and first order sideband nulls as function of the modulation index (*M*) and the modulating frequency (*Fm*) is given in Table 4.7-2.

Occupied Bandwidth

After examining the Bessel functions and the resulting spectra, it becomes clear that the occupied bandwidth of an FM signal can be far greater than the frequency deviation of the signal. In fact, the occupied bandwidth is infinite (if all sidebands are taken into account), so it is apparent that a frequency modulation system would require the transmission of an infinite number of sidebands for *perfect* demodulation of information. In practice, a signal of acceptable quality can be transmitted in the limited bandwidth assigned to an FM channel.

Effects of Bandwidth Limitation

Practical considerations in the transmitter RF circuitry and available spectrum make it necessary to restrict the RF bandwidth to less than infinity. As a result, the higher order sidebands will be altered in amplitude and group delay (time). Bandwidth limitation will cause distortion in any FM system. Consider the block diagram shown in Figure 4.7-5a, where a perfect FM modulator is connected to a perfect demodulator via an RF path of infinite bandwidth. The demodulated audio shown in Figure 4.7-5b contains no distortion components. In Figure 4.7-6a, a bandpass filter is inserted between the modulator and demodulator to restrict the bandwidth. Audio distortion products now appear at the output of our perfect demodulator, as shown in Figure 4.7-6b. These distortion products are due solely to the bandwidth restriction (300 kHz = 3 dB BW) imposed by the bandpass filter.

Figure 4.7-7 show the effects of a narrowband RF bandpass filter on the RF spectrum of a composite signal consisting of a stereophonic subcarrier modulated only on the left channel with 4.5 kHz plus a 67 kHz unmodulated Subsidiary Communications Authorization (SCA) subcarrier. In Figure 4.7-7a, through a wideband RF path, there are no baseband distortion products in the demodulated spectrum. Figure 4.7-7b shows the corresponding effects observed on the demodulated baseband spectrum for the same signal through a narrowband filter. The only distortion evident on the RF spectrogram is the loss of some sidebands greater than 150 kHz from the center frequency and some amplitude differences between the upper and lower sideband pairs. Note the creation of many undesired intermodulation terms in the demodulated baseband spectrum that cause crosstalk into both the stereophonic and SCA subcarrier bands. The change in the RF spectrum is subtle, but the resulting spectrum after demodulation is clearly affected. As a result, the distortion in any practical FM system will depend on the amount of bandwidth available as well as the transmitted modulation index.

Group Delay Symmetry *versus* Amplitude Response Symmetry

Although both the amplitude response and time response (group delay) across the FM channel have an effect on the amount of distortion added to the FM signal, the symmetry of the group delay response is more important than the total group delay variation or the amplitude response. Best FM performance is always obtained when the system is tuned for symmetrical group delay (time) response. Depending on the circuit

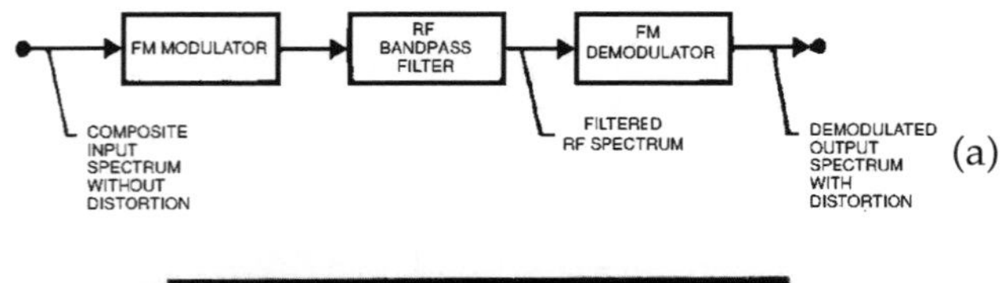

(a)

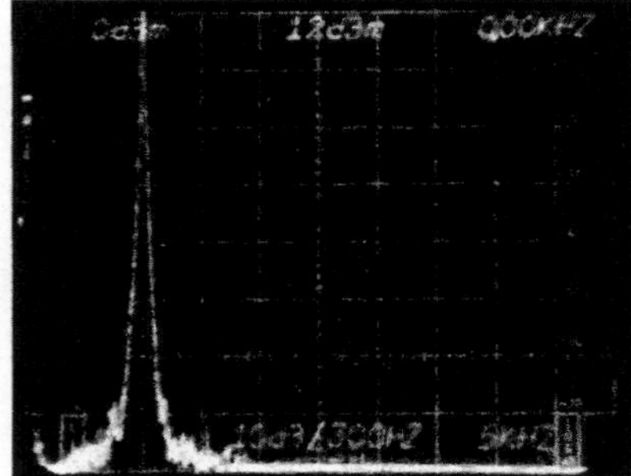

BASEBAND SPECTRUM TO FM MODULATOR

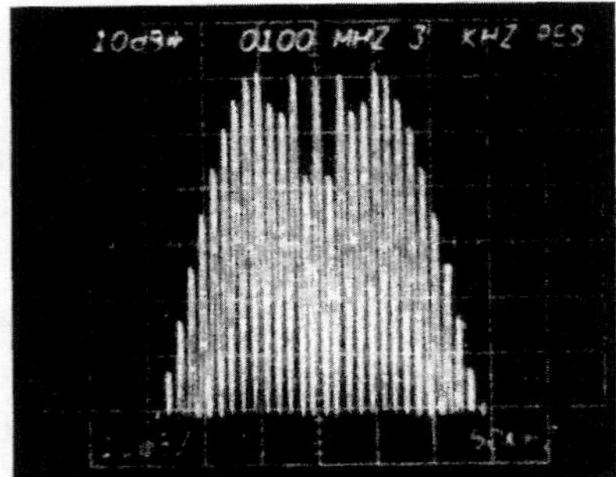

BANDWIDTH LIMITED RF SPECTRUM TO DEMODULATOR

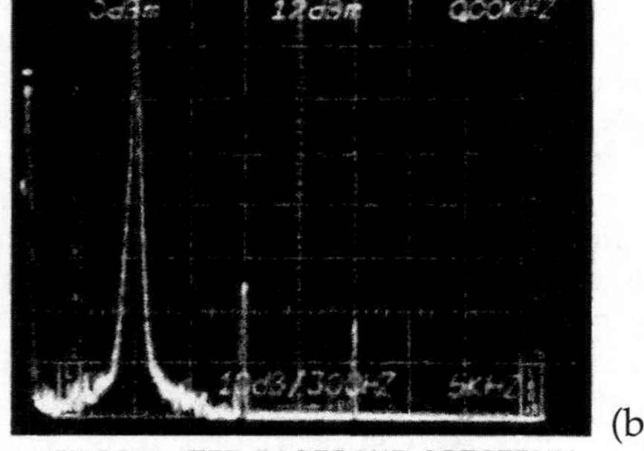
(b)

DEMODULATED BASEBAND SPECTRUM

FIGURE 4.7-5 (a) Bandwidth-limited RF path. (b) Single tone (10 kHz) modulation through narrowband RF path.

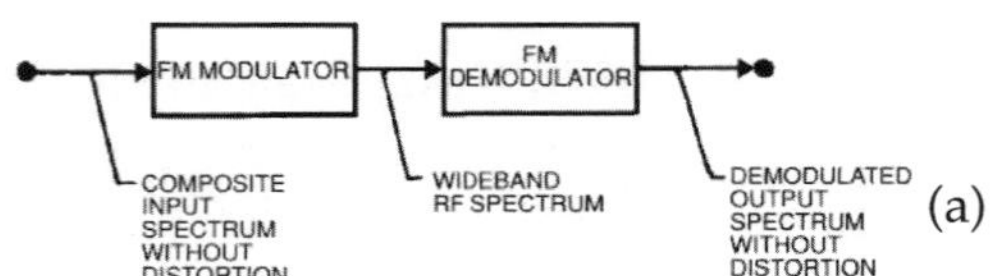

(a)

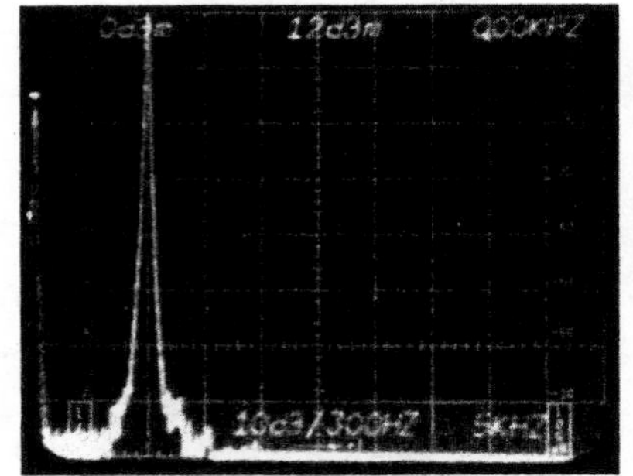

BASEBAND SPECTRUM TO FM MODULATOR

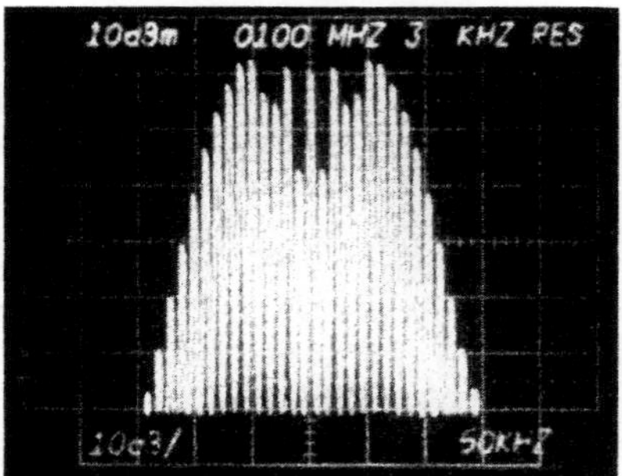

RF SPECTRUM TO DEMODULATOR

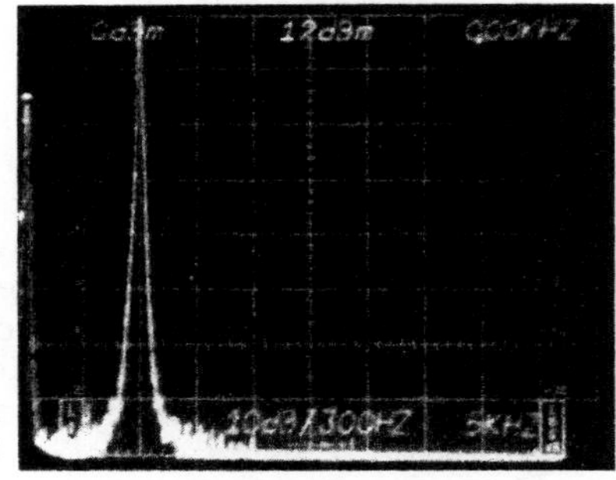
(b)

DEMODULATED BASEBAND SPECTRUM

FIGURE 4.7-6 (a) Wideband RF path. (b) Single-tone (10 kHz) modulation through wideband RF path.

topology, the tuning conditions for symmetrical group delay response may not coincide with the symmetrical amplitude response.

Limiting Factors within an FM Transmitter

Relating the specific quantitative effect of the bandwidth limitations imposed by a particular transmitter to the actual distortion of the demodulated composite baseband is a complicated problem. Some of the factors involved are:

- Total number of tuned circuits involved
- Amplitude and group delay response of the total combination of tuned circuits in the RF path
- Amount of drive (saturation effects) to each Class C stage
- Nonlinear transfer functions (AM-AM and AM-PM for HD Radio) within each amplifier stage

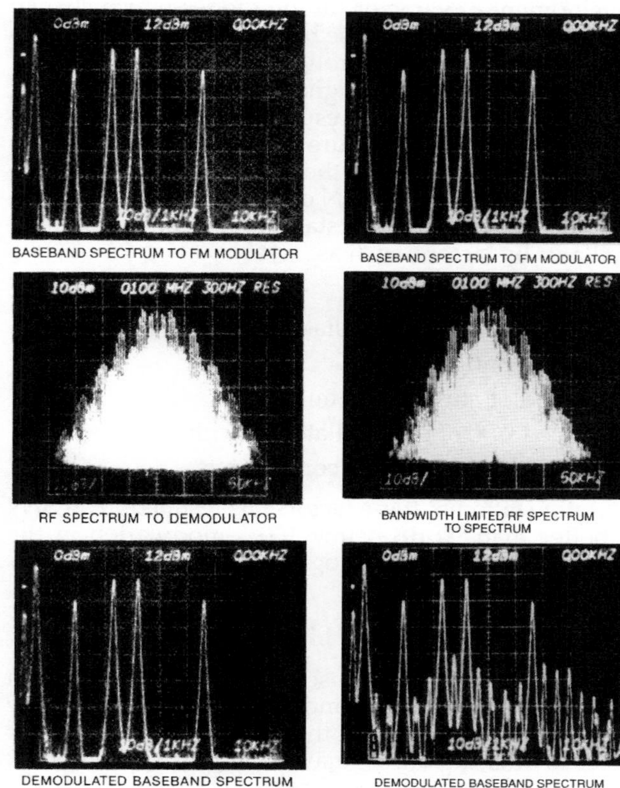

FIGURE 4.7-7 (a) Left group of images: stereo (L or R = 4.5 kHz) plus SCA (unmodulated) modulation through wideband RF path. (b) Right group of images: stereo (L or R = 4.5 kHz) plus SCA (unmod) modulation through narrowband RF path.

Improvement of the RF Path

The following design techniques can help improve the transmitter's bandwidth:

- Maximize bandwidth by using a broadband exciter and a broadband intermediate power amplifier (IPA) stage.
- Use a single-tube design or a broadband, completely solid-state design where feasible.
- Optimize both grid circuit and plate circuit of the tuned stage for the best possible bandwidth and symmetrical group delay response.
- Minimize the number of interactive tuned networks.
- Use a broadband antenna system that provides a low standing wave ratio on the transmission line.

For more information about FM and digital modulation theory, see References [2, 3, 4, 21, and 24].

FM AND DIGITAL RADIO TRANSMITTERS

The purpose of the FM transmitter is to convert a main channel audio signal and its associated audio or data subcarriers, or an AES3 serial digital audio data bitstream, into a frequency-modulated, radio frequency signal at the desired power output level to feed into the radiating antenna system. In its simplest form, the FM transmitter can be considered an FM modulator and an RF power amplifier packaged into one unit as shown in Figure 4.7-8. In fact, an FM transmitter consists of a series of individual subsystems, each having specific functions:

- The FM exciter converts the analog audio baseband or serial, AES3, digital audio data into frequency-modulated RF and determines the key qualities of the signal.

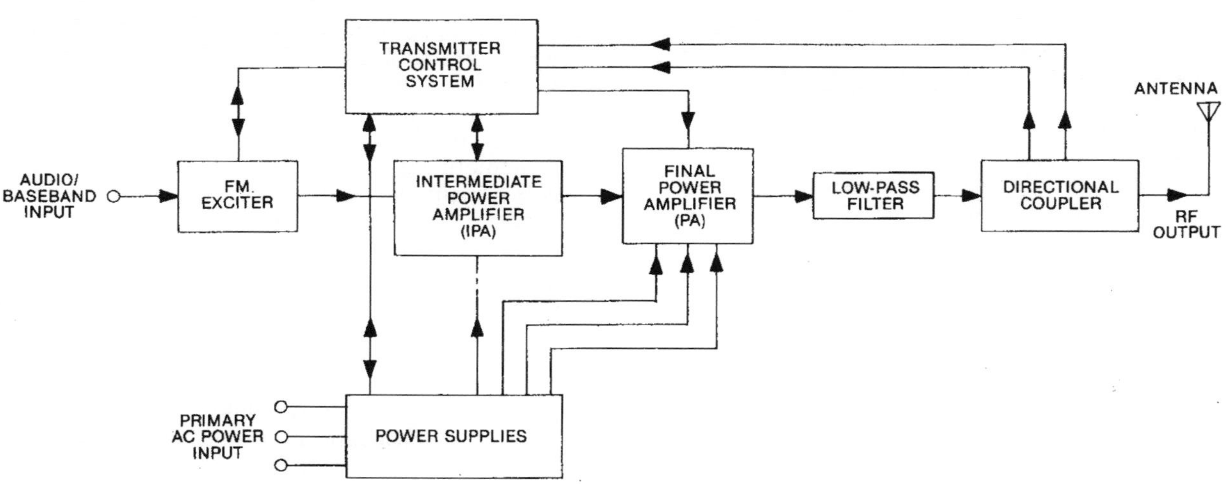

FIGURE 4.7-8 Simplified block diagram of an FM broadcast transmitter.

- An intermediate power amplifier (IPA) is required in some transmitters to boost the RF power level up to a level sufficient to drive the final RF power amplifier stage.
- The final power amplifier further increases the signal level to the value required to drive the antenna system.
- A transmitter control system monitors, protects, and provides commands to each of these subsystems so they work together as an integrated system.
- The RF lowpass filter removes undesired harmonic frequencies from the transmitter's output, leaving only the fundamental output frequency.
- A directional coupler provides an indication of the power being delivered to and reflected from the antenna system.
- The power supplies convert the input power from the AC line into the various DC or AC voltages and currents required by each of these subsystems.

Analog FM Exciter

The function of the exciter is to generate and modulate the carrier wave with one or more inputs (mono, stereo, SCA, data) in accordance with appropriate standards. Stereo transmission places the most stringent performance requirements upon the exciter. Because the exciter is the origin of the transmitter's signal, it determines most of the signal's technical characteristics including signal-to-noise ratio (S/N), distortion, amplitude response, phase response, and frequency stability. Waveform linearity, amplitude bandwidth, and phase linearity must be maintained within acceptable limits throughout the analog baseband chain from the stereo and subcarrier generators to the analog FM exciter's modulated oscillator. The introduction of AES3 [22] digital audio transport and digital FM modulation techniques such as direct digital synthesis (DDS), direct to carrier, and digital-to-analog conversion eliminate the distortions introduced by analog circuits. In a digital FM exciter, the left and right audio data is converted into a digital representation of stereo baseband by digital signal processing (DSP). This data is then further converted into a frequency-modulated carrier by a DDS numerically controlled oscillator (NCO). From here, the FM carrier is usually amplified in a series of Class C nonlinear power amplifiers, where any amplitude variation is removed. The amplitude and phase responses of all the RF networks that follow the exciter must also be controlled to minimize degradation of the signal quality.

Direct FM

Direct FM is a modulation technique where the frequency of an oscillator changes in direct proportion to an applied voltage. Such an oscillator, called a voltage tuned oscillator (VTO), was made possible by the development of varactor tuning diodes which change capacitance as their reverse bias voltage is varied (also known as a voltage-controlled oscillator, or VCO). If the composite baseband signal is applied to the tuning terminal of a VTO, the result is a direct frequency-modulated oscillator. Figure 4.7-9 is a block diagram representative of most of the modern direct FM exciters on the market. The S/N of an FM exciter is dependent on the short-term stability of the modulated oscillator by factors, such as

- Operating level
- Noise figure of the oscillator transistor
- Circuit configuration
- Method of amplitude limiting
- Loaded "Q" of the oscillator tank circuit
- Mechanical stability of components

Optimization of these factors has resulted in an S/N of better than 90 dB below 100% modulation in the current generation of analog FM exciters.

Analog FM Modulator Linearity

Nonlinearities in the FM oscillator alter the waveform of the baseband signal and create distortion in the demodulated output at the receiver. A secondary effect of this distortion may include stereo crosstalk into the SCA subcarrier signals [12]. The composite baseband signal is frequency modulated onto an RF carrier by the modulated oscillator. Frequency modulation is achieved by applying the composite baseband signal to a voltage tuned RF oscillator. The modulated oscillator usually operates at the carrier frequency and is voltage tuned by varactor diodes operating in a parallel LC circuit. To have perfect modulation linearity, the RF output frequency must change in direct proportion to the composite modulating voltage applied to the varactor diodes. This requirement implies that the capacitance of the varactor diodes must change as the square of the modulating voltage.

Unfortunately, the voltage *versus* capacitance characteristic of practical varactor diodes is not the desired square-law relationship. All varactor-tuned oscillators have an inherently nonlinear modulating characteristic. This nonlinearity is predictable and repeatable for a given circuit configuration, making correction by complementary predistortion of the modulating signal feasible. Suitable predistortion can be applied to the composite baseband signal by using a piece-wise linear approximation to produce the desired complementary transfer function. Figure 4.7-10 shows how the predistortion network is cascaded with a nonlinear voltage-tuned oscillator to produce a linearized frequency modulator.

It is also possible to improve both the linearity and S/N of the modulated oscillator by demodulating its RF output to baseband and then feeding some of this baseband with the proper phase relationship back into the composite input of the modulator. This configuration places the entire modulated oscillator within a negative feedback loop and transfers the responsibility

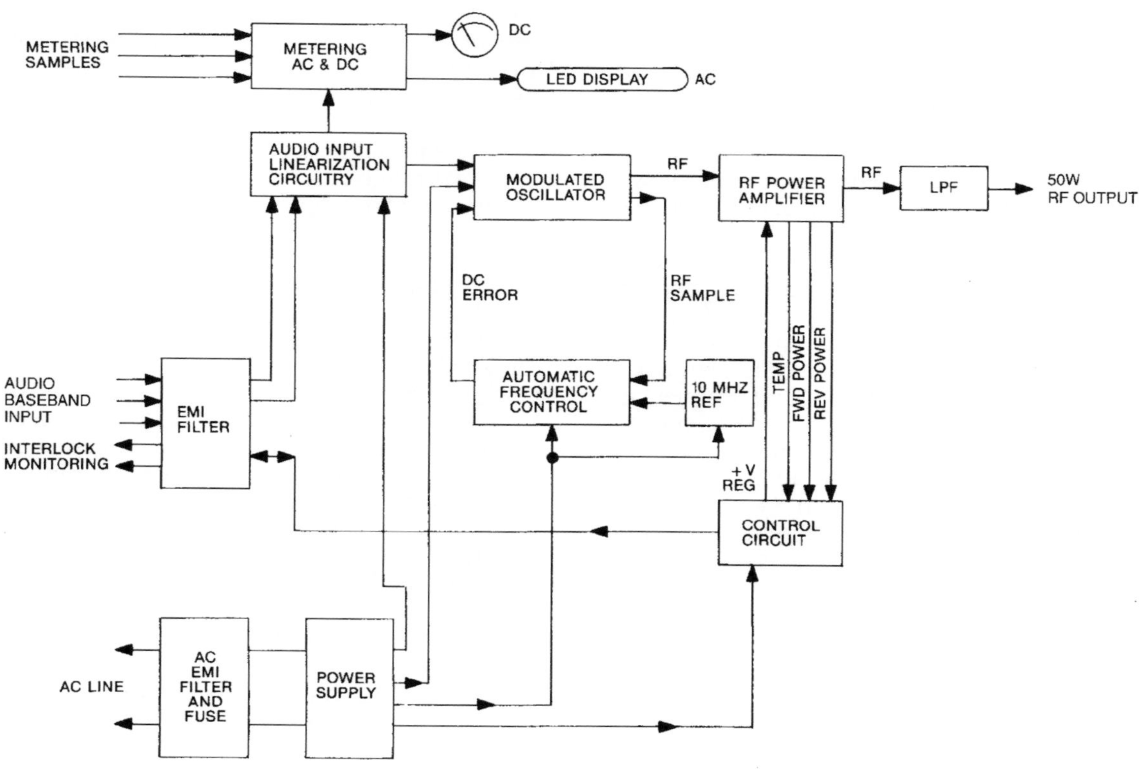

FIGURE 4.7-9 FM exciter block diagram using direct FM.

for maintaining linearity to the demodulator. Digital demodulation schemes can be made very linear, but the additional complexity and the potential problems with loop stability have limited the applications of this approach to linearization.

Analog modulator linearization has reduced harmonic and intermodulation distortion to less than .01% in the current generation of equipment. Any distortion of the baseband signal caused by the modulated oscillator will have secondary effects on stereo and SCA crosstalk, which are quite noticeable at the receiver in spite of the rather small amounts of distortion to the baseband. For example, if the harmonic distortion to the baseband is increased from .05 to 1.0%, as much as 26 dB of additional crosstalk into the SCA can be expected.

Transient intermodulation (TIM) distortion is usually not a factor in varactor-tuned modulated oscillators. The modulation bandwidth capability is generally more than ten times the composite signal bandwidth and no negative feedback is used to maintain linearity.

Ensuring that the composite baseband signal undergoes minimal distortion in the modulation process will reduce undesired harmonic and intermodulation products in the baseband, making the FM exciter transparent to the signals coupled into it. All exciter stages after the modulated oscillator operate as broadband amplifiers with minimal bandwidth limitations. Analog FM exciter technology is currently capable of transmitting near compact disc quality with less

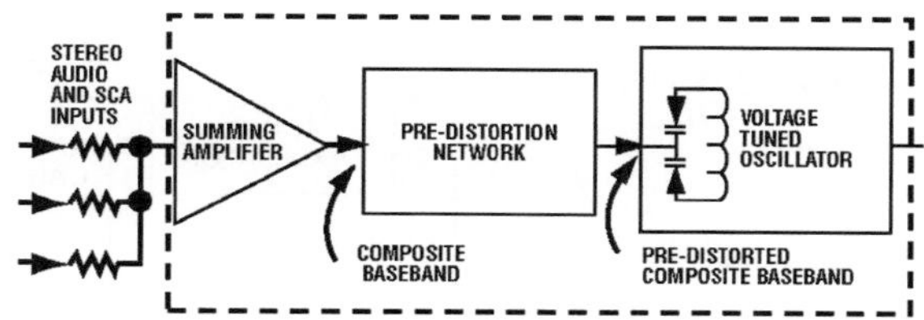

FIGURE 4.7-10 Linearized FM modulator block diagram.

than .01% distortion and an S/N of better than 90 dB. Digital FM exciter technology utilizing direct digital synthesis can exceed the limitations of analog modulators to provide the full 16-bit digital performance of a compact disc.

Phase-Locked Loop Automatic Frequency Control

The frequency stability of direct FM oscillators is not good enough to meet the FCC frequency tolerance of ±2000 Hz. This requires an automatic frequency control system (AFC) that uses a stable crystal oscillator as the reference frequency. The modulated oscillator need not have good long-term stability because the AFC feedback loop will correct for long-term drift to keep the average carrier frequency within limits. The modulated oscillator does need excellent short-term (less than 1 sec) stability because the control loop time constant must be long enough so the AFC circuit does not try to remove desired low frequency audio modulation. This means that the oscillator is essentially running open-loop at frequencies between 5 Hz and 100 kHz, so the noise performance of the modulator will also be determined by the short-term stability characteristics of the oscillator.

Phase-locked loop (PLL) technology has provided a means of precisely controlling the carrier's average frequency while permitting wide deviation of the carrier frequency at baseband modulating frequencies. This implies that a PLL system behaves like an audio high-pass filter with higher modulating frequencies being ignored by the control loop while lower frequencies are considered to be errors in the average frequency and are tracked out by the loop. An added advantage of PLL is the ability to synthesize the desired frequency from a single reference oscillator, thereby eliminating the need to change crystals when changing the frequency of the exciter.

The block diagram shown in Figure 4.7-11 includes the key elements in the PLL. The output of the modulated oscillator operating at the carrier frequency is digitally divided down to a frequency of a few kilohertz or less and is called the *comparison frequency.* Likewise, the reference crystal oscillator is also digitally divided down to the comparison frequency. The two frequencies are compared in a digital phase/frequency detector to develop an error voltage that corrects the carrier frequency of the modulated oscillator. The reason for dividing the modulated oscillator frequency so many times is to reduce the modulation index enough to limit the peak phase deviation at the comparison frequency to a value that will not exceed the linear range of the phase/frequency detector. If the linear range is exceeded, the loop will lose lock. This is why some exciters may lose AFC lock in the presence of low-frequency modulation components.

The phase detector output is integrated and low-pass filtered to remove the comparison frequency and all other frequency components above about 5 Hz so the AFC circuit does not try to track low-frequency modulation. Some FM exciters use a dual-speed PLL to keep the loop turnover frequency low enough to maintain good amplitude and phase response at 30 Hz while also providing quick lock-up time. The PLL error correction circuitry must respond quickly during the initial frequency scan of the FM band to achieve lock-up to the precision reference oscillator in a few seconds. The loop bandwidth is wide during acquisition and lock-up. After lock is achieved, the bandwidth is reduced to provide the optimum modulation characteristic.

The reference oscillator is usually temperature compensated and requires no warm-up to maintain ±3 parts per million (PPM) or better accuracy over the operating temperature range. A 10 MHz source may be used as the reference frequency for convenient

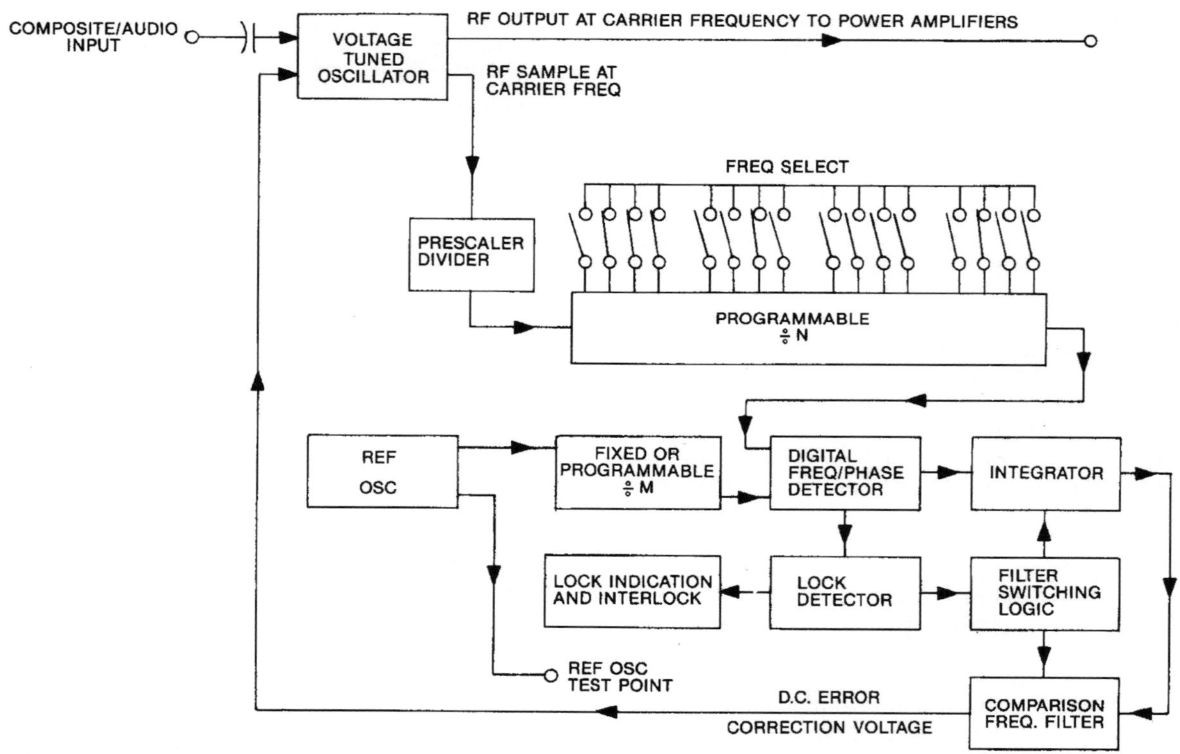

FIGURE 4.7-11 Phase-locked loop frequency synthesizer.

comparison to international or global positioning system (GPS) frequency standards. For more information about PLL frequency synthesizers, see reference [5].

FM by I/Q Modulation Techniques

Another method used to generate FM is the quasi-digital, in-phase/quadrature (I/Q) RF technique also known as an *RF modem*. Two identical analog RF mixers are fed in-phase (I) and quadrature (Q) analog baseband signals derived from two separate digital-to-analog (D/A) converters at the output of a DSP digital stereo generator. The I mixer is also fed the in-phase carrier frequency, while the Q mixer is fed the carrier frequency phase shifted by 90° from the I mixer. When the RF outputs of these two mixers are summed, the desired FM RF spectrum is produced if the system is perfectly balanced and matched. Practical limitations in the matching of these mixers require that feedback incorporating sophisticated nonlinear DSP adaptive correction be placed around these mixers and the I and Q D/A converters driving them. This technique has the advantage of producing FM at the carrier frequency instead of at an intermediate frequency.

Digital FM Exciter Using Direct Digital Synthesis

DDS eliminates the need for a PLL in the FM modulation process. It does so by directly synthesizing the carrier frequency, including FM modulation, from a sine wave look-up table in a programmable read-only memory (PROM) device operating in conjunction with a digital phase accumulator and a fast digital-to-analog converter. When this technique is combined with DSP technology, the entire process of generating stereo baseband with Radio Data System (RDS) and SCAs then frequency modulates this digital baseband information onto the RF carrier. This process is accomplished entirely in the digital domain. The cost-to-performance ratio of DDS/DSP technology has made it competitive with the analog technology. The full benefit of DDS/DSP technology requires digital transmission of audio information as an uncompressed, digital bit stream all the way from the digital audio source through a digital console, digital audio processing, and an uncompressed, digital studio-to-transmitter link (STL) to the AES3 digital input port of the DSP/DDS exciter [22,23]. This same technology is used in the fully digital audio broadcast (DAB) services, including HD Radio (IBOC), Digital Radio Mondiale (DRM), and the Eureka 147 (EU-147) transmission standards currently being implemented worldwide.

With DDS, the complete FM waveform is generated entirely in the digital domain. As digital modulation is an inherently linear process, no predistortion is required. The FM signal generated by a DDS device has low noise and distortion for true 16 bit digital audio quality (–96 dB FM S/N and 0.0016% harmonic distortion for ±75 kHz deviation and 75 μs preemphasis/deemphasis).

The current generation of DDS exciters uses a 32 bit NCO. The basic resting frequency of the NCO is set by a 32 bit tuning word. Frequency modulation occurs when modulation data varies the structure of the tuning word within the phase accumulator section of the NCO. The modulated output of the NCO is converted to analog FM, up-converted, filtered, and amplified to become the RF signal for a conventional FM broadcast transmitter RF amplifier chain. The recent generation of combined FM and HD Radio exciters uses direct-to-carrier, digital-to-analog conversion which directly converts the digital output of the NCO to the carrier frequency without the need for up-conversion or I/Q mixing at carrier frequency. A block diagram of a DDS digital FM exciter is shown in Figure 4.7-12.

DDS FM exciters also eliminate several basic limitations found in analog exciters using direct FM via the modulation of VCOs. Very low audio frequencies must be filtered from program signals feeding a VCO/PLL to avoid affecting the circuits of the analog exciter, which see very low modulating frequencies as an off-frequency condition that requires correction. A DDS-based FM exciter has no such limitation, and the modulation frequency response extends virtually to DC (zero hertz). These lower octaves of program material are important to accommodate digital audio source material and to preserve the phase correlation existing in the original program.

HD Radio Exciter

First-generation digital radio exciters do not include the analog FM function and generate the digital HD Radio (FM IBOC) signal separately from the station's existing analog FM exciter. If separate amplification is used, the digital radio exciter and transmitter operate independently from the existing analog FM transmitter. The separate digital HD Radio signal is either space-combined in separate antenna systems or is combined at a high level with the analog FM signal to be fed to a single antenna. The output from the HD Radio exciter may also be combined at low level with the output from the analog FM exciter, if common amplification of the two signals in a single, linearized transmitter is desired. First-generation HD Radio exciters also combine the functionality of the audio encoder (also called a "codec"), data multiplexing, and RF signal generation into a single unit; therefore, the audio coding and data multiplexing are located at the transmitter site and not the studio.

Digital FM + HD Radio Combined in a Single Exciter

There are several different HD Radio exciter RF operating modes. These are not to be confused with the iBiquity HD Radio digital operating modes MP1 through MP7. The exciter RF operating modes are:

- Analog FM only for separate amplification from the HD Radio signal or analog only broadcasting

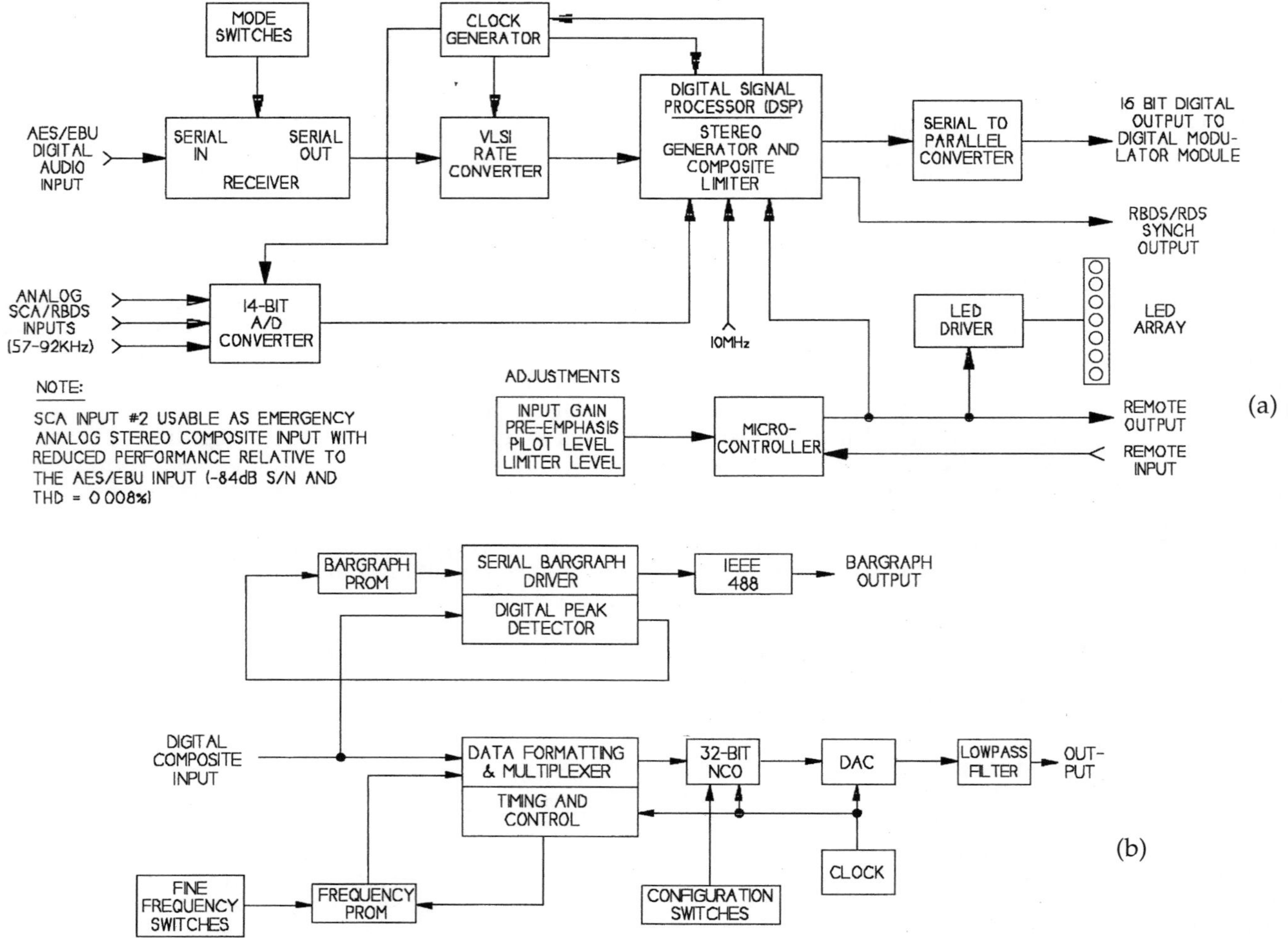

FIGURE 4.7-12 (a) Harris DIGIT-CD digital input module and DSP stereo generator. (b) Digital modulator in the Harris DIGIT-CD FM exciter.

- HD Radio only for separate amplification from the analog FM signal
- Analog FM + HD Radio for common amplification of the two low-level combined signals through a single transmitter
- Split-level analog FM + HD Radio for split-level, part common-amplification, and part separate-amplification described elsewhere in this chapter

Newer generation HD + FM exciters have the ability to instantly switch modes as required for emergency backup reconfiguration of the system.

HD Radio Exciter Functionality

Newer generation HD Radio exciters have incorporated the iBiquity "Exgine" architecture that separates the codec and data multiplexing functions ("Exporter") from the RF signal-generation function of the exciter subsystem ("Exgine" is a contraction of exciter and engine). The Exporter can be located at the transmitter site or the studio with only a single, unidirectional User Datagram Protocol (UDP) Ethernet connection between the two units. The unidirectional UDP stream carries the primary and supplementary audio and data services that have been multiplexed by the Exporter to the Exgine that creates the digital orthogonal frequency division multiplex (OFDM) signal for conversion to RF and amplification by the transmitter. Unlike the first generation exciters, which were based on personal computer technology, the new generation Exgine exciters use DSP and field programmable gate array (FPGA) hardware platforms that run on embedded, real-time, operating systems for greater reliability and much faster restart. Figures 4.7-13 and 4.7-14 illustrate the Exgine and direct to carrier exciter architecture.

Direct to Carrier Digital-to-Analog Conversion

The latest generation of HD + FM exciters incorporates high-speed digital-to-analog converters that can

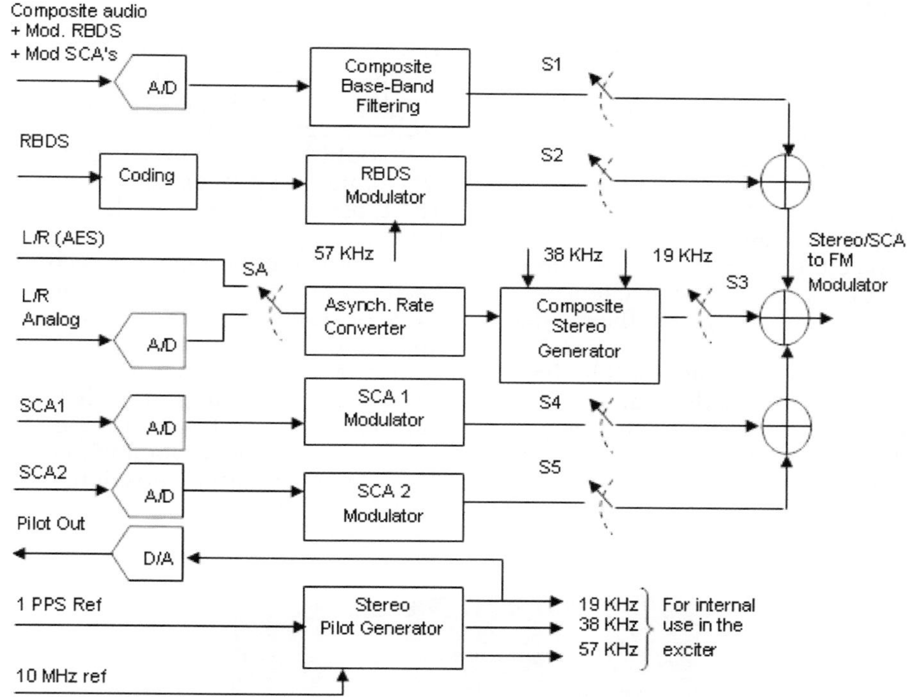

FIGURE 4.7-13 Flexstar HDx FM digital signal processing stereo generation and SCA modulation.

convert the digital representation of the HD plus FM signals directly to the carrier frequency in the 87 to 108 MHz FM broadcast band. This eliminates the need for modulation at a lower intermediate frequency and then up-converts the signal to the FM band. Direct to carrier digital A/D conversion provides a more accurate creation of the RF signal sideband structure than analog techniques. This technique also enables the creation of dual RF outputs from a single exciter that can have their phase relationships digitally controlled, thereby eliminating multiple exciters and external RF delay lines.

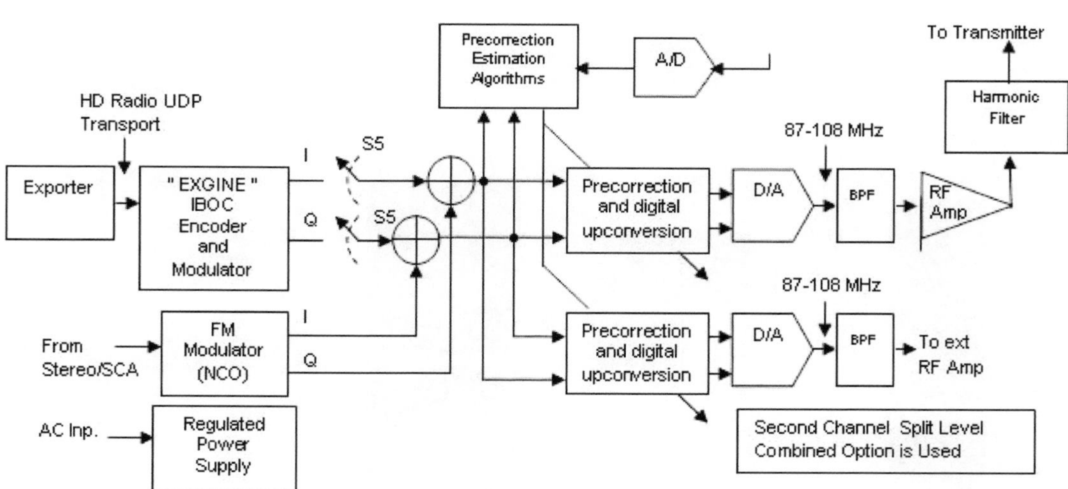

FIGURE 4.7-14 Flexstar HDx digital FM and IBOC and direct to carrier frequency digital-to-analog conversion.

Dual Outputs for Split-Level Combining

Separate amplification for high-level, split-level, or "antenna-space" combining normally requires two separate RF outputs from two separate exciters. As an example, the Harris FlexStar HDx FM exciter uses dual direct to carrier digital A/D converters to produce two separate RF outputs at the carrier frequency. These two independent RF outputs can have any combination of analog FM, digital HD Radio, FM + HD, and split-level modes on either output. This technique eliminates the need for two separate exciters and allows precise digital phase shifts between the two RF outputs for split-level combining. (See Figures 4.7-37 and 4.7-38 later in the combining portion of this chapter.) Figure 4.7-15a shows the HD Radio studio implementation of the Exgine architecture, and Figure 4.7-15b shows the HD Radio transmitter site implementation of the Exgine architecture.

Exciter RF Output Stage

The broadband RF amplifier in the exciter raises the output of the modulated oscillator from a power level of a few milliwatts up to the range of 5 to 50 W. The output stage is normally protected against damage that could be caused by an infinite voltage standing wave ratio (VSWR) on the output at any phase angle. The typical exciter RF amplifier is designed to have a bandwidth of at least 20 MHz, using successive broadband impedance matching sections for each stage. Each group of matching sections consists of micro-strip or lumped elements. The broadband performance of the RF amplifier eliminates the need for adjustments to any particular frequency within the FM band. The exciter output is transparent to the signal generated by the modulated oscillator, and the amplifier stability accommodates varying load conditions. A micro-strip directional coupler in the RF

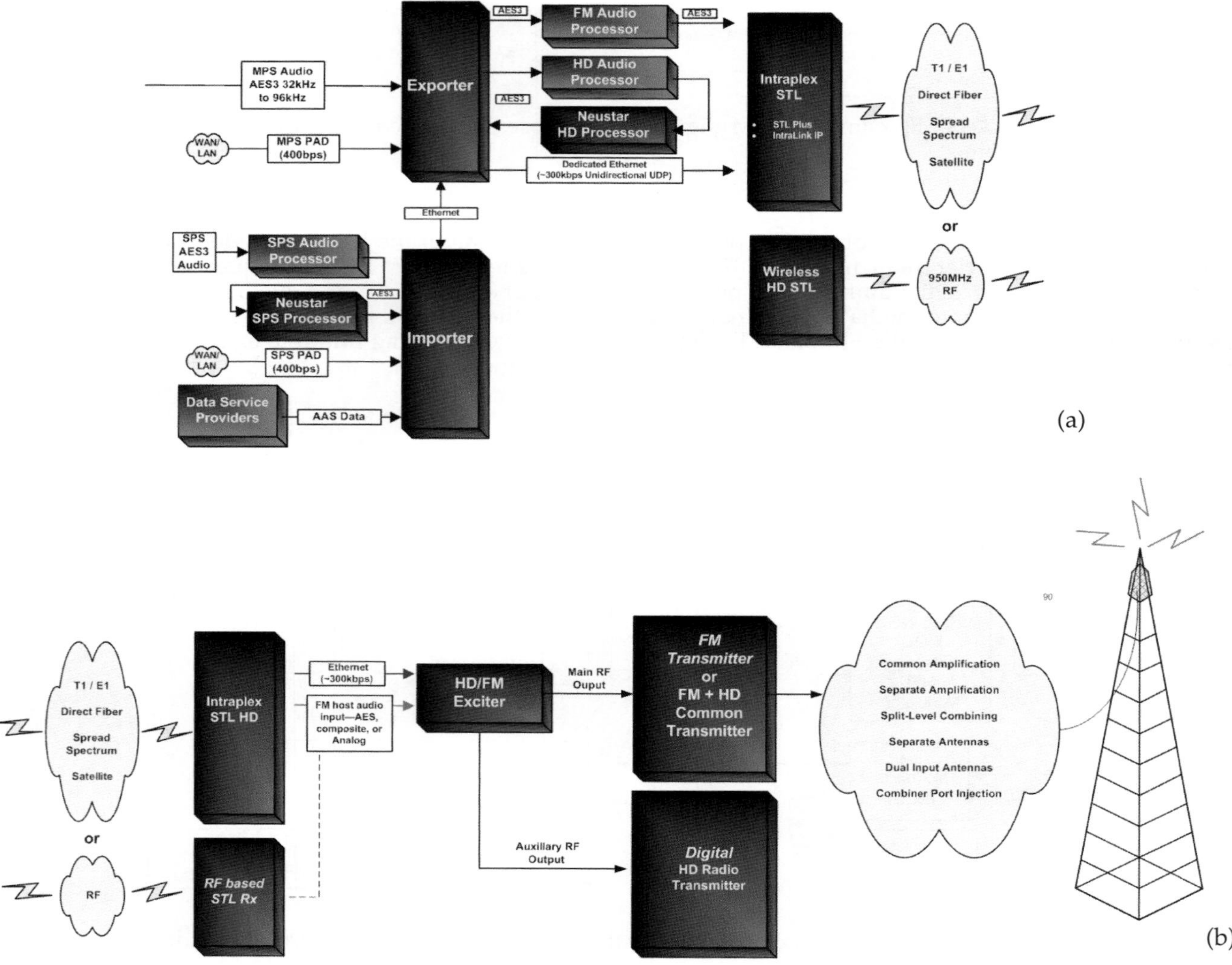

FIGURE 4.7-15 (a) Exgine studio system architecture. (b) Exgine transmission system architecture.

amplifier output network provides automatic control of power output level and protection against operation under high VSWR conditions. All current-generation FM exciters typically produce at least 50 W of RF output; thus, the exciter may be used as a complete transmitter for educational or low power stations with a harmonic low-pass filter at the output. For higher power level requirements, the exciter drives an external power amplifier.

Exciter Linear Amplification Requirements for HD Radio

The analog FM signal has a constant amplitude and does not require linear amplification. Typically, RF power amplifiers for FM operate in a nonlinear, saturated, Class C mode. When the FM signal is combined with the digital HD Radio signal, there is a moderate crest factor amplitude modulation component to the combined signal that requires linear amplification. The digital signal by itself has a high crest factor amplitude component due to the OFDM modulation that also requires linear amplification. Depending on the modulation mix requirements, the exciter RF amplifier can operate in three different modes:

- Class C, nonlinear for FM only
- Class AB, quasi-linear, for common amplification of FM + HD (moderate crest factor)
- Class AB, full linear operation for separate amplification of HD only (high crest factor)

Changes in operating modes are accomplished by changing the bias and drive level to the RF amplifier. There is typically a 1.6 dB back-off in power output from the saturated Class C power to accommodate common amplification and a 5 dB back-off in power to accommodate a digital only signal.

Adaptive Pre-Correction

Recent generation FM + HD exciters have digital adaptive pre-correction for both linear and nonlinear distortions in the high power portions of the transmission system. Linear distortions include amplitude and group delay variations in the transmitter and any filters following the transmitter. Nonlinear distortions include AM to AM and AM to PM conversion distortions in the transmitter RF power amplifiers. The exciter typically accepts signals from RF sampling points at the output of the power amplifier and after any RF filters at the feed point to the antenna. These RF samples are demodulated to digital baseband and compared with the original digital signal. The difference between these signals is used to drive the linear and nonlinear pre-correctors in the exciter to minimize undesired RF spectral components that would fall outside the RF mask limits. Real time, adaptive pre-correction automatically maintains the optimum spectrum and enables greater power utilization of the RF amplifiers. Figure 4.7-16 shows the RF spectrum before and after the application of adaptive pre-correction [32]. Linear, adaptive, pre-correction can also correct the analog FM signal for group delay distortions introduced by a multi-station filterplexer. Referring again to Figure 4.7-14, note the block diagram of the pre-corrector.

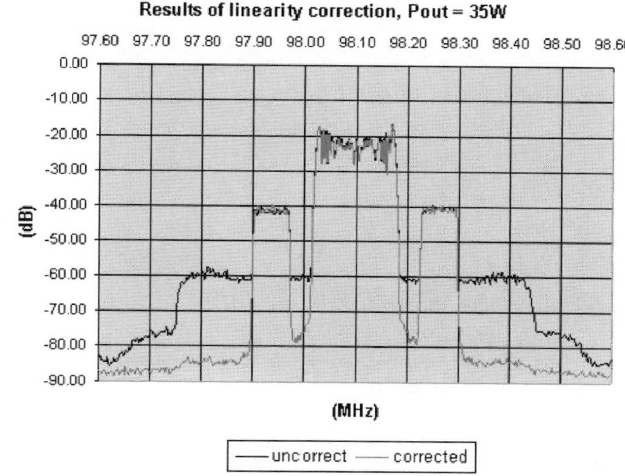

FIGURE 4.7-16 Adaptive pre-correction results. The light trace shows the corrected waveform.

Exciter Spectrum Analysis

The RF to digital baseband circuitry used for adaptive pre-correction allows the demodulated digital signal to be displayed as an RF spectrum on the front panel display of the exciter, as shown in Figure 4.7-17.

Exciter Control and Monitoring

The exciter control and monitoring system provides automatic regulation of the power output and protection against impedance mismatch of the RF output stage. The exciter's frequency control system is normally interlocked with the transmitter control system to prevent off frequency operation. Displays of important operating parameters are typically provided by a combination of analog metering, LED displays, or a graphical user interface (GUI). These parameters usually include supply voltages, RF power amplifier operating parameters, forward output power, reflected power, modulation level, and operating mode. A color-coded peak reading display may be provided to constantly monitor the peak FM deviation. A high-speed peak detector enables accurate peak modulaton readings. A peak hold function is often included to provide a clear indication of short transient peaks exceeding 100% modulation. Local control is provided through either front panel switches or a GUI. Remote control of the exciter is normally provided by a parallel hardware interface, serial port, or Ethernet connection. Most exciters provide both parallel and serial interfacing for flexible remote control.

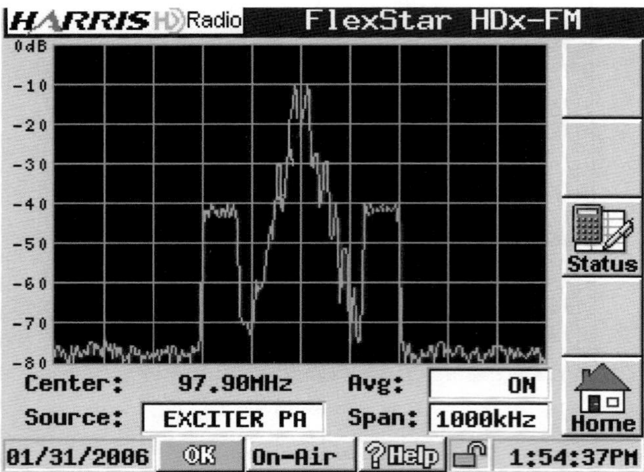

FIGURE 4.7-17 An example of front panel spectrum display on Harris FlexStar HDx exciter.

Exciter Packaging

Protection of sensitive circuits within an FM exciter from external electromagnetic interference (EMI) is important because the unit is often located in the near field of multiple broadcast antennas operating over a broad range of frequencies. The exciter should be protected from conducted EMI by use of RC and/or LC filters on all leads entering the cabinet, including the AC line. The power supply transformer may have an electrostatic shield between the primary and secondary windings. The modulated oscillator or frequency synthesizer is usually very sensitive to EMI, magnetic fields, and vibrations. It must be well RF shielded and is often shock mounted to prevent the transmission of mechanical vibrations from the transmitter's blower. This avoids microphonic pick-up that would degrade the FM S/N and HD Radio bit error rate (BER). Magnetic shielding of the oscillator or a hum-bucking circuit may be used to prevent hum pick-up from nearby transformers. The mechanical construction of most present day exciters incorporates a modular approach, which allows easy removal of subassemblies for repair or replacement.

FM Transmitter RF Power Amplifiers

The remainder of the FM transmitter consists of one or more power amplifiers, each having from 8 to 20 dB of power gain. Ideally, the transmitter bandwidth should be as wide as practical with a minimum of tuned stages. Broadband solid state amplifiers usually eliminate the need for tuned networks in the RF path. High-power transmitters in the multi-kilowatt range may use multiple tube stages each with relatively low gain such as a grounded-grid configuration or a single grid-driven power amplifier (PA) stage with high gain and efficiency. The cost, redundancy, and wide bandwidth benefits of solid-state transmitters make them attractive at power levels up to 20 kW. At higher power levels, the lower cost per watt of high-power, single-tube transmitters is still attractive even though the modulation performance and reliability are less than that of a solid-state transmitter. Design improvements in tube-type power amplifiers have concentrated on improving bandwidth, reliability, and cost-effectiveness while design improvements in solid-state amplifiers have focused on cost reduction to make them competitive with tube technology at high power levels.

RF Power Amplifier Performance Requirements

The basic function of the power amplifier is to bring the power of the exciter output up to the desired transmitter power output level. Most of the overall transmitter performance characteristics are determined by the exciter, but a few are established or affected by the following power amplifier characteristics:

- The RF output level at harmonics of the carrier frequency is almost completely a function of the attenuation provided by the power amplifier output matching circuit and output low-pass/notch filters. The FCC limit in decibels is (43 dB + 10 log [power in watts] dB) or 80 dB, whichever is lower. The specification is 73 dB for transmitters with 1 kW output increasing to 80 dB above 5 kW output power.
- The major source of asynchronous AM noise usually originates in the last power amplifier stage.
- The RF power output control system must keep the output within +5% and –10% of authorized power.
- Inadequate power amplifier RF bandwidth that affects phase linearity (constant time delay) across the signal bandwidth can reduce stereo separation and cause crosstalk to and from the SCA subcarrier.
- The presence of standing waves on the transmission line between the power amplifier and the antenna may also interact with the power amplifier to cause degraded stereo separation and SCA crosstalk.
- If the transmitter is amplifying an HD Radio signal, linear operation is required to prevent the generation of excess RF intermodulation products.

The power amplifier should provide trouble-free service and be easy to maintain and repair. Good overall efficiency is also desirable to reduce the primary power consumption and heat load released into the transmitter room.

Power Amplifier Bandwidth Considerations

As discussed earlier, the FM signal theoretically occupies infinite bandwidth. In practice, however, truncation of the insignificant sidebands (typically less than 1% of the carrier power) makes the system practical by accepting a certain degree of signal degradation; therefore, the transmitter power amplifier bandwidth affects the modulation performance. Available bandwidth determines the amplitude response and group

delay response. There is a trade-off between the bandwidth, gain, and efficiency in the design of a power amplifier [9,18]. The bandwidth of an amplifier is determined by the load resistance across the tuned circuit and the output or input capacitance of the amplifier. For a single tuned circuit, the bandwidth is proportional to the ratio of capacitive reactance to resistance:

$$BW \propto \frac{K}{2\pi f_c R_L(C)} = \frac{K(X_C)}{R_L}$$

where:

BW = Bandwidth between half-power points (BW 3 dB).

K = Proportionality constant.

R_L= Load resistance (appearing across tuned circuit).

C = Total capacitance of tuned circuit (includes stray capacitances plus output or input capacitances of the tube).

X_C = Capacitive reactance of C.

f_c = Carrier frequency.

The load resistance is directly related to the RF voltage swing on the tube element. For the same power and efficiency, the bandwidth can be increased if the capacitance is reduced.

Effects of Circuit Topology and Tuning on FM Performance

Analog FM broadcast transmitter RF power amplifiers are typically adjusted for minimum synchronous AM (incidental amplitude modulation) which results in symmetrical amplitude response. This ensures that the transmitter's amplitude passband is properly centered on the FM channel. The upper and lower sidebands will be attenuated equally or symmetrically which is assumed to result in optimum FM modulation performance. This is true only if the RF power amplifier circuit topology results in simultaneous symmetry of both amplitude and group delay responses [16].

The tuning points for symmetrical amplitude response and symmetrical group delay response usually do not coincide, depending on the circuit topology; therefore, simply tuning for minimum synchronous AM (symmetrical amplitude response) does not necessarily result in best FM performance. In fact, symmetry of the group delay response has a much greater effect on FM modulation distortion than does the amplitude response. Tuning for symmetrical group delay will cause the phase/time delay errors to affect the upper and lower sidebands equally or symmetrically. The group delay response is constant if the phase shift *versus* frequency is linear. In this case, all components of the signal are delayed in time, but no phase distortion occurs.

Measurements taken on a typical FM transmitter as well as computer simulations show that tuning the RF power amplifier for symmetrical group delay response resulted in minimum distortion and crosstalk and confirmed that group delay response asymmetry causes higher FM distortion and crosstalk than amplitude response asymmetry [17]. Therefore, the transmitter should be tuned for the symmetrical group delay response that results in best FM performance rather than the symmetrical amplitude response that results in minimum synchronous AM.

Intermediate Power Amplifiers

The IPA is located between the exciter and the final amplifier in transmitters that require more than about 50 W of drive to the final amplifier. The IPA may consist of one or more tubes or solid-state amplifier modules.

Interstage Coupling Circuits

The IPA output circuit and the input circuit to the final amplifier are often coupled together by a coaxial transmission line. Impedance matching is usually implemented by the input circuit to the tube. The interconnecting transmission line between the coupling circuits should be matched to avoid a high VSWR. Directional wattmeters may be placed in the line to measure forward and reflected power from which the standing wave ratio can be determined. The VSWR is established by the match at the load end of the transmission line. Solid-state RF power devices present a low load impedance at the device output terminal, so an impedance transformation is required to couple these devices into the relatively high impedance of the final amplifier grid circuit. Therefore, virtually all solid-state IPA systems have a 50 Ω impedance point within the system that can be used to feed the antenna should it be necessary to bypass the power amplifier. The tube in the final amplifier stage of most high-power transmitters requires between 150 and 600 watts of drive. This permits the use of solid-state, wideband power amplifier modules to boost the exciter output power to the level required to drive the grid of the final amplifier tube.

Linear Amplification Requirements for HD Radio

The peak-to-average ratio for the HD Radio signal is about 7 dB after crest factor reduction and causes the amplifier output level to go to zero when the vector sum of all the OFDM carriers is zero. This requires that the RF amplifier be linear throughout its entire dynamic range. Typically, a back-off of at least 5 dB from Class C operation is required to maintain enough linearity. The peak-to-average ratio for the combined analog FM + HD Radio signal does not require the RF amplifier output go to zero due to the presence of the larger analog FM signal in the mix which reduces the overall crest factor. Typically, a back-off of 1.6 dB from Class C is enough to obtain the required linearity at the upper end of the RF amplifier's power output

TABLE 4.7-3
Average Power for Different Modes*

Modes	Average Power (%)	Back-Off (dB)
FM only	100	0
FM + digital	70	–1.6
Digital only	32	–5.0

*As a percentage of Class C (FM-Only) operation.

range. Table 4.7-3 shows the average power (average of all the coded orthogonal frequency-division multiplexing [COFDM] carriers plus analog FM) available as a percentage of the rated analog-only transmitter power and the power back-off from rated analog power required (in dB) for each mode of operation.

High-efficiency amplifiers operate with less than 180° of a full RF cycle. This creates a nonlinear response of the envelope, specifically at low transition levels where the power output approaches zero. To correct for this, linear amplifiers are biased to conduct at least half the RF cycle, which is achieved by placing them in Class AB mode. Class A mode, the most linear operating mode, is not used in high-power transmitters due to its inherent low efficiency. The operating efficiency of the amplifier drops as much as 50% when changing from Class C to Class AB operation [27].

Common Amplification in a High-Power, Vacuum-Tube Amplifier

High-power vacuum tube amplifiers can also be operated in class AB linear mode to provide common amplification of the analog FM signal and the HD Radio signal. When operated in linear mode, the plate efficiency of the amplifier tube typically drops from approximately 80% to about 63%. The reduced operating efficiency of linear operation needs to be taken into consideration when sizing the AC power and building cooling requirements.

AMPLIFICATION OF HD RADIO SIGNALS

Separate Amplification

Figure 4.7-18 shows a block diagram of separate amplification of the FM analog and IBOC digital signals by the addition of new components from exciter to antenna.

Common Amplification

Figure 4.7-19 shows a block diagram of the FM analog and IBOC digital signals combined into a common power amplifier and using the same transmission line and antenna. For more information on HD Radio amplification, see References [33] and [34].

Solid-State RF Power Amplifier Systems

A solid-state RF power amplifier usually consists of multiple individual amplifier modules that are combined to provide the desired power output. The advantages of using several lower power modules instead of a single high-power amplifier include the following:

- Redundancy is provided by isolating the input and output of each module to permit uninterrupted operation at reduced power if one or more of the modules fail.
- Failed modules can be repaired without having to go off the air.
- More effective cooling of each power device is achieved by splitting the concentration of heat to be dissipated into several areas instead of one small area.
- Better isolation between the amplifier modules and the input circuit of the final power amplifier or antenna is provided by the combiner/isolator.
- Redundant power supplies and air cooling systems for each module improve overall reliability.

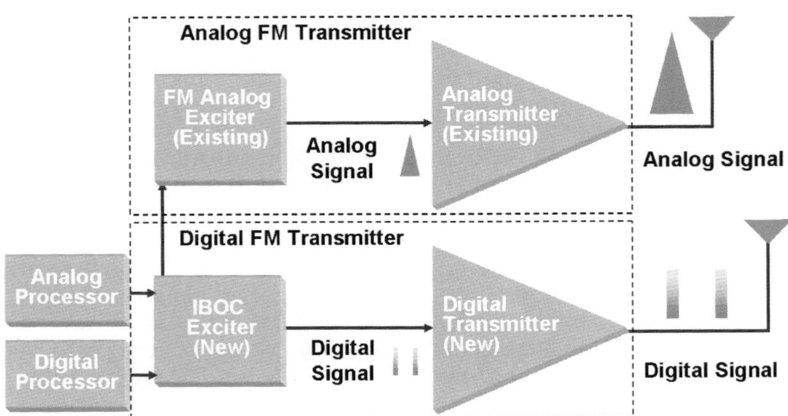

FIGURE 4.7-18 Simplified block diagram of separate amplification transmitters.

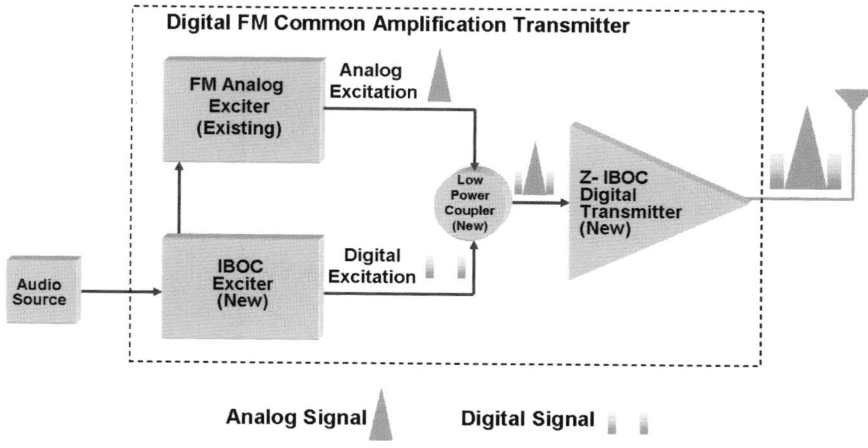

FIGURE 4.7-19 Simplified block diagram of common amplification transmitter.

Each RF power amplifier module consists of one or more solid-state devices with broadband impedance transformation networks for input and output matching. Vertical metal oxide semiconductor field-effect transistors (VMOSFETs) permit the design of broadband amplifier stages that exhibit both high efficiency and the wide bandwidth necessary to cover the FM broadcast band.

The input impedance to the solid-state device is always lower than the desired 50 Ω input impedance, so a broadband impedance transformation scheme is required. This is usually accomplished by using a combination of coaxial baluns and push–pull coaxial line sections that are cross-coupled to provide 4:1 or higher transformation ratios over the FM band.

By operating two devices in push–pull, the input impedance is double that of a single ended circuit, and the suppression of even order harmonics is obtained. Two devices fed in this manner also provide some degree of redundancy within the module itself, as partial RF output can be obtained with one failed device. In a similar manner, the low output impedance of these solid-state devices can be transformed up to the desired 50 Ω module output impedance where combining occurs. Figure 4.7-20a illustrates a simplified schematic of a broadband, 425 watt, VMOSFET RF amplifier module utilizing the push–pull configuration. Figure 4.7-20b is a photograph of this RF amplifier module.

Solid-State Amplifier Splitting and Combining

Two frequently used types of splitting/combining schemes are:

- A 90° hybrid splitter or combiner; $N - 1$ hybrids are required to split or combine N inputs (see the section on transmitter output combining)
- A Wilkinson N-way in-phase splitter or combiner

Either type of splitter/combiner must provide isolation between the individual power amplifier modules and low loss splitting or combining of the total power.

The cascaded 90° hybrid system shown in Figure 4.7-21 provides double isolation between the power amplifiers and the load by first combining the two pairs of amplifiers and then combining the outputs of the first two combiners. A portion of the reflected power, caused by a mismatch at the output, will be dissipated in reject loads so the power amplifier modules will see a lower VSWR than exists at the output.

The Wilkinson system shown in Figure 4.7-22 is a simple and effective way to split and combine modules operating in phase, but usually requires a balanced reject load that makes reject power measurements more difficult. By adding additional coaxial balun sections to the Wilkinson combiner (Wilkinson/Gysel), it is possible to use unbalanced reject loads. The unbalanced 50 Ω reject loads are accessible for monitoring of reject load power which is useful in determining the balance of the system. [6, 7]

Adaptive Control of the Combiner Configuration

Both the 90° hybrid and Wilkinson combining systems require resistive RF power reject loads to provide isolation between the amplifier modules in the event that one or more of the modules fail. A portion of the RF power from the remaining modules is wasted in the reject loads instead of being delivered to the output. A microcomputer can monitor the degree of imbalance in the system and adaptively change the configuration of the combiner to losslessly compensate for the failure of one or more power amplifier modules. This is accomplished by having the microcomputer substitute the appropriate reactances in place of the resistive reject loads to maintain enough isolation for the remaining power amplifiers to work efficiently. This technique is used in the Harris Z-plane combiner, as shown in Figure 4.7-23. Because most splitter/combiner systems are designed

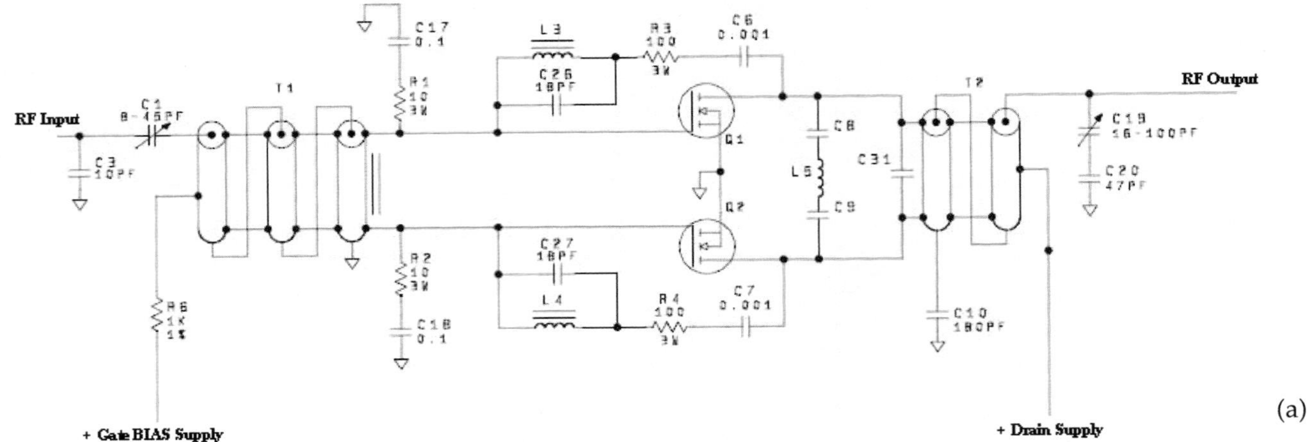

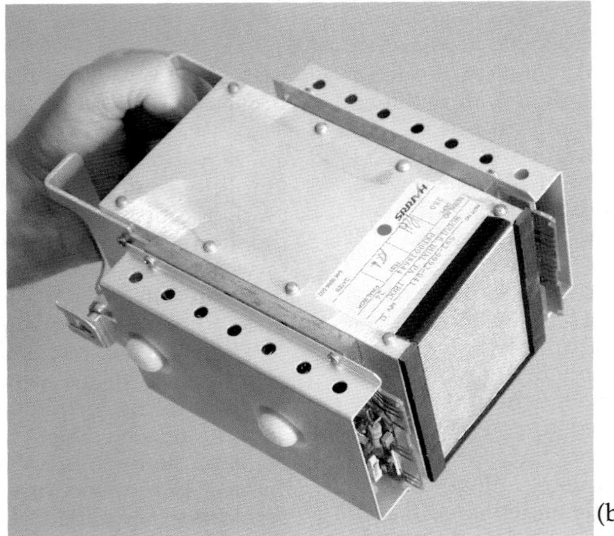

FIGURE 4.7-20 (a) Schematic of broadband, 425 watt, VMOSFET RF power amplifier subassembly. (b) Example of a broadband, 850 watt, VMOSFET RF power amplifier module. Note the relative size of the module compared to the hand holding it.

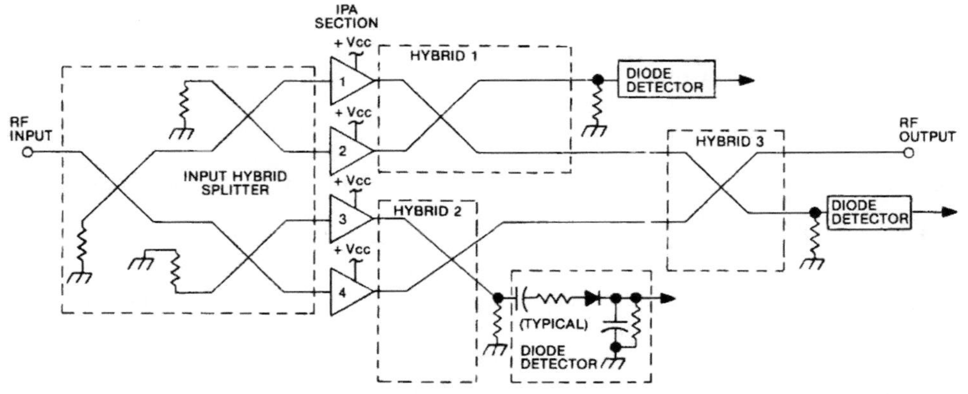

FIGURE 4.7-21 Cascaded 90° hybrid splitting/combining system.

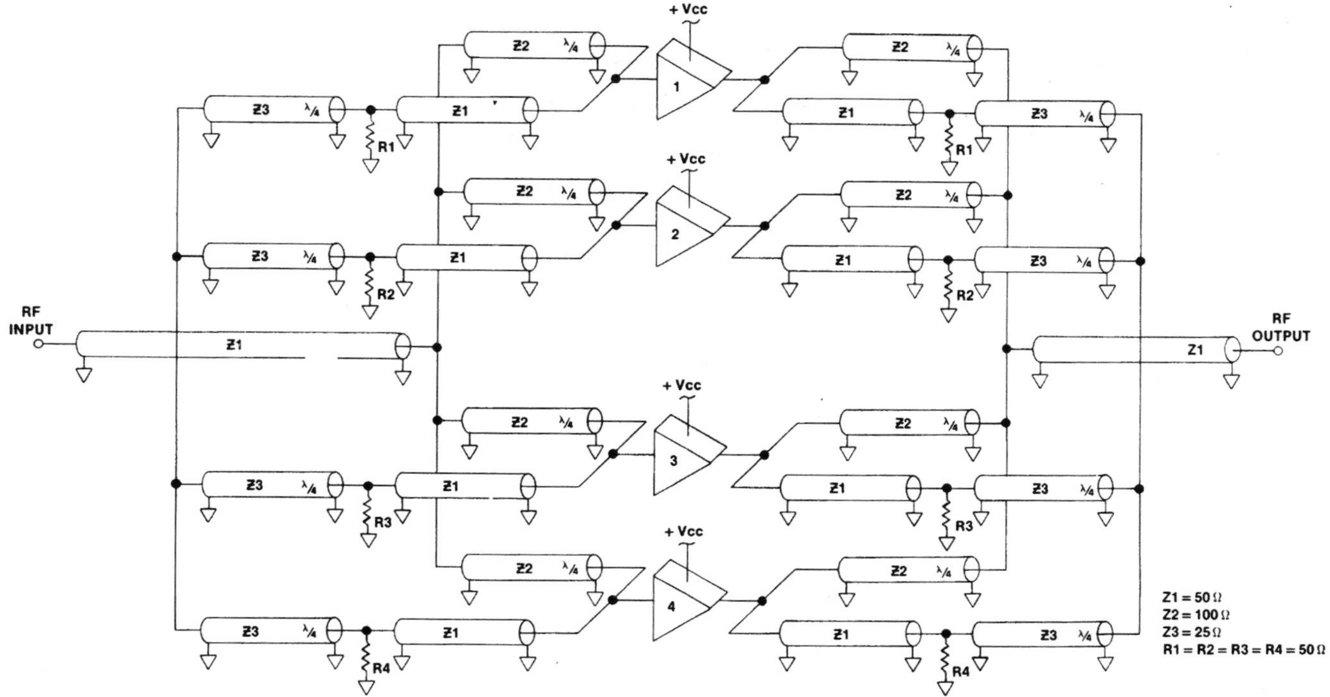

FIGURE 4.7-22 Wilkinson/Gysel in-phase splitting/combining system with unbalanced reject loads.

with 50 Ω input and output impedances, these systems can be easily used as low-power stand-by transmitters by routing the output to the antenna system. An RF low-pass filter (LPF) is required only when directly feeding the antenna system. The harmonic suppression of the IPA is not as critical when driving a nonlinear power amplifier that also generates harmonics, because the final amplifier stage will have its own LPF.

SOLID-STATE FM BROADCAST TRANSMITTERS

The techniques just described can be used to construct a completely solid-state transmitter using arrays of combined modules for the final output stage. An additional RF low-pass filter is usually required to meet FCC emission requirements.

Advantages of Solid-State Transmitters

The advantages of a solid-state transmitter when compared to a single tube transmitter, include the following:

- Built-in amplifier and power supply redundancy
- Improved FM performance
- The ability to cover the entire FM band without the need for retuning
- Elimination of tube replacement costs
- Less maintenance than tube ype transmitters

Solid-State Transmitter Design Considerations

Several manufacturers offer solid-state FM broadcast transmitters with power outputs ranging from 100 W up to 20 kW, but current economic factors still favor the single tube FM transmitter for power levels above 20 kW. In order for a solid-state transmitter to be cost and power consumption competitive with a single-tube transmitter, the efficiency of the solid-state RF power amplifiers and combining system should approach the 80% efficiency obtainable from tube-type RF amplifiers, a level that has been achieved with VMOSFET solid-state devices at VHF frequencies.

Solid-state designs can provide up to 80% DC to RF efficiency at the VMOSFET device level and over 62% overall efficiency, from AC power input to RF output. This is somewhat better AC to RF efficiency at the 5 kW level than a typical single-tube transmitter, which is normally less than 60%. Some solid-state designs have added a few percent to their overall AC to RF efficiency by optimizing their RF circuits over narrow-band sections of the FM band. This approach is beneficial if there will not be a need to change the transmitter's frequency.

Trends in the newest solid-state FM transmitters are to supply redundant RF amplifiers, power supply, and control circuits to keep the transmitter on the air at reduced power in the event that one or more

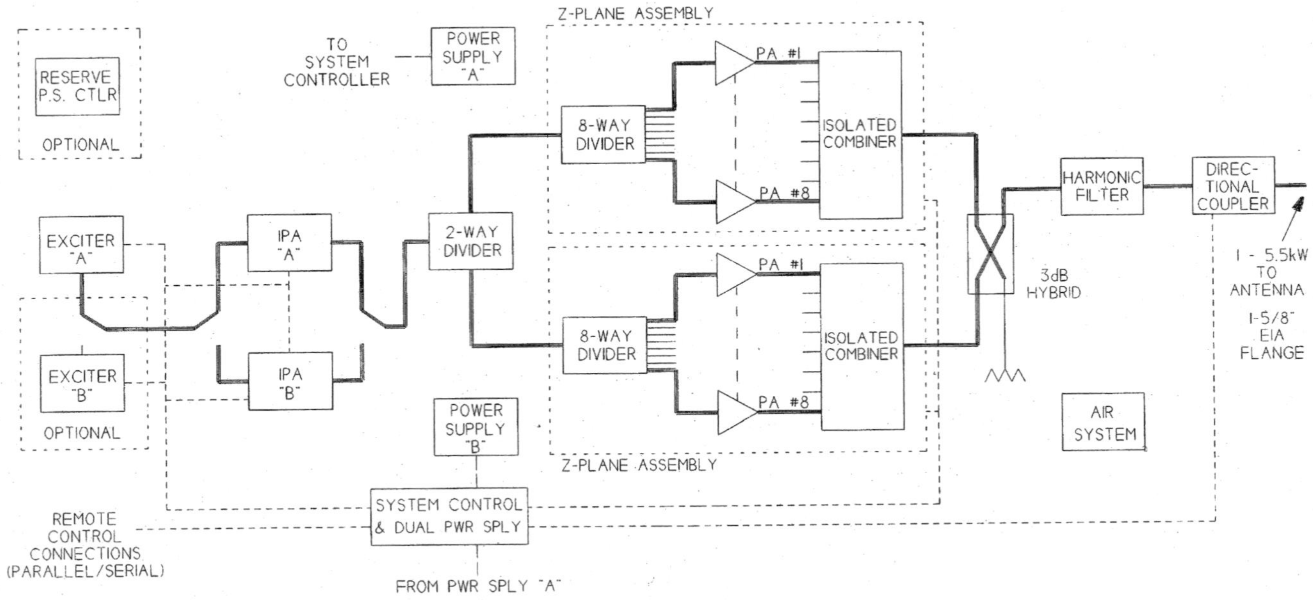

FIGURE 4.7-23 Harris Z-FM 5 kW solid-state transmitter block diagram.

components should fail. Identical and interchangeable IPA and PA modules offer additional redundancy. RF modules that can be removed and inserted in an operating transmitter also provide the advantage of not requiring an off-air period for some maintenance services.

Solid-state transmitter layouts using direct, cable-free connection of the RF modules to the RF combiner further enhance transmitter reliability and stability. Another enhancement provided in some current solid-state FM transmitters is a microprocessor-based control system that can monitor parameters within the transmitter and provide intelligent control of the transmitter system, including the RF combiner. This maximizes output power and minimizes reject load power under various combinations of active and inactive modules. Figure 4.7-23 shows a block diagram of a 5 kW solid-state transmitter.

VACUUM-TUBE POWER AMPLIFIER CIRCUITS

The amplitude of an FM signal remains constant with modulation so efficient, nonlinear, Class C amplifiers can be used. Vacuum-tube power amplifier circuits for FM service have evolved into two basic types. One type uses a tetrode or pentode tube in a grid driven circuit while the other uses a high-mu triode in a cathode driven (grounded-grid) circuit.

Grounded-Grid (Cathode-Driven) Triode Amplifiers

The high-mu, zero-bias triodes used in cathode-driven (grounded-grid) FM amplifiers were originally developed for linear single sideband (SSB) amplifiers. Their characteristics are well adapted to FM broadcast use because the circuit is simple, and no screen or grid bias power supplies are required. Figure 4.7-24 shows the basic circuit configuration. In this case, the grid is connected directly to chassis ground. In this case, the grid is connected directly to a chassis ground. The output tank circuit is a shorted coaxial cavity that is loaded by the tube output and stray circuit capacitance. A small capacitor is used for trimming the tuning, and another small variable capacitor is used for adjusting the loading. A pi-network matches the 50 Ω input to the tube cathode.

The triodes are usually operated in the less efficient, Class B mode to achieve maximum power gain, which is on the order of 20 (13 dB). They can be driven into high-efficiency Class C operation by providing negative grid bias. This increases the plate efficiency but also requires increased drive power and a bias power supply.

Most of the drive power into a grounded-grid amplifier transfers through the tube and appears in the stage's output. This increases the apparent efficiency so the efficiency factor given by the manufacturer may be higher than the actual plate efficiency of the tube. The true plate efficiency is determined by dividing the output power by the total input power, which includes both the DC plate input power (plate current times plate voltage) and the RF drive power. Because most of the drive power transfers through the tube, any changes in loading of the output circuit will also affect the input tuning and driver stage.

With RF drive voltage on the cathode (filament) of the tube, some means of decoupling is required to block the RF from the filament transformer. One method employs high-current RF chokes, as the induc-

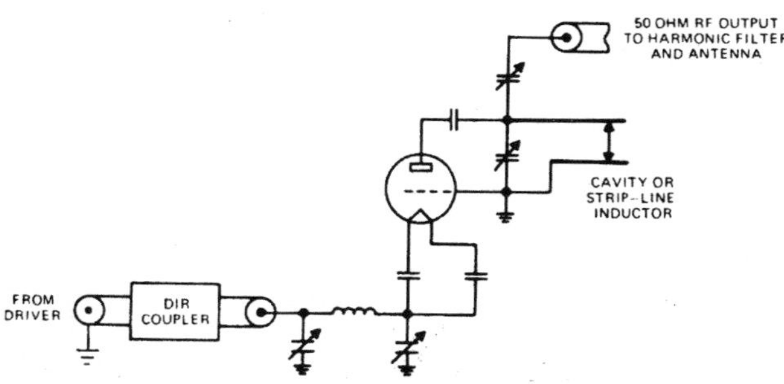

FIGURE 4.7-24 Cathode-driven, grounded-grid triode power amplifier.

tance can be very low at this frequency range. The other commonly used method feeds the filament power through the input tank circuit inductor. Cathode-driven stages are normally used only for the higher power stages. The first stage in a multi-tube transmitter is nearly always a tetrode because of its higher power gain.

Grounded-Grid *versus* Grid-Driven Tetrode Operation

There are several trade-offs between the performance of grounded-grid and grid-driven configurations of a tetrode PA with respect to gain, efficiency, amplitude bandwidth, phase bandwidth, and synchronous AM under equivalent operating conditions:

- When driving a grounded-grid PA into saturation, the bandwidth is limited by the output cavity. The PA bandwidth in the grid-driven amplifier is limited by the input circuit Q, which is determined by the input capacitance and the amount of swamping resistance.
- Output bandwidth under saturation can be improved in either configuration by reducing the plate voltage. This involves a trade-off in efficiency with a smaller voltage swing. The bandwidth improvement can be obtained with a loss of PA gain and efficiency.
- A grounded-grid saturated PA improves bandwidth over a grid-driven saturated PA at the expense of amplifier gain.
- The best performance for FM operation is obtained when the amplifier is driven into saturation where little change in output power occurs with increasing drive power. Maximum efficiency also occurs at this point.
- The phase linearity in the 0.5 dB bandwidth is better in a grid-driven configuration. The grounded-grid PA exhibits a more nonlinear phase slope within the passband yet has a wider amplitude bandwidth. This phenomenon is due to interaction of the input and output circuits because they are effectively connected in series in the grounded-grid configuration. The neutralized, grid-driven PA provides more isolation between these networks, so they behave more like independent filters.

Grid-Driven Tetrode and Pentode Amplifiers

Transmitters using tetrode amplifiers usually have one less amplifying stage than transmitters using triodes. Because tetrodes have higher power gain, they are driven into Class C operation for high plate efficiency. Against these advantages is the requirement for neutralization, along with screen and bias power supplies. Pentode amplifiers have higher gain than their tetrode counterparts. The circuit configuration and bias supply requirements for the pentode are similar to the tetrode because the third (suppressor) grid is tied directly to ground. The additional isolating effect of the (suppressor) grid eliminates the need for neutralization in the pentode amplifier.

Impedance Matching into the Grid

The grid circuit is usually loaded (swamped) with added resistance. The purpose of this resistance is to broaden the bandwidth of the circuit by lowering the circuit Q and to provide a more constant load to the driver. It also makes neutralizing less critical so the amplifier is less likely to become unstable with varying output circuit loading. Cathode or filament lead inductance from inside the tube, through the socket and filament capacitors to ground can heavily load the input circuit. This is caused by RF current flowing from grid to filament through the tube capacitance and then through the filament lead inductance to ground. An RF voltage is developed on the filament that in effect causes the tube to be partly cathode driven. This undesirable extra drive power requirement can be minimized by series resonating the cathode return path with the filament bypass capacitors or by minimizing the cathode to ground inductance by using a specially designed tube socket using thin-film dielectric "sandwich" capacitors for coupling and bypassing.

High-power, grid-driven Class C amplifiers require a swing of several hundred RF volts on the grid. To develop this high voltage swing, the input impedance of the grid must be increased by the grid input matching circuit. Because the capacitance between the grid and the other tube elements may be 150 picofarads (pF) or more, the capacitive reactance at 100 MHz will be very low unless the input capacitance is parallel resonated with an inductor. Figure 4.7-25 shows two popular methods of resonating and matching into the grid of a high-power tube. Both methods can be analyzed by recognizing that the desired impedance transformation is produced by an equivalent L network.

In Figure 4.7-25a, a variable inductor (L_{in}) is used to raise the input reactance of the tube by bringing the tube input capacitance (C_{in}) almost to parallel resonance. Parallel resonance is not reached because a small amount of parallel capacitance (C_p) is required by the equivalent L network to transform the high

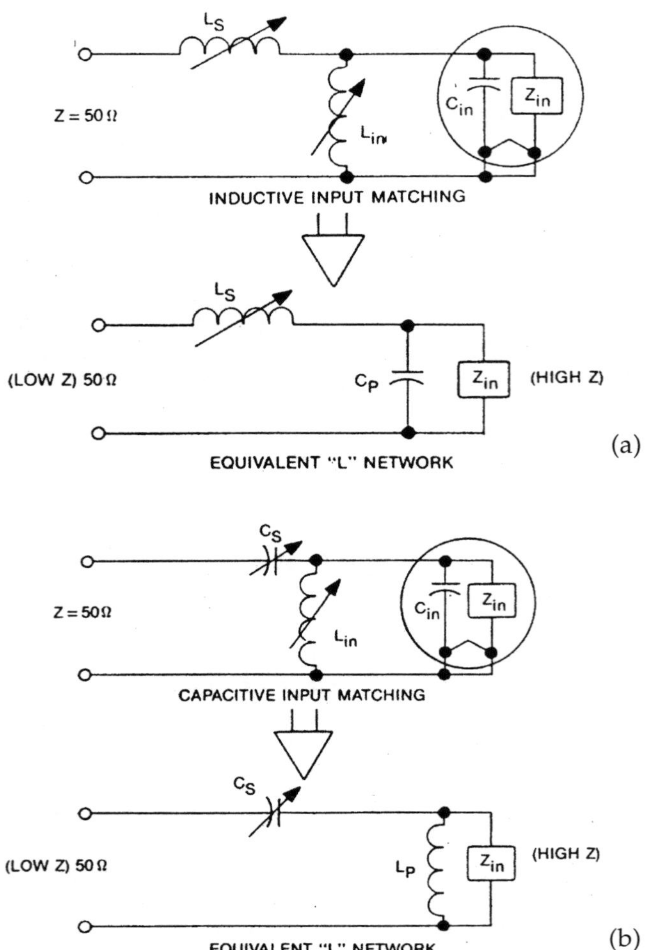

FIGURE 4.7-25 (a) Inductive input matching. (b) Capacitive input matching.

impedance (Z_{in}) of the tube down to a lower value through the series matching inductor (L_s). This configuration provides a low-pass filter by using part of the tube's input capacitance to form (C_p). Figure 4.7-25b uses variable inductor (L_{in}) to take the input capacitance (C_{in}) past parallel resonance so the tube's input impedance becomes slightly inductive. The variable series matching capacitor (C_s) forms the rest of the equivalent L network. This configuration is a high-pass filter.

Neutralization

Cathode-driven, grounded-grid amplifiers utilizing triodes do not require neutralization. The grid-to-ground inductance, both internal and external to the tube, should be kept low to maintain this advantage. Omission of neutralization will allow a small amount of interaction between the output circuit and the input circuit through the plate-to-filament capacitance. This effect is not very noticeable because of the large coupling between the input and output circuits through the electron beam of the tube.

Grid-driven, high-gain tetrodes require accurate neutralization for best stability and performance. This is particularly true if the amplifier is to be operated in Class AB linear mode for common amplification of FM + HD Radio signals. Self-neutralization can be accomplished by placing a small amount of inductance between the tube screen grid and ground that is usually in the form of several short, adjustable-length straps. The RF current flowing from plate to screen in the tube also flows through this screen lead inductance. This develops a small RF voltage on the screen, of the opposite phase, which cancels the voltage fed back through the plate-to-grid capacitance. This method of lowering the self-neutralizing frequency of the tube works only if the self-neutralizing frequency of the tube/socket combination is above the desired operating frequency before the inductance is added. Feedback neutralization utilizes a small coupling capacitor, usually in the form of a small plate located near the anode of the tube. The sample of the RF voltage from the anode intercepted by this plate is coupled through a 180° phase-shift network into the grid circuit. This technique has the advantage of providing neutralization over a very broad range of frequencies, if implemented correctly and stray reactances are minimized.

Special attention must also be given to minimizing the inductances in the tube socket by integrating distributed bypass capacitors into the socket and cavity deck assembly. Pentodes normally do not require neutralization because the suppressor grid effectively isolates the plate from the grid [8].

Power Amplifier Output Circuits

The output circuit usually consists of a high-Q (low loss) transmission line cavity, strip line, or lumped inductor that resonates with the tube output capacitance. A means of trimming the tuning and a means of adjusting the coupling to the output transmission line must also be provided by the output circuit. The tank circuit loaded Q is kept as low as practical to minimize circuit loss and to maintain as wide an RF bandwidth as possible.

Quarter-Wavelength Cavity

The quarter-wavelength coaxial cavity is the compact and popular PA output circuit illustrated in Figure 4.7-26. The tube anode is coupled through a DC blocking capacitor to a shortened quarter-wavelength transmission line. The tube's output capacitance is brought to resonance by the inductive component of the transmission line that is physically less than a quarter-wavelength long. Plate tuning can be accomplished either by adding end-loading capacitance at the high impedance end of the line with a variable capacitor or by changing the position of the ground plane at the low impedance end of the line. The plate tuning capacitor may be a sliding or rotating plate near the anode of the

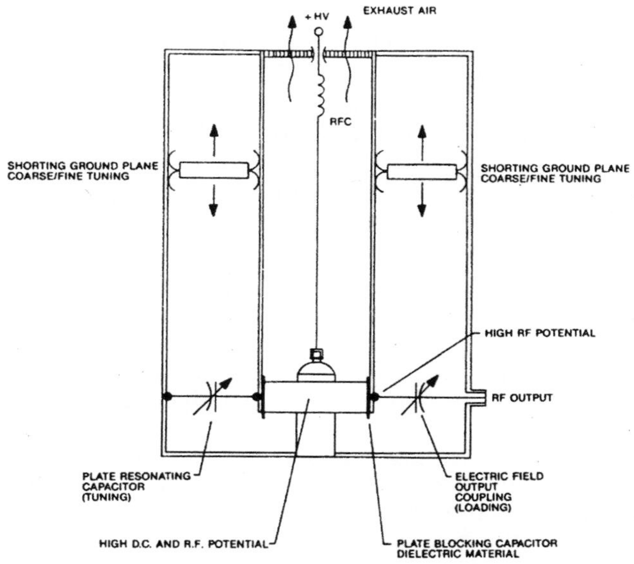

FIGURE 4.7-26 The quarter-wavelength cavity.

tube. The center conductor of the transmission line (air exhaust chimney) is at DC ground while the anode of the tube operates at a high RF and DC potential. High voltage is fed through an isolated quarter-wavelength decoupling network inside the chimney to the anode of the tube, while the plate blocking capacitor prevents DC current flow from the anode into the chimney. RF power may be coupled from a quarter-wavelength cavity to the transmission line by a capacitive probe placed at the high RF voltage point located at the anode end of the quarter-wave line. The loaded Q of this circuit varies with the degree capacitive coupling. Another method of coupling power from the quarter-wavelength cavity uses a tuned loop located near the grounded (high current) end of the line. In this case, the tuned loop operates as both an inductive and a capacitive pick-up device. The quarter-wavelength cavity has approximately twice the operating bandwidth of the half-wavelength cavity.

Folded-Half-Wavelength Cavity

Another approach to VHF power amplification uses the folded-half-wave cavity design illustrated in Figure 4.7-27. The DC anode voltage is applied to the lower portion of the plate line through a choke at the RF voltage null point. The half-wave line is tuned by mechanically expanding or contracting the physical length of a flexible extension (bellows) on the end of the secondary transmission line stub, which is located concentrically within the primary transmission line (air exhaust chimney). Coarse frequency adjustment is accomplished by presetting the depth of the top secondary section of plate line into the tank cavity. Power may be coupled from the half-wave line by an inductive loop located in the strong fundamental magnetic field near the center of the cavity.

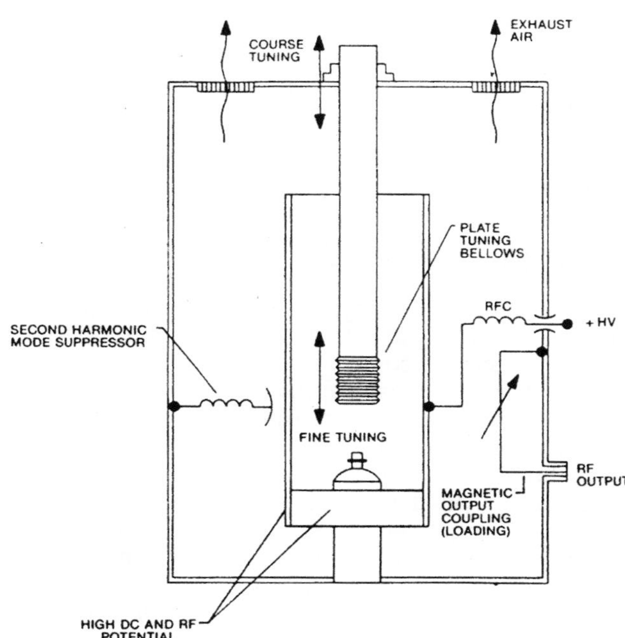

FIGURE 4.7-27 The folded-half-wavelength cavity.

Other power amplifier configurations may use lumped components or hybrid combinations with distributed transmission line elements to achieve similar results. The discrete circuit elements are chosen for their individual inductance or capacitance instead of being operated in a purely quarter-wave or half-wave mode. Stray inductance and capacitance add to the component values resulting in the hybrid nature of these circuits. Regardless of the specific configuration, the output circuit must transform the high resonant plate impedance down to the output transmission line impedance of 50 Ω. References at the end of this chapter give detailed information about the design of tube-type RF power amplifiers [8,13].

Power Supplies

Power supplies provide the appropriate DC or AC voltages to the various subsystems with the transmitter. In a typical FM transmitter, the voltages and currents can range from less than 5 volts at a few milliamperes to over 10,000 volts at several amperes. Safety must therefore be a prime consideration when working around potentially lethal power supplies [19]. Power supplies must be designed with adequate bleeder resistors and interlocks to discharge high voltages before an operator can come in contact with these circuits. The degree to which the AC ripple components are filtered out of the DC outputs of the power supplies will, in large part, determine the asynchronous (without FM modulation) AM noise of the FM transmitter. FM transmitters usually contain multiple power supplies for each of the functional blocks within the system. These power supplies fall into two general categories:

- Single-phase supplies (single input winding on the transformer)
- Polyphase supplies (three or more input windings on the transformer)

Single-Phase Power Supplies

Single-phase power supplies with conventional full-wave rectification and filtering are used most often for the FM exciter, the control system, bias supplies, and the IPA. A single-phase supply requires a larger filter choke to achieve the critical inductance requirement and a greater value of filter capacitance to maintain acceptably low ripple content compared to a polyphase supply. Large value filter components also mean that the greater stored energy in these components can have a more destructive effect if an arc-over occurs. Choke-input filter sections are normally used to help limit the in-rush current while the shunt capacitor is charging during turn-on. This reduces stress on the transformer and rectifiers by keeping the charging current nearly constant, producing the best filtering action. Choke-input filters have the undesirable characteristic of poor voltage regulation over a wide range of loads. The output voltage will rise well above the nominal value with no load unless there is enough current through the bleeder resistor to keep the choke in the constant current range. Fortunately, in an FM transmitter application, the load on the power supply is relatively constant because the power output of the transmitter does not vary significantly with FM modulation. In higher power transmitters with a three-phase main power source, it is important to balance each of the individual single-phase loads among the three phases so the total load on each of the individual phases is equal.

Polyphase Power Supplies

Polyphase power supplies are used for the final power amplifier, high voltage supply in most high-power transmitters. Sometimes they are used for tube or solid-state IPA supplies. Large blowers, used to cool transmitters, are usually operated from a three-phase power source. It is important to make the three-line connections to the blower motor in the proper sequence so the motor will turn in the proper direction. The most common type of polyphase supply is three-phase with full-wave rectification and LC filtering. Other polyphase systems encountered in broadcast equipment are usually multiples of the three phases, with 12 phase rectifiers becoming more popular. The main advantages of a polyphase power supply are:

- Division of the load current between three or more lines to reduce line losses and the size of each of the lines
- Greatly reduced filtering requirements after rectification due to the low ripple at the output of a polyphase full-wave rectifier
- Better voltage regulation with a choke input filter, with typically 6% or less variation from no load to full load
- Greater choice of output voltages from a given transformer by selection of either a delta or wye configuration

The main disadvantage of polyphase systems is their susceptibility to phase imbalance which causes degraded performance of the power supply. If significant imbalance exists in a polyphase system, ripple rejection will be reduced in the polyphase rectifier with a resulting increase in AM noise.

The broadcast engineer should determine that the local utility does, in fact, provide true three-phase power to the transmitter site. This can usually be verified by making sure that there are three transformers on the utility pole feeding the transmitter site. In many rural areas, the utilities are still synthesizing pseudo-three-phase service by providing the so-called open-delta (V–V) or Scott (T–T) connection with two transformers instead of true three-phase service. Operation on an open-delta service will degrade the transmitter's performance and increase the susceptibility of the transmitter to damage from transients on the line. Most transmitter manufacturers state that their warranty is void if the transmitter is connected to an open-delta system.

Regulated Power Supplies

In some cases, phase control switching regulation is applied to the high-voltage power supply feeding the final output tube. The regulation is accomplished by switching "thyristors" in the AC mains ahead of the primary winding of the transformer. As the switching duty cycle is reduced, the plate voltage is also reduced. It is important to protect solid-state devices connected to the main power line from transients. Heavy-duty transient suppressors are available for this purpose. Completely solid-state power amplifiers require lower voltages at much higher currents than tube amplifiers. Voltage regulation of these high-current supplies is necessary to suppress ripple, but the design of these specialized regulators is different from the typical high voltage power supply. Linear regulators are used at the lower power levels although they are low in efficiency, because they are simple and provide excellent ripple rejection without the need for suppression of switching transients. The linear regulators use series or shunt devices which change resistance dynamically to provide regulation with changes in load and therefore dissipate some of the power within the dynamic resistance.

Switched-mode regulators are used at higher power levels because they are high in efficiency, but they are more complicated and require additional suppression of the switching transients. The high efficiency comes from the digital "on" or "off" nature of the switching regulator that reduces resistive losses by using low-loss reactive components to store energy during switching. Switched-tap power supplies can provide good voltage regulation over a limited range at higher efficiency than a switched-mode regulated supply. A solid-state controller switches banks of silicon controlled rectifiers (SCRs) between several taps on the secondary of the power transformer to maintain a constant output voltage with changes in load. This type of power supply is simple and efficient and has an excellent power factor (PF).

Low-voltage, high-current power supplies contain extremely large amounts of stored energy. This can be dangerous due to the high peak currents that can occur during a short circuit across a component with high stored energy. For this reason, pay special attention to methods of safely discharging these circuits without damaging components or injuring the operator. The voltage regulator should provide short-circuit protection with some type of current limiting. The main danger to the operator from this type of power supply is burns due to the nearly instantaneous heating of metallic tools and other conductors (such as a metal watchband or ring) that accidentally get into a high-current path (such as a short across the filter capacitor or, in the case of tube-type amplifiers, the filament transformer secondary) [10].

Step-Start

Step-start is often used in large power supplies where peak in-rush currents become excessively high when the power supply is initially turned on at full power level. These peak currents are caused by the need to overcome the hysteresis effect (to initially magnetize the core of the transformer) when AC power is applied and to charge the filter capacitor. Step-start systems temporarily insert a resistance or reactance in series with the power lines to limit the current to a reasonable value until initial magnetization of the core and filter charge is completed.

Transmitter Control Systems

Transmitter control systems are often overlooked or given little priority in the selection and maintenance of a broadcast transmitter. The transmitter control system serves several important purposes including:

- Basic on/off control of the transmitter
- Overload protection to protect the transmitter from damage
- Safety interlock protection to prevent injury to people and accessory equipment such as RF switching equipment or RF loads
- A means of controlling the transmitter output power
- Remote control capability and interfacing at installations where the transmitter is not at the same location as the control operator
- Warm-up and cool-down timing sequences of filaments or other time-sensitive operations

Additionally, a transmitter control system may provide for:

- Status indications of overloads or other critical parameters
- Automatic regulation of the transmitter output power
- Local and remote diagnostic indications to aid in adjustment and maintenance
- Totally automatic operation of the transmitter plant
- Integrated remote control capability
- Control and monitoring via Internet connection

The transmitter's ability to stay on the air will only be as good as the reliability of the control system, so the selection and correct operation of the transmitter control system is important. Solid-state logic and microprocessor control systems are well protected against damage from high voltages by optical isolation, shunt protection techniques, and radio frequency interference (RFI) filtering. Operating experience with the current generation of transmitters has proven that a properly designed solid-state control system is far more reliable than older relay based designs.

Automatic Power Control

Many transmitters also provide automatic power control (APC) circuitry to maintain the transmitter's power output within preset limits by correcting for changes in line voltage, component aging, or small amounts of drift in operating parameters. The APC circuitry compares a sample of the transmitter output power to a reference and then adjusts the RF drive or other voltages within the transmitter to bring the output power within tolerance. Some of the more sophisticated APC circuits also utilize proportional VSWR fold-back of the transmitter output power. If a sample of the reflected power on the transmission line exceeds a safe limit, the transmitter output power is proportionally reduced to a safe level until the problem is resolved. This feature prevents lost air time during antenna icing or other limited VSWR situations. Of course, APC circuits should provide fast-acting shutdown of the transmitter during a catastrophic failure of the antenna system such as a short or open circuit.

Computer Control Systems

Most transmitters are now equipped with microprocessor-based control systems. Microprocessor technology lends itself well to industrial control applications like broadcast transmitters. The hardware can be made just as reliable as hard-wired digital logic. Changes and growth in the operational features are made by simple modifications in software instructions rather than a complete redesign of the hardware. Some of the features that distinguish these control systems from non-microprocessor systems include:

- Built-in *trouble tree* with fault location and diagnostic read-outs and user-friendly messages on GUI
- Simultaneous read-outs of all operating parameters
- Real-time calculation of efficiency, dissipation, VSWR, and other parameters requiring calculations
- Adjustment of parameters to maximize efficiency and minimize reject load dissipation
- Built-in clock/calendar for logging changes in operating status, power failures, and overloads
- Tolerance flagging on key operating parameters as warnings for logging and for preventive maintenance
- Ability to communicate with the outside world for remote control or logging purposes through a standard serial interface or Transmission Control Protocol (TCP)/Internet Protocol (IP)
- Integrated remote control capability without external remote control equipment and interfacing
- Provision to customize the system features to the station's individual requirements through the use of software menus
- Tuning aids that allow the operator to adjust the system for peak efficiency, minimum dissipation, and minimum VSWR by means of a real-time display of these calculated parameters

The method of communicating information to the operator varies among systems, but most use LED or LCD read-outs with codes or alphanumeric messages. Microprocessor controllers can also be equipped with GUIs so a large amount of information can be displayed in an easily read and understood format.

Controller Back-Up Systems

Some degree of redundancy is desirable in the transmitter control system so the transmitter can stay on the air if a portion of the system fails. There are several approaches currently in use to provide back-up systems. A multi-level hierarchy can be used that automatically hands over basic control functions to a primary life-support controller in the event of microprocessor hardware or software problems. Good system design separates diagnostic and supervisory functions from basic control functions so a failure in a higher level function will not affect the ability of the system to remain on the air without interruption. Watchdog circuits and software are embedded within the control system to detect failures and initiate corrective action before an interruption in service occurs. It is also possible to have distributed microprocessor systems with multiple processors that can automatically pick up the tasks of a failed processor without affecting the ability of the transmitter to remain on the air. The ability to quickly replace a controller subsystem while remaining on the air is a feature of the recent generations of transmitter control systems.

Remote Control Interfacing

Regardless of the type of control system used, the ability to interface easily with standard remote control systems is very important. Most transmitters have a parallel control interface with control lines for each individual function requiring a momentary contact closure of 24 volts DC or less at a current of 50 milliamperes or less. These levels are compatible with relay logic or optically isolated solid-state logic. Analog levels output from the transmitter for remote meter readings generally are fully buffered and fall into the range of 0 to 5 volts DC for a full-scale reading of a particular parameter at an impedance of less than 10 KΩ. The advent of microprocessor based control systems has been accompanied by a trend toward using a standard computer asynchronous serial interface in addition to the parallel interface. Serial interfacing reduces the number of connections to the transmitter and can carry both control functions and digitized meter readings through the same interface. By converting analog information into digital information before transmission of data to the remote control point, the need for calibration and recalibration of the remote metering point is reduced. The current trend is toward open, nonproprietary, serial data protocols and interface standards such as the Simple Network Management Protocol (SNMP) and TCP/IP. This gives the user maximum flexibility in choosing dedicated remote control equipment or using software based network management tools. Microprocessor-based control systems also allow the remote control system to have access to more in-depth information about the transmitter than is practical by parallel interfacing with an external remote control system. A personal computer or laptop computer can be used to control the transmitter through an ordinary dial-up phone line, Internet, or radio link.

Internet Protocol Remote Control and Web Access

Broadcast radio transmitters increasingly include IP for remote control and monitoring. The serial interfaces are usually either RS232 or Ethernet. When RS232 is used, an extra device is needed to convert the information to IP. Internet-based devices are the most extensive, and some generate GUIs on remote PCs via a web browser such as Microsoft Internet Explorer or Mozilla Firefox, as illustrated in Figure 4.7-28.

TCP/IP Protocol Suite

Internet-based services are built upon the TCP/IP protocol suite. TCP and UDP are both in this suite. UDP is known as a connectionless protocol. The connection between a server and a client is not guaranteed because there is no handshake. The protocol presumes one-way communication. A server knows where to send a message but its receipt is not acknowledged. UDP is therefore considered an unreliable protocol, but it is efficient when the network link is reliable.

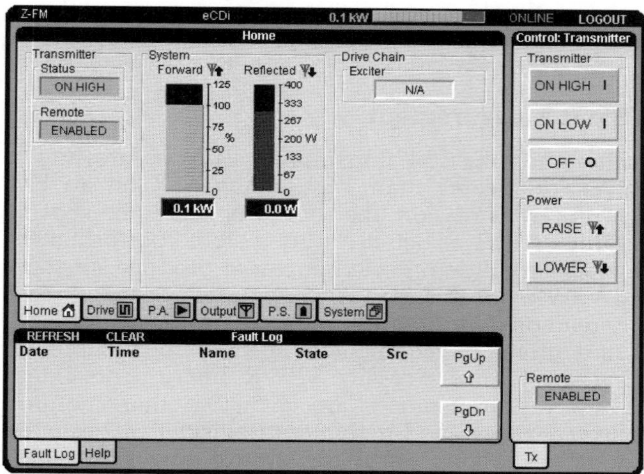

FIGURE 4.7-28 FM transmitter remote graphical user interface (GUI).

TCP is a connection-based protocol. Computers on each end of the communication link know about the other. As long as the physical link exists, TCP will ensure that messages are received reliably. It does this by acknowledging and requesting information multiple times if necessary.

HTTP and HTML

Hypertext Transfer Protocol (HTTP) is the basis of Internet-based web pages. Hypertext Markup Language (HTML) is a means of describing the layout and content of web pages and is the content of the files served to web browsers via HTTP. Web browsers decode HTML sent in these files to generate the proper display. HTTP and HTML work well for static graphical and textual images. A more powerful mechanism exists using Java Applets. Java Applets can be embedded within HTML and executed client-side, thereby relieving the server-side processing burden. This approach can be applied to radio transmitter systems because their embedded control systems are generally limited in order to be as cost effective as possible.

SNMP

The Simple Network Management Protocol was created to manage routers on a large network, but it has been applied to a multitude of other equipment, including radio transmitters. Data in the transmitter is identified with an object identifier (OID) that is unique; no two OIDs are the same. A management information base (MIB) describes the OIDs. An SNMP Manager is required on the remote computer, whereas the transmitter must be an SNMP Agent. An issue raised with the use of any of these protocols is that of setting up the gateway between the Internet and the control equipment. It is necessary to explicitly allow access through all the necessary ports; however, in general, the more ports that are open, the more vulnerable a network will be to hackers and viruses. For more information, see Chapter 9.5, "Transmitter Remote Control Systems."

RF Output and System Filtering

The high efficiency, nonlinear RF power amplifiers used in FM broadcast transmitters generate significant amounts of energy on frequencies that are integer multiples (harmonics) of the desired fundamental frequency. The output circuit alone does not provide enough harmonic attenuation to meet FCC regulations. To comply with Section 73.317 of the FCC Rules and to prevent interference to other services, a low-pass filter must be installed at the output of the transmitter. The FM band is narrow enough that one low-pass filter design can be used for any FM channel carrier frequency. These filters usually consist of multiple LC sections arranged so frequencies within the FM band are passed with typically 0.1 dB or less attenuation while frequencies above the FM band are attenuated 60 dB or more.

The most common type of filter used in this application is a reflective filter in which the frequency components outside the passband are reflected out of the filter and back toward the source because the filter exhibits an impedance mismatch at these undesired frequencies. The filter can be constructed using either lumped inductors and capacitors or by using a section of non-constant impedance transmission line to form distributed inductors and capacitors. The filters designed for low-power transmitters often employ lumped elements (coils and capacitors) because these elements are compact and can be integrated into the transmitter cabinet. The distributed type of filter is most often used with high-power FM broadcast transmitters because of its simplicity, extreme ruggedness, and ability to handle higher power levels. The distributed filter has the disadvantage of having larger physical dimensions than a similar lumped filter, which may require mounting the filter external to the transmitter cabinet. Figure 4.7-29 shows a cut-away view of a typical distributed low-pass filter. Note that the areas where the center conductor of the transmission line is smaller than that required for input Z_0 are inductive, while the areas where the center conductor is larger in diameter are capacitive.

When two filters (such as the output cavity and the harmonic filter) are connected together by a transmission line, the total harmonic attenuation will vary with

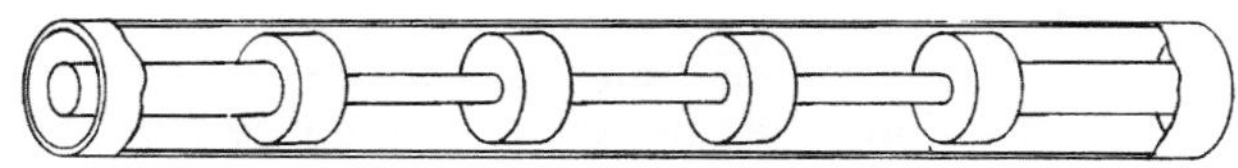

FIGURE 4.7-29 Distributed low-pass filter.

interconnecting line length. The attenuation characteristics of the harmonic filter are specified for the condition where both the source and load impedances are equal to the desired transmission line impedance. In practice, the source impedance at the output of the tank circuit is much less than the 50 Ω load impedance presented by a properly terminated filter. If an incorrect length of line is selected, the harmonic attenuation may be insufficient and the transmitter tuning may be affected. This undesirable condition can be corrected by changing the line length by approximately one-quarter wavelength. The line length between the transmitter output circuit and harmonic filter is usually supplied precut to the appropriate length by the transmitter manufacturer.

Harmonic Notch Filters

In some cases, a second -harmonic notch filter is required in addition to the low-pass filter if the second-harmonic component from the amplifier is high in amplitude and the cut-off slope of the low-pass filter is not steep enough to provide sufficient second-harmonic attenuation. The additional attenuation required (typically 30 dB) can be provided by a notch filter that places a short circuit across the transmission line at the second-harmonic frequency while exhibiting a high impedance at the fundamental frequency. A one-quarter wavelength (at the fundamental frequency) shorted coaxial stub is often used for this function. The second harmonic energy is partially reflected back into the power amplifier and partially dissipated in the equivalent series resistance of the series-tuned circuit formed by the stub. This shorted stub provides a very low inductance and a DC path from the center conductor of the transmission line to ground, providing a separate, protective advantage by shunting static discharges, such as lightning, to ground. Some transmitters have internal second harmonic suppressors that eliminate the need for an external notch filter.

Transmission Line Power and VSWR Measurements

Directional wattmeters are instruments that measure the forward power and reflected power in a transmission line. The net power delivered to the load (antenna) is the difference between the forward power and the reflected power. If the transmission line is perfectly matched, all the forward power will be absorbed by the load (antenna) and there will be no reflected power. The peak voltage at each point along the line will be the same value; similarly, the current at each point along the line will also have a uniform value. If the transmission line is mismatched, there will be reflected power with a resulting standing wave on the line. This means that the voltage and current distributions along the line will no longer be uniform, with high values at certain points on the line and low values at points one-quarter wavelength away. The ratio of the high-voltage value to the low-voltage value is the voltage standing wave ratio (VSWR).

VSWR Measurement

Although some FM transmitters can operate into a VSWR of greater than 1.8:1, the VSWR on an FM antenna transmission line should normally be kept to a value of 1.1:1 or better for good stereo performance. It takes very little reflected power to produce substantial VSWR. For this reason, the reflected power is usually read on a more sensitive meter position. Problems in the antenna system, such as loose connections or icing, may cause excessive VSWR. Instruments external to the transmitter are available that monitor reflected power and activate an alarm if it becomes excessive. As long as the transmitter power output is relatively constant, the use of reflected power to indicate excessive VSWR is simple and adequate [14].

Combined Transmitters

It is possible and practical to combine the output of two RF power amplifiers to obtain higher power output levels. The important advantage is that the broadcast signal is not interrupted should one amplifier fail. The radiated signal strength merely drops 6 dB until the failed amplifier is repaired and put back on the air. A dual amplifier system costs more than a single amplifier for a given total power output, but it offers the economic advantages of reducing lost air time and eliminating the need for a separate standby transmitter. Automatic or manual output switching can be used to route the full power of the remaining amplifier directly to the antenna.

Two methods may be used to bypass the output combining hybrid to allow 100% of the power of the remaining transmitter to be sent to the antenna if one transmitter of a combined pair should fail. The first method uses three motorized switches (or patch panels) to bypass the 3 dB hybrid combiner while connecting the operating transmitter directly to the antenna and the failed transmitter directly to the test load. This allows recovery of the 50% power lost in the reject load when one transmitter is off the air. A disadvantage is that the system must be taken off the air for several seconds for operation of the coax switches. A second method provided by some transmitter suppliers uses a pair of 3 dB hybrids interconnected with one fixed and one variable RF phasing section. The phasing section is constructed to operate while under RF power and can redirect the full output of either transmitter directly to the antenna and place the other transmitter into the test load without taking the system off the air. A dedicated system controller allows automatic or manual control. This so-called switchless combiner offers the highest possible on-air availability for combined FM transmitters. With complete redundancy in the RF power amplifier chain, some stations go one step further and also install dual exciters with automatic switching so if one exciter fails the other unit is quickly switched into service.

Hybrid Couplers

Hybrid couplers are reciprocal four-port devices that can be used either for splitting or combining RF sources over a wide frequency range. Figure 4.7-30 shows an exploded view of a typical 3 dB, 90° hybrid coupler. The coupler consists of two identical parallel transmission lines that are coupled over a distance of approximately one-quarter wavelength and are enclosed within a single outer conductor. Ports at the same end of the coupler are in-phase, while ports at opposite ends of the coupler are in quadrature (shifted by 90° in phase) with respect to each other. The phase-shift between the two inputs or outputs is always 90° and is somewhat independent of frequency. If the coupler is being used to combine two signals into one output, these two signals must be fed to the hybrid coupler in phase quadrature. The reason this type of coupler is also called a 3 dB coupler is that when used as a power splitter, the split is equal or half-power (3 dB) between the two outputs. Hybrid couplers can also be made with other coupling ratios. For example, a 6 dB coupler is often used for split-level combining of HD Radio and analog FM signals. A 10 dB coupler is used for high-level combining of HD Radio and analog FM signals.

Hybrid Combiners

The output hybrid combiner effectively isolates the two amplifiers from each other. Tuning adjustments can be made on one amplifier, including turning it on and off, without appreciably affecting the operation of the other amplifier. Good isolation is necessary so if one transmitter fails the other will continue to operate normally instead of in a mistuned condition. Two of the ports on the hybrid coupler are the inputs from the power amplifiers. The sum port is the antenna output terminal and the difference port goes to a resistive load called the *reject load*, as only the rejected power due to imbalance appears here. When the power fed to each of the two inputs is equal in amplitude with a phase difference of 90° the total power of both inputs is delivered to the sum port (antenna). Very little of the power appears at the reject load if the phase relationship and power balance are correct. If the phase relationship is reversed between the two amplifiers, all the power is delivered to the reject load, so it is important that the proper one of the two possible 90° phase relationships is used. When all the ports on the hybrid combiner are terminated, an isolation of 30 dB or more can be achieved between the power amplifiers. For perfect isolation between the amplifiers, the load impedance on the sum and difference ports must be exactly the same. This is approached in practice by providing a 1.0:1 VSWR with a resistive 50 Ω load for the termination (reject load) on the difference port and then reducing the VSWR on the antenna transmission line as low as possible by trimming the antenna match. This will prevent the input port impedances from changing significantly when one amplifier is not operating.

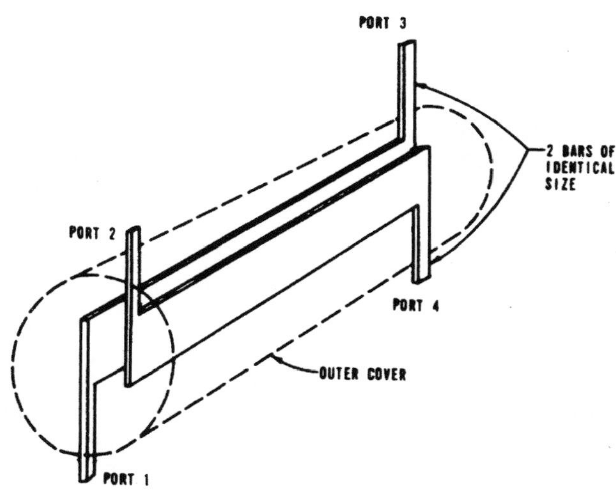

FIGURE 4.7-30 Model of 90° hybrid coupler.

The input ports will present a load to each transmitter with a VSWR that is lower than the VSWR on the output transmission line. This is because part of the reflected power coming into the output port will be directed to the reject load and only a portion will be fed back into the transmitters. The straight line in Figure 4.7-31a shows the effect of the output port $VSWR_3$ on the input port $VSWR_1$. The curved line in Figure 4.7-31a shows the change in isolation (dB) between the input port and output port as a function of the output port $VSWR_3$ for a 3 dB hybrid combiner. Figure 4.7-31b shows the reduction in isolation between coupler ports caused by antenna and/or reject load VSWR.

If the two inputs from the separate amplifiers are not equal in amplitude or not exactly in phase quadrature, some of the power will be dissipated in the difference port reject load. The match in input power and phase is not critical. Figure 4.7-32 shows how the percent of useful power available at the combiner output port is related to the square root of the power input imbalance between transmitter A (P_a) and transmitter B (P_b). Figure 4.7-33 shows how the percent of useful power available at the combiner output port is related to the phase error from the desired 90° between transmitter A and transmitter B.

The power lost in the difference port reject load can be reduced to a negligible level by touching up the amplifier tuning and by adjusting the phase shift. For example, if one amplifier is delivering only half the power of the other amplifier, only about 3% of the total available power remaining will be dissipated in the reject load; 97% of the remaining power, or 48.5% ($0.97 \times 0.50 = 0.485$), is still fed to the output transmission line. If one transmitter fails completely, half of the working amplifier's output goes to the antenna, and the other half is dissipated in the difference port reject load. This is why the radiated output drops by 6 dB or to one-fourth of the original combined power. The reject load must be rated to handle a minimum of

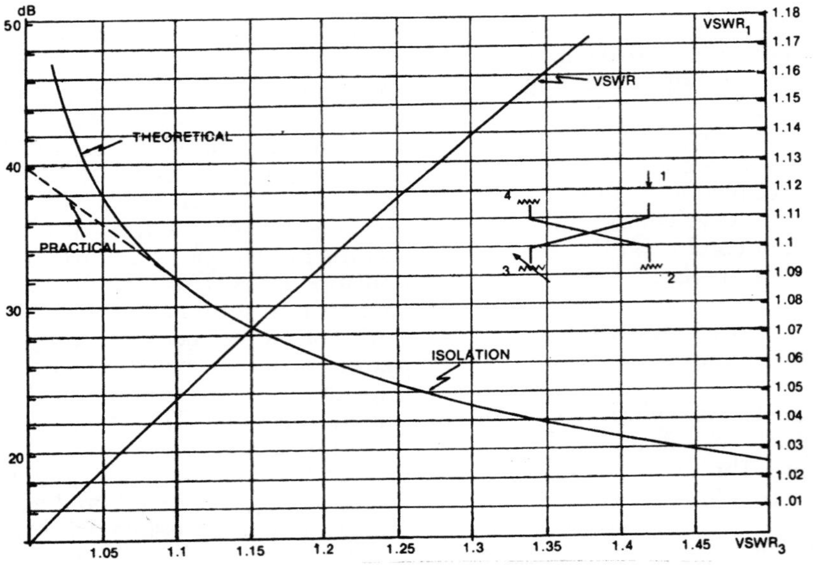

(a)

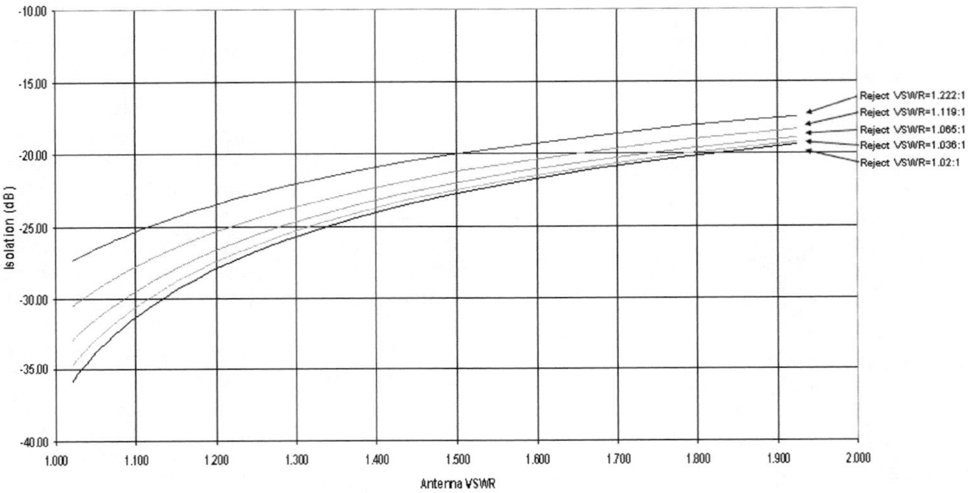

(b)

FIGURE 4.7-31 (a) Input port $VSWR_1$ as a function of output port $VSWR_3$ (straight line) and port-to-port isolation (in dB) as a function of output port $VSWR_3$ (curved line). (b) Coupler isolation (in dB) as a function of output port VSWR. (Traces are in same order as legend.)

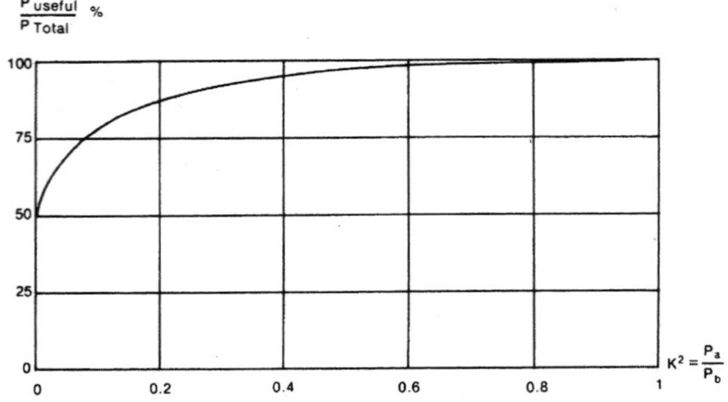

FIGURE 4.7-32 Power output as a function of imbalance in hybrid coupler.

one-fourth of the total combined power, but often the reject load is rated to handle one-half the total power so it can also be used as a test load for one of the transmitters.

Hybrid Splitting of Exciter Power

Figure 4.7-34 shows a block diagram of a pair of combined amplifiers with dual exciters. The exciters cannot be operated in parallel like the amplifiers because their RF outputs would have to be on exactly the same carrier frequency and almost exactly in phase under all modulation conditions. An exception would be to use a dual RF channel, digital exciter that can provide matched modulation on each of its RF outputs feeding

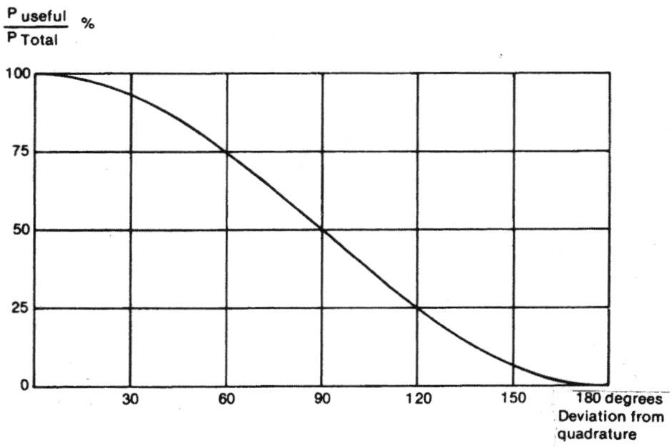

FIGURE 4.7-33 Power output as a function of phase error in hybrid coupler.

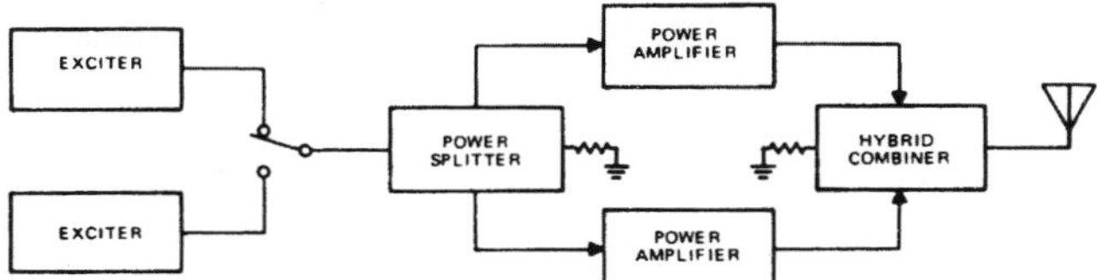

FIGURE 4.7-34 Block diagram of transmitter with two power amplifiers, a hybrid splitter, a hybrid combiner, and dual exciters with automatic switching.

each of the combined transmitters. An automatic or manual exciter switcher is used to direct the output of one exciter to the combined transmitter while the other exciter is routed to a dummy load. The one exciter in use feeds a hybrid splitter/phase shifter which transforms one 50 Ω input into two isolated 50 Ω outputs that have a 90° phase shift between them, with half the power going to each output. The operation of this hybrid splitter is the reciprocal of the hybrid combiner described above.

The exciter must have enough power output capability to drive both power amplifiers. In some cases, IPAs are inserted between the splitter and power amplifier to boost the drive level. The length of coax cable from the power splitter to each amplifier input must be cut to a precise length so the amplifiers will be fed in the proper phase relationship.

Each of the power amplifiers is assumed to have equal gain and phase shift. In practice, it may be difficult to get the amplifiers tuned so their gains and phase shifts are equal at the same time. For this reason, a *line stretcher* or variable phase shift network is usually included with the exciter splitter to adjust phasing independent of amplifier tuning. Recent generations of digital FM and HD Radio exciters have dual RF channels that digitally provide the necessary phase shift between the two outputs without external components. Additional information on the theory of hybrid couplers may be found in Chapter 6.2, "Analog Television Transmitters," and Chapter 4.10, "FM Combining Systems."

High-Level Combining HD Radio with Analog FM

High-level IBOC combining has been widely used since the beginning of HD Radio conversions. This method requires a new digital transmitter, about one-third the size of the main transmitter, whose IBOC output is combined at a high level with the FM signal, using a 10 dB hybrid coupler as shown in Figures 4.7-35 and 4.7-36. Two limitations of this method are the power lost in the reject load and the +10% additional power headroom demanded of the main analog FM transmitter to compensate for the loss of analog power into the reject load [35].

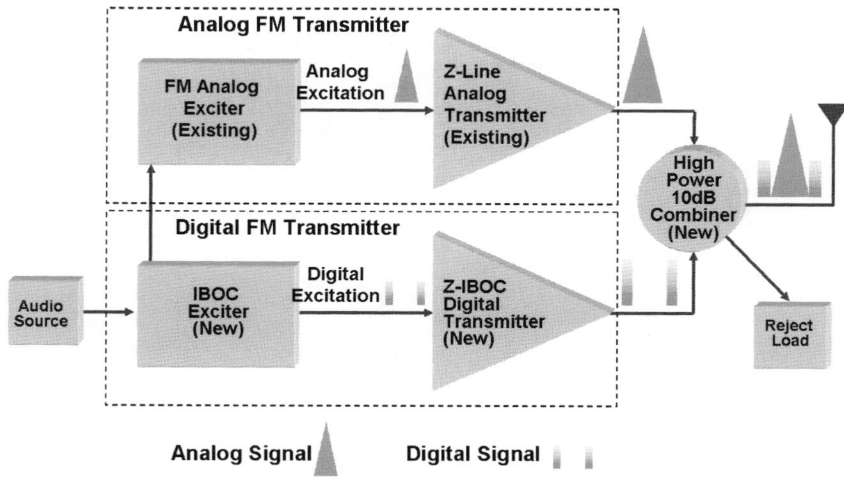

FIGURE 4.7-35 High-level combining with separate exciters.

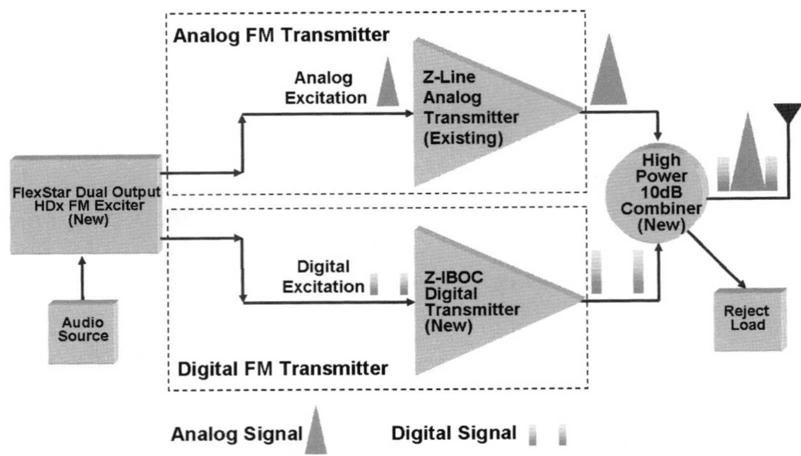

FIGURE 4.7-36 High-level combining with dual output exciter.

Split-Level Combining HD Radio with Analog FM

Split-level combining is an improved HD Radio combining scheme similar to high-level combining and is shown in Figures 4.7-37 and 4.7-38. The digital transmitter carries not only the IBOC signal, but also a portion of the FM signal that has been phased to add at the output combiner with the signal generated by the main FM transmitter. Split-level combining allows upgrading to HD Radio without increasing the power output requirement of the main FM transmitter. It also reduces the power lost in the reject load by about 50%.

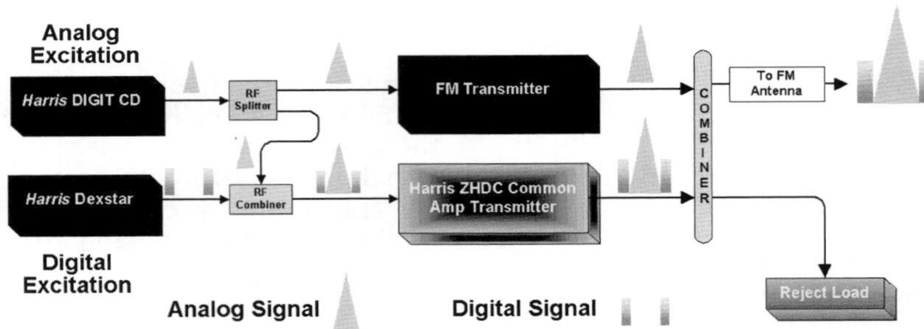

FIGURE 4.7-37 Split-level combining with separate exciters.

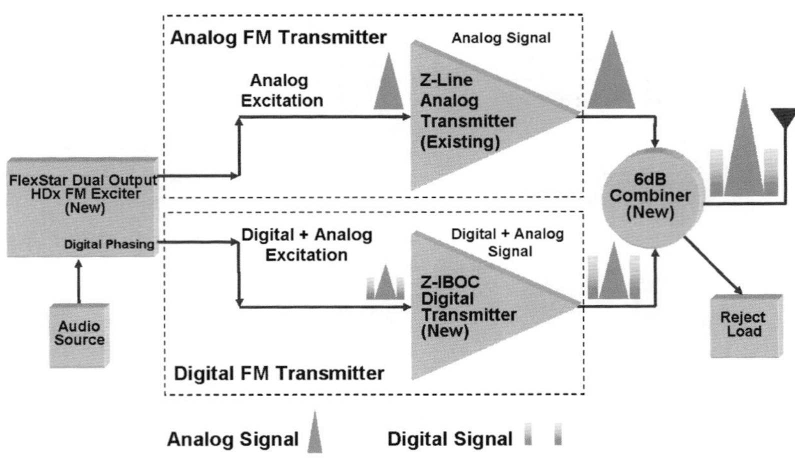

FIGURE 4.7-38 Split-level combining with dual output exciter.

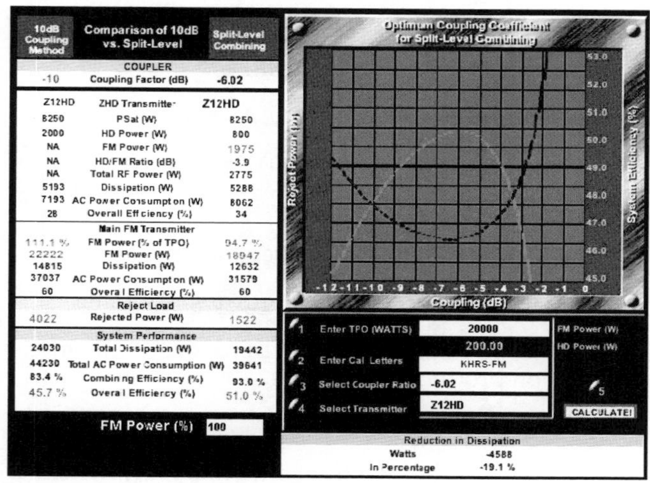

FIGURE 4.7-39 Split-level combining coupling coefficient calculator [36].

The coupling coefficient for this configuration is about 6 dB, instead of the 10 dB coupling used in high-level separate amplification. The coupling value is chosen to minimize the size required for the digital transmitter and to reduce the dissipation in the reject load while boosting the overall system efficiency by 3 to 4%. The total reject power is reduced by 50%. The digital transmitter power should be 35% of the main transmitter's power output to provide the secondary FM injection level that is required to make the main FM path lossless. Note that this value (35%) is comparable to the size required for separate amplification. Figure 4.7-39 is an example of a split-level system calculation [28–31,36]. A redundant FM + HD Radio combining system block diagram is shown in Figure 4.7-40.

COMBINING MULTIPLE FM STATIONS ON A SINGLE ANTENNA

It is common practice for several FM stations to share a single broadband antenna system by combining the multiple signals into a single, broadband antenna system.

Filterplexing

A device called a *filterplexer* (also known as an *RF multiplexer*) is used to connect several transmitters on different frequencies together onto one antenna system. The filterplexer provides isolation between the multiple transmitters while efficiently combining their power into a single transmission line. This is usually accomplished by a system of bandpass filters, band-reject filters, and hybrid combiners. The isolation is required to prevent power from one transmitter from entering another transmitter that can result in spuri-

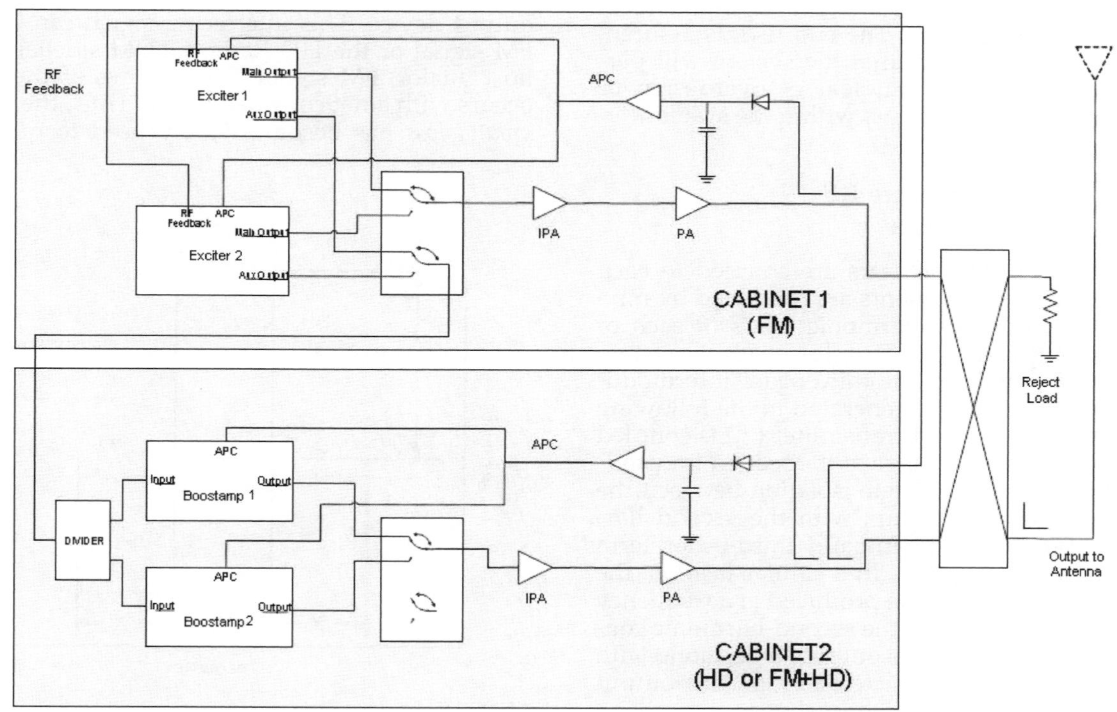

FIGURE 4.7-40 Redundant FM + HD Radio combining system.

ous emissions and to keep the rest of the system running in the event of the failure of one or more transmitters. The wideband port on a constant impedance filterplexer can also be used to sum multiple HD Radio signals into a dual-feed antenna system. An important consideration in the design of a filterplexing system is the effect on the phase response (group delay characteristic in the passband) of each of the signals passing through the system due to individual bandwidth limitations on each of the filterplexer inputs.

RF Intermodulation Between FM and HD Broadcast Transmitters

Interference to other stations within the FM broadcast band, as well as to other services outside the broadcast band, can be caused by RF intermodulation between two or more FM or HD Radio broadcast transmitters. Transmitter manufacturers have begun to characterize the susceptibility of their equipment to RF intermodulation so this information is becoming available to the designers of filterplexing equipment. The degree of intermodulation interference generated within a given system can be accurately predicted before the system is built if the actual mixing loss of the transmitters is available when the system is designed. Accurate data on *mixing loss* or *turnaround loss* not only speeds the design of filterplexing equipment but also results in higher performance and more cost-effective designs because the exact degree of isolation required is known before the system is designed. Filterplexer characteristics, as well as antenna isolation requirements, may be tailored to the specific requirements of the transmitters being used. The end user is assured, in advance of construction, that the system will perform to specification without fear of overdesign or underdesign of the components within the system.

Mechanisms That Generate RF Intermodulation Products

When two or more transmitters are coupled to each other, new spectral components are produced by mixing the fundamental and harmonic terms of each of the multiple output frequencies; for example, if two transmitters are involved, the third-order intermodulation (IM_3) terms could be generated in the following way. The output of the first transmitter (f_1) is coupled into the nonlinear output stage of the second transmitter (f_2). If there is not complete isolation between the two output stages, f_1 will mix with the second harmonic of f_2, producing an in-band third-order term with a frequency of ($2f_2 - f_1$). In a similar fashion, the other third-order term will be produced at a frequency of ($2f_1 - f_2$). This implies that the second-harmonic content within each transmitter's output stage, along with the specific nonlinear characteristics of the output stage, will have an effect on the value of the mixing loss. It is possible, however, to generate these same third-order terms in another way. If the difference frequency between the two transmitters ($f_2 - f_1$), which is an out-of-band frequency, remixes with either f_1 or f_2, then the same third-order intermodulation frequencies are produced. Empirical measurements indicate that the ($2f_2 - f_1$) type of mechanism is the dominant mode generating IM_3 products in modern transmitters using a tuned cavity for the output network. Figure 4.7-41 shows an example of how the intermodulation product frequencies may be calculated. Figures 4.7-42 and 4.7-43 show the resulting frequency spectra.

3rd ORDER INTERMODULATION PRODUCTS

f_1=100.3 MHz. f_2=101.1 MHz.

$2f_1$-f_2=[2(100.3)-(101.1)]=[200.6-101.1]=99.5 MHz.

$2f_2$-f_1=[2(101.1)-(100.3)]=[202.2-100.3]=101.9 MHz.

OR

[f_1-(f_2-f_1)]=[100.3-(101.1-100.3)]=[100.3-0.8]= 99.5 MHz.

[f_2+(f_2-f_1)]=[101.1+(101.1-100.3)]=[101.1+0.8]=101.9 MHz.

FIGURE 4.7-41 Calculation of third order RF intermodulation product frequencies.

Intermodulation as a Function of Turnaround Loss

Turnaround loss or mixing loss describes the phenomenon whereby the interfering signal mixes with the fundamental and its harmonics within the nonlinear output device. This interfering signal can be another FM signal or the HD Radio OFDM sidebands on the host analog FM signal. The mixing of these signals occurs with a net conversion loss. Thus, the term *turnaround loss* has become widely used to quantify the

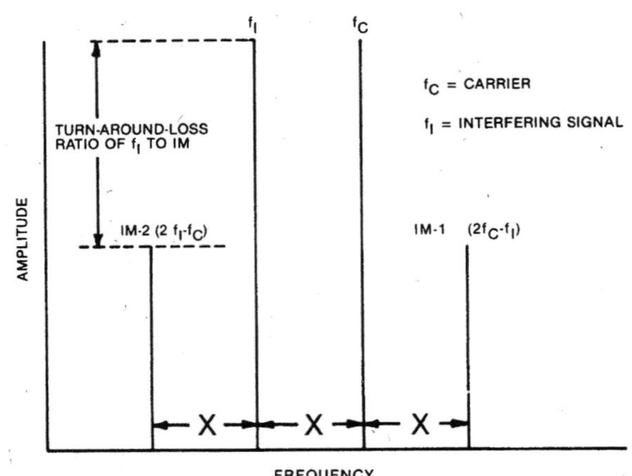

FIGURE 4.7-42 Frequency spectrum and turnaround loss ratio of third-order IM products with the interfering signal level equal to the carrier level.

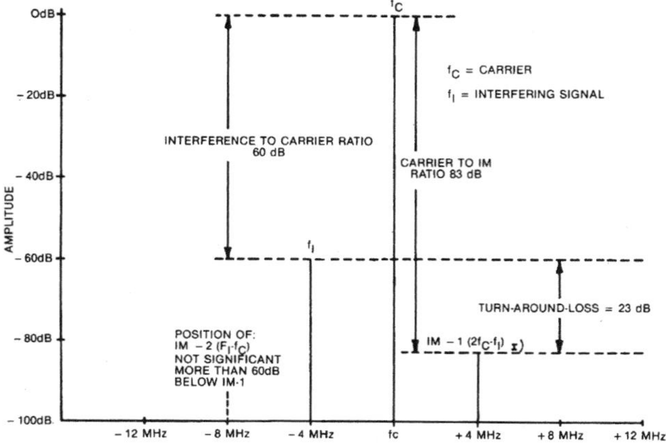

FIGURE 4.7-43 Typical frequency spectrum of third-order IM of a broadcast FM transmitter when interfering signal level is 60 dB below carrier level.

ratio of the interfering level to the resulting IM_3 level. A turnaround loss of 10 dB means that the IM_3 product fed back to the antenna system will be 10 dB below the interfering signal fed into the transmitter's output stage.

Turnaround loss will increase if the interfering signal falls outside the passband of the transmitter's output circuit, varying with the frequency separation of the desired signal and the interfering signal. This is because the interfering signal is first attenuated by the selectivity going into the nonlinear device; the IM_3 product is then further attenuated as it comes back out through the frequency selective circuit.

Turnaround loss consists of three components:

- Basic in-band conversion loss of the nonlinear device
- Attenuation of the out-of-band interfering signal due to the selectivity of the output stage (HD Radio interference is in-band and therefore not attenuated)
- Attenuation of the resulting out-of-band IM_3 products due to the selectivity of the output stage

As the turnaround loss increases, the level of undesirable intermodulation products is reduced, and the amount of isolation required between transmitters is also reduced.

The transmitter output circuit loading control directly affects the source impedance. This therefore affects the efficiency of coupling the interfering signal into the output circuit, where it mixes with the other frequencies present to produce IM_3 products. Light loading reduces the amount of interference that enters the output circuit with a resulting increase in turnaround loss. In addition, the output loading control setting will change the output circuit bandwidth (loaded Q) and therefore will also affect the amount of attenuation that out-of-band signals encounter passing into and out of the output circuit.

Second-harmonic traps or low-pass filters in the transmission line of either transmitter have little effect on the generation of intermodulation products. This is because the harmonic content of the interfering signal entering the output circuit of the transmitter has much less effect on IM_3 generation than the harmonic content within the nonlinear device itself. The resulting IM_3 products fall within the passband of the low-pass filters and outside the reject band of the second harmonic traps, so these devices offer no attenuation to RF intermodulation products. Figure 4.7-44 gives an overview of the various filtering options to prevent excessive IM_3 products.

RF Intermodulation between the FM and HD Radio Transmitters

Insufficient isolation between the analog FM transmitter and the digital HD Radio transmitter will cause undesirable RF intermodulation products. The required isolation can be difficult to achieve and depends on frequency in systems where space combining of the FM and HD Radio signals occurs in the antenna system. The desired isolation in a high-level or split-level system can be achieved by adjusting the impedances presented to the coupler ports.

Reduction of RF Intermodulation in Solid-State Transmitters

Depending on the topology of the combining system, broadband, solid-state transmitters are more likely to generate RF intermodulation products than single-tube transmitters, which have a narrowband selective cavity in the output stage; however, some solid-state designs use balanced *N*-way module combiners together with a conventional 3 dB hybrid as the final output combining stage. This topology tends to provide a uniform 15 dB minimum turnaround loss to incoming RF interference, regardless of the frequency separation of the solid-state transmitter's FM carrier from an interfering carrier. For frequency spacings closer than about 5 MHz from the interfering carrier, a solid-state transmitter using this combining technique actually produces fewer RF intermodulation IM_3 products than a single-tube FM transmitter in which close-in turnaround loss can be less than 6 dB. In some cases, a solid-state FM transmitter of this design may not require as much (or any) external RF filtering. In other cases, it may require more RF filtering than a tube/cavity transmitter. The turnaround loss of a narrowband tube/cavity transmitter is usually better than a broadband solid-state transmitter when there is a large frequency separation (>5 MHz) between the carrier frequency and the interfering frequency. This is due to the skirt selectivity of the narrowband final amplifier cavity. Co-sited FM transmitters are likely to require external RF filtering to prevent the generation or transmission of unacceptable RF intermodulation products, regardless of whether they are of tube/cavity or solid-state design. The turnaround loss of the

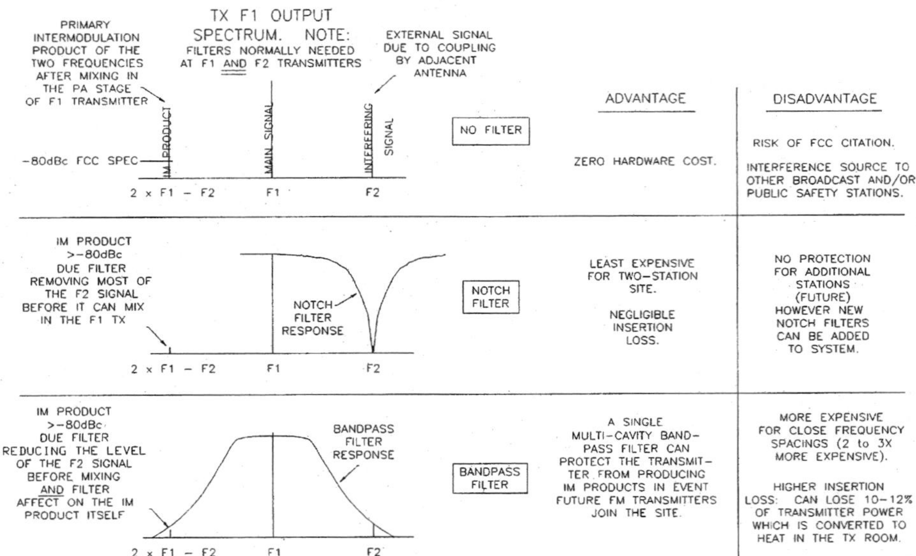

FIGURE 4.7-44 Three RF filtering options with advantages and disadvantages.

transmitter may also be improved by 20 dB or more with the addition of an external RF circulator to the output of the transmitter. The circulator is a unidirectional device that allows the transmitter output power to pass through it outbound with little attenuation to the antenna system while diverting incoming interfering RF signals to a reject load [11,25].

FM TRANSMITTER OPERATIONAL MEASUREMENTS

The FCC considers certain parameters important enough to justify almost continuous observation of modulation level, carrier frequency, and output power level.

NRSC-5A FM + HD Radio RF Mask Compliance

With the introduction of HD Radio, it is important to keep the transmission system operation within the RF emission mask prescribed by the NRSC-5 standard. This is due to the potential to generate RF intermodulation products between the host FM signal and the HD Radio sidebands. The NRSC-5 RF emission mask limits are shown in Figure 4.7-45. The spectrum analyzer setup parameters required to accurately measure the proposed NRSC-5B emission limits are listed in Table 4.7-4. The block diagram for the NRSC-5A test setup is shown in Figure 4.7-46. *Note:* Measurements are made by averaging the power spectral density (PSD) of the signal in a 1 kHz bandwidth over a 30-second period of time. The total FM power is defined as the reference at 0 dBc.

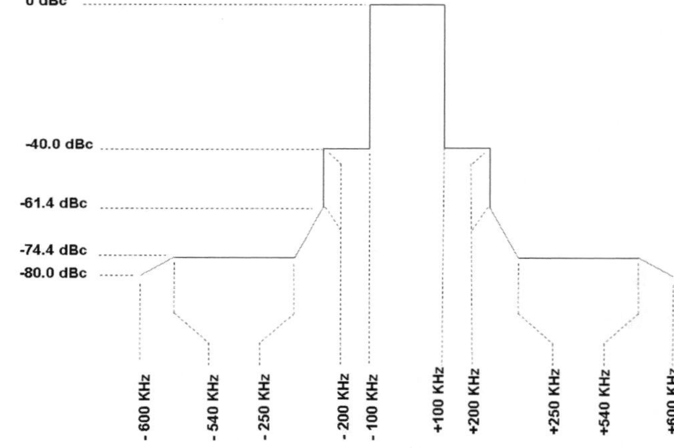

FIGURE 4.7-45 NRSC-5B RF emission limits.

FM Modulation Measurement

Measuring modulation levels can be accomplished with a broadcast-type modulation monitor or with a modulation analyzer test instrument. Most FM exciters have accurate built-in peak modulation displays for convenient setup and adjustment. Once the initial levels are correctly set, modern audio processing equipment will usually hold the modulation levels to the desired levels. Modulation monitors have a peak-indicating device that can be preset to flash at the

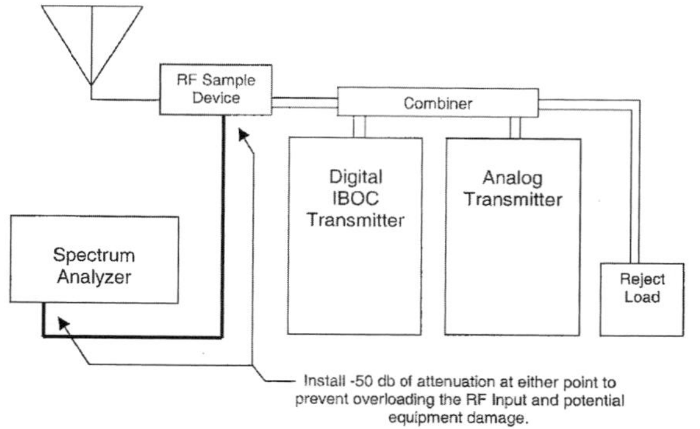

FIGURE 4.7-46 NRSC-5A RF emission measurement test setup.

particular level of interest. This device should be used instead of the meter to determine peak modulation conditions of the transmitter. See Chapter 8.3, "Radio Frequency Signal Analysis," for more details.

The reason for setting a peak deviation limit is so the related occupied bandwidth does not increase to the point of interfering with stations on adjacent channels and causing excessive audio distortion in the FM receiver by exceeding the receiver's IF bandwidth. The FCC currently enforces the modulation limit by monitoring the instantaneous peak deviation of the station as displayed on an oscilloscope. This method of measurement does not exactly correlate with the station's occupied bandwidth because the duty-cycle of the modulation peaks is not taken into account. As a result, many sophisticated peak-limiting and overshoot control devices have appeared on the market to maximize loudness without exceeding the peak deviation limit. These devices remove the low-energy peaks that would extend beyond 100% modulation. The use of these devices can cause some degradation of the audio quality, and they might not be used if the method of modulation measurement were changed to one based on occupied bandwidth. Recently introduced modulation measurement devices ignore short duration overshoots and provide modulation level indications that more accurately reflect the resulting occupied bandwidth. The FCC now enforces occupied bandwidth requirements for the FM band by measuring compliance with the FM emissions mask defined in FCC Rules Section 73.317.

TABLE 4.7-4
Nominal Settings of Controls on Spectrum Analyzer to Measure Emission Limits of Digital Radio Signal

Resolution bandwidth	1 MHz to set unmodulated FM carrier reference, then switch to 1 kHz
Span	2 MHz
Detect	Sample (not peak)
Video BW	1 kHz
Average	ON, at 100
Sweep	Leave in AUTO, sweep speed dependent on analyzer model
Sample points	400 minimum, use maximum points allowed
Marker	Set to peak
Marker	Reference level
Marker	Zero offset
Average type	Video

Carrier Frequency Measurement

The average carrier frequency must be measured with an accurate frequency monitor and maintained to within ±2000 Hz of the assigned channel. These monitors fall into two categories: analog display of the frequency error from the nominal carrier frequency and digital display of the absolute carrier frequency. The trend is toward the digital counter because of its accuracy and ease of use. Current generation exciters utilize a high-stability crystal oscillator or an external GPS frequency reference that can maintain the carrier frequency accuracy an order of magnitude better than the FCC requirement.

Historically, the FCC required FM stations to utilize type-approved modulation and frequency monitors, but this requirement has been eliminated. Each station is still required to maintain its frequency, modulation, and audio performance within the FCC limits defined in Part 73 of the Rules, but the responsibility for selecting the method of measurement and type of measuring equipment is now up to the licensee. Good operating practice is to have on hand the equipment necessary to accurately measure the broadcast signal; for example, modern modulation analyzers provide frequency agility as well as greater measurement capability than the more specialized modulation monitors. General purpose frequency counters are now available with sufficient accuracy to measure the carrier, subcarrier, and stereo pilot frequencies directly. Spectrum analyzers provide a wide range of capability, including the measurement of harmonic and spurious frequencies at the carrier frequency, composite baseband, Bessel nulls, occupied bandwidth, stereophonic and SCA crosstalk, and synchronous AM.

Measurement of RF Power Output

The methods for determining RF output power are specified in the FCC Rules. An accurately calibrated directional wattmeter provides an acceptable way of making a direct measurement of RF output power. Until recently, the directional wattmeter was seldom used as the primary RF power determining method because of the requirement for recalibration to a traceable standard at regular intervals. Use of the indirect method of power measurement avoided this requirement. The FCC permits the use of the transmitter power output meter directly, if it is periodically calibrated by comparison with the indirect method, instead of with a dummy load and standard wattmeter.

Using the indirect method, the output power is calculated from a measurement of the DC input power multiplied by the efficiency factor of the final amplifier stage. The efficiency factor is provided by the transmitter manufacturer on the final test data sheet or in the instruction manual and must be applicable to the particular frequency and power level in use. The power input to the final amplifier stage is normally defined as the product of plate voltage and plate current to this stage. Multiple output stages that are combined for the total power must have their individual DC power inputs arithmetically summed to obtain the total power input. The directional wattmeter can be used as a check when compared to the power output calculated by the indirect method to determine if the efficiency factor has changed due to incorrect tuning, changing antenna conditions, or a weak output device.

Measurement of AM Signal to Noise

The perfect FM transmitter will have a constant output, regardless of FM modulation or power supply variations. In practice, there will be some residual amplitude modulation of the FM transmitter. Two types of AM S/N are of interest to the FM broadcast engineer:

- Asynchronous AM S/N measured without FM modulation is required by the FCC Rules and is primarily related to power supply ripple.
- Synchronous AM S/N or incidental carrier AM (ICAM) measured with FM modulation is not required by the FCC Rules and is related to the tuning and overall bandwidth of the system.

Asynchronous AM

Residual amplitude modulation of the transmitter output without FM modulation, due primarily to power supply ripple, is measured with an AM envelope detector. Most FM modulation monitors include an AM detector for this purpose. The detector must include 75 μs deemphasis of its output. AM noise measurements must be made directly at the transmitter output (or an accurate sample of its output). No amplifying or limiting equipment may be used between the transmitter output and the AM detector, as this equipment would modify the residual AM noise level present. The FCC Rules require residual AM noise to be 50 dB below the level that would represent 100% amplitude modulation of the carrier. Because the transmitter cannot be amplitude modulated, this reference must be established indirectly by a measurement of the RF carrier voltage. (Refer to the instructions of the detector manufacturer to determine the reference level.) If the transmitter is unable to meet the 50 dB requirement, the problem can usually be traced to a power supply component or to line imbalance in a three-phase system.

Synchronous AM

Synchronous AM is a measure of the amount of incidental amplitude modulation introduced onto the carrier by the presence of FM modulation. Although this measurement is not required by the FCC Rules, it provides information about the amplitude response and tuning of the transmitter. Measurement of synchronous AM also gives the station engineer an idea of the overall system bandwidth and whether the passband is positioned correctly. Because all transmitters have limited bandwidth, there will be a slight drop-off in power output as the carrier frequency is swept to either side of the center frequency. This slight change in RF output level follows the waveform of the signal being applied to the FM modulator causing AM modulation in synchronization with the FM modulation. The concept is similar to the slope detection of FM by an AM detector used in conjunction with a tuned circuit.

Synchronous AM measurements are made directly at the transmitter output (or an accurate sample of its output). No amplifying or limiting equipment may be used between the transmitter output and the AM detector, as nonlinearities in this equipment could modify the synchronous AM level present. Because the transmitter cannot be fully amplitude modulated, an equivalent reference level must be established indirectly by a measurement of the RF carrier voltage. (Refer to the instructions of the detector manufacturer to determine this reference level.) Generally, the reference level is determined by setting a carrier level meter to obtain a specific DC voltage level at the output of the detector diode without modulation. It is important, when making these measurements, that the test setup does not introduce synchronous AM and give erroneous readings that would cause the operator to mistune the transmitter to compensate for errors in the measuring equipment.

The input impedance of the envelope detector must provide a nearly perfect match so there is a very low VSWR on the sampling line. A significant VSWR on the sampling line will produce synchronous AM at the detector because the position of the voltage peak caused by the standing wave moves along this line with FM modulation. A thru-line type of directional coupler normally used to drive the wattmeter movement has the envelope detector diode built into the sampling element. This provides a DC component that the meter movement responds to plus the demodulated synchronous AM component to which the meter movement does not respond. If the thru-line element output is fed to an oscilloscope instead of the wattmeter movement, the synchronous AM waveform can be accurately measured. This approach eliminates the errors due to VSWR on the sampling line, because the detector is located at the sampling point. The manufacturer of the thru-line coupler can supply the special connectors and/or cables to connect its output to the oscilloscope. It is important to avoid hum pick-up from AC ground loops while making these low-level measurements. Both the thru-line element detector and the precision envelope detectors have some residual RF on their DC output, so an RF filter network may be required between the detector and the input the oscilloscope.

Most FM demodulators cannot be relied upon to make accurate synchronous AM noise measurements, so it is necessary to cross-check the demodulator reading directly against the demodulated output of a precision envelope detector. This can be done by first measuring the DC component of the waveform with a voltmeter or by DC coupling the scope input. The scope is then AC coupled, and the input sensitivity is increased until an accurate peak-to-peak measurement of the AC modulation component can be made. The peak-to-peak AC voltage is then divided by twice the DC component to obtain the voltage ratio; 20 $\log_{10}$ of the voltage ratio is the actual synchronous AM noise level in dB below equivalent 100% AM modulation. Multiplying the voltage ratio by 100 yields the percent of AM modulation. Note that the peak detected value of the carrier must be doubled to convert it to the peak-to-peak value of the carrier. The ratio of the peak-to-peak modulation component to the peak-to-peak carrier is then used to calculate the percentage of synchronous AM modulation [15].

Acceptable Level of Synchronous AM

Synchronous AM of 35 dB or more below equivalent 100% AM is considered to be acceptable because the limited bandwidth of the IF filter in the receiver will reintroduce higher levels of synchronous AM to the FM signal before demodulation. Higher levels of synchronous AM can cause increased chopping of the signal at the receiver near the limiting threshold under weak signal fringe area conditions and can exacerbate multipath problems. Excessive synchronous AM is also an indirect indication of passband-induced distortion problems that degrade stereo performance and increase SCA crosstalk.

Many older multi-tube transmitter designs currently in use will have as much as 6% (–30 dB) synchronous AM when simply tuned for best power output and efficiency even though the asynchronous AM (without modulation) may be better than –50 dB. Some of the newer single tube transmitters can be adjusted for 50 dB or more suppression of synchronous AM. The synchronous AM level of virtually any FM transmitter can be improved by proper tuning techniques. An approximation to the overall system bandwidth can be related to the synchronous AM as shown in Table 4.7-5.

Limitations of Synchronous AM Measurements

Synchronous AM measurements are an indirect way of evaluating and optimizing FM performance. Even though synchronous AM measurements are a helpful aid to correctly tune an FM transmitter, these measurements tell only the amplitude response half of the total story. Transmitter tuning also affects the group delay (time) response which in turn affects the relative time delays of the higher order FM sidebands. Even though the amplitude response appears flat when the grid is heavily driven, the group delay (time) response still has a serious effect on the higher order FM sidebands.

Synchronous AM *versus* Symmetrical Group Delay Response

Computer simulations as well as empirical measurements made on FM transmitters showed that group delay asymmetry results in much more distortion than asymmetrical amplitude response [17]. As long as the group delay response is symmetrical, the amount of synchronous AM will have little effect on the FM modulation performance and distortion. Most FM transmitters will exhibit a significant increase in synchronous AM when tuned for symmetrical group delay response even though this results in the best FM

TABLE 4.7-5
Approximate System Bandwidth as Related to Synchronous AM

Synchronous AM Below 100% AM (+75 kHz at 1 kHz FM)	Approximate Bandwidth of Transmitters (–3 dB point)	RF Level Variation at Receiver Limiter	
		(%)	(dB)
–30 dB	410 kHz	6.32	0.57
–35 dB	550 kHz	3.54	0.31
–40 dB	730 kHz	2.00	0.18
–45 dB	1.00 MHz	1.12	0.10
–50 dB	1.34 MHz	0.64	0.06
–55 dB	1.82 MHz	0.36	0.03
–60 dB	2.46 MHz	0.20	0.02

modulation performance. Tuning for minimum synchronous AM is a good starting point, but it is more desirable to finish tuning at the symmetrical group delay point. Fine tuning the input and output for minimum even-order harmonic distortion will optimize the group delay (time) response. Transmitters that utilize wideband solid-state IPAs will add less distortion to the FM signal because both the amplitude and group delay (time) response will be better than systems utilizing several tuned stages.

Tuning the Transmitter for Best Performance

When properly adjusted as discussed in the preceding sections, modern power amplifiers can operate with high reliability and high-power efficiency without compromising subcarrier performance. All optimization should be made with the APC system disabled so the APC will not chase the adjustment to keep the output power constant. The transmitter should be connected to the normal antenna system rather than to a dummy load. This is because the resistance and reactance of the antenna will be different from the dummy load, and the optimum tuning point of the transmitter will shift between the two different loads. The tuning sequence is described below.

Initial Tuning and Loading

The transmitter is first tuned for normal output power and proper efficiency according to the manufacturer's instructions. The meter readings should closely agree with those listed on the manufacturer's final test data sheet if the transmitter is being operated at the same frequency and power level into an acceptable load.

Input Tuning and Matching

The input tuning control should first be adjusted for maximum grid current and then fine-tuned interactively with the input matching control for minimum reflected power to the driver stage. Note that the point of maximum grid current may not coincide with the minimum reflected power to a solid-state driver. This is because a solid-state driver may actually produce more power at certain complex load impedances than into a 50 Ω resistive load. The main objective during input tuning is to obtain adequate grid current while providing a good match (minimum reflected power) to the coaxial transmission line from the driver. In the case of an older transmitter with a tube driver integrated into the grid circuit of the final amplifier, the driver plate tuning and the final grid tuning will be combined into one control that is adjusted for maximum grid current.

Output Tuning

The output tuning control adjusts the resonant frequency of the output circuit to match the carrier frequency. As resonance is reached, the plate current will drop while both the output power and screen current rise together. Under heavily loaded conditions this dip in plate current is not very pronounced, so tuning for a peak in the screen current is often a more sensitive indicator of resonance.

Output Loading

There is a delicate balance between screen voltage and output loading for amplifiers utilizing a tetrode tube. Generally, there is one combination of screen voltage and output loading where peak efficiency occurs. At a given screen voltage, increasing the amplifier loading will result in a decrease in screen current, while a decrease in loading will result in an increase in screen current. As the screen voltage is increased to get more output power, the loading must also be increased to prevent the screen current from reaching excessive levels. Further increases in screen voltage without increased loading will result in a screen overload without an increase in output power.

Automatic Power Control Headroom

APC feedback systems are utilized in many transmitters to regulate the power output around a predetermined set-point with variations in AC line voltage or changes in other operating parameters. Modern FM broadcast transmitters may utilize a high gain tetrode as the final amplifier stage with adjustment of the screen voltage providing fine adjustment of the output power. For each power output level there is one unique combination of screen voltage and output loading that will provide peak operating efficiency. If the screen voltage is raised above this point without a corresponding increase in loading, there will be no further increase in power output with rising screen voltage and screen current. If the screen voltage is raised without sufficient loading, a screen current overload will occur before the upward adjustment in power output is obtained. To avoid this problem, tune the transmitter with slightly heavier loading than necessary to achieve the desired power output level and allow for about 5% headroom in adjustment range. The output loading can be adjusted for a peak in output power of 5% over the desired level, and then the screen voltage can be reduced enough to return to the desired level. This procedure will allow headroom for an APC system controlling screen voltage and will result in about a 1% compromise in efficiency, but it will ensure the ability to increase power output up to 5% without encountering a screen overload.

Centering the Passband

A simple method for centering the transmitter passband on the carrier frequency involves adjustment for minimum synchronous AM. If the bandpass is narrow or skewed, increased synchronous amplitude modulation of the carrier will result. A typical adjustment procedure is to FM modulate 100% at 1 kHz and fine-

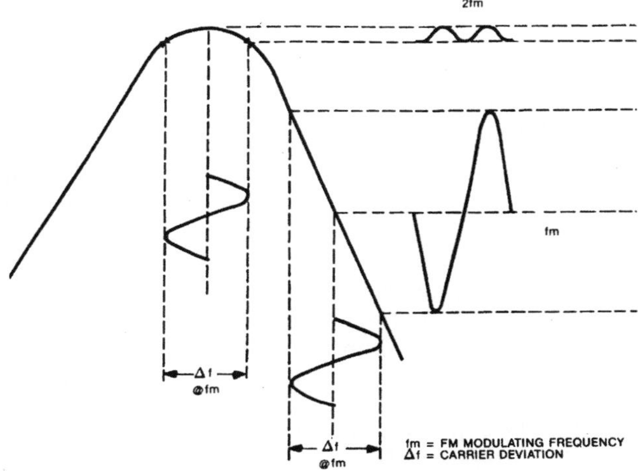

FIGURE 4.7-47 Synchronous AM waveforms.

adjust the transmitter's grid tuning and output tuning controls for minimum 1 kHz AM modulation as detected by a wideband envelope detector (diode and line probe). The FM modulating frequency of 1 KHz is used instead of the 400 Hz so the audio high-pass filter in the audio analyzer can be used to eliminate the AC line frequency related asynchronous component from the synchronous AM component. It is helpful to display the demodulated output from the AM detector on an oscilloscope while making this adjustment. Note that, as the minimum point of synchronous AM is reached, the demodulated output from the AM detector will double in frequency to 2 kHz. This is because the fall-off in output power is symmetrical about the center frequency, causing the amplitude variations to go through two complete cycles for every one FM sweep cycle as shown in Figure 4.7-47. It should be possible to minimize synchronous AM while maintaining output power and efficiency in a properly designed power amplifier. If an oscilloscope is not available for direct observation of the demodulated AM waveform, the 19 kHz bandpass filter and metering circuit used to measure pilot injection level in a stereo modulation monitor may be used as a tuning aid to center the passband. In this case, the main carrier is FM modulated with a 9.5 kHz tone (without pilot or any other modulation). If the transmitter is tuned for symmetrical AM response, the demodulated AM signal will have a strong second harmonic component at 19 kHz which falls within the passband of the pilot metering circuit. The output of the AM detector is fed into the composite baseband input of the stereo modulation monitor. The transmitter is then tuned for a maximum reading on the pilot injection metering position.

Effect of Transmitter Tuning on the FM Sidebands

The higher order FM sidebands will be slightly attenuated in amplitude and shifted in time (group delay) as they pass through the final amplifier stage. The alterations in the sideband structure that are introduced by the amplifier passband result in distortion after FM demodulation at the receiver. The amount of distortion depends on the available bandwidth *versus* the modulation index being transmitted. For a given bandwidth limitation, the distortion can be minimized by centering the passband of the amplifier around the signal being transmitted. This will cause the amplitude and group delay errors to affect both the upper and lower sidebands equally (symmetrically).

Tuning an amplifier for minimum plate current or for best efficiency does not necessarily result in a centered passband. One way to center the amplitude passband is to tune the amplifier for minimum synchronous AM modulation while applying FM modulation to the transmitter. Because the circuit topology of most transmitters exhibits a difference in tuning between the symmetrical amplitude response and the symmetrical group delay response, FM modulation performance can be further improved by tuning for symmetrical group delay rather than for minimum synchronous AM. The symmetrical group delay tuning point usually does not coincide exactly with the symmetrical amplitude tuning point but rather falls between the point of minimum synchronous AM and the point of maximum efficiency.

The transmitter may be tuned for minimum intermodulation distortion in left-only or right-only stereo transmissions. Stereo separation will also vary with tuning. For stations employing a 67 kHz SCA, transmitter tuning is important for minimizing crosstalk into the SCA. Modulate one channel only on the stereo generator to 100% with a 4.5 kHz tone. This will place the lower second harmonic (L–R) stereo sideband on top of the 67 kHz SCA. Activate the SCA at normal injection level without modulation on the SCA. Tune the transmitter for minimum output from the SCA demodulator. This adjustment can also be made by listening to the residual SCA audio while normal stereo programming is being broadcast.

A more sensitive test is to tune for minimum even-order harmonic distortion that will result in a symmetrical group delay response and will optimize distortion, separation, and crosstalk.

Modern power amplifiers have been designed to operate without compromising subcarrier performance. By providing broadband matching circuits, adjustment of these transmitters for optimum FM modulation performance (such as minimum distortion, minimum crosstalk, maximum separation) is repeatable and stable.

The following field adjustment techniques are listed in ascending order of sensitivity:

- Tune for minimum synchronous AM noise.
- Tune for minimum IMD_3 in the left or right channel only.
- Tune for minimum crosstalk into the unmodulated SCA subcarrier.
- Tune for minimum even-order harmonic distortion (symmetrical group delay).

In any of these tests, the grid tuning is frequently more critical than the plate tuning. This is because the impedance match into the input capacitance of the grid becomes the bandwidth limiting factor. Even though the amplitude response appears flat when the grid is heavily driven, the group delay (time) response has a serious effect on the higher order FM sidebands [11,12,15,16].

OPTIMUM TUNING *VERSUS* EFFICIENCY

VHF amplifiers often exhibit a somewhat unusual characteristic when tuning for maximum efficiency. The highest efficiency operating point does not exactly coincide with the lowest plate current because the power output continues to rise on the inductive side of resonance coming out of the dip in plate current. If the amplifier is tuned exactly to resonance, the plate load impedance will be purely resistive, and the load line will be linear. As the output circuit is tuned to the inductive side of resonance, the plate load impedance becomes complex, and the load line becomes elliptic instead of linear because the plate current and plate voltage are no longer in phase. Apparently, best efficiency occurs when the phase of the instantaneous plate voltage slightly leads the plate current. This effect is believed to be caused by the nonlinear gain characteristics of the power amplifier tube operating on an elliptic load line.

Care of Power Tubes

The operating life of high-power vacuum tubes can be extended by proper care. Most high-power tubes utilize a directly heated cathode composed of a thoriated tungsten filament structure. The key points to extending the life of RF power tubes include:

- Store tubes upright, along the axis of symmetry, not on their side, to keep the internal elements concentrically aligned.
- Use care when handling tubes to prevent mechanical shocks to the delicate internal structure. Do not set a tube on a hard surface without padding.
- Keep the tube seals and anode cooler free of dust and dirt by weekly cleaning even in a clean environment.
- Keep a spare tube on hand and rotate the tubes every few months to help keep the chemical "gas-getter" active so the tubes remain gas free.
- Keep a regular record of all tube operating parameters so any trend of changes will be noticeable. If a tube fails during the warranty period, this data will be necessary to receive credit on a replacement tube.
- Monitor the filament voltage on a true root mean square (RMS) responding instrument and log any changes for future reference. The sampling point for this voltage measurement should be located as close to the tube's filament contacts as possible to minimize errors due to voltage drops in the filament wiring.

A properly operated tube will gradually lose emission from the cathode until it is no longer useful because the emissive material has been consumed. The carcass of the tube can then be rebuilt with a new cathode and recycled back into service. Tube life is not directly related to plate dissipation (within the ratings) but is related to the filament operating temperature (filament voltage) and the current density (milliamperes per heater watt) emitted by a given size filament. This means that operating a given tube type at a lower filament voltage and plate current will proportionately increase the life of the tube. For directly heated thoriated tungsten filaments, the plate current should be less than 4 milliamperes per watt of heater power for extended life.

Normally a new tube will deliver full output at a reduced filament voltage. By operating the tube at the optimum filament voltage, the filament life can be significantly extended. The optimum value may be found by slowly reducing the filament voltage from the manufacturer's rated value until the RF power output drops about 2% and then increasing the filament voltage until the RF power increases back up 1%. Informational bulletins from tube manufacturers on extending tube life permit the filament voltage to be reduced more than 5% below the manufacturer's rating as long as operation is closely monitored to stay above the point where there is a 1% drop in power output [20]. A brand new tube should be operated at the full rated filament voltage for the first 200 hours before the voltage is reduced to the optimum value for long life. This will ensure that the "gas-getter" is properly activated. As the tube ages, the filament voltage will have to be increased to stay at the optimum value until RF output power cannot be maintained at or above the rated value of filament voltage. At this point, the tube's useful life comes to an end. Check the manufacturer's data sheets and application notes that are often enclosed with the tube. Guides to the proper care of power tubes and forced air cooling are listed in References [8] and [20].

INSTALLATION CONSIDERATIONS

Adequate planning and care in the installation of an FM + HD Radio broadcast transmitter and associated equipment will help avoid many problems that may be difficult and expensive to correct later; for example, poor grounds and ground loops may cause high electrical noise levels.

FM + HD Radio Transmitter Plant

The HD Radio transmission facility is usually an extension of the existing analog FM transmitter site. Consideration must be given for the additional floor space, AC power, and cooling required by the digital transmitter, RF power combiner, and RF reject load. Figure 4.7-48 shows a simplified block diagram of a

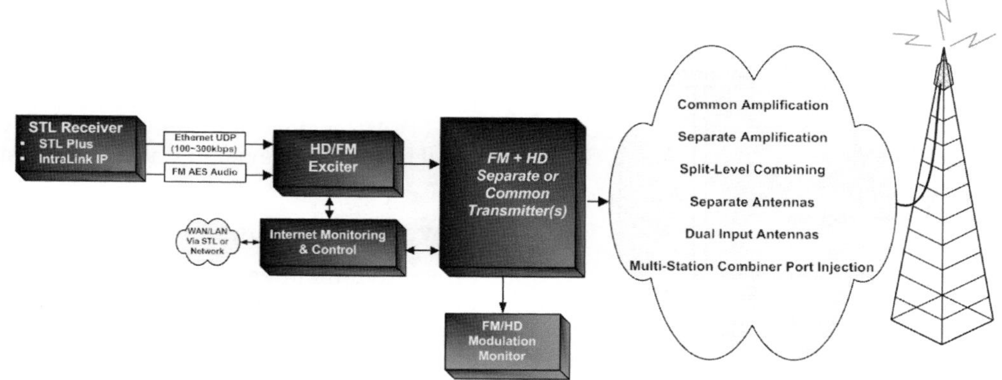

FIGURE 4.7-48 Simplified block diagram of a typical FM + HD Radio transmission facility.

typical FM + HD Radio transmission facility. For more information about converting a transmission facility to HD Radio, see References [26], [33], and [34].

Wiring the Transmitter Plant

Separate metallic shielded conduits or troughs should be provided for the audio and the AC wiring. A third conduit should be used if computer logic levels are employed for equipment control. High-voltage and high-current wiring should be isolated from low-level audio, control logic, and metering signals to prevent the coupling of transients that could produce fault conditions. Conduits or wiring troughs may be either overhead or below the cabinets. The AC wiring should be well separated from the audio pairs to prevent the induction of unwanted hum and noise into the audio circuits.

Audio shields should be grounded at only one point to prevent ground loops in the shields. This point may have to be found experimentally to give the lowest noise pick-up. The equipment racks and transmitter should be connected together by copper straps at least 2 inches wide tied to a good earth ground at one point. If a good ground screen is not available, a satisfactory ground can be provided by driving four or more copper ground rods 8 to 10 ft long into the ground spaced about 3 feet apart. These ground rods should be tied together with a wide copper strap. The straps connecting the equipment to the earth ground should be as short and direct as practical. See Chapter 9.3, "Facility Grounding Practices," for more information on grounding techniques.

It is often difficult to remove VHF RF from the equipment by grounding because, at FM carrier frequencies, nearly any connection to an earth ground has an appreciable impedance. The best way to keep RF out of sensitive low-level circuits is by keeping them enclosed within an RF shield and by filtering leads that enter the shielded unit. Filters in the audio lines may be appropriately sized shunt capacitors and small bi-filar RF chokes that add common mode inductance without adding differential inductance that would affect audio performance.

For stereo transmission, it is necessary to keep the L and R audio lines phased properly. To ensure proper monaural compatibility, correct audio polarity must be maintained throughout the station from the microphones and other audio sources through all of the audio equipment to the stereo generator audio input terminals. Stereo phone line pairs or separate RF studio-to-transmitter links should also be checked for correct polarity and equal phase delay. The transmitter equipment should be located and arranged to provide sufficient clearance around the front, sides, and rear for easy access during servicing and maintenance.

Transmitter Cooling

Almost all FM broadcast transmitters require forced air cooling to remove heat from the output stage and other assemblies within the cabinet. An important consideration in locating the transmitter is the provision for adequate cooling air. If the overall efficiency of the analog FM transmitter is about 50%, the transmitter will generate about the same number of kilowatts of heat as it does RF power output. HD Radio transmitter systems that require linear amplification will operate at lower efficiency than a comparable power FM transmitter. The extra heat load from the HD Radio transmitter and related signal combining reject load must be considered when planning the air conditioning system.

Figure 4.7-49a shows a transmitter located in an air-conditioned room. This type of closed-loop system requires no special ducting and has the advantage that the transmitter intake air is usually much cleaner than outside air. The transmitter exhaust air places a substantial heat load on the air conditioner during the summer, but it becomes a source of heat in the winter. The transmitter manufacturer can usually supply data on the number of cooling BTUs required, so the proper size of air conditioner may be selected. This method is frequently used with the lower power transmitters. A

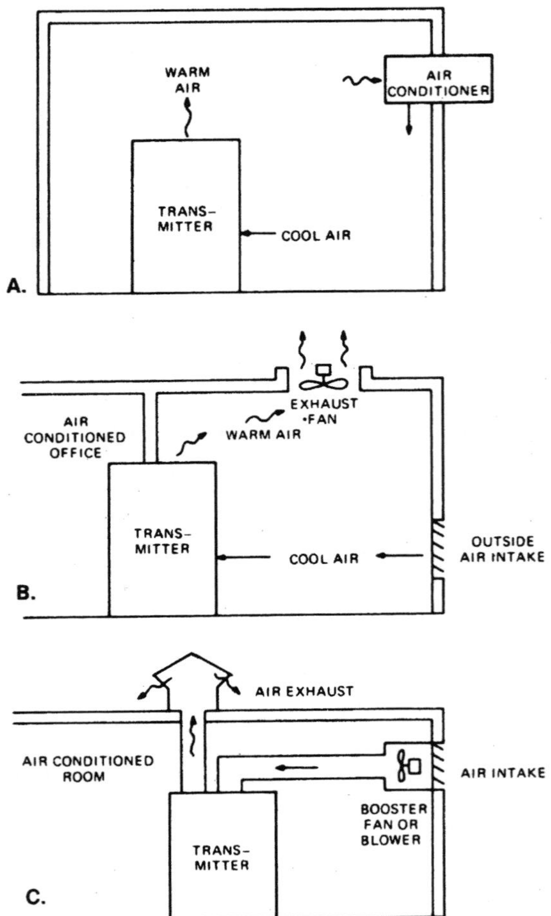

FIGURE 4.7-49 Three methods of providing cooling air for the transmitter (see text).

protective system should be provided to prevent overheating of the transmitter if the air conditioner fails.

Figure 4.7-49b shows a transmitter located in a wall separating an air-conditioned room and a ventilated but not air-conditioned room. A large exhaust fan is provided in the ceiling to remove the rising hot air, and an adequate cool air intake is provided in the lower portion of an outside wall. A filtering system is required to keep the transmitter interior clean.

Figure 4.7-49c shows a transmitter located in an air-conditioned room with intake and exhaust air ducts to the outside. An auxiliary blower or fan may be required to overcome pressure drop in the ducting. This type of system requires careful design to make sure the air flow through the transmitter is not impeded by the duct work. Additional air interlocks may be required to protect the transmitter from a failure of the auxiliary blower. The air intake and exhaust openings to the outside should be provided with rain shields, insect screens, and dust filters as dictated by the environment. The location of the air intake and exhaust openings should be arranged so wind pressure will not impede the air flow.

There is renewed interest in liquid cooling systems for high-power FM + HD Radio common amplification transmitters. Liquid cooling systems offer the ability to remove large amounts of heat from the transmitter room without significant changes to the existing HVAC system, and they are already widely used in UHF digital television transmitter systems.

Preventive Maintenance

Preventive maintenance is equipment inspection and maintenance performed at regular intervals before an operational problem develops. The long-term benefits are great because potential problems are discovered and solved while they are still easily manageable. A checklist of a few typical preventive maintenance items for an FM transmitter plant might include:

- Weekly overall internal and external cleaning and inspecting for damage or excessive wear
- Lubricating motors, tuning gears, and other moving parts at intervals recommended by the manufacturer
- Checking and logging all meter readings, including daily checks of filament voltage and comparing these readings with the previous set of readings as an aid to diagnosing a developing problem (this can be done automatically with digital transmitter control systems and remote control systems)
- Regularly exercising the automatic power control and any other mechanical servo systems
- Checking the antenna lighting and deicer systems
- Checking the transmission line pressurization and VSWR
- Checking all air filters in the transmitter plant and cleaning or replacing as required
- Checking the proper operation of all monitoring and remote control equipment

Air filters should be periodically cleaned or replaced according to the transmitter manufacturer's instructions. This is important because dust or insect clogged air filters may reduce the cooling air flow enough to cause overheating of some components. The probability of component failure increases rapidly when cooling is insufficient. Particular attention should be paid to removing dirt and dust from high-voltage components during regular maintenance after all power is removed and all components are discharged. Dust should be cleaned from the transmitter with a suitable brush and vacuum cleaner or as otherwise recommended by the transmitter manufacturer. Weekly cleaning is usually sufficient. In addition, good overall housekeeping will pay big dividends in the long run by keeping equipment clean and free of problems that would otherwise be caused by dirt build-up.

Maintenance Systems

The key to making any maintenance program work is to set up formal systems for checklists, logging, parts inventory management, and repair scheduling. These systems provide the discipline required to keep the maintenance routine accurate and complete. Each station should develop a system suited for the particular physical plant involved. When there is more maintenance and repair work requiring attention than there is time to do it all, set priorities for completing each item so no item is forgotten. Accurate notebooks, in either hard copy or software database format, describing all installation and maintenance work are a very helpful part of any maintenance system especially when work spans many years.

SUMMARY

Because of the importance of FM transmission, whether analog or digital, to radio broadcasting, this chapter has emphasized the need for a thorough understanding of the technical principles of frequency modulation theory, the methodology for implementing digital (HD Radio) systems, and how digital signal processing techniques have made significant improvements in both signal quality and reliability of broadcast transmitters.

References

[1] Lessing, L., *Man of High Fidelity: Edwin Howard Armstrong*, Bantam Publishing, New York, 1959.

[2] *Reference Data for Radio Engineers*, Fifth ed., Howard W. Sams and Co., Indianapolis, IN, 1970.

[3] Terman, F.E., *Electronic and Radio Engineering*, Fourth ed., McGraw-Hill, New York, 1955.

[4] Clarke, K.K. and Hess, D.T., *Communications Circuits: Analysis and Design*, Addison-Wesley, Reading, MA, 1978.

[5] Rohde, U.L., *Digital PLL Synthesizers: Theory and Design*, Prentice Hall, Englewood Cliffs, NJ, 1983.

[6] Krauss, H.L., Bostian, C.W., and Raab, F.H., *Solid-State Radio Engineering*, John Wiley & Sons, New York, 1980.

[7] Howe, Jr., H., Simplified design of high power, *N*-way, inphase power divider/combiners, *Microwave J.*, December, 1979.

[8] *Care and Feeding of Power Grid Tubes*, Eimac Division of CPI, Inc., Palo Alto CA, 2001 (www.cpii.com/eimac/).

[9] Hershberger, D. and Weirather, R., *Amplitude Bandwidth, Phase Bandwidth, Incidental AM, and Saturation Characteristics of Power Tube Cavity Amplifiers for FM*, Harris Corp. Broadcast Division, Quincy, IL, 1982.

[10] Hnatek, E.R., *Design of Solid-State Power Supplies*, Second ed., Van Nostrand Reinhold, New York, 1981.

[11] Mendenhall, G.N., *A Study of RF Intermodulation Between FM Broadcast Transmitters Sharing Filterplexed or Co-Located Antenna Systems*, Broadcast Electronics, Inc., Quincy, IL, 1983.

[12] Mendenhall, G.N., *The Composite Signal: Key to Quality FM Broadcasting*, Broadcast Electronics, Inc., Quincy, IL, 1981.

[13] Lyles, J.T.M. and Shrestha, M.B., *Transmitter Performance Requirements for Sub-Carrier Operation*, Broadcast Electronics, Inc., Quincy, IL, 1984.

[14] Bruene, W.B., An inside picture of directional wattmeters, *QST Mag.*, April, 1959.

[15] Mendenhall, G.N., *Techniques for Measuring Synchronous AM Noise in FM Transmitters*, Broadcast Electronics, Inc., Quincy IL, 1988.

[16] Shrestha, M.B., *The Significance of RF Power Amplifier Circuit Topology on FM Modulation Performance*, Broadcast Electronics, Inc., Quincy, IL, 1990.

[17] *FMSIM*, FM stereo simulation and analysis program, Quantics, Nevada City, CA, 1990.

[18] Anthony, E.J., *Optimum Bandwidth for FM Transmission*, Broadcast Electronics, Inc., Quincy, IL, 1989.

[19] Shrestha, M.B., *Personal Safety Considerations with Broadcast Transmitters*, Broadcast Electronics, Inc., Quincy, IL, 1989.

[20] Artigo, R., Extending transmitter tube life, *Broadcast Manage. Eng. Mag.*, March, 1982 (revised March 1990 and reprinted as Eimac Application Bulletin AB-18, Eimac Division of Varian Corp., San Carlos, CA).

[21] Twitchell, E.R., A digital approach to an FM exciter, *IEEE Trans. Broadcast.*, 38, 106–110, 1992.

[22] AES3-2003, *AES Recommended Practice for Digital Audio Engineering: Serial Transmission Format for Two-Channel Linearly Represented Digital Audio Data* (revision of AES3-1992, including subsequent amendments).

[23] Mendenhall, G.N., *Implementing an Uncompressed Digital Path from the Studio to the On Air Signal*, Harris Corp. Broadcast Division, Quincy, IL, 1997.

[24] Dittmer, T.W., *Advances in Digitally Modulated RF Systems*, Harris Corp. Broadcast Division, Quincy, IL, 1997.

[25] Agnew, D., *FM Harmonic Measurement Correction Factors*, Harris Corp. Broadcast Division, Mason, OH.

[26] Mullin, K., *A Planning Guide: Determining the Best IBOC Migration Path for Your AM or FM Radio Station*, Harris Corp. Broadcast Division, Quincy, IL, 2003.

[27] Cabrera, G., *Reducing FM IBOC Transmission Costs with the Proper Configuration and Linearization Techniques*, Harris Corp. Broadcast Division, Mason, OH, 2003.

[28] Cabrera, G., *Dual FM Injection Improves IBOC High-Level Combining*, Harris Corp. Broadcast Division, Mason, OH, 2004.

[29] Cabrera, G., *Improving Efficiency with Split-Level Combining*, Harris Corp. Broadcast Division, Mason, OH, 2004.

[30] Cabrera, G., *Understanding Split-Level Combining*, Harris Corp. Broadcast Division, Mason, OH, 2005.

[31] Fluker, S., *Split Level Combining Explained*, Cox Radio Corp., Orlando, FL, 2004.

[32] Mattsson, A., *Linearizing HD Radio Transmitters: A Technology Survey*, Harris Corp., Broadcast Division, Mason, OH, 2005.

[33] http://www.ibiquity.com/hd_radio/iboc_white_papers.

[34] http://www.broadcast.harris.com/support/papers.asp?cat=78.

[35] http://www.broadcast.harris.com/radio/hdradio/calculator/default.asp.

[36] http://www.broadcast.harris.com/radio/hdradio/calculator2/#.

CHAPTER

4.8

FM Stereo and SCA Systems

ERIC SMALL

Modulation Sciences, Inc.
Somerset, New Jersey

INTRODUCTION

A substantial portion of this chapter remains the work of John Kean, this chapter's original author. Although it has been revised and updated, much of the material is fundamental and remains current technology.

Stereophonic sound FM broadcasting and subcarrier services (commonly known as SCA) are a form of multiplexing that began over 70 years ago, when high-fidelity audio and facsimile messages were simultaneously transmitted from the Empire State Building in New York City to experimental receiving sites in New Jersey [1]. Those historic multiplexing efforts demonstrated the value of wideband frequency modulation, allowing modern FM broadcasting to provide services to the public of both quality and variety [2].

Despite its promising start, FM broadcasting in the 1950s and early 1960s was a marginal business. The number of operating FM stations declined from about 700 in 1947 to slightly over 500 in 1957. Most stations depended on additional income from a non-broadcast ancillary service. That service was called *Simplex*, and it allowed FM broadcasters to offer businesses "music only" receivers that muted when the station was not transmitting music. The station transmitted an unmodulated subcarrier when the microphone was active, thus eliminating, among other materials, commercials. These subcarrier-muted special receivers were rented from the station to businesses that wanted uninterrupted music. By 1955, the Federal Communications Commission (FCC) realized that allowing stations to broadcast signals that removed commercials was impeding the growth of the FM service. In 1955, the FCC banned Simplex and replaced it with Subsidiary Communications Authorization (SCA). SCA at that time was an FM subcarrier service that offered background music and other programs to stores and offices over medium-fidelity audio subcarriers. The quality of an FM analog audio SCA was, and remains, about that of an AM broadcast station.

The creation of SCA ensured the economic survival of FM broadcasting. Three more events, two technical and one regulatory, would set the stage for the emergence of FM as the dominant audio service. The first was the FCC's April 1961 Report and Order authorizing transmission of a stereophonic sound system that combined the system proposals of Zenith Radio and General Electric. At the time, FM broadcasters did not show much excitement for their new capability, but the growing availability of stereophonic LP records and home stereo equipment created a natural market for FM stereo. The second was the availability of affordable, solid-state automobile FM radios by the mid-1960s that offered improved fidelity and reduced noise as compared with AM. This brought the potential of a large drive-time audience for FM radio. The final and perhaps decisive event was the FCC action in 1964 (not implemented until 1967) that eliminated the practice of broadcasting the same material on FM stations that was broadcast on co-owned AM stations ("simulcasting"). The ban on simulcasting forced station owners to either sell their FM outlets or, as most stations did, make significant capital investments in them.

THE COMPOSITE BASEBAND

This section contains definitions of the terms commonly used in the FM stereo and SCA systems [3]. Because some of these terms are misused or ambiguous, the intention here is to establish meanings that will be used throughout this chapter [4]. The list was compiled to clarify certain terms and is not a complete glossary.

Key Terms for Stereo and SCA

- **SCA:** Until 1982, SCA stood for "Subsidiary Communications Authorization." It was clearly defined in the FCC Rules, and stations were required to apply for an authorization to operate the subchannel. After the deregulation of FM subcarriers in 1982, Subsidiary Communications Authorization ceased to be a defined term. Today, "SCA" generally refers to any subcarrier carrying information not related to the station's main program. The letters themselves no longer have any meaning.
- **Multiplexing:** In its simplest sense, multiplexing implies that two or more independent sources of information are combined for carriage over a single medium (namely, the radio frequency *carrier*) and then are separated at the receiving end. In stereophonic broadcasting, for example, program information consisting of left and right audio signals is multiplexed onto an FM carrier for transmission to receivers, which subsequently recover the original audio signals.
- **Channel:** A transmission path. The usage herein distinguishes between the concept of a channel and a signal (main channel, stereophonic subchannel, etc., and left and right audio signals).
- **Composite baseband signal:** A signal which is the sum of all signals that frequency modulate the main carrier. The signal includes the main channel signal, the modulated stereophonic subcarrier, the pilot subcarrier, and the SCA subcarriers.
- **FM baseband:** The frequency band from 0 Hz to a specified upper frequency that contains the composite baseband signal.
- **Main channel:** The band of frequencies from 50 (or less) Hz to 15,000 Hz on the FM baseband which contains the main channel signal.
- **Main channel signal:** A specified combination of the monophonic or left and right audio signals that frequency modulates the main carrier.
- **Stereophonic sound:** The audio information carried by two channels. These two channels may carry additional audio channels encoded by phase and amplitude, such as surround sound.
- **Stereophonic sound subchannel:** The band of frequencies from 23 to 99 kHz (53 kHz for two-channel transmission) containing sound subcarriers and their associated sidebands.
- **Subchannel:** A transmission path specified by a subchannel signal occupying a specified band of frequencies.
- **Subchannel signal:** Subcarriers and associated sidebands that frequency modulate the main carrier. It is synonymous with subcarrier, as in the stereophonic subcarrier or SCA subcarrier.
- **Frequency deviation:** The peak difference between the instantaneous frequency of the modulated wave and the average carrier frequency.
- **Percentage modulation:** The ratio of the actual frequency swing of the carrier to the frequency swing defined as 100% modulation, expressed in percentage. Although current FCC rules conditionally permit greater than 100% modulation when SCAs are transmitted, a frequency swing of ±75 kHz is still defined as 100% modulation.
- **Injection:** The ratio of the frequency swing of the FM carrier by a subchannel signal to the frequency swing defined as 100% modulation, expressed in percentage. The total injection of more than one subchannel signal is the arithmetic sum of each subchannel injection.
- **Crosstalk:** An undesired signal occurring in one channel caused by an electrical signal in another channel.
- **Linear crosstalk:** A form of crosstalk in which the undesired signals are created by phase or gain inequalities in another channel or channels. Such crosstalk may be due to causes external to the stereophonic generator; consequently, it is sometimes referred to as *system crosstalk*.
- **Nonlinear crosstalk:** A form of crosstalk in which the undesired signals are created by harmonic distortion or intermodulation of electrical signals in another channel or channels. Such crosstalk may be due to distortion within the stereophonic generator or FM transmitter; consequently, it is sometimes referred to as *transmitter crosstalk*.

Frequency Spectrum and Modulation Limits

The FCC's stereophonic transmission standards are contained in Section 73.322 of the FCC Rules, and the SCA transmission standards are contained in Section 73.319. Readers should refer to the Commission's rules [5] for specific rules and regulations regarding FM and SCA transmissions.

The composite baseband extends to 99 kHz and may be used in support of either stereophonic or SCA multiplex services. Within the frequency range of 23 to 99 kHz, any form of amplitude modulation (DSB, SSB, etc.), angle modulation (FM or PM), or frequency-shift keying of a multiplex subchannel is permitted.

Authorization was once required for a station to begin broadcasting a multiplex service. Although the familiar term SCA is still used, authorization is no longer required. Broadcast licensees may begin transmitting multiplex services without prior notification of or authorization from the FCC.

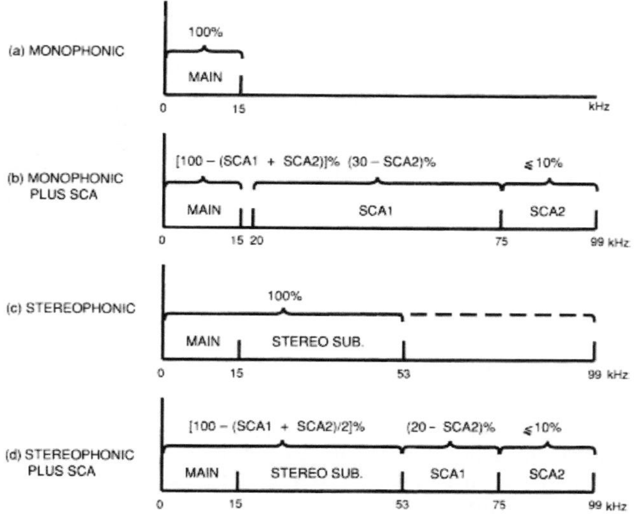

FIGURE 4.8-1 FM baseband scenarios with allowable modulation limits for monophonic, stereophonic, and SCA operation. Modulation percentages are referred to 75 kHz carrier deviation.

Under certain conditions when SCA multiplex subcarriers are operated, a total modulation of up to 110% is legal. Figure 4.8-1 shows the baseband frequency ranges and modulation limits for various modes, from monophonic to stereophonic plus fully loaded SCA operation. Figure 4.8-1a represents the basic monophonic program mode, where the baseband width is limited to approximately 15 kHz and no other signals are multiplexed. In this case, the main channel contributes all the modulating energy. No more than 100% modulation is permitted in this case.

Figure 4.8-1b shows the baseband with SCA operation in addition to monophonic main channel service. Total SCA injection up to 30% is permitted within the band from 20 to 99 kHz. This injection figure may be comprised of one or more SCA subcarriers. To ensure that the bandwidth of the main carrier (and its interference to other stations on adjacent and alternate channels) is not significantly increased, the arithmetic sum of all modulation must not exceed 100% plus 1/2 of the SCA injection. SCA injection between 75 and 99 kHz may not exceed 10% under any conditions. Note that the FCC Rules permit transmission of multiplex subcarriers when no broadcast program service is carried on the main channel, provided that the above modulation rules are met.

Figure 4.8-1c shows the basic stereophonic sound program mode, without SCA operation. As is the case for monophonic program operation, the modulation must be limited to 100%. Frequencies up to 99 kHz are available for multichannel sound program transmission.

Figure 4.8-1d shows the addition of a single band for SCA operation to the stereophonic mode. Although total SCA injection may be up to 20%, no more than 20% total injection may be employed within the frequency bands from 53 to 75 kHz and 10% from 75 to 99 kHz. The modulation contributed by the main channel and stereophonic subchannel signals may not be more than 100% minus 1/2 the total SCA injection. Because the total injection may be up to 20%, total modulation may be up to 110% [6].

A suggested method for adjusting modulation when SCAs are in use is as follows:

1. Remove all modulation from the baseband.
2. Apply one SCA signal.
3. Adjust its injection to the intended level, typically 10%.
4. Remove the SCA just adjusted and apply the next one.
5. Set this SCA to the intended level.
6. Remove this SCA and repeat steps 2 to 4 for any remaining SCA RBDS signals.
7. The sum of the values that each subcarrier was set to should be the intended value allocated for subcarrier operation.
8. Add the program signal. Adjust the program modulation for the desired peak total modulation.

Note that the reading on the total modulation scale of some modulation monitors will not be the sum of the injection of each SCA signal set individually. Typically, the reading will be greater. This is because the modulation monitor responds to the *vector sum* of the SCA signals, not their *arithmetic sum*. This characteristic can present a problem when program modulation is added to the composite baseband and an attempt is made to adjust the total deviation to the legal limit for whatever combination of subcarriers the station is employing. It has been shown [7] that the longer the peak response time of the peak indicator of the monitor, the more closely the indicated sum of the SCA subcarriers approaches the arithmetic sum of the subcarriers. When modulation monitors were deregulated in 1982, the FCC removed all guidance as to the response time of modulation monitor peak indicators. In a 1989 Declaratory Ruling [8] and a 1991 Public Notice [9], the FCC stated that a modulation monitor that complied with the pre-1983 technical requirements would produce valid readings of FM modulation under the current standard.

Generating the Stereophonic Baseband Signal

Figure 4.8-2 shows the composite baseband that modulates the FM carrier for stereophonic broadcasting. (SCA multiplex subchannels are not part of this band and will be discussed later.) The two channel stereo baseband has a bandwidth of 53 kHz and consists of:

- A main channel (L+R) consisting of the sum of left plus right audio signals. This is the same signal broadcast by a monaural FM station, but it is reduced by approximately 10% to allow for the stereo pilot injection.

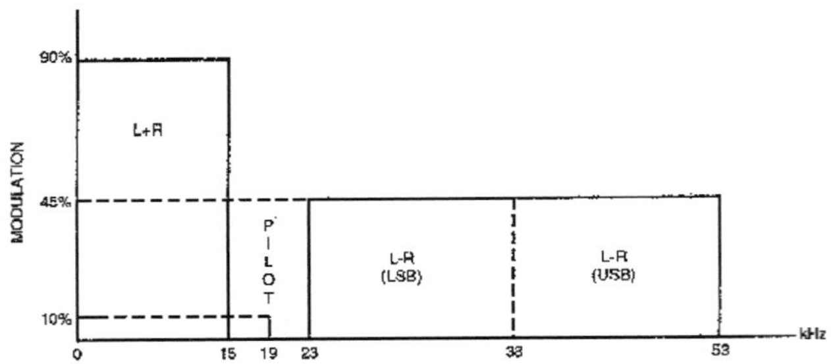

FIGURE 4.8-2 Two-channel (left and right) FM stereo baseband.

- A stereophonic sound subchannel (L–R) consisting of a double sideband amplitude modulated subcarrier with a 38 kHz center frequency. The modulating signal is equal to the instantaneous difference of the left and right audio signals. The subcarrier is suppressed to avoid wasting modulation capability. The pairs of AM sidebands have the same peak modulation potential as the main channel.
- A 19 kHz subcarrier pilot, which must be exactly one-half the frequency of the stereophonic subcarrier and very nearly in phase to it. It supplies the reference signal required to synchronize the decoder circuitry in receivers. The frequency tolerance is ±2 Hz, and the main carrier must be modulated (the injection level) between 8 and 10%.

Historically, two principles have been used to generate the stereophonic baseband: time-division multiplex (TDM), or switching method, and frequency-division multiplexing (FDM), or matrix method. Many stations still use these techniques. More recently, stereo baseband signals have been generated digitally.

Frequency-Division Multiplexing

A basic method for generating the stereophonic baseband involves the direct generation of the double-sideband suppressed L–R subchannel along with the L+R channel. A simplified block diagram of the FDM system is shown in Figure 4.8-3. Both left and right audio channels are pre-emphasized and low-pass filtered. In the matrix, the left and right audio signals are both added and subtracted. The audio signals are added to form the L+R main channel, which is also used as the monaural broadcast signal. The subtracted signals are fed to a balanced modulator, which then generates the L–R subchannel. Because a balanced modulator is used, the carrier at 38 kHz will be suppressed, leaving only the modulated sidebands. The 38 kHz oscillator is divided by 2 to make the 19 kHz pilot tone. Finally, the main channel, stereophonic subchannel and pilot are combined in the proper proportions (45 + 45 + 10) to form the composite output.

An examination of the composite stereo waveform in the time domain, as is displayed on an oscilloscope, is helpful, as shown in the three parts of Figure 4.8-4. First consider a 1 kHz sine wave applied equally to the L and R audio inputs. This is shown in Figure 4.8-4a without a pilot signal. The only frequency present in the spectrum graph is 1 kHz, because the matrix produces no difference signal necessary to generate sidebands in the stereophonic subchannel. Figure 4.8-4b illustrates the ideal composite signal when the same 1 kHz tone is applied to the L and R inputs but exactly out of phase. With the pilot still off, the symmetrical envelope shown represents a double-sideband suppressed carrier (DSBSC) AM signal. Finally, consider the waveform in Figure 4.8-4c, when the composite signal (still without pilot) is generated by applying a 1 kHz tone to the L input alone. The baseline of the waveform envelope will be a straight line if there is no amplitude or phase difference between the main channel and subchannel. Three frequency components are present: 1 kHz, 37 kHz, and 39 kHz. These sidebands are each one-half the voltage amplitude of the 1 kHz signal in the main channel; together, they equal the energy of the main channel in this instance. Figure 4.8-4c appears to be the same when an R-only 1 kHz tone is applied, but the phase of the two sidebands would be reversed with respect to the 38 kHz subcarrier (and the pilot). Adding the pilot at 8 to 10% produces similar waveforms, but an oscilloscope display of the waveform baseline is fuzzier. For this reason, most ste-

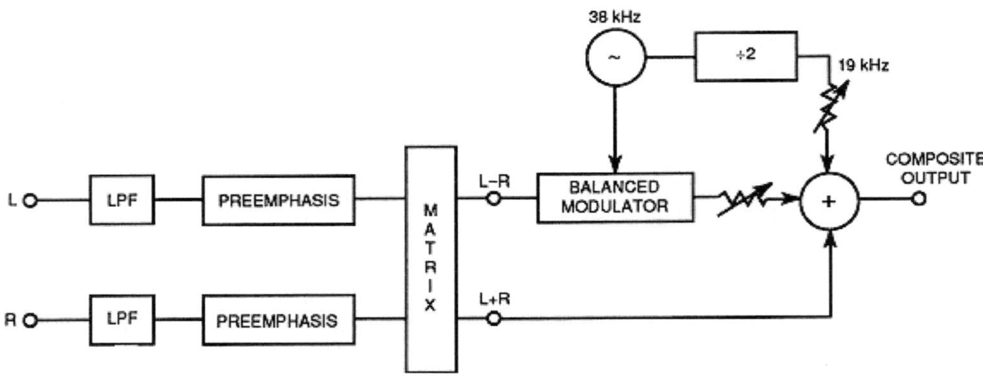

FIGURE 4.8-3 Functional block diagram of a frequency-division multiplex stereo generator.

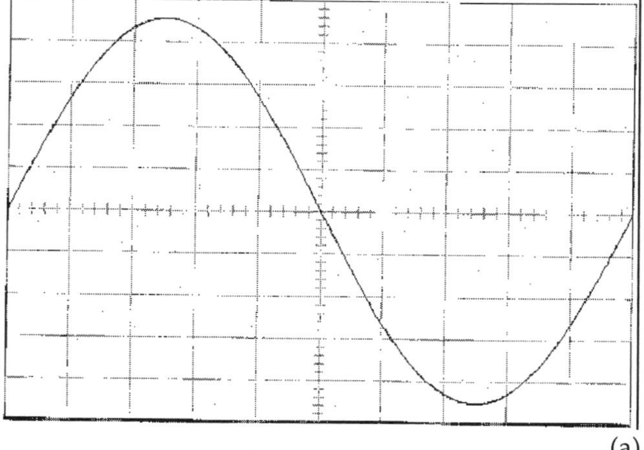

(a)

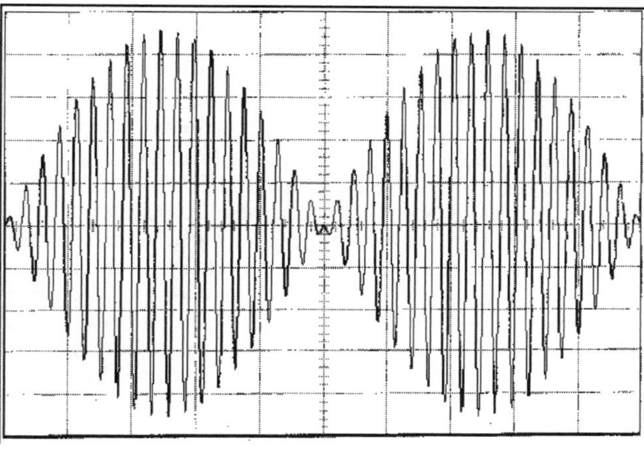

(b)

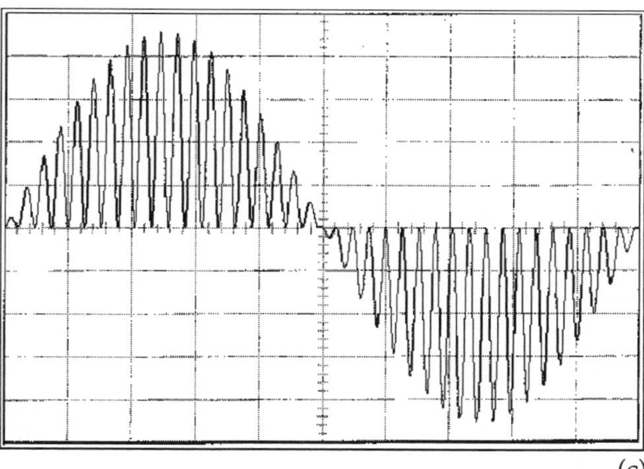

(c)

FIGURE 4.8-4 (a) 1 kHz left at 91%, 1 kHz right at 91%, 19 kHz at 0%; identical sinusoidal L and R inputs. (b) 1 kHz left at 91%, 1 kHz right at 91%, 19 kHz at 0%; identical, but out-of-phase sinusoidal L and R inputs. (c) 1 kHz left at 91%, 1 kHz right at 91%, 19 kHz at 9%; sinusoidal L input at 90% modulation. (Figures courtesy of Quantics; Nevada City, CA.)

reo generators allow the pilot to be turned off for baseline measurements.

Time-Division Multiplex

A different type of stereo generator is in use that produces a result similar to frequency-division multiplexing by using a switching technique. Generation of both the L+R and L–R channels is accomplished by an electronic switch that is toggled by a 38 kHz signal. The switch alternately samples one audio channel and then the other, as shown in Figure 4.8-5. According to Nyquist criteria, the original signal can be reconstructed from periodic samples, provided that the samples are taken at a rate at least twice the frequency of the highest audio frequency component (approximately 15 kHz in broadcast FM).

Figure 4.8-6 shows the output waveform for the TDM generator in the time domain (as an oscilloscope would display the signal) for a sequence of input signals. The diagrams at the right of the waveform show the same signal in the frequency domain (as would be displayed on a spectrum analyzer). In Figure 4.8-6a, no input signals are present. Ideally, no output signals are possible, and in practice only a small amount of leakage of the switching transients would occur. Because the transfer time of the switching signal is extremely quick, harmonics of the fundamental 38 kHz are possible. A 9 kHz audio tone is applied to the L and R inputs in Figure 4.8-6b. The 9 kHz input signals are combined at full amplitude (90% modulation), and no subchannel sidebands are generated.

In Figure 4.8-6c, only the left channel has a signal present. As the switch selects the L audio line, samples are passed along to the composite output; therefore, the output waveform shows the same signal, chopped into segments of 1/38,000 ($1/3.8 \times 10^5$) of a second. Because the total area under the waveform has been divided in half, it should be apparent that the energy of the 9 kHz signal in the L+R channel is only half the amplitude that it would be if an equal 9 kHz signal were also present at the right channel. The equation for the output signal e for an input signal σ at any instant t is:

$$e = 1/2 \sin \sigma t$$

(main ch. audio)

$$+1/\pi\left[\sin(\phi+\sigma)t + \sin(\phi-\sigma)t\right]$$

(DSBSC) (1)

$$-1/3\pi\left[\sin(3\phi+\sigma)t + \sin(3\phi-\sigma)t\right]$$

(3rd harmonic)

. . . etc. (higher harmonics)

Figure 4.8-6c shows the original 9 kHz signal (at half amplitude) and a pair of sidebands centered about the 38 kHz switching frequency. No 38 kHz signal is generated if the switching waveform has perfect symmetry—that is, if the switch is connected to the left and right channels for precisely equal periods. Note that a harmonic of the stereophonic subcarrier is

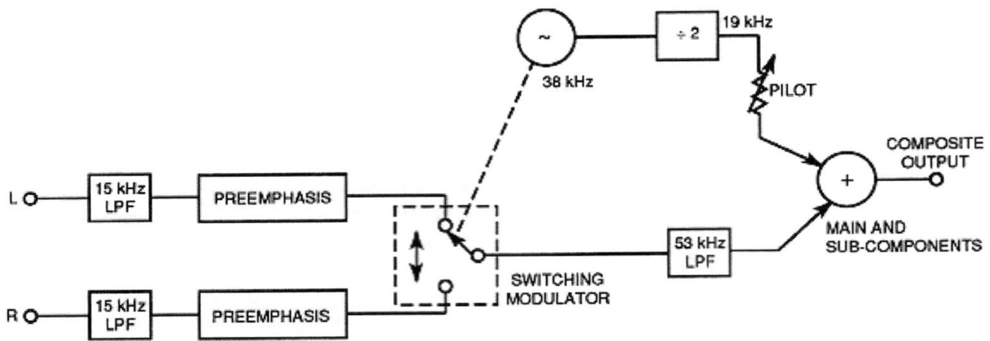

FIGURE 4.8-5 Functional block diagram of a time-division multiplex stereo generator.

shown, centered around 114 kHz, which is three times the switching frequency. Only one extra term was shown in the equation; however, other terms, at the fifth and seventh, are present. In addition to the odd-order harmonics of the 38 kHz subchannel, asymmetry in the switching signal or other circuit imbalances can create some sidebands centered at about the second harmonic at 76 kHz. All of these harmonics must be removed by filtering, as shown in the diagram. When the odd harmonics are filtered out, the proper DSBSC waveform results; however, it is slightly greater in amplitude than the L+R signal because the fundamental component of the square wave is $4/\pi$ or 1.27 times larger than the square wave amplitude. This is easily corrected by adding enough of the L and R audio to the output to equalize the amplitude. In Figure 4.8-6d, the TDM signal is shown when the L and R signals are equal in amplitude and exactly reversed in phase. This waveform matches the composite stereo signal shown earlier in Figure 4.8-4b.

The composite low-pass filter must have very steep cutoff characteristics but should have flat amplitude response and linear phase shift with frequency (equal time delay at all frequencies) below 53 kHz. Although this approach to stereophonic generation is simple and stable, the filter can degrade stereo separation, especially at higher audio frequencies.

The 19 kHz pilot square wave from the ÷2 digital divider must also be filtered to remove harmonics. This additional time delay (phase shift) of the pilot, with respect to the 38 kHz information, must be compensated to have optimum channel separation.

It is also critical that the left and right audio signals be sharply low-pass filtered at 15 kHz before being applied to the stereo generator. Any energy whatsoever above 19 kHz will violate the Nyquist criteria and result in aliasing distortion. Aliasing distortion is a very offensive signal that sounds like "monkey chatter" crosstalk. Avoiding any energy within ±500 Hz of the 19 kHz pilot is also desirable, as any signal there may confuse the stereo decoder and result in a sudden apparent rotation of the sound field.

A significant improvement on the original switching concept is shown in Figure 4.8-7. As mentioned earlier, the higher order terms of the square-wave-driven switch are responsible for generating the harmonics of the 38 kHz subchannel which must be removed by filtering. By using a *soft switch* to connect back and forth between the L and R channels it is possible to eliminate the low-pass filter and its side effects.

This is accomplished by using the electrical equivalent of a variable attenuator, shown in the diagram as a potentiometer. The slider is driven from end to end of the potentiometer by a sine wave. Because a sine wave is represented by a single, fundamental frequency, the signal output at the slider has the proper DSBSC characteristics without the harmonics. The equation for the composite signal generated in this manner is:

$$e = 1/2 \sin \sigma t \quad \text{(L+R audio)}$$
$$+ 1/\pi \left[\sin(\phi + \sigma)t + \sin(\phi - \sigma)t \right] \quad \text{(38 kHz DSBSC)} \tag{2}$$

As the equation shows, only the fundamental sidebands of 38 kHz are present in the sampled signal,

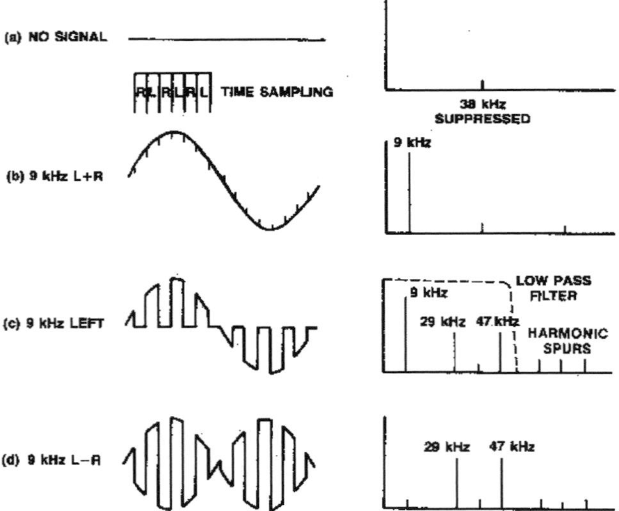

FIGURE 4.8-6 Time-domain and frequency-domain diagrams of stereo baseband signals.

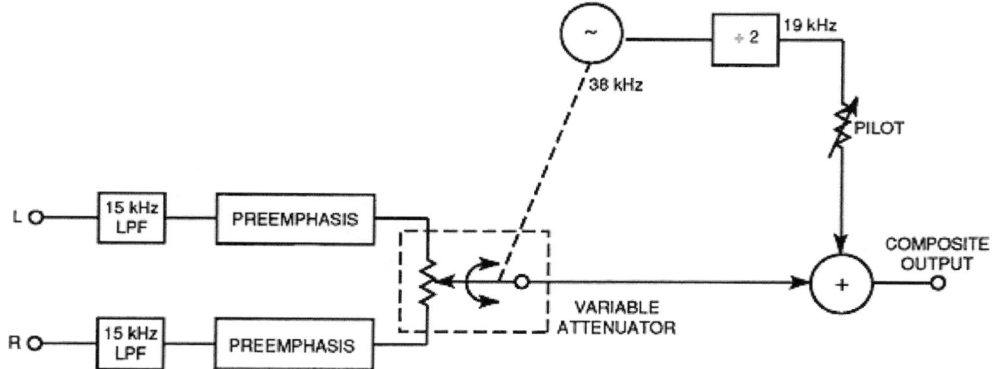

FIGURE 4.8-7 Functional block diagram of a time-division multiplex stereo generator using a variable attenuator.

along with the main channel component. Like the fast-switching TDM system, the L+R and L–R channels are generated in one operation so the circuit remains relatively simple. Filtering of the output is not required, provided that the 38 kHz sine wave is free from harmonics and the variable attenuator has good linearity.

Digital Stereo Generation

The major manufacturers of stereo generators have moved to all-digital composite generator designs. Initially, there were some designs employing hardwired logic, but that approach rapidly gave way to programmable state machines based either on processors designed for high-speed signal processing (DSP integrated circuits) or, more recently, programmable logic arrays. The advantages of a programmable state machine approach to stereo generation are manifold:

- A programmable state machine stereo generator is inherently stable, as it is a binary system and has only two states; thus, there are no adjustments that can drift.
- The hardware design is simple and generic. The two audio signals (stereo), if they are analog, are first digitized in A/D converters, and they then drive the digital state machine device, which may be a processor or a programmable logic array. If the audio is already in a digital format, such as AES/EBU, then it will connect directly to the state machine. The output is a stereo composite signal in digital form. If the FM exciter is a direct digital synthesizer (DDS), then it may accept the digital composite signal directly; otherwise, it will require a digital to analog converter (D/A) to create an analog composite signal. The same hardware design can realize an infinite number of different algorithms to generate the stereo baseband. Regardless of the algorithm that generates the stereo baseband, the hardware remains the same—only the executed software instructions change.
- The actual design of the stereo generator is realized in computer code, usually some form of the computer programming language C; therefore, modifying the design is no more complicated that editing and recompiling a computer program.
- Where the source program signal is available in digital form (usually AES/EBU), then the state machine stereo generator may be programmed to accept it directly. The same is true for the output signal, if the FM exciter is based on a direct digital synthesizer, then the digital form of the stereo baseband signal can drive it directly, again avoiding conversion to and from analog.

Because the computer instructions used to generate the stereo baseband in a digital generator are not contained in the product in source format, manufacturers have chosen to take advantage of this and keep their source code confidential. It is reasonable, however, to assume that computer code used to generate the stereo composite signal is based in some part on the principles discussed earlier for the analog generation of the stereo baseband.

Stereo Decoder Circuits

Stereo FM receivers include a circuit to convert the multiplexed composite signal at the FM detector into the original left and right audio channels transmitted by the FM station. There are at least as many ways to decode the stereophonic signal as there are ways to encode (generate) the composite signal. In practice, only one type of analog decoder is commonly used: the phase-locked loop (PLL) integrated circuit. Increasingly, stereo decoders employ digital techniques as well. There is at least one all-digital (including the stereo decoder) automobile receiver on the market; however, to avoid consumer confusion with Digital Audio Broadcasting (DAB) of various types, consumer manufacturers have chosen not to promote their DSP-based receivers as such. Again, as with the stereo encoders, it is reasonable to assume that the design of digital decoders began with the adoption of one of the basic analog designs for computer code.

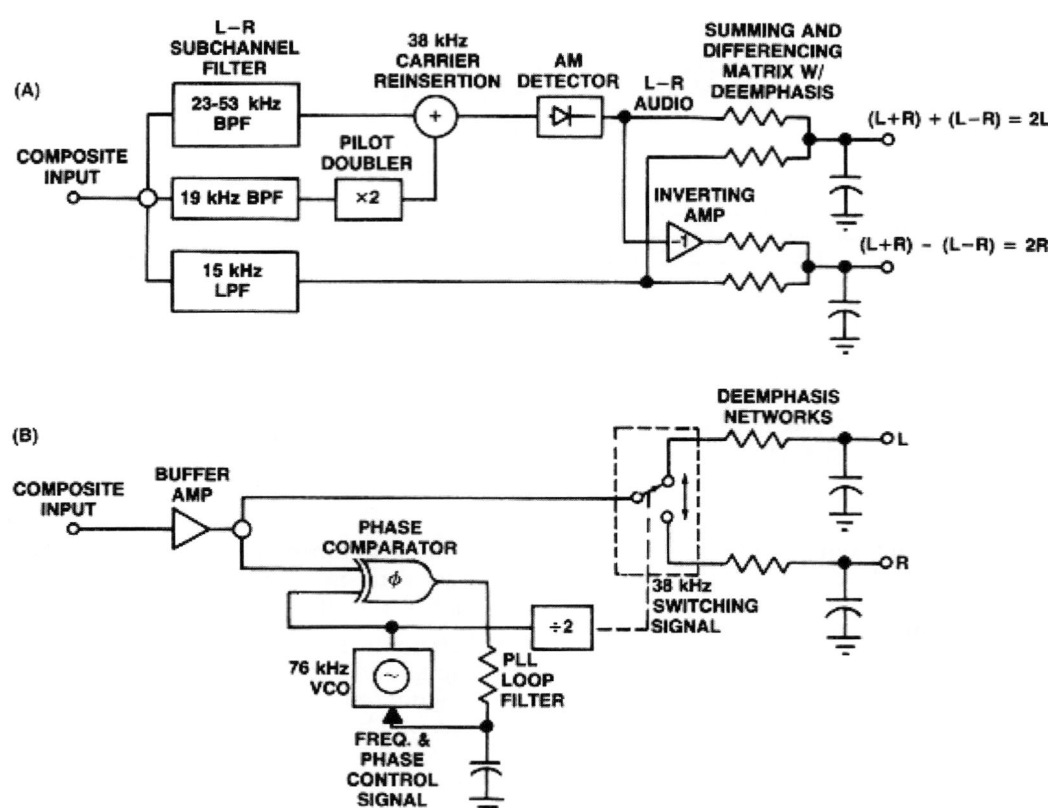

FIGURE 4.8-8 Functional block diagram of stereo decoders using (a) L+R and L–R matrixing, and (b) phase-locked time-division multiplexing.

The circuit in Figure 4.8-8a is seldom used but is shown for comparison. It is the closest complement to stereo generators using frequency-division multiplexing. At the input, the composite signal is split with three equal time-delay filters into the main (L+R) channel, pilot signal, and stereo (L–R) subchannel. Next, the pilot is doubled to 38 kHz and this regenerated carrier is reinserted into the double-sideband AM signal from the subchannel filter. This AM signal is demodulated to yield the L–R (difference) audio. Finally, the L+R and L–R signals are combined in a sum and difference matrix to produce original L and R audio channels. Because of the costly filters needed to separate the composite spectrum, the frequency-division multiplex circuit is not used in consumer equipment. Similar circuits have been used in broadcast modulation monitors, where metering of the separate channels is required. The circuit shown in Figure 4.8-8b is universally used, due to its simplicity, high performance, and low cost. Although this stereo decoder is commonly referred to as a PLL decoder, its performance is really distinguished by its time-division demultiplexer (shown in the dashed box as a toggle switch).

Following a buffer amplifier, the composite baseband signal is sampled by a PLL within the IC. A voltage controlled oscillator, usually running at 76 kHz (four times the pilot frequency), is held in phase with the pilot by a reference signal from the phase comparator and loop filter. When divided by two, the result is a square wave at 38 kHz having a nearly perfect duty cycle (high and low states have equal timing) and very fast rise and fall times. This signal is ideal for driving the output audio switcher (demultiplexer). This stage is a transistor matrix designed to rapidly transfer the composite baseband to the L and R audio output in time with the switch in the station's stereo generator. Fast, clean audio switches are relatively easy to make and do not drift.

Because PLL stereo decoders normally use square-wave switching, the circuit is able to demodulate baseband signals that are odd harmonics of 38 kHz. The third harmonic (114 kHz) is most troublesome, as noise and spurious signals near this frequency are shifted to the audio baseband, as is the 38 kHz stereophonic subchannel. Engineers should be watchful of the frequency band centered on 114 kHz in their transmitted signal because audible noise may occur in consumer receivers.

Recent stereo decoder designs have reduced sensitivity to energy outside the composite baseband. One approach utilizes a second composite audio toggle switch operated at 114 kHz. The demodulated product is inverted and mixed equally with the 38 kHz switching outputs, canceling the response to signals in the 114 kHz range. The other approach—Walsh demodulator—applies a properly timed stair-step imitation of a sinusoid to a digital multiplier in the output signal path [10]. This reduces sensitivity to third and fifth harmonics as well as adjacent channel noise and interference by up to 20 dB.

Centering the Transmitted Passband

A noise generator and synchronous AM detector form a synchronous AM tuning aid. This tool assists in adjusting the transmitter and related circuits for minimum synchronous AM, which tends to minimize crosstalk into the SCA. This is an indirect method for optimizing the bandwidth of the power stages of an FM transmitter. Mendenhall [11] provides a more direct and accurate method for minimizing synchronous AM. A simple method for centering the transmitter passband on the carrier frequency involves adjustment for minimum synchronous AM. If the bandpass is narrow or skewed, increasing synchronous amplitude modulation of the carrier will result. A typical adjustment procedure is to FM modulate 100% at 1 kHz and fine-tune the transmitter's grid tuning and output tuning controls for minimum 1 kHz AM modulation as detected by a wideband envelope detector (diode and line probe). One kHz is used as the FM modulating frequency rather than 400 Hz so the audio high-pass filter in the audio analyzer can be used to eliminate the AC-line-frequency-related asynchronous component from the synchronous AM component. It is helpful to display the demodulated output from the AM detector on an oscilloscope while making this adjustment. Note that, as the minimum point of synchronous AM is reached, the demodulated output from the AM detector will double in frequency to 2 kHz. This is because the fall-off in output power is symmetrical about the center frequency, causing the amplitude variations to go through two complete cycles for every one FM sweep cycle.

It should be possible to minimize synchronous AM while maintaining output power and efficiency in a properly designed power amplifier. If an oscilloscope is not available for direct observation of the demodulated AM waveform, the 19 kHz bandpass filter and metering circuit that are used to measure pilot injection level in a stereo modulation monitor may be used as a tuning aid to center the passband. The main carrier is FM modulated with a 9.5 kHz tone (without pilot or any other modulation). If the transmitter is tuned for symmetrical AM response, the demodulated AM signal will have a strong second harmonic component at 19 kHz which falls within the passband of the pilot metering circuit. The output of the AM detector is fed into the composite baseband input of the stereo modulation monitor. The transmitter is then tuned for a maximum reading on the pilot injection metering position.

FM SCA TRANSMISSION

From its beginning as a broadcast service, there has been recognition of the potential of FM for multiplexed services. As early as 1940, the FCC permitted multiplex facsimile transmission on FM stations, but not until much later did auxiliary FM services attain a wide acceptance among FM broadcasters. In 1955, the FCC established SCA and created a new business opportunity for broadcasters to permit programming such as background music to offices, stores, and restaurants. For many commercial FM stations, the SCA operation became a major source of revenue, which enabled them to survive economically in the 1950s and early 1960s. By the early 1980s, improvements in transmitter and receiver technology and the desire for new revenue prompted commercial and noncommercial broadcasters to seek changes in the SCA rules. In a series of rulemakings in 1982 and 1983, the FCC made numerous changes to expand technical opportunities and reduce regulation [12]. These changes extended the baseband frequency limit from 75 kHz to 99 kHz, allowed any type of subcarrier modulation to be used, changed the subcarrier injection requirements to permit multiple services, and increased limits for the total modulation during SCA multiplex operation to reduce main channel modulation loss.

FCC Requirements for SCA Operation

Section 73.319 of the FCC Rules sets forth the technical standards for FM multiplex subcarriers; however, the Commission does not set standards for minimum SCA subcarrier performance. This is left for the broadcaster or lessee of the service to determine. In its rules, the Commission defines the transmission conditions under which subcarriers may operate to minimize interference to the main channel and stereophonic subchannel and to other FM stations.

In this chapter, subcarrier injection and bandwidth are considered only with stereophonic operation, as this is the most common FM mode. Figure 4.8-9 shows the upper portion of the composite baseband, from approximately 50 to 100 kHz. This is the same as the upper portion of the spectrum of Figure 4.8-1 with the addition of two hypothetical subcarriers centered at 67 kHz and 92 kHz. Any number of SCA subcarriers may be operated in this frequency range provided that total bandwidth and injection limits are met. The vertical scale shows the level in decibels below 100% modulation of the main carrier; for example, –20 dB marks the injection at the center frequency of both subcarriers. Because 10% is the maximum injection permitted under the Commission Rules within each SCA subchannel:

$$\text{Injection} = 20\log_{10}(0.1) = -20 \text{ dB} \qquad (3)$$

At 53 kHz, an arrow marks a level of –60 dB. The FCC requires that any frequency modulation of the main carrier due to the SCA operation shall be at least 60 dB below 100% modulation in the frequency range of 50 Hz to 53 kHz when stereo is transmitted. This figure must include spurious and intermodulation products as well as subcarrier sideband energy. At 99 kHz, the level of –20 dB is marked, denoting the FCC requirement that instantaneous sidebands be

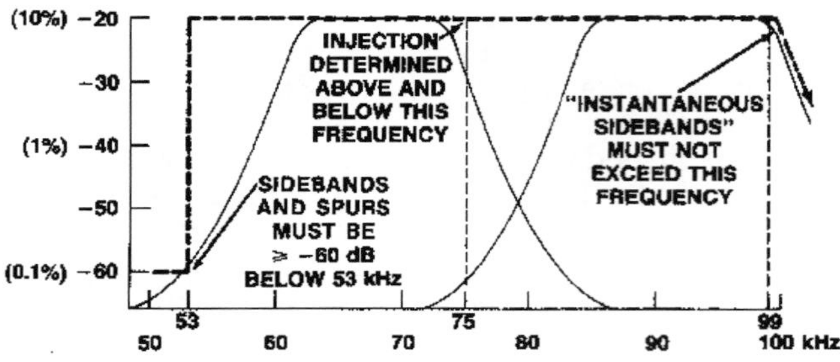

FIGURE 4.8-9 Injection, channel bandwidths, and spurs limits for SCA operation when stereo is transmitted. Two subcarriers (at 67 and 92 kHz) are shown.

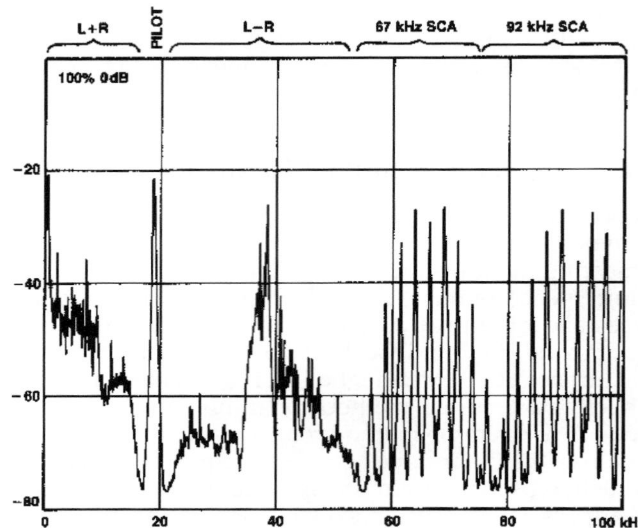

FIGURE 4.8-10 Composite baseband of FM station with 67 kHz and 92 kHz FM-SCAs, modulated with 2.5 kHz tone at 5 kHz and 7 kHz peak deviations, respectively. Note the overlap of sidebands between subcarriers. The station is carrying stereo programming.

restricted within this frequency limit. The Commission has not officially defined instantaneous sidebands, but they are normally considered to be the instantaneous frequency of the subcarrier at its peak deviation (for frequency-modulated subcarriers) or the highest sideband frequency (for amplitude-modulated subcarriers).

Crosstalk between FM SCA Subcarriers

In practice, two subcarriers should be separated as far apart in frequency as possible, while observing the limit of spurious energy below 53 kHz and the instantaneous sidebands at 99 kHz. Although the FCC Rules are silent on the choice of SCA frequencies, 67 kHz and 92 kHz have become the *de facto* standards for FM subcarriers. The first frequency was adopted when the original stereophonic standards restricted the subcarrier spectrum to between 53 and 75 kHz. The second frequency was recommended to place the instantaneous sidebands below 99 kHz while maintaining a safe separation from a 67 kHz subcarrier [13]. Some overlap of the subcarrier sidebands does occur, as depicted in Figures 4.8-9 and 4.8-10. This does not cause significant interference between the two SCA subchannels when the systems use frequency modulation. Laboratory tests of a standard table model SCA receiver show that crosstalk is greater from main-to-subcarrier than from a 92 kHz subcarrier into the 67 kHz subchannel demodulator. No multipath or other real-world impairments were added. In the real, over-the-air world of broadcasting, a typical main-into-SCA signal ratio would be 40 dB.

Examples of Analog SCA Operation

Figure 4.8-11 shows a functional block diagram of a popular aural SCA generator using frequency modulation. After an input amplifier, the audio is bandpass filtered. The low-pass portion cuts off the audio at 5 kHz. This limits the modulation sidebands that could interfere with the stereophonic subchannel or an upper SCA subchannel. The high-pass portion removes any energy from the audio below 50 Hz to allow for the possible use of subaudible telemetry signals.

For the protection requirements, equipment manufacturers often recommend 5 or 6 kHz peak deviation. An audio processor follows the bandpass filter. It is a three-part processor: a broadband compressor to maintain average levels, a broadband limiter to control overall peak deviation, and a high-frequency limiter to provide pre-emphasis and limit high-frequency peaks. The pre-emphasis is typically 150 µsec for music or speech SCA services.

The output of the audio processor is fed through another filter that removes any remaining out-of-band energy before the signal is applied to the modulator. The modulator accepts the processed audio and any low-frequency (subaudible) telemetry. The modulator, a voltage-controlled oscillator, is frequency locked to a quartz crystal to maintain a stable frequency. In this design, the frequency lock is a simple synthesizer that may be programmed to any frequency in the baseband by moving jumpers on the circuit board. A mute circuit is provided to remove the subcarrier entirely during absences of program material. This is an important function in background music operation but is rarely used today.

Two subcarrier outputs are provided. One is a high-level signal for connection to the SCA input of an FM exciter; the other is made up of a composite input and composite output. A mixer circuit adds the SCA to the composite at the proper level. This forms a composite loop-through for easy interfacing.

SCA DATA SYSTEMS

SCA subchannels have long seemed attractive for transmitting data. The first technology, and one still in

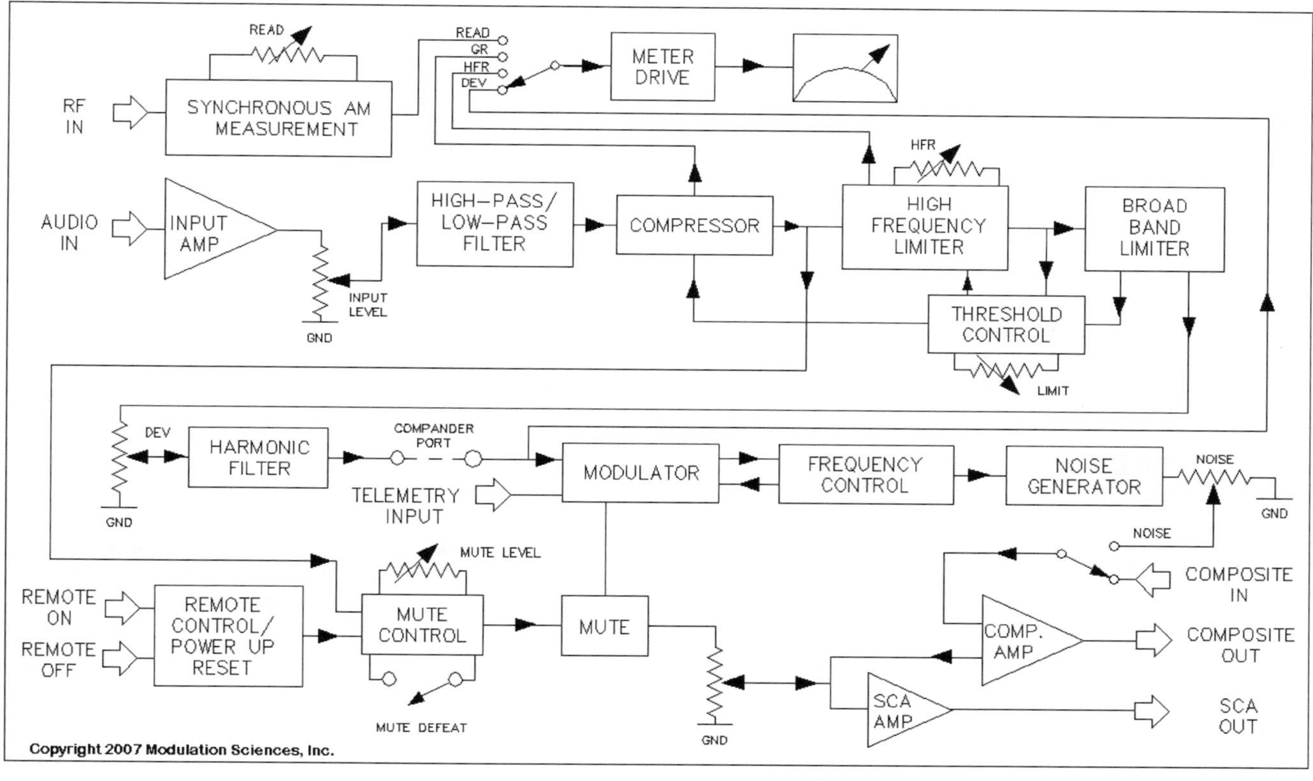

FIGURE 4.8-11 Functional block diagram of a frequency-modulated SCA generator.

limited use today, is audio frequency-shift keying (AFSK)—the connection of an old-fashioned Bell 202 type of telephone data modem to the audio terminals of the SCA generator. The cost and complexity of this form of SCA data broadcasting are low. This technique occupies the bandwidth of a standard audio subcarrier, and the highest practical speed is about 2400 bps; however, this technology results in an excessive bit error rate (BER) unless operated under ideal conditions. At the other extreme, proprietary systems exist that use multiple subcarriers on two or more FM stations to carry sophisticated data protocols with interleaved data correction schemes. This is a frequency diversity system that works well but at the cost of SCA spectrum on several stations. Between these two extremes are many systems, most of which are proprietary, that apply modern data communication technologies to SCA [14,15].

When deciding which SCA data system to use, two important issues should be addressed:

- What is the minimum acceptable bit error rate for the application under real-world transmission conditions?
- After all of the error correction coding, redundant transmission, and frequency diversity are taken into account, what is the bits per hertz efficiency of the system?

Once these issues are addressed, it becomes possible to estimate the real cost of a data SCA system and compare data SCA to alternative transmission systems.

Radio Broadcast Data System

The most successful SCA data system has been RBDS/RDS. The Radio Data System (RDS) was developed in Europe in the 1980s. The United States' version is called Radio Broadcast Data System (RBDS). Standards were first adopted by the National Radio Systems Committee (NRSC) in 1993. A revised edition of this standard was adopted in 1998. The RBDS standard uses an FM data subcarrier at 57 kHz, as shown in Figure 4.8-12 in the FM baseband to transmit information

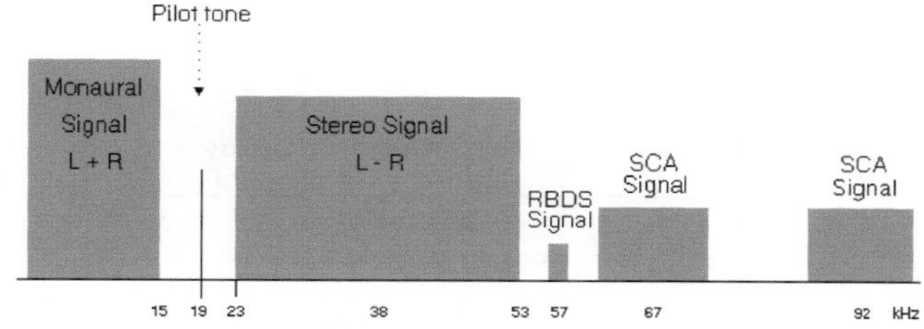

FIGURE 4.8-12 Baseband of FM station with RBDS subcarrier and two SCA subcarriers.

such as call sign, format type, song title, artist, and emergency information to RBDS-equipped receivers. This standard is explored in more detail in Chapter 4.14, "Radio Data Broadcasting."

SCA-to-Stereo Interference Considerations

Advances in consumer receiver technology over the past 20 years have removed interference from the SCA channel into the stereo as a concern unless specific transmission equipment is defective. The only other time that interference becomes an issue is when non-standard SCA parameters are employed. Overall, the "SCA birdie" in the stereo is a thing of the past.

Crosstalk from Main Channel Program into SCA Subchannels

Crosstalk remains a problem for audio SCA services. In the reference case illustrated in Figure 4.8-13, the baseband signal consists of one low-frequency sine wave of frequency F (1 kHz), which modulates the main carrier by deviation D (95%, or 71.25 kHz for broadcast FM), and a sine wave subcarrier at a variable frequency (f Hz) which modulates the main carrier with a very small deviation (d) having a modulation index of less than 0.3.

Second-order distortion will cause sidebands at frequencies of $(f - F)$ and $(f + F)$ with an output injection $d_2(f - F)$ and $d_2(f + F)$. The sum of the sideband amplitudes is:

$$d_2(f) = d_2(f - F) + d_2(f + F) \tag{4}$$

The distortion, defined as the ratio of the sum of the sidebands to the amplitude of the subcarrier d_1, is:

$$\delta = d_2(f)/d_1(f) \tag{5}$$

This distortion is easily measured as a function of the subcarrier frequency (f) with a spectrum analyzer. If the sidebands are equal, they may, in the extreme, either amplitude modulate or phase modulate the subcarrier. In general, there is a combination of amplitude and phase modulation. Third-order distortion is similarly defined by the sidebands created at frequencies $(f - 2F)$ and $(f + 2F)$.

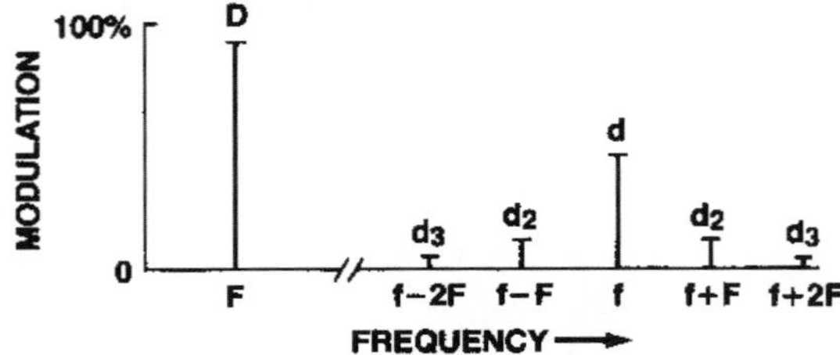

FIGURE 4.8-13 Spectrum of a subcarrier (f) shown with second- and third-order distortion products resulting from low-frequency modulation of f.

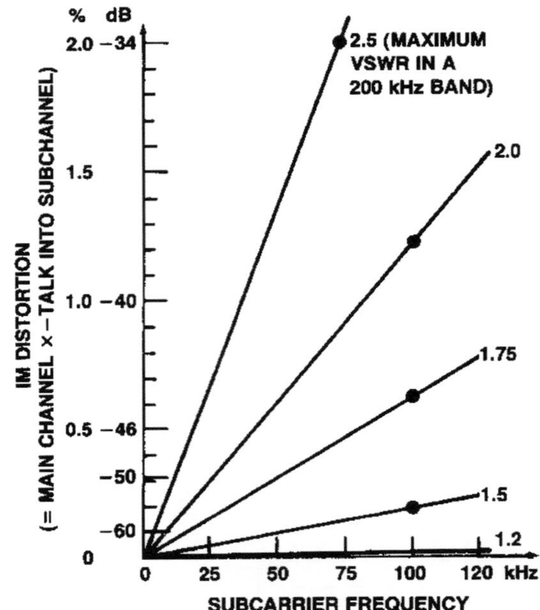

FIGURE 4.8-14 Distortion due to limited antenna bandwidth versus frequency of subcarrier with max VSWR in 200 kHz band as parameter.

Although FM SCA detectors usually include amplitude limiting, they are intended to convert any angular modulation, whether program audio or not, into an output signal. Thus, any phase nonlinearity in the radiofrequency system or multipath will generate second and higher order sidebands around the subcarrier, causing main-to-SCA crosstalk at audio-modulating frequencies.

Figure 4.8-14 shows the relationship between main-to-SCA crosstalk, subcarrier frequency, and maximum antenna voltage standing wave ratio (VSWR) in a 200 kHz band. It is evident that higher frequency subcarriers require somewhat larger bandwidth. It has also been determined that antenna matching must be improved as the transmission line becomes longer. This is especially important for higher frequency subcarriers because the phase error is compounded rapidly with an increase in subcarrier frequency. Distortion is proportional to the subcarrier frequency; thus, for equal distortion, doubling the subcarrier frequency requires halving the reflection coefficient [16].

PRACTICAL CONSIDERATIONS IN SCA

SCA has existed in its current form for more than 50 years. In that time, SCA has been applied to many communications needs. Some of these applications have been very successful, but many have failed. The notable successes were background music delivery, although over the past decade satellite delivery has largely replaced SCA. Reading services for the blind continue to serve millions of people who are vision

impaired in some way. Most cities have at least one SCA outlet serving the blind; many have two. In many parts of the country, ethnic broadcasting has become a major user of SCA and continues to grow. Ethnic broadcasting is a successful form of "narrowcasting." Small groups, without the numerical strength to support their own full- or low-power radio stations, can buy SCA generators, lease SCA channels, and purchase SCA receivers for a few tens of thousands of dollars, even in the largest markets. In some cities, the sale price of an FM radio station would be prohibitive for such narrow applications of broadcasting.

Several years ago the author conducted a survey of the SCA channels in the New York Metropolitan area. Fourteen different languages were being spoken on SCA channels, ranging from several dialects of languages spoken on the Indian subcontinent, to Serbo-Croatian, to Chinese, plus some French and Italian. SCA transmission, therefore, provide opportunities for broadcasting to limited audiences where conventional AM or FM stationswould not be practical because of cost. Low power FM stations cannot provide nearly the coverage area of an SCA signal on a full power FM station.

Effect of Multipath on SCA

At one time, broadcasters were expecting SCA-based paging to create a new profit center. Today, few or no commercial SCA paging systems are in operation in the United States. SCA paging failed because of multipath and very inefficient receiver antennas. Even personal FM radios employing headphones had a significant advantage over paging receivers, because the personal FM radios used the headphone cable as an antenna. These problems resulted in poor reliability, especially in urban areas, as compared with common carrier based paging systems with multiple transmitter sites and narrow bandwidth receivers. The slightly more reliable systems using multiple FM stations were hampered by their very poor spectrum efficiency and cost effectiveness. Another entrepreneurial idea was to deliver customized business news via SCA. That also failed due to excessive multipath in urban areas where the receivers were using simple whip antennas.

A key factor in the success of any SCA application is minimizing multipath. SCA channels are vulnerable to multipath. Because the SCA is typically 20 dB below the level of peak main channel modulation a small amount of nonlinear distortion caused by multipath will reduce the signal-to-noise ratio of the SCA channel to an unacceptable level.

Background music was technically successful due to the use of a roof-mounted directional or Yagi antenna often cut to the channel which resulted in optimum reception with little or no multipath. Reading services for the blind and ethnic radio on SCA rarely employ outside antennas and therefore are more prone to problems with multipath; however, the receivers are in a fixed location with built-in antennas. Here, success is derived from motivated listeners. The radios are moved about a room until acceptable reception is achieved. Also, because the SCA is delivering a service not available elsewhere and one that is greatly desired, the listener's tolerance of interference is much greater than it would be for a typical radio station.

Whether mobile or personal, portable (pagers and pocket) SCA receivers nearly always fail. The low antenna efficiency of personal portable receivers, combined with the inability to control the receiver location and orientation, ensures significant multipath and unreliable reception. Although automobile FM whip antennas can often be nearly ideal monopoles, the rapidly changing geometry of the reception path created by the car's motion causes constant and dynamic multipath. Consider that main channel FM reception is often compromised by multipath in automobile reception. Many business ventures based on SCA broadcasting have ended in failure. In most cases, these failures have stemmed, to a large degree, from a lack of understanding of the technical limitations of SCA.

SCA Receivers

The quality of SCA receivers often plays a role in the success of an SCA service. Because the number of SCA receivers purchased is such a small fraction of the volume of consumer FM receivers manufactured, they receive less of the engineering effort usually applied to consumer electronics and there are few if any custom integrated circuits available for SCA decoding. Further, because the quantities of sale of any one SCA receiver manufacturer are relatively low, optimum pricing on the parts or manufacturing processes is not achieved. The effect of these limitations on receivers is high cost and variable quality. One of the most expensive circuits unique to SCA receivers is the filter that extracts the SCA from the composite signal. Another is the filter that limits the audio bandwidth from the SCA signal to usually 5 kHz. Both of these filters are important to minimize interference to the SCA from other signals on the baseband. For 67 kHz SCA operation, interference from the RBDS signal at 57 kHz is a special concern. To reduce the cost of SCA receivers, the filters are almost always compromised. When SCA receivers are compared side by side, the range of performance can vary greatly. This wide variation in performance often traces back to differences in the filters.

Estimating SCA Coverage

The need to estimate the coverage achievable with an SCA often arises. The traditional methods of calculating monaural FM coverage are often unreliable because they fail to take into account many real-world factors such as urban terrain (building), diffraction, and multipath. Add to this the additional variability introduced by SCA receivers and traditional coverage calculations become ineffective. On the other hand, coverage models are available that have demonstrated significant accuracy in predicting SCA coverage. An example is the NTIA/ITS Communications Systems Performance Model (CSPM) [17]. When time and budget do not allow for modeling, there is a rule of thumb that often can be applied with surprising accuracy. Place a small FM stereo table radio in a location where a proposed SCA receiver would be employed. If the FM radio produces quality noise-free reception, then it is likely that the SCA receiver will work as well; however, if the FM radio exhibits noise or distorted sound, then SCA reception will be worse and ineffective. Note that many FM radios have a stereo blend circuit that under weak signal conditions will combine the stereo into monaural, thus greatly reducing the noise created by the weak signal. Such radios should be avoided for SCA coverage testing because they will mask the effects that suggest poor SCA coverage.

SUMMARY

FM Broadcasting is undergoing major changes during the transition to digital broadcasting and multicasting and data transmission. For more information on how digital signals are added to a standard FM broadcast channel, see Chapter 4.13, "AM and FM IBOC Equipment and Systems," and Chapter 4.14, "Radio Data Systems."

References

[1] Armstrong, E.H., A method of reducing disturbances in radio signaling by a system of frequency modulation, *Proc. IRE*, 24(5), 689–740, 1936. (Reprinted in Kapper, J., Ed., *Selected Papers on Frequency Modulation*, Dover, New York, 1970, pp. 3–34.)

[2] FCC BC Docket No. 82-536, *Notice of Proposed Rule Making* (August 19, 1982), paragraph 4, page 3.

[3] EIA, *Report on the National Quadraphonic Radio Committee (NQRC) to the Federal Communications Commission*, Vol. II, Electronic Industries Association, Arlington, VA, chap. 1, pp. 21–29.

[4] Federal Communications Commission, Rules and Regulations, 47 CFR, Section 73.310.

[5] The FCC Rules are embodied in the Code of Federal Regulations (CFR) Title 47, Chapter 1, Parts 0 to 199; http://www.access.gpo.gov/nara/cfr/cfr-table-search.html#page1/.

[6] National Association of Broadcasters/Westinghouse Broadcasting & Cable, Inc.; National Public Radio, *Increased FM Deviation, Additional Subcarriers and FM Broadcasting: A Technical Report*, August 30, 1983, pp. 5–6.

[7] Small, E., *Regaining Modulation Lost to SCA*, Modulation Sciences, Somerset, NJ, http://www.modsci.com/images/whitePapers/radioProducts/fmmm2/LOSTSCA.pdf.

[8] Declaratory Ruling letter (under 47 CFR 1.2) from Thomas P. Stanley, Ph.D., Chief Engineer, Federal Communications Commission, to Modulation Sciences, Inc., December 4, 1989.

[9] Modulation Measurement, Public Notice 77, Federal Communications Commission, Washington, D.C., January 31, 1991.

[10] Takahashi, S. and Iida, H., Application of Walsh functions to an FM stereo demodulator, *J. Audio Eng. Soc.*, 33(9), 1985.

[11] This technique was developed by Geoffrey Mendenhall, Harris Corporation, Mason, Ohio.

[12] FCC BC Docket No. 82-536, First Report and Order (April 7, 1983); Second Report and Order (March 29, 1984).

[13] Kean, J.C., *Laboratory and Field Tests of Several FM/SCA Frequencies*, National Public Radio Engineering Report, October 15, 1981, pp. 13–14.

[14] Yamada, O. et al., Traffic information services using FM multiplex broadcasting, *IVHS J.*, 1(1), 35–43, 1993.

[15] Mastrangelo, J.F. and Rust, W.R., *Testing and Evaluation of the Subcarrier Traffic Information Channel*, NTIA Report 96-333, Boulder, CO, 1966.

[16] *Ibid.*, NQRC/Gibson, p. 3/11.

[17] DeBolt, R.O. and DeMinco, N., *FM Subcarrier Assessment for the Intelligent Transportation System*, NTIA Report 97-335, Boulder CO, 1997, p. 21.

CHAPTER

4.9

FM Broadcast Antennas

PETER K. ONNIGIAN
Sacramento, California

ERIC DYE
Sacramento, California

With Updates by

DANE JUBERA AND SLAVA BULKIN
Jampro Antennas
Sacramento, California

INTRODUCTION

This chapter is for broadcast engineers, technicians, and station managers who must make important decisions regarding FM transmitting antennas. To ensure the best possible signal strength in the station's service areas, the site location, antenna height, antenna type, and propagation conditions must all be considered. The implementation of digital radio systems using in-band, on-channel (IBOC) techniques requires the use of separate antennas or higher performance common antennas to meet the more stringent digital signal requirements.

Most FM antenna radiation patterns are nonsymmetrical. That is, the antennas are often mounted on one side of a steel supporting tower or pole. FM antennas outside the western hemisphere on the other hand are usually symmetrical because they are generally installed on all faces (all around) of a tower. However, both methods are capable of providing acceptable omnidirectional azimuth patterns.

Antennas for FM broadcasting use horizontal polarization (H-pol), vertical polarization (V-pol), or circular polarization (CP). Cross polarization, that is alternate use of H and V, is used as a means to prevent cochannel interference in some European countries but not in the western hemisphere. CP, together with its special form, elliptical polarization (E-pol), was introduced in the United States in the early 1960s as a means to provide greater signal penetration into the many different forms of FM receiving antennas, which are now found in the service area. H-pol is the standard in the United States, but CP or E-pol may be used if desired. V-pol only is permitted for noncommercial FM stations seeking to limit interference to TV channel 6.

FM radio receivers use a variety of antennas including extendable monopoles (whips), dipoles, and capacitive coupling to power cords and headphone leads. Receiver antennas differ from their transmission counterparts, which have a fixed polarization.

Antennas for FM broadcasting must be chosen carefully in order to cover the service areas properly with adequate level and quality signals. For economic and technical reasons, the desired effective radiated power (ERP) should be produced with a balance between antenna gain and transmitter power. The height of the antenna over the service area, distances to areas of population, ERP, and economics are items that must be considered.

FM broadcasting antennas must meet the stringent requirements for FM stereo and subcarrier broadcasting and for digital broadcasting. Most FM stations in the United States are using CP antennas.

PROPAGATION

There is essentially no difference between day and night propagation conditions for FM broadcasting because of the nature of the spectrum employed (88–108 MHz). Therefore, FM stations have relatively uniform day and night service areas. FM propagation loss includes everything that can happen to the energy radiated from the transmitting antenna during its journey to the receiving antennas. That includes the free space path attenuation of the wave and such factors as refraction, reflection, depolarization, diffraction, absorption, scattering, Fresnel zone clearances, grazing, and Brewster angle problems.

Propagation is dependent on all these properties out to approximately 40 miles (65 km). Some additional factors enter the picture at greater distances. Radio wave propagation is further complicated because some of these propagation variables are functions of frequency, polarization, or both, and many have location and time variations.

It is the intent of the FM broadcast transmission system to put a signal into FM receivers of sufficient strength to overcome noise and to provide at least 20 dB carrier-to-noise ratio, which will provide at least 30 dB of stereo separation. The required RF signal level varies from about 2 μV/m (microvolt per meter) for high-sensitivity FM stereo tuners in the suburbs to about 500 μV/m for less-sensitive portables. Automobile receivers typically have wide-ranged sensitivity values.

FM antenna manufacturers do not guarantee coverage per se. They supply antennas that provide certain radiation pattern requirements and gain. Many antennas are assumed to have an omnidirectional pattern. Although achievable in free space, it is rarely fully achieved in practice due to sources of distortion, such as support structure and feed lines.

FM antennas are usually designed to provide a horizontal plane pattern circularity of about 3 dB when mounted on the side of a specific tower or pole. Note that specific radiation patterns and gain are initially designed for free-space conditions and may not relate directly to signal strengths measured at or near ground level, well away from the antenna.

Radiation pattern and propagation are two distinctly separate conditions. The pattern is the radiation that is transmitted by a given antenna in any given direction, without any propagation limitations, as measured on a good antenna test range. Propagation depends on path and environmental conditions existing between the transmitting antenna and the receivers.

The actual service area signal strength contours are based on two probability factors. Contours are not solid signal areas. For example, the FCC signal coverage charts referred to as the f(50/50) curves, are based on a probability of occurrence of certain voltage levels for at least 50% of the locations, at least 50% of the time. This means that at any given location within the predicted signal contour at 9 meters or about 30 ft. above ground, the signal has a 50% chance to meet the requirements. Furthermore, half the time at that location it may reach or exceed the level predicted while at other times it may be lower in strength.

These FCC signal propagation contour charts (FCC Rules Section 73.333) are based on the assumption that average propagation conditions exist. One or more of the propagation conditions mentioned earlier may reduce the measured signal strength from the predicted values substantially.

Propagation Loss

The power radiated from an FM transmitting station is spread over a relatively large area, somewhat like an outdoor, bare lightbulb on top of a tall pole. The power reaching the receiving antenna is a very small percentage of the total radiated power.

The formula used to compute free-space loss (FSL) is

$$\text{FSL(dB)} = 36.6 + 20 \log D(\text{miles}) + 20 \log F(\text{MHz})$$

At 100 MHz and a distance of 30 miles (48 km) the free-space path loss would be 106 dB [1]. Doubling the distance increases the space loss by one-half, or 6 dB. The path loss does not attenuate the signal as much as some other factors. Path loss between an Earth station and a satellite is a classic textbook example of a 6 dB loss every time the distance is doubled because the path is unimpeded. But a typical FM station signal travels through air with weather variables, over the imperfect earth's surface (ground), and through vegetation and around buildings, all of which can cause propagation problems.

Refraction, diffraction, and reflection from scores of objects, such as hills and buildings, may occur in the propagation path between the transmitting and the receiving antennas. These, along with absorption, scattering, lack of Fresnel zone clearances, and other factors, all can reduce the signal strengths.

Signal loss due to foliage has been well known to UHF TV broadcasters [2]. The same condition exists to a lesser degree for FM broadcasting. Trees, shrubs, and other foliage on hills or smooth terrain affect the reflected as well as the direct signal strength. With average values of permittivity and conductivity in both foliage and ground, a loss of about 2.5 dB was found to exist in a 10 mile path, at FM frequencies [3]. The height gain factor is increased with heights above the foliage.

Considerable depolarization takes place because the transmission through or reflections from ground foliage are a diffracted field contribution. Additional information on Propagation Characteristics of Radio Waves can be found in Chapter 1.8.

Multipath Problems

The ideal reception condition is a strong direct single source signal. When energy from two or more paths reaches the receiver (due to reflections), a condition called *multipath reception* occurs. Poor reception is experienced when there is insufficient strength difference between the direct and the reflected signals, because they can cancel each other where the geometry places them out of phase.

An important factor in achieving the goal of delivering an optimal signal to the listener is the location of the transmitting antenna. Great care must be exercised to find a suitable tower site because poor selection of the site can result in unfavorable signal propagation and poor signal quality. One very serious result of poor site selection is multipath propagation in some directions.

As an example, the transmitter should not be located so that strong reflections take place from nearby hills or mountains. This can happen when the transmitter is placed on one side of a large city and the other side of the city has a high mountain range. This is illustrated in Figure 4.9-1 where a mountain range causes reflections back into a large city. Radiation into the city directly from the transmitting antenna, as well as reflections from the nearby hills and mountains, will create two or more signal paths toward many receivers. These reflections can be strong enough that only a 10 dB difference may exist between the direct and the reflected condition, which causes severe multipath problems in receivers. A TV station at this same location would cause heavy ghosting, even with directional receiving antennas, which reduce signal pickup from their back.

A better FM transmitting site would be on the hills between the high mountain range and the city. Using a directional transmitting antenna with very little radiation toward the high mountains, reflections can be satisfactorily reduced, and the FM station will likely operate successfully.

Multipath reflections are easy to identify. On an automobile radio, the signal will drop out, sometimes abruptly, as the vehicle moves. This effect may be rhythmic with distance while traveling slowly. It is sometimes called the picket fence effect as it acts like a picket fence alternately blocking and letting the signal pass. A field strength meter will usually reveal great variations of signal when moving, say, 100 ft. (30 m) in a line with the transmitter. Cyclic variations over uniformly spaced intervals on the ground as great as 40 dB have been observed by the author.

This variation in signal levels is caused by the reflections adding and subtracting from direct and reflected signals caused by propagation problems existing in the path between transmitter and receiver. It usually has little to do with the qualities of the transmitting antenna. It is a function of site selection. This should not be confused with a similar effect observed near the base of the tower supporting a high-gain antenna. Nulls produced by stacking bays for gain are found near the antenna and may be reduced by null-fill techniques if needed. (See Beam Tilt and Null Fill sections later in this chapter.)

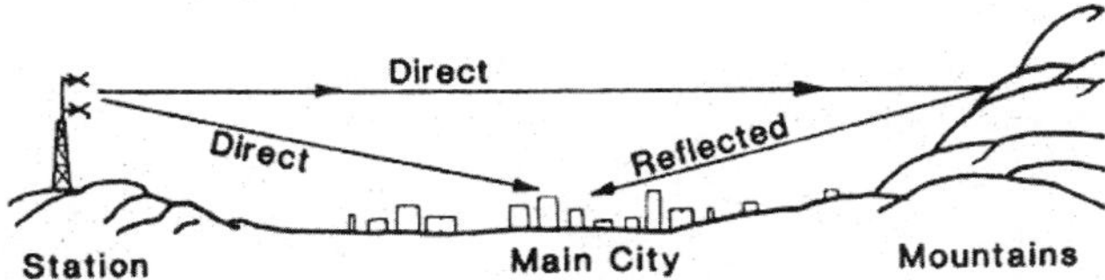

FIGURE 4.9-1 Example of how placement of a transmitting antenna can cause poor reception due to reflections.

Ground Reflections

In the elevation plane between transmitter and receiver, nearly all FM signal coverage lies between the horizon and 10 degrees below. Called the *grazing angle,* it lies between the horizontal plane and the earth's surface. Generally the higher the transmitting antenna above the service area, the greater this angle will be.

The angles of incidence and reflection are not the same, as shown in Figure 4.9-2. The depression angle and the grazing angle are not equal as would be the case for a flat earth. Reflections from these angles play an important part in the strength and the quality of the signal in FM broadcasting with circular polarization.

The ground, which causes reflections at these grazing angles, does not treat H-pol and V-pol in the same manner. The V-pol is attenuated considerably more than the H-pol, as shown in Figure 4.9-3. The phase of the V-pol changes substantially with angle, while H-pol remains nearly the same. At these useful low-propagation angles, there is considerably less V-pol signal reflected than H-pol, when grazing takes place. Field measurements confirm this fact [4]. For this reason, it is impossible to measure accurately axial ratios in the service area. To be meaningful, the H-pol and V-pol ratios must be measured on a good antenna test range.

It is quite difficult to predict accurately the reflection coefficient (efficiency), which varies considerably as a function of polarization, frequency, grazing angle, surface roughness, soil type, moisture content, vegetation growth, weather, and the season. There are complex formulas for predicting the ground conductivity at the frequency of interest. For 100 MHz, a value of 10 millisiemens per meter (mS/m) ground conductivity is often used, with a permittivity of 25, as being about the average for the continental United States [3].

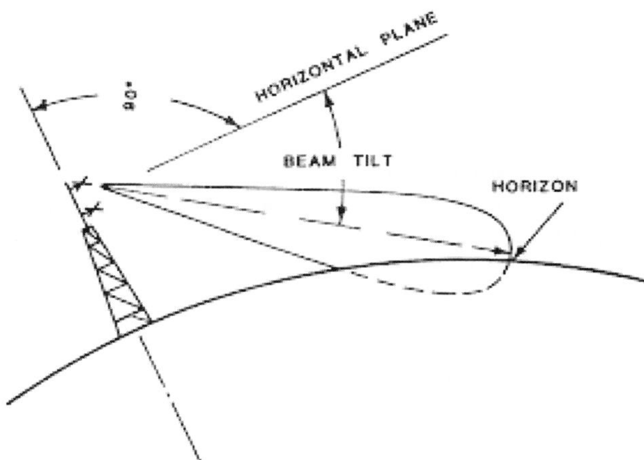

FIGURE 4.9-2 Illustration to show how beam tilt is used to radiate maximum ERP at the horizon.

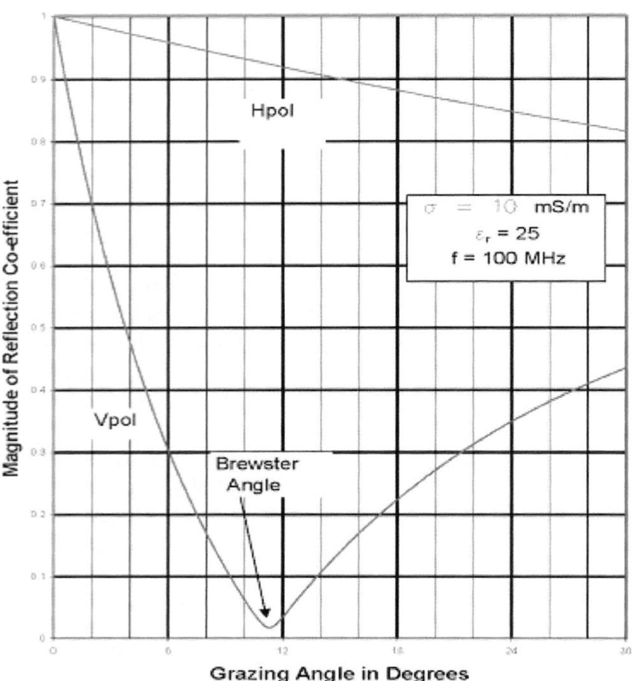

FIGURE 4.9-3 Magnitude of reflection coefficient versus grazing angle for horizontal and vertical polarization.

Brewster Angle

For polarization with the electric field normal to the plane of incidence, there is no angle that will yield an equality of impedances for earth materials with different dielectric constants but like permeabilities. An incident wave with both polarizations present will have some of the one polarization component but little of the other reflected. The reflected wave at this angle is thus plane polarized with the electric field normal to the plane of incidence, and the angle is the polarizing angle.

Notice that in Figure 4.9-3 the minimum reflection coefficient occurs at a grazing angle of about 11 degrees. Below this angle, the reflection coefficient rapidly increases to unity. The angle at which the minimum reflection coefficient occurs is called the *Brewster* or *polarizing angle,* after the English mathematician who first discovered this phenomenon.

For ground reflections occurring near the Brewster angle, the reflection coefficient is much smaller for V-pol than the H-pol. Therefore, the reflected V-pol signal component of CP is attenuated considerably. The greatest attenuation for V-pol from ground reflection occurs at this angle.

Field measurement of V-pol signals will usually show a significant variability of H-pol to V-pol ratios due to this Brewster angle phenomenon. The Brewster angle is also a function of soil conductivity and may change from place to place, as well as from season to season [5].

It is important, then, that the antenna height above the service area should produce grazing angles that are less than the Brewster angle. Otherwise the V-pol will be reduced and the radiation will be much more elliptical than circular in polarization.

Fresnel Zone Clearance

A much neglected consideration in FM transmitting antenna location and height is *Fresnel zone* radius clearance in the path to the service area. Microwave engineers always make certain that their signal paths have this important clearance.

The effect of clearance above ground or other obstacles was studied by August Jean Fresnel, a French scientist who first discovered this phenomenon in optics. Fresnel zones are circular areas surrounding the direct line-of-sight path of a radius such that the difference between the direct and indirect path length to the zone perimeter is a multiple of a half-wavelength longer than the direct path. This is illustrated in Figure 4.9-4. The zone diameter varies with frequency and path length. The greater the path length, the larger the required midpath clearance required for full signal.

Fresnel also discovered that the entire first zone radius is not required for full signal strength. Six-tenths of the first zone would suffice, which is fortunate since the radius is quite large at FM frequencies. The equation for determining the first Fresnel zone radius for 4/3 earth curvature is

$$R = 1140\sqrt{d/f}$$

where d is the path length in miles, f is in MHz, and R is in feet for the first radius.

In Table 4.9-1, the required 0.6 first Fresnel zone radii clearances at the middle of the path are shown for 98 MHz and service areas up to 52 miles (92 km) from the transmitter. The idea is to raise the height of the transmitting antenna so that the midpath height is as high as or higher than shown in the table. Due to the geometry of the Fresnel zone, if the terrain is relatively flat, the midpath radius will control and be larger than that required elsewhere along the path. If the midpath clearance is less than the values shown, the FM signal will be attenuated in accordance with the curve shown in Figure 4.9-5, presuming ideal reflection off the ground or obstructions.

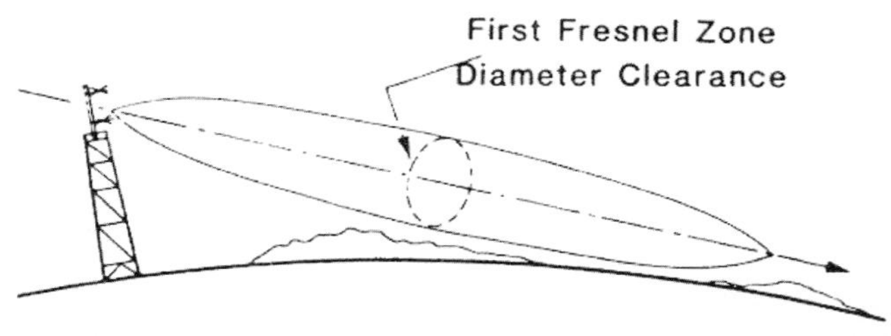

FIGURE 4.9-4 Fresnel zone at path midpoint.

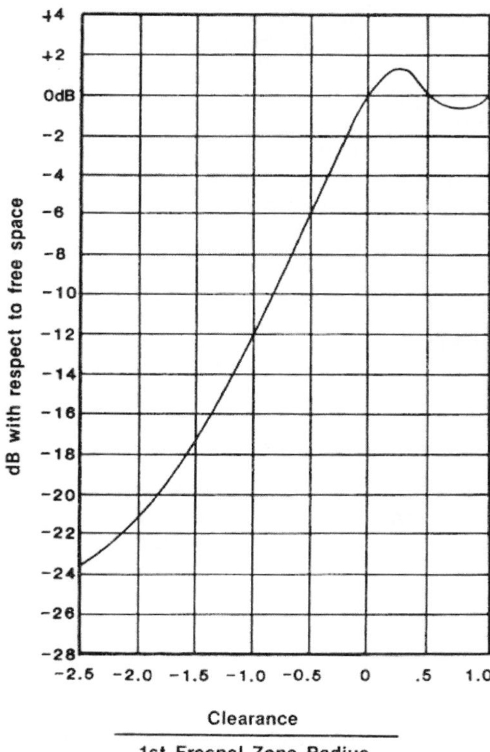

FIGURE 4.9-5 Attenuation of FM propagation when the path between transmitter and receiver lacks Fresnel zone clearance in the ratios shown.

The center-of-radiation heights of the antennas in Table 4.9-1 are actual and not height above average terrain (HAAT). Some of these recommended heights will reduce the allowable ERP in accordance with FCC Rules Section 73.211(b), depending on the class of station and the zone. However, it is better to have the Fresnel clearance than the maximum low height ERP values, as the higher heights will produce stronger signals.

It is a well-known propagation axiom that greater heights are more useful in producing higher signal strengths far from the antenna than ERP levels, everything else being equal. Without the first Fresnel clearance of 60%, the signal level at the distant point may suffer. This reduction will follow the curve shown in Figure 4.9-5 for different values of clearance and worst-case reflection conditions.

In order for the FCC prediction curves to be valid, the recommended minimum antenna heights should be employed. These heights not only provide line-of-sight conditions to the service limits but also proper Fresnel clearances. Both conditions are required for the FCC f(50,50) curves to be valid.

The values in Table 4.9-1 are for relatively flat terrain, but take into consideration the FCC suggested roughness factor of up to ±150 ft (50 m). Where the tower height is limited by HAAT values or other limitations, the signal strength will suffer due to those factors.

TABLE 4.9-1
Recommended Minimum Antenna Heights (for Flat Terrain and 98 MHz)

Service Area Radius Required (km)	Fresnel Zone Six-tenths Clearance (m)	Recommended Min. Antenna Height (m)	Probable FCC 80–90 Class
8	47	95	A
12	58	115	A
16	66	130	A
24	81	167	A, B, C-2
32	94	188	B-1, C-2
40	105	213	B-1
48	115	230	B
56	125	250	B
64	133	267	C-1
72	141	282	C-1
80	149	297	C-1
92	159	318	C

Linear Height Gain Effect

By raising the receiving antenna above the immediate effects of the soil, the signal level will be increased. Actual field measurements have shown a 9 dB increase in signal when the dipole was raised from 3.28 ft (1 m) to a level of 30 ft (9.1 m). This is due to reflection phenomena in the foreground of the receiver, not ground conductivity.

FCC Service Contours

From the FCC coverage prediction charts, it is possible to draw contours of the various grades of service for a given ERP and antenna height above average terrain. These predictions, for at least 50% of the locations at least 50% of the time, constitute the basis for the service contours. The city grade contour is 70 dBμV/m (3.16 millivolts per meter) and primary service contour is 60 dBμV/m (1.0 millivolts per meter). The FCC Rules Section 73.333 charts for these predictions have a built-in terrain roughness factor, as explained above.

GENERAL COVERAGE STANDARDS

There are certain height and power levels established by the FCC for various classes of stations. The United States has been divided into three geographical zones based on population density as well as propagation refractive index levels. For each zone ERP and height values have been set to prevent cochannel and adjacent channel interference.

TABLE 4.9-2
FM Station Classes, Zones, and ERP

Class	Zone	Max. ERP (kW)	Max. HAAT (Meters)	Distance to 60 dBuV/m (km)
A	I, I-A, II	6	100	28
B	I, I-A	50	150	52
B-1	I, I-A	25	100	39
C	II	100	600	92
C-0	II	100	300	83
C-1	II	100	300	72
C-2	II	50	150	52
C-3	II	25	100	39

Zone I, generally speaking, is the northeastern part of the United States. Zone I-A includes Puerto Rico, Virgin Islands, and that portion of California lying below the fortieth parallel. Zone II includes Alaska, Hawaii, and the remainder of the United States not in the above two zones. The zones are more fully described in FCC Rules Section 73.205 and figure 1 of Section 73.699.

Under the FCC Rules that resulted from Docket 80–90, *Modification of FM Broadcast Station Rules to Increase the Availability of Commercial FM Broadcast Assignments,* in 1983, new ERP levels and additional classes of stations were created. The distance to the 60 dBu (1 mV/m) signal contour is the controlling factor so that the ERP based on the HAAT is adjusted to produce that level and no more at a specific distance for a particular class station.

Table 4.9-2 shows for each FM class station, the zone, the maximum ERP, the maximum HAAT, and the distance to the 60 dBu contour calculated by using the maximum ERP and HAAT, and then rounding to the nearest kilometer and mile.

Stations may be upgraded using the easiest method, which is to increase existing location tower height. Such factors as local zoning laws and aircraft flight patterns may preclude this approach, however.

FM Signal Measurements

The signal strength received at 5 ft (1.5 m) above ground, which is about average for auto whip antennas, is several times lower in level than at the standard FCC measurement height of 30 ft (9.1 m). This fact should be taken into consideration when comparing low height measurements with the FCC Rules Section 73.333 prediction charts, which are based on a 30 ft receiving height.

Signal levels inside houses, apartments, offices, and other structures vary widely. Levels depend on the type of building construction, but in nearly all cases will be lower than those outdoors. Reflections inside the building reduce stereo separation, and cause crosstalk problems with SCA channels. Outside FM receiving antennas generally provide good reception.

Field strength measurements should not be used to determine the transmitting antenna radiation pattern or efficiency except under controlled conditions. The propagation factors discussed previously camouflage the true antenna performance. The only technically acceptable way to determine the antenna's characteristics is on an antenna test range.

This information may be used to determine the actual quality of service and the areas where useable signal levels in fact exist. Predicted contours may be considerably different from actual measured values.

Required Signal Strength

The history of FCC proceedings provides the rationale for the following levels for minimum satisfactory signal strength and maximum for the listed coverage areas:

34 dBu = 0.05 mV/m rural areas

60 dBu = 1.00 mV/m suburban areas

70 dBu = 3.16 mV/m principal community

82 dBu = 12.64 mV/m highest useful level

The first three levels were established by the FCC in the early 1950s when tube receivers and H-pol antennas were popular. Modern receivers have much better sensitivity. CP has added greater signal penetrating power than H-pol when the levels were first established.

The FCC defines two grades of signal contours on applications. The first is based on the 70 dBu contour (3.16 mV/m) required to cover the principal community of license. The second is the 60 dBu contour (1 mV/m), which defines the primary service area.

The FCC also stated that, in rural areas, levels as low as 50 µV/m were useful. Indeed current home stereo tuners and FM auto radios operate very well with only 25 µV/m. In practice 50 µV/m (0.05 mV/m) provides good quieting in nearly all automobile and portable radios receiving a stereo signal from a CP station antenna. Therefore, 50 µV/m should be considered the minimum useful signal level.

If the highest level of 3.16 mV/m is quadrupled, it will be 12.64 mV/m. This is a 12 dB increase, equal to increasing the FCC power level by more than 15 times. It can be safely said that this level of 12.64 mV/m is considerably more signal than necessary by any present-day working FM radio. Any signal level higher than this at the receiving antenna has not proven to be of significant value.

Blanketing

Excessive RF signals can overload the front end of receivers and make satisfactory reception impossible. The FCC Rules Section 73.318 defines the 115 dBu (562 mV/m) level as the *blanketing contour,* and adopted the free-space prediction method to predict how far this contour extends.

New or modified FM stations have the responsibility to satisfy all complaints at no cost to the complainant, of blanketing-related interference inside this contour within 1 year of commencement of operations.

The distance to the 115 dBu contour is determined using the following equation:

$$d\,(\text{km}) = 0.394\sqrt{P}$$

or

$$D\,(\text{mi}) = 0.245\sqrt{P}$$

where P is the ERP, in kilowatts of the maximum radiated lobe, irrespective of vertical directivity. For directional antennas, the horizontal directivity shall be used.

ANTENNA CHARACTERISTICS

Antenna gain can be increased by adding additional radiating elements (bays) to the antenna at the cost of narrowing the radiated beam. High-gain antennas concentrate the energy into such a narrow beam that often null fill must be employed to achieve the desired signal strength within the first few miles to the tower.

Directional antennas achieve increased gain over nondirectional antennas by limiting the radiated energy to specified directions. Directional antennas are useful when the tower is located near a large body of water, mountain range, or other areas where energy radiated in those directions is otherwise wasted. They are also employed to avoid interference where stations are not far enough apart.

Antenna gain is expressed in power ratio or in dB. For example, an antenna with a power gain of 2 is also said to have a gain of 3.0 dB.

FCC Rules Section 73.310(a) defines antenna gain as the inverse of the square of the root mean square value of the free-space field strength produced at 1 mile in the horizontal plane, in millivolts per meter for 1 kW antenna input power to 137.6 mV/m (in metric units, 1 km and 221.4 mV/m). Note that this gain is in reference to a horizontally polarized half-wave dipole. For a CP antenna, the gain is half for the same input power.

A two-bay H-pol antenna has a power gain of approximately 2. But a two-bay CP antenna in FCC terminology has a gain of about 1 because the other half of the power is V-pol and is not considered in the gain calculations. Only the horizontal polarization mode is used by the FCC. The vertically polarized energy must not exceed the H-pol (except for noncommercial, educational FM facilities attempting to minimize interference to TV channel 6 reception).

The power gain of an antenna is used with the transmitter power and transmission line and other losses when determining the ERP. Consider for example a 10 kW transmitter and an antenna power gain of 5. Neglecting transmission line loss, the ERP is 10 kW × 5 = 50 kW ERP. If the antenna gain were 10 and the transmitter power were 5 kW, the same ERP of 50 kW (5 kW × 10 = 50 kW ERP) would be obtained.

The FCC defines ERP to mean the product of the antenna input power (transmitter output power less transmission line loss) times the antenna power gain. Where circular polarization is used, the term ERP is applied separately to the H-pol and V-pol of radiation. For allocation purposes, the ERP is the H-pol component of radiation only. The V-pol component power normally must not exceed the H-pol power.

Beam Tilt

FM broadcasting antennas are normally mounted on towers that are plumb, so the peak power beam in the elevation pattern is perpendicular to the tower axis. A standard FM antenna without any beam tilt radiates more than one-half of the total radiated power above the horizon. All this power is lost.

The higher the antenna is above its average terrain, the larger the predicted coverage area. Since the earth is curved, the service horizon is bent lower than a perpendicular angle from the earth's surface. Thus, the strongest portion of the signal is aimed above the horizon. It also follows that the higher the antenna above the terrain, the greater the elevation angle down to the earth's horizon.

In order to strike the farthest service area from a high HAAT, the beam may need to be tilted down toward the earth. Electrical beam tilt lowers the beam angle equally in all azimuth headings and is chosen more frequently than mechanical tilting, which exhibits different effects in different directions. Choose enough tilt to position the center of the main beam on the farthest edge of the desired coverage area or just below the horizon, whichever is closer. Refer back to Figure 4.9-2 for an illustration of beam tilt.

For low gain antennas (two to four bays), the main beam is very broad, and if the antenna HAAT is less than 500 ft, there is little to be gained with beam tilt. On the other hand, beam tilt makes a large difference on high-gain antennas mounted on towers with a high HAAT.

Figure 4.9-6 shows the comparison between elevation angle path and coverage distance. It incorporates the curvature of the earth. This chart can be used to determine the optimum beam tilt. Follow the curve that is closest to actual HAAT, and mark the point where it intersects the horizon or crosses the distance of furthest desired coverage area (vertical axis). Read the beam tilt on the horizontal axis and round this value up to the nearest 0.25 degree.

Consulting engineers, familiar with this technique, can easily work out the required amount of beam tilt, if it is necessary. Typical values are 0.5 to 1.0 degree of tilt, depending on the antenna height, distance to the far service area, and the antenna elevation pattern.

Beam tilt is usually accomplished electrically, by delaying the currents to the lower bays, and advancing

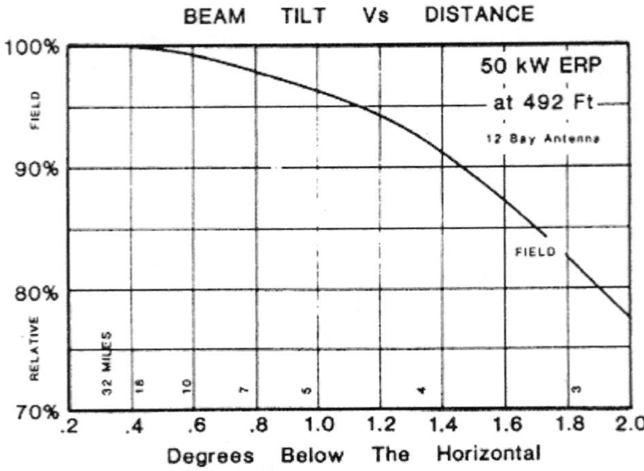

FIGURE 4.9-6 Twelve-bay C-pol antenna with 0.3 degree beam tilt showing ERP distribution with coverage from 492 ft (150 m) tower over flat land. Degrees below the horizon are based on 4/3 earth curvature. Horizon is –0.341 degrees at 32 miles (52 km).

the phase of the upper bay currents during the design and construction of the antenna at the factory.

Null Fill

While the beam tilt puts more signal into the far reaches of the service area, it does not solve the problem sometimes caused by high-gain antennas within several miles of the transmitter. Elevation angle nulls common to all antennas with two or more bays appear farther and farther away from the antenna as its gain is increased with more bays.

When multiple bay arrays are employed, lobes and nulls occur in the elevation pattern. As the number of bays increases, the main beam narrows and the first null radius increases. The advantage of beam tilt and null fill varies depending on factors such as tower height, site elevation, number of bays, and relative locations of communities to be served. A simple rule-of-thumb is that null fill is beneficial when there is desired service area within the radius of the first null.

In most FM applications, the null is relatively close to the antenna, thus a small amount of null fill (5–10%) takes care of the problem. Larger amounts of null fill are unnecessary and reduce the gain of the antenna. Note that null fill has no effect on distant coverage.

VSWR Bandwidth

According to theory, the bandwidth of an FM signal is infinite if all the sidebands are taken into account. Also, at certain modulation indices, the carrier amplitude goes to zero and all the transmitted power is on frequencies (sidebands) other than the carrier frequency. Practical considerations in the transmitter and receiver circuitry make it necessary to restrict the RF bandwidth to less than infinity.

Prior to 1984 the maximum deviation for FM stations was 75 kHz, representing 100% modulation. In that year the FCC changed the maximum deviation to 82.5 kHz (110%) for those stations with 10% injection of subcarrier channels. This additional deviation requires greater antenna system bandwidth than previously needed.

System bandwidth is measured at the point in the antenna system where the transmitter is connected. This usually includes the harmonic filter, the main coaxial transmission line, and the antenna.

The significant sidebands are usually considered to be those whose amplitude exceeds 1% of the unmodulated carrier. With 110% modulation (82.5 kHz deviation) these sidebands produce a bandwidth of 260 kHz.

The voltage standing wave ratio (VSWR) bandwidth is the range over which the system under consideration has a reflection coefficient of less than 5%; a VSWR of 1.1:1. Digital radio transmission systems using IBOC technology place greater emphasis on VSWR bandwidth as the sideband relationship to carrier is essential for optimum operation.

Checking System VSWR

The VSWR of the narrowband antenna system should be checked regularly and adjusted as necessary. One of several methods for checking VSWR in coaxial line systems using test equipment may be used. These include a signal generator test setup, an impedance test set, or a network analyzer.

The VSWR should be measured to ensure that the reflection response is balanced to 130 kHz on each side of the carrier frequency. With transmission lines longer than 300 ft (100 m), it is suggested that the VSWR bandwidth should be less than 1.08:1 for a bandwidth of 260 kHz. The additional delay due to increasing line length becomes more of a problem for digital signals, so the amplitude of the reflection must be reduced for best operational results.

Importance of Low VSWR

The VSWR shown by the transmitter reflectometer does not increase or decrease the range of the signal. It has nothing to do with coverage. But VSWR values above 1.1:1 may decrease the final amplifier efficiency. Other definite negative effects of VSWR are increased intermodulation products and AM synchronous noise. Stereo separation is also degraded with increased VSWR [6].

Intermodulation and SAM Distortion

Intermodulation distortion and synchronous AM (SAM) noise can be caused by narrow VSWR bandwidth in the antenna system, as well as by final amplifier circuitry [7]. SAM is important in FM transmitter facilities employing subcarriers. SAM is AM modulation of the carrier caused by frequency modulation of the carrier frequency in the VSWR notch. At the notch the reflected energy is the lowest. As the deviation takes place, the greater the frequency swing, the greater will be the reflections, due to the VSWR notch. With a flat VSWR curve, SAM does not take place. If the VSWR curve is skewed, SAM will occur and intermod and stereo crosstalk will increase.

Directional Antennas

The FCC sometimes requires that the azimuth radiation pattern be made directional to reduce the distance spacing of normally allocated ERP toward a given short-spaced station, or for other reasons. (See the FCC Rules Sections 73.213, 215, and 316(b) and (c).) To conform to these specifications, most broadcasters order antennas that are pattern adjusted, measured, and certified to the FCC's requirements.

Directional antennas are licensed for peak ERP values based on the azimuth pattern. The V-pol gain may not exceed the H-pol gain in a CP directional array nor may V-pol exceed the H-pol in the protected direction (except in the case of FM protection to TV channel 6, mentioned elsewhere). The amplitude away from the null cannot climb more than 2 dB per 10 degrees of azimuth.

Directional antennas are often mounted on poles on the side of a tower. Since the support affects the pattern, the antenna is specified and measured with the pole or tower on which the antenna is mounted. Most firms will make the antenna meet the specific pattern requirements.

Making a directional pattern is a combination of the natural pattern resulting from side mounting and the use of parasitic elements. Using these two factors, virtually any directional pattern can be produced.

Antenna gain is calculated differently for directional antennas. The azimuth directivity increases the gain value to correspond to the pattern. If all elements/bays are the same (the typical case) pattern multiplication can be used to determine gain. For linearly polarized antennas, the gain is simply the product of the azimuth directivity, the array factor, and the efficiency factor. The array factor is referenced to an ideal dipole. For directional CP antennas, the power distribution between polarizations must be taken into account.

$$\text{Antenna gain (H-pol)} = G_H \times \text{array factor} \times \text{efficiency}$$

$$\text{Antenna gain (V-pol)} = G_V \times \text{array factor} \times \text{efficiency}$$

where: $G_H = D_H \times D_V \times GA/D_H + D_V \times GA$, and $G = G_H/GA$; D_H and D_V are the directivities of the H-pol and V-pol azimuth patterns; while GA is the gain of the H-pol over the V-pol pattern.

Due to the gain of its azimuth patterns, directional antennas have gains that are typically 2–6 dB higher than their nondirectional counterparts of equal number of bays.

ANTENNA POLARIZATION

Radio waves are composed of electric and magnetic fields at right angles to each other and to the direction of propagation. When the electric component E is horizontal, the wave is said to be horizontally polarized, as shown in Figures 4.9-7(b) and (d). Such a wave is radiated from a horizontal dipole. References are with respect to the earth plane. If the desired electric component is vertical as in Figures 4.9-7(a) and (e), a vertical dipole could be used to produce the vertically polarized wave.

Circular Polarization

When the two plane waves are equal in magnitude, and if one plane wave lags or leads the other by 90 electrical degrees, the field will rotate as shown in Figure 4.9-7, at the speed of the carrier frequency and will be polarized circularly.

Only in the special case where the horizontal and vertical components are equal in strength with a 90 degree phase difference is the radiation said to be CP.

The direction of rotation shown by the vector arrows in Figure 4.9-7 depends on the relative phase of the two components. Thus, the polarization of the wave will appear to have either clockwise or counterclockwise rotation, as shown. The FCC has set clockwise rotation as the technical standard in order that similar sense of rotation antennas may be used for reception in the future.

Notice that in Figure 4.9-7 the polarization rotates as the field propagates in time and space. Importantly, vertical and horizontal components are in quadrature phase. It is this rotation that enhances the signal penetrating qualities of CP, so useful in FM broadcasting.

The axial ratio as shown in Figure 4.9-8 is that between the maximum and minimum voltage component at any orientation of the reference measuring test dipole that is placed perpendicular to the direction of propagation. An axial ratio of 1:1 (0 dB) is perfect. In practice, axial ratios of 2 dB or better are considered to be excellent and commercially available. Axial ratios over 4.9 dB (1.75 to 1 voltage ratio) are considered to be elliptically polarized, a hybrid form, and not as good in signal-penetrating qualities as CP.

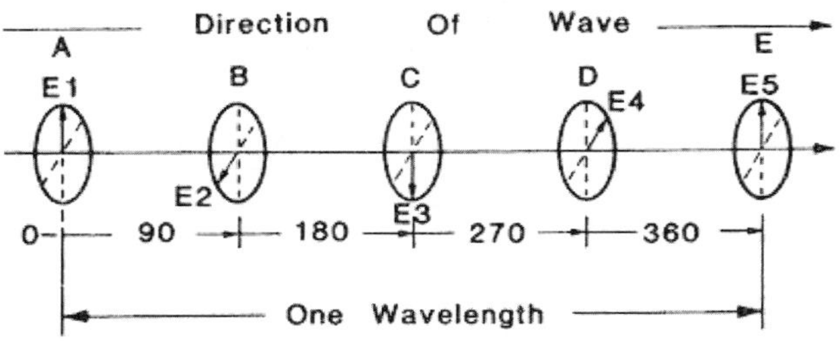

FIGURE 4.9-7 Circularly polarized wave propagation in one wavelength of travel, showing right-hand rotation. Note vector rotation with wave travel.

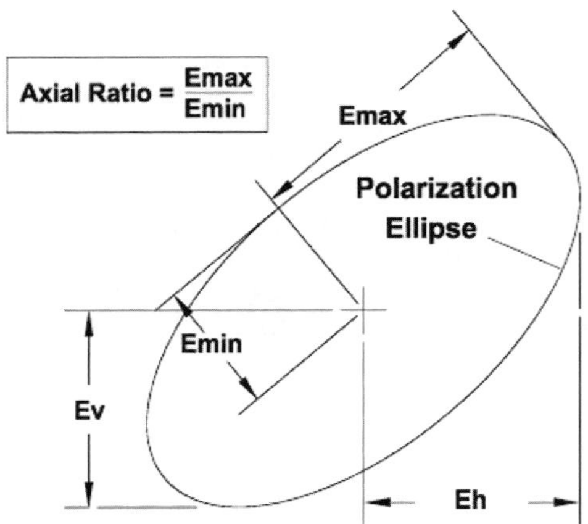

FIGURE 4.9-8 Axial ratio is the ratio of the larger polarized component divided by the smallest. This may be expressed in dB as 20log (Emax/Emin).

Since most receiving antennas are linearly polarized, the introduction of CP does not increase the net power received since the vertical and horizontal components never occur during the same instant. Even though CP requires nearly twice the transmitter output power, the signal does not propagate additional distance from the transmit antenna. Thus, using CP does not necessarily mean an increase in coverage. However, the introduction of CP eliminates the requirement of the receiving antenna to have a specific polarization (H or V). Thus, CP allows more consistent coverage within the contour. Its rotating vector can penetrate areas where linear polarization is stopped, shadowed, or cancelled due to out-of-phase reflections.

MATCHING COVERAGE AND ANTENNAS

Table 4.9-3 shows the FCC-predicted signal strengths for a typical class A facility on a relatively flat plane, with the antenna center at 328 ft (100 m) HAAT. A power of 3 kW is used. The first two columns show the distances, with the farthest being the horizon

TABLE 4.9-3
Transmitter Power versus Antenna Gain for a Class A 3kW ERP Station*

Service Distance			Signal Level in mV/m	
(miles)	(km)	Vertical Angle	7.5 kW Transmitter 1 Bay Antenna	1 kW Transmitter 6 Bay Antenna
1	1.6	3.58	275	210
2	3.2	1.80	88	81
3	4.8	1.21	42	40
4	6.4	0.92	24	22
5	8.0	0.74	16	16
6	9.6	0.63	11	11
7	11.3	0.55	8.5	8.5
8	12.9	0.49	6.2	6.2
9	14.5	0.44	5.0	5.0
10	16.1	0.41	3.7	3.7
12	19.3	0.36	2.5	2.5
14	22.5	0.33	1.8	1.8
16	25.7	0.31	1.4	1.4
18	28.9	0.29	1.1	1.1
20	32.2	0.28	0.85	0.85
22	35.4	0.28	0.70	0.70
24	38.6	0.27	0.55	0.55
26	41.8	0.27	0.40	0.40

* Showing the signal level at various distances for two transmitter/antenna combinations—zones 1 and 1-A maximum HAAT 328 ft (100 m)

from this height. The third column indicates the true earth angle from the antenna to the distances shown. From the elevation information the ERP from each antenna was determined at each vertical angle. This ERP value was used to find the signal strength from the FCC f(50,50) FM prediction chart, FCC Rules Section 73.333, figure 1. Under the signal level in millivolts-per-meter (mV/m) column, the predicted field strengths shown in Table 4.9-3 are based on the above procedure. From 5 miles (8 km) to the horizon, the signal strengths are identical. This is due to the shape of the antenna elevation pattern near the maximum.

Departure occurs as the depression angle to the receiver becomes larger. Beyond 4 miles (6.4 km), the one-bay antenna and the six-bay antenna produce nearly the same signal level.

Going toward the transmitter from 4 miles (6.4 km), the field increases in favor of the one-bay antenna. In this example, the table clearly indicates that the high-power transmitter, low-gain combination antenna does not improve the signal strength available to the receivers beyond about 4.5 miles (7.25 km). The signal level starts to increase between 4 and 5 miles (6.4 and 8 km). Any increase above this level is useless because full limiting has certainly taken place in even the poorest FM receiver.

In Table 4.9-3 the same signal strength of (16 mV/m) at 5 miles (8 km) and beyond comes from either transmitter-antenna combination. This is due to the fact that the ERP power at the vertical angle of –0.74 degrees is about the same from both antennas. The ERP at 0.0 degree elevation pattern will of course be exactly the same for both combinations. The field does not change measurably until observation is made beyond 1.5 degrees from the peak value in a six-bay antenna.

The signal strengths in Table 4.9-3 were based on relatively flat terrain for an antenna 328 ft (100 m) HAAT. The true earth curvature distance to the horizon is 25.56 miles (41.23 km). Therefore, the useful signal drops off very rapidly beyond this point for the typical class A station.

There are no nulls in a one-bay antenna pattern. In a six-bay antenna the first null occurs at about –10 degrees, which is approximately 0.37 miles (0.6 km) from the tower. Antenna arrays are never perfect so the null is never zero power. With a minimum radiation of 5 W in the first null, the predicted signal would be 31 mV/m. The second null is closer to the tower and with the same 5 W ERP would be even stronger in this example. So in practice there may be no need to fill in the nulls of the six-bay antenna.

Another consideration is that the nulls may fall close to the tower and the number of people within the null areas may be small. Thus, problems resulting from these close-in nulls would be minor.

Class B and C-2 Station Coverage

The same comparisons of transmitter-antenna combinations can be made for class B and the new class C-2 stations, operating with a HAAT of 492 ft (150 m) with 50 kW ERP. This is shown in Table 4.9-4. A 55 kW

TABLE 4.9-4
Transmitter Power versus Antenna Gain, Class B, C-2 50 kW EFP Zone 1, 1-A, and C-2

Service Distance			Signal Level in mV/m	
(miles)	(km)	Vertical Angle	55 kW Transmitter 2 Bay Antenna	10 kW Transmitter 10 Bay Antenna
10.	1.6	4.97°	900	140
1.5	2.5	3.25°	562	165
2.0	3.2	2.49°	310	230
3.0	4.8	1.67°	153	135
4.0	6.4	1.26°	92	88
4.6	7.4	1.15°	71	71
5.0	8.0	1.02°	57	57
7.5	12	0.70°	22	22
10	16	0.55°	13	13
15	24	0.41°	6.5	6.5
20	32	0.36°	3.1	3.1
25	40	0.33°	1.9	1.9
30	48	0.328°	1.1	1.1
35	56	0.332°	0.7	0.7

transmitter with high-efficiency coaxial lines and a two-bay CP antenna would provide the 50 kW ERP. It is compared with a 10 kW transmitter feeding a 10-bay CP antenna. The terrain flatness is assumed not to exceed ±150 ft (50 m).

Table 4.9-4 also shows that the signal levels are the same from 4.6 miles (7.4 km) to 35 miles (56 km) under similar columns as for the class A station comparisons. The FCC uses a receiving height of 30 ft (10 m) so the horizon is a bit farther away, at 31.3 miles (50.5 km).

From the transmitter out to about 2 miles (3.2 km), the signal rises much more rapidly in the two-bay antenna than in the 10-bay, the latter being somewhat similar to a cosecant curve. There is surplus signal close in and more than is needed or can be tolerated. This is one of several problems with high transmitter power and low-gain antenna combinations. With the two-bay antenna there is 900 mV/m at 1 mile (1.6 km) and 562 mV/m at 1.55 miles (2.5 km). This is above, or at, the blanketing level of 562 mV/m discussed earlier. The high-gain antenna does not cause this type of problem under identical conditions.

The signals from both combinations are much more than necessary for present-day FM receivers out to about 10 miles (16 km). There is no practical difference technically in useable signal strengths presented to receivers in the entire market area from either antenna. There is, however, a great deal of savings in capital costs as well as operating expenses between the two combinations.

One antenna factor is not clearly indicated in Table 4.9-4. The two antennas have elevation pattern nulls. The two-bay antenna null at –30 degrees falls 852 ft (260 m) from the base of the tower and can be disregarded. The 10-bay antenna null can be filled to as little as 2.5% field, which will not affect its gain. This would represent a minimum ERP at the nulls of 31 W. Although seemingly very small, it is effective, as shown in Table 4.9-5.

Note that the 10-bay antenna nulls can easily be filled to produce signal levels in excess of those required. If the transmitter is located in a populated area, these high levels prevent the loss of stereo separation and noise in the SCA (if there are reflections from high-level lobes in the built-up areas). This problem is common to TV transmitters that produce ghosts from high signal level lobe areas reflecting into null areas. This problem is greatly and satisfactorily reduced with null fill, as shown in Table 4.9-5.

TABLE 4.9-5
2.5% Null Fill-in, 10-Bay Antenna

Null	Angle	ERP	Distance	Field
First	–5.75°	31 w	4,800 Ft	31 mV/m
Second	–11.50°	31 w	2,240 Ft	70 mV/m
Third	–7.25°	31 w	1,512 Ft	109 mV/m

MATCHING TRANSMITTER POWER AND ANTENNAS

Several available combinations of antenna gain and transmitter power will provide the necessary ERP. The choice is complicated by the nature of the terrain in the service area such as flat land, some rolling hills, mountains, or perhaps a large valley. It now becomes necessary to understand the regulatory limitations on the tower height.

Important considerations when choosing the transmitter power and the gain combination to produce a given ERP are as follows:

- Transmitter
- Feed system
- Antenna
- Final amplifier
- Tower
- AC power consumption

The transmitter, antenna, tower, and coaxial feed line are one-time capital costs for the station. Tubes and commercial power, however, are continuing hour-by-hour cost factors. From Table 4.9-4, it is apparent that a low-power transmitter is much more economical than a high-power transmitter.

The ERP is the product of the antenna power gain and the antenna input power. Many different combinations of power gain and input power will yield the same ERP. The azimuth pattern will be quite similar for many different antenna power gains. The only difference in various combinations is the elevation pattern. As discussed previously, there is no significant or important difference in serving listeners from very different transmitter/antenna ratios.

The signal strength at any given location is a direct function of the ERP from the antenna elevation pattern angle to that location, the height of the antenna, and the propagation path. The ERP at the pertinent angle is the product of the elevation pattern relative amplitude at that angle squared, times the maximum ERP.

In practice there is no significant difference between a 3 kW ERP class A station using a 7.5 kW transmitter and a one-bay CP antenna, or, one using a 1 kW transmitter and a six-bay CP antenna, all other factors being equal.

Normally, all the power radiated above the antenna elevation pattern to the horizon is wasted. It is the radiated power below the angle to the horizon that reaches FM receivers. Therefore, only the radiated power toward the earth should be considered useful.

The ideal antenna system would put the same signal level from the base of the tower all the way out to the horizon. This requires an antenna whose elevation pattern is a cosecant curve, the normalized reciprocal of sine. It would be the most efficient antenna elevation pattern. Although this curve is impossible to achieve, it is approached as the antenna gain becomes greater.

ANTENNA SITE SELECTION

The transmitter location must be carefully chosen. Site economics should be secondary to the technical advantages of a particular site. Fresnel zone clearances and other factors outlined in this chapter should be considered. A site with an operating FM or VHF TV station makes an excellent source of signals to check propagation for a new station. If the existing station is FM, make certain that its antenna pattern has been optimized to provide as much circularity as possible.

A good field strength meter should be used to measure the actual signal from the existing station. Relative rather than absolute readings are important. Check for reflections as well as level changes within a short walking area of about 100 ft (30 m). Check for stereo separation. Using this information, the operation of a new station near the one being checked can be compared before moving or submitting the FCC application. The consulting engineer may find it useful to consider this information to evaluate the suitability of the new site.

High-Gain Antenna Contradictions

The advantages of high-gain, low-power transmitter combinations to produce the required ERP have been discussed. Their superiority in relatively flat land applications cannot be disputed. There is, however, the matter of unusual height over average terrain to be considered. As examples, if the transmitter is located on Mt. Wilson, in California, or on a very tall building in Chicago or New York, the elevation pattern issue can become a serious problem. This is true particularly when a significant portion of the audience is near the sites as is the case for these three locations.

Mt. Wilson, which serves the greater Los Angeles metropolitan area, is more than 1 mile (1.6 km) above most of its listeners. Coverage is required from 11 miles (17.75 km) out to the horizon, which is –0.57 degrees at 105 miles (168 km). Pasadena, the nearest city, is 13 degrees below the horizon. A high-gain antenna tilted down 0.5 degrees would serve the far reaches well, but would not lay down a moderate signal at –13 degrees.

FCC Rules Section 73.211 limits the ERP for overheight antennas such as those on Mt. Wilson with 2,900 ft (884 m) HAAT. New stations using that height must reduce ERP in accordance with the equivalence calculation, so that the predicted signal at the 1 mV/m contour does not extend beyond 32 miles (52 km) for class B stations.

In these situations a moderate (rather than high) gain antenna should be considered. From Mt. Wilson several existing four- and five-bay antennas now provide excellent service.

TV CHANNEL 6 PROBLEM

Television channel 6 occupies the band from 82–88 MHz with the sound carrier at 87.75 MHz. The FM broadcast band extends from 88–108 MHz. Noncommercial educational FM stations are assigned from 88–92 MHz. Interference can exist between the two, with the TV station viewers receiving sound and picture interference from the FM stations and channel 6 signals interfering with low-powered FM stations in the lower part of the FM band. The FM receiver is relatively selective with a response to about 200 kHz, but the TV receiver has a bandwidth of at least 6 MHz. (See FCC Rules Section 73.525.)

Three techniques can be employed to minimize channel 6 interference from FM stations: collocation, locating the FM station in an area of low population density, and antenna cross polarization.

Collocation

The purpose of collocation (that is, placing the FM transmitter at the channel 6 transmitter site) is to achieve the same propagation path for both TV and FM stations, thus maintaining a nearly constant desired-to-undesired signal ratio in the service area. If possible, both antennas should be mounted on the same tower. If not, a maximum separation of 0.25 miles (400 m) between the two is still considered as collocation.

The horizontal and vertical plane radiation patterns of both antennas should be similar because the objective is to maintain a near constant desired-to-undesired signal ratio. The HAAT should be similar, thus the desirability of collocating on the same tower. The maximum ERP of the FM stations operating on this basis is specified in FCC Rules Section 73.525(d), table B.

Alternate Locations

The FM station may not be intended to serve the same community as the TV station, or collocation may not be possible. In this event, the FM broadcaster should be located in an area of relatively low population density by imposing a limit on the population that may be included within that area where a particular undesired-to-desired protection ratio is exceeded.

Two ratios were proposed by a committee that studied this problem in 1983 [8]. Their recommendation varied according to the separation between the FM station from the channel 6 aural frequency of 87.75 MHz. In any event, the interference area should not have more than 3,000 people living in it. (See FCC Rules Section 73.525(c) and (e).)

Cross Polarization

Several organizations have made discrimination tests in the United States and in Europe with cross-polarized antennas from which it has been established that a discrimination of 16 dB can be expected in rural areas and 10 dB in urban areas between two stations with one using V-pol and the other using H-pol, and the receiving antenna being similarly polarized. This is sufficient in most cases to resolve the FM channel 6 problem.

While technically cross polarization will help solve the problem, the FCC rules do not require it. This is left as an option for the FM applicant to use. Most TV channel 6 receiving antennas will remain H-pol, while automobile FM antennas will stay V-pol. So if the TV station remains H-pol, this interference problem may be cleared up if the FM station switches to V-pol. (See FCC Rules Section 73.525(e)(4).)

COMMERCIALLY AVAILABLE ANTENNAS

There are several basic classes of antennas available for FM broadcasting. These and variations of them are made by several manufacturers in different models, gains, and input power ratings. They may be broken down into the following classes:

- Ring stub and twisted ring
- Shunt- and series-fed slanted dipole
- Multi-arm short helix
- Panel with crossed dipoles

These antennas have many things in common. For example, nonsymmetrical antennas are designed for side mounting to a steel tower or pole, as shown in Figure 4.9-9. Radiating elements are shunted across a common rigid coax line that eliminates the problems associated with the older corporate feed system using semiflexible solid dielectric low-power cables. Shunting elements every one wavelength across a transmission line makes impedance matching simple. Bandwidth is limited by the VSWR of the individual elements and the use of an internal transformer.

With more than about seven bays, the first three of the above antennas are more difficult to match and there is undesirable "beam squint," since the elevation beam angle changes with frequency deviation by the transmitter. Antennas with more than seven bays are fed from or near the center, thus dividing the phase change in one-half and effectively eliminating the beam squint. Center feeding the antenna also simplifies the VSWR matching.

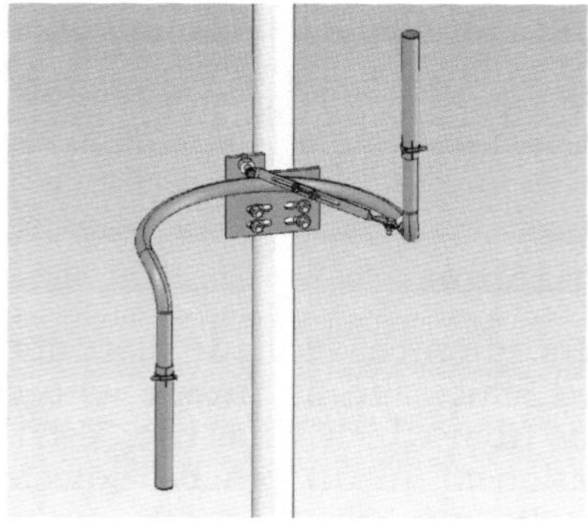

FIGURE 4.9-9 Example of ring-stub FM antenna element.

A means for tuning out reactance after the antennas have been installed on the tower is also common with all the antennas. Located at the input to the antenna, the VSWR tuner consists of adjustable location dielectric or metal slugs on the inner conductor of the main coax line. Several fixed-position variable capacitors, spaced one-eight wavelength along the main feeder near the antenna input, are also used on some side-mounted antennas to adjust the VSWR to low levels.

Another variety of antenna has curved radiating elements around a circumference whose diameter is determined by the number of element arms. Each radiator consists of two, three, or four such circular arms, depending on the model. Each element is fed through a shunt arrangement, and then shunted across the vertical rigid feed coaxial line.

Wideband panel antennas are becoming popular where high buildings, favorable mountain sites, or high towers are available. Several firms make wideband panel antennas. Some have very wideband VSWR features in each radiator. Others with not so broad VSWR use phase impedance compensation similar to the European scheme, which uses 90 degree phase quadrature impedance compensation.

Phase quadrature compensation makes it possible to cover the entire 88–108 MHz band with a VSWR under 1.1:1 while maintaining excellent elevation and azimuth patterns, together with very good axial ratios. This is especially important in IBOC operation where a single antenna is employed for both analog and digital signals. Power ratings up to several hundred kilowatts are offered so that many FM stations can be diplexed into one such antenna.

Only the wideband community FM antenna design now uses a corporate feed system, while the others are shunt fed from a common rigid coax line. This corporate feed system, using air dielectric semiflexible line at the lower power levels, is very successful. It splits the input power to many different dipoles at the correct amplitude and phase.

Standard Side-Mount Antennas

Standard side-mount antennas come in a variety of shapes and forms and are currently used in the majority of applications. Their chief advantages are low cost, easy installation, have relatively high gain, and low tower constraints. They are available in linear polarized configurations (H-pol or V-pol) or circularly polarized (CP).

Most side-mount antennas are comprised of a series of radiating elements, or bays, which are fed via a rigid inner-bay feed line. The most typical feed lines used are 1-5/8 inches for applications with less than 10 kW antenna input power, and 3-1/8 inches line for up to 40 kW. Most antenna elements come in high- and low-power versions. These antennas are mounted directly to the side of a tower or pole, as shown in Figure 4.9-10. Leg and face mounts are typical on tower structures.

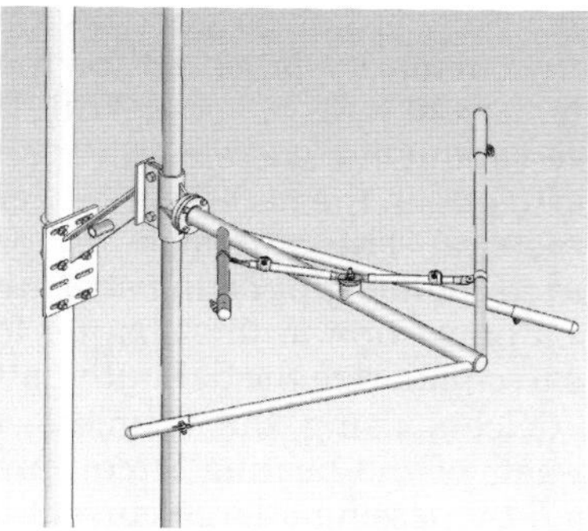

FIGURE 4.9-10 Shunt-fed slanted dipole antenna element.

Some manufacturers with test ranges offer side-mount antennas with custom directional patterns. The pattern shaping is accomplished by optimizing the mounting and adding parasitic reflectors, which are on the order of one-half wavelength. Repeated range tests have shown that side-mount antennas have largely distorted patterns due to feed lines, mounting structures, and other conductive items in or near the antenna aperture.

Side-mount antennas are inherently narrowband. The bandwidth of a single element rarely exceeds 1 MHz. Although diplexing two stations on a single side mount is practiced, the frequency spread between stations must be small (a few MHz), and compensation tuning (such as long stubs) must be used. In addition, none of these antennas are symmetrical, and each type has uncontrolled radiation from booms and feed lines in the aperture. This distortion deteriorates the antennas axial ratio and circularity.

Series-Fed V-Dipole Antennas

This antenna has similar bandwidth to its shunt-fed counterpart, but the array is typically intentionally tuned high in frequency. The combination of this tuning technique and the internal protection of its feed allows this antenna to be somewhat resistant to light icing.

This model is larger in size and heavier than other types of side-mount antennas so that tower constraints may become an issue. The antenna is typically field tuned for an optimized match. Careful placement of ceramic slugs can produce a good VSWR over the stations useful bandwidth.

Ring Radiators

There are several antennas that are simple adaptations of ring radiators and were designed and manufactured in the 1950s and 1960s for horizontal polarization. By adding vertical stubs to the ends of the radiator or twisting the ring, elliptical polarization (of sorts) is achieved.

Both the ring and the ring stub suffer from temperature variations that tend to change the spacing between the ring openings and thus the electrical capacitance and resonant frequency. The ring stub and the twisted ring are not really circularly polarized because the axial ratio varies considerably with azimuth. At best they may be said to be elliptically polarized.

The design has been improved by adding a second horizontal ring and improving the feed. Reducing bay spacing reduces high axial side lobes. The antenna has good circularity in free space, but like other types of side-mounted antennas, it is strongly affected by its support structure and feed line.

The radiation patterns are strongly affected by the tower mounting environment. Being of relatively high Q design, they are more susceptible to detuning because of icing. Radomes and electrical deicers are available to overcome this problem. While the icing problems may be overcome, pattern optimization is not offered for these antennas.

Ring-Stub Antennas

The H-pol radiation from these antennas comes from the ring portion whose plane is parallel with the earth. There is a minor lobe from each radiator, which is strengthened with vertical stacking for additional power gain. This nadir-zenith lobe is the result of 360 degree stacking on the rigid coax feed line. It reduces the gain and presents a lobe at the tower base that is detrimental to low-level audio equipment and personnel located in a building at the base of the tower.

In order to keep the cost down, like the twisted-ring antenna, the ring-stub is manufactured in several radiator-to-radiator spacings across the FM band. This results in some minor beam tilt up or down depending on the frequency. Most higher-priced slanted dipole and helix antennas are spaced exactly 360 degrees and are usually tested to assure this spacing during production.

Shunt-Fed Slanted Dipole Antennas

The slanted dipole antenna in its present configuration was developed and patented in 1970 [9]. It consists of two half-wave dipoles bent 90 degrees, slanted and fed in-phase. The slant angle is critical as it is the factor that determines the ratio of vertically and horizontally polarized radiated power. When fed through a vertical support pole on which the antenna was mounted during initial development tests, the axial ratio varied less than 1 dB.

The commercial adaptation uses a horizontal boom containing a step transformer. This boom supports two half-wave dipoles in which the included angle is 90 degrees, as shown in Figure 4.9-11. The two sets of dipoles are rotated at 22.5 degrees from the horizontal plane. Two opposite arms of the dipoles are delta matched to provide a 50 Ω impedance at the radiator input angle. All four dipole arm lengths may be adjusted to resonance by mechanical adjustment of the

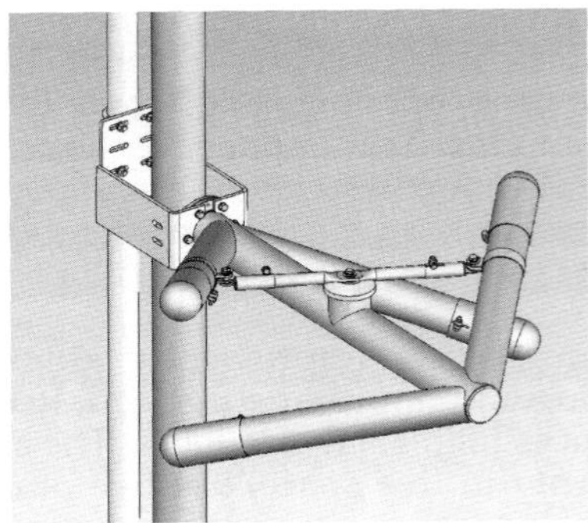

FIGURE 4.9-11 High-power shunt-fed slanted dipole antenna element.

end fittings. Shunt feeding, when properly adjusted, provides equal currents in all four arms resulting in excellent azimuth circularity.

Short Helix Antennas

An asymmetrical radiator is the four-arm shunt-fed helix. By using four dipoles, curved so that their circumference is about one wavelength, a CP antenna is produced, as illustrated in Figure 4.9-12 [10]. Each dipole is about one-half wavelength and is shunt fed. These are supported on a four-arm structure, one end of which is tied to the supporting structure. The dipoles overlap so that the current flow around the circumference is circular. The four feed arms are connected in shunt and the feed impedance is quite low, but may be improved with an internal step transformer.

The CP quality of the four-arm, side-fire, short helix is good. Three- and two-arm models are also available, but their axial ratio is not as good as the four-arm model. Pattern circularity is ±1 dB for the four arm, together with an axial ratio of about 3 dB. These radiators are stacked about one wavelength apart on a rigid coax feed line to obtain the necessary power gain. Like other asymmetrical FM antennas its patterns are strongly affected by the supporting structure. See Pattern Optimization section in this chapter for the need and methods to circularize the azimuth pattern.

Electrical deicers using the stainless-steel dipole arms as one-half of the heating circuit are available. Heat is created by passing a large current at low voltage through each arm from voltage-dropping transformers placed at each bay level. Plastic radomes are also available to keep snow and ice off the sensitive VSWR parts of the antenna.

Twisted-Ring Antennas

This type consists of one or more rings, which have been partially twisted so that the open ends of the ring are about 10 inches (25 cm) apart. One semicircular arm of the ring is fed with a small loop or by a direct tap on that arm, as shown in Figure 4.9-13. A number of these rings are fed in the same manner as the ring stubs, and have the same zenith-nadir lobe problem. The mechanical twist is not the same when viewed in all the azimuth directions. Therefore, the current is not the same, with the end result that in some directions there is much more elliptical radiation than in others.

These antennas are very simple and relatively inexpensive for single frequency use, but have some serious operational limitations for CP operation. They do not have the same signal penetrating effect as the slant dipole, short helix, or the flat panel antenna type of CP antennas.

Short Helix—Multi-Arm Antennas

The number of arms may be increased to four instead of the two in the slanted dipole variety. To provide CP, the arms are curved to form a one-wavelength circumference. These short multi-arm helices are also stacked

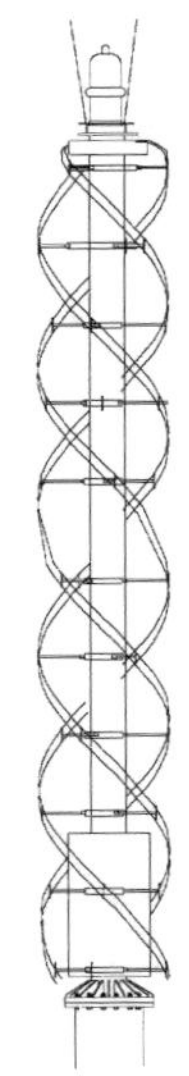

FIGURE 4.9-12 Multi-arm helical antenna.

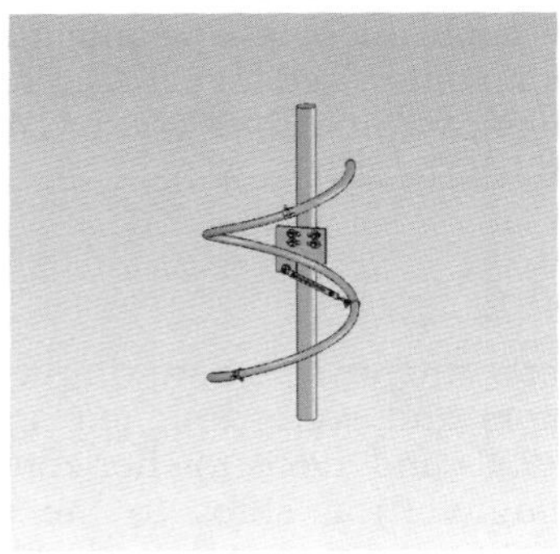

FIGURE 4.9-13 Twisted-ring radiator.

in the conventional manner, like the others in this series for power gains as desired. This design uses two-wavelength feed straps to feed all the elements in phase. This antenna is shunt fed, and is arrayed and mounted similarly to the slanted V-dipole antenna.

The azimuth pattern of all these nonsymmetrical antennas is affected by the supporting steel structure. With pattern optimization, the pattern can be made quite omnidirectional (see Pattern Optimization section in this chapter).

Series-Fed Antennas

A similar arrangement of arms supported by a T arrangement may be series fed. That is, part of the outer end is insulated from the rest of the dipole and fed across the insulated break as shown in Figures 4.9-14 and 4.9-15. To allow for adequate power-handling capacity and to increase the VSWR bandwidth, 3 inch (75 mm) diameter tubing is used. The antenna has a VSWR bandwidth of about 1%, so it makes an excellent single channel FM antenna. The antennas are usually mounted on the side of a supporting tower or pole, and stacked vertically to achieve required power gain.

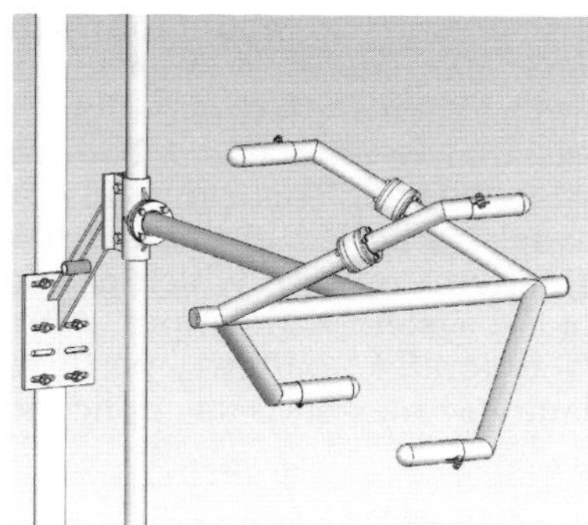

FIGURE 4.9-14 Side-mount dual-dipole FM antenna element fed across insulated breaks.

This antenna has greater wind loading than the shunt-fed version due to its larger element diameters necessary to achieve useable VSWR bandwidth. It requires large amounts of AC power for electrical deicing. Plastic radomes also present additional wind loading.

Flat-Panel Antennas

The panel antenna was developed in Europe to provide a wide bandwidth for several collocated FM stations without the need to change antennas when a new channel was added or if the operating frequencies were changed.

Panels are from 7–8 ft (2,100 to 2,450 mm) square in the flat configuration. In the cavity style they are about 8 ft (2,450 mm) in diameter and about 3 ft (1,000 mm) deep. A heavy metal frame is often used over which a large diameter wire mesh has been welded. The wire mesh screen openings vary from 4–12 inches (100–300 mm). Electrically they are considered nearly solid metal. These openings produce relatively low wind loads. The entire flat frame or cavity is strong enough to support a man on its mesh openings. Some manufacturers hot-dip galvanize their steel after fabrication; others use stainless steel construction.

For FM, two crossed dipoles are used as the illuminating source for each panel or cavity, as shown in Figure 4.9-16. Each dipole is fed in phase quadrature. That is, one dipole receives its peak current 90 degrees after the other, to produce CP. A typical set of electrical and mechanical specifications for a CP eight-bay cavity community antenna are shown in Table 4.9-6.

Flat-panel antennas are typically side mounted on large face–size towers. The screen panels greatly reduce interaction and distortion between the antenna and tower. The panels are directional, thereby requiring three or four panels to be mounted around the tower to achieve acceptable azimuthal circularity.

These antennas are usually branch fed, and often the array's top and bottom halves are fed separately. This allows operation of either half of the antenna separately when it is necessary for maintenance or repairs. Circular polarization is achieved on each

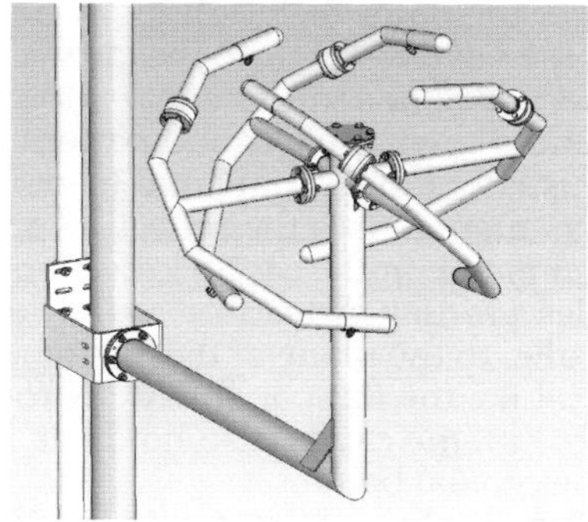

FIGURE 4.9-15 Side-mount four-dipole FM antenna element fed across insulated breaks.

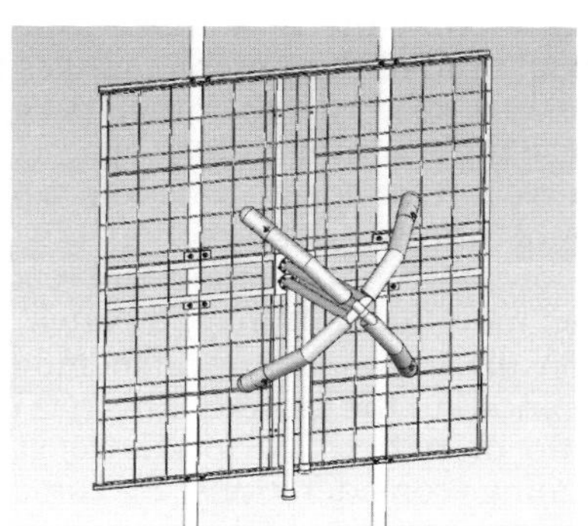

FIGURE 4.9-16 Dual-arrowhead dipole flat-panel antenna.

TABLE 4.9-6
Typical Measured Community Antenna Performance

Operational frequency range	88 to 108 MHz
Safe RMS input power rating	200 kW
Power gain ratio, each polarization	4.4 (6.43 dB)
Maximum VSWR any channel between 88–108 MHz	1.1:1
Elevation pattern beam tilt	–0.5°
Polarization	Right hand circular
Axial ratio	Better than 2 dB
Azimuth circularity V-pol or H-pol	Better than ±2dB
Antenna dead weight, less than	7,000 Lbs (3,183 kgs)
Active wind load, RS-222-C 50/53 PSF	8,000 Lbs (3,636 kgs)
Antenna input flanges, two, size	6-1/8 inch
Number of bays (stacks)	Eight
Radiator type	Circularly polarized cavity

panel by feeding two perpendicular dipoles 90 degrees out-of-phase. This phase offset helps this antenna achieve usable bandwidths on the order of 10 MHz.

By pulling the dipole back on its feed support arms, the arrowhead-shaped dipoles control both V-pol and H-pol azimuth patterns. Rotating the dipoles 45 degrees with the earth-ground reference further improves the polarization ratios.

Round dipoles made of tubing as large as 6-1/8 inches (155 mm) in diameter are used along with a single line quadrature feed. This combined arrangement makes an excellent wideband CP panel to cover the entire FM band. Power splitters, dividers, and cables, along with a number of these panels, complete the antenna design.

On large face towers, circularity in the H-pol can be quite good, on the order of ±2 dB. On standard configurations, the V-pol pattern is quite different. As a result, the axial ratio of this antenna ranges from good at some azimuth headings to rather poor at others. This is because the azimuth pattern of an H-pol dipole is like a figure eight, or cosine function, while the pattern for a V-pol dipole is not directional in its azimuth plane. Therefore, each polarization will react quite differently when mounted in front of a panel.

Dipoles on these panels are often mounted at 45 degrees referenced to the ground. This has no effect on the axial ratio or pattern performance; it instead is done for tuning considerations to compensate for mutual coupling.

Improvements to this design have reduced the differences between the patterns of the polarizations.

FIGURE 4.9-17 Broadband four-dipole flat-panel FM antenna.

These techniques are effective for applications requiring only a few MHz of bandwidth. One method optimizes the angle of the dipole bend as well as its distance to the panel. This design requires three panels to be mounted around a tower. Axial ratio and pattern circularity are improved at the cost of system bandwidth.

Another method uses four dipoles forming a square shape in front of the panel, as shown in Figure 4.9-17. By adjusting the spacing between the dipoles, the beam width of a panel can be controlled. Over a small bandwidth, the pattern performance is greatly improved. It is necessary to mount four panels around a tower for a circular pattern. A large amount of panel interaction and leakage are severe design limitations.

For projects that require wider bandwidth, skew mounting is often used. This physical configuration allows the panels to be fed in mode 1 (0°, 90°, 180°, 270° phase for four around), which can increase the bandwidth of the system at the input. Although skew mounting deteriorates pattern performance, the increase in bandwidth extends its applications.

Cavity-Backed Panel Antennas

The use of a cavity screen instead of a flat panel has greatly improved axial ratios. The cavity acts as a resonator with little leakage toward the tower. The shape of the azimuth pattern in each plane becomes both controllable and symmetrical. System bandwidth is improved over the flat-screen design. By adjusting the diameter of the cavity structure, beam widths can be altered to meet specific requirements. Mounting three cavities around a tower gives good pattern circularity. Axial ratios usually range from good to excellent.

The cavity antenna uses the reflective properties of the flat-screen panel. In the cavity however, the illuminating dipoles are flat instead of round and all four arms are parallel to the plane of the cavity, as shown in Figure 4.9-18. Like the flat panel with its round dipole supporting balun, the cavity also holds its flat dipoles with a double coaxial balun.

The dipoles in the cavity get their wide VSWR bandwidth through the sleeve dipole principle [11]. Capacity is provided by a metallic ring close to all four

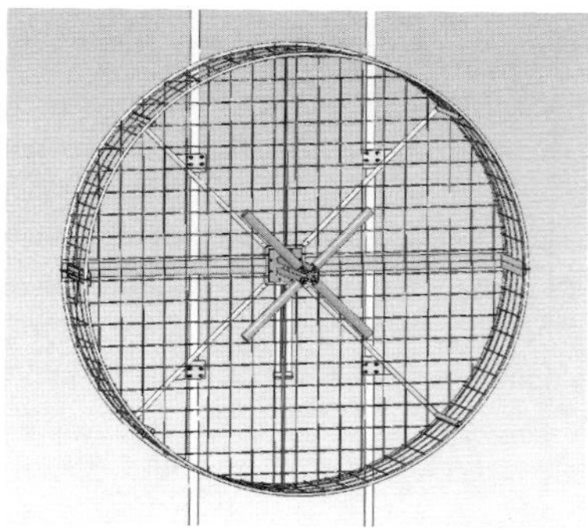

FIGURE 4.9-18 Dual-dipole cavity-backed panel antenna.

dipole arms placed between them and the back of the cavity. Circulating surface currents flow on the dipole arms, which result in evenly radiated patterns in all polarization planes. The bandwidth of a single cavity can cover the full 20 MHz band with a VSWR better than 1.1:1. Therefore, it is not necessary to skew mount these antennas for bandwidth considerations.

This antenna has the advantage over some other designs of greater VSWR bandwidth. It is considered closer to state of the art due to better elevation and azimuth pattern control of both planes of polarizations by the shape of the cavity.

Cavities and flat panels can be modeled using a computer. Factors, such as tower size and orientation, as well as the phase and skew of the elements, can be modeled to determine optimum mounting and feeding. This is useful in projects that require a directional pattern. A station should take the pattern constraints and gain requirements to an antenna manufacturer to determine what is feasible.

Crossed Dipole Theory

Common to the flat panel and the cavity is the operation of the dipoles that generate CP. The dipoles are fed currents in phase quadrature, through a coaxially balanced balun that provides equal currents to all four arms of the two dipoles. They excite the entire cavity or flat panel with a rotating radio frequency (RF) field in a plane parallel to the dipoles. The RF field is thus CP and may be ideally represented by a rotating vector of constant magnitude revolving one revolution per wavelength of propagation distance. It is right-hand polarized as the field rotation is clockwise as viewed from behind the screen, looking toward the direction of propagation, if the phasing between the two crossed dipoles is properly made.

Radiation patterns, associated beam width, and directivity are determined to a large extent by the size of the cavity or flat panel. The geometry of the dipole has less effect than the reflector size. The size and shape of the dipole control the antenna impedance and the VSWR. The screen panel, be it flat or a cavity, fulfills the following five important electrical functions:

- Isolates the radiating elements from the tower or the mounting structure, and reduces mutual coupling.
- Provides sharper beam width and more gain than achievable with the dipoles alone.
- Furnishes pattern control so that the beam width is nearly equal for both horizontal and vertical plane polarization.
- With an effective balun feed system, the crossed dipole radiated pattern phase is very uniform as the amplitude changes normally with azimuth.
- Computer-aided designs are easily achieved in production for various width towers because the pattern is simply pure electrical geometry.

Antenna Element Spacing and Downward Radiation

Most FM antennas have elements that are spaced one wavelength apart (9–11 ft) for reasons such as gain considerations, mutual coupling effects, and ease of feed design. There are cases, however, that require different element spacing. High levels of downward RF radiation are the most common reason, although considerations, such as aperture constraints and beam shaping, also utilize other than one-wavelength spacing.

Radio frequency radiation (RFR) safety levels must be considered in nearly all site locations. Power radiated in the lowest sidelobe can cause a variety of problems, of which human exposure levels are most critical. See Chapter 10.4, Human Exposure to Radiofrequency Energy, for further information on this subject.

When antenna elements are arrayed, the resulting elevation pattern contains lobes and nulls. The farthest sidelobe from the horizon typically peaks between 70 degrees and 90 degrees below the horizon for full-wave spacing. This lobe occurs since the physical path results in no phase cancellation in that direction, and thus the downward radiation of each element is additive. Shortening the spacing changes the difference in the elements physical path length, and results in some phase cancellation.

Antenna array pattern in elevation plane is a sum of multiple stacked sources as follows:

$$E(\theta) = \sum A_n(\theta) e^{-j2\pi/\lambda \cdot z(n)\cos(\theta)}, \quad (1)$$

where $0° < \theta < 180°$ is the elevation angle from zenith, $A_n(\theta)$ is a complex elevation pattern of a single radiating element, and $z(n)$ is a vertical location of radiating element ($n = 1, \ldots, N$).

For uniformly excited and linearly phased array of equally spaced elements the elevation pattern takes the form of

$$E(\theta) = A(\theta) \sum e^{-j2\pi d/\lambda \cdot n \cdot (\cos(\theta) - \Delta F \cdot \lambda / 2\pi d)}, \quad (2)$$

where ΔF is a phase shift between adjacent elements to tilt the beam, if required.

When spacing $d = \lambda$ and $\Delta F = 0$ the pattern of an array of horizontal dipoles (for which $A(\theta)$ = const) has three equal peaks: main beam at $\theta = 90°$ (horizon) and grating lobes at $\theta = 0°$ (zenith) and 180° (downward). For an array of vertical dipoles the upward and downward radiation is suppressed by element factor A_{max}/A_{min}. Most suppression on the order of –20 dB or better is for straight vertical dipoles, less suppression is provided by bent vertical dipoles, and no suppression due to element alone for horizontal dipoles.

Expression (2) is a finite geometric series and can be reduced to a single term:

$$|E(\theta)| = \left|A(\theta) \bullet \sin(\pi N d/\lambda \bullet \cos(\theta)) / N \sin(\pi d/\lambda \bullet \cos(\theta))\right|. \quad (3)$$

At $\theta = 0°$ and 180° the normalized pattern (3) represents additional suppression of upward and downward radiation by an array factor, and is a function of spacing d/λ and number elements N, as follows in (4):

$$|E(0°, 180°)| = \sin(\pi N d/\lambda) / N \sin(\pi d/\lambda). \quad (4)$$

Full suppression takes place at d/λ = **0.5** (for any even number N of vertically stacked radiating elements). See Figure 4.9-19.

Typical circularly polarized FM antennas are side-mounted slanted dual V-dipoles fed from a common point. One of the dipoles has sweptback arms, the other is swept-forward, both fed in-phase (refer to Figures 4.9-10 and 4.9-11). The dipoles in front view are in X form and therefore both contribute to both vertical and horizontal polarization components. Element factor A_{max}/A_{min} of suppression of downward radiation is different for H-pol versus V-pol in the same way as for vertical and horizontal dipoles forming a cross configuration.

The panel antenna is another common type of FM antenna. Its dual dipole is phased in quadrature and backed by cavity or screen (refer to Figures 4.9-16 and 4.9-18). With the size of screen and depth and size of a larger cavity, suppression (A_{max}/A_{min}) of downward radiation by the elements alone improves.

Calculation of power density at the base of a broadcasting facility can be approximated in (5) as

$$P = 1.6 \times \text{ERP} \times \left(\left|E(180°)\right| \times A_{max}/A_{min}\right)^2 / (4\pi R^2), W/m^2, \quad (5)$$

where factor 1.6 is used to account for ground reflection, and R is the height of the antenna above ground. Field strength at the base is $F = \sqrt{377 \cdot P}$ in V/m, and should be less than the FCC-mandated threshold for human exposure, which currently stands at 27.5 V/m (FCC OET Bulletin 65, Edition 97-10, August 1997, table 1).

Half-wave spacing, for example, greatly suppresses the levels of the side lobes, while increasing the width of the main beam. Despite the lack of power in the side lobes, the extra width of the main beam causes the pattern to be less directive. Thus, this exchange results in an overall gain reduction on the order of a third. Spacing the elements 0.8 wavelengths apart improves sidelobe suppression and the gain of the antenna is not greatly affected.

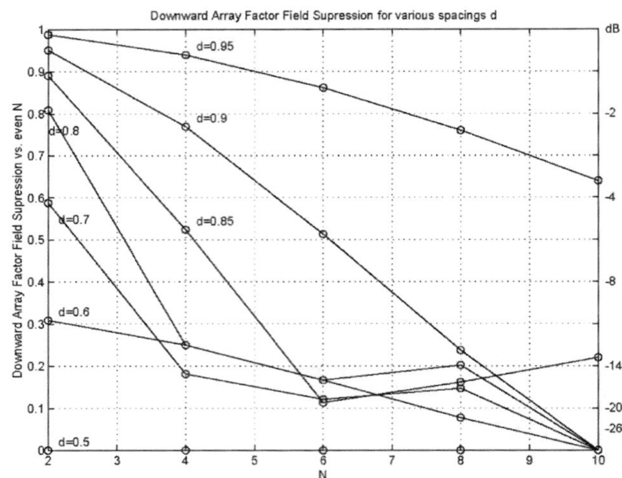

FIGURE 4.9-19 Downward array factor for number of bays.

Short-spaced antennas are fed either by a shunt or branch feed system. For half-wave spaced antennas, a shunt line delivers each element 180 degrees out-of-phase with the next element. This phase distribution problem can be overcome by flipping every other element upside down, thus inducing a 180 degree phase shift in the feed. Other spacings, such as 0.8 wavelengths, can be accomplished with a branch feed, which can deliver equally phased signals to each element regardless of spacing. Also, in the shunt feed case, an extra elbow complex in the form of U-link in between each pair of elements provides an additional path to deliver in-phase signals to elements spaced shorter than one wavelength.

Problems with Side-Mounted Antennas

Single-station FM antennas are typically side mounted on a pole or tower. Unlike panel antennas, the support structure greatly affects or distorts the radiation pattern. The resultant pattern may have large peaks and nulls that can result in coverage and reception problems.

In addition, the V-pol and H-pol patterns react quite differently to these distortions. Due to the geometric complexities of the CP radiating elements and tower structure, it takes computer modeling to accurately predict pattern effects. Therefore, the use of a test range is required to determine how an antenna behaves when mounted on a tower section similar to the one on which the antenna will eventually be used.

For nondirectional stations, a test range can determine the proper mounting of an antenna to achieve an acceptable circularity. Depending on the tower size, the depths of nulls can be greater than 10 dB. An optimized mounting configuration can make the nulls less significant and oriented in areas where service to the primary coverage area is not hurt. Parasitic reflectors are often used to improve the circularity of nondirectional antennas.

When the top spot on a tower or structure is available, pole mounting is often preferred. A pole provides a stable and symmetrical support structure that has low interaction with the horizontally polarized component. In combination with the feed line, the pole typically induces a null in the V-pol pattern, directly opposite of the elements. Proper orientation of the element can reduce the effects of pattern distortion.

Mounting an antenna on the side of a tower can produce unpredictable results. The positions of the peaks and nulls vary greatly from the orientation of the elements. As the face size of the tower increases, the distorting effects magnify. To compensate, many stations use smaller sections of tower at the top, where they plan to install the FM antenna. Eighteen inch and smaller face towers tend to produce good results. With careful planning, the use of a 24 inch or larger face tower can also be successful. Note that the patterns of each polarization react differently, and thus axial ratios can be quite poor.

PATTERN OPTIMIZATION

Single-station FM antennas are usually side mounted on a pole or tower. This is economical and it frees the tower top for other possible uses. Unfortunately, the pole or tower tends to distort the radiation pattern, seriously affecting station coverage in some directions [12].

This problem can be serious if the FM antenna has been randomly attached to a support tower. FM antenna makers do not manufacture and sell towers. A few have made supporting poles on which the FM antenna has been mounted, adjusted, and pattern tested. TV antenna makers, on the other hand, usually build the antenna as a complete self-supporting structure to be mounted on top of a support and are usually not faced with this side-mounting problem. The logical but more expensive solution would be to make the FM antenna a self-supporting structure just like TV antennas.

Improper FM antenna side-mount installation on a tower can cause serious pattern problems. Measured patterns have indicated that, in some cases, the maximum radiation can actually be in the opposite direction from the desired direction [13].

When Optimization Is Necessary

Side mounting an FM antenna on a tower may have serious consequences with an FM station's coverage. Nulls may be toward important service areas. Nulls as low as 1% of the RMS power have been measured with towers varying in width from 18–120 inches (0.5–3 m). Another problem is that with nulls come lobes. Lobes as great as 9.8 dB over RMS power have been found. When used without pattern optimization, this lobe would produce an ERP in a given direction nearly 10 times the FCC-licensed value. Translating this to a 50 kW ERP station there would be radiation in some directions of only 0.5 kW and others with 477 kW. This is a maximum to minimum ratio of 29.8 dB, and clearly not acceptable.

CP creates other problems, such as the H-pol and the V-pol ratios are not always the same and vary moderately in any given azimuth. This ratio can be as great as 15 dB and must also be addressed in order to resolve the horizontal plane circularity problem. The axial ratio could be degraded causing the V-pol radiation in certain directions to be much stronger than the H-pol. This violates FCC's requirement that with CP, the V-pol must not be stronger than the H-pol component.

Section 73.316 of the FCC Rules covers FM antennas but does not specifically address the problem of azimuth circularity. In fact the FCC assumes that FM nondirectional broadcast antennas have perfectly circular horizontal radiation patterns [14]. In actual practice, they seldom do.

In order to produce a horizontal plane pattern that even approaches a circle requires considerable work by the firm making the antenna. Since it is nearly impossible to produce a circular pattern with a nonsymmetrical side-mounted antenna, the term technique *optimization* (to do the best possible) is in common usage now.

Theory of Optimization

Figure 4.9-20 indicates how energy from the horizontal loop representing H-pol is intercepted by a pole and reradiated. Similarly, in Figure 4.9-21, energy from the vertical dipole is intercepted and reradiated as V-pol.

In the first case, the pole diameter is small in wavelength in the direction of the electric field, thus,

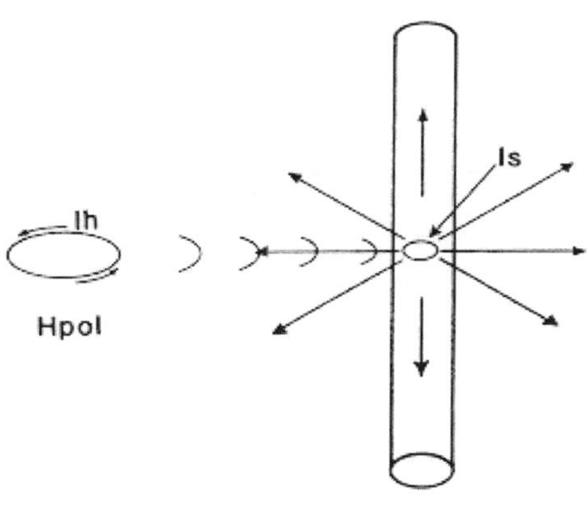

FIGURE 4.9-20 Influence of pole on H-pol signal.

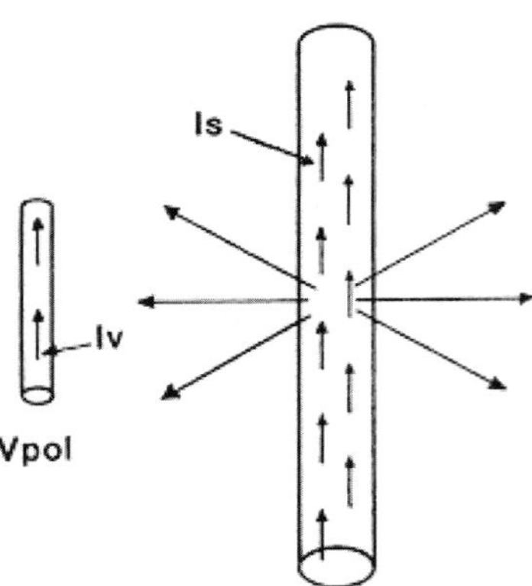

FIGURE 4.9-21 Influence of pole on V-pol signal.

scattering is minimal and not much H-pol radiation can be expected. However, when the V-pol dipole excites the pole, a large amount of energy is intercepted (Is) and reradiated because the large dimension of the pole (length) is parallel to the electric field. A similar effect is produced by the vertical transmission line that is common to the antenna itself (Iv in Figure 4.9-21). The result is appreciable distortion of the vertically polarized azimuth pattern.

Figure 4.9-22 shows the resulting H-pol and V-pol patterns. The pole and/or vertical coaxial line have transformed the V-pol pattern from circular to a cardioid, while the H-pol pattern remains essentially omnidirectional. The null of the cardioid can be more than 7 dB down from the RMS value. This phenomenon is well known, and as a compromise, broadcasters generally install the antenna on the side of the tower support structure facing the main service area. There are many exceptions to this, as some measured patterns on triangular towers of standard construction have shown.

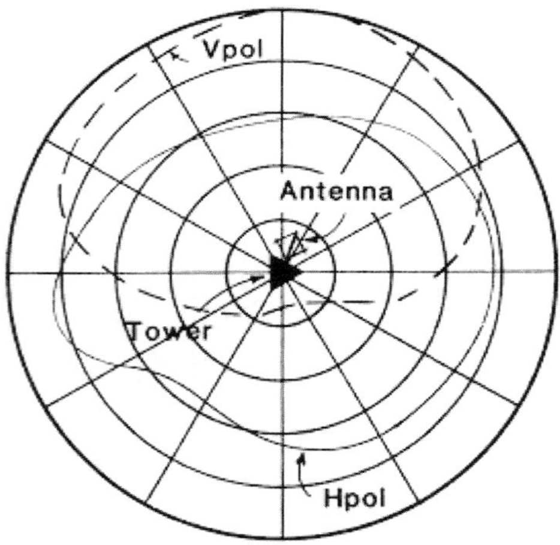

FIGURE 4.9-22 H-pol and V-pol azimuth patterns of antenna-mounted on a structure.

In Figure 4.9-22 the V-pol is much stronger than the H-pol in the favored direction. In the opposite direction, there is little V-pol. The power ratio of H-pol to V-pol is 16 times. This makes for a very poor CP antenna in some directions. Tower and poles under about 2 ft (0.6 m) in cross section will exhibit the same effects on the antenna patterns. Towers greater than this size obviously will increase the complexity of scattering effects. Three or four tower legs, the horizontal and diagonal cross members, transmission lines, ladders, tower lighting, and deicer conduits, all will be excited by the vertical and horizontal currents from the radiators, which in turn will reradiate and affect the horizontal plan patterns.

In contrast to the simplicity of the antenna on the side of a pole, the tower-supported antenna may be mounted on the face or on a corner, at or between horizontal cross members, or tilted at various angles compared to the tower—all multiplying the complex factors affecting the patterns.

Optimization Methods

The most popular technique to achieve the desired pattern is through the use of Yagi antenna principles, wherein parasitic elements are placed in the field of the radiator to modify its radiation pattern. For example, a shortened dipole (director) placed in close proximity to a radiator reinforces radiation in the forward direction and suppresses the signal in the opposite direction. If the parasitic element is longer than the radiator (reflector), the effect is reversed. The signal is suppressed on the side of the parasitic element and reinforced in the direction of the radiator. Similarly, parasitic elements can be used with FM antennas mounted on the sides of towers or poles to produce pattern changes. As discussed here, both directors and reflectors may be used. Both are frequency sensitive. The effects of the supporting structure are also frequency sensitive.

Therefore, an arrangement of parasitic elements for a given FM frequency will not necessarily be the same for another, nor will the pattern be the same for a given arrangement, if it is moved up or down the tower by as much as 1.5 ft (0.5 m).

The resulting patterns cannot be predicted. There are many factors that affect the horizontal plan pattern. Only by actual range testing can the patterns be adjusted and properly measured. Therefore, the cost for doing this is high, since it is time consuming and requires qualied antenna technicians and the use of an antenna range. In addition, the nal parasitic arrangement must be permanently fabricated and installed. However, the results are generally well worth the expense and time.

Pattern Service

There are two basic types of pattern service furnished by antenna manufacturers in the United States. FM antennas may be adjusted for the best omnidirectional pattern possible or they may be adjusted to proven minimum ERP values in particular azimuth directions. The minimum required values, plotted on a polar chart by the broadcaster, may be combined with the tower orientation. Using the customer's make and model tower, two or more bays of the antenna are fabricated, installed on a section of the tower, and put on the test range. Adjustments are made such as leg chosen, distance from the leg, and the orientation of the antenna with respect to that leg. Parasitic elements are then used to further improve and shape the pattern.

For example, a class A station may wish that a minimum of 3 kW be radiated from 90 degrees to 120 degrees and the remainder of the azimuth be no less than 1.5 kW. This would then require a pattern without any field voltages less than 70% and that the vectors between 90 degrees and 120 degrees be 100%. This sort of work has been done by many antenna manufacturers. Figure 4.9-23 illustrates a typical before and after optimized pattern.

Various methods have been used to optimize FM antenna azimuth patterns. Some firms use models at twice the operating frequency. Others use theoretical methods, backed up by experimental proof. The final optimized antenna is match marked on the tower sections so that it will be assembled exactly as it was made and tested at the fabricator's plant, and tested on the antenna range.

A complete set of installation prints must be provided so that the antenna is assembled exactly as tested, with all the correct locations, and angles of all the parasitic elements.

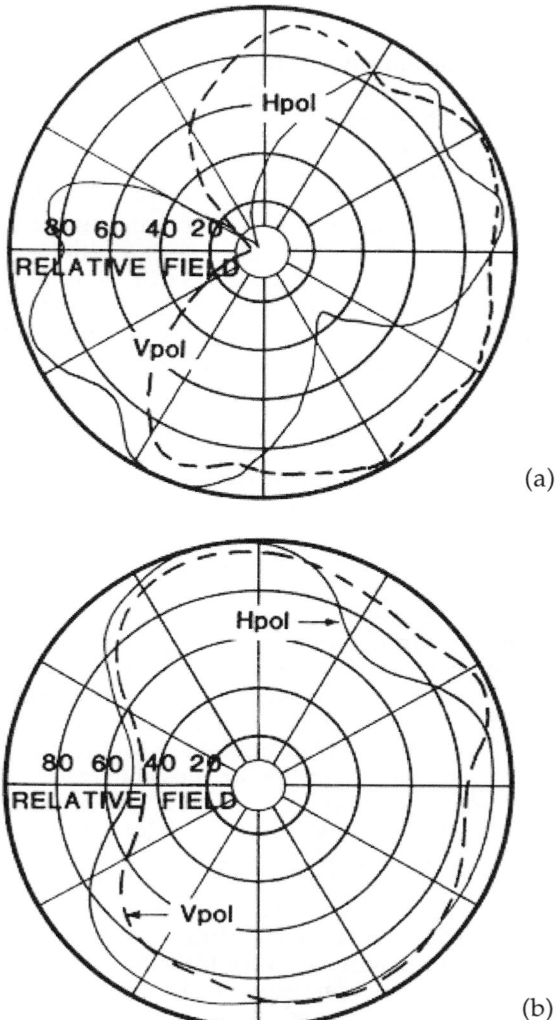

FIGURE 4.9-23 (a) Measured nonsymmetrical C-pol pattern of tower side-mounted antenna. V-pol variation is ±15 dB, while H-pol is ±12 dB. The axial ratio was 24 dB. Antenna patterns are considered to be poor. (b) Same antenna but much improved after pattern optimization. H-pol variation is ±3 dB and V-pol is ±3.4 dB. Axial ratio is quite acceptable at 2.9 dB.

WIDEBAND COMMUNITY AND MULTI-USER ANTENNAS

The most significant developments in FM antennas have come in broadband designs. Flat panels, cavities, helicals, and a side-mounted design have given broadcasters a range of choices to fit their application. Factors, such as pattern requirements, tower constraints, and system budget, can narrow the options for selecting the FM broadcast antenna.

In order for an antenna to be useful throughout the 20 MHz FM band, its operation must be the same on any frequency. The VSWR at 88 MHz, for example, must be just as good at 108 MHz. The CP azimuth pattern should remain the same on one end of the band as the other, as must the axial ratio. This is a much more rigorous requirement than placed on the single-channel slanted dipole or the ring stub.

In the wideband antennas, several factors go together in order to meet these important requirements:

- Basic wideband dipole radiators
- Screen-panel pattern control
- Quadrature phase distribution

By using these three principle parameters in a wide-band antenna the radiation pattern, VSWR, and the gain can be nearly the same on any channel within the FM band. Several methods are used to make the VSWR of the crossed dipoles as good as possible. The dipoles are usually fed with a folded balun or the split-tube–type balun [15]. This improves the impedance match, phase, and amplitude linearity of the resulting azimuth pattern.

The length to diameter (or width) ratio is usually about three. This not only reduces the Q but also increases the voltage flash-over levels. The low Q also increases the bandwidth by decreasing the rate of reactance change with frequency.

A natural factor aiding the VSWR problem is the fact that in order to obtain CP from two crossed dipoles, there must be a phase quadrature of the two currents feeding the crossed dipole. The two reflections, as a result of VSWR, return back to the phasing device 180 degrees out-of-phase with each other. Being the same amplitude, they cancel.

All of the above factors, plus two or three more levels of quadrature reflection cancellations, bring the overall system input at the antenna to under 1.08:1 across the band. This cancellation technique eliminates the need for electrical deicers or plastic radomes, as the VSWR is not affected by moderate ice coatings. However, radomes may be necessary with flat panels to physically protect some radiators from falling ice.

These and other factors all contribute to make the panel-cavity antenna the best possible for either single channel or community antenna use.

Top-Mount Antennas

There are a few broadband designs that require top mounting on a pole. The first incorporates a series of dipoles that are mounted on a pole. Each dipole is branch fed individually. By rotating the mounting scheme, circular patterns can be achieved. The rather small radius between dipoles keeps nulls from crossover points to an acceptable level. Large bandwidths and good patterns have been measured on this system. The considerable amount of feed cables that run through the aperture, however, can cause problems if not properly RF grounded or shielded. Line damage, failure, and pattern distortion are possible results.

The second, spiral (or normal mode helical) antenna design, is now being used for FM applications. This traveling wave–type antenna has bandwidth on the order of 15% to 20%. Fed from the top, a series of wires are wrapped around the conductive pole over an aperture of two to three wavelengths. The pitch angle of the wire wrap controls the pattern characteristics. Pattern circularities of ±1 dB over a 20 MHz band have been measured. Axial ratios better than 2 dB are considered typical. The feed system is enclosed by stainless-steel feed cans that eliminate pattern distortion by the feed system, a common fault of other top mounts. In regions where icing is a problem, heating elements inside the radiating wires will stabilize the antenna's performance.

Multichannel Side-Mount Antennas

Another broadband FM broadcasting antenna is the side mount. This design incorporates dual baluns, which ensure a balanced and symmetric current excitation. This balance eliminates spurious radiation from currents on booms, feed lines, and other nearby conductors. The feed arms are of the skewed V-dipole configuration. Excellent circularity and axial ratios have been measured in free space. Interaction with feed lines and support structures still distort the pattern, though to a lesser extent.

A single FM element has been measured to have a bandwidth of 5.5 MHz with a VSWR of 1.1:1. This bandwidth does not limit the spread of two combined stations. By centering the band edges, the antenna can tune two stations over 8 MHz apart with a VSWR of 1.1:1 over each channel. This comes at a cost of poor performance at center frequencies, thus three stations over 8 MHz are not achievable. As a side mount, this antenna has substantially reduced costs and tower constraints over a panel type.

Power-Handling Capacity

A multichannel antenna must be designed to not only accommodate the bandwidth requirement for all users, but also to handle the input power due to multiple users. Considerations for both peak power and average power must be made.

The average power capacity of an antenna refers to its ability to operate below some critical temperature. The antenna (including its feed system) will convert a certain portion of its input power into heat, and that heat must dissipate into the environment without raising the temperature of any critical components, such as insulators, to extreme levels. The generation of this heat is due to RF currents and the attendant resistive losses in the component materials. The average power of an analog FM transmitter may be considered constant with time and the average power due to multiple (N) FM transmitters each delivering P_o watts of average power to the antenna is simply $N \times P_o$.

The peak power capacity of an antenna refers to its ability to operate without producing excessive electric fields of such magnitude that arcing or corona discharge occurs. While average power capacity is about amperes and cooling, peak power is about volts and conductor geometries. Corona or arcing can occur where electric fields are concentrated by sharp points. On a dipole structure, the dipole ends are the location of the highest voltages, so larger radius parts are often used there to prevent discharge.

In multiple channel systems the voltage at a point in the system is equal to the sum of the voltages for each channel. In the absence of subcarriers or digital modulation, the peak power (as with the average power) of an analog FM transmitter may be considered constant with time. The peak power due to N nonsynchronous FM transmitters each delivering P_o watts of average power to the antenna is $N^2 \times P_o$.

Pressurization and altitude have an effect on both the peak and average power-handling capacity of an antenna. Generally at lower pressure the voltage at which corona discharge or arcing may occur will be lower, which bears on the peak power capacity. Also, since lower pressure reduces the cooling efficiency, the average power-handling capacity is impacted. Pressurization of feed system components may be used to increase the power-handling capacity of that portion of the antenna system. High relative humidity and fog can apparently precipitate corona discharge from the ends of dipoles, so these factors must also be considered in multiple-frequency antennas.

Community Antenna Economics

The community antenna fits best in multiple station service. This allows costs to be shared so all parties

benefit from a superior antenna that each station independently could not economically justify [16].

If enough planning is done in advance it may be possible to install all the FM stations in one community on one tower at considerable savings to all users. Some exclusions include lack of adequate mileage separations, the existence of excellent facilities, and FAA tower height limitations.

The break-even point appears to be with four stations. When five stations are involved, there is perhaps a 20% reduction in cost to each of them when compared to the costs of putting up their own individual single-channel antenna and tower at the same height.

One of the first large-scale community antenna projects in the United States was the Senior Road Tower. In 1984, a group of nine Houston, TX, broadcasters formed the Senior Road Tower Group and installed a 2,049 ft (625 m) tower supporting a 12-bay community FM antenna system, with an HAAT of 2,000 ft (610 m) [17]. This height permitted the maximum service allowed.

Two runs of 8-3/16 inches (208 mm) diameter coaxial lines are used to feed the antenna so that the power in both the lines causes right-hand CP from the antenna for nine FM stations. The nine stations use one diplexer each, all of which are housed in one 2,400 square foot (223 square meters) room. The 10-port modular diplexer has a total power-handling capability of 350 kW. The insertion loss for each station is 0.80 dB (17% loss). The isolation between the various transmitters meets FCC spurious emission (intermodulation) requirements. The diplexers are monitored at a central operating rack that displays each diplexer's forward, reflected, and rejected power. This permits troubleshooting in an orderly and rapid manner. Electrically operated coaxial switches permit each station to be connected to the dummy load for individual testing. Air conditioning and chilled water are used to remove heat produced during operation.

Technical Advantages

Besides the financial advantages cited under the economics heading, there may be the competitive advantage of protecting the channel classification and using the same height antenna as the competitor. Other advantages include its emergency upper-lower half feature for transmission line or antenna half backup. The flat VSWR curve is highly useful for SCA and IBOC operation. Stations sharing this type of antenna will all experience less intermodulation interference than if they had separate but closely placed antennas.

STATION DIPLEXING

Diplexers are passive devices used to combine the power of two or more stations and feed the combined power to a common transmission line and/or a common transmitting antenna. This system of utilizing one well-sited, high-quality antenna has become popular, convenient, and economical.

Wideband panel antennas, although expensive for use by one station, are cost effective for two or more stations. These antennas maintain their omnidirectional horizontal plane patterns and VSWR throughout the FM broadcasting band from 88–108 MHz. Thus, they make the ideal antenna for multistation diplexing. Chapter 4.10 discusses FM Combining Systems in detail.

FM ANTENNA INSTALLATION ON AM TOWERS

The current trend is to locate FM transmitters in places where the best service may be rendered to the most listeners. This usually permits the maximum possible height to be used. Sometimes, however, it may be economical and convenient to install the FM antenna on a tower used for AM broadcasting. If the steel AM tower is not base insulated but is grounded and shunt fed, the FM coaxial line may be connected to the tower, without any further problems.

TRANSFORMER ISOLATION

However, if the AM tower is insulated at the base, an isolation transformer may be used to couple the FM power across the base insulator without introducing objectionable mismatch into the FM antenna feed line. An isolation transformer is especially desirable for feeding high-impedance AM radiators or AM radiators that are part of an AM-directional antenna system that might be adversely affected by a quarter-wave isolation system.

These transformers have two tightly coupled RF coils that are resonant at the FM operating frequency. An air gap is provided for the AM power to pass through the two resonant loops. The capacity is quite low, resulting in a very high capacitive reactance placed across the tower base insulator. Figure 4.9-24 shows the internal basic construction of a typical isolation transformer. The insulation for AM under the top of the box may be high-density polyethylene, Teflon, or fiberglass. The metal top provides a rain shield as well as protection from dust, mud, or snow.

The use of these isolation transformers permits the AM tower to operate undisturbed by the presence of the FM antenna. It also allows the FM coaxial line to be connected in the usual manner, except for the placement of the isolation transformer. These have internal gas blocks and permit the passage of dry air pressure through the transformer via a plastic tube.

In addition to lower cost, the isolation transformer method has another advantage in directional AM tower use. It does not distort the AM radiator current distribution that may adversely affect the AM radiation pattern.

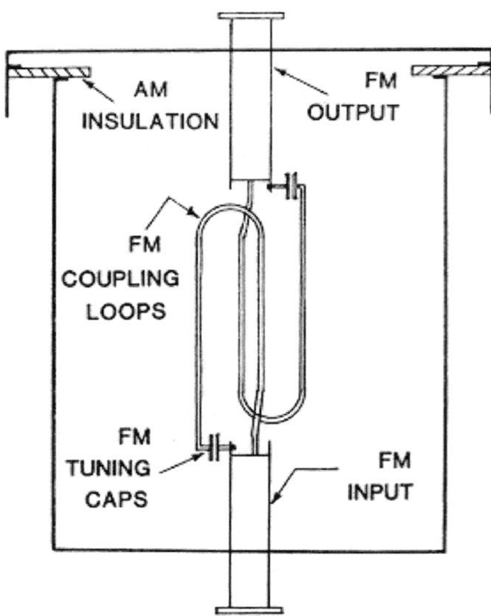

FIGURE 4.9-24 Example of AM-FM tower isolation transformer (isocoupler) used to decouple FM transmission line on a series-fed AM tower.

Quarter-Wavelength Isolation

A less popular and older method is to use the technique of quarter-wavelength transmission lines. Simply stated, the opposite end of a shorted quarter-wavelength line has high impedance. This high impedance is placed across the AM tower base and may be successfully used to provide the necessary isolation. It is more difficult to physically accomplish this as the tower should be at least one quarter-wavelength high, and the FM antenna coax line must be insulated all the way down the tower. In practice the insulated part may be as short as 75 degrees of line, as the line hangers and distributed capacity of the line tend to electrically increase the physically shorter line. For best results, the FM line should be placed within the tower body. The location of the shorting point of the outer copper sheath of the coaxial cable to the tower is experimentally found by first measuring the base impedance prior to cable installation and then choosing a point that produces the same impedance (the quarter wave short up the tower produces an open circuit at the base).

Guy Cable Considerations

The presence of continuous steel guy cables going through the FM antenna level on a steel supporting tower was studied by Jampro Antenna Company in 1968. It was found that guy cables had an effect of less than 0.6 dB in the azimuth pattern of an H-pol antenna. On V-pol the maximum variation was 1.8 dB on the azimuth pattern. The strongest effect is on CP where the azimuth pattern change was as great as 3.4 dB. The elevation pattern was also affected since the first and second nulls were filled as much as 4 dB.

In addition to pattern anomalies, the steel cables reradiate near the ground. This may cause RF problems in some high-power installations with low-level audio equipment located in a building near the base of the tower.

Strong currents may be induced when the steel guys are in the immediate vicinity of the radiators. If that guy passes close to the side of the radiator, the field on that particular side of the antenna element will induce currents in the guy wire. The radiator currents can become unbalanced and the impedance of the element is disturbed, changing its VSWR and radiation pattern. Because of these effects, it is common practice to break up the guys using insulators, fiberglass rods, or plastic guy cable, within 10 ft (3 m) of the antenna radiators.

When a CP antenna is side mounted on a guyed tower, the vertically polarized field will have an appreciable component parallel to the guy wire in its aperture and will induce currents in the wire. If the guys are continuous, a progressive wave traveling toward the ground will result, and will radiate most of its energy before reaching the ground. The energy will be radiated in cones concentric with the wire. A small amount of the V-pol power will thus be bled off.

Porcelain Insulators

If the FM antenna is side mounted on an AM tower, which usually will have metal guys and porcelain break-up insulators, those insulators will probably be spaced several FM wavelengths apart. The induced currents will form standing waves on the sections between insulators and radiate multilobed patterns into space at many angles from the wire axis. If the sections between insulators happen to be of a resonant FM wavelength however, currents in the guy wires and their radiated fields will be considerable. A single isolated piece of guy wire with its ends insulated can only resonate in multiples of one-half wavelength, so this spacing of insulators must be avoided. In fact, with the capacitive end loading of the insulators, the resonant length of wire will be somewhat less than one-half wave so three-eighths wavelengths should also be avoided. A one-quarter wavelength is much better, but this would be quite expensive as it requires insulators every 30 inches (762 mm).

Alternatively, the guys through the FM antenna aperture on the tower could be replaced with plastic cable, which is transparent to RF energy [18].

Nonconductive Guys

In order to eliminate FM pattern distortion, any guy cable going through the antenna level should be of nonconducting material. Plastic fiberglass (GRP) insulating rods as well as flexible plastic rope covered with a PVC plastic jacket can be used. The black jacket prevents deterioration due to ultraviolet sunlight radiation, which may be injurious to the plastic strands of the rope. Plastic rope has been successfully used for more than 25 years.

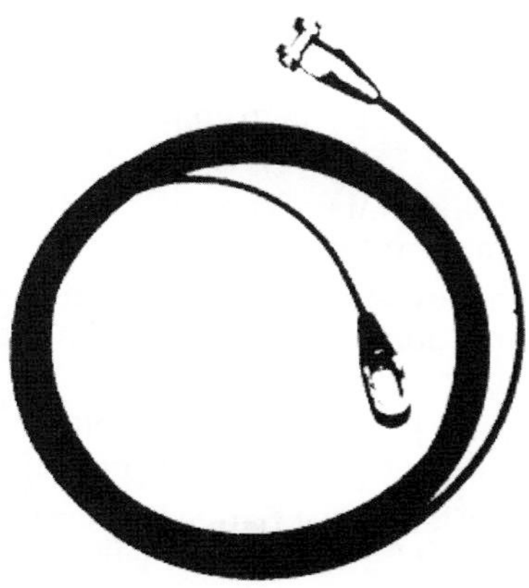

FIGURE 4.9-25 A roll of nonconductive guy wire with factory-installed fittings.(Courtesy of Phillystran.)

The idea is to remove metallic RF-conducting steel guy cables from within the antenna aperture. The rest of the guy may be of steel construction. The length of the steel guy from its attachment point near the antenna to a point well below the antenna is simply replaced with an equal length of fiberglass rods or plastic rope.

Plastic rope is available in continuous lengths of up to 1,000 ft (304 m) and kits are available for installing the end fittings. The cable is quite flexible as Figure 4.9-25 shows with a 225 ft (69 m) length coiled up with its end fittings installed at the factory.

The cable may be purchased in strengths exceeding similar diameter steel guy cable. These strengths are shown in Table 4.9-7 for corresponding size of commonly used extra high strength (EHS) steel guy wire. Sizes smaller and larger than shown in the table are available.

INSTALLATION PROCEDURES

If the installation is not properly planned and carried out, there may be unwarranted delay and cost associated with putting the FM antenna on its support tower. The following suggestions are offered to avoid unnecessary delays and expenditures.

Planning the Installation

The first step is to ensure the structural integrity of the tower. This should be performed by a qualified, licensed structural engineer. Some tower manufacturers offer this service. Because of the high cost of rigging services, it is essential to carefully plan the installation. Make sure that all parts are on hand. The installation of the antenna should be planned by a technically qualified person who must supply accurate tower construction information to the antenna manufacturer. If this information contains errors, these will be carried through the design and fabrication of mounting hardware, and finally show up in the field to frustrate and confuse the installing crew, wasting time and money at every stage of the process.

The station should consider hiring a tower rigging rm that is financially qualified and mechanically well equipped to do the work. A written contract should exist between the station and rigging firm, with a fixed price. The rigging contractor should be licensed as a contractor in the state, and should post a completion bond. The contractor should also supply an insurance policy holding the station harmless and making the station and its personnel coinsured. Only in this manner will the broadcaster be protected in the event of injury or property damage.

The riggers should be knowledgeable about antennas and coax line, and should inspect the tower and check out the mounting design of the brackets before the full rigging crew arrives. If errors are found, contact

TABLE 4.9-7
Phillystran-Type HPTG Plastic Guys

Outside Diameter		Break Strength		Jacketed Weight		EHS
Inches	mm	Pounds	kgs	1,000 Ft	300 m	Equivalent
0.20	5.1	4,000	1,815	18	8.2	3/16
0.29	7.4	6,700	3,039	31	14.1	1/4
0.42	10.7	11,200	5,080	55	24.9	5/16
0.46	11.7	15,400	6,975	69	31.3	3/8
0.53	13.5	20,800	9,435	93	42.2	7/16
0.58	14.7	27,000	12,247	115	52.2	1/2
0.63	16.0	35,000	15,876	142	64.4	9/16
0.68	17.3	42,400	19,235	167	75.8	5/8
0.73	18.5	58,300	26,445	195	88.5	3/4

the factory immediately. Particular attention should be paid to the following:

- Fit of mounting brackets to tower member.
- Freedom from interference of the mounts with gussets, leg flanges, guys, and their attachment points, tower face members, and obstruction lights.
- Compatibility of transmission line and antenna input coax terminals.
- Location of transmission line run relative to antenna input terminals.
- Use of fiberglass guys on AM towers in the immediate vicinity of the FM antenna (refer to the AM Guying section in this chapter).
- Availability of proper voltage, current, and cable size for deicers if required.
- Adequacy of tower to carry the wind load placed upon it by the antenna, particularly where radomes are used. This radome/antenna load should be checked by a competent structural engineer, as all antenna installations should be checked. This is usually required by the company carrying insurance on the tower.

Receiving and Unpacking

The shipping boxes are usually numbered and the total number is indicated on each box; contact the shipper if not all boxes are delivered, or if equipment is received damaged. If possible, do not store the material outdoors, boxed or otherwise.

As soon as the antenna is received, open and examine it for shipping damages so that any necessary claims may be filed with the shipping company immediately. Check the material against the parts list and installation drawing.

The box with the installation drawing and instructions is usually marked. Open it first, so that the balance of the items may be easily identified and counted. Contact the factory immediately if any material appears to be missing or is damaged during transportation.

Do not call the riggers until all antenna and coaxial line are at the site. Otherwise, unnecessary delays and costs may result.

Installation Tips

Broadcast antenna manufacturers furnish detailed installation instructions that should be closely followed by the rigger. Together they will ensure a perfect installation saving time and money.

The following items are specifically called to the attention of the broadcast engineer (in addition to all those stated before) to permit proper installation and good long-term performance:

- Follow manufacturer's instructions.
- Do not leave antenna parts where rain or moisture can enter. Store indoors and keep units capped as received.
- Do not allow dirt or other foreign matter to enter any coaxial part.
- Protect all antenna parts from physical damage and abuse.
- Hoist antenna members carefully, with a tag line to prevent damage by striking against the tower.
- Install on the tower as indicated by the manufacturer's instructions, remembering that bay number #1 is the uppermost top unit.
- Riggers should lubricate O-rings with a small amount of silicone grease before mating flanges.
- The full complement of flange bolts must be used and they should be as tight as instructed.
- Tuners or individual element devices, if used, should be adjusted only after the entire antenna and tower installation has been completed.
- Rigid transmission lines should be properly installed with two hangers per 20 ft (6 m) length, and with the inner conductor retaining pin on the top of each section.
- If semi-flexible cable, such as Heliax or Wellflex, is used, it should be firmly tied down at least every 5 ft (1.5 m) for 3-1/8 inch (76 mm) line, and every 3 ft (1 m) for 1-5/8 inch (43 mm) coax line. The line manufacturer's hangers should be used. The line should not be attached to the tower using plastic ty-wrap straps.
- After physical installation has been completed in accordance with the manufacturer's recommendations, the main transmission line should be pressurized with dry air through a dehydrator, air pump, or by using dry nitrogen gas. (See the Air Pressurization section in this chapter for more information.)
- Dry air or gas pressure should be maintained at all times. Most antenna warranties are not valid unless this is done. It is the rigger's responsibility to make certain that the entire coax and antenna hold air pressure.

The antenna system should be checked by a qualified rigger every time the obstruction lights are replaced, or if lights are not used, at least once a year. The rigger should look for vibration and storm damage, loose or broken coax hangers, and signs of arcing across exposed insulators. A dry rag soaked in 91% isopropyl or other solvent alcohol or equal should be used to wipe clean all exposed insulators in each antenna element. (Do not use carbon tetrachloride!) Chapter 7.1 discusses in detail the safety and OSHA rules that must be followed during antenna installation and maintenance.

STRUCTURAL CONSIDERATIONS

Most FM antennas in the western hemisphere are installed on the sides of a steel tower, between 18–60 inches (45–152 cm) wide. The antenna and its transmission line together with all mounting brackets introduce wind loading, in addition to their dead weight. The live wind loading is a result of the amount of physical surface presented to the wind. It is sometimes called the *wind catch area*. This consists of either at or round antenna members, coaxial lines, mounting brackets, and hardware, all represented as surfaces that are exposed to the wind.

The dead weight of the antenna system is fixed and is always present on the tower. The live load is a variable, depending on the wind velocity, and is added to the dead load for the total amount present.

The standard wind load starts with an assumed wind velocity of 87 mph (139 km/h), which will produce a push of 35 lbs per square foot. (170.8 kgs/m^2). With lesser wind speeds the wind push is less, and more with higher velocities. Various building codes determine the rated winds to be considered in the design of the tower system. While most of the United States has a 35 lb per square foot minimum rating, some parts of the country have higher requirements due to higher wind velocities. Some insurance companies may require even higher safe wind ratings.

ANTENNA POLE MOUNTING

Nonsymmetrical antennas may also be installed on a round pole, made of various diameters of steel pipe. Several antenna manufacturers supply these as a complete system and will optimize the horizontal plane pattern. The advantage of pole mounting on top of a tower or building is that the pattern may be more easily contoured. This provides more signal in the service area since the antenna orientation is not limited by a fixed triangle formed by a guyed tower.

TRANSMISSION LINE SYSTEMS

Two types of coaxial transmission lines may be used to feed FM antennas. One is rigid coaxial line sections, each 20 ft (6.09 m) long, and requires elbows, flanges, spring hangers, and other devices to attach the line sections to the tower.

The other has a semi-flexible coaxial line that is available with either air or foam dielectric and uses fixed hangers. Semi-flexible cable is available on a spool whose diameter depends on the line size. EIA end flanges mate to the antenna flanges as well as other RF equipment. Detailed information on FM RF transmission lines can be found in Chapter 4.11. Information on rigid transmission lines may be found in Chapter 6.6.

AIR PRESSURIZATION

If the antenna is operated without positive pressure of dry air or nitrogen, the manufacturer will not assume responsibility for failure under power. Moisture or the accumulation of water within the coaxial transmission line is a very serious matter. Its presence causes the VSWR to rise and corrosion to occur. When a sufficient amount of moisture is present, arcing will take place burning the line or antenna radiating elements. High humidity or moisture will cause the inside of the coaxial transmission line to corrode over time, thereby increasing the line loss. For this and other reasons, the entire antenna system must be dry-air pressurized.

After the antenna is installed and the transmission line connected, the system is purged with dry gas or dry air to remove trapped moisture before RF power is applied. A manually opened or pressure-actuated purge valve is installed in nearly all FM antennas made by American firms. When the gas pressure is raised to 10 psig (0.68 atmospheres) the automatic pressure relief valve will open up letting moist air out. The complete system purge requires a considerable volume of dry gas.

Before expending this amount, it is good practice to perform a quick check for major leaks. The system pressure is raised to a point below the relief valve setting, such as 8 psig (0.48 atmospheres) and the source of supply shutoff. A pressure gauge should be installed on the antenna side of the shutoff valve. The pressure, when corrected for temperature, should not fall to less than half its initial value in a 24 hour period. If the pressure loss is more than this, the system should be checked with a leak detector, or soap suds, to locate the leak. A pinched or missing O-ring is the most common cause for large leaks.

Once the system is known to hold pressure, it should be purged with dry air or nitrogen gas. Either must be dry enough to have a dewpoint well below the coldest temperature expected to be encountered. When using nitrogen, it should be of the oil-dried type, to remove nearly all the moisture from the gas.

Five to 8 psig (0.34–0.48 atmospheres) should be maintained in the system at all times to ensure that no moisture will be able to enter. Very small leaks will pull in moisture, if the pressure is lower than suggested above, when the transmitter is turned off nightly. This is due to the pumping action due to expanded dry air/gas pressure cooling down, and contracting below the outside air pressure, during cold ambient temperatures.

PROTECTION FROM ICING

High-Q antennas are subject to increased VSWR ratios as well as pattern distortion, with light to moderate coatings of ice. Low-Q antennas, such as the panel type, are usually not affected in this manner. Where climatic conditions cause sufficient ice or in some cases snow to affect the antenna's performance, there are two remedies. The radiating element may be covered with a plastic cover, or, it may be electrically heated to melt or prevent the formation of ice on its sensitive surfaces.

Electrical Heaters

By far the more popular method of deicing is to order electrical heaters at the time the antenna is ordered.

Electrical deicing equipment is supplied as an option and is factory installed. Kits are furnished for interbay connections, but the broadcaster must supply power from the building to the center of large arrays, or the bottom element on smaller antennas. Local electrical codes of course must be followed.

While a thermostat may be used with smaller deicer wattages, a power relay operated by the thermostat is required and furnished with most electrical deicer kits by the antenna supplier. Due to high power costs, a sophisticated deicer control, which operates when temperature and humidity conditions produce sleet or icing, is desirable.

Most deicers use a resistance heating element that is inserted inside the antenna radiator arms. One manufacturer, however, uses a different method, dropping the 230/240 volts to a few volts with a transformer located at each bay level. The low voltage is passed through the ice-sensitive arms of the radiator and connected to the far ends by a heavy Teflon-coated wire. The current return is by the stainless-steel antenna element, whose resistance is sufficient to produce enough heat to melt or keep the ice off. This method is becoming obsolete as the transformers are expensive and heating costs are rising, as hourly electrical rates go up. Further, the voltage dropping transformer technique is not as efficient as direct heaters.

A word of caution when selecting an FM antenna with electrical heaters. Some deicers use 1 kW of power for each bay as previously described and increase the wind loading by 225%. Others have a switchable power option feature using 125/500 W per element, with only a 15% increase in windloading, when compared to an antenna without electrical deicers. The continuing cost of electrical deicers is a consideration of the operational cost of the station and should not be overlooked.

Automatic deicers are those with a thermostat for mounting near the antenna for accurate temperature sensing of the actual ambient temperature. The temperature zone of +20° F to +35° F (–7° C to +2° C) is the most likely icing range, depending on humidity conditions. Deicers should be turned on at +35° F, prior to ice formation, because it is better to prevent icing than to remove it once it is formed. Power should be turned off when the temperature goes below +20° F since ice does not usually form below this temperature.

Radomes

A radome is a protective dielectric housing for an antenna-radiating element. Its function is to protect the antenna not only from ice, but snow and physical damage due to ice dropping from above. Radomes also help protect the radiating element from environmental corrosive atmospheres.

The primary purpose of using radomes on FM antennas is to prevent the VSWR from rising with the formation of ice, if the site and height cause icing to occur during the winter months. Ice formation detunes high-Q radiators, increases the VSWR, and causes vertical plane pattern changes. Figure 4.9-26 shows a typical radome enclosing a radiator.

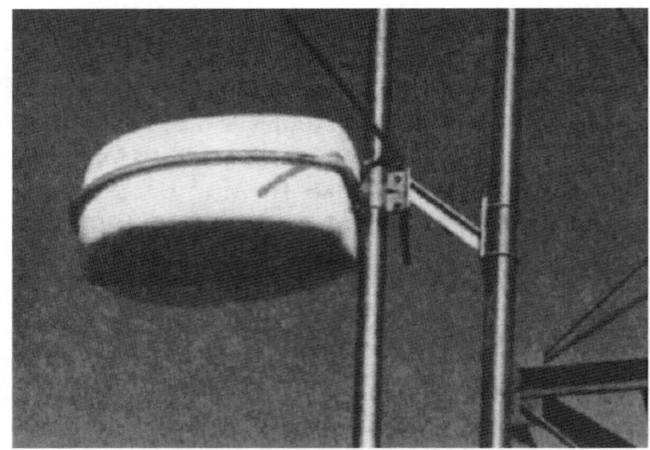

FIGURE 4.9-26 Typical radome installed on antenna bay.

Ice may form on the radome but does not particularly affect the operation of the radiator if that ice is kept at least 0.05 wavelengths from the sensitive portions of the antenna element.

Radomes are particularly desirable in heavy icing environments where deicers are not adequate even with very high heat density. They are also useful in protecting antenna elements from falling ice when they are so exposed.

Radomes are cost effective with single channel high-Q antennas where electrical deicer heating power costs are expensive. The deicer power cost is a continuing one, while radomes are a one time capital investment, which may be depreciated over time.

Radomes are generally composed of low-loss dielectric material with low values of dielectric constants and loss tangents. Laminated fiberglass, using glass cloth reinforcement, has a constant of about 4.1 and a loss tangent of about 0.15. Water absorption by the radome increases its dielectric constant and loss tangent. Materials that do not easily absorb water or those treated with a protective gel coat are often used to shed water and prevent the adhesion of ice.

Good radome designs take into consideration operating temperature, a relative humidity of 100%, safe wind pressures, ice, hail, snow loads, rain adhesion, wind, and supporting tower vibration. They are made from fire-retardant plastic, and must safely withstand air contaminants over the useful life of the antenna. All these factors increase the cost, but are necessary for a long, useful life. Radome shapes are dictated by the form of the radiating element in most instances.

In all cases radomes are supplied by the antenna manufacturer, and usually supplied in two pieces that are held together with stainless-steel fasteners.

LIGHTNING

Because FM towers are usually located on high ground, hilltops, or high buildings, they require lightning protection since they are likely recipients of lightning strikes. The type of damage that can be caused by lightning to a FM tower varies. Smaller coaxial lines will usually melt; larger coax (such as 1-5/8 to 3-1/8 inches) will also melt in some cases, and others will conduct the heavy current into the transmitter building to do damage there.

The FM antenna itself may heat, arc, melt, and otherwise be damaged. Holes in the outer conductor, burns, and melting at flanges are common. Teflon or polyethylene insulation will burn, depositing a film of carbon, causing further damage if RF from the transmitter continues after the strike.

Protection of the FM antenna system may be provided to some degree, by taking several precautions. The top of the tower should have a lightning rod, about 1 ft (0.3 m) higher than the uppermost obstruction light part. The FM antenna itself should be firmly grounded to the tower. If the coaxial cable is buried between the tower and the transmitter building, it must be at least 6 ft away from any tower base grounding system copper wire or strip [19].

A ground system should be located immediately around the base of the tower. This should have a direct current loss of less than 10 Ω to earth ground. This low resistance may be obtained by using ground wires buried in the soil. Six radials spaced at 60 degree intervals, buried as deep in the soil as possible and running out up to 150 ft (46 m) each, should provide a suitable ground of less than 10 Ω, even if the soil is shallow or rocky.

Guyed tower anchors should also be grounded. This is covered in Chapter 7.2, Design, Erection, and Maintenance of Antenna Structures. It is important to install the proper number of ground rods and/or copper wire radials in order to obtain a connection to earth ground of less than 10 Ω. In any event these ground rods or radial wires must be tied together with AWG 4 or larger copper wire, or 2 inch (5 cm) copper strap. This is to provide for thousands of amperes of current flow for less than one second in the event of a direct lightning hit.

If the FM antenna is located on an AM-insulated base tower, then the spark gap should be set at the lowest point that still provides protection for the highest AM modulation peak voltage.

Another way to protect the FM transmission line isolator (if one is used), as well as the tower and FM antenna, is to use an RF choke across the insulated tower base. This tends to reduce the static build-up voltages due to passing thunderstorm clouds, snow, hail, or dust storms. Arc-overs due to these less severe sources usually do not cause damage, but may trip the FM transmitter reflectometer since they will create a current flow through the reflectometer circuitry.

If the base-insulated tower supporting the FM antenna is located in an area of regular thunderstorms, a way to protect both antennas from lightning is to ground the AM tower at the base, and shunt feed it. Several excellent methods exist. The folded unipole method not only grounds the tower for lightning purposes, but may improve the VSWR bandwidth. See Chapter 7.3 for more information on lightning protection.

FM SCA MULTIPLEXING

With subcarriers in use and 110% overall modulation, intermodulation products may be created due to mixing of the various subcarriers with their own harmonics within nonlinear devices, one of which can be the antenna system.

Antenna linearity is determined by its VSWR response curve versus frequency. Phase delay in the antenna system is also important. In the past, with 67 kHz being the highest SCA frequency, the ±100 kHz bandwidth was considered sufficient. With subcarriers and 110% modulation (82.5 kHz deviation), the minimum bandwidth is ±130 kHz under 1.1:1 VSWR [20]. (See VSWR Bandwidth in this chapter for more specific requirements and recommendations.)

Tests have shown that 92 kHz is the frequency of choice for a new aural SCA service after 67 kHz [21]. The 92 kHz subcarrier produces lower intermodulation product levels and less interference to the main channel stereo service than 67 kHz [22]. It may be successfully operated in addition to stereo and existing SCA services.

Other nonlinearities in the exciter, and transmitter, plus multipath reception, receiver misalignment, and user mistuning are contributions to the received intermodulation distortion of the baseband signals. In addition, these products can cause small levels of audible swishing beat notes in some types of FM receivers.

Spurious Emissions

Interference to other stations within the FM broadcast band as well as to other services outside the band can be caused by RF intermodulation product energy developed between two or more FM broadcast transmitters. It may be due to coupling through a diplexer or coupling between two antennas. This phenomenon has been well documented [23].

When RF energy from two or more transmitters is combined, new spectral components are produced by mixing the fundamental and the second harmonic of each of them. The dominant intermod product generated by each transmitter is at twice the transmitter's frequency. For example, 101.1 and 102.7 transmitters would produce two intermod signals appearing on 99.5 and 104.3 MHz.

Second harmonic traps or low pass harmonic filters in the transmission line of either transmitter prior to the diplexer have little effect on the generation of intermod products. This is because the harmonic content of the interfering signal entering the transmitter output circuit has much less effect on intermod generation than the harmonic content within the nonlinear device itself. The resulting intermod falls within the passband of the low pass filters and outside the reject band of the second harmonic traps, so these devices offer no attenuation to intermod products.

Diplexers, however good, will reflect some of the undesired energy back to each transmitter, generating intermod products. The key to this problem is to keep that undesired power level as low as possible using proper transmitter output circuitry and tight diplexer specifications. Diplexed transmitter installations should be routinely checked for the presence of excessive intermod products.

HARMONIC FILTERS

FCC Rules Section 73.317(d) calls for the harmonics of FM transmitters to be as much as 80 dB or more below the transmitter output. This requirement is usually met by using a low pass filter that passes the station carrier frequency power but attenuates its harmonics.

The transmitter provides some harmonic attenuation of course, and is usually 25–38 dB for single-ended amplifiers. The worst-case harmonic is the third. Harmonic filters by several firms provide a minimum of 50 dB for harmonics from the second through the tenth. Adding the transmitter attenuation to that of the filter normally provides more than the required level.

The high level of rejection is made possible by using high-impedance (inductance) and low-impedance (capacity) coaxial sections for m-derived three to five section filters, with half-pi end sections. Harmonic filters are commonly made in three production schedules. Some are adjusted to the customer's operating frequency, so that there are no attenuation gaps in the higher harmonics. They are not tunable outside the factory as the insertion loss, and attenuation along with passband VSWR are closely related.

The insertion loss in the passband varies from 0.05–0.08 dB while the rejection for the second through the tenth harmonic is 50–60 dB, depending on the number of internal filter midsections. The VSWR in the passband varies from 1.05 to 1.1:1. Harmonic rejection is due to the very high VSWR on the harmonic frequencies, which may be as high as 15:1. This rejected power is passed back to the transmitter amplifier where it is absorbed.

Harmonic filters are available in straight rigid coaxial line sections and may sometimes be pressurized. Power capacity varies from 10 kW for 1-5/8 inch EIA line size to 50 kW for the 6-1/8 inch size.

IBOC ANTENNA OPERATION

There are three methods of transmitting FM IBOC (in-band on-channel). The first employs a single transmitter that generates both analog and digital signals and feeds them to a common antenna. The second method employs a separate analog and digital transmitter combined in the transmitter building and fed to a common antenna. The third employs separate analog and digital transmitters each feeding signals through separate transmission lines and antennas. The IBOC digital signals are lower in power than the analog signal so antenna and transmission line power ratings used exclusively for the IBOC signal may be smaller in size.

ANTENNAS FOR DIGITAL RADIO APPLICATIONS

There can be different types of antenna arrangements. The most important issue is isolation between analog and digital signals when combining them using shared or interleaved antennas. The analog signal (typically +20 dB above digital) will appear at the digital transmitter port as reflected or coupled signal, respectively.

Interleaved Antennas

Two antennas shown in Figure 4.9-27 are separate FM side-mount antennas sharing the same aperture and fed by separate transmission lines. Each antenna consists of wavelength-spaced elements sharing the same aperture at half-wavelength spacing of each other. Because of half-wavelength spatial and small frequency separation between the two antennas the coupled signal is large (typically –15 dB for H-pol signals). Therefore, the isolation at the digital input is poor. Coupling can be suppressed and isolation therefore improved by making one of the antennas a left-hand circularly polarized (LHCP) and the other right-hand circularly polarized (RHCP). Then the signals are effectively cross-polarized and the coupling between them is suppressed and isolation is improved to –30 dB or better.

Low-Level Injection

This technique, low-level injection, is used to combine analog and digital signals into a shared antenna and shared transmission line. To isolate the antenna-reflected analog signal from the digital transmitter a four-port 10 dB hybrid combiner is used for the two transmitters, as shown in Figure 4.9-28. The hybrid combiner dumps 90% of digital transmitted power and 10% of analog signal into a dummy load and passes 90% of the analog and 10% of the digital signals to the antenna. As the reflected analog signal travels back from the antenna (VSWR of 1.1:1) to the hybrid combiner, the 10 dB hybrid sends 10% of analog-reflected power toward the digital input and transmitter. If the antenna is tuned to have a VSWR of 1.1:1 (–26.5 dB) over an FM channel, then the reflected analog signal at the digital input (transmitter) is –36.5 dB below rated FM power.

High-Level Injection

A shared antenna and two separate transmission lines are practical for an array of panel antennas, as shown

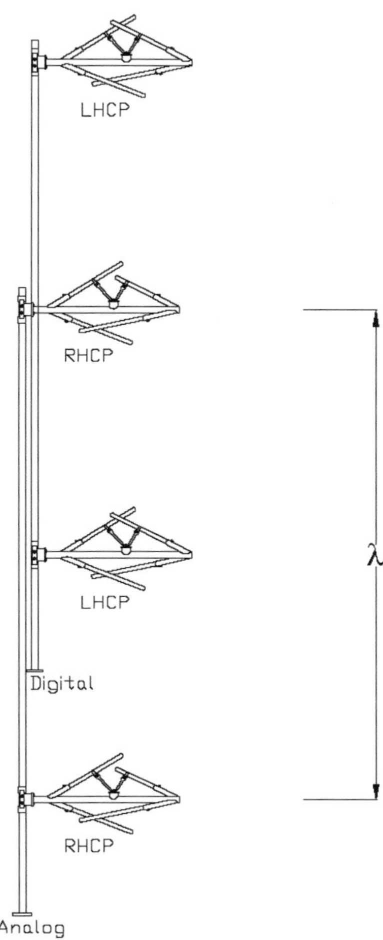

FIGURE 4.9-27 Two interleaved FM antennas fed by separate transmission lines and sharing the same aperture.

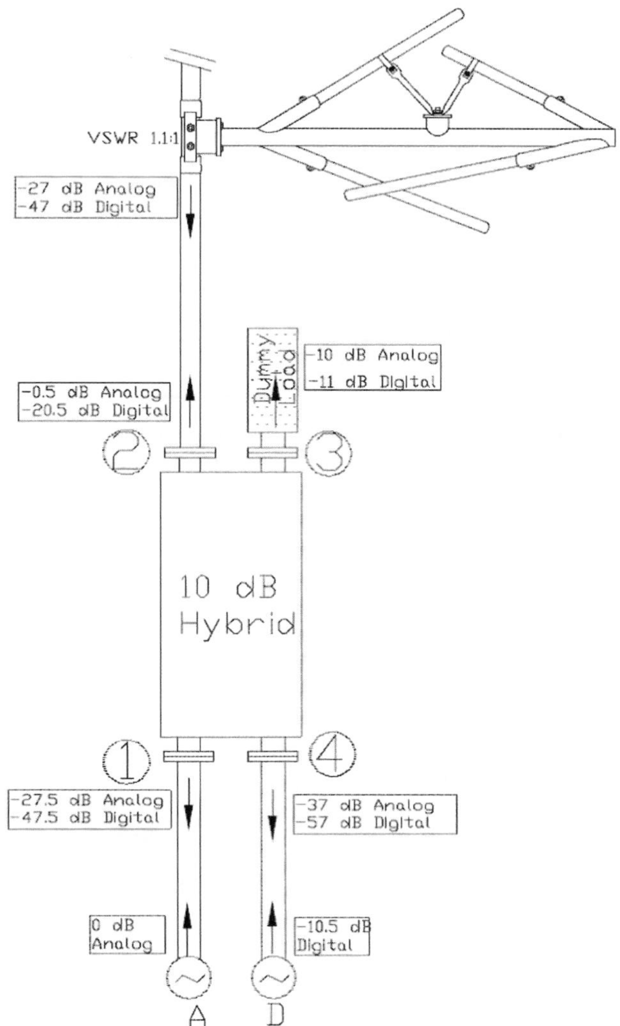

FIGURE 4.9-28 Example of 10 dB hybrid combiner used to couple analog and digital transmitters into a common transmission line and antenna. Note relative values for forward and reflected power.

in Figure 4.9-29. The typical FM panel antenna has dual inputs (two polarization ports) fed in quadrature phase (i.e., 0° versus ±90° for RHCP and LHCP, respectively). Polarization ports can be injected at high level with the digital signal via 3 dB hybrid combiners (at shared antenna) making analog and digital signals LHCP and RHCP, respectively. Two different arrangements of digital radio injectors are considered here (narrowband single FM channel and broadband).

Narrowband Digital Radio Injector

For narrowband, two 3 dB hybrid combiners are employed, one for each portion of the feed system (as shown in Figure 4.9-30). A 3 dB hybrid provides equal power split and 90 degree phase shift between output ports 2 and 3. Also, assuming that coupling between the cross-dipole is negligible, the hybrid's reject port isolation is limited to $S_{14} = S_{22} + S_{33}$, so when hybrid port 4 is for digital injection then the isolation of the analog signal from digital is determined by a close match of the dipoles at hybrid ports 2 and 3. Fine matching devices can be effectively used at ports 2 and 3 to improve isolation at digital port 4.

Broadband Digital Radio Injector

A schematic with hybrid combiners at every panel is shown in Figure 4.9-31. Broadband isolation of the analog signal at the digital input and vice versa is achieved by combining both analog and digital with feeds phased in a turnstile manner, that is, with incrementally increased phase-feeding panels of one bay.

Superturnstile Feed

Analog-reflected signals seen at port 4 are first suppressed by the close match of individual dipoles S_{22} and S_{33}. Then as reflections travel down toward the digital input they follow the turnstile route and there-

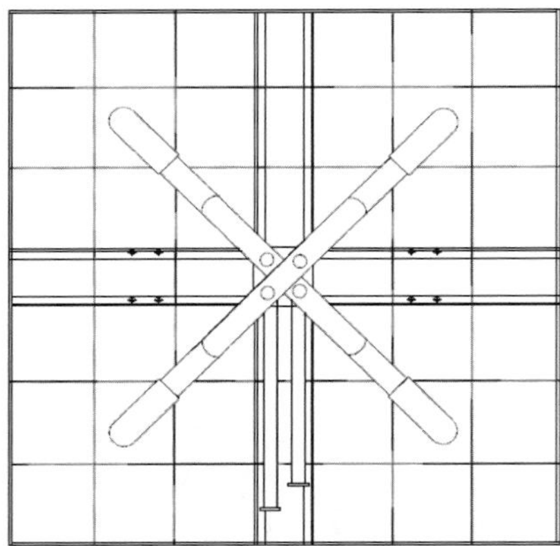

FIGURE 4.9-29 FM dual-dipole panel antenna, front view. Note two input transmission lines near bottom of image.

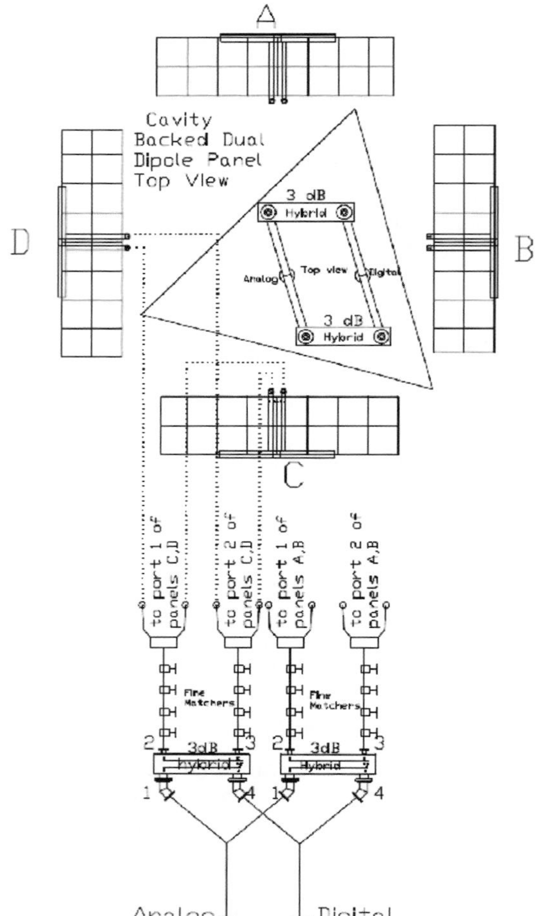

FIGURE 4.9-30 Schematic of high-level injection for shared antenna and separate transmission lines.

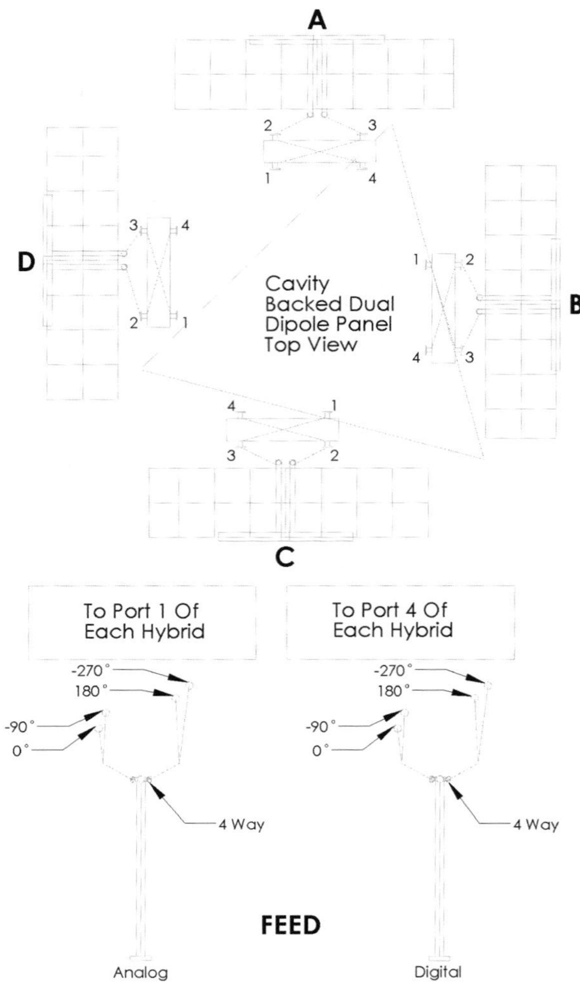

FIGURE 4.9-31 Diagram of high-level injection for shared antenna and separate transmission lines.

fore are greatly suppressed over a broadband by phase compensation at the digital input. With four-around panels per bay, each is fed with –90 degrees phase increments to produce an omnidirectional azimuth coverage. The panel lateral offsets are used to optimize circularity for both analog and digital azimuth pattern. Typical turnstile omnidirectional pattern circularity is within ±1.5 dB.

With three-around panels the bays produce a wide cardioid azimuth coverage either when fed in-phase or in 60 degree phase increments. Slight asymmetry of the latter can be compensated with lateral panel offsets of about 10 inches. Figure 4.9-32 shows the top view and a schematic of three-around panels with both analog and digital signal feeds to the panels in 60 degree phase increments. Reflected analog signals that show up at the digital input are effectively cancelled since the reflected signals from the three panels come back to the digital port in 120 degree phase increments.

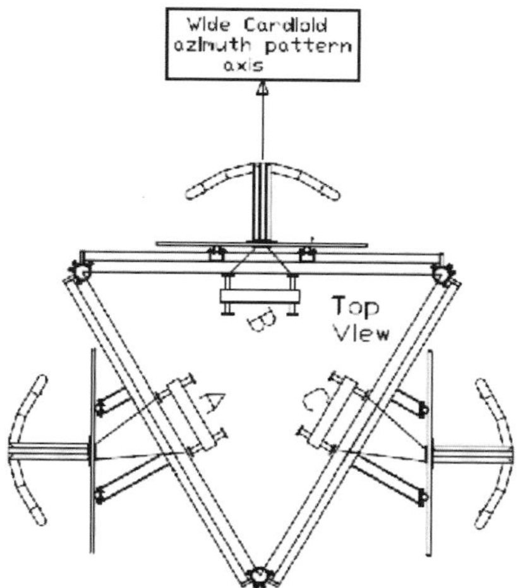

FIGURE 4.9-32 High-level injection for shared antenna and separate transmission lines (not shown). Superturnstile feed. Wide cardioid azimuth pattern, top view.

ACCESSORY ANTENNA SYSTEM EQUIPMENT

There are multiple devices and systems that support the operation of a broadcast antenna. The dry air pressurization of coaxial transmission line was discussed earlier in this chapter.

Reflectometers

The reflectometer is a device for detecting the ratio of power into the antenna (forward) and the rejected power from the antenna (reverse). It consists of a short coaxial line section about 12 inches (305 mm) containing diode detectors, coupling loops, and terminations that produce the DC current that drives a suitably calibrated VSWR meter.

Reflectometers are wideband devices and therefore must be placed *after* the harmonic filter. Putting them between the transmitter and the filter causes them to read the rejected harmonic power along with the reflected, thus giving an erroneous reading.

Dummy Loads

A very useful test device in an antenna system is the terminating (dummy) load. At least one is needed when two amplifiers are combined and fed to the antenna, or when a number of diplexers are used in a community antenna arrangement. Dummy loads are available in several power levels up to 50 kW or more and are cooled by air or water.

RF Switches

Often used with a dummy load, coaxial line switches are available to provide electrical or manual switching of transmitter power to diplexers, antennas, standby transmitters, etc. They typically are not pressurized.

SUMMARY

With the introduction of digital radio broadcasting using in-band, on-channel techniques it is essential that the FM broadcast engineer fully understand the concepts and practices of good antenna design, installation, and maintenance. Antennas deliver the signal to the audience and therefore are the single most important part of the transmission system. By understanding the interrelationships between antenna site, antenna design variables, propagation, and local terrain the broadcast engineer can develop the best configuration for a given FM station facility.

Selection of the proper antenna system for an FM station is essential to delivering optimum signal to the receiver. The antenna type, number of bays, mounting structure, height above ground, and antenna location relative to the desired audience are all essential factors in achieving that goal.

References

[1] Freeman, Roger. *Telecommunications Transmission Handbook.* New York: John Wiley & Sons, 1975, pp. 180–186.

[2] Head, Howard. Influence of Trees on TV Field Strengths, *Proceedings of the Institute of Radio Engineers,* 48, June 1960, pp. 1016–1020.

[3] Armstrong, A. Study of Electromagnetic Wave Propagation at 112 MHz, *Proceedings of the IREE Australia,* April 1969, pp. 105–110.

[4] Moeller, Adolph. Effects of Ground Reflections on Antenna Test Range Measurements, *Microwave Journal,* Mar. 1966, pp. 47–54.

[5] Reed, Russel. *Ultra High Frequency Propagation,* 2nd ed. Boston: Technical Publishers, 1964, pp. 223–238.

[6] Onnigian, Peter. Stereo Degradation as a Function of Antenna System VSWR, Audio Engineering Society Annual Meeting, Audio Engineering Society, New York, 1976.

[7] Mendenhall, Geoff. The Composite Signal—Key to Quality FM Broadcasts, *Technical Monograph.* Quincy, IL: Broadcast Electronics, Inc., 1984.

[8] Cohen, Jules. Proposed Solutions, Channel 6—Educational FM Broadcast Interference Problem, *Proceedings of the 38th Annual Broadcast Engineering Conference,* Washington, DC, NAB, 1984.

[9] Onnigian, Peter. Circularly Polarized Antenna, U.S. Patent 3,541,470.

[10] DuHamel, Ray. TV and FM Transmitting Antennas, in *Antenna Engineering Handbook,* 2nd ed., eds. R.C. Johnson and H. Jasik. New York: McGraw-Hill, 1984, chapter 28, pp. 8–9.

[11] Bock, E. Sleeve Antennas. *VHF Techniques,* vol. 1. New York: McGraw-Hill, 1947, chapter 5, pp. 119–137.

[12] Knight, Peter. Re-radiation from Masts at Radio Frequencies, *Proceedings of the IEEE,* 114, Jan. 1967, pp. 30–42.

[13] Jampro Antennas, Inc. International Technical Communication, Sacramento, CA.

[14] FCC Public Notice. Criteria for Licensing of FM Broadcast Antenna Systems, Notice 84–437 25004. Washington, DC: Federal Communications Commission, Sept. 14, 1984.

[15] Rudge, A. *Handbook of Antenna Design,* vol. 2. London: Peter Peregrinus Ltd., 1983, pp. 917–922.

[16] Onnigian, Peter. Multi-Station FM Antennas, Paper presented at the 23rd Broadcast Symposium, *IEEE Group on Broadcasting,* Washington, DC, 1973.

[17] Fisk, Ronald. Design and Application of a Multiplexed Nine-Station FM Antenna, Senior Road Tower Group, *Technical Monograph.* Quincy, IL: Harris Corporation, 1983.

[18] Gregorac, L. Electrical and Mechanical Analysis of Plastic Guys of Broadcast Towers, *Technical Monograph.* Radio-Television Ljubljana, Yugoslavia, 1973.
[19] Marshall, J.L. Lightning Protection, in *Canadian Broadcasting System.* New York: John Wiley & Sons, 1973, pp. 53–54.
[20] Kean, John. Distortion of FM Signals Caused by Mismatched and Limited Bandwidth Transmitting Antennas. Washington, DC: National Public Radio, 1984, pp. 37–42.
[21] McMartin, Ray B. Super Eight, *Proceedings of the 38th Annual Broadcast Engineering Conference.* Washington, DC: NAB, 1984, pp. 160–166.
[22] Denny, Robert. Report on SCA Operation, *Proceedings of the 37th Annual Broadcast Engineering Conference.* Washington, DC: NAB, 1983, pp. 187–196.
[23] Mendenhall, Geoff. Study of RF Intermodulation Between FM Broadcast Transmitters Sharing Diplexed Antenna Systems, in *Technical Monograph.* Quincy, IL: Broadcast Electronics, Inc., 1983.

Bibliography

Collins, G. W. "TV and FM Broadcast Antennas," in *Antenna Handbook,* eds. Y.T. Lo and S.W. Yee. New York: Van Nostrand Rienhold Co., 1988, chapter 27.
Jordan, E. C., and Balmain, K. G. *Electromagnetic Waves and Radiating Systems.* Englewood Cliffs, NJ: Prentice-Hall, Inc., 1968.
Kaluski, M., and Stasierski, L. "Electromagnetic Field Estimation in the Vicinity of Panel Antenna Systems for FM and TV Broadcasting," *IEEE Transactions on Broadcasting,* 41(4), Dec. 1995.
Kerkhoff, William. "A New Master FM Antenna Design," *2002 NAB Broadcast Engineering Proceeding,* National Association of Broadcasters, Washington, DC.
Pantsios, F. A. "Customized Pattern Applications of the FM CBR Antenna," *IEEE Transactions on Broadcasting,* 36(3), Sept. 1990.
Trainotti, V., and DiGiovanni, N. D. "FM Wide Band Dipole Antenna," *IEEE Transactions on Broadcasting,* 48(4), Dec. 2002.

CHAPTER

4.10

FM Combining Systems

ROBERT A. SURETTE
Shively Labs
Bridgton, Maine

INTRODUCTION

Transmitting several frequencies from a single broadband antenna system requires the use of a combining system, or combiner, composed of radiofrequency (RF) filters and interconnecting transmission line. In general, a combiner can be categorized as one of two types: branched (star point) or balanced (constant impedance). Any of these types may employ band-reject (notch) or bandpass filters. This chapter discusses the use of filters, other components in FM combiners, and the hardware used to combine an in-band-on-channel (IBOC) digital signal into an analog signal.

APPLICATIONS

For years, both the FM spectrum and the FM channel were straightforward and uncomplicated. Until the early 1980s, the number of stations on the air in all but the largest metropolitan areas was low by today's standards. In most areas, the frequency spacing between stations exceeded the 0.8 MHz minimum that is common in all parts of the country today. These wider frequency spacings, the relative ease of developing new tower sites, and the limited station ownership in any market worked against the economics of combining stations into common antennas; therefore, most stations operated on single-frequency antennas, and large, multi-station antennas were generally only found in a few of the largest markets.

In the late 1980s, with the arrival of Docket 80-90, the FM spectrum in the United States became increasingly crowded. Ensuing changes in ownership regulations which tightened zoning regulations changed the economics of combining. It has become increasingly common to combine stations in even the smallest markets, including stations with very low power levels. Further complicating the spectrum is the dramatic increase in auxiliary antennas that began as a result of the need to accommodate digital television (DTV) construction. Expansion of combined systems has not been limited to small and medium markets. Large metropolitan stations that rarely exceeded 10 combined stations in the 1990s are now routinely being replaced by systems with room for 20 stations or more.

At the same time that changes in the FM spectrum made combining attractive and increased filtration a necessity, the FM channel itself became increasingly complicated. In the 1980s, the 67 KHz Subsidiary Communications Authorization (SCA) became more widely used. This was quickly followed by the 93 KHz SCA, pushing critical information to the ±100 KHz fringe of the FM channel and closer to potentially interfering signals. With the introduction of IBOC in the early 2000s, the channel has increased in size to ±200 KHz from the center frequency, and its full channel width is being utilized. Even this enhanced channel is becoming more crowded as digital multicasting becomes commonplace.

The net result is that, as the FM channel becomes larger and more complex, filters and combiners have had to evolve to provide the necessary isolation between closer-spaced signals at the same time that their own pass bands must be more tightly controlled to pass the desired channel. Today's combiners are even being used to isolate separate signals on the same channel to facilitate the combining of analog and digital signals.

WHY COMBINERS ARE USED

As populations migrate to suburban areas, it has become more desirable to construct large broadcasting facilities that can reach these heavily populated areas from more central locations. Of course, these prime locations have become more valuable, so it important to use each location to its fullest potential. This can best be done by sharing a transmitter site and a common antenna among several users. To accomplish this, the broadcast industry uses combiners of various types and sizes. For example, in San Francisco (Mt. Sutro), Toronto (CN Tower), Montreal (Mt. Royal), New York City (Empire State Building), and Chicago (John Hancock and Sears Buildings), tall towers or towers on skyscrapers have been used to consolidate as many broadcasting facilities as possible, including VHF-TV, UHF-TV, FM, and land mobile communications services. This approach has proven very effective, not only by using real estate economically but also by spreading the tower costs over many users.

Shortage of Prime Locations

Group ownership of FM stations in a market has led to proliferation of combined stations, and, with the implementation of DTV systems, FM stations are being forced off existing towers, making it even more imperative that they share tower space, which increases the demand for combined systems.

FCC ISOLATION REQUIREMENTS

When more than one signal is broadcast over a single antenna, the signals must be combined in such a way that signals from one transmitter are not permitted to feed back into another transmitter. Failure to do so would cause intermodulation products to be generated within the final amplifier stages of the transmitters and subsequently radiated from the antenna. These intermodulation products are generally referred to as *spurs*. Spurs created between FM stations can occur not only in the FM band but also within the low-band VHF channels and above the FM band, causing interference to the aviation band. FCC Rule 73.317(d) specifies that spurs more than 600 kHz removed from the carrier must be attenuated below the carrier frequency by 80 dB or by 43 + $10\log_{10}$ (power in watts) dB, whichever is less. In practice, stations operating transmitter output powers of 5 kW or greater must usually meet the 80 dB requirement, while stations running lower transmitter power output (TPO) fall under the computational method.

Experience provides ways to prevent spurs. Each transmitter must be isolated from all others in the system by a minimum of 40 dB, with 46 to 50 dB ensuring regulatory compliance. Spur attenuation is accomplished by a combination of transmitter turnaround loss and filtering. Turnaround losses are inherent to the way spurs are created in the transmitter. These losses typically run in the 6 to 13 dB range for tube-type transmitters, while 15 to 25 dB is typical for solid-state units. An off-frequency signal is attenuated 40 dB as it passes through the bandpass filters of the combiner module toward the transmitter, with the spur it creates exiting the transmitter an additional 6 to 25 dB below the level where the signal entered. This spur is then attenuated 40 dB as it passes back through the bandpass filters. The result is spur attenuation of at least 80 dB, with 100 dB or more possible.

IN-BAND-ON-CHANNEL COMBINING

The IBOC signal is transmitted above and below the standard FM analog signal (Figure 4.10-1) and is discussed in Chapter 4.13 of this handbook. An IBOC signal can be combined in a modified analog transmitter. This is referred to as *low level combining* and is covered in Chapter 4.7. The IBOC signal can also be combined by using a dual input antenna or separate antennas; this topic is discussed in Chapter 4.9. The "Combining Digital and Analog Signals" section (below) discusses combining the digital and analog signals from separate transmitters into a common transmission line before sending them to the antenna.

COMBINER CHARACTERISTICS

Important characteristics of combiners are frequency response, insertion loss, group delay, impedance, physical size, and tuning compromises. Improving one parameter may result in a reduction in another.

Frequency Response

Energy transfer through the bandpass filter is highest, or least attenuated, at the resonant frequency and drops off at frequencies above and below that frequency. This frequency response is the fundamental property that enables a filter cavity to sort frequencies. If it were possible to design an ideal filter, its frequency response plot would be as shown in Figure 4.10-2. Response would be flat within the pass band, with a vertical roll-off at the edges of that band. Figure 4.10-3 shows the frequency response of a real world single-cavity bandpass filter. Note that the energy transfer is highest at the resonant frequency (f_0) and drops off gradually away from f_0.

Insertion Loss

Even at the resonant frequency f_0, energy transfer is not perfect; some energy is lost along the way which is expressed as insertion loss—that is, the loss of energy at the resonant frequency. The lost energy is converted to heat and dissipated in the metal surfaces

MP1 Spectral Mapping

Logical Channel	Throughput (kbps)
P1	98.4
PIDS	0.9

Lower Sideband
Primary
Main
PIDS
P1
Analog FM Signal
Upper Sideband
Primary
Main
PIDS
P1
-198 kHz
-129 kHz
Channel Center
129 kHz
198 kHz
Key: MPS Audio SIS Data
HD Radio
Pure Digital. Clear Radio.
Proprietary to iBiquity Digital Corporation

FIGURE 4.10-1 Spectral mapping of an FM channel. (Figure courtesy of iBiquity, Warren, NJ.)

of the cavity. A cavity that is larger in size is more efficient than a smaller sized cavity in that it will provide a lower insertion loss at the resonant frequency with comparable frequency response. Coupling efficiency also affects insertion loss; curve B of Figure 4.10-4 shows the effects of coupling adjustment. The theoretical ideal filter would show no insertion loss in the pass band.

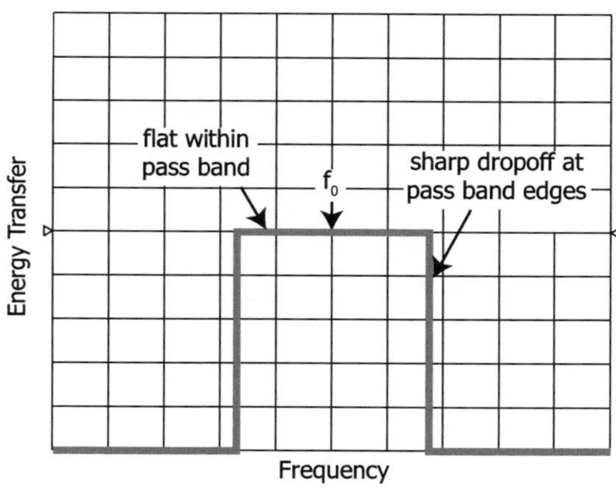

FIGURE 4.10-2 Ideal filter frequency response.

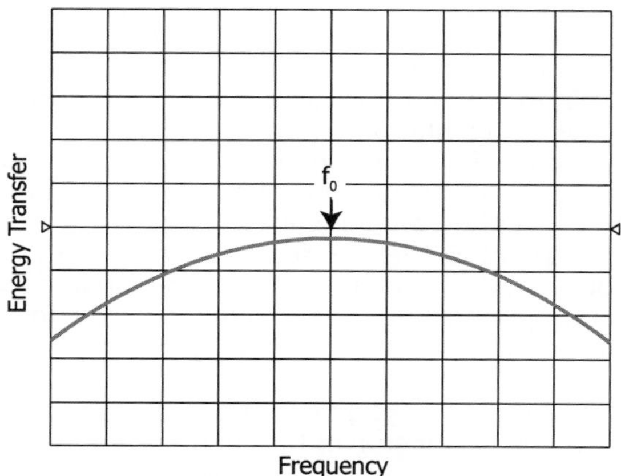

FIGURE 4.10-3 Frequency response for a single cavity filter.

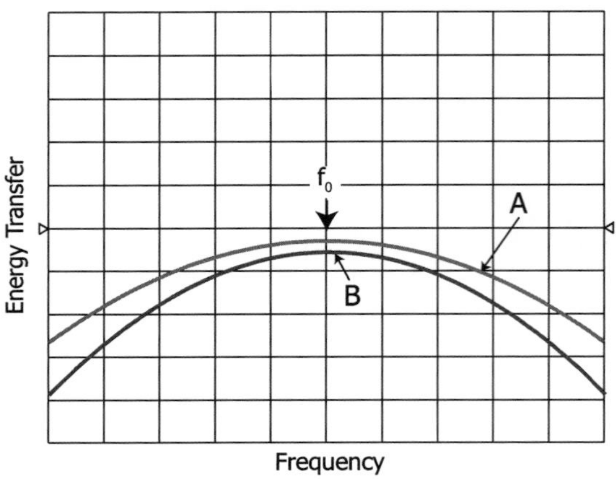

FIGURE 4.10-4 Insertion loss for a single cavity filter.

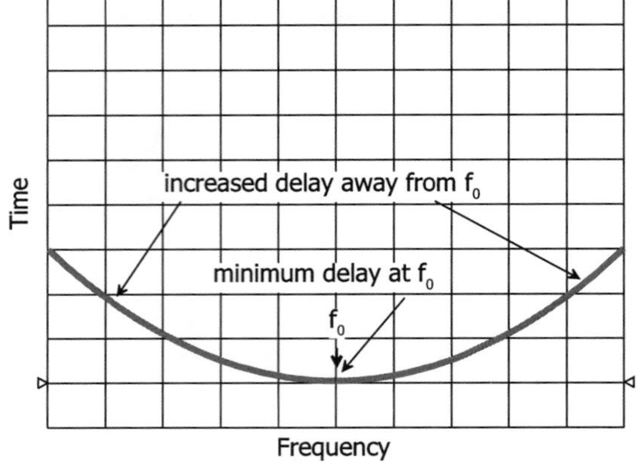

FIGURE 4.10-5 Group delay for a single cavity filter.

Group Delay

An RF signal takes a finite amount of time to pass through the cavity, and just as more energy is lost, more time is taken at nonresonant frequencies. Figure 4.10-5 shows a plot of time *versus* frequency and shows that, as the frequency changes further away from f_0, the signal takes more time to pass through the cavity. This is termed *group delay difference*, or *group delay* for short. Excessive group delay within the pass band can result in signal distortion. An ideal filter would have no group delay difference; that is, the curve would be a horizontal line, at least across the pass band. IBOC requires the full channel bandwidth, so it is important to limit group delay across the full channel.

Impedance

Current flow in any RF circuit must overcome resistance, capacitive reactance, and inductive reactance. The vector sum of these is termed *impedance*. Because this is a complex function, it may only be fully represented on a complex diagram known as a Smith chart. While a full discussion of Smith charts is beyond the scope of this chapter, a few features of the chart will aid in the understanding of filter performance and tuning. Figure 4.10-6 shows an expanded Smith chart. The center horizontal axis (A) represents a state of pure resistance. In a properly tuned system, this state exists at f_0, the resonant frequency, where the inductive and capacitive components cancel each other out. The center point on line A represents a resistive value of 50 ohms (50 Ω). To the left, the resistive value decreases, approaching a short circuit (0 Ω); to the right, it increases, approaching an open circuit (infinite Ω).

The region above the horizontal axis represents a state when the vector sum of the circuit is inductive in nature; conversely, below the axis the circuit is capacitive. Any point on the chart may be expressed as $R \pm jX$, where R is the resistive component, j is a constant, and X represents the magnitude of the net inductive or capacitive component of the circuit. A circle drawn around the center point would be a locus of points of equal voltage standing wave ratio (VSWR); for example, circle B in Figure 4.10-6 represents a VSWR of 1.1:1. Points within the circle then represent conditions of VSWR less than 1.1:1. Our ideal filter would be plotted as a dot at the center of the chart, representing a pure 50 Ω resistance throughout the pass band, with no capacitive or inductive components.

Figure 4.10-7 shows the Smith chart of a single cavity bandpass filter. At the resonant frequency f_0, the impedance is pure resistance and 50 Ω at the chart center. As the frequency changes away from f_0, the inductive and capacitive components grow, forming a vertical arc. The slight offset to the right of chart center represents insertion loss. The small circles (beads) on the curve indicate the pass band. A pass band of ±200 kHz is generally considered acceptable for a filter system.

Figure 4.10-8 is an impedance diagram showing manipulation of the coupling through the cavity. Curve A (truncated for emphasis) is a cavity with the

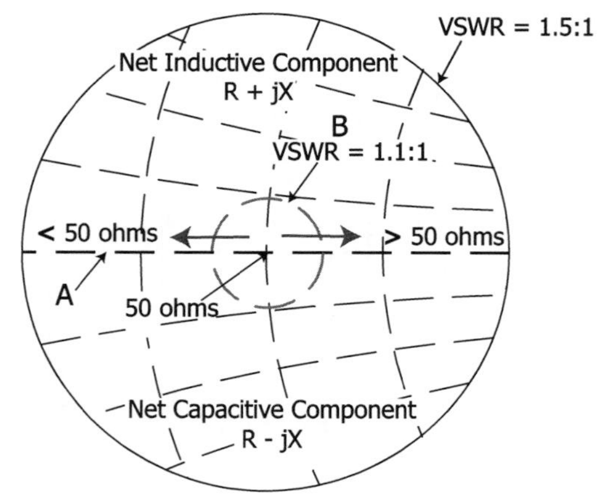

FIGURE 4.10-6 Smith chart components.

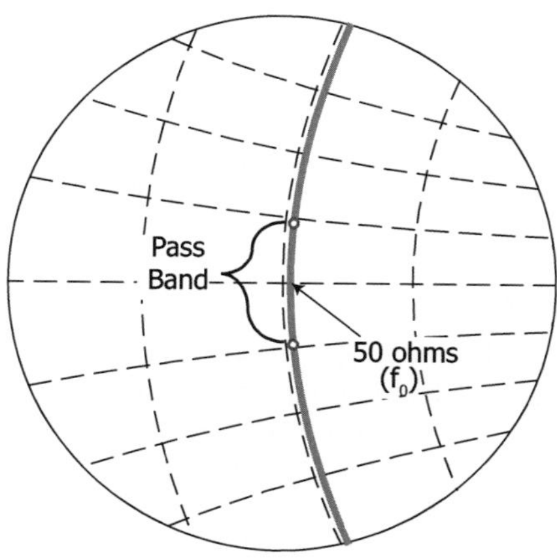

FIGURE 4.10-7 Smith chart for a single cavity filter.

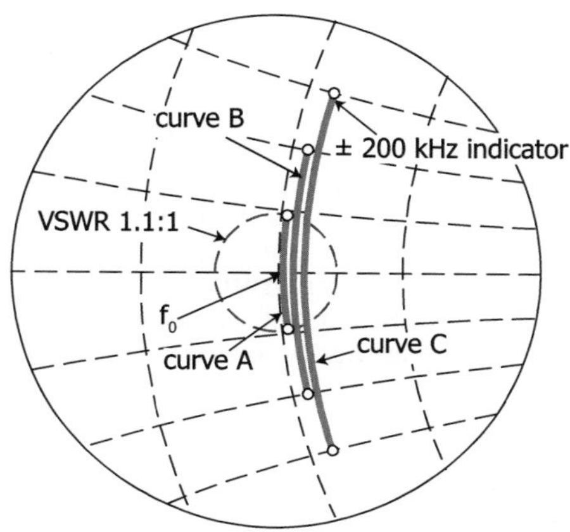

FIGURE 4.10-8 Tuning of a single bandpass cavity.

loops adjusted for maximum coupling. This curve almost passes through the center of the chart (R = 50 Ω) due to insertion loss, and the entire 200 kHz pass band (between the beads) is within the circle representing VSWR = 1.1:1. As the coupling is adjusted to achieve increased isolation (curve B) and extended for still more isolation (curve C), the center of the curve moves into the R > 50 Ω area to the right of chart center, an indication of greater insertion loss. In addition, the beads representing ±200 kHz move outward, well outside the 1.1:1 VSWR area. Again, this illustrates the tradeoff between increased isolation and increased insertion loss.

Physical Size

The physical size of the cavity is established for the purpose of power capacity and electrical performance. The cavity is then tuned to optimize the performance for a given application.

Tuning Compromises

Note that an ideal filter would have a 50 Ω impedance (unity VSWR), no insertion loss, no group delay, and flat frequency response within the pass band. Figures 4.10-3, 4.10-4, and 4.10-5 show that actual cavity type filters do not meet these ideal parameters. It is important to remember that filters are always designed for best real world overall performance and that, at times, a little performance must be sacrificed in each parameter to improve overall performance. To obtain increased isolation to meet today's standards, the number of cavities in a filter system must be increased, but this occurs at the cost of increasing group delay and insertion loss. In a four cavity system, the group delay curve becomes so steep as to be unacceptable (Figure 4.10-9); therefore, the tuning is modified to decrease group delay to an acceptable level, as shown in Figure 4.10-10. This adds some minor distortion to the frequency response (Figure 4.10-11). Although

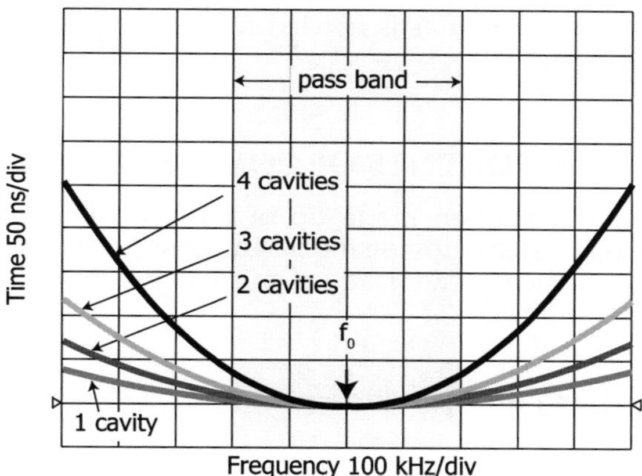

FIGURE 4.10-9 Group delay for one, two, three, and four cavity filters.

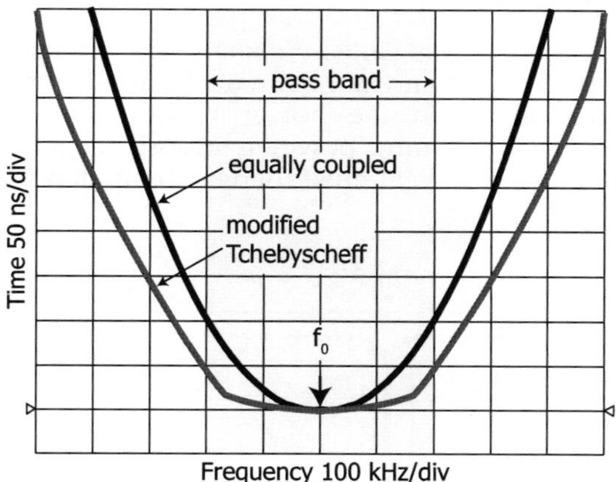

FIGURE 4.10-10 Group delay for a four cavity filter tuned for group delay.

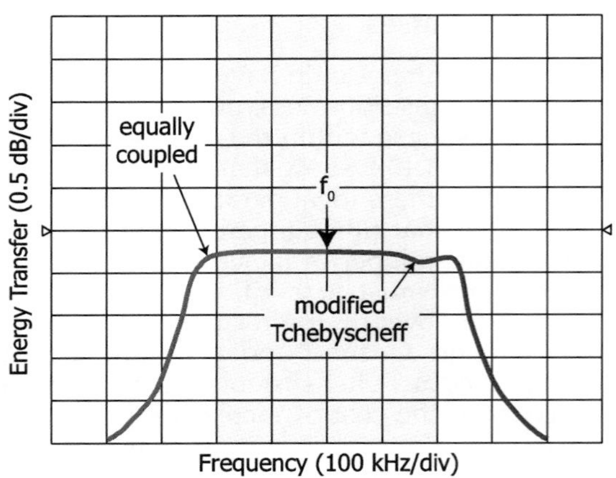

FIGURE 4.10-11 Frequency response for a four cavity filter tuned for group delay.

none of the individual parameters is optimized by itself, the overall performance of the filter is optimized and acceptable. A four-cavity bandpass filter is as large a filter system as is needed for most high isolation applications.

COMPONENTS OF COMBINERS

Combiners are often made up of multiple elements which as a system provide the desired results. Separating the elements can lead to mismatches or other incompatibilities.

Tee or Star-Point Junction

A tee junction, shown in Figure 4.10-12, is a coaxial component that allows two RF signals to flow into a common path; a star-point junction is a tee with more than two input paths. This basic coaxial component is one of the building blocks of a branched combiner.

Resistive Load

Resistive loads, often called *dummy* loads, are used in many applications and can be manufactured in many sizes depending on the power requirement. In a dummy load, incoming power is absorbed and converted to heat. The heat must then be dissipated to the surrounding air, so the power rating of a dummy load is determined by the size of the resistor and the amount of heat that can be dissipated before the resistor overheats and fails. If enough resistors can be chained together with enough cooling, they can dissipate almost an unlimited amount of RF energy.

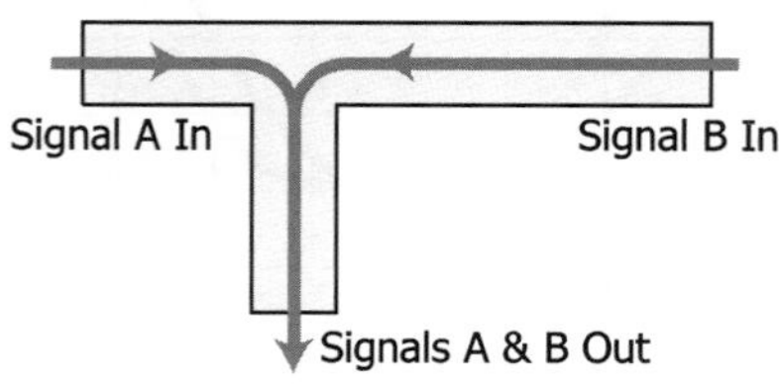

FIGURE 4.10-12 Tee junction.

Quadrature Hybrid

The heart of the modern balanced combiner system is the quadrature hybrid (usually just referred to as *hybrid*). A hybrid is a complex broadband device that has the ability to operate in various modes, either singly or simultaneously. The detailed mathematical explanation of a hybrid is beyond the scope of this work; this section covers only the use of hybrids in combining systems.

Hybrid as Signal Splitter

In Figure 4.10-13, the hybrid is acting as a power splitter and phase shifter. When an RF signal is applied to port 1 (TX1), the hybrid splits the signal in half, and the phase of port 4's output is delayed with respect to port 3's output by 90°. Port 2 is called the *isolated port*, because the isolation between ports 1 and 2 is approximately 35 dB and is usually terminated with a 50 Ω resistive load. If two inputs are required, port 2 can be used as an additional transmitter input (TX2). In this configuration, the output power levels are the same as above, but the phases are reversed (Figure 4.10-14).

Hybrid as Signal Combiner

A second mode uses a hybrid in reverse, for combining transmitters, as shown in Figure 4.10-15. If two equal RF signals, with the proper phasing, are introduced at ports 1 and 2, the combined signal exits the hybrid through port 4. If the phase of the two input signals is reversed, the signal will exit the hybrid through port 3. Again, the isolated port is usually terminated with a

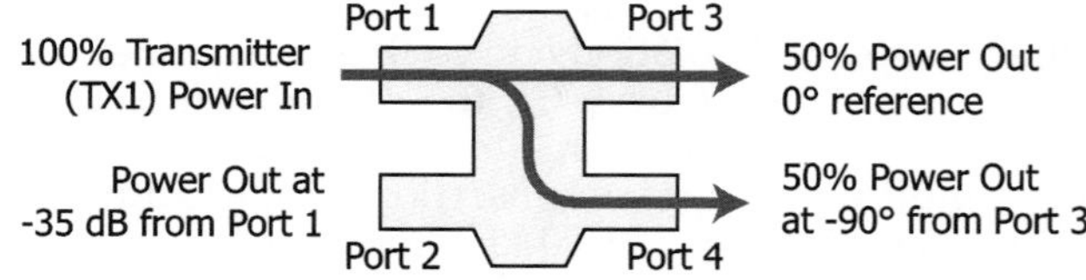

FIGURE 4.10-13 Hybrid as a signal splitter.

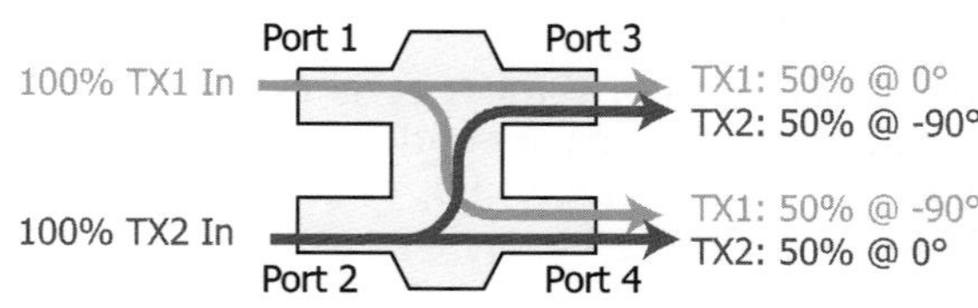

FIGURE 4.10-14 Hybrid splitting two input signals.

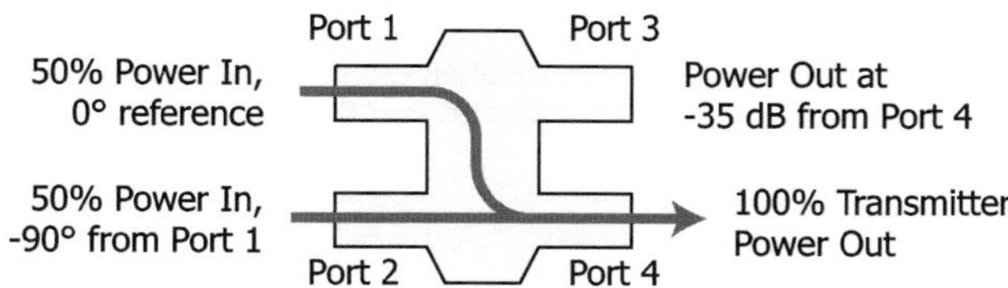

FIGURE 4.10-15 Hybrid as a signal combiner.

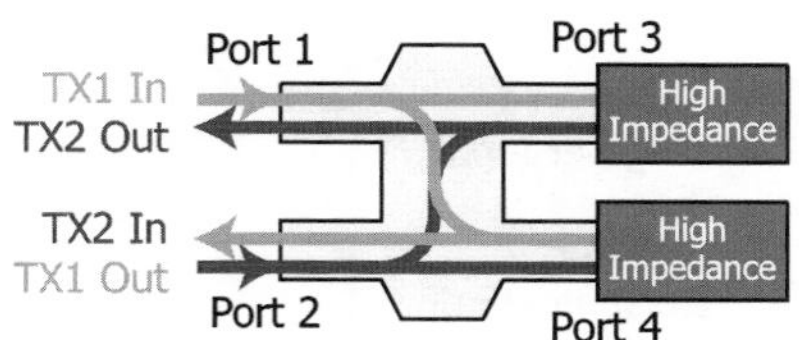

FIGURE 4.10-17 Hybrid as a signal reflector.

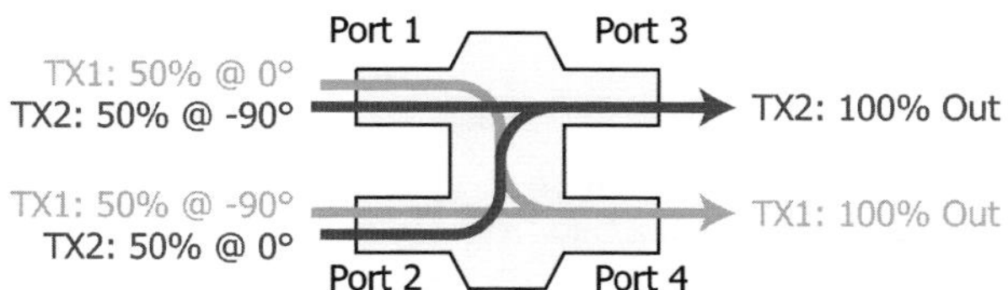

FIGURE 4.10-16 Hybrid combining two signals.

resistive load. The hybrid can be used to combine two incoming signals in the exact reverse of Figure 4.10-14. If two incoming signals with the correct phasing are present at port 1 and two at port 2, as shown in Figure 4.10-16, then port 4 is an output for one combined signal TX1 and port 3 is the output for the other combined signal TX2.

Hybrid as Signal Reflector

The third hybrid mode of operation is the *reflected mode* (Figure 4.10-17). When two identical devices with high impedance, such as bandpass filters tuned to another frequency or band reject filters tuned to the incoming frequency, are attached to ports 3 and 4 of the hybrid, the signal entering at port 1 is reflected and exits the hybrid through port 2. Again, the hybrid is symmetrical; if a second signal enters port 2 it will be reflected and exit port 1. The characteristics of this mode make the hybrid useful in conjunction with other hybrids and cavities in combining systems. A hybrid can operate in all three modes simultaneously.

With power moving in many different directions at once, it is imperative that the hybrid have good electrical characteristics and that it be as balanced and symmetrical as possible, both mechanically and electrically. Balanced and symmetrical hybrids show the same electrical characteristics through each port. The more identical the electrical paths through these ports are, the greater the isolation that can be achieved and the lower the VSWR at each port. Figure 4.10-18 shows the performance curve of a well-balanced and symmetrical hybrid.

Hybrid Ring

When two hybrids are used in a ring configuration (Figure 4.10-19) to both split and combine a single input signal, virtually 100% of the signal exits the ring through the hybrid leg opposite the input. In a balanced and symmetrical hybrid ring, if the signal is introduced at port 1, the outgoing signal will be at port 8, with isolation at ports 2 and 7. Likewise, if the signal is introduced at port 7, it will emerge at port 2, with isolation at ports 1 and 8, and if it is introduced at port 8, it will emerge at port 1, with isolation at ports 2 and 7. Energy can flow in all four directions at the same time without the signals mixing (Figure 4.10-20). The multiple flow paths of the hybrid ring make it the backbone of the balanced combiner.

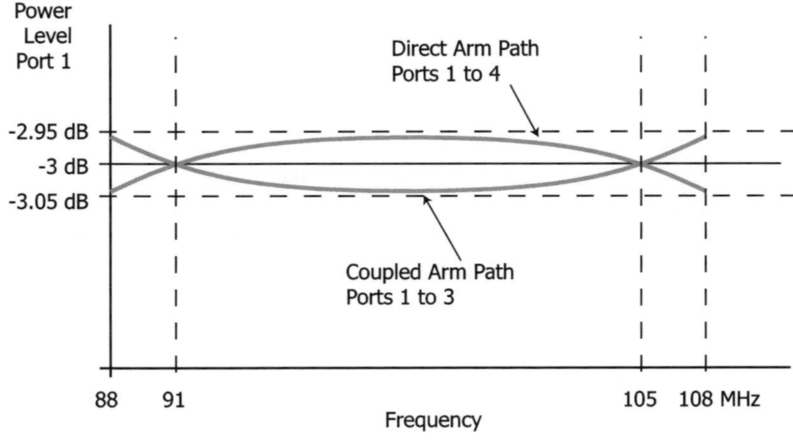

FIGURE 4.10-18 Hybrid frequency response.

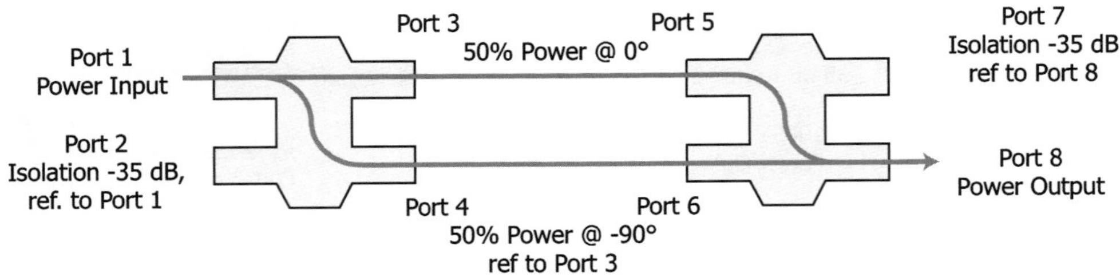

FIGURE 4.10-19 Hybrid ring.

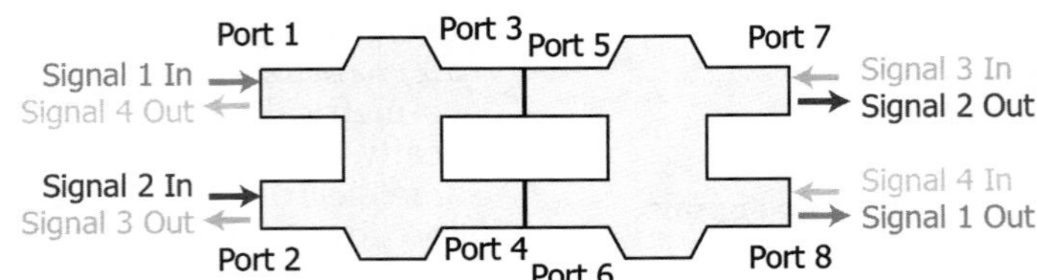

FIGURE 4.10-20 Hybrid ring multiple flow paths.

FILTERS

Filters sort RF frequencies, attenuating some while allowing others to pass readily. Depending on the design, a filter may either attenuate (band reject type) or pass (bandpass type) a relatively narrow bandwidth.

Band Reject or Notch Filter

There are several ways to design a band reject or notch filter (Figure 4.10-21), but they all accomplish the same purpose. In one form, a cavity with only an input coupling loop is mounted off the transmission line by means of a matched tee. This provides a path that removes the tuned frequency from the system, allowing other frequencies to pass with minimum loss. Other designs employ some form of capacitive coupling into the cavity.

Multiple Notch Cavities

The frequency response of a typical notch cavity is shown in Figure 4.10-22. When more isolation is required, two notch cavities are coupled in sequence. The resonant frequencies of the cavities may be identical, yielding a response curve with a very deep narrowband notch, as shown in Figure 4.10-23, or they may intentionally be staggered to give a broader notch response, as shown in Figure 4.10-24.

Performance and Limitations

The impedance plot of a typical notch cavity is shown in Figure 4.10-25. When a single notch cavity is used, an impedance matching network is added to the filter to improve the impedance bandwidth. The group delay plot of a notch cavity (Figure 4.10-26) distorts signal quality. No practical device has been marketed to equalize the group delay of a notch-cavity system;

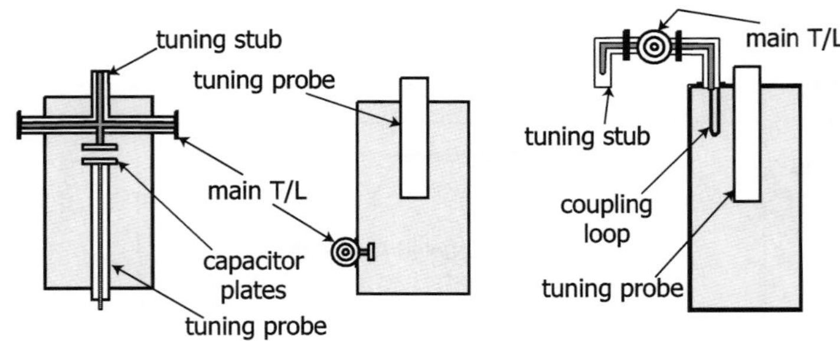

FIGURE 4.10-21 Notch filter configurations.

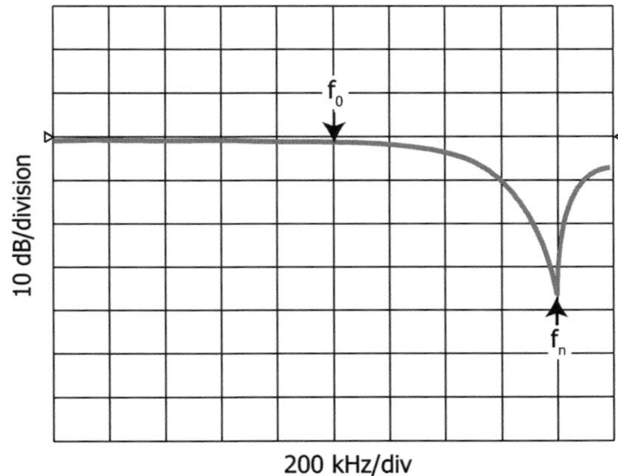

FIGURE 4.10-22 Frequency response for a single notch cavity.

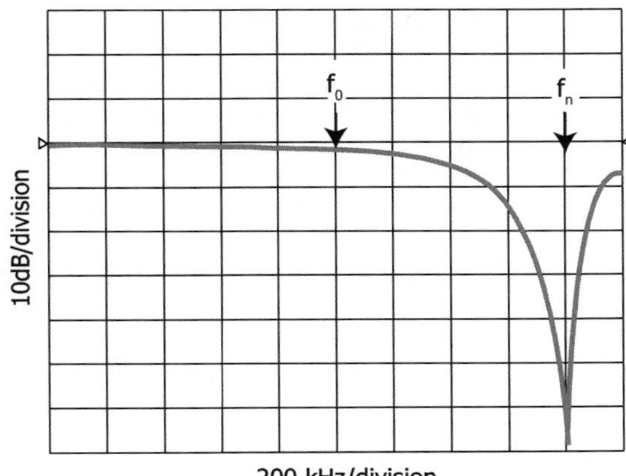

FIGURE 4.10-23 Frequency response for dual notch cavities.

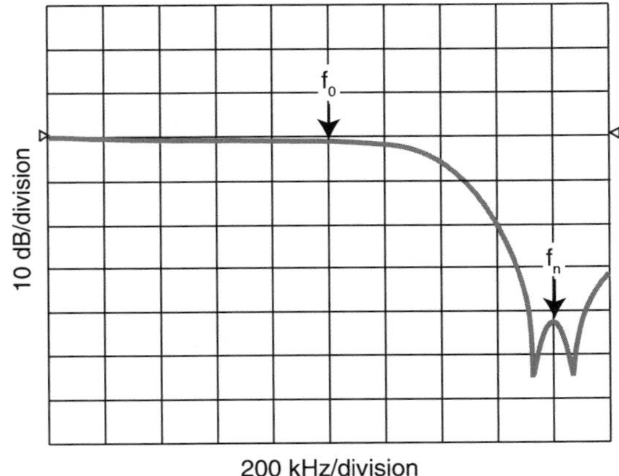

FIGURE 4.10-24 Frequency response for staggered dual notch cavities.

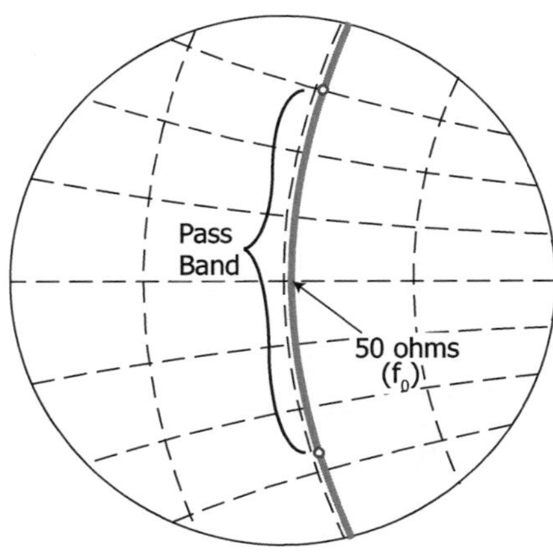

FIGURE 4.10-25 Impedance plot for a single notch cavity.

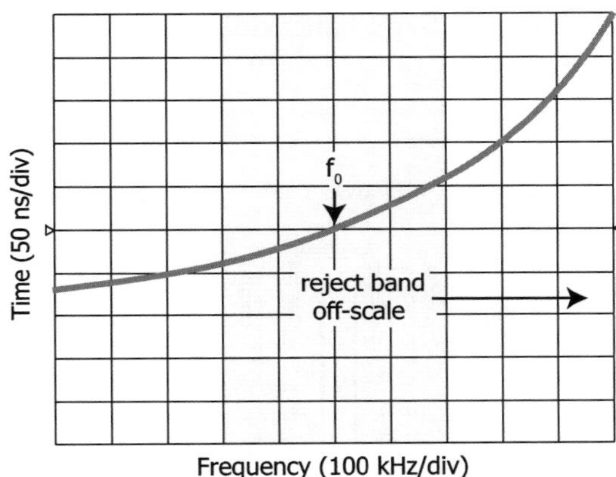

FIGURE 4.10-26 Group delay for a single notch cavity

however, this has not been a major issue, because about the same time (mid-1980s) that group delay was recognized as an issue the industry was turning toward bandpass filtering, anyway.

Bandpass Filter

Figure 4.10-27 shows the basic mechanical configuration of a bandpass filter cavity. When RF energy is applied to the input coupling loop, the loop inductively couples the energy into the cavity. Energy is transferred through the cavity and inductively coupled to the output coupling loop. The resonant frequency of the cavity is tuned by adjusting the tuning probe. The transfer of energy is maximized at the resonant frequency; therefore, a filter of one or more identical cavities can be used to attenuate frequencies other than the resonant frequency.

Multiple Bandpass Cavities

Generally, a filter system is considered adequate if it provides a VSWR of 1.1:1 over a frequency range of

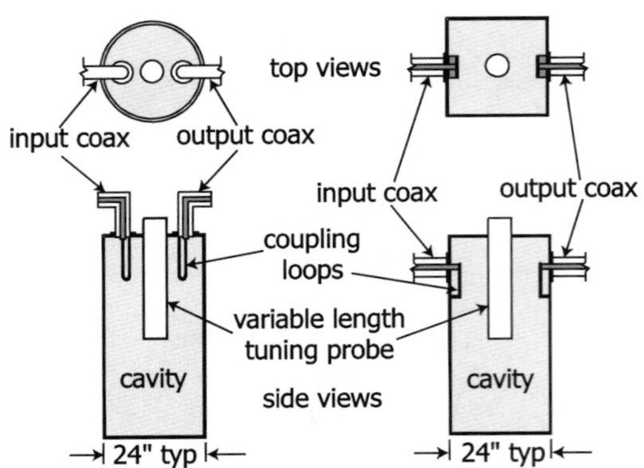

FIGURE 4.10-27 Bandpass cavity configurations.

±200 kHz. This is termed the *bandwidth* of the filter system. In most cases, a single bandpass cavity will not yield this much bandwidth. To increase the isolation and increase VSWR bandwidth, a second cavity may be added to the first, as shown in Figure 4.10-28.

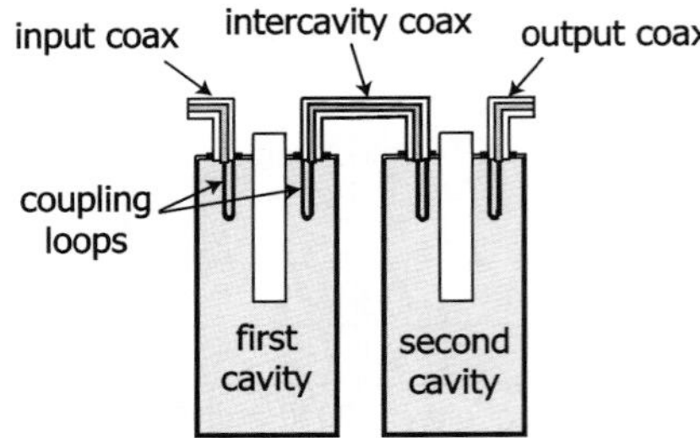

FIGURE 4.10-28 Two cavity bandpass filter.

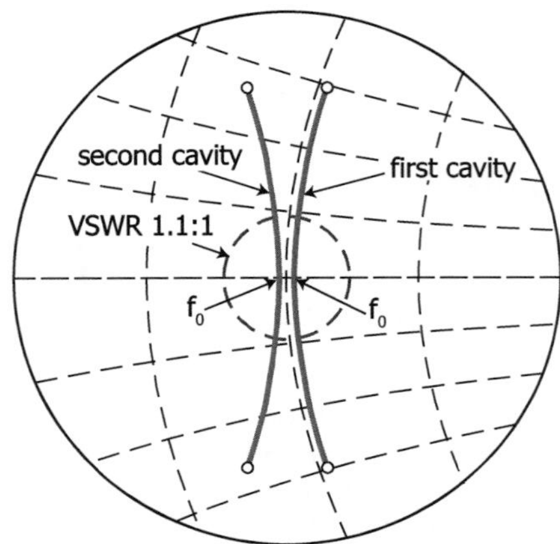

FIGURE 4.10-29 Superimposed impedance curves of a two cavity bandpass filter.

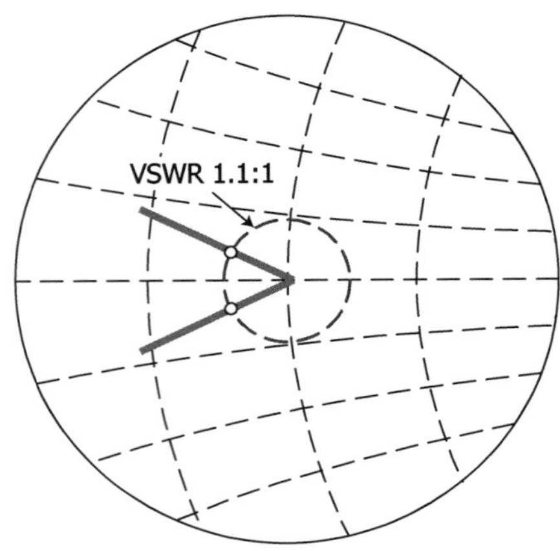

FIGURE 4.10-30 Impedance plot for a two cavity bandpass filter.

When two identical cavities are coupled a quarter-wave apart, the impedances superimpose themselves, as shown in Figure 4.10-29. Note that Figure 4.10-29 shows two Smith charts superimposed 180° apart. The small circles (beads) representing the ±200 kHz bandwidth fall on a VSWR circle of about 1.3:1. When their impedances are added together mathematically, due to phase cancellation, the VSWR bandwidth improves to about 1.1:1 (Figure 4.10-30). Curve A of Figure 4.10-31 shows the frequency response of the two cavity filter.

When still more isolation is required, more cavities can be added. Figure 4.10-31 shows the frequency responses of three, four, and five cavity systems. As more cavities are added, the curve becomes more square and flatter across the pass band, with a sharper roll-off; that is, it begins to approach our ideal filter shown in Figure 4.10-2. Consider, however, that the five-cavity filter does not show a great improvement over the four cavity filter, and, in fact,

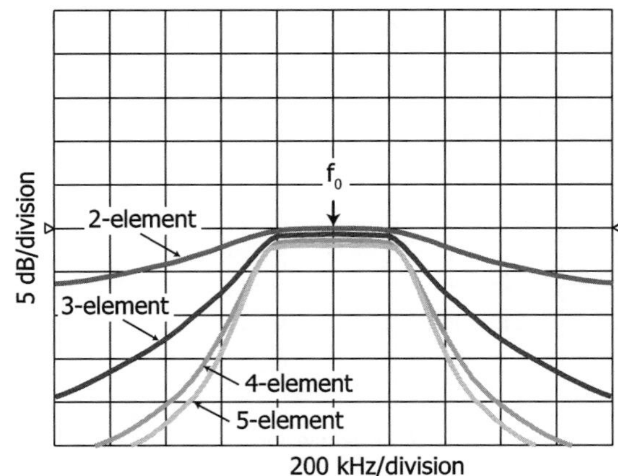

FIGURE 4.10-31 Frequency response for two, three, four, and five cavity bandpass filters.

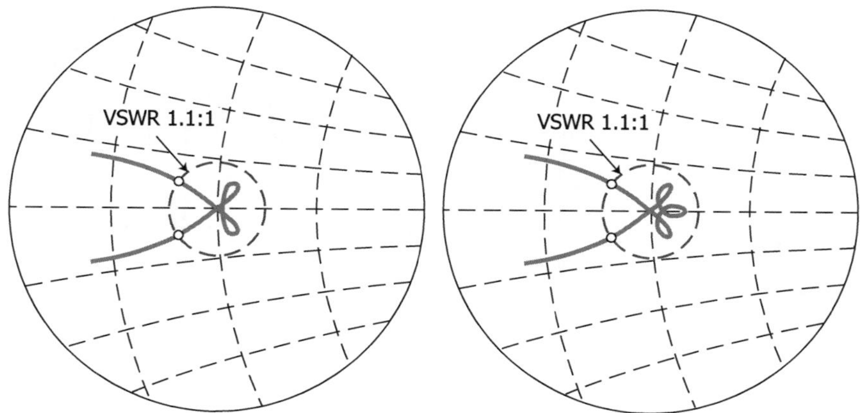

FIGURE 4.10-32 Impedance plots for three and four cavity equally coupled bandpass filters.

the four cavity filter represents the best compromise among isolation, insertion loss, and physical size for close-spaced stations transmitted through a combining system. Figure 4.10-32 shows Smith charts for a three cavity system and a four cavity system. Note that the beads indicating the ±200 kHz points are within the 1.1:1 VSWR circle.

Mechanical Constraints

To obtain the optimum mathematical cancellation shown in Figure 4.10-30, the cavities must be spaced at one-quarter electrical wavelength. As the frequency increases, the electrical wavelength decreases; therefore, the physical length of the intercavity coax must be shortened. At the higher frequencies of the FM band, the large cavities used for high power applications are difficult to link together, because the cavities themselves approach one-quarter electrical wavelength. As a result, when the intercavity coax is added, the electrical spacing is longer than one-quarter wavelength. In this case, the coupling loops must be manipulated to compensate for the extra length so the impedance bandwidth of the cavities is maintained.

Common-Wall Coupling

The spacing problem can be prevented by building the cavities contiguous to each other and coupling them through a tuned opening in the wall between them, as shown in Figures 4.10-33 and 4.10-34. The one-quarter electrical wavelength spacing is maintained by the coupled fields between the cavities.

Coupling Options

Although common-wall coupling can be accomplished using a true iris placed away from the top of the cavity (Figure 4.10-34), the size of the iris is difficult to control and adjust. Another method, shown in Figure 4.10-35, of coupling energy from one cavity to the next is neither an iris nor a slot but a trapezoidal opening designed so no adjustments are needed to couple the energy from one filter to the next across the

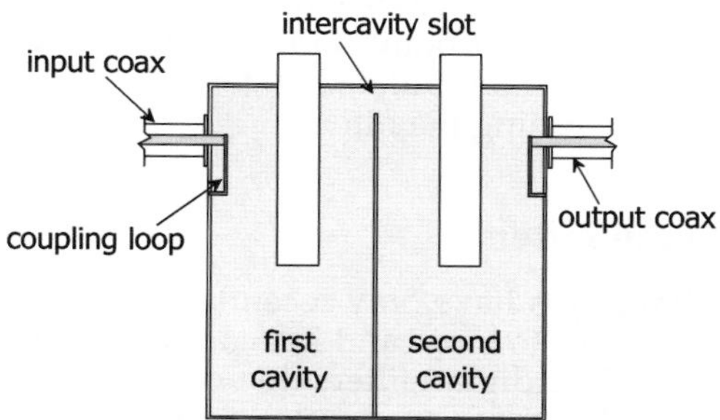

FIGURE 4.10-33 Slot-coupled cavities.

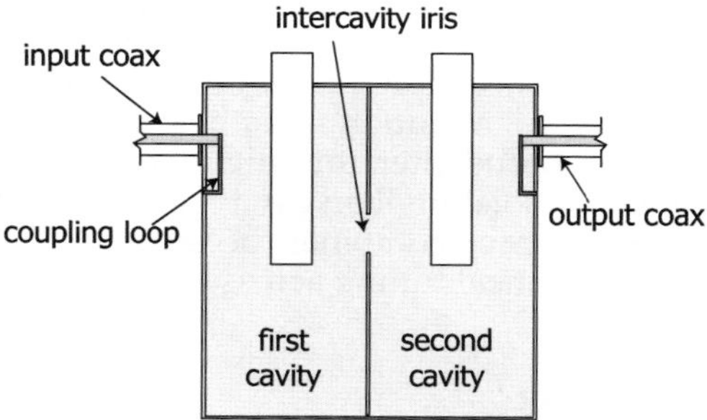

FIGURE 4.10-34 Iris-coupled cavities.

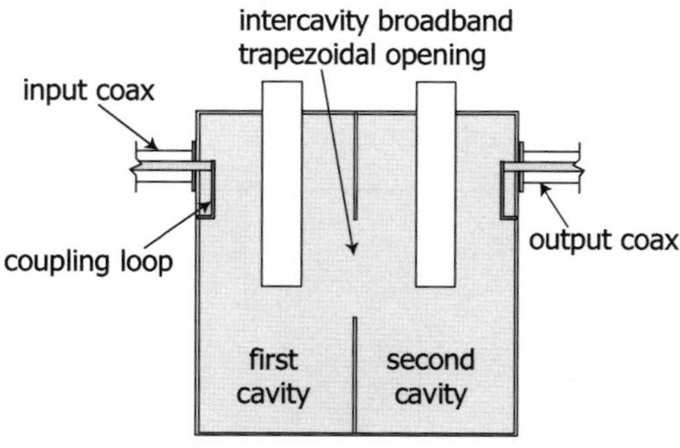

FIGURE 4.10-35 Cavities coupled by broadband trapezoidal opening.

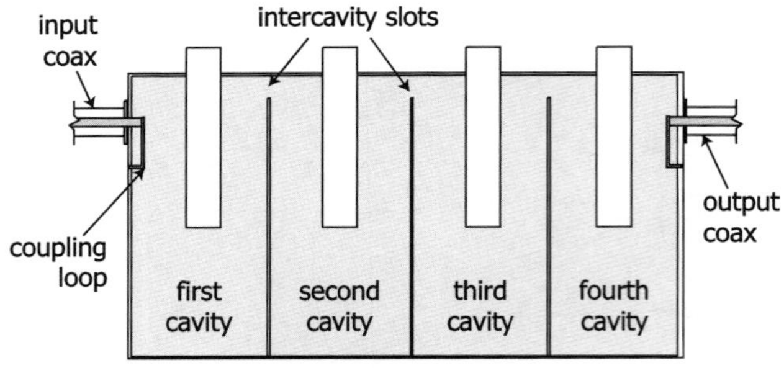

FIGURE 4.10-36 Four cavity slot-coupled filter.

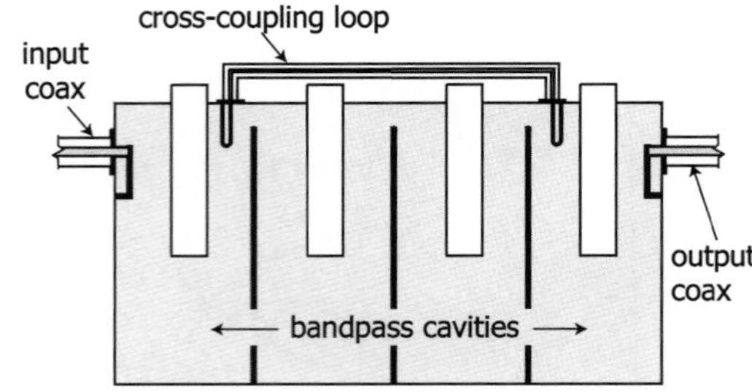

FIGURE 4.10-38 Filter with cross-coupling.

FM band. Like most broadband-tuned networks, however, it is difficult to optimize a filter set at any one particular frequency. Perhaps the best configuration, shown in Figures 4.10-33 and 4.10-36, is a slot at the very top end of the cavities where the magnetic fields are at their strongest point and the size and shape of the slot can be manipulated externally for ease of adjustment in tuning the filters.

Interdigital Filters

Interdigital filters have only recently been introduced as an alternative to loop- and iris-coupled filters at FM frequencies. Interdigital filters do not employ individual cavities that must be coupled together. As shown in Figure 4.10-37, the energy is directly coupled to the input and output tuning probes. Parts counts are minimized and interdigital filters are significantly smaller than iris-coupled filters. Because of their smaller size, interdigital filters have higher insertion losses than either loop- or iris-coupled filters of the same power rating, and careful attention must be paid to the thermal properties of the filter. Interdigital filters have better out-of-band isolation than cavity-style systems and are ideal for balanced combiners because of the ease of maintaining identical tuning across the channel.

Cross Coupling

If a transmission line segment is added between the first and last bandpass sections (Figure 4.10-38), a parallel transmission channel is created. This line segment is then tuned to achieve specific phase and amplitude characteristics, so unwanted frequencies at both ends of the filter cancel each other out. It therefore acts as a band reject component, creating notches at the edges of the pass band (Figure 4.10-39).

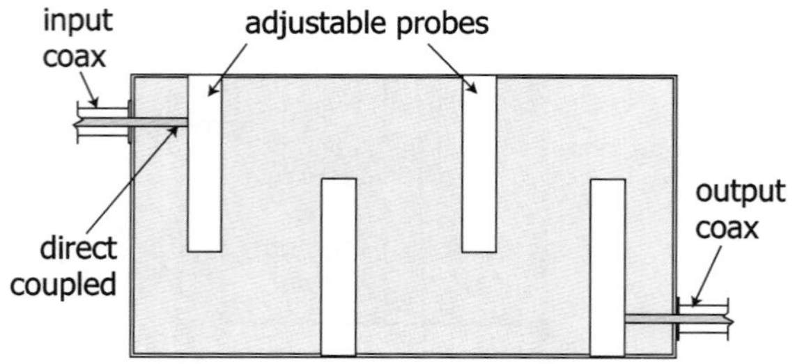

FIGURE 4.10-37 Four-pole interdigital filter.

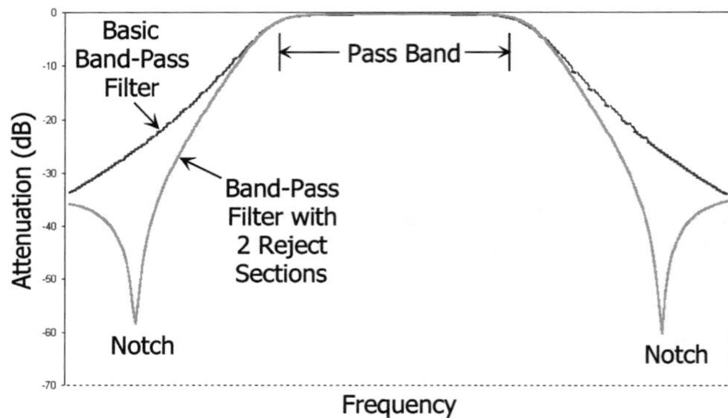

FIGURE 4.10-39 Frequency response for a filter with cross-coupling loops.

ISOLATORS

An isolator is comprised of a circulator and a load. Signals move between legs in only one circular direction, giving the device its name. Although it is theoretically possible for the signal originating at any given leg to reach any other leg, this is prevented by the existence of one high-impedance leg, which traps energy trying to move across it and shunts it off to a dummy load. Thus, it is possible to configure the circulator to allow the signal from the transmitter to flow freely out the adjacent antenna leg, but energy returning through the antenna leg is interrupted before it can reach the transmitter leg. This is shown in Figure 4.10-40. The signal from the transmitter is fed into the isolator at leg 1. It flows out leg 2 on the transmission line toward the antenna. At the same time, any signal from the antenna enters the circulator at leg 2 and is directed to the dummy load at leg 3. The actual isolation value is a function of the match of the dummy load and is typically 26 dB. This ability of isolators to divert on-frequency signals headed in the wrong direction is key to

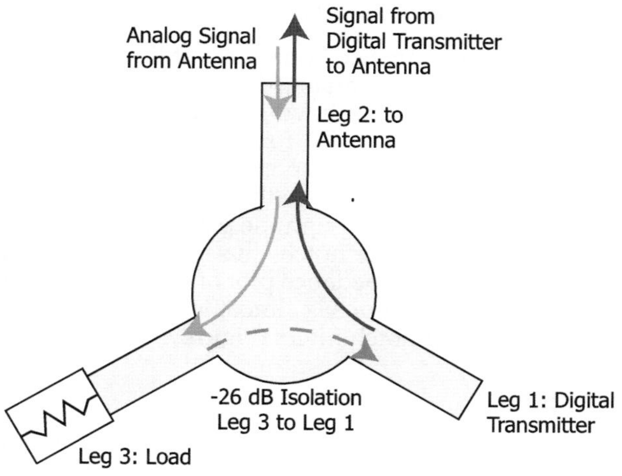

FIGURE 4.10-40 Isolator.

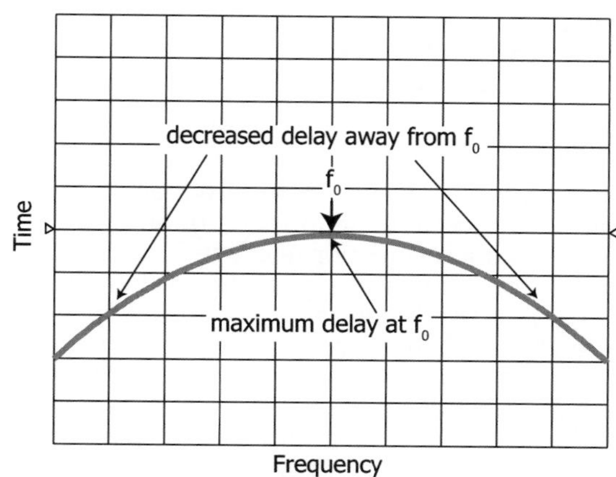

FIGURE 4.10-42 Group delay of group delay equalizer.

a number of modern combining strategies that employ separate digital and analog transmission paths and where the combining method does not afford at least 35 dB of isolation between the digital and analog transmitters.

DIRECTIONAL COUPLERS

Precision directional couplers are commonly found on each broadband output of a combiner system and are convenient ports for taking FCC-required test measurements, enabling diagnostics, and serving as a port for any protection and monitoring system the combiner may employ. Directional couplers located on the inputs to each module further enhance the versatility of the system.

GROUP DELAY EQUALIZER

A group delay equalizer consists of a quadrature hybrid and two identical bandpass filters that have only one coupling loop, so the energy is coupled in and out of the cavity by the same loop (Figure 4.10-41). The tuned frequency is delayed for longer than the off-resonant frequencies (Figure 4.10-42).

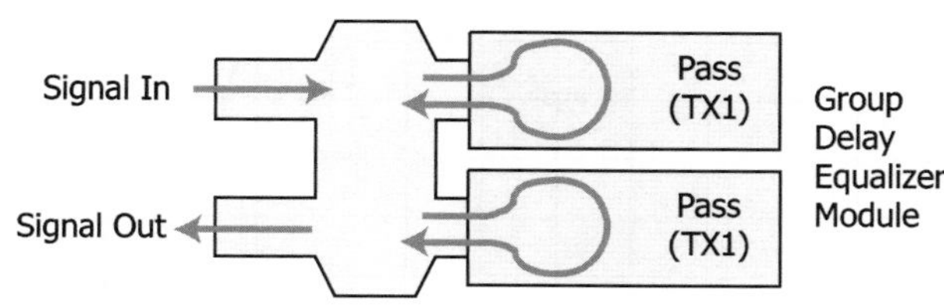

FIGURE 4.10-41 Group delay equalizer.

TYPES OF COMBINERS

In addition to using hybrids as combiners, a series of specialty devices are employed for combining multiple RF signals into a common path. These include the branched or star-point, balanced, notch filter, and band pass filter combiners.

Branched or Star-Point Combiners

A branched combiner is a simple combination of a tee junction and the required number of filters to ensure a sufficient amount of isolation to prevent spurs. For example, an FM branched combiner consisting of a three cavity bandpass filter in series with two band-reject cavities (Figure 4.10-43) may be used to provide the isolation required for two close-spaced frequencies 0.8 MHz apart. TX1 and TX2 are the signals from transmitters 1 and 2 as they enter the combiner. The signals pass through the notch and bandpass filters and arrive at the tee junction. The length of the coaxial line between each set of filters and the tee junction is adjusted to provide a very high impedance

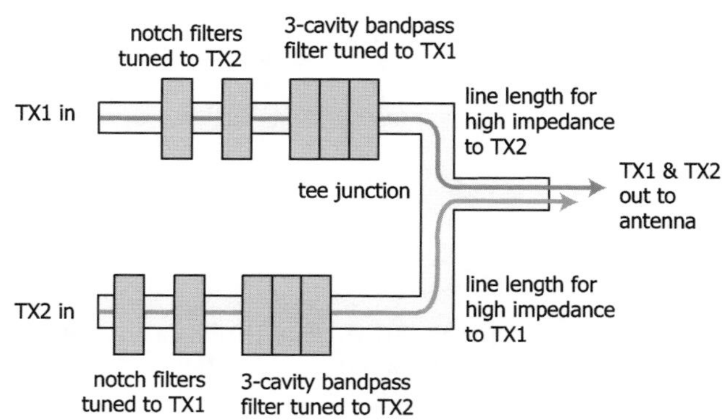

FIGURE 4.10-43 Branched combiner with notch cavities.

(approaching an open circuit) to the other frequency, so the power flow of each signal is through its own filter, out of the tee junction, and up to the antenna.

Performance

Refer again to Figure 4.10-30, the frequency response curve for a three cavity bandpass filter, and Figure 4.10-23, the frequency response curve for a two cavity, staggered-frequency band-reject filter. When these filters are used in combination, the resulting curve is as shown in Figure 4.10-44. Note that the insertion loss for the pass frequency f_1 is only about 0.25 dB and the isolation at the reject frequency f_2 is greater than 50 dB across the channel. The impedance plot, Figure 4.10-45, is likewise the combination of impedance plots for the same filter combination.

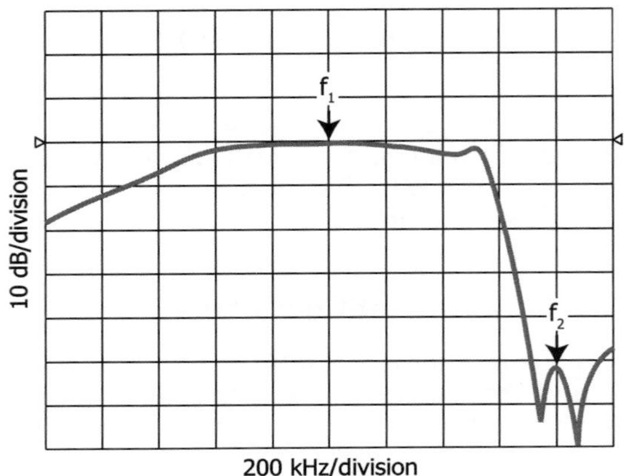

FIGURE 4.10-44 Frequency response of a three-cavity bandpass filter in series with two notch cavities.

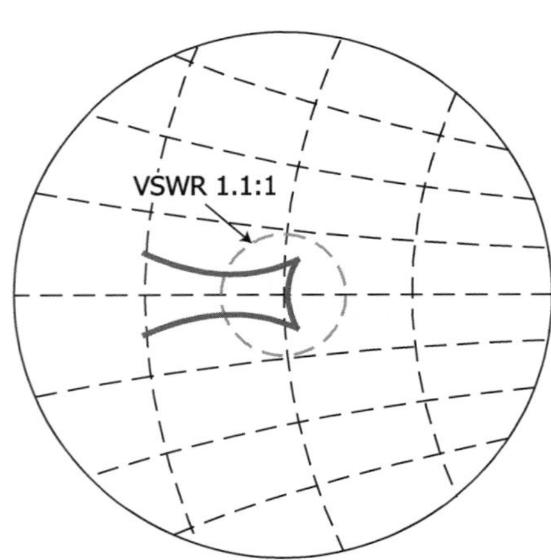

FIGURE 4.10-45 Impedance of a three cavity bandpass filter in series with two notch cavities.

Branched Combiners with Feedback Loops

Although many branched combiners still in operation use notch cavities for enhanced isolation, most modern branched combiners have gone to feedback loop technology (Figure 4.10-46) for this purpose. Figure 4.10-47 is the frequency response curve of the three cavity bandpass filter with feedback loops. Notice that the curve is smoother through the pass band and, even though it only has one notch, the isolation at f_2 still exceeds 50 dB. The impedance plot of a branched combiner with feedback loops is almost identical to that of the combiner with notch filters (Figure 4.10-45).

Limitations

A branched combiner is very efficient for a two station installation and has been used for as many as four stations, but a tee junction for more than four stations becomes impractically large, and adjusting the lengths of interconnecting coax becomes prohibitively complex as well. Also, a branched combiner cannot easily be expanded later to include more stations, although it

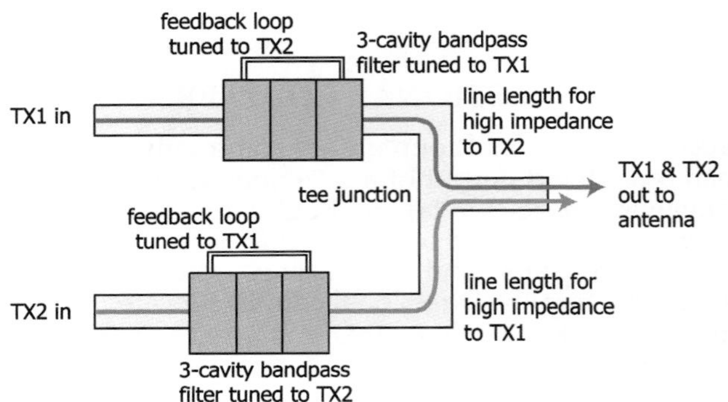

FIGURE 4.10-46 Branched combiner with feedback loops.

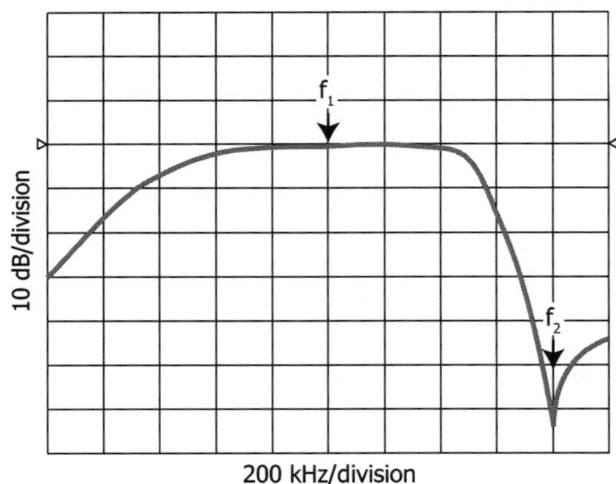

FIGURE 4.10-47 Frequency response of branched combiner with feedback loops.

can be expanded by integrating it with balanced combiner modules. To combine more than four stations, a balanced combiner becomes more practical and cost effective.

Balanced Combiners

The balanced combiner is based on a hybrid ring. Each leg of the ring contains an identical set of either bandpass or band reject filters, hence the term *balanced*. It is imperative that the filters of all modules be tuned to have as close to the same response characteristics as possible. The goal is to have the hybrids react identically to the filters. Small differences in electrical length through the hybrids quickly add up to an increased VSWR. For example, a phase difference of ±2° in the legs of a hybrid produces a VSWR of 1.07:1 (or a return loss of 29 dB). If that phase difference degrades to ±4°, the VSWR deteriorates to 1.15:1 (23 dB).

Most early balanced combiners used notch filters.

Notch Filter Balanced Combiners

In the notch-filter balanced combiner (Figure 4.10-48), both notch filters within the hybrid ring are tuned to reject TX1's frequency, which enters the combiner at port 1. That signal is reflected by the filters and exits at port 2. TX2 enters port 3, the broadband input port of the module, and passes through in the diagonal mode (shown in Figure 4.10-19) with minimal loss in the reject cavities.

Performance

The isolation of transmitter 2 from frequency TX1 is the sum of the hybrid ring isolation of 35 dB and the isolation of the notch cavities and can approach 35 to 40 dB. However, the isolation of transmitter 1 from frequency TX2 is only that of the hybrid ring—about 35 dB; therefore, additional filtering, either bandpass or band reject, is required to ensure that no spurs are generated within transmitter 1. This added filter is shown in Figure 4.10-49.

External Bandpass Filtering

A better way to reject multiple unwanted frequencies, of course, is to use a bandpass filter tuned to the desired frequency. For example, Figure 4.10-50 shows

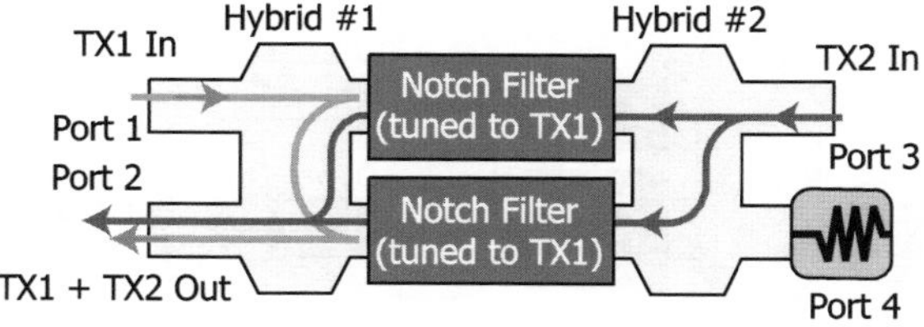

FIGURE 4.10-48 Two station, one module notch filter balanced combiner.

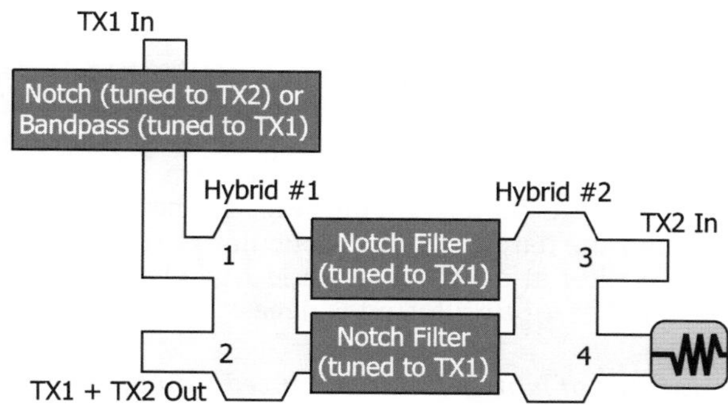

FIGURE 4.10-49 Two station, one module notch filter balanced combiner with input filter at TX1.

a five station, four module combiner. In this example, each input filter is a bandpass filter tuned to the frequency of that input. If reject filters were to be used at the various inputs, each input would have to filter all the frequencies previously introduced; therefore, port 3 of module 2 would have to contain two notch filters; port 1 of module 3, three filters; and port 3 of module 4, four filters. This proliferation is avoided by the use of input bandpass filters.

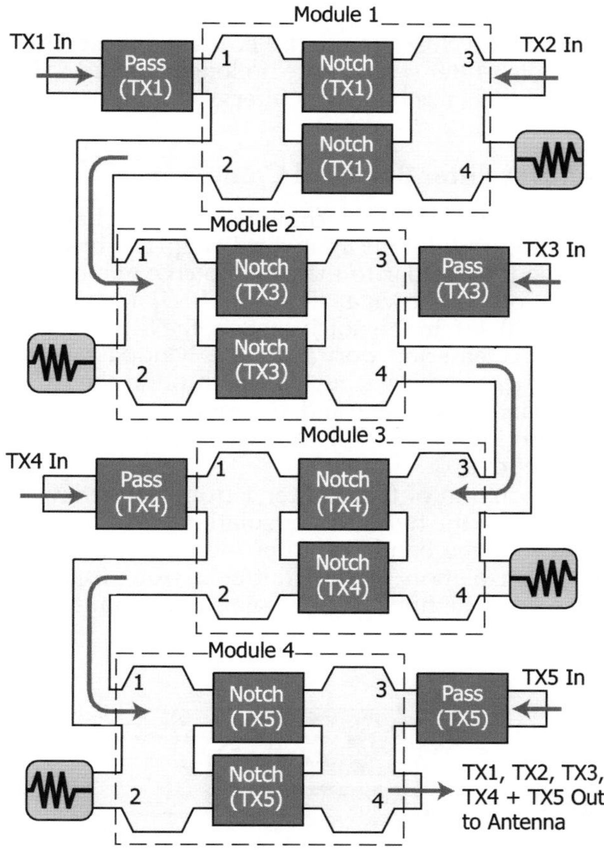

FIGURE 4.10-50 Five station notch balanced combiner with input bandpass filters.

Emergency Input Port

In some cases, instead of having a station located at port 3 of module 1, that port is terminated in a 50 Ω load and can be used as an emergency input for any station in the system. Providing an extra port in this way allows a damaged module to be bypassed. Because of the nature of that particular port, as long as the input filter at port 1 of module 1 is a bandpass filter, no further input filtering is necessary.

Limitations of Notch-Filter Balanced Combiners

A problem with using notch filters within the hybrid rings is that if the two filters in any module are not identically tuned, an imbalance occurs within the hybrid ring, thus reducing the isolation to a point where a spur can be generated within a transmitter. Once a spur has been generated, there are no filters within the system to reject that spur because the filters are tuned only to the expected frequencies; therefore, the spur is broadcast. A second disadvantage of using internal notch filters is that each module in turn has to conduct the accumulated power of all the previous modules. For a multiple high power system each module must be larger than the previous one, and the power rating of the system is limited by the size of the final module. Third, notch-filter combiners are impractically narrowband in nature for today's wideband IBOC channels, especially when the frequencies combined are closely spaced. Because of these limitations, notch-filter systems are no longer used. Modern FM combiners use bandpass filters.

Bandpass Filter Balanced Combiners

In a bandpass balanced combiner system, bandpass filters are used within the hybrid ring. The basic system layout is similar to that of a notch combiner. The power flow is shown in Figure 4.10-51 (compare to Figure 4.10-48). In the notch system, the filters rejected signal TX1 entering port 1. In the bandpass system TX1 also enters port 1 but passes through the hybrid ring's bandpass filters and out port 4, while signal TX2, entering at port 3, is reflected by the filters and exits at port 4.

The isolation of transmitter 1 from frequency TX2 is the sum of the hybrid ring isolation (35 dB) and the isolation of the bandpass filter (about 25 dB). However, the isolation of transmitter 2 from frequency TX1 is only the hybrid ring isolation of about 35 dB;

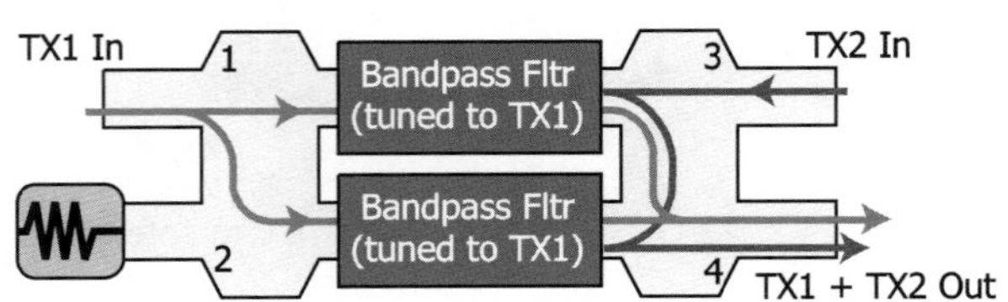

FIGURE 4.10-51 Two station, one module bandpass filter balanced combiner.

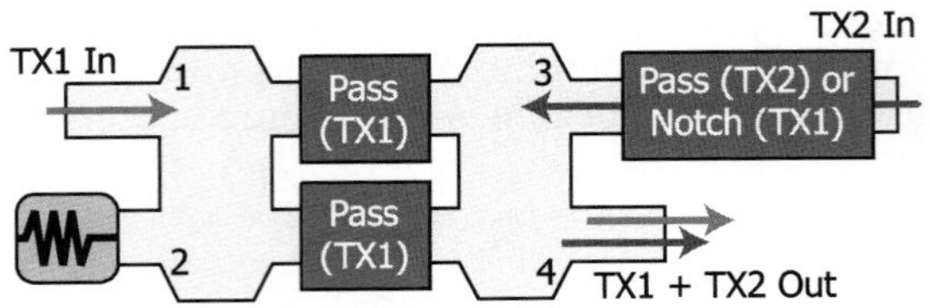

FIGURE 4.10-52 Two station, one module bandpass filter balanced combiner with input filter at TX2.

therefore, an additional filter must be added between transmitter 2 and its input port (Figure 4.10-52), similarly to the single module notch filter balanced combiner shown in Figure 4.10-49. Alternatively, a second module may be added to port 4 of module 1, and port 3 terminated in 50 Ω (and available as an emergency input port). Signal TX2 is then introduced at port 1 of module 2, as shown in Figure 4.10-53.

No input filter is necessary now for TX2, because it is isolated by the bandpass filters in module 2. The emergency input port now sees both frequencies TX1 and TX2, reduced 35 dB below each transmitter's power level. A multiple station bandpass balanced combiner (Figure 4.10-54) is an extension of the latter configuration, where each frequency has its own module. In a bandpass system, the accumulated

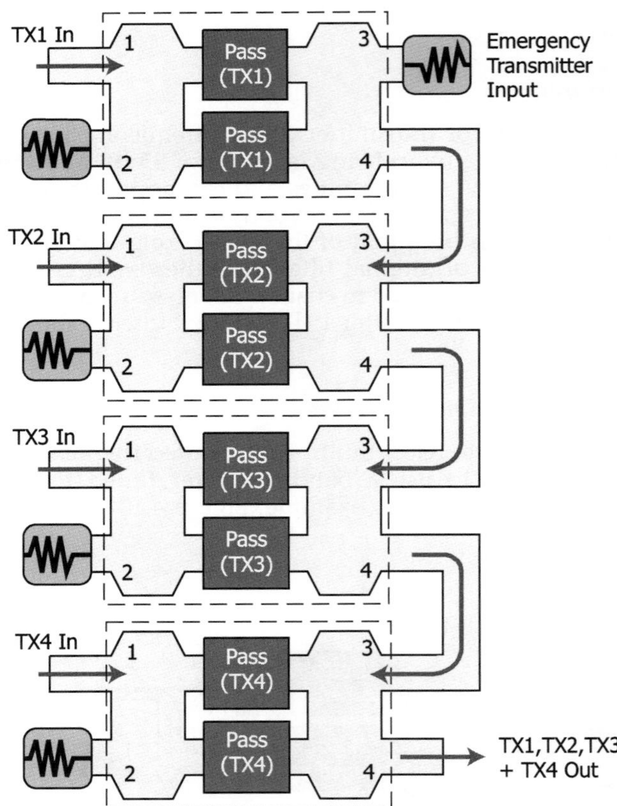

FIGURE 4.10-53 Four station bandpass filter balanced combiner.

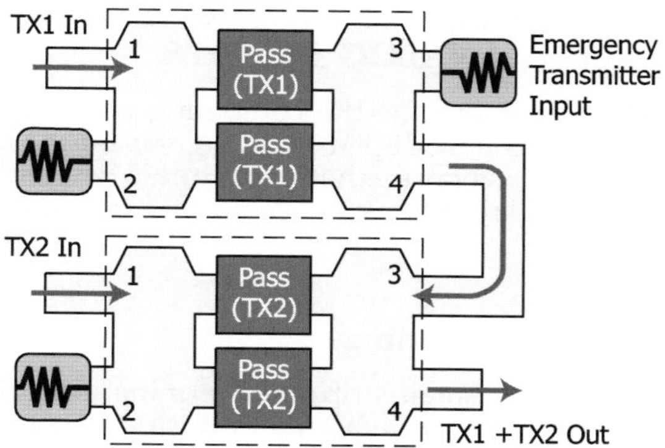

FIGURE 4.10-54 Two station, two module bandpass filter balanced combiner.

power entering each module flows only through the output hybrid, so the power handling capacity of the system is limited only by the size of the output hybrids and interconnecting transmission line, not the entire module.

Performance

The frequency response, the group delay, and the impedance diagram for this combiner are shown in Figures 4-10.55, 4-10.56, and 4-10.57, respectively.

Group Delay Effects

When two stations are 1.2 MHz apart or closer, the bandpass filter will not provide quite enough isolation and will allow a small amount of signal interaction. This affects the group delay curve of the module that is farthest from the antenna, as shown in Figure 4.10-58. A group delay equalizer can be installed either at the combiner input, using high power components (Figure 4.10-59), or between the transmitter's exciter and the IPA, using similar low power components (Figure 4.10-60).

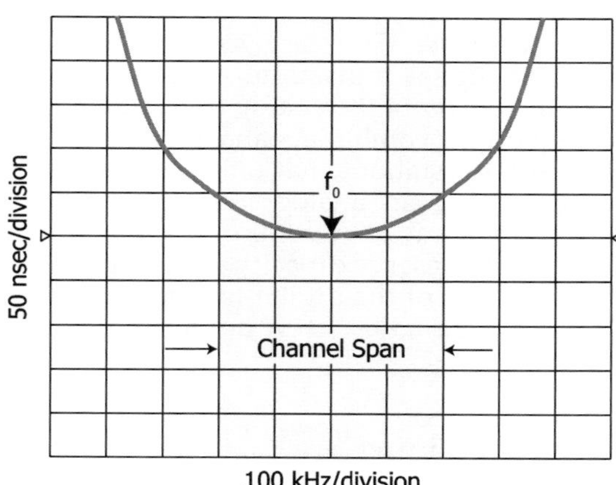

FIGURE 4.10-56 Group delay for a bandpass filter balanced combiner.

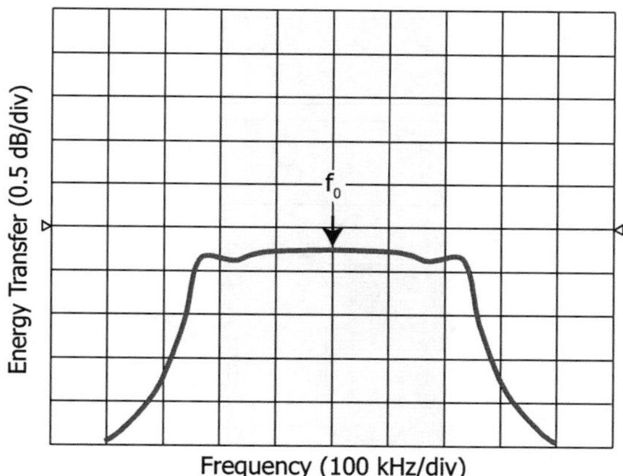

FIGURE 4.10-55 Frequency response for a bandpass filter balanced combiner.

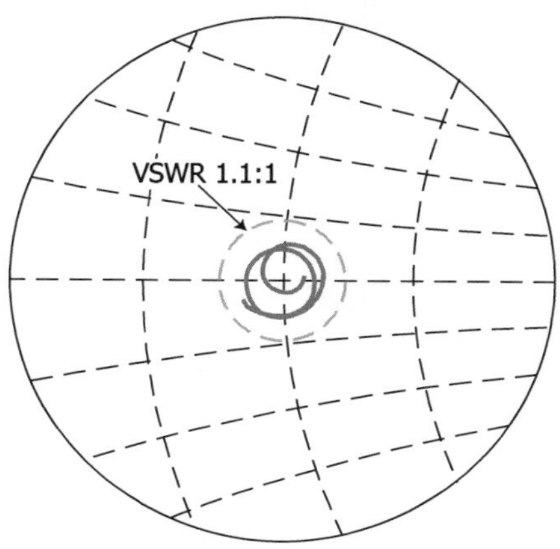

FIGURE 4.10-57 Impedance for a bandpass filter balanced combiner.

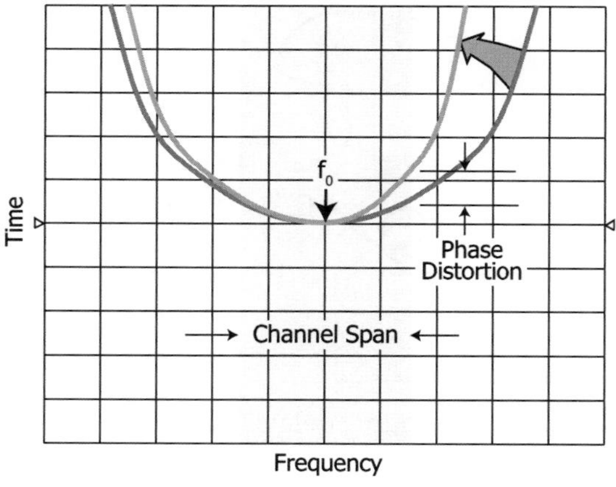

FIGURE 4.10-58 Distorted group delay.

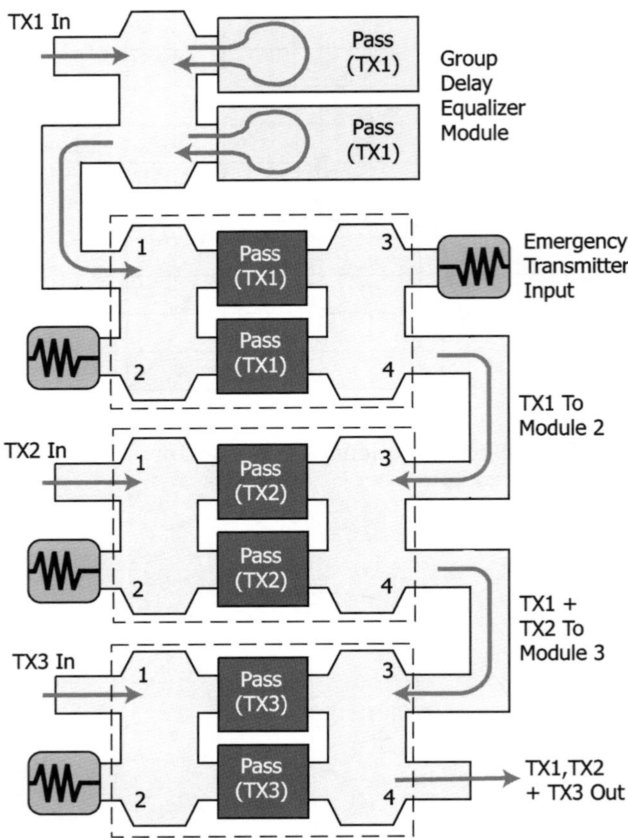

FIGURE 4.10-59 Balanced combiner with group delay equalizer at combiner input.

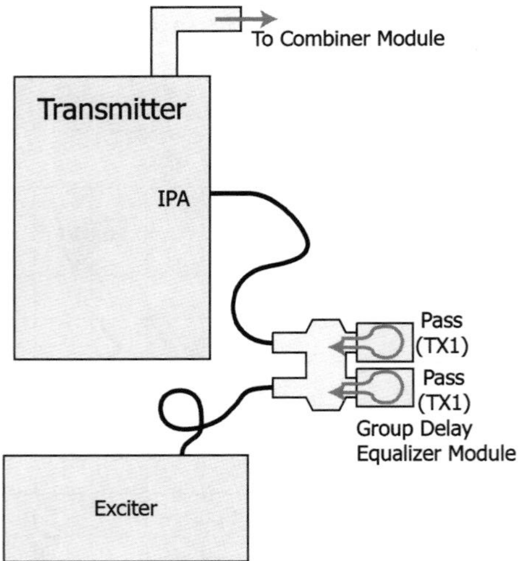

FIGURE 4.10-60 Group delay equalizer between exciter and transmitter IPA.

COMBINING DIGITAL AND ANALOG SIGNALS

Of particular interest is the combining of traditional analog FM and new digital IBOC signals on the same FM channel. Several methods to do this at high and medium levels and with separate antennas are described.

High-Level Combining

Small probes or small strip lines forming a precision directional coupler have been used historically to couple a small amount of energy out of the transmission line with coupling factors of anywhere from –40 dB to –60 dB from the RF power level being transmitted within that line. Typical uses of this level signal are to monitor or measure the high power signal with low signal level devices. Losses to the main path of the signal due to this sampling system are insignificant to the analog signal.

High-level combining uses a directional coupler (Figure 4.10-61) that has been mechanically enlarged to handle power levels in the kilowatt range, with a nominal coupling factor of –10 dB, and can be used in reverse to inject the digital signal into the analog RF stream. One strip carries the RF energy from the analog transmitter, which is considered the main line of the transmission system. The other strip is considered the coupling strip. The spacing of the strips determines the amount of coupling between the two signals. Increasing the coupling to –10 dB introduces a loss of 10% of the analog power, which is dissipated in a dummy load.

The digital signal enters the directional coupler at the reject port of the coupler, referenced to the analog input. Because it is a –10 dB coupler, only 10% of the digital signal is coupled to the main line. The remaining 90% flows to the dummy load. Several iterations of a high power combiner/injector have been tried. The –10 dB value is a good compromise for minimizing the loss to the analog transmitter while keeping the size of the digital transmitter to a reasonable level. An injector with a coupling factor smaller than –10 dB will increase analog losses, while a larger coupling factor will require a substantially larger digital transmitter. Note that the digital transmitter operates at only a fraction of the power of the analog transmitter.

This method of combining analog and digital is normally used for stations that only have one single-input antenna and an analog transmitter with the reserve capacity to make up for the 10% loss in power. Depending on the reserve capacity of the analog transmitter and the size of the digital transmitter, the coupling factor can be adjusted to optimize almost any installation.

Mid-Level Combining

Mid-level combining (Figure 4.10-62) was developed by incorporating a standard –3 dB quadrature hybrid and using two analog transmitters: one standard analog

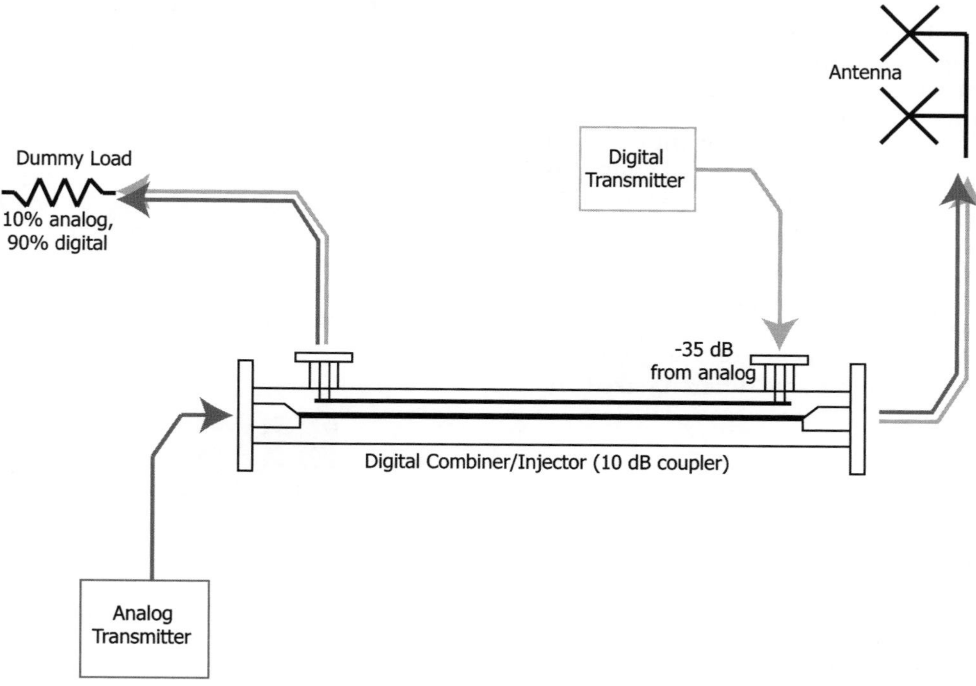

FIGURE 4.10-61 High-level combiner/injector.

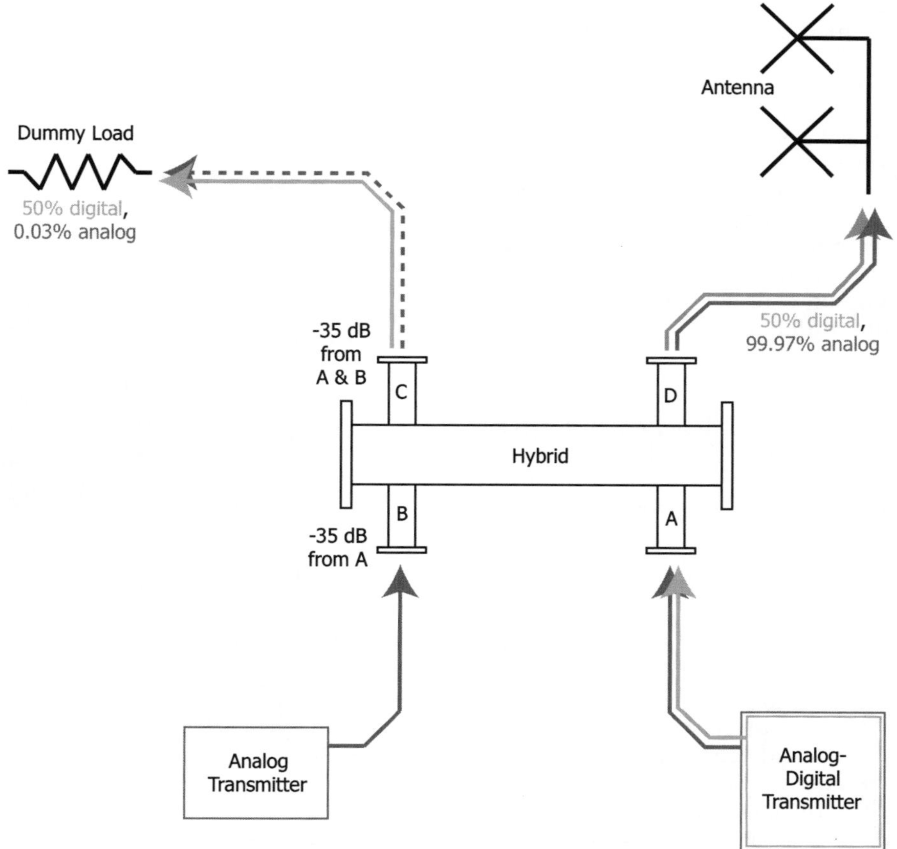

FIGURE 4.10-62 Mid-level combining.

transmitter and one linearized transmitter equipped to transmit digital along with the analog. It has been a well-established practice to combine two analog transmitters into a quadrature hybrid so most of the power goes to the antenna with minimal loss to the hybrid's dummy load. When the digital component of the linearized transmitter is turned on, the signal enters port A with its associated analog signal. Because no digital signal is entering port B, the digital signal is split in half. The benefit of this method over high-level combining/injection is that there is no significant loss to the analog signal and only a 50% loss to the digital signal rather than a 90% loss. As is the case with the high-level coupler/injector, the power split of the hybrid can be optimized to accommodate different-sized analog transmitters.

Combining Using Bandpass Balanced Combiners

Strategies that combine analog and digital signals in antenna radiators or use separate analog and digital radiators in close proximity are among the most popular IBOC implementation strategies because they minimize the size and cost of the digital transmitter and reduce the energy wasted.

Back-Feeding IBOC into a Balanced Module

The simple use of balanced combiner modules shown in Figure 4.10-54 is termed *single-feeding*. A variation on this configuration is called *back-feeding* (Figure 4.10-63), and is used for low level combining of analog and digital signals with minimal loss. Digital transmitters are fed through isolators into the hybrid ports opposite the analog transmitter ports, and a combined digital signal exits the wideband port (top left) normally occupied by the system reject load in a single-fed combiner. The only added hardware is the isolator in place of the dummy load of the single-feed combiner, to prevent analog on-channel signals from feeding back into the digital transmitter.

Purpose of the Isolator

Transmission systems that do not have enough isolation between the analog and digital components require isolators. When combiners are configured for back-feed operation, on-channel power is coupled from one transmission path into the other via the antenna elements and feeds back into the module through the opposite leg from which it exited. Although an efficiently operating antenna will minimize the energy coupled between paths, there still will be sufficient energy returned to require a dummy load for the port opposite the analog transmitter input. If a station runs an analog-only or high-level combined analog/digital signal, a stand-alone dummy load is used on this port. When the port is occupied by a digital transmitter, the dummy load becomes part of an isolator assembly. Isolators are not used where the analog and digital signals are already combined in the transmitter (low level), combined through a hybrid providing at least 35 dB of isolation (mid-level), or combined using a coupler/injector providing at least 35 dB of isolation (high-level).

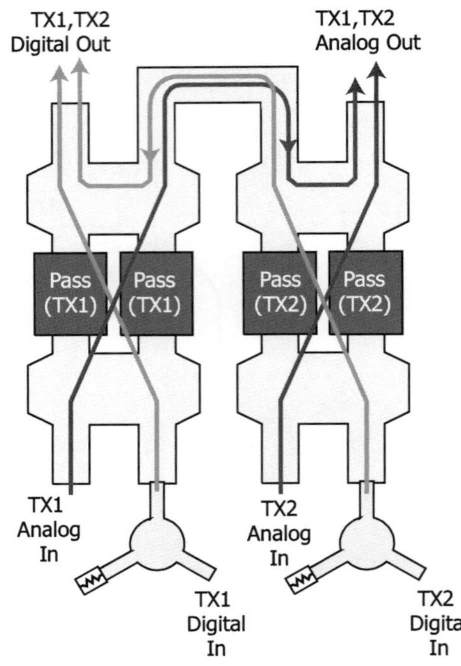

FIGURE 4.10-63 Back-feed configuration.

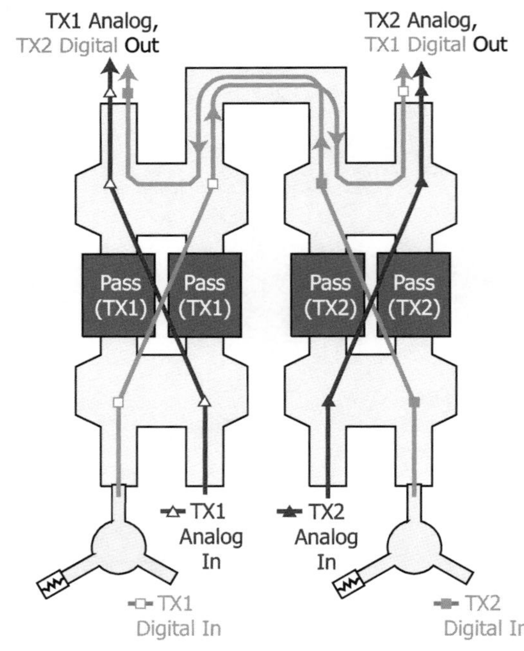

FIGURE 4.10-64 Cross-feed configuration.

Cross-Feeding IBOC into a Balanced Module

The cross-feed, or split-feed, configuration (Figure 4.10-64) is a further extension of back-feeding. Rather than segregating digital and analog signals into separate transmission lines, it combines the analog signals of some stations with the digital signals of others. Again, an isolator is used to provide additional isolation between the analog and digital transmitters. Usually, the analog power is split as evenly as possible, thus minimizing both the average and peak power any broadband line component carries. Thus, 9" components are eliminated in all but the largest systems. Using equal-sized transmission lines also provides redundancy. A failure in a transmission line or portions of the antenna feed system can be overcome by directing a station's primary transmitter (either analog or digital) over the remaining transmission line.

SUMMARY

Combiners are required when it is necessary to transmit multiple signals from a single antenna. Without proper combining, signals will interact in each other's transmitters, producing intermodulation products. This chapter has discussed the fundamentals of combining and the use of combiners in FM broadcasting. Several designs and many different components and configurations were described. The various types of combiners have their own advantages and disadvantages. The system designer must be aware of each so the appropriate filter system or systems can be selected for the specific combining application.

Books on Related Topics

Matthaei, G. L., Young, L., and Jones, E. M. T., *Microwave Filters, Impedance-Matching Networks, and Coupling Structures*, Artech House Books, Dedham, MA, 1980.

Smith, P. H., *Electronic Applications of the Smith Chart in Waveguide, Circuit, and Component Analysis*, McGraw-Hill, New York, 1969.

CHAPTER

4.11

FM RF Transmission Lines

KERRY W. COZAD
Dielectric Communications
Raymond, Maine

INTRODUCTION

Transmission lines are one of the main components in the RF transmission plant of a broadcast station. Acting as the connecting link between the transmitter and the antenna, the transmission line plays a critical role in both the quality and reliability of the broadcast signal; therefore, the proper choice of a transmission line type to be used can have a significant impact on the success of the station. With the development of digital formats for transmitting broadcast signals, it is important to review the existing transmission line performance specifications and any measured data on the installed line at the broadcast site prior to completing plans for implementing digital transmission. This review can prevent costly mistakes regarding long-term reliability and meeting necessary minimum transmission criteria for a robust hybrid analog/digital signal. See the Digital Transmission Requirements section in this chapter for additional information.

TYPES OF COAXIAL TRANSMISSION LINE

The choice of transmission line is typically decided based on the following criteria:

- Frequency of operation
- Power handling
- Attenuation (or efficiency)
- Characteristic impedance
- Tower loading (size and weight)

For most FM broadcast stations, these criteria result in the choice of either a semiflexible coaxial cable or rigid coaxial line for the connection between the transmitter and antenna.

Semiflexible Coaxial Cable

Semiflexible coaxial cables are designed with soft-tempered copper inner and outer conductors. The dielectric material may be solid (foam filled) or a spiral that has been wrapped in helical fashion around the inner conductor (air dielectric). A plastic jacket is applied to the outer conductor to resist abrasion. One advantage of semiflexible coaxial cables is that they can be fabricated in diameters up to 9 inches and in continuous lengths of hundreds and even thousands of feet. This type of transmission line is used extensively for radio and low power television broadcasting, as well as inter-element feeders for some antenna types.

Rigid Coaxial Line

Rigid coaxial lines are designed with hard-tempered inner and outer conductors. Discrete dielectric insulators are used to support the inner conductor within the outer conductor. Because it is rigid, it must be fabricated in defined lengths, typically no longer than 20 feet. The individual lengths are then attached to each other through the use of flanges and the inner conductors are typically spliced together using special connectors. Rigid coaxial lines can have diameters up to 14 inches. They have high power handling capabilities and low attenuation values. Special precautions must

TABLE 4.11-1
Common Types of Flexible RF Cables

Cable Designation	Nominal Z_0	Velocity of Propagation (Vp)	Diameter (inches)	Nominal Attenuation at 50 MHz (dB/100 ft)	Nominal Attenuation at 200 MHz (dB/100 ft)	Nominal Attenuation at 700 MHz (dB/100 ft)	Peak Power Rating (watts)
RG-8A/U	52	0.66	0.405	1.6	3.2	6.5	4000
RG-8/X	50	0.78	0.242	2.5	5.4	11.1	600
RG-213/U	50	0.66	0.405	1.6	3.2	6.5	5000
RG-58/U	53.5	0.66	0.195	3.1	6.8	14.0	1900
RG-58A/U	50	0.66	0.195	3.3	7.3	17.0	1900
RG-58C/U	50	0.66	0.195	3.3	7.3	17.0	1900
RG-11A/U	75	0.66	0.405	1.3	2.9	5.8	5000
RG-59B/U	75	0.66	0.242	2.4	4.9	9.3	2300
RG-62B/U	93	0.84	0.242	2.0	4.2	8.6	750
RG-71/U	93	0.84	0.245	1.9	3.8	7.3	750
RG-141A/U	50	0.69	0.190	2.7	5.6	11.0	1400
RG-178B/U	50	0.69	0.070	10.5	19.0	37.0	1000
RG6A/U	75	0.66	0.332	1.9	4.1	8.1	2700

be taken with the selection of rigid line section lengths, as standard available lengths (20', 19.75', 19.5') do not perform well at all FM channels due to flange reflections adding together at specific frequencies and line lengths. A rigid line section length of 17.5' allows the use of all FM frequencies between 88 MHz and 108 MHz. For purposes of this chapter, the focus here is on semiflexible coaxial cables as that is the transmission line used for the majority of FM broadcast transmission facilities. For additional information on rigid line and discussions on the choice of transmission lines and tradeoffs, both electrical and mechanical, refer to Chapter 6.6, "Coaxial and Waveguide Transmission." Before proceeding with a discussion on semiflexible coaxial cable, a short description of flexible RF cables will complete the descriptions of typical coaxial transmission lines used by broadcasters.

Flexible RF Cables

Coaxial transmission lines used for short interconnections between equipment, typically inside buildings, have traditionally been referred to as RF cables. They are small in diameter, making them very flexible and very useful in areas with minimum space. The inner conductor is usually solid copper or a copper-clad metal, and the outer conductor is a copper or aluminum braid. In order to maintain concentricity of the inner and outer conductors, the dielectric is a solid insulating material. The more commonly used flexible cables are shown in Table 4.11-1.

With the introduction of digital signals and multiple frequency grouping, the shielding performance of braided outer conductors may not be sufficient to prevent signal leakage. Interference between cables within an installation can be detrimental to signal quality, and it can be frustrating to track down the source. In such installations, cables with a solid outer conductor will improve the system performance. Typical solid outer conductor cables are listed in Table 4.11-2 and shown in Figure 4.11-1. Note that the velocity of propagation (Vp) is greater for these cables than the flexible cables shown in Table 4.11-1. Lower values of Vp will cause a given physical length of transmission line to behave as if it is electrically longer.

SEMIFLEXIBLE LINES

Globally, semiflexible coaxial lines are the most popular choice for broadcasting applications. It is ideal for a wide variety of low and medium power transmission systems. The line achieves its flexibility through the use of corrugated copper conductors and can be fabricated in extremely long lengths. This eliminates the need for interconnecting joints, thus minimizing possible installation and assembly problems. They are also easier to install because the corrugations eliminate most of the need for hanger designs that permit independent movement of the line from the tower (differential thermal expansion).

FOAM DIELECTRIC LINES

Foam dielectric lines are designed for systems that normally do not require a gas pressure path to the

TABLE 4.11-2
Common Solid Outer Conductor Cables

Foam Cable Size (inches)	Nominal Z_0	Velocity of Propagation (Vp)	Diameter (inches)	Nominal Attenuation at 50 MHz (dB/100 ft)	Nominal Attenuation at 200 MHz (dB/100 ft)	Nominal Attenuation at 700 MHz (dB/100 ft)	Peak Power Rating (watts)
1/4 SF	50	0.84	0.290	1.27	2.58	4.97	6400
3/8 SF	50	0.83	0.415	0.848	1.73	3.37	13,200
1/2 SF	50	0.81	0.520	0.73	1.50	2.97	15,600
3/8	50	0.88	0.44	0.736	1.50	2.93	15,600
1/2	50	0.88	0.63	0.479	0.983	1.92	40,000
1/4 SF, HT	50	0.82	0.29	1.27	2.56	4.89	6400
3/8 SF, HT	50	0.83	0.415	0.856	1.77	3.48	13,200
1/4 SF	75	0.78	0.29	1.3	2.68	5.30	3300
1/2 SF	75	0.81	0.52	0.673	1.39	2.75	10,000
1/2	75	0.88	0.63	0.435	0.896	1.76	26,000
Note: SF, superflexible; HT, high-temperature, high-power.							

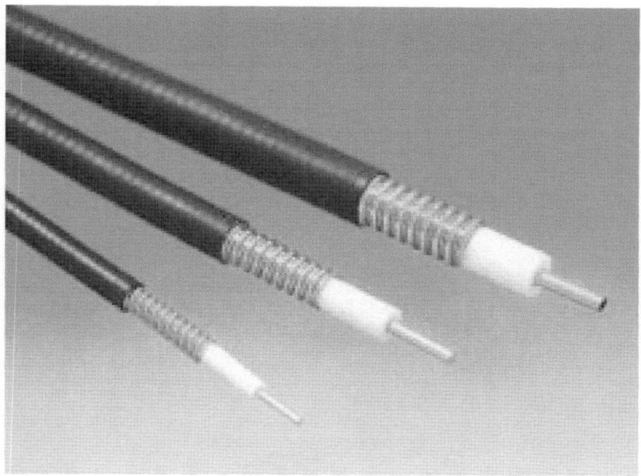

FIGURE 4.11-1 Typical small diameter solid outer cables in a cut-away view to show dielectric and inner conductor.

FIGURE 4.11-2 Typical foam dielectric cables in cut-away view to show dielectric and inner conductor.

antenna or other connected components. Typical applications include FM radio and low power television where the antennas do not require pressurization. Sizes range from 1/4" to 2-1/4". These cables are constructed with a closed-cell, low-density foam dielectric which prevents water penetration while providing low attenuation and high relative velocity of propagation. Examples of foam dielectric cable are shown in Figure 4.11-2. The maximum diameter of foam dielectric cables is limited by the manufacturing process of coating the inner conductor with the foam dielectric. The attachment process requires the foam to be applied to the inner conductor in a liquid state followed immediately by cooling the foam to prevent distortion of the foam shape and movement of the inner conductor within the foam. As the diameter of the foam and inner conductor increases, the outer portion of the foam hardens and acts as a thermal insulator for the inner portion of the foam. This action is much like the theory behind insulated coolers. Whatever is inside the cooler is insulated from outside forces. However, because the inner portion of foam is not hardened, the actual weight of the inner conductor can cause the conductor to sag in the liquid foam, resulting in a nonconcentric configuration of the conductors. As a result, the characteristic impedance can change and the electrical performance can be compromised. To produce larger diameter cables requires the use of spiral sections of dielectric wrapped around the inner conductor. This results in what is referred to as an air dielectric cable.

TABLE 4.11-3
Characteristics of Common Foam and Air Dielectric Coaxial Transmission Lines

Size (inches)	Maximum Frequency (GHz)	Velocity of Propagation (Vp)	Nominal Inside Transverse Dimension (inches)	Diameter Over Jacket (inches)	Minimum Bend Radius (inches)	Weight (lb/ft)
Foam dielectric cable						
7/8	5.00	0.89	0.83	1.09	10	0.33
1-1/4	3.30	0.89	1.22	1.55	15	0.66
1-5/8	2.50	0.89	1.59	1.98	20	0.92
Air dielectric cable						
7/8	5.20	0.92	0.80	1.11	10	0.54
1-5/8	2.70	0.92	1.57	1.98	20	1.04
2-1/4	2.30	0.92	1.95	2.38	22	1.16
3	1.64	0.93	2.50	3.02	30	1.78
3-1/2	1.43	0.96	2.96	3.50	30	1.98
4	1.22	0.92	3.37	4.00	40	2.50
5	0.96	0.93	4.45	5.20	50	3.30
6-1/8	0.86	0.97	5.79	6.73	79	7.33
9	0.65	0.97	7.68	8.90	98	12.50

AIR DIELECTRIC CABLE

Air dielectric cables utilize a spiral dielectric material to separate the conductors. Because the majority of the volume between conductors is now air, these cables have lower attenuation and higher average power handling than foam dielectric cables. They require pressurization to prevent water ingress and can also provide pressure to the antenna if needed. A benefit of the pressurization is that significant physical damage to the cable outer conductor will usually result in a pressure leak, allowing the damage to be detected before a more severe condition occurs as a result of voltage breakdown. Table 4.11-3 compares common foam and air dielectric transmission lines used in broadcasting. Spiral-wrapped air dielectric cables are shown in Figure 4.11-3.

FIGURE 4.11-3 Typical air dielectric cables in cutaway view to show spiral dielectric and inner conductor.

ELECTRICAL CHARACTERISTICS OF SEMIFLEXIBLE CABLES

For FM radio transmissions, the most significant factor driving the decision of the type of line needed is the average power rating. Both analog and digital FM transmitters are rated based on their average power output and this value corresponds directly to the catalog average power rating that is provided by the cable manufacturer. The average power rating is dependent on the physical sizes and materials of the cable conductors and insulators. Each manufacturer uses a slightly different configuration, resulting in differences between the published ratings for lines that use the same size designator (*e.g.*, 1-5/8", 3", 3-1/2"). It is important to check with each manufacturer to confirm the actual specifications for a line, including average

power rating, peak power rating (peak voltage rating), and attenuation (see the list of manufacturers at the end of this chapter).

PHASE-STABILIZED LINES

In some cases, such as AM directional antennas, it is necessary to use coaxial cables to sample current levels to confirm compliance with a broadcasting license. In this case, the phase stability of the cable over long periods of time is extremely important. Ordinary coaxial cable should not be used in these instances, as it is subject to significant phase variation with temperature change during its initial usage. Instead, phase stabilized foam polyethylene dielectric or air dielectric cables should be specified. These cables have been subjected to a process that cycles the cable through a wide temperature range to remove the phase instability caused by hysteresis (movement between the inner and outer conductors).

CONNECTORS, SPLICES, AND ADAPTORS

As described in the introduction, semiflexible coaxial cables can be fabricated in long continuous lengths. This is one advantage over rigid coaxial lines, which must be spliced approximately every 20 feet. The elimination of the numerous connections inherently improves the reliability and reduces the costs of the transmission line system when a cable can be used *versus* rigid line; however, it is still necessary to make connections to the output of the transmitter and input of the antenna through the use of connectors attached to the cable ends.

Usually, the connectors on the ends of the semiflexible cable are determined by the antenna and transmitter connection type and the power handling requirement of the transmission system. When the determination has been made, then a choice must be made regarding when and where to attach the connectors. If the total length is already known, the connectors should be attached at the factory by the cable supplier prior to shipment. Those connectors are attached in a controlled environment by highly experienced personnel, and the chance of a poorly made attachment is reduced considerably; however, if a connector must be assembled in the field, it is important to make the attachment properly. Close attention to the supplied assembly instructions, especially the cable trimming dimensions, will ensure a good connection. Clean mating surfaces and proper tightening are required for good electrical contact along the entire 360° of the cable edge.

It is important to make all connector and splice attachments with the cable ends in a horizontal position. This minimizes the chance of metal chips falling into the cable during the trimming process. This will normally mean making connections on the ground; however, there may be times when there is no choice but to make the attachment on the tower. In that case, it is important to take precautions, such as stuffing cloth into the open cable end to prevent chips from falling into the cable. Use care when removing the cloth so as not to drop any chips into the now open cable end.

Proper attachment of the finished connector to the antenna or transmitter prevents loosening of the connection from vibration or other stresses. It is important to recheck the cable assembly after attachment, specifically for connections using screw together interfaces. Twisting of the cable during attachment may cause the connection to loosen and result in poor performance and environmental damage to the cable. These connections should be securely taped to prevent loosening.

Electronic Industries Alliance (EIA)-style flange connectors must be carefully tightened to prevent distortion of the flanges. Tighten bolts that are opposite (not adjacent) to the proper torque specification. Overtightening can result in a warped flange. Application of silicone grease to the O-ring to hold it in place is a common practice; however, silicone is an insulating material, and excessive application of the grease can result in contamination of the contact surfaces. Excessive heating and damage to the O-ring can result.

DIGITAL TRANSMISSION REQUIREMENTS

The use of digital transmission for FM radio in the United States is underway. The in-band, on-channel (IBOC) system accepted as the U.S. standard for FM digital transmission can be implemented in several different ways. At the present time, two principal methods are employed. One is combining the analog and digital signals using a hybrid combiner at the outputs of the analog and digital transmitters and then using a single transmission line to a common antenna. The second involves installing a second antenna and transmission line to transmit the digital signal. In both of these cases, a review of the transmission lines that may be used is important to optimize performance and ensure long-term reliability for both the analog and digital transmissions.

In the introduction to this chapter, five criteria were presented as the primary factors regarding the choice of transmission line type for a specific installation. For FM radio transmissions, two of the five are effectively constant: The frequencies of operation in the United States fall between 88 and 108 MHz, and the characteristic impedance of the cable will almost always be 50 ohms. This leaves three other factors for discussion: power handling, attenuation, and tower loading. Note that these three factors are dependent on the size of the cable. Increased size is a positive for power handling (higher ratings) and attenuation (higher efficiency) but is a negative regarding line costs and higher loading on the tower. In the case of new tower construction, the choice to use a larger size line is determined based on the upfront costs of the tower structural design and the possibility of reduced ongoing operating costs due to lower attenuation (higher efficiency and subsequent lower transmitter power requirement) or the sharing of costs by combining multiple stations on a master antenna and transmission line system.

For an existing tower, the balance between the electrical and mechanical trade-offs can be difficult if the existing analog transmission line and antenna system were designed with minimal additional capacity from a power-handling perspective. Additionally, even though the digital signal is "on-channel" with the analog signal, many FM transmission line/antenna systems have been optimized at the center of the channel for best voltage standing wave ratio (VSWR) performance, and the VSWR at the edges of the channel, where the digital signals are located, may be greater than desired. This higher VSWR level will also de-rate the power handling of the transmission line for the digital signal if the station is planning to use a hybrid combiner and the existing line. Retuning the system for a lower average VSWR over the channel may be necessary for optimum digital performance.

It is important to verify the performance of an existing line that is going to be used under different conditions in the future (such as adding a digital signal, increasing analog power, or combining with another channel) by performing an electrical check of the line and a detailed mechanical inspection prior to making the change (discussed later) as well as a comprehensive specification review to update the performance expectations for the line. If the line has experienced problems in the past, it is not a good practice to increase the power levels, particularly if the power level will be near the catalog rated power for new line. The purchase and installation of new transmission line may save the station significant costs and air-time in the future.

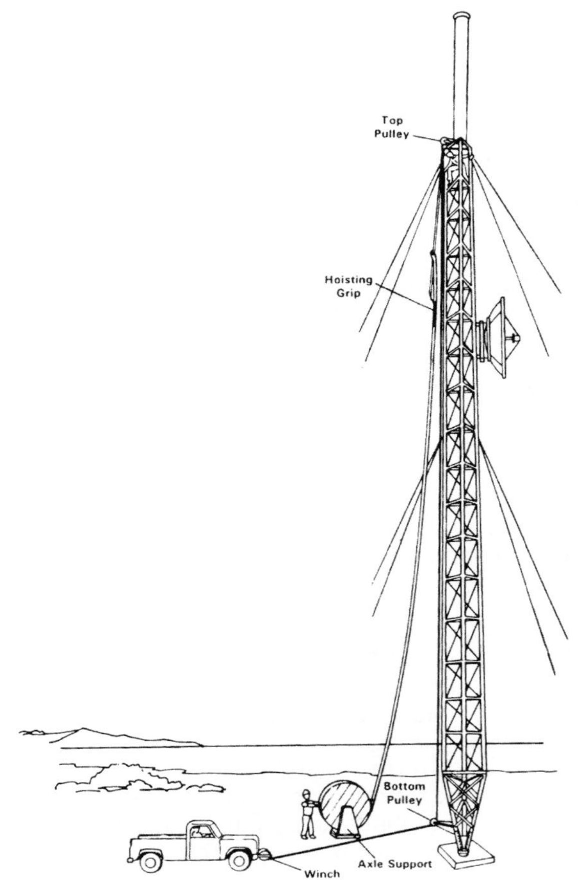

FIGURE 4.11-4 Example of rigging technique for installation of semiflexible transmission line.

INSTALLATION

A proper installation can be the difference between a smooth running transmitter site and a continuous effort to correct failures in the transmission line and antenna system. Note that the environmental forces acting on transmission lines installed outdoors and on tall towers are constantly changing. This results in substantial stress on the cable from corrosion, vibration, atmospheric contaminants, and thermal changes that are not present with indoor installations. Following the manufacturer's recommended procedures for installation and maintenance can prevent catastrophic failures.

General guidelines for installation of semiflexible cable systems generally emphasize the handling of the cable. Semiflexible cable is usually shipped from the factory on reels or in cartons and should be carefully inspected for shipping damage. In the case of air dielectric cables, check that the cables are still holding pressure. In storage, allowable pressure drop is typically 1 psig (6.9 kPa) per 24 hours. If there is excessive leakage, the cause should be determined prior to installation. Reels should always be stored and moved on the flange edges. Storing on a flange side or dropping a reel can cause damage to the inside lengths of cable due to crushing from the outside. Lifting and installation of the cable should only be performed by qualified installation crews. The proper use of lifting devices, such as hoisting grips, winches, and load lines is required for safety. Figure 4.11-4 shows a diagram of the installation technique used for semiflexible transmission line.

For long-term reliability, proper installation of cable hangers is essential. The purpose of the hangers is to support the cable's weight and to prevent movement from wind, rain, ice, and other forces. Excessive movement causes abrasion and damage to the outer conductor. Initially, the manufacturer's recommended hanger spacing should be considered. This spacing should be modified (typically twice as many hangers used) if the wind velocity will be consistently above 40 mph, peak winds exceed 125 mph, or heavy ice is expected to form.

A drip loop is recommended, as shown in Figure 4.11-5, for the horizontal run of semiflexible line to prevent moisture from using the line as a path to the building wall. Water can travel down the outside of the jacket and build up at the wall feed-through if a drip loop is not used and the line slopes toward the building. Water may also travel between the jacket

FIGURE 4.11-5 Note droop or drip loop in transmission lines to prevent moisture from entering transmitter building or collecting to cause other damage.

and outer conductor if damage to the jacket has occurred. If this occurs, a small notch cut into the jacket at the bottom of the drip loop will allow for drainage.

Because jacketed copper cables are corrosion resistant, it is also possible to bury the horizontal run either directly or pulled through a conduit. Sharp objects must not come into contact with the cable, or damage to the cable may occur. If installing within a conduit, do not exceed the tensile strength of the cable when pulling it through. Conduit pulling lubricants should be applied generously to the cable. Bends should be avoided if at all possible, as the corrugations may cause excessive friction when pulling around a corner. Because a buried cable may occasionally become submerged, any buried connectors or splices should be well protected with a good weatherproofing kit.

GROUNDING

Because the cable represents a direct connection between the outside and the transmitter equipment, proper grounding is necessary to protect the equipment from the large currents caused by a lightning strike. Minimum grounding requirements include grounding the cable at the top and bottom of the vertical run as well as just before entry to the building. For long runs, additional grounding in 200 ft increments is recommended. When mounting cables on insulated AM towers, bonding intervals should be no greater than 1/8 wavelength. Ground connections must be made with high-quality copper wire or straps. Braided copper straps and fine-stranded wire are not recommended, as they tend to deteriorate with time, reducing the effectiveness of the grounding connection. After installation, all ground connections should be weatherproofed to prevent degradation. It is best to

FIGURE 4.11-6 Example of neat installation of cables with grounding strap weatherproof kits and identification markers.

employ grounding kits supplied by the cable manufacturer. Figure 4.11-6 shows an installation for several cables. Note that the cables are labeled for ease of identification. Also note the "bulge" area just above the labels. These are the weatherproofing kits covering the grounding strap kits for grounding the cables at the base of the tower. More information on grounding may be found in Chapter 9.3, "Facility Grounding Practices."

ELECTRICAL TESTING

When the line has been installed and mechanically inspected, an electrical check is necessary for maintaining a reliable system. A more detailed discussion is included in Chapter 6.6, "Coaxial and Waveguide Transmission." For coaxial cable installations, the initial electrical tests for VSWR and time-domain reflectometry (TDR)/frequency-domain reflectometry (FDR) or network analyzer responses provide a baseline for future measurements that will allow the station engineer to identify changes that may be caused by damage, such as:

- Environmental factors such as falling ice

- Dents caused by personnel working on the tower
- Hangers and supports that have loosened over time

Identifying these potential causes early by performing annual inspections and electrical tests can prevent catastrophic failures due to continued wear on the cable over time. In most cases, metering that is used with the transmitter RF system for measuring forward and reverse power levels is not an adequate substitute for the details available from electrical testing using a network analyzer or similar instrument.

MAINTENANCE AND REPAIR

If the installation of the coaxial cable is done properly, then basic maintenance procedures will keep it operating trouble free for many years. Inspections made at least every 12 months, or at shorter intervals if the facility is located near bodies of saltwater or in environments that have extreme levels of wind, ice, snow, or temperatures, should look for loose or missing hardware on hangers and flanges, as well as areas where abrasion to the cable jacket is occurring due to vibration and movement and damage to the grounding straps. For air dielectric cables, a daily log and regular review of pressure levels can allow early detection of a future problem if a pressure leak is noticed.

Repair procedures usually involve tightening and replacement of hardware and damaged hangers. If found early, abrasion to the outer jacket can be halted through readjustment of the cable location or addition of hangers to stiffen the area of the cable that moves. If the jacket has been damaged so the copper outer conductor is exposed, it is necessary to seal the damaged area with a weatherproofing kit to prevent water intrusion between the jacket and outer conductor and protect the outer conductor from future abrasion that may result in a pressure leak or electrical failure. Figures 4.11-7, 4.11-8, and 4.11-9 show examples of proper installation and areas of concern to check during inspections and maintenance activities. Although these photographs of actual installations are of antenna feeder cables, the same principles apply for main transmission line runs and larger cables. If the outer conductor has been damaged so a pressure leak has occurred, it is important to repair this damage as soon as possible. Immediate action may be to apply a weatherproofing patch to prevent water ingress and to slow or eliminate the pressure leak. Long-term repair may require cutting and removing the damaged sec-

FIGURE 4.11-8 Example of incorrect method for securing semiflexible cable. Tape should only be used temporarily to secure cable during installation work.

FIGURE 4.11-7 Example of how clamps are correctly used to position semiflexible cable away from tower components to avoid abrasion and movement.

FIGURE 4.11-9 Cables in close proximity to tower structures without securing clamps will incur abrasion, punctures, and cuts that will degrade performance or cause premature failure of the line.

tion and inserting a splice. The manufacturer should be contacted for recommended practices to repair this type of damage.

SUMMARY

Subject to the necessary power handling requirements, semiflexible coaxial cable is an optimum choice for FM broadcasters. Its availability in long, continuous lengths eliminates the need for multiple connections, resulting in wideband electrical performance and high reliability. Also, it is typically easier and more cost effective to install than a similar-sized rigid coaxial transmission line. It is important that installers follow the manufacturer's recommended procedures to avoid damaging the line by denting or kinking and to correctly attach it to the tower using the appropriate hangers and spacing for long-term reliability. Finally, regular inspections, maintenance, and monitoring will help prevent minor damage from turning into a major repair or replacement problem.

Bibliography

Campbell, W.D., Troubleshooting transmission line using a time domain reflectometer, in *Proc. of the SBE Broadcast Engineering Conf.*, 1991, pp. 224–236.

Cozad, K.W., Coax/transmission line, in *The Electronics Handbook*, Whitaker, J.C., Ed., CRC Press, Boca Raton, FL, 1996.

Cozad, K.W., Coaxial transmission lines, in *NAB Engineering Handbook*, Ninth ed., National Association of Broadcasters, Washington, D.C., 1999.

Kerkhoff, W.A., Care and feeding of FM multichannel antennas, in *NAB Broadcast Engineering Conference Proceedings*, 2003, pp. 244–256.

Lampen, S.H., Gebs, B., Van Der Burgt, M.J., Impedance, in *NAB Broadcast Engineering Conference Proceedings*, 2002, pp. 311–315.

Margala, A.J. and Johnson, E.H., Transmission lines for broadcast use, in *NAB Engineering Handbook*, Fifth ed., Walker, A.P., Ed., National Association of Broadcasters, Washington, D.C., 1960, pp. 2-184–2-210.

Paris, D.T. and Hurd, K.F., Transmission lines and waveguides, in *Basic Electromagnetic Theory*, McGraw-Hill, New York, 1969.

Tellas, R., A new helically corrugated coaxial cable with a 6-1/8" coaxial cable performance, in *NAB Broadcast Engineering Conference Proceedings*, 1996, pp. 437–441.

Manufacturer Web Sites

www.andrew.com
www.belden.com
www.dielectric.com
www.eriinc.com
www.myat.com
www.rfsworld.com

CHAPTER

4.12

FM Translators and Boosters

RON CASTRO

Results Radio, LLC
Petaluma, California

INTRODUCTION

In the early days of FM, residents of communities that lacked adequate FM coverage due to terrain blockage, or simply because they were too far away from existing FM stations, could be served by applying to the FCC for a license to operate a low powered station located such that it could receive signals from an existing full-service station and retransmit or "translate" them onto another FM channel that covered the community. The FCC sanctioned these stations provided there was no financial connection between the translator station licensees and the full-service stations they carried. Today, in certain situations, the licensees of full-service stations are permitted by the FCC to use translators to fill in areas where terrain blocks predicted service contours.

TRANSLATOR COVERAGE METHODOLOGY

The method prescribed by the FCC to predict the service area of an FM station is based on a 1947 paper by Kenneth Bullington [1]. This was the basis of a practical method of estimating the usable coverage, as well as the potential interference of an FM station. Even though newer propagation-prediction methods, aided by today's powerful and affordable computers, have revolutionized the ability to predict signal coverage and interference, the FCC still uses the Bullington methodology as its primary standard.

The Bullington method is fairly accurate in predicting signals against the newer technology in relatively flat terrain situations, but its inability to account for terrain obstructions in the signal path causes anomalies such as the one illustrated in Figure 4.12-1. This shows a comparison of the 54 dBu coverage contour of a typical class B station serving San Francisco predicted by the FCC's method, and one using a more modern prediction method over the very diverse topography of the Bay area. In some directions, the coverage extends beyond the FCC contour, but in others it falls short. Often, as in this situation, a broadcaster finds that the FCC service contour includes many communities and highways where the actual signal is poor or nonexistent.

Fill-In Translators

FM licensees can remedy this anomaly by operating a "fill-in" translator to provide coverage within its licensed FCC-calculated service contour. For commercial stations, the service contour of the translator itself cannot extend beyond that of the primary station. Leeway is granted to noncommercial broadcasters to use non-fill-in translators to extend their coverage beyond that of their main stations' service contours, but in no case may any translator cause interference to other full-service stations or other existing translators.

FM TRANSLATOR TECHNOLOGY

A translator is a low power FM transmitter that rebroadcasts signals from another FM station on a different FM channel. This differentiates it from a station that gets its programming from an originating source

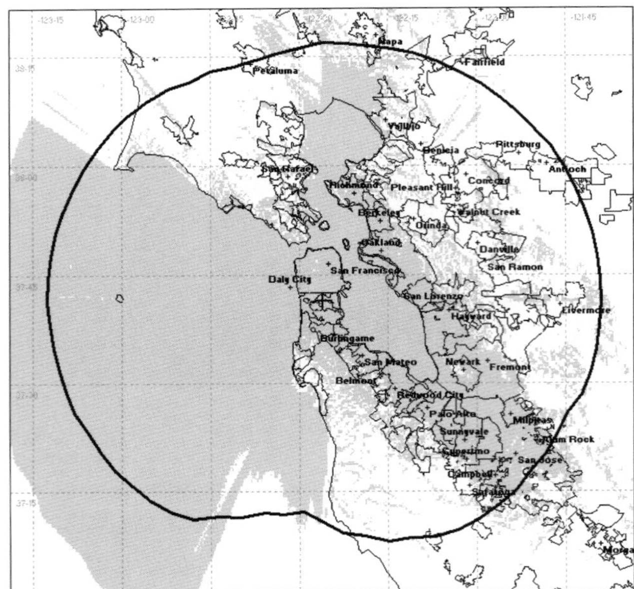

FIGURE 4.12-1 The 54 dBu FCC F(50,50) contour of a typical San Francisco class B FM station. The areas of gray show 54 dBu coverage using an alternative point-to-point terrain-sensitive methodology.

other than a licensed, full-service FM station, or from a "booster" that retransmits the signal of a licensed, full-service FM on the same channel. Figure 4.12-2 shows the most basic translator that consists of an FM receiver tuned to channel A, connected to a directional antenna pointed at the station to be rebroadcast. The audio output of the receiver is used to modulate an FM transmitter that rebroadcasts the audio from the receiver on channel B and is connected to an antenna that provides coverage to the intended community.

Noncommercial stations operating boosters in the nonreserved part of the FM band must use receivers to feed their translators, but in all other situations, alternate feed technologies may be employed.

FM BOOSTER TECHNOLOGY

FM boosters differ from FM translators in that they retransmit the received station on the *same* FM channel. The main advantage of this arrangement is there is less potential of interference to other full-service stations. Also, listeners don't have to retune their radios to a different channel (from the one announced on the station) as they move from one area to another. One disadvantage is that the booster transmitter can cause interference to the main transmitter (and vice versa) in areas where the signals of both transmitters are similar in strength. Another disadvantage is that the simple feed method used by the translator shown in Figure 4.12-2 quite often does not work due to feedback from the output to the input. Despite these challenges, boosters are often the preferred and sometimes the

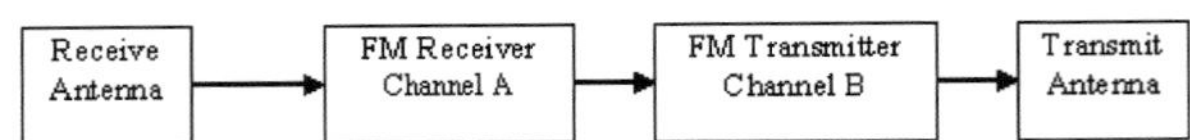

FIGURE 4.12-2 Block diagram of a basic translator.

only way to provide a usable signal to a blocked area in the main station's coverage contour.

TYPES OF FM BOOSTERS

There are three types of boosters that are generally in use today, each with its own unique advantages and disadvantages.

Amplified Band Pass Booster

The amplified band pass booster, which is considered to be the "original" type since it was the first to be authorized by the FCC, consists of a directional antenna connected to a band pass filter tuned to the primary station's frequency, followed by a high-gain amplifier that provides output power to a directional transmit antenna sufficient to provide fill-in service to the desired area. This arrangement is shown in Figure 4.12-3. The high-gain amplifier has an automatic level control that maintains the output power at a fixed level, and a device that shuts down the amplifier if the input signal drops below a certain threshold.

Besides being a simple design, the advantage of this type of booster is that the main and booster transmitters are always on the same frequency, preventing a destructive beat that could be caused if the two transmitters had independent frequency determining elements that were on slightly different frequencies.

There are several drawbacks to this system. One is that the input to the amplifier can often be as little as –60 dBm, and the desired output can be in excess of 40 dBm, requiring a system gain of 100 dB, plus the gain of the directional antennas. With such high gain, oscillation can result if the input does not have enough isolation from the output to prevent regenerative feedback that would result in wide areas of interference and no beneficial coverage. The other drawback is that the input can be subject to co-channel or adjacent channel interference or noise that will appear in the transmit signal if it gets through the band pass filter. Still another drawback that can be common to all boosters is that signals from the primary and booster transmitters can arrive at a given location with similar signal strengths, but at different times of arrival, resulting in a multipath effect that can cover significant areas, some of which experienced no interference before the booster was activated.

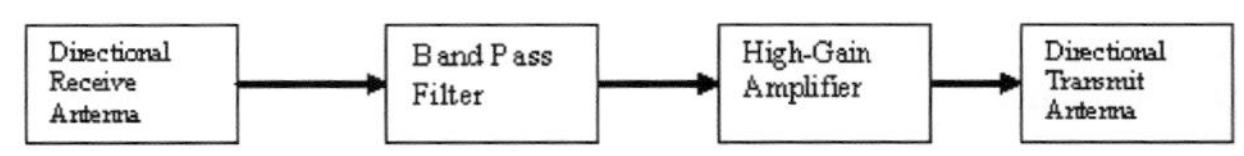

FIGURE 4.12-3 Simplified block diagram of an amplified band pass FM booster.

Alternate Feed FM Booster

While many boosters have been successfully constructed and operated over the years, many have also failed. The FCC recognized this and allowed broadcasters to mitigate the feedback problem by choosing alternate feed methods, leading to the second type of booster, one which is fed by the same audio as the main transmitters, but not directly from the main transmitter's over-the-air signal. There are many alternate feed methods that can be used with boosters as well as with translators. In this instance the transmitter is on the same frequency as the main signal but there is no receiver that can cause feedback or other drawbacks as described above.

Synchronous Booster

The third type of FM booster, the synchronous booster, is a relatively new concept that was described in a paper by Stanley L. Salek, P.E. [2] that addresses the problem of mutual interference created by the main and booster transmitters in "overlap" areas where the signals are similar in strength. Such areas of interference in conventional boosters were often so pervasive that many boosters were ultimately shut down because they caused more loss of coverage to the main transmitter's signal than they added.

Salek posited that these areas of interference were a result of the two transmitters' signals arriving at slightly different times where they were similar in strength, and that by delaying the signal of one of the transmitters so that both signals arrived at approximately the same time, the interference could be mitigated. His research further demonstrated that not only did the signals have to arrive at nearly the same time, they also had to have precisely the same carrier frequency, preferably phase-locked to a common source, and identical modulation characteristics. Although the first type of booster discussed, the band pass amplifier, inherently has identical frequency and modulation since it is a straight-through amplifier, there is no practical way to control the important time of arrival issue. Conventional alternately fed boosters can lend themselves to time of arrival control with the addition of equipment for that purpose, but the typical design does not allow for phase-locking the carrier to a common source or for perfectly replicating the modulation of the main transmitter. In the synchronous booster, all three factors are addressed.

POWER, COVERAGE PERMITTED, AND OWNERSHIP

The FCC has very specific standards for effective radiated power (ERP) and areas of coverage permitted for boosters and translators, as well as for their location and operation, all of which are covered under Part 74 of the Rules and Regulations [3]. The rules are nearly as complex as those for primary stations, and licensees considering a booster or translator would be well served by seeking advice from qualified consulting engineers.

The FCC recognizes two distinct types of translators: fill-in and non-fill-in. Fill-in translators can be owned and operated by the primary station licensee if its coverage does not extend beyond the FCC protected coverage contour of the primary station. Non-fill-ins are used to extend the coverage of a primary FM station beyond its protected coverage contour into communities that they were not predicted to cover. For commercial stations, this type of translator may only be owned and operated by an entity with no financial connection to, or support from, the licensee of the primary station. There is an exception to that rule for the few areas of the country where there is no coverage from any FM or AM station. Noncommercial stations may operate their own fill-in and non-fill-in translators. For all commercial licensees, only fill-in translators are permitted. Boosters are permitted for fill-in use only and may only be owned and operated by the licensee of the primary station. Persons or entities with no connections to primary stations may operate fill-in or non-fill-in translators for those stations, provided they have the express permission of the primary station.

Translator Power Level

Translators may operate with a power of up to 250 watts, with no restriction on the height above average terrain (HAAT). However, they must meet certain requirements regarding contour overlap that are covered in the FCC Rules [4]. Boosters are permitted to operate with an ERP of up to 20% of that permitted for the class of the primary station. As an example, a class B station is allowed a maximum of 50 kW, therefore, a booster for a class B would be allowed an ERP of up to 10 kW. As with translators, there is no restriction on the HAAT of a booster as long as the predicted protected contour of the main station (57 dBu for class B1 stations, 54 dBu for class B stations, and 60 dBu for all others) fully surrounds the same contour of the booster. There are no minimum power levels for boosters or translators.

Translators are not permitted to be used *solely* for the purpose of relaying a signal to another translator.

The FCC has provided an easy-to-read overview of the regulatory issues of boosters and translators, complete with hypertext links to the relevant regulations on its website at http://www.fcc.gov/mb/audio/translator.html. This is an excellent resource for potential applicants as well as current licensees.

COST ANALYSIS

Determining if a translator or booster is economically feasible is an important first step, and several factors must be carefully assessed before moving forward.

- *Population to be served and legal restrictions.* By referring to the FCC Rules and Regulations and previous paragraphs, the licensee can determine if it is entitled to a fill-in or non-fill-in station. Once that is known, the desired coverage areas can be determined and defined in terms of geographic area and population.

- *Revenue potential.* For commercial stations, potential revenue can be estimated with help from the sales and programming departments, which can determine how the added signal will enhance the ability to compete. The added coverage may be in the Arbitron market and add to existing ratings, hence adding to market revenue share. There may be potential advertisers in the new area. For noncommercial broadcasters, the mission must be considered and balanced against the cost. For those who are not licensees but are looking to import programming into a community, the cost to build and operate the facility must be determined.
- *Choosing the right method of coverage.* Once the desired coverage area has been identified, determine the best method for coverage, such as a translator or a booster, and if a booster, a synchronous or nonsynchronous type. Other criteria include the site and antenna elevation, power level, and directional characteristics of the receive and transmit antennas. Perhaps an existing site is available, and if not, the construction of a new facility introduces new potential problems. Plans may change as the project moves forward, but it is advisable to have a starting point to work from for the next steps.

Costs to Be Considered

The initial construction costs of the project and the ongoing expenses must be carefully considered. Start with a spreadsheet and get more specific with each item as the project progresses. Some of the costs to be researched include:

- *Engineering and legal costs.* These would include consulting engineers who will prepare the FCC application and its various exhibits, communications counsel, fees charged by the FCC for the construction permit (CP) and the license, and the engineering expenses related to the installation and testing of the equipment.
- *Equipment.* This may be difficult to estimate at this stage of the project since there may be uncertainty in the power level, program feed method, and antenna type. Equipment costs can range widely based on many factors. One way to make an estimate is to look at a high range, low range, and reasonable range in between. Continuously update the estimated costs as plans change.
- *Permitting and land use.* Every local jurisdiction will be different, and costs will range widely based on certain land-use factors even within a particular jurisdiction. The local zoning board might allow a large tower in an area that is zoned for industrial use with just a simple application, but it might require some very expensive environmental reports and special remediation work and perhaps a public hearing in an environmentally or esthetically sensitive area. It is advisable to seek the advice of a local land-use consultant. This may heavily influence where the station will be located, or if the project is feasible.
- *Construction.* The costs for utility installation and connection, and work that may be needed for access, such as roads; fencing and gates; the building or enclosure to be used; HVAC; electrical; tower costs including soils reports, foundation, and erection of a new tower or improvements that may be needed for an existing tower; and land improvement such as grading, trenching, drainage, tree removal, etc., must also be taken into account.
- *Ongoing expenses.* Recurring costs, such as rent, utilities, insurance, equipment maintenance, land and tower maintenance, local and state taxes, and FCC annual regulatory fees should be examined and worked into the budget.

Justifying the Initial Cost and Ongoing Expenses

Once a desired coverage area has been defined the next step is to justify the costs. For noncommercial operations, this may be as simple as showing the cost per additional potential listener that will be added to the service area.

For commercial operations, it is necessary to collect as much data as possible from sales, programming, and management about how improved coverage can be used to generate revenue. The departments should work collaboratively with the goal of estimating increased cash flow. There is going to be some estimation, but the final decision maker will need some well thought out projections on which to base a decision. To help each of the departments make their projections, prepare coverage maps and population data that depict the coverage before and after the addition of the translator or booster. If possible, these should be based on real-world signal projections rather than on FCC contours. (For a detailed look at real-world coverage maps, see Chapter 1.8.) Assemble the maps and other information into a presentation to help explain the benefits and drawbacks of adding a translator or a booster.

The final decision to go forward will most likely be made on the basis of a "capital budgeting project," which is a way of comparing the costs to the benefits in an objective manner. Some of these evaluations can be made on a simple determination of how long it will take to amortize the initial cost with the new cash flow generated. Another quick calculation is to determine the net annual cash flow increase, then multiply that by the "cash flow multiples" that are used in determining the valuation of a station. If that number is significantly greater than the initial costs, plus any negative cash flow in the early months or years of operation, the project will have a better chance of gaining approval.

A more formal method of capital budgeting used by some managers is "risk-adjusted net present value," which is based on the proposition that a dollar in hand today is worth more than a dollar expected in the future. A manager using this method will consider cash inflows and outflows, including such items as depreciation and taxes, for each of several years with the assumption that the project will end and the equipment will be liquidated after the final year. The net cash flow for each year is "discounted" along with the liquidation value of the equipment to make it comparable to the value of money today. This discount is determined by adding both the projected cost of borrowing money and a factor to adjust for the risk that the investment might not reach its goal. These adjusted cash flows are then added to find the risk-adjusted net present value of the project. If that value is greater than the initial investment, the project is considered to be viable.

PLACEMENT GUIDELINES FOR TRANSLATORS AND BOOSTERS

While boosters operate on the same channel and within the existing coverage of the primary station, finding a channel for a translator station can be a challenge in most populated areas of the country. The charts shown in FCC Rule Section 73.1204 show the various contours that must be protected by a translator station. There are several computer programs that can plot these contours and help determine if there is a channel available to serve a particular area, and if so, where it can be located and what the boundaries of the coverage contour will be. Boosters have few restrictions with regard to other stations, except that their signal contours must be 6 dB below the protected contours of any existing first-adjacent channel station.

Figures 4.12-4, 4.12-5, and 4.12-6 show an example of how a fill-in translator can be located to serve a community within the FCC protected contour of a primary commercial station. In this example, the gray areas in Figure 4.12-4 show the 60 dBu Longley-Rice projected coverage of the primary station, while the black line shows the FCC F(50,50) 60 dBu coverage contour. A highway running through the community is also shown.

A well-placed translator can be constructed by the licensee to serve this community if it does not extend the 60 dBu FCC coverage contour of the primary station, and does not cause a prohibited contour overlap with any existing station. In Figure 4.12-5, the 60 dBu F(50,50) FCC contour of a proposed translator on a nearby hill is mapped, along with the contour that could interfere with a first adjacent station. The first adjacent station's protected contour and the actual

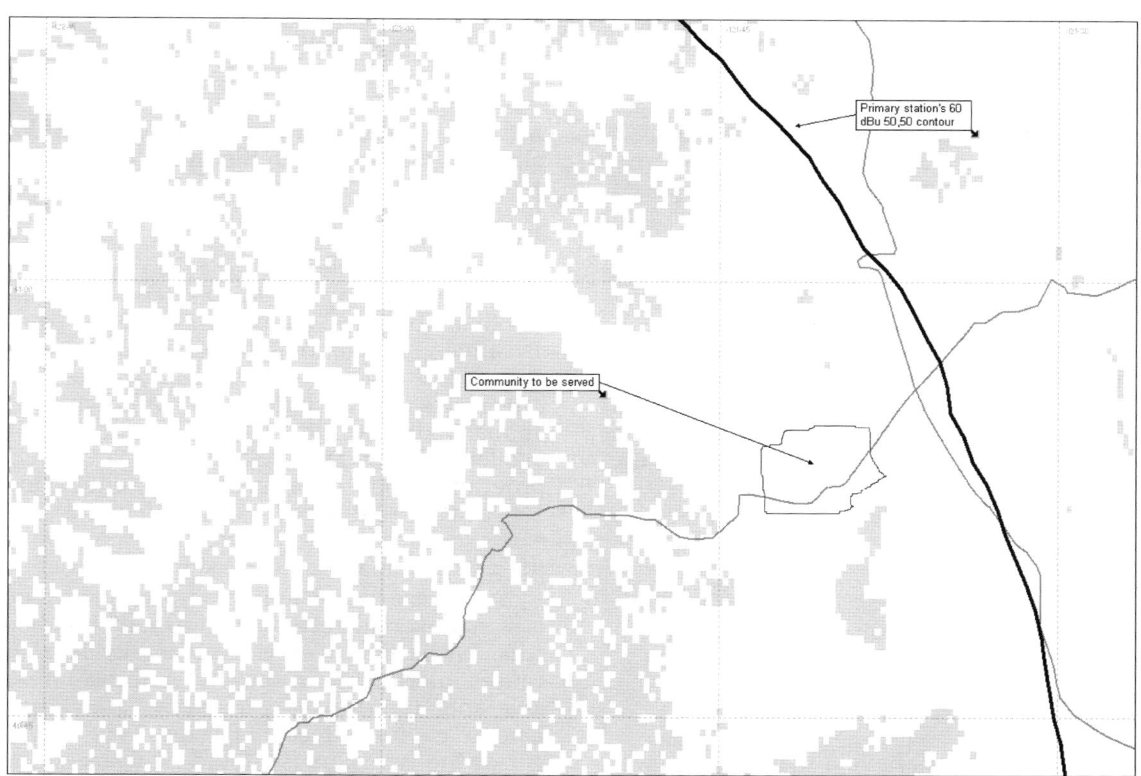

FIGURE 4.12-4 Longley-Rice (terrain sensitive) predicted 60 dBu signals (gray areas) do not cover community to be served while primary station FCC F(50,50) 60 dBu contour (heavy black line) shows apparent coverage.

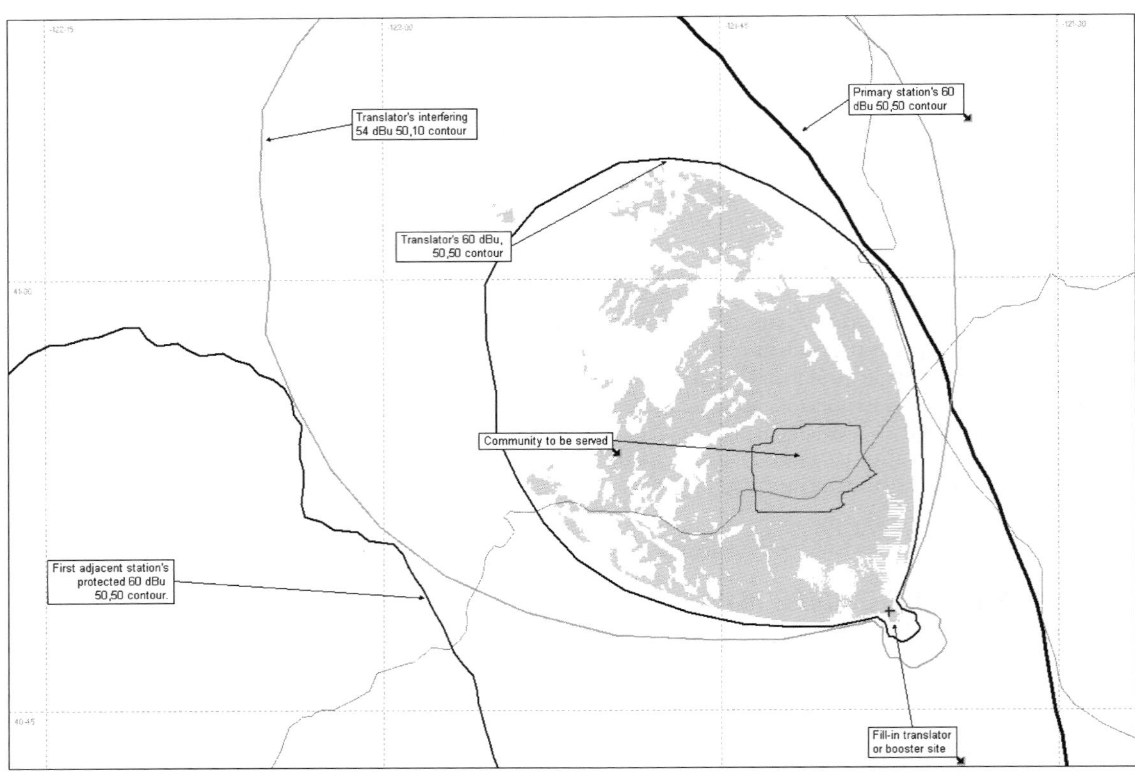

FIGURE 4.12-5 Map showing Longley-Rice predicted 60 dBu contour (gray areas) and FCC F(50,50) contour of proposed translator over the community to be served. Primary station FCC F(50,50) 60 dBu contour (heavy black line) and the translator F(50,10) interference contour are shown with relation to adjacent channel 60 dBu F(50,50) contour.

area of 60 dBu or better signal is determined using a terrain-sensitive, Longley-Rice–based mapping program. The FCC contour plots show that the proposed translator would meet regulatory requirements, and that the actual coverage would serve the community as well as its surrounding areas and portions of the highway. The combined coverage of the primary station and the translator is shown in Figure 4.12-6.

The selected site should mainly be line-of-sight to the community to be served and provide an appropriate signal intensity to the population to be served, but must not have prohibited contour overlaps. The parameters that need to be evaluated in reaching these three goals are:

- Topography of the area.
- Elevation of the transmission site.
- ERP of the translator.
- Directional patterns of the transmit and receive antennas.
- Co-channel and adjacent channel interference to be overcome.

In most cases, translators have a maximum power of 250 watts ERP and boosters may use as much as 20% of the ERP allowed for the class of the primary station. However, if there is a contour overlap, there will be a tradeoff between elevation and ERP that may play a significant roll in the signal intensity in the targeted area. Is some situations, a lower elevation site closer to the targeted area with a higher ERP may produce a more usable signal than a very high site with lower ERP even though the FCC coverage contour of the higher site may appear to extend the signal farther. In other situations, the opposite may be true. Terrain-sensitive propagation mapping programs are valuable, if not essential, in comparing sites that may be candidates for translators, and it is important to model as many diverse sites as possible (if the software is available) before making a final choice. Such computer programs (some of which are described in Chapter 1.8) can also be used to determine if a line-of-sight exists for any signal path that may be needed to feed the program material to the site.

In the case of boosters, an important area of concern is signal overlap between the primary and booster signals. While this overlap is generally legal (unless it causes interference in the city of license) it is undesirable and unfortunately, in many cases, unavoidable. Due to "capture effect" of FM receivers, only the stronger of the two signals should be heard by a radio in the overlap area. In theory, this is true, but in practice, one signal generally has to be 15–20 dB greater than the other for it to be "captured" to the point that the

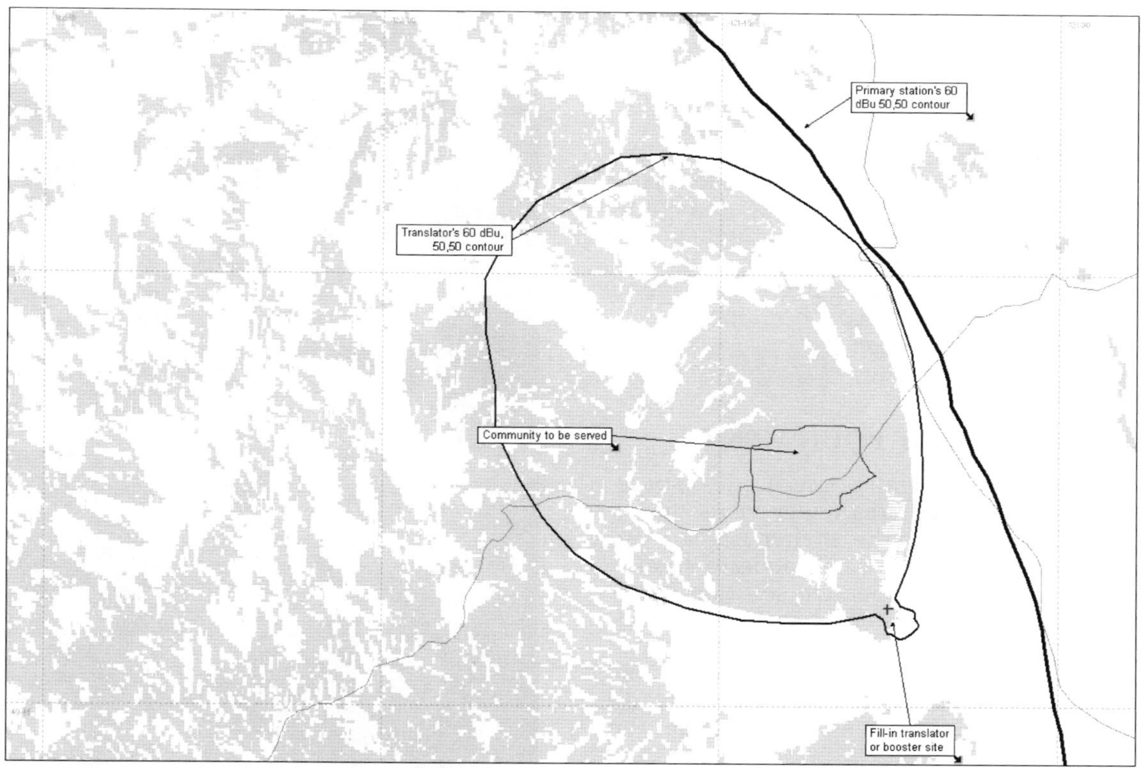

FIGURE 4.12-6 Map showing combined Longley-Rice predicted 60 dBu contours (gray areas) for primary station and translator compared with FCC F(50,50) 60 dBu contours (dark lines).

weaker signal causes no noticeable interference. The more similar in strength the signals are, the greater the potential for interference.

The best general advice is that the booster signal should be "cut off" from the primary signal by a combination of terrain, antenna directional pattern, and ERP that will result in as little overlap as possible, and that as much of those overlap areas as possible should be over sparsely populated or traveled areas. Interference in some small areas of overlap that fall over highly populated or traveled areas can, in many cases, be mitigated by use of a synchronous booster system as described earlier. Plotting the critical time of arrival versus signal-intensity difference issue is best done using terrain-sensitive propagation mapping techniques described in Chapter 1.8.

SITE SELECTION

Selecting a site for a translator or booster can be nearly as challenging as selecting one for a primary station. Important factors that should be considered include:

Obstructions. Unless planning to use a satellite or landline feed, a good line-of-sight path from the primary station transmitter is required, or from some alternate feed source such as a 950 MHz link. Avoid terrain features and trees or buildings that will cause blockage or multipath of the transmitted signal or of the received feed signal.

Accessibility and security. Unrestricted site access and security are important considerations along with methods to keep unauthorized people from the site, equipment, and tower.

Nearby radio frequency (RF) sources. Determine if there are any strong nearby RF sources that could prevent reception of the feed signal or mix with the transmitted signal to produce intermodulation distortion. If locating at an existing communications site, this could be a serious issue. Note that a newcomer to a multiuse RF site is responsible for solving problems created by the new facility. Use of 950 MHz links will require FCC-mandated coordination. However, a transmitter operating in the 850 MHz cellular band at the same site requires no coordination, but could require special filtering to prevent receiver overload.

Power and telephone service. Determine that adequate power is available at the site, and if not, investigate the costs and issues of installing it. The same is true of telephone lines if the system will require them.

Land-use issues. If proposing to build a new site in an environmentally protected area or in a residential area, there may be significant problems securing the permits needed from local or other governmental authorities. Many elevated areas that make excellent transmitter sites have been designated as "scenic resource areas" and may be off limits or require significant visual mitigation efforts and/or expensive environmental impact reports (EIRs). Check the applicable zoning regulations, or consider hiring a zoning consultant who has experience with the local authorities. Residents in the immediate vicinity may consider the project to be an eyesore. It may be necessary to secure variances that in turn may require hearings and notification of nearby property owners. In situations like this, a little PR goes a long way and both neighbors and community leaders are more likely to respond positively to the project if they have been included in the planning from the early stages.

HOUSING THE RF EQUIPMENT

A translator or booster generally occupies less space, gives off less heat, and requires fewer visits than a main transmitter. Therefore, the shelter needed can be smaller and significantly less elaborate than a main transmitter site. Many creative solutions have been used to house translators and boosters that represent a good compromise between utility and cost. If it is planned to locate at an existing communications site, most of the issues relating to shelter will already be addressed, but if building a new site, consider the following:

- *Type and size.* Buildings can be of cinder block, wood, fiberglass, or metal, such as prefabricated housings, cargo containers, or sheds. Don't overlook any existing structures, such as barns, garages, or other buildings. Depending on the local environment, only a small weatherproofed rack may be required, such as shown in Figure 4.12-7.
- *Temperature control.* If the climate is variable, heating or air conditioning may be needed in addition to reasonable ventilation to stay within the equipment manufacturers' recommended limits. If there will be more equipment added later, plan ahead for the additional heat.
- *Dust, dirt, and moisture.* These elements can cause failures and shorten equipment life.
- *Power issues.* Even if there is reliable power at the site plan for short-term outages or occasional "brown-outs" with a uninterruptible power supply (UPS) system, or longer-term outages with a generator.
- *Grounding.* Consider the three major types of grounding: safety, RF, and lightning. Safety grounding in AC power circuits is usually spelled out in local construction codes, which should be followed. RF grounding can be important if the equipment is located in a high RF field and especially if it is located at or near an AM broadcast site. Lightning grounding is an extremely important and often overlooked area that should be carefully researched, designed, and implemented. Surge suppression equipment should be installed on incoming AC power lines and telco lines, and transmission lines should be grounded at a common ground "window" at the building utility entrance that is connected to the racks and equipment. Chapter 9.3, Facility Grounding Practice, addresses this in detail.
- *Insects and rodents.* Pests can be avoided by using weather-tight, sealed entrances for transmission lines, power and telco, as well as for doors and windows. All air ventilation should be filtered.

FIGURE 4.12-7 A weatherproofed rack, surrounded by an 8 ft redwood fence, provides shelter for this synchronous booster and 250 watt amplifier. Since the station was located in an area with a mild climate, only the two small fans on the ceiling were needed to control the temperature.

FEED METHODS FOR TRANSLATORS AND BOOSTERS

Once the site has been chosen, the next challenge is to determine how the program signal will be fed from the studio to the translator or booster. There are a variety of methods available.

Direct Feed

If the chosen site is in the line of sight from the main transmitter, a feed system consisting only of a high-performance off-air FM receiver with its composite output connected to the translator transmitter's composite input is a good candidate. Nonsynchronous boosters can simply use a selective set of filters to feed a booster transmitter if there is adequate isolation between the receive and transmit antennas. This is the most economical method, but the main transmitter's signal at the site must be of adequate signal strength and free of multipath as well as co-channel and adjacent channel interference and noise.

950 MHz "Single-Hop" Analog or Digital Feed

If the studio has a clear line of sight to the translator or booster site, a feed consisting of a 950 MHz STL transmitter at the studio and a matching receiver at the site is a good choice. A composite analog system can be used, or if a stereo generator is installed ahead of the translator or booster transmitter, a discrete channel analog or digital system can be used.

"Multihop" Feeds

Often terrain or other obstacles will block a line-of-sight signal from the studio to the site, but if an intermediate site is available that has line of sight to both locations, the 950 MHz signal can be "translated" to another 950 MHz channel and sent to the second destination. These STL translators work well provided the transmitted signal does not overload its receiver. Another creative method is to locate a site with line of sight to the translator or booster site that also receives a good signal from the main transmitter or the studio. The main transmitter's signal or a studio STL signal can be received and retransmitted on 950 MHz to the translator or booster site.

IF Translators

A typical 950 MHz translator uses a receiver with a composite output feeding a 950 MHz transmitter on another frequency. This system works well, however, it can make subtle changes in the modulation that can be detrimental in the case of a synchronous booster system. This problem can be mitigated with the use of an STL translator that works by heterodyning the received signal down to an intermediate frequency (IF), which is then heterodyned back up the output frequency in the transmitter, thus producing an exact replica of the received signal.

Landline Feeds

Terrain, distance, or overcrowding may make use of the 950 MHz band impractical. In that case, a landline solution may be a viable alternative. Since equalized and phased pairs of 15 kHz analog phone lines are rarely available at a reasonable cost, digital transmission using ISDN, fractional T-1, or T-1 may be worthy of consideration. This method will cost more initially for equipment and will incur an ongoing cost, but both of these costs have been trending downward in recent years. Less-expensive alternatives, such as DSL and wireless broadband, have in the past been considered too unreliable for full-time delivery, however, reliability has been improving, and with adequate buffering to mitigate dropouts, they may be worthy of consideration.

Spread Spectrum RF Feeds

Point-to-point spread spectrum equipment operating in the unlicensed 2.4 GHz and 5.8 GHz UNII (unlicensed national information infrastructure) and ISM (industrial, scientific, and medical) bands are being successfully employed in many translator feed systems. However, since they are unlicensed, anyone can turn on equipment using the same frequencies and cause interference. Many consumer products, such as cordless phones and wireless LANs, share these frequencies as well as many Internet service providers (ISPs) that cover wide areas with service on the same channels. The advantage of using this method is that a very wide-band digital link, capable of delivering much more than just two uncompressed, high-quality digital audio channels, can be established with no need to obtain a license.

Satellite Feed

In cases where a single program may feed multiple translators that are separated from the main transmitter by long distances or challenging terrain, feeding by way of a satellite link may be an option. The cost of satellite transmission is high and varies widely based on many factors such as bandwidth and signal strength. It is rarely a cost-effective solution except where many translators are being fed from a common source.

RECEIVING THE FEED SIGNAL

Assuming that the translator or booster will be fed by an RF method, several important factors must be considered during the design phase of the project.

Line of Sight

While line of sight can sometimes be verified with visual observation, the typical distances involved usually require that computer mapping programs be used to verify that there are no obstructions. Adequate Fresnel zone clearance is an important factor as is allowance for annual tree growth.

Predicting the Signal Strength

Whether picking up the signal on the FM band or on a 950 MHz or other feed system described above, a study should be conducted to determine the predicted received signal strength. For the FM band, the existing signal at the proposed site can be measured with a field intensity meter, or it can be predicted by computer propagation analysis software. For feeds at 950 MHz and above, there are computer programs that can predict received signal strength based on variables, such as transmitter power, type and length of transmission lines, antenna gains, path distance, and earth curvature.

Antenna Types

For off-air signal feeds from the main transmitter, directional Yagi or log-periodic dipole antenna arrays are most often employed. Choose one with adequate gain, side and back rejection at the operating frequency to mitigate interference and front-end overload. Select one that is sturdy enough to handle the environment at the site and plan to place it at an elevation high enough to receive a solid signal, but not so close to the transmitting antenna that front-end overload becomes a problem.

Interference from Near and Far

A properly coordinated 950 MHz link should be free of interference from co-channel and adjacent channel signals if well constructed. However, nearby transmitters far removed from the receiving frequency can still cause front-end overload to the receiver. This is most prevalent when sharing a site with multiple users or when operating an STL translator, but can also be caused by paging transmitters or cell sites in the area. The potential for interference from these services can be reduced by using band pass filters on the receiving frequency alone or in combination with band reject filters on the frequencies of the offending transmitters.

Off-Air Reception

Off-air signals from the main transmitter can be affected by the transmitted signal from the translator output, and by strong signals from other nearby FM broadcasters as well as TV transmitters in the low-VHF-band channels. Band pass and band reject filters can cure this problem. Often the received signal from the main transmitter is weak enough to require a low-noise preamplifier. If a preamplifier is used in a high RF environment, it should always be placed *after* a set of filters, just ahead of the receiver.

Isolation

Wherever interference from nearby RF sources might be a problem, increased isolation of the receive antenna from the source of radiation can reduce problems. Consider both vertical and horizontal separation, positioning the antennas in the null of the other, cross-polarization, and placing them on opposite sides of obstructions such as buildings or hills if the situation allows. Avoid bundling receiving transmission lines with those used by transmitters on the same tower.

TRANSMITTING ANTENNAS FOR BOOSTERS AND TRANSLATORS

Once the proposed area of coverage has been determined and a nondirectional (ND) antenna or directional antenna (DA) has been selected, determining the specifics of the antenna is the next task. Some important criteria to be considered are discussed below.

Tower Space

The amount of "vertical real estate" available and the wind-load capacity of the tower are important issues to consider. If constructing a new tower, there will be more choices although economics may dictate the ultimate design of the antenna. If using space on an existing tower, there are likely to be some limitations. It is important to seek the advice of a competent, licensed structural engineer and to have the final design analyzed and "wet-stamped." Municipal jurisdictions, tower and property owners, as well as insurance companies may require such documentation.

Antenna Polarization

Boosters and translator transmit antennas may be vertically, horizontally, dual, or circularly polarized. There is no true consensus as to which type performs better, although the majority of systems is vertical. While circular polarization is now nearly universal for primary station antennas, choosing either vertical-only or horizontal-only antennas for translators and boosters will require less transmitter power, less antenna hardware, and less wind load than circular or dual polarization.

Stacking Antennas

Both DA and ND antennas can be "stacked" vertically to narrow the beam pattern in the vertical plane that increases gain and requires less transmitter power for a given ERP. Stacking also mitigates radiation toward the ground that may be an issue if the elevation above ground is relatively low or a multi-user site is already close to its radiation limit. ND antennas are not much different than primary station ND antennas in how they operate, although they can most often be designed to cost less by reducing their power-handling capability. Most antenna manufactures sell inexpensive, low power "educational" antennas that may suffice. Directional antennas that are most commonly used with boosters and translators are of two basic varieties described below.

Parasitic Array Antennas

One type is referred to as a "Yagi" or "Yagi-Uda" array that usually consists of a dipole element to which RF is fed, a passive "reflector" element on one side, and one or more "director" elements on the other, which when all combined, cause the antenna to radiate the signal in a unidirectional beam. All of the elements are mounted on a horizontal boom. The elements can all be vertical or horizontal, or both can be combined on the same boom for circular or dual polarization. Advantages of a Yagi include:

- Relatively high gain.
- A narrow beam pattern.
- Small amount of space.

Disadvantages include:

- It must be designed and tuned for a single FM channel.
- The front-to-back ratio of the pattern is not as high as log periodic directional antennas (LPDAs).
- That nearby conductors, such as other antennas sharing the tower, can cause the pattern to distort by detuning the parasitic elements.

Log Periodic Directional Array

Log periodic directional antennas (LPDA) look similar to their Yagi cousins, but they consist of a series of different length dipole elements on a boom, giving it a distinctive fish-bone look. RF is fed to the front dipole in the array, and each following dipole is tuned to a successively lower frequency and fed out of phase with the dipoles before and after it. The combination of parasitic reradiation and phase relationship of the dipoles that are nearest to resonance at the operating frequency, causes the array to have a unidirectional beam pattern. Characteristics of the LPDA include:

- Less gain for a given boom length than a Yagi.
- The design is inherently broadband.
- A typical design can cover the entire FM band with no tuning.
- Low voltage standing wave ratio (VSWR).
- Little change in gain or pattern over the design range.
- Not as susceptible to pattern distortion from nearby antennas as a Yagi.
- High front-to-back ratio, which is desirable in certain situations that require protection of another station.

LPDAs are made for vertical or horizontal polarization only, however, vertical and horizontal arrays can be arranged on a tower to deliver circular or dual polarization. Figure 4.12-8 shows the polar patterns of the Yagi and LPDA.

It is possible for multiple boosters or translators using the same site and covering substantially the same area to share a single broadband antenna with the use of a system of tuned cavities called a combiner. Conversely, a single booster transmitter can be used to serve more than one geographic area from a single site by splitting the signal into multiple beam antennas pointed in different directions.

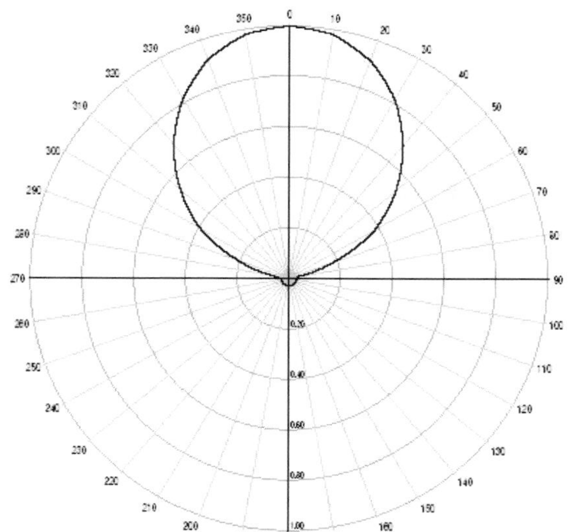

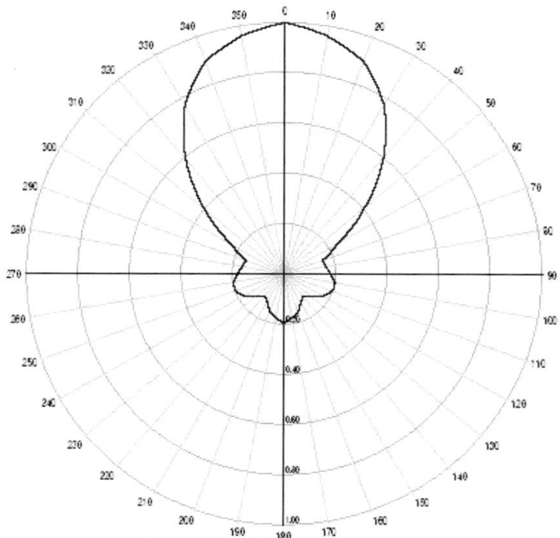

FIGURE 4.12-8 On the left is the azimuth pattern of a commercially available, eight-element, vertically polarized LPDA. On the right is the pattern of a circularly polarized five-element Yagi.

PROPAGATION AND COVERAGE MODELING FOR BOOSTERS AND TRANSLATORS

Longley-Rice–based, terrain-sensitive mapping computer programs have removed much of the guess work out of planning a successful translator or booster and a detailed discussion of these programs can be found in Chapter 1.8. Some specific applications of these programs useful to anyone planning a translator or booster project are described here.

Analysis of FCC Contours

When planning a booster or translator, it is necessary to determine if there will be any prohibited overlaps of coverage contours using the FCC's contour prediction method.

Analysis of Directional Antennas

By applying manufacturer's specifications of directional antennas being considered for a project, along with ERP and elevation at a particular set of coordinates, a pattern can be viewed both in terms of FCC-predicted contours and terrain-sensitive signal coverage methods.

Comparative Analysis of Sites

Often, several sites may be considered. Using a propagation analysis program, the best suited site can be chosen.

Preparation of FCC License Application Exhibits

When preparing the FCC application, it is necessary to demonstrate compliance with certain requirements, especially where there is potential for interference to another station. The FCC uses computer programs to check all applications, and inaccuracies or significant variations from the applicant's information may result in the application being returned or denied.

Evaluation of Interference Zones and Timing Issues for Boosters

It is important to obtain good performance and minimize interference in a booster system. Most computer programs will provide graphic representation of what will happen as the directional pattern and relative timing are adjusted as well as the signal between the primary transmitter and booster.

Checking for Line of Sight for the Input Feed

A path free of obstructions and having adequate Fresnel-zone clearance is necessary for any system using an RF feed for the program input.

Accuracy of Methods

There are many caveats to the use of Longley-Rice–based propagation programs, particularly in regard to the predicted signal strength once an obstruction has been encountered by the signal. Chapter 1.8 discusses propagation of radio waves in detail. Since the purpose of a translator or booster is to bring coverage to areas where coverage is less than desirable, determining just how poor existing coverage is may, in some cases, stretch the reliability limits of such programs. In the case of boosters, it is important to know where the signal strengths of the primary and booster stations overlap, and in cases where one or both signals traverse urban areas or hilly or mountainous terrain, it may be necessary to verify the signal conditions by field measurement, using a mobile, GPS-based data collection system as discussed in Chapter 4.15. Propagation analysis programs can help in displaying the collected signal data and comparing that to the predicted signals.

BOOSTER INTERFERENCE ZONES

A common problem that can accompany the use of a booster station is interference caused to the primary station. If all receivers had perfect capture ratios, this would not be a problem, since the receiver would demodulate only the strongest of the two signals on the channel and reject the other, but in fact, capture ratios on common consumer radios are less than perfect. Any area where the two signals are similar to each other in strength is likely to be an interference zone. Interference zones can be mitigated to some degree if the three parameters of the signal that cause interference can be controlled: carrier frequency difference, modulation difference, and time-of-arrival difference. If any of these three are not addressed in the design of the booster system, interference between the two transmitters is almost certain to occur at some location, particularly if the booster is desired to have seamless coverage between its service area and that of the primary transmitter.

The first task in mitigating interference zones in booster station design is to design the system from the beginning in such a way as to minimize them, or to cause them to fall over unpopulated areas. Careful use of Longley-Rice–based, terrain-sensitive propagation mapping programs can help in this process, as well as a thorough knowledge of the local terrain.

Terrain Shielding

The most effective way to separate the two signals is to use the terrain to block one signal from interfering with the other. This may mean locating the booster at a lower elevation to limit its coverage back into the coverage area of the main.

Directional Antennas

It may be necessary to experiment with several antenna designs and test them in the propagation software at various azimuth angles to get the null in a position that protects the primary station.

Power Levels

The tendency is to run the maximum amount of power that can be authorized, but in many situations, less is better. The converse can also be true—running a higher ERP can move an interference zone over a less important area.

Synchronization

If the interference zones have been minimized, precise synchronization of carrier frequencies can abate the undesirable heterodyne that results from two transmitters on slightly different frequencies. Synchronizing modulation such that each station's frequency swings precisely the same amount and in the same direction as the other at all modulation percentages will reduce the multipath-like noise that results from random differences in unsynchronized systems. A third parameter of synchronization involves controlling the time of arrival of the two signals such that they arrive simultaneously at a target area where the signals are nearly the same strength and where a good-quality signal is desired.

Synchronizing carrier frequency and modulation alone will reduce mutual interference problems. However, combing those with time-of-arrival control for full synchronization is the "gold standard" of booster systems. It is apparent why interference can result from two FM stereo signals of the same strength as they arrive at slightly different times and combine in a receiver. The top of the upper sideband of the stereo subcarrier is about 53 kHz. The time it takes for one cycle at 50 kHz is 20 µs, so a signal 90 degrees out of phase requires only a 5 µs delay, and is enough to cause slightly audible distortion in the stereo signal due to the effect of phase cancellation. At 10 µs, the 50 kHz signal is 180 degrees out of phase, the 25 kHz signal is 90 degrees out of phase, and the distortion gets progressively worse.

If a design goal is set of keeping the delay below 5 µs where the signals are equal or within ±5 dB, the distortion can be kept to minimal levels. As the signals depart in level, phase cancellation is reduced and the delay can increase on a sliding scale without causing significant interference until the signals are different by approximately 15 dB. In the range of ±5–15 dB, a reasonable design goal is to keep the delay to less than 10 µs. When the difference in signals is greater than 15 dB, phase cancellation is minimal and the capture effect of the receiver will reduce the interference to the point where time of arrival is no longer an issue. Note that every receiver will deal with the two signals differently due to differing capture ratios, but the above is a rule-of-thumb that has worked successfully in the field.

Time-of-Arrival Synchronization

It is possible to determine how much of an interference zone area time-of-arrival synchronization can mitigate. Consider that a radio wave travels just under 1 mile in 5 µs. The synchronization area for two signals that are within ±5 dB can be as small as ½ mile in either direction, or 1 mile total. On the other hand, if the primary transmitter and the booster transmitter both line up exactly in the direction of the interference zone to be synchronized and that zone is not located between the two transmitters, the synchronization zone can extend indefinitely. If the vector differential of the distances from the primary and the booster transmitter to a target are less than 1 mile, time-of-arrival difference can be synchronized to the point of insignificance.

Time-of-arrival synchronization is accomplished with digital delay devices that can be added on to analog or digital booster systems. The process can also be performed in software in some digital exciters. Determining the amount of delay for a specific system is an exercise in mapping and requires choosing a target location and determining the total difference in time-of-arrival delay and must include the delay caused in the STL paths in each signal. The delay is inserted at the transmitter with the shortest total path to the target, allowing the other transmitter to "catch up," and the amount of delay is generally set to 5.37 µs for each mile difference in the distances to the target. Fine tuning the delay time is done by tweaking it for minimum distortion on the stereo pilot at the target.

Frequency and Modulation Synchronization

Synchronization of the other two parameters, the carrier frequency and modulation, is inherent in an off-air fed system that simply amplifies the primary station's signal, making no significant changes to it. This can also be inherent in certain digital designs where identical, preprocessed digital audio streams are fed to identical digital exciters that are both phase-locked to a common source, such as GPS. With analog exciters, the most common method is to phase-lock the master oscillators of both the primary and the booster transmitters to the 19 kHz stereo pilot. If this method is used, it is important that the pilot frequency is always within ±2 PPM.

Modulation synchronization in the analog domain requires a complex system to ensure that both transmitters are fed with, and transmit, identical modulation. This can be achieved by feeding both exciters with 950 MHz signals from a single STL transmitter and then heterodyned down to the operating frequency. Alternately, a signal fed to either the primary or booster exciter by any feed method is extracted from the IF stage and heterodyned to 950 MHz for transmission to the other site, where it is then heterodyned back down to the operating frequency.

TRANSLATORS AND BOOSTERS USING IBOC

In Band, On Channel (IBOC) or HD Radio™ is the new digital system of broadcasting that promises to bring CD-quality broadcast signals to the audience. During the implementation period, IBOC will require that the analog signal continue to be broadcast with the digital carriers. Special designs will be required for translators and boosters that accommodate both signals. As with a primary station, any of the accepted methods of IBOC transmission (high-level combining, low-level combining, split-level combining, or use of a separate transmitter and antenna) can be employed, although the generally lower ERP of translators and boosters may point to some solutions that are more economic than their higher-power equivalents.

For a simple, low power translator that picks up the primary station with an off-air feed, the simplest implementation would be to shift the received signal to its new frequency and amplify it to the desired TPO. If the process was completely linear, theoretically, there should be an identical reproduction of the primary station's signal at the output of the translator. As of this writing, at least one manufacturer was testing an all-in-one unit that, when coupled with a truly linear amplifier, produces the desired signal on any other FM channel, including the first adjacent. According to the manufacturer, this method requires no additional HD Radio™ license fee (of course, an FCC license is required) and although it has not been tested, the method could theoretically work with an off-air fed booster as well, assuming adequate isolation between the input and output antennas.

An alternative is to apply existing HD Radio™ technology to an existing or proposed booster or translator, which requires use of an IBOC generator and exciter along with an appropriate feed method and combining scheme with the analog FM signal. One such system has been built and tested by National Public Radio at KCSN FM, Northridge, CA. The complete description of the system as well as the result of extensive testing can be found on the Internet at http://www.nprlabs.org/reports/KCSN_Report.pdf.

Although this was an on-channel booster system, the same technology could easily be applied to a translator. The system described in the report is fully synchronized and is fed using T1 landlines to both the primary and booster stations. The carrier frequencies are locked by GPS signals and time-of-arrival delay is accomplished in the codecs that are used to convert the processed analog audio at the studio to T1 digital signals. Bidirectional data circuits from the two transmitter sites ensure that the required delay does not change in the event the T1 service provider changes its network in such a way as to change the time delay in the delivery of the signals. This type of a system could be simplified if applied to a translator since time-of-arrival delay is irrelevant.

SUMMARY

Translators and boosters can bring important benefits to many FM broadcast stations but are often overlooked due to their complexity and their low priority in relation to overall station operation. However, as communities grow, populations move, and competition heightens, station managers are increasingly considering their value. The engineering department is usually charged with the responsibility of investigating, designing, and constructing a system, but often lacks the time, the tools, and the practical experience to maximize the benefits a booster or translator can bring. That is why it is important to seek advice from experienced technical advisors who can help organize the project and work with the local engineer and management to see the project through to a successful completion. As with any project, planning is the key to success, particularly in these areas:

- Regulatory
- Financial
- Environmental
- Structural design
- Electrical design
- Coverage prediction and maximization
- Interference prediction and mitigation
- Construction
- Final evaluation

The future of radio is as good as it has ever been, and translators and boosters will be an important factor in helping to bring new and exciting programming and technologies to millions of listeners and to better serve the broadcast audiences.

References

[1] Bullington, K. Radio Propagation at Frequencies above 30 Mc/s, *Proceedings of IRE*, 35, 1947.

[2] Salek, S. Analysis of FM Booster System Configurations, *NAB Broadcast Engineering Conference Proceedings*, 1992. See at http://www.h-e.com/pdfs/ss_nab92.pdf.

[3] A summary of FCC Rules can be found on the FCC's website at http://www.fcc.gov/mb/audio/bickel/part74rule.html.

[4] See FCC Rule Section 74.1204.

CHAPTER

4.13

AM & FM IBOC Systems and Equipment

JEFF DETWEILER
iBiquity Digital Corporation
Columbia, Maryland

INTRODUCTION

Enhanced performance, increased reliability, flexibility, and cost effectiveness are the principal motivations for industries transitioning from analog to digital technology. For many countries including the United States, insufficient spectrum exists to accommodate the transition of radio from analog to digital.

HD Radio™,[1] a digital radio broadcast technology, developed by iBiquity Digital, operates within the Federal Communications Commission (FCC) allocated Amplitude Modulation (AM) and Frequency Modulation (FM) bandwidths. iBiquity Digital's technology operates within the FM band (In-Band) while making use of existing broadcaster analog channel assignments (On-Channel), thus establishing the In-Band, On-Channel (IBOC) digital radio broadcasting (DRB) system.

The iBiquity HD Radio brand of IBOC technology allows broadcasters to add digital signals to their existing analog broadcasts, providing the capability for an eventual transition to digital-only transmission. Two modes define how audio and data are transmitted: *Hybrid* and *All Digital*. The Hybrid mode includes both the existing analog and the new digital services. Broadcasters are using this mode during rollout of Digital IBOC technology to permit analog-only radios continued operation. In the future, when the market is fully capable of receiving HD Radio digital signals, broadcasters can switch to the All Digital mode. Hybrid receivers employing HD Radio technology are backward and forward compatible, allowing them to receive current analog broadcasts in addition to all digital broadcasts.

[1]HD Radio ™ is a trademark of iBiquity Digital Corporation.

Conversion to Digital Technologies

HD Radio technology was designed to provide a unique opportunity for broadcasters and consumers to convert from analog to digital broadcasting without service disruption, while maintaining the current dial positions of existing radio stations. Consumers who purchase digital radios will receive AM and FM stations broadcasting with digital signals with digital sound quality. Additionally, consumers will have the capability to receive new multicast audio channels and wireless data services rendered on a radio's display screens, similar to those offered on web-enabled cell phones and pagers. Program Service Data (PSD) such as artist, title, program content, news, sports, local traffic, and weather are available today.

National Radio Systems Committee Test Reports

The NRSC has provided a forum for all parties interested in digital terrestrial radio to evaluate IBOC systems and to provide industry input to the FCC.

In 1999, the NRSC developed its first set of IBOC test guidelines, which were designed for testing prototype IBOC systems. On December 15, 1999, USA Digital Radio filed its Report on Laboratory and Field Testing with the NRSC. On January 24, 2000, Lucent Digital Radio filed its Test Report, which demonstrated system performance using the PAC™[2] audio CODEC technology. These reports confirmed the ability of IBOC technology to provide upgraded audio quality and robustness in both the AM and FM bands without causing harmful interference to existing analog broadcasts in those bands.

During 2000 and early 2001, the NRSC continued its analysis of IBOC technology. In December 2000, the NRSC adopted new FM laboratory and field test procedures to be used for final IBOC validation testing. In April 2001, the NRSC adopted new laboratory and field test procedures for AM IBOC testing.

During 2001, iBiquity completed an extensive test program of its FM and AM technology. These tests followed the NRSC test procedures and were conducted using independent laboratories in Alexandria, Virginia; Cincinnati, Ohio; and Austin, Texas. In addition, field tests were conducted using seven commercial and one experimental FM station and three commercial and one experimental AM station. NRSC observers actively participated in and oversaw all tests. The tests were designed to assess both the HD Radio system's performance in a variety of interference and impairment scenarios typically found in the real world. They also studied the potential impact of the IBOC system on analog operations of the host station as well as co- and adjacent channel analog stations. The tests involved the use of analog receivers representing OEM auto, aftermarket auto, home, and portable receivers. Simultaneous tests were conducted on the digital and analog receivers to allow for direct comparison of digital and analog broadcasting. In addition to objective measurements obtained in the lab and field, these tests also involved an extensive subjective evaluation of digital and analog audio samples. In the case of FM, more than 480 listeners subjectively evaluated thousands of audio samples.

Additional information about the NRSC and its evaluation of IBOC technology can be found at the NRSC website: http://www.nab.org/SciTech/nrsc.asp.

FCC Approval Process

iBiquity, as USA Digital Radio, initiated the current FCC consideration of IBOC by filing a Petition for Rulemaking on October 7, 1998. After a series of comments supporting the USA Digital Radio Petition, on November 1, 1999, the FCC issued a Notice of Proposed Rulemaking (NPRM) on terrestrial digital radio broadcasting. USA Digital Radio filed both Comments and Reply Comments on the NPRM. Comments of interested parties in reaction to the FCC's NPRM can be viewed at http://www.fcc.gov/mb/audio/includes/23-digital.htm.

On July 2, 2001, the Radio Board of the National Association of Broadcasters (NAB) filed with the FCC its resolution endorsing iBiquity's IBOC system and encouraged the FCC to take steps that will enable a fast rollout of the technology.

[2]PAC™ is a trademark of Lucent Technologies, Inc.

On November 29, 2001, the NRSC completed its evaluation of iBiquity's FM IBOC system and recommended that the FCC authorize the technology as an enhancement to the current analog FM broadcasting system in the United States. The NRSC's evaluation was based on extensive field and laboratory tests of the FM IBOC system conducted in accordance with the NRSC's test procedures. The NRSC completed its evaluation of iBiquity's AM IBOC system in April 2002.

On December 19, 2001, the FCC issued a Public Notice seeking comments on the evaluation of iBiquity's FM system by the NRSC. Supporting comments endorsing IBOC were filed by all components of the radio industry. The comments and reply comments for proceeding MM Docket No. 99-325 overwhelmingly supported FCC adoption of IBOC as the digital broadcasting standard for the United States. iBiquity filed comments and reply comments with the FCC.

On April 19, 2002, the FCC issued a Public Notice seeking comment on the NRSC's endorsement of iBiquity's AM IBOC system. Comments were due June 18, 2002.

Domestic Regulatory Status

On October 10, 2002, the FCC authorized broadcasters to commence digital broadcasts using iBiquity's HD Radio technology for FM and AM daytime broadcasting. This approval allowed broadcasters to move forward with the implementation of HD Radio technology and gave receiver manufacturers enough confidence in the systems' marketability to begin development of HD Radio products. Much progress has also been made on several open matters, including AM nighttime transmission, the use of separate antennas for FM analog and digital signals using an STA, and the formalization of the FCC's rules.

Further Notice of Proposed Rulemaking

On April 15, 2004, the FCC adopted a Further Notice of Proposed Rulemaking and Notice of Inquiry on IBOC technology. The Further Notice proposed to amend the existing rules for AM and FM radio to further the introduction of the IBOC system. The Further Notice also sought comment on several new initiatives including the introduction of nighttime digital AM broadcasting and multicasting of the digital service to allow for the introduction of multicast (supplemental audio) and datacasting services. The FCC intended to expedite its treatment of these issues to support the continued rollout of IBOC digital radio service.

AM Nighttime Transmission

When the FCC approved IBOC technology, it held off on its approval of AM nighttime transmission pendng completion of additional tests. iBiquity Digital subsequently carried out additional tests under the oversight of the NRSC and submitted the results for the

NAB's review. Based on the test results, on January 20, 2004, the NAB voted unanimously to approve IBOC AM broadcasting at night, and on March 5, 2004, the NAB submitted an endorsement of IBOC AM nighttime transmission to the FCC, asking for approval. In their letter, the NAB wrote that the dramatically improved audio quality from IBOC service is well worth the predicted and limited reductions in analog coverage. In support of the NAB's endorsement, iBiquity submitted a technical report to the FCC on March 5, 2004, consisting of an AM nighttime compatibility report and two field studies of HD Radio's AM nighttime performance. On April 14, the FCC issued a public notice seeking comment on the NAB's recommendations. Comments were due June 14 and reply comments due July 14. The NAB's recommendations and the AM nighttime reports are available on the FCC's website: http://gullfoss2.fcc.gov/prod/ecfs/comsrch_v2.cgi under proceeding #99-325.

Use of Separate Antennas for Analog and Digital Signals

On March 17, 2004, the FCC approved the use of separate analog and digital station antennas to initiate IBOC FM transmissions provided broadcasters apply for an STA. This separate antenna option will enable many FM stations to use existing equipment and reduce IBOC implementation costs. The FCC's decision was based on the recommendation and report filed by the NAB on July 24, 2003. This report included extensive field tests conducted by iBiquity that were specified and witnessed by the NAB. Effective with the release of the FCC's Public Notice, FM stations may file requests for STAs to begin IBOC transmission using dual antenna systems provided they meet the following criteria:

- Digital transmission must use a licensed auxiliary antenna;
- Auxiliary antenna must be within 3 seconds of latitude and longitude of the main antenna;
- Height above average terrain (HAAT) of the auxiliary antennas must be between 70 and 100% of the height above average terrain of the main antenna.

International Developments

In April 2001, the International Telecommunication Union (ITU) adopted Recommendation ITU-R BS.1514, which endorses iBiquity's AM IBOC technology as a standard for digital broadcasting in the broadcasting bands below 30 MHz. The ITU also endorsed iBiquity's FM IBOC system (referred to by the ITU as *Digital System C*) as a standard for digital broadcasting above 30 MHz in Recommendation ITU-R-BS.1114.

iBiquity also works with other international organizations to pursue international recognition of IBOC technology. iBiquity regularly participates in the meetings of CITEL, the Inter-American Telecommunications Commission of the Organization of American States (OAS). CITEL's Permanent Consultative Committee II (PCC II) has received regular briefings on the development of iBiquity's system. iBiquity also provides regulatory and technical briefings to broadcasting organizations in the United States and several countries abroad.

OVERVIEW OF HD RADIO IBOC CONCEPTS

In the Hybrid mode where both the analog and digital signals coexist, upper and lower low-level digital sidebands with multiple carriers are added to the analog spectrum, as shown in Figure 4.13-1 for FM and Figure 4.13-2 for AM. These carriers are modulated with redundant information to convey digital audio and data. Since the analog AM signal is amplitude modulated, a quadrature phase component can carry digital information. Thus, it can be placed directly beneath or in quadrature to the analog modulation, as shown in Figure 4.13-2. This additional information is transmitted at a lower power level to avoid increasing noise to the analog signal.

AM and FM Hybrid and All Digital Spectrums

In the FM All Digital mode, Figure 4.13-1, the analog signal is removed and additional data carriers are added. The main channel stereo audio and its associated data information are unchanged from the Hybrid mode; however, the power level may be increased to provide a more robust service. In the region between +100 kHz to +130 kHz, digital carriers are added to carry the digital audio backup and tuning channel and ancillary data. In the region bounded between +/–100 kHz, which previously carried the analog audio, additional carriers are added to carry new supplemental services like multicast audio channels, wireless data, surround sound, and more.

IBOC TECHNOLOGY BUILDING BLOCKS

Digital radio transmission systems are composed of three main building blocks:

- Audio coder/decoder;
- Modulator;
- Protocol stack.

iBiquity Digital's HD Radio modem technology utilizes patent and patent pending digital signaling techniques: QPSK modulation with Orthogonal Frequency Division Multiplexing (OFDM), Interleaving, Channel Coding/Error Correction, Time Diversity and Frequency Diversity, and others.

Audio Coder/Decoder (CODEC)

HD Radio technology is designed to deliver Compact Disc (CD)-like audio quality within the FM broadcast band and FM-like stereo audio quality in the AM band.

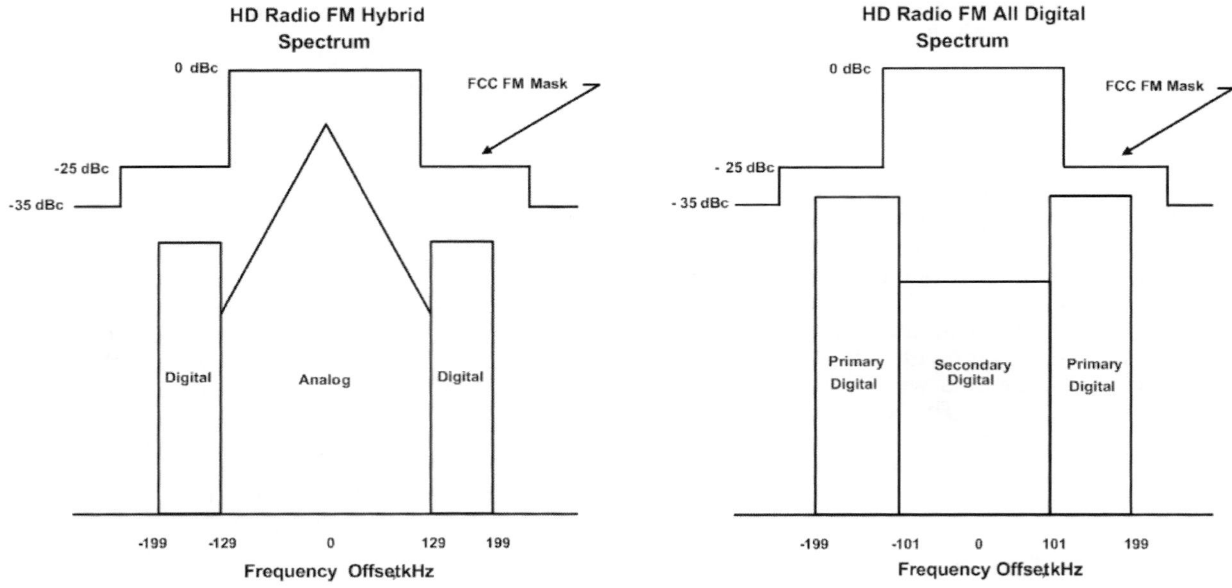

FIGURE 4.13-1 HD Radio Hybrid and All Digital FM spectrum.

iBiquity's HD Radio technology, like other digital broadcast technology systems, does not have sufficient bandwidth to carry the full digital data stream delivered by a CD player; defined as 16 bits per channel sampled at 44.1 kHz (44,100 × 16 bits × 2 channels = 1,411,200 bps). Therefore, some method of compression must be employed to reduce the audio data bit rate. The audio coder/decoder (CODEC) branded HDC™ is specifically designed to reduce the number of bits required to transmit a given quality audio signal.

CODECs use advanced signal processing and psycho-acoustic modeling to interpret human hearing and eliminate redundancies and irrelevancies in the audio signal. HDC employed in iBiquity's HD Radio technology for FM compresses the audio at a ratio of 15:1 enabling 96 bps to deliver CD-like quality audio. The AM CODEC delivers FM-like audio quality with a compression ratio of approximately 40:1. HDC offers compatibility with broadcast multichannel audio modes, e.g., 96 through 64 kbps for 5.1 channel audio configuration, built-in data channels, and transport features.

iBiquity's HDC is uniquely optimized for use in a digital audio broadcast system. In particular, it incor-

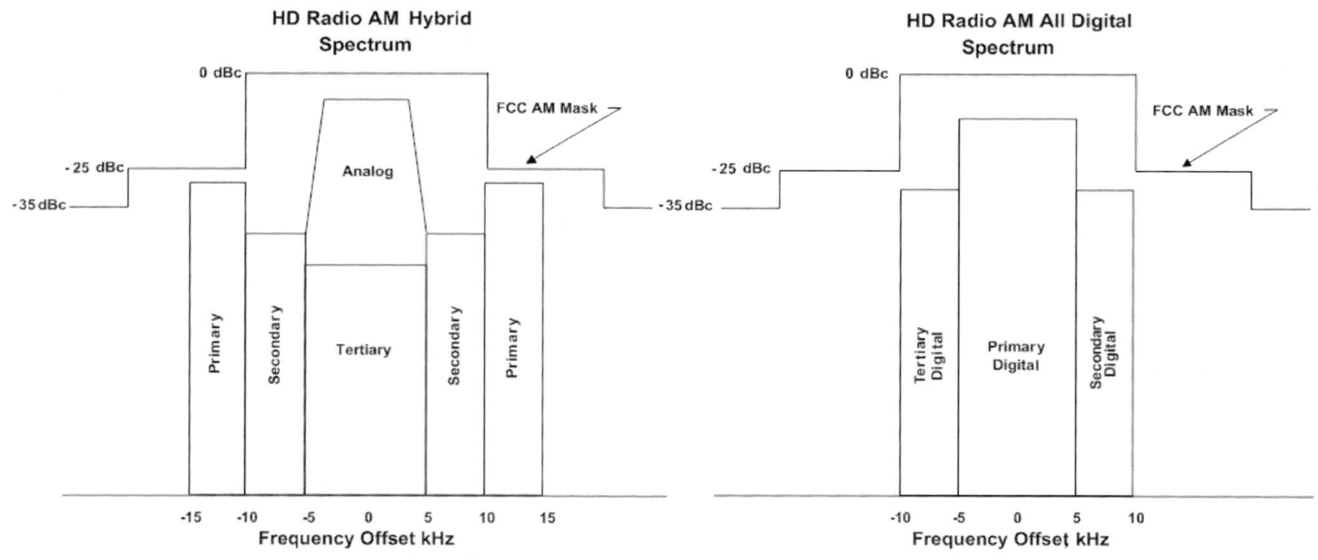

FIGURE 4.13-2 HD Radio Hybrid and All Digital AM spectrum.

porates powerful error-concealment techniques for mitigating the effects of channel errors. These features allow improved broadcast system design that is better matched to the prevailing channel conditions and interference.

Modulator

Unlike analog modulated audio, digital audio and data cannot be directly propagated over RF channels; therefore, a modulator is used to modulate the digital information onto an RF carrier. There are many signaling techniques used for digital modulation that vary in complexity and suitability to a given application. iBiquity's HD Radio technology employs Orthogonal Frequency Division Multiplexing (OFDM) to impress the data onto the RF channel, as shown in Figure 4.13-3. OFDM signaling employs multiple, overlapping, orthogonal subcarriers. Trade-offs between throughput and robustness can be made with an increasing number of phase states and signal levels, leading to more throughputs with less robustness in fading channels.

iBiquity's HD Radio technology uses OFDM techniques for robustness in the presence of multipath fading, interference, and noise. Additionally, OFDM is flexible for adding optional subcarriers and allows the placement of the most sensitive digital information in more robust portions of the spectrum. iBiquity's HD Radio technology for FM modulates the OFDM subcarriers with Phase Shift Keying (PSK) modulation. OFDM signals modulated with low complexity PSK are quite robust in the presence of interference and multipath fading.

HD Radio on AM operates in very narrow channels that are free of multipath fading. The narrow bandwidth leads to an OFDM modulation technique that is optimized for higher throughput. iBiquity's HD Radio technology for AM uses Quadrature Amplitude Modulation (QAM) on each subcarrier. In QAM systems, multiple phase and amplitude states are applied to each subcarrier in the OFDM waveform. Multiple phase and amplitude states allow greater data throughput in a given channel bandwidth. For AM systems, the timing of the symbols is optimized to ensure that duration of noise pulses is much shorter than the symbol duration of the QAM modulation, thus ensuring robust digital reception in the presence of static and noise that is prevalent in the AM channel.

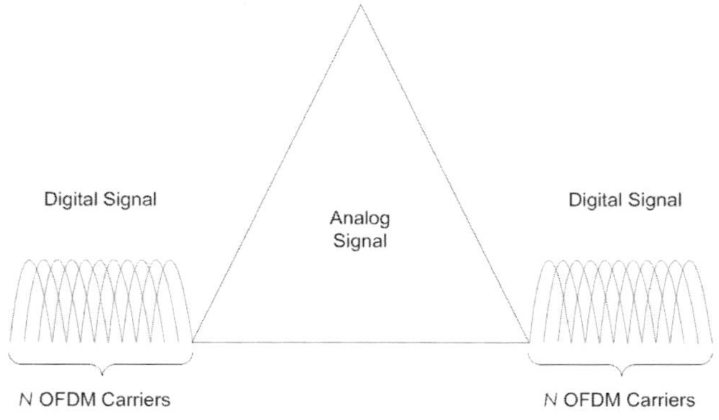

FIGURE 4.13-3 HD Radio spectrum depicting multiple overlapping OFDM carriers adjacent to FM signal.

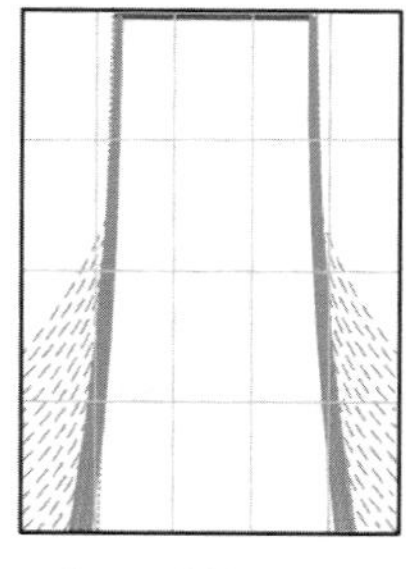

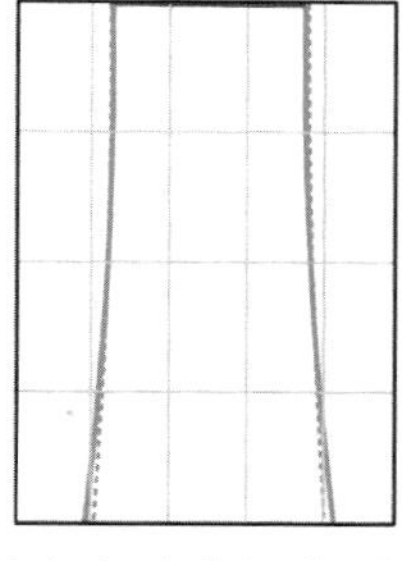

FIGURE 4.13-4 Side lobe reduction by Pulse Nyquist Root-Raised-Cosine Tapering.

iBiquity's HD Radio technology applies Pulse Nyquist Root-Raised-Cosine Tapering pulse shaping to its OFDM waveform. Pulse shaping aids in acquisition and relaxes frequency-tracking requirements while reducing spectral side lobes. The root-raised-cosine function creates a guard band between the symbols that improves intersymbol interference by reducing energy in the region where subcarriers overlap.

The benefit of side lobe reduction and its effect on out-of-band attenuation are clearly demonstrated in Figure 4.13-4.

The reduced side lobes result in less interference to adjacent channels, more robust performance in multipath, and less interference to the host analog signal.

OFDM is a highly flexible modulation technique that is ideally suited for digital radio broadcasting systems. This flexibility allows iBiquity's HD Radio technology to provide a high-quality digital audio broadcast, protect adjacent channels, and provide robust performance in the AM and FM bands.

Interleaving

Digital error correction techniques are enhanced if errors in transmission are spread in a manner that minimizes data loss in successive bits. Interleaving is a technique that scrambles the bits in a predetermined manner upon transmission and reassembles them in the receiver. The length of the interleaver is normally optimized to spread errors over a longer period of time than would exist in a channel fade. Figure 4.13-5 shows an example of bit interleaving.

In the example, the middle three bits—1, 7, and 5—are successive in the transmission path. If they were corrupted in the channel, the receiver de-interleaving process would despread the bits and reassemble the

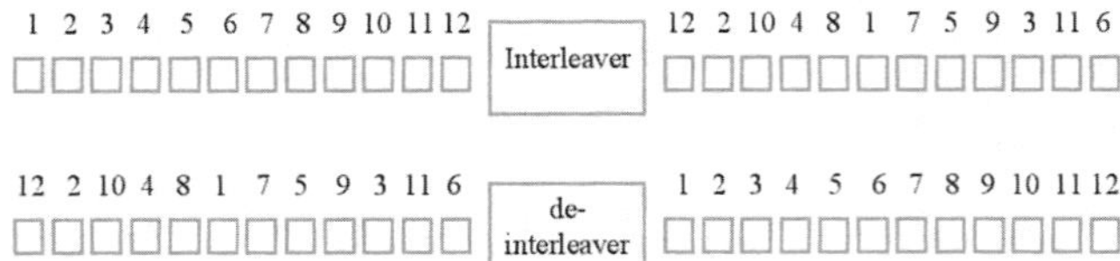

FIGURE 4.13-5 Example of interleaver and de-interleaver.

bitstream to ensure no two consecutive bits are in error. Digital systems typically employ interleavers to improve the performance of the system in the presence of channel impairments. Designers of digital systems must make trade-offs between interleaver length and system performance. Long interleaver lengths are desirable for providing robust reception at the expense of acquisition time. iBiquity's HD Radio technology for the Hybrid mode employs interleavers that are optimized for performance while using a backup channel for rapid tuning. The All Digital mode adds a backup digital audio channel with a short interleaver for rapid tuning. In each mode, the Main Program Service (MPS) has a rapid tuning channel provided as a backup signal service to ensure continuity of service in fading channel conditions.

Channel Coding/Error Correction

A digital bitstream, when passed through a transmission channel, is likely to encounter various forms of impairments including noise, multipath distortion, fading, and interference. Digital systems employ various error correction techniques to correct transmission errors. These algorithms improve robustness through the introduction of error correction bits. These error correction bits are used in the receiver to verify the accuracy of the recovered bitstream, detect errors, and provide restoration of the transmitted bitstream.

Error correction codes are typically specified as their coding rate (R = information bits/total bits). For example, a coder with R = 1/2 channel coding represents an algorithm in which half of the bits carry information and the other half of the bits carry the overhead of the error correction algorithm. Coding is sometimes called *Forward Error Correction* (FEC). With iBiquity's HD Radio technology, FEC and interleaving are closely tied together. In the presence of adjacent channel interference, the outer OFDM subcarriers are most vulnerable to corruption. The information, coding, and interleaving are specially tailored to deal with this nonuniform interference such that the communication of information remains robust in the presence of such interference. iBiquity's HD Radio technology for FM transmits a full-quality digital audio signal on each digital sideband (upper or lower) of the analog carrier. Each of these digital sidebands is detected and decoded independently with an FEC coding gain achieved by a rate 4/5 convolutional coder on each sideband. This redundancy permits operation on one sideband, while the other is corrupted. Using a patented technique known as Code Combining, the receiver combines the digital signals from both sidebands to provide additional signal power and coding gain. Similar techniques are employed in iBiquity's HD Radio technology for AM.

iBiquity Digital's HD Radio technology makes use of a patented technology known as Convolutional Punctured Pair Codes (CPPC). CPPC techniques take advantage of the redundancies in the upper and lower sidebands by allowing the error correction in these sidebands to be combined to create a more powerful error correction algorithm. In order to effectively achieve coding gain when the pair of sidebands is combined, the code on each sideband consists of a subset of a larger (lower rate) code. Each subset has been designed through complementary puncturing of the lower rate code. A simple way of constructing a code for this application is to start with an R = 1/3 convolutional code. This code can be generated as shown in Figure 4.13-6.

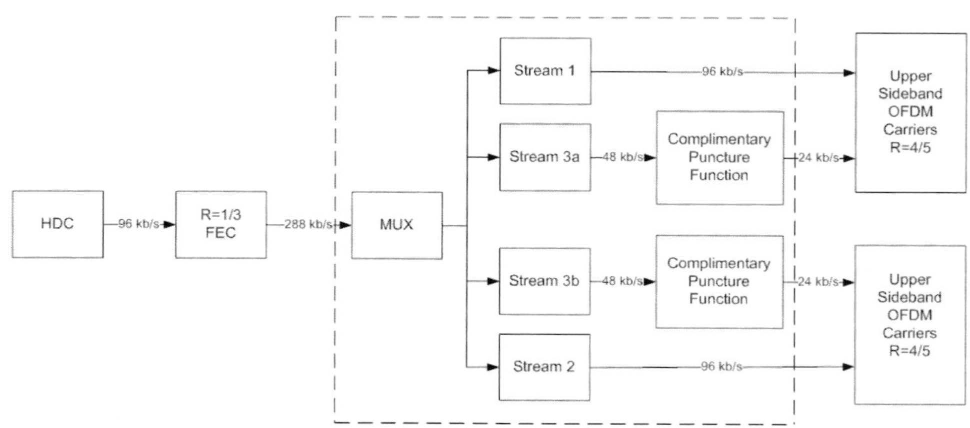

FIGURE 4.13-6 Example of an R = 1/3 convolutional encoder.

The R = 1/3 convolutional encoder produces three encoded bitstreams each at the same bit rate as the input. The combination of these three bitstreams produces the R = 1/3 code output. To create the complementary code pair, for example, a subset of the output code bits is assigned to the lower digital sideband and a different subset is assigned to the upper digital sideband. Each subset must contain at least the same rate of bits as the information input rate, plus some additional bits to provide some coding gain. An R = 4/5 code on each sideband requires 25% additional bits. The combined bitstreams of the two sidebands create a code rate of R = 2/5. An additional feature of CPPC codes is the capability to combine the power in the two sidebands and recover a signal that is 3 dB more robust than either sideband by itself.

Error correction algorithms add to the number of bits that must be transmitted to reliably recover digital data after passing through an RF channel. This extra overhead is necessary to ensure reliable reception. iBiquity's HD Radio technology uses error correction algorithms to provide a robust digital broadcast while minimizing the number of additional bits needed to reliably convey the digital information.

Time Diversity Backup Channel and Blend

An effective method for dealing with channel fading in a mobile environment is to provide a second channel conveying the same information. Transmitting the information on the second channel shifted in time can enhance the total system performance when the two channels are recombined at the receiver. This technique is called *time diversity*. iBiquity's HD Radio technology includes a time diverse backup channel in all AM and FM modes. If the diversity delay is sufficiently large such that the transmission outages are independent, then the probability of an outage after diversity is the square of the probability of outage without diversity. For example, if the probability of an outage in either channel is 1.0%, then the probability of outage after diversity delay is 0.01%, a significant improvement. The autocorrelation function represents the probability of channel outage after diversity improvement as a function of time offset. A sample autocorrelation function is shown in Figure 4.13-7; the bell curve shows a plot of the statistical distribution of outages where p^2 = Success Probability (reference for 4-second offset); R = represents the Probability of Outage, which increases as a function of offset time (t); and t = Time Offset in seconds. In other words, with zero second, time offset outages are significantly higher than with 3 or 4 seconds offset. In actual operation, however, an actual autocorrelation function depends on distance from the station, terrain, propagation conditions, and interference.

iBiquity's HD Radio technology takes advantage of time diversity by delaying backup transmissions by approximately 4 seconds and realigning the digital and analog signals in the receiver. Figure 4.13-8 shows the effect of the loss of information due to a signal blockage such as would be encountered when listening to AM in an automobile while traveling under an overpass.

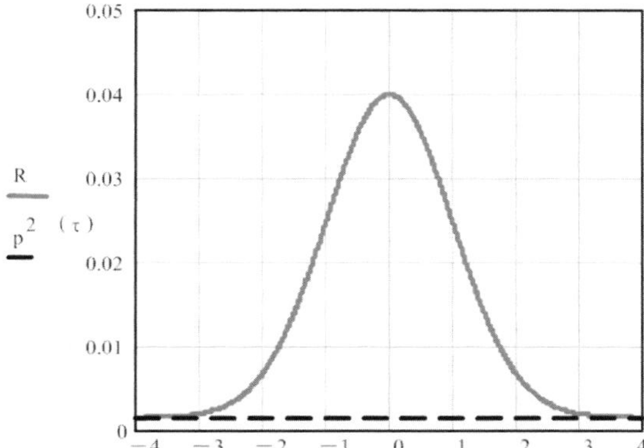

FIGURE 4.13-7 Autocorrelation function of channel loss due to blockage or severe impairment.

Time Diversity of Backup Channel

The top block of Figure 4.13-8 shows that the analog signal is delayed by a fixed amount from the digital signal at the transmission point. In a receiver with iBiquity Digital's HD Radio technology, the digital reception delay is set to be identical to that of the analog transmission. These two time delays allow for a seamless blend to occur between the two transmissions. The shaded blocks of Figure 4.13-9 depict what happens when such a blockage occurs.

In this case, the channel outage affects segments 2 and 3 of the analog and 6 and 7 of the digital signals, effectively decorrelating the outages. In this example, the receiver decodes the digital signal in time segments 1 through 5 and then seamlessly blends to the analog signal to recover segments 6 and 7 and then blends back to the digital transmission for segments 8 and 9, thus maintaining continuity of programming.

In the future, when the analog transmission is discontinued and transition to an all digital service occurs, iBiquity's HD Radio technology will continue to provide a time diverse backup channel. For the All Digital mode, this backup channel is a separate time-delayed digital signal. The use of iBiquity Digital's patented Time Diverse Blending technique provides continuity of service during channel blockages. For this reason, time diverse backup channels are an important part of iBiquity's HD Radio technology.

Frequency and Sideband Diversity

Transmission systems do not necessarily experience uniform fading across the channel bandwidth. In fact, frequency selective fading is common in the AM and FM bands. If the information is transmitted in two different parts of the spectrum, and there is sufficient frequency separations between these transmissions, it is possible to mitigate the effects of a fading channel. The

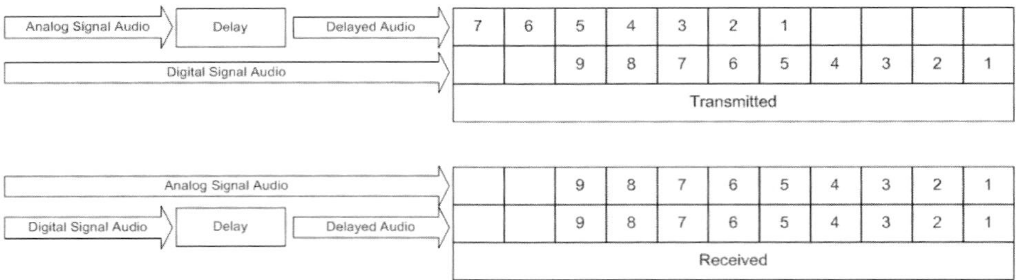

FIGURE 4.13-8 Diversity delay under normal reception.

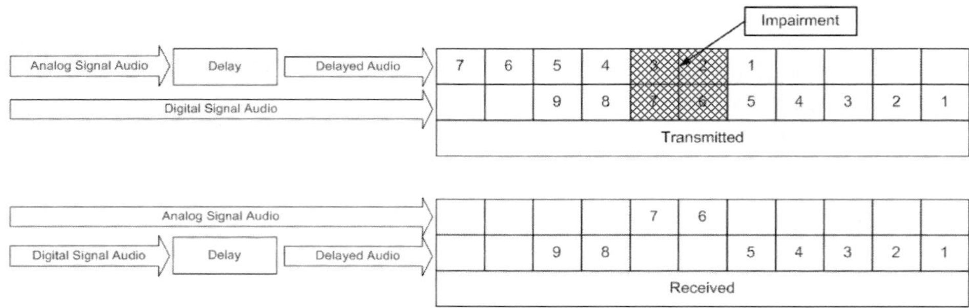

FIGURE 4.13-9 Blend transition during signal impairment.

iBiquity HD Radio technology employs sideband diversity as a form of frequency diversity. The system transmits identical digital information on both upper and lower sidebands. The data in the upper and lower sidebands is identical; however, the CPPC error correction codes are different. In the receiver, the sidebands are independently detected and decoded. Additional coding and power gain are achieved when both sidebands are combined. iBiquity's HD Radio technology essentially carries the same audio information in two different parts of the spectrum. Sideband diversity plays a key role in the system's robustness when either sideband experiences the effects of interference or multipath fading.

First Adjacent Channel Canceller

iBiquity's HD Radio technology signals operate at a lower power level than the analog signals in the FM band. The FCC allocations provide for a 6 dB desired-to-undesired ratio at the desired protected contour for adjacent channels. Therefore, at the protected contour the analog interferer is 19 dB higher in power than the digital sideband. iBiquity's HD Radio technology employs a patented First Adjacent Canceller (FAC) to mitigate the effects of first adjacent channel analog interference. The FAC, which is simultaneously active in the upper and lower digital sidebands, essentially tracks the instantaneous frequency of a first adjacent channel analog interferer and nulls out its effect on the digital information. Figure 4.13-10 shows two hybrid signals as they would appear at the protected contour. The arrow to the left (desired signal) depicts an analog carrier which is swinging back and forth with the analog modulation. The first adjacent channel interferer (undesired signal) is located to the right of the desired signal and is depicted as being about 6 dB lower than the desired analog signal. As shown in the left drawing, the analog interferer is higher in level than the digital subcarriers. The drawing to the right depicts the effect of FAC in eliminating the analog interference.

First adjacent channel interference from analog FM broadcasts is significantly higher in level than the digital signals. To ensure coverage of iBiquity Digital's HD Radio technology near the protected contour, an effective means of interference cancellation must be employed. FAC makes possible digital reception, in a mobile environment, in the presence of analog interference that is about 19 dB higher in power level than the digital signal.

HD RADIO DEVELOPMENT CONCEPTS

iBiquity's HD Radio broadcasting system represents the next generation of broadcasting in the AM/FM bands. Existing analog AM and FM broadcasts may be augmented and even replaced by digital broadcasting modes, resulting in improved and expanded services to the consumer. This section provides an overview of the iBiquity HD Radio system design concepts and what it means to the next generation of broadcasting systems.

The HD Radio concept encompasses all of the features currently envisioned for IBOC-based digital

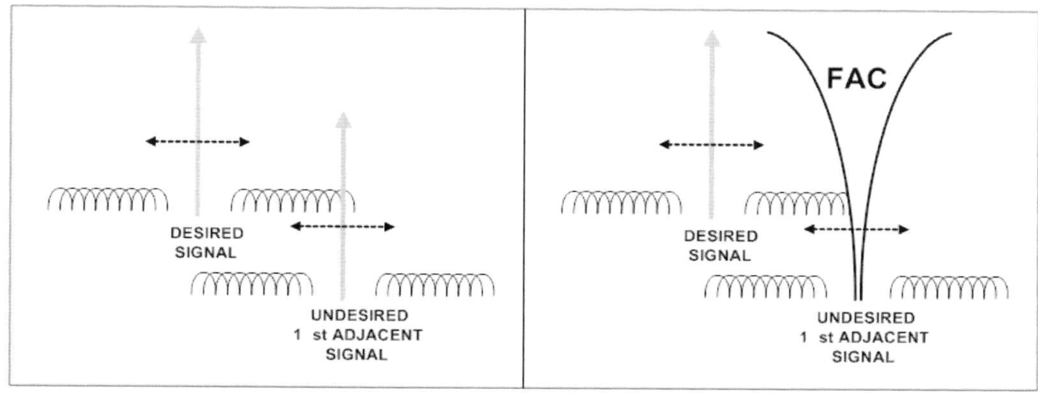

FIGURE 4.13-10 The left drawing shows digital sidebands impacted by FM carrier. The right drawing shows FAC cancellation of analog carrier interference.

systems for near-term and longer-range development. Features being developed are broad in concept and have a wide range of appeal to both the service provider and the consumer.

Initially, broadcast systems will support processing and transmission of analog and digital audio signals and digital data services. Analog processes will likely be phased out as the IBOC technology evolves and digital receiver penetration reaches critical mass. System design concepts have been modularized and grouped, for discussion, according to features or services to be offered. Those features are listed and described in the following paragraphs.

It is anticipated that broadcast system designers and developers will independently determine the features incorporated in the various HD Radio broadcasting equipment models offered based on the services supported by broadcasters and on consumer demand.

HD Radio System Features

Listener flexibility in the selection of services offered is a major objective of the HD Radio system. Individual broadcasters may elect not to offer all HD Radio services nor expeditiously update their capabilities as the system evolves. Therefore, stations will likely offer variations of the wide variety of defined and developed HD Radio digital data services such as multicast audio, on-demand audio, and traffic data, to name a few. The HD Radio broadcast system will support a wide array of receiver and broadcaster configurations to accommodate user preferences.

The HD Radio broadcasting system offers improved audio quality and services over existing analog systems. In the FM band, HD Radio technology can provide compact disc (CD)-like audio quality or multiple high-quality audio streams surpassing FM quality. In the AM band, it will provide audio quality similar to that of existing analog FM systems. HD Radio broadcasts will offer new audio features, such as surround sound, not available with existing analog services and will support multiple audio programs (multicast) on a single RF frequency assignment.

HD Radio technology will introduce new data services that greatly expand the range of broadcaster functions and applications. Digital audio will appear to the system as a source of data that may be traded off against data capacity to optimize both. Because of this, it will permit synergistic cooperation between the audio and data services to

- Permit the trade-off of digital audio fidelity (via bit rate adjustment) against data capacity;
- Permit the trade-off of analog audio fidelity (via bandwidth adjustment) against data capacity;
- Provide additional opportunistic variable data capacity based on audio activity;
- Provide various levels of data service quality (via adjustable error protection coding);

The HD Radio system will exhibit defined levels of audio fidelity and robustness and provide resistance to multipath, Doppler shift, adjacent channel, grounded conductive structures, and impulsive noise interference.

COMPATIBILITY WITH EXISTING ANALOG AM AND FM SERVICES

HD Radio broadcasting represents a major departure in the way future broadcasting equipment and receivers must be built and function. The technology can be introduced to the consumer and broadcasters gradually, in a way that does not immediately obsolete the current radio broadcast infrastructure (transmitters and receivers). The current population of radio receivers will continue to work without modification until such time as analog broadcasting is completely supplanted by the All Digital mode. This is made possible by the initial introduction of Hybrid modes consisting of analog signals augmented with digital signals.

HD Radio features are incorporated as the technologies to perform those features are developed and refined. Features development is based on both transmitter and receiver functionality. Some broadcast systems may continue to operate entirely in the analog environment, whereas others are converted, in stages, from analog, to hybrid, to the all digital HD Radio system. For those reasons, flexibility in design and performance characteristics is essential as broadcasters evolve from analog to hybrid to all digital technologies.

HD Radio broadcasting does not require new allocation of frequencies, and the HD Radio over-the-air waveform coexists within existing FCC spectral emissions masks. The impact on existing analog services is minimized. The HD Radio system

- Does not require a change in current FCC protected contours;
- Minimizes interference to the analog host in Hybrid modes;
- Avoids harmful interference to co-channels and adjacent channels;
- Avoids harmful interference to and remains compatible with existing Subsidiary Communications Authorization (SCA) services in Hybrid modes;
- Is compatible with the existing Emergency Alert System (EAS);
- Does not require changes to existing analog radiated power in Hybrid modes;
- Is compatible with existing radio frequency translators and boosters;
- Minimizes the interference of co-channels and adjacent channels on the digital signals, in both Hybrid and All Digital modes;
- Does not impair acquisition time of existing analog receivers.

On Channel Repeaters for FM Systems

The use of OFDM modulation in the FM HD Radio system allows on-channel digital repeaters to fill areas of desired coverage where signal losses due to terrain and/or shadowing are severe. A typical application occurs where mountains or other terrain obstructions within the station service areas limit analog or digital performance. On channel repeaters may take the form of hybrid and digital-only implementations.

Seamless Transition

The introduction of the HD Radio broadcasting system is incremental and evolutionary in nature. Market and regulatory forces will necessitate evolutionary changes to the system. Such changes will, in general, conform to the following principles to make changes as seamless as possible:

- Analog sunset (mandatory elimination of analog broadcasting modes) will not be required.
- The system exhibits forward and backward audio compatibility, supporting future audio CODEC enhancements while maintaining audio compatibility with earlier receivers.
- The system exhibits forward and backward data compatibility, supporting future data service enhancements while maintaining compatibility with earlier receivers.
- The system allows each broadcaster and consumer to upgrade according to individual needs.
- Initial HD Radio receivers provide analog, hybrid, and all digital functionality, so as to remain useful throughout the entire evolution of the broadcasting industry from existing analog service to digital service.
- The HD Radio broadcasting system provides a migration path for existing SCA services when their host stations transition to an all digital service.
- As the system evolves, it will seek to maximize use of existing electromagnetic spectrum.

Economical Implementation

The HD Radio system features a wide variety of options, not all of which are of interest to all broadcasters, all users, or to all user situations. Some features will interest travelers or mobile users but are of little interest in the home or place of business and vice versa. For that reason, manufacturers will likely offer specialty broadcast systems configured to the desired services for the various user situations or applications. That concept allows HD Radio broadcast system producers to economically produce systems that fit the needs of a broad range of users within a price range they are more likely to afford.

Conversion to the HD Radio broadcasting system is reasonable and affordable in cost to the broadcaster. Specifically, it accomplishes this by

- Minimizing new infrastructure requirements;
- Maximizing existing broadcaster infrastructure investment.

BROADCAST SYSTEM FUNCTIONAL TYPES

Implementation of HD Radio broadcasting systems is generally accomplished by one of the following strategies:

- Systems converted from "analog only" operation;
- Systems designed and constructed from the outset to be HD Radio compliant.

It is anticipated that most systems will follow the first (conversion from analog) strategy. HD Radio broadcasting systems, however implemented, are of the following defined types.

Commercial Hybrid Types

Hybrid HD Radio broadcasting systems transmit an analog carrier signal which is compatible with existing analog radio receivers and is substantially identical to the analog carrier signal of existing analog-only systems, augmented with digital sidebands which carry digital audio and data, in the manner prescribed herein.

Commercial All Digital Types

All Digital HD Radio broadcasting systems have been developed for future applications. The analog carrier portion of the broadcast signal is removed and secondary digital subcarriers utilized in the manner prescribed herein. The primary subcarriers are identical to those of hybrid systems.

COMMON AM/FM SYSTEM SERVICE DEFINITIONS

Functional Layer Definitions

The HD Radio broadcasting system functionality has been defined in terms of functional layers. These layers correspond in definition to those of the International Standards Organization Open Systems Interconnection (ISO/OSI) model. Layers communicate at defined interfaces through structured data blocks called Protocol Data Units (PDU). A PDU is the structured data block in the HD Radio system that is produced by a specific layer (or process within a layer) of the transmitter protocol stack. The PDUs of a given layer may encapsulate PDUs from the next higher layer of the stack and/or include content data and protocol-control information originating in the layer (or process) itself. The PDUs generated by each layer (or process) in the transmitter protocol stack are inputs to a corresponding layer (or process) in the receiver protocol stack.

Layer 1, or the Waveform/Transmission Layer, generally performs the functions of a modem. It receives several logical channels over which all digital data (including main audio and control and status data) is transferred, and its output is the modulated electromagnetic waveform broadcast. Logical channels are implemented as data streams internal to the processing functions of the broadcasting system. Organizing the Layer 1 input in terms of multiple logical channels, instead of a single bitstream, affords a convenient and more effective way of utilizing the data transmission resources of the HD Radio broadcasting system.

The upper layers (Layer 2 or higher) generally deal with generating and packetizing the data to be sent, including implementing the trade-offs between audio quality and data throughput.

Broadcasting Mode Definitions

Hybrid mode is defined as simultaneous broadcasting of analog and digital signals within the same channel, in such a way that existing analog receivers can satisfactorily receive the analog portion of the signal. HD Radio receivers will receive both the analog and digital portions of the signal. The analog portion of the signal functions as backup to the digital audio signal to mitigate outages and fades.

In the All Digital mode, digital audio and data occupy the entire HD Radio broadcast spectrum. For the FM mode, the primary digital portion of the spectrum is those carriers which are placed, in a noninterfering manner, on either side of the existing analog channel and are present in both Hybrid and All Digital modes. The primary extended portion of the spectrum is optional with additional carriers placed next to the primary carriers on either side of and adjacent to the analog channel. The secondary digital portion of the spectrum includes those carriers that are placed in the spectrum vacated when the analog carrier spectrum is removed for the All Digital mode.

For the AM mode, the digital carriers are grouped into mutually exclusive frequency sidebands designated as

- Primary subcarriers, occupying the outer part of the spectrum;
- Secondary subcarriers, occupying the middle part of the spectrum;
- Tertiary subcarriers, occupying the inner part of the spectrum located below the analog signal.

Primary AM subcarriers are further designated as core subcarriers, and collectively the secondary and tertiary subcarriers are designated as enhanced subcarriers. The tertiary subcarriers spectrally overlap the analog signal, but are phase orthogonal with it for minimal mutual interference.

Service Mode Definitions

The HD Radio broadcasting system functionality affords many degrees of freedom in the choices that can be made to optimize system end-to-end performance. These include

- Audio robustness versus audio quality and latency;
- Data throughput versus data robustness;
- Audio quality versus data throughput trade-offs (upper layers);
- Scrambling (Layer 1);
- Channel coding (Layer 1);
- Interleaver design (Layer 1);
- Allocation of the various subcarriers (Layer 1).

When these factors are chosen and combined in appropriate ways, a wide variety of information data rates, audio/data combinations, and bit error rate performance for given channel conditions can be achieved.

For the HD Radio broadcasting system, several such combinations have been defined, enumerated, and designated for the Hybrid FM system, the All Digital FM system, the Hybrid AM system, and the AM All Digital system. These designated combinations are the defined service modes of the HD Radio broadcasting system.

TABLE 4.13-1
Audio Service Classes

Class	Service	Number of Audio Channels	Minimum Audio Frequency Response	Minimum Stereo Separation (dB)	Dynamic Range	Quality Level
1	FM main	stereo	20–20,000 Hz	70 dB	96 dB	CD like
2	FM backup	monophonic	20–15,000 Hz	none	65 dB	FM mono
3	AM main	stereo	20–15,000 Hz	70 dB	72 dB	FM like
4	AM backup	monophonic	20–10,000 Hz	none	60 dB	AM mono

Audio Service Definitions

In Hybrid HD Radio broadcasting systems, the analog portion of the air interface is identical to the air interface of existing analog-only broadcasting systems. Any deviations from this requirement are compatible with existing analog-only broadcasting equipment, and the transmitted signal is compatible with existing analog-only receivers. Analog audio is the means by which the broadcast system maintains compatibility with analog-only legacy receivers. It is used with digital audio in Hybrid mode as backup audio, to provide improved robustness at the receiver under circumstances of signal fading and blockage. Multicast or Supplemental Program Services (SPS) are additional audio channels that may be added in addition to the Main Program Service (MPS). The exact bandwidth of the MPS and SPS channels will be determined by service mode and the total number of audio streams transmitted. Digital audio is the primary or main audio delivery mode in both Hybrid and All Digital modes. In All Digital mode, it is augmented with another lower latency and bandwidth digital channel which acts as backup audio in lieu of the analog channel. While the MPS channel is backed up by the analog signal, the SPS channels are stand-alone and mute at the point of failure (POF).

Audio Classes

The following definitions are applicable throughout this document.

In the Hybrid modes, Class 2 and Class 4 audios are accomplished by the analog channel. In All Digital modes, Class 2 and Class 4 audio are separate, possibly independent, encoded, low-latency, and low-bandwidth data streams. This relationship is defined in Table 4.13-1.

Audio Quality Definitions

Audio quality is measured as produced by the end-to-end HD Radio system, from unencoded source audio at the broadcasting system to the decoded audio output of the receiver. This is quantified in terms of two methodologies to test whether a system has reached a quality target: Degradation Category Rating (DCR) and Absolute Category Rating (ACR).

Digital Data Service Definitions

The HD Radio broadcasting system will offer extensive datacasting services. These services extend to both the HD Radio AM and FM systems. Six data transport services have been defined for AM and FM as follows:

- CODEC (HDC) digital audio transport;
- Text transport;
- Control channel transport;
- Packet/message transport;
- File transport;
- Generic streaming data transport.

To support these transport services, the HD Radio broadcasting system has defined the classes of data as shown in Table 4.13-2.

TABLE 4.13-2
Classes of Data Service

Class	Service	System	Minimum Rate	Maximum Rate
1	Dedicated fixed rate	Hybrid, All Digital	860 bps FM 430 bps AM	N/A
2	Adjustable rate	Hybrid, All Digital	0	Equal to at least the maximum rate of audio CODEC
3	Opportunistic variable rate	Hybrid, All Digital	0	Equal to maximum rate of audio CODEC

Dedicated fixed-rate services employ a fixed data rate that cannot be changed by the broadcaster. This data rate is allocated to the various data services defined by the upper layer functionality. It includes, among others that may be defined, the Station Information Service (SIS) that offers an array of RBDS-like services to the broadcaster.

Adjustable rate services operate at a constant rate that, unlike fixed-rate services, is selectable and changeable by the broadcaster by trading off audio quality or robustness for data throughput. To increase the data rate, the broadcaster reduces the audio bit rate (and therefore audio quality or robustness) and reallocates it to data. To decrease the data rate, the reverse occurs. The adjustable rate services operate by dynamically allocating digital subcarriers among error correction, audio, and data services.

Opportunistic variable-rate services offer data rates that are dynamically related to the complexity of the encoded digital audio. Simpler audio passages (e.g., simple tones, narrow bandwidth audio, and silence) require lesser bit rates, permitting the unused throughput to be used for data. The audio encoder dynamically measures audio complexity and adjusts data throughput accordingly, without compromising the quality of the encoded digital audio.

Station Information Service

The Station Information Service provides broadcast station identification and control information. SIS is transmitted in a series of SIS Protocol Data Units (PDUs) on the Primary IBOC Data Service (PIDS) logical channel. The PIDS channel is a fixed-rate channel that delivers basic control messages that carry service information. Service information is information about the services carried in real time, other information such as schedules or service event calendars, and station-related system broadcast information similar to existing RBDS services.

Datacasting System

Datacasting is defined as delivering content from a content provider to a receiver end user via the HD Radio system. Datacasting affords expanded data functions over those provided by the IDS channels. These include, but are not limited to, the following:

- Streaming perceptual audio CODEC (HDC) applications;
- Still and streaming video applications;
- Message/packet-based applications;
- File-based applications;
- Audio storage and retrieval applications;
- Billing and management;
- Text/XML (Extended Markup Language) applications;
- Specialized applications with specialized receivers;
- Datacasting services with various defined levels of quality of service for each;
- Control data beyond that provided by the IDS channel.

HD Radio system datacasting users are of three types:

- Content providers who create and package content for broadcast over the HD Radio system;
- Operations, Administrative, and Maintenance (OAM) users who manage the broadcast system for content delivery, billing, and other administrative tasks support;
- Receiver end users who make use of the content broadcast by the HD Radio system.

The HD Radio broadcasting system interfaces content providers to receive content for broadcast, with the receiver systems (via the air interface) to deliver the content and with OAM users to administer and maintain the data service. For purposes of datacasting definition and specification, the HD Radio broadcasting system consists of two parts:

- Broadcast network system that receives content from content providers and delivers it to individual broadcast station systems;
- Broadcast station systems that receive content from the broadcast network system or from local content providers for broadcast.

AM HD RADIO SERVICES

The AM IBOC system is capable of supporting the following services:

- Main Program Service (MPS): The Main Program Service preserves the existing analog radio-programming formats in both the analog and digital transmissions. In addition, Main Program includes digital data, which directly correlates with the audio programming.
- Personal Data Service (PDS): Unlike the Main Program Service, which broadcasts the same audio program to all listeners, the Personal Data Service enables the user to select the data services desired and when they are presented. This provides personalized, on-demand, user-valued information.
- Station Identification Service (SIS): The Station Identification Service provides the necessary control and identification information which indirectly accommodates user search and selection of IBOC digital radio stations and their supporting services.
- Advanced Application Service (AAS): This service allows a virtually unlimited number of custom and specialized IBOC digital radio applications to coexist concurrently. Advanced Applications can be added at any time in the future.

Support of the preceding services is provided via a Layered protocol stack, as illustrated in Figure 4.13-11.

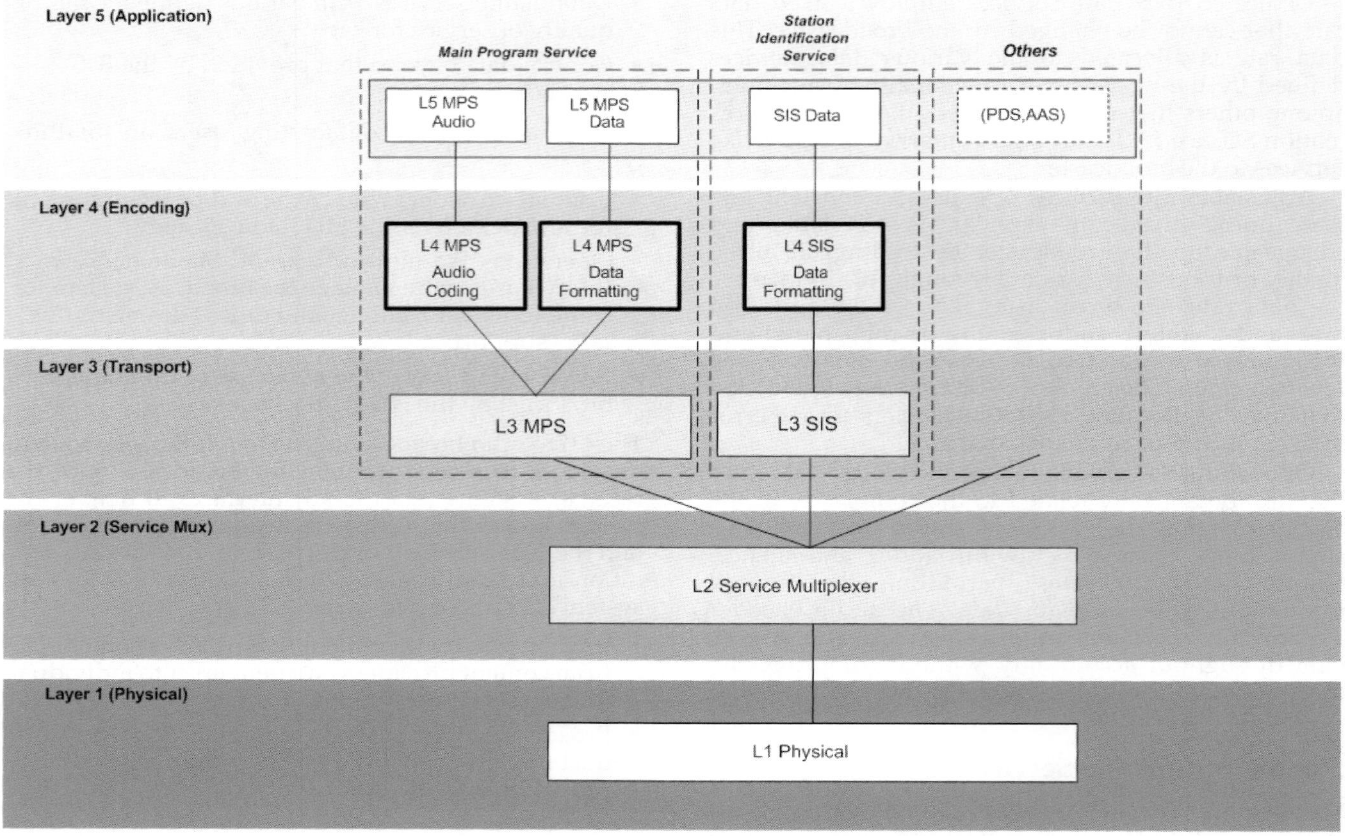

Layer 5 (Application)
Main Program Service
Station Identification Service
Others
L5 MPS Audio
L5 MPS Data
SIS Data
(PDS,AAS)
Layer 4 (Encoding)
L4 MPS Audio Coding
L4 MPS Data Formatting
L4 SIS Data Formatting
Layer 3 (Transport)
L3 MPS
L3 SIS
Layer 2 (Service Mux)
L2 Service Multiplexer
Layer 1 (Physical)
L1 Physical

FIGURE 4.13-11 International Standards Organization Open Systems Interconnection (ISO/OSI) Layered model.

This Layered protocol stack is based on the International Standards Organization Open Systems Interconnection (ISO/OSI) Layered model. Source material is received from the broadcaster in Layer 5, source encoded in Layer 4, multiplexed into logical channels in Layers 3 and 2, and formatted for over-the-air broadcast in Layer 1.

- Layer 5 (Application) accepts content from the user (i.e., program source).
- Layer 4 (Encoding) performs the necessary audio compression or data formatting of the various source materials.
- Layer 3 (Transport) provides one or more application-specific protocols tailored to provide robust and efficient transfer of Layer 4 data.
- Layer 2 (Service MUX) provides limited error detection and addressing. Its main function is to format the data received from Layer 3 into discrete transfer frames for processing by Layer 1.
- Layer 1 (Physical Layer) provides the modulation, FEC, framing, and signaling necessary to convert the digital data received from higher layers into an AM IBOC waveform for transmission in an existing allocation in the MF band.

Layer 1 can be thought of as simply a conduit for broadcast data with a specific grade of service; the source coding, formatting, and multiplexing of the program content are performed at the higher protocol layers. However, the AM IBOC system provides a number of different configurations, called *service modes*, in which the number, throughput, and robustness of the data pipes, called *logical channels*, can vary. Therefore, after assessing the requirements of their candidate applications, higher protocol layers select service modes that most suitably configure the logical channels. The number of logical channels and service modes reflects the inherent flexibility of the system, which supports simultaneous delivery of various classes of digital audio and data.

AM FUNCTIONAL DESCRIPTION

Several aspects of the AM HD Radio system are discussed in the following sections. First, the modulation and spectral occupancy of the waveforms are described followed by a description of the available service modes, as well as the various broadcaster options within each service mode. Next, the logical channels in each service mode are described in terms

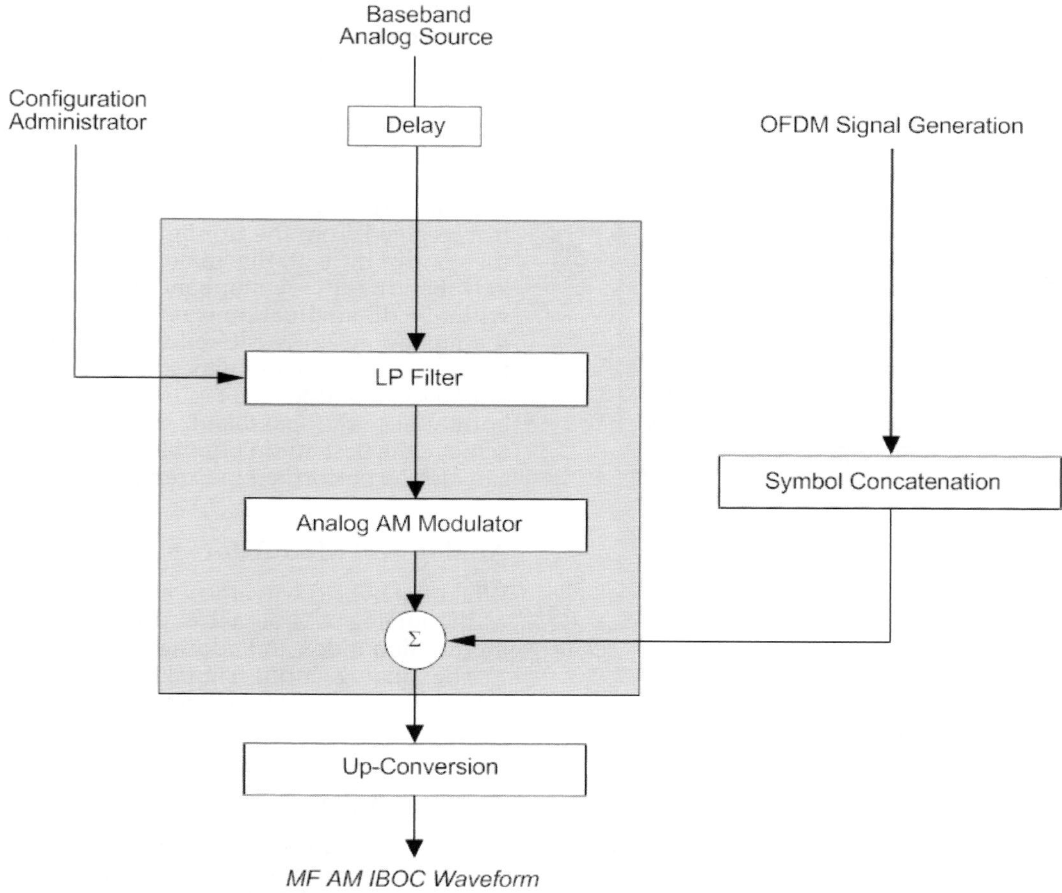

FIGURE 4.13-12 Hybrid transmission subsystem functional block diagram.

of throughput, robustness, and latency. Lastly, the functional components of AM waveform generation are presented, describing the processing necessary to convert the digital data in the active logical channels into an AM IBOC waveform.

Transmission Subsystem

The transmission subsystem formats the baseband AM HD Radio waveform for transmission through the Medium Frequency (MF) channel. Functions include symbol concatenation, precompensation, and frequency upconversion. In addition, when transmitting the hybrid waveform, this function filters and modulates the baseband analog audio signal before coherently combining it with the digital portion of the waveform.

The input to this module is a complex, baseband, time-domain OFDM symbol from OFDM signal generation. A baseband analog audio signal after application of diversity delay is also input from an analog source when transmitting the hybrid waveform. The output of this module is the MF AM HD Radio waveform.

Figure 4.13-12 and Figure 4.13-13 show functional block diagrams of the Hybrid and All Digital transmission subsystems, respectively.

HD RADIO FUNCTIONAL COMPONENTS

Precompensation

The pulse-shaping function used in the AM system guarantees that the frequency domain side lobes meet spectral limits imposed by the FCC, as outlined in FCC Section 73.44

Symbol Concatenation

The individual time-domain OFDM symbols output precompensation are summed to produce a continuum of pulses.

Diversity Delay

The first step in generating the HD Radio signal is the application of diversity delay to the baseband analog

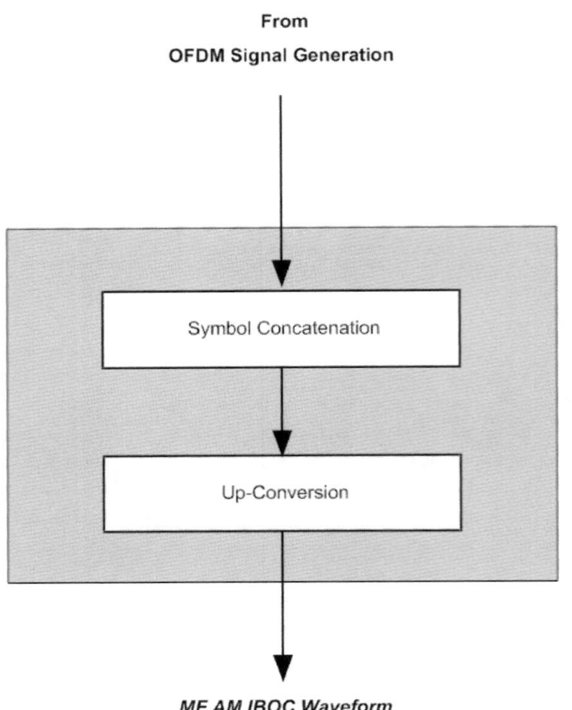

FIGURE 4.13-13 All Digital transmission subsystem functional block diagram.

audio signal. An adjustable delay is applied to the baseband analog audio signal so that, at the output of the analog/digital combiner, it lags the audio content of the corresponding digital signal. For example, if both the analog and digital signals carry the same audio program, the analog audio would be delayed from the corresponding digital audio at the output of the analog/digital combiner. The delay is adjustable to account for processing delays in the analog and digital chains.

Low-Pass Filtering

In Hybrid mode, this process low-pass filters the analog audio data according to the state of the AAB control received from the Configuration Administrator. If the control bit is 0, the analog audio is filtered to a 5 kHz bandwidth. A stopband frequency of 5116 Hz with >55 dB attenuation is used. If the control bit is 1, the analog audio is filtered to an 8 kHz bandwidth. Here, a stopband frequency of 9 kHz with >60 dB attenuation is used. See Figures 4.13-14 and 4.13-15. In both cases, the passband specifications are set to achieve the best audio possible. This low-pass filtering can also be performed in external audio processors.

Analog AM Modulator

When the hybrid waveform is broadcast, the envelope of the analog AM signal is processed by applying a modulation index and adding a DC offset.

The analog input signal must be preprocessed external to the AM HD Radio exciter, so negative peaks do not exceed 100%.

Analog/Digital Combiner

When the hybrid waveform is broadcast, the real analog AM baseband waveform is coherently combined with the digital baseband waveform to produce the complex baseband AM HD Radio hybrid waveform. OFDM subcarrier mapping appropriately scales the levels of the digital sidebands in the output spectrum.

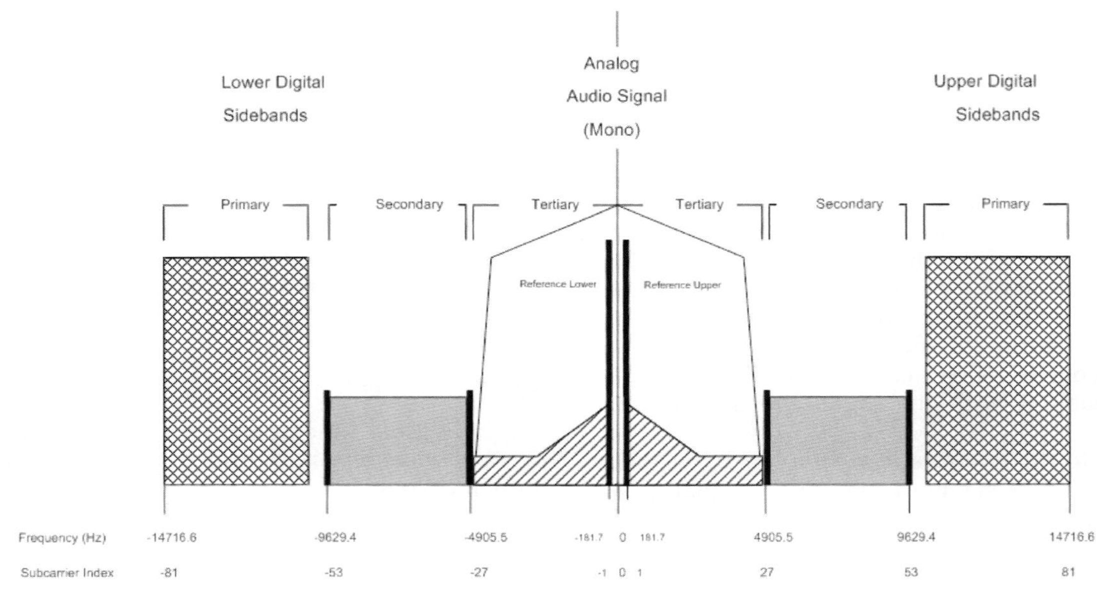

FIGURE 4.13-14 AM HD Radio hybrid waveform spectrum (5 kHz audio configuration).

Upconversion

The concatenated digital signal is translated from baseband to the RF carrier frequency.

The AM HD Radio waveform is broadcast in the current AM radio band, and its power levels and spectral content are limited to be within the spectral mask as defined in FCC Rules Section 73.44

The carrier frequency spacing and channel numbering schemes are compatible with FCC Rules Section 73.44. Channels are centered at 10 kHz intervals ranging from 540 to 1700 kHz. Both the analog and digital portions of the hybrid waveform are centered on the same carrier frequency.

GPS Synchronization

In order to ensure precise time synchronization, the transmitted signal for each station may be synchronized in time and frequency to the Global Positioning System (GPS). In the case in which transmissions are not locked to GPS, time, and frequency synchronization, the accuracy requirements are relaxed and transmissions cannot be synchronized with other stations.

RF Carrier Frequency and OFDM Symbol Clock

For Level I transmission facilities, all transmissions shall phase lock their L1 frame timing (and the timing of all OFDM symbols) to absolute GPS time within ±1 s.

If the preceding specification for a Synchronization Level I transmission facility is violated, due to a GPS outage or other occurrence, it shall be classified as a Synchronization Level II transmission facility until the preceding specification is again met.

GPS Phase Lock

For Level I transmission facilities, all transmissions will maintain phase lock to absolute GPS time within ±1 s.

If the preceding specification in a Level I transmission facility is violated due to a GPS outage or other occurrence, it will be classified as a non-GPS synchronization Level II transmission facility until the preceding specification is again met.

Waveforms and Spectrum

Digital data and audio cannot be directly propagated over RF channels; therefore, a modulator is used to modulate the digital information onto an RF carrier. The AM IBOC system employs Orthogonal Frequency Division Multiplexing (OFDM) for robustness in the presence of adjacent channel interference and noise. OFDM is a parallel modulation scheme in which the data streams modulate a large number of orthogonal subcarriers that are transmitted simultaneously. OFDM can be tailored to fit an interference environment that is nonuniform across frequency.

The narrow AM bandwidth leads to an OFDM modulation technique that is optimized for higher throughput. iBiquity Digital's IBOC technology for AM uses Quadrature Amplitude Modulation (QAM) on each OFDM subcarrier.

The current design calls for symbol durations of 5.8 ms and an OFDM subcarrier spacing of 181.7 Hz. In addition, the timing of the symbols is optimized to ensure that duration of noise pulses is much shorter than the QAM symbol duration, ensuring robust digital reception in the presence of static and noise prevalent in the AM channel.

The design of the AM IBOC system provides a flexible means of transitioning to a digital broadcast system by providing two new waveform types: hybrid and all digital. The hybrid waveform retains the analog AM signal, whereas the all digital waveform does not. The analog source must be monophonic, as the AM IBOC system does not support AM stereo broadcasts.

In the hybrid waveform, the OFDM subcarriers are located in primary and secondary sidebands on either side of the host analog signal, as well as underneath the host analog signal in tertiary sidebands, as shown in Figures 4.13-14 and 4.13-15. Each sideband has both an upper and lower component. Status and control information is transmitted on reference subcarriers on either side of the main carrier.

In addition to the primary, secondary, tertiary, and control subcarriers, there are two additional subcarriers between the primary and secondary and the secondary and tertiary sidebands on either side of the main carrier. These are known as IBOC Data System (IDS) subcarriers and are primarily used for low-latency, low-data-rate applications such as SIS or RBDS, which is currently being used in FM systems.

The number of OFDM subcarriers in the secondary, tertiary, and IDS sidebands is twice the number needed to transmit the QAM constellation values. The reason is that the overall digital signal must maintain a 90° phase relationship (quadrature) to the AM carrier, thereby minimizing the interference to the analog signal when detected by an envelope detector. Placing these subcarriers in quadrature to the analog signal also permits demodulation of the tertiary and IDS subcarriers in the presence of the high-level AM carrier and analog signal. The price paid for placing these subcarriers in quadrature with the AM carriers is that the information content on these subcarriers is only half of that for nonquadrature digital carriers.

During transmission, the phase relationship between the digital and analog signals must be maintained. Amplification of the analog and digital signals by a single transmitter is the most straightforward approach. Separate amplifiers can be used for the analog and digital signals and the transmitter outputs combined. However, proper phasing of the signals must be maintained.

The total power of all digital sidebands in the hybrid waveform is significantly below the power in the AM analog signal. The power level of each OFDM subcarrier in the primary sidebands is fixed relative to

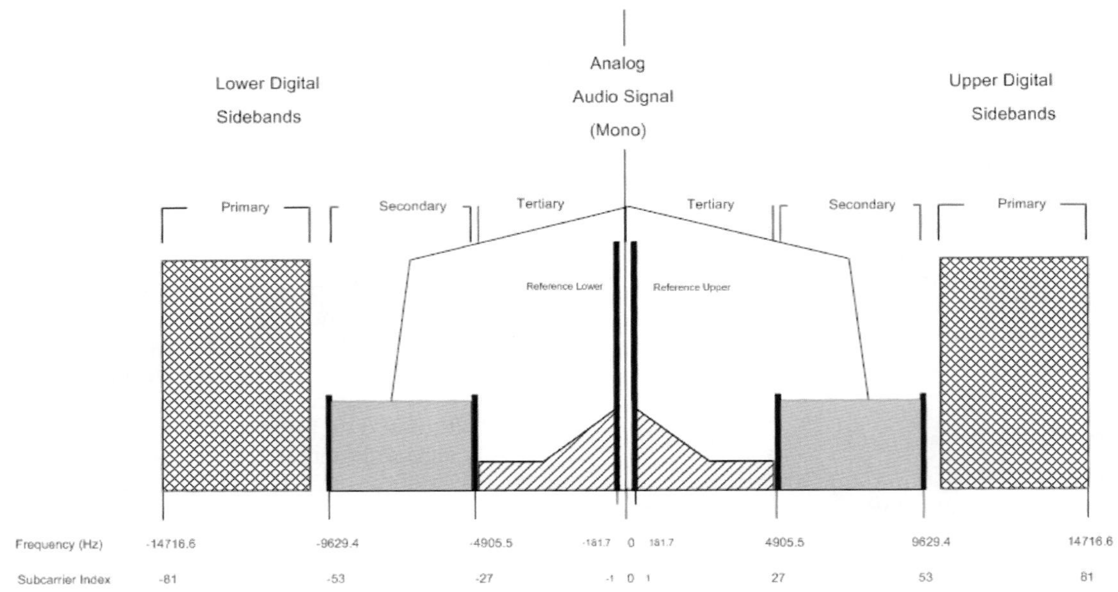

FIGURE 4.13-15 AM HD Radio hybrid waveform spectrum (8 kHz audio configuration).

the unmodulated main analog carrier. However, the power level of the secondary, IDS, and tertiary subcarriers is adjustable. Table 4.13-3 summarizes the spectral characteristics of the hybrid waveform. Individual subcarriers are numbered from –81 to 81 with the center subcarrier at subcarrier number 0. Table 4.13-3 also lists the approximate frequency ranges, bandwidth, levels, and modulation types for each sideband. Referring back to Figures 4.13-14 and 4.13-15, note that the subcarriers 54 to 56 and –54 to –56 are not represented because they are not transmitted in order to avoid interference with first adjacent signals.

The greatest system enhancements are realized with the all digital waveform. In this waveform, the analog signal is replaced with higher power primary sidebands. The unmodulated AM carrier is retained, and the secondary sidebands are moved to the higher frequencies above the primary upper sideband. In addition the tertiary sidebands are moved to the frequencies below the primary lower sideband. The sec-

TABLE 4.13-3
AM Hybrid Spectral Summary

Sideband	Subcarrier Range	Subcarrier Frequencies (Hz from Channel Center)	Frequency Span (Hz)	Power Spectral Density (dB/Subcarrier)	Modulation Type
Primary Upper	57 to 81	10,356.1 to 14,716.6	4360.5	–30	64-QAM
Primary Lower	–57 to –81	–10,356.1 to –14,716.6	4360.5	–30	64-QAM
Secondary Upper	28 to 52	5087.2 to 9447.7	4360.5	–43 or –37	16-QAM
Secondary Lower	–28 to –52	–5087.2 to –9447.7	4360.5	–43 or –37	16-QAM
Tertiary Upper	2 to 26	363.4 to 4723.8	4360.4	To Be Announced	QPSK
Tertiary Lower	–2 to –26	–363.4 to –4723.8	4360.4	To Be Announced	QPSK
Reference Upper	1	181.7	181.7	–26	BPSK
Reference Lower	–1	–181.7	181.7	–26	BPSK
Upper IDS1	27	4905.5	181.7	–43 or –37	16-QAM
Upper IDS2	53	9629.4	181.7	–43 or –37	16-QAM
Lower IDS1	–27	–4905.5	181.7	–43 or –37	16-QAM
Lower IDS2	–53	–9629.4	181.7	–43 or –37	16-QAM

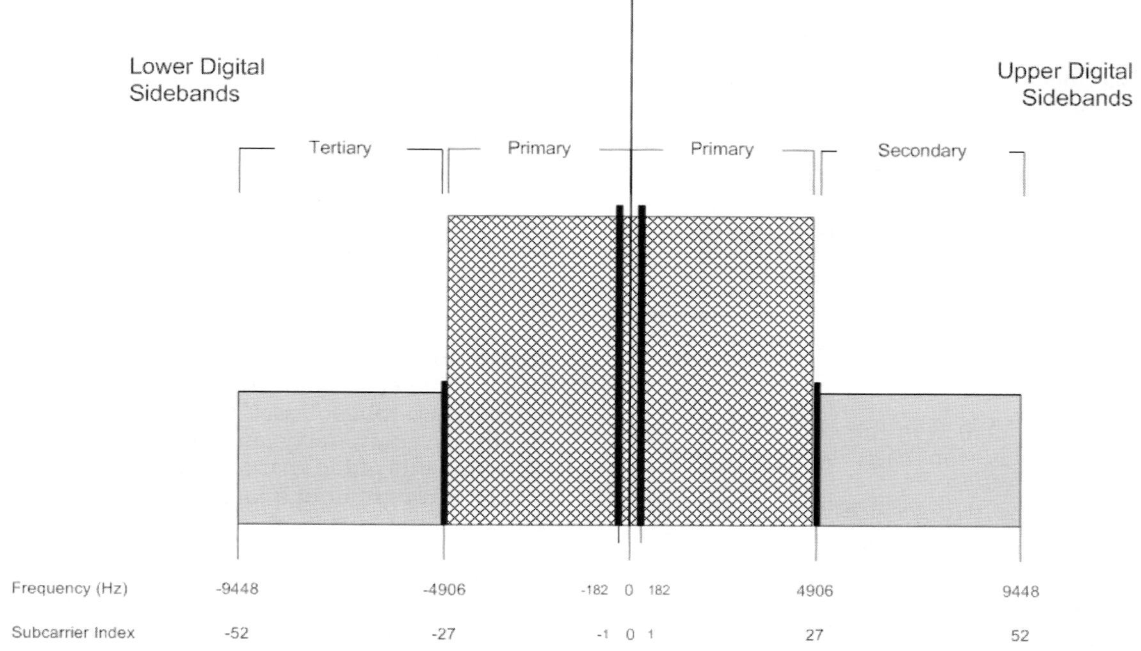

FIGURE 4.13-16 AM all digital waveform spectrum.

ondary and tertiary sidebands use half the number of subcarriers, as compared to the hybrid waveform, because there is no longer a need to place them in quadrature with the analog signal since it is unmodulated. As a result, the power of both the secondary and tertiary sidebands is increased.

These changes result in an overall bandwidth reduction, making the all digital waveform less susceptible to adjacent channel interference. The reference subcarriers are located on either side of the unmodulated AM carrier, as in the hybrid waveform, but at a higher level. The spectrum of the all digital waveform is illustrated in Figure 4.13-16. The power level of each of the OFDM subcarriers within a sideband is fixed relative to the unmodulated analog carrier. Table 4.13-4 summarizes the spectral characteristics of the all digital waveform.

Both the hybrid and all digital waveforms conform to the currently allocated emissions mask per FCC Rules Section 73.44 and summarized in Table 4.13-5. All measurements assume a resolution bandwidth of 300 Hz.

Service Modes and System Options

The AM IBOC system provides two service modes: MA1 and MA3. Service mode MA1 is used with the hybrid waveform, whereas service mode MA3 is used with the all digital waveform. Specifics of the various

TABLE 4.13-4
AM All Digital Waveform Spectral Summary

Sideband	Subcarrier Range	Subcarrier Frequencies (Hz from Channel Center)	Frequency Span (Hz)	Power Spectral Density (dB/Subcarrier)	Modulation Type
Primary Upper	2 to 26	363.4 to 4723.8	4360.5	–15	64-QAM
Primary Lower	–2 to –26	–363.4 to –4723.8	4360.5	–15	64-QAM
Secondary	28 to 52	5087.2 to 9447.7	4360.5	–30	64-QAM
Tertiary	–28 to –52	–5087.2 to –9447.7	4360.5	–30	64-QAM
Reference Upper	1	181.7	181.7	–15	BPSK
Reference Lower	–1	–181.7	181.7	–15	BPSK
IDS1	27	4905.5	181.7	–30	16-QAM
IDS2	–27	–4905.5	181.7	–30	16-QAM

TABLE 4.13-5
AM Analog (NRSC 4) Spectral Emissions Mask

Offset from Carrier Frequency	Level Relative to Unmodulated Carrier
10.2–20 kHz	–25 dBc
20–30 kHz	–35 dBc
30–60 kHz	–5 dBc–1 dB/kHz
60–75 kHz	–65 dBc
>75 kHz	–80 or [–43-10log(power in watts)] dBc, whichever is less

service modes are detailed in the next section, where the logical channels for each service mode are described.[3]

In addition to the two service modes, the broadcaster has the option of configuring service mode MA1 using two additional controls: power level control and analog audio bandwidth control.

The power level control selects one of two levels for the secondary, tertiary, and PIDS subcarriers, as shown in Figure 4.13-17. The higher power levels increase the robustness of the digital signal at the expense of decreasing compatibility with certain classes of existing analog radios, namely portable radios. As receiver manufacturers produce IBOC radios or if a broadcaster is not as concerned with compatibility with these classes of existing analog radios, this option allows broadcasters to increase the robustness of the digital signal.

The analog audio bandwidth control allows the analog audio to be broadcast using either a 5 kHz bandwidth or an 8 kHz bandwidth. Broadcasting 8 kHz analog reduces the robustness of the digital signal in the presence of second adjacent interferers.

To provide robust reception during outages typical in a mobile environment, the AM IBOC system applies time diversity between independent analog and digital transmissions of the same audio source. In addition, a blend function allows graceful audio degradation of the digital signal as the receiver nears the edge of a station's coverage. The AM HD Radio system provides this capability by delaying the analog transmission several seconds relative to the digital audio transmission. When the digital signal is corrupted, the receiver blends to analog, and by virtue of its time diversity with the digital signal, the audio is not significatly degraded.

Logical Channels

A logical channel is a signal path that conducts data through Layer 1 with a specific grade of service determined by service mode. Layer 1 of the AM IBOC system provides four logical channels to higher layer protocols: P1, P2, P3, and PIDS. P1, P2, and P3 are intended for general-purpose audio and data transfer, whereas the PIDS logical channel is designed to carry the IBOC Data Service (IDS) information.

The P1 and P2 logical channels are designed to be more robust than the P3 logical channel. Therefore, P1 and P2 typically transmit the "core" audio information, whereas P3 transmits the enhanced audio information (such as the stereo signal). Logical channels P1 and P3 are available for all services modes.

Logical channels are defined by their characterization parameters and configured by the service mode. For a given service mode, the grade of service of a particular logical channel may be uniquely quantified using three characterization parameters: throughput, latency, and robustness. Channel code rate, interleaver depth, diversity delay, and spectral mapping are the determinants of the characterization parameters.

The throughput of a logical channel is its allowable data rate and is typically defined in terms of kilobits per second (kbps). Latency is the delay that a logical channel imposes on the data as it passes through Layer 1. The latency of a logical channel is defined as the sum of its interleaver depth and diversity delay. It does not include processing delays in Layer 1, nor does it include delays imposed in upper layers. Robustness is the capability of a logical channel to withstand channel impairments such as noise, interference, and Grounded Conductive Structures (GCS). There are 10 relative levels of robustness designed into Layer 1 of the AM IBOC system. A robustness of 1 indicates a very high level of resistance to channel impairments, whereas a robustness of 10 indicates a lower tolerance for channel-induced errors. As with throughput and latency, higher layers must determine the required robustness of a logical channel before selecting a service mode.

Spectral mapping, channel code rate, and interleaver depth determine the robustness of a logical channel. Spectral mapping affects robustness by setting the relative power level, spectral interference protection, and frequency diversity of a logical channel. Channel coding increases robustness by introducing redundancy into the logical channel. Interleaver depth influences performance in GCS and impulsive noise, thereby affecting the robustness of the logical channel. Finally, some logical channels in certain service modes delay transfer frames by a fixed duration to realize time diversity. This diversity delay also affects robustness since it mitigates the effects of the mobile radio channel.

Tables 4.13-6 and 4.13-7 list the logical channels P3 and PIDS and multiple robustness values for the service modes. The reason is that there are two power levels associated with these logical channels. The lower relative robustness number, indicating greater robustness, is associated with the higher power level settings.

For a given service mode, each logical channel is applied to a frequency sideband. Figures 4.13-17 and 4.13-18 show the spectral mapping of each logical channel for each service mode.

[3]MA2 and MA4 are no longer a part of the system as they were specific to the PAC™ CODEC. The PAC CODEC was replaced by the HDC™ CODEC at the request of the NRSC.

TABLE 4.13-6
Logical Channel Characterizations—Service Mode MA1

Logical Channel	Throughput (kbps)	Latency (Sec)	Relative Robustness
P1	20.2	5.94	6
P3	16.2	1.49	7 or 10
PIDS	0.4	0.19	4 or 8

TABLE 4.13-7
Logical Channel Characterizations—Service Mode MA3

Logical Channel	Throughput (kbps)	Latency (Sec)	Relative Robustness
P1	20.2	5.94	1
P3	20.2	1.49	5
PIDS	0.4	0.19	3

Figure 4.3-17 reveals that the P1 logical channel is transmitted on both the upper and lower primary sidebands. These are redundant copies of the same information allowing the P1 logical channel to operate in the presence of a strong interferer on either the lower or upper adjacent channel. In addition to the frequency redundancy, the P1 logical channel in service mode MA1 also contains time redundancy. This is realized by transmitting redundant information that has diversity delay imposed at the transmitter. This redundant information uses a short interleaver so that the digital audio may be acquired quickly. It also serves as a backup to the nondelayed information, providing robustness to short-term outages such as those caused by GCS. This is why in Table 4.13-6 and Table 4.13-7 the latency of the P1 logical channel is larger than the other logical channels. The P1 logical channel, with its high degree of robustness, was designed to transmit core audio information.

Because of the requirement to minimize interference to the host analog signal, the carriers in the secondary and tertiary sidebands are placed at low levels and, as previously mentioned, maintain a phase relationship with the host analog signal. The P3 logical channel is transmitted on these carriers and therefore is less robust than the P1 logical channel. Because of this reduced robustness, audio enhancement information such as the stereo signal is typically transmitted on this logical channel. The P3 logical channel contains no time redundancy but contains frequency redundancy in the upper and lower secondary sideband, as long as the host analog signal is transmitted with a 5 kHz bandwidth or less. If the analog audio extends into the secondary sidebands, decoding may be impaired because both sidebands are needed to demodulate the digital signal. This is why the digital signal is less robust to second adjacent interferers when 8 kHz analog signals are transmitted.

Figure 4.13-18 shows that service mode MA3 is the all digital equivalent of service mode MA1. Since there is no analog signal to serve as a backup channel, the time diversity inherent in the P1 logical channel serves this purpose.

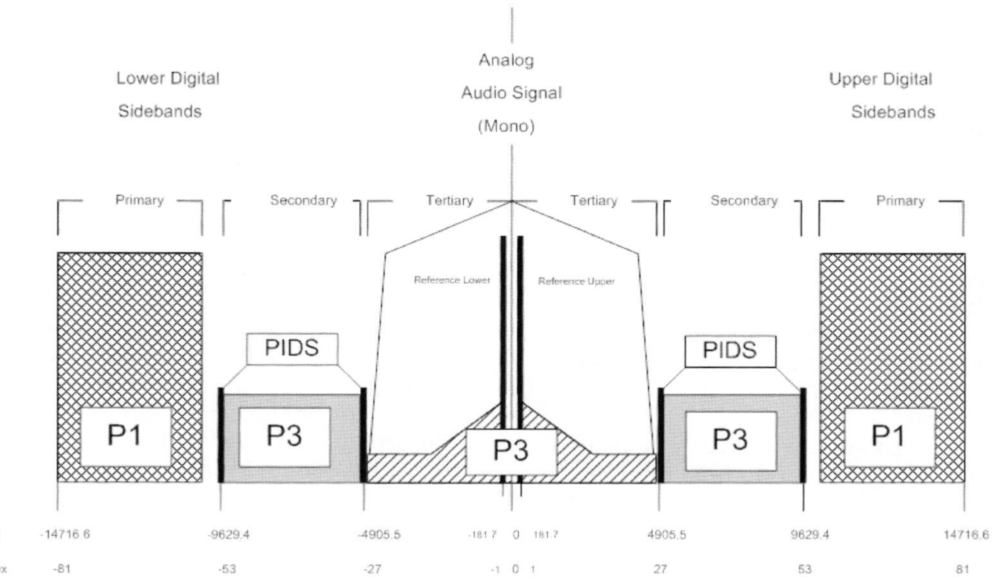

FIGURE 4.13-17 Logical channel spectral mapping—service mode MA1.

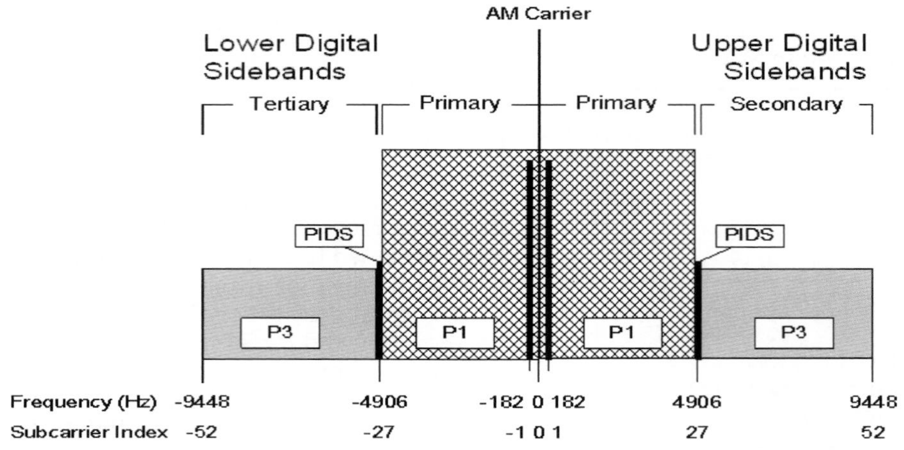

FIGURE 4.13-18 Logical channel spectral mapping—service mode MA3.

Layer 1 Functional Components

Several processing steps are necessary to convert the various logical channels into an AM IBOC system waveform. Figure 4.13-19 shows a functional block diagram of the Layer 1 processing. In this diagram, the logical channel name denotes that data is passed between the various functions as vectors. During the interleaving process, logical channels lose their identities as they are combined or split by the interleaving process. Of course, the identity is restored in the de-interleaving process.

Scrambling

The scrambling function randomizes (actually a pseudo-random sequence is employed at both transmitter and receiver) the digital data carried in each logical channel to mitigate signal periodicities and aid in receiver synchronization.

Channel Encoding

A digital bitstream, when passed through a transmission channel, is likely to encounter various forms of impairments including noise, distortion, fading, and interference. Digital systems employ various error correction techniques to restore transmission errors. These algorithms improve robustness through the introduction of error correction bits that are used in the receiver to verify the accuracy of the recovered bitstream, detect errors, and provide restoration of the transmitted bitstream.

Error correction codes are typically specified by their coding rate (R = information bits/total bits). For example, a coder with R = 1/2 channel coding represents an algorithm in which half of the bits carry information and the other half of the bits carry the overhead of the error correction algorithm.

The size of the logical channel vectors is increased in inverse proportion to the code rate. The encoding techniques are configurable by service mode. At the output of the channel encoder, the logical channel vectors retain their identity.

Interleaving

Interleaving in time and frequency is employed to mitigate the effects of burst errors. Digital error correction techniques are enhanced if errors in transmission are spread in a manner that minimizes data loss in successive bits. Interleaving is a technique that jumbles the bits in a predetermined manner upon transmission and reassembles them in the receiver. The interleaving techniques used in the AM IBOC system are tailored to the AM nonuniform interference environment and are configurable by service mode. In this process, the logical channels lose their identity. The interleaver output is structured in a matrix format. Each matrix consists of information from whole or partial logical channels and is associated with a specific portion of the transmitted spectrum. Diversity delay is also imposed on selected logical channels. It is through this function that the time redundancy of logical channel P1, in service modes MA1 and MA3, is created.

System Control Processing

The System Control Processing function generates a vector of system control data sequences that includes system control information received from Layer 2 (such as service mode and configuration options) and status for broadcast on the reference subcarriers. This information is used at the receiver to determine how to process the AM IBOC system waveforms.

OFDM Subcarrier Mapping

The OFDM Subcarrier Mapping function assigns the interleaver matrices and system control vector to OFDM subcarriers. One row of each active interleaver

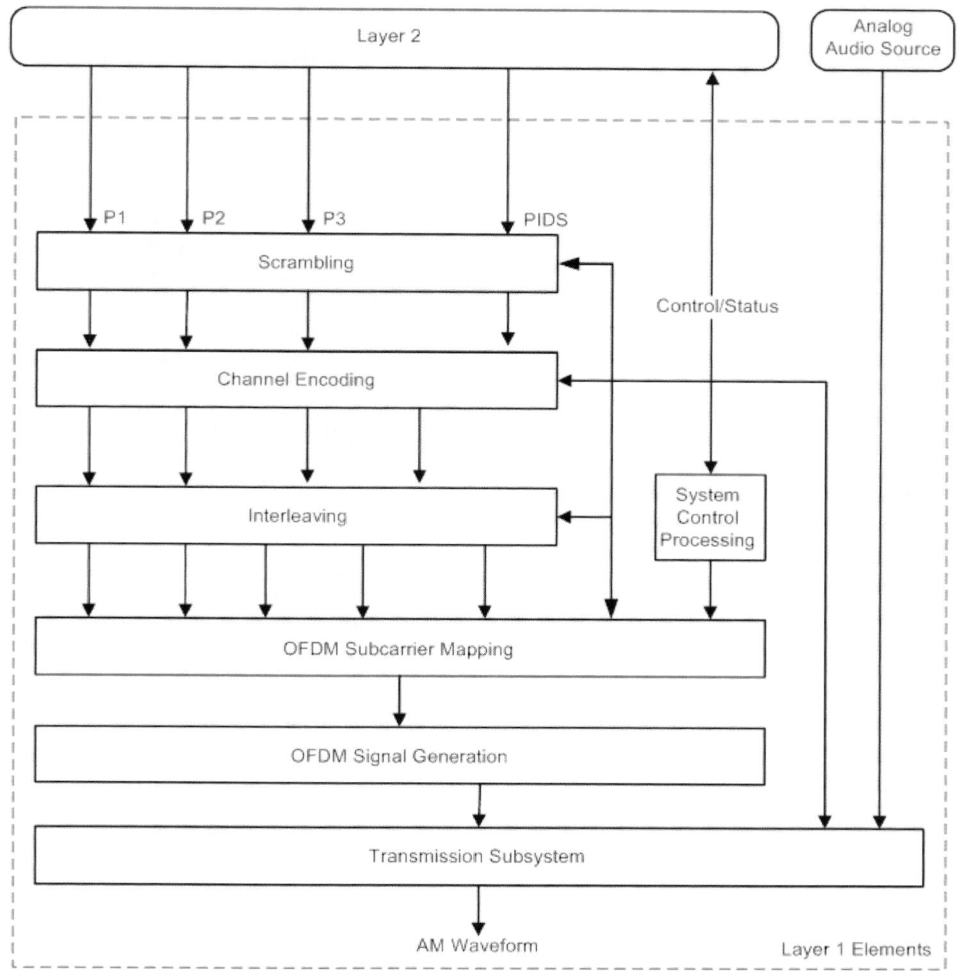

FIGURE 4.13-19 AM Layer 1 functional block diagram. The diagram illustrates the processing that takes place in Layer 1 after receiving data from Layer 2 at the top of the diagram.

matrix and one bit of the system control vector are processed for each OFDM symbol to produce one output vector, which is a frequency domain representation of the signal.

OFDM Signal Generation

The OFDM Signal Generation function generates the digital portion of the time-domain AM IBOC waveform. The input vectors are transformed into a shaped time-domain baseband pulse defining one OFDM symbol.

Transmission Subsystem

The Transmission Subsystem function formats the baseband waveform for transmission through the MF channel. Major subfunctions include symbol concatenation and frequency upconversion. When transmitting a hybrid waveform, this function modulates the AM analog audio source and coherently combines it with the digital signal to form a composite hybrid signal, ready for transmission.

FM HD RADIO SERVICES

FM Functional Description

The OFDM Signal Generation receives complex, frequency-domain OFDM symbols from the OFDM Subcarrier Mapping and outputs time-domain pulses representing the digital portion of the FM HD Radio signal. A conceptual block diagram of OFDM Signal Generation is shown in Figure 4.13-20.

The input to OFDM Signal Generation is a complex vector, representing the complex constellation values for each OFDM subcarrier. The output of OFDM Signal Generation is a complex, baseband, time-domain pulse, representing the digital portion of the FM IBOC signal.

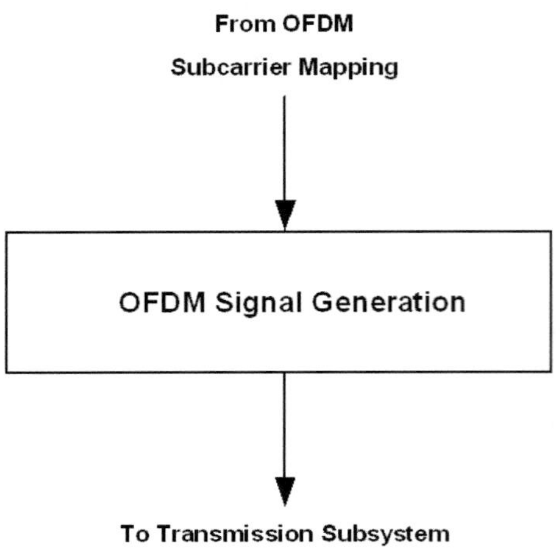

FIGURE 4.13-20 OFDM Signal Generation conceptual block diagram.

FM HD Radio Transmission Subsystem

The Transmission Subsystem formats the baseband FM HD Radio waveform for transmission through the VHF channel. Functions include symbol concatenation and frequency upconversion. In addition, when transmitting the hybrid or extended hybrid waveforms, this function modulates the baseband analog signal before combining it with the digital waveform.

The input to this module is a complex, baseband, time-domain OFDM symbol from the OFDM Signal Generation function. A baseband analog signal is also input from an analog source, after application of diversity delay along with optional Subsidiary Communications Authorization (SCA) signals, when transmitting the hybrid or extended hybrid waveform. The output of this module is the VHF FM HD Radio waveform. Refer to Figure 4.13-21 and Figure 4.13-22 for functional block diagrams of the hybrid and all digital transmission subsystems, respectively.

The FM HD Radio system affords broadcasters the ability to modify their digital audio broadcasts to meet their own specific needs. During the transition period to all digital broadcasting, each station will have the

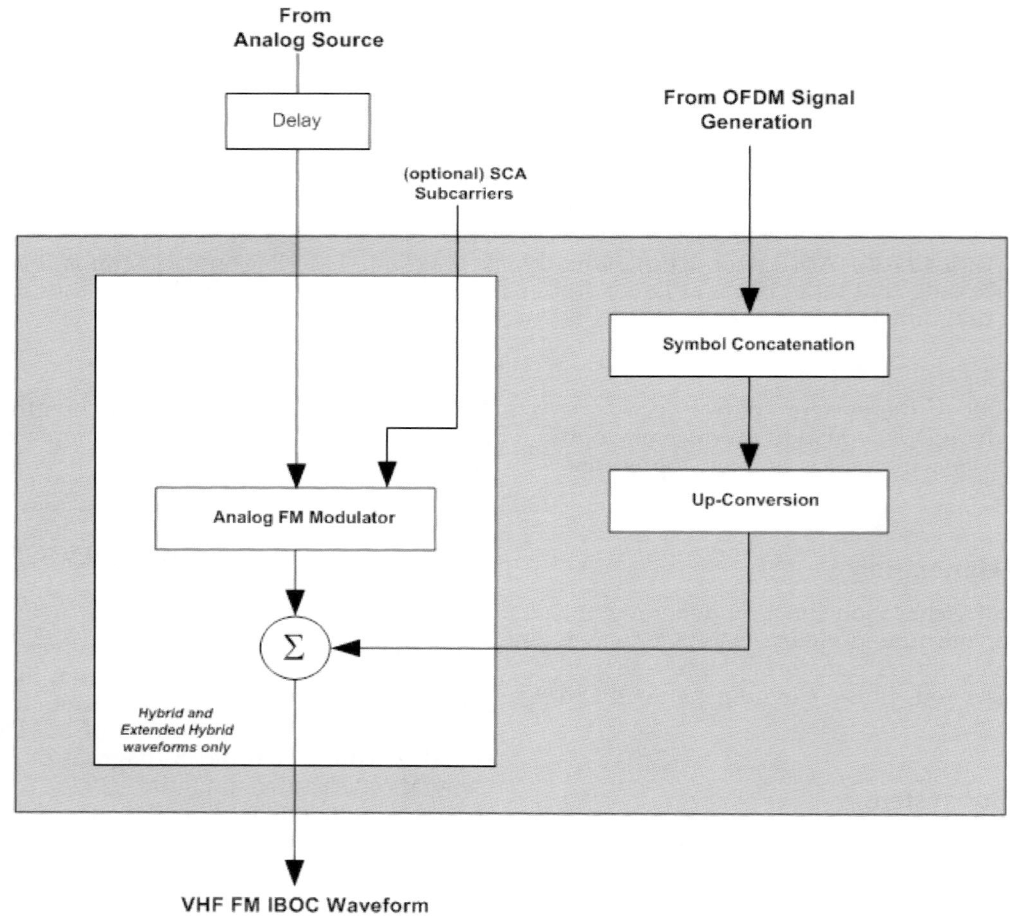

FIGURE 4.13-21 Extended hybrid transmission subsystem functional block diagram.

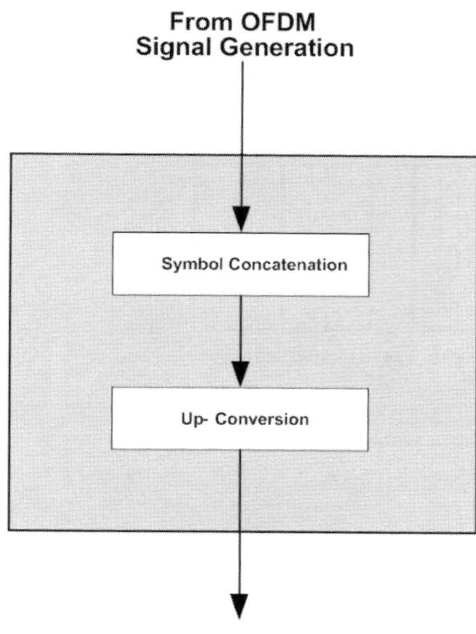

FIGURE 4.13-22 All digital transmission subsystem functional block diagram.

opportunity to convert at its own pace, beginning with a hybrid analog/digital waveform and eventually turning off the analog and broadcasting an all digital signal.

To support a wide variety of broadcast requirements, the FM HD Radio system was designed with a high degree of flexibility. All waveforms, hybrid or all digital, may be configured in a number of ways by sensibly adjusting the throughput, latency, and robustness of the audio and data program content as it is converted into an IBOC waveform. The following sections describe the structure and generation of the FM IBOC waveforms and present the various configurations from which a broadcaster can choose in order to transmit digital audio or data in a manner that best supports the broadcaster's needs.

FM IBOC SERVICES AND PROTOCOLS

In order to provide broadcaster flexibility and enhance the listening experience, iBiquity's FM HD Radio technology supports a variety of digital program services. They include a Main Program Service (MPS), Program Service Data (PSD), Station Information Service (SIS), and Advanced Application Service (AAS).

The MPS delivers existing programming formats in digital audio along with digital data that directly correlates with the audio programming. Whereas the MPS broadcasts a traditional audio program to listeners, the PSD enables listeners to select on-demand data services, thereby providing personalized, user-valued information. The SIS provides the control and identification information required to allow the listeners to search and select IBOC digital radio stations and their supporting services. The AAS allows a virtually unlimited number of custom and specialized IBOC digital radio applications to co-exist concurrently. Auxiliary applications can be added at any time in the future. Simultaneous support of these services is provided via the Layered protocol stack illustrated earlier in Figure 4.13-11. Source material (audio or data) moves down the protocol stack from Layer 5 to Layer 1 at the transmitter, is broadcast over the air, and upon reception is passed back up the protocol stack from Layer 1 to Layer 5.

At the transmitter, Layer 5 receives audio or data program content from the broadcaster. Layer 4 provides content-specific source encoding (such as audio compression), as well as station identification and control capabilities. Layer 3 ensures robust and efficient transfer of Layer 4 data, and Layer 2 provides limited error detection, addressing, and multiplexing.

Layer 1 receives the formatted content from Layer 2 and creates an FM HD Radio waveform for over-the-air transmission in the FM band. Since most of the digital signal processing required to generate an FM IBOC waveform occurs in Layer 1, it will be explained in greater detail.

Formatted program content is received from Layer 2 in discrete transfer frames via multiple logical channels. A transfer frame is an ordered collection of bits originating in Layer 2, grouped for processing through a logical channel. A logical channel is simply a signal path that conducts transfer frames from Layer 2 through Layer 1 with a specified grade of service. The service mode defines the active logical channels and their associated transmission characteristics.

The HD Radio system design provides a flexible means of transitioning to a digital broadcast system by providing three new waveform types: hybrid, extended hybrid, and all digital. The hybrid and extended hybrid types retain the analog FM signal, whereas the all digital type does not.

In all waveforms, the digital signal is modulated using orthogonal frequency division multiplexing (OFDM). In a single-carrier digital modulation scheme, the digital symbols are transmitted serially, with the spectrum of each symbol occupying the entire channel bandwidth during its appointed signaling interval. Conversely, OFDM is a parallel modulation scheme in which the data stream simultaneously modulates a large number of orthogonal subcarriers. Instead of a single, wideband carrier at a high signaling rate, OFDM employs a large number of narrowband subcarriers that are simultaneously transmitted at a much lower composite symbol rate. The long symbol times of OFDM provide superior robustness in the presence of multipath fading and interference. OFDM is also inherently flexible, readily allowing the mapping of specific logical channels to different groups of subcarriers.

The following sections describe the transmitted spectrum for each of the three digital waveform types. Each spectrum is divided into several sidebands, which represent various OFDM subcarrier groupings. All spectra are illustrated at baseband, with an upper and lower sideband centered around 0 Hz.

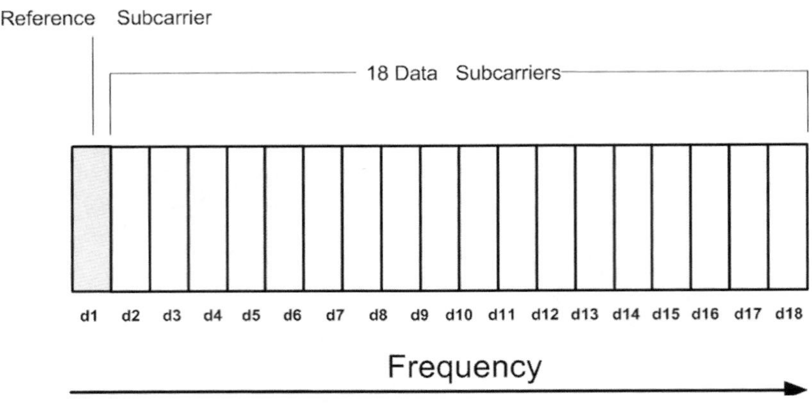

FIGURE 4.13-23 Frequency partition—ordering A.

Frequency Partitions and Spectral Conventions

The OFDM subcarriers are assembled into frequency partitions. Each frequency partition is composed of 18 data subcarriers and 1 reference subcarrier, as shown in Figure 4.13-23 (ordering A) and Figure 4.13-24 (ordering B). The position of the reference subcarrier (ordering A or B) varies with the location of the frequency partition within the spectrum.

For each frequency partition, data subcarriers d1 through d18 convey digital program content, while the reference subcarrier conveys system control. OFDM subcarriers are numbered from 0 at the center frequency to ±546 at either end of the channel frequency allocation.

Besides the reference subcarriers within each frequency partition, depending on the service mode, up to five additional reference subcarriers are inserted into the spectrum at subcarrier numbers –546, –279, 0, +279, and +546. The overall effect is the regular distribution of reference subcarriers throughout the spectrum. For convenience, each reference subcarrier is assigned a unique identification number between 0 and 60. All lower sideband reference subcarriers are shown in Figure 4.13-25. All upper sideband reference subcarriers are shown in Figure 4.13-26. The figures indicate the relationship between reference subcarrier numbers and OFDM subcarrier numbers.

Each spectrum described in the remaining subsections shows the subcarrier number and center frequency of certain key OFDM subcarriers. The center frequency of a subcarrier is calculated by multiplying the subcarrier number by the OFDM subcarrier spacing Δf 363.373 Hz. The center of subcarrier 0 is located at 0 Hz. In this context, center frequency is relative to the radio frequency (RF) allocated channel.

FM Hybrid Waveform

In the hybrid waveform, the digital signal is transmitted in primary main (PM) sidebands on either side of the analog FM signal, as shown in Figure 4.13-27. The analog signal may be mono or stereo and may include SCA channels. Each PM sideband is composed of 10 frequency partitions, which are allocated among subcarriers +356 through +545 or –356 through –545. Subcarriers +546 and –546, also included in the PM sidebands, are additional reference subcarriers. Figure 4.13-27 summarizes the upper and lower primary main sidebands for the hybrid waveform.

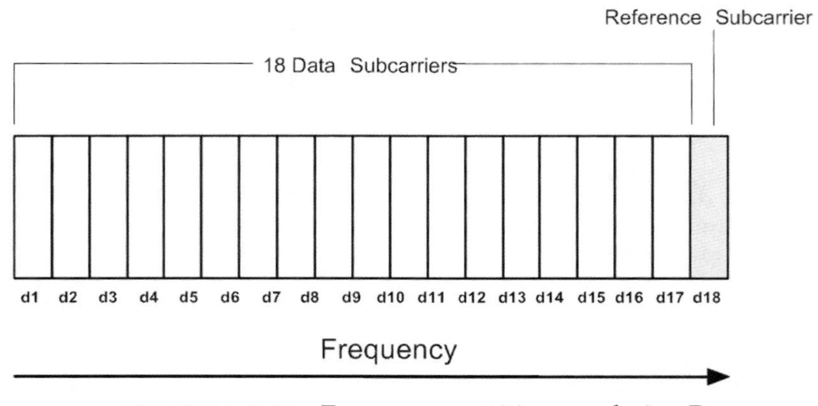

FIGURE 4.13-24 Frequency partition—ordering B.

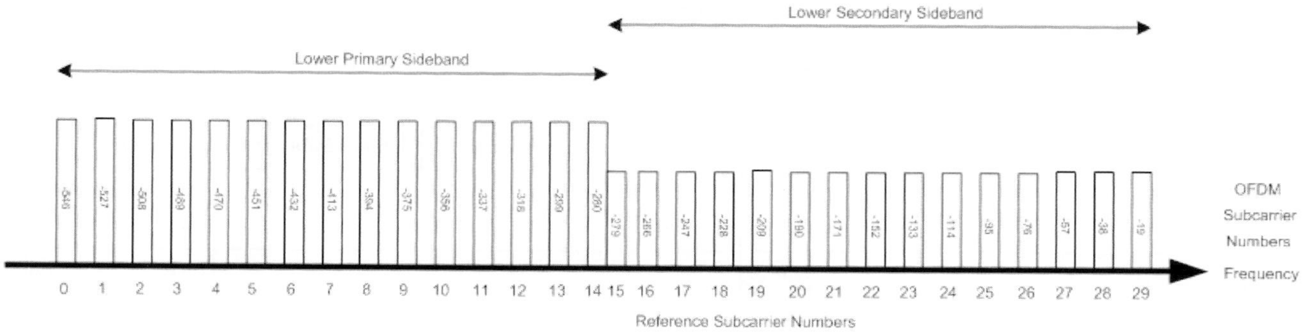

FIGURE 4.13-25 Lower sideband reference subcarrier spectral mapping.

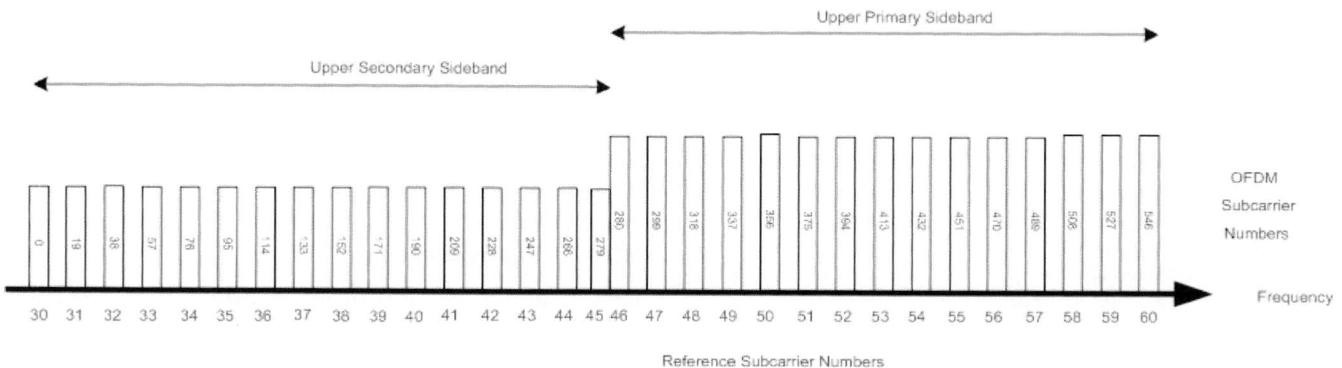

FIGURE 4.13-26 Upper sideband reference subcarrier spectral mapping.

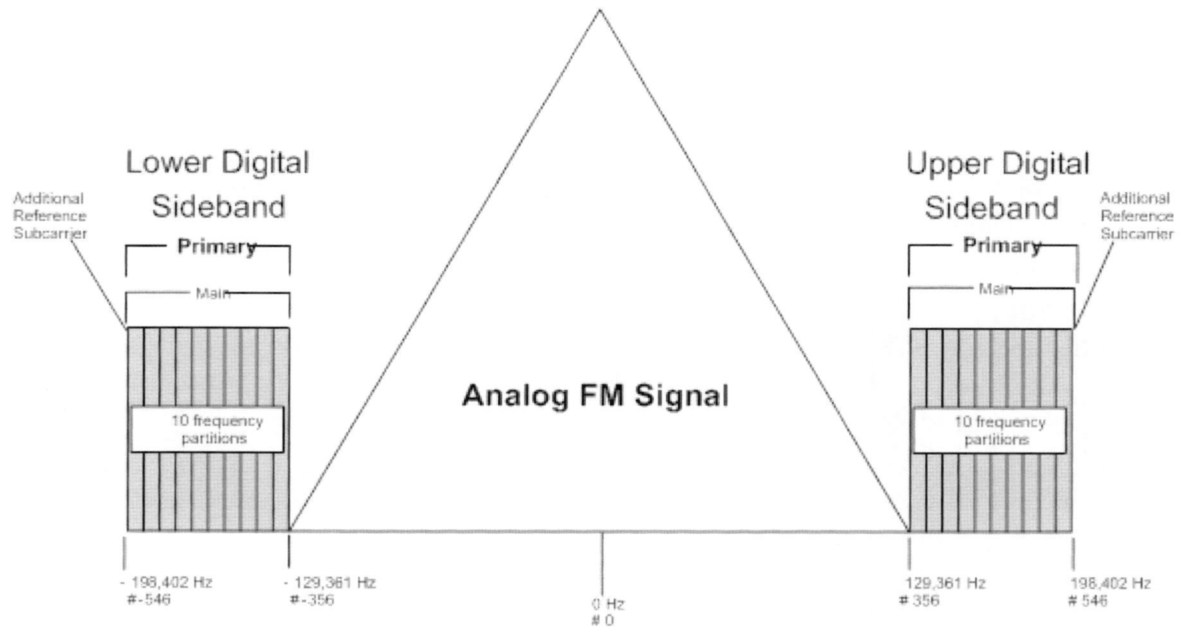

FIGURE 4.13-27 Spectrum of the hybrid waveform.

TABLE 4.13-8
Hybrid Waveform Spectral Summary

Sideband	Number of Frequency Partitions	Frequency Partition Ordering	Subcarrier Range	Subcarrier Frequencies (Hz from Channel Center)	Frequency Span (Hz)	Power Spectral Density (dB per Subcarrier)	Comments
Upper Primary Main	10	A	356 to 546	129,361 to 198,402	69,041	–45.8	Includes additional reference subcarrier 546
Lower Primary Main	10	B	–356 to –546	–129,361 to –198,402	69,041	–45.8	Includes additional reference subcarrier –546

The power spectral density of each OFDM subcarrier in the PM sideband, relative to the host analog power, and spectral summary are given in Table 4.13-8. A value of 0 dB would produce a digital subcarrier whose power was equal to the total power in the unmodulated analog FM carrier. The value was chosen so that the total average power in a primary main digital sideband (upper or lower) is 23 dB below the total power in the unmodulated analog FM carrier.

Extended Hybrid Waveform

The extended hybrid waveform is created by adding OFDM subcarriers to the primary main (PM) sidebands present in the hybrid waveform, as shown in Figure 4.13-28. Depending on the service mode, one, two, or four frequency partitions can be added to the inner edge of each primary main sideband. This additional spectrum is termed the primary extended (PX) sideband. Table 4.13-9 summarizes the upper and lower primary sidebands for the extended hybrid waveform.

The power spectral density of each OFDM subcarrier in the PM and PX sidebands, relative to the host analog power, is given in Table 4.13-9. Like the hybrid waveform, the value was chosen so that the total average power in a primary main sideband (upper or lower) is 23 dB below the total power in the unmodulated analog FM carrier. The level of the subcarriers in the PX sidebands is equal to the level of the subcarriers in the PM sidebands.

All Digital Waveform

The all digital waveform is constructed by disabling the analog signal, fully expanding the bandwidth of the primary digital sidebands, and adding lower-power secondary sidebands in the spectrum vacated by the analog signal. The spectrum of the all digital waveform is shown in Figure 4.13-29.

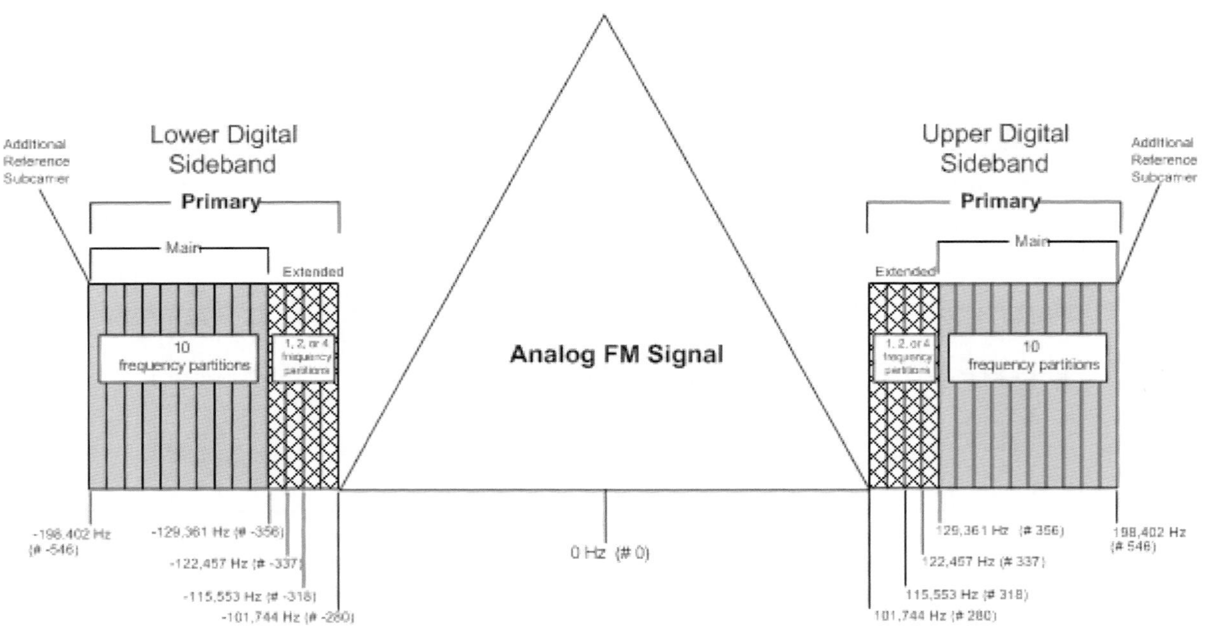

FIGURE 4.13-28 Spectrum of the extended hybrid waveform.

TABLE 4.13-9
Extended Hybrid Waveform Spectral Summary

Sideband	Number of Frequency Partitions	Frequency Partition Ordering	Subcarrier Range	Subcarrier Frequencies (Hz from Channel Center)	Frequency Span (Hz)	Power Spectral Density (dB per Subcarrier)	Comments
Upper Primary Main	10	A	356 to 546	129,361 to 198,402	69,041	–45.8	Includes additional reference subcarrier 546
Lower Primary Main	10	B	–356 to –546	–129,361 to –198,402	69,041	–45.8	Includes additional reference subcarrier –546
Upper Primary Extended (1 frequency partition)	1	A	337 to 355	122,457 to 128,997	6,540	–45.8	None
Lower Primary Extended (1 frequency partition)	1	B	–337 to –355	–122,457 to –128,997	6,540	–45.8	None
Upper Primary Extended (2 frequency partitions)	2	A	318 to 355	115,553 to 128,997	13,444	–45.8	None
Lower Primary Extended (2 frequency partitions)	2	B	–318 to –355	–115,553 to –128,997	13,444	–45.8	None
Upper Primary Extended (4 frequency partitions)	4	A	280 to 355	101,744 to 128,997	27,253	–45.8	None
Lower Primary Extended (4 frequency partitions)	4	B	–280 to –355	–101, 744 to –128,997	27,253	–45.8	None

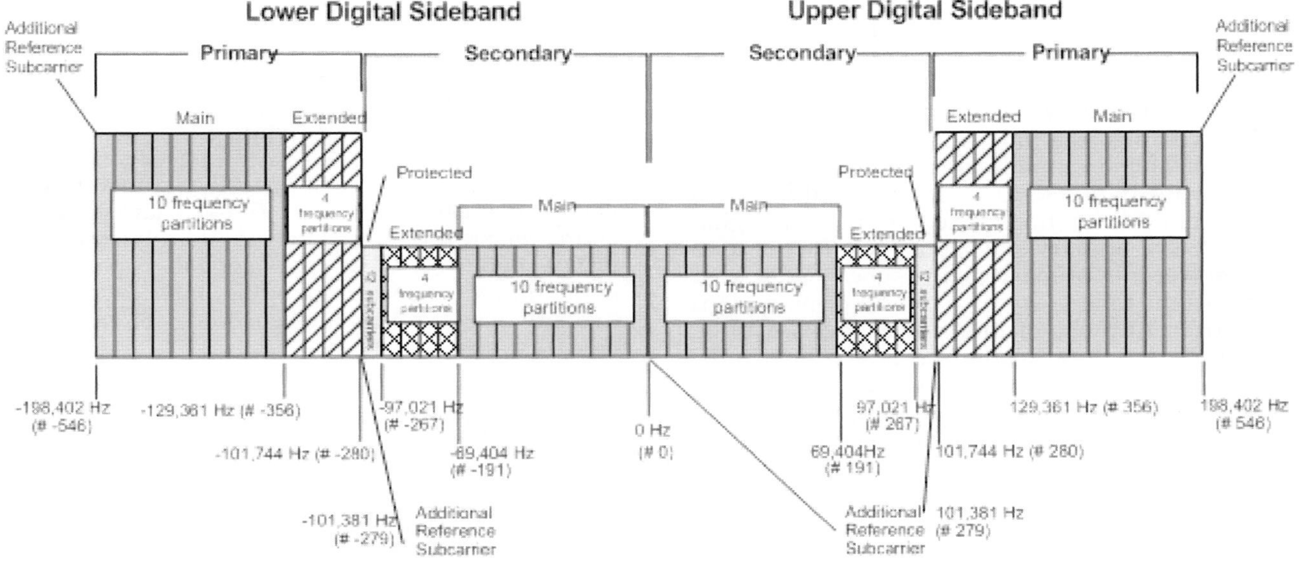

FIGURE 4.13-29 Spectrum of the all digital waveform.

TABLE 4.13-10
All Digital Waveform Summary–Service Modes MP5, MP6, MS1–MS4

Waveform	Service Mode	Sidebands	Amplitude Scale Factor Notation	Power Spectral Density dBc per Subcarrier	Power Spectral Density in a 1 kHz Bandwidth, dBc
All Digital	MP5–MP6	Primary	a_1	–27.3	–22.9
	MS1–MS4	Secondary	a_2	–32.3	–27.9
		Secondary	a_3	–37.3	–32.9
		Secondary	a_4	–42.3	–37.9
		Secondary	a_5	–47.3	–42.9

In addition to the 10 main frequency partitions, all 4 extended frequency partitions are present in each primary sideband of the all digital waveform. Each secondary sideband also has 10 secondary main (SM) and 10 secondary extended (SX) frequency partitions. Unlike the primary sidebands, the secondary main frequency partitions are mapped nearer to channel center with the extended frequency partitions farther from the center.

Each secondary sideband also supports a small secondary protected (SP) region consisting of 12 OFDM subcarriers and reference subcarrier +279 or –279. The sidebands are referred to as *protected* because they are located in the area of spectrum least likely to be affected by analog or digital interference. An additional reference subcarrier is placed at the center of the channel (0). Frequency partition ordering of the SP region does not apply since the SP region does not contain frequency partitions as defined in Figure 4.13-29. The total frequency span of the entire all digital spectrum is 396,803 Hz. Table 4.13-10 summarizes the upper and lower, primary and secondary sidebands for the all digital waveform. The power spectral density of each OFDM subcarrier is also provided.

For uniformity, as with the hybrid and extended hybrid waveforms, the values are referenced to the power level of the unmodulated analog FM carrier allocated by the station's FCC license (even though the analog carrier is not transmitted in the all digital waveform). The primary sideband level sets the total average power in a primary digital subcarrier at least 10 dB above the total power in a hybrid primary digital subcarrier. Any one of four power levels may be selected for application to the secondary sidebands. The four secondary power levels set the power spectral density of the secondary digital subcarriers (upper and lower) in the range of 5 to 20 dB below the power spectral density of the all digital primary subcarriers. A single secondary power level is evenly applied to all secondary sidebands.

For the all digital waveform, the value of a_1 was chosen so that the total average power of all the primary digital subcarriers combined is equal to 1. The values for a_2 through a_5 were chosen so that the total average power in the secondary digital subcarriers (upper and lower) lies in the range of 5 to 20 dB below the total power in the all digital primary digital subcarriers. The selection of one of the values a_2 through a_5 is determined by the amplitude scale factor select (ASF) received from Layer 2.

Logical Channels

A logical channel is a signal path that conducts program content through Layer 1 with a specific grade of service, as determined by the service mode. There are 10 logical channels, although not all are used in every service mode. The variety of logical channels reflects the inherent flexibility of the system.

There are four primary logical channels, denoted as P1, P2, P3, and PIDS. There are six secondary logical channels that are used only with the all digital waveform. They are denoted as S1, S2, S3, S4, S5, and SIDS. Logical channels P1 through P3 and S1 through S5 are designed to convey digital audio and data, whereas the PIDS and SIDS logical channels are designed to carry IBOC Data Service (IDS) information.

FM IBOC CHARACTERIZATION PARAMETERS

The performance of each logical channel is completely described through three characterization parameters: throughput, latency, and robustness. The service mode sets these characterization parameters by defining the spectral mapping, interleaver depth, diversity delay, and channel encoding for each active logical channel.

Throughput

Throughput defines the Layer 1 audio or data capacity of a logical channel, excluding upper layer framing overhead. The block-oriented operations of Layer 1 (such as interleaving) require that it process data in discrete transfer frames, rather than continuous streams. As a result, throughput is calculated as the product of transfer frame size and transfer frame rate. Spectral mapping and channel code rate determine the throughput of a logical channel, since spectral mapping limits capacity and coding overhead limits information throughput.

Latency

Latency is the delay that a logical channel imposes on a transfer frame as it traverses Layer 1. The latency of a logical channel is defined as the sum of its interleaver depth and diversity delay. It does not include processing delay or delays through higher protocol layers.

The interleaver depth determines the amount of delay imposed on a logical channel by its interleaver. Diversity delay is also applied to some logical channels to improve robustness. For example, in some service modes, logical channel P1 presents dual processing paths; one path is delayed and the other is not.

Robustness

Robustness is the capability of a logical channel to withstand channel impairments such as noise, interference, and fading. There are 11 relative levels of robustness in the FM IBOC system (but only 10 for the AM IBOC system). A robustness of 1 indicates a very high level of resistance to channel impairments, whereas an 11 indicates a lower tolerance for channel-induced errors. Spectral mapping, channel code rate, interleaver depth, and diversity delay determine the robustness of a logical channel. Spectral mapping affects robustness by setting the relative power level, spectral interference protection, and frequency diversity of a logical channel. Channel coding increases robustness by introducing redundancy into the logical channel. Interleaver depth influences performance in multipath fading. Finally, some logical channels in certain service modes delay transfer frames by a fixed duration to realize time diversity. This diversity delay also affects robustness, since it mitigates the effects of the mobile radio channel.

Characterization Parameter Assignments

Tables 4.13-11 through 4.13-20 show the active logical channels and their characterization parameters: throughput, latency, and relative robustness for a given service mode. A broadcaster might use these tables as a basis of comparison when selecting a service mode.

TABLE 4.13-11
Logical Channels—Service Mode MP1

Logical Channel	Throughput (kbps)	Latency (Seconds)	Relative Robustness
P1	98.4	1.49	2
PIDS	0.9	0.09	3

TABLE 4.13-12
Logical Channels—Service Mode MP2

Logical Channel	Throughput (kbps)	Latency (Seconds)	Relative Robustness
P1	98.4	1.49	2
P3	12.4	0.19	4
PIDS	0.9	0.09	3

TABLE 4.13-13
Logical Channels—Service Mode MP3

Logical Channel	Throughput (kbps)	Latency (seconds)	Relative Robustness
P1	98.4	1.49	2
P3	24.8	0.19	4
PIDS	0.9	0.09	3

TABLE 4.13-14
Logical Channels—Service Mode MP11

Logical Channel	Throughput (kbps)	Latency (Seconds)	Relative Robustness
P1	98.4	1.49	2
P3	49.6	0.19	4
PIDS	0.9	0.09	3

TABLE 4.13-15
Logical Channels—Service Mode MP5

Logical Channel	Throughput (kbps)	Latency (Seconds)	Relative Robustness
P1	24.8	4.64	1
P2	73.6	1.49	2
P3	24.8	0.19	4
PIDS	0.9	0.09	3

TABLE 4.13-16
Logical Channels—Service Mode MP6

Logical Channel	Throughput (kbps)	Latency (Seconds)	Relative Robustness
P1	49.6	4.64	1
P2	48.8	1.49	2
PIDS	0.9	0.09	3

TABLE 4.13-17
Logical Channels—Service Mode MS1

Logical Channel	Throughput (kbps)	Latency (Seconds)	Relative Robustness
S4	98.4	0.19	7
S5	5.5	0.09	6
SIDS	0.9	0.09	8

TABLE 4.13-18
Logical Channels—Service Mode MS2

Logical Channel	Throughput (kbps)	Latency (Seconds)	Relative Robustness
S1	24.8	4.64	5
S2	73.6	1.49	9
S3	24.8	0.19	11
S5	5.5	0.09	6
SIDS	0.9	0.09	10

TABLE 4.13-19
Logical Channels—Service Mode MS3

Logical Channel	Throughput (kbps)	Latency (Seconds)	Relative Robustness
S1	49.6	4.64	5
S2	48.8	1.49	9
S5	5.5	0.09	6
SIDS	0.9	0.09	10

TABLE 4.13-20
Logical Channels—Service Mode MS4

Logical Channel	Throughput (kbps)	Latency (Seconds)	Relative Robustness
S1	24.8	0.19	11
S2	98.4	1.49	9
S3	24.8	0.19	11
S5	5.5	0.09	6
SIDS	0.9	0.09	10

Spectral Mapping and Service Modes

For a given service mode, each logical channel is assigned to a group of OFDM subcarriers or frequency partitions. This spectral mapping contributes to the throughput and robustness of the logical channel.

Hybrid modes include MP1, MP2, MP3, and MP11 (replaced MP4).

All digital modes include MP5 and MP6.

Secondary modes that may be added to MP5 or MP6 are MS1, MS2, MS3, and MS4.[4]

Since this is a digital system, the various logical channels are simply conduits for the delivery of bits; the content of the bits is immaterial. However, the service modes were designed with specific services in mind for the active logical channels. As a result, although not strictly required, the recommended use of the logical channels is described along with the spectral mapping in the following sections.

Primary Spectral Mapping

The following sections describe the assignment of logical channels to the primary sidebands and describe the intended application of the logical channels for each primary service mode.

Service Mode MP1

The assignment of logical channels to OFDM subcarriers in service mode MP1 is shown in Figure 4.13-30. Both the P1 and PIDS logical channels are mapped to the upper and lower primary main sidebands. In service mode MP1, the P1 logical channel is designed to carry the MPS audio, whereas the PIDS logical channel would carry SIS data. Identical program material is carried on each sideband (upper and lower) so that the alternate sideband would be available if the other sideband were corrupted.

Service Mode MP2

The assignment of logical channels to OFDM subcarriers in service mode MP2 is shown in Figure 4.13-31. The transmitted spectrum for MP2 is identical to MP1, with the addition of a single extended frequency partition to each primary sideband. As in service mode MP1, the P1 and PIDS logical channels carry MPS audio and SIS data on each primary main sideband. In addition, the P3 logical channel is designed to carry additional data services, such as MPS, PDS, or AAS data, on the primary extended sidebands. Identical program material is carried on each sideband so that the alternate sideband would be available if the other sideband were corrupted.

Service Mode MP3

The assignment of logical channels to OFDM subcarriers in service mode MP3 is shown in Figure 4.13-32. The transmitted spectrum for MP3 is identical to MP1, with the addition of two extended frequency parti-

[4]MP4 had an error in carrier mapping that failed to have an all digital equivalent. As a result, a station implementing channels in hybrid MP4 would not be in a position to field the same services in MP6. The carrier groupings were harmonized and MP4 was replaced by mode MP11.

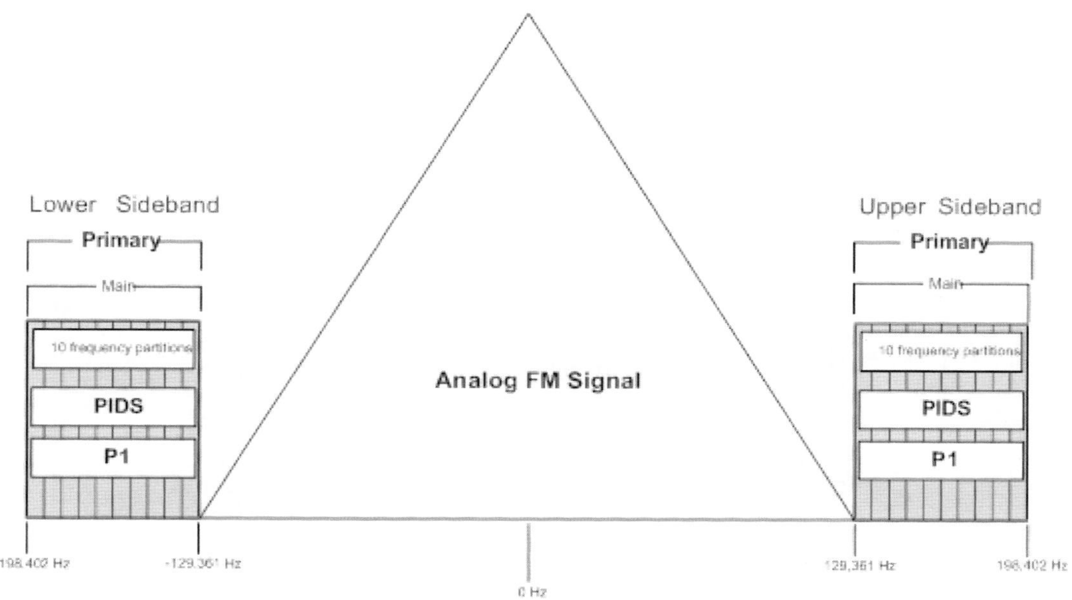

FIGURE 4.13-30 Spectral mapping—service mode MP1.

tions to each primary sideband. As in service mode MP1, the P1 and PIDS logical channels carry MPS audio and SIS data on each primary main sideband. In addition, the P3 logical channel is designed to carry additional data services, such as MPS, PDS, or AAS data, on the primary extended sidebands. Identical program material is carried on each sideband so that the alternate sideband would be available if the other sideband were corrupted.

Service Mode MP11

The assignment of logical channels to OFDM subcarriers in service mode MP11 is shown in Figure 4.13-33. The transmitted spectrum for MP11 is identical to MP1, with the addition of all four extended frequency partitions to each primary sideband. As in service mode MP1, the P1 and PIDS logical channels carry MPS audio and SIS data on each primary main sideband. In addition, the P3 logical channel is designed to

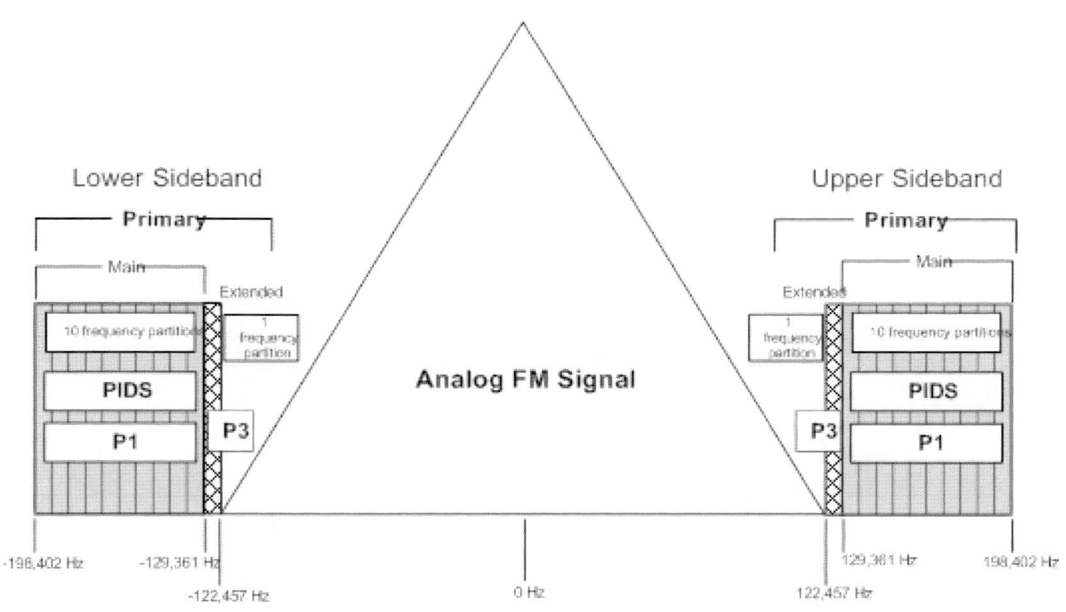

FIGURE 4.13-31 Spectral mapping—service mode MP2.

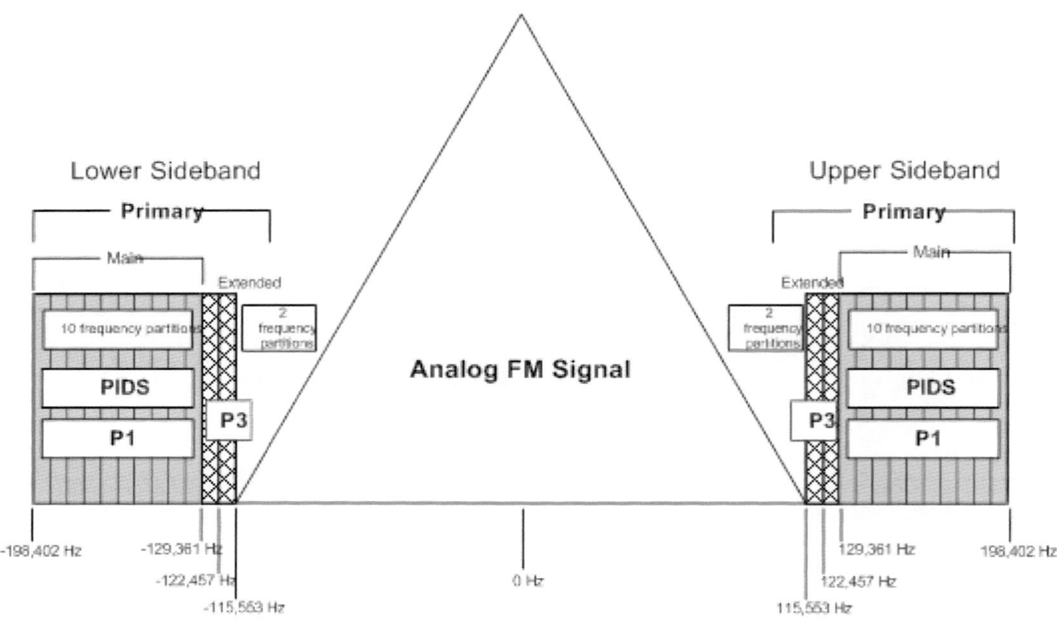

FIGURE 4.13-32 Spectral mapping—service mode MP3.

carry additional data services, such as MPS, PDS, or AAS data, on the primary extended sidebands. Identical program material is carried on each sideband so that the alternate sideband would be available if the other sideband were corrupted.

Service modes MP1 through MP11 provide essentially the same program services, with varying data capacity via the P3 logical channel on the primary extended sidebands.

Service Mode MP5

The assignment of logical channels to OFDM subcarriers in service mode MP5 is shown in Figure 4.13-34. The transmitted spectrum is identical to MP11. However, the spectral mapping for MP5 allows operation as either an extended hybrid or all digital waveform.

In service mode MP5, the MPS audio is divided into core and enhanced audio streams. The core audio is a stand-alone, low-bit-rate (~25 kbps), backup audio

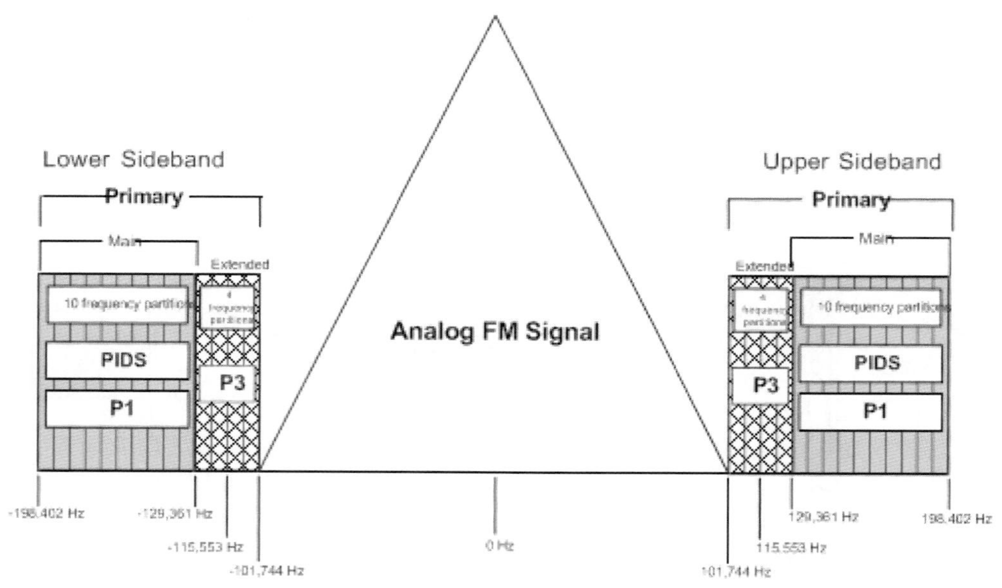

FIGURE 4.13-33 Spectral mapping—service mode MP11.

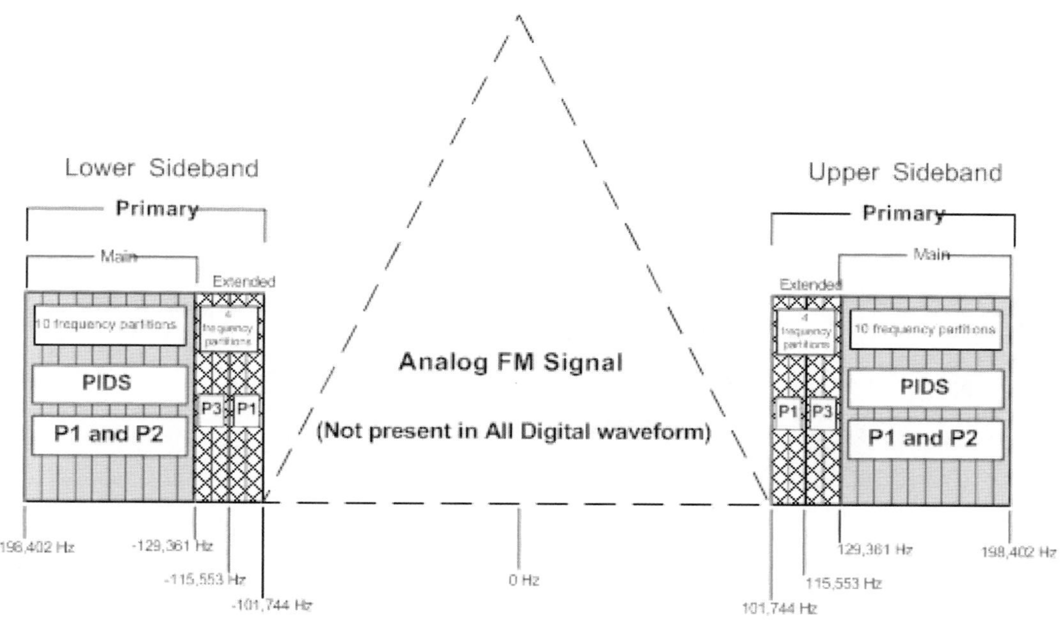

FIGURE 4.13-34 Spectral mapping—service mode MP5.

stream. When the core audio is combined with enhanced audio, the result is CD-like quality (~98 kbps) audio stream. The enhanced audio is not autonomous; it can be used in combination only with the core audio stream.

In service mode MP5, the core MPS audio stream is carried by the P1 logical channel, and the enhanced audio is carried by the P2 logical channel. Both P1 and P2 are mapped together in the primary main sidebands. In addition, the same P1 logical channel is diversity delayed and separately mapped to the inner two extended frequency partitions of each primary sideband. At the receiver, the two P1 channels are combined to form a more robust backup core audio stream.

In hybrid and extended hybrid waveforms, the analog host provides fast tuning and a diversity delayed backup channel for graceful degradation of audio near the edge of coverage. In the all digital waveform, the analog host no longer exists. In this case, the robust P1 logical channel, carrying core audio, acts as the backup for graceful audio degradation and fast tuning (since it is lightly interleaved). When the enhanced audio is not available, the receiver reverts to the backup core audio stream.

The P3 logical channel also carries additional data services, such as MPS, PDS, or AAS data, on the primary extended sidebands. As in service modes MP1–MP11, the PIDS logical channel carries SIS data over the primary main sidebands. Again, identical program material is carried on each sideband so that the alternate sideband would be available if the other sideband were corrupted.

Service Mode MP6

The assignment of logical channels to OFDM subcarriers in service mode MP6 is shown in Figure 4.13-35. The transmitted spectrum is identical to MP5. However, for MP6, the size of the core audio stream is doubled to a higher quality ~50 kbps. As a result, all four frequency partitions in the primary extended sideband are required to carry the backup core audio, and capacity is no longer available for data. Thus, the increased data capacity in service mode MP5 is traded for core audio quality in service mode MP6.

In service mode MP6, the core MPS audio stream is carried by the P1 logical channel, and the enhanced audio is carried by the P2 logical channel. Both P1 and P2 are mapped together in the primary main sidebands. In addition, the same P1 logical channel is diversity delayed and separately mapped to all four extended frequency partitions of each primary sideband. The PIDS logical channel also carries SIS data over the primary main sidebands. Identical program material is carried on each sideband so that the alternate sideband would be available if the other sideband were corrupted.

As in service mode MP5, the P3 logical channel carries additional data services, such as MPS, PDS, or AAS data, on the primary extended sidebands. The PIDS logical channel also carries SIS data over the primary main sidebands. As always, identical program material is carried on each sideband so that the alternate sideband would be available if the other sideband were corrupted.

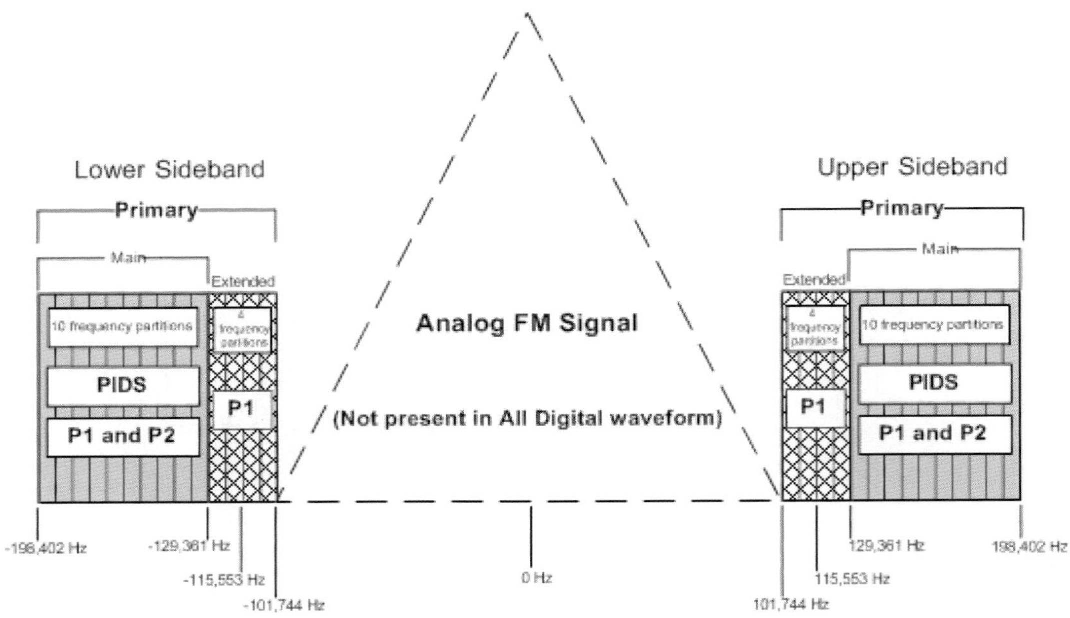

FIGURE 4.13-35 Spectral mapping—service mode MP6.

Secondary Spectral Mapping

The following sections describe the assignment of logical channels to the secondary sidebands and describe the intended application of the logical channels for each secondary service mode. Note that secondary sidebands are present only in the all digital waveform. Only the secondary sidebands are presented in the following subsections; the presence of primary digital sidebands in service modes MP5 or MP6 is implied.

By *implied*, it means that only the MS1–MS4 carrier groups are illustrated, and they are to be used in combination with the MP5 or MP6 modes. It is possible to run MP5 or MP6 with the analog signal present. If the analog signal is removed, there is room to add MS1 through MS4 secondary carriers in place of the analog signal (see Figure 4.13-29 earlier).

Service Mode MS1

The assignment of logical channels to OFDM subcarriers in service mode MS1 is shown in Figure 4.13-36. Service mode MS1 is intended for the transmission of secondary broadband data.

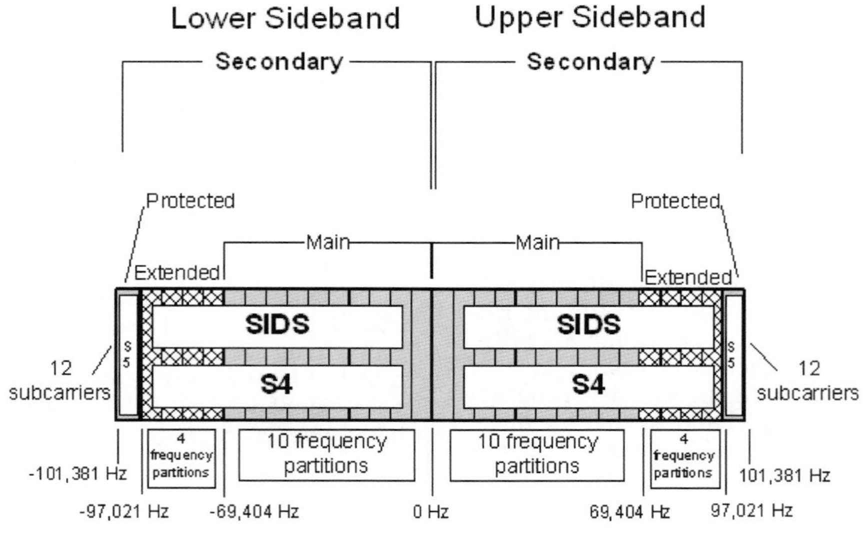

FIGURE 4.13-36 Spectral mapping—service mode MS1.

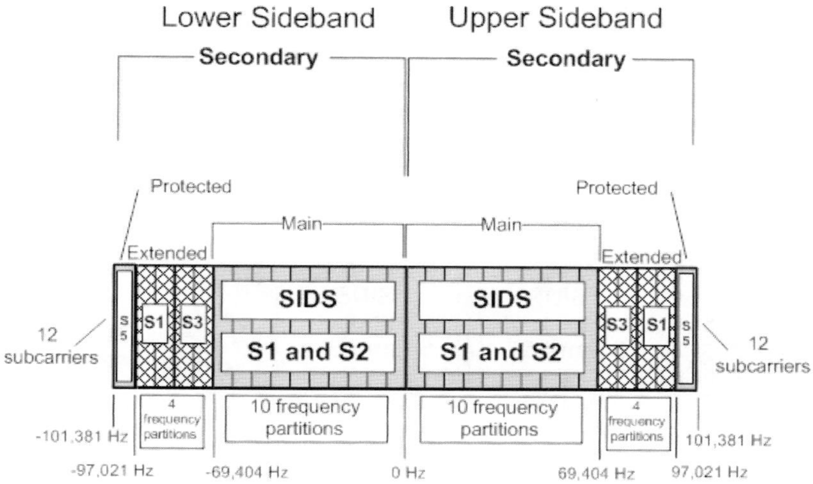

FIGURE 4.13-37 Spectral mapping—service mode MS2.

In service mode MS1, logical channel S4 carries MPS, PDS, or AAS data over the secondary main and extended sidebands. In addition, the SIDS logical channel also carries SIS data over the secondary main and extended sidebands. Finally, the S5 logical channel carries MPS, PDS, or AAS data over the secondary protected sidebands. As with the primary sidebands, identical program material is carried on each secondary sideband (upper and lower) so that the alternate sideband would be available if the other sideband were corrupted.

Service Mode MS2

The assignment of logical channels to OFDM subcarriers in service mode MS2 is shown in Figure 4.13-37. Service mode MS2 is the secondary equivalent of primary service mode MP5 in terms of throughput.

In service mode MS2, the S1 and S2 logical channels might carry core and enhanced auxiliary audio (such as surround sound), intended to enhance the MPS audio broadcast on the primary sidebands. Both S1 and S2 are mapped together in the secondary main sidebands. In addition, the same S1 logical channel is diversity delayed and separately mapped to the outer two extended frequency partitions of each secondary sideband.

The S3 logical channel carries additional data services, such as MPS, PDS, or AAS data, on the secondary extended sidebands. The SIDS logical channel also carries SIS data over the secondary main sidebands. Finally, the S5 logical channel carries MPS, PDS, or AAS data over the secondary protected sidebands. Identical program material is carried on each secondary sideband so that the alternate sideband would be available if the other sideband were corrupted.

Service Mode MS3

The assignment of logical channels to OFDM subcarriers in service mode MS3 is shown in Figure 4.13-38. Service mode MS3 is the secondary equivalent of primary service mode MP6 in terms of throughput.

As in service mode MS2, the S1 and S2 logical channels might carry core and enhanced auxiliary audio (such as surround sound), intended to enhance the MPS audio broadcast on the primary sidebands. However, in service mode MS3, the size of S1 is doubled and capacity is no longer available for S3 data.

Both S1 and S2 are mapped together in the secondary main sidebands. In addition, the same S1 logical channel is diversity delayed and separately mapped to all four extended frequency partitions of each secondary sideband. The SIDS logical channel also carries SIS data over the secondary main sidebands. Finally, the S5 logical channel carries MPS, PDS, or AAS data over the secondary protected sidebands. Identical program material is carried on each sideband so that the alternate sideband would be available if the other sideband were corrupted.

Service Mode MS4

The assignment of logical channels to OFDM subcarriers in service mode MS4 is shown in Figure 4.13-39. It is intended for broadcast of a single, low-bit-rate audio stream, with the remaining capacity reserved for data services.

In service mode MS4, the low-bit-rate audio is carried by the S1 logical channel, which is mapped to the outer two extended frequency partitions of each secondary sideband. Logical channel S2 carries MPS, PDS, or AAS data over the secondary main sidebands. As in service mode MS2, the S3 logical channel carries additional data services, such as MPS, PDS, or AAS data, on the secondary extended sidebands. The SIDS logical channel also carries SIS data over the secondary main sidebands. As always, identical program material is carried on each sideband so that the alternate sideband would be available if the other sideband were corrupted.

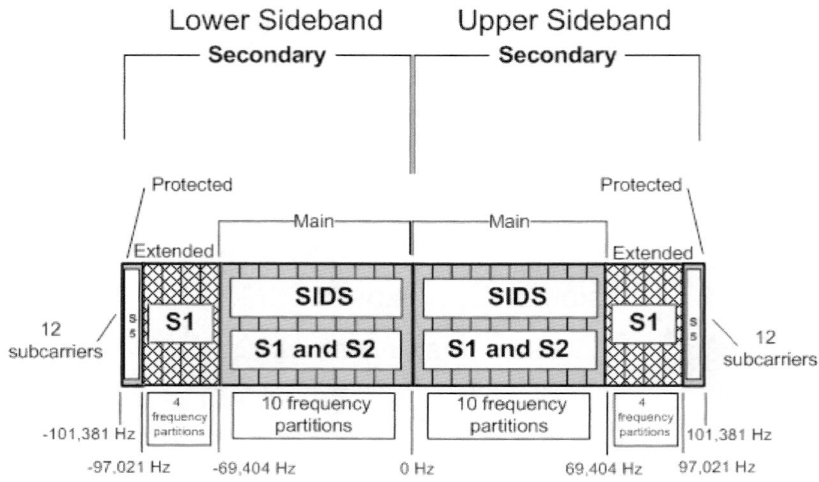

FIGURE 4.13-38 Spectral mapping—service mode MS3.

FM IBOC FUNCTIONAL COMPONENTS

The conversion of audio program content and data into the FM IBOC waveform is accomplished by the Layered protocol stack. Source material is received from the broadcaster in Layer 5, source encoded in Layer 4, multiplexed into logical channels in Layers 3 and 2, and formatted for over-the-air broadcast in Layer 1.

The following sections include a high-level description of each Layer 1 functional block and the associated signal flow. Figure 4.13-40 is a functional block diagram of Layer 1 processing.

Scrambling

The Scrambling function randomizes (using a pseudo-random sequence employed in the transmitter and receiver) the digital data in each logical channel to mitigate signal periodicities which could cause a higher peak to average power ratio resulting in transmission inefficiencies.

Channel Encoding

A digital signal, when passed through an RF transmission channel, is likely to be impaired by noise, fading, and interference. Digital systems employ error correction techniques to correct bit errors caused by these impairments. Forward Error Correction (FEC) algorithms improve signal robustness by adding error correction bits to the signal prior to transmission. These FEC bits are used by the receiver to correct bit errors and regenerate the transmitted bitstream. FEC codes are typically specified by their coding rate, which is

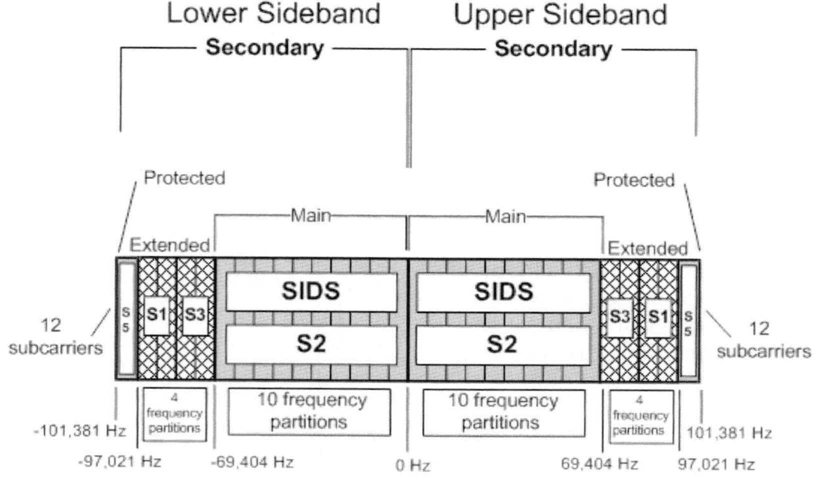

FIGURE 4.13-39 Spectral mapping—service mode MS4.

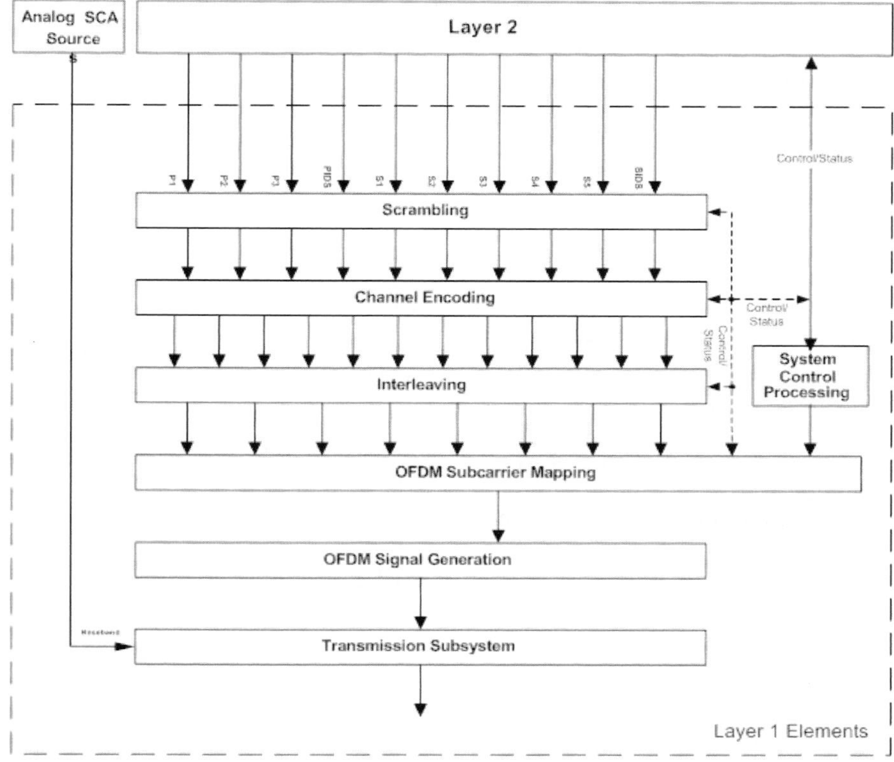

FIGURE 4.13-40 Layer 1 functional block diagram. The diagram illustrates the processes included in Layer 1 after the data is received from Layer 2 at the top of the diagram.

simply the number of information bits divided by the total number of transmitted bits. For example, in a rate 1/2 code, half of the bits carry information and the other half carry the FEC overhead. The channel encoding function uses convolutional encoding to add redundancy to the digital data in each logical channel, in order to improve its reliability in the presence of channel impairments. The size of the logical channel transfer frames is increased in inverse proportion to the code rate. The encoding techniques are configurable by service mode. Diversity delay is also imposed on selected logical channels.

Interleaving

Interleaving re-orders the transmitted bits to disperse burst errors typical of a fading channel. The FM IBOC waveform is interleaved in both time and frequency. The custom interleaving techniques are tailored to the VHF Rayleigh fading environment and are configurable by service mode. In this process, the logical channels lose their identities (which are restored upon de-interleaving in the receiver). The interleaver output is structured in a matrix format; each matrix is composed of one or more logical channels and is associated with a particular portion of the transmitted spectrum.

System Control Processing

The System Control Processing function generates a matrix of system control data sequences, which include control and status (such as service mode), for broadcast on the reference subcarriers.

OFDM Subcarrier Mapping

The OFDM Subcarrier Mapping function assigns the interleaver matrices and the system control data matrix to the OFDM subcarriers. One row of each active interleaver matrix is processed every OFDM symbol time to produce one output vector X, which is a frequency-domain representation of the signal. The mapping is specifically tailored to the nonuniform interference environment and is a function of the service mode.

OFDM Signal Generation

The OFDM Signal Generation function generates the digital portion of the time-domain FM IBOC waveform. The input vectors are transformed into a shaped time-domain baseband pulse defining one OFDM symbol.

Transmission Subsystem

The Transmission Subsystem function formats the baseband waveform for transmission through the VHF channel. Major subfunctions include symbol concatenation and frequency upconversion. In addition, when transmitting the hybrid or extended hybrid waveforms, this function modulates the analog source and combines it with the digital signal to form a composite signal ready for transmission.

FM HYBRID IMPLEMENTATION

FM hybrid transmission modes MP1, MP2, MP3, and MP11 all require the simultaneous transmission of an analog and a digital signal. The following methods can be used by the broadcasters to transmit the two signals at the proper analog-to-digital power ratios. Four methods exist for producing the HD Radio hybrid FM signal.

High-Level Combining

Initial station conversions utilized what is known as high-level combining or separate *amplification* shown in Figure 4.13-41. With this method, the existing station transmitter is combined with the output of a separate digital transmitter compatible with HD Radio technology. The combined signal is then fed to the existing station antenna.

HD Radio FM high-level combining uses two transmitters to produce the transmitted signal. This approach requires the addition of an HD Radio digital transmitter and the associated combiner, filter, and digital exciter. Since both an analog and digital transmitter will be operated at the site, power demands may require the upgrade of electrical service to the facility. Heat load will also increase and may require additional cooling to remain within acceptable limits.

The high-level combining method is inefficient due to the combining technique employed. In order to achieve the requisite isolation and linearity, combiners used for HD Radio sacrifice 10% (~0.5 dB) of analog power and 90% (~10dB) of digital power to the reject load. However, because the power requirements for the digital signal are low (–20 dB relative to analog power), this loss is tolerable. Additionally, because the digital signal varies in amplitude as well as frequency, the peak-to-average ratio (PAR) will vary about 5.5 dB. For example, in the case of an FM station with an analog TPO of 10 kW, the carrier power of the digital signal would be 100 Watts. Assuming combiner loss as listed above, the analog transmitter must be increased to 11.1 kW to overcome combiner insertion loss. The digital transmitter would require an average power of 1 kW to overcome the 10 dB combiner loss. The IBOC transmitter must be sized to accommodate 5.5 dB of additional overhead for PAR. This sizing for peak will amount to approximately three and a half times the average power. Chapter 4.10 explores FM combining systems in greater detail.

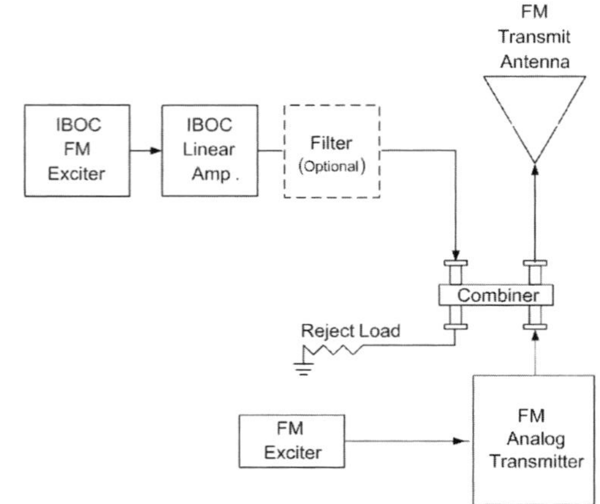

FIGURE 4.13-41 HD Radio high-level, separate amplification method.

Mid-Level Combining

A derivative of the high-level amplification technique is *mid-level* or *split amplification*, as represented in Figure 4.13-42. In this design, the FM analog signal generation is shared between a traditional Class-C FM transmitter and a transmitter that has been linearized for amplification of both the analog and digital components.

Low-Level Combining

The low-level combining or common amplification method is depicted in Figure 4.13-43. In this implementation, the output of an analog FM exciter is combined with the output of an HD Radio exciter both at low RF levels. The combined signal is fed to a common broadband linear amplifier to raise the power to the desired TPO. This method is both power and space

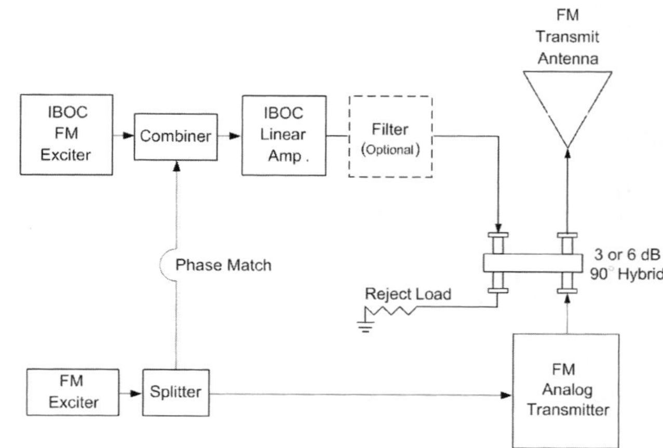

FIGURE 4.13-42 FM HD Radio mid-level, split amplification method.

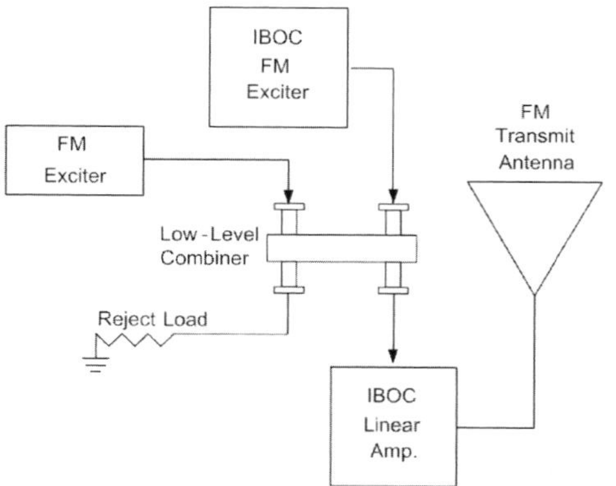

FIGURE 4.13-43 HD Radio low-level, common amplification method.

efficient and reduces the number of independent elements in the broadcast chain. Manufacturers are evaluating linearized versions of their transmitter design to determine optimal levels of headroom and linearity. This common amplification is both power and space efficient.

Low-level combining utilizes a shared transmitter power amplifier to boost the HD Radio digital signal and the host analog FM signal to desired output levels. This commonality reduces the demand on equipment space and may reduce power demands by increasing overall efficiency.

Separate and Dual-Input Antennas

The fourth implementation is the separate or dual input antenna method. This methodology takes one of two forms: a physically separate antenna, as depicted in Figure 4.13-44, or a dual-input antenna, as shown in Figure 4.13-45.

Separate antenna implementation routes the signal from independent IBOC and analog amplifiers to dedicated radiating elements for each signal. Two methods of separate antenna implementation are in use today. The basic form is an independent antenna, often previously installed as an analog backup. The second method, known as an interleaved antenna, places a digital bay at the midpoint of the analog radiating elements. In this design the digital elements phase is typically inverted (installed upside down), to provide additional isolation.

Regardless of which separate antenna method is employed, the FCC authorizes their use as follows:

- The digital transmission must use a licensed auxiliary antenna.
- The auxiliary antenna must be within 3 seconds of latitude and longitude of the main antenna.
- The height above average terrain of the auxiliary antennas must be between 70 and 100% of the height above average terrain of the main antenna.

Digital antennas require a minimum of 40 dB of isolation from the analog antenna in order to keep intermodulation products within acceptable limits. Careful placement and measurement of the antenna elements and RF isolator on one or both transmitters may be required to minimize mutual coupling.

Replicating radiating element placement with regard to tower leg and crossbar for both the analog and digital transmitting elements will ensure the patterns of the digital and analog signals are congruent. Since physically separate radiators are employed in this design, analog coverage may be superior to the digital due to the height difference between the centers of radiation. Despite this drawback, the advantage of separate antennas is the elimination of the combiner loss. This increase in system efficiency results in a significantly smaller IBOC transmitter required to develop the HD Radio carrier ratios.

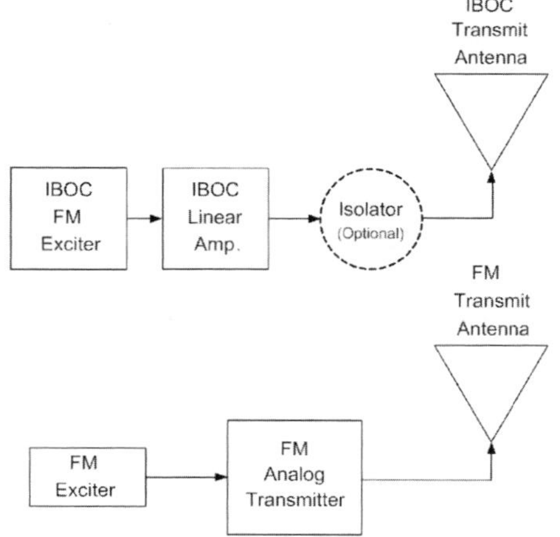

FIGURE 4.13-44 Radio separate antenna implementation.

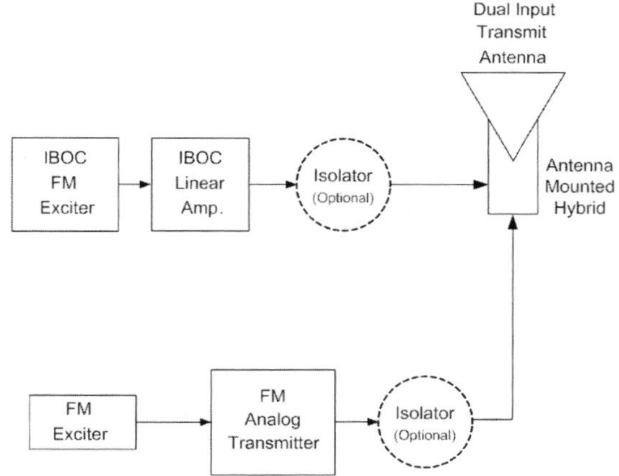

FIGURE 4.13-45 FM HD Radio dual-input antenna implementation.

Dual-input antennas utilize a hybrid to feed the two dipoles of the radiating element with independent analog and digital RF signals. In this implementation the resultant signal has right-hand circular polarization for the analog and left-hand polarization for the digital. Essentially, the dual-input antenna is a combination of free space signal combining in concert with the use of a 3 dB hybrid. Analog-to-digital coupling may be as high as 22 dB, typically necessitating the use of additional RF isolation. Dual-input antennas have the advantage that the center of radiation is identical for both the analog and digital aperture, resulting in near-identical analog and digital RF coverage.

Since system designs vary in dimension as well as configuration, the physical space and implementation constraints should be reviewed with equipment manufacturers to determine the appropriate solutions.

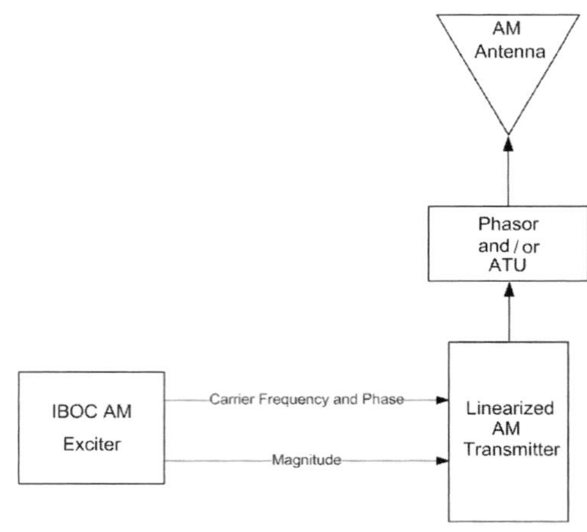

FIGURE 4.13-46 AM HD Radio implementation.

AM HD RADIO HYBRID TRANSMISSION

AM HD Radio requires the amplification of complex modulation. Unlike FM, AM IBOC requires strict phase coherency between the analog and digital signal. The most straightforward solution is to amplify both the analog and digital signals in a single transmitter. To accomplish this, AM HD Radio transmitters must provide ample bandwidth and minimize phase distortion.

The nonlineal transformation of complex I/Q to phase/magnitude requires infinite bandwidth to perfectly reconstruct the signal within spectral limits. Ideally, the transmission of discrete I/Q data would be desirable over a phase/magnitude transform, as it constrains frequency to the bandwidth of modulation. Transmitter bandwidth is a consideration, as negative modulation peaks reach pinch-off (100%). As these values approach the baseline, infinite bandwidth products are produced. Although these emissions are infinite, sample rates and filter characteristics employed in HD Radio transmitters constrain the products to acceptable emission limits.

HD Radio transmission requires similar response and phase characteristics from the antenna as did AM Stereo. State-of-the-art transmitter designs employ multiple solid-state amplifiers which are summed to obtain the required power. For optimal transfer, it is desirable to provide the appropriate match at the summing point of the final amplifiers. Multiphase PDM (Pulse Duration Modulation) transmitters with switching frequencies higher than 150 kHz and digitally modulated solid-state AM transmitters are generally compatible with minor input filter modifications. Figure 4.13-46 depicts the key transmission elements of an AM HD Radio implementation.

Traditional AM tube amplifier designs show insufficient phase and frequency fidelity to pass the HD Radio waveform. However, several tube manufacturers are exploring new methods of increasing tube frequency and phase performance.

AM Antenna System

Antenna characteristics should be viewed from the *driving point* of the transmitter, as this is where linear distortion characteristics will be introduced. Ideally, this driving point would be situated at the output of the final amplifier ahead of any matching network. Today's solid-state transmitters combine multiple amplifiers, each with different electrical lengths (delay) to the output network. It is therefore desirable to use the average of the electrical path lengths to determine the optimal point for achieving a match. This point in the circuit is known as the "amplifier summing point" of a transmitter that precedes the output matching network and sometimes has very low impedance, making it difficult to match to a higher load impedance. While no formal AM antenna specification exists, field trials indicate a sideband conjugate match to achieve optimal transfer with the fixed source impedance. The transmitter manufacturer can furnish this information as it is model dependent. Typically, the output network performs the functions of impedance matching and harmonic attenuation.

Hermitian symmetry[5] of the transmitted signal ±5 kHz is desirable to keep quadrature information in quadrature and to minimize crosstalk from the analog

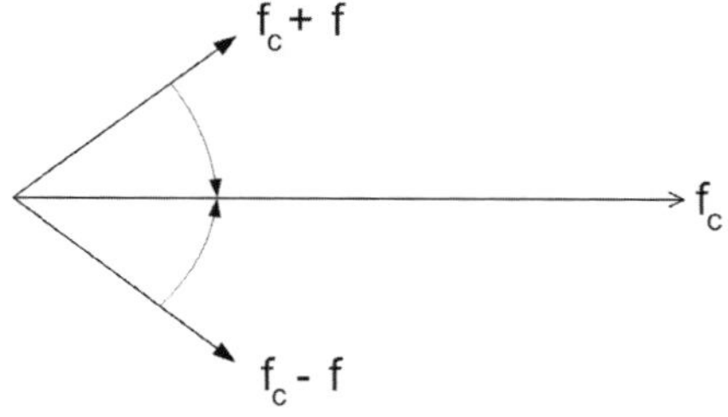

FIGURE 4.13-47 Hermitian symmetry.

[5]Named after Charles Hermite, French mathematician. See http://www-history.mcs.st-andrews.ac.uk/Mathematicians/Hermite.html.

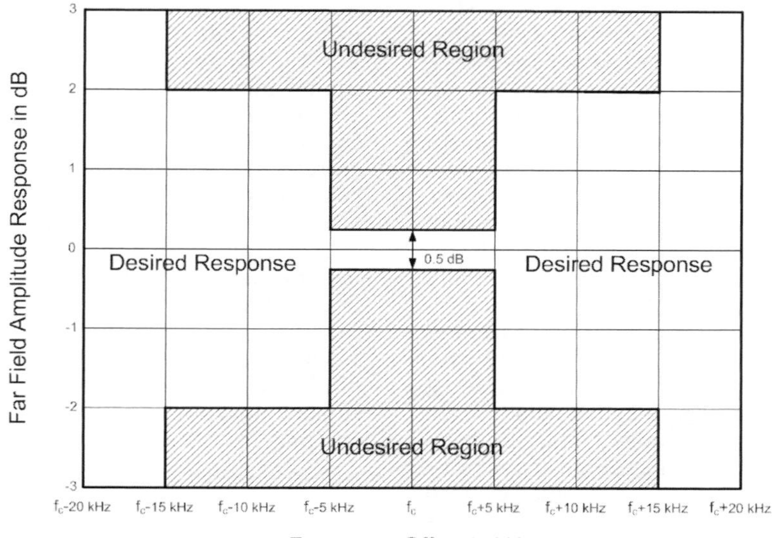

FIGURE 4.13-48 Desired HD Radio amplitude response limits.

into the digital and digital into analog. Figure 4.13-47 depicts the Hermitian symmetry vector relationship. The spectrum f_c to +5 kHz and f_c to –5 kHz is complementary Quadrature Phase Shift Keying (QPSK). If the carrier information is transmitted and then symmetrically filtered in the receiver, the signals will combine and null tertiary carrier interference.

iBiquity is exploring ways to specify the Hermitian symmetry in terms of a single quantity like Voltage Standing Wave Ratio (VSWR) or the magnitude of the reflection coefficient of one sideband normalized to the other sideband. While it is not traditional to calculate the VSWR of one sideband referenced to another, it allows more flexibility in defining a specification. A limited range of load impendence and a reactance could disqualify many designs in use today. This is the rationale for iBiquity suggesting a Hermitian symmetry: with a normalized sideband conjugate match within 1.035:1 VSWR +/– 5 kHz. This equates to an amplitude response variation of 0.5 dB across the region f_c to +5 kHz and f_c to –5 kHz.

Complex conjugate matching produces the maximum small signal transfer of power from a source that is not a transmission line to a load. Real-world anten-

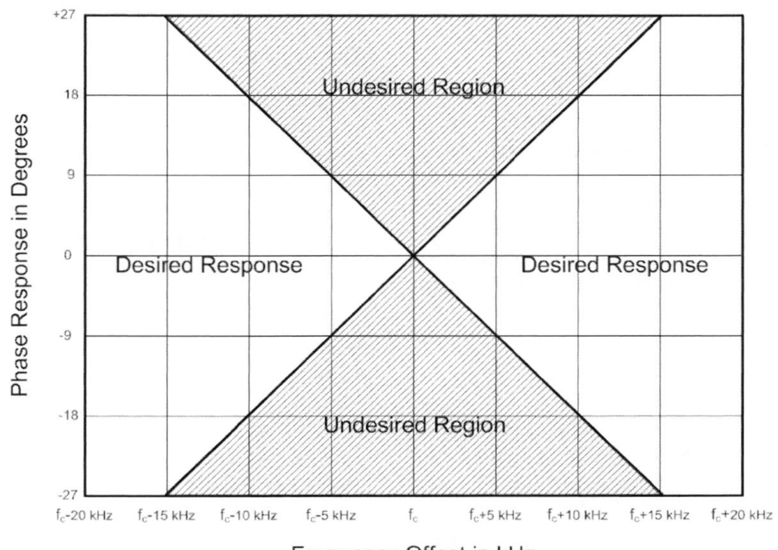

FIGURE 4.13-49 Desired HD Radio group delay limits.

nas do not have sideband impedance characteristics required for absolute Hermitian symmetry, making compromise in the resistance and reactance necessary. A conjugate match is typically employed when the situation involves source impedance that is stable. Making the phase angle of the load voltage and current equal is desirable, as complex load impedances produce an elliptical load line, which generates distortion. As long as a conjugate relationship is maintained, even high VSWR relationships work satisfactorily up to the point where an amplifier depletes headroom. The departure from a conjugate relationship is tolerable while the AM carrier angle-modulation is limited to avoid disrupting the HD Radio sidebands. Unfortunately, VSWR cannot directly be related to symmetry, as it is a relationship of two complex numbers to each other. The degree they depart from a conjugate relationship without regard to the AM analog carrier must be observed. In the far field of the antenna, pattern bandwidth effects move HD Radio sidebands just as the AM analog carrier. Since the antenna system is passive and linear, the equalizer in the receiver can track the sidebands and differentiate them from what is going on with the AM analog signal.

For acceptable IBOC reception, the following amplitude response and group delay limits are suggested:

Amplitude Response:

0.5 dB for the region f_c to +5 kHz and f_c to –5 kHz;

4.0 dB for the region f_c to +5 kHz to f_c to +15 kHz and f_c to –5 kHz to f_c to –15 kHz.

Group Delay:

5 microsecond for the region f_c to +15 kHz and f_c to –15 kHz.

See Figures 4.13-48 and 4.13-49 for a graphical representation of these limits.

Chapter 4.4 provides details of various matching networks employed in AM transmission systems.

HD RADIO NETWORKING

The advent of HD Radio Generation-3 hardware architecture and Advanced Application Service (AAS) has led the broadcast industry to re-evaluate the station network infrastructure. Minimizing network contention and resulting signal outages is crucial to delivering a quality HD Radio implementation. Network topology and protocol have a significant impact on overall system performance.

Hardware Topology

A discussion of HD Radio networking basics requires an understanding of the basic function and interface requirements of the transmission components.

The first and second generation HD Radio exciters are monolithic hardware platforms sourced with linear 44.1 kHz sampled AES-3 digital audio. This design was acceptable when only a single Main Program Service (MPS) stream was delivered to the exciter. When multiple Supplemental Program Services (SPS) were added, the audio payload requirements increased several fold. For a radio station to convey linear 20 kHz audio from the studio to the transmitter requires 1.4112 Mbps throughput for each stereo channel. If broadcasters were to connect multiple stereo audio channels to the transmitter site for MPS and SPS, it is efficient to employ the systems HDC™ bit-reduced audio. Transport efficiency is accomplished by bit reducing MPS and SPS audio and PSD at the studio end of the system and transmitting a single multiplexed data stream to the HD Radio exciter at the transmitter site. See Figure 4.13-50 for an overview of the system topology. This implementation known as the *Generation-3 architecture* has four major elements:

- Importer;
- Exporter;
- Synchronizer (EASU);
- Exciter Engine (Exgine).

Importer

The Importer contains the hardware and software necessary to deliver AAS. Data service providers use an Application Programming Interface (API) to pass service data to the Importer over the service link. The Importer establishes session connections between multiple service providers. Once a session is established, service providers can pass service data over the Importer-to-Exporter link (I2E), which in turn will be broadcast to HD Radio digital receivers. In addition to the AAS from data service providers, the Importer also accepts SPS and PSD. The Importer multiplexes all of the service provider data, multicast audio, and data streams into a full-duplex TCP/IP or bidirectional User Datagram Protocol (UDP) output.

Exporter

The Exporter contains the hardware and software required to generate the MPS and the Station Information Service (SIS). The SIS provides station information: call sign, station slogan, absolute time, and position correlated to GPS. The Exporter accepts digital MPS audio over its audio interface, bit reduces (i.e., compresses) the audio, and outputs that audio to the Exgine over the simplex Exporter to Exgine link (E2X). The Exporter applies preprogrammed delay to the analog MPS audio over its audio interface and broadcasts it as the backup channel for hybrid configuration. The delay compensates for the digital system latency, allowing receivers to seamlessly blend between the digital and analog program without a shift in time. In the FM system, the delayed analog MPS audio is returned to the synchronizer, which in turn is fed into the STL, and is stereo multiplexed and modulated by the analog exciter at the transmitter site.

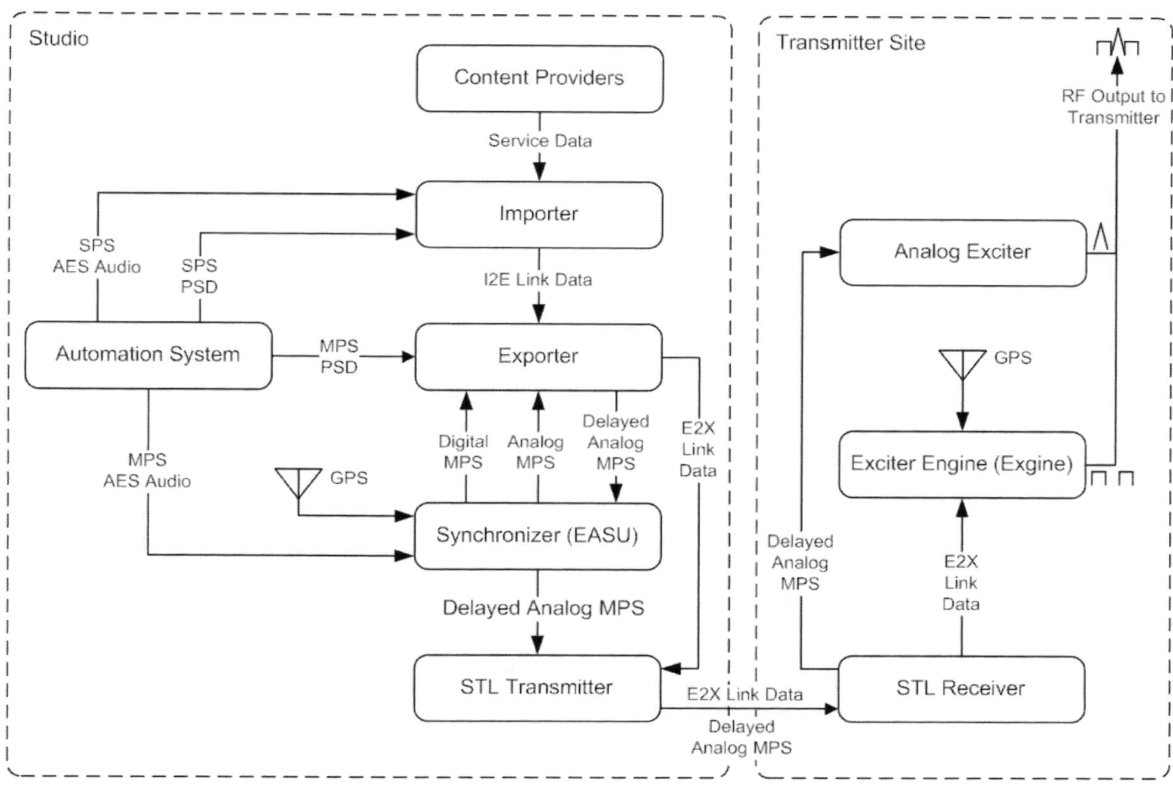

FIGURE 4.13-50 HD Radio Generation-3 system architecture.

Synchronizer

The Synchronizer, also known as the Exciter Auxiliary Service Unit (EASU), accepts MPS audio and rate-converts it to the proper system clock. It outputs two rate-converted and reference-clock-locked copies of the MPS audio to the Exporter. One output drives both the digital and analog MPS audio. The digital MPS audio will be bit reduced and modulated in the digital portion of the waveform and the analog MPS audio will be modulated in the analog portion of the hybrid waveform. In most installations, audio processing is employed on the analog and digital audio paths between the Synchronizer and the Exporter.

The GPS receiver in the synchronizer provides the master 10 MHz system clock used by the Exporter. This 10 MHz clock is also used as the reference input for the 44.1 kHz word clock used to synchronize the analog and digital AES audio streams.

Exgine

The Exgine (exciter engine) subsystem accepts the E2X data from the exciter's host processor and performs the Orthogonal Frequency Division Multiplexing (OFDM) modulation for the digital portion of the HD Radio waveform. The Exgine element is composed of a Texas Instruments C6415 processor, SDRAM, and Flash memory and enables the Layer 1 modulation to be executed on the Digital Signal Processor (DSP).

This configuration may be added to many manufacturers' digital implementations of their analog exciter offering an integrated solution for low-level analog and digital signal generation. In this distributed architecture, the HD Radio data stream is fed to the Exgine over a simplex UDP Ethernet connection.

To generate the multiplex over the E2X requires a new element in the system topology. The Exporter accepts the MPS AES audio and PSD from the automation server at the studio end as well as the multiplexed SPS audio, SPS PSD, and advanced application service data from the Importer. The Exporter may be visualized as the final multiplexer of MPS and all Advanced Application Service (AAS) data prior to developing the simplex UDP stream to the STL. An Exporter is essentially a Gen-2 exciter without the RFU and DUC subassemblies. Some manufacturers allow maximization of the original equipment investment by providing an upgrade path from existing Exciter to Exporter.

NETWORK IMPLEMENTATIONS

There are two distinct network implementations employed in AAS multicasting:

- Importer to Exporter/Exciter (I2E);
- Exporter to Exciter Engine (E2X).

The I2E network path connects an Importer to an Exporter or Exciter via either bidirectional UDP or

Transmission Control Protocol (TCP) over an Ethernet connection. In implementation, an AES digital audio path is required for the digital MPS as well as a bidirectional broadband Ethernet/IP connection for the supplemental programs and data between the Importer and Exporter/Exciter. Only the AAS signal components are transported by the I2E link and do not affect the main program digital service. The I2E configuration can be implemented only over a bidirectional network connection between the Importer and the Exporter/Exciter due to the inherent command-response design. Co-locating the Importer and Exciter within the same isolated local area subnet or over a dedicated WAN/WAN Extension can help constrain implementation costs. This can be accomplished either via wireless (RF) or wired data links.

Importer to Exporter/Exciter

Prior to Importer Version 1.2.1 and the iBiquity Reference System Software (IRSS) Exciter/Exporter software Version 2.2.5, the only network protocol available for I2E was bidirectional UDP/IP. UDP protocol's lack of error correction makes it susceptible to faults. The loss of one IP packet over the link will result in the loss of an entire frame, which represents 1.48 seconds of SPS audio or PSD. Importer Version 1.2.1 and the Exciter IRSS software Version 2.2.5 introduced TCP/IP as the default protocol. Using TCP/IP affords up to 20 frames of receive buffering on the I2E link, allowing sufficient time for retransmission of dropped packets. The low overhead and additional robustness offered by TCP leave little reason to use the UDP configuration on the I2E data link.

With up to 1% packet loss and 100 ms latency, the I2E path continues to perform well when configured for TCP. With TCP/IP adequate bandwidth is the key to system recovery from lost packets. Configuring to a maximum of 60% of the available WAN bandwidth will produce successful results under all but the most adverse network conditions.

The average bandwidth of the I2E link ranges from 17 kbps to 156 kbps dependent on the service mode and IP protocol employed. A WAN bandwidth of at least 90 kbps and latency below 100 ms are required for FM Hybrid mode MP1 operation with a single 32 kbps SPS over TCP. A 128 kbps LAN/WAN Extender or two DS0 data circuits will provide sufficient bandwidth for any MP1 configuration. For MP3 mode, the highest capacity Hybrid mode available commercially, a minimum bandwidth of 156 kbps is required. In practice, this would require three DS0 channels for a total of 192 kbps.

Exporter to Exciter Engine Link

Co-locating the Importer and Exporter at the studio enables bandwidth efficient communication across Exporter to Exciter Engine (E2X) link. This implementation requires a distributed system architecture consisting of the Exporter and Exgine. With this implementation, a single data stream is conveyed to the transmitter site containing the main program digital audio along with the supplemental programs and all of the associated data.

Typically, the Exporter and Importer along with the audio processing are located at the studio. The Exporter performs four primary functions:

- Input of AES MPS digital audio, MPS PSD;
- Input of Importer networked multiplex (SPS audio, SPS PSD, and AAS data);
- Diversity delay of analog AES audio path;
- Multiplexing of all services for transport over the STL.

The main program digital audio is delivered to the Synchronizer (some manufacturers incorporate this functional element in the Exporter hardware) that splits the audio into two streams which feed the analog and digital processing chains. Both streams are returned to the Exporter after processing. The audio destined for the legacy analog transmitter is time aligned with the digital audio and then sent to the AES digital input of the STL for transport to the legacy analog transmitter. The main HD Radio program audio is encoded into the MPS signal.

The Importer accepts AES audio and PSD of the Multicast programs. These services are encoded and sent from the Importer to the Exporter over a local network as bidirectional UDP or TCP data over Ethernet as the I2E configuration. Program Service and the Advanced Services (SPS and data) are then combined in the Exporter into a single data stream destined for the Exciter over the STL or WAN link.

Transport of the E2X data stream is currently supported only as simplex UDP and can operate over most unidirectional STL systems of sufficient bandwidth and robustness. IRSS Version 2.3.3 has provision for E2X over TCP, requiring a bidirectional link; however, the TCP connection is not fully supported in the IRSS Version 2.3.3 software load. Some manufacturers have independently enabled TCP service.

iBiquity's lab tests reveal the E2X configuration running UDP must have IP packet loss better than 10^{-5} for a successful implementation. It is not uncommon for WAN and STL systems to deliver only 10^{-3} performance (one dropped packet in every 1000), which will result in poor HD Radio system performance. Table 4.13-21 predicts the number of HD Radio dropouts that can be expected for a given network packet loss based on the number of packets sent per day when running UDP without error correction. Without error mitigation, 6 to 31 momentary dropouts per day can be expected, depending on the mode and configuration, when using UDP over a 10^{-3} link.

The TCP implementation of E2X can perform significantly better than UDP, tolerating packet losses of up to 0.03%. This level of performance is well within the capabilities of healthy WAN/STL systems.

E2X average bandwidth ranges from 120 kbps to 168 kbps, dependent on configuration and IP protocol. A WAN/STL bandwidth of at least 128 kbps with latency of less than 100 ms is necessary for MP1 mode UDP. A 128 kbps LAN/WAN Extender or two DS0

TABLE 4.13-21
Predicted Dropout at Percent Packet Loss

	Mode	Packets Per Day X1000	0.1% Packet Loss	0.01% Packet Loss	0.001% Packet Loss	0.0001% Packet Loss
			UDP—Expected Daily HD Dropouts			
I2E	MP1-w/ HD2 @ 24 kbps	6400	640	64	6	1
I2E	MP1-w/HD2 @ 48 kbps	873	873	97	9	1
I2E	MP2/3 HD2	1396	1396	140	14	1
I2E	MP2/3 HD2 & HD3	2036	2036	204	20	2

data channels will provide sufficient bandwidth for any MP1 configuration. For FM mode MP3, 256 kbps or four DS0 channels should be considered for UDP and 320 kbps or five DS0 channels should be more than sufficient for TCP. With UDP, packet loss across the link becomes a critical factor and must be kept below 10^{-5} for successful operation due to a lack of error recovery.

SUMMARY OF PROVISIONING

TCP/UDP

In order for a TCP data stream to function properly under adverse conditions, the link that carries it must have reserve bandwidth above and beyond the data rate of the stream. This is necessary to accommodate the higher data rate that occurs when the stream recovers from packet loss. Additional bandwidth beyond the recommended guideline allows operation under poorer conditions, but with diminishing returns. In general, bandwidth should not be used to adjust for a poor network.

When using TCP, the WAN link must have a minimum of 40% overhead (reserve bandwidth) in order to function properly. This overhead should be calculated on the total traffic through the WAN link, which can consist of the following components:

- IBOC data stream;
- Utilities such as VNC, telnet;
- Broadcast or multicast traffic.

The aggregate of this traffic should occupy no more than 60% of the available bandwidth. If other traffic is going through the WAN, the link should have class of service, QOS, or other prioritization techniques employed to ensure that the HD Radio traffic has the necessary bandwidth. For UDP, the total traffic can be no more than 75% of the provisioned bandwidth.

Traffic Control

Through field investigations, iBiquity found that the only sure way to prevent extraneous traffic from traversing the link is to make the link its own IP subnet. This separates the link from the production network and spares the link from having to carry any broadcast or multicast traffic from the production and/or office network. Traffic that switches or is directed to all ports should be kept to a minimum. In addition to broadcast and multicast packets, this category also includes Unicast packets sent to Media Access Control (MAC) addresses that are not in the switch's forwarding table. This is a situation that occurs when a device from outside the local network sends packets to a device that was recently operating in the subnet. The device is still in the router's Address Resolution Protocol (ARP) table, but has already "aged out" of the switch forwarding tables. The switch aging timer is typically shorter then the router's ARP aging timer, resulting in a discrepancy. There are steps that the network manager can take to prevent this situation. These steps include reducing the router's ARP aging timer to match the switch forwarding table aging timer or programming the switch to block Unicast traffic sent to unknown MAC addresses.

Using TCP, audio drops occur whenever the receive buffers are depleted. In most circumstances, the audio stream will not resume until the receive buffers are all restored. Since each receive buffer corresponds to 1.48 seconds of audio, depletion of 20 receive buffers will result in an audio drop of 30 seconds or longer.

UDP has the following advantages:

- Operates over a simplex STL or WAN;
- Has shorter broadcast delay due to fewer buffers;
- Operates with less overhead;
- Has shorter audio drops when they occur (although they may occur more frequently).

With UDP transmission, the loss of a single packet results in the loss of the entire audio frame of which it is a part. The resulting outage will last for the duration of that single audio frame –1.48 seconds. For any constant packet loss rate, one would expect fewer audio drops with TCP but of shorter duration with UDP.

Detailed resources on HD Radio network implementation may be found in the whitepaper section of the iBiquity website at the following URL: http://www.ibiquity.com/broadcasters/quality_implementation/iboc_white_papers.

SUMMARY

The concept of the IBOC system is to provide broadcasters a means to deliver robust, high-quality digital audio and data to receivers making optimum use of their FCC-assigned frequency without the need for additional spectrum. Additionally, data and other services may be added at the broadcaster's choice. IBOC is compatible with existing analog receivers and allows a means to an eventual all digital broadcasting format.

ACKNOWLEDGMENTS

The author wishes to acknowledge the contributions of Tim Anderson, Jeff Baird, Kathi Cover, Harvey Chalmers, Denise Cammarata, Ashruf El-Dinary, Steven Johnson, Brian Kroeger, Stephen Mattson, Marek Milbar, Russ Mundschenk, Paul Peyla, Al Shuldiner, Glynn Walden, Girish Warrier and all members of the iBiquity Digital Corporation team.

Bibliography

T. B. Anderson "HD Radio™ Data Network Requirements" iBiquity Digital Corporation, Oct. 2006.

S. A. Johnson "The Structure and Generation of Robust Waveforms for AM In-Band On-Channel Digital Broadcasting" iBiquity Digital Corporation, Mar. 2000

B.W. Kroeger, D.M. Cammarata, "Robust Modem and Coding Techniques for FM Hybrid IBOC DAB," IEEE Transactions on Broadcasting, vol. 43, no. 4, pp. 412-420, Dec. 1997

P.J. Peyla "The Structure and Generation of Robust Waveforms for FM In-Band On-Channel Digital Broadcasting" iBiquity Digital Corporation, Mar. 2000

"HD Radio™ Air Interface Design Description - Layer 1 FM", Doc. No. SY_IDD_1011s, Rev. E, iBiquity Digital Corporation, Mar. 2005.

"HD Radio™ Air Interface Design Description - Layer 1 AM", Doc. No. SY_IDD_1012s, Rev. E, iBiquity Digital Corporation, Mar. 2005.

"HD Radio™ Air Interface Design Description - Layer 2 Channel Multiplex", Doc. No. SY_IDD_1014s, Rev. F, iBiquity Digital Corporation, Feb. 2005.

"HD Radio™ Air Interface Design Description – Audio Transport", Doc. No. SY_IDD_1017s, Rev. E, iBiquity Digital Corporation, Mar. 2005.

"HD Radio™ Air Interface Design Description - Advanced Application Services Transport", Doc. No. SY_IDD_1019s, Rev. E, iBiquity Digital Corporation, Aug. 2005.

"HD Radio™ Air Interface Design Description – Station Information Service Transport", Doc. No. SY_IDD_1020s, Rev. E, iBiquity Digital Corporation, Sep. 2005.

"HD Radio™ FM transmission System Specification", Doc. No. SY_IDD_1026s, Rev. D, iBiquity Digital Corporation, Feb. 2005.

"HD Radio™ Air Interface Design Description - Program Service Data", Doc. No. SY_IDD_1028s, Rev. C, iBiquity Digital Corporation, Mar. 2005.

"HD Radio™ AM transmission System Specification", Doc. No. SY_IDD_1082s, Rev. D, iBiquity Digital Corporation, Feb. 2005.

"HD Radio™ Air Interface Design Description - Program Service Data Transport", Doc. No. SY_IDD_1085s, Rev. C, iBiquity Digital Corporation, Feb. 2005.

"Petition for Rulemaking to the United States Federal Communications Commission for In-Band On-Channel Digital Audio Broadcasting," Appendix C, p. 7, USA Digital Radio Corporation, Oct. 1998

Federal Communications Commission, Code of Federal Regulations, Title 47, Part 73

CHAPTER

4.14

Radio Broadcast Data System

SCOTT A. WRIGHT
Soneticom
West Melbourne, Florida

Updated for the 10th Edition by

THOMAS D. MOCK
WYE Consulting
Sheffield, Pennsylvania

INTRODUCTION

Radio Broadcast Data System (RBDS) is the official name used for the U.S. version of the Radio Data System (RDS) Standard, which was first adopted in 1993 by the National Radio Systems Committee (NRSC) and revised in 1998 and in 2005.[1] RDS is a standard originally adopted by the European Broadcasting Union (EBU) for transmitting low bit rate (on the order of one kilobyte per second) digital information to appropriate receivers using conventional FM radio broadcasts. The RDS system standardizes several types of information transmitted, including time and station identification. RDS has been a standard in Europe since the mid 1980s. The two standards (RDS and RBDS) are nearly identical, with only slight differences, mainly in which numbers are assigned to each of 31 musical and other program formats. Both use a 57 kHz subcarrier to carry data at 1187.5 bits per second. The 57 kHz frequency was chosen for being the third harmonic (3×) of the 19 kHz pilot tone for FM stereo, so it would not cause interference or intermodulation with it, or with the stereo difference signal at 38 kHz (2×). The data format utilizes error correction. RDS defines many features, including how private (in-house) or other undefined features can be "packaged" in unused program groups.

[1]When revised by the NRSC in 2005, the RBDS Standard was given the designation NRSC-4-A so as to bring it in line with the number scheme adopted by the NRSC for its standards. NRSC-4-A is now the official designation of the RBDS Standard.

THE RBDS AND RDS STANDARDS

The Radio Broadcast Data System (RBDS) Standard was first adopted in January 1993. It was developed by the National Radio Systems Committee (NRSC), a technical standards-setting body which is jointly sponsored by the National Association of Broadcasters (NAB) and the Consumer Electronics Association (CEA). Thus, RBDS is an agreed-upon standard between broadcaster and receiver manufacturer interests. The U.S. RBDS Standard is largely based on the European RDS Standard, which was first adopted in 1982.

Revised editions of both the European RDS Standard and the U.S. RBDS Standard were adopted in 1998. The RBDS Standard was revised again, in April 2005, adding an annex (Annex U) for harmonization of program-associated data (PAD) usage between RDS and the HD Radio in-band/on channel (IBOC) digital radio system (standardized by the NRSC in NRSC-5-A). As this handbook was being printed, the RDS Standard was in the process of being revised to incorporate a feature known as "Radiotext Plus," which links "metadata" information to the information in a Radiotext message, making the Radiotext information "machine readable." To obtain a copy of the current U.S. RBDS Standard (NRSC-4-A), visit the NRSC webpage (www.nrscstandards.org). The European RDS Standard was initially issued by the European Committee for Electrotechnical Standardization (CENELEC) under Document Number EN 50067 but is now an International Electrotechnical Commission (IEC) standard, IEC 62106, and may be obtained from the IEC website at www.iec.ch.

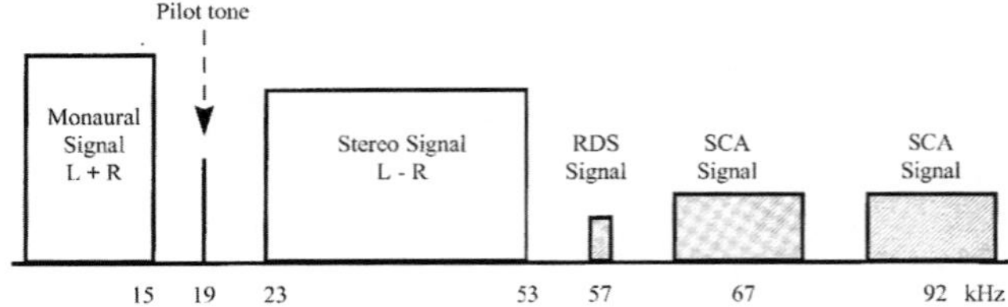

FIGURE 4.14-1 FM baseband spectrum showing main, pilot, stereo, RDS, and two SCA channels.

In the United States all references to the technology generally use the term "RDS" rather than "RBDS." RBDS is generally used only to refer to the actual NRSC Standards document itself. The NRSC felt that it was important to distinguish the U.S. Standard from its European counterpart by giving it a slightly different name. The U.S. standard includes a few features that the European standard does not. One of these is a specification for transmitting Emergency Alert System information via the RDS signal.

The RDS Subcarrier

The FCC allows almost any technology to be used for FM subcarriers as long as certain criteria are met. This means that analog or digital or a mixture of both may exist within the subcarrier frequency range as long as the total modulation does not exceed 110%. RDS has proven to be compatible with existing services due to its low injection requirements and narrow bandwidth. In addition, by phase locking the AM double sideband suppressed carrier RDS signal with the FM stereo pilot signal, there is little risk of interference to the main audio program, even under multipath conditions. RDS is carried on many classical networks throughout Europe and the United States with no degradation to the audio quality. Figure 4.14-1 depicts the baseband audio spectrum of an FM broadcast station. RDS nests into the 57 kHz position between the stereo multiplex and the commonly used analog 67 and 92 kHz subcarrier channels. Injection levels as low as 1.3 kHz allow RDS to be easily implemented without giving up program audio power. Since RDS lies at 57 kHz, it is compatible with nearly every subcarrier in use or development today. RDS was also shown, in the NRSC's evaluation of the FM IBOC system, to be compatible with the hybrid mode of operation of the HD Radio system.[2] Due to the low implementation cost of RDS, it is not likely to be displaced by another system for many years.

RDS Data Structure

The RDS signal is modulated and demodulated as a synchronous bitstream. The 57 kHz RDS subcarrier is

[2]See "DAB Subcommittee—Evaluation of the iBiquity Digital Corporation IBOC System, Part I—FM IBOC," November 29, 2001 (available on the NRSC web page at www.nrscstandards.org).

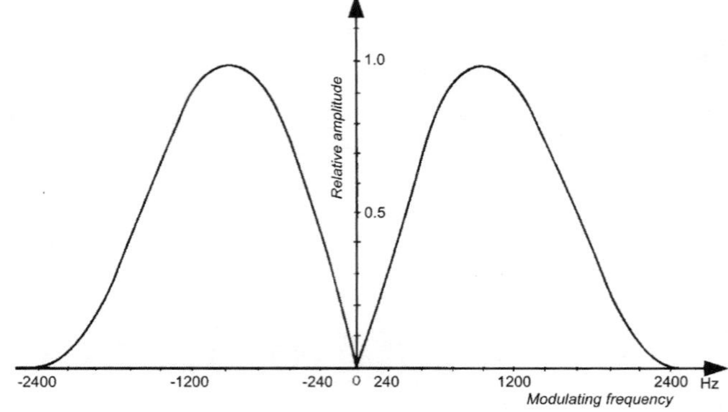

FIGURE 4.14-2 Frequency spectrum of RDS signal: this spectrum is centered at 57 kHz in the FM baseband.

amplitude-modulated by a shaped and biphase-coded data signal. The spectrum of this modulated signal is shown in Figure 4.14-2. The RDS data rate is 1187.5 bits per second. The data transmission is fully synchronous, and there are no gaps between the groups or blocks. The RDS group structure or baseband coding is depicted in Figure 4.14-3. This baseband structure is described as follows:

- The largest element in the structure is called a group (104 bits).
- Each group comprises 4 blocks of 26 bits each.
- Each block comprises an information word and a checkword.
- Each information word comprises 16 bits.
- Each checkword comprises 10 bits.
- All information words, checkwords, binary numbers, or binary address values have their most significant bit (MSB) transmitted first.

Information Word

The *information word* contains the actual data for each block. This information word is composed of 16 bits. The most significant bit (m_{15}) is transmitted first.

Checkword and Offset Word

The *checkword and offset word* consist of 10 bits. The most significant bit (c_9) is transmitted first. The checkword allows the receiver to detect and correct errors that occur during data reception. The error-protecting code has the following error-checking capabilities:

- Detects all single and double bit errors in a block;
- Detects any single error burst spanning 10 bits or less;
- Detects about 99.8% of bursts spanning 11 bits and about 99.9% of all longer bursts;
- The code is also an optimal burst error-correcting code and is capable of correcting any single burst of span 5 bits or less.

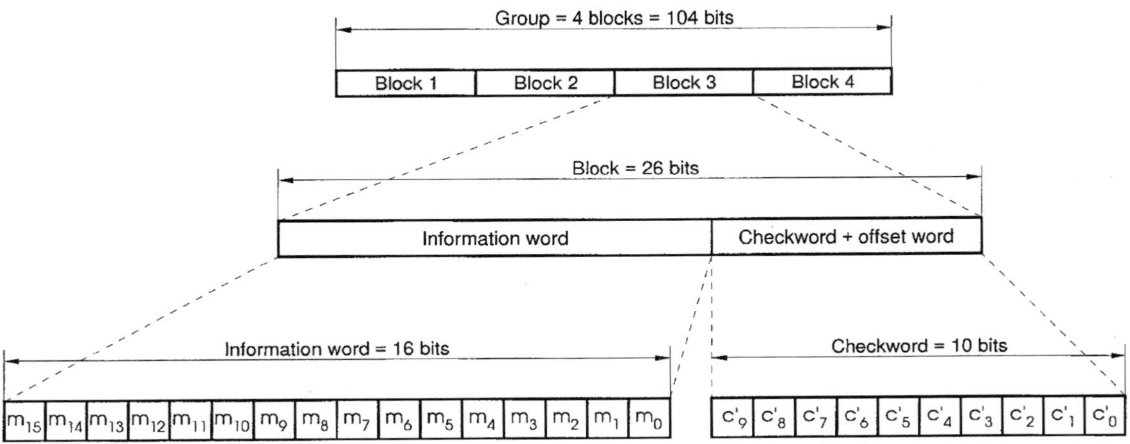

FIGURE 4.14-3 Structure of the RDS baseband coding.

Table 4.14-1 shows the possible offset words. The offset word allows detection of the block number within the data group. This allows critical data such as the PI, PTY, and TP codes (described later) to be decoded without reference to any block outside the one that contains the information. This is essential to minimize acquisition time for these kinds of messages and to retain the advantages of the short (26-bit) block length. To permit this to be done for the PI codes in block 3 of version B groups, a special offset word (C′) is used in block 3 of version B groups. The occurrence of offset C′ in block 3 of any group can then be used to indicate directly that block 3 is a PI code, without any reference to the value of B_0 in block 2. Offset word E is transmitted only by Mobile Broadcast System (MBS) or RDS/MBS time multiplexing (MMBS) stations. This allows receivers to maintain synchronization in MMBS applications.

TABLE 4.14-1
Offset Words*

Offset Word	Binary Value									
	d_9	d_8	d_7	d_6	d_5	d_4	d_3	d_2	d_1	d_0
A	0	0	1	1	1	1	1	1	0	0
B	0	1	1	0	0	1	1	0	0	0
C	0	1	0	1	1	0	1	0	0	0
C′	1	1	0	1	0	1	0	0	0	0
D	0	1	1	0	1	1	0	1	0	0
E	0	0	0	0	0	0	0	0	0	0

*Eight bits (i.e., d_9 to d_2) are used for identifying the offset words. The remaining two bits (i.e., d_1 and d_0) are set to logical level zero.

Synchronization of Blocks and Groups

The blocks within each group are identified by their offset words A, B, C or C′, and D. The beginning and end of the data blocks may be recognized in the receiver decoder by using the fact that the error-checking decoder will, with a high level of confidence, detect block synchronization slip as well as additive errors. This system of block synchronization is made reliable by the addition of the offset words (which also serve to identify the blocks within the group). These offset words destroy the cyclic property of the basic code so that in the modified code, cyclic shifts of codewords do not give rise to other codewords. A detailed explanation of a technique for extracting the block synchronization information at the receiver is given in Annex C of NRSC-4-A (the U.S. RBDS Standard).

Group Structure

The information coded into each group has a common fixed structure, as depicted in Figure 4.14-4.

All data groups share a common fixed structure of information coding to allow critical data to be transmitted in a fixed, highly repetitive pattern to ensure the best reception even under adverse reception conditions. This fixed structure is described as follows:

- *Block 1:* Block 1 of every RDS data group contains only the program identification or PI code.
  - — Block 1 is identified by offset word A.
- *Block 2:* Block 2 contains the group type code, version code, traffic program code, and program type (PTY) code. All codes are binary coded with the MSB transmitted first.
  - — Group type code: The data groups are identified through the 4-bit group type code identified as A3–A0, along with the 1-bit version code, B0. The bits A3–A0 yield decimal values 0 through 15.

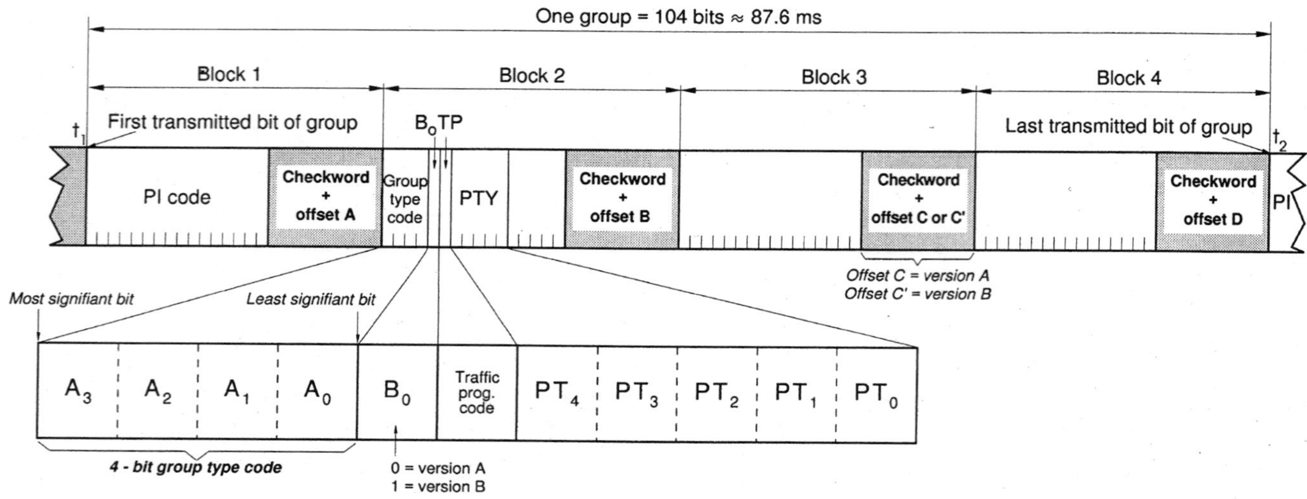

FIGURE 4.14-4 RDS group coding structure.

— Version code: Each group then has an "A" version and a "B" version based on the state of B0 where

- B0 = 0 is identified as group type A;
- B0 = 1 is identified as group type B;

— All B type groups repeat the program identification (PI) code in blocks 1 and 3.

— Traffic Program (TP) code: This single-bit code is utilized with the traffic announcement (TA) code to provide the traffic feature.

— Program type (PTY) code: This five-bit code describes the audio program type being aired. There are 31 possible PTYs available.

— Block 2 is identified by offset word B.

- *Block 3:* Dependent on the group version code (A or B) or offset word (C or C′), block 3 may carry two distinct types of data.

— Version code A: This block carries data defined by the group type code.

— Block 3 is identified by offset word C for version code A groups.

— Version B type groups carry only the PI code.

— Block 3 is identified by offset word C′ for version code B groups.

- *Block 4:* This block carries data defined by the group type code.

— Block 4 is identified by offset word D.

RDS Group Types

RDS data groups are referred to by both group type and version (0A, 14B). Taking into account both the group type and version codes, there are 32 defined data groups. Some of the group features are defined in their entirety and cannot be defined in any other way, whereas other data groups are openly defined such that the actual data contained in the group can be defined by the operator. The group type code defines the basic function of the data group. The "A" and "B" versions differ only in the coding of the data for the group. Because RDS data is transmitted in functional groups that share a common structure, the operator can select from these group types and transmit all or only a portion of them depending on the features in use. In this manner, there is no need to transmit data which is not in use, thereby maximizing data throughput. Table 4.14-2 summarizes the possible applications for all possible group types. Only group type 0A or 0B must be included in the transmission sequence.

RDS FEATURES

RDS provides primary and secondary features. Primary features are either contained in every information group or are included in type OA or OB groups, which are always transmitted. Secondary features are transmitted only when necessary or desired. This prevents data capacity from being wasted on unused features.

Primary Features

Program Identification (PI) Code

The PI code is a four-digit hexadecimal code that is unique for each station. In Europe, the PI code is assigned. In North America, the PI code is calculated from the station's call letters; thus, no two are alike. The receiver uses the PI code to identify a station rather than frequency. RDS allows frequency diversity such that if a station simulcasts on another frequency, the receiver can automatically tune to the strongest station. If a station simulcasts, then it must pick one station's PI code and use it on all of the simulcasting stations. While the PI code is transparent to the user,

TABLE 4.14-2
RDS Group Types

Group Type	Group Type Code/Version					Flagged in Type 1A Groups	Description
	A_3	A_2	A_1	A_0	B_0		
0A	0	0	0	0	0		Basic tuning and switching information only
0B	0	0	0	0	1		Basic tuning and switching information only
1A	0	0	0	1	0		Program Item Number and slow labeling codes only
1B	0	0	0	1	1		Program Item Number
2A	0	0	1	0	0		Radiotext only
2B	0	0	1	0	1		Radiotext only
3A	0	0	1	1	0		Application Identification for ODA only
3B	0	0	1	1	1		Open Data Applications
4A	0	1	0	0	0		Clock-time and date only
4B	0	1	0	0	1		Open Data Applications
5A	0	1	0	1	0		Transparent Data Channels or ODA
5B	0	1	0	1	1		Transparent Data Channels or ODA
6A	0	1	1	0	0		In-House applications or ODA
6B	0	1	1	0	1		In-House applications or ODA
7A	0	1	1	1	0	Y	Radio Paging or ODA
7B	0	1	1	1	1		Open Data Applications
8A	1	0	0	0	0	Y	Traffic Message Channel or ODA
8B	1	0	0	0	1		Open Data Applications
9A	1	0	0	1	0	Y	Emergency Warning System or ODA
9B	1	0	0	1	1		Open Data Applications
10A	1	0	1	0	0		Program Type Name
10B	1	0	1	0	1		Open Data Applications
11A	1	0	1	1	0		Open Data Applications
11B	1	0	1	1	1		Open Data Applications
12A	1	1	0	0	0		Open Data Applications
12B	1	1	0	0	1		Open Data Applications
13A	1	1	0	1	0	Y	Enhanced Radio Paging or ODA
13B	1	1	0	1	1		Open Data Applications
14A	1	1	1	0	0		Enhanced Other Networks information only
14B	1	1	1	0	1		Enhanced Other Networks information only
15A	1	1	1	1	0		Do not use; Fast PS being phased out
15B	1	1	1	1	1		Fast switching information only

its proper use is vital to proper receiver operation. Specialized PI codes are available for use for networked programming as well.

Program Service (PS) Name

The PS Name is whatever name the station chooses to present to its listeners. It can be call letters like "WXZH-FM," or a slogan like "X-100." The PS Name is displayed instead of frequency on an RDS receiver. When the listener tunes to an RDS station, the station's name will be shown on the radio display. The PS Name cannot be longer than eight characters. Any character referenced in NRSC-4-A may be used, but special characters might not be displayable on certain low-cost displays. The note appearing at the bottom of Table E.1 of NRSC-4-A details which characters are most commonly displayable.

Traffic Program (TP)

Stations offering listeners traffic bulletins should consider using this feature. The TP identifies the transmitting station to the listener as a station that offers traffic programs. RDS receivers can automatically tune to stations that offer traffic programs when the user turns on the traffic announcement feature of the radio. A broadcaster should not set TP high unless it actually uses the TA feature.

Traffic Announcement (TA)

When an actual traffic bulletin is broadcast, this information bit must be set to a logic "1." The RDS receiver detects this and will automatically stop any playback device that may be in use and will return to the FM tuner mode. Audio adjustments are also automatically made so that, if the user had the volume muted, for instance, the receiver will adjust the volume to a user preset level.

TABLE 4.14-3
North American Program Types

Program Type	Name	Description
1	News	News reports, either local or network in origin.
2	Information	Programming that is intended to impart advice.
3	Sports	Sports reporting, commentary, and/or live event coverage, either local or network in origin.
4	Talk	Call-in and/or interview talk shows either local or national in origin.
5	Rock	Album cuts.
6	Classic Rock	Rock-oriented oldies, often mixed with hit oldies, from a decade or more ago.
7	Adult Hits	An up-tempo contemporary hits format with no hard rock and no rap.
8	Soft Rock	Album cuts with a generally soft tempo.
9	Top 40	Current hits, often encompassing a variety of rock styles.
10	Country	Country music, including contemporary and traditional styles.
11	Oldies	Popular music, usually rock, with 80% or greater noncurrent music.
12	Soft	A cross between adult hits and classical, primarily noncurrent soft rock originals.
13	Nostalgia	Big band music.
14	Jazz	Mostly instrumental, includes both traditional jazz and more modern smooth jazz.
15	Classical	Mostly instrumentals, usually orchestral or symphonic music.
16	Rhythm and Blues	A wide range of musical styles, often called *urban contemporary.*
17	Soft Rhythm	Rhythm and blues with a generally soft tempo.
18	Foreign Language	Any programming format in a language other than English.
19	Religious Music	Music programming with religious lyrics.
20	Religious Talk	Call-in shows, interview programs, etc., with a religious theme.
21	Personality	A radio show where the on-air personality is the main attraction.
22	Public	Programming that is supported by listeners and/or corporate sponsors instead of advertising.
23	College	Programming produced by a college or university radio station.
24–28		Unassigned.
29	Weather	Weather forecasts or bulletins that are nonemergency in nature.
30	Emergency Test	Broadcast when testing emergency broadcast equipment or receivers. Not intended for searching or dynamic switching for consumer receivers.
31	Emergency	Emergency announcement made under exceptional circumstances to give warning of events causing danger of a general nature. Not to be used for searching—only used in a receiver for dynamic switching.

Program Type (PTY) Codes

The program type code is used to identify the type of program material being broadcast. There are predefined codes for Country, Rock, Top 40, etc. The RDS receiver can automatically search for stations by PTY, thus allowing listeners to find their favorite type of programming without tuning to all available stations. Advanced receivers can even interrupt listeners when, for example, news is broadcast. A definition of the terms used to denote North American Program Types is given in Table 4.14-3.

Alternate Frequencies (AF)

The alternate frequencies feature allows the RDS receiver to automatically tune to the best signal when multiple transmitters or translators are broadcasting the same program. Regional or national programs that are broadcast over large areas can even be linked together, providing the listener with the illusion of one very powerful transmitter. It is also possible to link stations only during specific times when the program material is common (perhaps during a sporting event) using the linkage actuation feature. Competitors cannot steal listeners away using this feature because an RDS receiver will look only for alternative frequencies that are specified by the station that is currently tuned in. An example of an AF network is shown in Figure 4.14-5.

Secondary Features

Program Type Name (PTYN)

The program type name feature allows additional flexibility for the broadcaster that wants to stand out from competitors. While the PTY codes are predefined, the PTYN can be any eight characters the broadcaster desires to further describe the current program. For example, a broadcaster that is currently using the PTY "Personality" may set the PTYN to "Limbaugh" to further describe the current program. A "Rock" station may set the PTYN to "BobTom" to describe their morning team. An RBDS receiver cannot search by PTYN but will display the PTYN in place of the PTY once tuned to a particular station.

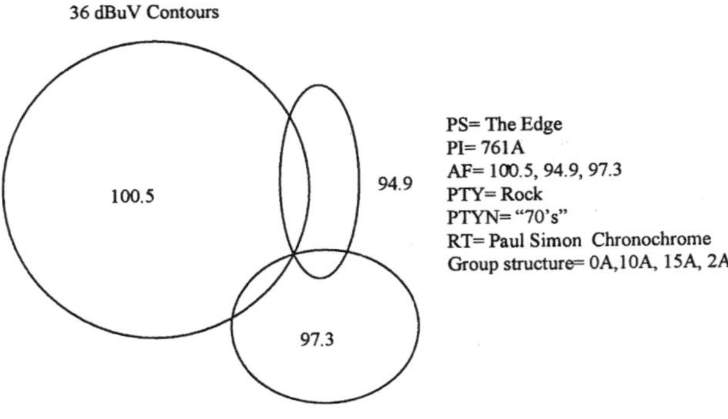

FIGURE 4.14-5 Example of an RDS alternate frequency network.

Radiotext (RT)

Radiotext allows a station to transmit up to 64 characters of information to the listener. Information such as the current artist and song title, station promotional information, local events, and even additional information about the advertiser whose commercial is currently playing can be sent. (Radiotext is currently being augmented with a feature called "Radiotext Plus," which will make the information contained in the Radiotext fields machine readable.)

Clock Time and Date

The clock time and date feature enables broadcasters to transmit the current date and time once every minute, thus enabling RDS-equipped receivers (clock radios, car radios, etc.) to be automatically set after a power failure or time zone change. Broadcasters using this feature should ensure the accuracy of the transmitted time to ±2 sec.

Emergency Warning System (EWS)

The Emergency Warning System feature allows the transmission of coded emergency information intended for specialized receivers. The FCC has recognized the capability of this feature and permits broadcasters to use RDS to transmit Emergency Alert System information. Broadcast EAS equipment can be linked to an RDS encoder for automatic retransmission of EAS information. (However, this does not eliminate the requirement to transmit EAS information in the main program channel as well.) An alert feature contained within consumer receivers will be activated when an alert code is transmitted. NRSC-4-A includes an entire annex (Annex Q) devoted to implementing EAS with RDS technology.

Open Data Channel (ODC)

Standards working groups were formed to determine the best way to add additional data services such as differential correction for the Global Positioning System (DGPS), and even the EAS to NRSC-4-A. These systems needed both public- (free) and private- (fee) based delivery. Using the remaining few undefined RDS data groups to perform these services proved to be too much effort, as coordinated systems could not be agreed upon. However, it was agreed that each data group left undefined was valuable, since once all the groups were defined, that meant the end of any future expansion to the system. Thus was born the concept of the open data channel.

The open data channel allows the definition of an unused RDS data group based solely on the Application Identification (AID) code. Thousands of AID codes are available, meaning that the remaining data groups could be defined thousands of different ways. Table 4.14-4 includes the AID codes issued as of September 2006. The AID codes are assigned by NAB in the United

TABLE 4.14-4
AID Codes (as of September 2006)

Application Name	Issue Date	ODA AID	
		Decimal	Hex
Normal features specified in standard	9-Apr-98	0	0000
Cross-referencing DAB within RDS	16-Nov-98	147	0093
Leisure & Practical Info for Drivers	22-Dec-99	3019	0BCB
ELECTRABEL-DSM 7	1-Jul-99	3108	0C24
Wireless Playground broadcast control signal	10-Oct-03	3265	0CC1
RDS-TMC: ALERT-C (for testing use, only)	20-Apr-01	3397	0D45
ELECTRABEL-DSM 18	1-Jul-99	3467	0D8B
ELECTRABEL-DSM 3	1-Jul-99	3628	0E2C
ELECTRABEL-DSM 13	1-Jul-99	3633	0E31
ELECTRABEL-DSM 2	1-Jul-99	3975	0F87
ELECTRABEL-DSM 1	1-Jul-99	7130	1BDA
ELECTRABEL-DSM 20	1-Jul-99	7262	1C5E
ITIS In-vehicle database	29-Jul-05	7272	1C68
ELECTRABEL-DSM 10	1-Jul-99	7345	1CB1
ELECTRABEL-DSM 4	1-Jul-99	7495	1D47
CITIBUS 4	10-Sep-97	7618	1DC2
Encrypted TTI using ALERT-Plus only (for testing use, only)	28-Feb-01	7621	1DC5
ELECTRABEL-DSM 17	1-Jul-99	7823	1E8F
RASANT	2-Mar-98	19105	4AA1
ELECTRABEL-DSM 9	1-Jul-99	19127	4AB7
RDS-TMC: ALERT-C with ALERT-Plus	11-Jun-97	19202	4B02
ELECTRABEL-DSM 5	1-Jul-99	19362	4BA2
RadioTextplus	11-Feb-05	19415	4BD7
Encrypted TTI using ALERT-C only (for testing use, only)	9-Apr-01	19428	4BE4
CITIBUS 2	10-Sep-97	19545	4C59
Radio Commerce System (RCS)	26-Mar-01	19847	4D87
ELECTRABEL-DSM 16	1-Jul-99	19861	4D95
ELECTRABEL-DSM 11	1-Jul-99	19866	4D9A
NRSC Song Title and Artist	20-Dec-04	50000	C350
Alertus Technologies, LLC	28-Feb-06	50388	C4D4
ID Logic	9-Apr-98	50531	C563
CITIBUS 1	10-Sep-97	52083	CB73
ELECTRABEL-DSM 14	1-Jul-99	52119	CB97
CITIBUS 3	10-Sep-97	52257	CC21
RDS-TMC: ALERT-C (for service use, only)	11-Jun-97	52550	CD46
ELECTRABEL-DSM 8	1-Jul-99	52638	CD9E
Encrypted TTI using ALERT-Plus only (for service use, only)	28-Feb-01	52843	CE6B
Stratos Audio	7-Feb-03	57635	E123

TABLE 4.14-4 *(continued)*
AID Codes (as of September 2006)

Application Name	Issue Date	ODA AID	
		Decimal	Hex
eCARmerce Inc.	1-Mar-01	57793	E1C1
ELECTRABEL-DSM 12	1-Jul-99	58137	E319
Cell-Loc	7-Nov-02	58385	E411
Encrypted TTI using ALERT-C only (for service use, only)	9-Apr-01	58391	E417
ELECTRABEL-DSM 15	1-Jul-99	58432	E440
ELECTRABEL-DSM 19	1-Jul-99	58534	E4A6
ELECTRABEL-DSM 6	1-Jul-99	58839	E5D7
EAS open protocol	9-Apr-98	59665	E911

States and by the RDS Forum for all other locations, and coordinated so that anyone can apply for an AID code and start a data service. The ODC is much more flexible than other data-only groups such as the transparent data channel because it can be automatically tuned (by use of the AID) and tracked regardless of any other data being transmitted. This is a dynamic, and potentially very profitable, data group.

Transparent Data Channel (TDC)

For specialized applications, information of any type can be transmitted by use of the transparent data channel data group. For instance, advertising messages could be sent to an electronic billboard using this feature. Virtually any way that data is sold can be supported by the TDC. Normal car and home receivers do not decode this information.

In-House Application

Data contained in the in-house application group is to be used only by the broadcaster. Remote control applications, station telemetry, or paging applications can be supported using this data group.

Radio Paging (RP)

Paging services including numeric or alphanumeric can be supported by RDS.

Traffic Message Channel (TMC)

The traffic message channel feature allows traffic information to be coded and displayed on specialized consumer receivers (i.e., navigation systems) so that normal audio bulletins are not required.

Changes Made in the 1998 Versions of the RDS and RBDS Standards

The 1998 versions of the European RDS and U.S. RBDS Standards offered new opportunities for FM data services. All changes were devised to be fully backward compatible with the older standards, while offering a substantial amount of growth.

While the original standards shared much commonality, the 1998 versions of the standards are nearly identical in form. The following features were modified from the original RBDS Standard:

- Program Type Name (PTYN)
- Fast Program Service Name (Fast PS)
- Location and Navigation (LN)
- Open Data Applications (ODA)
- Enhanced Radiopaging Protocol (EPP)
- Language Identification (LI)
- Decoder Information (DI)
- Extended Country Code (ECC)
- PI structure from B000–FFFF
- Analog SCA Cross-Referencing
- IRDS Updating
- PTY Code Table
- MBS EAS

Program Type Name (PTYN)

The PTYN contained within group 10A has been added to the RDS standard. This feature was previously defined only within the RBDS Standard and has been slightly modified with the addition of a text A/B flag, but is still backward compatible with the previous definition. The A/B flag is located in the unused 4-bit region and shares the same bit position as the A/B flag in Radiotext group 2A. The usage is the same in both groups in that the A/B flag should be toggled whenever the PTYN is changed. Within the receiver, the PTYN text buffer should be cleared when the A/B flag is toggled to prevent mixing of old and new PTYNs. The rationale is that the PTYN is designed to be semi-dynamic, potentially changing whenever the program changes from "news" to "weather" to "sports." Newly designed receivers can take advantage of the A/B flag without effects from existing encoders that do not incorporate this change.

Fast Program Service (PS) Name

The Fast PS is currently defined in group type 15A of the RBDS Standard and is not defined in the RDS Standard. This group was not adopted in Europe. In North America there was a planned phaseout of all transmissions. This group will not be defined for any future use and could later be reallocated as open data. The rationale for this change is that it will not cause significant performance degradation in the receivers, as the PS is also located in groups 0A and 0B.

Location and Navigation (LN)

LN was defined in Group 3A in the RBDS Standard only. This usage was deleted in the 1998 version and reassigned to the Open Data Application (ODA). This was recommended since there were no LN transmissions within North America and this feature was no longer required for ID Logic receivers. The definition as LN was found to be inadequate by the Differential Global Positioning System (DGPS) Working Group of the NRSC. This group, along with its European counterpart, determined that the LN structure could not meet current needs of private and public service providers. It was recommended that the ODA be used for LN and DGPS services.

Open Data Applications (ODAs)

Group 3A was redefined for ODA. Other, previously unused groups were assigned for ODAs as well. The use of an "A" type group allows low data capacity applications to exist entirely within Group 3A. The ODA allows multiple reuse of unused data groups rather than fixed definitions, ensuring a long future for RDS. The actual use of data can be either public or private. The ODA allows encryption of data for fee-based services. Data groups referenced via slow labeling (Group 1A) can also be used for ODA when not specifically referenced for use. Specific uses are identified by application identification codes (AID). Over 65,000 internationally allocated AID codes are available (see Table 4.14-4). This allows anyone to start a datacasting business locally, nationally, or internationally. ODAs are already defined or proposed for EAS, DGPS, DAB cross-referencing, and enhanced TMC.

Enhanced Radiopaging

To improve the existing paging capabilities of RDS, the enhanced radiopaging protocol was developed. This protocol is contained within Groups 13A and 1A. Annex M of the U.S. RBDS Standard contains the protocol definitions. With the enhanced protocol the possibility of international RDS paging has been added. The prior definition of RDS paging was based on the usage of PI codes only. This worked well in network-based broadcasts as found in many parts of Europe, but offered no solution where few network-based broadcasts are employed, such as in the United States. The enhanced radiopaging protocol opens local, national, and international paging opportunities.

Language Identification

To aid the international traveler, the language identification feature located in Group 1A Variant 3 has been expanded to include most world languages. The list of covered languages is located in Annex J of the RBDS Standard. The feature enables the broadcaster to indicate the spoken language being broadcast, thus allowing listeners to search for their native language on local broadcasts.

Decoder Identification (DI)

The decoder information feature has been modified to enable the identification of stations that perform dynamic PTY switching. This allows the broadcaster to indicate that it dynamically switches the program type (PTY) based on the current audio program, thus supporting the use of PTY-interrupts in consumer receivers. Prior to this, there was no way to tell if the broadcaster would ever change the PTY. The dynamic PTY indicator is analogous in use to the TP bit of the traffic announcement feature.

Extended Country Codes (ECC)

Extended Country Codes allow receivers to identify the country that the broadcast is coming from. Since PI codes are limited in number, they must be repeated throughout the world; when the PI code is received in conjunction with the ECC, the exact country of origin can be identified. The updated ECC code table has been expanded to be international in scope. It is contained in Annex D and Annex N of the RBDS Standard. ECC should be transmitted by all broadcasters, and it is recommended that it be a default automatic transmission in encoders.

Program Identification (PI) Codes

In the United States program identification codes are based on call letters rather than being assigned by any organization, as is done in Europe. A portion of the PI codes is reserved for network usage and also for assignment to stations in Canada and Mexico. During the 1998 upgrade to the U.S. RBDS Standard, several mistakes were discovered in the non-call-based PI codes which had to be corrected. The changes to the PI code assignment are summarized as follows:

- PI assignments below B000 will remain as is, allowing AF switching but no regionalization. PI codes "_0__" and "__00" are remapped into the "A" range of PIs.
- C000–CFFF assigned to Canada. Allows AF switching but no regionalization. PI codes C0xx and Cx00 are excluded from use.
- F000–FFFF assigned to Mexico. Allows AF switching but no regionalization. PI codes F0xx and Fx00 are excluded from use.

- B_FF, D_01–E_FF assigned for national networks in the United States, Canada, and Mexico. Regionalization allowed. NRSC to provide assignments for all three countries.

The ECC code table was modified for these changes as well. The rationale for these changes was that the current definitions for PIs above BFFF were contradictory. Also, the present PI code structure above BFFF was inadequate for the needs of Canada and Mexico because the existing PI structure allowed only 256 PIs for Canada and Mexico. The new PI assignments yield 3,584 possible nonregional PIs for Canada and Mexico, as well as 765 national network PIs for all three countries.

Analog SCA Cross-Referencing

The analog SCA cross-referencing feature was devised to allow RDS signaling to receivers equipped with analog subcarrier decoders to switch over automatically. This feature was defined in section 3.1.3.6, Note 3, as part of the transparent data channel (TDC), Group 5A Channel 2. This feature has been deleted from the revised standard because: (a) it is not in use; (b) it prevents harmonization with the European RDS Standard; and (c) it can be easily converted over to the ODA channel.

ID Logic RDS (IRDS) Updating

The ID Logic RDS updating feature was previously defined in section 3.1.3.6 of the Standard, Notes 1 and 2, as a reservation of the transparent data channel (TDC), Group 5A Channels 0 and 1. This was reserved for updating ID Logic receiver databases through RDS. The TDC reference was deleted from the revised standard. The rationale behind this change was that it was not currently in use and it prevented harmonization with the European RDS Standard. ODA will now be used for updating ID Logic receiver databases. The use of AID codes will provide better protection and international usage of the IRDS feature.

PTY Code Table

Both the RDS and RBDS Standards have included additional definitions for previously undefined PTY codes. RBDS has defined two new codes:

- PTY-29, Weather: It is defined for nonemergency weather-related information such as forecasts, watches, and advisories. Weather is a feature popular with listeners that can be supported with the PTY "Watch" mode to interrupt playback similar to traffic announcements and News PTY.
- PTY-23, College: This code is intended to be used to identify broadcasts oriented to college students; these broadcasts often contain a variety of formats and information. It is one of the top three noncommercial formats. The other two, Public and Religious, already had PTY definitions.

In the European RDS Standard all the PTY codes have now been defined. The main reason for this was to ensure coordination with the development of the digital audio broadcasting system being rolled out throughout Europe. These codes are all defined in the 1998 version of the RBDS standard.

DIFFERENCES BETWEEN RDS AND RBDS

With all the changes to the RDS and NRSC-4-A Standards, there still remain differences. These differences are mainly market driven. From a listener point of view, most of these differences are transparent. In exploring these differences, it will become evident that the receiver manufacturer must be well versed in the differences to ensure proper receiver operation. When these differences are properly employed by the receiver manufacturer, it is possible to construct a truly global receiver. If broadcasters and receivers make use of the ECC feature, it is possible that the receivers could self-configure automatically. The problem, of course, is getting broadcasters to utilize the ECC feature since it is not a required broadcast feature.

Summary of Differences between RDS and RBDS

- Program Type Definitions (PTY): To accommodate differing broadcast styles, the PTY code definitions are different. These differences may be accounted for through the use of a look-up table within the receiver. Annex F of NRSC-4-A includes both the European and North American PTY definitions.
- Program Identification Coding (PI): Due to the high penetration of network-based programming throughout Europe compared to the largely independent single station structure of North America, the derivation of PI codes is different. In NRSC-4-A, PI codes are based on call letters from 0000–AFFF and are network based from B000–FFFF. This means that PI codes above B000 in North America are treated by the receiver the same way that all PI codes are treated in Europe. Below B000, PI codes in North America do not employ the regionalization feature. While alternate frequency switching is still employed below B000, variants based on changes in the second nibble of the PI code do not exist and should be ignored. PI code definitions can be found in Annex D of the Standard.
- Mobile Broadcast System (MBS) and RDS/MBS time multiplexing (MMBS): The predecessor to RDS, MBS, was mainly used as a paging system through a network (now out of service) of approximately 500 stations within the United States. The MBS system utilizes the same modulation and data structure as RDS but employs a different data protocol. An MBS broadcast is identified through the offset word E. Since there are similarities between the two systems, it is possible to time multiplex MBS and RDS data. This time sharing is known as MMBS. Receiver manufacturers must be able to

differentiate between RDS and MBS, as well as accommodate MMBS broadcasts. Internal MBS/MMBS cross-references can be found throughout NRSC-4-A as a reminder of particular system requirements and as a possible alternative to RDS. A public domain EAS protocol is also contained within the MBS/MMBS annex.

- EAS ODA Protocol: Within NRSC-4-A, the EAS ODA protocol is defined for use in the United States. This optional feature set is constructed around the Federal Communication Commission's EAS protocol and is open for public use. RDS allows the silent (i.e., retransmission over the RDS data stream and not the station's main channel audio signal) retransmission of emergency information. This has been combined with existing consumer-oriented emergency features to allow additional feature functionality to consumer receivers as well. The EAS ODA can also accommodate private emergency systems.
- ID Logic (IDL): In-receiver database updates via RDS (IRDS) ODA protocol. The ID logic feature is a licensed technology that allows the incorporation of an in-receiver database that contains format type and call letters for all AM and FM stations. When combined with RDS, IDL can provide similar data and features for non-RDS FM and AM stations. The IRDS feature allows the database to be updated through an RDS ODA so that the information can be updated and maintained automatically. The introduction of ID Logic is contained in Section 7 of NRSC-4-A, and the IRDS ODA is described in detail in Annex R.
- AM RDS (Future System): Although a suitable data transmission system has yet to be developed that is compatible with the C-QUAM AM stereo system and/or the AM IBOC system specified for NRSC-5, there is a section in NRSC-4-A where such a system can be defined or referenced.
- Annex U, Open Data Application for Program Associated Data (added to NRSC-4-A in April 2005). Annex U describes an RDS/RBDS Open Data Application (ODA) for the transmission of Program Associated Data (PAD). This application ("ODA PAD") is in many ways analogous to the ID3v2 content "tagging" system developed around and popularized by "MP3" audio. The most significant benefit of ODA PAD over traditional forms of RDS datacasting is the clear delineation of individual data elements (title, performer, etc.) within the transmission. Additional benefits include a doubling of the available text message buffer over RDS/RBDS radiotext alone and several features intended to improve display update performance in enabled receivers. Another benefit provided by this ODA is that it facilitates "harmonization" between the PAD data features of IBOC digital radio and RDS. This is useful for IBOC digital radio systems utilizing "digital/analog blend" in that it allows for continuity of reception of PAD data as the IBOC receiver blends between digital and analog modes.

RDS DATA TRANSMISSION CAPACITY LIMITS

An analysis of the limitations of RDS transmission capacity limits is given here, and is based on the use of RDS/RBDS in North America. Due to differences in broadcasting requirements between North America and Europe, the importance of certain features is not the same. In Europe the networking of broadcasts is facilitated by the use of Enhanced Other Networks (EON). The use of EON consumes approximately 8% of the available RDS bandwidth.

The use of features to achieve automated tuning, essentially for mobile reception, requires a considerable portion of the available capacity. RDS offers a choice of some 20 well-defined features, but the limited data capacity restricts the number that can be implemented in a single channel.

Calculation of RDS Capacity

With a bit rate of 1187.5 bps, the RDS channel capacity is a rather limited resource. Since the four checkwords in each group of 104 bits occupy a total of 40 bits, and each group address needs 5 bits, the useful bit rate is

$$1187.5 - [(1187.5/104) \times (40 + 5)] = 673.7 \text{ bps}$$

Analysis of RDS Capacity

In order to better understand the usage of RDS capacity for each of the features, one can group them into categories according to their impact on the channel capacity. By examining the features required for RDS receiver tuning, then, one can identify the remaining capacity for non-program-related features.

As shown in Table 4.14-5, the program-related features can be divided into three categories:

- The primary features—AF, PI, PS, and TP/TA—mainly required for the automated tuning process;
- A group of features—CT, DI, MS, PIN, PTY, and PTYI—require relatively little RDS capacity to be implemented. Note that PIN requires a repetition rate of only one group type 1A per minute; however, in connection with RP it must be increased to one per second.
- Radio text

Furthermore, the non-program-related features—IH, RP, TDC, and TMC—are identified. The following can be seen from Table 4.14-5:

- AF, PI, PS, TP/TA: These primary functions of RDS, essentially supporting the automated tuning process, already require 48.35% of the available channel capacity.
- PTY, MS, DI, PIN, CT: All of these features require relatively little RDS capacity. Whereas the first three

TABLE 4.14-5
RDS Program-Related Features

Application	Feature	Group Types Containing This Information	Appropriate Minimum Group Repetition Rate per Second	Number of Occupied Bits per Group	Number of Occupied Bits per Second	Percentage of 673.7 Bits per Second	Accumulated RDS Capacity
Automated tuning	PI	All	11.4	16	182.4	27.07	48.35
	PS	0A	4.0	16	64.0	9.50	
	AF	0A	4.0	16	64.0	9.50	
	TP	All	11.4	1	11.4	1.69	
	TA	0A	4.0	1	4.0	0.59	
Various other program-related features	PTY	All	11.4	5	57.0	8.46	58.20/63.59*
	MS	0A	4.0	1	4.0	0.59	
	DI	0A	4.0	1	4.0	0.59	
	PIN	1A(B)	0.02/1.0* 37†	0.74/37*	0.11/5.5*		
	CT	4A	0.02	34	0.68	0.10	
Radiotext	RT	2A(B)	3.2‡	37	118.4	17.58	75.78/81.17*
Various non-program-related features	RP	1A/4A/7A/13A					
	TDC	5A(B)					
	IH	6A(B)					
	TMC	3A/8A					

*If Radio Paging is used, an increased repetition rate of 1 per second is necessary.
†Although 16 bits are actually used for PIN, the associated 16 undefined and 5 spare bits must also be taken into account since this represents used capacity.
‡A total of 16 type 2A groups are required to transmit a 64-character Radiotext message, and therefore, to transmit this message in 5 seconds, 3.2 type 2A groups will be required per second.

require 9.65% whether implemented or not, the latter two add an additional 0.21% (5.6% if RP is implemented), bringing the total required to 58.2% (or 63.59% with RP).

- RT: Radiotext requires an additional 17.58%, bringing the total required for program-related features up to 75.78% (or 81.17% with RP).

This then leaves 24% (18% with RP), for the implementation of the non-program-related features (IH, ODA, TDC, and TMC); that is 162 bps (or 121 bps with RP). Each of them will require, if implemented, a significant proportion of the remaining capacity.

However, due to the time-multiplexing possibilities of many of these features, the average capacity available for other features will be greater than the indicated "peak demand."

In conclusion, RDS has a finite capacity, but with careful planning, it can be optimized to provide many beneficial services to the broadcaster and the consumer.

SUMMARY

The NRSC-4-A Standard and its European counterpart, the RDS Standard, provide FM broadcasters and receiver manufacturers with the tools necessary to provide a meaningful supplement to FM audio programs. These standards continue to evolve with the goal of staying relevant as broadcasters transition to digital radio broadcast systems.

Bibliography

European Committee for Electrotechnical Standardization (CENELEC). *Specification of the Radio Data System (RDS) for VHF/FM Sound Broadcasting in the Frequency Range 87.5 to 108.0 MHz.* EN 50067:1998. Brussels, Belgium, www.cenelec.org

European Committee for Electrotechnical Standardization (CENELEC). *Specification of the Radio Data System.* EN 50067 Brussels, Belgium, www.cenelec.org

International Electrotechnical Commission (IEC). *Specification of the Radio Data System (RDS) for VHF/FM Sound Broadcasting in the Frequency Range from 87.5 to 108.0 MHz.* IEC 62106 Geneva, Switzerland, www.iec.ch

Kopitz, Dietmar, and Bev Marks. *RDS: The Radio Data System.* Artech House, 1998.

National Radio Systems Committee. *NRSC-4-A, United States RBDS Standard.* Washington, DC, www.nrscstandards.org

Wright, Scott. *The Broadcasters Guide to RDS.* Focal Press, 1997.

CHAPTER

4.15

AM–FM Field Strength Measurements

DONALD G. EVERIST
Cohen, Dippell and Everist, P.C.
Washington, D.C.

INTRODUCTION

Properly conducted field strength measurements can provide valuable insight on the performance of a broadcast transmission facility. Because of the differing characteristics of AM-band and FM-band propagation, the procedures used for taking AM measurements differ from those used for FM measurements. This chapter will explain both procedures.

SAFETY

Performing measurements requires that all safety considerations be made, including every facet of the measurement program. For FM measurement program, this includes proper mounting of all measurement equipment, raising and lowering of the calibrated antenna, point selection methodology, ensuring that there are no overhead obstructions, and navigating the measurement vehicle in traffic. Measurements of this type must never be taken at nighttime. In addition, if field strength measurements are going to be performed for extended periods, then appropriate rest intervals should be implemented so operator vigilance can be maintained. For more information on electrical shock safety, see Chapter 2.6 in this handbook.

AM FIELD STRENGTH MEASUREMENTS

AM field strength measurements are central to establishing and maintaining a directional array. They are performed to determine antenna radiation efficiency and propagation path characteristics. The requirements for U.S. broadcasters are given in Table 4.15-1. An AM directional antenna is composed of at least two or more radiating elements. The amplitude and phase relations of radio frequency (RF) energy fed to each element and the physical spacing between each element produce a predicted directional pattern shape. Each directional pattern is designed and constructed for a specific application to provide the optimum service to the desired area while providing adequate protection for other co-channel and adjacent-channel stations.

Proof-of-Performance

The proof-of-performance is used to establish the initial operation of the directional pattern at the time of licensing. It is a condition imposed by the Federal Communications Commission (FCC) before licensing can occur and is used to update the performance of the antenna system at such times as may be necessary or may be directed by the FCC. It is also used as a reference for subsequent partial proof-of-performance. Chapter 4.6, "AM Antenna System Maintenance," also includes a discussion of proof-of-performance procedures.

Reference Proof-of-Performance

A *reference proof-of-performance* is the latest complete proof-of-performance accepted by the FCC and is the proof to which partial proof measurements must be referenced. The reference proof specifies, for each

TABLE 4.15-1
Selected FCC Rules Applicable to AM Directional Antennas

Rule	Section
Antenna resistance measurements	73.54
Antenna Testing During Daytime Hours	73.157
Directional Antenna Monitoring Points	73.158
Directional Antenna System Parameter Tolerance	73.62
Emergency Antennas	73.1680
Equipment Test	73.1610
Operating During Modification of Facilities	73.1615
Operating Power	73.51
Partial Proof-of-Performance	73.154
Proof-of-Performance	73.151
Establishment of Effective Field at 1 km	73.186
Special Temporary Authority	73.1635

monitoring point, the point number, the distance in kilometers from the transmitter site, and the radial and point locations. FCC Rule Section 73.151 defines the general field strength measurement requirements to be made on the construction permit and non-construction permit radials for the reference proof.

A reference proof defines the nulls, suppression, and major radiation areas of the directional pattern. A reference proof requires taking nondirectional and directional measurements along each radial under similar environmental conditions. Although each proof has its own requirements, in general, nondirectional measurements begin (as specified by the FCC Rule Section 73.186) at a distance of five times the height of the nondirectional antenna and are to be made at approximately equal distance. Measurements shall be made on a minimum of six radials at the following intervals: 0.2 km up to 3 km, 1 km from 3 to 5 km, and additional measurements beyond 5 km as required. The goal is a minimum of 15 measurements that include a minimum of 7 measurements within 3 km.

Directional measurements are to be made under similar environmental conditions beginning at ten times the widest spacing between the elements of the antenna system. Section 73.151 of the FCC Rules specifies measurements shall be made as follows:

- Those radials specified in the instrument of authorization are required.
- A minimum of one radial in the major lobe is required.
- For a simple pattern, a minimum of 6 radials is required to establish the shape of the pattern; however, if adjoining radials are more than 90° apart, then an additional radial is required within that arc.
- For complicated patterns, up to 12 radials may be required to establish the pattern shape.

The ability to change patterns between nondirectional and directional at either a prearranged time or by two-way communications can be beneficial when taking measurements. This technique permits acquisition of the measurement data in the minimum amount of time and permits the continuous measurement observation without moving the field strength instrument. The accumulation of measurement data whereby both the nondirectional and directional measurements are made at a given point at the same time will eliminate relocation and time differences.

Partial Proof-of-Performance

FCC Rule Section 73.154 governs the undertaking of a *partial proof-of-performance* as currently required. A partial proof is required, for example, when:

- There is a change of directional operating parameters.
- Changes are made above the base insulator of the antenna (such as the addition or alteration of an FM antenna or transmission line mounted on the tower).
- There is a change in an existing monitor point value.
- Changes in the environment of the array dictate the necessity of demonstrating compliance with the station's instrument of authorization.

Partial proof measurements are to be made at the same locations as specified in the reference proof, but measurements are not required at all locations. The partial proof must contain a minimum of eight measurements and include a monitor point at points defined in the reference proof. If the directional pattern has less than four monitor point radials, the partial proof should include measurements on adjacent radials shown in the reference proof. The partial proof must contain an arithmetic or logarithmic analysis of the measurement data. It must demonstrate that the antenna system is operating within its instrument of authorization. Generally speaking, the measurements should be made between 3 and 15 km unless other physical constraints are present. A statement that the impedance of the common point has been measured and is unchanged from the licensed value prior to the making of the measurements should be provided. A change in common point impedance at the operating frequency requires an appropriate submission to the FCC using Form 302-AM.

In those areas where ground conductivity is not constant, the partial proof should be made under similar environmental conditions as the reference proof. More accurate information can be obtained when the directional partial proof is based upon current nondirectional measurements.

Measuring Instrument

When field strength measurements are made in support of the partial or reference proof-of-performance, professional specialized measuring instruments of known accuracy should be employed. For directional antenna systems, the taking of a proof-of-performance and the availability of the measurement instrument are conditions of the construction permit. For example, a typical granted authorization is subject to the following condition:

> A complete nondirectional proof-of-performance, in addition to a complete proof on the directional antenna system, shall be submitted before program tests are authorized. The nondirectional and directional field strength measurements must be made under similar environmental conditions.

General Requirements for Measurements

All field measurements should be made during daylight hours and in the absence of interference. Special temporary authority (STA) may be required prior to commencement of measurements for a new station. For established stations, the FCC Rules permit considerable flexibility in operation during periods of making field intensity measurements on the antenna system.

FCC policy has been that the measurement observation to be recorded and utilized as a basis of analysis of the inverse distance radiation is that observed with the field set oriented towards the station. Under some conditions, the field set maximum indication can have an orientation away from the transmitting source. This phenomenon can be caused by many factors, particularly when measuring in the depth of a null. Other factors include buildings and reradiators near the measuring point, nonuniform conditions inherent in the propagation path, and the position of the monitoring point in regions where there is a sharp change in the pattern.

A record, such as that shown in Figure 4.15-1, must be kept of the measurement data, including:

- Point number and description
- Field strength value measured
- Date and time
- Pattern under investigation
- Name of the individual taking the measurements
- General weather conditions
- Field strength instrument and date of last calibration

At least one equipment manufacturer[1] makes equipment to supplement its current professional manual measurement instrument that will automatically perform the functions for record keeping by recording the value of the field, date and time, and geographic coordinates. This instrument will also compute the distance between the measurement point and the transmitter site and put the data into a format that can be transferred to a computer (such as via a USB port). As an option for determining the measurement point location, the global position system (GPS) can be employed. Suitable GPS equipment can simplify determining the location of measurement points and compute the distance from the transmitter. The geographic coordinates obtained can subsequently be used to describe the measurement point location.

Monitoring Points

An important initial step in setting up a directional antenna is the selection of *monitor points* along the radials specified in the construction permit. These points must be reachable in inclement weather and free of possible reradiation objects such as underground pipes and overhead wires. Schoolyards,

[1]Potomac Instruments, Inc., Model PI 4100, http://www.pi-usa.com/4100.htm.

FIELD STRENGTH MEASUREMENTS

FOR________________

DATE: __/__/__

PAGE NO: ________

ENGINEER: ________

METER: ________

N ____ ° E

POINT NO.	DIST.	DATE	TIME	___kW ND MV/M	DATE	TIME	___kW DA-___ MV/M	RATIO DA___/ND	POINT DESCRIPTION

FIGURE 4.15-1 Sample field strength measurement form.

churches, and cemeteries often provide useful locations because the surrounding area is usually clear. Prior to submission to the FCC, data should be taken at the proposed monitoring points over a period of time until the accuracy and stability of measurements can be established. Readings must be made in the direction of the station. The direction can be established by using a U.S. Geological Survey (USGS) topographic map or by switching the antenna array to the nondirectional mode (if available).[2]

Look for New Construction

Once the directional array has been adjusted and its performance documented, an often overlooked item is monitoring the area for new construction. Vertical structure construction or new construction authorized by governmental authorities is often very difficult to detect in the advanced stage of planning. New power lines and other communication structures (cellular towers) can act as reradiators if located sufficiently close to the array. Such reradiated energy can affect the pattern and may affect interference protection to other stations.

Computer Analysis

Some measurement techniques employ a computerized analysis method for evaluation of the measurement data. As an example, the software[3] incorporates a mathematical[4] analysis method to analyze the measurement data. Such a technique can be used while performing measurements and can be a valuable evaluation tool when taking measurement data. Note that the FCC Rules still permit the manual method for the analysis of measurement data and the FCC staff will still analyze data in accordance with FCC Rule Section 73.186.

Graphical Analysis

The inverse distance field or attenuated field intensity at a reference distance (1 km) is that radiation predicted if the Earth were to behave as a perfect conductor. As the wave energy travels away from the antenna, this energy is reduced in value. The value of the radiated field is in inverse proportion to the distance from the antenna. For example, if the value of the attenuated field at 1 km is 100 mV/m, its value at 2 km will be one-half that value, or 50 mV/m; at 10 km, its value will be one-tenth, or 10 mV/m. After the distance from the transmitting antenna to each of the measuring points has been determined and tabulated along with the observed field strength values, the field strength can be plotted on log-log graph paper. The ordinate is expressed in mV/m and the abscissa is expressed in distance (km).

The FCC provides 20 frequency charts that encompass the frequencies from 540 to 1700 kHz. The uppermost portion of each chart contains conductivity curves *normalized* for 100 mV/m at 1 km from 0.1 to 50 km, and the bottom portion contains the same conductivity curves for distances from 10 to 5000 km. The chart for 670 kHz is shown in Figure 4.15-2.

For the logarithmic coordinate system, (log-log graph paper), the inverse distance line plots as a straight line as shown in the center of the figure. Each of the curves is drawn for the case of an inverse distance field of 100 mV/m at 1 km, and its use is not limited to that value. If an inverse distance field is 200 mV/m (twice the reference value) or 50 mV/m (one-half the reference number) or some other value at 1 km, and if all points on the curve are multiplied by that ratio, this would be the equivalent of moving the curve by that amount on logarithmic coordinate paper. This is the basis by which measurements are analyzed, and the appropriate graph for the frequency involved is made by matching the abscissa of the data with that of the FCC graph. By sliding the ordinate information data vertically, the best fit is obtained. By this method, both the unattenuated field at 1 km and the conductivity value along the radial path have been determined. The use of a light table will serve as a backlight for visual analysis of the data.

An individual attempting to analyze measurement data for the first time or not having benefit of supervision can find this to be a frustrating experience. One approach is to take log-log graph paper for the appropriate frequency (either the regular or expanded scale) and plot the measurement point values normalized to 100 mV/km. For example, if the nondirectional 0.25 kW operation is expected to possess an RMS field at 1 km of 91 mV/m, 70 degree electrical height tower (0.194 of a wavelength), with a normal ground system (see Figure 8, FCC Rule Section 73.190), then it has a field 91/100 less than the FCC log-log conductivity graph. Therefore, multiply all values (divide all values if the expected field is greater than 100 mV/m) of the measurement data by the ratio of 100/91 to normalize this data to 100 mV/m. Plot the normalized data. The plotted values can be viewed in relation to the conductivity values if the assumption of the inverse distance field is correct. If the normalized data appears to be over the inverse distance line, then the radiation value is higher than assumed; conversely, if the normalized data appears abnormally low, the assumed radiation value selected is too high.

This approach can be useful when the nondirectional measurements out to 3 km in the various directions have been taken and a quick evaluation of the conductivity values/radiation efficiency around the site is desired. It also will help to assess whether or not the nondirectional radiation pattern is being influenced by other adjacent towers in the directional antenna system.

[2]See Chapter 1.2, "Broadcast-Related Organizations and Information," of this handbook for information on obtaining USGS maps.

[3]Anderson, H. R., Systematic bivariate analysis of AM field strength measurement data, *IEEE Transactions on Broadcasting*, BC-32(2), 1986.

[4]V-Soft Communications, *Conductivity*, http://www.v-soft.com/Conductivity/Index.htm.

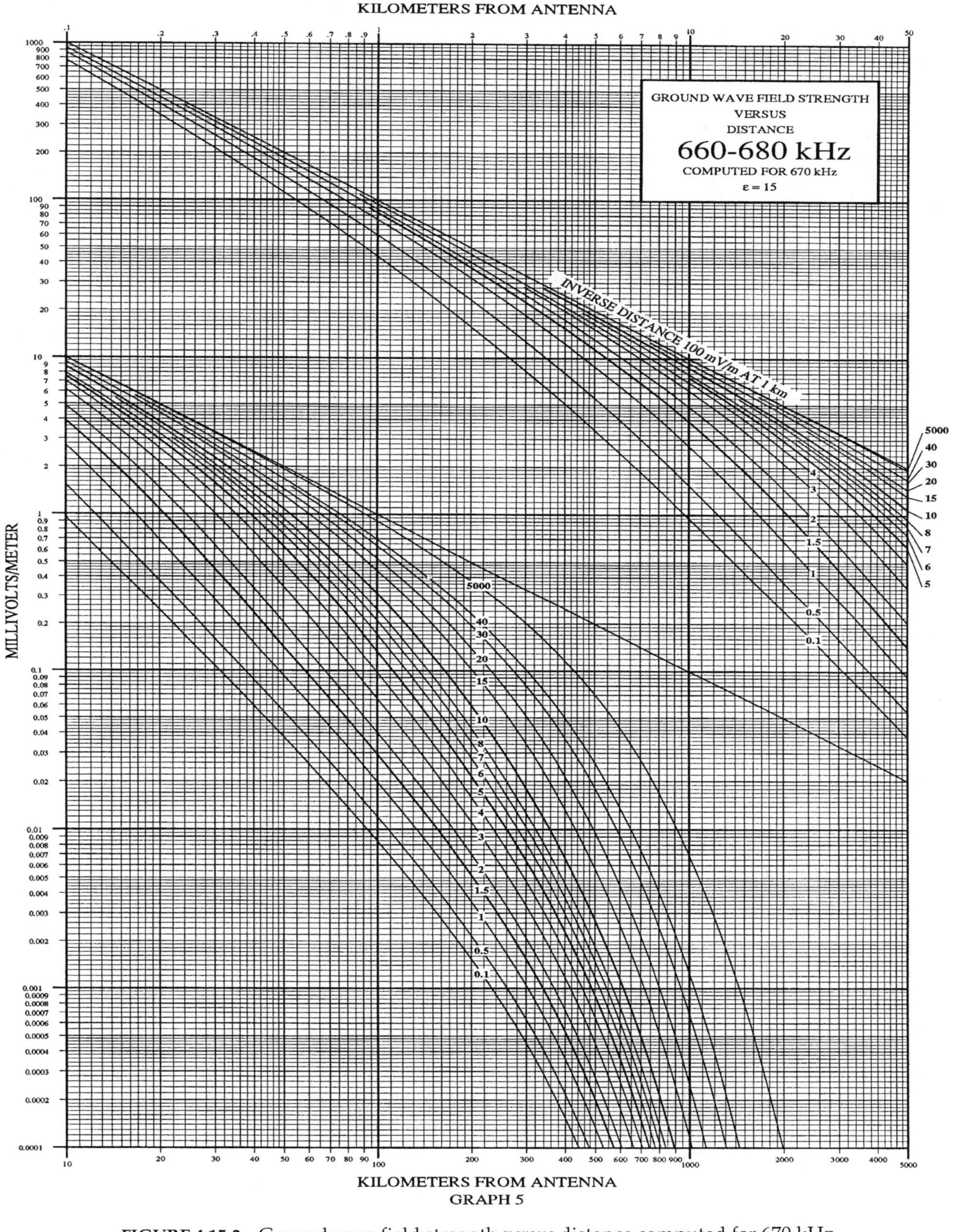

KILOMETERS FROM ANTENNA

GROUND WAVE FIELD STRENGTH
VERSUS
DISTANCE
660-680 kHz
COMPUTED FOR 670 kHz
$\varepsilon = 15$

INVERSE DISTANCE 100 mV/m AT 1 km

MILLIVOLTS/METER

KILOMETERS FROM ANTENNA
GRAPH 5

FIGURE 4.15-2 Groundwave field strength *versus* distance computed for 670 kHz.

Authorizations

Special temporary authority (STA) is usually required with any operation that has not received prior authorization from the FCC. FCC Rule Section 73.1635 provides that an STA must be requested in writing. The authorization request should include the station frequency, location, a complete description of the proposed operation and the necessity of the STA, the title of the requesting individual, and contact information should the FCC request additional information. For licensed operations, the FCC Rules provide that nondirectional power has been authorized for a proof-of-performance during daytime only. A station with a single, or more than one, directional pattern for day and night, etc., can utilize nondirectional power set forth in the latest proof-of-performance without further authorization from the FCC; however, this privilege is permitted only for the purpose of field strength measurements. In addition, the FCC permits without further authorization the nighttime pattern being operated during daytime hours when field strength measurements are being taken.

Authorization to Use an AM Directional Antenna System

The directional antenna system requires FCC approval before its operation can commence. Unlike nondirectional operation, a directional operation cannot be instituted upon a notice and a promise of a subsequent submission to the FCC to fulfill license requirements. The FCC program test authority requires that the FCC agrees that the antenna system is in compliance with the FCC Rules and that the permittee has met the basic obligations and authorizes that operation can commence as described in its licensed application. However, specific requests by the FCC may accompany the program test authority, and those requests must be satisfied before the FCC will issue a license.

Antenna Monitor System Approval

For each authorization for a new station, details of the antenna monitor system components and installation must be contained in the proof-of-performance in order to obtain the necessary recognition from the FCC that the antenna monitor system conforms to the FCC Rules. During construction or revision of the monitoring system, an STA for variance of parameters may be required for existing stations and may be obtained by submitting a request to the FCC. The purpose of the request as well as its duration should be provided. For specific situations, reference should be made to FCC Rule Section 73.68; however, with the revision in MM Docket 85-90, the FCC is less specific as to the details of sample-system construction. The FCC indicates as a matter of policy that the methods outlined in the Rules as modified by MM Docket 83-16 would receive continued FCC acceptance. Other less conventional methods may be subject to rigorous scrutiny including observations over a period of time and a partial proof-of-performance. For convenience, FCC Rule Section 73.68 prior to alteration by MM Docket 85-90 is provided in Appendix A to this chapter.

FCC License Application

The FCC license application is FCC Form 302-AM. Each proof-of-performance must include the information requested in Form 302-AM. It must comply with the provisions and intent of the FCC Rules, including Section 73.186 regarding the number of measurement radials, number of nondirectional and directional measurements made along each radial, mathematical or graphical analysis (if utilized), plot of the field *versus* distance measurements, or use of semi-log or log-log graph paper. Section 73.186 also describes form and substance concerning the submission of nondirectional antenna resistance data as well as the common point impedance measurements. Moreover, descriptions of the monitoring points as specified by the construction permit must be supplied (see Sections 73.152 and 73.158). The station must also comply with other provisions of the construction permit.

A diagram of the RF feed system as constructed (including the phasor, transmission lines, and the tower matching networks) is to be provided. A plot of the inverse distance field at 1 km for the nondirectional as well as each directional mode is to be supplied based on the interpretation of the measured data. The directional pattern must not exceed the authorized pattern in any direction and must have the requisite RMS. In certain situations, for new stations or revisions of existing facilities requiring a new reference proof-of-performance, the FCC will permit an adjustment of directional power which can be changed at the time of the license application.

For each directional operation, the parameters as indicated by the antenna monitor for both loop ratio value and phase (including sign) are to be furnished. Each field strength instrument and its type number, make and model, and date of the last calibration should be listed. If more than one instrument is utilized, a comparison of the accuracy observed for each instrument should be made.

The FCC has a long-standing policy regarding reduction of predicted radiation efficiency when abbreviated ground systems are found. A tabulation of that policy is included in Appendix B.

Information on the License

The FCC license provides, among other things, the licensee name, term of the license, station location, transmitter location and its coordinates, type of antenna and ground system (if nondirectional) frequency, nominal power, hours of operation, and any special conditions. For stations using a directional antenna system, the *second page* will provide a description of the directional antenna system; the spacing, orientation, and height of the towers; and a description of the ground system. Also provided are the theoretical and operating specifications determined from the most recent partial or full proof-of-performance.

The description of the monitor points and the maximum limits of the field strength are contained in subsequent pages of the license.

Changes and modifications of any of these items require appropriate notification to and approval by the FCC. When receiving a new license, it should be inspected for correctness by comparison with the license application. The operating parameters must be maintained in accordance with FCC Rule Section 73.62. The directional parameters must maintain the indicated relative amplitude of the antenna base currents and antenna monitor currents within 5% of the values specified in the license. In addition, the directional antenna relative phase angles must be maintained within 3° of the values specified in the license.

Monitor-point values must be maintained within the values specified in the license. An increase in an existing monitor point value can be accomplished by submission of a partial proof-of-performance along that radial to the FCC with FCC Form 302-AM. A change in monitor point location to a location measured in the last full proof-of-performance together with a recent measurement and appropriate description are to be submitted to the FCC also on FCC Form 302-AM.

The FCC, in a letter to the author on December 6, 1979, adopted a policy regarding the determination of monitor point limits and is included in Appendix C.

FM FIELD STRENGTH MEASUREMENT

FM field strength measurements performed for general determinations of service area require prior planning, including:

- Safety considerations
- Study of the topography of the propagation path
- Proper field strength measurement equipment
- Documentation and analysis tools

FCC Rules for FM Field Measurements

FCC Rule Section 73.314 outlines what is required when submitting FM field strength measurements in support of applications. The Commission addresses two methods of measurement: radial and grid. Either of these methods, when properly performed, can yield valuable information about a station's service area on a selective or overall basis. Further, a variant of this technique can be used when issues arise between neighboring administrations.

The field measurement data for the station can be helpful, for example, in verifying a site's propagation characteristics to an important service area[5] and in determining an antenna's horizontal and vertical performance characteristics.[6] Performing proper field measurements can be a daunting task because FM signal levels can vary dramatically over a short distance, particularly if the measurements are taken with the antenna located close to the ground. The FCC propagation curves F(50,50) and F(50,10) are referenced to a height of 9 m (30 ft). Examples of consecutive horizontal and vertical data recorded at 9 m in generally urbanized, hilly, and flat terrain are shown in Figures 4.15-3, 4.15-4, and 4.15-5. Although this measurement data is reasonably well behaved, it does provide insight as to the degree of difficulty in attempting to gather reliable and meaningful data under varying propagation paths.[7]

The FCC Rules provide guidance for the taking of field strength measurements, such as the following:

- Field strength measurements are to be taken with an antenna height of 9 m (30 ft) above ground.
- Field strength measurement data is to be continuously recorded (graphically and digital) over a path length of at least 30 m (100 ft).
- Field strength measurements are to be performed with an antenna and field meter designed to accurately measure the FM signal.
- The manufacturer's calibrated antenna is to be oriented for the maximum signal. Each field strength measurement point is to be made in an area that is representative of the area and free of overhead obstructions.
- Documentation should include GPS receiver coordinates or the location marked on the latest USGS quadrangle map with the date and time and a description of the point, including topography and the type of vegetation, buildings, obstacles, weather, etc.
- Analysis of the field strength data must be based on the median value of measured signal if it is to be referenced to the FCC F(50,50) propagation curve.

Planning Field Measurements

To minimize time and effort, determine the purpose for taking the measurements prior to going into the field. For an examination of the radiation performance of an aging antenna, this may require field strength measurements to be taken in the eight cardinal directions. Field strength measurements to characterize antenna radiation performance could require the use of a reference antenna with known gain characteristic and effective radiated power (ERP). This reference antenna could be another one on the same tower as the antenna under test and whose radiation performance is known, or it could be a top-mounted FM or VHF television antenna. The mathematical relationship between the desired FM signal and the signal

[5]Alternatively, properly conducted field strength measurements can help define an underserved area. Field strength measurements can provide important information if supplemental service via a booster or translator is being considered.

[6]Field strength measurements can also be made to determine radio frequency levels and spurious emissions.

[7]Measurements were performed with a specially designed vehicle that uses a method in which the chart recorder is dependent on the distance traveled. This allows field strength measurements to be repeated.

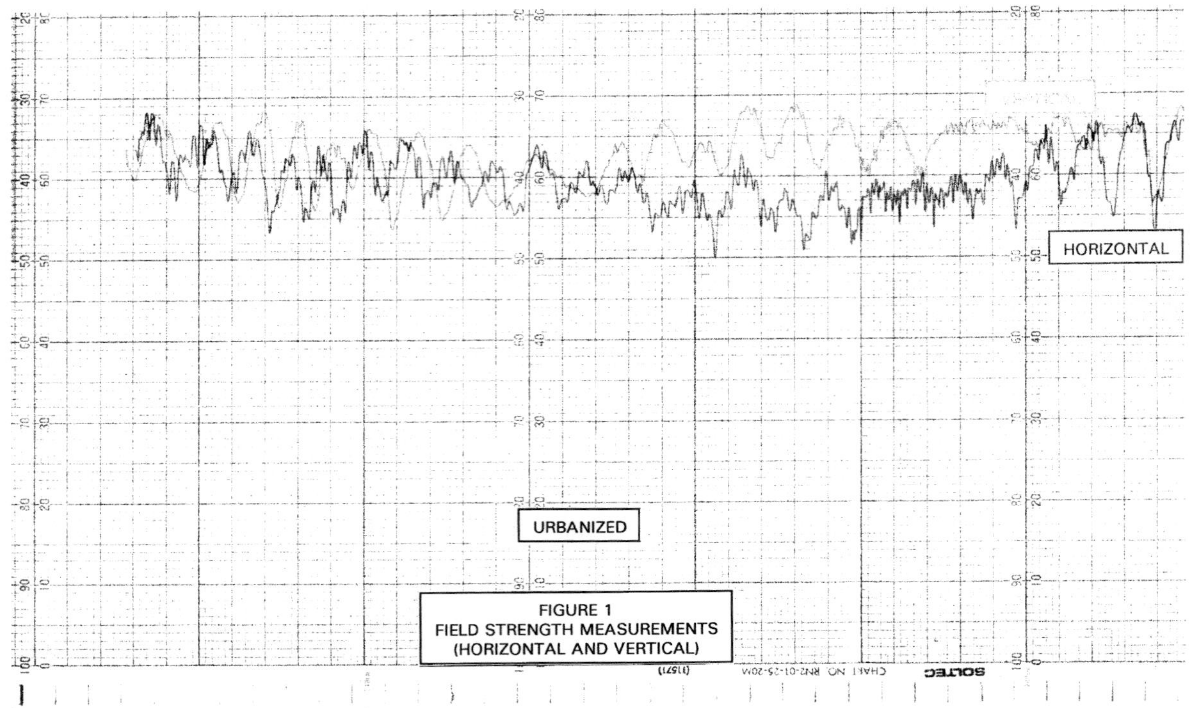

FIGURE 4.15-3 Field strength measurements (horizontal and vertical), urbanized terrain.

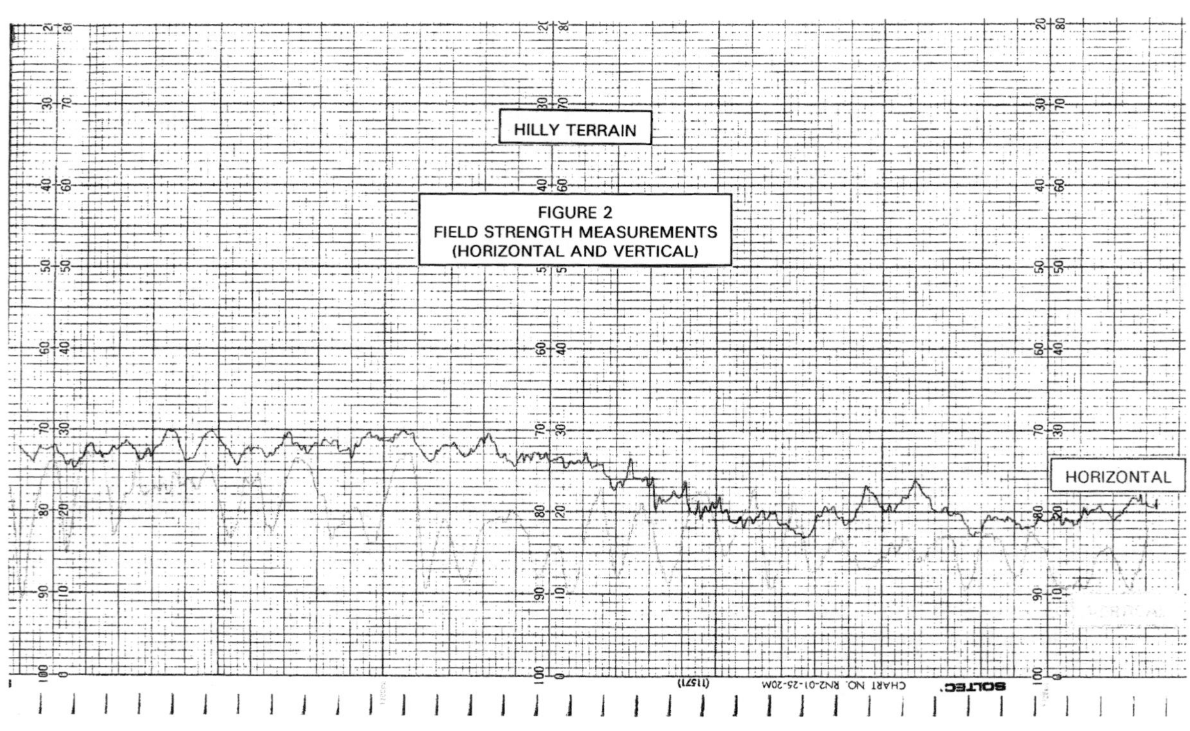

FIGURE 4.15-4 Field strength measurements (horizontal and vertical), hilly terrain.

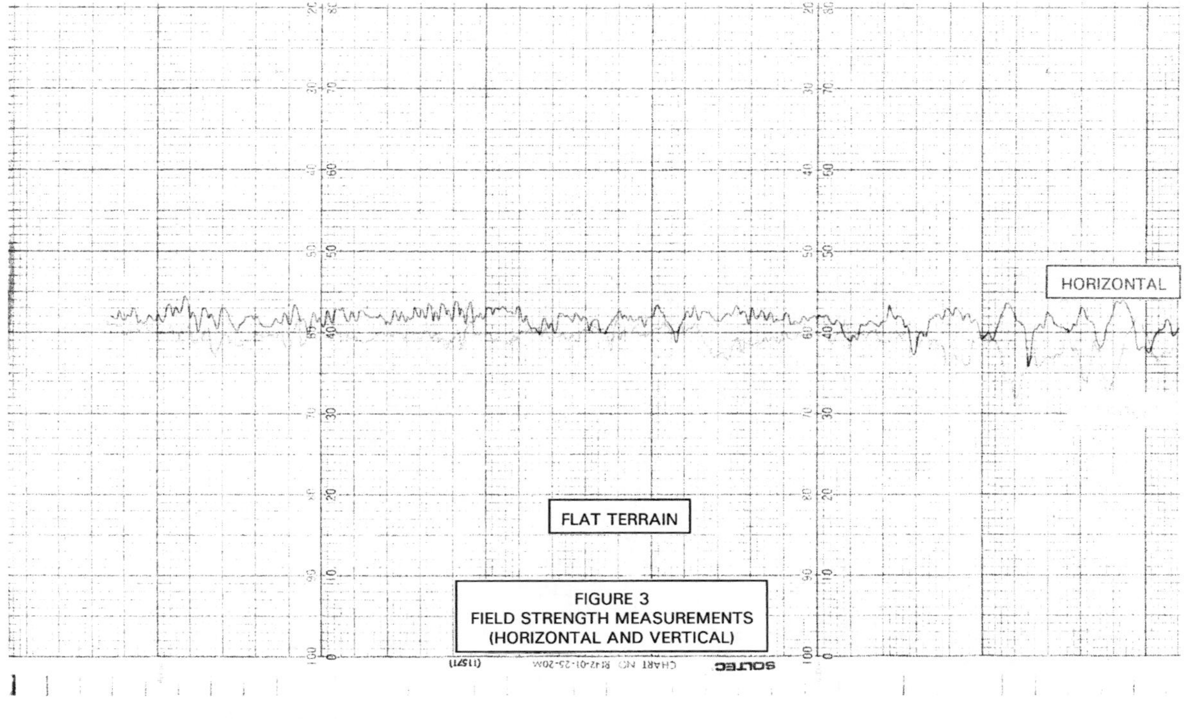

PERFORMED WITH POTOMAC INSTRUMENT MODEL FIM-71 METER
WITH MANUFACTURER'S SUPPLIED CABLE AND CALIBRATED ANTENNA AT 9.1 METERS

FIGURE 4.15-5 Field strength measurements (horizontal and vertical), flat terrain.

from the reference antenna can be determined and results compared with those anticipated. In any measurement program, measurement points located in clear and unobstructed areas are required. The number of measurements along each radial should be sufficient to establish the antenna's performance in a particular direction.

To investigate an area in which reception difficulties (low signal, multipath, or interference) are being experienced, organize the factors relevant to the field investigation. Service area concerns or locations, where possible, should be compiled by Zip Code and address. Also, additional information such as whether the complaint is from a stationary or mobile receiver, at what time of day, and other particular factors must be documented so they are available when performing the measurements.

When sufficient information is available to warrant further investigation, several approaches may be used. Initial physical investigation could be performed by documenting the area of concern by using several types of good-quality receivers. The results of this initial investigation could be augmented by manually plotting profile information or by using one of a variety of computer profile programs to ascertain the nature of the propagation path.

From this preliminary examination, if the cause of the service deficiency is not apparent, expansion of the study by making field strength measurements may be warranted. The preliminary examination, prior to measurements, must yield sufficient information to give focus to the nature of the further investigation. In essence, complaints by advertisers and the listening public must be verified before taking any measurements.

Data Analysis

Data analysis can be performed manually or through the use of computer software. Analysis techniques should be configured to adapt to the measurement program. Mathematical or graphical analysis may be appropriate. Figure 4.15-6 shows the FCC's F(50,50) propagation chart (from FCC Rule Section 73.333, Figure 1). This chart describes the FM signal levels *versus* distance at a particular average terrain value. The average terrain value is determined by an evaluation of the terrain along the radial from 3 to 16 km from the antenna location (see FCC Rule Section 73.313. The curves can be used to extract a set of data to be compared to the measurement data. For example, along a particular radial, a value can be determined that best describes the terrain in which the measurements will be performed. The second piece of information is the appropriate ERP in that direction in decibels—for example, 50 kW (17 dBk) or 100 kW (20 dBk). Next, select a distance. Now, using the F(50,50) propagation chart, a new curve can be generated from a set of values characterizing the propagation path in terms of predicted signal level. These steps are performed for arbitrary distances, such as 1 km, 5 km, or 10 km. Examples shown on Figure 4.15-6 are for a 50 kW ERP at a height above average terrain (HAAT) of 100 m for which the predicted signal level at 10 km is 87.5 dBuV/m and for 20 km is 75.4 dBuV/m. This data can be used at specific distances to provide a comparison to the measurement values. Alternatively, a set of points can be developed such that a curve can be plotted if a graphical approach is desired.

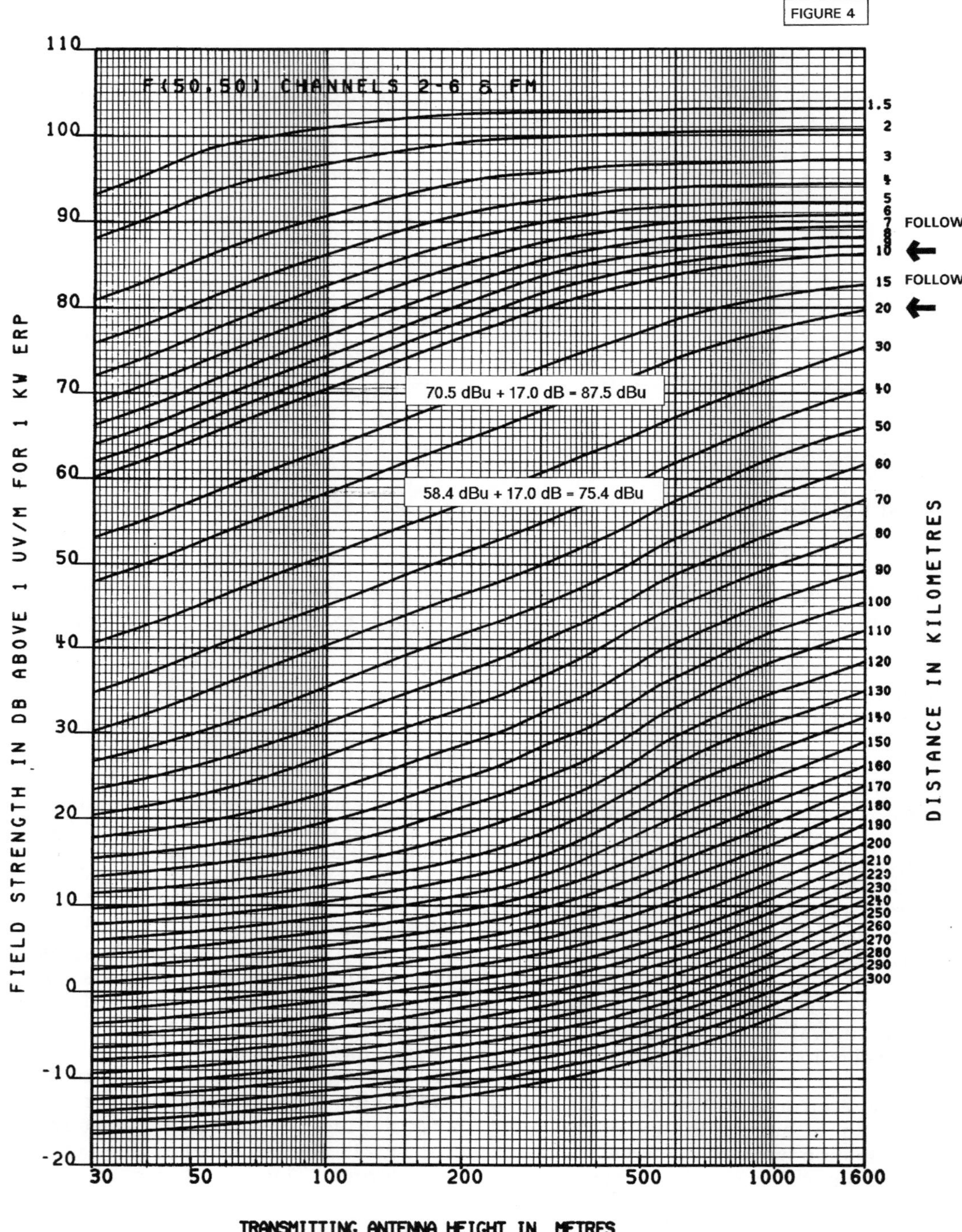

FIGURE 4.15-6 FCC F(50,50) propagation chart.

Field Measurements

Horizontal and vertical measurements performed at 3.0 m (10 ft) can be constructive in documenting difficult reception areas. This technique is used to gather general reception information so other measures of study can be implemented as warranted. For more difficult situations, enhanced measurement techniques using a receive antenna height of 9.1 m (30 ft) should be used. Note that, for a meaningful measurement program, appropriate planning, details of the transmission equipment, measurement equipment, geographic maps, and analysis tools are required.[8] In locations where other higher power transmitters are located, it may be necessary to employ a filter or attenuator at the input of the measurement instrument to eliminate overload in the measurement instrument. It is imperative that a shielded cable be used with good-quality connectors.

In low signal reception areas where intervening terrain may be a factor, an elevation profile study using 4/3 earth radius of the transmission path using 7.5 minute USGS geological quadrangle maps may be appropriate. For initial studies, profile radials using 4/3 earth radius generated by a computer database may be sufficient; however, the computer-generated profile data should be confirmed in critical areas by the use of current USGS maps. Plotting the center of radiation of the transmitting antenna above mean sea level and carefully drawing a straight line to each reception area under scrutiny may give insight if terrain is a factor. From this study, buildings or other manmade obstacles can be added. If a clear propagation path is evident, then the transmitting antenna and its placement should be examined.

Transmitting Antenna

A well-known phenomenon occurs when an FM transmitting antenna is side-mounted on a wide-faced tower.[9] Interaction by the FM signal with the tower members will modify the FM antenna radiation pattern. This can be confirmed by making a number of radial measurements at clear and unobstructed points. If this condition occurs, the horizontal and vertical signal behavior will typically be modified differently. The effect of the individual tower members upon the horizontal or vertical radiated signal is a function of the tower member dimension, position, and its physical relationship to the radiating element.

Field Measurement Configuration

The configuration of vehicle-mounted measurement equipment can be adjusted depending on the vehicle.

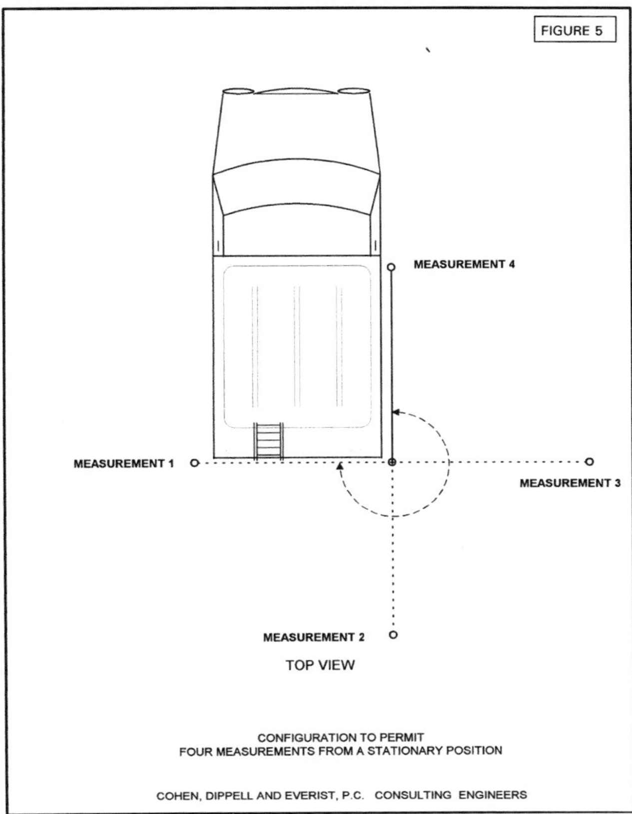

FIGURE 4.15-7 Configuration to permit four measurements from a stationary position.

It is essential to have all measurement equipment appropriately mounted in a stable configuration and secured. For all measurement programs, including those using non-dedicated vehicles, good documentation of the measurement configuration and the equipment is required so the measurement program can be re-established at a later date and time. A simplified method of making spot measurements is shown in Figure 4.15-7; for example, vehicles with a hinge-mounted rear tire support can be modified to permit four uniform height measurements at a single vehicle location. Figure 4.15-8 shows a sample measurement form that can serve as a guide for documentation.

SUMMARY

Field strength measurements are a valuable tool for broadcasters who want to determine exactly how their transmission systems are performing, and how their coverage is being affected by such things as their towers, propagation path, and other nearby signals. In the case of AM stations, these measurements are generally required as a condition of license. For FM, they are usually an optional tool available to the broadcaster. In either case, understanding the basic procedures for taking these measurements is a necessity for broad-

[8] It is recommended that a field meter of known accuracy with a manufacturer-supplied calibrated antenna be used. The antenna dipole length should be set to the length indicated by the instrument manufacturer's instructions. Data acquisition can be made using a chart recorder or digital storage device.

[9] Typically, pattern modification can occur when the tower face is greater than 3 ft or the antenna operation is in the presence of other antennas or guy wires.

FIGURE 6

FIELD STRENGTH MEASUREMENTS

FOR__________

N ° E

DATE: / /

FIGURE NO:______

PAGE NO: ______

ENGINEER: ______

METER: ______

POINT				FM HORIZONTAL				FM VERTICAL		RATIO	POINT DESCRIPTION/GEO COORDINATES
NO.	DIST.	DATE	TIME	dBu	mV/m	DATE	TIME	dBu	mV/m	V/H	

FIGURE 4.15-8 Sample field strength measurement form.

casters performing their own measurements and for broadcasters who will be hiring an expert to perform the measurements.

ACKNOWLEDGMENTS

The author acknowledges Ross Heide, PE; Ryan Felmlee, EIT; and Martin R. Doczkat, EIT; for their valuable contributions and review.

APPENDIX A

FCC Policy Statement Entitled "Criteria for Approval of Sample Systems for Directional AM Broadcast Stations," Dated December 9, 1985

On October 31, 1985, the Commission adopted a *Report and Order in MM Docket 85-90* concerning the antenna sampling systems and proof-of-performance for directional AM broadcast stations. The new rules are based on performance standards in terms of accuracy and stability rather than on construction specifications. This Notice clarifies the information required for directional AM sampling system approval under the new provisions of Section 73.68(a) of the Rules. As before, stations constructing new antenna systems pursuant to a construction permit must obtain approval of their sample system by informal request to the FCC in Washington, D.C.

To obtain antenna system approval, applicants may follow either of the procedures set forth in Paragraphs A or B below:

A. Demonstrate that the system complies with the provisions of Section 73.68(a) of the Rules in effect prior to January 1, 1986.

B. Demonstrate stability of operation by submission of the following information:

1. A detailed and complete description of the antenna monitoring system installation
2. Field strength readings taken on a monthly basis at each of the monitoring points specified in the instrument of authorization for a one year period prior to the date of the application
3. The following readings taken daily for each directional pattern used during the 30-day period prior to the filing of the application:
    a. Common point current
    b. Base currents and their calculated ratios
    c. Antenna monitor sample current ratios
    d. Antenna monitor phase readings
4. The results of either a partial proof-of-performance (Section 73.154) or a full proof (Section 73.186) conducted no longer than 3 months prior to the filing of the application and the common point impedance at the operating frequency measured at the time of the proof

Additional sampling system components and configurations found by the Commission to be accurate and stable over a wide range of environmental and operational conditions will be acceptable under Paragraph A above and announced periodically via FCC Public Notice.

APPENDIX B

AM Radiator Efficiency

The FCC indicates that per its current policy, adopted prior to 1974, for situations where 120 ground radials are used that are shorter than one-quarter wave, the Commission engineers are using as a guide the following table for reduction in radiation with reduction in length of radials:

Radial Length Reduction

Wavelength	mV/m at 1 Mile
.2401–.2500	No reduction
.2301–.2400	–2
.2201–.2300	–4
.2101–.2200	–6
.2001–.2100	–8
.1901–.2000	–10
.1801–.1900	–12
.1701–.1800	–14
.1601–.1700	–16
.1501–.1600	–18

APPENDIX C

FEDERAL COMMUNICATIONS COMMISSION
Washington, D.C. 20554

December 6, 1979

In Reply Refer to: 8800-DW

Mr. Donald G. Everist, Chairman
FCC Processing and Procedure Committee
Association of Federal Communications
Consulting Engineers
1015 15th Street, N.W., Suite 703
Washington, D.C. 20005

Dear Mr. Everist:

I have your letter of October 22nd, written on behalf of your committee, requesting modification of certain Commission engineering practices used in assigning monitoring point limits to AM directional broadcast stations. Your letter formalizes suggestions developed in a series of meetings, begun well over a year ago, between your committee and members of the Broadcast Facilities Division's engineering staff concerning the policies and procedures governing the preparation and processing of various types of applications. The interest shown throughout this period by your committee in helping improve our processing procedures has been helpful and is greatly appreciated.

Specifically, your committee feels that, under the present policy, monitoring point limits are often assigned which are unnecessarily restrictive and urges the adoption of a policy whereby the assignment of these limits is based on the "direct ratio" method. The committee also urges the establishment of a policy whereby stations subject to seasonal conductivity changes can achieve relaxed limits upon submission of "seasonal proofs." Additionally, the committee requests that the Commission refrain from altering monitoring point limits based on partial proofs of performance if "substantial conformance" of the radiation patterns is demonstrated and the antenna parameters are either essentially unchanged or, if changed, adequately justified.

In response to your first suggestion, I am pleased to announce that we have, on an experimental basis, adopted the policy of assigning monitoring point limits using the direct ratio method. Under the direct ratio method, monitoring point limits are obtained by multiplying the measured field strength at a monitoring point by the ratio of the authorized maximum radiation divided by the unattenuated radiation established in the proof of performance. This method simply restricts unattenuated radiation to within its maximum authorized value whereas the traditional method, in many cases, restricted radiation much more severely. Theoretically, objectionable interference is not caused if antenna radiation is maintained below its maximum authorized value. Assuming, therefore, that changes in monitoring point field strength correspond directly to changes in antenna radiation, monitoring point limits determined by the direct ratio method should be adequate to avoid interference. However, since the assumption of a linear relationship between monitor point readings and antenna radiation becomes somewhat questionable with excessive changes, we do not intend to assign limits higher than 200% above proof values. In addition, because operation with monitoring point field strength in excess of the direct ratio limit could result in objectionable interference, we will continue to deny requests to exceed those limits.

Your second suggestion addresses a problem encountered in many areas of the country where complete proofs of performance are done during the summer months when ground conductivity is significantly lower than during the winter months. Often monitoring point limits resulting from such summertime proofs are not sufficient to accommodate higher readings encountered during winter. In such a case increased limits are obtained by collecting supplemental wintertime data in the form of a partial proof of performance consisting of at least 10 measurements on each radial established in the complete proof (see Section 73.154(a) of the Rules). You suggest that the Commission accept "seasonal proofs" for this purpose in lieu of partial proofs. A seasonal proof would consist of "at least 20 field strength measurements, both nondirectional and directional, on each of the radials specified in the construction permit and at least one radial in the major lobe."

In responding to this suggestion, it is helpful to understand the approach used by Commission engineers in analyzing complete proofs of performance. These generally consist of 20 or 30 measurements per radial (see Section 73.186(a)(1)) and serve as the reference for all subsequent partial proofs. As you know, the fundamental problem is distinguishing between the effects of conductivity and antenna radiation. In making this distinction, we consider it imperative to establish, as conclusively as possible, the size and shape of the nondirectional radiation pattern. The nondirectional radiating system is simpler (fewer variables) than the directional system and its RMS (size) can be more accurately determined since each measured radial is of more or less equal significance, particularly if the radials are evenly spaced. With a directional pattern, many of the minor-lobe and null radials do not contribute significantly toward defining the RMS, leaving the remaining main lobe radials with a disproportionate influence on the determination of the pattern size. For these same reasons, the Commission relies entirely on nondirectional measurement data in determining the extent of seasonal changes in conductivity.

Because of the crucial role played by the nondirectional pattern resulting from a complete proof of performance, extreme care is used in analyzing the measurement data. Experienced engineers who have been carefully trained are used in this work. All known external factors such as terrain features, reradiating structures, pipe lines, etc., are taken into account. Each radial is repeatedly weighed against the others with constant attention to the resulting pattern shape and RMS and the analysis is not considered complete until the importance of each element of data is understood from the perspective of the whole. Of course, the more extensive and "well behaved" the measurement data, the more precise and confident the engineer can be with his/her analysis. Once the nondirectional pattern is established, analysis of the directional data can usually be done mathematically, rather than graphically, using either arithmetic or logarithmic averages. Any subsequent nondirectional partial proofs which are submitted to the Commission for the purpose of documenting suspected conductivity changes are mathematically analyzed, point for point along each radial, against the complete proof nondirectional data (see Section 73.186(a)(5)). If the possibilities of distortion and changed RMS can be eliminated from the partial proof nondirectional pattern, then the extent of conductivity change along each radial can be determined and applied to the directional partial proof data revealing whether, in fact, observed changes in directional field

strengths resulted from changes in the radiation pattern or simply from conductivity changes.

The notion of a seasonal proof, to the extent that some of the proof radials would be eliminated, strikes at the very heart of our approach which is an accurate determination of the nondirectional radiation pattern. Although, under the committee's suggestion, the minimum number of measurements on some radials would be raised from 10 to 20, we do not feel the value gained from additional data on these radials would be sufficient to offset the complete loss of data on the remaining radials. This is also the case for directional patterns where changes in radiation in some directions can affect radiation in other directions and assumptions of pattern symmetry are generally unreliable. The Commission encourages supplemental measurements in addition to the minimum of 10 per radial required by the Rules; this should not be accomplished, however, at the expense of fewer measurements on other radials.

Your last suggestion concerns the Commission's assignment of monitoring point limits in response to partial proofs of performance conducted following antenna repairs, refurbishment, construction or readjustment. Often such proofs result in a reduction in limits below those previously assigned because measurements were taken during periods of low conductivity or because antenna radiation in some directions was reduced. The committee suggests we not lower limits in such cases if the pattern remains in substantial conformance and the antenna parameters (phases and current ratios) are either essentially unchanged or, if changed, adequately justified. We believe this suggestion has merit and have, also on an experimental basis, ceased the practice of lowering limits based on partial proofs except when such limits would exceed measured values by more than 200%.

We feel that the current mandatory use of type-approved antenna monitors by directional stations and the widespread use of approved sample systems permit these changes in policy at this time without endangering in any way the technical integrity of our AM broadcasting system. Nonetheless, because of the significance of these changes, we intend to proceed on an experimental basis for at least a year, gaining the benefit of practical experience, before permanently adopting them. In addition, cases clearly falling beyond the scope of these policies will continue to be handled on a case-by-case basis.

We are hopeful that the changes we have initiated in response to your suggestions will provide many stations with operating tolerances sufficient to accommodate variations which, under our old policy, would have required a proof of performance and the filing of an application with the Commission. Again, I would like to express my sincere appreciation for the work done by your committee in bringing forth these suggestions.

Sincerely,
Richard J. Shiben
Chief, Broadcast Bureau

SECTION

5

VIDEO PRODUCTION AND STUDIO TECHNOLOGY

CHAPTER

5.1

Planning a Video Production Facility

RALPH S. BLACKMAN
Rees Associates Inc.
Dallas, Texas

INTRODUCTION

This chapter is intended to give a brief overview of key steps that must be considered when planning the construction or renovation of a video broadcast or production facility, taking account of some of the changes that the introduction of digital equipment and systems has brought about. This is not intended to be an all-inclusive approach, but it gives enough scope and guidance to the broadcast engineer to understand the principles involved. For anything but the smallest projects, it is suggested that specialist design professionals, who deal with these facilities on a daily basis, should be employed to help ensure the success of the new or renovated facility.

A methodical and logical approach is needed when planning and designing any technical facility. The process should address the project in terms of purpose and objectives, issues and planning criteria, quality of design and design intent, building systems, definition of spatial needs, conceptual planning, estimated cost, and schedule.

Broadcast video production facilities have needs that are different from most other building types. As well as conventional areas such as reception areas, offices and storage space, there are areas with special requirements, including studios, control rooms, edit rooms, and equipment areas. These all have particular acoustical, air conditioning, and electrical needs that should be addressed through efficient planning and construction techniques not usually used in other building types.

These specialized buildings typically require a large electrical service catering for two different types of distribution: technical and non-technical. The electrical supply, especially for the technical service, must be clean and reliable. This may require systems such as a backup generator, uninterruptible power supply (UPS), or flywheel to ensure clean and reliable power.

Video production facilities always require special air conditioning in sound-critical and technical areas. Noise and vibration from mechanical equipment and airflow must not effect the program production operations and cooling must be provided for electronic equipment. At the same time, temperature and humidity must be consistently maintained with widely varying loads in some areas.

A common thread through most broadcast facilities is the need to plan for the future when the future is not easily identifiable. This has never been truer than it is now with the advent of digital technology and it is necessary to address such concerns as:

- How will digital systems and equipment impact the new facility?
- What is the impact on an existing facility?
- What is the best way to plan for change?
- Who should be hired to help with this process?

The following sections are intended to give some insight into the obstacles ahead and show how to ease the process along the way.

GENERAL PLANNING CONSIDERATIONS

Building a Team

Once a video facility construction or renovation project is decided upon, it is never too early to start the selection of an architect and other specialist consultants. The first and foremost step to be taken is identifying a suitable design professional. It is important to select an architect with experience in this particular project type. There are many pitfalls and broadcasters do not have the time to educate the novice design professional to the particulars of a video facility.

Once selected, the architect should participate in the selection of qualified specialist engineers, acousticians, integrators, and other team members. It is the caliber of the design professionals, coupled with the selection of capable contractors, which ultimately ensures the success of the project.

Planning a New or Renovated facility

The design team should be able to assist the owners in determining whether to expand or renovate an existing facility or consider building a new facility. A *planning study* or similar exercise should be commissioned to determine whether to build new or renovate. A good example of a "planning study" is the *Facility Business Plan*®.[1] This process specifically assists in determining the following:

- Purpose and Objectives of the Project/Study;
- Project Issues;
- Project Planning Criteria;
- Identify applicable Zoning Ordinances and Building Codes;
- Functional Space Program;
- Probable Project Budget;
- Initial Building Systems;
- Schedule.

A typical flow of events for this process is illustrated in Figure 5.1-1.

Zoning and Building Codes

One of the first steps to be undertaken by the project design team should be to investigate the applicable *zoning ordinances* and *building codes* that will govern the project.

One aspect of zoning is to control which types of buildings can be located where in the community and their adjacency to other properties. Zoning ordinances also control issues such as exterior signage, where and when a new tower can be constructed, how many parking spaces must be provided, and landscape requirements, to name just a few. There are ways to gain exceptions to some zoning ordinances, called variances, but obtaining a variance approval usually requires a great deal of time and political know-how.

Building codes, on the other hand, drive the principles and details of design and construction for the project. Codes inform the design team of the type of fire-rated construction that must be used throughout the project, the number of exits required, and the number of restrooms that must be provided, just to name a few examples of building code requirements.

New construction is viewed differently from that of renovation and expansion. Most "grandfathering", which allowed an existing facility to get by with building design or systems that did not meet code requirements, is now being eliminated. In other words if any part of a facility is touched with a renovation project, then any previous deficiencies must be rectified and brought up to current code.

Building codes govern all areas of facility design, but one area that can have a major impact on the cost of the facility is *seismic constraints*. Every city in the country is classified by seismic zones. The degree to which a facility must be designed for seismic requirements is addressed in the current building codes. Primarily, anything that attaches to, or is part of, the building will require a seismic attachment of some type; this includes, but is not limited to, electronic racks, light fixtures, mechanical equipment, and the building superstructure.

Site Selection

The selection of a site and the way it is ultimately used can be a real asset to facility planning or a major challenge. The site can drive the location of key areas internally and externally. For example, if the site selected has a railroad adjacent, planning should place sound-sensitive areas away from the railroad, using areas with other functions as buffers from the source of noise.

The site layout can have a critical impact on workflows throughout the facility. Delivery of equipment, where employees enter, where visitors enter—all can be driven by site-related factors. If known up front, these challenges can be addressed with the appropriate planning.

For most sites, it is necessary to investigate what it will take to get line of sight to the tower or transmitter site for a studio to transmitter link and for a backhaul link for electronic newsgathering.

Other site-related issues to be considered include access to major traffic arteries, back roads (for when arteries are congested), flight paths, infrastructure needs (utilities, fiber, Telco, etc.), and space for future expansion.

Space Planning

Some of the issues that should be discussed and considered when planning the building layout include:

- Relationship of sound sensitive areas to other areas of the facility;

[1]Facility Business Plan®, is a registered trademark of Rees Associates, Inc.

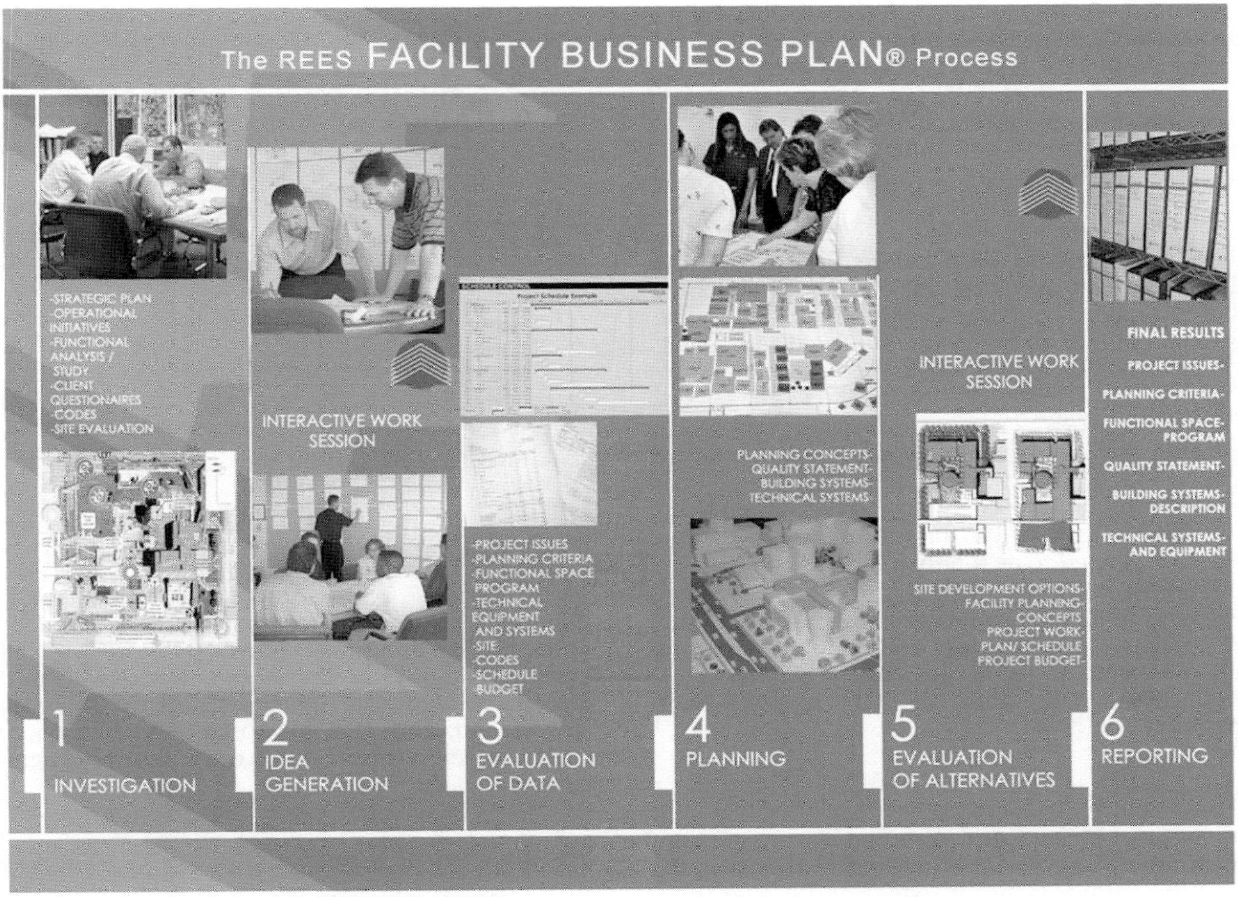

FIGURE 5.1-1 Facility Business Plan® process. (Courtesy of Rees Associates.)

- Relationship of studios to storage (primarily for sets and properties);
- Relationship of control and edit facilities to equipment and other technical areas;
- Relationship of technical areas to any and all areas that require technical support or cabling;
- Relationship of visitor and employee areas;
- Departmental relationships (Sales to Traffic to Business, Master control to Engineering, etc.);
- Security (now a very important consideration);
- Duopolies (multiple stations operating out of one facility).

DESIGN CONSIDERATIONS

Studios

Size and Height

Modern studios should be able to handle television production with widescreen pictures. Cameras that shoot 16:9 have a wider aspect ratio than the traditional 4:3, resulting in a need for wider sets, which ultimately can require wider studios. It might appear possible to reduce the height of the studio for a given horizontal angle of view when shooting 16:9. However, other considerations, such as the height of suspended lights and air conditioning ducts at high level, also come into play, as well as the need for shooting low angle shots. These factors indicate that the height of studios should not be reduced just because wide screen television is being introduced.

It should be noted that for high definition (HD) television, sets must be well defined and more detailed than in the past because HD will show up any and every imperfection of the set.

Floors

All efforts should be made to ensure that a super-flat concrete floor slab is placed. ASTM "F" numbers of F_L 55 (levelness) and F_F 35 (flatness) should be target levels.[2] When met, these targeted levels are more than capable of allowing the operation of pedestal-mounted cameras without visible camera-shake.

[2]ASTM E 1155-96, "Standard Test Method for Determining FF Floor Flatness and FL Floor Levelness Numbers," see http://astm.org/

Some studio floors are finished with a self-leveling epoxy-type surface treatment but, with a well-finished slab, all that may be required is a good concrete seal. This allows the production unit to paint the floor or add whatever surface is required for particular production needs.

Air Conditioning

Air should be delivered to the space with high volume and low velocity to provide the cooling required for the thermal load of lights and equipment with noise levels to meet the acoustic criteria for the studio.

Acoustics

Background noise levels and reverberation should be appropriate for the studio application. While all studios need to be quiet, exactly how quiet depends on the application, as does the amount of reverberation. Not all studios need to be totally dead.

Target acoustical noise criteria for different types of studios and methods of controlling reverberation with sound absorbers and diffusers are mentioned later in this chapter and discussed in detail in Chapter 3.2, Principles of Acoustics and Noise Control for Broadcast Applications.

Doors

Doors should be sized and of suitable construction (based on acoustic criteria) according to the ultimate usage of the studio. Typically, wider and higher than normal doors are provided for the movement of large scenery and set pieces. The opening itself is usually a minimum of 10 ft wide and high. Doors into studios, edit facilities, and other sound-sensitive areas should be coordinated with the sound transmission class (STC) rating of the surrounding construction (see Chapter 3.2).

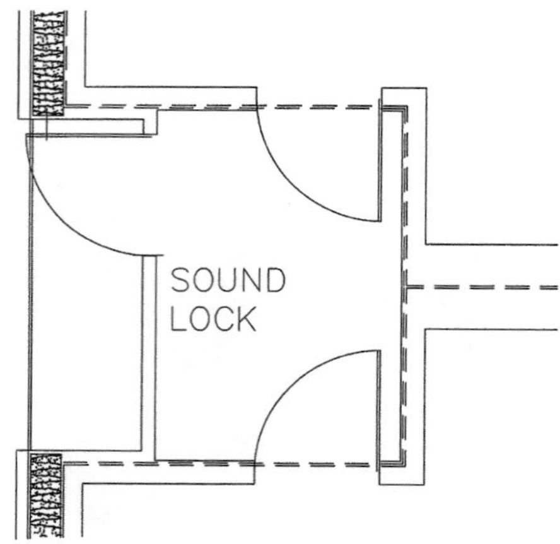

FIGURE 5.1-2 Studio vestibule/sound lock.

Sound Vestibules

To reduce noise ingress into a video production studio, *sound vestibules* or *sound locks* should be included in the facility design, as shown in Figure 5.1-2. These vestibules should be the main entrance into the studio for all audience members and, in some cases, crew, talent, and cast. In most cases, the doors will need to open out, as part of the egress requirements from the space.

Studio Control Rooms

Direct line of sight from the control room into the production studio is not as necessary as it may have once been. In fact, when control rooms are constructed with windows into the studio, they are often eventually obstructed with other materials or props. It is not usually necessary to provide windows for direct line of sight unless there is an over-riding functional reason for the productions being planned for the studio.

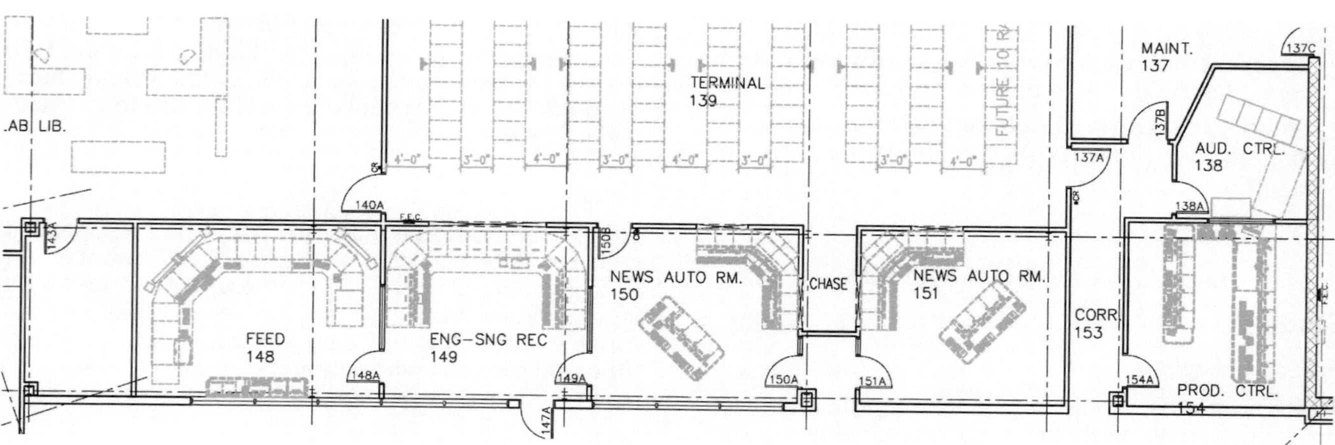

FIGURE 5.1-3 Plan of a typical television control room suite.

One valid reason for providing windows into control rooms, however, is to allow visibility for visitors touring the facility. In this case, observation windows can be strategically placed to allow viewing from an adjacent area in a manner that will not disturb the control room operations.

Inside the control room suite, it is still desirable to provide a visual link from the audio engineer to the remainder of the control. Often audio control will be in a separate room to provide acoustic isolation and a window between the two areas will be needed.

Figure 5.1-3 shows the layout of a typical suite of control rooms for a studio, with other rooms for news operations and feed control.

EFFECTS OF DIGITAL TECHNOLOGY

Changes to the "Tech Core"

"Tech core" for the purpose of this discussion includes all areas such as ingest control rooms, edit rooms, master control, production control, news control, audio control, electronic rack rooms, server rooms, IT and engineering areas.

A good design usually will address these areas by grouping rooms with related functions as close together as possible, which is generally good for station operations. In addition, placing as many technical rooms as possible on raised modular access flooring in turn addresses the primary cabling issues of most facilities.

The switch to digital technology has changed the size and quantity of racks in the tech core. There tends to be more racks that are more densely populated. While it is true that most pieces of digital equipment are getting smaller for similar or increased functionality, more equipment is often being populated within the racks. This has resulted in a deeper rack (36-inch depth is commonly used), with more power dissipation and a need for more air conditioning to keep the equipment cool.

Air Conditioning

The tech core requires air conditioning systems that maintain humidity and temperature within close tolerances. These conditioning systems require redundancy in order to prevent equipment from shutting down or being damaged due to adverse temperature conditions if an air conditioning unit fails.

Some digital equipment is more temperature sensitive than analogue and temperatures within the tech core should not rise above 85° F. Systems should be designed to provide uninterrupted temperature and humidity control. One approach that is employed to respond to this 365/24/7 criteria is the use of computer-room-type air-conditioning (CRAC) units with built in redundancy (dual compressors or with chilled water with direct expansion backup). Effective temperature and humidity control is easily achieved with these units when used in conjunction with a properly placed vapor barrier.

Technical Power

Just as digital equipment is sensitive to temperature, it is also power sensitive and tends to be unforgiving when it comes to power disturbances. Clean and reliable power is critical to a digital facility.

A site with two different sources of power is definitely a plus and this is a consideration when selecting a site for a new facility. Two power supply lines from the power company (if truly independent) would allow for a redundant system design with either an automatic transfer switch or a conveniently placed manual transfer switch. However, when this is not available some type of generator backup is usually essential. The generator should be sized to back up all critical loads of the facility, including those systems that serve the tech core.

An uninterruptible power supply (UPS) or electromechanical flywheel may be required to bridge the gap between power failure and the time when the generator comes on-line. In any case, given the sensitivity of modern equipment to power line disturbances, a large UPS supplying all tech core electronics is highly desirable. The alternative arrangement, of many smaller units distributed with individual items of equipment, is much less satisfactory and difficult to monitor and maintain.

Changes to Control Rooms, Edit Suites, and Post Suites

The advent of smaller equipment for digital production switchers and editing systems has reduced the spatial requirements for equipment within the design of these particular rooms, but the number of people wanting to occupy these rooms has not changed radically. So these areas should be designed around the number of people involved in the action, with less consideration needed for the amount of equipment than in the past.

The bulky consoles and huge walls of monitors are now largely obsolete. Today, many consoles are replaced with simple, functionally tiered desks (see Figure 5.1-4). Monitor walls are usually composed of a series flat panel displays or a projection system that allows multiple-source viewing.

Totally networked facilities can reduce the number of traditional viewing and editing bays. Initial video browsing and editing can now be performed at the desktop. This capability is becoming increasingly cost effective and therefore more staff may require quicker and easier access to these features. This implies a need for more working spaces with these facilities in offices.

Cable Management

Most facilities still require some traditional coaxial and twisted pair cable interconnections. However, nearly all modern equipment requires networking and copper- or fiber-based networks are now part of the landscape in digital facilities. Wireless network systems may be used in some limited applications.

FIGURE 5.1-4 Edit Room. (Courtesy of WJLA, Allbritton Communications, Arlington, VA.)

A properly designed facility will allow for multiple changes in technology, which inevitably will require change in equipment locations and replacement of wiring. In order to respond to these changes, a complete and cohesive cable management system is required, including all or some of:

- Modular access floor systems (where appropriate for the room;
- Overhead cable trays;
- In-floor duct banks and conduits.

Where multi-floor technical facilities cannot be avoided, fully accessible cable risers with cable support systems must be provided. How and where each of these methods for handling cables is incorporated into the design depends on the type and location of each room, the constraints of the overall building design, and, to some extent, how the end user sees the facility ultimately functioning.

Whichever methods are selected for running cables, part of the management system design is to separate different categories of cables in order to minimize interference between cables carrying different types of signals.

Audio

Quality acoustically treated spaces must be provided for audio production and monitoring. In the past, quality sound was not always a priority for television broadcasters. Today's viewer demands a higher quality of sound. Along with the great improvements in video, broadcasters and production facilities are expected to provide similar results in audio. Provision will need to be made for high-quality monitoring of stereo and, increasingly, 5.1-channel surround sound.

Technical Furniture

In today's environment, the more flexibility that can be designed into a facility the better. The use of manufactured moveable furniture instead of typically built-in millwork is one approach to ensure flexibility of space. Areas such as edit bays, engineering technical benches and audio consoles, to name a few, are areas to consider for movable modular furniture such as that shown in Figure 5.1-5. This will allow areas to be easily repurposed when the need arises. The use of premanufactured instead of built-in furniture will, however, have an impact on the budget of the facility.

Facility Monitoring Systems

Many facility support systems now have computer-based control systems, providing the ability for centralized monitoring. Most commonly monitored systems are those for energy management, security, phones, networks, and fire and safety. Discussions with the many different vendors regarding plans for using such monitoring will assist in selection of the

(a)

(b)

FIGURE 5.1-5 Modern modular premanufactured furniture (Courtesy of Winsted Corporation) appropriate manufacturer and system for each application. The level of monitoring of facility systems will also have a direct impact on the project budget.

Tape Storage

Servers are being used increasingly in throughout the broadcast and production industries. Many in the industry believe that ultimately such disk-based systems will replace tape-based technology completely. This may eventually happen, but until that time there is still a need for tape archives and storage space.

Most facilities will need some tape storage for the foreseeable future, both for legacy analog and digital tape material. Other digital media such as CDs and DVDs are also still going to need a home. The use of high-density storage systems, as shown in Figure 5.1-6, should be considered. Such systems may appear at first to be expensive but, when compared to construction cost, they may be cost effective, especially for a large library. High-density storage will also have an impact on the budget.

FIGURE 5.1-6 High-density storage systems. (Courtesy of Russ Bessett Corporation)

BUDGET DEVELOPMENT

When project budgets are developed, they should so far as is possible include all "probable" or "potential" project costs. These costs may include but are not limited to the following:

- Land cost;
- Land surveys;
- General construction of the facility based on a gross square foot cost. General construction costs cover all exterior and interior building costs for the facility. Renovation will be different from new construction. Gross square footage should include all the net square footage of assigned spaces in the facility, with a gross up factor, plus percentages for circulation, mechanical, electrical, plumbing, and building envelope;
- Moving or adding additional satellite dishes;
- Construction of a new tower, if needed;
- Any special construction required (such as dry pipe fire suppression, hurricane protection, etc.);
- Furniture, furnishings and accessories
- Fees for architecture, engineering, technical reports (such as geo-technical and environmental reports, etc.);
- Any radio frequency (RF) studies, if required;
- Any terrestrial studies, if required;
- Communication systems;
- Computer systems;
- Security system;
- Broadcast equipment systems;
- Bid escalation for time that the project maybe on hold;
- Design and construction contingencies.

Timing of construction is increasingly a major cost concern in the construction industry. Costs for new or renovated construction have escalated in recent years and, at the same time, the skilled labor force has dwindled. So the earlier the process can be locked down, the better chance there is of controlling the overall cost of the project.

The U.S. Green Building Council has introduced a "Leadership in Energy and Environmental Design (LEED) "green" building rating system.[3] LEED-friendly design and/or certification should be discussed in the initial stages of the project, primarily because of the cost implications required to accomplish true LEED certification. If LEED certification is desired, it must start at the beginning of the project because of the documentation required from initial design through the construction process. Buildings can incorporate green design techniques if desired and

[3]See http://www.usgbc.org/

not have to be certified. A few of the green techniques that can be employed include:

- Gray water irrigation;
- Lights-on motion sensors;
- Recycled materials and programs;
- Energy management systems that monitor carbon dioxide levels and that allow for economizer cycles. These allow fresh air to condition the space;
- Waterless waste technologies.

DESIGN AND CONSTRUCTION PROCESS

There are several different scenarios that may be used in delivering a project, including:

- Design, bid, build;
- Design/build;
- Fast track, or negotiated.

The following sections describe the process for each scenario.

Design, Bid, Build

The phases of a typical *Design, Bid, Build* process are as follows.

Schematic Design

During the schematic design (SD) phase, the architect will interpret the functional space program that is usually provided by the owner and reflect this interpretation in a floor plan as scaled representation of the new or renovated facility. During this time, aesthetics, materials, finishes, schedule, and budget will begin to be identified and discussed.

Design Development

After approval of the schematics, the design team will move into design development (DD). During this phase the project is further defined in terms of actual wall thickness, interior and exterior window systems, selection of all interior and exterior finish systems and materials; exterior elevations and building sections will be generated to assist in defining the building vertically as well as horizontally. All design decisions must be made prior to finishing this phase of the project. Also, at this point 3D sketches will begin to help clarify the overall aesthetic of the facility. Budget and schedule will again be discussed and updated.

Construction Documents

After approval of the design drawings, the design team will move into preparing the construction documents (CD). With all decisions made in DD, the project team now can build the project on paper. This is the technical exercise that allows the team to do all the final designing and documentation before putting the project out for bid.

Quality Assurance

Before any project goes out for bids, a third-party review or quality assurance (QA) review should be performed. This effort is the final step of coordination of all the disciplines—civil, structural, architectural, mechanical, electrical, and plumbing—that needs to take place before the CDs are given to contractors for bid.

Bid/Negotiation

With a bid/negotiation (B/N) type of project, this is the stage in which several pre-selected and pre-qualified contractors, usually no more than three, will be provided with the CDs and asked to formulate their bids (price) for the project that is defined in the documentation. Typically, a three to four-week period is allocated for the bid process, and, at the end of this time, bids are submitted. From the bid process, a contractor is selected and a contract for construction is negotiated. Once the contract for construction is executed, construction will usually start.

Construction Phase

Two points regarding the management of the construction process should be noted:

- *Point of Contact:* When construction starts, the owner should assign one individual as the main project contact. This individual will be the contact point between the design team and the contractor from day one through the end of construction. In fact, it is best if this individual is assigned during the initial planning or SD phase of the project. Direct contact by others with the contractor during this time should be discouraged and not allowed.
- *Change Orders:* No design team is perfect and there will inevitably be change orders in most construction projects. However, change orders add to the cost of the project and the key is to minimize them, if at all possible. Allowing a project to proceed in a methodical and logical progression is one of the steps that will help in the minimization of change orders. Well-coordinated and documented CDs will also minimize change orders.

The design process outlined in the preceding sections represents a conventional design approach for a "hard" or "competitive" bid project. While it is methodical and logical, the downside is that the contractor is not a design team member. The contractor comes in only at the B/N phase and then proceeds with construction. Waiting this late to bring a contractor into a project may create a more adversarial project environment.

Design/Build and Negotiated

An alternative approach for project implementation is known as *Design/Build*. In this arrangement, the contractor and architect, as a team, design and construct the project. A single contract is executed between the owner and the design/build team with an agreed price for the project. The team then proceeds to deliver a complete project within the budget and on schedule.

Another delivery method is that of the *Negotiated* approach for the construction part of the project. In this arrangement, the owner has already selected the architect and the owner and architect together pre-select a group of general contractors for interviews. These firms may be selected by a list of predetermined qualifications such as previous experience, personnel, and/or cost. The successful contractor is then included in the design team from the beginning of the project and through construction. The contractor firm is then available to monitor cost estimates and consult on constructability. Many customers have found this approach fits a broadcast facility project well and has the best track record of keeping a project within the initial budget and on schedule.

Selecting a General Contractor

The general contractor's expertise is a necessary part of the project's success, with skills that should include project scheduling, keeping with the budgeted dollars, material selections and constructability for the particular region of the country. This key player must also have a good knowledge of technical facilities and understand the importance of flexibility for the future.

The selection of the appropriate general contractor is one of the most crucial decisions a broadcast facility owner has to make when planning and constructing a broadcast facility. There can be serious challenges from the first day if the selected contractor is not suitable qualified.

When selecting a general contractor, there are four main areas to evaluate:

- *History:* A consistent management philosophy should have been established if the firm has been in business for at least ten years. This should be prevalent in the interviews of personnel during the selection process.
- *Reputation:* This consideration includes not only the firm, but its leaders and it relationships with other subcontractors. Does the reputation coincide with your company standards? Does the reputation reinforce the management philosophy? Does the reputation hold with subcontractors?
- *Financial stability:* The selected firm's net worth should be worth at least half of the projected project cost of the planned project.
- *Record:* What is the record of the firm concerning technical facilities? As part of the qualifications, request a list of projects the firm has completed, with total construction costs, sorted by year. Determine the bonding limit of the contracting firm and make sure this dollar amount is at least twice the construction budget. This guarantees the performance of the contracting firm, and the insurance company can provide valuable information on the company. Ask for detailed information of the last three technical facilities projects the firm has completed. The information should include each project's location, construction budget, completed cost, the year built, and a person to contact regarding the project at the facility. If there are no technical facilities then the contractor is not qualified.

CONSTRUCTION

Exterior Construction

The building exterior should be designed to adequately isolate extraneous noise at the site. In some cases, interior sound isolation construction should be resiliently decoupled from the exterior construction. Roofs over sensitive spaces should be designed to adequately attenuate rainfall impact noise and other environmental noise.

Selection of exterior cladding materials should take account of the area. For instance, the use of brick in northern areas is often undesirable because materials may be hard to come by and installation may be difficult due to the climatic conditions. In that case, a pre-manufactured panel of some type may be better value for money.

Selecting the proper building structure is critical to several areas of the project. This selection will have an impact on cost, schedule, and expandability of the facility. Whatever form of construcion is chosen should allow for ease of reconfiguration or future modifications. Steel or concrete framing will usually allow this. A load-bearing wall structure is much less flexible and does not easily reconfigure. Obviously, when remodeling an existing building there is not as much flexibility with the structure, so the interior planning must allow for ease of reconfiguration.

Interior Construction

Walls

Interior construction should usually be primarily based on partition walls built with steel studs. This type of construction allows for easy reconfiguration as the partitions are easily moved or removed and can be insulated for acoustic isolation when required. In areas such as studios, with critical acoustic requirements, grout filled concrete block walls are commonly used if dead load is not an issue. Otherwise, various designs of stud walls with multiple layers of cladding and insulation may be used as discussed in a later section.

Ceilings

Areas without special acoustic requirements may use conventional suspended ceilings with lay-in ceiling tiles. However, utilities such as water, electrical, fire suppression, wiring, communications, and data, are often run in the ceiling voids of corridors and other areas. In some areas, technical cabling is also run in overhead cable trays. To allow access to these overhead utilities and cabling, it is recommended that the use of suspended ceilings should be eliminated wherever possible. Clearly aesthetic considerations may require the use of ceiling tiles in some areas such as offices but their use should be kept to a minimum.

Areas that require acoustic isolation may require special sound isolating ceilings as discussed in a later section.

MECHANICAL, ELECTRICAL, PLUMBING, AND FIRE PROTECTION ENGINEERING

Main Service

A new main electrical service, typically fed from a utility company pad-mounted transformer will be required. Transformer secondary distribution voltage should be 480Y/277 Volts, 3-phase, 4-wire with a solid ground. The main switchboard and each level of distribution for the technical, mechanical, and studio lighting power systems should be equipped with separate transient voltage surge suppression devices in accordance with IEEE standards.

Secondary Services

Reliability of the secondary electrical distribution system should be a major design consideration to eliminate single points of failure in the critical power path. Separate panel boards should be provided for emergency, technical, mechanical, studio lighting, and normal power.

Emergency Backup Generator

A standby diesel-driven generator set should be provided, either located in a heated acoustical weatherproof housing or in an acoustically treated room within the building, with either a skid-mounted or external fuel tank sized for a minimum 24 hour run period.

UPS and Flywheel

All technical power systems that serve broadcast processing equipment, control equipment, technical equipment, and non-broadcast-related LAN equipment should be served by a central 3-phase static or rotary UPS system. UPS systems should be backed up either by lead acid batteries (sealed or flooded cell) with a minimum of 5 minutes supply duration or a rotary flywheel for 20 seconds duration, with a backup 24-volt supply from the UPS to start the standby generator.

Air Conditioning

A distribution of conditioned air should be provided to all occupied spaces in the building, with the exception of mechanical rooms, using either packaged rooftop units or air handling units located in mechanical rooms. Cooling may accomplished by central chiller units with a chilled water distribution around the building or by unitized direct expansion equipment located where required. Associated ductwork, terminal units, diffusers, registers, return fans, and associated controls are normally required for a complete system. Each rooftop unit should include air filters, hot water preheat coil, direct-expansion or chilled water cooling system, centrifugal supply and return fans, double-wall casing, stainless steel double-sloped drain pan, and vibration isolation roof curbs. Where required for noise considerations, duct silencers should be provided.

Each studio, control room, or other acoustically sensitive space should be provided with a dedicated constant-volume terminal unit served from a variable-volume air handling unit. Care must be taken in the location of the rooftop units, terminal units, selection of fans, and sizing of ductwork, to achieve the required acoustic and thermal goals for each space.

Technical equipment areas on raised access flooring should be served by cooling-only units. Air may be ducted from above ceiling to precision registers or through the raised floor air grilles. Redundant air conditioning units for these areas should be taken into account for both maintenance and failures. The technical area spaces should also be furnished with a humidification system to allow constant humidity to be maintained.

Technical Grounding

The facility should be provided with a low-resistance grounding system consisting of ground rods around the building perimeter. The grounding system should be connected to structural steel and to the lightning protection system. An isolated signal reference ground system should be provided for all broadcast and technical equipment, with a single point ground, as discussed in Chapter 9.3, Facility Grounding Practices and in Chapter 3.1, Planning an Audio Production Facility.

Energy Management

A central energy management control system should be used to monitor and control the air conditioning and heating systems. The system should utilize direct digital controls (DDCs), with variable volume terminals equipped with DDC controllers. This central system should monitor and alarm all critical temperature and humidity levels in technical broadcast and information technology equipment spaces. Such systems utilize remote paging options to alert personnel of approaching potentially serious conditions.

Fire Suppression

The facility should be furnished with a complete fire protection system that meets building code requirements. Sprinkler heads should be furnished throughout the facility to accomplish total coverage. The technical areas should be provided with a pre-action double interlock sprinkler system actuated by any two detectors (connected to the fire alarm system).

An addressable fire alarm system should be provided for a new facility. Smoke detectors should be provided throughout the building and also manual pull stations at egress points as required.

In the technical equipment spaces, an FM 200 or FE-25 gaseous extinguishing system should be considered for the areas above or below the raised floor.

ACOUSTICAL DESIGN CRITERIA

The following criteria can be used as a guide for the required maximum background noise levels for mechanical and electrical systems, in terms of Noise Criterion (NC) ratings. These criteria can also be used to characterize the sensitivity of each space to outside noise intrusion.

TV/Production Facilities

Large production studio	NC 20-25
Small production studio	NC 20-25
Production control rooms	NC 25-30
Audio control rooms	NC 20-25
Announce booths	NC 20-25
Master control rooms	NC 30-35
Instructional studios	NC 25-30
Editing suites	NC 25-30
Client editing suites	NC 20-25
Audio sweetening rooms	NC 20-25

Radio

Performance studio	NC 20-25
Announce booths	NC 20-25
Control room	NC 20-25
Editing rooms	NC 25-30

Miscellaneous

Offices	NC 35-40
Conference rooms	NC 33-35
Technical workrooms	NC 35-40
Data center/ core equipment rooms	NC 40-45

ACOUSTIC SEPARATION OF SPACES

Requirements for acoustic separation of the various program spaces should be established based on the sensitivity of the spaces, as defined in preceding sections, the sound-generating characteristics of each space, and the specific space adjacencies as developed in the building layout.

To the maximum extent possible, the building layout should take advantage of opportunities to minimize or eliminate expensive sound isolation construction—for example, decoupling certain elements of the building structure, providing buffer space such as corridors and storage rooms between sensitive spaces, locating noisy shop spaces remote from studios, and avoiding vertical adjacencies between the most sensitive spaces.

Most facilities have two general levels of physical acoustic separation required in critical spaces with requirements as follows:

Level I

Floor: Typically, there are no special isolation requirements for on-grade floors, unless there are special ambient circumstances such as an adjacent railroad or holding area for trucks. Above-grade floors may possibly require heavier than normal construction for insulation and isolation of airborne noise from vertically adjacent spaces.

Walls: Typical sound walls are of multiple layers of gypsum board on both sides of heavy-gauge steel framing, extending full height, and sealed to the structure above, with full-depth batt insulation in the stud cavities. Construction details are provided in Chapter 3.2.

Ceiling: Typical suspended acoustic tile ceiling, or no special requirement.

Level II

Floor: An isolated floor may be required within each room, consisting of a 4- to 6-inch concrete slab supported by a mat of fiberglass insulation with resilient isolators spaced on 12- to 16-inch centers atop the structural slab, as illustrated in Figure 5.1-7. The isolated floor should be decoupled from all other elements of the building construction.

The structural slab may require depression to resolve floor elevations between isolated and non-isolated spaces and cabling pathways will need to be incorporated into the isolated floor design as required.

If the actual structural *floor* slabs between adjacent sensitive spaces can themselves be resiliently decoupled, for example, with expansion joints, then the isolated concrete floor may not be required.

Walls: Walls should consist of separate multiple layer gypsum board assemblies for each space being isolated, supported on the respective isolated floors and decoupled from the structure above with resilient angle brackets. Construction details are provided in Chapter 3.2.

Ceilings: Ceilings should consist of multiple layer gypsum board assemblies supported with combination spring and neoprene isolation hangers, as shown in Figure 5.1-8, with all services surface mounted or suspended below the sound barrier. The ceilings should be decoupled from all wall and other surrounding constructions. When double ceilings are required for additional isolation, one ceiling will be of gypsum board construction as shown in the figure and

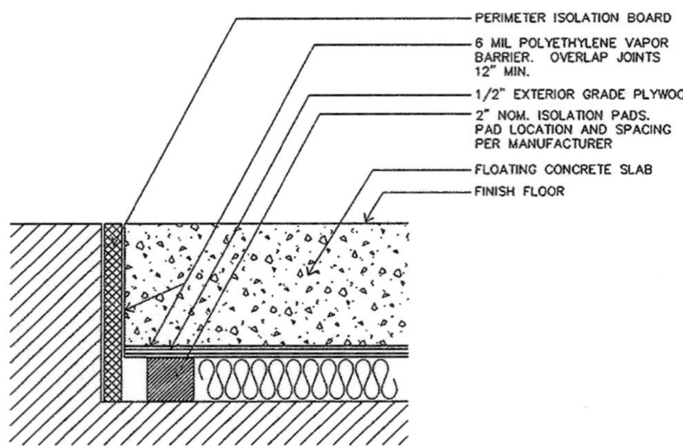

FIGURE 5.1-7 Sound isolation floor.

one will be lay-in. The lay-in ceiling is 1- to 1.5-inch thick glass fiber with a canvas face or other finish—not the typical lay-in ceiling found in offices.

Other acoustical treatment items for Level II areas include resilient details and seals at all duct, pipe, electrical/other wiring infrastructure, door, and window penetrations.

Doors

Three types of sound-reducing doors are typically used in sound-critical areas, depending on the use and acoustic isolation required.

- *Pre-engineered sound-rated door and frame assemblies.* These have integral doors, frames, and acoustical seals. They comprise heavy gauge, stiffened and insulated door panels, heavy gauge frames, magnetic head and jamb seals, cam-lift hinges, and fixed bottom seals. Glazed panels incorporate insulated laminated glass.
- *Standard doors with field-installed acoustical seals.* These assemblies typically have insulated door panels, standard frames, and field-installed adjustable perimeter gaskets, including an automatic door bottom seal.
- *Sliding glass doors.* These assemblies will typically be exterior weatherproof type with insulated glass assemblies incorporating laminated glass. In some cases, double sliding doors may be required.

Isolation for the most acoustically sensitive spaces will require sound locks—that is, two sets of doors separated by a vestibule. For sound locks, the inner door should be the sound-rated type and the outer door may be either a sound-rated door or a standard door with field-installed acoustical seals.

Acoustical Room Finishes

Substantial portions of the wall and ceiling surfaces in most of technical/production spaces will require sound absorption treatment of some type. Sound absorptive treatment may consist of rigid acoustic insulation between 1 and 6 inches thick, covered with sound transparent facing such as fabric, either in pre-manufactured panels or a field-fabricated system. Some spaces may be provided with special suspended ceilings as discussed previously, depending on the sound isolation requirements of the area.

Broadcast production studios should have sound absorptive acoustical treatment on the full extent of the wall, to grid height, and also on ceiling surfaces. The treatment can consist of 2- to 4-inch rigid acoustic insulation with protective sound transparent facing such as metal mesh or fabric on nailing strips, as shown in Figure 5.1-9. Studios will typically also require special units for absorbing low frequency sound and for sound diffusion. Further information on sound absorption and diffusion is provided in Chapter 3.2. Television studio floors are almost always left with a hard level surface finish to allow for smooth camera movements.

Critical audio listening and mixing rooms will require acoustic treatment with sound absorption materials and usually also require pre-fabricated sound diffusing elements on the walls and/or ceiling surfaces.

Floors in control room, master control, and technical operation areas are typically high-pressure laminate on computer access flooring systems. Although sometimes used in control rooms to assist with acoustic absorption, carpets should generally be avoided in these areas. Even with "static-resistant" carpet, there maybe some static discharge potential over the life of the material that can give rise to problems with equipment When planned correctly, these rooms should be located in close proximity so as to limit the amount of access flooring required and to minimize cable runs.

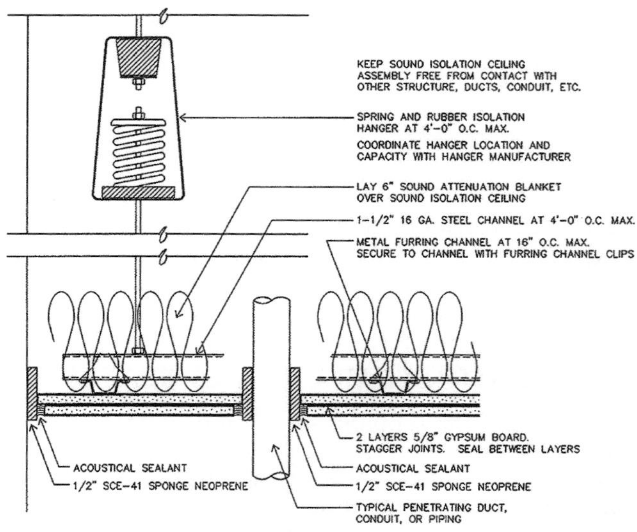

FIGURE 5.1-8 Sound isolation ceiling.

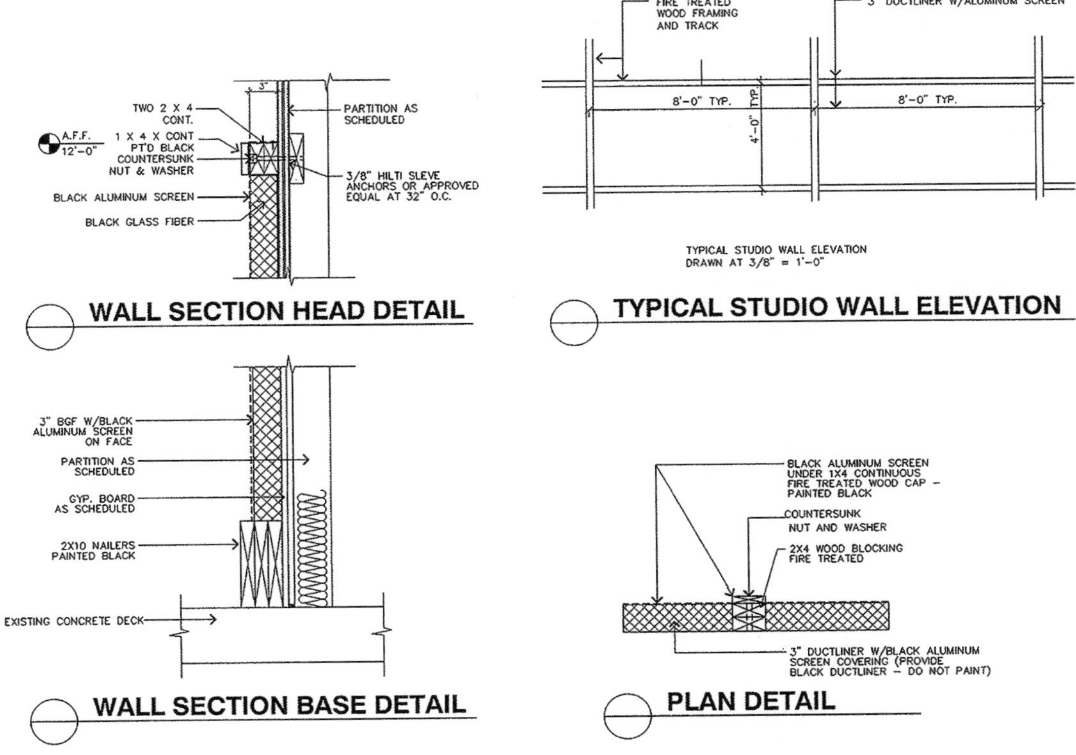

FIGURE 5.1-9 Wall acoustic treatment detail.

Office areas are treated very much in the manner of typical office space, with lay-in ceilings, paint wall finshes, and carpet or vinyl tile flooring. However, in sound-critical environments the use of hard reflective surfaces should be minimized. For example, there should not be a hard floor surface next to an edit bay or control room without allowing for proper acoustic isolation. The treatment would need to isolate the potential noise-generating surface from the interior of the room, to minimize any sound transference between spaces.

Mechanical Systems Noise Control

The noise and vibration control design for the building mechanical systems should anticipate and plan for the following features:

- Physical location of major equipment remote from sound sensitive spaces, on grade where possible;
- Selection of quiet type (low pressure) air distribution systems to serve the most sensitive spaces;
- Selection of quiet type equipment.
- Long lengths (30 to 60 ft) of duct run between air handling equipment and sensitive spaces, to allow dissipation of low-frequency noise;
- All ductwork serving critical spaces to have 1-inch or 2-inch internal acoustical lining;
- Ductwork to be sized and routed to minimize regenerated noise due to air turbulence;
- Ductwork to be designed according to guidelines for maximum air velocities, which in the most sensitive spaces will be as low as 300 or 600 feet per minute;
- All above-grade equipment to be vibration isolated from the building structure to minimize structure-borne noise and vibration. To the extent necessary, the pump and piping systems to be vibration isolated, and air handling units to incorporate flexible duct connections;
- Selection of quiet studio exhaust system equipment, with remote location to allow natural dissipation of noise in the ductwork.

Electrical Systems Noise Control

Power

Electrical power systems, including major substations and transformers, should be reviewed with respect to potential structure-borne noise excitation. Power transformers may require vibration isolation from the building structure.

Studio Lighting

Lighting systems should be reviewed with respect to the applicable background noise criteria for the studio and the noise characteristics of the fixture and control systems. Lamps and dimmer systems should be selected accordingly.

Unusual Ambient Noise Conditions

Sometimes a site is selected for reasons outside the control of the design team and they are left with site conditions that are less than desirable. An example would be a site that is directly in the flight path of an airport, or with an active railroad less than 30 feet away, or perhaps an active rock quarry not far away. Site conditions such as these clearly have major challenges for acoustic isolation and the project will require the expertise of an experienced acoustician. An acoustic consultant should be selected who is familiar with the requirements of a broadcast or production facility. The acoustician should be brought into the design process early and made familiar with the application of the facility and how it will be operated. Sometimes the smallest detail can lead to substantial cost savings.

SUMMARY

The intent of this chapter has been to give an overview of the design and planning process that should be undertaken once the decision is made to build a new video facility or renovate an existing one.

To ensure success, it is imperative that a team of experienced individuals be assembled in the early stages of the project to work through the details of programming, planning, designing, engineering, scheduling, and estimating for the new facility and its technical systems. The team should be composed of an architect, engineers, contractor(s), and system integrators that either have all worked together on a broadcast or production facility or at a minimum, has several members with experience in this type of facility.

ACKNOWLEDGMENT

Thanks are extended to Julian Rachman of DFW Consulting Engineers, Irving, TX, for his help and advice in preparing this chapter.

Bibliography

Gregg, Walter. Preparing Your Building for Digital Television, *Broadcast Engineering*, September 1988.

CHAPTER

5.2

Principles of Light, Vision, and Photometry

JERRY WHITAKER
Advanced Television Systems Committee
Washington, D.C.

INTRODUCTION

Television images originate as light coming from a scene being viewed by a camera. The television system ends with the presentation of those images on a display, which generates light that is picked up by the eye and perceived by the human brain as a representation of the original scene. The key to the whole process is the science of light and how the human visual system detects light and processes the signals that it receives. This chapter introduces some of the scientific principles of light, vision, and photometry that make television possible.

PRINCIPLES OF LIGHT

Sources of Illumination

Light reaches an observer directly from a light source or after being reflected from some object. The original source of such energy typically is radiation from molecules or atoms resulting from internal (atomic) changes. The exact type of emission is determined by:

- The manner in which the atoms or molecules are supplied with energy to replace that which they radiate
- The physical state of the substance, whether solid, liquid, or gaseous

The most common source of radiant energy is the thermal excitation of atoms in the solid or gaseous state, although light is also generated by other mechanisms.

The Spectrum

The electromagnetic spectrum comprises radiation of a wide range of wavelengths. As shown in Figure 5.2-1, light constitutes a small section in the range of electromagnetic radiation, extending in wavelength from about 400 to 700 nanometers (nm) or billionths (10^{-9}) of a meter. The wavelength of the light determines its color. When a beam of light traveling in air falls upon a glass surface at an angle, it is *refracted*, or bent. The amount of refraction depends on the wavelength, its variation with wavelength being known as *dispersion*. Similarly, when the beam, traveling in glass, emerges into air, it is refracted again (with dispersion). A glass prism provides a refracting system of this type. Because different wavelengths are refracted by different amounts, an incident white beam is split up into a number of beams corresponding to the many wavelengths contained in the composite white beam—thus is obtained the spectrum of many colors. If a spectrum is allowed to fall upon a narrow slit arranged parallel to the edge of the prism, a narrow band of wavelengths passes through the slit. Obviously, the narrower the slit, the narrower the band of wavelengths, or the sharper the spectral line. Also, more dispersion in the prism will cause a wider spectrum to be produced, and a narrower spectral line will be obtained for a given slit width. Note that purples are not included in the *spectral colors*. The purples belong to a special class of colors; they can be produced by mixing the light from two spectral lines, one in the red end of the spectrum, the other in the blue end. Purple (magenta is a more scientific name) is therefore referred to as a *nonspectral color*.

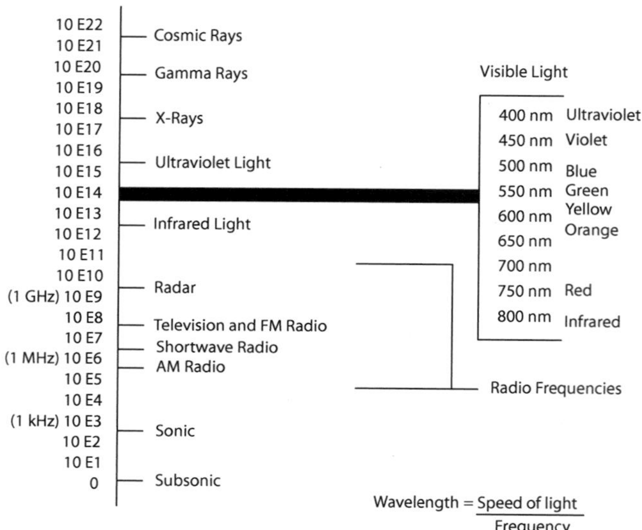

FIGURE 5.2-1 The electromagnetic spectrum.

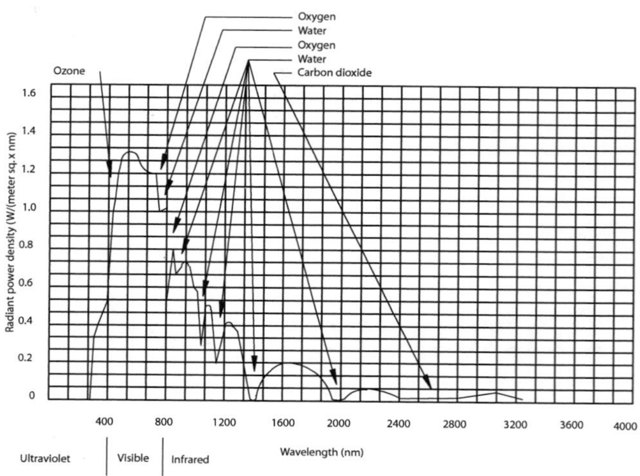

FIGURE 5.2-3 Spectral distribution of solar radiant power density at sea level, showing the ozone, oxygen, and carbon dioxide absorption bands. (Adapted from *IES Lighting Handbook*, Illuminating Engineering Society of North America, New York, 1981.)

Spectral Distribution

A plot of the power distribution of a source of light is indicative of the watts radiated at each wavelength per nanometer of wavelength. It is common to refer to such a graph as an *energy distribution curve*. Individual narrow bands of wavelengths of light are seen as strongly colored elements. Increasingly broader bandwidths retain the appearance of color but with decreasing purity, as if white light had been added to them. A very broad band extending generally throughout the visible spectrum is perceived as white light. Many white light sources are of this type, such as the familiar tungsten-filament electric light bulb for which the radiating characteristics are shown in Figure 5.2-2. Daylight also has a broad band of radiation, as illustrated in Figure 5.3-3. The energy distributions shown in Figures 5.2-2 and 5.2-3 are quite different and, if the corresponding sets of radiation were seen side by side, would be different in appearance. The light bulb would appear to have a yellow-orange tint, while the daylight would have a bluish tint. Either one, particularly if seen alone, however, would represent a very acceptable white. A sensation of white light can also be induced by light sources that do not have a uniform energy distribution. Among these is fluorescent lighting, which exhibits sharp peaks of energy through the visible spectrum. Similarly, the light from a monochrome (black-and-white) video cathode ray tube is not uniform within the visible spectrum, generally exhibiting peaks in the yellow and blue regions of the spectrum, as shown in Figure 5.2-4, yet it appears as an acceptable white.

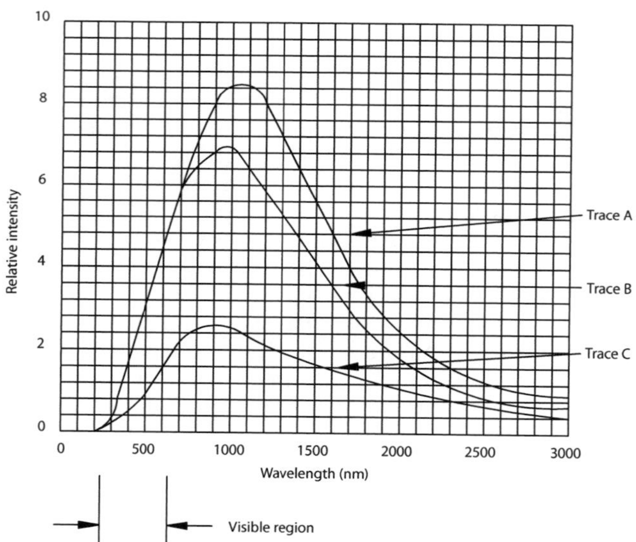

FIGURE 5.2-2 The radiating characteristics of tungsten: (Trace A) Radiant flux from 1 cm^2 of a blackbody at 3000 K. (Trace B) Radiant flux from 1 cm^2 of tungsten at 3000 K. (Trace C) Radiant flux from 2.27 cm^2 of tungsten at 3000 K (equal to curve A in the visible region). (Adapted from *IES Lighting Handbook*, Illuminating Engineering Society of North America, New York, 1981.)

Additive Color Mixing

It can easily be demonstrated using light sources of different colors that two or more colors can be added together to produce a mixture of light, which is perceived as a new color. Using the three colors of red, green, and blue, nearly all other colors can be reproduced. As shown in Figure 5.2-5 (when reproduced in color), mixing red and blue together produces magenta; red and green produce yellow; and blue and green produce cyan. Furthermore, when the correct

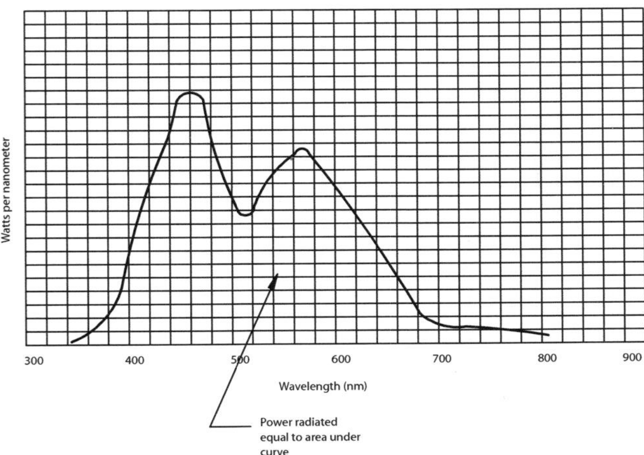

FIGURE 5.2-4 Power distribution of a monochrome video picture tube light source. (Adapted from Fink, D.G., *Television Engineering*, 2nd ed., McGraw-Hill, New York, 1952.)

proportions of red, green, and blue are mixed together, a white light is produced. The three colors red, green, and blue (RGB) are known as *primary* colors. The colors magenta, yellow, and cyan are known as *secondary colors*. One of the characteristics of primary colors is that two of them cannot be mixed to produce the third. Color television systems can only work because of this fundamental property of light and vision. It allows almost any color from a scene to be analyzed into values of red, green, and blue light by a camera pickup device and also enables those values to be added together on a display device to reproduce the original color.

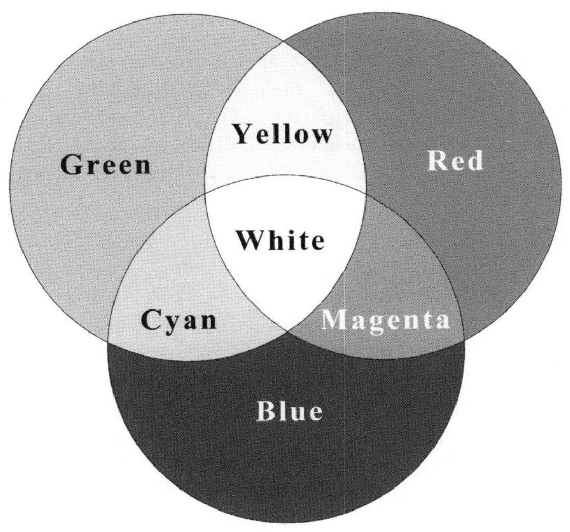

FIGURE 5.2-5 Additive color mixing.*

*Figures 5.2-5 through 5.2-8 are reproduced in color on the accompanying CD.

CIE COLOR SYSTEM

The International Commission on Illumination (CIE, the initials of its French name) is the international organization dealing with light, illumination, and color. In 1931, the CIE defined a set of color-matching functions and a coordinate system that have remained the predominate international standard method of specifying color since (see http://www.cie.co.at/).

RGB Color Matching

The color-matching functions of the initial CIE effort were based on experimental data from many observers measured by Wright [1] and Guild [2]. Observers had to produce a match for a series of test colors by mixing the appropriate amounts of red, green, and blue primary color light. Wright and Guild used different sets of primaries, but the results were transformed to a single set—namely, monochromatic stimuli of wavelengths 700.0, 546.1, and 435.8 nm [3]. The units of the stimuli were chosen so equal amounts were needed to match an *equienergy stimulus* (constant radiant power per unit wavelength throughout the visible spectrum). These tests resulted in a set of RGB color-matching functions, defining what is known as the "1931 CIE Standard Observer." One characteristic of the RGB functions was that they required some negative values in order to reproduce the full gamut of colors (that is, a primary color had to be added to change the test color before a match could be produced).

Figure 5.2-6 shows a chromaticity diagram based on these color-matching functions plotted on r and g axes with parameters that are calculated from R, G, and B. When reproduced in color, this diagram shows all visible colors plotted on the basis of the proportions of red, green, and blue required to match the color. It also shows the location of the monochromatic **R**, **G**, and **B** primaries and the spectrum locus. The straight line joining the extremities of the spectrum locus is called the *purple boundary*.

XYZ Color Space

At the same time it adopted these color-matching functions as a standard, the CIE also introduced and standardized another color space with a set of primaries that would encompass all the visible colors and allow the color-matching functions to be everywhere greater than or equal to zero. This involved some ingenious concepts. The set of real physical primaries was replaced by a new set of imaginary nonphysical primaries with special characteristics. These new primaries are referred to as **X**, **Y**, and **Z**, and the corresponding tristimulus values are *X*, *Y*, and Z. The chromaticity of the new **X**, **Y**, and **Z** primaries in the RGB system are shown in Figure 5.2-6.

Primaries **X** and **Z** have zero luminance, and all the luminance in a mixture of these three primaries is contributed by **Y**. This convenient property depends only on the decision to locate **X** and **Z** on the alychne,

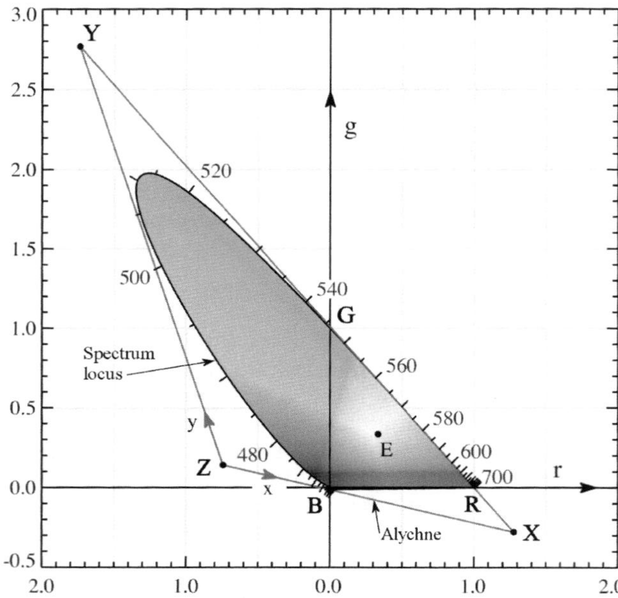

FIGURE 5.2-6 Spectrum locus and alychne of the CIE 1931 Standard Observer plotted in a chromaticity diagram based on matching stimuli of wavelengths 700.0, 546.1, and 435.8 nm. The locations of the CIE **R**, **G**, and **B** primaries and the imaginary primaries **X**, **Y**, and **Z** are shown

which is defined as the locus of colors with zero luminosity. It still leaves a wide choice of locations for all three primaries. The actual locations chosen by the CIE were based on the following additional considerations:

- The spectrum locus lies entirely within the triangle *XYZ*. This means that negative amounts of the primaries are never needed to match real colors. The color-matching functions are therefore all positive at all wavelengths.
- The line $Z = 0$ (the line from *X* to *Y*) lies along the straight portion of the spectrum locus. *Z* is effectively zero for spectral colors with wavelengths greater than about 560 nm.
- The line $X = 0$ (the line from *Y* to *Z*) was chosen to minimize (approximately) the area of the *XYZ* triangle outside the spectrum locus. This choice led to a bimodal shape for the color-matching function because the spectrum locus curves away from the line $X = 0$ at low wavelengths. A different choice of $X = 0$ (tangential to the spectrum locus at about 450 nm) would have eliminated the secondary lobe of *X* but would have pushed *Y* much further from the spectrum locus.
- The units of *X*, *Y*, and *Z* were chosen so that the tristimulus values *X*, *Y*, and *Z* would be equal to each other for an equienergy stimulus.

This coordinate system and the set of color-matching functions that go with it are known as the *CIE 1931 Standard Observer Colorimetric System*. The XYZ color-

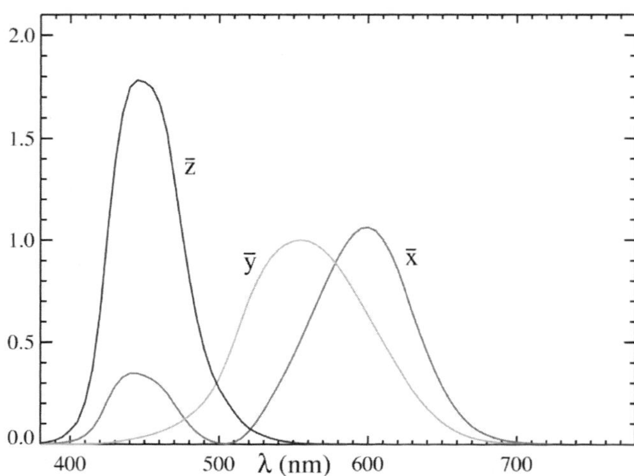

FIGURE 5.2-7 CIE 1931 XYZ color-matching functions

matching functions are shown in Figure 5.2-7, where relative tristimulus values are plotted against wavelength (λ).

CIE 1964 Supplementary Standard Observer

The color-matching data on which the 1931 Standard Observer is based were obtained with a visual field subtending 2° at the eye. Because of the slight nonuniformities of the retina, color-matching functions for larger fields are slightly different. In 1964, this prompted the CIE to recommend a second Standard Observer based on a visual field of 10°, known as the *CIE 1964 Supplementary Standard Observer*, for use in colorimetric calculations when the field size is greater than 4°.

Chromaticity Coordinates and Diagram

Chromaticity coordinates *x* and *y* may be calculated as functions of *X*, *Y*, and *Z* and a chromaticity diagram plotted as shown in Figure 5.2-8. This diagram (when reproduced in color) shows the color gamut and the **R**, **G**, and **B** primaries; it is commonly seen in the theory of color television.

Standard Illuminants

The CIE has recommended a number of standard illuminants, *E*(λ), for use in evaluating the tristimulus values of reflecting and transmitting objects. Originally, in 1931, it recommended three—known as A, B, and C. These illuminants are specified by tables of relative spectral distribution and were chosen so they could be reproduced by real physical sources. (CIE terminology distinguishes between *illuminants*, which are tables of numbers, and *sources*, which are physical emitters of light.) The sources are defined as follows:

- **Source A**—A tungsten filament lamp operating at a color temperature of about 2856 K. Its chromaticity

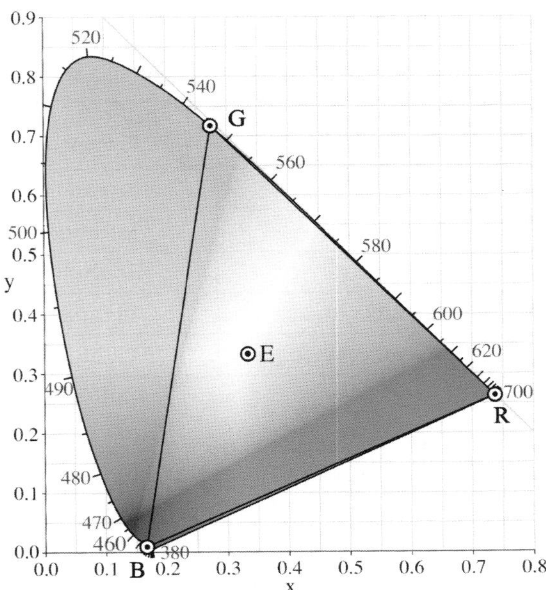

FIGURE 5.2-8 CIE 1931 Chromaticity diagram, showing RGB primaries.

coordinates are $x = 0.4476$ and $y = 0.4074$. Source A represents incandescent light.

- **Source B**—A source with a composite filter made of two liquid filters of specified chemical composition [4]. The chromaticity coordinates of source B are $x = 0.3484$ and $y = 0.3516$. Source B represents noon sunlight.
- **Source C**—This source is also produced by source A with two liquid filters [4]. Its chromaticity coordinates are $x = 0.3101$ and $y = 0.3162$. Source C represents average daylight according to information available in 1931.

In 1971, the CIE introduced a new series of standard illuminants that represented daylight more accurately than illuminants B and C [5]. The improvement is particularly marked in the ultraviolet part of the spectrum, which is important for fluorescent samples. The most important of the D illuminants is D_{65} (sometimes written D6500), which has chromaticity coordinates of $x = 0.3127$ and $y = 0.3290$. The relative spectral power distributions of illuminants A, B, C, and D_{65} are given in Figure 5.2-9.

Gamut of Reproducible Colors

In a system that seeks to match or reproduce colors with a set of three primaries, only those colors can be reproduced that lie inside the triangle of primaries. Colors outside the triangle cannot be reproduced because they would require negative amounts of one or two of the primaries. In a color-reproducing system, it is important to have a triangle of primaries that is sufficiently large to permit a satisfactory gamut of colors to be reproduced. To illustrate the kinds of requirements that must be met, Figure 5.2-10 shows the maximum color gamut for real surface colors and the triangle of typical color television receiver phosphors as standardized by the European Broadcasting Union (EBU). These are shown in the CIE 1976 u',v' chromaticity diagram in which the perceptual spacing

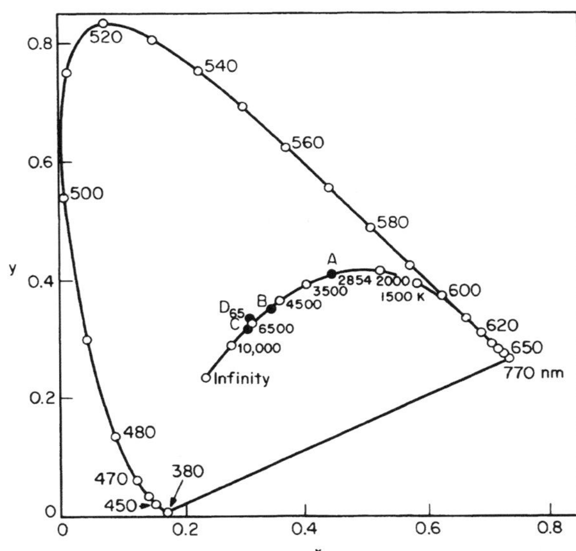

FIGURE 5.2-9 The relative spectral power distributions of CIE standard illuminants A, B, C, and D_{65}. (Adapted from Robertson, A.R. et al., in *Standard Handbook of Video and Television Engineering*, 4th ed., Whitaker, J.C., Ed., McGraw-Hill, New York, 2003.)

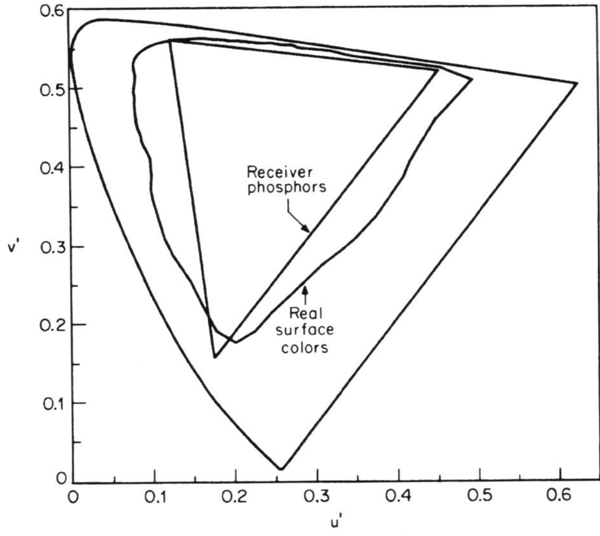

FIGURE 5.2-10 The color triangle defined by a standard test of color television receiver phosphors compared with the maximum real color gamut on a u',v' chromaticity diagram. (Adapted from Pointer, M.R., *Color Res. Appl.*, 5, 145–155, 1980.)

of colors is more uniform than in the x,y diagram. High-purity blue-green and purple colors cannot be reproduced by these phosphors, whereas the blue phosphor is actually of slightly higher purity than any real surface colors.

THE HUMAN VISUAL SYSTEM

The human visual system is powerful and exceeds the performance of artificial visual systems in almost all areas of comparison. Vision results from stimulation of the eye by light and consequent interaction through connecting nerves with the brain. Although the range of human vision is small compared with the total energy spectrum, human discrimination—the ability to detect differences in intensity or quality—is excellent. Under ideal conditions, the human visual system can detect:

- Wavelength differences of 1 millimicron (10 Å; 1 Angstrom unit = 10^{-10} cm)
- Intensity differences of as little as 1%
- Forms subtending an angle at the eye of 1 arc minute, and often smaller objects

Sensation and Properties of Light and Vision

The perceived (psychological) color vision sensation associated with a light stimulus can be described in terms of three characteristics:

- Hue
- Saturation
- Brightness

The spectrum contains most of the principal hues: red, orange, yellow, green, blue, and violet. Additional hues are obtained from mixtures of red and blue light—these constitute the purple colors. *Saturation* pertains to the strength of the hue. Spectrum colors are highly saturated. White and grays have no hue and, therefore, have zero saturation. Pastel colors are of low or intermediate saturation. *Brightness* pertains to the intensity of the stimulation. If a stimulus has high intensity, regardless of its hue, it is said to be bright.

The psychophysical analogs of hue, saturation, and brightness are:

- Dominant wavelength
- Excitation purity
- Luminance

These equivalent properties are listed in Table 5.2-1.

By means of definitions and standard response functions, which have received international acceptance through the CIE, the dominant wavelength, purity, and luminance of any stimulus of known spectral energy distribution may be determined by simple computations. Although roughly analogous to their psychophysical counterparts, the psychological attributes of hue, saturation, and brightness pertain to observer responses to light stimuli and are not subject

TABLE 5.2-1
Psychophysical and Psychological Characteristics of Color

Psychophysical Properties	Psychological Properties
Dominant wavelength	Hue
Excitation purity	Saturation
Luminance	Brightness
Luminous transmittance	Lightness
Luminous reflectance	Lightness

to calculation. These sensation characteristics as applied to any given stimulus depend in part on other visual stimuli in the field of view and upon the immediately preceding stimulations.

Color sensations arise directly from the action of light on the eye. They are normally associated, however, with objects in the field of view from which the light comes. The objects themselves are therefore said to have color. *Object colors* may be described in terms of their hues and saturations, as is the case for light stimuli. The intensity aspect is usually referred to in terms of lightness, rather than brightness. The psychophysical analogs of lightness are *luminous reflectance* for reflecting objects and *luminous transmittance* for transmitting objects.

Scotopic and Photopic Vision

Light entering the eye passes through a lens that focuses the image onto the *retina*, which has two types of light-sensitive elements known as *rods* and *cones*. Cones are able to respond to different colors of light, whereas rods respond only to luminance levels. At low levels of illumination, objects may differ from one another in their lightness appearances but give rise to no sensation of hue or saturation. All objects then appear to be of different shades of gray. Vision at low levels of illumination is called *scotopic vision*, as distinct from *photopic vision*, which takes place at higher levels of illumination. Table 5.2-2 on the next page compares the luminosity values for photopic and scotopic vision. Only the rods of the retina are involved in scotopic vision; the cones play no part. As the fovea centralis (at the center of the retina) is free of rods, scotopic vision takes place outside the fovea. Visual acuity of scotopic vision is low compared with photopic vision. At high levels of illumination, where cone vision predominates, all vision is color vision and visual acuity is high.

Trichromatic Color Vision

Color vision processing in the human visual system starts with the absorption of light by the light-sensitive cones. Based on research carried out during the 1970s and 1980s, three different classes of cones have been identified, each containing a different type of photosensitive pigment. These are short-wavelength sensitive (S-cones), middle-wavelength sensitive (M-cones), and long-wavelength sensitive (L-cones), all having different but overlapping spectral sensitivities. The spectral sensitivity of S-cones peaks at approximately 440 nm, M-cones peak at 545 nm, and L-cones peak at 565 nm, although different researchers have found slightly different results. Interactions between at least two types of cone are necessary to produce the ability to perceive color. In this way, the brain can compare the signals from each type and determine both the intensity and color of the light. Based on this *trichromacy* of color vision, it is apparent that many different physical stimuli, or mixtures of light, can evoke the same sensation of color. All that is required for two stimuli to be equivalent is that they should each cause the same number of light photons to be absorbed by any given class of cone. In that case, the message to the brain, and the thus the color sensations generated, will be the same even though the stimuli are physically different. This characteristic of the human visual system is fundamental to the science of colorimetry, and without it color television as we know it could not exit.

MEASUREMENTS AND PERFORMANCE

This section discusses some further aspects of vision and the visual system that can be quantified and measured.

Photometric Measurements

Photometry is the measurement of the properties of light. Evaluation of a radiant energy stimulus in terms of its brightness-producing capacity is a photometric measurement. An instrument for making such measurements is called a *photometer*. In visual photometers, used in obtaining basic photometric measurements, the two stimuli to be compared are normally directed into small adjacent parts of a viewing field. The stimulus to be evaluated is presented in the *test field*; the stimulus against which it is compared is presented in the *comparison field*. For most high-precision measurements, the total size of the combined test and comparison fields is kept small, subtending about 2° at the eye. The area outside these fields is called the *surround*. Although the surround does not enter directly into the measurements, it has adaptation effects on the retina and thus affects the appearances of the test and comparison fields. It also influences the precision of measurement.

Luminosity Curves

A *luminosity curve* is a plot indicative of the relative brightness of spectrum colors of different wavelength or frequency. To a normal observer, the brightest part of a spectrum consisting of equal amounts of radiant flux per unit wavelength interval is at about 555 nm. Luminosity curves are therefore commonly normalized to have a value of *unity* at 555 nm. If, at some other wavelength, twice as much radiant flux as at 555 nm is required to obtain brightness equality with radiant flux at 555 nm, the luminosity at this wavelength is 0.5. The luminosity at any wavelength λ is, therefore, defined as the ratio P_{555}/P_{λ}, where P_{λ} denotes the amount of radiant flux at the wavelength λ, which is equal in brightness to a radiant flux of P_{555}. The luminosity function that has been accepted as standard for photopic vision is given in Figure 5.2-11. Tabulated values at 10 nm intervals are given in Table 5.2-2. This function was agreed upon by the CIE in 1924. It is based upon considerable experimental work that was conducted over a number of years. Chief reliance in arriving at this function was based on the step-by-step equality-of-brightness method. Flicker photometry provided additional data.

TABLE 5.2-2
Relative Luminosity Values for Photopic and Scotopic Vision

Wavelength (nm)	Photopic Vision	Scotopic Vision
390	0.00012	0.0022
400	0.0004	0.0093
410	0.0012	0.0348
420	0.0040	0.0966
430	0.0116	0.1998
440	0.023	0.3281
450	0.038	0.4550
460	0.060	0.5670
470	0.091	0.6760
480	0.139	0.7930
490	0.208	0.9040
500	0.323	0.9820
510	0.503	0.9970
520	0.710	0.9350
530	0.862	0.8110
540	0.954	0.6500
550	0.995	0.4810
560	0.995	0.3288
570	0.952	0.2076
580	0.870	0.1212
590	0.757	0.0655
600	0.631	0.0332
610	0.503	0.0159
620	0.381	0.0074

TABLE 5.2-2 *(continued)*
Relative Luminosity Values for Photopic and Scotopic Vision

Wavelength (nm)	Photopic Vision	Scotopic Vision
630	0.265	0.0033
640	0.175	0.0015
650	0.107	0.0007
660	0.061	0.0003
670	0.032	0.0001
680	0.017	0.0001
690	0.0082	—
700	0.0041	—
710	0.0021	—
720	0.00105	—
730	0.00052	—
740	0.00025	—
750	0.00012	—
760	0.00006	—

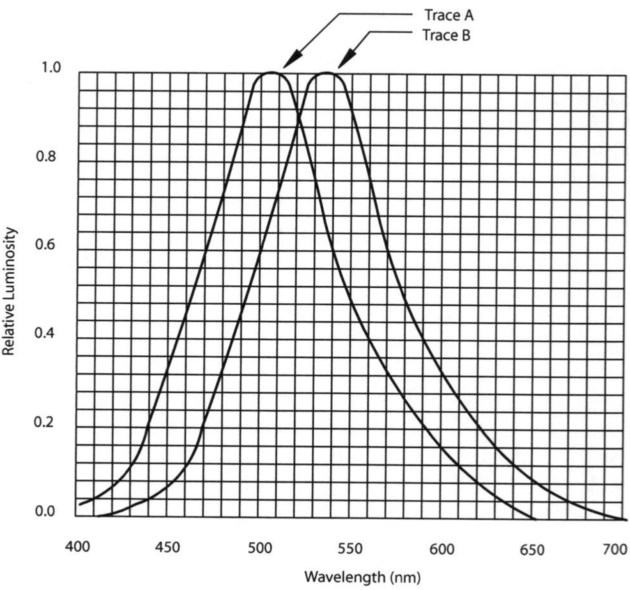

FIGURE 5.2-12 Scotopic luminosity function (Trace A) as compared with photopic luminosity function (Trace B). (Adapted from Fink, D.G., *Television Engineering*, 2nd ed., McGraw-Hill, New York, 1952.)

In the scotopic range of intensities, the luminosity function is somewhat different from that of the photopic range. The two curves are compared in Figure 5.2-12 and the relative values are listed in Table 5.2-2. The two curves are similar in shape, but there is a shift for the scotopic curve of about 40 nm to the shorter wavelengths.

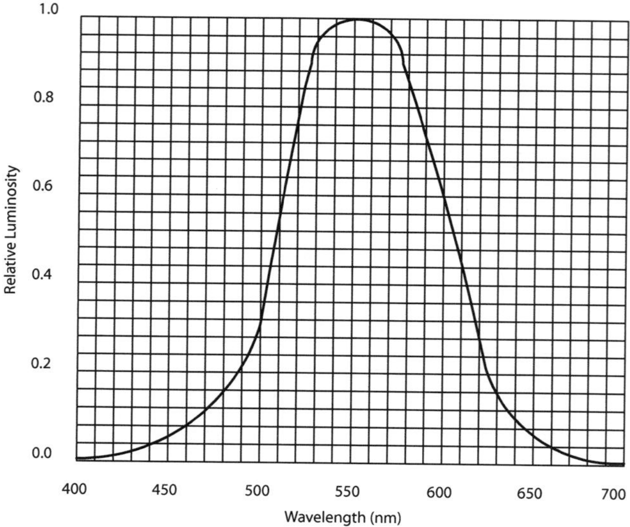

FIGURE 5.2-11 The photopic luminosity function. (Adapted from Fink, D.G., *Television Engineering*, 2nd ed., McGraw-Hill, New York, 1952.)

Measurement of Luminosity

Measurements of luminosity in the scotopic range are usually made by the *threshold-of-vision* method. A single stimulus in a dark surround is used. The stimulus is presented to the observer at each of a number of different intensities, ranging from well below the threshold to intensities sufficiently high to be definitely visible. Determinations are made of the amount of energy, at each chosen wavelength, that is reported visible by the observer a certain percentage of the time, such as 50%. The reciprocal of this amount of energy determines the relative luminosity at the given wavelength. The wavelength plot is normalized to have a maximum value of 1.00 to give the scotopic luminosity function. In the intensity region between scotopic and photopic vision, called the *Purkinje* or *mesopic region*, the measured luminosity function takes on sets of values intermediate between those obtained for scotopic and photopic vision. Relative luminosities of colors within the mesopic region will therefore vary, depending on the particular intensity level at which the viewing takes place. Reds tend to become darker in approaching scotopic levels; greens and blues tend to become relatively lighter.

Luminance

Brightness is a term used to describe one of the characteristics of appearance of a source of radiant flux or of an object from which radiant flux is being reflected or transmitted. Brightness specifications of two or more sources of radiant flux should be indicative of their actual relative appearances. These appearances will

depend in large part upon the viewing conditions, including the state of adaptation of the observer's eye. Luminance, as indicated previously, is a psychophysical analog of brightness. It is subject to physical determination, independent of particular viewing and adaptation conditions. Because it is an analog of brightness, however, it is defined in such a way as to relate as closely as possible to brightness.

The luminosity function is the best-established measure of the relative brightness of different spectral stimuli. In evaluating the luminance of a source of radiant flux consisting of many wavelengths of light, the amounts of radiant flux at the different wavelengths are weighted by the luminosity function, which converts radiant flux to luminous flux. As used in photometry, the term *luminance* is applied only to extended sources of light, not to point sources. For a given amount (and quality) of radiant flux reaching the eye, brightness will vary inversely with the effective area of the source.

Luminance is described in terms of luminous flux per unit projected area of the source. The greater the concentration of flux in the angle of view of a source, the brighter it appears. Luminance is therefore expressed in terms of amounts of flux per unit solid angle or *steradian*. In considering the relative luminance of various objects of a scene to be captured and reproduced by a video system, it is convenient to normalize the luminance values so the *white* in the region of principal illumination has a relative luminance value of 1.00. The relative luminance of any other object then becomes the ratio of its luminance to that of the white. This white is an object of highly diffusing surface with high and uniform reflectance throughout the visible spectrum. For purposes of computation, it may be idealized to have 100% reflectance and perfect diffusion.

Perception of Fine Detail

Detail is seen in an image because of brightness differences between small adjacent areas in a monochrome display or because of brightness, hue, or saturation differences in a color display. Visibility of detail in a picture is important because it determines the extent to which small or distant objects of a scene are visible and because of its relationship to the *sharpness* appearance of the edges of objects. *Picture definition* is probably the most acceptable term for describing the general characteristic of crispness, sharpness, or image-detail visibility in a picture. Picture definition depends on characteristics of the eye, such as visual acuity, and upon a variety of characteristics of the picture-image medium, including its resolving power, luminance range, contrast, and image edge gradients.

Visual Acuity Measurements

Visual acuity may be measured in terms of the visual angle subtended by the smallest detail in an object that is visible. The *Landolt ring* is one type of test object frequently employed. The ring, which has a segment cut from it, is shown in any one of four orientations, with the opening at the top or bottom or on the right or left side. The observer identifies the location of this opening. The visual angle subtended by the opening that can be properly located 50% of the time is a measure of visual acuity. Test object illuminance, contrast between the test object and its background, time of viewing, and other factors greatly affect visual acuity measurements. Up to a visual distance of about 20 ft, acuity is partially a function of distance, because of changes in shape of the eye lens in focusing. Beyond 20 ft, it remains relatively constant. Visual acuity is highest for foveal vision, on the center of the retina, dropping off rapidly for retinal areas outside the fovea. Normal vision, corresponding to a Snellen 20/20 rating, represents an angular discrimination of about 1 min. Separations between adjacent cones in the fovea and resolving power limitations of the eye lens give theoretical visual acuity values of about this same magnitude.

Resolution

The extent to which a picture medium, such as a photographic or a video system, can reproduce fine detail is expressed in terms of *resolving power* or *resolution*. Resolution is a measure of the distance between two fine lines in the reproduced image that are visually distinct. The image is examined under the best possible conditions of viewing, including magnification. Two types of test charts are commonly employed in determining resolving power: either a wedge of radial lines or groups of parallel lines at different pitches for each group. For either type of chart, the spaces between pairs of lines usually are made equal to the line widths. Figure 5.2-13 shows a test signal electronically generated by a video measuring test set.

Resolution in photography is usually expressed as the maximum number of lines (counting only the black ones or only the white ones) per millimeter that

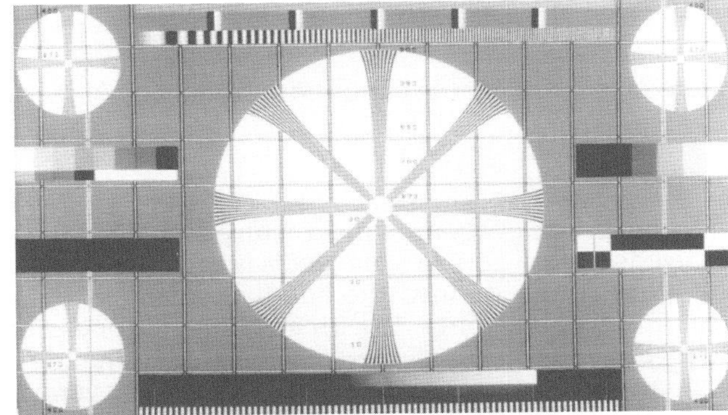

FIGURE 5.2-13 Test chart for high definition television applications produced by a signal waveform generator. The electronically produced pattern is used to check resolution, geometry, bandwidth, and color reproduction. (Courtesy of Tektronix.)

can be distinguished from one another. Measured values of resolving power depend upon a number of factors in addition to the photographic material itself. The most important of these typically are:

- Density differences between the black and the white lines of the test chart photographed
- Sharpness of focus of the test-chart image during exposure
- Contrast to which the photographic image is developed
- Composition of the developer

Sharpness of focus depends upon the general quality of the focusing lens, image, and object distances from the lens, as well as the part of the projected field in which the image lies. In determining the resolving power of a photographic negative or positive material, a test chart is generally employed that has a high density difference, such as 3.0, between the black-and-white lines. A high-quality lens is used, the projected field is limited, and focusing is critically adjusted. Under these conditions, ordinary black-and-white photographic materials generally have resolving powers in the range of 30 to 200 line pairs/mm. Special photographic materials are available with resolving powers greater than 1000 line pairs/mm.

Resolution in a video system is expressed in terms of the maximum number of lines, counting both black and white, that are discernible in viewing a test chart. The value of horizontal (vertical lines) or vertical (horizontal lines) resolution is the number of lines equal to the vertical dimension of the raster. Vertical resolution in a well-adjusted system equals the number of scanning lines, roughly 500 in conventional television. In normal broadcasting and reception practice, however, typical values of vertical resolution range from 350 to 400 lines. The theoretical limiting value for horizontal resolution (R_H) in a 525-line, 30 Hz frame-rate system is given by:

$$R_H = \frac{2 \times 0.75 \times \Delta f}{30 \times 525}$$

$$R_H = 0.954 \times 10^{-4} \times \Delta f$$

where Δf is the available bandwidth frequency (in Hz).

Sharpness

The appearance evaluation of a picture image in terms of the edge characteristics of objects is called *sharpness*. The more clearly defined the line that separates dark areas from lighter ones, the greater the sharpness of the picture. Sharpness is, therefore, related to the transient curve in the image across an edge. The average gradient and the total density difference appear to be the most important characteristics. No physical measure has been devised, however, that in all cases will predict the sharpness (appearance) of an image. Picture resolution and sharpness are to some extent interrelated, but they are by no means perfectly correlated. Pictures ranked according to resolution measures may be rated somewhat differently on the basis of sharpness. Both resolution and sharpness are related to the more general characteristic of picture definition. For pictures in which, under particular viewing conditions, effective resolution is limited by the visual acuity of the eye rather than by picture resolution, sharpness is probably a good indication of picture definition. If visual acuity is not the limiting factor, however, picture definition depends to an appreciable extent on both resolution and sharpness.

A Model for Image Quality

Researchers have studied the human visual system extensively to ascertain the most efficient and effective methods of communicating information to the eye. An important component of this work has been the development of models of how humans see, in an effort to improve image processing systems. The classic approach to image quality assessment involves the presentation to a group of test subjects visual test material for evaluation and rating. The test material may include side-by-side display comparisons or a variety of perception threshold presentations. One common visual comparison technique is the *pair-comparison* method. A number of observers are asked to view a specified number of images at two or more distances. At each distance, the subjects are asked to rank the order of the images in terms of overall quality, clearness, and personal preference. An image acquisition, storage, transmission, and display system need not present more visual information to the viewer than the viewer can process. For this reason, image quality assessment is an important element in the development of any new video system. Evaluation by human subjects, while an important part of this process, is also expensive and time consuming. Numerous efforts have been made to reduce the human visual system and its interaction with a display device to one or more mathematical models. (Some of this research is cited in References 6 through 10.) After the system or algorithm has successfully passed the minimum criteria established by the model, it can be subjected to human evaluation. The model simulation requires the selection of many interrelated parameters. A series of experiments is typically conducted to improve the model to more closely approximate human visual perception.

VIDEO SYSTEM REQUIREMENTS

The objective in any type of visual reproduction system is to present to the viewer a combination of visual stimuli that can be readily interpreted as representing, or having close association with, a real viewing situation. In order to achieve that, it is by no means necessary that the light stimuli from the original scene be duplicated precisely. There are certain characteristics in the reproduced image, however, that are necessary and others that are highly desirable. Only a general

qualitative discussion of such characteristics as they relate to video systems will be given here.

Resolution

In monochrome video, images of objects are distinguished from one another and from their backgrounds as a result of luminance differences. In order that detail in the picture is visible and that objects have clear, sharp edges, it is necessary that the video system be capable of rapid transitions from areas of one luminance level to another. This degree of resolution need not necessarily match that possible in the eye itself, but too low an effective resolution results in pictures with a fuzzy appearance and lacking fineness of detail. Matching the resolving capabilities of the human eye is a commendable engineering goal, but a difficult proposition in any practical imaging system. The move from standard-definition imaging (in particular, the constraints imposed by encoding systems such as NTSC, PAL, and SECAM) to high-definition imaging has led to enormous improvement in overall picture quality. This improvement has stretched from the camera through the entire transmission system to the display device. This trend will no doubt continue as imaging technologies continue to improve, particularly cameras and displays. Historically, improvements in one area tend to push improvements in another.

Dynamic Range

Luminance range and the transfer characteristic associated with luminance reproduction are likewise of importance in monochrome video. Objects seen as white usually have minimum reflectances of approximately 80%. Black objects have reflectances of approximately 4%. This gives a luminance ratio of 20:1 in the range from white to black. To obtain the total luminance range in a scene, the reflectance range must be multiplied by the illumination range. In outdoor scenes, the illumination ratio between full sunlight and shadow may be as high as 100:1 [11]. The full luminance ranges involved with objects in such scenes could not be reproduced in conventional video reproduction equipment. Modern imaging systems, with advanced cameras and displays, have made great strides in approximating this luminance range.

Monochrome video transmits only luminance information, and the relative luminance of the images should correspond at least roughly to the relative luminance of the original objects. Red objects, for example, should not be reproduced markedly darker than objects of other hues but of the same luminance. Exact luminance reproduction, however, is by no means a necessity. Considerable distortion as a function of hue is acceptable in many applications. Luminance reproduction is probably of primary consequence only if detail in some hues becomes lost.

Depending on the camera pickup element or the film, the dominant wavelength and purity of the light may be of consequence. Most films and video pickup elements exhibit sensitivity throughout the visible spectrum and, consequently, marked distortions in luminance as a function of dominant wavelength and purity are not encountered. Their spectral sensitivities seldom conform exactly to that of the human observer, however, so some brightness distortions do exist.

Scanning Frequency

Images in a video system are transmitted one point, or small area, at a time. The complete picture image is repeatedly scanned at frequent intervals. If the frequency of scan is not sufficiently high, the picture appears to flicker. At frequencies above a *critical frequency,* no flicker is apparent. The critical frequency changes as a function of luminance, being higher for higher luminance. The basic requirement for video is that the *field frequency* (the rate at which images are presented) be above the critical frequency for the highest image luminance.

Color Reproduction

Images of objects in color video are distinguished from one another by luminance differences or by differences in hue or saturation. A number of factors may contribute to color rendition in a practical system, categorized as follows:

- *Exact color reproduction,* where the reproduction is a metameric match to the original. Exact color reproduction will result in equality of appearance only if the viewing conditions for the picture and the original scene are identical. These conditions include the angular subtense of the picture, the luminance and chromaticity of the surround, and glare. In practice, exact color reproduction often cannot be achieved because of limitations on the maximum luminance that can be produced on a color monitor.
- *Colorimetric color reproduction*, a variant of exact color reproduction in which the tristimulus values are proportional to those in the original scene. In other words, the chromaticity coordinates are reproduced exactly, but the luminances all are reduced by a constant factor. Traditionally, color video systems have been designed and evaluated for colorimetric color reproduction. If the original and the reproduced reference whites have the same chromaticity, if the viewing conditions are the same, and if the system has an overall gamma of unity, then colorimetric color reproduction is indeed a useful criterion.
- *Corresponding color reproduction*, a compromise by which colors in the reproduction have the same appearance that colors in the original would have had if they had been illuminated to produce the same average luminance level and the same reference white chromaticity as that of the reproduction. For most purposes, corresponding color reproduction is a suitable objective of a color video system.
- *Preferred color reproduction*, a departure from the preceding categories that recognizes the preferences of

the viewer. It is sometimes argued that corresponding color reproduction is not the ultimate aim for some display systems, such as video games or even television programs, and that it should be taken into account that people prefer some colors to be different from their actual appearance. For example, suntanned skin color is preferred to average real skin color, and sky is preferred bluer and foliage greener than they really are.

Even if corresponding color reproduction is accepted as the target, some colors are more important than others. For example, flesh tones must be acceptable—not obviously reddish, greenish, purplish, or otherwise incorrectly rendered. Likewise, the sky must be blue and the clouds white, within the viewer's range of acceptance. Similar conditions apply to other well-known colors of common experience. The imaging system improvements brought into practical usage by high-definition video have given program producers considerable flexibility in using the video system for highly accurate picture capture and display, or for using the system creatively to convey a mood or effect.

SUMMARY

This chapter has introduced many of the basic principles of light and vision upon which color television systems rely. Other chapters in this book, particularly Chapter 5.3, "Television Camera Systems," build on this foundation. For more in-depth coverage of the subject, other sources of information are listed in the further information, bibliography, and references sections that follow.

ACKNOWLEDGMENTS

Portions of this chapter were adapted from Whitaker, J.C., *Video Display Engineering*, McGraw-Hill, New York, 2001. Used with permission.

DEFINING TERMS

- *Brightness:* A term used to describe one of the characteristics of appearance of a source of radiant flux or of an object from which radiant flux is being reflected or transmitted.
- *Critical frequency:* The rate of picture presentation, as in a video system or motion picture display, above which the presented image ceases to give the appearance of flickering. The critical frequency changes as a function of luminance, being higher for higher luminance.
- *Dispersion:* The variation of refraction as a function of wavelength.
- *Energy distribution curve:* A plot of the power distribution of a source of light giving the energy radiated at each wavelength per nanometer of wavelength.
- *Field frequency:* The rate at which images in an electronic imaging system are presented. In conventional analog television, two fields are presented each second to make up one *frame* (a complete picture).
- *Refraction:* The bending of light as it passes from one medium to another, such as from air to glass, at an angle.
- *Resolution:* The extent to which an imaging system, such as a photographic or a video system, can reproduce fine detail.
- *Sharpness:* The appearance evaluation of a picture image in terms of the edge characteristics of objects contained therein. The more clearly defined the lines separating dark areas from light ones, the greater the sharpness of the picture.

Further Information

The International Society for Optical Engineering (SPIE; Bellingham, WA, see http://www.spie.org/) offers a number of publications examining the characteristics of the human visual system. The organization also conducts technical seminars on various topics relating to optics and the application of optical technologies.

Bibliography

Boynton, R.M., *Human Color Vision*, Holt, New York, 1979.

Committee on Colorimetry, *The Science of Color*, Optical Society of America, New York, 1953.

Davson, H., *Physiology of the Eye*, 4th ed., Academic Press, New York, 1980.

Evans, R.M., Hanson, Jr., W.T., and Brewer, W.L., *Principles of Color Photography*, Wiley, New York, 1953.

Kingslake, R., Ed., *Applied Optics and Optical Engineering*, Vol. 1, Academic Press, New York, 1965.

Polysak, S.L., *The Retina*, University of Chicago Press, Chicago, IL, 1941.

Richards, C.J., *Electronic Displays and Data Systems: Constructional Practice*, McGraw-Hill, New York, 1973.

Schade, O.H., Electro-optical characteristics of television systems, *RCA Rev.*, 9, 5–37, 245–286, 490–530, 653–686, 1948.

Whitaker, J.C., *Video Display Engineering*, McGraw-Hill, New York, 2001.

Whitaker, J.C., and Benson, K.B., Eds., *Standard Handbook of Video and Television Engineering*, 4th ed., McGraw-Hill, New York, 2003.

Wright, W.D., *Researches on Normal and Defective Colour Vision*, Mosby, St. Louis, 1947.

Wright, W.D., *The Measurement of Colour*, 4th ed., Adam Hilger, London, 1969.

References

[1] Wright, W.D., A redetermination of the trichromatic coefficients of the spectral colours, *Trans. Opt. Soc.*, 30, 141–164, 1928/1929.

[2] Guild, J., The colorimetric properties of the spectrum, *Phil. Trans. Roy. Soc. A*, 230, 149–187, 1931.

[3] Robertson, A.R., Fisher, J.F., and Whitaker, J.C., The CIE color system, in *Standard Handbook of Video and Television Engineering*, 4th ed., Whitaker, J.C., Ed., McGraw-Hill, New York, 2003.

[4] Judd, D.B., and Wyszencki, G., *Color in Business, Science, and Industry*, 3rd ed., Wiley, New York, 1975, pp. 44–45.

[5] CIE, *Colorimetry*, Publ. No. 15, CIE, Paris, 1971.
[6] Grogan, T.A., Image evaluation with a contour-based perceptual model, in *Human Vision, Visual Processing, and Digital Display III*, Rogowitz, B.E., Ed., SPIE, 1992, pp. 188–197.
[7] Barten, Peter G.J., Physical model for the contrast sensitivity of the human eye, in *Human Vision, Visual Processing, and Digital Display III*, Rogowitz, B.E., Ed., SPIE, 1992, pp. 57–72.
[8] Daly, S., The visible differences predictor: an algorithm for the assessment of image fidelity, in *Human Vision, Visual Processing, and Digital Display III*, Rogowitz, B.E., Ed., SPIE, 1992, pp. 2–15.
[9] Reese, G., Enhancing images with intensity-dependent spread functions, *Human Vision, Visual Processing, and Digital Display III*, Rogowitz, B.E., Ed., SPIE, 1992, pp. 253–261.
[10] Martin, R.A., Ahumanda, Jr., A.J., and Larimer, J.O., "Color Matrix Display Simulation Based Upon Luminance and Chromatic Contrast Sensitivity of Early Vision," *Human Vision, Visual Processing, and Digital Display III*, Bernice E. Rogowitz ed., Proc. SPIE 1666, pp. 336-342, 1992.
[11] Brewer, X., Lyle, W., Morris, R.A., and Fink, D.G., Light and the visual mechanism, in *Standard Handbook of Video and Television Engineering*, 4th ed., Whitaker, J.C., Ed., McGraw-Hill, New York, 2003, pp. 1–13.
[12] IES, *IES Lighting Handbook*, Illuminating Engineering Society of North America, New York, 1981.
[13] Fink, D.G., *Television Engineering*, 2nd ed., McGraw-Hill, New York, 1952.
[14] Pointer, M.R., The gamut of real surface colours, *Color Res. Appl.*, 5, 145–155, 1980.

CHAPTER

5.3

Television Camera Systems

MICHAEL BERGERON AND STEPHEN MAHRER

Panasonic Broadcast and Television Systems Company
Secaucus, New Jersey

INTRODUCTION

This chapter discusses the inner workings and applications of modern broadcast cameras. Because digital standard definition (SD) and high definition (HD) technologies have increasingly become standard equipment in most production and broadcast facilities, descriptions of National Television System Committee (NTSC), tube-based, or otherwise analog systems are provided only as reference points in describing the newer digital, solid-state systems and standards. Many components of complete video systems are described in general terms so as to encompass the wider range of possibilities in today's camera systems. The intention is to bring the working broadcast engineer up to speed on the current technologies and to enable informed decisions when deploying these technologies. The chapter refers to 60 Hz-based television standards as used in the United States, but the principles and details of camera technology apply equally to 50 Hz-based systems used elsewhere.

At the time of this writing, the NTSC analog standard is still used for terrestrial broadcasting by most stations in the United States, in parallel with HD and SD digital television (DTV) broadcasting. With or without digital transmission, there are advantages with digital production, and many broadcasters began to implement the International Telecommunication Union, Radiocommunication Sector (ITU-R) BT 601 digital video standard [1] (referred to as "ITU 601" in this chapter) in the studio domain several years before the introduction of DTV transmissions. Others are in the process of doing so now. Cameras and downstream studio equipment were changed over incrementally to take advantage of many potential performance gains as it is a simple matter to take an SD digital program and convert it to an NTSC signal, either at the studio or just before the signal reaches the transmitter.

Similarly, it is straightforward to down-convert HD signals to SD (and NTSC if needed). The HD transition, therefore, appears to be following the same pattern as the move to SD systems, with cameras leading the way. At this time, few broadcasters or production companies are even considering standard definition cameras for purchase, simply because most of them will likely not have budget allocations for camera purchases again for several years, and an HD camera can produce either SD or HD programming as needed.

DIGITAL VIDEO STANDARDS

Before continuing a discussion on HD and SD digital video, we should first review what the standards are. The chart in Figure 5.3-1, based on Table A3 of Advanced Television Systems Committee (ATSC) A/53 [2], outlines the currently defined ATSC digital television formats for emission. The bottom two rows represent SD formats with 480 × 640 being closest in resolution to the quality of analog NTSC video, although this resolution actually corresponds to the computer VGA standard and not a video production format. The 480 × 704-line digital video formats are defined for production by the ITU 601 and Society of Motion Picture and Television Engineers (SMPTE) 125M [3] and SMPTE 293M [4] standards (where they actually have 720 pixels per line). ITU 601 defines

Scanning Lines	Horizontal Pixels	Aspect Ratio Pixel Shape	Picture Rate	Horizontal Frequency
1080 Active (1125 Total)	1920 Active	16:9 Square	60i, 30p, 24p	33.75 kHz (60i)
720 Active (750 Total)	1280 Active	16:9 Square	60p, 30p, 24P	45 kHz (60p)
480 Active (525 Total)	704 Active	16:9, 4:3	60i, 60p, 30p, 24p	15.734 kHz 30p/60i 31.5 kHz (60p)
480 Active (525 Total)	640 Active	4:3 Square	60i, 60p, 30p, 24p	15.75 kHz (60i)

FIGURE 5.3-1 ATSC-defined video emission formats.

spatial and temporal characteristics, colorimetry and transfer characteristics, and the sampling parameters for the images, while the SMPTE standards further define other aspects of particular formats. Transfer characteristics define the *gamma correction* applied to compensate for the nonlinear characteristics of traditional cathode ray tube display tubes (see later section in this chapter on gamma, under Signal Processing). The serial digital interface (SDI) for these formats is defined in SMPTE 259M [5].

The top two rows of the chart represent high definition formats. The 1080-line digital video formats are defined for production by the SMPTE 274M standard [6], and the 720-line formats are defined by the SMPTE 296M standard [7]. These standards define the spatial and temporal characteristics and the sampling parameters for the images. The colorimetry and transfer characteristics for the HD formats are defined by the ITU-R BT 709 specification [8] (referred to as "ITU 709" in this chapter), which also defines the spatial characteristics for international program exchange. The high definition serial digital interface (HD-SDI) for these formats is defined in SMPTE 292M [9].

To be clear, the above formats and standards describe digital video signals and are not specifications for imagers or signal processors. The purpose of the standards is to ensure that camera systems and other video equipment can work together; many cameras, however, tend to follow the standards internally from the beginning of the chain to simplify signal processing.

Regarding the frame rates of the above formats, although the chart shows 24, 30, and 60 frames per second (fps) and 60 fields per second, all the formats have equivalent versions with these frame and field rates divided by 1.001. These alternative rates are needed for compatibility with NTSC standard video (for which the frame rate had to be adjusted by that factor when color was introduced many years ago). This use of non-integer frame rates will be necessary as long as NTSC and DTV material coexists in the same plant and conversions between the NTSC and DTV video standards are required. For the alternative rates, 30 fps becomes 30/1.001 fps, which is usually referred to as 29.97 fps, and 60 fields per second becomes 59.94 fields per second. Similarly, 24 fps becomes 23.98 fps.

Broadcast Video Formats

Some common HD formats that current cameras are capable of shooting, with typical applications, include:

- 1080i59.94 (thirty 1920 × 1080 pixel frames every 1.001 seconds, transmitted as sixty 1920 × 540 pixel fields)—Used for production and HD broadcast; capable of displaying on most HD monitors and recording on most HD tape formats.
- 720p59.94 (sixty 1280 × 720 pixel frames transmitted every 1.001 seconds)—Used for production and HD broadcast; capable of displaying on many monitors and recording on many tape formats.
- 1080p23.98 (twenty-four 1920 × 1080 frames transmitted every 1.001 seconds)—This is primarily a production and mastering format. Many monitoring or recording systems require that it be changed to *progressive segmented frame* mode 24 PsF (twenty-four 1920 × 1080 pixel frames every 1.001 seconds, transmitted as forty-eight 1920 × 540 pixel fields).
- 720p23.98 (sometimes referred to as 720 24pN; twenty-four 1280 × 720 pixel frames delivered every 1.001 seconds)—Used in IT-based production and post-production to conserve bandwidth and storage space and take advantage of the variable frame rate capabilities of some cameras; not compatible with most HD monitors or tape-based recording systems unless inserted into a 720p59.94 signal (see below).

Some common quasi-formats used for production and broadcast include:

- 24p over 480i59.94[1]—This is another way of expressing the 2–3 pulldown arrangement used since the introduction of the telecine to transfer film material to video (see Chapter 5.23, "Film for Television"). It is how film-originated material is viewable on broadcast television. Some camcorders can now shoot in this format to simulate a "film look." To produce this quasi-format, the 24p video frames are converted to segmented frames (48 fields) compatible with fields in an interlaced video stream. By repeating fields from alternate frames (producing the 2–3 pulldown cadence), the 48 fields become 60 fields transmitted every 1.001 seconds, thus producing 59.94 fields per second interlaced video. Most IT-based editing systems have the capability to extract the original 24p frames so they can be cut on the correct frame boundaries.
- 24p over 1080i59.94[1]—This is the HD version of the above. Material from an HD telecine or 1080 24p native video must be converted to this format if it is to be broadcast in a 1080i system. 1080p23.98

[1]24p Advanced (24pA), for SD and HD formats, alternates 2-3 and 3-2 field cadences so that redundant fields are adjacent. This allows downstream systems to extract the original 24p frames without decoding compressed video.

imaging cameras can generate a signal that is 1080p23.98, 1080PsF23.98, or 24p over 1080i59.94.

- 24p over 720p59.94—This works similarly to the above two quasi-formats; however, because 720p HD does not have fields, the frames need not be segmented; they only need to be repeated following the 2–3 sequence. The original 24 frames can be extracted from this format to reduce the bandwidth, as the repeated frames are redundant; however, this changes the video back to 720p23.98, which is not compatible with some recording and monitoring systems.

Notwithstanding the above, the non-integer frame rates and associated formats are very often referred to in the literature by their equivalent integer numbers. For example, progressive video at 59.94 fps is often referred to as "60p" and interlaced video at 59.94 fields per second as "60i." Similarly, progressive video recorded or transmitted at 23.98 fps is very often loosely referred to as "24p" (as in the quasi-formats above). This needs to be kept in mind when discussing formats, as the precise meaning of the number may depend on the context.

Non-Broadcast Video Image Formats

Just as film has always been a part of the video production chain, several non-broadcast video format cameras are becoming significant to video broadcasters. Because digital cameras used for these formats are often the responsibility of the video engineering department, it is now necessary that broadcast video engineers become familiar with these alternatives. Dealing with 24p video is part of this but, as cameras continue to evolve with more formats, the engineering department will tend to be drawn into the creative process. Here are some other production formats that are beginning to gain acceptance.

- 1080p23.98 4:4:4—This is similar to 1080p23.98 already used for many film or HD masters; however, instead of consisting of a luminance signal (Y) and two color difference signals (Pb, Pr), with the color difference signals subsampled horizontally in a 4:2:2 ratio, 4:4:4 video does not subsample the color and usually retains the original RGB (red, green, blue) components rather than matrixing them into standard component video. The result is a higher quality master that is more easily integrated into computer-generated graphics and is better equipped to generate a clean chroma key against a blue/green screen. Because video cameras start in RGB space, this signal is easily generated from an HD video camera by bypassing some of the processing. Due to its increased bandwidth, 1080p23.98 4:4:4 is not compatible with standard HD-SDI, but a dual link system has been developed to handle the 4:4:4 HD interconnections.
- 2K and 4K—These formats were developed for the *digital intermediate* process and are designed to match the image capabilities of 35 mm motion picture film as scanned for digital manipulation. Developed by Kodak, the *Cineon* (www.cineon.com/) file format specifies a family of pixel densities, bit depths, and a log-based transfer function, as well as colorimetry parameters. Due to their high data-rates, these systems were not initially real time and were never considered useful for video-type imaging; however, a 23.98 frame rate 2048 × 1080 crop of a 2K frame provides a 4:4:4 moving image similar to HD and compatible with 2K, which is capable of being recorded and played back in real time. As imagers achieve the necessary pixel densities, it is becoming possible to digitally create Cineon-compatible images directly.

COLOR VIDEO

Color Perception

The human visual system is capable of not only interpreting varying intensities of light as an image but also analyzing the frequency characteristics of the light within the range of what is known as the visible spectrum. This spectral analysis is perceived as color. Laser or other coherent light sources produce a single wavelength; however, most naturally occurring radiated or reflected light sources emit a spectral response that includes components of many wavelengths (colors). The human vision system interprets the spectral makeup by analyzing it into three specific color components. The retina contains light receptors known as cones that have sensitivities centered on red–orange (580 nm) yellow–green (545 nm), and blue (440 nm).

A study conducted in 1931 by the International Commission on Illumination (CIE) analyzed and plotted the entire range of human color perception. One result of this work is the CIE chromaticity diagram shown in Figure 5.3-2. The range of all possible colors is referred to as the *gamut*. For a *tristimulus* (three-color) system, the gamut of colors that can be produced by any particular system is defined by the triangle whose vertices are the location of the three *primaries* on the CIE chart. Color video systems define color using red, green, and blue primaries of slightly different wavelengths depending on the particular standard being used. Like the human eye, color video cameras are tristimulus systems and they must separate the three primaries in order to capture a color image.

Color Space

For any video standard, the precise color (or placement on the CIE chart) of the primaries must be part of the specification. This is part of what we refer to as *color space*. The RGB primaries defined by the CIE set red at 700.00 nm, green at 546.1 nm, and blue at 435.8 nm. Because the human eye adds and subtracts colors in its system, pure red, green, and blue primaries are not suitable for a video system, and primaries must be chosen that can be mathematically combined more simply. These primaries are what define the gamut of

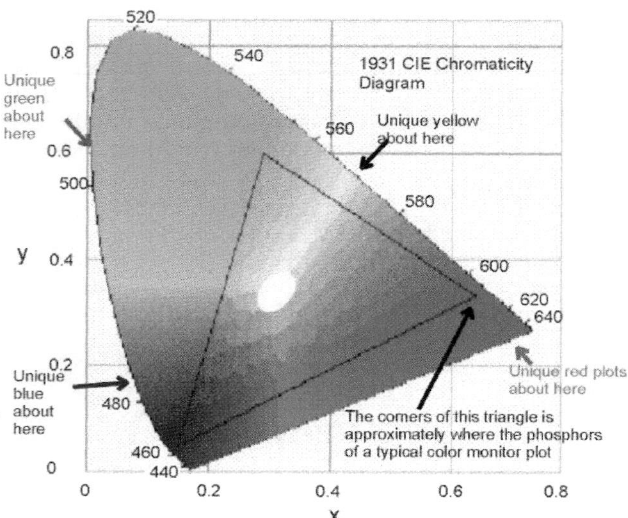

FIGURE 5.3-2 CIE chromaticity diagram.

each video standard, and the primaries defined for NTSC, ITU 601, and ITU 709 are all slightly different.

Red, green, and blue primaries are not the only three components that can be used to define a color image. Like any three-dimensional coordinate system, it is possible to transform RGB into an infinite number of three-variable bases, and these can be derived mathematically by using a 3 × 3 matrix to transform the components as follows.

$$[\mathrm{Y, Pb, Pr}] = \begin{vmatrix} a & b & c \\ d & e & f \\ h & i & j \end{vmatrix} \times [\mathrm{G, B, R}]$$

This matrix equation is a representation of the three equations that determine the values of Y, Pb, and Pr:

$$\mathrm{Y} = a\mathrm{G} + d\mathrm{B} + g\mathrm{R}$$

$$\mathrm{Pb} = b\mathrm{G} + e\mathrm{B} + h\mathrm{R}$$

$$\mathrm{Pr} = c\mathrm{G} + f\mathrm{B} + i\mathrm{R}$$

where Y is the *luminance* and Pb and Pr are the *color difference* signals.

These transforms form the basis of component video as it is usually managed in a broadcast environment. The actual values of the matrix terms/equation coefficients (*a*, *b*, *c*, *d*, *e*, *f*, *g*, *h*, and *i*) are specific to each video standard, and these parameters (along with the values of the RGB primaries) define the color space. It should be noted that the color space is likely to change when video is changed from one format to another, such as when it is downconverted from HD to SD.

Component Video

Cameras deal with color image acquisition in RGB; most monitors display images in RGB, and color imaging on most information technology (IT) platforms, including film post production, works in RGB space, so why do component digital video systems transform RGB to YPbPr? The answer is part legacy "baggage" and part smart engineering.

The NTSC color video standard was created with the mandate that earlier black and white equipment had to be compatible, so a transform was created such that one of the three primaries would be the black-and-white video signal. This component was, and still is, often (imprecisely) referred to as *luminance* after the CIE-defined term denoted as "Y." Two other *color difference* parameter signals were defined to carry the color information. These produced the "U" and "V" signals that were processed further, bandwidth reduced, and encoded on the color subcarrier, where the color information was referred to as *chroma* or *chrominance.* Those terms are not used in component video systems. The NTSC luminance signal by itself will provide a good monochrome image to a black and white monitor. Color difference signals are needed with the luminance in order to produce a color picture.

Because the luminance parameter used in video signals is *gamma corrected,* its value is not identical to the well-defined CIE term luminance, and it is correctly referred to as *luma*, denoted with the term "Y′." In this chapter we will use the precise term *luma*. As shown in the matrix equations above, luminance can be derived from R, G, and B. These equations relate to luma when the signals are gamma corrected.

It is the luma and color difference signals that are used in component video. As mentioned, the definition of luma is specific to the video standard. Looking at the ITU 601 specification, it can be seen that luma is defined as:

$$\mathrm{Y}' = 0.299\mathrm{R}' + 0.587\mathrm{G}' + 0.114\mathrm{B}'$$

and the color difference signals are defined as:

$$\mathrm{R}' - \mathrm{Y}' = 0.701\mathrm{R}' - 0.587\mathrm{G}' - 0.114\mathrm{B}'$$

$$\mathrm{B}' - \mathrm{Y}' = -0.299\mathrm{R}' - 0.587\mathrm{G}' + 0.886\mathrm{B}'$$

The terms used for analog color difference channels are Pb and Pr, which are directly derived from the R–Y and B–Y signals (indicating that they are essentially the remaining red and blue channels after the luma has been removed). Strictly speaking, digital color difference signals should be referred to as Cb and Cr. These conventions are not always followed, so if a schematic indicates Pb and Pr this does not necessarily mean that the signal is not digital; it may just be careless labeling on the part of the designer or writer.

Color Subsampling

The human visual system perceives significantly more detail in luminance (luminance is correct, because we are talking about human vision) than it does in color information. In fact, half of the color resolution in a video picture can be dropped without a perceptible difference in sharpness, as long as luma resolution is maintained. Based on this information, the horizontal sampling rates for digital video were chosen for digital video standards with a two-to-one ratio of luma to

color information. This is commonly referred to as color subsampling and is denoted by 4:2:2. The arrangement used in most component digital interface systems is 4:2:2. Other varieties of color subsampling, such as 4:2:0 and 4:1:1, are possible.

Y′ Cb Cr signals are subsampled as part of the digitization process, including those carried on SDI and HD-SDI interfaces. While the human eye may not notice the change brought about by color subsampling, it affects the robustness of the signal if it is processed downstream. To reduce the effects of color subsampling, image processing should wherever possible be carried out before the signal is transformed to Y′ Cb Cr component digital, which is one reason why the 4:4:4 format is almost exclusively used with RGB components.

In a video camera system design, it is up to the engineers to decide where in the signal processing chain to transform the RGB video signal into component Y′ Cb Cr, a process known as *matrixing*. Some signal processing takes advantage of the nature of luma and therefore can only be done after matrixing. Contouring and detail enhancement, for example, are applied only to the luma signal, as they seek to enhance the perceived sharpness, which is almost entirely a function of the luma signal.

VIDEO CAMERA SYSTEMS OVERVIEW

Turning Photons into Electrons

The human visual system is capable of capturing and interpreting an image focused on the image plane of the retina. Video cameras capture images similarly. Both must use some kind of lens to focus light bouncing from an object on to a light-sensitive surface. The light pattern, or image, must then be captured and interpreted. To electronically capture an image, it is necessary to represent varying light levels with analogous voltage levels. The voltage levels can then create a representation of the image focused on the image plane. Image capture is, in essence, an analog process, regardless of the technology involved. Early video cameras used an electron beam to scan lines on the imaging surface of a vacuum tube, which allowed detection of varying light levels on that surface. The resulting voltage variations were amplified and processed directly into an analog video signal.

Imagers

Modern cameras now use solid-state *imagers*, which are manufactured similarly to other semiconductor devices and as such are referred to casually as *chips*. Imaging chips comprise an array of *photoreceptors* or *photosites*, each converting photons into an electrical charge. When the charges from the individual receptors or *charge wells* are extracted sequentially from the imager, this creates a variable voltage, much like the variable voltage generated by the scanning in a tube camera. Although the solid-state imager architecture is based on individual pixel sensors, they are still analog devices. In most cases, the imager produces an analog signal, which must then be sampled. When the voltage values of the samples are converted to binary code words, the signal can be considered digital video.

Traditionally, in most solid-state imagers, the number of photoreceptors corresponded directly to the number of pixels required to produce a given video image format, so the photoreceptors on a solid-state imager are also referred to as pixels. When an imager's pixels match the final video format pixels one to one and are scanned similarly, this is referred to as *native capture*. Because of the need for cameras with flexible output formats, as well as the necessity to optimize the relationship between pixel size and pixel density, many current cameras do not employ native capture. Having a greater density of photoreceptors than output pixels allows *oversampling* of the image with potential quality improvements. Having fewer pixels than is required for native resolution is often a necessary compromise for smaller imagers that would otherwise lack sufficient sensitivity due to smaller pixel sizes.

System Architecture

A video camera system needs to accomplish a specific task, which is to capture an image and deliver the image data in a form compatible with downstream systems. Regardless of the technologies involved or the specific applications for the camera, this task is accomplished by the following components:

- *Optics*—Including lenses, filters, and prism block;
- *Imager*—Charge-coupled device (CCD) or complementary metal-oxide semiconductor (CMOS); 3-chip or single-chip; converts light images to video signals;
- *Processing*—Analog processing and digital processing;
- *Control circuitry*—Access to processing to adjust camera parameters and images;
- *Interfaces*—Necessary for reference, input, output, and control signals and essential for multiple cameras to work together;
- *Output or recording*—Camera must output a standard format.

Camera Components

The interrelation between these components is determined by the camera application. Components are combined in some systems and split in others to allows some components to be moved to a dedicated operation area.

A camera in a studio application usually has the camera optics and lens with its *zoom* and *focus* controls and the imagers and electronics in a *camera head* run by the camera operator, while control electronics, most interfaces, and often a good deal of processing, are in a *camera control unit* (CCU), which is installed in racks in an equipment room so it can be easily interconnected with other systems. *Lens iris* and camera *black level* and

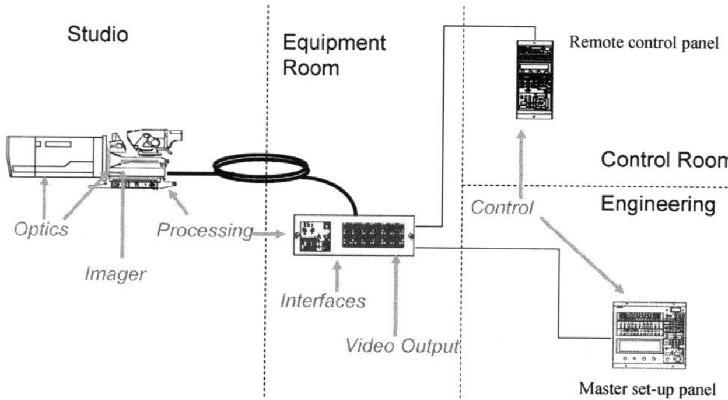

FIGURE 5.3-3 Studio camera system.

color adjustment (*paint*) controls are sent to a remote panel so an engineer or *shader* in a control room can watch exposure and other settings and match cameras. Fine image adjustments might be done in yet another area with another remote control unit.

Some point-of-view (POV) cameras have even more circuitry away from the imager to reduce the size of the camera head, while an electronic news gathering (ENG) camcorder will pack everything but the lens into one housing. Cameras without interchangeable lenses are now becoming available at the professional level, and these offer a true one-piece system.

Figure 5.3-3 shows a diagram of a typical studio camera system configuration and Figure 5.3-4 shows a typical ENG camera with a built-in recorder and also a radio link for a wireless microphone.

Camera and Imager Sizes

Historically, imager size has also been determined by the application. This is still a factor, but the lines have been blurred thanks to the ability of new solid-state devices to deliver good quality with smaller imagers. Full-size studio cameras are often referred to as *hard cameras*. They have fewer size and weight constraints than portable cameras, so camera designers would in the past often include larger imagers and more advanced processing in the camera. Currently available studio hard cameras, however, all use the same 2/3-inch 3-CCD image blocks found in ENG or electronic field production (EFP) cameras. The size of a hard camera is now driven by the need to include studio features and the ability to mount large box-type lenses, rather than the size of the imager, as the lens mount on the smaller cameras cannot support these heavy lenses.

Other than the potential for more sophisticated lenses, image quality available from even quite modest camcorders has largely caught up to that produced by the largest and most expensive studio cameras. Although the camera sizes still vary, nearly every broadcast application is currently employing 2/3-inch imagers. Until recently, imagers smaller than the 2/3-inch format were relegated to industrial applications; however, three CCD imagers of 1/2 inch and even 1/3 inch have begun to find a place in some broadcast applications. Most of these cameras are chosen based on the significant cost savings and an expectation that the image quality will be good enough for the particular application.

Achieving acceptable performance with small-format optics is particularly difficult, and many small-format cameras mitigate this by integrating fixed optics into the design. This dictates a fixed lens rather than the traditional professional design that allows for alternative lenses. That is not to say that the lens itself is fixed; zoom lenses are virtually universal for modern television cameras (although not for electronic cinema, where cameras that mount 35 mm film camera prime lenses are also available). Most producers have approached smaller imager cameras with the attitude that this allows the production to use more cameras for the same price and provide more content, albeit of inferior quality. The limitations of small-format imagers will be discussed in the next section.

Larger than 2/3-inch imager cameras have begun to make a comeback. These have primarily been introduced into production where the potential for better performance is combined with, and enhanced, by the desire to shoot with cameras more similar in operation to 35 mm film cameras.

FIGURE 5.3-4 ENG camera system.

BACKGROUND TO PIXEL-BASED FOCAL PLANE IMAGING ARRAYS

In 1985, the world's first solid-state imager-based camera made its debut in the world of television broadcasting. That early camera with its prototypical comparatively low-resolution CCD imagers, although primitive by today's standards, paved the way for a sea change in the methodology of television image capture. The benefits of solid-state imagers compared to tube-based cameras quickly became apparent to all. They were simpler and more robust, and they had no tube (target) burn, no highlight comet tail problems, and no requirements for tube registration and other complex alignment before use. Today, tube-based cameras seem archaic and quaint, a distant relic of a bygone age. Indeed, many of today's

broadcast engineers may have never seen one, let alone had to use or maintain one.

The advent of tubeless cameras was not, however, without a struggle. Device physics, analog signal processing, and technology limitations all conspired against the CCD pioneers; however, the relentless march of technology and the advent of digital computing power made it possible to overcome the problems. Today, CCD and CMOS video cameras are ubiquitous. They are used in broadcast and production cameras, consumer camcorders, cell phones, personal digital assistants (PDAs), point-and-shoot digital cameras, spy satellites, robotic vision, webcams, laptop computers, security cameras … the list is almost endless. They range from marginal to ultra-exotic in performance and from almost free to hugely expensive in cost. At the time of this writing, the state of the art for a CCD device is an amazing 100 million pixel array (*i.e.*, 10,000 × 10,000 pixels)! Such devices are obviously for ultra-specialized applications (e.g., satellite imaging), but even $100 digital point-and-shoot cameras now boast imagers of 5 million pixels or more. How those imagers are designed, manufactured, and integrated into a device sets the performance of the entire system.

CCDs were invented in the late 1960s; their initial application was to provide an analog delay function for computer applications. Their ability to move an electric charge in a controlled fashion soon led to other applications, such as audio "bucket-brigade" delay lines and as video imaging/storage devices. Bell Labs produced a prototypical CCD-based television camera in 1970. Device technology rapidly improved to the point that by 1975 a more advanced CCD camera was considered almost broadcast quality.

CCD Image Development

Figure 5.3-5 shows some examples of CCDs from the past 20 years of camera design. They range in size from an early 2/3-inch RCA Frame Transfer CCD to a 1/4-inch Interline Transfer device suitable for ultra compact point-of-view and closed-circuit television cameras.

It should be noted that the nominal size of a solid state imaging device is based on the diameter of an equivalent legacy camera pickup tube. The actual diagonal size of the active imaging area is approximately 16 mm for a 1-inch device, 11 mm for a 2/3-inch device, 8 mm for a 1/2-inch device, 6 mm for a 1/3-inch device, and 4.5 mm for a 1/4-inch device.

Early CCDs suffered from extremely low manufacturing yields, often less than 0.01% for broadcast-worthy devices, and were thus very expensive. As semiconductor production technology improved, it became possible to build CCDs with improved yields, or smaller CCDs for less demanding applications (e.g., the consumer and business/industrial markets). Those markets are considerably more tolerant of defects and appreciate the user benefits of less expensive, more reliable, and easier to operate cameras that the new technologies bring. The enormous consumer

FIGURE 5.3-5 A collection of CCDs from 1983 to 2006. The top left device is a 1/4-inch CCD from an inexpensive "lipstick" color camera; the lower right is a 2/3-inch high-definition 16:9 progressive HD imager.

camcorder market sector would not have been possible without solid-state imagers.

Current Imagers

Today's solid-state imagers are based on two technologies: CCDs and CMOS. Both are intended to fulfill the same basic function, namely to capture light on a pixel-based focal plane array and then to convert that light into a video signal that can be processed for later image display. Both device families have their own unique strengths and weaknesses, and the user should choose the best device for the intended application. As semiconductor physics and manufacturing processes for both device families improve, newer devices with improved resolutions or other specific parameters may become viable. Imagers are like most things in life in that there is no "one size fits all" solution.

CHARGE-COUPLED DEVICES

CCDs are a family of similar imaging devices:

- Frame transfer (FT) CCDs
- Frame interline transfer (FIT) CCDs
- Interline transfer (IT) CCDs

The FT, FIT, and IT imagers are basically analog devices. They all capture photons and accumulate an *electric charge* proportional to the incident light falling on the CCD's pixel-based *photosites.* Those charge packets are then moved in a conveyer-belt type of process, one packet at a time, to the CCD's output stage. Here the minuscule electric charge from each pixel is converted to a voltage proportional to the charge; this voltage is the output signal of the device. The CCD structure is fabricated by means of diffusion and doping of the silicon and by the addition of metal and polysilicon conductive bus wiring to provide the defined pixels areas with their interconnect and drive requirements. Although the basic imaging array works in a similar general fashion, the three device types transfer and store the imaged signal charge in different ways, which are now explained.

Frame Transfer CCD

The first CCDs were frame transfer (FT) devices. FT CCDs for broadcast applications comprise two basic sections. In an imaging array, light is converted by means of the *photoelectric effect* to an electrical charge. Light hits the silicon of the imager, and, in efficient devices, nearly every photon generates an *electron–hole* pair. The electron is captured under each pixel in a tiny electrically generated temporary storage area called a *charge well*. The CCD's photosensitive pixel array is normally exposed to light for the majority of the television field or frame period (1/60 sec), usually ~16 msec. The electrons present in each well represent the cumulative charge generated by the light falling on that pixel during that 16 msec period. CCDs are extremely linear devices; thus, more light equals more charge and thus more signal output.

In FT CCDs, this signal charge is rapidly moved from the imaging area to a similarly structured storage area during the TV signal's vertical blanking time. The only real difference between the imaging and storage areas is that the storage area is covered by an opaque mask of chrome or other such metal. This is necessary to avoid unwanted light from generating extraneous charge and corrupting the signal collected from the imaging array. Figure 5.3-6 details the basic structure of a frame transfer CCD. It is not necessary for the video engineer to be an expert on such devices; however, an appreciation of the differences may prove helpful in determining their best use in differing applications.

The white shaded imaging array (IA) consists of contiguous pixels; each pixel accumulates charge during the active integration period. During the vertical blanking interval, that charge is clocked at high speed to the chrome-masked storage array (SA). It is interesting to note that the FT imager uses the actual photosensitive pixels as vertical transfer columns. This is efficient and simplifies manufacturing but requires that the camera use a physical shutter to block light from impacting the CCD during this transfer period. If an optical shutter is not used, or if the shutter is inoperative or mistimed, transfer streaking will occur. This is a visible effect caused by extraneous light polluting the stored charge during its transfer to the masked storage area. The effect resembles a bright vertical highlight smearing artifact in the image. The use of a mechanical shutter is fairly simple but may have undesired and objectionable visual artifacts when viewing alternating current discharge lighting in a television scene, on

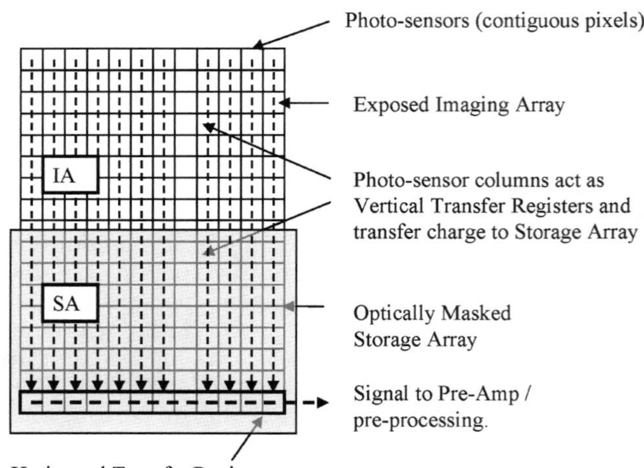

FIGURE 5.3-6 The typical frame transfer CCD is composed of a light-sensitive imaging array and an optically masked storage area. Charge is accumulated in the imaging array and then quickly transferred to the storage array during the vertical blanking interval. The stored charge is then clocked out of the storage array in real time by means of the vertical and horizontal transfer registers for subsequent video processing.

electronic scoreboards, or any devices that use scanned display arrays. Such devices may exhibit a strobing effect, even on still shots.

In the FT CCD, the stored charge is clocked out of the storage area by the horizontal shift register in a *real-time* line-by-line fashion; the charge for each pixel is then fed to the output stage of the CCD. This important part of the CCD is a tiny sample-and-hold circuit that converts each pixel's tiny electrical charge to a voltage proportional to that charge. It should be noted that the electric charge per pixel is extremely small, usually ranging from about 5 to 10 electrons for black (no light) to possibly 20,000 electrons for peak white. As can be imagined, such tiny charges and such sensitive circuitry require very specialized circuitry, which is critical to the performance of the CCD.

Techniques such as *correlated double sampling* are used to minimize various noise sources; for example, the N-MOS source followers in the CCD output stages have a significant flicker noise component. The processing at this stage is analog and determines the noise floor of the device; extreme care should be taken to provide the cleanest signal for subsequent video processing.

An example of a frame transfer CCD can be seen in Figure 5.3-7. Although an early device, it clearly shows the different imaging and storage areas of the imager. It is interesting to note that this apparently 16:9 widescreen imager is actually comprised of the two 4 × 3 imaging and storage sections sitting side by side. The lighter gray area is the chrome-covered storage area and the darker area is the imaging array.

FIGURE 5.3-7 An early RCA 535 × 480 pixel frame transfer CCD (circa 1984). Note the clearly demarked imaging and storage areas.

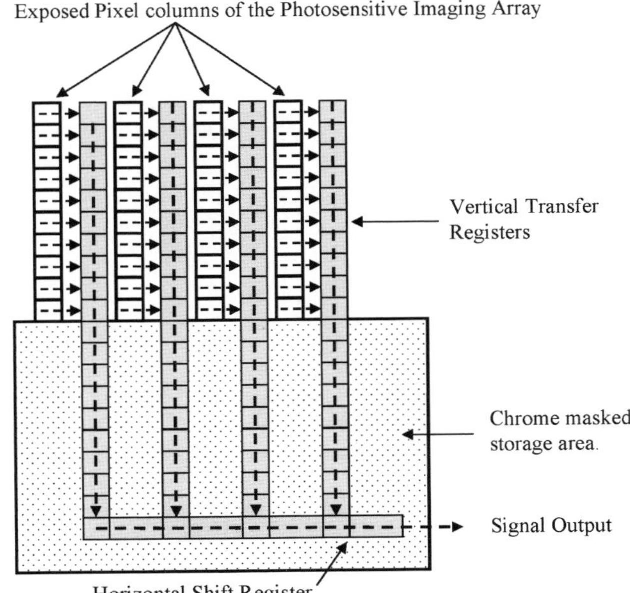

FIGURE 5.3-8 Frame interline transfer CCD basic structure shows an imaging array with a vertical shift register feeding the optically masked storage area. The signal charge is then clocked out of the storage area in real time.

Frame Interline Transfer CCD

The next type of CCD to consider is the frame interline transfer (FIT) device. This CCD was devised as a means of simplifying the structure and complexity of the FT imager and to obviate the need for a mechanical light-obscuring shutter. The FIT has the usual imaging array of pixels to capture the light and form the signal charge; however, in the FIT, optically obscured vertical transfer registers are added between the columns of pixels in the imaging array. After the imaging area's light integration period, the accumulated charge is quickly moved from the imaging section to these adjacent vertical transfer registers, then to the storage area.

The use of the metal-masked vertical transfer registers means that the CCD does not require a mechanical shutter during the transfer period, as the chromed vertical transfer registers are masked from seeing any extraneous light. The lack of a physical shutter means that the imager may be shuttered electronically and the shutter duration adjusted to eliminate any visual beating effects from strobing light sources. This feature is a significant benefit for both production and ENG cameras and is often referred to as "Synchro-Scan" or "Clear-Scan." It is used to eliminate strobing when shooting computer screens, televisions, and other non-constant light sources.

Figure 5.3-8 shows the basic structure of the FIT CCD. The imaging array is similar to the FT, but the readout/transfer is accomplished by the vertical transfer registers (shaded gray). The transfer from the imaging area to the storage area is very fast, usually less than 1% of the frame period. The fast transfer and chrome-masked transfer registers/storage array give the FIT CCD a transfer smear specification of about –130 dB, meaning that the classic transfer smear, visible as a vertical white line on highlights, is virtually eliminated.

There are some custom variations of the FIT CCD; one example is the M-FIT, a Panasonic CCD device introduced in the mid-1990s. This particular device was a 480-line 60 frames per second progressive 16:9 aspect ratio imager with a split storage area that provided options for clocking the image out as either an interlace or progressive signal. It was used in the AJ-PDW900P, a DVCPRO-50P 480p/480i camcorder.

Interline Transfer CCD

The third type of CCD imager is the interline transfer (IT) device. This CCD is of much simpler construction and therefore both easier and less expensive to manufacture than the FT and FIT devices. The IT device imaging array is very similar to that of the FIT CCD and, as in the FIT, vertical transfer registers are located between each column of the array. The IT CCD, however, does not use a separate storage area, as the vertical transfer registers perform both the charge transfer and storage functions. Figure 3.5-9 details the simple structure of the IT device. The imaging array accumulates charge during the active field or frame period; the charge is then moved laterally to the vertical transfer registers. These, as in the FIT device, are masked to eliminate extraneous light from contaminating the signal. During readout the charge is clocked in real time, both down and sideways, to the output section of the device.

The simplicity of the IT imager and low cost of production make this device ideal for consumer video cameras. Early IT devices, however, had an undesirable highlight overload effect that caused an ugly vertical

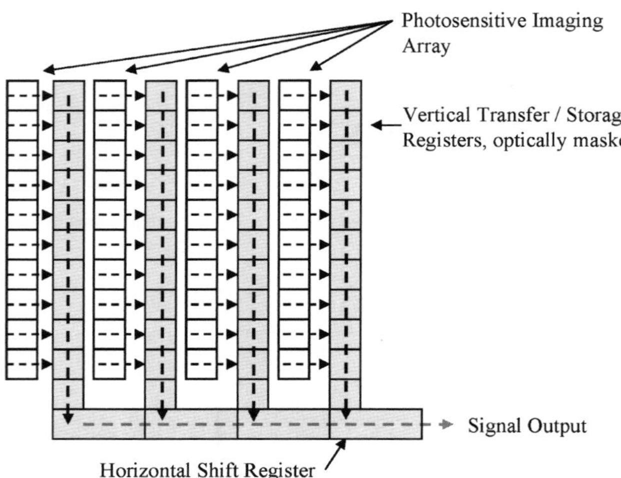

FIGURE 5.3-9 Interline transfer CCD basic structure shows an imaging array with vertical shift registers acting as both transfer registers and local storage for the charge from the imaging area. These dual-purpose registers then feed the horizontal shift register/output stage.

streaking artifact. This effect is widely seen on early ENG news cameras when shooting bright lights, most especially at night. The classic transfer smear effect is seen as a bright vertical line on picture highlights; in extreme cases it can be full picture height. This phenomenon gave IT imagers a bad name and relegated them to ENG and other low-end applications where such effects could be tolerated. The FIT imager with its much-improved highlight handling performance was usually specified for more serious EFP and production work.

As with most technological problems, time and advances in process technology came to rescue the lowly IT devices from the stigma of being suitable only for news applications. Current IT devices feature much more efficient blooming drains and very much improved highlight handling, so much so that current devices are equal to the more complicated and expensive FIT CCDs. An example can be seen in the Panasonic 1280 × 720 progressive HD imager, which has a transfer smear specification of –135 dB, basically on a par with current FIT devices.

A basic rule of thumb is that an HD FIT imager usually costs at least twice as much as a similar HD IT imager to produce, so if an IT device performs equivalently then the savings realized from using an IT imager mean more affordable products and a wider product choice for the end user. The era of IT cameras being subperformers is over; this can be seen in the marketplace today where most of the high-end production, studio, and EFP cameras feature IT imagers. The FIT imager is no longer worth the cost and most development work on CCD imagers is very much centered on IT devices.

COMPLIMENTARY METAL-OXIDE SEMICONDUCTOR IMAGERS

Complimentary metal-oxide semiconductor (CMOS) imagers, like CCDs, have variants and range in complexity from quite simple to amazingly complex. CMOS technology has the innate ability to mix both analog and digital processing functions on the same device. Thus, unlike CCDs, CMOS devices can place their digital clock generators, row and column drivers, and even complex analog-to-digital converters on the same chip as the analog photosensitive imaging array.

Passive Pixel CMOS

An example of a simple CMOS imager can be seen in Figure 5.3-10. This device, known as a passive pixel CMOS imager, is the least complex of any of the CMOS family. It is basically an array of photodiodes connected to MOS switches, which in turn connect to a series of horizontal and vertical bus lines that transport the pixel charge to the sample-and-hold circuit output stage. In passive pixel devices and other similar basic CMOS devices, the lack of any on-pixel buffering causes the tiny electric charge to encounter problems of the significant stray capacitance of the switched bus matrix. This results in noise problems and poor performance. It is, however, straightforward in construction and is often used for simple cameras (e.g., cell phones), where performance is less of a factor than cost and complexity. Figure 5.3-11 details enhancements to the PPS CMOS imager with the addition of buffering on the column busses; this helps reduce noise and improves performance. The next logical step in this process would be the addition of *on-pixel* sample-and-hold circuitry along with individual buffering. This raises the complexity of the

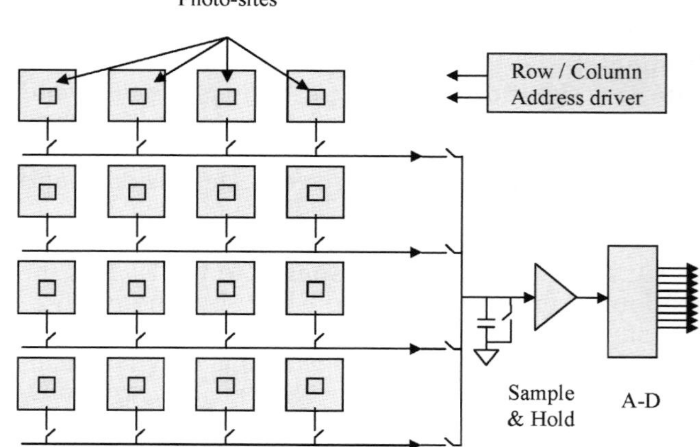

FIGURE 5.3-10 Block diagram of simple passive pixel CMOS imager. Row and column switches route charge to a sample-and-hold circuit followed by an analog-to-digital converter. This simple imager structure is used in applications such as cell phone cameras.

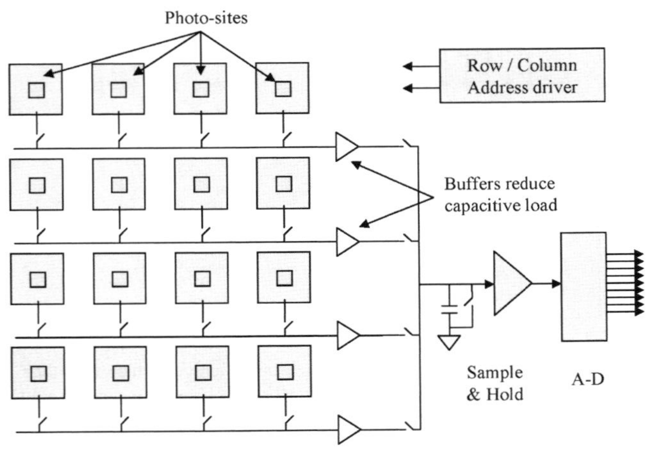

FIGURE 5.3-11 Block diagram of slightly more complex passive pixel CMOS imager with buffering on the column busses to reduce the stray capacitance. It increases the complexity but is worth the improvement in performance.

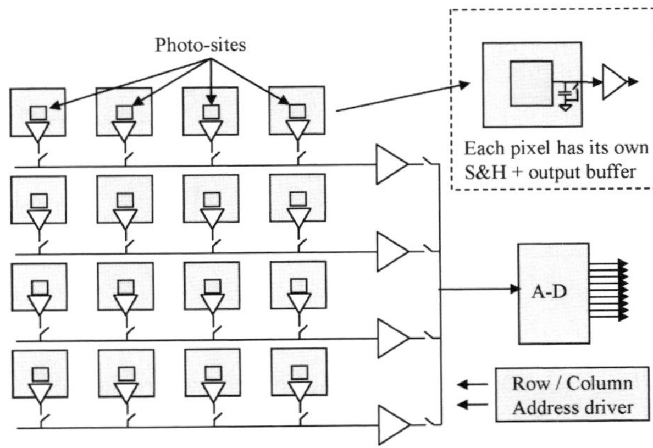

FIGURE 5.3-12 Block diagram of more complex active pixel sensor CMOS imager. In this device, each pixel has its own sample-and-hold and buffer amplifier; however, as each pixel's sample-and-hold circuit and buffer amplifier are unique and not 100% matched to adjacent pixels, fixed pattern noise can become an issue.

device considerably but results in marked picture quality improvement.

Active Pixel CMOS

Figure 5.3-12 shows details of a considerably more complex, but much improved, active pixel sensor CMOS imager. Although similar to the simpler passive pixel device, each pixel has its own sample-and-hold circuit and buffer amplifier to reduce signal deterioration when switched by the column/row switch matrix to the analog-to-digital converter. The introduction of a per-pixel sample-and-hold circuit and buffer can, however, lead to fixed pattern noise problems due to the individual pixel sample and hold circuits not being 100% identical. The fixed pattern noise comes from differences in both the electrical gain as well as DC offsets in the sample-and-hold circuits for each pixel. This gain/DC error signal causes a fixed pattern static modulation of the imager's signal and is often quite visible. Fixed pattern noise is not so much of a problem for digital still cameras, as it resembles random film grain in appearance; however, in television applications the human eye is very good at correlating this fixed pattern noise across time, and even small brightness differences (often less than ~1%) tend to become very visible.

CMOS Developments

CMOS imagers can be as simple as the passive pixel device or as complex as each pixel having its own sample-and-hold amplifier, buffer, *and* analog-to-digital converter. Such complexity, however, causes problems of cost, heat (from the many active devices on the wafer), and reduced fill factor of the pixel, due to the addition of more metalizing and polysilicon to provide power and signal distribution busses. More metal means less silicon is available to catch the all-important photons, and devices become less and less sensitive.

The pixels of CMOS imagers are addressed on an individual basis. Although this may initially be considered an inconvenience, it does permit the user to map the imager and download only the area of interest. This process has uses for applications where it may be desired to read only a certain area or zoom into an area of interest. CMOS imagers are often used for high-frame-rate cameras; for example, an imager may have a 1000 × 1000 pixel matrix and a maximum frame rate of 100 fps. It is possible, however, to address only the center 100 × 100 pixels of that imager and read them out at a much higher frame rate (e.g., 1000 fps). The same 10,000,000 pixels per second are being processed, so there is no issue of speed of the A/D or sample-and-hold circuitry. The user is simply choosing temporal over spatial resolution for the viewing application.

In broadcast applications, some of the current 1920 × 1080 pixel progressive CMOS imagers have the ability to be addressed as native 1280 × 720 imagers. In this mode, the user is choosing to address only the center portion of the imager. This is often done to permit operation at a higher frame rate or to switch the imager between the two main HD imaging formats (1920 × 1080 and 1280 × 720). This submapping process imposes a very high requirement for lens performance. It also has the additional unwanted side effect of the native 2/3-inch imager becoming windowed to the equivalent size of a 1/2-inch imager. This has an unfortunate twofold effect of increasing the effective focal length of the lens by about 22% and losing about one f-stop of sensitivity. In news applications, this can reduce a $20,000 8 mm

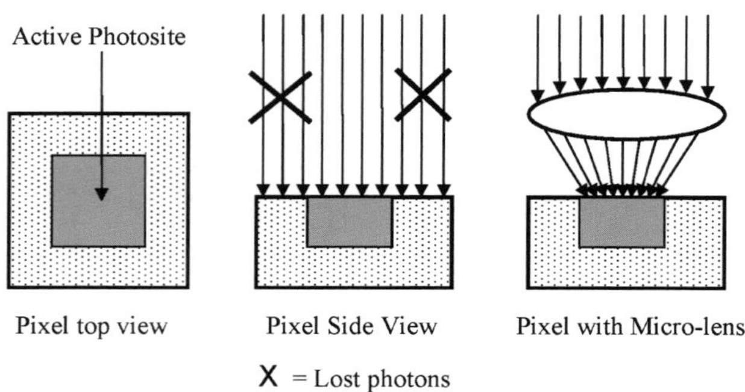

FIGURE 5.3-13 CCD imager with micro-lenses.

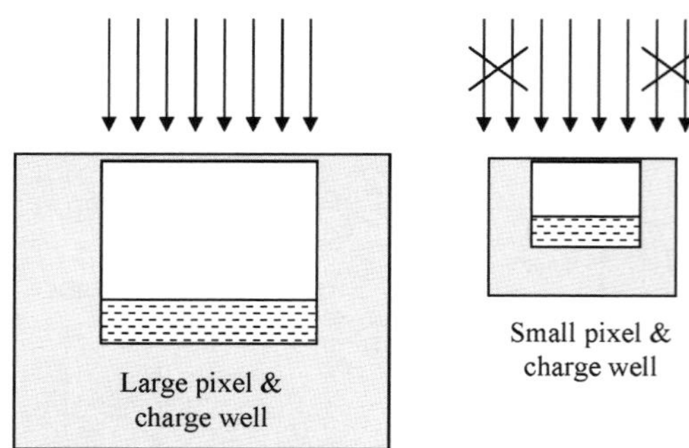

Large pixels are more sensitive due to their larger active surface area, they simply catch more photons. The charge well size is also large and can hold more electrons before the onset of blooming.

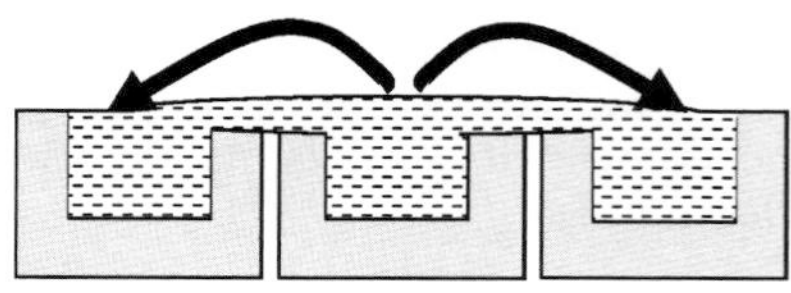

Small pixels are less sensitive and have smaller charge wells. When a pixel is overexposed, its charge well overflows and spreads the charge to the adjacent pixels in process know as highlight blooming.

FIGURE 5.3-14 CCD imager with micro-lenses.

wide-angle HD lens to a not-so-wide 11 mm lens. The modulation transfer function (MTF) requirements are also vastly increased as the effective 1/2-inch center portion of the lens is expected to provide the full resolution of the 2/3-inch format.

FILL FACTOR AND CCD MICRO-LENSES

Fill factor is the term used to specify how much of the surface area of a given pixel or imager is actually photosensitive and exposed to the incident light. By way of example, if an imager has a contiguous array of 10 μm square pixels, yet only the center 5 μm square of each is light sensitive, the fill factor for the pixel is 25 μm^2/100 μm^2, or 25%. This simple example shows that only 25% of the light actually hits the photosensitive portion of the pixels. This means that the imager will be some two f-stops less sensitive than an equivalent imager with a 100% fill factor. To mitigate that loss of sensitivity requires either a faster lens (usually very expensive) or having to add 12 dB of electronic gain to make up for the light loss; 12 dB of electronic gain is a significant amount of gain and will make any camera noisy.

Low pixel fill factors are not easy to correct. The surface of the CCD or CMOS imager is usually covered with numerous metalized busses, with additional metal plugs and via holes that move power or signals around the inside of the imager. The more complex the imager, the more circuitry it is likely to have and consequently the more likely it is to have opaque metalizing that contribute to the lower fill factor.

One means of correcting a poor fill factor to use a *micro-lens* to focus the light onto the photosensitive portion of the pixel. Figure 5.3-13 shows the concept of the micro-lens approach. Pixel-sized lenses above each active pixel focus the light onto the active photosite to greatly improve the sensitivity. Typically, micro-lens systems provide an increase in the light sensitivity of 200 to 300% (some 1 to 1.5 f-stops) and can bring the sensitivity back on par with a high fill factor imager.

SENSITIVITY AND DYNAMIC RANGE

Both CMOS and CCD imagers share the same basic imaging process and are covered by similar device physics. Figure 5.3-14 details large and small pixel devices. Usually, the larger the pixel is, the more photons it will catch, thus the more sensitive the imager is. These large pixels also have large charge wells and therefore are capable of holding a larger charge. An imager with large pixels at peak exposure may hold ~50,000 electrons in its charge well. That same imager with no incident light still may hold 25 electrons; the majority of these are caused by thermal activity and by chemical impurities causing leakage in the bulk silicon. Interestingly, one or two of those electrons may be caused by high-energy photons from cosmic rays and even radioactive isotope emissions from the CCD's ceramic packaging. Examining the minimum to maximum charge levels (50,000/25 or 2000:1), we can say that the imager has a useful dynamic range of about 2000:1, which represents some 11 f-stops.

Now consider a 2/3-inch high definition imager with 2 million pixels. In this case, the active pixel size will be about 5 μm square. Unfortunately, electrons do not change size just because we use smaller imagers; hence, the much smaller charge well of the HD imager may only hold 10,000 electrons at peak well level, so if the noise floor of the pixel is 10 electrons then we now have only a 1000:1 dynamic range (10 f-stops). This smaller HD pixel is also less sensitive compared to the larger pixel, as the photosensitive portion of the pixel may only be 1/4 the size of the larger pixel and thus able to only intercept 1/4 of the light.

In summary, the HD imager is two f-stops less sensitive and has half the dynamic range of the device

with larger pixels. This limitation can be seen with the current crop of digital point-and-shoot consumer camcorders; these usually utilize 1/4-inch or smaller imagers with around 5 million pixels. They have severe limitations with both sensitivity and dynamic range, and testing of these cameras has indicated only about 7 or 8 f-stops of latitude and poor low light performance—a testament to the fact that more is not always better!

INTERLACE AND PROGRESSIVE SIGNAL GENERATION

Most CCD and CMOS imagers have an inherently progressive imaging array; that is, the light hits the uniform array of pixels all at the same time. If the imager is required to output signals suitable for interlaced television applications, the stored image data will require processing to derive the required interlaced signal structure. This process is usually accomplished by a technique called *row–pair summation*. Figure 5.3-15 shows how the odd and even field signals are derived from the inherently progressive imaged data. The charge data from row 1 and 2 of the CCD is summed in the CCD sample-and-hold output circuit to provide a new pseudo-pixel of the required TV field.

This row–pair summation process has two interesting benefits. First, as the charge from the two pixels is summed, the signal value is doubled. This doubling increases the effective device sensitivity by 6 dB, or 1 stop. The second benefit of the row–pair summation is an increase in the device's signal-to-noise ratio by 3 dB. This occurs in the summing of the non-coherent random noise of each pixel; the differing noise of each pixel tends to slightly cancel the other, resulting in a useful 3 dB improvement in the signal-to-noise ratio.

The row–pair summation process of mixing charge from adjacent lines (*i.e.*, 1 + 2, 3 + 4, and so on) generates a new single pseudo-sample (pixel) sited between the two original pixel rows, thus generating the half-line vertical offset required to simulate the odd TV fields composed of lines 1, 3, 5, ... In the next field, pixels are summed from rows 2 + 3, 4 + 5, . . . to create the even TV field composed of lines 2, 4, 6, ...

Any imager used to create an interlaced signal must use a carefully optimized optical low-pass filter to limit the vertical resolution of the device/optics. If not, the row–pair summation process will result in unpleasant and very visible vertical alias artifacts in the video. This filtering may impose limitations of low vertical resolution upon the system if the camera is to be switched between interlace and progressive imaging modes. If optimally optically filtered for interlace, any progressive signals may appear as *soft*, or lacking in vertical resolution. Conversely, if optimized for progressive mode, the interlace signal may exhibit severe interlace *twitter.*

With the increases in large-scale integration (LSI) technology, a more recent approach to the problem of switchable interlace/progressive cameras is to generate a full-resolution progressive signal from the imager, then by means of filtering and spatial image interpolation derive an optimally filtered interlaced signal electronically. This approach, however, does lose an f-stop of sensitivity and the 3 dB reduction in the signal-to-noise ratio. As is usual in most things in life, there is no free lunch, and compromises abound.

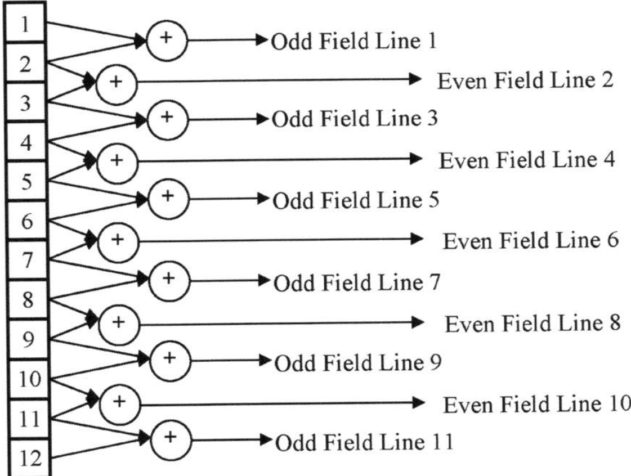

FIGURE 5.3-15 CCD row–pair summation generates interlace signals from progressive device.

DICHROIC PRISM BEAM SPLITTING ASSEMBLIES

Almost all ENG and production television cameras utilize a prism to split the light from the lens into the three constituent primary colors of red, green, and blue. This system is efficient in that none of the light is wasted by color absorption filters; instead, a three-channel R, G, and B prism beam splitter is used to channel the spectrally selected light to three dedicated imagers. Figure 5.3-16 details the structure and ray paths of the typical 3-port beam-splitting prism. The prism is assembled of three glass blocks glued together. *Dichroic coatings* on the prism faces are used to generate two *optical filters*, one low-pass (LPF) and one high-pass (HPF). Light enters the front prism port and proceeds to the dichroic coating on the back face of the first prism, which forms the high-pass filter; this filter permits green and blue light to pass through. The low-frequency red light is reflected and bounces to exit the lower prism exit port to the red imager. The back of the second glass prism block is coated with another dichroic layer to form a low-pass filter; this reflects blue light up and out of the top prism port to the blue imager. Green light is not affected by either the low- or high-pass filters and passes straight through the prism assembly to exit the rear port to the green imager.

Figure 5.3-17 shows an optical low-pass filter and infrared cut filter on the front of the prism front port. The optical LPF is a specially etched two-dimensional

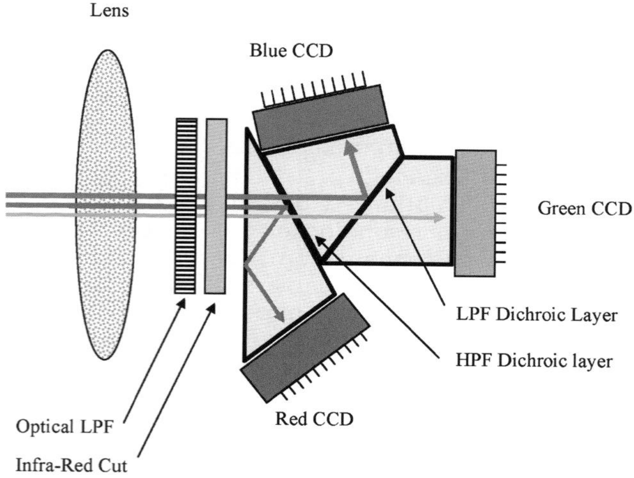

FIGURE 5.3-16 Basic three-port RGB prism and CCD optical assembly.

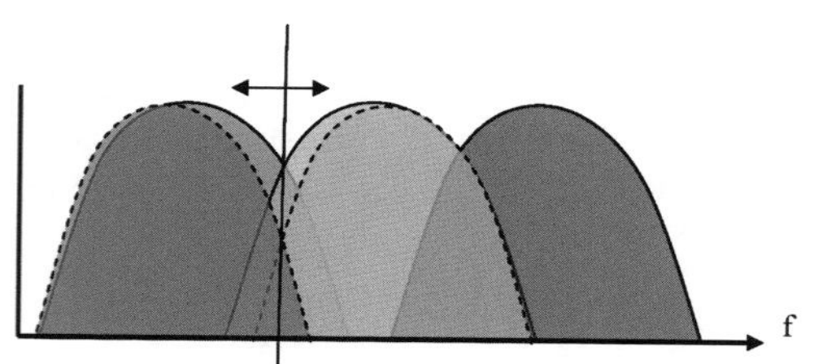

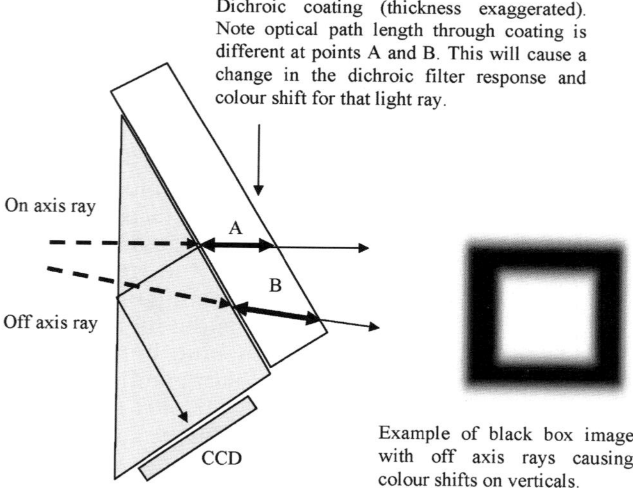

FIGURE 5.3-17 Off-axis light hitting the dichroic coating causes a shift in the low-pass filter and high-pass filter, thus causing a color shift.

filter chosen to limit the optical resolution of the light path to below that of the Nyquist limit of the imager. This will dramatically reduce aliasing on high spatial frequency information in the image.

The infrared (IR) cut filter is required to severely curtail the IR response of the optical path to avoid the red CCD imager from seeing this IR portion of the light. Infrared light will show up as a strange color shift in some colors due to some IR component of the subject's spectral emission or due to the different focus point of the lens to IR, as with an out-of-focus halo on street lights. CCDs are still relatively sensitive to IR light even at wavelengths approaching 1.0 μm. The IR cut usually starts at about 625 to 650 nm and will have response of less than 1% at 750 to 800 nm.

It is of particular importance that the ray path of the lens's exit pupil be parallel. If an optical mismatch occurs (e.g., due to the wrong lens on the camera, a misaligned rear extender, or the iris diaphragm opened too wide), there is a chance that the exit pupil light cone may diverge. This will result in an unsightly magenta/green shading effect on out-of-focus objects in the image.

The prism's dichroic filter is made by depositing finely controlled layers of metal oxide coatings that are adjusted in thickness to match the wavelength of the light being filtered. Figure 5.3-17 shows how an off-axis ray can cause a filter shift as it enters the dichroic, thus incurring a transit path length change. Off-axis light entering the dichroic coating at an angle will shift the filter response and thus cause a color shift for that ray. This problem is very obvious when using a long lens with a wide-open iris diaphragm. An external lens extender will exacerbate the effect.

Port Prism/CCD Assembly

An interesting variant of the common 3-port prism is the 4-port version shown in Figure 5.3-18. In this device, there are two green ports on the prism. This permits the attachment of two green imagers, G-1 and G-2, as discussed in the next section.

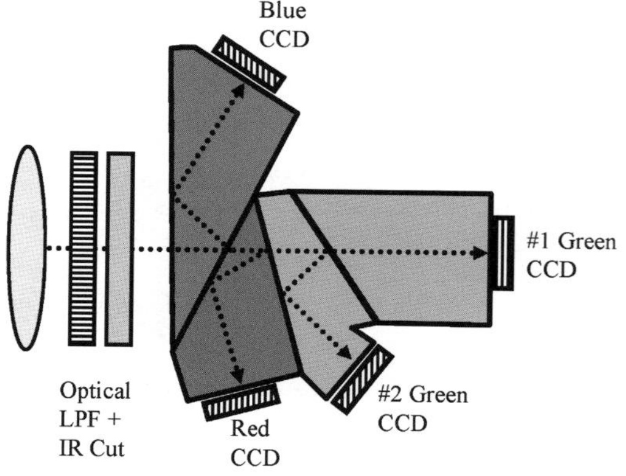

FIGURE 5.3-18 A 4-port prism assembly (G1 + G2 + R + B).

THE USE OF PRECISION SPATIAL OFFSET OF IMAGERS

Two Offset Green Imagers

If the alignment of the two green images is very carefully offset so there is a half pixel horizontal and vertical offset between the imagers, it is possible to generate a double-resolution image by merging the two green signals. CCDs and CMOS imagers have a defined pixel matrix for their imaging array. That X–Y matrix of pixels will determine the Nyquist limit for the imager in both the horizontal and vertical directions; for example, if an imager has a 1000 × 1000-pixel count, the Nyquist limit for both directions will be 1000/2, or 500 cycles. Any image resolution above that of the 500 cycle limited imager will result in an alias effect. It is possible to build imagers with higher pixel counts to improve resolution, but this often results in a loss of sensitivity and also lowers manufacturing yields. The spatial offsetting technique with two imagers is often used as a practical means of increasing the resolution of an imaging system; for example, using a 4-port prism assembly like that detailed in Figure 5.3-18 with two green channels. If these two imagers are aligned with a half pixel offset, it is possible to combine the signals from two green channels and double their combined resolution.

Figure 5.3-19 shows two CCD imagers with a half-pixel offset sampling a high-frequency sine wave chart. The imaged test chart is exactly at the Nyquist limit of each CCD imager, so if the imagers have a fairly narrow sampling aperture they will each integrate the image scene to a DC value of white or black depending on the phase of the pixel to that of the image. In a correlated double sampling system, the pixel samples from the two imagers are interleaved as CCD1, then CCD2, then CCD1, etc.:

Steady-state video values for CCD1 + CCD2 =
"white" DC levels + "black" DC levels

Correlated sampling of CCD1 + CCD2 provides an accurate reproduction of high-frequency image detail and doubles the pixel resolution of a single imager. Because green has by far the greatest contribution to luma, using two, offset, green imagers effectively doubles the system resolution.

Spatial offsetting can work in horizontal, vertical, and combined horizontal and vertical directions. This technique is becoming more common on higher resolution 4K and 8K HD cameras. An individual 4K or 8K imager would be prohibitively expensive and quite insensitive. The use of two 2K or 4K imagers, quincunx (biaxially) spatially offset, will produce the resolution of a 4K or 8K imager at much lower cost and with more practicality.

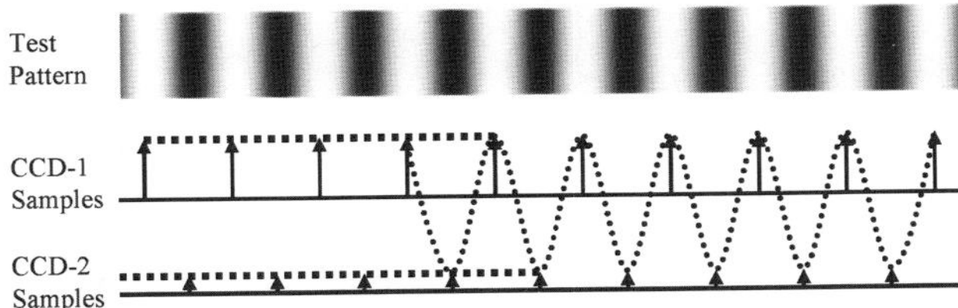

FIGURE 5.3-19 Spatial offsetting of imagers to increase resolution.

Offset Green with Red and Blue Imagers

A more common approach that uses spatial offset is a conventional 3-port prism camera with individual red, green, and blue imagers. Here, however, a half pixel horizontal offset is added to the green imager relative to the red and blue imagers. This spatial offsetting generates a "super-sampling" effect that occurs in the luma matrix when the red, green, and blue channels are combined. The ITU 709 luma matrix is described as Y′ = 0.701G′ + 0.212R′ + 0.089, and it is interesting to note that the 30% increase in effective luma response from this approach agrees closely with the approximate 30% contribution of the combined red and blue channel signals in the Y′ matrix. With the offset pixel imagers, the increase in resolution will only occur when there is coherent signal content in the green and one or more of the color channels. Should the system be imaging a primarily green image, there will be little or no red or blue signal content in the image and thus little or no additional sampling content to be added to the green, hence no increase in effective luma resolution.

An example of the benefits of the use of spatial offset in 3-CCD imaging systems can be seen in Figure 5.3-20. This image was taken from the Panasonic AG-HVX200, a small P2 memory-based SD/HD camcorder. This camera uses 3 × 1/3-inch CCD imagers in the usual R, G, and B prism configuration; however, the use of 1/3-inch HD CCDs has negative consequences for sensitivity and dynamic range if the pixel

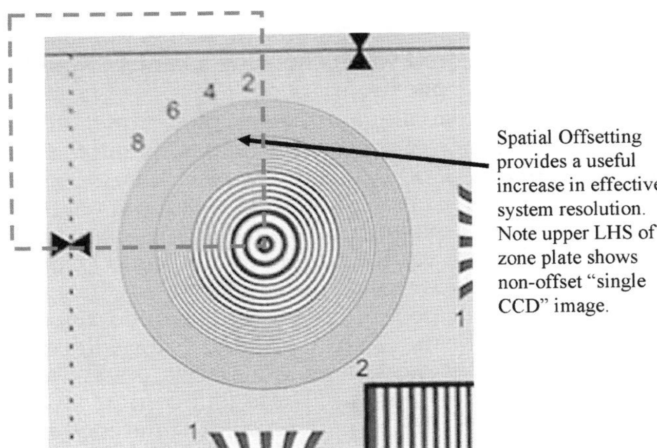

Spatial Offsetting provides a useful increase in effective system resolution. Note upper LHS of zone plate shows non-offset "single CCD" image.

FIGURE 5.3-20 Results of biaxial spatial offsetting of imagers.

density is too high. It was therefore decided that this camera would use a novel approach of biaxial spatial offsetting in combination with a new 1/3-inch 16:9 960 × 540 pixel progressive CCD. The choice of the imager resolution was a careful compromise between resolution and sensitivity/dynamic range. As the AG-HVX200 was intended for HD production applications, dynamic range and sensitivity were critical constraints. The use of a quincunx (biaxial horizontal and vertical) spatial offset greatly helps in raising the system resolution to about that of a native 1280 × 720 camera, while also maintaining the sensitivity and dynamic range benefits of the large pixel imagers. In the figure, note the marked increase in resolution, comparing the upper left corner (effectively that of a single 960 × 540 imager) with the rest of the image, which is effectively 1280 × 720. The significant increase in resolution results in a clean 600 TV lines-per-picture height resolution being quite visible in the combined luminance signal.

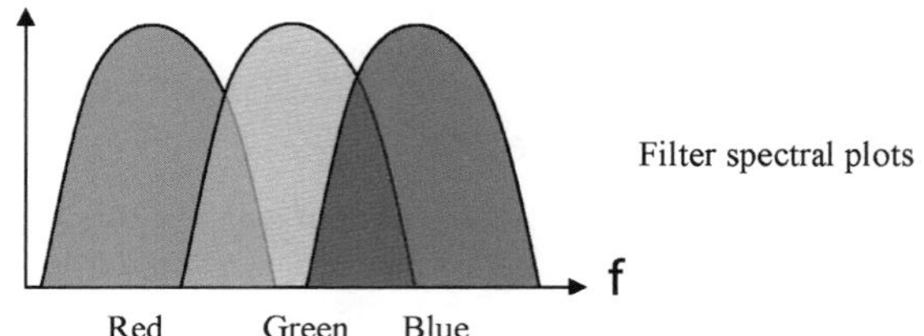

G	R	G	R	G	R	G	R	G	R
B	G	B	G	B	G	B	G	B	G
G	R	G	R	G	R	G	R	G	R
B	G	B	G	B	G	B	G	B	G
G	R	G	R	G	R	G	R	G	R
B	G	B	G	B	G	B	G	B	G
G	R	G	R	G	R	G	R	G	R

FIGURE 5.3-21 Beyer mask with R, G, B, G pattern. The mask absorbs about 67% of the light, and only 33% of incident light is captured.

Single Imager Color Systems vs. Three-Imager Systems

If a 35 mm film camera is considered as a model for a still or movie camera, it at first appears quite simple in concept. Light enters a lens system, passes through a shutter mechanism, and then falls onto the film plane. In this model, both electronic and film cameras can be considered pretty much identical. Both capture light and create an image at the focal plane imager, be it a piece of film or an electronic CCD or CMOS imager. There is no easy electronic analog to a piece of film; that simple square of film with three ultra-thin light-sensitive emulsions is still a wondrous achievement.

It is possible to make cameras using a single combined color imager that are smaller, simpler, and lighter than 3-CCD systems but suffer from being less sensitive and, depending on the type of imager/color separation filter system used, may have issues of resolution. Cameras with beam-splitting prisms and three or more CCDs or CMOS sensors do not absorb light by filtration as do single imagers; thus, they are more sensitive but also more costly, more complex, and physically larger. The relative merits of both approaches should be considered when choosing a camera system for a particular application.

Efforts to make single-imager color cameras go back many years. Ignoring CBS field sequential system of the early 1950s, which involved a large spinning color wheel, as being impractical, a more realistic approach was to use striped optical filters on a tube camera's pick-up tube. Diagonally opposed, thinly striped gel filters were carefully placed on the tube faceplate. The striped filters, when scanned by the tube's electron beam, would provide color information automatically modulated on subcarriers embedded in the video signal. These subcarriers were demodulated and the respective color difference signals produced and processed. Although of low color resolution and performance, they served well in millions of early tube-based VHS camcorders.

A simple yet elegant solution to the single imager requirement was developed for the digital still camera of today, of which hundreds of millions are sold annually. The most common approach to a single-imager color sensor is the Beyer color mask, which is composed of individual red, green, and blue color filters overlaid onto the CCD or CMOS imager's pixels, as shown in Figure 5.3-21. The interleaved R, G, B, G pattern has twice as many green pixels, because green is the color most sensitive to the human visual system. It is possible by means of a "de-Beyering" process to interpolate the Beyer pattern to individual red, green, and blue signal components. The Beyer mask approach is useful but results in a loss of light due to the absorptive optical filtering. It is obviously important that the filters are of the correct density and color response and also accurately matched to adjacent pixels, to avoid blotchy colored areas in the picture. It is a tribute to modern manufacturing technology that a tiny 1/4-inch CCD 5-million pixel imager can be covered with a Beyer mask and also with the tiny microlenses that are used to regain some of the lost sensitivity.

Large-Format Imagers

CCD and CMOS imagers range in size from tiny 1/10-inch or smaller imagers for cell phone cameras to the larger than 4-inch wafers for specialist large-field cameras, such as the Hubble Space Telescope cameras. Several companies are now producing 35 mm sized imagers for use in film-style production cameras. Most of these cameras use a single imager, with some form of Beyer or color mask. Very large imagers like these are difficult to make and are expensive, but such cameras are made in modest quantities, usually in the tens to one hundreds, so limited imager yields are not an insurmountable problem. The large-format imager means that the camera can make use of the large range of very high-quality 35 mm lenses used in motion picture film production. One interesting benefit is that large-format cameras are more tolerant of less-than-perfect lens specifications. A 35 mm sized imager may require a lens with only about 70 lines/mm response to provide a very high quality image. A much smaller 2/3-inch CCD requires a lens with almost three times that performance to provide equal quality, a very much more difficult task for the lens.

CCD Versus CMOS Quality

As of the time of this writing, CCD imagers have the advantage of superior image quality, especially in the broadcast size of 2/3 inch. CMOS, although an interesting technology, is still probably a few years away from viability in mainstream broadcast cameras. While offering some advantages, particularly lower power consumption, CMOS technology suffers from lower sensitivity, fixed pattern noise, and production manufacturing issues; little can be done to easily improve these parameters, although perhaps in due course they will be overcome. Figure 5.3-22 is an interesting testament to the relentless advances in technology, as it is an image taken by a current 1280 × 1024 pixel cell phone camera. Although optically challenged by cost and size constraints and lens physics, it has about twice the resolution of the 1984 RCA CCD1 professional ENG news camera with its 1/2-inch frame transfer 403 × 480 pixel imagers mentioned in the opening paragraph of this section. This cell phone camera has a lens that is about the size of the letter "O" printed here.

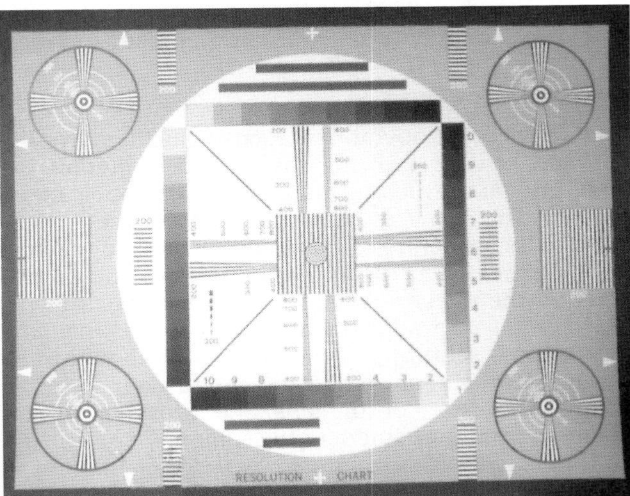

FIGURE 5.3-22 Test chart capture from a 1.3 mega-pixel cell phone camera with a 1280 × 1024 pixel native imager.

OPTICS

The optical system of every camera can be thought of as everything affecting the image before the light waves (photons) change to voltages (electrons). The optical systems of cameras possessing radically different image capture technologies often have a great deal in common. The core of the optical system is the lens; however, before the light reaches the lens, there are accessories, such as lens hoods and optical filters, that can improve the image by taking account of conditions in the scene, such as point light sources and other stray light, which may degrade the performance of the lens itself. As shown in Figure 5.3-23, behind the lens the optical system primarily involves built-in optical filters and the color separation system. Dichroic prism beam splitting with color separation was discussed earlier in the Imager section.

Optical Filters

Filters can be placed either in front of or behind the lens. Behind-the-lens filtering is common practice in video cameras because it keeps systems streamlined and there is less danger of a misaligned filter interfering with the imager. Filters behind the lens are usually of two kinds: permanent filters, which are part of the *optical block,* or those placed on a rotating *filter wheel.* Common fixed filters include:

- Infrared cut (IR) filters, which protect the imagers from infrared waves, to which solid-state imagers are very sensitive
- Anti-aliasing filters, which prevent high-frequency details that might cause aliasing in the video signal from being picked up

Other behind–the-lens filters are normally placed in a filter wheel. The filter wheel makes changing of filters easier. It is important to remember that placing any plane of glass behind the lens will change the back

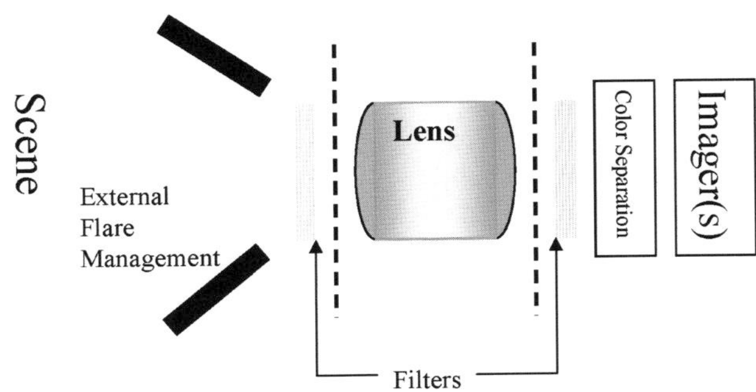

FIGURE 5.3-23 Camera optical system.

focus so most filter wheels include one clear filter to maintain consistent back focus when filters are removed. Filters on filter wheels include:

- *Neutral density* (ND) filters for adjusting the exposure without setting iris or gain
- *Color correction* filters for shooting outdoors with tungsten balanced imagers
- Effects filters including *cross* or *star* filters as well as *diffusion*.

Filtering is also used for color separation in the imager architecture, but this should not be confused with general optical filtering, which is no different than it has always been in motion picture or still-frame film photography.

Lenses

In principle, a lens can be as simple as a pinhole but it can also have almost unlimited complexity. Television lenses are always comprised of multiple *elements*, and the best quality lenses can easily outstrip the price of even an expensive HDTV or digital cinema camera. The two key parameters for a lens are its *focal length* (fixed or variable) and its *aperture*.

Focal Length

Lenses with a fixed focal length fixed are referred to by this parameter (see Figure 5.3-24). Variable focal length *zoom* lenses are referred to by their maximum and minimum focal lengths or by the minimum value and the *zoom ratio* (e.g., ×20). The actual measurement of the focal length (typically in millimeters) is the distance from the principle plane (optical center) of the lens system to the focal point (the point at which light rays converge when the lens is focused on a subject at infinity).

The focal length determines what viewing angle of the scene will be focused on an imager of a given size, so the choice of focal length, combined with camera placement, determines the point of view for the scene. Shorter focal lengths capture a *wider angle* of view, and longer focal lengths (tending toward *telephoto*) capture a narrower field of view. Increasing the focal length seems to bring the scene closer, as a smaller viewing angle is captured by the same size image plane. Adjusting focal length with a variable focal length lens is referred to as *zooming*.

An experienced camera operator will have an intuitive feel for what portion of a scene a given focal length captures; however, when the imager size is changed the relationship between focal length and field of view is changed. A smaller imager is looking at a fraction of the image a larger imager would capture. When these images are shown on the same size monitor, the smaller imager picture will capture a viewing angle similar to what the larger format system might have produced with a longer focal length lens.

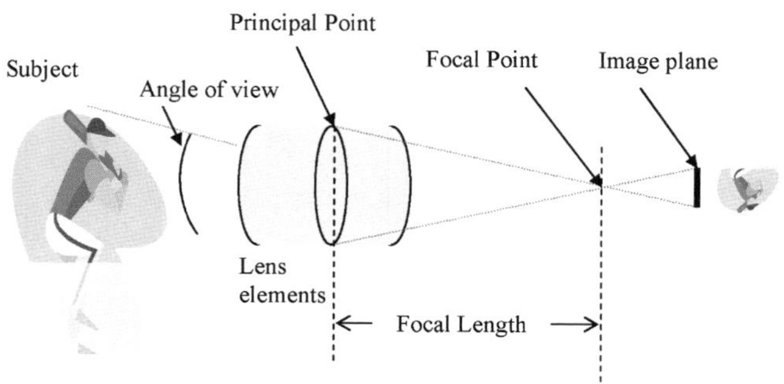

FIGURE 5.3-24 Focal length.

Aperture

The lens aperture determines the amount of light that passes through the lens and is a fundamental characteristic of the lens design based on the size and arrangement of the lens elements. In most lenses, the amount of light passed can be controlled with an adjustable *iris diaphragm* and is referenced by an *f-number* or *f-stop*. The term *t-number*, with slightly different meaning, is used on many lenses intended for electronic cinematography, as t-numbers are usually used in the movie industry. This adjustment is referred to as *aperture*, *iris*, or *exposure*. The f-numbers indicated on the lens are normally specified in one stop increments, and each change from one f-number to the next represents halving or doubling the amount of light passing through the lens. The smaller the f-number, the more light reaches the imager. When the lens is set to its lowest f-number, this is referred to as *full aperture* or *wide open*. Reducing the f-number by one increment (one stop) cuts the light in half. This is equivalent to reducing the camera gain by 6 dB or applying 0.6 of neutral density (ND 0.6) filtering. Camera sensitivity is often indicated by the f-number; in this case, the specification will normally specify the lux rating (light intensity), such as 2000 lux, that is needed for the camera to produce good-quality pictures at that f-number.

Front and Back Focus

Front focus is the operational focus adjustment for the lens and is used to bring the subject matter in the scene into sharp focus. This means the proper image plane in the scene is being focused on the imager plane in the camera. For this to occur the image plane has to be the correct distance from the lens and proper focus on the imager itself is set with the back focus, which is a setup adjustment (see below). When the back focus is set properly the front focus should track throughout the zoom range and the focus marks (the distance indicators on the lens) should represent real distances from the camera to the subject.

Depth of Field

Whereas front focus will set the focal plane, the *depth of field* determines if a given object will be in sharp focus. Depth of field is the range of distances from the camera that are simultaneously in focus. A shallow depth of field means that small ranges of distance, before and behind the focal plane (perhaps even a fraction of an inch), are in sharp focus. The depth of field is proportional to the f-number (F_{NO}) as well as the object distance (l), and is inversely proportional to the focal length (f). The formula for depth of field is a bit complicated for practical field use:

$$[\delta * F_{NO} * l^2]/[f^2 - \delta * F_{NO} * l] + [\delta * F_{NO} * l^2]/[f^2 + \delta * F_{NO} * l]$$

where δ refers to the "circle of confusion," which is determined by the smallest circle that can resolve as a point, which is important to know in terms of the near limit and far limit in relation to the object distance. The formulae for the far and near limits of depth of field are as follows:

$$D1 = [\delta * F_{NO} * l^2]/[f^2 - \delta * F_{NO} * l]$$

$$D2 = [\delta * F_{NO} * l^2]/[f^2 + \delta * F_{NO} * l]$$

Depth of field calculations in the field, if needed, are typically made using tables. If more depth of field is required, then it is necessary to shoot with a higher f-number or a wider focal length; if a shallower depth of field is required, then the iris must be opened if possible or a longer focal length selected. Iris settings can often be compensated for by adjusting gain, light levels, or neutral density filters; different focal lengths can achieve the desired framing by changing the position of the camera. It is worth remembering that, because different size imagers capture different angles of view with like focal lengths, a similarly framed shot will use a different focal length and therefore exhibit a different depth of field for a given f-number. For this reason, it may be difficult to achieve a shallow depth of field with a small imager.

Back Focus Adjustment

Back focus for film camera lenses has traditionally been set in the shop using a collimator, and these lenses are also designed to maintain consistent back focus as it cannot be checked in the field. Video lenses, on the other hand, have a user-adjustable back focus so they can be checked and adjusted in the field. The back focus adjustment technique normally used takes advantage of the difference in depth of field and depth of focus as focal lengths are changed. At longer focal lengths, front focus is critical due to the shallow depth of field; at wider focal lengths, back focus is more critical.

A long focal length with a shallow depth of field affords a more precise front focus setting and does not require precise back focus. A wide-angle focal length has a deeper depth of field but it requires a precise back focus setting. To set back focus in the field, the iris should be wide open to minimize the depth of field. After setting front focus zoomed in, the lens should then be zoomed out to a widest position. If the back focus is off, the image will go soft and it should be adjusted on the lens. At a wide angle, the back focus becomes critical, so a fine adjustment can be made. It is often necessary to repeat the process of focusing the front of the lens zoomed in and the back focus with the lens wide to achieve a good back focus.

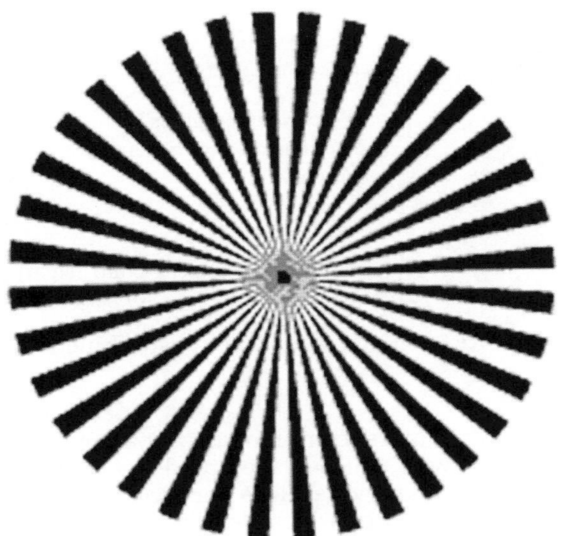

FIGURE 5.3-25 Siemens star test pattern used for adjusting back focus.

ENG cameras require frequent back focus checks because they are often handled roughly, but studio cameras also require a resetting of back focus from time to time, especially if they experience changes in temperature. A *Siemens Star* (see Figure 5.3-25) is a useful tool for setting back focus because sharp focus can be easily judged at many focal lengths. Any chart or object that has a variety of object sizes can be used for back focus as long as it provides multiple-sized high-contrast figures on the same focal plane.

Comparing Lens Quality

Camera lenses price ranges are a function of lens speed (or the wide-open f-number) wide angle capability; zoom ratio; MTF and overall image and construction quality. A lens capable of transmitting more light is clearly of more value, as is a lens capable of delivering a greater range of focal lengths. Achieving high specifications for all of these parameters simultaneously becomes increasingly difficult for lens designers, and small improvements can lead to large price changes. Comparing available lenses it is easy to see how the key lens specifications impact the cost. With the introduction of HD cameras, some confusion arises because lenses with similar focal length and speed specifications may vary greatly in price. This is a function of lens image quality, which is somewhat more difficult to measure.

Modulation Transfer Function

The *modulation transfer function* (MTF) (Figure 5.3-26) is a measure of an optical system's perceived sharp-

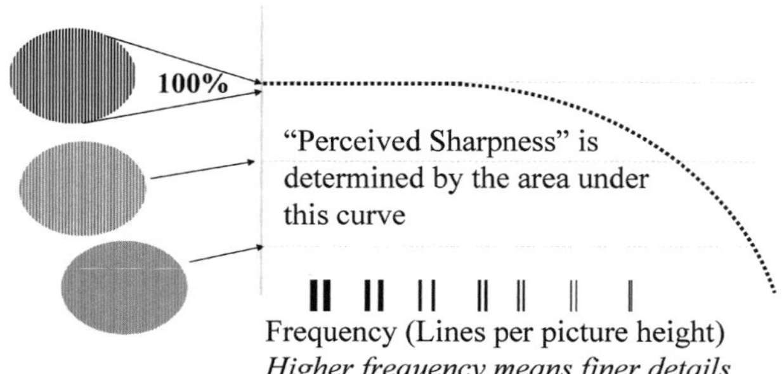

FIGURE 5.3-26 Modulation transfer function.

ness. It is a more comprehensive parameter than resolving power, which is measured optically as line pairs per millimeter or as lines per picture height on a video signal, both of which deal with the maximum frequency signal that can be resolved. MTF takes into account the fact that, even if an imager is capturing a high-frequency image, it may not be capturing it at its true contrast level. In other words, a series of black and white stripes might be resolved in a given system but it may be resolving as dark gray and light gray rather than black and white. The percentage of full contrast level for a given frequency is known as the modulation depth.

Modulation transfer function plots the modulation depth as a function of the frequency (size of the objects to be resolved). This parameter is called *frequency* because it can describe the frequency of a periodic signal on a horizontal scanning line imaging black and white line pairs. As is shown in the figure, the MTF exhibits a roll-off as the Nyquist frequency is approached. Sharpness is the measure of what a *human observer* will describe as *sharp* when viewing an image, and this is actually a function of the square of the area under the MTF curve. That means that the perceived sharpness of an image is determined by both the resolution and the contrast as combined in the MTF. The overall MTF of a displayed image is determined as an aggregate of the entire system, including the lens, camera, video format, recording algorithm, and display technology; therefore a high-quality lens can have a positive effect on any video system but it will have diminishing returns on a system where the imager recording format or display has a limited MTF.

Lens Performance

The stated MTF for a lens may not be useful for comparison unless something is known about the consistency of the lenses performance. Lens designers will always have more difficulty achieving good MTF at the extremes of the lens parameters, such as iris and focal length. At the extreme wide, long, or wide open settings, lenses may experience a reduced MTF, as well as a lesser degree of consistency across the image. Diminished MTF at the corners is a typical problem with zoom lenses, particularly at the widest focal lengths. The number and severity of lens artifacts are perhaps more important than MTF when determining lens performance in most applications. As with MTF, the artifacts will become better or worse depending on the focal length and iris setting, so time and care should be taken when looking for them.

There are five basic lens flaws known as aberrations (*Seidel's five aberrations*). *Spherical*, *chromatic*, and *astigmatism* aberrations appear as zones of the image that may exhibit color fringes or are not completely focused at some lens settings. *Curvature of field* means that the focal plane is not flat, and *distortion* is exhibited when an image that ought to be rectangular exhibits curved edges so as to appear like a *pincushion* (concave lines) or *barrel* (convex lines). All of these have to be corrected or reduced through complex lens design, and the level of success helps determine the quality of the lens. Failures contribute to lens inconsistencies and chromatic aberrations, which are observed as color fringing on contrasty images. Chromatic aberrations tend to occur near the edges of the image and at extreme settings. Another optical problem is known as *flare*, which is a catch-all term for unwanted light reflected off surfaces inside the optical system. Flare can prevent an imaging system from producing a true black and it adds a light bias because the reflected light typically has some tint.

In evaluating a particular lens, care should be taken to test it at focal lengths and exposures where it is most likely to be used. Some aberrations at very wide angles might be tolerable if the lens performs well in other circumstances. Lenses for ENG use can tolerate flaws at extreme settings (such as an inconsistent f-stop at long focal lengths) in order to have access to more focal lengths and exposure settings. Digital cinematography lenses may limit the zoom range to maintain consistent exposure throughout the range.

It may seem counterintuitive at first, but the larger the image format, the easier it is to manufacture a good-quality lens. This is because the large imager can function within wider optical tolerances. Line pairs per millimeter can only be compared for two lenses used on the same size imager, as a smaller imager will necessarily be enlarged more to be shown on the same display so any flaw will also be blown up. Small-image-format cameras typically use lower cost lenses, so it should come as no surprise that lens aberrations will be far more common in these cameras.

SIGNAL PROCESSING AND CAMERA SETUP

The raw video signal from the imager must be processed to make the camera output compatible with the other video systems in a broadcast plant or production chain. While this is happening, it is possible to compensate for many of the imperfections of the physical imaging system and to achieve the specific look desired by the director or producer. This is accomplished through *signal processing*. Signal processing can happen in either the digital or analog domain, and

camera manufacturers must make engineering decisions as to where specific processing is best accomplished. Since the advent of digital signal processing (DSP) technology, the trend has been to move more and more processing to the digital side. Analog processing is not limited by the potential quantization errors of digital systems, but digital processing is more versatile, and as DSP systems have grown from 8 bit to 14 bit and higher quantization errors have become less of a concern.

It is often apparent to the camera engineer whether a camera adjustment is utilizing digital or analog signal processing, but making assumptions regarding this can lead to unexpected results or problems. An experienced camera engineer might be tempted to open a camera to look for potentiometers that are no longer physically available. Analog or digital, solid-state systems have become more stable over time, and adjusting cameras has generally become less of a concern with maintaining legal signals and is more concerned with maintaining an excellent image. Camera setup has always been as much of an art as a science, and all of the basic engineering principles still hold true although more flexibility is often possible. The following adjustment parameters will be explained in the context of achieving proper settings. It is assumed that the engineer has access to an 11-step *gray-scale chart* (either front lit or back lit) and in some cases a *color chip chart* and a *vectorscope*.

Gain

To go from an imager with a charge well (holding just a few thousand electrons) to a legal video signal on the order of a volt will take a good deal of amplification. The 0 dB point (no gain) is a reference determined by the manufacturer based on the compromises made between the signal-to-noise ratio and sensitivity. This might seem arbitrary; however, it is also chosen to maximize dynamic range, which is measured between the imager's *saturation point* and the *noise floor*. The other internal camera settings are optimized around this 0 dB point. Gain, shutter speed, ND filtering, and iris adjustment all impact the overall output of the camera, and each has a side effect. For gain adjustments above and below 0 dB, the side effect is noise level. Depending on the camera, scene content, recording, or transmission method, noise will eventually become objectionable. Although noise can be reduced by running at negative gain, taking this too far may require some other sacrifice, such as a loss of dynamic range.

White Balance

Adjusting the gain of the R, G, and B signal channels individually can compensate for the *color temperature* of the light source. The proper balance between red, green, and blue gain levels depends both on the color temperature of the light in the scene and also on what optical color correction filter (if any) is used. Higher color temperatures such as daylight or halogen metal iodide (HMI) lighting are more bluish and require more red and less blue gain. Lower color temperatures, such as tungsten-based lamps, are more reddish and require more blue and less red gain.

With proper white balance, white surfaces will appear neutral white. If the light color temperature is changed then the R and B levels will have to be adjusted or the picture will develop either a red or blue *cast*. In the natural world, human vision automatically adjusts so a white object always appears basically white in anything from firelight to moonlight. Video camera systems have to balance the R and B levels each time the light source is changed. With earlier analog cameras, RGB black levels also tended to drift over time, but solid-state cameras rarely experience this problem. White balance can be set when shooting a gray-scale chart (see Figure 5.3-27) by adjusting RGB levels until the color subcarrier signals (if an NTSC output is available) or color difference signals (as viewed on a vectorscope or waveform monitor) are eliminated, leaving only luma. The scope trace in Figure 5.3-28 was generated by the 11-step gray scale. It shows still some color in the Pb and Pr signals (center and right traces). Red and blue gain should be adjusted until the color difference channels null out. Alternatively, a display of red, blue, and green can be

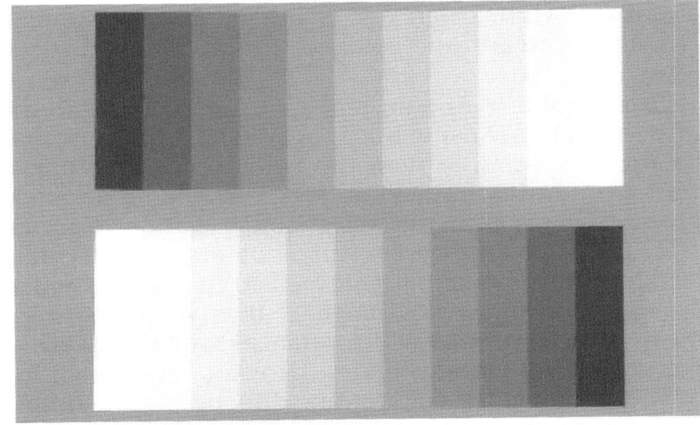

FIGURE 5.3-27 Eleven-step gray scale chart.

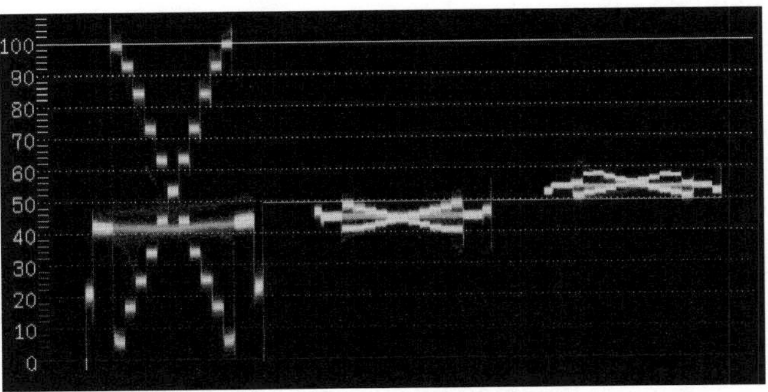

FIGURE 5.3-28 YPbPr scope trace of 11-step gray scale.

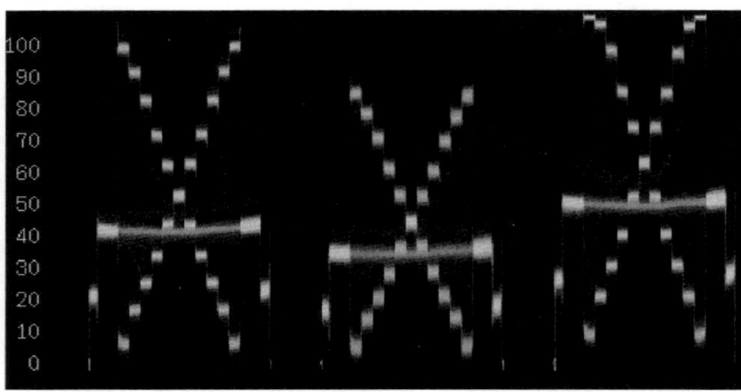

FIGURE 5.3-29 RGB scope trace of 11-step gray scale.

produced with a digital scope in either a parade mode, as seen in Figure 5.3-29, or overlaid. This signal should be white balanced until R, G, and B signals are even.

Camcorders will automatically white balance at the touch of a button if a white surface of sufficient size is in the center of the frame; however, there are many instances where white balancing in the field is not practical, and most cameras have at least one preset color temperature. The most common value is 3200° K, which corresponds to tungsten lighting. Many cameras also have higher color temperature presets for daylight (typically 5600° K) or a color correction filter to be used with a 3200° K preset. In some instances, electronic white balancing is out of the adjustment range of a camera, and color correction filters must be used.

Preset color temperatures are not usually as accurate as can be set using a scope, or even an auto white balance, especially given the color tint often introduced by individual lenses. It is possible to adjust a preset in many cameras so it is somewhat more accurate, and this is popular for shooters with a film background who are accustomed to film stocks with a single accurate color balance. It is worth noting, however, that precise white balance is not necessarily the goal in all circumstances. Directors may wish to purposely push the color in one direction or the other to create a mood or simulate an unusual lighting condition. A common practice in digital cinema is to fix white balance to a given color temperature and then allow color temperature changes, such as at sunset, to affect the image.

Gamma

The term *gamma* comes from the Greek letter γ (specifically, an exponent to the function $Y' = Y^{\gamma}$). It can cause considerable confusion because it can refer to different yet related meanings:

- The *transfer characteristic* is an imaging system's response, plotting video signal output against the light input. This is sometimes referred to as the gamma of the imaging system. Applying a look-up table (LUT) to an image affects the transfer characteristic and is said to change the gamma. Any change applied to a linear transfer characteristic might be referred to as a *gamma curve*.
- The *gamma of a D-Log E curve* (the transfer characteristic for a film stock) is the slope of the linear portion of the curve.
- The *gamma setting for a video camera* (master gamma, red gamma, or blue gamma) is the value of an exponent in a power law that determines the transfer characteristic. It is typically set between 0.35 and 0.75, with a standard setting close to 0.45. Adjusting this value will change the Y′ value of the "crosspoint" (the point where the top and bottom gray scale traces cross on a video waveform display of an 11-step gray-scale chart).

Gamma is the primary variable in the standard response function of a video camera. For most of the usable range of an imager, the actual output of the imager is raised to the power of γ so the end-to-end system response is substantially linear when viewed on a cathode ray tube (CRT) display. The complete gamma response is specified in video standards such as ITU 601 or ITU 709, including specifications to manage the step curve near the black point, but most of the gamma curve is a simple power law, which is the reciprocal of the characteristic response of the CRT (which was the only video display available when video standards were first developed). Today, non-CRT monitors employ look-up tables to mimic the behavior of a CRT. For a full explanation of gamma, including human perception, see reference [10].

Video cameras typically allow fine adjustment of the exponent (gamma), which has the effect of moving the middle range of the response. This changes the slope of the curve, which affects the relative contrast of the image. A lower gamma value such as 0.40 will increase the slope (and contrast) of the low end but decreases the slope in the middle and high ranges of luma. Higher gamma values such as 0.50 will decrease the slope of the low end and increase the slope of the middle and high ends. At normal exposures, a higher gamma value will create a more contrasty image. If an 11-step chart is exposed such that the white chip is 100% of the video level (assuming no knee function is used), then the cross point will typically be adjusted to 50% of the video level.

Shading and Gray-Scale Tracking

The imagers in video cameras will not always respond to light evenly across the image plane, and this is also impacted by the exit pupil of the lens. Adjustments known as *shading adjustments* must be made to compensate for this with any camera/lens combination. Black and white shading should be adjusted for red, green, and blue channels, both vertically and horizontally. Shading adjustments involve setting the *vertical saw correction* (top to bottom linear correction), *vertical parabola* (concavity of the vertical shading), and *horizontal saw* and *parabola* (see Figure 5.3-30). Poorly adjusted shading can lead to color shifts near the edges of an image. The characteristic response of

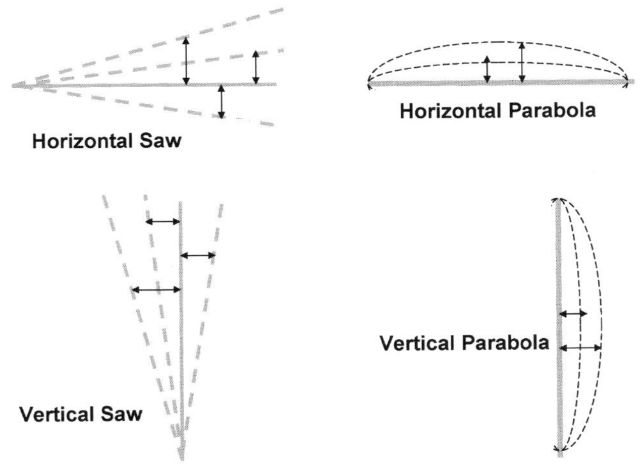

FIGURE 5.3-30 Shading adjustments.

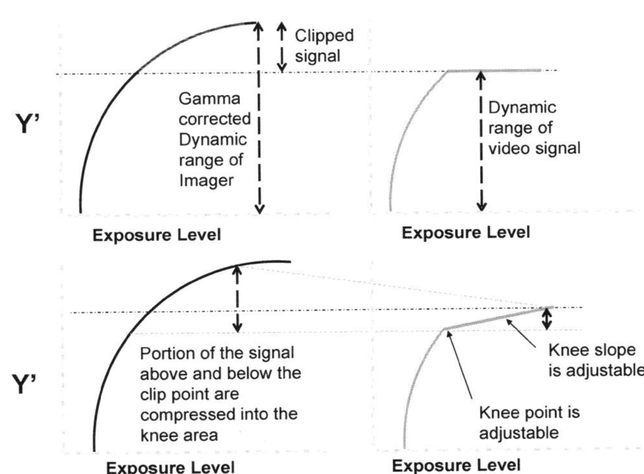

FIGURE 5.3-31 Illustration of knee.

individual red, green, and blue imagers may differ slightly, and for this reason, adjustments to shading and gamma will also need to be made independently for three color channels. If responses are not tracking identically a color shift may be observed as exposures or light levels change.

Knee, Black Gamma/Black Stretch

It is now possible to manipulate the response of a video camera in more specific ways than traditional gamma adjustments. Cameras can be adjusted to behave differently in the shadows and highlights using *knee adjustments* or *black gamma* (often called *black stretch*). Knee adjustments allow the camera to respond to light changes more gradually in the highlight region. This allows a camera to capture more contrast than a video signal can ordinarily carry by taking advantage of the full dynamic range of the imager. Signal above the normal clip point is combined with a portion of the signal near the clip point, and a very gradual response allows a significant range of illumination to be displayed in a highlight area near the clip point (see Figure 5.3-31).

It is possible to raise the black point of a video signal and reveal more shadow detail, but a more elegant means of bringing out detail in shadows is to use black gamma (black stretch). This works similarly to knee by establishing a "toe point" and pulling up or pushing down the response below that point. The effect of this is similar to adding fill light to the scene, but the electronic adjustment can introduce noise in the shadows. Conversely, crushing the blacks by applying a negative black stretch or *black press* can help reduce shadow noise and deliver darker blacks without necessarily pushing image details below the black point. Using knee can cause banding in the highlights because there are too few digital levels representing a significant range of illumination. Overexposed flesh tones cannot be corrected in postproduction if knee artifacts appear in highlights, so it is often desirable to increase the dynamic range of an imager without aggressive use of the knee.

Quasi Gamma

Another way to increase the dynamic range that can be carried with a standard video signal is to depart from the standard gamma function. Different manufacturers have implemented these kinds of techniques, usually in production-style cameras, as the resulting look is thought to resemble film-originated material. The Sony HDW-F900 CineAlta™ introduced a quasi-gamma like this in the /3 version of the product. Panasonic introduced CineGamma™, which works in the same way. In either case, the gamma response is flatter overall, and the middle exposure (gray-scale cross point) is set about 5 to 10% lower than standard gamma. Knee is still available, but the actual change in overall response increases the dynamic range below the knee point.

The Panasonic VariCam™ introduced "Film Rec" gamma, which does not use a knee and lowers the overall gamma response until the entire dynamic range of the camera corresponds to the video signal range. The overall slope of this new knee-less gamma is adjusted using the dynamic level, as shown in Figure 3.5-32; consequently, the raw image is quite dark and has low contrast until it is corrected in postproduction.

The Arriflex D-20 works similarly to this, offering a choice of knee-less transfer responses depending on the application. Cineon files do not apply a video-style gamma, either, but they have the option of applying a logarithmic response. The log response option was also adopted by the Thomson/Grass Valley Viper FilmStream™ camera. Standard gamma, CineGamma, and log exposure all apply the necessary nonlinearity to the video signal to fill a video signal or bit stream efficiently. Most of the more radical departures from standard video gamma assume a postproduction path

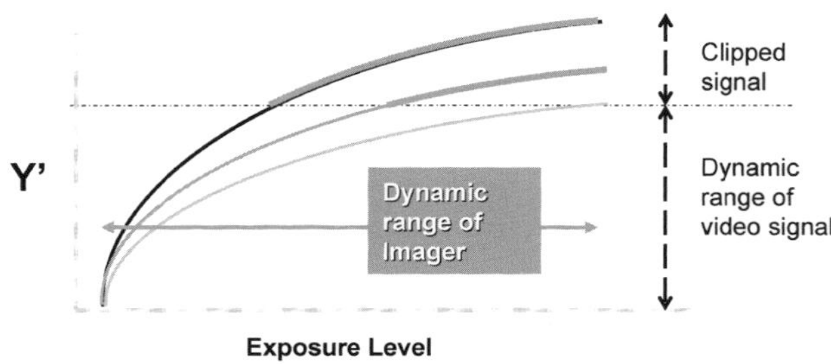

FIGURE 5.3-32 Overall response to gamma manipulation.

that will take advantage of the added range and optimize the image for the specific monitoring system.

Manipulating the response of a video system to maximize dynamic range helps to achieve some of the benefits of film production; however, it must be remembered that capturing a very wide range of tonal values in a recording system designed for a standard video signal (designed to capture only 5 or 6 f-stops of latitude) should be approached with caution. Most current digital video recording formats found in camcorders are 8-bit systems, meaning that they represent a video signal using only 219 discrete levels once signal overhead is accounted for. If an image includes over 44 dB dynamic range, or more than 10 f-stops of latitude, this means there are just about 22 levels per stop, which can easily lead to quantization errors. Recording systems designed for mastering such as D5 HD or HDCAM SR use 10 bit encoding and may be more appropriate for recording images spanning a very wide dynamic range.

Color Processing

If all lenses and color separation systems were ideal, properly white balanced cameras would show all colors identically on the same monitors. This is clearly not the case, and any one of the following will contribute to cameras showing differences in color reproduction:

- Any lens elements, prism, filter, or other optics will transmit light with a characteristic depending on the wavelengths found in the light. Filters do this intentionally, but no optical element passes all portions of the spectrum equally.
- Like optical elements, imagers also respond differently to variable wavelengths of light, and, although dichroic reflective coatings and mosaic filters attempt to perfectly separate R, G, and B, this is never perfect.
- Electronic manipulation of the color channels would ideally require negative gain on some frequencies from each channel so approximations have to be used.
- Many common light sources (e.g., fluorescent tubes) do not emit light with a continuous spectrum and exhibit spikes near specific wavelengths. These spikes can fall near nodes that exist in the color combination spectrum which will severely change the overall color gamut captured.

Color manipulation in the digital processing will seek to correct for the lack of ideal color representation, and can also be used to color match two cameras. This kind of color manipulation is more sophisticated than simple gain adjustment because it needs to be done without changing neutrals, which must remain black, white, and gray. To maintain gray scale, the changes must not impact the Y′ signal. To do so, changes are made to the matrix that transforms RGB to Y′PbPr, and, for this reason, these kinds of adjustments are referred to as *matrix* adjustments. To observe the effects of matrix adjustments on the color space, engineers will observe the signal on a vectorscope, which displays the values of Pb and Pr plotted in two dimensions (Figure 5.3-33).

Any neutrals will register at the center of the circle, and the Y1, R, Mg, B, Cy, G boxes show where yellow, red, magenta, blue, cyan, and green will appear if the vectorscope is fed color bars. Many color charts have been produced for this purpose, attempting to achieve perfect color chips. Charts will fade eventually so even the best chart must be checked periodically.

With a good chart and vectorscope an engineer can observe the changes in the hexagon as matrix values are changed. By shooting the same chart, two cameras can be made to match colors very closely. Some manufacturers provide a set of matrix tables to choose from, and others allow full adjustment to a set of variables. These kinds of adjustments are made possible by digital signal processing. Beyond general color space set-

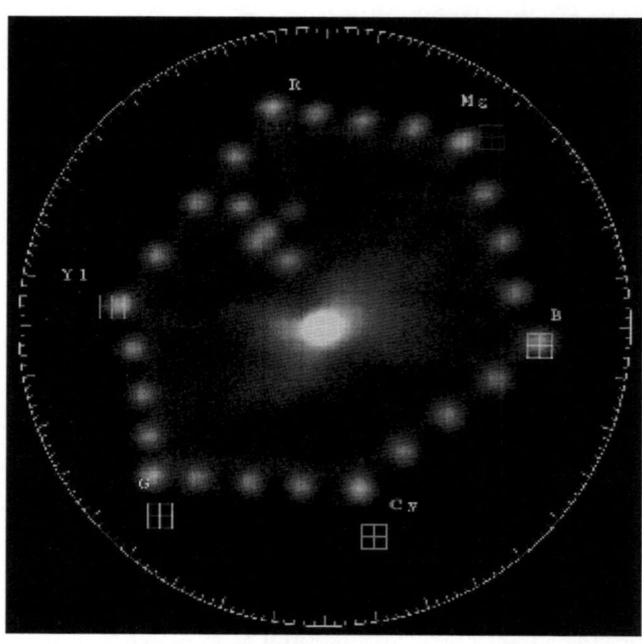

FIGURE 5.3-33 Vectorscope trace of a 24-color chart.

tings, camera manufacturers have begun to allow adjustment of the hue and saturation of individual colors; this is a very powerful tool and is a favorite when a client wants to see a specific product or logo shown on the monitor with a precise color.

As powerful as these controls are, care should be taken in using them. The human eye adjusts to color changes such that a major color adjustment may only appear to the engineer as a subtle one. The vectorscope will provide an unbiased view. Most importantly, these adjustments take signals from different color channels and remix them; if taken too far cross-color, noise can be introduced and other unexpected color artifacts may appear.

Detail Enhancement

As explained in the optics section, the perceived sharpness of an image is determined by the area under the modulation transfer function curve. Although it is not possible to increase the resolving power of a given video system without introducing aliasing, it is possible to improve the perceived sharpness. This is done in signal processing by increasing the modulation depth for the higher frequencies. One way of doing this is to create a *detail enhancement* signal. The detail enhancement signal is generated by applying an amplification to signals within a given frequency band near the Nyquist limit. When the detail enhancement signal is added to the luma signal, higher frequency signals are superimposed with the detail signal, increasing their modulation depth in the critical frequency bands, as shown in Figure 5.3-34. As can be seen in the figure, detail enhancement can also increase modulation depth beyond 100%; although this will mean a sharp picture, it can also be seen as an artifact. Increasing the modulation depth beyond 100% means that the Y′ will overshoot at a luma transition. An illustration of this can be seen in Figure 5.3-35. Depending on the aesthetics desired for the video image, this kind of hyper-sharp image may or may not be desirable. Extreme overshoots brought about by aggressive detail enhancement can also lead to *ringing*, which is a description given to a *ghost* edge.

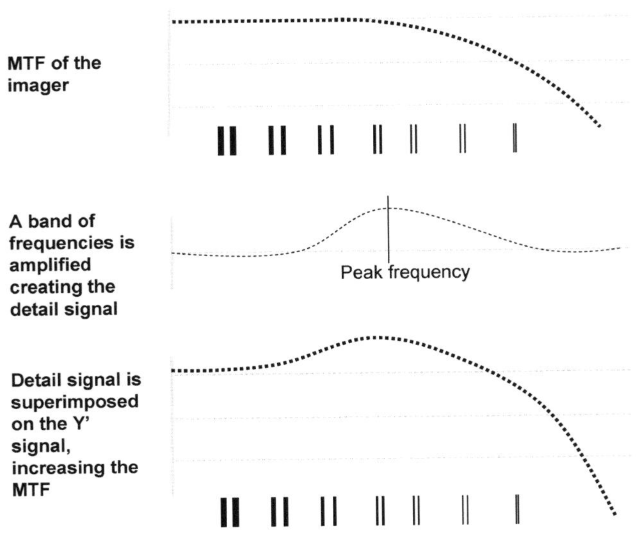

FIGURE 5.3-34 Detail enhancement.

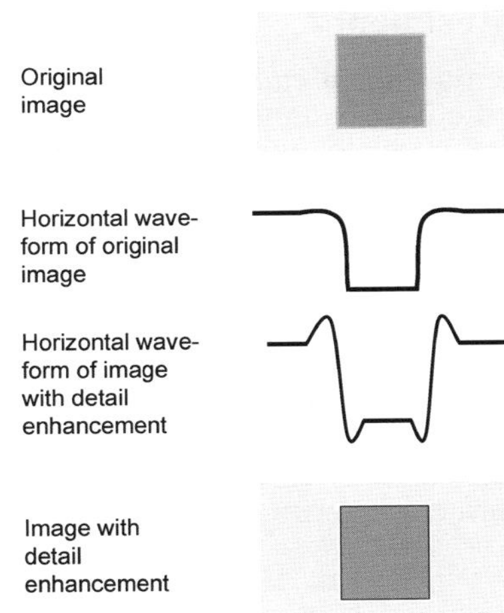

FIGURE 5.3-35 Illustration of detail enhancement with modulation depth >100%.

In professional broadcast cameras, detail adjustments can be made horizontally and vertically; the amplitude and peak frequency of the detail frequency can also be set. Most other adjustments to detail are involved with correcting for artifacts introduced by overly aggressive detail settings. Skin-tone detail is used to reduce, or dampen, the detail signals for colors deemed to be flesh tones, so these are usually very popular with aging on-screen talent.

Noise Management

Beyond the avoidance of extra gain and over-used matrix color correction, other steps can be taken to mitigate camera noise. To understand these, we need to determine where the noise is coming from. Detail enhancement amplifies a small band of frequencies, but because white noise (by definition) is spread over all frequencies, it inevitably amplifies some noise. Much of this increase can be avoided if we assume that most noise signals are low amplitude. This is the approach taken by *detail coring* or *crispening*. Before a detail signal is superimposed onto the video signal, it is subjected to a thresholding circuit that will reject any low-amplitude signals, most of which will be noise. Noise also tends to come from the low end of the luma scale. Shadow areas are closer to the noise floor, and items such as black gamma will apply more gain to these areas; consequently, noise can be reduced by crushing the blacks. A more elegant approach is to apply a coring to signals below a certain luminance level. This is sometimes referred to as *level depend*.

Camera Adjustment

Many problems with a camera's image may only be seen in specific circumstances, and the fix for it may cause another issue in very different circumstances. The task of a camera engineer can be formidable when circumstances are constantly changing. To help retain and reproduce camera settings, most modern cameras have begun to introduce scene files or user files that can save a complete table of camera settings in either internal camera memory or removable flash memory. This feature is also helpful when a station has only one fully trained camera engineer, who can save setup files for each camera so they can be called up by less experienced camera engineering staff.

CONCLUSION

The technology for acquiring moving images with electronic television cameras has advanced greatly over the years. The quality of pictures that is now available from a camera that can be held in the palm of the hand would amaze the broadcast engineer of 25 years ago, quite apart from the size of the camera itself. Cameras are now available in many video formats and form factors tailored to almost any application that can be imagined. Nevertheless, research and development continue, and it is likely that new device technologies will appear that will surpass today's CMOS and CCD imagers, providing increased performance and a new direction in camera technology. In parallel with this, new and refined systems for recording video signals at the camera will no doubt continue to be developed. Combined together, these advances will continue to provide the television producer with the tools demanded for that most fundamental part of the television process—acquisition of the moving image.

References

[1] ITU-R Recommendation BT 601-5, Studio Encoding Parameters of Digital Television for Standard 4:3 and Wide-Screen 16:9 Aspect Ratios, International Telecommunications Union.

[2] ATSC A/53E, ATSC Digital Television Standard, Revision E, with Amendments No. 1 and 2, Advanced Television Systems Committee, 2006.

[3] SMPTE 125M-1995, Television—Component Video Signal 4:2:2—Bit-Parallel Digital Interface, Society of Motion Picture and Television Engineers.

[4] SMPTE 293M-2003, Television—720 × 483 Active Line at 59.94-Hz Progressive Scan Production—Digital Representation, Society of Motion Picture and Television Engineers.

[5] SMPTE 259M-1997, Television—10-Bit 4:2:2 Component and 4fsc Composite Digital Signals—Serial Digital Interface, Society of Motion Picture and Television Engineers.

[6] SMPTE 274M-2003, Television—1920 × 1080 Image Sample Structure, Digital Representation and Digital Timing Reference Sequences for Multiple Picture Rates, Society of Motion Picture and Television Engineers.

[7] SMPTE 296M-2001, Television—1280 × 720 Progressive Image Sample Structure—Analog and Digital Representation and Analog Interface, Society of Motion Picture and Television Engineers.

[8] ITU-R BT 709-5 Recommendation, Parameter Values for the HDTV Standards for Production and International Programme Exchange, International Telecommunications Union.

[9] SMPTE 292M-1998, Television—Bit-Serial Digital Interface for High-Definition Television Systems, Society of Motion Picture and Television Engineers.

[10] Poynton, C. *A Technical Introduction to Digital Video*. Wiley, New York: Wiley 1996. Chapter 6, "Gamma," is available on line at: http://www.poynton.com/PDFs/TIDV/Gamma.pdf.

CHAPTER

5.4

Television Camera Supports and Robotics

PETER HARMAN

Vinten

Bury St. Edmunds, United Kingdom

INTRODUCTION

The methods used to support and move television cameras are constantly changing as technological advances influence and expand what can be created in the viewfinder. New generations of ultracompact cameras allow the camera operator to explore and get into the action as never before and longer lenses allow close-up views that were previously unobtainable. Virtual set and insertion technology in the studio offer opportunities for new production techniques.

These and other factors bring new challenges to program creators and those working behind the camera. They also place new demands on the devices used to support and move the camera. Of course, there will always be a place for handheld cameras for some types of productions, but for total control of the framing, picture stability, and producing seamlessly developed shots, a camera support of one sort or another will always be required.

Camera mounting products available today can be divided into several basic categories:

- Tripods
- Pedestals
- Cranes
- Jib arms
- Flying camera systems
- Body-worn stabilizing systems
- Robotic systems

Most of these systems have an associated *pan and tilt head* for pointing the camera.

There are other highly specialized gyro-stabilized support systems designed for airborne use, but they are outside the scope of this chapter. The following sections describe some of the products listed above and some of the factors that affect the choice for a particular application.

TRIPODS

Tripods are the most basic form of camera support and have been in use since photography began. The choice of tripod usually depends on the weight and size of the camera, where it's likely to be used, the speed of repositioning, and the types of shots that are likely to be needed.

Standard Tripods

Tripods are offered in a variety of carrying capacities; construction materials (plastic, carbon fiber, aluminum, and wood); configurations (single stage—one adjustable leg, two stage—two adjustable legs, or multistage telescopic); and with various accessories. They may have different interfaces for attaching the pan and tilt head. Typically this is a concave bowl feature at the top of the tripod with a diameter of 75 mm, 100 mm, or 150 mm. Other head interfaces are used for heavy-duty applications. Figure 5.4-1 shows an example of a 99 lb capacity carbon fiber professional tripod with a pan and tilt head fitted.

A clear understanding of the application is essential before a choice of tripod is made. There are several professional products available on the market, all

FIGURE 5.4-1 Vinten Fibertec tripod (shown unextended) fitted with a midlevel spreader, feet, and Vision 100 head. (Courtesy of Vinten.)

offering a variety of features aimed at the serious-minded camera operator. There are many more "low-end" tripods available, seemingly offering similar functions and features at a fraction of the cost of the professional models. However, few of these can achieve the performance needed for good camera work.

When choosing a tripod, the initial cost of purchase is only part of the decision. By going for the low-cost option, financial savings may be achieved, but it could be at the cost of creativity caused by restrictions due to physical limitations of the device. Such limitations may also add more cost and time in the edit process to correct shooting deficiencies. Some of the criteria to be considered include:

- Lightweight yet robust enough to take the knocks of the professional world.
- Torsional rigidity to resist the twist put into the tripod due to drag forces when panning. In other words, when the legs are fully extended and clamped with the spreader attached, the top moulding/casting should resist any twist. If present, this twist, which is known as *spring back* or *wind up*, may be seen in the viewfinder when the pan bar is released (the picture typically springs back slightly).
- The legs should not deform or the clamps slip when the tripod is loaded with the pan and tilt head and the camera. The tripod should also be rigid enough to resist movement due to wind buffeting.
- Simple, reliable leg locks that release the legs only when required and are easily maintained for optimum holding reliability. All fittings and clamps should be able to be used with gloved hands.
- Multistage legs (single stage, two stage, or multistage) that allow extremely low (e.g., as low as 16 inches and as high as 79 inches) shooting angles. Typically, the more stages, the greater the height range.
- Availability of *baby legs* for extremely low-angle shots (e.g., down to 9 inches). These are usually constructed in the same way as standard tripods but with reduced leg length; they usually accommodate standard pan and tilt heads.
- Robust stabilizing devices such as the ground or floor spreader (for hard, level surfaces) and midlevel spreader (for uneven surfaces) that add stiffness to the tripod by bracing the legs while allowing for a variety of terrains. Note that although a midlevel spreader seems to be the ideal solution, it doesn't add as much stiffness to the tripod as a ground spreader. The leg ends of the tripod should fit snugly into the ends of the floor spreader, rubber feet, or points of the triangle. Any clearance here will be noticeable during framing.
- Spiked feet, which allow the camera operator to push the feet into a soft surface such as grass, when a spreader is unavailable or unsuitable, or flat plastic/rubber feet that help to prevent slipping and damage to hard surfaces. Note that both spiked feet and plastic feet should be used wherever possible with a midlevel spreader to ensure overall stability.
- The tripod legs should have clearance in order for the leg tubes to move freely during setup and breakdown. However, these clearances must not be so large that they produce a "dead spot," known as *backlash*, which is noticeable when the direction of the camera pan movement is changed (the operator may feel the drag reduce and/or a positive "bump"). If backlash exits in any degree, it makes framing very difficult at the start and finish of a move, especially when using a long lens for a tight shot in adverse weather conditions.

Lightweight Tripods

Lightweight or electronic news gathering (ENG) professional tripods usually have either a 75 mm or 100 mm pan and tilt head interface. This is a fundamental feature that allows the pan and tilt head to be levelled accurately and locked off securely. Seventy-five mm tripods typically are used for small-bodied DV or HDV camcorders (typically weighing up to 15 lbs) and are popular with those working in event videography fields, such as weddings and corporate and industrial communications. They are also popular with camera operators and journalists working mainly on foot in the documentary field where portability is more important than stability and robustness.

One hundred mm or ENG tripods on the other hand are designed for fully featured ENG cameras (typically between 17 and 33 lbs) and this type is the choice of professionals working in news or similar areas. Of prime importance for these users is ease of use and maintenance, especially in the field where the

nature of their work often means that they are in difficult circumstances with very little time to react to an event as it unfolds. This type of tripod must be reliable and simple to fix or adjust using only the most basic of tools.

Medium-Weight Tripods

Medium-weight or electronic field production (EFP) tripods usually have a 150 mm pan and tilt head interface, suitable for a much larger camera configuration (typically up to 66 lbs). This design of tripod, although still relatively light in weight, uses much larger leg tubes and offers much more rigidity than could be achieved using an ENG tripod. Although speed and ease of setup is important, more emphasis is placed on stability and ruggedness. Typical configurations could either be a 16 mm film camera or an ENG or other portable style camera body with a large lens, viewfinder, matte box, and pan bar controls for zoom and focus. A teleprompter may even be used occasionally. EFP tripods are used extensively in sports applications such as at the touchdown line or corner position and are used in some television studios for fixed camera applications that don't need a pedestal.

Heavy-Duty Tripods

Heavy-duty tripods are a much more substantial construction and use either a standard four-bolt fitting, Mitchell screw, or other proprietary quick-release system to attach the pan and tilt head. Typically, these tripods are used to support full-facility camera systems weighing up to 200 lbs (studio camera with a box lens, viewfinder, pan bar controls, and possibly a teleprompter) and would be used either in the studio or on a remote (outside broadcast) where absolute stability (with long focal length zoom lens) is crucial. Figure 5.4-2 shows an example of a 300 lb capacity professional tripod with a pan and tilt head fitted.

Tripod Accessories

Accessories for the above-mentioned types of tripods range from simple *dollies* (also known as *skids*) with swivelling (castored) wheels, which allow the user to easily move the tripod and camera across a smooth surface from one position to another. These devices are not normally used live as the quality of movement and directional control is not on-air quality.

In order to make camera height changes, many tripods can be fitted with a simple pneumatic or crank-operated elevation unit, giving the advantage of adjustable additional height, such as looking over a crowd. These devices also are not normally adjusted on-air because the motion is not smooth enough.

As mentioned previously, it is essential to use a spreader to ensure stability. A variety of ground/floor or midlevel spreaders can be used to provide steady and predictable shots in a variety of locations where a tripod without a spreader could collapse.

FIGURE 5.4-2 Vinten HDT tripod with Vector head. (Courtesy of Vinten.)

PEDESTALS

Pedestals differ from simple tripods and skids principally in that they incorporate an assisted and balanced elevation unit (a telescopic column supporting the weight of the camera and the pan and tilt head at any position). Some pedestals are fully skirted to protect the wheels and some have an open design rather like a tripod in a skid. Pedestals are typically used in studios or on smooth and level purpose-built surfaces, and are designed for moving the camera during recording or live shows, so noise and smoothness of movement are critical factors to consider.

Some older pedestals use mechanical springs or counterweights to provide column balance, but these can be noisy. Also, due to the many moving parts, motion roughness may be seen in the viewfinder (and felt through the pan bars) if they are not maintained adequately. Modern pneumatic systems offer smooth movement and are almost silent in operation. Pneumatic pedestals are the norm in almost all professional studios around the world, enabling greater creativity and consistent performance. The number of stages of column elevation is an important factor as this will determine the height range (perhaps from a minimum of 18 inches up to 60 inches maximum). The studio pedestal should at least be able to place the camera lens below the eye height of a person sitting in a chair, and able to position the lens over the shoulder of a person standing.

Pedestals vary in sophistication from devices with simple castors to systems that offer both crab (wheels all point in the same direction) and steer capabilities. Once again, it is important to understand what the camera operator is required to perform before selecting a pedestal and whether it will be used only in the studio or also in other locations, and if it must support only a large camera or a variety of cameras. Finally the director may want to be creative with camera movement or it may simply be locked off.

Pneumatic Pedestal Balance

The pedestal column system should offer sufficient support so that the camera operator can raise and lower the camera with one hand. Pneumatic pedestals are based on the principle of using compressed gas as the means of support for the column. The pneumatic system is usually charged from a cylinder of compressed nitrogen gas or with compressed air from a pump.

To explain how this works, consider a simple bicycle tire pump. By covering the outlet of the pump and pushing down on the piston, the pressure inside the pump increases. As the pressure increases, so the effort needed to push down on the piston increases. Now imagine that the effort being applied to the pump is the constant force of a camera system sitting on top of a pedestal and that the pump body is the sealed pneumatic system in the pedestal. In this case, the column would drop to a point where the internal pressure is equal to the pressure being applied by the camera system—the system would be in balance at that particular point. This arrangement would not, however, be particularly helpful to the camera operator as the camera system may weigh up to, perhaps, 220 lbs and the operator would have to be strong enough to lift the camera from this point. Furthermore, if the camera needs to be lowered further, then the operator would need to lean on the steering ring to take it down by compressing the gas further.

Most modern pneumatic systems offer perfect balance by utilizing either tapered pistons or eccentric pulleys that overcome the problems associated with compressing a gas with a piston. They then can provide just the right amount of counterbalance throughout the stroke, such that the amount of effort required to raise and lower the camera remains constant and the camera can be left stationary at any point in the stroke.

Temperature Changes

Temperature changes can have an adverse affect on the pneumatic counterbalance system. For example, the temperature in the studio in the morning is likely to be cool. In that case, the operator setting up the pedestal may notice that there isn't enough pressure in the pedestal to support the camera (the camera drops when the lock is released). So the pedestal may then be pumped up until it balances perfectly (the camera remains stationary wherever the column is positioned). When the shoot begins the ambient temperature may gradually increase (due to lights, staff, and, perhaps, an audience), causing the gas pressure inside the pedestal to increase, eventually to a point where the pedestal does not balance and the camera rises. The operator may then let out a little of the gas too so that the camera remains in balance. This is not good practice, however, because after the shoot finishes, the studio will cool down again and the following day more gas will have to be added.

To overcome the effects of temperature variations correctly, most pedestals are supplied with trim weights, which should be added and removed as the temperature changes. These small weights are sufficient to overcome the changes in pressure so perfect balance can be maintained throughout the day without adding or releasing gas.

Portable Pedestals

Portable pedestals, as the name implies, are designed to be easily moved from one location to another. They vary in functionality and capacity and their choice depends on the camera system being used and the degree of on-air movement that the director needs. For example, a portable pedestal, as shown in Figure 5.4-3, provides perfectly balanced elevation for camera systems up to 66 lbs, but does not have a steerable skid, making it an ideal pedestal for a small studio using simple camera configurations. Other systems offer perfectly balanced elevation up to 165 lbs and a fully

FIGURE 5.4-3 Vinten Vision Ped Plus lightweight pedestal. (Courtesy of Vinten.)

steerable skid, providing both crab and steer functionality, making them suitable for most studios using a fully featured compact camera, lens, viewfinder, and possibly teleprompter. These lightweight portable pedestals can be separated into skid and column for breakdown, transport, and setup. Where larger cameras and/or large lenses are used, there are larger, stripped down versions of studio pedestals that support up to 220 lbs, provide a minimum height of 18 inches and a maximum height of 60 inches, and have both crab and steer functionality. This is the type of pedestal often seen at large concerts running along a level floor in front of the stage.

Studio Pedestals

A large studio television camera together with prompters, pan bar controls, and viewfinders can weigh nearly 220 lbs. By its very nature, a studio pedestal, as shown in Figure 5.4-4, is a robust product, typically utilizing a triangular base housing three sets of two steerable wheels and designed to carry loads of between 175 and 265 lbs. Both lightweight and heavy-duty studio pedestals can be single, two-stage, or four-stage versions offering height range anywhere between 18 and 60 inches.

On all studio pedestals, steering and elevation are controlled with a steering ring immediately below the pan and tilt head. A foot pedal controls how the three wheels are connected for steering. When all three wheels are locked together in parallel, the pedestal is said to be in *crab* mode and the pedestal travels in the direction of the steer spot or pip on the steering ring. When the pedal is depressed again, the mechanism locks two of the sets of wheels together (parallel) and the third set turn as the steering ring is rotated (pedestal is said to be in *steer* mode).

When the column is correctly balanced, the operator should be able to easily raise and lower the camera on shot, and skilled operators will be able to maneuver the pedestal almost as an extension of themselves, tracking in and out, crabbing left to right, or any combination, enabling the cameraperson to follow a performer walking around a set.

FIGURE 5.4-4 Vinten Quattro pedestal with Vector head. (Courtesy of Vinten.)

PAN AND TILT HEADS

A pan and tilt head is a precision device that allows the operator to move the camera horizontally and vertically in order to point the lens at a subject and compose the picture. Horizontal rotation is known as *pan* and vertical rotation is known as *tilt*. It is sometimes necessary to pan or tilt the camera in order to see a different part of the scene or to enhance the visual storytelling, but usually it is associated with following a subject as it moves.

Pan bars (either fixed length or telescopic) are attached to the sides of the head and are used to pan and tilt the camera. Typically, the operator would use one pan bar for ENG work where only the monocular viewfinder is used; this leaves the other hand free to operate the lens controls. Two pan bars are used when the camera has an overhead or side-mounted viewfinder screen and pan bar-mounted controls are typically used in this situation for lens zoom and focus.

For most program and creative purposes it is desirable to make camera movements as invisible to the viewer as possible. Jerky and unnecessary movements in the picture are disturbing to the viewer, and professional equipment is designed to help the camera operator overcome these unwelcome effects. Balance and drag characteristics are important functions that determine the usability of a pan and tilt head.

Camera Mounts

The camera may be fixed to the head in several ways. MiniDV and other compact cameras use the standard 1/4-inch screw and pin to attach the camera to a camera plate, which in turn either slides into a dovetail on the top of the pan and tilt head or into a quick release plate. Professional cameras use a quick-release camera plate that in turn attaches to the same plates as explained above, but use two standard 3/8-inch screws that offer a much more stable fixing. For larger cameras using large and heavy box lenses, it is usual for the camera to be fixed to a large V-shaped plate, called a *wedge plate*, which slides into a similarly shaped receptacle on top of the head, called a *wedge adaptor*.

Counterbalance Systems

Just as a pedestal needs a counterbalance system to assist with raising and lowering the camera, a pan and tilt head also has to provide counterbalance to overcome the falling away effect as the camera is tilted from the horizontal. This enables the operator to control the motion without having to compensate for an unbalanced head. There are many methods available to overcome this problem and four systems typically used in video production are described here.

Counterbalance Torque

First, it is worth explaining the physics behind a tilting camera. To help understand this, imagine someone holding a 2 lb weight in his or her hand. If lifted to be at shoulder level, with the arm tucked in to the body and bent completely closed so the hand is above the elbow, then it is relatively easy to support the weight. However, if the weight is lowered by extending the arm at the elbow, the further the arm extends, the harder the weight becomes to support. Now, the weight is not getting heavier, but the amount of effort needed to support it increases the farther the center of gravity (CoG) is moved horizontally away from the pivot point (elbow). The effort (torque) needed to keep the weight from falling to the floor follows the equation:

$$\text{Torque} = \text{Mass} \times \text{Radius (from the pivot to the CofG)} \times \text{Sin (angle of rotation)}$$

This is also true for a camera that is tilted when the center of gravity moves relative to the pivot point. The further the camera is tilted from horizontal, the more effort (torque) it takes to keep it from falling. The graph in Figure 5.4-5 shows this relationship. To prevent the camera from falling, the manufacturer may design a counterbalance system that generates a counteracting force.

Most torque-generating counterbalance systems generate a linear torque. As the equation and graph show, as the counterbalancing torque requirement is sinusoidal, a linear system will react rather like an unbalanced pedestal column; that is, the camera will fall away from horizontal to the balance point and will rise up from the bottom of its movement to the balance point. This is a poor characteristic and camera operators may be forced into either adding tilt friction to lessen the fall away or to dial in the amount of counterbalance that ensures the camera springs back to horizontal, neither of which is ideal.

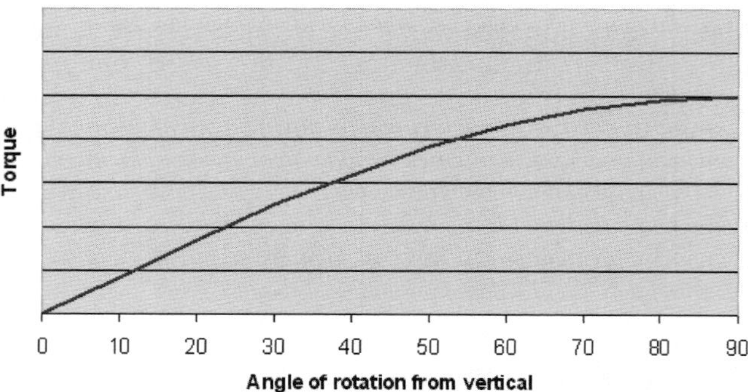

FIGURE 5.4-5 The relationship of counterbalance torque to angle of tilt.

A well-designed perfect balance system, however, will allow the operator to "dial in" exactly the right amount of counterbalance torque to match the camera system precisely throughout the tilt, resulting in even effort throughout the tilt and the ability for the operator to release the pan bar with confidence that the camera will remain stationary. It also allows the operator to use as much or as little drag as they choose without regard for counterbalance compensation.

An alternative design concept used for some pan and tilt heads, as described later, is to eliminate the need for counterbalance torque by an arrangement that rotates the camera about its center or prevents the camera's center of gravity from changing its height.

Drag Systems

In order to control the camera, not only is counterbalance necessary, but also resistance to movement (drag) is needed to help prevent overshooting and to smooth out transitional movement. For example, imagine the task of framing a tight shot when the camera operator is a long way from the scene. If there were no drag applied and the head were perfectly balanced, the cameraperson would find it nearly impossible to keep the shot steady, and minor framing adjustments would be difficult to make without the motion being obvious to the viewer. Therefore, drag is applied to help dampen out the unwanted movements and to help make minor adjustments unobtrusive.

There are several methods available to add drag. The easiest way is simply to apply friction to the moving axis, usually through the application of the brake. This method has inherent problems; the brake is typically either on or off and subtle adjustment to drag levels is difficult. The brake tends to stick and introduce jerking at the start or stop of a movement; this being particularly noticeable with light cameras.

A better drag method is the use of a lubricated friction system, a braking system immersed in a lubricant that works itself between the brake material and the moving surface. The lubricant prevents the sticking associated with the simple brake and the careful choice of brake material allows for fine adjustment of drag levels and both slow and whip (rapid) pan at any level of drag. Precise manufacturing techniques are necessary for this type of drag system so it is therefore more costly than nonlubricated drag systems, but smooth and reliable performance is achieved.

Newer developments in drag technology rely on the principles of achieving resistance to movement through fluid shear using a sandwich of fixed and moving plates that are separated by a thin film of grease. This type of sealed system can accommodate many types of fluids and wide operational temperature ranges can be accommodated. However, one undesirable characteristic of this type of system is that resistance increases proportionally to speed, which may result in a whip pan being impossible. Another problem is that infinitely variable adjustment is difficult to achieve. Some manufacturers approach this by making their systems modular, with levels of drag selected by switching in different combinations of modules. One manufacturer uses a fluid shear approach that is infinitely variable and has design features that modify the output/speed characteristics so that whip pans are still achievable.

Spring Balance Heads

Improvements in the design of springs have made it possible to design a head using spring balance systems for up to 90 kg of load. The main benefit of the spring-balanced head is the ease of adjusting the balance when the camera load is changed. This becomes important in the field when changing the eyepieces or viewfinder, or when batteries or recorders are added.

In this type of pan and tilt head, some form of spring mechanism (which may be fixed, stepped, or fully variable) counterbalances the weight of the camera. The amount of spring counterbalance (torque) required depends on both the overall weight of the camera and the height of its center of gravity. Therefore, a significant effect is felt when adding, for example, a viewfinder on top of the camera.

Post Heads

The simplest pan and tilt balance method is the post head. This is based on the alternative design concept mentioned previously and a means to produce counterbalance torque is not required. With this arrangement, the camera and lens can be balanced about its own center of gravity and therefore will remain where it is pointed until such time as the camera operator moves it to another position.

The head consists of a tilting L-shaped cradle attached to a vertical body, which is also able to pan. The camera is attached to the cradle and adjusted forward and backward until the camera sits horizontal with the center of gravity sitting central over the vertical axis of rotation. The camera is then raised and lowered until the cradle is able to support the camera at any angle (CoG is now over the horizontal axis of rotation).

Although the design of the post head now seems dated, its simplicity has a lot to commend it, and post heads have earned high reputations and have dedicated users. Variations on this design are often used in motorized robotic heads, as discussed later in the chapter and shown in Figure 5.4-12.

There are, however, two disadvantages with post heads. The first is that, in order to carry heavy loads, they need to be big and extremely robust and, therefore, are only practical for lightweight cameras. Tilt can be limited by the overall body design and some camera operators would say that unless the axis of the lens sits directly over the rotational axis, panning appears unnatural. Although this would appear to be so, tests have shown that, for most normal movement, being off axis by up to 15 cm is usually undetectable.

Cam Heads

For many years the cam head was the industry standard pan and tilt mounting for large television cameras and lenses, providing easy and precise movement. The cam head principle derives from a simple premise that if a mass moves such that its center of gravity remains in a horizontal plane, its potential energy does not change, and it will therefore remain in balance, again, without using counterbalance torque. A post head keeps the center of gravity of the camera in the same position throughout the tilt range. A cam head achieves this by the use of a developed cam, which allows greater tilt by letting the center of gravity of the load move forward and backward without moving up and down. In theory, every different payload requires a unique cam shape to achieve this balance relationship, but in practice, a compromise is obtained using an incremental range of cams. When tilted, the camera is, in effect, pivoting about its center of gravity, and although its mass has to be shifted backward and forward a few centimeters, this requires very little effort.

As with the other heads described, the camera will remain at any tilt angle and will only move to another if the operator moves it. However, if the payload changes significantly in use—say a lens is changed or the viewfinder repositioned—a different cam must be fitted to restore balance. Choosing the correct cam is a combination of guesswork and trial and error. Changing cams can be inconvenient and time consuming and is a major limitation to the design. This type of head is illustrated in Figure 5.4-6.

Pantograph Heads

The most recent counterbalance system comes in the form of a pantographic system, which works on the

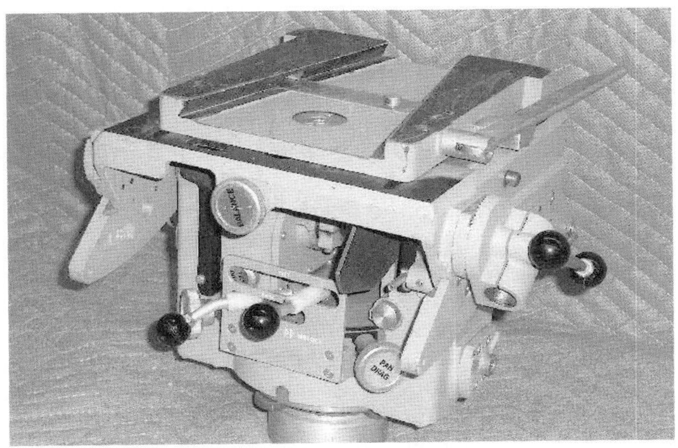

FIGURE 5.4-6 Vinten Mk 7 cam head. (Courtesy of the Broadcast Store, http://www.broadcaststore.com/.)

FIGURE 5.4-7 Vinten Vector pantograph balanced head. (Courtesy of Vinten.)

FIGURE 5.4-8 Vinten iScript prompting system. (Courtesy of Vinten.)

same principle as the cam system, keeping the center of gravity in a horizontal plane, but this time, the lift is generated by a series of mechanical linkages. The balance mechanism provides the lift necessary to maintain the camera center of gravity at the same height, as the head tilts from horizontal. The rate at which the mechanism lifts is infinitely adjustable to match any camera that falls within the head's center of gravity range and weight capacity. Therefore, this type of system doesn't suffer from downtime associated with cam heads if the camera system changes. Because the system is infinitely adjustable, the operator can trim or make major adjustments whenever necessary without having to strip the head down, saving time and improving camera control. This type of head is illustrated in Figure 5.4-7.

Teleprompters

On many occasions, it will be necessary for a *teleprompter* to be used with a camera. These come in a variety of forms, with CRT or LCD screens of different sizes, overslung or underslung mounting, and other variations. An example is shown in Figure 5.4-8.

The teleprompter device needs to move with the camera and must be securely attached to the pan and tilt head. For most prompters, brackets are attached to the camera plate (slide, quick release, or V) on top of which the camera is fixed. The prompting device is then attached to the front of the bracket and the whole assembly attached to the head, either into the dovetail groove, quick release plate, or wedge adaptor. Note that the teleprompter increases the mass of the overall camera system and its center of gravity is raised and moved forward considerably. This puts an extra load on both the pedestal and pan and tilt head counterbalance systems. To reduce the loading on the head, some manufacturers offer teleprompter fixings in the top platform of the pan and tilt head, making rigging and derigging faster and reducing the load on the balance system.

CRANES, JIB ARMS, AND FLYING CAMERA SYSTEMS

These categories of camera supports provide additional capabilities as needed by directors for some types of programs. They allow the camera to get a much higher point of view than tripods and pedestals and may enable closer views of the action without the intrusion of an operator. Many of the devices offer multiple axis movement and wide area coverage unavailable using traditional methods.

Platform Cranes

The Hollywood film industry created the ride-on platform crane, and the picture of one of these huge devices and the director with a megaphone shouting instructions to the camera operator and focus puller is a familiar sight.

Platform cranes traditionally not only supported the camera, but the operator also who sits at the end of the crane on a revolving platform, controlling the camera. These devices naturally found their way into the early television studios and have been utilized extensively. In television today, with camera size and weight decreasing continually and remote control commonplace, it is more likely that the camera operator will be controlling the camera and crane movement from the studio floor or from a track-mounted dolly. However, there are still many manufacturers making platform cranes and they are still very popular, but more so in film than in television.

The *Nike*[1] and *Tulip*[2] are examples of big, ride-on cranes still in use that cover a wide area and offer a huge range of creative options. The Tulip, shown in Figure 5.4-9, allows two operators at the camera end while a crew of *grips* would be responsible for controlling the

[1]Manufactured by Chapman/Leonard Studio Equipment, Inc.
[2]Manufactured by Egripment Support Systems.

FIGURE 5.4-9 Egripment Tulip platform crane. (Courtesy of Egripment Support Systems.)

arm and its rolling base. The longest version of this model can achieve a maximum height of 31 ft and a minimum height of –17 ft (assuming the ground contours allow) and it has a reach (swing) of 31 ft.

Remote Cranes

Remote cranes allow the director to cover an even bigger area by utilizing smaller cameras and lightweight materials. In this case, a remote control camera is situated at the end of the arm and controlled from the ground by a camera operator. A team of grips is employed to maneuver the crane either on track, a dolly, or in a fixed position. The *Xtreme T12*,[2] shown in Figure 5.4-10, is a telescoping remote crane that has a reach (sweep) of 33 ft, a maximum height of 38 ft, and a minimum height of –18 ft (assuming the ground contours allow). It is also able to telescope in and out and utilizes a series of sliding counterweights to keep the arm in balance throughout, offering an additional axis for even greater creative freedom.

Jib Arms

A jib arm, also known as a boom, is essentially a means by which an operator can freely maneuver the camera in all axes, utilizing a swivelling arm with a pan and tilt head on one end and counterweights on the other. Rather like a pedestal but with far less movement restriction, the operator has the freedom to elevate and track horizontally, vertically, and diagonally in addition to the pan and tilt function of the head. All this movement is controlled by the operator end of the arm. The jib arm may simply be attached to a tripod in a fixed position or may be attached to a tracking skid or pedestal for additional freedom of movement.

Flying Camera Systems

"Flying" camera systems, such as *Skycam*,[3] have become practical with the advent of ultracompact cameras. They allow the director to obtain dramatic

[3]Manufactured by Winnercom, Inc.

FIGURE 5.4-10 Egripment Xtreme T12 remote crane. (Courtesy of Egripment Support Systems.)

pictures from high above a venue. Such systems can be effective where cranes cannot reach and are used extensively for sporting events and concerts throughout the world. Utilizing a pan and tilt head with a moving cable system, flying camera systems provide high-angle shots covering a very large area. These systems offer an additional view to the director and create a feeling of involvement for the viewer. The camera is positioned with a series of cables, pulleys, and drive motors suspended above the venue with a computer-based remote control system. Pan and tilt and camera functions are managed via an RF link.

BODY-WORN STABILIZING SYSTEMS

Body-worn camera stabilizing devices, such as the *Steadicam*[4] and the *Artemis*,[5] offer complete freedom of movement, limited only by the stamina and capability of the operator. They aim to provide the stability of a tripod or pedestal with the freedom of a handheld camera. An example is shown in Figure 5.4-11.

These devices are supported by the camera operator through a body-worn harness. They remove unwanted motion disturbances by isolating the camera from the

[4]Manufactured by Tiffen-Steadicam.
[5]Manufactured by Sachtler.

FIGURE 5.4-11 Steadicam Clipper 2. (Courtesy of Tiffen.)

cameraperson's body by supporting the camera on a counterbalanced articulated arm housing a series of shock absorbing joints. In addition, the overall mass of the system (camera, battery, monitor, mechanics, etc.) is distributed widely so as to increase its moment of inertia, making it less prone to shock and consequently much more stable.

One disadvantage of body-worn devices is that they require a significant investment in training in order to operate them safely and reliably. It can also be tricky to reposition the camera with any degree of accuracy, relying on the expertise of the operator more than the support itself. As the operator supports the entire weight of the system, fatigue can be an issue, particularly with heavier cameras. However, the results obtained can be very good, relying perhaps on the immersive qualities of the picture rather than absolute picture stability. They allow the camera operator to explore many locations in ways that traditional devices would not allow, for example, stable pictures can even be obtained while running up a flight of steps.

Other simpler stabilizers have been introduced specifically for compact lightweight cameras. These are typically handheld devices using systems that may include special handles, gimbals, weights, balances, and distributed mass to stabilize the camera.

ROBOTIC SYSTEMS

Robotics can be utilized to control the functions of a pedestal (track, crab, and elevation); pan and tilt head (pan and tilt); and some of the camera functions (zoom, focus, and perhaps iris and black level), all without a camera operator being present in the studio. These systems can perform complex moves either from joystick controls or from a series of preset shot positions. Often, the robotic system operator will work in the studio control room alongside the production crew, but occasionally will be located on the studio floor. With the proliferation of computer networking, multiuser and multilocation control of pedestals, elevation units, and pan and tilt heads is now a reality.

Applications

Remote operation of cameras is becoming very important as new and unusual shots are always in demand. Robotic cameras are particularly useful where an operator's safety could be compromised, for example, in trackside coverage of high-speed motor sports. Robotic cameras also offer a cost-effective solution for a wide variety of studio programming and systems and are used extensively either in static locations or on robotic pedestals, offering unrivalled positional accuracy (especially important as the use of virtual sets increase), wide coverage, and unobtrusive operation.

Robotic cameras are particularly useful in predictable situations such as in a 24-hour news studio where repetition is all part of the look and feel of the program and where conventional operator costs can become a burden. They are also frequently used for coverage in legislative venues, where traditional manually operated camera methods are considered too intrusive.

There is little doubt, as technology improves and production demands increase, that the use of robotic products inside and outside the studio will increase as production companies realize its cost effectiveness and flexibility. Systems are now available with scalable system architecture. This enables a facility to purchase equipment to suit their current needs, while allowing future hardware and software upgrades, extra robotic heads, and pedestals to provide additional features and functionality as required.

Robotics Compared to Manual Systems

Robotic cameras have a number of operational advantages over manual production methods, which include the following:

- Repeatability, with the ability to reproduce the same camera shot time and time again, on- or off-air. Camera movement is as smooth and controlled as if being done by a skilled operator.
- The overall number and involvement of operational staff can be reduced as the robotic system does most of the camera work. Expensive and time-consuming rehearsals to block the shots are not required as

the preset camera shots only need to be set up once by the system operator.

- With broadcast-quality robotic cameras, more thought and creativity can be put into set design without fear of operator errors or the need for workarounds.
- With cheaper and smaller cameras, they can be placed in areas previously inaccessible to a manually operated camera—for example, in tight areas in a stadium or in dangerous areas by a racetrack. However, their positioning still needs to be controlled and small robotically controlled devices are the ideal solution. This proliferation of camera positions allows for greater and more creative coverage of an event.
- Robotic cameras can be set to work with and/or around complex set pieces, and repeat those same moves consistently.
- Reduction of costs. The system purchase price is a fixed cost whereas personnel costs are prone to escalate. Typical users have found that their robotic systems pay back in less than two and a half years.

Robotic Pan and Tilt Heads

Remote-controlled pan and tilt devices are usually designed around the post head principle as it allows for the inclusion of motors and control electronics in the space within the body that is not taken up by a complex counterbalance mechanism. Movement of the pan and tilt axis is by means of direct or indirectly linked servo motors. These motors, combined with their control electronics, must be able to offer on-air quality movement at any speed as any inaccuracy in the drive will be seen as a jerky movement, especially when using a long lens and making small framing adjustments. Smooth motion is achieved by the use of high-quality servo devices, which send many millions of data pulses to the electronic systems used to control the motion. In essence, the more pulses the servo generates per degree of movement, the smoother the motion will be. An example of such a head is shown in Figure 5.4-12.

Some manufacturers also offer the ability to operate these heads manually as a fall back, which means that a mechanical decoupling device between the motor and the driven axis is necessary. To make manual operation as controllable as possible, simple drag systems are also built in, allowing the user to set their preference and operate with a feel similar to that of a manual head.

Robotic Pedestals

Robotic pedestals, as illustrated in Figure 5.4-13, work in a similar way to the robotic pan and tilt head. They are essentially standard studio pedestals, but with precision servomotors driving the wheels, steering and column elevation.

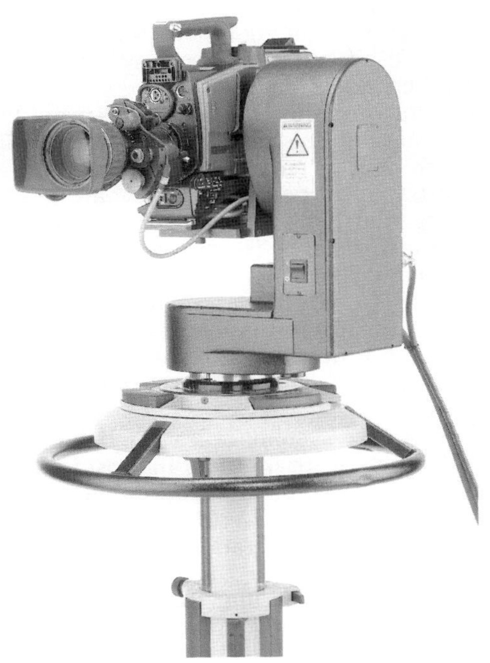

FIGURE 5.4-12 Vinten Radamec Broadcast Robotics pan and tilt head. (Courtesy of Vinten.)

FIGURE 5.4-13 Vinten Radamec Broadcast Robotics pedestal. (Courtesy of Vinten.)

All such pedestals use encoders on the column motor for vertical positional data. Most also use encoders built into the wheel and steering motors to provide the data needed to navigate from one preset position to another. Another system utilizes infrared devices to provide navigational data. There are advantages and disadvantages to both systems.

Wheel encoders, for example, count revolutions to give distance and this can sometimes provide misleading information. This is because no studio floor is perfect—there will be small peaks and troughs and chips in the surface and perhaps also a coating of dust. The floor irregularities give rise to distance-travelled errors (linear distance is not the same as the distance travelled) and dusty floors can allow the tires to skid, giving rise to additional distance-travelled errors. Over time, the accumulation of errors may mean that the preset shot framing must be adjusted or, if errors are serious, that the pedestal must be recalibrated back to its original datum.

The infrared system relies on being able to see enough targets to triangulate its position. Sometimes these targets may be moved, damaged, or simply obscured by sets, and therefore the pedestal becomes lost and requires recalibration back to its datum.

No manufacturer to date has come up with a design that does not suffer from one or both of these problems, so it is important when considering robotic pedestals to think about the state of the studio floor, make the necessary repairs, and have a regime of regular floor cleaning. If infrared-navigating pedestals are used, sets must be planned to leave space for the infrared targets. If all of these issues are considered prior to the installation, and manufacturer's guidelines adhered to, robotic pedestals should provide reliable service and consistent shots.

A robotic pedestal with pan and tilt head provides the director with full three-dimensional (X, Y, and Z) control that simulates a manually operated pedestal. Sometimes this sophistication is not necessary. In that case, a robotic elevation unit retrofitted to an existing pedestal is able to provide elevation only or the robotic head can be attached to a tripod for fixed position pan and tilt only.

Control Systems

Robotic camera systems typically consist of a user interface and control system connected to the robotic devices (pan and tilt head, pedestal, elevation unit, and camera lens controls) via a set of cables that carry the movement signals and power to and from the devices. An additional cable-carrying video also goes back to the control panel, which is used to set up shots, manage framing, and store shots to be repeated.

User Interface

The user interface typically uses simple joystick controls with one for pan, tilt, and zoom, and a second for x, y, and z, position in space, and focus via a panel-mounted wheel. Figure 5.4-14 shows the layout of a typical control panel, which would be used in conjunction with a touch screen, as shown in Figure 5.4-15. This panel layout shows, from left to right, rotary joystick pedestal control (x, y, and z); camera control for black and iris; lens focus; shot transition controls (cut, fade, stop, and transitional speed); camera switch bank (1–16); and function switches and rotary joystick control for pan, tilt, and zoom.

The speed of movement is usually controlled by the joysticks. A slow push of the joystick generates a slow movement from the device and a fast movement generates a rapid movement from the device, making acceleration, movement, and deceleration genuinely controllable, as if the operator were directly controlling the pedestal and pan and tilt head.

The function of the touch screen is to enable the operator to easily create and recall stored shots. The screen area itself acts as a preview of the camera the operator has selected, and when the shot is composed, an icon depicting that particular shot is generated for visual identification along with text identification.

In addition, the touch screen allows the operator to make changes to the system, create backups, and to generally manage the operating system and engineering functions.

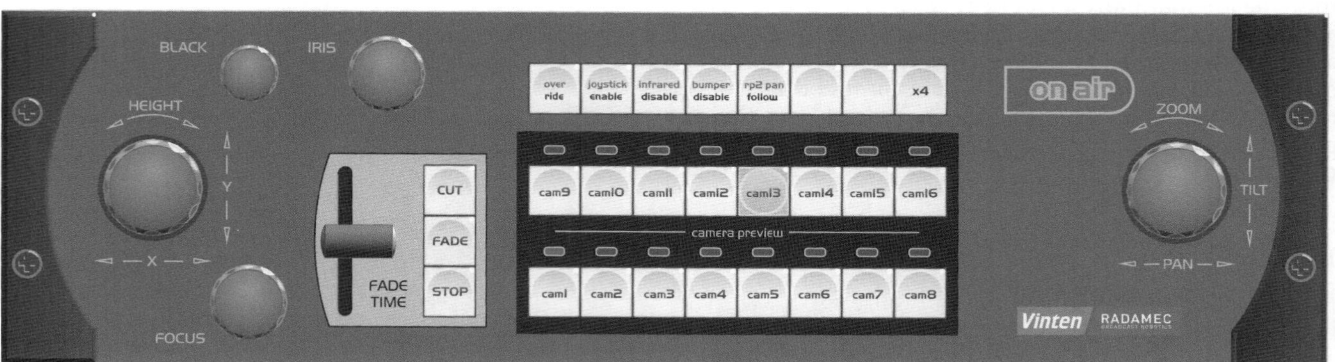

FIGURE 5.4-14 Vinten Radamec Broadcast Robotics Fusion control panel. (Courtesy of Vinten.)

FIGURE 5.4-15 Vinten Radamec Broadcast Robotics touch screen. (Courtesy of Vinten.)

System Calibration

Some systems rely on the controller hosting all the movement data, driving "dumb" devices that simply respond to the data as it arrives from the controller, while other systems rely on the devices storing the movement data (smart devices), which respond to much simpler data from the controller. Either way, all devices must be calibrated from a reference point as they use this point for navigational and reset purposes. For example, robotic pedestals are driven across the floor from a datum (zero X and zero Y), and as they do so, the encoders in their wheels measure the distance and bearing from the datum in terms of distance X and distance Y. If the pedestal is elevated, then its position will be stored as distance Z from a datum previously determined on the column (e.g., when the column is down fully). The same is true for robotic pan and tilt heads. The operator establishes a datum in pan and tilt (e.g., tilt could be when the platform is horizontal and pan could be when the camera is pointing at an object in the set) and movements from these datum points are recorded as angular data in both pan and tilt.

Having established these datum or reference points, the operator can then establish the shots to be used in the show and store each shot that can be retrieved using a touch screen or push-button identification.

Shot Setup

Setting up shots is a simple process. The operator selects the camera by choosing it from the control panel or touch screen and then manipulates the joysticks and focus controls until the picture is perfectly framed. This process generates positional data relative to the series of datum points established earlier. The operator then stores the shot (positional data) into a file. Most systems also allow the user to store, recall, and edit multiaxis motion, for example, for an opening shot.

During the show, the operator recalls these shots (either through a touch screen or through a series of buttons on the controller) as required by the director. Any reframing needed, for example, if the presenter moves, can be achieved using the joystick controls.

System Extensibility

Robotics control systems often come in several configurations—head control only, head and pedestal control, touch screen, simple pushbutton panel, and others—so it is wise to think carefully when selecting robotic camera equipment about what the requirements are now and how they may change in the future. However, some control systems now available have been designed with expandability and some can control more than one brand of device so an initial choice with this type of system gives maximum flexibility.

CAMERA TRACKING FOR TWO- AND THREE-DIMENSIONAL VIRTUAL SETS

Chroma key has been used for many years and is still in use today for video and graphics inserts for a variety of applications. Unfortunately, with conventional chroma key systems, the camera has to remain stationary because, if the camera moves from one side to the other, the perspective will change on the foreground (presenter), but not on the background (inserted graphic or other source). This is a problem with virtual sets, where a complete program may be created using computer-generated backgrounds with real actors or talent in the studio and camera movement is required. An example of such a studio is shown in Figure 5.4-16.

The camera movement and perspective problem can be overcome by fitting the pedestal, pan and tilt heads, and the cameras with positional encoders, rather like those used with robotic devices. These feed the camera's position in space back to the computer that is generating the virtual set graphics. By doing this, it is possible to change the perspective in real time, making the virtual set, and those within it, look completely lifelike.

Important to the process of feeding the camera position to the computer is the calibration of the devices on the set. Any errors in device position relative to the datum will be seen on screen. For example, the real world could move at a different speed than the virtual world or the perspective between the two worlds shifts as the camera moves.

An alternative to mechanical encoders is an infrared scanning system that uses special tracking cameras to identify the position of the studio cameras. This *Free-d* system, developed by BBC Research,[6] is used in the studio shown in Figure 5.4-16. The system has a number of passive coded targets placed in the studio

[6]See http://www.bbc.co.uk/rd/index.shtml.

FIGURE 5.4-16 Free-d camera tracking system in a virtual reality studio. (Courtesy of Vinten.)

lighting grid (visible as white circles above the set), each uniquely identified by a circular bar code. The targets are illuminated with infrared light and viewed by small CCD infrared cameras mounted on the studio camera and pointing upward. The use of narrowband LEDs for the infrared light and retro-reflective material ensures a sufficient number remain visible to the tracking camera under normal studio operating conditions.

The image from the tracking camera is processed by a Free-d unit and used to calculate the exact position and orientation of the studio camera. This is done by real-time analysis of the image to identify each target. Knowing the physical location of the targets, Free-d calculates the position of the camera to a high degree of accuracy. The zoom and focus axes are monitored by high-resolution optical sensors mounted on the lens. This data is combined with the video from the small camera CCD as ancillary data. The Free-d processor then calculates the Pan, Tilt, Roll, X, Y, Height, Zoom, and Focus parameters to accurately pinpoint the studio camera position. The complete data package is transmitted via a high-speed serial data link to the 3D virtual reality system. The computer can then control the virtual set image so its perspective matches the position and movement of the camera.

SUMMARY

No matter how good the camera or lens is, if the support system is inadequate or unstable, the quality of camera movement may be poor. In addition, if the camera support does not offer perfect balance and freedom in all axes, the camera operator will be distracted from the primary objective—that of composing the picture. Camera supports must be robust, reliable, easy to maintain, simple to set up and adjust, and safe to operate, and provide a stable platform for smooth, controllable, and quiet on-air movement. When properly implemented, such systems help the person behind the camera exploit emerging technologies while working with consistency and confidence to maintain creative production values.

Further Information

Further information on choosing camera support products for particular cameras and applications can be found in downloadable literature available at http://www.vinten.com/. Information on products from other manufacturers referred to in this chapter may be found at:

http://www.chapman-leonard.com
http://egripment.com/en/index.html.
http://www.skycam.tv
http://www.steadicam.com
http://www.sachtler.com
http://www.vintenradamec.com

CHAPTER

5.5

Video and Audio Switching, Timing, and Distribution

BIRNEY DAYTON

NVISION, Inc.
Grass Valley, California

INTRODUCTION

Since the last edition of this handbook was published, an immense change has overtaken the infrastructure of the typical broadcast facility. Only a few short years ago, most television plants had a largely analog infrastructure with islands of digital equipment. Today, virtually all facilities have converted to a fully digital core or are on a fast pace to get there. This chapter addresses the basis of the analog infrastructure for a broadcast facility, essentially as a historical reference. It then discusses various aspects of the contemporary digital plant, highlighting some of the issues and challenges for video and audio distribution, synchronization (sync), timing, and switching.

BACKGROUND

Analog Video

Analog video in North America conforms to the specification of the National Television Systems Committee (NTSC). The NTSC system has 525 total lines and 486 active lines of video. Typically about 483 lines are used for picture information and, in current practice, many images have only 480 active lines available because of MPEG compression limitations in the distribution chain.

The NTSC system operates at a field rate of 60/1.001 = 59.9401... Hz (usually described as 59.94 Hz) and a line rate of 15.750/1.001 = 15.7342... kHz. Color difference information is modulated onto a quadrature amplitude modulated, suppressed subcarrier (3.579545... MHz) at 455/2 times the line frequency.

Analog television signals are generated with sawtooth sweep waveforms at both the vertical and horizontal frequencies. This causes the video signal to contain both even and odd harmonics of the sweep frequencies. When the color subcarrier is placed at an odd harmonic of the line frequency, the spectral energy of the subcarrier and its modulation sidebands falls in between the line harmonics of the luma[1] signal, as shown in Figure 5.5-1. This characteristic prevents the subcarrier energy from being directly added to that of the luma signal and saves 3 dB in peak transmitter power. It also greatly reduces the visibility of the color subcarrier in the picture and allows the color information to be effectively separated from the luma with a comb filter in a NTSC decoder.

Line and Field Frequencies

At the time the NTSC color system was introduced, there were a number of receivers deployed with *intercarrier* sound that depended on a precise 4.5 MHz separation between the visual carrier and the aural carrier. For this reason, the video line frequency was changed to be 15.750/1.001 kHz to produce the following relationship: Fh = 4.5 MHz/286. This puts the aural intercarrier beat at a harmonic of the line frequency and causes the sidebands of the color subcarrier to straddle the sound carrier, thus minimizing color into sound interference.

To avoid naking obsolete about 100,000 deployed intercarrier receivers, this approach was chosen in

[1]Luma is an approximation of true luminance derived by combining gamma-corrected RGB signals rather than linear RGB signals.

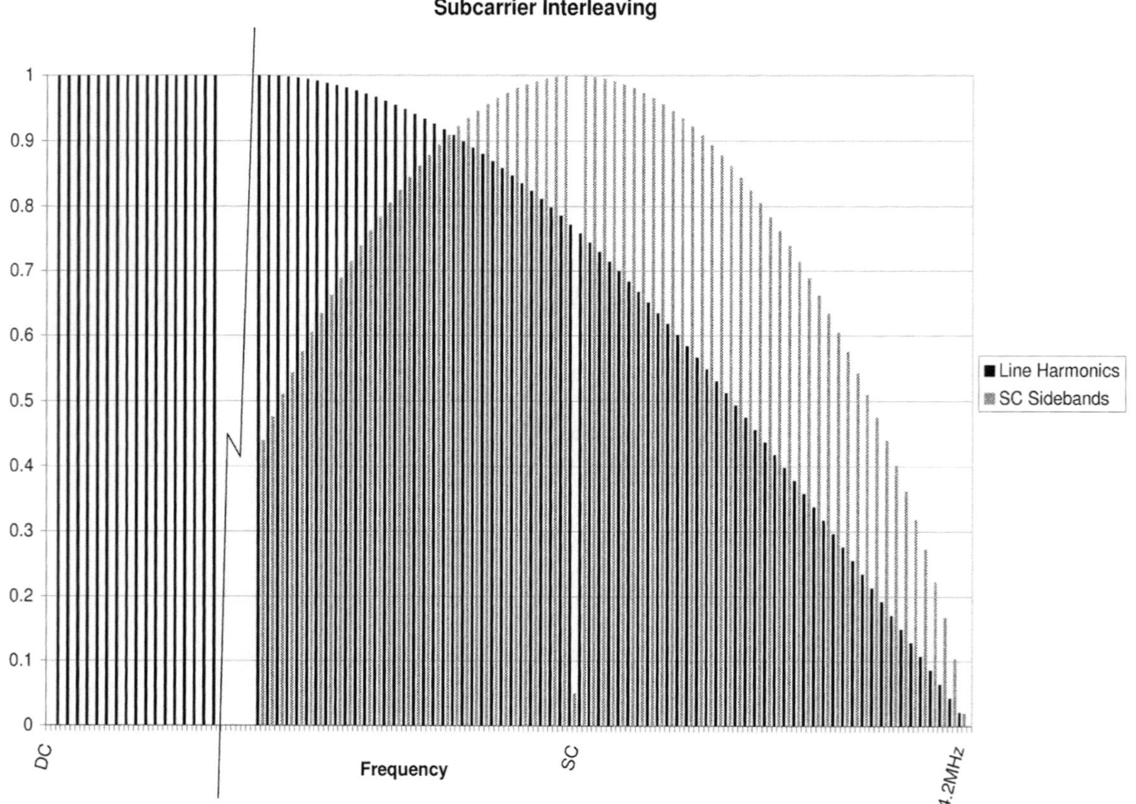

FIGURE 5.5-1 Illustration of luma/chroma interleaving.

preference to simply moving the sound carrier. In retrospect, many would say the choice was a poor one, but it is an important piece of broadcast history and it helps a new entrant to the business to understand why we still live with the complexity of 1/1.001 line and frame rates everywhere in the system.

Analog Audio

Television audio started as single channel monaural sound. The capability to broadcast stereo sound was introduced in 1984, but many broadcasters delayed converting to two-channel operations until the introduction of home video recorders and stereo television sets with high-quality stereo audio created increased consumer demand. The subsequent transition to stereo was very rapid.

Analog Sync

Early television systems used a synchronizing approach that simplified the design of synchronized equipment at the price of a rather complex sync distribution system. Synchronizing signals were produced by a central *sync pulse generator* and distributed around the facility to all cameras and other source devices. With monochrome television, the signals were *horizontal drive*, *vertical drive*, and *composite sync*. When NTSC color came along, *subcarrier and burst flag* were added.

As transistors replaced vacuum tubes and equipment evolved, it was realized that it should be possible and cost effective to distribute a single reference signal and for each device to internally produce all the references needed from it. In this way, they could back-time themselves and produce correct system timing. The signal chosen for this single reference was called *color black*, and that became the sole survivor for most sync distribution systems.

Color black is effectively composite sync with color burst, with 7.5% setup added to active video lines. This signal has all the information necessary to recreate timing references on a locked piece of gear. In spite of several issues, color black still prevails today as the most common timing reference for both analog and digital video systems.

Analog System Timing

For accurate switching and mixing to occur, analog video signals must be accurately *timed* so that all signals to be switched or mixed arrive at the switch point with their sync pulses and color subcarrier synchronized. The classic standard for 525-line analog systems has been to time signals within ±1 degree of subcarrier at routing switcher and production switcher inputs.

System	Scan	System Type	Standards	Field Rate	Fields/ Frame	Lines/ Frame	Active Lines/ Frame	Line Rate Hz	Line Integer Relation to 2.25 MHz. H=	Subcarrier	Luma Sample Rate	Active Y Samples/ Line	Color Difference Sample Rate	Bits/ Sample	Serial Data Rate	Field (V) to 48 kHz Audio (A) Relationships
NTSC	I	Analog	SMPTE 170M	60/1.001 Hz	2	525	483	15750/ 1.001	2.25 MHz/143	455H/2	Analog	Analog	Analog	Analog	Analog	V/5=A/4004
PAL	I	Analog	ITU-R BT.470	50 Hz	2	625	576	15625	2.25 MHz/144	25+1135H/4	Analog	Analog	Analog	Analog	Analog	V=A/960
525/59.94	I	Digital	SMPTE 125M, 259M	60/1.001 Hz	2	525	483	15750/ 1.001	2.25 MHz/143	Component	858H= 13.5 MHz	720	429H= 6.75 MHz	8 or 10	17160H= 270 Mb/s	V/5=A/4004
625/50	I	Digital	ITU-R BT.601, BT.656	50 Hz	2	625	576	15625	2.25 MHz/144	Component	864H= 13.5 MHz	720	432H= 6.75 MHz	8 or 10	17280H= 270 Mb/s	V=A/960
750/60	P	Digital	SMPTE 296M, 292M	60 Hz	1	750	720	45000	2.25 MHz/50	Component	1650H= 74.25 MHz	1280	825H= 37.125 MHz	8 or 10	33000H= 1.485 Gb/s	V=800A
750/59.94	P	Digital	SMPTE 296M, 292M	60/1.001 Hz	1	750	720	45000/ 1.001	2.25 MHz*20/ 1001	Component	1650H= 74.25/1.001 MHz	1280	825H= 37.125/1001 MHz	8 or 10	33000H= 1.485/1.001 Gb/s	5V=4004A
1125/60	I	Digital	SMPTE 274M, 292M	60 Hz	2	1125	1080	33750	2.25 MHz*3/200	Component	2200H= 74.25 MHz	1920	1100H= 37.125 MHz	8 or 10	44000H= 1.485 Gb/s	V=A/800
1125/59.94	I	Digital	SMPTE 274M, 292M	60/1.001 Hz	2	1125	1080	33750/ 1.001	2.25 MHz*15/ 1001	Component	2200H= 74.25/1.001 MHz	1920	1100H= 37.125/1.001 MHz	8 or 10	44000H= 1.485/1.001 Gb/s	5V=4004A
1125/60	P	Digital	SMPTE 274M, 424M, 425M	60 Hz	1	1125	1080	67500	2.25 MHz*3/100	Component	4400H= 148.5 MHz	1920	1100H= 74.25 MHz	8 or 10	44000H= 2.97 Gb/s	V=A/800
1125/59.94	P	Digital	SMPTE 274M, 424M, 425M	60/1.001 Hz	1	1125	1080	67500/ 1.001	2.25 MHz*30/ 1001	Component	4400H= 148.5/1.001 MHz	1920	1100H= 74.25/1.001 MHz	8 or 10	44000H= 2.97/1.001 Gb/s	5V=4004A

FIGURE 5.5-2 Television formats, system frequencies, and sample rates.

Analog audio timing is important to the extent of maintaining lip sync and stereo phase, the latter being a much tighter constraint.

HD and SD Video Formats

HDTV was standardized in the late 1980s and early 1990s via the Advanced Television Systems Committee (ATSC) and the FCC-driven Advisory Committee for Advanced Television Standards (ACATS) process. The final system finally agreed upon has a single transmission system but supports multiple HD and SD scanning standards (including a component video version of the NTSC system). The standards and video formats are discussed in Chapters 1.11 and 1.16.

The two most common HDTV formats are interlaced 1080 lines at 59.94 fields/sec (usually abbreviated to 1080i/60) and progressive 720 lines at 59.94 fields/sec (720p/60). In the production environment, the 1080i system has 1125 total lines (1080 active) and 2200 total horizontal pixels per line (1920 active). The 720p system has 750 total lines (720 active) and 1650 total horizontal pixels (1280 active). It may be noted that the 720p system has exactly two-thirds the pixel count in both directions as the 1080i system. This makes for a slight reduction in effective bandwidth (versus the nominal square root of 2) but allows simple scalers in cross converters and displays (ignoring the de-interlacing issue).

Figure 5.5-2 lists the main parameters and sample rates for the principal analog and digital video formats, both SD and HD.

Both 1080i and 720p are used with 2-3 pulldown to support 24 (or 24/1.001) frames per second film sources. In the case of 24-frame operation, the studio equipment operates as if it were handling a 60-field (or frame, for progressive systems) signal, and the transmission encoder may extract and encode only the 24-frame information.

When the HDTV system was being developed, there was a concerted effort to move back to the 60 Hz field rate, and the current high definition (HD) and standard definition (SD) digital systems are designed to be operable at both 60 Hz and 59.94 Hz. In practice, they are universally operated at 59.94 Hz because of the need to convert NTSC programming to digital and for simulcast NTSC/DTV transmission. Refer to SMPTE 170M and the various digital video standards [1].

BASEBAND DIGITAL VIDEO

Serial Digital Video Interface

Digital video equipment was first introduced more than 20 years ago, although the *serial digital interface* (SDI) standard, SMPTE 259M [2], was not published until 1993, followed by the *high definition serial digital interface* (HD-SDI) standard, SMPTE 292M [3], in 1996. For a number of years, digital systems were largely confined to production islands, and fully integrated digital plants were rare. However, the introduction of a wide range of digital equipment using SDI and HD-SDI interfaces, with higher performance at lower cost, coupled with increased demands from users, led to complete systems being designed using digital interconnections end to end.

Since the launch of DTV broadcasting in 1996, momentum has rapidly increased to build new facilities and to rebuild existing facilities to support both digital SD and HDTV programming. Many modern infrastructure products are designed to support both 270 Mbps and 1.5 Gbps operation, so a single switching and distribution fabric can be built to handle both SD and HD concurrently.

For facilities that have a large block of dedicated SD functions, at the time of writing (2006) some cost savings can still be realized by using interface cards (for routing switchers, for instance) that support only 270 Mbps and not 1.5 Gbps. This decision should be made with care, because even with equipment that supports upgrade by card replacement, the upgrade cost is significant, and if the operating mode of the block changes to include HD sooner than expected, the SD cards may have to be replaced before they are fully amortized.

Bandwidth

Serial digital video is a much different kind of signal from analog video. The spectrum of an NTSC analog video signal extends from DC to about 5 MHz. The serial digital signal spectrum extends from a low of a few kHz to at least the third harmonic of the data transition frequency. In the case of SD video, that is 405 MHz (3 × 135 MHz), and for HD it is just over 2.2 GHz (3 × 742.5 MHz). Some fifth harmonic will typically be present at driver outputs, but it is not critical to signal recovery.

As the preceding numbers show, serial digital HD video signals operate with roughly 500 times the bandwidth of analog NTSC and include spectral energy up to three times that of the highest UHF carrier frequency. Many of the rules that used to apply only at the transmitter are now very much part of building a studio. For example, as little as a quarter-inch stub on a cable or circuit board trace can cause a serial HD signal to fail due to signal distortions caused by reflections. This precludes the use of BNC "T" connectors and most other loop-through techniques that were commonplace in the analog world. In almost all situations, serial digital signals should be viewed as suitable for carefully matched point-to-point connections only.

Looking to the future, a number of cameras are currently being manufactured that scan at 1080p/60 rates, and SMPTE has now standardized a 3 Gbps interface [4] to support future applications. Already, there is infrastructure equipment appearing on the market to support rates from 270 Mbps all the way to 2.97 Gbps.

Ground Loops

Other factors have to be considered when moving from an analog infrastructure to SDI or HD-SDI signal distribution. A large percentage of analog equipment was supplied with differential input amplifiers that mitigated (at least somewhat) the need for rigorous system grounding. Serial digital signals, for electromagnetic interference (EMI) reasons, are typically grounded to the frame at both ends. The spectrum of the serial interface is such that hum-producing ground loops typically do not interfere with the signal, but due to stray currents, they can, and will, produce hot coaxial cables if not carefully managed. This makes careful single-point system grounding at least as important for digital video as it always was for analog systems.

DIGITAL AUDIO

AES3 Audio

The AES3 audio standard, discussed in Chapter 1.15, defines a stereo digital audio interface with up to 24-bit sampling. The interface is designed to operate at sampling rates from 32 kHz all the way up to 192 kHz and beyond, but in television broadcast facilities (except in special cases) only 48 kHz sampling is used. In most applications, the audio sampling must be kept synchronous with the video signal. The term *synchronous* here means frequency synchronous, because the relationship between 48 kHz and 59.94 Hz is somewhat less than "tidy." To be precise, there are 4004 audio samples in every five 59.94 Hz fields (or progressive frames). This relationship is clearly simpler at 60 Hz, 50 Hz, and 24 Hz, but for now, 59.94 Hz rules the day. Figure 5.5-2 lists the 48 kHz audio sample rate relationship for the principal video formats.

Embedded Audio

SMPTE standards 272M and 299M [5] define methods of multiplexing AES3 digital audio into the ancillary data area of SDI and HD-SDI digital video stream, respectively. Up to 16 channels of audio may be multiplexed with a video stream and, in situations where audio production requirements are limited, this can considerably simplify facility design, greatly reducing the amount of separate audio distribution.

Digital Versus Analog Levels

In the analog world, audio levels are measured above and below a nominal operating level, with some assumed headroom. For most purposes, analog broadcast operating levels in North America have been pegged at +4 dBu for unterminated voltage-driven audio systems (voltage equivalent to +4 dBm in a 600 ohm terminated environment). Headroom is typically 14 to 20 dB, giving a maximum level from +18 dBu to +24 dBu).

Digital systems inevitably exhibit the property of hard clipping at maximum level (when the analog to digital converter runs out of range, there are no more numbers to code), so general practice is to refer all levels to full-scale digital (FSD). The nominal digital broadcast operating level (in North America) is 20 dB below FSD or –20 dB FSD. With the advent of HDTV and multichannel audio, there are some real challenges to this practice. Movie producers, for instance, typically operate with considerably more headroom than broadcasters. Multichannel compressed audio streams (typically distributed with Dolby E coding, followed by Dolby Digital transmission) are designed to carry this additional headroom all the way to the consumer. There is no hard standard in the movie industry, but 27 dB of headroom is most common, so in any situation where mixing of broadcast audio with movie audio must occur (such as voice-overs), extreme care must be taken to assure that the levels are matched.

Level management is also a major issue when managing commercial inserts into movie-originated programming. The Dolby Digital transmission encoder includes a mechanism to manage source program level differences via *dialnorm* metadata. For this mechanism to work properly, all operating levels must be understood, labeled, and correctly communicated to the transmission encoder. SMPTE RDD6 [6] describes the metadata communication protocol and its use.

Channel Overhead Data

The AES3 standard carries up to 24 bits of audio per channel and has an additional 8 bits in each subframe. Four bits are used for the preamble, which can be made up of three specific biphase code violations to identify frame and block starts, and four bits are ancillary data, with the most important being channel status. The channel status bit is coded into 24 bytes of information over a block period (192 samples) with a number of useful bits of information, such as whether the audio signal was pre-emphasized before digitization and whether the signal is linear audio or some other form of data (for instance, compressed audio).

When AES3 audio is used, it is important to understand the limitations of devices in the chain with respect to carriage of the ancillary data. The majority of simple infrastructure devices will carry the data through unharmed, but recorders are particularly unaccommodating with carriage of ancillary data, and signal processing devices (such as sample rate converters) may change the data to reflect new conditions or choose not to pass the data at all to avoid confusion. An example is a simple audio mixer. At any point in a fade other than at the limits, deciding what to do with ancillary data is challenging. Often, devices of this sort will simply stripe new data onto the output stream. This situation makes the ancillary data bits difficult to use for any information that must follow the audio through the system from end to end.

Monaural and Multichannel Audio Issues

The AES3 standard defines a sample rate (typically 48 kHz for broadcast operations) and a block rate that is equal to 1/192 of the sample rate (at 48 kHz, that is a period of 4 ms). Re-aligning audio to the nearest block point at every processing station would cause an unacceptable amount of delay to be introduced into the signal, so typical practice is to re-align to the nearest sample. This means that at a device input—a routing switcher, for example—multiple audio streams may not (and typically will not) be in block alignment. If it is necessary to disassemble two or more AES3 streams and reassemble them into new streams, the channel status data, which is block aligned, must be delayed into alignment or restriped, with the former approach being highly preferable. A sample case is a routing switcher that switches at the individual channel (monaural) level.

Dolby E Issues[2]

Dolby E is a proprietary compression system designed for carriage of eight channels of audio through contribution, distribution, and production environments. The Dolby E coded signal is formatted exactly as an AES3 signal. Therefore, there are two ways of transporting it: either as a separate signal or embedded into an SDI or HD-SDI digital video stream. Some care must be exercised when processing the Dolby E signals, as it has certain properties that may be unexpected to some users.

Dolby E can be encoded at three different frame rates: 24 Hz (or 23.98 Hz), 25 Hz, and 30 (or 29.97) Hz. For North American television applications, the 30 (29.97) Hz rate is nominal. The encoder introduces two video frames of delay, so the video must be delayed to match.

For proper performance, the signal must be switched on frame boundaries in the appropriate vertical interval. This means that in interlaced systems, all switch strobes must be capable of and set to frame (not field) rate switching. For 720p systems, the switch strobes must be referred to a 30 (29.97) Hz reference (typically NTSC color black) and set to switch on every other frame of the 720p60 system, because the Dolby E frames span two 60p frames. There is no two-frame indicator in progressive systems, so there is also a potential frame offset ambiguity at start-up in a number of different pieces of progressive equipment (even if the equipment is locked to NTSC sync).

Care must be exercised to ensure that various system delays do not put the Dolby E frames excessively out of phase with any switch point. This applies to the signal both when formatted into an AES stream and when embedded into an SDI/HD-SDI stream. In the embedded case, a switch timing problem can develop fairly quickly due to repeated de-embedding/re-embedding delays as audio is bypassed around video processing equipment. In the embedded case, it is important to keep the total delay of the Dolby E audio frames closely matched with the delay of the video. A last point of care that must be taken is matching of audio and video delay when the Dolby E signal is decoded, as the decoding process introduces one (30 Hz or 24 Hz) frame of delay in the audio.

Note that 24 Hz Dolby E encoding is used only in 24 fps program generation and film transfer. Once the 24-frame video is inserted into a 60 Hz system with 2-3 pulldown, the audio is typically re-encoded in 30 fps Dolby E so the audio frames match the video frames. Alternatively, if the 24 fps source is speeded up to 25 fps, the audio must be decoded, sample rate converted, and re-encoded at 25 fps, since the audio sample rate must stay at 48 kHz for all video rates.

[2]Dolby is a registered trademark of Dolby Laboratories, Inc.

MADI

The AES10 [7] serial interface commonly known as MADI (Multichannel Audio Digital Interface) is an adaptation of the Copper Distributed Data Interface (CDDI), which is a coaxial implementation of the Fiber Distributed Data Interface (FDDI) computer LAN. The gross data rate is 125 Mbps, and the net rate is up to 100 Mbps. With the advent of ever-faster Ethernet interfaces, FDDI has gone by the wayside in the computer industry to the point that FDDI chips are no longer available, but broadcast manufacturers have effectively re-invented them using field programmable gate arrays (FPGAs), so the AES10 standard is alive and well. The MADI interface carries up to 56 channels of 48 kHz sampled audio with the option of 12.5% varispeed and up to 64 channels at a fixed 48 kHz. Since the interface packetizes audio data into an asynchronous channel, with bit stuffing to manage the average data rate, a separate AES sync channel is specified for synchronization. The MADI interface is commonly used on large audio consoles and multichannel audio recorders. Audio routing switchers are also beginning to support the interface.

VIDEO SYNCHRONIZATION FOR DIGITAL SYSTEMS

Color Black

As previously mentioned, NTSC color black is widely used as a timing reference for both NTSC and digital systems. However, two challenges are presented in using color black as a high-accuracy sync reference for digital video systems:

1. The signal is unipolar and is severely affected by the low-frequency group delay introduced by unequalized or poorly equalized coaxial cable. This sensitivity can cause timing jitter problems that, once introduced, are difficult to remove.
2. The repetition rate is quite low relative to current digital clock rates, so for use in digital circuitry, the signal must be multiplied in frequency many times.

Because of the first issue, it is particularly important when using color black as a reference for digital devices to take care in the distribution of the signal. All cable runs should be properly equalized and, since

the vast majority of devices to be synchronized do not have internal cable equalizers, this does mean installing equalizing distribution amplifiers near the devices to be locked. Also, signal looping should be limited to points close to the destination end of the cable. Looping near the source end of a long cable can produce large amounts of jitter in the recovered clock.

The second issue requires a trade-off between lock time and jitter. Several designs over the years have used ingenious two-speed phase locked loops to help overcome this trade-off, but for the most part, cost reduction has prevailed and the result is mostly a sea of compromise in equipment design.

Systems engineers are cautioned to pay careful attention to these issues.

Tri-Level Sync

The *tri-level sync* was developed for HDTV to minimize group delay sensitivity. It is still analog but is a bipolar signal where the timing edges are positive-going and always follow a peak negative excursion that can be used as a clamp point. This makes the signal much less sensitive to group delay introduced by poor cable equalization.

The tri-level signal is defined by SMPTE 240M [8] for 1080i and by SMPTE 296M [9] for 720P signals. The signals are identical to the appropriate Y channel but with video always at black level. Since the tri-level signal has much higher frequency energy than NTSC color black, wider bandwidth cable equalization is necessary, but some inaccuracy in equalization does not produce as much jitter in recovered clocks as when using NTSC color black. Since these reference signals operate at line frequencies roughly two and three times, respectively, the NTSC line rate, the frequency multiplication problem is reduced somewhat, but it is still an issue.

Many facilities use both NTSC color black and tri-level HD synchronizing signals in different areas of the plant. In particular, in the case of the 720p sync signal, there is no way to derive a 30 Hz reference with defined phase from tri-level sync, and this is an issue for certain editing constraints and for Dolby E audio (discussed later in the chapter). So 720p plants typically still either refer to NTSC sync or use both NTSC and 720p sync.

Master and Slave Sync Generators

Keeping the preceding points in mind, in order to minimize timing jitter, a master system timing generator should produce the appropriate reference signals referred to a common master clock.

With the large amount of delay inherent in digital image processing, slave sync generators that can offset timing by several frames are a necessity. Slave generators may be built into pieces of equipment, such as recorders, or may be separate units.

SYSTEM SYNCHRONIZATION AND TIMING ISSUES

Digital Audio Reference

The Digital Audio Reference Signal (DARS), which is simply an AES3 audio signal with specified stability, per AES11 [10], is also a necessary element of system synchronization. It is possible to reference the DARS signal to any video signal, but there is no guarantee of audio frame phasing due to the poor correlation between 48 kHz and 59.94 Hz (4004 audio samples in five video fields or progressive frames). For this reason, it is important to think of the DARS reference as a submaster reference (that is, referred directly to the master video reference at a single point) wherever possible. This is particularly important whenever multichannel audio phasing is being considered.

The DARS reference is best generated at the master sync generator and distributed in parallel with video sync to ensure accurate audio sample phasing.

Time Code Versus Time

Time code as defined by SMPTE 12M [11] is a label applied to a frame of video. Depending on the application, time code gets used in different ways. In a production situation, time code is really a frame counter that is used to assist in establishing accurate edit points. In a broadcast situation, *drop frame* time code is used to actually keep track of the length of a program in real time.

Drop frame might better be called "drop code," in that time code labels, not frames of video, are dropped in an attempt to keep the frame counter running at a 1000/1001 ratio to 60 Hz (59.94 Hz) and concurrently synchronized as closely as possible with a real-time clock. Drop frame is at best only an approximation and eventually requires a reset to get the time code generator back in sync with a real-time clock. A simple way to look at the issue is to recognize that the factors of 1001 (7, 11, and 13) are not shared with any of the integers in a clock, so it is possible to keep seconds aligned every 1001 seconds, and minutes aligned every 1001 minutes, and so on up to leap years aligned every 1001 leap years, but all the elements of time will not align past the reset point when using the offset counter.

The Society of Professional Audio Recording Services (SPARS) Time Code Primer publication [12] provides considerable insight into the different ways of using time code and, for the adventurous, even delves into drop frame arithmetic.

Frequency Versus Time

Television systems are based on frequency. To a first order, this is consistent with saying they are based on time, but on closer inspection, time is ultimately based on the rotation of the earth, which is gradually slowing. The official mechanism for resolving the difference between frequency and time is to add or subtract "leap seconds." For a frequency-based television system, this can be addressed by a reset or by an extended period of slightly off-frequency operation to align the frequency-based system with the "new" time. Leap seconds are not an everyday event (they occur about every 18 months on average), but with the prevalence of global positioning system (GPS)-based clocks, the concept is important to understand.

Frame Synchronizers

In 1980, the frame synchronizer was a new and wonderful tool for broadcasters. For the first time, remote video sources could easily be synchronized with local sources, allowing glitch-free switching and effects. Today, frame synchronizers are integrated into numerous pieces of equipment, and discrete units are ubiquitous and inexpensive. The upside of this evolution is that the concept of genlocking a local facility to a remote source has all but disappeared. The downside is that keeping audio and video aligned has become much more difficult.

Audio-Video Tracking

A frame synchronizer works its magic by periodically repeating or dropping a video frame. To the human eye, this is an insignificant event. The equivalent process in audio is often quite noticeable and annoying, and dropping or repeating whole frames of audio is generally considered unacceptable. Two approaches have been used to try to keep audio and video in alignment. One is to simply add one-half of the frame synchronizer delay to the audio all the time (typically 1/2 frame delay). With this approach, the maximum error will be plus or minus one field, but the delay will be perfectly matched only when the video frame buffer is half full.

The second approach is to *chase* the audio delay by slightly offsetting the internal buffer frequency until the audio is back in alignment with the video after a frame drop or repeat. In some systems, the pitch error is corrected, and in others it is not (typically dependent on the speed of the chase function). Interestingly, the peak error with the first approach is one field, and with the second approach, it is one frame (until the audio chase catches up). This is important to understand in a system with many (potentially dozens) of frame synchronizers in cascade.

Compression and Delay

Another source of delay in modern television systems is signal compression. Both video and audio are routinely compressed, and the delays are typically long (several to many frames) and not equal. It is therefore important to understand the amount and nature (fixed versus variable) of the delay in any compression equipment used and to develop delay compensation schemes to keep audio and video aligned. A couple of issues conspire to make this a difficult task, as follows.

Presentation Timing Versus Channel Timing

The only signal timing that matters to a viewer is presentation timing—the timing of the audio and video that is seen and heard. Many compression systems (typified by MPEG-2) have buffer systems feeding the compressed channel in order to accommodate peak signal complexity events (zooms, cuts, dissolves, and so on). This causes the channel timing to be only loosely coupled with presentation timing and adds considerable complexity to any process that attempts to switch compressed signals. It can also cause errors in presentation timing to occur if any switches are not carefully compensated.

The Computer Clock Model

Many compression decoders are built around the computer clock model, in which the presentation clocks (that is, the graphics card clock and audio card clock) are completely decoupled from the software clock (and in some cases from each other). The result is that the decoder (software) attempts to decode audio and video at the right time, but with no control over the presentation timing. The net effect is that frames can be dropped or repeated in the graphics card buffer in ways the decoder does not anticipate. This can lead to at least one frame of audio to video timing error, and sometimes more. When systems are built with multiple cascaded stages of compression and decompression, it is important for the systems engineer to understand the characteristics of the devices used in order to avoid accumulation of audio to video timing error.

Production Delay Management

Another potentially large contributor to audio/video delay mismatch is the dynamic insertion of digital video effects (DVE) engines into video production mixers. Some video mixers have the option of inserting as many as eight DVEs in cascade, with one to two frames of delay per DVE. Most production switchers do not have compensating delay built in, due to the inconvenience of working with long delays on a regular basis. In the case of multiple cascaded channels of digital effects, it is essential to include some form of audio tracking delay to avoid serious lip sync problems.

DIGITAL VIDEO ROUTING SWITCHERS

In modern television facilities, an increasing amount of video distribution and switching is accomplished via computer networks, but most real-time on-line distribution and switching is managed by full-bandwidth equipment capable of passing either or both SDI and HD-SDI signals, either with embedded audio or with associated separate audio.

In many facilities, there is still some analog signal distribution, although several modern digital routing switches have the capability of conversion from and to analog at their inputs and outputs. This feature allows legacy analog sources and destinations to have access to the entire facility connected to the routing switch.

Serial digital video routers must manage signals at data rates of 270 Mbps and 1.5 Gbps. New standards, SMPTE 424M and 425M [4], support signals up to 3 Gbps for signals such as 1080p/60, and an upcoming generation of routing switches is capable of handling all three rates.

Video Timing Considerations

For clean switch operation, all inputs to a video routing switcher must be accurately timed. In the analog world, "accurately" typically meant about ±1 ns. Digital systems are somewhat more forgiving, but a window of no more than a few microseconds is advisable, particularly in multirate facilities. At the time of writing, SMPTE RP168 [13] defines signal timing and switch points in the current standards. It is important to note that when inputs are aligned in the manner recommended in RP168, each signal type has a different switch point. Many modern routing switches will handle at least two different switch points concurrently, so a system can be set up per RP168 that will correctly switch SDI and one HD-SDI standard simultaneously. For example, 480i and 1080i or 480i and 720p can be switched correctly at the same time, but if all three formats are present, one of the HD signals will be switched at a time other than its nominal switch point. Since most broadcast facilities are dedicated to one or the other of the HD standards, this arrangement typically works out satisfactorily.

If a facility must handle multiple standards that must all be switched on their defined switch points, the plant timing can be adjusted so that all the switch points are aligned, and then only one switch point is needed. With this approach, the second available switch point in the router can be used to switch signals at a different vertical rate. For example, the first switch strobe could be referred to NTSC color black and be used to switch 480i/60 (59.94), 720p/60 (59.94), 1080i/60 (59.94), and even 1080p/60 (59.94). The second strobe could be referred to PAL color black and used to switch 576i/50, 720p/50, 1080i/50, and 1080p/50.

One additional consideration is that, in order to satisfy the constraints of Dolby E embedded audio, the video switch strobes need to occur at 30 (29.97) Hz and/or 25 Hz, even for progressive scan signals.

Video Conversion I/O

A number of modern video routing switches have the option of adding analog to digital (A/D) converters on inputs and digital to analog (D/A) converters on outputs. This feature allows composite NTSC (or PAL) sources and destinations to be attached to switch inputs and outputs. When this feature is used, some thought needs to be put into system timing, since conversion chips introduce delay. A/D converters typically include multiline comb filter decoders and will introduce several lines of delay, so analog composite sources need to be back-timed such that the decoded digital signal is in time with the rest of the digital inputs on the routing switch. The encoder function on a D/A converter typically introduces only a few microseconds of delay, but the timing accuracy of analog output signals is no better than the timing of the digital input signals. Thus, this type of analog output is suitable for feeding a monitor wall or any other function that does not require accurate analog timing, but is not suitable for a time-critical production application such as a feed to an analog production switcher.

DIGITAL AUDIO ROUTING SWITCHERS

Digital audio for broadcast is built around AES3 coding for distribution and uses 16, 20, or 24 bits of linear PCM to encode audio. The nominal sampling frequency for all broadcast-related applications is 48 kHz. Dolby E may be used to carry up to eight channels of audio compressed to fit into a single two-channel AES3 stream. When Dolby E, or any other audio compression system, is used, all parts of the infrastructure must be capable of passing bits accurately (that is, the AES3 signal must be treated as a data stream, not a linearly encoded audio signal) and switch points must be gated on video frames. Typically, in a digital audio switch, the actual switch point will be the first audio sample boundary following the video frame switch point. This allows for synchronous audio switching and stays within the constraints of the Dolby E specification.

Synchronous Versus Asynchronous Audio

Digital audio routing switches may be designed to manage audio streams synchronously or asynchronously. Since all audio associated with video typically must be synchronous with the video at most destination devices, synchronous switching is typically preferred for broadcast operations. However, in telecine operations, where the audio may have been captured separately from the film, asynchronous audio switchers are often preferred. For a synchronous audio routing switch to operate optimally, all the sources should be frequency locked to a common DARS reference, as described in the previous Synchronization and Timing Issues section.

Sample Rate Converted I/O

Many modern synchronous audio routing switchers include the option of sample rate conversion on inputs and, in some cases, outputs. This feature allows *wild* sources to be digitally synchronized with the video plant but does not address actual audio timing issues. One sample application would be the use of CD sources at 44.1 kHz sampling to feed audio clips into the broadcast plant.

Audio Conversion I/O

As with serial digital video routers, many digital audio routing switchers include the option of adding A/D converters on inputs and D/A converters on outputs. The audio case is typically somewhat simpler than video in that the conversion delay is short in relative terms, so synchronization is normally not an issue. Care should be taken to have A/D converters of one type referred to a common DARS reference for delay matching in the case of multiple microphones on a single set.

MADI I/O

MADI (described previously) is available on some digital audio routing switchers.

Sample Rate Conversion Back Timing

Some audio switchers with MADI outputs have the option of output sample rate conversion. Since MADI uses a separate AES signal for synchronization, it is possible to feed sync backward to the output card and thus have all the channels in the MADI signal converted to match. This feature is useful when feeding an audio console that is being run off-speed for timing correction or pitch correction. The output of the console can then be fed back to a sample-rate converted MADI input on the routing switcher, and all the channels will again be synchronous with the video plant.

Monaural and Multichannel Management

The AES3 interface is nominally a two-channel interface. It can be used to carry a stereo pair, a single audio channel, or two independent audio channels. In the last two cases, it is often necessary to switch individual channels at the monaural level. This requires management of the overhead data, such as channel status, separately from the audio in order to minimize delay in the audio. The overhead data block repeats every 192 samples and is typically not phase-aligned between different sources. By managing the overhead data with some latency, which is allowed in the AES3 standard, the audio delay can be minimized and the requirements of the overhead data can be met. There are digital audio routers available that accomplish this, and they are useful when managing multichannel audio through a production facility.

AES CABLING ISSUES

AES3 signals can be managed with two different connection methods. The first is via a 110 ohm balanced pair cable as defined in the AES3 standard, and the second is via 75 ohm coaxial cable as defined in AES3-id and SMPTE 276M [14]. Most broadcast operators prefer to use coax wherever possible, so most broadcast digital audio is distributed on 75 ohm coax, whereas most professional audio and many postproduction facilities use the balanced interface.

A key difference between the two methods is signal level. In the balanced case, the pp level is defined as between 2 V and 7 V but is typically about 3.5 V, and receivers are required to operate correctly down to 200 mV. In the unbalanced case, the level is fairly tightly defined as 1 Vpp, and receivers are required to operate correctly down to 320 mV. Transformers are available that will convert balanced signals to unbalanced signals. These transformers include an attenuator pad to correct for the level difference. Transformers without attenuators are also available to convert coaxial signals to 110 ohm balanced signals, but the level at the output is typically about 1.1 V. This process works once (because of the defined sensitivity of the balanced receiver) but cannot be cascaded. If a second conversion to unbalanced coax is needed, then a transformer without attenuation should be used. Any time an active device such as a distribution amplifier is introduced into the chain, the whole sequence starts over.

IT-BASED SIGNAL MANAGEMENT

Over the past 10 years, infrastructure products based on general-purpose computer technology have made their way into the broadcast plant. In areas such as postproduction and editing, they have largely displaced traditional systems. Information technology (IT)-based solutions are often perceived as more efficient, and they extend increasingly into all parts of the broadcast chain. One issue with IT-based sytems is their relatively short lifetime (often only 25% that of classical broadcast equipment). This brings to question the initial perception of lower cost when viewed in a life-cycle perspective. At this point in time, for several reasons, conventional full-bandwidth video and audio distribution systems continue to coexist with IT-based systems in various key areas of most broadcast facilities.

File-Based Systems

The single biggest conceptual change in making the transition from conventional infrastructure to IT-based infrastructure is the change to electronic file-based storage. With conventional storage systems (largely videotape-based), filing is a physical process involving an indexed storeroom, whereas in an IT-based storage system (typically using video servers), the filing system is electronic and virtual. The latter allows for access to stored material without a need for physical movement of storage media. This enhances opportunities for system automation and potentially reduces staffing requirements.

Header-Driven Control

Conventional systems are controlled by an external control system. That is to say, control information is

not embedded in the channel with the program stream. IT-based routing systems such as Ethernet, however, are typically two-way systems and are controlled in-band with control headers. They are also usually asynchronous packet-based systems, rather than continuous-framed systems, so synchronization information cannot easily be embedded in the packet stream with the kind of accuracy inherent in a conventional system.

Statistical Packet Switching

Conventional SDI video and digital audio switches are full bandwidth, meaning that, short of equipment failure, if a path is assigned, there is a 100% probability of the signal reaching its destination. Packet-based switching systems such as Ethernet are statistical in nature and may have more data assigned to a destination than the path to the destination can carry. In this situation, packets will be dropped. To avoid this situation, statistical packet systems must be run at a traffic level well below maximum capacity in order to approach 100% probability of signals reaching their assigned destinations. One tool that has been used to facilitate the reduction in required bandwidth is signal compression (carried out before packetization), which is also useful in conserving storage resources.

Ethernet, SCSI, FireWire, and Fibre Channel

Several different IT-based transport technologies have been employed in the management and routing of video and audio information. Early servers were small computer system interface (SCSI) based, and some effort was put into using the FireWire (IEEE 1394) interface for signal connections, but most systems today have migrated toward Fibre Channel for Storage Area Networks (SAN) and toward some level of Ethernet for larger networks. The FireWire interface is still used extensively for connection of portable devices to the network and Serial Attached SCSI (SAS) and Serial Advanced Technology Attachment (SATA) are rapidly making their way into newer storage systems.

Store and Forward Versus Real Time

In a store and forward situation, Transmission Control Protocol/Internet Protocol (TCP/IP) can be used to guarantee packet delivery. In a congested packet system, this may result in some of the packets arriving considerably out of order because TCP/IP ensures their delivery by requesting a retransmission. This is not an issue when a file is being transferred from one storage disk to another. In a real-time situation in which the stream is intended for immediate viewing (or listening), the amount that packets can be delayed is determined by the size of the storage buffer in the receiving device. With real-time systems, latency is very undesirable, so receiver buffers typically must be very short. Under these circumstances, the best performance is achieved by backing off considerably on the network utilization and using a protocol that doesn't guarantee delivery such as User Datagram Protocol (UDP) in conjunction with some form of forward error correction (FEC).

The Role of DVB ASI

DVB ASI [15] is an interface that uses the same gross data rate as SDI video (270 Mbps) and can be switched through digital video switches. ASI is used in many situations in which compressed packetized signals need to be reliably transported in real time. The control is out of band, as with SDI switching; and the delivery is not statistical as is the case with Ethernet.

SUMMARY

The introduction of DTV transmission and the steady migration from analog to digital video and audio equipment throughout the broadcast industry have resulted in a revolution in the facility infrastructure for signal switching, timing, and distribution that is still ongoing. The technical and potential economic benefits are many, but care is necessary to optimize the system design and installation to avoid the numerous pitfalls that can trap the unwary.

References

[1] SMPTE 170M-2004 – Composite Analog Video Signal – NTSC for Studio Applications, SMPTE 125M-1995 – Component Video Signal 4:2:2 – Bit-Parallel Digital Interface, SMPTE 259M-2006 – SDTV Digital Signal/Data – Serial Digital Interface, SMPTE 274M-2005 – 1920 × 1080 Image Sample Structure, Digital Representation and Digital Timing Reference Sequences for Multiple Picture Rates, SMPTE 292M-1998 – Bit-Serial Digital Interface for High-Definition Television Systems, SMPTE 296M-2001 Television – 1280 × 720 Progressive Image Sample Structure – Analog and Digital Representation and Analog Interface, *Society of Motion Picture and Television Engineers*

[2] SMPTE 259M-2006 – SDTV Digital Signal/Data Serial Digital Interface, *Society of Motion Picture and Television Engineers*

[3] SMPTE 292M-1998 – Bit-Serial Digital Interface for High-Definition Television Systems, *Society of Motion Picture and Television Engineers*

[4] SMPTE 424M-2006 – 3 Gb/s Signal/Data Serial Interface and SMPTE 425M-2006 Television – 3 Gb/signal/Data Serial Interface – Source Image Format Mapping, *Society of Motion Picture and Television Engineers*

[5] SMPTE 272M-2004 – Formatting AES/EBU Audio and Auxiliary Data into Digital Video Ancillary Data Space, SMPTE 299M-2004 – 24-Bit Digital Audio Format for SMPTE 292M Bit-Serial Interface, *Society of Motion Picture and Television Engineers*

[6] SMPTE RDD 6 – Description and Guide to the Use of the Dolby Audio Metadata Serial Bitstream, *Society of Motion Picture and Television Engineers*

[7] AES10-2003: AES Recommended Practice for Digital Audio Engineering – Serial Multichannel Audio Digital Interface (MADI), *Audio Engineering Society*

[8] SMPTE 240M-1999 (Archived 2004) – 1125-Line High-Definition Production Systems – Signal Parameters, *Society of Motion Picture and Television Engineers*

[9] SMPTE 296M-2001 – 1280 × 720 Progressive Image Sample Structure – Analog and Digital Representation and Analog Interface, *Society of Motion Picture and Television Engineers*

[10] AES11-2003: AES recommended practice for digital audio engineering - Synchronization of digital audio equipment in studio operations, *Audio Engineering Society*

[11] SMPTE 12M-1999, Audio and Film – Time and Control Code, *Society of Motion Picture and Television Engineers*

[12] Steve Davis, SPARS Time Code Primer, 1997, ISBN 0-9658309-0-X, published by *The Society of Professional Audio Recording Services*, see http://www.spars.com/

[13] SMPTE RP 168-2002 Definition of Vertical Interval Switching Point for Synchronous Video Switching, *Society of Motion Picture and Television Engineers*

[14] SMPTE 276M-1995 – Transmission of AES-EBU Digital Audio Signals Over Coaxial Cable, *Society of Motion Picture and Television Engineers*

[15] ETSI TR 101 891, *European Telecommunications Standards Institute*

CHAPTER

5.6

Workflow, Metadata, and File Management

BRUCE DEVLIN
Snell & Wilcox
Petersfield, United Kingdom

THOMAS EDWARDS
Public Broadcasting Service
Alexandria, Virginia

BRAD GILMER
Gilmer & Associates
Atlanta, Georgia

PHIL TUDOR
BBC
Tadworth, United Kingdom

INTRODUCTION

From acquisition through transmission, technical facilities in television plants are changing at a rapid rate. One of the greatest areas of change is in the use of computers and computer infrastructures. Over the years, computers have replaced purpose-built systems for the storage and manipulation of images and sound. At first, only audio and still images were involved because the sheer volume of data created by moving images overwhelmed early computer systems. Today, computers are employed in almost every aspect of the broadcast industry. This chapter discusses some of the latest developments in program workflow and the use of metadata (data about data—in this case, information about the program video and audio content), and in particular, it:

- Describes how the change to computer-based technologies has impacted the program production and distribution workflow;
- Provides an example of a computer-based workflow from the Public Broadcasting Service;
- Describes how metadata can be associated with video and audio;
- Introduces the Advanced Authoring Format (AAF) and Material eXchange Format (MXF), file wrappers that carry video, audio, and metadata;
- Lists some of the standards that enable this new workflow process.

NEW INFRASTRUCTURES—NEW OPPORTUNITIES

One of the fundamental differences between the traditional broadcast domain and the computer domain is in the infrastructure. In the video domain, the traditional storage and distribution infrastructure technologies have been videotape, analog video over coax, and digital video over coax. In the broadcast plant, the predominant method for switching video signals from one place to the other has been a video router.

In the computer domain, early infrastructures for storage and signal transmission were punch tape, large removable disk platters, proprietary serial interfaces, and rudimentary networking technologies. As broadcasters started to use computers, they frequently kept metadata completely separated from the media it was associated with, except for a tape label and perhaps some text on a slate at the head of the tape. The slate was the only link between computer records and the images on tape.

As technology improved, compression reduced the bandwidth and storage requirements for video. At the same time, computer networking and processor speeds increased to the point where television engineers began to work with video in the computer domain. This new infrastructure has led to changes in workflow and new benefits to end users. Some of the benefits include:

- Faster than real-time transfers;
- The ability to share content across multiple users at the same time;

- Eliminating or reducing rekeying of data;
- Attaching metadata to the images and sounds so critical information remains with the content throughout the workflow process;
- Eliminating compression effects from multiple compression/decompression steps (interchange in compressed domain rather than baseband video).

Engineers began to imagine a world in which all video and audio would be carried as bits and metadata could move along with the video and audio without the possibility that it would be misplaced. Users could exchange files just as they exchange videotapes, but the files could be moved across a computer network. Content stored on a central server would be accessible by several users at the same time. Not only would it be possible to access and transfer entire programs and commercials over a network, but it would also be possible to play back a stream of video from a remote server just as if the content was playing back locally. Engineers expected to leverage much of the economies of scale that are present in the computer industry to lower the overall cost of these facilities. Some hard problems had to be solved—the major issues were capacity, bandwidth, speed, and cost. Substantial resources were brought to bear on these problems, and information technology-based systems are now able to handle the considerable demands of networked video and audio.

In the broadcast industry, the issues to be tackled were largely identified in the EBU/SMPTE Task Force report released in 1998 [1]. The task force identified several areas that required further study—systems, compression, wrappers and metadata, and networks and transfer protocols.

Wrappers and their associated metadata can enable significant changes in workflow. The development of standardized wrappers and metadata means that, for the first time, combined video, audio, and metadata can be exchanged between different vendors using nonproprietary methods. To the user this means that content can be uniquely identified, once in the chain, using a technology such as the SMPTE Unique Material Identifier (UMID) specified in SMPTE 330M [2]. Information about the content can be kept in an organized and predictable structure based upon a standardized data model, and this information can stay with the content as it works its way through the creation and emission workflow.

PBS VIDEO WORKFLOW WITH AAF AND MXF

To understand how wrappers and metadata can enable new workflow possibilities, consider a specific example as a case study. This is an automated file-based workflow system introduced by the Public Broadcasting Service (PBS) in 2006 as an aid to the creation and airing of content. Figure 5.6-1 shows a simplified diagram of the system. The workflow is based on the use of AAF and MXF open standards; however, some aspects of the system utilize proprietary interfaces and protocols, and therefore, the following description includes names of particular equipment and systems. Some of the technical terms used in this example are explained in the later sections describing the MXF and AAF wrapper systems. Note that at the time of writing, this system is in an early stage of deployment and details may be subject to change.

Production

The video content workflow at PBS begins when programs are "greenlighted" for production go-ahead. PBS Program Management establishes a contract with the producer, and this information creates the initial "Deal" within the PBS BroadView Traffic System for a number of episodes of a program.

The producer then uses a PBS Web interface called Orion to begin entering key program metadata, including program title, video format, and frame-accurate segment timing (known as the media inventory). Media inventory is added to and revised during the production process. Upon PBS giving approval for a program episode, based on the information provided, the producer prints a bar-code identification label for the program, obtained through the Web interface. The producer affixes this label to the finished program videotape for delivery to PBS.

Ingest

Upon arrival of the tape at the PBS Media Operations Center (MOC), the tape bar-code is scanned in, which leads the BroadView system to initiate two processes. The first uses the Avid application programming interface (API) to automatically create an AAF file on the Avid Unity system used by PBS; this contains the program metadata in a media object (MOB), including the segment timings from the media inventory. Simultaneously, the ScheduALL scheduling system is instructed (using Microsoft BizTalk) to schedule an ingest work order.

Technicians then open the ingest work order, which provides them with relevant instructions on how to perform the simultaneous technical evaluation and ingest of the program on an Avid Media Composer Adrenaline. During the ingest process, an automated HD/SD test and measurement console[1] is used to ensure proper video levels and color gamut of the program, and a broadcast loudness meter[2] is used to monitor loudness levels. Logs generated by these devices become part of the metadata associated with the ingested material, so that technical discrepancies can later be investigated and corrected.

Segment lengths are indicated on the Avid timeline through locators that are defined in the AAF file using metadata from the original media inventory.

[1]VideoTek VTM 440.
[2]Dolby LM100.

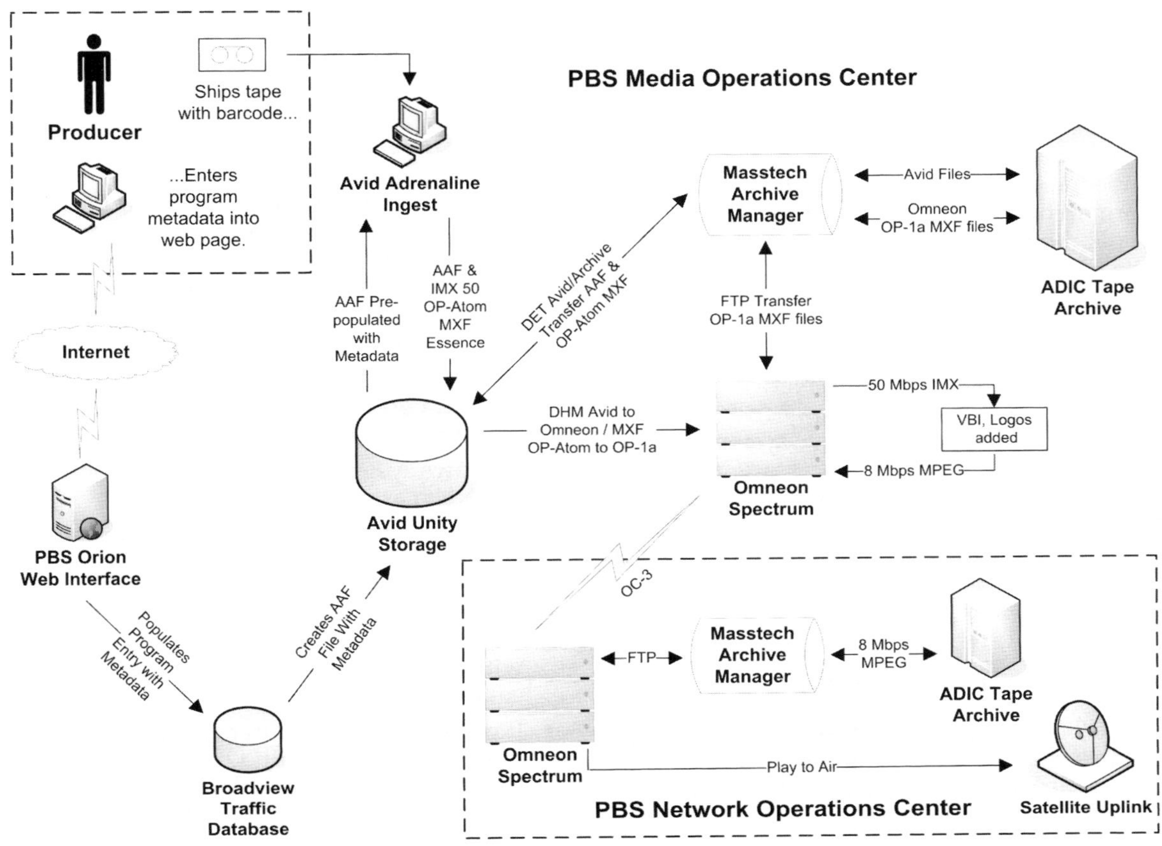

FIGURE 5.6-1 PBS ingest and playout workflow.

The technician confirms these timings, and if they are off, the technician can move the locators slightly if needed. Having clean segment breaks marked allows segments to be automatically added and removed in the future should there be a change in program elements, such as an underwriter change or schedule-specific promos. Once the program has been ingested, Broadview extracts the locator information from the Avid AAF file to update its segment timing database.

Compression

During ingest, standard definition (SD) video is compressed using a 50 Mbps 4:2:2 I-frame IMX codec, and HD video is compressed using the 145 Mbps 8-bit 4:2:2 DNxHD codec. The ingested essence (program video and audio) is wrapped as OP-Atom MXF files.

Archiving

After ingest, the technician uses the "Send to workgroup" function on Media Composer to initiate a Masstech Smart Avid Interface (SAvI) that uses the Avid Dynamically Extensible Transfer (DET) API to transfer the ingested program Avid media objects to an ADIC Scalar 10K tape-based video archive system. The file transfer protocol (FTP) transfer to the archive takes place at about three times faster than real time for SD material. About 200 terabytes of data per year are expected to be archived on the Scalar 10K, which is expandable to hold over 4 petabytes.

Conversion from OP-Atom to OP1a MXF Files

After the Avid native files are archived, the Omneon AO Transfer tool is used to convert and transfer the OP-Atom MXF essence files for standard definition programming on the Avid Unity system into OP1a MXF files on the PBS Omneon Spectrum video server system at 50 Mbps. This is an intermediate step in an automated process that archives the OP1a MXF files on the ADIC Scalar 10K archive system, usually weeks or months ahead of air time.

Final arrangements for the use of HD MXF files, and for HD bit rate reduction and playout in this new workflow, are still in development.

Bit Rate Reduction and Playout

As the time for airing of the program comes near, BroadView generates automation playlists for the "flattening" process to prepare programs for distribution to PBS stations. The OP1a MXF files are retrieved from the ADIC archive and transferred back to the Omneon Spectrum server. These SD 50 Mbps IMX files are played back under automation and external baseband devices are used to add elements to the video signal, such as V-chip, Nielsen data, and in-picture logos. The processed video is simultaneously recorded on the Omneon server as an 8 Mbps 4:2:0 long-GOP MPEG-2 format in an OP1a MXF wrapper. The flattening process is monitored with automated test equipment[3] to ensure that all video, audio, and VBI signals are present.

A few days before transmission, the flattened files are sent over a network connection from the MOC to an Omneon Spectrum server at the Network Operations Center (NOC) about 10 miles away and, for security, archived on an ADIC i2000 archive system. Before air time, the flattened files are, if necessary, retrieved from the archive and transferred back to the server for final playout over the PBS satellite interconnection system. The play-to-air Omneon server is controlled via an automation system,[4] following playlists created by the Broadview traffic system.

Future IP-based Network Distribution

In the future, PBS will distribute these MXF files direct to PBS member stations using Internet Protocol (IP)-over-satellite technology. PBS is working with the key vendors of video servers to public television stations to develop an Application Specification for the MXF file delivery to stations (AS-PBS). This AS will ensure that the MXF files distributed by PBS will play back on station servers.

Benefits of the AAF- and MXF-based Systems

The use of the Avid AAF and MXF-based system for ingest provides three benefits. First, Avid files can be retrieved from the archive and reedited without the need for another ingest. For example, PBS occasionally needs to repackage programs with updated underwriting spots in the program. A second benefit is the possibility of direct file transfer from producers. Although tapes are still ingested at PBS today, it is expected in the future that producers using Avid systems will contribute files to PBS directly using Avid Transfer Manager over a network connection, avoiding the tape shipment and ingest process altogether. Finally, Avid can provide the flexibility to ingest both SD and HD material using the same tool set.

The close relationship between AAF and MXF and the use of MXF for the archive and playout systems allows PBS to exchange metadata and essence between disparate systems from multiple vendors. Without such defined standards, gluing together an efficient, file-based workflow would be difficult, and involve multiple proprietary format conversions.

[3]Evertz AVM 7760.
[4]Omnibus.

TECHNOLOGY FOR WORKFLOW IMPROVEMENT

With standards for wrappers and metadata in place, as well as those for digital video and audio formats and file compression, when a receiving device gets a file, it will know how to unwrap the contents and quickly determine what is contained within the file. Perhaps more significantly, by reading the metadata the application will determine if it can play the contents, without having to actually load the video or audio file. Furthermore, it will know how to interpret the metadata that is contained in the file or stream. This is significant for several reasons:

- It allows users to count on delivery of metadata along with their content.
- It allows system designers to build on the original metadata associated with a file.
- Because of the designed-in interworking between current exchange formats, metadata created in the content creation process (primarily the AAF application area) can flow into the distribution/emission environment (primarily the MXF application area) without rekeying, which is a manual process and subject to error.
- It gives the manufacturers the opportunity to design systems that find out what is contained within the wrappers without having to dive deeply into the files, thus saving time and computational resources.

As the PBS example demonstrates, it is now possible to build systems that can automate workflow. By standardizing on wrappers and essence formats, content can be automatically moved from one location to another when needed. By using unique identifiers contained at a specific location in the essence wrapper, systems can be designed that maintain the link between data about a piece of the content and the content itself. Furthermore, the combination of unique identifiers and a metadata model that allows a system to describe the heritage of a piece of content allows systems to trace the path of a piece of content back to its origin. There are systems being built now that sift through a storage system, extracting metadata, which can be used later for search and retrieval.

Other users are creating their own custom metadata sets by building on existing standards. These extensions may describe non-video sources such as infrared or radar imaging that is part of an overall data set. By using standardized technologies, these systems can employ off-the-shelf components reducing the overall cost of the project. The power of these technologies is that they are extensible—they can be expanded beyond what they were originally designed to do. This is possible because the standards upon which these technologies are based clearly define how extensions are made. By allowing flexibility, these technologies have a longer life and may be used in applications beyond their original scope.

The next two sections introduce the GXF, MXF, and AAF interchange formats and explain some of the

differences between them. Later sections provide more details of the functionality and structure of AAF and MXF.

GENERAL EXCHANGE FORMAT

The General eXchange Format (GXF) was originally developed to transfer audio and compressed video on Fibre Channel data networks for the Grass Valley PDR-200 servers. It was first used in on-air facilities in the mid-1990s. Additional compression families, more audio types, user-defined metadata, and editing capabilities were added as newer products with these features were made available.

GXF was designed for finished material transfers and introduced several ideas such as bit-perfect file format transfers, embedded user-defined metadata, interleaved audio and video content in a file, "partial restore" from data storage device archives, and many others. GXF established the baseline for future finished material file formats. It is supported by several organizations and is used in a wide variety of products. However, it is somewhat limited in its capabilities and applications and has increasingly been superseded by the more recently developed MXF.

GXF became a SMPTE standard in 2001. It was later updated to add new features and the revised standard SMPTE 360M-2004 was issued in 2004.

BACKGROUND TO MXF AND AAF

In simple terms, the Advanced Authoring Format (AAF) is intended for use during program production and postproduction while the Material eXchange Format (MXF) is intended for the exchange of completed programs or program segments. The background to the development of these two complementary file formats is as follows.

The Professional-MPEG Forum [3], often known as Pro-MPEG, is an association of broadcasters and program-makers, equipment manufacturers, and component suppliers with interests in realizing interoperability of professional television equipment, according to the implementation requirements of broadcasters and other end users. In 1999, Pro-MPEG started work to define a file format for program interchange. Existing (proprietary) formats were reviewed to determine if these might be suitable to build upon, but there did not appear to be one that met all the necessary requirements, in particular for the carriage of metadata with program material. Fundamental initial requirements for this file interchange format include:

- The format should be aimed at exchange of completed programs or program segments.
- It should be able to carry program-related metadata components as well as the video and audio components.
- It should be possible to start working on the file before the file transfer is complete. This is particularly important when sending large files over slow networks.
- It should provide mechanisms to decode useful information from the file even if parts of the file are missing. This might happen, for example, if a file transfer (e.g., over satellite) is picked up halfway through the transfer.
- The format should be open, standardized, and independent of the particular compression format used for the video and audio.
- Above all, the format should be simple enough to allow real-time implementations.

At the same time that Pro-MPEG started to work on MXF, the manufacturer Avid was in the process of launching an association to support the AAF as a file format for use in the postproduction environment. AAF was intended to allow interchange between different production and editing systems, and had the support of several manufacturers and end users, but Avid decided to also contribute to the Pro-MPEG work on MXF, as it was intended for a somewhat different environment.

AAF did not meet all the requirements intended for MXF, in particular the requirements (a) to be able to start using a file before transfer is complete, and (b) to be able to form up and read files simply in real time. However, there was a clear end-user call for MXF to be easily interoperable with AAF, with the particular desire that it should be possible to carry metadata transparently from postproduction through to the distribution/exchange of the completed program. Consequently, Pro-MPEG and the AAF Association [4] decided to work closely together on the project to define MXF.

In the early days of MXF development, the different perspectives of the engineers coming from a background of video engineering and the engineers coming from a software background created some difficulties as the participants interacted on the project. AAF was already defined in the form of open-source software known as an SDK, or Software Developers Kit, maintained by the AAF Association. However, this SDK was fairly opaque to the video engineers coming from Pro-MPEG. Also, video engineers traditionally look toward the Society of Motion Picture and Television Engineers (SMPTE) as the place where their standards are documented and these standards are traditionally described in words, not software.

MXF was not designed to support all the rich functionality provided for postproduction in AAF. For example, MXF does not provide the capability to describe complex transitions between two clips in a file. It was a fundamental requirement within the MXF development work to keep the structure as simple as possible while meeting defined user requirements. However, different users have requirements differing in complexity and therefore MXF was designed to support separate operational patterns with different levels of complexity. For example, the simplest generalized operational pattern "OP1a" carries a single program or program segment within a file. A more complex operational pattern can contain several program segments, with the ability to do cut edits between them (useful, for example, when distributing one copy of a program to different countries that may require some scenes to be removed). Other more complex patterns may contain more than one version of the same program—for example, full-quality and browse-quality versions.

Whatever the complexity of MXF, it is a fundamental requirement for metadata transfer between MXF and the AAF file formats to be simple and straightforward. In order to achieve this, it was important that the data models of the two formats should be as closely aligned as possible. Therefore, Pro-MPEG and the AAF Association agreed between themselves on a Zero Divergence Doctrine (ZDD) that enshrined the principle that all MXF developments should be aligned with the AAF data model. Also, if new requirements emerged during the development of MXF, then the AAF data model would be extended in ways that were compatible with MXF. This ZDD continues to exist today, even though some of the later developments in MXF have been done within the SMPTE.

ADVANCED AUTHORING FORMAT (AAF)

AAF is a standard file format for interchanging essence (video and audio content) and metadata between many kinds of authoring tools—for example, between systems for editing and functions such as compositing and rotoscoping.

Purpose and Functionality

AAF has a comprehensive and extensible interchange data model specifically designed to meet the complex requirements of real-world audio, video, and film authoring in the production and postproduction environments. AAF can be implemented by different manufacturers, within different types of authoring tools, and on different underlying computer platforms. An open-source software reference implementation (SDK) is available that is suitable for use in products.

The AAF Association ensures that AAF is not controlled by any one manufacturer and cannot be changed unilaterally. AAF carries a royalty-free license for implementers and requires no license for end users.

Types of AAF Material

AAF is structured around the notion of interchanging pieces of *material*. An AAF file can interchange *metadata* for the following types of material:

- Physical source—where the essence is held on physical media such as tape or film;
- File source—where the essence is held in a file on a computer file system;
- Clip—the authoring tool notion of a *clip* that references sources;
- Composition—the authoring tool notion of an edit *composition* that references sections of clips.

An AAF file can interchange the *essence* for file source material only. Table 5.6-1 gives some examples

TABLE 5.6-1
Examples of AAF Metadata and Essence for Different Material Types

Material Type	Examples of Data in AAF File	
	Metadata	**Essence**
Physical source	For tape media—tape name, tape format, video signal standard, and time code. For film media—reel name, film format, frame rate, aspect ratio, edge code, manufacturer, and type.	Essence not in AAF file.
File source	Picture size, frame rate, compression type (if any), and location (e.g., network URL).	Video, audio, graphics, animation, text or MIDI, and any compression type or uncompressed.
Clip	Clip name, track layout, source, and alternatives.	No essence.
Composition	Track layout, chosen sections of clips, alternatives, and layering. For effects—effect type, location, key frames, and parameters.	No essence.
All material types	Annotations such as operator comments, user-specific data, and descriptive schemes.	
	Extensions—standardized, manufacturer-specific, and user-specific.	

of AAF metadata and essence for each type of material. Although the AAF notion of material is defined to include both its metadata and essence, it can be seen that the existence and storage of essence depends on the material type.

AAF Material Identification

AAF assigns a unique identifier to each piece of material that identifies both the metadata and essence parts of the material. AAF uses the 32-byte SMPTE 330M UMID [2] for this purpose.

A SMPTE UMID has the useful property that it may be locally generated by a device and yet be globally unique; that is, there is practically no possibility of the same identifier being generated again by that device or by another device. This is achieved by using an identifier with a sufficiently large range of values and generating the identifier from elements based on date, time, device identifiers, and random numbers. The idea of locally generated globally unique identifiers originates in the computer industry, where it is widely applied as the Universally Unique Identifier (UUID) or Globally Unique Identifier (GUID) [5].

When AAF references material, it does so using this unique identifier. The referencing mechanism is unlikely to fail, because the identifiers are so unlikely to be accidentally duplicated and are not changed once they are assigned. Traditional means of referencing material, such as tape names and computer file names, can be less robust because it is relatively easy to create duplicate tape names and change computer file names. Note that tape names and computer file names may still be included in the AAF source metadata, but are not the primary means of identification.

Methods of AAF File Interchange

AAF is a file interchange format. This means that data interchange occurs by writing a file in one authoring tool and reading it in another. AAF supports two different methods for using AAF files to interchange data between multiple authoring tools: *export/import* and *edit in place*. Figures 5.6-2 and 5.6-3 show file interchanges from tool A to tool B and then tool B to tool C, using each method.

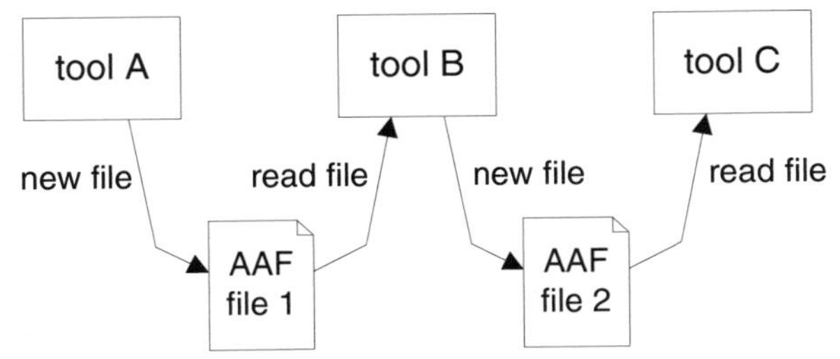

FIGURE 5.6-2 AAF file interchange—export/import method.

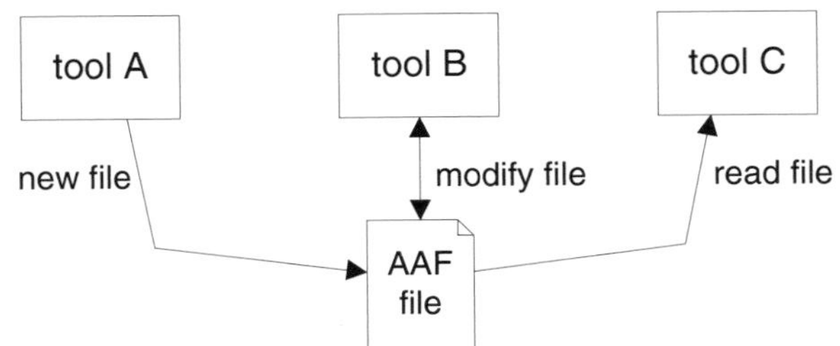

FIGURE 5.6-3 AAF file interchange—edit in place method.

With the export/import method, tool A creates a new AAF file that is read by tool B, and tool B creates a new AAF file that is read by tool C. The two interchanges are independent. The transfer of any data from tool A to tool C relies on tool B to import that data and subsequently re-export it. If tool B is a different kind of authoring tool than tool A and tool C, (e.g., an audio tool between two video tools), then this transfer may not be reliable. Export/import is appropriate for a simple interchange between two tools but has limitations for moving data between multiple tools.

With the edit in place method, tool A creates a new AAF file, which is then modified by tool B before being read by tool C. Any data from tool A intended for tool C can simply be ignored by tool B. Edit in place allows data to pass reliably between different kinds of authoring tools without requiring them to read and write data structures that are potentially outside their scope.

When AAF files are created or modified, a record of the application used is added to the file to maintain an audit trail.

Methods of AAF Essence Interchange

When interchanging file source material, AAF supports two different methods for interchanging the essence. The essence may either be internal to the AAF file, or held in an external file and referenced by the metadata. In both cases, the metadata for the file source material is in the AAF file. Figures 5.6-4 and 5.6-5 show a file interchange from tool A to tool B, using each method of essence interchange.

With the internal essence method, tool A creates an AAF file containing both the file source metadata and the essence. Tool A copies the essence into the AAF file and tool B copies it out again. Clearly, as the essence is being copied, the AAF file will potentially be large and the time to create and read the AAF file may become appreciable. Internal essence is an appropriate choice for interchanging data between tools that do not have any other common essence storage beyond whatever mechanism is being used to transfer the AAF file between the two systems. For audio and compressed video proxy essence, internal essence is quite workable. The authoring tool creating the AAF file will typically allow control over precisely what essence is copied into the file—for

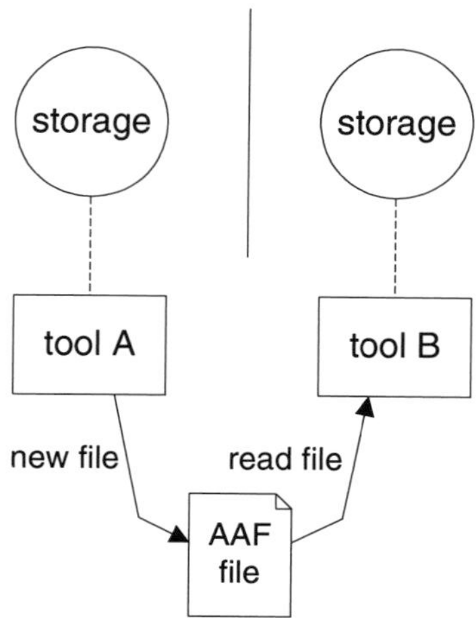

FIGURE 5.6-4 AAF essence interchange with essence internal to the AAF file.

example, all source essence or only source essence referenced by the composition.

With the external essence method, tool A creates an AAF file containing the file source metadata but not the essence—the essence remains external. The file source metadata includes a location (e.g., a network URL) for tool B to follow to read the external essence. External essence is an appropriate choice for interchanging data between tools that have common essence storage and can therefore share essence files. It is efficient because it avoids copying large essence files and the AAF file remains small. Note that the external essence file may itself be another AAF file acting to contain the essence, another container format such as MXF, or a plain essence file such as a DV DIF or an MPEG-2 Elementary Stream.

In some workflows, there is no requirement to interchange file source metadata or essence from tool A to tool B at all. For example, the tools may be working at

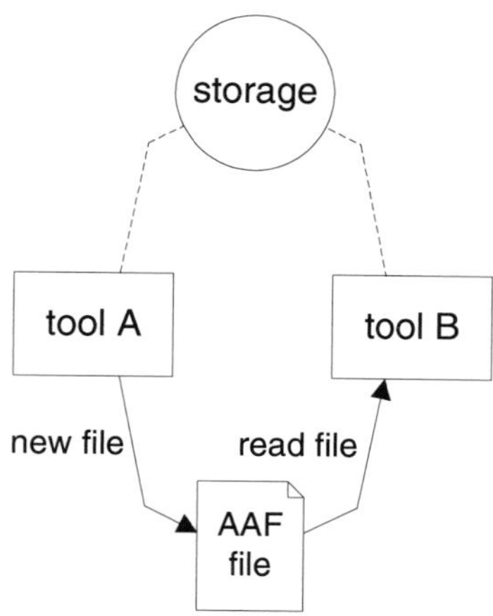

FIGURE 5.6-5 AAF essence interchange with essence external to the AAF file and referenced.

TABLE 5.6-2
Layered AAF Specifications

AAF Specification	Scope
AAF Object Specification	Defines data structures that can be interchanged by AAF, in terms that correspond to the authoring domain. AAF data structures are *objects*, defined by a *class model*.
AAF Stored Format Specification	Defines mapping of objects into a low-level container.
AAF Low-Level Container Specification	Defines mapping of low-level container into a file (or other persistent storage).

different picture resolutions and may each have transferred the material in from physical source tapes. In these cases, the source metadata may describe only the physical sources, such as tape or film.

Structure

The AAF file format is defined by three AAF specifications [6], organized into a layer structure. The scope of each layer is described in Table 5.6-2.

Organizing the specifications into cleanly separated layers improves clarity and allows for the independent extension or substitution of layers. For example, the Object Specification could be extended to support new types of objects without affecting the lower layers. Alternatively, the Stored Format and Low-Level Container Specifications could be substituted to provide an additional storage format for the objects.

Note that the relationship between AAF and MXF can be viewed in these terms—MXF reuses a subset of the AAF Object Specification and maps it onto a different storage format based on SMPTE 336M Key-Length-Value encoding [7].

AAF Object Specification

The AAF Object Specification defines the data structures that can be included in an AAF file. An instance of a data structure in an AAF file is an *object*. An object has *properties*, which have a *type* and a *value*. The Object Specification defines objects by specifying a *class model*. It is expressed in terms that closely correspond to the authoring tool domain. For example, it defines classes to hold metadata for the different material types, and to hold essence for file source material. The mapping of these objects into a file (or other persistent storage) is specified by the lower layers of the AAF Specifications.

Viewed in terms of the Object Specification, the objects contained within an AAF file are depicted in Figure 5.6-6, with a brief description of their purpose.

An important concept in the construction of an AAF file is *object containment*, where one object logically contains, and is the owner of, another object or a

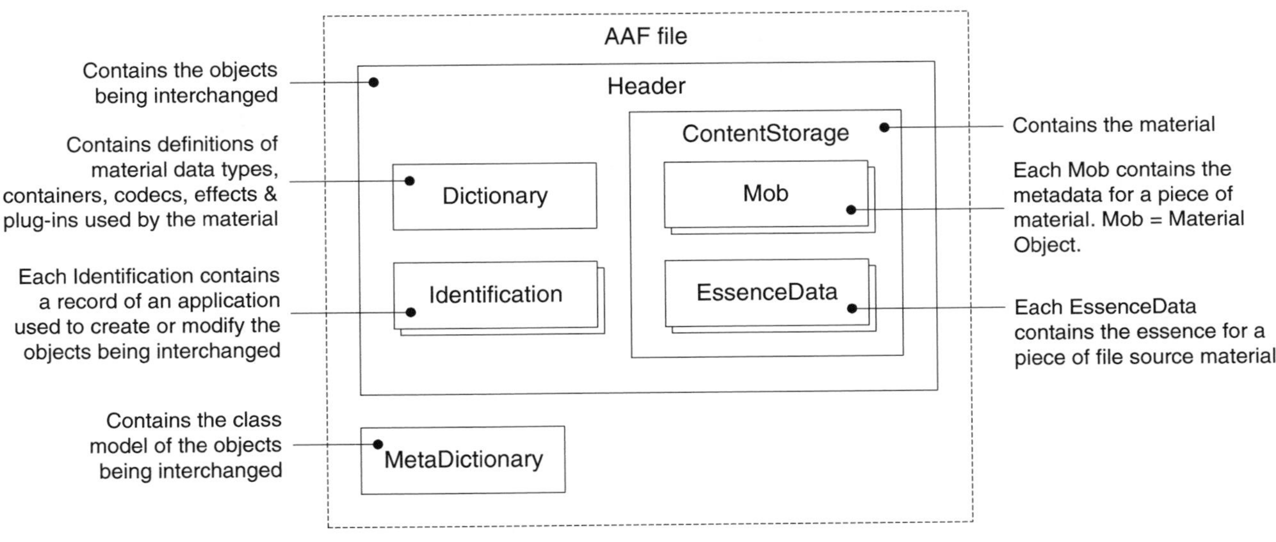

FIGURE 5.6-6 Objects contained within an AAF file.

collection of objects. An object or a collection of objects can only be owned by a single object at a time, which allows treelike structures to be constructed.

Object containment can be seen in Figure 5.6-6. At the highest level, there are two objects in the AAF file—Header and MetaDictionary. Contained within Header, there are Dictionary, ContentStorage, and a collection of Identification objects. Contained within ContentStorage, there is a collection of Material Objects (Mobs) and a collection of EssenceData objects. The MetaDictionary, Dictionary, and Mob objects also contain objects, and so on. Object containment is also known as *strong object reference.*

In contrast to object containment, an object may also reference another object or collection of objects without specifying ownership. This is known as *weak object reference*. An object can be the target of weak object references from more than one object. A target of a weak object reference must have an identifier so a reference can be made to it. In an AAF file, the Dictionary and MetaDictionary contain objects to which weak object references are made.

An object reference, strong or weak, may be made to an individual object or a collection of objects. Collections are divided into two types: *sets* and *vectors*. Objects in a set have a unique identifier and are not ordered. Objects in a vector do not need a unique identifier and have an order—the class model defines whether the order is meaningful for each vector.

The data interchanged in an AAF file is no more and no less than the tree of objects contained within it.

COMMON ELEMENTS OF AAF AND MXF

MXF reuses a subset of the AAF Object Specification and maps it onto a different storage format based on SMPTE 336M Key-Length-Value (KLV) encoding. The parts of the Object Specification dealing with clips and source material are reused in MXF, and the parts dealing with compositions, effects, and the in-file Dictionary and MetaDictionary are removed.

The motivation behind the MXF Key-Length-Value encoding was to create a simple, standards-based, stream format that could be processed by embedded systems such as videotape recorders and camcorders as well as PC-class devices. The MXF encoding has the capability to provide *play during record* and operate with isolated sections of an MXF stream, but it does not have the capability to provide efficient *edit in place.*

By using a common data model, the metadata in an MXF file is directly compatible with AAF, which allows AAF and MXF to work well together. For example, an AAF authoring tool can create an AAF file containing a composition that references file source material in an external MXF file.

During the development and standardization of MXF, the names of certain classes and properties taken from AAF were changed, although their meaning and unique identification were not changed. As a general guide, an MXF *Package* is an AAF *Mob*, an MXF *Material Package* is an AAF *MasterMob*, and an MXF *Track* is an AAF *MobSlot*.

MATERIAL EXCHANGE FORMAT (MXF)

MXF Files

A very simple MXF file is represented in Figure 5.6-7. At the beginning of the file, there is the file header, in the middle is the file body, and at the end is the file footer. The fine detail of the physical structure of an MXF file is discussed later. Discussed initially is the structural metadata that is stored in the header to describe the structure of the MXF file and the multimedia content contained within it.

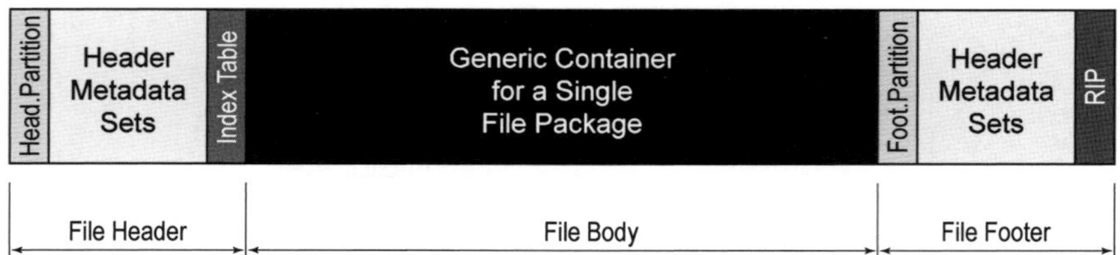

FIGURE 5.6-7 Basic structure of an MXF file.

Structural Metadata

The *header metadata* of an MXF file is a comparatively small amount of data at the beginning of the file. It may be repeated at points throughout the file and contains a complete description of what the file is intended to represent. This structural metadata does not contain any of the video, audio, or data essence, it merely describes the synchronization and timing of that essence.

Everything that is to be synchronized within an MXF file is represented by a *track*. An MXF track is a representation of time that allows different essence components to be synchronized relative to each other. Figure 5.6-8 shows a representation of a *timecode track*, a *picture track*, and a *sound track*. The metadata that describes these tracks can be found in the header metadata area of the MXF file. The structural metadata is very compact and allows the picture, sound, and timecode to be synchronized during playback or during capture. In the example shown, the picture track would describe the duration of the long-GOP MPEG video, the duration of the Broadcast Wave audio would be described by the sound track, and a timecode track would describe any timecode that was present in the original content (e.g., that which was generated when the content was captured).

MXF is intended to be used for more than just the simple playback of stored essence. The intention has been to provide a mechanism for simple editing or selection of material from content stored within the file. Therefore, it is necessary to provide tracks that describe the required output timeline as well as tracks that describe the material as it is stored. Figure 5.6-9 shows an example of two collections of tracks within an MXF file. The upper collection of tracks is the *material package,* which represents the required output timeline for the material as it is intended to be played. The lower collection of tracks is the *file package,* which describes the content as it is stored within the MXF file.

Constraints are placed by the MXF Specification on the possible relationships between an MXF material package and an MXF file package. These constraints are known as *operational patterns.* A file with the simplest standardized operational pattern is known as an OP1a file (see Operational Patterns section later).

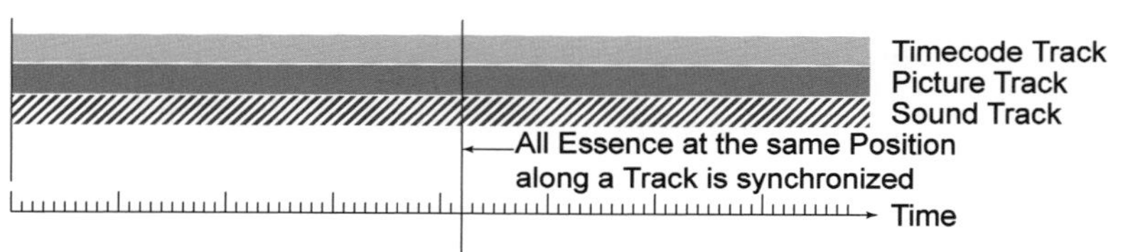

FIGURE 5.6-8 Header metadata synchronizes essence using tracks.

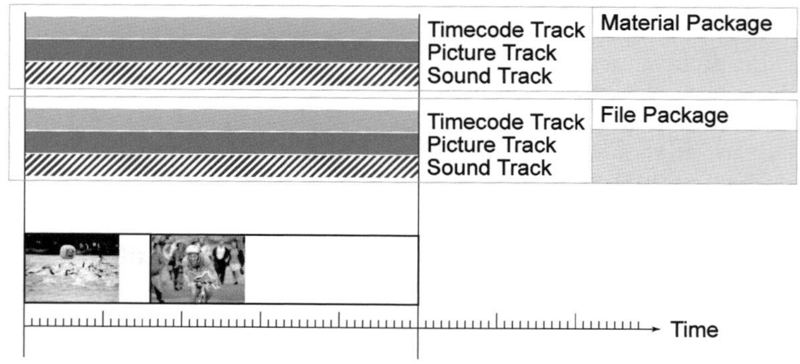

FIGURE 5.6-9 OP1a material package and file package relationship.

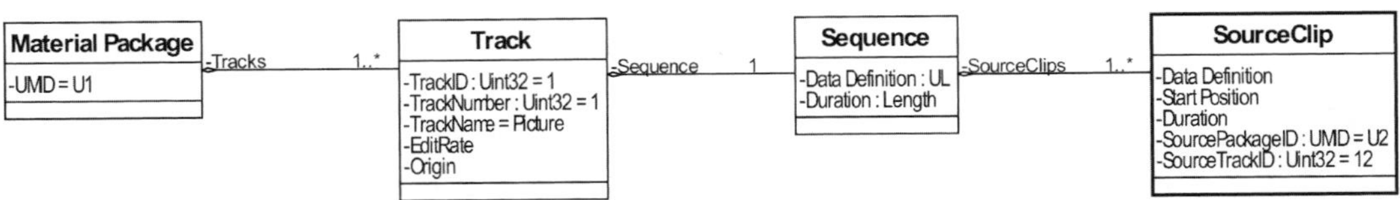

FIGURE 5.6-10 MXF metadata sets controlling synchronization.

Figure 5.6-8 showed a conceptual model of several tracks being synchronized for a given position along the timeline, and Figure 5.6-10 shows the actual MXF data sets that make this possible.

The material package has one or more tracks, each of which has a TrackID, an Edit Rate, and an Origin. The track has a sequence of one or more SourceClips, each of which has a Start Position, Duration, and the PackageID and TrackID of the file package track where the actual essence can be found. Using the SourceClips on the material package, any position on a material package track can be converted into a position on a file package track. Then *index tables* can be used to convert these time-oriented positions into byte offsets within the stored picture, and the sound essence used to play back synchronized material.

Figure 5.6-9 shows an OP1a representation of the example file where the material package *start* and *duration* are identical to the file package *start* and *duration*. In other words, the output timeline is equal to the timeline of the entire stored content; that is, what is played is equal precisely to what is stored.

In some applications, it may be desired to play out only the central contribution portion of a file. In the case of the example mentioned above, a material package is used that describes only this central portion of the file, not the color bars, slate, or black portion of the stored content. The operational pattern that describes this functionality is OP3a, as shown in Figure 5.6-11. To play back this file, random access functionality is required in a decoder, to skip over the unplayed portions of the file. In this way, operational patterns are used to describe the complexity required in coder and decoder functionality. Other representations of the file and more operational complexity will be covered later in the chapter.

Tracks

A *track* is the basic structural metadata component that allows the representation of time in MXF. The *duration* property of a track describes the duration, in terms of *edit units*, of the corresponding essence. The position along a track is used to describe the expected playout time of the start of a given sequence, and in this way synchronization between two different tracks is obtained.

The *origin* of a track is a property that describes the start point of the content (e.g., the sound track may start at a different time from the video track). The tick marks that are shown along the track in Figure 5.6-12 correspond to the edit units that have been chosen for that track. Typically, edit units will correspond to the video frame rate of the combined audio/video file and will usually have a duration of 1/25 or 1/30 of a second.

A collection of tracks is known as a *package*. As previously explained, the material package describes the output timeline of the file, and the top-level file package describes fully the stored content of the file. Lower-level file packages may be present, but these are only used to carry historical metadata about the "genealogy" of the stored content. There will be a track for each component of the output timeline. Typically, there will be a single picture track and a sound track for each of the audio channels. Stereo audio may be described by a single sound track.

Timecode is also represented by a track. A timecode track is actually stored metadata that describes timecode as a piecewise linear representation of the actual

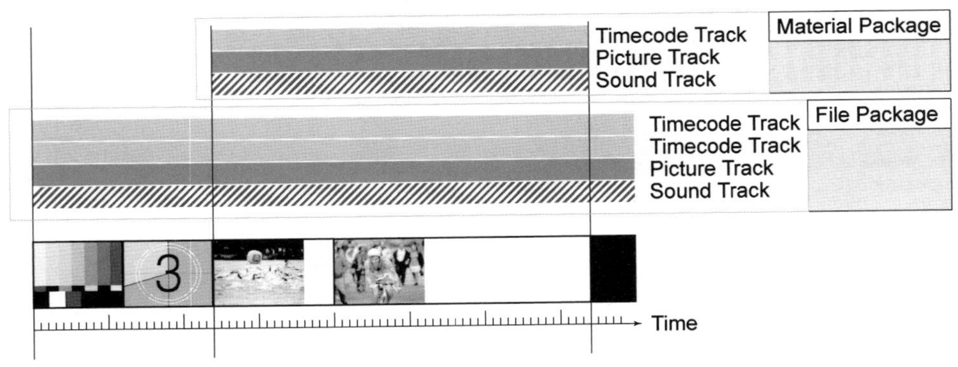

FIGURE 5.6-11 OP3a material package and file package relationship.

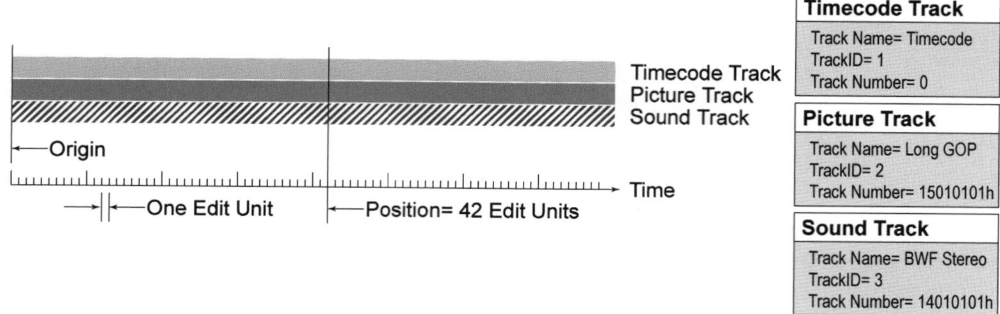

FIGURE 5.6-12 Properties of a track.

timecode value. The timecode is calculated by an MXF playout device when the file is played and would normally be reinserted into the output of the player. For example, Figure 5.6.13 shows a file package with several tracks, two of which are timecode tracks. This file was created as an ingest process from tape where there was a requirement to maintain the timecode that was originally found in the vertical interval timecode (VITC) of the tape. The second timecode track was added at ingest, corresponding to the linear timecode (LTC) of the tape. MXF allows as many timecode tracks to be stored in the file as are needed. The only hard requirement is that a material package must contain at least one timecode track that is linear and monotonic.

Timecode is stored this way because, in some applications, timecode can become discontinuous. This in turn can lead to synchronization errors between the various elements of a stream. MXF uses the position along a track as its internal synchronization mechanism, not the timecode value. This means that synchronization within a file is always preserved, even when an MXF file must re-create a discontinuous timecode at its output. The goal of MXF is to faithfully describe content while preserving synchronization between the tracks.

Timecode is defined by the SMPTE specification SMPTE 12M [8]. That document defines a counting mode called *drop frame* mode, which is provided for use in countries where TV signals work at the 29.97 frames per second rate. The important point about drop frame is that it is a counting mode and does not actually change the number of frames of video that are stored. To quote from SMPTE 12M:

> Because the vertical field rate of an NTSC television signal is 60/1.001 fields per second (≈59.94 Hz), straightforward counting at 30 frames per second will yield an error of approximately +108 frames (+3.6 seconds) in one hour of running time.
>
> To minimize the NTSC time error, the first two frame numbers (00 and 01) shall be omitted from the count at the start of each minute except minutes 00, 10, 20, 30, 40, and 50.
>
> When drop-frame compensation is applied to an NTSC television time code, the total error accumulated after one hour is reduced to --3.6 ms. The total error accumulated over a 24-hour period is --86 ms.

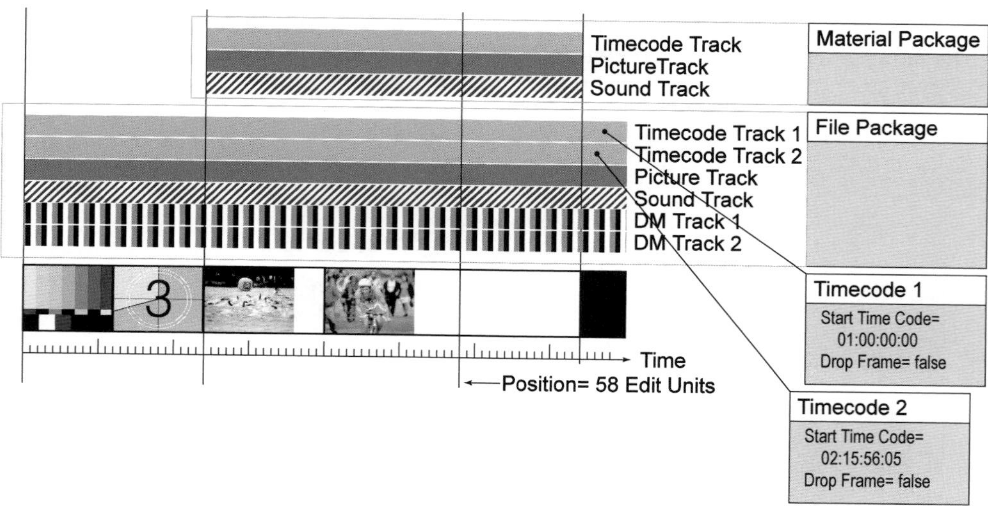

FIGURE 5.6-13 Timecode track.

Descriptive Metadata

To this point, this section has considered structural metadata—that is, metadata that binds the file together, controls synchronization, and identifies the various tracks and packages. In general, the structural metadata is machine-created and is intended to be read by another machine, such as an MXF player. The two lower tracks in Figure 5.6-13 are labeled *descriptive metadata* (DM) tracks. Descriptive metadata is usually created by humans for human consumption.

Descriptive metadata is represented by a track. In the design of MXF, it was realized that creating a single vocabulary for the description of broadcast and film content was neither achievable nor really desirable. However, without a standardized structure for the interchange of descriptive information, the power of MXF could not be realized, so an attempt had to be made to create a structure that was good enough for a large number of applications. The result was SMPTE 380M—MXF Descriptive Metadata Scheme 1 [9]. This scheme divided metadata descriptions into three broad categories:

- Production Framework—descriptions that relate to an entire production;
- Scene Framework—descriptions oriented towards a scene/script or what the content is intended to show (e.g., nightfall over Moscow);
- Clip Framework—descriptions oriented towards how the clip was shot (e.g., shot in a back lot in Hollywood).

Descriptive metadata may not be continuous along a track in the way that video and audio essence tends to be. A description may last only for the portion of a timeline where the information is valid. For this reason, different types of tracks are defined within MXF:

- Timeline Track: This track is not allowed to have any gaps or overlaps. This is the sort of track used to describe the video and audio essence.
- Event Track: This track can have components on it that overlap or are instantaneous; that is, they have zero duration.
- Static Track: This track is independent of time and is used to describe static content. Note that this is not the same as an event with zero duration that occurs at a specific time. A static track contains metadata that is always valid for the entire length of a sequence.

In Figure 5.6-14, the validity of the metadata has been shown on the metadata tracks. The track labeled DM Track 1 is valid from the end of the slate until the start of the black content at the end of the example. This might be a production framework with a producer's name and contact details. The track labeled DM Track 2 has a framework that lasts for a small portion of the middle of the clip. This might be used to indicate a shot list on a scene framework highlighting some significant event in the clip, such as cyclists cresting a mountain.

Finally, descriptive metadata tracks are able to describe one or more essence tracks in a package. For example, DM Track 1 describes the picture, sound, and timecode essence tracks. This has been shown in the figure by the arrows between tracks. DM Track 2, however, is an annotation of only the picture track, as shown by the single arrow in the figure.

Source Reference Chain

This is a grand title for a simple concept. One of the goals of MXF was to fully describe the content and where it came from, or in other words, the derivation or "genealogy" of the content. This means that the MXF file should be able to store the relationship between the output timeline and the stored content, as well as information about what source file and source tapes were used to make the content in the first place. The system designers recognized that a full, in-depth history of every frame was not needed for every application, but there were some applications for which it was vital. This area of functionality is extremely well designed

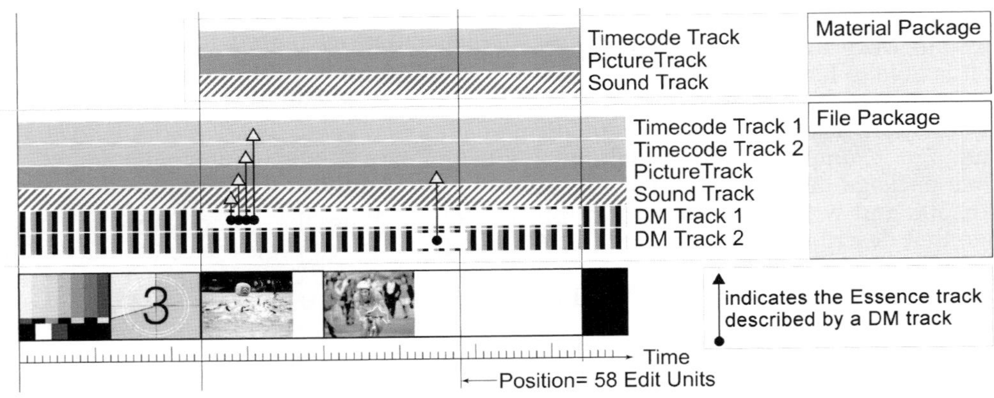

FIGURE 5.6-14 Descriptive metadata tracks describe other tracks.

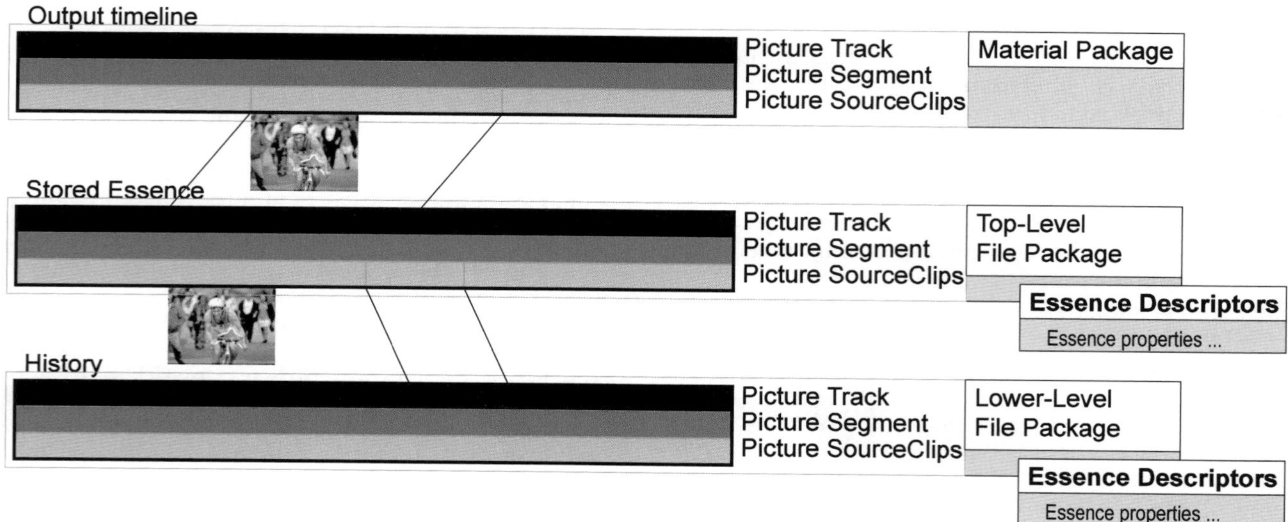

FIGURE 5.6-15 Source Reference Chain.

within the AAF data model and is one of the reasons why MXF designers put the AAF model at its core.

The *source reference chain* is the linking mechanism between different packages (see Figure 5.6-15). When several packages are linked together, the links form a chain from the top-level material package down to the lowest-level file package. In an MXF file, the material package is responsible for synchronizing the stored content during playout. The top-level file packages describe the stored content within the MXF file, and each of the lower-level source packages describes what happened to the content in a previous generation. This metadata allows an audit trail to be built up of the processes that the content has undergone. Each package is identified by a SMPTE UMID. Each of the UMIDs associated with the previous generations is preserved, along with a description of the essence that was contained at that time.

Each of the essence tracks in a top-level file package has an *essence descriptor* associated with it that describes the parameters of the stored essence. Lower-level source packages may also contain essence descriptors that describe the physical parameters of previous generations of the material. These essence descriptors allow an MXF application to determine whether or not it is able to handle the stored content. They also allow the source reference chain to be "mined" for information relating to the current pictures and sound. For example, a file containing a lower-level source package that describes the content as DV, 720 × 576, 50i, and which also contains a top-level file package that describes the content as MPEG long-GOP, 720 × 480, 59.94i must have undergone a frame rate standards conversion and a DV to MPEG transcode at some stage.

Essence descriptors fall into two broad categories: *file descriptors* and *physical descriptors*. A file descriptor is basically a description (e.g., resolution, sample rate, or compression format) of the stored content of an MXF File. It may be attached to a top-level file package, in which case it describes the content that is actually in the file. It may be in a lower-level source package, in which case it describes the content as it was stored in some previous generation of the file.

The file descriptors are intended to provide enough information to be able to select an appropriate codec to decode, display, or reproduce the content. They are also intended to provide enough information to allow an application to make decisions on how essence might be efficiently processed. An example of typical parameters is given in Figure 5.6-16.

A physical descriptor is a description of *how* the content entered an MXF-AAF environment. This may have been as the result of a tape digitization, in which case a tape descriptor may be in the file, or it may have been a result of an audio file conversion operation, in which case an AES audio physical descriptor may be in the file.

The physical descriptors are intended to give enough information about the original source of a file so the content can be appropriately processed. When the physical descriptor defines another file format, the physical descriptor is often the place where extra metadata is placed to allow transparent round-tripping to and from the MXF environment—for example, MXF to BWF to MXF.

Operational Patterns

Controlling the complexity of the source reference chain also controls the complexity of the MXF encoder and MXF decoder required to generate or play an MXF file. In the design of MXF, several attempts were made to categorize applications in order to simplify the vast flexibility of the MXF format. In the end, the approach chosen was to control the relationship between the material package(s) and the file package(s) in an MXF file.

Generic Sound Essence Descriptor		
Name	**Type**	**Meaning**
Instance UID	UUID	Unique ID of this instance
Generation UID	UUID	Generation Identifier
Linked Track ID	UInt32	Value of the Track ID of the Track in this Package to which the Descriptor applies.
SampleRate	Rational	The field or frame rate of Essence Container (not the audio sampling clock rate)
Container Duration	Length	Duration of Essence Container (measured in Edit Units)
Essence Container	UL	The UL identifying the Essence Container described by this descriptor. Listed in SMPTE RP 224
Codec	UL	UL to identify a codec compatible with this Essence Container. Listed in SMPTE RP 224
Locators	StrongRefArray (Locators)	Ordered array of strong references to Locator sets
Audio sampling rate	Rational	Sampling rate of the audio essence
Locked/Unlocked	Boolean	Boolean indicating that the number of samples per frame is locked or unlocked.
Audio Ref Level	Int8	Audio reference level which gives the number of dBm for 0VU.
Electro-Spatial Formulation	Uint8 (Enum)	E.g. mono, dual mono, stereo, A,B etc
ChannelCount	Uint32	Number of Sound Channels
Quantization bits	UInt32	Number of quantization bits
Dial Norm	Int8	Gain to be applied to normalize perceived loudness of the clip, defined by normative ref 0 (1dB per step)
Sound Essence Compression	UL	UL identifying the Sound Compression Scheme

FIGURE 5.6-16 Properties of the generic sound essence descriptor.

During the design of MXF, the words *templates*, *profiles*, and others were used to describe the constraints on the file. Most of these words already had various meanings coming from the video and IT industries. The phrase *operational pattern* was chosen as this was reasonably descriptive and did not carry any historical baggage. Figure 5.6-17 shows the 3x3 matrix of standardized, generalized operational patterns. The different constraints on MXF functionality can be described by looking at the columns and rows independently. There are also operational pattern qualifiers (not shown in the diagram) that carry extra information about the operational pattern, such as whether the bytes are physically arranged to make the file streamable.

The operational pattern columns differentiate the time axis complexity of an MXF file. The first column constrains a material package to play out the entire timeline of the file package(s). This means that everything that is stored is played out. The second column constrains a material package to play out the file packages one after the other according to a playlist. Each of the file packages is played out in its entirety. The third column requires some kind of random access in an MXF player. Here, small portions of essence within an MXF file can be played out one after the other. This functionality allows cut-edits to be expressed using the source reference chain.

The operational pattern rows differentiate the package complexity of an MXF file. The first row constrains a material package to have only a single file package active at any one time along the output timeline. The second row allows two or more file packages to be synchronized using the material package to define the synchronization. The third row allows more than one material package in a file. This allows the selection of different output timelines in a file to cover versioning and multilanguage capabilities.

Heading upward or leftward in the diagram of Figure 5.6-17 indicates decreasing MXF encoder or MXF decoder complexity. Any encoder or decoder must support the functionality of operational patterns that are above or the left of its stated operational pattern capabilities. The MXF rules state that a file must be labeled with the simplest operational pattern required to describe it. This is to ensure maximum interoperability between all applications.

Specialized Operational Patterns

The specialized operational pattern, OP-Atom, is not shown in Figure 5.6-17. OP-Atom defines a tightly constrained file structure for a single track of essence. It is targeted as a material format for nonlinear editing applications requiring high-performance access to individual tracks of essence. The main features of an OP-Atom file are:

- Within each OP-Atom file there is one top-level file package, which references the essence in the single essence container within the file.
- There is only one essence track in the top-level file package.
- There is exactly one essence container containing essence from a single instance of an MXF essence mapping. This essence container is internal to the file.
- Within each OP-Atom file there is one material package, which may have an unconstrained number of tracks and an unconstrained number of SourceClips within each track. This allows synchronization metadata to be held in the material package for a group of OP-Atom files. The grouping may be across the timeline (such as separate audio and video files generated from a synchronized audio-video source), along the timeline (such as where the entire essence will not fit on a single disc), or both. Using this approach, each OP-Atom file within a group contains the same material package, describing the structure of the complete audio-

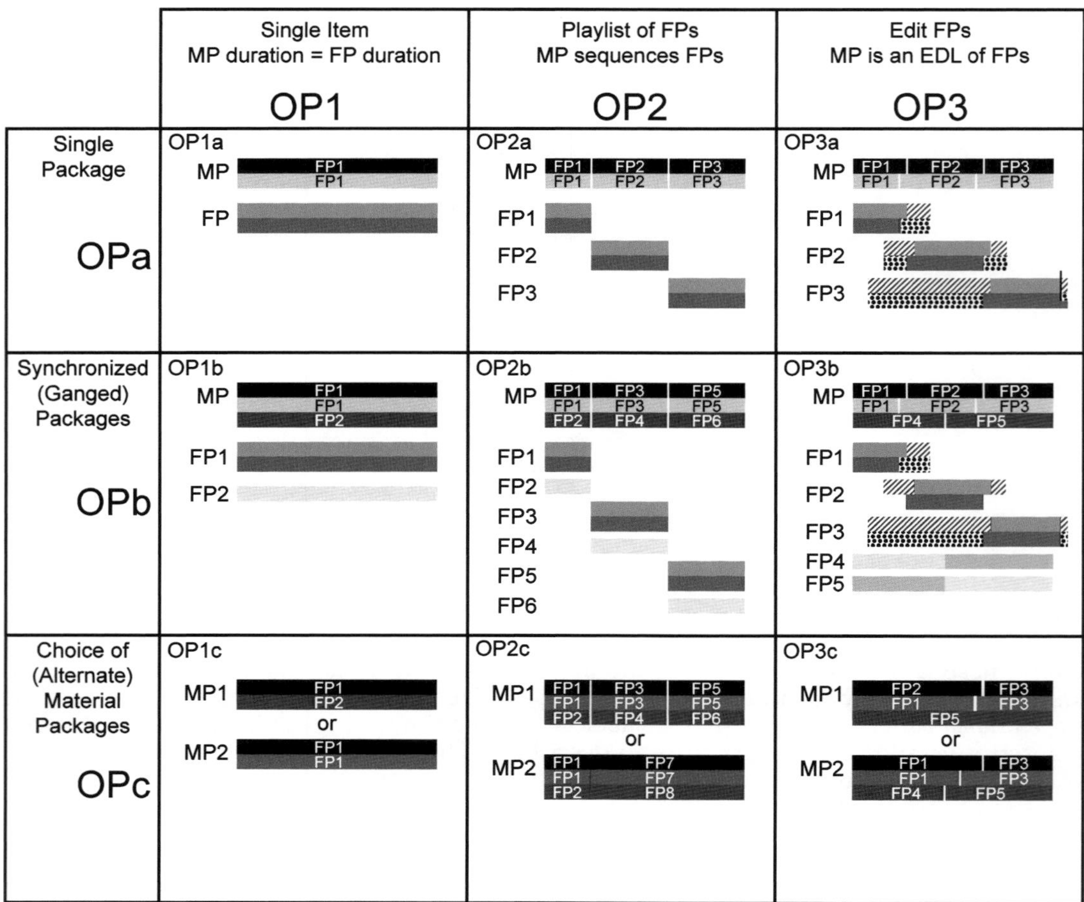

FIGURE 5.6-17 Operational patterns.

visual clip of which each OP-Atom file is a part. By including this material package in each OP-Atom file, the original clip metadata is always available with any individual essence file. This avoids the necessity for an additional metadata-only master MXF file, although one could be used to define new audiovisual clip associations.

Physical Storage

So far, in this section the concentration has been on the logical view of an MXF File—that is, what the file is intended to represent. Next is how MXF arranges the bytes on a storage device. One of the goals in the design of MXF was to create a format that was extensible. This means that a device built in accordance with revision 1.0 of the specification must be able to parse and decode a file that is version 1.2 or 2.0 without modifying the device. A device must be able to ignore elements of the file that are unknown to it and yet still be able to decode known elements.

The low-level mechanism for achieving this is key length value (KLV) encoding. Every property, set, and chunk of essence in an MXF file is wrapped in a KLV triplet as shown in Figure 5.6-18. The key of the KLV triplet is a 16-byte number that uniquely identifies the contents of the triplet. If an MXF parser does not recognize the key, then the length property can be used to skip the value bytes. If the MXF parser does recognize the key, then it is able to route the KLV payload (the value field) to the appropriate handler.

Comparing 16-byte keys can consume time and storage space when a large number of small values need to be KLV-wrapped. In addition, it is useful for the KLV structure to be able to group together a number of different KLV properties as a single set. For this reason, KLV allows *local set* coding, which is a mechanism whereby 2-byte tags are substituted for the 16-byte keys as shown in Figure 5.6-18. To ensure that a decoder is able to associate all the 2-byte tags with the 16-byte keys, a lookup table to convert tags to ULs called the *primer pack* is included in every file.

Generic Container

Essence is placed in an MXF file using KLV encoding. The *generic container* was created to provide a mechanism for encapsulating essence that was fast, could be frame-oriented, and provided easy implementation

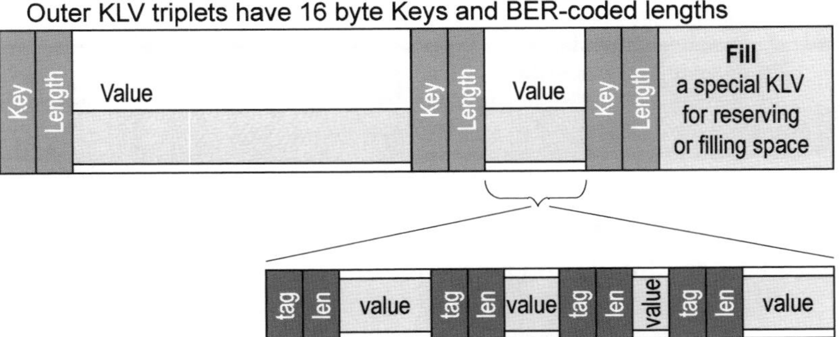

FIGURE 5.6-18 KLV coding and sets.

for streaming or file-bridging devices. Inside the file, "essence in a generic container" is often referred to as the "essence container" and MXF allows several essence containers to be stored in a single file. Each essence container in a file is associated with a single top-level file package that holds the metadata description of the stored essence.

Figure 5.6-19 shows the basic structure of the generic container. It divides the essence up into *content packages* that are of approximately equal duration. Many (in fact probably most) MXF files are divided up into content packages that are one video frame in duration. These files are called *frame-wrapped* files. When the duration of the content package is the same as the duration of the files, these are called *clip-wrapped* files (because the entire clip is wrapped as an indivisible lump).

Each content package is categorized into five different essence categories called *items*. Within each item, there may be a number of individual *essence elements*. For example, the content package in Figure 5.6-19 has five items, of which the sound item contains two sound elements. Each content package may only have a single essence item from each category and the order of the items in the file should be the same in each and every content package. This makes it easier for parsers to work out where one content package starts and another ends. As well as keeping the order of the items constant in the file, it is also important to keep the order of the elements within the item constant. Each essence element is linked to a unique track in the top-level file package using the KLV key of that essence element. It is therefore important to ensure that the order of the elements remains constant *and* that the element in any given place within the content package has a constant KLV key, because this is the binding that ties the logical description to the physical stored content. Essence categories are as follows:

- System Item—used for physical-layer-oriented metadata. System elements are used to provide low-level functionality such as pointing to the previous content package for simple reverse play. There is no system element in the example shown previously.
- Picture Item—contains picture elements, which in turn contain essence data that is predominantly picture-oriented (e.g., the long GOP MPEG samples in the example).

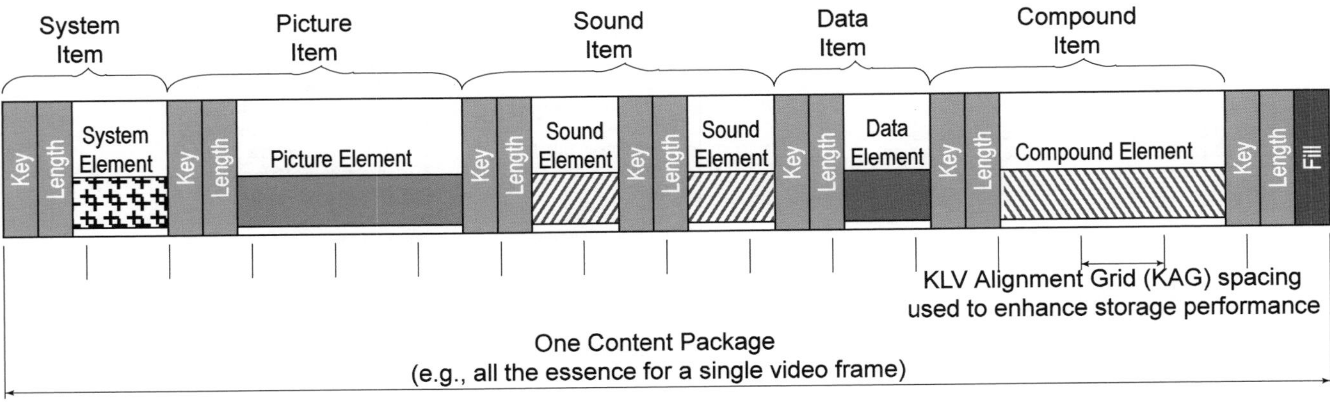

FIGURE 5.6-19 The generic container.

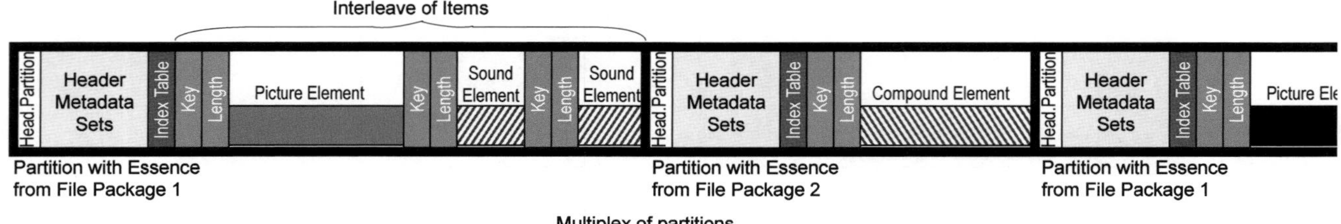

FIGURE 5.6-20 Interleaving, multiplexing, and partitions.

- Sound Item—contains sound elements, which contain predominantly sound essence data bytes (e.g., the BWF sound samples in the example).
- Data Item—contains data elements, which are neither picture nor sound, but continuous data. There are no data elements in the example.
- Compound Item—contains compound elements that are intrinsically interleaved. DV essence is a good example of this. The stream of DV block contains an intrinsic mix of picture, sound, and data.

Tracks, Generic Containers, and Partitions

Within the content package, the essence items and elements are said to be *interleaved*, as shown in Figure 5.6-20.

A single file package describes a single generic container. If the generic container has a single stream of sound essence elements, then the file package will have a single track to describe them. If the generic container has a single stream of picture elements, then the top-level file package will have a single track to describe them.

If a file has more than one file package, there will be more than one generic container. It is necessary to multiplex together different file packages, which themselves may be interleaved. In order to separate the different stored generic containers, MXF inserts *partitions* between them. Each partition marks a point in the file where header metadata may be repeated, index table segments may be inserted, and data for a different generic container may start.

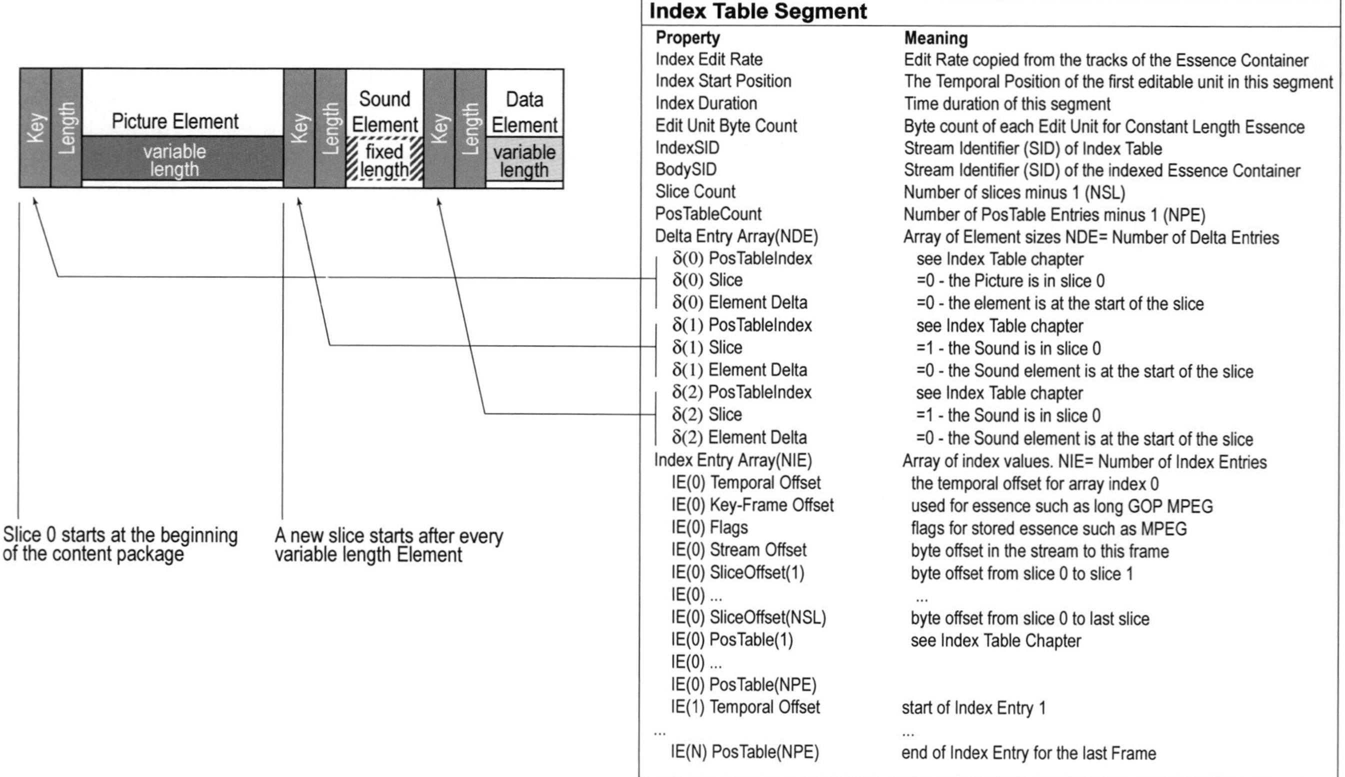

Index Table Segment

Property	Meaning
Index Edit Rate	Edit Rate copied from the tracks of the Essence Container
Index Start Position	The Temporal Position of the first editable unit in this segment
Index Duration	Time duration of this segment
Edit Unit Byte Count	Byte count of each Edit Unit for Constant Length Essence
IndexSID	Stream Identifier (SID) of Index Table
BodySID	Stream Identifier (SID) of the indexed Essence Container
Slice Count	Number of slices minus 1 (NSL)
PosTableCount	Number of PosTable Entries minus 1 (NPE)
Delta Entry Array(NDE)	Array of Element sizes NDE= Number of Delta Entries
δ(0) PosTableIndex	see Index Table chapter
δ(0) Slice	=0 - the Picture is in slice 0
δ(0) Element Delta	=0 - the element is at the start of the slice
δ(1) PosTableIndex	see Index Table chapter
δ(1) Slice	=1 - the Sound is in slice 0
δ(1) Element Delta	=0 - the Sound element is at the start of the slice
δ(2) PosTableIndex	see Index Table chapter
δ(2) Slice	=1 - the Sound is in slice 0
δ(2) Element Delta	=0 - the Sound element is at the start of the slice
Index Entry Array(NIE)	Array of index values. NIE= Number of Index Entries
IE(0) Temporal Offset	the temporal offset for array index 0
IE(0) Key-Frame Offset	used for essence such as long GOP MPEG
IE(0) Flags	flags for stored essence such as MPEG
IE(0) Stream Offset	byte offset in the stream to this frame
IE(0) SliceOffset(1)	byte offset from slice 0 to slice 1
IE(0) ...	...
IE(0) SliceOffset(NSL)	byte offset from slice 0 to last slice
IE(0) PosTable(1)	see Index Table Chapter
IE(0) ...	
IE(0) PosTable(NPE)	
IE(1) Temporal Offset	start of Index Entry 1
...	...
IE(N) PosTable(NPE)	end of Index Entry for the last Frame

FIGURE 5.6-21 Indexing content.

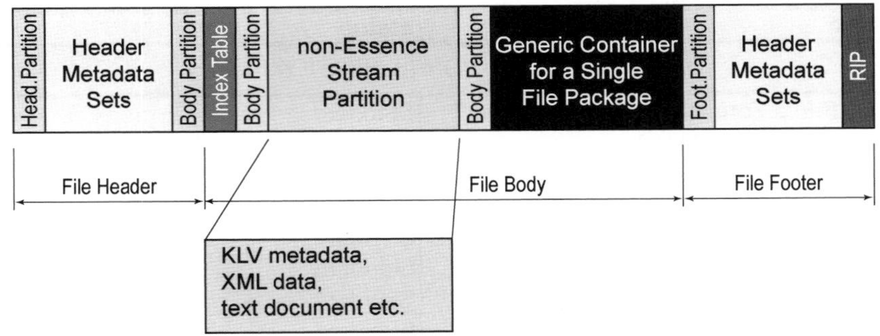

FIGURE 5.6-22 Non-essence stream container.

Since the original design of MXF, it has been recognized that it is good practice to place only a single item in a partition. Although the MXF specification allows each partition to contain header metadata *and* index table *and* essence, current wisdom advises that a partition should contain header metadata *or* index table *or* essence.

Index Tables

With the essence data mapped into the generic container and described by the header metadata, it is useful to be able to perform random access within the file so that efficient partial restore (that is, picking out a portion of the stored material), scrubbing (moving quickly forward and backward through the material), and other operations can be performed. Figure 5.6-21 shows the basic *index table* structures that are provided within MXF to allow a time offset to be converted to a byte offset within the file. The richness of MXF makes the index table design rather complex.

The index tables have to cope not only with an MXF file containing a single generic container with simple constant-sized elements but also with MXF files containing multiple multiplexed generic containers with interleaved elements of variable length. This gives rise to the structures shown in the figure.

Non-Essence Streams

There are certain types of data and essence that do not fit well in the generic container. These types of data are referred to as non-essence streams and may include things such as bursty KLV data streams, text documents, and other large miscellaneous lumps of data, which need to be included in a file (see Figure 5.6-22).

A stream container has been defined that allows this sort of data to be included in an MXF file and associated with the header metadata. One of the key features of this sort of data is that the MXF index table structure either does not work or is extremely inefficient for indexing the content.

Data Model

The term *data model* has been mentioned previously. The structural metadata forms part of the MXF data model, which is basically the definitive set of relationships between the various metadata sets and properties used in MXF. The data model for MXF is derived from the AAF data model described earlier.

MXF Documents

Figure 5.6-23 shows the documents that had been standardized, and those that were in progress at the time of this writing (mid-2006). The status of these documents will change over time.

SUMMARY

This chapter has provided an introduction to the subject of AAF, MXF, file formats, and metadata in the broadcast environment. It is a complex subject and those wishing to learn more will find in-depth information in the references, including the MXF Engineering Guidline [10], and in the books listed in the bibliography.

ACKNOWLEDGMENTS

Thanks are extended to Bob Edge of Thomson Grass Valley for providing the description of the GXF format. The section on Advanced Authoring Format (AAF) is adapted from the *File Interchange Handbook*, Chapter 6, "Advanced Authoring Format," by Phil Tudor, published by Focal Press in 2004, and used with permission of the publisher. The section on Material Exchange Format (MXF) 9 is adapted from *The MXF Book*, Chapter 2, "What Is an MXF File?" by Bruce Devlin, published by Focal Press in 2006 and used with permission of the publisher.

Number	*doc*	*status*
	On the SMPTE CD-ROM in SMPTE directory	
SMPTE 377M	MXF File Format	Standard
SMPTE 378M	Operational Pattern 1a	Standard
SMPTE 379M	Generic Container	Standard
SMPTE 380M	Descriptive Metadata Scheme 1	Standard
SMPTE 381M	Mapping MPEG into MXF	Standard
SMPTE 383M	Mapping DV (&DV-based) into MXF	Standard
SMPTE 384M	Mapping Uncompressed into MXF	Standard
SMPTE 385M	SDTI-CP compatible system Item	Standard
SMPTE 386M	Mapping D10 into MXF	Standard
SMPTE 387M	Mapping D11 into MXF	Standard
SMPTE 388M	Mapping A-law audio into MXF	Standard
SMPTE 389M	Reverse Play	Standard
SMPTE 390M	Operational Pattern Atom	Standard
SMPTE 391M	Operational Pattern 1b	Standard
SMPTE 392M	Operational Pattern 2a	Standard
SMPTE 393M	Operational Pattern 2b	Standard
SMPTE 394M	GC System scheme 1	Standard
SMPTE 405M	GC System scheme 1 Elements	Standard
	On the SMPTE CD-ROM in EG directory	
EG41	Engineering Guideline	Engineering Guideline
EG42	DMS Engineering Guideline	Engineering Guideline
	At www.smpte-ra.org	
RP210	Metadata Dictionary	Standard
Class 13-14	Private number spaces	Registered UL number spaces
UMID	UMID registries	Registered UMID number spaces
	In the Final Proof Reading stages of publication	
-	-	-
	With the Standards Committee	
SMPTE 382M	AES - BWF audio	
-	-	-
	In Trial publication on the SMPTE website - subject to change	
-	-	-
	In Technical Committee Ballot - subject to change	
SMPTE 407M	OP3ab	-
SMPTE 408M	OP123c	-
SMPTE422M	Mapping JPEG2000	-
-	Generic Stream container	-
-	XML Representation of Data Models	-
RDD9	Sony Interop Spec - MPEG Long GOP	-
-	MXF- XML Encoding	-
-	MXF Mapping for VBI Ancillary Data	-
-	MXF on Solid-State Media Card	-
SMPTE423M	MXF Track File Essence Encryption	-
SMPTE416	D-Cinema Package Operational Constra	-
-	dCinema Track File Specificaiton	-
	Unballotted Working Drafts - subject to change	
---	---	---

FIGURE 5.6-23 MXF documents at time of publication.

References

[1] EBU/SMPTE, Task Force for Harmonized Standards for the Exchange of Program Material as Bitstreams, *Final Report: Analyses and Results*, 1998, (http://www.smpte.org/engineering_committees/pdf/tfrpt2w6.pdf)
[2] SMPTE 330M, *Unique Material Identifier (UMID)*, http://www.smpte.org/smpte_store/standards/.
[3] Professional-MPEG Forum, http://www.pro-mpeg.org/.
[4] AAF Association, http://www.aafassociation.org/.
[5] ISO/IEC 11578-1, *Information Technology—Open Systems Interconnection—Remote Procedure Call (RPC), Annex A, Universally Unique Identifier.*
[6] AAF Specifications, http://www.aafassociation.org.
[7] SMPTE 336M, *Data Encoding Protocol using Key-Length-Value*, http://www.smpte.org/smpte_store/standards/.
[8] SMPTE 12M, *Time and Control Code*, http://www.smpte.org/smpte_store/standards/.
[9] SMPTE 380M, *Material Exchange Format (MXF)—Descriptive Metadata Scheme-1*, http://www.smpte.org/smpte_store/standards/.
[10] SMPTE EG 41, *Material Exchange Format (MXF)—Engineering Guideline*, http://www.smpte.org/smpte_store/standards/.

Bibliography

Gilmer, B., et al., *File Interchange Handbook.* Focal Press, 2004.
Devlin, B., Wilkinson, J., Beard, M., Tudor, P., Wells, N., *The MXF Book: An Introduction to the Material eXchange Format*. Focal Press, 2006.

CHAPTER

5.7

Video Compression

PETER SYMES

Grass Valley Inc.
Nevada City, California

INTRODUCTION

The electronic transmission or recording of moving images has always presented challenges because of the sheer volume of information that must be handled. From the early days it has been necessary to find compromises between bandwidth and quality. The challenge has always been to find the techniques that provide the greatest savings in bandwidth for the least loss in delivered quality.

In the analog world, the first major development in this direction was interlaced scanning. For a given picture size and static resolution, and a chosen repetition rate sufficient to prevent large-area flicker, interlace permits halving the bandwidth that would otherwise be required. But, there is a cost; segments of vertical and temporal spectra are overlapped. This results in a loss of vertical resolution on moving objects and interlace artifacts caused by vertical energy appearing as small-area flicker.

The other technology that provides dramatic saving of bandwidth is the National Television System Committee (NTSC) color system. This brilliant development interleaved color information into the luminance spectrum so as to permit color transmissions in the same bandwidth as monochrome NTSC (and maintained monochrome compatibility). Again, there were consequences: some degradation of the monochrome signal, cross-channel artifacts, and a barely adequate chroma bandwidth. However, few would argue today that the compromise was a bad one or that a better solution was available at the time.

Digital video brings many benefits, but, in the formats used for production, it requires vast quantities of data. Standard-definition 525 line component 4:2:2 video at 29.94 frames/sec, as used for NTSC coding, is carried in the studio at a rate of 270 megabits per second. This is more than needed for transmission—about 23% is dedicated to blanking and carries no picture information. Also, for delivery to the home, 10 bit precision is unnecessary, and 4:1:1 or 4:2:0 coding would be adequate. Taking all of these factors into account, it would be possible to save about half of the original 270 Mbps; however, even the remainder would require ten or more 6 MHz broadcast channels using conventional transmission systems.

High definition (HD) is, of course, much more demanding. In the studio, HD signals are transmitted at almost 1.5 gigabits per second, five and a half times the standard-definition rate.

Today, digital transmission of high-definition signals is accomplished in a single 6 MHz channel, sometimes less. Advanced modulation techniques provide one part of the solution, but the most important enabling technology is video compression. MPEG-2, the compression standard used for almost all digital video broadcast systems (in 2006), can provide acceptable picture quality while reducing the required bit rate by a factor of about 50 to 1. The latest compression systems can provide even higher ratios.

Delivery to the home is the obvious imperative for video compression—without this technology digital transmission would be totally impractical—but compression is also a vital technology for storage of digital video. A two-hour movie, even in standard definition, represents well over 100 GB but can be compressed to less than 9 GB for a DVD. Personal video recorders (PVRs) use compression to store many hours of video

on standard PC hard drives at very low cost. In the professional areas of production and program storage, lower compression ratios are used, but video compression is still the enabling technology for all current digital videotape formats, nonlinear editors, playout servers, and digital archives.

Compression systems really are systems: complex combinations of a variety of tools that operate in quite different ways. This chapter will discuss the various approaches to compression, the tools that implement these, and the ways that tools are combined into complete systems. In most applications, video compression is used to provide an appropriate balance between compression ratio and picture quality. The term "appropriate" is very important, as diverse applications and the various steps in the production/delivery chain call for very different approaches to compression. In all cases, the objective is the same as discussed at the beginning of this introduction: maximum savings in bandwidth/bit rate for a loss of quality that is acceptable in the specific application.

OVERVIEW

Terminology

In a practical application, there are three major elements to a system that uses compression. The *encoder* takes a standard video signal as its input and provides a compressed *bit stream* that is sent to a transmission or storage system. The bit stream is then sent to a *decoder* that reconstructs the video. The video from the decoder will likely be of lower quality than that input to the decoder. Terminology varies, but most people differentiate between picture *losses*, such as a loss of sharpness, and *artifacts* that are visible effects in the output picture that were contributed by the compression system. The most obvious example of an artifact is the "blockiness" often seen in stressed MPEG systems.

Lossy and Lossless Compression

There are two fundamentally different parts of practical compression systems. The first is to ensure that only necessary information is transmitted (or stored); the second is to code that information in the most efficient way possible. In the video world, compression is usually thought of as a lossy process; it is expected that the output of an encoder will not be an exact match to the input. Lossy compression works by approximating the image in a way that can be transmitted with fewer bits while minimizing the loss *as perceived by the human visual system* (HVS). This is a critical factor. An image compression system must be designed to produce the desired quality as perceived by a viewer under a particular set of assumed viewing conditions. This is the first step described above—ensuring that only necessary information is transmitted. This part of the process may also be described as *eliminating visual redundancy.*

Lossless compression refers to techniques that reduce the number of bits required by coding the information in more efficient ways but always guaranteeing that the output of the lossless decoder will be absolutely identical to the input to the lossless encoder. Lossless compressors can themselves be systems employing a number of tools. Well-known examples are the *PK Zip* and *StuffIt®* programs used on personal computers. This process of lossless compression, also known as *entropy coding*, can also be viewed as eliminating redundancy. It substitutes non-redundant coding (or less redundant coding) for inefficient (redundant) coding of information; however, within this article the term "redundancy" will be used to refer to visual elements; improving coding efficiency will be referred to as just *lossless compression*.

Lossless compression is rarely used alone for video because it can provide only a relatively small degree of compression (2:1 is regarded as typical); however, after lossy techniques have been used, lossless compression is generally applied to the resulting bit stream to improve the efficiency of the overall system. The distinction is very important. As described above, the first step is to remove, as far as possible, all redundancy in the representation of the image. The necessary principles and techniques will be discussed later in this chapter. However, if this step is performed efficiently, any subsequent error will remove non-redundant information with unpredictable results, so it is essential that the following steps be, so far as is practical, lossless.

Spatial and Temporal Compression

Video is a sequence of fields or frames, each of which is a still image. Spatial compression treats each image separately and compresses it without reference to any other image. This is straightforward, relatively simple to implement, and the preferred solution for some applications. However, video is not a sequence of unrelated images—in any real video sequence there is obviously a great deal of similarity between adjacent or nearby frames. This means that a spatial-only system will necessarily send similar or identical information many times and cannot be the most efficient approach. There is still redundancy in the signal. Temporal compression takes advantage of the real-world similarity of frame sequences to improve the compression ratio.

Image Types

Compression systems used for video are based on the assumption that the information to be coded represents real-world "photographic" images. This is important in a number of ways. As explained below, the initial conversion from analog to digital must be performed correctly to avoid aliasing. In fact, all processing steps must be designed carefully to prevent artifacts such as excessive overshoots and clipping. Noise is inherently unpredictable and, therefore, generally uncompressible. High-quality encoders will

often include sophisticated processing to remove as much noise as possible prior to compression. Graphics systems must be selected carefully. Simple systems can produce very fast edges that can exceed the bandwidth of the television system. Filtering these edges can result in clipped overshoots that can substantially impair the performance of a compression system.

The Human Visual System

The human visual system, or HVS, is the term used to describe the combination of eye, nerves, and brain that allows us to see. It is a remarkable and complex system (still not fully understood), but it does have quantifiable limitations, and these are exploited by image compression systems. The most obvious parameter is limited acuity. There is a limit to the smallest details we can see—generally about 20 cycles/degree; however, the design of television systems already takes this into account. The number of lines in a television system is chosen so the line structure is not visible at the design viewing distance; for example, NTSC is designed for viewing at a distance of 6 to 7 times picture height, and the 480 active lines correspond to about 25 cycles/degree at this distance.

Compression systems use another, related, limitation of the HVS. The figure quoted above, 20 cycles/degree, represents (approximately) the absolute limit of vision. These spatial frequencies can be resolved only under optimum conditions and at high contrast. The HVS has a characteristic known as the *contrast sensitivity function*, shown in Figure 5.7-1. The peak of HVS sensitivity is around 4 cycles/degree. Below this value, sensitivity decreases somewhat, but this effect is not generally useful for compression systems. Sensitivity falls off rapidly at higher spatial frequencies, and this effect is the most important element of lossy image compression systems. This characteristic of the HVS means that image elements that have high spatial frequency and low contrast do not need to be transmitted, because they will not be perceived by the viewer. More importantly, the decreasing sensitivity with increasing frequency means that the higher the spatial frequency the lower the precision that needs to be transmitted. Less precision means fewer bits and more efficient transmission.

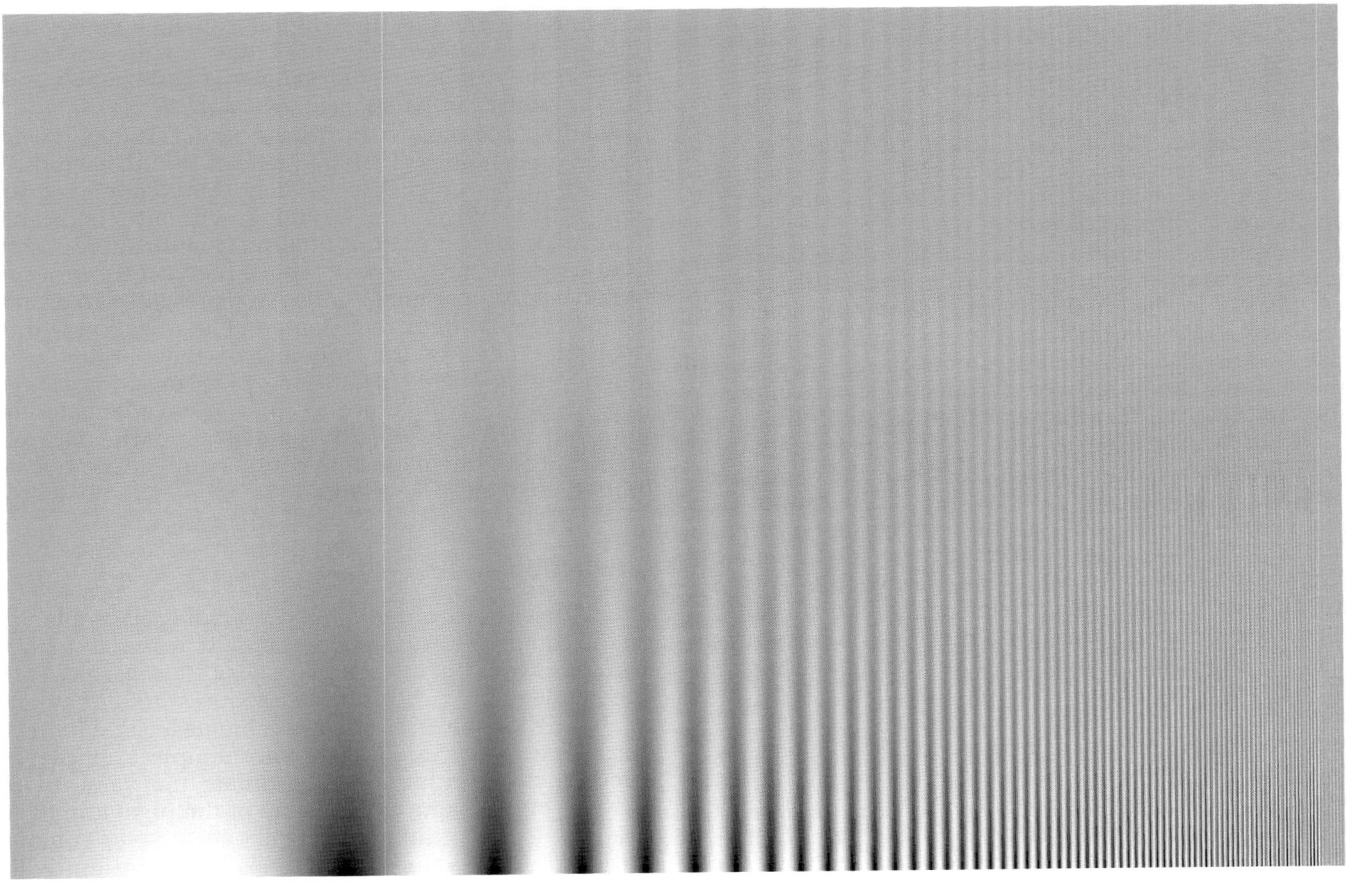

FIGURE 5.7-1 This illustration shows the increasing spatial frequency (left to right) and decreasing contrast (bottom to top) to demonstrate the contrast sensitivity function of the human visual system.

TOOLS FOR SPATIAL COMPRESSION

Spatial compression is the compression of the data representing a single static image. Originally developed for still images (see JPEG section below) the techniques may be used to compress each single field or frame of a video sequence. Simple video compression systems, such as Motion JPEG, use only spatial compression.

Sampling

Sampling is not really part of the compression process but is closely linked to it. Sampling is the conversion of a continuous analog signal into a sequence of discrete values, each representing an instantaneous measure of the signal. Sampling can be performed spatially and/or temporally; for example, a film movie camera samples a complete scene temporally by imaging it onto a photosensitive emulsion 24 times per second.

The sampling mechanisms are not always simple or immediately obvious. A non-shuttered camera tube produces an almost continuous analog signal that approximately repeats every frame. If this signal is sampled for digital conversion, each sample represents a unique point in space–time. In a shuttered charge-coupled device (CCD) camera, the shutter samples the scene temporally into a sequence of frames. The physical cell structure of the CCD samples each frame horizontally and vertically, but all these samples represent the same instant of time. The cell charges are then read out sequentially from the sensor in the form of an analog signal that is then sampled in an analog-to-digital converter!

Sampling is a critically important process because, if it is performed improperly, errors and artifacts will be created in the sampled signal that can never be removed. An important concept in sampling theory is the Nyquist theorem, which can be stated: "To ensure that a signal can be recovered from a series of samples, the sampling frequency must be more than double the highest frequency in the signal." This statement suggests that frequencies higher than half the sampling frequency (known as the Nyquist limit) cannot be recovered. This is true, but the problem is far worse than that. If a signal is sampled that contains frequencies above the Nyquist limit, this energy will reappear as different, lower frequencies when the signal is recovered from the samples. This effect is known as *aliasing*. A classic example of temporal aliasing is filmed wagon wheels—rotation that is too fast (frequency too high) for the sampling is reproduced as slow rotation in the opposite direction. Spatial aliasing can often be observed as moiré-like patterns in the image.

The relevance of this discussion is that the compression process will start with a sampled signal. The design of the compression system will assume that the signal has been correctly sampled; any aliasing present will, like any other "improper" signal content, impact performance. Particular care must be taken to ensure that signals to be compressed have been properly filtered prior to sampling.

Quantization

Quantization is the process of setting a value to one of a predefined set of possible values and is important in a number of ways. First, quantization is the tool used with sampling to perform analog-to-digital conversion. The sampling process was described above. Sampling alone produces a series of analog values—the values are constrained only by the precision of the sampling process; however, to generate a digital signal it is necessary to limit the possible values to a set that can be represented by the chosen number of bits. For example, with an 8 bit analog-to-digital converter, the output value of each sample must be represented by a single 8 bit word that can represent one of 256 possible values. Quantization in this example is the process of choosing one of these 256 possible values for each sample. Usually the quantizer will choose the value closest to the actual analog value measured by the sampler. So, quantization is fundamental to creating any digital signal. The number of bits used determines the precision with which the original signal can be represented. It also determines the best possible signal-to-noise ratio of the digital signal, because the quantization process allocates values that are not identical to the original signal values. This deviation is a form of noise known as *quantization noise*.

In the discussion of the human visual system, it was stated that high spatial frequencies could be represented at lower precision than the frequencies to which the HVS is most sensitive, without significant loss to perceived picture quality. Quantization is the tool used for this purpose. The section on transforms will describe how images may be represented in terms of their spatial frequency content. Quantization is the tool that permits transmission of the various frequency components at differing precisions; for example, it may be determined that it is adequate to transmit only eight possible values of a particular frequency component, so this component can be transmitted using only three bits.

Prediction

Prediction is one of the most powerful tools available to the compression system designer. It is used in many different ways but always plays a substantial role. The technique of prediction is to use some information that is already known and coded to estimate a value for the next item to be coded. As a very simple example, a prediction algorithm might estimate the value of a pixel to be the same as the value of the pixel immediately to its left, or immediately above, or some combination of these values. Sometimes the algorithm will predict correctly; more typically, the estimate will be wrong to some degree. The estimate must be compared with the actual value to determine the prediction error; however, if the prediction algorithm is a

good one, it will be more efficient to code the prediction errors than to code the actual values.

Some spatial predictors are very simple, as described above. Others are more complex, using adaptive algorithms that choose a predictor based on the context created by the surrounding image. Modern codecs such as MPEG Advanced Video Coding (AVC) provide multiple predictors, and it is possible for the encoder to test all of the possibilities and choose the one that represents the lowest number of bits for the image segment in question. Most spatial predictors operate on actual pixel brightness values, but some operate on coefficient values after a transform. Prediction is also the key tool in providing temporal compression, but this application will be discussed separately.

Transforms

As discussed above, a key element of compression is the exploitation of the limitations of the human visual system, but this requires knowledge of the spatial frequency content of the image. This may be derived by examining the brightness values of multiple pixels, but individually the pixel values provide no spatial frequency information.

A *transform* is a mechanism for changing the representation of information. The most familiar transform to television engineers is the conversion between RGB and YUV color spaces. Each representation consists of three values that together specify a single color. RGB and YUV are two different representations of the same thing, and with sufficient precision the transform is reversible to any chosen degree of accuracy; however, the two representations have substantially different properties. The UV signals may be frequency limited without substantial damage to the image, and this is a fundamental requirement for color systems such as NTSC and Phase Alternation Line (PAL).

For video compression, there is a need to represent image information in a form permitting exploitation of the contrast sensitivity function of the HVS. The transform must take pixel brightness values for all or part of an image and transform the representation of that information into a set of coefficients that represent the horizontal and vertical spatial frequencies in the image. In this case, a block of brightness values is transformed into a block of frequency coefficients.

Most compression systems use some version of the *discrete cosine transform* (DCT). The principle exception is the JPEG2000 Standard, which uses the *discrete wavelet transform* (DWT), and this transform will be discussed in the JPEG2000 section. DCT is a derivation of the Fourier transform, long used for frequency analysis. The DCT used by compression systems is a two-dimensional transform, usually operating on an 8 × 8 block of pixels. The transform takes 64 pixel brightness values and returns 64 spatial-frequency coefficients. Each coefficient represents the value of one combination of horizontal and vertical frequencies. The possible combinations are shown in Figure 5.7-2. The top-left component in the illustration is used to represent the DC (average) value of the pixel block. To the right of this are seven components representing various horizontal frequencies ranging from 1/4 cycle to about 3-1/2 cycles across the block of pixels. Down the left-hand side of the illustration the same vertical frequency components appear. All of the rest of the frequency components represent some combination of horizontal and vertical frequency.

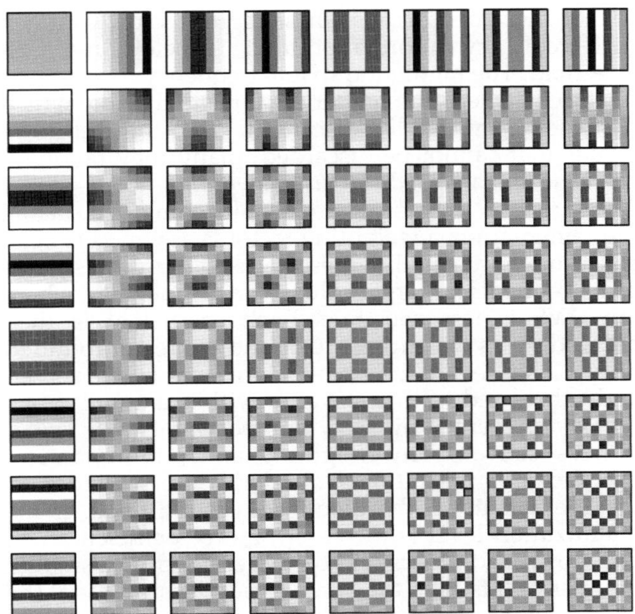

FIGURE 5.7-2 A representation of the 64 basis functions of the DCT transform.

The transform is reversible. Each of the spatial frequency components, known as *basis functions*, can be represented by a block of 64 pixel values forming a pattern; for example, all 64 values for the top-left DC component are the same; the 64 values of the bottom-right form a checkerboard pattern. The coefficients generated by performing the DCT transform represent the quantity (positive or negative) of each of these patterns that must be added to get the original 8 × 8 pattern of brightness values.

As stated above, a transform creates a new representation of the same data. The DCT transform takes 64 brightness values and returns 64 coefficients. With adequate precision, these 64 coefficient values may be reverse-transformed to yield the original set of brightness values. "Adequate precision" means generally that the coefficients must be represented with about three more bits than the brightness values; transforming 8 bit brightness values requires 11 bit coefficients for reversibility. So, the transform has actually created more data! However, when the coefficients resulting from the transform of a real image are examined, it is found that many of the coefficients are zero, or very close to zero, and can be discarded without significant impact.

Because of the HVS contrast sensitivity function (CSF), low values of high-frequency coefficients can also be discarded because these represent low-contrast, high-frequency information that is below the CSF threshold. The higher the spatial frequency represented by the coefficient, the higher the threshold below which that coefficient may be discarded. Even for coefficient values that cannot be discarded, many may be quantized more coarsely, thus using fewer bits.

In practice, these steps are usually combined. In one implementation, a weighting factor is assigned to each basis function; small values are used for very significant frequencies, large values for frequencies where the HVS is less sensitive. Each coefficient is then divided by its weighting factor and the result rounded down. This results in small coefficients being zeroed and others reduced to a value within a small range that is represented by a low number of bits.

Irrespective of any weighting, an overall quantization factor (an additional divider, uniform across all the coefficients in the block) may be applied. If the quantization level is increased, more coefficients will be zeroed, and others will be represented by a smaller number of bits. The total data needed to represent the block will be decreased but at the expense of accuracy and quality.

Following all these steps, each block of the image is represented by a block of 64 coefficients, many of which are zero and some others coarsely quantized. The compression system must take the data from all of the blocks and code the information in the most efficient way possible. Many tools are used in this process, which is described under the Entropy Coding section.

Integer Transforms

The DCT transform described above is a complex calculation requiring many cosine values and floating-point arithmetic. During the development of advanced compression systems, an alternative was developed that implements a rough approximation to DCT but uses just integer values in the calculations. It was found that such a transform could be used to provide compression efficiency very similar to that of true DCT. Integer transforms are incorporated into the MPEG AVC and VC-1 systems (see below).

TOOLS FOR TEMPORAL COMPRESSION

Prediction

As discussed above, video is not a sequence of unrelated images. The content of a frame is usually very similar to the preceding frames. If one frame of a sequence has been compressed and transmitted, this frame is already present at the compression decoder when we send subsequent frames. It should be possible to predict the content of some areas of a subsequent frame and to use the image content already transmitted. For static scenes, this is simple, but it is more complex if objects in the scene move or if the camera moves, thus changing the position in the image of background content. Even when the camera pans and objects or actors move, however, there is typically considerable content in common, albeit in a different location in the image. It should still be possible to benefit from prediction if we can gather information about where content has moved. This is the role of *motion estimation*.

Frame Types

Most compression systems that perform temporal prediction use *reference frames*, or *intra-frames*. An intra-frame is one that is coded without reference to any other image, using just the spatial compression tools described above. Reference frames may be intra-frames, but predicted frames may also be used for reference (see below). A *predicted frame* or *inter-frame* is coded, so far as possible, by prediction from one or more reference frames. The reference frame is transmitted, and at the receiving end it is decoded and stored. The encoder also incorporates a decoder so it can store a reference identical to that stored in the decoder (not the same as the original image because the compression is lossy). The new frame to be coded is compared to the stored reference frame. When a valid prediction is found, the decoder is instructed to use the appropriate data from its copy of the reference frame.

Motion Estimation

The most obvious case where prediction is possible is in a sequence where the camera has not moved so the background is static from frame to frame. In this case, the encoder can instruct the decoder to extract unchanged blocks from the stored reference frame and place them in the same position in the new frame. More generally, of course, there may be content common to the reference frame and the frame to be predicted, but likely not in the same position in the image. Objects in the scene may have moved, or the camera may have moved. Prediction using these elements is much more difficult but is essential for efficient temporal compression.

Motion estimation is the process of searching a reference frame for image content that matches the content in the predicted frame that needs to be encoded. This is the most complex part of an encoder, and the range and accuracy of motion estimation are major factors determining encoder performance.

Motion estimation is generally based on the *macroblock*—usually 16 × 16 pixels. A motion estimator takes a macroblock and attempts to find an area in the reference frame that is a close match in content. When a match is found, a *motion vector* is generated. The motion vector is a pair of values representing horizontal and vertical offsets; it is used to instruct the decoder where the matching content is to be found in the stored reference frame. Although the content that is being searched *for* is determined by the macroblock boundaries on (usually) a 16 × 16 pixel grid, the matching content generally is not found on this grid. For efficient motion estimation, the system must search to very high precision, varying from 1/2 pixel

resolution in most systems to 1/4 pixel in more advanced systems. An exhaustive search is, therefore, an enormous computational task.

The performance of an encoder is closely linked to the efficiency of its motion estimation. It is obvious that the more content that can be matched to image information already at the decoder, and the more accurate the match, the better the system will perform. Less obvious is the fact that it is important to find the right match. There may be many places in the stored reference frame that are a good match for the macroblock being coded, particularly for image content such as blue sky; however, once the match is found, the motion vector has to be transmitted to the decoder. Each motion vector is a pair of numbers, each perhaps 10 or 12 bits long, so these must be coded in an efficient manner or much of the benefit of temporal compression will be lost.

In image areas where an object is moving (or the background is apparently moving because of a camera pan or tilt), there should be a high correlation between motion vectors of adjacent macroblocks; in fact, for linear motion, the motion vectors will be identical. This is an ideal situation for predictive coding (again). After initialization, instead of sending the motion vector values, the coder will send just the differential from the previous macroblock, quite likely zero. However, if the motion estimator were to select suitable matches (for blue sky, for example) from anywhere in the reference frame, all correlation would be lost, and the predictive coding would not be effective.

Residual Coding

In practice, the motion estimator will usually find excellent matches for some macroblocks, marginal matches for others, and no useful match for the remainder. Where the match is sufficiently good, only the motion vector needs to be transmitted; where there is no useful match, that macroblock has to be coded using just the spatial tools. When the match is marginal, some benefit can still be obtained by subtracting, pixel-by-pixel, the reference macroblock from the macroblock being coded. This array of error values is known as the *residuals* and can be encoded using the same spatial tools. If the match is reasonably good, it is likely that the total bits required for the motion vector and residuals will be less than required to code the entire macroblock spatially.

I-, B-, and P-Frames and the GOP

Temporal compression requires image content to be stored at the decoder to be used as a reference. Usually complete frames are used for reference. A system generally will use three types of frame. The simplest type of frame is the I-frame, or *intra-frame*. Intra-frames are coded using only spatial tools and the information within that frame; no use is made of any information outside the frame being coded. An I-frame will generally be used as the first frame of a sequence, providing a reference frame. I-frames are also inserted, usually at regular intervals, to prevent the propagation of coding errors.

A P-frame, or *predicted frame*, is coded using predictions from a previous I-frame or P-frame. Macroblocks that cannot be predicted are coded using the spatial tools. B-frames are coded using two reference frames, one from earlier in the sequence and one from later in the sequence. This can only be implemented if both reference frames have been transmitted to the decoder, so the use of B-frames requires reordering of the frames for transmission and introduces significant delay. The reference frames may be I-frames or P-frames; in most systems, B-frames are not used as reference frames.

The *group of pictures* (GOP) represents the structure or cadence of the coded sequence. In typical applications, a GOP may have a length from a few frames to a few tens of frames. For transmission channels such as terrestrial broadcast with a relatively high error rate, a GOP length of about 1/2 second is common. A typical GOP structure for such an application might be:

I-B-B-B-P-B-B-B-P-B-B-B-P-B-B-B-[I-B-B-B-P-...]

and is described as *long GOP*. Systems may use a *closed GOP* structure, where all references (for temporal compression) are within the same GOP, or *open GOP*, where references are permitted across GOP boundaries. GOP structures are usually flexible; the chosen structure is used when there is no reason to change it, but an event such as a scene change may cause the encoder to insert an extra I-frame or restart the GOP cadence (but see the Rate Control section below).

TOOLS FOR ENTROPY CODING

The final step in a compression encoder is to take the result of all the processes described above and to reduce this information to the minimal number of bits for transmission. This process has to be lossless; otherwise, the data fed to the decoder would be meaningless. The entropy of a set of data is a measure of its lack of predictability. If the coding is to some degree inefficient, this means that there is some predictability (or redundancy). The coding will use more bits (per pixel, per frame, for example) than the calculated entropy of the data. In general, entropy is expressed in bits per symbol, where a *symbol* is any one of the possible values that might be transmitted (for example, for an alphanumeric data stream, a letter or number would be a symbol). Shannon, the mathematician responsible for much of modern communication theory, stated that data cannot be represented in any coding scheme by fewer bits than its entropy. The job of the entropy encoder is to approach this limit as closely as possible, within appropriate constraints of cost, delay, and computational resources.

Run-Length Coding

This is the simplest tool. If a value is repeated many times, it is obviously more efficient to code this information by a scheme that says, in effect, "Take value *m* and repeat *n* times." Sometimes the data must be manipulated to maximize the gain available from run-length coding. During the development of the original JPEG system, it was found that, after quantization, the transform coefficients of a typical block included a large number of zeros. It was discovered that reading out these coefficients in a diagonal pattern (rather than horizontally or vertically) resulted in a stream that usually contained long runs of adjacent zeroes—suitable for run-length coding.

Variable-Length Coding

Many coding schemes such as ASCII use a fixed number of bits for each item to be coded. Variable-length coding is used when the statistics of a data set are known; it uses small codes for common values and longer codes for infrequent values. A simple example of a variable-length coding scheme is the Morse code, which is (crudely) optimized for English language text. Common letters such as *e* and *t* are represented by short codes; less frequently used letters such as *q* and *w* are allocated much longer codes. Variable-length coding is particularly useful following prediction. In a predictive coding system, the prediction error (rather than the actual value) is transmitted. If the prediction is good, there will be many more small errors than large ones, and the most common errors can be represented by very short codes. A system used in several standards that allocates variable-length codes in an efficient manner is referred to as *Huffman coding*.

Arithmetic Coding

Techniques such as Huffman coding can usually compress losslessly to a number of bits close to the calculated entropy of the data, but are limited by the fact that each symbol coded has to be transmitted using a whole number of bits. Some systems mitigate this effect by coding combinations of symbols, but at the expense of much greater complexity.

Arithmetic coding is a family of technologies originally developed by IBM for facsimile transmission. Like Huffman coding, the techniques are based on an analysis of the statistics of the data and the probability of occurrence of each of the possible different symbols. Arithmetic encoding transmits a code stream that represents a binary fraction of the interval between zero and one. The operation is recursive: When each symbol is received, the precision of the fraction is increased by adding more bits to the code stream, effectively coding the cumulative probability of the input sequence. Decoding is also recursive and is performed by analyzing what the encoder must have received to generate the successively finer granulation of the interval.

The principle advantage of arithmetic coding is that the techniques effectively permit fractional bits to encode symbols. Also, arithmetic encoders can adapt on the fly to the statistics of the data, rather than relying on a prebuilt table of codewords. The principal disadvantage is complexity; the techniques are relatively demanding in hardware implementations and are difficult to implement efficiently with general-purpose processors.

RATE CONTROL

Rate control is the science and art of using all of the tools discussed above to achieve the best compromise of bit rate *versus* quality. In most systems, the actual control mechanism is very simple—it is just the quantization level applied to each block of data being encoded. The algorithms used to decide the appropriate quantization at any point, however, may be extremely complex. The objective is generally to compress the input video to a given number of bits per second while maintaining the highest possible subjective quality. The term "subjective" is very important; heavily compressed images will contain artifacts, but a good compression system will ensure that the artifacts are as invisible as possible.

Many factors affect visibility of artifacts. Many artifacts are less visible in busy parts of the picture—areas with considerable detail or of rapid motion. The human visual system is very good at integrating over time, and it is generally possible to quantize B-frames more coarsely than reference frames without significant perceived impairment. Sophisticated encoders will use "look ahead" techniques to determine the complexity of upcoming frames in advance of starting the actual compression process. This allows the encoder to use fewer bits on simple frames, reserving more for the complex frames.

Statistical Multiplexing

Conventional rate control mechanisms aim to achieve the best possible quality within a given permitted bit rate. In television distribution and delivery, several program streams are commonly combined into a *multiplex*. Early systems used a fixed allocation of bits from the total stream for each program, but efficiency can be improved to a remarkable degree by statistical multiplexing. Video programs vary enormously in their bit-rate requirements. "Talking head" static shots contain little information and may be heavily compressed, whereas a sports event such as basketball contains a great deal of complex motion and is a severe challenge to compression systems. Bit-rate requirements vary not only by program genre but also with time in a single program. Statistical multiplexing uses these variances to fit more programs into a multiplex while maintaining adequate quality. It relies on the fact that not all programs need their maximum number of bits at the same time. In a statistical multiplexing (*statmux*) system, one encoder is provided for each program; each encoder measures the complexity of its video and requests an appropriate bit rate from the statmux controller. The controller aggregates the requests, applies any priority rules that may be in

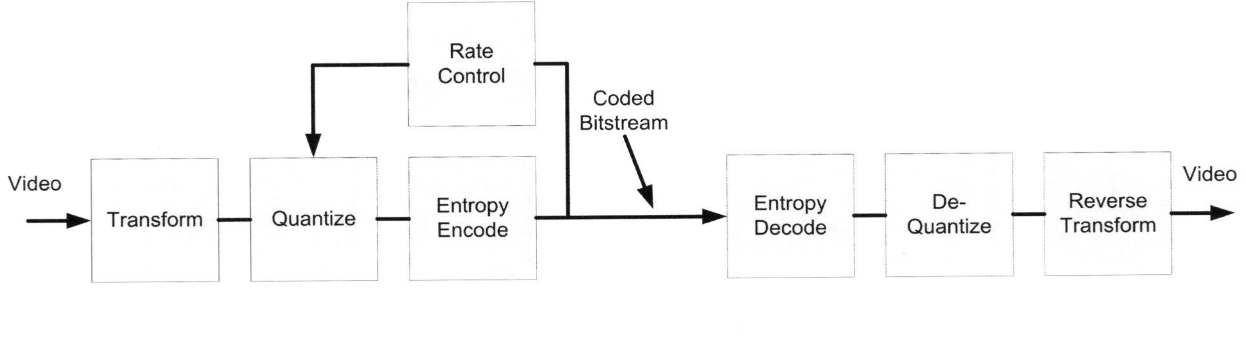

FIGURE 5.7-3 A simple spatial encoder and decoder.

force, and tells each encoder the actual number of bits it will be allowed.

PUTTING IT ALL TOGETHER

Figure 5.7-3 shows the elements of a spatial encoder and decoder. Only the luminance signal path is shown. Each frame is split into blocks, and the blocks are transformed, producing transform coefficients. These are weighted and quantized, and the resulting sparse set of coefficients is scanned, run-length encoded, then passed to the (lossless) entropy coder. The stream is monitored by the rate controller, which determines the appropriate level of quantization.

In the decoder, the entropy encoding and run-length coding are reversed, generating a set of values identical to that output by the quantizer. The coefficients are then de-quantized. (Note that this creates an approximation to the original magnitudes, but the precision discarded by the quantizer cannot be recovered.) The recovered coefficients are then subject to the reverse transform, resulting in a close approximation to the original image.

Figure 5.7-4 shows a simplified diagram showing the basic elements of an encoder with motion estimation and temporal compression, such as an MPEG-2 encoder.

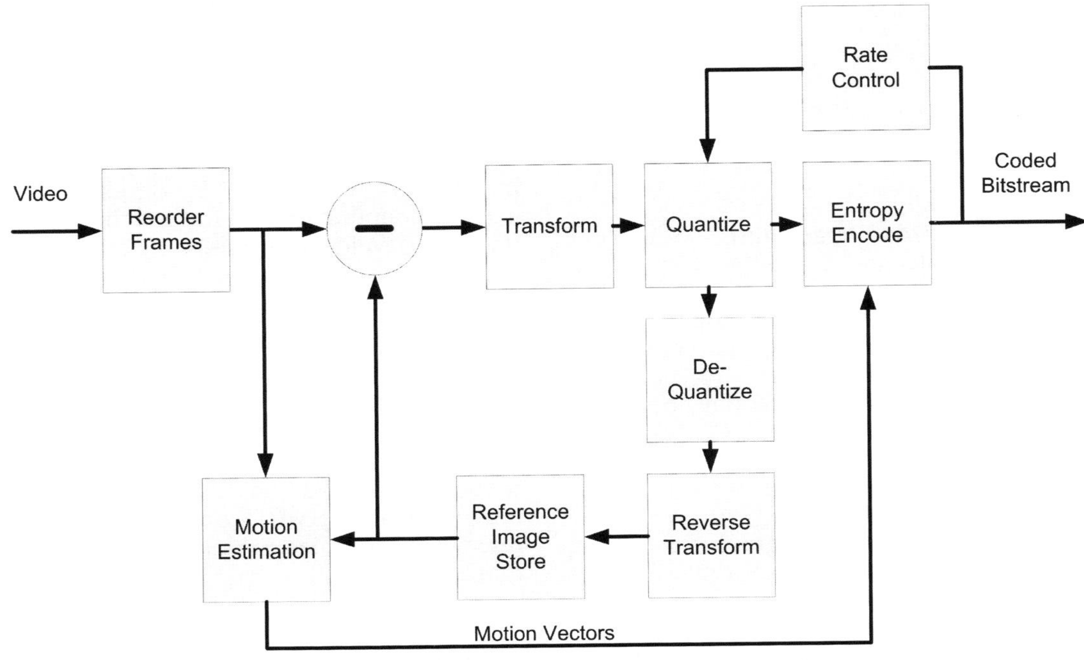

FIGURE 5.7-4 A simplified encoder with motion estimation and prediction.

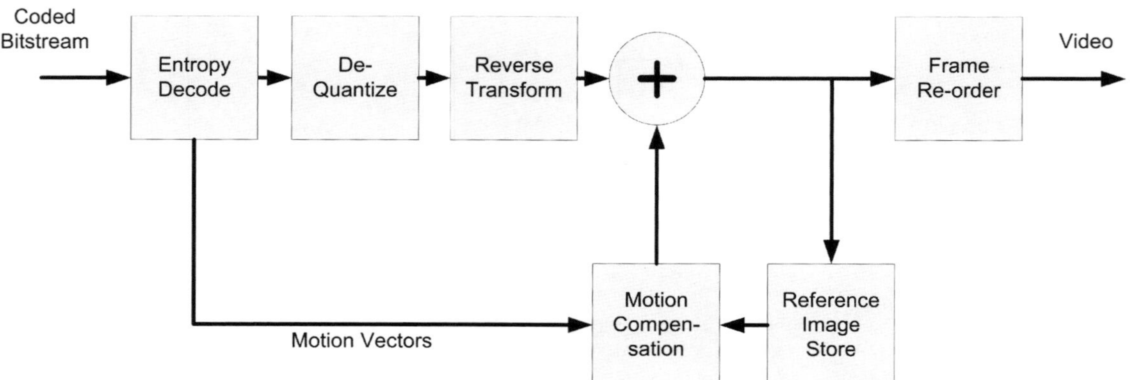

FIGURE 5.7-5 The simplified decoder corresponding to the encoder of Figure 5.7-4.

Many of the elements are the same as for the spatial encoder, but a prediction loop is added. Quantized coefficients are de-quantized and reverse transformed. This generates the same image (an approximation to the original) as will be generated within the decoder. This reference image is stored, and is available for comparison when subsequent images are coded. When a match is found, the reference block is subtracted from the block being coded, and any residuals are transformed and quantized. A motion vector is generated to point to the block used for prediction, and the motion vectors are multiplexed with the rest of the encoded bit stream. Where no useful prediction is found, blocks are encoded spatially, just as in the previous description. In the decoder (Figure 5.7-5), reference frames are stored, and motion vectors are used to fetch predicted areas from the store. Where there is no motion vector, the transmitted values are treated as spatially encoded, just as in the spatial decoder.

THE STANDARDS BODIES

ISO/IEC

The International Standards Organization (ISO) and the International Electrotechnical Commission (IEC) formed a Joint Technology Committee (JTC-1) to develop standards for information technology. One of the subcommittees of JTC-1 is SC29, charged with image technology. Two Working Groups (WG1 and WG11) are responsible for much of the development of modern compression systems.

JPEG

JTC-1/SC29/WG1 is the Joint Photographic Experts Group (JPEG). The development of the original JPEG standard for still pictures created the basis for most modern image compression schemes. Although the ISO version of JPEG was not published until 1994 (one year after MPEG-1), the early work begun around 1983 created the spatial tools used in both JPEG and MPEG standards.

MPEG

JTC-1/SC29/WG11 is the Motion Picture Experts Group (MPEG). This group was established in 1988, and the MPEG-1 standard was published in 1993.

ITU

The International Telecommunication Union (ITU) is a treaty organization covering all aspects of telecommunications; it is divided into ITU-R for radio (wireless) communications, and ITU-T covering (generally) wired communications. Part of ITU-T is the Video Coding Experts Group (VCEG), which has developed video compression systems for video conferencing, etc. ITU-T has also published standards such as JPEG and MPEG-2 as ITU-T Recommendations.

JVT

In 2002, ITU-T and ISO/IEC agreed to formalize the existing cooperation and formed the Joint Video Team (JVT) to develop compression standards for publication by both organizations. JVT developed the standard now known as H.264 (ITU-T) and (within MPEG) as MPEG-4 Part 10, or the MPEG AVC.

SMPTE

The Society of Motion Picture and Television Engineers (SMPTE) is a body that focuses on standards for moving images. SMPTE does not generally engage in research to optimize compression systems but ensures that coding specifications are accurate and complete to permit interoperability. SMPTE standardized the VC-1 coding scheme proposed by Microsoft, and at the time of this writing is processing other schemes such as the BBC-developed *Dirac* and *DNxHD* from Avid.

COMPRESSION SYSTEMS

JPEG

JPEG (now often referred to as JPEG-1 to distinguish it from JPEG2000) was intended for static images. It is a block-based system using DCT and Huffman variable-length coding.

Motion JPEG

The compression provided by JPEG was more efficient than anything else available at the time, and a number of manufacturers recognized the benefit of coding video as a sequence of static images, each JPEG coded. This technique became know as *Motion JPEG*. However, the JPEG committee did not provide any mechanism for rate control in this application or any other tools for coding a sequence. Unfortunately, this resulted in each manufacturer developing a proprietary scheme, and Motion JPEG streams are not generally interchangeable.

DV

This compression system was developed by a consortium of manufacturers for use in consumer camcorders. It uses the DCT transform followed by quantization and a complex system of priorities and ordering to ensure that standard-definition video can be encoded at 25 Mb/sec, with a constant number of bits per frame, suitable for recording on tape. Encoding and decoding are performed by (the same) single chip. The performance of DV exceeded expectations, and it has been widely applied in professional environments, particularly in camcorders for news broadcasts. A professional variant was created, using two chips and 50 Mb/sec, for applications requiring higher quality.

MPEG-1

The first compression standard specifically intended for video was developed between 1988 and 1992 and was aimed specifically at the Video CD application. MPEG-1 encodes CIF Video (352 × 240) at 30 Hz and two-channel audio all within a bit rate of 1.5 Mb/sec. MPEG-1 uses motion prediction and then DCT for transforming blocks that could not be predicted, or residuals of predicted blocks. After quantization, Huffman coding is used for entropy coding. MPEG-1 also established a new standardization philosophy. The standard defines the *syntax* used in the bit stream, and exactly what the decoder must do in response to the bit stream. The standard does *not* specify the encoder—anything that produces a legal bit stream that can be decoded according to the standard is a legal encoder. This approach permits a great deal of evolution in encoder design and substantial improvements in performance, without changing the standard or the decoders in the field.

The importance of MPEG-1 should not be underestimated. Although designed for an application that had only a short life, it established the foundation for the video compression systems in use today. (It also included the MP3 audio compression scheme that is still dominant in many fields.) The syntax created for MPEG-1 was so robust that, prior to the availability of MPEG-2, it was extrapolated for use in satellite television and in an early high-definition system.

MPEG-2

Work started on MPEG-2 even before completion of MPEG-1. It was realized that there were important applications for compression in broadcast television, but that these applications required coding of larger pictures and provision for coding interlace with reasonable efficiency. The standard is based heavily on the concepts of MPEG-1, but with many extensions and improvements. MPEG-2 introduced *profiles* and *levels*. Profiles define a set of tools that may be used—more precisely, syntax elements that a compliant decoder must be able to process. A level defines a set of parameters such as maximum bit rate, decoder memory requirements, etc., that size a decoder for a specific application. A *conformance point* is a combination of a profile and a level. In MPEG-2, a decoder for a particular conformance point must decode bit streams from all lower profiles and levels. For example, a common conformance point is the *main profile at main level* (expressed as MP@ML), suitable for standard-definition television and used in DVD players. The *main profile at high level* (expressed as MP@HL) may be used for high-definition television, as in the ATSC television system. An MP@HL decoder must be able to decode MP@ML bit streams, but not *vice versa*.

MPEG-2 is the dominant video compression scheme at the time of writing. It is the basis of the current digital television transmission systems (ATSC, DVB, and ISDB) and is used for satellite and digital cable television distribution. It is also used for video distribution on DVDs, and a decoder is present in almost every personal computer. There are probably well over 1 billion MPEG-2 decoders, MP@ML or higher, in the field.

The MPEG-2 standards also included very robust systems-level tools that have been used by later standards and diverse applications. The *MPEG-2 transport stream* is the most widely used mechanism for transmitting synchronized audio, video, and ancillary data in a single bit stream.

MPEG-4

Many people consider that MPEG-4 is where the committee lost its way. The set of standards included an enormous array of techniques such as facial and body animation schemes, but none of these has achieved widespread employment. MPEG-4 audio has been adopted in some areas, most notably in the Apple iPOD®. MPEG-4 did include a range of new video compression schemes, collectively known as *MPEG-4*

Part 2, and offered some performance improvement over MPEG-2 (perhaps 15%), but in most applications this improvement was not sufficient to overcome the benefit of the large number of MPEG-2 decoders already deployed.

MPEG AVC (H.264)

Both MPEG and ITU recognized that there was a need for a video compression system substantially more efficient than MPEG-2 and that they should focus their efforts in this direction. The two organizations formed the Joint Video Team (JVT) to produce such a standard, to be published by both organizations. This accounts for the many names of the resulting standard. Officially, in ITU-T it is H.264; in MPEG, it is MPEG-4 Part 10 or the MPEG Advanced Video Codec (AVC). The standard is also known to many by the moniker of the committee that produced it, JVT.

AVC is very efficient. It has been shown that it can achieve up to roughly twice the efficiency of a sophisticated MPEG-2 encoder (and this will likely improve with further encoder developments in the future). This is sufficient advantage to make it worthwhile to upgrade, when the possibility arises to change decoders. In the United States, there is no immediate possibility to use AVC for terrestrial transmission because all of the ATSC decoders in the field are MPEG-2 MP@HL decoders, ready for high definition but not capable of decoding AVC. In countries such as Australia, where the Digital Video Broadcasting (DVB) transmission system has been adopted for high definition, using MPEG-2, the same applies. In Europe, however, existing DVB decoders are standard definition only. High definition will be a new service, requiring a new generation of receivers and decoders, and it is very likely that AVC will be used for compression. Brazil has adopted the Integrated Services Digital Broadcasting (ISDB) transmission system but will use it in a unique mode, employing AVC for compression.

AVC faced a challenge in its first versions from VC-1 (see below) that appeared to be more efficient with large images such as HDTV. JVT moved on and produced an amendment that added *Fidelity Range Extensions* (FrExt), which provided better tools for large images and again is becoming the preferred choice for most applications. AVC has been adopted in both of the high-definition DVD systems (Blu-Ray and HD-DVD).

AVC uses tools similar to MPEG-2, but with many refinements and improvements. The transform is an integer approximation to DCT, and spatial compression is aided by a sophisticated spatial predictor. Motion estimation is much more sophisticated. One major improvement is the ability to define a global motion vector, improving efficiency considerably for camera pans and tilts. The basic unit for prediction is still the macroblock, but to track accurately the edges of complex moving objects a motion vector may be applied to blocks as small as 4 × 4. AVC improves on the prediction of B-frames and can use multiple reference pictures for even more capable prediction. The system can use an entropy encoder called *Context-Adaptive Variable-Length Coding* (CAVLC), an enhancement of Huffman coding, but, for greatest efficiency, an alternative entropy encoder, *Context Adaptive Binary Arithmetic Encoder* (CABAC) is employed. CAVLC is a significant improvement on Huffman coding, but CABAC improves by a further 10 to 15% in many applications. Finally, AVC includes a de-blocking filter that masks the very obvious blocking artifacts seen when MPEG-2 systems begin to fail.

The greatest challenge of AVC is its computational complexity. Decoding is about three times more complex than for MPEG-2. Encoding, as with any MPEG standard, is greatly variable, but a high-efficiency AVC encoder may be perhaps ten times as complex as a good MPEG-2 encoder.

VC-1

In 2003, Microsoft Corporation decided to freeze the bit stream definition for the video coding in Windows Media 9 and publish this as a standard to permit implementation by multiple vendors. VC-1 is similar in many ways to AVC, except that the tool set is less comprehensive and was chosen to reduce computational complexity. VC-1 also uses an integer transform; it has a de-blocking filter, but it is of a simpler design than in AVC, and it uses variable-length coding rather than an arithmetic coder. The result is a compression system that is not quite as efficient as AVC but with a decoder that is much easier to implement in software on a general-purpose processor.

VC-2

VC-2 was originally a proposal from the British Broadcasting Corporation to SMPTE for a simple compression system specifically designed to compress 1920 × 1080 50 or 60 Hz progressive (requiring 3 Gbps uncompressed) by a factor of 2:1, permitting transmission over a standard 1.5 Gbps high-definition serial digital interface. The original integer-DCT-based design has been abandoned, and the latest proposed variant uses a wavelet transform. The full proposal now includes a wide range of capabilities, including the combination of block-based motion prediction with a wavelet transform. The intent is to have an "open source" (and believed royalty-free) compression system suitable for use throughout the production and delivery chain. At the time of this writing, it is not clear exactly what will be offered for standardization.

VC-3

VC-3 is the compression system proposed by Avid Technology, Inc., to SMPTE, based on its DNxHD compression schemes for use in editing environments. It is a spatial compression system that uses a DCT trans-

form, and standardization is in process at the time of this writing.

JPEG2000

JPEG2000 is the successor to the original JPEG standard and is also intended for still images; however, this time provision has been made for a standardized, interoperable mechanism for rate-controlled compression of a video sequence.

This system used a wavelet transform rather than a transform from the DCT family. The wavelet transform is essentially a set of complementary high- and low-pass filters that operate on the image, both horizontally and vertically. This produces high (H) and low (L) horizontal frequencies, and high and low vertical frequencies. The filtered image is separated into four parts or quadrants, one with just high vertical and horizontal frequencies (HH) and the other combinations (HL, LH, and LL). Because of the filtering, alternate samples from each of these parts may be discarded, so each of the four quadrants has 1/4 of the number of pixels of the original image.

Three of the image parts (HH, HL, and LH) are wavelet transform coefficients and are quantized in a manner similar to that used in DCT systems. The LL image is treated differently. It is a quarter-sized (1/2 × 1/2) version of the original and can be filtered again by the same filters, resulting in four new quadrants. Again, the resulting LL is a smaller version of the original (now 1/4 × 1/4 size). This process, called *decomposition*, may be repeated—a typical JPEG2000 compression may use five iterations. The JPEG2000 coder uses sophisticated techniques to predict values in one iteration from those in others and a complex arithmetic encoder as a final step.

JPEG2000 is remarkable in that it is scalable in many dimensions. The bit stream that represents the compressed version of the full-size, full-quality image may be ordered in many ways. Using these techniques, it is possible to decode many possible subsets of the complete image. Using data from less than the full number of decompositions will yield a smaller image. Other techniques may give just chosen areas of the picture, or a monochrome image, or lower signal to noise. Often these techniques are used in a progressive manner to provide a small, low-quality image immediately, building to a large, high-quality image as more data is decoded. JPEG2000 has been adopted by the Digital Cinema Initiative (DCI) and SMPTE for digital cinema and has been used in a number of television products.

INTELLECTUAL PROPERTY AND COMPRESSION

Compression systems depend on very sophisticated technology, resulting from very expensive research in many companies around the world. This expertise is brought together in committees such as MPEG, but the contributing companies need to see some return on their research investment. This results in the requirement that most compression systems may be legally implemented only with the appropriate patent licenses.

MPEG-2 resulted in a new concept—the patent pool. Each company that believes it has patents essential to implementing the standard may submit the patents for examination by independent experts. If a patent is found to be essential, it may be included in the pool. Companies wishing to build systems may take one license for all the patents in the pool, and the royalties are shared among the patent owners.

This system has been controversial, particularly with MPEG-4 and later systems. The original MPEG-4 license required payment of royalties, not only on equipment but also on each actual use of the compression (for example, a television transmission). Many believe that this license was a contributing factor to the poor reception of the original MPEG-4. When AVC was first completed, the proposed license terms were similar to MPEG-4, and there was a similar reluctance to adopt the standard. Since then, the license terms have been modified substantially and are now widely accepted.

JPEG2000 is unusual in that the committee adopted a policy that the only technologies adopted within Part 1 of the standard would be those that were contributed on a royalty-free basis. It is believed, therefore, that no royalties are payable for basic JPEG2000 implementations—one of the reasons that it was attractive to the Hollywood movie studios and a factor in its adoption for digital cinema.

SUMMARY

Compression is a factor in almost every aspect of the television system. The technology has evolved rapidly over the last 20 years—today the most efficient systems can transmit good-quality color images at an average of about 1/10 of a bit per pixel. This is made possible partly by the development of flexible and extensible standards, with contributions from hundreds of experts from all parts of the world. The other major factor is the continuing effort in many companies to exploit these standards and the availability of more and more computational power, to provide ever-improving performance, often beyond the expectations of those who created the standards. Like interlace and NTSC in earlier times, compression always represents a compromise between bandwidth and performance. Like many compromises, it can produce disastrous results if applied incorrectly or without understanding. Compression is an art as well as a science!

Bibliography

Books on Compression

Hubbard, B.B., *The World According to Wavelets*, 2nd ed., A.K. Peters, Ltd., Wellesley, MA, 1998.

Mitchell, J.L., Pennebaker, W.B., Fogg, C.E., and LeGall, D.J., *MPEG Video Compression Standard*, Chapman & Hall, New York, 1996.
Pereira, F. and Ebrahimi, T., *The MPEG-4 Book*, Prentice Hall, Upper Saddle River, NJ, 2002.
Rabbani, M. and Jones, P.W., *Digital Image Compression Techniques*, SPIE Optical Engineering Press, Bellingham, WA, 1991.
Richardson, I.E.G., *H.264 and MPEG-4 Video Compression*, John Wiley & Sons, New York, 2003.
Symes, P.D., *Digital Video Compression*, McGraw-Hill, New York, 2004.

Useful Websites

http://www.chiariglione.org/MPEG/
http://www.jpeg.org/
http://mpegif.org/
http://www.symes.tv/

CHAPTER

5.8

Video Recording

STEVE EPSTEIN
Broadcast Buyers Guide
Columbia, Missouri

INTRODUCTION

In the early days of television, film provided the only method for recording and storing video images and by 1954 American television operations used more raw film than all of the Hollywood studios combined. Following the introduction of the first practical videotape recorder (VTR) in the 1950s, there was a rapid migration from film to videotape, with its advantages of instant replay and reusable media. Tape-based systems revolutionized and dominated television recording for more than fifty years, although film continues to this day to be the first step in the production chain for some types of programming both in the studio and on location. Since the 1990s there has been a new migration, this time from videotape to new methods of recording using a variety of IT-based magnetic, optical, and solid-state media.

Today's recording systems are based on the pioneering work of numerous individuals and companies. The principles of magnetic recording were discovered and developed into primitive recorders in the later 1800s. Practical high quality recorders were developed first for audio recording in the 1930s (see Chapter 3.4) and then for video in the 1950s. The breakthrough for video recording came in 1956 with the demonstration of the quadraplex video recorder, developed by a team from Ampex led by Charles Ginsburg [1]. Since then, video recording media and systems have undergone continual evolution, with the most significant changes including the introduction of color, the change from transverse to helical scan devices, size reductions (both of machines and media), increased media packing density, and the transition from analog to digital recording. More recently, these changes have been accompanied by an increasing usage of sophisticated compression systems and the move to record data, as opposed to a defined video or tape format. This has enabled a file-based approach for storage and networking—another major revolution for the industry (see Chapter 5.6 on workflow and file management).

This chapter examines aspects of professional video recording systems using tape, optical discs, and other removable media.[1] Video servers based on non-removable hard disk drives are covered in Chapter 5.9. The path from quadraplex VTRs to portable handheld camcorders has been covered earlier in Chapter 5.3 on television camera systems. The theory of analog video recording has been covered exhaustively in other texts and the reader is referred to the literature in the bibliography. For a detailed analysis of digital video recording, the reader is referred to *The Art of Digital Video* by John Watkinson, and other texts as listed.

FUNDAMENTALS

For any video recording method, some basic design choices must be made. The first is whether the recording is to be analog or digital. Virtually all of today's

[1]This chapter draws extensively on material first published in the 9th Edition of the *NAB Engineering Handbook*: Chapter 5.5, "Magnetic and Optical Recording Media," by Joe Grega, and Chapter 5.7, "Video Recording Principles," by Steve Epstein. It also draws on material from Chapter 9 of *An Engineering Tutorial for Non-Engineers*, by Graham Jones, Focal Press, 2005; used with permission.

systems record digital information. Most modern systems record component video based on a luminance channel and two subsampled chrominance channels, with a resolution of 8 or 10 bits. The critical parameter is the bit rate that can be recorded since, this, in conjunction with the compression system used, largely determines the quality of the video recording.

The next choice is media type. Whether the medium is magnetic tape or disk, optical disc, or solid-state devices, physical size and storage capacity are often the determining factors. The capability to separate the media from the recorder can also be a deciding factor in the choice of technology. If a tape machine goes bad, the tape can readily be moved to another machine and the same is true of optical discs. This is generally not true of conventional hard disk drives where, if the drive or the internal media fail, separating them and recovering the data is, at best, an expensive and time-consuming endeavor and at worst an impossibility. Regardless of the storage method, the capability to store video information as a file instead of a linear recording is becoming increasingly important.

When a technology is chosen for recordings, several parameters should be considered to ensure the system chosen is appropriate for the application.

For tape machines, these factors include:

- *Physical tape size:* Tape widths have ranged from 1/4 inch up to 2 inches. Current systems largely use 1/4-inch and 1/2-inch tape. Lengths, rather than being expressed in terms of feet or inches, are expressed in minutes or hours, with 2–3 hours being the typical maximum.
- *Tape style:* Styles include open reel and cassette. Today's machines are almost exclusively cassette based.
- *Scanning method:* All current machines use *helical scan*. With helical scan, the tape wraps a rotating drum that contain the recording head(s). The tape falls in the shape of a helix as the wrap progresses around the head drum. This results in a long slanted track being recorded on the tape.
- *Recording format:* Nearly all current formats are component digital machines. It is important to realize the format discussed here is how the signal is recorded, not what the machine can input and output.
- *Input/output connections:* Most professional machines provide input and output (I/O) connections for analog composite as well as component video and two to four channels of balanced audio. Some modern digital machines provide eight or even twelve audio channels. Digital video and audio connections are becoming popular as most facilities have some digital infrastructure in place.

Disc recorders for video, whether optical or magnetic, have similar criteria including

- *Maximum record time:* The record time is based on the amount of storage within the device and the amount of compression used. Many devices can be expanded using external drives. Storage time may be for video and two audio channels, or for video only.
- *Compression requirements:* Compression may or may not be required, depending on internal bandwidth issues. Some units have selectable compression rates but require that all clips to be played in a sequence are compressed to the same level. Others can play back a variety of compression levels as needed.
- *Input/output connections:* Units may have video I/O, LAN connections, or disk drive connections such as SCSI or Fibre Channel. Some units provide all of these and more. In addition, some units can handle multiple I/O streams simultaneously.
- *Random access:* One of the primary features of disc-based recordings is the capability to randomly access clips. A few units can truly play any frame straight after any other frame; however, most require either some cueing or rendering time for real-time playback. One parameter that affects this is the rotational speed of the drive; another is whether the drive system has been optimized for streaming-style playback.
- *Metadata and MXF:* Another factor, of increasing importance in the modern video environment, is the support for metadata and MXF (Material eXchange Format), discussed later.

Solid-state systems have a few factors to consider, but at the time of writing, the number of solid-state systems used for recording is limited. Factors today are largely based on capacity, portability, and price. As these systems follow the road to commoditization, additional factors such as interchangeability will grow in importance.

Videotape Recorders

VTR basic building blocks include:

- *Transport and related control systems:* The transport is an electromechanical assembly that provides a precise path through which the tape moves. Included are the various motors and guides used to move and position the tape since the tape transports require precise mechanical alignment. Tape must be capable of moving at various speeds, forward, still, or reverse, without damage to or distortion of the tape surface and edges.
- *Helical Scan:* Rotating head system used to achieve the necessary head to tape speed.
- *Servos:* These devices are used to control the head drum assembly, capstan, and, in many units, the reel motors.
- *Video and audio circuitry:* Basic video circuits consist of input, modulator, demodulator, and output stages. Some newer machines include bit rate reduction (compression) circuitry as well as A/D, D/A converters, and/or composite-to-component

encoders/decoders. Linear audio tracks in most analog VTRs are comparable to professional audio recorders. Some units record audio as digital bits using the rotating video head.

- *Additional circuits:* Many units have additional circuitry for timecode, editing, and other special functions such as dynamic tracking.
- *True digital machines:* These machines have signal paths that are entirely digital. They may also act on the data and provide a lossless compression to reduce the amount of data recorded to tape. In addition, these units include error correction circuitry to ensure error-free playback.

Helical Scan

With the exception of the obsolete quadraplex VTR, all practical analog and digital VTRs rely on helical scan to achieve the high head-to-tape speed needed for video recording. Details vary, but the basic principle is the same for all machines, as shown in Figure 5.8-1.

One, two, or more record and playback heads are fixed to the outside of a rapidly rotating drum, around which the tape is wrapped at an angle, like part of a spiral helix (hence the name). The tape moves quite slowly past the drum, but each head moves rapidly across the surface of the tape, laying down a series of diagonal tracks. The angles shown in the figure are exaggerated, so the tracks are much longer than shown in the figure.

Because it is not practical to record video signals directly onto tape, various modulated carriers are used for different VTR formats, digital and analog being quite different. The video information on the slant tracks, recorded by the rotating heads, takes up most of the tape area. There may also be fixed heads that record and playback audio, control, and timecode signals on narrow longitudinal tracks along each edge of the tape (not shown in the figure). These tracks vary with different machines and, in some cases, the audio and/or timecode signals may also be recorded on the slant tracks using the rotating heads. In the case of digital VTRs, the audio is embedded with the video bitstream and recorded to tape with the rotating heads.

The principal videotape formats are listed later in the chapter.

Timebase Correctors

Videotape recording for broadcasting would not have become possible without the invention of the timebase corrector (TBC). Because a VTR is a mechanical device, there are slight variations in the speed of the tape and the rotation of the drum. These produce irregularities in the timing of the replayed video signal, and the sync pulses and picture information may arrive at slightly the wrong time. The timing variations are not important if the signal has only to feed an analog television set direct (as a consumer VCR does). In that

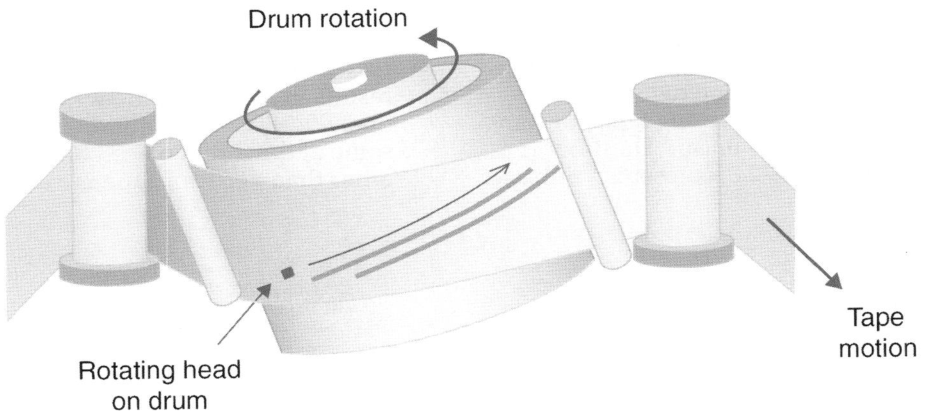

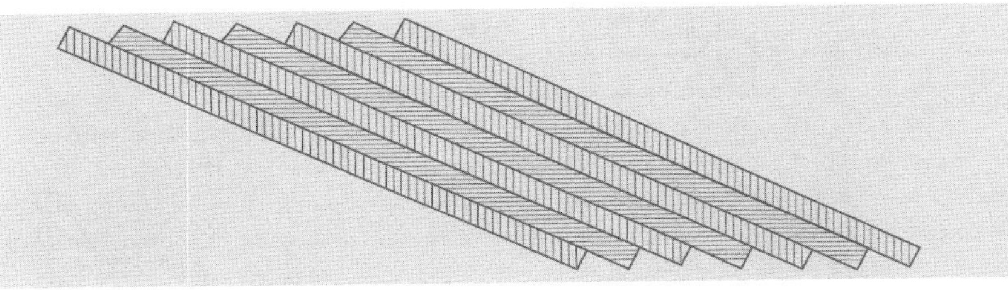

FIGURE 5.8-1 Principle of helical scan recording.

case, the set will remain locked to the sync pulses even as they vary and will produce a basically stable picture. Such timebase errors are disastrous in a broadcast station, however, where the signal has to be mixed with other signals, perhaps edited, and re-recorded. To make VTR output signals usable, they are passed through a TBC that comprises an electronic buffer store. This removes the input irregularities and sends out video with stable timing restored.

Digital TBCs offer a much greater correction range than analog and are universally used, even with VTRs that otherwise record an analog signal on tape. However, VTRs that record a digital signal on tape do not actually have a separate TBC function, because the techniques required to recover the digital data from the tape also take care of correcting any timing irregularities.

MAGNETIC TAPE

Since its introduction, magnetic tape has evolved and diverged into a variety of different types. As recording technology has evolved, magnetic tape packing density has increased (more information/data per square inch). This is possible through improvements in record head technology and improved tape formulations. Consistent magnetic particle orientation also contributes to the recording properties of magnetic tape. The direction of magnetization in a standard iron oxide particle is parallel to its long axis. The more uniform the orientation, the more discrete the transitions are on playback, giving better signal-to-noise properties. Today's metal particles are considerably smaller and have much less surface area, allowing a greater number of individual particles to be located within a given area. This, along with the higher coercivity and retentivity, allows a much greater packing density rate.

Manufacturing of magnetic recording tape is a complex process that is essentially linear. The flow diagram shown in Figure 5.8-2 represents a basic magnetic tape manufacturing process. The entire coating of one batch or mix may take one week, depending on the amount of time required for milling, heat curing, and final finished goods assembly.

Formulation Components

Magnetic tape is typically made up of three parts or layers: backcoat, magnetic coating, and base film. The formulation of any magnetic layer contains several components including magnetic pigments, binders, plasticizers, and head-cleaning agents. The final formulation in its wet, coating state has a consistency similar to that of fingernail polish and typically includes

- *Magnetic particles:* Generally, magnetic particles are acicular in shape (similar to a sewing needle). The smaller the size, the larger the quantity that can be packed into a given surface area. Magnetic material used today includes gamma ferric oxide, chromium dioxide, and magnetic particles. A typical gamma ferric oxide particle is 25 to 35 microinches in length and about 7 to 12 microinches in width. For comparison, a cigarette smoke particle is about 25 microinches in diameter.
- *Binders:* These can be likened to paste, as they provide the cross linking necessary to keep all of the ingredients together (cohesion) and to ensure the coating does not shed or flake off the base film (adhesion). Binders also add considerably to the stiffness of the finished tape product. Stiffness is required so that the tape wraps the helical scanner and transport assemblies properly and adequate head-to-tape contact is maintained.
- *Plasticizers:* Transport design geometry typically results in tape-to-guide wrap angles of greater than 90°. Because of these guiding requirements, some of the forces placed on the tape are in a plane other than vertical, Tape must be flexible yet have

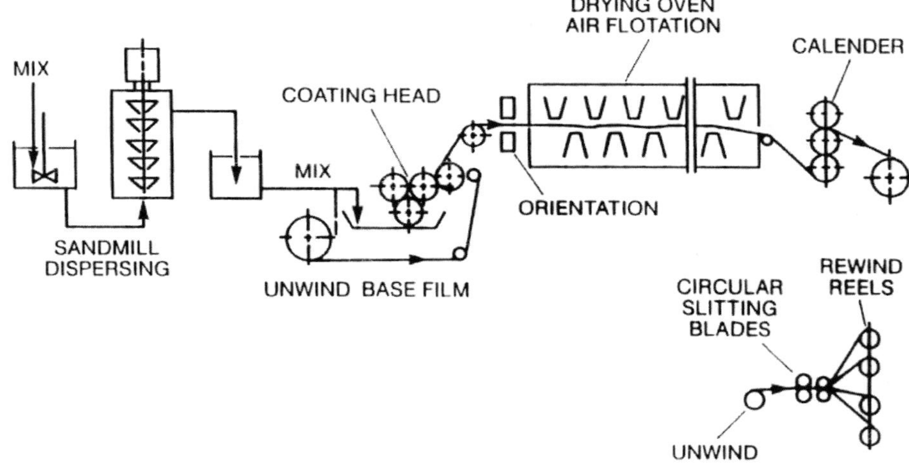

FIGURE 5.8-2 Schematic of mix preparation, tape coating, and processing.

a relatively high coating modulus. Plasticizers typically perform this function. Without plasticizers, tape would be somewhat brittle and unable to survive the high physical demands and speeds of the transport.

- *Head-cleaning agents:* Typically, head-cleaning agents are added to formulations in the form of an alumina particle of a specific size. Their purpose is to ensure the heads of the machine remain clean. In addition, these cleaning agents keep the remaining portions of the tape path free from dirt and contamination. Too much alumina increases abrasivity and head life deterioration. Too little alumina may result in dirty transports, poor head-to-tape contact, and poor-quality recordings. The head-cleaning agents added to a tape's formulation are not a substitute for routine cleaning and transport maintenance, which will help reduce the incidence of dropouts and machine-induced physical damage.
- *Solvents:* All of the mix ingredients are added together to form a slurry, and part of the formulation will include some solvents. These are used to dissolve other mix chemicals such as binders during the milling process. While the solvents are needed during the mixing and coating of the formulation, they are not required in the final state of the magnetic tape. Solvent-laden air can be collected during the drying phase of manufacture and routed to a solvent recovery system.
- *Dispersants:* Because of their magnetic nature, particles tend to clump together in their raw form. Dispersants are added to aid in dispersing the particles evenly throughout the mix.
- *Fungicides:* Fungicides are added to the mix formulation to help prevent fungus growth, which may occur on the tape surface. Fungus problems are almost entirely confined to storage or operating environments with very high humidity.
- *Antistatic Agents:* One of the causes of dropouts (the momentary loss of head-to-tape intimacy) is static electricity buildup. The buildup causes dust and dirt particles to attach themselves to the tape surface. Antistatic agents such as carbon black are added to prevent this problem.

Mix Preparation

Once all of the formulation ingredients have been combined, it is necessary to mill or disperse the slurry so that two conditions are met. First, the magnetic particles and other mix ingredients must be thoroughly dispersed throughout the slurry. Second, all ingredients must be dissolved so that no particles larger than a given size remain. This is achieved in a milling process that may take from 72 to 96 hours and depends on the chemistry of the formulation. The mix is normally filtered after milling to remove unwanted materials such as broken pieces of the milling media which were introduced during that process. At the end of the milling process, the slurry is ready for coating.

Base Film and Coating

When magnetic tape was first coated, it was constructed on kraft paper backing. This was later replaced with a more resilient cellulose acetate and then with today's range of polyester base films or backings. The current polyester base film includes Polyethylene Terephthalate (PET), Polyethylene Naphthalate (PEN), and Polyaramide type films. Polyester-type base film can be produced in a variety of widths, lengths, thicknesses, surface smoothness, and tensile properties. The base film is electrically inert and provides no electrical or magnetic function. Base films are becoming thinner with newer tape formats, presenting a unique challenge to tape manufacturers, as it is essential that tensile properties are maintained despite the reduced thickness. Coating can be achieved by using a variety of coating technologies.

Knife Coating

Knife coating is probably the oldest type of coating method employed. The coating mix (wet) is fed onto the base film and then past a knife blade that spreads it evenly and removes the surplus material. The blade height (with respect to the base film) determines how much mix is left on the film. Knife coating is particularly good for relatively thick coatings.

Reverse Roll Coating

Reverse roll coating is more complex than knife coating and is capable of much thinner coatings. In the reverse roll coating method, the mix is picked up from a mix pan by the pickup roll and transferred to the applicator roll. A metering roll meters the amount of mix on the applicator roll, which is determined by the distance between the two rolls. The mix is then transferred to the base film from the applicator roll. The applicator roll is turning in the reverse direction or against the motion of the base film, hence the name. Reverse roll coating was the mainstay of the tape industry for many years and is still in common use today.

Gravure Coating

Gravure coating (see Figure 5.8-3) was adapted to the tape industry from the printing industry. It is the most accurate means of coating thin thickness where tight tolerance control is required. The gravure roll is pitted with cells, small grooves, or cuts of given width or diameter and depth. The diameter and depth of the cell, as well as the viscosity of the mix, together determine how much mix is retained by the gravure roll and thereby deposited onto the base film. Gravure coating is an ideal method for thinner coatings such as consumer VHS and the backcoat coating found on most professional videotapes.

Metal Evaporation Process

The process or coating technique of depositing a metal layer onto a flexible substrate through evaporation of a

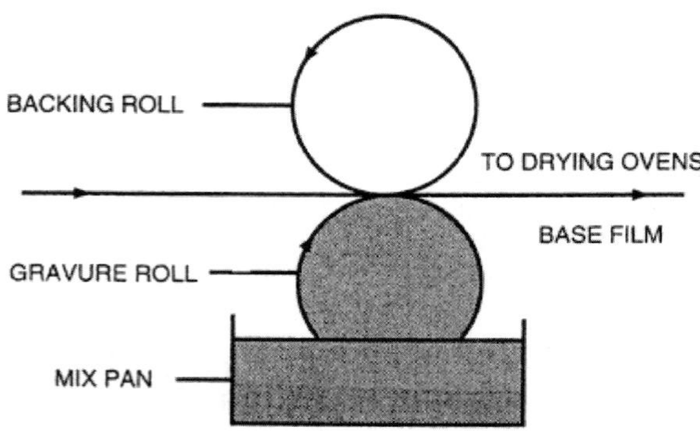

FIGURE 5.8-3 Application of gravure coating.

metal in a vacuum is referred to as metal evaporation (ME) coating (see Figure 5.8-4). The most direct way to maximize the magnetic volume of the recording layer is to remove all nonmagnetic components (oxygen) of the particle itself and by removing binding and resin matrix from the formulation. An ideal magnetic recording medium is achievable through this physical deposition of magnetic material on the base film through metal evaporation. The magnetic coating thickness of a metal-evaporated tape is typically 1 or 4 microinches. This is more than 20 times thinner than the metal particle layer of D2 tape. This practically means that the entire thickness of the tape is basically that of the base film being used.

Slot Die Coating

Slot die coating uses a coating head that has a slot cut into it. The geometry of this slot and its position relative to the base film determine the parameters of the coating surface. Slot die coating is one of the newest of coating technologies and is typically used for the thinnest and most critical coatings. The slot die head usually has one slot but is capable of two or three slots, which results in multiple and different layers being applied to the base film. These layers can be applied wet on wet or wet on dry.

Drying, Solvent Evaporation, and Calendering

After leaving the coating area, the web, as it is now called, passes through the orientation process. This process physically moves the individual particles into a specific direction. The direction desired is always in the direction of head travel across the tape or as near to this direction as possible.

Solvents are removed during the drying process by heating, causing evaporation. After leaving the orientation stage, the coated web passes through an oven with a series of different temperature zones. The temperature of each zone is usually matched to the evaporation temperature of one of the solvents, which may be recovered. Within the ovens, the web typically travels over a series of air plenums, which suspend the tape on a cushion of hot air.

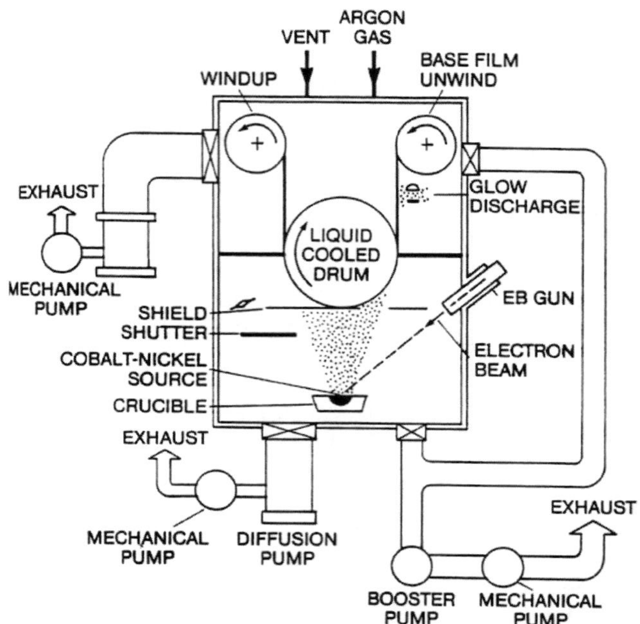

FIGURE 5.8-4 The vacuum metal evaporation deposition process.

After passing through the ovens, the coated web then passes through a series of calendars. The calendaring process can be roughly likened to ironing of clothing. During the process, the tape surface is smoothed and compressed. The smoother tape surface tends to give higher electrical outputs than rougher ones and better signal-to-noise ratio.

Slitting and Finishing

A "jumbo" roll of tape as manufactured is up to 3 meters wide and must be slit and packaged into a usable form. Slitting takes the jumbo roll and cuts it into strands of the appropriate width for the given format. Each strand is then wound onto a hub, where it is commonly referred to as a pancake. The final task in the magnetic tape manufacturing process is to load and assemble the cassettes or to add flanges to the hub to make a reel. In cassette format products, manufacturers have a choice between spool loading and in-cassette loading. Spool loading is a process in which the spools are wound with the correct amount of tape prior to being placed into the unfinished cassettes. The cassettes are then assembled around the spools. With in-cassette loading, the cassettes are assembled first, including the leader material. The tape is then wound into the cassettes by cutting and splicing the leader material from outside the cassettes. Large-scale assembly of cassettes and their component parts is largely achieved using relatively simple robotics and industrial controls. The assembled cassette or flanged open reel is then placed in the appropriate unit container, labeled, and placed in the shipping carton and ready for sale.

OPTICAL MEDIA

The use of optical media has become increasingly widespread since the consumer introduction of the audio compact disc (CD) in 1984. Optical recording media read and write information by different means. Reflectivity and Kerr effect are the basic underlying principles of optical media. Recorded media are read as the pickup reads the intensity of a reflected laser beam as it is changed by indentations in the surface or changes in a dye layer. The signal received from the disc depends on both changes in reflectivity and the polarization angle of the reflected light.

Optical Discs

There are three basic types of optical media: read-only-memory (ROM), write-once read-many (WORM), and rewritable or write-read-erase memory (WREM).

ROM devices cannot be rewritten or erased. Prerecorded audio CDs, computer software CDs, prerecorded Mini Discs, and prerecorded DVDs are examples of these types of media. ROM media consist of a series of pits of varying length that are formed as the polycarbonate substrate disc is molded. Once a CD-ROM disc has been molded, the data cannot be altered. A thin reflective film is coated onto the pit surface of the substrate. When light is focused on the pit, it is scattered by edges of the pit. Light reflected from the area between (the land) the pits is scattered very little. The difference between the nonscattered and scattered light is sufficient to be read by the detector.

WORM discs are best known as the CD-R, DVD-R, and DVD+R formats, which have become popular for computer storage applications. These discs are ideal for distributing large amounts of information. WORM discs have a special reflective layer that is melted by the laser to form a pit or hole in the reflective layer. During the forming of the pit or hole, the laser is operating in a high power mode. Chemical and physical changes in the dye layer result in changes in reflectivity contrast. The reflectivity difference between the hole or pit and the reflective surface is read by the detector. The Double Layer DVD+R system uses two thin organic dye films for data storage. These films are separated by a spacer layer. Heating with a focused laser beam changes the structure of each layer. These modified areas have different optical properties than those of their unmodified surroundings, causing a variation in reflectivity as the disc rotates. The principles of optical disc recording are covered in detail in Chapter 3.4.

Write-read-erase memory (WREM) is also called erasable or rewritable. Data can be written, erased, rewritten, and erased as many times as desired. One form of WREM media that has become popular in recording systems is magneto-optical (MO) media because it can be erased and rewritten over one million times. Rewritable optical discs can be written on and erased just like floppy discs. Rewritable MO discs are ideal for a variety of uses, ranging from personal computing to video applications such as short-form segment storage.

Digital Versatile Disc

The various varieties of CD and Digital Versatile Disc (DVD) discs have data capacities ranging from 650 MB for a CD-R up to 8.5 GB for a double layer DVD+R. DVD developed out of CD recording to provide greater data capacity. This format actually represents a family of discs using the DVD format and performing different functions. A basic DVD-ROM disc is constructed in the same basic way as any ROM disc and has a capacity 4.7 GB per side. DVD-R, +R, –RW, and +RW (rewritable) writers and discs are all available and can be used for a variety of portable storage applications, including video.

In its initial version, the DVD-R disc was capable of recording 3.95 GB per side, or six times the capacity of a CD-R disc. Data is written at 11.08 Mbps, or nine times a CD-R's 1X speed. The write technology employed is similar to that used by CD-R except that the data can be written at a higher speed and the recording density is much higher. To achieve the six times increase in density, two components of the recording hardware were changed. The wavelength of the laser was changed to red laser with a wavelength of 635 nm. In the case of CD-R, the laser is infrared with a wavelength of 780 nm. The numerical aperture of the lens was improved to 0.5 from 0.6. These changes allow a DVD-R disc to record marks as small as 0.44 μm as compared to 0.834 μm for CD-R. The recording action happens by momentary exposure of the recording layer to a high-power laser (approximately 10 mW) beam that is tightly focused onto the surface. The heated dye polymer layer is permanently altered so that microscopic marks are formed in the pregroove. The light intensity of the recording laser is tuned to a wavelength of light that is not affected by exposure to ambient light or playback lasers. Playback occurs when a low-power laser of approximately the same wavelength is focused onto the disc. Today, DVD Jukeboxes are available in a variety of formats that can read and write on 600 disks and store 5.6 TB and more.

More recently, the capacity of optical disk recording has further increased greatly with the introduction of HD DVD and Blu-ray discs, providing storage capacities up to 25GB, 30 GB, 50 GB, or even more, depending on the system and configuration. These devices are intended for high definition video and mass data storage. Fundamental principles are similar to conventional CD and DVD but with several variations—in particular the use of a shorter wavelength blue laser to enable increased storage density. The reader is referred to the website reference in the bibliography for more information.

Magneto-Optical Discs

MO discs have a number of layers bonded to their surface during manufacture (see Figures 5.8-5 and 5.8-6). In these disks, magnetic and optical principles work

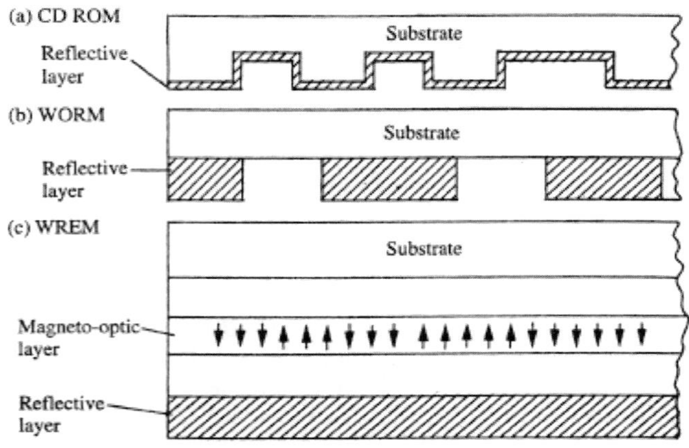

FIGURE 5.8-5 The differences between ROM, WREM, and MO discs.

together to enable the storage of hundreds of megabytes. The physical recording process is based on two principles: (1) a magnetic material's coercivity (its resistance to changing magnetic orientation) drops when heated, and (2) laser light polarization is rotated when exposed to a magnetic field. The signal received from the MO disc depends on both changes in reflectivity and the polar Kerr rotation. High data storage is achieved because each data spot is identified by the narrow wavelength of laser beam. At the focus of the beam, the local temperature of each spot increases to 200° C, and its coercivity drops from 8,000 Oe to less than 200 Oe. At such a reduced coercivity level, it needs the influence of only a low-intensity magnetic field to alter the magnetic polarity of each spot. The information to be stored on the disc is coded by the sequence of north and south polarities, each representing a binary state of "0" or "1."

During MO recording, data is written onto the disc when a laser beam is focused to a spot on the spinning disc. The laser heats the magnetic material to the Curie point temperature, the threshold at which its coercivity drops. A magnetic head on the opposite side of the disc is driven by the digital signal to be recorded. The spot heated by the laser takes on the magnetic orientation given by the head. The laser beam in MO does not encode the magnetic material; it only heats the medium to make it more easily magnetized. A switchable electromagnet is used to record the data as either 1's or 0's" In conventional MO drives, the laser is pulsed while the magnetic bias field, which determines the magnetization of the recording layer, is held constant. During playback, a lower power laser beam is reflected from the spinning disc where it is exposed to the magnetic fields on the recorded disc. These magnetic fields rotate the laser beam's polarization (Kerr effect). The reflected beam can be of two alternating polarizations as a result of the north-south or south-north recorded magnetic fields. A polarizing beam splitter directs the two polarizations to photodetectors. Separate erase and verification passes may be required before rerecording can be done on a previously recorded MO.

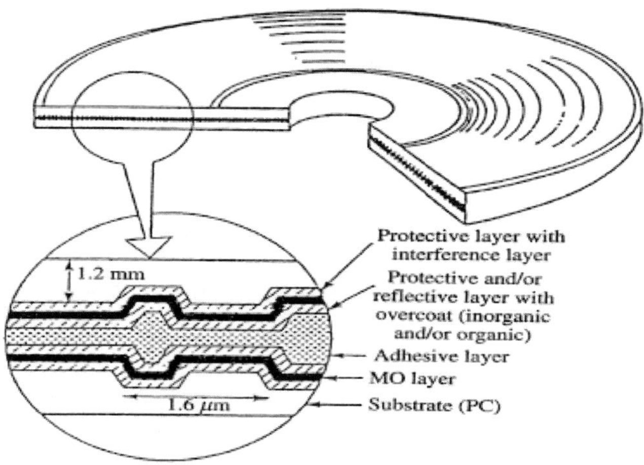

FIGURE 5.8-6 A conventional MO disc.

VIDEOTAPE FORMATS

In approximate historical order of introduction, we list below the principal videotape formats, with their main characteristics. Some of the machines mentioned are now obsolete and may be found only in specialist facilities for dubbing archive material onto more modern formats, but they are included for completeness. There are other less well-known formats that are not listed. With the exception of Quadraplex and Type C, all VTRs listed here use tape on two reels enclosed in a cassette for protection, so this is not mentioned each time.

Analog Tape Formats

- *Quadraplex (1956):* Developed by Ampex Corporation in the United States, and also known as *Quad*, this format is now completely obsolete and is listed here only because it was the first successful VTR format. The name comes from the wheel with four heads, which rotated across the width of the tape to produce the video tracks. The system used open reels of two-inch-wide tape, which were extremely heavy and cumbersome but at one time were the means by which nearly all television program material was shipped from one place to another.
- *U-matic (1970):* The U-Matic format, developed by a Japanese consortium, uses 3/4-inch tape. Video quality is not particularly high, but it improved considerably with the later High Band version. The format is still in regular use by some stations and production houses and machines are still manufactured on a limited basis. Portable versions were once widely used for news gathering, as a replacement for 16-millimeter film. U-matic is a color-

under format that records chrominance and luminance on separate carriers (see glossary).

- *Type-C (1976):* The Type C format, developed by Ampex in the United States and Sony in Japan, used one-inch tape on open reels to record composite video, with two channels of audio. It was the first machine to provide usable still frame, slow- and fast-motion playback, and pictures in shuttle. At one time, this was the most widely used VTR format. It has now largely disappeared but remains in limited use. Its single-field per scan design and sophisticated transports provided high-quality, uncompressed analog video recordings.
- *VHS (1976) and S-VHS (1987):* Developed by JVC, VHS (Video Home System) is the very successful standard format for consumer VCRs, using a 1/2-inch tape. It is used in homes and businesses throughout the world, although largely replaced by DVDs for movie playback. Both VHS and S-VHS are color-under formats. The S-VHS version has higher resolution, delivers better color performance, and can record timecode. It is also very cost effective, so it has been used by some small-market broadcasters, but it has largely been superseded by other, higher-quality professional formats.
- *Betacam (1982) and Betacam SP (1986):* The Betacam format, from Sony, uses 1/2-inch videotape to record analog video. Input and output interfaces use composite video, but the signal is actually recorded on tape in analog component form. This considerably improves picture quality. Betacam SP (superior performance), with metal particle tape and higher recording frequencies, further improves the quality. The format has up to four channels of audio (two longitudinal and two FM) and quality multigeneration editing. Most of these decks have a built-in TBC, and several offer variable speed dynamic tracking. Betacam SP became the industry workhorse for portable recording and studio applications before digital formats were introduced and many of these recorders are still in use today.
- *M (1983) and M-II (1985):* These formats from Panasonic, using 1/2-inch tape, were developed for professional use from the VHS format. They were direct competitors to Betacam and Betacam SP but were not widely adopted and are now obsolete.
- *Video8 (1983) and Hi8 (1989):* These formats, using 8 millimeter tape, were developed by Sony for consumer camcorders, in conjunction with several other manufacturers. Both are color-under formats. Hi8 was occasionally used for professional assignments requiring a very small camcorder.

Digital Tape Formats

There are many legacy analog VTRs still in use but digital VTRs have largely replaced analog machines for new systems due to there superior video and audio performance and other advantages. One crucial improvement of digital VTRs is the ability to record many *generations* of video without degradation. This is important with editing and post-production, where the original material needs to be re-recorded many times. Analog recordings show a buildup of video noise and other degradations after a very few generations, which does not occur with digital. Figure 5.8-7 shows an example of a standard definition digital VTR, the Digital Betacam.

FIGURE 5.8-7 Digital Betacam Studio VTR. (Courtesy of Sony.)

Video Compression for Recording

It is possible to record digital video signals onto tape in uncompressed form. However, uncompressed video has a high data rate (270 or 360 Mbps for SD, 1.485 Gbps for HD), so many digital VTRs reduce the amount of data by using *video compression*, while trying to maintain quality as high as possible (see Chapter 5.7 for more on video compression). This allows smaller tape sizes and smaller head drums to be used, which can make the machine more compact, and also allows less tape to be consumed. MPEG compression is used for some VTRs but with lower compression ratios than used for transmission. Several other compression systems are also used for recording.

The digital VTRs listed first are all designed to record standard definition video only.

- *D1 (1987):* These 8-bit, uncompressed, component digital recorders were manufactured by Ampex and Sony and used 19 mm oxide tape and three different cassette shell sizes. The L-size can record 76 minutes of CCIR-601 video. The M-size can record 34 minutes; and the S-size, only 11 minutes. For many years, D1 was regarded as the top of the line digital recording standard and was used throughout the postproduction industry for multigenerational editing. It is no longer widely used.
- *D2 (1989):* Produced by both Sony and Ampex, the D2 format recorded composite NTSC (or PAL) video in digital form. This 8-bit uncompressed format used 19 mm metal tape with up to three hours of record time. While allowing easy integration into existing analog facilities, it was soon overtaken by other component video recorders for new installations.
- *D3 (1991):* This format came from Panasonic and also recorded composite video in digital form. It

uses 1/2-inch metal particle tape and provides for two cassette sizes; the small cassettes hold a maximum of 125 minutes of tape, and the large hold up to 4 hours of tape. With the small cassette alternative, it was possible to make equipment suitable for electronic news gathering (ENG) and other portable applications.

- *DCT (1992):* The Ampex DCT component recorder used 19 mm tapes in specially designed shells. A mild 2:1 discrete cosine transform, essentially lossless, compression scheme was used. Ampex claims that the use of error correction rather than error concealment, makes the deck robust enough for data applications and allows perfect duplicates to be made from a master tape. The DCT video recorder is now obsolete but a parallel product, DST, is a still-current data storage unit based on the same technology. The units are similar, with the most obvious difference being the I/O modules. Numerous large-scale tape archive systems are available today using the DST format, which can store up to 660 GB per tape.
- *Digital Betacam (1993):* Developed by Sony, this component video format uses 1/2-inch tape with 2:1 DCT compression and is capable of 10-bit recording. Quality is excellent, and the machine has become the industry standard for high-quality standard definition recording for portable and studio applications. Some models are backward compatible with analog Betacam SP.
- *D5 (1994):* D5 was produced by Panasonic as a backward compatible development of D3. D5, however, records component video with four audio channels onto 1/2-inch tape. It uses a higher data rate than Digital Betacam and records uncompressed video at up to 360 Mbps for the very highest quality, but it has the disadvantage of not being available in a portable recorder version.
- *DV and MiniDV (1995:* Developed by a consortium of ten companies, this format was intended for consumer camcorders. It records video and two audio channels on 1/4-inch metal evaporated tape using the DV codec and a recording track pitch of 10 microns. Maximum recording time in standard mode is 80 minutes. No linear tracks are supported, as everything (video, audio, and data) is recorded digitally by the rotating heads in the scanner assembly. The signal structure used in the DV format is 4:1:1, 8-bit recording and, when combined with a compression ratio of 5:1 and the necessary data and error correction, produces a data rate of 25 Mbps This format is now also widely used for professional products.
- *DVCPRO (1995) and DVCPRO 50 (1998):* DVCPRO from Panasonic was a development of the DV format for broadcast applications, using a wider 18 micron track pitch and a metal particle tape material to increase robustness and quality. It also records longitudinal cue and control tracks for improved editing performance. This format is widely used by broadcasters, particularly for ENG, and has a good range of portable and studio recorders, and editing facilities. A small cassette records 63 minutes, while standard size is 123 minutes (184 minutes for some models). The later DVCPRO 50 format further improves quality by recording 4:2:2 video at a higher bit rate (50 Mbps), at the expense of reduced recording time. Figure 5.8-8 shows the tape footprint parameters for the DVCPRO video format and Figure 5.8-9 shows a DVCPRO tape transport with, for comparative scale, a MiniDV tape and coins. Notice that the DVCPRO headdrum is about the same diameter as a US 25 cent piece!
- *D9 (Digital-S) (1995):* This format, developed by JVC, uses a 1/2-inch tape cassette, similar to VHS, but is, in fact, a much higher-quality professional machine, rivaling Digital Betacam. It uses a variant

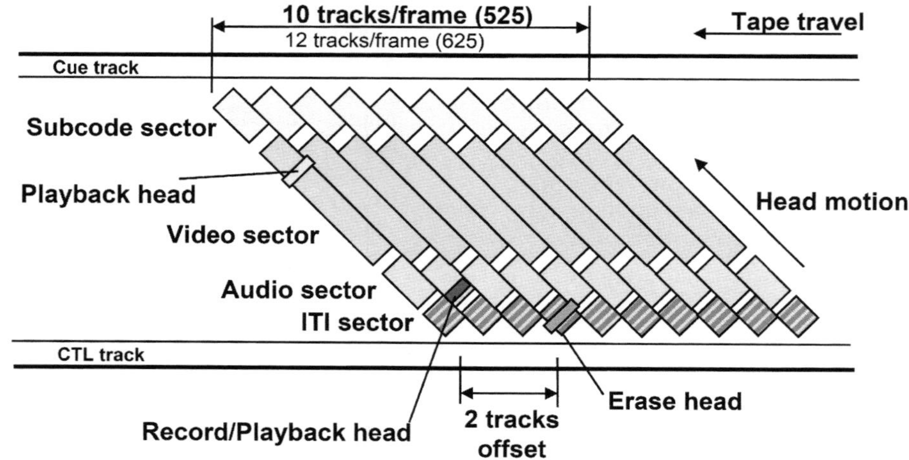

FIGURE 5.8-8 DVCPRO tape format. (Courtesy of Steve Mahrer, Panasonic.)

FIGURE 5.8-9 DVCPRO transport and minidv tape cassette. (Courtesy of Steve Mahrer, Panasonic.)

of the DV codec, similar to that used for DVCPRO 50, but with a different tape format, and is available in both portable and studio versions.

- *DVCAM (1996):* This is the Sony professional version of DV. It uses metal evaporated tape, with a 15 micron track pitch. No longitudinal tracks are recorded but a small integrated circuit within the shell is used to record cue points and other information to simplify the editing process.
- *Betacam SX (1996):* This format from Sony was targeted for ENG and newsroom applications. It provides 8-bit recordings on 1/2-inch tape, using approximately 10:1 compression with MPEG-2 compression and 4:2:2 Studio Profile. Some decks include hard disk drives, which allow a single machine to be used as an editor. Betacam SX VTRs were the first to allow video to be sent back to the studio over video links at twice the normal speed.
- *Digital8 (1999):* This consumer format from Sony records the same digital signal as DV onto less expensive Hi8 tapes, and can play back analog Video8 and Hi8 tapes.
- *IMX (2000):* This Sony format has some similarities to Digital Betacam but uses MPEG compression and is able to record eight audio channels. As well as regular playback of video and audio, the recorded compressed data can be output directly from the tape, for transferring to video servers.

HD Digital VTRs

Nearly all the modern tape formats offer versions capable of recording compressed HD signals. These VTRs are intended for recording high definition video signals with either 1080 or 720 lines of resolution. Figure 5.8-10 shows an example of an HD studio digital VTR, the D5.

- *D5 HD (1994):* This HD format from Panasonic is based on the SD D5 recorder and uses 4:1 Motion JPEG video compression to record video and four audio channels (eight channels on later models). It provides high-quality HD recordings and is widely used as a mastering machine, but is available only as a studio recorder.
- *HDCAM (1997):* This 1/2-inch format from Sony is backward compatible with Betacam and uses a proprietary compression technology with a data rate of 140 Mbps. Compared to D5, this format has more aggressive compression, of about 7:1 and, in order to reduce the data rate sufficiently for the tape format, it also subsamples the video to a horizontal resolution of 1440 pixels for the 1080-line format, compared to 1920 for D5 HD. HDCAM is switchable among a variety of frame and scan rates. It is available in both studio and portable

FIGURE 5.8-10 D5 multi-format, multi-standard HD/SD studio mastering recorder with full bandwidth 4:2:2 10-bit recording. (Courtesy of Panasonic.)

versions and is widely used for high definition program acquisition.

- *DVCPRO HD (2000):* The HD version of the DVCPRO format records video with considerably higher compression and more aggressive subsampling than used for either D5 HD or HDCAM, to achieve a bit rate of 100 Mbps recorded on tape. Video is subsampled to a horizontal resolution of only 1280 pixels for the 1080-line format.
- *D9 HD (2000):* The HD version of Digital-S records video and up to eight channels of audio at a data rate of 100 Mbps, using a proprietary JVC codec.
- *HDCAM SR (2004):* This version of HDCAM from Sony records HD video at a data rate of 440 Mbps with full 1920 × 1080 resolution, using mild compression of only 2.5:1, and with up to 2 hours of recording time. It carries 12 channels of uncompressed audio and is intended for the highest quality production and distribution applications. The format is capable of recording 1080p60 video signals. There is a high speed mode available providing a data rate of 880 Mbps that allows full-bandwidth 4:4:4 recording with 2:1 compression and an option for recording two separate video channels for 3D television
- *HDV:* Sony, JVC, Canon, and Sharp agreed to an HDV standard for consumer and professional use. HDV records an HD signal to MiniDV tapes using heavy MPEG-2 compression and aggressive subsampling to permit recording at 25 Mbps, with a recording time of 60 minutes. The format is based on technology previously developed by JVC. Recorders and playback systems are available from a variety of manufacturers.

OPTICAL, SOLID-STATE, AND HARD DISK RECORDERS

Recording technology has once again moved on and tape-based recording in now giving way to IT-based devices that record simply data. These devices leverage technology from the IT industry to provide large data storage volume at low cost. Examples are hard disk drives, optical disks and flash memory. The ability to record data as opposed to a defined video or tape format is a great advantage, it permits flexibility to record what ever is needed. Video formats can be mixed with impunity and the recording technology is computer friendly and IT accessible. Files recorded on any of these units can be transferred at high speed from the recorder to a video server or nonlinear editing system, with no requirement for the time-consuming digitizing procedure needed when working with conventional video recorders.

The most widely accepted formats include:

- *XDCAM:* Sony's XDCAM uses an optical disk for recording digital video and audio. The disc has a capacity of 23 GB and is similar to (but not compatible with) Blu-ray DVD. It can be recorded, erased, and rerecorded. SD and HD signals are recorded with a choice of compression formats, including DVCAM and IMX, with up to 85 minutes duration. Files can be accessed and transferred direct from the disc to other systems such as video servers. Figure 5.8-11 shows an XDCAM HD optical disc camcorder. Studio recorders are also available.
- *P2:* The P2 format from Panasonic uses solid-state memory for recording either SD or HD digital video and audio and offers the advantage of containing no moving parts. Storage is on 2, 4, or 8 GB secure digital flash memory cards. Signals are recorded with DV or DVCPRO compression, giving

FIGURE 5.8-11 XDCAM HD camcorder. (Courtesy of Sony.)

FIGURE 5.8-12 DVCPRO HD/50/25 P2 portable recorder/player using solid-state memory for recording. Six slots may be populated with 8 GB compact flash cards, giving a total storage of approximately 48 GB. (Courtesy of Panasonic.)

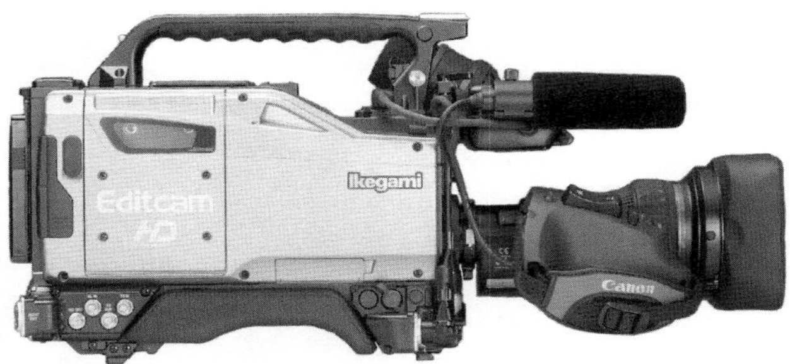

FIGURE 5.8-13 HD camcorder with hard disk drive recorder. (Courtesy of Ikegami.)

up to 32 minutes of video per card in standard definition and 20 minutes in HD. P2 is available in a variety of hardware and software configurations from different manufacturers with integrated camcorders and stand-alone recorders. The memory card is in a standard PCMCIA form factor and can be plugged into a laptop for playback. Figure 5.8-12 shows a portable P2 solid-state recorder.

- *Hard Disk:* Several camera manufacturers have produced portable camera systems that can record digital video data direct to a hard disk drive (HDD). Current systems include the Editcam developed by Ikegami and more recently adopted by Avid, and the Infinity from Thomson Grass Valley. Typically these systems can use either a removable hard drive media or solid state flash memory recording. Figure 5.8-13 shows a camcorder with removable hard disk drive. Figure 5.8-14 shows the removable field pack hard drive, and Figure 5.8-15 shows a portable disk recorder that can also accept the removable hard drive.

Codecs, MXF, and Metadata

One of the great advantages of the IT-based storage systems is that they can readily support a variety of codecs, which may include DV, MPEG-2, JPEG-2000, IMX, DNxHD, JPEG 2000, and others, at different bit rates to meet particular applications. The use of many compression standards can present a problem for interchange of file-based material between different users in the production, postproduction, and distribution environments. This issue has been addressed with the SMPTE MXF suite of standards, which are supported at various levels by virtually all modern IT-based recording systems.

Not only does MXF provide an interchange solution, it also enables the use of rich metadata, to be captured, added to, and carried with the video and audio essence. For more information on MXF, see Chapter 5.6.

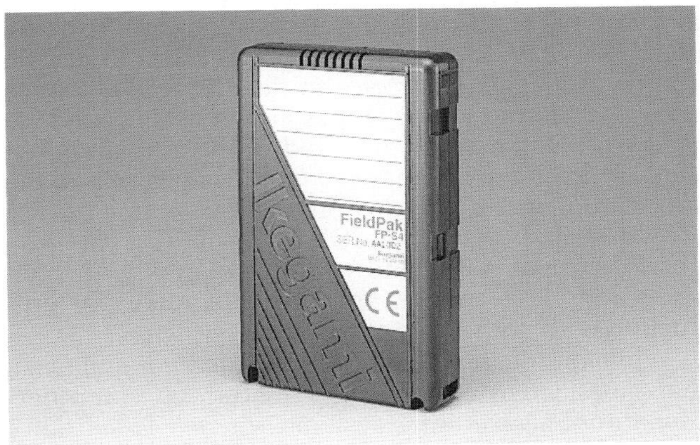

FIGURE 5.8-14 Removable hard disk drive with 40 GB storage capacity. (Courtesy of Ikegami.)

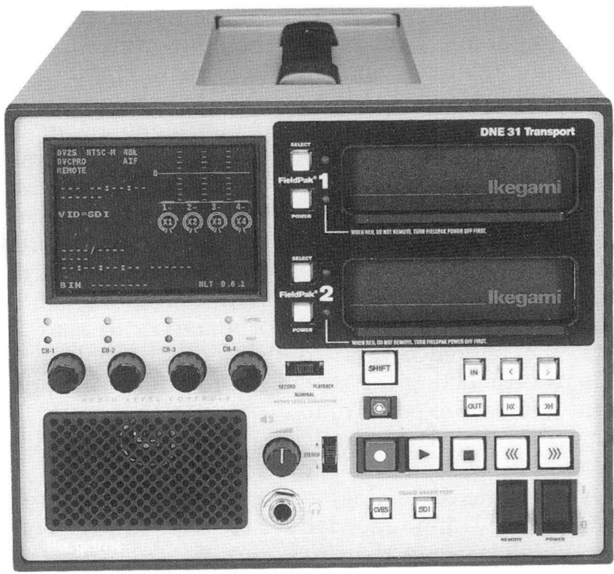

FIGURE 5.8-15 Portable disk drive recorder. (Courtesy of Ikegami.)

HOLOGRAPHIC RECORDING[2]

Developments in holography have enabled a new type of recording system with potential for reliable, low-cost data storage and archiving. Holography breaks the density limits of conventional storage by recording through the full depth of the medium. Whereas typical storage systems record data a single bit at a time, holography allows a million bits of data to be written and read in parallel with a single flash of light, enabling significantly higher transfer rates. Holographic data is recorded by splitting the light from a single laser beam into two beams: the signal beam (which carries the data) and the reference beam. Data is encoded onto the signal beam using a spatial light modulator (SLM). The SLM translates the electronic

[2]This section is based on information from InPhase Technologies (http://www.inphase-technologies.com).

data of 0s and 1s into an optical "checkerboard" pattern of light and dark pixels. The hologram is formed where the reference and signal beams intersect in the recording medium. At the point of intersection of the reference beam and the data-carrying signal beam, the hologram is recorded in the light-sensitive storage medium. A chemical reaction occurs in the medium when the bright elements of the signal beam intersect the reference beam, causing the hologram to be stored. When the reference beam angle, wavelength, or media position is varied, many different holograms can be recorded in the same volume of material. To read the data, the reference beam deflects off the hologram, thus reconstructing the stored information. This hologram is then projected onto a detector that reads the data in parallel. This parallel readout of data provides holography with its fast transfer rates.

At the time of writing, a practical holographic system, recently announced by two companies,[3] combines high storage densities and fast transfer rates using removable media. It is expected to provide a cost-effective, tapeless solution for archiving large video files. The initial product will be a 300 GB external holographic drive associated with a PC, while second- and third-generation drives are in development with capacities of 800 GB and 1.6 TB. The holographic drive for use with PC systems employs a removable 130 mm disk-based cartridge. The storage disc media is composed of two substrates with 1.5 mm of recording material between them. Data is recorded between the substrates, with no surface recording and this use of the full depth of the recording material contributes to the robustness of the holographic media. Data is recorded at 160 Mbps, using a blue laser from 405 to 407 nm in wavelength. The new holographic recorder and media are shown in Figures 5.8-16 and 5.8-17.

FIGURE 5.8-16 Holographic data recorder. (Courtesy of InPhase Technologies.)

[3]InPhase Technologies and Ikegami.

FIGURE 5.8-17 Holographic disk media with 300 GB capacity. (Courtesy of InPhase Technologies.)

EDITING VIDEOTAPE

In the early days of quadraplex VTRs, videotape was edited by actually cutting and splicing the tape. This crude system soon gave way to editing by copying (*dubbing*) portions of the source VTR material onto a second VTR. This method provides *cuts-only* edits. While cuts-only edits have served news departments well for many years, commercial production requires more capabilities. VTR editing suites developed that typically have several playback VTRs for the program material sources (known as *A-roll*, *B-roll*, and so on) fed through a video switcher to allow transitions with mixes, wipes, and effects to be carried out. Graphics, captions, and other elements may also be added at the same time. The output of the switcher is recorded on another VTR. All of the machines, the switcher and audio mixer and possibly other devices in the suite, are controlled by an *edit controller*.

The edit controller uses an edit decision list (EDL) to locate the right program material. It then instructs the VTRs when to start and stop and when to record. The controller uses *timecode* to determine the precise points to cue the various tapes and effects needed for the edit. EDLs may be prepared in the edit suite, but they are frequently prepared *offline*, using copies of the program material recorded on a low-cost tape or disc format such as VHS, MiniDV, or on DVD. It is also possible to use some edit control systems with both VTRs and video servers, enabling integrated postproduction with both types of recording systems.

While online, tape-based editing facilities still exist, they are increasingly being replaced by computer-based nonlinear editing systems where VTRs are not required except for ingest of content that may be delivered on tape or to produce a tape version of the finished product (see Chapters 5.6 and 5.11). Nonlin-

ear systems also enable many of the effects previously done with a video switcher to be achieved using the editing system software.

SMPTE TIMECODE

For video production, editing, and transmission, it is necessary to time program segments accurately and also to be able to identify any single frame of video. For this purpose, the Society of Motion Picture and Television Engineers (SMPTE) developed a system called *SMPTE Time and Control Code*, usually known simply as timecode and specified in SMPTE 12M [2]. This digital code is recorded onto videotape to identify how many hours, minutes, seconds, and video frames have passed since the beginning of the recording. On early analog VTRs, timecode was recorded on a separate track along the edge of the tape; this is known as *longitudinal timecode* (LTC). The disadvantages of LTC are that it takes up extra space on the tape and cannot be read when the tape is stationary or moving slowly. A later development puts timecode in the vertical interval of the video signal, where it is known as *vertical interval timecode* (VITC). VITC can be read at most tape speeds (whenever the video is locked), which is a great advantage during editing or when cueing a tapes. For maximum flexibility, some VTRs and editing systems can use both LTC and VITC.

SMPTE timecode is recorded when the video is first recorded; it can then be used to identify edit points during postproduction. If necessary, a new continuous timecode is laid down for a finished program. This can then be used for timing and machine control purposes, in conjunction with an automation system, when the tape is played out for broadcast.

Timecode is now used with video servers and nonlinear editing systems. It also has other applications; for example, it may be used to synchronize a separately recorded audio recording with a video recording, to control station clocks, and to act as the time reference for station automation. *Burnt-in timecode* is used to place the timecode frame numbers in the visible area of the picture where an editor, producer, or other viewer can see the timecode. This make it easy to determine edit points and create scene logs. Dubs using burnt-in timecode are often made on low-cost video formats that allow for viewing almost anywhere.

Glossary

Coercivity: A measurement of the magnetic characteristics of a given particle. Coercivity refers to the amount of magnetic energy required to bring a saturated magnetic particle to a zero, or demagnetized, state.

Color-under recording: A recording process that adds an AM chrominance subcarrier centered near 650 kHz to an FM luminance carrier for recording on tape. This process was used for some analog professional video recording formats and most analog consumer formats.

Control track (CTL): One of several longitudinal tracks laid down during the recording process. Information from the CTL is used to position the tape for proper playback. In many ways, the control track is the electronic equivalent of sprocket holes in film.

Curie point (or Curie temperature): The temperature at or above which the coercivity of magnetic material decreases substantially, permitting a change in orientation by a weaker magnetic field.

Drum (assembly): The drum assembly in helical scan machines typically consists of an upper and lower drum. The upper drum contains the head and preamp assemblies. The lower drum consists of the reference edge guide (helix), rotary transformers, and the servo motor used to rotate and position the video heads. The lower drum also contains a pulse and/or frequency generator that provides rotational velocity and position information to the drum servo.

Helical scan: A recording method in which the tape is wrapped around a rotating drum in the shape of a helix. Various methods exist which vary drum diameter, number of recording heads, wrap angle, and direction of tape travel (either the same or opposite direction as drum rotation).

Kerr effect: The phase change exhibited by certain substances when exposed to a magnetic field, due to rotation of the plane of polarization of light reflected from them.

Laser: A device that uses the natural oscillations of atoms to amplify or generate electromagnetic waves into an intense beam of coherent optical radiation. (Acronym for *light amplification by stimulated emission of radiation.*)

LTC: Longitudinal timecode.

Quadraplex: Videotape machines that used 2" tape and the transverse scan recording method pioneered by the Ampex Mark IV in 1956. The format was referred to as quadraplex because four heads, mounted 90° apart on a drum assembly, were used to record the video signal on tape.

Timecode: A system of numbering the individual frames of a video recording. Valid timecode numbers run from 0 to 23:59:59:29—a 24-hour clock. Timecode can be generated as an audio signal and recorded longitudinally (LTC) on audio tracks, or can be inserted into video on a single horizontal line of the vertical interval (VITC). Many small format recorders insert two identical lines of timecode due to the possibility of a dropout making a single line unusable.

Transverse scan: A recording method in which the recorded track is perpendicular to the direction of tape travel.

VITC: Vertical interval timecode.

References

[1] Ginsburg, C.P,. *The Birth of Video Recording.* Society of Motion Picture and Television Engineers Convention, 1957. See http://www.labguysworld.com/VTR_BirthOf.htm

[2] SMPTE 12M-1999, for Television, Audio and Film—Time and Control Code, Society of Motion Picture and Television Engineers.

Bibliography

NAB Engineering Handbook, 6th ed. (1975), 7th ed. (1985), 8th ed. (1992), and 9th ed. (1999). Readers are referred to these earlier texts for in-depth information on analog videotape and videotape machines.

Jones, G., *A Broadcast Engineering Tutorial for Non-Engineers*, NAB and Focal Press, Burlington, MA. 2005

Tozer. E.P.J., *Broadcast Engineer's Reference Book*, Focal Press, Burlington, MA. 2004.

Watkinson, J, *The Art of Digital Video, Third Edition*, Focal Press, Oxford, UK. 2000.

Useful Websites

http://ampexdata.com/
http://www.blu-ray.com/
http://www.dvdforum.org/
http://www.ikegami.com/
http://inphase-technologies.com
http://jvc.com/
http://panasonic.com
http://www.sony.com

CHAPTER

5.9

Video Servers

KARL PAULSEN
AZCAR USA, Inc.
Canonsburg, Pennsylvania

INTRODUCTION

Portions of this chapter were derived from the NAB Engineering Handbook 9th edition chapter on the subject by Jerry Whitaker. It has been updated and expanded to include the burgeoning area of storage technologies. It covers primarily video servers, but some of the principles apply to media servers in general so the chapter is complementary to Chapter 3.6, Radio Station Automation, Networks, and Audio Storage.

Professional broadcast video servers today are available with many storage configurations, various input-output configurations, and sophisticated capabilities for the storage and presentation of video and audio. Early video disk recorders evolved into the highly capable server systems of today by combining fundamental core technologies with modern-day networking technology to create flexible, extensible, scalable, and interoperable platforms that generated a paradigm shift for the industry in the way that television program material is captured, stored, distributed, and presented.

The first core technology is linear magnetic recording, which was applied in the development of the spinning magnetic disk drive. The second technology is digital image compression, beginning with JPEG, followed by Motion-JPEG, and the advanced predictive encoding technologies of MPEG and beyond. The third component, networking, consists of real-time operating systems, network topologies, file systems, and content management principles. All three of these technologies contribute to the makeup of video server systems.

The video server is a descendent of the *digital disk recorder* or DDR. The earliest DDRs, employing similar recording principles to the D-1 videotape format, sampled video as eight-bit (CCIR-601) data, and recorded individual paired sets of Y frames (luminance) and Cb/Cr frames (based on R-Y and B-Y color difference signals) onto predetermined track locations stored progressively and continuously around the magnetic surface of each respective Y and Cb/Cr disk drive pair. For example, frame 39's Y component always could be found in the thirty-ninth position of the Y drive, with the corresponding Cb and Cr color difference frame always mapped to the thirty-ninth position on the Cb/Cr drive. The two sets of disk drives, running in exact synchronization, were either written to (during record) or read from (during playout), one frame at a time, for repetitive stop motion, or sequentially as video clips. One early DDR recorded precisely 750 frames of luminance data on one drive, and 750 frames of color difference data on the other, for a total of 25 seconds (NTSC 30 frames per second) or 30 seconds (PAL 25 frames per second). Using frame buffer technologies, a frame could be read into memory, manipulated (e.g., *painted* or *layered* onto another image) externally, and then written back to the same position on the same disk drives. The result was early frame-by-frame graphic image manipulation, the concept leading to the early days of digital or computer animation.

Combining multiple sets of these digital disk recorders and using external devices, such as compositors and digital video effects units, the age of digital video graphics, animation, and special effects was born. Video compression technology aided in allowing many more frames to be recorded on the disk drives, leading to the development of nonlinear editing. The concepts extended to the real-time recording, storing, and playing back of multiple sets of finished clips (e.g., commercials or promotional material) that could be rearranged without disturbing the original image and played back live to air or to videotape in a fashion that heretofore could only be done on linear videotape. Larger-capacity hard disk drives increased the amount of storage for these images and applications software allowed the shuffling of segments during playback. In parallel with the nonlinear editing work, file server technologies from the computer data storage domain were worked into the video disk platform and the video server became a reality.

The paradigm shift from transport-based linear tape recording to professional video server and nonlinear digital disk recording is now well into its third decade, although historically the video disk recorder has been around for much longer than that.

MAGNETIC RECORDING—HISTORICAL PERSPECTIVE

Valdemar Poulsen, a Danish telephone engineer who started work at the Copenhagen Telephone Company in 1893, began experimentation with magnetism to record telephone messages. Poulsen built and patented the first working magnetic recorder called the Telegraphone. However, it was not until the early 1950s that there was commercial development for storing data in a semipermanent format. Figure 5.9-1 lists disk drive developments from the 1950s to present.

The initial method for data storage employed cylindrical drums, whereby magnetic patterns were deposited and then recovered by a device that would later become the magnetic *head*. First-generation disk drives had their recording heads physically contacting the surface, which severely limited the life of the disk drive. IBM engineers later floated the head above the magnetic surface on a cushion of air, a fundamental principle that would become the mainstay methodology for magnetic disk recording technology for decades.

The first commercially manufactured hard disk drive was introduced September 13, 1956, as the IBM 305 RAMAC, which stood for Random Access Method of Accounting and Control. With a storage capacity of five million characters, it required 50 24-inch diameter disks, with an areal density of 2 kilobits per square inch (that compares to gigabits per square inch in the drives of the early 2000s). The transfer rate of the first drive was only 8.8 kilobits per second. The IBM model 355-2 single-head drive, at that time, cost $74,800, which is equivalent to $6,233 per megabyte.

A need for removable storage followed. IBM assigned David I. Noble the job of designing a cheap and simple device to load operating code into large computers. Called the Initial Control Program Load, it was supposed to cost only $5 and have a capacity of 256 KB. During 1968, Noble experimented with tape cartridges, RCA 45 rpm records, dictating belts, and a magnetic disk with grooves developed by Telefunken, but finally created his own solution—the *floppy disk*.

Called the "Minnow," the first floppy was a plastic disk 8 inches in diameter, 1.5 mm thick, coated on one side with iron oxide, attached to a foam pad, and designed to rotate on a turntable driven by an idler wheel. It had a capacity of 81.6 KB. A read-only magnetic head was moved over the disk by solenoids that read data from tracks prerecorded on the disk at a density of 1100 bits per inch. The disk was *hard-sectored*, meaning the disk was punched with eight holes around the center marking the beginning of the data sectors.

By February 1969, the floppy was coated on both sides and had doubled in thickness to a plastic base of 3 mm. In June, the Minnow was added to the IBM System 370 and used by other IBM divisions. In 1970, the name was changed to Igar, and in 1971, it became the 360 rpm model 33FD, the first commercial 8 inch floppy disk. With an access time of 50 milliseconds, the 33FD was later dubbed the Type 1 diskette. The eight hard sector holes were later replaced by a single index hole, making it the first soft sector diskette; it contained 77 tracks and was referred to as "IBM sectoring." A double-density, frequency-modulated 1200 kilobit model 53FD floppy, introduced in 1976, followed the 43FD dual-head disk drive, permitting read and write capability on both sides of the diskette.

Ironically, the floppy disk emerged from IBM at the same time that the microprocessor emerged from Intel; with disk-based recording technology arriving some 20–25 years prior to the August 1981 debut of IBM's first personal computer, the PC.

By contrast, in 1962 IBM had introduced its model 1301, the first commercially available 28 MB disk drive with air-bearing flying heads. The 1301's heads rode above the surface at 250 microinches, a decrease from the previous spacing of 800 microinches. A removable disk pack came into production in 1965 and remained popular through the mid-1970s. A year later, ferrite core heads became available in IBM's model 2314, to be introduced later with the first modern PCs.

The IBM Winchester drive, introduced in 1973, bore the internal project name of the 30-30 Winchester rifle and employed the first sealed internal mechanics. The IBM 3340 Winchester drive had both removable and permanent spindle version, each with a capacity of 30 MB. The drive's flying head height had now been reduced from the original 800 microinches to 17 microinches.

Seagate introduced its 5-1/4-inch form factor ST-506 in 1980, featuring four heads and a 5 MB capacity. Once IBM introduced the PC/XT, it would use a 10 MB model ST-412 drive, which set the standard for the PC-compatible future.

The 3-1/2-inch form factor RO352, introduced in 1983 by Rodime, remained the universal size for modern hard disk drives through the early development of modern personal computers until the 2-1/2-inch disk

HALF CENTURY OF DISK DRIVES
DEVELOPMENT FROM 1956 TO 2000

1956 FIRST DISK DRIVE
IBM 350 RAMAC
5 MB TOTAL (50 - 24" DISKS)

1961 FIRST USE OF ZONED RECORDING
BRYANT 4240
90 MB (TWENTY FOUR - 39" DISKS)

1962 FIRST DRIVE WITH AIR BEARING HEADS
IBM 1301 "ADVANCED DISK FILE"
28 MB (TWENTY FIVE - 24" DISKS)

1963 FIRST REMOVABLE DISK PACK
IBM 1311 "LOW COST FILE"
2.69 MB (SIX - 14" DISKS)

1965 FIRST VOICE COIL ACTUATOR
FIRST SINGLE DISK CARTRIDGE DRIVE
IBM 2310 "RAMKIT"
1.024 MB (SINGLE - 14" DISK)

1966 FIRST DISK DRIVE WITH FERRITE CORE HEADS
IBM 2314
29.17 MB (ELEVEN - 14" DISKS)

1971 FIRST TRACK-FOLLOWING SERVO SYSTEM
IBM 3330-1 "MERLIN"
100 MB (ELEVEN - 14" DISKS)

1971 FIRST FLEXIBLE DISK DRIVE (READ ONLY)
IBM 23FD "MINNOW"
0.0816 MB (SINGLE - 8" DISK)

1973 FIRST FLEXIBLE DISK DRIVE
SETS INDUSTRY STANDARD FOR 8" DISK
IBM 33FD "IGAR"
0.156 MB (SINGLE - 8" DISK)

1973 FIRST LOW MASS HEADS, LUBRICATED DISKS
FIRST SEALED ASSEMBLY
IBM 3340 "WINCHESTER"
35 or 70 MB (TWO or FOUR - 14" DISKS)

1976 FIRST 5.25 INCH FLEXIBLE DISK DRIVE
SHUGART ASSOCIATES SA400
0.2188 MB (SINGLE 5.25" DISK)

1979 FIRST 8 INCH RIGID DISK DRIVE
IBM 62PC "PICCOLO"
64.5 MB (SIX - 8" DISKS)

1980 FIRST 5.25 INCH RIGID DISK DRIVE
SEAGATE ST506
5 MB (FOUR – 5.25" DISKS)

1981 FIRST 10.5 INCH RIGID DISK DRIVE
FUJITSU f6421 "EAGLE"
446 MB (SIX – 10.5" DISKS)

1981 FIRST 3.5 INCH FLEXIBLE DISK
SONY OA-D3OV
0.4375 MB (SINGLE – 3.5" DISK)

1982 FIRST 9 INCH RIGID DISK DRIVE
CONTROL DATA 9715-160 "FSD"
150 MB (SIX - 9" DISKS)

1983 FIRST 3.5 INCH RIGID DRIVE
RODIME RO 352
10 MB (TWO – 3.5" DISKS)

1985 FIRST DISK DRIVE MOUNTED ON A CARD
QUANTUM HARDCARD
10.5 MB (SINGLE – 3.5" DISK)

1988 FIRST ONE INCH FORM FACTOR, 3.5 INCH DISK
CONNER PERIPHERALS CP3022
21 MB (SINGLE - 3.5" DISK)

1990 FIRST DISK DRIVE WITH PRML ENCODING
IBM 0681 "REDWING"
857 MEGABYTES (TWELVE – 5.25" DISKS)

1991 FIRST USE OF MAGNETORESISTIVE HEADS
IBM 0663 "CORSIAR"
1.004 MB (EIGHT – 3.5" DISKS)

1993 FIRST 7,200 RPM DISK DRIVE
SEAGATE TECHNOLOGY ST 12550 "BARRACUDA"
2,139 MB (TEN – 3.5" DISKS)

1997 FIRST 10,000 RPM DISK DRIVE
SEAGATE TECHNOLOGY ST19101 "CHEETAH 9"
9,100 MB (EIGHT - 3.5" DISKS)

1999 FIRST ONE INCH DISK DRIVE
IBM "MICRODRIVE"
340 MB (SINGLE - 1" DISK)

2000 FIRST 15,000 RPM DISK DRIVE
SEAGATE TECHNOLOGY ST318451 "CHEETAH X15"
18,350 MB (THREE – 2.5" DISKS)

FIGURE 5.9-1 A half-century of disk drives.

was introduced for portable applications in 1988. A diagram of a modern IDE magnetic hard disk is shown in Figure 5.9-2.

Video Image Recording

Recording video images onto both linear and rotating magnetic storage surfaces had a similar and parallel development. Alongside the development of magnetic recording tape, random-access video-on-demand, predicted in 1921, was demonstrated in principle as early as 1950. John Mullin at Bing Crosby Enterprises in 1951 demonstrated an experimental 12-head VTR that ran at 100 inches per second, while the concept of recording video onto a spinning platter was demonstrated shortly before in the late 1950s, about the same

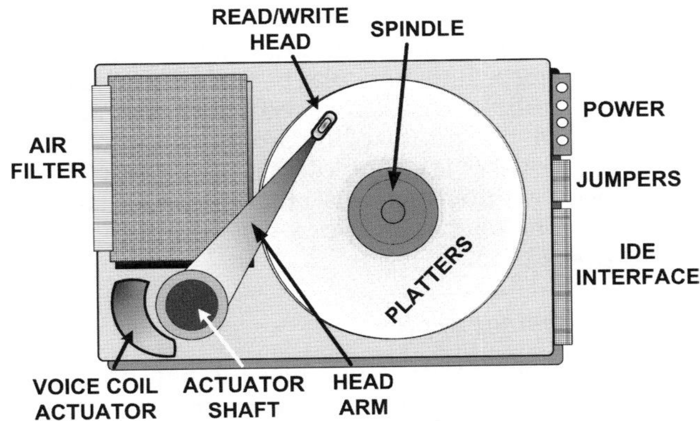

FIGURE 5.9-2 Components of the modern IDE magnetic hard disk drive.

time the first disk drive and NTSC were introduced, which was adopted in 1954.

A rudimentary plastic video disk was demonstrated at the Salone Internazionale della Tecnica in 1957 by Antonio Rubbiani; a few years later, technologists at CBS developed a procedure for a video disk recorder.

Videotape recording was developed by a team of engineers from Ampex Corporation led by Charles Ginsburg, who began work on the videotape recorder (VTR) in 1951. Ampex demonstrated the first three-head system in November 1952 and, in March 1953, a second system, using four heads, was shown. However, problems continued with the "Venetian blinds" effect due to discontinuous recording from one head to the next. In 1954, Charles Anderson and the Ampex team, including Shelby Henderson, Fred Pfost, and Alex Maxey, were working on an FM circuit that debuted in February 1955. Later, Ray Dolby designed a multivibrator modulator, Maxey discovered how to vary tape tension, and Pfost developed a new sandwich-type magnetic head. A half-century later, in 2005, the original team was awarded, some posthumously, a Lifetime Achievement in Technology Emmy for their contributions.

In preparation for the first public demonstration of video recording at the Chicago convention of the National Association of Radio and Television Broadcasters (NARTB) in April 16, 1956, an improved VTR, later to become the Ampex Mark IV, was shown to Bill Lodge of CBS and other TV people. The Mark IV, later renamed the VRX-1000, used 2-inch wide tape, which ran at 15 inches per second past a transverse track, rotating head assembly and employed FM video and AM sound recording. Ampex took out a trademark on the name *videotape* for its recorder, and in 1959 color videotape debuted during the Nixon-Khrushchev Kitchen Debate in Moscow.

Dawn of the Video Disk

The Minnesota Mining and Manufacturing (3M) Company, which produced the first 2-inch wide videotape for the VRX-1000, showed a 1964 video noise-plagued video disk, publicly demonstrating that this new *disk* format had a future. Although less than a year before the first demonstration of a random access, still-image generating, recording, and playback device, the demonstration utilized disk drive recording technologies and changed the future of recording in a profound way. Twenty years later the concept of a "video serving device" would emerge, a breakthrough that took its roots from television sports—in the form of the *instant replay.*

Television Instant Replay

At the July 1965 SMPTE conference in San Francisco, MVR Corporation showed a 600 frame (20 second), black-and-white video recorder, the model VDR-210CF, that recorded individual frames. CBS used as a freeze-action video disk around August of that year. The MVR, with its shiny aluminum, nickel-cobalt–coated magnetic disk, was used in football telecasts to instantly play back short-action sequences in normal motion or freeze the motion on a single frame.

Ampex took a different approach creating an 1800 RPM spinning metal disk with a series of stepper motor-driven recording heads that moved radially across the platters, creating 30 seconds of normal video using analog recording technologies. It recorded 30 video tracks per second, with each track holding one NTSC frame, giving a total of 1800 NTSC fields. The heads could be rapidly moved to any location on the disk for replay at normal speed or, when the heads were slowed down and the same frame repeated in multiple sets, for *slow-motion* playback. When the playback stepper heads were stopped, and the platter continued to spin, the same frames were repeated with a *freeze frame* continuously displayed.

In March 1967, the first commercially available video magnetic disk recorder with true slow and stop motion, the Ampex HS-100, was placed into service. Effects included rapid playback in normal, slow, or stop action. With the World Series of Skiing program from the U.S. Ski Championships in Vail, Colorado, the disk recorder marked the dawn of instant replay for television.

Early Video Disk Recording

The developments for recording moving images to spinning disks continued with overall disk drive advances. Graphic arts applications came with the 1981 introduction of the Quantel Paintbox. With no digital tape recorder available at the time for storage, digital images from the Paintbox were offloaded onto either 8 inch floppies or the FSD removable hard drive, a transportable hard disk drive from Control Data/Hitachi/NEC, permitting exchange between proprietary Quantel systems only. In 1986, Quantel's Harry became the first high-quality, integrated, and true nonlinear editing (NLE) system to use disk-based digital technologies to create multilayered moving video and effects. Harry remained the benchmark for television graphics (through Henry and Hal, circa 1992) and, with the 1993 introduction of their Dylan fault-tolerant drives, set the stage for RAID-like protected video storage systems going forward, albeit entirely for proprietary-dedicated systems.

Following the developments at Quantel, the need for a simpler, stand-alone disk recording device with component digital quality surfaced. NTSC (composite) recording, onto less than full-resolution rigid and semi-rigid media was already available, but that was constrained to short segment recording and playback with stunt features, such as reverse or stop motion, faster than real-time playback, and for the general-purpose storage of graphic images as electronic still stores (ESS).

The Abekas A62/A60/A64 series of disk-based production systems began a new trend for production-quality digital video recording. Available in a single disk chassis, or as pairs of disk sets with compositing or layering engines, Abekas digital disk recorders worked with both PAL and NTSC standards in either composite analog, composite digital, or component digital video. Animation, graphics special effects, and early 3D computer graphics drove single-frame recording and short segment clips. However, another important and significant technology also emerged from the concepts of the Abekas designs.

The stand-alone Abekas A60 disk recorder accepted and produced files that were data representations that could be transported (via TCP/IP) to other devices, in particular 3D and 2D computer-generated imaging (CGI) platforms. The files were produced in two formats: 24-bit RGB and 16-bit YCbCr.[1] Both image formats were 720 pixels wide by 486 pixels high; almost precisely what the first component digital video format would become for television moving images. The pixels were not square, and an aspect ratio conversion factor of either 0.9 or 1.111 needed to be applied depending on whether squaring or unsquaring the pixel was needed for the proper resolution and scanning rates of finished video. The Abekas A60 also featured control and transfer capabilities, such as Telnet, that allowed files to be exchanged between other computer-centric devices. Using simple commands such as this single line of ASCII code:

```
a60> play # play from the current frame to the
end (frame 749)
```

would cause the device to play from the current frame to the end frame.

The Abekas A60 stored 750 PAL frames (25 f/sec) or 750 NTSC frames (30 f/sec) of broadcast standard video in the digital storage format specified by (at that time) the CCIR-601 specification. The disk format's native YCbCr format also had encode or decode firmware that processed 3 byte Red-Green-Blue (RGB) images on-the-fly, permitting TCP/IP file transfers to and from external devices. This relatively new concept became the de facto process for computer graphics 3D and animation, given there were, at the time, no standards for file formats or interoperability between platforms. Facilities often developed their own structures for dealing with the individual frames as discrete files. The precise video storage format created files that had no options, no header, and were always 720 pixels per line, 576 lines per image.

Abekas (and its predecessor Accom) developed other disk-based products using its concepts for digital disk recorders, including still store and clip management servers, thus effectively spawning the age of the digital video recorder.

VIDEO COMPRESSION AND SERVERS

Need for Compression

Videotape systems have utilized compression techniques for many years to reduce the bit rate of the signal to be recorded. This is necessary due to limitations on the highest frequency (and hence data rate) that can be recorded using magnetic medium based on the characteristics of the record/playback head and the tape formulation. It is also used in order to increase the length of program that can be recorded on a given length of tape. In the late 1980s, the D-1 format used 4:2:2 coding to reduce chroma channel bandwidths. Digital Betacam followed with its mild 2.5:1 compression algorithm. Many other formats of video compression, all the way through DV, MPEG-2, and MPEG Advanced Video Coding (AVC), have followed. Similar constraints to videotape apply for recording on magnetic disks and several of the same codecs have been employed for video server technologies as the markets demanded.

Codecs

Professional broadcast video operations have stabilized on MPEG-2 as the predominant compression coder-decoder (codec) format. However, early video servers employed codecs, which used the tool kit developed by the committee that wrote the JPEG (Joint Photographic Experts Group) recommendations, coining the term *Motion-JPEG* (MJPEG), with its many flavors of implementation. Techniques evolved into a set of solutions that met the demands of both early nonlinear editing and video storage for editorial and commercial broadcast applications. Motion-JPEG is an image compression mechanism utilizing an all intraframe discrete cosine transform (DCT) coding foundation, designed for compressing full-color or grayscale images of natural, real-world scenes.

Like MPEG-2, JPEG is considered a lossy compression scheme, meaning that the decompressed image is not quite the same as the original. JPEG is designed to exploit known limitations of the human eye and brain, notably the fact that small color changes are perceived less accurately than small changes in brightness. JPEG is therefore intended for compressing images that will be looked at by humans, and thus became acceptable for applications related to moving images in the video domain. An aspect of JPEG that lends itself to video servers is that JPEG decoders can trade off decoding speed against image quality by using fast but inaccurate approximations to the required calculations.

As development of both editors and video servers continued, the introduction of MPEG-based compression and other codecs largely superseded Motion-JPEG. More recently, JPEG2000 has been introduced, designed primarily for digital cinema applications.

Data Rates

During early development of nonlinear editing platforms, primarily Avid Technology's NLE products, the Motion-JPEG data rate and the quality of the encoder drove image quality (in particular resolution), which Avid would refer to as the *AVR* (Avid video resolution) rate. The higher the AVR number, the better the

[1]Abekas adopted the file extension ".yuv," leading to some confusion, as U and V are the color components for NTSC and PAL analog video. Signals used were correctly scaled as digital YCbCr components.

image quality and the more storage space the images required on the disks.

In order to achieve broadcast-acceptable pictures from Motion-JPEG codecs, data rates would range between 24 and 48 Mbps. However, once codecs employing the ISO-13818 standard—the Moving Pictures Expert Group (MPEG) compression format—were developed for professional video servers, observations showed that the same images when compressed using MPEG could utilize data rates between 12 and 18 Mbps and achieve the same image quality as Motion-JPEG images at nearly twice the data rate. Today's broadcast applications employing MPEG operate in the 8–12 Mbps range for standard definition images, providing Betacam-quality resolution, the benchmark for the late 1980s and 1990s.

Data rates and image resolution are closely linked together, which complicates video server development. High definition (HD) MPEG-2 compressed imaging employs typical data rates in the 45–50 Mbps range, in video contribution and distribution, with upward of 100 Mbps and beyond for video recording. Full-bandwidth baseband high definition servers and associated storage components may employ data rates in excess of 1.5 Gbps, especially where used in applications for high-end motion picture and commercial production.

Motion picture film images at 2K and 4K resolution are often scanned into disk arrays at upward of 3–5 Gbits per image, yet there are no commercially available playout systems for real-time playback of images of this size and scale. Thus, when scanned images of this size are captured, viewing of finished production sequences is done by down converting to HD resolution and playing out from HD production-quality video servers. At the time of this writing, servers for digital cinema and their projection systems work with very large-scale images, at very high resolution, and digital cinema standards have settled on JPEG2000 compression, for which specialized playback video servers will be deployed. MPEG-2 remains the de facto choice for consumer-released products, such as DVD, although HD DVD and Blu-ray DVD formats also allow for AVC and VC1 coding.

VIDEO SERVER PLATFORMS

The continued and successful growth of video server technologies relies on the principles of scalability, extensibility, and interoperability. Video server platforms have various models based on these principles for expansion or to service other needs. Part of the video server selection process depends on how these principles are supported, making it essential to understand which set of features makes best sense from both a cost and functionality perspective.

Scalability

Scalability is the capability of a system to increase performance under an increased load when resources are added. A system whose performance improves after adding hardware proportionally to the capacity added is said to be a *scalable* system.

A system should be easy to expand based on the loading placed on the system, that is, the number of input or output ports, the number of streams being ingested or played back, the size and dimension of the files stored on the system, and the amount of storage that can be incrementally added without excess reconfiguration or complete obsolescence.

Extensibility

Extensibility is a system design principle where the implementation takes into consideration future growth. It is a systemic measure of the ability to extend a system and the level of effort required to implement the extension.

Extensions can include either the addition of new functionality or, through the modification of existing functionality with a central theme, provide for change while minimizing impact to existing system functions. Obsolescence is a fact—as technology continues to evolve, so comes obsolescence. Thus, for a product or a feature set to be extensible means that it can continue to be of value and will not be rendered useless.

This is no easy task and requires that the system be capable of accommodating data that can be interpreted and used at that instant, while ignoring other information for which it has no use or value. It may seem difficult to apply the concepts of extensibility to a video server system without knowing a great deal about not only the technical inner workings, but also the marketing forces behind the product. The general concept to understand is that the devices used in the server, built before later definitions were established, will accept and process variations in the materials that they have been designed to handle.

In short, extensibility is the ease with which a system or component can be modified to fit the area it addresses now and in the future.

Interoperability

A goal since the first introduction of video servers has been the ability of systems to provide services to, exchange services between, and accept services from, other systems, that is, interoperability. For a video server to be most efficient, services must be exchanged in a fashion that will enable them to operate effectively together.

A service may include the transfer of data between systems at its root form, that is, the file level. Or, it may include established video standards, primarily SMPTE 259M and SMPTE 292M, for the interchange of baseband video and audio information in standard and high definition television.

The ultimate goal for video servers is to exchange information transparently between systems, regardless of the native format of the video server itself, and regardless of how the services are rendered. For a server to be future-proofed from an interoperability perspective, it should be capable of conveying its information over interfaces it currently uses and those that have not yet been defined.

As evidenced by the lack of interchange on Motion-JPEG files, the marketplace and industry both recognize that future video servers need to establish a common set of standardized interchange formats and protocols with which to carry and translate data from manufacturer X to manufacturer Y.

The development of the Material eXchange Format (MXF) opened the door to this solution. As these new standards unfold, the ability to interchange content at a file level, and transport any compressed digital video packet over existing networks and cabling infrastructures, becomes closer to reality.

Inputs and Outputs—Video or Data

For the most part, the structure of a video server is based on digital video. The serial digital interface (SDI) is the standardized methodology used to transport baseband digital video around a studio environment and to present that data to such real-time operating devices as video switchers, display monitors, and video servers. Once thought of as the only methodology (beyond analog video/audio) for getting real-time program content into and out of the server, the baseband technologies of video and audio are still in use, but are being rapidly superseded by file formats that wrap the digits of a compressed stream of video, audio, and metadata into a compacting structure that can avoid using the coaxial BNC inputs and outputs via video codecs. Thus, video baseband connections have evolved into a method based on *file transfer*.

To understand further how these technologies are being enabled, a simplistic overview of how a video server platform is constructed is presented.

PRIMARY SERVER COMPONENTS

In its elementary form, the video server is composed of two major components: a record/play unit that is the interface to and from the second component, the storage system. As illustrated in Figure 5.9-3 this includes inputs and outputs (I/O), encoders and decoders (also referred to as a codec), and an interface for machine control (usually RS-422 or network-based interfaces). Storage may be integral to the server, as in a stand-alone videotape transport replacement or smaller dedicated server, or as is the case with large server systems, external. A final storage element is the system drive, frequently redundant and separate from the media storage. Not shown in the figure are power supplies, RAID or other storage controllers, or the user interfaces.

The record/play unit, or video engine as it is sometimes called, consists of the elements described in the following sections, and illustrated in Figures 5.9-4 and 5.9-5. It should be noted that the structures of the I/O and control components are often combined onto a

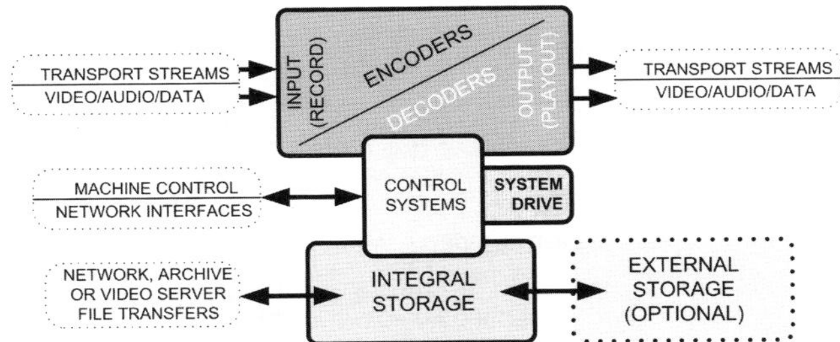

FIGURE 5.9-3 Major system components of a video server.

single codec board, allowing the video server's port configuration to easily convert from an ingest device to a playout device. Also, some vendors provide access to the data/metadata only over a network interface or not at all, relying on a third-party application to input or extract that information at the file level, or during ingest.

Selectable Video Inputs and Outputs

Video inputs and outputs may comprise serial digital SMPTE 259M for standard definition, SMPTE 292M for high definition in 8- or 10-bit, or analog composite video and, in some systems, component analog (Y, Pb, Pr) video. Analog video is becoming less prevalent as digital video becomes more predominant. Many professional video server systems manufactured after 2000 have excluded most analog I/O (audio and video), moving toward an all-digital environment.

Most servers produced since the 2003 timeframe include embedded (as well as discrete AES) audio capabilities, whereby the video signal carries multiple tracks of audio in the same transport.

ASI Input–Output

ASI inputs and outputs are used on most professional video servers to allow input and output of compressed bit streams. In this arrangement, the full high definition bandwidth of 1.485 Gbps is precompressed using an external encoder and fed to the ASI port, capable of directly storing MPEG as either single or multiple transport streams (with an equivalent decoder on the output).

Selectable Audio Inputs

Similar to the video inputs, these may be either a single stereo-analog pair or generally two or more sets of AES digital audio. Individual AES interface options include the 75 ohm BNC coax (unbalanced) input or the three-pin XLR input for 110 ohm twisted pair (balanced). Most servers nominally accept 48 kHz AES sampling at 16-, 20-, or 24-bit resolution. With the emphasis on multichannel and 5.1 Dolby surround sound, most servers now handle three sets of AES as

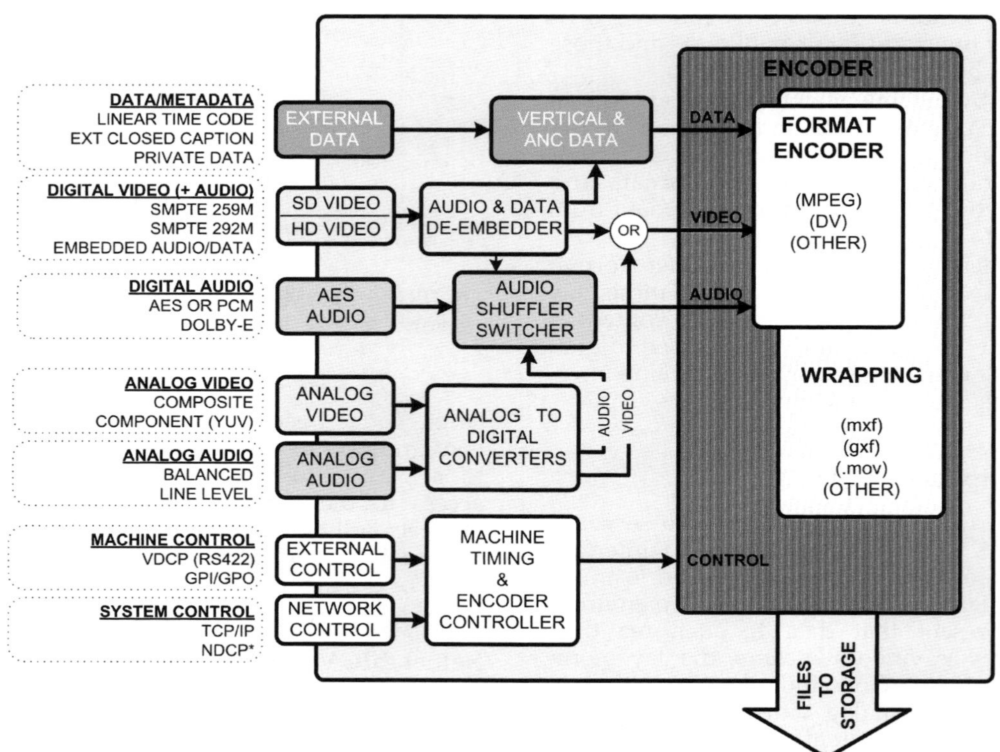

FIGURE 5.9-4 Typical input, control, and encoder components.

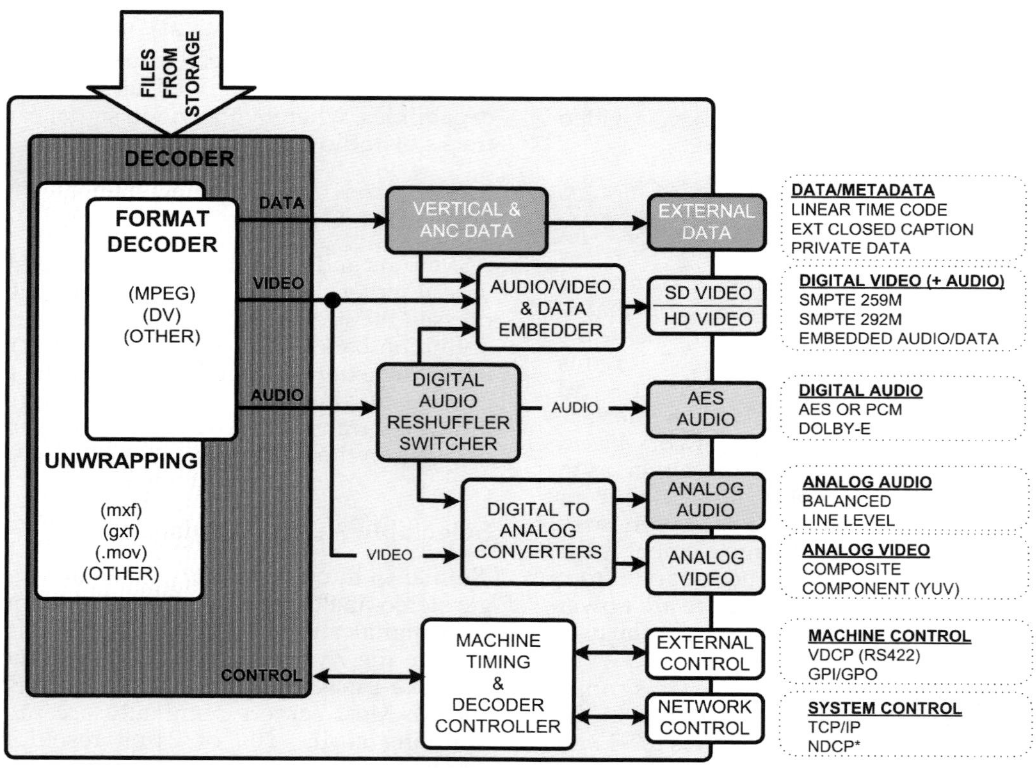

FIGURE 5.9-5 Typical decoder and output components.

either discrete or embedded signals. The ability to input and preserve Dolby E signals is almost universal on professional video servers; extending the ability to store many more audio tracks in the same space as an AES set.

Analog Video Reference

Video black-burst is fed from the facility's reference generator. The user may typically select between external reference, internal reference (free run), or one of the video inputs (either digital or analog).

Control Interfaces

An RS-422 machine control, for emulation of common VTR modes such as the BVW-75 protocol, is provided. Most servers will supply at least one RS-422 machine control port for each channel of the server. The terminology *channel* has no common definition and may vary depending on the manufacturer's implementation. The control protocol usually is VDCP (video disk control protocol), but there is a trend toward using Ethernet-based interfaces as more sophistication in control and interchange is developed.

GPI/GPO

General-purpose input (GPI) or general-purpose output (GPO) interfaces usually have at least the same number of inputs and outputs as there are channels. GPI/GPO functions may be dedicated or set into software and assigned by the user. GPIs allow a server operating without facility automation to start, stop, re-cue, or advance clips during playout, provided the server had the ability to generate its own playlist internally.

Timecode

External linear digital timecode, per SMPTE 12M, for use in systems at 30, 25, and 24 frames per second, is invariably provided. Timecode will generally be referenced to the facility house clock. Timecode data, specific to the video clip recording, may be carried and conveyed to the video server from the source VTR on the RS-422 serial control interface.

Vertical Interval Blanking

Digital video has little use for the traditional synchronization-related blanking information of analog television and effectively does not encode that portion of the analog picture. The information needed to reconstruct the vertical blanking interval (VBI) portion of the picture can be carried in the digital bit stream without consuming the nearly 8% of a video frame that is used for blanking in the analog domain.

However, because broadcasting still has requirements for carriage or retention of certain lines in the VBI, such as closed caption data, teletext, rating, and content validation or audience measurement information, the data originally held in the VBI of the original source material may need to be preserved. Some nonbroadcast digitization processes may obliterate portions of the VBI, so another method of retaining data such as CEA 608 (line 21) closed captioning data is provided so that it may be converted and stored on the storage system as ancillary data. The number of lines to be retained is generally selectable at a setup level on the server's configuration menus. Digital video equivalents for this data are often carried in the ancillary data portions of the digital transport signal, for both HD and SD applications.

Network Connection

Servers generally have at least one 10/100baseT network port, used for Ethernet connectivity to or from the facility's broadcast LAN. Some vendors have applications that permit access to these ports for diagnostics and software uplifts. Other vendors have employed control system capability permitting native level control interfaces to the servers.

Gigabit Ethernet Interfaces

Ultra-high-speed network interfaces may be employed for everything from control, to data transfer, to storage area networking. Like the 10/100baseT connectivity, this port may be used for whatever the manufacturer intends, such as FTP of MXF files to/from the server, and access to storage in next-generation platforms. Gateways are often interfaced through Gigabit Ethernet ports that buffer or transfer data to/from peripheral interfaces including WAN, remote storage sites, and system archives.

Terminal or Control Ports

RS-232 or USB ports supplied for connection to a dedicated PC terminal are used for the setup, interrogation, and diagnostic monitoring of the server. In today's networked environments, this information may be transmitted via TCP or similar protocols direct to the facility's broadcast LAN or other network.

Other interfaces to or from the server are most likely manufacturer specific. These connections might include data drives, controllers, or possibly IEEE 1394 interfaces between servers and media I/O ports.

System Drives

System drives include the internal hard disks that are dedicated to the operating system that manages all the functions of the video server. Often they are mirrored (RAID 1) for the protection of the critical system functions and should never hold content of the actual program essence or metadata.

Storage Interfaces

Depending on their drive structure, the physical drives may be mounted internally or may be in secondary enclosures. Most drive systems use RAID configuration and, in early servers, were deployed as simple *direct-attached storage* (DAS).

When the server's data drives are internally mounted, there may be no external *small computer systems interface* (SCSI) connections. Should the server use an external drive chassis, the number of SCSI ports on the primary chassis will depend on the number of SCSI buses that the system supports (sometimes several) or is handled by the RAID controller, if internal, or other methods.

Fibre Channel Interface

In the early phases of external Fibre Channel (FC) interfaces, many servers were using FC over twisted copper pairs. The need for an optical fiber connection did not surface until larger-scale drive arrays or common shared file systems (NAS or SAN, discussed later) came about. Today, most FC interfaces are on optical fiber.

Alarm Ports

Connections for alarms are in the form of either GPO contact closures or other independent ports, which allow users to configure signaling as they desire. The signals that trigger alarms might include primary power supply failure, fan or airflow failure, over temperature, or a fault in the computer or compression engines themselves.

User Interface

If the server platform uses a PC-based circuit board set it is most likely to have an array of the traditional PC human interfaces or I/Os. At the very least, the dedicated server chassis will utilize a computer VGA connection for display, a keyboard, and a mouse port. These three items are lumped into the term *user interface*, are coupled with applications software, and become the means by which a human communicates with the various levels of the server system.

Peripheral Ports

Generally these comprise at least a 9-pin serial port and/or a 25-pin parallel port, although the 25-pin implementation is dwindling. Some servers made use of the 25-pin port to drive an external CD-ROM drive, used for loading software. Early units typically had built-in floppy or CD-ROM drive for loading software or archiving system data. Today that is being replaced by the USB port. Seldom were these peripheral drives used for the interchange of video content data.

Internal Storage Expansion

Many servers, including self-contained storage and I/O chassis, offer some degree of internal drive expansion and in certain cases provide a variety of upgrade paths for the platform. The upgrades might be an additional bank or group of drives to fill out the integral storage, or as an uplift with larger drives when, or if, qualified during the life of that particular server or storage platform. However, many of the integrated server/storage products that began their life as 4 GB or 9 GB drives abandoned the ability to grow beyond those drive platforms for a variety of manufacturing obsolescence and other technology reasons. Current data storage drives typically range from 73 GB to upward of 300 GB, with new drives being qualified routinely.

Internal Codec Expansion

Server platforms nearly always provide for at least a few additional card slots destined for new or additional codecs. As server products have progressed, some of the new codecs feature additional value, such as playout only, Motion-JPEG to MPEG conversion, bidirectional input/output offerings, ASI, and more.

Evolution

As systems became more advanced and technology began to shrink the footprint of the silicon on these add-in cards, feature sets, and capabilities grew and the size of the physical server engine began to diminish. Some manufacturers began to break away from the integrated codec/storage platform and replaced the physical storage space with more codec offerings—sending all storage to external drive arrays.

Eventually, the actual physical architecture of the server became the distinguishing benchmark of each manufacturer's product line, setting each system farther apart from its competition. Careful analysis of the long-term needs of the facility began to make the difference between selecting one server platform versus other choices.

Once self-contained or integral storage migrated to a network-centric topology, the entire perspective of the video servers distinguished by those early 1990's platforms changed dramatically. Today, nearly all video server manufacturers have adopted a network-like design philosophy, with only a handful of vendors playing into the more traditional simple VTR-replacement model.

VIDEO SERVER STORAGE SUBSYSTEMS

Early video servers consisted of internal or dedicated external storage arrays, principally around the era when individual drive sizes were confined to 4 GB or 9 GB standard Fibre Channel (FC) drives. As storage systems evolved and the introduction of MPEG began to overtake Motion-JPEG, external server storage

subsystems arranged in modular form factors began deployment.

In advance of network-attached storage (NAS) or storage area network (SAN) storage technologies, this next-generation server was limited by its overall system bandwidth—the rate at which data could be exchanged between server engine and storage subsystem. To overcome this limitation, another change was necessary.

Storage drives, often referred to as *storage arrays* or the *RAID chassis*, became a problem. In some cases, the development of the physical drives themselves outpaced the manufacturers' abilities to employ them in the systems they provided. Once the server's integral, internal storage boundary was crossed, the only choice was to use a specific external RAID chassis for all media storage. This concept had its own problems. In particular, the ability to upgrade drive sizes to accommodate storage growth became locked by the very structure of the RAID controller, the physical drive and media interface, and the availability of qualified FC drives, all of which created a complex balancing act for overall server performance.

The prominent form for external RAID storage is the FC disk array, which originally employed dual serial connections of shielded twisted pairs set in a loop arrangement from drive to drive. Optical fiber, initially quite expensive, was an alternative for up to a few arrays. In the early years of FC deployment, around the mid-1990s, the use of intelligent switching, that is, FC switches, was cost prohibitive for all but the largest or highest profiles of systems. Today, as server storage systems have grown beyond terabyte capacities, and the number of I/O channels exceeds dozens, FC switches are now an integral part of most storage networks.

Storage system components need to be fully qualified and are generally supplied directly by the video server vendor. With storage technologies advancing faster than video server manufacturers can develop their new products, the effect is that servers lag behind the size of the storage drives that could be employed, and when those systems could theoretically reach the market.

Dedicated external storage platforms have reached a point where performance cannot be improved by simply adding more or larger external arrays. Hence, the rationale for NAS or SAN came into being.

EXTERNAL ARRAY COMPONENTS

The components and capabilities found in pre-NAS/SAN storage architectures include:

- *RAID controller:* The hardware-based circuit card that manages the disk drives to make them appear as one volume (i.e., "drive") to the system. RAID may also be implemented in software, in which case that application may also be referred to as the "controller."
- *Backup RAID controller:* A secondary, standby card, typically an option that can be placed into operation either manually or automatically.
- *Power supply systems:* At least a primary and one or more backup supplies are specified in most server systems so that the failure of one does not incapacitate the entire system. Generally the supplies are hot standby with automatic fault tolerance that provides performance status back to the system controller and the RAID controller.
- *Cooling assemblies:* Multiple fans designed so that a failure of one or more fans would not take the drive system off-line.
- *Drive carriers:* Depending on the chassis, generally single slots that house 3-1/2-inch form-factor Ultra SCSI or Fibre Channel, high-performance disk drives.
- *Fault tolerance:* The ability of a system to respond gracefully to an unexpected hardware or software failure. Levels of fault tolerance range from being able to continue operation in the event of a power supply or drive failure, to fully redundant data and/or system drive, server engine, and operating system changeover.

Part of the RAID 3 specification includes automatic detection and identification of failed or failing drives, plus, upon replacement, automatic rebuilding of the data on the new drive from information remaining on the other sets of drives. Usually a transparent *failover*, there sometimes is a slight yet generally unnoticeable degradation in performance during drive reconstruction. The RAID 5 specification offered an alternative that was amplified by RAID 6, with the addition of an extended parity drive, sometimes offered as an option.

RAID

In their landmark 1988 paper, "A Case for Redundant Arrays of Inexpensive Disks," David Patterson, Randy Katz, and Garth Gibson of the University of California at Berkeley describe the concepts for five methods of data protection on disk arrays [1]. The methods were aimed to guard against data destruction due to failure of drives, and described techniques for mapping (or striping) data across arrays and their associated performance benefits. The commercial development of the concept of RAID began in 1989 when the first nonmirrored version with commercial viability was announced by Compaq. By the middle of the 1990s, the concept had become a virtual household word. When first introduced, RAID stood for "redundant arrays of inexpensive disks," but the word "inexpensive" has since been modified to "independent," presumably because the word inexpensive is relative, and the cost of all storage has literally become comparatively inexpensive over time.

The numeric numbering system for RAID was an outgrowth of the first publication and was adopted over time to indicate certain types of protection and data mapping schemes. In the researchers' first publication, only five levels were described, but a sixth form of protection was later added.

The principle advantage behind the RAID concept is that data is spread across a number of individual disks in such a manner as to provide additional security for the data compared to using a single disk. Data protection is provided by the way that redundancy data is stored separately from the primary data. The redundant data may be a complete copy of the data or the addition of information that can be used to reconstruct the primary data, should some element of the hardware fail.

RAID technology can be applied to stand-alone independent subsystems for computer data, image storage, media servers, and systems designed for video-on-demand. With the cost-effective availability of small form-factor disk drives, promoters of RAID claim significant improvements in reliability and maintainability. RAID subsystems also provide higher performance than individual magnetic disk drives. Early opposition to the concept stated that RAID can be expensive and unnecessary, and that there are alternative methods to similar performance at a better cost point. Those arguments have subdued significantly, as the size and depth of arrays and storage on disks have become universally accepted.

Thus, the recognized architecture for the storage of data on magnetic spinning disks, including media content, became characterized by its RAID number. This well-recognized, but often misunderstood, term was well entrenched in the computer data industry long before it was used for video servers. RAID terminology, when used in the broadcast video server context, may still be a source of confusion brought on by marketing hype and continual fear of failure resulting in loss of data.

During early video server roll outs, the term RAID was both widely used and abused. RAID numbering was used as much as a marketing tool as it was a technology—especially when trying to understand what the numbering system meant as it was applied to RAID levels. This further resulted in the myth that "the higher the RAID number, the better the performance." Unfortunately, this misconception about RAID numbering continues even after several years of explanation and implementation.

RAID LEVELS AND ARCHITECTURE

The term *architecture* is meant to describe the physical structure of the elements in a RAID storage subsystem. The term *subsystem* is used because, for the most part, RAID is intended to be an element of the overall system that is comprised of a host (typically a computer and system drive); an input/output engine (that is, a video encoder or decoder); and a storage array of some size significance.

To distinguish between different architectures of RAID, a numbering scheme was developed that was intended to identify the various combinations of drives, striping, and parity. In the early development of RAID, some expressed concern that the use of the term *RAID X*, sometimes referred to as *Level X* (where the X is generally a number from 0 to 5) was an attempt to capitalize on the incorrect notion that the higher the RAID level, the higher the reliability or performance.

RAID levels that are recognized by the original Berkeley papers and the RAID Advisory Board (RAB) have been categorized as RAID Level 1 to Level 6. Other levels recognized by the Berkeley papers but not by the RAB include Level 0 (disk striping), where data is mapped in stripes across the entire array.

A drive array cannot be created by simply connecting a series of SCSI drives to a single SCSI controller. Arrays generally consist of electronics that format, code, and distribute data in some structured form across all of the drives. A RAID system consists of a specialized set of electronics and instructions that operate in conjunction with the various drives to perform protective and fault-tolerance functions necessary to meet the level of RAID designated. Some video server manufacturers have developed their own RAID architecture in software, thus eliminating the dependence on a physical set of circuitry that perform the RAID operations.

Combinations of existing RAID levels, sometimes called hybrid RAID levels, include Level 53, a combination of Level 0 and Level 3 that combines disk striping and the features of Level 3 parity. Level 10, another combination, mixes disk striping and mirroring resulting in excellent I/O performance and data reliability. Some people suggested that this numbering system could lead to confusion among the end users or that it is contrary to generally accepted industry practice. The *RAIDbook* and the RAB suggested that, "RAID levels be chosen with numbers that impart meaning, not confusion." Regardless, the confusion surrounding RAID nomenclature continued.

By 1997, the RAB had recognized nine RAID implementation levels. Of the nine levels, five conformed to the original Berkeley RAID terms developed from the 1988 researcher's efforts, with four other RAID terms that are used and acknowledged by the RAB.

RAID 0, RAID 6, RAID 10, and RAID 53 were developed through committee work anchored by manufacturers, suppliers, and consumers. The technique (or techniques) used to provide redundancy in a RAID array is a primary differentiator between RAID levels. Redundancy is provided in most RAID levels through the use of mirroring or parity (which is implemented with striping).

RAID Level 0

Also referred to as *striping*, where all data is striped (that is, distributed randomly) across an array of disks without providing redundant information or parity. The striping of an array, although enhancing performance especially in high transfer rate environments, has one serious downside. Since there is no redundant or parity information recorded and there are no redundant disks, the failure of any one drive results in the complete loss of all data on the drive array.

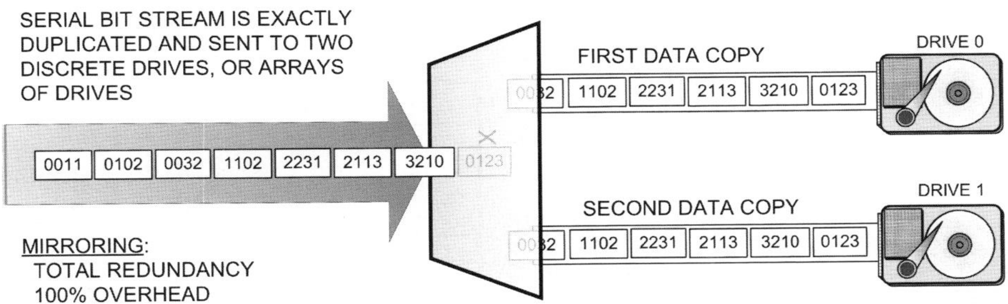

FIGURE 5.9-6 Bit mapping of data string to drive 0 and its "mirror" drive 1 is referred to as RAID 1 or mirroring. All data is duplicated for 100% redundancy. RAID 1 provides good reliability for data, does not require a complicated RAID controller, and is relatively inexpensive to implement.

The term RAID 0 is seldom used even though the principle is commonly practiced. In earlier times, RAID 0 referred to the absence of any array technology. The RAB states that the term implies data striping—a means to evenly distribute data so that during the read request, the blocks can be rapidly recovered at random with a minimum of latency.

RAID Level 1

Level 1 is the simplest, most reliable, and easiest of the RAID architectures to implement and understand. All data is continually duplicated and managed as a redundant copy, residing on at least one separate drive, as shown in Figure 5.9-6.

RAID 1 is referred to as disk-mirroring, shadowing, or duplexing. When data is stored on two or more separate drives, total redundancy is achieved. This approach further provides for a continual backup of all the data, thus improving reliability while providing a high level of availability, especially when reading data. If one disk component fails or is replaced, all of the information can be restored without any down time because it has already been backed up.

RAID 1 writes as fast as a single disk. The systems are typically single I/O and nonscalable, which means that performance is not increased by adding disk elements to the array. RAID 1 is not prominent in servers designed for video purposes only and the term *mirroring*, in the world of video, is typically accomplished by providing two complete and independent video server systems (including I/O, processing engines, controllers, and storage) that operate with an automation system that monitors the health of the overall system. These mirrored video server systems nearly always run in parallel, with a failure of any component in the primary system resulting in a switch over to the mirrored system until the primary system can be repaired.

For highly reliable systems, the RAID 1 is sometimes extended to DVD archives, data tape systems, and other large-scale spinning disk arrays that keep content online, near-line, or off-line.

RAID Level 2

Level 2 is the only RAID level that does not use one or more of the techniques of mirroring, striping, and/or parity. RAID 2 was developed for those early, very large diameter disks that needed to operate in 100% synchronization. In those early implementations, it was assumed that these large mass storage devices would be prone to constant disk errors, so RAID 2 employed error checking, such as Hamming codes, to produce higher data transfer rates. The technique used RAM to detect and correct errors.

RAID 2 interleaves bits or blocks of data, with the drives usually operating in parallel and employing a special form of striping with parity that is similar to error-correction code (ECC) coding, the process of scrambling data and recording redundant data onto the disk as it is recorded. This redundant information helps to detect and correct errors that may arise during read cycles. In a RAID 2 system, the drive spindles are fully synchronized. The system will use redundant disks to correct single bit errors and detect double bit errors. Error correction algorithms determined the number of disks required in the array.

RAID 2 was very expensive, and there were virtually no commercially available implementations of RAID 2. This method has all but evaporated from the traditional storage environments.

RAID Level 3

Level 3 is an array of N drives with bit or byte striping across all but one of the drives, as shown in Figure 5.9-7. A single dedicated drive contains redundancy data that is used to mathematically reconstruct the total data stream in the event any one member drive fails. The redundancy data is kept separately on a parity drive.

RAID 3 employs parallel data access and has the advantage of high data rates, which increases performance over single-disk structures. Each disk in the entire array is used for each portion of the read/write operation.

In a RAID 3 array, all of the disks are synchronized so that both read and write performances cannot be

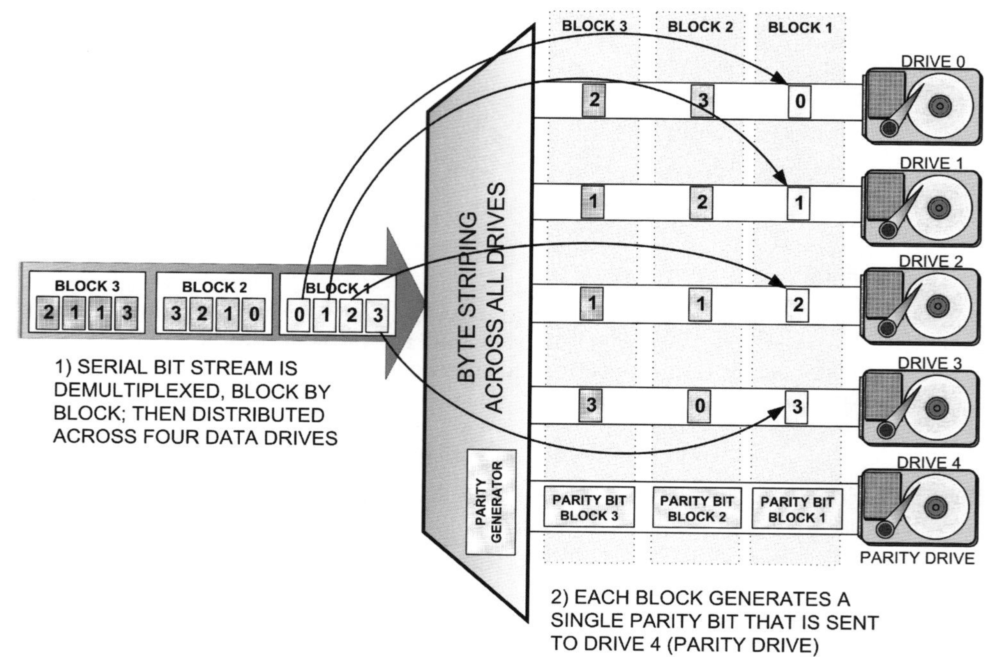

FIGURE 5.9-7 In RAID 3, data is distributed across all four primary drives. A parity bit is built from each block of data and stored on the fifth drive.

hampered. Each logical record from the data stream is broken up and interleaved between all of the data drives except the parity drive. Each time that new data is written to the data drives, a new parity is calculated and then rewritten to the parity drive.

RAID 3 is often a preferred choice when large blocks of sequential transfers are required. Applications in imaging, CAD/CAM, and digital video or media servers may select RAID 3 because of its ability to handle large data block transfers. In streaming video applications contiguous blocks of data are written to the array so that during playback only a minimal amount of searching is required, increasing efficiency and throughput.

One disadvantage that results, which is common to all striped arrays, is a poorer level of write performance when compared with single or duplex/mirrored drives, a drawback that can be controlled by proper buffering and sectoring during the write process.

By definition, the entire RAID 3 array can execute only one I/O request at a time, referred to as *single-threaded I/O*, which may or may not be important, depending on the application. Some controllers and smart arrays have minimized this impact by providing intelligent algorithms and larger disk caches to buffer data temporarily while being written to the drive. With the use of a discrete parity drive, if a drive goes down protection is temporarily lost until that drive is replaced and the parity information is reconstructed. Vendors configure mission-critical arrays with two parity drives as a second level of protection. The cost of this implementation must be weighed against the volume of data to be stored per unit dollar and physical space.

RAID Level 4

Level 4 is characterized by block-level striping with a dedicated parity drive. RAID 4 improves performance by striping data across many disks in blocks and provides fault tolerance through a dedicated parity disk.

The significant difference between RAID 4 and RAID 3 it that it uses blocks instead of bytes for striping. It is like RAID 5 except that RAID 4 uses dedicated parity instead of distributed parity. Block striping improves random access performance compared to RAID 3, but the dedicated parity disk remains a bottleneck, especially for random write performance.

RAID 4 lets the individual member disks work independently of one another. Benefits include good input and output performance for large data transfers and good read performance, fault tolerance, format efficiency, and other attributes similar to those found in RAID 3 and RAID 5.

As with most striping, write times are extended because the data is dispersed across several drives in segments.

With RAID 4, if a block on a disk goes bad, the parity disk can rebuild the data on that drive. If the file system is integrated with the RAID subsystem, then it knows where the blocks of data are placed and a management control scheme can be implemented such that parity is not written to the hot disk.

Another drawback to RAID 4 is extra steps become necessary to update check data and user data. Furthermore, if the parity disk fails, all data protection is lost until it is replaced and the parity drive data is rebuilt.

RAID 4 storage systems are uncommon in most storage architectures and almost never implemented in video server applications.

RAID Level 5

Level 5 is characterized by high transaction throughput with support for multiple concurrent accesses to data. RAID 5 is employed when independent data access is required. High read-to-write ratios are most suitable for RAID 5. Typically, the more disks in the array, the greater the availability of independent access.

RAID 5 uses block or record striping of the data stream, as shown in Figure 5.9-8. The serial digital data word is spread across all of the disks in the array in block form. Independent access is available because it is now possible to extract the entire specific data block from any one drive without necessarily accessing any other. Latency and seek times are effectively reduced, resulting in performance increases.

As each block of data is written to the array, a rotating parity block is calculated and inserted in the serial data stream. The parity block is interleaved throughout all the disks and can be recovered from any of the drives at any time. Because parity information is rotated over all of the drives in the array, the I/O bottleneck of accessing the single parity disk (as in RAID 3) when concurrent accesses are requested is significantly reduced.

When a single drive in a RAID 5 array fails, the read and write operations will continue. Recall that data is block striped over all of the drives, so if the data to be read resides on an operational drive, there are no problems, yet if data is to be written to the array, the controller simply inhibits writing to that particular failed drive.

If data resides on a failed drive, the read process uses the parity information interleaved on the remaining drives to reconstruct the missing data. The algorithms that determine where the data resides are quite sophisticated, statistically tailored to allow nearly transparent operations during failure modes. The host never sees the failed drive and the entire array remains transparent.

RAID Level 6

For Level 6, two sets of parity values are calculated for each parcel of information. Often a second parity drive is added to the structure of a RAID 5 array. The benefit is that two drives may fail without a loss of data, provided they are both not the parity drives. In the event one drive fails, the entire array can continue with a second level of protection while a replacement drive is found and installed.

With two parity drives, every write operation requires that two parity blocks be written, thus write performance is lower, yet the read performance of RAID 6 still keeps on a par with RAID 5.

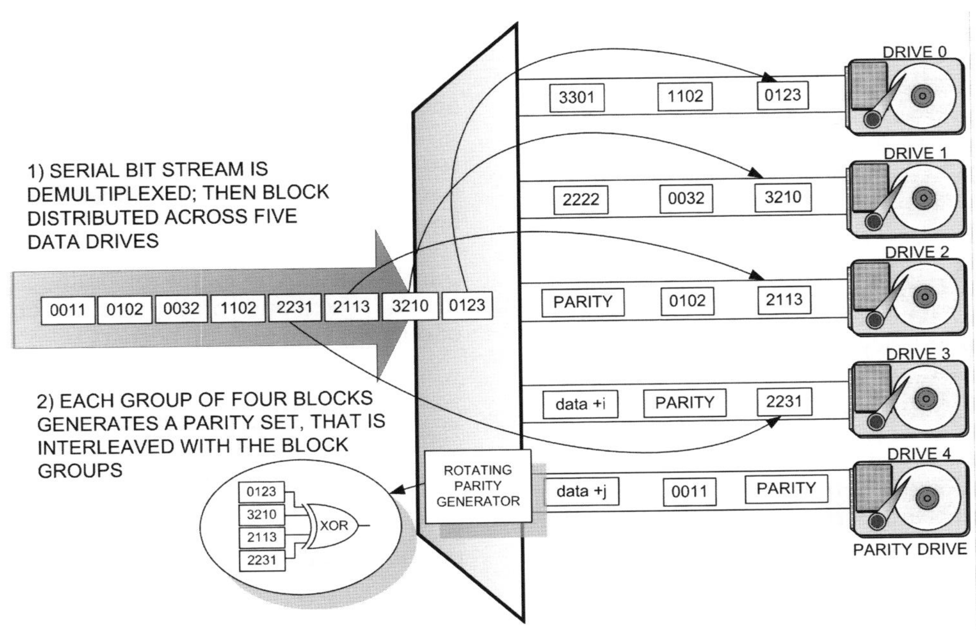

FIGURE 5.9-8 In a RAID 5 configuration, should any single drive fail, the read and write operations may continue. The parity bits generated are interleaved throughout the entire set of disks, allowing for continued operations and rapid recovery. RAID 5 (like RAID 3) is used in high-performance applications.

MIRRORING & STRIPING
RAID 0+1 (1+0, 10) HYBRID OF STRIPING AND MIRRORING
RAID 0+3 (35) STRIPED RAID 3 SEGMENTS
RAID 0+5 (5+0, 50) STRIPED RAID 5 SEGMENTS

FIGURE 5.9-9 Common nomenclatures for multiple RAID levels, generally employed for additional protection and drive redundancy. The numbers indicated in parentheses are alternative expressions for that particular RAID level.

Multiple RAID Levels

In some instances additional levels of RAID are layered in order to improve performance, access time, overall bandwidth, and data throughput. See Figure 5.9-9 for common nomenclatures for multiple RAID levels.

Mirroring

Mirroring is where multiple copies of a given RAID level are employed. RAID Levels 0+1 and 1+0 (i.e., RAID 10), represent a layering of the two elementary levels of RAID. In this example, RAID 10 then combined data striping (from RAID 0) and equal data splitting (RAID 1) across multiple drive sets. Duplexing is a variant of RAID 1 that includes mirroring of the disk controller as well as the disk.

Striping with Parity

Striping with parity is where single RAID Levels 2 through 7, and multiple RAID Levels 0+3 (e.g., RAID 53), 3+0, 0+5, and 5+0 use parity with striping for data redundancy.

Mirroring and Striping with Parity

When multiple RAID levels, such as 1+5 and 5+1, are employed for redundancy protection, it is both mirroring and striping with parity.

RAID Level 7

RAID Level 7 is not an open industry standard, but a trademarked marketing term from a computer storage vendor, used to describe the proprietary RAID design—described as a hybrid of RAID 3 with a write cache and RAID 4.

Hot Standby Drives

Some video server vendors will provide a second, hot standby disk drive that becomes active should more than one primary drive fail, or when certain other conditions prevail. This hot standby is not generally a part of the operating storage environment. Hot standby drives provide another level of protection, minimizing the requirement that a technician be immediately available to replace a failed drive. Servers located in remote environments or where not readily accessible may employ this additional protection.

STORAGE ALTERNATIVES

Deep archives are traditionally employed for storing massive amounts of inactive data. This data is usually kept on linear magnetic data tape or optical storage, such as DVD-RAM. Deep archives are used for off-site storage, backup protection, or disaster recovery.

When access to data is infrequent, yet that data may need to be readily accessible, in whole or in part, another method of near-line storage may be employed. A *massive array of idle disks* (MAID) is a near-line, short- or long-term storage technique, that retains accessible data on spinning disks but allows those disks to be placed into an idle state, either in groups of drives or individually.

MAID utilizes a massive number of inexpensive disk drives, arranged in a configuration that literally shuts down the spindles when the data on that drive isn't being accessed. Configuration and an intelligent, active file system management application will suspend only those drives that are not expected to be used for many hours or days. MAID powers down the head motor and parks the heads until the file system calls for a particular segment of data to be accessed. When access is requested, only those appropriate disks are restarted, and the data is transferred to another higher bandwidth store or sent by FTP to another server for appropriate use.

The value in MAID is that it increases longevity, reduces power consumption, and mitigates the negative issues associated with linear tape, robotic libraries, or optical drives. MAID will require a few more seconds to return to an active state, but the principle purpose of MAID is not for online activity; MAID is for fast access and rapid return to service once blocks of data or large files are needed.

FIBRE CHANNEL

Fibre Channel (FC), differentiated from "fiber" by its British spelling, is a standardized generic interface with multiple uses and support for many different protocols, such as:

- SCSI
- IPI-3 disk and tape
- Link encapsulation
- Internet protocol
- ATM

FC has become the de facto implementation standard for a majority of video server storage architectures in the professional and broadcast space.

Fibre Channel Loop

Initially, Fibre Channel storage arrays utilized the Fibre Channel Arbitrated Loop (FC-AL) principle, introduced in 1995 and employed by video server manufacturers around the 1996–1998 timeframe.

FC-AL was designed for new mass storage devices and other peripheral devices that require very high bandwidth. It uses loop architecture as opposed to the buslike standard SCSI or Intelligent Peripheral Interface (IPI). FC-AL is compatible with, and is promoted to eventually replace, SCSI for high-performance storage systems.

A Fibre Channel loop can have any combination of hosts and disks up to a maximum of 126 devices (implementation has shown that not more than 40 devices may be effectively used), with a data transfer rate of 100 MBps (1.062 GHz using an 8B/10B code). Fault tolerance was achieved by dual porting each drive, and with hot plugging capabilities.

FC-AL made it possible for FC drives to be used as a direct disk attachment interface, opening a new level of performance for high-throughput, performance-intensive systems. SCSI-3 defined the disk protocol, which is also technically referred to as the SCSI-FCP (Fibre Channel protocol) for FC-AL.

Fibre Channel has evolved to include both optical and electronic (nonoptical) implementations, with the ability to connect many devices to a host port, for example, disk drives, and storage arrays, in a relatively low-cost manner.

Manufacturers employing the dual-loop principles use dual-ported nodes and interfaces, with automatic failover, that protects the storage system should one direction on the loop open. In this mode, each node utilizes a second port that may be connected to an entirely electrically isolated and separate loop. Both loops are concurrently active and are designed for the automatic transfer of data traffic between them. Losing one port on the node does not cause a system failure.

FC Loop Resiliency Circuits

Loop resiliency circuits (LRC) are high-speed bypass circuits used in arbitrated loop hubs. LRC detects the presence or loss of the connected port, acting as a switcher by placing the port online, if active, and switching it out if inactive or nonresponding. This LRC system is designed to eliminate a total system crash in the event a node, port, or a loop failure occurs.

Fibre Channel Topology

Video server storage expanded geometrically once the marketplace recognized the value and reliability of server versus videotape technologies. This growth forced the requirements for multiple combinations of storage arrays, and for a topology that would be both resilient and reliable. FC-AL eventually became inadequate to deal with the growth of storage for video servers.

To facilitate the next generation of storage, an important key technology had to be employed. Fibre Channel Switch Topology (FC-SW) and the Fibre Channel Fabric were developed, and have essentially replaced FC-AL in today's advanced storage area network configurations. See Figure 5.9-10.

Fibre Channel connections are selected based on their application and the size or scale of the storage subsystem. Fabric Switching, utilizing a *node* interface, is the preferred choice for many large-scale video server storage subsystems because of its bandwidth and throughput properties. External Fibre Channel switches are necessary, adding to cost and configuration efforts. When only a single or small set of drives are necessary, an N_PORT will be used for what is considered a P2P (point-to-point) connection. Arbitrated Loop was commonly employed in the early development of Fibre Channel uses for video servers.

The Fibre Channel switch became one of the components in this new switched/fabric topology for these relatively new "networked-storage" applications.

Fibre Channel Switch

FC switches are responsible for passing Fibre Channel packets to the target port regardless of which Fibre Channel loop or switch the port physically resides on. Multiple switches may be connected to create large networks with up to 224 addressable ports. The Fibre Channel switch is a storage management device that allows the creation of a Fibre Channel fabric.

Fibre Channel Fabric

Defined as a network of Fibre Channel devices, a fabric utilizes Fibre Channel switches that enable many-to-many communications, device name lookup, security, and redundancy. The mesh like structure of a fabric consists of paired links of unidirectional transmissions in opposite directions between transmitters and receivers. The Fibre Channel fabric becomes that network of Fibre Channel devices, enabled by those FC-switches that communicate utilizing FC-SW topology. The FC-switches allow for multiple paths to interconnect many devices and provide for both redundancy as well as greater system bandwidth. Fibre Channel mesh networks are subdivided into zones, with each fabric having a name server and capabilities for providing additional services.

Fibre Channel Zoning

Only applicable to FC-SW, zoning is the partitioning of a fabric or SAN into subsets in order to restrict interference. Zoning is also a method in which to add security. The fabric can be configured with hard or soft zoning, and with two sets of attributes, name, and port.

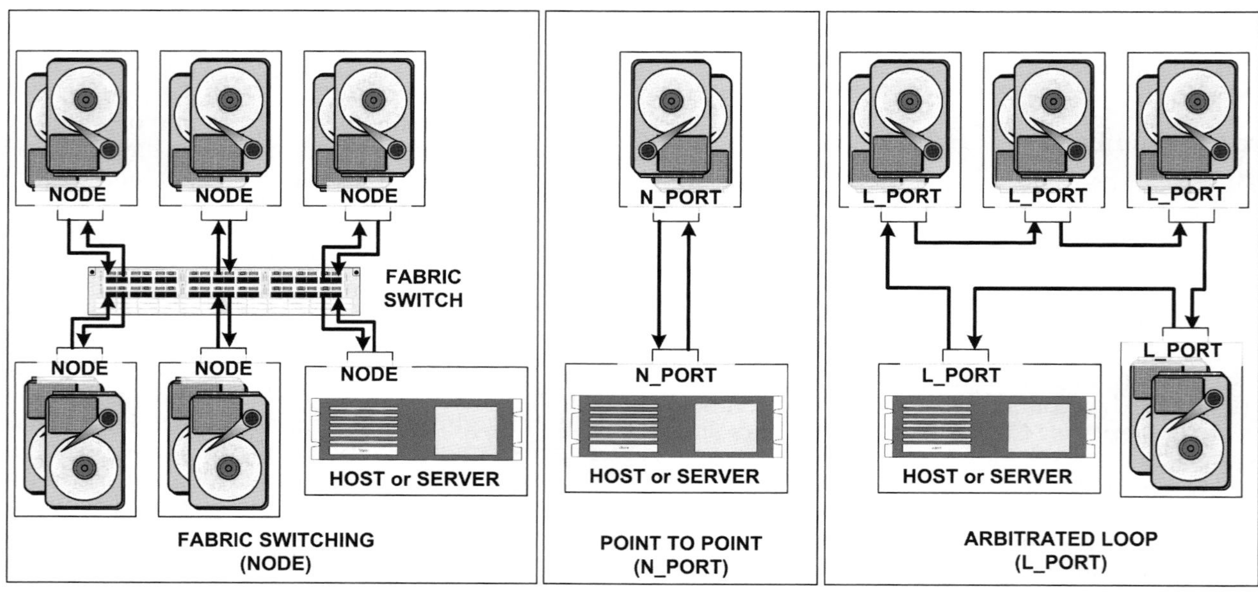

FIGURE 5.9-10 Three types of Fibre Channel connections.

Soft zoning restricts the name services of a device to reflect only those devices it is configured to see. Thus, when a server addresses the content of the fabric, it can only see the device it was configured to reach. Like the computer concept of security through obscurity, soft zoning still allows any server to attempt contact with any device on the network by address.

Contrarily, hard zoning restricts actual communication across a fabric, creating a much more secure environment by utilizing the expanded resources in the fabric switch. Zoning may be applied to either switch ports or end-station names. Here, port zoning restricts ports from communicating with unauthorized ports (and requires a heterogeneous SAN if it is to be used beyond a single switch). Name zoning restricts access by World Wide Name (WWN), the 64-bit address used in FC networks to uniquely identify each element in the network. While more flexible, the WWN of a device is a user-configurable parameter and can be spoofed, thus reducing security.

GROWTH IN VIDEO SERVER STORAGE

The wide availability of cost-effective disk storage continues to provide incentives for the deployment of video server platforms into new environments. Editorial applications, including collaborative postproduction, news clips, short-form content creation, nonlinear editing, and long-form content storage, are all candidates for the tapeless production environments shaping up industry wide.

Rich media server architectures for production activities need to include a set of network storage systems to satisfy the evolving workflow issues of continual content development and distribution. The volumes of media as data created in these newest working environments have nearly eliminated integrated and dedicated storage systems for broadcast news, on-air ingest or playout, and production-editorial server installations. Servers with specific dedicated functionality, such as just a VTR replacement or sports replay drives, are limited in their abilities to increase their I/O capabilities or their total storage. Expansion of these platforms necessitates adding another chassis in order to gain additional I/O ports. Storage could be expanded, but sharing that storage between multiple chassis becomes complicated and slow, often requiring an actual transfer of files from one chassis to another in order to play out the transferred content from a decoder. Over time, the dependence on FC-based transfers between the various sets of servers and their storage arrays moved storage management to a level of impracticality.

Once storage reached the prescribed capacity in terms of physical drives, an additional set of drives and/or chassis would be added, but eventually a limit for this type of expansion would be reached. The overtaxed server system would then need a major upgrade to a network-based storage topology.

Complex media server architectures are now configured with large sets of shared-storage arrays connected in a managed, network-like architecture, and serving multiple access points in an intelligent networked storage model.

ADVANCED, INTELLIGENT NETWORKED STORAGE

Early digital disk recorders and some basic video server deployments were configured with simple dedicated directed attached storage. When compressed motion imaging was limited to Motion-JPEG file for-

mats, the limitations on run times for these devices became directly proportionate to the physical storage capacity of the drives.

As multichannel video server products were introduced, their storage systems' capacities grew up to the limit of the RAID controller, which governed how many and which type of hard drives could be added to the storage subsystem. Video server performance was limited by the level of continuous I/O and data transfer activities, restricted by the bandwidth necessary for transfers between the encoders and storage (ingest), and the decoders and storage (playout). Any overhead activity required for operational purposes, including external file transfers to archive devices and to/from other servers' storage subsystems, was secondary to getting media into or out of the server.

At these early stages of deployment, video servers generally had just one or two inputs with one or two output complements available. As uses continued, more I/O was required, and more storage to support the additional I/O was required. To facilitate the sharing of the data across many I/O ports, the concept of a *centralized storage system* was recognized. Only a select few video server manufacturers fully embraced the concept of true centralized or shared storage. However, the growth of Fibre Channel technologies supported high bandwidth data transfers between multiple server chassis, storage arrays, and other devices. This permitted the desired shared storage and Fibre Channel, therefore, to grow into the predominant storage technology for video servers, where it remains today.

The size and types of media files continues to grow. High definition and compressed multiprogram transport bit streams, increased ancillary requirements for proxy generation, near-line archive management, and external transfers of data (via FTP) between systems all necessitate more advanced means of addressing storage and a continual movement toward higher bandwidth systems.

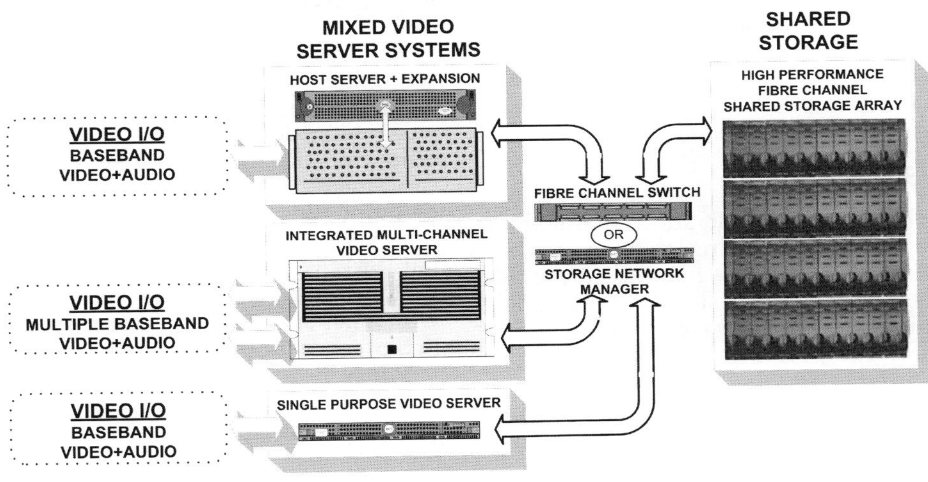

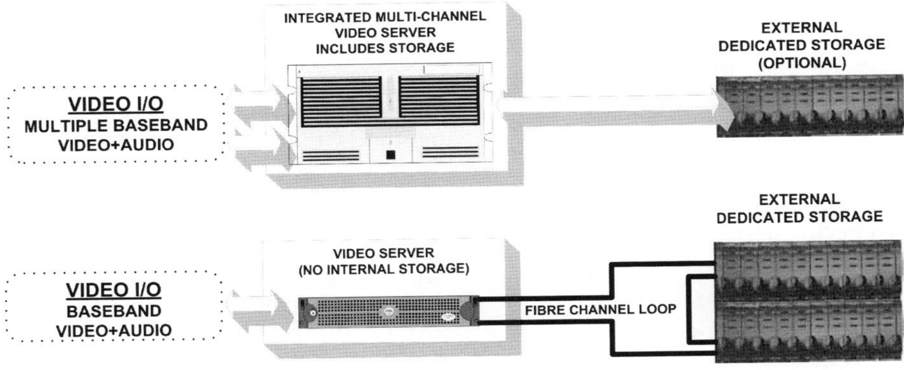

FIGURE 5.9-11 Traditional storage and the current network storage system models for broadcast operations.

Video media server systems must process massive amounts of contiguous data. To satisfy the management of that data, a higher degree of intelligence is required for the storage architecture. This intelligent storage must appear to function like the elements of a common network environment. Storage intelligence, as it applies to networked storage, grew out of the increasing requirements for high availability, security, and a measurable quality of service.

Network storage architectures, as shown in Figure 5.9-11, provide a path for disaster recovery and aid in the prevention of data loss. Network storage systems enable a means for consolidated backup and the archiving of data assets. Employing networked storage to a media server system offers that consolidation of storage, with improved capacity, utilization, and a unification of storage management.

Early video server systems either handled all the media storage in the same chassis as the codecs or were directly attached to storage consisting of a single or small group of arrays over an SCSI or FC interface. As server systems have grown to address the tapeless working environments of modern production, editorial systems, and play-to-air, a networking solution for storage was required.

The influence of network-centric systems has brought three terminologies to defining storage architecture: *direct-attached storage* (DAS), *network-attached storage* (NAS), and *storage area networks* (SAN).

Direct-Attached Storage

Direct-attached storage was the first, and remains the most common, approach for storing data on magnetic disk drives. Evolving from the five megabyte to present-day multigigabyte hard disk drives, DAS is the easiest and simplest of storage architectures still employed. Direct-attached storage describes a storage device that is directly attached to a host system. DAS devices housed in external chassis, such as the early SCSI-transportable storage devices, are still available.

DAS may utilize FC, IDE, ATA, and other forms of SCSI drives, but will be limited in capabilities and performance by the controllers that interface the bus to the drive. There are a finite number of devices that can usually be directly attached to the server or computer bus, and expansion is limited by such systems as the type of connector or cabling, the number of command channels available to the operating system, and the actual operating system itself.

DAS drive systems are generally found on smaller, dedicated, stand-alone computer platforms, such as a Windows or Mac PC, that do not require external storage subsystems or high performance, high bandwidth, or high throughput.

Network-Attached Storage

Network-attached storage detaches storage access and management from the primary server, as illustrated in Figure 5.9-12. By definition an NAS device is a server dedicated to file sharing. The NAS *head* consists of both the file system and the drives themselves. NAS is typically simple to implement, and is found deployed on various video server systems depending on the applications of that server—often a choice of the manufacturer and based on several factors including cost, throughput, performance, and bandwidth requirements.

Following a traditional client/server design, NAS is a group of one to many disk drives with its own network address. NAS removes storage management from the main processor's activities; thus, storage applications run independent of the main server, which results in both file management and run time applications that can be served faster. In an NAS environment, storage and computer processing no longer compete for the same processor resources.

Network-attached storage devices are attached to the local area network (LAN) typically via an Ethernet network. Requests for files are then mapped by the main server to and from the NAS file server.

The NAS does not need to physically reside with the server. NAS consists of hard disk storage, including multidisk RAID systems, and software for configuring and mapping file locations to the network-attached device. NAS allows for the incremental addition of storage without the requirement for shutdown or rebooting. Network-attached storage can be included or may be seen as part of a SAN.

An NAS device will generally contain its own operating and file system. Most NAS systems communicate over TCP/IP, with I/O requests support by common file sharing protocols, such as the Unix-based Sun Network File System (NFS), Samba, and the original IBM/Microsoft developed file-sharing support for DOS, formerly known as Server Message Block (SMB), which has since become Common Internet File System (CIFS) for Windows machines.

An NAS identifies data by file name and byte offsets, transfers file data or file metadata (i.e., the file's owner, permissions, creation dates, etc.), and handles security, user authentication, and provisions for file locking.

Storage Area Networks

The counterpart to network-attached storage is called a storage area network or sometimes simply a *storage network*. Storage networks are distinguished from other forms of network storage by the low-level access methods that they use. Most storage networks will communicate between servers and devices using SCSI protocol, although the low-level physical interfaces (e.g., the parallel cabling between drive devices) will not be used. SANs employ block storage, where it is the server that initiates a transfer request for specific blocks (referred to as data segments) from specific drives, addressing that data by disk block number, and then transfers raw disk blocks of data.

In a SAN, storage administration is simplified, adding flexibility because physical cables and storage devices are not moved one server to another. However,

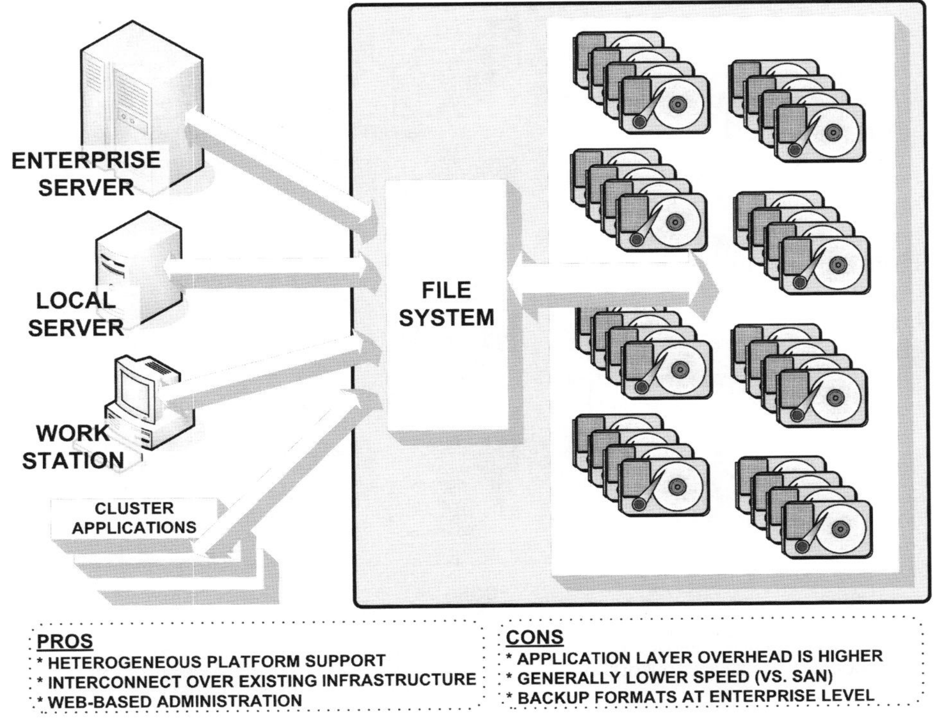

FIGURE 5.9-12 Network-attached storage systems utilize a shared file system.

with the exception of SAN file systems and clustered computing, each logical unit number (LUN) on the SAN is owned by a single computer host, called the initiator, and thus storage in a SAN is still on a one-to-one relationship. By contrast, NAS allows many computers to access the same set of files over a network.

SAN Types

SANs tend to provide faster and more reliable access than the higher level protocols employed in an NAS because the SAN will generally be built on a specifically designed infrastructure that handles storage communications only. Over its relatively short history, the SAN has undergone the highest degree of architectural change, from a mostly exclusive to a Fibre Channel system to a hybrid of iSCSI, Gigabit Ethernet, and more.

Fibre Channel networking is the most common SAN deployment technology, whereby a SAN would be comprised of a number of Fibre Channel switches connected in a fabric or network structure. A SAN may have one or more server hosts and one or more storage structures, as shown in Figure 5.9-13. Storage may be comprised of disk arrays, tape libraries, or optical-based storage devices. When large contiguous blocks of data must be transferred at high throughput data rates, such as in a video server, the SAN can provide greater bandwidth and thus exchange data between the store and the server with much improved performance.

Early in its development, the SAN developed a bad reputation, initially in part because Fibre Channel was rushed to market without thorough compatibility testing. The lack of compatibility between components, servers, operating systems, and storage arrays added to the cost of deployment, and became a significant factor as FC fabrics and associated drives were far more expensive than the Ethernet and conventional drive components of an NAS.

Interoperability problems surfaced even though basic storage protocols were formally standardized. Certain higher-level functions did not work with different devices even though they met current standards compliance. Most of these early incompatibility issues have since been resolved, but it is for these reasons that mission-critical professional video server manufacturers will seldom supply storage systems that have not been thoroughly qualified by themselves and, therefore, will only provide SAN (or NAS) storage directly.

A SAN includes a collection of control and management systems that collectively provide for the connections, data transfers, and other block-based services. In this context, services are defined as the input and output operations for data movement between servers and storage systems. Note that a SAN may provide file-based services, yet, for media-centric applications, SANs are generally configured to meet the demands of large continuous blocks of data that must move predictably and efficiently between I/O and storage.

In the 2003 timeframe, a nonfabric network SAN protocol referred to as iSCSI took the existing SCSI command sets and implemented them (typically over Ethernet) using TCP/IP protocol, thus creating a SAN

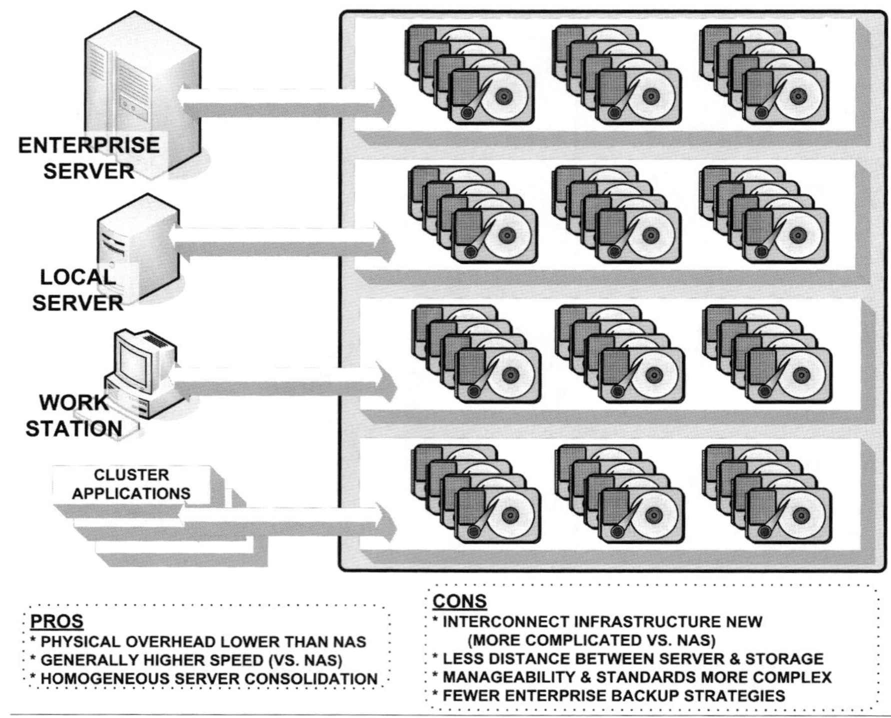

FIGURE 5.9-13 Storage area networks have all the disks shared across a network interface. They are typically used when large contiguous blocks of data (as in media content transfers) are required.

using Ethernet switches. iSCSI has only recently moved into the video server space, following its development in the data-server space.

Elements of a SAN

There are specific functional elements required in a SAN. The first is the device, which can be a collection of storage elements or storage systems. Storage devices may be a single JBOD (just a bunch of disks) array, a series of RAID chassis, or a massive array of FC drives arranged, for example, in a split-bus configuration that provides high system bandwidth and high availability storage. The device can also be an archive system driven by a gateway server that buffers data flow and interruptions from the tape drive mechanics while maintaining a high throughput, constant performance archive (e.g., a transfer to data tape), or restore (a transfer from data tape) operation.

The second element is connectivity, consisting of components such as routing, switching, the physical media (copper or optical cabling), and the appropriate protocols for the exchange and transport of data between those components. Connectivity interfaces are at the PHY (physical) layer and are administered through specific standardized protocols that provide for compatibility between media and the physical elements on the network.

The third element is control, the management of the data paths, transfers, resources associated with those devices (e.g., storage arrays), and the regulation of actual data within the SAN. For the control element, network management is the process by which a stable transport of data across a network's infrastructure is maintained. In the case of the SAN, control must be maintained such that peak limits are obfuscated, that server requests are handled according to their preset hierarchy for delivery, and that backup or protection paths are enabled and ready to take over when or if needed. Additional elements of control include volume management, data resource and data backup management, file access, and the reliable transport of data between storage elements and servers when called upon.

Comparing NAS and SAN

Traditional SANs differ from the NAS in several ways, even though on the surface they might appear quite similar. Both SANs and NAS generally employ RAID-protection schemes for storage, are connected in a networklike environment, and at a high level, can serve the same purposes, yet there are important differences that can affect the way data is utilized, as listed in Figure 5.9-14.

From a connection perspective, NAS employs TCP/IP networking, including Ethernet, FDDI, and ATM, while a SAN will employ Fibre Channel. The NAS protocol will use TCP/IP and NFS/CIFS/HTTP, with a SAN using encapsulated SCSI.

NAS	SAN
Almost any machine that can connect to the LAN (or is interconnected to the LAN through WAN) can use NFS, CIFS, or HTTP protocol to connect to a NAS and share files	Only server class devices with SCSI Fibre Channel can connect to the SAN. The Fibre Channel of the SAN has a limit of around 10 km at best
A NAS identifies data by file name and byte offsets, transfers data or file meta-data (e.g., owner, permissions, creation, description), and handles security, authentication, file locking	A SAN addresses data by disk block number and transfers raw disk blocks
A NAS allows greater sharing of information especially between disparate operating systems, such as UNIX and NT	File sharing is operating system dependent and does not exist in many operating systems
File system managed by NAS head	File system is managed by servers
Backups and mirrors are done on files, not blocks, for a savings in time and bandwidth. A snapshot can be very small compared to its source volume	Backups and mirrors require a block by block copy, even if the blocks are empty. An entire mirror machine must be equal to or greater in capacity compared to the source volume

FIGURE 5.9-14 Comparison of principal differences between NAS and SAN.

The lines between NAS and SAN are blurring, as evidenced by the crossover in transport, topologies, and even networks. SANs can now use Ethernet, NAS systems will sometimes use Fibre Channel, and NAS systems may now incorporate private networks with multiple endpoints. Still, the principal differentiator between NAS and SAN deployments depends most on the choice of the network protocol.

For example, a SAN system will transfer data over the network in the form of disk blocks (fixed-sized file segments, using low-level storage protocols like SCSI), whereas the NAS system operates at a higher level with the file itself.

STORAGE MANAGEMENT

Continued growth in media storage is demanding that storage itself be managed. Storage management is essentially a hierarchy of processes associated either with discrete applications or, on an overall sense, an integrated storage platform at an organization-wide perspective. The highest level of the hierarchy is the enterprise, ranging in size from the office or studio, to a campus, to an entire set of users and locations spread geographically over great distances. As one moves from the highest level of the storage management hierarchy to the lowest level, the physical model contains devices, such as the Gigabit interface connector (GBIC), hubs, switches, and even fabrics, as shown in Figure 5.9-15.

A principal benefit to a SAN is that it provides a much higher degree of control over the overall storage network environment. Applications in which SANs are deployed generally expect a high degree of availability, must be completely predictable in performance, and will not tolerate wide fluctuations in overall system performance. Fibre Channel is the connectivity component that aids in consistently meeting this level of performance, and is why most video server systems continue to employ it.

One of the values in the SAN approach is that it requires less processing overhead when interconnecting nodes among servers. A SAN performs best when data isn't broken into small segments, but is transferred in large blocks, thus making it ideal for media-centric operations. Furthermore, SANs operating over FC protocol are arguably most effective when delivering large bursts of block data and provide little benefit for occasional desktop work, such as word processing.

Media content delivery applications move data between storage and server in large contiguous chunks (that is, *blocks*). In live-to-air playout, as a file is called from storage to the server's decoder, data transfer must be steady and isochronous such that decoding results in a noninterrupted stream for a continuously moving image. A SAN's performance is tuned for this kind of delivery, balanced for the server's internal data plane, overall system bandwidth requirements, and prioritized to deal with delivery over a finite and usually predetermined or predictable time period. SANs further provide these high-availability applications with additional features, such as hot standby or protected switching, multiple server connections, and ease of scalability, because the barriers of managing direct-attached storage to servers are removed.

The design of a properly functioning SAN is no trivial task. For these reasons, the selection of the components necessary to make up a SAN is best left to the vendors of video server systems themselves. The qualification of switches, Gigabit interface converters, hubs, and storage arrays takes considerable time and effort. Most end users would find little reward for the effort it would require to select a proper mix of specific elements for the storage network and a wrong choice could pose considerable risk to the owner's investment for a small return in the cost/benefit ratio.

ENCODING AND DECODING

Motion image capture and playout devices, those outside the production level digital disk recorders, began as units that could string together a 30 (or 25) frame per second set of JPEG files into a coherent set of frames called a *clip*. As MPEG became an ISO/IEC standard, servers moved from the Motion-JPEG file format to the MPEG-2 Main Profile at Main Level (MP-ML) encoding and decoding platform. For standard ingest and

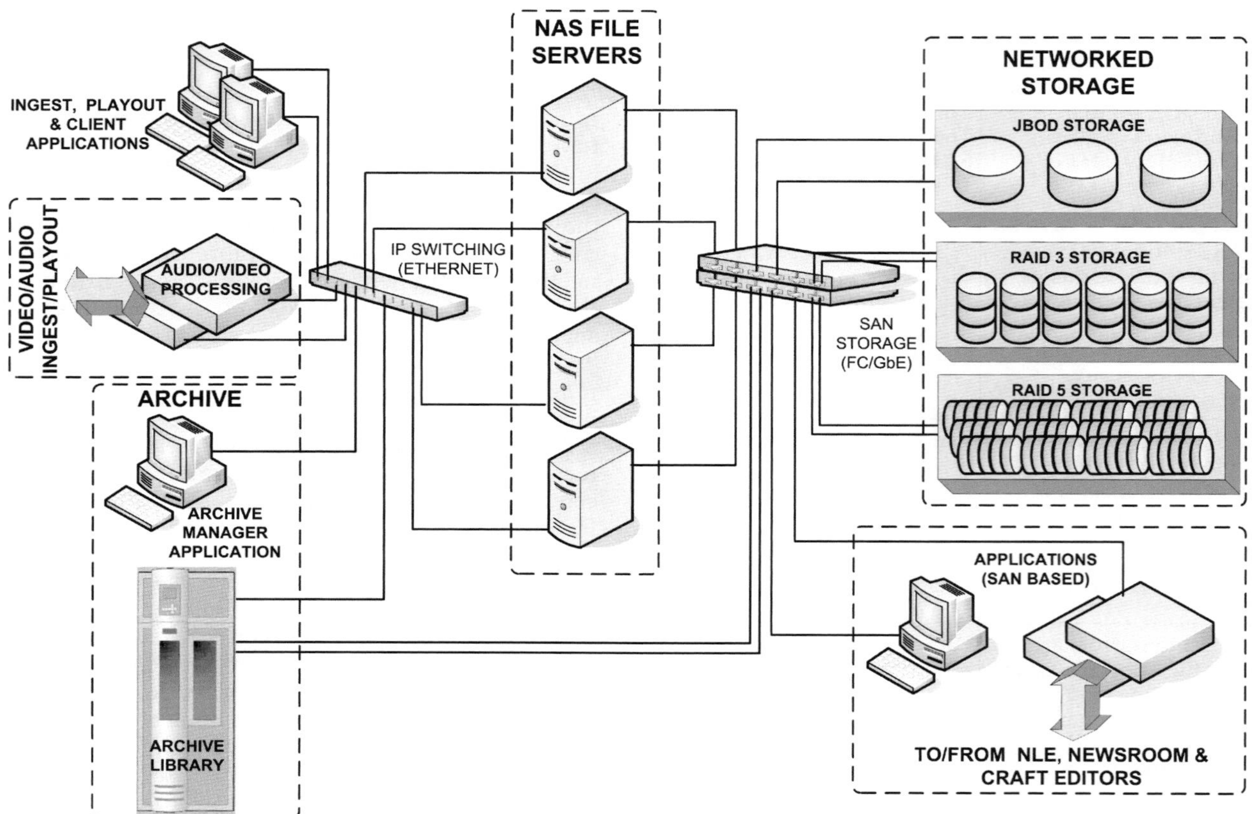

FIGURE 5.9-15 This diagram depicts a complete server, client/application, storage, and archive system that might typically be deployed throughout a broadcast facility. The architecture has a mix of both NAS and SAN systems, A/V-I/O, and serves as a central store comprised of individual (JBOD) stores, RAID 3 and 5 drives, and interfaces to various applications both as clients and as dedicated processing subsystems.

playout, MPEG-2 made more practical sense as it was not only a recognized standard, it consumed less physical storage space when compared to comparable image quality in Motion-JPEG.

A typical frame rate for standard definition, MPEG-2 peaked at 15 Mbps (long-GOP), yet many vendors benchmarked 8 Mbps as the reference level image resolution that was equivalent to Betacam-SP, the then predominant metric for image quality in a broadcast facility.

As the various flavors of MPEG began to mature, the professional MPEG-2 4:2:2 profile thrust the video server into the I-frame–only realm (for editing) and kept the long-GOP-IBP coding format for playout. Server bit rates next began a movement toward 25 and 50 Mbps, around the time that DVCPRO-25 became a viable format for professional video applications, both on videotape and in video servers.

With these encoding formats well established in the broadcast and professional marketplace, additional advancements have surfaced, including (and by no means limited to) MPEG-4 part 2 and part 10 (part 10 is now usually referred to as AVC or H.264), and other video-encoding platforms, including Windows Media 9 and its recently standardized counterpart, VC-1. While the later two principal encoding formats (AVC and VC-1) have not yet made their way with significant strides into the over-the-air terrestrial broadcast marketplace, AVC is certainly finding its way into cable, satellite, and video over Internet protocol and other content-delivery facilitators.

With the advent of solid-state–based image capture (e.g., Panasonic P2), the imminent standardization of other encoding and image storage formats, and the emergence of handheld and other mobile devices, this segment of the video server technology space will certainly remain the most fluid and dynamic. Heretofore, developed video serving architectures will soon need to provide a format-agnostic platform in order to deal with the ongoing evolution.

FILE FORMATS AND MATERIAL INTERCHANGE

No discussion of video servers would be complete without mention of file interchange and interoperability, although to thoroughly do justice to this topic would require considerably more space than is available for this chapter. The adoption of both MXF (with its many SMPTE standards) and its earlier predecessor, GXF (SMPTE 360), is changing the means by which recordings and files are stored and exchanged between serving platforms.

MXF, AAF, and GXF are the results of industry professionals and users in search of a common denominator by which essence can be configured or wrapped for interchange between comparable image storage, encoding, and decoding devices—principally for media-serving platforms. Significant work by vendors, manufacturers, users, and creators has been put into these standards, with the intention that it will tie together and provide a common link between files and formats that will survive the test of time.

The extensibility of MXF will be in what its users make of it. Certainly the flexibility and usability of the myriad hooks and placeholders for metadata, unique identifiers, descriptors and syntaxes, etc. are there so that a uniform structure for files can be created, maintained, and utilized. MXF is in its infancy and together with the many standards and engineering guidelines that SMPTE and others have developed, will find its way into all video-serving platforms in much the same fashion that MPEG has over the past half decade.

More information on MXF, AAF, and GXF and how they work is provided in Chapter 5.6.

CONTROL PROTOCOLS

As discussed in Chapter 5.13, broadcast facility automation systems provide a wide set of control mechanisms especially tailored to video servers. As both servers and automation systems have become critical entities in broadcast television facilities, a high degree of flexibility has ensued. Device controllers and server interfaces now incorporate the management of the content on the server, manipulate the locations of the data on the storage platforms, control the encoders and decoders during ingest and playout, handle the migration of data in native and standard file formats (MXF, GFX, and others), provide the tools for segmenting and timing of video clips, and a number of other features sets such as addition of metadata, wrapping of elementary essence to containers, and the extraction or importation of ancillary data such as closed captioning. While automation systems frequently interface to routing switchers, machine control systems, and signal-processing devices, their highest level of active intelligence occurs at the interface to the video server.

Servers can be controlled in a number of variant operational modes. In the simplest of functionality, a GPI may trigger a string of predetermined individual clips to begin playout. At the higher level, control of the server can include the control for disassembly of an MPEG multiple program transport stream (MPTS) into individual program streams, the disembedding and shuffling of audio tracks from the video stream, the reembedding of those audio tracks, and the cross converting from 720p to 1080i for playout. Most of the software intelligence for the sequencing of these functions stems from the video servers' architecture, yet the hooks between the server and the automation control systems are thought of as a joint effort between vendor camps. There are yet no general standards for the control of these processes, given that each vendor will generally devote significant engineering effort to creating those capabilities in a method that gives the server vendor a marketing advantage on price and performance.

Video Disk Communications Protocol

The most common of the control interfaces for video servers in professional broadcast television applications is the *video disk communications protocol* (VDCP) (also known as the Louth protocol after the company that developed it, since acquired by Harris). VDCP is a proprietary communications protocol that uses a tightly coupled master-slave methodology, where the controlling device takes the initiative in communications between the controlling device and the controlled device.

The VDCP topology is point to point, usually carried via a nine-pin D-subminiature connector utilizing EIA RS-422A protocol over a balanced full duplex link as an asynchronous bit serial, word serial 38.4 kbps data stream. VDCP conforms to the OSI (open system interconnection) reference model. A timeline command set has been included for systems that must use the protocol over a nondeterministic network environment. One RS-422 interface, per server channel, is generally required, although in some applications it is possible to change the functionality of that control port to address an encoder during one function and a decoder in another.

Network Device Control Protocol

Other, mostly proprietary, schemes for non-RS-422 interfaces have been proposed or employed by others. For example, in early 2001, Harris Corporation introduced its own *network device control protocol* (NDCP), which it aimed at becoming the successor to VDCP. NDCP took the dedicated, serial RS-422 concept and moved it to an open, network-based standard for control of audio/video devices. NDCP is built on TCP/IP using XML and, like VDCP, is an asynchronous, time-based protocol that uses a master-slave model. The model further allows for multiple devices to be linked via a network, and in turn can manage the transfer of media assets between devices. Other machine control protocol (e.g., Odetics, BVW-75, etc.) continue to be available as legacy interfaces for the numerous installed systems in the field.

Other Control Systems

As the applications for video server controls continue to expand, and the connectivity among systems grows more complex, a network-controlled environment

begins to make practical sense. With the additions of MOS protocol, high-demand FTP transfer activities between edge and catch servers, task-based workflow, and the hybridization of both baseband and compressed bit streams with those of MPEG-transport streams, the older legacy methods, such as VDCP, will become less important to the future of media content management, control, and store and forward activities. Newsroom computer and editing systems have become predominant influencers in this arena, constantly adding feature sets to online and off-line server applications, in both dedicated server/storage platforms and other large-scale server systems.

Advances at the newsroom production and on-air level have brought the *media object server* (MOS) protocol[2] into the server realm as an evolving communications protocol for newsroom computer systems (NCS) and media object servers. The binding intent of MOS is to spawn the development and implementation of a common communications protocol that will bridge the integration of diverse NCS and MOS equipment. The interaction between video servers and other broadcast devices is at MOS level, and will occur at the points in the workflow of ingesting, editing, storing, or broadcasting of those media objects. Included in the MOS set are video and audio servers, digital still stores, clip stores, and character generators. MOS has a widely accepted feature set for many server-to-news and server-based production systems, and may very well be on its way to becoming a de facto standard for these operations (see also Chapter 13).

Servers are also being controlled at a variety of different levels and through an equal variety of interfaces. At the desktop level, based on a distributed network model, multiple operators can be given access to server ports and streaming proxies without the need for a direct physical connection. Broadcast automation, newsroom, and content management products all contribute to the future growth and variances in server interfaces and controls, thus reducing the requirements for expensive refits and significant technology updates.

ANCILLARY VIDEO SERVERS

Ancillary video serving devices are taking on the appearance of "appliances," meaning that the size, scale, and complexities of those servers are configured for specific functions other than those typically found in the high-end, mission critical, and full-featured video servers found in pure on-air applications.

Catch Servers

Serving platforms now appear in a variety of formats, architectures, and purpose-built configurations. For example, when a server is dedicated to collecting data or files from a unicast or multicast distribution model, they are often referred to as *catch servers*, a smaller-scale, single-input (usually IP-based) device that is set to retrieve on a prescribed schedule-specific content for use by a specific contractual arrangement. Alternatively, a catch server might collect the daily commercial feeds, the weekly set of syndicated programs for air, or a set of preordered clips that were first reviewed by browsing on a *proxy* and then later delivered to the catch server during off hours or nonpeak periods.

[2]See http://mosprotocol.com

Edge Servers

Video servers that deliver content fed from a central serving platform to a local or distant server are generally referred to as *edge servers*. Often the edge server acts first as a catch server, then during a playout operation becomes an edge server. Edge servers typically have less storage and dedicated purposes, yet may also be part of a distributed broadcasting platform that sits at the ends of a hub-and-spoke centralized content delivery network.

Proxy Encoding Servers

Video servers that collect and make available for preview functions, media that was initially captured at a high-resolution level (e.g., MPEG-2, 15 Mbps) while at the same time, or shortly thereafter, those same images are also encoded to a lower bit rate (e.g., MPEG-1 at 1.5 Mbps or Windows Media at 350 kbps), are considered to be *proxy encoding servers*. Proxies permit desktop platforms to preview, scan, review, or approve the content without burdening the online full bandwidth servers employed in the mission-critical on-air environment. Typically the proxy server is a smaller form-factor storage and server platform, with its own database and its own set of disk drives, which can, and usually do, operate in an autonomous domain. They may be *clients* added to desktop PCs or dedicated library servers for metadata entry platforms.

Proxy servers offer feature sets that include the applications and interfaces for off-line editing, clip trimming of tails and tops to match-frame accuracy, and the conveyance of those decisions back to the primary server's database for playout under third-party automation interfaces. The proxy server is often where metadata is added, edited, or validated, once again without disturbing the original material essence on the primary server platform. Media asset management systems frequently make use of the proxy servers so that the primary media content is left undisturbed.

Transport Stream Servers

Transport stream servers are another form of server that extend the capabilities of ingest and playout, specific to MPEG or other transport stream files or formats. Typically an ASI server's input and output ports accept DVB-ASI signals. These input ports do not accept SMPTE 259 or SMPTE 292 baseband digital video, but are purpose-built to manage the data as a compressed video (e.g., MPEG-2) transport stream. Features in the transport stream server platform may selectively segment MPTS streams (or files) into one or more program streams, either during the ingest process or once stored on the server, for discrete program stream playout. Typical applications for ASI servers include digital turnarounds for satellite distribution, delay servers for time shifting (such as the high definition feed from a broadcast or cable network), catch servers for playout to an external decoder, or a storage system for externally encoded high definition program streams.

Some broadcast server manufacturers have built-in HD-encoding capabilities, some in software and some in silicon. Other manufacturers elected not to provide baseband high definition inputs. The latter determined that users may prefer to provide their own "best of breed" external encoders, typically with ASI outputs that are recorded on the video server as an MPEG transport stream, as an ASI stream, or in an MXF compliant wrapper. The most recent generation of video servers is providing internal HD decoders that can produce both an HD and a down-converted SD copy simultaneously during playout, thus retaining the original higher bit rate, HD resolution, and MPEG-encoded content on the server and/or on the archive.

Philosophically, down-conversion and cross-conversion functions when integrated onto video servers' decoders aid in the design and may further promote the advancement of an all HD broadcast infrastructure. These conversion concepts make the video server somewhat agnostic to formats and provide a high degree of flexibility and extensibility to the end user.

ARCHIVE FUNCTIONS

Throughout these many server topics and discussions, archive and content management has been frequently mentioned. The process of archiving valuable assets, such as media content, becomes a complex and site-specific and enterprise policy-based activity. Archives are not for every operation, and deciding on an archive involves assessment of the operation, workflow needs, market size, value of the assets, etc. Archives may consist only of backup videotapes or they could involve linear data tapes, optical and DVD-RAM, spinning magnetic disks (e.g., MAID), and even remote offsite storage delivered to mirrored servers via WAN.

Nearly all recent video server platforms provide the flexibility to transfer native files from the storage platform to another media. The methodologies surrounding those techniques and technologies are varied and complex, typically involving a sophisticated hierarchical media asset management (MAM) or digital asset management (DAM) application. Independent solution providers often team up with automation vendors, server vendors, and/or archive medium (such as robotic tape systems) to select a comprehensive set of components—both hardware and software—that are end user– and workflow-specific applications.

The products that interface between the server/automation solutions and the physical robotic disk or tape libraries are referred to as *archive managers*. As mentioned, these archive managers are tightly integrated into each of the components in the system. They become the handlers of the assets, making certain the assets are cataloged, protected, secured, backed up, and ready for use with sufficient notice that the content appears always available.

SUMMARY

In the past the ability to replicate a video recording in a nonlinear way was a luxury afforded only by the largest program distributors. What grew from an early sub-one-minute, instant-replay, disk-based recorder into modern-day server systems has generated yet another paradigm shift in how motion images for broadcast and other purposes are collected, manipulated, and distributed. Video server technology is the driving force that has permanently changed video storage, ranging from the personal video recorder (PVR), now in widespread use in the home, to the myriad professional/storage delivery systems throughout the world.

Reference

[1] Patterson, D., Katz, R., and Gibson, G. "A Case for Redundant Arrays of Inexpensive Disks," *Proceedings of the Association for Computing Machinery*, SIGMOD International Conference on Management of Data, 1988.

Bibliography

Benner, A. *Fibre Channel–Gigabit Communications and I/O for Computer Networks*. New York: McGraw-Hill, 1996.

Kovalick, A. *Video Systems in an IT Environment*. Boston: Focal Press, 2006.

Wells, N., Devlin, B., and Wilkinson, J. *The MXF Book: Introduction to Material eXchange Format*. Boston: Focal Press, 2006.

Paulsen, K. *Video and Media Servers: Technology and Applications*, 2nd ed. Boston: Focal Press, 2001.

Paulsen, K. Media Server Technologies, monthly column, *TV Technology*, Falls Church, VA: IMAS Publishing, various 1995–2006.

CHAPTER

5.10

Video Production Switchers and Special Effects Systems

JEFF MOORE*

Ross Video
Iroquois, Ontario, Canada

JEFF MAZUR**

Disney-ABC Television Group
Los Angeles, California

INTRODUCTION

This chapter covers the video production switcher (or vision mixer as it is called in Europe), which is used to mix between and layer video sources during a live production. It includes a description of the video effects that are provided by the switcher and also covers new developments in production automation systems. The second part of the chapter provides more details on special effects techniques, with an emphasis on systems used in the postproduction process for recorded programs.

Production switchers are widely used in television studios and production facilities to mix video for newscasts, dramatic productions, game shows, talk shows, and live comedy programs. They are installed in television remote trucks and used for every type of production from news gathering through sports to major features and special events. They are also used in a variety of nonbroadcast applications.

One of the fastest growing fields in entertainment production is special visual effects. Every movie has them, and the proliferation of music videos, commercials, and even video games has created an insatiable need for newer and better effects. Once used only for truly *special* effects, or impossible live shots such as magic or sci-fi environments, they are now routinely used to make subtle changes to brand names and lighting, and to create realistic scenes that might be mistaken for live action.

PRODUCTION SWITCHER CONFIGURATIONS AND APPLICATIONS

Figure 5.10-1 shows a typical switcher installation in a television station production control room.

Due to the demanding nature of live production, production switchers have evolved into complex devices with numerous buttons, knobs, fader handles, and usually a system menu screen and a 3-axis joystick. This control panel is the tactile user interface that a technical director (TD) manipulates.

The control panel is typically separate from the processing electronics with commands and status passed back and forth on a high-speed serial or network link. The processing electronics contain dedicated hardware capable of performing the video processing functions in real time. This could be as simple as selecting a different camera to the program output or as sophisticated as running a complex timeline involving digital video effects (DVEs) and device control.

Many of the functions of a production switcher can now be emulated in software using computer-based systems and these have become ubiquitous in postproduction. In the live production environment, however, switcher control panels provide the optimum interface. There are some variations between manufacturers but the basic design features of most switchers are similar to those described here.

All production switchers have a number of video inputs that are mixed down to a single program

*Jeff Moore wrote the first part of this chapter, on video production switchers.
**Jeff Mazur wrote the second part of this chapter, on special effects.

FIGURE 5.10-1 Production control room at WXXI, Rochester New York.

stream output (although there can also be split feeds produced at the same time on the same production switcher—more on that later). They usually include a number of video outputs for preview functions and auxiliary outputs to provide separate feeds for recording, monitors, and return video to cameras.

These systems come in a range of sizes based on the number of inputs and the number of Mix-Layer-Effects (MLE) units, otherwise known simply as Mix-Effects (ME). Each MLE is a discrete layering engine allowing mixing and keying, and producing a composite output. The more MLEs, the more layers of mixing and keying available, and the more complex productions that are possible. Figure 5.10-3 shows a 32 input, 4 MLE switcher capable of handling the most complex of productions. Similar units but with 2 or 3 MLEs are more common in news rooms and mid-sized mobile trucks. Figure 5.10-4 shows a smaller switcher with 16 inputs and 1 MLE, suitable for small scale production use, while Figure 5.10-5 shows a more compact 1 MLE units with 8 inputs.

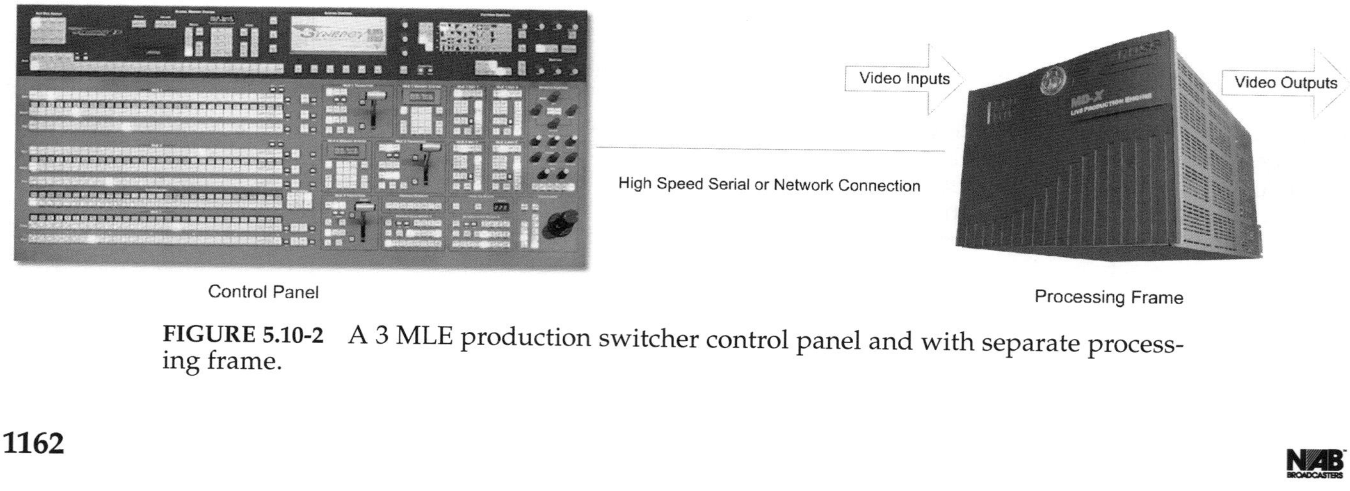

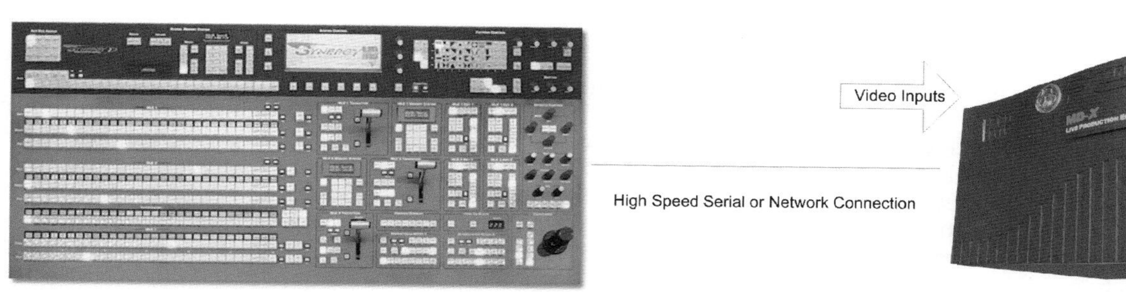

FIGURE 5.10-2 A 3 MLE production switcher control panel and with separate processing frame.

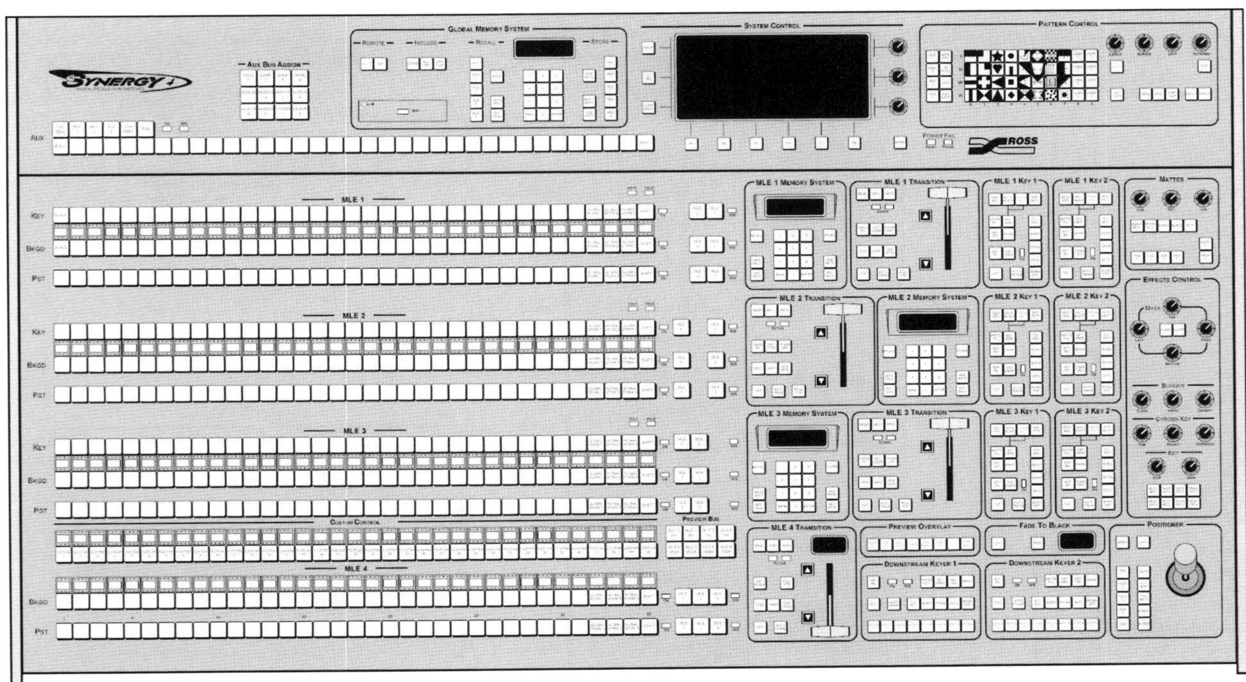

FIGURE 5.10-3 A 4 MLE production switcher control panel, typically found in large studio control rooms and large mobile trucks.

SWITCHER SYSTEM ARCHITECTURE AND CAPABILITIES

The system architecture of a large production switcher is illustrated in the block diagram shown in Figure 5.10-6. The functions of the main elements are described in the sections following. It should be noted that the basic production switcher functions were first implemented in analog systems, which are still available today. These functions are also implemented using digital processing in modern production switcher designs that also include enhanced capabilities. The theory of the analog and digital signal processing involved is beyond the scope of this chapter and the reader is referred to other literature listed in the bibliography.

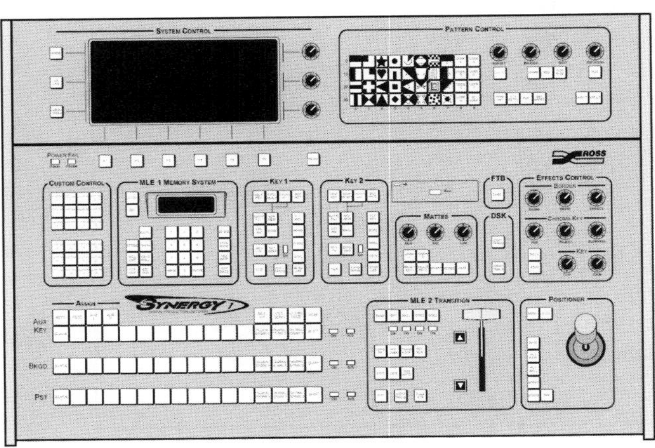

FIGURE 5.10-4 A full-sized 1 MLE production switcher control panel, typically found in small controls rooms and linear edit suites.

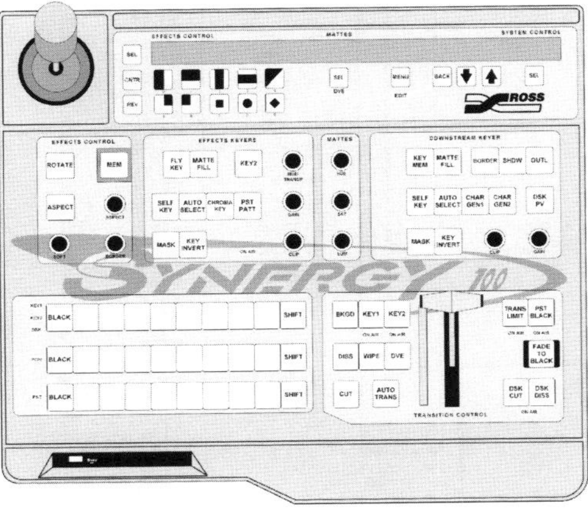

FIGURE 5.10-5 A compact 8 input, 1 MLE production switcher control panel, typically found in small control rooms with fewer sources, small ENG and SNG trucks, and linear edit suites where space is limited.

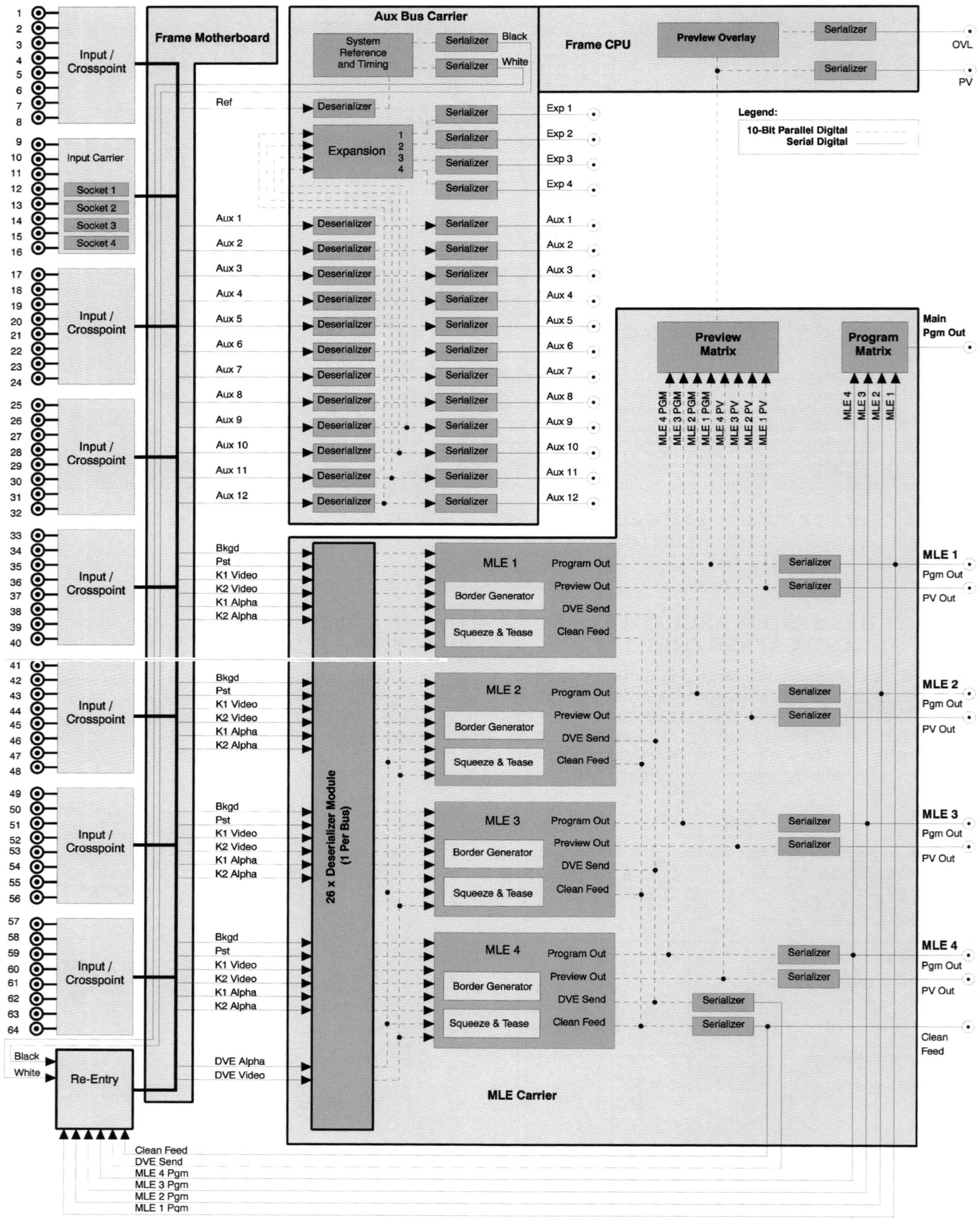

Input / Crosspoint (1–8)
Input Carrier (9–16): Socket 1, Socket 2, Socket 3, Socket 4
Input / Crosspoint (17–24)
Input / Crosspoint (25–32)
Input / Crosspoint (33–40)
Input / Crosspoint (41–48)
Input / Crosspoint (49–56)
Input / Crosspoint (57–64)
Black
White
Re-Entry
Clean Feed
DVE Send
MLE 4 Pgm
MLE 3 Pgm
MLE 2 Pgm
MLE 1 Pgm

Frame Motherboard

Aux Bus Carrier
System Reference and Timing
Serializer Black
Serializer White
Ref
Deserializer
Expansion 1 2 3 4
Serializer Exp 1
Serializer Exp 2
Serializer Exp 3
Serializer Exp 4
Aux 1 … Aux 12: Deserializer → Serializer → Aux 1 … Aux 12

Frame CPU
Preview Overlay
Serializer OVL
Serializer PV

Legend:
10-Bit Parallel Digital
Serial Digital

Bkgd
Pst
K1 Video
K2 Video
K1 Alpha
K2 Alpha
(for each of MLE 1–4)
DVE Alpha
DVE Video

26 x Deserializer Module (1 Per Bus)

MLE 1 / MLE 2 / MLE 3 / MLE 4
Border Generator
Squeeze & Tease
Program Out
Preview Out
DVE Send
Clean Feed

Preview Matrix
MLE 4 PGM, MLE 3 PGM, MLE 2 PGM, MLE 1 PGM, MLE 4 PV, MLE 3 PV, MLE 2 PV, MLE 1 PV
Program Matrix
MLE 4, MLE 3, MLE 2, MLE 1
Main Pgm Out

Serializer
MLE 1 Pgm Out, PV Out
MLE 2 Pgm Out, PV Out
MLE 3 Pgm Out, PV Out
MLE 4 Pgm Out, PV Out
Clean Feed

MLE Carrier

FIGURE 5.10-6 Production switcher block diagram.

Inputs and Autotiming

Production switchers have a number of inputs ranging from 4 or 8 inputs on small, single MLE units up to 96 inputs on the larger 3 or 4 MLE units. The more inputs available, the more direct video sources are available to the TD in putting together a production without having to access an external router. One might think that 96 is an overly large number of inputs; however, in sports like golf or car racing, these inputs can become occupied quickly, as there can be up to 20 or 30 camera sources alone.

Digital production switchers have auto timing inputs typically in the +/–1/4 to +/–1/2 horizontal line range (see Figure 5.10-7). This eliminates the need for fine timing adjustments such as cutting video cables to a precise length, as was necessary with analog production switchers. Sources must still be timed to coincide with the input timing window of the production switcher. On some production switchers, synchronous out-of-time sources can still be used, as the timing circuits simply drop the image down the screen by one or more horizontal lines, depending on how far out of time the source is.

When a video system is designed, a timing diagram for the facility should be prepared that shows the timing relationships between the sources and destinations to ensure that sources can be properly timed to all the various destinations they are required (Figure 5.10-7). This is especially important when dealing with hybrid analog/digital plants in which sources are encoded and decoded. A quality decoding process has a processing delay of a little more than one horizontal line, so if the source is required in both analog and digital destinations that are timing sensitive, this delay must be taken into account.

In addition to the external sources feeding the production switcher, a number of sources can be generated internally, including black, color bars, color background, and internal still and clip stores. Inputs to the switcher can be assigned to the buttons on the control panel in the order of preferences of the TD.

MLE or ME

Production switchers are referred to by size in terms of the number of effects banks they are equipped with; for example, "I have a 3 MLE production switcher in my main control room and a 1 MLE production switcher in my linear edit suite."

Each MLE is a self-contained effects bank that has the capability to mix and layer sources and, on some production switchers, access digital video effects units (DVEs), as shown in Figure 5.10-8. An MLE is made up of the areas described in the following sections.

Crosspoint Area (Source Buttons)

The crosspoint area is the place where the source is selected and contains Program, Preset, and Key selection button rows. The Program row is where the background source is selected; for example, if Camera 1 is wanted at the output of the MLE, Camera 1 should be selected on the Program row. The Preset row is the bottom row of buttons and is the place where the next background is selected. When a background transition

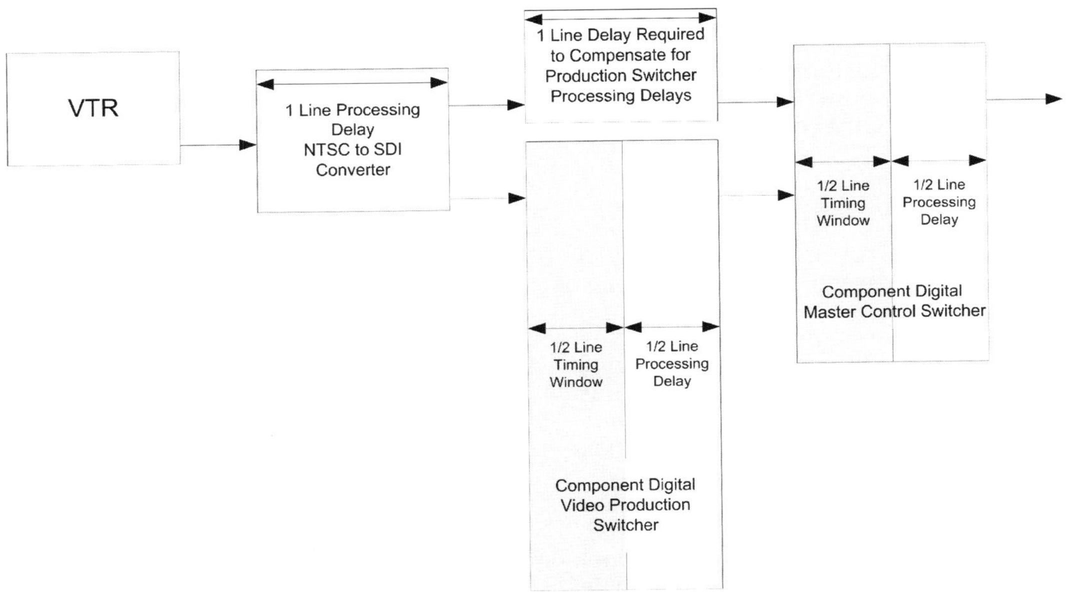

FIGURE 5.10-7 Simplified timing diagram example showing timing relationships from an analog source through digital production switcher and component digital master control.

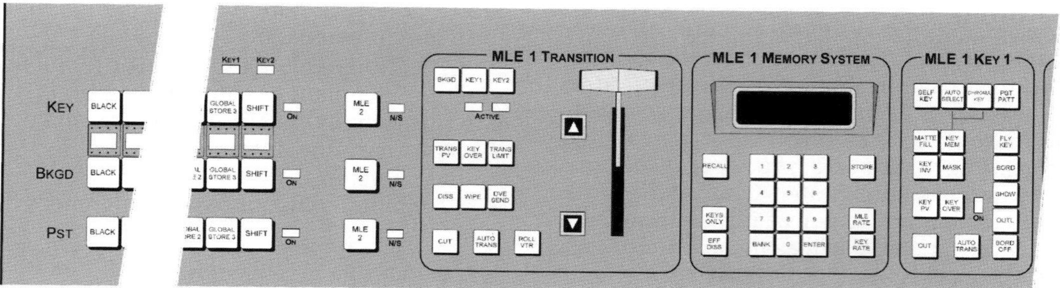

FIGURE 5.10-8 MLE bank showing crosspoint area (left), transition area with fader handle, memory system area, and keying area (right).

is performed, selecting a source on Preset—for example, Camera 2—allows the TD to see the source on the preview monitor for that MLE and then mix between the Program Source (Camera 1) and the Preset Source (Camera 2) by moving the fader handle or pressing the Auto Transition Button.

Above the Program row is the Key row. This row of buttons is the place where the key sources are selected. The Key row is either a single or dual row of buttons. In the case of a single row, this button row is shared between keyers but controls two separate pairs of crosspoint busses. Each pair consists of a Video Fill Source and a Video Key Source (or Alpha Channel). The button automatically follows the selected keyer, allowing the TD to route the desired key source pair to the keyer for processing.

Source buttons are labeled with the source names by inserting a printed Mylar insert under the button cap. Alternatively, many production switchers offer LCD mnemonic displays that are installed in parallel with the source button rows and offer a corresponding mnemonic display to each button. This eliminates the need for insert labels and increases the flexibility of the system, as different productions can have different source-to-button mapping with the mnemonic displays automatically showing the correct source names.

Transition Area

The transition area is the area that is used to set up and initiate transitions between video sources and bring keys on and off (see the following section on keying). The transition area consists of the "Next Transition Buttons" labeled BKGD, Key1, and Key 2. These buttons are the "what's going to change buttons?" and are pushed individually or in combination to select which layers will be affected by the next transition.

A fundamental group of buttons is the "Transition-Type" buttons or "how is it going to change buttons?" These buttons determine what the transition will look like when it happens; with the choices being dissolve, wipe (geometric pattern transition), or DVE transition (page turn, push, etc). A dissolve changes gradually from one picture to the other by mixing the two signals together and changing the proportion of each as the transition proceeds. A wipe uses a dynamic geometric pattern as the boundary between the two sources, with patterns ranging from simple vertical or horizontal lines (giving a split screen effect) to complex patterns.[1] The different types of DVE effects are covered later in the chapter.

The transition is actioned by moving the fader handle or pressing the Auto Transition Button, which triggers the transition at a predetermined rate. Alternatively, pressing the Cut button switches instantly between the layered program output of the MLE and the layered preview output of the MLE.

Keying Area

Keying is the process of layering text, graphics, or other video over a background source. The keying process is akin to cutting a hole in the background and dropping the new video into that hole. Thus, keys have both a video fill source as well as a key signal that cuts the hole, as mentioned earlier in the "Crosspoint Area" section. The primary key types are described in the following sections.

Self Key

A self key is a key in which a video source (selected on the key bus) is used both to cut the hole and fill the hole. As an example, a video camera can be used to shoot a graphic, which is then used as a key source. The keyer utilizes the luminance (brightness) value to key. Clip and Gain controls are used to set the luminance value at which the key "cuts the hole."

Auto Select or Linear Key

An Auto Select Key is a key in which there are separate externally generated video fill and key sources. These sources are mapped together when the production switcher is configured and are automatically selected together, feeding the keyer when the source is selected on a key bus. This is the most common type of

[1]Illustrations of standard wipe patterns, with assigned code numbers, may be found at: http://www.w3c.rl.ac.uk/pasttalks/slidemaker/XML_Multimedia/htmls/transitions.html and in SMPTE 258M (see bibliography).

FIGURE 5.10-9 Auto select key layered over a background camera source.

FIGURE 5.10-10 Chroma key green screen camera image.

key used with graphics equipment. Most character generators and still stores have both video fill and key (or alpha) outputs. The alpha output is generated to give the key a nice clean edge and also supports semi-transparent keys and keys with gradient transparencies. An example of auto select keying is shown in Figure 5.10-9.

Chroma Key

Chroma keys are keys in which a specific color or color range in the video signal are used to "cut the hole in the background." Chroma keys are commonly used in keying a weather person over a weather map and are also used in virtual sets and other applications. The most common colors used in a chroma key background are green and blue. Figure 5.10-10 shows an example of a subject with a green chroma key background. Figure 5.10-11 shows the same person with a world map keyed behind the subject.

Lighting is very important in establishing a good chroma key since the quality of the color backing determines the quality of the key that will be produced by the keyer. Having a high saturation and low signal-to-noise ratio in the color backing will make it a lot easier to produce a good key. Lighting should also be set to minimize backing spill onto the subject as any spill of the backing color onto the subject will result in that area being keyed out.

Camera detail circuits should be turned off on the camera used for chroma keying, as the pre-emphasis that detail circuits add on the edges of dark-to-light and light-to-dark transitions can place an unwanted edge around the chroma key subject.

Preset Pattern Key

Preset Pattern Keys use the production switcher's internal pattern generator to generate a geometric shape that is used to generate the key signal. An example of its use is on split screens in which the video is split between the key and background sources with a vertical line.

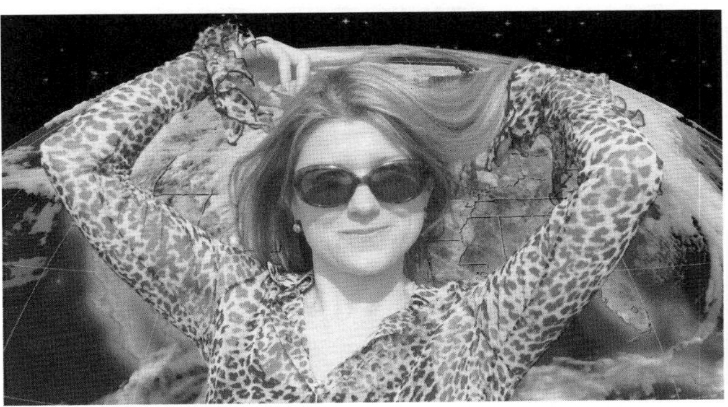

FIGURE 5.10-11 Chroma key composite image with replacement background source.

Digital Video Effects

Real-time digital video effects, or DVEs, were first introduced in the early 1980s. Initially, these devices were large units with separate control panels that were used in conjunction with production switchers either as a background source or key source. As time went by, the integration between the production switcher and DVE improved. Effects send outputs, from the switcher MLE to the DVE, and control over the DVE from the switcher fader handle now make external DVE transitions work like a wipe effect. Today, the DVE is inside the production switcher and is fully integrated into the unit. DVE effects fall into two broad categories: planar effects and warp (or curvilinear) effects.

Planar effects include the ability to size and position an image and fly it in either 2 or 3 dimensional space. They are used to produce multiple on screen "boxes" (see Figure 5.10-12), size and reposition keys, as well as many more creative effects. Most modern DVEs provide 3D space manipulations that keep track of the images being flown. These provide a much more sophisticated look as the subtleties of perspective, lighting, shadows, motion paths, which image is

FIGURE 5.10-12 DVE two-box planar effect with lighting and picture frame.

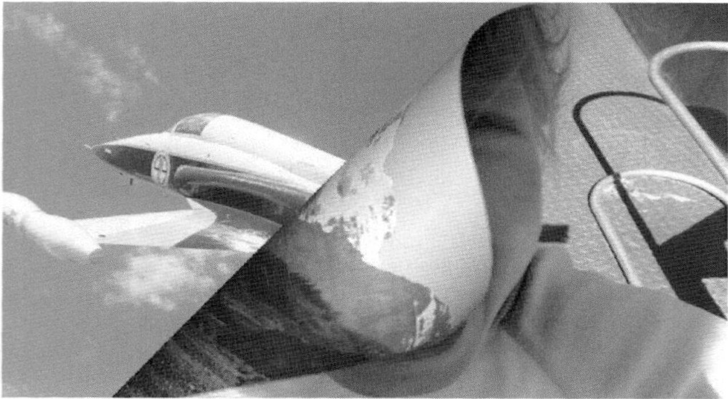

FIGURE 5.10-13 DVE single-channel warp effect—page turn with lighting.

in front of another, and so on are all tracked in the DVE and present in the output image.

Warp effects describe the ability to bend, break apart or otherwise distort an image for creative effect. Warp effects include things like page turn, stretch, split, slats, sphere, magnify, swirl, and so on. Warp effects are often used for fancy transitions, for example from one news story item to the next (see Figure 5.10-13).

Key combiners are the part of the DVE system that allows the TD to combine multiple DVE channels together inside the effects unit. The combiner puts two or more channels together and allows them to be flown in one combined 3D space. This feature is used to create and manipulate slabs, cubes and other multichannel objects.

DVEs are usually provided as a single channel or as a channel pair. A single channel can fly a single video image for an effect like an over the shoulder box. Two channels are required to put up a two-box interview. Two channels are also required to fly an auto select key as the video and key channel must both be flown simultaneously with the same translation being applied to both. More information on how DVEs work is provided in the special effects section later in the chapter.

Working with Multiple MLEs

Production switchers contain one or more MLEs. Each MLE is a self-contained layering device. On larger mixers, additional MLEs can be used by the TD to recall effects that are coming up next in a production. Then the TD will transition to that MLE in the Program/Preset area to bring it to air. Often the TD will mix between MLEs, for example, one MLE with the Weather set chroma key and another one with the sports DVE setup. More MLEs give the TD more power and flexibility in both building effects and recalling effects during a live production.

Additional MLEs can also be used for adding layers. Production switchers allow MLEs to be re-entered from one to another cascading through the production switcher, adding layers along the way. Every MLE is a self-contained layering device whose output is a composited series of layers, Background +Key1+Key2. This composite can then be re-entered into another MLE on any of the background, preset, or key busses. An example of this use would be to produce a chroma key of the weather set on one MLE and then squeeze it into an over-the-shoulder box on another MLE.

Downstream Keying

Downstream Keys, or DSKs, are keyers that are available at the output of the production switcher in the Program/Preset area. These keyers are typically used for adding titles, logos, time clocks, and other text information. The name *Downstream Key* originated because the keying device is downstream in the signal chain from the main MLE units and is separately controllable apart from the transition in the Program/Preset Area. In modern production switchers, the DSKs can typically be transitioned in both the P/P area as well as through separate, dedicated DSK transition buttons.

Memories, Macros, and Timelines

All modern production switchers incorporate a built-in memory system to allow previously built effects to quickly and easily be recalled. The memories can recall static settings of an MLE, combination of MLEs, or the entire switcher. The TD will typically have all of the effects required in a production setup in memories and simply recall them during a show.

Macros are buttons that can be used to record and then recall a series of keystrokes. The macro can be used to create a button to do something very simple or quite complex. For example, a macro could be used to initiate a general purpose interface (GPI) trigger to an automation system to run a commercial stack while another macro could run a complex timeline involving memory recalls, keystrokes, and external device controls.

A timeline is a memory system that allows the recording, editing, and playback of a sequence of production switcher events. Timelines can be useful for creating complex effects sequences like show openings that require tight timing between elements.

Memories, macros, and timelines can be saved to a hard drive and/or portable media, such as a USB flashdrive, so that the switcher settings for that production can be recalled in the future.

EXTERNAL INTERFACES AND ASSOCIATED SYSTEMS

Device Control

Initially, production switchers were used only to switch, mix, and key video; however, over the past several years, as the demands of the broadcast and production industries have changed and as technological capabilities have increased, production switchers have taken on the job of managing an increasing number of the control room devices. These capabilities are implemented with the goal of giving the technical director more control over the end product, improving the tightness of the production timing, and increasing operational efficiencies. Controlled devices include video servers, audio servers, still stores, digital disc recorders (DDRs), video tape recorders (VTRs), robotic camera systems, routing systems, and even audio mixers.

Device control is typically via an RS-422 serial connection between the production switcher and controlled device although network-based control schemes are becoming more common and GPIs are often used in cases in which a simple trigger is required. RS-422 is a robust, dedicated connection between two devices that can run up to 1000 meters. Both transmit and receive lines employ differential transmission on two wires each for noise immunity.

Network control schemes are usually based on TCP/IP over Ethernet. In this case, the network can be shared between multiple users. It is important when using a network-based control scheme to take special care in designing the system to ensure that sufficient bandwidth is available to allow the commands to be transmitted in a timely fashion from the controlling device to the controlled device. In many cases, a separate, dedicated network is best for this type of control system, ensuring that general office traffic or movement of large files does not impede the operation of the on-air control system.

There are a few industry standard protocols such as the Sony VTR Protocol that is widely used for VTR and DDR control and the VDCP protocol established by Louth Automation (now part of Harris) for video server control. Most other devices have manufacturer-specific protocols, although some emulate other devices.

When a modern production studio is being planned, device control is now an important issue. It is best to have a comprehensive plan regarding the devices the production staff wish to control and the capabilities required. An example of such a list for planning purposes is shown in Table 5.10-1.

The GPI is a simple on/off control that uses TTL-compatible input and output connections to connect devices together for control or status signaling. It is a simple way to trigger devices without requiring a complex protocol or control software to be written. An example of GPI control is a server being triggered to play a clip. GPIs are generally tied into macro buttons so that the TD can select the desired controls when needed.

TABLE 5.10-1
Device Control List

Device	Manufacturer/ Model	Protocol
VTR	Panasonic D5	Sony VTR
Video Server	GVG Profile	VDCP
Audio Server	360 Systems Digicart/E	ES-Bus
Routing Switcher	NVISION	Native
Character Generator	Avid Deko	III (Chyron) serial
Still Store	Chyron Aprisa	Chyron
Audio Mixer	Yamaha DM1000	Native
Robotic Camera System	Sony BRC-300	VISCA
Virtual Monitor Wall	Miranda Kaleido-K2	Native

Tallies

Tallies are connections to the monitor wall or studio camera that cause a red light to turn on so that the talent and crew know which video source or sources are on the air or currently being recorded. Two types of tallies are provided on production switchers: parallel tallies and serial tallies.

Parallel tallies are simple controls that provide a contact closure for each device connected. The contact closure is made when the source mapped to that contact closure is routed through to the program output. This may be a complex path through the production switcher involving one effects bank cascading into another. The production switcher incorporates logic that follows the source path backward through the switcher to determine what sources are contributing to the on-air output and instantly makes the appropriate tally indicator as the sources feeding the output are changed.

Serial tallies are transmitted over a serial RS-422 link. They are useful in systems in which a virtual monitor wall system is used. Wiring is simplified, as a single RS-422 cable run is required as opposed to a large multicore cable.

On some production switchers, tallies can be mapped for outputs other than the program output. As an example, split feeds can have their own tally mappings that are tallied from MLE 1. Another appli-

FIGURE 5.10-14 Dedicated aux bus control panel, controls a single aux bus output.

cation is to provide tally for ISO (isolated) record sources, in which, for example, a camera following a specific hockey player is routed through an auxiliary bus to a VTR or server to capture all of the movements of that player. In this case, the camera should be tallied to make the camera operator aware that it is being recorded.

Auxiliary Busses

Auxiliary (aux) busses are extra outputs from the production switcher that can be used for a variety of purposes, including camera switching for the camera shader operator, feeds to on-set monitors, feeds to VTRs for ISO-camera recording, engineering source monitoring, extra program or MLE outputs, and feeds to external devices such as DVEs and graphics equipment.

Auxiliary busses are controllable from the main switcher control panel and, in the case of most manufacturers, from other locations within the facility via a separate auxiliary bus control panel, as shown in Figure 5.10-14. It is also possible for an assignable panel to control multiple aux busses. Some aux bus control panels allow a contact closure from camera paint control panels to connect into the control panel for the purpose of selecting that camera to the aux bus output. This allows the person shading the cameras to simply press a button on the camera paint panel and have that camera appear on the monitor in front of the shader.

Auxiliary bus keying provides the capability to mix and key on aux bus outputs. This facility can be useful for branding on split feeds or prekeying of incoming sources.

Graphics Networking

Many production switchers now incorporate networking capabilities. Ethernet ports are commonly available and can make it possible to transfer graphic elements, stills, animations, and even short clips directly into digital storage in the production switcher. These on-board graphic stores allow the TD to easily incorporate the graphics into transitions, DVEs, and other show elements for a particular "look."

DEALING WITH ASPECT RATIOS AND HD/SD FORMATS

One challenge in modern productions is dealing with the different source and destination aspect ratios in standard definition (SD) and high definition (HD) formats. This is especially significant in a live production, as there can be very little time to deal with problems as they arise. It is very important to have a strategy in place to deal with aspect ratios; otherwise, the production staff will be faced with decisions for trade-offs that they were not expecting to have to make.

Needless to say, the producers and operations staff should be involved early and often in the planning of any new production facility, as they are the ones that will have to use it and live with the constraints of the design. Aspect ratio planning must now be an important component of these discussions.

Following are a few questions to keep in mind when designing a production system for multiple aspect ratios:

- What output formats are necessary? For example, is it HD or 16:9 SD for a DTV transmitter and 4:3 SD for the NTSC transmitter?
- What sources are to be used, and what are their output formats?
- How will the graphics be produced, and how will they be compatible in the different output formats? HD and 16:9 are a real issue here, as up-converted SD graphics tend to look soft, and designing for 16:9 and 4:3 together needs to be carefully managed. For a discussion on this topic, please see Chapter 5.15 on Television Graphics Systems.
- For signals that are to be converted from a source format to a different destination format, what quality is required? Is the source a road traffic camera in which a lower quality HD up-conversion might be acceptable, or a camera that is used regularly in the studio that should have a high-quality conversion?

Some production switchers have special features to be able to automatically convert between aspect ratios on the fly and even provide up- and down-conversion. These features can be crucial in certain production scenarios in which a mix of source formats is a requirement.

SAFE ACTION AND SAFE TITLE AREAS

Most production switchers provide a preview output with SMPTE standard Safe Action and Safe Title area overlay on the video (see Figure 5.10-15). The Safe Action zone is a white box that shows the area that is viewable on most TV sets, accounting for the typical horizontal and vertical overscan of the traditional picture tube. The Safe Title zone is somewhat smaller and shows what should be visible under all circumstances. A further guide may be provided for minimum text

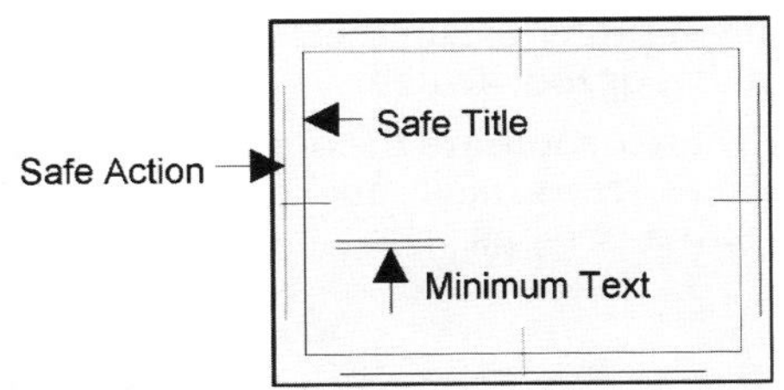

FIGURE 5.10-15 SMPTE Safe Action and Safe Title area display, as seen overlaying the switcher preview output.

size, which consists of two horizontal lines, the separation of which represents the height of the smallest recommended text size.

In addition to Safe Action and Safe Title, some production switchers are capable of displaying additional information overlaying the preview output, such as count up/down timers and source identification.

SYSTEMIZATION

The diagram in Figure 5.10-16 shows a simplified view of video and control connections in a typical production studio. Video and Alpha (or Key) Sources are wired to the inputs. Outputs are fed to monitors and on to the master control, plant router, and other destinations. Note the monitors for MLE 1 and MLE 2 program and preview outputs. These monitors are used by the TD to get these MLEs ready to go to air. The auxiliary busses are utility outputs and have a wide variety of uses, from feeding camera shading and on-set monitors, to ISO VTRs, and other destinations.

PRODUCTION CONTROL ROOM STAFF

Understanding the job functions of the various staff involved in the production is important. For those who are new to a television production facility, following is a list of job titles with brief descriptions of their job functions. As every production organization is dif-

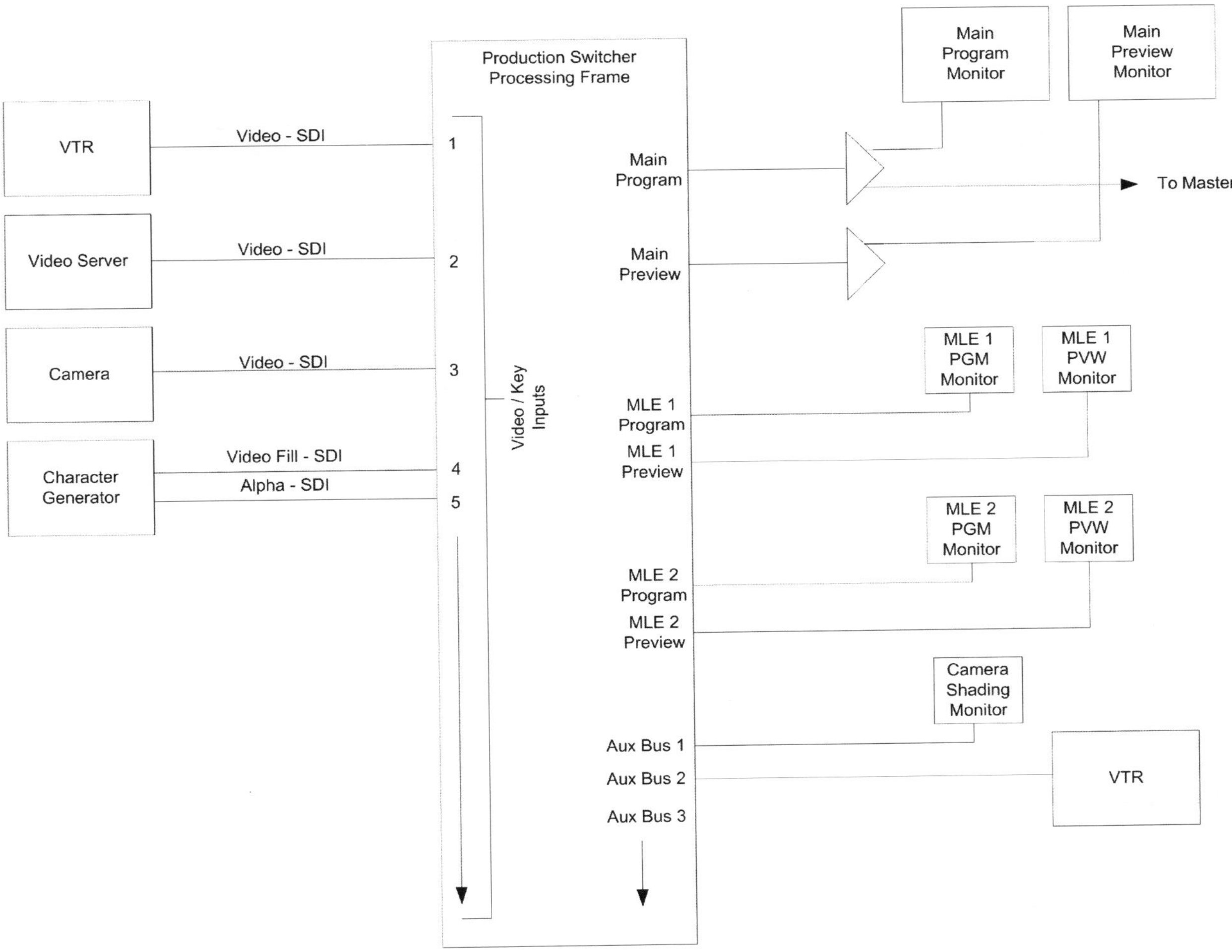

FIGURE 5.10-16 Simplified production control video flow diagram.

ferent, these roles can vary from place to place or, as is increasingly the case, can be combined.

- *Technical Director:* This person runs the production switcher and may also direct the production.
- *Director:* This person directs the production, calling the cameras, graphics, and so on, and may also act as the technical director.
- *Producer:* This person oversees the production and decides the content and its running order.
- *Production Assistant:* Production assistants have various duties. This person is often responsible for timing the show with a stop watch.
- *Character Generator Operator (CG Op):* This person runs the character generator and sometimes other graphics devices during a production.
- *Audio Operator:* This person runs the audio console, ensuring optimal audio levels during a production and is responsible for the overall audio quality of the production.
- *Floor Director:* This person directs the talent and activity in the studio.
- *Teleprompter Operator:* This person runs the teleprompter. This job function is less common now, as most talent prefer to run their own teleprompter using a dial or foot switch.
- *Talent:* These people are the ones in front of the camera, such as actors, news anchors, sports casters, and weather persons.

PRODUCTION AUTOMATION SYSTEMS

Computer-based production automation systems are a relatively new development that allows semi-automation of a production. The goal of these systems is to reduce the staff required, reduce mistakes, tighten timing, and increase the overall quality and repeatability in producing a sophisticated production such as a newscast. Production automation systems allow as few as one person to control all of the devices in a live production, including the production switcher, audio mixer, video servers, graphics systems, robotic cameras, and many other device types.

Production automation systems operate by defining a series of templates and custom control macros, which define the various "looks" that will be used in a given production. The various settings are prepared in advance of the production and then recalled by an

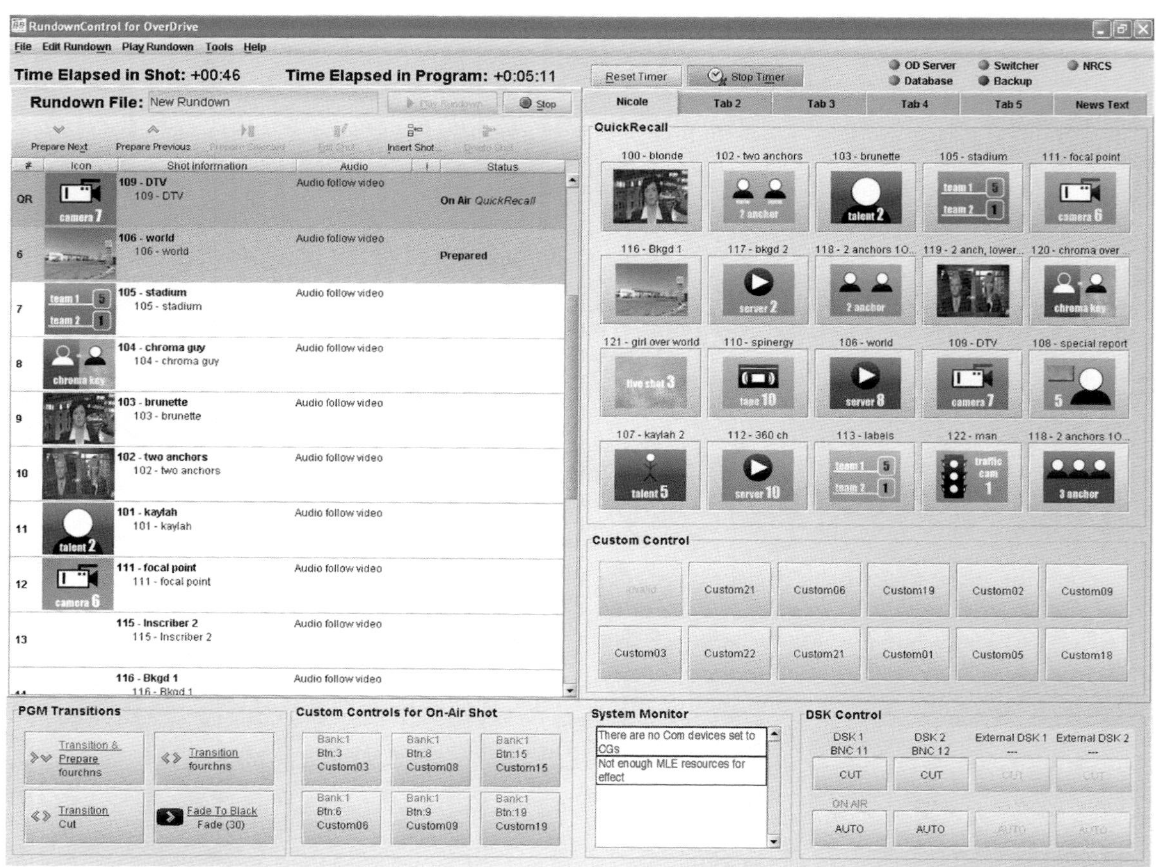

FIGURE 5.10-17 Touch screen control for production automation system.

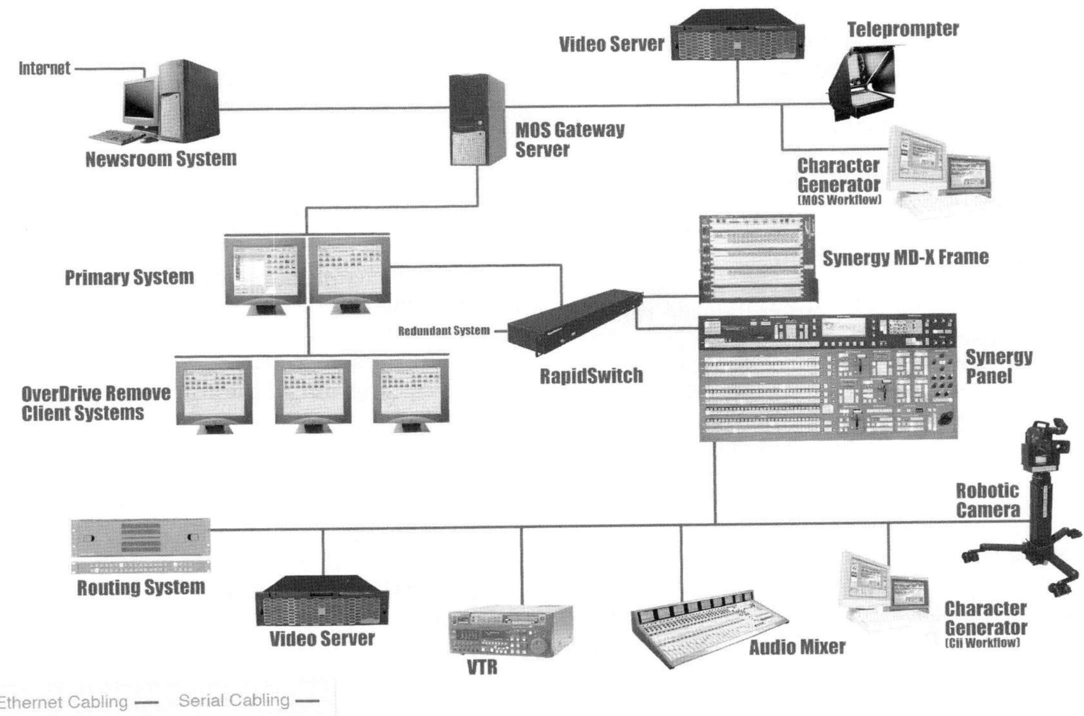

FIGURE 5.10-18 Connectivity for production automation system.

operator as the show is being produced. These systems tend to require more advance planning as to the various looks and elements that will be needed in a production; however, once they are configured, they make it easy to get complex effects to air repeatedly.

Figure 5.10-17 shows a touch screen control panel for a production automation system. This shows a program run-down on the left of the screen, with prebuilt setups for each sequence. The right of the screen shows various other set-ups that can be instantly recalled, allowing for changes on the fly.

Some production automation systems have integration with newsroom editorial systems—the system that is used by journalists, assignment editors, and the news director to write the scripts and build the rundown for a newscast. In this case, it is possible to embed the segment "look" right into the story. Through the industry-standard Media Object Server (MOS) protocol, this information can then be transmitted to the production automation system. Thus, the producer's vision of the show is directly connected through to the system that will be used to take it to air. (for more information on MOS, see Chapters 5.9 and 5.13.)

Production control automation is sometimes confused with facility automation, as used for master control operations. There are, however, several differences. Production automation systems are designed to easily change complex live production elements at any time, even seconds before taking shots to air. Also, the MOS protocol link to the newsroom system allows for complete synchronization between rundowns at all times. In production automation, the management and control of all the on-air devices is centralized in one GUI operated by the TD or Director. The system then typically allows for operator intervention, or control, at a level that suits the facility's requirements, from traditional manual operations, to semi-automated or "production-assist" to full system control of all devices and shot elements. Typical interconnection arrangements are shown in Figure 5.10-18.

BRIEF HISTORY OF SPECIAL EFFECTS

There are many techniques that go into the making of special effects, from mechanical (wires) and chemical (pyrotechnics) to motion control, camera tracking and virtual sets. Some of these effects are created in real time, during original camera shooting. Other visual effects such as 2D and 3D computer generated images (CGI) and digital effects are added later during postproduction. This section focuses on the use of digital video effects (DVE) devices and compositing systems.

The history of video effects dates back to the early days of television when the only effect that could be performed was superimposition (mixing two video feeds) similar to making a double exposure on film. Normally used to add graphics or credit rolls to a live video feed, the video mixer was simply dissolved halfway between two cameras. Although the effect was crude—and lowered the brightness level from each camera—it was also possible to create composite images for effects such as ghosts and giants. For this effect to work, one subject had to be shot against a black background to avoid introducing unwanted elements and spoiling the other image.

Later, video switchers developed the capability to actively switch between two cameras, thus creating a wipe. Different signal waveforms can create horizontal, vertical, circular, diamond, star, and many other shaped wipe patterns. With the advent of the VTR and with careful planning, it became possible to wipe between a live camera and tape playback, allowing one actor to play against herself, much like the traditional optical effect used in film.

As discussed earlier, switching between two video sources can also be done by a keyer. The switching signal is derived from a video source rather than a fixed pattern generator. An internal key typically uses the luminance level of the video to create the switch; this is practical for superimposing black-and-white graphics or text.

For effects work, the next advance took place with the introduction of the *chroma keyer.* A chroma keyer generates a key signal based on the color of objects in the scene. Thus, an actor posed in front of a strongly lit blue or green screen can be isolated from that picture and placed over, or into, a completely different scene from another camera.

All these effects were carried out using analog signal processing, and, for more than 20 years, this was about the extent of the effects that could be performed on video signals. In the early 1970s, however, a revolution took place in the field of electronics. In the broadcast arena, it began with the digital time base corrector (TBC). When a small portion of the video signal was stored in random access memory (RAM), a digital TBC could eliminate jitter from the raw output of a VTR. As memory capacities grew, it was soon feasible to hold an entire frame of video in RAM, thus creating the *frame store*. At this point, many inventive engineers realized the potential this had for creating digital video effects, which revolutionized video special effects with new capabilities as described in the next section.

Subsequently, all the traditional dissolve, wipe, and key effects were also implemented using digital processing in video production switchers as described earlier in the chapter. In due course, these capabilities were also replicated in software with computer-based systems for postproduction as described later.

DIGITAL VIDEO EFFECTS SYSTEMS

Principles

The techniques of sampling, digitizing, utilizing RAM storage, and converting from digital to analog video were honed on the TBC and frame store. With an entire frame of memory, video could be stored into RAM at one rate and read out at another. This is the basic operation of the frame store—allowing asynchronous video signals to be combined without the need for genlock. In this scenario, the digitized video samples from the incoming signal are stored more or less sequentially in a RAM buffer. They are also read out sequentially although the offset between the read and write addresses will change depending on the difference in frequency and phase of the incoming video versus the station reference.

However, it should be obvious that there is no restriction on how the video samples can be retrieved from memory. For example, starting with the assumption that each digital sample represents one pixel of video.[2] If the read addressing is then altered so that the video pixels are read out from RAM in reverse order, this would create an upside-down and mirror image picture coming out of the frame store. That's a digital video effect!

When one is trying to define the various effects capabilities of a DVE, it is helpful to think in terms of source and destination spaces. Imagine the original video signal which is fed into the DVE as represented by a flat piece of paper; that would represent the two-dimensional source space. Since the output is also a video signal that could be displayed on a monitor, it too can be represented by the two-dimensional face of the monitor screen; this is the destination space. The relationship between these two spaces can be altered in various ways within the DVE including separation into an imaginary third dimension going into the screen. These operations are usually performed using a 3-axis joystick.

Figure 5.10-19 shows a basic block diagram of a typical DVE. First, if necessary, the analog video and/or key signals are digitized by an A/D converter. The video samples are written directly into the frame buffer. The Reverse Address Generator (RAG) then determines how these video samples will be read out. The RAG is controlled by a CPU, which calculates the sequences necessary to give the desired effect. The output from memory is passed through an interpolator and one or more filters to create the final digital output frame. If necessary, the output is converted back to analog. This process continues for each field of video passing through the DVE; thus, the CPU must be fast enough to calculate the RAG sequences in real time.

Translation

The simplest effect that can be performed involves *translation*, or movement in the horizontal and/or vertical direction. It is easy to see that these effects can be accomplished by altering the sequence in which the stored pixels are read out. If the sequence itself is then manipulated in real time, it is possible to have the DVE smoothly move the picture left or right, up or down. When trying to read out past the edges of the source space, the DVE will usually substitute black or a fixed background color.

[2]In reality, the luma and chroma data from each pixel are stored and processed separately through different sections of the DVE.

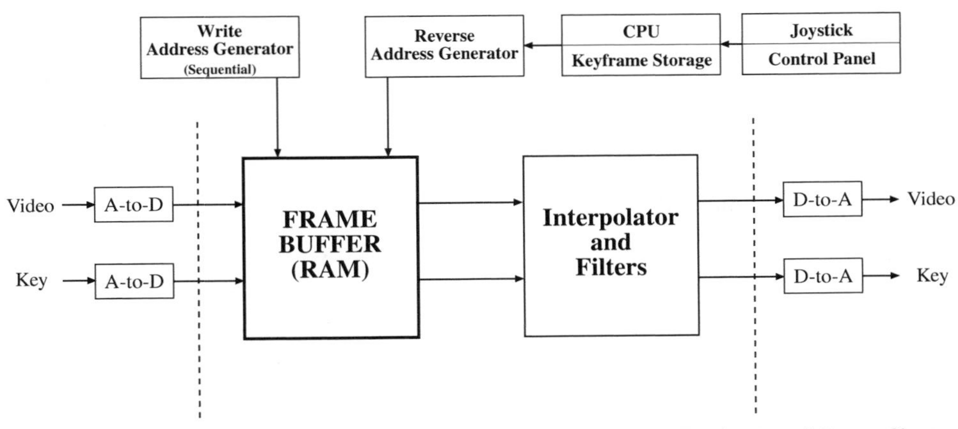

FIGURE 5.10-19 Block diagram of a typical stand-alone DVE device. Most effects are created by the Reverse Address Generator (RAG), under control of the CPU.

Interpolation

The previous examples did not alter the original video samples. They were simply read out in a different order. Almost all effects, however, will require some manipulation of the stored samples to create the pixels in the output image. This is the job of an *interpolator*. For each output pixel, the interpolator examines the surrounding pixels stored in memory. Then, using a precisely weighted average, it calculates (or interpolates) what the final output pixel should be.

Because the NTSC video signal is an interlaced format, finding an adjacent pixel is somewhat ambiguous in the vertical direction. For example, consider what would be called the pixel directly above. If the pixel from the previous line (within the same field) is used, then this ignores half of the vertical resolution contained in the other field. Looking farther back into memory to the previous field gives the true pixel that is on the line directly above. This restores the full vertical resolution but introduces another potential problem—motion artifacts. If the pixel in question represents a part of the scene that is in motion, then looking back at the previous field represents what the pixel looked like 1/60 of a second ago.

In a nutshell, if the video stream is processed as fields, then half of the vertical resolution is given up, but fluid motion is retained; with frame processing, full resolution is retained at the expense of possible jittery motion or *judder*. For this reason, the frame mode is usually used only for still, high-resolution images such as computer-generated graphics. For most other video sources that contain any amount of movement, the field mode will usually produce better results. If the motion is great enough, it will mask much of the loss of resolution.

Most interpolators also have an *adaptive* mode, which can select the optimum processing for each pixel of the video image. By comparing every pixel with its counterpart in the previous frame, the interpolator can determine whether there is motion occurring in that particular spot on the screen. If motion is detected, then that pixel is processed using field interpolation. If not in motion, then frame interpolation is used for that pixel. The motion detector can sometimes get fooled, but this mode usually produces excellent results.

Rotation and Size Manipulation

Although slightly more complicated, rotation of an image is accomplished in a way similar to translation. The read addresses are calculated using standard polar to Cartesian coordinate transforms. The center, or axis, about which an image is rotated can also be easily altered.

Size manipulation is likewise performed by mathematical operations. Shrinking a picture involves tossing away information (a one-quarter size image can be created by eliminating every other pixel and line). But to blow an image up greater than full size requires more information than is available in the original video signal. Thus, the interpolator is relied upon to create the missing pixels. At best, the image will appear slightly fuzzy, and at large magnifications the loss of resolution may not be acceptable.

Digital Optical Effects

If the computer generating the read addressing is fast enough, then more complicated functions can be performed. Generally referred to as *warps*, these effects look like page turns, rolls, ripples, or wrapping the video onto a solid object such as a sphere.

It is also quite an easy trick to manipulate the pixel size to give a mosaic look, add multiple images on the screen, or add video effects such as a drop shadow or glow around an image. To further enhance the illusion that a video image has been moved through three-dimensional space, a DVE can add perspective size and defocus to the video or even bounce an imaginary light source off the source video plane.

Latest Computer Effects

Going even further, computerized alteration of the video can produce morphs (changing one image into another), distort (mapping the video onto a sheet of rubber and then deforming it), and painterly effects (changing the video to look like an oil painting, for example). Because of the mathematical processing necessary to produce these effects, they usually cannot be created in real time. Thus, they may not be found on a dedicated DVE device such as those used for live broadcasts.

For effects work done in postproduction, non-real-time compositing workstations with built-in DVE functions are used to create such effects. Dedicated compositing software (along with numerous third-party plug-ins) offers a multitude of effects that can be applied. Many are programmed with real-world physics formulae to simulate particles, gravity, collisions, fire, etc. Although processing times vary, a typical workstation can take up to several seconds to render each frame.

Recursive Effects

Other interesting effects can be generated by feeding all or part of the processed video back through the DVE. Much like pointing a camera at its video monitor to produce video feedback, the DVE can recycle its output video to provide recursive functions. The most popular uses for this mode are to create decays or trails from moving parts of the image.

Key Channel

Although the primary function of a DVE is to manipulate the video coming into it, most DVEs also keep track of the original source rectangle or an external input key through a separate *key channel*. This channel undergoes all of the transformations along with the video so that it can be used to key the output from the DVE onto another video signal. The key channel can also be used to create a drop shadow of the processed video onto the background.

Motion Keyframing

Sometimes a DVE is used to simply perform a single, stationary effect on an incoming video stream. The real power of the DVE, however, comes with its capability to alter its effect over time. A simple effect, for example, might be to move a shrunken source image from off-screen left, through the target space, and then completely off-screen to the right. It would also be necessary to specify the length of time that this move should take place.

To set up such an effect, *keyframes* are created where absolute parameters are to be entered. In this case, it would be necessary to specify the picture size parameter and then create a starting and ending keyframe. These keyframes would then be set with the beginning and ending x-coordinates, respectively. The DVE will then calculate the x-coordinate for all of the other frames of the effect; this is called *in-betweening*. Most DVEs have a timeline display that shows the location of each keyframe within the total length of the complete effect.

Types of Motion

Adding the fourth dimension of time to the DVE also allows manipulation within this new dimension. The simplest type of motion between keyframes is linear motion, where each frame moves the same fraction of the difference between keyframes. Thus, the video moves with constant velocity, in a rather mechanical fashion. Another type of motion is S-linear, which adds a small amount of acceleration at the start of the move and deceleration at the end. This usually gives a more natural feel to the motion and is sometimes referred to as Ease In and Ease Out. Other variations such as smooth or curve motions also try to mimic more closely the movement of actual objects in nature. Very specific motion paths, such as that of a bouncing ball, may require many keyframes or finer control over the motion parameters.

One scheme for setting precise control over keyframe motion is the use of motion vectors. Three parameters—called tension, continuity, and bias (TCB)—determine how the video moves into and out of each keyframe. Figure 5.10-20 shows the effects of various motion types on a simple video clip that is moved across the screen.

For the ultimate in motion keyframing, some systems present one or more graphs for each parameter that changes. The curve representing how the parameter changes from one keyframe to the next can then be manipulated to give the exact motion desired. Another graph can represent the speed at which the parameter changes.

COMPOSITING

While some visual effects can be added during production (and, indeed, *must* be for a live broadcast), most effects work today is done during postproduction. Here, at a much more leisurely pace, the full force of the postproduction tools can be focused on perfecting each effects shot. Many devices and techniques can be combined to achieve the desired look. In most cases, the final effect will come together in a compositing suite where all of the various elements are combined. A typical effects shot might consist of 5 to 10 layers: an original background plate with one or more keyed-in elements, plus shadow passes, reflections, highlights or glints, and possibly a CGI or hand-animated element.

In its simplest form, *compositing* involves the layering, or keying, of one video image over another. This multiple image is then recorded off to tape or disk where it can be recalled later as the background image for the next layer. Thus, an entire image or video clip is built up one layer at a time. In a linear compositing

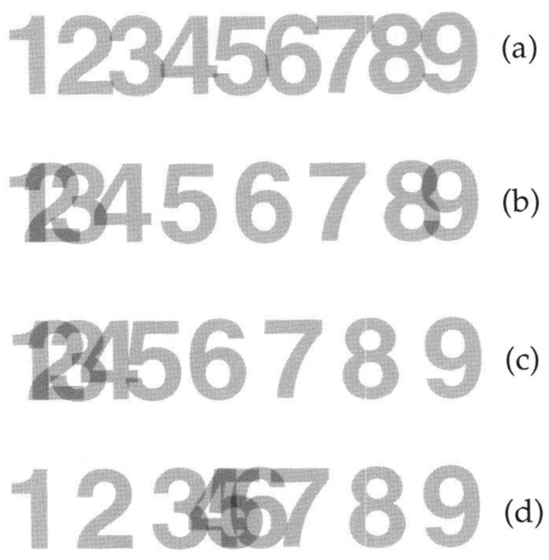

FIGURE 5.10-20 Various motion paths that can be applied to an effect. In all of these figures, the same nine-frame video clip is moved from left to right across the screen (a superimposed function has been applied to show all nine frames in the same figure). In (a), a *linear* motion gives rise to equidistant spacing and thus constant velocity; in (b), the effect is given an s-linear motion which starts off slower, accelerates through the move, and then decelerates at the end; by varying TCB values, many other motion paths are possible such as (c) constant acceleration or (d) deceleration in the middle of the effect.

suite, clips would be laid off to a component digital VTR or DDR.

Today, however, falling prices and increased capacities have made complete compositing workstations much more popular. An entire edit suite can now be built into a single box with nonlinear editing, keying, and digital video effects all being controlled by a single operator. The increased efficiency and ease of operation now make them the logical choice for most effects work.

With early compositing suites, changing something on the first or second *bed* in a 30-layer composite was very time consuming. It could take almost as long to rebuild the composite just to make a small change to one of the first layers. It was like peeling off the layers of an onion to make a change inside and then carefully reassembling the pieces back together. Modern compositing workstations now allow the operator to work with many layers concurrently, making changes to any layer at will. Other features, such as the capability to marry a matte channel to a video clip, can drastically improve the efficiency of a compositing system.

When one is compositing, there are basically two ways in which video sources can be combined: *mixing* or *keying*. Mixing can be performed in either an additive or nonadditive fashion. Keying—also referred to as matting—can be performed using one of the existing sources as a key signal (internal key) or using a third, separate source for the key signal or matte (external key).

Additive versus Non-Additive Mixing

When mixing two video signals, the usual operation is to add a portion of each signal such that the total percentages add up to 100. This is referred to as an *additive* mix. This duplicates the function of the dissolve lever on a switcher. Inside the switcher or compositing computer, the mix is performed by mathematically adding the two signals in this fashion:

$$Vo = M * V1 + (100 - M) * V2$$

where Vo is the video output signal from the mixer, V1 and V2 are the video sources, and M is the mixing percentage.

For certain operations, however, it may be desirable to combine the signals in such a way that the output video represents the level from the video source with the higher luminance. This is called a *nonadditive* mix and can be used, for example, to combine mattes from multiple objects to make a single combination matte. Another common use is to add one or more garbage mattes to eliminate areas where the camera shoots past the area of interest. Mathematically, the nonadditive mix can be expressed as

$$Vo = V1 \text{ where } V1 >= V2$$

$$Vo = V2 \text{ where } V1 < V2$$

Figure 5.10-21 shows the difference between an additive and nonadditive mix using two *hicon* (high contrast, 0–100 IRE) matte signals.

Keying

As previously mentioned, keying involves the combination of two video sources by selectively switching between them. A separate, monochrome key signal is used to determine when to switch. A simple keyer switches between the two sources based on the level of the key signal in relation to key level and/or clip controls. The key signal can be derived from either the overall brightness level (luminance key) or from the color hue information (chroma key), or possibly a combination of both.

The output from a keyer can be expressed mathematically by the formula

$$Vo = Vf \text{ for } K >= Kth$$

$$Vo = Vb \text{ for } K < Kth$$

where Vf and Vb are the foreground and background video signals, respectively; K is the key signal level; and Kth is the key threshold level set by the operator.

The preceding statement describes the operation of a hard clip keyer that switches instantaneously between video sources. Such a switch often creates undesirable artifacts that are visible in the output video. To reduce this problem, most keyers can set two thresholds over which a linear dissolve between signals is produced. This is often referred to as *softening* the key. However, the most precise control can be achieved by using a linear key.

The linear keyer, with its full range of transparency, has allowed much more natural and pleasing compos-

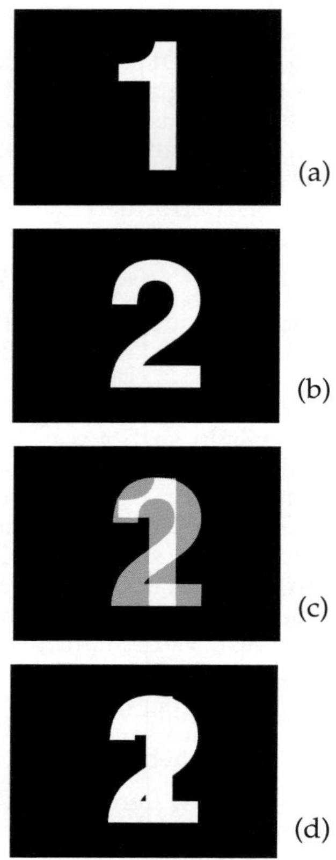

FIGURE 5.10-21 Additive versus nonadditive mixing. The two hicons shown in (a) and (b) are combined in (c), a 50% additive mix, and (d), a 100% nonadditive mix.

iting of images. With a linear keyer, the key signal is used to effectively dissolve between the two video sources. Where the key signal is zero, or black, the foreground image will be totally transparent and thus does not show through over the background. Where the key is 100 units, or white, the foreground image becomes completely opaque and thus covers up the background. In between, as the key signal takes on other values, the foreground becomes as opaque as the absolute value of the key signal, and the background shows through accordingly. This is expressed mathematically as

$$Vo = K' * Vf + (1 - K') * Vb$$

where the key signal K is normalized to vary between 0 and 1 (i.e., $K' = K/100$).

One of the great advantages of a linear keyer is that it can maintain the proper levels of anti-aliased images such as graphics when creating the composite.

Chroma Key

In a chroma keyer, the key signal is extracted from the foreground video source. Thus, the quality of the key will depend on the video source as well as the chroma keyer's capability to discern colors. For a live production, the RGB outputs from a camera would be used; for later use in postproduction, the blue (or green) screen shot must be recorded. Although it is possible to pull a key from composite video, this should be done only as a last resort. All blue screen source work should be shot on 35 mm film and/or recorded directly onto a full-bandwidth component VTR using 4:2:2 or 4:4:4 color subsampling. This will preserve the greatest detail in the color signals and eliminate artifacts that are hard to remove from lower sample rate and NTSC composite signals.

Most chroma keyers work in the RGB domain. Thus, the foreground video signal must first be transformed into an RGB signal. The keyer will then analyze the individual red, green, and blue levels and isolate a small range of color values to represent the key signal. It can take a lot of patience to set up the keyer properly, especially if there are other production or lighting problems. A bad key is quite noticeable; there may be parts of the foreground subject that start to disappear, or often there will be a blue or green halo around the keyed subject.

Edge effects around the key are the number one giveaway that a chroma key has been used. Unfortunately, even if everything is set up correctly, there may still be a colored halo around the foreground subject. Most often, this is caused by light from the backing material reflecting back onto the subject. It can also be caused by reflections or flares within the camera lens or camera itself. To reduce these effects, many keyers have the capability to *hue suppress* the foreground image before keying it over the background. Hue suppression selectively eliminates the backing color from the foreground image. At worst, any remaining halo would be monochrome and would blend into the background more naturally.

For the ultimate in chroma keying, an additive mixing process can be used. First, the foreground image (against the colored screen) is processed to produce both a matte signal and a hue-suppressed foreground image (against black). Then the two signals are combined in a fully additive mix according to the equation

$$Vo = Vf' + (1 - K) * Vb$$

where Vf is the preprocessed (hue suppressed) foreground signal.

Note that there is no attenuation of the foreground video, other than removal of the backing color. With this technique, very high quality keys can be achieved despite the usual problem areas such as fine details (wisps of hair) and transparent objects such as glass. Even reflections from shiny surfaces and shadows cast onto the backing wall can be re-created in the final composite.

Effective Chroma Keying

Since many effects rely on chroma keying, it is worth discussing the various parameters that go into making a believable composite. Blue (or green) screen effects can basically be divided into two categories, depending on whether or not the foreground and background

cameras move. Early chroma keys were mostly performed with locked-down camera shots. Since the purpose of the chroma key is to place a new object into an existing background scene, there could be no relative motion between the two without destroying the illusion. Later advancements with motion control rigs, motion tracking, and virtual set technologies have allowed moving chroma keys to be created. Since most people are accustomed to seeing such effects shots as stationary, these new effects seem even more believable.

Whether the shots are stationary or moving, many tricks can be used to make a successful chroma key. First, some decisions need to be made during preproduction. This would include the choice of backing color and the type of keying required. Knowledge of the keyer that will be used for compositing and whether the original shoot will be captured on film or video will aid in the effect's planning.

Backing Color

Although technically a chroma key can be pulled from any solid background color, blue and green screens are used almost exclusively. In many cases, the choice between the two is very subtle and may be dictated solely on the availability of a suitable stage. When given the option, however, there are several factors to consider. The most common concern is with colors that are to be used in the foreground scene.

Conventional wisdom implies that there shouldn't be any strong blues on a subject that is to be shot against a blue screen. This is especially true with older keyers that are not very selective (the notorious background peeking through a spot of blue on a subject's tie, for example). Modern keyers, however, are much more accurate and flexible, and this is not such a great problem today.

When live actors are being shot, a green screen is often preferred. This is due to the fact that any green spill on the subject is much less objectionable than blue. Again, however, most modern keyers offer some degree of hue suppression, which can greatly reduce this problem. The higher resolution and lower noise of a camera's green channel also favor the use of green screens. Other factors such as the effect of detail-correcting circuits in video cameras should also be considered.

Linear and Shadow Keyers

Depending on the effect required, it is important to choose the appropriate type of chroma keyer. A simple keyer can be used when the subject simply needs to be isolated from the backing and inserted into the composite image. This is often the case when the composite is not meant to be realistic (e.g., when the subject will be flown around in the composite using a DVE).

When a live chroma key where the foreground subject contains translucent objects such as glass or smoke is being shot, a linear keyer must be used to preserve the illusion. Otherwise, the object simply disappears or, worse yet, shows the backing color behind it. A linear chroma keyer reduces the level of the background scene according to the level of the backing color as seen through the translucent object. This makes the composite much more realistic.

Another way to add realism to live chroma keys is to let the foreground subject cast shadows onto the backing (or other keyed-out objects in the foreground). As with translucent objects, a linear keyer will recognize the reduced level of the backing color and darken the background video accordingly. This creates a very real illusion of the subject casting shadows into the background scene.

When the chroma key compositing is to be performed later in postproduction, keep in mind that some compositing workstations have relatively simple keyers. Fortunately, however, numerous tricks can be used to enhance these effects. For example, it may not be possible to pull shadows directly from the foreground video. To get the same result, it may be necessary to create a separate shadow matte, possibly requiring the use of rotoscope. Another trick—although somewhat of a cheat—is to take the shadowless matte and pass it through a DVE to lay it down in the proper perspective; when it is mixed with the original matte signal, the very effective illusion of a shadow can be obtained.

Once again, the foremost goal when compositing with a chroma key is to reduce any edge effects. This may take a careful balance between all of the keyer and hue suppression controls. Another technique for improving chroma keys in postproduction is to add a slight blur to the matte signal. Even better, some devices have the capability to add or subtract one or more pixels around the edge of a matte signal. Such a grow (or shrink) function can be very effective in creating the perfect composite.

Matching Foreground and Background

After edge effects, nothing spoils a chroma key shot more than mismatched foregrounds and backgrounds. The single most important consideration in matching both foreground and background scenes is *lighting*. Obviously, the intensity and direction of key lighting in both scenes should be similar. Color temperature and atmospherics should also be matched whenever possible. For example, when the background scene is an outdoor shot on a sunny day, it might be best to shoot the foreground blue screen outdoors as well under similar lighting conditions (paying close attention to shadows).

Other camera parameters such as depth of field may also have subtle effects on the composite image. If the foreground subject is to be placed well in front of the background scene, consider shooting the background with the camera focused onto the area where the foreground subject will later be placed. If this is not possible, the background may be defocused later during compositing to give a similar effect. Once again, pay attention to the details.

Another quality that must be matched between foreground and background is *colorimetry*. Since each scene might be shot with a different camera, lighting setup, film/video setup, etc., it will be necessary to color correct each scene so that it matches. Sky colors, skin tones, and shadows are usually first to be tweaked. Color correction capabilities vary widely from system to system, but they are very important in creating that seamless composite.

Film versus Video

Film shoots introduce several other factors to consider. Most importantly, when the telecine transfer of either background and/or blue screen film to video is made, a pin-registered gate must be used to eliminate film weave. *Film grain* is another quality that must be matched if the foreground and background scenes are to appear as if they are parts of the same scene. Fortunately, software tools and filters now exist which can match clips, even if one is shot on film and the other on video.

Motion Considerations

When the effect calls for a moving chroma key, there are even more details to watch out for. Even though a tracking keyer or motion control rig will be used to keep the foreground image in the right place with respect to the background, optical parameters such as perspective and motion blur must also be considered. This is especially true when either the foreground or background scene is shot with miniatures. In this case, camera height and distance to subject should be scaled accordingly.

When an object is moving fast enough so that it would not appear sharply defined in a single frame, there should be some degree of blurring in the direction of travel. Many video cameras have a high enough shutter speed to mask this effect. When called for in the composite, or if the foreground is moved around with a DVE, it may be desirable to add a degree of motion blur electronically to enhance the effect.

Other Effects

Full-featured compositing systems will also have the capability to process video clips in various ways. This might include the capability to de-interlace/re-interlace or cine compress/expand a clip. The first function creates two frames from each original frame—one for each field. The latter is used to remove or replace the 2–3 pulldown used in a telecine to convert 24 fps film to 30 fps video. Both of these functions can be useful for rotoscope.

It is also possible to stretch a video clip to make it fill any given time period. Much like an interpolator, the output frames are calculated from the input clip by mixing adjacent frames in proportion to their proximity to the desired time slice. This is often used to add extra speed to a fast scene or to present something in slow motion. There is a limit, however, to how much a clip can be slowed down. Just as the DVE loses resolution when blowing up an image, a slowed-down clip may look strange if the added dissolves between frames become noticeable. Newer "time-warp" software has become available that mathematically examines every pixel to see how it moves from frame to frame. This can be used to predict how various objects in the picture are moving and thus create in-between frames from scratch. However, this is very processor intensive (i.e., slow) and may produce new artifacts, such as stationary parts of the image being "pulled" along with the moving areas. If very clean slow motion is required, then the original footage must be shot at a higher than normal frame rate (either high-speed film or super slo-mo video).

Motion Tracking

Another useful tool for creating special effects is *motion tracking*. This can be used to composite a new image onto an existing object in a scene that is moving or in a scene where the camera moves. The computer uses a high contrast edge or point to track the movement over a series of frames. This *motion data* can then be used to stabilize the image. This is an effective way to remove film weave from non-pin-registered film transfers or to eliminate hand-held camera shaking. This process can even be repeated a second or third time to remove any residual motion.

The motion data can also be applied to another image, which is to be keyed into the original scene. When the motion of the background scene is tracked, the new keyed-in foreground element appears to blend into that picture. When done correctly, the tracking motion reinforces the illusion and makes for a great effect; it only takes a small amount of mistracking (wandering), however, to give the effect away.

Motion tracking with one point only lets a key follow horizontal and vertical movements. If the object being tracked rotates or changes size, then the keyed-in video must be adjusted accordingly. This can be accomplished using *corner pinning* and *four-point tracking*. With corner pinning, a roughly rectangular area is defined on both the tracked background object and the insert video element. The computer then uses its built-in DVE to match the perspective of the insert video by pinning its corner points onto the corresponding points in the background image. If four-point tracking is then applied to independently and simultaneously track all four corner points, the result can be quite spectacular. This technique is often used to track images onto a sign or video screen on the set.

Rotoscoping

Shooting an object or person against a blue screen makes it very easy to generate a matte for the object. This matte can then be used to composite the image over a different background. However, it is not always possible to have the subject shot blue screen. Sometimes, it is necessary to generate a matte for one object within a busy picture. In such a case, present computer tools are usually unable to perform this separation. It then becomes necessary for a human artist to painstakingly create the matte on a frame-by-frame basis. This is called *rotoscoping*. While mostly used for generating a traveling matte, roto can also be used for other effects such as wire removal and adding hand-drawn objects such as sparks, pixie dust, or laser beams.

Virtual Sets and Ad Replacement

One technology that appears to hold great promise for the future of video special effects is that of *virtual sets*. A virtual set is basically a CGI background scene created entirely within a high-speed supercomputer. Foreground subjects are then shot blue screen and chroma keyed into the virtual background. The real trick is to then have the CGI background automatically track movements by the foreground camera in real time. This is where the supercomputer comes in; it must calculate the position and field of view of the foreground camera and then render the background scene at the proper perspective. And, of course, it must do this at 30 or 60 frames per second.

Current computer technology limits the complexity of the background that can be rendered in real time. Thus, early virtual sets were somewhat surreal. But with sufficient computing power, it will become possible to create truly photorealistic imagery (current computers often require up to several minutes to create just one such frame).

Creating a virtual set requires two basic components: *camera tracking* and *real-time rendering* of a 3D graphic database. Camera tracking—including position, height, pan and tilt angles, lens focal length, and even focus—is derived either by position sensors or via pattern recognition. The former method involves the physical attachment of various devices to the camera to relay pertinent data to the render computer. These devices must be calibrated in advance to give the computer a starting frame of reference. For more information, see Chapter 5.4 under Tracking for Two- and Three-Dimensional Virtual Sets.

Pattern recognition offers an alternative technique for gathering the camera parameters. With this system, the blue screen is crosshatched with various-sized lines. These lines are painted in the same hue, but slightly darker than the regular background. The colors are close enough to be completely keyed out together, but different enough that they can be detected. The camera parameters can then be extracted by analyzing the camera's video signal. By detecting the presence of and the relative relationship between various lines of the crosshatch pattern, the computer can determine all of the needed parameters. Even the fuzziness of the lines can be used to determine the relative focus needed to render the background correctly.

The camera tracking technology of virtual sets was first used for advertisement replacement during sporting events. It is now quite possible that a banner or logo seen in the stands (or even on the field, court, rink, etc.) of a televised sporting event might not actually be visible to a spectator at the same event. And pattern recognition can also be used to create other interesting effects (tracking the motion of a single player or other object in a live sports production). The virtual first down line has become so commonplace that it's hard now to watch a football game without it.

Some believe that the next great visual effect will be the use of *virtual actors* (dead actors brought back to life on screen or new, totally computer-generated actors). Currently, live actors wearing body suits with motion sensors can be used to animate CGI characters. Higher resolution, however, is needed to capture slight nuances such as facial expressions. In the future it may be possible to eliminate the actor altogether and have a realistic character completely controlled by computer (à la the movie *Simone*).

VIDEO TO COMPUTER INTERFACES

The power and low cost of personal computers have not gone unnoticed by video professionals. Some of the most exciting and lucrative fields of software development are in the area of video creation and processing. Many tasks that were traditionally performed on dedicated video production equipment or 3D supercomputers can now be executed on low-cost PCs.[3] Of course, it may take much longer to perform the same task on a PC, but often this is well worth the trade-off of much lower equipment costs. And as PCs get faster and more powerful every year, the gap between them and dedicated equipment continues to narrow. With the addition of one or more specialized plug-in boards (or an external video processing unit), this gap can even be eliminated; indeed, many new video production tools are now based on, or *driven by*, PCs. With the advent of digital television (DTV), this approach will undoubtedly be used even more.

Video Format Conversions

To process video in a computer requires some manipulation of the video signal. Analog composite or component video must obviously be digitized; composite signals must also be decoded into RGB or YUV components. Even component digital video signals may require YCbCr to RGB color space conversion and/or pixel aspect ratio correction. For compressed video systems, conversion to an M-JPEG, MPEG, or similar bitstream must also be accomplished. And these processes will likely need to be reversed when bringing the video back out of the computer realm.

To understand all of the conversions necessary, it is helpful to examine each of the standards currently in use. Broadcast engineers should be familiar with the NTSC composite, RGB analog, and perhaps the ITU-601[4] digital video standards. Inside computers, video is most likely to be represented by 24-bit RGB values.

[3]The use of the term *PC* (personal computer) here is not platform specific. This includes Windows and Mac OS-based computers as well as high-end, UNIX-based workstations.

[4]Formerly CCIR 601, now completely specified as ITU-R BT.601-1994.

An optional 8-bit alpha channel is sometimes added to ease keying.

The NTSC composite video format is described in detail in Chapter 1.9, "NTSC Standard." The RGB[5] analog format utilizes three separate monochrome video circuits to convey the complete video signal. Sync and setup are optional on each signal, and obviously there is no color burst. If sync is not present on any of the RGB signals, it must be carried as a fourth signal and the format is referred to as RGBS. Since this adds to the complexity of the overall system, most equipment is designed to add/extract sync from the green channel, otherwise known as *sync on green*. Sync may also be added to the red and blue channels to prevent discrepancies in the way the three channels are processed; however, this sync is never used.

Component digital (ITU-601)—also referred to as SDI, per ANSI/SMPTE 259M when carried over a single coax—is based on YCbCr[6] component signals. These signals are derived from the standard, gamma-corrected R′G′B′ signals according to the following equations:

$$Y = 0.257R' + 0.504G' + 0.098B' + 16$$

$$Cb = -0.148R' - 0.291G' + 0.439B' + 128$$

$$Cr = 0.439R' - 0.368G' - 0.071B' + 128$$

By definition, the Y signal has a range of 16–235, and the Cb/Cr signals have a range of 16–240.[7] This leaves some digital headroom for over- and under-shoots of the video signal. Even so, much of the YCbCr color space is outside the standard RGB gamut, as shown in Figure 5.10-22. This is a result of restricting the RGB values to a range of 0–255.

To make matters worse, some RGB combinations translate into illegal colors when encoded to NTSC composite (due to excessive chroma levels). For this reason, it is imperative to check levels on all signals brought in from YCbCr space (e.g., from a digital paint system). Any objects created on the computer should be drawn using an NTSC legal palette. Compositing programs should have an NTSC filter function to ensure that their rendered output is within spec. For a detailed discussion on this topic, including issues related to HD conversions, see Chapter 8.2 under Verifying HD Color Gamut.

Picture Resolution and Pixel Aspect Ratio Conversions

Picture resolution (both horizontal and vertical) and pixel aspect ratio must also be considered when interfacing video signals with a computer. This is most likely to be an issue when dealing with SDI signals. The ITU-601 standard specifies a horizontal resolution of 720 pixels by 486[8] lines. This would represent a picture aspect ratio of 1.48. However, given the 4 × 3 (or 1.33) aspect ratio of NTSC, this effectively means that each pixel must be rectangular (with a *pixel* aspect ratio of 0.9). Computer graphics adapters, however, almost exclusively use square pixels.

This leads to three basic ways of working with SD video in a computer. If the need is to make only minor changes to the video and then lay it back out to SDI, it is possible to work at the video's native resolution of 720 × 486. This will incur the minimum amount of processing and conversion, so it will yield the cleanest result. It does mean, however, that images will be horizontally stretched on the computer screen, and drawing tools will not work correctly when constrained to squares and circles, etc. The latest versions of many graphics programs now have preview modes, which can compensate for the nonsquare pixels, but this usually comes at a price of reduced speed and picture quality. Another option is to run a second video monitor off an SDI output, which enables working on the video monitor in true aspect ratio.

More commonly, the computer graphics/editing program will be set to work at either 648 × 486 or 720 × 540 resolution. This restores the pixels to a square format, and all of the tools will work correctly. When it comes time to output the final result, the video is then resized to 720 × 486. Working at 720 × 540 is probably more common since the final step involves shrinking the 540 vertical pixels to 486. However, in many cases, working at 648 × 486 may be more desirable. For one thing, the storage requirements go down by 10%. Plus, the computer's CPU has 10% fewer pixels to manipulate. But more importantly, there is no need to do any vertical interpolation, which often makes dealing with rectangular objects and masks more difficult. The downside, of course, is that the 648 horizontal pixels will have to be stretched to 720 pixels, which is less than optimum. All high-definition formats use square pixels, so this issue will eventually disappear.

When the final result is to be laid off to film, the computer must render a higher resolution image. Typically, film images will be handled at either *2K* or *4K resolution,* which implies 2048 or 4096 pixels per line. Since film aspect ratios vary considerably, this could represent a frame size up to 4096 × 2990 (35 mm @ 1.37:1). At 24 bits per pixel,[9] that translates into a file size of 22–35 MB per frame. This compares to approximately 1 MB for a single frame of 720 × 486 SDI video.

Interfacing Video with Computer Workstations

For offline quality editing, composite video will probably suffice. Even online editing is possible thanks to

[5]RGB may alternatively be denoted as GBR.

[6]YCbCr is the correct terminology for this luminance (Y), R-Y (Cr), and B-Y (Cb) component format. This is sometimes incorrectly referred to as YUV (which also is related to the unscaled and offset signals upon which YCbCr is derived).

[7]When 10-bit levels are used, a 2-bit fractional value can be added, or the levels can be expressed as 64–940 and 64–960, respectively.

[8]Technically, the standard allows for 487 active lines in NTSC. When the 525 lines per frame are divided into two fields, the standard does not allow half-lines. Thus, an extra line is added to field one. Whether this line is used for "active" video is somewhat debatable, but the majority of ITU-601 devices specify an image size of 720 × 486.

[9]Newer systems extend the color depth up to 32 bits per color or even 128 bits per pixel; this increases the file sizes proportionately.

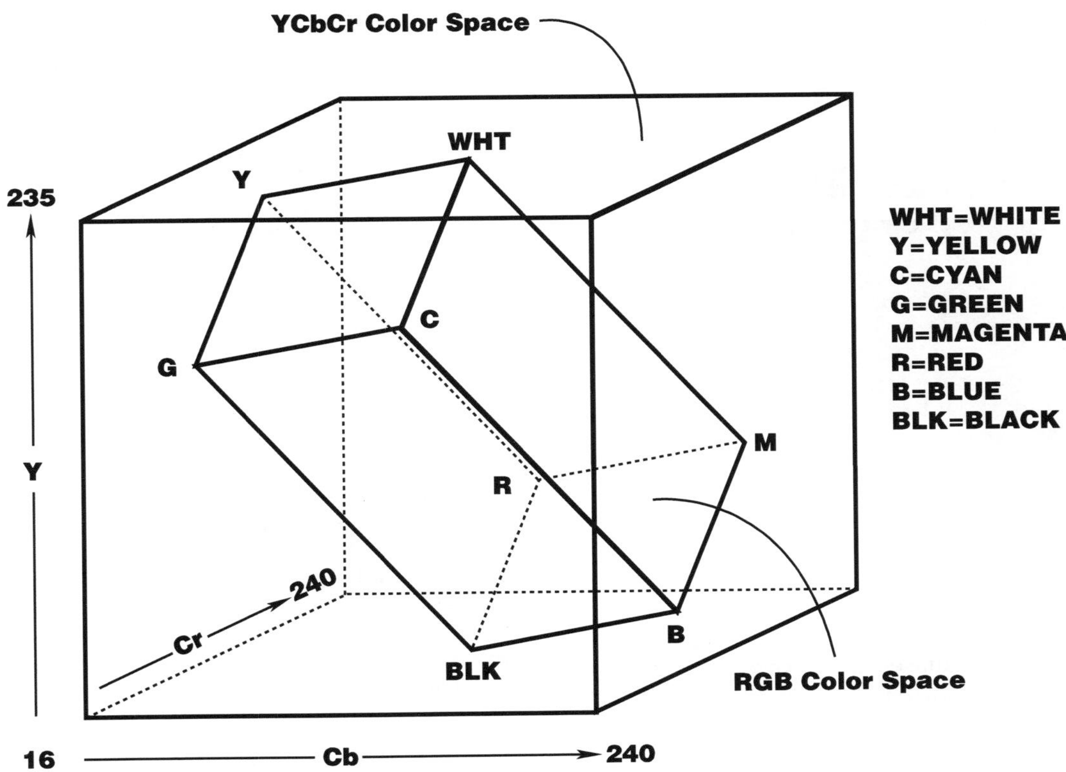

FIGURE 5.10-22 Comparison between the YCbCr and RGB color spaces. Note that about one-half of the YCbCr values are outside the RGB gamut.

faster computers and hard drives which allow for lower compression ratios. Special effects and compositing, however, are usually done in an uncompressed, component video domain. For most television work, this will be based on the ITU-601 Standard or SMTPE 274 M for HD.

Images created solely within the computer, however, can be rendered to any resolution. Higher resolutions, of course, will require corresponding increases in rendering time. For this reason, long video clips (or short film-resolution scenes) can require rendering overnight. While this may make good use of computing time, it also carries the danger that something will go wrong which halts the process. Or, after the final render is seen, there may be a problem that requires making a change. In either case, another long render may be required. For those working under tight deadlines, this can be disastrous. To avoid this, rendering can often be distributed over multiple computers set up as a *render farm* to cut down the overall time needed.

Film resolution jobs will usually be laid off from the computer to some form of digital media, or possibly transferred directly over a network to the digital film recorder. Most video work, however, is done by transferring component video in and out of the computer. This can be accomplished in one of three ways: using *analog component video, digital component video,* or via *data transfer.*

Analog component video usually requires an add-on video card that feeds RGB (or YCrCb, Betacam™, etc.) to the computer. This card performs sampling, digitizing, and possibly color space conversion on the input side. For output, it must perform D/A conversion and filtering. Make sure that the card is set up correctly for the type of component signals to be used (e.g., whether or not 7.5 IRE setup is present on RGB signals). Also, be careful when feeding the analog output from a computer into other ITU-601 devices. Images created on the computer often have risetimes exceeding the standard NTSC bandwidth. When these signals are fed into a digital device, they can cause ringing on sharp edges. Some computers also have built-in S-video ports; while certainly better than standard composite, they are still not as clean as full-component video.

Feeding component digital to the computer requires a high-end video card. SDI most likely will be color space converted between YCbCr and RGB. The levels may also be scaled to match the computer (Y=16 black would scale to Y=0, or R=G=B=0; Y=235 white becomes Y=R=G=B=255). Color bandwidth is expanded from 4:2:2 to the computer's equivalent of 4:4:4 sampling. Although these operations are mostly transparent, beware of illegal colors and optional filters that can cause slight variations in color. This could make the processed frames stand out if they are edited

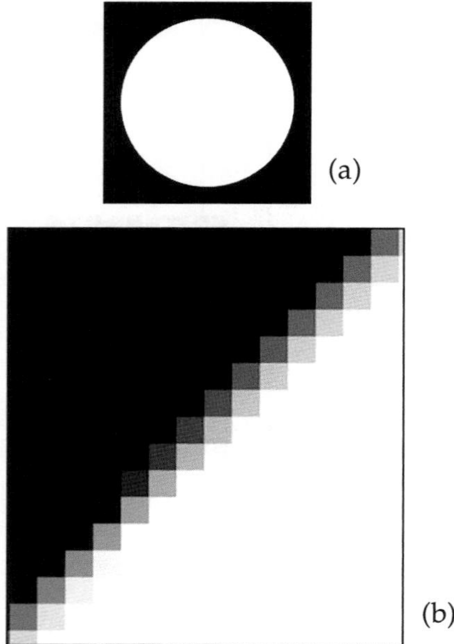

FIGURE 5.10-23 To create smooth edges of the circle in (a), a computer anti-aliases the nearby pixels using various levels of transparency as shown in the enlarged view (b).

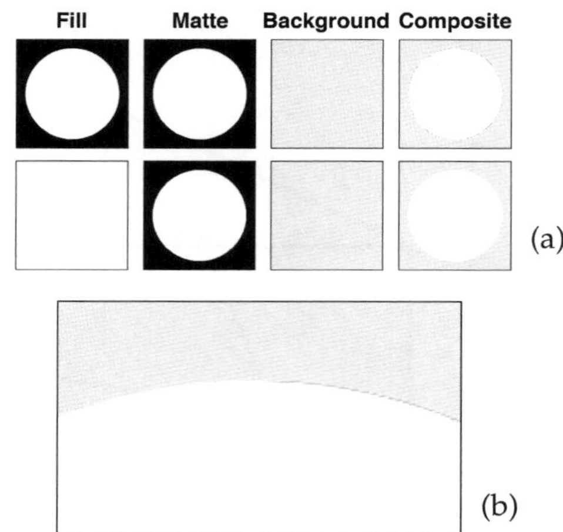

FIGURE 5.10-24 In (a), when a premultiplied (anti-aliased) fill image is used with a separate matte image, undesired artifacts can occur around the edges. This can be avoided by using a straight (non-anti-aliased) fill or, in this case, substituting a completely white image as the fill. In (b), an enlarged view clearly shows the difference between the correct (left side) and incorrect (right side) composites.

back into the original clip. The best advice is to work in YCbCr space whenever possible.

Video frames can also be passed as data files. These files can either be written to a standard CD or DVD, placed on a removable hard drive, or transferred over a network to another digital device. Sometimes, the other device is a DDR. In this case, the video is rendered and transferred to the DDR one frame at a time over Ethernet. When this process is complete, the entire clip can then be played back in real time on the DDR.

When creating keyable graphic elements, pay close attention to how the fill (color) and matte (alpha) channels are created. Since computers generate images within a fixed digital resolution, they use anti-aliasing techniques to smooth out edges that cross over pixel boundaries (see Figure 5.10-23). As with an interpolator, this is done by altering nearby pixels with various percentages of foreground and background.

When the alpha channel is anti-aliased, this will produce the correct transparency levels for each pixel. The color channels are anti-aliased by mathematically multiplying each pixel by its transparency, that is, the value of the corresponding pixel in the alpha channel. These are then called *premultiplied* channels. However, this assumes that the graphic will be keyed over a black background, which usually is not the case. Therefore, the color channels must be unmultiplied before compositing them over a background using the alpha channel.

Information on whether the color channels have been premultiplied is not always carried correctly from the graphics creation program to the compositing device. This is especially true when the fill and matte images are sent as separate files or sequences. Therefore, it is usually better to create unmatted or *straight* color channels. This will give the fill image a "chunky" look, but it will key correctly without additional artifacts. Figure 5.10-24a shows how to avoid this problem. When the image from Figure 5.10-23 is used as both the fill and matte to key over a gray background, the result is a white circle over the gray background, but with a tiny dark edge around the circle. When the fill is replaced with a completely white frame, the edge disappears and the composite is created as it should be. Figure 5.10-24b shows a close-up of the difference.

The same holds true for solid objects that are not meant to be fully opaque. Place the transparency information in the alpha channel only; keep the color channels at full levels. Of course, full-screen graphics do not need an alpha channel and thus should be rendered with premultiplied (anti-aliased) color channels.

Video Editing/Compositing Programs

Once video has been digitized and stored within the computer, it is up to a software program to perform whatever manipulation is required. This could be as simple as cuts-only editing or as complicated as full multilayer compositing with DVE moves, morphs, motion tracking, etc. Most programs offer an intuitive *user interface*, often representing clips as reels of video frames (like filmstrips) and using menus and buttons to operate on these clips. One of the main advantages of general-purpose PCs is their capability to run several different programs on the same hardware. There-

fore, it is possible to capture a video clip with one program, send the video to a stand-alone morphing program, and finally composite the morph using yet another piece of software.

Video editing programs are most likely to use compression techniques to store the lengthy video clips used in long-form production. High-end systems will use the least compression (usually 2:1) for online work. With additional hardware, it is possible to process two or more video streams simultaneously, allowing dissolves and simple effects to be created in real time. These programs also handle multiple channels of digital audio as well.

For effects work, 2D and 3D animation programs are often used in conjunction with compositing software. These programs tend to work with highest quality, uncompressed video. As such, they often do not operate in real time. Although this type of work generally deals with short video clips, even a few seconds of complicated compositing can take an hour or more to render.

Glossary

CGI: Computer Generated Images. Still pictures or complete motion video created solely within a computer (i.e., without a camera).

Compositing: The layering of multiple video elements to create a montage of motion video.

DVE: Digital video effects creator. May be a standalone unit, combined with a switcher, or part of an integrated video workstation.

Field dominance: The specification of which field (one or two) is to denote the start of a new frame. Most facilities adopt the standard of field one dominance. Incorrect field dominance can result in juddery motion and "flicker frames" at edit points.

Fill: The foreground video signal that is keyed over a background using a matte.

Hicon: High contrast matte. A monochrome key signal ranging from black (making the fill totally transparent) to white (making the fill totally opaque).

In-betweening: Creating parameters for those frames "in between" keyframes of an effect.

Interpolation: Calculating the value of a pixel based on its surrounding pixels. Also used to describe the creation of frames when altering the speed of a clip.

ITU-601: Formerly CCIR-601, the standard for component digital video. Specifies a resolution of 720 × 486 (NTSC), 4:2:2 sampling (chrominance sampling at one-half that of luminance), and 8- or 10-bit signaling. Also the format used internally by many digital video devices, most notably the D1 VTR.

Matte: Also referred to as a key (or alpha channel when part of a computer image file); the signal used to determine which parts of the foreground and background images appear in the final composite. Can also be used to describe to the keying process itself.

Keyframe: A specific frame of an effect where certain parameters are specified.

Premultiplied: An image whose color channels have been multiplied by an anti-aliased alpha signal to create transparency and smooth edges. In making some pixels transparent, an implied background color (usually black) is assumed.

Rotoscoping: The frame-by-frame hand drawing of a video clip. The drawing may be used to create a moving matte or to add special hand-drawn effects to a clip.

Straight: An image whose color channels have not been premultiplied by an anti-aliased alpha signal. Any pixels with a transparency level less than 100% are represented as 0% transparent in the color channels, and the transparency level is carried in the alpha channel to create a correct composite.

Bibliography

A60 Interface Handbook, Abekas Video Systems, 1992.

"GREEN OR BLUE—Selecting a Backing Color for an Ultimatte Composite," *Ultimatte Technical Bulletin No. 2*, Ultimatte, Chatsworth, CA.

ITU-R Recommendation BT.601-1994, "Encoding Parameters of Digital Television for Studios," ITU.

Jack, Keith. *Video Demystified, Second Edition*, HighText Books, Solana Beach, CA, 1997.

"Matching Foreground & Background in a Composite," *Ultimatte Technical Bulletin No. 7*, Ultimatte, Chatsworth, CA.

Mazur, Jeff. "Compositing and Effects," *Broadcast Engineering*, Intertec Publishing, Overland Park, KS, pp. 48–54, November, 1996.

Pank, Bob. *The Digital Fact Book, 8th Edition*, Quantel, 1996.

"Shooting Film for Ultimatte," *Ultimatte Technical Bulletin No. 3*, Ultimatte, Chatsworth, CA.

SMPTE 125M-1995, "Television – Component Video Signal 4:2:2 – Bit-Parallel Digital Interface," Society of Motion Picture and Television Engineers.

SMPTE 258M-2004, "Television – Transfer of Edit Decision Lists," Society of Motion Picture and Television Engineers.

SMPTE 259M-1993, "Television – 10-Bit 4:2:2 Component and 4fsc NTSC Composite Digital Signals – Serial Digital Interface," Society of Motion Picture and Television Engineers.

Tozer, E.P.J., *Broadcast Engineer's Reference Book*, Focal Press, Burlington, MA, 2004.

"Ultimatte vs. Chroma Key," *Ultimatte Technical Bulletin No. 5*, Ultimatte, Chatsworth, CA.

Watkinson, J., *The Art of Digital Video, Third Edition*, Focal Press, Oxford, UK, 2000.

C H A P T E R

5.11

Postproduction and Nonlinear Editing

TIM CLAMAN

Avid Technology, Inc.
Tewksbury, Massachusetts

INTRODUCTION

The craft of making film and television programming has evolved dramatically over the past century, both aesthetically and technically. During this time, the working practices and tools used to create programs have evolved in parallel, constantly expanding the boundaries of creative possibilities. Film and television postproduction has become a diverse discipline, blending the crafts of storytelling, visual artistry, and sonic virtuosity with the hard science of motion imagery and sound.

Perhaps one of the most notable developments in the history of postproduction is the advent of the digital nonlinear editor (NLE), a tool that has fundamentally changed the way that programs are made. Rooted in the film and video editing tools that preceded it, the nonlinear editor offers content creators complete creative control, while automating the management of technical information critical to the integrity of the process. Nonlinear editors have transformed what had been a laborious and time-consuming process into a highly dynamic, creative process in which changes can be instantaneously explored without penalty. Now a mature technology, the nonlinear editor occupies the nexus of modern postproduction workflows.

This chapter covers the subject of postproduction in general, discusses the craft of editing, and briefly covers the history and development of the different techniques. It then provides information on modern NLE systems and examines some trends for the future.

POSTPRODUCTION DEFINED

Postproduction is the process of assembling raw images and sound into finished programs. As its name implies, the postproduction stage sits downstream of the planning and acquisition stages, but upstream of distribution and archive. Figure 5.11-1 shows where postproduction fits into the program production chain.

Postproduction encompasses far more than technical assembly. It represents an important phase in the creative process during which raw elements are given meaning, final decisions are made, and the original concept is fully manifested in a continuous stream of images and sound.

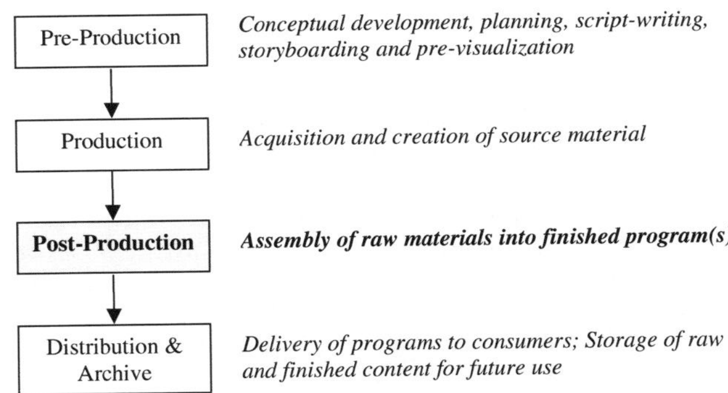

FIGURE 5.11-1 Basic program production chain.

The majority of television programming goes through some form of postproduction, although live television events are often produced and broadcast in real time. However, many live broadcasts also include previously assembled segments designed to supplement the live content.

When one works with film, particularly for movie production, the medium necessitates a postproduction process that is typically more complex than for video. This is true partly because the images must be processed before they can be viewed, but also because of the wide range of steps involved, both creative and technical.

A BRIEF HISTORY OF POST

The genesis of postproduction, also called simply *post*, can be found in the silent films of the early twentieth century. Before the emergence of motion picture technology, continuous visual performances could not be captured and replayed. While many early films were little more than a literal record of a single, staged performance, filmmakers quickly recognized the enhanced creative control they could exert in postproduction. Film editing enabled filmmakers to synthesize an idealized single performance by combining snippets from many disparate performances and viewing angles. One could say that without the film medium, postproduction would not have become necessary, but without postproduction, filmmaking would never have become a true art form.

Before the advent of television broadcasting, postproduction changed the way that news information was disseminated. Newsreels gave civilians a firsthand, if delayed, glimpse of warfare and other newsworthy events. By the 1930s, postproduction had also become a powerful propaganda tool as Nazi Germany pioneered new techniques in film and sound editing to create forceful, nefarious films such as *Triumph of the Will*.

Since the early days of filmmaking, public awareness of postproduction has steadily risen. Today's consumer can often choose between different edited versions of the same program, such as the "Original Release" and the "Director's Cut." The inclusion of supplemental material in DVD releases often gives the viewer direct visibility into the postproduction process. For example, the supplemental material included in the DVD for David Fincher's *Panic Room* provides a detailed look at the postproduction process used for the film. The emergence of interactive content has transformed the passive viewer into a virtual participant in the postproduction process. In the DVD for James Wong's *Final Destination 3*, for example, the viewer can enable a "Choose their fate" option, which occasionally pauses playback of the movie to offer choices to the user that affect the resulting action.

Furthermore, inexpensive consumer camcorders and personal computer based software packages intended for consumer use[1] have provided the creative power of postproduction to the general public.

Despite the wide proliferation of basic tools, professional postproduction workflows continue to expand in complexity and diversity, stretching the capabilities of the creative teams and driving the continuous advancement of professional systems.

POSTPRODUCTION WORKFLOW

Today, professional postproduction workflows range from very simple tasks such as "topping and tailing" a clip for a news program to highly complex workflows in which dozens of specialists collaborate for several months to craft an elaborate feature film. The broad diversity of workflows makes it impossible to abstract a universal postproduction workflow process. For instance, the postproduction workflow for an animated feature film looks very different from that of a news magazine program. However, a very basic postproduction workflow for a typical live action program can be outlined roughly as shown in Figure 5.11-2.

Although this diagram depicts a linear progression through different phases, postproduction is a highly iterative and often chaotic process. Workflows must be carefully tailored dynamically to meet the needs of the program.

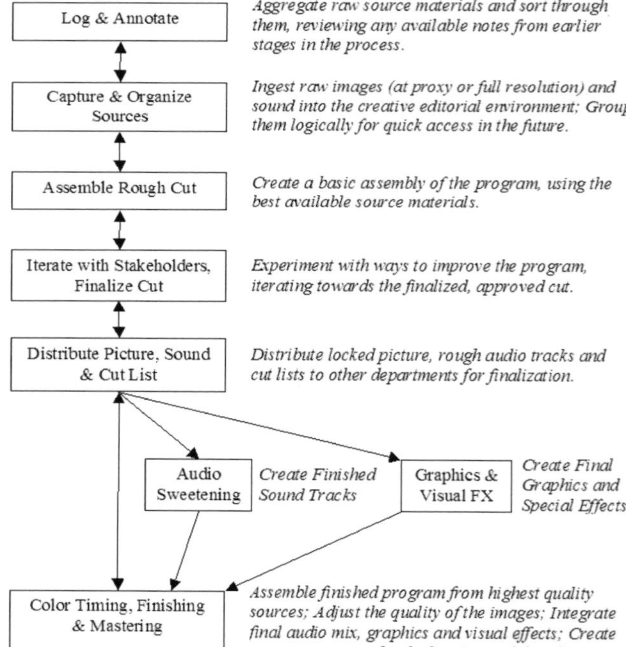

FIGURE 5.11-2 Postproduction workflow for live action program.

[1]Examples are Pinnacle's *Studio* and Apple's *iMovie*.

TABLE 5.11-1
Factors Affecting the Postproduction Workflow

Scope	What kind of program is to be made? What is the aesthetic vision of the creative team? What technical challenges will the team encounter?
Delivery Requirements	What exactly needs to be delivered? How will this program be distributed? How many versions must be created? How will the program be stored for future re-use?
Available Resources	How much budget is apportioned to postproduction? What personnel, equipment, and facilities are available?
Delivery Schedule	When is the finished product(s) to be delivered? What are the interim milestones?
Leadership & Authority	Who has ultimate creative authority for this program? Whose sign-off is required for each milestone? How decisive and well aligned are the stakeholders?
Talent Preferences	Who has the required skill and talent to fulfill each role? What are the team's preferences regarding workflow, process, and tools?

Designing Postproduction Workflows

Many factors influence the structure of a postproduction workflow, as shown in Table 5.11-1.

The answers to these questions vary broadly from one production to another. As such, perhaps the most important factor in the success of a postproduction team is its ability to anticipate the specific needs of that production during the planning and budgeting stages.

Postproduction Roles

The successful fulfillment of a postproduction workflow depends on the team and its "chemistry." While there are many different specialized roles in the postproduction process, there is no universal naming convention for these roles. Generally speaking, postproduction roles can be classified into roughly six functional groups, as shown in Table 5.11-2.

TABLE 5.11-2
The Six Functional Groups of Postproduction Participants

Stakeholders	Provide creative leadership and direction Make decisions and provide sign-off at milestones *Sample roles include the Director, Studio or Network Executives, Investors, Producers, and Writers.*
Process Managers	Coordinate the team, manage resources, define and execute the process Act as communication link between stakeholders and the team *Sample roles include the Postproduction Supervisor, Line Producers, Postproduction Coordinators and Assistants.*
Creative Editorial	Carry out the assembly of the program Guide creative decisions regarding which elements will be used in the program as well as their relative timing *Sample roles include Editors and Assistant Editors.*
Sound Department	Create finished soundtracks from a variety of disparate sources *Samples roles include Sound Supervisor, Sound Editor, Sound Designer, Music Supervisor, Music Editor, Rerecording Mixer.*
Visual Effects & Graphics	Create visual elements for inclusion in the program, including graphics and special effects *Sample roles include Visual Effects Supervisor, Artist.*
Finishing	Adjust the visual quality of the finished images Build the finished program masters Create multiple versions and deliverables *Sample roles include Colorist, Online Editor, Negative Cutter.*

Team size can vary significantly from one production to another. An independent documentary, for example, may require only a half-dozen roles including writer, editor, sound editor, music editor, colorist and rerecording mixer. Due to budget constraints, the documentary filmmaker himself or herself may play virtually all of these roles. In contrast, a high-budget feature film typically has a large number of highly skilled specialists involved in different aspects of postproduction, as can be seen from the number of credits listed at the end of a motion picture, although not all of those involved are typically listed. Television programming often falls somewhere in between, although there are great variations depending on the material.

THE ROLE OF EDITOR

The role of editor lies at the heart of all postproduction workflows. The primary function of the editor is to execute and refine the vision of the program's stakeholders. In many cases, the editor may play a pivotal part in the creative process, acting as a consultant and advisor to the stakeholders. Editing is a craft that balances aesthetic sensibilities, technical knowledge, diplomatic skill, experience, and technique. The editing process is the place where all timing decisions are made, as well as all decisions about which material will be used in the finished program.

In many production schedules, the editor begins work while shooting is still in progress, reviewing takes (often called *dailies*) as they become available to gain familiarity with the variety of performances and camera perspectives captured for each scene. Early assemblies of scenes may also reveal inadequacies in the source material, which may trigger additional shooting, often referred to us *pick-ups*.

The editing process is highly dynamic, iterative, and at times experimental. The editor works with the stakeholders to explore the performances and visual imagery available for use in each scene, piecing elements together in different combinations. In many cases, the plot line may be altered significantly during editing as scenes are removed, truncated, or rearranged in the cut. Entire characters and subplots may be eliminated or enhanced in the cutting room.

Rhythmic pace and flow of the program are explored by the editor early in the postproduction process. It is the editor's job to painstakingly adjust each transition in a program to elicit the desired reaction from the viewer. In some cases, this requires careful attention to "continuity," the degree to which a series of edits appears as continuous action. For example, it may be disruptive to the viewer to see a character reach for the doorknob and then instantly appear in the open doorway on the other side of the edit. To create the illusion of one organic performance, the editor watches each frame of footage and carefully adjusts transitions between takes by adjusting their relative timing on a frame-by-frame basis. Of course, sometimes continuity is intentionally broken by the editor to accelerate the pace of the program or to intentionally jar the viewer, evoking a sense of displacement and disorientation.

The first assembly of the program is typically the "Editor's Cut," which may be built by the editor with or without input from the stakeholders. For programs that include a script, screenplay, or storyboard, the editor will usually follow the original vision closely, selecting the takes preferred by the director, as indicated in shot logs—notes taken by the production team during the shoot.

In film projects, the primary creative stakeholder is the director, so after the editor's cut, the director may be permitted to complete a "Director's Cut." In television, the producer may be entitled to create a version of the cut. At this point, the cut is viewed by other stakeholders, such as executives and investors. The editor may then receive additional direction in the form of *change notes* that are executed to produce the final cut. In reality, this evolution process is anything but linear. There are typically many iterations and retrogressions. It is the editor's job to track and maintain many alternate versions as they branch from one another so that different cuts can be compared, merged together, or reverted to as necessary.

Through all these gyrations, the editor must remain responsive, flexible, collaborative, creative, and well organized. But editing can be a very stimulating and rewarding endeavor. Stanley Kubrick said, "I love editing. I think I like it more than any other phase of filmmaking. If I wanted to be frivolous, I might say that everything that precedes editing is merely a way of producing film to edit." (See *Stanley Kubrick Directs* listed in the Bibliography.)

EDITING STYLES AND TECHNIQUES

Different programs require different styles of editing. Over the past century, the art of editing has evolved in conjunction with the art of filmmaking. Although many approaches and techniques have emerged, they can be grouped into three basic editing styles:

- Documentary;
- Montage;
- Continuity.

Documentary Editing Style

As its name implies, *documentary-style editing* is meant to simply convey the on-screen action in a highly literal, nonfictional manner. The earliest films on record, dating to the turn of the twentieth century, simply documented a real event. These "actuality" films were typically very brief and often contained only one continuous take, shot from one camera angle. But as film editing technology evolved, the films themselves became more coherent, if at times less authentic. Travelogue films and newsreels of the early twentieth century appeared to the viewer as unadulterated accounts of unstaged events, even though the action they depicted was often completely staged.

Subsequent decades witnessed the emergence of several different movements within the documentary filmmaking community, including *Cinéma Vérité* and the *French New Wave*, which tested the outer boundaries of realism in film. Today, documentary-style editing is used on television and in film. So-called reality shows and news magazines dominate program schedules, while documentary-style films[2] have yielded great commercial success at the box office and in DVD distribution.

The emergence of inexpensive high-quality cameras, such as DV camcorders, and powerful but readily available software editing tools has helped to spark a new wave of documentary filmmaking during the past decade. More independent films (as well as home moves) are being produced now than ever before.

Editing documentary-style programs is a highly challenging process, as the raw material may have no structure of its own to begin with. The story itself is literally created during the editorial process, elevating the role of editor to principal storyteller. Documentary-style programs often begin with large amounts of raw footage that must be viewed, categorized, culled from, and ordered into a coherent whole.

Montage Editing Style

In the French language, the word *montage* simply means *assembly*. While this word may be used by French filmmakers to refer to the general process of editing, it has also come to connote a style of editing that is more subjective and evocative. *Montage editing* does not attempt to depict a literal, coherent scene. Instead, montage juxtaposes disparate images together to evoke a more symbolic, impressionistic meaning.

As film editing evolved in the 1920s, early filmmakers like Lev Kuleshov and Sergei Eisenstein began to examine the discipline in a more analytical, academic manner. For Kuleshov, editing embodied the unique, defining essence of filmmaking. His influential experiments provided evidence that the basic act of editing—assembling different shots together—could lead the viewer to infer coherent new meaning from the program. Later, Sergei Eisenstein stretched the boundaries of montage editing by intentionally contrasting disparate shots together to trigger subjective associations in the viewer that may evince a more abstract meaning.

Today, montage is often used to convey a mood or to condense the narrative progress of a program. In montage sequences, the passage of time becomes nonliteral and subjective. In place of on-screen dialogue and narration, music is often employed as the primary sound element to spark the user's emotions and to suspend the sense of literal time. Examples of montage can be found in feature films, such as the inspirational training sequences in *Rocky*, as well as in television, such as the opening sequence of dramatic television programs such as *CSI*.

Continuity Editing Style

The goal of *continuity editing* is to create an apparently seamless, continuous scene. Continuity editing techniques are used widely in dramatic programming, where the goal of the editor is to present a natural flow of events that is perceived by the user as continuous and natural. In this style of editing, the editor desires the work to be as invisible as possible—the viewer should feel as though it is continuous, real-time action and should not notice that any editing has occurred.

Continuity editing has its roots in early Hollywood films. American directors like D.W. Griffith pioneered the art of continuity editing in films,[3] evolving new techniques that became a fundamental part of the editing vocabulary, including

- *Establishing Shot*: Typically, a brief static shot employed at the beginning of a scene to convey the location and/or relative position of key objects and characters.
- *Cut In:* An instantaneous cut from a distant shot to a closer view of the action, used to quickly draw the viewer into the scene.
- *Reverse Shot*: A method of alternating camera angles used typically for conversations between characters.
- *Crosscutting*: Alternating between two or more lines of actions such that the action appears to be occurring simultaneously, but in two locations.
- *Eyeline Match*: A sequence of shots that begins with a shot of a character looking at something, followed by a shot of what he or she is supposedly looking at.
- *Action Match*: Editing so that the action appears to be continuous and seamless.
- *The 180 Degree Rule*: A method of shooting and editing whereby camera angles are confined to one side of the action, enhancing the viewers' sense that they are physically present, observing the action.

Successful execution of continuity editing requires careful planning and *storyboarding* before the shoot to ensure that the proper camera angles are covered. Special attention to continuity is required during shooting, typically the responsibility of the script supervisor. Continuity editing is perhaps the most common of editing styles and is used widely today in dramatic films and television shows.

THE EVOLUTION OF EDITING TECHNOLOGY

While the craft of editing transcends the tools, the introduction of new editing tools and technologies has

[2]Recent examples include *Fahrenheit 9/11, Super Size Me,* and *March of the Penguins.*

[3]Examples are *The Birth of a Nation* and *Intolerance.*

clearly influenced the editorial discipline. For example, the fast pace of editing in MTV-era music videos was enabled by the widespread use of nonlinear editing tools in the late 1980s and early 1990s.

The history of editing tools can be segmented into roughly four time periods that are exemplified by the defining technologies of each era:

- Film editing
- Manual videotape editing
- Electronic videotape editing
- Nonlinear proxy editing
- Nonlinear finishing

Film Editing (ca. 1900 to Present)

Editing film is a highly tactile process involving cutting and splicing pieces of film together with film cement or a form of adhesive tape. Film editors cut a *work print*, a positive copy of the original camera negative. The work print allows the editor to experiment with editing the program together without touching or damaging the original camera master. In this sense, film editing is a proxy workflow in which the work print acts as a surrogate for the original camera negative until all the creative decisions have been finalized.

Because the work print is positive, it can be viewed easily on an optical viewing device such as the traditional Moviola editor that was widely used in Hollywood, or flatbed editing machine, as shown in Figure 5.11-3.

At any time, the edited work print can also be viewed with a projector in a screening room or theater.

Film Editing Process

The television and film industries have long used film stock that includes *visible ink image edge numbering*, unique number codes that are applied to the film stock by the manufacturer. These *edge code* numbers, which can be seen on the edge of the film stock and which survive film processing, are used to uniquely identify each frame of source material.

FIGURE 5.11-3 Flatbed film editor.

Once the creative editorial decisions have been finalized, a *cut list* is created that specifies which source frames have been sequenced together to create the final cut. This cut list contains instructions to the negative cutter, a list of edge numbers, which prescribe how to create the same edits using the original camera negative. The negative is then physically cut and spliced together according to the cut list, typically in two "A" and "B" rolls of film, from which the final film prints are printed optically.

The film medium is inherently nonlinear. The editor can add or remove any frames from any part of the program at any time, enabling complete editorial flexibility and creativity. But the physical nature of film means that making changes can be very labor intensive. Small clips of film called *trims*, sometimes as small as a single frame, must be stored and organized in such a way that they can be found and used in the future at the whim of the editor. To help with this process, a film editor typically has more than one assistant actively managing the physical film media using synchronizers, pen and paper, and large canvas bins.

Sound was also traditionally edited on magnetic film stock, also known as *mag*. Layering of sounds in the soundtrack was accomplished by creating many parallel tracks of edited mag (often on separate lengths of film), which would be played back simultaneously and summed through a mixing console.

Today, virtually all film workflows involve digital tools for picture and sound editing as well as visual effects creation. With the move to *digital cinema*, film finishing and distribution are increasingly accomplished in the digital domain. The introduction of ultra-high resolution digital cameras has begun to reduce the number of productions shooting on film. Still many cinematographers continue to shoot on film, and a few prominent editors continue to edit on film.

Manual Videotape Editing (ca. 1956 to 1970)

In the late 1950s, television broadcasters were introduced to videotape, a new technology that could record programs for future playback. Videotape had several benefits over film: it stored picture and sound on a single reel and could be erased and re-used. Furthermore, videotape captured and generated signals that were designed to be directly compatible with television broadcast systems.

Before the introduction of videotape, television was a predominantly live medium, for which different input signals were switched on the fly in the control room and transmitted instantly to viewers. Videotape was initially attractive to American broadcasters as a means of delaying the broadcast of programs for different time zones, but quickly became the standard physical medium for storing and distributing television programs. Unfortunately, the relatively high expense and also the lack of durability of early videotape stock mean that very few early programs have survived.

The first successful videotape machines, which gained widespread use by broadcasters, were the *quadraplex* (or simply *quad*) machines introduced by Ampex in the United States in 1956. The quad format, which would dominate the broadcast world for roughly two decades, utilized magnetic tape stock that was 2 inches wide, running at 15 inches per second and scanned transversely by a combination of four rotating video heads as well as linear heads for the audio signals. Quad was inherently somewhat fragile but delivered to the viewer for the first time an image quality that was largely indistinguishable from live broadcasts.

Quad was somewhat film-like in that it was stored in open reels and could be cut and spliced in a splicing block or *jig*. However, editing quad tape was much more difficult than editing film due to the lack of a visible image on the medium itself and the fact that still frame playback was not possible with this tape format. Edit points could therefore only be selected on the fly during playback with cut points physically marked on the tape. To join two pieces of videotape together, a solution of iron filings suspended in carbon tetrachloride was applied to each end of tape to be joined to expose the magnetic tracks so they could be visually aligned in the jig. It was difficult to edit the tape without disturbing the odd/even field ordering inherent to the interlaced scanning of television signals, and the physical splices were unreliable.

Another significant drawback of manual videotape editing was the lack of *edge codes* (or an equivalent method) for uniquely identifying individual picture frames. This meant that there was no feasible proxy workflow for videotape editing, rendering it cumbersome for creative editorial applications, adequate only for basic compilation of complete programs.

Therefore, although manual videotape editing was theoretically nonlinear like film, it was much more cumbersome. In light of the workflow disadvantages, manual tape was not widely adopted and most non-live television shows continued to be shot and edited on film for decades after the introduction of videotape.

Electronic Videotape Editing (ca. 1970 to Present)

In the early 1960s, new techniques emerged for assembling programs on tape by cueing and playing back videotape machines sequentially, recording the result on another videotape machine. However, the results were often poor, producing audible audio artifacts and visible color shifts at the edit points. And the edits were not frame-accurate as they were with film. Broadcasters and videotape machine manufacturers recognized that a more sophisticated method for synchronizing videotape machines was required.

In the late 1960s, a method for applying frame-accurate codes to videotape was created, similar to the edge code concept of film. The Society of Motion Picture and Television Engineers (SMPTE) standardized a form of *timecode* that enabled precise synchronization of videotape machines and reliable, frame-accurate assembly of programs from cut lists. Today, SMPTE timecode[4] remains ubiquitous in most postproduction workflows.

TABLE 5.11-3
Simple Edit Decision List

Edit	Source	Start Timecode	End Timecode
1	Source Tape 1	1:30:43:25	1:30:47:29
2	Source Tape 2	4:23:37:18	4:23:44:12
3	Source Tape 1	1:42:21:06	1:42:28:00
4	Source Tape 3	6:10:12:11	6:10:19:17
5	Source Tape 2	4:15:06:10	4:15:11:12

Timecode enabled accurate, repeatable videotape editing. Linear, electronic editing suites were typically equipped with at least one machine to play back source tapes, one machine to record the assembled program (on the edit master tape), and an *edit controller* for managing the synchronization and offsets between machines.

As computer technology evolved in the 1970s, computer-based editing systems emerged. Vendors such as CMX, a collaboration between CBS and Memorex, created software programs that would automatically store the edit timing information in *edit decision list* (EDL) files so the program creator needed only to focus on the visual imagery at each edit point. Although there are several different specific EDL formats, an EDL is basically a set of instructions for how to re-assemble a program from the original sources. The example shown in Table 5.11-3 is a basic EDL that describes how to assemble a 30-second program containing five edits, compiled from three different source tapes.

With the introduction of SMPTE timecode, videotape editing was finally as accurate and repeatable as film editing. More sophisticated EDLs include other information such as the type of transition between clips (wipes, dissolves, and so on) and separate audio sources.

Proxy Tape Editing

Electronic editing with timecode also introduced the concept of *proxy editing* for videotape, which was possible because timecode signals could be recorded along with the program during videotape duplication. With this system, a copy of the program was recorded on a low-cost tape format such as U-Matic or VHS, usually with timecode *burned* into the video so it could be seen on screen. The time-consuming creative process of reviewing and making all the edit decisions was then done *offline*, without tying up an expensive

[4]SMPTE 12M for Television, Audio and Film, Time and Control Code, Society of Motion Picture and Television Engineers.

broadcast-quality edit suite. A resulting EDL was then used in an *online* edit suite to edit the original source tapes to produce the finished program. This process is known as *conforming*.

One drawback of electronic editing was that it required another generation of recording. Given the quality limitations of analog tape recording, this placed a limit on how many passes through the editing process could be made before the quality became unacceptable.

Limitations of Tape Editing

Unfortunately, in moving to electronic editing, the process of editing videotape had also become completely linear, a step backward for the creative editorial craft. With linear electronic tape editing, it is impossible to go back and make changes in a program without having to reassemble the remainder of the program. If the editor wants to add or remove frames at the beginning of a program, there is no way to "push" the rest of the program downstream as with physical media—it must be reassembled or rerecorded onto a *submaster* for re-inclusion in the master tape, incurring time and effort and a significant loss in image quality.

With the advent of computer-based nonlinear editing in the late 1980s, linear tape-based editing would be relegated gradually to the domain of program finishing (online editing) where few, if any, timing changes take place.

Nonlinear Proxy Editing (ca. 1989 to Present)

In the early 1970s, although linear videotape editing was becoming a mature technology, there was a growing awareness of its drawbacks, which included

- *Speed:* Source material could not be accessed quickly. Tapes had to be constantly loaded, cued, rewound, and unloaded, slowing down the creative process.
- *Change Management:* It was difficult to experiment with changes to the program after the initial assembly because the "downstream" remainder of the program had to be re-assembled.
- *Nonsequential Editing:* It was not possible to edit different portions of the show out of order without introducing cumbersome "submasters" that would eventually have to be tied together.

As early as 1972, nonlinear computer-based editing systems were being developed to provide random access to source material and the ability to produce trial edits that could be reviewed and changed.

Early systems, while providing some of the required features for the creative process, were not capable of producing broadcast-quality video. They were therefore used as *proxy* systems, in which the program was first edited to produce an offline version of the end result, together with an EDL. The EDL was then used to produce the finished program using the original source material in an online tape edit suite.

Systems such as the CMX *600* stored video on magnetic discs for random access during creative editing session. But high prices (roughly $250,000 per system), poor image quality, and small storage space (27 minutes maximum) hindered adoption. It was not until the late 1980s that advances in computer technology made digital nonlinear editing both practical and economical.

During the 1980s, several hybrid editing systems were brought to market. In 1983, the Montage Group's *Montage Picture Processor* system utilized a graphical user interface (GUI) in combination with timecode-based synchronization of up to 17 videocassette decks to assemble a videocassette master in quasi-real-time. For editorial flexibility, each video deck was loaded with an identical copy of a single source tape, providing some semblance of nonlinearity, and a computer kept track of where the machines were. Black-and-white head and tail frame images were digitized from tape and displayed on screen, providing the editor with a way to identify clips visually without rolling tape. The Montage system also utilized physical controllers designed to emulate the tactile interaction of flatbed film editors.

In the mid-1980s, the Cinedco *Ediflex* system enjoyed some success in episodic television production. The Ediflex combined a bank of 12 VHS tape cassette decks with a script-driven software interface, making it ideal for dramatic program editorial work. Roughly concurrent with the Ediflex, Bell & Howell introduced *TouchVision*, a similar system with a more intuitive icon-driven user interface driven by touch-sensitive screens. While short-lived, these hybrid computer/tape systems began the evolution of the nonlinear editing graphical user interface.

But hybrid systems were clumsy, complex, and expensive. Aimed at delivering true nonlinear editing, laserdisc-based systems such as the LucasFilm *EditDroid*, Amtel *E-Pix*, and the CMX *6000* were introduced in the late 1980s. These pioneering systems were rather expensive, typically costing more than $100,000, but gave the industry its first taste of the possibilities of true digital nonlinear editing.

In late 1988, Editing Machines Corp. (EMC) introduced an editing system that used magnetic hard drives and IBM-compatible personal computers (PCs). Although image quality was poor (single field, black-and-white images at 15 frames per second), the *EMC2* was considerably more affordable than laserdisc systems and provided easy re-use of the magnetic disks.

Macintosh and PC-Based Systems

In early 1989, Avid Technology introduced *Avid Media Composer*™, built on the Apple Macintosh II platform, which enabled true 30 fps editing on magnetic disks for the first time. While still being a proxy offline editing system, the Avid Media Composer was widely adopted and quickly evolved the standard three-point editing paradigm used in most modern nonlinear editors today.

During the early 1990s, the Media Composer application was expanded and customized for other

applications such as film editing at 24 fps (*Film Composer*™), multicamera editing, and news editing (*NewsCutter*™).

Since that time, many nonlinear editing applications have proliferated and extended the seminal Avid Media Composer user interface, including inexpensive applications like Adobe *Premiere*™, Apple *Final Cut Pro*™, and Avid *Xpress Pro*™, running on both Macintosh and PC platforms.

Nonlinear Finishing (ca. 1998 to Present)

In the late 1990s, computer technology had advanced to a point where nonlinear editing was no longer confined to proxy workflows. Systems such as Avid *Symphony*™ could capture, edit, process, and output multiple streams of uncompressed standard definition (SD) digital video (ITU-R BT.601) in real time.

Today, a variety of nonlinear editing products from Adobe, Apple, Avid, Autodesk/Discreet, Canopus, Quantel, Sony, and others work natively with SD video and uncompressed high-definition digital video (SMPTE 174M [2], SMPTE 296M, ITU-R BT.709) and, in some cases higher resolution formats like 2K and 4K (film-derived formats with 2000 and 4000 horizontal pixels; see Chapter 5.23, "Film for Television").

NLE SYSTEM COMPONENTS

Historically, the components of an NLE edit suite typically included the following:

- A computer with keyboard, mouse, and monitor(s);

FIGURE 5.11-4 Modern nonlinear edit suite. (Courtesy of Fotokem Film and Video.)

- Nonlinear editing software;
- An input/output interface which converts between baseband signals and digital data streams;
- A calibrated viewing monitor;
- An audio monitoring system;
- Audio mixing system (optional);
- Videotape source and record machines (as required).

Figure 5.11-4 shows these components installed in a modern NLE suite, and Figure 5.11-5 shows a typical system arrangement.

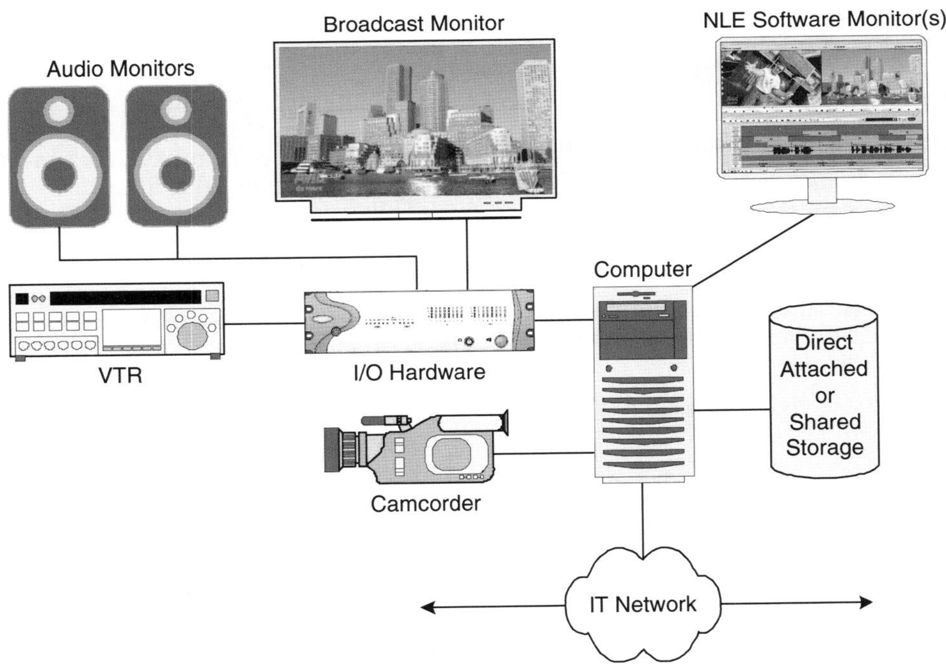

FIGURE 5.11-5 Nonlinear editing system diagram.

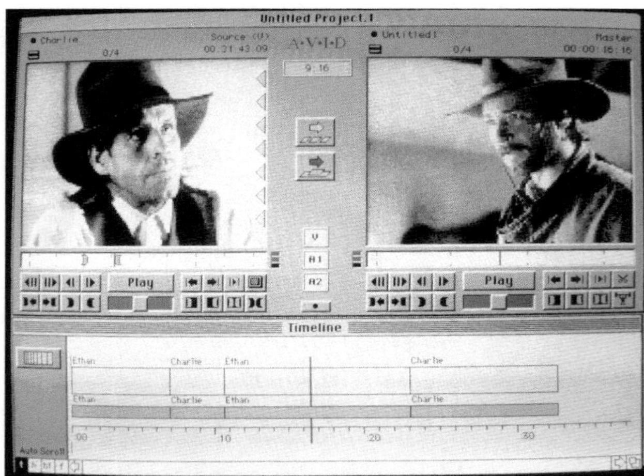

FIGURE 5.11-6 Screen shot of Avid Media Composer/1™.

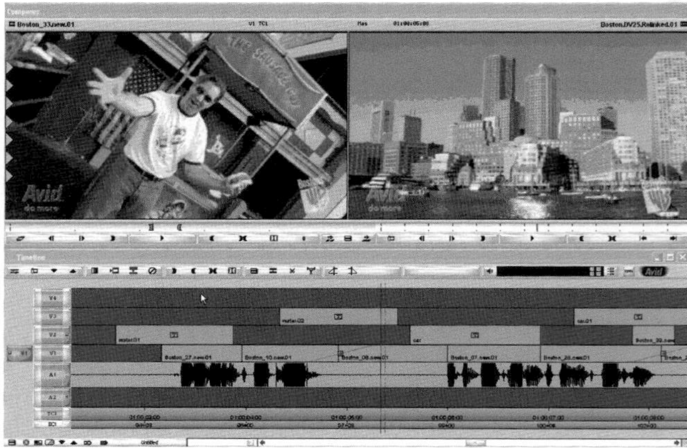

FIGURE 5.11-7 Screen shot of modern videotape editor.

Today, the only absolute requirement for nonlinear editing is a computer with NLE software, although some of the peripherals mentioned in the preceding list may be required in a system environment. The advancements in computing power, efficient codecs, and editing software have enabled even laptop computers to be fully capable of professional nonlinear editing.

NLE USER INTERFACE BASICS

During the past two decades, nonlinear editing applications have matured into highly sophisticated and capable tools with many different modes, menus, and windows. Yet most NLEs continue to employ an editing interface with four primary windows as introduced in the original systems:

- *Project Folders or Bins:* Virtual containers used for organizing material
- *Source Viewer:* Used for reviewing material and identifying which frames of content will be used in the program
- *Timeline Window:* A horizontal representation of the program which depicts the relative timing and layering of the source elements
- *Master Viewer:* Used for navigating through the program as currently assembled

The image in Figure 5.11-6 shows the original Avid Media Composer/1™ user interface from 1989.

The upper-left area on the screen is the Source Viewer, the upper-right area is the Master Viewer, and the Timeline is along the bottom. This basic three-pane layout became the industry standard because it combined the best elements of film editing and electronic videotape editing. The Source and Master Viewers closely emulate the two-monitor configuration used in videotape editing bays, allowing the editor to see the incoming frames (Source) and outgoing frames (Record) simultaneously while editing.

The program tracks in the timeline are laid out horizontally—virtual media on a virtual flatbed editor or synchronizer. However, unlike film or videotape, the Timeline provides the editor with a graphical representation of the edit and tools for directly manipulating the edit at random, without affecting other areas of the program.

This basic setup is still used today, as shown in Figure 5.11-7. Products from different vendors have some variations, but most are based on similar principles. Common NLE terms and concepts are outlined in Table 5.11-4.

NLE PROXY WORKFLOW

As mentioned previously, nonlinear editors were designed originally for proxy offline editorial workflows. Typical NLE proxy workflows follow roughly the path shown in Table 5.11-5.

This legacy proxy workflow is still in wide practice today, despite the fact that NLE systems can now work with full-resolution images, graphics, effects, and sound. Some of the reasons why offline editing has failed to disappear from postproduction workflows include

- *High Definition (and beyond):* HD images can contain six times the number of pixels of standard definition video. And image sizes larger than HD are becoming increasingly common, including 2K and 4K. Because many programs contain many hours of source material, it is often most practical to work with reduced-resolution proxies of the high-resolution sources to save on the disk capacity required for storage.
- *Economics:* High-resolution playback sources, such as high-end HD videotape recorders, are expensive to own and rent. It is often more practical to distrib-

TABLE 5.11-4
Common Nonlinear Terms

Clip	A virtual strip of film or videotape May contain multiple parallel streams of video (e.g., multicamera takes) and/or audio Contains time-based metadata which describes the source used, annotation information, and any other known technical parameters
Sequence	An edited assembly of one or more clips Sequences may be very short and simple or very long and complex, representing an entire edited program Sequences vary in horizontal density (e.g., slow- versus fast-paced edits) and in vertical density (e.g., one versus many visual layers)
Bin or Folder	A virtual container used for organizing clips, sequences, and other application objects (e.g., titles, effects presets, etc.)
Project	A virtual collection of bins, folders, and settings associated with a particular project
Media Files	The files stored on disk that contain the frames of picture and samples of sound that are used in a program, often referred to as "essence" Each clip may be "linked" to a set of media files at a given image quality Unlike the other concepts outlined above, Media Files are *not* virtual application objects; they are actual files on a hard drive and must be managed carefully

ute proxy files or tapes than it is to make expensive decks available to all editors.

- *Speed:* Proxy files are much smaller and therefore faster to move across networks. File-based camera workflows are greatly accelerated by the use of high-quality proxy files that contain metadata links to high-resolution files.
- *Specialization:* Creative offline editorial, color grading, visual effects creation, and finishing are different disciplines which require different skills and in many cases different specialists. It is not always beneficial for the offline editor to work with full-resolution images, since the editor may not be responsible for the visual look of the program.

NLE WORKFLOW TRENDS

Over the past decade, advancements in several areas of technology have exerted a profound influence on NLE workflows, including

- *Proliferation of Tools:* The dramatic decrease in prices for professional cameras and NLEs has drastically

TABLE 5.11-5
NLE Proxy Workflows

Transfer & Duplication	Create proxy tapes from original camera masters Film sources and original video masters are transferred to inexpensive tape, such as ¾" U-Matic or DV Audio sources are synchronized and recorded onto proxy tapes Source metadata such as timecode and edge code are logged and preserved during the transfer
Capture Sources in NLE	A/V signals with timecode are "digitized" into the NLE Compression algorithms are used to reduce the amount of storage bandwidth and capacity required for editing Source metadata is imported (e.g., FLEx file, ALE file) or manually entered during capture sessions File-based sources (e.g., graphics) are imported into the NLE for integration into the program
Creative Editorial	Source material is evaluated and pieced together in rough assemblies, or "cuts" Visual effects, titles, and graphics are previsualized and integrated with source material in the NLE Iteration eventually results in an approved final cut
Dissemination of Edit Decisions	The final cut is manifested in an Offline Master tape or file, which acts as a reference for other collaborators Edit decisions are exported in an electronic file—EDL, OMF, AAF, XML, or Cut List Rough audio edits are delivered to audio sweetening on tape (e.g., DAT, DA-88) or as files (e.g., WAV, OMF, AAF) In some cases, changes to the program continue after initial dissemination, necessitating careful change management
Conform & Finish	The edit decisions established in the creative editorial process are used to reassemble the program from the original full-resolution sources (camera masters or film negative) Finished graphics, audio sound tracks, and special visual effects are integrated into the program Finished masters are created on tape, film, or as files

increased the quantity of unique content created by the postproduction industry.

- *IT Interconnection:* Application of standard information technology protocols such as TCP/IP, IEEE 1394, and USB2 have made it possible to transfer professional-quality video between systems without special-purpose hardware.
- *Codecs:* Industry-standard video compression coding systems, such as MPEG, JPEG, and DV, have made it possible to store high-quality compressed images economically on hard drives, as well as exchange them between different systems.
- *Processors:* The continuing growth of processing power found in commodity central processing units and graphics processing units has enabled NLEs to support higher resolution content, manipulate more simultaneous layers in real time, and generate complex visual effects.
- *Storage:* As IT-based storage solutions have grown in storage capacity and bandwidth, NLEs have been given access to larger stores of higher resolution content. The development of custom file systems and the adaptation of IT storage servers have created cost-effective pools of storage that can be shared simultaneously by multiple NLEs.
- *Internet:* The ubiquity of broadband Internet connectivity has enabled distributed workflows where project files are moved between collaborators and approval files are easily transferred over the wide area network (WAN) for review by stakeholders.
- *File-based Cameras:* The development of file-based cameras such as Panasonic P2™, Sony XDCAM™, Ikegami EditCam™, and Thomson Infinity™ have enabled faster than real-time ingest of source material, while enhancing the metadata link between production and postproduction.
- *On-air Servers:* The proliferation of video servers in on-air applications has eliminated the need for videotape masters in many workflows. Finished programs can be sent to air by simply transferring a rendered file over a standard IP network. All these advancements have created new opportunities and challenges for postproduction teams.

POSTPRODUCTION TRENDS

The needs and processes of postproduction have also evolved, and will continue to do so, based on changes in technology and the availability of higher performance systems at lower cost. Driving forces for change also include new business models based on the economics of the media business and the changing expectations of program producers, broadcasters, and viewers.

Perhaps the most significant trend in postproduction is the rapid transition away from physical media such as film and tape. This evolution towards digital source and delivery formats along with the adoption of IT technology is sparking the innovation of new, nonlinear workflows in an enterprise environment, and promises many potential benefits to content creators, including:

- Higher team efficiency, productivity and creativity;
- Higher resolution content with higher production values;
- Lower production costs;
- Access to more distribution channels;
- Content that can be flexibly customized and repurposed for different outlets.

So far, however, this transition has been anything but painless for postproduction professionals. The recent explosion of different file formats, codecs, color spaces, aspect ratios, and metadata schemas has created confusion and disarray in the postproduction industry. The costs of managing additional complexity have coupled with an increase in delivery requirements, effectively offsetting productivity gains while destabilizing historically robust workflows.

Postproduction workflows are evolving in real time to adapt to this changing landscape. One of the key challenges of this transition is managing larger quantities of metadata throughout the production chain: both technical and user metadata.

Historically, technical metadata was relatively easy to manage. Film and video tape were the dominant media, with very few technical permutations. The transition to digital file-based media has created more flexibility, but also more complexity. A much larger range of technical metadata must now flow organically through the postproduction process, including:

- *Aspect Ratio:* 1.25:1 (LCD), 1.33:1 (4:3), 1.43:1 (IMAX), 1.5:1 (35 mm still), 1.56:1 (14:9), 1.66 (Super 16), 1.78:1 (16:9), 1.85 (widescreen theatrical), 2.39:1 (scope);
- *Image Size/Raster:* 480i, 480p, 720p, 1080i, 1080p, 2k, 4k, 8k, etc.;
- *Frame Rates:* 23.98, 24, 25, 29.97, 30, 50, 59.94, 60, and the complication of pulldown cadence for 24-frame material;
- *CODECs*: DV25, DV50, DV100, MPEG-1, MPEG-2, MPEG-4, AVC/H.264, VC-1, and their many variants;
- *Colorimetry:* YUV, RGB, color sampling (4:2:2, 4:4:4, 4:2:0, 4:1:1), color spaces (601 & 709), non-standard gamma, look up tables (LUT), arbitrary white and black points;
- *Source Metadata:* source time code, edge code, file names, URI file locators, frame counts, UMID;
- *Compositional Metadata:* EDL, AAF, XML metadata containing: information about how sections of program material are combined and modified;
- *Camera Metadata:* lens, camera;
- *Temporal and Geospatial Metadata:* acquisition date, time, location, camera direction, temperature.

The transition towards electronic workflows has also enabled broader exchange of user metadata, which also must be carefully stored and tracked, including:

- *Production Metadata:* production name, act, scene, take, director, director of photography, journalist, camera operator;
- *Annotations:* circle take, points of interest, text comments, graphical annotations;
- *Source Metadata:* camera roll, tape name, sound roll, mark in/out points.

Ideally, all this metadata must be aggregated intelligently even as teams of collaborators share the same content, adding new metadata in parallel in an ad hoc fashion. Making matters more complex, users may work simultaneously on multiple instances of the same content, in different formats. And for the aggregated metadata to be meaningful, system designers must be able to structure and validate the metadata, enforcing custom metadata schemas and dictionaries.

The explosive growth of technical and user metadata has triggered evolution in the tools themselves. It is no longer adequate for nonlinear editors to track time code, tape name and edge code information, eventually producing an EDL or cut list. Now more than ever, NLEs must occupy the very nexus of metadata management: capturing, transforming, and disseminating technical and user metadata to the greater creative team.

To manage today's larger more dynamic creative teams, production management systems are becoming essential. Built around core IT database and services infrastructures, production management systems form the centralized data hub for distributed teams, aggregating and transforming metadata. Production management systems are innovating along several vectors including:

- *Asset Management:* Central storage of all electronic production assets and metadata;
- *Secure Production Environment:* Centralized user management and authentication; policy-based watermarking and DRM;
- *Dependency Tracking:* Graphing and management of the interrelationship between assets;
- *Scalable Storage Management:* Bridge online, nearline and archive stores; policy-based aging of assets;
- *Content Distribution and Data Replication:* Enable geographically dispersed teams by automating the flow and sharing of data;
- *Workflow Status Reporting:* Track tasks, resource allocation and due dates;
- *Policy-based Workflow:* Changes in workflow status trigger actions and processes, based on customizable rules;
- *Version and Revision Control:* Automated change management, unlocking iterative workflow; streamlined content distribution and repurposing.

With metadata-aware creative tools interconnected by an intelligent workflow engine, postproduction teams will gradually reap the benefits of the transition to data-centric file-based workflows.

SUMMARY

From the early beginnings of splicing two pieces of film together to the complexities of sophisticated integrated editing and effects systems for television and feature films, the science and art of editing programs have been key parts of the business of movies and television. Even as the technologies continue to advance, the basic principle of taking raw program material and preparing it for the viewer to see and enjoy will remain. However, to reap the maximum benefits of the transition to a data-centric file-based workflow, post production teams will increasingly require metadata-aware creative tools interconnected by intelligent workflow engines.

Bibliography

Dmytryk, Edward. *On Film Editing: An Introduction to the Art of Film Construction.* Boston: Focal Press, 1984. ISBN 0-240-51738-5.

Nuffer, Eberhard. "Filmschnitt und Schneidetisch" ("Film Editing and Editing Equipment"), in *Weltwunder der Kinematographie—Beiträge zu einer Kulturgeschichte der Filmtechnik* (vol. 7), Joachim Polzer, ed., Potsdam: Polzer Media Group, 2003 (available through amazon.de). ISBN 3-934535-24-0

Ohanian, Thomas A. *Digital Nonlinear Editing.* Butterworth-Heinemann, 1998. ISBN 0-240-80225-X

Read, Paul. "A Short History of Cinema Film Post-Production (1896–2006)," in *Zur Geschichte des Filmkopierwerks/Weltwunder der Kinematographie—Beiträge zu einer Kulturgeschichte der Filmtechnik* (vol. 8), Joachim Polzer, ed., Potsdam: Polzer Media Group, 2006 (available through amazon.de). ISBN 3-934535-26-7

Walker, Alexander. *Stanley Kubrick Directs.* Harcourt Brace, 1972. ISBN 0156848929

CHAPTER

5.12

Format and Standards Conversion

JED DEAME
Teranex Division, Silicon Optix, Inc.
Orlando, Florida

INTRODUCTION

Now that the digital television (DTV) era has arrived, most broadcasters and facility operators have to deal with several different television formats. In the analog to digital transition, video transport mechanisms improved while image formats stayed largely the same. In the DTV era, transport mechanisms continue to evolve, but now the image formats in almost every facility vary in line rate and often in frame rate as well. Preserving image integrity in this multiformat environment presents a unique challenge for the television engineer.

Format conversion is used in many production, postproduction, and broadcast workflows. It is required any time the source format and destination format differ. Whether it is tape-to-tape conversions, broadcast conversions, or conversions for monitoring, this conversion capability is required for proper interoperability with the myriad of video formats in use today.

BASIC PRINCIPLES

In this chapter, a *video format* is defined as a particular number of *lines* and *pixels* that make up the image (and determine its spatial resolution) and a particular *frame rate* at which the picture information is repeated. Other parameters that are part of the format definition are whether the image is *interlaced* or *progressive*, and the picture *aspect ratio* (usually 4:3 or 16:9).

This chapter adopts the commonly used shorthand where, for example, 720p60 means a format with 720 active lines, progressively scanned at 60 frames per second (fps), and 1080i60 means 1080 active lines, interlaced scanned at 60 fields per second. It is noted that some literature always states the frame rate with the last number, while others, as here, state the field rate when describing interlaced formats.

Resolutions

There are a large number of resolutions used in digital television. The formats in Table 5.12-1 are known as the 18 ATSC Table 3 formats and are used for emission. The table shows the active lines (vertical size) and pixels (horizontal size), the frame aspect ratio, and the frame rate, also showing progressive (p) or interlace (i). Each frame rate has an alternate 1000/1001 frequency change to accommodate compatibility with NTSC color signals (24 to 23.98, 30 to 29.97, 60 to 59.94), resulting in a total of 36 formats. The fractional frame rate will remain part of the DTV landscape until NTSC is no longer needed.

The 480-line standard definition formats have 704 pixels for emission but the production standards for these formats typically have 720 horizontal pixels in accordance with ITU-R BT.601. The formats with 1080 and 720 lines are high definition (HD) and the formats with 480 lines are standard definition (SD).

Table 5.12-2 lists the main current digital video production formats. Format and standards conversion usually takes place between formats listed in this table. The number of lines of active video in HD production formats is the same as for the emission formats. The 525-line SD production formats have 483 or

TABLE 5.12-1
ATSC Table 3 Formats for DTV Transmission

Vertical Size Value (Active)	Horizontal Size Value (Active)	Aspect Ratio Information	Frame Rate and Scan
1080	1920	16:9 (square pixels)	24p, 30p, 30i
720	1280	16:9 (square pixels)	24p, 30p, 60p
480	704	4:3 (nonsquare pixels)	24p, 30p, 30i, 60p
480	704	16:9 (nonsquare pixels)	24p, 30p, 30i, 60p
480	640	4:3 (square pixels)	24p, 30p, 30i, 60p

486 active lines and the extra lines are discarded when video is coded for emission with 480 lines.

For convenience, in this chapter all SD formats intended for emission with 480 lines are referred to as having "480" lines, even if the actual production format has 483 or 486 lines.

Interlaced and Progressive

Most video sources today, including standard definition DV, Betacam, Digital Betacam, and high definition HDV, HDCAM, and HD-D5, produce predominantly interlaced images (although most HD recorders can also handle 720p progressive scan images). Instead of transmitting each video frame in its entirety with each line in order one after the other (as in progressive scan), interlaced video sources transmit only half of the image in each frame at any given time. This concept also applies to recording interlaced video images where the device records only half (one field) of the image in each frame at a time.

The words *interlaced* and *progressive* arise from CRT-based television displays, which form the image of each frame on the screen by scanning an electron beam horizontally across the picture tube, starting at the top and working its way down to the bottom. Each horizontal line "drawn" by the beam includes the part of the picture that falls within the space occupied by that line. If the scanning is interlaced, the electron beam starts by drawing every other line (all the odd-numbered lines) for each frame; this set of lines is called the *odd field*. Then it resets back to the top of the screen and fills in the missing information, drawing all the even-numbered lines, which are collectively called the

TABLE 5.12-2
Digital Television Production Formats

Vertical Size Value (Active)	Vertical Size Value (Total)	Horizontal Size Value (Active)	Horizontal Size Value (Total)	Aspect Ratio Information	Frame Rate and Scan
1080	1250	1920	2376	16:9 (square pixels)	50p, 25i
1035	1125	1920	2200	16:9 (nonsquare pixels)	30i
1080	1125	1920	2640	16:9 (square pixels)	25p, 25i
1080	1125	1920	2200	16:9 (square pixels)	60p, 59.94p, 30p, 29.97p, 30i, 29.97i
1080	1080	1920	2750	16:9 (square pixels)	24p, 23.98p
720	750	1280	1650	16:9 (square pixels)	60p, 59.94p
483	525	720	858	16:9 (nonsquare pixels)	59.94p
486	525	720	858	16:9 (nonsquare pixels)	29.97i
486	525	960	1144	16:9 (nonsquare pixels)	29.97i
576	625	720	864	4:3 (nonsquare pixels)	25i
486	525	948	1135	4:3 (nonsquare pixels)	29.97i
576	625	948	1135	4:3 (nonsquare pixels)	25i
486	525	768	910	4:3 (nonsquare pixels)	29.97i

even field. Together, the odd and even fields form one complete frame of the video image. A similar concept was used for interlaced scanning when video was originated in tube-based cameras.

Interlacing provides one major advantage over progressive scan—it halves the bandwidth required to carry the video signal while maintaining the image display rate sufficiently high (50 or 60 Hz) to avoid flicker and without substantial reduction in spatial resolution.

Converting an interlaced video signal from 480i, 576i, or 1080i sources into progressive format is required as the first step in DTV format conversion, even if the final output is to be an interlaced format. Until the image is in the progressive domain, it is not possible to scale it to the desired output resolution without creating unwanted image artifacts. This interlaced to progressive conversion, discussed later, is the most important step in the format conversion process and determines the overall quality of the output video signal.

TYPES OF CONVERSION

Moving images exist in three dimensions. In the horizontal direction they are made up of individual pixels. In the vertical direction they are made up of the lines contained in the field or frame. This is referred to as the *spatial domain*. Finally, there is the number of fields or frames per second, which is referred to as the *temporal domain*.

The process of format or standards conversion is a form of sample rate conversion in two or three of the above dimensions. It consists of expressing moving images sampled on one three-dimensional sampling grid to a different grid.

The process of *interpolation* is used to convert between these various spaces. Interpolation is defined as computing the value of a sample or samples that lie outside the sampling matrix of the source signal. In other words, it is the process of computing the values of output samples that lie between the input samples.

Format conversion typically refers to conversions where the frame rate does not change, whereas standards conversion refers to conversion between frame rates. However, conversions from 24 fps to 30 fps and from 25 fps to 50 fps are considered to be format conversions as they may be accomplished by repeating fields or frames. Finally, there is aspect ratio conversion, which may take place in conjunction with format or standards conversion, or as a stand-alone process.

Format Conversion

The term *format conversion* is generally associated with changing the number of pixels and lines in an image format. Most format conversions do not involve temporal frame rate changes. Examples of format conversion include:

- 480i59.94 to 720p59.94

FIGURE 5.12-1 Line structure for conversion from 480i to 1080i.

- 480i59.94 to 1080i59.94
- 720p59.94 to 1080i59.94
- 576i50 to 1080i50

When converting from 480i59.94 to 1080i59.94, two of the three parameters of the signal are changed. The first is the number of pixels in each line. A 480i signal has 720 pixels per line while the 1080i signal has 1,920 pixels. The second parameter to be changed is the number of lines in each field (or frame). The 480i signal has 240 active video lines per field while the 1080i signal has 540 lines, as illustrated in Figure 5.12-1. The third parameter, the number of fields (or frames) per second, does not change between the 480i signal and the 1080i signal in this example.

Standards Conversion

Standards conversion is generally concerned with changing the number of lines and fields (or frames) per second in an image. The process requires intelligent Invention (temporal interpolation) of new frames in between the input frames in order to provide the desired output frame rate. It is an imperfect process and usually generates artifacts of some kind.

Examples of standards conversions include:

- 480i59.94 to 576i50
- 720p59.94 to 1080i50
- 1080i59.94 to 1080i50

To convert from 480i59.94 to 576i50, two of the three parameters of the signal must be changed. The first is the number of lines in each field. A 480i signal has 240 lines per field while the 576i signal has 288 lines, as illustrated in Figure 5.12-2.

The second parameter that needs to be changed is the number of fields per second. The 480i signal has 59.94 fields per second while the 576i signal has 50, as illustrated in Figure 5.12-3.

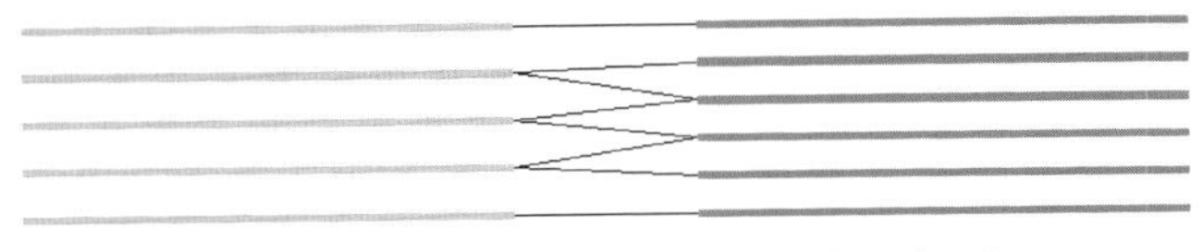

FIGURE 5.12-2 Conversion from 480i59.94 to 576i50 line structure.

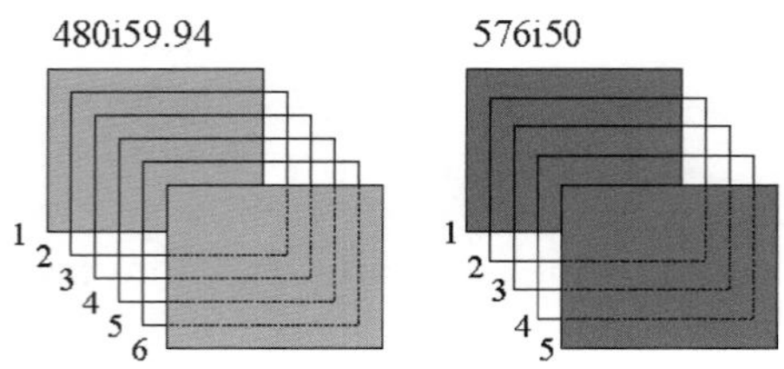

FIGURE 5.12-3 Conversion from 59.94 to 50 fields per second.

The third parameter, the number of pixels on each active picture line, does not change between 480i and 576i.

There are also cases where all three domains must be changed. Examples include:

- 480i59.94 to 1080i50
- 1080i59.94 to 576i50
- 480i59.94 to 1080psf23.98

Aspect Ratio Conversion

The change in aspect ratio is one of the more challenging aspects of the DTV transition. Standard definition video is typically captured at an aspect ratio of 4:3, but may also be captured in 16:9 aspect ratio (SD 16:9 is common in Europe and beginning to be used in the United States). High definition, on the other hand, is exclusively captured in 16:9. DTV transmissions may use either 4:3 or 16:9 aspect ratio, depending on the source material. A further complication is that such transmissions may be viewed on displays that are either 4:3 or 16:9 aspect ratio.

Movies may be shot in a wide variety of aspect ratios, commonly 1.85:1, which is quite close to 16:9, but also in wider aspect ratios, such as 2.35:1, using an anamorphic lens to allow the image to be captured in a conventional aperture on 35 mm film. In the professional environment, such widescreen film images may be transferred to video without cropping or letterboxing by recording an anamorphic image in the video frame, which must then be processed further before transmission.

Aspect ratio conversion, therefore, becomes a critical component of any format conversion process. Various techniques may be used, and some or all of these techniques may take place in a professional format or standards converter, in a dedicated aspect ratio converter, or in a DTV receiver and/or display.

4:3 to 16:9 Mappings

If the source material is 4:3 and the output needed is 16:9, there are several possible image mappings to align the aspect ratios.

The first and most common broadcast mapping is called *pillarbox*. In this case, the 4:3 source image is centered in the 16:9 output frame with black (or sometimes gray or other colors) bars placed along the sides to fill the rest of the frame. This technique avoids any distortion of the image, but can sometimes result in burn-in of the bars when such images are displayed on CRT or plasma televisions.

Another technique is called *cropped* or *common sides*. In this case, the 4:3 image is cropped on the top and bottom and the remainder of the image zoomed to provide an undistorted 16:9 image, albeit losing some of the original image. This technique is undesirable unless it is known that the critical action in production was kept within a *protected* area that will not be cropped but it may be available in receivers.

A third variation, called *14:9*, was pioneered in the United Kingdom and combines pillarbox with cropping. It assumes that in production all the critical action is kept within the protected area of the frame. The active image can then be cropped to an aspect ratio of 14:9, resulting in narrower pillarbox bars lines at the expense of some loss of lines at the top and bottom of the picture.

These mappings are illustrated in Figure 5.12.-4, which also shows the mapping of a widescreen movie into a 16:9 frame with letterbox bars. This figure also shows *active format description* codes that may be used to signal the various formats with this frame aspect ratio.

A fourth technique is called *anamorphic stretch*. This technique stretches the 4:3 input image horizontally to fill the 16:9 output image. This is an undesirable technique to use in broadcast as it introduces significant geometric distortion to the image, but it is commonly used in television receivers because it avoids any potential display burn-in problems.

A more recently introduced fifth technique is *nonlinear stretch* mode. This maintains correct image geometry in the center of the image, but progressively increases the horizontal stretch toward the edges of the image. It is not a panacea, but does provide the linearity of pillarbox in the critical center area of the screen combined with the screen-filling characteristics of the anamorphic stretch mode. This technique is available with some television receivers and is also available in the professional environment with systems such as the Turner/Teranex Flexview™ aspect ratio converter.

16:9 to 4:3 Mappings

If the source material is 16:9 and the output needed is 4:3, the solutions are similar but inverse. In this case, the techniques are *letterbox*, *cropped* or *common top and bottom* (also known as *center cut*), and *anamorphic squeeze*. 14:9 again is sometimes used, this time resulting in narrow letterbox bars.

These mappings are illustrated in Figure 5.12.-5, which also shows the mapping of a widescreen movie into a 4:3 frame with letterbox bars. Again, this figure shows *active format description* codes that may be used to signal the various formats with this frame aspect ratio.

Aspect Ratio Signaling

Various techniques have been developed to tag the content with the image format in order to automate

Active Format	Illustration in a 16:9 coded frame	Description
AFD = '0100' Box >16:9 (center)	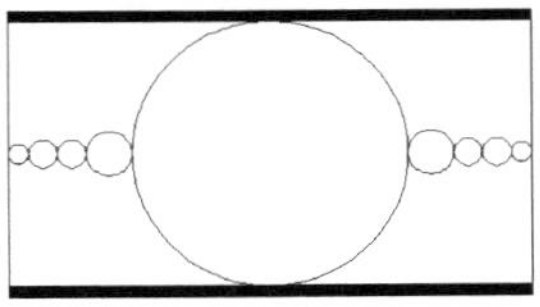	Image with aspect ratio greater than 16:9 as a vertically centered letterbox in a 16:9 coded frame.
AFD = '1000' Full frame	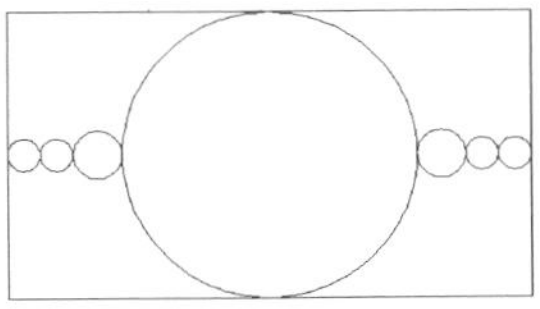	Image is full frame, with an aspect ratio that is the same as the 16:9 coded frame.
AFD = '1001' 4:3 (center)	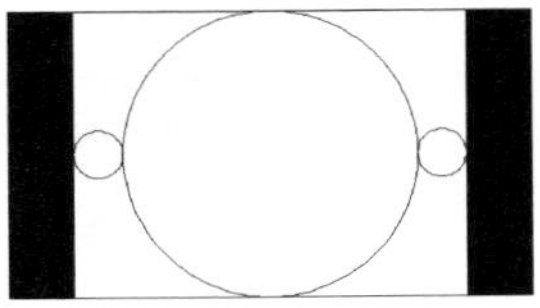	Image with a 4:3 aspect ratio as a horizontally centered pillarbox image in a 16:9 coded frame.
AFD = '1010' 16:9 (with complete 16:9 image protected)	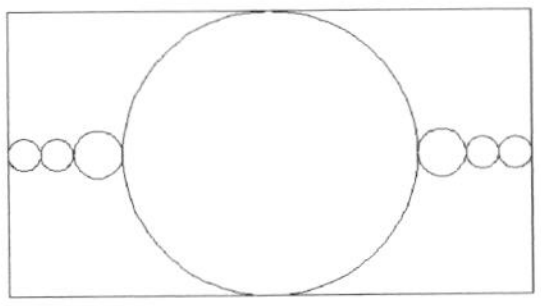	Image is full frame, with a 16:9 aspect ratio and with all image areas protected.
AFD = '1011' 14:9 (center)	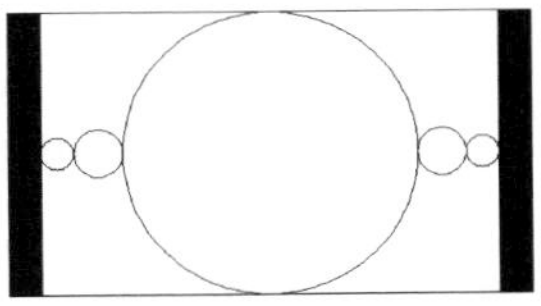	Image with a 14:9 aspect ratio as a horizontally centered pillarbox image in a 16:9 coded frame.
AFD = '1101' 4:3 (with alternative 14:9 center)	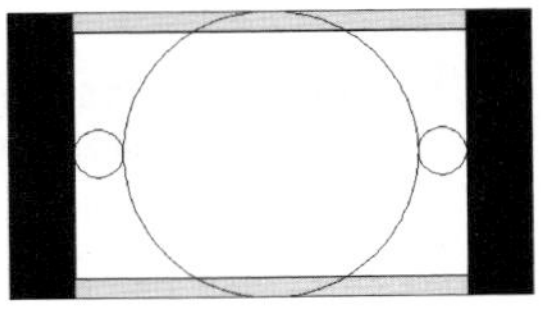	Image with a 4:3 aspect ratio and with an alternative 14:9 center as a horizontally centered pillarbox image in a 16:9 coded frame.
AFD = '1110' 16:9 (with alternative 14:9 center)	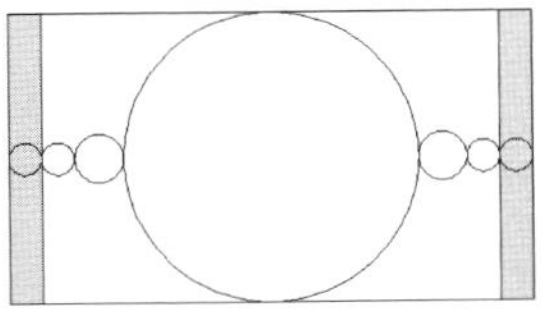	Image with a 16:9 aspect ratio and with an alternative 14:9 center in a 16:9 coded frame.
AFD = '1111' 16:9 (with alternative 4:3 center)	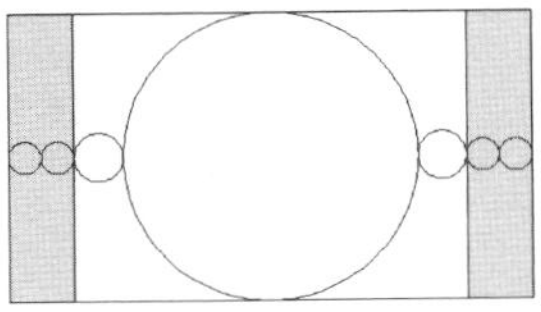	Image with a 16:9 aspect ratio and with an alternative 4:3 center in a 16:9 coded frame.

FIGURE 5.12-4 Image formats in a 16:9 frame. (Courtesy of SMPTE.)

the aspect ratio conversion process. Tagging standards include Wide Screen Signaling (WSS), which is standardized in ETSI EN 300 294; Video Index, which is standardized in SMPTE RP 186; and Active Format Description (AFD) and Bar Data, which is standardized in SMPTE 2016. These tags can be read by the format conversion equipment enabling automated changes to the aspect ratio with frame accuracy.

In addition to the AFD codes assigned to various active image formats. A further parameter, Bar Data, is

Active Format	Illustration in a 4:3 coded frame	Description
AFD = '0010' Box 16:9 (top)	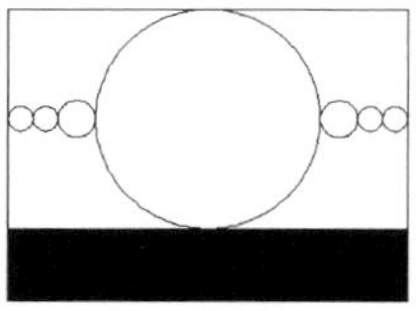	Image with a 16:9 aspect ratio as letterbox at the top of a 4:3 coded frame.
AFD = '0011' Box 14:9 (top)	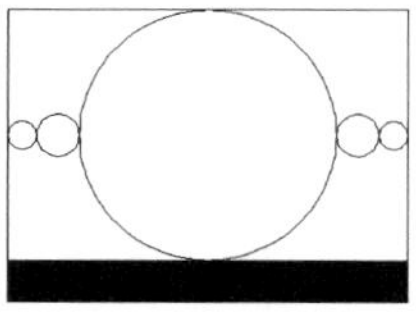	Image with a 14:9 aspect ratio as letterbox at the top of a 4:3 coded frame.
AFD = '0100' Box >16:9 (center)	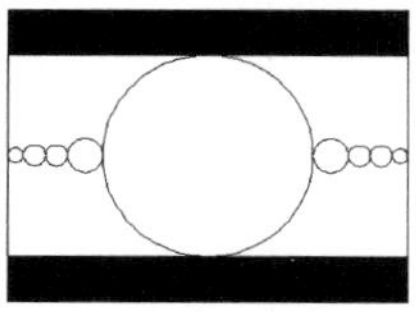	Image with aspect ratio greater than 16:9 as a vertically centered letterbox in a 4:3 coded frame.
AFD = '1000' Full frame	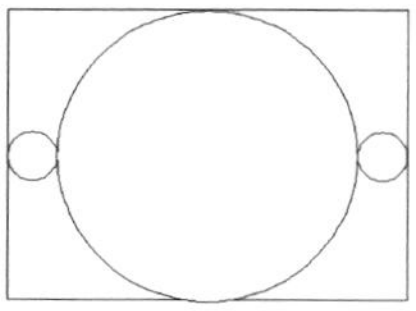	Image is full frame, with an aspect ratio that is the same as the 4:3 coded frame.
AFD = '1010' 16:9 (center)	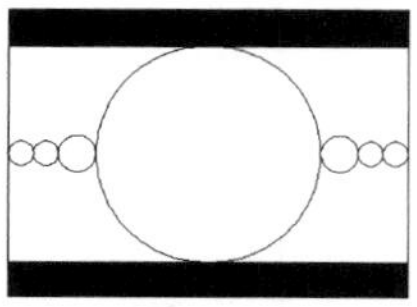	Image with a 16:9 aspect ratio as a vertically centered letterbox in a 4:3 coded frame.
AFD = '1011' 14:9 (center)	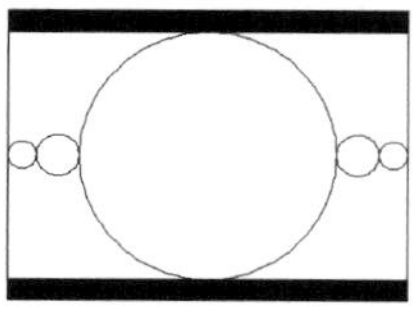	Image with 14:9 aspect ratio as a vertically centered letterbox in a 4:3 coded frame.
AFD = '1101' 4:3 (with alternative 14:9 center)	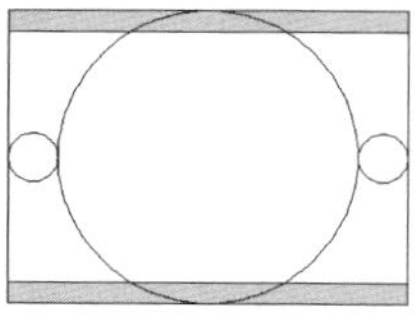	Image with a 4:3 aspect ratio and with an alternative 14:9 center in a 4:3 coded frame.
AFD = '1110' 16:9 (with alternative 14:9 center)	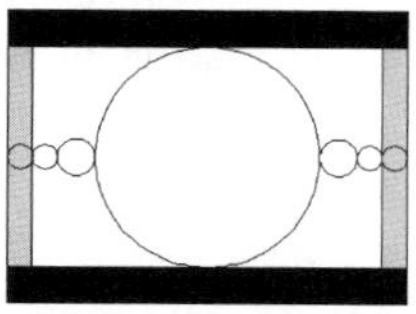	Image with a 16:9 aspect ratio and with an alternative 14:9 center as a vertically centered letterbox in a 4:3 coded frame.
AFD = '1111' 16:9 (with alternative 4:3 center)	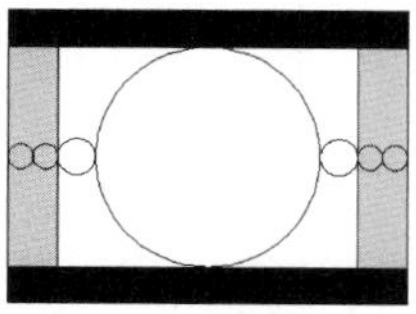	Image with a 16:9 aspect ratio and with an alternative 4:3 center as a vertically centered letterbox in a 4:3 coded frame.

FIGURE 5.12-5 Image formats in a 4:3 frame. (Courtesy of SMPTE.)

defined in SMPTE 2016. This exactly defines the width of the letterbox bars (when used with AFD 0100) or pillarbox bars (with AFD 0000) since, in those cases, the AFD code alone is not sufficient to fully describe the active area of the frame. Bar Data may be used for transmission with ATSC DTV signals.

When such tags or codes are broadcast, the television receiver may use them to optimize the image for showing on the particular display associated with the receiver.

Pan and Scan

For broadcasting in a video 16:9 or 4:3 frame, widescreen movie images with aspect ratios greater than 16:9 must be adjusted to fit the frame, either by adding bars to produce the well-known letterbox format or by cropping portions of the original picture. The technique, known as *pan and scan* is used to allow selected portions of the original image to be shown, depending on the action in the frame, rather than just using a fixed center cut. Further information on this technique may be found in Chapter 5.23.

APPLICATIONS FOR FORMAT AND STANDARDS CONVERSION

Until the arrival of digital television broadcasting there were only a few types of conversions necessary for program interchange, in particular NTSC to PAL or SECAM or vice versa. DTV broadcasting worldwide has brought with it the need for a wide variety of conversions at various stages in the television production and distribution process.

In the United States, the ATSC standard allows both SD and HD images to be broadcast. The majority of broadcasters have opted for the latter for transmission, sometimes also with other SD minor channels. In the early days of DTV, HDTV format converters were essential in order to create an HD version of existing SD programming. Even as primetime fills up with native HD programs, these converters are still in active use, converting SD content for other parts of the day to HD.

Production

Format conversion is increasingly used in the production process as the demand for HD acquisition grows. In many live HD venues, it is impractical to use HD cameras exclusively. Remote RF cameras are often SD due to wireless bandwidth limitations, and some locations require the size and portability of SD cameras. In these applications the SD camera feeds are upconverted on the way into the HD switcher and are mixed with the HD camera shots. Another application for format conversion in production is down conversion of HD camera feeds for low-cost SD confidence monitoring.

Postproduction

Postproduction facilities have many uses for format conversion. Program producers must deal with many common acquisition formats in use today. For example, an editor producing an HD program may have the majority of the source material in 1080i, but may need to use some stock footage that only exists in 480i. In that case, the editor would upconvert the 480i material to 1080i and ingest it into the editing system for incorporation into the HD program. Many editing programs offer the option of software conversion as well, but the image quality is typically not ideal (often half resolution) and the conversion times can be slow. In addition, program producers often need to deliver their programs in multiple formats. They may deliver a 1080i59.94 version, say, to NBC, a 720p59.94 version to ABC, and a PAL or 1080i50 version to the BBC in the United Kingdom. Format and standards converters enable the program to be converted to these different formats.

Postproduction facilities usually provide conversion services to outside program producers as well since they typically have a variety of videotape recorders (VTRs) and real-time format conversion capabilities.

Broadcast

Broadcasters use format conversion in a variety of ways. Many initially used SD to HD upconverters primarily for their SD master control output to feed an HD program service. With the growing demand for HD news, some local broadcasters are now finding that they must include a bank of format converters on their HD switcher in order to accommodate legacy SD sources. Also, HD news producers use format converters to incorporate SD stock footage into HD news clips. This is typically done in nonlinear editing (NLE) facilities, which may include external palm-sized converters, capture cards, or software conversion inside the NLE program.

Display

As the venerable CRT slowly disappears and is replaced by digital display devices for monitoring, high-quality format conversion becomes necessary to obtain an accurate representation of the image. Digital displays (LCD, LCOS, DLP, plasma, and other) are natively progressive and are thus unable to display an interlaced image without format conversion. Many digital displays rely on simple non-motion adaptive *field scalers* rather than true format converters. The result is imagery with half the resolution because a single field has only half the information in a video frame. For high-quality monitoring digital displays, it may be desirable to use an external format converter or scaler to convert interlaced input signals to the native progressive format of the display.

International Exchange

Standards conversion is needed for international program exchange between countries with television systems based on 50 Hz and 60 Hz standards. In addition to the traditional NTSC–PAL conversions, there is an increasing need for 50 Hz to 60 Hz standards conversion for HD 1080-line and 720-line programming.

System Design Considerations

In designing multiformat systems, it is important to understand the types of formats that are ingested as well as played out of the system and the limitations of the intermediate processing equipment. Most multiformat facilities have an SD routing plane and an HD routing plane with a series of format converters connected between them. This enables SD and HD sources to be available at different resolutions by simply routing through one of the format converters.

In broadcast production, and postproduction applications, it is recommended to install a series of upconverters in front of at least some of the HD switcher inputs. This provides a seamless interface between the SD sources and the HD production switcher.

INTERLACE TO PROGRESSIVE CONVERSION

The process of interlace to progressive conversion consists of several complex image-processing steps. It is important to understand the details of the processing steps in order to appreciate the many places where quality may be lost in the conversion process.

Automatic Video/Film Detection

The first step in the interlace to progressive conversion process is to detect whether the content is of video origin (either interlaced fields or progressive); film-based video (originating from 24 fps material, with 2:3, 2:2, or other assorted cadences); or mixed video and film in the same program. The processing requirements are very different depending on the content type and it is imperative that this detection logic be accurate in order to ensure that the proper conversion algorithms are used. It is also important that the video/film detection be fast and automatic. If it takes too long for the logic to make a decision, significant conversion artifacts may result. If it is not automatic, the operator must constantly switch the processing mode to reflect the content type being processed. In some postproduction applications where the content type is known and fixed, this may be desirable, but for broadcast or display applications when the content type changes often, automatic video/film detection is key.

Deinterlacing

Deinterlacing is the process of transforming video-based content with interlaced fields into a progressive

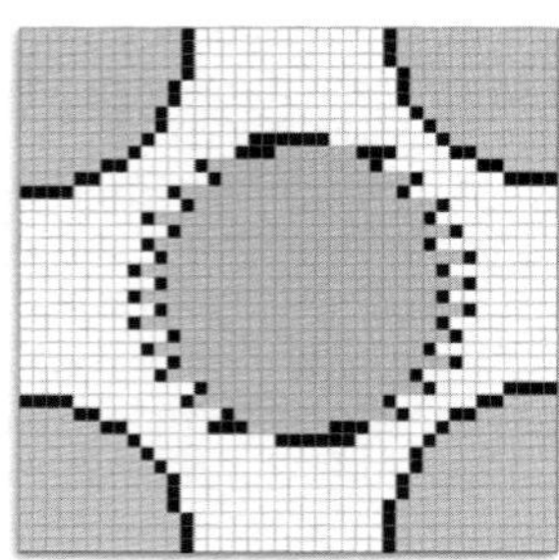

FIGURE 5.12-6 Illustration of the effect caused by combining two interlaced fields with motion into one progressive image.

format. There are many techniques available, each with its own advantages and disadvantages, but all tend to fall into the three classes below.

Basic Field Weave

The field weave technique was one of the first deinterlacing techniques attempted by converter designers. It was initially used in personal computers for display of interlaced video clips or DVDs. The simple premise behind the field weave technique is to weave the two fields together and combine them to form a complete frame. This works well if the objects in the video image are not moving. However, if there is motion in the image, this technique can generate significant artifacts.

When recording is performed in an interlaced manner, the two source fields that make up a complete frame are not recorded at the same time. Each frame is recorded as an odd field at one point in time, and then as an even field recorded 1/60 or 1/50 of a second later. Hence, if an object in the scene has moved in that fraction of a second, simply combining fields causes the errors in the image called *combing* or *feathering* artifacts, as illustrated in Figure 5.12-6.

Non-Motion Adaptive

The simplest approach to avoid these artifacts is to process a single field only using the *non-motion adaptive* approach. In this method, when the two fields reach the processor, only data from the first field is used and the data from the second field is not taken into consideration.

The video-processing circuitry recreates or interpolates the missing lines by averaging pixels from above and below. While there are no combing artifacts, image quality is compromised because half of the detail and resolution have been discarded as illustrated in Figure 5.12-7, which shows:

(a) Only half of the frame is used for interpolation.

(b) Zoom view of a section of the field.

(c) The missing pixels are recreated by averaging.

(d) The new interpolated frame (half resolution).

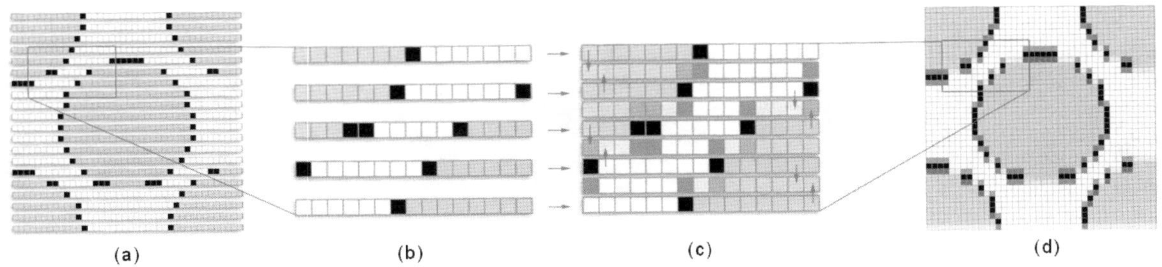

FIGURE 5.12-7 Interpolation technique where only one field is used resulting in loss of resolution (final image on right).

More advanced techniques have been adopted by virtually all SD video processors, but some manufacturers still use this basic approach for HD signals, due to the increased computational and data-rate requirements of higher video resolution. Therefore, in some processors only 540 lines from a 1080i source are used to create the image that makes it to the screen.

Frame-Based Motion Adaptive

More advanced deinterlacing techniques include a frame-based, motion-adaptive algorithm. By default, these video processors use the same technique described above. However, by using a simple motion calculation, the video processor can determine when no movement has occurred in the entire picture.

If there is no motion in the image, the processor combines the two fields directly. With this method, still images can have full vertical resolution, but when there is motion, half of the data is discarded and the resolution drops to half. With this technique, static test patterns and images look sharp, but moving images tend to look soft.

Pixel-Based Motion Adaptive

Pixel-based motion adaptive processing represents one of the highest performing deinterlacing techniques available. In this technique, motion is identified at the pixel level rather than the frame level. While it is mathematically impossible to avoid discarding pixels in motion during deinterlacing, this technique discards only the *pixels* that would cause combing artifacts, as illustrated in Figure 5.12-8. Everything else is displayed with full resolution.

Pixel-based motion-adaptive deinterlacing avoids artifacts in moving objects and preserves full resolution of nonmoving portions of the screen even if neighboring pixels are in motion.

Motion Compensated

Motion-compensated deinterlacing is often referred to as the "Holy Grail" of deinterlacing. In this technique, pixels in the first field are shifted to align with corresponding pixels in the second field in order to preserve the full resolution of objects in motion. This technique is extremely computationally complex and requires significant processing power to do properly. Although it has the capability of providing full-resolution conversions of objects in motion, there are some practical limitations to how well this can be achieved. For instance, as the motion between fields increases, the search area increases exponentially, as does the probability that the objects will change in shape. This creates a much larger possibility for artifacts due to false matches in the search process.

It is important that motion compensation be performed on a per pixel basis and be utilized in conjunction with a high-quality per pixel motion adaptive deinterlace framework in order to avoid the pitfalls and realize the potential benefits. Although region-based motion compensation techniques may be used to reduce the computational complexity, the benefits are often indistinguishable from pure motion adaptive approaches and much more prone to image artifacts. Due to the high computational requirements and associated costs of properly performing motion compensation on a per pixel basis, this technique is often used in systems serving the high end of postproduction.

Directional Filtering

To recover some of the detail lost in the areas in motion in the pixel-based motion adaptive approach, it is beneficial to employ a second-stage multidirection diagonal filter that reconstructs some of the lost data at the edges of moving objects, filtering out any *jaggies*

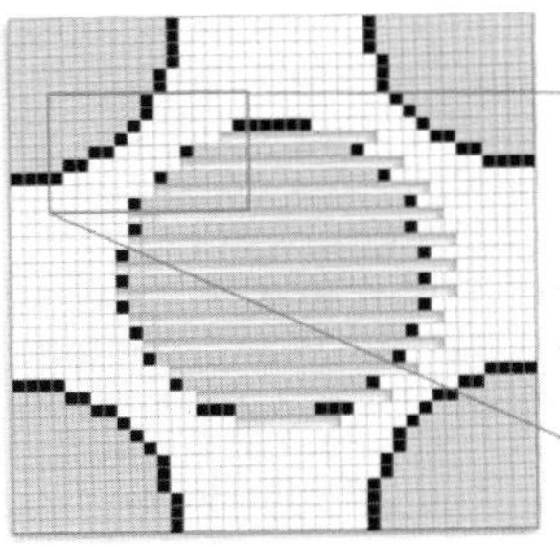

FIGURE 5.12-8 Pixel-based motion adaptive technique removes only the pixels that would cause combing.

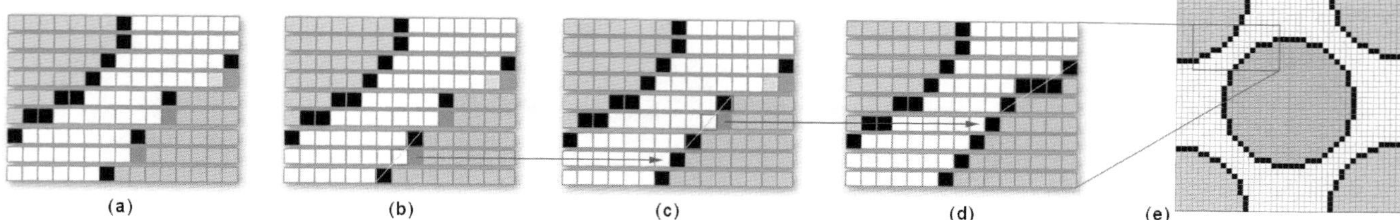

FIGURE 5.12-9 Diagonal interpolation removes jaggies from the reconstructed image.

(jagged edges), as illustrated in Figure 5.12-9, which shows:

(a) Diagonal lines resulting from moving object interpolation (green edges).

(b), (c), (d) Highlighting the jaggy section of the line, the missing detail is recreated by averaging along the diagonal lines.

(e) The new interpolated frame (full resolution).

This operation is called *second-stage* diagonal interpolation because it is performed after the deinterlacing. By interpolating along the diagonal axis, significant detail may be recovered. High-performance diagonal filters are capable of performing this selective filtering on very low-angle lines (close to horizontal).

Motion Detection

As discussed previously, automatic video/film detection and per pixel motion adaptive deinterlacing are key to providing high-quality deinterlaced images. The engine that enables these is the motion detector. Motion detectors are available in 2-, 3-, and 4-field varieties. In general, the larger number of fields used, the more accurate the motion detection will be. This additional accuracy comes at the price of additional latency through the system.

2:3 Film Mode Detection

Most motion picture films are shot at 24 fps. When the film is converted to video for DVD or television broadcast, those 24 frames must be converted into 60 interlaced fields. The actual frame rates are 23.98 and 59.94, but round numbers are used for clarity. The process is as follows. Consider four frames of film—A, B, C, and D—as shown in Figure 5.12-10.

The first step is to convert these four frames into eight fields. This transforms 24 frames per second into 48 interlaced fields per second. Then, to account for the faster rate of the NTSC standard (roughly 30 frames per second or 60 interlaced fields per second), it is necessary to repeat certain fields. This is done by repeating a field every other frame. That is, both fields of frame A are recorded (A-odd, A-even), but three fields of frame B are recorded (B-odd, B-even, B-odd). The cycle repeats with frames C and D. This is called a *2:3 pull-down cadence* because two fields of one frame are output followed by three fields of the next frame.

When this sequence is played back on a progressive-scan video display, it is desirable to implement the same deinterlacing techniques described earlier (non-motion adaptive versus motion adaptive, etc.). However, it is possible to perfectly reconstruct the original frames without losing *any* data. Unlike interlaced video, in which the two fields were recorded a fraction of a second apart, these fields were recorded at the same time in the same film frame and later separated into fields.

To display a video signal that originated as 24 fps film, a video processor must analyze the fields and determine that there is a regularly alternating pattern of two fields followed by three fields, etc. This recognition and reconstruction is called *2:3* (or *3:2*) *pull-down detection*, and it is found in all but the most basic deinterlacers.

Unfortunately, the 2:3 pattern is not always as regular and consistent as described above. Electronic editing is the most common cause of discontinuities in the 2:3 sequence. If all edits occurred on an A frame there would be no problems. However, it is not uncommon for edits to be made without regard to the underlying 2:3 sequence. Whenever an edit is made on something other than an A frame there will be a discontinuity in the 2:3 sequence. These discontinuities make it difficult to detect film sequences.

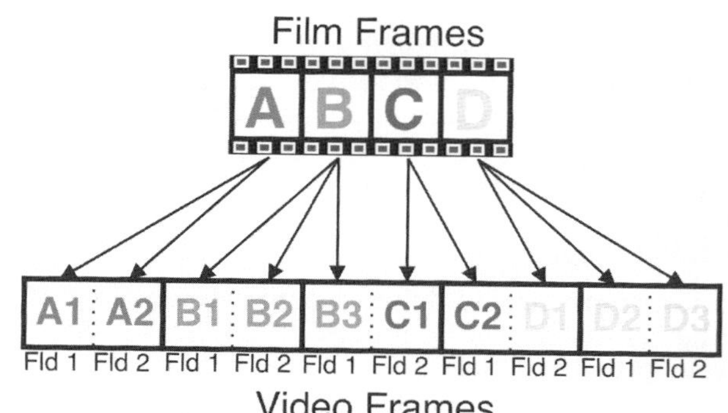

FIGURE 5.12-10 Conversion of 24 frames to 60 fields using 2-3-2-3 technique.

Other Cadences

Although 24 fps film and its associated 2:3 video cadence is the most common format, it is not the only cadence used today. Sometimes, broadcasters accelerate their film-based movies and TV shows by playing a VTR or server 4% faster to better fill the time slot available. The dynamic tracking VTR accomplishes this by dropping fields as needed to keep up with the output rate. This speedup is usually too small to be noticed by the average viewer, but these *vari-speed* broadcasts end up having unusual cadences such as 3:2:3:2:2. If the video/film detector is unable to detect this sequence, it has no choice but to treat it as video resulting in a loss of half the resolution.

The variety of cadences does not end there. DVCPRO camcorders are increasingly used in television and film production. In order to simplify computer-based extraction of the original 24p progressive content, these camcorders use a 2:2:2:4 cadence or a 2:3:3:2 cadence to store a 24p progressive source signal as 30 fps on the tape.

Animation is often rendered at 12 fps. Two pull-down cadences can be used to convert this to the 30 fps broadcast standard. Doubling every frame, and then applying 2:3 pull-down to the resultant fields will generate a 4:6 (or 6:4) cadence, as illustrated in Figure 5.12-11. Applying 2:3 pull-down to the frames (rather than the fields) will generate a 5:5 cadence. The Japanese *Anime* format is often rendered at 8 fps. To convert this to 30 fps, each frame of animation is repeated three times, and then 2:3 pull-down is inserted, resulting in an effective cadence of 7:8 (or 8:7).

The problem that arises with these other cadences is that many format converters rely on counting of incoming fields and attempting to match them against known sequences, such as 2:3 or 2:2, in order to select the right decoding. This works in the simplest cases, but there may still be a short delay before the processor is able to *lock on* and determine the right cadence. In addition, when the video processor encounters an unusual sequence, such as animation or DVCPRO, it may never lock and will resort to discarding half the data until it can find a known sequence.

It is important for format converters to have film mode detection logic that is insensitive to the various cadences described above. This will ensure that they will provide full-resolution conversions of all types of content.

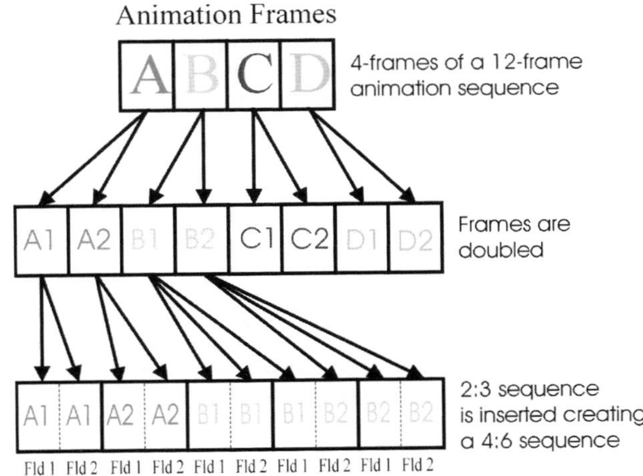

FIGURE 5.12-11 Conversion of 12 fps animation material to 60 fps using frame doubling and 2.3 pull-down insertion.

Mixed Video and Film Handling

Further editing and postprocessing is often required on film that has been converted to video. This includes adding titles, transitions, and other effects, which may result in frames that contain both film-based and video sources. As a result, simply reconstructing full frames using 2:3 frame matching results in combing artifacts because parts of the image are best processed using a standard deinterlacing approach, while other parts will look better by detecting the right cadence and reconstructing the original frames.

Like the various approaches to standard deinterlacing, there are many approaches to dealing with mixed video and film. If the processor interprets the entire frame as film, feathering artifacts will appear around the video portion; if the processor interprets the entire frame as video, the film portion will be displayed at half of its resolution. Some processors determine whether there is more film or more video content and choose the approach with the greatest benefit. Since this usually means film, this can result in feathering artifacts. Other processors are designed with the idea that these artifacts should never be seen and use the video deinterlacing techniques in all cases, at the expense of half the resolution of the film material.

The most advanced techniques make per-pixel decisions. These systems implement cadence-detection strategies for the pixels that represent film content while implementing pixel-based, motion-adaptive deinterlacing for the video content that has been superimposed.

Noise Reduction

Random noise is an inherent problem with all recorded images. Noise can enter the system in many places along the path from acquisition to consumption. Not only does noise get introduced during post-production, compression, and transmission, but it is also present at the source in the form of film grain or imaging-sensor noise. Noise-reduction algorithms can minimize the noise in a picture.

Spatial Filter

The simplest approach to noise reduction is to use a *spatial filter* that removes high-frequency data. In this approach, only a single frame is evaluated at any given time, and structures in the image that are one or two pixels in size are nearly eliminated. This does remove the noise, but it also degrades the image quality because there is no way to differentiate between noise and detail. This approach can also cause an artificial appearance in which people may look as though their skin is made of plastic. Spatial filtering is the most widely used noise reduction approach.

Temporal Filter

A *temporal filter* takes advantage of the fact that noise is a random element of the image that changes over time. Instead of simply evaluating individual frames, a temporal noise filter evaluates several frames at once. By identifying the differences between two frames and then removing the spurious data from the final image, visible noise can be reduced effectively. If there are no objects in motion, this is a nearly perfect noise-reduction technique that preserves most of the detail. Temporal filtering is used in many high-end noise-reduction processors.

Motion Adaptivity

The problem with the temporal approach is that if there are moving objects in the image, which also cause differences from one frame to the next, these differences should be retained. If moving objects are not distinguished from noise, they will be filtered and a ghosting or smearing effect is seen.

High-end format converters use advanced per-pixel, motion-adaptive, and noise-adaptive temporal filters to avoid the artificial appearance and artifacts associated with conventional noise filters. To preserve maximum detail, moving pixels do not undergo unnecessary noise processing. In static areas, the strength of noise reduction is determined on a per-pixel basis, depending on the level of noise in the surrounding pixels as well as in previous frames, allowing the filter to adapt to the amount of noise in the image at any given time. The end result is a *natural*-looking picture with minimal noise and grain and maximum preservation of fine details.

Detail Enhancement

Detail enhancement, also called *sharpening*, is an important component in format conversion, both SD and HD. Unfortunately, due to the historically poor implementations of sharpening algorithms, this process has received a reputation as something to avoid.

All digital video goes through a low-pass anti-aliasing filter to prevent false color and moiré effects that can occur during the digitization process. The filter improves overall image quality, but it necessarily blurs some of the detail. The data compression stage can also remove some detail. Fortunately, much of the lost detail can be mathematically recovered.

Because the human visual system perceives sharpness in terms of apparent contrast, exaggerating the differences between light and dark can produce what appears to be a sharper image. Unfortunately, due to rudimentary implementations of sharpening in the past, this process has been associated with artifacts known as *ringing* or *halos* in which objects are surrounded by a bright white edge. The resulting image appears harsh and does not reflect what was originally captured. The halos can sometimes be more distracting than the softness from the uncorrected image. For that reason, to avoid these kinds of artifacts it is important to use high-quality sharpening algorithms in the format conversion process.

High-end format converters use more conservative algorithms along with edge adaptivity to avoiding halo or ringing artifacts. Proper detail enhancement processing can enable high-quality SD video to be converted and delivered at quality that compares well with HD sources.

IMAGE SCALING

Once the input image has been converted from interlace to progressive (if not already progressive), the new output format is created through the process of image scaling. The scaler increases a low-resolution image to higher resolution or reduces a high-resolution image to a lower resolution, it may convert a 4:3 image to 16:9 or vice versa, and may correct keystone distortion resulting from off-axis projection. In each case, the processing is similar, although these functions represent increasingly sophisticated scaler operations.

Before any actual scaling can be performed, each frame must be stored in a series of line buffers, where the scaler analyzes each pixel in the frame. It is then processed based on various algorithms.

Scaling Control

The scaler analyzes the input frame to determine its format, or takes its settings from local or remote operator controls or presets. The desired output format is determined from operator controls or presets. Where appropriate, the scaler will also take input from metadata that may be associated with each video frame, such as WSS or AFD. Based on all this information the scaler determines the processing that is required for the conversion.

When aspect ratio conversion is required, bar widths, luma, and color are determined; any image cropping is calculated; and stretch or squeeze determined—all to meet the output formats described in the earlier section on aspect ratio. Operator controls also provide settings for keystone correction and other parameters.

FIR Filtering

This analysis is performed by digital FIR (finite impulse response) filters, which are mathematical algorithms that manipulate numbers. FIR filters include what are called *taps*, which, in a scaler, correspond to the YCbCr digital component-video values (that is, the intensity values of Y, Cb, and Cr) of individual pixels.

As a scaler analyzes each input pixel, it takes into account some of the pixels surrounding it. The number of neighboring pixels used in the analysis depends on the number of taps in the filters: the more taps, the more neighboring pixels are considered and the higher the quality of the scaled image.

Scaling normally adds pixels to (or removes them from) the input image—a process that is called *interpolation*. The goal of interpolation is to add or remove pixels so that the final image looks as if the new pixels had been there all along. The more that is known about the pixels in the neighborhood of each input pixel, the better the scaler can interpolate what the output pixels should be.

The scaler measures the YCbCr values for each input pixel and the surrounding pixels. It then multiplies those values by *weighting factors* assigned to each tap and combines the results together to calculate the YCbCr values for the final output pixels that are sent to the display. The weighting factors define each input pixel's importance in determining the final output pixel. In general, the closer a tap is to the location of the pixel being analyzed, the higher will be the weighting factor, because closer pixels have more relevance to the final output pixel than those that are farther away. The distribution of weighting factors among the taps is not simple or straightforward; this is the art of digital filter design.

Separable versus Non-Separable

In many scalers, the filters analyze the surrounding pixels in the horizontal and vertical directions sequentially. These so-called 1D (or separable) filters are relatively simple and inexpensive, and they work well enough for rectangular resizing. Non-Separable 2D filters analyze pixels in both directions simultaneously and produce a much better result, preserving more of the high-frequency components.

Taps versus Performance

Most scalers use a fixed number of taps for the horizontal and vertical filters; typically between 8 and 24 taps, depending on performance requirements. Some of the more advanced scalers use a single-pass 2D filter that can vary in size and aspect ratio for each pixel, up to a maximum of 32 × 32, yielding a total of 1,024 taps.

STANDARDS CONVERSION

Standards conversion is the process of converting image sequences between differing frame rates. The various standards conversion techniques can be broken down in to two main categories: non-motion compensated and motion compensated. The category of non-motion-compensated converters can further be broken down to linear and motion adaptive.

Non-Motion Compensated

Standard conversion is inherently an imperfect process as entirely new frames of information must be created. Non-motion–compensated standards converters must make a choice between *judder* and *blur* when creating the new output information. If an object is moving, it will be in a different place in each successive field. Interpolating between four fields gives four images of the object on the output field. If the position of the dominant image does not move smoothly, it will be seen to judder. The judder can be explained in terms of sampling theory. If the field rate is 60 Hz, then by the Nyquist sampling theorem, the maximum movement frequency allowable in the signal being sampled is 30 Hz. Unfortunately, objects frequently move at a rate that exceeds this, so temporal *aliasing* almost always occurs. When this temporal aliasing is presented to a converter, the converter cannot tell the difference between the original object and the ones created by the aliasing, so it resamples both. These multiple alias images are the cause of the perceived judder.

The alternative to judder is blurring. In an effort to minimize the judder in the output, the interpolation aperture can be changed to take a greater contribution from the two inside fields of the four-field sequence and less from the two outside. This will effectively soften the image and create a blurring effect on motion as less temporal information is being used than before. Note that even motion-compensated systems must rely on non-motion–compensated techniques as a fallback for the areas of the image where no image correlation exists, which can be a significant percentage of the image frame in practice.

Linear

A linear-based system uses a fixed temporal interpolation process where the weighting factors, or coefficients, are dependent only on the relative position of the line being synthesized. Typical linear systems will offer a number of different coefficient sets to allow for the handling of both moving and nonmoving images. The selection of a particular set is left to the user, with no one set being optimal for all conditions. This then requires the user to change between different sets as the material in a given input changes in order to maintain optimal performance.

Motion Adaptive

A motion-adaptive system tries to compensate for the limitations in a linear system by using an aperture function, which is adjusted to suit the needs of different areas of the image. The adjustment to the aperture is made based on the presence of motion.

The input image is analyzed pixel by pixel to determine if motion is present. Unlike the motion-compensated systems described below, the adaptive system does not attempt to track motion; it simply knows whether motion is present or not. This motion information is then fed to the interpolation process and used to select the appropriate weighting factors for that particular area of the image. In a typical 4-field system, when no motion is present the aperture will remain wide, using a contribution from each of the four fields. When motion is present, however, the aperture narrows and the greatest contribution occurs from the two closest fields. It is this ability to change the aperture size and weighting of the samples based on image content that allows a motion-adaptive system to balance the effects of blur and judder.

The motion-adaptive system, while eliminating many of the problems of a linear system, has one main fault. If the motion content of the scene changes quickly, for example, two people talking in one scene followed by a car chase in the next, the motion-detection system may have problems reacting fast enough to change the aperture and weighting. This has the result of the wrong coefficients being applied until the system catches up with the image content. The solution to this problem usually involves two-part motion detection, one looking for field-based changes while the other looks for smaller changes within the field. The other option is to provide a user control to restrict the range of aperture changes available to the system.

Motion Compensated

As described above, a non-motion–compensated system must make a choice between blur and judder. Whether that choice is made globally as in a linear-based system or on a pixel basis as in a motion-adaptive system, the choice must still be made because of the spatial and temporal offsets between lines and fields. The concept behind a motion-compensated system is to eliminate these offsets by analyzing the input video signal to identify each object in the scene and figure out how it is moving. From this information, it is able to theoretically calculate where all the objects will be at the time that it wants to generate each new output field.

By knowing the location of objects and to where they are moving, the input fields on either side of the required output field can be shifted to eliminate any offsets. Thus, when the interpolating temporal filter is applied, there is little possibility of any blur or judder in areas where object correlation is high. In challenging scenes, where many objects are moving in different directions (such as zooms), it is very difficult to find field-to-field correlations and the benefit of motion compensation is much reduced. It is important to acknowledge that only the areas of an image with valid temporal object matches will benefit from motion compensation. In practice, this is limited to only a portion of typical imagery, those regions with regular non-complex movement, such as pans and title scrolls. In such areas, the improvement is significant.

There are three basic types of motion compensation: block based, pixel based, and phase correlation.

Block Based

In the block-based (or block-matching) approach a block of pixels is compared, a pixel at a time, with a similarly sized block in the same place in the next image. If there is no motion between fields, there will be high correlation between the pixel values. However, in the case of motion, the same, or similar, pixel values will be elsewhere and it will be necessary to search for them by moving the search block to all possible locations in the search area. The location that gives the best correlation is assumed to be the new location of the moving object.

Block matching requires an enormous amount of computation because every possible motion must be tested over the search range. Thus, if the object is assumed to have moved over a 16-pixel range, then it will be necessary to test 16 different horizontal displacements in each of 16 vertical positions, in excess of 65,000 positions. At each position every pixel in the block must be compared with every pixel in the second picture.

Pixel Based

The pixel-based solution consists of a motion estimator using a full exhaustive search on 1 × 1 block sizes, generating a motion vector for every pixel in the image. This allows motion of any kind to be detected at pixel granularity. The pixel motion vectors are then used to precisely position the objects in the image to the correct spatial and temporal position, maintaining the full vertical resolution of the imagery even when significant motion is detected. Pixel-based solutions require advanced techniques for filtering motion vectors to ensure their accuracy. As a result of these filtering requirements along with the per-pixel motion vector generation, this technique is highly computationally complex and requires supercomputer performance to execute in real time.

Phase Correlation

Phase correlation takes advantage of the fact that displacement from one temporal sample to another is proportional to the phase difference from one temporal sample to another divided by the frequency of the component being sampled.

A fast Fourier transform (FFT) is performed on each field, mapping it into a two-dimensional frequency domain represented as amplitude and phase. The amplitudes are normalized to avoid any reliance on lighting levels, and for each frequency component, the phase values are subtracted to find the phase differences. Once the inverse FFT is performed, this gives a correlation surface in which the peaks represent movement.

If there is no movement, a single large peak results in the center of the correlation surface. If there is movement this results in peaks displaced from the center point. The distance from the center is the distance moved between fields. The direction of the peak from the center is the angle of movement. The size of the peak is proportional to the number of pixels that have movement. These are the candidate motion vectors together with a value that determines their significance.

The phase correlation process produces the correlation surface with peaks representing movement; it then uses an image correlator to figure out which pixels each movement belongs to. Candidate motion vectors are examined one at a time, starting with the one that corresponded to the largest number of pixel movement. The entire input field is shifted by an amount specified by the candidate vector and then is compared with the next input field in the sequence. Any pixels that are found to now be in the same place in each of the two fields therefore must have the motion described by that vector. After the appropriate discarding of spurious pixels, the pixels with the same movement are grouped together as an object and assigned the motion vector.

This matching process is not looking for motion, since the motion is known from the phase correlator, but instead is looking for the outline of objects that have known motion. The process is repeated for each of the candidate vectors. Some candidate vectors will be spurious ones and will not produce any pixel matches, so they will be discarded.

Once all of the vectors are correlated with pixels in the image, the objects are shifted to their new positions. The result is that each object is perfectly lined up in the current input field, the required output field, and the next input field. Another way to look at it is to say that the objects are all aligned to the time axis. Any objects not identified in this process must be handled with a linear or motion-adaptive fallback approach.

There are many techniques available for standards conversion, each with its own performance level and quality. The motion-compensated techniques have been shown to outperform the linear and motion-adaptive systems, particularly in sports, where there is significant motion. The largest problem in performing a high-quality motion-compensated frame rate conversion is in the determination of the motion characteristics of each object and avoiding false correlations. The traditional motion-compensated systems all use some form of estimation or intelligent guessing to help in determining where an object has moved. This estimation is used to help limit the number of searches because of the restrictions in the performance of the hardware used.

2:3 Sequence Insertion Removal versus Temporal Rate Conversion

When 29.97 fps (59.94 Hz U.S. standard) material is standards-converted to 25 fps (50 Hz European standards) there is always the possibility that the source video may have originated as film and have a 2:3 field sequence. One method used to overcome any possibility of artifacts created in the conversion of 2:3 pull-down material is to first remove the 2:3 sequence. Once the sequence is detected, it is removed and the original 23.98 fps progressive frames are recovered. This 23.98 fps video may then be spatially converted, say, from 480 lines to 576 lines (for SD material) and recorded at 23.98 fps on a specially modified *Slow PAL* VTR. The 23.98-frame, 576-line video may then be played back at 25 fps in a standard PAL VTR.

This is similar to the process whereby 23.98 fps film is converted to video for European distribution by running the telecine 4% fast to arrive at 25 fps. This process eliminates the need for a temporal rate conversion and thus eliminates the motion artifacts and mixed frames that can occur when converting this type of material.

This same approach can also be used when converting film-originated material (with 2:3 pull-down) to 1080p23.98. Once the 2:3 sequence has been removed from the 480i59.97 film original, the resultant 480psf23.98 (segmented frame) signal may then be spatially converted in the horizontal and vertical domain to 1080p.

SELECTING A FORMAT CONVERTER

Format converters come in a variety of shapes and sizes ranging from chipsets integrated into other devices, through handheld devices, to modules, and stand-alone 19-inch rack mount units. In general, the larger, more expensive solutions provide greater performance than the smaller, low-cost units. Due to the advances in ASIC technology driven by the consumer HDTV market, high-performance format converters are now available in small, affordable packages.

Integrated Format Converters

The need for integrated format conversion is growing at a rapid pace in order to meet the needs of the multiformat broadcasting world. Cameras, VTRs, servers, encoders, switches, routers, and displays all have options for integrated format converters. It is important that the chips used in these converters meet the high-quality professional standards described above. If a particular device is known to have substandard internal format conversion, the converter can often be bypassed and/or supplemented with higher quality outboard converters. For example, HD switchers often offer integrated format conversion capabilities, but due to the density and number of channels required, it is often better to use outboard conversion gear.

Modular Format Converters

Modular format converters are available in 1RU, 2RU, 3RU, and 4RU configurations, each holding numerous converter modules and sharing redundant power supplies and system reference inputs. They also provide a convenient mechanism for unified control via SNMP, Web, automation system, or discrete control panels. Remote control and status monitoring make modular converters good choices for large facilities with many channels, particularly when the converters will be left in a "set and forget" configuration. Note in this configuration it is imperative to have converters with robust automatic content detection to ensure high-quality operation regardless of source material.

Utility Format Converters

Utility format converters, available in handheld boxes, are useful as supplements to integrated format converters in cameras, VTRs, displays, and nonlinear editing (NLE) systems. Their small size and low cost make them affordable alternatives to the integrated solutions.

Stand-Alone Format Converters

Stand-alone format converters are available in 1RU through 6RU configurations and usually have elaborate front panel controls. These are typically used in postproduction houses and network feed centers. These units typically have the highest quality available and also have the largest array of I/O options. It is important to have intuitive user interfaces in these devices to prevent on-air glitches and ensure that the state of the conversion is as desired.

Quality Considerations

Quality is an important consideration in format and standards conversion. The primary goal of the conversion process is to maintain as much of the original image detail as possible throughout the conversion process and avoid the introduction of artifacts. Due to the complex nature of the processes involved, if appropriate care is not taken by the converter manufacturer, it is very easy to end up with significant artifacts or half-resolution images. Traditional static image test patterns such as color bars and multiburst are not sufficient to properly evaluate converter performance. Dynamic motion sequences are required to properly characterize performance. There are industry-accepted measurement test sets available that can help the system designer evaluate the quality of competing format converters. One such test set is the HQV Benchmark DVD (details at http://www.hqv.com).

SUMMARY

Format and standards converters have evolved significantly from early analog designs to become sophisticated digital systems using complex mathematical algorithms and powerful computer processors. The selection of a format or standards conversion system is a complex process because of the many choices that must be made regarding input, output, level of quality, flexibility, and applicability to the specific conversion tasks at hand. To make a selection, the user should understand the technical processes involved and how they affect the images to be converted. Making the right selection will result in better conversion quality for the end user whether it is broadcast, postproduction archives, or confidence monitoring.

Bibliography

Deame, Jed, and Ackerman, Scott. "Advanced Techniques for Conversion to and from p24." *SMPTE Journal*, June 2002, pp. 265–268.

Ackerman, Scott. "Film Sequence Detection and Removal in DTV-Format and Standards Conversion." From the Teranex website (http://www.teranex.com).

Watkinson, John. *The Art of Digital Video*, second edition. Oxford, England: Focal Press, Burlington, 1994.

CHAPTER

5.13

Television Station Automation

CHRIS LENNON

Harris Corporation
Colorado Springs, Colorado

INTRODUCTION

Television station automation has evolved over the years from rather humble beginnings into the highly sophisticated systems that exist today. Such systems are capable of controlling large amounts of equipment for the assembly and playout of content for multiple program channels and affect nearly every aspect of the on-air broadcast process. They provide functionality and reliability not previously possible, while also improving overall business efficiency. This chapter summarizes some of the developments that have taken place and introduces the technologies that comprise such systems in a modern broadcast plant.

BACKGROUND

To help understand today's automation systems, it is beneficial to first gain an appreciation of where they came from.

Before there was any type of automation in a master control room (MCR), there was "assisted" manual switching that consisted of simple machine control associated with a master control switcher. Tapes were loaded and cued manually on the playback devices, but at least the operator had a single interface with which to roll machines and switch *events* (the elements that make up a continuous program output). It was an improvement over having to control each individual device using its own control panel, but everything was still essentially manual, albeit with centralized control.

Early automation systems included dedicated mechanical devices that could play out a sequence of events, such as the Ampex ACR-25 and the RCA TCR-100 videotape cartridge machines (see Figure 5.13-1). These could automate the cueing and playing of a series of spots and also load the tape cartridges (*carts*) from a magazine, but they rarely handled programs, and they did not control other devices.

FIGURE 5.13-1 RCA TCR-100. (Courtesy of Lytle Hoover, at www.oldradio.com.)

Software-Controlled Cartridge Machines

The beginnings of master control automation capabilities were seen in sequencers, such as the original Sony Betacart machines (see Figure 5.13-2). What made these revolutionary was that they were the first widely deployed cart machines that were run by a software control system. Traffic systems sent *playlists* that were used to control the playout of material contained on videotape cartridges. *As-run files* were generated automatically, enabling automatic reconciliation of *spots*. This took an operation that had formerly required considerable operator effort and automated it. Video content no longer had to be sequenced manually. Commercials played to air automatically in precisely the manner dictated by the station's traffic department. This is also the point at which limited multichannel systems first became available, but there was minimal sophistication in terms of what could be automated.

There were limitations with this generation of devices. Typically, only short-form content was played back from them, meaning that long-form program material still had to be played manually from videotape recorders (VTRs). A one-to-one relationship existed between a piece of material and a video cart, and these machines had a limited capacity of carts. While most provided some general-purpose interface (GPI) capabilities, meaning that external devices could be triggered, for all intents and purposes, the Betacart served as a spot playback device.

Following early cart machines came so-called "smart cart machines" that included the Sony LMS, Panasonic MARC, Ampex ACR-225 (see Figure 5.13-2), and others. These expanded the tasks of the Betacart. Smart cart machines were capable of playing single- or multisegment material. GPI control was expanded, allowing for the control of more external devices. The concept of *media management* was born, and was offered as a core part of the smart cart machine's software suite. Schedules could be acquired from traffic and played back, and as-run files generated. In many cases, the media management software could accept *dub lists* and *purge lists* from traffic, further automating media management.

FIGURE 5.13-2 Ampex ACR-225 (left) and Sony Betacart (right). (Courtesy of Tim Stoffel.)

True Automation Systems

If a broadcaster wanted to do more than just sequence and play out tape-based material, a more sophisticated system was needed. Such systems were envisaged to allow the master control operation to move from a hands-on environment with highly-skilled operators to essentially a hands-off operation with the operator monitoring the system for problems and dealing with them as they arose. Broadcasters saw real benefit in these systems and, in response to the demands of users, true master control automation systems were developed and became commonplace during the 1990s.

In addition, around this time, hard disk-based devices were starting to become an economically viable means of storing and playing back broadcast material. Video servers began to replace cart machines, offering significant advantages. Perhaps most significant among these was the dramatic reduction in the number of mechanical parts, which resulted in greatly increased reliability and reduced maintenance costs. Automation systems quickly adapted and supported the array of new video servers that came on the market. With this came a new responsibility for automation systems—managing the movement of material among storage devices. No longer did a tape or cassette have to be physically moved from one device to another. With content stored as *files*, it could be moved electronically from *off-line* to *near-line* to *online* storage.

Paralleling the evolution of devices used to store and play content, a movement began in the area of master control systems, from primarily proprietary, closed approaches to those that were more standardized and open. The fact that content was now on disk-based storage, and was being managed by computer-based systems, led to the introduction of information technology (IT) standards into the master control environment, an evolution that continues.

INDUSTRY EVOLUTION

Capability and Efficiency

Along with the technological advances taking place in the master control area, broadcasters began to see new opportunities for further operational improvements with increased efficiency.

Without automation, master control operators were overloaded with responsibilities, and station management did not want the operators to take their attention off their duties to manually process the complex control sequences needed to generate *spots* and *interstitials* between programs, requiring custom tags, graphics, and voiceovers. Preproduction compilation for these items was therefore a necessity. As automation systems

became more prevalent, the need for preproducing spots was reduced because the new systems had expanded capabilities to manage complex simultaneous events. Now, the *secondary events* needed for such events could be programmed into the system, to be triggered automatically, in effect producing the on-air look for spots on the fly. There were real benefits associated with doing this. First, the staff's need to preproduce broadcast material was reduced, resulting in substantial cost savings. Second, there was no need to maintain multiple versions of essentially the same spot. Third, the on-air *look* improved with tighter transitions and better quality. Master control operators could use their time to monitor quality and assist in handling news feeds and maintain the automation equipment.

Automation systems continued to evolve to meet the expanding needs of broadcasters. An ever-increasing array of software device drivers became essential as broadcasters required automation to control new devices, and in some cases, entirely new classes of devices. As broadcasters moved away from the one-to-one operator to channel relationship toward a one-to-many relationship, automation systems became better at this as well. When coupled with the trend toward centralization of broadcast operations, this also meant that scalability became a very important feature, as did the ability to customize the master control operator's control interface so that multiple channels could be easily viewed and managed on a large single or multiple screen array.

Operating Systems

The progression of software operating systems (OS) paralleled the development of automation systems. As operating systems added capabilities, so did automation systems. The Microsoft Disk Operating System (DOS) was a step up from earlier options and Microsoft Windows brought a new level of operational capability and features of its own to automation, some good and some bad.

The introduction of the graphical user interface (GUI) and control using point and click with a mouse seemed a great advantage for the operator, but it was quickly learned that there were downsides to this as well. Operators accustomed to jog/shuttle wheels for data and control input found that performing the same task on a Windows-style GUI was often cumbersome, slow, and prone to errors. Being able to drag and drop items on screen seemed easier, but also could be very dangerous for the schedule, such as when a moved event did not end up in the intended place or time.

Basing the automation system on a standard Windows OS platform certainly offered advantages, but performance and reliability issues that came along with it were significant and had to be considered. These included such issues as the time required to boot up a Windows-based personal computer (PC), the mean time between the need to reboot it, and the fact that Windows does not lend itself well to deterministic processes that must be executed at a specific time.

As a result, there was a balance to be found between putting standard Windows-style functionality into automation systems and maintaining a reliable and user-friendly architecture. Different approaches were used, but the most successful systems achieved a workable balance compromise between all of these issues.

Integration Between Systems

Automated or semi-automated transfer of information into and out of the automation system have been available for some time, through the importing of logs, dub and purge lists, and as-run logs. However, the next logical step in this evolution is full integration of automation with the rest of the broadcast operation. In order to accomplish this, automation systems have to implement rich, dynamic interfaces with other systems, including programming and traffic (upstream) and distribution and monitoring (downstream).

As this occurs, it will not only allow all the actions of the master control operations to be fully automated, but it will also enhance the interaction between personnel with different, interdependent tasks, responsibilities, and interests in the operation of the channels. The goal is to position automation as the fulfillment of the upstream planning, the source of downstream deliverables, and a mechanism for coordinating metadata, actions, and responses in the real world. This topic will come up as a recurring theme as this chapter continues.

PLANNING CONSIDERATIONS

There are numerous considerations to take into account when planning for and selecting an automation system. Some of the key issues are outlined below.

Benefits of Automation

When stations make significant capital investments in automation systems, those making the purchasing decision must see a clear return on investment (ROI). While headcount reductions appear to drive the returns, what is perhaps more significant is that automation allows the broadcaster to do more with the same staffing level. Personnel who were formerly occupied with mundane, repetitive tasks, such as material management and on-air transmission responsibilities, can be freed up to address other needs within the facility. Another factor contributing to the ROI of automation is that less specialized (and theoretically less expensive) personnel can be used to run a master control operation. Fewer devices that are mechanical in nature, coupled with more reliable devices and systems, also means a reduced need for highly skilled maintenance staff in those areas.

It is clear that many of today's broadcast channels could not be run without automation. The nature of the services being presented and their complexity make it impractical for human control to keep up with the numerous and complex events in a broadcast day.

Presentation Quality

One of the tougher-to-quantify advantages of an automated operation over one running manually is presentation quality. A well-designed and properly configured automation system will switch events perfectly and consistently, eliminating up-cuts and intervals of black between events. Transitions, effects, and the display of graphics over primary video will be seamless and pleasing to the viewer. Performing tasks in a repeatable and consistent manner is something that machines inevitably are capable of doing better than humans, and achieving a clean on-air look necessitates that all of this happens perfectly, every time.

Operational Requirements

There are several important operational requirements that a broadcast operation in search of the right automation system must consider. These include the following.

Channel Count and Complexity

The number of channels to be controlled is a major factor. Some automation systems are very capable when managing, say, one to four channels, but when stretched beyond that, show that they were not designed for higher levels of performance. This can be due to core architecture restrictions; many systems were designed when single-channel operations were the norm, and there may be limitations in hardware, operating system, or the automation software itself. When considering a particular automation system, it is often helpful to look at existing deployments in other broadcast facilities. Evaluating how well the system is handling the requirements at other sites will help to guide local decision making.

Channel count alone is not the only factor to consider. One site might be running five very simple channels with minimal secondary events, and perhaps no commercials. Another site running five fully advertising-supported channels, with a complex array of devices to control, could place much greater demands on the automation system. Yet another five-channel site, with channels joining and leaving network feeds at various times of day, broadcasting a large amount of live programming, such as news and sporting events, would require even more from an automation system. The demands placed on the automation system in each of these scenarios are substantially different, although each is a five-channel system. Just as the channel count can stretch a system to its limits, channel complexity can test those boundaries even further.

Type of Operation

An important and often overlooked consideration when examining automation systems is the type of operation to be automated. A single-channel standalone operation has very different requirements from a centralized multichannel facility, and other types of operations may fall between these two in its requirements. A centralized ingest operation with individual playout facilities is quite a common arrangement. Distributed operations with regional hubs are another. Each model presents its own challenges for automation and often different systems are optimized for different models. Because there are so many operational models (with new ones emerging constantly), it is important that a thorough understanding of the model to be used is established between the broadcaster and the automation vendor.

Equipment and Devices

New devices, and even entirely new classes of devices, appear on a regular basis. Unfortunately, true plug and play of broadcast devices has yet to become a reality. Standards do exist, but even a group of devices that all adhere to a certain control protocol can differ in significant ways, and require custom device drivers in the automation system. New drivers can often be costly and time consuming to develop. For this reason, it is necessary to carefully research the extent of the device support available within the alternative automation systems. Many automation vendors have established close ties with a collection of partners who sell devices controlled by the system. When examined superficially it may seem that many automation vendors claim a close partnership with a certain equipment provider but closer examination often exposes important differences in the level of control possible over the devices in question.

Expandability and Adaptability

While a facility may be of a certain size and scope today, that does not mean it will be that way forever. Broadcasters who purchase automation systems based only on current requirements may find that, as their needs grow, the system they purchased is unable to grow to meet the new demands. Consider what the operation will look like in 1, 3, 5, or more years. Of course, some of this is impossible to anticipate, but much of it is predictable. Taking the time to think through the requirements in detail, and mapping the requirements onto the systems under consideration, can help ensure that the right purchasing decisions are made, so that the system acquired will be just as suited to future operations as it is today.

The purchase of an automation system is a significant capital expenditure, and in most cases, is written down over a period of many years. This means that the financial decision makers must fully understand the need to purchase a new system before the current one is fully depreciated.

Bandwidth Utilization

For digital terrestrial broadcasting in the United States, each broadcaster is allocated 6 MHz of spectrum. The

ATSC system allows this to carry a digital bandwidth of 19.39 Mbps but there is a certain amount of latitude available with regard to how this bandwidth is utilized. Traditionally, broadcasters have thought of the broadcast commodity they are selling as time. While this is still true, another commodity being sold in the digital world is bandwidth. Thus, inventory has moved from being a one-dimensional object to two dimensional. The goal is to now to generate as much revenue as possible across that 19.39 Mbps bandwidth every minute of every day. Automation systems that work in concert with systems that manage bandwidth, such as scheduling systems, encoders, and multiplexers, can get the most out of the commodity being sold.

HARDWARE DEPLOYMENT

As with any equipment to be installed in a broadcast facility, the physical installation of the hardware, wiring, and other infrastructure requirements to support it, must be considered. This is particularly true in cases when the system must be integrated into an existing facility; where space must be found for equipment, keyboards, mice, and monitors; and the proper connections to every controlled and interconnected device and system ensured. The degree to which equipment can be physically distributed may be as important as the actual space required. While some parts of the system might be safely tucked into an obscure rack, other equipment must be placed according to operator efficiency and convenience, or in proximity to related equipment. As with any equipment, consideration must be given to the power and cooling requirements and access for maintenance as well.

Modern automation systems are invariably comprised of multiple computers networked together. While the number of computers and the specific distribution of functions between them can vary, based both on the particular system chosen and the scale of the installation, certain components exist in virtually every design, as shown in Figure 5.13-3. The key elements that make up a typical automation system include the following devices.

Device Control Server

This computer handles the actual control of the broadcast devices that play, record, switch, and apply effects to the channel. The majority of broadcast devices are controlled via an RS-422 connection, so some means of providing a large number of serial ports for these individual connections is necessary. However, as network control of devices becomes more prevalent, it will supplant these multiple point-to-point connections. A method of synchronizing the device control server computer to the broadcast plant's timecode and house sync must be provided.

List or Application Server

This computer handles the centralized management of the processes throughout the system, and serves as the overall administrator for the system.

User Workstations

These computers support the operator's console, allowing monitoring and control of the system. While some systems require separate workstations for different functions such as playout, recording, and ingest, in many cases, any workstation can be utilized for any task by running the appropriate application. In more widely distributed installations, it is possible to access these applications through existing desktop computers or even remotely, although primary control room operation almost always maintains a dedicated workstation.

Depending on the particular system and installation, certain functions may be coresident on the same computer. As system size increases, not only are functions typically distributed among computers, but multiple computers are deployed for each function. In addition to scaling, multiple computers are often deployed to provide redundancy within the system. As well as these core functions, automation installations also often utilize additional machines for such tasks interfacing to other systems, translating and transcoding data, and hosting dedicated services.

File Server

This computer provides storage for the data utilized by the system, such as content and media metadata, logs and schedules, and configuration files. This important element is discussed in more detail later in the chapter.

SECURITY

In deploying an automation system to run on-air broadcast channels, a major consideration must be the security of the system. There is an inevitable tradeoff between providing convenient access and preventing unauthorized access. Traditionally, master control security was maintained by controlling physical access to the equipment itself. While this could be accomplished when operator and equipment were located in a machine room or control booth, the ability to access a modern automation system from networked computers limits this approach.

Operator Authorization

At the very least, any workstation not in a secure location should be protected by a password login. Note that this requirement is not limited only to workstations that can directly affect playout. As data-driven as modern systems are, improper modification of any data can have unforeseen and often unpleasant effects

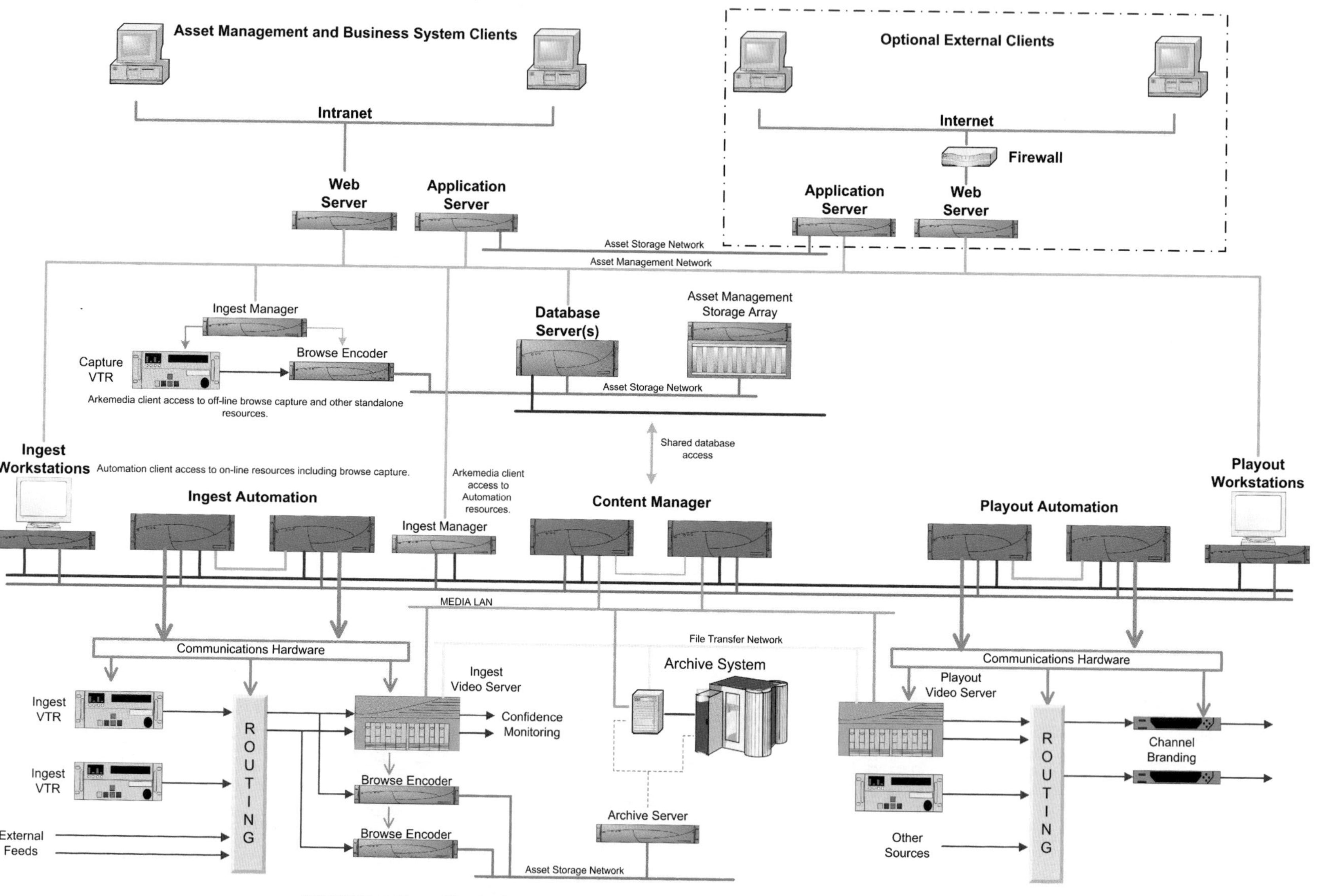

FIGURE 5.13-3 Typical automation system. (Courtesy of Harris Corporation.)

on broadcast operations. For this reason, it is desirable that log ins control the ability to access the system as well as the specific actions that a user can take. This allows access to be granted to users of varying skill and responsibility. The same approach should be taken when other systems, such as traffic, are directly interfaced. A balance must be struck between the advantages of unrestricted access to the schedule and other resources and the need to ensure the integrity of the actual execution of the channel. In many cases, there will be additional complications stemming from operational or contractual constraints regarding who or what may accomplish certain tasks. Broader access to an automation system should be provided only if limits are enforced on the different groups of users and their ability to act on the system.

From the standpoint of physical implementation of the network, larger systems, or systems with higher security requirements, can be implemented on multiple levels. This not only provides better control of the bandwidth available to different processes, it also allows critical functions to be segregated while other network segments are made more widely accessible. This is particularly important if an automation system is interconnected with a house or corporate network, and even more so if there is a path to a public network such as the Internet.

Data and System Integrity

Another security consideration is data and system integrity. The most obvious aspect of this is protection against viruses and other malicious software. It is essential that any point in the system that can be externally accessed, whether by a network connection or removable media, be protected. In most cases, commonly available antivirus software can be deployed. In some cases, there may be limits to the deployment of protection software on certain timing-critical computers, although newer such applications tend to be less intrusive than in the past. Also, in some cases, dedicated platforms that are not generally vulnerable to virus attacks are utilized for certain critical components. It is important that any protective software be regularly updated. While IT departments may be responsible for maintaining protection on the majority of the personal computers in an organization, computers in dedicated broadcast roles may have to be explicitly managed to maintain protection.

EQUIPMENT INTERFACES AND CONTROL

Types of Interfaces

In order to execute events, an automation system must control a wide range of devices. The traditional approach to this was the use of deterministic commands (e.g., "play this clip now"). It was the automation system's responsibility to determine device and system latencies, and adjust for them, so that these deterministic commands would occur as expected, resulting in a clean on-air product. As IT infiltrates the master control suite, an increasing number of devices that are capable of deferred command interfaces have emerged. The major influence in this area is the emergence of IP-based connection and control of broadcast devices. Whereas the traditional deterministic approach was based around point-to-point connections with virtually no latency, such as serial cables, deferred command interfaces are based on point-to-multipoint connections such as a LAN (local area network).

Determining which approach to use deserves consideration. In most broadcast facilities, both have their place. An automation system should be capable of controlling devices using whatever approach is required by the devices involved. Some will require serial deterministic control, while others may only be controlled via IP-based deferred command. Still others may offer both interfaces. Any automation system that takes an all-or-nothing approach on one side of this or the other is unlikely to be able to control all of the devices in place in a typical broadcast environment.

Drivers and Protocols

Since the primary task of any automation system is to control, an important factor in the applicability of any particular system to a facility is its ability to control existing and future equipment. The control of devices, regardless of the physical means of connection, typically requires that the automation system implement a driver specific to that device. In some cases, controlled devices may emulate other devices for control purposes and use the same protocol. A third possibility is for the device and automation system to communicate with one of the various standardized device control protocols. In general, the first option yields the best results, as a device's native protocol is the most likely to best exploit the device's capabilities. The disadvantage is the variability of support for a particular device, which often is a factor of the device's popularity. On the other hand, the use of standard protocols brings a relatively high level of support to every device that uses it, with the downside that these "generic" protocols rarely fully support the capabilities of each broadcast device that may use it. Of the three, emulation is typically the least successful, as a protocol developed for a single device will often not only fail to fully support all the functions of another device, but may in fact exhibit unexpected behavior unless the emulating device closely matches the functionality and behavior of the original.

Program Chain

Up to this point, the critical parts of the broadcast program chain have been purposely omitted when discussing the automation system. This includes such items as the devices that produce, switch, modify, and deliver the broadcast streams; the physical media and files that contain the material to be broadcast; the broadcast streams themselves; and the cabling and transmission paths by which streams are transported and delivered. However, in the increasingly integrated and network- and file-based environments found in broadcasting, the automation system is increasingly acting on more elements of the program chain, and acting more directly on parameters previously available only through device control.

For instance, the movement of media files, such as between video file servers, can in many cases be more efficiently accomplished by the direct manipulation of files by the automation system. This also permits the tight integration of lower cost mass storage, such as off-the-shelf disk array solutions, without having to make use of a proprietary intermediate system that makes it appear to be a broadcast device. Additionally, there are efficiencies in embedding, transporting, and extracting metadata within or closely associated with the media itself. An example of this is integration with digital content distribution systems, where items can be taken into a facility along with timing and other metadata, reducing or even entirely eliminating preparation tasks.

Emerging technologies, ranging from real-time viewership data from set-top boxes to direct user input from wireless network devices, will make use of the automation system to provide real-time feedback to upstream programming and traffic systems, and allow last-minute changes and even personalized delivery of content to be returned. In this role, the automation system becomes the mechanism by which the goals of upstream decision makers, and the expectations of the viewers, are fulfilled.

Operator Workstations

If one considers the ubiquitous master control switcher, all of the audio and video processing is accomplished in the rack-mounted frame, but the control surface is what almost any user would point to as being the actual master control switcher. In the same way, while the critical functions of the automation system will in most cases be handled by an anonymous rack-mounted chassis, the user workstations present the visible manifestation of the system.

In addition to controlling the execution of channels, the user interface must provide tools to manage the schedule, content and media metadata, and the configuration itself. Particularly significant in the deployment of workstations is the overall trend toward file-based playout from video file servers that reduces or eliminates tape handling, and where appropriate, the use of low-resolution proxy copies directly on computer desktops. This reduces the requirement to locate workstations adjacent to specific pieces of broadcast equipment, allowing both the distribution of tasks throughout a facility or, conversely, the performance of multiple unrelated tasks at the same workstation. The majority of workstation user interfaces utilize Microsoft Windows or other mouse-oriented platforms, which further serves to streamline operation by reducing the training requirements and making systems more approachable by a greater number of potential users, although there are potential perils here, as mentioned earlier in this chapter. This important element of the system is discussed in more detail later in the chapter.

STAFF REALIGNMENT

While, at its simplest, automation can be used merely to add precision and convenience to the operation of a television channel, there is almost always an expectation of both labor savings and an increase in the complexity of the operations that an operator can handle. An important potential remaining in many cases is furthering the integration between personnel, particularly as decisions and status are passed between different tasks, departments, and systems.

Master Control and Traffic

One of the fundamental ways an automation system must be integrated into a facility is the acquisition of schedules. In the past, it was acceptable for an operator to take the next day's printed log and manually type each event into the automation system. All of that effort, however, simply served to duplicate work that had already been accomplished by programming, public affairs, promotions, and traffic personnel in producing the program log. Thus, middleware was put into place to automatically convert a traffic log into an automation playlist.

This capability, along with the creation by the automation system of an as-run log indicating the successful execution, or failure, of each event, to be returned to the traffic system for reconciliation, greatly streamlines the interface. Not only does this result in a labor savings, the chance of error is reduced. However, once a log has been translated and passed to the automation system, the inevitable changes must typically be managed manually. This not only requires the efforts of both traffic and master control personnel and increases the chance of error, it forces the personnel involved to divert their attention from their primary responsibilities. In addition, the time involved limits the changes that are practical or even possible.

The solution is for the automation system to have a dynamic interface with which changes can be made directly in upstream systems, such as traffic, and, subject to user rights intended to maintain on-air integrity, immediately be reflected in the automation system with the minimum of user interaction. For instance, if made sufficiently far in advance of air time, a change might be automatically made to the schedule by a traffic operator, while a change made closer to air might require no more than the on-air operator's approval directly from the master control workstation to be applied. Similarly, as-run information can be immediately returned to upstream systems, such as traffic, by means of the same interface as each event completes. This could be used to more efficiently react to the changes made of necessity as a channel executes.

FILE SERVER

Of the relatively invisible components of most automation systems, the file server is most often overlooked but is in many ways the most critical. As the amount of data, both related to content and metadata, and the schedule itself increase, and as this data becomes more and more dynamic, the file server goes from being merely a repository to an active broker of the data it stores. In addition, the increased integration of systems means that data stores may be called on to also manage related files such as low-resolution proxy copies of media, related deliverables such as web pages, and in some cases, the broadcast media files themselves.

Distributed Systems

Some facilities have implemented massive storage systems in which this broadcast data might physically share space with everything from payroll data to a digital media archive. While conventional IT management tools can be utilized to present discrete storage to each system, the architecture of an organization's network and storage infrastructure becomes an increasingly important component of long-range planning.

Modern computer hardware and software serves to provide substantial capability in a single computer. This allows the potential for an entire automation system to be contained in a single device, encompassing all of the individual functions described. This can not only reduce the cost of the system, but also simplify implementation and management. However, as the system is scaled in terms of the number of channels, the complexity of operation, or additional users, distribution of the system across multiple computers becomes preferable or even necessary. The most common candidate for separation is the user workstation. This not only provides greater flexibility in their placement, but also reduces the risk of failure. By separating the automation control functions from the workstation, workstations can be safely rebooted, or the automation workstation application stopped and started, without directly affecting on-air playout. This is particularly important in cases where general-purpose computers with other applications are utilized as automation workstations. This also permits the use of more robust hardware for the critical functions, with such features as watchdog systems, redundant power supplies, and high mean time between failures (MTBF) components.

As systems become increasingly large and complex, and particularly as additional dynamic interfaces to external systems are implemented, the separation of real-time device control from the remainder of the system's processing becomes important. In order to deliver reliable, frame-accurate control, those portions of the system that are dedicated to real-time control must ensure their availability of every frame, as derived from house timing and sync sources. Doing so can have an impact on the processing capabilities of that machine. Additionally, separating the device control allows it to be deployed in a distributed manner. This not only reduces the effect of any disruption to the device control, it also allows support for geographically distributed control.

INGEST AND MANAGEMENT OF MATERIAL

Ingest Process

Content must be prepared or *prepped* before it can be played to air under automation control. Material preparation historically has involved the manual process of timing the material, that is, locating and marking the start of material (SOM) and end of material (EOM) points, entering content metadata (e.g., title and house ID), and then copying from the source, such as tape or video file server (VFS), to a destination device. Quality control (QC) of the audio and video information is also part of the preparation process.

Ingest stations provided by automation vendors allow media prep operators to perform these functions quickly and accurately. These stations should not be thought of as editing workstations but as points where SOM and EOM are checked. Editing stations allow the user to add such things as transitions between scenes, voiceovers, and sophisticated graphics to the final production, but the material prep operator in master control is primarily concerned with speed, accuracy, and quality. The task at hand consists of marking the correct in and out points on tape or VFS and entering the associated metadata.

While broadcasters have enjoyed the efficiency benefits of playout automation for many years, material preparation has been largely a manual process. Another inefficiency has been that material preparation is often performed at each station facility even though other stations within the same group might receive and air the same spots and programs.

Automated Ingest

Several technological trends emerged to help alleviate these problems. The first is the advent of digital content and the second is lower-cost, high-speed, secured electronic data distribution channels or networks.

Long-distance content distribution systems are a good example of this. In the United States, the prime players are DG FastChannel, Pathfire, and Vyvx. It is true that in the beginning, these systems aided the media providers more than the broadcaster since the process of transferring media from these *catch servers* was largely manual in nature. However, automation vendors and other broadcast equipment suppliers have stepped in to automate this workflow.

Since content delivered to these servers is pre-prepped with accompanying metadata files (content and timing metadata), there is no reason the ingest process cannot be automated. A typical workflow is outlined below:

- Traffic sends a dub list containing the Industry Standard Coding Identification (ISCI) code, house identification (ID), and other metadata to the automation system.
- The automation system checks for available content on the content delivery servers using the ISCI code as the unique identifier.
- When content is found, a transcode operation is initiated from the catch server to the broadcast VFS and the content is transferred via file transfer protocol (FTP) over the data distribution channel.
- The automation system pulls in the content distribution metadata and the *traffic dub list* data to create a database record.

In reality, the process is not quite as simple as outlined in the steps above. Flexible workflow configurations based on various business rules must be allowed. Some of those rules might include the following:

- Should a low-resolution proxy copy of the media be created?
- Can some material prep/QC be performed outside traditional channels using proxy material on a desktop?
- Should the process be initiated with a traffic dub list or when material arrives on the catch server?
- Should a manual validation task be inserted in the workflow to allow an operator to verify timing and material quality?
- Should a "trusted source" be allowed to go to air without this validation step?
- Can refeeds that arrive on the catch server be automatically detected and handled without operator intervention?

State-of-the-art ingest systems must allow for these and other workflow scenarios to optimize the efficiencies within the broadcast facility.

Other related trends in the broadcast industry are the migration from videotape to digital content and from a video/audio-based network to an IT network that offers several other benefits, including:

- Material can be easily shared on multiple channels and at multiple facilities.
- Improved utilization of media assets.
- Metadata can be shared among various systems.

Digital content (and the associated database records) can more easily be created, cataloged, stored, and shared on a LAN/WAN network. Material can be prepared once (or efficiently located and repurposed) and utilized on one or more channels and among multiple facilities. In effect, facilities can duplicate the content distribution workflow described above on their secured, private network.

Material Management

Almost every broadcast facility has a large tape library dating back decades. The problems of deterioration of physical media, knowing what is in the library, and in some cases, finding a specific VTR format for playback, can be substantial. Of course, existing tape libraries can be digitized and cataloged, albeit involving significant investments of time and money. More importantly, moving forward, modern facilities can take advantage of available technology and system applications to improve efficiencies and protect valuable assets.

Metadata

Of equal importance as the rich content assets themselves is the metadata that must exist to describe each piece of content. Metadata defines the instances of material (there may be many including VFS, archive, and proxy copies) that exist within a facility, various descriptors (title, producer, agency, and various alphanumeric IDs), and timing data for each spot and program segment.

Sharing content also implies sharing this metadata, and perhaps even a common database. For example, material may be initially prepared by a news editor with an emphasis on speed with minimal metadata. That same database record (with the associated content) can be enhanced at a digital asset management (DAM) workstation. This operator can add many more descriptors for more elaborate database searches so the material can be more easily found and used in the future.

Automation systems have taken two different approaches to managing metadata. The more traditional approach is to rely on a database, which resides on a computer accessible by the automation system. Records are maintained in this database, relating to each piece of broadcast material that is managed by the automation system.

The second approach is that of embedded metadata. Rather than store metadata in a separate database, this approach involves the recording of the metadata along with the essence, and storing that data along with the essence on broadcast devices.

Each approach has its pros and cons. Storage in an external database enables that data to be accessible to applications that do not necessarily have access to broadcast video storage devices. It also has the advantage of not requiring the accessing of these broadcast devices every time metadata needs to be retrieved. On the downside is the fact that the metadata is decoupled from the essence. This means that it is possible for the two to become out of synch. When dealing with a database, the potential also exists for database corruption across a wide array of material metadata records. Finally, the reliance on additional pieces of technology (database server, database software, etc.) that may fail could endanger the on-air product.

Embedded metadata can only be retrieved by systems capable of communication with the broadcast devices on which the material resides. It can also imply a single point of failure, wherein when the essence is lost, the metadata is lost as well. There is also the question of what to do when encountering devices that are incapable of storing metadata along with essence.

Fortunately, the two approaches are not mutually exclusive. Some successful implementations involve a hybrid of the two, with metadata residing both in an external database, and embedded with the material itself. This is desirable, leveraging the advantages offered by both approaches, and mitigating the downsides of each.

In summary, material preparation and management, like playout automation, has evolved based on enabling technologies. The modern facility can ingest and prepare media assets with efficiency. Digital content available on a secured network can be efficiently utilized among various departments and facilities, thus reducing costs and potentially adding new revenue streams.

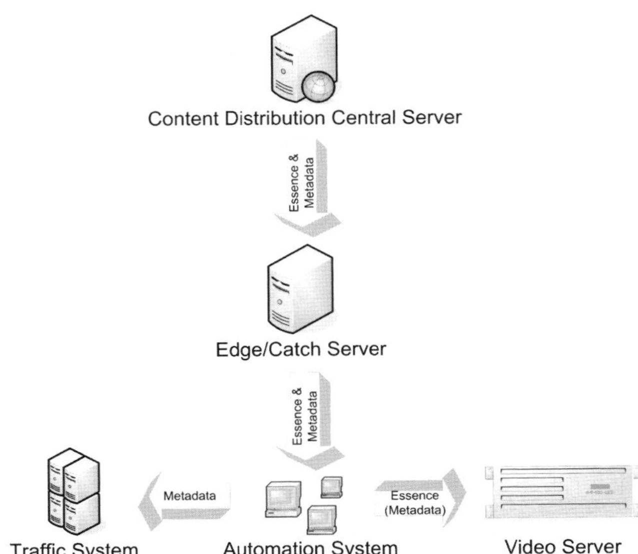

FIGURE 5.13-4 Essence and data flow among systems.

INTEGRATION WITH CONTENT DISTRIBUTION SERVICES

Legacy Procedures

As discussed earlier, the way in which content is sent to broadcast facilities has changed substantially. In the past, commercial content and some program content arrived on tapes, which could be in a variety of formats. The tapes were loaded onto the appropriate tape machine, reviewed, and ingested onto the media from which they would be played to air. Some program material often arrived via a scheduled satellite feed. Dishes had to be pointed, transponders set, and VTRs set to record the correct signal. If a feed was missed, phone calls were made, and hopefully a later feed arranged. Metadata typically consisted of information on a slate prior to the beginning of the content and/or on a piece of paper that accompanied the tape, or in the form of a fax received from the program distributor. The entire process of content distribution to the broadcast facility could fairly be characterized as labor intensive and prone to error.

Modern Distribution Systems

Today, much of the content received at a broadcast facility is in the form of digital files. It is not unusual to have 75% or more of the commercials arrive this way, and this is rapidly moving toward 100%. Some programming is moving from scheduled satellite feeds to file-based distribution. Some news content is following this trend.

File-based distribution of digital content can travel over a variety of transports before arriving at the broadcast facility. Some services utilize satellites, some use dedicated circuits, and still others use the Internet. However, a common feature of most of these services is that a *catch server* is needed in the broadcast facility to receive this content.

Having the content transcoded and moved from the catch server(s) to broadcast server(s) is a significant time-saver and having the metadata that accompanies the content sent to the automation systems in the broadcast facility is of great value.

Figure 5.13-4 illustrates the relationship of content distribution, automation, and traffic, and how the essence and metadata typically flows between each.

Material eXchange Format

The Material eXchange Format (MXF) will play a major role in the content distribution and ingest area. MXF has the potential to simplify the exchange of essence (the audio/visual component) and embedded metadata among these systems by using a standardized wrapper. Prior to the introduction of MXF, transcode engines handled the conversion of video formats, and metadata was extracted and exchanged using largely proprietary means. How this evolves depends largely on the pace of MXF essence and metadata adoption by content delivery and automation systems. For more information on MXF see Chapter 5.6.

PLAYOUT

This section discusses some of the considerations for control of on-air playout and the desirable characteristics of automation systems that achieve this successfully.

Staffing

Master control operators' responsibilities have changed dramatically with the introduction of increasingly powerful automation systems. The capability of early automation playout systems was more a function of assisting the operators than actual automation. Multiple operators spent a part of their time performing manual activities like pulling, loading, and cueing tapes. As described earlier in this chapter, automation systems, along with devices being controlled, became more sophisticated, moving the operator responsibility away from many manual tasks toward system monitoring.

Master control now requires much less human involvement, lowering the cost of operation while improving quality of the on-air product. However, master control operators must increasingly take on more responsibility as more channels are added to the operation. Many broadcasters cannot justify additional staff to support these new channels, and a single operator may have responsibility for multiple on-air channels. The task is how to manage a growing number of channels simultaneously. To accomplish this, the operator must move away from monitoring the whole list to a management by exception model. To support this playout model, the automation system must provide the operator with a new set of management tools.

Operator Graphical User Interface

The graphical user interface for the operator is perhaps the most important element in the new playout model. How an operator is informed of the status of the system directly affects the success of the operation. Messaging should clearly designate the nature and urgency of the situation. Since the operator must now monitor multiple channels, the interface must allow the operator to instantly identify and gain access to the channel and event or operation that requires attention.

The layout of the GUI should be easily configurable and easy to use. Each list may be different enough to require different event elements to be displayed. The system should be flexible enough to highlight the important elements for each event on each list. When an operator needs to make a change to the list, the interface should allow that change to take place at any time right up until the point of going to air.

Modern business systems (such as traffic) are moving toward support of dynamic interfaces with the automation system. An interface should be supported that will free the operator from the manual operation of editing the list based on updates from traffic and documentation of changes on the paper log. This will also benefit an operation that does not operate from a single location. This topic will be dealt with in more depth later in this chapter.

Multiple Schedules, Multiple Locations

In addition to managing multiple schedules, the application should allow the system to operate from multiple locations. Any transmission list should be able to be run from any location in the system, no matter if the operator sits in the next chair or in another facility. The workflow of a particular operation may require that certain portions of the schedule be run by a different operator at a different location. The system should instantly and seamlessly share or transfer control of any particular list at any time to any operator. In this openly flexible environment, access to the system and level of control should be restricted with a combination of system configurations coupled with user management to ensure that playout is not interrupted and that the transfer of control is predictable and does not adversely affect the operation.

Device Allocation

With multiple playlists, run by multiple users, residing in multiple locations, all needing access to the same pool of playout devices, some element of intelligence must be used to ensure enough playout devices are available and allocated in a priority that meets the business and service needs of the operation.

Most early automation systems were used in a single-channel environment. Thus, all available devices were assigned to one list. As operations evolved into multichannel environments, a need emerged to efficiently employ all available equipment across different lists and functions. An automation system should have the ability to manage these resources so that an operation can run with a minimal amount of equipment, yet use the equipment available in the most efficient way possible.

Device assignment should be prioritized based on the importance of the operation. A material transfer operation would carry a lower priority than playing a clip to air, for example. The system should anticipate this need before assigning devices to operations. In the event of a resource conflict that cannot be automatically resolved, the system should clearly identify the problem to the operator. The operator should then have the tools necessary to manually prioritize the system to resolve the conflict. Once devices are assigned to the lists, the system needs to ensure that those devices have the media required.

File Movement

Material must be available on the available devices as needed by the transmission list. The system must be intelligent enough to analyze the playlist far enough in advance to determine what material is needed and ensure that it finds its way to an appropriate playout device in time for air. Because the majority of content on a playlist originates from a video file server, this process may be as simple as a file-based transfer from another video server, near-line, or archive storage device. In some cases, this transfer may require a baseband video transfer, or even a transcoding operation from one file format to another. Regardless of the source, the list should automatically initiate the material request and keep the operator apprised of the progress. In the event that a piece of material appears on the list but is not available on a resource that can be accessed by the list,

the system must provide sufficient warning, so that the operator can manually resolve the issue.

RELIABILITY AND REDUNDANCY

In order for any system to be entrusted with the operation of a broadcast facility, it must be able to provide a high degree of reliability. In addition, since failures must be accounted for regardless of the reliability, some form of system redundancy is typically provided. The overall reliability of any system is based on the combination of its ability to avoid errors whenever possible, and to recover from errors that do occur as quickly and transparently as possible.

Broadcast equipment and facilities are typically designed for high availability. Redundant, hot-swappable components, such as power supplies, fans, and drives, limit the impact of the most common failures of the hardware that make up the automation system and the devices it controls. Also, since modern automation systems are primarily software products, attention must be paid to general computer security issues. This is particularly important as these systems are increasingly interconnected. Virus protection, user security, and regular data backups are necessary elements of the overall system reliability. While maintaining security, care must be taken to avoid doing anything that affects the system's operational capabilities. It should be noted that there are often constraints placed on what other processes can be run on the automation system's computers, particularly those that support real-time functions.

Redundancy

Redundancy must accommodate a failure of either a component of the automation system itself, or one of the systems or devices under its control. In both cases, there are several redundancy models available, each with costs, benefits, and drawbacks. The likelihood of a particular failure and its impact need to be weighed against the costs involved. Within a single system, and certainly within a facility, different techniques can be applied as required.

Popular redundancy models involve a backup device continuously shadowing the operation of a primary device, ready to take over immediately. This arrangement is equally applicable to protecting automation components as it is to controlled devices. For automation redundancy, two computers execute continuously, with user interaction transparently controlling both. A watchdog processor monitors the execution and determines which computer controls the actual playout. Where this approach is used for device redundancy, often referred to as air/protect, both devices cue and play simultaneously, but only one is switched on-air. If a fault is detected in the primary device, the backup is available to be switched on-air, with minimum disruption.

In certain cases where full redundancy cannot be justified, an "N + 1" model may be applicable. In this case, a single device is designated as the backup for multiple (N) identical on-air devices. In the case of failure of a primary device, the backup is automatically assigned to replace it. Obviously, this cannot be a seamless process, nor can it handle multiple or catastrophic failures. In some cases, the backup device can be used for some lower priority function when not needed as a backup. N + 1 solutions are ideally suited to devices such as video file servers with shared storage, as the backup device has immediate access to all of the media used by any of the devices it might be called upon to replace.

Certain software processes can benefit from clustering and load balancing. In this process, two or more computers, networked together, combine to provide processing capacity. Processes are split and their execution is spread out across the cluster. This allows the system to be scaled to support increased demands by increasing the number of machines in the cluster. Furthermore, the cluster can be sized in an N + 1 model, with sufficient excess capacity that upon a failure of a server within the cluster, the remaining machines can continue to service the system's demands. These techniques require specific programming techniques, but are increasingly supported by enterprise platforms.

Similar to N + 1 redundancy is *pooling*. In this approach, a number of similar devices are grouped together and jointly made available to multiple channels or other processes. As long as the number of devices required at any moment does not exceed those available, the devices can be freely shared among a large number of channels. This makes very efficient use of available resources, but requires a specific N + 1 process to handle failures. In addition, there may be operational constraints, and some flexibility is required.

The simplest form of redundancy is to have a spare unit in house that can be installed in place of a failed unit. If full redundancy is considered a *hot standby*, and N + 1 is a *warm standby*, this would be a *cold standby*. Using cold standby spares results in operational disruption, but is often sufficient for devices or components with a very low chance of failure or whose absence is of little impact.

Data Backup

In addition to the real-time control of devices, the preservation of a facility's data is essential to continued operation. This is one area where multiple complementary methods of redundancy are used simultaneously. The first priority is regular backups of automation, media, and content databases. As the sophistication of the captured metadata increases, the value of this data, both in terms of potential disruption and cost to recover, becomes significant. Simultaneously, mirrored file servers can provide seamless operation in cases of the failure. Use of *RAID* arrays can provide an additional layer of security. An important consideration is that the implementation of any data-mirroring technique cannot be used to justify the elimination of regular backups. Although the primary consideration is the loss of data, if by some chance the data is corrupted, without a recent backup, a facility may be left with two perfectly mirrored copies of unusable data.

Cost and Service Considerations

Obviously, the cost of implementing full redundancy can be high and can more than double the cost of both implementation and maintenance. In some cases, different levels of redundancy can be applied to various parts of the operation. Critical devices are obviously likely to be made fully redundant, while auxiliary functions might be candidates for N + 1 configurations. In addition, a lower service level might be applicable to certain channels.

One often overlooked aspect of a redundancy scheme is how normal operation is resumed after a failure condition. Ideally, any failed device can be restored and resume operation seamlessly. Often, it is desirable to allow the restored device to be put back into service in a standby or backup mode, to allow testing. In addition, data and media often must be restored to a device before it can resume operation. A mitigating factor is that while failures are unpredictable and can happen at any time, recovery can be scheduled and managed. Nonetheless, no redundancy mode is better than its recovery process.

Regardless of the precautions taken to ensure continued operation, the possibility that the broadcast facility itself might be affected must be considered. Extensions of these same redundancy techniques can be applied to off-site disaster recovery systems as well. An important decision is whether a set of relatively static programming will be utilized; if not, some continuous process must be put in place to copy schedules and material to the disaster recovery site. Another factor is determining how control will be passed from the primary facility. This is significant if the sites are geographically separated from one another. If the automation system is capable of remote operation, it might be possible to take control across a network connection, at least until operational personnel can travel between the sites. In sum, reliability and redundancy are two issues that must be carefully considered when dealing with any broadcast facility.

AUTOMATION INTEGRATION WITH BUSINESS SYSTEMS

Traffic and automation systems must exchange data in order for the overall station system to function. Currently, these two functional areas still largely operate independently from each other. Files are handed off periodically between the systems, but they are typically proprietary, inflexible, and difficult to work with.

Batch versus Interactivity

The batch files that have traditionally been exchanged between traffic and automation represent an entire broadcast day's worth of events, whether that be a schedule or an as-run log. Once a day's schedule has been sent to automation, any changes from that point on are typically hand entered, introducing the issues of extra time spent and the inevitable errors that occur when human input is involved.

Recent developments in this area have been in the area of supplementing these daily batch downloads with updates on the day of air. This offers several benefits, including the cost savings associated with automating a formerly manual process, the elimination of human error, and the enabling of new business workflows. This can all potentially result in new revenue for the station. When day-of-air changes are easily and accurately made, selling day-of-air spots is more easily handled. Potentially, lower value spots can be replaced with higher value last-minute additions. *Avails* formerly filled with *promos* or public service announcements because they were unsold have another day when they can be sold. Day-of-air program changes and sudden increases in ratings due to breaking news can be easily handled and even leveraged.

Static versus Extensible Data Models

Another limiting feature of existing traffic–automation interfaces is the fact that they are largely based on static data models. A format is agreed on and defined, then "cast in stone." If new requirements arise after this point, they must be accommodated within the current framework. In some cases, new data fields can be concatenated to the end of an existing record, but this often results in a patchwork of fields that becomes increasingly complex and difficult to use.

Proprietary versus Standard

The acceptance of XML (eXtensible Markup Language) for exchange of data means that it is now possible to update and extend as necessary the information that is passed between systems or applications when the need arises.

In the past, virtually all traffic–automation interfaces were proprietary in nature. System X could talk to system Y using protocol A, but if system X wanted to talk with system Z, protocol B would have to be used. This resulted in a large collection of vendor and system-specific interfaces that, at a high level, all did the same thing, but were incompatible with each other. The net result was that broadcasters and vendors spent considerable amounts of money not only developing support for these various protocols, but also in supporting them.

Application to As-Run

It is easy to fall into the trap of considering all of these advances purely in terms of broadcast schedules, but they apply equally to as-run reporting. Dynamic interfaces enable immediate notification to the proper business staff of spots that did not run successfully. Previously, this notification was often not provided

until the next day, at which point, the process of making good spots was more difficult, and in some cases, impossible, resulting in lost revenue. With traffic, accounting, and sales staff having immediate knowledge of as-run discrepancies, action can be taken to make spots good quickly and efficiently, recouping otherwise lost revenue.

Standardization

Considerable effort has gone into enabling much-improved integration of broadcast systems. Among those efforts is the BXF (Broadcast eXchange Format), developed under the auspices of the Society of Motion Picture and Television Engineers' S22-10 Data Exchange Working Group. BXF is designed to not only replace existing proprietary interfaces in place today between systems such as traffic, program management, automation, and content distribution, but also to enable the new workflows discussed in this chapter, as illustrated in Figure 5.13-5.

BXF is the result of the efforts of over 150 individuals, representing over 80 broadcast organizations. It is expected that BXF will achieve status as an SMPTE standard soon after the publication of this book. By replacing hundreds of proprietary interfaces with a single, extensible protocol, true integration between these systems can become a reality, and new workflows realized.

Benefits of Dynamic Interaction

The benefits of new dynamic open protocols for the exchange of data among broadcast systems are both cost savings and increased revenue. Cost savings result from elimination of formerly manual processes, saving on labor as well as eliminating errors. Increased revenue results from enabling new business models, such as selling spots up until airtime, and immediately making-good spots that were missed. There are peripheral benefits as well, with reduced development and support costs for interfaces. Building and supporting products that support one standard for all systems is considerably cheaper than writing a separate interface for every combination of systems.

Material Metadata

Recent standardization efforts are not limited to improving schedule and as-run data exchange. The other two common data exchanges, dub and purge lists, have been addressed as well. These benefit from the same advantages that schedules and as-runs do, with a tighter integration between the material databases of traffic and automation systems.

Integration with Asset Management Systems

In recent years, the terms MAM (media asset management) and DAM (digital asset management) have come to define separately deployable sets of products. This obscures the fact that automation systems have from the outset provided media asset management capabilities. These took the form of media prep stations that allowed for the timing and dubbing of programs and interstitials, and the storing of some amount of metadata, such as timing, either on a barcode label or in a database. Typically, this functionality allowed the capture of only a limited amount of metadata, in most cases little if anything more than the automation system required to subsequently play the material out.

The primary information, provided by a traffic system, was a house ID that was used to identify the content and correlate the stored data with events subsequently placed on a log. In addition, this house ID was also used to identify the media itself, both tape and later file based.

As automation systems have evolved, there is an increasing demand for additional metadata related to content and metadata to be stored and subsequently accessed. This metadata can be divided into two broad categories: that which the automation system has some understanding of and can act on, and that which is merely stored and returned on demand. In the former category are such elements as the house ID, cueing and duration data, media identifiers, and playout parameters such as audio formatting. The most obvious examples of the latter include such human readable labels as the title, although some systems have limited ability to extract meaning from these if specifically formatted. Increasingly, however, the automation system is presented with data that it is expected to deliver to some external system when the associated content is acted upon. For instance, a block of data might be associated with an item, to be delivered to the transmission stream when the material is played out for synchronized delivery to the viewer.

Another increasingly important set of requirements revolves around ratings and rights management. As regulatory restrictions are put into place, content is increasingly repurposed and delivered via multiple means; the automation system needs to be able to either limit the playout and delivery of material to appropriate channels or *dayparts*, or pass ratings and rights to downstream devices.

Where dedicated MAM systems are in place, or in any case where material data is managed by another system other than the automation system, a means of integration must be provided. In the simplest case, the automation system simply makes use of house IDs provided by a traffic system, but all data is manually entered, or copied from a dub list, into a separate automation database. More sophisticated techniques include processes whereby frame-accurate timing data is downloaded subsequent to prep by a business system.

However, the large-scale integration of these systems requires that there be some means for these systems to either share a common metadata store or dynamically share any updates. The former has the potential to provide the highest level of tight integration, and has the

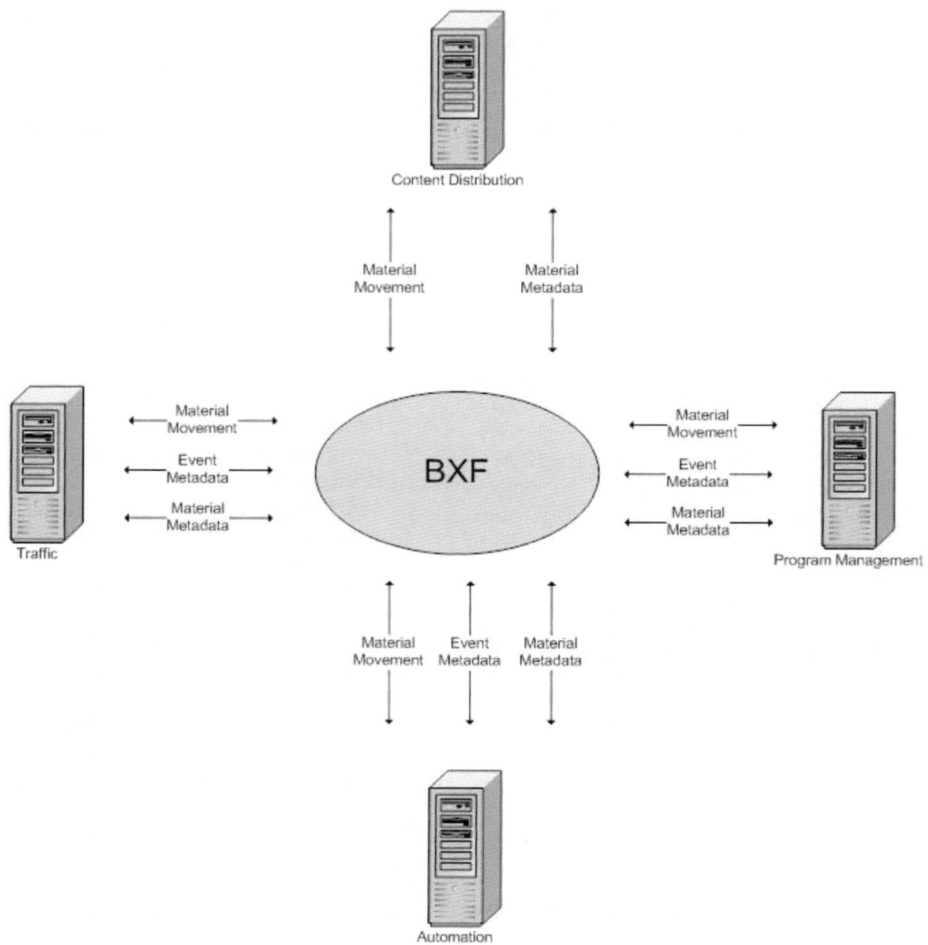

Material Movement:
- Push of material from Content Distribution to Automation
- Request for material from Traffic, Automation, Program Management to Content Distribution
- Purge list from Traffic to Automation

Material Metadata:
- Dub list from Traffic to Automation

Event Metadata:
- Playlist from Traffic to Automation
- As run from Automation to Traffic
- Record list from Program Management to Automation

FIGURE 5.13-5 Simplified data flows using the BXF protocol.

potential for the lowest latency between the systems. However, the level of cooperation necessary would most likely necessitate that all systems be provided by a single vendor, or a very close partnership relationship. Additionally, the risk is created that playout has a greater dependence of other systems. The alternative approach involves messaging between the systems. While there is some additional overhead inherent in this approach, it allows each system to retain autonomous, independent operation.

One critical factor that cannot be overlooked, however, is that one system must have final authority if multiple systems are to share any common data store. This is particularly important in cases where communications between such systems is lost, and restored only after both systems have updated the same items.

Tight integration of systems requires cooperation beyond mere translation of data. For systems capable of this, however, the major benefits include not only the ability to much more broadly share both metadata and the effort of acquiring it, but the ability for data to be made directly available to those systems that have interest in it. These same admonitions can be applied to not only content and media metadata, but to the schedule itself.

NEWSROOM AUTOMATION

The dynamic collaborative work environment in which news journalists, photographers, producers, directors, copy runners, and a host of other support staff run their entire news creation, production, and newscast enterprise presents particular challenges for automation systems, which are quite different from those for master control automation.

Manual Production

In a manual system without automation, when a newscast has been ordered in the rundown and is ready for air, scripts are printed and distributed among all the production control room (PCR) staff with a role in playing out the content in all the devices involved in the production. Typically a director makes all the calls for the character generators, still-store, and camera operators to take their cues. The technical director executes these cues through the production switcher, the audio engineer follows, mixing microphone and line outputs from the audio console, and the prompter operator follows the talent reading the script on-air. Manual production can require anywhere from a few to a dozen staff in the PCR.

Drive for On-Air Efficiency and Lower Costs

Lowering costs has been an ongoing theme for broadcast news and current affairs producers for years. This is the same issue that drives playout automation. The opportunity for automation in news does not avail itself until one particular digital technology is in place at the station—video servers. In addition to video servers, nonlinear editing (NLE) systems in a shared storage environment are helpful to establishing a digital workflow.

Digital Workflow

Once a newsroom has made a commitment to digital workflow, only then is news automation of value, with system elements and capabilities that typically will include:

- Digital video servers for content creation in shared collaborative environments
- Nonlinear video editing
- Digital low-resolution proxy browsing, shot listing, and editing
- Integrated digital graphics production
- Robotic cameras

News Automation

News automation oversees these islands of digital production technology and brings their control and playout under one user interface (UI) that requires as few operators as possible. *News automation* is a loose term with little resemblance to its sister automation system in the master control room. It is described as loose because, unlike program playout automation, which follows a fixed playlist, news production is live and unpredictable and better described by the term *live assisted* news production.

Newsroom Computer System

At the center of the newsroom universe is the newsroom computer system (NRCS or NCS). The work done within this system includes:

- Planning and assignments
- Messaging and mail
- Web browsing and research
- Wire service input
- Ingest scheduling
- Searching
- Low-resolution proxy browsing, shot listing, and editing
- Scripting
- Voiceover recording
- Show timing
- Graphic selection and content creation
- Teleprompting
- Subtitling
- Rundown management and production

Media Object Server Protocol

The NRCS uses the *Media Object Server* (MOS) protocol to communicate with the production devices as to the order of their media in the script, their status, and availability. This is done over a TCP/IP network using socket communication (see http://www.mosprotocol.com).

First organized in 1998, the MOS development group described the media object server communications protocol as enabling the newsroom computer systems to communicate with other production systems via a standard computer language/protocol. This protocol enables the exchange of the following types of messages:

- *Descriptive data for media objects:* The MOS *pushes* descriptive information and pointers to the newsroom computer system as objects are created, modified, or deleted in the MOS. This allows the newsroom computer system to be "aware" of the contents of the MOS and enables the newsroom computer system to perform searches on and manipulate the data the MOS has sent.

- *Playlist exchange:* The newsroom computer system can build and transfer playlist information to the MOS. This allows it to control the sequence that media objects are played or presented by the MOS.
- *Status exchange:* The MOS can inform the newsroom computer system of the status of specific clips or the MOS system in general. The newsroom computer system can notify the MOS of the status of specific playlist items or running orders.

Typically, each NRCS and each production device has its own *MOS gateway* (MOS server) that interprets the MOS commands as they are exchanged between MOS compatible systems. As script order often changes in the NRCS rundown, the playlists for these devices are updated and stay synchronized with the rundown via MOS. News automation takes all this MOS data and organizes it into a UI that can be dynamically controlled for playout by only a few operators.

Master control automation providers who also offer news automation can offer their customers some economy when the device drivers and control layer for the MCR can be also used for the PCR. This does not mean that the same device can necessarily be used for both master control automation and news automation simultaneously. Both automation environments require instantaneous device control and no port or device conflicts are allowed.

User Interface

The UI used by news automation venders is typically expressed in either a timeline or in a grid format. The most contemporary UI view is in a scripted format that is familiar to journalists. Some automation systems offer just video server and graphic control playout. More complex news automation systems will put every possible PCR device under the control of one operator including robotic cameras, studio lighting, audio, effects and switcher control, stage monitors, teleprompters, etc.

News automation is adventurous in that it all happens live in an environment where production staff are challenged every day at the five, six, and eleven o'clock newscast. News automation systems have to be well designed and implemented to accommodate the rigors of live news production.

The Bottom Line and Future Developments

Making more money from all the work news staff put in to produce a live newscast is the goal of a new generation of newsroom computer system production tools focused directly on applications of Internet protocol television (IPTV).

A news producer can take the current content creation models and inventory and repurpose them to meet the needs of viewers who in the future may receive their news via cell phones, PDAs, laptops, PCs, and similar devices. More importantly, the broadcast news producer can reduce costs or even generate revenue. Efficient operations are key and this opens up the need for news automation tools in the IP world.

There are some fundamental techniques for the news broadcaster to adopt in adapting to this new world order. While easy to describe, they are a challenge to implement:

- *Digital asset management:* DAM, at minimum, tells the content owner what they have so they can find it and repurpose it.
- *Digital rights management:* DRM, at minimum, identifies who owns what and who is allowed to see what.
- *Subscription-based news distribution:* This requires changes in habits by both the news producers and their audience. Billing and payment systems are needed that reimburse the content and copyright owner at a price the consumer is willing to pay.
- *Multiple IP-based automated publishing tools:* This is where the real value is added and if done well, where news consumers will pay a premium for the news they want, when they want it, and in a published format that they can easily navigate through.

Automation tools, as they evolve to serve broadcast news producers to meet this challenge, will become more tightly integrated into the news production process than ever in the past.

SUMMARY

Automation encompasses far more aspects of a broadcast operation than most people realize, and is a technology that is undergoing constant evolution. What was considered *automation* 5 years ago differs from that which exists today. Undoubtedly, automation systems 5 or 10 years from now will be quite different again from the systems of today. This chapter was intended to provide readers with a good grounding in where automation came from, what it is today, how it can help to improve broadcast operations, and what factors are important to consider when selecting a system for their facility.

ACKNOWLEDGMENTS

The author would like to thank several colleagues at Harris for the contributions they made to this chapter. Ben Peake, Chris Jones, Peter Lotz, Virgil Moore, Chris Reynolds, and Harn Soper provided a significant amount of the source material. Thanks are due as well to the many exceptional individuals I have had the privilege of learning from over the past 20+ years—you know who you are.

Bibliography

The following reference white papers are available on the Harris Corporation website, at http://www.broadcast.harris.com/support/papers.asp?cat=79.

Advantages of a Platform Approach, Moving Rich Media across a Hybrid Network.

H-Class Content Delivery Platform: The New Media Business Opportunity: The Road Ahead for Originators, Aggregators, and Broadcasters.

Lennon, Chris. The Bottom Line on Automation, at http://www.broadcast.harris.com/television/automation.

Lennon, Chris. New Directions in Transmission Automation H-Class Platform Impact on a Model Broadcaster: A Day in the Life of a Model Broadcaster.

CHAPTER

5.14

Weather Radar Systems

ROBERT BEACH

Radtec Engineering, Inc.
Deephaven, Minnesota

INTRODUCTION

Radar was invented in the 1930s and has been used extensively for air traffic surveillance and control. Beginning in the late 1940s, it was determined that radar could also be useful for detecting and measuring precipitation. Since then, radar has become the preferred technology to detect and track severe weather. More recently, Doppler capability has been added to many weather radar systems to permit measuring the wind velocity associated with the precipitation. The U.S. National Weather Service (NWS) has upgraded the U.S. National Weather Radar network to WSR-88D (NEXRAD) radar systems. WSR-88D is sophisticated, fully coherent Doppler weather radar with true three-dimensional volume scanning and processing capability. The U.S. Federal Aviation Administration (FAA) has installed Terminal Doppler Weather Radar (TDWR) systems at selected airports in the United States. TDWR is also a sophisticated, fully coherent Doppler radar system whose primary function is to detect low-level wind shear in the approach and departure flight path of an airport runway. Low-level wind shear has proven to be extremely hazardous to aircraft during takeoff and landing. Dual-polarization technology is now being introduced that provides increased accuracy of precipitation information.

BASIC RADAR THEORY

The term *radar* is an acronym for *radio detection and ranging*. The basic principle of weather radar is very simple: Transmit a pulse of radio signal, form that pulse into a very precisely focused beam, and listen for reflections that bounce back from any targets the beam happens to come into contact with. For weather radar, the desired target is precipitation: raindrops, hail stones, snow flakes, etc. The larger and/or more numerous the precipitation particles, the stronger the reflection. The reflections are typically displayed on a Plan Position Indicator (PPI) display as colored areas where the color represents the strength of the reflected signal. The PPI display is essentially a map with the radar at its center. Thus, intensity of precipitation, in a given geographical location within the radar's range, can be estimated quickly.

A radar system with Doppler capability also measures the speed at which the precipitation is moving toward or away from the radar. Moving precipitation implies wind, thus a Doppler radar can also estimate wind velocities. In a PPI Doppler velocity display, the wind velocity is displayed with the color indicating the wind velocity in each geographical location. In addition to forming the transmitted signal into a very precisely focused beam, the radar antenna also aims that beam into the area of interest in the atmosphere. Typically, radar antennas will be set to scan the atmosphere at the rate of two or three revolutions per minute. If the radar beam were visible, it would look very much like the beam of light from a lighthouse sweeping around through the atmosphere.

Wavelength

A very narrow range of radio wavelengths may be used for weather radar. If the wavelength is too long, there are no useful reflections from the precipitation; if

TABLE 5.14-1
Wavelengths Used for Weather Radar

Wave-length	Band	Typical Radar	Comments
10 cm	S	National Weather Service WSR-88D (NEXRAD)	Good penetration of torrential rain storms but requires very high power and a very large antenna.
5 cm	C	FAA TDWR and most commercially available weather radars	Reasonable trade-off of operating characteristics; works well with modest antenna size.
3 cm	X	On-board aircraft weather radar systems	Small antenna compatible with aircraft space limitations, but absorption of signal by intense precipitation can produce misleading images.

the wavelength is too short, the signal is absorbed rather than reflected by intense precipitation. The range of useful wavelengths is between 3 and 10 cm, as shown in Table 5.14-1.

Locating the Reflection

Radar systems locate the target with respect to the radar antenna (*i.e.*, the direction and distance from the radar antenna to the target). The distance is typically referred to as the *range*.

Direction

The radar signal processor monitors the azimuth position of the antenna as the antenna rotates; thus, the direction of transmitted and received pulses and consequently the direction to each target are known.

Range

The range (distance) from the radar antenna to the target is determined by measuring the length of time that elapses between transmitting the pulse and receiving the reflection. Because the propagation velocity of the signal is a known constant, the distance from the radar antenna to the target can be accurately calculated. Many radar systems use *range bins* to determine the range of target reflections. The radar's range is divided into some number (typically somewhere between 100 and 1024) of *bins* of equal size. Each bin has a time, corresponding to its distance from the radar antenna, associated with it. The receiver's signal is sampled at each range bin time, and the strength of the signal at that time is stored in that bin. The resulting array of data represents the target reflectivity *versus* distance along the radar beam in the direction the antenna is pointed at that moment.

Pulse Rate and Range

All the reflections of interest must get back to the radar before the next pulse is transmitted; therefore, the pulse rate and the radar's maximum range are related. The pulse rate, in pulses per second, is often referred to as the *pulse repetition frequency* (PRF). That relationship is expressed by the following formula:

$$R_{\max} = \frac{c}{2(PRF)}$$

where $R_{\max}$ is the maximum unambiguous range; c is the speed of light (typically 300,000 km/sec); and PRF is the pulse repetition frequency (pulses per second). For example, the maximum unambiguous range at a PRF of 1000 is:

$$R_{\max} = \frac{300{,}000}{2(1000)} = 150 \text{ km}$$

From this relationship it is apparent that longer ranges require lower PRFs.

If a reflection from a previous pulse returns to the radar after the next pulse has been transmitted, it is frequently referred to as a *second trip echo*, and the range at which it is displayed is said to be *folded*. The term *unambiguous range* means the range at which all reflections are first trip echoes with no range folding. In the above example, with the radar set to a pulse rate for 150 km range, a target located at 200 km range producing a very strong reflection might appear to be at 50 km range. Range folding can be confirmed by setting the radar to its longest range and observing the data for one or more scans. Targets producing strong reflections that appear to move out to the longer range are almost certainly folded when they appear at shorter ranges.

Reflection Strength

The strength of the reflected signal is proportional to the number and size of precipitation particles within the radar beam and to the type of precipitation—for example, rain, snow, or hail. The reflection strength is also proportional to certain parameters in the radar system such as transmitter power and pulse width. Typically, the weather radar's signal processor categorizes the strength of the reflected signal into one of 16 levels. Each level is displayed in a different color; therefore, it is easy to look at a display and quickly determine the strength of the reflection in any given location. Typically, the weakest reflections are shown in light blue, with a spectrum of colors up to bright red for the most intense reflections. For meteorological purposes, these levels are usually measured in units of Z. Z is the radar reflectivity factor for precipitation. Z

is usually displayed as decibels of Z. This is done because going from the weakest to the strongest Z routinely covers six or seven orders of magnitude of signal strength. The decibel scale provides a convenient way of expressing such a wide variation. The reference standard for Z is the signal reflected by precipitation with a single 1 mm drop of rain per cubic meter of space.

A signal strength of +10 dBZ means the reflected signal is 10 times stronger than the signal reflected by the 1 mm drop/m^3 standard; +20 dBZ is 100 times stronger; and +30 dBZ is 1000 times stronger. A −10 dBZ signal is 10 times weaker than the standard, −20 dBZ is 100 times weaker, etc. A 0 dBZ signal is equal to the standard. Figure 5.14-1 provides a comparison of the sensitivity *versus* range of a typical commercial weather radar with the sensitivity of both the TDWR and WSR-88D radars.

The actual amount of power reflected back to the radar can be calculated using the Probert–Jones equation. In practice, the radar set measures the power reflected back (P_r, the received signal) and solves the Probert–Jones equation for the Z value that caused that reflection. The precipitation intensity is then estimated from that Z value:

$$P_r = \frac{P_t G^2 \theta^2 H \pi^3 K^2 L Z}{1024(\ln 2)\lambda^2 R^2}$$

where:

P_r = Reflected signal power received at the radar antenna.

P_t = Transmitted power (fixed by transmitter design).

G = Antenna gain (fixed by antenna design).

θ = Antenna beam width (fixed by antenna design).

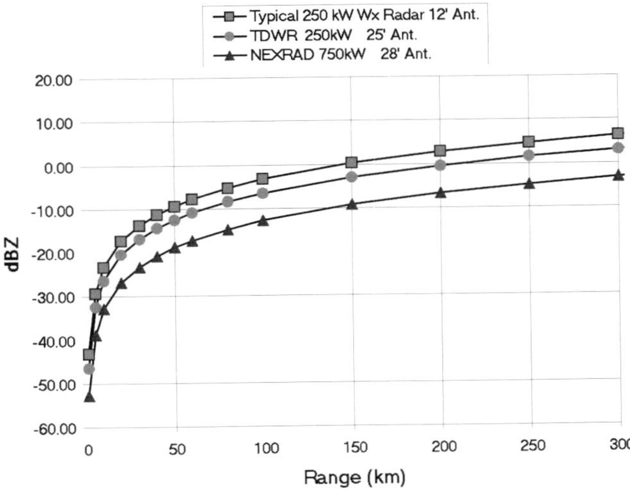

FIGURE 5.14-1 Sensitivity *versus* range.

H = Pulse width (limited to certain values by transmitter design).

K^2 = Physical constant (typically 0.93 for rain or 0.197 for ice).

L = Loss factors (determined by radar installation and atmospheric factors).

Z = Target reflectivity (determined by the type and intensity of precipitation).

λ = Transmitter wavelength (fixed by transmitter design).

R = Range (distance from radar antenna to target).

To be strictly accurate in calculating the reflected power, raindrops would have to be perfectly spherical. Because raindrops are not perfectly spherical, there is a small difference between the theoretical and actual results. This difference is accounted for in the empirical nature of the value of Z.

Pulse Width and Resolution

Weather radar systems typically transmit pulses with widths in the range of 1 μs to 5 μs. Some systems are capable of narrower or wider pulses. The advantage of a wider pulse is that more signal power is reflected from the target; therefore, it is easier to see weak targets with wider pulses. Doubling the pulse width is equivalent to doubling the transmitter's peak power. This is expressed by the H term in the numerator of the Probert–Jones equation. The advantage of a narrower pulse is the ability to see the separation between closely spaced targets. To be separately distinguishable, targets must be separated by more than 1/2 the pulse length. Because the propagation velocity of a radio signal is approximately 1000 ft/μs, a 1 μs pulse will be approximately 1000 ft from the front edge to the back edge in the direction of its travel. Therefore, with a 1 μs pulse, multiple targets that are separated by less than approximately 500 ft will appear on the radar display as one single target. For many meteorological situations, targets of interest are frequently several miles wide and several miles apart. In this situation, the ability to use wider pulses to see weak targets may be more important than the ability to distinguish between several closely spaced targets.

Doppler Effect

The *Doppler effect* is named after Christian Johann Doppler, an Austrian physicist, who became interested in why the sound of a train whistle seemed to change to a lower pitch as the train passed by where he was standing. He determined that, by measuring the change in pitch, it would be possible to accurately calculate the speed of the train. This principle applies to radio waves as well as sound waves. One of the more familiar examples of the Doppler effect is a police radar used to measure the speed of vehicles from a distance.

A Doppler weather radar measures the speed of the target as well as the target's location and reflectivity. The speed of the target (precipitation) implies a wind velocity. The wavelength of the reflected radar signal is slightly compressed in front of the moving target and stretched by a corresponding amount behind the target. The radar will measure either a slightly shorter or slightly longer wavelength reflection, depending on whether the target (precipitation) is moving toward or away from the radar antenna. After the reflection is received, the radar's signal processor compares the wavelength of the reflected signal to the wavelength of the transmitted signal. The velocity of the target is calculated from the difference in wavelengths.

Because the Doppler wavelength shift is very small, obtaining accurate Doppler velocity data requires a radar system with a high degree of frequency and phase stability. If the frequency or phase of the transmitted signal changes, even a little bit, during a pulse, there is no way for the signal processor to determine if the difference it is measuring is due to the real velocity of the target or to changes in the transmitted pulse.

Figure 5.14-2 illustrates the manner in which the radar signal pulses in effect sample the Doppler shift wavelength. The Nyquist data sampling principle requires that there must be a minimum of two samples per wavelength for the samples to accurately measure the wavelength. As a result, the maximum Doppler velocity any pulsed radar can measure is a function of the PRF (Nyquist principle) and the wavelength (Doppler principle). This is expressed by the following formula:

$$V_{max} = \frac{PRF\lambda}{4}$$

where:

V_{max} = maximum unambiguous velocity.

PRF = pulse repetition frequency (pulses per second).

λ = transmitted wavelength (0.0535 m for C-band, 5600 MHz).

For example, for a C-band radar with a PRF of 1000, V_{max} is:

$$V_{max} = \frac{(1000)(0.0535)}{4} = 13.375 \text{ m/sec}$$

Many radar systems provide velocity unfolding capability. These techniques often use staggered pulse rates and a processing algorithm, which can multiply the maximum velocity that can be measured by a factor of 2. Some radars also offer a 3 times velocity unfolding capability. From this relationship, it is evident that higher velocities require high pulse rates.

Dual Polarity

Dual polarity operation, also sometimes referred to as multi-parameter radar, is an important new technology that is becoming available for weather radar systems. Dual polarity gives a weather radar the ability to make accurate rainfall measurements anywhere within the radar's prime coverage area. Figure 5.14-3 illustrates the operation of a dual-polarity radar. Two signals are transmitted, one horizontally polarized and the other vertically polarized. Research has demonstrated the ability to use the difference between the horizontally and vertically polarized reflections to determine both the type of precipitation and the rainfall rate.

Small raindrops are nearly spherical and reflect horizontally and vertically polarized signals nearly equally. As raindrops become larger, they tend to flatten on the bottom, into a shape something like a hamburger bun. This flattening is the result of airflow pressure on the underside of the drop as it falls; therefore, nearly all drops will fall with the widest dimension of their shape aligned horizontally. The widest dimension (horizontal) will reflect signals more strongly than the narrower dimension (vertical). The relationship between the size and shape of raindrops is well established, thus the difference between the horizontally and vertically polarized reflections can provide an estimate of the size of the raindrops.

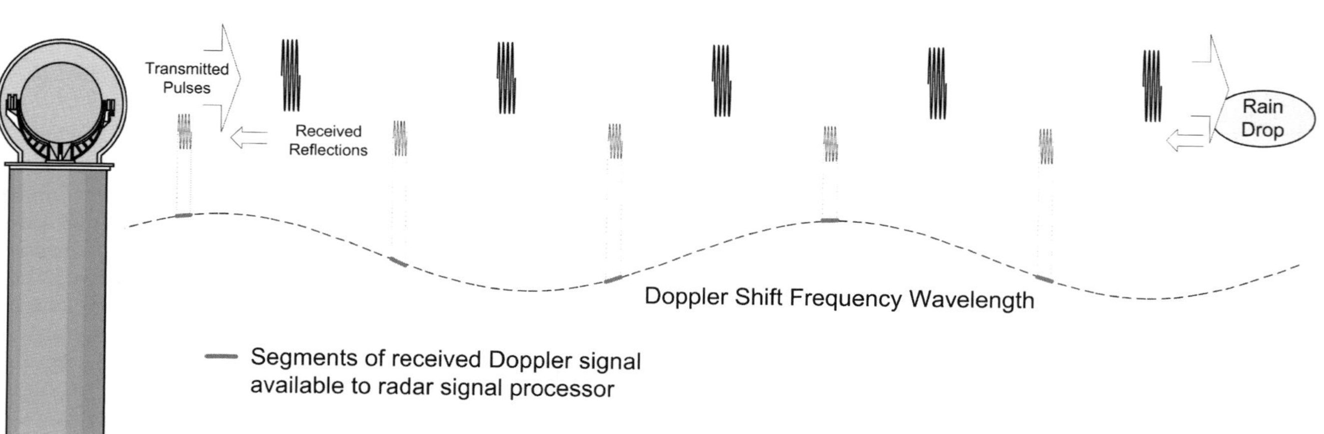

FIGURE 5.14-2 Doppler shift sampling technique.

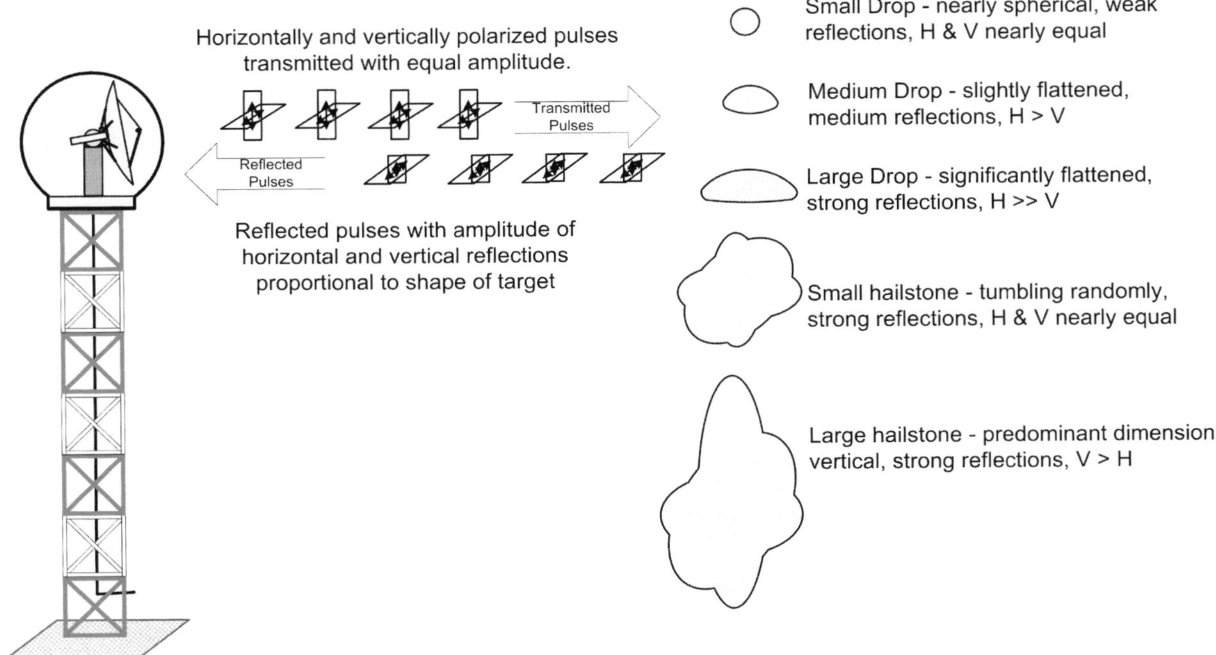

FIGURE 5.14-3 Basic concept of dual polarity.

Dual polarity radar operation requires the following items:

- An antenna that has the ability to transmit and receive both horizontally and vertically polarized signals.
- A transmitter that can provide two outputs. This is typically done in one of two ways: (1) using a power splitter to divide the transmitter's output power equally between the horizontal and vertical feeds and transmitting the two pulses simultaneously; or (2) using a switching arrangement to switch the transmitter's power output between the horizontal and vertical feeds on alternate pulses, such that every other pulse has opposite polarity.
- Receivers and a signal processor with the ability to receive and process the horizontally and vertically polarized signals.
- Analysis software with the algorithms required to process the horizontal and vertical data into results that display rainfall rates.

Maximum Unambiguous Range and Maximum Unambiguous Velocity

In a pulsed Doppler radar, the PRF, the maximum unambiguous velocity, and maximum unambiguous range are interrelated. The graph in Figure 5.14-4 expresses that relationship. Measuring high velocities requires a high pulse rate, and measuring long ranges requires a low pulse rate; therefore, there is a trade-off between the maximum velocity and the maximum range. This situation is sometimes referred to as the *Doppler dilemma*. The end result is that, at any given range, there is one unique PRF that will provide the maximum velocity measurement capability at that range.

Coherence

To obtain Doppler velocity data, the frequency and phase of the reflected pulses that are received must be compared to the frequency and phase of the transmitted pulse. There are two challenges in making this comparison:

- The Doppler frequency shift is very small, so this comparison requires a high degree of stability and accuracy to get reliable velocity data. A great deal of precision is required in the radar's transmitter and signal processing circuits.
- The received pulse arrives back at the radar long after the transmitted pulse has ended; therefore, it is impossible to make a direct comparison between the transmitted and received pulses.

There are two commonly used ways of making an indirect comparison between the transmitted and received pulses:

- A *coherent-on-receive* radar typically uses a magnetron tube in the transmitter. The magnetron is a power oscillator that starts up, transmits a pulse, and then shuts down. This cycle is repeated for each pulse. Thus, each pulse can have a slightly different frequency, depending on magnetron voltage,

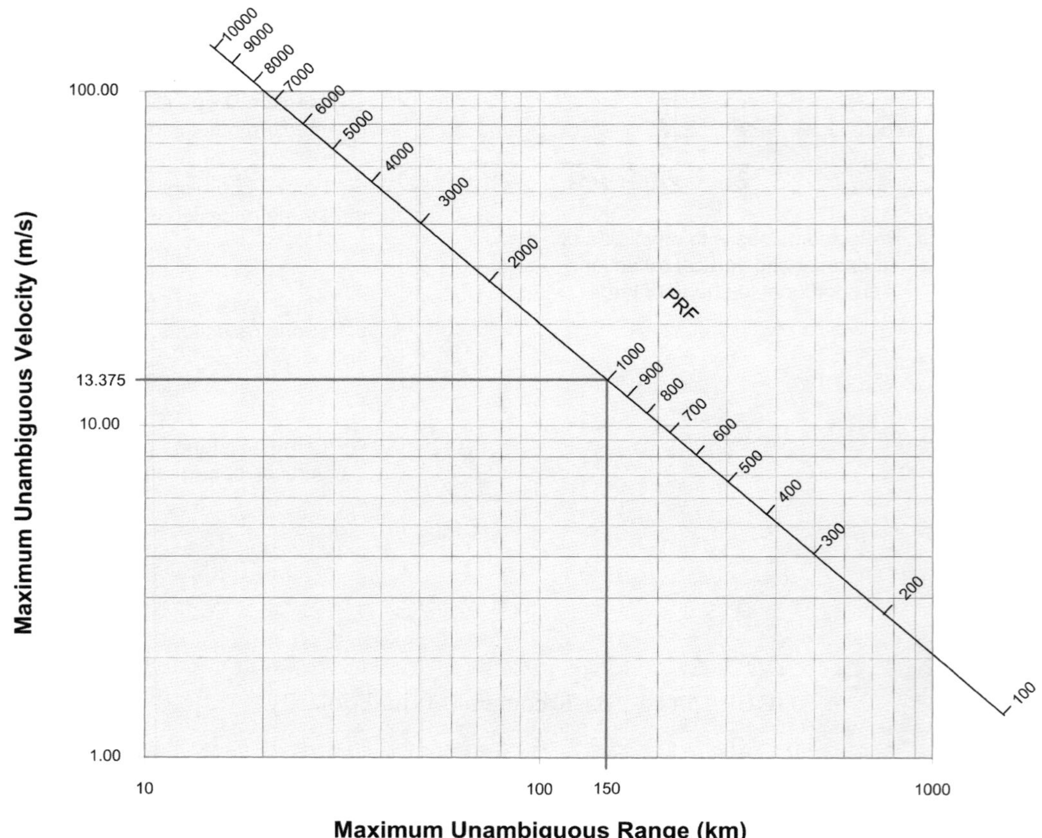

FIGURE 5.14-4 C-Band Doppler weather radar PRF *versus* range and velocity.

current, temperature, magnetic field, etc. The phase of each pulse from a magnetron is random. Circuits typically referred to as *STALO* (stable local oscillator) and *COHO* (coherent oscillator) sample and remember the frequency and phase of the magnetron's last pulse. The output of the COHO is used as the reference for Doppler processing of the received pulses. Many commercially available weather radar systems are coherent-on-receive. As a result of the small variations between pulses, these systems are typically limited to around 30 dB of ground clutter suppression by clutter filters.

- A *fully coherent* radar has a precision oscillator that operates continuously. This oscillator serves as the source of a master reference signal for all circuits in the radar. The transmitter is a true amplifier that typically uses a klystron tube, traveling wave tube (TWT), or solid-state device. Each transmitted pulse is a small portion of the reference signal that is amplified; thus, the frequency and phase of transmitted pulses are accurately known and are consistent from pulse to pulse. The signal processor uses the precision oscillator as the reference for determining Doppler shift in the received pulses. As a result of the phase-locked consistency of all signals, fully coherent radars can typically achieve around 50 dB of ground clutter suppression from clutter filters. WSR-88D, TDWR, nearly all research radars, and a few commercially available weather radar systems are fully coherent.

BASIC RADAR METEOROLOGY

Radar meteorology is a science in itself. The information presented here contains only the most basic concepts. The bibliography at the end of this chapter provides references to far more detailed and comprehensive information.

Reflectivity and Precipitation Intensity

Interpreting the meaning of radar reflectivity displays requires meteorological knowledge plus experience with local conditions. As an extreme example of local conditions, there are places where, at sunset, large numbers of bats leaving their daytime roosts may appear on the radar as a strong gust front line. The amount of signal reflected is determined by the number, size, and type (rain, snow, hail, ice crystals, etc.) of precipitation. Light snow is a very weak reflector of radar signals, hailstones that are wet on the outer surface are very strong reflectors, and liquid raindrops are somewhere in between depending on the size of the drops. Research and practical experience have shown that precipitation tends to reflect horizontally polarized signals best, and nearly all weather radar systems in use today have antennas that transmit a single polarized signal in the horizontal plane.

The reflectivity of raindrops is approximately proportional to the diameter of the drop raised to the sixth power, while the volume of water in the drop is approximately proportional to the diameter of the drop raised to the third power; thus, the relationship between reflectivity and actual water content of a raindrop is highly nonlinear. The reflectivity is also proportional to the number of drops; twice as many drops reflect twice as much signal. Because neither the drop size nor the number of drops can be readily determined, calculating the actual water content of precipitation from radar reflectivity data alone, using only a horizontally polarized signal, is at best an estimate. One commonly used method of estimating rainfall rate from radar images is to use the formula:

$$Z = aR^b$$

where:

Z = radar reflectivity factor of the precipitation.

a = an empirically determined constant.
R = rain rate in mm/hr.

b = an empirically determined constant.

It should be emphasized that this formula provides only an estimate of precipitation. Over the years there have been a large number of experimental data sets collected to attempt to refine the values of a and b in this relationship. Table 5.14-2 provides rainfall rates in inches/hour for various commonly used Z/R relationships.

Many radar systems allow the operator to enter values for a and b and display reflectivity directly as estimated rain rates. Some radar systems also provide a capability to total the estimated rainfall over a period of time. As a point of reference, for a number of years the National Weather Service used the values in Table 5.14-3 to correlate display color, estimated rain rate,

TABLE 5.14-2
Radar Rainfall Precipitation Estimates

dBZ	$300R^{1.4}$ Stratiform and Convective Rainfall (in./hr)	$200R^{1.6}$ Stratiform Rainfall (in./hr)	$486R^{1.37}$ Convective Rainfall (reduces hail effect) (in./hr)
20	0.02	0.03	0.01
30	0.09	0.12	0.07
40	0.48	0.47	0.36
50	2.5	1.9	1.9
60	12.9	8.1	10.3
70	67.0	34.1	55.4

TABLE 5.14-3
Display System Color Definitions

DVIP Level	Rain Rate (in./hr)	Reflectivity (dBZ)
1, Light green	0.1	0–30
2, Dark green	0.25	30–41
3, Yellow	0.5	41–46
4, Orange	1.25	46–50
5, Red	2.5	50–57
6, Dark red	4.0	57 or more

and reflectivity dBZ. It should be cautioned that these relationships only provide estimates of the water content based on radar reflectivity. In any given storm, the actual water content may vary significantly from the values indicated by the formulas and data shown here. Estimating the actual amount of precipitation using this Z/R relationship is a classic case of one equation with three unknowns (number of drops, size of drops, and whether the drop is liquid or frozen).

Precipitation Measurements and Flood Warnings

The problem with using conventional rain gauges to predict flooding is that there typically are not enough rain gauges to provide the required resolution within the coverage area. Radar improves the situation by providing essentially continuous coverage throughout its primary coverage area (typically about a 100 mile radius around the radar antenna's location). In order to improve the accuracy of precipitation measurements, much of the weather radar research for the last few years has focused on radar with dual-polarization capability. Dual-polarity radar-based rainfall measurements now offer the potential to significantly improve the accuracy and timeliness of flood warnings, particularly for flash floods. As explained in the Basic Radar Theory section, with dual polarity the differences between horizontally and vertically polarized reflections are analyzed to extract more information about the precipitation. Simply stated, dual-polarity capability provides the additional equations required to make accurate estimates of actual rainfall from radar measurements.

The end result of this is that a dual-polarity radar, with suitable hydrology software, has the ability to predict, with confidence, a flood warning with the flood crest height and the time of arrival of the crest at locations downstream in the watershed area. Recently, three separate groups of researchers have demonstrated the ability of a dual-polarity radar to produce rainfall measurements that compare favorably with the accuracy of actual rain gauge data. Typically, the radar measurement is within 10 to 15% (and, better in some cases) of the rain gauge amounts at the corresponding location.

To make full use of the radar rainfall data, the radar software should be able to segregate the data into individual watersheds within the radar's coverage area. A time-based histogram of the rainfall data can be input into a hydrology model of that watershed. The output of the hydrology model will reflect the flood potential for that watershed. The output of the hydrology model is the basis for flood warnings.

Reflection Shape

The shape of the area of precipitation is also significant; for example:

- If a line of precipitation begins to bulge out, the bulge (known as a *bow echo*) may indicate the presence of strong winds, possibly the outflow from a downburst.
- A hook-shaped echo may indicate the presence of a tornado.
- A long thin line ahead of an area of precipitation may indicate a gust front.

Non-Meteorological Reflections

Weather radar systems from time to time will also display reflections from non-meteorological targets, such as dust, insects, birds, and even bat migrations. Correct interpretation of these factors is usually a matter of local experience.

Velocity

A Doppler radar measures the velocity of the precipitation's movement. That velocity implies a wind blowing the precipitation; thus, a Doppler radar can measure wind velocity. The velocity the radar measures is either directly toward or directly away from the radar antenna. This is called *radial velocity*. For winds blowing at an angle to the radar beam, the velocity measured by the radar will be the vector component of velocity along the center line of the radar beam. In the wind velocity display, velocity is usually indicated by color. Typically, winds blowing toward the radar antenna are shown in shades of blue and green, while winds blowing away from the radar antenna are shown in shades of orange and red. Each color indicates a range of velocities; typically, the darker or brighter the color, the higher the velocity.

Velocity measurement capability has dramatically improved the accuracy of tornado warnings. Research has shown that most tornadoes are produced by mesocyclones (medium-sized rotating thunderstorms). The Doppler velocity measurements can usually identify the rotation within a storm cell many minutes before a tornado actually forms. This rotation is typically identified by winds within a storm cell blowing in opposite directions in close proximity to each other.

The velocity measurement capability can also detect winds associated with low-level wind shear. Low-level wind shear has been determined to be extremely hazardous to aircraft during takeoff and landing. Some Doppler radar systems can be programmed to automatically detect the patterns of storm-related wind shear and produce an alarm when those wind shear patterns are detected.

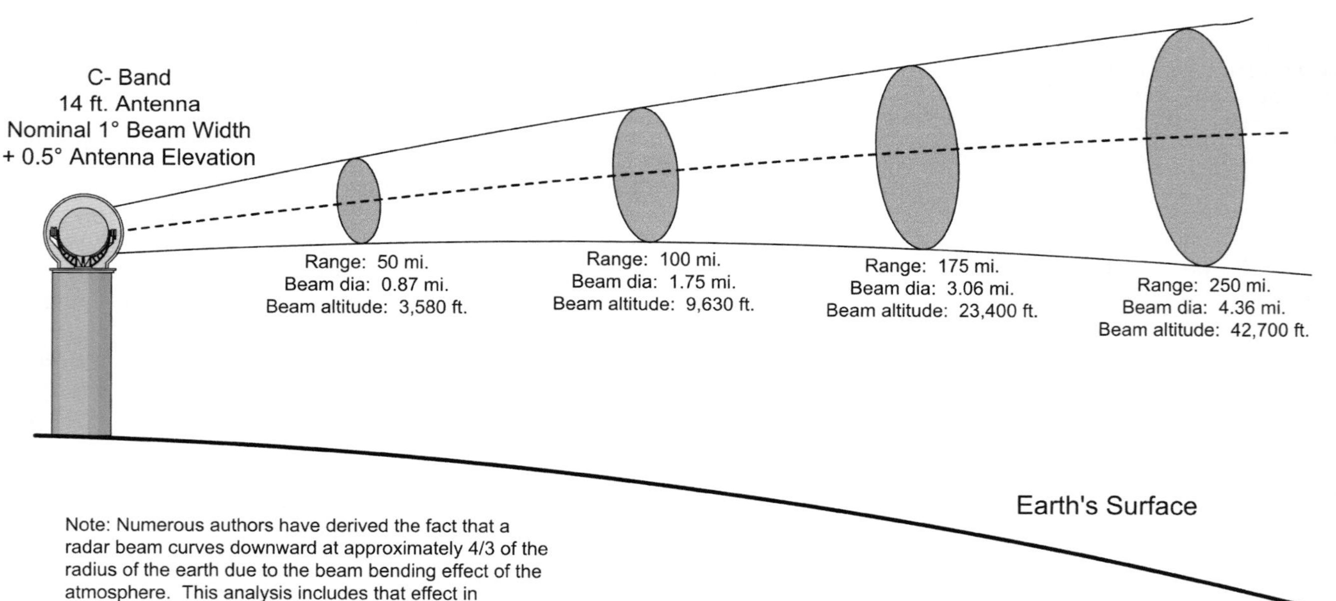

FIGURE 5.14-5 Radar beam size and altitude *versus* range.

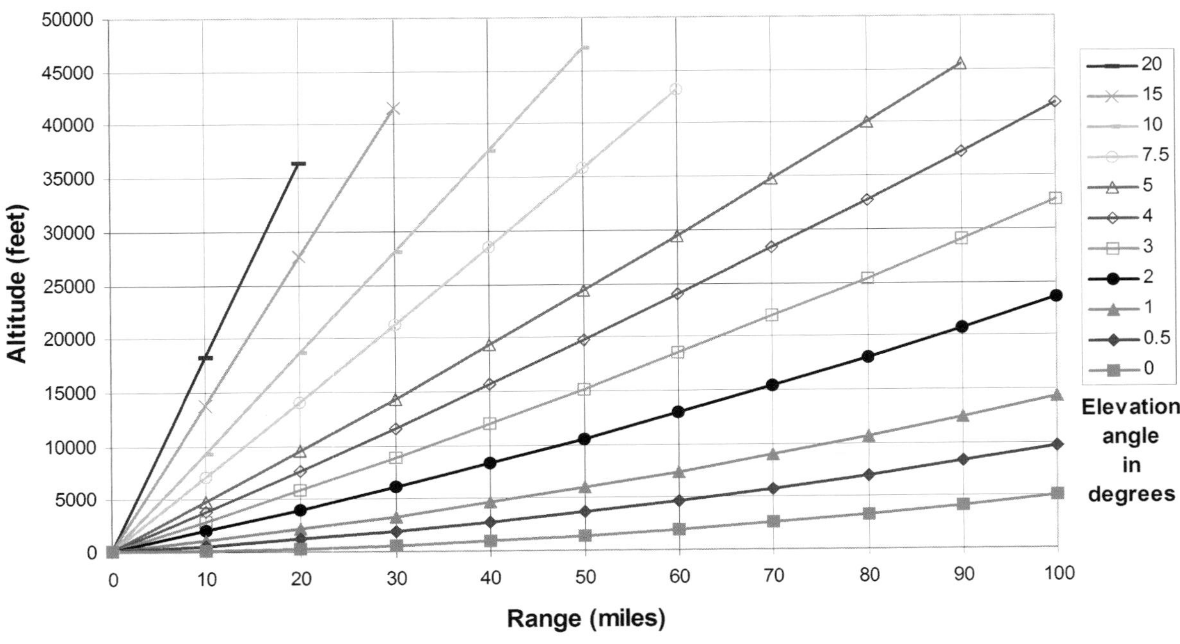

FIGURE 5.14-6 Beam altitude *versus* range at various antenna elevation angles.

Spectrum Width

Spectrum width is the difference between the maximum and minimum wind velocities in close proximity to each other. If the wind were perfectly uniform, all velocity measurements in the same area would be identical and the spectrum width would be zero. In essence, spectrum width is a measure of turbulence in the atmosphere. Many radar operators will use the spectrum width display to locate areas where the development of severe weather is most likely. Areas of high spectrum width are also of interest to pilots, as they are areas of turbulence to be avoided.

Beam Width and Beam Altitude

There are several factors related to the antenna beam width and altitude:

- Radar theory requires that, for accurate measurement of precipitation intensity, the precipitation must completely fill the pulse volume of the radar beam—that is, the height, width, and length of the volume of space the transmitted pulse occupies. For many meteorological purposes, precipitation areas more than 3 km (1.86 mi) wide are of most interest. To ensure beam filling, and therefore accurate measurement of precipitation, the radar beam should be no more than 3 km wide in the areas of greatest interest.
- As a result of the Earth's curvature, the altitude of the radar beam increases with distance from the radar antenna. Most precipitation occurs at relatively low altitudes, typically below 10,000 ft and nearly always below 20,000 ft. At very long distances, the radar beam is too high to provide any meaningful data. A useful rule of thumb is that, with the antenna elevation set at the normal +0.5° tilt, the beam will rise approximately 1000 ft for every 10 miles of range.

The diagram in Figure 5.14-5 indicates the beam width and altitude for a radar with 1° beam width (C-band with a 14 ft antenna at +0.5° elevation angle). Figure 5.14-6 illustrates the altitude of the beam centerline as a function of range at various antenna elevation angles. Figure 5.14-7 illustrates two situations in which the radar can be misleading because of beam altitude. The first shows virga, which the radar detects; however, the virga evaporates, such that no rain actually reaches the ground even though it is displayed on the radar. The second situation, which is typical at longer ranges, shows the radar beam well above precipitation; thus, that precipitation will reach the ground but will not appear on the radar.

Cone of Silence

The radar antenna directs its beam largely horizontally rather than vertically; therefore, as illustrated in Figure 5.14-8, there is an area directly above the radar antenna where there is no coverage, usually referred to as the *cone of silence*. The size of this blind zone is affected by the elevation angle of the antenna—the greater the angle, the smaller the cone of silence. There is also a small region very close to the radar antenna where there is no coverage, typically within 1 mile or less from the weather radar's antenna. This blind area, also illustrated in Figure 5.14-8, is due to the time

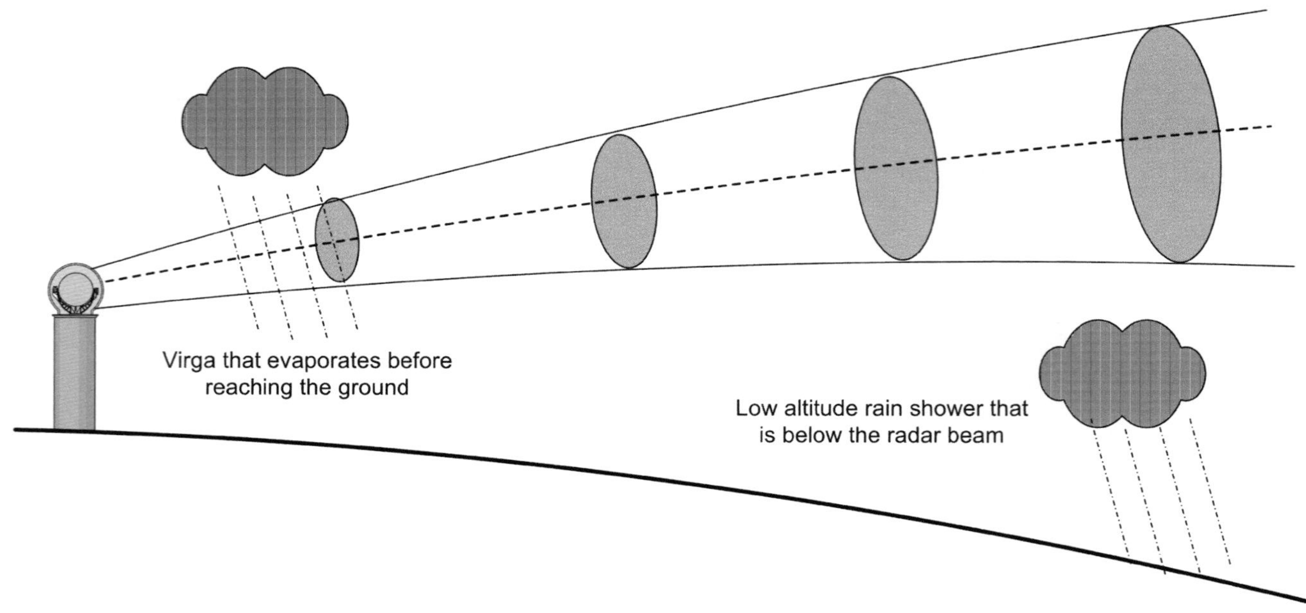

FIGURE 5.14-7 Rain detection considerations.

required to switch the radar from transmitting to receiving.

Ground Clutter

Ground clutter is one of the largest single problems for a weather radar. It often masks real precipitation. For a radar covering a metropolitan area, ground clutter is likely to be worst in the area where radar data is of most interest (i.e., directly over the metropolitan area). Unfortunately, some of the transmitted pulse leaks out of the antenna at angles away from the main beam. These leaks are called *side lobes*. Figure 5.14-9a illustrates how these side lobes cause reflections from buildings, trees, vehicles, etc. These reflections are referred to as ground clutter. Figure 5.14-9b illustrates how reduced side-lobe energy avoids much ground clutter. The antenna's first side lobe is usually the primary cause of ground clutter. Stationary ground clutter from buildings, terrain, etc., can be filtered out either by using a *clutter map* to subtract known clutter sources from an image or by using *clutter filters* in a

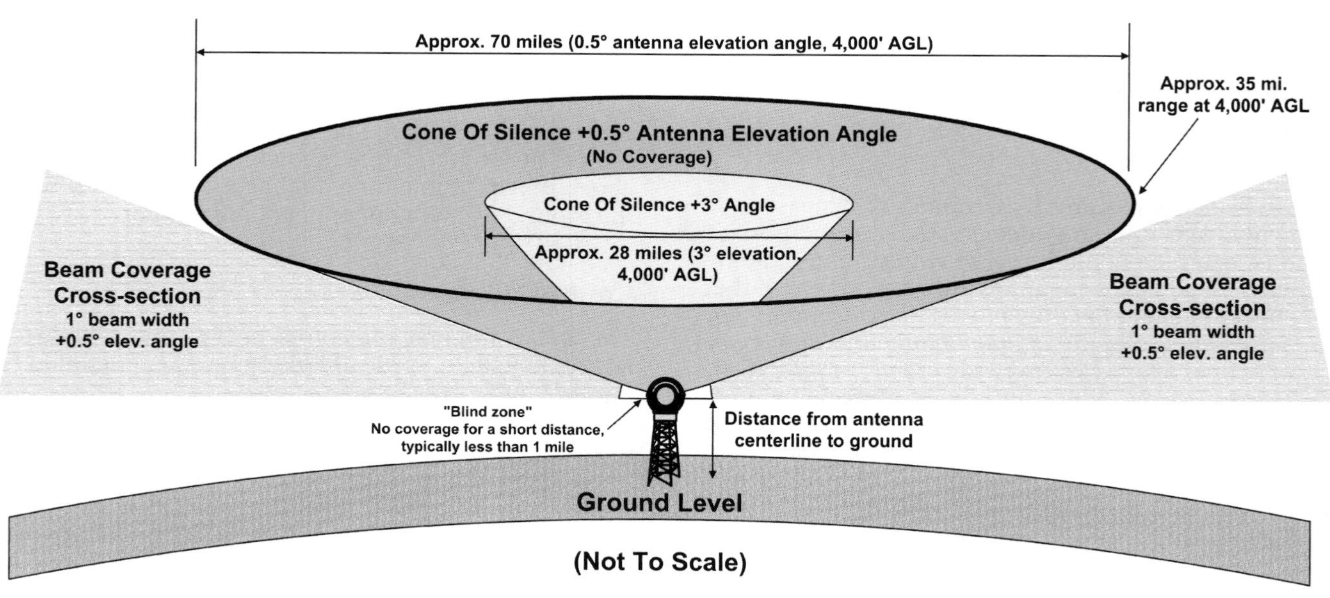

FIGURE 5.14-8 Cone of silence illustration.

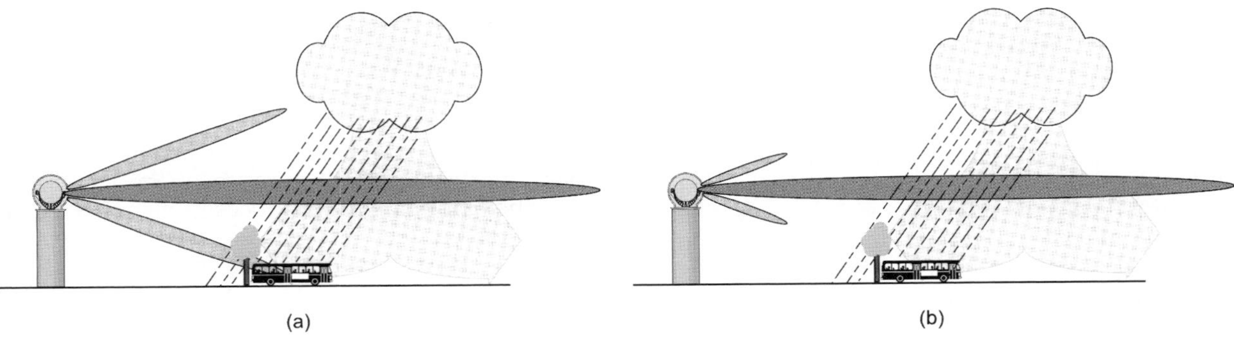

FIGURE 5.14-9 The effects of side lobes on radar performance: (a) large side lobes, (b) small side lobes.

Doppler radar. The effectiveness of Doppler ground clutter filtering is heavily dependent on the stability of the radar transmitter. Filtering moving clutter caused by vehicles, sea surface waves, etc., is an extremely difficult problem. If moving clutter is expected to be a problem in a given location, selecting a radar antenna with the lowest possible side-lobe energy is likely to be the most effective way to suppress that clutter.

Anomalous Propagation

Anomalous propagation (AP) is caused by atmospheric effects bending the radar beam downward so far that it actually hits the ground. The reflections travel back via the same path so the image displayed is actually the ground. This is illustrated in Figure 5.14-10. AP is typically a very low altitude situation caused by the radar beam passing through a thermal inversion layer. Raising the antenna elevation angle from the normal 0.5° to 1.0° or even 1.5° (*i.e.*, getting the beam above the inversion layer) may help to identify AP. If raising the elevation angle causes a reflection area to move away from the radar or disappear entirely, that reflection is likely to be AP.

RADAR SYSTEM COMPONENTS

Antenna

The *antenna* is a critical element in the overall performance of a weather radar system. The antenna performs two functions: forming the transmitted pulse into a very narrow beam and aiming that beam in the desired direction. The antenna rotates horizontally and, if the weather radar beam were visible, it would look very much like the beam of light from a lighthouse: a very powerful, narrowly focused beam sweeping around through the atmosphere.

Weather radar antennas are almost universally parabolic. The one exception is small *slot array* antennas, sometimes referred to as *pizza pan antennas*. For most weather radar purposes, the performance of a properly designed slot array antenna is equivalent to a parabolic antenna of the same diameter. For weather radar, the general premise is that the narrower the beam the better; however, basic physics requires a progressively larger antenna to obtain a progressively narrower beam. The costs of a radar antenna, radome, tower, etc., increase dramatically for larger antenna sizes.

Figure 5.14-5 illustrated the relationship between range, beam width, and beam altitude. From this data, it is apparent that a weather radar antenna with an angular beam width of approximately 1° will provide a beam width of less than 3 km at ranges and altitudes that are often of the most meteorological value. Side lobes are caused by a combination of basic physics plus very small imperfections in the shape of the antenna plus the waveguide, support struts, and feedhorn being located in front of the antenna reflector where they are in the main beam.

Side-lobe suppression is typically measured as decibels of side-lobe signal strength, with the center of the main beam as the reference. This is typically

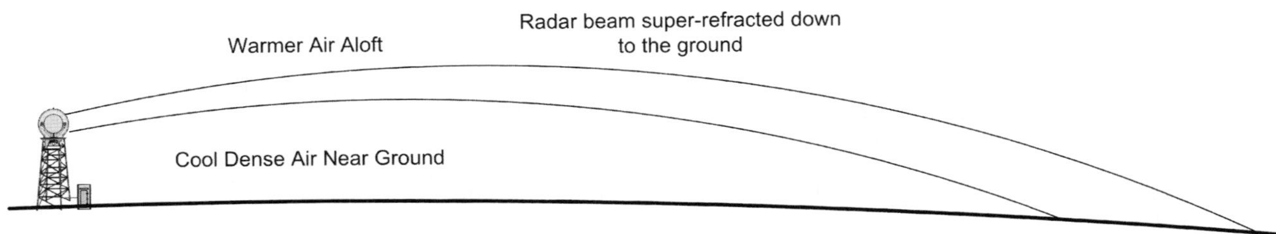

FIGURE 5.14-10 Radar beam super-refracted down to the ground.

quoted as the one-way value. The actual signal path is a round trip (transmitted signal out and received signal in). The round-trip attenuation is double the normally quoted value. Center feed parabolic antennas typically achieve side-lobe suppression in the range of –25 dBc (25 dB down from the carrier) to –30 dBc one way. (The larger the negative number, the greater the attenuation.) Precision antennas with an offset feed arrangement have been reported to have side-lobe suppression of –40 dBc or better one way.

For dual-polarity operation, the antenna must be able to transmit and receive both horizontally and vertically polarized signals. In addition, the antenna pattern for the two polarities should be as nearly identical as possible. Side-lobe suppression becomes very critical for dual-polarity operation. Ground clutter tends to have random polarization characteristics and appears as noise in the dual-polarity processing algorithms. This ground clutter noise reduces both the sensitivity and accuracy of the received data.

Pedestal

The weather radar antenna is typically mounted on a device called a *pedestal*. The job of the pedestal is to accurately turn and elevate the antenna to aim the antenna's beam in the desired direction. Typically, the pedestal can turn the antenna continuously a full 360° either clockwise or counterclockwise in the azimuth direction at a specified rate. The elevation (up and down) angle is often limited by drive mechanics to something like –2° to +60°, with horizontal being the 0° reference. This creates the *cone of silence* area directly above the radar that cannot be scanned. Because most weather occurs at lower altitudes (*i.e.*, at lower antenna elevation angles), the cone of silence is not a significant limitation in most weather radar installations.

The pedestal sends both the azimuth and elevation angles to the radar processor. Frequently ±0.1° (1/10 of the antenna's beam width) is specified as the required accuracy for the antenna position and position measurement. Improving the pedestal's angular accuracy can be done, but the cost increases dramatically and typically does not significantly improve the meteorological quality or usefulness of the data. Most radar systems permit adjusting the speed at which the antenna turns. Turning the antenna faster provides more frequent updates for data at a given location; however, it also puts greater demands on the radar signal processor (*i.e.*, one revolution's data must be processed in less time). If the antenna turn rate is too high, the accuracy of the Doppler velocity data is reduced. Typical scan rates are often in the range of 2 or 3 revolutions per minute. Many antennas are capable going as fast as 5 or 6 rpm. Many older designs of radar antenna pedestals use DC motors, which require periodic replacement of motor brushes, to drive the antenna. Newer designs often use brushless motors. Because the pedestal is typically on top of a tower where it is not readily accessible, changing the brushes can be a significant undertaking.

Transmitter

A radar transmitter typically has three major functional areas; the high-voltage power supply, the modulator, and the radio frequency (RF) power output stage. In a Doppler radar, stability of the transmitter is particularly important because of the small changes that the Doppler shift produces in the reflected signal. If the transmitter causes small changes in the signal, it is impossible to separate those changes from the Doppler shift changes. When evaluating radars, there is a tendency to place heavy emphasis on the peak power, with more being better. Peak power is only one factor. A radar having a more capable signal processor and an antenna with better side-lobe attenuation may outperform the more powerful radar in many situations.

High-Voltage Power Supply

The high-voltage power supply typically produces a DC voltage between 10,000 and 75,000 volts, depending on the design of the particular radar. Peak pulse currents, at those voltages, may be 20 amps or more. Although the peak power is very high, the average power is typically quite low, often not much over 100 watts. For the best accuracy of data, it is very important that the high voltage should be well regulated and stable. If the transmitted power, which is a function of the high voltage, varies, the reflected power will also vary, giving erroneous reflectivity readings. Variations in voltage may also affect the transmitted frequency or phase during a pulse, particularly with a magnetron transmitter. This variation in frequency will in turn appear as a Doppler shift in the reflected signal, giving erroneous Doppler velocity data.

Modulator

The modulator forms the high voltage into pulses of the proper duration and pulse repetition frequency (PRF) to drive the RF power output stage. Many radars use a pulse-forming network to do this. The pulse-forming network typically uses some combination of capacitors and inductors to store energy such that a high-energy pulse is available at the right time. Pulse-forming networks are typically designed around a particular pulse width and PRF. This factor limits a radar system that is based on pulse-forming networks to a very narrow range of choices of pulse width and PRF. Some research radars and a few commercially available radars use DC-switch-type modulators. A DC-switch-type modulator does not have the timing restrictions of a pulse-forming network; therefore, a radar based on a DC-switch-type modulator typically allows the selection of pulse widths and PRFs over a wide range that is limited only by the duty cycle of the components involved. Many older radars use thyratron tubes to conduct the pulse from the pulse-forming network to the RF power output stage. Newer designs often use solid-state devices, which tend to be much more reliable than thyratrons.

RF Power Output Stage

There are two categories of RF power output stages: *coherent-on-receive radars*, which typically use magnetron tubes, and *fully coherent radars*, which use a true amplifier that is often based on a klystron tube, although some commercially available fully coherent radars also use traveling wave tubes or solid-state devices. A magnetron can be described as a power oscillator. Every time a pulse arrives from the modulator, the magnetron starts up, begins to oscillate, produces a pulse at the RF and power level, and then shuts down. This sequence is repeated for every pulse.

A magnetron operates by causing a beam of electrons to sweep around the interior of a cylindrical cavity. That cavity typically has a series of slots around the outer surface of the cylinder. The number of slots and the speed at which the beam sweeps around determine the operating frequency. The beam speed is determined by a combination of the physical geometry of the tube and slots, voltage, current, magnetic field strength, and load on the magnetron's output. As a result, the frequency of oscillation for the magnetron is likely to be slightly different for each pulse. The phase of each pulse is random.

Coaxial magnetrons have an additional cavity that couples the slot output to the waveguide output. This additional cavity tends to isolate the beam from output loads and makes the operation of the tube more stable and less affected by changes in load.

A fully coherent radar has a master oscillator that is always running. The RF power output stage amplifies that signal during the pulse from the modulator; thus, the frequency and phase of each pulse are very stable, accurately known, and essentially identical to every other pulse. Klystrons and TWTs are both linear electron beam tubes. In both, the input signal is injected into the tube and the beam energy is used to amplify the input signal. The frequency and phase of the output signal are determined by the input signal and are largely independent of the tube and circuit parameters in the RF power output stage. For dual-polarity operation, there must be a method to either split or switch the transmitter's output to the horizontally and vertically polarized antenna feeds. Splitting the power output is often accomplished with a waveguide device known as a *Magic T*.

Receiver

The reflections returning to the radar antenna are extremely weak; therefore, the receiver must be sensitive and have a low noise level such that weak signals can be detected and are not masked by noise. In addition, the receiver must be able to receive a strong reflection from a nearby target without overloading. Meeting both objectives is a challenging engineering problem. Receiver sensitivity is typically referred to as minimum detectable signal (MDS) and is measured in dBm. Commercially available radar systems typically have an MDS in the range of –108 to –115 dBm (–115 is more sensitive than –108). A few systems have MDS values of –120 dBm or better.

Dynamic range, the difference between the weakest and strongest signal the receiver can handle without overloading, is typically at least 90 dB and some can exceed 100 dB. In many systems, this dynamic range is achieved by using an intermediate frequency (IF) amplifier with logarithmic rather than linear response. Many recently designed weather radar systems use receivers with a digital IF amplifier. The digital IF amplifiers typically have dynamic ranges of at least 80 or 90 dB, such that a separate logarithmic amplifier is no longer necessary.

Classical radar theory indicates that the IF bandwidth is related to the transmitted pulse width. Narrower pulses require wider IF bandwidths. Some radar systems automatically adjust the IF bandwidth when different pulse widths are selected. For dual-polarity operation, many systems require two independent receivers, one each for the horizontally and vertically polarized signals. These receivers must be identical.

Signal Processor

Most radar systems today use digital signal processors. The received signal is analog, so it must be converted to digital format. This is done with an analog/digital (A/D) converter. In many radar systems, the A/D converter has 12 bits of precision, and 14-bit precision is available on a few signal processors. The A/D process is critical in the overall performance of the radar system. The conversion process introduces noise and is subject to a number of factors that may reduce the quality of the signal. The signal processor must perform several functions:

- *$1/R^2$ compensation.* Radio signals propagate according to the inverse square law; that is, the signal has 1/4 the strength at twice the distance. If this were not compensated for, precipitation would always appear to be weaker at greater distances. The compensation causes precipitation of a given intensity to be displayed at the correct intensity, no matter what its range.
- *Doppler velocity processing*. The signal processor compares the frequency and phase of the signal in each range bin with the frequency and phase of the transmitted pulse and computes the radial velocity from that difference.
- *Spectrum width processing*. The signal processor computes the difference between the maximum and minimum velocity within the same range bin.
- *Clutter filtering*. In a Doppler radar, a ground clutter filter assumes that targets with zero or near-zero velocity are buildings, smokestacks, etc., and suppresses those reflections.
- *Dual-polarity processing*. Dual-polarity processing provides accurate rainfall data throughout the radar's primary coverage area based on horizontally and vertically polarized reflectivity values.

If the radar is to produce true real-time images, the signal processor must have the capacity to process all

the data immediately as the antenna scans. Processing capacities of 500 MIPS or more are readily available.

Radar Data Product Generator and Volume Scanning

The signal processing referred to in the previous section applies to data from the radar at a single antenna elevation angle. Many state-of-the-art radars have the ability to do true three-dimensional volume scans. This is typically done by taking successive scans of one revolution each with the antenna at successively higher elevation angles and then repeating the sequence from the ground up. In many radar systems, the volume scan data is processed by a separate workstation. That workstation will generate many different radar image products from the volume scan data. Typical volume scan products are:

- CAPPI (constant-altitude plan position indicator) is a horizontal slice through the atmosphere at an arbitrarily specified altitude.
- VIL (vertically integrated liquid) is the integrated value of the total amount of liquid in the atmosphere above a specified point on the ground.
- XSECT (cross-section) is a vertical slice through the atmosphere above a line between two arbitrarily located points on the ground.
- VAD (velocity azimuth display) is a display of a time series of wind velocity *versus* direction and altitude parameters.

A number of other volume scan products have been devised for various purposes. Radar manufacturers can provide details of the products their particular equipment supports.

Display

The basic images a Doppler weather radar system displays are reflectivity, velocity, and spectrum width. Some systems provide both raw and filtered reflectivity. The operator controls which image is displayed. The reflectivity, velocity, and spectrum width displays are typically shown with specific colors representing calibrated data values; thus, precipitation intensity, wind velocity, and atmospheric turbulence at a given location can be quickly estimated. The basic radar display contains a great deal of information but is typically not very useful by itself because the location of the data is not clear. Adding a map overlay to the radar image is one of the most common display graphic functions. A map permits users to quickly identify where the precipitation, velocity, etc., areas are with respect to known locations on the ground. Many radars provide the ability to blink selected colors in a display. This makes it easy for users to identify areas of particular interest. Some radars provide the ability to generate additional graphic overlays that can be created by the user for specific purposes. If a radar is to be used for live on-air presentations, it goes without saying that it should have the ability to produce broadcast-quality genlocked video output. For on-air presentations, having a radar system with a powerful graphics processor is a significant advantage. That graphics processor can be used for the creation of sophisticated weather products that have the look the station desires.

Weather Graphics Software

In a typical TV station, weather radar images and other weather images are created in a format that is compatible with broadcast video standards. Several vendors provide graphic workstations and software packages with the capability to do that. Figure 5.14-11 illustrates how a weather radar system is configured with a typical weather graphics system. The TV viewer is arguably the toughest critic of the technical accuracy of the weather broadcast. If the TV weather display does not agree with what the viewer sees out the window, the station's credibility is impaired. Several considerations need to be taken into account for a weather graphics system:

- Can the different types of data in different formats be made compatible? Many suppliers of graphics software provide data (for a monthly subscription fee) in formats compatible with their software. In addition, the software must be able to accept the radar data format.
- Are the different types of data located and scaled correctly to match the map being used?
- Is the detail level of the background maps appropriate? Do simple line maps show state lines, cities, etc.? Do detailed topographic maps show the terrain? Do satellite/aerial photographs show actual land features with streets, buildings, etc.? In general, increased level of detail adds both cost and complexity to the system.
- Does the on-air appearance of the final product include the desired level of detail, animation, etc.? Also, does the degree of customization accommodate local requirements?
- How much time, effort, and expertise are required to manipulate the data and the graphics to produce the final product?
- How much control over the displays during the on-air broadcast is possible?

Uninterruptible Power Supply

Severe weather is one of the most frequent causes of power failures. Having an uninterruptible power supply (UPS) for all radar equipment is good insurance against having the radar go off just at the moment when it is of most value and interest. A typical UPS system will have both a battery supply and an engine-driven generator. The installation is typically designed such that the batteries will keep the minimum essential items running for a few minutes until the engine starts and can take the full load. Some jurisdictions have special building codes that apply when large arrays of bat-

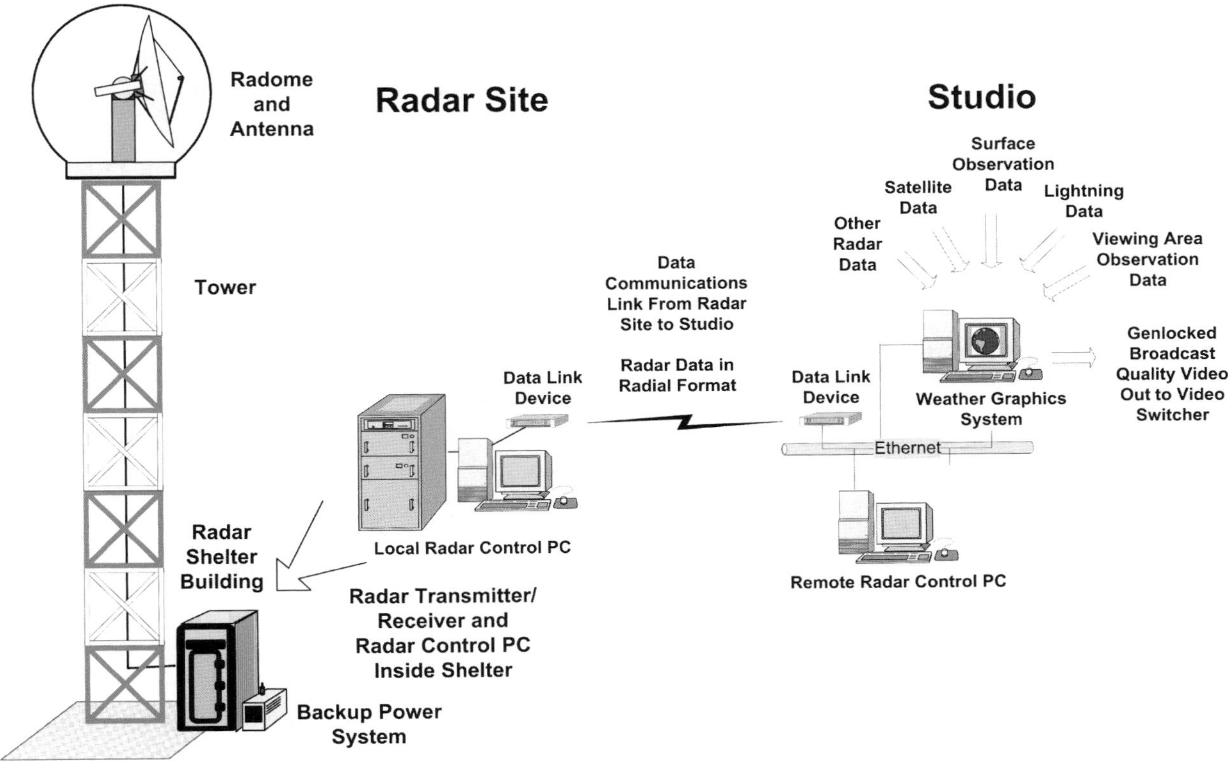

FIGURE 5.14-11 A weather radar system configured with a typical weather graphics system.

teries are used inside of a building. The general concern is that, under certain conditions, the most commonly available types of batteries have the potential to produce hydrogen gas. In a small enclosed area, such as a radar transmitter building, that hydrogen can accumulate to an explosive concentration very quickly. If not covered by specific building codes, the UPS batteries should be in an area that is well ventilated, with that ventilation going directly to the outside and not connected to the rest of the building's ventilation system. In addition, the amount of electrical equipment in the battery room should be minimized, and equipment that is in the battery room should be explosion proof and should also be on the UPS.

LOCATION OF THE RADAR SYSTEM

Many factors affect the choice of location for a weather radar system and its antenna; however, the practical reality is that, in many cases, only a certain location is available, and that is where the radar must be placed. In these cases, effective use of the radar requires understanding the limitations imposed by the specified location.

Define Coverage

Determining the nature of coverage to be provided and the primary area of interest for that coverage require addressing these questions regarding the primary objective of the weather radar:

- Is it to provide severe weather warnings for a metropolitan area?
- Is it to provide rainfall data for a particular watershed?
- Is it to detect potential aviation hazards?
- Is it to provide scientific data for a specific application?

Selection Factors

Meteorological Factors

If providing warnings of approaching severe weather is one of the objectives, more advance warning time is usually desirable. That dictates locating the radar system between the area to be protected and the usual direction of approach of severe weather. In many locations in the United States, severe weather approaches from a generally westerly direction; therefore, the radar in those locations should ideally be to the west of the area to be protected. Placing the radar elsewhere would typically reduce the warning time. The radar beam from the antenna should have a clear view of the area to be protected. Locations where buildings, hills, etc., block or obstruct that view should be avoided. Trees also can block the view, and trees grow. One common practice is to place the radar antenna on a tower that is tall enough so that it will still be comfortably above the tops of the trees after 20 years of tree growth.

Electromagnetic Compatibility Considerations

The radar must be in an area that is free from interference on its operating frequency. Other radar systems are a common source of interference. Interference includes both other sources interfering with the radar's operation and the radar interfering with other equipment. Non-RF sources can also be significant; for example, normal corona leakage from very high-voltage power lines near the radar may raise the background noise level to the point that very weak radar reflections will be masked.

Site Access

The radar site location should have access 24 hours per day, 12 months of the year. Selecting a site that is known to be periodically inaccessible as a result of snowfall, flooding, etc., is not recommended.

Radar to Control Location Data Link

In many situations, the radar system must be placed in a location different than the location where the radar data is needed. In many broadcast situations, the radar system may be located at the transmitter site, and the radar data and control must be at the studio. Weather radar systems produce large volumes of data. Having real-time coverage means getting that large volume of data from the radar to the studio in real time. That requires high-bandwidth data circuits. It is not uncommon for a real-time radar installation to require a T1 (1.544 Mbps) data circuit between the radar and the data/control location. The availability and cost of high-speed data circuits or a private microwave link between the radar site and the studio are factors that should be considered in the site selection process.

INSTALLATION PLANNING

Planning the installation of a weather radar system involves a great many details. Some of the issues to be considered include the following.

Lightning Protection

Lightning is one of the most frequent and costly causes of damage to weather radar systems. Systems from different manufacturers vary in their resistance to lightning strikes; however, nearly any radar installation will benefit from lightning protection that includes a grounding system suitable for the purpose. See Chapter 7.3, "Lightning Protection," and Chapter 9.3, "Facility Grounding," for more information on these topics.

Operating Frequency

An integral part of the plan should be the process of determining an operating frequency for the radar. That process is usually easier if an existing radar is being replaced. If the radar is new, a determination of a clear frequency must be made. Some radar manufacturers will require that the operating frequency be specified very early in the order process.

FCC

The radar system includes a transmitter, which requires an FCC license to operate in the United States, just as do most other transmitters.

FAA

A tower for the radar antenna may be subject to FAA regulations and require registration as an aviation hazard and an approved lighting system, based on its location and height.

Local Permits

Construction of a radar tower is likely to require local permits, variances, and zoning changes.

Checklist for Weather Radar Site

A checklist for planning and documenting some of the many other aspects of a weather radar is provided as an appendix to this chapter. The checklist demonstrates the complexity of a weather radar installation and serves to provide insight into the extent of planning, maintenance, and operations issues involved.

SAFETY ISSUES

Radars are complex and powerful systems that should be treated with great respect. Personnel who will operate and maintain a radar system should be trained in the safety aspects of the radar's operation as well as in the operational details of the radar.

Microwave Hazards

The effect of microwave radiation on the body has been the subject of a great deal of study. Of the hazards investigated to date, the heating effect on tissue is the only one that has produced general agreement and is readily replicated under controlled scientific conditions. The joint ANSI/IEEE Standard for Safety Levels with Respect to Human Exposure to Radio Frequency Electromagnetic Fields, 3 kHz to 300 GHz (C95.1-1991) is based on tissue-heating effects. In 1997, the FCC's Office of Engineering Technology issued some guidelines for levels of RF emission fields in OET Bulletin 65. These guidelines are neither laws nor regulations, but it is clear that the FCC will review any transmitter

license application with respect to them. The location of the radar beam with respect to areas where people live and work must be considered. The radar antenna should be located such that no persons are within the area where the Maximum Permissible Exposure (MPE) limits are exceeded. The MPE limit typically results in a minimum safe distance in front of the radar antenna given the transmitter power, antenna gain, frequency, and duty cycle. See Chapter 10.4, "Human Exposure to Radio Frequency Fields," for more information.

Electrical Hazards

The high-voltage, high-energy circuits in most radar transmitters present an extreme hazard to personnel. Because of the high voltage, normal skin resistance is ineffective at limiting current. Due to the ability to deliver large quantities of energy very quickly, often in the submicrosecond range, large amounts of energy will be conducted before even the fastest protective devices can operate. Only personnel who are thoroughly trained should be permitted to work on a radar transmitter. Even then, strict safety procedures must be observed at all times. There may not be a second chance. See Chapter 10.6, "Electrical Shock," for more information.

Mechanical Hazards

The motors that move the antenna are powerful and are typically geared down to provide the massive amounts of torque necessary to move a heavy antenna with precise accuracy. Typically, the antenna is in a location out of sight of the circuits that control its movement. The antenna is often in a very confined space inside a radome. The end result is that personnel near the antenna have a high risk of being injured or crushed if the antenna moves unexpectedly. Again, only personnel who are thoroughly trained should be permitted to work on the radar system, and strict safety procedures must be observed at all times.

X-Ray Hazards

Some radar transmitter tubes may produce x-rays in certain operating modes. The radar transmitter should not be operated if x-ray shields provided by the manufacturer have been removed.

SUMMARY

The addition of a weather radar system to a station's facility is a complex and potentially expensive endeavor. Careful planning and a thorough understanding of weather radar technology are essential to successful use of the system.

Bibliography

ANSI, *IEEE Standard for Safety Levels with Respect to Human Exposure to Radio Frequency Electromagnetic Fields, 3 kHz to 300 GHz,* Institute of Electrical and Electronics Engineers, New York, 1994.

Doviak, R.J. and Zrnic D.S., *Doppler Radar and Weather Observations,* Academic Press, Orlando, FL, 1984.

FCC, *Evaluating Compliance with FCC Guidelines for Human Exposure to Radio Frequency Electromagnetic Fields,* Bull. No. 65, Office of Engineering & Technology, Federal Communications Commission, Washington, D.C., 1997.

Rinehart, R.E., *Radar for Meteorologists,* University of North Dakota, Fargo, 1994.

APPENDIX: CHECKLIST FOR WEATHER RADAR SITE

1. **Area to Be Covered:** Population centers, watershed area, airport, etc., are prioritized in order of importance. A map with coverage areas outlined is helpful.
2. **Type of Coverage To Be Provided** (arrange list in order of importance):
   - General purpose meteorological data/forecasting
   - Severe weather detection/tracking
   - Rainfall measurement for hydrology application
   - Aviation safety hazard detection
   - Other ______________________
3. **Antenna Site Location:** Lat. _______ Long. ______

   Brief description of physical location: ___________
4. **Radar Beam Obstructions:** Type, direction, approximate distance; consider coverage priorities.
5. **Access to Site:** Road, trail, cross-country; whether special vehicles and equipment are required; whether access is subject to being cut off (snow, floods, etc.).
6. **EMC Factors:** Other radars, radio transmitters/receivers that may be an interference problem and their location; also, consider radar beam safety (*e.g.*, personnel who may be exposed to beam during normal operation).
7. **Tower Height:** Radar must see over tops of trees and buildings and may be restricted by proximity to an airport.
8. **Soil Conditions at Tower Site:** Sand, clay, rock; whether soil test data is available.
9. **Electrical Service to Radar Site:**
   - Voltage, frequency, single- or three-phase _____
   - Reliability of source: voltage surges, sags, interruptions, or seasonal problems.
   - Number and duration of interruptions in the past 12 months:
   - Less than 1 second ______
   - 1 second to 5 minutes _______
   - More than 5 minutes ______

10. **Uninterruptible Power Supply:**
    - Battery capacity ______ Watts/VA ______ Minutes
    - Generator capacity ______ Watts/V/A ______ Hours fuel capacity
    - Testing/maintenance schedule established
    - Items powered by UPS
    - Radar transmitter/receiver
    - Antenna pedestal
    - Radar data communications
    - Radar workstation
    - Tower warning lights
    - Radar air conditioning
    - Security system
    - Work lights
    - Other ______________________
11. **Data Communications** (more than one system may be employed):
    - Telecom service:
      - — Nearest connection point ________________
      - — Type of service ________________________ (analog, ISDN, T1, E1, etc.)
      - — Data rate required/available _____________
      - — Interface required to connect to service ____ ____
    - Private microwave:
      - — Single-hop or multi-hop ________________
      - — Obstructions in path ____________________
      - — Multi-hop tower site locations ____________ ____
      - — Rain fade potential ______________________
      - — Radio interference ______________________
      - — Data rates required/available ____________
      - — Interface required to connect to microwave ____
    - Private fiberoptic:
      - — Right-of-way for cable route _____________
      - — Interface required to connect to fiberoptic __
    - Satellite data link:
      - — Antenna locations _______________________
      - — Antenna look angles _____________________
      - — Type of service required (SCPC, TDMA, etc.)
      - — Data rates required/available ____________
      - — License requirements, restrictions _________
      - — Rain fade potential ______________________
12. **Tower Grounding:**
    - Required ground resistance ________ ohms
    - Type of grounding system required: _________

13. **Installation Schedule:**

Item	Date	Initials
Site location selected		
Site location approved		
Contractor selected		
Drawings approved		
Permits, licenses applied for		
Permits, licenses obtained		
Tower and building foundation done		
Radar site electrical service		
Crane to erect tower		
Pull cables		
Install radar		
Installation complete		

CHAPTER

5.15

Television Graphics

JIM MARTINOLICH
Chyron Corporation
Melville, New York

INTRODUCTION

Television graphics covers a broad range of products and applications that include character generators (CG), clip players, still stores, virtual studios, three-dimensional (3D) graphics systems, logo generators, and other similar devices. Each of these categories can be further differentiated into distinctly different subcategories. Today these products are differentiated more by their traditional workflows and interfaces than by their hardware differences. In fact, with the substantial growth in the graphics-rendering capability of commodity personal computer (PC) platforms many of these devices now share similar if not identical hardware architectures. While in the past, "big iron" products used specialized hardware to perform the required signal processing, today's feature sets are mostly implemented in software.

That isn't to say that there are no differences in the hardware. Even with a similar architecture, components need to be optimized for a particular application. A clip player will have more storage than a logo generator, a 3D graphics device will have a more powerful rendering system than a still store, and so on. But the architectures are remarkably similar, and the components are usually based on commonly available generic PC hardware.

The benefits of an open platform are clear. First of all, there is no longer a need for very expensive custom hardware that was typical of the last generation of graphics products. This speeds the time to market; new features can be implemented faster and the industry can be more responsive to users' needs. Open platforms lower the cost of development and ultimately the cost to the end users.

The other major benefit is that these open platforms inevitably grow in performance at a rapid rate that represents a "free ride" to better performance for a properly designed graphics system.

This chapter will describe a generic graphics device model and analyze the various components. It describes some of the technical issues, specifications, interfaces, and existing workflows that must be considered. By understanding the generic model, today's video engineer is in a good position to deal with a wide assortment of graphics devices.

TYPES OF GRAPHICS SYSTEMS

Before describing the generic graphics model, a review of how these devices are used in television production is provided along with an analysis of system requirements. The first important differentiator is that between off-line devices and online devices.

Off-Line Devices

Off-line device generally do not need to create a real-time video signal; they are file based instead. Therefore, off-line systems do not need to be as powerful as an online system, or given the same power, they can render much more complex graphics. Also, since off-line systems are used in postproduction environments they do not have to be as rugged as their online

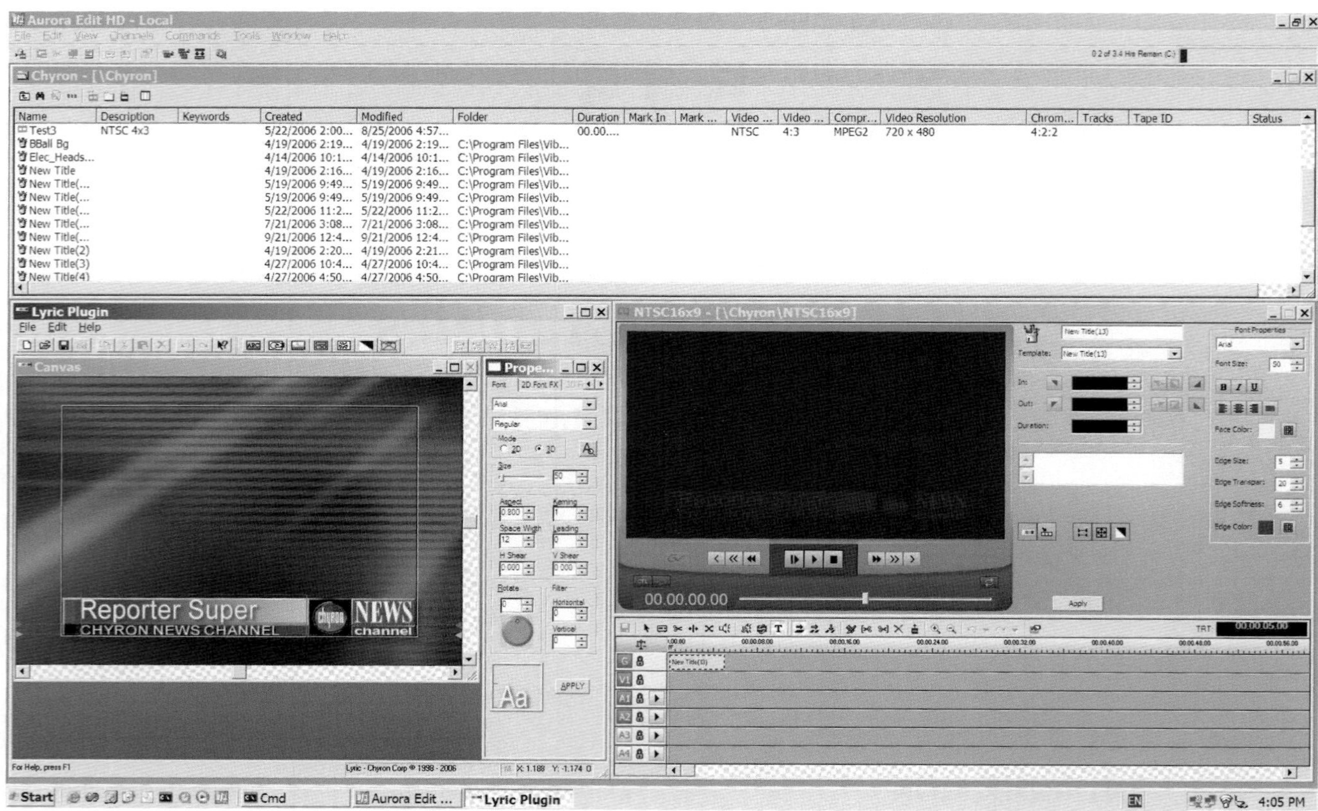

FIGURE 5.15-1 Nonlinear editor with graphics plug-in. (Courtesy of Grass Valley.)

cousins. One simple example of off-line graphics is a character generator plug-in to a nonlinear editor (NLE) (see Figure 5.15-1.

Off-line devices are often generic PC graphics applications with a few basic additional requirements unique to television. For example, they must create images with the correct number of pixels, such as 720 × 486 pixels per frame for standard definition (SD) 525-line systems. Pixel aspect ratio differences between digital television and computer displays must be taken into account. Each pixel should have at a minimum 4 bytes of data (8 bits each for the red, green, blue, and alpha components).

Alpha, or key, is important because television graphics are almost always meant to be "keyed over" a program video signal in a mixer somewhere. A key signal should have 256 levels to give the linear keying required to blend the graphics in smoothly with the program video and reduce keying artifacts.

Table 5.15-1 shows a list of commonly used file formats for saving still images to disk. Some of the formats are marked to show that they save a linear alpha signal along with the image. Some image formats, such as JPEG, are compressed, but uncompressed image formats, such as TGA, are usually preferred to minimize unnecessary image artifacts. With so many formats to choose from, graphics devices should avoid using proprietary formats whenever possible.

Animation is where television graphics software starts to deviate significantly from more generic graphics applications, especially when working with interlaced video systems. Generating an interlaced image and keeping track of the field sequence can be challenging. Some of the issues are discussed later in this chapter. Reversing the order of the interlaced fields either spatially or temporally can lead to unpleasant visual artifacts.

The simplest way to save animating video to disk is as a numbered sequence of still images. Sequences of still files are often called "flipbooks" or "cell animations." Video can also be saved as clip files in AVI or MOV wrappers. Since clip files can be large, some compression is usually necessary, though not desirable from a quality point of view, and care must be taken to minimize compression artifacts.

Table 5.15-2 shows a list of commonly used file formats for saving animations to disk. Again, with so many formats to choose from, graphics devices should avoid using proprietary formats whenever possible.

Note that that there are some off-line devices that output video rather than files and so are more like online devices. One example is the use of video graphics devices in traditional linear editing applications. While linear editing has been replaced by NLEs in most applications, it still has its place, especially in high-definition (HD) finishing work, such as adding a credit roll to long form content.

TABLE 5.15-1
Commonly Used Still Image File Formats

File Extension	File Type
.bmp	Windows Bitmap Graphics*
.cal	CALS Bitmap Graphics
.chy	Chyron Infinit Image File
.clp	Windows Clipboard File
.dcx	Multipage PCX Bitmap Graphics
.ica	IOCA File
.iff	Amiga Graphics File
.imt	IMNET Graphics File
.jpg	JPEG/JFIF Image*
.mod	MO:DCA IOCA
.pcd	Kodak PhotoCD Image
.pct	Macintosh Drawing Format File
.pcx	Paintbrush Bitmap Image
.png	Portable Network Graphics File
.psd	Photoshop Image File
.ncr	NCR Image File
.ras	Sun Raster Graphic
.rgb	Silicon Graphics RGB Bitmap
.sgi	Silicon Graphics Image File
.tga	Targa Graphic*
.tif	Tagged Image Format File*
.vpb	Quantel VPB Graphic
.xbm	X Bitmap Graphic
.xwd	X Windows Dump

*Most commonly used File Formats.

TABLE 5.15-2
Commonly Used Animation File Formats

File Extension	File Type
.avi	AVI is a file wrapper that can contain many types of a video content; Radius Cinepak, M-JPEG, MPEG-1, MPEG-2, MPEG-4 Part 10 (H.264/AVC), uncompressed, and many more
.mov	Quicktime file format—MOV is a file wrapper that can also contain a wide variety of video content
.mp4	Special case of a MOV file and which was standardized in MPEG-4 Part 14
.mpg	MPEG-1 File format

Online Devices

Online devices have much tougher performance requirements. They need to generate a genlocked broadcast-quality video in real time without missing a frame. Because they are generally used live on-air, they must be reliable. Mobile production equipment also must be ruggedized to withstand the stresses of transportation. Online devices can be further differentiated between upstream and downstream applications.

Upstream Devices

These devices (Figure 5.15-2) are fed into a production switcher to be mixed into program video. The genlock circuitry must allow the upstream device to be timed to the switcher's input requirements. Since graphics are usually layered over other video content, the upstream device must output the key, or alpha, signal precisely in time with the video signal. Upstream devices are usually involved in producing a live program and are manually controlled or controlled by specialized automation systems and playlists. Examples of upstream devices are traditional character generators, still stores, and clip players used in news and sports production.

Downstream Devices

These devices (Figure 5.15-3) are called downstream because they are typically placed downstream of the switcher. They fit in the program video path and mix the graphics directly into the program video. Because they are in the video path, downstream devices have the most stringent broadcast requirements. They must perform the mixing process without compromising the signal quality. They must pass ancillary data and audio. They should have program video bypass in case of system failures. Downstream devices are usually controlled by station automation systems. Examples of downstream devices are stock tickers, channel branding (logo) devices, weather information, and promo inserters used in master control applications.

GENERIC GRAPHICS DEVICE MODEL

Figure 5.15-4 shows a simplified block diagram of a typical modern broadcast graphics device. It makes a good framework for discussing the many technical issues faced by the developers and users of graphics systems.

Many graphics systems are built on PC platforms for all the advantages described earlier, but even non-PC devices share most of the same building blocks. The block diagram breaks the system down into the following components:

- Central processing unit (CPU)
- Graphics processing unit (GPU)
- Hardware clip player
- Control interface
- Data interface

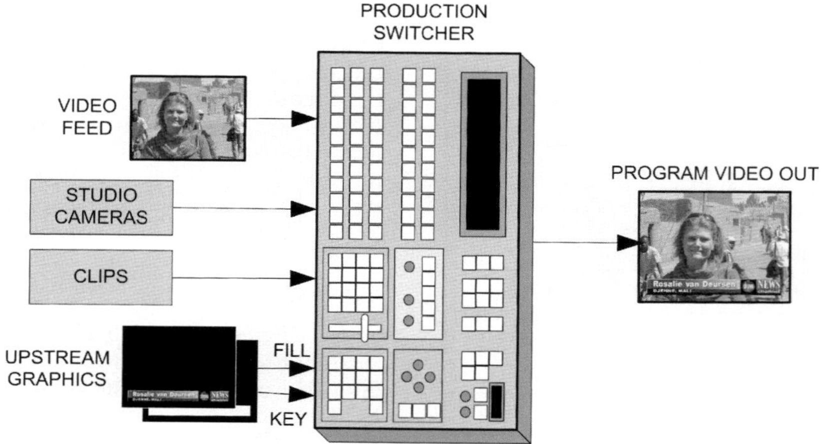

FIGURE 5.15-2 Upstream graphics system diagram.

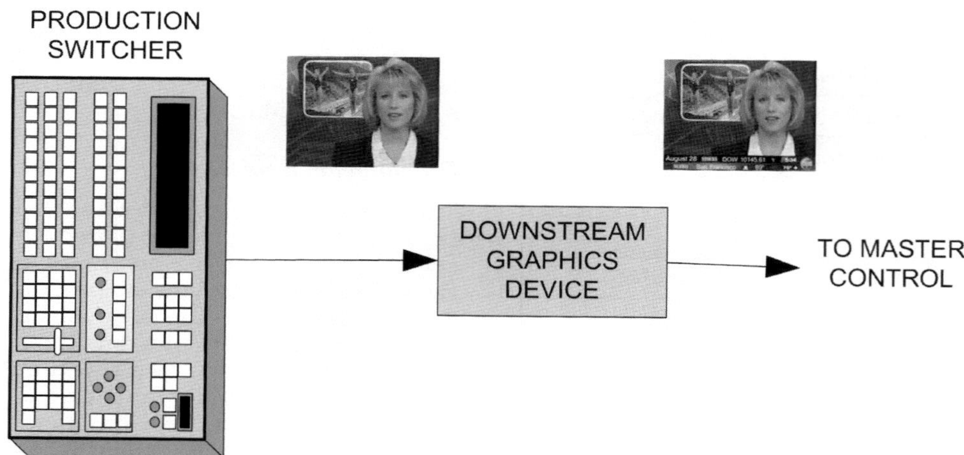

FIGURE 5.15-3 Downstream graphics system diagram.

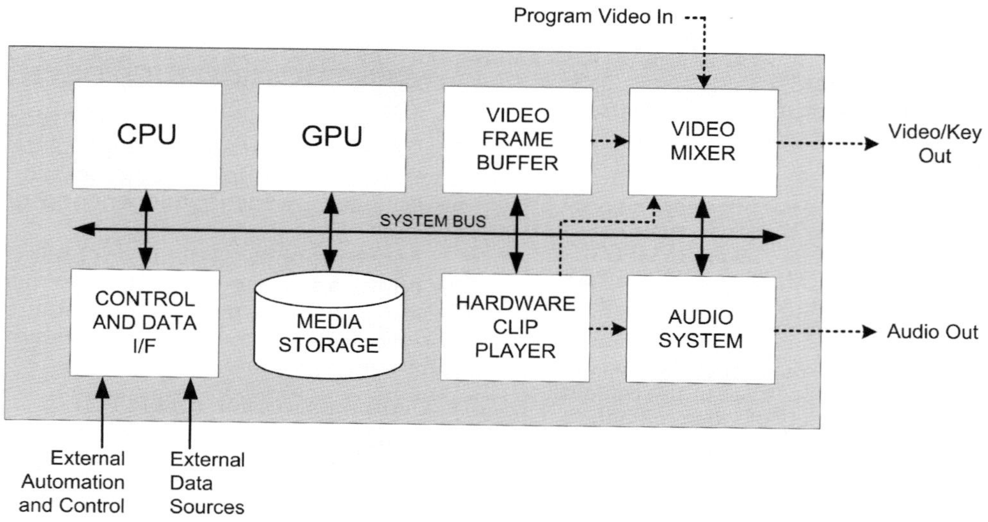

FIGURE 5.15-4 Generic graphics device block diagram.

- Media storage and pipelines
- Audio system
- Video frame buffer
- Video mixer

The CPU handles much of the non-graphics-related processing, such as managing the system, communication with data sources, and calculating the data to be displayed. For the most part, modern PC platforms exceed the CPU power needed for the non-graphics processing needs. But, sometimes the CPU is asked to handle some graphics-rendering loads. Modern graphics systems will typically use multiprocessor motherboards when there is a significant amount of CPU loading required.

Creation of the graphics image can be accomplished using one or more of the following three methods: the image can be rendered by the CPU, rendered by a separate device (i.e., the GPU), or played back directly from a file using a hardware clip decoder. A still store will typically only use the first method, a clip player will use the third method, while a character generator will use all three.

The GPU is the workhorse that generates the graphics image in cases where the image is created on the fly rather than played back from a file, such as in live CG applications. The GPU performance often defines the overall performance of the system. In a PC-based system, the GPU is the graphics board used to generate the video graphics adaptor (VGA) display. These boards are developed by companies like nVidia and ATI. Their technology is driven by the high-volume home computer and gaming market. The GPU in a current high-performance computer has more than twice as many transistors as the CPU, and they are doubling in performance about once a year.

Some performance metrics of GPUs are given in Table 5.15-3. How they affect the quality of the broadcast video signal may be hard to quantify. In terms of raw numbers, the modern GPU can generate more pixels, deal with higher frame rates, and handle more flying 3D objects than is needed in a typical television application. On the other hand, it does not natively want to be genlocked nor does it generate interlaced images or a key signal. Net performance depends more on how well the vendor's application manages the GPU and ultimately has to be evaluated from the output itself. However, since GPU performance is doubling about once a year, upgrading the GPU frequently can be a simple way to increase system performance.

TABLE 5.15-3
GPU Performance Metrics

Year	Vertices/Sec	Fill Rate pixels/sec)
1997	3 M	.033 B
2004	600 M	6.4 B
2006	2000 M	24 B

The *hardware clip player* is a hardware decoder used to play back full-resolution compressed video clips. In a dedicated graphics clip player, this is the only video-rendering device. In a character generator, a clip may be played in the background and mixed with the rendered graphics coming out of the GPU. In either case, graphics clip players are differentiated from generic video servers by several requirements.

In a graphics server there is always the need for two output streams per channel: video and key. Video and key must play back exactly in sync. Also, graphics systems do not need or perform well with a large amount of compression. Graphics clips are only measured in seconds and minutes, not hours, so high compression is not needed. Graphics images readily show compression artifacts because of their large flat color areas and sharp edges, so it is better to keep the level of compression low.

The compression algorithm should use only intraframe compression, that is, the compression of each video frame as a static image. It must be possible to splice the stream at any point for looping and to decode any frame independently of its neighbors. This means that intraframe-only MPEG-2 compression is preferred to long-GOP MPEG-2.

If possible, the compression algorithm should not undersample the chroma component. This will preserve color resolution and minimize keying artifacts. 4:2:2 profile compression is preferred to Main Profile (4:2:0) compression. Typical compressed video data rates for graphics applications are 10 Mbytes/sec for SD video and key, and 30 Mbytes/sec for HD video and key, which is a much higher data rate than is typically used in other applications.

The hardware clip player may soon be obsolete as the trend in the industry is toward software decoding of clip files. Software decoding is more flexible to changing compression standards and saves the additional hardware costs. As the CPU power increases, the CPU will be able to handle clip decoding without compromising its other functions. When evaluating this option, note that the key channel usually requires a second decoder, and that two complete channels, for a total of four output streams, are usually preferred in most broadcast applications. This is more of a concern in HD applications than for SD.

The *control interface* of the graphics playback is wired into the studio through automation interfaces or manual control via keyboard, general-purpose interface (GPI), or custom hardware control panels. A properly designed control interface is almost more important than the graphics capabilities since in broadcast television workflow is king. New products often try to emulate older product interfaces in order to ease the transition in the production suite. Automation interfaces are usually serial RS-232 or RS-422 with a gradual change over to newer Ethernet networking protocols. Hardware control panels can have a wide variety of functions and protocols.

Different graphics devices traditionally have different automation interfaces. Character generators use the Intelligent Interface protocol originally developed

FIGURE 5.15-5 Character generator custom keyboard. (Courtesy of Chyron Corp.)

by Chyron and adopted by most character generators. Graphics clip players can use many automation protocols borrowed from video servers and even videotape recorders (VTRs). Examples are the Odetics Protocol; the Video Disk Control Protocol (VDCP) originally developed by Louth Automation, now part of Harris Corp.; the PBUSII switcher control protocol; and the Sony RS-422 VTR control protocol, often called the BVW-75 protocol.

Hardware control panels are designed to simplify the operation of a graphics device by giving the operator fast-action control of special functions. Character generator control panels, such as the one shown in Figure 5.15-5, are custom QWERTY keyboards that include shortcut keys that perform simple common composition and file-handling functions. A clip player hardware control panel, such as the one shown in Figure 5.15-6, is designed to simplify searching for and controlling playback of video clips.

For the data interface, in many cases, especially in character generator applications, the purpose of the product is to display information that comes from some external source of data. Time and temperature bugs, stock quotes, news tickers, and sports scores are all examples of a data interface requirement. Traditional news feeds have used proprietary serial data feeds between the graphics device and some middleware that managed the database. Today the Internet, database technology, and open platform systems provide an almost limitless variety of options for accessing data. Data can be gathered from a remote database directly using ODBC or parsing it from an XML data source such as RSS news feeds. Data can be "scraped" from a website using smart HTML parsing. Software plug-ins can share complex data intelligently between applications by helping to filter and process data before sending it to the graphics device.

For *media storage* and pipelines, all graphics devices must store and manage substantial amounts of media content. Some applications, such as multichannel HD character generators, push the limits of PC storage and bus speed technology. Designing a storage subsystem is a major issue in selecting components for a graphics system. Requirements that must be considered include storage size, access speed (both for streaming and random access data), and redundancy in case of failure. Also, the system backplane should be capable of delivering the data to the GPU or frame buffer at the appropriate rates.

FIGURE 5.15-6 Clip player hardware control panel. (Courtesy of DNF.)

Table 5.15-4 shows the data rate needed to move a rendered 4:4:4:4 HD image in real time, compared to the bus speeds of modern computers. The PCI bus standard used in most desktop computers is not fast enough to move full high-definition red-green-blue-alpha (RGBA) graphics data in real time. The PCI-X standard, on the other hand, is fast enough and the next-generation PCI-Express bus will prove to be even faster.

Obtaining the overall size of storage needed is simple, as commodity drives start in the hundreds of gigabytes. However, access speeds can still fall short in multichannel applications. Random access or seek speeds can vary greatly in drives with similar streaming data rates. Careful testing is required. Striping disk drives (RAID-0) may be needed to meet the access speed requirements, but the striped pair is now twice as likely to fail as a single drive.

Since they are one of the few moving parts in a PC, disk drives are notoriously unreliable. Graphics systems in a critical application should have redundant hard drives in a configuration that tolerates the loss of

TABLE 5.15-4
Video Data Rates versus Current Bus Speeds

SD Data Rate (4:4:4:4)	42 Mbytes/sec
HD Data Rate (4:4:4:4)	250 Mbytes/sec
PCI Bus Speed	132 Mbyte
PCI-X Bus Speed	400 Mbytes/sec and up
PCI Express Bus Speed (per lane)	250 Mbyte/sec (each direction)
PCI Express Bus Speed (16 lanes)	4 Gbyte/sec (each direction)

at least one disk (RAID-5). Creating a RAID that meets the redundancy requirements and the access speed requirements can be difficult and costly.

The audio system is a recent addition to the basic block diagram of a graphics system, and audio has become part of the graphics event, whether it is a "whoosh" tied to a graphics animation or a voice over tied to a clip. The audio signal is typically mono or stereo and stored uncompressed as a 48 kHz WAV file. Audio file size and processing load is trivial compared to video so compression is rarely used.

Many graphics systems simply use the motherboard audio device and jacks. This analog audio is subject to hum and, more annoyingly, also subject to an unexpected Windows sound effect. Separate digital audio processing is preferred. In a downstream configuration, the audio generated by the graphics events must be mixed into the program audio in a controlled way to ensure proper levels. This means the addition of multichannel audio mixing circuitry under control of the graphics event. Digital audio inputs and outputs should conform to AES3 serial transmission format, or be embedded in the serial digital interface (SDI) program video stream.

The *video frame buffer* is where the rendered image is converted to a true broadcast video signal. This is what differentiates a video graphics system from other graphics devices. The video frame buffer performs many functions that will be discussed in more detail in the next section. A rendered image is transferred frame by frame from the GPU or CPU to the video frame buffer as an RGBA signal. The data is then retimed, converted into a YUV video signal and a Y-only key signal, and embedded into the SDI digital video-timing structure.

To some extent, video frame buffers have become commodity items like the rest of the generic model. They are available for purchase from a variety of vendors at competitive pricing and with published software development kits (SDKs). However, features vary widely and integrating a particular graphic application to a video frame buffer can be a difficult and costly process. It is unusual to see a given application support more than one hardware frame buffer.

A *video mixer* will be needed in downstream graphics devices to mix the graphics into the program stream. Because the video mixer is in the program video stream, it is a critical component and must be aligned and tested carefully for proper timing and levels. It must not disturb the program video, vertical blanking interval (VBI) ancillary data, and embedded audio. The generated graphic must be correctly aligned in H and V and must respect the blanking areas. It should have a relay bypass to pass through program video in case of system failure. A later section will look at some of the details of video mixing in both upstream and downstream configurations.

GRAPHICS SIGNAL PROCESSING

The CPU, GPU, video frame buffer, and video mixer need to work together to make a compliant broadcast video signal. This section describes some of the technical issues that need to be considered when generating a video signal and the problems that can arise.

Genlock Performance

The first step in making a video signal useful, whether inserted upstream or downstream, is to genlock to an external reference sync. In both SD and HD systems, the genlock reference is almost always an analog black-burst signal. (HD systems can also lock to the analog trilevel sync defined in the HD specifications, but this is rarely seen in use.) Analog black-burst is the same reference used in the days of analog production so the standard was chosen to ease compatibility with existing infrastructure. HD and SD systems will usually lock to a digital video input as well if there is no black-burst available.

Two parameters are important when genlocking. First, the allowable jitter is only 20% of one SDI clock cycle, which for SD is 740 psec and for HD is 135 psec. Jitter should be carefully specified and tested often. Secondly, a wide range of alignment in both directions are needed so the advance or delay of the video output can be timed accurately at the switcher.

Pixel Aspect Ratio

Since PC technology is so ubiquitous, television graphics content often is created on a desktop computer. When sharing image content between computer graphics systems and television graphics systems a common problem is the incompatible pixel aspect ratios and the resultant distortion of the image.

The pixel aspect ratio is defined as the ratio of the height to the width of the pixel. In computer graphics the pixel has a simple square 1:1 aspect ratio, so that a 640 × 480 display has a 4:3 scene aspect ratio. In standard definition digital television the pixel aspect ratio is not 1:1 as might be expected but is rectangular and slightly higher than it is wide. Since a 720 × 486 pixel display has a 4:3 scene aspect ratio the pixel aspect ratio can be calculated as 1.111 to 1, or simple 1.1 to 1. The difference is small but can be significant with some images. Figure 5.15-7 shows the image distortion that would result from moving computer images to video and back without compensating for the aspect ratio.

Note that high definition standards use a square 1:1 pixel aspect ratio. Since scaling is required when upconverting or downconverting the image, that is a good time for performing pixel aspect ratio conversion as well.

Image Dimensions and Placement in the Video Raster

Graphics systems have become so complex, and the process that maps an image into the video signal has become so abstract that it is sometimes difficult to predict where exactly an image will display in the video raster. This can result in simple alignment problems to

FIGURE 5.15-7 Rectangular versus square pixel image distortion. Left–distorted image; right–with correction.

more complex problems that cause significant interlace and motion artifacts.

Alignment problems are most noticeable when components of the same image are played back on different graphics systems, for instance, CG lower-third text against a clip player background, or against a branding bug inserted downstream. Also, some modern flat-panel monitors that do not have overscan may show artifacts that occur on the edges of the raster, if full-screen graphics are not aligned correctly. It is, therefore, important to occasionally calibrate this process and check for these problems.

A quick test of the alignment is to create a simple test pattern with crosshairs in the center. Place it as a full-screen image in the graphics system and check the position on video output with a calibrated digital video waveform monitor. For a more complete test, inspect the edges of the raster with the waveform monitor and check to make sure that the edges of the image are falling where expected and that they are all there.

Start with the basic dimensions of the video image. A SD video frame has 720 × 486 pixels, the 1080i HD system has 1920 × 1080 pixels, and the 720p HD system has 1280 × 720 pixels. When mapped into the video raster, these pixels should fall precisely into the video pixels shown in Tables 5.15-5, 5.15-6, and 5.15-7, for each of the three scanning systems.

TABLE 5.15-5
720 × 486 Image Alignment in a 525i SD Video Raster

	Image Line	Video Line	
		F1	F2
Top	0		283
	1	21	
	...		
Center	242		404
	243	142	
	...		
	484		525
Bottom	485	263	

	Image Pixel	Video Pixel
Left	0	138
Center	359	497
	360	498
Right	719	857

TABLE 5.15-6
1920 × 1080 Image Alignment in a 1080i HD Video Raster

	Image Line	Video Line	
		F1	F2
Top	0	21	
	1		584
	...		
Center	539		853
	540	291	
	...		
	1078	560	
Bottom	1079		1123

	Image Pixel	Video Pixel
Left	0	280
Center	959	1239
	960	1240
Right	1919	2199

TABLE 5.15-7
1280 × 720 Image Alignment in a 720p HD Video Raster

	Image Line	Video Line
Top	0	26
	...	
Center	359	385
	360	386
	...	
Bottom	719	745

	Image Pixel	Video Pixel
Left	0	370
Center	639	1009
	640	1010
Right	1279	1649

Note that the top line of the 525i image maps to line 283 of field 2. This is referred to as *even field dominance*. On the other hand, in the 1080i system the top line of the image maps to line 21 of field 1, or *odd field dominance*. Even if all the correct lines are populated it is possible to map the image to the raster with the wrong field dominance. In that case interlace structure gets inverted. The resulting artifact can be easily recognized as a field dominance problem, as shown in Figure 5.15-8.

In the case of a moving image, another artifact can occur when there is a field dominance problem. If a frame of video is rendered showing an object moving from left to right, field 2 is supposed to occur later in time than field 1 so the object is shifted to the right. If this frame is mapped to the raster incorrectly, the field 1 information may end up on field 2, and vice versa. During playback this object will appear to have jumped backward for a moment and the overall motion will appear to stutter.

Another type of field dominance problem occurs in progressive segmented frame (24 psf) systems in which the video is interlaced but there is no motion between the two fields in the frame. By specification, field 1 and the next field 2 are supposed slices of the same static frame. If the graphics system inadvertently renders a frame into field 2 first and then the next field 1, the image will look identical on interlaced display monitors because it is not possible to tell which field displays first. However, signal processing downstream can cause problems. For instance, if the misaligned 24 psf signal converted to 30 frames per second using a common 2:3 pulldown technique, as shown in Figure 5.15-9. The resultant video will stutter in an unexpected way. If 24 psf is deinterlaced (e.g., when down converting to SD), the resultant signal will have needless interlacing artifacts because each frame is comprised of fields that have motion between them.

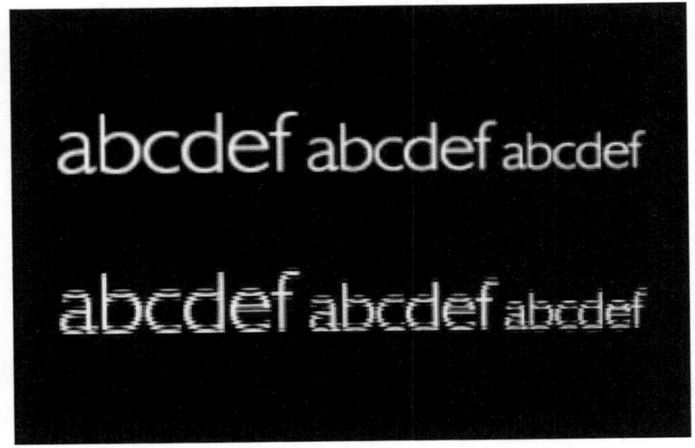

FIGURE 5.15-8 Field dominance artifacts.

Safe Title Area and 4:3 Protect

Television graphics design tools will usually show a safe title area. This is meant to keep important graphic information out of the area that might not display on an overscanned picture tube monitor or TV receiver. SMPTE RP 218 defines the safe title area as the central 80% of the screen height and width. In theory, modern flat panel displays do not need to allow for this overscan area and this should be less of an issue today. In practice, many new flat panel consumer displays available in the marketplace still exhibit significant amounts of overscan and it must still be expected that the whole raster will not be displayed. Some broadcasters, especially graphics-intensive news channels, are pushing their graphics well out of the safe title area. RP 218 also defines a safe action area as the central 90% of the screen.

Figure 5.15-10 illustrates 4:3 protect, safe action, and safe title areas. A 16:9 1080i raster is shown with a 4:3 protect area inside for down-conversion to SD. The edges of the different zones are given in pixels relative to the upper left corner of the 1920 × 1080 image.

As HD is phased in, broadcasters often need to create content for both HD and SD simultaneously. And, as much as possible, they would like not to have to create every graphic twice. When repurposing graphics between HD and SD, the biggest problem is not the

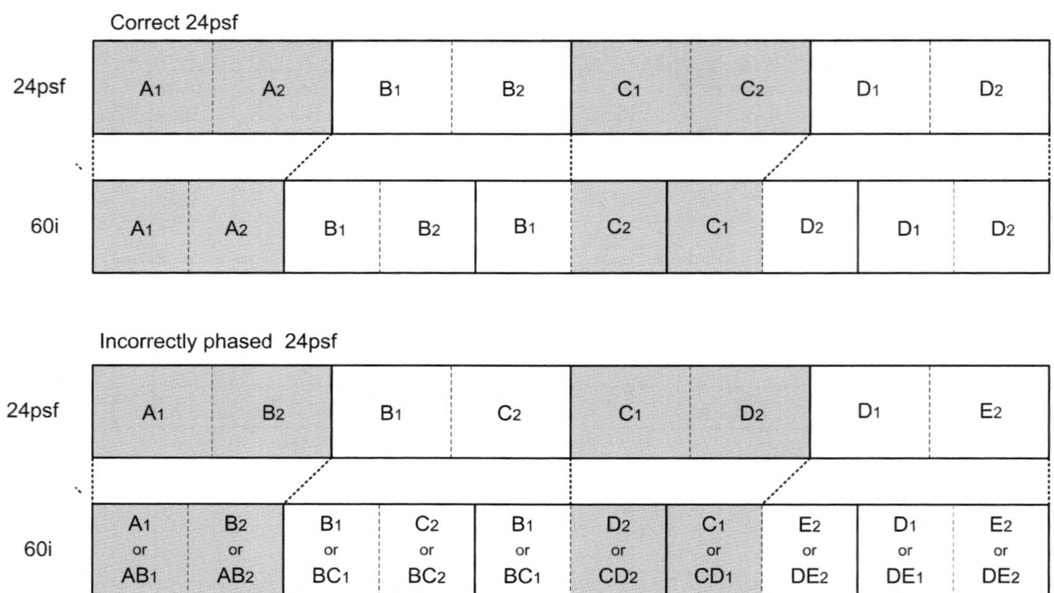

FIGURE 5.15-9 Misaligned 2:3 down artifacts.

resolution, since most graphics applications can easily scale an image up or down. The main problem is the different image aspect ratios: 4:3 for SD and 16:9 for HD.

A popular solution to this problem is to create the graphics templates in 16:9 with a 4:3 protect region. All important information, images, and text, are kept in a 4:3 area in the center. Backgrounds, lower-third banners, and other similar graphic elements can be extended to the end of the 16:9 area. When displaying in 4:3, the CG is programmed to crop the unused side panel area. A similar technique is used when HD camera shots will be downconverted for SD use.

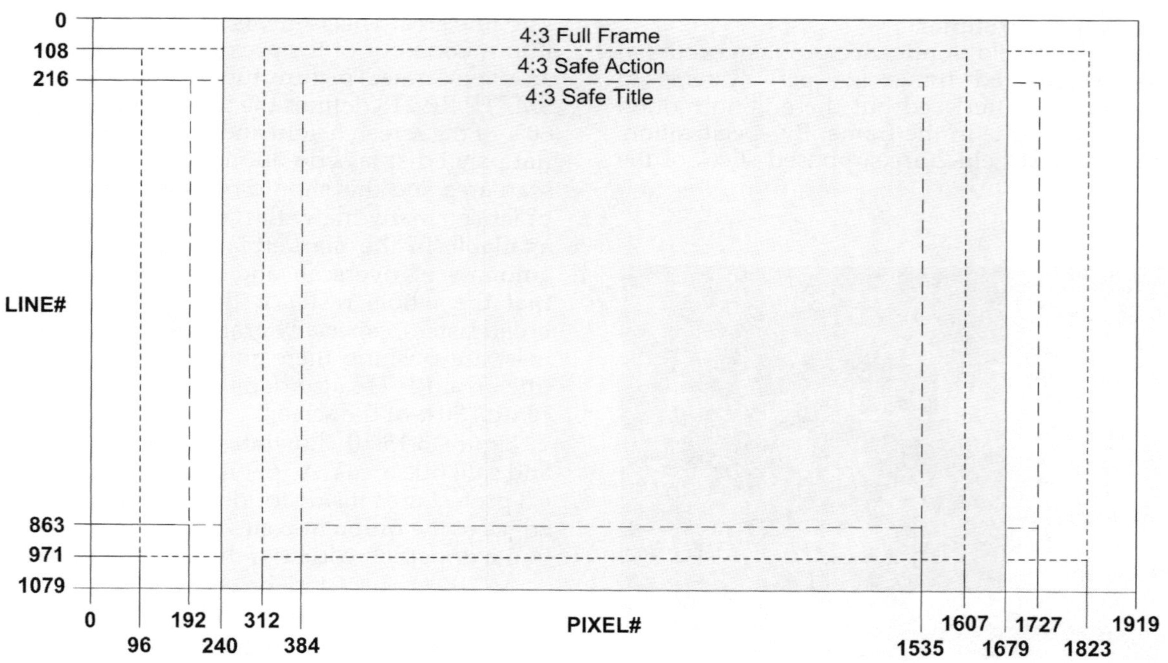

FIGURE 5.15-10 4:3 protect, safe action, and safe title areas.

Video Mixing and Shaped Video

A common problem that occurs in graphics production is the handling of premultiplied (or shaped) video in a downstream switcher. Since most switchers normally assume that video is not shaped this can lead to muddy or washed-out colors and incorrect blending of transparent backgrounds. Why this happens and some tips on correcting it are discussed below.

Figure 5.15-12 shows the functional blocks typically used for keying and mixing in a production switcher. Assume the white character is to have 50% transparency against the background. The key signal multiplies the fill (CG output), and the program is multiplied by (1 – key). The outputs of the two multipliers are then added together. The assumption here is that the video level is not changing at all, but that the blend is controlled only by the key level. This assumption was true in the past, but for reasons described below, many modern graphics devices output video that is already premultiplied by the key signal as shown in Figure 5.15-11.

In this case, the production switcher should not multiply the fill by the key. Most production switchers allow for *shaped fill* in their setup menus, though each manufacturer calls it something different. It can be referred to as clean mode, linear key, or additive key mode. The correct block diagram is shown in Figure 5.15-13.

Unfortunately, some older production switchers and many master control switchers do not have this mode. If the production switcher does remultiply the shaped fill signal the resultant effect is a reduced amplitude fill that is often described as looking "muddy." This is shown in Figure 5.15-14. The muddiness is due to a double multiplication of the fill by the key signal, once in the graphics device and again in the switcher.

If the production switcher does not allow for a shaped video input it is necessary to "unshape" the video signal. This is done by multiplying the video signal by 1/(1 – key signal) with an external unshaper, as shown in Figure 5.15-15.

The fill is premultiplied because in high-quality character generation and other graphics devices, linear keying has another important use, which is anti-aliasing character edges. This technique has been used since the late 1980s. Edges of characters are blended into the program to give the illusion of higher resolution than are possible by the number of pixels in the

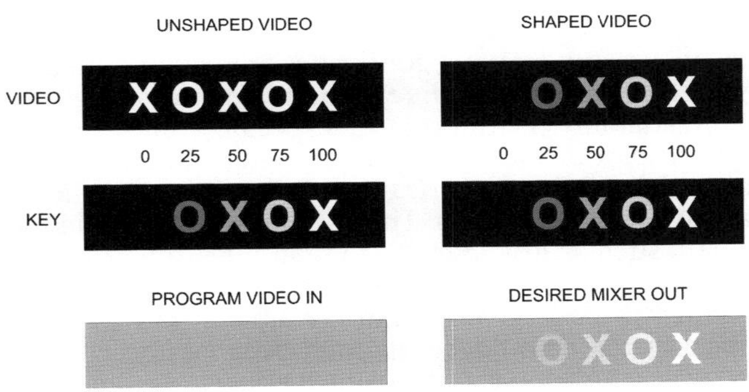

FIGURE 5.15-11 Shaped video.

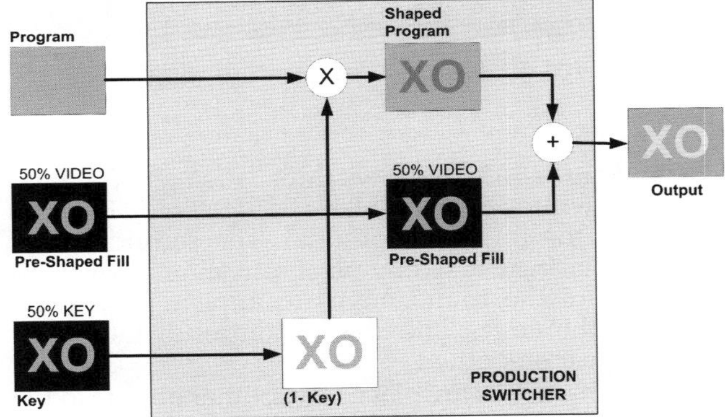

FIGURE 5.15-13 Correct mixing of shaped and unshaped video.

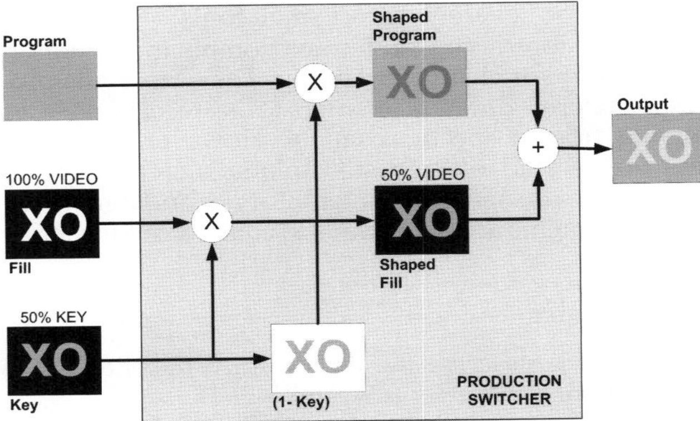

FIGURE 5.15-12 Production switcher keying/mixer functional diagram.

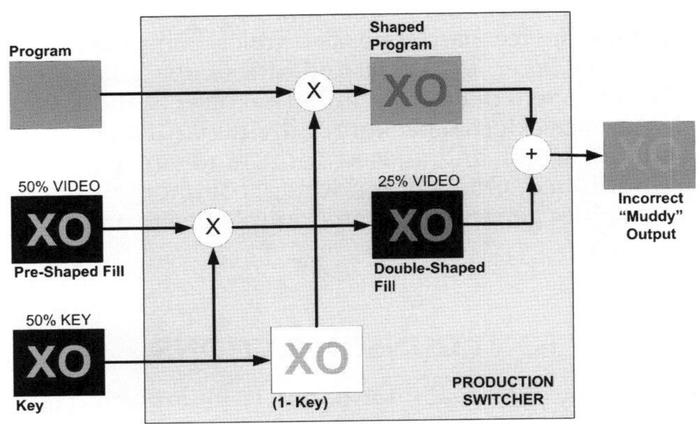

FIGURE 5.15-14 Incorrect mixing of shaped video.

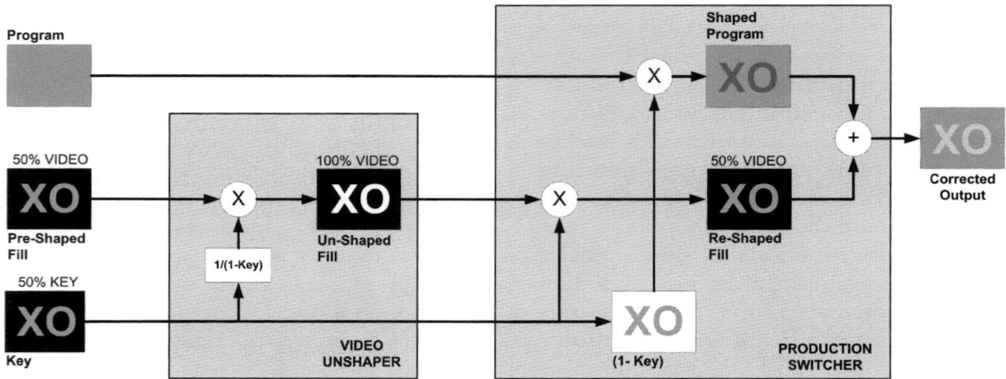

FIGURE 5.15-15 Video unshaper.

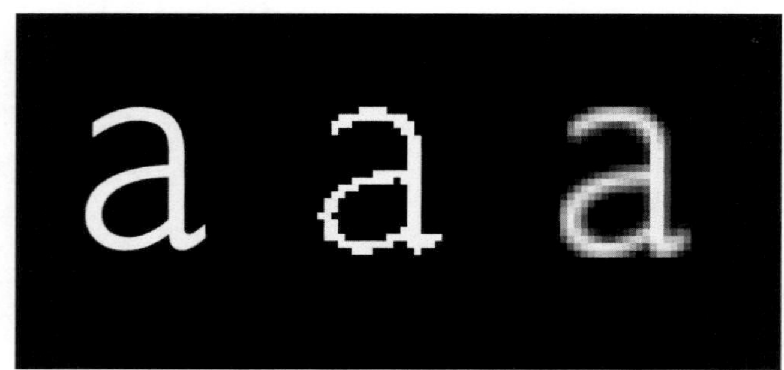

FIGURE 5.15-16 Font anti-aliasing.

format. Figure 5.15-16 shows three versions of the letter "a" in white against a black background. The first version is the unaliased original. The second shows extreme aliasing caused by pixel sampling. Pixel sampling causes the round edges to appear jagged. By blending the edges into the background, the object edges appear smoother.

When characters are placed over other objects in the CG scene, the blending is done internally in the frame buffer. But, when the characters are placed over live video, the blending must be done in the production switcher. The graphics device will generate the appropriate key signal to make this happen.

Unfortunately, the scene building process in the graphics device hardware assembles components of the scene layer by layer and does not distinguish between layers placed over internal objects and layers placed over external background video. All objects are rendered in an anti-aliased form, and are therefore premultiplied. OpenGL rendering, commonly used in 3D graphics generators, is an example of a process that does this.

FONT CONSIDERATIONS

The display of textual information is an important task of broadcast graphics systems. Character generators are highly specialized devices that have evolved over the years. Much of the concepts and terminology in character generation come from the print industry, which has been around since the fifteenth century. Today, the print industry, like broadcast, has moved to software-based platforms so there is more overlap than ever. This has greatly increased the choices available to the broadcaster.

Fonts are sets of alphanumeric characters of a particular design. Arial and Times New Roman, the basic Windows sans serif and serif fonts, are examples familiar to PC users. A simple search on the Web can find thousands of font styles. New fonts are being designed every day for a variety of applications. In television applications, font styles are usually selected with two goals in mind: readability and branding.

Readability is important, especially in relatively low-resolution SD systems. Television fonts are usually plain and geometrically simple. They are almost always sans serif fonts. They should not be made too small, such as less than 10–15 scan lines, and are usually larger. The color of the font should contrast well against the background and certain saturated colors, such as red, should be avoided because the chroma crawl distortion of composite displays would make the font unreadable. Very thin fonts, on the order of one or two pixel widths, should be avoided because they will not display well on a soft display and will suffer from interlace artifacts. Video compression artifacts can also severely distort small fonts.

Unique font styles and color combinations are an opportunity for channel branding. They can make program material different enough to be recognizable as coming from a specific channel. However, finding a font that is readable and still unique is a challenge to the graphics designer. Mixing font styles and distinctive edge treatments are other ways to create a unique look.

Font Metrics

Font metrics describe the important dimensions of a font and are illustrated in Figure 5.15-17.

Point size in print is the overall height of a character set, given as the distance from the highest ascender including any diacritics (i.e., the top of a capital Ž) to

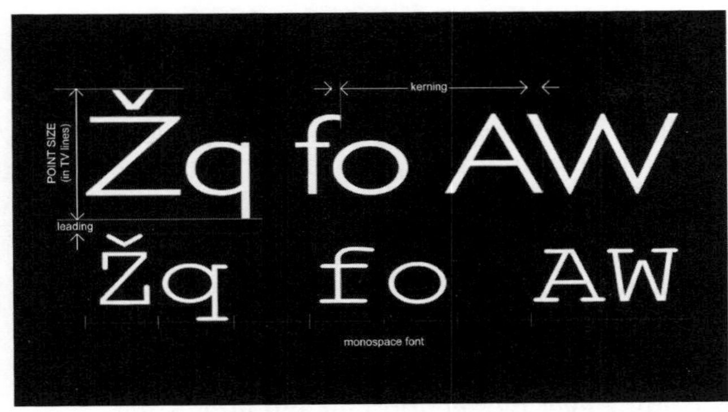

FIGURE 5.15-17 Font metrics.

the lowest descender (i.e., the bottom of a small q). Each point is approximately 1/72 of an inch when printed on paper. In television systems, the size is given in scan lines rather than points and is independent of the actual display size. Note that a given font size in scan lines will look smaller on an HD screen because of the number of scan lines per picture height.

Leading describes the spacing between lines, and the term is derived from lead strips that were placed between rows of text on a printing press. Adjusting the leading will change the spacing between the rows of text.

Kerning is the distance between characters in a word. Kerning is a complex topic because of the different widths and shapes of different characters. To make the printed text look balanced, characters are often nested together by different distances depending on their relative shapes. For instance, in the character pair "fo," note that there is a slight overlap of the character f over the o. The kerning of the "AW" character pair is even more pronounced. Kerning tables are used to determine the amount of kerning between certain character pairs. A monospace font, such as Courier, is an example of a font that does not account for the width of a character (that is, i versus m) and the space allowed for each character is the same. Character generator software will usually allow the operator to change the kerning between characters to achieve a more pleasing look.

Fonts used by most character generators are supplied as OpenType, TrueType, or PostScript files. These font file types have minor differences but all describe the font using geometric equations that can be scaled to smooth bitmaps of any size. They also contain hints for optimizing the rendering at particular sizes, and the kerning tables described above. The rendered character image is often referred to as a glyph.

FOREIGN LANGUAGE SUPPORT

Foreign language support is increasingly important for globally minded broadcasters and certainly cannot be ignored by the graphics systems vendors. It is worth reviewing character encoding, the Unicode standard, and other terminology surrounding extended foreign language character sets.

Character Encoding

The simplest and most familiar character encoding system is ASCII (American Standard Code for Information Exchange). ASCII was originally limited to seven bits, or 128 codes, the eighth bit being reserved for parity. These 128 codes were sufficient to represent the uppercase and lowercase letters, numerals, and punctuations used in the English language, plus the 32 control characters used by early printers and teletypes, but little else.

By enabling the eighth bit, IBM created the Extended ASCII character encoding set. The extra 128 codes (128 to 255) were used to support accented characters for European languages, mathematical symbols, and simple graphic elements. Other mappings for the upper 128 codes were then developed by different vendors that supported other languages. These are referred to as *code pages* and are identified by their code page numbers, such as code page 850 for multilingual Latin and code page 737 for Greek.

In DOS machines, OEM code pages were literally loaded into the output display hardware as bitmaps. Later, in Windows, ANSI code pages were used to render the fonts in software. ANSI code pages are still used today in non-Unicode applications. The correct code page must be installed and selected in order to display text properly.

For Asian languages, often referred to as CJK languages (Chinese–Japanese–Korean), even 256 character codes are not nearly enough to represent the tens of thousands of characters needed. Regional double-byte character sets (DBCS) were developed that use two bytes per character code, for a total of 64,000 codes. Common DBCSs are Big5 and GB. Big5 is the "traditional" Chinese character set used in Taiwan and Hong Kong. GB, or Guobiao, is the "simplified" Chinese character set used in mainland China.

Another related issue to foreign language support and extended character sets is the keyboard input method used. The world has standardized on the 101-key PC keyboard (more or less). For "western" character sets that fit in a 256 table, it is possible to assign each character to a key plus some control characters (shift, alt, control, etc.). Keycaps can then be made that represent all the characters, and most languages have fairly standard keyboard layouts that correspond with their character sets. CJK languages, however, cannot possibly have a unique key assigned to each character. Therefore, keyboard input methods have been developed that make it possible to enter any character using a standard keyboard and some logical combination of keystrokes. For instance, the *Pinyin* keyboard input method uses the English keyboard to phonetically spell a Chinese word. An alternate input method, called *Wubi*, is stroke based. Keys on the keyboard represent shapes and pen strokes that would be used to build the character when drawing it.

Unicode

The net result of all the incompatible extensions to ASCII was that it became difficult or even impossible to get a document to display correctly if you did not have the correct font. It may not even be obvious which language or character set the document calls for. The Unicode standard was developed to solve this problem by uniquely representing all characters for any language in a single character-encoding set. Most modern operating systems and software will support Unicode character encoding from input to output. Unicode fonts are available, which in theory support every character you need. There are, however, even in Unicode, some limitations and incompatibilities that one should be aware of, especially when passing data from one system to another.

Unicode comes in three encodings: UTF-8, UTF-16, or UTF-32. UTF-8, which can have from one to four bytes per character, is commonly used in web pages. For English characters only one byte is used, so UTF-8 is compact and compatible with legacy ASCII-based software. UTF-16 has a minimum of two bytes per character and is commonly used where a fixed character width has more benefit than a smaller size. UTF-16 is most common in Windows and Java-based applications. UTF-32 is very rarely used.

When transmitting double-byte Unicode data from one system to another, you must consider the *endianness* of the data. Big endian systems store or move the high-order byte first. Networks usually are big endian. Little endian systems store or move the low-order byte first. Personal PCs are little endian. This incompatibility makes it possible to switch the order of the bytes and leads to bad data. UTF-16 Unicode data has a special sequence at the beginning of every string called the Byte order mark (hex FE-FF) to check the endianness of the data.

In theory, a Unicode font could have a character glyph corresponding to every Unicode code point. But most do not, usually specializing instead in a particular language. They will generally leave the unused code points blank, so if you use the wrong font, you will see a lot of blank text. The CJK languages share many of the ideographs, or Han characters, but draw them differently. When Unicode was developed, the common CJK Han characters were given overlapping code points. One font set cannot properly display all CJK characters as they should look in each language. For these reasons, it is still important to have the correct font for the language you are using.

Many other languages present other challenges. Right-to-left languages, such as Hebrew and Arabic, require bidirectional text processing, especially when mixed with numbers and English characters that are typed left to right, A standard crawl, common in television, needs to crawl from the left edge of the screen to the right in order to reveal the text correctly. Complex script languages, such as Arabic and Indic languages, have ligatures that connect characters in a word and each character can change its shape depending on its location in the word.

SUMMARY

Today, television graphics sit at the intersection of the two rapidly changing technologies of computer graphics and television. Computer graphics technology changes daily, being driven by consumer video gaming and imaging products and the creativity of the graphics artists. Television technology, which was once relatively stable, is going though dramatic changes due to new compression and distribution techniques, from very high-definition electronic cinema to very low-definition podcasts and mobile television. One thing is certain, that the role of graphics is becoming ever more important and interesting.

Through all these changes, the broadcast engineer must ensure that these new graphics devices meet the high level of quality and precision that the viewer expects, that they fit well in the production workflow, and that they interoperate correctly with the other equipment in the broadcast plant. This chapter has sought to give a brief but useful overview by focusing on some key concepts—the basic architecture of the various graphics devices, their similarities and differences, how they interface with other devices in the production chain, many of the common terms that you will encounter, and how to spot some of the pitfalls when integrating graphics systems into broadcast facilities.

CHAPTER

5.16

Digital Asset Management

ROD FAIRWEATHER
Harris Corporation
Thames Ditton, United Kingdom

INTRODUCTION

Digital asset management (DAM) systems come in a range of sizes and capabilities. This chapter discusses some of the options that should be considered when deciding to implement a DAM system and ways long-term benefits can be achieved.

A digital asset consists of the core content (*essence*) plus additional metadata. In the broadcasting industry digital assets are rapidly growing. Program video and audio files, low-resolution copies, scripts, subtitle files, audio files, press photos, logos, and lower third *supers* are examples. From a computer's point of view, each of these is just a digital file.

Where DAM differs from most sections of this book is that the topic is not related to broadcasting alone. Many other industries are facing similar problems that need to be addressed. As examples, the medical world has massive and growing amounts of digital assets, including diagnostic records, patient records, drug advice for medical professionals, training videos, and so on. The aerospace industry has a mounting archive of design drawings, manufacturing records, photographic references, training videos, trial results, and other records. It would be difficult to find a major industry sector that does not have to deal with growing volumes of digital assets.

Broadcasting has moved steadily toward digital assets since the early 1990s, as it became faster and more cost efficient to work within the digital domain using new digital video and audio technologies. Traditionally, expensive dedicated hardware-based systems such as edit suites moved to computer-based nonlinear editing, not just because of the benefits of working in a nonlinear fashion, but because the reduced infrastructure costs became a more compelling factor even as the technology improved.

Creatively the digital domain has opened up many more possibilities that artists have exploited, despite being traditionally technology adverse. Once the capability and speed of operation of graphics devices became known, the graphics community drove the industry forward to developing more capable applications at lower costs.

Print-based organizations have used asset management systems to manage their documents and image archives, along with rights, for many years. Since print preparation moved to computer applications, it became a much more efficient process particularly when used in conjunction with DAM systems, speeding production and raising technical quality. One result of this transition was that it helped the magazine market to rapidly grow, as short-run magazines could be produced at a market-sustaining cost.

Developments in the print media world also helped focus DAM manufacturers' effort on the broadcasting industry. When we look to the future, there may be other industries and new markets that will capture the attention of the broadcast industry's favorite suppliers for DAM systems.

DRIVERS IN THE BROADCAST INDUSTRY

The rapidly growing numbers of digital assets are causing organizational problems for broadcasters. Research has shown that creative people in the industry may look for media files perhaps 80–100 times a week and, without proper support systems, they may fail to find what they are looking for as much as a third of the time.

However, simply managing these assets so they can be easily searched and located is not the only driver for the movement to enterprise DAM systems. Business opportunities for broadcasters are changing rapidly as delivery channels increase and consumers adapt to an increasingly digital world with different expectations from traditional television and radio for program delivery.

Figure 5.16-1 outlines some of the major changes taking place in the industry and identifies the future direction we are traveling. Each of these developments requires changes to our current working practices, which will not survive in the new competitive environment without rapid evolution.

With the move toward more distribution paths that are now available at lower costs, the sheer number of channels will likely increase further. Broadcasters will need to identify specific markets, deliver content with pinpoint accuracy to those niche sectors, and be able to prove to advertisers that the targeted audience was achieved. This means having greater knowledge about the audience and their behavior patterns.

Being able to achieve these tasks also means tailoring the content to those audiences. This is not just the primary content (traditionally regarded as programming), but also the revenue drivers, particularly commercials and convenience of delivery to the preferred platform.

The expanding number of platforms is changing the business of broadcasting in two ways. Not only is there a need to deliver differing formats of content, but there is a fundamental shift in the broadcasters' portfolio. The lines between telcos, Internet service providers, broadcasters, and content owners and creators is blurring rapidly, as these traditional businesses merge or collaborate to create combinations of content and control of the method of distribution. In particular the move is toward multiple "play" services, where a single provider offers packages of TV, Internet, landline communications, mobile and handheld devices, and also the program content carried.

Methods of streaming will continue to change as new technologies become available, in particular local storage capabilities. End users are not concerned whether a program arrived as a trickle feed, real-time feed, or faster than real-time file transfer. They simply want the right content on the right device at the right time.

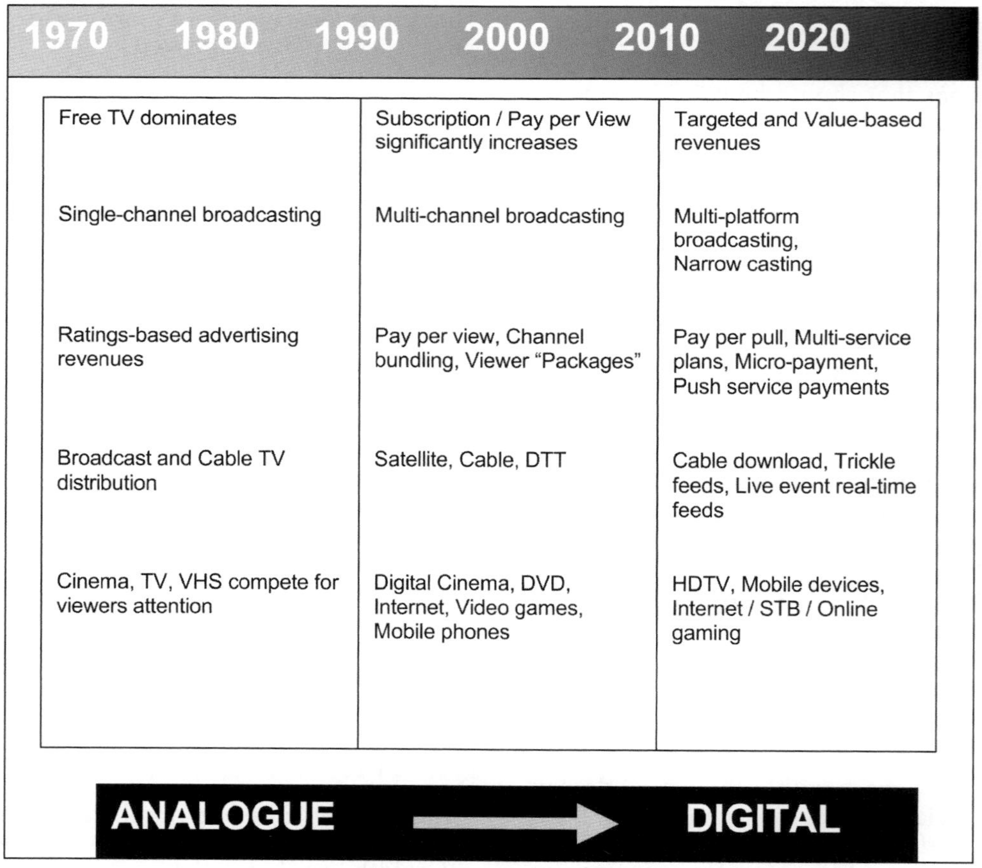

1970 1980 1990 2000 2010 2020

Free TV dominates	Subscription / Pay per View significantly increases	Targeted and Value-based revenues
Single-channel broadcasting	Multi-channel broadcasting	Multi-platform broadcasting, Narrow casting
Ratings-based advertising revenues	Pay per view, Channel bundling, Viewer "Packages"	Pay per pull, Multi-service plans, Micro-payment, Push service payments
Broadcast and Cable TV distribution	Satellite, Cable, DTT	Cable download, Trickle feeds, Live event real-time feeds
Cinema, TV, VHS compete for viewers attention	Digital Cinema, DVD, Internet, Video games, Mobile phones	HDTV, Mobile devices, Internet / STB / Online gaming

ANALOGUE → DIGITAL

FIGURE 5.16-1 Changing market conditions and opportunities in broadcasting.

DIGITAL ASSET MANAGEMENT SYSTEMS

In this changing and challenging world, there are two main types of digital asset management systems: smaller, local workgroup systems and systems spanning a whole enterprise.

Workgroup DAM

Workgroup DAM systems tend to concentrate on a limited range of benefits, as shown in Figure 5.16-2. Typically, they remain within a single working group that needs to share or access common material. As a project, workgroup DAMs are fairly straightforward to implement, since they have to handle only a limited range of requirements.

In the example of a workgroup DAM system shown in Figure 5.16-3, there is a limited range of potential users, and a defined range of assets is being managed, usually within a small number of operational groups. There are real advantages in working with limited scope projects, particularly in areas that are well understood by niche suppliers.

Workgroup applications usually require limited external assistance for successful implementation, primarily because limited workflow changes are achieved. Commonly, it becomes possible to share material or have faster access to newly received content, which is particularly useful in a news environment where several editing suites may be working on the same story or different packages for major breaking news.

However, workgroup solutions rarely address many of the changing business requirements that broadcasters face. To tackle these issues and to make significant changes to the economics of the upcoming requirements require an enterprise approach to digital asset management.

Workgroup DAM Benefits	
Project	• Easy to Define Scope • User roles clearly understood • Limited interaction with adjoining departments • Application(s) can be seen complete before purchase, so high confidence in suitability • Material quantities understood and trends identified
Price	• Predictable • Usually very little custom development (which is expensive) • Cost of scaling well known • Training costs low
Implementation	• Whole project duration predictable • Limited impact on physical installation • Training is usually short due to limited functional scope • Benefits easy to identify and quantify

FIGURE 5.16-2 Workgroup DAM benefits.

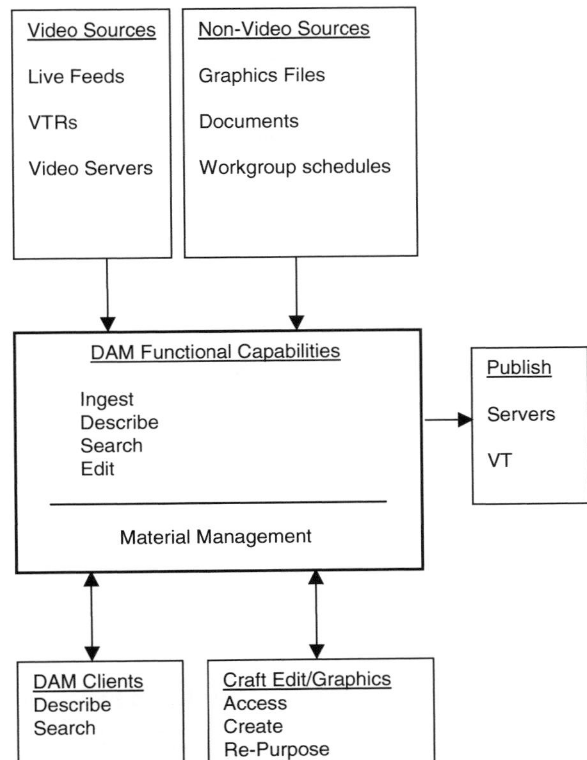

FIGURE 5.16-3 Typical Workgroup DAM system.

Enterprise DAM

Enterprise DAM systems can be complex and expensive—a disincentive for introduction at a time when many broadcasters are already making significant investments for the digital transition across much of their material gathering and infrastructure capabilities. However, a successful DAM project can drive recurrent costs down and increase revenue, enabling some of those infrastructure replacement costs to be recovered much sooner.

Time is a big cost factor for broadcasters. Whenever humans have to spend time performing a function, major costs are incurred. With the increase of channels and platforms and the fragmentation of the market (often resulting in declining viewership per channel), broadcasters are looking to see how they can achieve "more with less."

A common error when looking at an enterprise solution is to start from the wrong place. Suppliers often hear that a broadcaster client is looking for a technology solution that will "support and distribute all its digital assets." While not a surprising viewpoint because this may be the primary problem that triggers the desire for a DAM solution, the result may be less than perfect. A full analysis of requirements is needed.

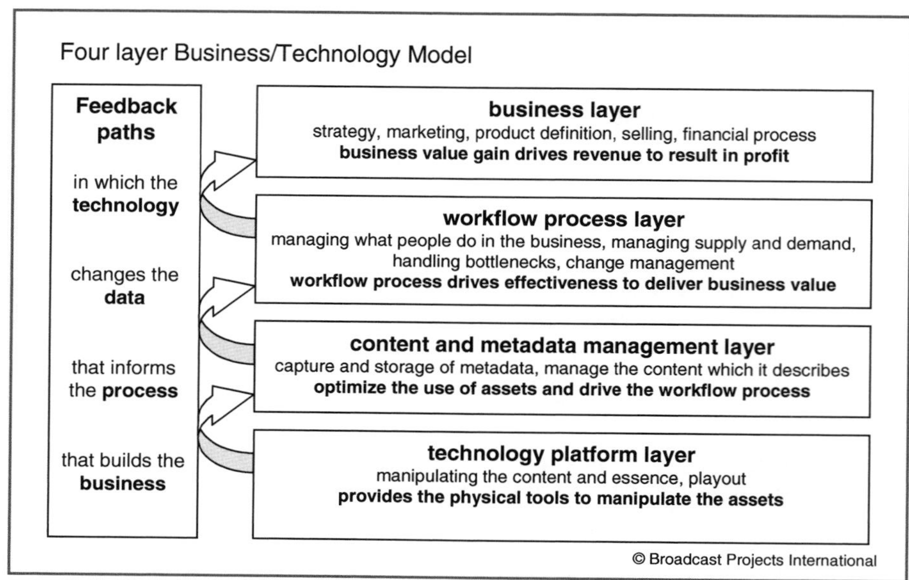

FIGURE 5.16-4 Four-layer DAM model. (Courtesy of Broadcast Projects International.)

Broadcast Projects International[1] has developed an interesting four-layer model that has proved helpful for a number of major clients looking to achieve a step change in efficiency and capability. This model, shown in Figure 5.16-4, helps clients think first about who they are and where they really want to be. What are the services they intend to supply? How will they make money? Who are their markets? What is the compelling content that will support their long-term business proposition? Understanding at an early stage how revenues will be realized and processed is vital before designing enterprise systems, and will help identify applications and skill sets necessary to achieve the outcome.

When we examine the business layer in more detail, a number of benefits will be achieved using a DAM system. They should go well beyond achieving workflow efficiencies of current production processes. Figure 5.16-5 outlines some examples of the benefits that broadcasters may expect to realize.

The advantage of thinking top-down is that it helps define a system that delivers the overall business objectives, many of which reach beyond single department views or technology benefits.

Content that is already owned by a client can be reused either to create additional content (new programs) or repurposed for different platforms. For this to work, content must be thoroughly described and managed within the DAM system. One of the fundamental concepts behind DAM systems is the mantra: If it can't be found, the content has no value. This concept is discussed further later in this chapter.

Content is also of value to other content providers. If another program maker can reduce its production costs by using high-quality material that already exists, its cost and time to delivery are reduced. Therefore, visibility to the external world is obviously important, as is an efficient and secure process to deliver the material with usage rights. Reduced program acquisition costs benefit all broadcasters, who are constantly looking to drive down their programming production costs.

The newer delivery platforms are a potential revenue source for personalized services. Perhaps the most

[1]Broadcast Projects International website: http://www.bpi.uk.com/home.asp

Business Benefits to Be Achieved	
Revenues	Increased by re-selling of content to other content suppliers
	Increased by providing new platform services that provide compelling value to the consumer
Costs	Reduced by improved workflows on production processes
	Reduced by effective re-purposing and re-use of content on in-house production
Flexibility	Ability to respond to new distribution opportunities
	Ability to reach targeted market
	Ability to integrate to additional revenue applications (e-commerce)
Competitive Advantage	Integrated systems working across multiple revenue streams enable more agile response to revenue opportunities

FIGURE 5.16-5 Examples of desired business benefits to be achieved.

obvious example is sports highlights, where consumers can request rapid delivery of sports action from a broad range of sports, or perhaps highlights of their own team. With the increase of the range of mobile devices and the quality of the consumable product (improved video quality), this has been identified as a major opportunity for rights holders. News and reality shows are also proving to be useful revenue streams when personalized services can be implemented in a cost-effective fashion.

Recent years have seen rapid technology changes. While the "killer" technologies and applications won't be known for a few years, it is certain there is more change on the way. This drives the requirements of the DAM system toward open architecture using industry standards since additional applications will almost certainly need to be integrated over time.

As distribution technologies open up, the trend is from a broad audience for a program to a "market of one." This will require broadcasters to know a great deal more about each individual audience member, how to deliver directly to each consumer, and what the consumer regards as compelling content. This may be at odds with concerns about privacy and consumer protection groups that are trying to prevent companies and organizations from holding too much personal data. However, broadcasters operate in a competitive environment, and DAM systems will initially become a separator, providing significant competitive advantages to early adopters.

WORKFLOW

Workflow is an area where significant business improvements can be made but is also where many broadcasters do not yet capitalize effectively on their DAM systems. The primary reason is that very few personnel work across multiple business units, so benefits of enterprise workflow improvements are rarely understood from the outset. To benefit from workflow efficiencies, new workflows must be designed and implemented. "Workflow engines" help drive new workflows, by being aware of the status of tasks and schedules, and automatically triggering next actions based on predefined decision criteria. Such workflow engines can use DAM systems as an infrastructure to drive the processes.

When discussing requirements for a DAM system and workflow definitions, users tend to ask for an electronic version of what they do today. Incorrect definition of system requirements is not the fault of the users, but system analysts should be looking to see where overall system benefits can be achieved.

This discussion raises an important question of when broadcasters should bring in external assistance for major projects. DAM projects affect business, technology, change management, and project management, which are transformation resources that many (even large broadcasters) do not have in-house. While DAM suppliers have technical and application knowledge, business transformation experience may be better provided by independent specialist consultants, who may be better equipped to help identify and implement major workflow improvements. Clearly, there will be different workflow benefit expectations identified in different organizations for the introduction of a DAM system, but the examples in Figure 5.16-6 give guidance as to the sort of benefits that need to be analyzed before a system is designed.

Material movement through a production process is clearly easier to achieve in the digital domain. Not only does quality remain consistent (assuming the engineering of the system is of a sufficiently high standard), but speed is a great deal faster than with traditional tape-based analog methods. More important than the electronic transfer of the digital content is the alignment of the work to be completed. Ideally, when logging on to a work station, an operator should be able to immediately identify the most urgent work. Typically, this is where integration with scheduling systems (either channel management or resource management) is useful, so the highest priority work can be completed first.

Defined workflow processes, particularly when combined with a workflow engine, would then see the automatic movement of material to the next person in the production line. This might be, for example, a senior graphic artist checking and accepting the work of a staff graphic artist. This is similar to the principles employed in document management systems, which have been implemented across many enterprise organizations.

Workflow Benefits to Be Achieved	
Reduced operational costs	Smooth and Efficient material movement throughout a production and transmission process
	Reduction of bottlenecks through clear reporting
	Reduction of volatility
	Removal of expensive, time consuming non-productive processes
	Exception based reporting
	Single sided Quality Control flows
Reduction of Errors	Removal of data re-entry
Improved time of delivery	Agile capability enabling later / quicker decision making
	Visibility of material enabling collaborative working practices
	Parallel working enabling reduced timelines

FIGURE 5.16-6 Examples of workflow benefits to be achieved.

DAM systems can be designed to track both quantity of work and progress through workflow processes. This helps identify bottlenecks and trends toward upcoming problems. For example, it might be found that one designer is able to consistently get through 10 tasks a day, while another achieves only 5 tasks. Alternatively, a trend over time may be found that an increasing amount of graphics work is awaiting completion, which would give a clear indication to management that additional resources will be required to successfully deliver all the items, or action must be taken to reduce the number of graphics requests.

Volatility is the biggest obstacle to efficient resource planning. If a production operation must have sufficient technical and staffing resources to cope with spikes in workload, then there may be considerable periods of time when those resources are not going to be needed. Process modeling has been successfully used by many broadcasters to predict where and when spikes will occur. If they know in advance when the peaks are likely to be, it may be possible to move other work through the system in advance to make resources available for the predictable surges.

However, there are always occasions when unpredictable and large spikes occur in broadcasting, particularly in news. The death of a world leader or a catastrophic geological event cannot be planned, and there must be processes available to cope with these situations. It will usually be more cost effective to have workarounds planned to cope with these events (for example, arrangements with rental companies for additional equipment) than to have invested in custom development to be ready for an event that happens once every three years.

Organizations that have not moved to enterprise DAM systems have spent a great deal of money on nonproductive processes. A clear example is the copying of material onto VHS or DVD for use by external companies or for internal approvals. Not only does this copying tie up expensive resources (VTR/video servers), but it is time consuming (human cost), has to be organized (additional logistic costs), and adds extensively to the delivery time scales (material has to be delivered and the results returned). Workflows should be identified that are able to achieve the same process result in a fraction of the time and at a fraction of the cost. Typically, this solution might be making a low-resolution copy of a piece of material and having it sent to a subtitling house, with the resulting subtitles returned as a digital file.

Current processes add zero value to the final content, and it's through the use of workflow improvement such as the one described above that the cost of DAM systems can be quantified and justified.

Exception-Based Reporting

The move to exception-based reporting is another fundamental shift in working operations. As an example, when checking that a series of items are ready for transmission, the operator will typically get a list of all the material that is required and start checking-off each item. The majority of items will probably have been processed properly, but because of the impact of the failure to have an item ready in the right place at the right time, everything has to be checked. If one in a hundred items is discovered to be missing, then the work currently being done to check everything is worthwhile—finding out that a program is missing in sufficient time to correct the error can save a lot of problems. However, the overhead of having to check all items that have gone through the workflow process properly is massive and adds zero value to the end product.

DAM systems can offer the benefit of "exception only"-based reporting, when combined with effective workflow management. In other words, the systems can be defined to report exactly what is *not* in the right place rather than what is. If material has flowed correctly through an entire workflow process, then there is no need to know about it. Human effort should be concentrated on the problem areas only. Exception reporting simply reports about the things that aren't right, not the ones that are.

Quality Control

Throughout production and transmission process chains, there are consistent repeated efforts at maintaining quality control. DAM systems can remove a number of these repeated steps, since it is possible to use technology to confirm material has arrived in exactly the same condition that it left the supplier or source. If the provider guarantees the quality of the product as sent, then this quality of service is inherently still there at the point of reception. Repeated checking of incoming material is therefore redundant. Avoiding the redundancy enables the broadcaster to remove a series of quality checks and to put the responsibility on the provider as part of the contract or service level agreement (SLA), resulting in significant savings of both hardware and human effort.

Error reduction is also a major saving through a DAM system. Errors are most commonly introduced by the inaccurate re-entry of data, which then causes incorrect processes to be triggered or for the wrong material to be accessed. Having data flowing without being re-entered all the way through the entire production and transmission chain reduces the likelihood and cost of data errors.

This discussion raises the importance of data entry at the earliest possible time, preferably electronically, and the benefit of data validation at the time of entry. If data can be rejected because it does not conform to system design, this can be flagged at the initial entry point, which is where the correct data is probably still available.

Process Visibility

An increasing amount of data is being captured and delivered from the front of the delivery chain. Ideally, cameras are connected to input devices that are controlled by production assistants, who can add data about material as it is shot in the first place. Date and time of the shots can be provided by the camera itself, but if additional data can be wrapped into the video (project, location, camera operator, shot number, good/no good, color filters, exposure, and other parameters), this information can be used and reused by everyone all the way down the production line. It is common for editors to look for cut-away shots to cover an edit, and shots classified as no good (N/G) are often a source for this material. Being able to quickly search for alternative shots reduces editing time and cost.

Improvements in time of delivery are usually based around improved visibility and consequent decision making. For instance, by being able to easily see the work that a graphics artist is doing, the editor can make rapid and better decisions about what work can be done at any particular time. If a graphic is almost complete, an editor may decide to wait a few minutes and get it dropped straight into the part of the production being worked on, enabling that section to be passed on for approval. However, if the graphic hasn't been started yet, then the editor would probably decide to move on to another completely different section of work that has all the components available to make effective use of the time and resource.

For productions with time constraints, having the state of the process visible can also help with decisions about how to complete the work within the allotted time. News is a classic example, where a graphic sequence may have been requested to help illustrate a story. If the item is due to go to air within a few minutes and the graphic sequence is not ready, an editor may decide to lay down general pictures underneath the voice track so the material can go to air even if it is not in the desired state. A program editor can then look at the edit and decide whether the urgency of the story justifies the less-than-perfect edit, possibly with a view to the graphics being dropped in later on for subsequent transmissions.

This improved visibility of work also helps creative resources to be used more effectively. Producers can monitor how edits, graphics, and music, for example, are progressing. If the work being done is not going in the direction originally envisaged, creative teams can receive feedback that too much effort has been spent producing unwanted work.

Workflow processing helps where projects involve many people working on the same piece of work. Consistent visual styles throughout a piece are important, and if a number of graphic artists are enhancing raw footage, it's helpful for them to see how the rest of the graphics teams are treating the work. The result is more consistent color balancing, detailing, and visual look throughout an extended piece of work.

Parallel working enables reduced timelines by running processes that would traditionally have been run in a linear fashion to happen at the same time. Typically, this may be color balancing, adding visual effects, and laying music on a piece of work at the same time. Also where one creative area would benefit from slight changes in others (e.g., background music might prefer a slightly longer shot), these changes can be requested at the same time that the music is being worked on. This collaborative approach results not only in a shorter time to delivery, but also an improved end result.

Other Benefits

A DAM will provide various workflow improvements based around the earlier identified business requirements. Another example would be defined workflow processes to enable delivery of content onto a series of different platforms.

In this case the broadcaster will also be looking for an automated process, using available metadata, that copies original material onto a different format and delivers it to an appropriate place for the alternative delivery methods.

This benefit also ties in with revenue processes. Consider a business plan to make program music available for downloading from the Web. In this case a process would be identified to capture some suitable images from the program, along with a high-quality version of the music, and have it delivered to an e-commerce system that is able to interpret the core content (name of music, composer, etc.) and the right metadata that travels with the content.

RIGHTS MANAGEMENT

The difference between broadcast rights and digital rights management (DRM) needs to be made clear from the beginning. When broadcasters buy the rights to a program, many conditions can be attached. They include the exhibition window, the number of runs, the territories that the rights are bought for, and fast rerun rights. The fast rerun enables many broadcasters to repeat a program within a short period of time (typically 24 hours) without incurring additional costs.

At the time of purchase, broadcasters also agree to the payment terms, which may vary from outright up-front payment, to staggered payments upon exhibition, and many other alternatives. Additionally, internally, broadcasters may value their remaining asset (that is, the rights to broadcast) according to how they run their business. For example, a broadcaster may value the first run of a program as the bulk of the contract value, with remaining runs decreasing in value for each play. Others may prefer to amortize the contract evenly over each exhibition. Due to the complexity of all these variations, most broadcasters use a program scheduling system (also known as a channel management system) to manage these complex rights and to maintain payment schedules as well as their inventory value.

It is possible to embed rights into a digital file—they are simply more metadata. However, since program files are typically copied to different servers, the program may play from one server, while a different copy is played from a different server to another channel. This can create a situation in which different rights are used up by different files (which are oblivious to the other transmissions going on).

Consequently, it makes more sense to manage the broadcast rights centrally, using the unique references back from the playout systems, with consumer rights embedded in the distributed content in circumstances in which the consumer platform is capable of interpreting this data.

DRM is a broad subject in its own right, but normally it concerns the management of distributing and copying digital content. It goes well beyond simple anticopying protection, providing information to the consumer regarding identification of the work and the rights holders and authorized use. It often controls single and multiple playing of the content, time windows in which the content can be played, as well as whether the content can be copied. DRM can also be linked to conditional access systems, where the purchase rights of the consumer may be checked and verified against the content.

The DRM rights embedded in digital content may well be derived from the rights purchased by the broadcaster as they apply to the content being delivered to the individual consumer. For example, in video-on-demand (VOD) distribution, an operator may sell to the consumer the ability to play a piece of material for the next seven days, which would need to be within the exhibition rights the operator bought from the original content provider.

Some DRM also deals with collection of play data, reporting back to central systems the usage that the content has received.

METADATA

Metadata is all the data that describes the program *essence*. There are many different ways to describe and organize metadata. Broadcasting is slightly unusual in that the core program data is often separate from the metadata, although there will always be a metadata reference to the storage of the essence. The reason is that the video files involved are large, and with high definition, larger still. The technology that stores these large files is not always appropriate for storing metadata, which commonly needs to be available quickly and across an entire enterprise.

Metadata has two main properties, *fixed* and *flexible*, that are important to understand, along with the benefits of each.

Fixed metadata is a schema that is predefined and used throughout an organization. The main benefit of a fixed schema is that it is easy to apply business rules, since specific data is stored in predefined places. In other words, when many applications are reaching into the metadata, they know exactly where the data is stored.

The downside is that it is difficult to change the metadata schema for additions that are found necessary over time, as more platforms, transactional methods, and other changes come along. Changing fixed metadata schemas may involve having to go back to the application development house to have the extra fields added.

In DAM systems that are based around "flexible" or "extensible" metadata, users can add extra fields at any time, but it is harder to put a fully integrated flexible metadata system together, as the fields where users store data may change over time. Rigid administration control is required to prevent the system from growing in a noncohesive fashion.

Agreeing on an enterprise metadata schema is an important stage during the development of a DAM project. Digital assets that are received into many departments must be mapped onto the internal schema. Achieving these mappings is not complex so long as the content of the mapping is correct. This means that even if metadata received from an external supplier is referenced by another name, it should be easy to copy the data into a field within the DAM system. However, the data must be consistently correct and match the purpose of the internal field.

There are many different structures that can be used for organizing metadata, and many standards have been, and continue to be, published. Adherence to industry-standard schemas makes passing to and receiving assets from external suppliers easier, but there are no standards yet that comprehensively cover all the needs of broadcasters.

One simplified method of thinking of the organization of metadata is shown in Figure 5.16-7. Organizing metadata in different business areas helps in understanding the purpose of the data. For example, a broadcaster may buy the rights to the use of a program; the core metadata would contain the basic information about the content, which will not change from that point on since it is the essence itself.

Additional metadata may be applied. For example, there may be an edit list to create a version of the program that is suitable for family viewing, or programs may be split up into different parts to allow for commercials to be played. All of these are applied to the core essence to make the applied version.

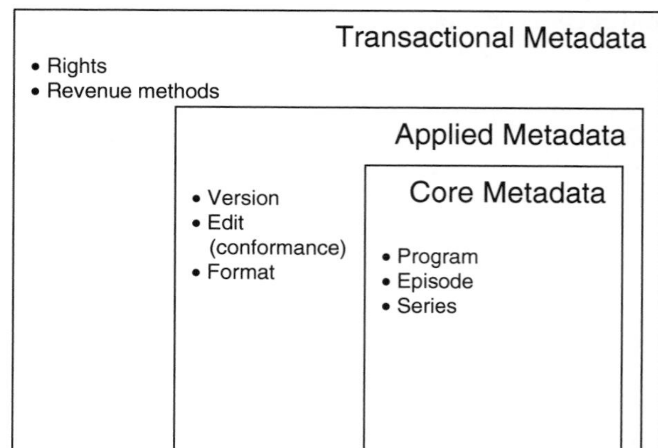

FIGURE 5.16-7 Simplified metadata model.

At a business level, all transactional data must be tracked and used to trigger various processes. Broadcasters may have the VOD rights for a particular window (the date range during which they can make the material available to consumers). The transactional rights (VOD with window conditions) must be carried around with the content and be available when the material is placed at locations where it can be "pulled" by the consumer. When the rights are finished, the material can then automatically be prevented from being downloaded by end users.

Metadata is not inherently complicated, but a structure is required for entering the information. From a user point of view, this can either be a list of fields that can be filled in or, if the DAM system offers a hierarchical structure, a selection of major headings with subheadings and details. This simplifies the organizing of descriptive data about different types of content. Opening up a heading in a hierarchical system shows the subheadings appropriate for that branch, and so on.

The following is an example of a hierarchical and "flexible" metadata structure. Suppose there is a library that catalogues sports feeds. If it receives, say, a soccer game, then the metadata might have a structure that looks like Figure 5.16-8.

This example is obviously a very basic version of the type of data that may need to be logged. Typically, this would be set up in a way to match the types of searches that a sports department would want to run. For example, "show me all the matches where Beckham is the scorer between date A and date B."

This sort of structure is fine, until the sports department starts receiving material from a completely different sport. If motor racing started to be shown, a completely different data structure, allowing the entry of data about the drivers, the final results, or perhaps major crashes, would be needed.

Being able to create new branches at any point in the descriptive metadata structure without having to

```
Football

    Team 1      ______
    Team 2      ______
    Stadium     ______
    Date        ______
    Result
        Team 1 __
        Team 2 __
    Goals
        Scorer      ________
        Timecode    __:__:__:__

    Major Events
        Players     ________
        Timecode    __:__:__:__
```

FIGURE 5.16-8 Simplified example of a subset of metadata structure.

go back for additional development is obviously helpful in this situation and is not a significant problem from a business rules point of view. It is also easy from an IT standpoint to generate XML files that show all the data that has been stored, even if it is not yet understood by any other application. At least this way, the data is being captured correctly from day one, even if work has to be done to be able to effectively use the data at a later time.

Data Types

Numerous different types of data can be entered into metadata fields, and the type is usually part of the definition of the field. Most users are familiar with the standard range, and limiting the data to specific types can have great benefits to an organization over the long term.

The most common type is "freeform," in which any letter, number, or character can be used. When used for large descriptions, which can be indexed to create search indexes, this format should be used sparingly.

Wherever possible, fixed lists should be used. An example would be the names of team members when entering data to describe a football game. If all the team names have been entered into a list (an administration task), then users select the one they want. This prevents spelling errors, abbreviations, etc., from being used, which becomes important later when the material is being searched.

Boolean values are usually available, which makes it easy for business rules to be applied from other applications or for workflow control. Date data generated from calendars should ensure adherence to proper date formats. Dates and time codes in particular formats are used in broadcasting. Dates must work across the world for multigeographical organizations, and time codes in particular must be handled carefully and validated upon entry. This has the obvious complication of different recording formats in different parts of the world.

Sensible labeling disciplines need to be applied. If a field description of digital content is "New Graphics," this will inevitably cause confusion in years to come. Descriptions must be specific and, if at all possible, unique. Figure 5.16-9 shows a screen shot of DAM metadata for a particular asset.

DAM CAPABILITIES

Having established the business requirements and the workflows that must be supported to achieve these business needs, it is necessary to look at the capabilities that are inherent to DAM systems and how external applications interact.

It is important to understand that the core of a DAM system doesn't do much by itself. Its purpose is to support operations, based on metadata, making content and data easily visible throughout an enterprise operation. DAM systems usually have applications (for

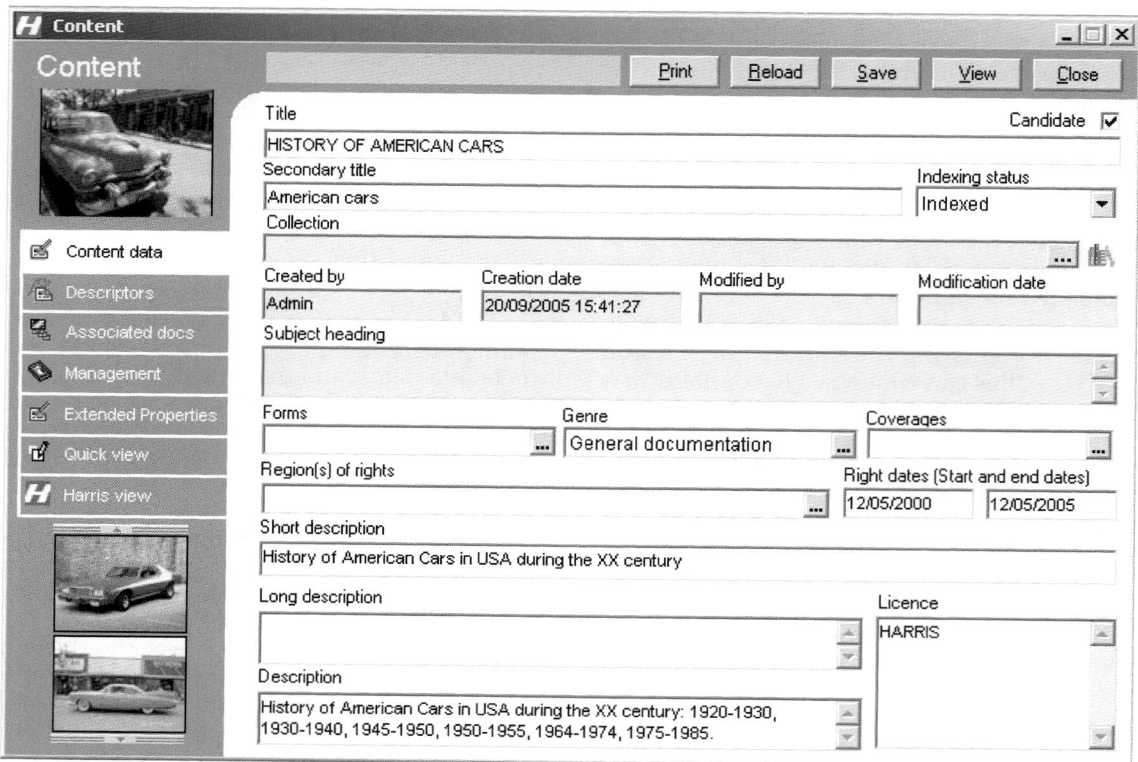

FIGURE 5.16-9 Sample representation of data within an asset.

example, edit suites, graphics machines, and other systems) attached that perform the functions required. The level and range of functions that can be supported will depend on the capabilities of the integrated system and their ability to work with the metadata as presented by the DAM system.

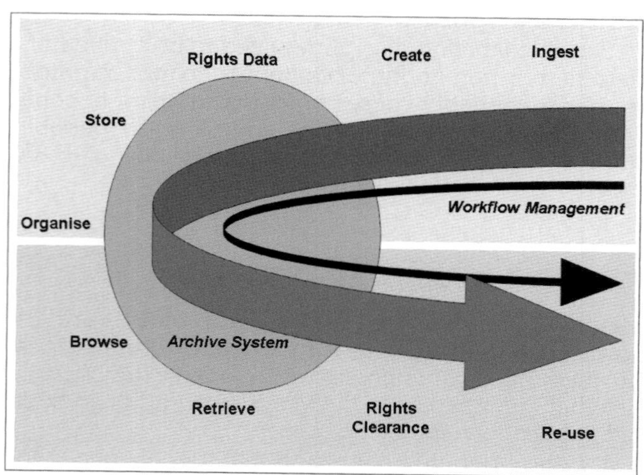

FIGURE 5.16-10 DAM helps the asset supply chain work effectively throughout the production process. (Courtesy of Harris.)

Figure 5.16-10 illustrates how DAM helps the asset supply chain work effectively throughout the production process. The description works whether looking at production from a single program point of view (rushes are shot, individual graphics commissioned, music selected, and all combined to form a final program), or looking at DAM from a broadcaster's point of view, with channels made up of programs, channel graphics, promos, and so on. At both macro and micro levels, the same efficiencies generate benefits through reuse, parallel workflows, and *pull* rather than *push* processes.

INGEST

For most industries *ingest* means uploading material, either directly in a digital format, or by digitizing an analog format. As well as obvious material such as analog video, an example of an analog format might be an old document that has to be scanned to transform it into a digital format. The core digital content will usually then have metadata attached with fields such as "Title" and "Author."

The core video content used by broadcasters is massive, and with current technologies it is impractical to store the vast quantities of data in central networks or to pass the high-quality versions of data around from person to person for browsing. Additionally, large quantities of video material are sitting on library

Video Ingest Quantities	
Initial State	How much material needs to be ingested? Do you need to digitize your entire back library? Can this be done over time, or does it need to be completed for go-live?
Steady State	How much material will be coming in on a daily/weekly basis? What format will it arrive in? What levels of volatility can you expect to face?
Longer term	How will the steady state change over time, and in particular how will the delivery technology be changing as far as you can see?

FIGURE 5.16-11 Video ingest quantities.

shelves in a wide variety of videotape formats. Consequently, ingest for broadcasters has slightly different requirements from most enterprise operations in other businesses. When the requirements for a DAM system are put together, there must be a clear grasp of material quantities, in three specific time stages, as shown in Figure 5.16-11. Additionally, the three main technology delivery methods need to be clearly understood and appreciated for how they fit into the enterprise technology strategy, as illustrated in Figure 5.16-12.

The *file transfer* ingest method reduces the storage requirements inside an organization and provides efficiencies in both the production and transmission processes. Figure 5.16-13 shows how the file transfer process with automatic delivery to the playout server benefits a transmission operation; similar benefits can apply in production.

Many news organizations have traditionally stored final story packages onto videotapes in sequence, with a period of black recorded between each item. Some manufacturers have developed black space identification tools, allowing a tape to be played, and automatic identification of the *in* and *out* points of the actual items. These features can save considerable time if the original time codes of the finished items were not properly recorded.

Video Delivery Technologies	
Video Tape	Large libraries available. Low long-term storage costs. Substantial asset values. High cost to ingest (real-time operation, expensive hardware).
Direct to Disc/ Solid State	Much faster to ingest, and at lower cost, this type of video material usually has a range of metadata that is already available and can be imported with the core essence. The storage medium for solid-state is currently high, and usually requires material to be ingested quickly to make medium available for re-use. Particularly suitable for new production.
File transfer	For finished material this will become the dominant delivery mechanism, often in a "just-in-time" approach. Content that is scheduled for delivery to the consumer will be pulled from a central repository, and received by the required device along with its core, applied and transactional metadata.

FIGURE 5.16-12 Video delivery technologies.

Sports logging of live feeds has particular requirements; for instance, highlights are often needed while the material is coming in. Specialized sports ingest tools have been created to help turn material around quickly. Typically, this will include the capability to play back material from the ingesting device while still recording and the capability for producers to use an extensive range of shortcut keys, enabling the sport to be accurately described while the feed is being received. This makes it easy for a director to call for a particular clip (for example, the first goal) to be found and replayed. All this descriptive metadata is then stored for future reference.

Metadata Storage

Automation systems (playout) have been using ingest systems since the early 1990s and have always used some method of attaching metadata to the video files.

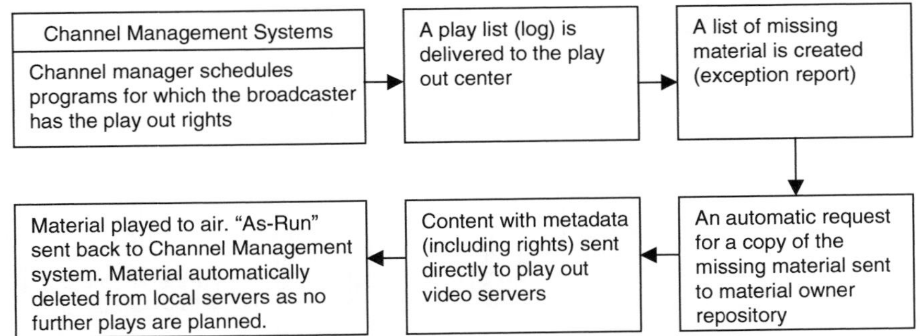

FIGURE 5.16-13 Automatic file transfers for program delivery and ingest reduce the need for long-term storage and material preparation.

While some suppliers chose to wrap this metadata into the file itself (typically in the file header), others preferred to use separate databases to hold the additional data.

There are different benefits to each method. Simple databases are easy to change, whereas data wrapped up in file headers is much more complex to open up and edit. However, having metadata wrapped into the essence means the metadata can never be lost, avoiding the situation of having a piece of video and not knowing the information about it. Losing metadata is equivalent to having an unlabeled tape in a box.

Metadata Enrichment

Describing and enriching metadata have two main purposes for broadcasters. In the traditional library function, data is added to content to enable it to be found efficiently. With hundreds of thousands of digital assets, cataloguing of material must be specific and done in a method that will allow people searching to find what they want without having to wade through copious returns on a search.

Major cost efficiencies can be achieved using automatic data enrichment. The most obvious example is perhaps storyboarding. When material is ingested, a low-resolution *proxy* copy is usually made for use by DAM users on their office computers. This proxy copy may use a variety of formats including, for example, MPEG-1 and Windows Media. However, when users are searching or browsing content, it is time consuming for them to look through complete items, so storyboards (selected key frames) may be created, showing the sequence of shots that forms the material. The number of frames created in the storyboard can be chosen either by fixed times (for example, one every five seconds) or automatically by detecting shot changes.

After a storyboard is created, each frame can have additional data entered. This helps very rapid searching of material, since a user can search for "Venice" in the example shown in Figure 5.16-14 and immediately be shown the shot in this piece of material that has been catalogued to this level of detail.

DAM systems themselves cannot necessarily do this sort of automatic data enrichment, but they can be used to send material to applications that have these

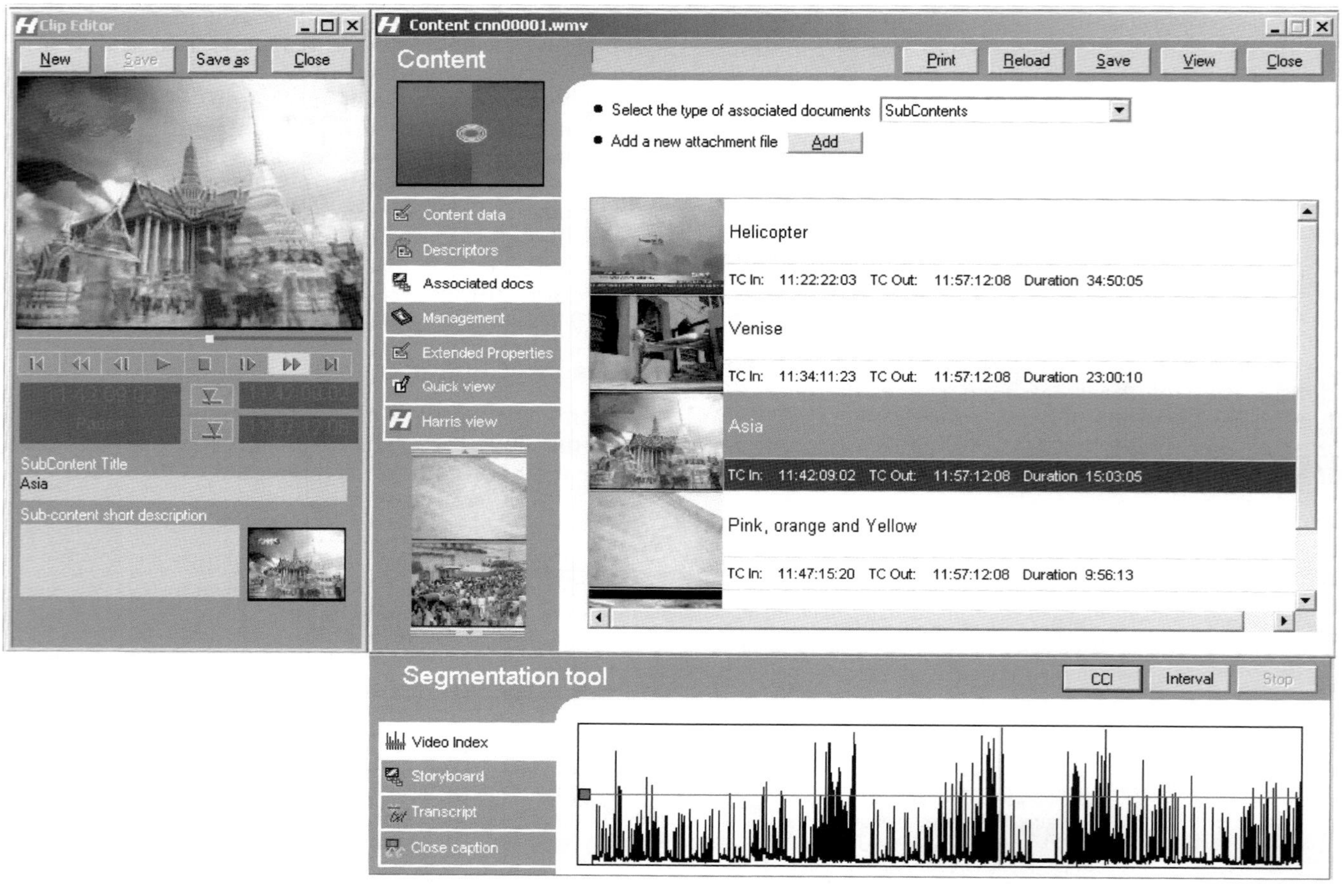

FIGURE 5.16-14 A segmentation tool with controllable scene detection threshold being used to create a storyboard. (Courtesy of Harris.)

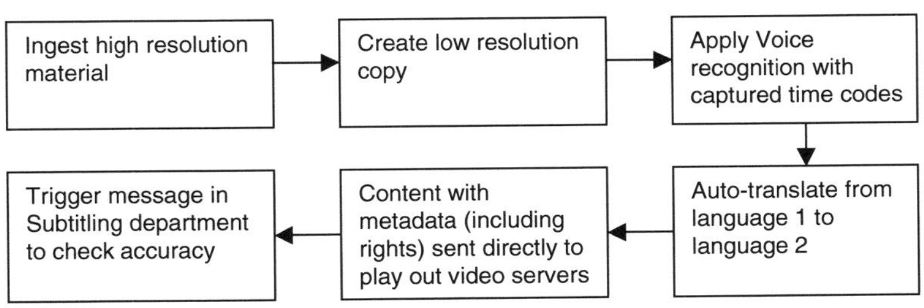

FIGURE 5.16-15 Example of a workflow using a DAM infrastructure system to create a new language subtitle.

capabilities and then store the resultant data along with descriptive metadata, enabling re-use.

When multiple technologies are integrated, there is potential for significant time and cost benefits. In the example shown in Figure 5.16-15, a (simplified) workflow process has been defined to create a new language subtitle. Using a workflow engine in conjunction with a DAM system, work that would traditionally have taken several people many days to achieve can be done within a matter of minutes. With current technology, the end results will not always be of sufficiently high standard for on-air use, but if even 75% of the work effort can be removed, leaving a person to correct the automatic results where the machine is not yet capable of perfection, major savings can be made.

These types of automatic data enrichment will be particularly useful when large quantities of poorly catalogued material exist. Face and voice recognition processors that run across large quantities of older news rushes may enable those expensively acquired assets to be reused. Without having useful, searchable data, producers may never find them; hence, they have little value.

SEARCH TOOLS

A core component of a DAM system is the search engine. Searching vast quantities of data has driven the need for a range of search tools. The number of items and range of searchable data are only going to increase, making this capability even more important. Complex searches can be built up, using ranges across a series of metadata fields such as catalog roots, subheadings, dates, descriptors, and so on, to help users come up with a sensible range of results.

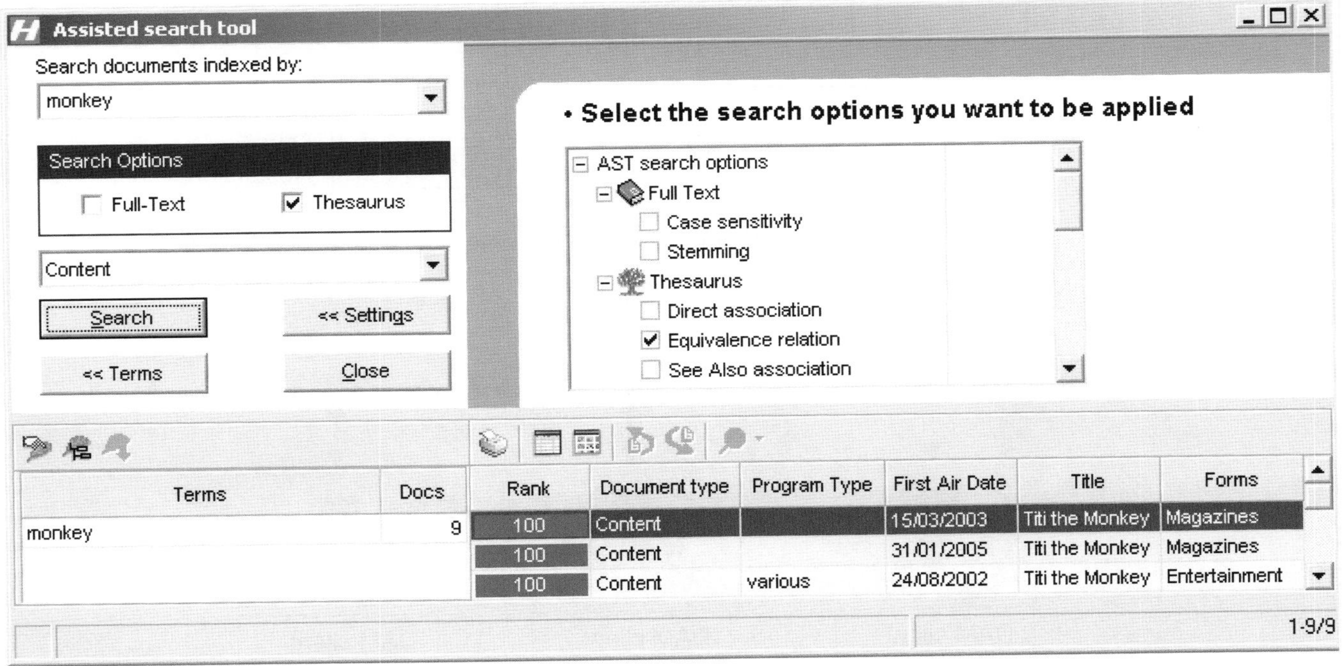

FIGURE 5.16-16 Advanced search techniques have been implemented in some DAM systems. (Courtesy of Harris).

Some search systems are built from scratch by the DAM vendor, whereas other suppliers prefer to use powerful "off-the-shelf" engines that are integrated into the DAM system. The capabilities of these search engines vary greatly, and the suitability of different systems depends on the quantity of data, the number of users, and the range and formats of the data. Search capabilities vary greatly across different DAM systems, with some providing more advanced searching techniques. Figure 5.16-16 shows the user interface for an advanced search engine.

MATERIAL MANAGEMENT

Material management is central to any broadcast DAM system. It is so important that some manufacturers refer to their systems as media asset management (MAM) systems. This area of the system primarily covers discovery (knowing what essence you have and where) and managing that material (moving, copying, deleting).

Very large media files are stored on a range of technologies, including video servers, storage area networks (SANs), network attached storage (NAS), and deep archives (including off-line tape-based systems). It is common in broadcast systems to have a central unit, sometimes referred to as a *Content Manager*, that can look into the different storage devices and present the contents in a common format to the rest of the DAM system (see Figure 5.16-17).

The content manager can present different technologies in a similar way. For example, it could indicate what material is sitting in a video server from the file

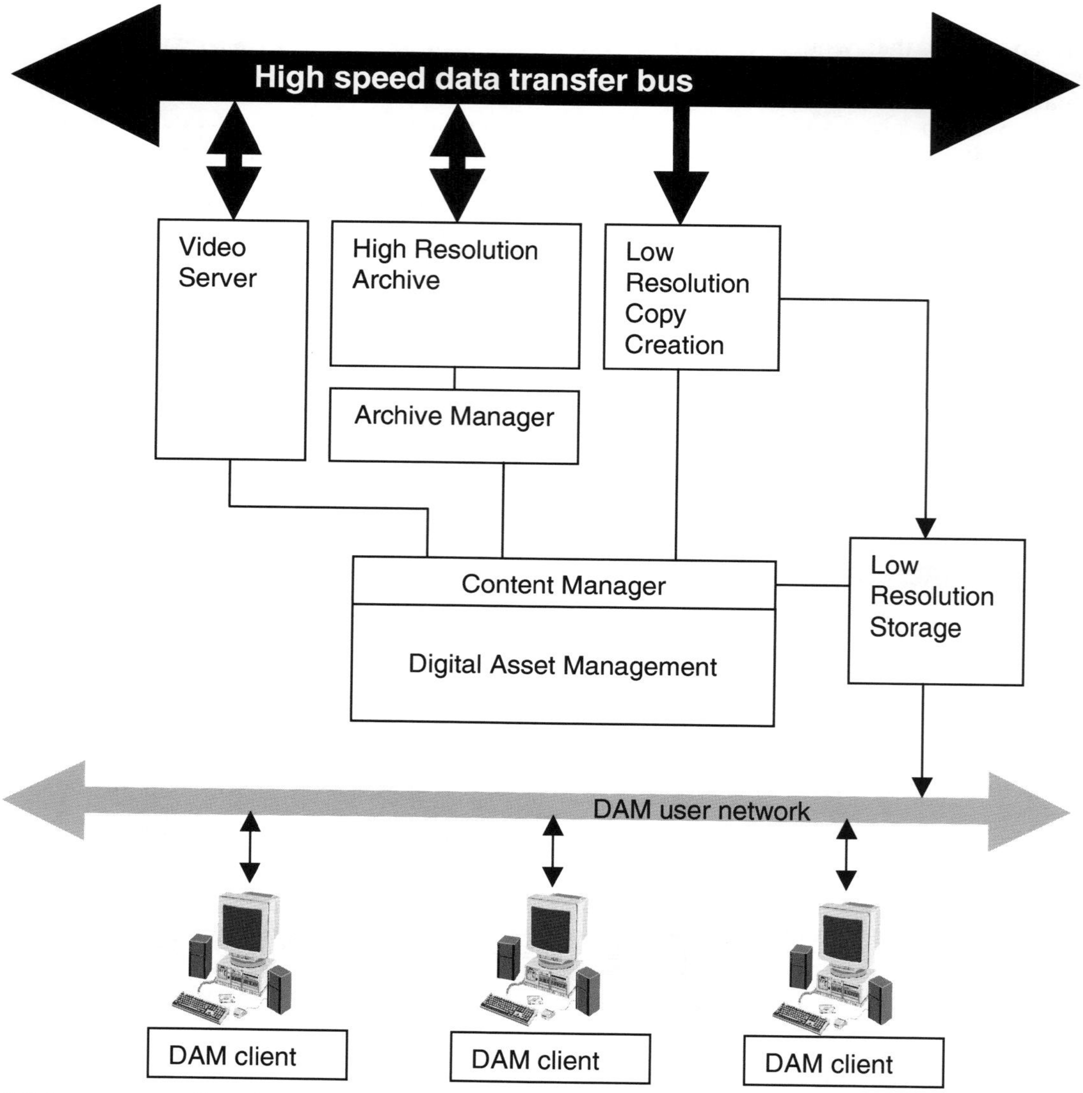

FIGURE 5.16-17 Simplified schematic showing a Content Manager reading data from and sending instructions to video storage devices.

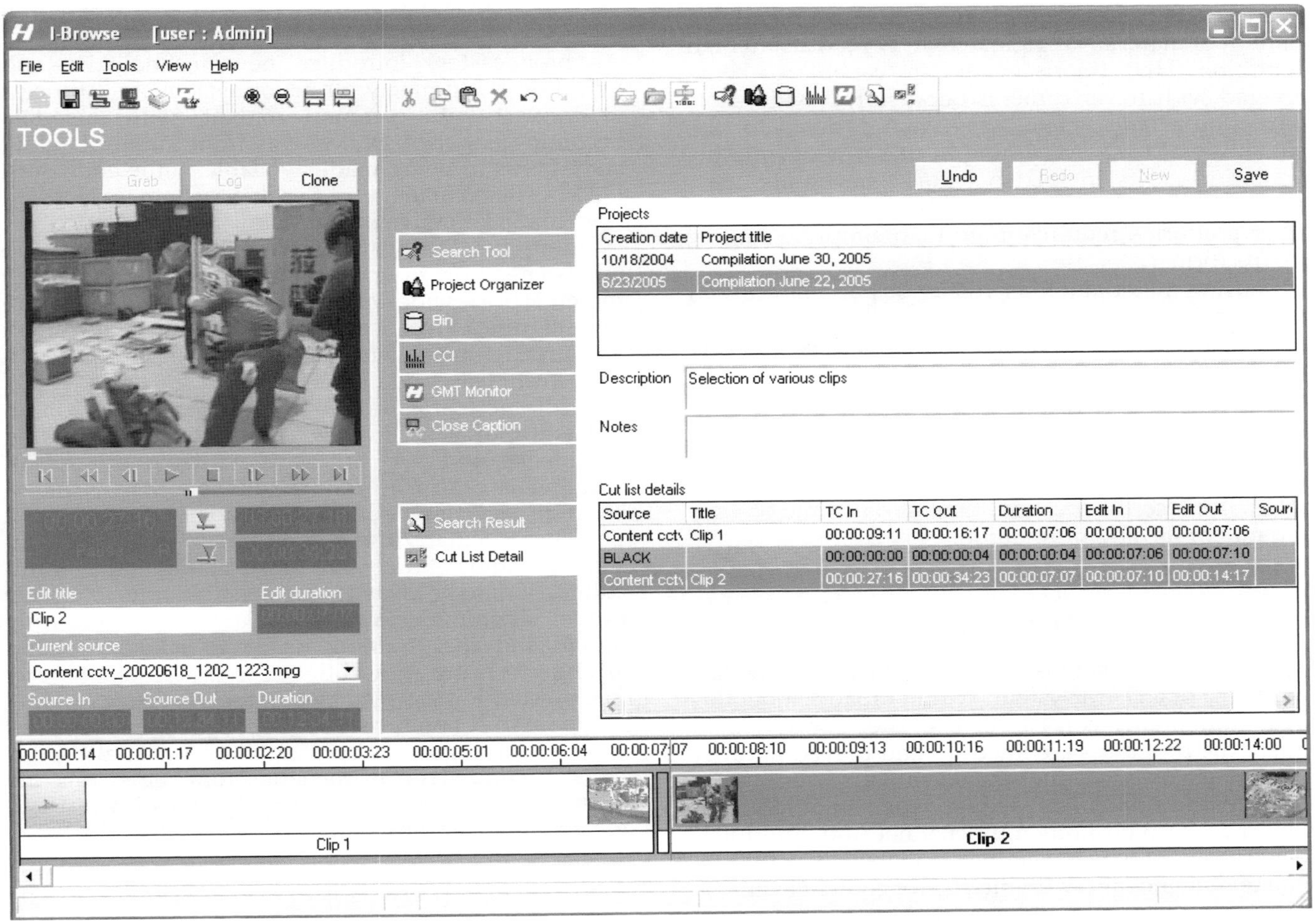

FIGURE 5.16-18 DAM browse mode user interface. (Courtesy of Harris.)

records and also report what material is sitting in a *Flexicart* by reading data from barcodes or user bits off the tape. A high-speed data bus is used to send material between the various storage devices and also to reformatting machines. The most common example is browse copy creation, in which a full-resolution file is sent to a device to make the proxy copy. Normally, this proxy copy is stored in a separate storage device, and those copies are made available to the DAM users on their office network. See the example in Figure 5.16-18.

An *archive manager* is used to remove the complication of managing a deep storage archive from the content management system. Typically, these deep archives use a lower cost storage medium, but it takes longer to restore material to a video server than to move it from a near-line storage system.

Repurposing material goes beyond simply reusing a piece of material in new programs or finding a piece of raw footage that can be used in a news report. Large, multinational broadcasters often own outright the usage of a major library of programs. As they seek to exploit these rights by delivering services to other parts of the world, there is a need to localize material. Typically, this will include *compliance editing* to remove material that may be offensive to local customs, *subtitling* or *dubbing soundtracks*, and replacing any burnt-in graphics with localized versions.

ALTERNATIVE PLATFORMS

With the newer distribution platforms, such as mobile phones and handheld devices, material is rarely transmittable in its original format, for both technical and editorial reasons. Obviously, the screen size of the reception device may be a different shape (aspect ratio), size, and resolution. Delivery bandwidth may also be an issue, so material may have to be heavily compressed to suit the delivery mechanism.

Editorially, many of the shots that work on traditional televisions simply do not work for small screens. Wide shots do not work effectively, and highly compressed pictures rarely handle rapid movement well. Typically, service providers looking to achieve additional revenue streams by offering premium content through new media platforms must edit the source material to find pictures and sound that work well for that end device.

Editing becomes more of an issue as an increasing amount of material is originated and broadcast in high definition (HD). Sports events are capable of being covered with fewer cameras because HD wide shots shown on a large screen have enough detail for the human eye to see everything at one time. While the resolution exists in HD to extract the tighter view required for devices with smaller screens, reuse on other platforms requires more than simply changing the file format.

Reusing material (either rushes or parts of finished programs) in new programming has already been discussed and is an obvious area where an expensively acquired asset can be reused in later programs. For effective reuse, the material has to be thoroughly described, or else much time will be wasted searching material that is inappropriate. The description therefore has to cover not only the subject in considerable detail, but also the location, time of day, time of year, and other relevant information.

DELIVERY

Delivery or *publishing* is normally considered the last main area of a DAM system. Within broadcasting, this area traditionally consists of a master control room where the programs are lined up and played in order as planned on a channel management system. More recently, the method of delivery for some organizations has become both multichannel and multiplatform, with content becoming available to "pull" by viewer demand, and with interactive content available on the Internet or through set-top boxes, all of which must be coordinated by a channel management system that may be integrated with the DAM.

As time goes on, more "schedule" information may be delivered to DAM systems to manage each of these distribution platforms, with the metadata gathered throughout the entire broadcasting chain used to automate many of these deliveries.

JUSTIFYING THE COST

It is difficult to quantify the return on investment (ROI) for enterprise projects, particularly for a DAM system, which is fundamentally an infrastructure technology. However, it may be instructive to look back to the 1980s, when technical teams put together ROI predictions to justify office networks enabling PCs to be connected together. Such infrastructures are now understood to be a basic requirement by anyone putting together a modern office environment, and in the near future no doubt the same will happen with DAM systems.

One critical point is that to achieve a successful ROI, projects have to be completed on time. Late completion usually has a greater impact on the ROI figures than the additional overrun costs, so tight project management is necessary to achieve overall success.

SUMMARY

As broadcasters introduce new technologies and workflows, ultimately all major systems will process or interact with common content, with metadata becoming enriched as it moves from initial planning stages toward being made available to the end users. DAM systems manage and enable this process for efficient and productive operations.

Enterprise DAM systems are major projects and take some effort to implement successfully. However, the enterprise-wide implementations are able to produce significant cost savings through reuse of content and process improvements, some examples of which have been discussed in this chapter.

Most successful projects are carried out in a multiphase approach, limiting disruption and building confidence and support as the project progresses. Training is central to any technology change and is often difficult to schedule in a broadcast environment. This is particularly the case in news, where the constant demands of daily live broadcasting make it difficult for staff to be away for an extended time to receive proper training. However, many managers fully understand that training needs to take place and arrange to bring in additional resources to properly free up staff for this vital step in the project.

Successful enterprise DAM implementation not only returns significant cost savings, but it also positions broadcasters to benefit from new revenue streams and provides the capability to readily respond to new opportunities.

CHAPTER

5.17

ENG, SNG, and Remote Video Production

JIM BOSTON
DTV Engineering
Auburn, California

GEORGE HOOVER
NEP Broadcasting, LLC
Pittsburgh, Pennsylvania

INTRODUCTION

On April 20, 1939, David Sarnoff of RCA introduced television from the field to the few viewers of an experimental television broadcast system. He took a camera to the World's Fair in Flushing Meadows, NY, and relayed the signal to a New York City transmitter for reception of the broadcast by television sets in department store windows. From this humble beginning, television production originating outside the studio, or a *remote* as it is known, has become a staple of programming produced by broadcast and cable networks and television stations throughout the world.

Today, technology has made the delineation between low-end personal camcorders and suitable for professional use equipment a matter of just a few thousand dollars. The World's Fair remote in 1939 required an entire truck to carry the gear for a single camera, along with several engineers to make the equipment function. Current high definition (HD) electronic field production equipment can fit in a backpack and requires a crew of just one. Remote television grew and gained utility because of advances in technology as well as the demand for programming. The size of this part of the industry has reached a point where a significant amount of new technology is being developed and marketed specifically to meet the needs of every type of television programming produced on location, from news, sports, awards shows, and concerts to nature programs, documentaries, and reality shows.

A television segment or program generated on location may be produced by a single person with a camera or camcorder and perhaps a microwave link or Internet connection to send the content to the station. Or, it may be a sophisticated multi-camera production costing tens of millions of dollars in equipment and involving hundreds of people. This chapter introduces the various types of television remote operations and the equipment and facilities that are used. It provides guidance on planning a remote event and discusses some of the issues to be considered in procuring or building a truck for ENG, SNG, or remote video production.

FIELD PRODUCTION

Video production work performed beyond the confines of the studio is considered *remote* or *field production*. Television remotes can be grouped into a number of subsets with categories that often overlap: *electronic news gathering* (ENG), *satellite news gathering* (SNG), *electronic field production* (EFP), and the specific term *remote*. The Europeans often use the term *outside broadcast* (OB) for a remote.

Electronic News Gathering

Historically, prior to electronic cameras, images captured inside and outside the studio were shot on film. There were studio productions and shoots on location. Film productions shot on location tended to be of three categories: news, documentaries, or movies. In the early days of television news, the station sent out a photographer with a 16 mm film camera to shoot a story. The film was brought back to the station, developed, mechanically edited, and converted into a television

FIGURE 5.17-1 Handheld camera operator and talent doing a "stand up." Tripod-mounted camera on the right is shooting cut aways.

signal using a *telecine* machine for integration into the newscast.

With the advent of electronic cameras and recorders, producers immediately took them outside of the studio for remote production. As the cameras became smaller and more reliable, they were placed in service for news gathering by the broadcast industry, and the term *electronic news gathering* (ENG) was born in the early 1970s. ENG is typically a "shoot it as it happens" activity. Crews are minimal, retakes are few, and time constraints are tight. The ENG category is defined by the lack of control the production crew has over the subject of the story. As a result, ENG equipment must be adaptable to function in a variety of climatic, shooting, and lighting situations and must be able to do so with battery power.

Today, an ENG shoot usually involves a *camera operator* (or *shooter,* as they are known in the news business) and a *reporter,* as shown in Figure 5.17-1. Sometimes an *audio operator,* who handles the audio portion of the shoot, will be part of the team, as shown in Figure 5.17-2. If the assignment is a news feature or part of the production of a larger show, a *producer* or an assistant, often called an *associate producer* (AP), will orchestrate or direct the shoot.

On ENG shoots, a camera operator will often bring the tape back to the station where an editor turns the raw footage into a one or two minute story. Sometimes, especially if the story is a big one and will command more airtime, the reporter might direct the editing process. In many stations, the camera operator puts on an editor's hat back at the station and edits the piece shot in the field.

Live Reporting

To provide the capability for immediate (live) reporting from the field, a mobile microwave link is used to send the remote program back to the studio. The microwave equipment, with an antenna mounted on a hydraulic mast, is usually installed on an ENG truck or van (see Figure 5.17-3), which may also be used for transporting the other ENG equipment. The receive point for the microwave link is often on a tower or high building at the studio center. Alternatively, to provide greater coverage, it may be separately located on the station's transmitter tower, which may be used as an intermediate relay point with a further *backhaul* microwave or fiberoptic link to the studio. Further arrangements for microwave links are discussed later in the chapter.

Multi-Camera ENG

Multi-camera ENG shoots are possible and usually involve multiple trucks, each feeding a single camera via multiple microwave paths back to the TV station. There they are fed through frame synchronizers so they can be switched and mixed into a composite show. Some ENG vehicles have video switchers and

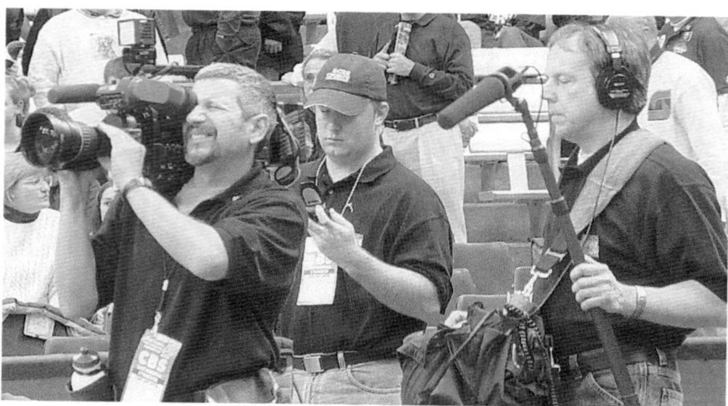

FIGURE 5.17-2 Camera operator, audio operator, and associate producer (AP; center) to direct the activities.

FIGURE 5.17-3 Typical ENG microwave vehicle. (Courtesy of Frontline Communications.)

audio mixers so multiple video and audio sources can be handled by a single truck.

ENG Style

Through necessity, ENG is usually produced with a single camera, often handheld, and with heavy use of ambient light and minimal or harsh lighting augmentation. ENG shoots allow less-than-perfect framing, and, for single-camera productions, style is limited to moving the camera to change the angle. The change may be live or edited in later with the addition of cutaway shots. These restrictions tend to produce an *ENG style* characterized by a live or reality feel. This ENG style is now often used in drama productions when a producer wants to evoke a feel of live action with a sense of immediacy and importance. In that case, the ENG style may in fact be part of a more elaborate EFP shoot or a studio production with multiple cameras. ENG style should not be confused with the real use of the ENG remote category.

Extending the Range—SNG

Terrestrial microwave links are generally limited to operation with line-of-site transmission back to the central receive site. This restriction on range for live news gathering can be overcome by using a truck-mounted or portable satellite link for the ENG camera. If a satellite truck is used to uplink to a satellite for sending the program to the studio, then the activity is known as *satellite news gathering* (SNG). Arrangements for satellite links are discussed later in the chapter.

Electronic Field Production

If the activity on location is not news related, then it is generally referred to as *electronic field production* (EFP). Generally speaking, EFP refers to productions with one or two cameras in the field, with the recording device contained in the camera. Images and sound are usually captured for later post-production. Commercials, reality shows, documentaries, and travel programs all fit into the EFP category.

Unlike ENG, where the most important factors are the news value of the content and the portability of the equipment, for EFP the picture and sound quality and production values have a high priority, as reflected in the choice of equipment and the style of working. Like ENG, EFP shoots, at their simplest, involve a camcorder, a mic, a camera operator, and usually a tripod for the camera. Depending on the production requirements, ancillary equipment such as lights and more advanced camera platforms, monitors, audio mixing, and processing are often added. Typically, EFP shoots have much more control over time, staging, lighting, and artistic elements than the news shoot. All of these factors contribute to the larger size of a typical EFP crew compared to the one- or two-person ENG crew.

A subset of EFP motion picture production that utilizes electronic cameras is called *electronic cinematography*. For some applications, particularly for documentaries but also increasingly for drama, high-definition electronic cameras have replaced image capture using traditional 35-mm film cameras. The cameras may be very similar to those used in broadcasting, but the workflow, lighting, and artistic conventions tend to be different and follow the language and workflow of film.

Remotes

When the added capabilities of cutting between cameras, more sophisticated audio systems, and the playback of recorded elements in a switched program are implemented, a level of production is reached that is called a *remote*. Remotes can be live, live-to-tape, or recorded for further editing in post-production. For this type of remote production, multiple cameras, recorders, and a production control area are housed in a vehicle of some kind. Trucks used to facilitate these activities can take many forms, from small ENG vans to large tractor-trailers (see Figure 5.17-4). As more equipment is added, the size of the vehicle to transport it grows. The largest remote trucks top out at 53 feet long with expanding sides ("expandos") on both sides and the rear, and they weigh in at just under 80,000 lb. For the largest remotes or even just the "A" game (the football game that the network selects as the national game), one of these trucks is not enough. Often the graphics, audio, or replay requirements are so complex that they are housed in separate trailers. Remotes can require many trucks tied together to provide the necessary gear and facilities at the site to accomplish the shoot. These types of remotes often require multiple days to set up and, obviously, are more expensive to produce.

ENG LINKS AND BACKHAUL

If a news, sports, entertainment, or other event is to be transmitted live, the signal will be sent from the remote site to the station or network studio or *network operations center* (NOC) via a circuit called a *backhaul*. It could be by *microwave*, *landline*, or *satellite*. A combination of all three is shown in the diagram in Figure 5.17-5. At most sports venues, permanently installed landlines are available. Often a satellite path is used as a backup. Landlines, almost exclusively fiberoptics, provide the best quality and are the most reliable method. The use of landlines requires a common carrier, such as the phone company or other service provider, to have a *point of presence* (POP) at the venue. If none is available at the venue, one alternative is to use a local station's ENG truck to get the signal back to a local TV station's master control and then hand it off there to a long-haul service such as Vyvx or other service provider. In addition, some local stations have their own fixed C-band satellite uplink facilities or a path to someone else's uplink facility. In that case, the ENG microwave truck could be used as a terrestrial link to a satellite uplink.

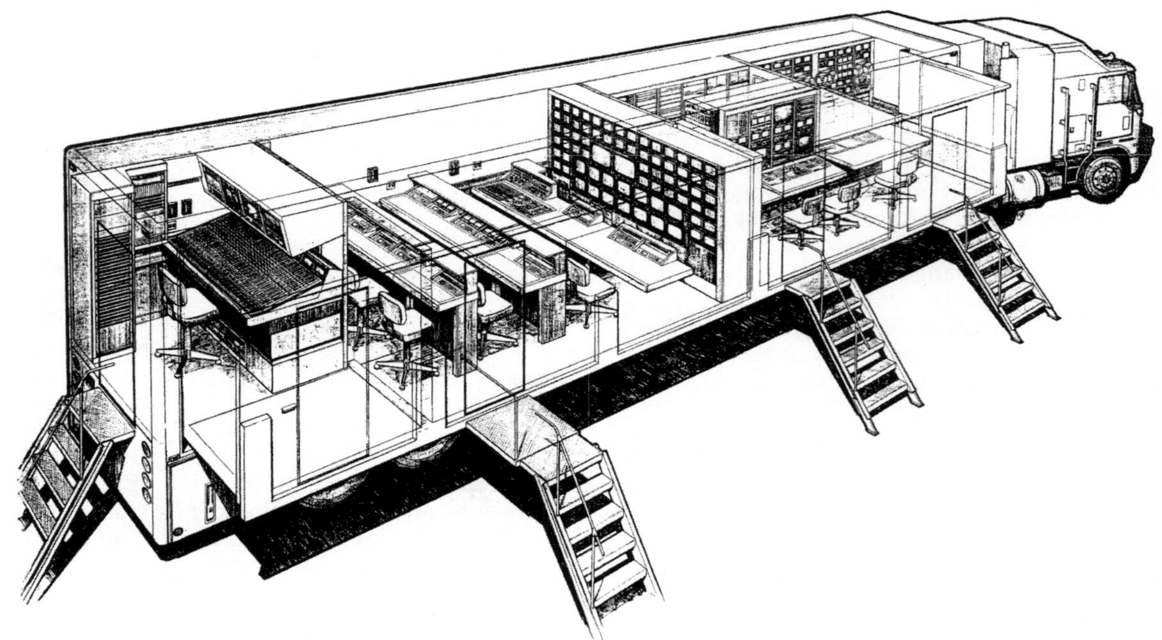

FIGURE 5.17-4 Artist rendition of full-featured, "expando" TV remote production vehicle. (Courtesy of Wolf Coach.)

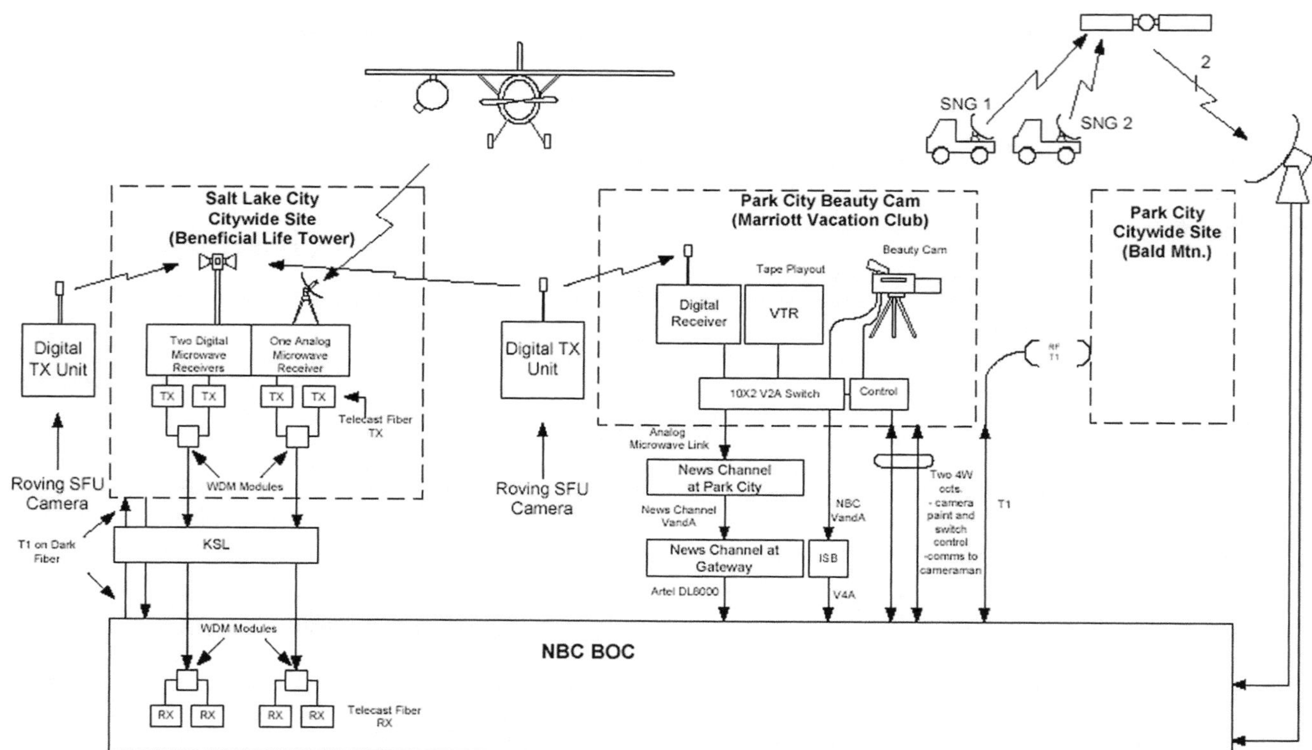

FIGURE 5.17-5 Diagram of complex interconnection system for multi-camera, multi-venue production at the 2002 Winter Olympics in Salt Lake City. (Courtesy of NEP Broadcasting.)

Between main and backup backhauls a wide combination of paths and links may have to be used. The usual path for a live news link (ENG) is via microwave. If the venue is outside the range of the station's microwave receive sites, then a station with a satellite truck might use it to get from the venue to the station via a satellite uplink, or SNG, truck. Although microwave frequencies may already be assigned in a metropolitan area, a check on frequency coordination with the local Society of Broadcast Engineers (SBE) frequency coordinator (http://sbe.org/freq_index.php) is advisable to make certain that interference can be avoided. Often, cooperation between stations is needed to ensure interference-free operation for special events.

Establishing a Microwave Path

After the truck operator has determined that the mast can be safely raised, the microwave dish at the top is pointed toward a receive site. The truck will then contact the station operator responsible for establishing feeds. If the operator is expecting a feed and the intended receive dish is not already in use, then the receive side antenna will be aimed in the general direction of the truck awaiting contact from the field (see Figure 5.17-6). Even if the station is expecting a feed, the operator still should contact the station before powering up to confirm that plans have not changed. Once powered up, the truck will pan the transmit antenna until the best signal is received at the station. Then the receive end will also pan the receive antenna to optimize the signal. This might go on for a few iterations until the best, most reliable signal is received.

FIGURE 5.17-6 ENG remote-controlled receive antenna on tower.

Often the path will end up bouncing off one or more objects, such as hills and buildings. Sometimes the receive site will find that the best reception from the truck is from a direction far different than the truck's true bearing. Although this usually works for the short duration of an ENG shoot setup and on-air live shot, it is not recommended if the path is required for long periods, especially if the entire show, not just a segment in a news show, is counting on that path. A bounce path might work earlier in the day, but not later as atmospheric conditions change or buildings used to bounce the signal heat up or cool down. A 30 mile 2 GHz microwave shot will typically have a line-of-sight path loss of approximately 130 dB. Any reflections or other obstructions will obviously lower the range and stability of a path. Where a single hop is not feasible due to intervening terrain, a multi-hop microwave relay may be used, as shown in Figure 5.17-7.

FIGURE 5.17-7 Microwave relay facility on mountain top.

Satellite Links

Closely related to ENG operations are SNG remotes. The major difference is the signal path to the station. Instead of a microwave path between truck and station, the signal is uplinked to a satellite. In the case of news, it is usually a satellite that operates in the

FIGURE 5.17-8 SNG uplink vehicle. (Courtesy of Frontline Communications.)

Ku-band. SNG or satellite uplink vehicles require several hundred pounds of additional weight in the form of a high-gain 2 to 3 meter transmit dish, more complicated radio frequency (RF) distribution paths, and antenna controller hardware. Satellite transmitters require more power, with maximum power for Ku band operations around 300 watts *versus* 20 watts for ENG. This additional hardware requires a larger and heavier truck than for ENG, as shown in Figure 5.17-8. The Ku-band satellites used in the United States have a directional pattern that covers most of the continental United States or a "conus" pattern, as illustrated in Figure 5.17-9.

In addition to being expensive and complex, satellite uplink operation is also potentially dangerous, and not something to be undertaken without proper training. Many transmitters have lethally high voltages and the current to cook as well as shock. But, more dangerous, because it is easier to get in its way, is the RF power. Hundreds of watts of RF power can be coming out of a transmitter and channeled down a waveguide only a centimeter or two wide. Waveguides can spring leaks. If the transmitter is on long enough, the leaks manifest themselves by heating up the area around the leak. Leaks are found by checking for hot spots only after the transmitter is turned off. If a waveguide is dangerous, then so, too, is what is reflected off the dish. A satellite dish takes the power that the waveguide delivers and directs that energy in a single concentrated high-power beam (transmitter power times the antenna gain) that may be the equivalent of hundreds of thousands of watts. This amount of power is extremely dangerous to any human flesh within that beam.

Satellite transmission levels are at least 50, and usually 100, times more powerful than any ENG microwave transmission. Whereas microwave shots often intentionally bounce their signals, satellite uplink transmissions must be aimed directly at the satellite. A Ku dish can emit enough RF energy that it is unsafe within almost 200 meters from the dish. The dish must not attempt to shoot through any obstructions. Shooting through leaves can scatter RF energy, so instead of illuminating the intended satellite the signal may be hitting many obstacles. A Ku dish with no scattering has a beam width of hundreds of miles when it reaches the satellite. A 2° spacing between the many

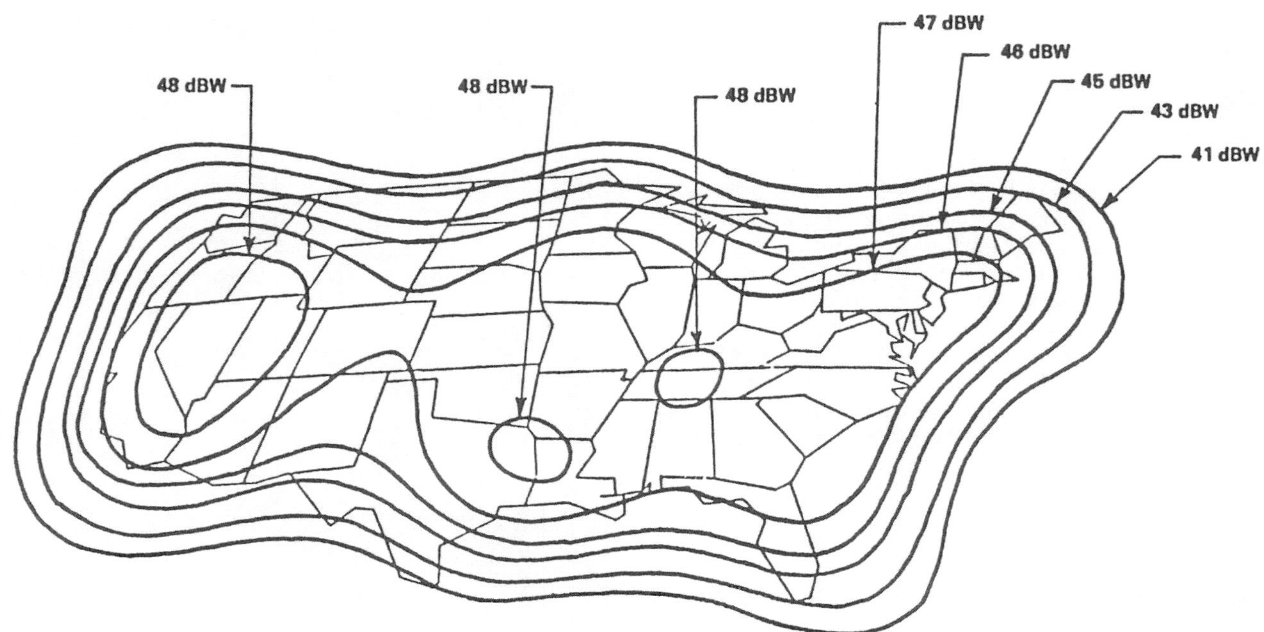

FIGURE 5.17-9 Example of "conus footprint" coverage pattern of Ku-band satellite used for SNG operations.

satellites in orbit means that they are only a little over 900 miles apart, which leaves little margin for error.[1]

SNG Operating Practice

Satellite operation is more complex than a simple microwave setup. The equipment required is larger and thus heavier. The transmit dish must be aimed accurately so stabilizing levelers (or landing gear, as they are often known) must be deployed first. Next, many transmitters have a warm-up time, so often the transmitter is put into warm-up mode before anything else is done. The required satellite or "bird" is then found with a monitor receiver. Automatic locating systems that are part of the antenna controller can be used to find the correct satellite, but verification by the operator is important to ensure that no interference will occur by firing up at the wrong satellite or transponder.

The geosynchronous satellites used by television uplink industry, both Ku- and C-band, are all in the same orbit approximately 23,000 miles above the equator. They actually range across 140 miles from the lower orbit to the highest due to differences in satellite mass. By using a spectrum monitor and looking at the spectra associated with various satellites, it is possible to recognize a particular satellite's spectrum. This spectrum is the mix of analog and digital services across all the transponders on a particular satellite. Often, satellites will have special pilot signals and other fixed services that make its spectrum easily identifiable. Here is where the accurate logging of dish azimuth, elevation, and location is important. Determining the dish's elevation and azimuth from a nearby location gives an experienced operator a place to start if the automatic satellite location equipment is not working.

With practice, an operator can locate a satellite in the arc of satellites above the equator and recognize its spectrum. Then, using a satellite chart, the operator can work across the arc until the required satellite is found. Also, use of the satellite receiver or integrated receiver–decoder (IRD) to look at decodable services will identify what satellite is being observed. Once the correct satellite is found, it is important to contact the satellite operator and confirm the set of services that are in use on the selected satellite.

The next step is to adjust the rotation of the feed horn, which is known as *cross-pol* (for dual or cross-polarization capability). Adjacent transponders alternate between horizontal (H) and vertical (V) polarization. On the receive side, half will disappear when switching between H and V. Using either H or V receive polarity, adjust the feed horn polarity until the opposite polarity downlink signals are minimized. Always use this minimizing approach. The satellite operator will expect this to be done when it is time to transmit and may require the adjustment to be made again when "up" on the satellite.

[1]The NAB provides an in-depth satellite uplink operators training seminar on a regular basis; for details, see the NAB website.

It should be stressed here that, during news *blocks* (the common newscast times, generally 4 to 7 p.m. and 10 to 11:30 p.m.), the demand for transponder time is at a premium, and news departments usually buy only the time they think they need. Thus, it is important to have talent and camera person in position and ready to go at the proper time, in addition to having the satellite transmission facility ready to go. If the news show runs late or the SNG operator is not ready for the time slot booked, the satellite operator may require shutdown whether the live shot is complete or not.

With cross-pol adjustment completed and the transmitter warmed up, it is important to feed the output to the dummy load before allowing the transmitter feed RF power to the antenna. Next, establish an interruptible fold-back (IFB) connection with the station. Some satellite providers have a *phone switch* in the satellite to connect to the station IFB bridge (see Chapter 5.19 for more about IFB). At this point, the facility is ready to go live.

At the proper time, the satellite operator is called, either via traditional methods or via the satellite's phone switch (if the SNG unit is out of cell phone range). When the operator allows illumination of the satellite, then switch from the dummy load to the antenna. Use low power at first so that if a mistake is made with regard to transmitting to the wrong transponder then the low power will minimize interference to the existing service on that transponder. Also, if a transponder is being shared by others, the satellite operator will have the power increased slowly to a point where maximum saturation of the transponder occurs without bringing another user on the same transponder out of saturation.

Ku satellites also allow a transponder to be split up, with two or more trucks using the same transponder at the same time. With analog transmission, the maximum number of trucks that can use the same transponder is two. With digital transmission and accompanying compression and modulation techniques, multiple transmissions can pass through the transponder simultaneously. Note that interference to analog transmissions usually results in picture and sound impairments that are obvious. With digital transmissions, interference results in a complete loss of picture and sound and no indication of whether the problem is caused by interference or an equipment problem.

With analog, the split transponder is known as a *half-transponder*; its use results in slightly noisier signals due to the decrease in modulation index. This is generally acceptable for news applications, as the cost for transponder time is correspondingly less. Often, on large stories where a news site has many Ku trucks on site, half-transponder use is all that is available due to the demand, especially during the news blocks mentioned earlier.

Digital Satellite Transmission

The early adopters of digital satellite transmission have been station and network news departments.

Digital SNG (DSNG) can be superior to analog SNG in many ways. To move to this new technology, a new mindset is required to understand what is happening in the digital satellite transmission process. It is helpful to think of a digital video stream as data. In a transmission path, this becomes imperative. Digital satellite encoding and transmission can be compared to telephone modems. Modems come with multiple transmission protocols, which specify various modulation techniques and constellations along with baud rates, among other things. Baud rate is also referred to as the *symbol rate*. A common misunderstanding is that baud rates and bits per second are synonymous. In the case of RF, one carrier cycle is its symbol or baud rate; thus, the baud rate must be less than the channel bandwidth. Advanced modulation techniques allow the bits per second rate to be higher than the channel bandwidth.

Digital signals suffer in split transponder use in the sense that the smaller slice of the transponder bandwidth used means higher compression, with more compression artifacts. Although the usual analog noise *per se* is absent, compression artifacts can often resemble noise. Digital transmission is a trade-off. It allows the use of less bandwidth, and bandwidth is what is bought, but too little bandwidth and the correspondingly high compression can result in poor quality. Compression and modulation techniques continue to improve, and what was marginal at one bandwidth a year ago might be acceptable this year.

Satellite Link Budgets

One measure of the health of a satellite link is the parameter called *received carrier-to-noise ratio* (C/N). In the case of digital satellite transmission, another specification is often stated, *energy-per-bit* versus *noise* (Eb/No) (pronounced "ebno"). Eb equals the time allotted to transmitting each bit multiplied by the signal power. This means that, as the modulation constellations get more complex (such as 16QAM *versus* 8 QAM, or 16PSK *versus* QPSK), the required C/N to maintain a given received error rate increases. The C/N and Eb/No ratios have a one-to-one relationship. The difference between the transmitted Eb/No and the received Eb/No is known as the *link margin*.

Satellite transmission paths commonly have losses in the 200 dB range. With satellite receive and transmit amplifiers having gains in the range of 50 to 70 dB each, and send and satellite and earth receive dishes having gains of 40 to 50 dB each, there is not much room for unplanned degradation. Ku-band *rain fades* alone can absorb 10 dB on the uplink or downlink side. This means it is necessary to start with as high a transmission effective isotropic radiated power (EIRP) and receive signal as needed, as well as the lowest possible noise power.

In conjunction with the received signal and noise, there is a parameter called *antenna gain to system noise temperature ratio*. A common figure for this is 30 dB/°K or more. Noise will increase in northern latitudes, as the dish must be pointed closer to the horizon (to see the satellite that is over the equator), which means increased atmospheric absorption and increased noise radiated from the earth. Lower link margins can occur at the edges of the United States, as satellite transmit/receive antennas commonly have more gain at the center of the mainland than near the borders or coasts, with as much as a 6 to 8 dB difference (see Figure 5.17-9). For more information on satellite systems, see Chapter 6.11, "Satellite Earth Stations and Systems."

Transponder Use

Common satellite transponder bandwidths vary from 24 to 110 MHz. One hundred ten MHz transponders are really two transponders feeding the same antenna. The most common transponder bandwidths are 24, 36, and 54 MHz. Half-transponder usage is normal for analog transmissions that are frequency modulated for their trip through the satellite path. There are no standards for the bandwidth use of half-transponders, so it is up to the individual satellite operator to decide. Some operators specify both normal and peak deviation rates, while others specify peak deviation. Peak deviation of ±7.50 MHz seems to be common, with normal white levels around ±6.85 MHz. Satellite operators sell small slices of a transponder for digital transmission, but *spectral occupancy* by the SNG facility is based on where the skirts at the edges of the signal fall. One satellite operator considers occupancy to be the –26 dBc falloff point.

Satellite spectrum and time are precious commodities. The cost of launching a satellite is in the $50 million to $100 million range. The cost of the satellite itself can rival the cost of its ride into space. On-board systems, including receivers, transmitters, and sophisticated control equipment, consume approximately 1 to 5 kW of total power, which generally comes from solar panels. Efficient use of all these resources is required and careful management and use by the satellite operator, and by users, is essential. For more information on satellites see Chapter 6.11.

Satellite Receive Issues

Many modulation schemes used for digital signals use no carrier component in the spectrum; therefore, local carriers must be derived at the receiver to recover the information. Ku-band DSNG systems must use *digital* low-noise block converters (LNBs) with low phase noise at the receive end. *Analog* LNBs can have local oscillators (LOs) that drift as much 2 to 3 MHz, and phase noise is less important. Some digital decoders need 70 MHz intermediate frequency (IF), or in some cases L-band, outputs that drift no more than 100 kHz, although some newer ones can stand drifts of hundreds of kilohertz. Good LNBs have temperature-compensated crystal oscillators in their phase-locked loop circuits. LNB LO stability is also measured by a parameter called *phase noise*. This is a measurement of how much energy is found at various frequencies away from the desired LO frequency. An example of a good phase noise measurement would be –65 dBc at 1 kHz.

Another consideration in the digital realm is *spectrum inversion*. The signal, just like analog signals, is upconverted at least once on the uplink side, downconverted in the satellite, and downconverted at least once on the receive side. These conversions are usually accomplished through a heterodyne process. The product of this process is the original signal, the local oscillator carrier, and the sum, and differences of the first two. The sum product is a replica of the original signal, but at a new frequency. The difference signal has a spectrum at its new frequency that is a mirror image of the original. Some frequency converters use the sum product (filtering out all the other products), and others use the difference product. If an even number of these difference products is used in the path, there is no problem because the double inversion cancels out. But, encountering an odd number can cause problems. Digital receivers must cope with this situation. Analog FM satellite signals are not affected by this because once the signal is detected a simple inversion of the baseband signal is all that is needed, if it is required at all. Most new digital receivers sense the inversion and correct for it automatically; older models require manual switching. When quadrature phase-shift key receivers are used for digital signals, the I and Q signals must be able to be inverted to solve this problem.

2 GHZ BAS RE-ALIGNMENT

Most broadcasters using ENG or digital ENG (DENG) facilities are familiar with the new FCC rules (FCC 04-168) requiring 2 GHz broadcast auxiliary service (BAS) operations be moved to new 12 MHz bandwidth channels in the revised 2025 to 2110 MHz band. To make this possible, and in consideration for both returning 8.5 MHz of spectrum in the 700 MHz and 800 MHz bands as well as funding the 800 MHz band reconfiguration, Sprint Nextel was allocated 10 MHz of spectrum, 5 MHz of which the FCC had originally licensed to certain mobile satellite service (MSS) providers. The FCC has mandated that BAS licensees using the current 1990 to 2110 MHz band convert from the current seven-channel 17 MHz channel plan to a new seven-channel 12 MHz channel plan, which extends from 2025 to 2110 MHz, and replace or upgrade their equipment to operate on the new frequencies. The FCC deadline is September 2007.

To meet this requirement Sprint Nextel will replace all current analog equipment used in this band with DENG facilities. Sprint Nextel and the broadcasters will first agree on what needs to be replaced, the equipment is then ordered and installed. The new equipment creates narrowed-in-place microwave channels that allow all new installed equipment in a market to have their center channel frequencies changed or reprogrammed via software, freeing up the bandwidth that Sprint Nextel has been allotted.

DENG provides some built-in benefits such as the possibility of high definition video. Using a higher data rate modulation and line-of-sight (LOS) paths over distances up to 20 miles, HD can be fed from the field to stations, or a number of programs in standard definition, such as multiple-camera shots, can be sent in the same transport stream. The transport stream consists of data representing video and audio data, metadata, and even Internet Protocol (IP) or UDP data. These attributes may apply to DSNG as well.

Technical Differences between Analog and Digital ENG

Digital techniques have added several levels of complexities to DENG and DSNG operations that will require further training and education for operators of this equipment:

- A compression layer is added. The narrow spectral bandwidth allowed requires a large-scale bit reduction.
- The modulation layer is made more complex. Coded orthogonal frequency-division multiplexing (COFDM) is the most popular modulation scheme used to combat multipath problems.
- The RF amplifier layer must be more linear. Multiple carrier COFDM uses QPSK, 16 QAM, or 64 QAM.
- The RF path layer has relaxed requirements. FM may be used for path alignment, with a switch to digital for program transmission.

Drawbacks include the encoding time delay, which can be several seconds, so IFB delay coming back to the truck can be twice what a single hop satellite link would be. The normal encoding choices found in other MPEG encoders offer a trade-off of delay *versus* bandwidth, compression, and quality. Modulation constellations, carrier guard band intervals, and forward error correction all affect video quality and reliability. These choices add up to a useful data payload ranging from just under 5 Mbps to just over 30 Mbps in an 8 MHz channel, depending on parameter settings, which is also the useful channel width using DSNG. In time, operators will find the optimum settings for specific types of shoots under specific circumstances. To address some of these issues, in 2006, the Advanced Television Systems Committee (ATSC) published its candidate standard for automatic power control of ENG transmitters over data return link bands created as part of the re-organized TV broadcast auxiliary service between 2025 MHz and 2110 MHz, and at the time of this writing, is seeking comments on the specification.

DRL Channels

As part of the relocation of TV BAS operations to 2 GHz, the FCC allocated two 500 kHz spectrum segments for digital return link (DRL) use. Each is divided into 20 25 kHz wide channels. The DRL channels allow communications between ENG receive-only sites and ENG trucks for automatic transmitter power control. Additionally, the channels can be used for two-way data communications, IFB, camera intercom, and control functions.

WORKING IN THE FIELD

People working in this area of the broadcast industry need certain sets of skills and traits. Remote production staff must work well under pressure, in harsh environments at times, and in different settings and situations. Surprises and last-minute changes are routine. The staff must also be able to get along with people who are strangers or who might only work occasionally. Remotes are a jumble of equipment, cables, and interfaces that usually have to come together quickly and then, more often than not, are changed at the last moment. More so than in a fixed studio facility, personnel on a remote must prepare for the possibility that anything that can go wrong probably will. The key to a technically successful remote video production is to rehearse, prepare for the worst, and always have a backup plan.

Personnel Required

Many different disciplines are required for on-location production. All shoots always have a camera person. On small-scale shoots, the person running the camera has other duties, as well. On large-scale shoots many different jobs are involved.

Camera Operators

Operating a camera can require long stretches of concentration and spending a long time on one's feet. On a multiple-camera shoot, a good camera operator not only gets the shots assigned to that camera position but also anticipates what the director is trying to show and is thinking ahead about show continuity. In a live event, the announcers are often directing as much as the director, so a good camera operator listens to the program audio and offers shots that fit the audio dialog. Also, good camera operators keep their heads out of the viewfinder and are looking at the action that is outside their shot so as to react faster and more smoothly to action that might intrude into their shot. On a large-scale shoot, the director usually has a camera meeting before the show to let the camera operators know the key stories and what areas of the action they are expected to cover. For large-scale events, the number of cameras has greatly increased. Producers and directors often use miniature point-of-view (POV) cameras to capture unique views unavailable to even the best seats at the venue. Cameras over baskets and over goals, in race cars, in blimps, on cranes, and on helmets have transformed sports and entertainment to the extent that the home viewer sees and hears much more than the live audience. Camera angles are often lower and closer to the action, or may be high up over the action, or any number of places—allowing the director to weave a story out of any drama arising from the event.

Technical Director

For a large-scale event, turning the producer's and director's vision into a video story is up to the technical director (TD). The TD operates the video switcher and usually any associated video effect devices, such as digital video effects (DVEs) and digital memory effects (DMEs). A good TD can quickly develop a good rapport not only with the director but also with the entire crew. An important factor is how well the TD can "fly" or operate the switcher's control panel—that is, how well the TD can recall an effect on the switcher when required. Most directors will appreciate any comments that TDs relay to them about any limitations they have regarding a specific switcher/ DVE package ahead of time.

Video Operators

Multi-camera shoots mean that all the cameras used must maintain a constant look. While white and black balancing cameras on a common object would get most cameras looking relatively the same, cameras on larger remotes are often in diverse locations apart from one another and often under different light levels and color temperatures. Switching between these various cameras could produce annoying changes in appearance from one camera to the next. The video operators are charged with maintaining the correct video levels and using camera setup control to ensure that all the cameras have the same *look*. The director often only notices video operators when it is obvious that they are doing a good job under adverse conditions, or if they are doing a poor job.

Replay Operators

In most sports, but also in some news situations, video replays are an important part of telling the story. Multiple replay activities often take place at once. While the director is concerned about covering the drama of the moment and the producer about telling the overall story, a replay strategy is also required. Often assistant directors direct instant replay coverage and let the director know what replays are available immediately after the fact. At the same time, associate producers could be at work assembling highlight packages. Replay operators, like camera operators, must be aware of their assignments and be prepared to present replays nearly instantly when called upon. This job has become easier with video servers replacing videotape recorders (VTRs). In this role, operators no longer have to make split-second decisions as to when to stop recording the action and rewind to be ready for a replay, as server channels can be set up to continue to record while a second channel instantly starts playing out the captured action from a set *in point*. Conversely, the job has become tougher because operators now usually handle more channels of replay input and output than when VTRs were solely in use.

Assistant Directors

Large-scale remotes will find assistant directors (ADs) involved in various roles such as moving talent through the production by relaying commands and providing cues. ADs in the truck are often used for commercial and feature coordination.

Audio Operators

Some personnel are devoted solely to the audio effort even in small-scale shoots, but when the shoot is of any substantial size the audio effort can become daunting. These remotes will have a person known as the "A1"—the audio person who actually does the main mix for the show and is the lead audio engineer on the shoot. The A1 sets up the sub- or group mixes, understands the general audio needs of the event, and takes most cues from the video, with only generalized direction from the director. The director and A1 will often discuss specific audio production goals and philosophies before the show. The A1 can have a number of assistants (known as "A2s") who also do sub-mixes. If an event has effects mics scattered in diverse locations, often a sub-mix is done by an A2 in a location where he or she can see the action. The A1 and crew are usually charged with running and setting up intercom and IFB drops.

Spotters

Spotters are often found near the field of play, as well as in the truck, and are the eyes and ears of the director and producer, letting them know when something is happening that might not be obvious on-camera. Spotters take up places in the booth, at corners on racetracks, in dugouts, at the sidelines, or other places. Often one is in the truck as another set of eyes to observe overall action and alert the director if something interesting or important is happening. Spotters prove useful because the crew might be concentrating on a current story line and not notice that a new and perhaps more interesting story is unfolding.

Producer

The producer generally establishes the concept for the event and the director carries it out with the equipment at hand. When the operations and engineering staff associated with a remote are able to anticipate and react to the requirements and concerns of the producer and director before and during a shoot, everyone will be less stressed, resulting in a better show.

PLANNING FOR A REMOTE

This section contains information about working outside the studio on facilities that will be used for live on-the-air operations. Who does what, the physical and communications arrangements are all important details that must be worked out in advance. In general, fairly early in the preplanning process, a producer and a director are selected for the remote. The major responsibility at this early stage is to define and communicate their vision of how the production will look and sound. This vision and style of production required will determine the deployment of facilities and personnel—ultimately driving the budgeting process and the final look and feel of the program. For a series of events, such as a network's coverage of the NFL season, the style and format of the production will be set by a team comprised of the executive producer, coordinating producer, coordinating director, and generally a member of the organization's graphics and operations departments. In this example, the producer and director are responsible for carrying out the vision and ensuring that their production maintains the desired look and feel.

Steps Required for a Remote Production

As an example, we will consider a parade to be televised by a local television station in the middle of town. Following are the main steps in the process leading up to the actual setup of the remote.

Securing Rights

First and foremost, the station's management must determine who is organizing the parade and approach the organizers about securing rights to broadcast the event. The broadcast of the parade may be viewed as a great idea by the organizers; in fact, the organizers might view it as a way to recoup some of the cost of putting on the event and require a payment from the station in the form of a rights fee.

Development of the Initial Production Plan and Budget

The station management in conjunction with the sales department will conduct a few quick calculations on the projected revenue to be made by the sale of advertising time and subtract the rights fee, if any, to determine roughly the amount of money available for the production. Note that this is not a simple "income minus expenses equals profit" equation for production. Advertising sales revenue is not the only factor used to determine the value of the broadcast. Goodwill, community service, cross-promotional opportunities, and prestige also factor into the equation. The major broadcast networks consistently report losses on their coverage of the NFL but continue to produce the games because the carriage contributes to their status as major networks and allows cross-promotion of other programming.

Selecting the Team

The producer and director may already be employed by the television station, in which case they may have been involved in the earlier rights discussions. Or they may be hired from the community of freelance producers and directors based on their expertise in this type of programming. Usually at this point a technical manager (TM) is brought into the team to assist with planning the implementation details, and a production manager (PM) is brought in to deal with logistics. The team has probably determined whether the station has the necessary in-house technical facilities to successfully pull off the remote. Equipment might include a combination of the station's news vans and camera crews. The team may also determine that an outside facilities vendor is required to bring in a larger, more sophisticated mobile production unit. Again, budget constraints and audience potential will be big factors in this preliminary decision. Depending on the need for lighting and the complexity of the audio requirements, a lighting director and an audio mixer, or A1, may be added to the team at this point. Once this core group is assembled, it is time for the team to meet with the parade's organizers and survey the parade route.

The Survey: Gathering On-Site Information

A remote survey will be conducted by going to the site, walking around and taking a look at the parade route. The initial survey should be attended by at least the producer, director, technical manager, and members of the event's organizing group. More complex events, such as a parade, may require several surveys, often involving lighting designers, audio, lead camera, and other disciplines. For the first survey, it is important to walk around the venue's exterior, checking for building access, loading docks, fire zones, etc., before proceeding to the actual event site. The mobile unit will have to park somewhere and many cables will be running to the cameras and announcer locations, so options for parking must be determined early on. Often, the technical manager will arrive early to get the preliminary exterior walkaround out of the way before initiation of the official survey. A remote production can involve extended areas. When surveying for a parade remote, a block or two on either side of the route where the anchor coverage will be located should be checked. Pictures should be taken of possible parking sites and other points of interest. When the production coverage requirements are decided, the TM will confirm the place to park the mobile unit. This location could turn into a TV compound containing the TV trucks, generators, office trailers, catering tents, and a satellite uplink. Ideally, the mobile unit would park as close to the action as possible, eliminating long cable runs. A modern expanding trailer with tractor attached will require a reasonably level area approximately 75 by 25 feet for operation and setup.

Placing Talent and Equipment

The most important part of the survey is identifying where the cameras will be located. In many cases, platforms or scaffolding must be erected to support the cameras and their operators (Figures 5.17-10 and 5.17-11). Special mounts such as jibs and pedestals will be identified (Figure 5.17-12). Preliminary lens needs will be noted. The TM should take photos of the camera locations, including shots showing the cameras' points of view of the action. When production has identified all the camera locations, cable routes to the truck will be plotted out. Production's second priority will be the location of the announcers and any special platforms and tables required. This first survey will include an initial check of light levels and possible lighting positions. Depending on what is found, additional surveys with a lighting designer's services may be required.

FIGURE 5.17-10 Camera on scaffolding behind spectator seating area.

The TM should also determine what cables to run and the paths the cables must take. Often, special equipment, such as cherry pickers, or union or other venue-specified crews must be used to install and remove cables. Some venues have the necessary cables in place, but in those cases the TM should make sure

FIGURE 5.17-11 Camera on motorized platform behind spectator seating area.

FIGURE 5.17-12 Over and under coverage—camera on jib and camera and operator on ground.

which cables will be available for their use, as others might be using the installed cable at that time.

Incidentals

Additional items that the survey team must consider during the initial survey are the type of peripheral support equipment and supplies that will be required and where they will be housed. Such additional equipment and supplies might include portable bathrooms, telephones, copiers, furniture, drinking water, soda, food, and ice. If these support items will require office trailers or perhaps tents, it must be determined early on where they will go.

Power

The units or single truck must be parked as close as possible to the power connection. Most trucks carry a hundred feet or so of power cable for the three phases, plus ground. Power requirements of the truck should not exceed 65% of the source circuit breaker to ensure that it will not trip during the event. When possible, it is best to use separate circuits for electronics and lights. Lighting equipment has a tendency to generate noise spikes on the AC power line, and most electronic equipment is susceptible to interference. Quartz lights, when powered over time, tend to heat and trip circuit breakers when operating near their maximum capacity. Keeping the lights on a separate circuit will help keep lighting and electronic problems separate.

There are two philosophies involving truck grounding. A ground cable is necessary for safety. With no ground cable, a voltage source shorted to the trailer frame could make the whole trailer rise to the shorted potential. Some argue that if the truck is grounded it and everything connected to it become lightning targets. Also, in theory, an ungrounded truck would have no ground loops. But, invariably, something external will ground the truck—a monitor plugged into a wall outlet in the venue, a camera pedestal leg touching a railing, or connection to another truck. Ground loops will also occur if power legs are not balanced. On trucks that do not use a ground, it is possible for an isolated coax cable (to a grounded monitor, for example) to be carrying many amps of current if it has inadvertently become a ground path for an ungrounded truck. In general, grounding the truck is the best policy.

Generators

If it is determined that the remote will require backup power or that venue power is inadequate, then provisions for positioning a generator must be made. Rent a unit that can supply at least double the total load the production equipment will draw. Look for a well-maintained generator, inquire about its condition, and ask to see its maintenance log. Do not rely on a generator that is nearly due for scheduled maintenance or one that has no maintenance record. Some generator systems can weigh many tons and can reach high sound levels of over 70 dBA at 50 feet. On the other hand, many generators have sound-baffling enclosures that can lower noise levels by 30 dB or more. Such units (Figure 5.17-13) should be selected when noise levels are an issue.

Radio Communications

Unless the remote is in the same locale as the station where frequency coordination is not a problem, it is important to check with the SBE frequency coordinator to ensure that the production will neither be subject to interference nor cause interference with anyone else. Frequency-agile RF equipment helps avoid interference problems. Note that radios must have proper FCC licensing. The use of raised directional antennas,

FIGURE 5.17-13 Example of generator with acoustic cabinet to reduce noise so it may be placed closer to the vehicles it is powering.

FIGURE 5.17-14 Example of antennas for wireless devices (such as mics, intercom, IFB, two-ways) at a remote venue; shown on raised platform to improve coverage.

as shown in Figure 5.17-14, helps improve reception but also introduces the potential for interference.

Contingency Planning

The performance of ancillary systems not directly responsible for capturing audio or video can have a profound effect on how smoothly a remote production goes. In particular, tallies and intercom systems can wreak havoc if they are not working correctly. Probably the greatest confusion that could arise in the middle of a production would be the result of a loss of the intercom system. The orchestration of a production without the ability to communicate with everyone involved is almost impossible. Expect failures and have contingency plans available for backup and maintenance when problems arise.

Logistics: Firming Up the Plan

When the preliminary survey has been completed, a review of potential expenses will take place. Usually prior to the initial survey, production has a general idea of the required equipment and staff. The survey should have refined those assumptions; for example, maybe an extra camera is needed for a special shot. Perhaps there are special labor requirements with the venue's electricians or stage crew or additional lighting may be required. These items all need to be included in a budget based on that initial survey.

The Crew

A television remote crew is made up of a mix of creative and technical staff members. Beside the technical and operations staff required, there is a cadre of production support that may include writers, researchers, editors, and statisticians. For a leased truck, the mobile unit vendor will supply an engineer-in-charge (EIC), engineering maintenance, and a driver. Staff for utilities, runners, and possibly riggers round out the crew, but do not forget the announcers, caterers, generator operators, and security personnel. A production manager serves as the logistics coordinator for the remote. Most of the crew may be local and live in the area, but others may be flown in for the production, in which case the PM coordinates travel arrangements. The out-of-town crew will require hotels, transportation, and food. The survey will have identified any specialized equipment necessary, such as jib arms. Typically, the PM will make arrangements for any extra gear not supplied by the mobile unit. The PM also arranges for office trailers, catering, telephones, and just about everything else. Particularly big shows will have multiple TMs and PMs concentrating on specific areas.

Bag of Tricks

Just as the PM and TM plan a remote from a high level, most of the crew need to plan for remotes at an individual level and prepare to come to the site with the proper tools to do the job. Most people who work in remote production for any length of time build a remote "bag of tricks." A typical remote bag contains a variety of disaster-averting tools and useful gadgets. Items found in many remote bags can range from gaffer's tape and permanent ink markers to volt-ohmmeters and portable test generators. The bag usually also includes various audio and video cable adapters and turn-around devices that can rapidly solve a problem. Other items might include extra audio and video cables, work gloves, and prepackaged wet handwipes. It is standard procedure for people who shoot ENG to rely on their remote bags for quick access to supplies that may be difficult or impossible to find on location. Note, that the use of three-way to two-way electrical outlet adapters is not recommended, as removing the ground connection from equipment can be a real safety hazard.

The Budget

Once initial costs are determined, cuts are often made to bring the project and budget into line. The total number of cameras is a natural target for cuts. Large numbers of sources, whether cameras, VTRs (or video servers), or graphics devices, mean more trucks, more support pieces, more space, and more time for setup, all of which equate to greater cost. The PM and TM must be able to provide the required production tools in the most economical way, but, as is often the case, they must balance this against the expectations of the production staff and client.

SAFETY

Remotes are, by their very nature, involve electrical systems in which lethal voltages and currents are always present. Death and mayhem have occurred on location as a result of tragic accidents caused by a lack of training, by inattention to the subject at hand, or by a rush to get on the air. Most deaths have occurred on ENG vans, but shooting on location in general provides for greater exposure to danger. ENG operators have been electrocuted by raising antenna masts into power lines. Some of those operators were even aware of the power lines above but miscalculated the tilt caused by the crown in the road. Standard power lines along residential streets may be at 19,900 volts or higher. Survivors of contact with such potential will often be minus arms and legs, as these often act like fuses when encountering this kind of electrical power.

In the case of a truck whose mast has become entangled in power lines, there are varying opinions as to what to do. An argument is made that remaining in the truck should prevent providing a path to ground, because of the truck's rubber tires, but, if equipment and air bags are exploding, a natural reaction might be to get out. Generally, advice on leaving the truck is to jump out to prevent contact with the truck and ground at the same time. Some operators have survived the initial energizing of the vehicle caused by contact between the truck mast and a power line because a power company circuit breaker tripped, but some have mistakenly thought the emergency was over and were caught in the circuit because many breakers automatically reset and re-energized the line a few seconds later.

Federal and state regulations, along with common sense, are appropriate guidelines. Stay at least 10 feet away from all overhead lines, as the warning sign says in Figure 5.17-15. If the lines are large, stay farther away. Staying across the street from wires is appropriate. It is important to have a spotlight in a truck at night when lines are not easily visible.

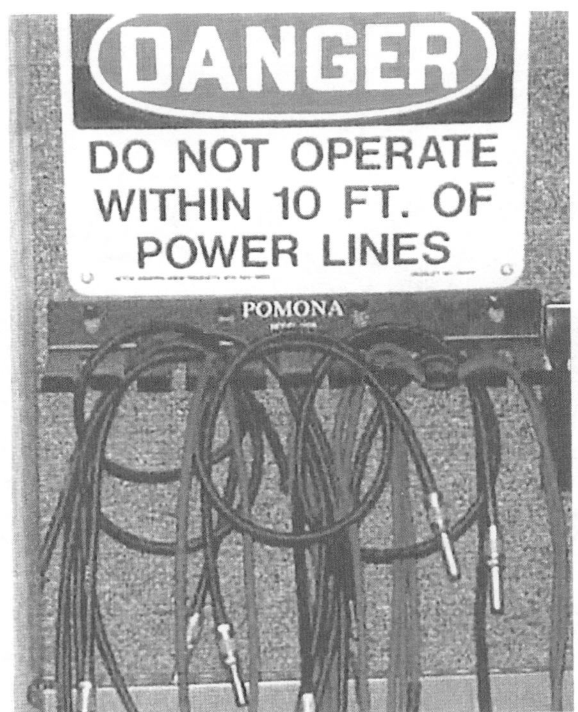

FIGURE 5.17-15 Warning sign posted in a conspicuous location in an ENG vehicle.

General Safety Guidelines

When arriving at the remote site, look around, walk away, and look up. This is an important discipline for safety. There likely will not be a power line accident if the view from 50 to 100 feet away has not revealed any danger. After the event, take extra and specific time to check that the microwave antenna mast has been properly stowed. Walk around twice to double check that the expensive mast-top equipment is properly lowered and safe. Inspect safety devices and write up any problems found so they can be rectified as soon as possible—those devices are there to save lives. Make sure clearance and danger signs are clearly visible to all operators. Signage saves lives and property. Be sure all operators have read and reviewed operator's manuals and safety literature for the equipment they are to operate. Make sure drivers realize that, even with the mast down, the equipment extends several feet above the roof of the vehicle and caution must be exercised when driving into areas without enough clearance, as shown in Figure 5.17-16. Remote trucks and facilities are not always comfortable environments. Because of space and weight restraints, many things around, in, or attached to trucks are as utilitarian as possible and may have hard, blunt, or sharp edges that can cause injury. Care is needed at all times. Establish safety priorities; the following are guidelines from a remote truck operator but are appropriate for all cases:

- Priority 1. Safety of the spectators (audience)
- Priority 2. Safety of the participants (athletes, performers)
- Priority 3. Safety of the crew and talent (you, me, and the reporters)
- Priority 4. Safety of the gear and mobile unit

Before the setup of any piece of gear or running any cable or connecting any wires, review these priorities. Ask what can go wrong and who could be hurt. Check with others to be sure.

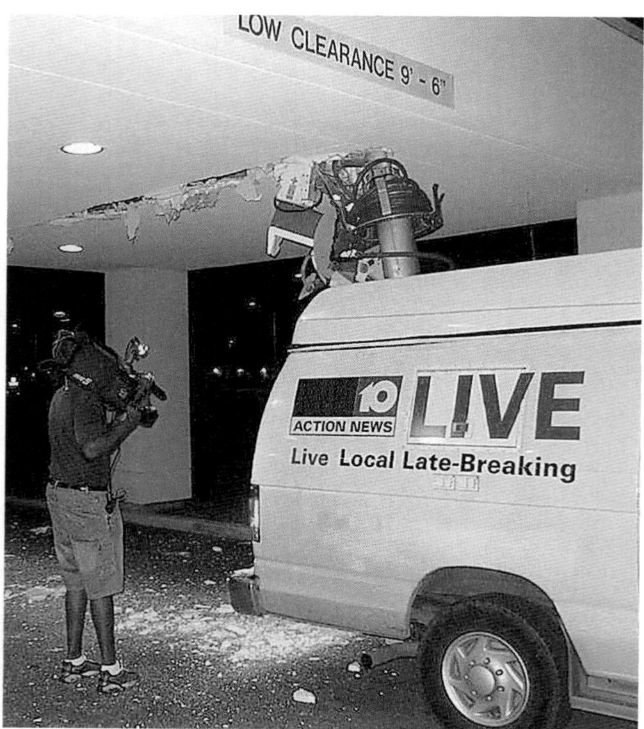

FIGURE 5.17-16 What can happen when the driver forgets that the mast (loaded with equipment) is still higher than the roof of the van, even when lowered.

ON-SITE SETUP

Setup for a remote production event can take a few days or a few months depending on the size of the event. With SNG remotes, setup on-site to air can be under 15 minutes and ENG often under 10. It is these short setup times, however, that increase the chances of mishaps, as described in the previous section.

Power

SNG and ENG units usually have on-board power generators. Larger trucks rely on local power sources or bring separate generators. Large trucks require upwards of 100 amp per leg of three-phase power; thus, the cables are thick and heavy with *camlok* type connectors to ensure that the connection is tight and has very little resistance. A general rule is that the male connectors go toward power, so, in general, cables running to the truck have the male (or pins) ends toward the truck. Truck power cables have their male connectors toward the breaker box. The exception is the neutral (if one is run); its male connector is toward the truck (this actually is in keeping with the "toward-power" rule).

Opening Up the Truck

Besides powering up the truck, the EIC and any assistants available must also provide access to the vehicle and its equipment. If the truck has only an EIC and no other help, the stairs are not installed until other crew members arrive. The unit's truck driver will often help with stair installation if it is too much for a single person to handle. Because many trailers today have expandable sides, if necessary the EIC will also bring the expandable sections out via a hand crank or via motorized system, when available.

Crew Call

It takes a number of different skills to produce a television show. The need for these skills is emphasized when the show is produced in the field. Often the pace of the show is much quicker than in a television studio. The amount of equipment, even for straightforward remotes, is far greater than is typically used inside a studio. In a studio, the use of four cameras would be considered a large production. That number in the field today would be considered on the small side for most non-ENG/EFP/SNG remotes. Many operators in a typical television station have never seen a studio camera broken down into a lens, camera head, pan/tilt head, and dolly or pedestal, as separate pieces sitting on the floor. Remote personnel are expected to gather all the pieces required to build a camera from the trailer bays, haul them out to their required locations, and assemble the pieces into a complete working system. Venues that see a lot of television coverage usually have preinstalled cables that run between the deployed equipment and the truck location. Trucks working in venues that do not have these cables must run their own. When cables are to be run, most of the crew usually helps with the task. The camera operators will generally work on running the triax cable out to the various camera positions, along with any other video coax necessary to support the production.

Early Production

Often, the needs of the production take some of the crew from cable-running and equipment-building duties. The producer may want to start pre-production in the truck immediately. Pre-production might entail producing the show's open or close or sending a handheld camera and VTR out to gather local color material, which is edited in the truck into a package for playback during the show. A major part of pre-production involves loading and creating all the graphics needed for sports productions.

The TD must often support the pre-production efforts by loading the transition moves into the effects devices (DVEs or DMEs) and the video switcher. These effects are most notably evident when switching between live action and replays in sporting events. The truck EIC generally assists the setup crew by showing the crew where equipment is stored. Good crews, especially ones that have worked on a particular unit before, will need very little help from the EIC. They know what parts are needed for their task and where they can be found. An important duty of the

EIC is as the onsite representative of the truck vendor to the client. EICs need to handle change well because things often change once on location, in spite of a good initial survey. The numbers of cameras and intercom drops and monitor placements often increase, and ones originally planned for can often change.

Pre-production demands can start as soon as the truck arrives. Most EICs have stories of clients with tapes in hand waiting impatiently as the truck is still being parked, thinking that as soon as the stairs are up pre-production will commence; however, a trailer must undergo an environmental stabilization process before equipment can be turned on.

SNG or ENG operators often race to the scene of breaking news and are expected to be feeding a signal back to the station within minutes of arriving on site. These trucks usually are concerned with delivering a single camera's video and the output of a single mic back for broadcast. As long as the SNG or ENG transmitters work and the camera and microphone are operational, their show goes on as expected.

Different crews, clients, venues, and trucks make for a different mix of connectivity and equipment topology each time. Larger productions not only demand that more equipment be spread through a venue, but they can also mandate larger crews with more complex communications requirements, resulting in more cabling throughout the site.

Engineering Setup

While the cables are run, cameras built or assembled, announcers' facilities assembled, and microphones placed, setup in the truck is also taking place. Just as cameras and mics are placed throughout the venue to accomplish what the client wants, the infrastructure in the truck is also configured to meet the client's needs. The facilities in a remote truck must be flexible. The video switcher and its associated effects—along with the audio, graphics, tallies—must work in a way that allows the client (or more exactly, the client's producer and director) to capture its vision for the production.

The technical manager, who works for the client and not the truck vendor, oversees the technical aspects of the shoot to be sure that the show is produced according to the client's wishes. The technical manager works with the truck's EIC and engineers so the truck vendor's equipment is configured as needed. Sometimes this requires a little subtle negotiation and tact on the part of the EIC, as the client's needs are not always in sync with what is best for the truck. Because most technical managers were, and usually still are, part of the truck crew, viewpoints usually come into reasonable alignment during setup.

If the show has a large production component—multiple producers, assistant producers, and editors, among others—the client might provide a production manager. Here, more coordination is required as the pre-production needs often interfere with technical setup and *vice versa*. Technical and production managers work together to balance the needs of each.

Truck Configuration and Facilities Check

During setup, the truck's engineers program a number of core systems that tie racks of equipment and scattered gear into a cohesive production system. These systems include monitor walls, routers, and intercom. After the completion of the technical setup, the system fax occurs. In this case, *fax* is short for *facilities check*—it is the television version of an aircraft take-off checklist. During the fax, all the sources are checked to confirm that they will work as part of the overall system, that accepted technical parameters are met for each, and that the various source(s) can be seen at every operator's position and be received by every piece of equipment that needs it. The fax also determines that the tallies that accompany each source and the control of sources such as TD control of a video server are exactly where the client desires them to be.

A transmission fax must also be done. Here, the main and backup feeds must be checked back to the broadcast operations center (BOC), network operations center (NOC), or other destination. In these cases, the truck's EIC must work with both the receiving television NOC and with the company providing the backhaul circuits to ensure transmission paths are working within required technical parameters.

When the technical fax is complete, the director's fax will occur. All operators and production people will take their operational positions and recheck all facilities, particularly checking all aspects of communications, IFB, and audio levels and confirm that they can perform the desired transition effects. Once both faxes are complete, then rehearsal can commence. For shows, this often involves talent and crew. For sports, rehearsal often just involves the director going over shot assignments with the camera personnel, and some practice with opens, closes, or other production items that will occur during the game.

As mentioned previously, many backhauls are a combination of fiber and satellite paths, but even for large events ENG paths are often used. If the backhaul is via satellite, at least to begin with, the EIC must coordinate the transmission fax with the BOC and the satellite truck uplink operator.

Ancillary Setup Activity

Large entertainment shows often require many lights and possibly the installation of lighting grids. This requires the use of *riggers*, a term originally coined for people who worked with sailing ships rigging but which now includes people who work at heights. For sports shoots, lighting often is just a matter of lighting the announce booth which can be handled by the production truck vendor. Entertainment shows often have a lighting director in the production truck in communication with the director. The lighting director also communicates with the people who control the dimmers and any computer-controlled movable lights in use.

While the truck and its equipment are set up other components of the overall production are being set up in parallel. An entertainment shoot often requires a large amount of lighting, expanded audio operation, or specialized camera gear such as stabilized body-mounted cameras, cameras on jibs and cranes, and even skycams (either suspended from wires or actually flown on a blimp). Such specialized platforms often require special setups. They often come with their own crew and can have setup times lasting longer than that for the main remote, especially in a venue where the system has not been used before.

There is a cadre of vendors who specialize in handling the requirements that are outside the usual scope of the remote production truck. Sometimes the truck vendor will hire these specialists vendors as part of an overall package for the client, similar to a general contractor hiring subcontractors with specialized skills. Or the client itself will hire the truck company and the other vendors needed to complete the production. Many of these specialty vendors bring trucks and crews of their own, while some involve only a single person. Besides the main production facility, there is an interconnection with other specialty vendors providing, for example, additional lighting, projection displays, audio recording facilities, or enhanced graphics.

The Audio Effort

The personnel assigned to audio, including the person who actually runs the audio mixer (the A1) and one or more assistants, work on installing the required audio lines. The audio crew also usually handles the intercom and IFB drops where needed outside the truck. This task is often as large as the actual audio requirements for the show. Audio mixers are so flexible that most likely no two A1s would organize and mix a show the same way. Some boards have multiple sets of faders and allow faders to control different sources with the push of a button. Most high-end boards assign often-used sources to close faders and the "set once and forget" sources, often part of a sub-mix, to faders not as conveniently located. Many audio consoles display their computer underpinnings. Some boards have motorized faders and "soft" buttons, as well as rotary pots. This allows a particular setup to be saved and recalled.

The A1 and all the A2s on many shoots can have their hands full, depending not only on the size but also the type of shoot. On an entertainment remote, the placement and number of microphones can be a large effort, and having more than 60 audio sources on a sports shoot is not unusual. Many A1s place mics so the audio coverage equals the video coverage. Coverage does not mean just capturing the sounds but also building a sound image for the viewer. Many viewers now have more sophisticated television listening environments than simply the one or two speakers that are inside their television receivers. The television audio experience is catching up with the enhancements that multi-channel audio brings. The audio production effort must result in all the sounds as well as their placement in space for the viewer. An audio image must be created to accompany the video image. Even with a monaural feed, poor mic placement can result in inadequate pickup from desired directions and at some frequencies. With stereo, the problems are compounded as phase of the audio signals becomes important. Simple stereo mic placement for football would entail mics on each side of the field at both 20 yard lines and the 50 yard line. In addition, four portable parabola mic operators and an umpire mic to catch backfield and quarterback audible calls would be needed. A football field is about four times the size of a basketball court, so a mic set-and-forget strategy is not possible. Field coverage must move as the game moves up and down the field. Thus, a number of operators operating parabolic microphones are often used. Many football games are also open to the elements and RF interference is often significant, all adding to the challenges.

Microphone manufacturers have helped the stereo effort, placing two mikes in the same package. When placed on a camera and on-air, they can produce stereo sound coincident with the observed video. If stereo were to be generated from two mics not collocated with the camera, the sound image could be dramatically wrong. Parabolas and shotgun mics are tools usually only found on sport shoots where a mic cannot be placed in the middle of the action.

Entertainment shows usually call for a different scheme of mic placement. For concerts, be they classical or rock, the audio becomes equally as important as the video. If the show is being taped, a separate sound truck might be called in to record the audio separately, often on as many as 48 tracks. Many, if not all, of the individual audio sources will be laid down on separate audio tracks to be mixed down later in postproduction. Usually a working mix is recorded live during the show to be used as a reference. This kind of production usually calls for different mic placement, one that puts the mics close to the instrument or person being recorded. This might mean that, in the post-production mix processing, reverberation will have to be added to give warmth back to the overall mix.

Generally the A1 will set up a number of sub-mixes for the show. Instead of treating every source individually, they will group types of sources (also known as *stems* in the film world) to be sub-mixed. Stems can be announcers, effects, replays, etc. These sub-mixes are then finally mixed into complete output channels, of which there can be many. Mono, left, right, center, rear, natural sound, and IFB could all be separate audio output channels.

Surround Sound Mixing and Encoding

For NFL games and other events produced in Dolby® Pro Logic® or Circle Surround™ sound, a surround sound image is built for the viewer. The field sounds are in front and the crowd noise is in back. Actually, there are four audio channels to be created: left, center, right, and rear surround. These are encoded in a Dolby four-channel matrix encoder, which produces two signals: left-total and right-total. The left-total has in- and out-of-phase audio. The in-phase is left and center audio, and the out-of-phase is the rear. The right in-phase is right and center audio, and out-of-phase audio is surround. The rear channel is delayed slightly to mask front channel leakage, based on the fact that if two similar sounds arrive at slightly different times, the listener will not perceive the later sound. To stop rear surround sounds from leaking into the front channels, the surround audio is run through a low-pass filter. These two channels can be decoded by any Dolby Pro Logic decoder back to the four original channels. Anyone listening just in stereo or mono would hear normal stereo or mono audio. Dolby surround encodes four-channel analog into two-channel analog.

More recently, Dolby has introduced Dolby E, which lightly compresses and encodes eight channels of audio into one AES3 digital audio stream. Six of these channels are 5.1 channel surround sound and the other two channels usually carry left-total and right-total encoded signals. The availability of Dolby E now allows full 5.1 surround sound with discrete channels to be remotely produced, recorded, and distributed. Unlike the earlier surround system, the rear channels are now stereo and full-bandwidth audio, whereas the surround rear channel was mono and of limited bandwidth.

A lot must be done before the audio gets to the encoders. Many sub-mixes must be done first, and it often takes the coordination of multiple people. When multiple parabolic mics are used for NFL football, an A2 often sub-mixes these from the announce booth to coordinate with the action. This mono mix is then sent down to the truck where it is turned back into pseudo-stereo and becomes part of the main mix.

Some audio mixing consoles today have joystick or track-ball controls so the A1 can quickly move left, right, and rear mixes in the listener space to capture the sense of listener perception location changes. In essence, left, center, right, and rear (left and right in the case of 5.1) mixes can be made to shift clockwise or counterclockwise from one channel to the next adjacent to give the illusion of the listener turning his or her head in different directions. Often, however, the nature of the event will require that, besides the fully mixed four channels of surround sound, additional instances of background or effects audio must be created. Often a show will be live-to-air or tape (show is recorded but shot as if it is live-to-air) but will have to be edited later for reuse, where the announcer's commentary is not desired.

In the case of stereo or four-channel surround sound on two analog channels, two additional channels would have two additional channels for the "clean audio." But it isn't as simple as that. If something has to be modified afterward—for example, the announcer's track is to be replaced with only natural sound or a new voice-over by the announcer is desired—there would be a noticeable change in the sound, as there is some natural sound leakage into the announcer's track. Switching away to just natural sound or to new announcer's tracks without this leakage is usually evident. To mask this, an extra mic is sometimes hung somewhere in the announce booth to pick up additional natural sound heard in the booth. This is then added to the natural sound tracks.

Other Audio Concerns

Two common audio problems are dead mics and hum. Due to the popularity of the condenser mic, dead mic batteries have become a problem, particularly if batteries remain in mics for long periods of time. At best, the batteries should be replaced before a production begins. At the very least, the operation of all mics should be checked and extra mic batteries placed on hand. Hum can be a more difficult problem to resolve. In addition to the potential for ground loops, the production in the field introduces many variables into the audio and AC wiring not found in studios. Grounding problems are most likely. All audio circuits should be checked for hum and ensure that there is no AC leakage to the audio grounding system. Not all field locations are ideal for all types of mics. Lapel or clip-on mics are popular for field work because they are less visible and automatically follow the speaker. They also tend to eliminate background noise due to their close proximity to the body. However, these mics tend to muffle the sound as compared to a headworn boom mic, which sounds better and can be controlled. Wind is another common source of problems on location. A good microphone windsock will reduce or eliminate wind noise. Windsocks are made from a variety of materials such as fur.

DIFFERENT TRUCKS FOR DIFFERENT EVENTS

Whether considering the design of a new vehicle or preparing to lease a vehicle, it is important to establish what the truck will be used for and the expectations of the users and program makers. Equally important is to understand the different types of trucks and what their capabilities are. Besides specifying a truck for purchase by the station, a truck may be rented or leased for a one-time event or extended use. For high-end extended productions (such as covering a season of sports), a leased truck will be configured to conform to the client's operational requirements. An engineer or operator must understand the considerations and trade-offs involved in the integration of a vendor's production unit. Trucks can be designed mostly for sports, news, entertainment, post-production, or all of these and the specifications will be dictated by the particular applications.

ENG and SNG

Most television mobile units on the road are ENG or SNG vans used for live shots involved in local newscasts. With a small vehicle, a mast-mounted transmitter, a camcorder, and optional onboard editing, virtually any news event can be quickly transmitted to the public live. For larger and more sophisticated productions, equipment packages typically grow larger, requiring larger vehicles.

Sports

Large remotes are often referred to as entertainment or sports shoots. Sports shoots may be as different as the various sports themselves. It has been observed that football is a made-for-television event: a play and then 30 seconds to regroup or tell the story along the sidelines or to capture some of the spirit and spectacle that can be part of the college game. Some think that golf and motor racing are a couple of the most difficult sports to cover. In both of these sports there are no time-outs. In motor racing, even when there is a yellow flag, there is usually activity in the pits to be covered as cars take advantage of the situation. Golf can be thought of as a number of matches, a different one at each hole, all being covered at once. The producer and director must have good knowledge of these sports to be able to determine where the most compelling stories are on venues that can be miles long and when a commercial break can be taken without losing too much of the live action.

Entertainment

Entertainment shows present a different set of challenges for the director and crew. Whereas the sports crew can anticipate some of the action, a lot of the activity cannot be predicted—thus camera moves and switching can be a little looser and less precise to allow for rapid response as events unfold. With most entertainment shows, however, the program is scripted and usually rehearsed at least once; therefore, the producer and director are expected to create a show that has smooth, calculated, and deliberate camera moves and switching that happen at precise moments during the production.

Side by Side

Many times two sets of trucks are parked side by side to cover an event for different outlets; an example would be baseball coverage for the home and away teams. This is known as a "side-by-side" setup. One truck may take a clean feed (a feed without graphics) from the main truck and supplement it with additional cameras and its own graphics to produce a separate program output. This is often the case where a host broadcaster in one country provides the base coverage and a visiting broadcaster from another wants cameras to shoot its booth talent and concentrate on its participants, plus do its own branding for the show sent back home. Two of the largest side-by-side remotes are the Indy 500 and the Super Bowl. Besides the fact the approximately 50 cameras are shared between the various separate productions (pre-game, game, and post-game shows as well as international feeds in the case of the Super Bowl), both events have side-by-side shows and multiple rebranded feeds going out with additional trailers providing central support and program release for the simultaneous shows. In the case of the Super Bowl, the amount of gear used is so vast that 30 miles of cabling is typical. The setup starts almost a month in advance, even before the playoffs commence.

TRUCK DESIGN

The Planning Process

Unlike most other TV plant projects, remote production vehicles cannot be designed to allow for expansion in size. When a basic chassis is ordered, a number of aspects of the unit become fixed. Purchasing a remote production truck is a major investment which, if not planned carefully, will adversely affect the station's bottom line. It is usually a more straightforward process when it comes to an ENG or SNG truck where the choices are perhaps more limited and the investment less. Most trucks are designed and built by systems integrators, with broadcasters or other customers specifying the attributes that are required for the finished truck. A remote news or production truck, of whichever type, must have the following attributes:

- High quality
- High reliability
- Operational simplicity
- Ease of maintenance
- Compatibility with other station equipment
- Reasonable price

Whether considering a truck purchase or just the rental of one, a good grasp of the truck's purpose and workflow is necessary in order to determine the number of staff and what equipment is needed, as this fundamentally affects the size of the vehicle. This process is, in fact, far more complex than it might at first appear and requires addressing many questions and decisions. These include issues such as how many cameras, tape decks, servers, monitors, and other items of equipment are needed. Parallel technical issues include what formats and systems to use (480i, 1080i, or 720p, tape formats, server compression systems, and so on). Most new equipment is digital but there may be some economies achieved with certain analog equipment or some sort of hybrid (particularly for audio). Another factor is the skill level of the crew and how large a crew will be needed to produce the type of productions planned. The engineering contractor or consultant assists with these issues, but there will have to be a general idea of what direction the project is headed in before a consultant can be of much real assistance.

Building a truck involves coping with a long list of engineering trade-offs. Many of these are fairly clear and will be dictated by the truck's intended use, but a number are not. Subjective decisions that have to be grappled with include which way to orient the production compartment and what the general layout of the truck should be. Intended customers will often

express their desires and concerns early on, but these may not align with a truck that is suitable for general use.

Vehicle Size and Layout

ENG and SNG Trucks

ENG trucks may be as small as a converted sport utility vehicle, although vehicles based on somewhat larger vans or a small-box construction on a chassis are commonly used. The challenge is usually squeezing the equipment required into the small amount of space available. Two examples of well-equipped ENG truck interiors are shown in Figures 5.17-17 and 5.17-18. SNG vehicle tend to be somewhat larger than those for ENG, depending on the size of the dish that is mounted on the roof. Figure 5.17-19 shows an example of the layout of an SNG truck.

FIGURE 5.17-17 View of video and audio operating areas in a well-equipped ENG vehicle.

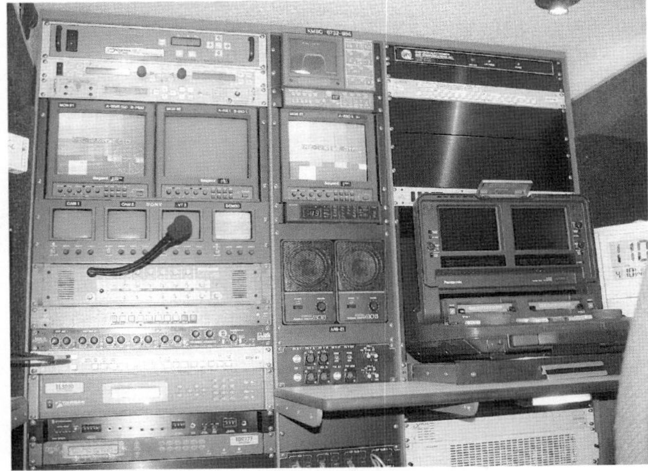

FIGURE 5.17-18 View of equipment wall in a well-equipped ENG vehicle.

FIGURE 5.17-19 Model showing cutaway view of a large SNG truck. (Courtesyof Wolf Coach.)

Mid-Sized Trucks

For many stations, the step beyond the typical ENG vehicle is a cutaway van chassis with a cube-shaped box attached to the frame directly behind the driver. The height of the box offers adequate headroom and is commonly about 14 ft long and 7 ft wide. Approximately 100 ft^2 of usable area is available for work and storage space in such a vehicle. Single-unit vehicles have drawbacks. One is a 29 ft maximum box length and resulting storage space limitations. Because of the drive shaft connecting the engine to the rear driving wheels, under-truck storage is reduced. Belly box storage depth is usually limited to about three feet, which prohibits its use for long tripods, light kits, cables, and bulkier items. For somewhat more room, a stretch van, a recreational or ambulance type of vehicle, or a straight truck can be considered, each of which may be outfitted as a very serviceable unit for two- to four-camera productions with onboard graphics. A straight truck, with a 29 ft box, provides approximately 200 ft^2 of deck space (see Figure 5.17-20).

FIGURE 5.17-20 Mid-sized remote truck. (Courtesy of Gerling and Associates.)

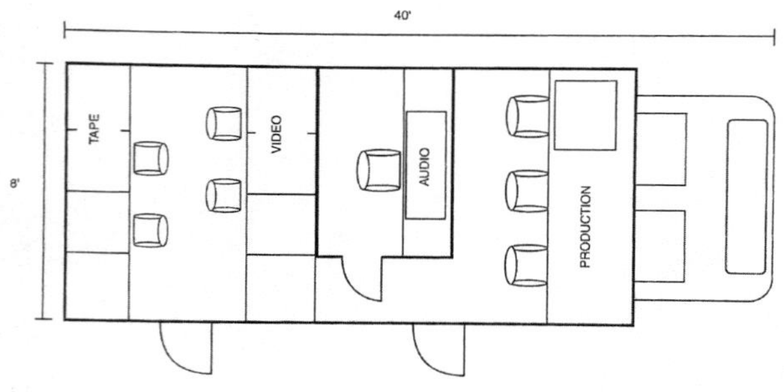

FIGURE 5.17-21 Typical layout of medium-sized remote production vehicle with room for eight. (Courtesy of AMV – All Mobile Video.)

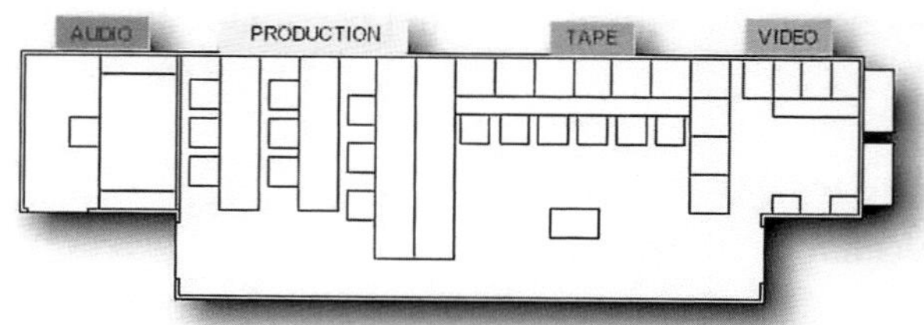

FIGURE 5.17-22 Typical layout of large "expando" remote TV truck showing the locations of the functional areas. (Courtesy of AMV – All Mobile Video.)

The size and layout of the truck will depend not only on the amount of equipment in the truck but also on the operational philosophy of the truck's owner and customers. If the director will be doing the video switching and running audio at the same time, a layout must be implemented to accommodate that workflow. Such a production compartment will be laid out much differently than one where a TD, director, producer, and multiple production assistants are used during the show. Figure 5.17-21 illustrates a typical remote van for modest sports or field production operations with room for eight staff.

Large Trucks

Truck layouts vary greatly, depending on the size of the vehicle and its intended use, but large remote trucks have two basic configurations for the main production area. An almost equal number of trucks have production compartments with monitor walls running street-to-curb facing the front or rear of the truck or, alternatively, with monitor walls parallel to the side of the truck, facing the side of the truck.

In both cases, layouts that facilitate scanning the monitor wall and making eye contract with other crew members and that minimize surrounding commotion and confusion are required. Monitor walls parallel to the side of the truck facilitate the large number of sources needed for some entertainment shows. Trucks with this orientation are often referred to as *show trucks*. Walls facing the front or back of the truck allow for more benches and consoles for additional producers and graphics operators, as shown in Figure 5.17-22 and earlier in the chapter in Figure 5.17-4. These trucks are often referred to as *sports trucks*. There is no clear consensus as to which orientation is best, and trucks laid out in a sports configuration are often used for entertainment shows and *vice versa*.

Weight

The weight load and its distribution in the vehicle must be carefully considered in relation to axle loading. A large concentration of weight over or behind the rear axle will create problems in steering. Ideally, the greatest concentration of weight should be fairly far forward; however, if the weight between the front and rear axles is too great, a tandem axle may be necessary to relieve some of the strain. Some modifications to the suspension may have to be made including heavier springs and shock absorbers along with electrical or hydraulic stabilizers.

It is important to know where the truck will operate and what the weight restrictions are in that area. As well as the weight of the basic vehicle, the equipment package can be considerable. ENG equipment packages can weigh in at a couple thousand pounds, SNG can easily double that. High-end box trucks can have equipment that weighs from 6000 to 8000 lb. At the very high end, equipment can push 15,000 lb or more. As a rough rule of thumb, the cable weight in a truck is often nearly equal to the equipment weight. At the high end, besides a large amount of required space and power, most trailers will be near the U.S. bridge restricted weight of 80,000 lb.

Cables constitute substantial weight for a remote production vehicle to carry. The combination of hundreds of feet of power cables and thousands of feet of camera, video, audio, and intercom cables may weigh several thousand pounds and may cause the truck to be overweight when it hits the scales on the highway. Although miniature cables may be lighter, they are more prone to damage and signal degradation. Careful selection of cables is important to maintain a balance between weight and performance and ruggedness. In some instances, a separate vehicle may be required to transport some of the cables to the shoot location.

Power

Power consumption rises rapidly with the use of digital equipment, which tends to be denser than analog. More computing power can be put in the same amount of space as earlier equipment, but it draws more power. Whereas a high-end standard definition truck might require 80 kW of power, a comparable high definition truck can draw well over 100 kW. At that consumption level, a generator or venue will require three-phase service with 208 volt circuits at 150 amp. These high current requirements mean that running the truck on single phase is no longer a viable

option. The biggest consumers of power are the on-board HVAC systems, as essentially all the power consumed by the electronics and lights ends up as heat. For heat and cooling considerations, a rough conversion factor is to multiply total equipment and lighting load current by 400 to calculate BTU/hr and then divide total BTUs by 12,000 to get the HVAC capacity required in tonnage. 500 BTU/hr per person inside the unit should also be added to the total heat load.

Driver Requirements

The size of the truck will also affect the requirements for the driver. Reliability and experience are very important. Requirements for drivers of high-end vehicles include a commercial driver's license (CDL), a spotless driving record, and experience with all types of driving conditions in all four seasons in all areas of the country. Because the driver is in charge of a multimillion dollar package of equipment, a very professional attitude toward the job is important. The Department of Transportation (DOT) has requirements for drivers of vehicles with a gross vehicular weight of more than 10,000 pounds. DOT regulations require a record of duty status reports to be kept by the driver (Title 49 CFR, Section 395.8). Requests by DOT inspectors to inspect the record could result in substantial monetary fines and a forced out-of-service period for a driver if rules have been violated. In some areas, additional local requirements more strict than those of the federal government will also apply. The driver must know all such rules for areas where the truck will travel and must meet medical physical requirements specified by the DOT.

Planning the Budget

Rough initial budget guidelines must be set before any contractor can provide meaningful help. It is important to match the needs with the budget in the early planning stages. There are huge differences in the costs of, say, a straight 32-foot truck and a 53-foot "expando," even with the same equipment in each. And those additional costs extend far beyond the difference in price of the basic box. Also worthy of consideration are the operating costs of the unit. Whereas the "expando" requires a tractor and driver with a CDL, the small straight truck requires no tractor and may or may not require a CDL driver, depending on several factors. All the above factors are heavily interdependent. Several iterations of the above steps are usually required before building a truck within budget that will serve its intended use. Vehicle type, size, equipment complement, layout, cost, and schedule all have a major impact on each step in the process.

TRUCK CONSTRUCTION

When the decision to build a new truck has been made, it is time to begin looking at the contractor or contractors required to get the truck on the road. When constructing the truck, there are four major components to consider:

- Engineering design
- Mechanical construction
- Technical construction
- Equipment suppliers

There are companies or contractors that specialize in one, several, or all of these aspects. When a decision is made as to how to proceed, most truck owners, especially in the ENG and SNG range, take the "turnkey" approach; that is, they hire just one company to provide all of the services, including financing. Others wish to have closer control over individual portions of the project. At least one mobile unit company manufactures its own units, other than the base trailer and HVAC unit. This decision is obviously based in large part on the level of expertise of truck owners and whether they have the personnel willing and able to be involved with some or all of the four major areas.

Turnkey Approach

In the turnkey approach, a full-service company is contacted and initial conversations are conducted between this firm and the client's in-house financiers, engineers, and other trades people. The rest of the decisions that the truck owner will need to make will be brought to their attention by the turnkey contractor.

Individual Contractor Approach

Up-front engineering is the glue that binds a vehicle and a list of equipment into a useful production facility. In the individual contractor approach, the engineer or consultant is probably going to be the initial and primary contact for the project. Regardless of who builds the truck, or how, the importance of sound engineering cannot be overstated. In the end, the engineer—be it one of the staff or an outside contractor—will ultimately be responsible for the truck's proper operation. That person is the architect of the truck, and it is that person's job to make sure all the pieces fit together properly. The engineer will have to individually select each of the contractors or suppliers. Even in this approach, one contractor will often handle a couple of the areas, such as mechanical and technical construction or engineering and technical construction.

Building the Box or Trailer

The mechanical contractor is responsible for the trailer body and running gear, the environmental systems, power systems, racks, consoles, lighting, walls, and floors to produce an empty "rack ready" trailer. The engineering usually has begun and at least a rough floor plan has been developed before the mechanical contractor is brought in. The truck owner and engineers will need to go over every detail of the physical truck with this company. The goal is to make sure that when equipment and the associated wiring are installed everything fits. Lighting needs to be over the consoles where it belongs. Wall switches should not be blocked by racks. There needs to be enough power and cooling in the proper racks. The power system needs to be large enough for the truck as currently conceived and for the future. The racks must be tall and deep enough for the equipment that will be installed. Raceways and conduits must be installed where needed before walls and ceilings are closed up. These are all inputs that the mechanical contractor needs from engineering.

Turning the Box or Trailer into a Production Facility

The technical systems contractor will take the empty rack-ready truck and turn it into a television production truck by following the plans drawn up by the engineers. This contractor will begin by studying the engineering drawings and discussing the project with the engineers. This contractor actually makes the cables and cable harnesses, wires the patch panels, puts connectors on all the wiring, installs the wiring and equipment, and coordinates the entire installation process to get the truck out the door on time. This will involve what is known as *commissioning*—making each piece of equipment and the entire unit as a whole work properly. This includes initial programming of the routers, intercom, switchers, setting up distribution amplifier levels, and labeling everything.

Testing of every wire and system will also be performed before the truck rolls out. There is usually a great deal of pressure on this contractor because, as the last link in the chain, he or she are expected to somehow make up for all the earlier delays. Allowing enough time up front in the project can make this contractor's job easier and will usually end up providing a better built truck. Most installation contractors have stories about the 12 week project that had to be done in 6 because of accumulated delays and opening day at the ballpark will not wait. Although the truck will probably get to the field for the first pitch in a situation like this, the installation will be lower quality than the full 12-week build would have allowed. Often the shakedown crew will not know what was left out until they try to use it and find it is not there.

SUMMARY

Like all areas of the broadcast and television production industry, digital technology has transformed the capabilities for news gathering and video production in the field. This has affected how remote operations are run and provides greatly enhanced production capabilities. Much of the evolution of equipment and operational facilities has been necessitated by the demands of television production outside the studio. With broadcasting's heavy emphasis on news and sports, this area of the industry continues to be an important and challenging one. Remote programming requires versatility, reliability, and quality in both equipment and personnel. Today's television engineers and technicians have the tools available to respond to the demands presented by remote productions in whatever situations and environments they may occur.

Bibliography

Remote news and production, in *NAB Engineering Handbook*, 9th ed., Whitaker, J., Ed., National Association of Broadcasters, Washington, D.C., 1999, chap. 3.9.

Boston, J. and Hoover, G., *TV on Wheels: The Story of Remote Television Production*, 2003.

Bennett, B., Trends in trucks, *Broadcast Eng.*, January, 2005 (http://broadcastengineering.com/mag/broadcasting_trends_trucks/index.html).

ENG Safety, http://www.engsafety.com.

Ericksen, D., Playing both sides of the street, SBE, 2006.

Maier, G., HD ENG: Is it ready?, *Broadcast Eng.*, May 2006 (http://broadcastengineering.com/mag/broadcasting_hd_eng_ready/index.html).

2 GHz Relocation Project, http://www.2ghzrelocation.com.

CHAPTER

5.18

Audio for Digital Television

TIM CARROLL
Linear Acoustic
Lancaster, Pennsylvania

JEFFREY RIEDMILLER
Dolby Laboratories
San Francisco, California

INTRODUCTION

Audio is an area in television broadcasting and production that causes apprehension with some engineers because of its complexity and the care needed to get it right in design, installation, and operations. With the transition to digital television (DTV), there is a new level of uncertainty because the requirements for audio are far more advanced and complex than before. Yet digital audio provides significantly more sophistication, artistic creativity, greater processing power, and higher quality than its predecessor. This chapter provides a general overview of the key features and requirements of audio for DTV, a detailed review of one of the largest problems today (loudness), and some real-world suggestions on making it all work. By no means an exhaustive discussion of what is involved, the material here will at least bring to light what must be done, and a view of what is possible with television and digital audio beyond just getting a signal on-air.

TELEVISION AUDIO SYSTEM OVERVIEW

Analog Audio

Television audio has been, until quite recently, a largely analog medium. It began as a single monaural frequency-modulated carrier at 4.5 MHz in the 6 MHz TV channel, as defined by the NTSC specifications. Stereo was superimposed on this carrier by the BTSC standard in much the same way as stereo was introduced to FM radio (see Chapter 6.3). However, while most FM radio stations transmit stereo, only about half of all television stations in the United States have implemented stereo transmission.

Compared to methods of delivering audio to consumers by the many digital means that are available, even the most carefully aligned BTSC system is not comparable in quality. Some of the issues include the side effects of high-frequency preemphasis (far less than FM radio, but still present), somewhat limited frequency response, low channel separation, and elevated levels of noise and distortion. For subcarrier channels that carry auxiliary programs, such as second audio program (SAP) or descriptive video service (DVS), the comparison is worse as these are monaural channels, limited to 7.5 kHz frequency response, and noisy due to limits on modulation.

Digital Audio

When the process for standardizing a digital television system began, one of the requirements was that the audio system should be enhanced well beyond the capabilities of the BTSC system. It was decided that stereo, then later multichannel, audio would be delivered digitally. To conserve data bandwidth, audio data rate reduction, or data compression, would be employed, with the objective of maintaining performance that was as measurably and audibly as close to the original as practical.

Multichannel Sound

The first actions toward achieving this goal focused on MPEG-compressed delivery of two-channel audio. The need for supporting multichannel surround sound was introduced because a substantial quantity of theatrical motion pictures had been mixed with four or more channels of audio: left front, right front, center, low-frequency effects (for driving the subwoofer), left surround, and right surround. With 70 mm films, the magnetic striped soundtracks were capable of carrying all of these channels, while 35 mm used matrix techniques such as Dolby Pro Logic to level- and phase-encode them into the standardized stereo variable area soundtrack.

Matrix techniques were briefly also considered for digital television, as MPEG could simply carry the audio in the same way the stereo variable area (SVA) track does for 35 mm film. This approach was not pursued, as Dolby Laboratories introduced a new system called AC-3 (Audio Coder 3), also now called Dolby Digital, which delivered nonmatrixed 5.1 channels of audio via a composite bit stream that fit into precisely the same space as the stereo MPEG audio. Today, Dolby Digital (AC-3) is a part of the ATSC standard, the European DVB standard, the Open Cable standard, and various DVD standards.

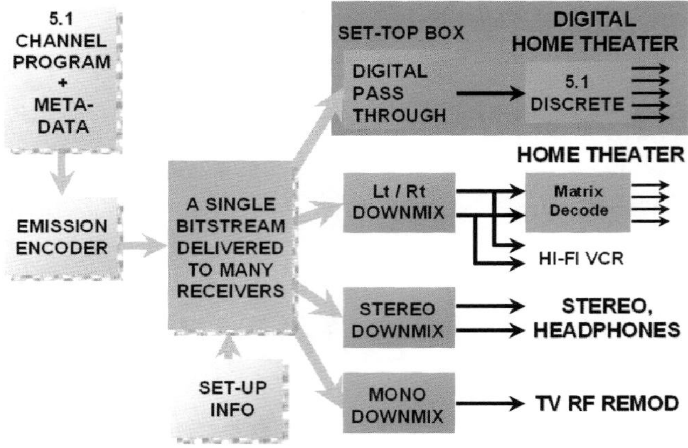

FIGURE 5.18-1 Single bit stream delivered to many different types of decoders; any bit stream can be decoded by any decoder under the guidance of metadata.

AC-3 Audio

ATSC standard A/52B[1] describes in detail the AC-3 audio coding system, and readers are encouraged to download a copy for reference purposes. This document also describes E-AC-3, the enhanced version of the coder for use with E-VSB (enhanced VSB) mode of the ATSC system. Enhanced AC-3 offers increased efficiency, but broadcasters should be aware that transmission of a standard AC-3 stream is still required to maintain compatibility with the large installed base of receivers.

AC-3 is an efficient audio coding system capable of carrying from 1 to 5.1 channels of audio at bit rates from 32 to 640 kbps with high quality 5.1 audio typically coded at data rates of 384 to 448 kbps (the maximum rate for ATSC emission). The sample rate for television uses is 48 kHz, and the system is capable of 24-bit audio resolution. Importantly, the AC-3 stream carries a parallel audio control data path called *metadata* (described in detail below), which provides additional important features used by decoders to optimize reproduction of the audio program for a given receiver in a given listening environment. The system is designed such that a single encoded bit stream can be reproduced by any consumer decoder, regardless of whether the source content is encoded as 5.1 channels or as mono, and regardless of whether the decoder is capable of reproducing 5.1 channels or only a single channel. Programs of any number of channels can be reproduced by *any* AC-3 decoder. Figure 5.18-1 shows an example of how a single bit stream can serve all types of decoders.

Note that multichannel decoders also contain a matrix decoder of some type (usually Pro Logic or Pro Logic II) to handle surround sound content that is carried in two channels. This audio will be matrix decoded using level and phase information and will then have the capability of providing audio to all speakers when appropriate. Unless care is taken, it is possible to cause audible clicks or pops when transitioning between 5.1- and two-channel material.

A local television station will need at least one AC-3 encoder per audio program. These encoders are often integrated into a full ATSC encoder, which, in addition to audio encoding, contains video encoding and multiplexing capabilities. Until recently, most of these encoders contained one or more two-channel AC-3 encoders, but for 5.1-channel programming an external encoder was required. Newer encoders may also contain 5.1-channel encoding. Note that it is likely that several audio encoders will be required for *each* video program in order to carry all relevant program audio.

Audio that is transmitted using the ATSC or OpenCable standards must be AC-3 encoded. Uncompressed audio is fed to the AC-3 encoder and an audio packetized elementary stream (PES) results. It is this PES data that will be multiplexed with video and other data into an MPEG-compliant transport stream for transmission to consumers.

AUDIO METADATA

An essential part of digital television audio is its metadata, or data about the audio data. Metadata conveys information such as the loudness of a program, how many channels have been encoded, how to downmix those channels if the bit stream is decoded by a two-channel decoder, and dynamic range control values to help match the audio to the listening environment.

Dialog Loudness

Based on research performed by Bell Laboratories, the BBC, Dolby Laboratories, and others, it has been determined that dialog provides the most common

[1]Available at www.atsc.org/standards.

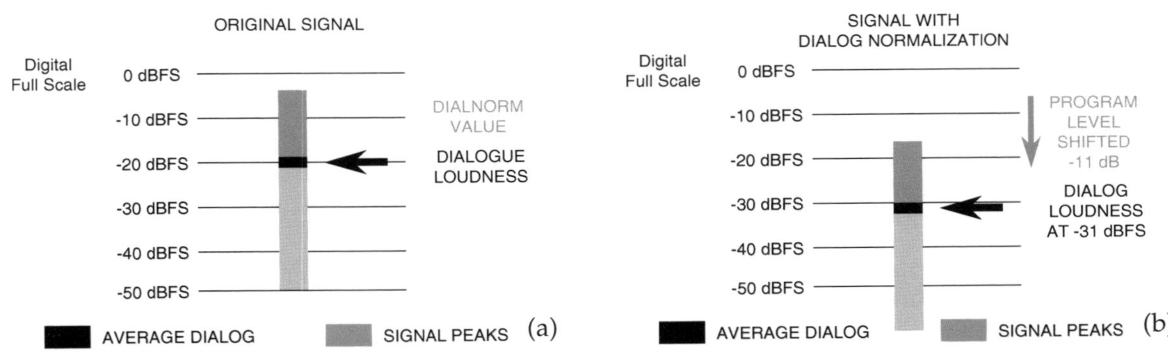

FIGURE 5.18-2 (a) Dialog loudness measurement and (b) application to signal.

loudness "anchor" of a program. It is what most listeners will use to judge the relative loudness of one program versus another. The *dialog loudness* (also known as *dialnorm*) metadata parameter is used to indicate the long-term, A-weighted loudness of a given program with respect to 0 dBFS (full scale). Each program requires its own measurement and the assignment of a unique dialog loudness value. This value directly controls a 1 dB per-step attenuator present in all AC-3 decoders, allowing programs to be scaled to the internal target for AC-3 of –31 dBFS.

Figure 5.18-2 shows how a typical program would be analyzed and dialog loudness parameter determined, and the results of applying this value after decoding.

The usefulness of this approach is illustrated in Figure 5.18-3(a), where multiple programs having different dialog loudness values are applied to the encoder, and upon decoding, are then scaled by the proper amount at shown in Figure 5.18-3(b).

It should be immediately apparent that while the average dialog loudness of each program in the figure is matched, the signal peaks have not been affected. Audio metadata delivers the unique ability to separate loudness matching from dynamic range control. This means that it is possible to more closely preserve content as originally produced, and leads to a discussion of the *requirement* for some sort of dynamic range control that supports this concept.

Dynamic Range Control

Television audio has traditionally kept dynamic range tightly controlled to ensure that loudness is consistent and that programs are intelligible in as many listening environments as possible. Unfortunately, this has meant that processing has been adjusted for the lowest common denominator, the 3-inch speaker on the side or front of a TV set, to the detriment of larger, higher-quality reproduction systems. With digital audio, loudness can now be matched without the need for severely impacting a program's dynamic range, and it should be apparent that far less dynamic range control is necessary and it can be different than the traditional approach.

Audio metadata contains Dynamic Range Control (DRC) gain words that can be generated by the AC-3 encoder and applied to the audio at the time of *decoding*—the original audio is not affected. Many Dolby Digital decoders offer the consumer the option of defeating the DRC metadata, but some do not. Decoders with six discrete channel outputs (full 5.1-channel capability) typically offer this option. Decoders with stereo, mono, or RF-remodulated outputs, such as those found on DVD players and set-top boxes, often do not. In these cases, the decoder automatically applies the DRC metadata associated with the decoder's selected operating mode.

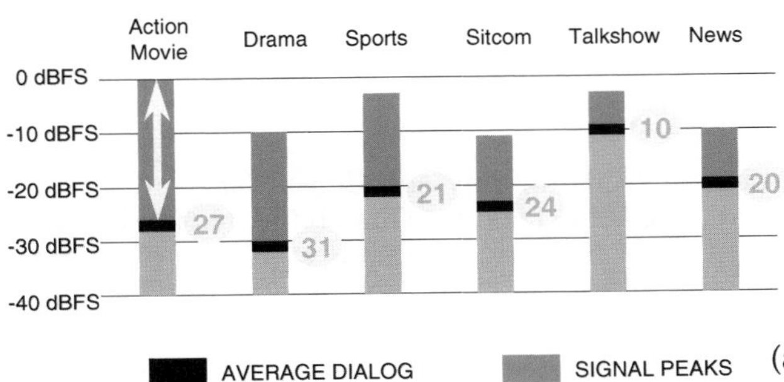

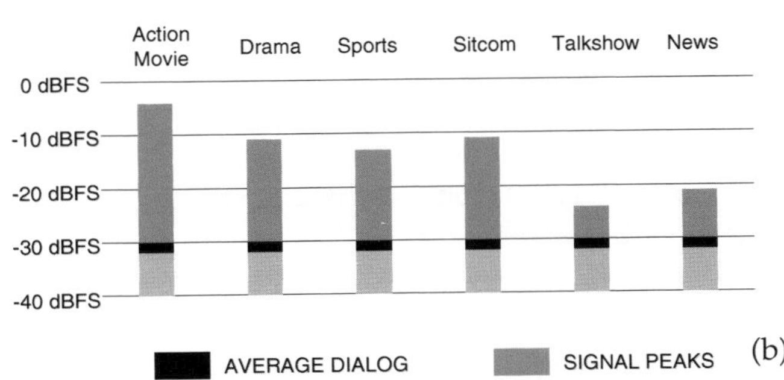

FIGURE 5.18-3 (a) Dialog level assignments for each program prior to encoding and transmission and (b) results after reception and decoding.

The DRC parameters that are part of the AC-3 bit stream are called *Line Mode* and *RF Mode*, also known as "dynrng" and "compr." Line Mode has ±24 dB of range in 0.25 dB steps, while RF Mode has ±48 dB of range in 0.5 dB steps. These modes are described in greater detail in the Loudness section below.

The system relies completely on the dialog loudness parameter being measured and set correctly or inappropriate DRC values will be generated, which will result in level control problems. Also, while the control loop used for generating these values is frequency weighted, the gain control is wideband and is applied to all channels simultaneously. This places practical limits on the degree to which dynamic range can be controlled without causing objectionable audible artifacts.

Figure 5.18-4 illustrates the transfer function for the DRC subsystem of Dolby Digital (AC-3). It shows that although it is constrained by the fact that is it wideband, the sophisticated structure of the compressor can yield reasonable results, again assuming that the dialog level parameter has been correctly set. Due to inherent natural delays in the audio encoding process, look-ahead processing is possible that further improves the audible performance of the DRC system.

No knowledge of transfer functions is required for users to generate dynamic range control parameters. Five presets available in all Dolby Digital (AC-3) encoders make the task straightforward. The available selections and corresponding key differences are listed below.

- *Film Light*
  - — Max Boost: 6 dB (below –53 dB)
  - — Boost Range: –53 dB to –41 dB (2:1 ratio)
  - — Null Zone Width: 20 dB
  - — Early Range: –26 dB to –11 dB (2:1 ratio)
  - — Cut Range: –11 dB to +4 dB (20:1 ratio)
- *Film Standard (Default)*
  - — Max Boost: 6 dB (below –43 dB)
  - — Boost Range: –43 dB to –31 dB (2:1 ratio)
  - — Null Zone Width: 5 dB
  - — Early Range: –26 dB to –16 dB (2:1 ratio)
  - — Cut Range: –16 dB to +4 dB (20:1 ratio)
- *Music Light*
  - — Max Boost: 12 dB (below –65 dB)
  - — Boost Range: –65 dB to –41 dB (2:1 ratio)
  - — Null Zone Width: 20 dB
  - — Early Range: None
  - — Cut Range: –21 dB to +9 dB (2:1 ratio)
- *Music Standard*
  - — Max Boost: 12 dB (below –53 dB)
  - — Boost Range: –55 dB to –31 dB (2:1 ratio)
  - — Null Zone Width: 5 dB
  - — Early Range: –26 dB to –16 dB (2:1 ratio)
  - — Cut Range: –16 dB to +4 dB (20:1 ratio)

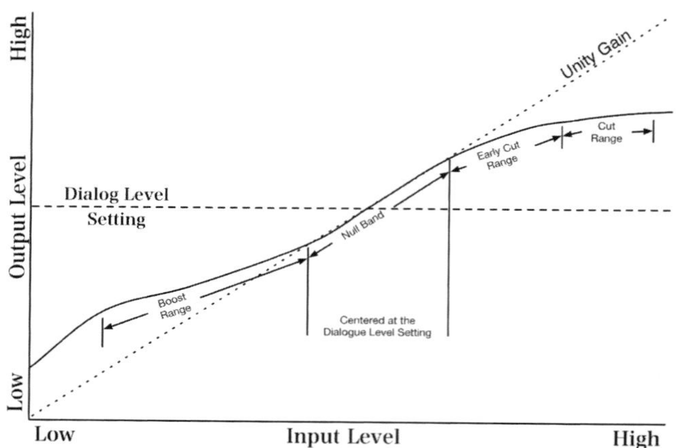

FIGURE 5.18-4 Transfer function of the DRC system in Dolby Digital (AC-3). Note that the minimal action null zone is roughly centered around the dialog level.

- *Speech*
  - — Max Boost: 15 dB (below –50 dB)
  - — Boost Range: –50 dB to –31 dB (5:1 ratio)
  - — Null Zone Width: 5 dB (–31 dB to –26 dB)
  - — Early Range: –26 dB to –16 dB (2:1 ratio)
  - — Cut Range: –16 dB to +4 dB (20:1 ratio)

The default selection is Film Standard, and is applicable to most programming commonly found on television. There is also the capability for selecting a preset that will not generate any DRC values. This selection is appropriately called "None," but note that while it will not act under most circumstances, it will do so in cases where dialog level has been set incorrectly and downmixing the audio channels in a decoder could cause overload. This so-called "Protection DRC" is automatic and cannot be bypassed. For example, consider a 5.1-channel program with signals at digital full scale on all channels being played through a stereo, downmixed line-level output. Without some form of attenuation or limiting, the output signal would obviously clip. Correct setting of the dialog level and DRC profiles normally prevents clipping and unnecessary application of overload protection. It is good engineering practice to avoid the "None" setting.

Downmixing

In order to handle any encoded bit stream, decoders must have some facility to match the decoded audio to the number of output channels available. This is particularly important in cases where the number of output channels is less than the number of encoded audio channels. *Downmixing* is the technique that makes this possible. Note that the LFE (low-frequency effects) channel is not included in any downmix but instead discarded.

Decoders generally exist in two forms: two-channel and 5.1-channel. This means that the two-channel decoder must then be able to downmix 5.1 channels to two, and this is further subdivided into two types. The first is for a standard stereo output and is called Lo/Ro (Left only/Right only) and is created by the following formula:

Lo = Lf + C*cmixlev + Ls*surmixlev

Ro = Rf + C*cmixlev + Ls*surmixlev

where Lf/Rf are left and right front channels, C is the center channel, and Ls/Rs are the left and right surround channels. Note the metadata parameters *cmixlev* and *surmixlev,* which determine the contribution of center and surround channels. The default setting is 0.707 (–3 dB), but can be adjusted to –4.5 dB or –6 dB in all decoders, and in the newest decoders can also be adjusted to off and 0 dB.

The second type of downmix is by far the most common found in the field as it produces a matrix surround compatible output. The resulting signal is fully appropriate for matrix decoding by systems such as Dolby Pro Logic, Pro Logic II, DTS Neo:6, and others. Note the differences in the following formula, where Lt/Rt represent left total/right total:

Lt = Lf + C*cmixlev – (Ls+Rs)*surmixlev

Rt = Rf + C*cmixlev + (Ls+Rs)*surmixlev

This formula is important. Note that channels are being added *and* subtracted and, therefore, can cause cancellations and other undesired effects. For example, it would be improper to place dialog equally in the center and surround channels, as upon downmix it would cancel almost completely in the Lt signal.

Proper downmixing relies on a 90-degree phase shift being applied to the left and right surround channels by the Dolby Digital (AC-3) encoder. Normally, it is enabled by default and it is necessary that it remains that way. It will produce little or no audible effect in a 5.1-channel environment, but is required for successful downmixing.

PROGRAM PRODUCTION

Now that the capabilities of the system have been defined, program producers can begin supplying audio content that will make it intact all the way to consumers. This is substantially true if certain rules that should be common to all delivery specifications (if not present already) are applied. Important aspects are monitoring, channel layout, synchronization (that is, lip sync), and audio upconversion also known as *upmixing*.

Monitoring

Proper configuration of 5.1-channel monitoring systems has been well defined by the ITU,[2] and a repre-

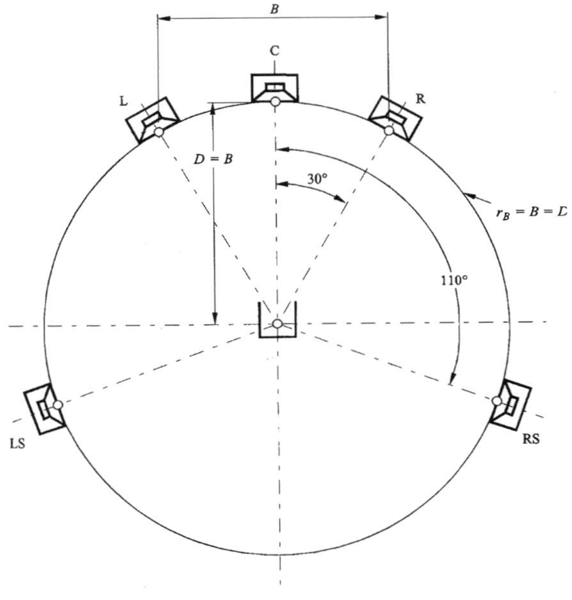

FIGURE 5.18-5 ITU-R BT.709-4 recommendation for multichannel speaker setup.

sentative speaker layout is shown in Figure 5.18-5. The subwoofer is generally placed on the floor near the front channel speakers.

Although beyond the scope of this book, room alignment is critical for creating a reference monitoring environment and a brief overview of the measurement and adjustment procedure will be presented.[3] It is impossible to know if a problem truly exists with the audio if the monitoring system is uncalibrated. Alignment is best done with a real-time analyzer (RTA) and calibrated microphone, but can also be accomplished with an inexpensive sound pressure level (SPL) meter. A suitable digital instrument is available from Radio Shack (see http://www.radioshack.com) and an analog version from ATI (see http://www.atiaudio.com). Meters from other manufacturers can, of course, also be used.

All channels should have an individual speaker. While some monitor systems allow for so-called "phantom center" operation, this is not an ideal scenario. In some cases, such as remote or outside broadcast vans, it cannot be avoided, but every effort should be made to have one speaker per channel, including a subwoofer for the LFE channel.

Once the speakers have been physically aligned, electrical alignment is next. Set the real-time analyzer or SPL meter to apply a C-weighting curve and a slow response. A reference listening level of 79 dB/C/Slow

[2]ITU-R BT.709-4

[3]An excellent reference called "5.1 Channel Production Guidelines" is available for download from www.dolby.com, and provides a comprehensive discussion of monitor setup and calibration.

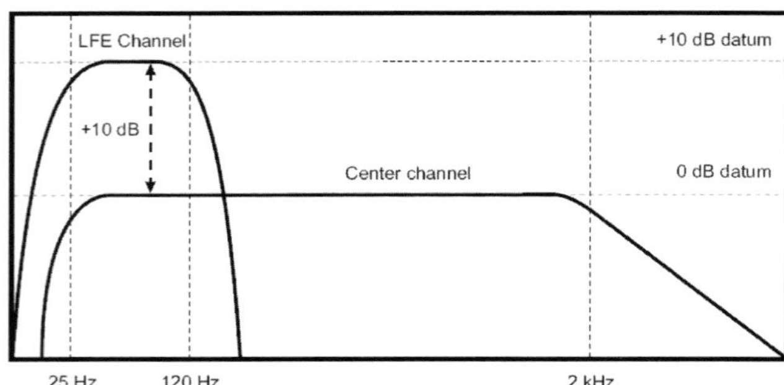

FIGURE 5.18-6 Real-time analyzer display of pink noise as reproduced by the center and subwoofer channels.

is recommended as it most accurately matches the average listening level for most viewers. Generate pink noise at the reference level as shown on the meters of a console or other metering device used during mixing (i.e., set the level of the pink noise so that it averages around 0 VU). Route the pink noise to the center channel and adjust the monitor volume control on the console or the monitor controller until the SPL meter reads 79 dB/C/Slow. Mark the position of this volume control and keep it there for the remainder of the calibration. Apply pink noise individually to each of the remaining speakers except for the subwoofer and trim their gain (not the master gain that is set at reference) until each reads 79 dB/C/Slow. When pink noise is panned to each speaker, they should all reproduce at the same level of 79 dB/C/Slow with no need to adjust any levels.

The subwoofer requires a slightly different alignment, as it needs an additional 10 dB of gain as compared to the other main channels in order to match consumer reproduction standards. Figure 5.18-6 shows how this would look on an RTA.

Note that the figure shows the response of a typical film mixing stage. Due to the large distance from the speakers to the mixer, there is a natural roll-off above 2 kHz (described in SMPTE 202M, and sometimes called the "X-curve"). In small mixing and control rooms, this response would be flat. The important point of this figure is to show that the subwoofer has 10 dB of additional gain as compared to the reference center channel.

The net result of this extra subwoofer gain is that it will cause operators to mix the sounds to the subwoofer channel 10 dB quieter. When this audio then reaches the consumer system, the subwoofer is again boosted by 10 dB and the net of the process is unity.

Channel Configuration

SMPTE 320M specifies the following track layout for multichannel audio media:

- Track 1: Left front (L)
- Track 2: Right front (R)
- Track 3: Center (C)
- Track 4: Low frequency effects (LFE)
- Track 5: Left surround (Ls)
- Track 6: Right surround Rs)
- Track 7: Left total or Left only (Lt or Lo)
- Track 8: Right total or Right only (Rt or Ro)

It is helpful if the channels are specified in this order. However, other channel configurations exist in film and music, so a facility and its operators must be prepared to shuffle individual channels.

DISTRIBUTION

While ATSC specifications fully define how to send audio from the broadcaster to the consumer, an area subject to wide variability is network distribution, or routing the signal from the program distributor to the broadcaster. There are several approaches that can be used, and a logical combination of different techniques may yield the best results.

Baseband Methodology

Keeping audio in the uncompressed PCM domain allows straightforward access to the audio and perhaps easier control of audio/video synchronization. Audio can either be separate AES pairs or can be embedded along with video in an SD or HD SDI serial digital interface stream.[4] Note that a separate time-aligned path for audio metadata storage and routing is required for baseband systems.

One audio metadata distribution approach inserts it, among other things, into the vertical ancillary data (VANC) space of the SDI or HD-SDI signal.[5] Done properly, this method allows for video, audio, and all associated metadata to be stored, routed, and switched within one signal path, and synchronization between the audio and its metadata can be accurately maintained.

Metadata is video synchronous so as to provide data "gaps" at video frame boundaries. This allows switching to take place without interruption of the metadata stream. When storing metadata in VANC, it is important to maintain video synchronization. Asynchronous insertion of metadata may cause problems at switch points that will be heard by consumers. Information about the structure of audio metadata is available from Dolby Laboratories.

Compression Methodology

Using audio compression systems, such as Dolby E[6] or the Linear Acoustic StreamStacker-HD e^2 system,[7]

[4]Described in SMPTE 259M/272M (SD) and 292M/299M (HD).

[5]Described in SMPTE 334M.

[6]Dolby E carries up to eight channels of PCM audio and metadata over a single 48 kHz/20-bit AES pair (see http://dolby.com/).

[7]The e^2 format generated by StreamStacker-HD carries up to 16 channels of audio, metadata, and auxiliary data over a single 48 kHz/20-bit AES pair or via TCP/IP (see http://linearacoustic.com/).

allows multiple channels of audio to be carried along with metadata over a single bit-accurate path. Noting that the status bits of the AES frame are often not preserved in the recording process, the data rate is confined to the audio and aux portions of the frame, equaling approximately 1.92 Mbps. Although compressed systems impart some amount of latency during decoding, they also guarantee that audio and metadata remain tightly synchronized. Any VTR or server capable of storing uncompressed 20-bit 48 kHz AES audio can carry these lightly compressed "mezzanine" compression formats.

When dealing with compressed audio, it is important to choose equipment that has known timing and performance characteristics. While Dolby E is video frame-based, different equipment can cause timing shifts. Just like video, this timing must be known when designing a system. Dolby Laboratories tests and certifies equipment through its Dolby E partner program and maintains a list of those models on their website. *StreamStacker*-HD is timed to AES reference and is not video frame rate dependent.

Transport Stream

Several U.S. terrestrial broadcast networks have chosen to send a precompressed, ready-to-air ATSC transport stream to their affiliate or member stations. This was done initially as a cost-saving method in the early days of DTV, which allowed local stations to simply feed the transport stream direct to the transmitter, getting them on the air with no need to purchase an expensive local ATSC encoder. Of course, this meant a lack of local programming or branding, but it did help stations obey FCC rules while simultaneously getting SD and HD content on the air.

A local station has two choices when local content must be intercut with a transport stream from the network: decode to baseband audio and video or splice in the compressed domain. Decoding to baseband audio and video produces signals that can be routed and switched with locally generated audio and video signals, then re-encoded for transmission. However, this results in some quality degradation due to the decode-encode cycle. For this reason the preferred method is usually splicing the transport stream, in which case original quality can then be maintained all the way to the consumer.

Technologies now available for processing and splicing transport streams have overcome most of the early drawbacks of transport stream distribution and allow many operations, such as local logo insertion, to be performed in the compressed domain. They also allow the output of local ATSC encoders to be seamlessly spliced in place of the network transport stream.

It is also possible to apply splicing techniques directly to audio that has been Dolby Digital (AC-3) encoded. The benefit of allowing preencoded content to pass through a facility and straight to transmission to the consumer is that it eliminates all other coding steps, and the ultimate quality is maintained with no need for local personnel to worry about metadata or other settings. This is useful also in situations that require the video to be decoded to baseband for local processing and logo insertion through traditional means. Using AC-3 bit stream splicing allows the encoded audio to pass through to consumers except when local content is switched in. The AC-3 bit stream is particularly well suited to this type of operation and responds well to splicing so long as timing constraints are obeyed. Techniques and products to support all types of local station needs, such as voiceover and program insertion, exist and are successfully being employed by many stations.

AUDIO/VIDEO SYNCHRONIZATION

Audio Synchronization Issues

Most film editors are able to detect audio/visual (A/V) sync errors as short as ±1/2 film frame. At 24 fps this equates to approximately ±20 ms. It is claimed that some editors can detect even smaller errors, but this might be more accurately attributed to their familiarity with the material being viewed. Other figures for A/V sync include ±1 video frame or ±33–40 ms. Dolby Laboratories specifies that any Dolby Digital decoder must be within the range of +5 ms audio leading video to –15 ms audio lagging video. This is because human perception of A/V sync is weighted more in one direction than the other due to our experience in the real world.

Light travels much faster than sound. For example, the sound of a baseball bat hitting the ball in a large stadium would appear relatively in sync to a viewer sitting in the first few rows of seats, but the further back a viewer gets, the more the sound lags behind the sight of the ball being hit. Because we are used to this common phenomenon, except in extreme cases over very long distances, it does not seem to be wrong.

Now, imagine if the audio/video timing was reversed. If, while watching a baseball game, the sound of the bat hitting the ball arrives before the bat looks like it makes contact. This would be an unnatural sight and would seem incorrect even to those in the first few rows. The point is that the error is in the "wrong" direction. In summary, human perception is much more forgiving for sound lagging behind sight, probably because this is what we are used to naturally observing.

The International Telecommunications Union (ITU) released ITU-R BT.1359-1 in 1998. It was based on research that showed the reliable detection of A/V sync errors fell between 45 ms audio leading video and 125 ms audio lagging behind video. That was just for detection, while the acceptability region defined by ITU, and therefore the recommended maximum, was quite a bit wider. In summary, the recommendation states that the tolerance from the point of capture to the viewer and or listener shall be no more than 90 ms audio leading video to 185 ms audio lagging behind video. This range is probably far too wide for truly acceptable performance. More recently, the ATSC has published documentation[8] stating a goal of 15 ms audio leading to 45 ms audio lagging at the input to the DTV encoding devices. The ATSC document acknowledges that there are various sources of different delay throughout the broadcast system. It recommends that designers should correct these delays at each step in the chain and strive for 0 ms offset as the goal for every step.

Delays in the Television Plant

While A/V sync issues within the TV plant are not new to digital television, they have become more noticeable in recent years. Some basic points to keep in mind are that in general, audio operations are very low latency. Compression, equalization, mixing, and processing can typically be accomplished in under 1 m in the digital domain, falling to microseconds in the analog domain. Generally, no compensating video delay needs to be added for these operations, as the latency is so low.

Video processing, on the other hand, takes substantial amounts of time, usually no less than one video frame. Similar to audio, any time a video signal is digitized, operations upon that signal will take longer. As most video effects are unable to be performed in the analog domain, delay is inevitable.

It is interesting to note that processing delay of audio and video signals has the opposite of the desired effect on each. As video processing takes longer, the video signals will be delayed with respect to the audio signals and A/V sync will seem incorrect much sooner. It is important that compensation is provided for any video device that has a delay in excess of a few milliseconds. An equal amount of delay should therefore be applied to the audio path.

The ITU recommends in ITU-R BT.1377 that audio and video apparatus should be labeled to indicate the amount of processing delay. This delay should be indicated in milliseconds to avoid any frame rate discrepancies, and if the delay is variable the range should be stated. In the case of variable delay, a signal that can control an audio delay should also be provided. By following these recommendations, it is apparent that regardless of the actual delay, compensation can be made and A/V sync ensured.

Some typical operations are presented below. Two are common to any video facility and have simple, logical solutions. The third is a somewhat surprising source of sync error that should be taken into account during facility design and troubleshooting.

Video Frame Synchronizers

A video frame synchronizer causes between one and two variable frames of delay. In this case, a special audio delay, able to track the variable delay of the frame synchronizer, is required. Most video frame synchronizers are available with matching and tracking audio delays and should always be purchased as a set as there is currently no standard interface that represents A/V sync values.

[8]ATSC IS Finding IS-191, available at http://www.atsc.org/.

Digital Video Effects

Digital video effects (DVE) can add from one to many video frames of delay. As the delay of a DVE is generally a fixed value, a fixed value audio delay can also be used. Devices with fixed delays are easier to deal with if they are always kept in line, or if they must be removed then a fixed video delay equal to that of the device is inserted in its place. This will prevent having to dynamically adjust audio delay and create an audible disturbance.

Cameras and Displays

Traditional tube-based television cameras generally have no intrinsic delay due to video processing because the timing of the video signal out of the camera is directly synchronized with the physical scanning of the image by the pickup tube. Because of the process in which the picture is scanned from top to bottom over a one-frame (or two-field) interval, there is some small variability of A/V timing depending on where the source of the sound is positioned in the video frame. Similarly, for traditional cathode ray tube (CRT) monitor displays or televisions, the displayed picture is directly synchronized with the input video, with no intrinsic delay. Also, because the CRT display will usually replicate the scan of the original camera image, the small variability due to the sound source position in the frame is removed.

Modern cameras using solid-state image sensors and picture displays based on new technologies (such as LCD, plasma, and micro-mirror) are fundamentally different from scanned tubes, being based on pickup and display of complete images one whole frame at a time. These processes mean that the devices all have significant internal storage and processing delays. The video signal out of, or into, such devices is, therefore, no longer synchronous with the image being viewed by the camera or shown on the display. Delays will be different, depending on the particular technology and processing, and may be of the order of one frame up to several frames. Professional picture monitors for quality control purposes may also be used with external video processors that introduce further video delay.

These video delays should be taken into account when designing or troubleshooting a television plant. Consumer television manufacturers and system designers also need to take them into account so that signals transmitted by broadcasters are correctly presented to viewers.

A/V Sync in the MPEG-2 System

The MPEG system provides the proper tools to make A/V sync correct through the transmission system. Each audio and video frame has a presentation time stamp (PTS) that allows the decoder to reconstruct the sound and picture in sync. These PTS values are

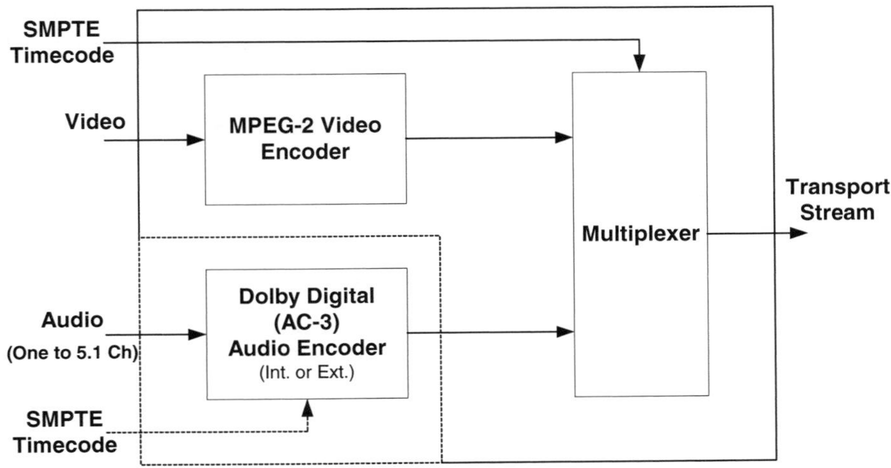

FIGURE 5.18-7 Simplified MPEG-2/Dolby Digital (AC-3) encoding system block diagram.

assigned by the multiplexer following the MPEG encoder. The decoder receives the audio and video data ahead of the PTS values and can therefore use these values to properly present audio and video in sync. Figure 5.18-7 shows a simplified block diagram of a typical MPEG-2 video encoder, Dolby Digital (AC-3) audio encoder, and a multiplexer. Note that the Dolby Digital (AC-3) encoder can be either internal or external to the video encoder and multiplexer.

Aligning the MPEG-2 Encoding System

Video and audio encoding take some time to accomplish, and the multiplexer must know exactly how long. This delay depends on the manufacturer of the equipment, but the value is necessary for the PTS values to be correctly assigned. Many of the A/V sync problems encountered in the field can be attributed to these delays not being properly accounted for or just not set at all.

In practical terms, this simply means that if the transmission system uses an external Dolby Digital (AC-3) encoder, there is a known, fixed audio encoding latency that must be entered into the MPEG-2 encoding system. There is usually a setting called MPEG-2 Encoder Audio Delay, or possibly AC-3 Delay. Once set, it need not be changed unless either the audio or video encoder latency is reset. In many cases, SMPTE timecode can be applied to the audio and video encoders and can be used by the multiplexer to calculate exact PTS values, thereby removing encoder delay as a source of error.

Testing the MPEG-2 Encoding System

In its simplest terms, testing entails feeding typical A/V sync test material such as beep/flash (audio pip with simultaneous video flash) to the encoder, capturing the resulting transport stream from the multiplexer, and using analysis software to determine compliance. It is best to avoid using a consumer set-top box to verify the performance of the MPEG-2 encoding system. There are commercially available tools that make the measurement easier and more accurate by directly analyzing the MPEG-2 transport stream.

The latency of the Dolby Digital (AC-3) encoding algorithm is fixed regardless of the number of encoded audio channels. Therefore, a two-channel signal is adequate to test the A/V sync relationship of an MPEG-2 encoded video signal and a Dolby Digital (AC-3) encoded audio signal. This means that a test tape can be easily created with a video flash and an audio beep, verified with an oscilloscope, and used as the source for A/V sync testing.

Testing the MPEG-2 Decoder

Testing the MPEG-2 decoder is a straightforward process. It requires that a reference transport stream be applied to the decoder under test. This transport stream is a beep/flash-type signal that has been encoded and verified for proper synchronization as described above. The audio and video outputs are then displayed on a dual-trace oscilloscope and compared as shown in Figure 5.18-8.

It is necessary to perform this testing with different video scanning formats and at different frame rates. This is due to the use of video format converters after the MPEG-2 decoder that may respond differently to native rates than to rates that must be converted. It is also important to test the decoder response to bit stream discontinuities as errors and splices are handled differently from one decoder to the next.

LOUDNESS

This section briefly discusses and provides a high-level overview of loudness estimation as it applies to digital broadcasting. It by no means attempts to give

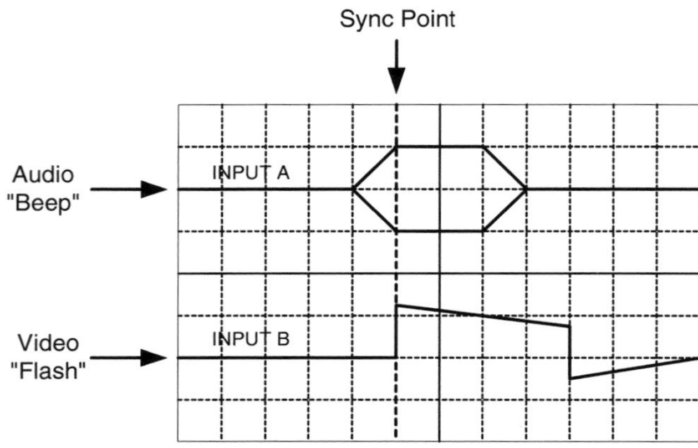

FIGURE 5.18-8 Audio and video outputs of a decoder under test as displayed on a dual-trace oscilloscope. In the display, audio and video are exactly in sync. Note the ramp-up and ramp-down of the audio beep. This is due to the windowing function present in the Dolby Digital (AC-3) process. The measurement point is after the windowing, or when audio reaches maximum.

the reader a full treatise on this subject. There has been, and continues to be, extensive research and development in the area of loudness estimation and devices to measure this subjective quantity. The reader is encouraged to follow the publications of several professional societies and standards bodies dedicated to work in this area, including the Audio Engineering Society (AES), Acoustical Society of America (ASA), and standards of the International Telecommunications Union (ITU).

It should be noted that the following sections on loudness and the digital set-top box refer to Dolby Digital (AC-3) bit streams. However, all of the principles and background also apply to Dolby Digital Plus (E AC-3) bit streams.

Background

The term *loudness* is generally defined to be the attribute of auditory sensation in which sounds can be placed on some scale extending from quiet to loud (corresponding to a ratio of intensities of 1,000,000,000,000:1). Loudness itself is also a highly subjective quantity (and as such, cannot be measured directly) that involves psychoacoustic, physiological, and other factors still under investigation. Hence, this highly subjective quantity (loudness) often results in substantial differences in loudness perception between listeners, making a single measurement method that considers all of the above factors—for all individuals—a complex problem. This is proven by real-world experience, as there is often no single loudness level that will satisfy all listeners (or even a single listener) all of the time.

At best, the loudness of sounds can only be approximated by artificial means. One study performed by Dolby Laboratories concluded that even when audio programming has been "normalized" by a group of people "by ear," the "normalized" programs do not completely satisfy a different group of listeners 100% of the time. In fact, the different groups only agreed approximately 86% of the time. Given this level of uncertainty among groups of listeners, any loudness measure utilized in broadcast will not guarantee that we satisfy 100% of the listening audience.

Loudness Perception

A brief recap of what is known about the science behind loudness perception is in order at this point. First, the human auditory system is nonlinear with respect to frequency. Thus, perceived loudness is dependent on the frequency content of a sound. For example, a person with normal hearing would perceive a very low-frequency sound, such as a 20 Hz tone at 40 dB SPL, to be quieter than a 1 kHz tone at 40 dB SPL. If this process is repeated for various frequencies (with the 1 kHz tone still fixed at 40 dB SPL), a 40 phon equal-loudness contour is created. (The term *phon* is defined as a unit of loudness level. For example, if a given sound is perceived to be as loud as a 40 dB SPL sound at 1,000 Hz, then it is said to have a loudness of 40 phons.)

Readers may also be familiar with the equal-loudness contours that were first developed by Fletcher and Munson in 1933; approximations of these contours have been utilized in sound level meters for several years and are commonly referred to as *frequency weighting networks*. In such a network, the intensity of each frequency is weighted according to the shape of the equal-loudness contour—and for a particular loudness level in phons (e.g., A-weighting approximates the sensitivity of human hearing similar to the 30 phon loudness contour)—before summing the energy across the entire frequency range, and devices of this type perform quite well at estimating the *relative* loudness of signals with similar spectra, such as dialog.

However, calculating the loudness of more complex groupings of sounds (sounds with heterogeneous spectra) requires further thought, as something called the "critical bandwidth" comes into the picture. Critical bandwidth is a measure of the frequency resolution of the ear. For example, when two sounds of equal loudness, when sounded separately and are close together in pitch (narrowband), their combined loudness when sounded together will be perceived as only slightly louder than one of them alone. Hence, they are probably in the same critical band where they are competing for the same nerve endings on the basilar membrane of the inner ear. However, if the two sounds are widely separated in pitch (wideband), the perceived loudness of the combined tones will be considerably greater because they do not compete for the same nerve endings. Third-octave frequency bands can, and have been, used as an approximation to the critical bands in some standardized methods of calculating loudness (namely ISO 532-1975 Method B). As a side note, the critical band is about 90 Hz wide below

200 Hz, and increases to approximately 900 Hz for frequencies around 5 kHz [1].

Because loudness perception is partially dependent on whether the signal is wideband or narrowband, it is challenging to design a measurement system that can detect, and subsequently apply a specific loudness measurement function for, each of these signal types on a continuous basis.

Given this brief overview, it is understandable how the development of a measurement system that factors in even these few characteristics of human hearing (as well as numerous others not described here) would be quite complex, and yet it still wouldn't provide a measurement method perfect for every individual.

Reference/Line-Up Levels

There have been several attempts to standardize a common analog and digital reference level that is intended to facilitate the seamless exchange of programming among broadcasters or other program users. In the United States, SMPTE RP155 defines a reference or *line-up* level of –20 dBFS (at 1 kHz) to be used for the calibration of audio level indicators and to be recorded on digital VTRs. It is common to find (but not always) that this –20 dBFS reference level represents +4 dBu in the analog domain. In contrast to RP155, EBU R68 defines a reference or line-up level of –18 dBFS (at 1 kHz) that often represents 0 dBu in the analog domain. Thus, 0 dBFS is equivalent to +18 dBu analog within facilities that follow EBU R68, but 0 dBFS is equivalent to +24 dBu analog in facilities that follow SMPTE RP155. This discrepancy has caused problems and confusion for years, not only with the exchange of programming but with the manufacturers of broadcast equipment as well.

It is not uncommon to find variations among different facilities (and, at times, within the same facility) as to what analog level indicators are adjusted to indicate with standardized reference signals. For example, some facilities in North America have calibrated their VU (volume unit) indicators (or similar devices) so that a 1 kHz steady-state sine wave at 0 dBu indicates 0 VU, instead of the more usual setting where +4 dBu indicates 0 VU. Given this condition, programming that has been adjusted to average 0 VU and then recorded onto a digital VTR will be recorded 4 dB lower than expected (i.e., at –24 dBFS) since the recorder's gain controls at the default "detent" position will only line up a +4 dBu analog input signal with –20 dBFS (per SMPTE RP155). If this program is then passed through a contribution or distribution circuit that follows the EBU R68 recommendation it will emerge at the opposite end 6 dB lower than expected.

With situations like the one just mentioned it is quite clear that SMPTE RP155 and EBU R68 only work well for calibrating level indicators to a known reference and determining the amount of headroom through a given system. However, they each assume that every facility abides by these practices. Since this has not been proven to be true in all cases, measured level discrepancies will continue to exist. This can be summarized by acknowledging that reference or line-up levels can be arbitrary from facility to facility and that measurements made relative to them (in the analog domain) are prone to significant errors.

Fixed Loudness Reference

To address the measurement problems experienced with these various, and often arbitrary, analog reference levels and calibration philosophies, there are recommendations that the industry should move toward the use of loudness measurement devices and methods using fixed references. For loudness measurements in the digital domain there is a convenient "fixed" reference of 0 dBFS. This fixed digital reference of 0 dBFS is universally agreed upon and is identical everywhere throughout the industry. Hence, a program produced in Europe with an average loudness of –18 dBFS can be verified (measured) in the United States to be at –18 dBFS. For analog television broadcast measurements there is also a fixed reference for measurement, 100% modulation. In the United States this equates to 25 kHz peak deviation for monophonic audio and in most PAL and SECAM countries 100% modulation equates to 50 kHz.

The arguments between geographic regions, broadcast, and postproduction facilities over the proper line-up and level indicator calibrations may never be solved. Therefore, fixed measurement references that relate directly to innate channel-coding properties (analog or digital) must be utilized when controlling and verifying a program's loudness value.

To summarize the arbitrary relationship between the loudness level and line-up level of broadcast programming, consider the following situations. If several different voices are adjusted in level so that they all deflect VU or PPM meters to the same mark, they may still sound somewhat different in loudness to the listener. Both peak program meters (PPM) and VU meters are also frequently used to measure and/or align to a predetermined *house reference* level, and thus produce an arbitrary relationship to the dialog or speech loudness within a given program. For example, if a VU meter and a PPM meter are calibrated to display a reference tone equally, and speech that averages 0 VU is applied to both, the PPM meter will indicate levels considerably above its reference level and possibly above the maximum permitted level. On the other hand, speech that averages at the PPM reference will most likely indicate many dB below 0 VU. This confirms the concept that the reference/line-up level is *not* the same as the dialog level of a program.

AUDIO METERING AND MEASUREMENTS

This section briefly discusses and provides a high-level overview of a few audio level and loudness measurement devices and practices. It by no means attempts to give the reader a full treatise on this subject. As mentioned in the previous section, there has been and continues to be research in the area of loudness estimation and devices to measure this subjective quantity.

Historical Milestones of Loudness Meters

In the 1960s, the former CBS Laboratories developed a Loudness Level Monitor (meter) for use in broadcasting, which, in the early 1980s, was improved based on further psychoacoustic studies. The CBS Loudness meter functions by dividing the signal into eight bands, each weighted according to a 70 phon equal loudness contour [3]. This equal loudness contour is similar to the Fletcher-Munson contour below 1 kHz and the Robinson and Dadson contour above 1 kHz. According to research, the output of each of the filters should respond with a 10 ms attack and 200 ms decay time constants. The filter outputs are then summed and applied to a 200 ms time-constant before being displayed. During listener (validation) experiments, CBS found that the meter performed very well and the indicator was never more than 2.5 dB from the median listener judgment. The CBS meter is no longer made, but the loudness work of CBS Laboratories continues to be applicable today being utilized in multiband processors to control program dynamics and overall level.

Another method for measuring loudness is defined in ISO 532 [4], which specifies two methods for computing the loudness of complex groupings of sounds. Method A (Stevens) utilizes an analysis of the spectrum in one octave band, where method B (Zwicker) uses 1/3 octave bands. The 1/3 octave bands are utilized to approximate to the critical bands. Zwicker's method also takes into account level dependence, nonlinear frequency response, and the masking effects of the human auditory system. The procedure for calculating loudness found in method B involves three steps based on a set of graphs included in the standard. The graphs themselves provide the user a manual means of plotting and combining the physical spectrum (in 1/3 octave bands) of the sound being measured then converting them to yield the total loudness. Different graphs are provided for frontal sounds and diffuse fields. While ISO 532 showed promise, it was complex and difficult to implement onto real-time hardware platforms that were available during the mid- to late 1970s.

VU and PPM

VU meters and PPMs were originally developed for measuring signals in order to best match the signal level to audio production, reproduction, or transmission equipment. This is a different process from matching audio levels to a desired perceived loudness. Nonetheless, these and other signal level meters are often used to estimate subjective loudness or to level programs to a reference level.

The VU meter has considerably slower ballistics than the PPM and will indicate somewhere between the average and peak values of a complex waveform. Moreover, the VU meter only approximates momentary loudness changes in program material and can indicate moment-to-moment level differences that are greater than what our ears perceive. The VU meter also incorporates a relatively flat frequency response over the entire audio spectrum and, therefore, does not address the nonlinear nature of the human auditory system. This can result in large meter deflections that do not correlate with a change in perceived loudness. Perhaps most important, these types of devices can lead to subjective interpretation errors among operators.

Even to the experienced operator, the VU meter is often very difficult to interpret due to its dynamic characteristics and small useable dynamic range. The useable dynamic range is approximately 13 dB, where the top 6 dB of this range is dedicated to 50% of the meter's overall scale. Thus, with uncompressed material the indicator tends to fluctuate more than the perceived loudness change, therefore, making it difficult to assess the subjective loudness of broadcast programming with this type of device. Experienced operators generally use their ears to balance the different elements of a program and to set the overall loudness, while watching the PPM or VU meter to make sure that signal peaks are not overloading the equipment.

As implied by the name, a PPM responds very quickly to changes in the signal level. It was designed to help identify peaks or transients that may exceed distortion limits in a device or system. However, the human ear is not particularly sensitive to instantaneous peaks in signal level. While peaks of short duration may be present in a signal, the perceived loudness of the overall signal is typically not significantly affected. This is why a PPM is less effective in indicating loudness. Psychoacoustic experiments have shown that for short intervals of time, perceived loudness is less for shorter sounds, but that at some time interval, somewhere around 100–200 ms, increasing the duration of a sound doesn't make it any louder to the listener.

Equivalent Loudness

In order to help measure loudness, the loudness equivalent (Leq) is defined as the level of a constant sound, which in a given time period has the same energy as does a time-varying sound. The equivalent loudness measure itself is often coupled with a frequency-weighting network (a filter) to better approximate the frequency sensitivity of human hearing at different loudness levels. Common (and standardized) filters include A, B, and C weighting, which approximate the equal loudness contours at increasing loudness (playback) levels, where A-weighting approximates the 30 phon equal loudness contour, B-weighting approximates the 70 phon contour, and the C-weighting approximates the 100 phon contour. Equivalent loudness measurements utilizing one of these weighting networks is often referred to as Leq(A), Leq(B), and Leq(C), respectively.

In mathematical terms, Leq is computed as the mean-square energy within the measurement interval where, x_w is the frequency-weighted signal, and x_{ref} is a reference level as follows:

$$\text{Leq}(w) = 10\log_{10}\left[\frac{1}{T}\int_0^T \frac{x_w^2}{x_{ref}^2}dt\right] \text{dB}$$

The label (w), if present, represents the type of frequency-weighting network utilized for the measurement. For example, a measurement expressed in Leq(A) indicates that the signal being measured was passed through an A-weighting filter, as shown in Figure 5.18-9, before the energy was summed and the average overtime was computed. Therefore, any equivalent loudness method (e.g., Leq, Leq(A), Leq(B), Leq(RLB), etc.) provides the operator with a single loudness value that represents the entire measurement period (program) on a long-term average basis.

Loudness Meters

Leq(A) has been a standardized loudness measure since the mid-1980s and is defined in IEC 60804–Integrating-Averaging Sound Level Meters. The use of Leq(A) is also referenced in several other standards, including ATSC A/53E and CEA-CEB11–NTSC/ATSC Loudness Matching. In ATSC A/53E, Leq(A) is recommended for use in measuring the average level of spoken dialog within broadcast programming to determine the proper dialog normalization (dialnorm) value within the AC-3 bit stream. This is an important and often overlooked step in the encoding process that contributes directly to the reproduced level of the decoded stream. The CEA-CEB11 document provides guidance to set-top box manufacturers on maintaining uniform audio loudness (via the internal gain structure relationships) between NTSC audio services and ATSC audio services while preserving the dynamic range of the ATSC audio service. Importantly, Leq(A) is the metric utilized to quantify the dialog levels throughout the CEB11 documents.

In July 2006, the ITU-R approved and published a new recommendation for estimating the loudness of broadcast programs, ITU-R Rec. BS.1770, Algorithms to Measure Audio Programme Loudness and True-peak Audio Level. It is the intent of this new algorithm to estimate the overall loudness of an audio program by computing the frequency-weighted energy average over time. Hence, it also provides (as does Leq(A)) a single measurement value that represents the overall loudness of an entire program.

The ITU-R Rec. BS.1770 algorithm was derived from the Leq(RLB) algorithm described by Soulodre [5] to support mono, stereo, and multichannel audio signals while retaining a low computational complexity. This in turn allows it to be easily implemented and/or adopted by many equipment manufacturers at

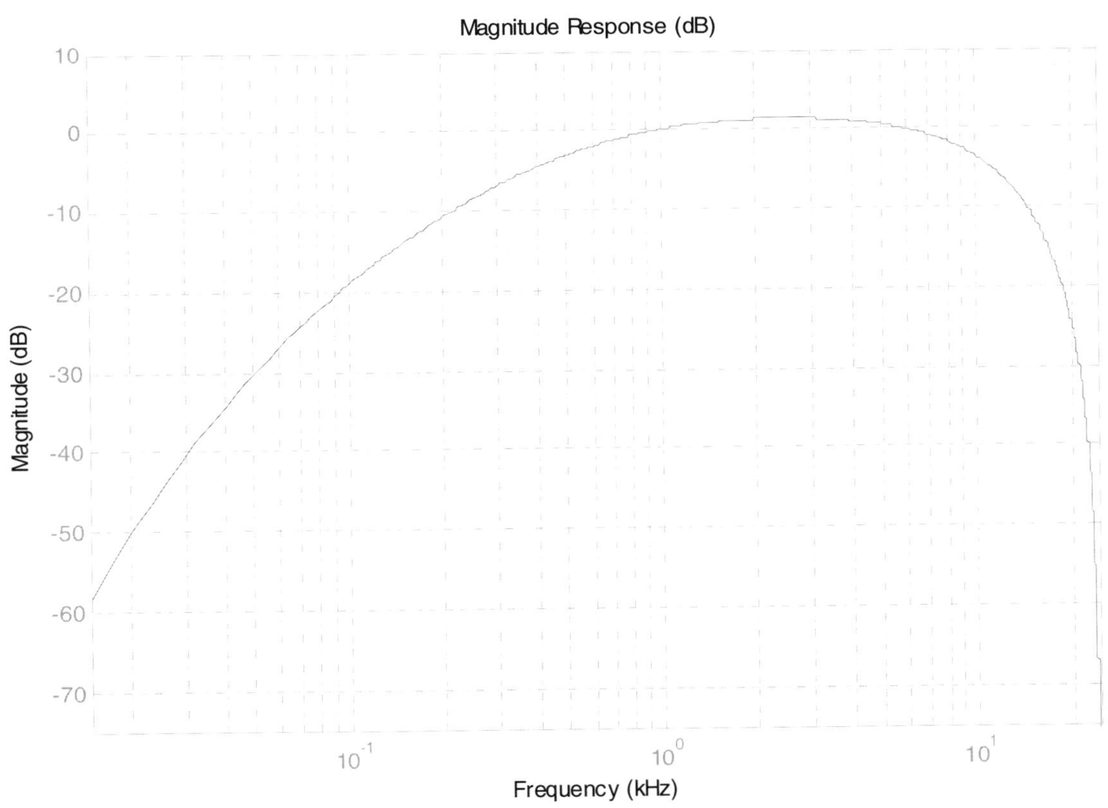

FIGURE 5.18-9 A-weighting filter contour utilized in Leq(A).

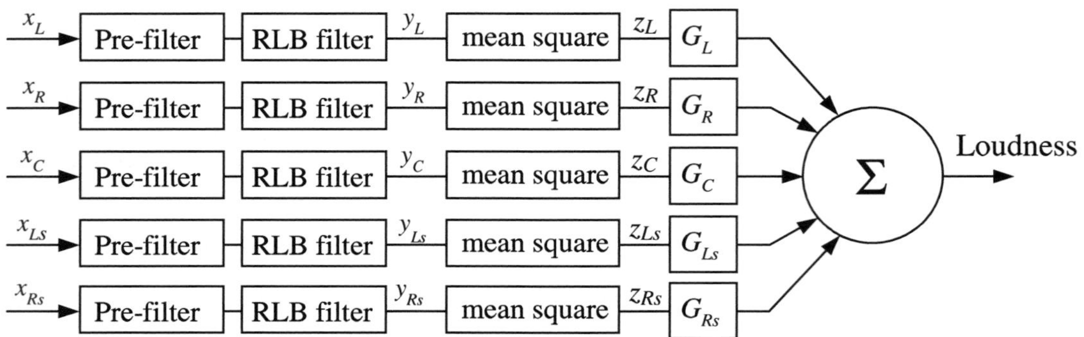

FIGURE 5.18-10 ITU-R Loudness meter/algorithm functional block diagram.

a low cost. During the ITU-R study, subjective testing showed that the BS.1770 algorithm yielded the best performance among several other methods, including algorithms based on psychoacoustic models. Figure 5.18-10 shows a high-level functional block diagram of the ITU-R meter taken from ITU-R Rec. BS.1770.

The BS.1770 algorithm can accommodate any number of channels for multichannel audio programs. However, it is assumed that the multichannel signals conform to the ITU-R BS.775-1 5.1-channel configuration. Note that the LFE channel is not included in the measurement.

Within the BS.1770 algorithm, each of the individual audio channels being measured is first passed through two filters in cascade. The prefilter, which has a shelving characteristic, shown in Figure 5.18-11, is to account for the acoustical effects of the human head and the RLB (revised low-frequency B) filter, shown in Figure 5.18-12, is a modified version of the standard B-weighting curve with a low frequency response that falls between the C- and B-weighting curves.

Once the input signal for each channel(s) is filtered, the mean-square energy for each channel is computed for the measurement interval. The individual channel loudness results are then weighted in accordance to the angle of arrival (refer to table 3 in [6]) and then linearly summed to provide the overall loudness value as follows:

$$\text{ITU Loudness} = -0.691 + 10\log_{10}\sum_{i}^{N} G_i \times z_i \text{ in dB.}$$

The weighting factor (G_i) (shown as blocks in Figure 15.18-10) is always 1.0 (0 dB) for left, center, and right channels and the weighting for the left surround and right surround channels is always 1.41 (+1.5 dB). The emphasis on the surround channels acknowledges the fact (based on the research of the ITU-R group) that sounds arriving from behind the listener are perceived as being louder relative to those arriving from the frontal direction.

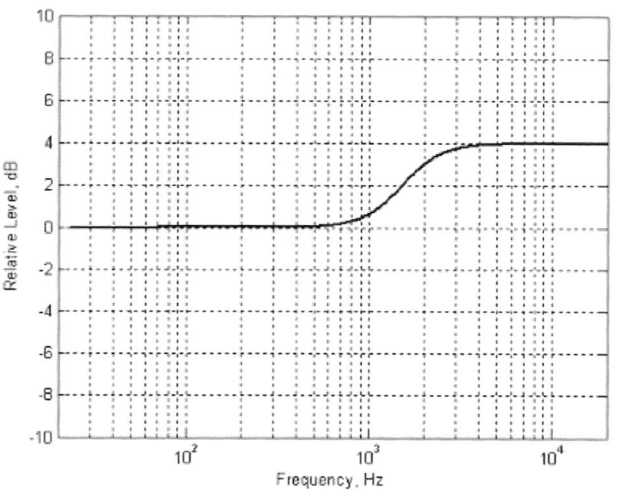

FIGURE 5.18-11 ITU-R Rec. BS.1770 prefilter characteristics.

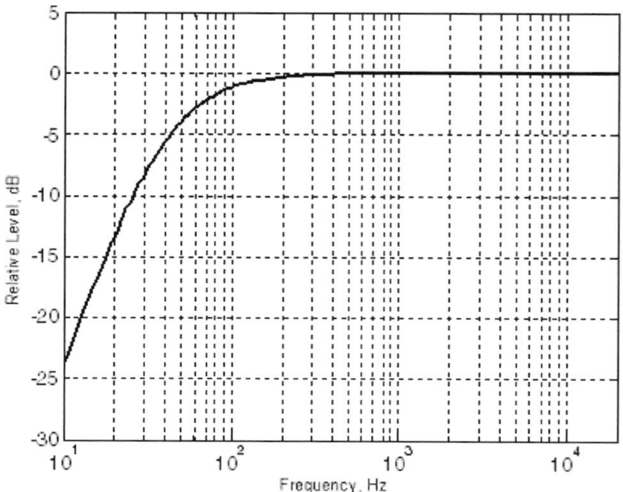

FIGURE 5.18-12 ITU-R Rec. BS.1770 RLB filter characteristics.

The constant value (–0.691) in the equation above is a calibration constant that addresses the combined effects of the prefilter and RLB filter at 1 kHz (note the filter *gain* at 1 kHz in Figure 5.18-11).

A final important note: the definition for ITU-R Rec. BS.1770 described above utilizes a full-scale square wave as the 0 dBFS reference point. However, it is common broadcast practice for meters to utilize a full-scale sine wave as the 0 dBFS reference point. Hence, a 1 kHz sine wave in either left, right, or center channel of a program at 0 dBFS (i.e., whose peaks reach full scale) would measure and display –3 dBFS with ITU-R Rec. BS.1770, but will measure and display 0 dBFS on other Leq meter types commonly found throughout the broadcast industry. As an example, consider an operator checking a 1 kHz line-up level in a single channel as per SMPTE RP-155. The RP specifies that the reference signal shall be a 1 kHz sine wave at 20 dB below the system limit (0 dBFS). A signal of this type and a display that follows this convention would tell the operator that 20 dB of headroom exists. However, if the measurement and display device were following ITU-R Rec. BS.1770, the display would read –23 dBFS with the same signal source, but 23 dB of headroom does not truly exist, only 20 dB as before. It is clear that displaying measurement results relative to a full-scale sine wave will lead to reduced uncertainty about the available headroom when measuring tones. Therefore, this practice has been utilized widely in broadcast meter implementations.

All this being said, while both ITU-R Rec. BS.1770 and speech-based Leq(A) have been effective in estimating the subjective loudness of typical broadcast programming, they do not claim to effectively estimate the subjective loudness of tones. It is important to be aware of this difference.

Leq(A) and ITU-R Rec. BS.1770—Dialnorm Measurement

This section briefly discusses the importance of speech-based measurement and its application within the North American DTV system. Also provided is a high-level comparison between Leq(A) and the BS.1770 algorithms and their use for provisioning the dialog normalization value within the North American DTV system. The information is also relevant to DTV systems outside of North America as a guideline for normalizing the loudness of broadcast programming based on speech (speech-based normalization).

It is useful to measure speech loudness levels. Studies have found that listeners are generally more satisfied when program leveling is based on the dialog segments within programs. For many television programs, speech, as the information bearing portion of the signal, is most important to the listener. Evidence also suggests that television listeners make adjustments to their volume controls in an effort to create consistent speech levels within their own listening environments. One study conducted by Benjamin [7] concluded that television viewers in a typical living room environment preferred the dialog level to be at a

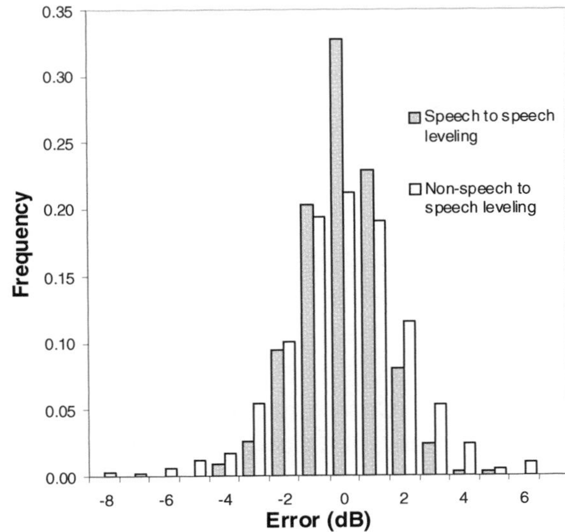

FIGURE 5.18-13 Frequency of listener leveling error.

mean sound pressure level of 57.7 dBA, where listeners in a typical home theater setting preferred a mean dialog level of approximately 64.8 dBA. In another independent study, Pearsons et al. [8] determined that speech levels range from 55 dB to 66 dB at conversational distances. Thus, television viewers choose to set the listening levels such that the dialogue level is consistent and that the dialogue level chosen is closely related with ordinary conversation levels.

A second benefit of loudness normalization based on dialogue is that measuring within a specific content type yields a higher correlation among listeners. Listeners more closely agree on relative loudness when comparing two dialogue segments than when comparing arbitrary audio signals. This is not necessarily the case for general audio content.

Figure 5.18-13 is a histogram from a study [10] that shows the frequency of leveling errors between speech and nonspeech broadcast programming. The dashed line shows the frequency of errors made when the listeners were comparing speech to speech, and the solid line shows error frequency when they were comparing nonspeech to speech. As the figure shows, errors when comparing speech to speech are more closely clustered around 0 dB with approximately one-third of the listeners agreeing to within 0.5 dB.

The correlation histogram in Figure 5.18-14, taken from [9], compares the results of only two specific samples evaluated by the listener panel. One of the samples primarily contained dialogue, and the other contained a portion of a program where only footsteps were heard (such as a scene from a movie). According to the correlation histogram, it is apparent that there was general agreement among the listeners when they leveled the dialogue item, where 19 out of the 21 listeners agree with each other to within 1 dB. However, there was a large disagreement within the group when they attempted to level the "footsteps" piece to the reference. Indeed, one person indicated a need to adjust

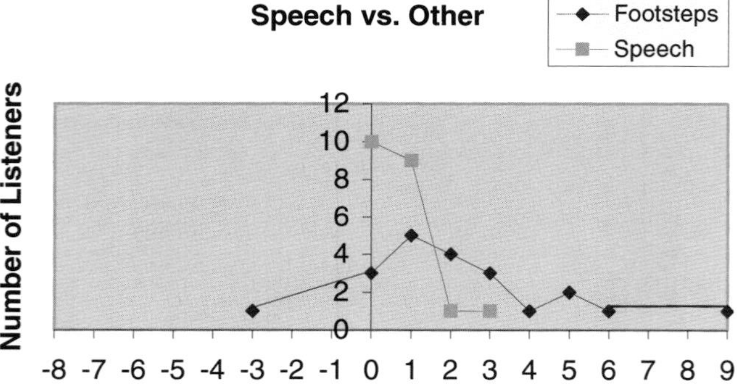

FIGURE 5.18-14 Agreement among listeners when evaluating speech items versus other signal types.

the footsteps up by 3 dB, while another indicated a decrease of 9 dB to make it agree to the reference.

Given the results of the studies outlined above, focusing loudness measurements on the speech portions of the content has been successfully utilized to improve the performance of objective loudness measurements (even simple methods such as weighted and nonweighted Leq types) for use in broadcast applications.

DIALNORM

The Dialogue Normalization (dialnorm) parameter within the Dolby Digital (AC-3) system is defined to indicate the long-term average level of spoken dialogue to the decoder (DTV receiver). Hence, the decoder utilizes this information (carried in the AC-3 bit stream) to normalize (scale) the reproduced audio level at the decoder output to a consistent (i.e., normalized) level. Therefore, by properly utilizing the dialnorm parameter within the AC-3 system, it will allow audio programs that have their overall dialogue level produced at differing levels (due to different headroom requirements and production philosophies) to be transmitted without any change to their levels or dynamic range, and let the decoder in the set-top box normalize the level with guidance from the dialnorm value carried in the bit stream. The use of dialnorm does not require the broadcast industry to agree on any one production philosophy or standard practice.

Dialnorm Facts

- Dialnorm is an audio metadata parameter carried within the AC-3 bit stream that *only* indicates (and matches, when set correctly) the long-term average spoken dialogue level of the audio program.
- The dialnorm value is placed into the bit stream every 32 ms by the encoder, either under local manual control, or via external metadata. Hence, this value is present in every AC-3 frame.
- The dialnorm value is only validated by measuring the long-term average level of dialogue within the program to determine whether it agrees with the transmitted dialnorm value.
- The dialogue level can be measured by utilizing integrating-averaging measurement devices that conform to IEC 60804 or ITU Rec. BS.1770. Note, both standards make no specific mention or recommendation on the measurement of speech levels. Hence, it is up to the operator of these devices to ensure that the results are in agreement with the actual long-term dialogue levels for the program. Note, a commercial device is available that can automatically detect and measure only on the dialogue segments of the measured program (discussed later).
- The dialnorm value has a finite range defined in ATSC A/52B of –1 to –31 dB relative to 0 dBFS.
- Dialnorm is provisioned in the encoder and always applied in the STB or home theater decoder.
- The dialnorm value within the AC-3 bit stream does not (typically) change on a frame-by-frame basis throughout the program to achieve normalization. Hence, it is only a single (scalar) value that represents the overall dialogue level of a program.
- The dialnorm value is also utilized by the encoder dynamic range control subsystem to "calibrate" the position of the "null band" within the chosen compression profile. Hence, an improper dialnorm setting in the encoder could produce adverse effects, including an unintended shift in audio levels, on the decoded audio.
- For DTV, the FCC mandates the use of ATSC A/53D [11], which in Section 5.5 states: "The value of the dialnorm parameter in the AC-3 elementary bit stream shall indicate the level of average spoken dialogue within the encoded audio program." The word "shall" denotes a mandatory provision of the ATSC A/53D standard.

The previous section briefly discussed two measurement methods recommended for use in estimating either the overall loudness or the long-term dialogue loudness of broadcast programming. Since both Leq(A) and ITU BS.1770 are defined to estimate the overall loudness of a program with a single value—by computing the weighted energy average over time (via integration)—it is important to consider and recognize several key points that could impact measurements whether they are speech based (to use for dialnorm) or not. For this discussion, the focus will be primarily on speech-based measurements.

Automated Speech Detection

Utilizing either Leq(A) or ITU BS.1770 to estimate the long-term average level of speech requires the operator to manually pause the integration function of the meter during moments of silence and nonspeech-

based portions of the programming. Acknowledging this, it can be cumbersome and time consuming for personnel throughout the broadcast delivery chain to manually select and measure only the dialogue sections of the programming to use for normalization, leveling, or for use as the metadata parameter dialnorm. An alternative to manually selecting a speech segment is to use an automated speech detector to isolate and measure program segments that are dominated by speech. A system that has been developed that meets these requirements is described by Vinton [12] and has been implemented and widely deployed into a commercially-available loudness meter for broadcast applications.[9] In this system, the speech/nonspeech discrimination works by computing seven features from the input signal, based on well-known speech characteristics.

Measurement Period

In some cases, taking speech measurements (to come up with a dialnorm value) on shorter "representative" portions of programming (and not computing the overall speech loudness value over the entire program) can be successful. However, the success of this practice is highly correlated with how dynamic the particular program is. For example, with a sitcom that has a very limited dynamic range, measuring over only a portion of the entire program can yield an accurate dialnorm value that represents the entire program well. On the other hand, a television drama can exhibit large differences in dynamics throughout the program. In general, taking a single shorter measurement in hopes to come up with an accurate dialnorm value that represents the overall speech loudness of the entire program has often met with little success. Moreover, this practice can lead to confusion among different stages throughout the broadcast production and delivery chain. Dialnorm based on short measurement dwell times cannot be easily validated by individuals involved in downstream processes because these individuals do not know what sections were actually measured. It is clear that measuring the entire program reduces the uncertainty and inherently fosters more consistency among different production processes.

Short-Term Loudness

One of the caveats of computing the long-term average speech loudness that represents an entire program is its inability to effectively and immediately report (display to the operator) sudden changes in loudness that may have taken place during the program. To overcome this, some measurement devices[10] currently available have the ability to simultaneously compute (and display) a short-term speech loudness value along with the long-term value. In this case, the long-term value (at the end of the measurement period or program being measured) is utilized as the dialnorm value and the short-term values (plotted against time) represent the "dynamic" speech loudness history of a given program. Current practice has shown that being able to visualize this short-term information is very useful during the quality check process, postproduction, and for monitoring within a cable head-end facility by allowing the operator to easily determine whether or not a particular program or service maintains consistent loudness levels within and from program to program.

[9]Dolby Laboratories LM100 Broadcast Loudness Meter with Dialogue Intelligence.

[10]Dolby Laboratories LM100 Broadcast Loudness Meter with Dialogue Intelligence.

Other Considerations when Setting Dialnorm

For various reasons, in the early days of DTV many broadcasters set their dialnorm value to a fixed default value. Depending on the actual speech levels and dialnorm setting, this can result in the Dynamic Range Control subsystem within AC-3 exhibiting "brute force" normalization, by constantly being driven into compression and/or limiting. While this will certainly reduce any significant level differences from program to program, the penalty is reduced dynamic range. For example, if the dialnorm setting is –31 but the actual speech level average is –20, then the DRC subsystem will frequently request (via metadata) gain words that indicate "cut" in the decoder, thereby reducing the level fluctuations from moment to moment but with the penalty of reduced dynamic range.

While it is not always possible to generate and provision a specific dialnorm value for each broadcast program, a first step to improving this situation is to set an appropriate dialnorm value based on an actual long-term speech average of the station programming (averaging the speech measurements over many different programs, perhaps at different levels, to come up with a typical value). Setting dialnorm to this long-term average of the program speech level will reduce the amount of time that the actual speech levels are on the "cut" portion of the DRC curve (see Figure 5.18-4). That is, overall, the programming will be more true to its original dynamic range and not subjected to significant amounts of compression/limiting via the DRC subsystem. DRC is then only used when needed to control loud or soft events that exceed the null-band limits).

While this measured average setting for dialnorm is in improvement on a perhaps arbitrary default setting, clearly the ideal situation is to for the system to have an appropriate dialnorm value for each program.

Speech Loudness Measurement

The ITU-R Rec. BS.1770 study group recognized the need to harmonize (so far as possible) speech-based measurements between BS.1770 devices and Leq(A) meters that display measured values relative to a full-scale sine wave (as discussed earlier). Upon investigation, it turns out that the difference between BS.1770 devices that properly display measured values relative to a full-scale square wave and Leq(A) devices that display measured values relative to a full-scale sine wave will generally agree on speech-

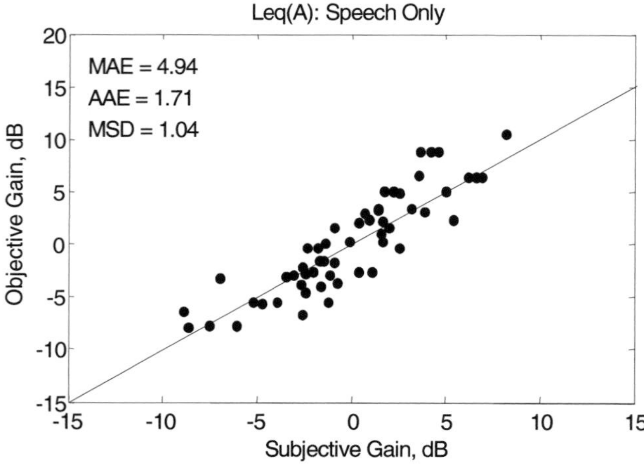

FIGURE 5.18-15 Leq(A) versus subjective loudness measurements for speech only.

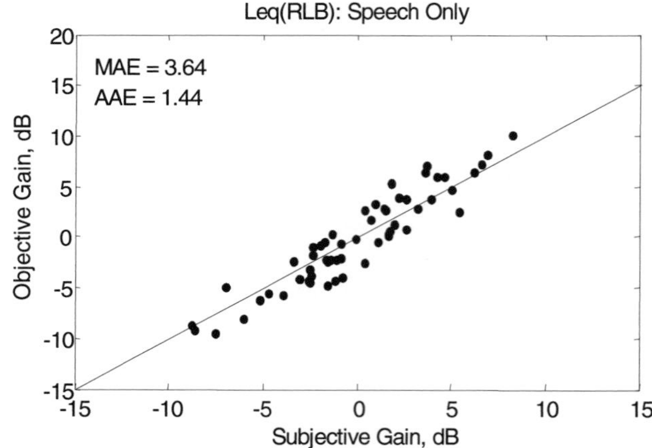

FIGURE 5.18-16 Leq(RLB) versus subjective loudness measurements for speech only.

based material. It is important to understand how the two methods relate to each other with respect to their accuracy against subjective test results.

Figure 5.18-15, taken from [9], depicts Leq(A) measurements plotted against mean subjective measurements of loudness for a database of speech segments taken from broadcast programming. The average absolute error (AAE) between the objective (Leq(A)) measurements and subjective (listener) opinions is only 1.71 dB, and the maximum absolute error (MAE) across the database of signals is 4.94 dB.

In contrast, and as suggested by Soulodre [5], loudness measurements performance may be improved by using a revised low-frequency B-weighting (RLB), which is the basis of the ITU-R Rec. BS.1770 algorithm. Figure 5.18-16 shows results for the same tests as above, but using Leq(RLB). In this case, the AAE between the objective (Leq(RLB)) measurements and subjective (listener) opinion has decreased to 1.44 dB, and the maximum absolute error (MAE) is 3.64 dB.

In summary, speech-based measurements taken with commercial meters using Leq(A) (common prior to the standardization and implementation of ITU-R Rec. BS.1770) will not be significantly different or more accurate (on average) than speech-based measurements taken with BS.1770. The data suggests that the AAE has only decreased by 0.27 dB and the MAE has decreased by 1.3 dB for the samples utilized in the comparison above. Hence, if a broadcaster has been utilizing speech-based Leq(A) to set dialnorm, it is unlikely that remeasuring and resetting dialnorm as per BS.1170 is necessary.

Note that the data above compares Leq(A) with Leq(RLB). Leq(RLB) is the basis of the ITU-R Rec. BS.1770 algorithm but does not include the prefilter (HF shelf) specified in the final version of ITU-R Rec. BS.1770.

To assist in determining the effectiveness of a particular objective measure or approach to loudness normalization, the authors would like to share the following piece of information on a somewhat new term called the *comfort zone*.

The comfort zone, first described in [4], is the range in which a typical listener will accept loudness changes within and among broadcast programs. Assuming further that the nonspeech portions of the program(s) have been "balanced" (in an appropriate manner by the program producer) around the speech portions, listeners will not be annoyed by the natural changes in loudness that typically occur during programs if the speech elements fall within their individual comfort zone.

Figure 5.18-17, generated by Lyman and taken from [9], shows the results of the comfort zone tests. The results indicate that an increase of 2–3 dB in subjective loudness is enough to move a program out of

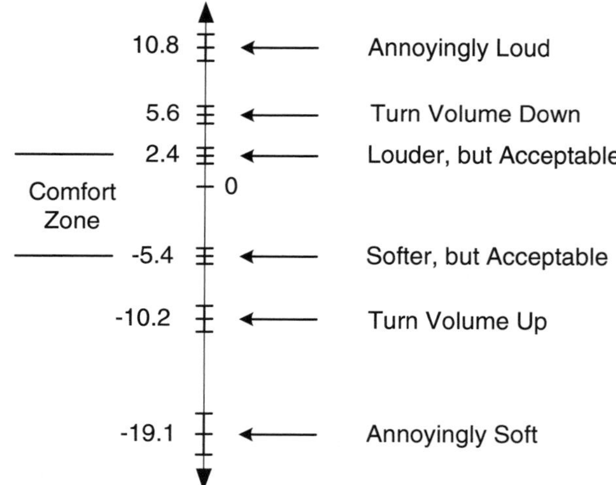

FIGURE 5.18-17 Comfort zone.

the typical listener's comfort zone, and toward the point at which they would like to turn the volume down. In contrast, there is much more latitude (that is, listeners are a bit more forgiving) available on the softer side of the "comfortable volume" point (shown in the figure as 0). As a side note, Reidmiller et al. [9] stated that "the ambient noise level in the listening room used for the tests was quite low; similar to a suburban living room on a tranquil evening. Since the 'annoyingly soft' point can reasonably be expected to fall somewhere above the ambient noise level in the listening environment, the figure of –19.1 dB may depend on the ambient noise level. The other points are far enough above the ambient that they should not be affected."

DIGITAL SET-TOP BOX GAIN CONSIDERATIONS

Many television viewers are presented with programming from both analog and digital sources via a single piece of hardware, the digital set-top box (STB). As a result, one of the most important and challenging goals of everyone involved in delivering content, including the broadcaster, cable operator, and set-top box manufacturer, is to provide viewers with a seamless listening experience as they switch between digital and analog channels. At first glance, this level of operability may seem impossible to achieve since it requires lining up two types of signals (analog and digital) that can differ greatly in areas such as available headroom above dialogue peaks and dynamic range. However, with a thorough understanding of the Dolby Digital system and current NTSC practice, and some knowledge of set-top box internal gain structure, the source of the problem can be addressed in a given situation.

Many broadcasters do not realize that the digital set-top box has been designed with several assumptions regarding analog broadcast levels, digital decoder operating mode, the validity of the digital audio metadata, and the level relationships within the STB itself. For this discussion, the focus is on three considerations:

- Digital set-top boxes assume that while tuned to an analog service (either off-air or via cable) the average dialogue level is ~–17dB Leq(A) below 100% modulation.
- While tuned to a digital service (either off-air or via cable) the transmitted dialnorm value (in the AC-3 bit stream) is assumed to be correct for that program.
- While utilizing the channel 3/4 remodulated RF output of the set-top box, the AC-3 decoder must default to the RF operating mode.[11]

[11]The dialogue normalization value in an AC-3 bit stream should be set by the program originator to indicate the dialogue level of the program relative to 0 dBFS. The valid range for this value is from –1 dBFS to –31 dBFS. Also note that the dialnorm value is utilized by the decoder (within the set-top box) to normalize the programming to a consistent level.

Significantly, all of these assumptions are referenced in the Open Cable Host Device Core Functional Requirements (i.e., for Open Cable digital set-top boxes) and in a bulletin issued by the Electronic Industries Association (EIA) and the Consumer Electronics Association (CEA) entitled, EIA/CEA-CEB-11 NTSC/ATSC Loudness Matching [13]. This document provides guidance to digital set-top box manufacturers on how to maintain uniform audio loudness between existing NTSC programming and digital television services while simultaneously preserving the dynamic range capability of the digital services. The bulletin also addresses the capabilities of consumer broadcast products to match loudness from the listener's perspective, internal gain structure, and output specifications.

As mentioned earlier, Dolby Digital decoders found in consumer products (including DTV set-top boxes), in general, can operate in two modes (line and RF mode). Each of these modes has a specific application, and care *must* be taken when the set top is designed and deployed to ensure that the intended mode is used by default. Decoders can be found in many places, including digital cable set-top terminals, home theater systems, consumer satellite integrated receiver decoders (IRD), DBS receivers and commercial IRDs that are used in cable headends, and cable turnaround uplink facilities. It is important to note that the default operating modes in each of these cases may vary and are based on that particular device's function within a given system.

AC-3 Decoder—Line Mode

Line mode operation generally applies to the baseband line level outputs from two-channel set-top decoders, two-channel digital televisions, and multichannel home theater decoders. Line mode operation is a requirement for all digital set-top boxes that have analog baseband (line level) outputs. With respect to consumer-type applications, a decoder's outputs operating in this mode will usually be connected to a higher-quality sound reproduction system than that found in a typical television set. In this mode, dialogue normalization is enabled and applied in the decoder *at all times*. Furthermore, in this mode the normalized level of dialogue is reproduced at a level of –31 dBFS Leq(A), but *only* when the transmitted *dialnorm* value has been correctly adjusted/provisioned for the particular program.

In general, with the reproduced dialogue level at –31 dBFS, this mode allows wide dynamic range programming to be reproduced without any peak limiting and/or compression applied as may be intended by the original program producers. And since the AC-3 system can provide more than 100 dB of dynamic range, there is no technical reason to encode the dialogue peaks at or near 100% as is commonly practiced in analog television broadcasts. This allows the AC-3 system to meet the goal of being able to deliver high-impact cinema-type sound to the digital subscriber's living room.

AC-3 Decoder—RF Mode

RF mode is intended for products such as terrestrial, cable, and satellite set-top boxes that generate a monophonic and/or downmixed signal for transmission via the channel 3 or 4 remodulated output to the antenna input of a television set. This mode was specifically designed to match the average reproduced dialogue levels and dynamic range of digital sources to those of existing analog sources such as NTSC and analog cable TV broadcasts. In this operating mode dialogue normalization is enabled and applied in the decoder *at all times*. However, the dialogue level in this mode is reproduced at a level of –20 dBFS Leq(A) *only* when the transmitted *dialnorm* value is valid for a particular program. In this mode, AC-3 decoder introduces +11 dB gain shift and thus the maximum possible peak-to-dialogue level ratio is reduced by 11 dB. This is achieved by compression and limiting internal to the AC-3 decoder (but is calculated in the encoder). Digital set-top boxes that include an RF modulator are required to provide and default to RF mode operation.

Figure 5.18-18 shows the signal relationships in the decoder for both line and RF operating modes. Note the reproduced dialogue level and dynamic range of each mode (assuming dialnorm was set correctly).

The audio subsystems within the digital set top typically include:

- NTSC tuner
- Two-channel Dolby Digital decoder IC and associated digital-to-analog converter
- Monophonic channel 3/4 FM modulator

Figure 5.18-19 shows the proper level relationships between the output of the Dolby Digital decoder and the NTSC tuner/demodulator for the modulated RF output of a typical digital set-top box used in North America for DTV and digital cable.

Note that the NTSC scale has a maximum value of 0 dBFM and that this level is equivalent to 100% modulation as per FCC rules (that is, 25 kHz peak deviation) and the Leq(A) dialogue level is shown to be 17 dB below 100% modulation. This value (–17 dBFM) indi-

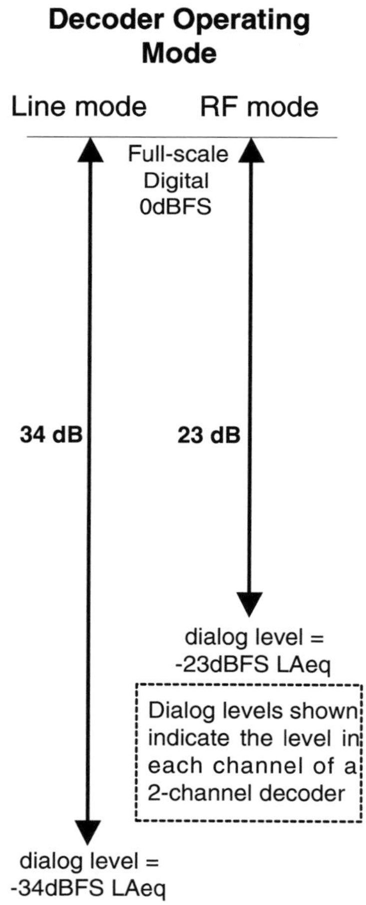

Decoder Operating Mode

Line mode RF mode

Full-scale Digital 0dBFS

34 dB

23 dB

dialog level = -23dBFS LAeq

Dialog levels shown indicate the level in each channel of a 2-channel decoder

dialog level = -34dBFS LAeq

FIGURE 5.18-18 AC-3 decoder operating mode levels—line versus RF.

Digital Set-Top Box
Ch. 3/4 RF Output Level Relationships

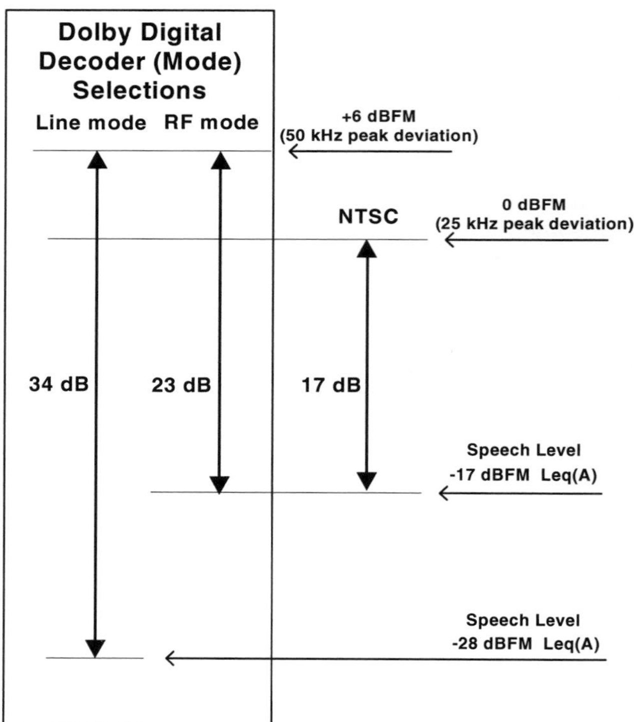

Note: Dolby Digital decoder mode selections affect digital service levels only. Digital and analog sources match in RF mode only.

FIGURE 5.18-19 Analog adjusted so that the level of speech is ~–17 dB FM Leq(A) (that is below 100% modulation 25 kHz peak deviation).

cates the ratio between the maximum program peaks and the average Leq(A) dialogue level. Measurements taken over several years show that the equivalent loudness of dialogue (A-weighted) for NTSC broadcasts is typically 17 dB below 100% modulation. Therefore, a properly designed digital set-top box assumes this condition is always true (for any analog channel) and requires it in order to provide a level match to digital sources.

For digital sources, the Dolby Digital decoder can typically be operated in two modes (as discussed earlier): line and RF. In many cases, these modes can be selected by the viewer. Referring to the decoder mode selections in Figure 5.18-18, note that the maximum permissible level of +6 dBFM (equivalent to 200% modulation for 50 kHz peak deviation) for Dolby Digital sources is available at the channel 3/4 RF remodulated output. This level relationship is intentional since the Dolby Digital signal being decoded by the set-top box potentially has 6 dB more headroom above dialogue peaks than NTSC analog audio (that is, while the decoder is operating in RF mode). Furthermore, since the BTSC system leads to a maximum peak deviation of 73 kHz, most television tuners can accept up to 8 dB above 25 kHz peak deviation, in the absence of pilot and subcarriers, without distortion.[12]

In Figure 5.18-19, with the digital set-top decoder operating in RF mode, the decoded dialogue level for digital sources will match the dialogue level of analog sources *if and only if* the analog source has its dialogue level provisioned at ~–17 dBFM *and* the dialnorm value within the Dolby Digital bit stream is set properly (that is, the dialnorm value represents the long-term A-weighted level of spoken dialogue in the program). Also note that if the viewer unknowingly switches the Dolby Digital decoder into line mode, the decoded level of dialogue will be reproduced at –28 dBFM Leq(A), or 11 dB lower than the analog source. This could potentially generate viewer complaints that could lead to hours of wasted time trying to find a problem within the broadcasters' facility, when in reality the set-top box is in the wrong operating mode.

This leads the discussion to an important (and often overlooked) point for set-top boxes. What is the audio decoder operating mode default and what mode selections are available to the user?

Differences exist between manufacturers of digital set-top boxes as well as program guide application providers (used in cable) as to the nomenclature they utilize to indicate line or RF mode operation to the viewer.

The differences in nomenclature have led to situations where viewers can easily create a level mismatch even between properly transmitted analog and digital programming. This can generate complaints to the broadcaster, cable, or satellite provider when, in fact, the level mismatch was generated by the viewer inadvertently selecting the wrong operating mode.

There have been instances where digital set-top boxes were deployed into viewer's homes defaulted to the wrong audio decoder operating mode. This may have been due to a lack of understanding as to what these user-selectable parameters do or what their impact would be on the reproduced audio. Therefore, it is in their best interests for broadcasters, cable, and satellite providers to fully understand what modes their audio decoders (both set top and IRD) default to, and which viewers can select.

Related to this is another point frequently raised by cable operators across North America, where digital channels (even with the correct dialnorm value) seem to often sound quieter when compared to their analog services. The source of this problem is that the average dialogue levels on analog channels of cable television systems may be significantly higher than –17 dBFM, either due to modulator misadjustment or the use of overly aggressive AGCs built into the FM modulator. In cases like these, the digital set-top box (even when in RF mode) can never provide a match between analog and digital sources. To correct the situation, the deviation on the analog modulators must be adjusted so that the average dialogue level is truly –17 dBFM Leq(A) (which is the dialogue level that the digital set top expects for analog services). However, in many cases the cable headend will pass through the local off-air analog broadcast signals at IF and, therefore, cannot make any adjustments locally. Therefore, in these cases, it is up to the local terrestrial broadcaster themselves to ensure that the correct levels are set.

Three recommendations are suggested for the television production industry:

- Provision any NTSC analog audio modulation equipment (off-air or cable) so that the A-weighted average dialogue level is ~17 dB below 100% modulation.
- Digital set-top boxes that include an RF remodulator must default to RF mode. Also recognize that viewers who want full dynamic range will probably use the S/PDIF output of the STB to feed their 5.1 home theater systems.
- Properly provision the dialnorm value to match the long-term dialogue level for programming.

AUDIO PROCESSING FOR DTV STATIONS

Audio processing for NTSC television has evolved from a necessity to guard against overmodulation, to guarding against annoying loudness shifts, to a way to create a so-called "signature sound" for a station, to combinations of all of these things. The net result has been a permanent reduction of the peak-to-average ratio of audio signals, and a trend toward higher, denser modulation.

With the advent of digital television and the uncoupling of loudness and dynamic range controls, the elusive goal of transmitting audio as it was intended by producers, while simultaneously protecting viewers from overly dynamic or objectionably loud programs, might be within reach.

[12]The FCC limits are of no significance to RF modulators in set-top boxes, VCRs, and PVRs.

It might seem to some that metadata provides all of the answers to this problem, while others believe that the old way of doing things is more reliable. It is very likely that a combination of both approaches may be the answer.

Under ideal conditions, dialogue loudness metadata is set correctly and program producers tailor their program to provide dynamic range that is acceptable under many listening environments, using mixing techniques and the DRC system inside of AC-3. In that case, the system will work as designed and viewers have a pleasant experience. However, if metadata is incorrect or simply run in default on any channel or on any program, then all of this tends to fall apart. Metadata must be correct everywhere, or unacceptable loudness shifts are guaranteed to creep in.

Unfortunately, everything is not always ideal, and experience has shown that, in these early days of 5.1-channel DTV audio, delivery of proper metadata with programs is still somewhat erratic. Local broadcasters must have fallback audio processing systems to protect viewers from objectionable loudness shifts. The same FCC regulations adopted decades ago concerning loudness still apply, but compliance is made more difficult now due to the 5.1 channels of wide dynamic range audio in the AC-3 system. Certainly an increase in quality can be obtained, but attention must be paid to the powerful capabilities afforded by the system.

The use of traditional two-channel broadcast processors that are likely limited to 15 kHz, and, more importantly, without linkage between the channel pairs for 5.1 channels may result in erratic behavior and be rather unpleasant to listen to, so this approach is to be avoided.

An alternative solution would seem to be the use of a simple six-channel processor where all of the channels are linked. On the surface, it seems that this approach would work. In reality, there are several problems. In particular, in controlling loudness, this type of unit will permanently reduce the dynamic range of the audio and remove the ability for consumers to enjoy the original audio if they so desire. All viewers are affected by the processing and it cannot be undone.

The answer is a processor that accepts metadata and checks the audio against the incoming metadata settings, correcting it only if necessary. This system might also include the optional ability to upmix legacy two-channel programs to 5.1 in response to incoming metadata. This type of processor should allow local stations to decide how closely the audio should match the incoming metadata, correcting if outside the limits, and would allow the station to decide what happens in the absence of metadata.

It is a fact of life that abuses of the advanced features provided by the metadata portion of the AC-3 system are possible and that parameters may be set incorrectly. In both cases, the results can be disturbing to viewers and should be minimized. Doing so in a way that supports the goals of metadata is good engineering practice, and the tools to make this happen are available.

SUMMARY

Audio for digital television needs to be considered carefully from the initial planning stages through to installation and calibration. Knowledge of not only the production and transmission side of the system is necessary, but also a thorough understanding of what happens at the consumer side is required. Knowing how audio will be received will help understand how it should be transmitted. Audio for television is still adjusting to the DTV environment, and certain details vary considerably outside of the United States, but the basic underlying principles remain the same worldwide. Good engineering practice still applies.

References

[1] ATSC A/52B, Digital Audio Compression (AC-3, E-AC3) Standard, Rev. B. Advanced Television Systems Committee, Washington, D.C., June 2005.

[2] Moore, B. *An Introduction to the Psychology of Hearing*. San Diego: Academic Press, 1997.

[3] Jones, B. L., and Torick, E. L. A New Loudness Indicator for Use in Broadcasting, *SMPTE Journal*, Sept. 1981.

[4] ISO 532, Acoustics: Method for Calculating Loudness Level, International Organization for Standardization, Geneva, Switzerland, 1975.

[5] Soulodre, G. A., and Norcross, S. G. Objective Measures of Loudness, 115th AES Convention, Oct. 2003.

[6] ITU-R Rec. BS.1770, Algorithms to Measure Audio Programme Loudness and True-peak Audio Level.

[7] Benjamin, E. Preferred Listening Levels and Acceptance Windows for Dialog Reproduction in the Domestic Environment, 117th AES Convention, Oct. 2004.

[8] Pearsons, K.S., Bennert, R.L., and Fidell, S. Speech Levels in Various Noise Environments, Report No. EPA-600/1-77-025, EnvironmentalProtection Agncy, Washington, D.C., 1977.

[9] Riedmiller, J. C., Robinson, C. Q., Seefeldt, A., and Vinton, M. Practical Program Loudness Measurement for Effective Loudness Control, 118th AES Convention, May 2005.

[10] Riedmiller, J. C., Lyman, S. B., and Robinson, C. "Intelligent Program Loudness Measurement and Control: What Satisfies Listeners?" 115th AES Convention, Oct. 2003.

[11] ATSC A/53E, ATSC Digital Television Standard, Revision E with Amendment No. 1. Advanced Television Systems Committee, Washington, D.C., September 2006.

[12] Vinton, M., and Robinson, C. Q. Automated Speech/Other Discrimination for Loudness Monitoring, 118th AES Convention, May 2005.

[13] CEA CEB11, NTSC/ATSC Loudness Matching, Consumer Electronics Association, Arlington, VA, 2002.

CHAPTER

5.19

Intercom and IFB Systems

ANDREW MORRIS
Consulting Engineer
Denver, Colorado

INTRODUCTION

Communications is an often underappreciated but altogether necessary component of broadcast television production. Director communication with camera operators, equipment operators, and stage managers, as well as producer communication with talent, are all required to put a show on the air. Communications between broadcast centers and remote sites for news coverage, sports, and remote entertainment events is also an indispensable production requirement.

The heart of any broadcast communication system is the *intercom system*. Wired intercoms come in two basic types: *matrix* and *two-wire*. Wireless communications are also used but, in the context of a broadcast environment, usually connected to work with a wired system. Many intercom systems incorporate matrix, two-wire, and wireless in one fully integrated communications system.

Intercoms based on matrix routing systems offer substantial operational capabilities and flexibility. As they have decreased in cost and physical size, they have become the dominant form of intercom deployed throughout broadcasting and production facilities. Matrix intercoms offer the advantage of four-wire communication, where send and receive audio are provided on separate and isolated audio circuits or "pairs." The result is cleaner audio that does not suffer from the crosstalk and artifacts introduced by the hybrid technology that is used in two-wire systems.

Traditional two-wire intercoms offer the advantages of quick set up and distribution of DC power over standard microphone cabling for the powering of beltpacks and speaker stations. With these unique capabilities, two-wire systems still have an important niche in broadcast communication. Increasingly sophisticated, cheaper, and smaller digital matrix systems have, however, replaced two-wire systems in many broadcast operations.

Wireless intercom systems offer production personnel the ability to communicate without being physically tethered to physical cables. For simple communication applications, this freedom from wires is of great benefit for those who need to be mobile.

Interrupted foldback (IFB) is the system that provides communications to production talent in the studio or remote location, and is typically integrated with the rest of the intercom system for control room and technical personnel.

NOMENCLATURE

A problem with understanding intercom functionality and operational capability is the confusing use of identical terms that have different meaning in different contexts. In addition, intercom manufacturers have their proprietary systems and they often use different terms to describe identical operational and technical functions.

Party Line

The term *party line* can refer to any of the following:

- A two-wire intercom system developed by Clear-Com.
- A conference on a two-wire system.
- A conference on a matrix Intercom system.

A party line intercom is a *two-wire* system. A two-wire system refers to an intercom system that carries both intercom audio and DC voltage for powering belt-packs and speaker stations over a single twisted audio pair. The term two-wire refers to the two wires that make up the audio conductors of a twisted pair. Bidirectional audio and DC power are carried over the two wires plus shield in a two-wire system.

Clear-Com developed the Party Line and RTS developed the TW (for Two-Wire) systems that are the most common two-wire intercom systems. Differences between them will be explained later in this chapter.

PL is an abbreviation for party line but in the vernacular can refer to an intercom system, a user station on an intercom system, as well as any kind of communication conference (including a teleconference).

Four-Wire

Four-wire refers to systems that have separate transmit and receive audio pairs. The two audio pairs each have two signal conductors for a total of four wires. Matrix intercoms are inherently four-wire systems since each port on a matrix intercom has separate transmit and receive audio pairs.

Digital systems do not use separate audio pairs for transmit and receive audio yet the term four-wire persists to describe transmit and receive audio that is of high quality and is isolated from each other. Matrix intercoms, even though they are digital, are often referred to as four-wire intercom systems to distinguish them from the Party Line and TW systems, which are referred to as two-wire intercom systems.

Mnemonic, Alpha, or Alphanumeric

These terms refer to the name applied to a variety of functions in a matrix intercom system. The intercom panel used by the director in control room A might, for example, have a mnemonic of "ADIR." In addition to panels, individual party line conferences and IFBs are generally assigned mnemonics that are meaningful to the users of an intercom system.

IFB

Interruptible foldback (IFB) refers to providing a program feed to production talent, usually via an earpiece. This program feed can be interrupted by production and engineering personnel whenever there is a need to speak directly to the talent. Pressing a key on an intercom panel that has been assigned an IFB will result in that user's voice interrupting the program feed. The program can be muted or dimmed (attenuated) to a predetermined level. The program feed is generally a feed of the program mix without the talent's voice and is therefore referred to as a mix-minus feed. An IFB can be provided to people other than those designated talent and it can also feed a loudspeaker as well as an earpiece.

ISO

ISO stands for "isolated" and is an intercom function that permits a user on a conference to be isolated from that conference for a private conversation. ISOs are typically used by a video engineer to isolate a camera from the camera PL. The video engineer can then work directly and privately with the camera operator on camera setup without disturbing the other camera operators on the camera PL.

Group Call or Special List

The *group call*, or *special list*, function allows for multiple participants in a special group communication circuit. For example, a group call can be an "all" call for a director to get the attention of everyone involved in the production.

Point-to-Point

Point-to-point refers to a communication between one user with an intercom panel to another user with an intercom panel. The talker presses a key that has been assigned a mnemonic for another user. The panel that is being called will hear the caller out of the intercom panel speaker or headset and will also receive a tally (a flashing mnemonic or lighting of an LED) that indicates the identity of the caller.

INTERCOM FEATURES

GPI Control Interface

Intercom systems typically have GPI (general-purpose interface) inputs and outputs for control purposes. A common use of an output is the triggering of a two-way radio base station. GPI inputs are used to trigger an intercom event from an external contact closure. An example of the use of a GPI input would be an external button push that triggers the intercom to change the program source of an IFB.

Telephone Interface

Telephone interfaces allow dial-up telephone system access to an intercom. These interfaces are typically integrated into the intercom control system allowing a phone call to be initiated from and received by an intercom panel. Telephone interfaces typically offer auto-answer capability, password protection, and the ability to remotely signal specific ports via DTMF (dual-tone multifrequency) tones.

Telephone interfaces are often used to communicate with a camera operator or talent at a remote location.

AES Interfaces

AES-3 digital audio interfaces have become important as digital audio systems have entered broadcast facilities. Matrix intercom systems often have a substantial

number of ports directly connected to digital production audio consoles for program audio and mix-minus feeds. In addition, many operations use the intercom as a monitoring device with remote feeds, talent microphones, and program feeds serving as sources to the intercom. A digital AES interface eliminates the need for external digital-to-analog converters.

Internet Protocol Interfaces

Internet protocol (IP) interfaces have recently become an integral part of intercom systems. Manufacturers offer voice over IP (VoIP) cards that are designed to work over private IP networks. Intercom control data and voice traffic can be sent over these IP networks.

Hybrids

The term *hybrid* refers to a device that converts two-wire to four-wire audio and vice versa. Analog hybrids initially used transformers, as shown in Figure 5.19-1, and later used op-amps to convert between two-wire and four-wire audio. Digital hybrids with digital signal processing (DSP) chips are most commonly used to perform two-wire/four-wire conversion.

One problem with hybrids is poor trans-hybrid loss, which is a measure of the loss or isolation between the transmit and receive ports on the four-wire side of the circuit. Trans-hybrid loss depends on signal cancellation accomplished by defining the line impedance and mirroring it in a balance network. The variations in impedance presented on the two-wire side of the hybrid make balancing difficult and often result in poor trans-hybrid loss.

Nulling refers to adjustments made in a balancing network to achieve greater trans-hybrid loss by making resistive, capacitive, and inductive adjustments to match the impedance of the two-wire side of the circuit. Inductive, resistive, and capacitive nulling affect

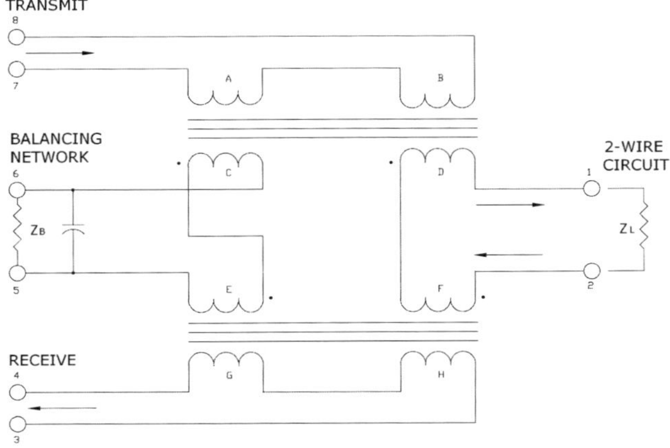

FIGURE 5.19-1 An analog hybrid circuit using transformers.

low-, mid-, and high-frequency bands. Modern hybrids are digital and are capable of auto-nulling.

Hybrids are used in intercom systems as interfaces to telephone networks and as interfaces between two-wire and four-wire intercom systems. The inherent problems associated with two-wire to four-wire conversion explain why it is better to use the four-wire interface that is available on most camera control units (CCU) instead of a two-wire interface when connecting CCUs to a matrix intercom system.

MATRIX INTERCOMS

Matrix intercoms are analogous to crosspoint switchers for audio or video, where one or many inputs are switched to one or more outputs. Some intercom systems use a time-division multiplex (TDM) bus to accomplish this same type of crosspoint switching. By mapping individual time slots in a TDM bus any talker can be routed to any listener, as shown in Figure 5.19-2. In addition to mapping time slots, mixing is also performed in the digital domain to dynamically control a mix of multiple talk ports to a given listen port.

While the audio internal to many matrix intercoms is digital, the audio input/output (I/O) for some system is analog. Some matrix intercoms are digital all the way to the intercom panel. Most intercom manufacturers have a variety of I/O cards offering analog, AES, and IP-based audio interfaces. The major distinction is the intercom panel itself. Some manufacturers offer panels with analog audio I/O while others offer panels with digital audio I/O. Panels with VoIP interfaces are also available.

Digital matrix intercoms (see example in Figure 5.19-3) pack a large amount of I/O in a relatively small physical space. Intercom manufacturers have been able to offer up to 208 ports in a single six-rack unit frame. Manufacturers have developed methods of extending the TDM bus to additional frames to create intercoms as large as 1024 ports. Coaxial and fiber-based interconnect between frames permits the creation of such large intercoms. The frames for these bus-extended intercoms are generally collocated but in some cases can be a mile or more apart.

Trunk Systems

Two or more intercom systems can be combined by what is known as *trunking* or *intelligent linking* to create what is, in effect, one large virtual intercom system. In the case of trunked intercoms, a controller, generally external to the intercom, receives control data from all of the interconnected systems. A predetermined quantity of four-wire audio trunks provides audio paths between the trunked intercoms. When a user on one intercom wants to talk to a user on another system, the trunking controller makes the necessary crosspoint connection that enables a trunk to be used between systems. The trunking controller also communicates with each intercom's control system so

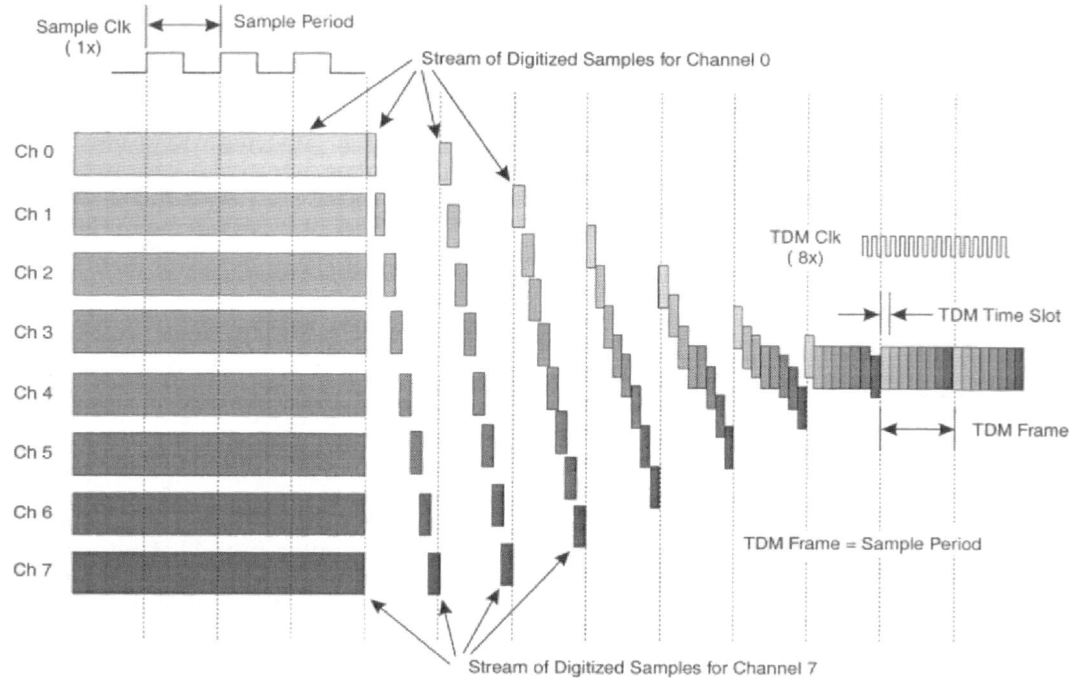

FIGURE 5.19-2 An example of "time-slicing" in a time-division multiplex system. (Courtesy of Telex Communications, Inc.)

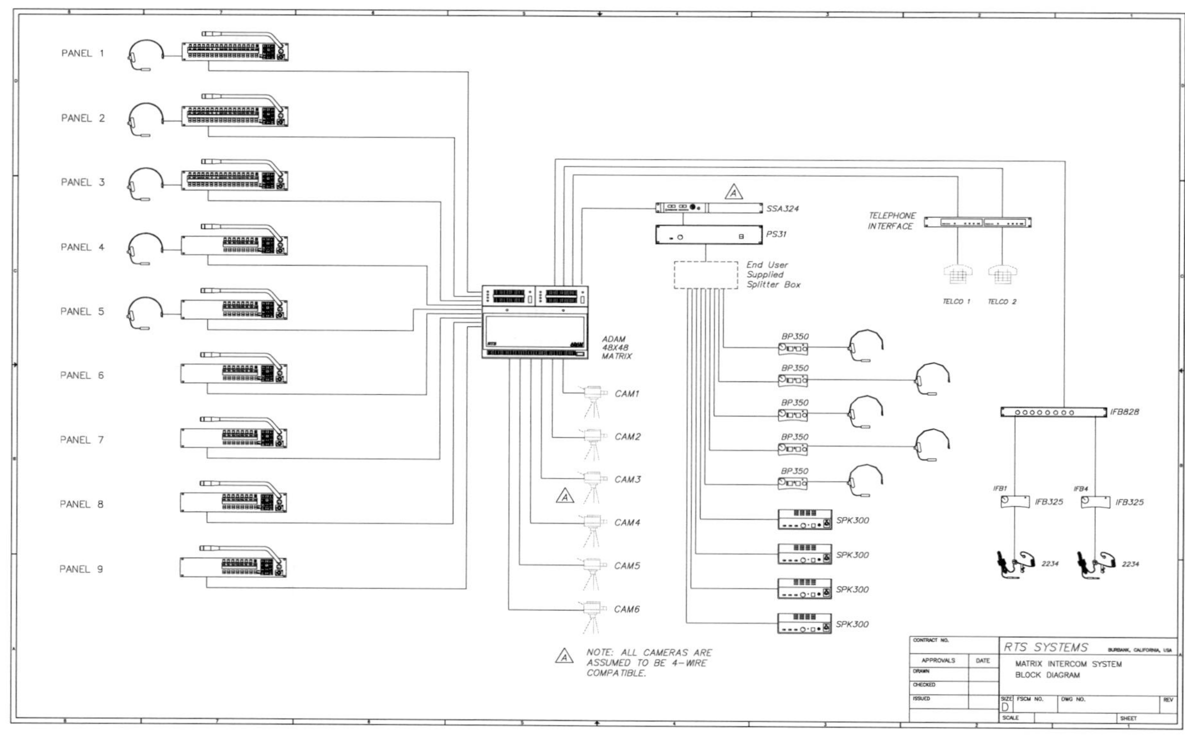

FIGURE 5.19-3 Example of a digital matrix intercom system. (Courtesy of Telex Communications, Inc.)

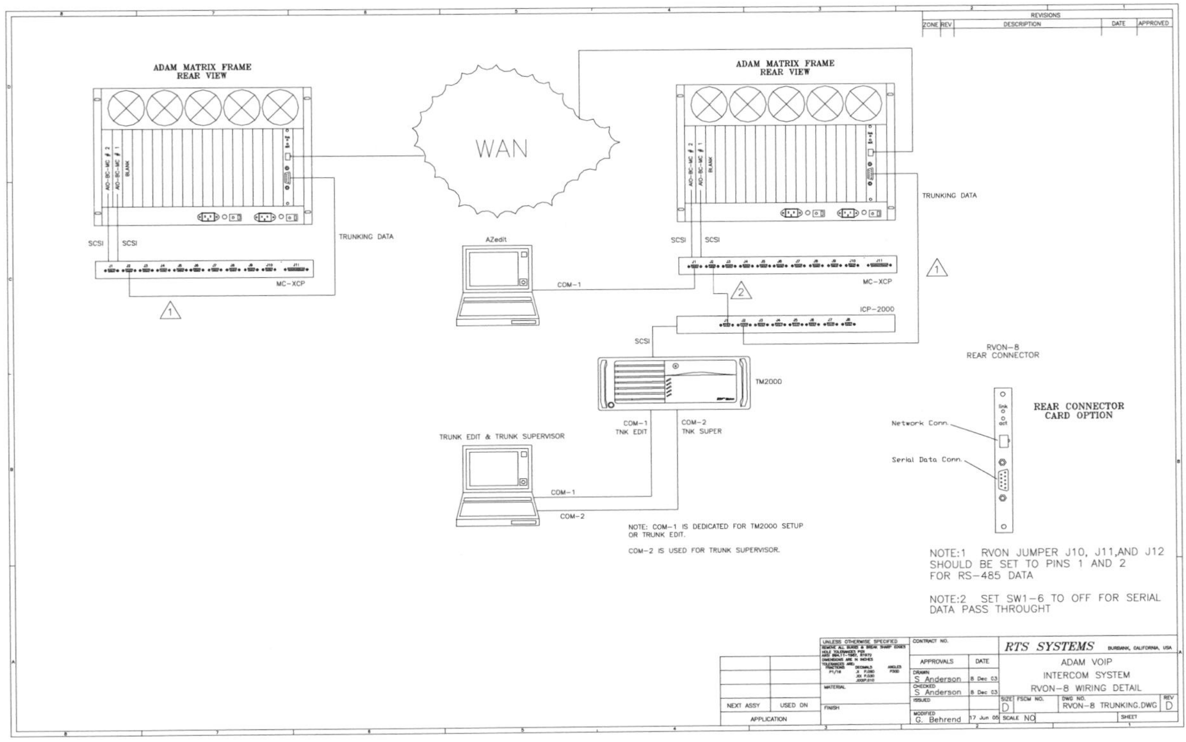

FIGURE 5.19-4 Example of system with trunked matrix intercoms using VoIP technology. (Courtesy of Telex Communications, Inc.)

that appropriate crosspoints are made in each intercom to enable a communications path between a user on intercom A with a user on intercom B. An example of an intercom system with interconnecting trunks using VoIP technology is shown in Figure 5.19-4.

Trunked intercoms are often used to connect intercoms that are geographically diverse. They may be in different parts of the country or even in different parts of the world. Trunking has also been used in large systems to connect multiple intercoms that serve different areas within a given facility.

One intercom manufacturer has implemented a number of methods for interconnecting multiple intercoms using fiber-optic cabling in order to provide varying degrees of fault tolerance and redundancy. A *meshed decentralized system* provides connection between each and every intercom in a system. If one intercom or its connection fails, all of the other intercoms are still able to communicate with each other. With a *decentralized ring system* a dual-fiber ring also allows for communication between all functioning matrices should one intercom fail. This method allows for continued operation in the event of a failure of one of the fiber-optic cables on the ring.

Matrix Intercoms Operations

Point-to-point communication is one of the most basic and powerful operations provided by a matrix intercom. Keys on intercom panels can be programmed either from the intercom configuration computer or from the panel itself to allow communication with any other user on the system. In the case of trunked intercoms, users from one intercom can be programmed to appear on keys on intercom panels attached to a a different system. This simply and easily enables communication between users that might be as much as half a world apart.

It is worth noting that point-to-point communication may not be as private as it might seem. Other users of intercom panels on a system can turn on a listen key to eavesdrop on one or both sides of a point-to-point conversation. The configuration software may offer a privacy feature that blocks listen key access to ports that have this *privacy* feature activated.

With matrix intercoms, PLs are conferences that can be joined by any user. In the configuration software, a PL is defined and given an alphanumeric name. "DIRPL" might, for example, be the name of the Director's PL. Anyone with an intercom panel can have that PL assigned to a key so that a user of that intercom panel can talk or listen on that PL. For nonintelligent devices, such as two-wire and wireless beltpacks, the configuration software can permanently assign their intercom ports to a PL so that they are always talking and listening on that particular PL.

IFB Feeds

IFB feeds are created in the intercom matrix. Typically, an audio console provides a number of mix-minus feeds that are inputs to the intercom. Outputs from the intercom feed the IFB receivers used by talent and others needing to hear the IFB.

The intercom configuration assigns one of the mix-minus feeds (also called a program source) as the input to an IFB. The intercom configuration software also assigns the output from the IFB to an intercom port that feeds an IFB receiver. The IFB is given an alphanumeric name that is assigned to a key on intercom panels.

Talent will hear a mix-minus program source until a user of an intercom panel presses the key that has been assigned that IFB. Upon a press of that key, the intercom interrupts the program source that is being fed to the talent earpiece with the audio from that intercom panel. The program source is dimmed (attenuated) or muted so that the talent can hear the user of the intercom panel at full level. When the IFB key on the panel is released the program source returns to full level. The audio routing, mixing, dimming/muting, and control of the IFB all occur in the intercom matrix itself.

The intercom configuration software can prevent intercom panel keys from being latched. *Latch disable* of IFB keys is used to prevent a producer from mistakenly latching a key and having the program source continuously dimmed or muted while undesirably putting his or her intercom panel's output full time into the talent's ear.

Voice Over IP

Voice over Internet protocol (VoIP) technology has been incorporated into many matrix intercom systems. Panels and intercom I/O cards are available that offer an IP connection to the intercom and an IP interface to the outside world. Both intercom audio and control data are transmitted over an IP network to and from the intercom.

In the case of an IP-based intercom panel, the connection between the intercom and the panel is over CAT5 cable and can travel over a private network that is dedicated to intercom use or over a corporate network. Note that the Internet itself is not conducive to reliable, mission critical, high-quality communication. The risk of latency and packet loss is too great to run a critical communication application over a public network.

Other uses for VoIP include trunking and four-wire interconnect between different intercoms. VoIP cards and a wide area network (WAN) can replace, at substantially lower cost, common carrier–provided four-wire trunks over a T1 data connection. Similarly, with VoIP cards and a local area network (LAN) there is no need to pull audio pair and control cable between intercoms in a broadcast facility. Interconnected intercoms on a network equipped with VoIP cards can send audio traffic and trunking data over that network. Physical interconnect is provided by CAT5 cables that connect to the network via a hub, switch, router, or gateway. Proper IP addressing and network setup parameters are required for the setup of the router or switch that is the interface between an intercom VoIP module and the network.

VoIP cards generally offer different versions of codec (coder-decoder). These codecs, with names such as G.711, G.722, G.723, and G.729 A/B, offer varying degrees of audio quality, bandwidth requirements, and delay. Choosing an appropriate codec is based on tradeoffs between those factors according to user requirements and the bandwidth available on the network.

In an effort to conserve bandwidth, most manufacturers offer optional voice activation detection so that bits are not wasted coding low-level noise during moments of relative silence.

Some intercom manufacturers are using VoIP technology to offer virtual intercom key panels. These are panels that exist in software and are accessed via an application that runs on a personal computer (PC). The connection between the PC and the intercom is via Ethernet with both the audio and the control data run over the network connection using IP protocols. The PC needs to be equipped with a headset or a microphone and speaker.

Matrix Control Systems

Matrix intercoms have three components: input audio (audio to the matrix), output audio (audio from the matrix), and control data. Different intercom manufacturers use different physical and electrical topologies to connect their intercom panels to the intercom matrix. Physical topologies include twisted pair, coaxial cable, and optical fiber to provide bidirectional audio and control data between the matrix and the user panel. Electrical topologies include serial data in the form of EIA-422, EIA-485, as well as AES-3 audio with user bits containing the control data.

It is the configuration software that makes matrix intercoms so powerful. The software that runs on a PC connected to the intercom allows for the creation of mnemonics, the programming and set up of intercom control panels, and party line conferences. It also configures IFBs, relay outputs, and GPI inputs, and the set up of ISOs, group calls, and other special features.

All of the above parameters can be set up and dynamically changed via the intercom configuration software. This is a significant change from the older two-wire intercom systems. Two-wire systems use cabling, jumper plugs, and DIP (dual in-line package) switches for configuration of many of the same parameters, which are not easily changed during production operations.

Matrix intercoms offer the ability to store and load setup files to customize the intercom according to different shows or day parts. Partial setup files can be created and sent to the intercom and this provides a broadcast operation the ability to change intercom settings for one section of an intercom. For example, loading a partial setup file can reconfigure the intercom panels, IFB, and PLs for a control room according to the production that is active in that control room

while the intercom settings for the rest of the broadcast plant remain unchanged.

The control software also offers extensive diagnostic and maintenance capabilities, including the ability to monitor real-time performance of intercom hardware including I/O cards, controller cards, and intercom panels. It also gives a user the ability to monitor crosspoint status and to enable and disable crosspoints from the control system itself.

The intercom control software can also set audio levels for specific crosspoints and intercom ports as well as for party line conferences.

Most intercom configuration software features programmable logic statements. An example is programming a key closure on an intercom panel to send a partial setup file to the intercom. This would be a quick and simple way for an operator, with the press of one key, to reconfigure a control room for its next show without having to use the intercom configuration computer.

Matrix Intercom Panels

The most visible part of the intercom, and the one piece of equipment that most users associate with an intercom, is the intercom panel, an example of which is shown in Figure 5.19-5. Most of these panels can be programmed from either the intercom control system or at the panel itself. Panels typically use LED, LCD, electroluminescent, or fluorescent displays to show intercom mnemonics and tallies. Tallies take the form of flashing mnemonics or either steady-state or flashing LEDs to indicate such things as incoming calls or whether the talk or listen function assigned to an intercom key is active.

Keys and Buttons

Matrix intercom panels typically have keys or buttons that activate talk or listen functions. An intercom key can be held in position to function as a momentary button push or it can be latched via a quick push up or down. Some panels have separate listen and talk keys but most panels use keys that share talk and listen functions. An example of intercom key usage is a user pressing a key down and enabling the talk function; at the same time an LED lights red to tell the user that the talk function is active for that key. Press that same key in the up position and the listen function becomes active and the LED turns green.

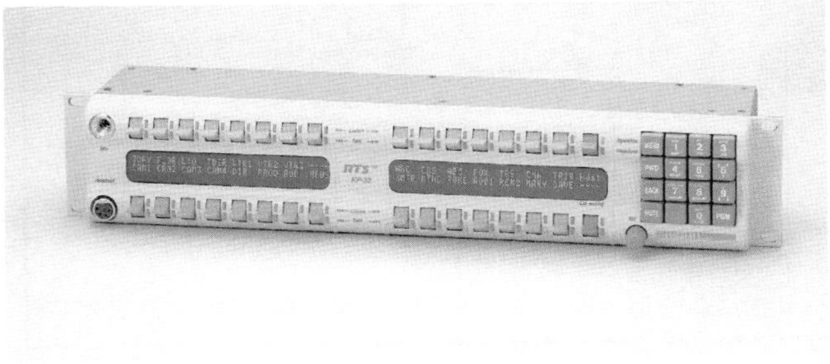

FIGURE 5.19-5 An RTS KP-32 matrix intercom panel. (Courtesy of Telex Communications, Inc.)

Latching of talk keys is generally discouraged as a latched talk key will pick up ambient room noise, which can be annoying to those required to listen to that intercom panel. This is especially true in the case of a PL, which by definition will have many listeners. A latched talk key can be annoying to the primary user of a PL, such as a production control room director. Generally, talk keys are only latched by important users, such as the director who will be talking to many stations continuously throughout a show. Listen keys are often latched so a user can monitor specific intercom sources, such as a party line conference.

Most intercom configuration software has the ability to disable the latching of keys. This is important with IFB usage where an accidentally latched IFB key could be disruptive to the talent on the receive end of that IFB circuit.

Footswitches are often used in control rooms and at graphics workstations to allow hands-free operation of the intercom panels.

Panel Audio Inputs and Outputs

Intercom panels offer the ability to switch between the panel microphone and panel speaker and a headset, and the ability to mute the panel microphone.

Intercom panels typically have a rear connector panel that provides access to a variety of features. These include auxiliary (*aux*) audio inputs, an external speaker output, an external headset connector, a microphone preamp output (generally referred to as *hot mic*), a footswitch input, GPI inputs, and relay outputs.

An example of the use of an intercom panel aux input is a program audio feed. This gives the user of that panel the ability to monitor program audio as well as any intercom communication.

Another method of accomplishing the same monitoring capability is to provide program audio as an input source to the intercom itself. Program audio can then be assigned to a key on an intercom panel, and by activating the listen function on that key an intercom panel user can monitor program audio.

An intercom panel's microphone preamp output can be used to distribute the talk audio of key production personnel around a facility. For example, if it is deemed critical for production personnel throughout a plant to monitor all communication from a show director, that director's hot mic audio may be distributed to speakers around a broadcast plant. Anything the director says (whether or not a talk key has been activated) will be heard through those speakers. This gives production personnel some ability to anticipate the director's commands.

Hot mic audio from intercom panels can also serve as a backup should the intercom itself fail in whole or in part. Since it is the audio from the panel microphone that is distributed, and not audio that travels through the intercom, that audio will remain available even if the intercom system itself experiences a failure.

Most modern matrix intercoms and some intercom panels give the intercom panel user the ability to adjust listen levels. An intercom panel user can therefore equalize the levels of soft and loud talkers and this contributes substantially to the overall intelligibility of all talkers on a given intercom panel.

Setup

Many matrix intercom panels can store multiple setup pages. A 16-button panel may, for example, have four setup pages. Each button on the panel on each setup page can be given an assignment. This provides the ability to have panel assignments preset for four different productions. In a matrix system these setup pages can be created from either the panel or from the configuration computer. Similarly, a given setup page can be activated from either the panel or the configuration computer.

PARTY LINE/TWO-WIRE INTERCOMS

Two-Wire Intercom Systems

Two-wire intercoms provide communication by offering a conference on one or more channels with each channel functioning over a single pair of wires. Generally speaking there is no point-to-point communication in a two-wire system. Multiple channels can be implemented so that a number of different party line conferences may be used, as shown in Figure 5.19-6. Thus, there may be, for example, a camera PL, a director PL, and a production PL.

Clear-Com calls its system a Party Line intercom. RTS refers to its as a TW (for Two-Wire) system. Both the Clear-Com and RTS systems offer bidirectional audio over a single shielded twisted audio pair. The shielded twisted pair also carries a nominal 30 volts DC that powers beltpacks and remote speaker stations. Clear-Com's Party Line system and the RTS TW system put a nominal 26–32 volts DC between pin 2

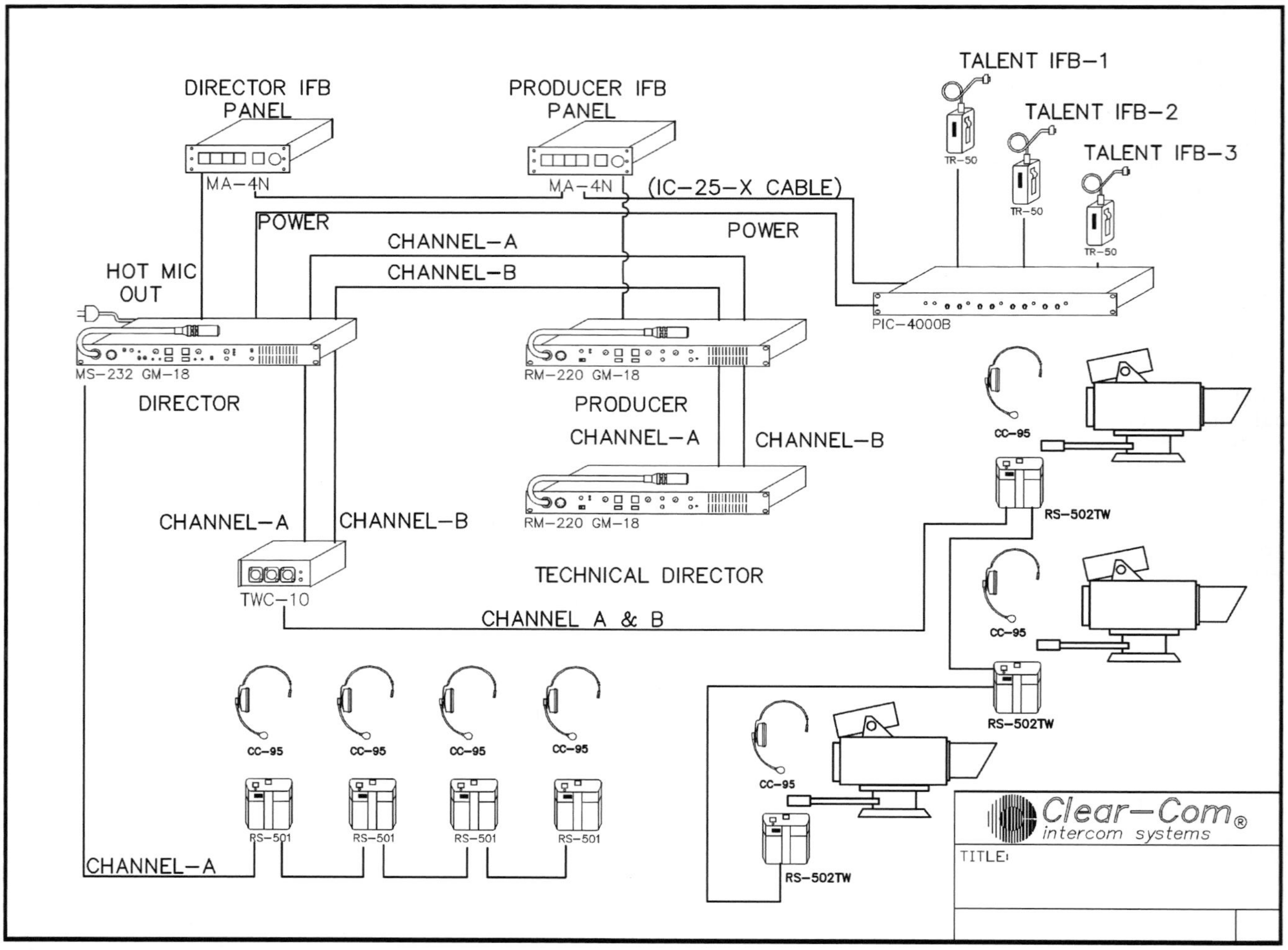

FIGURE 5.19-6 Party Line system drawing. (Courtesy of Clear-Com Vitec Group Communications.)

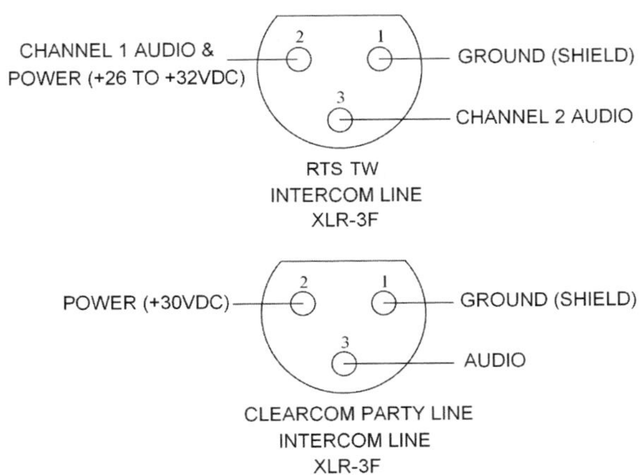

FIGURE 5.19-7 Clear-Com and RTS pinouts.

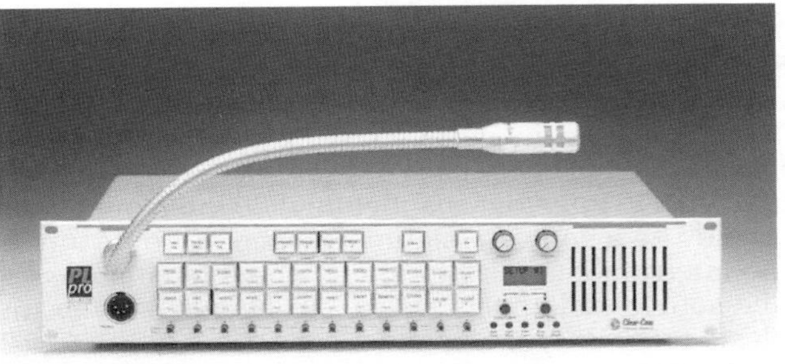

FIGURE 5.19-8 Clear-Com MS-812A party line master station. (Courtesy of Clear-Com Vitec Group Communications.)

and pin 1 (shield). The TW system allows for two channels of audio with audio channel 1 on the pin 2 conductor and audio channel 2 on the pin 3 conductor, as shown in Figure 5.19-7.

Both Clear-Com's Party Line and RTS's TW operate in an unbalanced mode where the audio signal is developed between a signal conductor and ground.

Party Line and TW intercom systems are typically designed around a power supply or (as is often the case with Clear-Com) a main station that provides the DC voltage applied to the pin 2 audio conductor, and which also provides a 200 ohm termination to the system. User stations bridge the intercom line at impedances of greater than 10,000 ohms and, therefore, only minimally load the system. Audio levels remain constant as stations (within the specification of the system) are added to or subtracted from the intercom line.

Two-wire systems are referred to as distributed amplifier systems since they do not have central electronics as does a matrix intercom system. Advantages of these systems include the ability to power beltpacks and speaker stations without external AC power supplies or batteries and the ability to use microphone cable for interconnect between power supplies, user stations, beltpacks, and speaker stations.

Party Line intercom systems are well suited for remote television production since they are relatively easy to set up and tear down. Providing power to beltpacks over the intercom line is an important advantage in this type of situation.

Each user station in a Party Line system has its own electronics and a power supply for powering microphone preamps, headset and speaker amplifiers, as well as for providing power to signaling electronics such as call lights.

Two-Wire User Stations

While two-wire user stations vary from matrix intercom panels, they have many of the same features, including hot mic outputs, program inputs, relay outputs, and logic inputs. The major difference between a two-wire user station and a matrix intercom panel is the data that is transmitted between a matrix intercom panel and the intercom matrix. This data includes mnemonic and tally information as well as intercom panel status. Figure 5.19-8 shows an example of a typical party line, two-wire, master intercom control panel.

Since a two-wire system is a distributed system it does not have a central intercom unit. Two-wire intercom panels can display mnemonics but these are created in the panel. Tally information is transmitted as a call light from a calling panel to a called panel.

Some of the smaller two-wire user stations, such as wall mount stations, are powered by the DC voltage on the intercom line and do not require an external power supply.

Signaling

Call lights allow user stations to generate a visual indication for cueing purposes and for providing a visual indication that someone is trying to talk to his or her station or beltpack. Clear-Com uses a DC voltage to activate call lights; RTS uses a super-audible 20 kHz tone. Call signals can also be used to trigger two-way radios and other devices. Devices are available that will detect both types of call light signals and generate a relay closure.

Remote Mic-Kill

Remote *mic-kill* is a useful feature of many two-wire intercom systems. An open, unattended microphone is an annoying and disruptive aspect to an intercom system. Extraneous noise picked up by open, unattended microphones can be heard by everybody on the intercom channel and can make communication difficult.

Clear-Com systems accomplish remote mic-kill by momentarily interrupting the DC on an intercom line. RTS TW systems use a super-audible 24 kHz tone to signal user stations to shut off their microphones.

Source Assignment

Source assignment panels function as a means of routing two-wire intercom channels to intercom busses. These are used in more complex systems that have many intercom channels. For example, a two-wire system with 12 intercom channels and 20 intercom busses would use a source assignment panel to route any of the intercom channels to any of the intercom busses. The source assignment panel can be used, for example, to route intercom channel 1 to busses 1, 2, 3, 5, and 7, and intercom channel 2 to busses 4 and 6. Users can plug their beltpacks into the appropriate bus to be on the desired intercom channel. The source assignment panel offers flexibility in assigning intercom channels to intercom busses and provides the ability to easily reassign channels to busses on the fly.

Two-Wire IFB Systems

Two-wire IFB systems generally require a central electronics control unit to allow multiple users to select one of many IFB channels to talk to different talent. An example of a two-wire IFB system is shown in Figure 5.19-9.

The circuitry in the central electronics unit allows for the selection of program sources and the switching of those sources to an IFB channel. When a user talks on an IFB channel, that interrupt audio is mixed with the program source audio, which is dipped in level according to a user-defined setting.

The IFB control unit also superimposes a DC voltage on the IFB line to drive the talent receivers. These IFB receivers can, according to model, be either a two channel model that can provide noninterrupt audio in one ear and program with interrupt (IFB) audio in the other ear, or a single channel model that provides only interrupt audio.

The interrupt audio is generated at IFB control panels that are interfaced to the IFB controller or from a user station. User stations can also generate a local IFB where the program audio that is fed into the user station is interrupted by the talk at the user station and is sent down an intercom channel. A locally generated IFB does not have the flexibility of a centrally based IFB system since only that user station can talk on a

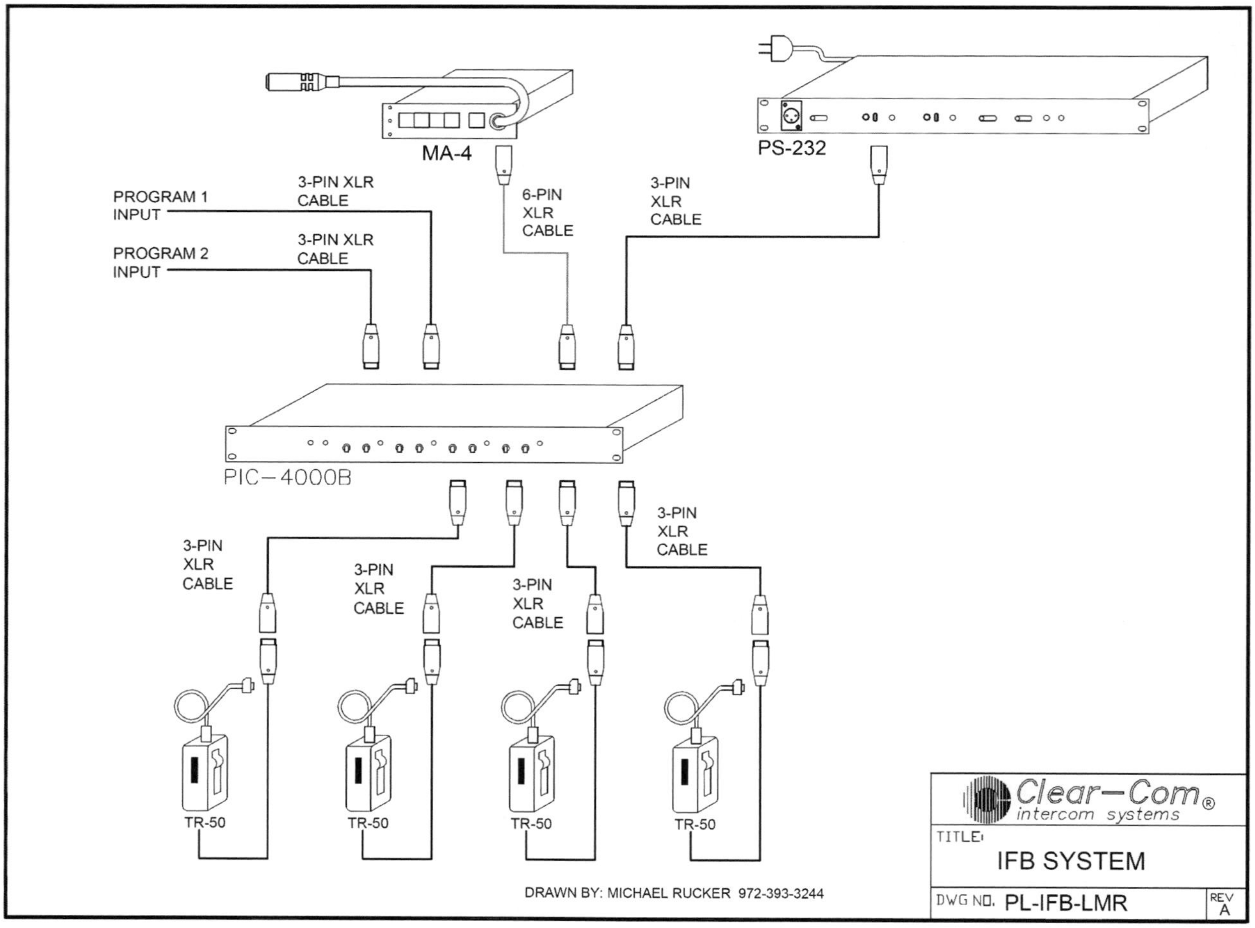

FIGURE 5.19-9 Party Line IFB system. (Courtesy of Clear-Com Vitec Group Communications.)

given IFB channel. With a central system any number of users can potentially cue talent on the same IFB channel.

In a two-wire system, IFB priorities can be set so that one station has priority over another station for talking on an IFB channel. User panels can be programmed with different priority levels to ensure that the most important producer in a production will always have communications access to talent.

The central electronics unit provides tally information so that an IFB button assigned to a channel will illuminate to indicate when that channel is in use. For example, if a user talks on IFB channel 1, the IFB 1 button on every user station will illuminate to indicate that IFB channel 1 is in use.

IFB control panels and appropriately equipped user stations have an IFB all call button that will allow a user to speak on all IFB channels simultaneously. This is a handy feature when a producer needs to provide the same cue to multiple talent on a set.

Interfaces to Two-Wire Systems

Interfaces exist for using two-way radios and telephones in Party Line/TW-type systems. For two-way radios, user stations are set up so that a call light signal is generated along with the talk audio in order to key the radio base station.

Two-Wire System Issues

Crosstalk between intercom channels can be a problem with two-wire intercom systems. Using appropriately sized low-capacitance cable and wiring techniques can help solve or minimize crosstalk problems. Running all cable in a star configuration to a suitable central location, such as the system power supply or main station, can help reduce system ground resistance.

When multiple channels are run in a multipair cable the resistance of the ground return can cause crosstalk. Tying all the individual shields and drain wire together will reduce DC ground resistance and help minimize crosstalk.

Excess cable capacitance can also cause oscillation and sidetone instability and act as a low pass filter that reduces frequency response on an intercom channel.

It is important that each intercom is properly terminated in 200 ohms at the power supply or main station. An unterminated line will cause excessive level and can cause oscillation and squealing. Intercom lines with double terminations will cause low level on the intercom lines and can also cause problems with the nulling of headset hybrids.

Depending on the capacity of the power supply or main station and the cable lengths deployed, there are limitations on how many user stations, beltpacks, and speaker stations can be used on an intercom line. Manufacturer specifications should be studied to determine how many stations can be used with a given power supply on a system.

Digital Party Line System

One intercom manufacturer provides a digital party line system. Similar in concept to the Party Line and TW systems, it offers interconnections via standard shielded twisted pair mic cable. Instead of a single DC voltage applied to pin 2, this implementation applies a 30 to 48 volt DC phantom power to both pins 2 and 3 (with the shield as return) to power beltpacks and speaker stations. The DC voltage is generated from a power supply or a master station. Instead of analog, the audio is in a modified AES-3 format and all signaling, such as remote mic-kill and call signaling, is transmitted as user bits in the modified AES-3 bitstream. The digital audio is two channels each at 16-bit resolution sampled at 48 kHz.

WIRELESS INTERCOMS

The obvious benefit of a wireless intercom system is the freedom it gives those who need mobility within a production area in order to do their jobs. Stage managers, lighting technicians, and A2s (also known as audio assists) are examples of personnel who move around a studio environment as they perform their jobs and would prefer the benefits of untethered communication.

Wireless systems may be used alone or connected with a matrix or two-wire intercom system. In broadcasting applications they are almost always connected to a matrix or two-wire intercom system and function as an extension of a wired intercom system. In addition, wireless systems commonly offer an ISO (isolate) option that allows wireless users to communicate between beltpacks without tying up an intercom channel.

A wireless IFB system employs a transmitter whose source audio is the IFB output of a matrix or two-wire intercom system. The talent wears a wireless receiver to hear the IFB. The IFB itself is created upstream of the wireless transmitter in either the two-wire or matrix intercom system.

Interfaces

Intercom wireless transceivers typically have both two-wire and four-wire interfaces. The two-wire interfaces generally have options for interfacing to both RTS TW and Clear-Com Party Line systems. The four-wire interface is merely an audio interface that provides transmit and receive audio and that typically serves as an interface to a four-wire port on a matrix intercom.

A program input connector is generally available at the wireless base station to feed program audio (or any audio input) to the wireless beltpacks.

Stage Announce

Stage announce (SA) is another wireless intercom system feature. In this case audio from a beltpack appears

on an XLR connector on the base station. Audio from this connector is typically wired to an amplifier and speaker in order to put the wireless audio on a speaker for the benefit of anyone on a studio set. A typical application is the stage manager announcing to everyone how much time remains until a show is on the air.

A relay can be associated with the SA output so that the activation of the SA will key a two-way radio or turn on an on-air light.

Radio Frequency Features

Wireless intercoms have advanced dramatically in the last few years. While VHF bands continue to be used, frequency congestion has increased and the UHF band has become increasingly popular. Digital wireless systems are now being introduced that use or plan to use frequencies in or around the 2 GHz range.

An important feature of some digital wireless intercom systems is the ability of base stations to scan and find available channels within a frequency band. This is especially useful in temporary setups, such as sports remotes, where it is difficult to know in advance which frequencies in a band will be available.

One vendor offers a system based on the DECT (digital-enhanced cordless telecommunications) standard. This is a multicarrier TDMA (time division multiple access) technology that operates over 10 carrier frequencies in the 1880–1900 MHz range.

This system offers security through encryption and the registration of each beltpack with the base station. Each beltpack has a unique ID assigned to a particular TDMA time slot. Communication routing can be set up in the base station and any changes made at the base station are transmitted to the beltpacks. Features include adding sources to a group and the reassignment of a talk/listen button on a beltpack in real time.

SUMMARY

With the introduction of digital technology, intercom systems have become increasingly powerful, flexible, and complex. This is particularly true for matrix and wireless intercoms.

The size and features of modern intercom systems allow more users on a communications system and the high-quality communication between those users can take place from disparate and far away locations. The added complexity can make these intercoms more difficult to set up and use so installation and configuration often require specialized knowledge and skill. The end result is worth the added effort as these communication systems have helped make increasingly complex broadcast productions possible.

Bibliography

The following publications were referred to by the author in preparing this chapter and may provide the reader with additional useful information:

Christensen, Steven G., and Strader, Ralph G. *Design and Implementation of a Time Division Multiplexing (TDM) Communications System*, rev. 3.0. Focal Press, Burlington, MA.

Clear-Com Party Line and Digital Matrix System Installation Manuals, at http://www.clearcom.com/support/manuals.html.

Telex Handbook of Intercom Systems Engineering and RTS Basic Intercom Application Guideline, at http://www.rtsw.com/application_guidelines_handbooks.php.

CHAPTER

5.20

Lighting for Television

BILL MARSHALL

Harvey, Marshall, Berling Associates
New York, New York

INTRODUCTION

Lighting is a fundamental and critical component of the television process. It requires a unique blend of optics, mechanical, electrical, and electronic engineering coupled with a major artistic element. Lighting systems, both for studios and stage, have increased greatly in complexity and sophistication over the years and new technologies continue to be introduced in light sources, robotics, and lighting control systems. This chapter covers the various parts of the system, including basic requirements, lighting fixtures, suspension systems, dimmers and control systems. It also mentions some of the developments in control system protocols and standards.

Lighting techniques in television have become as deliberate and carefully crafted as in film production. Many of the compromises of real-time multicamera production are no longer necessary due to lower light level requirements, sophisticated control systems, and enhanced editing capabilities.

Television is a two-dimensional medium, and therefore lighting of the subject is critical to the suggestion of three-dimensional forms. Even the impressive resolution of high-definition television can only replicate an image as it is revealed by light and shadow. Good lighting will both model the form of the subject and its surrounding and maintain its relative balance of intensity to the rest of the image. Lighting design can be most accurately defined as *constructive* use of controlled light for a predetermined objective.

BASIC PRINCIPLES OF LIGHTING

Many of the shots used in television are close-ups; therefore, well-composed lighting of faces is probably the single most important task to be accomplished. The techniques for modeling faces with light, as used in still photography and later adopted for motion pictures and television, were simply a modern adaptation of the same approach used by portrait painters for centuries and can be summarized as follows.

Three-Point Lighting System

The basic arrangement of lights for television lighting is commonly referred to as the *three-point system.*

Key Light

The first *point* is the *key light*, which is the principal illumination of the subject's face. It is placed on an axis or slightly to one side of the camera and generally at an elevation of about 30° above the horizontal line between the subject's face and the camera lens. The elevation should be adjusted relative to the particular structural elements of each subject's face such as length of nose or depth of eyes relative to the brow. The location in practical terms will obviously be determined by where or how the fixture can be supported. In studio work, this means the key light is very often suspended much lower than the general grid height to gain precise positioning.

Backlight

The second point of this lighting approach is *backlight*. Located above and to the rear of the subject, the backlight creates a glow on the hair and a highlight on shoulders, separating the subject from the background.

The combination of a camera-mounted key light and stand-mounted backlight is the most prevalent arrangement used in electronic news gathering (ENG) and electronic field production (EFP). The use of the camera-mounted key light often creates a somewhat less than ideal picture, but has come to have direct association with realism.

Fill Light

The third element of the classic three-point lighting system is the *fill light*. A single key light can often provide modeling that is too severe in the absence of other ambient light sources. Fill light generally softens and blends the back and key light accents while maintaining their purpose to highlight and separate the form.

Production Complexities

While a setup such as described here is relatively easy with one camera and one subject, the reality of television production is that there is generally more than one camera and multiple subjects. To light an actual production, the lighting director will expand the basics of the three-point system to cover all the subjects and for all the camera angles. This often requires clever fixture arrangements wherein the function of a fixture has multiple uses. For example, in a typical two-person interview, the lighting design might be arranged such that one subject's key is simultaneously serving as the other subject's backlight.

Simultaneous with careful positioning of light fixtures for modeling, the lighting director must also carefully consider the control of the shadow the light sources will create. For each camera angle, the key light and backlight relationship must continue as well as control of the shadow they create. Other common shadow problems, which are generally within the lighting director's control, are elongated nose and chin shadows that occur from improperly positioned key or backlight fixtures.

It should be noted that projected shadows, shapes, or patterns can be some of the most powerful techniques available to a lighting director in suggesting time, place, mood, or even pure background decoration. The character and quality of light are often more apparent from the shadows it casts. Shadows can be hard, soft, transparent, or even a different color than the apparent source. Because shadows are so important, major lighting arrangements should be carefully planned and drawn to scale. Section views in scale will help to predetermine where the shadows will fall, if there will be any scenic conflicts, and most of all, this preplanning will save time in the studio or on location.

Once the lighting of the faces is established, the lighting director plans the background lighting. While a TV program can occur in almost any conceivable background, the lighting of faces continues to determine the relative brightness of the backgrounds. The reflective value of a face ranges between 28% and 41% while a white wall can be 96% reflective. A lighting director must therefore be careful to use lighting to maintain a proper balance since the viewer's attention will be naturally drawn to the brightest spot of a television picture. Lighting intensity, as well as scenic element selection, is critical to controlling the viewer's focus.

The cyclorama (cyc) is a generic background found only in studio lighting. A cyc can easily be over lit and then in contrast overpower and appear to darken the skin tones. Projection of light patterns on the cyclorama is often an inexpensive way to create a setting with light, but again, the patterns that decorate a long shot should not appear undesirable or too bright relative to natural skin tones in the close-ups. Elaborate scenic treatment may require all types of built-in specialty lighting and a whole assortment of fixtures for each unique requirement. The cyclorama is discussed in more detail later in this chapter.

LIGHTING LEVELS

There is no one proper level of light, except as required to allow a particular camera to make a good picture under specific conditions. The level required will vary by camera type, lens, type of action, quality of teleprompter glass (mounted in front of the camera lens), existing ambient lighting conditions, and other purely aesthetic considerations. The lighting director must control the balance of the lighting for every given situation whether it is in brilliant sunlight or in a dark dramatic stage setting. Although newer technologies have increased the viewable contrast range, the enormous base of tube-based displays with their limited contrast range cannot be ignored. Therefore, it is also the lighting designer's responsibility to lead the director and design team away from these difficult conditions.

As with any camera lens system, the amount of light required is a direct relationship between the pickup medium, lens aperture, and depth of field desired. Long focal length lenses used in sports events have smaller apertures by design. Since the fast-moving action demands a high depth of field, the lighting level must therefore be high enough to satisfy this combination of factors. Conversely, the same camera in a dimly lit studio close-up may require a fraction of the illumination level of the sporting event. Good picture quality and proper lighting level can be determined only through the lens of the video camera.

Controlling Light Levels

Many devices are available to the lighting director to control the levels of illumination including a wide

variety of television studio and location lights. The distance between the lighting fixture and the subject will directly affect intensity. Intensity can also be controlled by *scrim*, which is actually a wire screen inserted in front of the lamp. The greater the density of the screen, the more the light level is reduced. A wide variety of available sheet *diffusion* materials will reduce intensity and reshape the pattern and quality of the light. Diffusion materials soften the light and shadows produced.

Color filters also reduce light output relative to the coefficient of transmission of the particular color. For example, a blue filter passes very little light, whereas a yellow-green filter will reduce the output light level very little.

The most precise and convenient level control of incandescent lighting is *dimming*. Compact portable dimmers are also used outside a studio or theatrical environment, but do not necessarily eliminate mechanical means of level control. While dimming does change the color temperature of the lamp, it can still be used effectively.

Color Temperature

For accurate color rendition, the television system must operate within a consistent range of *color temperature*. Color temperature is measured in degrees Kelvin (°K) and varies dramatically from type of artificial light to time of day and atmospheric conditions of daylight. A television camera must be white-balanced for the same lighting conditions as the scene that will be shot. Segments shot separately, which are to be seamlessly edited together, must be shot under consistent color temperature light for accurate color rendition between segments. Recalibration of the camera for varying lighting conditions has somewhat the same effect, yet it is often important to retain some of the original lighting color and character. The television system is actually quite flexible, permitting color temperature swings of ± 300°K without major effect to visible color rendition.

Outside light from exterior windows into a set with artificial lighting presents the greatest color temperature problem because exterior daylight is a substantially different color temperature and is extremely bright relative to the artificial interior lighting. Large sheet filters that correct both color temperature and intensity are available for this purpose; however, if the window area is extremely large and daylight dominates the scene, it is generally easier to work with high color temperature light sources rather than incandescent.

LIGHTING FIXTURES

Lighting equipment differs considerably between the studio and on location. On location, equipment is chosen for its light weight and efficiency. The lighting designer on location must cope with varying natural light conditions from full sunlight to little or no light. Conversely, studios are designed and equipped to give the lighting designer complete control over all aspects of lighting: intensity, color, and placement.

There are two basic categories of lighting *fixtures* (also known as *luminaires*): those, such as *spot lights*, with directional, hard-edged, light beams that produce distinct shadows; and those that produce a softer light with diffuse shadows. There are numerous varieties of each for different applications, and the nature of the light can usually be adjusted, sometimes by moving the internal position of the lamp or lens and always by using external accessories for particular applications.

Location Lighting

In some instances, good lighting design may not even require the use of lighting fixtures per se. In many daylight on-location (remote) situations, a lighting director may choose to simply use reflectors and scrims to achieve the desired effect. Movement of the sun is predictable, so a survey of the site must be made at the same time of day of the event shoot in order to select workable camera positions and to determine the proper equipment.

The sun as a key source is harsh and requires intense fill light for the talent. Reflectors are very effective for this purpose. They require no electrical power but must be attended by a person at all times to keep them oriented correctly to the moving sun. Lightweight folding reflectors are convenient to pack and set up but do not present the necessary hard, stable surface needed to provide a smooth field of light. The heavy solid panels are less convenient but are stable.

Another way the harsh light of the sun can be controlled in intensity is by using large silks or nets to soften or to shadow the talent. Nets and silks under 6 ft × 6 ft attached to frames are called *butterflies*, and those up to 20 ft × 20 ft are called *overheads*. Wind is an obvious factor in the decision to use these tools.

Small-Scale Setup

For smaller-scale remote setups, the lighting director's selection of fixtures is often governed by weight and portability relative to lumen output. A wide variety of ingenuous and lightweight fixtures are available in kits that have an assortment of stands and grip hardware, as shown in Figure 5.20-1, to accommodate any situation which may occur while on location. Many of these kits are designed to incorporate filters to match the 3200°K incandescent lighting to the 5600°K daylight. This filter reduces the light output quite noticeably.

Large-Scale Setup

For larger location work involving daylight conditions *HMI® lighting* has become the standard of both the film and television industries. HMI® is a registered trademark of OSRAM, although it is often used (or misused) in referring to lamps of other manufacturers. The name refers to the basic elements that are combined in the lamp's quartz envelope to create the

FIGURE 5.20-1 For remote lighting, compact kits containing a large assortment of fixtures, stands, reflectors, and grip hardware are available. (Courtesy of Lowel-Light Manufacturing, Inc.)

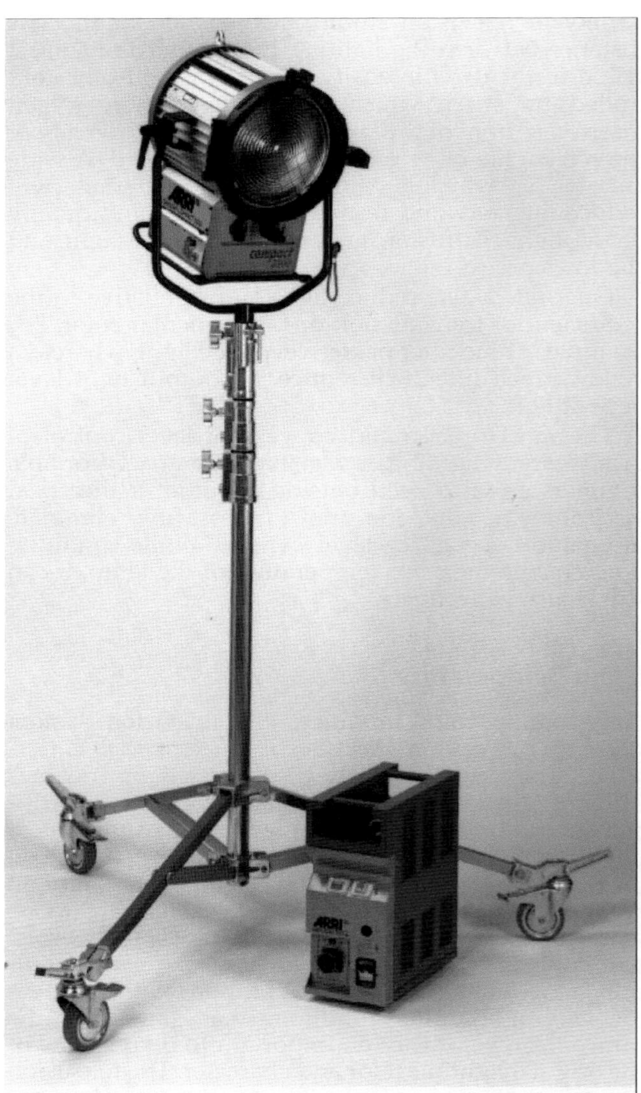

FIGURE 5.20-2 HMI lamps are most often used in *Fresnel lens* housings and are available in a wide variety of sizes. The associated lamp ballast is often a separate unit from the lamp housing for high wattage fixtures. (Courtesy of Arriflex Corporation.)

unique color temperature: "H" for mercury (Hg), "M" for the various metal halide rare earths, and "I" for the halogen iodine and bromine.

These high-efficiency ballasted arc source lamps come in a wide variety of sizes from 200 watts to 20,000 watts. They provide about 80+ lumens-per-watt at 5500°K (which is four times the output of an incandescent light). HMI® lighting is often the best solution for interior situations with substantial window exposure. As in any exterior situation, the lighting director must carefully monitor the changing color temperatures of daylight and remember that the light of a clear sky late afternoon is quite a different color and character than mid-morning when the sky is overcast. Figure 5.20-2 shows a typical HMI® device.

Dealing with Fluorescent Lights

Often a lighting director has to shoot in places where the existing lighting is a low ceiling covered with fluorescent fixtures, and they must be used because they characterize the space. Common architectural fluorescent lights do not provide a full spectrum of light, and the most common type cool white is very short of red energy. While it is possible to color correct each fluorescent lamp using a minus green color correction *gel,* it is often impractical. By color balancing the camera under a representative mix of the fluorescent and the 3200°K talent's lights, the lighting director can produce acceptable skin tone while the background may still appear greenish. Today, a wide variety of fluorescent lamps are available, and some offer quite a high color rendering index. Fluorescent studio fixtures have become very popular but are somewhat limited in control capability due to the relatively large size of the light source and the diffuse nature of the output.

Nighttime Location Lighting

Exteriors at night for TV require great simplicity to look real. Large HMIs® are very useful because one large source can supply the basic illumination for a wide area. Within an exterior scene there are always elements supposedly lit by artificial sources, either seen or imagined. With the camera balanced for 3200°, quartz lights can be used for the people areas while the overall HMIs® at 5500° will seem very blue. If the HMI® light is too blue, it can be corrected slightly until it seems to be the proper gray-blue of moonlight. Using smaller HMIs as back and rim light will further enhance the moonlight effect.

In attempting to provide realism, lighting must suggest the correct mood. Lighting is essential in creating the feeling or mood the production team envisions.

Studio Lighting

In a studio the television lighting director has the greatest control over lighting quality. A TV studio is by definition an idealized environment for production of television programming.

In a well-equipped studio the television lighting director uses a variety of lighting fixtures or luminaries to accomplish the basic objectives of TV lighting:

- Separation
- Modeling
- Accent
- Illumination
- Directing viewer attention

This variety of fixtures and the qualities of light that they produce have been developed to enhance the efficiency of studio operation. The careful manipulation of light quality by external diffusers, cutters, and reflectors for each shot is still not as common in video production as film but has increased greatly. A complement of fixtures with given characteristics plus the mechanical light control techniques of the film industry provide a television lighting designer with a versatile palette of choices to design the lighting.

Spot Lights

In a studio, the most common and useful fixture is the *Fresnel lens spotlight,* as shown in Figures 5.20-3 and 5.20-4. The light from a Fresnel lens[1] is a diffuse but directional beam of light and has a smooth, even field. When the position of the lamp is adjusted behind the lens, the fixture will produce either a narrow spot or wide flood beam of light. Equipped with *barndoors,* this beam can be further shaped to virtually any pattern. Slots in front of the lens allow *color filter, diffusion media,* or *screens* to be inserted for intensity control. A complement of single and double thickness and half- and full-frame screens should be provided for all

[1]French physicist Augustine Fresnel in 1822 invented a lens consisting of concentric rings of glass prisms to make a more efficient lighthouse beacon.

FIGURE 5.20-3 Studio-quality Fresnel spotlights are more durable and have superior optical systems compared to some inexpensive theatrical units. (Courtesy of De Sisti Lighting.)

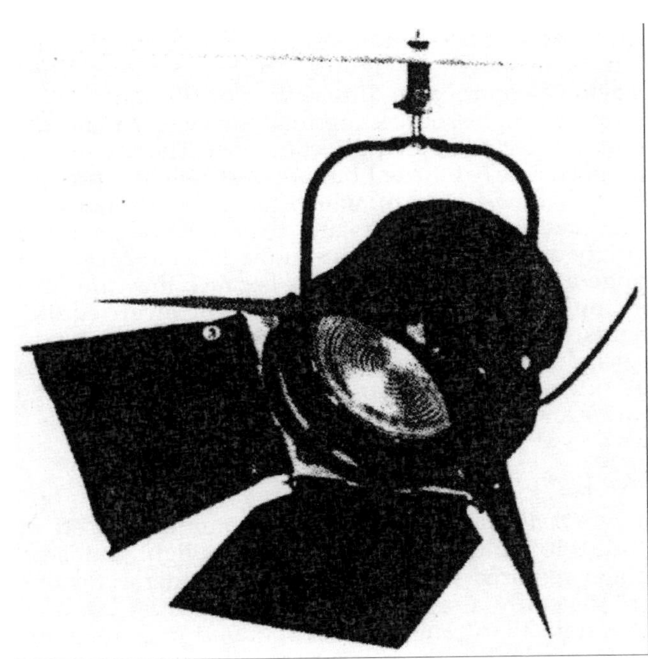

FIGURE 5.20-4 Studio-quality Fresnel spotlights must have long-leaf barndoors as an essential accessory. (Courtesy of Strand Lighting.)

Fresnel lights even when dimmers are available. Fresnel lights are commonly used for key and backlights or where direct lighting is required.

For most 14 ft high grid studios, Fresnel spots in 4", 5", 6", and 8" sizes, with wattages from 300 to 1000 watts, are the most common units. In smaller studios or tight applications, smaller fixtures are often employed. These units enclose the same wattage lamps in small housings. A number of people prefer these baby-size units for their easier handling and somewhat different optical characteristics. In larger studios with grids higher than 14 ft, fixtures with higher wattages are required. Therefore, in higher studios, the standard complement of fixtures might be 2 and 5 kilowatt units or even up to 10 kilowatts. Conversely, Fresnel lens units are also available in sizes down to a 2-inch lens with a 100 watt lamp.

In selecting a line of Fresnels, the studio lighting designers must weigh both the optical and mechanical features including:

- *Stability.* Poorly made fixtures will not focus properly once they are hot.
- *Balance.* Fixtures should be well balanced, even with barndoors to remain in focus and not change position.
- *Accessories.* These features include barndoors, screen sets, diffusion frames, and stand and grid mounting hardware.

Soft Lights

Soft lights are the second most common studio lighting fixture, as shown in Figures 5.20-5 and 5.20-6. They provide a controllable diffuse base or fill light. In this fixture the light sources are totally concealed and the light is reflected in an indirect manner. These units are not particularly efficient but are unequaled in providing a shadow-free light, which can be controlled with *egg-crate*-like attachments and they are available in a variety of sizes. Typical units in the 1,000 to 2,000 watt range are popular as fill lights because they are easy on the talent's eyes. With lower light level requirements, softlights have become the preferred broad baselight source. Color corrected flourescent fixtures are often used in similar applications and offer large power and heat savings.

Area Lighting

Scoop floodlights were once very commonly used in studios to build up high levels of base light that is no longer required with more sensitive cameras. Scoops are generally used with diffusion media, so it is important that frames be incorporated into the lighting fixture. While the majority of lighting can be done with Fresnels and some type of diffuse source, various fixtures produce a different quality of light for special applications. For a harder, rectangular field of fill light, a "broad" may be used; this is similar to a scoop in concept, except its housing/reflector creates a somewhat less diffuse light. It too is always used with diffusion media. These fixtures are now most commonly limited to lighting backgrounds.

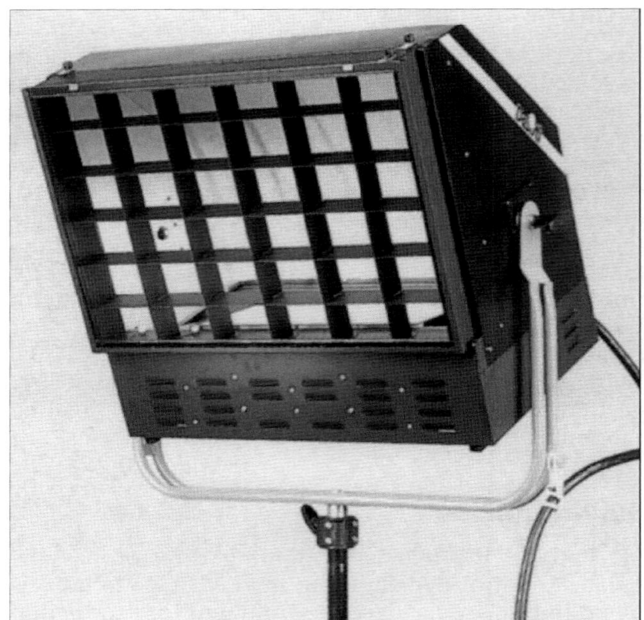

FIGURE 5.20-5 Spotlights fitted with egg crate louvers offer controlled, smooth, and even lighting.

Other Fixtures

A popular and efficient incandescent fixture used in television studio lighting is the *PAR light* (parabolic aluminized reflector), as shown in Figure 5.20-7. The lamp itself contains both the reflector and the lens, which is available in five different beam spreads. The PAR is an inexpensive and powerful tool for the lighting director, and while it has little in the way of adjustments, the application of barndoors and holders for color filters makes this fixture popular for key and backlighting. A new variation of this project, called the ParNel, allows adjustment of the beam spread without changing lenses, acting much like a fresnel.

The *ellipsoidal reflector* spotlight shown in Figure 5.20-8 is a unique effect light in a studio. The optical system of this unit allows the beam of light to be hard or soft edged, and the projected pattern of the beam can be precisely shaped by *internal shutters.* It is most commonly used to project patterns on the cyclorama. The modern ellipsoidal spotlights with dichroic reflectors and sophisticated optics offer bright, clear projections of *patterns,* which can be made from photographic half-tone etched metal or lithographed glass slides. These sophisticated patterns are available in a variety of standard and custom designs. Ellipsoidal spots come in a variety of beam spreads, but the wide-angle short-throw sizes are most useful in small and medium studios.

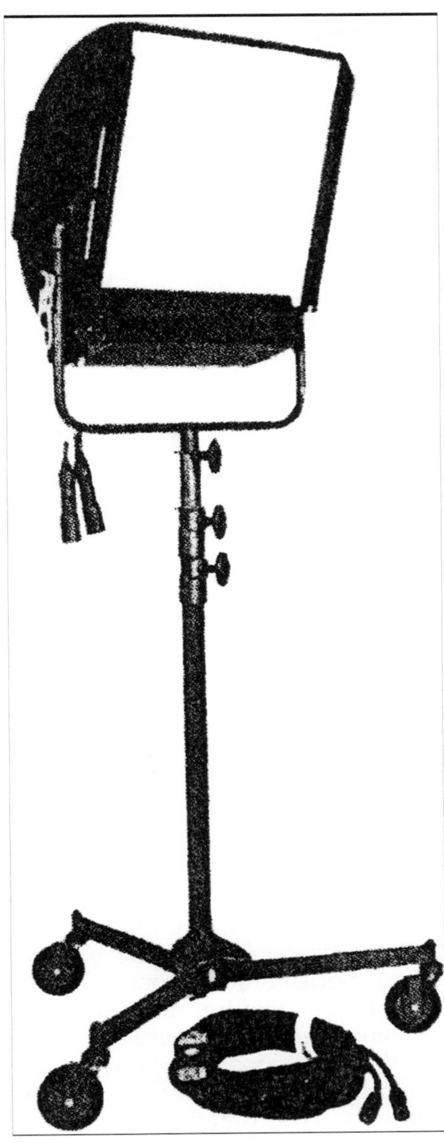

FIGURE 5.20-6 Soft lights produce a diffuse source and have switches to select as many lamps as necessary. (Courtesy of Mole-Richardson.)

Robotic Fixtures

Robotic fixtures have become commonplace in theatrical and television lighting. These lights fall into two categories. The first group can position its light by means of a motorized yoke, which therefore requires a sturdy mounting position that does not sway from the inertia of the moving unit. The second group utilizes a motorized mirror that reflects the output beam of light into the correct position. Within these two mechanical solutions, there are *wash* lights with a soft-edge beam of light and *hard* lights, which project a defined beam of light and can project many types of patterns. These lights are primarily used for concert lighting and theatrical productions where movement of the lights and

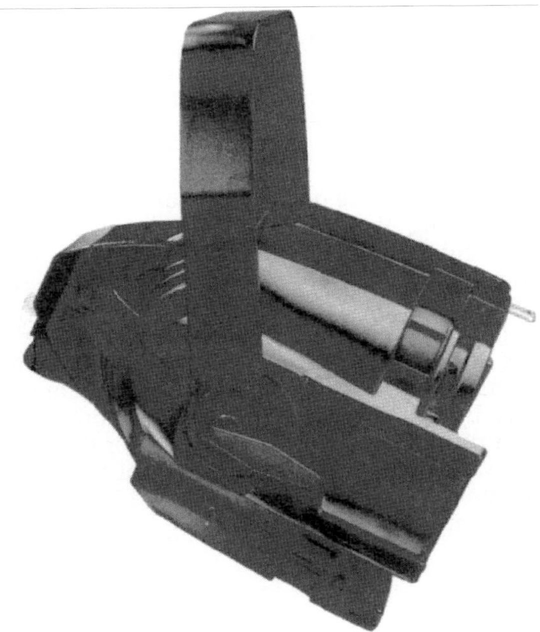

FIGURE 5.20-7 The Source 4 PAR provides similar output as a standard PAR 64 but at a lower wattage. It has interchangeable lenses for a variety of beam spreads.

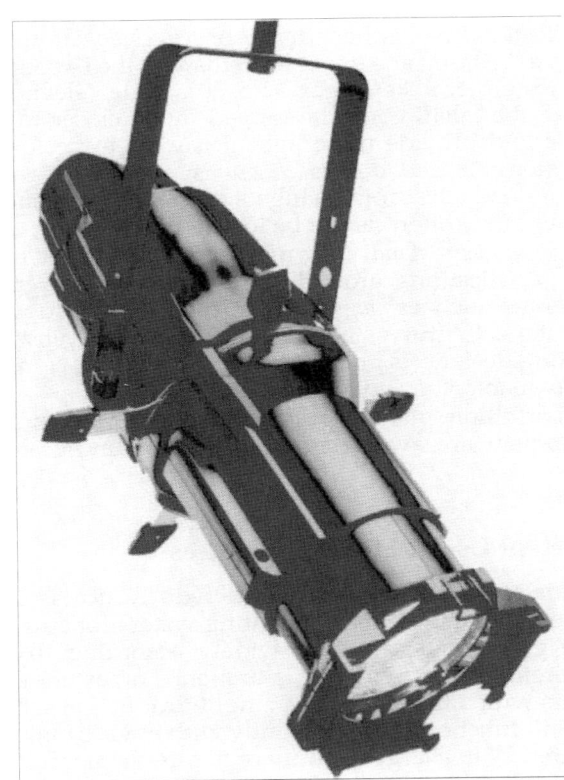

FIGURE 5.20-8 The Source 4 ellipsoidal lighting instrument can produce sharp-edged patterns for a variety of special effects.

changing colors provide dramatic effects under computer or manual control. These lights are used for face lighting only when they present the only practical alternative because of the useful remote focus capabilities.

Fluorescent Fixtures

Fluorescent fixtures are popular in many studio installations. Utilizing the improved color temperature of compact fluorescent lamps, several manufacturers have developed a variety of useful fixtures. These systems offer the compelling features of low energy consumption and minimal heat production. They can provide soft or fill light but are somewhat limited in throw distances. Fluorescents are most commonly employed in news and other continuously on-air forms of production to take advantage of their energy efficiency. Fluorescents are another tool for the lighting director and can be effectively used in conjunction with standard incandescent fixtures.

Accessories

There are numerous accessories that make positioning light fixtures easier and more accurate. Despite the expense and care expended on planning the lighting grid system, it cannot satisfy every fixture mounting requirement.

Rolling floor stands to position lighting fixtures in locations that are not easily reached from the lights on the grid are essential for every studio.

Extension rods, either *straight* or *telescopic,* allow fixtures to be hung at a lower level than on the fixed grid. For example, a key light at a 30° angle (elevation) above the talent could be moved much closer to the talent and provide more light if it were hung lower. Counter-balanced devices such as pantographs and spring-load telescoping hangers are not recommended because they often cannot be locked in place.

The *Century stand,* shown in Figure 5.20-9, has multiple applications around the studio. It can grasp materials such as *flags, cutters, cookies,* and support reflectors. Century stand use is limited only by the lighting designer's imagination. No matter how many are available, they will all be put to good use.

In addition, many types of *clamps* and general grip equipment are available for special mounting situations.

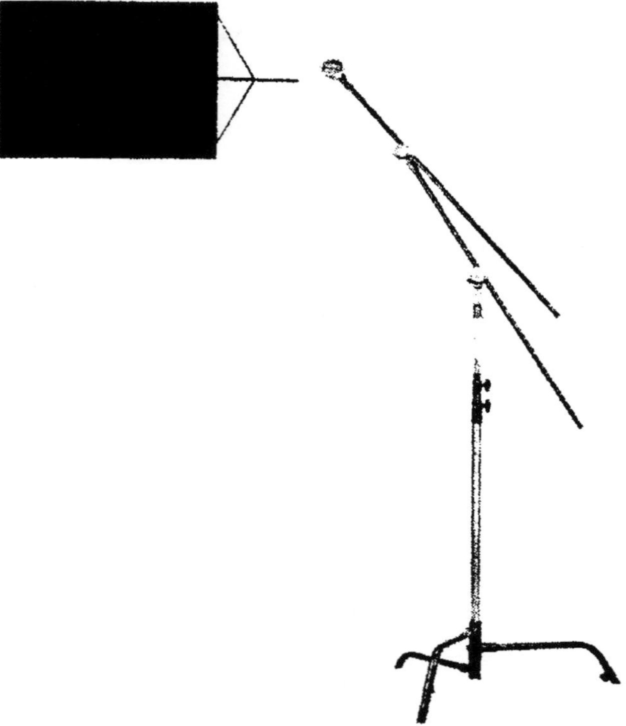

FIGURE 5.20-9 The three level legs allow the "C-stand" to fold flat for storage. The head can grasp flags, cutters, cookies, and other accessories.

Efficient Use of Lighting Fixtures

Ultimately, the proper quantity of fixtures, accessories, and other components of a lighting system for a studio of a given size can vary widely according to the requirements of a specific situation. For example, a studio with a fixed grid and a tight production schedule will function more efficiently with a wide range of fixtures. With a large quantity of fixtures, major relocation of the units is minimized. When sophisticated motorized grid systems are available, the ease of relocating fixtures will reduce the total quantity of fixtures for the same size studio. The choice of either of these two approaches is also dramatically affected by labor costs within a particular facility. The expense of a few extra fixtures is not significant compared to the additional time and labor required to fully utilize a minimal complement of equipment or the cost of elaborate rigging. Final discussions on equipment purchase require a careful analysis of a station's specific production requirements in coordination with all the financial ramifications of various approaches.

It should be evident from the preceding general discussion of lighting problems that for any production, close cooperation between the lighting designer for the show and the video shader in charge of the cameras is important. The video shader controls a versatile visual system whose limits are stretchable in the name of art. Great television pictures are the result of careful teamwork by the lighting designer and the video shader (see Chapter 5.3 on television cameras).

STUDIO DESIGN

In the design of a television studio, the lighting system is a major consideration because it is so closely interrelated to the physical size and shape of the room. While the television industry has seen major technological improvements throughout its evolution, the basic physics of light and its ability to describe three-dimensional forms cannot change. Consequently, while

many older studios have upgraded virtually every piece of electronic equipment, the basic components of an older, well-designed lighting system may still function extremely well.

Studio Size

Ideally, a studio provides the optimum environment for any type of production. In practice, however, studios of various sizes tend to function best for particular types of production. The typical broadcast plant requires several sizes of studios to most effectively service its programming schedule. Typically, a small studio (1200 sq ft) is dedicated to news, interviews, and public affairs, whereas a larger (2400 sq ft) studio is the most common size of small program production facility. For general production, a studio of 5000 sq ft or more will offer fewer limitations. The size of the studio must be carefully determined by existing and future programming requirements, keeping in mind that the larger the studio, the fewer the limitations (see Figure 5.20-10). These are important decisions, since the lighting system must be planned relative to the size and specific requirements of the studio.

Studio Height

When determining the height of a new studio, the lighting suspension system must be considered. The grid height for small- and medium-size studios is a function of the TV aspect ratio and the normal wide-angle zoom lens. Most zoom lenses can cover approximately a 45° field. When the actual width of this maximum horizontal dimension is calculated, the height of the suspended lights can be determined by applying the aspect ratio, as illustrated in Figure 5.20-11, which shows the calculated heights using a 4:3 aspect ratio camera. In theory, the use of 16:9 cameras could reduce the studio height requirement, but, despite widespread acceptance of 16:9, it will still be some time before it becomes the universal aspect ratio. Note that the studio lighting fixtures will hang at least two feet below the grid, possibly more, depending on the suspension system. In addition, a normal studio pedestal and camera operator will prevent the lens from getting closer than approximately five feet in from a wall or any other obstruction, further reducing the maximum coverage of the lens. This method describes the theoretical maximum picture possible and the height the fixtures need to be mounted to stay out of the picture.

However, television in the studio is primarily a close-up medium, and limitations in grid height can often be overcome by various camera angles and special fixture-mounting systems.

The suspension system for the studio lighting fixtures is an important factor in determining studio height. The system must allow lighting fixtures to be

FIGURE 5.20-10 This large multipurpose studio has a flexible lighting system equipped with motorized battens and hoists that allow fixtures to be easily adjusted and located for different productions. (Courtesy of Bavaria Studios/De Sisti Lighting.)

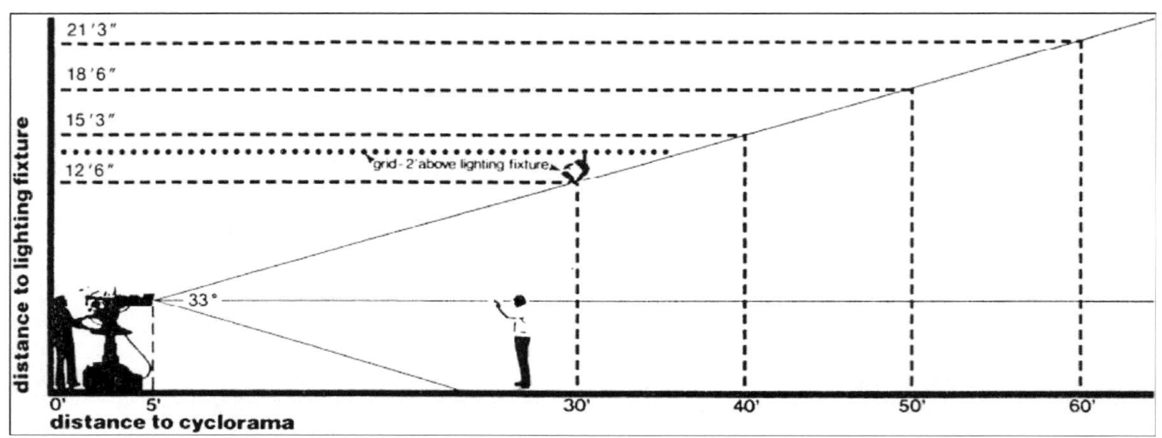

FIGURE 5.20-11 This drawing illustrates how the vertical aspect of a camera lens at eye level determines the height of a studio grid.

hung anywhere over the entire studio, and several arrangements are possible, ranging from simple fixed pipe grids to systems with adjustable hangers and manual and motorized hoist systems, sometimes with several different types or combinations.

Fixed Grid

A *fixed grid* as shown in Figures 5.20-12 and 5.20-13(a) is the most common and least expensive system for mounting lighting fixtures. The pipes are generally

FIGURE 5.20-12 This fixed grid studio employs two levels to create a higher apparent background. Multiple pigtail outlets are positioned for cyc lighting units. (Courtesy of WFAA-TV.)

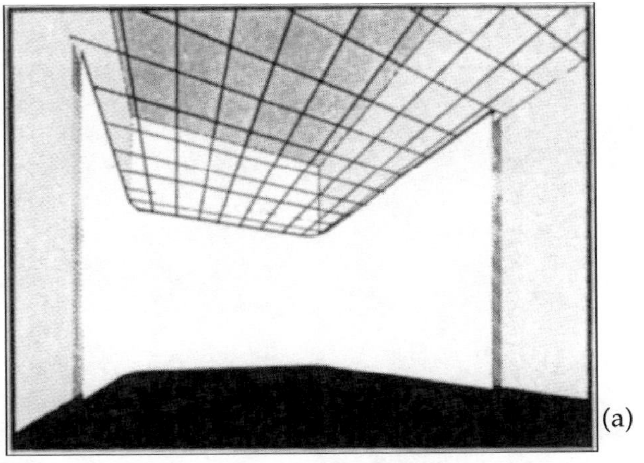
(a)

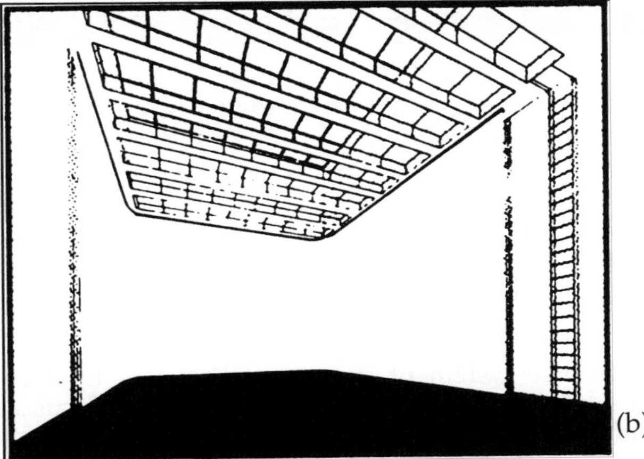
(b)

(c)

FIGURE 5.20-13 (a) Fixed grid plan. (b) Catwalk grid plan. (c) Batten plan.

laid out with a 4 ft × 4 ft grid spacing that provides adequate flexibility in hanging positions. The most common fixed grid height is 14 ft because it offers a good compromise between easy ladder reach and adequate clearance for wide shots. A fixed grid that is much higher than 14 ft increases the amount of labor needed to install and adjust fixtures. Bi-level fixed grids have been utilized to create a higher apparent background without additional cost and to keep the major portion of the grid at a reasonable height.

Fixtures may be attached directly to the pipes of a fixed grid, but when lights are needed at a lower level for particular effects, then various fixed- or adjustable-length suspension rods or poles and sometimes spring-loaded adjustable *pantographs* are used.

Self-powered manlifts can be used to provide reasonable access to higher fixed grids, but in a large multipurpose studio, a fixed grid can be a limitation. Several other systems have been developed that are more flexible and efficient to use, albeit at a higher initial cost.

Catwalks

Catwalks, as shown in Figure 5.20-13(b), are generally also built at a fixed level above the studio floor and need sufficient height above them to allow personnel access. Catwalks provide a measure of increased efficiency over a fixed grid, but they are considerably more expensive. They allow studio electricians to work on lighting at the same time carpenters are handling the scenery. This can be an important time savings on a tight production schedule. Catwalks are usually arranged to create a fixed grid utilizing handrails in conjunction with an extension rod for suspending each lighting fixture. When hung too low, catwalks create the same height limitation as a fixed grid, but generally they are employed at heights greater than would be reasonable for a fixed grid. This arrangement primarily saves time in setup and strike and offers the lighting crew a safe work platform above the studio. In recent practice their high cost has prevented many installations. Most shows set up and stay in a studio for their entire run. With only one major set up per production, the cost of catwalks is hard to justify.

Adjustable Hoist Systems

The larger and higher the studios become, the more complex the lighting and its suspension system become, and various alternatives to fixed lighting grids may be used. Figure 5.20-13(c) illustrates a plan for suspended battens.

Battens

For very large studios, in addition to the TV aspect ratio, a basic theatrical staging factor—"flyout clearance"—determines the grid height. Just as in a theater stagehouse, flying scenery is a common requirement in larger-scale TV production. Flown scenery commonly requires a minimum grid height of 40 ft. To raise and lower the scenic units and the lighting fixtures, a regular pattern of long battens or pipes is suspended across the width of the studio. These battens move up and down on steel cables and may be manually balanced by cast iron weights. While counterweighted battens solve the height problems, there is additional work in rebalancing the counterweight arbors when fixtures are added or removed from the batten. Counterweight rigging mounted along the studio wall encroaches on valuable floor space.

Winches

To solve the problems of counterweight systems, many installations use electric winches to raise and lower the battens. These winches can be operated by sophisticated control systems that will allow the battens to be lowered for service or adjustment and then returned to an exact preset trim or level. The motors will easily lift any variable load within their designed capacity.

Many variations on the motorized batten system have been developed. In order to increase flexibility in the height of particular fixtures, the battens have become shorter with fewer fixtures on each one. This increases the number of battens and the number of motors required and hence the cost.

In studios that have either counterweighted or motorized battens, a full walk-over grid should cover the entire studio. This grid-iron is usually made of steel grating or channels, as in a theater stagehouse, and provides the support for the adjustable rigging. This walk-over grid provides easy access for most overhead suspension tasks a production may require.

Slotted Grid with Monopoles

A further variant of a walk-over grid is for the grid to have a series of slots on a regular spacing through which adjustable-length monopoles are suspended, with one lighting fixture attached to each pole. The pole can be moved by personnel working on the grid to any position along the length of the slot. This system is popular in Europe but has had almost no acceptance within the U.S.

Self-Hoisting Battens and Modular Grids

To eliminate the complexity of installing separate winches for each batten, self-hoisting short battens (also known as "self-climbing hoists") are available that have their own built-in electric motor, as shown in Figure 5.20-14. Alternatively, to take advantage of the convenience of electric winches without excessive cost, several modular grid systems have been developed. These modules are essentially small adjustable pipe grid sections, also with self-hoisting motors. They offer much of the flexibility of battens, but since they cover a larger area, fewer motors are required.

Self-hoisting battens and modular grid systems typically simplify the overhead steelwork and structural requirements of the studio (see Figure 5.20-15) and reduce the cost of installation.

In the development of a fully integrated lighting system, it is essential to coordinate all the building's structural, electrical, and mechanical systems in relationship to each other and the grid. Whereas in normal construction many of the mechanical elements in a ceiling are placed where convenient, it cannot be overemphasized that, in a TV studio, improperly planned

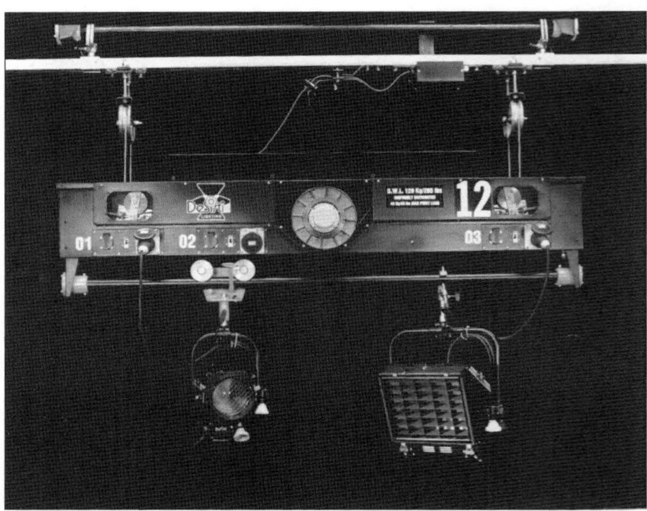

FIGURE 5.20-14 Manufactured self-hoisting battens are easy to install and efficient to use. (Courtesy of De Sisti Lighting.)

or installed air conditioning ductwork and electrical conduit runs can be a hindrance to production.

Pole-Operated Fixtures

Whichever system of suspension is used, light fixtures have to be adjusted for direction and focus. In studios with fixed sets such as news, adjustments may be rare, but for some general-purpose production studios, frequent adjustments may be needed. To avoid having to use a ladder or man lift to reach a fixture to adjust it, most studio fixture manufacturers offer pole-operated yokes for their instruments. Such fixtures have sockets to which a pole crank can be attached, with adjustments for each of the basic functions of pan, tilt, and spot-flood (for adjustable spot lights), as shown in Figure 5.20-16. There is a reasonable limit to how long a pole can be easily manipulated, but this feature considerably enhances most types of suspended light fixtures, whether attached direct to a fixed grid or on a hanger. This feature is especially useful for adjusting otherwise unreachable fixtures blocked by scenery or other obstructions and, although more expensive, is seriously worth considering when purchasing light fixtures.

CYCLORAMA

Designing the cyclorama is integral in planning the studio lighting and grid systems. The grid or suspension system must provide a mounting position for the cyc lights located at the proper relationship to the cyc. Also, the total area of cyclorama will affect the calculation of the studio power service because of the number of lights needed to illuminate it evenly.

FIGURE 5.20-15 Self-hoisting battens suspended from a grid structure, equipped with a variety of spot and soft lights. (Courtesy of NFL Films/De Sisti Lighting.)

FIGURE 5.20-16 When a long pole-mounted crank is inserted into the adjustment cups, this fixture can be panned, tilted, or spot/flood focused from the floor. (Courtesy of Mole-Richardson.)

Hard Cyc

Small studios often incorporate *hard* cycloramas that are smooth, plastered surfaces that actually blend flush into the floor. They provide the ideal infinity effect that draperies cannot equally simulate. Hard cycloramas can be easily painted any color as needed and are especially effective for certain chroma-key techniques that can be spoiled by even the most invisible cloth seams. Hard cycloramas are generally limited to small studios since their hard surface area invariably creates acoustical problems.

Cyc Pit

In some large studios a *cyc pit*, illustrated in Figure 5.20-17, is a compromise solution for creating the infinity effect. The pit contains and conceals the bottom cyclorama lighting, which is essential for a tall cyc and the bottom of the drapery. When it is shot from the proper angle, the pit will help to simulate a background without a horizon. Cyc pits are more often provided in studios primarily dedicated to film production and are rare in television studios.

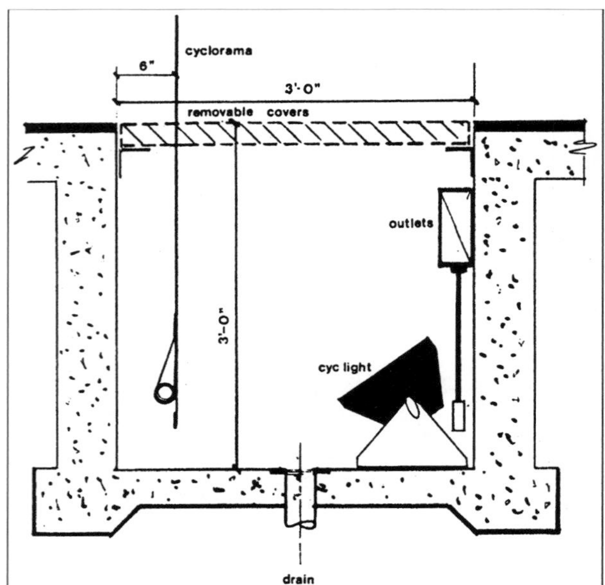

FIGURE 5.20-17 Cross-section view of a typical cyclorama lighting pit.

FIGURE 5.20-18 Cycloramas and draperies can be shifted to various track configurations by utilizing transfer switches (CBS-NY). (Courtesy Peter Albrecht Corporation.)

Drapery Cyc

For most studios, a *drapery cyclorama* is the most convenient solution. This seamless drape is hung on carriers which roll along a track and allow the cyc to be positioned anywhere around the perimeter of the studio. Two parallel tracks permit another type of background to be pulled in front of the stretched cyc, as shown in Figure 5.20-18. Switches on the track system allow the draperies to be easily transferred to the front or rear tracks. *Leno* or *filled scrim* is the most common drapery material, although seamless muslin is an inexpensive alternative material. Translucent plastic rear-projection screen–type material is also a popular material because it can be lit from either front or rear.

In large studios where experienced lighting directors are available, true white cycloramas are used to achieve greater color intensity on the cyc. However, in smaller studios, where the talent occasionally must work very close to the background or where limited control equipment is available, a 60% TV white cyc should be used for better control of contrast.

Drapery cycloramas are generally furnished with jack chain weights. Removable pipe weights bent to match the shape of the track should also be provided to create a wrinkle-free background. One of the most common errors in cyclorama design is an insufficiently large radius at the corners of the cyc. No matter what material the cyc is made of, the larger the radius, the easier it is to light evenly and accomplish the desired effect.

Cyclorama Lighting

Generally, the arrangement of doors into the studio will define the most functional area for the cyc to be positioned. In the basic design of the grid or suspension system, the type of cyc lighting system should be predetermined and the proper hanging system provided.

There are several types of cyclorama lighting fixtures, including striplights and a fixture commonly known as a *Far Cyc,* shown in Figure 5.20-19. Strip-

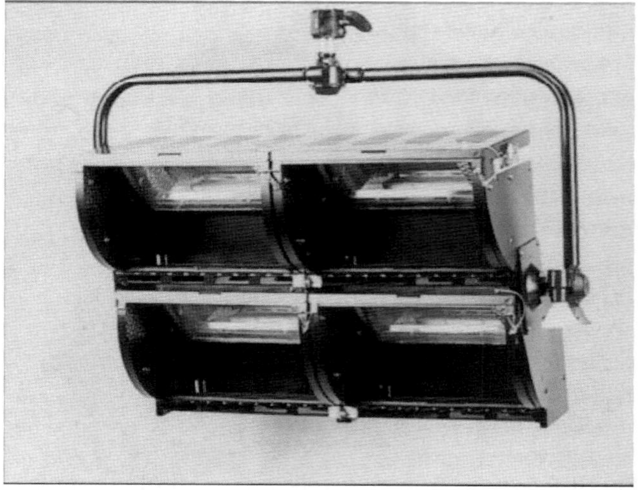

FIGURE 5.20-19 Four-unit "Far Cyc" fixtures are very effective in lighting tall cycloramas. (Courtesy of Colortran.)

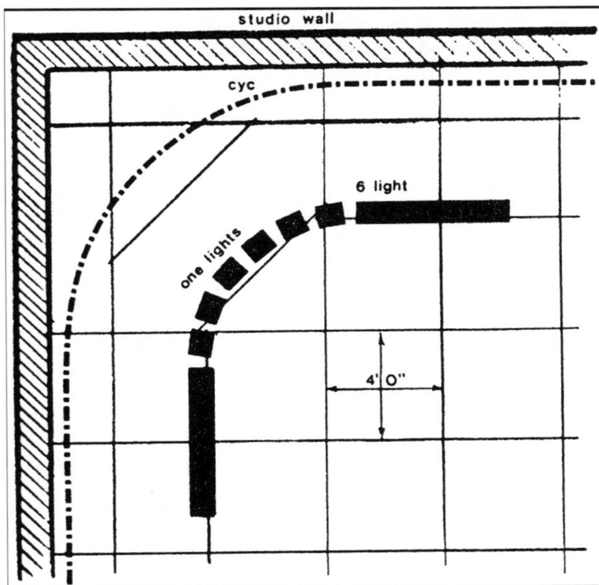

FIGURE 5.20-20 Cyc light layout: typical arrangement of strip cyc lights relative to the cyc curve.

lights are continuous rows of quartz halogen, MR-16, or PAR lamps, which, for an 18 ft high cyclorama, should be mounted 5 to 6 ft from the cyc, as illustrated in Figure 5.20-20, whereas the Far Cyc units should be mounted 7 to 8 ft from the cyc. Another type of cyc fixture is a version of a striplight made up of individual cells that can be placed to match any curve or straight surface. The entire suspension system should be designed around these dimensions. Far Cycs generally will light the cyc as evenly as striplights but with less wattage. Because Cycs are mounted a greater distance from the cyc, they force the talent farther from the background. Generally, striplights should be limited to three colors; otherwise, the separation between alternately colored lamps is too great to provide even coverage.

ARCHITECTURAL CRITERIA

Electrical Power Requirements

With the basic studio size determined and the net production area (NPA) defined by the cyclorama, it is possible to determine the power requirements for any given studio. The power requirements for studio lighting are a direct function of area and the required level of illumination. This power requirement remains consistent regardless of the grid height. For a lower grid, a greater quantity of smaller wattage fixtures is used, whereas a higher grid will require fewer fixtures of increased wattage. In either case, the total watts per square foot will remain roughly the same.

An average of 55 watts per square foot of NPA has been proven in production to provide sufficient power for any normal television lighting requirement. This method of calculating the studio load will provide sufficient power for virtually any situation. This is generally more power than is actually required for the average studio with typical television camera equipment, but it allows for over-lighting by novices and higher levels necessary for special situations. It also provides sufficient power for an average cyclorama as well as lighting the entire studio as required for audience participation. Very tall cycloramas, which may require double rows of cyc lights at the top, bottom, or both, should be calculated as an additional power requirement based on the wattage per lineal foot of cyclorama according to the lighting system chosen.

This calculated power service describes a real maximum probable load, and the feeders must be able to supply this full amount of power. Only certain limited productions will ever require this full amount. Also, note that for dimmer-per-circuit systems, the dimming capacity will be far greater than this calculated power service. The full dimming capacity need not be fully serviced, since the larger number of dimmers is a matter of convenience and will never be fully loaded beyond the maximum probable load. Large production studios will require a variety of additional power services in various voltages for special effects, rigging, and motorized lights.

Calculating Air Conditioning and Electrical Power

Because the maximum lighting load seldom occurs, it is unnecessary to use that maximum load as the basis for the air conditioning capacity. Production practice has shown that a diversity of 60% can be applied to the maximum load and still provide sufficient capacity for full period shooting. Of course, any other heat-generating devices and the population of the studio should be included in the air conditioning calculations. A properly-designed air conditioning system will require very large ducts to meet stringent acoustical requirements. These large ducts can often interfere with a grid system and should be closely coordinated.

Determining studio power and air conditioning requirements according to the factors discussed earlier:

- Start with the net production area (NPA), which is the usable studio area in sq ft minus the area behind the cyc and other areas that are unusable for production.
- To determine the number of outlets required at the patch panel, divide NPA by 18 (one outlet for each 18 sq ft).
- Then divide the outlets into 20 amp and 50 amp circuits to determine how many dimmers will be needed of each variety.
- To determine the Studio Lighting Load, multiply the NPA by 55 watts to find the maximum possible load in watts on the air conditioning system.
- To determine the Studio Lighting Power Service required, divide the maximum load power in watts by 120 volts to obtain the maximum current in amps.

- To determine the Studio Lighting Power Service, divide the total amps by 3 (for three-phase, 4-wire 120/208 volts) to find the maximum load in amps per phase leg and then round off to the next larger standard panel size.
- To determine the Studio Lighting Heat Load for Air Conditioning, multiply the Studio Lighting Load in kW by 60% (diversity).
- To determine the Dimmer Room Heat Load for Air Conditioning, multiply the Studio Lighting Load by 5%.

Electrical Distribution

For a studio to be flexible, lighting equipment power must be distributed uniformly throughout the studio. At the grid level, power is commonly distributed through prewired plugging strips. These strips are mounted directly to the grid, catwalks, or on fly-in-and-out battens or movable grid sections.

Each circuit terminates in the studio in a pigtail/outlet. There are two types of connectors in common use: *stage pin connectors* and *twist lock*. Stage pin connectors are less expensive and more common in rental equipment. If additional fixtures are rented on occasion, this may be an important consideration. In addition, the cost savings of stage pin connectors recur with each fixture and cable purchase. Twist lock connectors have a positive locking feature. The final choice should be based on the studio's specific requirements.

Wall-mounted outlet boxes should be provided at 30 inches above floor level, around the perimeter of the studio, as shown in Figure 5.20-21. Generally, the governing factor for their placement is relative to a layout of floor-mounted cyc strips. Otherwise, these outlets are used for miscellaneous lights on floor stands and practical lighting fixtures on the set.

The number of circuits and their capacities are also related to studio size. The actual number of circuits is based initially on the square footage (approximately one outlet every 18 sq ft of the NPA) and then altered as necessary to conform to the particular grid system and the cyclorama layout, as illustrated in Figure 5.20-22. Dedicated circuits for the cyclorama are often overlooked. The cyc lights require a large number of circuits, and once they are hung in place, they will seldom, if ever, be moved. Because of the rather wide spacing of the cyc units, it is sometimes more efficient to feed these lights from individual grid-mounted junction boxes rather than a plugging strip. Also, when Far Cyc lights are used, it is often convenient to double up the outlets to take full advantage of the 20 amp capacity of the dimmers.

Additional circuits should be located around the perimeter of the studio at grid level. This is a natural backlight position for a set facing away from the wall. Properly located circuits save considerable time in running jumper cables.

For most studios the majority of circuits will be rated 20 amp. However, the larger the studio, the greater the quantity of 50 amp circuits. In a medium-size studio (3500–5000 sq ft), 50 amp circuits are generally located in a regular pattern throughout the center area of the grid and slightly more frequently around the perimeter backlight position. In larger and higher studios the density of 50 amp circuits must be increased, although the overall outlet density should not be significantly decreased. On adjustable height grids, a full complement of both 20 amp and 50 amp outlets should be provided for use of

FIGURE 5.20-21 Wall-mounted outlet boxes are generally located around the perimeter of the studio. (Courtesy of Electronic Theater Controls.)

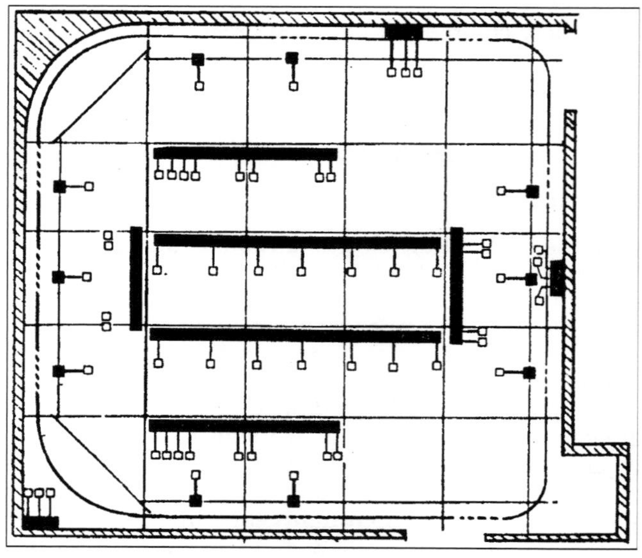

FIGURE 5.20-22 Typical studio lighting outlet distribution plan for a small industrial studio.

appropriate fixtures at varying heights. Large studios (over 10,000 sq ft) with high grids must be furnished with 100 amp circuits.

Lighting Control

The dimming and control system is an important part of the lighting system. Dimmers allow easy control of numerous fixtures, balancing and recording of levels, and the blending of colors. A dimmer system frees the lighting director of the unnecessary burden of calculating and controlling the loads through more labor-intensive mechanical methods. Electronic dimming and control allow the execution of complex lighting cues, which are a very effective production element.

Most studios are outfitted on a basis of one dimmer per circuit in which every circuit terminates in its own dimmer, with the integral circuit breaker protecting both the dimmer and the circuit. The dimmer-per-circuit system gives the lighting director individual control of each individual fixture, or group of fixtures, plugged into that circuit. Normally, the studio is outfitted with 20 amp (2.4 kW) and 50 amp (6 kW) dimmers. Figure 5.20-23 illustrates a typical studio lighting control dimmer system arrangement.

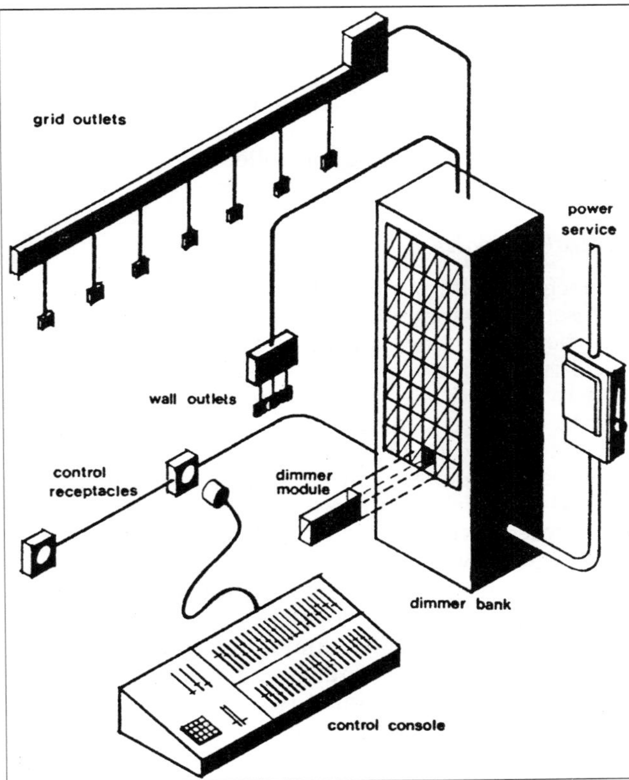

FIGURE 5.20-23 Diagram showing the basic components of a studio lighting control and dimmer system.

Dimmer Bank

The individual dimmer units, as shown in Figure 5.20-24, generally plug into electronic equipment racks to form the dimmer bank. This modular system also allows quick plug-in substitution of faulty dimmer modules. Depending on the manufacturer, up to 96 individual 2.4 kW dimmer modules can fit into a single rack. From the 1970s until recently, most studio-quality electronic dimmers have utilized silicon controlled rectifiers (SCRs), although a few have used MOSFET. These units are reliable and are universally available in 2.4 kW, 6.0 kW, and 12.0 kW (20 amp, 50 amp, and 100 amp, respectively, at 120 V) ratings. Only dimmers that have sufficient filtering to prevent unwanted RF interference and excessive filament vibration should be considered for use in television studios. Toroidal chokes are utilized to control the rise time of the dimmer and the associated electrical noise.

More recently, sine wave dimming has become commercially available. These dimmers, described in the next section, are more expensive but have several advantages compared to normal SCR dimmers.

The location of the dimmer bank is an essential part of the initial studio space planning. The dimmer room should be centrally located to minimize the length of all the wiring to avoid voltage drop and excessive installation cost. This room should be sized to allow sufficient space for required conduit radii, access to the feeder lugs, and adequate front clearance as specified by the local code.

Most SCR dimmers are 95 to 98% efficient. Therefore, they could create heat perhaps up to 5% of the energized lighting load. Since the maximum lighting load is an infrequent occurrence, the dimmer room cooling should more reasonably be based on 5% of the diversified load on which the studio air conditioning system is based.

Centralized and Distributed Dimming

SCR dimmers tend to be both acoustically and electrically noisy, which is one of the reasons why in the past they have always been located in a separate dimmer room—an arrangement known as *centralized dimming*. New-technology dimmers such as the *sine wave dimmers* described below, are acoustically silent and do not create electrical interference, enabling an alterna-

FIGURE 5.20-24 Tray-mounted dual dimmer modules allow high density and quick replacement. (Courtesy of Electronic Theater Controls.)

tive *distributed dimming* arrangement to be used for all or some of the studio lighting channels. Because these new-technology dimmers can be located in sensitive areas, they can be located closer to the fixtures and loads, on the lighting grid or elsewhere in the studio at a convenient location. This capability provides increased flexibility in design and can simplify installation, which may be useful, particularly when upgrading or expanding an existing installation.

Sine Wave Dimmers[2]

Although SCR dimmer designs have improved over the years, they have several flaws due to the basic characteristics of their design; they tend to generate acoustical and electrical noise, and may also induce noise in the lights that they control.

SCR dimmers work by varying the switch-on point of the lamp current each half cycle. These dimmers slice the waveform in half and, using chokes, increase the rise time and use the inertia of the filament to smooth out the switching change (see Figure 5.2-25). This causes the filaments in the lamp to buzz (creating noise and shortening lamp life). In addition, the electronic spikes that occur in these dimmers create radio frequency interference. The high frequencies created by the fact SCR switching waveforms may also be radiated or carried through electrical power circuits. The unfortunate results of this cross-interference may be picked up by other electronics (such as amplifiers and wireless devices). Although lighting equipment manufacturers have improved these negative artifacts with advanced technology and high quality chokes, it is impossible to completely silence these dimmers.

In noise-sensitive environments such as television studios, modern SCR dimmers may just be too noisy. If complete silence from the lighting system is desired, sine wave dimming may be the solution. Unlike SCRs, sine wave dimmers produce a pure sine wave output with variable amplitude to control lighting levels (see Figure 5.2-26). Through the use of transistors to slice the mains into pulses, these dimmers vary the current using pulse-width modulation, average the result, and produce a continuous, variable-amplitude smooth sine wave. This, in turn, eliminates the noise to the filament. In addition, this technology lowers the operating cost as it uses less reactive power (produced by harmonics in SCR dimmers), lowers the maintenance cost as lamps last longer, and eliminates radio cross-interference.

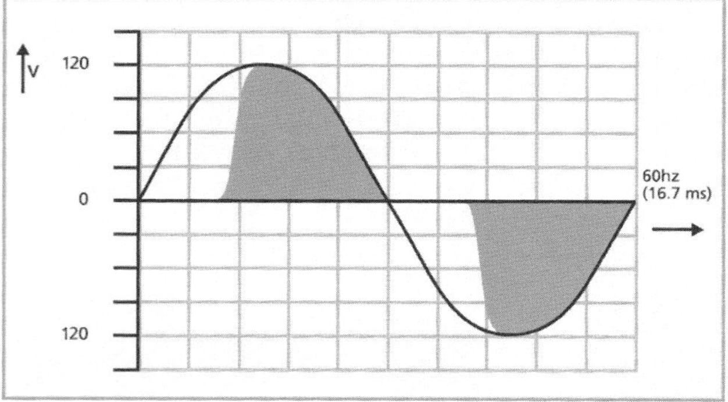

FIGURE 5.20-25 The voltage waveform of an SCR-dimmed circuit showing the sliced waveform characteristics. (Courtesy of Electronic Theater Controls.)

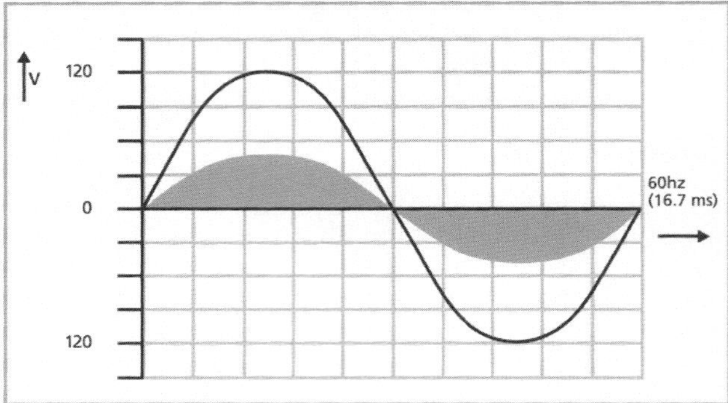

FIGURE 5.20-26 The voltage waveform of sine wave dimming technology showing how the waveform does not change regardless of the output voltage. (Courtesy of Electronic Theater Controls.)

Sine wave dimming also has the advantage of allowing control of almost any kind of load. This includes neon, HMI, and LED, lights, and even some motors, which cannot be controlled with conventional SCR dimmers

Control Consoles

In electronic dimming systems, each dimmer circuit represents a channel that can be programmed to retain a particular dimmer level setting. A group of preset levels is called a *scene*. While manual control panels with two scenes are sufficient for some productions, even a relatively simple production could require many more. While the physical size of manual multi-scene control panels can be cumbersome, numerous methods of storing and recalling the level settings for each preset have been developed.

Even the most modest control consoles offer enough control channels for individual settings of every dimmer, although it is possible to electronically patch dimmer circuits to a smaller number of control channels. Although the simplest multiscene systems physically resemble a standard manual two-scene preset system, they have virtually all the capabilities of the largest systems. Each slider can represent any channel or group of channels, or act as a submaster rather than a hardwired individual dimmer. The specific number of control channels is a matter of the physical electronic design of the console. Obviously, more channels afford greater control within a single scene.

[2]This section is adapted from *Modern Dimming: Choosing the Right Technology* by David Martin Jacques, published in Church Production Magazine, November/December 2006. Used with permission.

In a dimmer-per-circuit system, it is not unusual for a medium-size studio to have in excess of 200 dimmers and large studios may have a thousand or more. To control this large quantity of channels, it is possible to patch the dimmers to control channels. The patching function allows any dimmer to be controlled by any channel. For example, the blue cyc lights load may require 12 separate dimmers. If they are patched into the same channel, they will operate together in perfect unison. All this patching occurs within the console and does not require any cords, plugs, or diode pins. For small to medium-size studios, consoles with over 100 miniature potentiometers in two or three presets are available to provide the operational convenience and simplicity of manual systems.

After the dimmers are patched to channels and the desired levels are set, the entire preset can be stored in a memory by assigning a memory number. This preset can then be recalled by keying the appropriate number or operating the slider to which the preset has been assigned. This type of system is very economical and is suitable for most small- and medium-size studio situations.

For studios that encounter more complex production requirements, there are a number of systems that resemble a personal computer with specialized keypad and controls with much greater capabilities. One type of system that is common to several manufacturers utilizes a video display to show the various functions. In operation, the screen displays the channel numbers, and below each channel is a two-digit number for the intensity. At the bottom of this field of numbers is the *cue sheet*, which displays various operational functions such as cue numbers and fade times. The most complex systems employ dual displays to actively display more information. These systems offer many features and provide an indication if the lamp is burned out. Examples of control consoles are shown in Figures 5.20-27 and 5.20-28.

Computerized lighting systems store their memory on one of several types of removable storage media for reuse. This permits a complete copy of all the settings in memory. Most larger systems are available with varying degrees of internal backup memory systems and can be networked to interconnect various types of remote controls and peripherals. The operational software is permanently stored in ROM within the machine.

While many computer lighting control systems were designed for theatrical shows, where the easy daily repetition of very complex multipart cues is the primary goal, some have modified their programming to be more sympathetic to television's somewhat unpredictable demands. In selecting a system with this caliber of sophistication, the lighting director must carefully determine which features are really necessary.

Computerized lighting systems have greatly simplified the installation of the control wiring. While analog systems required at least one wire for every dimmer, these computerized systems use standard computer network wiring. An advantage to this simplified cabling is that the console can be easily relocated to any of the plug-in stations. In addition, when the main computer is located outside the studio, a small remote control the size of a handheld calculator can be used in the studio to activate fixtures as necessary for focusing or other simple operations. The technology of computer networking is now used on lighting control systems to add greater flexibility and multiple control locations. This allows the console to be located in the most convenient position for a particular phase of the production. It also allows several studios to share a more sophisticated system as required, since the largest systems can plug into the same control wiring as the smallest.

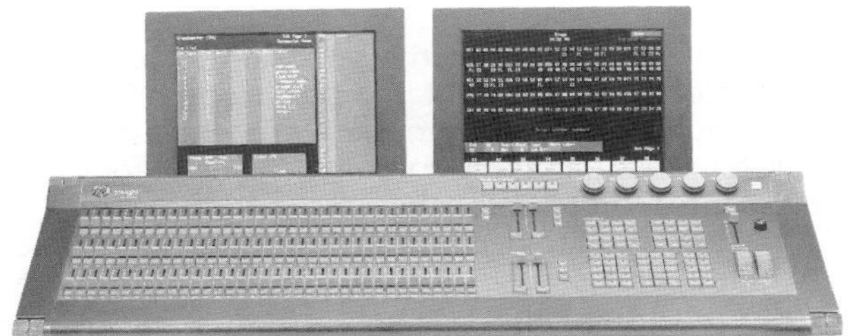

FIGURE 5.20-27 For medium-size studios, lighting consoles offer both manual and memory control. (Courtesy of Electronic Theater Controls.)

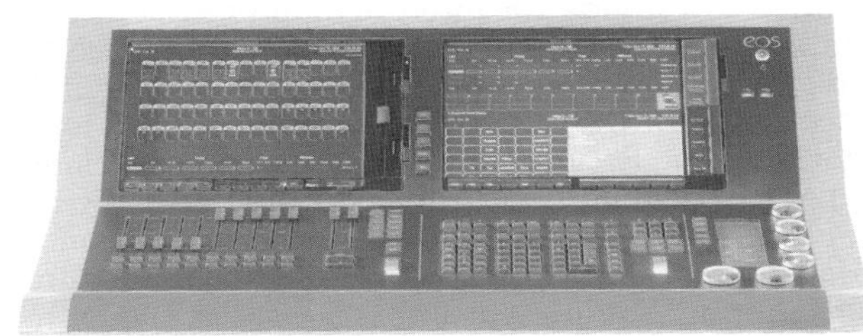

FIGURE 5.20-28 The most sophisticated consoles can control up to thousands of channels, with advanced sequencing and memory capabilities for all types of robotic fixtures, using both tactile and touch screen control user interfaces. (Courtesy of Electronic Theater Controls.)

Computerized control has made sophisticated remotely controlled (robotic) fixtures possible. Most consoles now have the capability to control moving lights as well as the dimmer system. A single console can be used to program and record the intensity, position, speed of movement, iris, internal pattern selection, color mixing, pattern rotation, as well as soft or hard beam patterns. The console records and replays all these functions as desired by each cue programmed.

Control Protocols and Standards

Early lighting control systems used analog control signals, with multicore cables and one wire-per-channel to connect control consoles with lighting dimmers. As systems became more sophisticated, most manufacturers developed their own dimmer control protocols with coded signals to eliminate the bulky multicore cables, but protocols were proprietary and incompatible with other manufacturers' equipment. As the industry grew and systems became more complex, the need for cross–manufacturer compatibility became critical.

In response to the need for standardization, in 1986 the United States Institute for Theater Technology (USITT) published the DMX512 Digital Data Transmission Standard for Dimmers and Controllers (usually known simply as DMX), later revised in 1990. This standard describes a method of digital data transmission between controllers and lighting equipment and accessories. DMX covers electrical characteristics, data format, data protocol, and connector type. It is intended to provide for interoperability at both communication and mechanical levels with controllers made by different manufacturers. In 1998, maintenance of the DMX512 standard was transferred to the Entertainment Services and Technology Association (ESTA). Subsequently, a further revision of DMX was adopted as ANSI standard E.1.11-2004, Asynchronous Serial Digital Data Transmission Standard for Controlling Lighting Equipment and Accessories. DMX has been widely adopted throughout the entertainment industry and the majority of current control systems and dimmers use this system.

ESTA coordinated development of further standards for control of integrated systems for the entertainment industry and a new standard ANSI E1.17-2006, Entertainment Technology – Architecture for Control Networks, better known as ACN, has recently been published. ACN will not replace DMX but is a suite of documents that specifies an architecture, including protocols and language, which may be configured and combined with other standard protocols to form flexible, networked audio, lighting, or other control systems. Wireless implementation of DMX (W-DMX) is possible and has been implemented by various companies; a standard has been proposed but is not finalized.

Sources of information on DMX, W-DMX, ACN, and other standards for studio and staging systems are listed in the Resources section.

SUMMARY

This chapter has introduced the subject of television lighting, describing the various parts of the system, with some guidelines on their use, including basic requirements, lighting fixtures, suspension systems, dimmers and control systems. New technologies and techniques will no doubt continue to be introduced and the reader is encouraged to refer to the resources provided here for more information.

Most television production professionals have the opportunity to be involved in planning a new studio only once or twice in their entire careers. Even years of experience in studio production are not necessarily the best preparation for coordinating studio requirements into a construction process. Often, a new studio provides an opportunity to acquire a complement of new equipment which is more sophisticated than the existing staff's level of experience. In this situation, an experienced lighting designer should be consulted to assist while evaluating the requirements and in the preparation of the equipment purchase orders. While manufacturers are sometimes helpful in this area, they are still primarily interested in selling their product, and no single manufacturer offers a full line of suitable equipment in every area. Working with a consultant with experience in the planning and design of the lighting system for a television studio will provide a better perspective and a more reasoned approach to designing the overall lighting system.

RESOURCES

Contact information is given below for a selection of resources related to studio lighting:

- United States Institute for Theater Technology: http://www.usitt.org/usittHome.html
- Information on DMX512 and other standards: http://www.usitt.org/standards/DMX512.html
- Summary of DMX512:
  http://www.ubasics.com/DMX-512
  http://www.artisticlicence.com/app notes/appnote004.pdf
- Entertainment Services and Technology Association (ETSA):
  http://www.esta.org
- The ACN standard is available from the ESTA Foundation website:
  http://www.estafoundation.org/pubs.htm
- A summary of ACN is also available at the ETSA website:
  http://www.esta.org/tsp/news/news-details.php?newsID=
- Information about wireless DMX solutions with numerous downloadable fact sheets, is available at: www.wirelessDMX.com
- Information on lighting fixtures, grid and suspension systems, control systems, and dimmers is available from various companies, including:

  Electronic Theater Controls (ETC):
  www.etcconnect.com

  Entertainment Technology:
  http://etdimming.com/

  Leviton Lighting Management Systems:
  http://lms.leviton.com

  Strand Lighting:
  http://www.strandlight.com

  Lowell-Light Manufacturing:
  http://www.lowel.com/

  Arriflex Corporation:
  http://www.arri.com/

  De Sisti Lighting:
  http://www.desisti.it/

  Mole-Richardson:
  http://www.mole.com/

CHAPTER

5.21

Television Master Control Systems and Network Distribution

JOHN LUFF
Media Technology Consultant
Pittsburgh, Pennsylvania

INTRODUCTION

Master control operations in broadcast stations take many forms, but all are intended to accomplish similar goals. Commercial group broadcasters often have tried to standardize their station operations around workflow and style, as well as hardware. Cable network operations, though they perform essentially the same tasks of concatenating content into one coherent stream, often operate with different workflow and stronger ties to automation. Small-market broadcast stations in many instances still operate with largely manual systems. And many variations exist in different types of stations including noncommercial operations, noncommercial educational (NCE) licensees, and public and religious broadcasters, all driven partly by the unique nature of their output.

While this chapter is not intended as a complete recitation of all possible permutations and combinations of approach, content, workflow, and hardware for master control, it provides an overview of the essential elements of modern broadcast operations and the different topologies in which stations have chosen to combine operations over wide areas into a centralized location.

Also included, in an Appendix, are descriptions of several major television network digital distribution systems. These descriptions include information on some fundamental changes in master control arrangements at local stations that are being introduced along with the new concepts in network distribution. These descriptions are by Greg Coppa of CBS Television, James M. DiFilipis of FOX Television, and Thomas Edwards of the Public Broadcasting Service (PBS).

MASTER CONTROL FUNCTIONALITY

Content Assembly

One common element ties all of the variations of different stations and implementations together. As illustrated in Figure 5.21-1, the prime function of master control is to assemble elements of content together into a predefined format for emission to consumers. That content may include many different types of media: short- and long-form video; sound and graphic elements; ratings and emergency alert messages required by regulation; and weather, school closings, and other data useful to the viewing public. The methodology for accomplishing this is usually tied directly to the workflow of the complete station.

The master control room (MCR) can be thought of as the "broadcast factory" which simply fills the orders sent from the Traffic department. The *air log* delivered from Traffic to MCR is the shorthand version of the final output from the business operation departments of the station. It instructs MCR to sequence and/or overlay video and audio elements into the output stream, which fulfills the commercial or noncommercial goals of the station management. The air log is the blueprint for the delivered product, which must be faithfully reproduced if the consumer is to receive what station management has determined to be the desired product.

The delivery of the air log can be either in paper form or a database file with individual records detailing the content and time it is to be displayed to the consumer, as well as any voice-over or graphic elements that must be combined at the time of transmission. If the station is automated, electronic delivery is

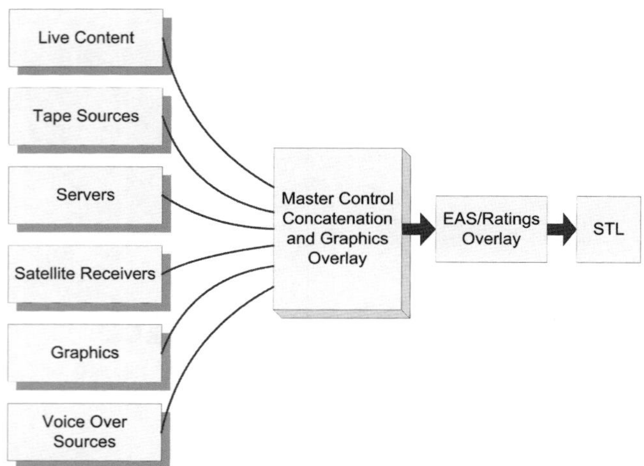

FIGURE 5.21-1 Master control basic functionality.

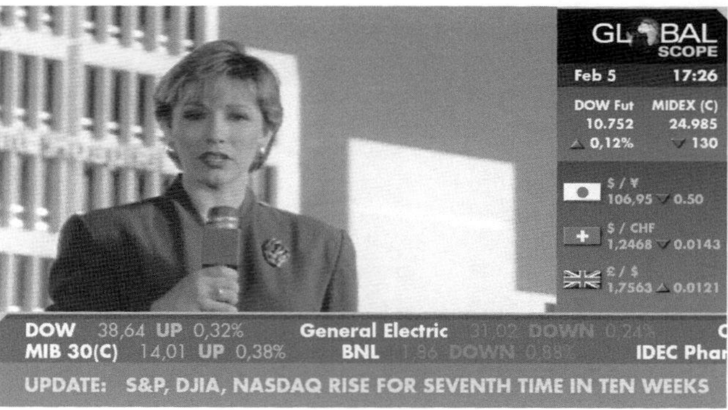

FIGURE 5.21-3 Modern screen graphics. (Courtesy of Miranda.)

preferred, though many stations still take a paper log and input automation commands into the MCR system. Partial automation, which runs only interstitial elements, can more easily be done in this manner. The partial automation approach, however, is more labor intensive and requires MCR operators who are more highly trained than for a fully automated system.

In markets where labor costs are low, it is not uncommon for management to believe that automation cannot save sufficient labor to pay for the investment. However, as digital television (DTV) increasingly becomes a multichannel environment, it may be difficult to accomplish the error-free switching on multiple simultaneous streams without the support of automation.

Clearly, there is a need to concatenate both program material and interstitials, as illustrated in Figure 5.21-2, which of course includes commercials in North America and other areas where commercial broadcasting is predominant. Note that in the United States both commercial and noncommercial broadcasters maintain an air continuity look that is similar, with interstitial messages filling time between long-form programming. Therefore, it is appropriate to consider all broadcasters as a homogenous set for the purposes of this chapter.

In addition to full-screen interstitials, MCR continuity increasingly includes sophisticated graphic elements, often preproduced and stored in still stores or other clip player products. Some manufacturers have designed template-based products that allow the format for promotional messages to be stored and the content to be populated at the time of air insertion. The content can be retrieved from information sent via the air log, or triggered to be retrieved from other data sources at the time of playback. An example of a finished screen from this process is shown in Figure 5.21-3.

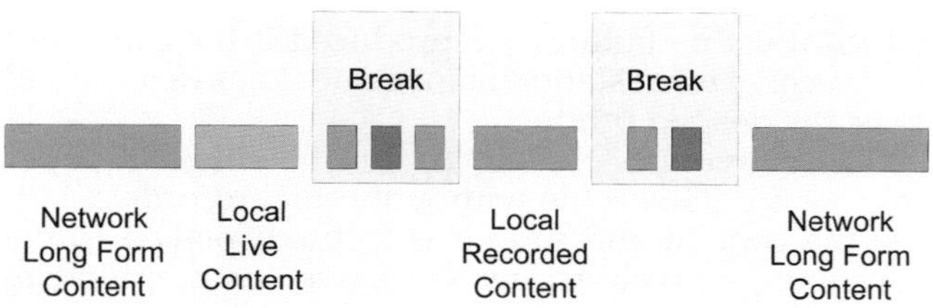

FIGURE 5.21-2 Concatenation of content segments.

Quality Control and Monitoring

A critical function of the MCR operation is to maintain control over the quality of the air signal and maintain suitable content in the event of failures. With multichannel sound, data broadcasting, and, in the future, interactive applications being delivered by DTV broadcasters, it is necessary to extend the conventional video and audio monitoring systems to include quality control for media types that are not currently monitored in many facilities.

In the past simple waveform monitoring, audio monitoring, and picture monitors have sufficed, but sophisticated new instruments will be needed to adequately test and monitor the complete DTV signal and the media it contains. Primary instruments must include vector error analysis, spectrum analyzers, MPEG and ATSC bitstream analysis, and picture and audio monitoring systems that are flexible for use with HDTV and SDTV signals, as well as for multichannel sound. Section 8 of this handbook contains several chapters devoted to test and measurement of baseband and compressed digital bitstreams and broadcast RF signals.

As illustrated in Figure 5.21-4, modern master control monitoring systems are able to integrate many of these test displays with the picture monitoring that is required, providing both operational and economic efficiencies.

MCR monitoring positions often are assigned the responsibility for transmitter operation monitoring and initial fault chasing. Sophisticated transmitter site remote control and monitoring systems greatly ease this task.

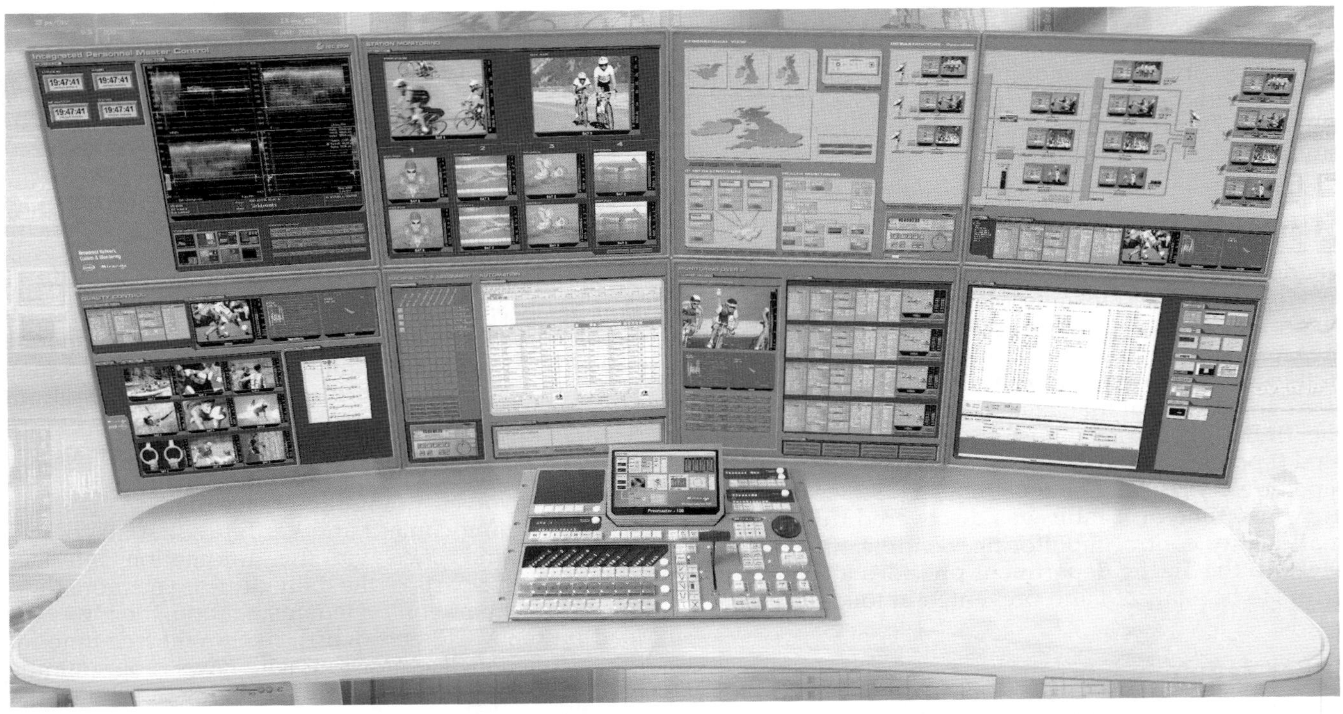

FIGURE 5.21-4 Master control panel and monitoring screens. (Courtesy of Miranda.)

Content Verification

A critical function for all broadcasters, commercial and noncommercial alike, is verification that the content scheduled to go to air actually makes it to air. This is important with commercials and other forms of paid interstitials, but of course includes other paid long-form programming as well. The use of paid underwriting also requires many noncommercial broadcasters to have verification of actual content run. This is often done by both recording the air signal on time-lapse videotape (VHS) or, increasingly, on hard disk as searchable archive files such as MPEG-4, WM9, or other low-bandwidth compression formats.

This air log is in fact the front end for the closed-loop sales process. Though MCR does not control the process of reconciling the air log with the Traffic log and sales orders for the day, it must provide a data path that begins the reconciliation process.

The *air log* (paper or electronic) is annotated with details of air content, and the log or log file is passed back to the Traffic department.

The combination of the as-run air log (or file) and the visual recording is used by Traffic and sales (underwriting in the case of noncommercial broadcasters). It is important that the records be available on a shared network for all to use in a nondestructive manner. Recording as-run data in a computer as a searchable file clearly enhances the utility of this historical record. It can be copied to a DVD or CD-ROM for delivery to others outside the station easily. In the future, some electronic method of confirming the delivery of data and other services sold in the DTV stream will need to be developed and deployed in every station offering such services.

Off-Air Monitoring

It is important that the MCR operation monitor the off-air signal delivered to the consumer. However, for reasons of complexity and cost, cable, DBS, and other carriers whose facilities deliver the broadcaster's air signal to end users are generally not actively monitored. Therefore, broadcast stations may not actually know if the signal was received in homes served by these other carriers. Leased circuits and remote receivers not under the control of the originating station are commonly used by these carriers. Inexpensive remote monitoring devices can allow the station to monitor the signal as received at a remote location if Internet or dial-up access can be provided to the monitoring device. Those remote sensing devices can all be displayed in part of a diagnostic system within MCR.

In the future, TV distribution may be by way of Internet Protocol Television (IPTV) providers. In many markets in the United States and Europe, providers offer national and local feeds in much the same way as cable and satellite. Some stations may prefer to have a web browser looking at live feeds delivered by web interface for the same purposes. With the beginning of distribution to mobile service providers such as handheld DVB-H devices and cell phones, it may be necessary that the monitoring devices will need more than simple web browser interfaces.

In stations that are staffed after business hours, it is common for the building security system, perhaps including surveillance monitoring, to be duplicated in MCR where personnel are present 24/7. These systems can be integrated into the overall monitoring environment.

Maintenance of Logs

In addition to the air continuity log, the MCR operator usually keeps the transmitter log for the official station records.

An important business function is the maintenance of the *discrepancy log*. This record allows station operations and technical management to understand where the Traffic log or other portions of the workflow created problems. Perhaps a piece of content delivered on tape or electronically on a commercial service did not arrive on time or a fault was discovered during normal quality control monitoring. Sending timely messages to the Traffic department provides a chance for errors to be analyzed and prevented in the future.

Other Requirements

The FCC requires the carriage of various supplementary information (such as closed captioning and program rating information) for virtually all programming, as well as insertion of emergency alert messages when they occur. MCR must be able to both insert the required data and video and audio information into the NTSC and DTV signals and monitor the success of the insertion. As with other monitoring requirements, the sophistication need not necessarily be high, but the effectiveness of the monitoring must be such that both the content and presence of ancillary signals can be verified. Part of monitoring the DTV bitstream includes the presence of the descriptors and data. However, only on-screen verification that the data is what was intended is sufficient to meet the needs of MCR workflow.

Monitoring devices are available that can serve both the functions of verifying the presence of the data and demonstrating proper decoding, displayed on the screen. These devices can achieve many of the desired monitoring functions in MCR. However, there is no effective substitute for a consumer-grade monitor (TV set) that "certifies" that ordinary consumer devices also properly process the information. Consumer devices often are produced with a minimalist approach to standards compliance, and some DTV consumer devices may have trouble decoding what broadcast test equipment deems to be a perfectly legal signal. In an age of technical complexity, using both professional and consumer reception equipment provides a level of certainty that the broadcast signal is appropriate to the target audience.

Monitoring must extend to Program and System Information Protocol (PSIP), closed captioning signals (DTV format and conventional NTSC format), and, in the future, any conditional access information needed in the DTV transmission.

Role of Automation

In the past, a trained operator could perform nearly flawless switching with only manual operations. Experienced staff could effectively spot problems as they occurred, take corrective action, and maintain on-air continuity, with acceptable results. However, in the fast-paced, multichannel environment that DTV may become, it is no longer practical or cost effective to attempt to run manual operations. There is simply too much going on. That is not to say that once automation has been installed and debugged in a station, all problems will cease or that operators will no longer be needed. However, the number of operations personnel required for master control in a multichannel environment can be reduced with automation.

When automation is introduced, the role of the operator becomes more supervisory and analytical and less involved in minute-by-minute decisions and actions. Timing content and verifying that all content is loaded and cued via the automation interface become major parts of the daily duties. The more thorough the automation system installation, the lower the load on the MCR staff, which likely will free time for other duties such as assisting with news programming and processing incoming programming.

One important issue is how Traffic systems, automation, DTV encoding, and content management systems communicate with each other. The ATSC and SMPTE standards organizations have been working on strategies to facilitate that communication across multiple manufacturers' platforms. ATSC's PMCP standard,[1] based on XML communications, provides the tools for multiple systems to communicate with PSIP generators for populating the numerous PSIP tables that are required to be transmitted. At the time of writing, SMPTE has a working group actively developing interoperability standards for business systems (programming, sales, and Traffic), and the MCR content delivery engine, especially automation and content management systems. The SMPTE S22-10 Working Group on Data Exchange is developing interoperability standards for this purpose. For more information on this BXF data exchange standard, and automation systems in general, see Chapter 5.13.

When a new MCR system is being planned, it is important to pick an automation system that is extensible to more channels for the future, as well as one that will adopt the newly developed data interchange standards.

CENTRALIZING BROADCAST OPERATIONS

While most stations operate as stand-alone operations, increasingly MCR systems can be implemented that allow varying degrees of centralized control over content and automation operations. In the context of this chapter, it is valuable to review some of the reasons for

[1]ATSC Standard A/76, available at http://atsc.org/.

the move toward centralization and the implications that centralized operations have for the operation of the station and workflow.

The principal factor driving the industry toward *centralization* is economics. In some instances, lower cash flow, resulting from increasing costs or falling revenue, has made it difficult to run a local television business. When a technological solution is presented that has the possibility of reducing costs and increasing cash flow, it must be considered seriously. Recently, broadcasters have been experimenting with, and implementing, centralized operations of varying sizes and levels of technical and operational complexity.

Conceptually, the approach is straightforward. If some of the operations are moved from geographically disbursed local broadcast stations to a central site, it should be possible to reduce the number of personnel involved in the day-to-day air operations at each of the stations served by the central operation. In the end, it is a balance between reduced cost for personnel who are no longer needed and the increased operating expenses due to depreciation, interconnection cost, and personnel staffing the central operations site. The intention is to utilize best practice approaches to the assembly and distribution of the centralized operations site, and interconnect the sites at the minimum cost commensurate with the risk the owner is willing to take for quality and reliability of service.

Economic Justification

While it is beyond the scope of this chapter to review the economic calculations in detail, it is relevant to note that the majority of companies that have rejected centralized operations has done so due to the cost of interconnection bandwidth and personnel issues related to the assembly of a new workflow. One issue is the need to retain some local staff to support emergency operations, sales, and maintenance.

Over time the cost of interconnections has dropped substantially, and with advanced compression such as AVC/H.264 on interconnection links, it should be possible to provide adequate quality of service, reliable interconnection, and sufficiently low cost to make centralized operations an increasingly important part of the toolset available to develop successful broadcast operations.

A major driving force is the substantial improvement in the capability and diminished cost of automation. At one time, the cost of adding full automation to a local station often exceeded $250,000, whereas today the cost per site is likely to be approximately one-third of that. Besides the slow and steady decline in cost per stream, the capability of automation, especially when coupled with advanced servers, has made it much more feasible to successfully automate a station. The added capability of automation software, intended for this specific market niche of centralized operations, is a major factor as well. Although some large-market stations would still find remote operations unacceptably risky due to the high value placed on every commercial, in mid-market stations it is easy to make a credible case that centralization presents little, if any, technical risk.

There are many different models for centralized operations, which are often misunderstood as being part of one strategy. The topology of interconnection, the services to be centralized, the back-up and disaster recovery strategies, and the portions of the workflow that are centralized—all are variables in the planning of a centralized operation system.

Centralized Traffic

Some station groups have chosen to centralize Traffic, leaving a token presence in the station and running the Traffic department for the entire group in the central location. This approach can be quite attractive, and the interconnection cost is low, especially when *thin clients* are used for the remote Traffic computers in the station. There is no requirement that the quality of service (QoS) for the interconnections achieves extremely high values, as short outages do not affect the capability of the department to continue to produce logs and service customers. Virtual private networks (VPN) or dial-up restoration circuits can replace missing network segments for short-term recovery operations after a system failure.

Centralized Production of Promotions

Some station groups have chosen to centralize promotional activities. If a particular program is purchased for many stations in a group, assembling the promotional spots in a central location can be efficient. *Donuts* (partially completed templates for items, with a hole to be filled with station-specific program information) can be created and *tags* inserted quickly and efficiently instead of duplicating the effort in many sites, with resulting higher labor costs. Delivering the *promos* to the stations can be via the public network in near-real time at low cost using a number of different commercially available store and forward systems. If the group stations already have dedicated interconnection bandwidth for Traffic, and perhaps telephony and business computer networks, adding the bandwidth to deliver promos is cost effective.

TOPOLOGY AND CENTRALIZATION STRATEGIES

The most difficult objective is the centralization of the actual air operations. While some topologies carry significantly different requirements, it is possible to categorize all centralization efforts into three basic categories:

- Remote monitoring and control from a central site;
- Centralized stream assembly and delivery to emission location;
- Distributed media and control.

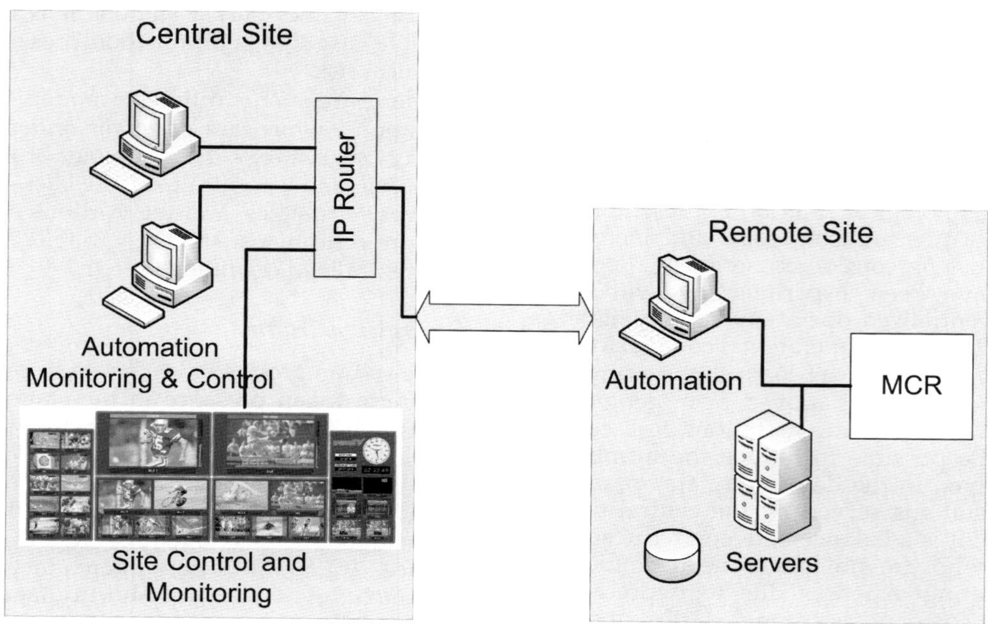

FIGURE 5.21-5 Centralized remote monitoring and control.

Remote Monitoring and Control

In this context, *Remote Monitoring and Control* means that no full-bandwidth video is shared in the WAN environment between the central location and the remote site, only status information and control. In this approach, illustrated in Figure 5.21-5, a local automation system runs in the station, and the central site monitors the automation status using tools which may be as simple as Microsoft's *Remote Desktop* utility.

In this model, video and audio can be monitored using low-bandwidth, low-latency streaming codecs with reasonable success. Technical monitoring can be achieved by the use of both streaming codecs and remote *monitoring probes* installed at the station which communicate over the wide area network (WAN) to the central site. Such an approach is sometimes called the "NY Times Model," named after the New York Times broadcast group that first utilized this topology.

This approach is not perfect. There is some latency in the monitoring feeds, and the full quality of the video and audio cannot easily be assessed over low-bandwidth circuits. Probes, which are available from a number of manufacturers, can provide good data about relatively static conditions and can trap errors such as missing video and audio and note the time that they occurred. They cannot substitute for full-bandwidth oscilloscopes and other test equipment in front of trained operators.

This approach shares an important characteristic with *Distributed Media and Control,* discussed in a following section. In both architectures, the local stations are able to continue operations in the event of a WAN interconnection failure. Remote Monitoring and Control is superior because audio and video systems in the station are not disrupted and are only enhanced in the process of centralized implementation.

This approach is relatively low cost and offers significant features. In the event of a WAN failure, the station continues to run under local automation, and all events already entered into the automation play list should run as scheduled without intervention. When the WAN failure is restored, the system should return to normal with a minimum of effort.

Centralized Stream Assembly

The second topology, *Centralized Stream Assembly,* is the variant most broadcasters think of as *CentralCasting*™, a term trademarked by Ackerley Broadcasting. This approach, illustrated in Figure 5.21-6, moves the complete broadcast MCR of each station to a central site. Economies of scale can be achieved by using scalable systems of routing and branding engines, effectively saving a portion of the investment that might be made at each station individually to achieve the same results.

In the case of a station that continues to produce local content, including news, two different approaches might be combined together to integrate the local production into the stream assembled in the central location.

Alternative Implementations

In the most expensive approach to Centralized Stream Assembly, sufficient bandwidth is needed to deliver the local content to the central site. It is then switched into the return broadcast feed to the station.

In the other approach, a small switcher at the local station is controlled from the central site (or from the production control room), which switches the locally

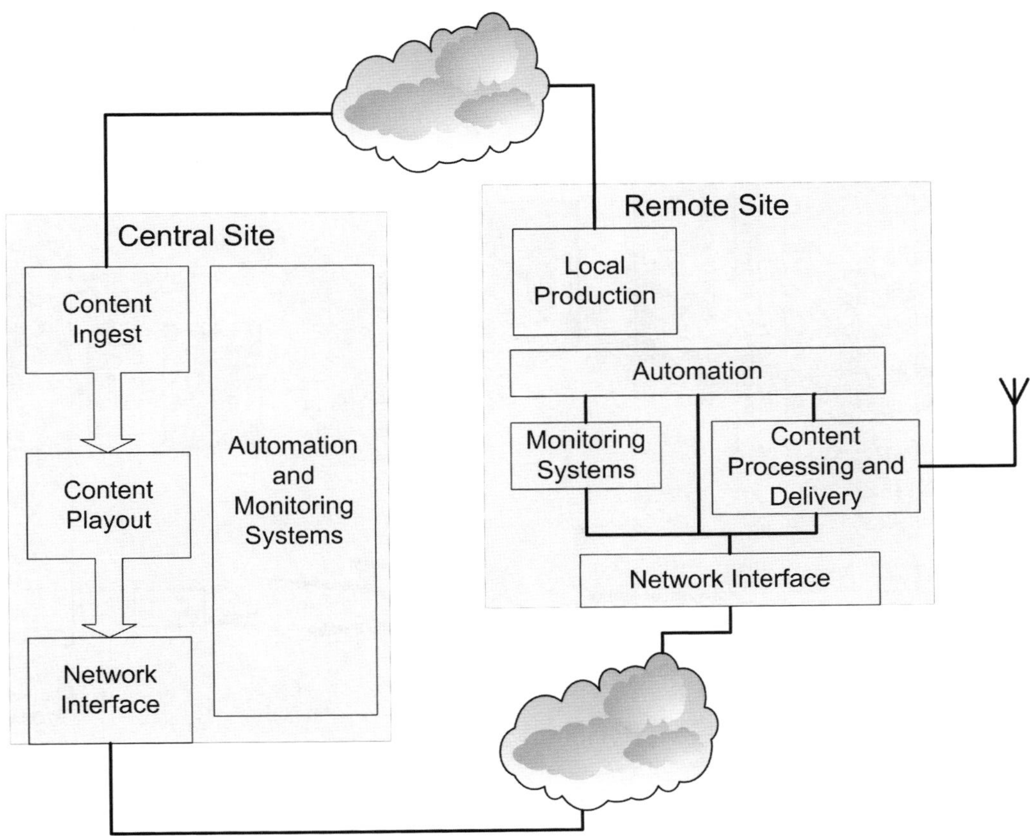

FIGURE 5.21-6 Centralized stream assembly concept; note that local production is fed to the central site where it is integrated into the program stream fed to the local site.

produced content into the broadcast stream from the central hub. In this approach, the timing of commercials is complicated since the commercials are typically rolled remotely from the central site before being switched in locally, a problem exacerbated by transmission latency.

For either approach, there must be sufficient WAN bandwidth with a high QoS to deliver the complete broadcast stream reliably to the station. In the general case, this is the most difficult part of the system, as the cost of interconnection is highly dependent on the specific locations, the bandwidth chosen, and the QoS requested. Assuming that the ATSC encoder is located at the central location, a continuous 19.39 Mbps segment must be available to feed the DTV transmitter, but additional bandwidth is needed for status and monitoring circuits, transmitter remote control, telephony, and other business systems like Traffic and promotions. The total bandwidth required most likely approaches DS3 (45 Mbps), which can be expensive in the WAN, and requires expensive local loops as well.

The major advantage this approach offers is the assurance that the stream was complete when it left the central hub, plus the ease of monitoring the quality of the signal as sent, though not necessarily as delivered. A complication is that some networks have remote control over the integrated receiver-decoders (IRDs) used for network backhaul and may well regionalize commercials or sports content. Moving this reception equipment to the hub may be possible, but it must coordinated with the network operator.

Latency in this approach may be a difficult problem to solve as well. If a journalist is on location watching the off-air DTV signal for cueing information, DTV latency is already an issue. Adding the latency of backhaul and WAN delivery circuits to the signal could end up giving the journalist several seconds of delay, which would not be acceptable. An alternative cueing system would have to be employed, adding more cost to the overall system.

Using this approach, locally-delivered commercials must be transferred to the centralized site before they are cut into the composite program stream. This presents an operational issue that could require new constraints on the Sales and Traffic departments' workflow.

Distributed Media, Control, and Operations

The third topology, *Distributed Media, Control, and Operations*, is a set of variations on one theme. The simplest explanation of this approach is that the media and the control over it are all distributed. As shown in Figure 5.21-7, the central broadcast facility receives,

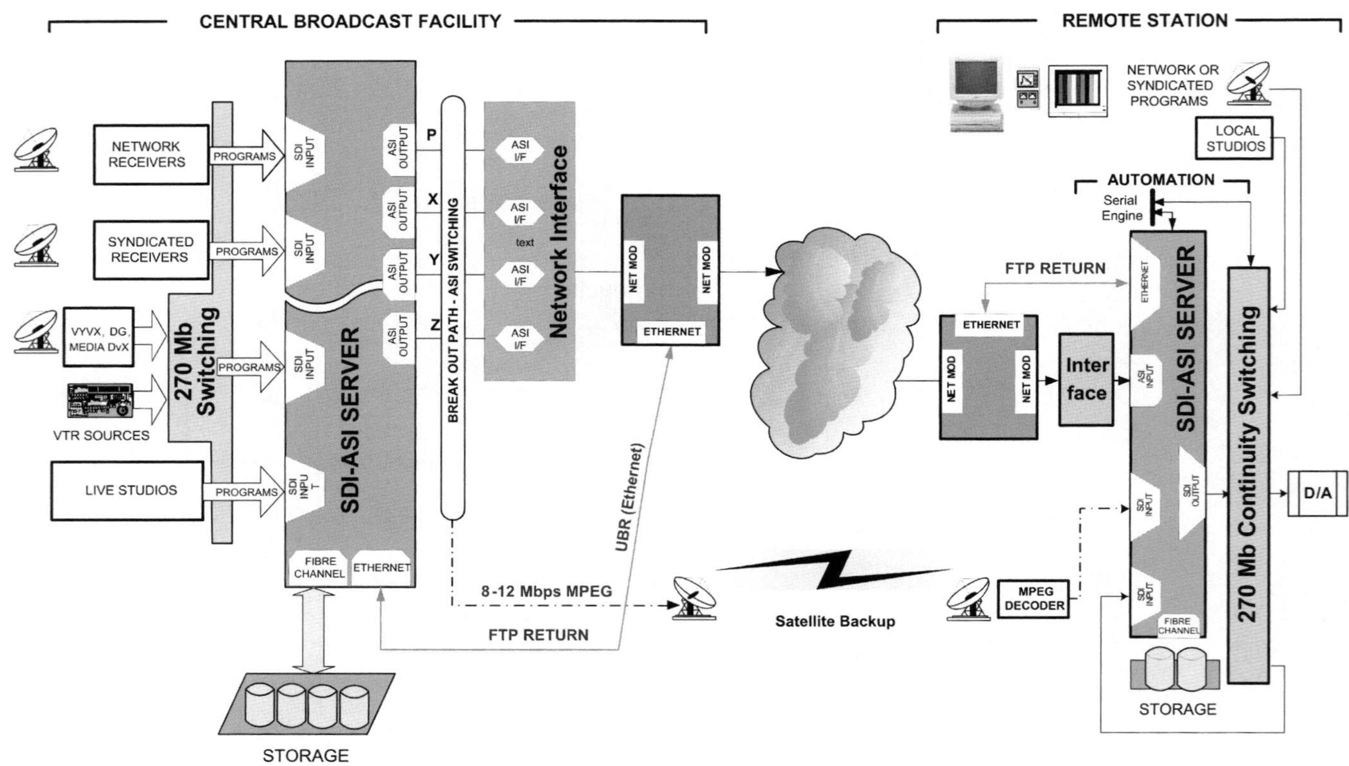

FIGURE 5.21-7 Distributed media, control, and operations. (Courtesy of AZCAR.)

processes, and forwards some content, whereas other content is received directly at the remote station and switched to air locally. The automation system runs locally, but under the control of a "supervisory" centralized automation system, which controls several (or many) stations.

The intent of this approach is to take advantage of store and forward techniques in which content need not be delivered in real time, but only that it arrives before its scheduled playout. In this respect it might be thought of as similar to the systems built for delivering commercials and programs nationally, which are discussed later in this chapter.

In this approach the automation system runs locally, as it does in the Remote Monitoring and Control model. This provides the same level of protection for WAN failures and assures that a viable disaster recovery system can be crafted. As with the "NY Times Model," the hardware in the station is kept intact. Unlike the fully centralized operations approach, this technique allows local content to be switched in locally and commercials run from local playback equipment, eliminating the need to move locally delivered commercials off site and providing a solution to the problem of spots delivered at the last second.

Advantages

There are important advantages for such a system. Since video is not delivered at high bit rates, but rather delivered in non-real-time, the bandwidth is much lower than for a live distribution system. It is practical and more affordable to build a WAN connection with a very high QoS and guarantee of delivery as long as latency in the circuit is not constrained to values needed for real-time delivery. Thus, practical WAN protection and restoration are more affordable, and it is practical to use a high-speed VPN over the public network for disaster recovery.

In addition, because the media are distributed across the local station sites, content that is easy and cheap to receive at the station need not be moved to the central site to be incorporated into a composite program stream for delivery back to the station. For instance, if the station already has network reception hardware in place, it need not be duplicated or moved to the central site. It suffices to gain control over the switching of such content to air from the central site.

Syndicated programming is more likely to be moved to the central site since it must be prepped for air and metadata distributed to more than one station. This allows syndicated content to be handled once, eliminating duplicate effort at multiple sites. If the distribution system is capable, IP multicasting can be used to deliver such content to several sites in one transmission, cutting network bandwidth substantially.

Complexity

Because hardware systems at the station are augmented and not replaced, this is not a simple system to install. New automation and distribution approaches are seldom possible without complexity, and this approach is the most complex of the three. There is an inextricable link between the station and the central site. Neither can complete the processing of content alone. IP multicasting, store-and-forward, WAN management of content, bandwidth management, and effective monitoring are all challenges, but this approach, when properly implemented, permits the most synergy between a local operation and centralized management of content. It is particularly true that, like the fully-centralized operations hub, this approach works very well with centralized Traffic. Both databases live in one location, making communication easier. When Traffic sends a log entry to the central automation system, that entry is entered into the automation play list, which is forwarded to the station. The play list lives in both locations, allowing for a degree of redundancy.

Cost of Deployment

In terms of cost of development, this approach is the middle ground, with less cost at the central site than is required for a fully centralized operation, but somewhat more hardware at each station. The cost of design and implementation is approximately the same. Since all stations would receive similar racks of hardware to connect to the distributed broadcasting hub, including servers, automation systems, IRDs, and monitoring systems, an efficient assembly process can be taken to building multiple sites before delivery to each station for final connection and checkout. The existing automation system at the station can continue to operate until cutover, and the new automation can operate in "shadow" mode, thus removing a major source of stress at the time of implementation.

Variations on this approach have been used for a number of implementations, notably NBC-owned and -operated stations. The degree to which the content and automation is distributed, the method of interconnection, and the intended workflow can vary widely.

Factors to Be Considered

All three of these approaches offer benefits and come with pitfalls that must be avoided. The final analysis of costs to benefits is a complex process that requires a team approach including all operational departments. Change itself is disruptive, and it may be that the largest hurdles are not technology. Rather, the problem lies as much in changing the culture to believe that centralization is an effort to allow the best of all business practices to prevail, while eliminating redundancy where practical. If the staff adopts a positive attitude toward change, the implementation of new workflow approaches, like centralized operations, can be a smooth though complex process. If the entire concept is met with great reluctance, no technology solution can be smoothly implemented. The desired outcome is, after all, the reduction in labor in at least some part of day-to-day operations.

Some station groups, particularly those in public broadcasting, offer particularly attractive targets for centralization efforts. Their content, while not identical from station to station, is more similar than different, and the number of sources of programming is small. For example, the Public Broadcasting Service (PBS) is in the process of designing a "Next Generation Interconnection System, (NGIS)"described in the Appendix, which bears close resemblance in many ways to portions of a distributed broadcasting approach.

Other practical implementations include group broadcasters with several stations affiliated with one network, particularly when they are in one time zone or a large group of stations with a significant amount of syndicated programming.

All centralized operations approaches must deal with common problems. Live programming represents one difficult issue because of the timing of commercials and promos. Exacerbating the problem is the latency of the WAN circuits used for distribution, monitoring, and control. The control issue can be minimized with each of the three main approaches; however, monitoring latency can create problems. A centralized operations center that is monitoring stations affiliated with several networks in more than one region may find that each station has truly unique programming, with differing commercial content. Remotely monitoring and switching content of this type may be difficult, even with automation assistance. As a result, it is often part of the plan that content of this type should be switched locally, with control returned to the central site at other times.

Monitoring for Content and Quality

A critical element of all centralized operations schemes is effective monitoring of both content and technical quality. Monitoring circuits must be of minimum latency and maximum information content, which of course is often mutually exclusive with cost. At a minimum, two feeds should be returned as a continuous stream of "thumbnail" pictures: off-air and MCR output. In a multichannel environment, now common for many DTV transmission systems, each station output feed should be continuously monitored, while the off-air feed could be remotely switched between the various feeds in the multiplex. In addition, transmitter remote controls must be monitored as well as the status of the automation system.

It may be appropriate in many stations to provide each monitoring feed with a small and independent routing switcher at the station intended for remote monitoring only. This allows the central operations supervisor to choose sample points within the station—for instance, the output of network receivers, input to the studio-transmitter link (STL), TSL return, and another output which links to the main station router. This would allow the central site to view any feed available in the remote station. These sample points could be returned with streaming media encoders but also should be monitored with devices that sample video and audio for technical parameters and for closed captioning information.

Part of any centralized monitoring scheme in the DTV environment must be the return of MPEG stream analysis data to the central site, including analysis of PSIP information. Other important systems to be monitored include encoder management systems, routing switcher control system diagnostics, and status monitoring available from servers and other complex computer-based systems.

CENTRALIZED OPERATIONS INTERCONNECTION

Each of the three general topology models discussed previously requires some form of WAN interconnection. In general three physical methods exist for such networks; microwave, satellite, and terrestrial leased data circuits.

Microwave

Private microwave interconnection over long distances is, in general, not practical due to cost and licensing issues. However, in certain cases, it should be considered. In the intermountain western states, where interconnection distances are long and there is less interference with other terrestrially licensed services (such as cell phone interconnection, intercity relay links, and satellite transmit and receive services), it may be practical to set up moderate-distance interconnections for networks within one state or between adjoining states. Using digital radios for the microwave circuits, it would be possible to craft a DS3 link that would have low operating cost and high quality of service (QoS).

Satellite

Satellite interconnection is practical if used in an asymmetrical fashion. This provides high-bandwidth TCP/IP data outbound and lower bandwidth terrestrial return for data acknowledgments and services needing full-duplex interconnectivity. A group broadcaster might choose this method to feed all of its stations, particularly in the distributed central operations topology. The bandwidth of the return circuit may, however, exceed T1, or multiple T1 bandwidth through the use of frame relay or other cost-effective products available from the regional telephone companies and competitive access providers (CAPs). The return data path need not have a high QoS unless isochronous services are needed, but it must be protected against lengthy disruptions that could complicate full-time operations.

Satellite services could be set up in either the Ku or C band. C band can be less costly and is less affected by rain fade, but C band satellite dish sizes are large and frequency coordination may be complex, making them impractical on some urban sites. An initial computer study can establish the likelihood of completing interference coordination before unnecessary funds are expended for the actual coordination. Ku services do not require frequency coordination, licensing is straightforward, and implementation is sometimes more rapid. The Ku and C bands are equally applicable to the interconnection strategy, so whichever scores highest during evaluation of financial, legal, and technical aspects will be appropriate to the task.

Terrestrial Data Circuits

Terrestrial leased data interconnections for centralized operations are the most common approach.

Factors to Be Considered

It is important in all three cases to create a "bit budget" analysis to establish the bandwidth in both directions that can sustain full-time operations. A hypothetical case showing required bit rates (BRs) is shown in Table 5.21-1.

TABLE 5.21-1
Sample Interconnection Bandwidth Requirements

Service	Central Site to Stations			Stations to Central Site		
	Lowest BR kbps	Target BR kbps	Maximum BR kbps	Lowest BR kbps	Target BR kbps	Maximum BR kbps
Video File Transfer to Stations	4000	6000	10,000	0	100	150
Streaming Video to Stations	3000	6000	8000	0	100	150
Traffic System Interconnection WAN	64	128	256	64	128	256
MCR Intercom	32	64	128	32	64	128
Automation Interconnection	128	224	384	32	64	128
Status and Monitoring	32	64	128	128	256	512
Total Required Bandwidth	7256	12,480	18,896	256	712	1324

The success of the project will depend on the viability of the interconnection strategy, which must be chosen to cover the required services and QoS. Setting an upper bound and finding the target bit rate for each direction may limit the choices of interconnection. For instance, if more than T1 is required in the return direction, an alternative may be frame relay, which offers more flexible bandwidth utilization and committed data rate options. If the aggregate bandwidth for the stations to be covered exceeds 45 Mbps in total, other product offerings should be considered.

Some types of interconnection are more appropriate to centralized operations, such as isochronous video in the general case. Frame relay is applicable to the best-effort return path used with distributed central operations topology. However, since the data may arrive with packet jitter exceeding that needed for smooth decoding of streaming content, it is not applicable to the circuit outbound from the central site to the stations. In the past, the jitter and packet loss of IP appeared to be unsuitable for video services. In more recent trials and implementations, broadcasters have found IP to be both suitable and reliable, and capable of providing the needed QoS for video services. IP can also be ordered with variable bandwidth from some carriers, allowing for scalable requirements at each site covered.

It is important to find a carrier that can provide end-to-end service. It is not sufficient to buy circuits from multiple carriers in the local and long-distance business. Implementation should proceed with only one carrier taking responsibility for the entire network implementation. QoS guarantees should be provided that give the bandwidth and reliability needed to service the complex needs of the centralized operation. When trouble occurs, the operations staff should not be troubleshooting a complex network of interconnected carriers. They should be implementing the carefully rehearsed disaster plans needed to keep the stations on the air. It must be the job of the carrier to analyze the problem and restore service, not restore a portion of a circuit at a time.

Today, it is tempting to buy a large block of Internet bandwidth and set up VPN services to handle some portions of a centralized operation on a routine basis. Though tests might show this to be a valid strategy, in some cases the entire bandwidth of the Internet has been absorbed due to malicious software attacks on a prime target, so use of the Internet for broadcast links is very risky. However, as a disaster recovery approach, it is practical and cost effective.

PLANNING FOR CENTRALIZED OPERATIONS

Any large project requires careful analysis, but none is more important to a broadcaster than a project that will fundamentally alter the workflow and financial results of the business. A complete business case and technical analysis is the first step in planning such an ambitious project. A flow diagram for this process is shown in Figure 5.21-8.

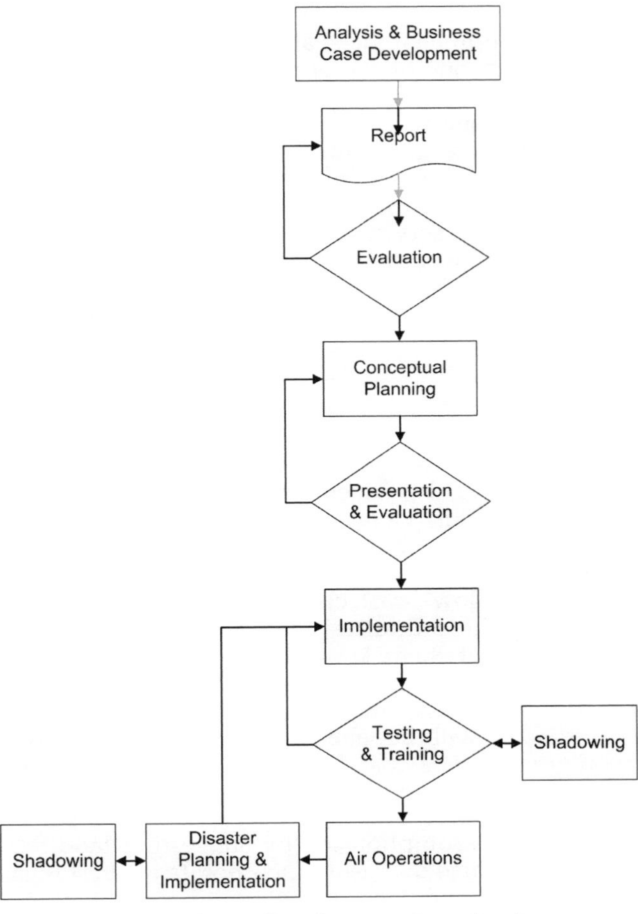

FIGURE 5.21-8 Centralized operations business case process flow diagram.

Analysis and Business Case Development

The process of planning a large project is often foreign to broadcast organizations due to the mature nature of the business. Chapter 2.3 discusses project management, which is particularly important to this process.

In the context of centralized operations, it is necessary to set goals that the implementation is intended to achieve. For example, the goal may be a financial target, or a reduction in staff, or perhaps simply to facilitate future growth in services provided to the public. These goals and the implications of them must be thoroughly understood and agreed upon by the team carrying the project forward. If the goals cannot be reasonably achieved by known methods, note the challenges and put special effort into problem solving.

Conceptual Planning

Once the overall goals are in place, the implementation approaches should be established. Use the three generalized topologies presented previously as test cases to determine if the goals of the project can fit logically within one of the approaches. A consultant is often called to assist in project conceptual planning, particularly one who has experience in both the process the team is going through and the implementation of centralized operations in another project. A system integration firm may also be consulted if it is likely that such services may be used for constructing the facility. The firm's experience may prove to be invaluable and help in moving the project to completion with a minimum of backtracking and unnecessary changes later.

Presentation and Model Development

Once the conceptual plan is in place, it should be presented both to the development team and management. The goal is not necessarily to win their support, but rather to provide an understanding of the entire picture. Dropping a fully-developed plan on staff whose workflow must change is both stressful and perhaps could lead to resistance to the project implementation. Let them know how it will likely affect their part of the operation. Also have a plan for how to handle objections and skeptical questions about whose jobs will be eliminated. Have information about the success that has been achieved by other broadcasters by doing what is contemplated. Take their input where it has been constructive, either by challenging assumptions or by suggesting modifications to optimize the result. Incorporate the best ideas and be ready to defend in constructive and nonjudgmental ways decisions not to use their ideas. This modified plan becomes the core for development of the final implementation plan.

Approval and Implementation

Once approval and funding are in place, carefully outline the implementation plan. Review each aspect of the technical implementation and the changes in the workflow that will be needed to achieve the project goals. Lay them out on a timeline and look for conflicts. Rushing to implementation is never wise, but neither is delaying it unnecessarily, since while the planning and implementation is going on, costs accrue without any offsetting benefits.

Training, Testing, and Shadowing

The plan should include time for testing and training and "shadowing" (running the new system simultaneously with) the existing operation. When the new workflow is run in parallel to the largest extent possible, technical and operational issues will be found that must be addressed before going live. Tests need not extend for months, but if five stations are to be implemented, a three-month test period would be in order before beginning the implementation of the rest of the sites. Doing so will help avoid pitfalls seen in the first case and allow time to find solutions and optimize the approach prior to cutover. Of course, avoid cutting over the week before the "May Sweeps" (when audience ratings are used to establish advertising rates). While there may be discrepancies on air, the goal of the shadowing period is to avoid serious issues affecting air operations. When the time feels right, cut over to the new system and work through the issues quickly to keep the process in motion.

Disaster Planning

Though most stations have back-up plans for many of the failures that can be predicted, the loss of capability from the centralized site is a different class of problem. A disaster plan should be created that identifies known ways for the system to break down. It should be a formal document, an Operations Manual, with tabbed chapters clearly labeled for the classification of failure. One tab, for example, might be the loss of the WAN connection. A hot list of initial actions on the first page should include

- Who takes command
- What actions each person is required to take
- What the expected response should be, such as calling the carrier to report the problem, obtaining a trouble ticket, and the name of the person working on the solution
- What interim actions need to be taken to avoid compromising the air signal

The book for the receiving stations and the centralized site will be different but coordinated to avoid duplication of effort that might slow down effective solutions. Failures are not all hardware. For example, a log might fail to convert and load into automation, or an operator might not show up for a shift. The Operations Manual should be a living document that is reviewed regularly with all concerned. Discrepancies should be reviewed against the plan and, if appropriate, the plan should be changed. Over time, the document will become more thorough and effective at laying out the best route to solving problems. It also becomes a valuable training aid when new personnel are hired.

A disaster recovery plan is insufficient if personnel are not run through drills to help them understand the process. Some large operations with many personnel have implemented full training rooms that include the actual hardware used for air operations. New staff are assigned normal shifts and given training in simulated emergencies using the manual they will be expected to use when they are in on-air operations. While this level of training may not be appropriate for small operations, the concepts should be considered carefully. A training room might actually be set up to serve as a second operations center and disaster recovery site if it is located away from the main operation.

Recognizing the signs of impending or actual problems may be difficult to train for, but when the station is running a late-night program without need of services from the central hub, the central site and local station working together can conduct live training with simulated emergencies. Supervisors might have the interconnection carrier pull a patch cord on command, or the air monitor might be turned off without warning. Drills should be designed to replicate situations that operators will see in practice and which will help them learn how to use the Operations Manual the way it is intended. If the manual is insufficient for operators to take their input seriously, change the manual to reflect the best solutions possible.

VARIATIONS ON MCR HARDWARE APPROACHES

Until quite recently, MCR switching systems did not evolve in any conceptually significant manner. Features have been added by manufacturers to differentiate products from the competition, such as more video keyers, internal *squeeze back* digital video effects (DVE), more channels of audio, and others. However, no radical changes occurred in MCR design for many years other than the development of digital electronics to replace aging analog designs.

Recently, manufacturers saw a need for a new style of "MCR-like" image and sound processing. It may be that this need was first identified in Europe because cable operations there often customized feeds with localized graphics and separate commercials, often requiring separate graphics overlays and keys. Some of the same needs were developing in the North American market, where localized commercial content had been a fact for a long time, but graphics overlays were generally not done on nationally distributed feeds.

It is important to understand the differences in approaches that are now possible and where each alternative offers advantages to the operator and designer of MCR systems.

Conventional Baseband Systems and Switchers

A conventional switching system, as illustrated in Figure 5.21-9, has internal, or dedicated external, crosspoints in a plant routing system configured to provide the functionality of a very small production switcher. In general, this is a stand-alone system that is configured at the time of ordering to include the options needed to satisfy the workflow requirements of the output channel as they are known at the time (such as the number of inputs, keyers, DVE channels, bypass switching, redundant power, control panel options). Often one control panel is assigned to control a number of frames of image and sound processing, either ganged together or assigned individually to the control panel. However, once the system is installed, it is difficult to change its functionality in any significant manner. These systems, still available today, are manually-driven systems that can be controlled by automation, and not the other way around.

Many such systems are similar in design. They typically have a hardware panel that enables manual intervention if automation is not sending the right commands. Though these systems could have been designed with software control panels, in general they were not, and indeed most predated the development of modern touch-screen control techniques. Likely their designs were conceived before modern web browsers with interactive JAVA/Flash code existed. This does not mean that the systems are unsuitable for

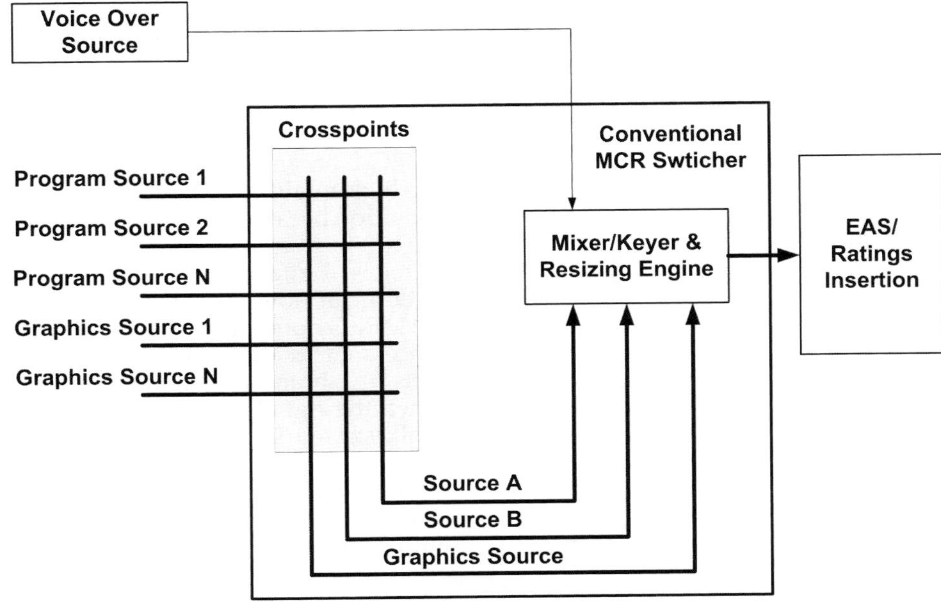

FIGURE 5.21-9 Conventional MCR switcher simplified block diagram.

critical modern applications, but it should be noted that an operator was an important factor in the design of the product.

In some baseband MCR switcher designs, internal crosspoints are provided in the processing frame so that MCR can function, if necessary, without the plant routing switcher. Most were designed with either dedicated or optional external passive bypass switching to further protect the air signal against failures. Other systems took advantage of using an external crosspoint routing system for source selection, often a router designed by the same manufacturer, and delivering a small number of signals to the MCR switcher itself for processing as Air, Next, and Key signals. The MCR switcher requires a small internal router to transition between the presets and perform takes and overlays. Usually, these systems are designed as AFV (audio follow video), with audio-over channels as secondary inputs.

Systems matching this description have been available for decades as analog MCR switchers, with serial digital and HDTV versions available since the early 1990s. In many cases the processing electronics for both analog and digital video, as well as HDTV, are available from the same manufacturer, allowing all emission output formats to originate in a similar manner.

Branding Engines

In contrast to conventional MCR switchers, *branding engines* started life with only air and preset busses, and perhaps overlays, as illustrated in Figure 5.21-10. Although, over time, control panel options have been developed to allow the products to move into conventional MCR environments, they were originally intended to give limited processing power, under automation control alone, to simply locally brand a feed delivered from a program producer. The first branding engines were small, limited capability systems without the options that conventional MCR switchers would be assumed to have, such as redundant power supplies. These products have now taken advantage of the growth of DSP processing and ASIC design approaches to allow more functionality to be inserted into smaller and less expensive products.

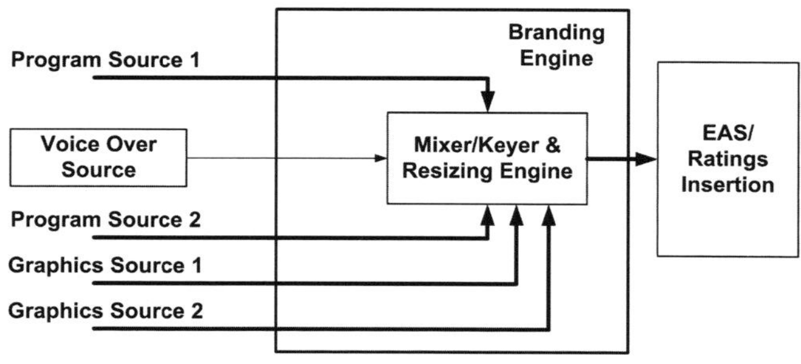

FIGURE 5.21-10 MCR branding engine.

Fundamentally, they still operate under computer control, with software control panels often available. However, most manufacturers have added optional manual control panels of significant complexity, thus narrowing the gap between these products and full-blown baseband MCR switchers.

Recently, a second generation of branding engine has appeared on the market, based on plug-in modules, which can accept either high definition (HD) or standard definition (SD) inputs (an option at the time of installation). This capability allows significantly enhanced system flexibility and, as a bonus, the capability to utilize in the future assets bought today, when perhaps an SD channel will convert to HD content. The modular approach also allows many channels to be grouped in a small amount of space. Some manufacturers have provided for both crosspoints and processing modules to be mounted in the same chassis, further improving design and growth options.

Software MCR

At one time, it was assumed that a high-powered personal computer could be configured to process video, given sufficient processing power and a fast graphics engine. This would provide all the functions of a hardware-based system, but with a processor-driven approach. Recent demonstrations have shown the functionality of a branding engine, MCR automation, and clip playback within one self-contained computer. There are appropriate places for such technology including disaster recovery, secondary channels, and small-market unattended operations. Other choices include public venue displays that can run high-quality outputs under automation and are monitored remotely from a centralized point.

COMPRESSED BITSTREAM MCR SYSTEMS

Until the late 1990s, the only compressed bitstreams in the broadcast plant were usually locked inside video servers and were specific to each manufacturer. When MPEG-2 became the predominant method of delivering high-quality content to stations, a new mode of operation became a practical option. Instead of decoding the bitstream of a network provider or the commercials digitally delivered by service providers, a station can now potentially keep all of the signals in the MPEG domain and perform bitstream splicing and processing to affect most, or perhaps all, of the functions of a branding engine and clip playback system. When the signals are kept in the compressed domain, as illustrated in the system shown in Figure 5.21-11, quality loss during repeated decoding and encoding is minimized.

Bitstream processing, however, has important implications for MCR usage. Although technology exists to do partial decoding of the signal for the insertion of keys, it is complicated and expensive. In addition, not all signals exist in the MPEG domain for direct input to such a system. To overcome this, one

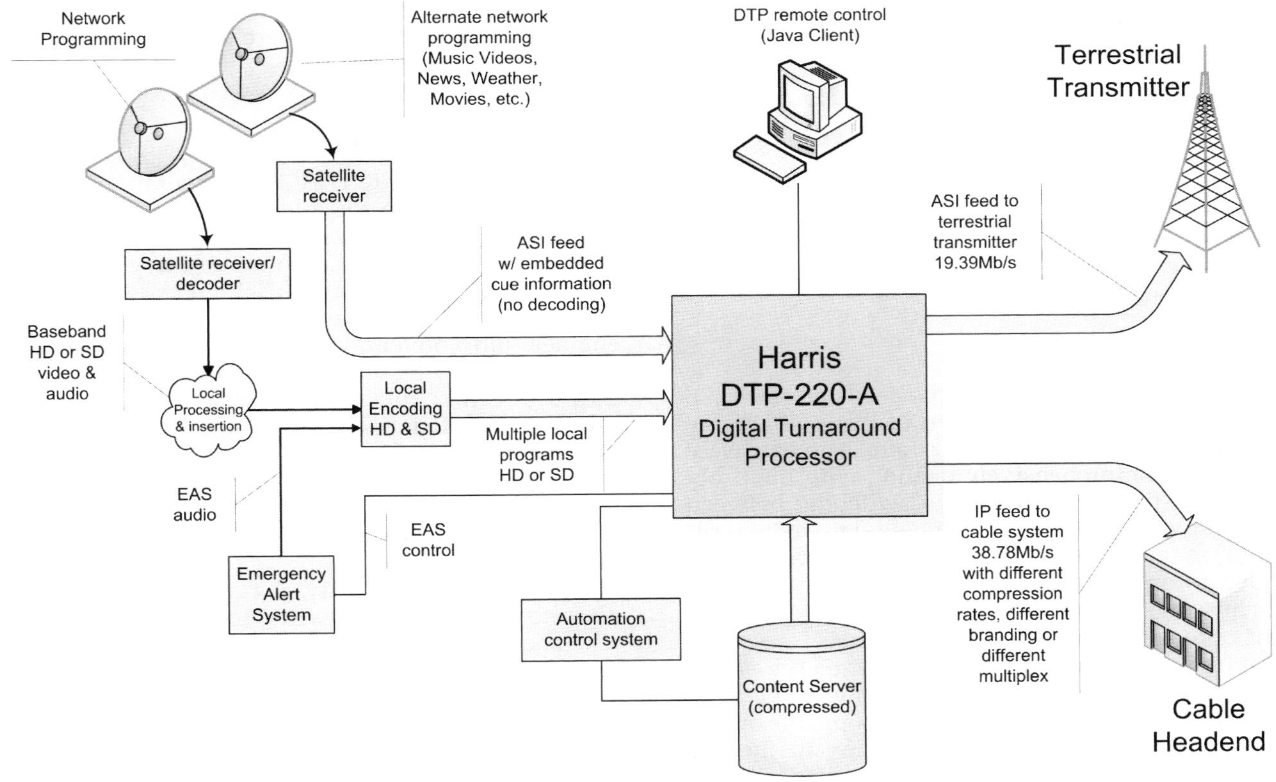

FIGURE 5.21-11 Harris DTP compressed bitstream diagram. (Courtesy of Harris Broadcast Communications.)

technique for implementation might be to take live feeds that do not already exist in MPEG and use a "utility encoder" of sufficient quality to allow feeds to be input directly as if they had been received by satellite or other means. The MPEG MCR engine then treats this as any other streaming feed and incorporates it into the outgoing stream. While technically feasible, the economics of such a system need to be considered carefully since, for live operation, at least two such utility encoders would be needed, driving up the cost. There is no doubt, however, that the capability to process MPEG material without any decoding is a powerful tool for MCR operations.

Some processors for compressed bitstreams with full MCR capability also allow *transrating*, or changing the bit rate of signals for different programs to be placed into the multiplex. This capability could have various applications, for example, when the multiplex will hold SD and HD feeds that must be changed in bit rate so as not to exceed the DTV bandwidth.

Such systems hold great promise for many applications in the future, as their capability expands further and cost drops. FOX Television has already adopted this model for use in FOX local stations and affiliates. A description of this system is included in the Appendix.

AUDIO FOR DTV

Audio transmitted with DTV has a unique characteristic not available in analog transmission, even when digital techniques are used to handle the signal in the studio, and that is *audio metadata.*

Metadata

Metadata can be added to describe various characteristics of the audio signal. This metadata is needed for emission, such as *dialnorm* (dialog normalization) and *DRC* (dynamic range control). It is associated with the audio stream right from its source and, to comply with the intent of the DTV system specifications, must be transitioned between sources so that the applicable values for each program are carried through from the point of creation to the point of emission. Every step in the production process, including backhaul to network origination centers and their output to stations, must carry the original data related to each composite program.

However, unlike video and audio, there is no easy way to transition the audio metadata intelligently when audio sources are mixed in master control. If a voice-over is performed in MCR, the new composite metadata must be authored and passed to the emission encoder. This requirement is illustrated in Figure

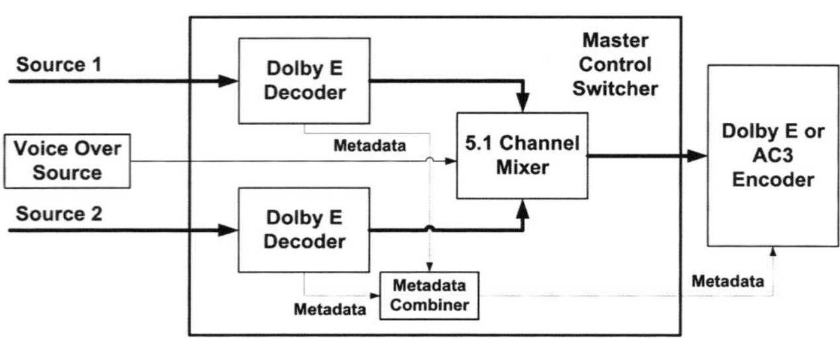

FIGURE 5.21-12 DTV MCR audio with Dolby E-encoded incoming audio sources.

5.21-12, where the incoming audio feeds are Dolby E encoded. The system implications of this problem are substantial, and as of this writing there is no hardware intended for comprehensively dealing with the problem. The existing solutions allow switching between data from two sources or forcing new data to be written, but as an integral portion of a system often intended to operate under automation, an important tool is missing.

Part of the solution to these audio issues is to maintain high-quality sound monitoring of the DTV signal in MCR. When audio problems are noticed, perhaps by ear more likely than by a meter or scope, the operator must adjust the emission multiplex to include more appropriate values for the audio metadata. This is, however, easier said than done. Fortunately, hardware now exists to do this job of monitoring that the operator must otherwise learn.

Audio Modes

A second complication of DTV audio comes from the very nature of the Dolby AC-3 signal. It can operate in many modes, including stereo and surround sound. Commercials are often stereo (called 2/0 mode in AC-3) or even mono (called 1/0 in AC-3), and when a stereo commercial is inserted into a 5.1 channel surround program (called 3/2 in AC-3), the correct control of the emission encoder would be to change the data that tells the decoder in the home how to decode the audio. Unfortunately, this leads to undesirable effects in many consumer decoders (clicks, pops, or even seconds of silence—something advertisers would certainly notice). There is no immediate practical solution that fits all applications except to avoid switching the audio back and forth. However, that is not easy to do when the content is not all under the control of the station emitting the final mix. A practical solution is to use a device that synthesizes surround sound when the input is stereo and then leave the encoder in 5.1 mode (3/2). Surprisingly good results can be obtained by using such an approach, and consumers are unaware of the switching that other approaches would require.

The DTV standard allows multiple 5.1 channel streams to be transmitted with each video program (e.g., for multiple languages), each of which may require six baseband channels, or three AES pairs. As a result, it is important that servers and routing and on-air switching equipment are able to pass no fewer than three AES pairs.

Embedded versus AES Audio

MCR systems prior to DTV implementation generally used discrete audio for interconnection. With digital SD and HD video systems, SMPTE 259M and 292M interconnections offer the option of carrying 16 channels of uncompressed audio within the serial digital interface (SDI) bitstream. Many modern branding engines and conventional MCR switchers allow the option of receiving embedded audio, which for MCR purposes can be quite convenient. In that case, fewer cables are needed, and matrix routing is held to a single level instead of multiple levels. However, embedded audio also has drawbacks, including cost and latency of embedding and disembedding. In particular, voice-over feeds can be difficult to handle in this manner.

Inside the DTV facility, it is possible to utilize a discrete audio system with no embedded signals whatsoever, and most plants have at least some analog audio. It is reasonable to use an audio routing switcher at the input of the MCR processing engine, allowing flexible selection of analog and AES inputs and internal conversion between them, in order to facilitate MCR needs such as voice-over or other special audio applications.

Network Distribution

Dolby E encoding allows eight lightly compressed audio channels to be carried in a single AES audio channel for intra-plant connections or for carriage over longer distances. It is currently used for network distribution by at least one North American network (CBS).

AC-3 was the choice of the ABC TV network for audio distribution for HDTV, but using 640 kbps instead of 384 kbps as specified for ATSC emission streams. This reduces the quality impairments caused by decoding AC-3 at the affiliate station to baseband for master control switching and subsequent re-encoding for emission. FOX and PBS use AC-3 for audio encoded at emission bit rates, as they do for HD video distribution.

Another arrangement for network distribution and routing is to use multiple stereo audio channels, as previously used for the digital distribution of NTSC program audio.

The Appendix to this chapter includes further information on network distribution systems for FOX, PBS, and CBS.

SERVERS FOR CONTENT ORIGINATION: THE TAPELESS FACILITY

The cost of video storage has been reduced to the extent needed to make a tapeless facility possible. If

sufficient low-cost storage is included in MCR server systems, it is possible to transfer incoming material directly into the server during the *ingest* process. For the most efficient operations, this process uses incoming data files, and not baseband video.

Servers for MCR can take several forms. A modest facility can utilize a single chassis with internal storage with well over 200 hours of content. Larger facilities, with more extensive needs, may move to servers in which the storage is expanded in a variety of ways. Video servers and storage systems are covered in detail in Chapter 5.9 and this section will only highlight strategies for implementation in the context of MCR.

Storage should be thought of not so much in terms of monolithic blocks of storage, but rather as *pools* that can be *on-line*, *near-line*, or *off-line*, as appropriate to the task and the cost of software management systems, as illustrated in Figure 5.21-13.

Near-line storage is particularly attractive due to low cost per hour stored and low technical complexity. Such storage is sufficiently fast when connected with high-bandwidth networks to the main server system. Automation usually can control the content in both the near-line and on-line pools transparently to the operators. The speed of disk access may be close to that needed for direct play-out to air, though it should be thought of as an intermediate cache in replacement of expanded on-line storage.

Off-line storage may be held on media including DVD RAM, linear archive tape, volumetric optical storage, or others. This storage approach allows lower cost per hour and, since the media are removable, it can be expanded nondestructively to an arbitrary size. This type of storage is, however, more costly to implement and slower to access in both read and write modes. The speed deficit is most commonly mitigated by using more than one transport to access media, permitting simultaneous read and write, or parallel read operations. Often a cache is built into such systems to move the media to an intermediate location and free up the power of the on-line server systems. Some automation systems can directly control archives, though it is most often the case that *Media Asset Management* and archive are implemented together.

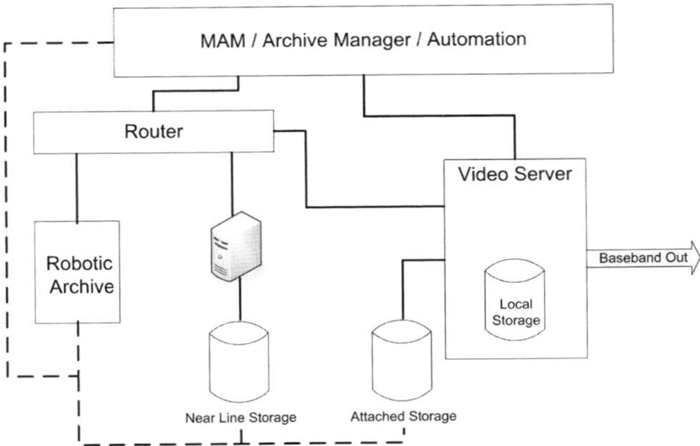

FIGURE 5.21-13 Tiered storage arrangements for efficient automation systems.

Compression Approaches

Servers for MCR usage today use primarily MPEG-2 (at various bit rates) for storage, though DV format (in all its bit rates) is often also available. MPEG-2 is particularly attractive due to the ability to transfer files between commercial servers and air playback with a minimum of complications.

MPEG-4 Part 10, also called AVC or H.264, is a likely candidate for future server implementations due to the high-bandwidth efficiency it offers.

Other compression systems, including JPEG 2000, may see applications in the future, particularly due to the development of servers for digital cinema that will use this codec. As with AVC, there is no current deployment for master control.

Audio is obviously an important part of the total system, and with DTV it has become significantly more complicated for servers as well. Most servers offer the option of using either uncompressed AES or one of the MPEG audio compression variants. It is important that servers installed for MCR operations have the capability to pass uncompressed AES *data* mode of AES, since that is needed for Dolby E and Dolby AC-3 signals. Server manufacturers are almost universally capable of handling this requirement today.

CONTENT DELIVERY SERVICES

Rather than being delivered on tape as in the past, advertisements and, increasingly, full-length programs are now often delivered as files shipped to *catch servers* located at stations. PathFire, Media DVX, Vyvx, and others have contracts to deliver such content to broadcasters and, frequently, the hardware needed is installed at the service provider's cost. The terminal devices often have both control and baseband video ports to play out content for rerecording in the station's servers or on videotape. As discussed earlier, it is desirable to have the video delivered as a file to the playback server at the station. Networking connections on the catch servers facilitate this, with of course the proviso that there is no guarantee that the internal file structure of the delivered file matches that used by the station.

MXF Standard

The solution to the problems of incompatible file formats is now provided by SMPTE in the Material eXchange Format (MXF) standard (SMPTE 377 and multiple other documents). MXF defines the wrapping and communication of information needed so it is possible to transparently exchange files between servers. It is a rich and complex set of standards, beyond the general reading of specialists in MCR implementation. However, content delivery providers are now starting to provide terminal devices with a networking port and MXF compliance to permit transfer of files directly without going to baseband.

An important characteristic of MXF is the transfer of metadata with the content file, which describes the content in detail. Because it has no physical media to fall back on, it is important that the file contains a full description of the material. As a minimum, metadata includes the length of each segment, starting time code, and descriptive information such as (for advertisements) the ISCI code (12-digit coding system promulgated by the Association of National Advertisers), spot title, and other information necessary to be sure the content is correct.

With long-form content, both the segment times and data on spots run in *barter positions* are often included. This information saves considerable effort in prepping content for playback by automation, since timing information does not need to be re-created. Passing this metadata directly from the delivery service to automation and Traffic software is a valuable part of the service from the station's viewpoint.

MXF is covered in more detail in Chapter 5.6, which includes a case study of content ingest by a network.

Transcoding between Formats

Another class of products, *content transcoding engines,* is also related to the content delivery process. These products allow content created on one manufacturer's server to be converted to use for another server, often automatically. Operationally, content arrives in the catch server and is placed in a directory that is visible to a network port. An external conversion system, which is in fact nothing more than a computer with specialized conversion software, determines that content has arrived and pulls it in, converts it to the output format, and either places it on another external server or puts it in a directory that acts as an "outbox." It does this without operator intervention, and content is converted seamlessly to the format required by the station. These products can be included in the catch server itself, permitting content to be delivered to the station's air server ready for air, although that software and hardware (if needed) are generally not included in the system as installed without cost to the station. This process is often referred to as *flipping*, a name derived from one manufacturer's product name.

ANALOG TECHNOLOGY IN THE TRANSITIONAL FACILITY

While many stations have converted their facilities to digital audio and video, some portions of the facility continue to employ analog technology. With proper conversion equipment, analog devices can, in fact, be useful long into the future although there are limitations that must be considered when judging the applicability of analog techniques.

Analog audio systems will have a longer life in all probability than analog video devices. High-quality audio analog-to-digital conversion hardware is not expensive, and back-and-forth conversions introduce few distortions. In some cases existing and perfectly serviceable analog hardware can fit well into the digital world by using converters. Audio cart systems, with analog outputs, can easily be used to feed digital air switcher and/or branding engine inputs. This saves unnecessarily replicating working hardware when implementing new systems. As noted previously, a mixed environment with analog and digital signals can easily be handled in a routing switcher with internal analog-to-digital and digital-to-analog conversion capabilities.

The same is not necessarily as true for video for two important reasons. First, high-quality video analog-to-digital conversions are relatively expensive, and low-quality conversions are obviously undesirable. Second, NTSC, with its interleaved color subcarrier and luma information, does not lend itself well to compression for emission where all MPEG encoding is based on component digital video. All NTSC signals can be decoded, as they must be for display, but the *footprint* of NTSC is visible on the component video, making the results of the conversion less than desirable. This is significant when up-conversion for transmission is needed, as it frequently is for HDTV. As a result, it is more appropriate to use analog video only in utility paths, such as monitors, rather than in the program path.

Another class of analog video hardware that can survive the transition so long as it is serviceable is the analog component videotape recorder with analog inputs and outputs. It is quite easy, and not expensive, to convert from component analog to component digital video, and the quality is acceptable. If the signal has never been encoded previously as composite NTSC, subsequent up-conversion and MPEG encoding can be done with little more loss of quality than if the digital signal were carried as serial digital SMPTE 259M.

SUMMARY

When taken as a whole, the topics relevant to master control operations add up to a confusing picture of an industry that has been mature for some years but has begun a forced transition. Because of the need in over-the-air broadcasting to separate its product from those of Cable, DBS, Mobile TV, and IPTV, broadcasters have been looking for those applications which could provide both differentiation and a "hook" to keep consumers tuned to broadcast television. HDTV and multicast DTV may provide the platform that can sustain broadcasters for some time, just as analog broadcasting has provided nearly 70 years of service to the television viewer. It is likely, though, that future broadcast systems will have to provide for the delivery of content to all comers and to all relevant delivery media. Not doing so would limit the size of the broadcaster's slice of the economic pie.

The implementation of DTV comes with the potential of technological obsolescence for analog techniques. Digital technology and, in particular, compressed systems hold the promise of new economical and multifaceted delivery systems to the consumer. Responding to this challenge will require an open mind as to just what master control needs to become in the future.

Bibliography

Broadcast Engineering Magazine: http://broadcastengineering.com/searchresults/?terms=master+control&ord=r

Kovalick, Al. *Video Systems in an IT Environment*, Burlington, MA: Focal Press, 2006.

Luff, John. *Centralized Broadcast Operations*, Burlington, MA: Focal Press, 2007.

Paulsen, Karl. *Video and Media Servers*, Burlington, MA: Focal Press, 2001.

Poynton, Charles. *Digital Video and HDTV.* New York: Morgan Kaufmann, 2003.

Poynton, Charles. *A Technical Introduction to Digital Video*, New York: John Wiley and Sons, 1996.

TV Technology Magazine:
http://www.tvtechnology.com/features/index.shtml

Whitaker, Jerry. *DTV: The Revolution in Electronic Imaging*, New York: McGraw Hill, 1998.

APPENDIX

TELEVISION NETWORK DISTRIBUTION TECHNOLOGIES

The following contributions from FOX, PBS, and CBS provide descriptions of the three main architectures currently adopted or planned by U.S. broadcasters for network television program distribution to local television stations. They cover methods for distributing standard definition and high definition video service and their associated audio. Some information is provided on arrangements at local stations where this differs significantly from traditional systems for NTSC.

FOX NETWORK PROGRAM DISTRIBUTION AND FOX SPLICING SYSTEM

Contributed by James M. DiFilipis

FOX Television

Introduction

In 2003 FOX began preparations for the delivery of high definition (HD) programs to its terrestrial affiliate network. Many factors went into the decision on delivery architecture, including the existing FOX Digital Satellite system, MPEG encoding equipment and MUXes, the existing satellite integrated receiver-decoders (IRDs), and the capability to overlay an HD infrastructure at the FOX Network Center in Los Angeles and to fully support the distribution patterns that the current satellite system supported. There also was concern about the cost borne by the affiliate stations so as to maximize the number of affiliates converting to HD.

It was determined that conventional baseband delivery of HD was not going to satisfy all of these requirements. The satellite system could support only 4 HD feeds in addition to the 16 SD feeds. The requirement was to support 16 HD and 16 SD feeds with the 4 C-Band transponders (at 72 Mbps per transponder). In addition, a minimum HD side chain for FOX affiliates was neither readily available nor cost effective. Further, flexibility was needed in the use of the ATSC multiplex, including full AC-3 5.1 audio, the addition of branding and local ID graphics, and the capability to add SD secondary programs (both local and network) and to integrate network and local PSIP information.

Because of these constraints, FOX decided that the system should deliver to the affiliate a "ready to air" emission-level HD MPEG-2 signal format. This approach had the advantage of reducing the bit rate per program, and with some reduction of the SD program bit rate, allowed for 16 program streams for both HD and SD.

MPEG Splicing Technology

The key technology to enable the HD pass-through at the affiliate station is MPEG splicing.

FOX researched MPEG splicing technology and wrote a state-of-the-art specification for MPEG splicing of video and audio, including additional features such as PSIP merge, data download, in-band splice control signals, and logo insertion (for branding bugs). No one company had the solution in hand, so FOX worked with a select number of manufacturers to refine the requirements to develop the appropriate technology.

In December of 2003, a contract was awarded to the Thomson/Terayon team. Thomson was the prime contractor and system integrator/installer, and Terayon was the splicer software/hardware developer. Triveni Digital was contracted to develop the head-end control/download software systems and the download software module for the Terayon splicer. Installations began in March 2004, while the software was still being refined and tested. An additional requirement came in February 2004 due to the announcement that FOX would broadcast the 2004 NFL (National Football League) season in HD. This required a change to the specifications to include dynamic program selection (in-band), which required modifications to both the BP5100 (Terayon splicer) and the current Wegener Unity 5000 IRDs. In addition, FOX replaced the then current MPEG-2 multiplex systems with more advanced MPEG-2 encoders (HD and SD) from Tandberg.

On September 12, 2004, FOX launched its HDTV service with a full day of eight NFL telecasts and three hours of prime-time programming. Sixty-nine stations were ready with the BP5100 splicing system and aired FOX in HD that day. Currently, there are over 135 FOX affiliate stations on the air in HD using the FOX Splicing System, as illustrated in Figure 5.21-A1.

Compression and Delivery

At the FOX Network Center in Los Angeles, there are five state-of-the-art MPEG-2 multiplexers, each of which can encode four SD and four HD programs.

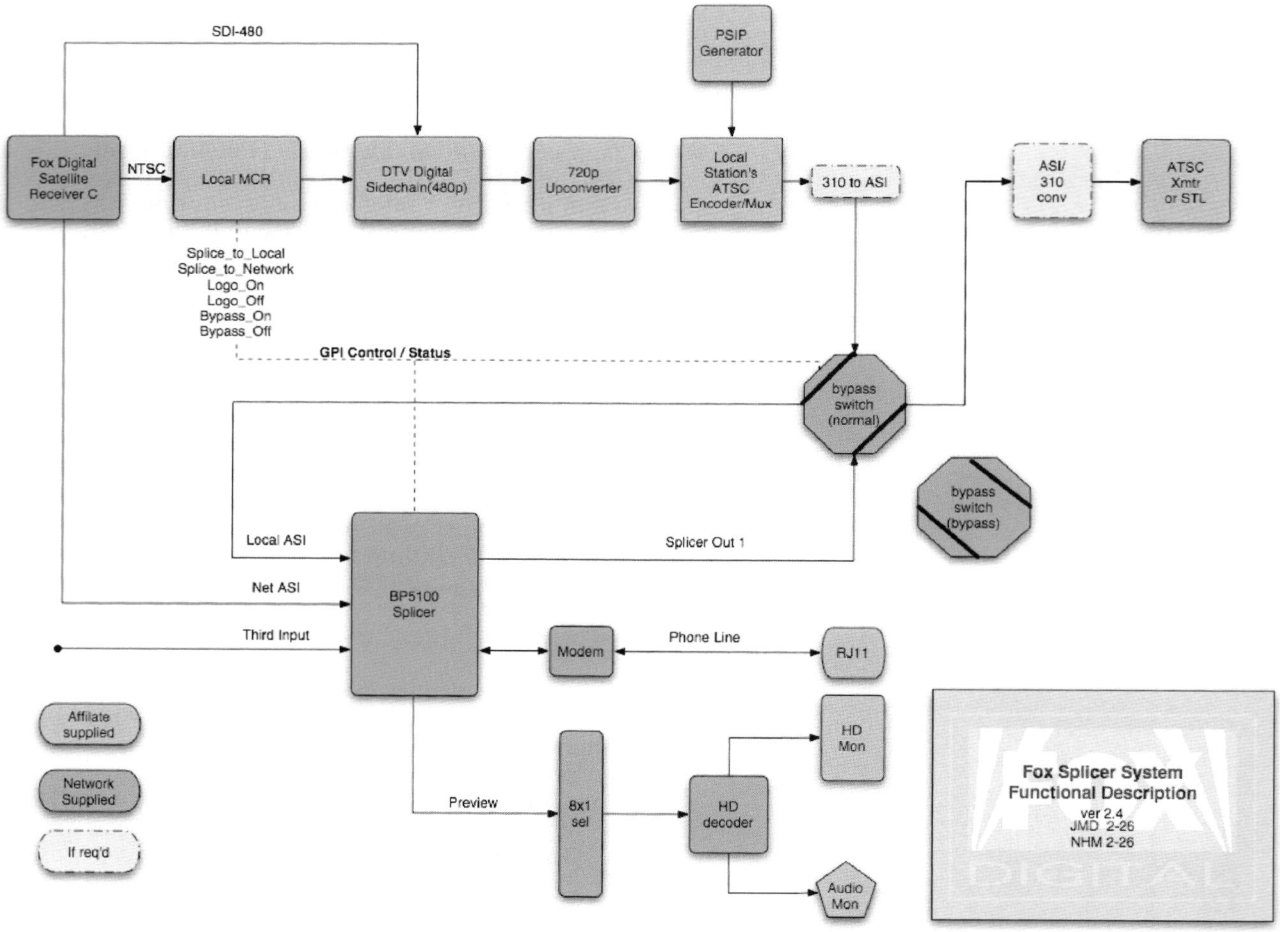

FIGURE 5.21-A1 FOX Splicing System diagram.

These multiplexers, supplied by Tandberg Television, use advanced statistical multiplexing as well as dual-pass MPEG-2 encoding for the best compression quality at a minimum bit rate. FOX distributes programming in both SD and HD with a maximum of 16 pairs of program content at the same time. The suite of four SD and four HD signals is statistically multiplexed, sharing the available bandwidth. The HD signal's maximum video bit rate is limited to keep it within the available ATSC payload. However, on average, the bit rate is far below the maximum, thus freeing up bandwidth for secondary services to be added by the affiliate station.

Each of four multiplexers (the fifth is used for testing and spare capacity) is assigned to a standard C-band satellite transponder (36 MHz bandwidth). Using advanced 8PSK modulation and precorrection/equalization, the transport payload is 72.73 Mbps. Each affiliate earth terminal consists of two 5 meter Andrews antennas using prime focus LNBs with low-phase noise. The IFL is engineered to minimize group delay and return loss from the LNBs to the satellite integrated receiver-decoders (IRDs). The minimum Eb/No for the system is 11.3 dB for quasi-error-free reception. Typical installations have Eb/No performance 3 dB above the minimum Eb/No.

At each affiliate, there are three Wegener Unity 5000 digital IRDs, which, under Compel (Wegener's network management system) control, select one of four transponders and one of four SD programs. They also output the complete MPEG-2 transport stream over the ASI output port, including the FOX special SCTE 35 extension "select_now" commands to inform the BP5100 splicer which one of four HD programs to "groom." The splicer "grooms" the selected HD program and, when commanded to go to network via a locally generated GPI contact, splices that FOX HD program into the transport that feeds the DTV transmitter or, in some cases, that is connected via fiber to a local cable company. The affiliate's local HD encoder (720p) is connected to the second input port of the splicer. The local input contains the local HD program, any SD programs (secondary), as well as PSIP data. The splicer is programmed to groom the SD programs to the main output while rate-shaping them to fit within available bandwidth.

The FOX Network SD program is decoded within the Unity 5000 and output in baseband (analog and SDI). The HD and SD program signals are in time with each other although the SD is baseband and the HD is compressed.

Bitstream Splicing

Splicing is controlled via the GPI inputs at the affiliate station. Typically triggered either by local tally from the MCR switcher or via station automation, the FOX BP5100, when commanded to splice, looks at the groomed stream for an exit point (P frame or truncated B frame) as well as an entry point in the new stream (I frame/start of GOP). The splicer adjusts the audio streams such that there is a minimum audio gap (<2 ms) at the video splice point (thus ensuring A/V synchronization). The splice time occurs within plus or minus half a GOP (group of pictures). For a GOP of 15 of an MPEG-2-encoded 720p60 (mp@hl), the splice occurs within 1/4 second (15 frames).

In order to synchronize the network and local streams as well as to fine-tune the splice time, there are delay buffers within the BP5100 on the network input. Also, the GPI controls the inputs to compensate for the delay of the local encoder/multiplexer path. Typically, the delay is 700–1200 ms, with a maximum delay of 2000 ms.

Logo Insertion

To provide local branding and station identification, the BP5100 can insert static graphic logos on-screen using partial decoding. The logos are controlled both by the local affiliate using a general-purpose interface (GPI) as well as via in-band commands (using a FOX extension to SCTE-35) from the network center. The BP5100 partially decodes the video only over the area of the image where the logo is to be inserted. The logo image is then "keyed" into the stream, and the BP5100 rate shapes the bit stream to prevent exceeding the available bit rate of an ATSC payload. There are two keyers in the splicer, and 16 logo files are available. Each keyer can select any of the 16 logos. The placement, transparency, and on/off times are set within the splicer configuration file or can be modified on the fly via the FOX SCTE-35 commands.

HD Audio

The FOX BP5100 can splice up to four audio services. Typically, FOX distributes two audio programs: a main audio service, which is a full AC-3-encoded, 3/2 (5.1) surround service at 448 kbps; and a secondary language service as an AC-3-encoded 1/0 mono at 64 kbps.

A special feature of the splicer called Audio Replication duplicates the main audio service at the output of the splicer if the stream being spliced to has fewer audio services supplied on the current output stream. Thus, if a local affiliate does not have a secondary audio service, when splicing from the network stream which has a secondary language service, the main audio of the local affiliate is replicated on the same PID assigned to the secondary language. Thus, viewers who are listening to the program's secondary language continue to hear audio with no loss of continuity.

The splicer does not alter any parameter of the AC-3 stream. The mode and audio metadata are transparently passed on to the consumer decoder unaltered.

Another audio feature is local audio breakaway. The audio breakaway allows an affiliate to splice to its audio program stream while still grooming the network video stream to the output.

PSIP

Program and System Information Protocol (PSIP) is an important ATSC feature that provides information to a DTV receiver for tuning of the station's virtual channels and programs as well as to provide the program guide information that is important for the viewer's experience of DTV. FOX wanted the guide information to be as fresh as possible, so Terayon was asked to add a feature called PSIP Merge. PSIP Merge processes the local PSIP data and can add information from the network PSIP tables following a set of rules. Primarily used only during network time, PSIP Merge adds information from the network TVCT (descriptors) as well as the EIT and ETT information to the PSIP tables. The BP5100 rebuilds the PSIP tables and outputs them in conformance with the ATSC specifications.

Control and Download

As part of the splicing system, it was necessary to provide an in-band control system as well as the capability to download files and new software. Working with Terayon and Triveni Digital, FOX designed an in-band control system that is based on the SCTE-35 Splicing Command standard used in cable distribution for control of splicing at head-ends. These commands provide dynamic control of the BP5100 such as selecting which network program to splice (select_now), turning on/off logos (keyer_on, keyer_off), and overriding commands (bypass_on). Terayon implemented these commands inside the BP5100, while Triveni Digital provided the head-end software (Station Manager). There is a dynamic addressing scheme that targets each splicer individually as part of the address field in the FOX SCTE-35 command set.

The download functionality is based on the ATSC A/90 download protocol. Triveni Digital provided the receiver code module inside the BP5100 as well as the head-end data download software (SkyScraper). Each file is encapsulated with the proper header and delivery information (address) and then put in a catalog for download via data carousel over the satellite MUX (1 Mbps bandwidth). FOX can target multiple stations for new software or individual stations for logo download.

MUX Secondary Programs

Another feature of the BP5100 splicing system used by FOX is the capability to insert additional program channels (nonprimary) into the output stream. In addition to grooming these channels, the splicer rate shapes them so as to fit within the existing residual bandwidth left over from the primary program (HD). Rate shaping is a process which alters the MPEG-2 quantization parameters to reduce the total bit rate of the encoded video. The BP5100's internal MUX processes and removes all null data, both within the null PIDs as well as within the video PIDs. This capability allows for up to two SD programs to be added during FOX prime-time programming and up to one SD and a static graphic channel (weather information, for instance) during sports programming.

The BP5100 maintains the correct ATSC transport bit rate (19.39 Mbps) at the output of the splicer, including PSIP and any other data services present.

THE PBS VIDEO INTERCONNECTION SYSTEM

Contributed by Thomas Edwards

Public Broadcasting Service (PBS)

Introduction

PBS operates the interconnection system used for the distribution of programming to noncommercial public television broadcast stations and networks in the United States. PBS delivers a variety of different real-time video streams to stations over satellite and is in the process of implementing non-real-time distribution of video files to stations using IP-over-satellite. See Figure 5.21-A2 for an illustration of the PBS Interconnection System.

The real-time feeds distributed by PBS include both packaged and nonpackaged feeds. Packaged feeds are intended for end-user continuous viewing, complete with interstitials filling all time between programs. These packaged feeds can be taken directly to air as-is. Nonpackaged feeds, such as the National Program Service (NPS), deliver programs individually with black, test color bars, and information slates carried between shows. Stations are expected to fill in the space between the shows in the unpackaged feeds with their own interstitial material.

Public television stations are independent organizations and thus can schedule their programming to meet their local needs. There are some requirements for "common carriage" of shows by PBS member stations at particular times, especially during prime-time hours. But outside those few requirements, stations have significant flexibility in their schedules. This means that stations often record programming to videotape or video servers for time-delayed playback. Rights windows on public television programming can be quite long, so rebroadcast of programs also adds to the heterogeneous nature of schedules across public television stations.

PBS Satellite Operations

PBS primarily uses four 36 MHz Ku-band transponders on the geosynchronous SES Americom AMC-3 satellite located at 87 degrees West longitude to deliver content to public television stations. Ku-band distribution was chosen based on the ease of siting and operating the smaller Ku-band dishes compared with larger C-band receive dishes used in the past. Today, it is difficult to obtain the physical space and zoning permission to install new C-band receiving dishes. C-band satellite reception also faces significant

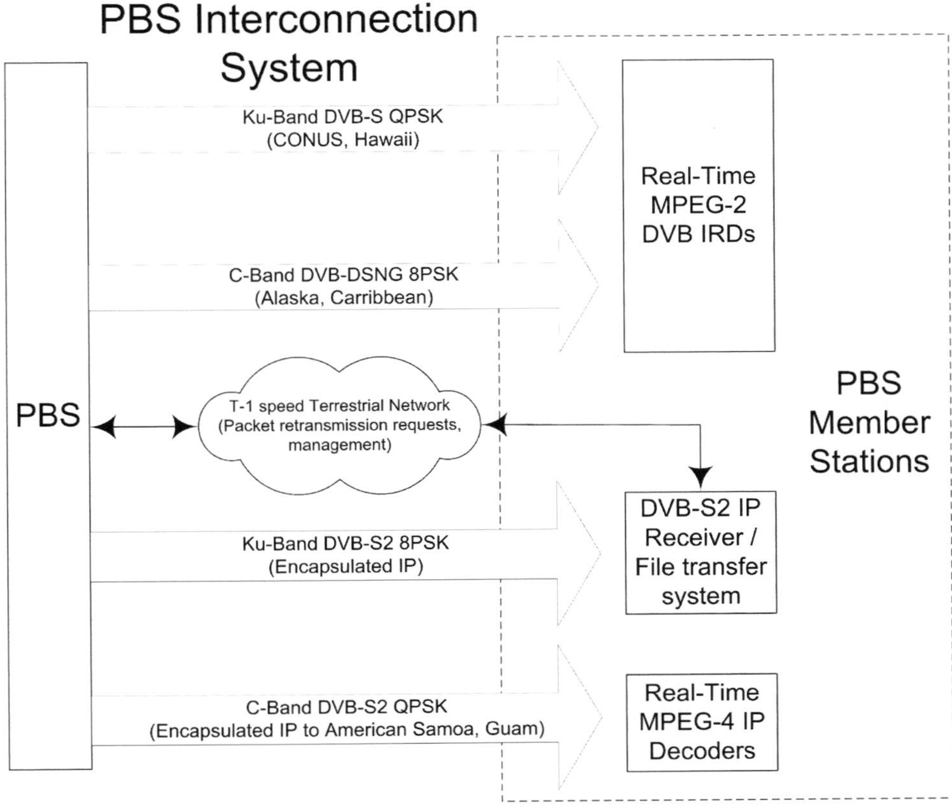

FIGURE 5.21-A2 PBS Satellite Interconnection System.

amounts of terrestrial microwave interference, especially in urban areas.

The disadvantage of Ku-band distribution is the susceptibility of the signals to attenuation due to precipitation, known as "rain fade." Extreme rain events may cause stations to lose the satellite signal briefly. Appropriate choices of downlink power, dish size, and modulation technique can make such outages relatively rare, but they do occur from time to time.

PBS uses DVB-S QPSK 3/4 rate modulation to deliver its real-time programming streams on Ku-band. DVB was chosen as the modulation technique because of its flexibility and the availability of transmit and receive equipment from multiple competitive vendors.

Program streams that originate from the PBS uplink in Alexandria, Virginia, are placed into a single multi-program transport stream (MPTS) carrier per transponder where possible. Having multiple carriers on a single transponder requires more power back-off from the transponder saturation point to avoid intermodulation noise. Thus, uplinking a single carrier per transponder provides for the highest power downlink signal and maximizing the rain fade margin.

The current PBS Ku-band satellite footprint does not cover stations in the Caribbean (including Puerto Rico and the U.S. Virgin Islands) or much of Alaska. Those areas outside the AMC-3 Ku-band footprint receive content from PBS using two 36 MHz C-band transponders on the SES Americom AMC-1 satellite located at 103 degrees West longitude. These C-band real-time program streams are modulated using DVB-DSNG 8-PSK 2/3 rate because of the reduced need for rain fade margin in C-band.

In 2008, it is expected that stations in all areas of the United States, with the exception of the Pacific Territories, will be served by four 36 MHz Ku-band transponders on a new SES Americom satellite with a wider footprint, AMC-21, which will be located at 125 degrees West longitude.

Public television stations in American Samoa and Guam currently receive programming from PBS on DVDs sent to them using standard commercial international shipping services, which take many days to arrive. PBS expects to begin delivering a portion of PBS video content to these stations on SES Americom AMC-23 located at 172 degrees East longitude using 3 Mbps of IP-over-satellite. It is expected that this link may utilize a video codec more advanced than MPEG-2 due to the bandwidth restriction.

In addition to the PBS main satellite uplink facility in Alexandria, Virginia, there are four additional regional satellite uplinks and several other occasional uplinking stations that also use the interconnection system to share SCPC carriers on PBS managed satellite transponders.

PBS Digital Distribution: Real-Time

On one of the 36 MHz Ku-band transponders, PBS uplinks up to nine SD channels in a single carrier MPTS with a data rate of about 40 Mbps. The SD programs are statistically multiplexed ("stat-muxed"), with variable bit rate (VBR) ranging from 1.2 Mbps to 8 Mbps per channel. A second transponder carries another single carrier MPTS with a constant bit rate (CBR) HD program of 18.8 Mbps, and up to four additional stat-muxed SD programs. A third transponder carries three single program transport stream carrier slots for use by non-PBS uplinks.

SD programs carry two 192 kbps AC-3 stereo pairs to include both main program stereo audio, second language, and descriptive video. HD programs carry a 384 kbps AC-3 5.1 multichannel primary audio stream and a 192 kbps AC-3 stereo pair for second language and descriptive video. Both HD and SD programs use MPEG-2 for video coding. Although newer video codecs may be capable of higher levels of compression than MPEG-2 at equivalent bit rates, these systems were not available during the initial deployment of DVB IRDs to public television stations.

The other advantage to the use of MPEG-2 video coding is its use in the ATSC digital television system for broadcast to viewers. New technology can provide transport stream splicing, branding, and digital video effects in the "compressed mode" of the MPEG-2 transport stream without requiring the full decoding of the video. It is also possible for the MPEG-2 transport streams to be carried within a station's plant in MPEG-over-IP using Gigabit Ethernet. This technology may provide a level of cost savings for some stations. At this time compressed-mode technology cannot currently provide all of the possible video effects processing that can be achieved by uncompressed video technology.

PBS Future File-Based Interconnection

A fourth Ku-band transponder will have a single carrier for non-real-time file transfer using IP-over-satellite. It is expected that this carrier will use either DVB-S2 8-PSK 5/6 or DVB-DSNG 8-PSK 2/3 rate modulation, depending on the availability and cost of DVB-S2 receiver technology during deployment of receiving equipment in 2007. The use of higher order modulation (and thus reduced rain fade margin) is possible due to the non-real-time nature of IP file transfer.

The majority of PBS program content is completely produced weeks or months before initial air date. Only a small number of programs requires real-time interconnection because these programs are produced live or arrive only a day or two before their initial air date. Therefore, most PBS content can be delivered in a non-real-time fashion using IP file transfer.

IP encapsulators that use multi-protocol encapsulation (MPE) place IP packets into an MPEG-2 transport stream for use in DVB applications. The IP packets used by PBS for file transfer are multicast UDP packets for the one-to-many unidirectional distribution. IP encapsulators typically have an Ethernet input for IP packets and deliver an MPEG-2 transport stream over ASI as an output. IP receivers are available as stand-alone units with L-band RF input and Ethernet outputs. Other IP receivers are simply PCI cards that reside within a PC and have an L-band RF input.

Note: Some care must be taken with IP multicast when it is carried over wired networks, as there is risk of accidental denial-of-service from improperly configured IP routing systems flooding the network with packets. Isolating LAN segments that carry multicast IP and having a good understanding of multicast IP routing are precautions to minimize this risk.

In IP multicast file transfer systems, file data is sent out as IP multicast UDP packets from a file transmitter system, and the packets are accumulated by file receiver systems over time. Once all required packets have been received successfully, the file is complete and can be played back by a video server.

It should be noted that the file transfer system "packet" may encompass one or more UDP/IP multicast packets, and each of those may be fragmented into multiple MPEG-2 188-byte transport stream packets during IP encapsulation.

On Ku-band, rain fade is a particular problem. Depending on modulation technique, rain fades can range from a few seconds to several minutes. To provide protection from packet loss, additional forward error correction (FEC) data packets calculated from the original file data are sent to provide replacement packets for those that might be missed due to rain fade or noise. This "file-level FEC" of IP multicast file transfer systems is different from and in addition to the FEC used in DVB modulation that is applied on individual MPEG-2 transport stream packets. The DVB FEC cannot correct data lost in long periods (several seconds) of rain fade, whereas the file-level FEC of the IP file transfer system can repair such loss, up to the limit of the number of additional FEC packets generated. Generally, additional FEC packets numbering from 2% to 10% of the number of packets in the original data file are sent for Ku-band satellite file transfer.

If packet loss in a multicast file transfer is beyond what can be corrected using the file-level FEC data, the file receiver system may automatically send a request to the file transmitter for the retransmission of missed packets through a lower bit rate back-channel. PBS will use a low-speed (DS-1 or slower) terrestrial back-channel connection for this purpose. This connection will also be used for SNMP monitoring and management of station-deployed file transfer equipment.

An advantage of sending files in non-real-time ahead of the initial broadcast date is that the FEC and packet retransmission mechanisms can assure the delivery of 100% correct files. Thus, video quality using a non-real-time distribution system is improved compared to that of a real-time distribution system that cannot correct for rain fade outages. In addition, higher order modulation techniques (such as 8-PSK) can be used for non-real-time distribution, as rain fade is less of a risk to the system. This allows for more data bits to be sent per symbol, reducing the amount of transponder space required for interconnection.

PBS intends to wrap video files in the SMPTE Material eXchange Format (MXF) wrapper, which will provide the capability to deliver rich metadata along with audio-video essence and also will enable the files to be more interoperable with video servers from a range of MXF-compliant vendors. If a video server is not MXF-compliant, the open and fully documented nature of the MXF file format will make it easier for video server vendors to specify file translation and transcoding solutions. PBS is currently producing an MXF Application Specification (AS-PBS) so that video server vendors can know what elements of the MXF standard will be utilized in the video files.

Non-real-time file distribution results in higher video quality and greater efficiency of satellite bandwidth use. It also provides an opportunity for stations to engage in efficient file-based workflows (if metadata that travels along with video essence files is accurately generated upstream in the production chain). Furthermore, file distribution provides the opportunity for stations to make greater use of less-expensive COTS (commercial off-the-shelf) IT elements for working with files within their plant for proxy viewing, archiving, and other processes.

CBS TELEVISION NETWORK DISTRIBUTION

Contributed by Greg Coppa

CBS Television

Introduction

CBS Television Network programming is distributed from its New York–based Broadcast Center where prime-time, news, live sports, and entertainment programs originate. Commercial and interstitial material are integrated with the programs at the Broadcast Center, and the resulting integrated programming is delivered via satellite to the CBS Network of owned and affiliated television stations for either immediate or delayed broadcast.

CBS distributes nearly its entire network schedule (news and some entertainment programming being the exceptions) in high definition (HD, 1920 × 1080I per SMPTE 274M) while simultaneously distributing its programming in standard definition (SD).

Distribution is digital, with MPEG compression and digital modulation techniques used to achieve efficient, reliable, and robust transmission performance.

Standard Definition Program Signal Description

A standard definition program delivered to the CBS SD Network is a high-quality, 4:2:2 chrominance-sampled video signal, which includes stereo and Secondary Audio Program (SAP) signals. The SD program also includes ancillary data services, EIA-608 captions with extended data services (e.g., V-Chip, Nielsen rating information, and internal network data with network and satellite switching schedules) carried in the Vertical Blanking Interval (VBI).

High Definition Program Signal Description

A high definition program delivered to the CBS HD Network is also a high-quality, 4:2:2 chrominance-sampled video signal and includes a Dolby E and stereo audio signal. The Dolby E signal contains two audio programs; the first is a 5.1 surround sound service, and the second is either a SAP signal or a stereo down-mix of the 5.1 surround signal.

Ancillary data services such as EIA-708 captions audio metadata (used to dynamically control the performance of an AC-3 audio decoder, typically found in a viewer's home), and, possibly, the Broadcast Flag are included with the HD program and are carried in the Vertical Ancillary Data (VANC) portion of the HD serial digital signal.

Standard Definition Program Multiplex

CBS delivers four SD programs in a digital multiplex which is distributed via satellite to the CBS SD Network. The four programs share a total bit rate of 40.5 Mbps, with the average bit rate for each program (video, audio, and data) equal to approximately 10 Mbps.

Each SD program's video signal is MPEG-2 compressed (ISO/IEC IS 13818-2, Main Profile @ Main Level) to reduce the video bit rate while maintaining high quality. Audio is sampled at 48 kHz using 20 bits and is compressed to a bit rate of 256 kbps. The compressed video and audio together with the ancillary data signals of each program are multiplexed using Motorola's DigiCipher™ product to form an MPEG-2 transport stream (ISO/IEC IS 13818-1) that is transmitted to satellite uplinks for distribution.

High Definition Program Multiplex

CBS can deliver up to six simultaneous high definition programs via satellite to the CBS HD Network. A single program is delivered at a bit rate of 45 Mbps, which includes the compressed video, audio, and ancillary data signals.

Each HD program's video signal is MPEG-2 compressed (ISO/IEC 13818-2, 4:2:2 @ High Level) to approximately 41 Mbps. The Dolby E signal is "passed through" at a bit rate of 2.3 Mbps, and the stereo audio signal is compressed to a bit rate of approximately 256 kbps. The ancillary data signals are multiplexed with the compressed video and audio signals to form an MPEG-2 transport stream (ISO/IEC IS 13818-1) that is transmitted to satellite uplinks for distribution.

Satellite Distribution

CBS uses C-Band satellite transmission to deliver the SD and HD multiplexes to its network of owned and affiliated television stations. The transmissions are uplinked to Intelsat Americas satellites 5 and 6 (IA5 and IA6), where CBS leases 36 MHz transponders.

The SD multiplex is transmitted using DigiCipher™ QPSK modulation at a symbol rate of 29.26 Msps and occupies an entire 36 MHz transponder. The HD multiplex is transmitted using DVB-S QPSK modulation at a symbol rate of 32.36 Msps and also occupies an entire 36 MHz transponder.

The SD and HD transport streams are delivered to two satellite uplinks that are located at geographically diverse sites. The two uplinks are connected to the Broadcast Center using fiber and microwave transmission paths.

Transmission Redundancy

CBS can simultaneously deliver up to four separate multiplexes with a distribution of up to 16 separate programs. Typically, a redundant operation is employed in which a main and back-up multiplex containing identical programs is delivered to two separate satellites and is received using separate antennas.

SD and HD audio compression encoders are configured with N:1 redundancy, and the multiplexers are 1:1 redundant, providing a highly reliable distribution architecture.

Receive System Description and Operation

The CBS Television Network consists of approximately 200 TVRO sites which receive programming distributed via satellite and are controlled by the New York–based Satellite Management Center (SMC). Of these 200 downlink sites, approximately 30% are not collocated with the TV station's studio. SD and HD signals from the non-collocated sites are delivered from the TVRO to the studio using digital microwave and digital or analog fiber interconnects.

In addition to the primary downlink sites, CBS services an additional 30 to 35 TV stations where the TVRO is not under SMC control. Finally, CBS services approximately 35 TVRO sites located in Canada, the Caribbean, and Mexico that are within the satellite's transmission footprint.

TVRO earth stations are equipped with two satellite receive antennas for primary and secondary program reception. The primary antenna is a steerable, 7.3 meter dish which receives the IA6 satellite-delivered programming. The secondary antenna is a fixed 4.5 meter dish, receiving signals from the IA5 satellite. Both antennas can receive horizontally and vertically polarized feeds.

Each earth station is equipped with a micro-controller-based Digital Earth Station Controller (DESC) system. The DESC is a custom software application that schedules and controls both SD and HD receive system functions such as steering antennas, tuning receivers, and switching baseband audio and video. Additionally, the software monitors equipment status and, in the case of equipment failure, will provide an alternate path for audio, video, and data signals.

The DESC receives its schedule and control information from ancillary data that is transmitted as an MPEG private data service contained in the SD multiplex. The data originates from the CBS SMC and is available on all SD-transmitted satellite signals. An SD satellite receiver supplies the data to the DESC controller via an asynchronous serial connection.

SMC can access a station's DESC system through a dial-up modem connection. This allows SMC personnel to interrogate equipment status and to download schedules in the event of a problem with the satellite-delivered data.

A control terminal, located at the receive station, allows local personnel direct access to the system. This can be used for maintenance purposes or to generate and execute, during non-CBS Network programming periods, a local schedule to receive syndicated or other non-network programming as required.

Real-Time Operation: Transponder and Satellite Switching

Operationally, two receivers each for SD and HD are always active. One is tuned to a primary transponder, whereas the other is tuned to a secondary transponder.

The SD and HD systems are controlled using a schedule that originates from SMC. These schedules contain information that allows different program streams to be selected as required for regional commercial sales, program time shifting, or regional sports coverage. For the most part, transponder or satellite switching is scheduled and downloaded in advance. In addition, the system is capable of real-time *direct* command switching, which is used to reconfigure the distribution patterns during live programs.

Switching of transponders may occur at a local station break during programming between a group of commercials. Tuning of the receivers to change transponders is controlled by the DESC.

The distribution network is broken up into basic regions, with regions added or subtracted according to programming and sales requirements.

Generally, a station's receivers will be tuned to one transponder for the duration of a program. In this case, receivers are tuned during local station breaks, which is time allocated to local station's material during network programs. This operation is referred to as a "Station Break" switch. After the change is initiated, the primary and/or back-up receivers are instantly tuned to a new transponder. Since this switch typically takes place during a 64-second station break, the glitch or momentary video discontinuity generated will not be seen on air.

The other type of switch occurs within a program segment between a group of commercials. This operation is referred to as "Clean Switch." In order to perform this switch, the back-up receiver is first tuned to the new transponder, and the SD and HD video

switcher is switched in the vertical blanking interval to select the program signal to be delivered from the secondary receiver. The primary receiver is retuned to a new transponder, and the video is again switched to re-establish the program signal from the primary receiver.

The CBS Satellite Management System allows control of the primary TVRO earth station either based on a schedule downloaded in advance or under real-time control. Typically, a 24-hour schedule is downloaded the night before it is required. Dynamic changes to the switching schedule are permitted, thus allowing the schedule to be updated or held minutes before a switch.

CHAPTER

5.22

ATSC Encoding, Transport, and PSIP Systems

MATTHEW GOLDMAN
TANDBERG Television, Inc.
Bedford, New Hampshire

RICHARD CHERNOCK
Treveni Digital Inc
Princeton, New Jersey

CHRIS LENNON
Harris Corporation
Colorado Springs, Colorado

INTRODUCTION

This chapter covers the implementation of equipment and systems for producing compressed digital television (DTV) bit streams[1] in accordance with the ATSC Digital Television Standards (see Chapter 1.11). The first part of the chapter covers the video encoder, the transport multiplexer, system solutions, and next generation video coding technology. Next, there is discussion of issues related to the implementation of Program and System Information Protocol (PSIP), followed by the use of Programming Metadata Communication Protocol (PMCP) and the role of the various sources of PSIP metadata. The chapter ends with a summary of the PSIP requirements stated in the FCC rules.

Other issues related to DTV encoding and transport systems are covered elsewhere in this handbook. Video compression is covered in Chapter 5.7, while AC-3 audio is discussed in Digital Audio Data Compression, Chapter 3.7, and Audio for Digital Television, Chapter 5.18. Closed captioning is covered in Chapter 5.24 and data broadcasting standards and applications are discussed in Chapters 1.17 and 5.25, respectively.

OVERVIEW

Encoding and multiplexing are central parts of the ATSC digital television (DTV) system, which is described in the ATSC Digital Television Standard, A/53 [1]. A generic block diagram of the encoding and multiplexing subsystem in a typical broadcast station is shown in Figure 5.22-1, with basic components comprising the video and audio encoders, a transport stream multiplexer, a PSIP generator, and (optionally) a data server. The output of the multiplexer is fed to the 8-VSB modulator that is associated with the over-the-air transmitter.

The main interface between encoders and the transport multiplexer conforms to the ATSC transport layer

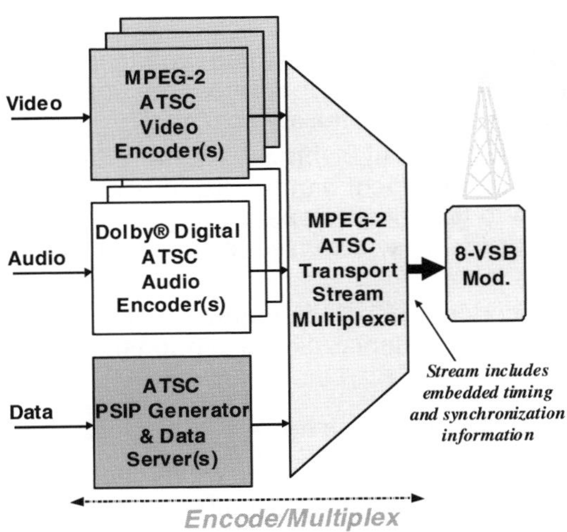

FIGURE 5.22-1 Encoding and multiplexing subsystem in a typical broadcast station.

[1]This chapter shows "bit stream" as two words, following ATSC A/53 convention.

as defined in A/53. The ATSC *transport layer* is based on the MPEG-2 *Transport Stream* (TS) format, as defined by the MPEG-2 Systems standard [2]. The link and physical layers are defined by the *DVB Asynchronous Serial Interface* (ASI) [3]. In typical implementations, the video and audio encoders and the ATSC transport multiplexer each create output bit streams in the TS format. The TS format provides a mechanism to encapsulate and multiplex coded video, coded audio, and generic data into a unified bit stream. It includes timing information in the form of *time stamps* in order to enable the real-time reproduction and precise synchronization of video, audio, and data (as necessary).

In order to facilitate parsing of the information contained within the bit stream, in-band control information, known as *Program Specific Information* (PSI), is also defined. PSIP expands upon the PSI to provide comprehensive channel "tuning" and program guide information as discussed later in this chapter. The TS format also was designed to facilitate real-time transmission and reception of DTV over error-prone physical transmission paths, in particular, over-the-air broadcasting.

For a detailed explanation of the theory behind the ATSC transport and compression layers, see *Proceedings of the IEEE, Special Issue on Global Digital Television,* January 2006 [5].

VIDEO ENCODER

When specifying or reviewing an MPEG-2 video encoder, there are many factors to consider:

- What is the image quality, particularly at the lower bounds of practical bit-rate usage?
- Are the necessary input and output interfaces supported?
- Is multichannel audio encoding supported internally or is an external audio encoder required?
- How many audio services are supported (to be encoded or pass-through from an external audio encoder)?
- Will the encoder be used in a single broadcast service solution or will multiple services be broadcast in the same 6-MHz transmission channel?
- If multiple services will be broadcast, does the encoder support efficient bandwidth utilization; in particular, will the encoder operate as part of a statistical multiplexing system (see section later in the chapter)?
- Does the encoder support the required ancillary data carriage (for instance, closed captions)?

Video Compression Performance

Arguably the most important factor when choosing a video encoder is the picture quality it produces. Unlike reviewing a checklist of, for example, input/output interfaces, the factors that impact video compression performance are often the least understood and, to many people, the most mysterious. Particular areas to consider are preprocessing (including noise reduction), motion estimation, and rate control [4].

Since bandwidth is a limited commodity, the performance of an MPEG-2 encoder is usually tested at the lowest possible bit-rate. While it is true that some encoders fail in more noticeable ways than others at the low end of practical data rates, differences between encoders may show up under a number of different test conditions. Some encoders might have difficulty with particular types of motion, others with noise, others with scene cuts, and so on.

The picture quality produced by an MPEG-2 encoder depends on many factors that include:

- The quality of the original video source material;
- Preprocessing operations;
- Encoding architectures and algorithms;
- System parameters (pixel rate, bit-rate, and number of services in the transmission channel, in the case of multiservice systems);
- Size and quality of the display device.

The encoder, however, has to produce the highest picture quality of the compressed video bit stream for any set of operating parameters and for any given source material.

Preprocessing

The preprocessing functions of an encoder typically include:

- Picture resizing;
- Noise reduction;
- Noise level detection;
- Film mode (or 2-3 pull-down) detection;
- Forward analysis.

Picture Resizing

At low bit rates, reducing the detail (pixel rate) of the source is a commonly used mechanism for trading between image sharpness that contains visible MPEG *coding artifacts* and a softer picture that reduces or eliminates visible MPEG coding artifacts. The most common MPEG-2 coding artifact that occurs when there are insufficient bits to encode the detail is known as *blockiness* or *macroblocking* (see Figure 5.22-2).

Most encoders provide a large number of resolution subsampling options. Common picture resizing is to reduce the horizontal resolution by one-quarter or one-third, so:

- 1920 × 1080 (1080i) → 1440 × 1080 or 1280 × 1080
- 1280 × 720 (720p) → 960 × 720
- 720 × 480 or 704 × 480 (480i/p) → 544 × 480 or 528 × 480

If too much detail is removed, the picture will lack contouring and edge detail, commonly known as being "too soft."

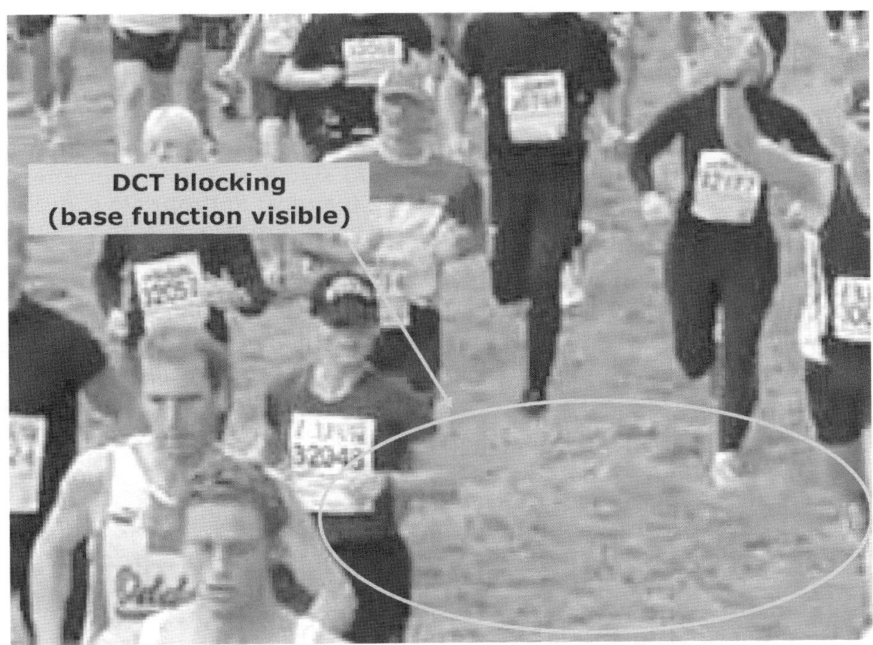

FIGURE 5.22-2 An example of the macroblocking artifact. (See CD for color version of this and other figures.)

As such, the *coding efficiency*—picture quality versus bit rate used—is one of the most important parameters that need to be tested and established for an encoder.

Noise Reduction

Most video sources contain a small amount of noise. While it may not be visually very annoying, even a low level of source noise may add a significant encoding overhead that increases the required bit rates. This is because noise, by its nature, is random and usually contains substantial high-frequency components that do not compress well. When an encoder does not recognize noise as such, it will treat it as a high-detail component of the image and will waste valuable bandwidth by attempting to encode it properly. This behavior becomes even more significant in cases where the compressed bit stream is statistically multiplexed with other compressed video sources, as discussed later in the chapter. In that case, instead of reducing its bit rate for input sequences that should be simple to code (say, with little motion and low detail), an encoder with a noisy source will continue to require relatively high bit rates even if the picture material itself (without the noise) could have been compressed further.

Measurements have shown that even low levels of source noise can increase the bit rate demand of encoders almost by a factor of two [6]. Figure 5.22-3 shows the bit rate demand of a noisy source (30 dB SNR) with and without noise reduction as a probability density function. In this example, noise reduction reduces the most likely bit rate from 5 to 3 Mbps.

Properly designed noise reduction algorithms reduce the visibility and the effects of noise without introducing picture artifacts themselves. As a preprocessing function to compression, it is particularly important that the processes of noise reduction, followed by compression, do not in themselves lead to undesirable artifacts. Examples of such artifacts are *motion blur* or *double imaging* (both commonly caused by excessive temporal noise reduction). In the case where the source material contains high levels of noise, it is generally better to add an element of spatial noise reduction while keeping the amount of temporal noise reduction at a moderate level. If such a filter is overapplied, the image will become noticeably soft and lacking in detail.

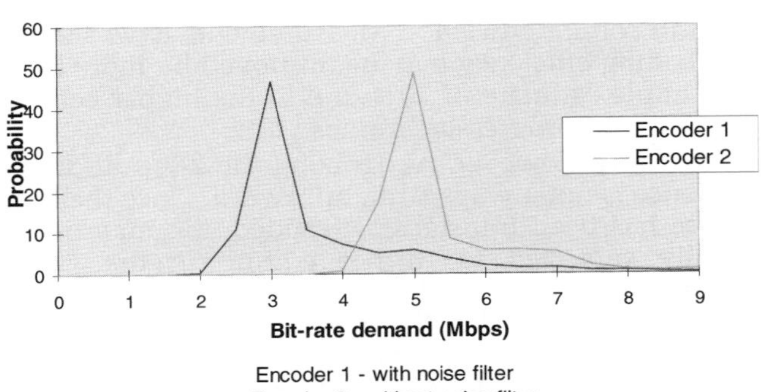

FIGURE 5.22-3 Bit rate required with and without noise reduction for a noisy source.

Noise Level Detection

As noted, an encoder must be adaptive in how it applies the noise filters or poor image quality will result. Therefore, even more important than the basic capability to reduce the effects of source noise, is the requirement to measure the amount of noise present on the source material so the proper type and amount of filtering may be applied. Noise levels may vary over a wide dynamic range, from extremely low levels of noise (typically expected at the output of a digital studio) to relatively high levels (such as from older, archived material or material that has already undergone a series of recordings or compression/decompression cycles).

A studio or broadcast center typically contains or passes content material from a myriad of different sources. Most noise-reduction algorithms, when applied to clean sources in full strength, will reduce picture quality. Since, in real-time broadcast operation, it is not practical for an operator to manually change the configuration of an encoder for each input source, the encoder itself must be adaptive and vary the level of noise reduction in accordance with the level of source noise present at the input. The flexibility and effectiveness of noise processing are therefore very important functions to test and verify.

Film Mode Detection

A considerable proportion of broadcast material originate from film. Since the compression efficiency for film material is much higher than for interlaced material (due to the lower frame rate, the fact that progressive frames are easier to code than interlaced, and for 60-Hz systems, the elimination of duplicate video fields), it is important to detect the presence of film-originated input and change the compression algorithm accordingly.

For film transfer to 60-Hz video, the well-known 2-3 pull-down method is used to generate 60-Hz video fields from 24-Hz film frames. In this method, two and then three fields are pulled down from successive film frames, thus generating a video sequence that contains two repeated fields for every five video frames. By detecting and eliminating the repeated fields, the encoder is able to reconstruct and compress the original film sequence, resulting in a significant improvement in coding efficiency. Measurements have shown that coding efficiency may be improved by more than the simple estimate of 20% due to the higher coding efficiency on progressive frames.

In reality, however, recognizing the 2-3 pull-down sequence is not always straightforward. Once the film source has been transferred to video, the material is usually subjected to further editing in the video domain. As a result, the 2-3 pull-down sequence, or *cadence*, which would otherwise be readily detectable, is disrupted and encoders must implement a more involved algorithm to reliably detect the sequence. An encoder's ability to detect and eliminate repeated fields in heavily edited film-mode material is therefore another aspect to test and verify.

Forward Analysis

In addition to the noise level and film mode, there are a number of other parameters that may be automatically measured at the input of the encoder to help with the compression process. Statistical analysis of the input signal assists the compression engine in selecting the most appropriate coding modes for a particular piece of content. A superior encoder will measure field dominance as well as field and frame picture activity to select between field and frame picture coding. It will also detect scene cuts and fades to select coding modes and to give advance warning to the rate control algorithm.

An extreme case of forward analysis involves doing a full encoding pass, measuring the efficiency of the encoding, and passing parameters such as image complexity and picture type to a subsequent compression engine. Typical industry terminology for this is "look-ahead processing" or "multipass encoding." This is not only helpful for single service encoding, but also greatly improves the efficiency of statistical multiplexing for multiservice encoding.

Motion Estimation

One of the most crucial components that determine the overall quality of a real-time MPEG-2 encoder is its motion estimation engine. Performance criteria of motion estimation algorithms encompass more than just the total search range explored for the motion displacement vectors, and the ability to track true motion is less significant than one might expect. What is important is the reliability of the motion vector and prediction mode to produce an overall (global) minimum prediction error.

Motion estimation is computationally intensive. In order to achieve a large enough search range, therefore, motion estimation is often carried out in a hierarchical manner using a down-sampled, lower resolution picture. Down-sampling reduces computational complexity because pixel displacements in the down-sampled image correspond to larger displacements (and therefore a larger search range) in the original image. Hierarchical motion estimation is not without its disadvantages, but the details are beyond the scope of this book.

Exhaustive motion estimation is an alternative to hierarchical, while still keeping computational complexity in check. A smaller search range is used but assuming the search range is still adequate for all prediction modes, exhaustive motion estimation results in the best possible prediction as it always finds the global minimum of prediction errors. The search range required depends on the specific content, with high-motion-sports programming having the largest potential range. However, the vast majority of motion vectors has been shown to be in a relatively small search area [7]. Therefore, the reliability of the motion estimation is more important than the size of the search area. Note also that fast motion is typically less difficult to compress because of camera blur and the human

visual system's lack of fine detail response during fast motion.

An encoder that does a good job at motion estimation will maximize picture quality for a given bit rate; hence, this is another important factor that impacts the overall performance of an MPEG-2 encoder.

Rate Control

Without rate control, the bit rate generated by an MPEG-2 encoder may vary over more than three orders of magnitude. Black frames require less than 70 kbps to code whereas white noise may generate instantaneous bit rates exceeding 100 Mbps. Rate control is required to convert this very large variation into a compliant constant bit rate (CBR) or variable bit rate (VBR) bit stream, with a known (and often restrictive) bit rate cap. To achieve this, the rate control algorithm must react quickly on both ends of the scale. At low bit rates, stuffing bits must be put into the bit stream to prevent buffer underflow. At the maximum allotted bandwidth, the rate control algorithm must be able to reduce the bit rate such that buffer overflow is prevented under all conditions.

In practice, the large variation in bit rate is due to a number of factors, including:

1. The type of picture material (film, sports, etc.)
2. Temporal correlation and predictability (fades, scene cuts, etc.)
3. Global scene criticality
4. Type of motion (transversal, zoom, rotation, etc.)
5. Picture type (I, P, or B frame)
6. Spatial complexity variation within pictures
7. Macroblock coding type (intra, forward predicted, etc.)

To achieve a fast response to changes in the video signal, the encoder needs to analyze the input signal at the earliest possible stage and forewarn the rate control about significant statistical changes in picture criticality, noise level, and other parameters. This is often done in high-end encoders by using dedicated forward analysis processing. The most accurate results are achieved by using a full-function "look-ahead" or multipass encoder before the actual encoding of the video signal. Information obtained from both the initial forward analysis of the video signal and the "look-ahead" encoder is passed on to the final encoder in order to optimize compression parameters and to obtain an accurate estimate of the required bit rate.

Audio Encoding

Two-channel (stereo) Dolby® AC-3 audio compression is typically included within a video encoder unit because it greatly simplifies system configuration and audio/video sync. In most cases, multiple stereo services (for example, multiple languages associated with the same video service) may be supported through the addition of optional additional audio cards. Some manufacturers are also beginning to offer 5.1-channel AC-3 audio compression as an option in an integrated video/audio encoder unit.

In most cases, however, where local 5.1-channel audio encoding is needed, an external multichannel audio encoder is used. In such situations, the audio encoder must be locked to the video encoder in order for A/V sync to be maintained. This is typically done by sending an AES/EBU clock reference from the video encoder or the station clock source to the external audio encoder (note that this is separate from locking the video encoder to the station reference clock, which is often done as well). In addition, the delay through the external audio encoder must be compensated for in the video encoder unit so that A/V sync will be aligned. Typically both the external audio encoder and the video encoder have configuration adjustments so that proper lip sync may be obtained.

Interfaces

The following is a summary of typical interfaces on a video encoder:

- *Video Input*
  - — Serial digital interface (SDI) (SMPTE 259M)
  - — High-definition serial digital interface (HD-SDI) (SMPTE 292M)
  - — Analog component/composite
- *Audio Input*
  - — Analog balanced
  - — Analog unbalanced
  - — AES-3 digital
  - — Note: Audio may be embedded in the video SDI/HD-SDI (as per SMPTE 291M)
- *Caption Input* (CEA 608 and CEA 708)
  - — Serial SMPTE 333M

    Note: "Grand Alliance" interface also supported on some equipment

    Note: Captions may be embedded in the video SDI/HD-SDI (SMPTE 334M)

    Note: Analog SD captions (CEA 608) may also be carried in video Line 21
- *Other Inputs*
  - — H-Sync (for synchronization with a station master clock)
  - — Asynchronous Serial Interface (ASI) (ETSI EN 50083-9 Annex B), if an internal remultiplexer is included
  - — Serial interface (RS-232 or similar) for control system
  - — 10/100Base-T Ethernet or similar for control system
  - — Voltage-level sensitive general purpose interface (GPI) for control system
  - — Feedback control channel for statistical multiplexing (typically Ethernet)

- *Outputs*
  - — Asynchronous Serial Interface (ASI) (ETSI EN 50083-9 Annex B) (typically more than one for redundancy support and confidence monitoring)
  - — AES/EBU clock reference (for connection to external multichannel audio encoder)
  - — Synchronous Serial Interface (SSI) (SMPTE 310M)
  - — IP/Ethernet (for direct connection to an IP network, for scenarios that do not use ASI)
  - — Control channel for statistical multiplexing (typically Ethernet)

TRANSPORT MULTIPLEXER

When the ATSC system was first envisioned, it was assumed that the DTV channel would comprise a single high definition (HD) service. However, standard definition formats were included in the standard, and their reduced bit rate requirements allow multiple SD services to be carried simultaneously. Today, various combinations of HD and SD (and also data) services are being delivered by broadcasters in a single DTV channel. This is made possible with the transport multiplexer.

The *transport multiplexer* (also known as a *mux)* is responsible for receiving the compressed video bit stream, compressed audio bit streams, PSIP bit stream, associated data (synchronous, asynchronous, and synchronized) bit streams, and independent data bit streams, packaging them up into a single multistream transport, and ensuring that streams that require synchronization among them are correctly aligned and do not exceed established bit rate or buffer size parameters, and then deliver (in a real-time stream of bits) the multistream transport to a channel modulator or other network interface.

The transport multiplexer's role can be best illustrated by understanding the ATSC transport subsystem. Figure 5.22-4, from ATSC standard A/53, illustrates that the ATSC transport subsystem resides between the video and audio encoders (application encoders in the figure) and the video and audio decoders (application decoders in the figure), both at

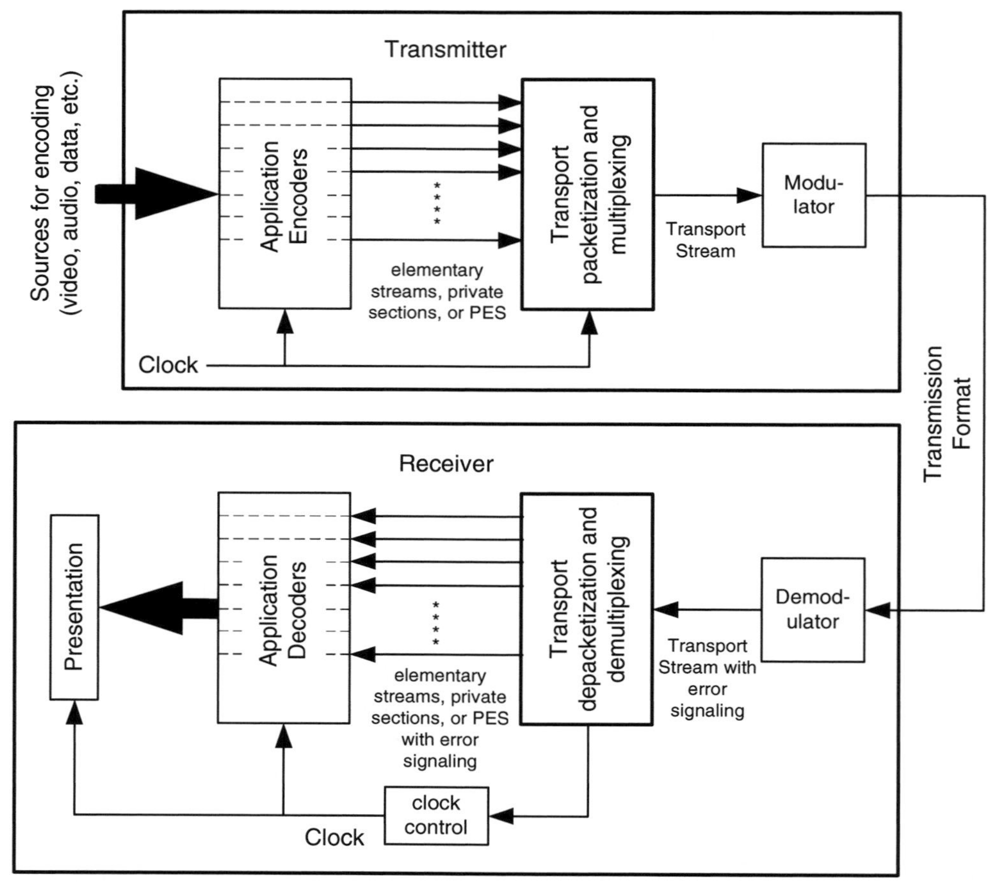

FIGURE 5.22-4 A functional overview of the ATSC transport subsystem. (Courtesy of ATSC.)

the transmission site and the consumer receiver location. The transmit site's transport subsystem is responsible for formatting the coded bit streams and multiplexing the different components of the program for transmission. The receiver's transport subsystem does the inverse function, recovering the coded bit streams to pass them to the appropriate decoder and for the corresponding error signaling.

The transport multiplexer is a physical device that incorporates the functions depicted as transport and packetization multiplexing in the figure. Note, however, that transport packetization and multiplexing are also implemented in a typical video encoder because the TS format is the *de facto* standard for interconnections between encoders and transport multiplexers. The difference is that the transport multiplexer receives TS-formatted bit streams from many sources. In essence, a transport multiplexer is both a multiplexer of data to be put into the TS format and a remultiplexer/resynchronizer of TS-formatted data from encoders and other devices that choose to use the TS format as an interconnect.

In a simple system, a remultiplexer function may be implemented on a single card and that card may be placed inside a video encoder. In such cases, the video encoder may perform almost all the functions of a "station in a box": video encoding, multiple audio encoding, and remultiplexer for data and PSIP.

Two-Layer Multiplexing

One way to visualize the TS multiplexing function is to consider it as a combination of multiplexing at two different layers. In the first layer, a *single service Transport Stream*[2] is formed by multiplexing packets from audio, video, and in-band control information, known as *program specific information* (PSI). This typically occurs within a video encoder. In the second layer, one or more single service Transport Streams are combined to form a *service multiplex* of Programs (also known as a Multi-Program Transport Stream (MPTS) in the MPEG-2 Systems standard, and a Digital Television Standard *multiplexed bit stream* in the ATSC standard). This occurs in the transport multiplexer (remultiplexing).

Data streams—such as a data application in its own right or PSIP data—may enter the transport mux in one of two ways: either as real-time streaming bit streams (already packetized into the TS format by the data server or PSIP generator) or as a typical file transfer. In the latter case, the transport multiplexer is responsible for packetizing, multiplexing, and synchronizing with other services within the multiservice TS, if applicable. In the former case, the data stream is remultiplexed into the multiservice TS in a similar fashion as done for video/audio services entering the transport multiplexer. The transport multiplexer is responsible for handing timing/synchronization of packets, the correctness of the stream syntax (including in many cases the content of the PSI, although in some cases this is done by the PSIP generator) and the overall output rate control.

[2]It should be noted that a "service" in MPEG-2 terms is known as a "Program," but this is different from a television or network program; see the ATSC and MPEG standards [1] and [2] for a complete description.

Conceptually, the TS may be represented as a large communications pipe containing one or more smaller pipes, as discussed in [5]. Each smaller pipe represents a single service (such as a DTV minor channel—see PSIP description later in the chapter). Figure 5.22-5 shows an example where the TS carries a single service, but this concept can easily be extended to multiple services within the TS by adding additional smaller pipes (each one with a different program number from program number 2000 shown in the figure). Each service (MPEG-2 Program) comprises one or more elements, which may include video, one or more audio (such as multiple languages), and data streams.

The program specific information (PSI) provides information about the contents of the Transport Stream. For ATSC streams, the PSI comprises the Program Association Table (PAT), which lists all the Programs (services) in the multiplex; the Program Map Table (PMT), which identifies the elements that make up each program; and the Conditional Access Table (CAT), which provides information relating to scambled programs.

Basic Transport Stream Structure

Figure 5.22-6 illustrates the packet structure hierarchy of an MPEG-2 Transport Stream. A video encoder that encodes both video and several audio services produces a set of elementary streams, one for each video and audio. Each elementary stream is segmented into a series of packetized elementary stream (PES) packets as shown in the figure. The PES packets, in turn, are further segmented into fixed-length TS packets to facilitate multiplexing and transmission in real time. Multiplexing of an MPEG-2 Program is the process of interleaving the TS packets of all the elementary streams (and other section data produced by the encoder, such as PSI) that make up the MPEG-2 Program into a single unified bit stream, while maintaining timing synchronization of each elementary stream. At the next layer, the multiplexing of multiple MPEG-2 Programs into a single multiservice multiplex is the process of interleaving the TS packets of more than one MPEG-2 Program into a unified bit stream, called a *multiprogram* TS (MPTS). The MPTS construct enables the deployment of practical, bandwidth-efficient digital broadcasting systems, with each service capable of being delivered at an independent, variable bit rate from other services within the overall fixed bit rate of the MPTS.

The packet identifier (PID), contained in the header of each TS packet, is the key to sorting out the components or elements in the TS. The PID is used to locate the TS packets of a particular component stream within the service multiplex in order to facilitate the reassembly of the payload of each TS packet back into its higher-level constructs; that is, TS packets into PES packets and PES packets into an elementary stream. A series of TS packets containing the same PID include

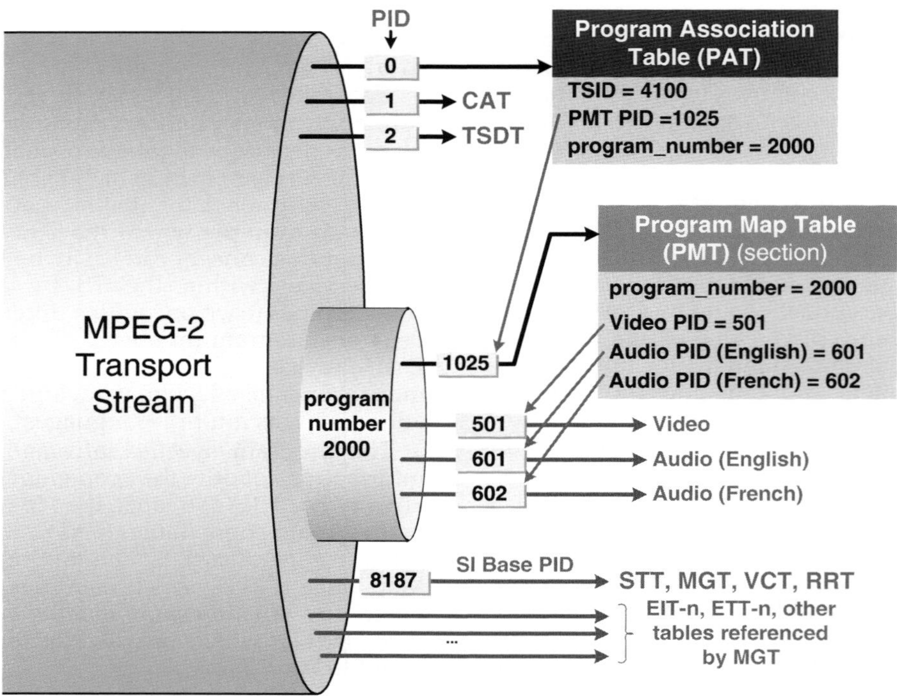

FIGURE 5.22-5 An illustration of a Transport Stream showing two-layer multiplexing. (Courtesy of IEEE.)

either a single program element (for instance, a video elementary stream), or descriptive information about one or more program elements (for instance, a PSI table).

An MPEG-2 Transport Stream is a continuous series of TS packets as shown in Figure 5.22-6. A TS packet is 188 bytes in length and always begins with a 4-byte (including the synchronization byte) TS packet header. The remaining 184 bytes are available to carry up to 184 bytes of TS packet payload.

A transport multiplexer may be configured for a fixed set of services (MPEG-2 Programs), with a fixed set of streams for each service. However, in practice, encoders and multiplexers are controlled by a sophisticated control and management system, so that high-availability solutions (automatic switching to backup or redundant components in case of a failure) and service flexibility (such as the changing of one HD service to multiple SD services) are possible. In addition, in

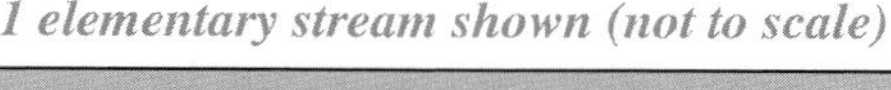

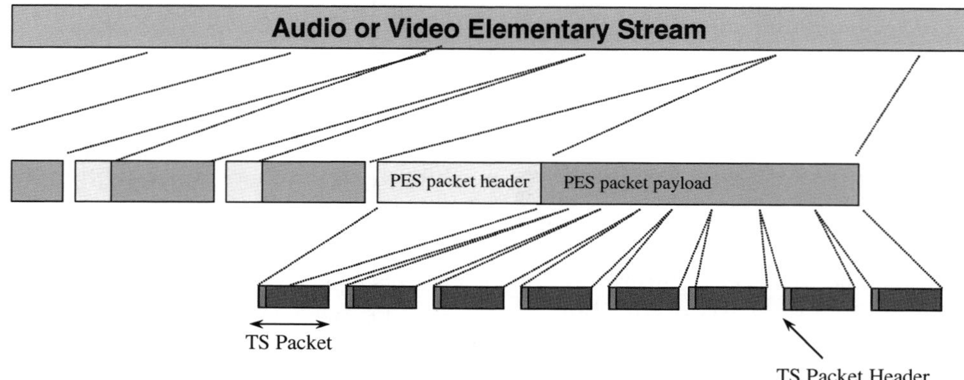

FIGURE 5.22-6 Packet structure hierarchy of a Transport Stream. (Courtesy of IEEE.)

the case of multiservice systems, there could be hundreds of parameters that need to be provisioned and using a control and management system to handle these operations reduces human error and eases changes in configurations.

Statistical Multiplexing

The ATSC transport is limited to 19.39 Mbps in a 6-MHz transmission channel, which was originally intended to provide sufficient bandwidth for an HDTV service. However, today many stations are either broadcasting or planning to broadcast multiple services in a single transmission channel—either using multiple SD services or a combination of HD and SD. As such, a mechanism is needed that will use the available bandwidth in the most efficient manner. Closed-loop *statistical multiplexing* is the most efficient mechanism for combining multiple services into a single transport (refer to Figure 5.22-7).

The quality measure is a very important part of a statistical multiplexing system. It should reflect the picture quality that will be perceived by the viewer, and this can be difficult to achieve. The quality measurement in the encoder and the bit rate allocation algorithm in the multiplexer form a closed-loop system. This must be capable of responding quickly to changes in the criticality of the video input, for example, at a scene change. Using a variety of parameters, each encoder computes the video quality and complexity and forwards this information to the transport multiplexer. The multiplexer, in turn, informs each encoder how much bit rate is available to it over a specific time period (typically adjusted on a video frame-by-frame basis) based on an evaluation of each encoder's parameters and service priorities defined. Each encoder then sets its output bit rate to the exact amount allocated for exactly the period allocated. The multiplexer then combines all the streams together to create an MPTS that, in the case of an ATSC 6-MHz transmission channel, is 19.39 Mbps.

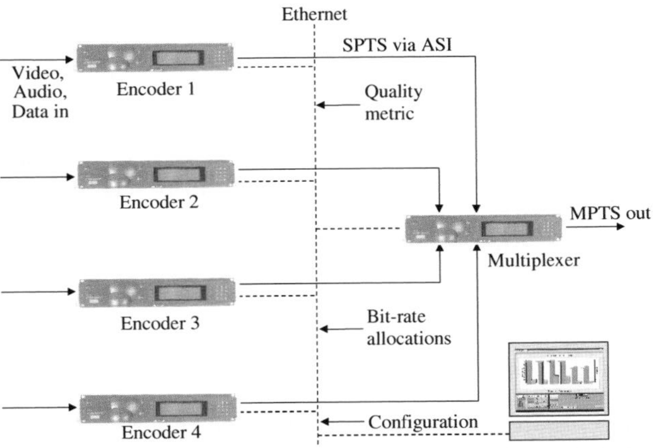

FIGURE 5.22-7 A 4-channel statistical multiplexing system. SPTS = Single Program (service) Transport Stream. MPTS = Multi Program (service) Transport Stream.

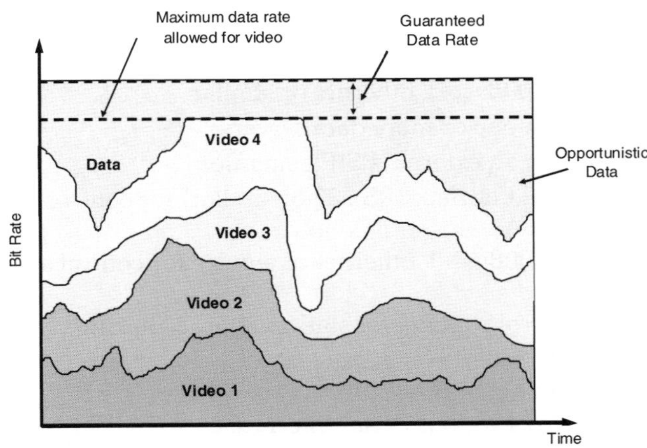

FIGURE 5.22-8 An example of statistical multiplexing within an MPEG-2/ATSC transport multiplex. (Courtesy of IEEE.)

Data services may be combined with the TV services in the same manner. Both "best effort" (*opportunistic*) and fixed data rates may be assigned. See Figure 5.22-8 for an example of a multiservice statistical multiplex that includes four video services, and opportunistic and fixed data services.

Statistical multiplexing works because MPEG-compressed video, by its nature, is variable bit rate for each frame. In addition, different bit rates are required to encode different types of content. Mostly still pictures that consist of large areas with relatively little motion and/or little detail (such as scenes of landscapes with blue sky or news commentators) are easiest to code, whereas fast sports material is hardest. There are also short-term variations due to scene changes. By statistically taking advantage of the mixture of low complexity and high complexity content, instantaneous bandwidth may be reallocated to the service that needs it. Statistical multiplexing systems typically also allow priority to be placed on certain content to prevent high complexity in a secondary service from "stealing" bandwidth from a primary service when both call for additional bit rate.

Statistical multiplexing is difficult to do well. If multiservice systems are highly desired or a requirement, the resultant picture quality of all the encoders in a statistical multiplex should be evaluated closely.

Interfaces

The following is a summary of typical interfaces on a transport multiplexer:

- *Inputs*
  - —Asynchronous Serial Interface (ASI) (ETSI EN 50083-9 Annex B) (multiple)

    From one or more video encoders

From one or more data servers
From an external PSIP generator
- 10/100Base-T Ethernet or similar
From one or more data servers
From an external PSIP generator
- Serial interface (RS-232 or similar) for control system
- 10/100Base-T Ethernet or similar for control system
- Control channel for statistical multiplexing (typically Ethernet)

- *Outputs*
  - Asynchronous Serial Interface (ASI) (ETSI EN 50083-9 Annex B) (typically more than one for redundancy support and confidence monitoring)
  - Synchronous Serial Interface (SSI) (SMPTE 310M)
  - IP/Ethernet (for direct connection to an IP network, for scenarios that do not use ASI or SSI)
  - Feedback control channel for statistical multiplexing (typically Ethernet)

ENCODING AND TRANSPORT SYSTEM SOLUTIONS

Broadcast requirements differ from station to station, and there are many possible encoding and transport solutions or system architectures to address those needs. In general, however, these may be grouped together into a few common system solutions that will address most of the station's needs. Variations on these common scenarios are possible. The three main types of system architectures are:

- *Basic:* Single HD service, no redundancy, simple control and management
- *Mid-level:* Multiservice, automated redundancy, dedicated control and management system
- *Advanced:* Multiservice with statistical multiplexing, automated redundancy, and flexible/high-function control and management system

The basic architecture covers the simplest of broadcast scenarios, where there is no requirement for system redundancy (controlled fail-over to backup hot standby components) and there is a strong desire to keep the system simple to operate and maintain.

The mid-level architecture is typically used by small or mid-sized stations that want to broadcast a combination of HD and SD services, and have a requirement for system redundancy (automated fail-over to a hot standby component).

The advanced architecture is typically used by large stations or those that want the highest level of flexibility and performance. It supports a combination of HD and SD services, automated redundancy for high availability, and closed-loop statistical multiplexing for the most efficient use of service bandwidth possible.

Basic System Architecture

The basic system architecture as shown in Figure 5.22-9, is for the simplest scenario where there is no requirement for system redundancy or statistical multiplexing. Configuration is handled using front-panel controls or a simple Web browser loaded on a colocated personal computer (PC); a dedicated control and management platform is not needed or desired in this scenario.

The system solution supports one encoder (HD or SD) with a built-in multiplexer card that supports an external PSIP generator. Optionally, two additional SD encoders can be supported as well since the multiplexer card has multiple inputs. A SMPTE 310M (SSI) output is also available in addition to ASI.

An Ethernet switch is used to interconnect all the devices to a local PC so that users can launch the graphical user interface (GUI) for any Web browser available on each device.

A separate AC-3 encoder has been included in the diagram to represent the case where the video encoder does not include an integral multichannel (5.1 surround) audio encoder and local multichannel encoding is required. The audio encoder is fed into the audio input interface of the HD encoder.

For closed caption handling, a translator unit is shown. This is to illustrate the required configurations at encoder level for closed captions handling in its possible standards and formats, when such a device is required; that is, for CEA-708 DTV captions via a separate SMPTE 333M or "Grand Alliance" interface, or in any case where captions are not embedded in the HD-SDI or SDI using SMPTE 334M.

Mid-Level System Architecture

The mid-level system architecture as shown in Figure 5.22-10, is designed for a small or mid-sized station that wants to broadcast a combination of HD and SD channels, has a requirement for automated redundancy, but does not need statistical multiplexing. It provides easy-to-use configuration and alarm monitoring via GUIs using a dedicated control and management platform.

The system solution supports 1+1 (one on-line and one hot standby) HD encoder, 2+1 SD encoders, 1+1 transport multiplexer with PSIP insertion and SMPTE 310 (SSI) output, and associated redundancy switching. A dedicated control and management system is used to configure and monitor the system, and automatically controls redundancy.

An Ethernet switch is used to interconnect all the devices to a local PC. The dedicated control system GUIs are launched from the PC.

A separate AC-3 encoder has been included in the diagram to represent the case where the video encoder does not include an integral multichannel (5.1 surround) audio encoder and local multichannel encoding is required. The audio encoder is fed into the audio input interface of the HD encoder.

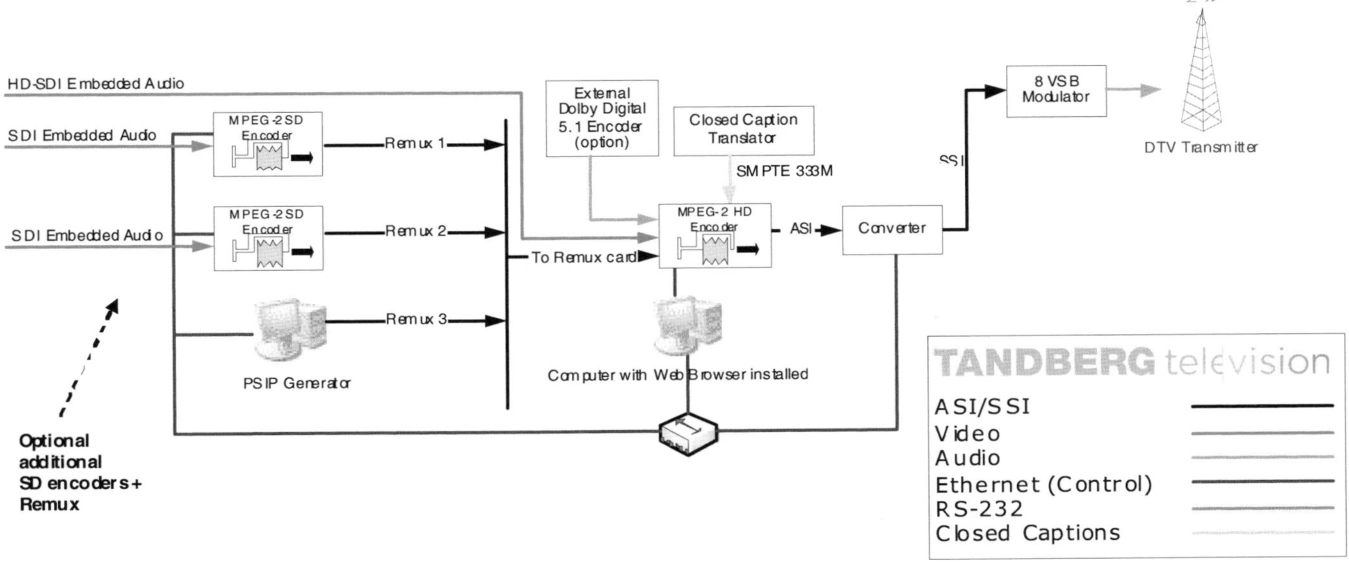

FIGURE 5.22-9 Basic encoding/transport system architecture. For color figure detail, please refer to the included CD. (Courtesy of TANDBERG Television.)

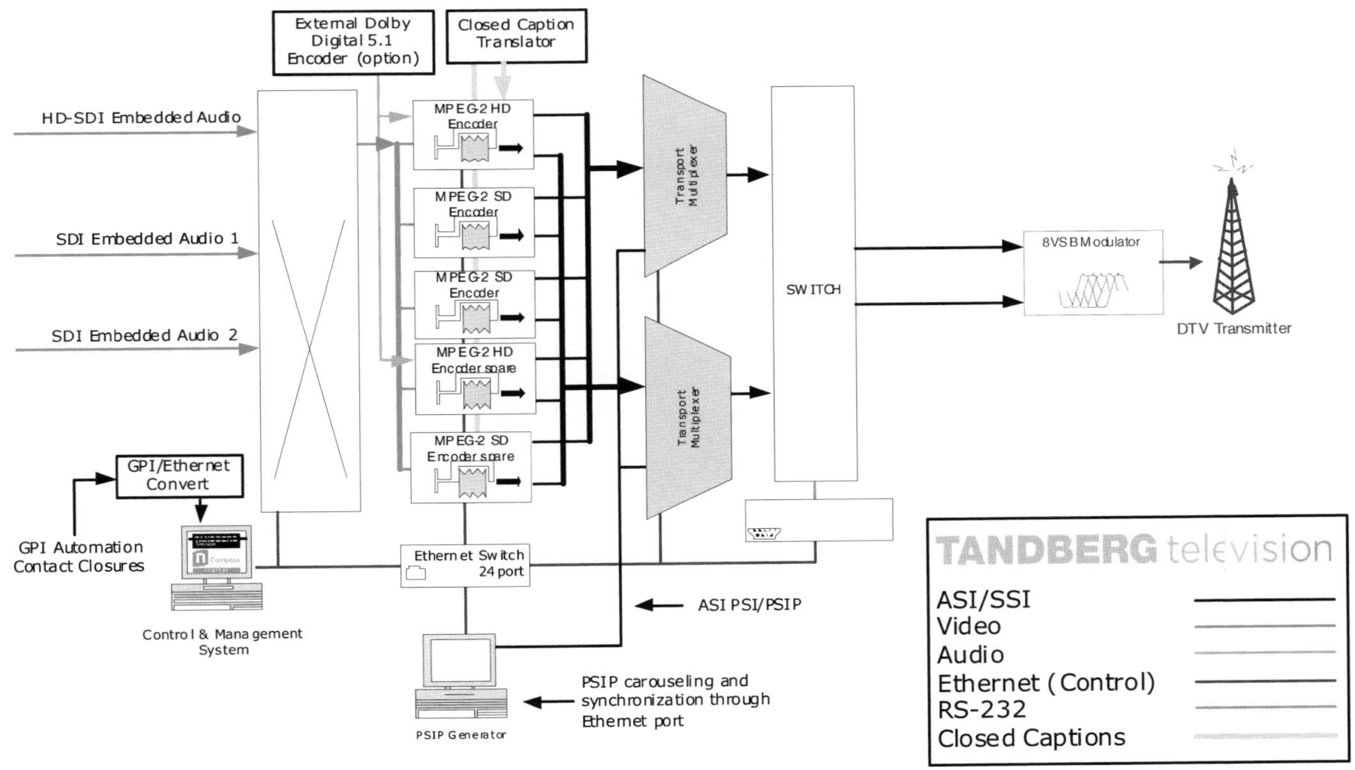

FIGURE 5.22-10 Mid-level encoding/transport system architecture. For color figure detail, please refer to the included CD. (Courtesy of TANDBERG Television.)

For closed caption handling, a translator unit is shown. This is to illustrate the required configurations at encoder level for closed captions handling in its possible standards and formats, when such a device is required; that is, for CEA-708 DTV captions via a separate SMPTE 333M or "Grand Alliance" interface, or in any case where captions are not embedded in the HD-SDI or SDI using SMPTE 334M.

Advanced System Architecture

The advanced system architecture as shown in Figure 5.22-11, uses a dedicated comprehensive control and management system to handle and configure the devices involved in the Transport Stream processing. An advanced control system typically provides intuitive interfaces, sophisticated "n+m" redundancy capabilities, easy-to-download and back up physical maps representing equipment interconnections, profile files that represent TS configurations (for example, PID assignments, stream types contained within the TS, and other configuration data), and the ability to run the configurations live (on-line) or in a test/demo mode.

The advanced system architecture includes the highest-end solution for video compression, which may include features such as multipass or "look-ahead" encoding and statistical multiplexing of the services/channels in order to optimize picture quality per bit rate across all services.

The inputs to the encoders are handled by a router that is both HD-SDI and SDI compatible with embedded SDI audio. However, in situations in which external digital or analog audios are a requirement, an audio router could be included to handle their input to the system. The encoders are set up as 1+1 HD and 2+1 SD, although many other configurations are possible.

A separate AC-3 encoder has been included in the diagram to represent the case where the video encoder does not include an integral multichannel (5.1 surround) audio encoder and local multichannel encoding is required. This encoder could also be the input to a digital audio matrix if needed.

For closed caption handling, a translator unit is shown. This is to illustrate the required configurations at encoder level for closed captions handling in its possible standards and formats, when this device is required; that is, for CEA-708 DTV captions via a sepa-

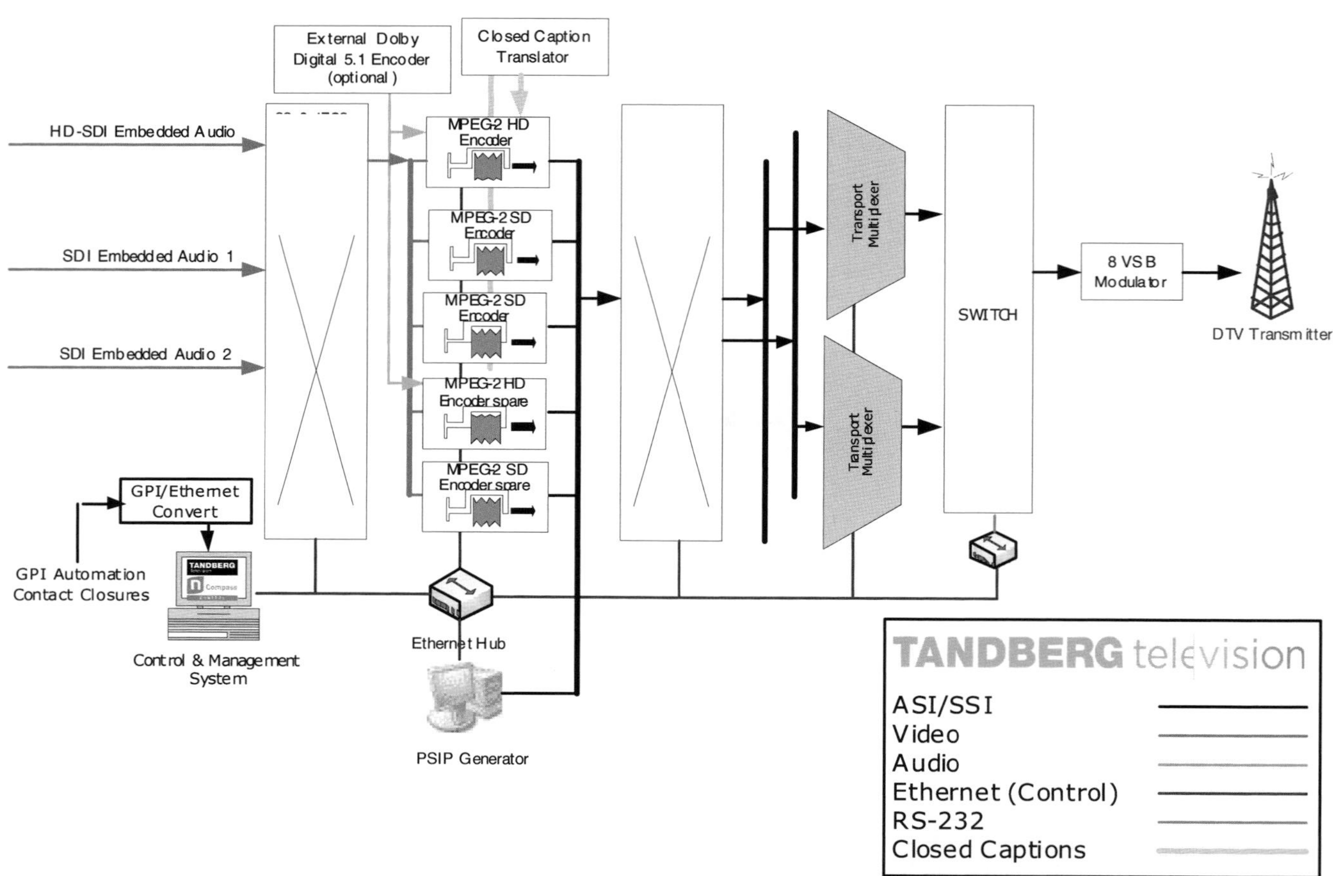

FIGURE 5.22-11 Advanced encoding/transport system architecture. For color figure detail, please refer to the included CD. (Courtesy of TANDBERG Television.)

rate SMPTE 333M or "Grand Alliance" interface, or in any case where captions are not embedded in the HD-SDI or SDI using SMPTE 334M.

For multiplexing purposes, two transport multiplexers are shown in a 1+1 redundant configuration. Each transport multiplexer contains two multiport ASI input cards and one SMPTE 310M SSI output card.

The PSIP generator could either feed the multiplexers via Ethernet or via one of the ASI ports of each multiplexer.

At the output stage, a switch is used to control the SMPTE 310M Transport Stream outputs to the RF system section.

The Ethernet switch keeps all the devices connected in the same network, and the translator server converts the RS-232 control protocols into TCP/IP for those devices that require it.

NEXT GENERATION VIDEO COMPRESSION

The MPEG-2 Video standard [8] is the ubiquitous video compression technology used today for digital terrestrial broadcasting, digital cable, direct-to-home satellite, and digital versatile disc for video (DVD-video). The standard was finalized in October 1994, and was the state-of-the-art at that time when memory was approximately $50 per megabyte, silicon wafer fabrication technology was greater than 1 micron, and typical microprocessor clock rates were well below 500 MHz. While many of the mathematical algorithms used were known years earlier, the technology for practical encoder signal processing and compact, cost-effective receiver implementation limited what could be achieved.

More than 12 years have passed and the state of technology has leapt ahead. At the time of writing (2006), memory is now less than $50 per gigabyte (and still falling), silicon wafer technology is less than 0.1 micron, and processor clock rates now exceed 2 GHz and perform many more functions in parallel. For the first time since MPEG-2 Video, two new systems are now available that significantly decrease the amount of bandwidth needed for broadcast-quality digital video: MPEG-4 AVC [9] and SMPTE VC-1 [10].

Application and Usage

At the time of writing, the new codec technologies are not approved by the FCC or the ATSC for use in the main, free, over-the-air television broadcast service in the United States. MPEG-2 Video is still required because existing digital TV receivers do not support the next generation compression technologies. However, there is interest in using next generation coding for backhaul, contribution, robust VSB transmissions (such as Enhanced VSB or A-VSB), alternative pay TV services, and mobile applications. As such, future digital TV receivers may include these new compression technologies.

Because of the different organizations involved, the industry is using various terms to describe the same technology standards. Figure 5.22-12 explains the names in use.

• **MPEG-2** • **H.262**	The ubiquitous video codec standard used digital television today – Terrestrial Broadcasting, Cable, DBS, DVD-V
• **MPEG-4 Part 2** • **MPEG-4 SP/ASP**	A follow-on video codec standard – not widely used for DTV (*therefore not discussed further here!*)
• **MPEG-4 AVC** • **H.264** • **MPEG-4 Part 10** • **AVC** • **"JVT"**	Advanced Video Coding (AVC) – A next generation video codec standard jointly developed by ISO/IEC MPEG and ITU-T VCEG
• **SMPTE 421M** • **SMPTE VC-1** • **Windows Media™ Video 9 (WMV9)**	A next generation video codec initially developed by Microsoft and now a SMPTE standard

FIGURE 5.22-12 Terminology primer for various compression technologies.

Improvements over MPEG-2 Video

In a similar way to MPEG-2 Video, both next generation codecs are organized into Profiles and Levels to define specific interoperability points. A Profile is used to specify the exact set (or subset) of algorithmic tools used and a Level defines constraints on those tools. As Figure 5.22-13 shows, only certain Profiles are applicable for broadcast-quality video such as that required for terrestrial and satellite broadcasting and cable television. Both the terminology itself and the Profile usage have caused some industry confusion as potential users attempt to compare video quality of what they believe are encodings made by the same technology but in fact are not. Examples include digital cameras and World Wide Web video streaming applications to PCs.

First-generation implementations of both MPEG-4 AVC and SMPTE VC-1 have realized 30% to 50% compression efficiency gains over MPEG-2 Video. As with MPEG-2 Video, continual refinements of real-time

Application	*MPEG-2 Video (H.262)*	*MPEG-4 AVC (H.264)*	*SMPTE VC-1 (Windows Media™ Video 9)*
Mobile Devices Video Conferencing Internet Streaming	-	Baseline	Simple
Broadcast Quality (TV, Cinema, IPTV)	Main Profile SD: Main Level HD: High Level	Main Profile SD: Level 3 HD: Level 4	Main (progressive displays only)
		High Profile SD: Level 3 HD: Level 4	Advanced Profile SD: Level 1 HD: Levels 2-3

FIGURE 5.22-13 Profiles and Levels versus application for video codecs. (Courtesy of IEEE.)

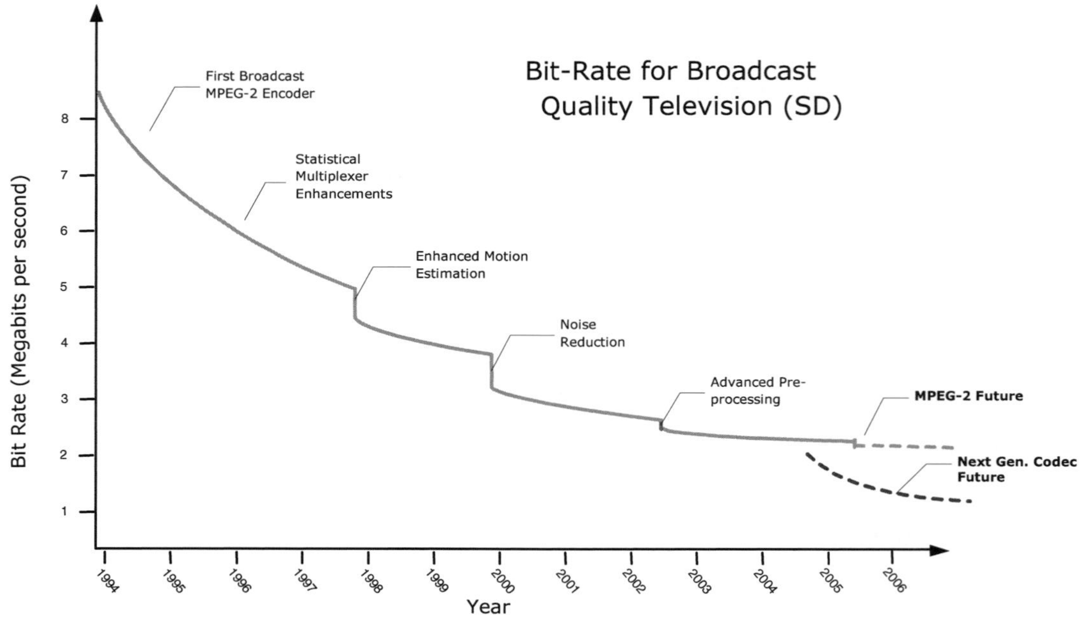

FIGURE 5.22-14 Comparison of SDTV picture quality for MPEG-2 and next generation codecs. (Courtesy of IEEE.)

implementations will almost certainly occur over the next few years. In 1994, the state-of-the-art for real-time full ITU-R SD resolutions was 8 to 8.5 Mbps (see Figure 5.22-14). With refinements in algorithmic implementations, advanced preprocessing, technology advances, and statistical multiplexing, this has been reduced to under 3 Mbps for the same picture quality. Most experts believe that the MPEG-2 Video improvement curve is near its asymptotic theoretical minimum. For next-generation compression technologies, SD rates for similar picture quality start at under 2 Mbps today and may drop below 1 Mbps within the next few years.

With HD content, bit rate reduction is even more dramatic as an amount of consumed bandwidth per

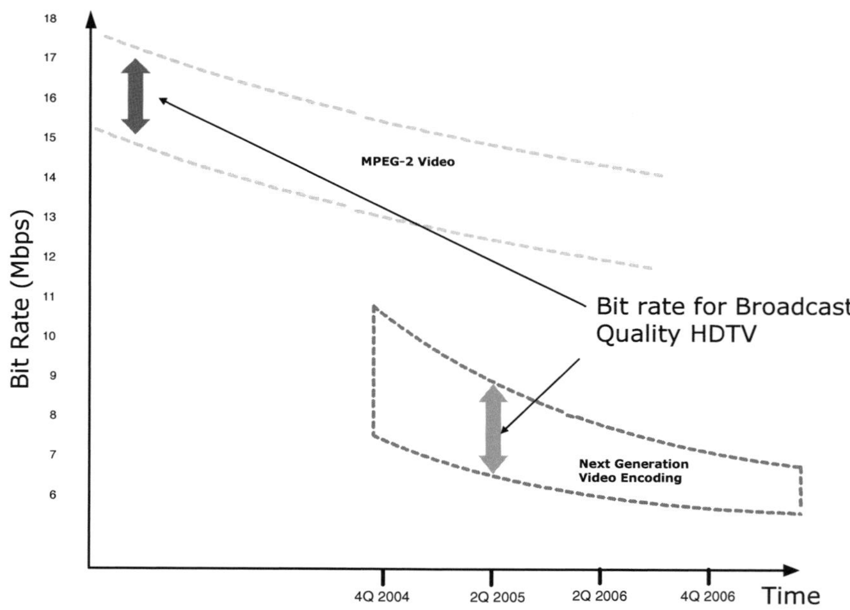

FIGURE 5.22-15 Comparison of HDTV picture quality for MPEG-2 and next generation codecs. (Courtesy of IEEE.)

Tool	MPEG-2 Video (H.262)	MPEG-4 AVC (H.264)	SMPTE VC-1 (Windows Media™ Video 9)
Intra Prediction	-None: MB encoded -DC predictors	- 4x4 Spatial - 16x16 Spatial - I_PCM	- Frequency domain Coefficient
Picture Coding Type	**- Frame - Field - Picture AFF**	**- Frame - Field - Picture AFF - MB AFF**	**- Frame - Field - Picture AFF - MB AFF**
Motion Compensation Block Size	- 16x16 - 16x8, 8x16	- 16x16 - 16x8, 8x16 - 8x8 - 8x4, 4x8 - 4x4	- 16x16 - 8x8
Motion Vector Precision	**- Full Pel - Half Pel**	**- Full Pel - Half Pel - Quarter Pel**	**- Full Pel - Half Pel - Quarter Pel**
P Frame Feature	- Single Reference	- Single Reference - Multiple Reference	- Single Reference - Intensity Compensation
B Frame Feature	**- 1 Reference Each Way**	**- 1 Reference Each Way - Multiple Reference - Direct & Spatial Direct Modes - Weighted Prediction**	**- 1 Reference Each Way**
In-Loop Filters	- None	- De-Blocking	- De-Blocking - Overlap Transform
Entropy Coding	**- VLC**	**- CAVLC - CABAC**	**- Adaptive VLC**
Transform	- 8x8 DCT	- 4x4 Integer "DCT" - 8x8 Integer "DCT"	- 4x4 Integer "DCT" - 8x4,4x8 Int "DCT" - 8x8 Integer "DCT"
Other	**- Quantization Scaling Matrices**	**- Quantization Scaling Matrices**	**- Range Reduction - In-Stream Post Processing Control**

FIGURE 5.22-16 Comparison of algorithmic tools used in MPEG-2 Video, MPEG-4 AVC, and VC-1.

service. Only a few years ago, HD content required over 19 Mbps. While today's MPEG-2 HD content is being compressed at rates between 12 to 18 Mbps, next-generation compression rates, at similar picture quality, are approximately 6 to 10 Mbps today, and will likely be reduced further in the next few years (see Figure 5.22-15).

As with MPEG-2 Video, obtainable bit rates for a particular overall picture quality vary greatly with content, with real-time encoded high-motion sports being one of the most difficult classes to code.

The coding gains come from the ability to perform more parallel processing and select better matches (that is, better results on the rate-distortion curve) in real time in the signal processing stages, a more efficient bit stream syntax, and more computationally intensive entropy coding resulting in fewer bits used in the stream processing stage.

Figure 5.22-16 contains a summary of the algorithmic tool differences among MPEG-2 Video, MPEG-4 AVC, and SMPTE VC-1.

WHAT IS PSIP?

Program and System Information Protocol (PSIP) is an essential set of broadcast metadata that is required under the FCC rules. It has three functions. First, it allows end users to access the DTV broadcast in a manner similar to the way they are used to with analog broadcasts. Second, it provides critical information for the DTV receiver so it can correctly decode the signal. Finally, it provides promotional information, which allows the station to brand itself and advertise present and future services.

PSIP helps to make DTV as simple to access as analog TV, hiding the complex interrelationships between digital components of the system from the end user. PSIP provides capabilities familiar to analog TV users such as channel numbering, up and down tuning, and on-screen program guide.

In addition, PSIP provides the information necessary to tune and decode DTV signals. Decoders depend on the PSIP to help them differentiate different services in the digital stream. While MPEG defines digital services (with the PSI tables mentioned earlier in the chapter), it is the PSIP that defines *virtual channel numbers*, and then associates that information with a station name.

The PSIP in a DTV stream also serves an important marketing function. PSIP allows the DTV broadcaster to brand its signal with promotional information such as the *electronic program guide* (EPG). At its simplest, PSIP provides signaling information about what is playing now, in the future, and where the user has to go in order to find the virtual channel he or she is looking for.

PSIP Generator

The PSIP information is collected and processed in a device known as a *PSIP generator*, which typically is a dedicated personal computer running a specialized software package that provides the required functionality. The PSIP generator has network and other interfaces to allow it to communicate with other systems for receiving source information and for feeding the PSIP output to the multiplexer.

PSIP IN THE DTV STREAM

PSIP consists of a number of MPEG-2 tables, whose structure and usage are defined in ATSC standard A/65C [11]. The tables are carried in MPEG-2 transport packets in the DTV Transport Stream. To provide for random access tuning, the PSIP tables are repeatedly cycled and updated according to a specific schedule, also defined in A/65C. The following is a list and brief description of the most important tables.

- *Master Guide Table* (MGT): The MGT serves as a directory of all PSIP tables. It lists the locations, sizes, and versions of each table so the receiver can extract them from the stream.
- *Virtual Channel Table* (VCT): The VCT defines each of the virtual channels within the DTV stream, defining major and minor channel numbers as well as which program elements to assemble to reconstruct the television program.
- *System Time Table* (STT): The STT carries current time information from the broadcaster to the receiver.
- *Event Information Table* (EIT): The EITs carry the main program guide information (event titles, times, and information about captioning, ratings, audio, and broadcast flag).
- *Extended Text Table* (ETT): ETTs carry extended text descriptions of individual events within the EITs.
- *Rating Region Table* (RRT): The RRT defines the rating parameters used for the region.

For a more detailed explanation of the PSIP tables, see *Proceedings of the IEEE, Special Issue on Global Digital Television, January 2006* [5].

PLANNNG FOR PSIP

Some of the information that the PSIP generator needs is relatively static and should be configured at installation. Static information includes items such as the station's Transport Stream Identifier (TSID, identified by the FCC), virtual channel number, and packet IDs (PIDs) associated with audio and video streams.

To facilitate a smooth and quick transition when implementing a PSIP system at the station, it is important to generate a list of this information before beginning installation. The information should be collected in a *Digital Site Survey* document (DSS).

Preparing a Digital Site Survey

The DSS is a crucial step that should be pursued prior to implementing PSIP in the TV station. The information to be collected is listed in Table 5.22-1.

The finished site survey may look something like Table 5.22-2.

Many stations undergo changes in the virtual channel configuration during the broadcast day, known as *day part changes*. Day part changes are transitions from multiple standard-definition virtual channels to a single high-definition channel and back. Because the virtual channel configuration is signaled by PSIP, the PSIP data must be modified to match the actual configuration dynamically, as the day part change takes place. If the station has implemented or plans to implement day part changes, the site survey should contain information for both time periods (see Table 5.22-3).

Sources of Schedule Data

In addition to the information required at startup, the PSIP generator requires a great deal of information or *metadata* to create the EPG. This information includes:

- Branding;
- Characteristics: Ratings, captioning, broadcast flag (at the service level);
- Announcement;
- What programs are coming up and when;
- Descriptive information;
- Characteristics: ratings, captioning, broadcast flag (at the program level).

TABLE 5.22-1
Digital Site Survey for PSIP Implementation

Type of Information	Where to Get This Information
Transport stream ID	The unique identifier for a specific station's stream, provided by the FCC
Packet IDs (PIDs) for all audio and video streams	Mux user interface, or from the encoder
Program number for all digital services	Comes from the source for the station MPEG tables. See the next section for more details
Major channel number	The channel that users will use to enter into their remote control in order to tune to the station's DTV signal. These numbers are typically between 2 and 69 and are usually (but not always) based on the station's NTSC channel number (which may be historical).
Minor channel numbers	For stations that broadcast multiple services, the minor number differentiates the different virtual channels. For example, if a station is broadcasting one high-definition service and one standard-definition service: 39-1 High-definition service 39-2 Standard-definition service The minor channel number is the -1 and -2

TABLE 5.22-2
Digital Site Survey Results Example

Station TSID:	1432					
Number of Services	3					
	Service Name	**Virtual Channel**	**Program ID**	**Video PID**	**Audio PID**	**Audio PID 2**
Service 1	WEXA-HD	31-1	3	49	52	53
Service 2	WEXA-SD1	31-2	4	65	68	69
Service 3	WEATHER	31-3	5	81	84	85

TABLE 5.22-3
Digital Site Survey Results with Day Part Changes

Station TSID:	1432					
Number of Services	Daytime: 3	Evening: 1				
	Service Name	**Virtual Channel**	**Program ID**	**Video PID**	**Audio PID**	**Audio PID 2**
4:00 AM—11:00 PM						
Service 1	WEXA-HD	31-1	3	49	52	53
Service 2	WEXA-SD1	31-2	4	65	68	69
Service 3	WEATHER	31-3	5	81	84	85
11:01 PM—3:59 AM						
Service 1	WEXA-NTL	31-1	3	49	52	53

All of this information is dynamic; it requires updating on a regular basis to remain current. Dynamic information is available within systems already in use in broadcast operations:

- *Listing services (traditionally provide guide information for print and other media):* Channel name/number, program name, time, descriptions, CC, ratings, audio types.
- *Traffic/program management (traditionally used for business operations—ensuring that content is available when needed and advertising revenue):* Program name, time (optionally: channel name/number, program descriptions, CC, ratings, audio types).
- *Automation (controls actual play out of content and switching between feeds):* Program ID, accurate time (optionally: channel name/number, program name, descriptions, CC, ratings, audio types).
- *Human operator (ultimate control of operations):* Subset or all of the above.

As one moves down the list of sources above, the information becomes more chronologically accurate but less informative. At one end of the spectrum, the listing services have a considerable amount of detail for each event (such as the title, full description, actors, and other details that are useful). However, due to the long lead times characteristic of these services, the schedule information might be inaccurate. At the other end, with the automation system the actual event timing is exact (because the automation system is responsible for starting and stopping the playout of content). However, the automation system typically does not have much detail about the events themselves and in many cases does not differentiate between shows and interstitials. In order to formulate the most accurate and informative guide, it is necessary to draw information from multiple sources over time and intelligently merge them.

When selecting an input source for the scheduling information, cost and time both play a role. A listing service may provide accurate and detailed information, but also typically requires a yearly subscription fee. Furthermore, listing services typically require an Internet connection to the PSIP generator. If the PSIP generator is located at a transmitter site, it may not have an Internet connection. When evaluating vendors for PSIP generators, it is important to check that the device supports the schedule import method that is being planned to use. More information on the ATSC Programming Metadata Communication Protocol (PMCP) standardized method of communicating this information is provided later in the chapter.

PSIP INSERTION

Once the programming information is encoded into the binary PSIP tables, the PSIP data needs to be injected into the broadcast transport through the station multiplexer. Either the PSIP generator will have an ASI output, which will feed directly to the mux, or it will communicate with the mux via IP over a network connection. The actual configuration depends on the interoperability between the mux and the particular PSIP generator.

ASI PSIP Insertion into the Transport Stream

The simplest method for doing the PSIP data injection is to utilize an ASI connection from the PSIP generator into one of the input ports of the multiplexer, in a similar manner to the connection from an encoder, as shown in Figure 5.22-9. In this arrangement, the PSIP generator creates an MPEG-2 packetized stream with all of the PSIP tables encoded and scheduled to meet the required (or desired) cycle times. The multiplexer is provisioned to allow sufficient bandwidth to accommodate the PSIP stream (typically, less than 250 kbps is necessary).

PSIP Insertion Using Mux Carouseling

Some muxes have the capability of storing encapsulated PSIP tables in internal memory and playing them out according to a predetermined schedule; this is known as *carouseling*. For carouseling, an Ethernet connection is used between the PSIP generator and the mux. The PSIP generator uploads the encapsulated PSIP tables into the mux using User Datagram Protocol (UDP) over an IP connection, along with instructions on how to schedule the playout. The tables only need to be refreshed when there is an update, typically at the 3-hour EIT boundaries or when an operator manually changes schedule information. Carouseling offers a form of error resiliency in case of communication problems between the PSIP generator and mux. If the PSIP generator fails to communicate, the MUX will continue to carousel the information it has indefinitely.

Multiplexer Configuration Synchronization

Some PSIP generators can synchronize directly with the encoder/multiplexer systems to automatically download static data and deal with day part changes. If day part changes are being planned (see above), then a *mux/PSIP generator pair* should be used, which automatically synchronizes this data. Otherwise, manual intervention will be required every day to switch between the two different program lineups.

PSIP AND PSI CONSIDERATIONS

One point, which is often overlooked, is the coordination of MPEG PSI (program specific information) and ATSC PSIP. Some of the information carried in the PSIP is also signaled in the PSI. In order to guarantee that services will be decoded by receivers, the MPEG-2 program information (from the PSI) and the PSIP virtual channel information (from the PSIP) must be consistent. Besides the necessary matching between the MPEG-2 programs and the PSIP virtual channels, some program-specific signaling information is carried in both PSIP and PSI structures (examples include

the *caption_service descriptor* and *redistribution_control descriptor*). Only the PSIP generator has access to the schedule data required for proper insertion of these descriptors.

The links between PSI and PSIP present a potential problem for broadcasters. If the PSI information does not match the information generated by PSIP, there is no guarantee that decoders will function. Furthermore, if more than one source of PSIP and/or PSI resides on the system, they may collide, resulting in intermittent or total decode failure.

PSIP and PSI conflicts remain among the most common problems encountered when implementing PSIP in a digital television system. To help avoid these problems, two important rules should be followed:

1. Use only *one* source for PSI information into the multiplexer. All other PSI sources should be blocked or disabled. The same is true of PSIP information—only *one* device in the system should supply encoded PSIP data to the multiplexer.
2. Regardless of the device that generates the PSI tables, the configuration of the PSIP generator must *match* the information in the PSI tables.

Rule 2 is automatically satisfied if PSI information is generated in the PSIP generator, and blocked or disabled elsewhere on the system. For this reason, it is recommended that PSI information come directly from the PSIP generator.

Time Information

Accurate time information is critical for PSIP. The system time table (STT) carries the broadcaster's notion of current time to the receiver, where it is used for the EPG and increasingly for the start and stop of recording through digital video recorders (DVR). ATSC standard A/65 requires that the time carried in the STT be within 1 second of GPS time to avoid confusion during channel tuning.

It is relatively straightforward to broadcast an accurate STT. Many stations utilize SMPTE time code, carrying time of day, in their operation. It is a simple matter to equip a PSIP generator with a SMPTE time code reader, which will automatically synchronize to the station time. The station clock should itself be synchronized to a standard time reference—typically GPS. Obtaining time reference direct from a GPS receiver is another relatively low-cost option and easy to implement with a connection into the PSIP generator. If neither of these references is available, accurate time information can be obtained from the Internet using the Network Time Protocol (NTP). NTP servers are available on many office networks or on the Internet.

If the station in question utilizes a UDP connection from the PSIP generator to the mux rather than an ASI input, the STT timing information is usually the responsibility of the mux. In this case, the mux generates the STT based on an internal clock, which should be synchronized to an external source. Specific arrangements about SMPTE, GPS, or NTP time sync are usually determined in consultation with the PSIP generator and mux vendors.

PSIP ARCHITECTURES

There is a wide range of PSIP architectures in use for broadcast television stations, groups, and networks. The simplest (and most common) configuration is for each station to be considered a separate entity that gathers all of the information required for PSIP and then injects the PSIP data tables into the broadcast Transport Stream, within the station. At the other end of the spectrum are "central casting" operations—large station groups that may be geographically distributed but with a centralized location for managing and creating PSIP.

Individual Stations

For an individual station, the architecture will resemble Figure 5.22-17; however, typically only some of the schedule sources will be involved. Most often, the PSIP generator and the encoding/multiplexing equipment will be collocated, allowing direct connections from the PSIP generator.

Central Casting

For large station groups, economies of scale can be achieved via central casting—especially in terms of operations. Commonly, schedule management and PSIP-related operations are performed at a central location, with the resultant information carried to the edges for broadcast. In some cases (as discussed in Chapter 5.21 on master control systems), master control has also been centralized. Central casting PSIP arrangements involves a single PSIP generator at the central location (network operations center or NOC), which feeds the individual remote stations/transmitters. Schedule aggregation and PSIP table generation takes place at the NOC. The method for carrying the

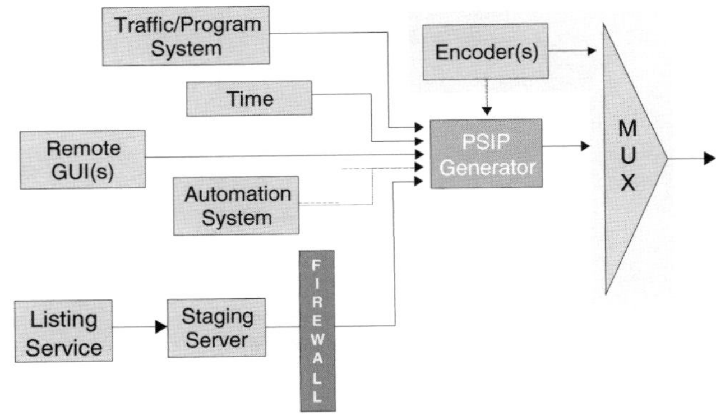

FIGURE 5.22-17 PSIP data flows.

PSIP tables to the network edges for broadcast varies depending upon the central casting design, as discussed below.

The architecture used for central casting depends upon a number of factors:

- Common programming at all stations versus varied programming: Can the same PSIP information be used at different stations with only changes to certain fields (TSID, major/minor channel numbers, and others) or does the PSIP information need to be generated for each station?
- Content distributed in compressed (MPEG) form to remote stations (remote encoding) versus analog form: Can encoded PSIP be carried in-band (part of the transmission) versus out-of-band, typically over a wide area network (WAN)?
- Network connectivity to remote stations: How is the PSIP data carried to the remote station—over a WAN? Over POTS (Plain Old Telephone Service)? As part of a digital (MPEG) transmission?

Figure 5.22-18 shows a generalized architecture for central casting, with three possible connections illustrated. Path A illustrates a connection for remote emission sites with broadband WAN connectivity. While it is possible to stream the MPEG-2 packetized PSIP tables across the WAN connection, uncertainty in packet loss and latency suggests that a carousel connection is best for this situation. Path B illustrates a similar situation (remote emission sites), but with no WAN connectivity available. For this situation, carouseling is the only reasonable approach, using IP modems across POTS connections. Path C illustrates connections to local multiplexers, where LAN connectivity is available. In this situation, it is reasonable to either stream MPEG-2 packetized PSIP tables or to utilize carouseling. In many instances, the actual solution will involve a mixture of these types of connections.

Central casting typically requires close coordination between the station information technology (IT) and engineering departments, as well as consideration of a number of IT policies: security (firewall, antivirus, VPNs, etc.), services that are allowed/disallowed (for example, SNMP auto-discovery), and naming and addressing schemes such as static IP addresses and dynamic domain name servers (DDNS).

PSIP AND PMCP

As previously mentioned, it is necessary to draw programming information, or metadata, from multiple sources and intelligently merge them to generate the PSIP tables. To try to avoid a multitude of proprietary interfaces to achieve this, the ATSC developed the Programming Metadata Communication Protocol (PMCP), published in 2004 as ATSC standard A/76.

What is PMCP?

PMCP is a common language that systems can use when exchanging PSIP-related metadata. It is based on XML (eXtensible Markup Language), and supports exchanges of data via files or sockets-based connections.

PMCP is used for communication between the PSIP generator and other sources of PSIP-related information, which may also communicate among themselves

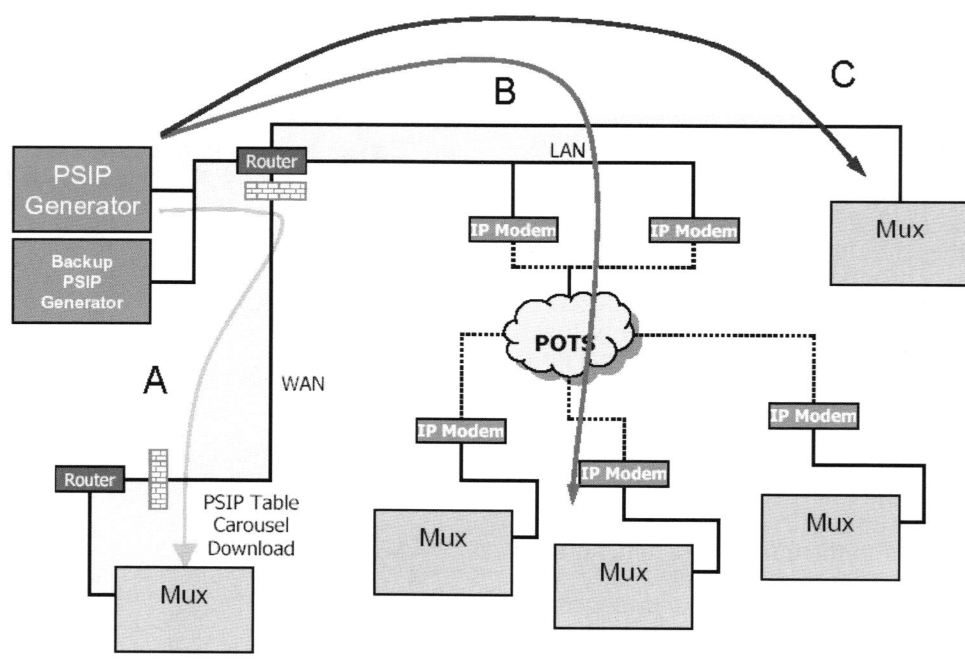

FIGURE 5.22-18 Example PSIP central casting architecture.

using PMCP. Data transfer will typically take place over a network, although any means of transport may be used for file transfer.

What Are the Benefits?

PMCP is clearly of benefit to manufacturers of systems that are involved in the PSIP production chain. A manufacturer can build a single PMCP interface and use it for communications with multiple partners' systems. Reduced development time, reduced costs, improved reliability, and simpler implementation are just a few of the benefits.

Broadcasters are likely to benefit from the adoption of PMCP, with an assurance of compatibility between systems that need to exchange PSIP data. If any of these systems in a facility is replaced at a later date, providing the new system also includes PMCP compliance, there should be few problems when connecting with other systems already in place at the station.

There are other benefits. Because PMCP is based on XML it is quite human-readable, even by those without an extensive technical background. XML support is easily incorporated into products using most modern development environments, and is already well known to most of the manufacturers of broadcast systems. It is by its very nature, extensible, which means that it can be extended and updated as needs evolve. In addition, provision has been made for carriage of private information within PMCP so that systems can enhance their PMCP interfaces with heretofore unthought-of data elements.

Static Versus Dynamic PSIP

In the first years of digital television in the United States, many over-the-air broadcasters sent out *static PSIP* (or, in some cases, no PSIP data at all). Static PSIP provided enough information for receivers to tune to the broadcast signal, but not much else. FCC rules now require transmission of *dynamic PSIP*, which is of far more use to viewers. Dynamic PSIP provides viewers with an accurate listing of what is scheduled to air on each channel, permitting them to tune to what they want to watch at the time the program will actually air, and to know how long it will be on air.

Dynamic PSIP is important for other reasons. The most obvious are recording devices that use PSIP, such as DVRs. In order for such devices to record what is intended, PSIP must accurately reflect what is going to air, and at exactly what time. Less obvious, but still important is rating (V-Chip) information. If PSIP for the next program comes on before the program begins, or if incorrect PSIP is sent, and the rating differs from what it should be, viewers who have set their receiver to prevent viewing of programs with certain ratings may be unable to view an otherwise suitable program.

Typical Systems Involved

One of the things that make PMCP so helpful is that it can be used to connect a wide variety of systems. Because PSIP data typically resides on several different systems, in order to have accurate, dynamic PSIP, this data must often be obtained from multiple sources.

A typical broadcast station may obtain PSIP data from some or all of the following systems:

- *Listing service:* A few companies provide services that aggregate program listing information for stations across the country. Similar services are also available outside of the United States. Unlike the other systems involved in PMCP, these systems are typically not hosted in the broadcast facility.
- *Program management system:* Broadcasters who themselves originate a good percentage of their programming lineup may have a formal program management system. These typically manage program contracts, rights, and program schedules, as well as financial aspects of programming.
- *Traffic system:* Virtually every broadcaster has a traffic system. These systems manage spot sales contracts, commercial material management, assembly of the complete broadcast schedule, reconciliation of what ran, and invoicing of spots, among other things.
- *Automation system:* Since the early 1990s, broadcasters have increasingly moved to systems that manage their on-air operations. These systems control the ingest of broadcast material, management of material and devices, on-air playback of that material, and a variety of other tasks.

Challenges: Data Ownership and Granularity

When PSIP data resides in multiple systems, it may be asked, why not just choose the one with the highest quality and quantity of data, and use that as the single data source for the PSIP generator? That would be wonderful if such a system existed, but it does not.

The PSIP paradox is that as accuracy of PSIP data increases, the richness of that data decreases, as illustrated in Figure 5.22-19.

A system-by-system examination of the problem follows.

Listing Service

Because the job of these services is to maintain rich viewer-friendly program listings, the depth of the data found here is excellent. Many details for the event information table (EIT), as well as much information for the extended text table (ETT), can be found here. These systems typically have databases with full and consistent information for every program that could be scheduled, removing any worry about typos or incomplete information. The problem lies in the fact that listing services' view of the program schedule is typically frozen in time about 2 weeks prior to air. This means that any program schedule changes occurring within this window are not reflected in these systems.

Program Management System

In terms of depth of information, in many cases these systems can challenge listing services. Program man-

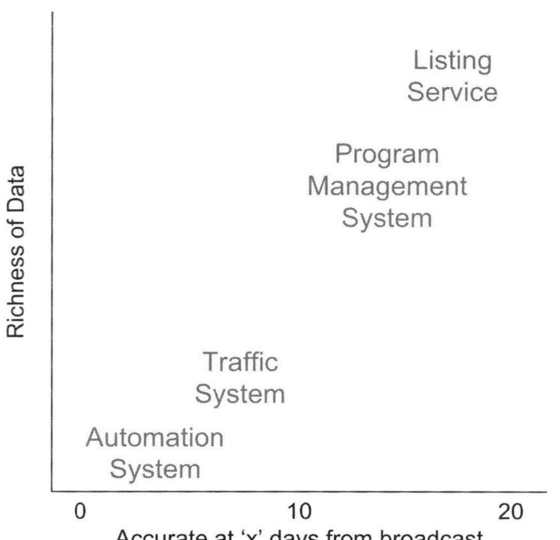

FIGURE 5.22-19 Richness of data versus accuracy at 'x' days from airtime.

agement systems have the capability to track virtually every detail concerning programs. However, there are some issues. The first point is that data must be entered into the system. Unlike listing services, which have centralized databases containing details relating to virtually every program ever produced, program management systems typically rely on personnel at the station or station group to either hand-enter this information, or capture it from external source(s). As a result, information held here can be incomplete, and is more likely to contain typographical errors than that which is found in listing services. However, program management systems have one significant advantage. Being in-house systems, they are typically updated as the schedule gets closer to air, and often contain accurate representations of the on-air schedule up until a day or two before broadcast.

Traffic System

At this point the depth of program-related metadata often degrades significantly. Because traffic systems are primarily concerned with the management of commercials, they do not require the depth of programming metadata that the other system mentioned thus far do. The typical depth of information that can be expected from a traffic system is generally limited to date, time, channel, show name, episode/title name, and show duration. In some cases, it may be possible to get more, but those will be the exception. Although the data is not rich, it is normally far more accurate than that found in either the program management system or listing service. Traffic systems have an accurate and complete view of the entire broadcast schedule up until a few days, or in some cases, a few hours prior to air. Changes to the program schedule made in this time window will normally be reflected in the traffic system.

Automation System

In terms of richness of metadata, automation is not typically any better than traffic. In fact, in many cases, automation may know even less than traffic when it comes to program material. It normally does not have visibility of the program schedule more than 1 to 3 days in advance. So, what makes it an essential PMCP source? In a word, *accuracy*. Automation's primary responsibility is to control what airs, and at precisely what time it is to air. So, if accurate PSIP is the goal—to reflect exactly what is airing and at what time it is airing—the information must come from the system responsible for the on-air product, which is automation. In addition, automation is probably the only source of PSIP data that is truly mission-critical. As such, it is typically built with a great deal of redundancy and reliability, in most cases far more so than any of these other potential PMCP sources.

Clearly, ownership of PSIP data changes as a function of time. Although there can be multiple sources, the authoritative source of this data will vary as the time gets closer to broadcast.

Which system should be relied upon for PSIP data? Ultimately, the decision is up to the broadcaster, and different broadcasters will make different choices. However, two factors are key in determining which systems to get PMCP data from:

1. How close to air does your program schedule generally get "locked in"?
2. How accurate do you want or need your PSIP to be?

Another Challenge: Accurate Names

As discussed, it may be desirable to have the ability to update PSIP from the traffic system, and possibly from the automation system. As outlined earlier, although the information available in these systems is very basic (typically, start time, duration, and show name), the accuracy is very good. An important thing to consider before putting in place any PMCP interface from either of these systems is that the show names in use in these systems may not be in a form that is acceptable for display to viewers. Some broadcasters have show names such as "Friends (Barter Version)" or "Late Night Movie (Weekend)" in their traffic and automation systems. These names are not appropriate for the EPG. So, before putting these interfaces in place, it is good practice to spend some time considering how to clean up these names and make them presentable for the public.

PMCP: A Use Case

To better understand the system interactions involved, a use case is presented.

1. Months in advance, a blockbuster movie is scheduled. This is done in the program management system.

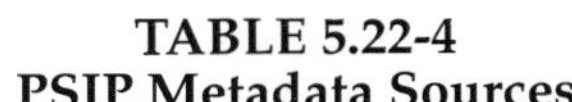

2. Two weeks prior to broadcast, the program management system sends this information to the listing service.
3. Shortly thereafter, the program management system sends this information to the traffic system.
4. The day prior to air, the traffic system sends details of this movie to the automation system.
5. On the day of air, a breaking news event preempts the movie at the last minute. Automation must communicate this to the PSIP generator.
6. Eventually, the live event ends, and the movie is started later than expected. Again, the automation system must communicate this to the PSIP generator, as well as any shifts of programming that follow the movie and whose start times are affected.

This is a fairly simple, common, and straightforward illustration, and it alone has six exchanges of PSIP data taking place. It can be far more involved than this. Because of this, it becomes clear that standardized communications are of great benefit in helping to solve this problem.

Recommended Implementations

How PMCP is implemented in a particular facility is up to the broadcaster. Hopefully, the information provided here is helpful in understanding the topic well enough to make the right decision for each particular situation. As a guide, Table 5.22.4 offers some possible implementations based on different types of programming.

SUMMARY OF THE FCC RULES ON PSIP

The entire Report and Order issued by the FCC in September 2004 concerning PSIP [12] is quite lengthy, but perhaps the most important section relating to PSIP and PMCP is the following.

> We conclude that adoption of ATSC A/65B (PSIP) into our broadcast transmission standards will serve the public interest. As pointed out by commenters, during the development of PSIP, the ATSC carefully considered which elements of PSIP should be mandatory and which should be optional...We therefore require that broadcasters fully implement PSIP to the extent that ATSC A/65B requires. In order to give broadcasters adequate time to come into compliance, this requirement shall take effect 120 days after publication in the Federal Register. We expect broadcasters to populate the required tables and descriptors with the proper information to help receivers assemble functioning guides. *All tables and descriptors that require one time setup should be set correctly, including TSID, Short Channel Name, Service Type, Modulation Mode, Source ID, and Service Location Descriptor. ATSC A/65B also requires that broadcasters send populated EITs covering at least a 12 hour period. These EITs should be populated with the correct information, so that the user knows what programs are on for this 12 hour period.* Also, we expect that manufacturers will have every incentive to build equipment that looks to PSIP for its basic functionality, but we will revisit the issue if necessary. Standardized use of the data transmitted through PSIP will ensure that the full benefits and innovations of the new digital system will be available to the public.

TABLE 5.22-4
PSIP Metadata Sources

Program Schedule is Locked Down	Is Programming Mainly Local Origination?	Suggested PMCP Sources
2+ weeks out	Yes	Listing service/program management
2+ weeks out	No	Listing service
1 week out	Yes	Listing service, then program management
1 week out	No	Listing service, then traffic
A few days out	Yes	Listing service, then program management, then traffic
A few days out	No	Listing service, then traffic
Never—lots of live programming	Yes	Listing service, program management, traffic, then automation
Never—lots of live programming	No	Listing service, traffic, then automation

A few portions of this are particularly important, and those have been highlighted. An examination of these passages is helpful in understanding the core of the rule.

> *All tables and descriptors that require one time setup should be set correctly, including TSID, Short Channel Name, Service Type, Modulation Mode, Source ID, and Service Location Descriptor.*

The implication here is that all of these attributes must be set up correctly in the PSIP generator. Since these are "static" elements of PSIP, they should only have to be set up once, saved, and rarely revisited. This is the easy part. The rest of what the FCC requires implies "dynamic PSIP," which is not quite so straightforward.

> *ATSC A/65B also requires that broadcasters send populated EITs covering at least a 12 hour period.*

EITs contain the basic program guide information that allows the consumer to see what programming currently being shown and what is coming up. These tables include information such as show name, start time, and duration of the entire program. ATSC A/65B actually requires that the first four EITs (EIT-0 through EIT-3) are sent, each containing 3 hours of events. However, as programs are completed in the current EIT (EIT-0), it will reduce from 3 hours, eventually down to zero, meaning that right at the end of EIT-0, only 9 hours of EITs (basically EIT-1 through EIT-3) will be sent out. Although the FCC Report and Order specifies a minimum of 12 hours, the actual A/65B standard that they cite really

only requires this 9- to 12-hour range covered by EIT-0 through EIT-3.

> *These EITs should be populated with the correct information, so that the user knows what programs are on for this 12 hour period.*

The key word in this sentence is "should," which may be considered as meaning, "we highly recommend you do this, and if you don't, we'll make you." So, it would be prudent to make sure that the information sent out in the EITs is accurate.

Required Accuracy

If EITs are based on information from a listing service that might be 2 weeks old, this may not be accurate enough. This leads to the question of where to get more accurate information regarding program names, start times, and durations. Traffic and automation systems certainly have this information.

If the goal is to have all of this information as accurate as possible, an interface from the automation system to the PSIP generator is required. However, some believe that getting this information from traffic may be good enough. To resolve this, the following question must be answered.

How Many Hours of EIT Information Should Be Sent?

Although the minimum requirement in A/65B is the next four EITs, covering 9 to 12 hours of programming, will that be sufficient? To give viewers a good experience, probably not. The ideal amount of EIT information to send out may be more along the lines of a week's worth. As mentioned in the dynamic versus static PSIP section, there are several reasons for this, relating to DVR use, rating information, and the viewer experience.

However, there is another compelling reason to send out this much EIT data. PSIP is sent out by each broadcaster individually, and it will typically only provide information about their own services. This means that a receiver, when tuned to a particular channel, must acquire the PSIP data relating to that channel, and make it available to the viewer. This may take several seconds. To permit quicker tuning from channel to channel, some receivers perform a scan of PSIP available from all available services, periodically when not in use (perhaps in the middle of the night). This information is stored and displayed to the viewer very quickly when they tune from channel to channel, giving a channel surfing experience similar to that available in the analog world, without a potential delay of a second or two while changing channels. This helps provide an improved viewing experience.

Rating (V-Chip) Information

The FCC order also requires that rating information be included in PSIP, and be accurate.

The pertinent passages from the FCC Report and Order regarding V-Chip are as follows:

> ...the Event Information Tables ("EITs") defined within PSIP will contain any available Content Advisory Descriptors ("CADs") for broadcast programming
>
> ...we believe it is reasonable to provide an 18 month transition period. After the transition period, all digital television receivers will be required to provide v-chip functions following the regulations that we adopt in this proceeding.

This means V-Chip data must be provided to the PSIP generator. Traffic and automation systems typically either do not track this information, or if they do, they often do so in nonstandard ways. One broadcaster may include this information on the schedule as secondary events, another may do so in text-based comments, while yet another may find other ways in which to include this information. As a result, there is no single straightforward way in which to extract this information from traffic and automation systems for the purposes of providing it to the PSIP generator.

For this reason, it may make more sense for PSIP generators to get this information from systems such as program management and listing services, which are more geared toward dealing with this parameter.

SUMMARY

Encoding, multiplexing, and the resultant Transport Stream are fundamental parts of any DTV system. This chapter has discussed the principles of operation of the equipment that broadcasters need to use for this vital part of their operations. The discussion about PSIP should help ensure that broadcast stations comply with the FCC rules in what is often a confusing area. PSIP compliance should also be made easier as equipment manufacturers adopt the PMCP standard for metadata communications. While MPEG-2 encoding will be with us for many years to come, it may be expected that the new advanced codecs discussed earlier in the chapter will inevitably become a part of the toolkit that broadcasters have to draw upon to provide high quality yet highly efficient coding for broadcast programming.

ACKNOWLEDGMENTS

Figure 5.22-4 is taken from the ATSC Digital Television Standard A/53 published by the ATSC and used with permission. Figures 5.2-5, 5.22-6, 5.22-8, 5.22-14, and 5.22-15 are based on figures in "ATSC Video and Audio Coding" and "The ATSC Transport Layer, Including Program and System Information Protocol (PSIP)," first published in *Proceedings of the IEEE, Special Issue on Global Digital Television*, January 2006, ©2006 IEEE, used with permission of the publisher.

References

[1] ATSC Digital Television Standard, A/53 Revision E with Amendments No. 1 and No. 2, September 2006. Advanced Television Systems Committee. See http://www.atsc.org/.

[2] ITU-T Rec. H.222.0 | ISO/IEC 13818-1:2000 Coding, Information Technology—Generic Coding of Moving Pictures and

Associated Audio. Part 1: Systems. Information: Systems, December 2000.

[3] ETSI EN 50083-9: Cabled Distribution Systems for Television, Sound and Interactive Multimedia Signals. Part 9: Interfaces for CATV/SMATV Headends and Similar Professional Equipment for DVB/MPEG-2 Transport Streams, (DVB Blue Book A010), Annex B, Asynchronous Serial Interface.

[4] Bock, A., "What Factors Affect the Coding Performance of MPEG-2 Video Encoders?," *DVB '99* conference proceedings, March 1999.

[5] *Proceedings of the IEEE, Special Issue on Global Digital Television,* January 2006.

[6] Jordan, J., and Bock, A., "Analysis, Modelling and Performance Prediction of Digital Videostatistical Multiplexing," *Proceedings of 1997 International Broadcasting Convention*, September 1997, Amsterdam, Netherlands, pp. 553–559.

[7] Bolender, S., Hackett, A., Heimburger, C., and Knee, M., "Motion Content of Transmitted TV Images: A Statistical Survey," *Proceedings of 1994 International Broadcasting Convention,* September 1994, pp. 405–410.

[8] ITU-T Rec. H.262 | ISO/IEC 13818-1:2000 Coding, Information Technology—Generic Coding of Moving Pictures and Associated Audio. Part 2: Video, February 2000.

[9] ITU-T Rec. H.264 | ISO/IEC 14496-10 |:2005, Information Technology—Coding of Moving Pictures and Associated Audio-Visual Objects. Part 10: Advanced Video Coding, December 2005.

[10] SMPTE 421M-2006, VC-1 Compressed Video Coding 1. Bitstream Format and Decoding Process, February 2006.

[11] ATSC Program and System Information Protocol for Terrestrial Broadcast and Cable, A/65C Revision C with Amendment No. 1. May 2006. Advanced Television Systems Committee. See http://www.atsc.org/.

[12] FCC 04-192 Second Periodic Review of the Commission's Rules and Policies Affecting the Conversion to Digital Television. September 7, 2004. Federal Communications Commission

CHAPTER

5.23

Film for Television

DAVID J. BANCROFT

Thomson
Reading, United Kingdom

INTRODUCTION

Film has been a technology for moving image capture and reproduction for over 100 years—far longer than television. As a result, vast libraries and archives of material have accumulated all around the world, covering an extremely broad range of content from news footage to major feature movies, in addition to all the non-entertainment uses. Estimates vary, but probably somewhere between 1 and 2 million hours of filmed content exist in the world today.[1] In addition, new content is constantly being created. Despite the increasing inroads of video capture and digital cinematography techniques, film remains the medium of choice for most directors of full-length feature movies as well as for many directors of documentary, wildlife, TV miniseries, and other genres. This chapter covers some of the basic characteristics of film and describes the systems that may be used for transferring film material into the television medium. This process is quite complex, and the place of transfer has accordingly migrated from the broadcaster to specialized facilities houses; however, networks and stations may still maintain the necessary equipment in house for short-notice requirements such as airing library footage that exists only on film.

[1]One real-time film scanning machine (telecine) would take 100 to 200 years to transfer this material, on a 24 × 7 continuous basis, not counting film reel changing times, cleaning, defect remedy, and set-up.

FILM-TO-TELEVISION INTERFACE

In the past, film was shot exclusively with theatrical projection in mind, implying a large dynamic range appropriate to a dark viewing environment. In comparison, television was designed as a low dynamic-range medium for the comparatively well-illuminated consumer viewing environment. This caused some difficulty in interfacing the film medium to the television system and was exacerbated by the design of early film scanning devices, which could not convert the full density range of a release print into a television image with correct gray scale.

An early solution to this problem was the use of special low-contrast film print stock, designed to make a print from the original negative compatible with television but unsuitable for direct projection. This method was assisted by exhorting the director of photography to limit scene contrast, with television, rather than the movie screen, specifically in mind. This approach is still valid for television-only projects such as episodics and documentaries but does not help when the production is intended for movie theaters but a version is also required for television.

The solution today lies in improved telecine design. Not only can modern telecines scan the original negative directly, but they can also reproduce its entire density range. In the case of negative stock, they can do this without recourse to any incremental adjustment during the scan of entire camera rolls. This has been made possible largely through the replacement of analog with digital image processing.

Film Ingest for Television

Film is a physical, photochemical medium that requires a significant transformation process for ingest into the electronic/digital domain of television. Before considering the different techniques that are discussed later for film transfer using *telecine* machines and other types of film scanners, some fundamental factors should be considered. In the earliest years of telecine development, film content was ingested using the telecine machine to transform the finished film release print into a television video signal. As the sophistication of telecine equipment improved, it became possible for the transfer to video to be taken from an earlier point in the film production chain—from an intermediate version of the content known as the *interpositive print*—that did not have adjustments for the display environment (the movie theater) embedded into it and which was thus much more amenable to the rather different settings needed for optimal television reproduction. This is now the preferred method for transferring completed motion picture titles from film libraries. A third ingest method is to take raw footage from the film camera,[2] transfer it directly to video or image data, and perform in the electronic domain *all* of the processing steps required for the final output: editing, special effects and compositing, color correction, and other adjustments that may be needed. This method has become the preferred practice for new productions destined predominantly for broadcast television distribution.

[2]In practice, a duplicate copy of the precious camera original may be used instead. However, it will still not have any image modification decisions embedded into it (contrast range, color balance, etc.).

NATURE OF FILM

Physical Structure

A piece of film is basically comprised of a light-sensitive emulsion that is applied to a much thicker transparent base material. Early film used a base of cellulose nitrate, which was extremely flammable. After about 1940, the industry gradually transitioned to using a much safer acetate plastic base, and, more recently, there has been an increasing use of a stronger polyester film stock for postproduction use and release prints.

Layout and Dimensions

Raw film stock is supplied in a number of gauges, ranging from 8mm to 70mm in width [1]. Some of these are shown in Figure 5.23-1. The most commonly encountered gauges are 16mm and 35mm. Perforations along one or both sides of the film (depending on the format) enable film transport and picture registration with various mechanisms. The image areas [2] may be in "Academy" format, which has the image frame size reduced to leave room for an optical sound track, or in "Super" format, taking up the full available width. Each frame's height may span one perforation (8 and 16mm); two, three, or four perforations (35mm); or five perforations (65 and 70mm). Other variants are found in special-effects work.

The image may be in a *flat* or in an *anamorphic* format. "Flat" means that only spherical lenses are used throughout the chain; "anamorphic" means a cylindrical distorting lens element was added to the film camera lens [3]. This has the effect of optically compressing the image in the horizontal direction (2:1 is a typical ratio), allowing a wider aspect ratio image to be accommodated in the same width on the film stock. A complementary adaptor is fitted to the projector to

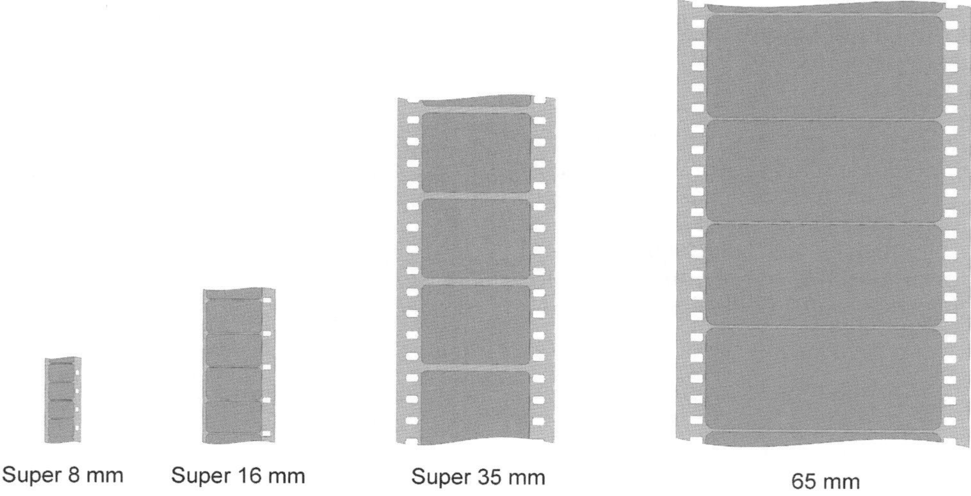

FIGURE 5.23-1 Film gauges.

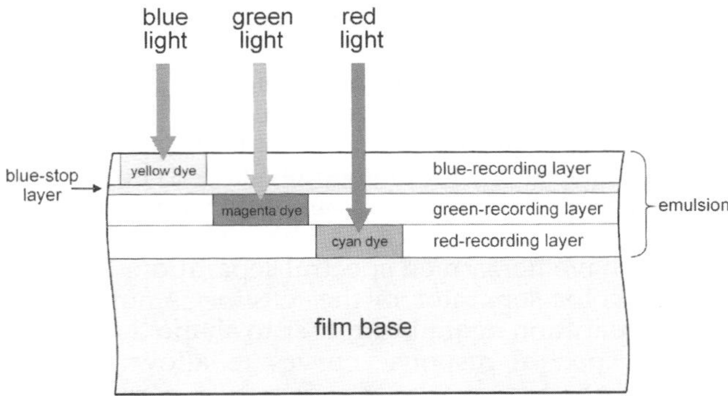

FIGURE 5.23-2 Color negative film dye layers.

restore geometry; a telecine will perform the equivalent function via signal processing.

Color Dye Layers

Today's color negative film uses multiple dye layers coated on the film base, as shown in Figure 5.23-2. Each layer responds to one of three primary portions of the visible spectrum by producing a density record from dye of complementary color. To register the blue component in the incident scene light, the blue-recording layer produces yellow dye. The yellow dye thus subtracts blue from the incident light before passing on the remainder to the next layer down. This process is repeated in subsequent layers. The densities of the dyes produced are in proportion to the intensities in the original light (brighter scene makes darker film).

In some cases, this *negative* film, once exposed and processed, will be scanned directly by the telecine. In other cases, an intervening printing operation will occur, to create a normal tonescale orientation. Print stock employs a similar structure of multiple dye layers, but the green layer is interchanged with blue (brought to the top) to minimize losses in contact printing.

There are two main kinds of prints. An *interpositive* print has a unity gamma or transfer characteristic, resulting in an image that is tonescale mirrored with respect to the negative, but which still preserves the full exposure latitude of the negative. The other is a *release print*, which employs a much higher gamma in the emulsion to increase the apparent contrast. An *internegative* is physically identical to an interpositive but achieves the opposite tonescale orientation because it is created by printing from an interpositive instead of a negative.

In telecine operations, it is preferable for quality reasons to reproduce an interpositive print or a negative, rather than a release print. The negative and interpositive retain the full exposure range but low contrast, making it much easier for the telecine operator to repurpose the reproduction for television's different viewing environment. Generation losses, such as modulation transfer function (MTF) loss, are also lower in these upstream stages in the film production chain.

The Picture Image on Film

Compared to most images captured by direct electronic means (for example, a charge-coupled device camera), film images have some important embedded differences, principally as follows.

Inherently Progressive Scan Images

Regardless of whether a telecine is required to output a progressive or interlaced video signal, a film frame is effectively a progressive frame, because all the information on it has been captured by the film camera in one exposure.

Different Frame Rate

Most film is shot at 24 frames/second. This differs from North American television's (and that of some other countries) nominal 30 frames/second (60 fields) rate and 25 frames/second (50 fields) in many other world regions. Together, the progressive scan factor and the frame rate difference necessitate special processing in the telecine for compatibility with television image standards. In addition, progressive scan at the slow rate of 24 frames/second, as opposed to, say, 60 frames/second for one of the high definition television formats, contributes to the film "look" that is distinctly different from television. Although perfectly acceptable most of the time, this low frame rate does require careful control of camera movement in shooting in order to avoid excessive motion judder. Because of the considerable temporal undersampling that occurs, it also produces a familiar artifact: the apparent contrarotation of wagon wheels. In addition, the 2-3 pulldown technique used for converting 24 frames/second film to 30 frames/second video can result in a characteristic loss of resolution on moving objects and a slightly irregular motion judder. This is due to the way that alternate film frames are displayed in a different number of video fields (see later section on Signal Processing).

Different Image Aspect Ratios

Television images have aspect ratios of either 4:3 (most standard definition television) or 16:9 (widescreen for both standard and high definition television), but film has a wide variety of aspect ratios, nearly all of them considerably wider than 4:3. Although 16:9 television is a closer fit to film, some film formats (for example, CinemaScope at 2.39:1) are still wider than the television image. The result is that there is a frequent requirement for a telecine transfer to include either resizing or controlled cropping of the film image. In resizing, the film image is digitally zoomed out by the telecine until it fits within the television image limits. This has the advantage that the full extent of the film image is seen at all times. The disadvantage is that black bars will appear at the top and bottom of the screen, leading to the term *letterbox mode*.

In controlled cropping, the film image is kept full height, but a portion of the image, selected by an operator, is digitally shifted to the left and right in a spatial processor to keep the most important action or scene content within the television frame limits. As the action moves within the film frame, the shifting function follows it dynamically. This mode is known as *pan-scan*. Because of the impracticality for an operator to perform this continual adjustment in real time, pan-scan transfers first require a *learning pass* in which the telecine is stopped whenever the center of action moves, and the pan-scan control factor is adjusted and recorded in a list synchronized by time code or a film counter. For the subsequent transfer, an *execution pass* is then carried out; the telecine's control system continually reads the pan-scan list and applies the changing stored control values in real time to the spatial processor. Management of the pan-scan list is commonly one of multiple functions provided by the telecine's color corrector accessory (discussed later).

Different Requirements for Color Spectral Separation

As the front end of a three primary color image reproduction system, a video camera is required to generate the three drive signals for the display device at the end of the chain. These are known as *primaries* or *tristimulus values* (usually red, green, and blue), because, for any color occurring in the original scene, a triplet of drive signal values can be derived that, to the human vision system, will subjectively match that color when added together or seen together on a display. This principle applies to both film and television color reproduction systems, but the film system has the additional complication of the negative–positive subsystem at its front end.

The three color tristimulus values are derived in a video camera by spectral separation of the band of visible colors impinging on the camera's sensor from the lens. These tristimulus values have to be derived in a telecine, too, but there is an important difference: The negative film stock in the film camera will already have performed spectral separation. The task for the color separator in the telecine is not to perform separation again but rather to shape the overall system spectral response curves to allow accurate reading of each dye layer's density record, with a minimum of crosstalk from the other layers, as shown in Figure 5.23-3. The filter curves are designed to capture the maximum spectral width of each dye's response for a good signal-to-noise ratio but with minimum response in the spectral crossover zones between dyes. Video camera curves applied to a telecine would not achieve good color reproduction because their shaping would not have this requirement in mind.

Impairments

Since they are carried on a mechanical and photochemical medium, reproduced film images can contain impairments not found in video sources. Some of these accumulate quite rapidly with repeated handling and playing of a section of film. Often, correction or compensation as part of the telecine film-to-video

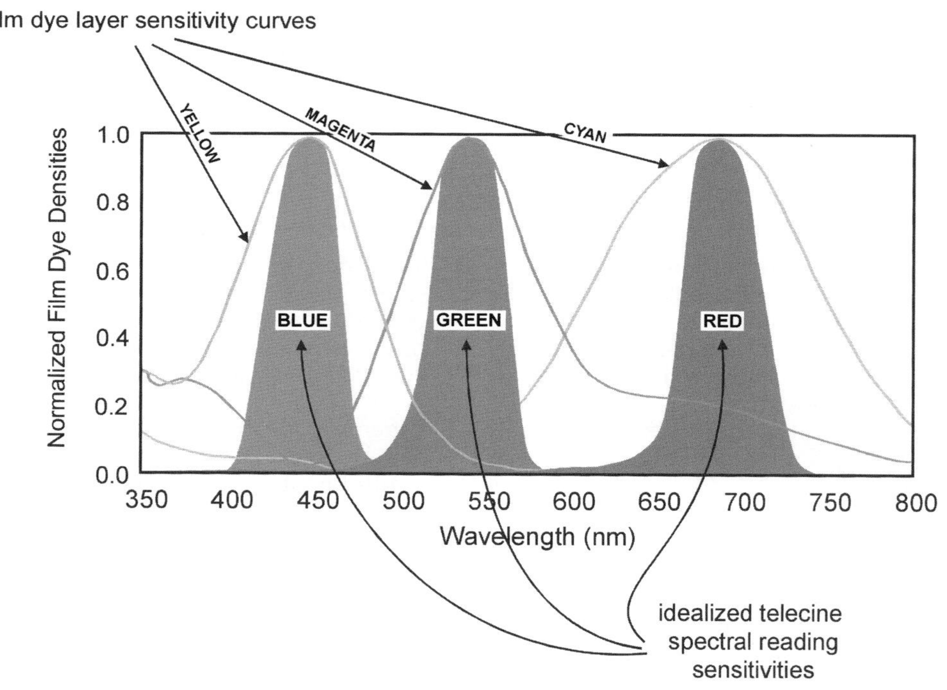

FIGURE 5.23-3 Desired relationship of telecine color channel responses to film sensitometric curves. Refer to the CD for a color version of this figure.

transfer process will be required. The major impairment categories are:

- Scratches and dirt, either on the base side of film (can be concealed) or on the emulsion side (more difficult to correct); loose dirt is removed by film cleaning, and embedded dirt may respond to electronic detection and concealment.
- Image instability; this will be far less in the original camera negative than in a release print; it can be minimized by telecine film transport design and further reduced electronically. It normally only requires special attention if the film image is to be composited with non-film-originated images such as computer-generated imagery (CGI).
- Other deteriorations due to long-term storage, including fading of film dyes causing hue and saturation changes, and fading of density; fading may not be constant across the film frame—a left-to-right differential due to film can storage orientation may be present.

SCANNING FILM

All methods of conversion from the optical images residing on a succession of film frames to the generation of a video signal have five basic components in common:

- A light source
- A transport system to move each film frame in succession into a gate in front of the light source
- A method of scanning the film frame to produce a series of image samples that can be made compatible with the television scanning system in use
- A detector that converts light modulated by the film image information into an electrical signal
- A method of separating the color information in the film into the color channels of television

Historically, there have been several completely different physical implementations of these common components. A key distinction is how these components are physically combined. Three telecine implementations in substantial use today will be described based on film scanning using *area array, flying spot*, and *line array charge-coupled device* (CCD) techniques.

Another determinant in classification is real time *versus* non-real time. A telecine is designed for real-time operation; that is, 24 frame film will be transferred at 24 frames/second. Conversely, a so-called film scanner will generally be a non-real-time device. A few machines, such as the one described at the end of this chapter, combine both modes of operation.

Area Array Telecine

As shown in Figure 5.23-4, the area array telecine is, in essence, a color television camera looking at film frames, one after the other. It could also be considered a film projector with a television camera in place of the theater screen and has the similarity that intermittent film motion is required. The method of operation is as follows:

- A broad-spectrum light source is used, so as to illuminate all three dye layers in the color film sufficiently. This will generally be a xenon type, similar to those used in optical film projectors, although of much lower wattage. An intermittent-motion film

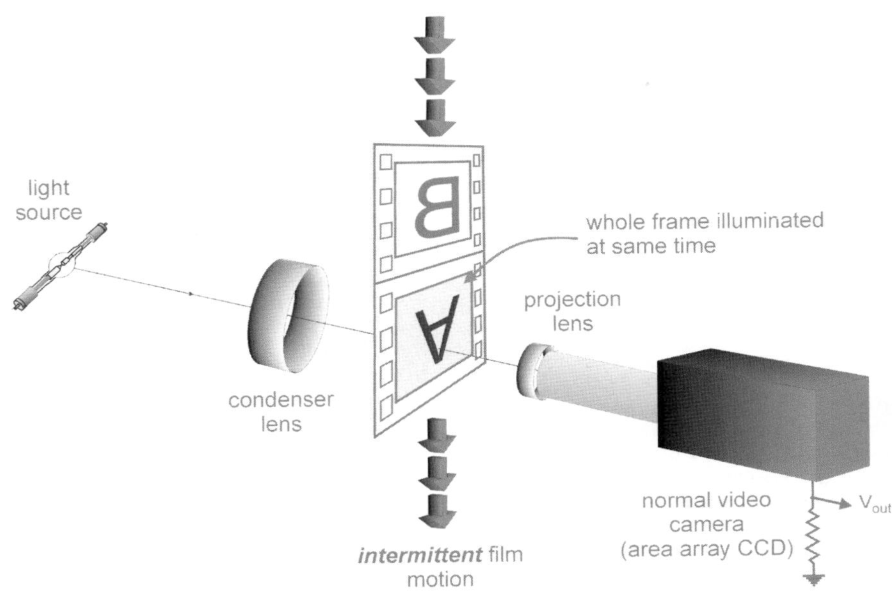

FIGURE 5.23-4 Area array telecine.

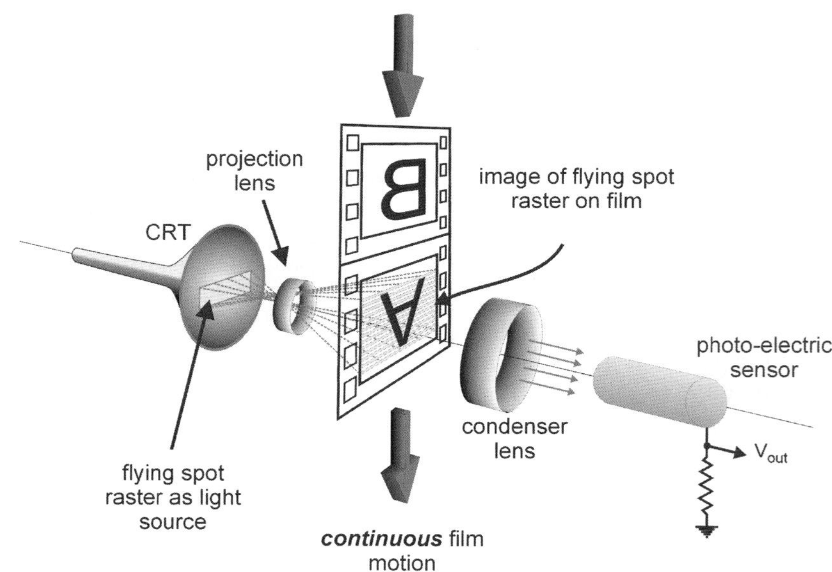

FIGURE 5.23-5 Flying spot telecine.

transport, similar to that found in a film projector but with much faster "pulldown" from one frame to the next, positions each film frame in sequence in a gate and holds it still for scanning.

- A projector lens focuses the light modulated by the film density information in the gate onto area array CCDs, similar to those found in a broadcast video camera. The red, green, and blue CCDs produce a video output in the usual way, although if interlaced video is required an interlace structure will have to be added by subsequent processing.

The distinguishing characteristics of the area array telecine are:

- The light source is a simple projector lamp.
- The transport has to provide intermittent motion of the film.
- The scanning for television is performed in the CCD.

The transport must hold the film frame sufficiently steady during the CCD's exposure and charge integration period; otherwise, smeared images will result. In addition, rapid advancement of each new film frame is required to allow a sufficiently long integration period for the CCD.

Flying Spot Telecine

In the flying spot telecine, shown in Figure 5.23-5, the scanning is part of the light source. An unmodulated raster on a cathode ray tube (CRT) is produced by a discrete spot of light deflected horizontally and vertically into a rectangular patch, hence the term "flying spot." A broad-spectrum CRT phosphor is used so as to illuminate all three dye layers in the color film. Image sensing then requires only a simple photoelectric cell or photo diode to measure the amount of light passing through the film frame and modulated by the film density at a given instant in time; the output of the photocell will then be a video signal.

The much-simplified illustration in the figure shows the light output after passing through the film being detected by a single photoelectric sensor. In practice, the light goes through a color beam splitter to three separate sensors for red, green, and blue. The distinguishing characteristics of the flying spot telecine are:

- The light source is a very high precision, continuous phosphor cathode ray tube (not a color tube).
- The film transport is a simple continuous motion type.
- The sensor is a simple photoelectric cell.
- Scanning the image into a video signal is a combination of the motion of the spot on the CRT with the continuous vertical motion of the film through the gate; all of the horizontal scan component is provided by the spot, but the total effective vertical scan is the summation of the spot's vertical deflection and the film's motion.

Although both transport and image-sensing sections of the flying spot telecine are much simpler than those of the area array CCD, the CRT as a light source is a more complex subsystem. The light spot must be very bright for good signal-to-noise ratio yet remain tightly focused for good resolution (especially for high-definition television). The position, dimensions, and linearity of the CRT raster must be very precisely controlled and stable; this is achievable only with elaborate sensing and feedback circuitry for the scan-generating electronics. The CRT also has a short operational life, with realignment of the telecine scanning

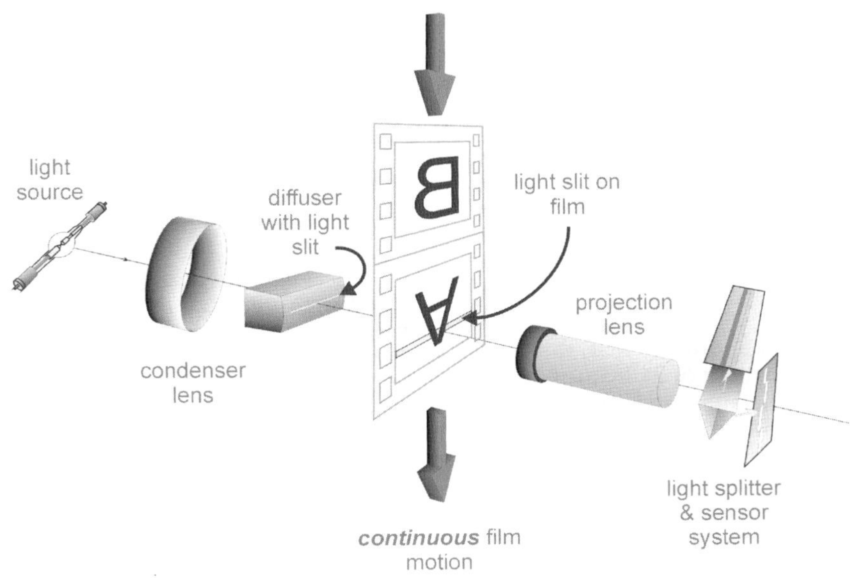

FIGURE 5.23-6 Line array CCD telecine.

section required at each replacement. On the positive side, the flying spot telecine is very flexible in adaptability to different film formats.

Line Array CCD Telecine

As shown in Figure 5.23-6, the line array CCD telecine combines the simple light source of the area array telecine with the simple continuous motion film transport of the flying spot telecine. This combination is made possible by using a one-dimensional, or line array, CCD sensor rather than a two-dimensional area array. Scanning the film frame's image into a video signal is then divided between horizontal scanning from the clocking and shift register action of the CCD subsystem and vertical scanning from the continuous vertical motion of the film. Scanning just one line at a time, the illumination is confined to a narrow slit of light impinging on the CCD. As shown in the illustration, after passing through the projection lens, the light goes through a color beam splitter and then to three separate CCD sensors for red, green, and blue. The line array CCD telecine shares the advantage of the flying spot telecine in accommodating multiple film formats easily. It also has the advantage that its sensor system does not wear out, so less maintenance is required.

Non-Real-Time Film Scanners

Achieving high image reproduction performance (resolution, dynamic range, signal-to-noise ratio, and other parameters) in real time can result in a complex and comparatively expensive telecine; great precision in mechanical, optical, and electrical engineering is required. If there is less urgency in a film transfer project or if only relatively short runs of film are to be scanned, the non-real-time film scanner may be a viable alternative. Without the need for real-time operation, there is more scope for variation in the scanning mechanism; for example, in a line array film scanner, the extra time allows the vertical component of scanning to be obtained without dependence on the film motion. Each film frame can be advanced into a gate and then held stationary while the line array CCD is mechanically swept from top to bottom of the frame. Further, rather than needing a color spectrum splitter and three CCD sensors (one per primary color), the vertical sweep can be repeated three times, using successive red, green, and blue filters on a rotating wheel for each sweep of the single CCD sensor. Such measures simplify and reduce the cost of the device but result in a transfer speed several times slower than a conventional telecine.

Non-real-time scanners produce *image data files* rather than a television signal. These will have parameter values (*e.g.*, numbers of pixels, transfer characteristics, color characteristics) quite different from those of a broadcast television signal. This means that subsequent conversion into the required television standard will be required before the scanned images can be played to air.

TELEVISION SCANNING STANDARDS

Legacy Standards

One major advantage of film is that the changes in a telecine required to switch the scanning characteristics between different world television standards are quite small because of the global consistency of film standards. For example, the frame rate of film is a nominal 24 Hz, regardless of whether the local television standard is based on 50 Hz or 60 Hz. In 50 Hz regions, the

TABLE 5.23-1
Related SMPTE and ITU-R HDTV Production Standards and Recommendations

1080i60(50)	SMPTE 274M [4] ITU-R Rec. BT.709 [5] (for international exchange)
720p60(50)	SMPTE 296M [6] (no ITU-R international exchange format)

telecine transport is simply run about 4% fast, to play the film at 25 frames/second instead of 24 frames/second. In 60 Hz regions, the telecine runs the film as shot, at 24 frames/second, but in order to bridge to the 60 Hz television rate additional duplicate television fields are inserted into the output video, following a sequence commonly referred to as *2-3 pulldown*, as described later. The accommodation of the different scanning line numbers (525 *versus* 625 for standard definition television) may be achieved by switching the parameters of the actual physical scan, but a more modern method is to oversample the film image in the scanning, then downconvert to the desired values in an onboard signal processor.

HDTV Standards

For high definition applications, the film, of course, remains the same. The output television standards, however, are different. Since the 1990s, the Advanced Television Systems Committee (ATSC) in North America (DVB and ISDB in other world regions) has set transmission standards for delivery of digital television (DTV) to consumers, including both standard definition television (SDTV) and high definition television (HDTV). In North America and some other countries, HDTV is transmitted in 1080-line and 720-line variants with nominal frame rates of 30 frames/second and 60 frames/second. In other parts of the world, these same variants are employed, but with different frame rates based on 50 Hz, according to region. Although ATSC, DVB, and ISDB define the world's DTV transmission formats, the feeder formats for content production are set by the Society of Motion Picture and Television Engineers (SMPTE) and the International Telecommunications Union, Radiocommunications Sector (ITU-R). The respective HDTV formats that the telecine engineer needs to be aware of are shown in Table 5.23-1.

SIGNAL PROCESSING

Signal processing in a telecine can be broadly divided into that which is common to all telecines and that which is required to deal with the specifics of a particular type of film image sensing and capture (for example, area array *versus* flying spot or line array CCD). Examples of the latter are covered in the section at the end of the chapter on a specific telecine implementation. The signal processing common to all telecines is considered in this section, which examines some particular areas of interest.

Processing Generics

The response of film to light is not linear and, in addition, telecines are frequently required to reproduce images from film in its negative form. These two factors determine much of the special nature of the signal processing required in a telecine. This can be contrasted to the simpler situation in television cameras, where the solid state imagers have an intrinsically linear response to light and this transfer characteristic always has a positive orientation.

Film Gamma versus *Television Gamma*

Although related, the term *gamma* has a different meaning in film than in television. Instead of referring to a display characteristic or its associated precorrection, gamma in film refers to the slope of the linear portion of the film density to exposure characteristic, as shown in Figure 5.23-7. In camera negative film, the slope, or gamma, is generally about 0.6 (*i.e.*, less than unity), giving the well-known large exposure latitude of negative film. Increasing amounts of exposure cause smaller amounts of incremental density increase than a linear response would produce. At the opposite extreme, release print film stock can have a gamma greater than 3.0, giving the stock its characteristic "contrasty" look. Intermediate (interpositive or internegative) stock is in between, having a unity gamma. A telecine must process all of these variants to a common linear gamma before applying television formatting.

Negative Inversion

The telecine allows the cost and performance losses of multiple film processing steps to be avoided by scanning the negative directly. Inversion of tonescale polarity in the output from the telecine's sensors (black-to-white and white-to-black) must therefore be

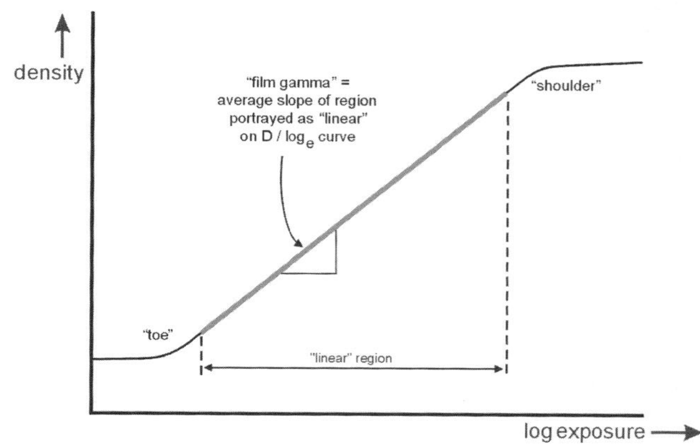

FIGURE 5.23-7 Definition of film gamma.

performed in signal processing. Complementing the color hue is also required (cyan to red, magenta to green, and yellow to blue).

Aperture Correction and Detail Enhancement

Any scanning operation involves a sampling aperture—whether a flying electron beam spot exciting a phosphor or a photon-sensing pixel in a CCD sensor—that is not infinitely small in relation to the spatial information being scanned. Aperture loss therefore occurs, and the necessary sin x/x aperture correction must be performed in the signal processor. Detail enhancement is slightly different, being more of a creative or production tool than a technical correction. It is therefore optional and more adjustable in its implementation.

Shading Correction

Film that has spent a long time in storage may have accumulated differential fading across its width or from end to end of the reel. This may take the form of variations in density or in color balance between the dye layers. The shading corrector is needed to add or subtract corrections in the appropriate video signal parameter to make the correction differential along the chosen axis.

Spatial Sizing and Positioning Functions

These functions are required both for technical reasons (for example, compensating for different image sizes and locations on film stock) and for creative reasons (for example, to assist in compositing a film-derived image with an electronic or CGI image). They are normally executed in a frame store. A related function that may be supplied is the horizontal-only scaling required for anamorphic film reproduction. A further related function that may be provided is z-axis rotation.

Television Signal Processing

Some of the processing adjustments just described are executed in a section of the telecine where the signals are not in standardized broadcast video form. The signals must therefore be converted for compatibility with the studio or network standard. Functions that are performed in television processing include:

- Conversion from linear RGB 4:4:4 to TV gamma-corrected R'G'B', then to Y'Cr'Cb' 4:2:2 components, with matrixing to ITU-R colorimetric specifications and filtering to SMPTE or ITU-R specifications
- Conversion from progressive to interlaced scanning structure
- For 60 Hz regions, 2-3 pulldown insertion
- Formatting for desired output interface—for example, SMPTE serial digital interface for standard or high definition

The 2-3 pulldown insertion process requires special mention. Film runs at 24 frames/second but most television in North America is 30 frames/second (60 fields/second). In the time that 4 film frames pass through the telecine transport, 5 television frames (10 fields) are required at the telecine's interface to the broadcast plant. This difference is accommodated by modifying the interlacing function; instead of always making two television fields out of one film frame, a sequence is followed in which every other film frame yields three fields, as shown in Figure 5.23-8. It is important that this 2-3 sequence (sometimes known as

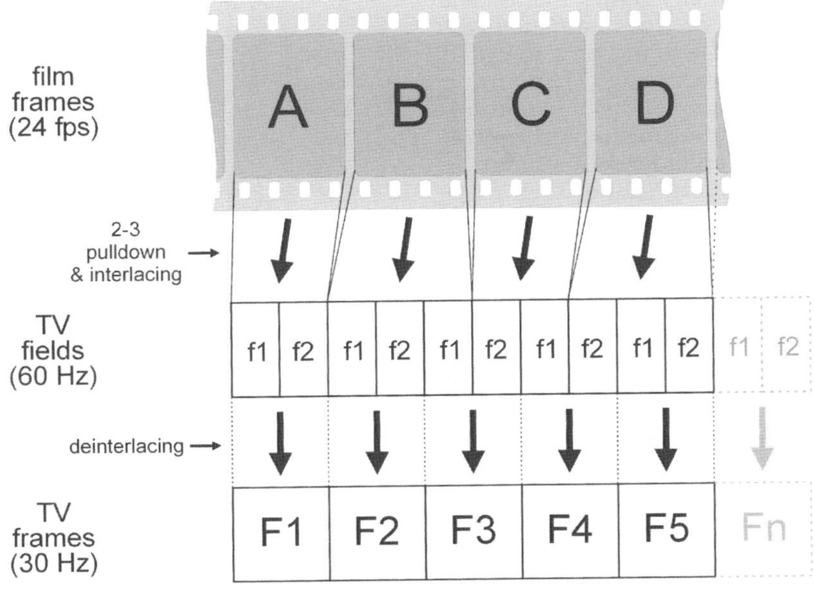

FIGURE 5.23-8 Film 2-3 pulldown sequence.

cadence) be maintained throughout the video recording made of the movie and not be broken by subsequent non-sequence-aware editing. Otherwise, deinterlacers found in devices such as consumer flat-panel displays will show a blurred blend of two different film frames instead of a clear image of the original film frame in the reconstructed progressive scan image. This distortion will continue from the point of discontinuity until the end of the movie or the next 2-3 discontinuity is reached.

SOUND FOR FILM

Sound for film is treated somewhat differently from sound for electronic television production. It is recorded separately and stays separate through more stages in post-production. For telecine users in post-production involved in new projects, sound is therefore rarely an issue, because the production sound (sound captured during shooting) will have used its own special recording formats (tape or disk) and been handled by a separate department. For the television broadcaster, it is more likely that the telecine is required to transfer a finished film production from an archive, which by definition comes complete with finished sound tracks. This section discusses the various different sources of these sound tracks and how they are handled in each case.

Types of Sound Track

Although movies are distributed to theaters in release print format, the broadcaster transferring a feature film or a documentary film from an archive will hopefully have been supplied with an interpositive or original negative as discussed earlier; however, unlike theatrical release prints, interpositive prints and negatives do not usually come with any sound tracks. The sound will have been recorded on a separate medium. With older movies, sound will be found on a so-called *SEPMAG* (separate magnetic) recording [7], requiring a separate reproducer—a telecine follower, sound follower, or "mag dubber"[3]—to be run in synchronization alongside the telecine. Even older movies may be accompanied by a separate optical (rather than magnetic) sound track, sometimes called a *SEPOPT* (separate optical) track, the optical negative or print that was used to add sound to the final release prints. In some cases, film material will have magnetic sound in the form of magnetic strips attached to either side of the perforations of the same film that carries the picture. This is known as *COMMAG* (combined magnetic) [8]. All the sound systems described so far employ analog mono, stereo, or multi-channel surround sound. The tracks may also be encoded using Dolby® A or SR [9] noise reduction, but still remain analog. Today, many movies are archived with the sound tracks on digital media—a set of CD-ROM data disks or R-DAT tapes, for example, preserving the highest quality sound source for the film transfer.

[3]A machine with a transport like a telecine but equipped to read sound tracks from sprocketed, magnetically coated film.

Sound on Release Prints

In instances where the only copy of the movie available is in theatrical release print format, the sound may be on the same film reels as the picture. In this case, there are again several possibilities as to where the telecine operator will find the sound tracks and in what form. Traditionally, sound has been recorded optically on an analog track or tracks in a space to one side of the image frame, as shown in Figure 5.23-9. Because these tracks are carried on the same film as the picture, they are referred to as *COMOPT* (combined optical) [10]. Telecines can be equipped with a special head to read these tracks. More recently, digital multi-channel sound tracks have been added to the remaining spaces on the release print. As shown in Figure 5.23-9, these include Dolby® Digital, Sony® Dynamic Digital Sound (SDDS), and the sync track for Digital Theater System (DTS™), for which the audio is carried on a separate compact disc. A telecine would not be equipped to recover these digital tracks; they are designed to be read by special readers attached to a theater film projector. They would have to be read by separate equipment, either via a theater projector in a separate sound-only transfer or by using a telecine follower equipped with special heads.

Sound Synchronization

COMOPT and COMMAG sound tracks achieve automatic lock to the picture by sharing the same recording medium; however, the sound information is

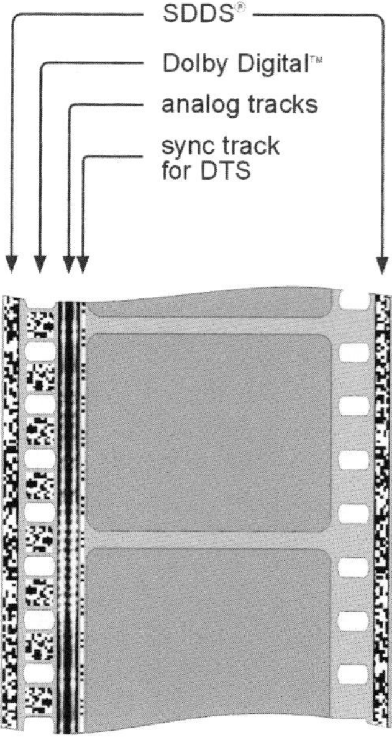

FIGURE 5.23-9 Sound tracks on a release print.

deliberately offset in timing from the corresponding picture frame. This allows the sound reproducing head to be positioned away from the picture gate and also enables adjustment of sound synchronization in theaters. On the telecine a similar adjustment of sound-to-picture synchronization will be required. This used to be performed by adjusting the effective length of film between sound head and picture gate by moving a film path roller. On modern telecines, it is done by programming an adjustable audio delay line. SEPOPT and SEPMAG sound tracks must be externally synchronized to the picture track: A sync mark placed on the sound track (a short burst of 1 kHz tone) is aligned in the sound reproducer with a corresponding sync mark (a punched hole) on the picture track in the telecine. A system of interlock pulses exchanged between the two transports then retains that synchronization. When the sound tracks are on separate digital media, a more elaborate time code system is used to synchronize sound and picture.

Sound Playback Processing

In many cases, the telecine will have sufficient built-in processing to handle simple analog COMOPT tracks, but some tracks will have been recorded with noise reduction and spectral filtering, as well as matrixed multi-channels. The best known is Dolby® Surround. Playback of these requires a special processor to reverse the characteristics applied in recording. This may be a separate external unit or in some cases a module built into the telecine.

Cyan Dye Sound Tracks

There is an issue with COMOPT sound tracks on release prints that can affect their playback on a telecine. Relying on the same chemical processing formulation used for picture information on color film has never been successful for sound, because the dominant light from the tungsten exciter lamps in theater projector sound heads penetrates right through a color emulsion, reducing the sound modulation. The traditional solution was to add an early processing stage to the film print that fixes a black-and-white silver image in the sound track area before all the silver is removed from the color picture area; however, the cost and environmental impact of this complex laboratory process have led to the development of an alternative arrangement. With this method, the broadband tungsten lamp is replaced by a narrow-band red LED, and the sound track area is printed cyan in color (maximum blocker of red) for maximum modulation of the LED and thus best signal-to-noise ratio. Using a cyan dye track in this way avoids special chemical processing requirements. If such a print is to be reproduced on a telecine, the machine must be equipped with a similar red LED to read the sound tracks properly.

Operational Philosophy for Telecine Sound Reproduction

From the previous discussion, it is apparent that a large number of variants in coupling sound to picture have been tried over the years and will therefore be reflected in the contents of a significant film archive, as well as in new productions. A considerable arsenal of replay equipment and interconnection support is required to be able to reproduce them all.

TELECINE PERIPHERALS

Control Peripherals

Like videotape recorders (VTRs), the telecine generally has its transport functions under remote control. Commonly, the telecine is in a separate machine room visible behind a glass door and accessed only for loading film and cleaning the film path. The machine is then run from a remote control panel that integrates telecine control with that of associated facility machines, such as a VTR or disk-based video server assigned to record the telecine's output. The remote controller may also include basic list management for cueing the telecine rapidly between preselected takes or positions on the currently loaded reel in accordance with a *pull* or *transfer list* (or this function may be performed by the *color corrector*, as discussed next).

Signal Processing Peripherals

The most significant of the signal processing peripherals is the color corrector, which is often a separate external module supplied by a different manufacturer. At its most basic, it provides lift, gamma, and gain adjustment in red, green, and blue, as well as a "master" (R, G, and B channel controls ganged together). This basic level performs what is generally referred to as *primary* color correction. The next level is *secondary* color correction. Corrections here can also be made to the color complements cyan, magenta, and yellow; a further feature is *hue isolation and replacement*. The corrector searches for a particular color hue in the image and changes the color of each pixel where it is detected to any other color hue, saturation, or luminance value. Any hue can be set as the target for replacement, while the detection window, in hue terms, can be made narrow or broad. A good secondary corrector will provide multiple parallel channels of such color replacement. An example of an application would be to change the color of an actor's eyes, the color of the sky, and so on.

The color corrector functions described apply to the entire image frame. A third level of sophistication allows the detection and application of corrector functions to be confined to just a part of the image frame—for example, to make the sky appear more blue but not to change the landscape. A combination of key masks and other processing tools makes these effects possible.

Many ancillary features will be found in a sophisticated telecine color corrector. One of the most important is list management, which allows the creative task of color correction to be divided into an adjustment pass and an execution or "transfer" pass. In the adjustment pass, color corrector settings are optimized in non-real time and stored in a color decision list, rather like an edit decision list. These settings will change shot by shot or scene by scene. When all the decisions have been made, the completed decision list is recalled for execution during film transfer. The telecine and color corrector signal processor will then switch on frame boundaries from one list entry to the next, causing the telecine output to stay correct as it flows in one continuous transfer to the VTR or other recording device. A supporting feature is a reference frame storage and management system. This allows the telecine colorist to assign a color-corrected frame from a scene as the reference frame for the entire sequence or project being transferred and load it into a frame store. When color correcting other scenes, the colorist can then quickly retrieve this reference frame and use it to achieve a matched look throughout the project. Integrating these multiple functions requires coordinated control; the sophisticated color corrector will therefore also perform the remote control function described earlier.

Recording the Telecine Output

The importance of recording the telecine output is specific to the film medium. Many archival films, as well as new productions, are of high intrinsic value, and their owners do not want them to stray far from their normal storage home, preferring to release a copy obtained by telecine transfer. Film is also physically fragile, its mechanical/analog format being vulnerable to damage such as scratches and dirt. Further, the cost and size of telecines, the associated cleaning and other labor-intensive attributes, as well as limited shuttle speeds discourage their direct use as on-air sources. Instead, the telecine output is invariably recorded to tape or disk first, and this recording is then used for on-air playout.

As described previously in the Television Signal Processing section, the functions in the onboard output processor of the telecine result in video outputs that are completely compliant with the SMPTE and ITU-R image structure and interface standards quoted earlier; for example, 2-3 pulldown insertion to accommodate the frame rate difference and conversion from progressive scan to interlace will already have been performed if the local studio standard requires it (*e.g.*, 1080i60). The telecine is therefore no different from other facility equipment items in being able to benefit from the wide choice of recording options available today. Random access disk recorders and file servers are the alternative to the digital VTR, although the latter is still favored for interfacility exchange because the removable tape cassette allows secure, long-term storage with no parts in constant rotation and is fairly immune to shock during transportation. For standard definition outputs, Digital Betacam® is the dominant videotape format of choice. For high definition, HD-D5 and HDCAM are widely used.

TRENDS IN FILM TRANSFER

Single HD Film Transfer Scan for All Resolution Levels

Rather than performing separate film transfers for standard-definition and high definition video applications, modern telecines and post-production plants make it practical to transfer the film content just once at HD resolution and store it at that resolution for all purposes. A standard definition version can then be made later via a downconverter.

Film Scanning and Digital Image Processing as Separate Activities

This is an extension of the first trend. Traditionally, the telecine has converted the signal derived from the film into a broadcast-standard video signal as soon as possible, processed it, and delivered it at its output interface in that particular standard. However, this meant that the film-to-video transfer, as well as all of the downstream production processes, such as color correction, had to be repeated for each television standard if standards conversion artifacts were to be avoided. This was a source of considerable extra time and expense.

An elegant solution to this problem is the *standard- and resolution-independent transfer.* Parameters of the original film image that are generic to all possible wanted output standards are used as the basis for an image representation that is not a video standard, but an image data format. It is then used internally in the telecine's processing path and made available at the machine's output as a file or stream. This cannot be recorded on a broadcast VTR, and computer disks and/or data tape recorders must be used for this purpose. A suitable image data format might include the following characteristics:

- Frame rate 24 frames/second (generic to both 50 Hz and 60 Hz TV regions, via output conversion)
- Spatial resolution of approximately 2000 pixels/scan line (referred to as *2K resolution*)
- A digital coding of the film's dye layer densities

This image data is a digital representation of what is on the film, rather than what is to be delivered to any particular one of the multiple broadcast platforms. As such, it carries the flexibility of the original film image into the digital domain. Time-consuming and expensive operations such as creative color correction can then be performed on this single image data representation and locked into it just once, leaving the delivery stage as a semi-automated digital conversion process from the single image data file into whichever broadcast standard is required. A telecine that can operate this way is the Thomson Spirit DataCine® system.

Telecine operations could then flow as follows:

Step (a) In one facility, film reels are transferred via the so-called *one-light process* (telecine settings constant throughout the transfer) to digital files on the DataCine®. Essentially, the digital coding range (up

to 16 bits per color) simply spans the film's density from minimum to maximum. No interpretation is made at this stage of the meaning, in terms of output luminance level or any other quantity, of any density level thus encoded. That is left until step (b). The film reels are then returned to a safe archive.

Step (b) The digital files from (a) are edited, color corrected, composited, and processed in other ways as needed. This has the advantage that the digital files cannot get scratched or dirty as the original film would with repeated access; also, modern disk-based file servers provide much faster random access than shuttling film, and no repeated unloading and reloading is required where the content is spread across multiple reels (sometimes hundreds of them).

Step (c) The common digital files are converted into the desired broadcast transmission format or formats for as many as are needed.

The major advantage is that steps (a), (b), and (c) can be separated not only by time (night shift and day shift), but by facility: One facility could perform step (a) and own telecines but no post-production plant, while facilities responsible for steps (b) and (c) need not own telecines.

This method is rapidly becoming established in the post-production of feature films; the digital stages subsequent to scanning are now referred to as the "digital intermediate" process, in part because compared to the photochemical intermediate (interpositive) produced in the conventional film-making process, the digital files produced have similar flexibility: They can be converted into video for broadcast or DVD, as well as recorded back to film to make conventional release prints. The method might also be justified eventually for broadcasters, but there does have to be a certain minimum level of need for such a multiformat release to justify the increased costs; a telecine running in conventional video-out mode will produce a broadcastable transfer of film material directly without any conversion.

EXAMPLE FILM SCANNER/TELECINE

The Thomson Spirit 4K DataCine® system, shown in Figure 5.23-10, is described here in some detail as an example of a current state of the art film scanner and telecine. This machine scans most film image formats in 35mm and 16mm gauges at a resolution up to 4096 × 3124 pixels (4K), at a rate of 6 frames/second. The scanning section may alternatively be switched to 2K resolution mode (up to 2048 × 1562 pixels), providing real-time operation at 24, 25, or 30 frames/second. The type of output is also selectable: Spirit DataCine® system modes provide image data files over a Gigabit System Network (GSN) interface [11]; conventional telecine modes deliver SDTV and HDTV video in most world standards over SMPTE 259M [12] and 292 [13] serial digital interfaces.

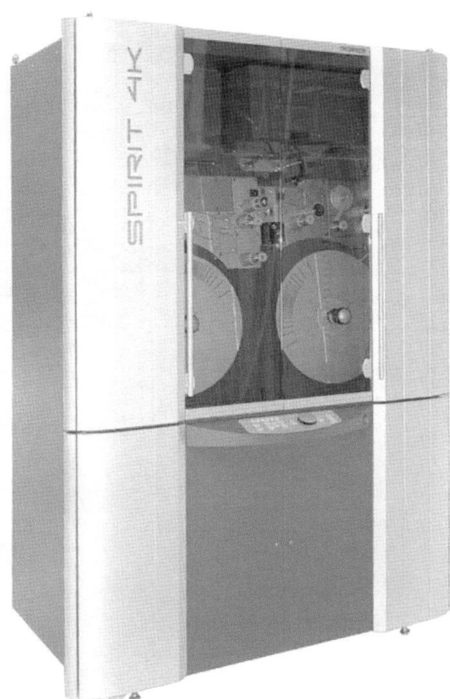

FIGURE 5.23-10 Example of telecine machine: Thomson Spirit 4K DataCine® system. (Courtesy of Thomson.)

Scanning System

The Spirit 4K DataCine® system operates on the line array CCD principle described in the Scanning Film section. A refinement is the provision of separate 4K and 2K resolution imagers, allowing both non-real-time scanning at 4K resolution and real-time scanning at 2K resolution. In the real-time mode, the 2K image data can be internally downconverted on-the-fly to HDTV and SDTV resolutions. In a third mode, the machine can scan at 4K resolution and internally downconvert to 2K resolution. This gives an advantage in reproduced image sharpness and reduced aliasing compared to native 2K scanning, but at the expense of non-real-time operation.

Film Transport

Figure 5.23-11 identifies the principle film transport components of the Spirit 4K DataCine® system. The machine uses a continuous motion transport, with the film's perforations being used only to generate servo feedback pulses for capstan and framing control purposes. Play mode traction is provided by a large wrap angle capstan assembly. By changing rollers and gate assemblies, both 16mm and 35mm (full aperture and Academy aperture) film gauges can be played. Provision on the transport is also made for 8mm and 65/70mm gauge film gates.

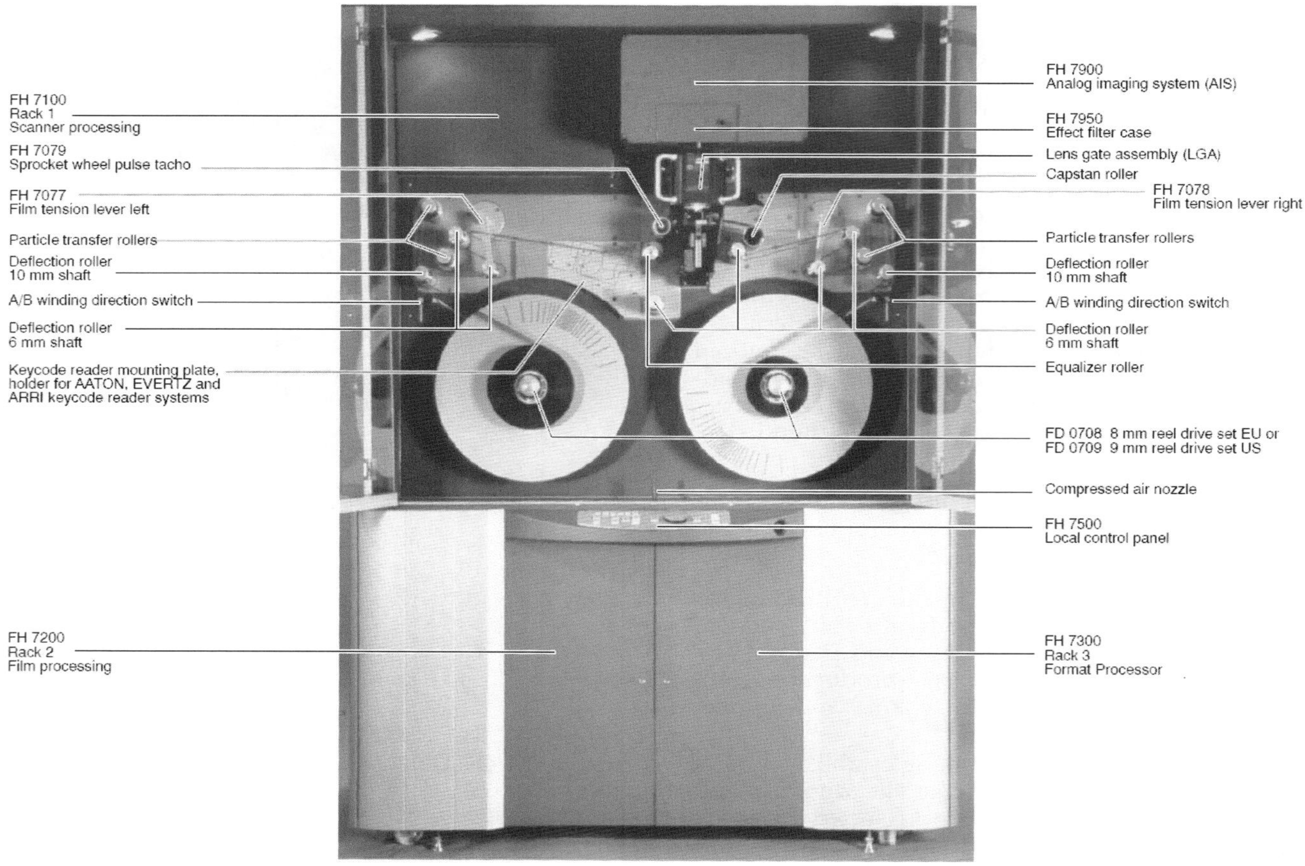

FH 7100
Rack 1
Scanner processing
FH 7079
Sprocket wheel pulse tacho
FH 7077
Film tension lever left
Particle transfer rollers
Deflection roller
10 mm shaft
A/B winding direction switch
Deflection roller
6 mm shaft
Keycode reader mounting plate, holder for AATON, EVERTZ and ARRI keycode reader systems
FH 7200
Rack 2
Film processing
FH 7900
Analog imaging system (AIS)
FH 7950
Effect filter case
Lens gate assembly (LGA)
Capstan roller
FH 7078
Film tension lever right
Particle transfer rollers
Deflection roller
10 mm shaft
A/B winding direction switch
Deflection roller
6 mm shaft
Equalizer roller
FD 0708 8 mm reel drive set EU or
FD 0709 9 mm reel drive set US
Compressed air nozzle
FH 7500
Local control panel
FH 7300
Rack 3
Format Processor

FIGURE 5.23-11 Front view of the Spirit 4K DataCine® system showing the film transport components. (Courtesy of Thomson.)

Light Path

Figure 5.23-12 shows the light path in the Spirit 4K DataCine® system. The light source is a 700 watt xenon lamp, chosen for its broad spectral power distribution, especially in the blue spectral region. This is important for reproducing negative film, which has an orange mask that attenuates blue light. A multi-position filter wheel balances the spectral power distribution of the light source according to the type of film. The different positions cater to negative, intermediate, and print stock. A fast-response aperture allows dynamic control of illumination intensity, allowing a close-to-real-time response to execution of color correction decision lists, where illumination has to switch quickly from one setting to another. Before reaching the film gate, the light passes through an integration cylinder to reduce the visibility of any base film scratches or dirt. The light then impinges on the film through a narrow slit, giving maximum efficiency of the transfer of light energy through the film and onto the line array (one-dimensional) CCD sensor assemblies. The slit image of the film is sharply focused onto the sensors by the objective lens, which is integrated with the gate in a lens/gate assembly. Different assemblies are used for different types and formats of film: 16mm, Super 16mm, 35mm full aperture, and 35mm Academy. A complete gate assembly is shown in Figure 5.23-13.

Imaging System

A dichroic filter-based RGB beam splitter separates the image from the film into three spectral parts, each directed to a CCD sensor. A CCD sensor assembly is shown diagramatically in Figure 5.23-14. Two separate line arrays are mounted integrally on each sensor integrated circuit package: one for 4K resolution (4096 pixels along the line) and the other for 2K (2048 pixels). The output of each red, green, or blue CCD sensor goes to an individual analog-to-digital converter (A/D). The A/Ds deliver the digitized raw image signals to the 3 × 16-bit wide internal signal processing system.

Signal Processing System

Figure 5.23-15 is a block schematic diagram of the all-digital signal processing system, containing a number of modular subsystems. Some key stages are described here in functional terms.

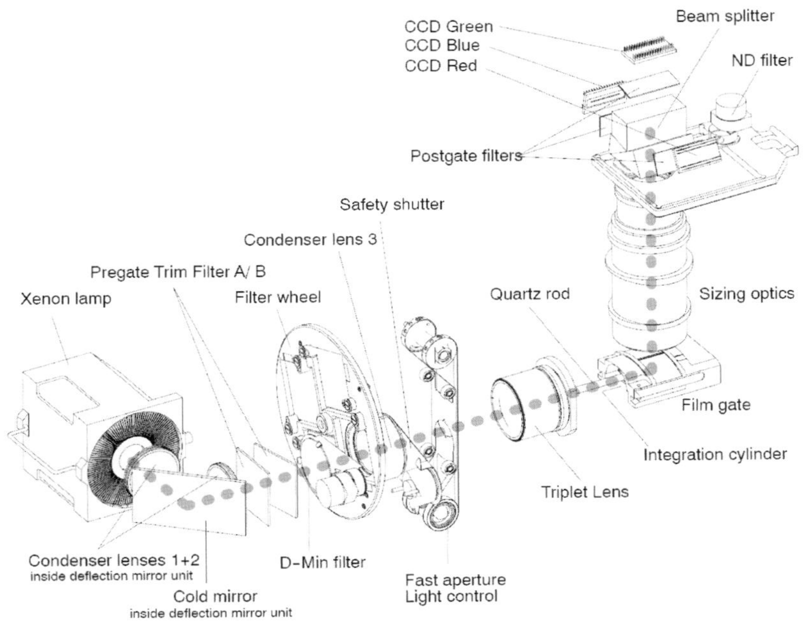

FIGURE 5.23-12 Light path of the Spirit 4K DataCine® system. (Courtesy of Thomson.)

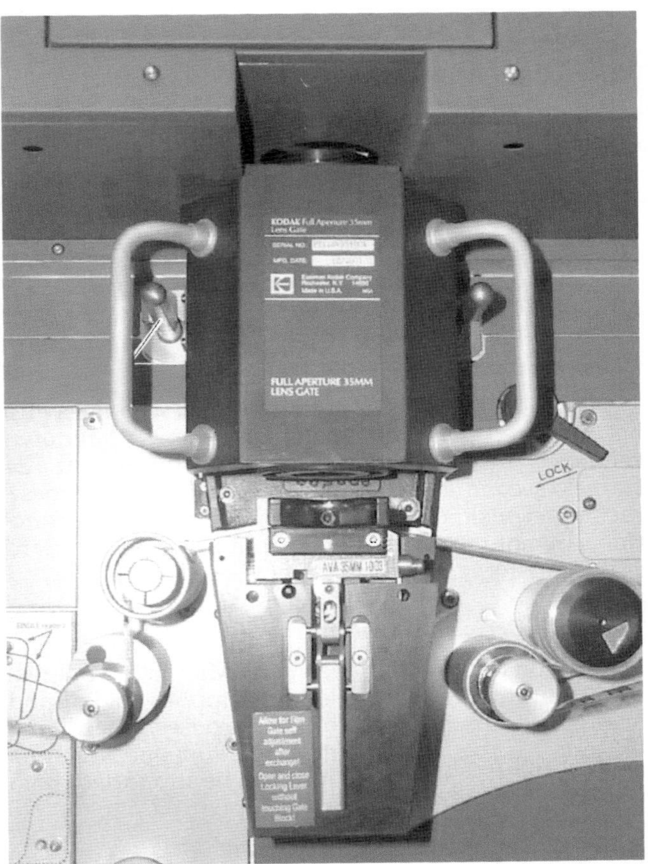

FIGURE 5.23-13 Film lens/gate assembly of the Spirit 4K DataCine® system. (Courtesy of Thomson.)

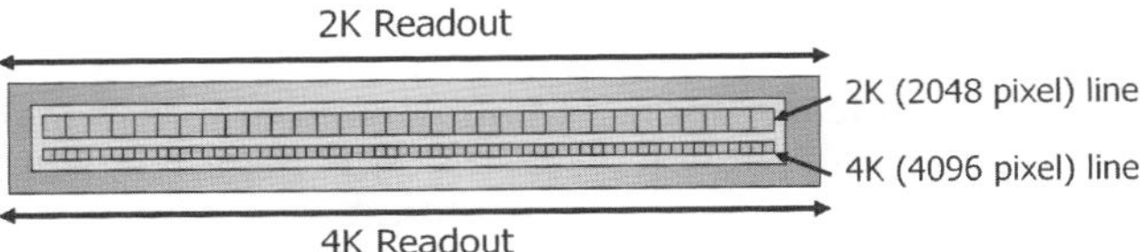

FIGURE 5.23-14 Dual 4K/2K CCD sensor assembly.

Film Image Data Frame Store

A line array telecine uses the motion of the film past the sensor to perform the vertical component of video scanning. When the film is stationary, this component disappears. To maintain a video output under these conditions, a film image data frame store is used. A high-capacity store is required because some film formats such as VistaVision extend over a large area of film, creating a very large number of pixels (up to 4096 × 16,384) when scanned at full resolution.

Aperture Correction and Film Primary Color Correction

The aperture correction function that compensates for the finite size of the CCD photosites (light-sensitive diode elements) allows adjustment of both correction amount and the spatial frequency of its peak. The color correction function refers to matching and primary color correction, with lift, gamma, and gain adjustments to each channel. This is also the stage where the different gammas of the straight-line portion of different films' transfer characteristics are equalized (gamma of 0.6 for negative, 1.0 for intermediate, and 2.0 to 3.0 for release print stock), to give an

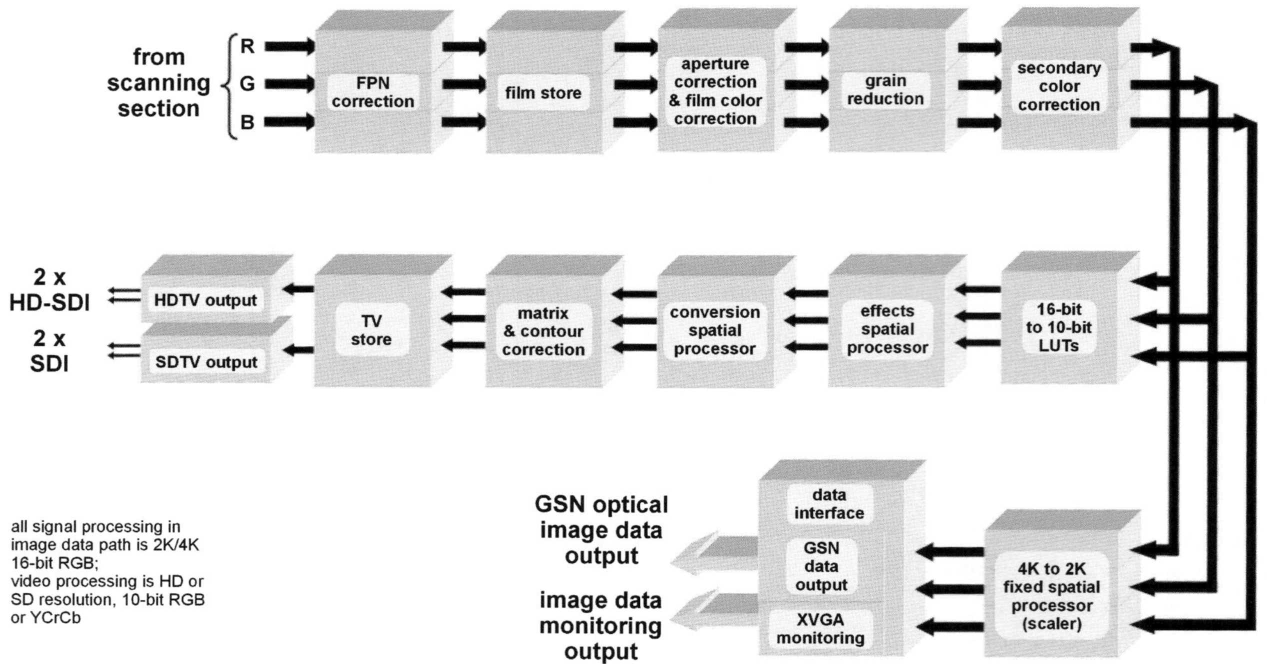

FIGURE 5.23-15 Spirit 4K DataCine® system signal processing system schematic. (Courtesy of Thomson.)

essentially linear response to light, comparable to that from a video camera's sensor. For negative film, matching of slightly different gammas in the film's color dye layers is also applied here, as well as the necessary signal inversion.

Secondary Color Correction

This is a six-sector implementation of the secondary color correction principle outlined earlier, meaning that the full system color hue range of 360 degrees is divided into six segments, with separate hue replacement actions possible in each. Following secondary color correction, signal processing branches into a video path and a data path.

Bit Depth Conversion, Gamma Correction, and LUT Insertion

The bit depth upstream of this point is 16 bits per R, G, and B component. This is maintained in the continuing data branch of the processor. In the video branch, however, for compatibility with studio video interface standards, television gamma correction is applied and the bit depth reduced to 10 bits per color component. Other modifications to the dynamic transfer characteristic can be made in this module for the data path, by means of so-called *look-up tables* (LUTs). These provide the operator with the means to compensate for factors such as particular film characteristics.

Video Path Modules

Spatial Processor for Effects

This module performs image repositioning and resizing (*e.g.*, for pan-scan) and *z*-axis rotation.

Spatial Processor for Video Downconversion

This module downconverts image data from 2K resolution to HDTV and SDTV resolutions.

Video Frame Store

This module converts each film frame's scan lines into an interlaced pair, and also creates the 2-3 pulldown sequence necessary to translate between the 24 frames/second film frame rate and 30 or 60 frames/second television frame rate. Some or all of these functions are not required when the output video standard is progressively scanned or is not based on 60 Hz.

Standard Definition Output Interface

In addition to providing the main SMPTE 259M digital serial outputs, this module also supplies digital and analog outputs for monitoring purposes. The digital outputs are available in both single-link and dual-link form, so RGB as well as YCrCb can be supported. Both 525/60 and 625/60 systems are supported.

HDTV Output Interface

This module provides SMPTE 292 HD digital serial outputs. Again, both single- and dual-link options support RGB modes as well as YCrCb modes. 1080i and 720p HD standards are supported at both 50Hz and 60Hz field/frame rates. A digital monitoring output is also provided.

Image Data Path Modules

4K to 2K Image Spatial Processor (Scaler)

This module performs the following functions:

- Formatting of scanned image to the desired pixel resolution and aspect ratio
- Image standard conversion from 4K to 2K when requested
- Image zooming and vertical/horizontal positioning

Data Interface

This module's function is to make the internal film image data available on an external interface for post-production in IT-based data formats as an alternative to video standards. In data output mode, this module will encode each film frame as a set of image data records in R, G, and B, accompanied by descriptive and functional metadata, and place all of this in a DPX [14] file wrapper. Digital Picture Exchange (DPX) wrappers are then transmitted over a GSN optical interface. The interface speed exceeds 4 Gbps (over 500 megabytes per second), allowing data transfer at 6 frames/second in 4K resolution mode, or in real time in 2K resolution mode.

Accessories

Several accessories are available, both from the telecine supplier and third parties. These include:

- Remote-controlled film transport controller, linkable to include VTRs and other output capture devices in control sync with the telecine
- Film defect management system for the concealment of scratches and dirt (those not already removed by the diffusion chamber in the light path), fading, flickering, and other impairments; operator intervention is possible, allowing selective acceptance, modification, or skipping of suggestions made by the impairment detection engine
- External version of the grain reduction module
- Full-featured color correctors, offering primary and secondary correction, with selective application of correction by operator-defined and image-defined masks, as well as remote transport control and color correction list management
- Time code and edge code readers, with associated synchronization systems
- Optical sound track readers and processors

SUMMARY

This chapter has provided an introduction to the current technologies used for the optimum transfer of film material into the television medium. Compared to images captured with broadcast television cameras, film images require a number of processes of conversion or modification before they can be freely intermixed with other image sources. Some of these processes are television standard dependent, but, as described in the previous sections, they include adding interlace and 2-3 pulldown, resizing the image, color correcting the image for the television viewing environment, compensating for impairments, and so on. Many of these functions are built into today's telecines and film scanners, but others require peripheral equipment, expanding the role of the responsible engineer to the system level.

References

[1] SMPTE Standards 149, 109, 139, 93, 145, 119, others.

[2] SMPTE Standards 157, 201M, 59, 195, 96M, 215, 152, others.

[3] *American Cinematographer Manual*, Eighth ed., ASC Press, Hollywood, CA, 2001, pp. 21–43; 44–51.

[4] SMPTE Standard 274M-2005: 1920 × 1080 Image Sample Structure, Digital Representation and Digital Timing Reference Sequences for Multiple Picture Rates.

[5] Recommendation ITU-R BT.709-5: Parameter Values for the HDTV Standards for Production and International Program Exchange.

[6] SMPTE Standard 296M-2001: 1280 × 720 Progressive Image Sample Structure—Analog and Digital Representation and Analog Interface.

[7] SMPTE Standard SMPTE86-2005 (for motion-picture film): Magnetic Audio Records—Two, Three, Four, and Six Records on 35-mm and One Record on 17.5-mm Magnetic Film.

[8] SMPTE Standard SMPTE177-1995 (for motion-picture film; 35-mm): Four-Track Magnetic Audio Release Prints—Magnetic Striping.

[9] http://www.dolby.com/professional/pro_audio_engineering/technologies.html.

[10] SMPTE Standard SMPTE203-2003 (for motion-picture film; 35-mm): Prints—Two-Track Photographic Audio Records.

[11] Gigabyte System Network (HIPPI-6400); specifications are maintained by the High-Performance Networking Forum (http://www.hnf.org).

[12] ANSI/SMPTE Standard 259M-1997: 10-Bit 4:2:2 Component and 4fsc Composite Digital Signals—Serial Digital Interface.

[13] SMPTE Standard 292-2006: 1.5 Gb/s Signal/Data Serial Interface.

[14] SMPTE Standard 268M-2003: File Format for Digital Moving-Picture Exchange (DPX), Version 2.0.

CHAPTER

5.24

Closed Captioning Systems

JEFF HUTCHINS*
Accessible Media Industry Coalition
Pittsburgh, Pennsylvania

ALAN LAMBSHEAD**
Evertz Microsystems
Burlington, Ontario, Canada

INTRODUCTION

Captioning, according to the Accessible Media Industry Coalition, is "the process of converting the audio content of a television broadcast, webcast, film, video, CD-ROM, DVD, live event, or other productions into text, which is displayed on a screen or monitor." Captions not only display words as the text equivalent of spoken dialogue or narration, but also include speaker identification and sound effects. It is important that the captions be

- Synchronized and appear at approximately the same time as the audio is available;
- Equivalent and equal in content to that of the audio, including speaker identification and sound effects;
- Accessible and readily available to those who need them.

Captions may be displayed on or adjacent to the video image. *Open captions* are a permanent part of the image and cannot be turned off. *Closed captions* are transmitted with the audio and video, but require a *decoder* to detect, decipher, and display the captions so that only viewers who wish to see the captions will do so. The data for closed captions is transmitted with the television program, either on line 21 in the vertical blanking interval (VBI) of an analog program, or in a separate data packet accompanying the audio and video of a digital program.

Captioning is distinct from *subtitling* in that captioning includes aural information other than just dialogue, such as sound effects, music effects, and indications of who is speaking, all intended to aid the viewer who is unable to hear the soundtrack. The original target audience for captioning was primarily people who are deaf or hearing impaired (about 20 million in 1980). The market has since expanded to include people learning English as a second language; those learning to read, especially students with reading disabilities; people in noisy places (like bars and airports) or quiet places (like hospitals and spas). Captions generally appear in the lower portion of the television screen and vary in size in proportion to the size of the television screen. The caption characters are sized to be easily visible, typically white letters against a black background (for analog television), and usually do not obstruct essential parts of the picture. Captions for digital television (DTV) can have a wider range of font styles, sizes, and colors than those for NTSC transmissions (see Figure 5.24-1).

Closed captioning may be added in real time to a live program or may be added before transmission as part of postproduction for a prerecorded program.

*Jeff Hutchins updated the analog captioning portion of this chapter drawing on material first published in the 9th edition of the *NAB Engineering Handbook*, Chapter 5.13, "Closed Captioning and Extended Services," by Amnon Salomon and Gerald Freda.

**Alan Lambshead adapted the DTV captioning portion of this chapter from material first published in the *NAB Broadcast Engineering Conference Proceedings 2004*, "Implementing Closed Captioning for DTV," by Graham Jones.

FIGURE 5.24-1 DTV caption data processed by a decoder and displayed as text across the lower portion of the picture.

A BRIEF HISTORY OF CLOSED CAPTIONING

The Public Broadcasting Service (PBS) developed closed captioned technology for NTSC analog television during the period 1973–1979 with funding support from the federal government (Department of Health, Education, and Welfare). Field test transmissions were conducted on all aspects of caption generation, encoding, decoding, and display features of the service.

During those years, a small number of programs were open-captioned by The Caption Center at PBS station WGBH and carried on PBS. The newly created National Captioning Institute (NCI), in cooperation with the ABC, CBS, NBC, and PBS networks, launched the closed captioning service in March 1980 with approximately 16 hours per week of captioned programming. The first consumer product containing the decoding feature, called *TeleCaption®*, was sold by Sears, Roebuck and Co. and was a separate box that worked with a standard television.

FCC Rules and CEA Standards

The line 21 captioning data signal of NTSC analog television signals is protected from interference from any other VBI service, test signal, or spillover from active video under FCC Rules and Regulations, Section 73.682(a)(22), adopted in 1976. These rules also established the transmission standards for captioning and list the uses of the data channel.

Following are highlights from rules associated with the Television Decoder Circuitry Act of 1990 that amended Part 15 of the FCC rules (Radio Frequency Devices, which relate to television receivers).

Television Decoder Circuitry Act of 1990

The U.S. Congress passed the *Television Decoder Circuitry Act of 1990* (Pub. L. 101-431, 104 Stat. 960 (1990)). This act required that, effective July 1, 1993, all television receivers with picture screens 13 inches or greater must be equipped to display closed captioned television transmission, and it required the FCC to enact rules to implement this requirement.

FCC Report and Order 6 of General Docket 91-1 was adopted April 12, 1991, and released to the public on April 15, 1991. The Order became effective July 1, 1993. The Order amends Part 15 of the FCC's rules by adding a new Section 15.119 to set out the FCC Rules for captions and caption decoders. The highlights of this section as it relates to NTSC analog television captioning are as follows:

- Effective July 1, 1993, TV receivers 13 inches or larger must have caption decoders;
- Closed caption information to be carried in line 21 of field 1;
- Decoders to have user-selectable Caption display modes. Text display mode is optional;
- Caption and Text modes may contain data in either of two operating channels, usually referred to as C1, C2, T1, and T2;
- Receivers must decode at least C1 and C2 captions;
- Captions to be displayed in "boxes" on the screen within the Safe Title Area defined by SMPTE 27.3 (now replaced by SMPTE RP218-2002);
- Caption decoding circuitry must be tolerant of cable security systems that alter line numbering, etc.

The complete text of the *FCC Report and Order* (R&O) may be obtained from the U.S. Government Printing Office. The text was also published in the *Federal Register,* Vol. 56, No. 114, p. 27200. The FCC Rules and Regulations are contained in the *Code of Federal Regulations*, Part 47, Telecommunications. See Chapter 1.2 of this handbook, "Broadcast-Related Organizations and Information," for information on contacting the U.S. Government Printing Office and other organizations mentioned in this chapter.[1]

Update Since Passage of Decoder Circuitry Act

Since the passage of the Decoder Circuitry Act, the FCC has ruled several additions. These additions are documented with other technical details in a Consumer Electronics Association document CEA-608, titled *Line 21 Data Services* (formerly EIA-608). At the time of writing (2006), the current revision of this document is CEA-608-D.

The regulation and use of line 21 for captioning has been expanded to include both fields. Additional features have been added to deal with additional languages. Also, within field 2 is a capacity to deliver additional captioning services (C3, C4, T3, T4) in addition to extended data services (XDS). XDS includes,

[1]The entire Part 15 regulations, as updated February 16, 2006, may also be downloaded as a PDF file from http://www.fcc.gov/oet/info/rules/part15/part15-2-16-06.pdf.

but is not limited to, content advisory (V-chip), Time of Day, and source of transmission.

For NTSC television, the Accessible Media Industry Coalition recommends using C1 (also called CC-1) for the primary captioning service, and C3 (or CC-3) on field 2 for any secondary captioning service. Bandwidth limitations make it extremely difficult, if not impossible, to use C1 and C2 simultaneously for two separate captioning services.

Digital Television Captioning

In June 1997, the next generation of closed captioning for digital television was adopted. The new Consumer Electronics Association caption standard, CEA-708, titled *Digital Television (DTV) Closed Captioning*, was developed by an industry group composed of caption service providers, receiver manufacturers, broadcasters, and encoder manufacturers. It enables improved closed captioning to be provided for digital television transmissions, including both standard definition (SD) and high definition (HD) television, while continuing to provide a method of delivering CEA-608 captions to analog receivers.

DTV closed captions (known as DTVCC) have greatly enhanced formatting and display capabilities compared to line 21 captions, with up to 63 services per program; eight independently controlled display windows; and an extended range of characters and multiple fonts, sizes, and background and character colors and edges. The standard specifies how caption information is to be coded and processed, minimum implementation recommendations for DTV closed caption decoders, and recommended practices for caption encoder and decoder manufacturers. It also requires that decoders give users control over caption font, color, size, and location that may override the parameters as transmitted.

During the transition from analog to digital broadcasting, analog NTSC and DTV transmissions will coexist, as will analog and digital cable distribution systems. In addition, analog and digital television production and distribution systems may coexist in many facilities and networks, frequently with both SD and HD video formats. This situation creates added complexity for closed captioning as well as other aspects of system implementation. A discussion of how to implement DTVCC, together with a summary of the relevant FCC rules, is in a later section of this chapter.

DISPLAY FORMAT

Analog CEA-608 captions can operate in any of three different styles: *pop-on*, *roll-up*, and *paint-on*. DTV CEA-708 captions also allow text to scroll in multiple directions.

The *pop-on* style is used when the captions are prerecorded. Captions are transmitted to a nondisplayed memory in advance of the time they are to appear. The caption data stored in this memory is then displayed by transmission of a single display code.

The *roll-up* style is most often used for captioning of live programs. From one to four rows of text display the captions in a scrolling manner.

The *paint-on* style is sometimes used in analog systems because it allows characters to be displayed immediately, avoiding the transmission lag of pop-on style that occurs thanks to the 60 characters per second data rate of line 21. Paint-on can also be used to create special effects, such as a simulated "reveal" of text added to a displayed caption.

CEA-608 Captions

CEA-608 captions (as well as material displayed in line 21 Text Services) may have any of several attributes in addition to block monochrome characters. Attributes include upper- and lowercase, six different colors, italics (or slanted text), underline, and flash. The caption attributes are determined during caption or text authoring and are communicated to the decoder using special control codes in the data stream.

The display of CEA-608 captions consists of 15 rows with up to 32 monospace characters per row. Decoder controls permit the viewer to select Captions or Text. In the Captions mode, a maximum of four rows is used onto which each caption pops on when prerecorded captions are received and rolls up when the captions are live. The four rows can appear anywhere on the 15 displayable rows which occupy most of the screen area. The first row starts at line 43 in each field. Each row occupies 13 lines of a field scan or 26 lines of the 525 lines. Each row of characters is displayed within a black surround box to enhance the readability against the normal video background. The box extends one character position to the left of the initial character in each row and one character to the right of each row.

CEA-708 Captions

The display of one CEA-708 caption service consists of up to eight windows of variable dimensions and positions that contain the text. Caption text may be proportional or monospace, with or without serifs, in up to three character sizes. Up to 64 colors are allowed, as well as a range of transparency levels, for characters, character edges, and window backgrounds. Support for these features among decoders varies, however, because FCC rules require decoders only to support a subset of the features defined in CEA-708.

CAPTION DATA ENCODERS

Encoding is the process of inserting the caption data into the VBI of an analog television signal or into the appropriate data packet of a DTV signal. The data contains the caption text in addition to positional instructions and display attributes (for instance, color and italics). The encoder is placed in the video path of the program to be captioned. There are two

types of the line 21 VBI encoders: *smart encoder* and *simple encoder.*

Simple Encoder

The simple encoder was the first type of line 21 encoder. It unconditionally strips any data already encoded on line 21. It generates the line 21 data signal, to be inserted on a video signal, from caption data received at an RS-232 serial input. This encoder is used exclusively in postproduction applications for inserting pre-authored captions on program material where there is no previously recorded caption data. The simple encoder cannot add captions to a video signal already containing line 21 data.

Smart Encoder

The smart encoder is used to insert caption data into line 21 of fields 1 or 2 of the VBI. Under user controls, it can strip all previously encoded data contained in the video, or intelligently leave one or more data channels intact while inserting new data. (Note that all previously encoded data that is to be passed through intact will actually be stripped and reinserted typically two video frames later.)

The smart encoder typically receives new data for encoding through a serial data port or via telephone line. Using a smart encoder, locally produced Caption and Text Mode Services may be added to an already closed captioned video program or if noncaption text is contained on line 21. If captions are present on the incoming program, noncaption data may be interleaved into the gaps between the captions by the smart encoder.

Smart encoders are used in the process of inserting real-time captions, live-display captions, and captions derived from scripts generated by newsroom computers (also used to feed teleprompters), and can also be used by off-line captioning systems that control the timing and preparation of data (including the generation of all control codes needed to manage the caption display).

Digital Encoders

Since the advent of DTV captions, encoder manufacturers have produced a new line of encoders to deal with analog and digital broadcasts. See the section on implementing DTVCC for more information on various types of digital encoders.

AUTHORING CAPTION DATA

Several techniques are employed to create caption data based on whether the broadcast is live or prerecorded. There are also variations within the live and prerecorded captioning technologies.

Prerecorded Captioning

Prerecorded captioning (off-line or nonlive captioning) involves the preparation of closed captions for programs that have been recorded prior to their telecast. The captions are created and then merged with the video signal and a closed captioned videotape version of the program is made. A typical block diagram of the prerecorded captioning process is shown in Figure 5.24-2. The main steps in the process are as follows.

A captioning facility receives a time-coded copy of a program master either on videotape or as a digital-video file. A caption author reviews each scene, transcribes the soundtrack, and formats the transcript into discrete captions using a personal computer. Several quality-control checks are performed, including spelling, grammar, syntax, timing, and screen positioning of the captions. The captions are saved as a file with their corresponding display time codes. The file is then transferred by any of the usual means for file transfer (LAN, WAN, modem, or by e-mail, disk, or flash memory) to the location where the program is to have the captions added or "encoded."

The recorded program and caption data are then merged, using the time codes to trigger the captions, to produce a closed captioned master video signal, which is typically recorded but may be transmitted live to air. Captions that are produced in advance are displayed on a caption-equipped receiver as pop-on captions, each one complete and timed to coincide with the spoken dialogue on the screen.

Live Captioning

Live captioning involves the addition of caption data to the television signal at the time of a live transmission or broadcast. Examples of live captioned programming include news programs, sporting events, live events (such as meetings, news conferences, and award shows), and special bulletins or reports. There are three methods of captioning live programs: real-time, live-display, and newsroom computer. A typical block diagram of the live captioning process is shown in Figure 5.24-3.

Real-Time Captions

Real-time captions are created and transmitted simultaneously. There are several methods of generating real-time captions: (1) by a court reporter (often called a "stenocaptioner") using a stenotype or similar machine that deciphers phonetic shorthand typed on a special keyboard; (2) by a "voicecaptioner" who listens to the soundtrack and simultaneously repeats each word into a speech-recognition system that has been modified to generate a formatted caption output; or (3) by a speech-recognition system attempting to transcribe directly from program audio.

Stenocaptioning

With a stenotype machine, the real-time captioner can key in up to 260 words per minute. The phonetic codes

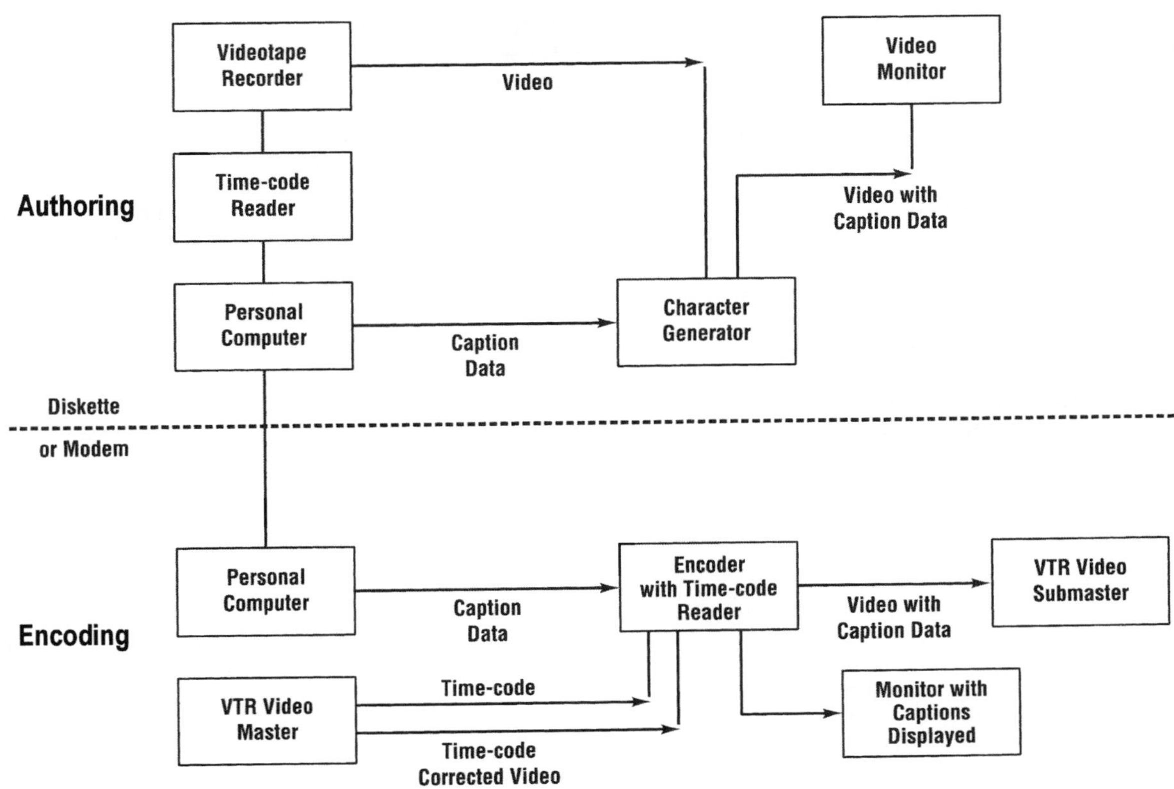

FIGURE 5.24-2 Prerecorded captioning system. Top: caption authoring; bottom: encoding.

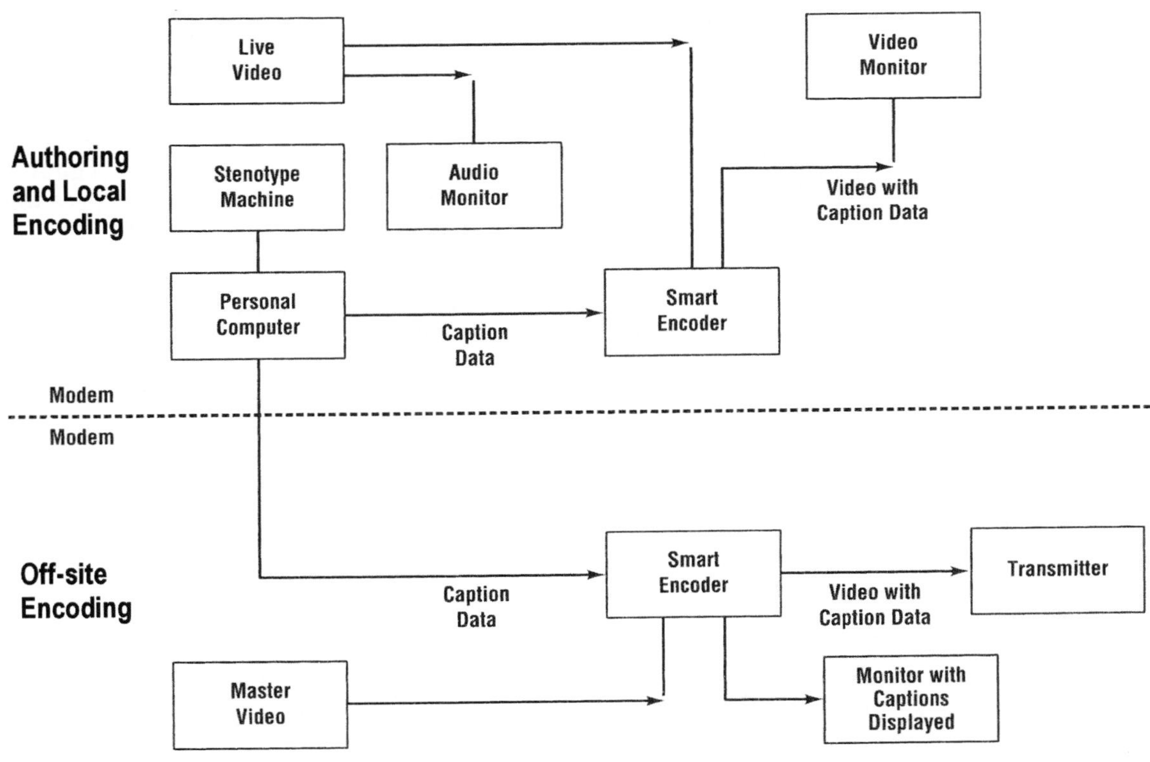

FIGURE 5.24-3 Live captioning system. Top: authoring and local encoding; bottom: off-site encoding.

are translated into English words by a computer that has been programmed with the phonetic codes and a dictionary customized to the individual court reporter. From the computer, the words and caption control codes are sent via a data circuit to a caption encoder where they are encoded into the television signal as real-time, or roll-up, captions.

A typical lag time between the time a word is spoken and the same word appears in a caption on a viewer's screen is two to three seconds.

Errors in real-time stenotype-generated captions are generally the result of one of four factors: (1) the captioner may hit the wrong keys; (2) the captioner may be unable to keep pace with the speakers; (3) the captioner may not hear the spoken word correctly; or (4) the computer may not have a particular word in its "dictionary," which is often the case when unfamiliar proper nouns (such as the names of people or places) are spoken.

Voicecaptioning

Captioning by a voicecaptioner is similar to stenocaptioning except that input is by voice rather than keyboard. The voicecaptioner trains a customized speech-recognition system to understand his/her voice accurately and quickly. Ideally, the speaker uses a high-quality microphone in a controlled, sound-isolated environment. The speech recognition software must be "trained" to the voice of the voicecaptioner. A system that must resolve only one voice has many advantages over a system that must recognize multiple voices, especially if those voices are recorded or delivered via low-fidelity methods.

Errors in real-time voice-generated captions are generally the result of one of three factors: (1) the captioner may say the wrong word; (2) the captioner may say the wrong form of a word (failing to distinguish between conflicts such as "their" and "they're" and "there"); or (3) the computer may not have a particular word in its "dictionary," which is often the case when unfamiliar proper nouns (such as the names of people or places) are spoken.

Speech-Recognition Captioning

At the time of this writing, speech-recognition systems that attempt to transcribe program audio directly with no human intervention are beginning to appear. Although it is anticipated that such speech-to-text systems will improve greatly in coming years, at this time these systems are generally far less accurate and lag farther behind program audio than the stenocaptioning and voicecaptioning systems.

Live-Display Captions

Live-display captioning is used when an accurate script is available in advance of a televised live event such as a speech or newscast. The scripted words are converted to captions by an editor and stored on a computer disk. When the live event is televised, the editor manually triggers each caption for insertion into the video signal. With live-display captioning, the words are timed to appear on the television receiver as they are being spoken. If last-minute changes are made in the script, it may be necessary to switch to real-time captioning.

Newsroom Computer Captions

The third method of live captioning is the generation of captions through the use of *newsroom computer* equipment. Many television newsrooms convert news stories from their word processors into data for use on a prompting machine for the news reporter to read while on the air. At the same time the reporter reads the script from the prompter, the computer controlling the teleprompter passes the data through a serial interface to the caption encoder that inserts the captions into the video signal (see Figure 5.24-4).

There are some dangers in this process. It is not as automatic as might be assumed. Some newsrooms add cues to the prompting text designed for the on-air personnel, some of which might be inappropriate to appear as captions. Program breaks or changes of scene may require changes in the prompting operation that could interrupt the captioning process. Finally, it may be necessary for an operator to signal the captioning encoder to add captions at the beginning of the local program and to revert to the pass-through of upstream captions at the end so that previously encoded captions (for example, on a network newscast that follows local news) will be transmitted intact. It is desirable for the operators and program producers to have an off-air monitor to ensure the integrity of the captioning process.

Note that FCC rules prohibit ABC, CBS, Fox, and NBC and their affiliates in the top 25 television mar-

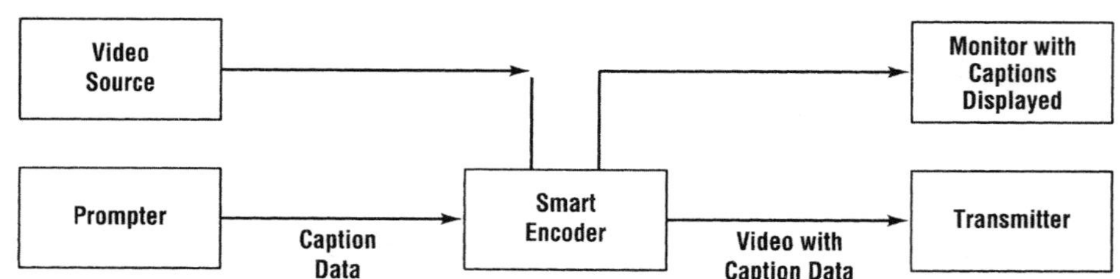

FIGURE 5.24-4 Newsroom captioning with teleprompter.

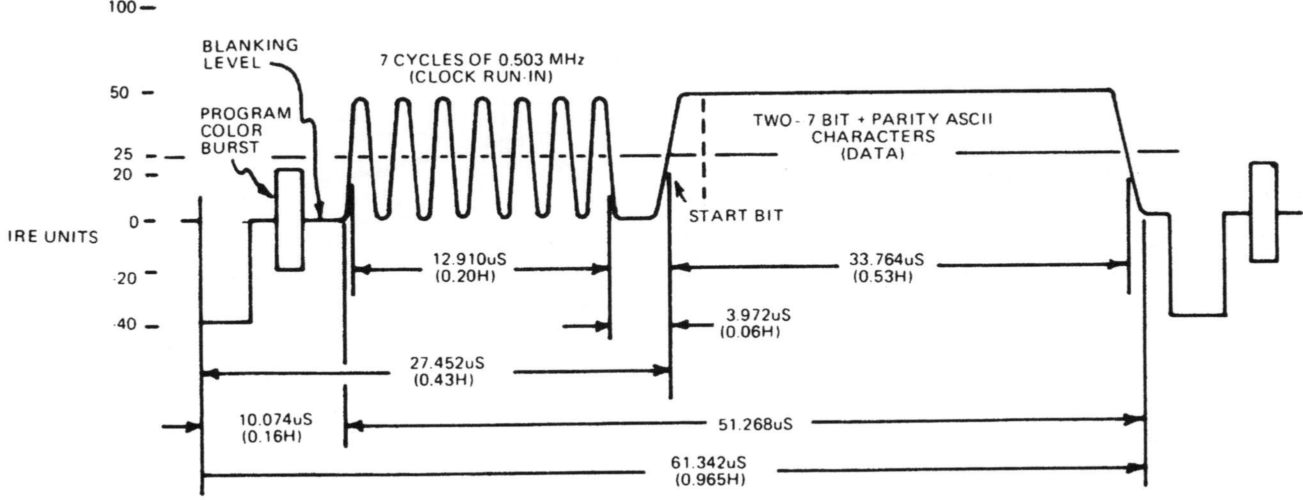

FIGURE 5.24-5 Line 21 data signal waveform specification.

kets from counting news programming using electronic newsroom techniques toward their captioning requirements. Rather, these networks and stations must provide real-time captioning using a stenographer to convert the entire audio portion of the live program to captions. Others may use electronic newsroom techniques to meet captioning mandates.

LINE 21 DATA FORMAT

Captions associated with an analog NTSC television program are transmitted as an encoded composite data signal during line 21 on field 1 of the video signal, as shown in Figure 5.24-5. The signal consists of a clock run-in signal, a start bit, and 16 bits of data corresponding to 2 bytes of 8 bits each (7-bit character code plus 1 parity bit). Therefore, transmission of actual data is 2 bytes every 1/30 of a second, equivalent to 60 bytes (characters) per second. The data stream also contains control codes that provide the instructions for the timing, screen position, and attributes of the data.

The clock run-in consists of a seven-cycle sinusoidal burst that is frequency and phase locked to the caption data clock. The frequency of 32 f_H (32 × 15,734.26 Hz = 0.503496 MHz) is twice that of the data clock and provides synchronization for the decoder clock. The clock run-in signal is followed by the equivalent of 2 data bits at logical zero level and then a logical one start bit. The last two cycles of the clock run-in, the 2 logical zero bits, and the logical one start bit constitute an 8-bit frame code signifying the start of data, as shown in Figure 5.24-6.

The 7-bit character data is coded in a nonreturn-to-zero (NRZ) format. An 8th bit is added to each character to provide odd parity for error detection.

The sequence of identification, control, and character code transmission is shown in Figure 5.24-7 and Figure 5.24-8. Each caption transmission is preceded

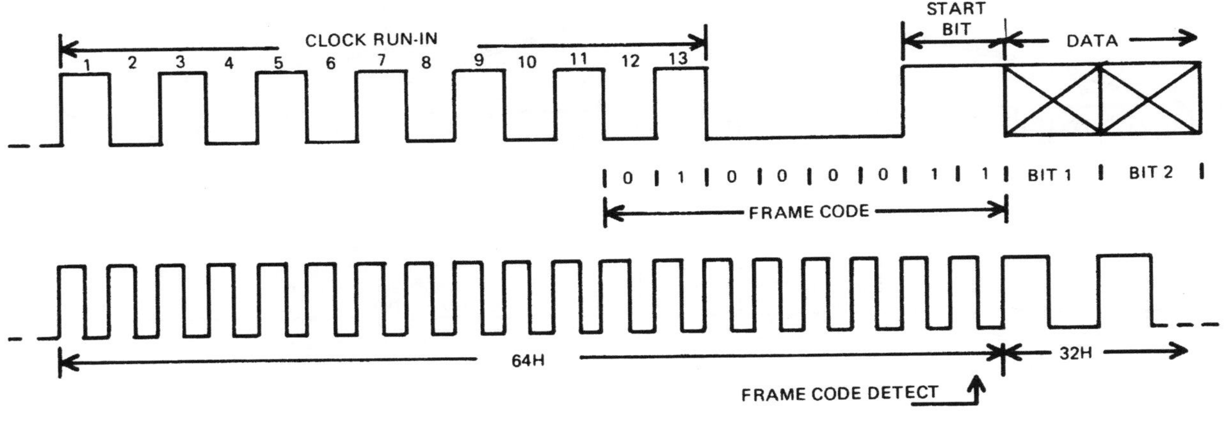

FIGURE 5.24-6 Line 21 data timing.

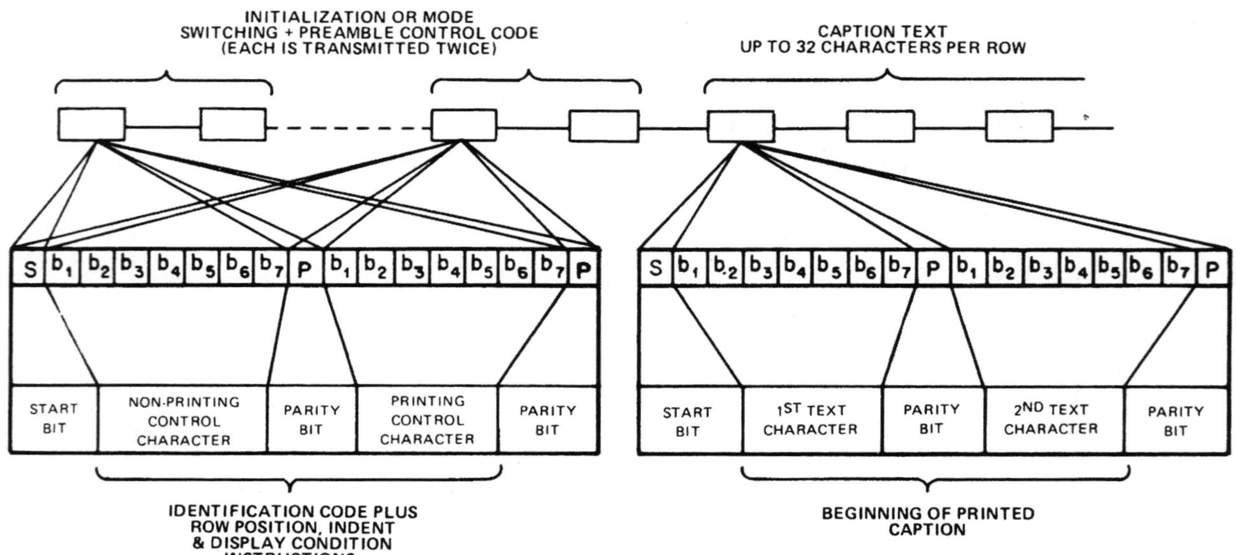

FIGURE 5.24-7 Beginning of caption sequence.

by a preamble code, which consists of a nonprinting character followed by a printing character to form a row address and display color code. Both characters of all control codes are transmitted within the same field of line 21 and twice in succession to ensure correct reception of control information. A transmitted caption may be interrupted by a mid-caption control code between two words in order to change display attributes such as color or italics. At the completion of a caption transmission, an end of caption control code is sent.

Codes are composed of 2 bytes. The first byte is a reserved code, in the range of 0x10 through 0x1F. This is followed by a printing character, in the range of 0x20 through 0x7F. All characters that are received after a valid control code are in the range of 0x20 through 0x7F. These are interpreted and loaded into the decoder memory as printing characters. Character codes with bad parity result in the display of a white box (the delete symbol) in place of the character.

The data rate is 480 bps or 60 bytes per second (8 bits per byte including parity). At an average of 5 bytes or letters per word, an average maximum word rate of 12 words per second or 720 words per minute can be achieved. In practice, the maximum word rate is somewhat less than this, about 600 per minute, due to the time required for transmission of control codes. Because most speech is much slower than this, there is adequate time for additional data services in each field.

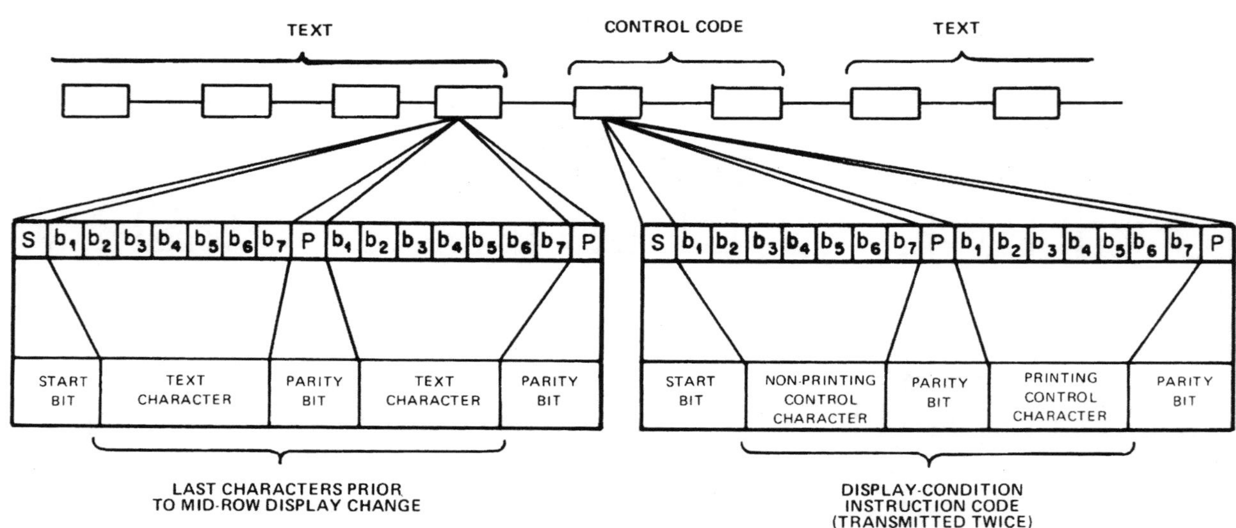

FIGURE 5.24-8 End of caption and display sequence.

LINE 21 TECHNICAL ADVISORY

The FCC requires that any local TV station or cable headend that redistributes a signal that contains closed captions *must* preserve and forward that caption data.

Regular observation and monitoring are essential in order to verify the presence and proper location of the data signal. Equipment throughout a video facility should be routinely checked and, if necessary, adjusted to pass the line 21 caption data signal.

Maintenance personnel and control room operators should be made aware of the importance of the line 21 signal and adjust all applicable equipment accordingly. A waveform monitor with line-select capability is required to view the line 21 caption data to determine that it is on the correct line and field. For example, a Tektronix 1480 series, VM 700A, 1700 Series, 1780R, or Hewlett-Packard equivalent with line selector will easily locate and display the signal. In addition, caption decoders can be used to verify that the caption content is being encoded or preserved correctly on the outgoing feed.

The following video processing equipment may adversely affect the line 21 signal:

- *Video processing amplifiers:* Certain types may delete the line 21 signal, move it to another line, exchange the fields, or partially blank the signal.
- *Time base correctors (TBCs):* Some may blank the line 21 signal or advance or delay line 21 by a field or a line.
- *Videotape machines:* Certain types contain a TBC that may blank or move line 21.
- *Video switchers or digital store devices:* May blank or move the line 21 signal.
- *Frame store or frame synchronizers:* May blank or distort the line 21 signal.
- *VITS generators or inserters:* Should be checked to determine whether they are programmed to pass the line 21 signal correctly.
- *Digital effects and text generators:* May delete or distort the line 21 data signal.
- *Digital-to-analog composite encoders:* May add setup to the line 21 data signal.

The line 21, field 1 data signal may be partially, but inadvertently, blanked. This can occur when the active video line is advanced or delayed relative to horizontal sync. The decoding system is more tolerant of this problem when the data signal is early and causes part of the clock run-in sinusoidal burst to be blanked. On the other hand, if the data signal is delayed relative to horizontal sync, loss of the last data bit (parity bit) may occur, producing decoding errors that in turn will cause the characters to be displayed as white boxes. In an extreme case, too many errors will cause the decoder to disable all closed captioning functions.

Some video processing amplifiers have sensitive vertical phase blanking and horizontal blanking controls. These may also adversely affect the line 21 signal. All signal paths, backup transmitters and their VITS generators and processing amplifiers, and tape-delayed broadcast installations should also be routinely checked to ensure passage of the line 21 signal. It is important to note that if the overall signal level is allowed to drop substantially below the normal 50 IRE level, captions could become garbled. Also, the normal 7.5 IRE setup is not present on the line 21 signal.

Cautionary Notes

Network Origination/Local Origination

Network programs and commercials prior to, during, and after locally originated captioned programs may also be closed captioned. Therefore, at the end of captioning local programming, the local smart encoder must be sent a command to re-enable passage of upstream line 21 data. If both network and local programming are passing through the smart encoder, the line 21 signal should be checked to ensure its integrity.

Newsroom Computers

Newsroom computer captioning typically uses the smart encoder in a newswire/real-time mode. When this mode is invoked, incoming line 21 data is automatically blocked, preventing any previously recorded caption information from being passed through and displayed. In newswire/real-time mode, the captions are encoded in roll-up style.

Other systems may use the smart encoder in the "line 21 direct-entry" mode rather than the newswire/real-time mode. Direct entry offers more complete control of the encoder, allowing captions to be authored in paint-on or pop-on styles.

In either captioning mode, the newsroom computer must issue a command to end captioning at the conclusion of local live captioning to re-enable passage of upstream line 21 data.

Using Captioned Excerpts

Excerpts of programs received or recorded off-air or from external satellite or network feeds, for use in other programming, may contain line 21 closed captions. In these cases, the captions should be stripped if the excerpted material audio will be edited or replaced because the captions may no longer be germane. A time base corrector or video processing amplifier can be used to remove or move the caption data signal from line 21 prior to inserting new captions.

Broadcast Decoder

In addition to consumer models, professional broadcast-type decoders are available for stations to use to monitor the line 21 data signal on the incoming network feed, videotape machines, or the broadcast signal. Broadcast models accept a standard baseband video signal with captioning data on line 21 and produce a video output with the captioning data decoded and displayed as open captions for viewing on a standard video monitor.

Emergency Information Messages

The FCC adopted rule 79.2 in 2000 to require broadcasters, cable operators, and other multichannel video programming distributors to make local emergency information accessible to persons with hearing disabilities. Emergency information not provided through closed captioning must be provided through some other method of visual presentation, such as open captioning or text crawls.

Of particular note to broadcasters, this FCC rule requires that the presence of captioning and emergency alert services must not disrupt each other. As already noted, text generator devices may delete or distort the waveform that carries caption data on line 21. Additionally, captions displayed on a television set may appear over the emergency message text. Either condition would violate the FCC rule.

Manufacturers of closed caption encoding equipment responded to these requirements by creating features to reposition existing captions to prevent overlap with emergency alert messages, and to bridge the caption data around emergency alert text inserters and reinsert it onto line 21.

LINE 21 TEXT AND EXTENDED DATA SERVICES

Typically, captions do not use all the available data capacity or "bandwidth" of the line 21 system, and the unused bandwidth may be used to transmit additional services. These services are time multiplexed between caption data.

Text Mode Services

The *Text Mode Services* consist of four data channels (field 1 includes T1 and T2; field 2 includes T3 and T4). The text service channels may be used to transmit program notes, news, weather, sports, farm, and financial reports and other information that may be independent of the program with which it is transmitted.

Text Mode Services are accessed by the user using the receiver's decoder controls. There are two text "channels" for each field. Unlike the captions, the text service appears as rolling text that obliterates the entire image or just the top or bottom half. Caption data is time sensitive and is given higher priority by encoders, which means that the display of Text mode data will pause during caption transmission.

Extended Data Services

Extended Data Services (XDS) is another data channel that is time-multiplexed with the caption and text channels on line 21 of field 2. XDS is used to convey ancillary data such as broadcaster identification, program description, content advisories (V-chip), Time of Day, and copy protection state.

Each data type is encoded in an XDS "packet" with a unique identifier and an 8-bit checksum. Caption data services on field 2 have higher priority than XDS and can interrupt insertion of an XDS packet. CEA-608 defines a mechanism to resume XDS packets that have been interrupted by caption data. However, not all XDS decoders have implemented this XDS resume feature, and will process only XDS packets that are encoded uninterrupted.

CEA-608 states that a caption waveform shall be present on field 1 whenever there is a caption waveform present on field 2. Consequently, many devices that include an XDS decoder will not enable it unless a caption waveform is present on field 1, even if the program is not being captioned. Broadcasters should be cautioned to always include a valid caption waveform on field 1 for this reason.

Note that XDS data carried with CEA-608 captions is intended for use only by analog NTSC television receivers. While the XDS data may be transcoded into the section of the DTV caption distribution packets reserved for CEA-608 captions, this data is intended only for reinsertion onto line 21 by DTV receivers with analog NTSC outputs. DTV receivers should use content advisories, copy protection guidance, and other ancillary data from the PSIP tables that are broadcast along with the DTV video content. See the section on DTV captions later in this chapter and also Chapter 5.22 for more details on PSIP.

A large number of XDS packet types have been defined in CEA-608. A subset of the packet types most commonly used by broadcasters and those with certain caveats are described herein.

Content Advisory Packet

The content advisory packet is also known as the "V-chip" packet. This packet provides a rating for the associated video program. Consumer electronics that support the content advisory packet may be configured to block content that exceeds a particular rating level. FCC rules state that the data within the content advisory packet shall not change during the course of a program, which includes program segments, commercials, promotions, station identifications, etc.

Four content rating systems have been defined: TV Parental Guidelines (TVPG), MPAA, Canadian English, and Canadian French. These four systems are mutually exclusive, so if one is included, then the others shall not be.

FCC Report and Order 98-36 required half of all new television models 13 inches or larger manufactured after July 1, 1999, and *all* sets 13 inches or larger manufactured after January 1, 2000, to have V-chip technology to allow parents to block the display of violent, sexual, or other programming they believe is harmful to their children. While support of the content advisory packet by content creators is voluntary in the United States, *FCC Report and Order 98-36* requires that TV stations and cable operators may not alter rating information inserted by the network or program provider.

FCC rules stipulate that the default state of a receiver should be to *not* block unrated programs, but users may be provided with the option of blocking programs that are not rated. The wording of the FCC rules does not make a distinction between programming with a content rating of NOT RATED and programs that have no content advisory packet encoded at all.

In Canada, the CRTC mandated that Canadian broadcasters must carry the Canadian content rating systems starting in March 2001. Programs originally produced for American audiences must have new ratings encoded in one of the two Canadian rating systems when they are broadcast in Canada. Decoders on many older V-chip-capable television sets do not support the Canadian rating systems, however.

Time Packets

Time information can be broadcast using the Time of Day (TOD) and time zone and daylight saving time use (TZ) XDS packets. This data can be used by VCRs and other consumer electronics devices to automatically adjust their internal time clocks. Whenever they are encoded, the time packets must always be inserted at broadcast time, never subjected to tape delay; otherwise, the data will be delayed, and receiving devices will adjust their clocks to the incorrect time.

Every PBS affiliate station has an encoder to insert time packets. Other broadcasters in the United States are discouraged from inserting TOD and TZ packets. Some consumer electronics will behave erratically when time packets are available on different channels in the broadcast area.

The TOD packet data includes the Coordinated Universal Time (UTC) time and date, and a flag to indicate whether daylight saving time is presently in effect. The data in this packet is independent of the broadcast area.

The TZ packet describes the time zone of the broadcast area and a flag to indicate whether daylight saving time is observed in the broadcast area. The time zone is encoded as an integer number of hours to be subtracted from the UTC time. Consequently, the time zone in some regions cannot be represented in the TZ packet. Examples include Guam and Newfoundland, Canada. Additionally, the TZ packet must not be encoded when the intended broadcast area encompasses multiple time zones, or across regions that do and do not observe daylight saving time. In these areas, consumers must manually set the time zone and daylight saving modes in their VCRs.

Copy Generation Management System

The Copy Generation Management System – Analog (CGMS-A) packet is a copy protection mechanism for analog television signals. It is supported on a voluntary basis in many consumer electronics video recording devices such as personal video recorders and DVD recorders.

The CGMS-A packet can specify that a program can be freely copied, copied once, or not copied at all. The packet also contains controls to specify that subsequent copies of the program are to be artificially degraded. Another field in the packet called the Analog Source Bit (ASB) specifies whether the source is prerecorded material.

CGMS-A data is also specified by IEC 61880 to be transmitted on VBI line 20 of NTSC and PAL transmissions. When it is encoded in this manner, CGMS-A is encoded independently of closed captions.

STANDARDS AND TERMINOLOGY FOR DTV CAPTIONS

Although broadcasters may not need to know every detail of the complex standards involved for DTV captioning, it is necessary to be aware of the terminology used. Knowledge of the standards will certainly be useful in troubleshooting and helping determine compliance of installed equipment and systems.

CEA-708

CEA-708, *Digital Television (DTV) Closed Captioning*, defines the coding of DTVCC as it is delivered in an ATSC emission bitstream as specified in ATSC A/53D, and is applicable equally to high definition and standard definition video formats, and to both terrestrial broadcasting and cable distribution. The captioning data is carried in the video user bits of the MPEG-2 bitstream.

Captions generated in accordance with CEA-708 standard are generally referred to as "708" captions. At the time of writing, CEA-708-C is the current revision of this standard.

CEA-608 Legacy Data

In addition to the actual 708 caption data required for use by a DTV set-top box or integrated receiver to display DTV captions, the CEA-708 standard requires carriage of equivalent data for a subset of the captions coded in the 608 format. This legacy 608 data, also known as the *608 compatibility bytes,* is required for use by DTV set-top boxes so that line 21 data can be inserted in an analog composite video output (if provided) to feed a legacy NTSC TV set. Some DTV receivers may use the 608 data when 708 data is not available for providing closed captioning on the DTV display, but this is not mandatory and is not implemented in many receivers.

SMPTE 334-1 and SMPTE 334-2

SMPTE 334M defines a method of coding which allows data services to be carried in the vertical ancillary data space of a bit-serial component television signal. At the time of writing, SMPTE 334M is being separated into two separate parts, namely SMPTE 334-1 and 334-2. SMPTE 334-1, *Vertical Ancillary Data Mapping of Caption Data and Other Related Data*, defines the coding and mapping of caption data and other related data into the vertical ancillary space (VANC) of baseband digital video signals. SMPTE 334-2, *Caption Dis-*

tribution Packet (CDP) Definition, defines the caption distribution packet (CDP).

Caption Distribution Packet

SMPTE 334-2 defines the caption distribution packet (CDP) consisting of a specific sequence of bytes that can hold the actual 708 DTV caption data, the legacy 608 caption data, caption service information, and optional time code for synchronization (not needed with transport methods where the CDP is directly associated with video frames). The CDP is the basic unit of data that is transported through the professional portion of a DTV caption distribution chain.

The CDP was previously defined in CEA-708-B. The specification of this data structure was handed off to SMPTE in 2006 because the CDP is used exclusively by professional equipment and is not broadcast in the transmitted DTV bitstream.

ATSC A/65

ATSC A/65, *Program and System Information Protocol for Terrestrial Broadcast and Cable*, defines information in the Program and System Information Protocol (PSIP) that describes the contents of an ATSC broadcast. It defines the *caption service descriptor* (CSD) to be carried in the transmitted bitstream to announce the presence and format of captions being carried. The CSD provides metadata that may be used by the DTV receiver to display information about captioning in an electronic program guide and may be needed to properly decode and display the closed captions.

A/65 states that for terrestrial broadcast transport streams, the CSD must be present in the Event Information Table (EIT), and that for transport streams delivered on cable, the CSD must be present in the Program Map Table (PMT) for captioned programs and, if an EIT is sent, must also be present in the EIT. At the time of writing, the current revision of this document is A/65C with Amendment 1.

While CEA-708 captions may use up to 63 caption services numbers, only 16 services are carried simultaneously and the CSD describes only up to 16 services.

Native Captions, Translation, and Transcoding

Native Captions

Captions that are encoded and transmitted in the 608 or 708 format in which they were authored may be referred to as "native 608" or "native 708" captions.

Translated 708

It is possible for captions to be encoded in the 708 DTVCC format by conversion from 608 legacy captions already encoded on line 21 of an analog NTSC video feed. This method can be used when the program was created primarily for NTSC transmission but is being encoded for DTV transmission, either as an SD program or with up-conversion to HD. For new programming this is expected to be an interim solution until native 708 caption authoring is generally adopted, but may persist indefinitely for legacy material that is already captioned.

There is no standardized algorithm for 608 to 708 caption translation, but the goal is to produce captions with an appearance as close as possible to the appearance of the original data on a CEA-608 decoder. Several manufacturers have implemented 708 encoders fulfilling this goal.

The resulting DTV captions are usually known as "translated 708." They are also referred to in the FCC R&O as "upconverted captions," but this term should be deprecated because of the connotation that the process is related to video up-conversion, which it is not. Some manufacturers use the term "transcoded" in this regard, but that should preferably be used only to refer to the 608 compatibility bytes transcoded from 608 caption sources.

Transcoded 608

Where 708 captions are produced by translation from existing 608 line 21 data, the same data may be "transcoded" to fill the required 608 compatibility bytes. Because the 608 data itself is unchanged when it is transcoded into CDPs, it can be subsequently read and transcoded back onto line 21 of an NTSC or SDI signal later, if so desired.

FCC RULES AND REGULATIONS FOR DTV CAPTIONS

Rules

Since July 1, 2002, U.S. broadcasters have been required by FCC 00-259–2000, *Report and Order, Closed Captioning Requirements for Digital Television Receivers*, to provide closed captioning on at least some of their programs transmitted on DTV channels.

The FCC R&O amended Part 15 of the FCC rules, adopting technical standards for the display of closed captions on digital television receivers. It adopted into the rules Section 9 of industry standard CEA-708, specifying the decoding and display of closed captioning for DTV systems. For informational purposes, it also incorporated by reference the remaining sections of the standard, which specify the encoding, delivery, and other aspects of DTVCC.

The R&O amended Part 79 of the rules to require transmitted closed captioning to reflect the changes in Part 15 and to require all caption data to be passed intact through program distribution facilities in a format consistent with Part 15, unless such programming is recaptioned or captions are reformatted.

The R&O also clarified the schedule for including closed captions in digital programming.

No distinction is made in the FCC rules between DTVCC for high definition or for standard definition programming; the requirements are the same for both. The R&O makes it clear that during the transition period to digital television, DTVCC in accordance with CEA-708 may be derived from legacy CEA-608 (analog) captions as well as from native 708 authoring. It also confirms that to count captioned DTV programming hours toward the captioning total for each channel, the broadcast DTV signal must include both CEA-708 and CEA-608 caption data.

The FCC 04-192A1 – 2004, *Report and Order, Second Periodic DTV Review,* confirmed the requirement to carry both 708 caption data and 608 caption data in both SD and HD DTV programming. It also confirmed the requirement to carry the caption service descriptor in both the PSIP Event Information Table and the Program Map Table.

Receivers

The rules require all DTV set-top boxes and DTV receivers with a screen height of at least 7.8 inches (the height of a 13-inch 4:3 display) manufactured after July 1, 2001, to include a caption decoder complying with Section 9 of CEA-708. Such devices shall provide the user with control of caption font, size, color, edges, and background. Decoders must be able to decode the six standard services in CEA-708 and allow users to choose at least one for display. Set-top boxes that have an analog NTSC output shall insert 608 caption data carried in the DTV signal into line 21 of the NTSC video output.

Cable Carriage

The rules require digital television programming with CEA-708 caption data services to be delivered to digital cable subscribers with captioning intact.

Programming Requirements and Schedule

The FCC rules allow for a phased introduction of captioning as described in the following sections (note that at this time the term "channel" has not been clarified in relation to multicast broadcasting).

New Programming

New digital programming is that "prepared or formatted for display on digital televisions that was first published or exhibited after July 1, 2002." Each DTV channel is required to have 100% captioned content, with some exceptions, for *new,* nonexempt, English language programming broadcast after January 1, 2006. For Spanish language programming the schedule is somewhat delayed, leading to 100% by 2010.

Prerule Programming

Prerule nonexempt English language digital (and analog) programming is defined as that "first published or exhibited before July 1, 2002." The broadcast schedule requiring captioning on *prerule* programming is as follows:

1/1/03 to 12/31/07	30% of programming
After 1/1/08	75% of programming

Equivalent dates for Spanish language programming are 30% at 1/1/05 and 75% at 1/1/12.

Thus, DTV services now have the same hourly captioning requirements as NTSC services. The programming requirements were reconfirmed in an FCC Public Notice issued January 6, 2004, titled *FCC Public Notice DA-04-2, Notice to Video Programming Distributors and the Public of the January 1, 2004, Requirements for the Closed Captioning of New Nonexempt English and Spanish Language Video Programming and Reminder Regarding Other Captioning Requirements.*

Exempt Programming

There are exemptions to the preceding captioning requirements for some types of programs; examples include

- Most programs shown between 2 a.m. and 6 a.m.;
- Local non-news programming with no repeat value;
- Commercials no more than five minutes long;
- Some instructional programming that is locally produced by public television stations;
- Programs in languages other than English or Spanish;
- Programs shown on new networks for the first four years of the network's operations;
- Public service announcements and promotional announcements shorter than 10 minutes, unless they are federally funded; and
- Programming provided by program providers with annual gross revenues under $3 million (although such programmers must pass through video programming that is already captioned).

In addition, a video programming provider or distributor may file with the FCC a petition for an exemption for specific programming.

DTV CAPTIONING SYSTEM IMPLEMENTATION

In accordance with CEA-708, ATSC A/65, ANSI/SCTE 43, and ANSI/SCTE 54, DTV signals for broadcast and cable distribution, both for HD and SD program material with closed captions, need to carry

- 708 DTV caption data;
- 608 compatibility bytes;
- Caption service descriptor in the EIT and PMT.

To enable ATSC encoders and PSIP generators to generate signals that meet the preceding requirement, 708 caption encoders should produce the 708 DTV caption data and 608 compatibility bytes. Caption service information is typically inserted into the ATSC video stream by a PSIP generator device, which may obtain this data from a number of sources.

Because of the varying requirements of different networks and production facilities, and the expanding capabilities of new equipment, it is not practical to describe every possible system implementation in detail. The following sections provide an overview of typical captioning applications.

CEA-708 VANC Caption Encoding

Where 708 captions need to be carried in a video signal, as shown in Figure 5.24-9, the architecture for encoding is similar to the 608 arrangement. This technique is independent of production format and equally applicable to captioning HD and SD program material. The program video is HD-SDI or SDI.

For native 708 captioning, the data from the captioning computer is processed and inserted as caption distribution packets (CDPs) embedded in the VANC of the video signal in accordance with SMPTE 334-1. Alternatively, as indicated by the dotted line input in the drawing, 608 captions from an NTSC video feed may be used as a source of captions to be translated to 708. The two video sources must, of course, have identical program content.

The VANC packets are transported as an integral part of the digital video signal; however, special arrangements have to be made at any point in the distribution chain where VANC data may be removed.

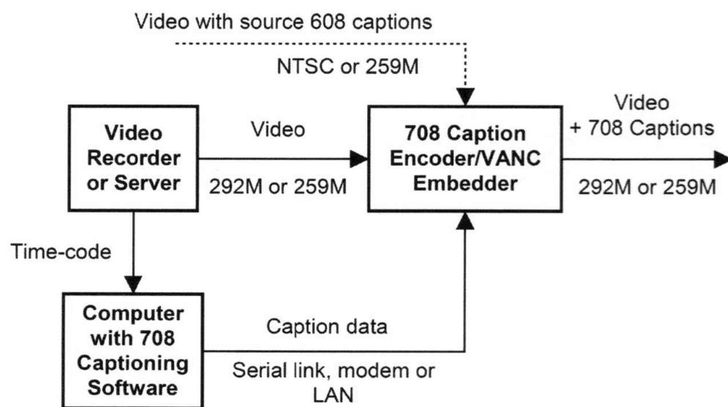

FIGURE 5.24-9 Typical 708 caption encoding.

Authoring of 708 Captions

At this time, work is in progress by U.S. national captioning organizations for an agreed 708 caption intentions format. In the meantime, some 708 caption encoders can accept data from files configured with data in native 708 format—that is, exactly as defined in CEA-708. It is also possible for 708 caption encoders to accept data derived from 608 "caption intention" files, which would then be translated to 708 as described previously. The resulting DTVCC are limited to the authoring capabilities of CEA-608 and cannot exercise all features of a 708 caption decoder.

For full implementation, it is recommended that CEA-708 caption authoring systems should produce both 708 and the associated 608 caption data types and provide appropriate data for the caption service information.

Recording, Processing, and Distribution

Embedding DTVCC in VANC, whether for HD or SD program material, allows them to automatically follow the video through most standard routing and switching, and some types of processing and recording equipment, similar to the principle of carrying 608 captions in line 21 of an analog signal. However, several issues must be considered, as discussed in the following sections.

Recording

For recording systems, the integrity of the DTVCC data should be ensured by selecting a tape format and equipment or video server technology that preserves the SMPTE 334-1 caption data in VANC. This function may be integrated into the recording system; in other cases it may be by use of external bridges. Current recording equipment is generally not capable of recording and passing all VANC data from input to output, and care is needed to ensure that the relevant data packets needed to implement DTVCC are preserved by the selected equipment.

Processing

Various types of video processing equipment in the broadcast plant may delay, repeat, or drop video frames and/or delete the VANC part of the video signal. Wherever possible, system design should minimize the use of equipment that disrupts the flow of VANC data, or a special provision should be made for bridging around it. The bridging equipment should preserve the relative timing between captions and video as far as possible.

Compression Distribution Systems

If the program is distributed through a medium that involves MPEG compression, such as a satellite link, VANC data is not carried directly with the compressed video; therefore, the CDPs must be bridged from the VANC of the input HD-SDI or SDI video signal to a data channel in the MPEG multiplex. In general, there are three different ways to do this:

1. The VANC data is packaged into the transport stream multiplex as a private data service on dedicated packet IDs (PIDs).
2. For captions generated direct from a 708 caption encoder, a serial representation of the captions can be packaged into a private data service.
3. The uplink MPEG encoder takes the caption data and packs it into the video user data according to CEA-708 and ATSC A/53.

It is possible for all three of the above methods to be implemented at the uplink system, and the most useful one(s) selected for use at the downlink site. All will equally support 708 captioning data and the 608 compatibility bytes at the same time.

All these methods are supported with MPEG transmission at bit rates including, and higher than, the 19.39 Mbps ATSC emission standard, such as the 45 Mbps rate typically used for network distribution. Method 3 is similar to the arrangement used for emission station captioning and ATSC encoding. It is usually used for network distribution at 19.39 Mbps as implemented by some networks, but may also be used when the bit rate differs from the ATSC nominal, using the video outputs from the receiver/decoder with caption data in VANC or feeding caption data via serial link to an MPEG encoder.

In each case the equipment bridging the caption data through the MPEG process should preserve the relative timing between captions and video.

Local Station Arrangements

Figure 5.24-10 shows a possible configuration of captioning-related equipment in a local station with ATSC and NTSC transmission. This is intended only to illustrate some of the captioning arrangements discussed later and is not a full system diagram.

The drawing shows a DTV system for HD only. A similar arrangement could be used for SD DTV using SMPTE 259M in place of SMPTE 292M. In that case, up-conversion to and down-conversion from the NTSC feeds are not required. It should be noted that some equipment manufacturers have combined several functions shown in the figure into one unit.

NTSC Station Output

For the NTSC station output there are two main arrangements; it may have

1. A legacy NTSC station master control with legacy network distribution and local sources; or
2. A standard definition (possibly down-converted) feed from one of the DTV program services.

For the first case, NTSC caption arrangements will be unchanged from the legacy arrangement. For the second case, 608 caption data will usually need to be transcoded from VANC CDPs onto line 21.

ATSC Station Output

For the ATSC station output there are three main arrangements; it may have

1. A dedicated DTV (possibly HDTV) station master control or simple switcher with a (H)DTV network distribution and local live and recorded sources. The DTVCC should already be present on the live network and recorded sources but need to be added to the live local sources;
2. A feed (possibly upconverted) from the NTSC network distribution and master control. In this case the 608 line 21 captions will need to be translated to 708 as well as transcoded;
3. A bitstream received from the network already encoded in ATSC transmission format (at 19.39 Mbps) and ready to send to the DTV transmitter with local content spliced in and new PSIP information added. In this case, DTVCC should already be present in the incoming bitstream.

Some stations may have combinations of these arrangements, depending on network DTV distribution arrangements and the number of DTV program services.

Local Live DTV Captions

Live programming may be captioned using a 708 caption encoder with VANC CDP embedding, as discussed previously but with live video and with caption intentions generated in real time, with authoring from a stenographer or with electronic newsroom techniques, as mentioned previously (where allowed). This is shown at the upper-left side of Figure 5.24-10.

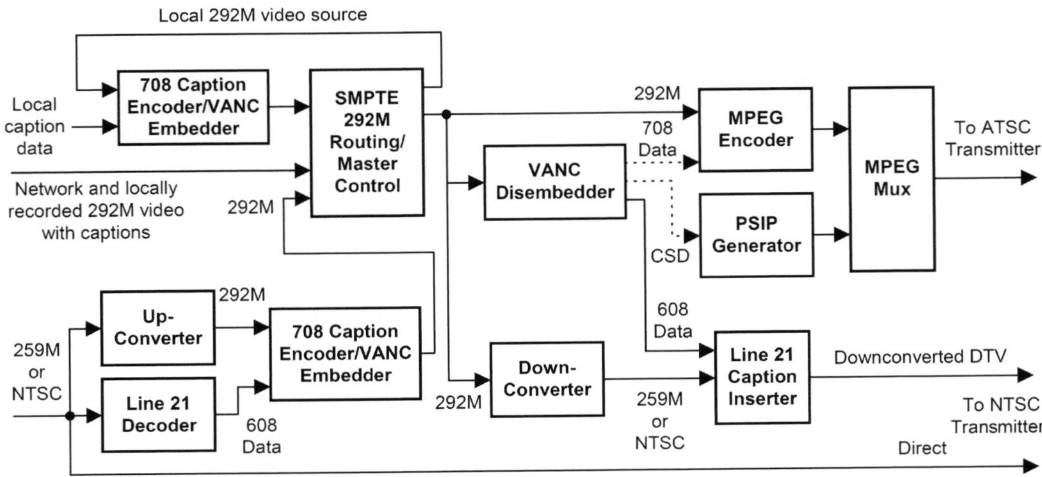

FIGURE 5.24-10 Emission station captioning.

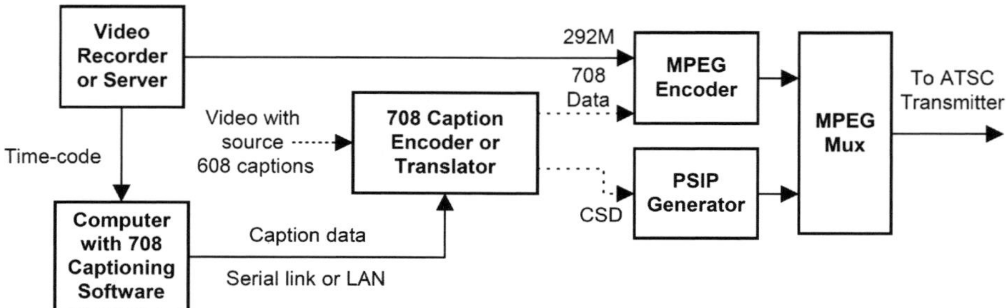

FIGURE 5.24-11 Alternative captioning feed to emission encoder.

Alternatively, the 708 caption encoder may feed the ATSC encoder directly, as shown in Figure 5.24-11. If the local programming originates as an NTSC source, existing legacy arrangements for live 608 captioning arrangements may be used, with bridging to DTVCC.

NTSC–DTV (Up-Conversion)

Some HD DTV programs may be produced by up-conversion from an analog or SDI program intended for NTSC transmission. In this case, the captions must be bridged from the analog NTSC or SDI line 21 to CDPs in the VANC of the HD-SDI stream. This can be accomplished by having a decoder extract the 608 data from line 21 and feed it to a 708 caption encoder/VANC embedder, as shown at the lower-left side of Figure 5.24-10. Similar techniques can be used to insert DTVCC in the VANC of SD programs for DTV transmission, without up-conversion. Some companies offer equipment that integrates the line 21 decode, 608 to 708 caption translation, and VANC CDP embedding functions in a single device.

DTV–NTSC (Down-Conversion)

There may be a need to derive an NTSC broadcast signal from an HD-SDI stream. The CDPs carried in the VANC in the HD-SDI stream should contain CEA-608 data, which can be embedded into line 21 of the analog NTSC or SD-SDI video. This can be achieved by having a VANC de-embedder extract the data from the HD-SDI stream and feed the 608 caption data to an analog NTSC or SDI line 21 closed caption inserter, as shown in the lower-right portion of Figure 5.24-10. Some companies offer equipment that can transcode 608 data from VANC CDPs directly onto line 21 in a single device.

Server-Based Captioning

The techniques described previously for carrying captioning data in VANC of the video signal are appropriate for many applications of DTV caption distribution. An alternative server-based captioning architecture, where the caption data is distributed separately from video, may have advantages for some station operations, and is particularly applicable in facilities using video-servers for program origination.

In this type of system, instead of inserting caption data into the video at some upstream point for distribution with the video program, caption files are generated and then distributed separately from the video, typically over a computer network. A separate caption file for each program is stored on a captioning computer (server) at the emission station that runs software to manage the various caption files and playout. Software running on the captioning server, or on other networked computers, allows a program playlist to be established. As programs are played out, time code from the video server triggers captions from the caption server to be fed to the caption encoder and inserted into the transmission feed. Information on specific features and arrangements for server-based captioning is available from the manufacturers.

Server-based captioning is applicable to both 608 captioning for NTSC and 708 captioning for DTV. Such systems use basically the same interfaces between the captioning computers and caption encoders and between caption encoders and ATSC encoders as systems using captions embedded in video streams.

Caption Input to the ATSC Encoder

The task of embedding the caption data into the ATSC stream in accordance with A/53 and CEA-708 falls to the ATSC encoder. Several different methods may be used for getting the caption data into the encoder, depending on system configuration and manufacturer and the source of the caption data, as follows:

1. As shown in Figure 5.24-10, a VANC CDP de-embedder can extract the CDP stream from VANC in an HD-SDI or SDI video feed and transform it to the data format expected by the ATSC encoder. Most ATSC encoders have a caption input serial port. This port accepts data in one of two documented formats: SMPTE 333M, which is known as the "pull" protocol; or the so-called "Grand Alliance" format, also known as the "push" protocol.
2. For ATSC encoders that themselves have the capability of extracting CDPs from CDPs located in the VANC, the "708" connection shown dotted in Figure 5.24-10 is not required.

3. Alternatively, as shown in Figure 5.24-11, a local 708 caption encoder without VANC embedding can be used to feed the ATSC encoder using a serial connection and taking caption data from a captioning computer or live authoring workstation. Figure 5.24-11 does not show the video routing and arrangements that may be needed for switching captioning data signals for different sources.
4. As shown in Figure 5.24-11, a 608 to 708 translator/transcoder unit without VANC embedding can also be used to feed the ATSC encoder, with captions originating from a legacy line 21 caption source.
5. Some ATSC encoders can accept analog NTSC or SDI video and generate the 708 DTVCC captions for transmission from 608 legacy captions present on line 21 of the video, without using an external 708 caption encoder. Note that some early ATSC encoders inserted only the 608 compatibility bytes.

A combination of the preceding methods may be used. This requires that special switching (not shown in the drawings) is used to route to the ATSC encoder the appropriate caption data for DTV network, local sources, or NTSC converted sources.

PSIP AND THE CAPTION SERVICE DESCRIPTOR

Caption service information is required to be carried in the DTV bitstream as set out in ATSC A/65, ATSC A/53, CEA-708, and ANSI/SCTE 54. Methods for originating caption service information are not standardized, and different users and manufacturers may have different arrangements for generating the caption service descriptor (CSD). It cannot be assumed that all caption encoders generate complete and correct caption service information for the caption services as required for insertion in the PSIP tables.

For systems using CDPs carried in VANC, the CSDs may be present in the CDP. SMPTE RP207-2002, *Transport of Program Description Data in Ancillary Data Packets,* describes another means of encoding CSDs into VANC. In either case, the CSD data may be extracted when the VANC data is disembedded and fed to the PSIP generator or, where appropriate, the ATSC encoder. That equipment should enter the CSD into the PMT and current program EIT table entries.

Ideally, local 708 caption encoders feeding the ATSC encoder and PSIP generator directly without using VANC, as shown in Figure 5.24-11, should also generate the caption service information. Caption service descriptors to be entered in PSIP tables for future events have, of necessity, to be generated and distributed separately from the captioned video service. Arrangements for this will vary depending on the equipment and system design.

Data Interface to the PSIP Generator

Methods for conveying CSD data from the VANC data de-embedder to the PSIP generator vary between manufacturers. A proprietary message format has been developed by one manufacturer to allow the CSD and also the content advisory descriptor to be provided from a caption encoder or VANC de-embedder to a PSIP generator upon request.

However, ATSC standard A/76, *Programming Metadata Communications Protocol*, defines an XML-based open standard for communication of PSIP-related metadata between equipment in the facility. This facilitates transfer of both current and future program metadata, including the caption service descriptor, to the PSIP generator. This is the preferred method of PSIP metadata communications.

MONITORING OF DTV CAPTIONS

Monitoring of 608 and 708 caption encoding and content should be carried out at appropriate points in the distribution chain to check for

- Presence of captioning information;
- How captions will be displayed on a consumer receiver (basic caption functionality);
- Video/caption synchronization;
- Regulatory compliance.

Caption data should be checked when the captions are generated and when programs are delivered. Master control facilities should be provided with monitoring for closed captioning data and functions, both for incoming program material and for transmitted broadcast signals. Points to note for 708 caption monitoring include

1. Where 708 captioning data is carried in VANC, a VANC de-embedder with appropriate caption decoders should be used to examine the 708 and 608 captions and associated data. Some 708 encoders themselves provide the capability for monitoring and verifying embedded caption data.
2. Verify correct insertion of the 708 and 608 caption data in the outgoing DTV bitstream. Presence of the caption data and the caption service descriptor in the ATSC transport stream should be checked using an MPEG stream monitor.
3. Monitor the broadcast 708 captions using an off-air DTV receiver with 708 closed captioning decoding capability. Professional receivers usually have more comprehensive functionality for this than typical consumer receivers and may also provide a bitstream output (typically ASI) derived from the demodulated RF, which can feed an MPEG stream analyzer. In addition, professional receivers may have a transport stream input that can be used to monitor the bitstream output from an ATSC encoder prior to sending to the transmitter. This capability will assist in system troubleshooting.
4. DTV caption monitoring should include the associated 608 caption data using the derived NTSC signal generated by the DTV receiver to feed 608 monitoring equipment.

5. Monitoring of cable television captioning will require use of a suitable cable television interface unit or integrated receiver able to select and display the DTV captions.

Verification

Consideration needs to be given to arrangements for proving captions were correctly broadcast in the event that a complaint of noncompliance is received. One possible arrangement would be to maintain an off-air recorded log of station output with a bitstream recorder—disk-based server or tape-based system.

SUMMARY

For nearly three decades, closed captioning has been an important enhancement to television programming for a large number of deaf, hearing-impaired, and other television viewers. FCC-mandated requirements to caption an increasing amount of material have made the service indispensable to many viewers. Technological developments have made it easier for broadcasters to comply with the FCC requirements and improve the service being offered, and such enhancements are still ongoing. Meanwhile, the migration to DTV captioning has created a new set of opportunities and challenges, but it is comparatively recently that standards have been completed that properly address all aspects of the caption production and distribution chain. Equipment manufacturers, program makers, and broadcasters need to work together to implement complete systems that exploit the capabilities of the CEA-708 standard to provide even better captioning services in the years ahead. The reader is referred to the bibliography and other sources of information listed below for more comprehensive data.

ACKNOWLEDGMENTS

Thanks are extended to Graham Jones of NAB, who wrote the NAB 2004 Broadcast Engineering Conference Paper that forms the basis of the DTV captioning portion of this chapter. The authors extend thanks to SMPTE for permission to quote from and summarize parts of EG 43, System Implementation of CEA-708-B and CEA-608-B Closed Captioning.

Bibliography

Society of Cable Telecommunications Engineers. See http://scte.org/home.cfm

ANSI/SCTE 43 2004, Digital Video Systems Characteristics Standard for Cable Television

ANSI/SCTE 54 2003, Digital Video Service Multiplex and Transport System Standard for Cable Television

Advanced Television Systems Committee. See http://www.atsc.org/

ATSC Standard A/53E: Digital Television Standard, Revision E, with Amendment No. 1, December 2005, as amended April 2006.

ATSC Standard A/65C: Program and System Information Protocol (PSIP) for Terrestrial Broadcast and Cable, Revision C, with Amendment No. 1, January, 2006, as amended May, 2006

Consumer Electronics Association. Documents available from Global Engineering Documents. See http://global.ihs.com/

CEA-608-D – 2006, Line 21 Data Services

CEA-708-C – 2006, Digital Television (DTV) Closed Captioning. Consumer Electronics Association.

Federal Communications Commission. See http://www.fcc.gov/

FCC 00-259 – 2000, Report and Order, Closed Captioning Requirements for Digital Television Receivers

FCC Public Notice DA-04-2, January 6, 2004, Notice to Video Programming Distributors and the Public of the January 1, 2004, Requirements for the Closed Captioning of New Nonexempt English and Spanish Language Video Programming and Reminder Regarding Other Captioning Requirements

FCC 04-192A1 – 2004, Report and Order, Second Periodic Review of the Commission's Rules and Policies Affecting the Conversion to Digital Television

National Center for Accessible Media. See http://ncam.wgbh.org/

Delivering Captions in DTV, DTV Access Brief, October 2002

Society of Motion Picture and Television Engineers. See http://smpte.org/

SMPTE 333M-1999, Television—TV Closed-Caption Server to Encoder Interface

SMPTE 334-1-2006, Television—Vertical Ancillary Data Mapping of Caption Data and Other Data

SMPTE 334-2-2006, Caption Distribution Packet (CDP) Definition

SMPTE EG 43-2004, System Implementation of CEA-708-B and CEA-608-B Closed Captioning

CHAPTER

5.25

Data Broadcasting Systems and Applications for Television

PETER LUDÉ
Sony Electronics, Inc.
San Jose, California

INTRODUCTION

The past decade has seen an explosion in the electronic delivery of content. Music files, television programs, and feature films are now routinely downloaded to consumer devices over broadband connections to homes and businesses. However, in situations in which large files need to be widely distributed within a limited time window, broadband Internet connections exhibit significant shortcomings. For such applications, data broadcasting has several distinct benefits, and broadcasters are poised to leverage these advantages.

The transition from analog to digital terrestrial broadcasting affords a new opportunity to launch data broadcasting services as an adjunct to primary programming channels. Since the early 1990s, there has been vibrant experimentation with various data broadcast applications, business models, and technologies. While most early datacasting services failed to become sustainable businesses, many valuable lessons were learned during the numerous trials and tests conducted. These findings have served as a solid foundation for the second generation of successful data broadcast services operating in the United States, and for future services certain to take advantage of the strategic benefits of data broadcasting. The purpose of this chapter is to review the proven and potential applications for data broadcasting, and to explore the practical aspects of implementing such a network. Information on the data broadcasting standards that enable these services is provided in Chapter 1.17.

DATA BROADCASTING

Data broadcasting, sometimes referred to as *datacasting*, is the mechanism for propagating data from one point to many, usually over a wireless network. In this context, *data* has traditionally referred to anything other than audio and video content. But the definition of video has been recently blurred with the advent of new digital animation formats such as Flash®, and of video codecs now widely used in nonbroadcast applications, including QuickTime™ and SMPTE VC-1. For this reason, many broadcasters have come to use the term *data broadcasting* to refer to the transmission of any content other than that intended for reception on a standard digital television receiver.

Under this expanded definition, data broadcast applications might incorporate audio and video content as well as other file formats, and may be broadcast either as a file download or a live stream. For example, an episode of *CSI: Miami* [1] broadcast by a local terrestrial DTV station is, of course, not considered data broadcasting. This program would be readily viewed on any ATSC-compliant [2] DTV receiver within the station's coverage area if fitted with a suitable receiving antenna. However, if the same episode is instead packaged as a QuickTime encoded file, broadcast within the same MPEG-2 transport, and cached within a specially designed set-top box for later viewing on the same consumer receiver, it would be an example of data broadcasting. From a user perspective, these two methods appear very similar—both present the same captivating program using the same receive antenna, broadcast channel, and

TABLE 5.25-1
Summary of Data Broadcasting Attributes

	ATSC Program Broadcast	ATSC Data Broadcast
Content File types		
Video with audio	YES	YES
Audio only program	NO	YES
Executable software	NO	YES
Still image graphics	NO	YES
Text document	NO	YES
Other file types	NO	YES
Codec type		
Video	MPEG-2	Any
Audio	AC-3	Any
Transmission mode		
Streaming	YES	YES
File download	NO	YES
Examples of Metadata		
Descriptive	PSIP (A/65C) V-ISAN (A/57A)	Any (User Defined)
Parametric	DIALNORM	Any (User Defined)

video display. However, there are important technical differences, as summarized in Table 5.25-1 [3,4,5].

Types of Data

In some applications, the data being transmitted is associated—or coupled—with the primary audio and video programming services. An example of *tightly coupled* data is where JavaScript™ or ECMAScript files are broadcast synchronously with a primary program to support interactive television (ITV) applications. In other applications, data may be associated with a program but not closely synchronized with it. An example of this *loosely coupled* data is sending a portable document format (PDF) file of a product brochure during a commercial for that product.

But most deployments of data broadcasting carry content unrelated to the primary television programs, that is, *noncoupled.* This would include transport of movies, music, e-books, videogames, executable software, map images, and many other file types that are intended to be independent of the television viewing experience.

Three factors have contributed to widespread content distribution through digital networks rather than physical media:

- Consumer acceptance and preference for convenience;
- Proliferation of consumer media players;
- Maturing infrastructure technologies.

These trends are likely to continue as consumer habits continue to evolve, creating more demand for digital content distribution.

ADVANTAGES OF DATA BROADCASTING

In applications requiring mass distribution of large content files, data broadcasting provides substantial advantages over broadband Internet distribution; these advantages include:

- All locations will typically receive the broadcast data at the same time, making it well suited for time-sensitive data.
- Bandwidth costs are comparatively inexpensive because one transmission channel can serve thousands or even millions of receivers.
- Network operation is greatly simplified due to the multicast architecture.
- The population of potential users is greater, since more households are able to receive (or have access to) a DTV broadcast signal than have broadband Internet access.
- Deployment will generally have only modest impact at many broadcast stations where the DTV bitstream may be underutilized during large portions for the day, with excess capacity lying fallow.

The most dramatic benefit of data broadcasting derives from its use of multicast protocols that allow simultaneous transport of a data file from the broadcast site to an unlimited number of receivers. By contrast, content distribution over broadband Internet connection almost always is accomplished over a unicast (point-to-point) transport. This topology stems from the common use of Transmission Control Protocol (TCP), one of the core transport techniques used on the Internet. In TCP/IP, a unique connection is established by the host for each receiver. As more receivers are added, the bandwidth requirement quickly multiplies. Since the identical data is being transported separately for each viewer, a high degree of redundancy occurs. For purposes of content distribution, this results in TCP being highly inefficient with respect to network resources.

A Case Study: Internet Content Distribution

During the NCAA Basketball ("March Madness") tournament in March 2006, CBS Digital Media (a division of CBS Corporation) set an Internet record [6] by serving over 268,000 [7] simultaneous video streams, and over 14 million streams during the first two days. The audience size for this streaming broadcast was constrained by bandwidth capacity. During the first day of the tournament, 700,000 unique visitors entered the "waiting room," with 479,000 of them eventually accessing the live video feed. Presumably, the rest gave up.

Typical Internet video streaming bandwidth is in the range of 256 kbps, supporting an image of 320 × 240 pixels at 30 frames per second. Assuming all 268,000 simultaneous viewers were able to access this bandwidth, the content delivery network hosting this historic transmission would have required over 68 Gbps of backbone connectivity. This represents a substantial portion of the capacity available at an average Internet service provider (ISP), which will typically have about 175 Gbps of traffic at any given time [8]. The content delivery network that provided services for the 2006 March Madness tournament subsequently announced plans to increase its raw bandwidth capacity to 1,000 Gbps, or one terabit per second, to accommodate this increasing demand. Such extraordinary bandwidth access is very costly. Furthermore, the popularity of the streaming feed created troublesome network congestion on many workplace local area networks (LANs), since nearly three out of four viewers were enjoying the tournament from their office computers. Corporate information technology (IT) departments are now implementing Internet-filtering appliances to block employees' access to such streaming content, preserving the limited corporate LAN bandwidth for more appropriate business use.

Internet versus Broadcast

Of course, by broadcast television standards, a nationwide audience of 268,000 viewers is miniscule. The television coverage of the same tournament, broadcast by CBS in high definition, captured an audience of 15.5 million households—60 times greater than the Internet streaming audience. In addition, since the television program was propagated in a multicast (or broadcast) topology, one stream could serve all viewers within a station's coverage area. The high-definition broadcast over an ATSC channel required only 19.39 Mbps—about 0.03% of the capacity used to host the much smaller Internet streaming audience.

The popularity of video distribution websites is increasing rapidly, offering consumers programming that ranges from amateurish home videos to live sports coverage to high-definition versions of major Hollywood feature films. There is no doubt that over the next decade, consumers will continue to expand their use of such Internet-delivered video. The growth will accelerate even more when viewing this Internet Protocol (IP) content on the living room television is as easy as tuning to a traditional broadcast channel. This is far from the current situation in which complex integration between broadband modems, computers, home networks, and television IP hubs is required. However, ultimately, this Internet video distribution is best suited to niche programming, in which a vast number of choices are available but only a small number of consumers are interested in any particular title. For highly popular programs, multicasting the files over a data broadcast architecture will serve as the ideal complement to the Internet's one-to-one connection.

TRANSPORT ALTERNATIVES

The principle of data broadcasting has been applied to many types of transport networks. The most promising is carriage on digital terrestrial television (DTT). The data broadcast system architecture can be used with virtually all worldwide digital television (DTV) standards including ATSC, DVB-T, ISDB-T (in Japan), or DTMB-T (adopted in China). Although these standards differ in some aspects, all share the use of MPEG-2 (ISO/IEC 13818-1) transport stream as a basis of service multiplex. In order to fully understand the specific benefits of data broadcast using DTT, it is useful to consider alternate transport networks that have been deployed, or considered, for various data broadcast applications.

Cable Television

In the United States, cable television service is available to 99% of households with televisions [9], with nearly 72 million of these households subscribing to the service [10]. Due to this substantial penetration, cable is an attractive transport for data broadcasting. Like digital terrestrial television broadcasts, digital cable systems typically use a transmission infrastructure compatible with digital multiplex bitstreams constructed in accordance with MPEG-2. The SCTE has published a standard [11] for such in-band data broadcast applications, including the capability to signal service availability through out-of-band announcements. In this context, out-of-band (OOB) refers to data paths, typically limited to a relatively narrow 256 kbps, carried on frequencies outside those used for video services.

One widespread application of data broadcasting on cable systems is the transmission of program guide information to set-top boxes. In the past, system vendors have used their own proprietary approach to broadcast program guide data for hundreds of television channels to a software application usually residing within the middleware in a set-top box. However, the cable industry is in the process of moving these OOB signals from the proprietary channels to a standards-based DOCSIS Set-top Gateway, or DSG [12]. This move will allow expanded bandwidth for OOB signals and encourage broader interoperability through the use of open standards.

Cable operators and vendors have also conducted trials of downloading video programs to the hard drive cache storage in digital video recorder (DVR)-equipped set-top boxes. In these cases, the programming content is multicast over an in-band channel within the cable system but can be viewed only after being downloaded into the specially equipped set-top box. Encryption of the digital file and digital rights management software prevents unauthorized viewing.

Multicast-Enabled Internet

Since nearly all of the data broadcast applications being discussed here rely on Internet Protocol, or IP, one might assume that the Internet provides a workable infrastructure for data broadcast. Unfortunately, there are significant technical issues. As described earlier, most Internet traffic is over the TCP/IP protocol, which is inherently point-to-point. TCP was created to provide reliable, error-free delivery of data by controlling packet ordering and retransmission of dropped packets. Multicast networks must instead employ User Datagram Protocol (UDP), which provides fewer functions. Instead of guaranteeing packet delivery, UDP provides only error detection. If a packet is detected in error under UDP, it is simply discarded. Other mechanisms must therefore be used to provide error correction and recovery.

In the mid-1990s there were experiments in providing multicast services over the public Internet. Internet pioneers at the Xerox Palo Alto Research Center developed the MBone (short for Multicast Backbone) in 1992, which led to many promising experiments in file sharing, content distribution, and videoconferencing. By October 1996, the IP Multicasting Standards Initiative (IPMSI) was formed [13], soon followed by Networked Multimedia Connection (NMC), launched by Microsoft, Intel, and Cisco to promote multimedia and multicast networking. A series of multicast addresses were established in the range 224.0.0.0 through 239.255.255.255 to support such services [14].

By 1997, there were more than 3000 servers on the experimental MBone infrastructure. It was hoped that the birth of IPv6 (the latest version of Internet Protocol, adopted by the Internet Engineering Task Force in 1994) would lead to widespread deployment of multicast network services, including data broadcasting applications.

Despite these ambitious plans, as of May 2006, IPv6 connections accounted for only a tiny percentage of the live addresses in the publicly accessible Internet, which is still dominated by IPv4. The IPMSI, NMC, and MBone appear to be largely dormant. Enabling a true multicast-enabled Internet met with unanticipated deployment challenges in conversion of private networks, scalability, and security. For the near-term, IP multicast is likely to be found only in special applications on privately controlled networks.

Mobile Device Networks

Wideband wireless networks intended for transporting video programming to mobile devices are now emerging in the United States and elsewhere worldwide. In the United States, three such services were announced in the mid-2000s, as spectrum became available. These nascent wireless offerings are MobiTV™, a technologically agnostic video service provider working with several wireless carriers in 3G, WiFi, and WiMax spectrum; Modeo™, a service launched by Crown Castle International using the DVB-H (Digital Video Broadcasting–Handheld) standard [15]; and MediaFLO™, a mobile TV technology from QUALCOMM Incorporated. Each of these services can be used for data broadcasting, restricted only by bandwidth availability and by the capabilities and features of the handsets. The DVB-H standard, similar to DVB-T and ATSC, supports data broadcast through multiprotocol encapsulation and forward error correction (FEC) [16].

Of these three services, MediaFLO has perhaps been most accommodating in regard to plans for data broadcast applications. MediaFLO is designed for one-way digital transmission (Forward Link Only, hence FLO), to hybrid voice and data handsets and other mobile devices. The network uses spectrum in the 700 MHz band, previously occupied by UHF channel 55, under Part 27 of the FCC Rules. In addition to live audio and video streams, the MediaFLO network has demonstrated the broadcast of data including stock market quotes, sports scores, and weather reports [17]. Suitably equipped handsets can access the continuous stream of low-bandwidth data and present subscribers with real-time updates. At the CTIA Wireless [18] conference held in Las Vegas in April 2006, MediaFLO data broadcast applications were demonstrated. These applications included graphics representing play-by-play updates of Major League Baseball games and a stock portfolio tracker capable of alerting subscribers when major activity is occurring with the stocks in their portfolio. Such services provide an enticing view of data broadcast applications that may become popular over the next decade.

Direct Broadcast Satellite

Direct Broadcast Satellite (DBS) services, like cable, have become a mainstay of multichannel video delivery, with over 26 million U.S. households subscribing to DBS services as of June 2005 [19]. DBS, when using preferred orbital slots, has the benefit of achieving coverage of the entire continental United States (full-CONUS). Like other transmission systems described previously, digital broadcasts on these DBS services also use MPEG-2 transport. This readily enables data broadcast on the same transponder multiplex that carries digital television channels. In DBS networks, data broadcast technology is typically used for transmission of electronic program guide data, set-top box software updates, interactive services and, recently, download of video-on-demand content to DVR-equipped satellite receivers.

VSAT

Very Small Aperture Terminal (VSAT) systems are two-way satellite data networks in widespread use for business communications since the late 1980s. The technology has long been attractive to enterprises needing to reach hundreds or thousands of locations, such as retail groups. As one example, VSAT antennae can be found on the roofs of most gas stations throughout the United States. Contemporary VSAT services utilize IP network standards to provide inexpensive connectivity to geographically dispersed locations—often thousands of sites. A typical VSAT network might support 128 kbps upstream and 500 kbps downstream bandwidth that could be used to speed credit card authorizations, manage inventory flow, and consolidate payroll data. VSAT networks are increasingly being used for data broadcast of training videos, corporate messages, and digital signage content. Such a network is particularly attractive in cases in which the VSAT infrastructure has already been

installed (and funded) for other uses, and the data broadcast content can "piggyback" on existing applications. Since the available bandwidth is limited, VSAT is a good fit where files to be broadcast are of modest size, they are updated infrequently (weekly or monthly), and the delivery time window is flexible.

FM Radio SCA Subcarriers

As described in Chapter 4.8, "FM Stereo and SCA Systems," the FCC originally enacted the Subsidiary Communications Authority (SCA) in 1955, allowing insertion of subcarriers into FM radio broadcast channels. However, it was not until the deregulation in 1983 that the licensing requirements were relaxed by the FCC to permit radio stations to use their subcarriers for essentially anything. Using this system, data transmission of up to 12 kbps is possible.

In early implementations, companies such as Cue Paging and Bonneville Data used SCA subcarriers to broadcast custom news, sports, weather, stock information, and paging data to end users on a subscription basis. FM subcarriers were also used in some more novel experiments in data broadcasting. In 1984, video game maker Atari used this low bit rate subcarrier to transmit new video game software to its Atari 2600 game console. Unlike modern video games, a typical Atari 2600 game contained just 4 kilobytes of data. These files were transmitted in a repetitive sequence through FM radio stations, received on a small demodulator built into a game console cartridge, and stored in ROM. Field trials were successful, but the service was never widely deployed. Nearly two decades later, data broadcasting was again used to transmit video game files, but this time using terrestrial DTV channels to multicast multimegabyte games to personal computers. This service, launched in 2002 by iBlast, Inc. under the name *GameSilo* (see description later in the chapter), also never made it past the trial stage.

In late 2002, Microsoft launched its Smart Personal Object Technology (SPOT) initiative [20]. This technology has been deployed as a subscription-based data broadcast service marketed under the name *MSN Direct*. The service also utilizes SCA subcarriers, leveraging a nationwide network of radio stations to send information to tiny receivers built into wristwatches.

A new development by Digital Radio Express called *FMeXtra* [21] uses digital modulation to enable much higher bit rates to be carried using the SCA subcarriers. Data rates of up to 64 kbps are claimed when used with stereo audio and 128 kbps when a mono FM audio signal only is present.

FM Radio RDS and RBDS

As described in Chapter 4.14, Radio Data System (RDS) is the standard from the European Broadcasting Union for sending small amounts of data over these SCA subcarriers within conventional (analog) FM radio channels. Radio Broadcast Data System (RBDS) is the name of the similar standard in the United States, published in 1993. The primary application of this data broadcast system is to transmit text information to RDS/RBDS-capable radio receivers, such as station call letters, program format, clock time, and traffic messages. Provision also exists in these standards to transmit a continuous stream of binary data using a Transparent Data Channel (TDC). While the data rate is slow (typically less than 200 bps) for a typical TDC application, it is still possible to send 1 or 2 megabytes of data each day through an RDS data network.

IBOC Radio

As described in Chapter 4.13, the In-Band On-Channel (IBOC) system is a method for simultaneously transmitting digital radio and analog radio in the same radio frequency channel. For FM broadcasts, the HD Radio system developed by iBiquity Corporation, in addition to high-quality digital audio programming, provides the opportunity for broadcasters to add data channels for new services—and the potential for new revenue streams. The capacity available for ancillary data broadcast within FM IBOC is determined by several variables, but is in the range of 12.5 to 64 kbps, enough to download several hundred megabytes per day.

Analog Television Vertical Blanking Interval

The vertical blanking interval (VBI) is part of the analog television transmission left clear of viewable content—originally to leave time for the electron gun of the cathode ray tube in early television sets to move from the bottom to the top of the screen between picture frames. For many years, broadcast engineers have used VBI capacity to transmit program information, teletext, test signals, and closed caption information. As mentioned in Chapter 5.24 on closed captioning, also included is the Extended Data Service (XDS), a scheme that allows broadcasters to send information about the station and its programming to consumers. XDS data is interleaved with the closed captioning data on line 21 and documented within the CEA-608 standard [22]. In 1989, Nielsen Media Research launched its proprietary *Automated Measurement of Lineup* (AMOL) service, using line 22 of the vertical blanking interval. This service supplies metadata about programming to aid in ratings measurement.

In the early 1990s, the VBI was used for broadcast of generic data content as well. The *North American Broadcast Teletext Specification* (NABTS) is a protocol utilized for encoding such digital data into an analog television channel's blanking interval. The standard, classified under standard EIA-516, supports a data rate of up to 15.6 kbps per line of video, or up to 268.8 kbps in aggregate over 16 video lines. However, given available VBI capacity and forward error correction requirements, broadcasters could practically accommodate only about 128 kbps for these services. This capacity was put to use for a number of forward-looking applications.

One of the U.S. pioneers in data broadcast is National Datacast, Inc., a for-profit subsidiary of the Public Broadcasting Service, formed in 1988. The National Datacast digital datacasting network grew from PBS's early work in closed captioning for the hearing impaired on line 21 and employed other vertical interval lines for data transmission. The organization has supported data broadcast for such firms as Merrill Lynch, IBM, Microsoft, Visa, and TV Guide On Screen.

In 1996, Intel launched a short-lived service called *Intercast*. Using VBI capacity and the NABTS standard, this service included transmission of web pages and computer software to compatible TV tuner cards in personal computers, using Intel Intercast Viewer software. The service was used for trials with NBC, CNN, and The Weather Channel, often with web pages providing interactivity with concurrently transmitted television programs.

About the same time, WavePhore, Inc., launched a similar service, using data broadcast over VBI to deliver customized news and information. Partner content creators included CBS SportsLine, Quote.com, The Weather Channel, and *Time* magazine. Despite the *WavePhore* software and service being embedded into the release of Microsoft Windows 98 operating system [23], it never attracted a large user base. By 1999, WavePhore was working with General Instrument on creating a DTV version of its content delivery service, to replace the constrained capacity of VBI. The company ceased operations a few years later.

Analog Television Channel Injection

Prior to the widespread adoption of digital terrestrial television, and mindful of the limited capacity of VBI data broadcasting, there were several attempts to transmit high-speed digital data within analog TV broadcast signals [24]. For example, one approach placed a quadrature phase-shift keying (QPSK) carrier into the analog television channel at a less sensitive spectral position, such as the space between the color subcarrier and the aural carrier, or at the end of the vestigial sideband area. Another solution borrowed spread-spectrum techniques by distributing the inserted digital data across multiple carriers at an insertion power below the noise floor of the television signal. However, these techniques were of limited use, since the injected data was often visible to the viewer as interference artifacts in video. Attempts at lowering the data power to the point where interference was imperceptible had the undesirable side effect of preventing reliable data reception under normal circumstances.

In the early 1990s, Dotcast, Inc., began tests and deployment of its proprietary *dNTSC* (for data NTSC) technology. By inserting data as a quadrature component to the television signal, the system allows insertion of between 1.0 and 4.2 Mbps of data payload into a 6-MHz NTSC analog television channel, depending on the targeted coverage area. An abatement filtering process is used to counteract the effect of the Nyquist filter in the television receiver. Experimental results allowed injection of data into both the visual and aural carriers, with robust reception of a 1 Mbps data transmission within the broadcast station's Grade A coverage contour.

At the receiver, data is extracted by recovering the quadrature component of the composite video signal and applying signal processing to mitigate multipath artifacts and video leakage into data. The company reports that since the data is channel coded using Reed-Solomon and Trellis codes, error rates better than 10^{-8} can be achieved without visible picture interference.

MovieBeam

In early 2006, the dNTSC technology was used for one of the most ambitious data broadcasting services to date. The technology was deployed for the launch of the *MovieBeam* movies-on-demand service in 29 major metropolitan areas throughout the United States, potentially reaching nearly 50% of U.S. television households [25] (over 40 million homes). MovieBeam is a venture backed by Disney, Cisco, and Intel, with additional funding from several Silicon Valley venture capital firms. The service distributes up to 10 feature films per week by encoding the content into encrypted data files and broadcasting them over the dNTSC quadrature carrier within the visual band on selected PBS analog broadcast stations. Files are then received on set-top boxes equipped with a simple VHF/UHF antenna, analog NTSC tuner, dNTSC data extraction circuit, and 160 GB hard drive. The player supports Windows Media 9™/SMPTE VC-1 video codec, and Dolby Digital® 5.1 audio, allowing approximately 100 film titles to be stored within the player at any given time, in either standard-definition or high-definition (720p) format.

Analog television channels, of course, will cease operation when the conversion to all-digital broadcasting in the United States is completed. Plans call for these analog channels, including their dNTSC data, to go dark as of February 17, 2009. At that time, it is likely, at least technically, that the MovieBeam service will convert from dNTSC analog to ATSC datacasting.

DTV DATACASTING SYSTEMS

Over a decade's worth of early data broadcast experiments has proven the utility of this promising new type of broadcast service. However, with the vertical blanking interval or other communications channels described previously, payload bandwidth was limited—often to a few hundred kilobits per second or less. With the advent of the ATSC digital television standard (as well as other DTV standards), the potential data rates are greatly increased. It is now possible to deliver as much as 75 GB of content a day from each digital broadcast station, while simultaneously transmitting regular high-definition or standard-definition video programming.

DTV data broadcasting provides the flexibility to support virtually any data type, with multicast coverage over a large audience area. The receiver components, particularly the ATSC receiver chip, are inexpensive due to the mass adoption of consumer DTV television sets and the FCC mandate that all television sets with a 13-inch screen size or larger contain ATSC receivers by 2007. Reception, while not perfect, has proved to be sufficient to support consumer services seeking mass audiences.

DATA BROADCAST APPLICATIONS

In recent years, data broadcast systems have been put to use for a wide variety of purposes. The most appropriate applications are those with five key attributes:

- File sizes are large, making the service cumbersome to transport over broadband networks.
- The service is primarily for downstream transport. For symmetrical two-way services, datacasting is not a good fit.
- Many users require access to the files at about the same time—a requirement easily accomplished using multicast (or broadcast) protocols but difficult if using multiple point-to-point connections.
- The target audience is large. If only a few recipients need the file, other transport options may work better, but to reach thousands or millions of users, datacasting is ideal.
- The applications are independent from broadband networks. Datacasting, being a wireless transport serving a wide geographic area, can reach locations where broadband penetration is low or wired services are unavailable.

Entertainment

Music Video and Movies

Transmission of media files to consumers for general entertainment has been among the most practiced applications for datacasting. Although music, movies, and games are available via file download from many Internet sites, the process—especially for large video files—is slow and cumbersome. A full-length feature film, encoded in MPEG-2 at DVD quality, will occupy in the range of 4 GB of data. Such a file will require more than eight hours of downloading time over a broadband network at 1 Mbps. For this reason, many Internet websites offer movies with higher compression ratios. This results in smaller files that will download more quickly, but at the sacrifice of video quality. The challenges facing Internet-based download services will become even more pronounced as movies are increasingly released in high definition. New consumer optical disc formats will accommodate HDTV movie files of up to 15 GB (for HD DVD) or 25 GB in the case of a Blu-ray Disc™. Transporting such large files will prove to be extremely difficult over typical Internet connections.

On an ATSC digital television channel, a substantial portion of the available 19.39 Mbps transport can be allocated for data broadcast of movie files. For example, during overnight hours, a DTV broadcaster may schedule just one standard-definition program, using 7 Mbps. This leaves at least 11 Mbps to be allocated for data broadcast. A DVD-quality 4 GB file can now be sent simultaneously to all households in the coverage areas in less than one hour. A high-definition DVD might take three to four hours. Due to the substantial marketing campaigns launched by Hollywood studios, there is a peak consumer demand for feature film titles when they first enter the home video sell-through window. On average, the movie studios release between 10 and 20 titles per week. Therefore, it is entirely feasible to transmit all new releases as they become available to the home video window over one DTV broadcast station in just a few hours each night. The transmissions would be received over the air on a DVR equipped with large hard disk drive storage capability, an ATSC tuner, and applications software to manage the file caching, user interface menus, and digital rights management systems.

As described previously, the MovieBeam service, relaunched in early 2006, provides such a movie download service. However, the MovieBeam data channel is contained in the NTSC analog visual carrier rather than the DTV channel. iBlast, a short-lived data broadcast company based in Beverly Hills, provided a movie-download service on a trial basis in Los Angeles and San Francisco markets during 2003. The iBlast service utilized an average of 7 Mbps on the DTV channel and transported six movies per week. The receivers were prototype units based on an inexpensive IPTV set-top box running the Windows operating system, with a digital rights management system from Macrovision and equipped with a PCI card ATSC tuner module and internal hard drive.

Music files could also be data broadcast in a similar manner. However, the business case is less compelling, since music files in popular file formats (such as MP3) are small enough to be easily transported on a broadband network.

Games

Video games offer another interesting application. Since PC-based game files may be as large as 4 GB, the capability to leverage large data rates over DTV channels is attractive. Like movies, new video game titles are often in high demand upon their initial introduction, and are too large to be quickly downloaded on typical broadband connections. In 2002, the GameSilo service was launched by iBlast, using commercial DTV stations in six U.S. markets (see Figure 5.25-1). GameSilo provided up to five new video games per week, along with demo video clips, patches, tips, and "cheats" to the video game enthusiast community. Subscribers to the GameSilo service purchased an inexpensive ATSC tuner with tabletop UHF antenna and USB port to connect with their PC. Game files were stored in a designated file space on the PC's hard

FIGURE 5.25-1 The GameSilo service provided videogames and game related content through data broadcast to personal computers, during a 2002 trial. (Courtesy of iBlast, Inc.)

drive, and made available for limited free trial or for purchase using a digital rights management system.

The challenges in operating these entertainment services—whether games, movies, or music—were found to be of a commercial nature rather than technical. From an engineering perspective, the capability to send files within the ATSC transport to many users proved to be reliable and relatively inexpensive to implement. Off-the-shelf systems were, and still are, available for content management, performance monitoring, and digital rights management. However, securing rights to high-quality content can be expensive. Marketing campaigns and equipment subsidies to attract new subscribers are high, often running into hundreds of dollars per subscriber. Further, operating a customer support desk to provide quality customer service is costly. For these reasons, an entertainment-based datacasting service may need to attract hundreds of thousands or even millions of subscribers to sustain positive operating cashflow.

News and Weather

Many commercial television stations have invested heavily in staff and infrastructure to produce local news programming. Broadcast stations are uniquely positioned to serve their community with coverage of local events and to generate advertising revenue needed to support the significant cost of new production. As an added service, television broadcasters have experimented with making news segments available over datacast. An example was the *TotalCast* service, inaugurated in 2000 by WRAL-DTV, a pioneering digital broadcaster in Raleigh, North Carolina, owned by Capitol Broadcasting. Among other services, TotalCast offered subscribers access to local news segments on demand. News clips were reformatted for viewing on the subscriber's personal computer, which was equipped with a PCI card ATSC tuner. The service helped promote the WRAL brand and build affinity with the community. Another application that has been implemented by some digital television broadcasters is the datacast of weather information, including continuously updated Doppler radar and satellite imagery. Kentucky Educational Television (KET) has included transmissions of weather information and alerts as part of its service.

Education

From the start, data broadcasting has been embraced by educators looking for more efficient methods of communicating with students and with each other. From primary education through colleges and universities, there is a need to transport multimedia content for distance learning and supplemental classroom materials. To accommodate real-time distance learning, large files can be distributed in advance to students' PCs over a data broadcast network. Then, live interaction with the instructor can take place during the lesson over an Internet connection of modest bandwidth. One such software application is the *vLearning* system marketed by Vistacast, LLC. To implement this system, Vistacast has announced collaborations with Triveni Digital, a prominent data broadcast technology provider, and SpectraRep, a subsidiary of BIA Financial Networks involved in providing data broadcast bandwidth from commercial DTV stations.

Data broadcast systems have also been successfully deployed to deliver educational courseware, including full-resolution video segments, lesson plans, and photos, from institutional archives. Educational download services have been launched by KLCS-DT, a television station owned and operated by the Los Angeles Unified School District serving over one million students and teachers by DTV datacasting, and by Nebraska Educational Telecommunications (NET), a network of 9 transmitters and 14 translators covering the state of Nebraska [26].

Software Updates

Datacasting has been used for the important task of distributing software patches and upgrades to consumer devices. Consumer electronics manufacturers have the daunting task of updating television sets, set-top boxes, and other devices. Sending new software loads through the mail and requesting the consumer to self-install is, of course, slow and unreliable. Dispatching service technicians is reliable but expensive. In early 2006, UpdateLogic, a Massachusetts-based firm, launched the *UpdateTV*™ service to deliver and install software updates to digital television sets in consumers' homes [27]. The company has secured contracts with at least six prominent consumer electronics manufacturers. For the trial, software upgrades were transmitted over the data broadcast capacity provided by National Datacast over four PBS digital television stations (in Boston, Denver, Indianapolis, and San Diego.) This promises to be a beneficial and effective use of data broadcasting for commercial as well as residential applications.

Digital Signage

As advertising dollars continue to move from broadcast television to Internet and out-of-home media, digital signage networks have become prominent. Flat-panel displays are rapidly being installed in shopping malls, retail outlets, sports venues, and even gas pumps for the purpose of exposing mobile audiences to sponsored advertising messages. Distributing the advertising content, however, can be a complex challenge. In the digital signage network, media files have become larger as more HDTV content is used. Screen locations may be widely dispersed over a geographic area, sometimes in hard-to-reach locations. The size of a typical network, often covering thousands of locations for a chain retailer, demands efficient multicast delivery. Sales promotions and inventory conditions drive rapid reprogramming of the advertising message. For these reasons, data broadcast is now being utilized for these retail digital signage systems. In one example, Scala, a pioneering provider of digital signage systems, announced a partnership with Microspace Communications [28], a subsidiary of Capitol Broadcasting, to provide targeted delivery of content to locations nationwide over a satellite datacast network. At about 50 cents per megabyte delivered during prime-time hours, the transport cost is relatively economical compared to Internet distribution. For deployments within a single market area, DTV datacasting could provide an attractive solution and prove more cost effective than a full-CONUS satellite channel.

Emergency Information

With heightened concerns for domestic security, there is an increasing demand for methods of quickly and reliably distributing emergency information. During a crisis, it is imperative that disaster alerts, graphs, maps, and operational communications be quickly disseminated to emergency responders. Within a year after the tragic events of September 11, 2001, the Association of Public Television Stations (APTS) proposed a way to meet this challenge to the Federal Emergency Management Agency (FEMA): the use of data broadcasting on PBS digital television stations [29]. Data broadcasting provides an attractive solution due to the wide signal coverage, simplicity, and bandwidth availability. A datacast network is well suited for transmission of real-time video and telemetry data to emergency crews situated over a wide area. In addition, DTV data broadcasting channels can potentially be used for sending emergency information to the general public, including Amber Alerts, broadcast Emergency Alert System (EAS) notices, and government communications.

Secure emergency network systems were introduced by at least two firms. Triveni Digital offers its *SkyScraper ESN* (Emergency Services Network) system to allow multiple agencies within a communications network to prepare, schedule, and insert emergency data into the DTV broadcast stream [30]. SpectraRep has developed the *AlertManager* critical event messaging system, also intended for deployment on datacast channels [31].

Emergency data broadcast systems were used in trials in Washington, DC, and other municipal areas, and in 2006 the APTS and the Department of Homeland Security announced the deployment of the new Digital Emergency Alert System (DEAS) with roll-out nationwide [32].

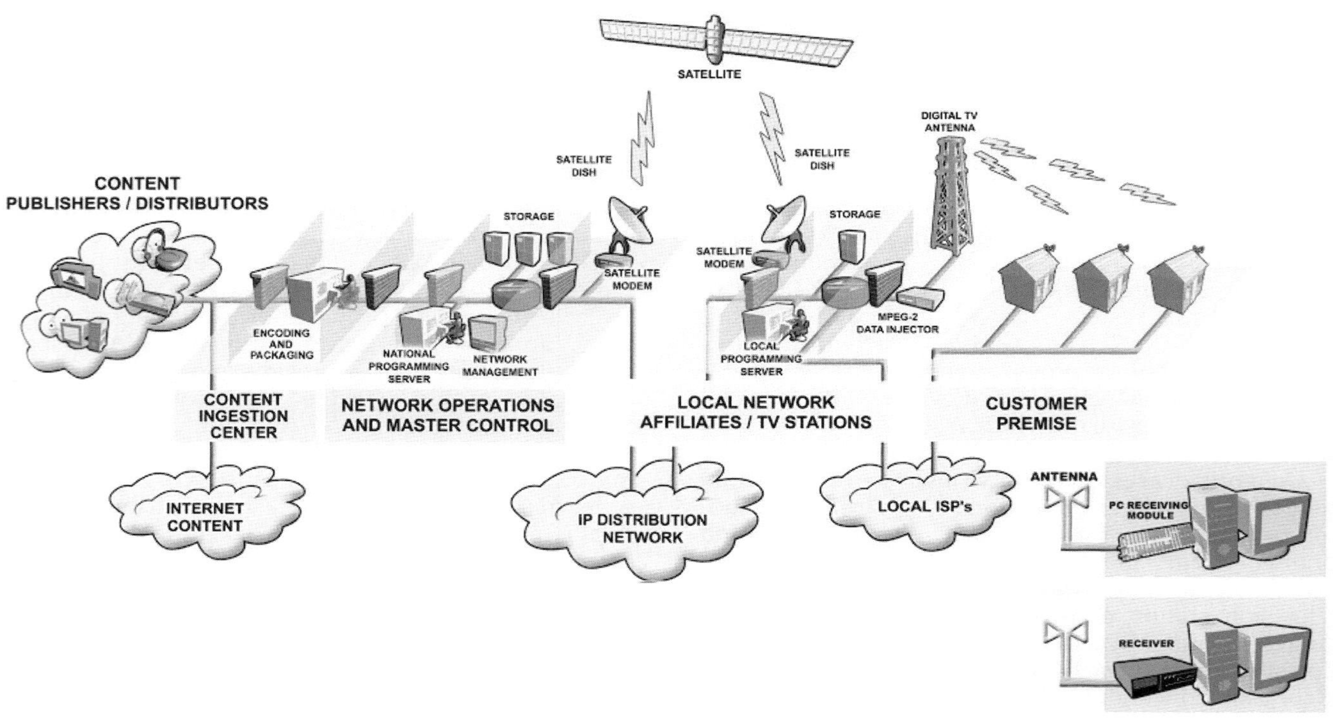

FIGURE 5.25-2 System architecture for a typical two-hop data broadcast network. (Courtesy of iBlast, Inc.)

SYSTEM ARCHITECTURE

A typical data broadcast network involving two or more DTV stations will utilize a two-hop architecture. As illustrated in Figure 5.25-2, the system consists of four primary subsystems: content ingest, network operations center, local television station node, and customer premises equipment. The first "hop" is used to transfer data files from the central *Network Operations Center* (NOC) to each affiliated DTV station, each of which is considered a *network node*. This connection can use terrestrial data circuits, such as T-1 or DS-3 lines, or a satellite interconnect. The second hop is from the broadcast station node to the customer.

Content Ingest

In most cases, content files will require some conversion or processing prior to being transmitted across the network. Since a television broadcast station is subject to license restrictions, it is imperative that no inappropriate or malicious content is inadvertently datacast. In addition, high-value content, such as popular movies, games, music, or e-books, will have defined user rights, which must be tracked and enforced. Examples of common requirements for the content ingest process include:

- Cataloging and tracking files, using a database or digital asset management system;
- Confirming content ID, version, and broadcast rights;
- Converting or re-encoding to match the desired output format;
- Confirming and appending metadata descriptors;
- Performing technical quality assurance checks;
- Running virus and software worm detection applications;
- Encrypting and generating security keys to support a digital rights management system;
- Packaging and creating file associations with related material.

This ingest process may be simple, in the case of a tightly controlled content workflow, or complex if material is arriving from many sources with unknown internal process controls.

Network Operations Center

The Network Operations Center is a central facility equipped to manage real-time data broadcast operations (see Figure 5.25-3). Typically, the NOC performs three primary functions. First, the NOC staff will oversee the scheduling of data files to be transmitted. In the case of a network of multiple DTV broadcast stations, there may be unique transmission schedules for each station. This will accommodate different time zones and variations in the availability of bandwidth. Schedules may be customized to avoid data-

FIGURE 5.25-3 The Network Operations Center for a data broadcast service enables monitoring, scheduling, and control of the network from a central location. (Courtesy of MovieBeam Inc.)

casts during prime-time hours or at times when increased bandwidth is allocated to high-definition primary programs.

Second, NOC personnel will manage the health and welfare of the end-to-end network. This includes monitoring of data error rates, MPEG transport stream insertion, and availability of backbone connections. Network management tools such as OpenView™ or Tivoli™ are often used to automate this function.

Finally, the NOC is equipped to take corrective action in the event of failure in any portion of the transmission chain. This might include operation of a satellite uplink for data transport to the affiliate stations, routing of backup data circuits, or switchover to redundant hardware components in the event of equipment failure.

In the case of a small data broadcast operation from a single DTV station, the NOC functions might be easily handled as part of the station master control operations.

DTV Station Node

The network node is the key subsystem for injection of data files into the ATSC (or other DTV standard) bitstream. The node consists of a *data server, RAID storage* of suitable capacity, an *IP encapsulator* (see Figure 5.25-4), and an *emission multiplexer.* Data files may be received from the NOC or, in the case of a single location service, may be generated locally at the broadcast station. In the case of a multinode network, each node will have provision for receiving data from the NOC, such as satellite data receiver or fiber-optic network interface, and a router. A scheduling software application may be used to assign the transmission sequence of each data file. In some applications, the node will add a separate announcement stream of metadata to the emission multiplex. This announcement stream will be used to inform each data receiver as to what data objects are being transmitted for purposes of targeting selected receivers, populating a data program guide, or managing the cache memory within the subscriber's receiver device.

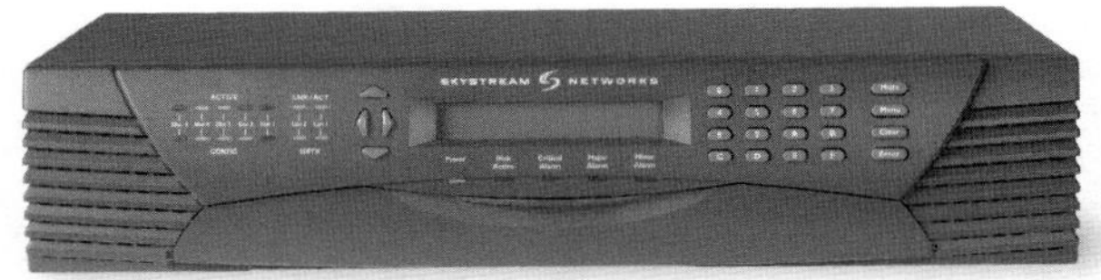

FIGURE 5.25-4 An IP encapsulator is used at the DTV station to receive IP data and inject it into an MPEG-2 transport stream. (Courtesy of Tandberg Television.)

Physically, the data broadcast node will generally occupy about one equipment rack at the broadcast station and is installed near the existing MPEG-2 transmission encoders, ATSC transport stream multiplexer, and PSIP generator.

Interconnections

A 100BaseT local area network is the appropriate interconnect between the data receiver, router, and IP encapsulator. Either a UDP or TCP/IP session can be initiated to transport files from local cache memory to the encapsulator. Connection from the IP encapsulator to the multiplexer may use either the DVB-ASI or SMPTE 310M interfaces. These standards are designed to transmit MPEG-2 transport streams over a 75-ohm coaxial interconnection. DVB-ASI is a packet asynchronous fixed-frequency serial interface with a clock rate of 270 Mbps and usable payload of 216 Mbps. SMPTE 310M is a packet synchronous interface that operates at a fixed frequency of either 19.39 Mbps or 38.78 Mbps.

Opportunistic Data and Flow Control

With an emphasis on improving the information throughput of networks and maintaining high image quality, variable bit rate (VBR) encoding is becoming increasingly widespread. VBR coding assigns compression ratios according to the frame content complexity. This results in significant savings over constant bit rate (CBR) coding techniques. With VBR, as picture complexity decreases, more bandwidth becomes available for injection of data files. This nondeterministic data channel is known as *opportunistic data* and will vary over time. For this reason, opportunistic data is not appropriate for certain applications, such as tightly coupled data used in interactive programs. However, opportunistic data is ideally suited for transport of large files with less stringent delivery deadlines.

In order to take advantage of opportunistic data, flow control messages between the data server and multiplexer are required. SMPTE 325M is a standard that facilitates the interoperable controlled delivery of MPEG-2 transport stream packets for this purpose.

Alternatively, a system can be designed to use the inherent flow control within the TCP/IP protocol. In this case, the functions of the data server and multiplexer are integrated, and data can be delivered to the emission multiplexer in a guaranteed flow-controlled manner over a TCP/IP connection.

Data Receiver

At the customer's premise, a data receiver equipped with an appropriate antenna is used to capture, store, and use the data files. The data receiver may be a dedicated device, such as a purpose-built set-top box, or a PC with datacast hardware and software added.

For deployments covering a modest number of users, a PC provides a convenient and economical means for receiving data broadcasts. Low-cost ATSC receiver modules, as shown in Figure 5.25-5, are available on PCI cards that can be fitted to most PCs. Software is required to capture the data broadcast stream according to the PID value assigned to the program element of the MPEG-2 transport stream, or by using a separate announcement stream to identify the target data packets.

Figure 5.25-6 is an example of a software stack for PC applications. Software drivers are installed to communicate with the PCI data broadcast receiver module. The MPEG-2 transport stream is extracted, and a TCP stream is further unwrapped. At this point, the data may be decrypted and presented to the middleware layer, where metadata handling, memory cache management, diagnostics, and antenna positioning aids are provided. At the top of the "stack" is the user application that will be created for the specific service, such as video on demand, HTML browser navigation, and program guide services.

While using a general-purpose PC is a convenient means of launching a data broadcast service, there are

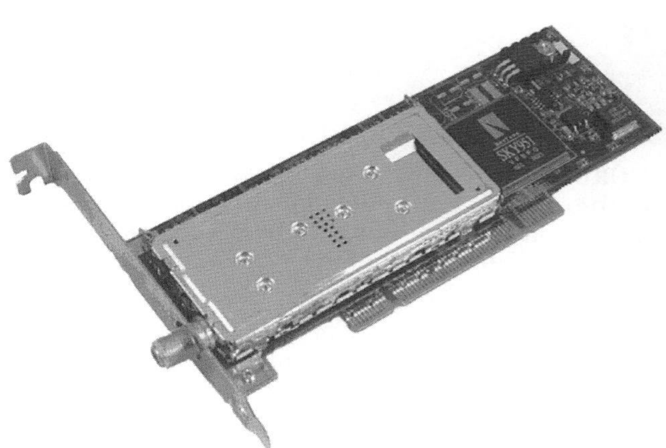

FIGURE 5.25-5 Model DTA-150 Cat's Eye—an inexpensive ATSC receiver card may be fitted into a PCI slot of a personal computer to create a data broadcast receiver. (Courtesy VBox Communications.)

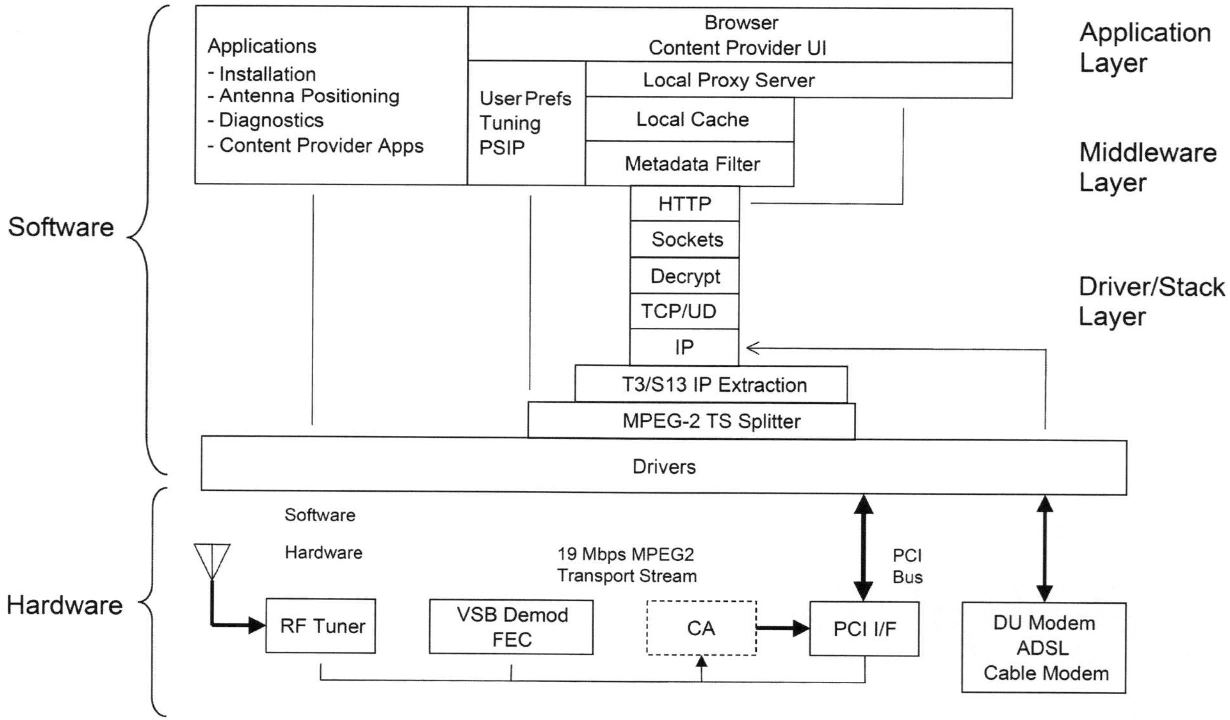

FIGURE 5.25-6 Example of a software stack for a PC-based data broadcast receiver.

several disadvantages to this approach. First, it requires the user to add new hardware and software, which requires a customer support function to assist. Second, the complexity of computers and the large number of possible configurations make system testing difficult. Unexpected incompatibilities may occur. Third, the PC-based system will generally be more costly when compared to a purpose-built hardware device, assuming the device is manufactured in large quantities. For this reason, services targeted at mass audiences, such as the MovieBeam service, use specially designed receivers that are inexpensive and simple to install and use.

OPERATIONAL CONSIDERATIONS

The specific design and operation of a data broadcast network, as with any broadcast system, will depend on the performance expectations, programming requirements, and budget. Special attention should be paid to the issues described in the following sections.

Security

Data security has become a critical design attribute in all IT systems. Security strategies should be carefully considered in at least four areas. First, high-value content must be protected from piracy in accordance with business rules dictated by the content owner and service programmer. Fortunately, digital rights management systems are available to facilitate content protection. Second, the data broadcast network must be protected from malicious attacks. The capability to transmit data files to thousands or millions of devices in an instant will undoubtedly be discovered by criminal programmers intent on propagating viruses or spyware. Third, wherever the business model relies on subscription or pay-per-use, the on-line financial transactions must be protected from tampering. Finally, customer privacy must be considered, and any data about end users should be strongly protected from unauthorized access.

Forward Error Correction

As with any data channel, DTV data broadcast is an imperfect conduit. For primary program channels, using MPEG-2 coding, the viewing experience is protected by multiple layers of error correction and concealment built into the ATSC standard. Up to a reasonable limit, lost packets are invisible to the viewer. However, when sending media files or executable software over a datacast channel, even one missing packet could render the file unusable. The counterstrategy used is to transmit redundant information to compensate for the inevitable lost packets, a method known as forward error correction (FEC). On the most basic level, this can be accomplished through use of a data carousel, wherein data files are repeated multiple times. Repeated retransmission of the file can be continued until all (or nearly all) receivers have successfully captured every packet. A more sophisticated means of FEC is through the use of algorithms such as *Tornado codes*, which allow reconstruction of lost packets with only minimal error correction packets. A third strategy uses a return path, such as a broadband Internet connection, to replace specific packets on a receiver-by-receiver basis. Fortunately, telecommunications systems providers offer many commercial products to address FEC in any of these modes.

Monitoring and Quality Assurance

Monitoring a data broadcast service is more complex than a primary DTV program. When a new service is designed, it is important to consider how failures at any point in the chain can be quickly identified, diagnosed, and rectified. Consideration should be given for the heterogeneous set of receive devices that may be attached to the service. Different users may be using different operating system software, antennas, receiver boards, and hardware configurations. MPEG transport stream monitors are available, which are necessary for finding and solving problems with the multiplex or flow control systems.

SUMMARY

Data broadcasting on DTV channels can benefit from well-established standards and commercially available equipment. Data broadcasting is an ideal solution for many applications, but it is important to consider the specific benefits over alternative transport means, particularly broadband Internet connections. The system architecture requires careful planning to accommodate needs of metadata injection, security, and reliability. For some services, back-office functions including subscriber management, billing, and customer are important components and must not be overlooked. Receiver devices are now widely available, but for mass deployments to thousands or millions of users, a purpose-built receiving device will prove more reliable and user-friendly than general-purpose PC receivers. In any case, ease of use is critical to the success of any service.

Many applications have been launched on data broadcast networks, with varying degrees of commercial success. More innovations will undoubtedly unfold as the technology is better understood and infrastructure expanded.

References

[1] *CSI: Miami* is a popular CBS prime-time program.

[2] ATSC Standard A/53E: Digital Television Standard, Revision E, with Amendment No. 1, Advanced Television Systems Committee (ATSC), December 27, 2005, as amended April 18, 2006. See http://www.atsc.org/.

[3] ATSC Standard A/65C: Program and System Information Protocol (PSIP) for Terrestrial Broadcast and Cable, Revision C, with Amendment No. 1, Advanced Television Systems Committee (ATSC), January 2, 2006, as amended May 9, 2006. See http://www.atsc.org/.

[4] ATSC Standard A/57A: Content Identification and Labeling for ATSC Transport, Advanced Television Systems Committee (ATSC), July 1, 2003. See http://www.atsc.org/.

[5] R. Chernock and F. Schaffa. "Issues in DTV Broadcast-Related Metadata," IBM, 2002.

[6] "CBS NCAA March Madness on Demand Shatters Record," CBS Press Release, March 20, 2006.

[7] G. Keizer. "March Madness Webcast Set Internet Record," *TechWeb Technology News*, March 17, 2006.

[8] "Must Stream TV? Bosses Worry," *Red Herring*, March 20, 2006.

[9] NCTA, *Industry Statistics*, Cable Developments 2005. According to the NCTA, at the end of 2004, cable systems passed 108.2 million occupied homes with a television set, and 109.6 million had a television set.

[10] MAGNA Global Research, *On Demand Quarterly*, Multichannel News, July 10, 2006.

[11] ANSI/SCTE (Society of Cable Telecommunications Engineers) 80-2003: In-Band Data Broadcast Standard Including Out of Band Announcements. See http://www.SCTE.org/.

[12] Brian Santo. "Out of Band, Out of Mind," *CED Magazine*, July 1, 2006.

[13] The IPMSI was originally known as the Internet Protocol Multicast Initiative (IPMI).

[14] Internet Engineering Task Force (IETF) RFC 1112

[15] Digital Video Broadcasting (DVB) Transmission System for Handheld Terminals (DVB-H), ETSI EN 302 304 v1.1.1 (2004-11).

[16] G. May. "The IP Datacast System—Overview and Mobility Aspects," IEEE International Symposium on Consumer Electronics 2004, Proc. pp. 509–514, September 2004.

[17] "MediaFLO USA Demonstrates Live Datacasting Application on FLO™ enabled 3G Handsets at CTIA Wireless 2006," QUALCOMM Incorporated press release, April 5, 2006.

[18] See http://www.ctia.org/.

[19] Twelfth Annual Report, "Annual Assessment of the Status of Competition in the Market for the Delivery of Video Programming," Federal Communications Commission, March 3, 2006.

[20] "Microsoft Launches Smart Personal Object Technology Initiative," Microsoft press release, November 17, 2002.

[21] "FMeXTra Multi-Channel Digital Programming FM Subcarrier System," Digital Radio Express. See http://www.dreinc.com/.

[22] Kelly Williams and Neil Mitchell. "Television Data Broadcasting," *NAB Engineering Handbook* (9th ed.), National Association of Broadcasters, Washington, D.C., 1999, pp. 1163–1168.

[23] Microsoft to Include WavePhore's WaveTop Software in Windows 98," Microsoft press release, October 8, 1997.

[24] Wonzoo Chung. "Datacasting over Analog Television: A Tutorial on Dotcast's dNTSC Technology," Dotcast, Inc.

[25] MovieBeam, Inc. Launches Across the Country, Movies-on-Demand Service Dramatically Improves the Movie-Rental Experience, Company press release, February 1, 2006.

[26] R. Crinon, D. Bhat, D. Catapano, G. Thomas, J.T. Van Loo, and G. Bang. Data Broadcasting and Interactive Television, *Proceedings of the IEEE*, vol. 94, no. **??**, January 2006.

[27] Update Logic Launches Software Distribution Service for Digital TV Devices, Company press release, March 28, 2006.

[28] Microspace and Scala Join Forces to Offer Shared Network Solutions for Digital Signage, Company press release, June 8, 2005.

[29] Letter to FEMA from J.M. Lawson, M.D. Erstling, and L.M. Thompson of APTS to FEMA, September 30, 2002.

[30] Triveni Digital Introduces Solution for Secure Data Broadcasting of Emergency Services Network Information, Company press release, February 7, 2006.

[31] SpectraRep Announces the Release of AlertManager Version 1.2, Company press release, October 19, 2006.

[32] APTS and Department of Homeland Security-FEMA Announce National DEAS Rollout, Press conference July 12, 2006, at public television station WETA in Arlington, Virginia. See http://www.apts.org/news/deas.cfm.

SECTION

6

TELEVISION TRANSMISSION

CHAPTER

6.1

Planning a Television Transmitter Facility

JOHN M. LYONS
The Durst Organization
New York, New York

INTRODUCTION

The planning of a television transmitter facility requires that many variables be considered before decisions can be made. Planning ahead can save time and money as the project proceeds toward completion. Consulting with experts during the early design stages can save time and money in the long run and make the facility perform as anticipated. Described in this chapter are many of the complexities associated with this type of project, whether big or small, solo or community based (see Figures 6.1-1 and 6.1-2). With proper preparation, the project can be accomplished within budget and on schedule.

Many questions come to mind when designing a new facility from scratch or redesigning an existing facility for updated or additional equipment:

- Is a completely new facility being built?
- Is equipment being replaced in an existing one?
- Is it standalone or on a community or master antenna system?
- Is the facility in an inner city area, an urban area, or a suburban community?
- Is the builder the site owner or a tenant/licensee?

Planning a television transmitter facility is not something that is done often, but careful planning will result in obtaining the desired functionality and appearance of the facility and will determine how it works when completed. The time frame to plan and construct a transmitter facility is often a year or more from start to finish; the procedure must take into account the following:

- Permitting process
- Zoning processes
- Time to manufacture equipment
- Delivery of major equipment components
- Negotiating the lease or license agreement

The process begins with planning the facility, finding the location, checking zoning, and buying the property or negotiating the lease/license agreement. The next step is the hiring of an architect and the necessary mechanical, structural, electrical, and other consultants to design the facility using the original design concepts as the basis, as well as assembling a project design team. After the design team is satisfied with the facility design and its development, specifications are then completed to be sent to bidders, the eventual award of the contract and, finally, building the facility. Although this all might seem to be a rather simple process, it can be very complex.

FACILITY DESIGN SCHEDULING

Problems with the design of a TV transmitter facility may be discovered only after construction is completed, when it may be too late to solve them economically and in a timely manner. Detailed advance planning will make the project run smoothly. An important concern is the weather. Adverse weather can seriously affect the construction schedule and substantially affect the costs. If downtime is necessary when the project has started, the trades involved may require the owner to pay for their time, even though

FIGURE 6.1-1 Standalone Class A TV transmitter installation (WEBR-CA, New York).

FIGURE 6.1-2 Multiuser tower site at 4 Times Square, New York, showing four master antennas and a single channel antenna.

there may be little or no production from them. Tower crews are usually tightly scheduled and must be scheduled well in advance of the work; for example, the tower working season in the north is much shorter due to the winter cold and snow than in the south, where other weather conditions must be anticipated.

Long-lead-time items, such as the transmitter, antenna, or radiofrequency (RF) system, can affect the schedule if the arrival time is not accommodated. Note that it is not the date that the item leaves the factory that is important, but the date that the item will arrive on the job site. Written commitments from the tower company and the equipment manufacturers should be obtained early in the project. Zoning approvals, tower design, analysis, and modification all take time, and changes in the schedule could affect the anticipated on-air date. Moratoriums on construction around holidays or other events can affect the schedule and should be known early in the planning. The time taken in advance to develop a realistic delivery and work schedule will result in fewer unanticipated problems.

ZONING PROCESS

Whether planning a new facility in a new area or joining an existing system, knowledge of the zoning regulations for that area is important; for example:

- Are there any deed restrictions or other covenants on the land?
- Can a zoning variation or a special-use permit be secured to use the land as a transmitter site, if not already so designated?
- What are the Federal Aviation Administration (FAA) regulations for the parcel of land that is contemplated?
- Could any environmental problems cause ownership liability?
- Are any wetland issues involved?

Consultants know how and where to find answers in an expeditious manner, and involving them in advance can save time and money later.

BUILDING CODES

Ordinances in place in most cities and towns specify construction standards that must be met during construction of the site. These ordinances include controlled inspections at specific times during specific parts of the project, including such items as electrical, environmental, and structural inspections for welds and bolt torque. Most communities require the issuance of a building permit from their building department or other issuing authority before work can begin. Because this can be a lengthy process, the use of an expediter who is familiar with the permit system process can shorten the length of time from application to issuance of the building permit.

Meeting ADA Requirements

It is important for the proposed installation to be in conformance with Americans with Disabilities Act (ADA) legislation. Enacted in 1992, the ADA was intended to protect disabled persons from discrimination in public accommodations and commercial facilities. New construction and existing facility renovations must adhere to the ADA provisions in every regard. Conformance to the Act includes having doorways and paths with sufficient room for wheelchair passage and replacing doorknobs with lever-type handles. Restrooms must also comply with the Act; fixtures must meet stringent requirements for wheelchair access, even if no disabled or wheelchair-bound persons are currently working at the facility.

LEASE/LICENSE IMPLICATIONS

If construction is in an existing building, the lease/license agreement may determine what can or cannot be done, who can or cannot be hired as a contractor, and when the construction work can be performed. Most major sites have a standard set of building regulations, which generally include a listing of approved contractors who have demonstrated that they can perform the assigned tasks in a workmanlike manner, that they do not cause undo disruption to the building operations, and that they have the appropriate kinds of insurance, both general liability and workers' compensation (these can be quite expensive in major cities). The lease/license agreement may also have clauses as to the hours of work, noise requirements, shutdowns for condensers and chilled water, sprinkler system tie-ins, elevator usage and charges, plan review and permit processing, work letters or allowances, tap-in charges, and window coverings, among other issues that can impact the schedule and costs. Being aware of them in advance can eliminate surprises when an invoice comes from the owner or management company. Other issues that may appear in leases or license agreements are RF radiation (RFR), tower light monitoring, and interference resolution at multiuser sites.

Radiofrequency Radiation Issues

RFR is a concern wherever a facility is planned. Studies that should be performed in advance include computer modeling of the expected RF profile. At a multiuser site, the lease or license agreement will spell out the site owner or manager's responsibilities with regard to making the RF exposure from the site conform to, for example, FCC Rules and Occupational Safety and Health Administration (OSHA) regulations. In some instances, the reduction of RF power necessary to protect workers will be spread across all the stations using one master antenna if that is the section of spectrum that is over the limit. This procedure allows the other users on other antennas (whether they are single antennas or another master antenna) to remain at full power. There are also ways to monitor

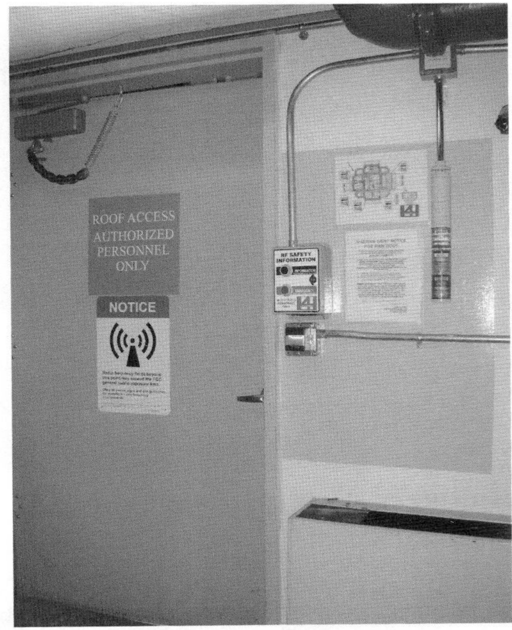

FIGURE 6.1-3 Roof access warning light and signage.

an active site, such as positioning area RF monitors at the hot spots of the site where maintenance personnel are likely to be located. Warning lights and audible alarms (see Figure 6.1-3) can warn personnel when they are entering an area on a rooftop where the RF exposure may exceed or exceeds the safe human exposure levels. See Chapter 2.4, "Human Exposure to Radio Frequency Energy," for more information on this important issue.

Lockout/Tagout

Lockout/tagout procedures must be in place as part of RF control and monitoring. Lockout is secured by the turning of a switch (*e.g.*, the shorted position) and a lock being placed through the control to prevent unauthorized tampering. Tagout is a procedure whereby the person locking out the switch places a tag through the lock shackle with information as to who performed the lockout and who has the key. Lockout/tagout monitoring should also be a part of site management controls. Figure 6.1-4 shows the switch with locks in place and a microswitch with cables going to the monitor system with position information.

Tower Light Monitoring

Tower light monitoring techniques vary from site to site (see Figures 6.1-5 and 6.1-6). Usually the site owner or manager is responsible for monitoring, logging, reporting outages to the FAA, and having repairs performed. In some cases, the user of the top antenna takes responsibility for all monitoring obligations.

FIGURE 6.1-4 Lockout/tagout switches.

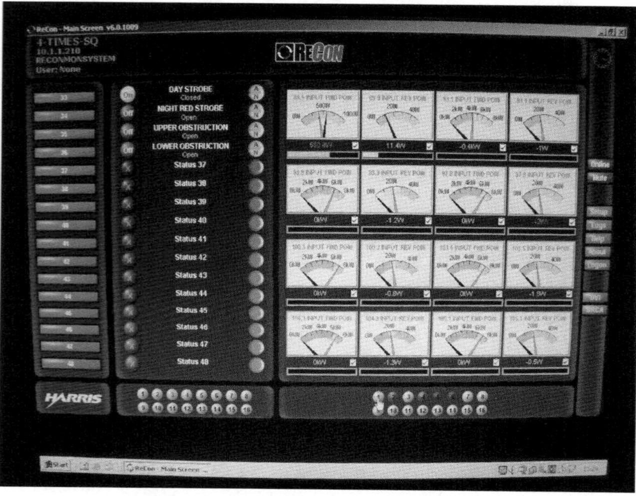

FIGURE 6.1-5 Site tower light monitoring.

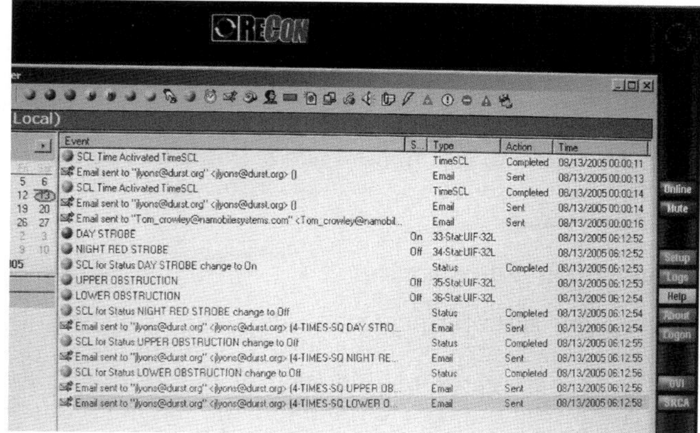

FIGURE 6.1-6 Monitor log shows tower light change of state.

Interference Resolution

Resolution of interference between multiple users at a common site is generally the responsibility of the most recent user at the site. Most lease/license agreement language specifies this fact and can force the assumed offender to lower power to a point where there is no interference or, if this cannot be accomplished, to shut down until the problem can be resolved. In an effort to make this work for all users, leases/license agreements usually specify the frequencies of all existing users at the site for analysis so there are no surprises when the newest user turns on the transmitter. Included in the frequency list are the new user's services, main television or radio frequencies, and all ancillary channels for STL, TSL, ENG, RPU, spread spectrum, or any other services at the common site.

Common Pressurization System

Pressurization of transmission lines increases their lifetime and adds a safety factor through the use of dry air or preferably nitrogen. Nitrogen keeps lines dry inside and, because oxygen in the lines is removed, oxidation is deferred for a longer period of time. Moisture is also removed to reduce the potential of arcing. An example of a community pressurization system is shown in Figure 6.1-7.

FIGURE 6.1-7 Pressurization system showing nitrogen generator, manifold, and vacuum system.

ASSEMBLING THE PROJECT TEAM

The project team includes the facilities manager, architects, engineers (electrical, structural, mechanical), expediters, project managers, contractors and subcontractors, fabricators, and suppliers (see Figure 6.1-8). Weekly design meetings should be held and turnaround deadlines for modifications or other changes tightly controlled. If the design process is allowed to start slipping, additional slippage may occur once the construction starts. Selection of the team relies on the planner's knowledge.

Role of Consultants

The consultants should be chosen from among known entities and from recommendations by friends or business associates and organizations. A request for proposal (RFP) should be submitted for their response and contact made with their references; also, a sit-down interview should be conducted. It is important to determine that consultants can meet the requirements of the RFP and that they have worked on similar types of projects. Also, find out if their costs are in line with others in the discipline, that they have demonstrated that they know how to meet schedules and control costs, that they will sign a contract, that they have professional liability insurance, and that they are registered and in good standing with the state where the work will be performed. After all the above has been satisfied, the planners should check to make certain of the chemistry between them and the consultants. Although a consultant may be the best available, if the parties cannot work together, they are off to a bad start, and the situation may not improve.

The consultant should be asked to answer questions in the RFP such as:

- Other projects of similar size and scope they have done
- How they will staff the project
- How they will develop the project (in the case of the architect) from design stage to the punch list
- What back-up capabilities they have
- How they will deal with client changes to the project once it has started
- How they will charge for services that are above and beyond the original scope of work

They should provide a detailed schedule for the anticipated project through all stages of the project through completion and turnover to the client. If the project is being done in phases, it should be determined how that will impact the project costs. Also ask what the consultant expects from the client. These are all things that must be known in advance, before the project begins; there should be no surprises when the project is at a point of no return and unanticipated funds must be spent to keep the project on track and on schedule. The drawings and illustrative documents must be prepared in sufficient detail so bids can be solicited from all the parties involved in the project.

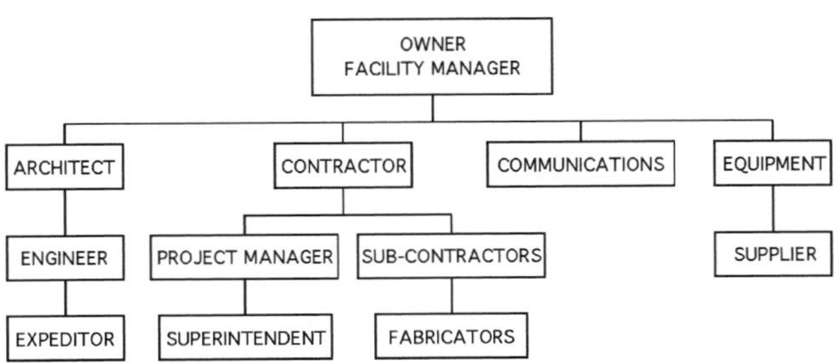

FIGURE 6.1-8 Project team.

This is the point in the process where the architect should submit an estimate of the costs of the construction for the project. Communication with contractors is essential in all phases of the project. If changes are necessary, they should be in writing and all involved parties should sign off and date their confirmation on paper or by e-mail.

DESIGN DEVELOPMENT

After the owner's written approval of the schematic design documents and any changes have been made to the drawings, budget, or scope of the project, the architect prepares the drawings and document set describing the entire project, including all appropriate materials and other project elements. The set then goes to the other consultants so they can prepare their drawing sets for bidding and should include all preliminary drawings, floor plans, reflected ceiling plans, power and lighting plans, floor and ceiling loading drawings, specifications, and any other documents necessary for construction. The owner and the architect must work closely to coordinate the design with the equipment suppliers. They should set the schedule, identify long-lead-time items for preordering, and review recommendations for space, finish, and light fixtures, as well as other elements that make up the overall project.

Construction Documents

Once the architect has prepared the drawing sets and has sent them to the list of prospective contractors, the architect then helps the owner to evaluate the received proposals to find the best fit for the project. The bids should be solicited with an emphasis on long-lead-time items and the impact they will have on the schedule. The architect coordinates with the consultants to produce complete construction drawings and specifications (*e.g.*, coordinating with the electrical, mechanical, structural, plumbing, fire protection, and equipment suppliers). Together, the owner and architect will determine the submittal dates for drawings and the time allowed for approval, redraft, and resubmittal to keep the project on schedule. Weekly project meetings should be held to coordinate drawings among the trades and to find conflicts, errors, omissions, or ambiguities in the trade submittals (*e.g.*, the steam fitter and the electrician could have the water lines for the IOT transmitter and the transmission lines occupying the same space). Good communication and the use of coordinated drawings (discussed below) can help avoid such conflicts.

Building Owner Drawing Review

In most jurisdictions, the building owner must sign the permit applications and is entitled to a review of the drawings for code compliance, errors, omissions, conflicts, and any other items that could affect the owner's building. Ample time should be allowed for this review. The drawings may be accepted on the first pass, or they may come back from the owner's consultants requiring a resubmittal to correct deficiencies. It is important to specify the owner's drawing turnaround time, as it can affect the project schedule.

TABLE 6.1-1
Types of Typical Permits

Construction	Concrete Design Mix
Architectural	Blasting
Structural	Sewer
Mechanical	Asbestos
Sprinkler	Site safety plan
Standpipe	Fire protection plan
Generator	Paving plan
Fire alarm	Zoning
Public assembly	Egress
Excavation	Hydraulic
Foundation	Fencing
Borings	Department of Transportation
Subgrade	Department of Buildings
Plumbing	Sidewalk closing
Equipment use	Sidewalk shed
Piles	Street closing
Underpinning	Weekend work
Welding	Night work
High-strength bolts	Netting
Masonry	Containers
Fire stopping	Trucking
HVAC	Exterior hoist
Shoring	Temporary Certificate of Occupancy
Spray-on fireproofing	Certificate of Occupancy

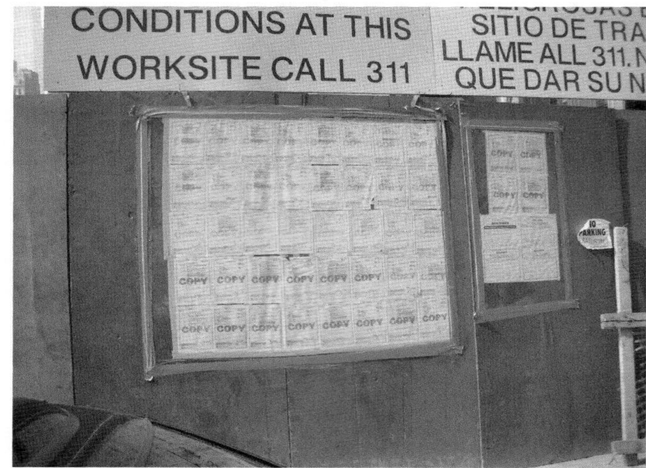

FIGURE 6.1-9 Posting of permits as required at job site.

Permits

Permits are required by most communities to make sure all construction is in conformance with local codes and zoning requirements. Many types of permits are needed, and the requirements vary greatly by municipality. Some are for new construction, some are for renovation or modification of an existing facility, and some cover both types of construction. Table 6.1-1 shows examples of permits needed at various sites.

In some cases, self-certification by licensed professionals is acceptable for filing for the permit. Different types of permit applications may require different individuals to sign off on them. In larger cities, this process can be complex and time consuming and is often assigned to experts. In most areas, the permits are only issued for a specific length of time and in most cases the expiration date is tied to the expiration date of the contractor's insurance (see Figures 6.1-9 and 6.1-10). If changes are made to the drawings after the permitting process has begun, an amendment is usually required by the building department.

SELECTING A CONTRACTOR

The criteria for choosing a contractor are essentially the same as choosing the architect. Look for known entities, talk with friends and associates, ask the architect, and contact references. Conducting a face-to-face interview with a new contractor may be in order. Determine that the contractor has worked on this type of specialty project. Make certain a contractor's cost submittals are clear and the proposed schedule is consistent with the proposed project. Determine that a good project manager is assigned to the job. Check that the contractor knows how to meet schedules and limit change orders. Some contractors underestimate the initial price of the job and charge excessively for

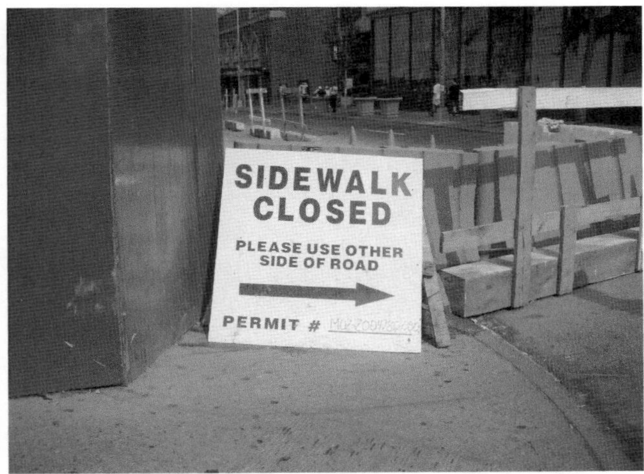

FIGURE 6.1-10 Sidewalk closing notice with permit number posted.

change orders. Make certain that those bidding on the project have read the specifications documents as well as the drawings and hold them responsible for errors or omissions caused by not being fully aware of the written specifications.

Construction Team

The construction team encompasses all the trades working the project along with the owner, the owner's architect, and consultants. The general contractor's project manager is critical to the job. This person must thoroughly understand the project from start to finish to meet the schedule; for example, the transmitters should not arrive on site prior to installation of the sheet rock. If the ceiling grid contractor has installed the grid before the plumbing or sprinkler lines are run, extra costs result for extended installation time. A concise and well thought out sequence of events is necessary for a transmitter installation project. The project manager is the main contact, and the owner must plan to be on site regularly during the entire project. In this way, when questions arise, the owner will be able to respond quickly to keep the job moving. Downtime can be expensive. It is important to take the time to brief the construction team on what is expected to be accomplished and the time frame.

Building Knowledge

A contractor who has worked in the building in which the project is to take place is well suited for the project and will know when the loading dock is empty, when the freight elevators can handle the larger deliveries, and where the electric service panels are located. This knowledge can lead to fewer change orders and keep costs in line with the contract.

Value Engineering

If the project is about to exceed budget, a contractor that is willing to prepare a value engineering report can reduce projected overruns. With value engineering, the contractor can recommend a less expensive ceiling tile or wall finish, different floor tile, or other alternatives that can cut costs from a project without making the project look substantially different from what was planned.

Building Owner Drawing Review

The building owner or their consultants may question an item and send the drawings back for an explanation; therefore, allow time for resubmittals. Seldom will the drawing set pass review on the first try. The owner's cooperation is essential, as the owner will be a signatory to the permit application.

BASIC CONSTRUCTION

Coordinated drawings (encompassing all trades) detail the placement of all material that is to be installed. The process of coordinating the drawings begins with the architect and the consultants. The contractor should also prepare coordinated drawings using the ductwork drawing as the basis, as the ducts are generally the first materials to be affixed to the ceiling. The contractor should be responsible for measuring the freight elevators to make certain all job-related material fits. If the material cannot fit in the elevator or make turns within the building, it must be broken down at the manufacturer's location or on the job site or be hoisted from the outside, at extra cost. Chillers, generators, and combiners are examples of items that must be shipped broken down for reassembly on the job site. If a tower is being erected, be sure

FIGURE 6.1-11 Standby power generator at 4 Times Square, New York.

FIGURE 6.1-12 Antenna hoisted to the tower. (Photograph courtesy of John Fink.)

FIGURE 6.1-14 Combiner module being reassembled at the job site.

the ironworkers or riggers and the electricians have a time plan for installing the transmission lines and antenna so the line is installed as the tower is constructed. Note that safety issues preclude ironwork occurring above workers who are performing installations at lower levels.

Union Issues

In major markets and many others as well, trade union members may be required to perform all the tasks associated with transmitter/antenna installation. Their costs are often higher than non-union work and their workday shorter. For electricians performing work on the tower, or places on the site that are 16 feet or more above the floor, a premium known as "high time" may be charged above the basic hourly rate. Overtime hours and weekends carry heavy premiums, but may be necessary if the project is behind schedule.

FIGURE 6.1-13 Tower with transmission lines. (Photograph courtesy of John Fink.)

Inspections

Periodic inspections must be performed before the next step is undertaken; for example, welds or high-strength bolts must be inspected before the contractor can spray on fireproofing. Be ready for the scheduled appointment dates. If a missed appointment holds up the other trades on the job, such delays could affect the schedule and be quite costly (see Figures 6.1-15 and 6.1-16).

Advance Equipment Purchase and Preassembly

Long-lead-time items must be purchased and scheduled for later delivery to arrive on time. Lead times vary (*e.g.*, antennas, 120 to 180 days; tower analysis, 45 to 60 days; tower modifications, 30 to 120 days). Add the time required for zoning approval; a new tower may require up to a year. Transmitters may take 120 to 150 days after receipt of order. STL and ENG equipment can take 90 to 120 days for manufacture. If installing a new tower or modifying an existing one, plan on a 6 month time frame for FAA clearance, in addition to the installation time at the site. Where

FIGURE 6.1-15 Beams with high-strength steel bolts ready for inspection.

FIGURE 6.1-17 Tower section preassembly. (Photograph courtesy of Tom Silliman.)

conditions permit, major complex sections of tower and antennas can be preassembled at the factory or off-site to reduce installation time on the site (see Figures 6.1-17, 6.1-18, and 6.1-19).

Hazardous Material Removal

Remedial work at a work site, which is necessary to remove hazardous material previously stored there, can be time consuming, as contractors must be hired to properly remove and document the material.

FIGURE 6.1-16 Spray-on fireproofing.

FIGURE 6.1-18 UHF antenna preassembly.

FIGURE 6.1-19 VHF antenna preassembly.

Approved Shop Drawings

It is important to have the contractors work from approved shop drawings. The general contractor issues copies of the approved shop drawings to other subcontractors involved in the project. Additional work required to conform to the latest drawing set can cost the project money and time. Any subcontractor using unapproved shop drawings whose work impacts other trades may be required to pay for any delays that are caused by use of the unapproved shop drawings.

Change Orders

Change orders usually take three forms: time and materials, lump sum, and unit pricing. Spell out in the contract how change orders are to be handled.

LIEN WAIVERS

A lien is a legal right to hold another's property to satisfy a debt. A mechanics lien is usually filed by a subcontractor for nonpayment of bills; it is filed against real property (*i.e.*, the building in which the project takes place) and may be filed if there is a dispute with the contractor or subcontractor. After the first payment is made to the contractor, the contractor should submit a partial waiver of liens when submitting an invoice for the second payment and after each subcontractor has been paid. The partial waiver of lien should indicate the amount of money that the subcontractor has received from the contractor's payment from the owner. After the final payment has been made, a final waiver of lien should be requested and should include a stipulation that, if a mechanic's lien is filed, the contractor will post a bond to protect all relevant parties. Partial lien waivers should be supplied from all suppliers of transmitters, antennas, and towers as individual payments are made.

Retainage

As contractors perform work they submit requisitions. The owner should hold back some amount of retainage. The industry standard for retainage is 10%. Included in the requisition is a description of the work performed, scheduled value, percentage completed, and retainage. Retainage essentially is insurance to make sure that the punch list (described below) is completed.

Finishing

Finishing includes setting the light fixtures, polishing the walls (final paint coat), fixing dings in the walls, removing scuffs on the equipment, and general cleanup.

Punch Lists

Punch lists are items that have not been completed according to the plans, specifications, and other contract documents. A suggestion is to make drawings of where the corrections are required and to reference drawing numbers in the punch list. A punch list walk-through should be scheduled, and a final punch list sign-off meeting should conclude the project. If it becomes necessary to hire outside contractors to complete the work, the cost of this effort should be deducted from the contractor's requisition. Prior to this kind of action, the contractor should be given a time period, such as one week, to complete the work.

Sign-Offs

There should be an agreed-upon process established for signing and dating each completed item by the authorized representatives.

AS-BUILT DRAWINGS

As-built drawings should reflect all modifications and change orders made after the construction drawing set has been released. An as-built drawing set should be kept on-site and copies given to the building management office. The contractor is responsible for supplying and checking as-built drawings for accuracy.

MANUFACTURER'S CERTIFICATIONS

The certifications are performed after complete installation. Equipment is calibrated to match the data achieved at the factory; for example, transmission lines must be fine-tuned and the antenna certified that its specifications are the same as the factory tests.

CERTIFICATE OF OCCUPANCY

A certificate of occupancy is issued by the Department of Buildings or other appropriate municipal agency. All inspections and sign-offs must be completed before the final certificate of occupancy is issued, then work on transmission system equipment installation can begin in earnest (see Figure 6.1-20).

REDUNDANCY, SECURITY, AND DISASTER PREPAREDNESS

Engineers nearly always want backup facilities. Planning of new or modified facilities should also include a plan for redundancy and security. A disaster recovery plan with scheduled reviews, updates, and practices is a necessity. Vulnerability assessment should be part of transmitter facilities planning. Site physical security, video surveillance, and access card readers are some of the means of securing the facility. For

FIGURE 6.1-20 Completed transmitter equipment space.

additional information, see Chapter 12.7, "Broadcast Facility Security, Disaster Planning, and Recovery."

ACKNOWLEDGMENTS

The author would like to acknowledge the instruction of Richard Lambeck, P.E., Professor in the Graduate Construction Management Program at New York University.

C H A P T E R

6.2

Analog Television Transmitters

HARRIS CORPORATION ENGINEERS*
Quincy, Illinois

Revised for the 10th Edition by

JAY C. ADRICK
Harris Broadcast Communications Division
Mason, Ohio

INTRODUCTION

Significant advances continue in TV transmitter technology. New technology and ideas have been introduced to continue to provide high-quality TV signal transmission while improving reliability, reducing maintenance, and lowering overall cost of ownership. These new technologies include solid-state power amplifiers for VHF up to 100 kW and UHF up to 30 kW, plus improvements in UHF high-power tube transmitters. Further, the Federal Communications Commission (FCC) continues its policy of technical deregulation, which allows more flexibility in transmitter design and system operation. This chapter discusses the relevant technology and provides the information needed for selection, installation, operation, and maintenance of TV transmitters in today's environment.

TV transmitters may be considered to comprise two essential components: the exciter and the RF power amplifier (PA). The exciter provides the signal-processing functions required to convert a baseband TV signal into a modulated RF signal on the assigned channel. These functions include baseband signal processing, modulation, precorrection and equalization, upconversion, band limiting, and amplification to a relatively low-power RF signal. While different methods are used for different signals, all functions must be performed whether the baseband signal is video or audio. Because the output of the exciter is a modulated RF signal, most commercially available exciters may be considered low-power transmitters.

The PA provides the muscle to amplify the modulated RF signal to the desired level for transmission. Thus, the PA technology to perform this function is key. Both solid-state and tube devices are available in commercial equipment. For VHF channels, solid-state devices have been the exclusive amplifying technology utilized in all products produced for over 15 years. For UHF, solid-state devices are used at low- and medium-power levels, while tube devices are used for high-power levels. Other key functions in the PA common to all amplifier technologies include AC distribution, AC to DC power conversion, cooling, and control.

TV transmitters are unique in that no other application requires such high levels of linear radio frequency (RF) power generation while operating virtually uninterrupted. This has led to the development of specialized techniques to assure highly efficient and reliable operation. The need for high efficiency has led to the near universal use of partially saturated class AB final power amplifiers. This, in turn, has resulted in the development of precorrection and equalization techniques required to compensate for nonlinearities inherent in this class of operation. To achieve the levels of reliability required, redundant system architectures that minimize single point failures are used.

*Gerald W. Collins, retired, Robert J. Plonka, retired, George Maclean, Dana Myers, Anders Mattson, Ph.D., David Danielsons, and Randy Restle.

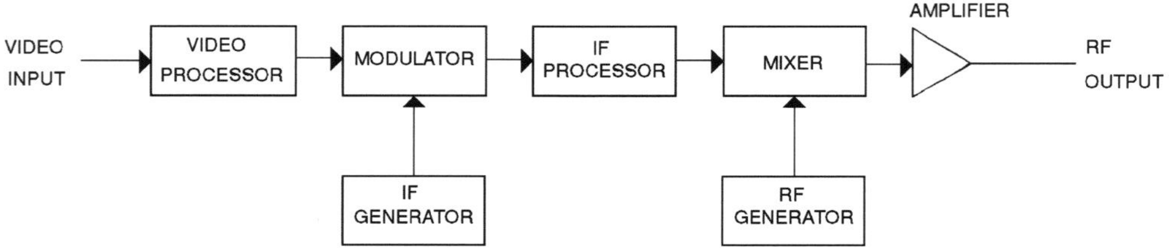

FIGURE 6.2-1 IF-modulated transmitter.

EXCITERS FOR ANALOG TV

Exciter performance is the key to excellence in TV transmission. The exciter performs baseband video and audio processing, modulation, precorrection, equalization, and upconversion functions. Some of these functions are common with exciters for digital transmission.

Visual Modulator

The visual modulator receives a video baseband signal, and processes and converts it to a fully modulated vestigial sideband (VSB) signal. Since intermediate frequency (IF) modulation is used, most of the signal processing occurs in the video and IF stages. The basic block diagram of an IF-modulated transmitter is shown in Figure 6.2-1. For this discussion, the modulator is considered to include all blocks through the IF processor.

Video Processing

For transparent transmission of the incoming signal, it is important to optimize the incoming signal. It is often difficult to define where the station processing ends and the transmitter video processing begins. Typically, some video processing is done external to the transmitter by a processing amplifier.

The main functions of the exciter video processing circuitry are:

- Obtain proper sync-to-video ratio.
- Remove common mode signals.
- Provide overall video level control.
- DC restoration.
- Prevent overmodulation.
- Frequency response correction.
- Provide chroma phase and level control.

Direct current restoration (clamping) is important because picture brightness information is contained in the DC component of the video signal. If AC coupling is used in the video circuitry, the DC level will tend to vary as the capacitors in the coupling circuits charge and discharge. Alternating current coupling is convenient since differences in AC and DC ground can be allowed without introducing distortion. By clamping a particular level (such as sync tip or back-porch) during each line to a fixed voltage, the correct DC level is applied to the modulator.

Common mode signals, such as noise or AC hum, can be removed by using a differential input for the video input stage. It is desirable to use an RF choke at the video input to prevent rectification of ambient RF and remodulation of the main signal.

White peak limiting is used to prevent modulation from reaching 0% or to keep the carrier from being pinched off, as shown in Figure 6.2-2. When an intercarrier TV receiver encounters a carrier that has been pinched off, it temporarily has no signal to receive and the automatic gain control (AGC) adjusts to maximum gain. Since there is nothing but noise, the noise is greatly amplified, transferred to the aural intercarrier, and becomes audible as a "buzz" in the sound output.

High-frequency peaking circuits are often used to compensate for roll-off due to long video signal path runs in a transmitter plant or STL system.

In UHF exciters, the horizontal and vertical sync portions of the TV signal may be detected and utilized to drive klystron pulsers.

Precorrection can be accomplished by changing the gain of a video amplifier over a portion of the total video signal. The threshold of the gain change is set at a certain level of the video signal. Differential phase can be corrected by splitting the video signal into two paths that are in quadrature at the color subcarrier. Phase changes can be obtained by changing the gain of only one path over a portion of the video waveform.

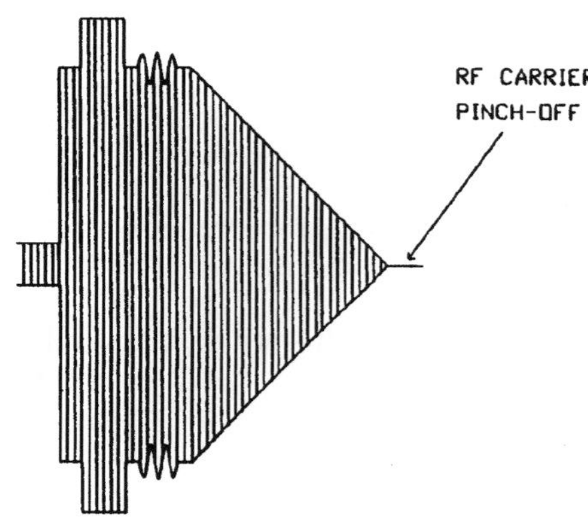

FIGURE 6.2-2 RF carrier pinch off.

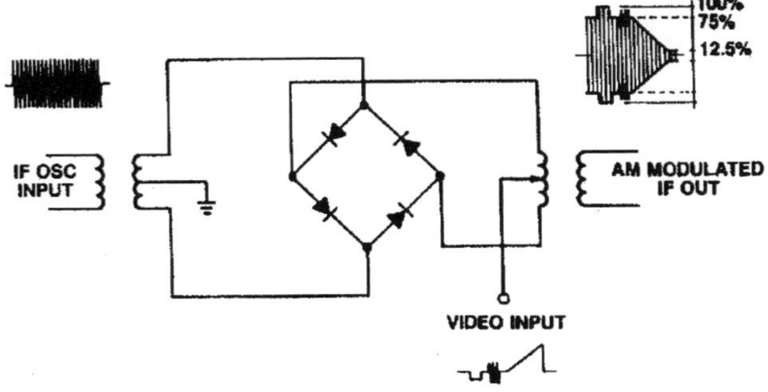

FIGURE 6.2-3 Balanced mixer.

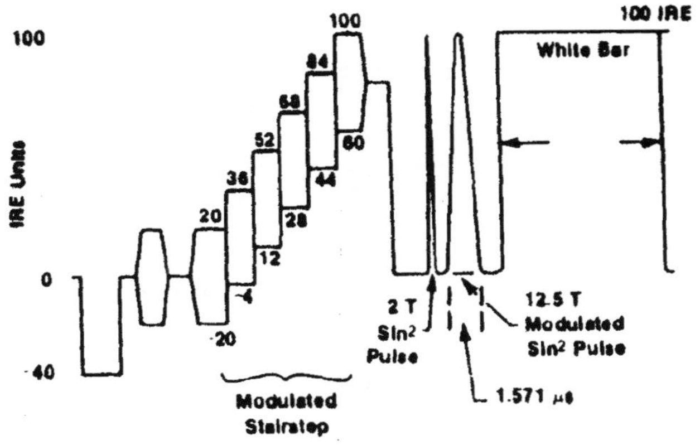

FIGURE 6.2-5 Composite test signal.

Modulator

In nearly all IF-modulated transmitters, the modulator is a broadband, balanced diode mixer. It is configured for maximum rejection of the local oscillator signal and biased so as to provide excellent linearity, low noise, and capability to achieve carrier cutoff. A schematic of a typical balanced mixer is shown in Figure 6.2-3. The video signal is DC offset to provide the proper modulation level and is used to control the attenuation of the diodes. Peak of sync corresponds to maximum IF envelope output and white corresponds to minimum IF output. The output signal of the modulator is a double sideband AM signal having the proper depth of modulation (12.5%).

Visual Group Delay Distortion

Group delay distortion appears as color smear and halo effects on edges. On a waveform monitor the effects may be seen using the 2T and modulated 12.5T contained in the composite test signal (see Figure 6.2-4). Pulse responses with unacceptable group delay may include exaggerated pre-or postringing on 2T and modulated 12.5T baseline disturbance. Low-frequency group delay and amplitude errors are referred to as short-time waveform distortions. Figure 6.2-5 shows typical distortions of a 2T pulse.

Intermediate Frequency and Video Delay Compensators

The techniques used to accomplish group delay equalization at baseband or IF are similar in concept. Both active and passive equalizers may be employed. Intermediate frequency group delay correction is necessary to correct group delay errors below visual carrier while not affecting the signal above visual carrier. Above visual carrier, both IF and video correctors are effective. If the group delay error affects only the high-frequency side of the passband, equalization can be accomplished at video. This is the case for most visual-aural notch-type diplexers.

The FCC requires the TV transmitter to predistort group delay according to the curve shown in Figure 6.2-6. The purpose is to compensate for the delay caused by discrete LC aural notch filters in TV receivers. The notch filter in TV receivers is necessary to prevent the aural carrier from mixing with the detected video and chroma subcarrier signals and producing visible spurious beats.

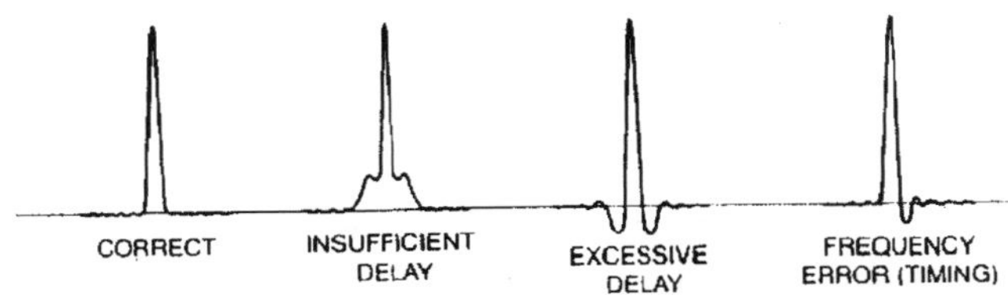

FIGURE 6.2-4 Typical distortion of a 2T pulse.

Passive Group Delay Equalizer

A common form of a passive group delay equalizer is shown in Figure 6.2-7. It provides a flat frequency response and a nonlinear group delay that peaks at resonance. This circuit is referred to as a passive all-

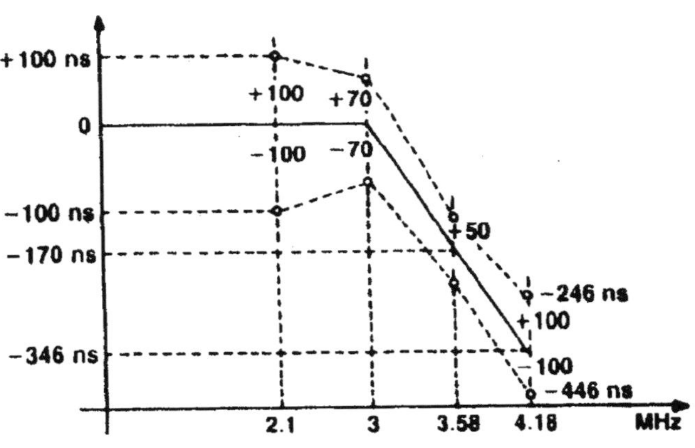

FIGURE 6.2-6 Predistorted group delay curve.

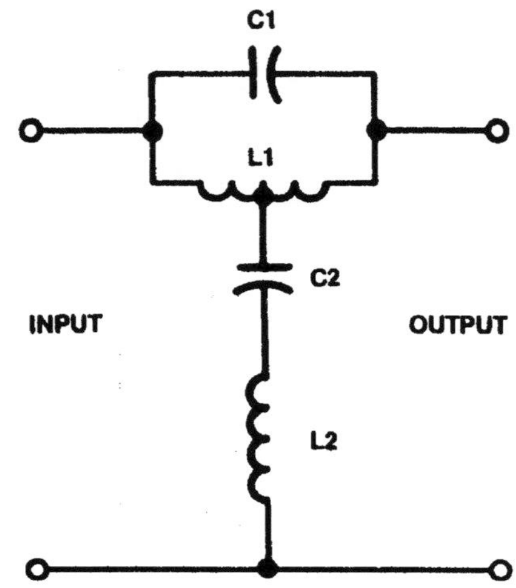

FIGURE 6.2-7 Passive all-pass network.

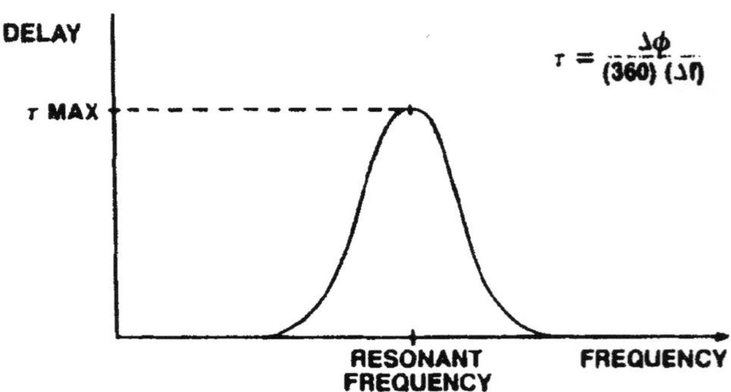

FIGURE 6.2-9 All-pass network group delay.

pass network. It can be understood by examination at frequencies well below and above resonance. At low frequencies the network can be approximated by a low series inductive reactance due to L1 in which the output voltage leads in phase. At high frequencies well above resonance, the circuit can be approximated by a series capacitive reactance due to C1 in which the output voltage lags in phase. The output is constant across the band. The phase of the network is plotted in Figure 6.2-8. A plot of the group delay is shown in Figure 6.2-9.

Active Group Delay Equalizer

As in the passive type, this network has constant amplitude versus frequency response and a nonlinear phase response. The action of an active all-pass network can be best seen using the simplified schematic of Figure 6.2-10. The phase of voltage (e1) plotted as a function of frequency traces out a circle with maximum amplitude and zero phase at resonance. Voltage (e2) has a constant amplitude and phase.

When voltage (e1) is equal to twice (e2), the output voltage of the summing amplifier, (e3), has the characteristics of an ideal all-pass network. The output amplitude is constant and the resonator tuning determines the frequency of maximum group delay. The resonator "Q" determines the magnitude of delay. For the case where voltage (e1) is larger than twice (e2), the output has an amplitude peak at resonance. If (e1) is smaller, the output has a dip at resonance.

Vestigial Sideband Filter

The FCC requires that the radiated signal has a major portion of the lower sideband (vestigial sideband) suppressed. In addition, the upper sideband signal must be contained within 4.75 MHz of the visual carrier. With IF modulation the filtering is in the low-power stages. Transmitter manufacturers use solid-state filters using surface acoustic wave (SAW) technology to accomplish the filtering requirements.

The electrodes in a SAW filter act like tapped delay lines and are designed to scale the amplitude and delay the signal, as shown in Figure 6.2-11. The length, spacing, and number of the electrodes determine the wave-shaping properties of the filter. The output is selectively attenuated depending on the time delay and signal frequency relationship. This type of filter is called transversal because the attenuation is controlled by delay lines rather than resonators. The wavelength of an IF acoustic signal is approximately 0.003 inches. The short wavelength allows the filter to be very compact. Because the transducers are on the surface, pho-

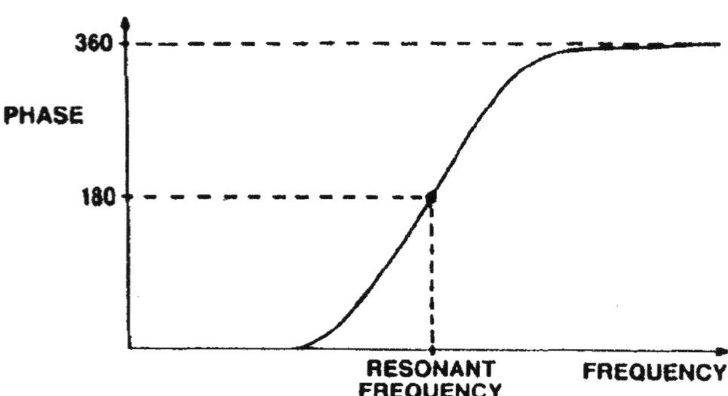

FIGURE 6.2-8 All-pass network phase response.

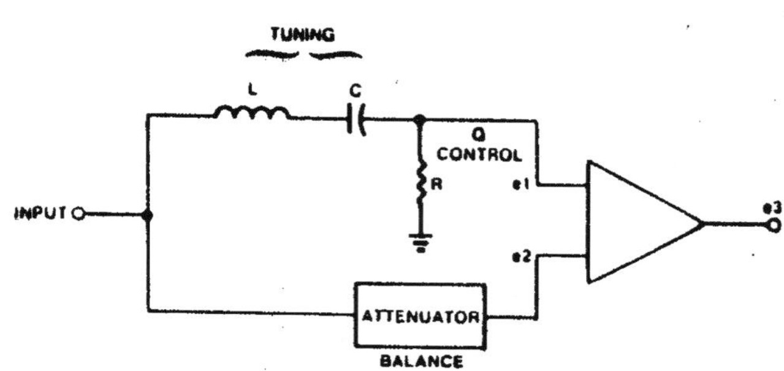

FIGURE 6.2-10 Active all-pass network.

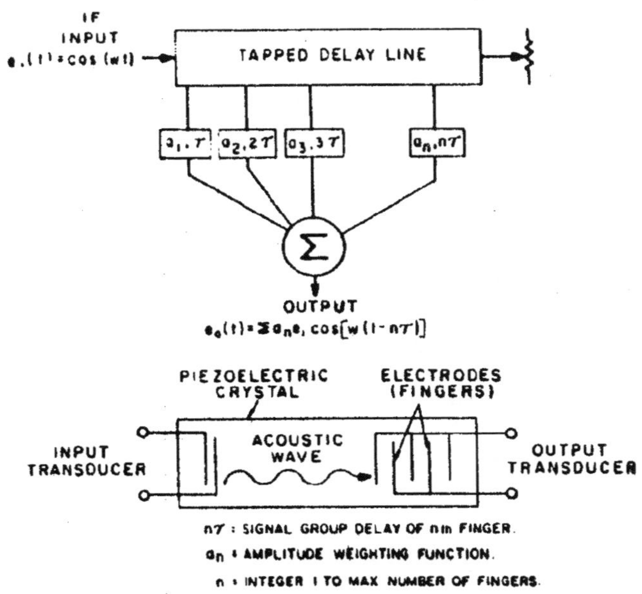

FIGURE 6.2-11 SAW filter.

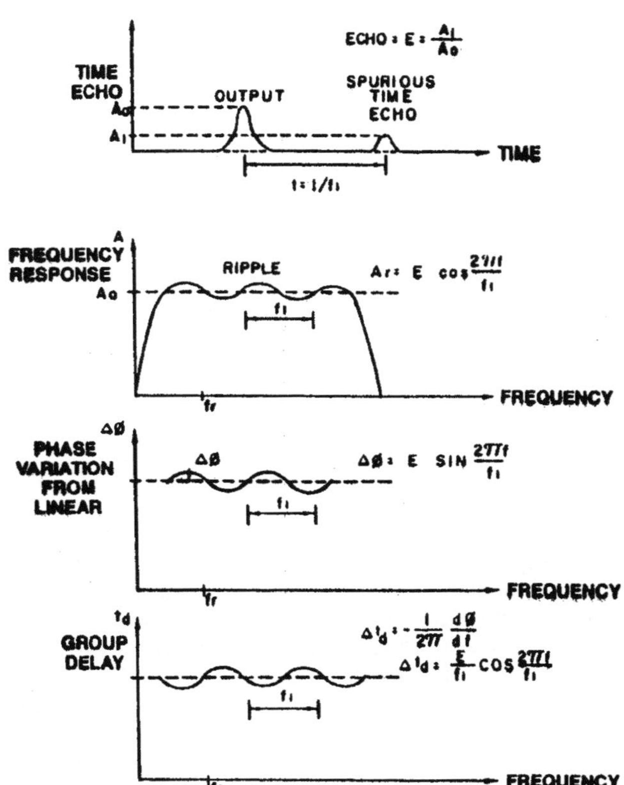

FIGURE 6.2-12 Response, phase, and delay ripples caused by time echoes.

tographic masks can be used to accurately control their physical dimensions. The photographic mask lends itself to modern manufacturing techniques and ensures a reproducible filter with permanent amplitude versus frequency response and group delay.

A characteristic of a transversal filter is that group delay can be set independently of the amplitude. The independent nature of group delay and amplitude allows a manufacturer to provide almost any type of group delay curve across the passband, such as a constant delay or the TV receiver group delay equalization curve.

Although the FCC does not require attenuation of visual signals in the aural passband, video signal components can cause interference called visual-to-aural crosstalk. Reduction of visual-to-aural crosstalk is essential to the proper transmission of stereo sound and other subcarrier broadcast services. High-resolution cameras and character generators can produce spectral components at 4.5 MHz and beyond. Without sufficient attenuation these signals can cause distortion, especially in aural subcarriers that are more sensitive to visual crosstalk than the main aural signal. SAW filters thus are able to perform many functions simultaneously (bandwidth shaping, group delay equalization, and visual-to-aural crosstalk reduction).

Distortion products can be created by SAW filters due to signal reflections (triple transit) and direct feedthrough. These distortions cause time echoes displaced either before or after the desired responses. In the frequency domain these echoes show up as ripples in the passband (see Figure 6.2-12). A given echo will contribute uniquely to the amplitude, phase, and group delay characteristics by the addition of a ripple component in the passband. The amplitude of the ripple is proportional to the echo level and its period is proportional to the reciprocal of the SAW time delay.

SAW group delay varies sinusoidally with the same period as the passband ripple. Magnitude, however, is a function of echo level and is inversely proportional to time delay. It is, therefore, not a sure test of signal distortion. The group delay ripple for long delayed echoes will give peak-to-peak values that have no correlation with conventional signal distortion estimates. For that reason, fast peak-to-peak group delay errors that occur at greater frequency than the reciprocal of the filter time delay can usually be ignored.

Intermediate Frequency Linearity Precorrection

An alternative to video correction is to provide the correction at IF frequencies. There are advantages to correcting at IF. Since most distortions are caused in the high-power RF amplifiers after vestigial sideband filtering, a corrector placed after the vestigial sideband filter can more accurately predistort the modulated signal. Any precorrection spectra generated at IF after the VSB filter will produce energy components that can cancel intermodulation products generated in the final amplifier stage. This is particularly important in the cases of pulsed klystron transmitters or common amplification.

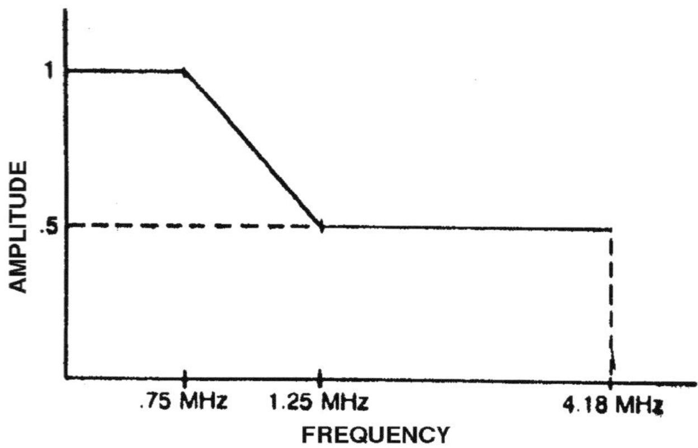

FIGURE 6.2-13 Frequency response of ideal diode detector.

Correction of distortion at IF is particularly helpful for chroma distortions. Chroma spectra at 3.58 MHz have only single sideband information and thus less energy than equivalent luminance signals. An ideal diode detector frequency response of the NTSC modulation signal is plotted in Figure 6.2-13. Note that beginning at 0.75 MHz the video begins to fall to 6 dB. For video signals lower than 0.75 MHz, the RF spectrum is double sideband and has twice the peak RF voltage.

Intermodulation products are caused in high-power amplifiers by a nonlinear transfer function. As the power output increases toward saturation, amplitude compression and phase lag occur. The nonlinear transfer function gives rise to mixing products that occur at sum and difference frequencies around the visual carrier. This process creates the frequency spectra of the lower sideband, commonly referred to as *lower sideband reinsertion*.

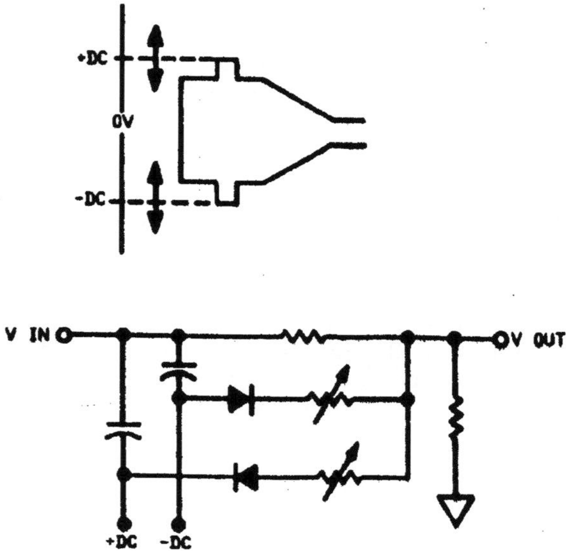

FIGURE 6.2-14 Basic gain expansion circuit.

Intermediate frequency correctors for amplitude distortion are usually similar in concept to video differential gain correctors. Linearity correctors generally use diodes that are set to conduct at a specific level of the IF-modulated signal. When the diodes conduct, the gain or attenuation is reduced as needed.

An example of a basic gain expansion circuit is shown in Figure 6.2-14. The signal is normally attenuated a fixed amount by using a resistive L-pad. The diodes are normally reverse biased by equal DC voltages. Reducing the DC voltage permits the diodes to conduct on the signal peaks. This inserts additional resistance in parallel with the series arm of the L-pad attenuator thereby decreasing the attenuation. Varying the resistance in series with the diodes provides for variable gain expansion.

Incidental Carrier Phase Modulation

Phase distortions in high-power amplifiers produce incidental carrier phase modulation (ICPM) or spectral components in quadrature with the modulation signal. Fast video amplitude changes, such as a step or pulse, will cause larger incidental phase spectral components than slow changes. Receivers make this condition worse by attenuating the lower sidebands below 0.75 MHz. The receiver then responds to the extra sidebands created by the phase modulation as if they were amplitude-modulated single sidebands, producing spikes. The faster the rise time of the signal, the more high-frequency energy is present, resulting in edge distortions in the displayed picture.

The picture impairment is similar to simultaneous group delay and differential phase errors in that edges are less sharp and color hue changes with brightness. On a waveform monitor, overshoots are visible as trailing edges and as rounding on leading edges. These overshoots vary in severity depending on how close to saturation the power amplifier is driven.

Audio impairment is produced by ICPM in receivers employing intercarrier conversion. Intercarrier receivers use an AM or synchronous detector to produce a 4.5 MHz aural IF from the composite video IF. Any phase modulation present on the visual carrier is then transferred to the aural intercarrier. In monaural baseband audio, the increasing amplitude versus frequency effect of ICPM is nullified by deemphasis to some degree. With multichannel sound, however, there is no deemphasis applied to the baseband stereo signal, and thus the distortion is more pronounced at the stereo subchannel and pilot frequencies. To counteract the effects of ICPM and other noise sources on the stereo subchannel, audio companding is employed. Although the audio companding process can reduce some of the effects of ICPM, correction is essential in delivering clear, low-noise audio to intercarrier receivers.

There is no defined level of ICPM for a given stereo performance level since the signal-to-buzz ratio is highly dependent on the picture spectral components.

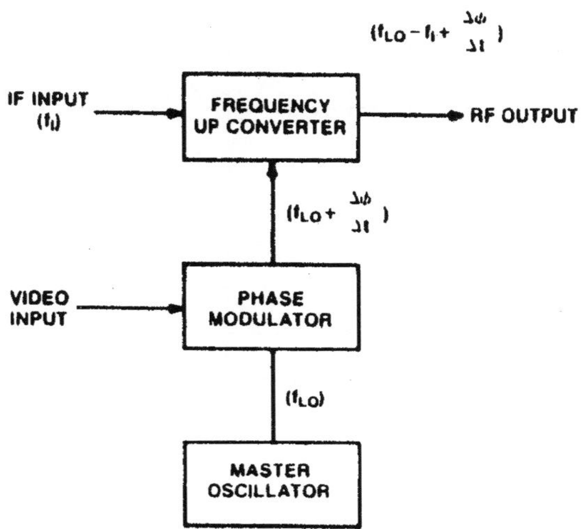

FIGURE 6.2-15 Simplified block diagram of a master oscillator phase modulator.

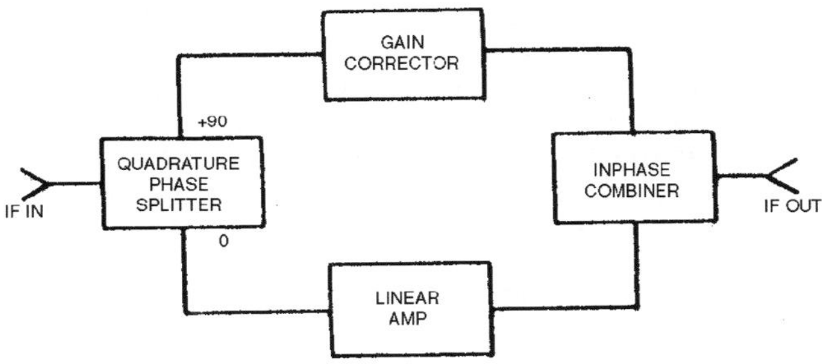

FIGURE 6.2-16 Direct IF ICPM corrector.

ICPM precorrectors can be grouped into two types: those using a phase modulator and those operating on the signal directly. The phase modulator uses video to modulate the IF or master oscillator with a phase characteristic opposite that of the nonlinear amplifier. A phase modulator can also operate on the IF signal directly using a video signal to set the amount of modulation. A block diagram of a master oscillator phase modulator is shown in Figure 6.2-15.

ICPM precorrectors operating directly on the IF signal can be implemented through several methods. Direct precorrection at IF is similar in concept to baseband differential phase precorrection (see Figure 6.2-16). In both cases, the visual signal is split into two paths that are in phase quadrature. In the IF corrector, the entire channel of frequencies is in quadrature, whereas in the video precorrector only the chroma band is in quadrature. One method of implementation is to modify the quadrature signal gain function with level dependent diode expansion or compression circuits. This can be done using the same techniques as in the linearity corrector.

The vector diagram in Figure 6.2-17 illustrates nonlinearities and the operation of the ICPM and linearity correctors. The desired output signal is represented by the left vector. Uncorrected, the transmitter output signal is phase shifted and compressed. To compensate, the signal in the exciter is precorrected by an amplitude expansion and a correction in quadrature. When the resultant signal is amplified, the output signal is the desired TV signal.

Aural Modulator

In its most basic form, the aural modulator consists of some audio processing and an FM-modulated IF oscillator.

Audio Processing Circuits

To ensure that the transmitter is not the limiting factor in audio (monaural and stereo) reproduction, the transmitter should add as little distortion to the incoming signal as possible.

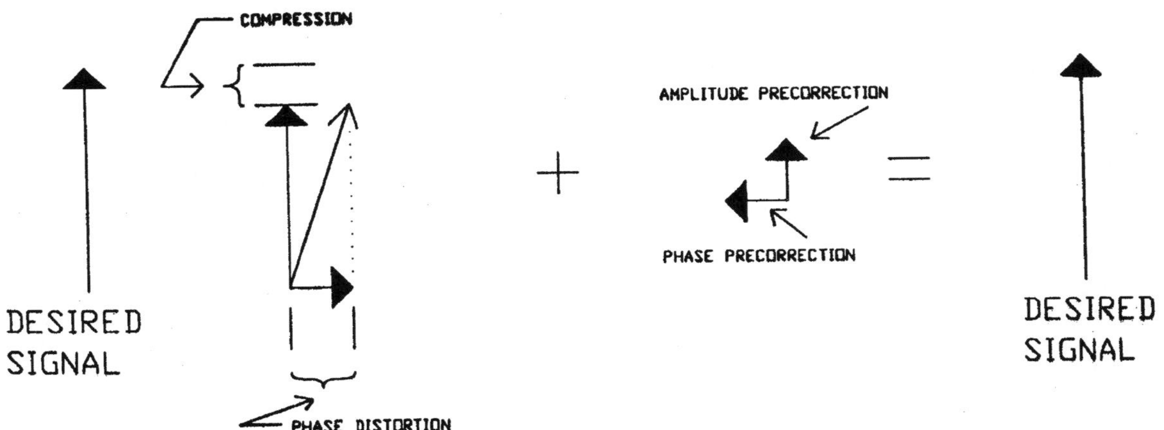

FIGURE 6.2-17 Vector representation of precorrection.

Baseband audio of the BTSC Multichannel Television Sound (MTS) system including mono, stereo, second audio program (SAP), and professional channels (PRO) contains frequency components to 105 kHz. Emphasis must be placed on phase linearity, low distortion, reduction of any amplitude ripples, and roll-off over the stereo passband to achieve good stereo separation and minimum crosstalk between the main stereo and the SAP channels.

While unbalanced coaxial inputs are used for the MTS input to the aural exciter, it is necessary to use some form of common mode rejection to reduce the possibility of hum and noise from entering the audio stages.

All errors in phase linearity and amplitude response within the audio circuitry contribute to stereo separation degradation. As a general rule, amplitude roll-off should be less than 0.1 dB and departure from phase linearity should be less than 1 degree for good quality stereo.

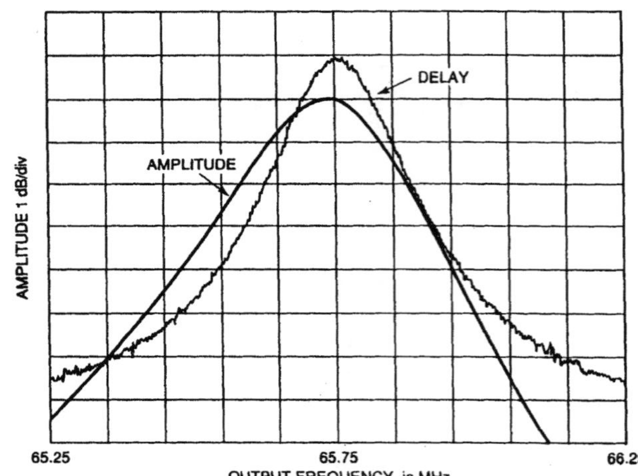

FIGURE 6.2-18 Typical single-cavity diplexer amplitude and group delay.

IF Modulation

With the MTS system, intermodulation (IMD) and harmonic distortion products lie in the stereo channel or the SAP channel and will degrade stereo separation or crosstalk into the SAP channel. In addition, IMD products generated in the stereo channel may lie in the mono or SAP channel. Modulated oscillator linearity requirements include flat modulation sensitivity versus frequency up to 47 kHz minimum, and typically, out to 120 kHz.

As the MTS signal is upconverted to the aural RF carrier frequency, the residual FM of the local oscillator signal becomes a determining factor. The level of FM produced by the local oscillator should be 10 dB lower than the modulated oscillator. Synthesized sources should be tested for spurious frequencies that may show as FM noise. To ensure low-power stages do not contribute any group delay or amplitude roll-off, wideband amplifiers should be used.

Compensation of Aural Passband for Optimum Stereo Performance

Group delay equalization for the aural transmitter can be introduced in the IF section of the aural modulator to effectively correct the adverse group delay in the diplexer. The end result is improved stereo separation. Equalization of the FM bandpass allows the use of lower cost, single-cavity notch diplexers. Stagger-tuned dual-cavity notch diplexers have been used to provide the broad bandwidth desired for good stereo separation and negligible crosstalk between the different MTS components. However, dual-cavity notch diplexers introduce more group delay in the visual path and are more expensive than single-cavity diplexers.

The notch diplexer is a passive device that can introduce distortion and stereo separation errors. The basic problem is that the FM stereo signal is sensitive to the notch diplexer group delay and amplitude response over the occupied bandwidth of the FM signal.

The group delay and amplitude response of a single-cavity diplexer is shown in Figure 6.2-18. It is a typical notch diplexer optimized for minimum aural reject power. The bandpass is somewhat narrow and the group delay is steep. Fortunately, the response curves show a high degree of symmetry that makes equalization possible.

The group delay equalization concept as applied here is essentially a feed-forward scheme that is a technique of generating equalization early in the RF lineup for the purpose of correcting distortions occurring downstream in the system. Feed-forward correction is operated open-loop (without feedback) and is manually adjusted for optimum operation.

Aural IF Group Delay Equalization

The equalization technique of adjusting the circuit to produce an inverse curve of that of the diplexer is the same as that used on the visual signal. There is, however, a significant difference between group delay correction on an FM system versus an AM system. The aural transmitter must incorporate group delay correction at RF or IF to compensate for group delay distortions occurring in the diplexer. Group delay correction before the FM modulation process is not effective for equalizing group delay occurring at RF. The reason for this is that the occupied bandwidth of an FM signal increases or decreases as a function of baseband signal level. Figure 6.2-19 shows the occupied bandwidth when the carrier is deviated 25 kHz compared to a typical notch diplexer group delay curve. When the baseband signal level is increased, the FM deviation increases and a number of additional significant sidebands are generated beyond the acceptable group delay curvature region. The result is distortion in the demodulated FM signal.

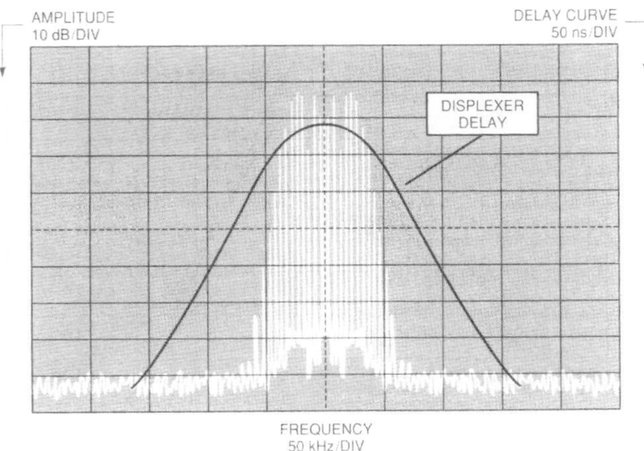

FIGURE 6.2-19 Occupied bandwidth with 25 kHz deviation with overlaid notch diplexer bandwidth.

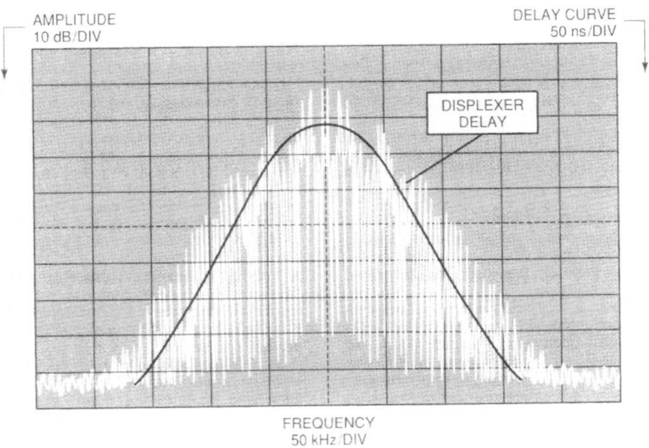

FIGURE 6.2-20 Occupied bandwidth of TV stereo sound signal with 55 kHz deviation and overlaid notch diplexer bandwidth.

In addition, as the baseband high-frequency content increases (such as with a stereo, SAP, or PRO signal) significant sidebands extend further increasing the demodulated distortion. Figure 6.2-20 shows the spectral content of a TV stereo signal with notch diplexer group delay curves overlayed. The equalizer location in the aural transmitter system is shown in Figure 6.2-21.

The circuit configuration is shown functionally in Figure 6.2-22. The measured amplitude and group delay are shown in Figure 6.2-23. An ideal precorrection circuit without any circuit losses would have a flat response without a dip. The response dip is very useful, however, because it also provides a first-order correction to the notch diplexer amplitude response.

Figure 6.2-24 shows the system equalization when the delay corrector is switched in and switched out. A

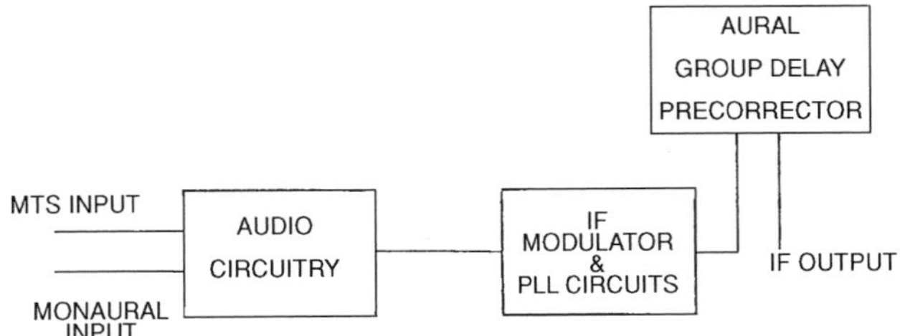

FIGURE 6.2-21 Aural exciter with group delay precorrector.

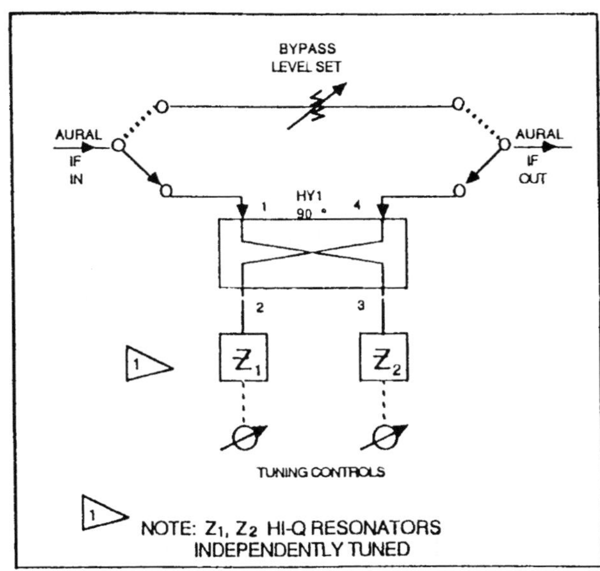

FIGURE 6.2-22 Functional block diagram of an aural group delay corrector.

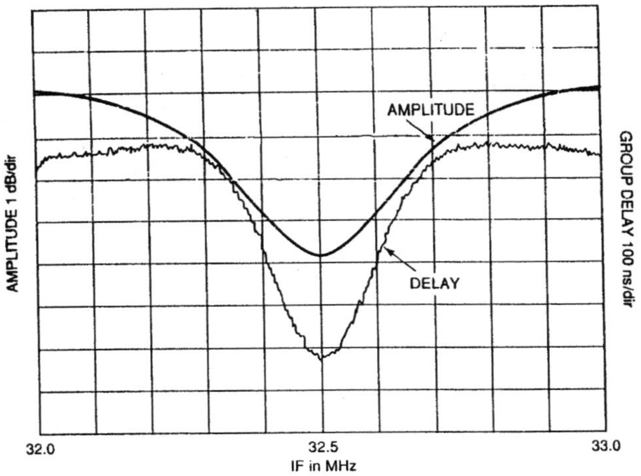

FIGURE 6.2-23 Measured results of an aural group delay corrector.

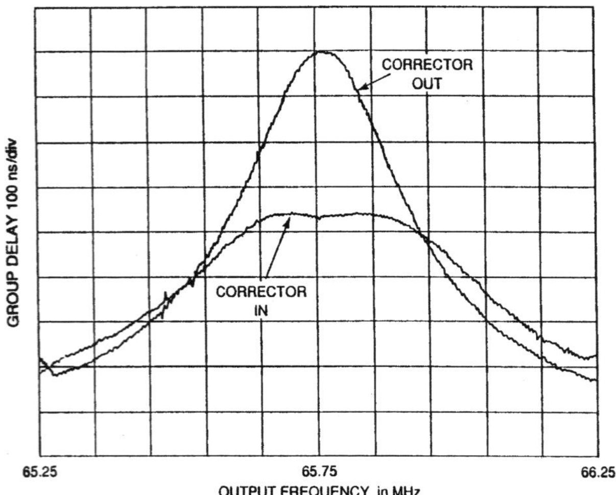

FIGURE 6.2-24 Overall transmitter delay with aural group delay corrector switched in and out.

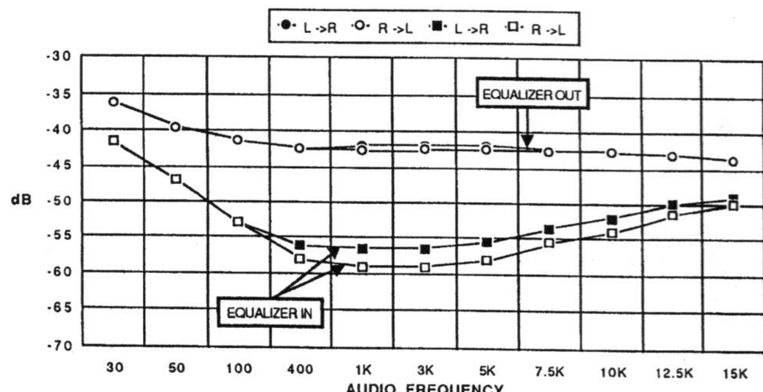

FIGURE 6.2-25 TV stereo sound separation with aural group delay corrector switched in and out.

significant amount of equalization is achieved over the occupied bandwidth of a stereo signal.

The effect on stereo separation is shown in Figure 6.2-25. More than 10 dB stereo separation improvement can be obtained over midband audio frequencies.

Common Amplification

Traditionally, TV transmitters have used separate RF amplifier chains for the visual and aural signal paths. Assuming a single antenna and feed line, this requires a high-level diplexer to combine the visual and aural signals prior to transmission. With the introduction of inductive output tubes as final amplifiers, it has become popular to combine the visual and aural signals in the exciter and amplify them together in the following stages.

Even though very linear amplifiers are used, some residual nonlinearity remains. This has the effect of mixing the visual, color, and aural signals to produce in-band as well as the out-of-band IMD and cross-modulation products. Out-of-band products that are sufficiently removed from the channel of operation are attenuated by the high-level filter. However, in-band products can be removed only by making the transmitter sufficiently linear to reduce these products to required levels. This requires the use of effective correction circuits in the IF and/or RF signal paths. For example, the IMD at ±920 MHz can be precorrected by low-level IF circuitry.

Upconversion

Transmitters employing IF modulation generate the following frequencies: visual IF, aural IF, and master oscillator signals for translating visual and aural IF to the final carrier frequencies. These oscillators have been implemented with either digital synthesizer techniques or crystals. An advantage of the synthesizer is that only one crystal is needed at a single standard frequency for all TV channels. The crystal oscillator approach, however, may involve simpler circuitry.

The two commonly used IF frequencies are 37 MHz and 45.75 MHz. There are many reasons for selecting one IF frequency or the other. One advantage of 37 MHz is that the temperature drift sensitivity of most IF components, such as the SAW filter, is related directly to carrier frequency. Thus, the lower IF has a 12% less drift sensitivity than components at 45.75 MHz. The second harmonic of 37 MHz falls in between channels 4 and 5 so as not to cause interference. On the other hand, 45.75 MHz is a common demodulator IF that can be useful for IF troubleshooting. Temperature drift may be minimized at either IF by maintaining the SAW filter at a stable temperature.

The important performance characteristics of an oscillator are phase noise, frequency stability over time, temperature variations, and the level of microphonic response.

When replacing a crystal, it is important to follow the recommendations of the oscillator manufacturer to ensure proper operation. Synthesizer performance should be properly maintained to prevent inadvertent phase noise and spurious frequency generation.

Offset Frequency Control

The limited number of available channels for TV broadcasting makes it necessary to assign the same carrier frequencies to many stations. To avoid interference between stations operating on the same frequency (cochannel interference), geographical separation and radiated powers are carefully selected.

Cochannel interference between television stations appears to viewers as a horizontal pattern of alternating light and dark bars on the viewing screen—very much like the shadows cast by venetian blinds. It has been demonstrated that the visibility of these bars varies cyclically as a function of the difference in frequency of the interfering carriers (see Figure 6.2-26). The interference is least visible when the carriers are offset by odd multiples of one-half the video line frequency (15,734.264 Hz). Fine-grain maxima and min-

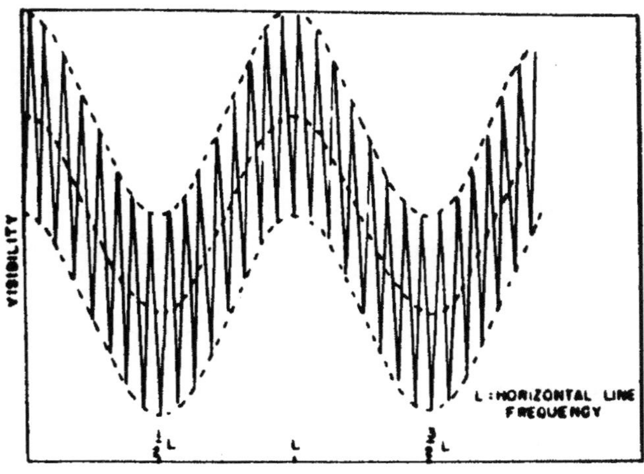

FIGURE 6.2-26 Cochannel interference.

ima occur when the frequency offset is an integer multiple of the video frame frequency (29.97 Hz).

Ideally, stations would be offset by odd multiples of one-half the line frequency to provide minimum interference visibility. However, a third station in the same area would be offset from one of the other stations by an even multiple of the line frequency. Hence, maximum visibility of the interference would occur. Therefore, 10 kHz offsets currently are used in the United States to provide approximately equal reduction of the interference patterns for any number of stations in geographical proximity (see Figure 6.2-27).

For DTV transmitters operating on channels immediately adjacent to (above) and within 88 km of a transmitter on an NTSC channel, the pilot frequency must be maintained 5.082138 MHz above the visual carrier frequency of the NTSC transmitter within a tolerance of ±3 Hz. This requires that both carriers be locked to a common, highly stable source.

Precise Frequency Offset

Although it is not practical to utilize the gross minima occurring at odd multiples of one-half the line frequency, use of the fine-grain minima, occurring at even multiples of the NTSC video frame frequency, is advantageous in reducing the visibility of cochannel interference.

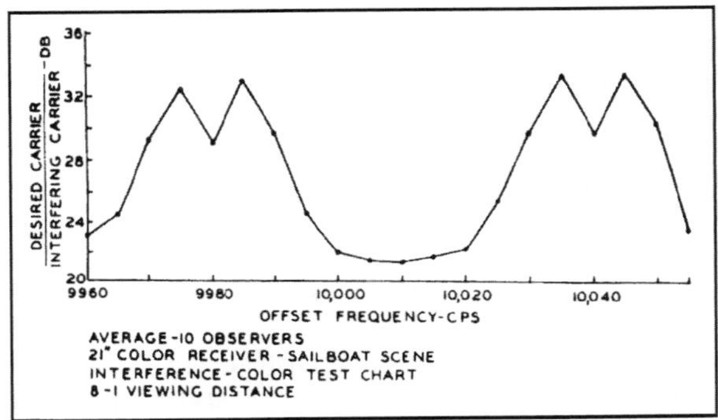

FIGURE 6.2-27 Ten kHz offset pattern.

The nearest even multiple (334th) of the frame frequency to the 10 kHz offset is 10,010 Hz. In a three-station arrangement, one station will have zero offset and the other two stations will be offset by +10,010 Hz or −10,010 Hz. Changes in the frequency differences of 5 Hz have a negligible effect on the reduction of the interference visibility.

To maintain the precision offset within 5 Hz requires maintaining each visual carrier frequency within 2 or 3 Hz. Maintaining a television transmitter to such tight frequency tolerances requires an extremely stable frequency source. Note that 3 Hz stability is also required in the case of an adjacent DTV signal.

In exciters using two independent oscillators (one for IF and one for the local oscillator), the visual carrier signal may be derived from mixing the oscillators together and comparing that signal to the reference signal in the comparator. The resultant error signal can be used to adjust one of the oscillators, preferably the local oscillator. For exciters using synthesizers, the synthesizer reference oscillator is compared to the precise frequency standard in the phase detector and the resultant error voltage is used to adjust the synthesizer oscillator.

By phase locking the visual carrier to a stable reference oscillator, the master oscillator acquires the stability of the reference source. Typically, this is accomplished by using a conditioned oscillator that is phase locked to the GPS satellite system. The conditioned reference oscillator is used to provide a 10 MHz reference signal to each of the master oscillators in the TV exciters.

This method of frequency reference allows a transmitter to be maintained within ±1 Hz of a desired frequency for an indefinite period of time.

RF Amplifiers

The last active stage in a typical exciter is the RF amplifier. For visual signals in separate amplification, combined aural and visual signals in common amplification, and for digital signals, it is important that this amplifier have a linear transfer characteristic in amplitude and phase, and a flat symmetric frequency response with minimum group delay variation across the modulation passband. For visual-only signals, the required bandwidth is at least 4.5 MHz. For common amplification and digital signals, at least 6 MHz bandwidth is required.

For optimum stereo performance a nonlinear amplifier with flat response and group delay across the modulation passband is required. Since FM modulation and demodulation are nonlinear processes there is not a one-to-one correspondence between RF amplitude/phase response and baseband stereo separation and crosstalk. A 3 dB bandwidth of 1.5 MHz provides excellent stereo and SAP performance.

Digital Signal Generation Technology

All RF signals are analog signals, albeit that they might contain analog (NTSC) or digital (ATSC/8-VSB) information. Sooner or later, any exciter will have to create an analog signal.

In theory, analog and digital signal generation is equivalent. An advantage in one method over the other is purely in the implementation. The trend in television, as well as all broadcasting, is that more and more of the processing is done digitally. In television, as well as in radio, the signal generation is done in an exciter, which takes the input signal, modulates it, and puts it out on channel. Originally, this was done using analog means only. Since a TV signal has a bandwidth of around 6 MHz, it wasn't until digital signal processors (DSPs) and field programmable gate arrays (FPGAs) could handle the necessary data rates that digital processing of a TV signal became possible. Hence, most exciters tend to generate a digital signal at IF that is subsequently upconverted to RF using an analog mixer.

There are some inherent limitations in a digitally implemented exciter. First and foremost is the bandwidth limitations set by the sampling rate. The maximum sampling rate is always hardware dependent and cannot be changed. Even if a digital exciter is completely software configurable, it can never handle a signal that requires more bandwidth than what it was designed for. The second issue also relates to the sampling rate, more precisely different sampling rates: If the exciter and the digital input are at different sampling rates, the data rate needs to be changed. In some systems, it is preferable to lock the sampling rate to the incoming signal. In other cases it is preferable to lock the sampling rate to an external reference.

The main advantage with digitally generated TV signals is the flexibility offered and the possibility for adaptive correction. The only thing that is needed to change from one standard to another, say from NTSC to DVB-T, is to change the software in the exciter—provided that the exciter is capable of such change.

Analog TV/NTSC Processing

In a digital implementation, the first step is to sample the analog inputs unless the signals are already in a digital format. The exciter will have to create the chroma, luma, and, possibly, audio signals, and insert the synchronizing signals.

It is also possible for a digitally implemented analog exciter to generate internal test signals that simplify testing and maintenance and eliminate the need for some test equipment.

Linear and Nonlinear Correction

Traditionally, correction was broken into two parts: linear and nonlinear. The former being the correction for the high-power filter, which generally causes most of the distortion. Since a filter only adds linear distortion, the correction is fairly straightforward. Note that it is necessary to only correct for the in-band distortion caused by the filter.

Nonlinear correction is achieved mainly through AM-AM and AM-PM correction for the transmitter itself (excluding the filter).

Baseband versus RF Predistortion

With baseband predistortion the interest is in the envelope and phase of the RF signal. The actual shape of the RF carrier is ignored. This is the same as ignoring the harmonics at multiples of the carrier frequency. When examining an amplifier's AM and PM response curves, baseband predistortion is implicitly assumed. Consequently, baseband predistortion will not get rid of harmonics at multiples of the carrier frequency. The reason is that the extra bandwidth needed is too great—an RF predistorter will need about 1 GHz. With today's technology, this excludes any digital implementation of RF predistorters. In addition, RF predistortion is seldom needed as harmonics are easily handled by filters at the output of the transmitter.

Note that the actual principles for predistortion are the same in both cases. It is only the aim of the predistortion that differs.

Correction Methods

In one way or other, the predistorter should be the inverse of the distortion produced in the transmitter. The theory behind nonlinear systems is much less developed than for linear systems. There are presently no known methods to analytically find inverse function to nonlinear systems except for the simplest cases such as AM-AM and AM-PM curves. This makes the modeling of the transmitters and their inverses more difficult.

AM-AM AND AM-PM CONVERSION

A fundamental limit is imposed by the saturated power, P_{sat}. No predistorter will be able to push beyond P_{sat} without damaging the transmitter. When predistorting, adjust the amplifier to be linear up to P_{sat} and then saturate. This means that for maximum output power from the transmitter, some clipping of the signal must be accepted. Since the clipping will result in distortion, this is equivalent to accepting some spectral regrowth. Therefore, further increases in amplification will result in compression and distortion, some of which can be corrected. When the point is reached where no further correction is possible some backoff is often necessary. Figure 6.2-28 shows input versus output as the output reaches saturation. At some point along the curve predistortion is no longer possible and that establishes the maximum power.

For example, consider an amplifier with a P_{sat} of 10 kW and a signal with 10 dB peak-to-average. Assume that the signal is clipped at 6 dB peak-to-average and still has acceptable distortion. Then the output power

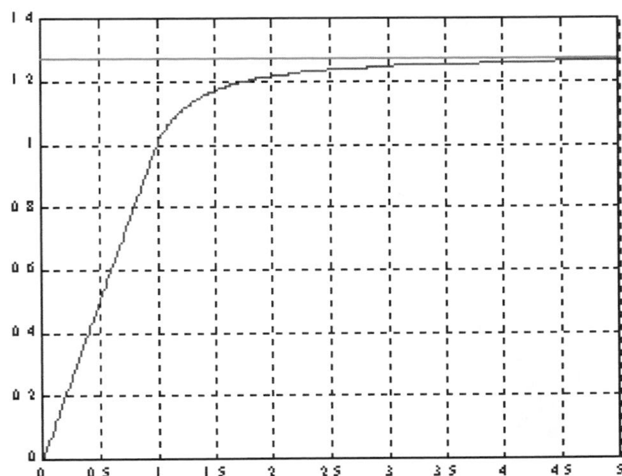

FIGURE 6.2-28 Example of a transmitter with asymptotic P_{sat}.

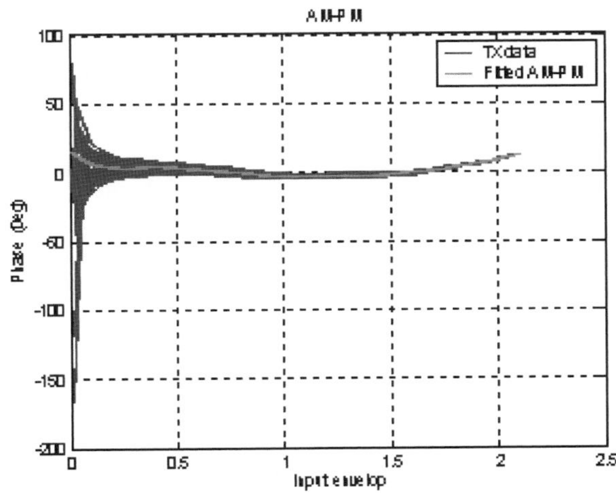

FIGURE 6.2-30 Typical AM-PM curve.

can be no more than 2.5 kW (6 dB below 10 kW). For less distortion, the output power might only be 1 kW (10 dB below 10 kW).

An example of an amplifier that is fairly well described by its AM-AM/PM curve is shown in Figures 6.2-29 and 6.2-30. The figures also show that it is not being driven into saturation, so it should be possible to linearize it. Since it isn't driven into saturation, it is impossible to estimate P_{sat} from this plot. The AM-PM curve also shows hysteresis, very noticeable around zero. To some extent this is due to the chosen format, the phase gets sensitive to small changes for a small envelope.

However, it also shows that the transmitter is not completely described by its AM-AM. This is clearly seen in Figure 6.2-31, the AM-AM/PM curve. Note that the envelope follows different paths when ascending and descending. Since this behavior resembles that of a transformer hysteresis curve, this term is often used to describe the memory effect. This type of behavior can't be described by an AM-AM/PM curve.

Precorrection

One simplified implementation is to use multiple AM-AM/PM curves. This can be seen as a special case of a Volterra series or neural network. In a sense, all precorrection methods are ad hoc, since there is no known way of directly finding the inverse from the amplifier. In addition, this assumes that the amplifier is completely known, which is seldom the case.

A predistorter is based on the inverse of the error of the amplifier. One advantage is that the error can include any particular frequency weighting. Another is how the predistorter follows from the amplifier, showing in a more direct way how the predistorter

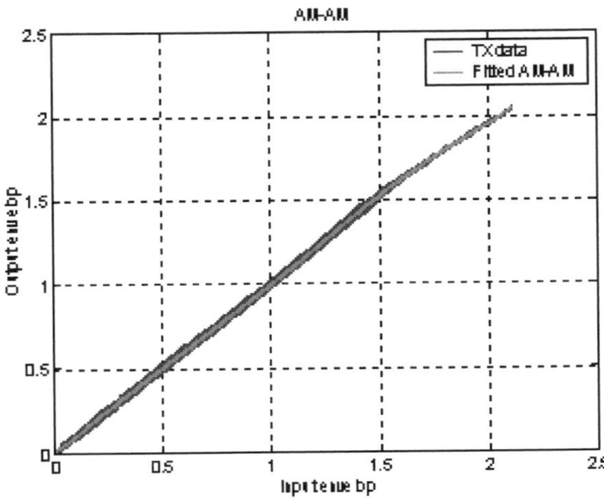

FIGURE 6.2-29 AM-AM curve.

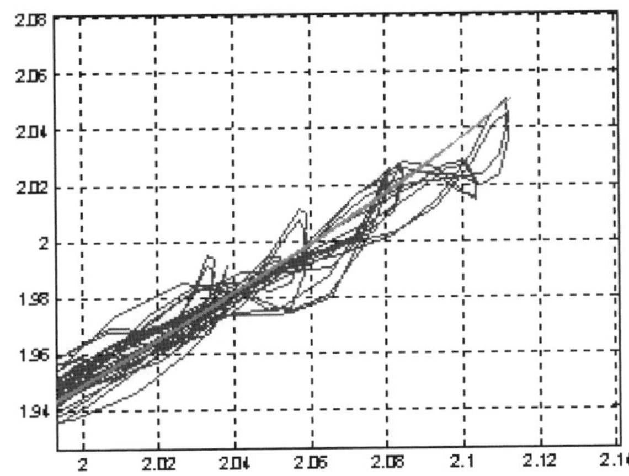

FIGURE 6.2-31 Detail of the AM-AM/PM curve showing hysteresis/memory.

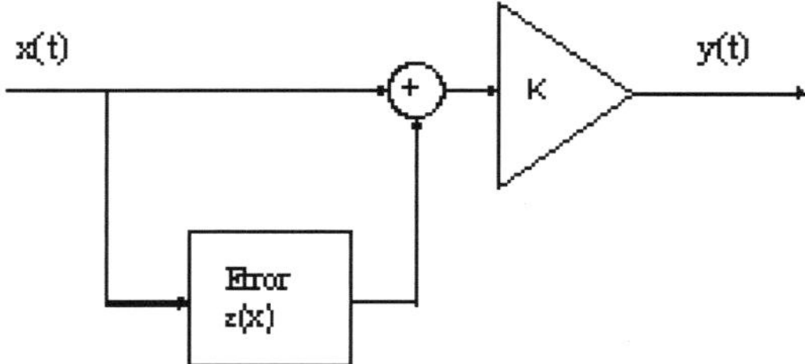

FIGURE 6.2-32 Transmitter model.

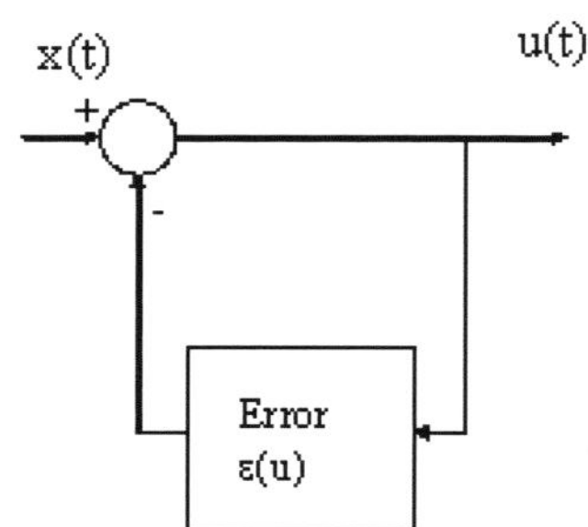

FIGURE 6.2-33 Predistorter.

depends on the amplifier. Given the amplifier model shown in Figure 6.2-32, it is not too hard to show that the inverse will be as seen in Figure 6.2-33.

The mathematical intricacies of the predistorter are complex, but there are some characteristics that a more careful analysis will reveal. The main one is stability. The predistorter is only stable if the error is sufficiently small. For typical broadcast transmitters that are linear to start with, this requirement should not be too hard to meet. At least up until saturation. A second one in is causality of the inverse, which is easily accommodated. The predistorter in Figure 6.2-33 can also be implemented as in Figure 6.2-34. In this implementation the issue is the number of stages needed. This depends on the error; the smaller the error, the fewer stages are needed. For typical broadcast transmitters, more than three should seldom be needed. Quite often one is sufficient.

It is possible to implement this predistorter, as well as any other, either digitally or analog. A modular transistor-based transmitter is particularly suitable for an analog implementation since an actual module can be used to get the error shown in Figure 6.2-35.

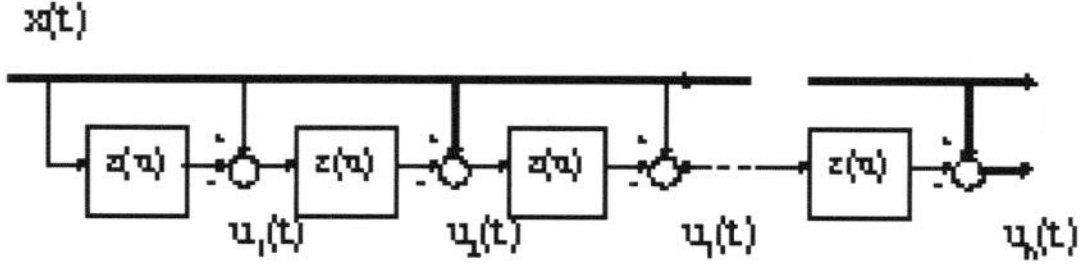

FIGURE 6.2-34 Alternative realization of the predistorter.

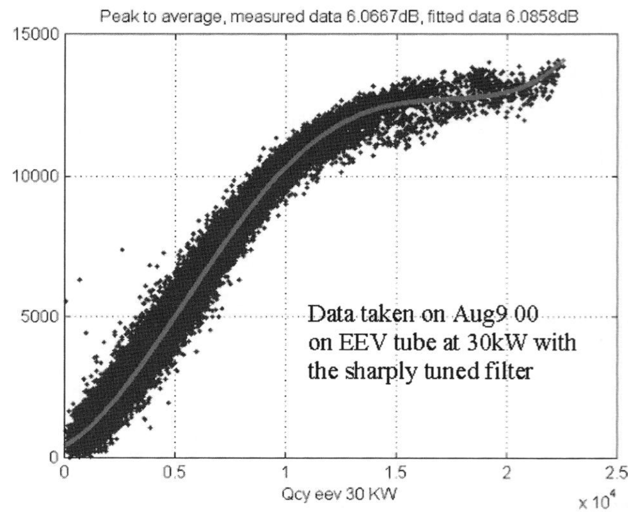

FIGURE 6.2-35 Actual AM-AM.

Amount of Correction

The amount depends on how much clipping is expected. This in turn depends on the desired output power and P_{sat} of the transmitter. If the clipping itself creates distortion/intermods products that are, say, 30 dB down, the correction only needs to make the amplifier linear to 35–40 dB. Then the distortion from the clipping will dominate that from the remaining nonlinearities. As a general rule: The correction should be about 5–10 dB better than that from the clipping. Consider this rule an approximation; spectral differences between the distortion from the transmitter and the clipping might possibly require the correction to be slightly different from what the rule suggests.

The acceptable amount of clipping is important for the designer of a corrector in that it is easier to design one that gives 40 dB linearity than 70 dB.

Adaptive Correction

Transmitters tend to change over time and temperature and the use of an automatic corrector to constantly adjust to current conditions is better than one that is set manually under only one set of test conditions. But a corrector that tries to compensate beyond the ability of the transmitter to sustain corrected conditions may actually cause damage. These limitations are important when designing adaptive correction that is reliable, and involves tradeoffs between correction capabilities and stability/performance. Best performance is generally found when the behavior of the likely transmitters is known.

POWER AMPLIFIER DESIGN CONSIDERATIONS

Aside from near-linear, transparent performance, it is essential that TV transmitters operate efficiently with

minimum lost air time. Efficient power conversion, while important at VHF, is more important for UHF transmitters. Many stations operate continuously and unattended, making reliability a key requirement.

Reliability

There are many factors that can affect the reliability of a TV transmitter. Overall design philosophy, device technology, module design, control architecture, power supplies, cooling, and cabinet design are critical areas that must be considered. If a transmitter design that uses circuits in series with no system redundancy, when one device fails, the entire transmitter fails. Figure 6.2-36 shows a system of three series devices with no redundancy. If each device (a, b, c) has a probability of survival (P) of 0.5 over some time interval the probability of the system surviving $P(s)$ is given by the formula:

$$\begin{aligned} P(s) &= P(a) \times P(b) \times P(c) \\ &= 0.5 \times 0.5 \times 0.5 \\ &= 0.125 \end{aligned}$$

If three identical devices are operated in parallel (Figure 6.2-37) and only one is required for adequate operation of the system, the probability of the system surviving is greatly enhanced. The overall system survival probability now becomes:

$$\begin{aligned} P(s) = {} & P(a) + P(b) + P(c) \\ & - P(a)P(b) - P(a)P(c) - P(b)P(c) \\ & + P(a) \times P(b) \times P(c) \\ = {} & 1.5 - 0.25 - 0.25 - 0.25 + 0.125 \\ = {} & 0.875 \end{aligned}$$

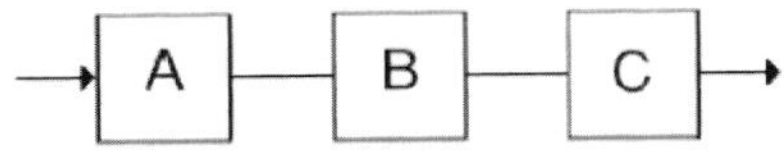

FIGURE 6.2-36 Circuits in series.

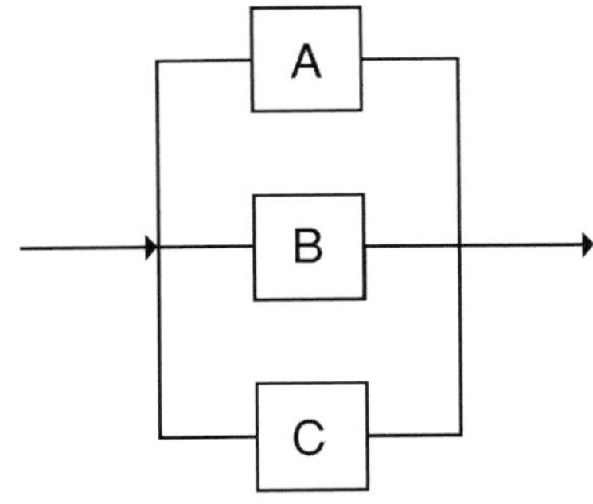

FIGURE 6.2-37 Circuits in parallel.

On-Air Availability

Related to reliability, but perhaps even more important, is on-air availability. On-air availability is the percentage of time the transmitter is available for service, defined by the following equation:

$$\text{On-air availability} = \frac{\text{MTBF}}{\text{MTBF} + \text{MTTR} + \text{MPMT}} \times 100\%,$$

where:

MTBF is mean time between failures (hours).

MTTR is mean time to repair (hours).

MPMT is mean preventative maintenance time (hours).

It is apparent there is little point in designing a transmitter that has a high MTBF figure if, due to poor design and mechanical packaging, it takes an inordinate length of time to make repairs, or the transmitter has to be shut down frequently for routine preventative maintenance.

Many stations have very short sign-off windows or operate 24 hours a day. This often results in a less-than-optimum maintenance schedule that can lead to premature failure or out-of-tolerance operation. One way to reduce the amount of off-air maintenance time is by making provisions for on-air maintenance or to have redundant transmitters. This significantly reduces the MPMT.

Several design factors should be considered for optimum on-air availability:

- High reliability for the fundamental circuits.
- Provision for fast, easy access to all subassemblies.
- Maximum use of like parts and subassemblies. Because fewer items are needed, this allows a TV station to maintain a full inventory of spares. If spares are on hand, it follows that the repair time is much shorter.
- Repair of transmitter at subassembly level. A subassembly that has been removed can then be repaired by station personnel or returned to the manufacturer for exchange.

Efficiency

When the fundamental concept of efficiency is applied to TV transmitters several factors must be considered. For purposes of determining total power consumed, the AC to RF conversion efficiency is the parameter of interest. For systems with unity power factor, determining the AC input power is relatively straightforward. For power factors less than unity, the relative phase of the fundamental voltage and current must be determined. In addition, in systems generating significant line frequency harmonics, the relative level of these harmonics must be known. Power factor is expressed either as displacement power factor or total or true power factor.

Displacement power factor is the cosine of the phase between the voltage and current at the fundamental frequency. It is equal to the total power factor only for an undistorted sine wave.

Total power factor is the ratio of total power to the apparent power, given by:

PF = AC power input P(watts)/V(volts) × I(amps).

Determination of input power is somewhat simplified by considering only the DC to RF conversion process. In this case it is necessary to determine the voltage and current provided by the power supply(s) for the final amplifying devices. While this is a useful tool for comparison purposes, it has the disadvantage of ignoring the power consumed elsewhere in the transmitter, such as driver stages, cooling systems, filament, and magnet power and control. If these powers are to be included in the efficiency calculation, they must be determined separately.

For NTSC transmitters, determination of output power is equally complex. Transmitters are commonly rated in terms of peak sync visual power. Exclusive use of this number neglects the aural output. In addition some amplifier technologies may exhibit efficiencies greater than 100% using visual peak power. This has given rise to the use of figure of merit defined as:

Visual peak sync output power to DC input power at 50% APL.

This definition is valid for transmitters using separate amplification. For common amplification, the aural output and input powers must be added to the numerator and denominator, respectively. Typical figures of merit for common tube amplifiers operating in separate amplification are shown in Table 6.2-1.

Consider a typical klystron amplifier in pulsed operation. Direct current input power is calculated as follows:

DC input power = beam voltage × [(beam I_{sync} × sync duty cycle) + (beam I_{video} × video duty cycle)].

For 60 kW output, DC input power = 24 [(5.5 × .08) + (3.7 × .92)] = 92.2 kW

Figure of merit = 60/92.2 = 0.65.

For digital transmitters, the situation is somewhat simpler. The output power most readily measured is the average level, which is constant. Since the audio is encoded with the video, there is no separate aural amplifier. Thus, the DC to RF efficiency is readily calculated as a ratio of the average RF output to the DC input.

TABLE 6.2-1
Typical Figures of Merit for Common Tube Amplifiers Operating in Separate Amplification

Amplifier Device	Figure of Merit
Tetrode	0.9–1.0
Integral cavity klystron	0.65–0.75
External cavity klystron	0.65–0.75
Klystrode or IOT	1.1–1.3
Depressed collector klystron	1.2–1.3

SOLID-STATE TRANSMITTERS—VHF AND UHF

Technological advances in bipolar and field effect transistors (FET) have made the development of solid-state high-power, linear amplifier modules for TV applications both practical and cost effective. By combining RF modules, it is practical to create transmitters at any power range up to 100 kW with cost rather than technology being the limiting factor. Solid-state transmitters maintain their performance over extended periods of time due primarily to the fact there are no tuning controls nor filament emission degradation with time. No warm up time is required—solid-state transmitters produce full-rated power within seconds of activation and provide a high degree of reliability as a result of their modular design.

Solid-State Devices

Both bipolar and FET technology exist today as suitable RF amplification devices. Power amps are operated in class AB for the best tradeoff of efficiency and linearity. Driver stages usually contain class A amplifiers.

Although both device types have merit, FETs have some advantages over bipolar devices. FETs have a higher amplification factor than bipolar transistors, helping to reduce the number of driver stages. Higher-supply voltages help to reduce the current capacity of the power supply. Simpler bias circuitry minimizes parts count.

RF Amplifiers

Combining several RF power modules to achieve the desired transmitter output power increases the parallel redundancy and the on-air availability. Output power of 1–2 kW per module has been adopted by nearly all manufacturers based on overall cost, practical weight, and size limitations.

Self-protection of each PA module against various fault conditions is good engineering practice. By using self-protecting modules, the cabinet control logic and overall transmitter control logic can be kept simple, thus improving overall reliability. Self-diagnostics for the module aid in minimizing time to repair. Protection from overvoltage, overdrive, VSWR, overtemperature, and ensuring proper load sharing among devices, is essential to maintaining amplifiers for long life. It is desirable that one subassembly failure not cause another subassembly to fail.

Modular amplifiers that can be removed while hot (powered) improve overall on-air availability. If an amplifier module fails, the transmitter continues to function indefinitely without disrupting transmitter operation. If a spare PA amplifier is on hand, it can be used while the failed unit is repaired.

Temperature-compensated regulated supplies for the amplifiers are important. Otherwise, power output varies as temperature or supply voltage changes.

Combiners and Dividers

There are several methods used for dividing and combining RF power for the solid-state amplifier modules. An effective method is the use of in-phase N-way combiners. Three common examples are described.

Microstrip Wilkinson Combiner

Figure 6.2-38 shows an example of a Microstrip Wilkinson Combiner. Microstrip is used to carry the RF power. When all amplifiers are operating, equal voltages are presented to each node of the load resistor so that no power is dissipated. When an amplifier failure occurs, the power is distributed between the load and the output. The impedance of the transmission lines and the length of the lines are selected to achieve the desired impedance transformation. Balanced reject loads are used to absorb RF power in case of amplifier failure and to provide isolation between amplifiers.

The combiner is housed in a shielded casing so that outside fields cannot alter the balanced configuration. The reject loads are cooled by conducting heat through the flange to the heat sink where the heat is exchanged to the moving air stream. This type of combiner is used generally for lower numbers of amplifiers (two to six).

Ring Combiner

Figure 6.2-39 shows an example of a ring combiner. The higher power-handling capability of the coax lines used in this combiner allows a range of amplifiers to be combined (2–20). It also provides isolation between amplifiers using reject loads that are not in the direct RF path to the output. The operation of the multiport combiner is easily understood by considering a two-way version (see Figure 6.2-40). Each transmission line is one-quarter wavelength. When equal voltages are applied to both input ports (both amplifiers operating), the combined signal arrives at the output. This is true because the distance from each input port to the output is electrically equal whether the signal follows the shorter or longer path, the signal from one amplifier arrives at the load port out of phase with that from the other amplifiers so no power is dissipated, and the signals from amplifier 1 arriving at input port 2 via the short and long paths are out of phase and vice versa. Under normal conditions, all power appears at the output, none is absorbed in the loads, and there is complete isolation between amplifiers.

Power is absorbed in the load resistors only when an amplifier is not operating. Assume that only amplifier number 1 is operating. The signal path is electri-

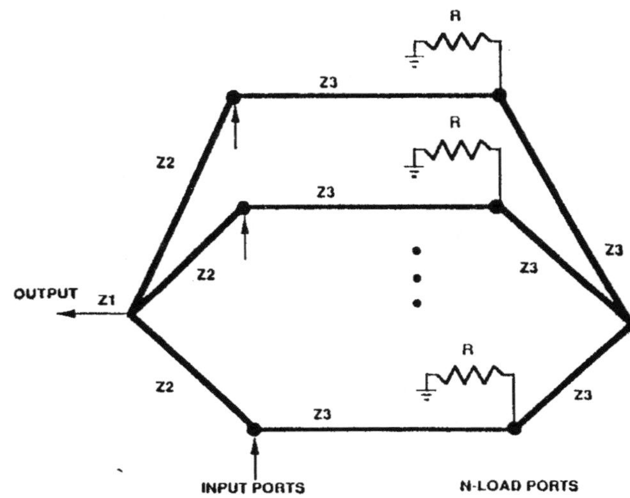

FIGURE 6.2-39 N-way ring combiner.

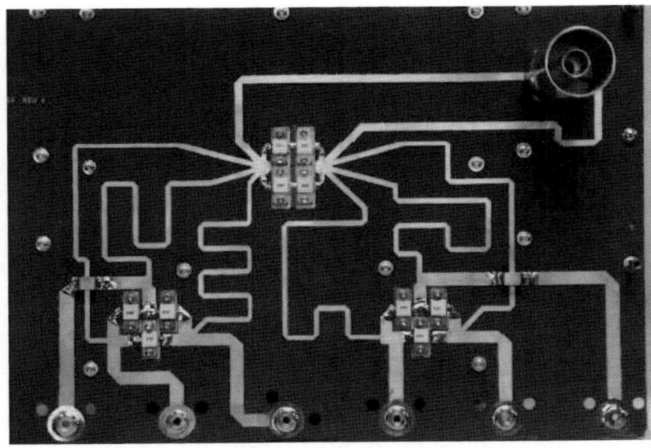

FIGURE 6.2-38 Microstrip Wilkinson combiner. (Photo courtesy of Larcan.)

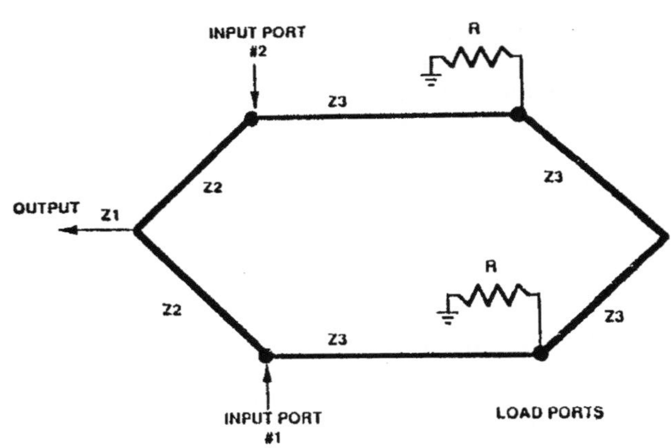

FIGURE 6.2-40 Two-way ring combiner.

cally equal not only for the long and short paths to the output, but also, for the right and left paths to either load. The power from the operating amplifier is split equally between the output and the isolation loads. Due to the isolation inherent in the network, the input ports remain matched even when one or more amplifiers are removed.

Since the transmission line that connects the amplifier port to the output port is one-quarter wavelength long and 75 Ω transmission line is used, the 50 Ω amplifier impedance is transformed to:

$$(75/50) \times 75 = 112.5\ \Omega.$$

For combining N amplifiers, the impedance at the output combiner junction is 112.5/N. This impedance is then matched to 50 .

The reject loads are mounted to a grounded structure. One method of cooling the loads is to use a vapor cycle heat pipe for the mounting structure. In case of amplifier removal, the reject load temperature rises, the fluid in the lower section of the heat pipe heats until it vaporizes. The vapor rises to the finned area where the heat is exchanged to the moving air stream. The vapor condenses as it releases its heat and returns to the bottom of the pipe to repeat the cycle.

Starpoint Combiner

Starpoint combiners are simple low-loss devices and operate in the following manner. Consider parallel operation of four sources of equal phase and amplitude, each with a source impedance of 50 Ω and no interaction. The resulting source impedance is 12.5 Ω. All that is needed to complete the combining process is to transform this impedance to 50 Ω. Obviously, this concept can be extended to any value of "N." The "Q" of the impedance transformation determines the combiner bandwidth. For optimum bandwidth it may be necessary to combine and transform in several corporate steps. In practice, the combined amplifiers have slightly differing phases and amplitudes. As in any combiner design, this represents a combining loss that must be accounted for. A flat air-spaced strip line structure may be used to provide a combiner with extremely low-resistive loss.

This simple combining technique works best only if all inputs are present. If a failure occurs the isolation may be insufficient to prevent interaction between modules, especially for small values of "N." Using circulators at the output of each module resolves this, producing a combiner with excellent isolation and bandwidth. Circulators not only provide port-to-port isolation but also protect against high VSWR due to icing on the antenna and other transmission line problems. The combiner bandwidth should be consistent with module and circulator bandwidth.

Combining Multiple Amplifier Cabinets

When combining RF power amplifiers, they must be matched in phase and gain for maximum power to the antenna and minimum power to the reject load. Electronic phase shifters and attenuators should have the capability for remembering their settings in case of AC power failure.

Cooling System

Proper cooling of the solid-state modules is important for high MTBF. The MTBF of a transistor essentially doubles for every 10°C drop in the junction temperature. Distributed cooling systems employing more than one fan offer good redundancy. Current motor/fan technology has matured to the point where a few larger direct drive fans are as reliable as many smaller fans. Since many RF power amplifiers may be employed, a large volume of air is needed to adequately cool the heat sinks. Low-pressure fans or blowers may be used if heat sink fin density is not high. This aids in reducing audible noise. The heat is distributed over a large volume of air and the temperature rise is relatively low.

Automatic Gain Control

Automatic gain control (AGC) is used to maintain constant power output from the transmitter. Ambient temperature changes will cause gain changes in a solid-state amplifier. RF drive power must be changed to maintain constant power output. A detected RF sample of the PA output may be fed to an input of a comparator. The exciter output sample or a voltage proportional to the exciter output is applied to the other input of a comparator. The DC output is then integrated and fed to an attenuator that varies the RF drive level at a low power level. Alternatively, the RF sample may be taken from an intermediate stage. In this case, other means must be used to temperature compensate the output stage or some power reduction with temperature must be tolerated.

Power Supplies

Power supply design is critical to the reliability of a solid-state transmitter. Since FET and bipolar devices are low-voltage devices, the power supplies that serve them must provide low voltage and high current. High-reliability connections must be guaranteed in the DC distribution. Since available power output from a transistor varies as

$$\text{Po} = \frac{(V)^2}{2R_L}$$

it is desirable that the supply remain very tightly controlled over incoming AC line variation. Since the amplifier current demand varies with picture level (for NTSC), the power supply output voltage must remain stable from low load (white picture) to high load (peak sync output). Efficiency of the power supply is important since the lost power results in heat as well as high utility costs. Any anomaly present, such as voltage or current transients or voltage sags at the AC input, should be significantly suppressed before

reaching the amplifier transistor device. Transmitters should successfully pass the applicable portions of the ANSI/IEEE C62.41 Transient Testing Standard (also referred to as the IEEE-587 Standard).

Power supply architecture for solid-state modular transmitters generally follows one of two designs. One involves a common power supply powering a group of modules. This typically results in a high-current DC supply, and wiring and connectors to distribute the DC to a bank of modules, typically in one cabinet. In some cases, a second supply is available to provide redundancy. The second approach is to dedicate a single power supply to a single RF amplifier. Many of the early designs had the power-supply module mounted directly behind the RF module resulting in a short DC and control path. More recent designs now have the power supply integrated into the RF amplifier module. High-efficiency switching power-supply designs have made this self-contained design possible.

Alternating Current Distribution

A reliable method of AC distribution provides power to modular RF amplifier cabinets through a parallel system. Each cabinet is protected by a separate AC circuit breaker external to the transmitter. This concept allows a cabinet to be safely serviced while the remaining cabinets are operational. Phase monitors guard against low voltage, loss of phase, or reversal of phase.

Control Systems

If individual amplifier modules and power supplies are self-protecting, control and monitoring functions can be simple and straightforward. One approach for the control system is to use a single controller to control and monitor all the functions of the transmitter. Another approach is to distribute the control system throughout the transmitter. The distributed control

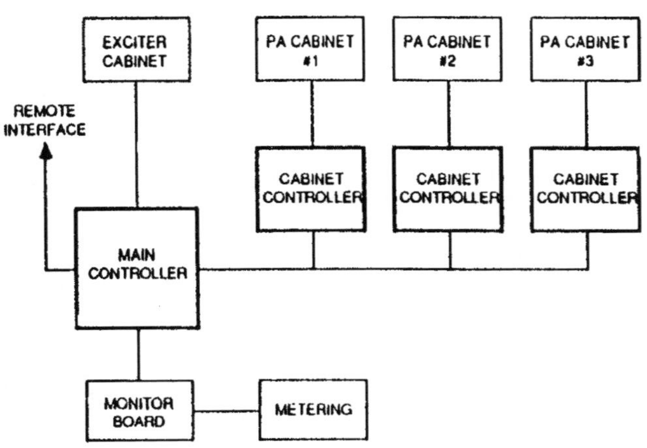

FIGURE 6.2-41 Distributed control and monitoring.

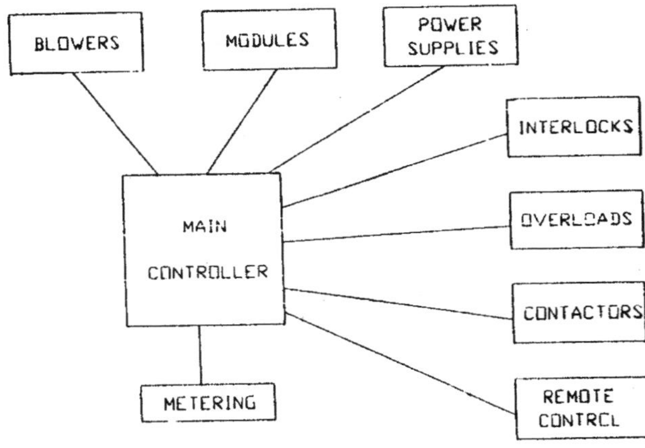

FIGURE 6.2-42 Centralized control system.

system can be designed so that the failure of any individual controller does not affect the operation of the others.

Both distributed control and centralized control systems are shown in Figures 6.2-41 and 6.2-42. After an AC power failure, the controller should have backup memory to restore the transmitter to the same operating condition as before.

A system of indicators is essential to quick fault diagnosis. Typical status conditions that may be displayed are exciter fault, VSWR fault, VSWR foldback, power-supply fault, controller fault, air loss, door open, fail-safe interlock, phase loss, module fault, visual drive fault, aural drive fault, and external interlocks.

VSWR foldback reduces power during high-VSWR operation, such as antenna icing, and restores RF power to normal when difficulties are removed. Other options used to maintain on-air capability may include dual exciters, aural power reduction, and redundant drive chains. A block diagram of a solid-state transmitter with VSWR foldback is shown in Figure 6.2-43.

UHF TUBE TRANSMITTERS

With high-UHF transmitter power levels, power consumption and efficiency are important parameters. A variety of tube technologies are available to address this requirement. Some are most suited for lower-power transmitter designs; others are more appropriate for the highest power requirements. These technologies include:

- Tetrodes
- Klystrons
- Multiple depressed collector klystrons (MSDC)
- Inductive output tubes (IOT)
- Multiple depressed collector inductive output tubes (MSDC IOT)

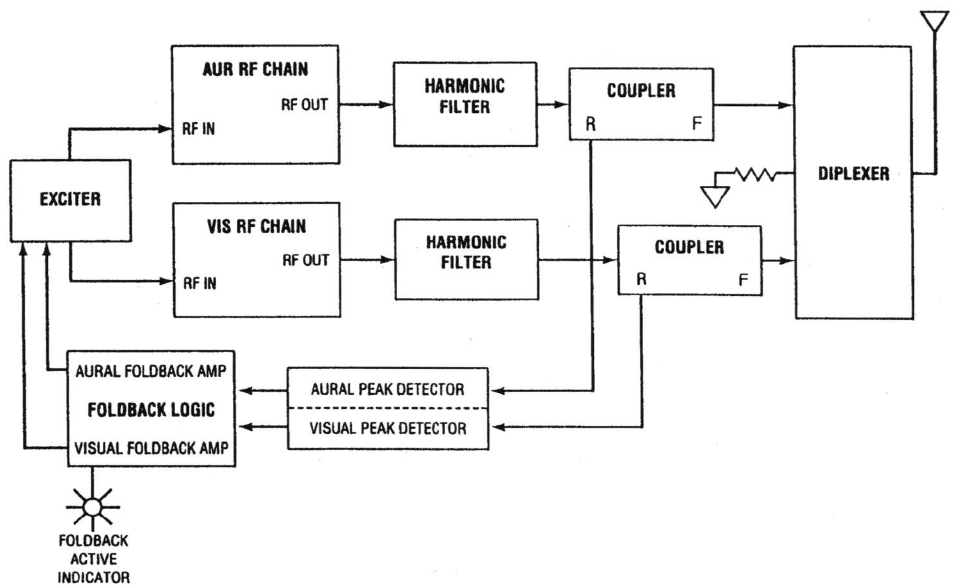

FIGURE 6.2-43 VSWR foldback block diagram.

Tetrodes

Tetrode refers to a generic category of four element tubes suitable for the linear amplification of RF signals to the power levels required for broadcasting. This discussion is limited to water-cooled UHF TV tubes made available in recent years. These tetrodes are of cylindrical construction. The inner diameter of the anode of a typical 30 kW peak sync tetrode is barely 2 inches in diameter. The anode is domed at the upper end to make it watertight. The water jacket is a watertight container around the anode. Element spacing is quite close, especially the spacing between the control grid and the cathode, which is about 0.003 inches. Gain is an inverse function of the control grid to cathode spacing while a direct function of control grid to anode spacing. Figure 6.2-44 is a horizontal cross-sectional sketch of typical power tetrodes.

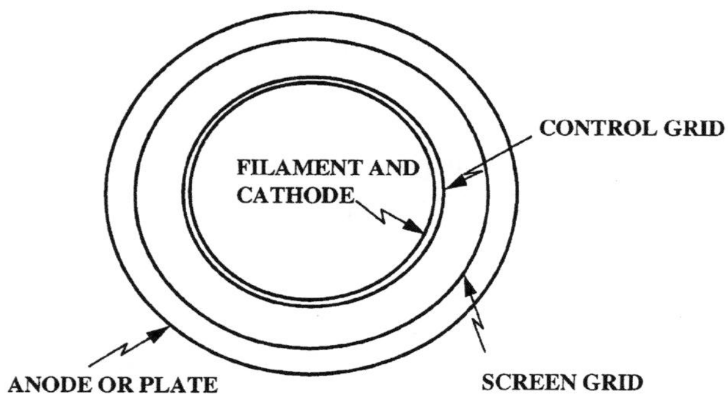

FIGURE 6.2-44 A horizontal cross-section of a typical power tetrode showing the relative placement of the four elements.

Tetrodes can be biased for class AB operation and are more efficient than the class A klystron. Tetrodes operating at UHF at higher power levels are made possible due to advances in ceramics and ceramic-to-metal bonding, pyrolytic grid structures, and the method of anode cooling. They exhibit excellent envelope and RF linearity requiring only the degree of correction normally seen in VHF transmitters. The tradeoff for performance in these areas is lower power gain than most other amplifiers. For example, the gain for a tetrode is about 15 dB.

Modern UHF power tetrodes used for linear television applications are capable of combined visual and aural amplification. Ten percent aural power is typical at UHF and tubes are rated according to peak sync power with the aural carrier. For example, a typical UHF tetrode may be rated at 30 kW peak sync. Ten percent aural power is assumed giving a PEP rating of 52 kW. It is capable of 43 kW peak sync if the aural carrier is amplified separately.

Connections to these tubes are made at the base by means of concentric ring conductors, staggered in height and mating to a socket that sits above the tuning sections of the input and output networks. This allows nearly all the RF current to flow in the lower two-thirds of the tubes' vertical structure, the upper end being open circuited. Since the tube sits above the cavity sections, tuning is not disturbed when the tube is removed or inserted. Because inner electrode capacitances vary from tube to tube, some touch-up retuning is required. Tuning is not done at full power but at milliwatt levels.

Typical tetrodes are operated in grounded grid. Neutralization is required, but it is a one-time physical setting and does not change from tube to tube. The tetrode is cathode driven with a low "Q" input tuning section. Output power is taken from the anode

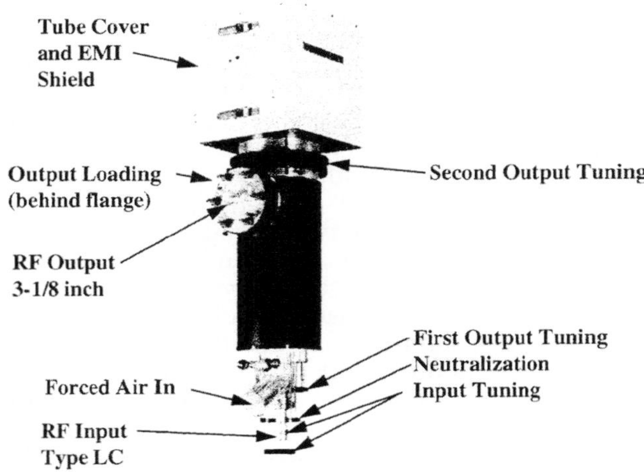

FIGURE 6.2-45 A representative cavity for the tetrode is shown. RF drive connects to the bottom while the output is taken from the side through the 3-1/8 inch EIA flange.

through a double-tuned cavity section with loading control. The grids are grounded for RF. The tubes are cooled with distilled water. Cooling water must be nonconductive since it comes in contact with anode voltage. A representative cavity is shown in Figure 6.2-45.

The tetrode node voltage is much lower than that of the klystron. No special standoff requirements are necessary and the negative supply line of the power supply is grounded. The cathode is at DC ground potential.

Contributing to the service life of modern tetrodes is adequate water cooling of anode and screen grid. The method used is a process in which the water coming in contact with the anode surface is allowed to vaporize to the nucleate boiling state within tiny pits on the anode surface as illustrated in Figure 6.2-46. As soon as the vapor enters the main water stream a few thousandths of an inch above the surface, it condenses, releasing its heat content to the main stream. Nucleate boiling is enhanced by deliberately roughening the anode surface providing millions of boiling sites uniformly spread over the surface. If the anode surface were visible and not hidden under a water-tight cap, nucleate boiling would be recognized by the formation of nearly microscopic bubbles, which quickly condense and disappear. Film boiling is maintained in between nucleate boiling sites. The screen grid is also water cooled, but no boiling takes place.

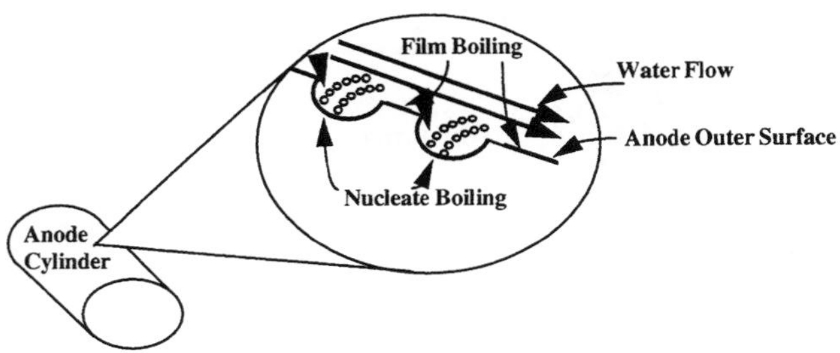

FIGURE 6.2-46 Cooling of the anode surface.

TABLE 6.2-2
Typical Power Tetrode Electrical Characteristics

P_{out} PEP	52 kW
P_{out} peak sync	30 kW
10% aural?	Yes
One dB bandwidth	12.8 MHz
Circuit configuration	Grounded grid
Integral to the cavity?	No
Power gain	15 dB
IMD	–48 dB no correction
Neutralization	Yes
Sync compression	5 IRE no correction
Plate (anode) voltage	8,500 V
Anode current idle	2.5 A
Anode current black	5.6 A
Screen grid voltage	650 V
Screen grid current black	40 mA
Control grid voltage	–120 V
Control grid current black	20 mA
Filament voltage	5.1 V AC or DC
Filament current	180 A

The important electrical characteristics of a 30 kW peak sync tetrode are shown in Table 6.2-2. The basic structure of these tubes is silver-plated copper and ceramic with pyrolytic graphite grids and a thoriated tungsten directly heated cathode. Filament currents are high to minimize cathode current modulation of the cathode temperature.

Drive power is in series with the input/output circuits that are in series. Since the cathode is the driven element, its current is nearly identical to the anode current. Thus power gain is realized from the voltage gain.

Tetrode Transmitter

A block diagram of a NTSC tetrode transmitter using a 30 kW tetrode is shown in Figure 6.2-47. It is an example of an IF-modulated combined amplification transmitter requiring almost 1 kW of drive power. Drive is supplied by two solid-state class AB amplifier drawers having gain of about 23 dB. Correction circuits include low-frequency linearity, differential gain, differential phase, and sync compression, all performed at video, and ICPM, stereo pilot protection, in-band

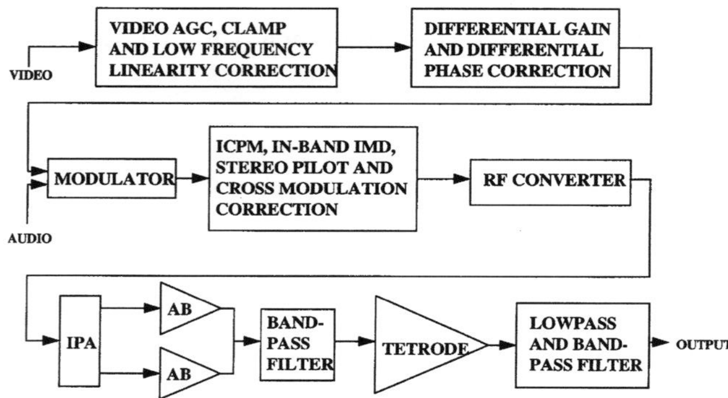

FIGURE 6.2-47 Block diagram of a 30 kW common amplification tetrode transmitter.

intermodulation distortion, and cross-modulation, all performed at IF.

For good linearity the control grid and screen grid power supplies must be well regulated. These supplies are capable of modulating the amplified signal if allowed to shift with modulated drive level. It is not necessary to regulate the anode supply since the tetrode is not saturated and is relatively insensitive to anode voltage variations. Both oil-filled and dry supplies may be used for the anode voltage source. Oil has a cooling advantage over convection or forced-air cooling.

The filament supply must also be well regulated and should be capable of two different voltages: full voltage according to the value in Table 6.2-2 and about 20% of full voltage. The latter is known as "black heat" voltage and is used when the tube is in off-air, standby mode. No cooling of any kind is needed when the tube is operated in the black heat mode.

Black heat is used to reduce the time to on-air, since the filament is already warm, and to reduce the thermal stress of going to full filament power. The filament voltage should be gradually increased from the black heat value to full value over a period of 60 seconds. Likewise, it should be gradually decreased at the end of the broadcast day for longest life.

It is imperative that the control circuitry be designed so that faults, overvoltage, and overcurrent situations quickly disconnect the power supplies from the tube. It is also imperative that the cooling system be maintained in proper operating condition. Blocked air filters or reduced water flow due to blockages will shorten tube life even though the control system judges conditions that are adequate to operate. The control system should be designed so that the transmitter will remain on-air as long as minimum cooling requirements are met. The transmitter should be designed to have warning indicators posted on the front panel, available by remote monitoring. When a trip occurs, the high-voltage power supply should be removed within 15 msec and the DC output of the screen grid power supply immediately shorted to ground while being disconnected from the AC mains.

Klystrons

The klystron uses velocity modulation in the amplifying device. The electron beam emitted from the cathode is accelerated to high velocity by the electric field between the cathode and anode and is directed into the RF interaction region, as shown in Figure 6.2-48. An external magnetic field is employed to

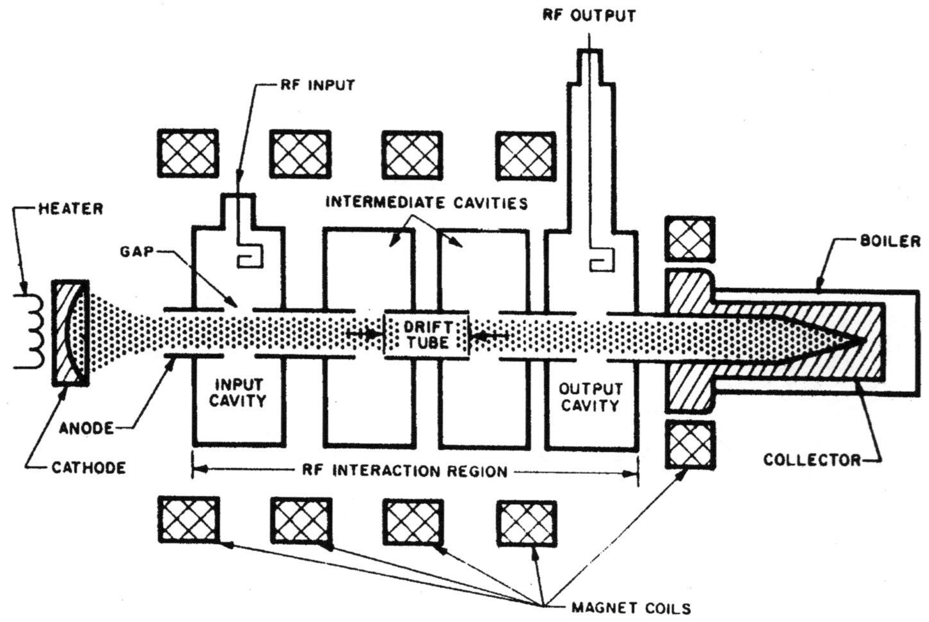

FIGURE 6.2-48 Principle elements of a klystron.

focus the beam as it passes through the tube. The electron beam impinges on the collector, which dissipates the beam energy and returns the electron current to the beam power supply.

The RF interaction region, where the amplification occurs, contains resonant cavities and field-free drift space. The first resonant cavity (the input cavity) is excited and an RF voltage is developed across the gap. Since electrons approach the input cavity gap with equal velocities and emerge with different velocities, the electron beam is said to be velocity modulated. As the electrons travel down the drift tube, bunching develops, and the density of electrons passing a given point varies with time.

The RF energy produced by this interaction is extracted from the beam and fed into a coaxial or waveguide transmission line by means of a coupling loop in the output cavity. The DC beam input power not converted to RF energy is dissipated in the collector. The cavities can be mounted external to the klystron or can be included in the vacuum envelope. The resonant frequency of each of the cavities is adjusted by changing the volume of the cavity in external cavity klystrons, or the capacitance of the drift tube gaps is changed in integral cavity klystrons.

The RF energy is fed through a coaxial line with its center conductor inserted into the cavity. The end of the center conductor is formed into a loop. This is a simple one-turn transformer that couples energy into the cavity as shown in Figure 6.2-49. Intermediate cavities may have their loops coupled into RF loads to vary the "Q" of the cavities thus to change the overall bandpass characteristics.

All cathodes have an optimum operating temperature that must be high enough to prevent variations in heater power from affecting the electron emission current (beam current). However, the temperature of the emitting surface must not be higher than necessary, since excessive temperature can reduce cathode life.

There are two electrodes that may control the beam current: the modulating anode and a lower voltage (0–1,400 V) electrode typically used for pulsing the beam current. If the low-voltage electrode is connected to the cathode, the modulating anode voltage controls

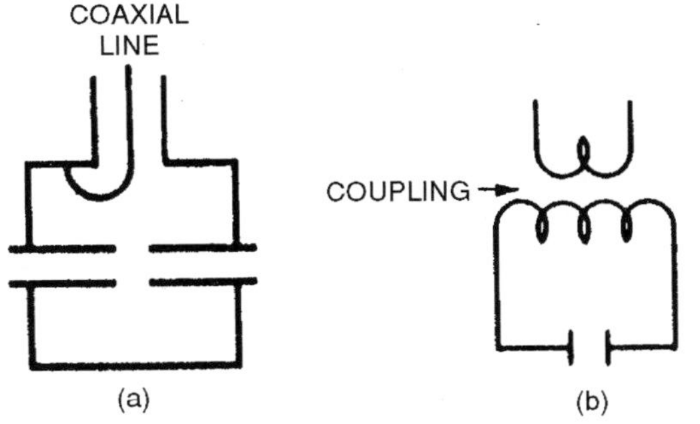

FIGURE 6.2-49 Loop coupling and equivalent circuit.

beam current. Perveance is a function of the geometry of the cathode-anode structure. This can be calculated using the following equation:

$$I_b = KV_b^{3/2},$$

where:

K is perveance.

I_b is beam current in amperes.

V_b is beam voltage.

Figure 6.2-50 shows the relationship between beam current and voltage described in the equation. Operating condition A on the chart with a modulating anode at 4,000 V, with respect to the cathode beam voltage, produces a beam current of 0.6 A. The intersection point lies at a perveance expressed as 2.4×10^{-6} or 2.4 micropervs. Operating condition B illustrates a practical television transmitter situation in which a common beam supply of 18 kV is used to power both the visual and aural klystrons. The visual tube operates at a beam current of approximately 5.0 A if the modulating anode is connected (through an isolating resistance) to the body of the tube, and the perveance is 2.1 micropervs. Since the aural output power required is much less, the DC input power is less than that required to operate the visual tube. Points B′ indicate that if the modulating anode is supplied with only 8 kV (through a voltage divider), the intersection with the 2.1 microperv line yields a beam current of only 1.5 A, thus accomplishing the necessary reduction of input power for aural service.

Electromagnets are placed around the klystron to develop a magnetic field along the axis of the electron beam. This controls the size of the electron beam and keeps it aligned with the drift tubes. If the magnetic field is interrupted or insufficiently controlled, the electron beam will land on surfaces other than the collector and may destroy the tube.

The output cavity is generally tuned to the carrier frequency. It is important to operate with the coupling loop adjusted so the output cavity is slightly overcoupled. Figure 6.2-51 shows the relationship of output power to proper coupling loop adjustment. If the coupling loop is adjusted so that the cavity is undercoupled, arcing and ceramic fracture resulting in klystron failure may occur.

Figure 6.2-52 shows output power as a function of drive power. As expected, when drive power level is low, output power is low. As the level of drive power increases, output power increases until an optimum point is reached. Beyond this point, further increases in drive power result in less output power. In the underdriven zone, output power increases when the input power is increased. The point labeled "optimum" represents the maximum output power obtainable. The tube is said to be saturated at this point, since any further increase in drive only decreases the output power. The overdriven zone is formed at the right side of saturation. To obtain maximum output power, sufficient drive power must be applied to the tube to reach the point of saturation. Operating at

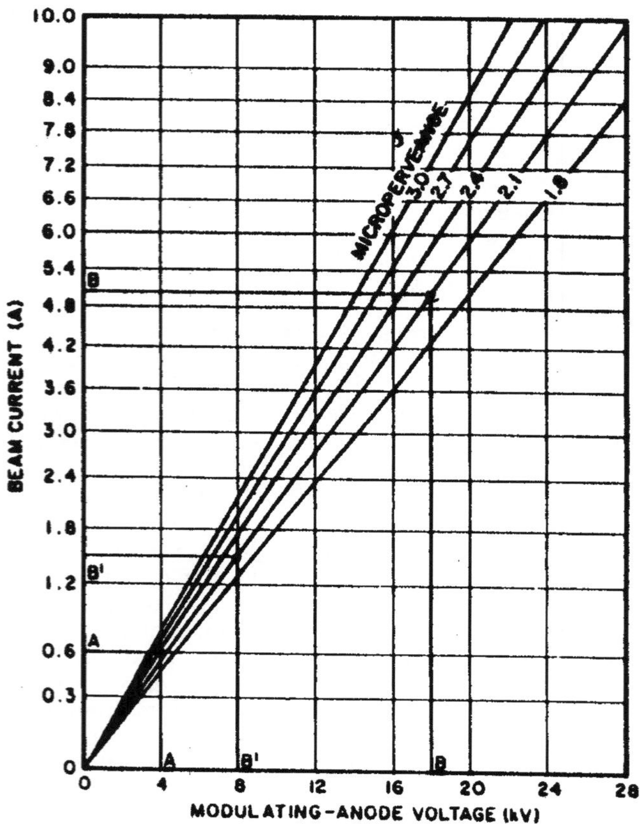

FIGURE 6.2-50 Beam current variation with modulating anode voltage.

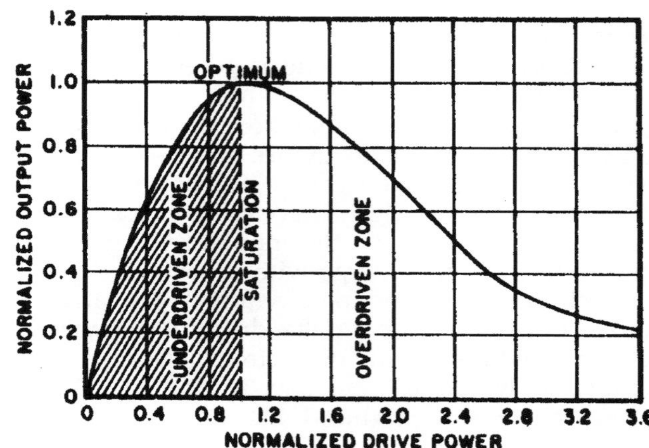

FIGURE 6.2-52 RF output power as a function of RF drive power.

drive levels beyond saturation point only overdrives the tube, decreasing output power, and increasing the amount of beam interception at the drift tubes (body current). Klystrons tuned for TV service are operated within the underdriven zone.

Figure 6.2-53 shows how output power changes with various levels of drive power applied under different tuning conditions. Curve A represents saturation for a synchronously tuned tube. Curves B and C show saturation reached by tuning the penultimate (closest to the output cavity) cavity to higher frequencies. Increasing the penultimate cavity frequency beyond curve D no longer increases output power. Instead, it reduces the output power as shown with curve E. The input coupling is adjusted for the best tradeoff of minimum reflected power and best overall bandpass.

Operating a klystron at saturation improves efficiency but requires more linearity and phase compensation. By using a pulser to switch to a higher beam current during sync and back to a lesser current during video, the average beam current is significantly reduced. The practical limit of reduction of the video

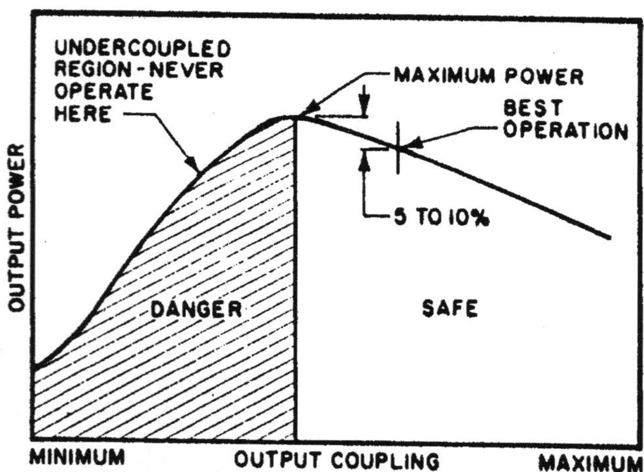

FIGURE 6.2-51 Adjustment of output coupling control.

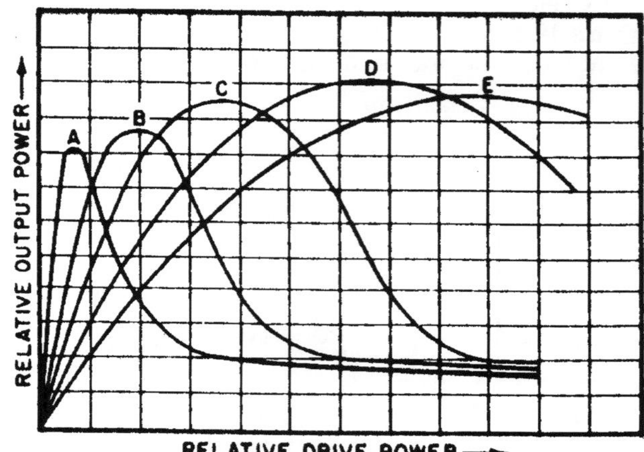

FIGURE 6.2-53 Output power variation with drive power under different tuning conditions.

beam current is the point at which tip of burst and back porch distortion are not correctable.

With beam current pulsing, the effective "Q" of the electron beam and cavity combination is altered. This results in passband tilt from visual carrier down to the upper edge of the passband. This requires readjustment of the cavities to obtain a flat response. Tilt of as much as 7 dB has been observed.

In static (nonpulsed) operation, the tube is supplied with enough beam current to saturate at 100% power with a small amount of headroom for changes in beam voltage. During pulsed operation only enough beam current is supplied to saturate at 100% power during sync. The amount of beam current reduction during the video picture is dependent on operating channel, perveance, available drive power, and precorrection capability.

The optimum value of beam current must be experimentally determined. However, figures of merit of 77% have been achieved. The absolute value of the pulsed efficiency is directly proportional to, and therefore limited by, the efficiency obtained for nonpulsed operation. In order to prevent severe burst and back-porch distortion, a small amount of amplifier headroom is needed, approximately 2–5%.

Pulsed operation for klystrons is accomplished by connecting the beam control electrode to a voltage source of 0 to –1,400 V with respect to the cathode. During sync, the pulser operates at 0 V. During the video portion of the signal, values of –400 to –800 V are used to achieve reliable high-efficiency operation.

In some transmitters, sync is actually reduced or removed from the input video. As the beam current is pulsed, the klystron gain change increases the RF power to produce the proper sync level.

Effect of Pulsing on Transmitter Precorrection

Very little can be done to reduce ICPM by klystron tuning or selection of magnet current. Also, ICPM increases rapidly near saturation of the klystron.

The phase shift through a klystron changes due to the beam current (when the klystron is pulsed during sync). When the klystron switches back to the video current level, the previous value of phase returns. ICPM correctors operating at the exciter IF introduce a correction signal equal and opposite to the distortion produced by the klystron. A phase-modulation stage in the exciter may be keyed by sync and adjusted to precorrect for incidental phase distortions caused in the klystron during pulsing.

When amplifiers are operating very close to saturation during the color burst, more differential gain and differential phase correction may be needed. Modern exciters can fully precorrect these conditions.

Sync pulse oscillations may produce "ringing" when the klystron is operated at saturation. This may exhibit itself as a tearing of the picture. This ringing is believed to be caused by secondary electron feedback enhanced by the reverse gain of the klystron cavities. Rebiasing the tube slightly out of saturation will eliminate the ringing.

Multistage Depressed Collector Klystrons

Klystron amplifiers operate by converting energy from a beam of electrons to RF output power. At full output power, about half of the beam power is converted to output power. Correspondingly, half the DC input power remains on the beam as it exits the cavity region. In a standard klystron this spent electron beam energy is dissipated as heat. The MSDC klystron, however, uses the spent electron beam, recovering its energy to reduce the dissipation. In the depressed collector technology, power recovery in the collector region is accomplished by decelerating the electrons in the spent beam. This is done by providing an electric field in the collector such that the electrons are slowed before they strike the collector wall. A collector composed of multiple elements is utilized, with each element operated at a negative potential with respect to ground. Consequently, the collector potentials are referred to as being depressed below ground.

The interaction process produces a wide range of velocities for individual electrons. The collector geometry is carefully selected to provide an electric field shape that sorts the impinging electrons according to their velocity, reducing their energy as much as possible, yet ensuring that all electrons strike one of the elements and are not reflected into the electron beam.

Figure 6.2-54 shows the resulting collector configuration. The collector is composed of five elements and is designed to operate with equal voltages between elements. For a 60 kW klystron, the voltage per collector stage is typically 6.25 kV. This means that element 1 is at ground potential, element 2 is at –6.25 kV, element 3 is at –12.5 kV, element 4 is at –18.75 kV, and element 5 is at full cathode potential of –25 kV. To make sure that the electron beam is optimized before

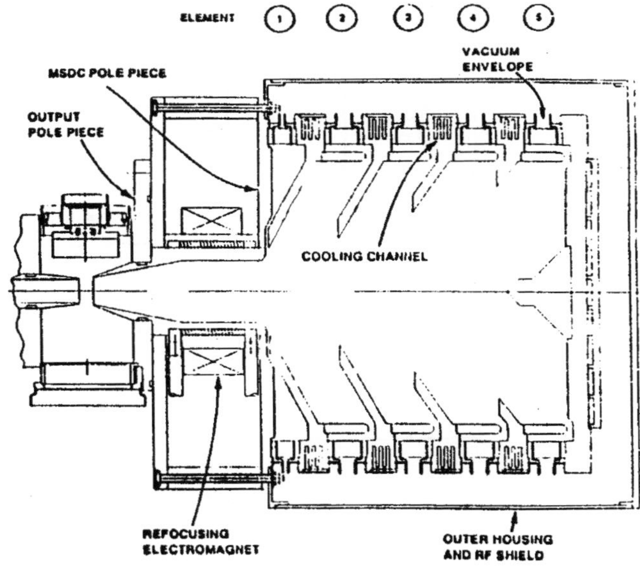

FIGURE 6.2-54 UHF-TV MSDC klystron tube design.

entering the collector region, a refocus coil is provided just ahead of the collector.

The significant benefit of the depressed collector klystron is the reduction in power consumption for a given power output. The individual collector beam currents vary depending on the picture level. Since recovery of the spent beam takes place in the collector, it has essentially no impact on the tuning and normal precorrection of the klystron. The depressed collector technique can be applied to either external or integral cavity klystrons.

Since the power dissipated in the collector is reduced, only six gallons per minute of high-purity water coolant is needed. The efficiency of the MSDC is dependent on the transmitter configuration. Figures of merit of 1.2–1.6 have been obtained in transmitter installations in the field.

Transmitter Design Using MSDC Klystrons

The primary differences in this transmitter from a standard klystron transmitter are in the beam supply and the cooling system. Figure 6.2-55 shows the power connections made to an MSDC klystron. Since the RF performance of the MSDC klystron is the same as the standard klystron, there are no differences in the RF driver chain.

The cooling system of an MSDC klystron transmitter uses a two-stage heat-transfer system as shown in Figure 6.2-56. A two-stage system is chosen to allow outside heat exchangers to be used. The cooling system consists of a high-purity water loop and a glycol-water mixture loop. High-purity water (resistivity of 200,000 -cm or more) must be used with the cooling system to prevent current flow between collectors. To remove ions, free oxygen, and other possible contaminants from the water, a three-stage purification loop is used. The filter cartridges sample part of the water flow so the whole system is continuously cleaned. The filter cartridges can be replaced without taking the transmitter off the air.

Separate beam supplies for each klystron are frequently used. Since the currents from the beam supplies change dependent on the power level, there are video frequency currents present on the power-supply leads. Therefore, sufficient bypassing and power-supply high-voltage wire shielding are required. Monitoring each section of the beam supply current is required to obtain the currents for power dissipation calculations.

Protection circuitry should include

- Magnet overcurrent and undercurrent trip points.
- Beam supply overcurrent and overvoltage sensors (for each collector).
- Water flow sensors.
- Arc detectors within the third and fourth cavities.
- Sufficient interlocks to prevent personnel from accidentally coming in contact with high voltage.

Inductive Output Tube

The IOT combines features of a tetrode and a klystron. The tradename Klystrode® is also used to describe this tube. The electron beam is constrained similar to that of a klystron—that is, by electromagnets. The modes of operation of the IOT are similar to that of a tetrode. However, there are significant differences due to the different geometries.

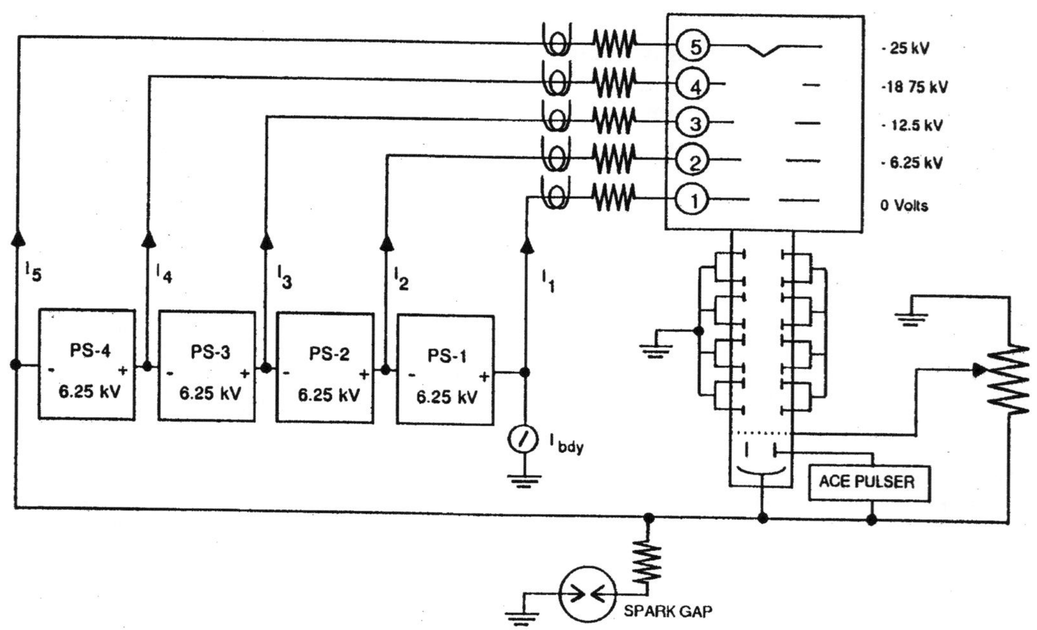

FIGURE 6.2-55 MSDC high-voltage connections.

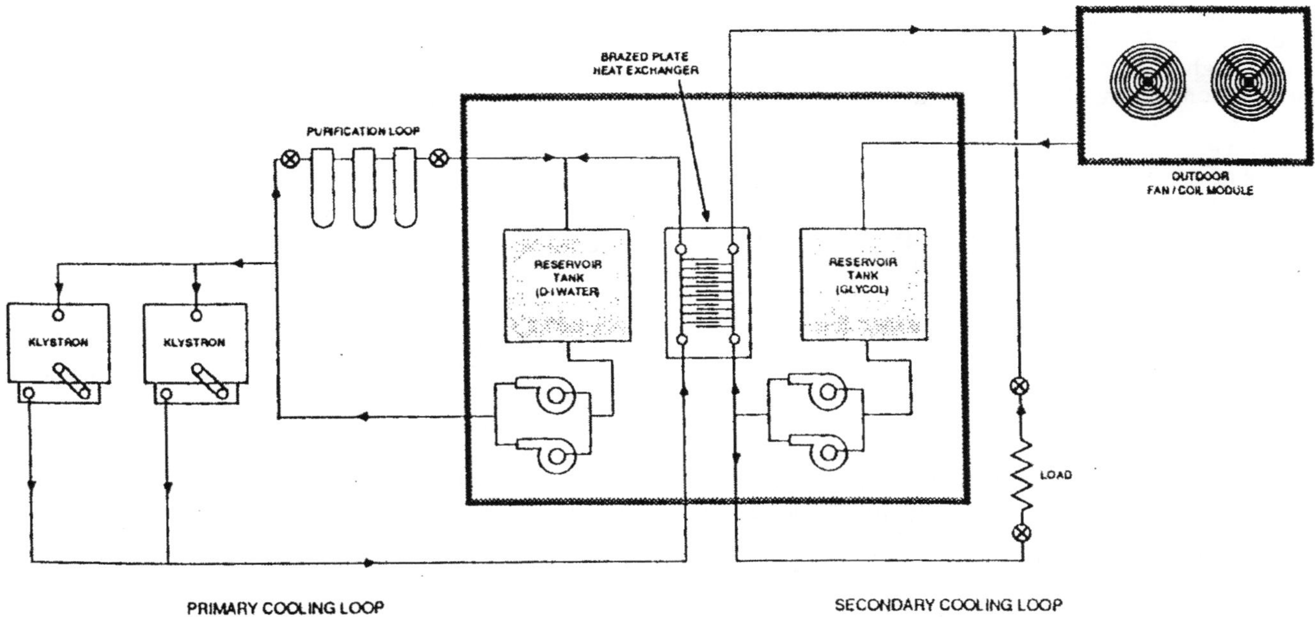

FIGURE 6.2-56 Dual-loop heat exchanger system.

An IOT is shown schematically in Figure 6.2-57. It is composed of an electron gun—very similar to a klystron—a control grid, an input cavity, accelerating anode, drift tube, output cavity, and collector. The electrodes are arranged linearly, unlike the concentric configuration of a tetrode. It is physically smaller than a klystron, which weighs approximately 400 lbs including the tuning cavities and magnet assembly. The tube itself weighs about 60 lbs.

The electron beam is formed at the cathode, density modulated with the input signal applied to a grid, and then accelerated through the anode aperture. In its bunched form, the beam drifts through a field-free region and then interacts with the RF field at the drift tube gap in the output cavity. Power is extracted from the beam in the output cavity in the same manner as a klystron. The grid structure may intercept some electrons causing a small amount of grid current. This will increase if the tube is overdriven.

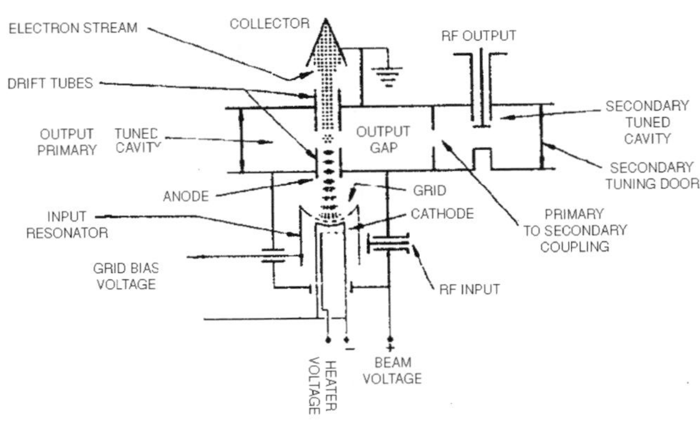

FIGURE 6.2-57 IOT or Klystrode schematic.

Input power is applied to the control grid via a resonant cavity. The grid is biased negatively near cutoff as a class B or class AB amplifier. The first part of the tube may be thought of as a triode with a perforated anode through which the electron beam is guided by electric and magnetic fields. The beam is bunched at the radio frequency rate and is accelerated by the high anode potential. It passes through the anode extension cylinder, which is an electrostatic shield, and then interacts with the RF field in the output gap. The spent beam is dissipated in the collector, which is separate from the output RF interaction circuit.

The tuned input and output circuits are external to the vacuum envelope. The input circuit consists of a stub tuner and resonant coaxial cavity, which match the drive source to the high-impedance grid. Also included is a circuit that provides a DC block for high voltage. The output cavity is clamped to the body of the tube with techniques similar to those used for external cavity klystrons. The output circuit is double-tuned to achieve the bandwidth required for visual service. The double-tuned output circuit consists of a primary cavity clamped to the body and an iris-coupled or loop-coupled secondary cavity. The output transmission line is probe coupled to the electric field or loop coupled to the magnetic field in the secondary cavity. Coupling methods are different for different manufacturers. Variable controls adjust primary and secondary cavities to the same resonant frequency. Additional controls are used to adjust the intercavity and output couplings. Because the tube has only two cavities, it is much shorter than a klystron. The magnetic field requirements of the tube are produced by a solenoid powered by a supply of about 6–7 V at 25–30 A.

The high-voltage circuitry is contained within a shielded compartment on top of the input circuit. This circuitry consists of various filters to contain the fields and prevent instabilities at video frequencies. High-voltage grid bias and filament power enter this section via high-voltage cables.

The fundamental benefit of the IOT is that it may operate as a class B amplifier, resulting in high efficiency. Thus, the beam current (I_b) is proportional to the RF drive signal (V_g), following the modulation envelope according to the three halves law:

$$I_b = K(V_g + V_a/\mu)^{3/2},$$

where μ is the amplification factor. The perveance (K) is proportional to the cathode area and inversely proportional to the square of the grid to cathode spacing. The drive voltage is not normally high enough to cause the instantaneous grid voltage to become positive. Although the efficiency of the IOT is dependent on the transmitter configuration, figures of merit of 1.2–1.4 have been obtained for 60 kW visual service from transmitter installations in the field. In aural service, the IOT is tuned the same as for visual service. A single tube covers the entire UHF operating band, although two slightly different input cavities are required.

Power gain in either visual or aural service is about 21 dB. Thus, drive power is about 500 W for the visual and 50 W for the aural (assuming 10% nominal aural power). A typical transfer curve is shown in Figure 6.2-58. The amount of nonlinearity is similar to a tetrode; the transfer function is S shaped. The nonlinearity of the beam current causes distortion of the fundamental UHF component and generation of signals at multiples of the UHF frequency. Harmonics do not excite the output cavity and, therefore, do not affect the quality of the transmitted signal. The linearity improves as the quiescent current is increased from zero, becoming optimum when at 8–9% of the maximum value. The transfer characteristic eventually saturates at high-power levels, as is the case with all power amplifier devices. However, in an IOT, the grid voltage may be driven positive with respect to the cathode for short periods of time. This enables the tube to amplify the high power, accommodate short duration signal peaks present in digital TV signals, and also to transmit the high-peak envelope power that occurs during common amplifier operation in an analog transmitter. The nonlinearity increases as the load impedance is increased to boost efficiency. Limited bandwidth of the input circuit also increases nonlinearity.

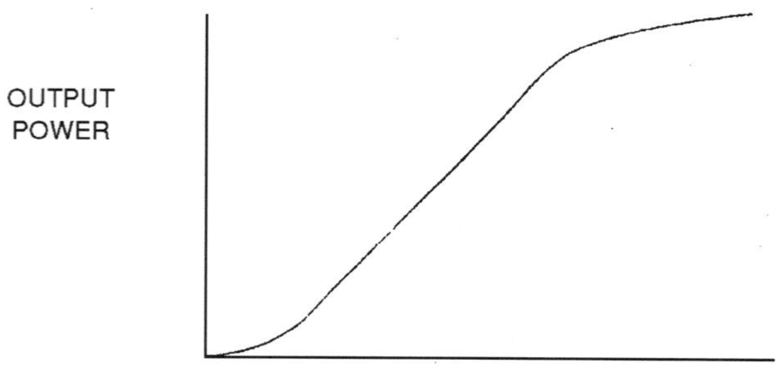

FIGURE 6.2-58 IOT transfer curve.

The IOT may also be used in multiplexed or common amplification mode. For example, a 60 kW tube may be used at 40 kW visual with a 10% aural signal. Intermodulation products at +920 kHz from visual carrier generated by the tube can be precorrected by low-level IF and/or RF circuitry.

Cooling of the IOT at the 60 kW power level under maximum ambient temperature conditions requires about 25 gallons per minute of water for the collector. The body of the tube is also water cooled. A 50/50 water/glycol solution is typically used in cold climates without any special water purification. The input and output cavities are air cooled. The 15 kW and 40 kW IOTs may be air cooled. The volume of air and air pressure needed is comparable to a tetrode of similar power.

Transmitter Design Using IOT

Support circuitry for the IOT consists of the necessary drivers, precorrection, power supplies, and protection circuitry. An IOT transmitter block diagram using common amplification is shown in Figure 6.2-59.

The IOT uses a beam voltage of 32–36 kV. Since the beam current changes with modulation, video frequency currents are present in the beam supply. The beam supply must be designed to provide excellent regulation from no load to full load and to also provide a low source impedance for all video frequencies. In the event of a high-voltage failure, the conventional beam supply should limit the energy dumped into an arc. The supply, being stiffer, requires a triggered *crowbar circuit* to limit the beam supply arc energy. A block diagram of a crowbar circuit is shown in Figure 6.2-60.

The grid bias supply of –10 to –70 V with respect to the cathode floats with the beam voltage. A simple method of developing the grid voltage is to use zener diodes connected between the power supply and the tube cathode connection and tap the grid to the appropriate zener to obtain the desired bias current. As with klystrons, the magnet power supply must have sufficient energy storage that the beam remains focused until the beam decays. A power supply for the ion vacuum pump is needed with appropriate sensors for the protection circuitry. Other protection circuitry provided in the IOT transmitter is similar to that needed in the klystron transmitter.

Multistage Depressed Collector IOT

The efficiency of a single-collector IOT power amplifier can be improved by employing multiple collectors. But there is a tradeoff in beam power efficiency

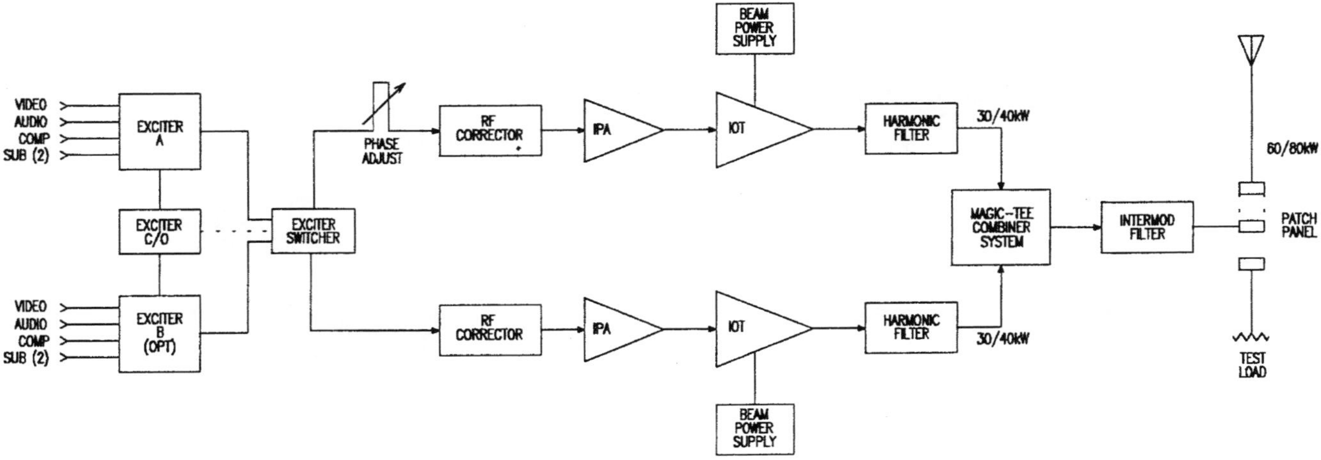

FIGURE 6.2-59 IOT transmitter block diagram.

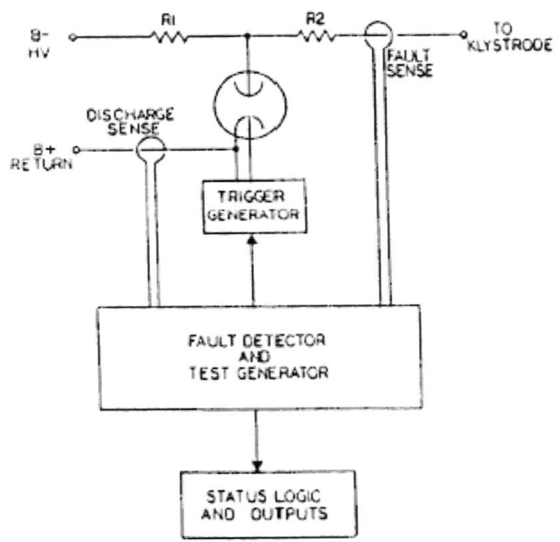

FIGURE 6.2-60 Triggered crowbar circuit.

and the cost of adding collectors and their supporting circuits. The optimum point is typically reached when four or five collectors are employed.

A five-collector IOT is shown schematically in Figure 6.2-61, which shows collectors connected to successively higher negative voltages. The gun of the tube is exactly the same as the IOT. The cathode operates at approximately –36 kV relative to the anode, which is grounded. A heater element raises the cathode temperature; electrons are emitted by the cathode, accelerated toward the anode because of the voltage gradient, then pass through it to the output gap. A grid near the cathode is connected to a regulated DC supply and is typically set to –100 V relative to the cathode, thereby reducing cathode current almost to cut-off for class AB operation. External solenoids and a magnet frame create a magnetic field to confine the electron beam from the time it enters the anode until it leaves the output gap.

An RF signal is applied between the grid and cathode via a tuned circuit that also isolates RF from the high negative DC potential. The RF signal is superimposed on the DC grid voltage resulting in electron beam density variations (bunches).

The IOT output gap is enclosed by a resonant circuit (cavity). As bunches of electrons pass the output gap they induce high-RF voltages in the cavity. Electrons are slowed by giving up energy to the cavity; the more RF they induce in the output gap, the lower their residual velocity. A high-power RF signal is coupled from the cavity to a secondary cavity then to the amplifier output. The two cavities and their couplers are adjusted for substantially flat amplitude over the desired bandwidth. The input cavity, set for best match, has a broad bandwidth relative to the output.

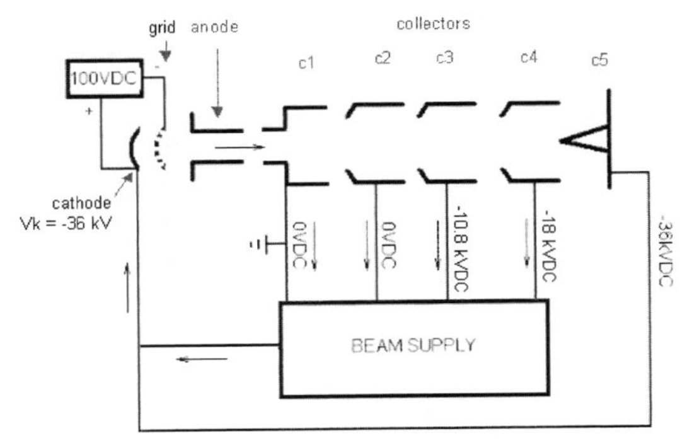

FIGURE 6.2-61 Simplified schematic of MSDC IOT power-supply configuration.

The "spent" electron beam spreads as it enters the collector region. Electrons with greater residual velocity penetrate further into the collector region and encounter a progressively higher repelling potential. Reduced velocity at the moment of contact with a collector results in less waste heat generation. Depending on tube power rating and design, waste heat may be conducted away from the collectors by water, forced air, or oil.

Efficiency (η) is calculated by dividing RF output power (Wout) by the total collector powers; collector power = current × difference between a collector's voltage (Vcn) and cathode voltage (Vk):

$$\eta = \text{Wout}/(\text{Ic1}(\text{Vc1} - \text{Vk}) + \text{Ic2}(\text{Vc2} - \text{Vk}) + \text{Ic3}(\text{Vc3} - \text{Vk}) + \text{Ic4}(\text{Vc4} - \text{Vk}) + \text{Ic5}(\text{Vc5} - \text{Vk}).$$

Typical beam efficiency improvement of a five-collector IOT over a single-collector type is a factor of 1.5 (e.g., 57% versus 38%) for 8 VSB. Figures 6.2-62 and 6.2-63 depict type ESCIOT 5130 W rated for 130 kW peak output. Note, ESCIOT is a registered trademark of e2v technologies (www.e2v.com).

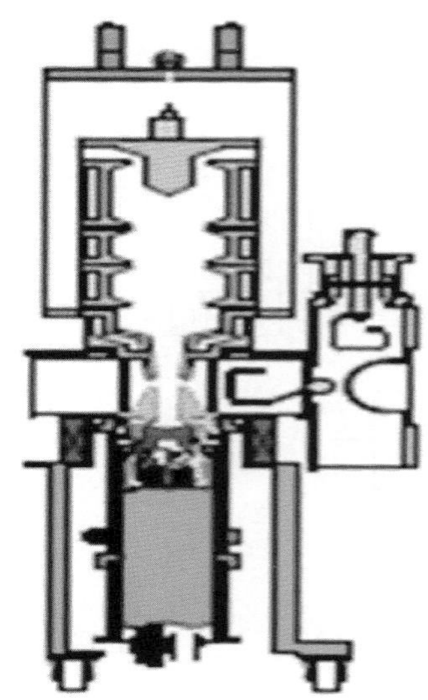

FIGURE 6.2-63 Cutaway view of ESCIOT 5130W. (Courtesy of e2v.)

Transmitter Design Using IOT and MSDC IOT

The single collector IOT operates with one high-voltage value at –36 kV whereas the MSDC can have up to five values of high voltage. These high voltages are generated in a multi-output beam supply. Each collector supply must be monitored for voltage and current for protection of the tube and all high voltages must be grounded for access to the power supply and main cabinet of the transmitter. While this multi-output supply arrangement makes packaging of components a problem, the increased cost of it is offset by use of a smaller beam supply, as the tube operates at a higher efficiency than the single collector IOT.

Because the anode of the IOT is at ground potential, glycol can be used as a coolant. However, in the MSDC IOT, the collectors are at high voltage and, therefore, must be cooled by pure water or oil. This means a secondary cooling system is necessary with redundant pumps that must be used to prevent freeze up. Water purity must be maintained by use of a polishing loop. The conductivity is required to be in the ranges of 1.25–3 megohm-cm.

FIGURE 6.2-62 e2v ESCIOT 5130W™. (Courtesy of e2v.)

TRANSMITTER COOLING SYSTEMS

Transmitter cooling may take a variety of forms in air- and liquid-based cooling systems according to the power level required, the devices used, and the architecture of the system in general.

Modern transmitters are grouped as

- Solid-state air cooled
- Solid-state liquid cooled
- IOT or klystron air cooled
- IOT or klystron liquid cooled

Air Systems for Transmitters

Transmitters dissipate large amounts of heat. Air that has already passed through the equipment and has picked up this heat must be removed from the immediate vicinity to prevent the hot air from being recirculated through the transmitter cooling system. In addition, provisions must be made for sufficient intake air to replace that which has been exhausted.

To provide adequately for hot air exhaust and fresh air intake, and the maximum and minimum environmental conditions in which the equipment may operate, the following information about the equipment and the installation location should be known:

- Altitude (feet)
- Maximum temperature (°C)
- Minimum temperature (°C)
- Total air volume through the transmitter (CFM)
- Pressure drop across the transmitter tube (inches of water)
- Air temperature rise through the transmitter (°C)
- Air exhaust area (square feet)

Equipment layouts usually provide an area for heated air to exit from the top of the cabinet. The size and location of this exhaust area are usually shown on a manufacturer-supplied outline drawing.

Most broadcast equipment internal air systems are designed to be operated into free space (back pressure of 0.0 inches of water) so exhaust ducts attached to the cabinet must have minimum loss. Good practice is to design for no more than +0.1 inches of water pressure in the duct close to the exhaust area of the transmitter.

Any exhaust installation other than a large cross-section duct (equal to the cross-section of the transmitter exhaust port) with no bends and with a long radius turn outside the transmitter building for weather protection, will need an auxiliary exhaust blower or fan.

The recommended system is sized only for cooling the transmitter. Any additional cooling load in the building must be considered separately when selecting the air system components. The transmitter exhaust should not be the only air exhaust in the room as heat from the peripheral equipment may be forced to go through the transmitter.

The *sensible-heat* load is the sum of heat loads, such as solar radiation, heat gains from equipment, and lights and personnel in the area that are to be cooled.

The following exhaust duct design illustrates the key cooling concepts:

- Air volume through transmitter = 325 CFM (ft^3/min)
- Air exhaust area (A) = 3.4 sq ft
- Air exhaust velocity (A) =

$$\frac{325\ \text{ft}^3/\text{min}}{3.4\ \text{ft}^2} = 94.5\ \text{ft}/\text{min}$$

Air velocity of 94.5 ft/min is relatively low allowing a transition to a smaller diameter pipe if desired. Assume a transition to a 10 inch diameter pipe.

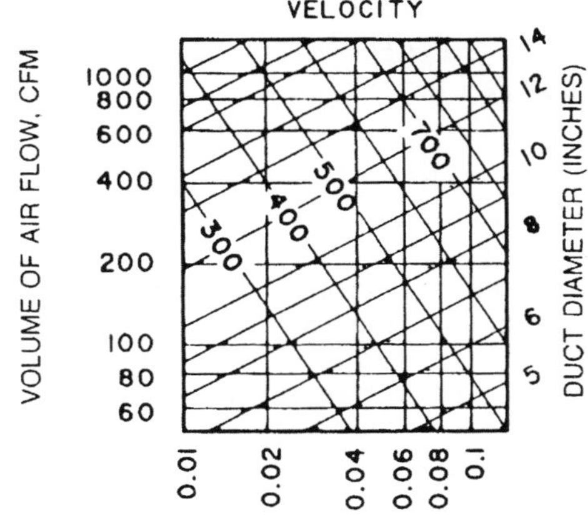

FIGURE 6.2-64 Friction loss in pipes.

$$A = \frac{\pi R^2}{(12)^2} = \frac{\pi(5)^2}{144} = 0.545\ \text{sq ft}$$

$$V = \frac{325\ \text{ft}^3/\text{min}}{\text{Pipe Area } 0.545\ \text{ft}^2} = 596\ \text{ft}/\text{min}$$

The air friction in a 10 inch diameter pipe is 0.06 per 100 ft of pipe at a flow of 325 CFM (see Figure 6.2-64). Assuming there is 20 ft of straight pipe to a roof, with two 90 degree elbows to turn the pipe down for weather protection, the total loss of the exhaust is estimated as follows:

$$\frac{20}{100} \times 0.06 = 0.012\ \text{inches of water}$$
(pressure drop in 20 ft of pipe).

From Figure 6.2-65 a 10 inch elbow at 900 ft/min gives less than 0.01 pressure drop. However, there is only 596 ft/min in this system. Conservatively, use

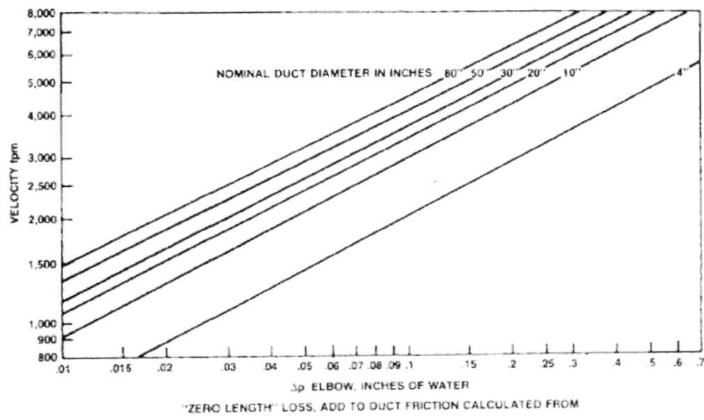

FIGURE 6.2-65 Friction loss in 90 degree elbow.

0.01 for each 90 degree elbow and 0.12 for the 20 ft section of 10 inch pipe. Thus,

0.012 inches + 0.01 inches =
0.032 inches of water (total pressure drop).

Therefore, no exhaust fan is necessary since 0.032 inches is less than the 0.1 inch pressure that requires a fan.

In a different situation, the installer may have a problem exhausting the transmitter if a roof exit is not available. It may be required to add two additional 90 degree elbows and a straight run of 10 inch pipe 100 ft long. The pressure drop in the exhaust system is then:

Friction loss in 20 ft of 10 in diameter pipe	0.012
4 elbows × 0.01 in of water (each)	0.04 in
Friction loss/100 ft of 10 in diameter pipe	0.06 in
	0.112 in of water

The 0.112 inch of pressure now exceeds the level at which a fan is needed.

The performance curve for the fan, shown in Figure 6.2-66, indicates that it will deliver about 350 CFM into 0.1 inches of water. This is sufficient to handle all transmitter air and overcome all estimated duct losses in the example given.

The outline drawing in Figure 6.2-67 shows a typical exhaust duct and blower system. The recommended minimum ceiling height to properly handle exhaust air as shown is 12 ft. The outline drawing also indicates a typical air intake and prefilter system.

These calculations are for a very simple air system. Most television transmitters have multiple enclosures with large volumes of air required. In these cases, the equipment user is required to combine the heat exhaust air from several enclosures into a complex duct system. When these circumstances occur, the safest practice is for the user to contract for the services of a heating, ventilating, and air conditioning (HVAC) consultant. The cost involved in having the best possible air exhaust and supply system will pay in extended transmitter life and lower service costs.

The intake vent and blower should be sized to provide a slight positive room pressure. Installing a manometer to detect pressure drop across the filters helps determine the replacement interval of prefilters.

If existing space will not permit the construction of the transmitter manufacturer's recommended air system, it may be necessary to modify the design to fit the available space and still properly cool the transmitter.

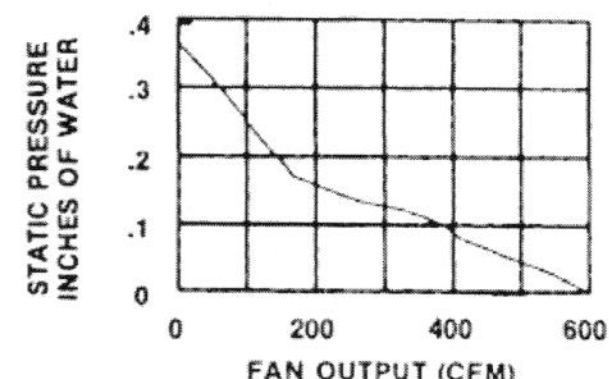

FIGURE 6.2-66 Fan performance curve.

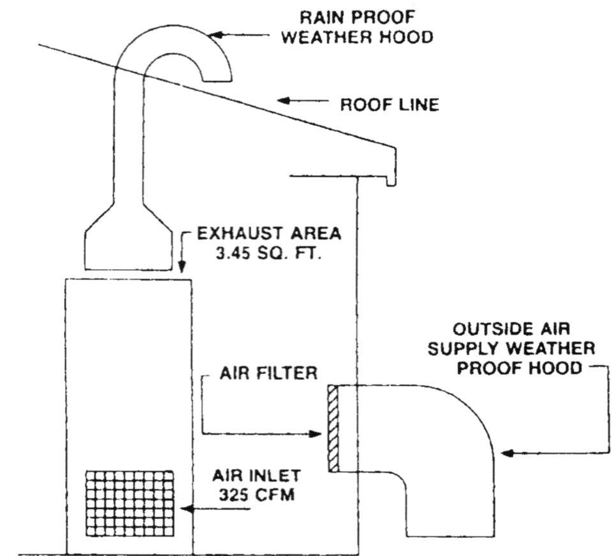

FIGURE 6.2-67 Suggested intake and exhaust duct installations.

Additional flushing air is recommended for the removal of heat from any surrounding equipment that shares space with the transmitter. Input air should be no greater than 5°C above ambient.

Air Conditioning

It is a common practice to set the transmitter into a sealed wall forming a plenum chamber behind the transmitter. The plenum is supplied with outside air and separate air conditioning is provided for the front side to cool personnel and source equipment.

In areas with severely polluted air, it may be necessary to run the transmitter on air-conditioned air to avoid bringing in corrosive salts or gaseous contaminants.

The amount of air conditioning will depend on several factors. The air-conditioning system should be shared by a number of units rather than one large central system so that operation can continue in the event of the failure of one unit. Air-conditioning units are usually specified in tons of cooling capacity with one ton equal to 12,000 btu per hour.

Air systems may consist of ventilation only, 100% air-conditioned spaces, and split rooms with ventilation for power amplifiers, and a smaller air-conditioned area for the operator, low-level drive, and test equipment. Solid-state transmitter air cooling will require a large volume of air at low velocity and pressure. Air cooling of a tube device generally requires a lower volume of air but with much greater velocity.

Liquid systems may use water and glycol mix, high-purity water, or split systems with an intercooler between the pure water side and the outdoor glycol

mix to prevent freezing the cold climates. Some equipment providers use oil to cool IOT and MSDC IOT tubes.

Liquid-cooled systems radiate a small amount of heat into the room, which must also be air conditioned or removed by ventilation.

Klystron/IOT Water/Glycol Cooled

Early klystron tubes with a single collector required a flow of about 30 gallons per minute of water.

There was a time period during which vapor phase cooling for G, H, and S series klystrons took advantage of the latent heat of water as it turned to steam that allowed greatly reduced flow rates of the order of 3 gallons per minute to the collector.

Split vapor phase shell and tube systems allow indoor gravity-fed water to cool the collector and a glycol mix to cool the shell as well as body and magnets. This provided freeze protection and reduced the number of pumps required.

Single-collector IOTs are cooled with glycol-based liquid, or, in warm climates, pure water with a small amount of inhibitors and additives. Typical flow rate is on the order of 18 gallons per minute. Such a system is illustrated in Figure 6.2-68.

The MSDC klystron uses a pure-water forced liquid cooling system with a glycol-based plate heat exchanger.

The ESCIOT and MSCD collectors are similarly cooled with pure water and an intercooler with glycol mix that is cooled by a second pump and outdoor fan cooler.

"Pure water" is considered as distilled, deionized, or demineralized water. Each coolant and equipment provider has detailed specifications defining what is required as suitable water for the application. The typical pure-water specification would include these factors:

- A pH of 7 is desired, with 6–8 generally accepted for grounded collector cooling.
- Resistivity/conductivity in megohms-cm.
- PPM-particulate dissolved oxygen or other contaminant.

The typical glycol mix specification is an alkaline pH of 9–10 and not neutral 7 as is desired for water. The presence of inhibitors contributes to the alkaline pH and it will become lower as the inhibitors are consumed. The coolant must be tested and monitored over time. Fluid replacement in total or renewal of the inhibitors is a typical annual maintenance activity.

IP-BASED CONTROL OF TELEVISION TRANSMITTERS

Television transmitters increasingly include protocols compatible with the Internet in order to control and monitor them remotely. Remote connection to a transmitter is through serial or parallel interfaces. The most popular interfaces are parallel but serial interfaces are more powerful and continue to be embellished. The serial interfaces are usually either RS232 or Ethernet. When RS232 is used, an extra device is often used, like Harris' eCDi™, to convert the information to Internet-based protocols (IP). IP devices are the most extensive and some generate graphical user interfaces (GUIs) on remote PCs via web browsers, such as Microsoft's Internet Explorer or Mozilla's Firefox, as shown in Figure 6.2-69.

Parallel interfaces contain limited information because there is only a one-to-one correspondence between parallel wires and signals and no multiplexing is employed to reduce the number of conductors. These connections are compatible with standard interface levels like TTL but the amount of control is modest. Commands through parallel interfaces are often limited to on/off, raise/lower power level, and activate exciter A/exciter B.

TCP/IP Protocol Suite

Internet-based services are built on the TCP/IP protocol suite. Transmission control protocol (TCP) and user datagram protocol (UDP) are both in this suite. Each of these protocols has 65,536 ports numbered 0 to 65,535. Each port distinguishes a specific service like HTTP and SNMP (discussed later). The set of ports from 0 through 1,023 is so-called "well-known ports."

UDP is known as a connectionless protocol. The connection between a server and a client is not guaranteed

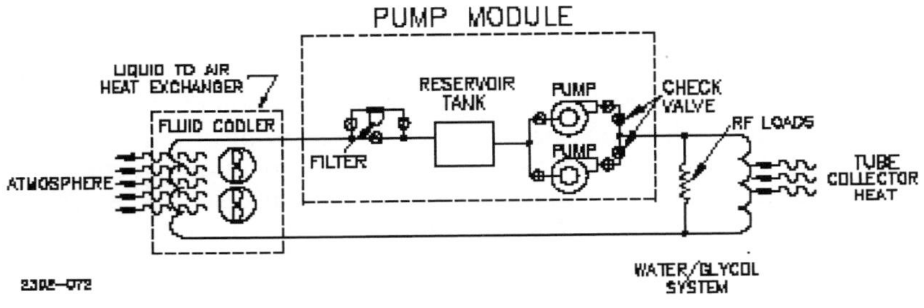

FIGURE 6.2-68 Water/glycol cooling system.

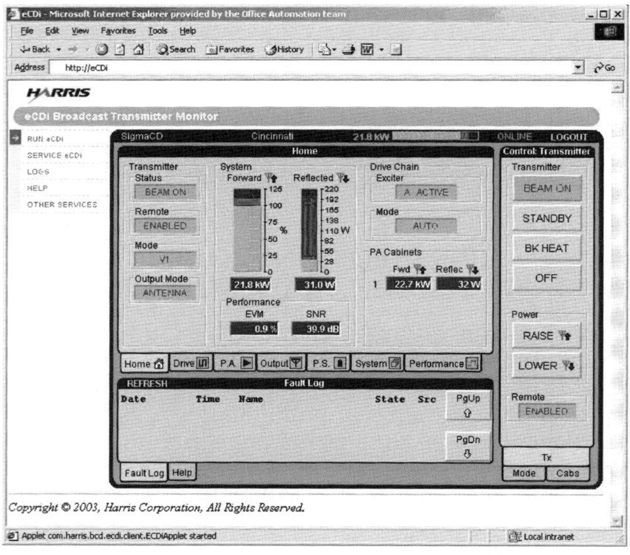

FIGURE 6.2-69 eCDi™ GUI screen.

because there is no handshake. The protocol presumes one-way communication. A server knows who to send a message to but its receipt is not acknowledged. UDP is, therefore, considered an unreliable protocol but it is fast and efficient when the network link is reliable.

TCP is a connection-based protocol. Computers on each end of the communication link know about the others. As long as the physical link exists, TCP will assure that messages are sent and received reliably. It does this by acknowledging and requesting information multiple times if necessary.

HTTP and HTML

Hypertext transfer protocol (HTTP) is port number 80. This protocol is the backbone of Internet-based web pages. Hypertext Markup Language (HTML) is a means of describing the layout and content of web pages and is the content of the files served to web browsers via HTTP. Web browsers decode HTML that is sent in these files and generates the proper display.

HTTP and HTML work well for static graphical and textual images, but they do a poor job with dynamic data. HTTP and HTML put the entire processing burden on the server and not the client (the local PC). Screen images can be generated dynamically by the server. Server-side programming languages like PERL, Active Server Pages (ASP), and PHP are popular. All things being equal, server-generated dynamic pages limit the number of connections a server can maintain without noticeable and undesirable delay. It is entirely possible to overload a connection between a server and its clients to the point that no time is left to respond to the user activating the control buttons.

A more powerful mechanism exists using Java Applets. Java Applets can be embedded within HTML and executed client-side, thereby relieving the server-side processing burden. This approach is beneficial to television transmitter systems since their embedded control systems are generally limited in order to be as cost effective as possible.

SNMP

Simple network management protocol (SNMP) is a protocol served on ports 161 and 162. SNMP was created to manage routers on a large network but it has been applied to a multitude of other equipment including television transmitters. Data in the transmitter is identified with an object identifier (OID) that is totally unique—no two OIDs are the same. Extensive information is available with SNMP, but it is terse and using it is tedious. For example, floating point values cannot be sent. Scalar information must therefore be sent as an integer and scaled before it is used. A television transmitter power of kilowatts might be sent in deciwatts or centiwatts in order to properly report power accurately over its full dynamic range. A management information base (MIB) describes the OIDs. This file is both machine and human readable. Scaling factors are found here among other things.

To make use of SNMP, an SNMP manager is required, such as Hewlett-Packard's OpenView™, running on the remote computer whereas the transmitter or a device attached to it must be an SNMP agent.

An issue using any of these protocols is the process of setting up the gateway between the Internet and your equipment. Access through all the necessary ports must be explicitly given. However, the more ports that are open, the more vulnerable a network becomes to hackers and viruses.

PERFORMANCE MEASUREMENTS

The quality of the broadcast television transmitted signal is the responsibility of the broadcaster. IF modulation, solid-state SAW vestigial filters, and sophisticated precorrection circuits make transparent performance possible. Precision demodulators with synchronous detection and SAW filters are capable of near ideal detection. Test waveforms and digital signal synthesis provide accurate test signal generation to complement transmission facilities.

The Electronic Industries Association Standard RS-508 takes into account empirical quality factors and reflects a common denominator for new transmitter performance. This standard is a valuable reference document that describes performance parameters, standards, and methods of measurements.

While a proof-of-performance at the time of installation is an invaluable record of normal operating performance, it also serves as verification of proper signal quality and emission standards. The number of detailed measurements required after the proof usually can be limited since several performance characteristics are a measure of the same impairment. Also, some test waveforms are more useful for transmitter

adjustment than others. Transmitter measurements can be considered in two primary categories: frequency response and linearity.

In the time domain, 2T-modulated sine-squared pulses and multiburst waveforms are used in identifying and correcting linear distortions. In the frequency domain, swept amplitude and group delay measurements can be used, but are normally reserved for out-of-service testing.

The multipulse signal shown in Figure 6.2-70 may also be used to analyze group delay. The multipulse signal consists of a gray flag (80 IRE), 2T pulse, and five sine-squared pulses (one 25T pulse and four 12.5T pulses) modulated with five discrete frequencies.

The modulated pulses contain low-frequency and high-frequency information, as illustrated in Figure 6.2-71. If the low-frequency spectra and the high-frequency spectra are delayed equally, the results will be a symmetrical modulated pulse with the same shape as the input. The gain versus frequency distortion will alter the baseline flatness but will not change pulse symmetry. Delay errors will result in an asymmetrical pulse baseline. Combinations of group delay and gain errors are shown in Figure 6.2-72.

A method of quantifying these distortions uses the waveform graticule shown in Figure 6.2-73. This graticule was arrived at empirically and represents constant perceptible distortion levels. This shows that overshoots closer to the desired pulse are not as perceptible as ringing further away. The desired K-factor graticule (2%) is overlayed on a waveform monitor and the group delay corrector is adjusted until the 2T waveform lies entirely within the graticule. This technique is often preferred to swept group delay measurements because the results are in terms of perceptibility.

To ascertain the carrier frequency of a station operating under precision frequency control, measurement equipment should meet the following accuracy requirements:

VHF low band	$+2.5 \times 10^{-9}$
VHF high band	$+1.0 \times 10^{-9}$
UHF	$+3.0 \times 10^{-10}$

Frequency counters with the above accuracy and calibration traceable to the National Institute for Standards and Technology (NIST) may be used for making these measurements. A reference 10 MHz signal from a suitable GPS receiver may also be used to provide the reference signal source for modern counters and signal analyzers.

To test for nonlinear distortions as a function of average picture level (APL), one line of the video waveform is alternated with four lines containing a static luminance level. Figure 6.2-74 contains a modulated staircase with three average picture levels.

Modern transmitters are designed for unattended operation for extended periods of time. A properly adjusted transmitter operating with adequate cooling and regulated power lines may need only be checked in detail every 3 or 4 months or whenever a major component is replaced or repaired. Weekly routine

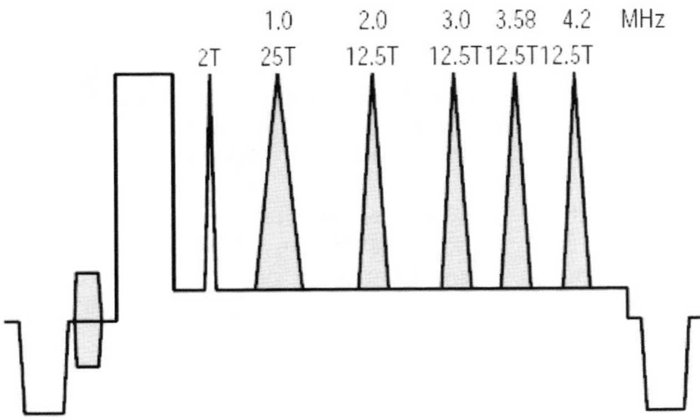

FIGURE 6.2-70 Multipulse test waveform.

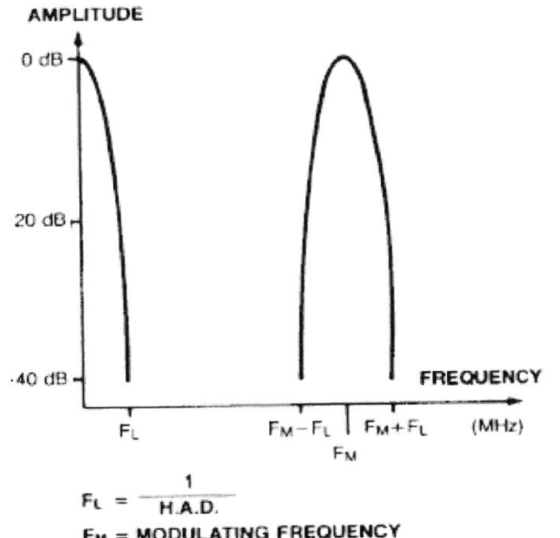

FIGURE 6.2-71 Energy spectrum for modulated pulses.

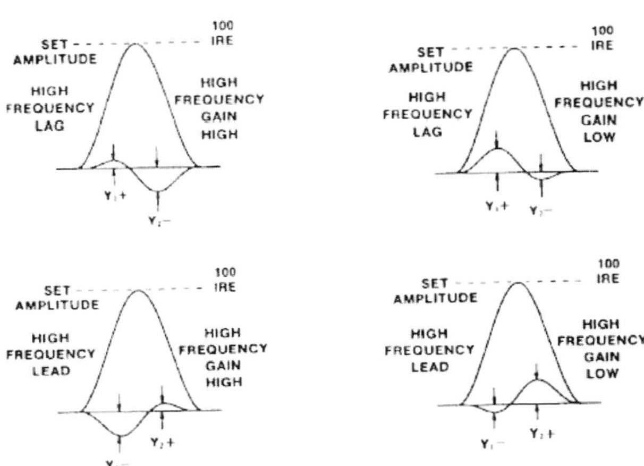

FIGURE 6.2-72 Modulated sine-squared pulses with gain and phase errors.

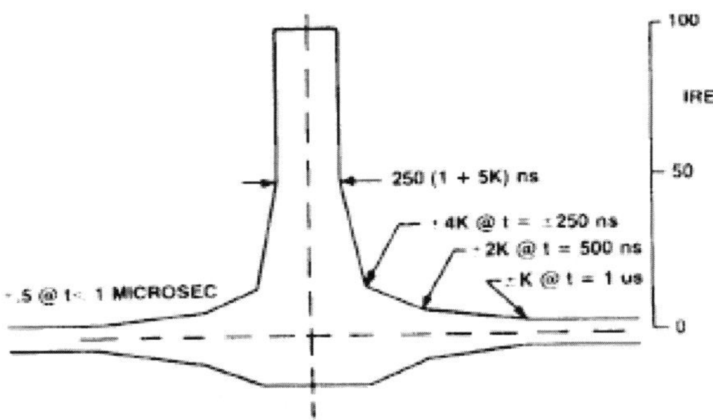

FIGURE 6.2-73 2T sine-squared pulse graticule.

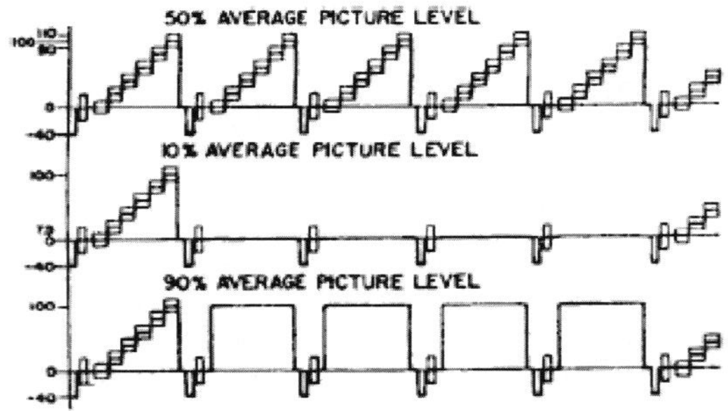

FIGURE 6.2-74 Modulated staircase waveform.

inspections to discover performance trends will help avoid more serious problems that could lead to downtime.

The following pretest checklist and sequence of tests are presented as a general guide for a properly adjusted transmitter. Many adjustments are interrelated and require returning to previous tests to verify proper performance. The test sequence is intended to minimize the number of adjustments.

Pretest Checks

- Check online test equipment.
- Check input video signal parameters.
- Check output transmission line, station load, and antenna VSWR, and for hot spots.
- Check AC mains input voltage, phase, and symmetry.
- Record operational meter readings, adjustment settings, and key performance parameters.

Transmitter Test Sequence

- Exciter linear and nonlinear performance.
- Intermediate and driver linear amplifier performance.
- Swept frequency response.
- Modulation depth (includes UHF pulser and sync ICPM adjustments).
- Power output meter calibration.
- Nonlinearity response (differential gain, ICPM, differential phase.)
- Linear response checks (group delay, pulse tests, composite waveform K-factors.)
- Record meter readings, adjustment settings, and key performance parameters.

During periodic measurement and adjustment, it is a good practice to record meter readings and adjustment settings before and after the test. These recordings will indicate normal setting range and can aid in getting the transmitter to near normal in the event of mistuning errors during adjustments.

In order to make satisfactory measurements of video signals, adequate test equipment is required. Measuring the output of a television transmitter at baseband requires a precision television demodulator. For greatest accuracy, this demodulator should provide a zero carrier reference pulse for determining percentage of modulation and have both synchronous and envelope detection modes.

For most baseband measurements, a video waveform monitor is all that is needed. The waveform monitor should include a provision for making measurements on vertical interval test signals (VITS) and provide filtering to allow separate examination of the luminance and chrominance components of color signals.

In addition to the waveform monitor, a vectorscope must be used for making differential gain and phase measurements. If a greater level of accuracy is desired in timing measurements, a conventional oscilloscope with an associated digital counter/timer is a great asset. For off-air monitoring, a picture monitor is convenient for quickly identifying major video degradation.

Maintenance logs should include date, time duration of outages, corrective action taken, and if possible, cause of failures.

Monitoring TV Multichannel Sound

Maintaining a high-quality aural signal requires high-quality monitoring facilities. To be able to perform the proof-of-performance measurements is another important reason to have a monitor that demodulates the RF signal and separates the components in the composite MTS signal for analysis.

An ideal modulation monitor should:

- Demodulate the composite signal from the aural carrier.
- Separate the components in the composite MTS signal for measurements.
- Provide in-service monitoring capability.
- Be suitable for proof-of-performance measurements.
- Contain a precision (BTSC) expander.

PREVENTATIVE MAINTENANCE

A good preventative maintenance program includes periodic inspection and cleaning of the equipment. A vacuum cleaner is preferred to remove dust instead of compressed air, which will blow the dirt onto something else. A paintbrush can be used to dislodge dust from delicate circuit boards. Avoid using a nylon-bristled brush with a plastic handle as the static charge may damage CMOS or other static-sensitive components. A natural bristle brush with a wooden handle and metal binding is recommended.

High-voltage wires and insulators must be cleaned with denatured alcohol, or another cleaner capable of removing the dirt without leaving any residue. Meter cases are cleaned with glass wax or other nonstatic cleaner.

Air filters should be replaced or cleaned to maintain adequate air flow to the equipment. A second set of washable filters save time when using a single transmitter in critical service by quickly switching the clean filters and washing the dirty ones later. Blowers should be inspected to see if the curved fins are filled with debris that would reduce air flow. Motor windings may collect a layer of dirt and interfere with the cooling of the motor itself. The fins of high-power tubes must be cleaned of debris that may have passed through the air filters. Bearings should be lubricated and checked for excessive noise.

Color change in silver-plated cavity parts is a sign of overheating and may require the disassembly of the cavity to check for obstructed air passages of loose connections. Set screws in gear and chain drive-tuning mechanisms should be checked for tightness. Black silver-oxide is a good conductor and need not be removed. Small parts should be dipped in Tarnix® for cleaning. Be sure to flush the parts after cleaning to remove any residue. Scotch-Brite® is a good nonmetallic cleaning pad for silver-plated parts. Do not remove the plating when cleaning these parts.

High-current wires may move during turn-on surges and can suffer abrasions that may cause an arc if not properly dressed away from sharp edges. Wiring on terminal boards may loosen through thermal cycling and vibration. All connections must be checked to be sure they remain tight. If wires need to be replaced, the correct gauge, voltage rating, and temperature rating must be considered when selecting the replacement.

Edge connectors on printed circuit boards should be cleaned with Cramolin® or other cleaner. A small amount is applied to the connector and then removed with a lint-free cloth. Do not use pencil erasers as this will remove gold or silver plating from the edge traces, and could degrade the connection or create an intermittent **????**. The sulfur in the eraser causes chemical reaction to the edge connector material.

Backup systems or emergency modes of operation should be checked periodically. Relays should be exercised to keep contacts polished.

Transmitter site cleaning should include a check of the building for such things as leaks in the roof that may cause damage to the transmitter, and the presence of insects or small animals.

Intake blowers with filters capable of creating positive pressure in the room can minimize the need for cleaning.

Careful records must be maintained in order to establish a good history for future reference. Such a log will include a description of what was done, when it was done, and the name of the person who performed the work. Seasonal events such as harvesting in farm locations, severe weather, and construction projects in and around the transmitter building can require special action, but usually a pattern will emerge that allows the maintenance to be scheduled on a regular basis.

A complete set of meter readings taken when the equipment is working properly and updated weekly or monthly can greatly assist problem diagnosis.

Create a maintenance program with weekly, monthly, quarterly, semi-annual, and annual tasks evenly spread throughout the year.

Maintenance Items

The following list of items can be used to develop a maintenance routine:

- Prefilter and postfilter manometer readings.
- Inspect/replace prefilters.
- Cabinet input air manometer reading.
- Inspect/replace transmitter air filters.
- Vacuum cabinets.
- Clear tube fins.
- Measure blower currents.
- Clean fan blades and motor windings.
- Check connections for tightness.
- Inspect MOVs (transient protectors).

Recommended Data to be Recorded

- All parameters on meters or user displays.
- DC input power and calculate dissipation.
- Transmitter currents in black picture and at idle (no RF drive).

Visual Performance Checks

- Ensure proper video and sync levels.
- Optimize differential gain/phase.
- Optimize ICPM.
- Optimize group delay using T pulses.
- Assure proper power calibration.
- Optimize swept response.

Aural Performance Checks

- Check for proper audio processor setup.

- Ensure proper modulation levels.
- Establish proper SCA input levels.
- Assure proper aural power calibration.
- Optimize audio-frequency response.
- Check for and minimize distortion.
- Optimize stereo separation.
- Check and minimize crosstalk between MTS channels.

Control System Checks

- Check for proper operation of all interlock circuits.
- Verify proper operation of all overload circuits.
- Confirm proper operation of all control processes (VSWR foldback, filament timing, coax switches, etc.).

Key Terms

Baseband: The video, audio, or digital signal supplied from program source equipment. For analog TV in the United States, these signals are defined by the NTSC standard. For digital TV in the United States, the input signal is defined by the Advanced TV Standards Committee Standard A/53; see www.ATSC.org.

Common amplification: Use of a power amplifier with sufficient bandwidth and linearity in which the visual and aural signals are combined at low level and amplified together.

Differential gain: Nonlinear chroma gain as a function of luminance level. A change in the color saturation results.

Differential phase: Nonlinear chroma phase as a function of luminance level. A change in hue results.

Efficiency: The ratio of RF output power to input power, expressed as percent. For TV transmitters, special conditions apply (see section on TV transmitter efficiency).

Equalization: A technique introduced to compensate for linear distortions. The objective is to provide a complementary transfer function that, when multiplied with the frequency response function, will minimize linear distortion. Equalization may be introduced in the baseband, IF, or RF sections of the system and may be manually or adaptively adjusted.

Group or envelope delay: Nonuniform delay of different frequencies over the signal bandwidth; the first derivative of phase with respect to frequency. It is caused by nonlinear phase as a function of frequency inherent in RF amplifiers, filters, combiners, and other output devices. In general, the closer the amplitude roll-off is to the passband, the higher the group delay distortion.

Linearity: Refers to the degree with which the transmitter output voltage is directly proportional to the input. Common terms used to quantify the degree of transmitter nonlinearity include low-frequency or *luminance nonlinearity, differential gain*, and *AM to AM conversion*. Output phase may also be a function of input level. The deviation from ideal linear phase is often quantified as incidental carrier phase modulation (ICPM), differential phase, and AM to PM conversion.

Linear distortions: Distortions to the transmitted signal that are not level dependent. They are introduced by linear components in the transmission path. These components include any device with a nonconstant frequency response, such as matching networks, cavities, filters, diplexers, and other tuned circuits. Variations in amplitude and phase are included.

Low-frequency or luminance nonlinearity: The change in luminance gain as a function of brightness level.

Microphonics: The susceptibility of a circuit or system to mechanically induced phase and frequency shifts.

Nonlinear distortions: Distortions to the transmitted signal introduced by nonlinear components in the transmission path. These components include any device whose output voltage is not directly proportional to input, such as power amplifiers operating near compression.

Precorrection: A technique or circuit introduced to compensate for nonlinear distortion. The objective is to provide a complementary transfer function that, when multiplied with the nonlinear transfer function, will minimize nonlinear distortion. Precorrection may be introduced in the baseband, IF, or RF sections of the system and may be manually or adaptively adjusted.

Reliability: The degree to which equipment will operate without failure or off-air time. Quantitatively, reliability is often expressed as the *mean time between failures* (MTBF); alternatively, availability is a useful measure.

Surface acoustic wave (SAW): The propagation of waves on the elastic surface of a piezoelectric crystal. The wave propagation is roughly the speed of sound, and therefore called acoustic.

Bibliography

ANSI/IEEE C62.41 Standard (sometimes referred to as the IEEE-587 Standard).

Electrical Performance Standards for Television Broadcast Transmitters, Electronics Industries Association, RS-508.

Electronic Industries Association. BTSC System Recommended Practices, EIA Systems Bulletin No. 5, July 1985.

Engineering Design Manual for Air-handling Systems, United Sheet Metal Bulletin 100-5-178.

Ennes, Harold E. *Television Broadcasting: Equipment, Systems, Operating Fundamentals*. Howard W. Sams & Co., Inc., 1979.

Ennes, Harold E. Television Broadcasting: Systems Maintenance. Howard W. Sams & Co., Inc., 1978.

Gonzolez, Guillermo. *Microwave Transistor Amplifiers*. Englewood Cliffs, NJ: Prentice-Hall, 1984.

Gysel, Ulrich. A New N-Way Power Divider/Combiner Suitable for High Power Applications, IEEE-MTTS-5, International Symposium Digest, 1975.

Lin and Costello. *Error Control Coding: Fundamentals and Applications*. Englewood Cliffs, NJ: Prentice-Hall, 1983.

Mark's Mechanical Engineers Handbook. New York: McGraw-Hill.

Plonka, Robert. Group Delay Corrector for Improved TV Stereo Performance, NAB Broadcast Engineering Conference Proceedings, 1989.

Rhodes, Charles. The 12.5T Modulated Sine-Squared Pulse for NTSC, *IEEE Transactions on Broadcasting*, vol. 18, March 1971.

Weirather, Robert. A Distributed Architecture for a Reliable Solid State VHF Television Transmitter Series, NAB Broadcast Engineering Conference Proceedings, 1989.

Weiss, Merrill S. *Issues in Advanced Television Technology.* Boston: Focal Press, 1996.

Whitaker, Jerry C. *Maintaining Electronic Systems*. Baco Raton, FL: CRC Press, 1991.

ADDITIONAL INFORMATION

The original information on tetrodes for UHF transmitters was supplied by Timothy P. Hulick, Ph.D., Acrodyne Industries, Inc., Blue Bell, PA.

CHAPTER

6.3

Multichannel Television Sound for NTSC

EDMUND A. WILLIAMS
Editor-in-Chief, 10th Edition NAB Engineering Handbook

Updated for the 10th Edition by

ERIC SMALL
Modulation Sciences, Inc.
Somerset, New Jersey

INTRODUCTION

Multichannel television sound (MTS) is a generic term for the process of adding subcarriers to the aural carrier of a National Television System Committee (NTSC) television station. The development of a two-channel stereo MTS system was accomplished through the Broadcast Television Systems Committee (BTSC) of the Consumer Electronics Association (CEA), formerly known as the Electronic Industries Association (EIA). The resulting system is known as the BTSC system for television stereo sound. Subcarriers in the BTSC system are designed to be received by the public and are used for stereophonic sound and, simultaneously, a second audio program (SAP) channel for a second language or other program services intended for reception by the public. A third subcarrier in the BTSC system may be used by the broadcaster for professional applications such as news crew cues (IFB), transmitter telemetry, or other digital/analog or aural/data services. The BTSC stereo system is completely compatible with monophonic receivers.

Non-BTSC subcarriers may be used for a different stereo system or for non-program-related material; however, such signals must not put any energy on or near the 15,734 Hz aural pilot or crosstalk into any of the protected BTSC subcarriers, if they are in use. Modern broadcast television transmitters built after 1984 are designed to accommodate multichannel sound. The introduction of BTSC multichannel sound as an all-industry recommendation allowed receiver manufacturers to develop new models that offer good stereo performance.

By 1997:

- Over 50% of all television receivers sold in the United States had BTSC stereo.
- Nearly 50% of all U.S. homes with televisions had one or more stereo television receivers.
- Nearly all television programs were produced in stereo.
- More than 90% of all television stations in the United States had BTSC transmission capability.
- Nearly all video services available to the public (VCR, cable TV, satellite, and DVD) contained stereo sound tracks, and many of these sources include surround sound material encoded into stereo (two audio channels).

The adoption of the BTSC system accelerated the use of stereo sound tracks for television program production. Handling the stereo sound throughout the broadcast facility initially challenged audio design and production, installation, distribution, operation, and maintenance. Today, a combination of experience and the conversion of most television plants to digital transmission internally makes the correct handling of stereo sound a routine matter.

During the introduction of the new stereo service, many stations employed stereo synthesizers to provide an immediate sensation for viewers with newly acquired stereo television receivers; however, the use of any stereo synthesizer or stereo synthesized material should be strictly avoided. Stereo synthesizers often create substantial spatial errors in surround sound decoders. Acoustic positional errors, many dynamic with program content, are common effects of synthesized stereo audio later processed by consumer surround sound decoders.

BACKGROUND

The CEA (then, EIA) established a laboratory and conducted transmission tests on several multichannel television sound systems. The system selected by the Committee uses the transmission parameters developed by Zenith Electronics Corporation and the noise reduction system developed by dbx® [1] Incorporated. This combination, called the BTSC Multichannel Television Sound system, was recommended to the FCC as an industry-wide standard and adopted in March 1984 by the FCC [2]. While not mandating the exclusive use of the BTSC system, the FCC provided protection if the BTSC system is employed [3]. Some provisions of the FCC rules include:

- The BTSC aural pilot is protected from any interference.
- FCC Bulletin OET-60 [4] is the document that contains the BTSC specifications described in FCC Rules.
- Stations are permitted to implement any other MTS system of subcarriers desired if no significant energy is transmitted in the aural channel on or near the BTSC pilot frequency [5].

Although the BTSC system specifications are not embodied in the FCC Rules, the Commission provided the next best thing—protection; therefore, transmitter and receiver manufacturers can produce equipment conforming to the BTSC system with full knowledge that a compatible, protected MTS system is recognized by both FCC and industry. The CEA also developed and published "BTSC System: Television Multichannel Sound Recommended Practices" [6], a set of recommended operating practices aimed mainly at transmitter and receiver manufacturers. Television stations may employ aural subcarriers for virtually any purpose as long as they do not interfere with the BTSC pilot or normal monophonic receiver operation. All stations operating or contemplating operating in stereo should obtain a copy of FCC Bulletin OET-60 [4].

It should be noted that the multichannel sound system adopted for use in the United States is unique. Multichannel sound transmission schemes used in other countries have different configurations. The Japanese MTS [7] carries either a stereo sound channel or an additional second language channel but not both simultaneously. Germany adopted a two-carrier sound transmission system providing either stereo or a second language, sometimes referred to as the "German IRT System." The BBC developed an MTS for the U.K. Near Instantaneous Companded Audio Multiplex (NICAM) that is widely used outside the NTSC countries. It is a digitally modulated second sound carrier for stereo sound that is inserted just above the normal monaural FM carrier. Brazil and Argentina, whose television systems are PAL-M and PAL-N, respectively, both employ BTSC for MTS.

Except for Japan, all countries that employ the NTSC system have adopted the BTSC MTS system. In comparison, the U.S. system, with its combination of AM and FM subcarriers, can simultaneously transmit stereo and a separate program or second language and a subcarrier for professional or non-public uses. Further, the U.S. system allows stations to employ other subcarrier arrangements, when not using the BTSC system, to serve specific station requirements.

A noise reduction or companding (compression and expansion) system for both the stereo and separate audio program channels is an integral part of the BTSC system and not simply an add-on feature. A high-performance, very precise compressor is built in on all BTSC encoders. A matching expander is built in on all BTSC decoders. The compressor or encoder is a specified part of the BTSC documents and is described in precise detail in FCC OET-60 [4].

It is important not to confuse the BTSC noise reduction compressor that is used to encode the stereo signal for noise reduction with the audio processing compressor that is used for automatic gain-riding and loudness control. Often, both of these subsystems are included in what is broadly termed the "BTSC Stereo Generator." The BTSC compressor, as part of the BTSC compandor system, employs a precisely defined algorithm with very strict and tight tolerances. In contrast, the audio processing compressor and associated limiter are highly idiosyncratic devices, varying greatly in design philosophy from one manufacturer to another and offering user controls able to greatly alter the characteristics of its operation.

THE BTSC SYSTEM

The BTSC MTS encompasses several major performance objectives and considerations, including:

- Full sound fidelity
- Compatibility with monophonic sound television receivers
- Ease of implementation in transmitters
- Ability of the MTS signals to pass through cable television systems, master antenna systems, and TV translators
- Ruggedness to withstand typical transmission impairments (noise, multipath, interference) in the path between the transmitter and receiver

The BTSC MTS is designed to be compatible with monophonic television receivers and provide high-quality stereo sound transmission. The system requirements and performance capability are as follows:

- The BTSC MTS is compatible with monophonic television receivers. Establishing the correct operating levels in the stereo encoder is essential for proper stereo and monophonic operation. Monophonic performance is also a function of the balance of the original audio program material and the distribution and processing within the broadcast plant.

- The system provides both a stereo and a SAP channel simultaneously. The SAP channel is available on most BTSC-equipped television receivers. The quality of the SAP channel is nearly comparable to that of the monaural channel. The only significant specification difference is that the monaural channel has an upper frequency response of 15 kHz, but the SAP channel is limited to 10 kHz. Psychoacoustically, this is not a large difference, especially because the signal-to-noise ratio and distortion of the SAP channel are comparable to those of the monaural main channel.
- There are provisions for a professional-use (PRO) or station-use subcarrier. The PRO channel is used for a variety of purposes for non-program-related station operations. It cannot be received on consumer television receivers.
- A noise reduction (companding) system is employed for both stereo and SAP channels. Nearly 30 dB subjective reduction in noise is achieved. The decoders used in all consumer BTSC-equipped television receivers are matched to the encoders used in the transmitter. The CEA published a detailed description [6] of the compandor system.
- The aural FM carrier frequency deviation capability of the aural transmitter exciter must be able to accommodate the additional BTSC subcarriers. The peak deviation for monophonic transmission remains at ±25 kHz. With stereo, SAP, and PRO, the main FM carrier peak deviation will increase to ±73 kHz. When BTSC MTS was first introduced, there was concern that transmitter bandwidth issues, particularly in high-level diplexers, would set a limit on MTS performance, especially stereo separation. This never proved to be a significant problem [8].

Excellent stereo audio quality and stereo separation can easily be obtained from modern aural exciters if care is taken in the setup and maintenance of the modulation sensitivity. Table 6.3-1 shows the typical MTS performance available with BTSC. These numbers are representative of professional broadcast encoders. Minimum performance requirements are provided in FCC OET-60 [4]. Performance levels for these parameters depend primarily on the BTSC encoder hardware. Stereo separation depends on the adjustment and maintenance of the aural exciter modulation sensitivity. Incidental carrier phase modulation (ICPM) correction is a function of the visual exciter compensation circuits and transmitter adjustment.

Degradation of any of the performance parameters will occur under the following conditions:

- Loss of stereo separation will occur if the matching of the BTSC encoder reference to absolute modulation is in error by more than +/− 0.05 dB.
- Noise reduction system artifacts such as pumping or changes in level will occur with an error greater than 1 dB in matching the BTSC encoder reference to absolute modulation.
- Buzz in the received audio may be due to excessive ICPM in the transmitter or video over-modulation.

The BTSC System Transmission Format

The transmission format of the BTSC baseband configuration is shown in Figure 6.3-1. The monophonic component (L+R) is transmitted as a standard FM signal as in the past. The stereophonic component is transmitted as a difference, or left minus right (L–R), signal and is based on the current U.S. FM stereo

TABLE 6.3-1
Typical MTS Performance

Channel	Frequency Response	Noise (dB)	Distortion (%)
Main (monophonic)	50 Hz to 15 kHz ± 0.5 dB	> −60	< 0.1
Stereo (L–R, 75 µsec equiv.)	50 Hz to 15 kHz ± 0.5 dB	> −60	< 0.1
Stereo (L–R, dbx)	50 kHz to 14 kHz ± 0.5 dB < −2.2dB @ 15 kHz	Dyn*	< 0.5
SAP channel	50 kHz to 10 kHz ± 2 dB	Dyn*	< 1.0
PRO channel	300 Hz to 3 kHz ± 3 dB	> −40	< 3.0
Other Parameters			
Stereo separation > 30 dB (50 Hz—14 kHz)			
Crosstalk into SAP > −50 dB			
Crosstalk into main > −60 dB (> −40 dB for stereo into main)			
Crosstalk into stereo (L–R) > −60 dB (> −40 dB for main into stereo)			

*The noise varies dynamically with program content and the characteristics of the noise reduction system. Subjectively, the noise will be greater than 60 dB below peak operating levels.

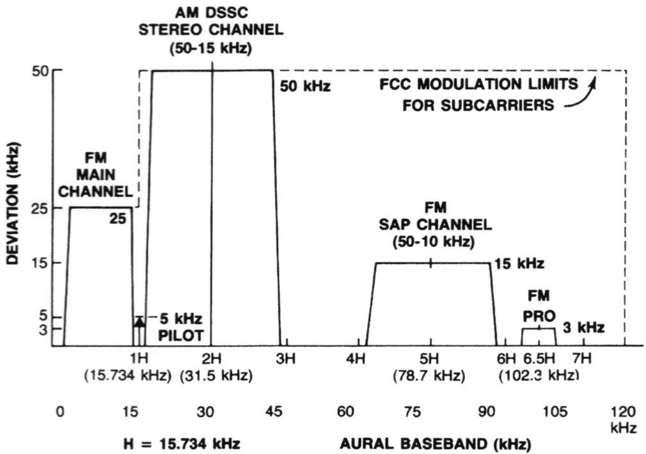

FIGURE 6.3-1 BTSC MTS baseband.

broadcast model, which is familiar to most broadcast engineers. Subcarriers for cues (IFB), data, telemetry, or other station-related purposes may also be employed. Space for professional use, non-program-related subcarriers is provided in the aural baseband.

Figure 6.3-2a shows how the BTSC signal is configured at the transmitter. Figure 6.3-2b shows how BTSC is handled in a typical receiver if the decoding were done by discrete analog circuits. Today, virtually all consumer BTSC decoding is accomplished algorithmically by some type of digital signal processor.

Monophonic Channel

The monophonic, or sum (L+R), channel is maintained at its current place in the aural baseband and continues to have ±25 kHz peak deviation and the standard 75 μsec pre-emphasis. The addition of BTSC or MTS subcarriers, while increasing the total deviation of the main FM carrier to ±73 kHz, does not result in any degradation to the normal monophonic sound in existing receivers. The monophonic channel is maintained at ±25 kHz deviation, unlike FM radio broadcasting, where total deviation is reduced when transmitting the stereo pilot and any SCA channels. This would result in lowering the signal-to-noise ratio of the aural channel and the loudness of the sum channel compared with stations not operating with the BTSC system. Increasing the total deviation provides the needed deviation for the stereo pilot, SAP, and PRO subcarriers to be added to the aural baseband.

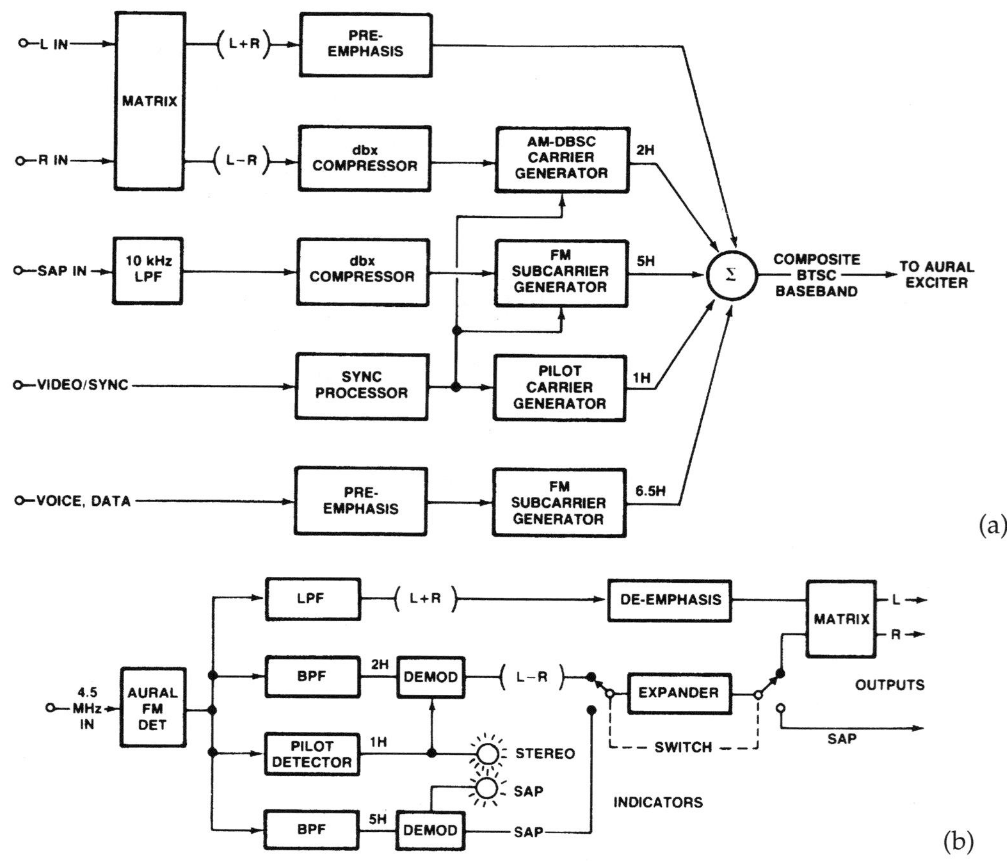

FIGURE 6.3-2 (a) Typical MTS transmitter encoder; (b) typical MTS receiver decoder.

The fidelity of the main channel monophonic audio is not adversely affected by the addition of the BTSC subcarriers. However, because the sidebands of the stereo subcarrier are close to the main channel audio and to prevent crosstalk between channels, BTSC stereo encoding equipment incorporates sophisticated low-pass filters in the audio circuits. These filters limit the frequency response of each audio channel to 15 kHz but are not required to maintain stereo separation above 14 kHz; however, most broadcast-quality BTSC encoders maintain some separation to 15 kHz.

Stereo Subcarrier

The stereophonic subchannel is an amplitude-modulated, double-sideband (DSB), suppressed-carrier subcarrier. The carrier is suppressed to avoid wasting modulation capability. The subcarrier operates at twice the NTSC video horizontal scanning rate of 15.734 kHz, or 31.468 kHz. Station video is fed to the encoder where the horizontal sync is extracted to frequency lock the pilot and stereo subcarrier. The subcarrier is modulated with the stereophonic difference, or left minus right (L–R), signal. Maximum peak deviation of the main carrier (injection) by the stereo subcarrier energy (not the carrier itself, which is suppressed) is ±50 kHz.

The modulation level is controlled by the companding system algorithm and is normally quite high during stereo program material due to the compression. However, because of the nature of stereophonic sound (significant correlation between the sum and difference channels), a form of interleaving takes place that, in effect, prevents the deviation of the main carrier from becoming simply the sum of the two components that would appear to produce a total deviation of ±75 kHz. Instead, the maximum rarely exceeds ±50 kHz. This is because a left-only or right-only signal originating from the studio would produce only a 50% modulated signal in the mono channel, although it produces a high modulation level in the difference channel. It should be noted that this interleaving effect is not as pronounced in BTSC as in FM radio because of the decorrelating effect of the compandor. This interleaving breakdown takes place mainly on high-frequency program material. The potential over-modulation can be prevented by either audio clipping or dynamic high-frequency reduction.

The difference channel of the BTSC system employs the dbx BTSC noise reduction circuit (the compression portion of the compandor) which precisely controls the maximum modulation level of the difference signal. Note that the difference channel is nonlinear by design; that is, the compandor processes the audio signal differently at low levels than at high levels. When conducting certain tests on the difference and SAP channels, a 75 μsec pre-emphasis network is substituted for the compressor in order to make measurements in a linear system. This is called the "equivalent mode" and is discussed in detail in FCC OET-60 [4] and the CEA's Recommended BTSC Practices [6]. Equivalent mode is little used today at the station level. It is mainly a tool for circuit testing the encoder and qualifying transmission hardware following the encoder. Its one application in station maintenance is in measuring signal-to-noise after the point where the signal would otherwise be companded.

Selection of the center frequency of the stereophonic subcarrier is based on its relationship with the 15.734 kHz horizontal frequency (H) component of the video in both the transmitter and the receiver. Crosstalk caused by video signal circuits, power supply, and radiofrequency (RF) paths in the transmitter generally occurs at multiples of H and therefore is canceled or at least substantially reduced in the receiver without significantly affecting general stereo audio quality.

The fundamental of H (15,734 Hz), present to some extent in many transmitter and receiver audio circuits, is above the hearing range of most viewers and not normally audible. Receiver audio circuits generally employ notch or low-pass filters to eliminate all but vestiges of H at the audio output. The second harmonic falls at 2H (31,468.6 Hz), which is the frequency of the BTSC stereo subcarrier. By locking the stereo subcarrier frequency to the video sync, the harmonic and subcarrier remain zero beat and no beat note occurs. It is critical that the video from which the sync is derived for lock to the BTSC signals be the same video signal that is being broadcast.

Modulation of the 2H harmonic by 59.94 Hz field rate sync components of the video signal can also occur in either the transmitter or receiver. The sharp rise time of the 60 Hz component causes the 60 Hz sound to have the characteristic buzz that is heard on some receivers. A related *buzz-beat* effect is reduced with high-pass filters in the stereo subchannel in the receiver to remove unwanted audio components below about 100 Hz that can beat with the 59.94 Hz buzz. While this has virtually no effect on the stereo audio quality, it substantially reduces the low-frequency buzz and hum caused by the field rate modulation.

Placing the stereo subcarrier, with its 15 kHz wide sidebands, so close to the upper end of the main audio channel invites potential crosstalk problems; therefore, it is essential that the subcarrier frequency be an integer multiple of H, as previously described. Placing the stereo subcarrier at higher or odd multiples of H is not desirable due to the much higher noise levels encountered in the upper portion of the aural baseband and the potential for problems with phase-locked loop audio detector circuits in the receiver. A higher subcarrier frequency would also have made the addition of the SAP channel more difficult to implement as well.

The buildup of noise in the aural baseband, shown in Figure 6.3-3, illustrates why the stereo subcarrier is placed as low as possible in the baseband and indicates the necessity for the audio companding system. In frequency-modulation systems, noise increases at a rate of 6 dB for each doubling of the bandwidth. As a result, the noise the stereo subcarrier would add to the received dematrixed sound is about 23 dB. This is reduced about 6 dB because the stereo subcarrier

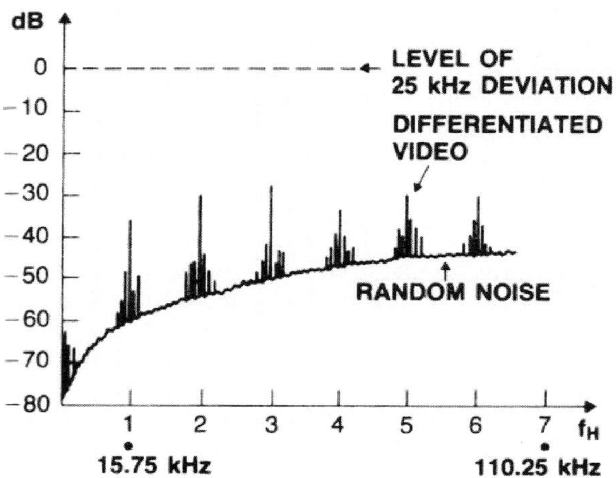

FIGURE 6.3-3 Spectrum of noise on the received aural baseband. (Figure courtesy of J.J. Gibson, RCA.)

deviates the main carrier by ±50 kHz compared to the main channel deviation of ±25 kHz. The resulting degradation in signal-to-noise is only about 18 dB. Further, modulation components caused by the horizontal and vertical sync signals also add to the noise buildup. Both interference sources are overcome by the 30 dB improvement provided by the BTSC noise reduction system.

Using an AM-modulated, double-sideband, suppressed-carrier subcarrier for the stereo channel also provides several other significant advantages over an FM subcarrier. AM offers lower theoretical distortion levels than would be possible for FM given the limited spectrum available (30 kHz) for transmission of the stereo subcarrier. The 15 kHz frequency response of the stereo subchannel also properly complements that of the main channel.

Pilot Carrier

The amplitude modulated stereo subcarrier is transmitted with the carrier suppressed. To provide a reference for the AM detector, an unmodulated pilot signal is transmitted at the TV horizontal line rate of 15.734 kHz, which is one-half the subcarrier frequency. The pilot is used in the receiver to reinstate the carrier on the exact frequency and phase as the original. This is similar to the technique employed in broadcast FM stereo transmission systems, except that in radio a 19 kHz pilot tone is employed. For BTSC stereo, the pilot is locked to the video horizontal sync rate of 15.734 kHz. The pilot modulates the main aural carrier (injection) to a deviation of ±5 kHz. The stereo encoder requires a feed from the companion video signal being broadcast to obtain its frequency reference.

Pilot Carrier Protection

The pilot must be protected from extraneous signals near the pilot frequency in the aural baseband. Principally, this protection is needed to ensure that there is no energy near or above the pilot frequency. Since the stereo encoding process can be seen as a switched, sampled system, it is subject to aliasing distortion if any energy is present at greater than one-half the sampling frequency. The switching frequency is twice H sync. Any energy above the 15,734 Hz pilot will produce highly objectionable aliasing distortion (which sounds like "monkey chatter" crosstalk). Another reason to limit energy around the pilot (below as well as above) is to simplify the circuitry needed to extract the pilot and to improve the reliability of the stereo presence detectors in consumer receivers. FCC OET-60 [4] also requires that a 1 kHz (±500 Hz) window around the pilot be protected to better than 30 dB below the pilot. As can be seen in Figure 6.3-4, a sophisticated filter is required to fully meet the requirements of FCC OET-60.

For stations *not* operating under the BTSC standards, including monaural, the FCC Rules [9] require that modulation components around the BTSC pilot frequency of 15.734 kHz ± 20 Hz be attenuated to a deviation of ±0.125 kHz (46 dB below maximum modulation of ±25 kHz).

For some monaural stations operating with older equipment, problems may arise with acoustic pickup of H sync from studio monitors. Often, aided by the 17 dB pre-emphasis boost in the transmitter, this will exceed the FCC limit on energy around 15,734 Hz. The solution is to install a 15,734 Hz notch filter just before the transmitter. Acoustic pickup of 15,734 Hz sync is rarely a problem for stations operating in BTSC stereo, as all broadcast quality stereo generators include very effective low-pass filters that will remove any incoming H sync.

Operators of monophonic transmitters should also check their facilities for residual horizontal sync signals in either the aural transmitter or in the combining network at the output of the transmitters. The monophonic audio input to the transmitter should incorporate a 15 kHz low-pass filter to maintain the FCC-

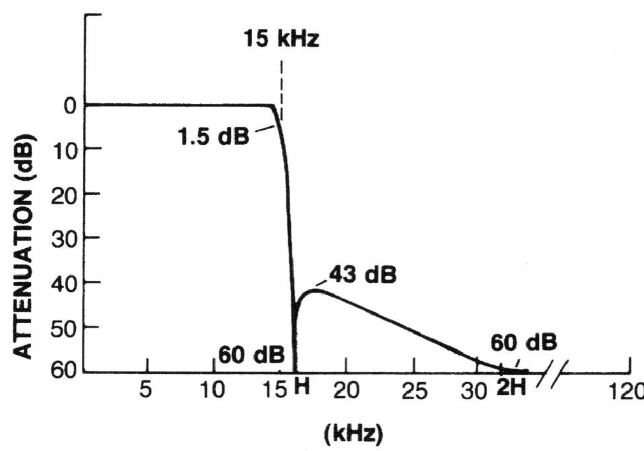

FIGURE 6.3-4 Audio input low-pass filter.

required 46 dB protection of the pilot frequency, thus preventing false triggering of the stereo detector on television receivers.

The Second Audio Program Channel

BTSC MTS transmissions may include a SAP channel in addition to stereo. Although the SAP channel may be program related or not, as the station chooses, it is nevertheless designed for reception by the public. The SAP channel subcarrier, an FM subcarrier, is located at exactly 5H (78.671 kHz) in the aural baseband (see Figure 6.3-1). The same BTSC noise reduction system on this channel provides an excellent signal-to-noise ratio. The audio frequency response is limited to a maximum of 10 kHz. The use of the companding system eliminates the need for separate pre-emphasis in the SAP channel. The SAP subcarrier is frequency modulated by program material to a maximum deviation of ±10 kHz. The subcarrier injection into the main carrier is limited to a peak deviation of ±15 kHz. That is, the subcarrier modulates the main carrier by ±15 kHz. The subcarrier center frequency is locked to 5H by reference to H sync. The SAP channel uses the same dbx noise reduction scheme as the stereo subcarrier. This permits a single expander decoder to be used in the receiver for either stereo or SAP. Most receivers provide either stereo or SAP output but not both at the same time. To make both available simultaneously a receiver would require two expanders.

The Companding System

Companding is a term used frequently in the telecommunication industry to describe what broadcasters refer to as *noise reduction*, a process well known to studio audio technicians. Its use is restricted to systems where there is control over both ends of the circuit. Examples include magnetic analog audio recorders, microwave and satellite circuits, and telephone lines used for program transmission. Simply stated, noise reduction is compressing the dynamic range at the sending end and expanding the signal an equal amount at the receiving end of the circuit. In the process of expansion, the noise is reduced. Until the development of the BTSC MTS system, noise reduction techniques could not successfully be used for over-the-air transmission because a complementary expander is required in the receiver. BTSC MTS required the use of expanders in the receiver from the beginning of the service. As a result, BTSC MTS is the first broadcast service to use companding as an integral part of the transmission system. dbx, Inc., developed the BTSC companding system. Figure 6.3-5 graphically depicts how a typical noise reduction system alters the dynamic range of the program material during transmission according to the content. Dynamic range is reduced from 70 dB to about 40 dB. A matching expander in the receiver restores the dynamic range.

The companding system employed by the BTSC MTS system was specifically designed to operate in the comparatively hostile environments presented by

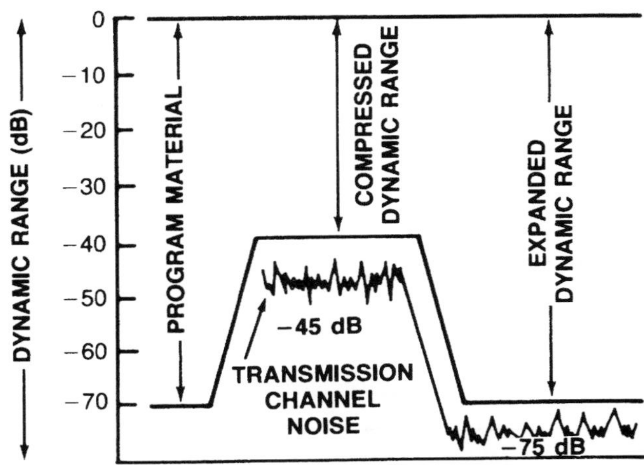

FIGURE 6.3-5 Typical noise reduction action.

television transmitters, signal propagation, retransmission systems (cable TV, MATV, translators), and the receiver itself. Thermal and impulse noise, multipath distortion, and intercarrier buzz generated by the transmitter and receiver all combine to present formidable obstacles for the companding system to overcome.

The companding system used in the BTSC system employs a combination of fixed pre-emphasis, spectral compression, and amplitude compression. The fixed pre-emphasis combines the familiar 75 μsec rising frequency response with a 390 μsec network. The resulting curve is shown in Figure 6.3-6a. By itself, this extremely steep pre-emphasis curve would cause problems with high audio frequencies. To avoid this, a dynamic spectral compressor (variable pre-emphasis) circuit is employed to increase the gain of low-level, high-frequency material and reduce the gain of high-level, high-frequency audio. Only frequencies above about 200 Hz are affected. High-frequency material is increased during transmission by as much as 30 dB at low levels but virtually not at all at high levels. Figure 6.3-6b illustrates the dynamic characteristics of the spectral compressor. A complicated algorithm controls the fixed and variable pre-emphasis to produce the sharply rising frequency response shown in Figure 6.3-6c. The third feature of the companding system is the amplitude compressor, which reduces the dynamic range of the input signal by a factor of 2:1 for low frequencies and 3:1 for high frequencies. In other words, a 40 dB dynamic range audio signal is reduced to 15 to 20 dB or less for transmission as shown in Figure 6.3-7. In addition to dynamic range compression, the maximum input level applied to the stereo subcarrier is reduced to just below the maximum modulation of the subcarrier. This feature provides some headroom to accommodate instantaneous peaks without resorting to clipping. High-level transients are clipped to avoid severe distortion during reconstruction of the compressed signal in the receiver expander.

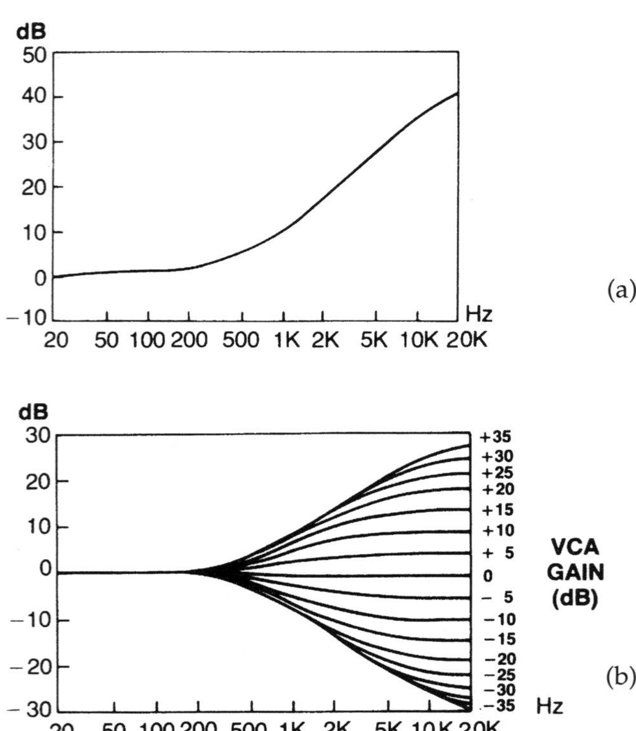

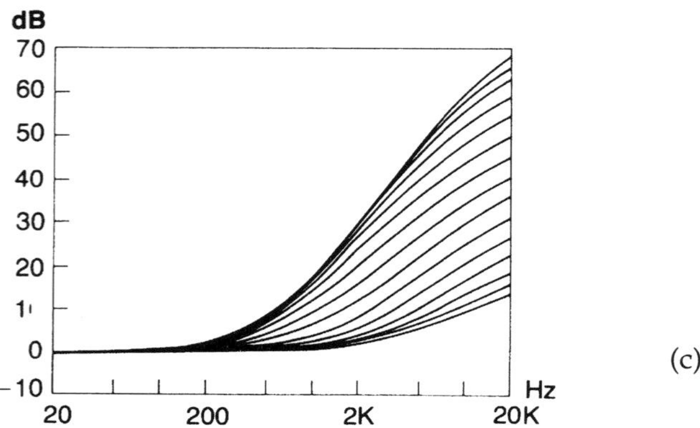

FIGURE 6.3-6 (a) Fixed pre-emphasis frequency response; (b) dynamic characteristics of spectral compressor; (c) frequency response range of spectral compressor and fixed pre-emphasis.

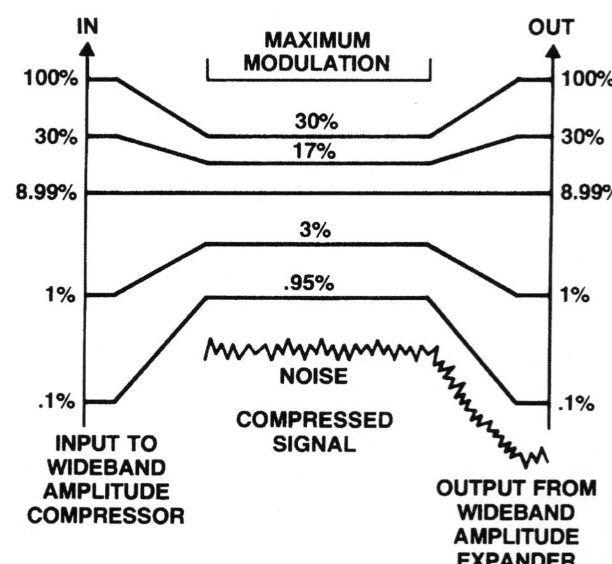

FIGURE 6.3-7 Companding of (L–R) and SAP. (Figure courtesy of dbx, Inc.)

The *equivalent mode* in the BTSC system uses a precision 75 μsec network in place of the nonlinear dbx compressor when tests for separation, crosstalk, and noise are conducted. In this mode, the BTSC encoder contains only linear elements, which allows some adjustments on the transmitter and measurements of the main, stereo, and SAP channels to be performed more accurately and very much as conventional FM broadcast stereo exciters. Overall, this mode is not often used by television stations except as a troubleshooting tool.

It should be noted that the normal BTSC stereo mode includes a 6 dB boost in the L–R channel as compared with FM radio stereo. This 6 dB boost is *not* removed when the BTSC-specified equivalent mode is employed. The significance of this is that the familiar stereo setup patterns, easily observed on an oscilloscope (see Chapter 4.8, "FM Stereo and SCA Systems"), cannot be used to set up or test the linear portion of a BTSC stereo generator. At least one manufacturer of professional broadcast BTSC stereo generators makes provision to remove the 6 dB offset so conventional stereo generator patterns may be employed.

PRO Channel

The PRO channel, which is not designed to be received by the public, can be used by broadcasters for many purposes, such as voice cues to remote news crews, IFB, signaling, and data transmission. The PRO channel is located at 6.5H (102.28 kHz) in the BTSC baseband, as shown in Figure 6.3-1. The carrier need not be locked to H. The peak deviation of the frequency-modulated subcarrier (modulation) is not specified in FCC OET-60 but is typically ±3 to ±5 kHz. The peak deviation (injection) of the main carrier by the PRO subcarrier is specified at ±3 kHz. There has been some confusion between these two parameters. It is important to distinguish between the deviation of the main channel by the PRO channel (injection), which is a static parameter set by FCC OET-60 at ±3 kHz, and the deviation of the PRO subcarrier by audio (modulation), which is a dynamic parameter *not* specified by the FCC. When used for voice or audio transmission, heavy audio processing (compression and limiting) and 300 Hz to 3 kHz band-limiting filters and pre-emphasis are normally employed to provide a usable subjective signal-to-noise ratio. Because of sync buzz issues with intercarrier aural

detectors (to be discussed later in this chapter), intercarrier detection of the PRO channel often seriously limits the utility of this channel. True non-intercarrier (split) detection of the aural signal has proven effective at providing robust, low noise recovery of the PRO channel.

Other Subcarriers

In addition to providing for stereo and SAP subcarriers, the FCC permits television stations to use any other scheme of subcarriers on the aural carrier as long as they do not interfere with the normal operation of monophonic or BTSC television receivers. When not transmitting stereo or SAP, broadcasters may elect to use the aural baseband for subcarriers that are not program related. Virtually any kind of subcarrier for program material or data may be transmitted using AM, FM, or other analog or digital modulation methods. Subcarriers may be placed in the aural baseband between 16 and 120 kHz when not transmitting BTSC stereo but must avoid the BTSC pilot frequency. Although any number of subcarriers may be used, the maximum deviation of the main carrier by these subcarriers (excluding the main or L+R channel) is limited to a total of ±50 kHz. The total modulation of the aural carrier including main channel sound and non-BTSC subcarriers may not exceed ±75 kHz.

Modulation Summary

The FCC permits the BTSC MTS signal to modulate the aural carrier a total peak deviation of ±73 kHz. This is produced by the combined main monophonic and stereo subcarrier deviation of ±50 kHz, the pilot subcarrier deviation of ±5 kHz, the SAP subcarrier deviation of ±15 kHz, and the PRO subcarrier of ±3 kHz for a total of 73 kHz [10]. Table 6.3-2 lists the peak deviations of each of the components of the MTS signal. Experience has shown that when a professional broadcast BTSC stereo generator is aligned to its companion aural exciter with sufficient accuracy to maintain at least 30 dB of stereo separation, the metering on the generator will be sufficient to accurately set the deviation levels shown in the table. Other subcarrier arrangements may modulate the main carrier to an additional peak deviation of ±50 kHz. The combined main monaural channel of ±25 kHz and non-BTSC subcarriers of ±50 kHz must not exceed ±75 kHz deviation [11].

TABLE 6.3-2
Typical Modulation Monitor Functions

Signal	Monitored Parameter (Peak Deviation, in kHz)
Main channel (L+R)	25
Stereo channel (L+R)	50
Main and stereo	50
Pilot carrier injection	5
SAP modulation	10
SAP injection	15
Other subcarrier	3
Total composite	73

TRANSMISSION REQUIREMENTS

Many of the technical rules in Part 73 of the FCC Rules that described quality characteristics of the transmitted signal were deleted by 1984. Those rules, requiring stations to meet certain distortion, frequency response, and signal-to-noise performance levels, were eliminated in favor of marketplace pressures on stations to maintain high quality levels; thus, there are no FCC rules that require minimum audio quality performance levels in the AM, FM, or TV broadcast service. The FCC Rules that permit stations to transmit MTS also do not specify performance objectives; however, rules that serve to control interference have been maintained.

The BTSC MTS performance objectives achievable under normal operating conditions are shown in Table 6.3-1. Although most of the objectives are easily achievable with modern broadcast BTSC equipment, the stereo separation performance is very much in the hands of the user. A small error in the matching of the absolute modulation to the reference point of the BTSC stereo generator, less than a few tenths of a decibel, will effectively render the transmission monaural. This adjustment, modulation sensitivity, is the only truly critical user adjustment necessary to maintain BTSC stereo performance. When adjusted carefully and regularly, it is possible to transmit a stereo signal with 40 dB of separation across most of the audio range.

While it may be argued that so much separation is not necessary, it is desirable to maintain as much separation as practical to provide a margin for degradation due to less than ideal receiver decoders, losses in propagation, and losses in cable systems. The nonlinear circuits in the companding system and the critical adjustments that must be made in the BTSC encoding system may result in less than the desired 40 dB due to tracking errors in the BTSC compandor. The setting of absolute modulation to match the BTSC stereo encoder is the single most important adjustment that determines this tracking, thus setting the maximum stereo separation, as shown in Figure 6.3-8a.

Phase shift between the sum and difference channels after BTSC encoding also reduces stereo separation. Only a few degrees' difference can reduce separation to 30 dB or less, as shown in Figure 6.3-8b. By locating the BTSC generator physically near the aural exciter, there is little opportunity for phase errors of this type to occur. The problem sometimes occurred in the early days of BTSC transmission when aural exciters that were not designed for stereo operation

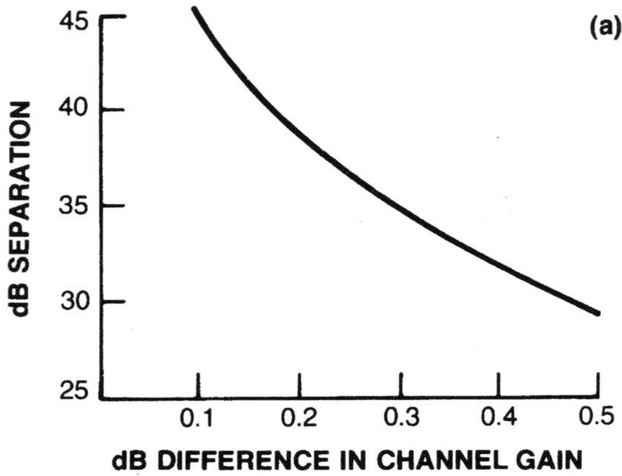

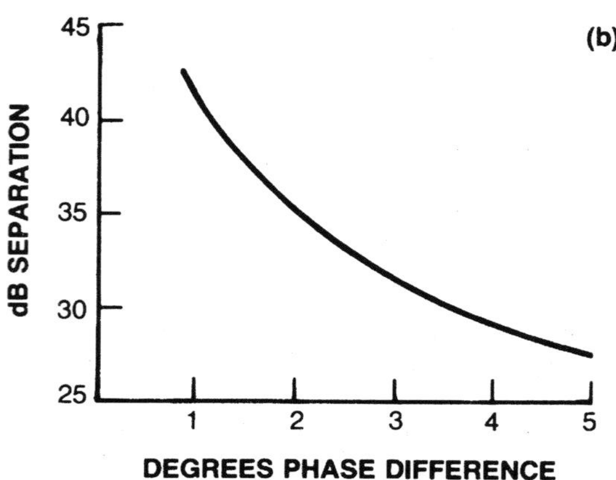

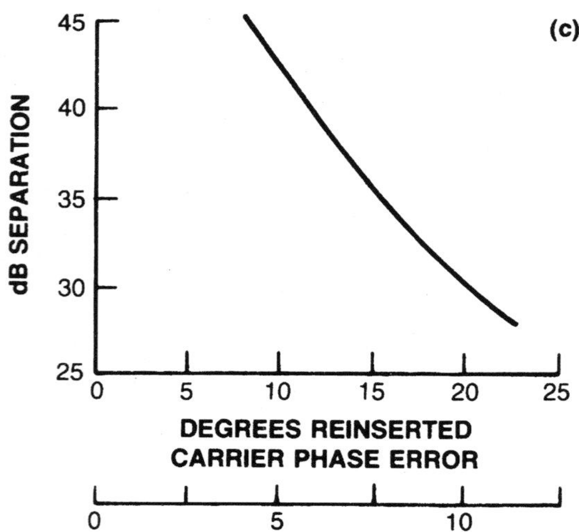

FIGURE 6.3-8 Stereo separation *versus* (a) amplitude difference between main and stereo channels, (b) phase difference between main and stereo channels, and (c) error in phase of pilot or reinserted carrier.

were "converted" to wideband operation, but today the issue rarely arises.

Another way that both phase and amplitude errors are introduced into the transmission system is when the BTSC stereo generator is located remotely, usually at the studio, and the composite signal is sent to the transmitter via a wideband subcarrier on the visual STL. Although this technique has been employed with great success in FM stereo broadcasting for many years, the phase and amplitude accuracy and stability demands of BTSC are at least an order of magnitude greater than linear FM broadcast stereo. In FM radio composite STL service, the stereo composite signal is the direct modulation of the RF carrier. When a visual STL carries BTSC, the stereo signal first modulates an aural subcarrier, usually at a frequency well above the upper limit of the video channel. This adds another modulator/demodulator to the transmission path that must be at least as accurate and stable as the aural exciter. With careful adjustment and maintenance it is possible to meet BTSC performance standards with a composite STL, but too often this mode of transmission becomes the weak link in BTSC stereo performance. Put another way, the use of left and right stereo audio is more robust and easier than using the BTSC stereo composite signal.

It is important to distinguish between amplitude and phase issues before and after encoding. Correct amplitude and phase between the left and right stereo is important but no more critical than in any other stereo audio application. What is unique to BTSC stereo, and demanding, is the balance that must be maintained between the BTSC encoder reference point and the absolute modulation of the transmitter. Absolute modulation is the common reference between the compressor of the BTSC encoder and the expander of the decoder. If this reference is in error by even a few tenths of a decibel, the compandor will not track correctly, resulting in an error between the reconstructed (expanded) L–R signal and the linear L+R signal. As described earlier, this will rapidly degrade the stereo signal; however, and importantly, the signal is only "delicate" between the output of the BTSC encoder and the point that it is RF modulated (typically the varactor diode of the aural exciter).

A major feature of the BTSC MTS system is that there is no stereo noise penalty as is the case with FM stereo radio. Because the FM stereo radio subchannel has no companding it is subject to the increased noise in the channel baseband by as much as 23 dB. In FM radio, the main or L+R channel must also be reduced by 8 to 10% to accommodate the addition of the pilot subcarrier, as well as by 50% of any SCA injection. With BTSC MTS, the aural carrier deviation is increased when adding the pilot and stereo subcarriers; thus, there is no reduction in deviation of the mono signal during BTSC stereo transmission as there is when adding stereo in broadcast FM radio operation.

The stereo separation performance values are the most important technical parameters with which the broadcaster must be concerned for the BTSC system. If the stereo separation performance is good, experience has shown that all the other parameters of the BTSC system will be acceptable.

Setting Modulation Levels

The most important means for maintaining good stereo separation is to correctly set the absolute modulation of the aural exciter to agree with the internal reference of the stereo exciter. Errors of less than 0.2 dB in setting the composite stereo level will result in a reduction of stereo separation to 30 dB, the FCC minimum. For best performance, levels must be set to an accuracy of +/– 0.05 dB to help account for errors in other parts of the system. The only method for setting levels to this accuracy is to use the Bessel null procedure with an accurate, low-distortion audio test signal and an RF spectrum analyzer. Modulation monitors have insufficient accuracy and resolution to align a stereo generator with its aural exciter directly. Tweaking the modulation while watching a modulation monitor, a practice nearly universal in FM radio broadcasting, must be avoided in BTSC. A change of more than a few tenths of a decibel will place the station effectively in monaural mode.

Incidental Carrier Phase Modulation

ICPM in the visual transmitter is also an important characteristic of the transmitter that will affect the quality of multichannel sound. ICPM must be adjusted and maintained to less than 3° during the luminance portion of the video and less than 5° during sync in order to keep buzz to acceptable levels as shown in Table 6.3-3. Excessive ICPM in the transmitter will cause an audible and annoying buzz in intercarrier sound television receivers. The buzz will be modulated by the action of the BTSC expander in the receiver and increase the level of annoyance. A more undesirable effect of excessive ICPM is to reduce the operating margin of the BTSC noise reduction system, thus reducing over-the-air coverage and increasing the likelihood of buzz problems on cable TV systems.

TABLE 6.3-3
Intercarrier Buzz and Buzz-Beat Levels*

Degrees ICPM ->	1°	2°	3°	4°	5°
Baseband buzz level below 25 kHz deviation (worst case) (dB)	–50	–44	–40	–38	–36
Stereo SBR (dB)	56	50	46	44	42
SAP buzz-beat total harmonic distortion (dB)	–49	–43	–39	–37	–35

**Note:* Nyquist slope equivalent ICPM is 2.4° at (L–R) subcarrier, 6° at SAP subcarrier.

Consumer television aural detectors have improved significantly since the simple intercarrier detector of early vacuum tube television receivers. Such detectors offered no protection against ICPM. Since then, aural detectors have grown increasingly resistant to ICPM interference; however, all consumer television receivers employ aural detectors that to some degree depend on the visual carrier to recover aural modulation. This is necessary in order to cancel phase noise from the receiver's local oscillators. The aural and visual carriers beat together in some manner prior to the video detector to produce the 4.5 MHz aural intermediate frequency (IF) signal. Because common mode phase modulation affects both carriers, the phase modulation is factored out of the 4.5 MHz carrier. Therefore, ICPM that occurs on the visual carrier in the transmitter, which is independent of the aural carrier, will produce undesired phase modulation of the 4.5 MHz aural carrier in the receiver, and the result will be the familiar buzz. Reducing sensitivity to phase noise is a significant factor in controlling the cost of TV receivers; however, the same principles that help visual carrier-based detection of the aural be immune to phase noise in the receiver make it susceptible to ICPM.

Intercarrier receivers carry the aural signal to the detector as a subcarrier of the visual carrier. Translators and analog cable television systems carry the aural signal through head-end processors as a subcarrier; therefore, if phase modulation is introduced in these systems, it is common mode and will be rejected by the receiver. Intercarrier-type sound detectors cannot reject phase modulation that is introduced independently into the visual or aural carriers at the transmitter or in subsequent equipment or systems.

Measuring and Correcting ICPM

Digital signal processing (DSP)-based transmission parameter measurement techniques have simplified ICPM adjustment significantly and result in a number on a digital display. It is no longer necessary to use a special waveform monitor graticule to make careful adjustment of the gain of the I and Q channels, or to take subjective readings of a distorted waveform monitor display in an attempt to put a number on the ICPM. To further improve matters, modern television visual exciters include effective circuits for canceling transmitter-generated ICPM; however, most require periodic adjustment to compensate for component drift and changes in transmitter tuning.

Compatibility with Monophonic Receivers

Although stereo television has been in use for some time, a significant number of monaural TV sets remain in use and are still being manufactured and sold. During development of the BTSC MTS, considerable effort was spent evaluating the effect of a fully loaded BTSC signal (stereo, SAP, and PRO) on existing monophonic television receivers. Extensive compatibility testing of a wide range of receiver makes and models, under adverse transmission conditions, revealed that the BTSC system caused no significant degradation to either the sound or picture of the desired or adjacent channel.

Compatibility with Cable Television Systems

After BTSC MTS was introduced, several difficult years followed for broadcasters and their viewers while the cable industry learned how to deal with BTSC and cable equipment manufacturers developed the needed hardware to process and transmit BTSC via cable. Today, the major issues with cable carriage of BTSC material have been resolved through the use of improved analog equipment and large-scale conversion of cable systems to digital transmission. A few issues remain that bear discussion.

Although a few cable TV systems (generally smaller systems) take broadcast signals off-air, most obtain video and audio signals directly from the studio. Techniques include microwave, dedicated fiber, or backhauling on the cable system itself. The signal is in the form of either baseband video, with discrete left and right audio, or video with a 4.5 MHz composite stereo audio carrier. In either case, the actual generation of the BTSC signal takes place in the cable system and is not under the broadcaster's control. Problems arise from two sources: inferior cable BTSC modulators and a lack of proper audio processing.

Inferior cable BTSC modulators nearly always have the same problem: insufficient audio filtering to prevent aliasing distortion, which can be objectionable, and is often mistaken for crosstalk. Although the quality of cable TV type of BTSC generators has improved with the introduction of DSP-based generators, some of the older, inferior equipment remains in use. Audio processing is important in maintaining a consistent-quality audio signal. Insufficient processing will result in wide swings in audio level, loudness complaints, and the risk of driving the stereo encoder into overload distortion; however, most modern, high-quality, cable TV BTSC modulators do not have sufficient audio processing. One solution to both problems is to provide the cable modulator an audio feed that is demodulated from the station's transmitted signal. This signal has passed through a broadcast BTSC-compliant generator, and the audio has been filtered and processed.

Digital transmission of cable television has greatly improved many aspects of cable performance, but it has introduced a new problem for BTSC stereo. The audio is transmitted to the viewers digitally, without any reference to BTSC. If the viewer has a stereo analog TV, the digital stereo audio must be converted to BTSC by a single-chip BTSC integrated circuit generator in the set-top box. Although there is some variability in the performance of various manufacturers' BTSC integrated circuits, the greatest source of variable quality performance is the large differences in the quality of the setup of the various integrated circuits. There is no simple solution for this problem, but changing cable boxes, especially if the cable company has set-top boxes from several manufacturers, may help.

Compatibility with Television Translators

Many television broadcast signals are carried to remote communities by the use of translators that use RF power amplifiers ranging in size from 1 to 1000 watts. Television translators that use heterodyne or direct RF conversion techniques normally do not cause degradation to BTSC MTS signals if the translator is properly maintained. Some translators, however, receive little attention unless they fail completely or viewers report poor performance.

Modulation Monitoring

There is much less need for a traditional FM radio-style modulation monitor in BTSC MTS. The greatest limitation of any modulation monitor is that none is sufficiently accurate to be useful in making the critical match of the BTSC encoder to the absolute modulation sensitivity of the transmitter. For maximum separation, that adjustment must be done to within +/− 0.05 dB. Such accuracy in the direct measurement of deviation is not obtainable in any practical modulation monitor.

Operating Practices

The most important adjustment in maintaining a BTSC stereo transmission facility is matching the BTSC encoder reference level to the absolute modulation of the aural exciter using the Bessel null technique. The reasons for this have already been discussed in the section "Setting Modulation Levels"; however, there are some practical aspects to this adjustment that more than two decades of experience with BTSC stereo have proven must be taken into account.

Without question, failure to routinely check this critical adjustment degrades more BTSC stereo transmission than all other causes combined. This is not because modern broadcast-quality BTSC stereo encoders are unstable; quite the contrary. The issue is the nature of the majority of TV aural exciters. With only a very few exceptions, TV aural exciters employ direct frequency modulation in which a single element controls the frequency of a free-running RF oscillator. An automatic frequency control (AFC) circuit maintains the RF oscillator on precisely the desired frequency. The AFC correction voltage varies slowly, often only over days or even weeks, just to compensate for minor drift in the free-running oscillator. The baseband modulation also changes the frequency of the free-running oscillator but at much faster rate and usually over a wider frequency range than does the AFC. The problem arises when both of these frequency changing signals are applied to the same frequency determining component—a solid-state varactor diode. These

diodes are reverse biased and vary their capacitance with varying voltage.

To maintain the required match between the BTSC stereo encoder reference point and the absolute modulation of the aural signal, the modulation sensitivity of the aural exciter must be equally stable. The problem is that, when the AFC voltage changes to correct the free running oscillator's frequency, it moves the operating point of the varactor diode, thus changing its modulation sensitivity somewhat. It takes a change of only a few tenths of a decibel to significantly degrade the stereo separation of the signal.

In an experiment, a BTSC stereo transmission system consisting of a stereo generator, aural exciter, demodulator, and precision BTSC decoder was adjusted for maximum separation using the Bessel null method. The exciter AFC was turned off and the frequency of the free running oscillator measured. It was 5 kHz below the frequency when the AFC was on. This was a reasonable value and well within the lock range of the AFC. Then, the free running of the oscillator was adjusted, using its capacitor trimmer, to 5 kHz above the desired frequency, for a total readjustment of 10 kHz. This is a very small percentage change for a free-running LC oscillator operating in the 40 MHz range. When the AFC was switched back on, the exciter locked to the desired frequency immediately; however, the measured stereo separation was now less than 10 dB! That is 20 dB less than the FCC standards in OET-60 and was caused by a minor change in the operating point of the varactor diode.

As a practical matter, the free-running frequency of an LC oscillator will not shift by 10 kHz overnight, but such a drift amount is entirely realistic over a few months. This sensitivity points to the need to check the Bessel null alignment of the stereo generator to its exciter every few months. In addition, the AFC voltage, which is usually metered on the aural exciter, should be logged regularly. If any noticeable change is noted, the Bessel null should be checked promptly. Stereo synthesizers, used by some stations to provide a sense of stereo during monophonic programming, have no place in contemporary BTSC MTS broadcasting. Use of stereo synthesizers invites severe spatial distortion for viewers equipped to decode surround sound (see the later section on surround sound).

It is important that the video sync used to sync lock the BTSC encoder be the same video that is being broadcast. A frame synchronizer or a sync-regenerating proc amplifier located between the BTSC stereo generator and the video input of the transmitter can cause serious degradation of the audio signal to all stereo viewers. Even if the processing device does not normally replace the sync in the video, if it has the capability to do so automatically then that feature should be disabled.

A difference in H sync frequency between the sync supplied to the stereo generator and the sync on the video will defeat all the carefully designed features intended to prevent buzz-beats. In one incident in a major market, the BTSC stereo generator was inadvertently locked to a sync generator at the transmitter, while the video was locked to network sync. The difference in frequency was only 0.5 Hz, but it caused the stereo present lights on consumer televisions throughout the market to flash twice a second, as well as the stereo decoders to switch on and off at that same rate. It is left to the reader to imagine how this sounded. …

In another television plant, the signal routing between the studio and transmitter was so multiply redundant and the plant so redundant—with four stereo generators and four aural/visual exciters—that it was impossible to be sure that the stereo generator on the air was sync matched with its video signal. In that plant, until a switcher could be installed and programmed to control the video to the stereo generators, sync for the stereo generators was derived from a demodulator that sampled its RF from the antenna transmission line. ICPM, visual carrier over-modulation, severe group delay, and phase distortion in the visual amplifier will cause buzz in the received audio and loss of stereo separation. These two characteristics are most sensitive to problems with the RF systems.

FCC Rules require all stations to provide protection to the pilot frequency by limiting modulation in the vicinity of the 15,734 Hz pilot (±20 Hz) to no more than 46 dB below ±50 kHz deviation. The BTSC pilot frequency is immediately adjacent to the upper end of the frequency response for the main L+R channel. It is essential that monophonic as well as stereo stations ensure that the BTSC pilot is protected to avoid causing improper operation of stereo receivers and false stereo indications. All broadcast BTSC stereo exciters incorporate low-pass filters with a sharp cutoff above 15 kHz, and monophonic stations should install such a filter if adequate protection is not otherwise provided (see Figure 6.3-4). For a monaural station, measuring the pilot protection can be performed with a low-frequency spectrum analyzer observing the aural baseband using a wideband aural demodulator. RF spectrum measurements of the aural carrier are of little value in determining excessive 15,734 Hz energy.

Many stations prefer to locate the stereo encoding equipment at the studio and transmit a composite MTS signal to the transmitter over a studio-to-transmitter link (STL). The STL in effect becomes another section of the RF path where degradation to the BTSC signal can occur. Stereo separation, a good indicator of overall performance, will suffer if the baseband BTSC signal is degraded. To maintain at least 30 dB separation at the output of the transmitter, the STL should have at least 40 dB separation to compensate for degradation in the transmitter and elsewhere in the RF path. This means that the phase and amplitude frequency response across the 73 kHz baseband of the STL must be maintained to less than 0.1 dB combined with no more than 1° phase error, as shown in Figure 6.3-8. This level of performance, achievable with modern BTSC encoding equipment, can exceed the capability of many STL systems. Placing discrete left and right stereo audio channels on the STL remains popular because most stereo encoders are located at the transmitter. Variations in the performance of discrete audio channels will produce significantly less degradation of the stereo signal than will the same performance STL characteristics on the BTSC composite signal. The BTSC stereo system has not become attractive for use on satellite transmission circuits for the reasons cited above for composite STL systems.

Monitoring the Off-Air BTSC Signal

Off-air monitoring is essential to maintaining good stereo and monophonic sound. Reversed channel polarity (sometimes called *phase reversal*) on one channel may cause the stereo sound to be slightly degraded on stereo receivers, but severe degradation will result with regard to the sound on monophonic receivers. By using a consumer receiver for audio monitoring, buzz caused by over-modulation or ICPM can be detected at the station. Many stations employ both stereo and monophonic monitoring of the transmitted audio to ensure that the stereo imaging and channel polarity are correct. Another technique for monitoring mono and stereo simultaneously is to employ three loudspeakers in the control room. Two speakers carry left and right, respectively, and a third is located in the center to carry left plus right (monaural). See Figure 6.3-9 for one method to implement this approach. If the relative level of the center speaker to the stereo speakers is set carefully and is fixed, then any polarity reversal or loss of correlation affecting the monaural channel will be immediately apparent to the operator as a "hole" in the center of the sound field.

A third method to monitor BTSC stereo is to install a surround sound, two-channel decoder system and speakers in master control. Not only will such a system allow for monitoring stereo and mono at the same time, but it will also allow the operator to note conditions that will cause tracking errors in home surround sound systems. Because surround sound speaker placement is not extremely critical, setting up a surround sound monitoring situation in a typical control room is not difficult. Note that television station master control monitoring is concerned with major errors. More careful placement of speakers in control rooms that deal with programming quality control is the subject of various manufacturers' application notes, which discuss the positional accuracy required for doing original production.

Monitoring discrete 5.1 surround sound as opposed to surround sound recovered from two channels will *not* allow detection of most audio transmission defects. Discrete 5.1 is transmitted as six independent digital audio channels and is therefore very robust. It is also recovered only by digital reception equipment where downward compatibility to mono and stereo is generally not an issue. Multichannel-sound-equipped television receivers have substantially improved audio systems that will reveal audio faults often concealed by most television receivers. As a result, special attention must be devoted to the audio transmission facilities of all television stations, whether transmitting monophonic or multichannel sound.

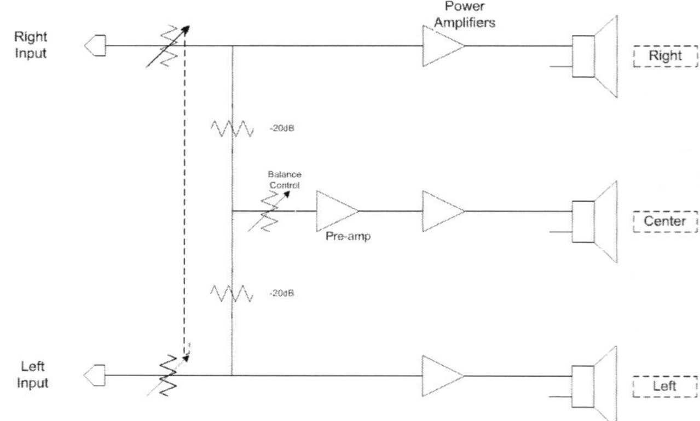

FIGURE 6.3-9 Three-channel monitoring system.

SURROUND SOUND

The introduction of surround sound to television broadcasting is becoming more prevalent. Although surround sound is more closely associated with digital television (DTV; both standard-definition and high-definition), the vast majority of broadcast viewers experiences encoded surround sound via NTSC/BTSC transmission. According to Dolby® Laboratories, Inc., a major licensor of surround sound technology, as of December 2003, there were about 50 million two-channel Dolby® Digital (surround sound) decoders licensed for TV and satellite receivers.

The impact of surround sound can be dramatic. Sound is delivered with a degree of positional resolution previously heard only in electro-acoustically well-designed motion picture theaters. In some cases, the impact of surround sound in a home can be more impressive that the experience in a movie theater because of the closer proximity of the speakers to the listener. For sporting events, the enhancement offered by creatively mixed surround sound often rivals the experience of being there.

Two principle transmission modes are used for surround sound. Most digitally based systems carry discrete 5.1 surround sound as six channels of audio. The 5.1 designation stands for five channels of full-bandwidth audio and one channel (0.1) of frequency-response-limited, augmented bass, sometimes called the *low-frequency enhancement* (LFE) channel. The 5.1 surround sound system is discussed in Chapter 5.18, "Audio for Digital Television." The BTSC system cannot transmit discrete 5.1 surround because it is a two-channel system.

A critical part of the surround sound system is compatibility with two-channel stereo transmission systems. Several manufacturers have developed systems for encoding the six discrete channels of 5.1 surround into two discrete stereo channels. These systems employ different, but largely compatible, phase and amplitude encoding methods. The encoding reduces the six channels of surround sound to two. Literature that discusses these encoding techniques in detail is available elsewhere [12].

At issue is the possible impact of the BTSC stereo broadcasting process on surround sound reproduction. Two-channel stereophonic transmission remains, at least for the time being [13], the dominate television broadcast sound delivery system. Most programs produced in surround sound and heard by viewers with surround sound decoders have been transmitted to the viewer by a two-channel link. Few television stations originate surround sound; most simply pass it along and believe they do no damage to it. The problem is that, when a two-channel stereo signal is passed through a surround sound decoder (a device that takes in stereo and decodes it into the six audio channels of 5.1 surround), any phase or amplitude errors earlier introduced to the two-channel stereo signal will translate into positional (spatial) errors. Errors of this type are readily apparent to the audience.

It is important to note that surround sound decoders are typically designed to remain active in the consumer audio system at all times. They are programmed to determine the type of signal being received (mono, stereo, or surround sound) from the phase and amplitude signature of the two incoming audio channels. Based on that decision, the surround sound decoder routes the audio to the appropriate speakers. Problems arise when errors disturb the expected phase and amplitude relationship of the two channels and create unpleasant sounding spatial effects on all six speakers, as well as on material transmitted as conventional monaural or stereo.

The following is a list of errors common to television stations operating in BTSC stereo and their impact on surround sound reproduction:

1. *Stereo synthesizers*—Stereo synthesizers that switch in automatically when monaural program material is detected can have a variety of adverse effects on surround sound; however, the most common effect of the synthesizer is to take monaural program material that should reproduce from the center channel speaker, directly under or behind the picture, and place it in "full surround" mode. This is a full immersion effect and can be unpleasant.
2. *Automatic tape azimuth correctors*—Although most of these devices have disappeared from television station air chains along with analog audio tape machines, some remain in service. Their impact on surround sound is also unpleasant, as they create the impression of a constantly and randomly rotating sound field.
3. *Automatic phase flippers*—During the introduction of television stereo, polarity reversals (often improperly called "phase reversals") were common. The penalty for true polarity reversal is the cancellation of audio to monaural listeners, which were in the majority when television stereo was first introduced. Devices were also introduced that looked for the anti (negative) correlation condition between left and right channels that might indicate a polarity reversal and then switched the polarity of one of the channels in response. A problem with that approach is that surround sound frequently uses high levels of negative correlation to encode certain effects. It is done in such a way so as not to detract from the quality of the program for anyone listening in monaural (L+R), but it will often trigger automatic polarity error detectors that then reverse the polarity of one audio channel in an attempt to fix a problem that does not exist. When the polarity is reversed, the entire sound field inverts. The program material that was in front, including the critical center (usually dialogue) channel, trades position with ambient or surround material. For example, consider a baseball game where suddenly the announcer and the crack of the bat are behind and around the viewer, while the crowd noise and the echo of the stadium announcer are now directly front and center.
4. *Failure to couple the left and right channels of program audio processors*—Some audio processors, intended for television stereo sound, offer uncoupled operation as user selectable mode. This defect causes the sound field to sway back and forth with the average balance between the left and right channels.
5. *Errors in matching modulation of the transmitter to the BTSC stereo encoder*—The effect on stereo separation of failure to make this match to the required degree of accuracy was discussed earlier, but the loss of stereo separation also affects two-channel encoded surround sound as a proportional loss of the surround sound effect. It is as though the station is transmitting a monaural signal.

SUMMARY

Multichannel television sound is an important feature of a modern television system, especially in the age of the "home theater" where newer expectations demand a multichannel experience. The two-channel stereo sound features supported by NTSC television, using the BTSC standard, and the surround sound features of the new ATSC DTV system, can adequately fulfill that expectation if utilized by the broadcaster to their full potential.

References

[1] dbx is a registered trademark of Harmon International, Sandy, UT.

[2] Second Report and Order of FCC Docket 21323.

[3] 47 CFR 73.682(c).

[4] FCC Bulletin No. 60, Revision A, February 1986, Office of Engineering and Technology, http://www.fcc.gov/Bureaus/Engineering_Technology/Documents/bulletins/oet60/oet60a.pdf.

[5] 47 CFR 73.682(c)(3)

[6] EIA, Multichannel television sound BTSC system recommended practices, in *EIA Television Systems Bulletin No. 5*, Engineering Department, Electronic Industries Association (now CEA), Arlington, VA; available from Global Engineering Documents under a royalty agreement with CEA as document CEA TVSB5 (www.global.ihs.com).

[7] MTS employed for analog television (NTSC) in Japan is based exclusively on FM subcarriers rather than the DSB and FM subcarriers of BTSC. This was done to enhance the multilanguage capability of the system.

[8] See *Is Your Transmitter Stereo-Ready?*, by Eric Small, Modulation Sciences, Inc., Somerset, NJ, www.modsci.com.
[9] 47 CFR 73.682(c)(3).
[10] 47 CFR 73.682(c)(3).
[11] 47 CFR 73.682(c)(9).
[12] http://www.dolby.com/; http://www.dtsonline.com/; http://www.srslabs.com/; http//www.neuralaudio.com.
[13] At the time of this writing, the shut-down of analog NTSC broadcasting is scheduled for February 17, 2009.

CHAPTER

6.4

Digital Television Transmitters

BRETT JENKINS

Thomson Broadcast & Multimedia, Inc.
Southwick, Massachusetts

Appendix material provided by

JOHN FREBERG

Freberg Engineering, Inc.
Homewood, Illinois

INTRODUCTION

Digital television (DTV) transmitters began appearing in widespread commercial use in 1998 shortly after the Federal Communications Commission (FCC) created rules adopting the Advanced Television Systems Committee's standard for digital transmission [1]. The introduction of DTV has required television transmitters to change to accommodate the new standard. At the same time, technology has advanced at a rapid pace over the past decade. Standardized network protocols are available for communication between devices, even when those devices are geographically remote. RF amplification technology continues to advance as new devices and techniques attempt to provide improved performance in cost and efficiency. Broadcast engineers seek to take advantage of these advances as their day-to-day activities become more complicated and the transmitter becomes a relatively smaller part of the overall complexity of a broadcast station. This chapter covers the basics of digital television transmitters:

- Characteristics of the DTV signal
- Types of amplifying devices
- Support systems needed in a DTV transmitter
- Installation considerations
- Signal performance and quality metrics
- Recommended maintenance practices for various transmitter types

This chapter also contains an appendix, available only on the CD version, with a tutorial, "Understanding DTV Transmission," by John D. Freberg (Freberg Communications Corp.; Homewood, IL) which describes the 8VSB transmission system in detail.

THE DTV SIGNAL

The DTV signal in the United States is defined by the Advanced Television Systems Committee (ATSC) in ATSC Document A/53 [2]. Unlike an analog television transmitter, which accepts a composite video signal and a baseband audio signal as its main inputs, a digital television transmitter receives a digital stream of ones and zeroes called an *MPEG-2 Transport Stream*. The specific electrical and mechanical characteristics of this input signal are defined by SMPTE 310M. This interface is the one most commonly used for ATSC transmission. The SMPTE 310M signal is a serial stream of ones and zeroes delivered over a 75 ohm coaxial cable. The data is transmitted at a rate of 19.392658 Mbps, and the clock is transmitted along with the data. Other interfaces, including Digital Video Broadcast (DVB) Asynchronous Serial Interface (ASI), can be used as a modulator input only if the modulator itself is capable of reclocking the data to ensure that the specified data rate is met. The SMPTE 310M standard specifies that the incoming data rate must be maintained to ±2.8 ppm, which translates to about ±54 Hz on the 19.39 Mbps signal [3].

Once the serial stream is received by an ATSC modulator, it must be formatted according to the standard. Digital processes that enable the digital signal to be modulated on a radiofrequency (RF) carrier and received reliably after transmission are shown in the block diagram in Figure 6.4-1. The type of modulation

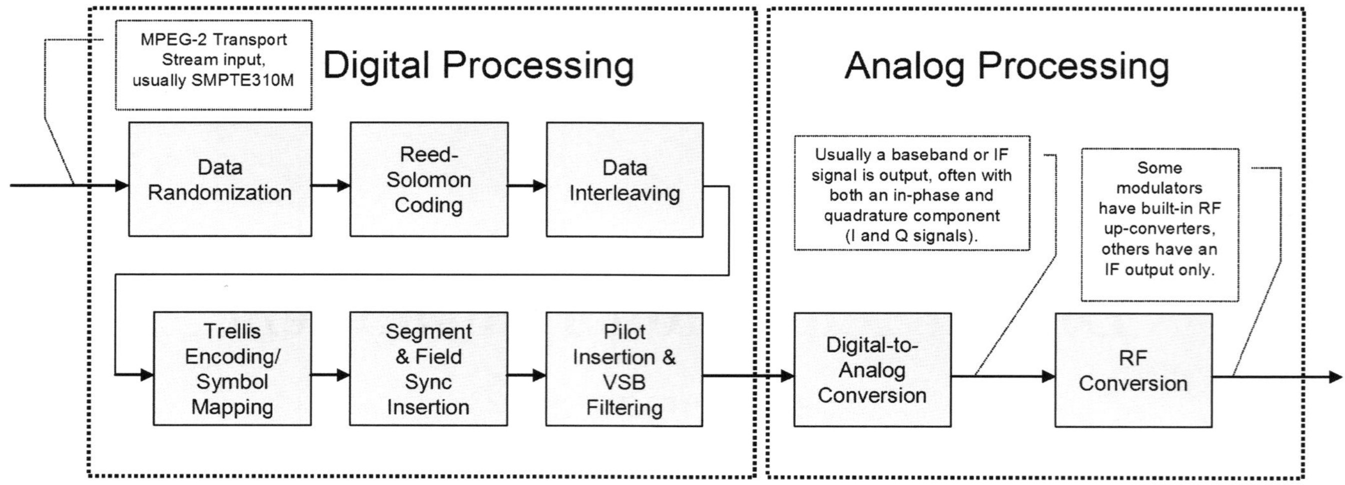

FIGURE 6.4-1 Block diagram of an ATSC modulator.

used in the ATSC system is called 8VSB for reasons that will become apparent in the next few paragraphs.

The first process is a data randomizer. This process takes the input stream and randomizes it according to a polynomial function. If the processing was truly random, there would be no way for a receiver to recover useful information, but the receiver also knows the polynomial function so it is able to remove the randomization by performing a reverse process. The randomizer ensures that there will be a random mix of ones and zeroes at the output regardless of the input signal. Randomization helps give the signal its noise-like characteristic; for the transmitter, this is important because it keeps the average power of the signal constant.

The next process is the insertion of 20 bytes of Reed–Solomon error correction code. For each 187 bytes of payload data, 20 bytes of error correction are generated. Each MPEG-2 packet contains 187 bytes of payload plus 1 sync byte, for a total packet size of 188 bytes. This additional information is used for error detection and correction at the receiver. The codes work by adding some redundant information in such a way that allows the receiver to reconstruct data in a packet that may have become corrupted during transmission [4].

After the additional bytes are added, the data is interleaved, meaning that the order of the bytes is changed in time. This helps protect against burst noise interference, which may degrade a section of data for a short duration. When the signal is de-interleaved at the receiver, these burst errors get distributed over a longer period of time, each with a smaller duration, allowing the error correcting codes to recover the data.

After interleaving, the data is sent to a Trellis encoder and symbol mapper. The digital stream is mapped into one of eight discrete levels, called *symbols*. Each three bits are taken and mapped to one of these levels, and each three bit combination can be given a unique symbol value because there are only eight possible sequences of three bits ($2^3 = 8$). The Trellis encoder adds one bit for every two input bits. The bit is added in such a way as to control the transitions from one symbol to the next. The receiver can then fix errors when it sees that an unlikely symbol transition has occurred. The eight symbol levels are the reason why the name 8VSB modulation contains the number eight.

At the output of the Trellis encoder, the data is grouped into data segments with 828 symbols of payload data plus error correction in each segment. A four-symbol segment sync is added to the data to make a complete data segment of 832 symbols. Data segments are further grouped in data fields. A data field consists of 313 data segments, of which one segment is inserted by the modulator as the data field sync. This field sync contains some useful information about the signal, but most importantly the field sync contains a predefined sequence that can be used by receivers to equalize channel distortions that may have occurred during propagation. The final symbol rate (the rate at which the eight-level symbols are output from the modulator) is given by the equation $F_{sym} = (684/286) * 4.5$ MHz, which gives us 10.762238 MHz rounded to the nearest hertz. This is the symbol rate, 10.762238 Msymbol/sec. This rate is matched exactly to the incoming data rate, such that no over- or underflow conditions exist within the modulator.

The next step in the modulation process is to add a low-level digital pilot to the signal just before filtering and conversion to an IF or RF signal. The filter used is a vestigial sideband filter, which gives the name of the type of modulation: VSB. This filter removes most of the lower sideband of the signal, leaving only a vestige. The filter shape is defined as a root raised cosine filter with an excess bandwidth of 11.5%. This filter shape is common in digital communication systems and is chosen to allow recovery of the symbols without intersymbol interference (ISI). The final RF signal

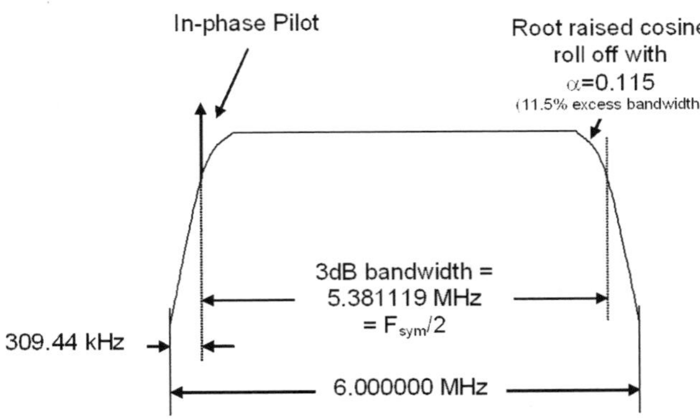

FIGURE 6.4-2 Spectral characteristics of the ATSC signal.

occupies a 6 MHz channel bandwidth with the pilot approximately 309 kHz from the lower band edge. The filtering is performed in the digital domain unlike analog broadcast where a surface acoustic wave (SAW) filter was typically used to shape the transmitted signal.

The final step in an 8VSB modulator consists of converting the digital signal back to analog using D/A converters. The signal is then upconverted to the final RF channel frequency. In some modulators, an IF signal is fed to an external upconverter and exciter assembly for upconversion to RF for final amplification. The ATSC signal occupies a bandwidth of 6 MHz. Figure 6.4-2 shows the general shape of the signal with an exaggerated roll-off at the band edges. The pilot signal is shown approximately 309 kHz from the lower band edge.

Because of the noise-like characteristics of the DTV signal, the appearance of the signal on a spectrum analyzer will vary depending on the resolution bandwidth settings. An example plot of a DTV signal viewed with a narrow resolution bandwidth is shown in Figure 6.4-3. Notice that, even though the pilot power is below the average signal power, it can still be seen at the lower edge of the channel when the resolution bandwidth is narrow enough.

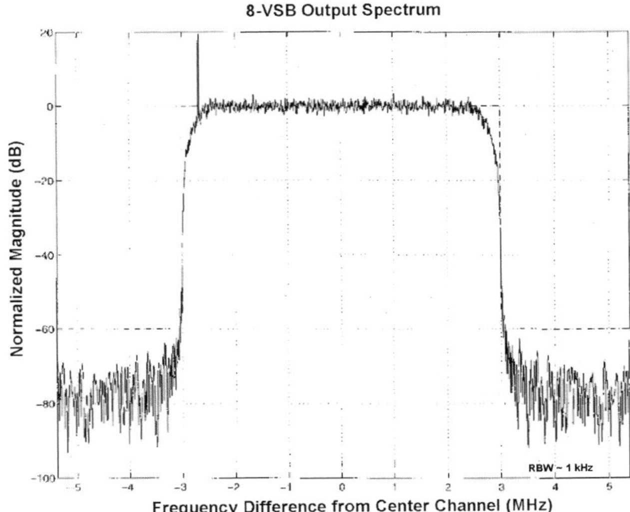

FIGURE 6.4-3 Spectral plot of the ATSC signal.

A major difference between analog and digital transmitters is that there is no need for a separate amplification chain for the aural carrier. The digital signal carries video and audio, as well as any other data required. This simplifies the design of the amplification chain and eliminates the need for any external diplexers in the RF system. For a different perspective on the DTV signal, the reader is encouraged to look at the appendix in the CD version of the handbook.

DTV SIGNAL POWER

A significant effect the digital signal has on a broadcast transmitter is its power characteristic. Recall that an analog TV signal is defined at fixed powers for certain periods of time (*e.g.*, during horizontal or vertical sync) but has variations in power during the video portion of a line. The power of the signal during a line of video depended on the video content. The digital signal is completely different. The power at any given time is randomized and does not depend upon the content of the video. There are no defined time intervals where the power is constant such as during the analog sync pulse; however, the average power of the DTV signal is constant regardless of the input signal. Because the average power is constant and the instantaneous power is random, a way to express these power characteristics of the DTV signal is needed. A cumulative distribution function (CDF) plot is used for this purpose. The CDF shows the probability that the signal is a certain amount above the average power. In other words, it defines the percentage of time that the signal spends above the average power. Figure 6.4-4 shows a cumulative distribution function for the DTV signal [5].

Peak-to-Average Power

From the curve on the figure, the peak power of the signal is 5.3 dB or more above the average power only 1% of the time. Note that the curve intersects the point where 1% and 5.3 dB meet. It is this curve that gives us the description of peak-to-average power. Peak-to-average ratio is defined as the 0.1% point on the cumulative distribution function. For 8VSB, this power is approximately 6.4 dB. Restated, with a peak-to-average ratio of 6.4 dB, the DTV signal is 6.4 dB or more above the average power only 0.1% of the time. Transmitter and amplifier designers often use this value to determine how much peak power capability an amplifier must have to accurately reproduce the input signal without significant distortion. If an amplifier does not have sufficient peak power capability, the result will be high levels of intermodulation products appearing both in-band and out-of-band. This impact

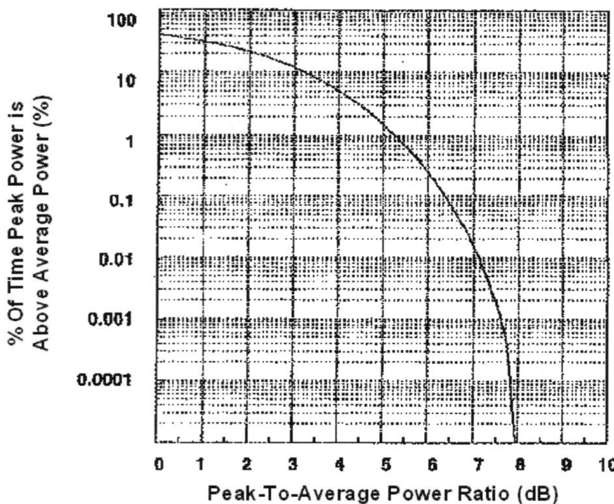

FIGURE 6.4-4 Cumulative distribution function of the DTV signal from ATSC Recommended Practice A/54. (Reproduced with permission of the Advanced Television Systems Committee, Inc.)

of intermodulation distortion will be discussed in detail later in the chapter.

The peak-to-average ratio characteristic of the digital signal has important implications for a digital transmitter. The power characteristics affect the total amount of average power that can be delivered by a transmitter. It is usually the peak power capability of an amplifier that limits the total amount of useful power. Thermal dissipation or electron density will limit the absolute average power of an amplifier, but often this limit cannot be approached because a peak power limit is reached first. The power characteristics also impact the way power measurements are made in a transmitter as well as how the automatic gain control (AGC) or automatic level control (ALC) circuits function. Detectors used in these circuits must be usable over the entire dynamic range of the signal to avoid false readings.

DIGITAL TRANSMITTER AMPLIFIER TYPES

There are several amplifying devices commonly used in digital transmitters. For VHF frequencies, solid-state transmitters utilizing metal-oxide semiconductor field-effect transistor (MOSFET) devices are used almost exclusively. For UHF transmitters, the most common devices are lateral diffused MOS (LDMOS) transistor-based amplifiers and inductive output tube (IOT)-based amplifiers. Klystron amplifiers that were common in UHF analog transmitters are not practically suitable for digital transmission. Klystrons do not have a wide enough dynamic range to pass the high peaks of the digital signal at an operating point that would yield reasonable efficiency. As an example, a klystron amplifier might be able to deliver peak powers of 60 kW. This would correspond to a DTV power of less than 12 to 15 kW after RF system losses. Running at this power, the device would be less than 10% efficient.

The choice in amplifier technology is based on which technology yields the most cost-effective implementation at a given frequency and power level. Additional considerations include the level of redundancy given by a technology choice. Solid-state based transmitters offer a greater level of redundancy than tube-based transmitters because the amplifiers can be designed with relatively low power modules operating in parallel to produce high power. If any single module fails, the transmitter output power is lowered only by the amount being contributed by that module. Each amplifier in a tube transmitter contributes a larger amount of power. Moreover, if the transmitter has only a single tube to achieve its rated power and that tube fails, the transmitter is off the air. Redundancy in a tube transmitter is achieved only when multiple tubes are operated in parallel. Efficiency is another consideration when choosing an amplifying technology. Tube-based amplifiers are more efficient than their solid-state counter parts if the tube is operated near its maximum rated power. A recent advance in IOT design has been introduced that allows an even greater gain in efficiency. The improved design is the multi-stage depressed collector (MSDC) IOT, which is a standard IOT with a modified collector that allows the main electron beam in the tube to be collected at different voltage potentials. A similar concept was used in MSDC klystron transmitters which were used in analog operation; however, the MSDC klystron is still not a good choice for digital operation because of its lack of ability to pass high peaks in the digital signal at a power that would yield reasonable efficiency. MSDC IOT transmitters were first used in broadcast service in 2002 and are becoming a state-of-the-art choice for high power UHF digital transmitters, especially in areas where the cost of electricity is high.

Transmitter efficiencies are described in terms of amplifier (or amplifying device) efficiency and plant efficiency. In DTV service, the amplifier efficiency is measured by taking the output root mean square (RMS) RF power delivered by the amplifier and dividing it by the input power consumed by the amplifier. There are different ways of measuring the input power, so care must be taken when looking at amplifier efficiencies; for example, the efficiency measurement of a solid-state amplifier module might take into account the AC input required or the DC power consumed. Any loss in the AC/DC power supply or rectifiers would not be reflected in the latter case. Plant efficiencies take into account not only the amplifier but also all of the equipment required to run the transmitter such as cooling systems, power supplies, and exciters. The output RF power also takes into account any losses due to combining, filtering, or other RF system components; therefore, plant efficiency of a DTV transmitter is the total RF power out divided by the total AC input power consumed.

In practice, digital VHF transmitters are made exclusively with solid-state MOSFETs. Digital UHF

TABLE 6.4-1
Comparison of Amplifying Technologies Used for DTV Transmitters

	Band			
	VHF	UHF		
Technology available	Solid-state MOSFET	Solid-state LDMOS	IOT	MSDC IOT
Best power ranges (DTV RMS power)	0–20 kW	0 kW–20 kW	10–100 kW	10–100 kW
Plant efficiencies	10–18%	10–18%	22–30%	40–50%
Main cooling system type	Air or liquid	Air or liquid	Liquid (glycol/ water typically)	Liquid (pure water or oil)
Ease of maintenance	Excellent	Excellent	Good	Good

transmitters below 10 to 15 kW are usually made with solid-state LDMOS transistors. Higher power UHF transmitters are usually made with IOT amplifiers. Table 6.4-1 presents some general guidelines on which technologies are suitable for which kinds of applications.

Solid-State Digital Transmitters

Solid-state digital transmitters are usually built with field-effect transistors (FETs). Some older solid-state transmitters used bipolar junction transistors (BJT). Newer designs take advantage of the FET's higher gain and better thermal stability. Solid-state transmitters in the UHF range use LDMOS. In the late 1990s, another transistor technology, called silicon carbide (SiC), showed promise, but the transistors were never commercially realized at UHF frequencies. At the time of this writing, SiC remains an interesting subject of research but likely will not have an impact on television transmitter design for the next several years.

Solid-state digital transmitters are built using amplifier modules operated in parallel. This architecture allows for soft-fail redundancy and increases the availability of the on-air signal. Figure 6.4-5 shows a simplified block diagram of a digital solid-state transmitter. The typical architecture allows for a digital exciter to feed a splitter network which in turns feeds each power amplifier (PA) module. Most designs allow for a standby exciter to be switched in if there is a failure in the main exciter. The outputs of the PA modules are fed into a combining network, which in turn is fed to a bandpass filter, sometimes called a *mask filter*. This filter removes a majority of the out-of-band energy and allows the transmitter to meet the performance criteria established by the FCC emission mask (more on this topic can be found in the Digital Signal Performance Measurements section).

Power Amplifier Module Protection

Protection of the PA modules against high levels of reflected power is critical in the design of a system. The simplest approach to this problem is to include a high power circulator at the output of the transmitter, before the mask filter and other RF components. This protects against any reflected power coming from the RF system. Another method is to include lower power circulators at the output of each PA module. While sometimes more expensive, this method has the added advantage of being able to individually protect each module from failures in both the RF system and the combining system. A third method is to provide a sensing circuit that removes power from the system when a high reflected power condition occurs. This method requires careful design to ensure that it operates fast enough to protect the PAs before damage can occur. All three of these methods are currently in use in today's digital transmitters.

Amplifier Operating Class

The key element in the design of a solid-state transmitter is making the amplifier capable of passing the digital signal in terms of bandwidth and power capability with the lowest distortion and highest efficiency. The final amplification stages in solid-state amplifiers are operated in class AB, allowing a good trade-off between power handling capability and linearity. Operating in class A would allow more linear operation but only over a smaller dynamic range. In class A, the amplifier would not be able to supply the high peak power required for a digital signal.

Power Amplifier Module Design

Figure 6.4-6 shows an example of a solid-state amplifier module. Other individual design architectures will vary by manufacturer. Two low-level amplifier stages are shown driving six final stages in parallel. The transistors are mounted to a large heat sink, making up the bulk of the amplifier. The heat sink fins are on the back, and the module is cooled by forcing air through the fins from bottom to top.

Amplifiers are often designed to operate within particular frequency bands. Most transmitter manufacturers have an amplifier design that covers the low VHF channels (2 to 6) and a different amplifier design to cover high VHF channels (7 to 13). To cover the UHF channels (14 to 69), some manufacturers offer a

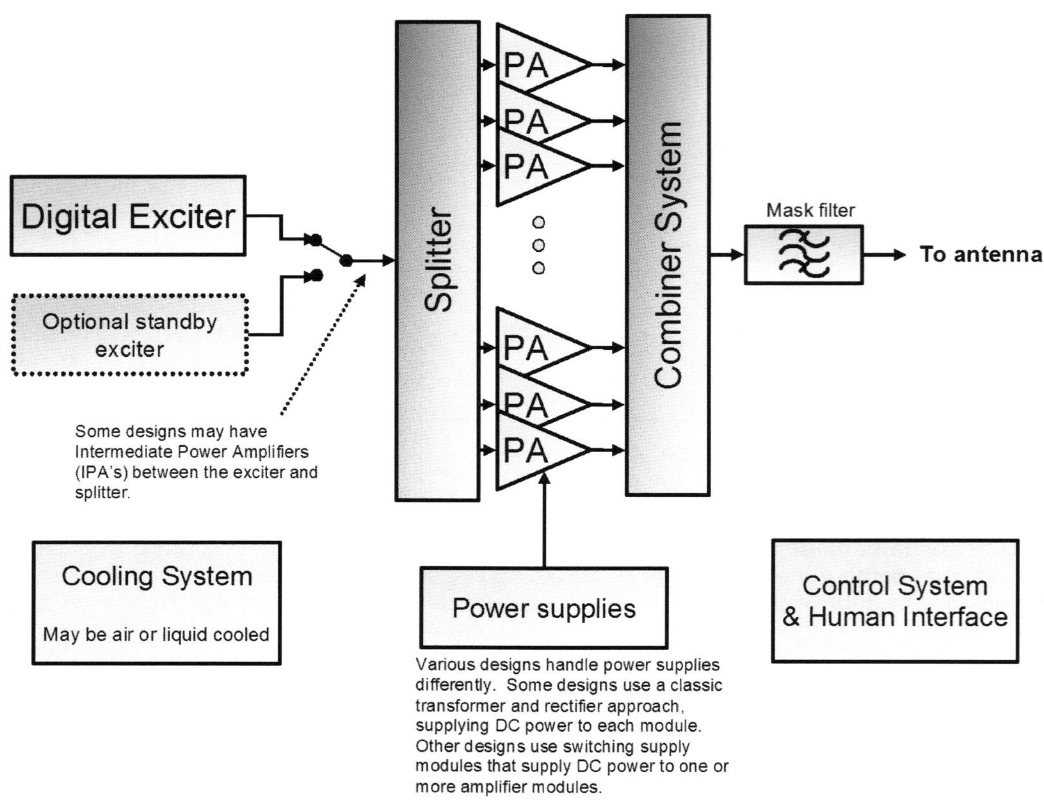

FIGURE 6.4-5 Solid-state DTV transmitter block diagram.

single broadband design to cover any channel, while other manufacturers optimize (or tune) their amplifiers to cover specific channel ranges. Having a broadband design covering the entire range of channels may be beneficial for station groups or multiple users at a common site using the same transmitter that wish to share a common pool of spare modules.

Different transmitters have different methods of supplying DC power to the amplifier modules. The classical approach is to use a single large transformer with rectifiers and capacitors to create the DC voltage required which may be done on a per cabinet basis rather than a per transmitter basis. This method is well known and reliable; however, it does not offer redundancy in the case of a power supply failure because all of the modules in the transmitter (or cabinet) are fed from a single supply. Other designs take advantage of a modular approach by providing a single switching mode power supply on a per module or two PA module basis. Figure 6.4-7 shows an example of a liquid-cooled power supply module used in a solid-state transmitter.

Solid-state transmitters provide a high reliability and availability. In large part this is due to the fact that solid-state transmitter architectures take advantage of combining multiple transistors and multiple power amplifier modules such that if one device or module fails, the entire transmitter is not taken off the air. Instead, it continues to operate at a slightly reduced

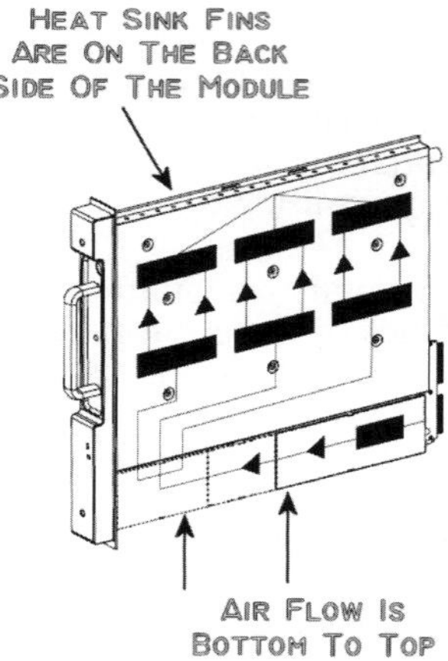

FIGURE 6.4-6 Typical solid-state amplifier module (air cooled).

power. This is the concept of *graceful degradation*. An important metric to consider for a solid-state transmitter is how much the transmitter power will be reduced in the event of a failure in one, two, three, or more devices in the amplifier modules. The reduction in power will depend on various design considerations and the type of combiner employed.

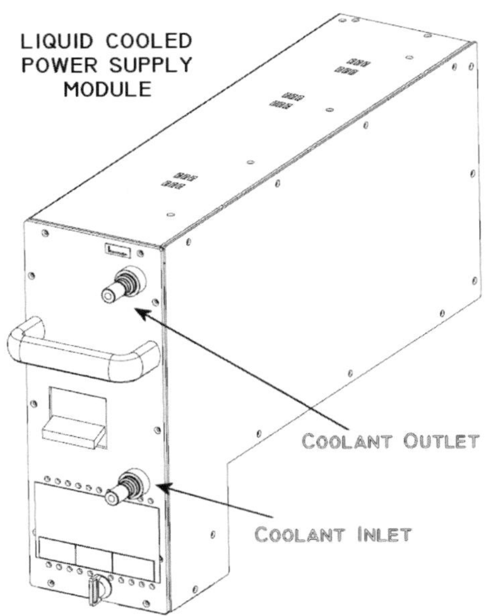

FIGURE 6.4-7 Typical solid-state transmitter power supply module (liquid cooled).

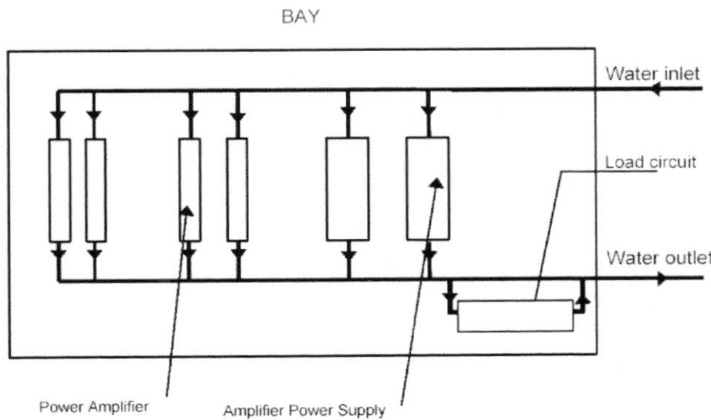

FIGURE 6.4-8 Schematic representation of a liquid-cooling system for a solid-state DTV transmitter.

Power Amplifier Module Combiners

There are many different types of N-way combiners in use in DTV transmitters such as the Wilkinson, conical waveguide, and radial line combiners [6]. The impact of device failures will depend on the impedance of the failed devices and the s-parameters of the type of combiner used. In a well-designed combining system, the expected reduction in power is proportional to $(1 - m/N)^2$, where m is the number of failed devices and N is the total number of devices in the system [7]. For example, if one-half of the devices in a system have failed, one might expect to be at half power, but the formula indicates $(1 - 1/2)^2$ power, or 25%. The expected power is less than the maximum capable of being delivered by the remaining devices (50% in the example) because of the impedance mismatches that occur when devices have failed. A portion of the remaining power ends up being reflected back into combining structures and dissipated into their isolating loads, if the combiner provides such loads. Furthermore, if the combiner has poor or no isolation, a portion of the remaining power is reflected back to the devices, affecting their original dynamic range. Transmitter designers attempt to minimize the total amount of power lost due to a failure (beyond theoretical values) and the adverse impact of any resulting impedance mismatch through careful design techniques.

Power Amplifier Module Cooling

Cooling for solid-state transmitters has been an area for innovation with new designs using water to cool the amplifier modules and the power supplies. In air-cooled systems, comparatively large blowers force air through PA module heat sink fins. This requires a large amount of heated air flow that either has to be ducted out of the transmitter building or cooled with air conditioning units. Large blowers also create a noisy environment, which can be detrimental to workers. In a liquid cooled design, the heat-generating components of the amplifiers and power supplies are mounted to mechanical structures (bays) through which liquid is circulated to remove the heat. The schematic representation is shown in Figure 6.4-8. Liquid is circulated through the three major heat-generating components of the system: power amplifiers, power supplies, and load networks. Figure 6.4-9 shows an inside view of a liquid cooled solid-state transmitter. The liquid, usually a glycol/water mix-

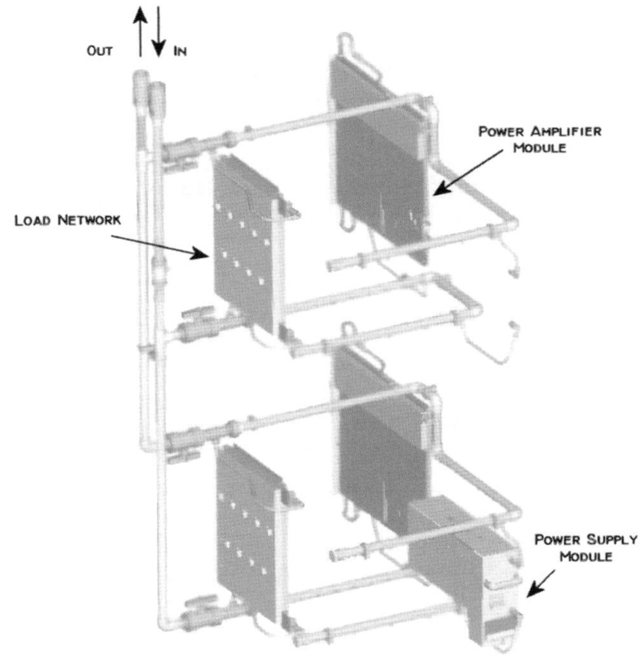

FIGURE 6.4-9 Inside view of a typical liquid cooled solid-state DTV transmitter.

ture, is pumped through the cabinet and modules. Hoses with self-sealing disconnects bring the water to each module. To perform maintenance, the hose connections allow for a single module to be removed without disturbing the overall operation of the transmitter—that is, without turning off the coolant flow. Hot water is pumped out of the cabinet, and heat can be removed outside the transmitter room using heat-exchanger units. Exhausting the heat into outside air can substantially reduce the load on a building air conditioning system.

Another reason why water cooling of solid-state transmitters has become popular is the promise of increased device reliability. The average life of a high-power RF transistor is linked closely to the temperature of the junction of the device. The heat capacity of the water allows more heat to be extracted from the device compared to most air-cooled amplifiers. This lowers the junction temperature and often increases the life of the device. Individual device data sheets will often have curves showing the effect of junction temperature on mean time to failure (MTTF). Note that water cooling technology is not limited to digital transmitters, but because this is relatively new to broadcast transmitters it is more common in digital transmitters than analog ones.

Once solid-state technology has been selected for the power amplifier, the type of cooling to be used is the next most important decision to make. The different methods of liquid or air cooling have unique installation requirements. In the case of liquid cooling, the installation must include provisions for heat-exchanger equipment, which is most often installed outside of the transmitter building on concrete pads. For air-cooled systems, the installation must have provisions to duct air (and heat) outside or to handle hot air from the transmitter blown into the room using large air conditioning systems. Ducting systems can be complex and sometimes include various blowers, mixers, and plenums. These systems must ensure that the intake cooling air stays at a reasonable temperature and that heat is exhausted in a way that controls the temperature of the transmitter room. Provision must be made for the altitude at which the transmitter will be installed. As altitude increases, the effect of air cooling decreases, requiring a larger air flow to obtain the equivalent cooling effect. To properly size the cooling system, the engineer must know the heat load produced by the transmitter. Table 6.4-2 provides some typical values of heat loads for various solid-state transmitters. The values shown are approximate and will vary depending on the specific transmitter model.

Inductive Output Tube Digital Transmitters

A popular technology choice for digital transmitters with average power greater than 10 to 15 kW is the inductive output tube. The IOT was popularized in the 1990s in analog service as a more efficient alternative to the klystron. An IOT is operated in class AB, so the current flowing through the tube varies with the input drive level.

An IOT consists of an input cavity and a gun section with a filament, cathode, grid, and anode, as shown schematically in Figure 6.4-10. A relatively short drift region of the tube has focus magnets to keep the electron beam from spreading and hitting the body of the tube. Finally, there is a collector sec-

TABLE 6.4-2
Approximate Heat Loads for Various Types of Solid-State DTV Transmitters*

Transmitter Output Power (kW)	Air-Cooled		Liquid Cooled	
	Indoor Sensible Heat Ducted Outdoors (kBTU/hr)	Indoor Sensible Heat with Indoor Exhaust (kBTU/hr)	Indoor Sensible Heat (kBTU/hr)	Outdoor Sensible Heat (kBTU/hr)
	VHF Transmitter			
2.5	15	55	5	40
5	30	125	10	80
10	65	260	25	200
20	125	500	45	340
	UHF Transmitter			
2.5	15	55	5	40
5	30	110	10	80
10	60	220	20	160
15	90	320	30	240
20	120	430	45	320

*For illustrative purposes only.

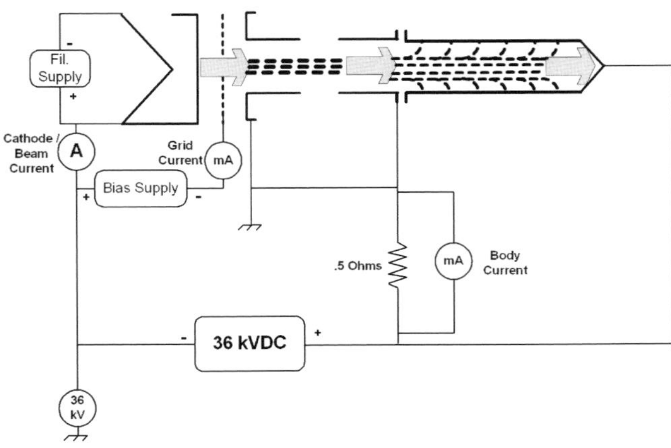

FIGURE 6.4-10 Schematic diagram of an inductive output tube.

tion. Between the anode and collector there is an output gap from which RF energy is extracted. The IOT works as an amplifier by directly modulating the grid of the tube with a DTV RF signal. This modulation allows the current in the electron beam of the tube to vary based on the input voltage. This is known as *density modulation* because the density of the electron beam changes with power level. (*Note:* Klystrons work on the principle of *velocity modulation*, and the total beam current stays constant regardless of the signal input.) As the modulated electron beam passes through the output window, it excites a cavity tuned to resonate at the channel frequency. In practice, two output cavities are used in series to achieve the 6 MHz bandwidth required for digital broadcast operation.

An IOT amplifier requires several power supplies and support systems that typically include:

- High voltage power supply (beam supply)
- Grid bias power supply
- Focus coil power supply
- Filament (or heater) power supply
- Vacuum ion supply (for the ion pump)
- Crowbar or other subsystem to limit the energy dissipated in a high voltage arc
- Water (or water/glycol) cooling for the anode and collector of the tube
- Air cooling for the input and output cavities and electron gun of the tube
- Control system for proper application and monitoring of the various supplies and cooling systems

Inductive Output Tube Protection Systems

IOT amplifiers require that the energy dissipated in a high voltage arc in the tube be limited because of the delicate nature of the IOT gun structure, which can be damaged by an uncontrolled arc event. IOT manufacturers have established a *wire test* that specifies that a length of 37 gauge wire must not be damaged

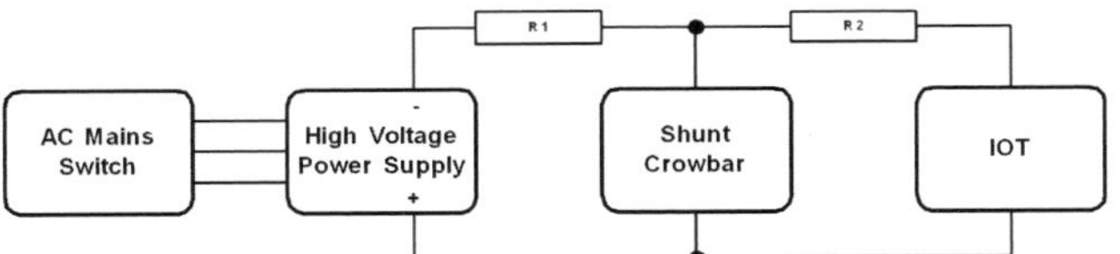

FIGURE 6.4-11 Schematic of a shunt crowbar protection system.

when the high voltage is shorted through it. There are various ways to meet this specification, all of which involve finding a way to quickly remove the high voltage from the arc (or the 37 gauge wire in the test) by either switching it off or by rerouting the power through a path that is electrically closer to the high-voltage power supply. This method of bypassing the energy into a secondary path is called a *shunt crowbar,* as shown in Figure 6.4-11. The term *crowbar* refers to the concept of taking an actual crowbar and using it to short circuit the high voltage. A crowbar is thick enough so that it would be able to conduct the current without vaporizing. Most commercially available IOT transmitters use a thyratron-based crowbar IOT protection circuit. The thyratron is a vacuum tube which in its normal mode acts as an insulator but can be triggered to conduct and is not damaged by the large amount of energy when it does conduct. Some manufacturers use a device called a *triggered spark gap* to accomplish the same thing. In these designs, a small air gap works like the thyratron, normally holding off the high voltage but triggered to conduct in the event of a detected arc in the tube.

In addition to the protection mechanism, the AC mains to the high voltage power supply should be removed quickly. This leads to another method of protection that involves limiting the stored energy in the high voltage power supply combined with a fast switch on the AC mains input to ensure that any arc within a tube is extinguished before damage can occur. These "crowbar-less" systems are available in commercial transmitters. Some of these systems use switching-type high voltage supplies to limit the stored energy in the supply.

Inductive Output Tube Transmitter Configuration

IOT transmitter systems are made up of one or more IOT amplifiers operated in parallel along with appropriate cooling systems and a passive RF system. The simplified IOT system block diagram in Figure 6.4-12 shows a system that includes two high-power amplifiers (HPAs). Each HPA in an IOT transmitter is usually capable of delivering around 25 kW of average DTV power. These HPAs are used as building blocks to create transmitter systems from 25 to 100 kW, depending on the number of HPA cabinets (from one to four). The passive RF system design will vary depending on the number of cabinets being combined and switching complexity desired.

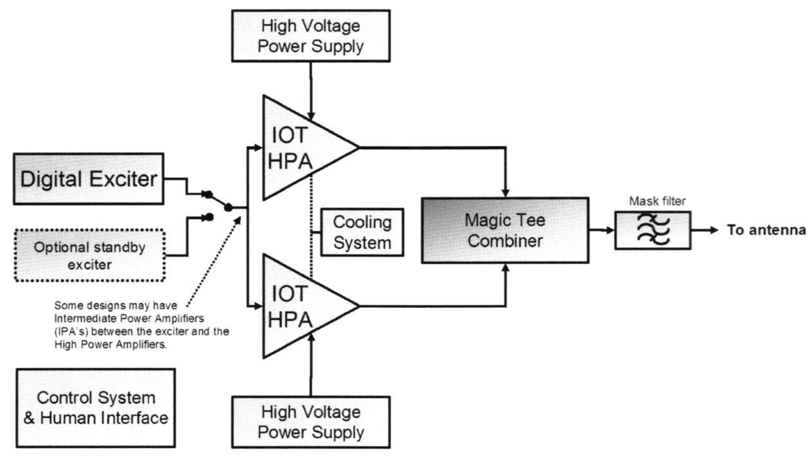

FIGURE 6.4-12 IOT DTV transmitter block diagram.

Inductive Output Tube Transmitters: Analog and Digital

An IOT transmitter in digital service is almost identical to one in analog service, making the conversion of an IOT amplifier from analog to digital service a simple process. Some characteristics of operation in digital service are worth noting. In analog service, the beam current of IOT varied depending on the average picture level (APL) of the input video signal. In digital service, due to the nature of the signal, the beam current for a given power level will remain relatively constant. This makes calculating the transmitter efficiency quite simple and eliminates the so-called "figure of merit" calculations that were often used in analog service. In digital service, the efficiency of the IOT can be calculated by its *beam efficiency*, given by the equation:

Beam efficiency (%) = [Average DTV power out/(beam voltage × beam current)] × 100

An example calculation is shown in Figure 6.4-13. Actual values will vary depending on the output power, the channel frequency, and a number of other factors.

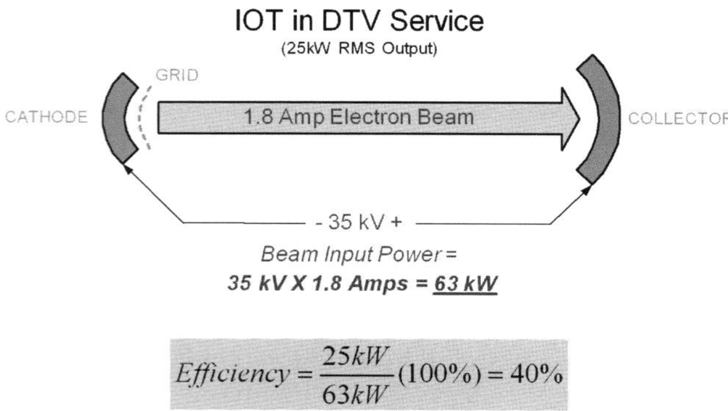

$$Efficiency = \frac{25kW}{63kW}(100\%) = 40\%$$

FIGURE 6.4-13 Calculating the beam efficiency of an IOT.

Other aspects of the analog transmitter impacted by digital operation are similar to those that affect a solid-state design. Power metering is done at average power rather than peak sync, and AGC is performed at average power. There is no longer a separate aural carrier, so any differences between externally diplexed and common amplification systems are no longer important.

MSDC Inductive Output Tube Technology and Systems

The latest technology to affect UHF tube-based transmitters is the MSDC IOT. This new tube was developed because it has the capability to be twice as efficient as a standard IOT in DTV service [8]. It accomplishes this increased efficiency by adding multiple collector sections that operate at various depression voltages. The depression voltages on each collector segment allow the electrons in the beam to be drawn to a segment with a voltage potential close to their excitation potential. Electrons traveling in the beam essentially sort themselves and are collected at the various depression stages, limiting the amount of power that would otherwise be wasted as heat. Figure 6.4-14 shows a schematic diagram of a five-stage MSDC IOT. In this tube, the various collector stages are supplied with 36 kV, 28 kV, 21 kV, 19 kV, and 10 kV, respectively. For example, collector 2 is supplied with 28 kV; this collector is depressed 8 kV from what would normally be thought of as the collector in a standard IOT, collector 1. For another example, collector 4 is supplied with 19 kV and is depressed 17 kV compared to the first collector. The input section and electron gun of the tube are identical to those of a standard IOT; it is only the collector design that has changed.

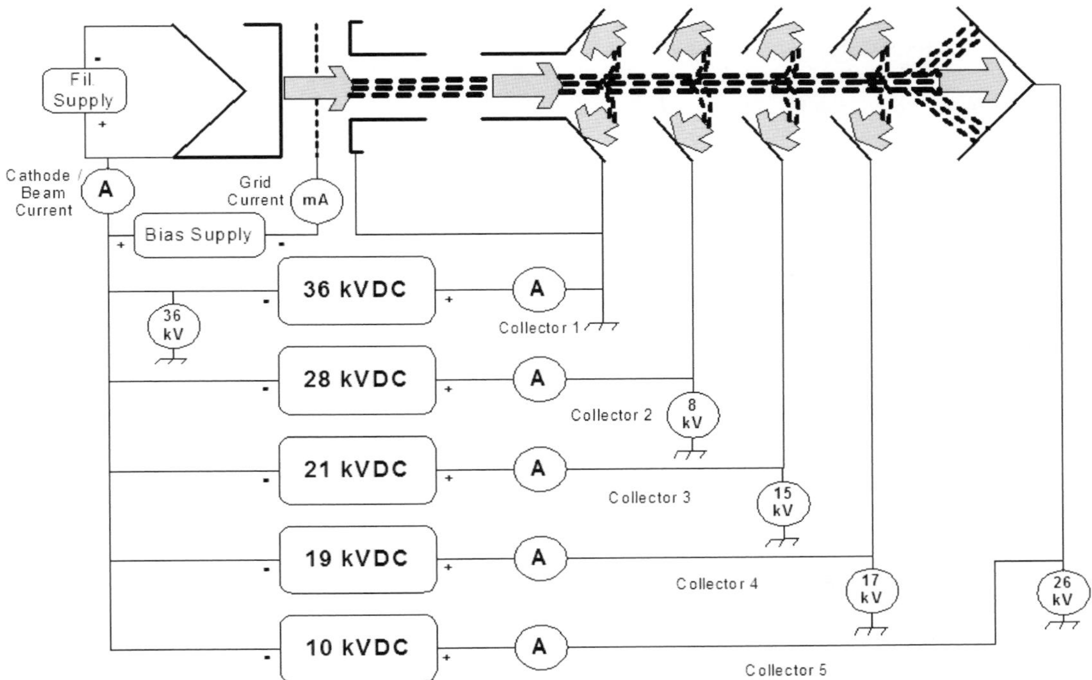

FIGURE 6.4-14 Schematic diagram of a five-stage multi-stage depressed collector IOT.

One way to think of an MSDC IOT is to think of it as five beams (or even five different tubes) operating within one envelope and one gun structure. (This is not to be confused with a multi-beam tube which would have a gun for each beam.) When different power levels are required due to the modulation of the signal, the tube in effect self-selects the most efficient electron beam path. Because the electrons are not collected at a single voltage potential, calculating the power needed for the beam is not as simple as multiplying a single current by a single voltage. In an MSDC tube, calculating the beam efficiency involves taking the current drawn into each depressed collector stage and multiplying it by the corresponding voltage supplied to the collector. This is done for each of the collector stages. The total power in the beam is the sum of all these products. The beam efficiency is the RF output power extracted from the tube divided by the total power in the beam. A sample calculation is shown in Figure 6.4-15. In this figure, the cathode is at –35 kV and the total current through the tube is 1.8 amps. The output power (DTV RMS power) is 25.0 kW. For example, collector 1 has 0.48 amps supplied to it with the main beam voltage of –35 kV. This means the total power in this circuit is 16.8 kW. Collector 2 has a current of 0.44 amps with a supplied voltage of 26.6 kV. Collector 2 is 8.4 kV depressed compared to collector 1, so it is supplied with 35 kV – 8.4 kV = 26.6 kV. The power in collector 2 is therefore 11.7 kW. This calculation is repeated for all the collector sections. The beam input power is the sum of the power dissipated in all the collector segments. In this example, the total beam input power is 42.3 kW, giving a calculated beam efficiency of 25 kW/42.3 kW which equals 59%.

The improvement in efficiency above a standard IOT is made possible because of the peak-to-average characteristics of the DTV signal. If the modulated signal passing through the tube is often at or near the highest power, most of the electrons will still be collected at the highest voltage potential, resulting in roughly the same efficiency compared to an IOT with a single collector. As shown previously in the cumulative distribution function in the section on the DTV signal, the signal spends much of its time at or even below the average power. It is during these portions of time that electrons will be drawn to the collectors with higher depression voltages.

MSDC Inductive Output Tube Cooling Systems

The MSDC IOT is a more complex tube than its predecessor. The multiple collector stages must all be cooled. This task is complicated by the fact that the collectors are in relatively close proximity and that they are all at different voltage potentials. MSDC tubes cannot be cooled with standard water or glycol/water mixtures, as these liquids are conductive and do not allow the collectors to remain electrically isolated from each other; therefore, a liquid with reasonably high dielectric properties must be used in MSDC IOT cooling. Two choices have emerged as a solution for this difficulty, and both are available in the marketplace. The first is to use deionized (DI) water as a coolant. This requires filtering the water in the cooling system such that the water is purified to a level where

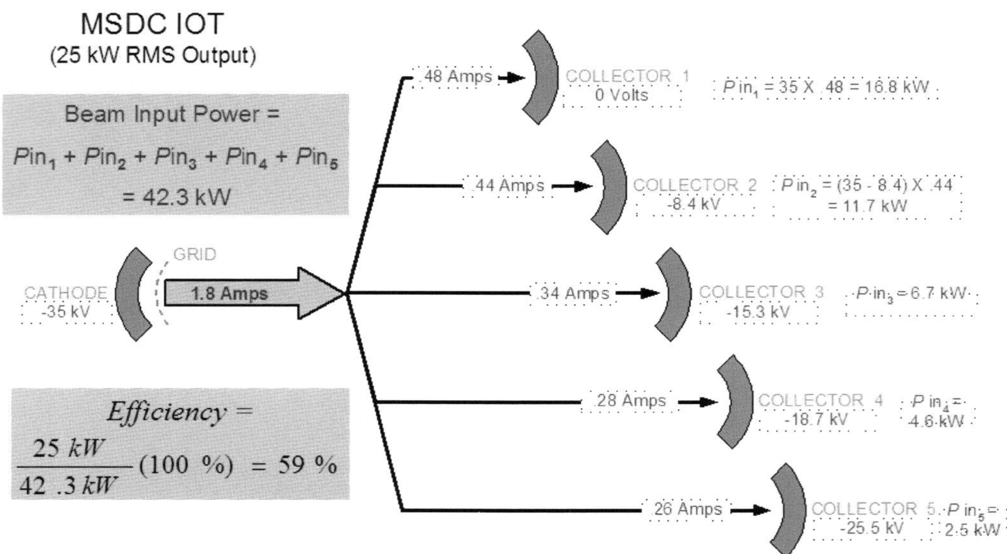

FIGURE 6.4-15 Calculating the beam efficiency of an MSDC IOT.

it can stand off the voltage differences between the segments. The tube or transmitter manufacturer will specify what level of resistivity is required, and the system must ensure that the water maintains this level of purity. The resistivity of a material is measured as its resistance (ohm) to electric current per unit length (meter) for a uniform cross section (for example, ohm-meter). A typical specification for purified water is 1 MΩ-cm. A deionized water cooling system must include the filter to purify the water along with purity sensors to make sure that resistivity is being maintained. A deionized water system cannot take advantage of the use of glycol. In normal liquid cooled transmitters, water is almost always mixed with glycol

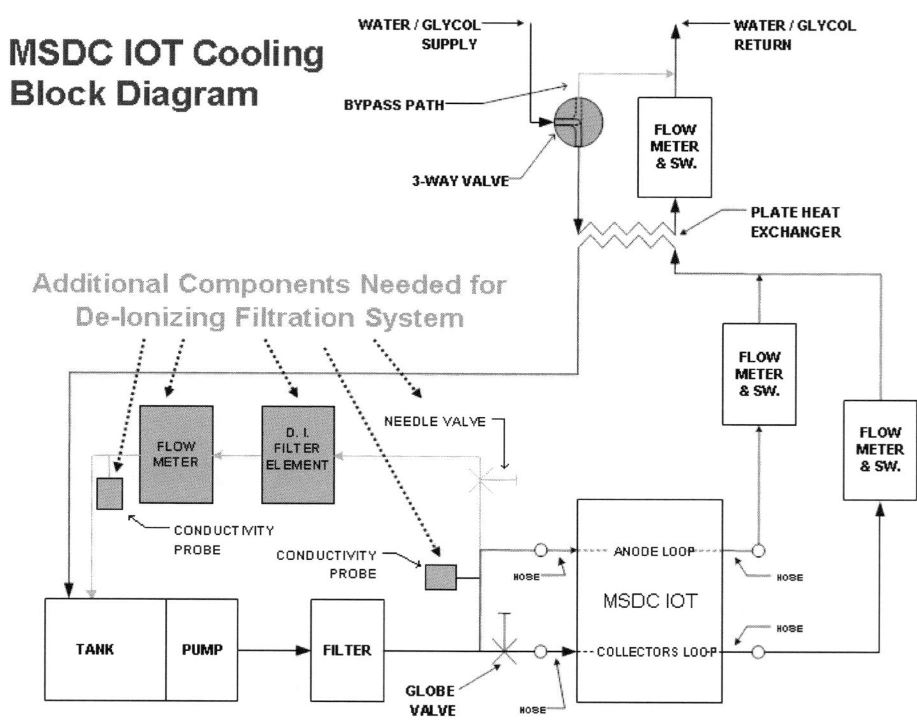

FIGURE 6.4-16 An MSDC IOT DTV transmitter cooling system.

to prevent freezing. Because this is not possible with a pure water system, some other method must be used to make sure the water does not freeze when operating in cold environments.

The alternative to deionized water is to use dielectric oil as the coolant. Dielectric oil is less efficient at extracting heat away from the collectors. Oil has both a lower specific heat and lower thermal conductivity compared to water; however, this can be overcome with appropriate collector mechanical design. Oil is naturally a good insulating material, so it is capable of standing off the voltage between the collector segments. It is also less corrosive than pure water.

Regardless of the coolant chosen, most MSDC IOT transmitter designs simplify the issues of dielectric coolants by providing a primary cooling loop (either deionized water or oil) and a secondary cooling loop. The secondary loop is operated with a standard glycol/water mixture and the same components that are common in standard IOT transmitters (plumbing, heat exchangers, etc.). Heat is passed from the primary loop to the secondary loop via a small liquid-to-liquid plate heat exchanger. Having a secondary cooling loop also allows for glycol/water to be used as the coolant for the anode cooling loop which remains separate from the collector cooling loop. A schematic diagram of this kind of arrangement is shown in Figure 6.4-16. This diagram shows a possible configuration for both oil and deionized water systems. The difference between the two types of systems is that the pure water system requires the addition of the DI filter element and conductivity probes. This diagram also shows a bypass path in the secondary loop. This path allows the water/glycol coolant to be rerouted away from the plate heat exchanger during times when the temperature is low. This makes the primary loop thermally closed. This mode can be used to prevent freezing of the pure water coolant even when the transmitter is not producing RF power.

Several different versions of MSDC IOT transmitters are available today. The major differences between manufacturers' tubes and transmitters are related to two variations in the tube design, the number of collectors used, and the liquid used for cooling.

SUPPORT SYSTEMS

Control systems and remote access to digital transmitters are important aspects of broadcast transmission systems. Standardized wide-area networking protocols have made it possible to more easily access and present transmitter operational data to a remote operator or maintenance engineer.

Remote Control Systems

Most digital broadcast transmitters provide a variety of ways for monitoring and control. These include but are not limited to:

- Dry loop contacts and 0 to 5 V metering outputs for interface to a traditional remote control system
- A serial connection (RS-232 or RS-485) with documented protocols for monitoring and controlling
- Embedded web servers to allow a browser to connect directly to monitor and control functions
- An Ethernet connection with a connection to a proprietary application running on a PC
- A Simple Network Management Protocol (SNMP) interface

When Internet-based technologies are used for remote interfacing, it is important to maintain secure access to the equipment. Most solutions will not implement their own security systems but rather will rely on security measures that should be implemented on the network being used. Never allow the equipment to be accessed through the public Internet without implementing network security. Make sure that the equipment and its associated servers are located behind firewalls. For remote access, use virtual private networking (VPN) or some other secure access method. If a network interface is used, the engineer should consider the possibility of a network failure. Backup methods must be in place if regular access is not available or bandwidth is limited for an extended period. Dry loop interfaces, which almost never fail, should be used as a backup for at least a minimum number of functions. Often, resolving these issues will require broadcast engineers to work with the Information Technology personnel in their stations.

The technologies used to provide control and monitoring can vary between manufacturers. Some systems are built around discrete logic control. More advanced systems might use embedded microprocessors or programmable logic controllers. Most modern digital transmitters include some kind of local graphical user interface (GUI). The amount of data monitored and presented to the user can vary greatly depending on the transmitter type (solid-state or tube) and the transmitter manufacturer. Generally the control system will present the overall operating state of the transmitter, along with any alarm conditions or module faults. Minimum controls include the ability to turn the transmitted signal off or on and adjust the output power. Chapter 9.5 discusses transmitter remote control systems in greater detail.

Cooling Systems

Cooling systems are an extremely important consideration in digital transmitters. As previously discussed, solid-state transmitters can use either air or liquid as their primary cooling method. IOT transmitters use liquid as a primary coolant. MSDC IOT transmitters can use pure water or oil as a primary coolant which interfaces to a secondary liquid cooling loop. Transmitter system engineers must take into account the method of cooling and design a way for the heat to be exhausted. In liquid cooled systems, the most common way is to use heat exchangers that exhaust the heat into outside air. Some buildings have chilled water available, in which case a liquid-to-liquid exchange can be made. In the case of air-cooled solid-state transmitters, it is important to exhaust the hot air through ducting to the outside or to use air conditioning systems large enough to handle the heat load.

INSTALLATION CONSIDERATIONS FOR DTV TRANSMITTERS

The first consideration for the installation of a transmitter is the building size. Different transmitter types will have different requirements in terms of floor space. Various system options can also have an effect on required floor space. Most transmitter manufacturers will provide guidelines for indoor and outdoor space requirements once system options are selected. In general, transmitter buildings should have a ceiling height of at least 12 feet, and large systems using WR1800 waveguide may require a 14 foot ceiling height. Most transmitters are designed to fit through standard door openings, 8 ft high by 5 ft wide, and some large RF systems may require wider door openings. Outdoor equipment can include heat exchangers and IOT high voltage power supplies. This equipment is designed to operate in all kinds of weather conditions; however, it can be damaged by falling ice from towers. In those cases, provisions should be made to shield the equipment. Sun shields are usually recommended as well. Shielding should be designed in such a way as to prevent exhaust air from heat exchangers from being trapped under roofing.

Building HVAC systems will be affected by the presence of a transmitter cooling system. In the case of liquid cooled transmitters, there is still a significant amount of heat exhausted into the transmitter building; however, this heat is usually not sufficient enough to be the only source of heat for the building, especially in colder climates. For air-cooled transmitters, it is helpful to subdivide the type of systems into closed systems, where air is exhausted into the transmitter room, and open systems, where fresh inlet and exhaust air comes from and goes outside the room. In closed systems, it is imperative that the room air conditioner be sized appropriately to handle the transmitter heat load (measure in kBTU/hr). In open systems, the heat load is less important because most of the heat is exhausted outdoors. Of more concern is that the supplied airflow rate in cubic meters per minute must be sufficient and also balanced between the air inlet, transmitter, and exhaust duct. An engineer should compensate for what might happen when an HVAC system fails by including a redundant system or a method to bypass the system in the event of a failure. Transmitters may be able to operate with an air bypass, even if it requires operating at reduced power.

Most manufacturers recommend that a transmitter building should be maintained at room temperature (nominally 75°F). To maintain this temperature, the following factors should be considered:

- Sensible heat load of the transmitter system
- Sensible heat of other equipment such as microwave links, test/monitoring equipment, lighting, etc.
- Local climate conditions (extreme heat or cold during summer or winter months)
- Solar heating of roofs

Radiofrequency Support Systems

RF systems also impose certain constraints on the transmitter building design and can usually be divided into two types: floor-mounted and ceiling mounted. Floor-mounted systems consist of RF components (filters, switches, combiners, loads) mounted to a self-supporting angle-iron frame. The frame is usually placed on the floor near the transmitter amplifiers and therefore requires additional floor space, which needs to be planned. Ceiling mounted RF systems are suspended by a rod-and-grid system (usually threaded rod and metal strut) and place an additional weight load on the ceiling of a building. This load can be significant, so it is necessary to check with a structural engineer to make sure that the building is able to support such a system. Some additional building considerations include:

- Space and proper construction to support wall mounted components
- Clearance spaces around transmitter cabinets for maintenance access
- In IOT transmitters, additional space in front of amplifier cabinets for installation and removal of the IOT and its cavities which are supplied on a rolling cart

Electrical Systems

Another major installation consideration is the electrical hookup. In most high power transmitter installations, the main electrical feed will be 480 V. Some lower power installations are done with a 208 V feed. This is much more common in older VHF sites that run at lower power levels. In all cases, the transmitter manufacturer should supply a power distribution drawing. Note that electric power utilities are vested with a great deal of authority and codes vary from area to area; therefore, it is important to check with local experts to make sure that the transmitter system as installed will meet all necessary local requirements.

As part of the electrical design, an engineer should consider the ramifications of possible problems with the AC feed to the system. The addition of transient protection, voltage regulation, and generators to an electrical system can certainly add cost to a system. At the same time, these components provide important protection against common problems with the AC mains. Some common conditions and various possible solutions that should be considered are listed in Table 6.4-3. See Chapter 9.2, "AC Power Conditioning," and Chapter 9.4, "Standby Power Systems," for more information.

Proper electrical grounding is required for any transmitter system. Most transmitters are required to

TABLE 6.4-3
AC Power Disturbances That Can Affect DTV Transmitters

AC Power Condition	Causes	Protection
Over-voltage	Lightning, power company	AC mains voltage regulator
Under-voltage	Power company	AC mains voltage regulator
Phase loss	Power company, lightning	Standby generator, transmitter protected by three-phase monitor
Energy surge	Lightning, load switching, power company	Transient protection on AC mains
Induced transient	Lightning, nearby motor operation, power company	Transient protection on AC mains

have cabinets and enclosures bonded to a single common point. This greatly reduces any problems with radiation or ground loops. In addition, proper grounding is a basic safety requirement that should command a great deal of attention. In all cases, engineers are advised to follow the recommendations of the equipment manufacturers. See Chapter 9.3 for more information on facility grounding practice.

Lightning strikes can be devastating to most electrical equipment, and a DTV transmitter is no exception. Lightning is extremely unpredictable and very powerful, so it is really impossible to completely prevent all problems from strikes. Some areas of the United States are particularly prone to electric storms, such as Florida, Colorado, and Louisiana. Transmitter installations in these states should take additional precautions to prevent problems from lightning. Some basic precautions to help minimize any problems with a transmitter include following good grounding principles, using underground feeders on the input power lines, and avoiding running underground wires or cables in close proximity to tower bases or guy anchors. Doing the latter may allow induced currents to enter through these cables.

Power line regulation is a requirement in electrical installations to ensure good transmitter performance, stability, and reliability. Most transmitters have a tolerance for reasonable fluctuations in the AC line (*e.g.*, ±3%). Although most transmitters will handle certain power line fluctuations, no transmitter can withstand all impairments. If the area has a history of power line disturbances, it is essential to use power conditioning techniques to protect the transmitter against variations. This will ensure higher reliability and less maintenance. Large fluctuations will often cause alarms in the transmitter control system to occur or breakers to trip. Various protection methods can be used for various problems. Large electromechanical regulators can provide a reasonable level of transmitter stability with changing line conditions. These regulators can substantially increase the reliability and long-term life of a digital television transmitter. In installations where the input power may be interrupted, a stand-by generator may be required to maintain the availability of the on-air signal.

DTV SIGNAL PRECORRECTION METHODS

Regardless of the type of amplifying technology employed, some distortion will be introduced as the signal passes through the digital transmitter, but the goal of the transmitter is to transmit a clean signal, because the channel itself will degrade the signal. Distortion introduced by the transmitter will have an impact on the ability of an ATSC receiver to correctly recover the signal. To compensate for distortions in the transmitter, DTV exciters have built-in precorrection circuits. The purpose and general principle behind precorrection in DTV is the same as it is in an analog transmitter. The exciter introduces a distortion in the low-level signal that is opposite from the distortion that will be introduced in the high power amplification stages. The effect is that the distortions cancel and the final output signal is close to ideal.

DTV signal distortions fall into three categories, two of which are correctable: linear distortions, nonlinear distortions, and noise. Noise is introduced into the signal from typical well-known sources: the thermal noise of the electronic components, the noise floor of an amplifier, AC ripple or switching noise from a power supply, and phase noise of the local oscillator used for RF upconversion. These noise sources are not considered correctable; rather, they become design criteria in the DTV exciter and transmitter. All of these sources must be considered and dealt with in transmitter design. In practice, all good-quality commercially available equipment take these factors into account and ensure that they are minimized to the point that they do not enter into consideration for the transmitter engineer or operator. In contrast, both linear and nonlinear distortions are usually dealt with using precorrection techniques.

Linear Distortion

A linear distortion by definition is a distortion that is not dependent on the amplitude of input signal; that is, the distortion does not change when a higher power or lower power signal is fed into the system. The mathematical definition of linearity relies on a principle called *superposition*. This principle shows that if a signal, $x(t)$, is fed through a linear system and gives a response, $y(t)$, then the same signal scaled by a factor A, $A * x(t)$, will give response $A * y(t)$. The principle of superposition also applies to additive signals. So, if $a(t)$ gives a response of $b(t)$, then $x(t) + a(t)$ will give a response of $y(t) + b(t)$. In other words, linear distortions are ones that do not change characteristics based on the input signal amplitude. In practical terms, linear distortions are those that affect frequency response and group delay response. Any deviation from a flat response across the 6 MHz channel is a linear distortion. In a transmitter system, the most common sources of linear distortion are filters. In an IOT system, linear distortion is introduced in the tuned response of the output cavities. In both solid-state and IOT systems, linear distortion is introduced in the mask filter and any channel combiners that might be present in the system.

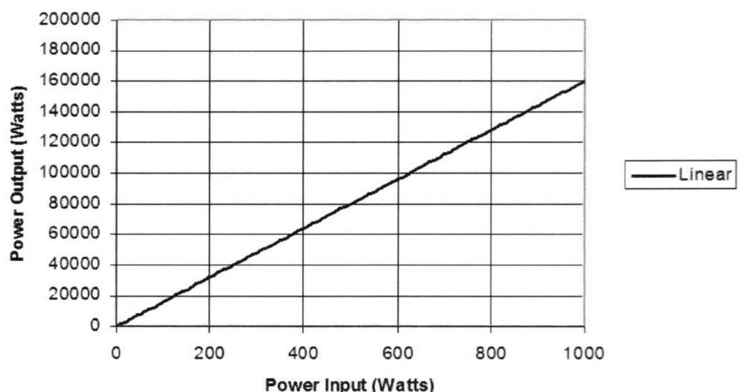

FIGURE 6.4-18 Idealized transfer curve showing power out *versus* power in for a perfectly linear amplifier.

Nonlinear Distortion

In contrast to linear distortions, nonlinear distortions are variations that occur when the signal input changes in amplitude. Nonlinear operation in an amplifier is described by the Power_{in} *versus* $\text{Power}_{\text{out}}$ curve. If an amplifier were perfectly linear, then the gain, for example, would remain constant for any input power, but actual amplifiers cannot be linear over an infinite range. If this curve were a perfect straight line, then the amplifier would be linear. When the gain of the amplifier changes at different power levels, nonlinear distortion occurs. Nonlinear phase distortion can also occur. This occurs when the phase of the output signal is dependent on the amplitude of the input and varies based on the input amplitude. Figures 6.4-17 and 6.4-18 show transfer curves (power out *versus* power in and gain *versus* power in) of an amplifier that would be linear. If an amplifier exhibited these characteristics, it would introduce no nonlinear distortion.

Real-world amplifiers are always nonlinear to some extent. Figure 6.4-19 shows an example of the transfer curve of an IOT amplifier compared with a linear transfer curve. It is this characteristic curve that causes nonlinear distortions. These transfer curves are one way to characterize nonlinear distortion of a transmitter, and DTV test equipment will often include a plot that shows output amplitude *versus* input amplitude and output phase *versus* input amplitude. These curves are typically referred to as AM-to-AM (amplitude distortion) or AM-to-PM (phase distortion) curves. Another way to measure the amount of nonlinear distortion in a system is to plot the cumulative distribution function (peak-to-average measurements) and compare it to an ideal plot. The deviation in the measured response gives an indication of nonlinearity. Figure 6.4-20 is an image from a Tektronix model RFA-300A that shows both linear (light portion of the curve) and measured response (dark line).

In a transmitter system, nonlinear distortion is introduced in the final amplification stages of the transmitter, whether the transmitter is based on an IOT or solid-state device. Nonlinear distortion can also be introduced in the intermediate power amplification stages. Both linear and nonlinear distortion have the potential to create problems for a DTV receiver, which is why minimizing distortions in the transmitter is important. With a DTV signal, there is a single quality metric which takes into account all sources of distortion and presents a result representative of the several distortions. The metric is called *modulation error ratio* (MER), but it is nearly equivalent to the more common metric *signal-to-noise ratio* (SNR). The MER or SNR measurement is a ratio of the magni-

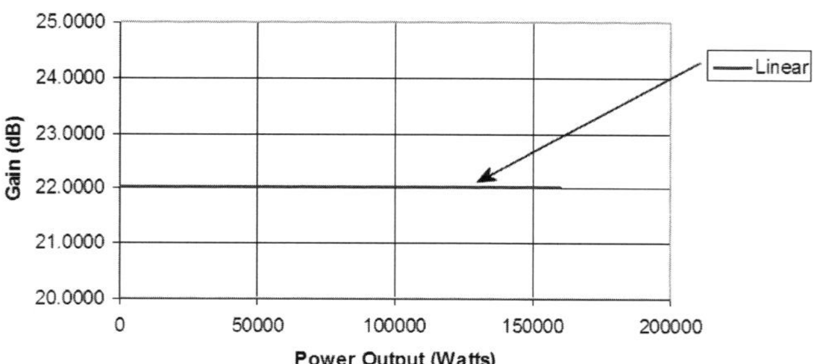

FIGURE 6.4-17 Idealized transfer curve showing gain versus power in for a perfectly linear amplifier.

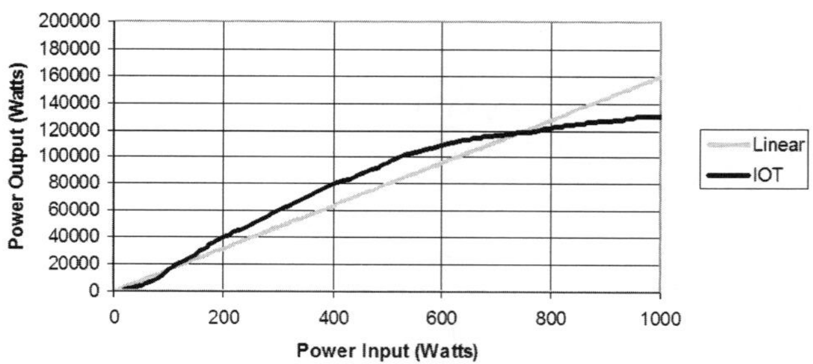

FIGURE 6.4-19 Transfer curve of an IOT amplifier compared with a linear transfer curve.

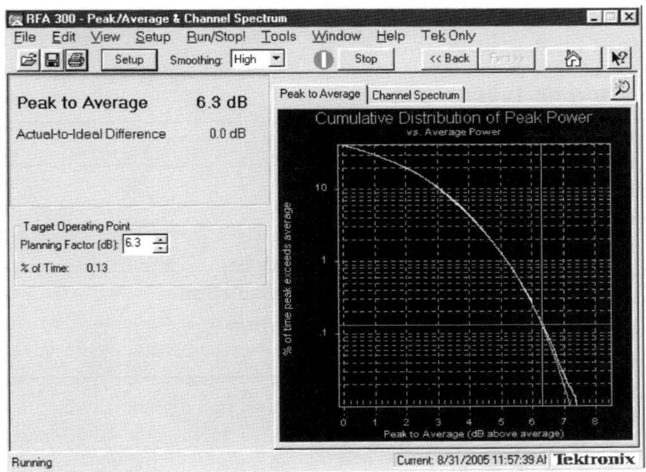

FIGURE 6.4-20 Measured CDF of a DTV transmitter compared to an ideal DTV signal.

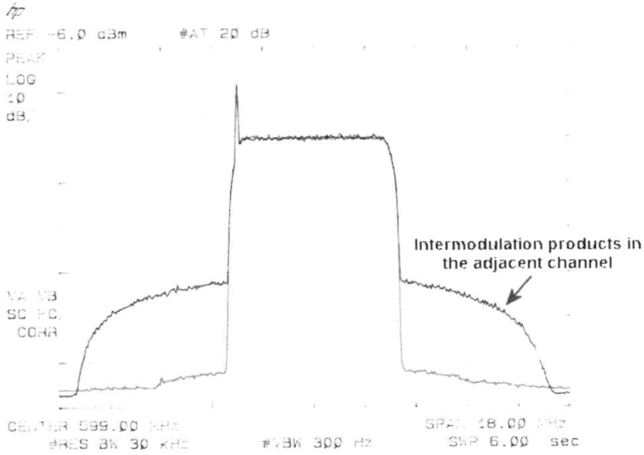

FIGURE 6.4-21 Intermodulation products in an adjacent channel, shown without the FCC mask filter.

tude of all the errors in a measured signal compared to the ideal signal and is usually expressed in decibels. The larger the SNR number, the greater the signal power compared to the distortion power and therefore the better the signal quality. The ATSC recommends that a DTV transmitter maintain an MER or SNR of 27 dB or greater.

Another metric is the *error vector magnitude* (EVM), which is expressed in percent. Signal quality improves as the EVM becomes smaller (*i.e.*, as the percentage of error magnitude decreases). The EVM ratio is opposite the SNR in that it expresses the magnitude of the error compared to the signal rather than the magnitude of the signal compared to the error. The equivalent EVM for an SNR of 27 dB is roughly 4%. For further information, refer to Chapter 8.4, "Digital Television Transmitter Measurements."

Linear and nonlinear distortions impact the SNR of the DTV signal itself (in-band), but nonlinear distortions can create out-of-band intermodulation (IM) products. IM products have the potential to cause interference with other broadcast signals. Adjacent channels are of particular concern because the amplitude of IM products decreases as the frequency delta from the main signal increases. The level of the out-of-band products can be easily viewed and measured on a spectrum analyzer, as shown in Figure 6.4-21. The section on performance measurements discusses the requirements for out-of-band IM products and methods for making compliance measurements. It is important to note that when IM products are present out-of-band they are also present in-band even though they are not easily viewed on a spectrum analyzer. Table 6.4-4 provides a quick reference summary of linear and nonlinear distortions.

There are three methods used in digital transmitters for performing precorrection, all of them implementing the "opposite distortion" principle described previously. The first and simplest method is the use of analog correctors. The second method involves using manually adjusted digital correctors. The last method uses adaptive or automatic digital correction. This last method has become the most common in DTV transmitters. Regardless of the method used, linear and nonlinear corrections are performed independently using two different corrector functions.

Analog Correctors

Analog correctors function essentially the same way as in analog transmitters. In a nonlinear corrector, a sample from the signal is fed through a series of diodes and resistors which clip the signal at an adjustable threshold, vary the amplitude, and then add it back into the main signal. Multiple corrector cells with independent threshold and amplitude controls are grouped together. These adjustments allow the ability to create a piece-wise linear approximation to the AM-AM curve of the amplifier. This is illustrated in Figures 6.4-22 and 6.4-23.

The ideal precorrection curve is the opposite curve, a reflection of the amplifier transfer curve around the linear curve as shown in Figure 6.4-22. The analog linearity corrector creates an approximation to this curve with straight line segments as shown in Figure 6.4-23. The more correction cells supplied, the more straight-line segments can be used and therefore the more accurately the curve can be approximated.

Phase correction can be accomplished using the same method. The only difference is that the clipped signal pieces are shifted 90° before being added back into the main signal. This has the effect of changing the phase of the signal while only changing the amplitude a small amount. As long as the phase correction required is relatively small, correctors implemented in this way can be effective.

Analog correction of linear distortions is also the same as it is in analog transmitters. The goal is to create an inverse filter that exhibits the opposite frequency response and group delay characteristic. This

TABLE 6.4-4
Summary of DTV Signal Distortions

	Distortion Type	
	Linear	**Nonlinear**
Typical causes	Mask filter Channel combiner IOT output cavities	Final amplifier linearity Intermediate power amplifier linearity
Effect on signal	Frequency response Group delay response	Output amplitude *versus* input amplitude distortion Output phase *versus* input amplitude distortion Intermodulation products
Measurement	SNR (equalizer off)	SNR (equalizer on) Out-of-band intermodulation products
Effect on receiver	Can usually be compensated by receiver equalizer	No compensation possible. Impacts receiver C/N threshold and ultimately signal coverage area.
Analog signal equivalents	Frequency response Group delay response Chrominance–luminance delay	Low-frequency linearity ICPM Differential gain and phase Intermodulation (in common amplification systems) Visual–aural cross-modulation (in common amplification systems)

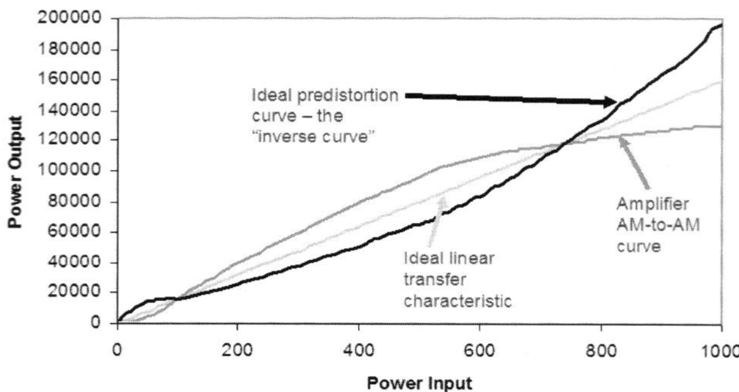

FIGURE 6.4-22 An amplifier transfer curve and the corresponding precorrection curve.

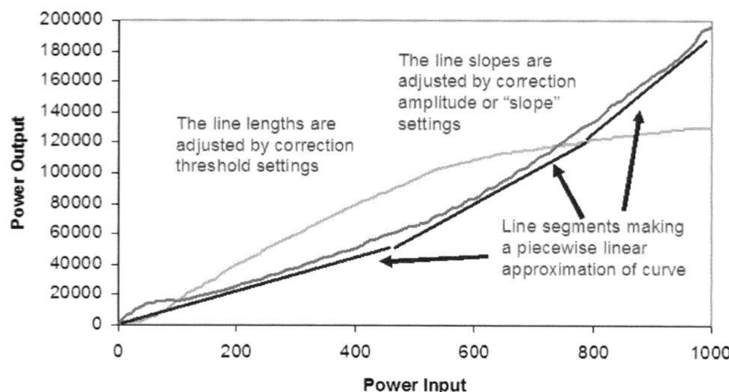

FIGURE 6.4-23 A piece-wise linear approximation of a precorrection curve.

can be done using lumped element cells that can vary the amplitude or group delay response at specific frequencies. Multiple cells centered at various frequencies across the channel can be used to create the overall desired response.

One difficulty with analog correction of DTV signals is the lack of a training signal—that is, a signal that easily gives a display of the parameters being adjusted. In National Television System Committee (NTSC) transmitters, video test signals serve this purpose; for example, a stair-step signal is used to look at low-frequency linearity and incidental carrier phase modulation (ICPM). Linearity corrections were performed while looking at the test signals on a calibrated measurement instrument which gave the operator guidance about which portion of the signal needed to be adjusted and how much correction was required. DTV has no such training signals. Another drawback of analog correction is the fact that the inverse correction can only be approximated with line segments. This limits the precision of the correction and in practice makes it difficult to achieve required specifications.

Fixed Digital Correction

An improvement over traditional correction is fixed digital correction. Fixed digital correction operates on the same principles as analog correction, but it allows for more precision because correction is applied in the digital domain. The signal must be digitized in order for digital correction to work. In other words, the sig-

nal must be in a digital format to be passed through digital processing steps. With a DTV exciter, this is easy because all of the modulation is accomplished in the digital domain; therefore, it is relatively easy to place precorrection functions after the modulation functions but before the conversion of the signal to analog. First consider nonlinear digital correction. In this form, the signal can be passed through a look-up table that takes each input sample and changes its amplitude and phase according to a fixed AM-to-AM and PM-to-PM curve. This is superior to the piecewise linear approach described earlier because it allows each sample to be precisely adjusted to the needed precorrection value. So, rather than line segments approximating a curve, the curve is drawn with a series of dots. And, because these dots can be placed completely independently through the use of a look-up table (as opposed to, say, an equation), the correction can be accurate and precise. The only problem in using fixed digital precorrection for nonlinear distortion is filling the look-up table with the correct values to approximate the inverse curve. Typically, this is done through trial and error, and it can take many hours of work to achieve correction values that are acceptable.

A look-up table is used to correct for nonlinear distortions, but a digital filter can be used to compensate for linear distortions. As it is with analog correction, the goal is to create in inverse filter that compensates for the linear distortions in the transmitter system. Implementing a digital *finite impulse response* (FIR) filter is a straightforward task [9]. The response of an FIR filter, $y[n]$, given an input $x[n]$, is given by:

$$y[n] = \sum_{m=0}^{N-1} h[m]x[n-m]$$

In this equation, N is the total number of taps in the filter. The more taps the filter has, the more precisely it can estimate an inverse filter. For each tap, the filter has a coefficient, shown as $h[m]$ in the equation. These coefficients are really the required inputs to the filter. The difficulty with fixed digital linear precorrection is determining these coefficients in an easy manner. One way to get the required coefficients is to use DTV test equipment that has an adaptive equalizer built in. Such equipment can be used to calculate the tap coefficients required, which can then be input manually or via a computer interface into the exciter. Another common way is to use a GUI to draw the desired inverse filter response on a computer screen. A computer algorithm can then calculate the coefficients needed to create the filter response.

Adaptive Precorrection

The final precorrection method is adaptive or automatic digital precorrection. This method is almost the same as the fixed digital methods described. The difference is that the exciter provides a means to automatically calculate the look-up table values (in the case of nonlinear correction) and the filter coefficients (in the case of linear correction). This is usually accomplished by taking a sample of the transmitter system output and feeding it back into the exciter. The exciter then digitizes the feedback signal and compares it to the reference ideal output of a clean exciter. Computer algorithms then calculate and load the appropriate information into both the nonlinear and linear correctors. If the algorithms operate continuously, the correction is said to be adaptive because the correction will automatically adjust the correction values to compensate for changes in the closed loop system. Adaptive precorrection is now considered state of the art in digital television transmitters. It has the advantage of reducing set-up and maintenance time, thus ensuring that operating parameters are continuously maintained. Because the correction is digital it is more precise than realizable analog correctors. This means that amplifiers can be made to deliver more power for the same performance, thus decreasing cost and increasing efficiency.

DIGITAL SIGNAL PERFORMANCE MEASUREMENTS

NTSC transmission required many different measurements to ensure that an acceptable quality video signal could be reproduced by a receiver and that the transmitter was not interfering with other transmitted signals. Performance measurements for the DTV signal are much fewer, and modern test equipment has made these measurements simple. Performance measurements can be broken into two categories: in-band and out-of-band. In-band measurements are made to determine the acceptability and quality of the transmitted signal itself, while out-of-band measurements ensure that the transmitted signal will not interfere with other signals. It is interesting to note that the FCC does not have requirements on the quality of the radiated in-band signal, as long as the modulation format is correct. The ATSC published a standard that has been widely adopted by the industry to determine what an acceptable quality signal is at the output of a transmitter. These guidelines are contained in ATSC Standard A/64A, Transmission Measurement and Compliance for Digital Television [10].

Before beginning the transmitter measurements, the engineer should first verify that the input signal stream is within tolerance. The specified bit rate for the transport stream input is ~19.392658 Mbps. This should be maintained to within ±54 Hz (2.79 ppm). It is important that this rate be accurate because the modulator symbol rate is normally locked to the incoming rate. Significant variations in the input data rate can prevent some receivers from demodulating the signal.

Requirements for the in-band signal are straightforward. First, the frequency of the signal should be checked. The pilot carrier allows a quick check of frequency. The pilot should be 309.4405 kHz from the lower band edge (assuming no assigned offsets). A digital spectrum analyzer with markers, along with a good 10 MHz reference, such as an internal high-precision reference or from a global positioning system (GPS) receiver, allows an accurate measurement of the pilot signal. If the exciter has a monitoring output for the pilot, a frequency counter can be used. The power of the total signal must be measured using an RF power meter that is able to measure the true RMS power of a digital modulated signal with a high peak-to-average ratio. Calorimetric methods may also be used, although these methods are becoming less common. The final in-band measurement necessary to give an indication of the quality of the signal is the SNR (expressed in dB). It is a measure of all sources of disturbance in the signal, including noise, intermodulation, and amplitude or group delay variations across frequency. Typically, a specialized piece of test equipment is required to make this final measurement. The equipment will usually display a number in decibels (or percent) along with either a constellation or eye diagram, as shown in Figure 6.4-24.

Figure 6.4-24 shows the SNR and EVM measurements of the output of a digital exciter. The measured SNR of 39.2 dB indicates a clean signal. The constellation diagram shows eight lines corresponding to the eight levels of the VSB signal. The sampled symbols are shown as dots along the lines. In this case, the samples are almost exactly on the lines (making them difficult to discern). As the SNR decreases, the samples will begin to spread away from the eight lines. The recommended value for SNR is 27 dB or greater. The higher the SNR is, the better the signal quality. An SNR reading lower than 27 dB can degrade coverage of the transmitted signal.

Test equipment with built-in SNR measurement is usually capable of distinguishing between a low SNR due to linear distortion and a low SNR due to nonlinear distortion or noise. This is accomplished through the use of an adaptive equalizer. Enabling an adaptive equalizer effectively removes the linear distortion from the measurement, giving the engineer an SNR due to nonlinear effects. This allows the engineer to diagnose problems with a transmitted signal. If the SNR improves significantly by using the equalizer on the test equipment, then the transmitter has a high amount of linear distortion. The engineer can usually solve this by adjusting the linear correctors (amplitude response and group delay). If the problem is severe enough, the engineer should look for the cause of the distortion, such as filtering or transmission line problems, excessive voltage standing wave ratio (VSWR), or tube cavity tuning (as in the case of an IOT transmitter). If the SNR does not improve when the equalizer is activated, then the engineer knows the SNR is being limited by a nonlinear distortion. The engineer can usually improve this by adjusting the transmitter's nonlinear correction. Other sources of this include phase noise of the local oscillator used for upconversion, thermal noise of the electronics, or sometimes a problem within the ATSC modulator itself.

Figure 6.4-25 shows a plot of the theoretical degradation in carrier-to-noise ratio (C/N) that will be caused by the transmitted SNR. The plot assumes that all sources of distortion are not capable of being removed by a receiver equalizer. To use the plot, allow the equalizer on the test equipment to remove the linear distortion from the measurement. Look up the resulting SNR on the x-axis of the plot. The corresponding value on the y-axis indicates the amount of lost C/N due to the transmitter SNR.

Linear distortion effects are more difficult to describe. To find the impact on the receiver, it is necessary to know the values of the tap weights in the receiver equalizer given a particular linear distortion characteristic. The tap weights determine the calculation of *white noise enhancement* [11]. This is a measure

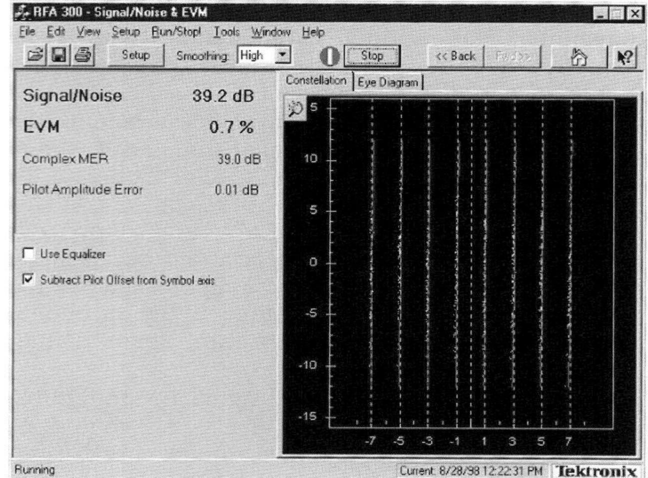

FIGURE 6.4-24 Measurement of SNR from a DTV exciter output.

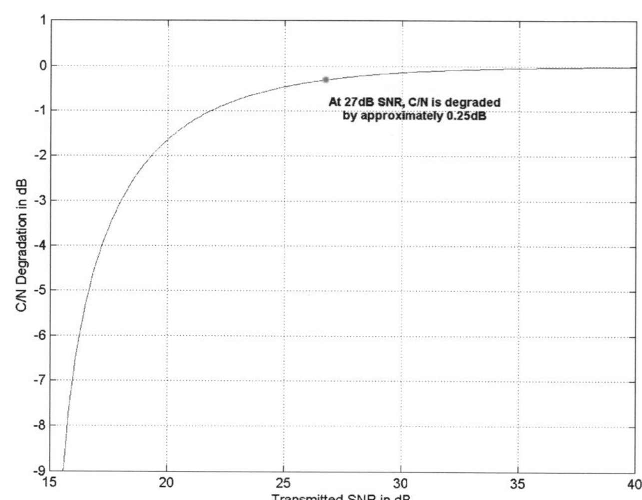

FIGURE 6.4-25 Theoretical degradation in C/N resulting from transmitted SNR.

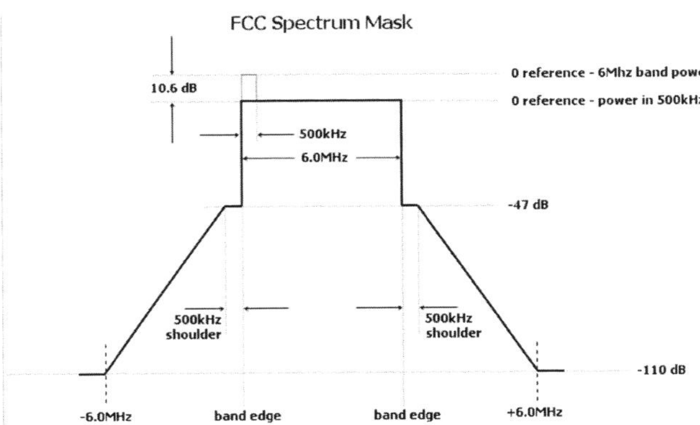

FIGURE 6.4-26 The DTV out-of-band emission mask described in FCC Rules, Section 73.622.

of the amount of effective noise that will be added into the signal as a result of the receiver equalizer. Although it is possible to use the white noise enhancement calculation to find the theoretical effect of the linear distortion, it is not recommended. In general, the linear distortion caused in a transmitter system will have little impact on the receiver threshold, which is about 15 dB, especially if the measured SNR without the equalizer is better than 27 dB.

Out-of-band measurements are made looking at the "shoulders" of the transmitter, as well as any other unwanted emissions including harmonics. Unlike in-band measurements, out-of-band emissions are regulated by the FCC. The rule for out-of-band emissions is given in the FCC Rules Section 73.622. This rule defines a mask for levels of out-of-band emissions as shown in Figure 6.4-26. The rule states that [12]:

> (h)(1) The power level of emissions on frequencies outside the authorized channel of operation must be attenuated no less than the following amounts below the average transmitted power within the authorized channel. In the first 500 kHz from the channel edge the emissions must be attenuated no less than 47 dB. More than 6 MHz from the channel edge, emissions must be attenuated no less than 110 dB. At any frequency between 0.5 and 6 MHz from the channel edge, emissions must be attenuated no less than the value determined by the following formula:
>
> $$\text{Attenuation in dB} = -11.5(\Delta f + 3.6);$$
>
> where Δf = frequency difference in MHz from the edge of the channel.
>
> (2) This attenuation is based on a measurement bandwidth of 500 kHz. Other measurement bandwidths may be used as long as appropriate correction factors are applied. Measurements need not be made any closer to the band edge than one half of the resolution bandwidth of the measuring instrument. Emissions include sidebands, spurious emissions and radio frequency harmonics. Attenuation is to be measured at the output terminals of the transmitter (including any filters that may be employed). In the event of interference caused to any service, greater attenuation may be required.

Shoulder Measurements

Industry practice is to break up the out-of-band measurement into three parts: shoulders (the first 1 to 3 MHz adjacent to the channel), out-of-band intermodulation products, and harmonics. For shoulders, it is convenient to use a digital spectrum analyzer with power band markers to measure each 500 kHz segment and compare each to the in-band signal power in 6 MHz. One way to measure the out-of-band products directly uses the observed level of the in-band signal on a spectrum analyzer as a reference. To do this, calculate a correction factor and apply it to the mask. The correction factor can be shown to be 10.6 dB by taking into account the difference in the resolution bandwidth between 6 MHz and 500 kHz [13]. To account for the pilot power in the calculation, 0.3 dB is added to the total power in the signal. The flat portion of the 8VSB signal spans 5.38 MHz. When viewed at a 500 kHz resolution bandwidth, this flat portion will appear 10 log(0.5/5.38) dB, or 10.3 dB, below the total power without the pilot. The pilot adds another 0.3 dB making the flat portion appear 10.6 dB below the total power in the 6 MHz signal. This correction factor can then be applied to the out-of-band signal as viewed on a spectrum analyzer at any resolution bandwidth assuming both the in-band signal and out-of-band signal exhibit noise-like characteristics. This is a good assumption, as most if not all of the energy being measured out-of-band is the result of intermodulation distortion. The 10.6 dB factor means that, instead of looking for 47 dB on a spectrum analyzer, an engineer would look for the energy in the first 500 kHz from the channel edge to be 36.4 dB below the flat portion of the signal. This is useful for making a quick determination of the performance of a transmitter by simply viewing a sample on a spectrum analyzer; however, this method is not advised for making precise measurements as it can be difficult to exactly define the flat portion of the 8VSB spectrum.

Out-of-Band Measurements

The out-of-band intermodulation products more than 4 to 5 MHz away from the channel edge can be difficult to measure if the dynamic range of the spectrum analyzer is not more than 110 dB, the FCC requirement for emissions at 6 MHz away. Most spectrum analyzers available at transmitter sites do not have this dynamic range. An alternative method to direct measurement involves taking measurements in two parts. The first part is to measure the transmitter emissions at a point before the final output mask filter. The second part is to measure the response of the mask. These two sets of data can be superimposed to obtain the final output response of the transmitter. Taking the measurement in two parts effectively doubles the dynamic range of the measurement and easily allows standard test equipment to be able to demonstrate compliance to the FCC mask. This method of measurement has become standard practice in the industry and is used in almost all DTV transmitter proofs.

Harmonic Measurements

The FCC requirement for emissions also applies to harmonic energy. Demonstrating compliance that these emissions are 110 dB below the transmitter output power at these higher frequencies presents some measurement challenges. The high power levels of broadcast transmitters do not allow termination directly into measurement equipment so directional couplers are used to obtain samples. The directional couplers used with waveguide transmission line provide a flat accurate representation of the signal in the main transmission line over a narrow band. As the band is extended, a +6 dB per octave correction is applied to account for the change in sample coupling to the main line as it varies with frequency. This +6 dB per octave relationship is accurate only if the transmission media stays in the same operating mode. Unfortunately, it often does not. Rectangular waveguide transmission line operates under a single octave in its principle operating mode. Harmonic measurements made from a coupled sample in waveguide rely on a reference level measured at the fundamental prior to taking the reading at the harmonic frequency. These two measurements will always be made with the transmission line operating in different waveguide modes. As a result, the +6 dB per octave relationship is questionable. Coaxial transmission line couplers are also prone to inaccuracies. The power levels required in many UHF transmission plants require physically large transmission line sizes. The transmission lines are always designed to operate in the TEM mode at the fundamental frequency; however, the harmonic frequencies may be well above the frequency where higher order waveguide modes start to develop in the transmission lines. For example, EIA 3-1/8" 50 ohm transmission line, a relatively small transmission line size, will develop waveguide modes above 1800 MHz. This is still in the range of three times the fundamental frequency for the upper UHF band. Measurements made above these cutoff frequencies could deviate from the +6 dB per octave relationship. This measurement challenge is something that engineers tend to deal with on a case-by-case basis because alternative modes of propagation do not always manifest themselves.

PROOF OF PERFORMANCE

As is the case with analog broadcasting, every DTV transmitter should have documented performance measurements in the form of a proof-of-performance document. This document serves a dual purpose of being able to demonstrate compliance to the FCC and other regulatory authorities and being able to capture a baseline transmitter performance that should be maintainable during the life of a transmitter. The format of proof of performance of a DTV transmitter may vary depending on the engineer making the measurements; however, all DTV transmitters require the following measurements in a proof of performance:

- Output power measurement
- Frequency measurement (made by verifying the pilot frequency)

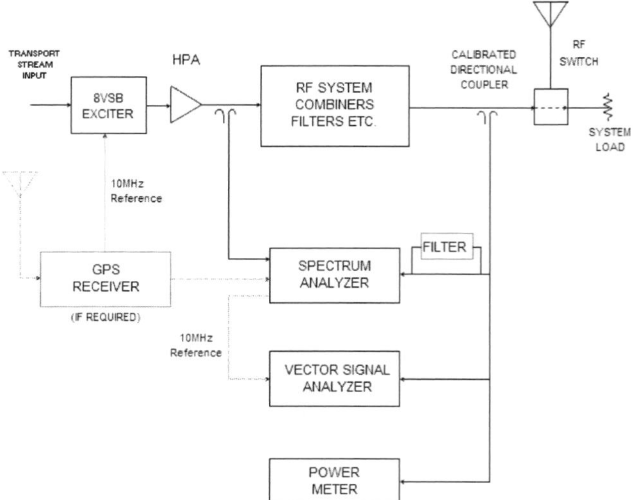

FIGURE 6.4-27 Test equipment setup for measuring a DTV transmitter.

- In-band signal performance (such as SNR, EVM, MER)
- Out-of-band shoulder measurement
- Out-of-band intermodulation measurement (sometimes this is combined with shoulder measurement)
- Harmonic measurements

With the exception of phase noise, these few measurements are required to demonstrate compliance to FCC regulations and ATSC transmitter specifications. It is not necessary to measure phase noise on a regular basis, as this is a performance parameter that rarely changes over time. Once the equipment has been checked, it is unlikely that there will be degradation in performance. Figure 6.4-27 shows a block diagram of a test equipment setup for making proof-of-performance measurements. In many of the measurements, a specialized 8VSB measurement set can be substituted for the vector signal analyzer shown. The figure is intended to show examples of how DTV proof-of-performance measurements can be taken. The examples are measurements from an actual DTV solid-state transmitter system.

Output Power

Output power can be measured using a power meter that is capable of measuring the true RMS power of a digitally modulated signal. RF power meters used in NTSC service may not be accurate enough to give a true reading. A power meter measurement can be made by taking a sample from a directional coupler at the transmitter system output, usually just after the mask filter. The coupler must be characterized so the exact coupling value at the channel frequency is known. This value is then taken into account when the power meter is read. To calculate the average power in kilowatts, use the formula:

$$Power = \left(mW \times 10^{\left(-1 \times \frac{CV}{10}\right)} \right) \times 10^{-6}$$

where *Power* is in kW, *mW* is the reading in milliwatts from the power meter, and *CV* is the coupler value in dB. For example, if the meter reading is 250 mW and the coupler value is –40 dB, the output power is 250 × $10^4 \times 10^{-6}$ (this factor converts milliwatts to kilowatts), or 2.5 kW. Newer digital power meters often allow the coupling value to be entered as an input so the meter displays the transmitter power output with the correct scaling.

Calorimeters may also be used to determine output power if the system is equipped with a liquid cooled dummy load. For this measurement, the transmitter is operated into the dummy load and a formula is used to calculate the power output by measuring the rise in temperature of the coolant as it passes through the load. The formula for calorimetry is:

$$Power_{kW} = 0.264 \times P \times CP \times \Delta T \times FC \times Flow$$

where the output power is given in kilowatts of average power, *P* is the density of coolant in g/mL, *CP* is the specific heat of the coolant in BTU/lb, ΔT is the rise in temperature of the coolant between the inlet and outlet in degrees Celsius, *FC* is the flow meter correction factor, and *Flow* is in gallons per minute. The flow meter correction factor is needed because the density of a glycol/water mix is not the same as water. It is calculated from the formula:

$$FC = \sqrt{\frac{7.02 \times P}{8.02 - P}}$$

Flow is the coolant flow as measured in gallons/minute. This formula is valid as long as the inlet and outlet temperatures stay in the range of 43 to 71°C. The coolant parameters, *P* and *CP*, will depend on the type and mixture amount of glycol used. Various flow meters may require different correction factors, so it is necessary to consult the manufacturer to verify the correct factor.

Pilot Frequency

The pilot of the DTV signal is an excellent way to verify the frequency of the transmitter. The exact frequency of the pilot carrier depends on whether or not there is an offset required. The ATSC recommended several possible offset scenarios for minimizing interference, but the FCC only mandated the use of an offset in the case of a DTV assignment in an upper adjacent channel to an analog assignment. These assignments are known as "N+1" assignments, where the N signifies an NTSC assignment. These stations are clearly identified in the DTV table of allotments (FCC Rules Section 73.622) with a "c" designation following the channel assignment. In these cases, the rule states that the station "must maintain the pilot carrier frequency of the DTV signal 5.082138 MHz above the visual carrier frequency of any analog TV broadcast station that operates on the lower adjacent channel and is located within 88 kilometers. This frequency difference must be maintained within a tolerance of 3 Hz."

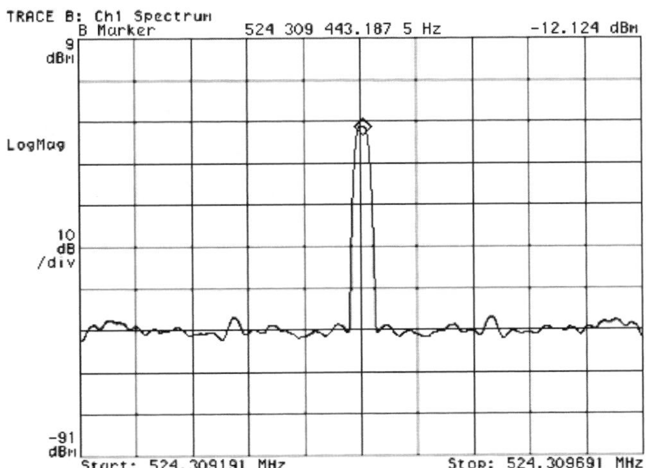

FIGURE 6.4-28 Measurement of the DTV pilot frequency using a vector signal analyzer.

In cases with no offset, the DTV pilot should be +0.309441 MHz from the lower band edge. The frequency tolerance is ±1000 Hz. In N+1 cases, the pilot carrier is offset from this frequency by +22,697 Hz plus or minus any positive or negative offset on the analog station. Note that the NTSC station may already have a ±10 kHz offset that must be taken into account, as the FCC requirement specifies an offset from the NTSC visual carrier. The N+1 offset must be maintained to within ±3 Hz. Figure 6.4-28 shows an example of the pilot frequency measurement on a DTV transmitter without an offset. This measurement was taken on a vector signal analyzer configured to look at a spectrum measurement. A spectrum analyzer could also be used. A marker was used to take the measurement. The instrument was locked to a 10 MHz reference from a GPS receiver to ensure accuracy.

In-band signal performance is measured with an HP-89440 series vector signal analyzer or another piece of test equipment with the capability to measure SNR. An example measurement is shown in Figure 6.4-29. The SNR in dB is automatically measured and printed as text in the lower left quadrant of the display. Figure 6.4-30 shows the measurement of the shoulders of a DTV transmitter using power band markers. Notice that the 6 MHz power is measured with the first set or markers. The second and third sets of markers are positioned to measure the lower and upper 500 kHz shoulders, respectively. The shoulder power relative to the channel power is calculated by subtracting the channel power reading from the shoulder power reading. In this example, the calculation would be (–54.909 dBm) – (–0.32 dBm) = –54.589 dBm. Note the sign of the measurements when doing the calculation of shoulders. Also, the pilot of the signal is

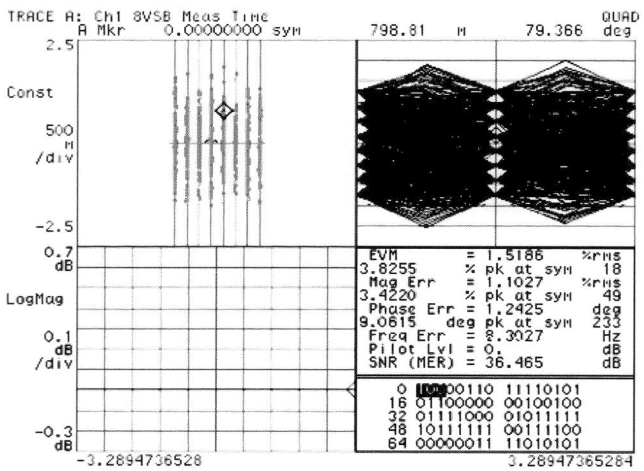

FIGURE 6.4-29 Constellation, eye diagram, and SNR measurement using a vector signal analyzer.

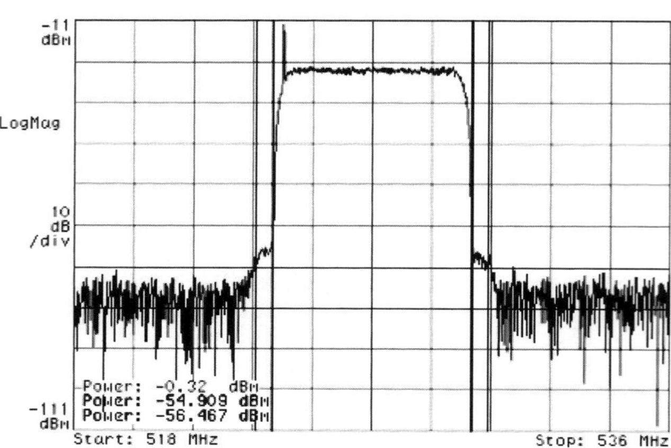

FIGURE 6.4-30 Measurement of DTV transmitter out-of-band intermodulation products using power band markers.

easily seen above the flat top of the signal because the resolution bandwidth (RBW) of the analyzer is set at 30 kHz. As the RBW decreases, the pilot will appear larger in amplitude compared to the signal flat top despite the fact that the power is not changing. This may be confusing at first, but using power band markers takes the guesswork out of the measurement. The marker function automatically integrates the power in the band regardless of the RBW used.

Intermodulation Products

Measurements of the out-of-band intermodulation products that are further away than 500 kHz are shown in Figures 6.4-31 and 6.4-32. These plots were taken using the method described in the previous section on measurements where the data is taken before the output mask filter, and the filter response is then added to the measured data. This method is used to overcome limitations in the measuring equipment.

The plots shown were created using a Microsoft® Excel® spreadsheet. This allows an engineer to concatenate the data measured on a spectrum analyzer with the swept filter response from a network analyzer. For the resulting measurement to be accurate, the spectrum analyzer must have consistent frequency spans and number of data points. As long as this is the case, the data can be imported into a spreadsheet with three columns, the first column containing the frequency vector (span), the second one containing the measured spectrum analyzer data (in dB), and the third containing the swept filter response from a network analyzer. The response of the full system is the sum of the first and second columns. (This works because the filter is considered a linear system.) Excel can be used to plot the data as shown in the two figures.

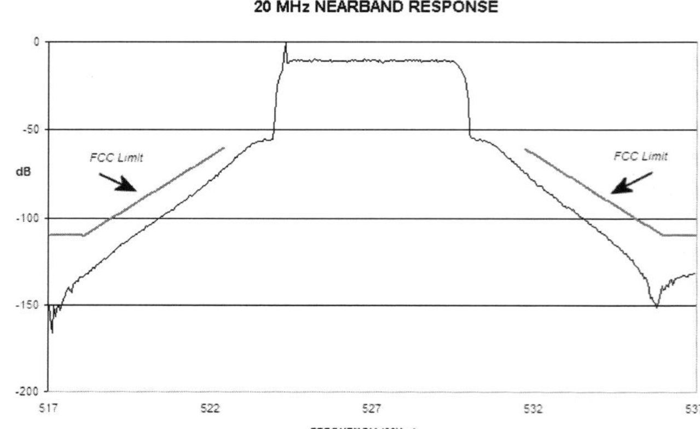

FIGURE 6.4-31 Plot of intermodulation products over 20 MHz span using measured data and superimposed filter rejection data.

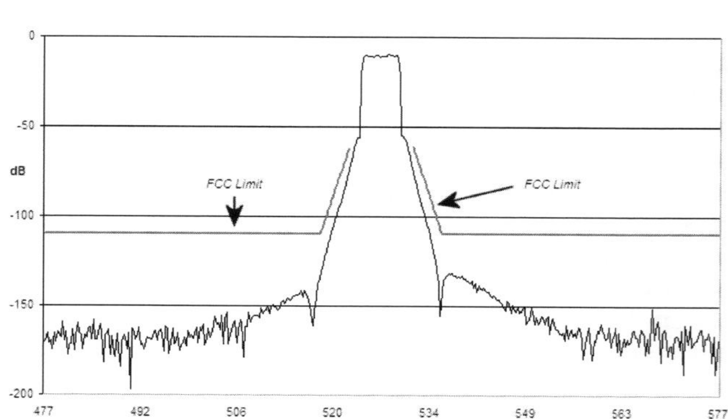

FIGURE 6.4-32 Plot of intermodulation products over 100 MHz span using measured data and superimposed filter rejection data

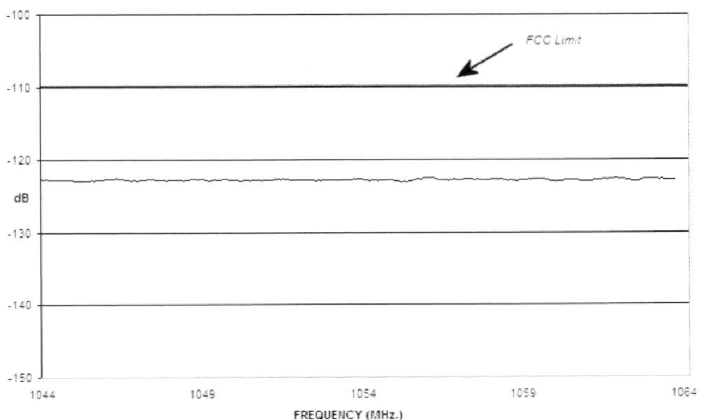

FIGURE 6.4-33 Plot of measured emissions at two times the fundamental frequency.

Figures 6.4-33 and 6.4-34 show the measurement of the harmonics at two and three times the fundamental frequency, respectively. In order to make these measurements, a reference level is set by tuning a spectrum analyzer to the fundamental frequency. Then, a high-pass filter with known characteristics is used to block the fundamental signal but pass the harmonics. In this manner, the reference level of the spectrum analyzer can be lower without overloading the front end of the instrument. Measurements are made at the harmonic frequencies of interest, and the response of the output coupler is taken into account. It is important to make sure the output coupler response is known at all frequencies of interest. These two figures were also created by importing spectrum analyzer data into Excel and then plotting. Having the data in Excel is useful because an engineer can write a formula to automatically check if any limits are exceeded.

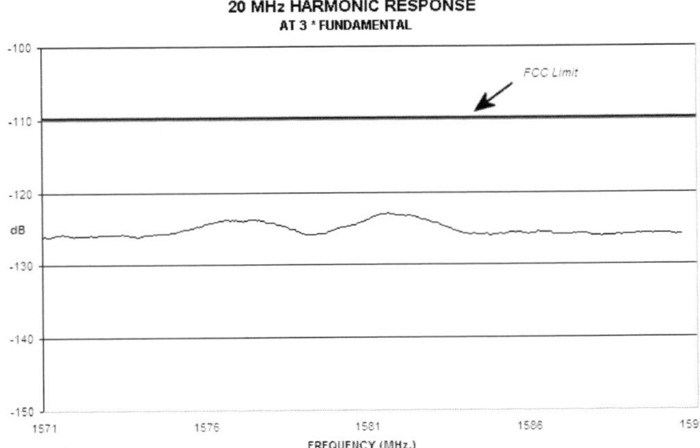

FIGURE 6.4-34 Plot of measured emissions at three times the fundamental frequency.

PREVENTATIVE MAINTENANCE

A preventative maintenance routine is essential to the efficient and reliable operation of any digital transmitter. Always follow the maintenance guidelines provided with the transmitter equipment.

Inspection and Cleaning

A good program begins with a regular inspection of the transmitter and its ancillary equipment. Electrical connections should be checked periodically. All wire connections can be susceptible to loosening due to thermal cycling or mechanical movement or vibration. In particular, wires carrying high currents such as the AC mains feeds (or shunt crowbar wires in the case of IOT transmitters) can move during turn on surges due to the high magnetic fields produced during these times. Strands of heavy-gauge wires can deform during thermal cycling, causing a loosening of the wires from the connector. Connections such as screw terminals should be inspected and tightened periodically, particularly within the first few months of transmitter operation. Replace or repair connections that are loose that cannot be tightened. Loose connections can cause intermittent failures that can be difficult to diagnose. An infrared (IR) sensor can assist in finding potential troublesome connections before they become real problems. Passive RF components should also be inspected for any discoloration and routinely checked for hot spots.

Regular cleaning of the equipment is required especially in installations where there may be significant build up of dust and dirt. Use a vacuum to remove dirt and dust. Dirt can build up on parts in such a way as to prevent efficient cooling. High voltage compartments of high power tube transmitters tend to attract dirt and dust more rapidly than other areas. This can cause corona and arcing because the dirt makes a low-resistance path across insulators and wiring. This may impact the operation of the transmitter in intermittent and unpredictable ways. These areas should be cleaned periodically with denatured alcohol. Where possible, use lint-free cloths to clean dirty surfaces of the equipment. Analog meters wiped with a cloth may cause a static charge to be built up. Use of an antistatic cleaner will prevent this.

Use this quick checklist when inspecting and cleaning the transmitter equipment.

- Inspect electrical connections; use IR sensor to find hot spots.
- Tighten screw terminals.
- Vacuum dust and dirt.
- Clean high voltage areas with denatured alcohol.
- Wipe surfaces with lint-free cloths.
- Clean meter faces; use antistatic cleaner.

All air filters should be cleaned or replaced on a regular basis. Even a small amount of dust build-up will begin to limit the efficiency of the cooling by blocking air flow. Fans must also be cleaned regularly, as dirt on the impellers will limit the amount of air the fan can move. Inspect and clean the windings and bearings of motors, as accumulation of dirt can cause overheating and decrease the life of the motor. Liquid-cooling systems also have filters or screens that must be cleaned or replaced. Pumps in these systems should be lubricated according to the manufacturer's recommendations. Be attentive to the flow rates through a liquid cooling system through the various components. Keep in mind that RF system components such as loads may also be liquid cooled and may have screens in the liquid inlet.

Record Keeping

Beyond performing routine maintenance and repairs, station engineers should establish a system of record keeping. Good records can help reduce maintenance time and show trends that may allow a fix even before a problem or failure occurs. Record keeping should include logs of daily meter readings. Most manufacturers recommend daily or weekly logs of meter readings. The specific readings will depend on the type of transmitter but at a minimum should include any adjustable operating parameter. Meters should be calibrated on a regular basis according to the manufacturer's recommendations. Note the calibration in the maintenance log when it is performed. Any unusual event or maintenance item should also be recorded. A repair should be recorded with information on the problem, the solution, and any parts replaced.

Checklist

Use the following checklist to assist in developing a regular preventative maintenance routine:

- Take meter readings.
- Inspect and clean equipment.
- Inspect and clean or replace filters or screens (air and liquid cooled transmitters).
- Check pH of liquid and flush cooling system if required (liquid cooled only).
- Inspect, clean, and lubricate fans, blowers, and motors.
- Check indicators and meters: calibrate, if necessary.
- Replace any components that are known to wear out (fans, batteries on CPU cards, etc.).
- Record all maintenance actions in a log.

Safety

Virtually all television transmitter maintenance involves working with electrical components. Electrical maintenance should only be performed by qualified personnel according to established specifications and in compliance with local and state electrical codes. When operating on tube transmitters in particular, engineers must exercise caution due to the presence of high voltage. Signs with safety guidelines (or requirements as determined by station management) should be posted in conspicuous locations at the transmitter site and can include the following:

- Allow only qualified individuals to work on or around high voltage equipment.
- Never work on or around high voltage alone.
- Never work on or around high voltage when impaired, tired, or otherwise not fully alert.
- Post emergency procedures and telephone numbers in a highly visible and accessible location.
- Always completely de-energize equipment on which work is to be performed.
- Always carefully follow any instructions or precautions provided by the equipment manufacturer.
- Disconnect, lock-out, and tag-out the high-voltage disconnect switch associated with the equipment being worked on.
- Be sure that all high voltage interlock and grounding switches are functioning properly.
- Always use a grounding stick prior to coming into close proximity of or contact with high voltage apparatus, even if it is believed to have been de-energized. Capacitors may still be storing potentially lethal voltages. Keep the grounding stick in contact with the high voltage circuit at all times when working on or near the apparatus.
- Inspect the grounding stick(s) frequently to ensure that there is no wear or damage.
- Reinstall any interlocked covers before re-energizing the equipment.
- Never disengage or disable interlocks or other safety features.
- Never wear jewelry, watches, or conductive material while working on or near high voltage equipment.
- Be sure that the work area is dry and has adequate lighting and space in which to work.
- Use only tools and equipment specifically intended for working on high voltage.
- Consult with a local utility company or a licensed electrician before working on main building power circuits.
- Be familiar with first aid procedures such as CPR. (See Chapter 2.6 for a discussion of electric shock.)

The Occupational Safety and Heath Administration (OSHA) requires that all work places use lockout/tagout procedures to protect employees from unexpected activation of equipment when it is being serviced. When service on a transmitter requires access to areas that are blocked or interlocked, disconnect the equipment from its energy source (circuit breakers) and place a lock on the source so it cannot be energized. The lock should be tagged with information about who has locked the equipment and why. Make sure that all stored energy sources are discharged before working on the equipment, even if the energy source has been locked out. For more information, refer to the OSHA website (www.osha.gov).

Annual Tests and Inspection

At least once a year, it is good practice to complete tests to verify that the transmitter is operating within all the parameters established by regulations and recommended practices. Use the original proof of performance as a guide along with the information in this chapter. The method of measurement is extremely important when performing these tests. Poor measurement methods will lead to incorrect results, which in turn can lead to wasted time and frustration. Keeping good records and following documented procedures will help keep a digital television transmitter operating in good condition for many years.

Common Acronyms

8VSB (8-level vestigial sideband)—The name given to the type of modulation used in the ATSC digital television broadcast standard.
AGC (automatic gain control)—A circuit used to regulate the gain of an amplifier, usually so the output power stays constant. This is often used interchangeably with ALC.
ALC (automatic level control)—A circuit used to regulate the gain of an amplifier, usually so the output power stays constant. This is often used interchangeably with AGC.
ATSC (Advanced Television Systems Committee)—The name of the industry group responsible for the creation and maintenance of digital television standards and practices used in the United States (see http://www.atsc.org).
BJT (bipolar junction transistor)—A transistor consisting of a base, emitter, and collector.
D/A (digital to analog converter)—An electronic component used to create an analog output from a digital input.
DTV (digital television)—The industry term used to describe digital television terrestrial broadcast service, primarily in the United States. Digital Terrestrial Television Broadcast (DTTB), is more common in other parts of the world.
DVB-ASI (Digital Video Broadcast Asynchronous Serial Interface)—A standardized interface used to carry MPEG-2 transport streams.
EVM (error vector magnitude)—A metric that expresses the total amount of distortion in a digital communication system by measuring the average magnitude of errors at each symbol instance and comparing that to the average magnitude of the ideal symbols (usually expressed in %).
FCC (Federal Communications Commission)—The governmental body in the United States that is responsible for regulating both wired and wireless communications (see http://www.fcc.gov).
FET (field-effect transistor)—A transistor consisting of a source, drain, and gate.
HPA (high power amplifier)—Refers to the final amplifier stage in a transmitter system.
IF (intermediate frequency)—A modulated signal frequency used as an intermediate step before the final upconversion to the channel frequency.
IM or IMD (intermodulation distortion)—Unwanted products produced in nonlinear systems that are the result of various combinations of sums and differences of frequencies in the fundamental signal.
IOT (inductive output tube)—A type of gridded vacuum tube used in high-power UHF amplifiers.
IPA (intermediate power amplifier)—An amplifier stage that precedes the final. These are typically used in IOT based transmitters.
LDMOS (laterally diffused metal-oxide semiconductor)—Refers to a particular arrangement of the elements of a MOSFET transistor where the current flow is lateral or across the die (as opposed to vertical).
MOSFET (metal-oxide semiconductor field-effect transistor)—A type of transistor named for the materials used in its construction.
MPEG (Moving Picture Experts Group)—An industry association that created standards for video and audio compression. Also, generic expression for data streams with compressed video and audio (see http://www.chiariglione.org/mpeg).
MSDC IOT (multi-stage depressed collector inductive output tube).
PA (power amplifier)—Often refers to the higher power modules used in solid-state transmitters.
RF (radiofrequency)—In television, RF is normally used to refer to the electromagnetic energy on the sets of frequencies used for transmission. More generally, this can refer to any frequency used for wireless transmission.
SAW (surface acoustic wave)—A technology used in electronic circuits (filters and oscillators for example) where electrical energy is converted to acoustic energy and back again.
SMPTE (Society of Motion Picture and Television Engineers)—An industry association that has established many standards and recommended practices used in television broadcasting (http://www.smpte.org).
SNMP (Simple Network Management Protocol)—A protocol used in networks for the control and monitoring of devices attached to the network.
SNR (signal-to-noise ratio)—Generally refers to a measurement of noise compared to the signal level. In DTV, SNR is a specific measurement of the power of all signal distortions compared to the average power of the signal computed at the symbol instances. It is expressed in dB.
UHF (ultra high frequency)—The frequencies in the electromagnetic spectrum ranging from 300 MHz to 3 GHz. In typical DTV usage, UHF refers to television channels 14 through 69 (470 to 806 MHz).
VHF (very high frequency)—The frequencies in the electromagnetic spectrum ranging from 30 to 300 MHz. In typical DTV usage, VHF refers to television channels 2 through 13 (54 to 216 MHz).

References

[1] Fourth Report and Order, *DTV Standards Adopted*, FCC 96-493; Fifth Report and Order, *DTV Rules*, FCC 97-116; Sixth Report and Order, *DTV Allocations*, FCC 97-115.
[2] ATSC Standard Document A/53, *Digital Television Standard*, Advanced Television Systems Committee, Washington, D.C., 2004 (http://www.atsc.org).
[3] SMPTE310M Standard, *Synchronous Serial Interface for MPEG-2 Digital Transport Stream*, http://www.smpte.org.
[4] Wicker, S.B. and Bhargava, V.K., *Reed–Solomon Codes and Their Applications*, IEEE Press, Piscataway, NJ, 1994.
[5] ATSC Recommended Practice Document A/54A, *Guide to Use of the ATSC Digital Television Standard*, Advanced Television Systems Committee, Washington, D.C., 2003.
[6] Change, K. and Sun, C., Millimeter-wave power combining techniques, *IEEE Trans. Microwave Theory Tech.*, MTT-31(2), 91–107, 1983.
[7] York, R.A., Some considerations for optimal efficiency and low noise in large power combiners, *IEEE Trans. Microwave Theory Tech.*, 49(8), 1477–1482, 2001.
[8] Symons, R.S., The constant efficiency amplifier, in *Proc. of NAB Broadcast Engineering Conf.*, 1997.
[9] Oppenheim, A.V. and Schafer, R.W., *Discrete Time Signal Processing*, Prentice Hall, Englewood Cliffs, NJ, 1989.
[10] ATSC Digital Television Standard Document A/64A, Advanced Television Systems Committee, Washington, D.C., May 30, 2000.
[11] Sgrignoli, G., DTV field test methodology and results and their effect on VSB receiver design, *IEEE Trans. Consumer Electron.*, 45(3), 894–915, 1999.
[12] FCC Rules, Section 73.622, http://fcc.gov.
[13] Gumm, L., Measurement of 8VSB DTV transmitter emissions, *IEEE Trans. Broadcasting*, 45(2), 234–245, 1999.

APPENDIX: UNDERSTANDING DTV TRANSMISSION (CD VERSION ONLY)

John Freberg

Freberg Engineering, Inc.
Homewood, IL

This Appendix is contained in the CD that accompanies this edition of the *NAB Engineering Handbook*.

C H A P T E R

6.5

DTV Single-Frequency Networks and Distributed Transmission

S. MERRILL WEISS
Merrill Weiss Group, LLC
Metuchen, New Jersey

Editor's note: The maps in this chapter are best viewed on the CD version of the handbook, in color.

SINGLE FREQUENCY NETWORKS

A single-frequency network (SFN) consists of several transmitters, operating on a single channel and covering a service area to deliver a unitary program service or complement of such services. The signals from the various network transmitters overlap each other and have the potential to interfere with one another. Steps are taken in the design of the network to maximize the service and to minimize the interference within the network.

In some ways, SFNs can be thought of as being akin to cellular telephone networks. They use multiple transmitters spread throughout a service area to place those transmitters closer to receivers. They are described in terms of "cells" when discussing the areas served by each of the transmitters and are sometimes called cellular television systems. But, they are radically different from cellular telephone systems in that cell phone systems use different frequencies for transmission and reception in each direction within each of the sectors associated with each cell in the network. The frequencies are re-used in the network at locations sufficiently separated from one another that there will be no interference between transmitters on the same frequencies. In SFNs, all the transmitters are on the same frequency; interference between transmitters will exist and must be treated.

The reasons for implementing SFNs range from service improvement to economic, regulatory, and interference considerations to spectrum efficiency. Although theoretically possible with any form of modulation, SFNs only become really practical when digital techniques are used for transmission. They depend, in particular, on techniques routinely applied in receivers to enable operation of digital transmission systems. They come in two basic configurations, having different philosophical approaches and different design objectives. They often involve trade-offs between improvements in service and interference caused within their service areas.

Benefits of SFNs

SFNs enable service to be provided in places that cannot be reached using single, high-power transmitters alone. If there are areas within a station's service contour that would not receive adequate signal levels due to terrain considerations, additional transmitters in an SFN configuration can be used in conjunction with a high-power transmitter to fill in the gaps in coverage. The high-power transmitter in such a case becomes part of the SFN and serves as a main facility that is supplemented with smaller, extension transmitters, often called *gap fillers*. In cases where it is desirable not to have a high-power main transmitter, the service areas of the multiple transmitters in an SFN can be tailored to achieve the coverage and signal levels desired, while doing so with lower overall power emitted.

Because they do not necessarily require the tall towers and large antennas that single-transmitter operations do, transmitters in SFNs often can be installed on existing towers used for other purposes, such as cellu-

lar telephone and other wireless communication services. They sometimes can be installed on the roofs of buildings. If there is not already an existing tower on which to install the antenna for a single-transmitter operation, these alternatives can avoid the need for zoning and planning approvals that sometimes delay or preclude progress on the otherwise necessary tower construction. In circumstances that require construction or reinforcement of a large tower and depending on the particular situation, using alternative locations may be less expensive and may result in a lower overall cost of implementation.

Because they permit transmitters to be located closer to receivers, SFNs reduce the amount of fade margin that must be provided to achieve reliable communications. As a consequence, they can be operated at lower power levels to obtain a given signal level over a particular service area. With lower power operation comes less interference caused to neighboring stations, both in frequency and geography. At the same time, the shorter distances from transmitters to receivers, combined with the increased number of transmitters, can result in a more uniform signal level being delivered to the area to be served by the station. The more uniform signal levels can enable the use of indoor receiving antennas in larger portions of the service area.

If a transmission system is to be used to deliver signals to pedestrian and mobile receivers, the use of SFNs permits signals to be delivered to those receivers from multiple directions. Doing so helps to minimize the impact of buildings and terrain on reception as the receiver moves. When the signal field from one transmitter is cut off by virtue of the receiver moving behind an obstruction, it is more likely that a signal field will be available from another transmitter when a sufficient number of transmitters are in use.

Digital Differences

Digital transmission involves the communication of data through naturally occurring channels[1] and using forms of modulation that are inherently analog in nature. To recover the data at a receiver, a variety of steps must be taken to retrieve the transmitted bits and to overcome the analog impairments to the signal that occur along the path of the transmission channel. The types of impairments caused by the path from transmitter to receiver include the addition of noise, variations in the amplitude response across the spectrum of the signal, variations in the time delay across the spectrum, and dynamic variations in those characteristics with time.

Depending on the type of modulation in use and the encoding of the data prior to modulation, different techniques are applied in receivers to recover the data and to overcome the channel impairments. To receive the trellis-coded, eight-level vestigial sideband (8VSB) modulation specified by the Advanced Television Systems Committee (ATSC) in its Digital Television Standard (A/53), as adopted for use in the United States and several other countries, receivers routinely incorporate adaptive equalizers to mitigate channel impairments other than noise. (Noise only can be overcome through improvements in the link budget between transmitter and receiver, which is one of the reasons for using SFN techniques, as described below.) Other modulation approaches use similar methods, although they may not be based on the same types of adaptive equalizers.

Adaptive equalizers are applied to 8VSB receivers in recognition of the fact that channel impairments behave very much like the application of linear filters (albeit complex ones) to the signals that traverse the channels from transmitters to receivers. The filters represented by the channels cause intersymbol interference (ISI) in the received signals and could make it impossible to recover the data carried in the modulation because of the interference from one symbol to another. By characterizing the channels (often through determining their channel impulse responses), it becomes possible to fashion filters the characteristics of which are the converse of those of the channels. Applying such filters to the received signals reverses the effect of ISI and allows the data carried in the modulation to be accurately retrieved. The filters used can operate either in the frequency domain or in the time domain, although time domain filters are most prevalent in consumer receiver designs as of this writing.

The channel impairments treated by adaptive equalizers largely derive from multipath transmission of the signals; that is, the signals propagate over more than one path from a given transmitter to a given receiver. Multipath causes the receiver to experience the effect of echoes in the received signals. With analog transmission, these echoes were called *ghosts* because of the ghost images they produced in the displayed pictures that were recovered from signals experiencing multipath transmission. With digital transmission through multipath channels, there are no ghosts—only a complete failure of reception when the ISI caused by the echoes crosses a threshold beyond which the error correction system in the receiver becomes incapable of recovering the data. With adaptive equalizers, the threshold is much higher than it would be without them.

SFNs take advantage of the techniques built into receivers for treating multipath by making the signals from the multiple transmitters in a network appear to receivers as echoes of one another. The result is that receivers then act on the multiple received signals just as they would on signals from a single transmitter received through multipath channels. Because different receiver designs vary in their ability to deal with multipath, the extent to which SFN designs can depend on receivers to treat their multiple signals as echoes of one another also varies. As described in detail below, the characteristics of receivers with

[1]The term *channel* can have several different meanings. As used in this chapter, it can refer either to the path through the environment taken by a signal or to the portion of the electromagnetic spectrum used to transmit a signal. Efforts have been made in the text to clarify which meaning applies to each use.

respect to their handling of multipath put limits on the designs of SFNs.

SFN Prerequisites

To enable receivers to treat the signals from multiple SFN transmitters as echoes of one another, those signals must have the characteristics of *echoes* when they arrive at receiver inputs. In other words, the signals must appear to the receiver as though they emanated from a single transmitter, modified only by the effects of the environmental transmission channel along the path from transmitter to receiver, including multipath.

Achieving the effect of single emitted signals even though the signals are originating from multiple transmitters, SFN signals must be on essentially the same frequency. To the extent that they are on slightly different frequencies, they will have the characteristics of echoes that have been reflected by moving objects; that is, they will seem to have had a Doppler effect added to one or more of them. Receiver adaptive equalizers are designed to treat signals with Doppler-impacted multipath, but the extent of any Doppler shift reduces their ability to handle other forms of multipath effects. Thus, the frequency difference between transmitters must be kept to the minimum practical level.

For receiver adaptive equalizers to determine the channel characteristics and to compute a filter that can correct for the channel, the multipath signals must contain the same data symbols, which must be separated in time from one another by no more than an amount determined by the receiver design. Thus, the transmitters in an SFN must emit the same symbols at approximately the same time. This means that the transmitter outputs must be synchronized with one another in terms of what they emit and when. As described below, it may be desirable to offset the output timing of some transmitters to aid in achieving the required time relationships at receiver inputs in certain geographic areas.

Types of SFNs

There are two fundamental approaches to the design of SFNs: a *small-cell* scheme and a *large-cell* scheme. The primary difference between them is in the way the overlap of signals from the several transmitters in the network is handled. The small-cell method is implemented using lower power transmitters that cover smaller areas and are in closer proximity to one another than in the large-cell method. In the small-cell scheme, there is intentional signal overlap to a very great extent. As discussed below, the close physical transmitter spacing has the benefit of keeping the time spacing of signals from nearby transmitters within the delay spread windows of receiver adaptive equalizers. The large-cell method is implemented using higher power transmitters that cover larger areas with greater separation from one another than in the small-cell method. In the large-cell scheme, the signals from the several transmitters are separated to the extent possible using terrain shielding, more than normal beam tilt and other elevation pattern techniques, azimuth pattern shaping, and transmitter power-level choices. As discussed below, the wide separation of the transmitters tends to stress receiver adaptive equalizers and leads to portions of service areas where signals are likely to fall outside the delay spread windows of receiver adaptive equalizers.

As of this writing, all of the SFN implementations built have been large-cell designs. Small-cell designs currently are too expensive to implement in terms of the capital costs of the transmitters and antennas and the infrastructure to interconnect them. With the declining cost of electronics and the plummeting cost of terrestrial data distribution bandwidth, the economic situation for the small-cell scheme may change. If it does, the small-cell method will be the more preferable of the two approaches, for reasons that will become apparent in later sections of this chapter. Because of the current greater practicality of implementation of the large-cell method, it will be the principal focus of the remainder of this discussion.

Large-Cell SFN Trade-Offs

Implementing large-cell SFNs implies design trade-offs or compromises with respect to the coverage and performance of the network. Because of the larger separation of the transmitters in the large-cell scheme, there are likely to be areas in which the combined signals from two or more transmitters fall outside the range in which receiver adaptive equalizers can compensate for the effective channel and accurately recover the data. It becomes part of the design process to minimize such areas and to place them where they will have the least effect on service to the public.

The design trade-offs implicit in large-cell SFN designs, then, are between increased signal levels and service in areas where there should be no difficulty caused to receivers in recovering the transmitted data and internal network interference (INI) having characteristics such that it will create difficulties for receivers in data recovery. The trade-offs are made so as to maximize the areas where signal levels and service are increased without INI causing problems, while minimizing those areas where INI is predicted to make reception more difficult. The maximization and minimization described generally are determined based on counting populations affected rather than geographic areas. The use of population counts for making decisions tends to place any deleterious INI in regions with low population densities. The methods applied in the design process for making such trade-offs are discussed below.

Resources, Abbreviations, and Terminology

- *A/110* is the document number of the ATSC Synchronization Standard for Distributed Transmission, which is available in Revision A as of this writing. It documents the DTx method that uses a centralized data processing model in a DTxA to embed timing and control information contained in DTxPs in the data stream to DTxTs. [1]

- *A/111* is the document number of the ATSC Recommended Practice: Design of Synchronized Multiple Transmitter Networks. It describes in detail methods for design of networks using DOCRs, DTxTs and DTxRs, the RF watermark, and related concepts. [2]
- *ATSC* is the Advanced Television Systems Committee, which documented the 8VSB Digital Television Standard and has adopted several documents on the subject of SFNs.
- *DOCR* is a digital on-channel repeater, which receives signals over the air from a main transmitter or from another type of transmitter closer to the signal source in a network and re-transmits them on the same channel as the one on which they were received.
- *DTS* is the nomenclature for distributed transmission systems technology used by the U.S. Federal Communications Commission (FCC), which it applies to SFN transmitters involving synchronized transmission.
- *DTx* is the abbreviation for distributed transmission used in the ATSC documents.
- *DTxA* is the abbreviation for distributed transmission adapter used in ATSC documents. DTxAs insert DTxPs into transport streams to carry the information necessary to synchronize DTxTs and/or DTxRs.
- *DTxN* is the abbreviation for distributed transmission network used in ATSC documentation; DTxN is a network of transmitters fed through studio-to-transmitter links (STLs) and synchronized with one another using DTxPs inserted into the transport stream by a DTxA.
- *DTxP* is the abbreviation for distributed transmission packet used in ATSC documents; DTxPs are inserted into transport streams by DTxAs to carry the information necessary synchronize transmitters in a DTxN.
- *DTxR* is the abbreviation for distributed translator used in ATSC documents; DTxRs receive signals over the air and translate them to another channel, where their outputs are synchronized with one another in similar fashion to the synchronization of DTxTs.
- *DTxT* is the abbreviation for distributed transmitters used in ATSC documents; DTxTs produce outputs that are synchronized with one another from transport stream inputs that contain DTxPs carrying the necessary synchronization information.
- *E-DOCR* is the abbreviation for equalization DOCR, which adds an adaptive equalizer to a DOCR to overcome many of the limitations of conventional DOCRs; in particular, the signal distortion resulting from signal coupling from the transmitting antenna to the receiving antenna can be significantly reduced.
- *INI* is internal network interference, which is the interference that occurs between transmitters operating on the same channel in an SFN.
- *ISI* is intersymbol interference, which is the interference that occurs in a transmitted data stream between symbols nearby in time because of the effects of transmission through a propagation channel and the resulting echoes. ISI occurs in the signals from single transmitters due to the natural effects of signal propagation, and it occurs in SFNs as a result of reception of signals from multiple transmitters.
- *SFN* is the abbreviation for single-frequency network, a network of transmitters collectively serving an area by transmitting on the same channel.

Example Designs

To help in visualizing and understanding many of the concepts involved in the design of SFNs, two examples are included in this chapter. One is of an actual system that has been on the air for over 3 years as this is written. There is severe terrain blockage from the main transmitter to the principal population centers on the southern extreme of the service contour of the main transmitter. Without additional transmitters acting as gap fillers, there would be little or no service to the bulk of the public that the station is intended to serve (although it would more than adequately serve its community of license, which is located near the main transmitter). The SFN is designed for a total of four transmitters: one main and three lower power gap fillers. Of these, the main transmitter and one gap filler are the ones currently on the air; the other two gap fillers are soon to be installed.

The second SFN example involves eight transmitters. It is a design developed to overcome the problem of a station not being able to construct a tower where it needed to do so within a reasonable period due to local governmental restrictions. The station has a construction permit for maximized facilities, but the site intended for its tower was withdrawn from consideration by the same community leaders who had first suggested it. It also happens that the station is on a channel adjacent to an in-market neighbor, the reference points of the two stations being separated by about 50 km. To complicate the design further, there are co-channel neighbors in two adjacent markets on either side of the station's allotted service area. The SFN generally is designed to use low-power transmitters, with one moderate-power transmitter required to take advantage of the well-known benefits of collocation to the extent possible.

TYPES OF SFN TRANSMITTERS

Three basic transmitter types can be used in designing an SFN. They differ primarily in their spectrum utilization, in the way signals are delivered to them, and in their configurations. Each has advantages and disadvantages that make it most applicable for solving par-

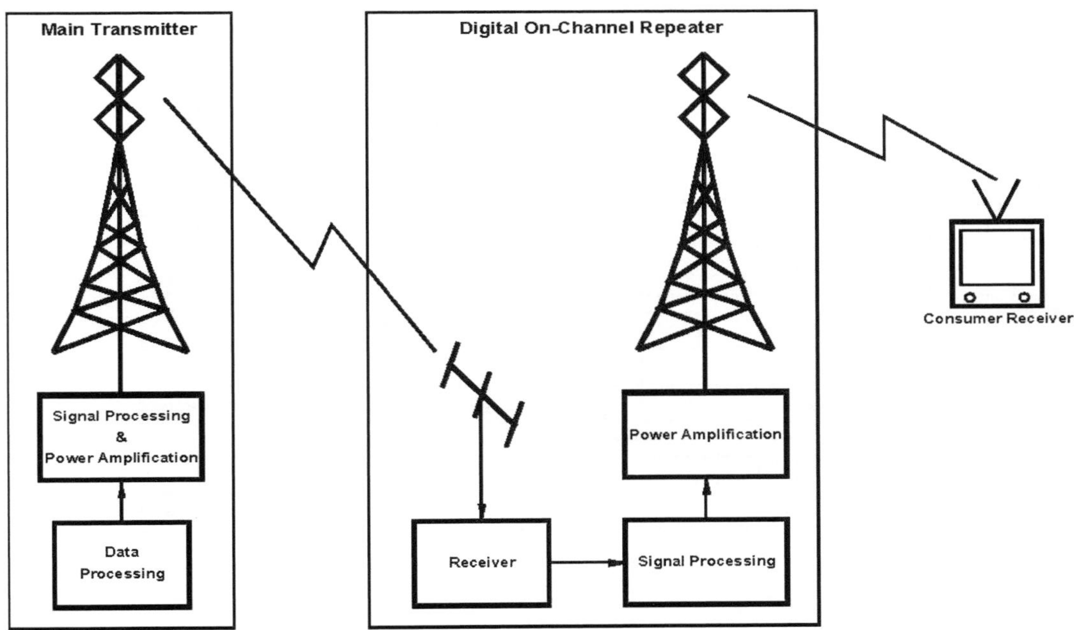

FIGURE 6.5-1 Basic digital on-channel repeater (DOCR) configuration.

ticular issues in network design. They are digital on-channel repeaters (DOCRs), distributed transmitters (DTxTs), and distributed translators (DTxRs). Their basic characteristics and designs are considered next, followed by a comparison of their respective advantages and disadvantages, which can help in determining where they are best applied. Details of their designs are presented in later sections of this chapter. It should be noted that, although the transmitter types are considered separately here and below, they can be combined into SFNs having composite characteristics of the various methods.

Digital On-Channel Repeaters

Digital on-channel repeaters are the digital equivalent of the boosters that have been used in analog services for many years, but they can take advantage of several digital processing techniques that are not applicable to analog signals. They range from very simple combinations of a receiving antenna, a channel filter, an amplifier, and a transmitting antenna to quite sophisticated combinations of receiving antenna, receiver, digital signal processing, transmitter, and transmitting antenna. To operate within SFNs, DOCRs must be designed to meet the SFN prerequisites discussed above. The basic configuration of a DOCR is shown in Figure 6.5-1. Designs appropriate for SFN use are described in detail below in the major section on DOCRs.

DOCRs provide the simplest method for establishing an SFN. They receive signals over the air from a main transmitter and retransmit them on the same channel. Because they are receiving and transmitting on the same channel, isolation between the transmitting and receiving antennas becomes a critical factor in the installation of a DOCR. To the extent that high levels of isolation are not maintained, the performance of a DOCR can be degraded. In the extreme, with a simple DOCR design the feedback from transmitting to receiving antenna can make a DOCR oscillate if the amplifier gain is too high. Because of the antenna isolation factor, the power output of a DOCR is limited, and there are limitations on the locations where they can be used. DOCRs of a particular design also have a fixed time delay that depends on the design, which further limits the locations where such a DOCR can be used. These limitations are discussed below in the sections on network designs. Certain new techniques (*e.g.*, equalization DOCRs) that have been applied in relatively sophisticated DOCR designs can mitigate some of the limitations in DOCR application. These recent developments also are discussed below in the major section on DOCRs.

Distributed Transmitters

Distributed transmitters are used in distributed transmission (DTx) networks (DTxNs). The primary characteristic of a DTx system is that it uses a studio-to-transmitter link to deliver signals to each of the transmitters in a network. The inclusion of the STL in the design avoids the issue of antenna isolation that is so limiting in the DOCR approach. As a result, DTxTs are not inherently limited in the power levels they can transmit or in the relative emission timing of their outputs. Because of the use of an STL to deliver signals to the transmitters in a DTxN, however, each transmitter must include its own exciter, of one form or another, depending on the DTx method used. The exciters of

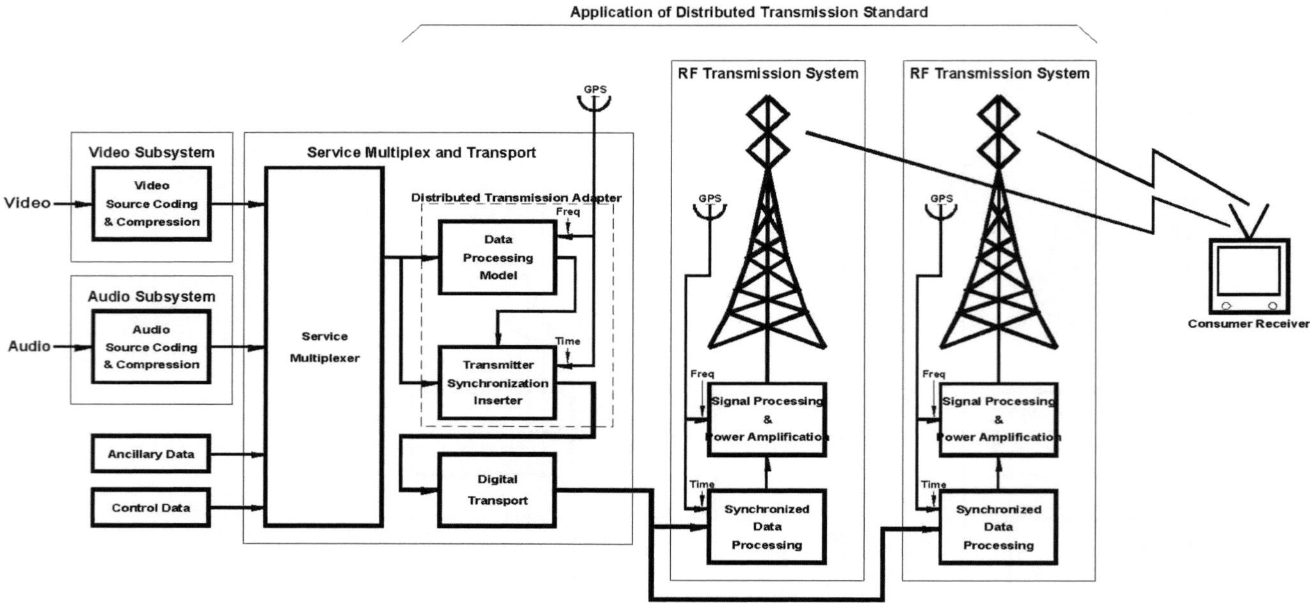

FIGURE 6.5-2 Conceptual view of a distributed transmission (DTx) system, showing a distributed transmission adapter (DTxA) and multiple distributed transmitters (DTxTs).

the several transmitters must be properly synchronized with one another in order to comply with the SFN prerequisites discussed above. Several schemes for synchronizing transmitters in DTxNs are described in detail below in the major section on DTxTs.

DTxTs represent the most complex method for establishing an SFN. They require both an STL and a means of sending synchronization information through the STL to all of the DTx transmitters in the network. Because they are not limited in power or timing, however, they provide the greatest flexibility in achieving any particular network design goals. In fact, they permit solutions not possible with DOCRs in some instances; they similarly can provide solutions not possible with DTxRs. A basic DTx system is shown at the conceptual level in Figure 6.5-2.

Distributed Translators

Distributed translators are a hybrid of DOCRs, DTxTs, and translators. They receive their signals over the air from a broadcast transmitter the output of which also can be received by the public; thus, they do not require STLs to feed signals to them, just like DOCRs. Under the right conditions, they are capable of virtually unlimited power and timing flexibility, similar to DTxTs, but they require additional radiofrequency (RF) channels, just like translators, which may not be available in many locations. A conceptual DTxR network involving a single tier of translators and two channels is shown in Figure 6.5-3.

DTxRs essentially receive signals over the air and repeat those signals on different RF channels. Because of the different input and output frequencies, many of the issues related to coupling between transmitting and receiving antennas can be avoided if the channels are separated widely enough in frequency to permit avoiding desensitization of the receiver by the transmitter. DTxRs do have the issue of requiring an adequate received signal level from the preceding transmitter in the network, thereby, to some extent, limiting their placement geographically, but they are not limited in the way that DOCRs are.

Advantages and Disadvantages

The relative advantages and disadvantages of the respective types of SFN transmitters are summarized in Table 6.5-1, which considers limitations in emitted power, emission timing, and geographic placement, plus the need for an STL to each transmitter and the need for an additional RF channel. This table also shows the relative costs of the several types of solutions. As can be seen in the table, DOCRs have several design limitations, do not require an STL or more spectrum, and are lowest in cost. DTxTs have the greatest design flexibility, do require an STL but no additional spectrum, and are highest in cost. DTxRs have medium design flexibility, do not require an STL but do require additional spectrum, and are intermediate in cost.

NETWORK DESIGN OBJECTIVES

Designing a single-frequency network requires careful balancing between competing objectives. On one hand, it is desirable to cover the largest area and to

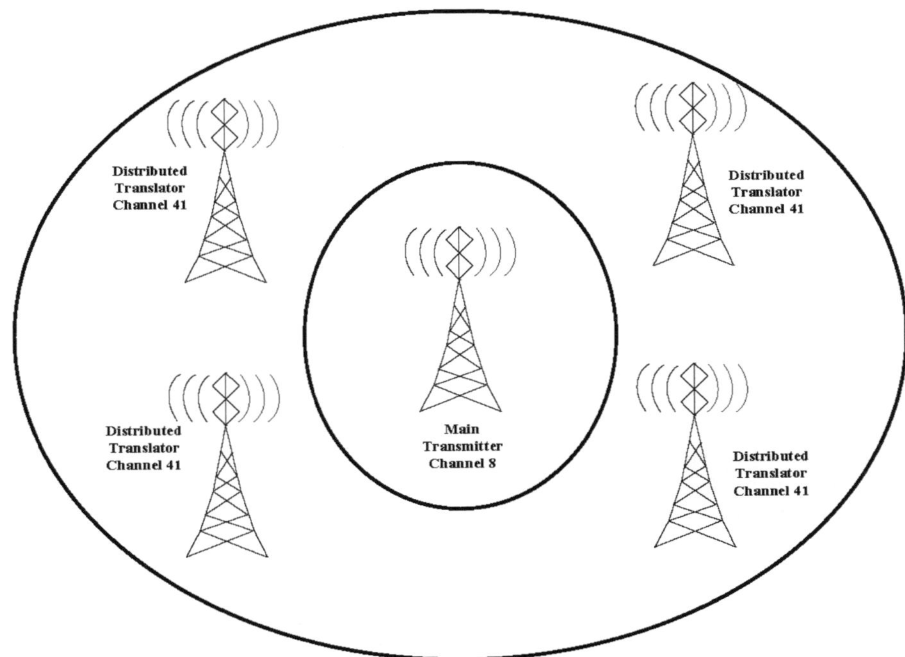

FIGURE 6.5-3 Distributed translator (DTxR) network with a main transmitter on one channel and a tier of DTxRs on a second channel. (Courtesy of Advanced Television Systems Committee, Washington, D.C.) [2]

provide the highest reasonable signal level over the coverage area. On the other hand, higher signal levels may result in more interference to other stations, both within and outside the station's market, and they may result in more INI for the station for which the network is being designed. To optimize the coverage and service, it helps to begin with the characteristics of receivers and to design the network to accommodate their capabilities.

TABLE 6.5-1
Relative Advantages and Disadvantages of SFN Transmitter Types

Characteristic	DOCRs	DTxTs	DTxRs
No power limits	–	++	+
No timing limits	–	++	++
No geographic limits	–	++	+
No STL needed	++	–	++
No additional spectrum needed	++	++	–
Cost	$	$$$	$$

Note: A minus sign is used for a beneficial property that does not apply to a particular transmitter type, a plus sign indicates that the beneficial property does apply to that transmitter type, and two plus signs indicate that the property strongly applies; the number of dollar signs indicates relative cost.

Coverage and Service

Before attempting to design a network to cover a particular area, it helps to determine what received signal levels will be considered as providing service under various conditions. The classic regulatory threshold for service, while appropriate for application and licensing purposes, may not have much meaning with respect to practical service.

Using the values for UHF operations (the values for VHF can be substituted into this discussion with the same results), the field strengths specified in, for example, the rules of the FCC in the United States, are often held to be about 10 dB below what is necessary for semi-reliable service when an outdoor antenna is used. Thus, a target threshold value for outdoor antenna reception (at 9.1 m, or 30 feet, above ground level) might be a field strength of 51 dBu (dB microvolts/meter) instead of the 41 dBu included in the rules.

When indoor service on a set-top or similar antenna is considered in a suburban or rural setting, several adjustment factors must be applied to the threshold for reliable outdoor reception. First, the antenna generally will be at a lower elevation above ground level, resulting in lower field strength. Next, the antenna gain (10 dB) assumed in arriving at the outdoor threshold likely will not be available. Then, there will be additional attenuation of the signal from passing through the roof and floors or the walls and windows of the structure. All of this leads to the need for a signal at the standard measurement elevation above ground level about 30 to 40 dB stronger than required for reliable outdoor reception—that is, according to a consensus among experts on the subject, having a field strength of 80 to 90 dBu at UHF.

When the area to be served is in a downtown canyon or a high-rise area, the signal may need to be another 10 dB stronger under standard measurement conditions in order to achieve reliable set-top reception inside apartments and similar abodes in such areas. Thus, according to the expert consensus, the signal level required for reliable reception may be as high as 100 dBu in such areas.

Using Receiver Performance Characteristics

After providing adequate received signal level, the next most important factor in designing an SFN is the expected capability of receiver adaptive equalizers. Typically, adaptive equalizers are characterized in terms of the length of the echoes (*i.e.*, the time delay) they can correct and the amplitudes of those echoes with respect to the strongest signal arriving at the receiver input. From an SFN design standpoint, the most desirable adaptive equalizer would accommodate the longest expected delay spread (*i.e.*, time from the earliest leading echo to the latest trailing echo to which the equalizer can adapt) and the smallest expected echo amplitude separation at any particular delay value.

Because an ideal equalizer cannot be expected to be found in consumer-grade receivers, it helps to have some concept of what level of performance practical equipment may be capable of achieving. To that end, the ATSC has produced a Recommended Practice: Receiver Performance Guidelines (A/74) document that contains sufficient information to guide SFN design in the near term. In particular, Figure 4.3 of A/74 shows the relationship between time delay and amplitude that can be expected in consumer receivers in the years following its publication. That chart is reproduced here as Figure 6.5-4.

Figure 6.5-4 contains a mask in which conditions falling above the line drawn on the chart can be expected to be in the range that receiver adaptive equalizers will be able to treat, while conditions below the line may not be adequately handled by some receivers. The conditions shown are the separation in amplitude (on the vertical scale) and in time (on the horizontal scale) between a pair of arriving echoes of the same signal. One echo is considered to be the main signal path, as determined by its being the strongest received echo, and the other either leads or lags the main echo. A DTx network design that keeps signals from the several transmitters in the space above the line should reasonably ensure reception on relatively modern receivers. In those parts of a DTx network service area in which the conditions fall below the line, it is far more likely that reception will be impaired.

A few observations can be made about some significant points in the chart of Figure 6.5-4. The best adaptive equalizer performance is represented by values near the bottom of the chart and near the sides. These are the values with the lowest amplitude separation between echoes and with the longest time delays. The closer an adaptive equalizer comes to the bottom corners, the better it is from the standpoint of flexibility of SFN design.

Above the level where the line in the chart flattens out to the sides, adaptive equalizer performance does not matter; the adaptive equalizer essentially is not needed above that level of amplitude separation between echoes. The flat top level represents the

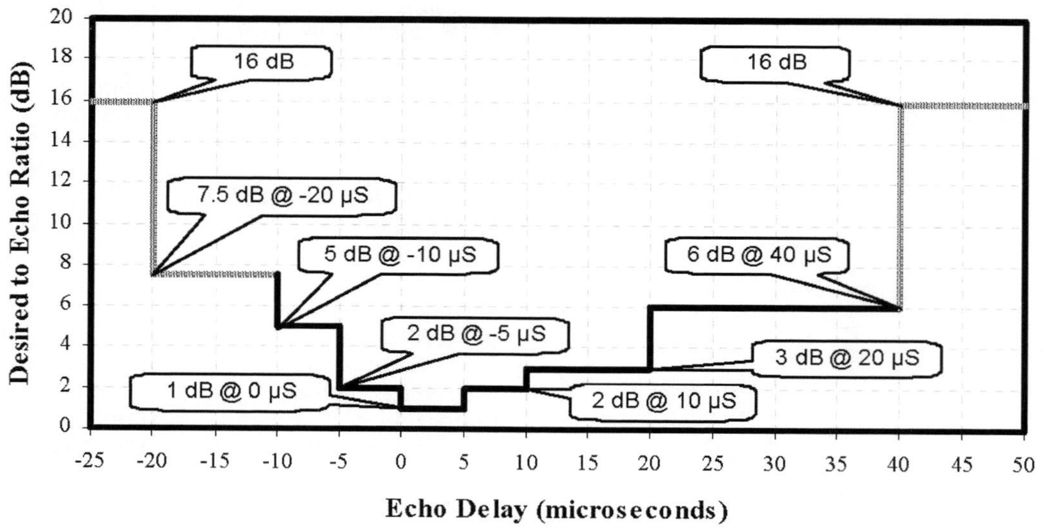

FIGURE 6.5-4 Minimum adaptive equalizer performance recommendations from ATSC A/74. (Figure courtesy of Advanced Television Systems Committee, Washington, D.C.) [3]

desired-to-undesired (D/U) ratio between signals from different transmitters at which one of them becomes dominant and network timing adjustments become irrelevant. This relationship is discussed in detail below in the section on network design techniques.

Note that, for a given amplitude separation between echoes, there is more time delay capability for echoes trailing the main signal (*i.e.*, arriving at the receiver after it) as opposed to the time advance capability for echoes leading the main signal (*i.e.*, arriving at the receiver before it). This relationship results from the tailoring of adaptive equalizer designs to approximate what occurs most often in nature and from the structure used in adaptive equalizers, with infinite impulse response (IIR) filters used to compensate leading echoes and finite impulse response (FIR) filters used to compensate trailing echoes. It means that there is more design flexibility in SFNs with apparent trailing echo relationships between received signals from multiple transmitters than there is with apparent leading echoes.

Minimizing Leading Echoes

To put the least stress on receiver adaptive equalizers, thereby leaving the maximum margin for treating path impairments, it is important to design SFNs to minimize the leading echoes they create. Determining where leading echoes are likely to occur and their expected extent is readily done when the relationship between relative levels and timing of the signals arriving at a receiver input is understood with respect to geography. This relationship can be seen in Figure 6.5-5. The figure shows a generalized map with two SFN transmitters (Tx-1 and Tx-2). Each transmitter has a contour surrounding the region in which its signal exceeds that of the other transmitter by a ratio of field strength of 10-to-1, which corresponds to 20 dB or close enough to the flat top of the mask in Figure 6.5-4 to be suitable for purposes of discussion. (In a real design effort, the value of 16 dB from Figure 6.5-4 would be used, and the ratio of the signal levels from the transmitters would be predicted at each study point on a grid surrounding all the transmitters in the network.)

Two other features of significance to the evaluation of a network design are shown in Figure 6.5-5, both represented by dashed lines. One is the line representing the locations where the field strengths from the two transmitters are equal; the other is the line representing the locations at which the arrival times of the signals from the two transmitters are equal (*i.e.*, zero delay between the two signals). The positions of the two lines can be controlled by adjusting the relative power levels of the two transmitters (for the position of the equal signal level line) and the relative emission times of the two transmitters (for the position of the equal arrival time line).

Yet another feature of Figure 6.5-5 is the gray area between the equal field strength line and the equal arrival time line. In that gray area, the first arriving signal (from Tx-2) will be weaker than the second arriving signal (from Tx-1). In other words, the space between the two lines represents the area with leading echoes. As previously discussed, leading echoes put the greatest stress on adaptive equalizers. Thus, the optimum adjustment of the network parameters generally will be one that minimizes or eliminates the gray area and, with it, the leading echoes. Depending on the choices of basic transmitter types, such an adjustment may or may not be possible.

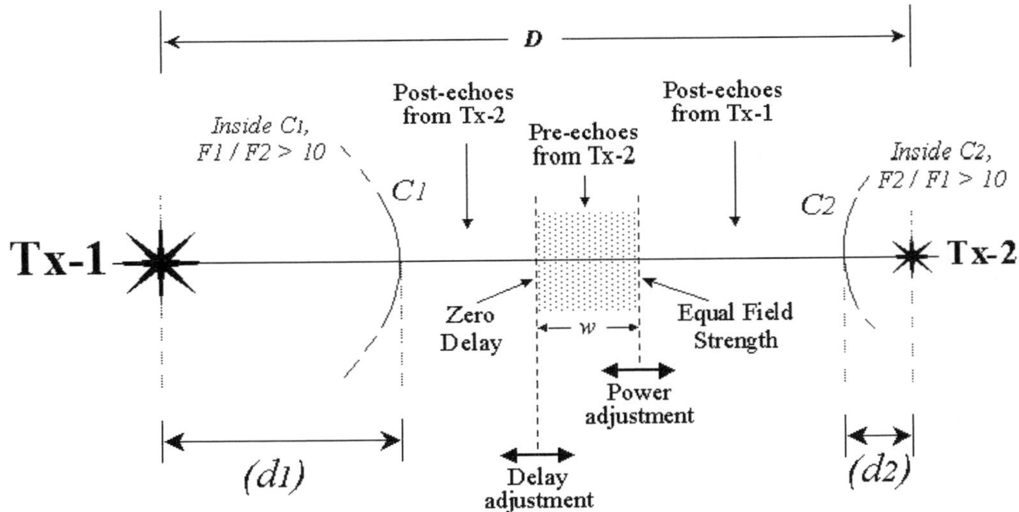

FIGURE 6.5-5 Relationship between signal levels and arrival times from a pair of transmitters at a receiver. (Figure courtesy of Y. Wu and K. Salehian, Communications Research Centre Canada, Ottawa, Ontario.)

Avoiding Internal Network Interference

In the small-cell scheme described previously, there is intentional overlap of the signals and consequent creation of internal network interference. The effect of INI can be minimized by placing the transmitters close enough to one another that their signals will fall within the delay spread windows of receiver adaptive equalizers. Controlling the areas in which the signals from the transmitters can be received, so there is overlap of signals from only a small number of transmitters at any one location, helps further to reduce the stress on the adaptive equalizers.

For large-cell systems, the avoidance of INI is a much more challenging design factor. Combinations of power level selection and antenna pattern control, along with terrain shielding, can be used to minimize the overlapping of signals that produces INI. In laying out the service areas for each of the transmitters in a network, it is desirable to minimize the width of the bands between the transmitters where the signals will overlap without one of the transmitters being dominant. Reduction of the width of the overlap zones can be achieved through selection of antenna elevation patterns with characteristics that are rather different from the norm used for stations covering large areas from a single transmitter. In particular, greater than conventional beam tilt, combined with reduction of the energy radiated from the top of the main beam, can minimize the width of the overlap region.

Use of elevation pattern control of the overlap regions between cells, as well as for control of adjacent channel interference as discussed below, implies the use of larger antenna vertical apertures than often would be required for the power levels and service areas involved. The increased antenna size tends to drive up the cost of using elevation pattern control in two ways: the cost of the antennas themselves and the cost of the tower space on which to install the antennas. Nevertheless, when the lower elevations of the antennas above ground level often possible are taken into account, substantial savings frequently can be achieved with respect to the cost of building a single, tall tower with a high-power antenna when one does not already exist.

Minimizing Interference to and from Other Stations

Two types of interference to neighboring stations are of primary concern when designing a single frequency network: co-channel interference to stations in neighboring markets and adjacent-channel interference to stations in the same market. Co-channel interference actually can be easier to control through the use of SFNs than with a high-power, single-transmitter design, while in-market, adjacent-channel interference can become more problematic when using SFNs. Both are manageable if attention is paid to the differing characteristics of each type of interference.

Co-Channel Interference to Neighboring Markets

Treatment of co-channel interference is made manageable by the ratio between the radii of the interference zone around a transmitter and the service zone around the same transmitter, which is on the order of 3-to-1 at UHF, for example. This relationship has the effect that a smaller interference zone is created when a smaller transmitter service area is used, and, consequently, the transmitter can be placed closer to the co-channel neighbor without causing unacceptable interference. The relationship is shown in Figure 6.5-6. The figure shows a very simplified example of the same coverage as that of a single, central transmitter being provided by a group of four SFN transmitters. In the diagram, the large star represents a high-power transmitter at the center of a circular coverage area, having a radius R, and the four smaller stars represent four lower power SFN transmitters, each having a coverage area of radius $r = R/2$. Point A is at a distance of $3 \times r$ from the closest SFN transmitter and point B is at a distance of $3 \times R$ from the single, central transmitter. Substituting $R/2$ for r, the distance of point A can be shown to be $2 \times R$ from the single central transmitter.

Disregarding the aggregation of signals from the more distant SFN transmitters, these computations result in point A, located a distance of $2R$ from the center of the larger circle, being the limit of the interference zone of the SFN in the particular direction shown. Because the limit of the interference zone of the single, central transmitter is point B, located a distance of $3R$ from the transmitter, it is apparent that, in the case described, using an SFN results in reducing the radius of the interference zone to a distance of approximately 2/3 that of the single, central transmitter.

In this simplified example, if an SFN transmitter were moved toward the edge of the coverage area and its effective radiated power (ERP) were reduced in the direction in which it was moved, its interference to a station in a neighboring market in that direction would be further reduced. It is worth noting that substituting an SFN for a single, central transmitter also

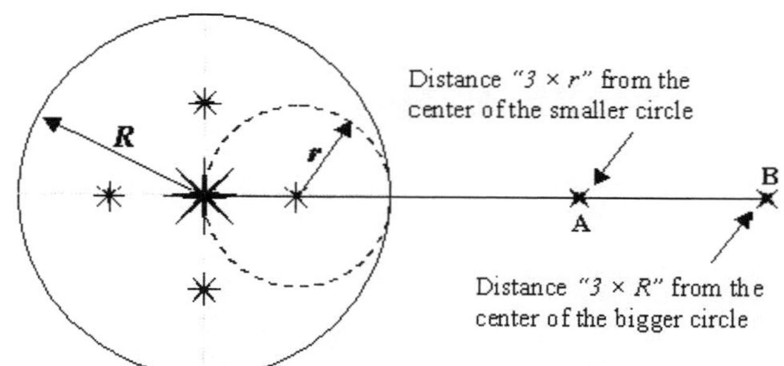

FIGURE 6.5-6 Relationship of service and interference zones of a single, central transmitter and of a group of SFN transmitters. (Figure courtesy of Y. Wu and K. Salehian, Communications Research Centre Canada, Ottawa, Ontario.)

has the effect of making reception in most parts of the service area more robust against interference from neighboring stations. Distribution of power among the SFN transmitters makes the received signal strength more uniform across the coverage area and more tolerant to stronger unwanted signals from neighboring stations.

The example given demonstrates that, when an SFN replaces a single, central transmitter, its design can have the effect of reducing interference to stations in neighboring markets. It is important to recognize, however, that pre-existing DTV allotment plans are based essentially on single, central transmitters that are assumed to replicate the coverage of existing analog stations. In such a situation, design and implementation of an SFN actually constitutes a replacement for an already-planned single, central DTV transmitter and may have to respect its protected contour and coverage area. In the planning for such a single, central transmitter, adequate separation distances from neighboring market stations already would have been provided. When the single, central transmitter is replaced by an SFN, the separation distances already provided likely will be more than those required by the SFN alone.

It is likely that there will be cases in which an SFN is designed independent of a planned single, central DTV transmitter. In such circumstances, the impact of the SFN on neighboring market stations can be predicted from the required protection ratios, the ERP of the transmitters, and the configuration and topology of the SFN. For a more precise determination, the cumulative effect of all the network transmitters should be taken into account. Based on the results of the calculations, SFN design parameters can be adjusted to avoid causing unacceptable effects on neighboring-market stations. It is to be expected, however, that network transmitters in an SFN service area replicating the coverage of a single, central transmitter can be closer to neighboring market stations than could the single, central transmitter itself.

In-Market Adjacent-Channel Interference

Providing protection to and obtaining protection from adjacent-channel neighboring stations can be markedly different from the case with respect to co-channel neighbors. Adjacent-channel stations may be in geographically neighboring markets, they may be in the same market as the one in which an SFN is to be built, or there may be partial overlap of the coverage areas of the respective stations. When adjacent-channel neighbors are in separate markets, the analysis and process become essentially the same as with co-channel stations in the same places; only the required D/U protection ratios are different. When there is overlap of the service areas, whether partial or complete, a number of additional considerations come into play.

The primary difference when there is overlap of service areas between stations on adjacent channels (whether first-adjacent or other channel relationships) is that transmitters usually have significantly stronger signals in the immediate areas surrounding them than they do in the far distance from their locations. This phenomenon is often described as the presence of "hot spots" around transmitters, in which D/U ratios to other stations can become quite low (*i.e.*, the undesired signals can be quite a bit stronger than the desired signals). It has long been understood that the best way to reduce or eliminate interference between adjacent-channel stations is to collocate them with one another, taking advantage of similar radiation patterns from nearby locations to keep the D/U ratios between the stations relatively constant. Placing transmitters on adjacent channels at substantial distances from one another seems to contravene the optimum solution.

Two ways are possible to provide protection to neighboring stations from non-collocated transmitters: reduce power of the separated transmitters to the point at which the hot spot signal levels are predicted not to cause unacceptable interference or adopt techniques that avoid creating hot spots in the first place. The power reduction method is straightforward and relatively inexpensive to implement, but it results in very significant reduction in the service that can be obtained from a transmitter so constrained. Avoidance of creating hot spots can be accomplished using specialized, but relatively expensive, antenna elevation pattern control techniques, which are described below in the major section on network design methods (in the subsection on making trade-offs). Fundamentally, the method for avoiding creating hot spots requires the use of a pattern that creates uniform field strength from the base of the tower on which the antenna is installed out to some distance from the tower at which the field strength begins to fall off. With such a pattern in use, it becomes possible to set the D/U ratio between the stations over the service area of the transmitter so as not to exceed a value selected in the design process.

Obtaining protection from adjacent-channel neighbors, especially those operating at high power levels on tall towers, often requires more of a brute-force approach. Ideally, adjacent-channel stations would build SFNs together, collocating their transmitters at all sites and thereby minimizing interference to one another (and obtaining economic efficiencies, too). When such a coordinated design is not achievable, then the next best approach is to place a moderate-power transmitter in the SFN nearby to the high-power, adjacent-channel operation to serve the area where its signal is strongest. Doing so will allow setting the D/U ratio from the other station to the SFN signal at a selected value in the strong-signal region of the adjacent-channel transmitter. The moderate-power SFN transmitter presumably would operate at lower power than its neighbor, but the D/U ratio would be selected to maintain a workable value over the service area of the SFN transmitter. Then, smaller transmitters could be situated throughout the remainder of the service area allotted to the station using an SFN, and those smaller transmitters could actually deliver stronger signals in their regions than would the high-power, adjacent-channel neighbor in those same areas. Then, of course, protection would have to be provided in the opposite direction—from the SFN to the other station.

Regulatory Considerations

Any SFN will be designed and built in an environment constrained by what the spectrum regulatory body having local jurisdiction will permit and license. Regulatory requirements determine technical matters such as the power levels and antenna heights permitted to be used for each of the transmitters, service matters such as any requirements to serve particular areas or to provide certain minimum signal strengths in specific areas, and the methods to be used in calculating the accumulation of interference from multiple transmitters to neighboring stations.

In the United States, for example, the FCC permits (on an interim basis, as of this writing) use of the same power levels and antenna heights for SFN transmitters as are allowed for single transmitter facilities.[2] Additional limitations occur through operation of the interference protection requirements of the rules in combination with limitations on where transmitters are permitted to be located and the areas they are permitted to serve. The Commission requires that stations using DTS technology provide service to essentially all of the areas they have served from their analog single-transmitter operations (known as *replication*) and allows them to serve all of the areas that they may have authorized in construction permits for their digital facilities (known as *maximization*). The Commission also requires that stations using DTS technology deliver signal levels to their communities of license equal to or greater than would be required from a single-transmitter facility. Under its interim policy, the FCC requires that interference to neighboring stations be calculated using an aggregation of the populations in study cells (2 km or less on a side and used to evaluate interference) predicted to receive interference, while under the expected final rules the Commission likely will require that the predicted signal levels from the several transmitters in an SFN be aggregated before the interference calculations are performed.

NETWORK DESIGN METHODS

Designing an SFN requires careful consideration of all the factors that will impact upon or be impacted by its operation, selection of the optimum set of techniques for each application, and frequently making trade-offs with respect to optimization of various system characteristics. It is important to recognize that the technologies of SFNs are effectively toolkits that provide a variety of solutions in most cases; the design task becomes one of finding the optimum combination.

Choosing Among Techniques

The first step in designing an SFN involves making choices of the types of transmitters to be used and the cell design approach to be applied. These choices will depend on an analysis of the environment in which the network will be built and the objectives of the network design. Among the characteristics requiring analysis are the presence of neighboring stations that must be protected from interference, whether there already is a single-transmitter facility the service of which is to be enhanced or extended by the SFN, the presence of terrain features either that must be overcome or of which advantage can be taken in the network design, the availability of tower space and its geographic relationship to the areas to be served, and the like.

Although this discussion treats each of the various characteristics as independent of the others, in fact they interact with one another, and systems can be designed with combinations of the several techniques. Thus, a service area could have portions that use large-cell techniques and other portions that use the small-cell approach. Similarly, a single, central transmitter could be converted to one of the transmitters in a DTxN, with several, most likely smaller, transmitters added to form the network. It also is possible to use a DOCR to repeat the signals from one of the transmitters in a DTx network or to use a DTxT to feed one or more DTxRs.

Transmitter Types

Each of the transmitter types has situations in which it can be optimally applied. The detailed technical characteristics of each type of transmitter are treated below in sections devoted to each type. For purposes of network design, it suffices to be aware of certain application implications of the differences between the transmitter types. These differences have to do with the ways in which signals are delivered to the several transmitter types, the spectrum requirements of the various types, and the geographic limitations on placement of the transmitters.

For example, assuming the existence of a single transmitter facility that does not provide adequate field strength throughout its FCC-allotted service area, any of the three types of SFN transmitters might be added to provide service in the under-served areas. The possibility of using DOCRs would depend on the power levels that are required of the added transmitters, the geometry of the places available to locate those transmitters relative to the locations of the areas to be served and of the original single transmitter, and the signal levels from the original transmitter that can

[2]As of this writing, the FCC policies described are those in an Interim Policy established to permit the authorization of distributed transmission system (DTS) technology for use by stations choosing to implement SFNs. The FCC has initiated and partially completed a rulemaking proceeding to adopt permanent rules for the licensing of facilities based on DTS technology. The statements made in this section about the Commission's policies are based on a combination of the provisions of the Interim Policy, the Commission's proposals in the rulemaking proceeding, and the supportive comments made by others in the rulemaking proceeding. It should be recognized that, when the Commission does adopt permanent rules, some may be different in particular details than represented here. Nonetheless, this discussion is intended to indicate the type of regulatory framework to be expected.

be received at those potential DOCR locations. The potential for use of DTxRs would depend on the availability of an additional channel in the areas in which service is to be improved. If higher power is needed than available from DOCRs, if additional spectrum is not available for the use of DTxRs, or if timing adjustments are required to minimize INI, then DTxTs, fed through some form of STL, would be necessary.

If the design is not based on inclusion of an existing, central transmitter, then either a single transmitter can be used to feed a group of DOCRs and/or DTxRs or a network of DTxTs can be established to cover all or part of the desired region. A network of DTxTs, in turn, can be extended by DOCRs and/or DTxRs. The DTxTs would provide complete flexibility in the placement of the transmitters, the power levels at which they can operate, and the relative timing at which they can emit their signals. DTxRs, of course, would require a separate channel on which to emit their signals, but they provide the same flexibility on that second channel as DTxTs have on the primary channel. DOCRs would operate on the same channel as the DTxTs, but, when compared to the DTxTs, they would be limited in power output, placement, and relative timing.

Cell Arrangement

Whether to use the small-cell or large-cell approach in a given area depends on a number of factors. Principal among these is cost. The small-cell scheme provides the optimum technical solution, but it generally costs more than the large-cell method. It involves many more transmitters, requires a means to deliver signals to each of those transmitters, and may involve lease payments at all of the transmitter sites for space to mount antennas and in which to locate equipment. Large-cell systems have similar types of costs, but the multiplier for the number of transmitters will be much lower. It also is possible to combine large-cell and small-cell regions within the service area of a station; for example, the large-cell method can be used in rural and suburban areas, and the small-cell method can be used in urban areas and high rise canyons.

In choosing between the two approaches, it is important to start with the locations of the population centers and an understanding of their spread across the service area. Ideally, large-cell transmitters should be placed adjacent to or within isolated population centers, with the regions between population centers served from the transmitters associated with the population centers. Depending on interference considerations, this layout may allow sufficient signal levels in the population centers for reception with indoor antennas, while areas away from the population centers might require outdoor antennas. When extended urban areas exist within the service area, especially when there are high-rise canyons, use of the small-cell scheme likely is a better choice and can be justified economically by the higher population density to be served.

Important factors in deciding which type of cell to use are:

- The level of service intended to be provided (*e.g.*, to outdoor or indoor receiving antennas)
- The interference protection that must be provided to neighboring stations

Indoor reception typically requires considerably higher field strengths (typically measured at 9.1 m, or 30 feet, above ground) than does reception using outdoor antennas. The difference is on the order of 40 dB higher field strength required for indoor reception, which includes the effects of reduced receiving antenna height, reduced receiving antenna gain, and losses from penetrating into a building to reach the receiving antenna. Placing transmitters closer to the receivers reduces the necessary fade margin required to maintain reliable service and provides more uniform signal levels throughout the service area. At the same time, if there is an adjacent-channel station with a service area that overlaps the service area of the network being designed, then special efforts must be taken in the design of individual transmitters to avoid creating signal hot spots around those transmitters that will cause interference to the adjacent channel neighbor. The special efforts usually involve use of well-controlled elevation patterns with uniform field strengths surrounding the transmitters.

Interference Aggregation

When multiple transmitters are used, it becomes necessary to aggregate the impacts of the signals from the several transmitters in a network on the signals of neighboring stations. There are two fundamental ways in which to carry out the aggregation. In either case, it is assumed that interference is determined by computing the population receiving interference when signals from the SFN are treated as the undesired signals and those from the neighboring station being studied are treated as the desired signals. Interference is studied in each of an array of cells, called *study cells*, within a geographic grid, and all of the population within each study cell is treated as receiving interference if there is determined to be interference predicted at the reference point for the respective study cell. In the first case, the interference is determined in the normal way for the particular channel relationship being studied, and the impacted populations receiving interference from each network transmitter are totaled. In the second case, the received signal levels (RSLs) from the several transmitters in the network are aggregated, and the total power obtained is used in calculation of the interference caused to the neighboring station.

Aggregating the population affected by each transmitter in an SFN is a relatively straightforward exercise, termed *cell aggregation*; however, it requires avoidance of double counting of the population receiving interference, as described in the next subsection on population counting. All of the normal factors included in an interference analysis, such as pointing of the receiving antenna toward the desired station and applying the off-axis antenna discrimination to the signal from each SFN transmitter, are computed individually for each SFN transmitter with respect to each desired station studied. The process is repeated for each SFN transmitter, and the results are accumulated. To make this method work, it is necessary to collect data on the locations and populations of the study cells receiving interference from each of the SFN transmitters.

Aggregating RSLs prior to determining whether interference occurs in each study cell is somewhat more complicated than the cell aggregation method and is termed *signal aggregation*. In the signal aggregation method, the predicted RSLs from each of the SFN transmitters, after consideration of antenna discrimination in the direction of each SFN transmitter, and from the desired station studied are calculated at the reference points of each of the study cells. For each study cell, the RSLs from the SFN transmitters are combined into a single value, which is used as the undesired signal level in a calculation with the RSL of the desired station studied to determine whether interference is predicted to occur in the respective study cell. If there is interference predicted in a cell, its population is included in the total population predicted to receive interference from the SFN signals. In the case of signal aggregation, there is no need to avoid double counting because each study cell of a desired station is analyzed only once.

There are several methods by which RSLs could be added together in the signal aggregation method. These include simple addition of the RSLs of the signals, a root-mean-square (RMS) calculation of the combined RSL, and a root-sum-square (RSS) calculation of the combined RSL. For several reasons, too mathematically complex for treatment here but having to do with the fact that the signals from the several transmitters cannot be coherent with one another by the time they reach a receiver, the RSS summation method is the appropriate one for use in all cases in which it is necessary to combine the RSLs from multiple SFN transmitters into a single value.

Population Counting

A significant aspect of all determinations of interference, at least as practiced in the United States, is the counting of the population predicted to receive interference above a threshold D/U RSL ratio, where the value of the threshold depends on the frequency relationship between the channels used by the desired and undesired stations. The D/U ratio thresholds for various channel combinations are published in the rules of the FCC and have the force of law. Interference is considered to be permissible when the population affected by that interference falls below a specified percentage of the population potentially served by the station receiving the interference. Populations receiving interference below the specified percentage are considered to be *de minimis*.

When counting populations in evaluating interference involving an SFN, there are several possible cases that may require consideration. Earlier sections of this chapter focused on interference from SFNs to single-transmitter operations. Interference in the reverse direction, from a single transmitter to an SFN, also must be evaluated, as must interference from one SFN to another. When the interference to be evaluated is from a single transmitter to an SFN, study cells in the service area of the SFN may be served by more than one desired SFN transmitter. In such instances, the appropriate evaluation method is to consider service to the study cell to be provided by the strongest of the SFN transmitters, as seen from the study cell and evaluated by field strength over the cell, and to aim any directional antenna used for modeling toward that strongest transmitter. Without such an approach, it would be possible for interference to a particular cell to be counted as being caused with respect to the signals from more than one desired SFN transmitter, resulting in double- or multiple-counting of interference to the study cell and an incorrect total population being predicted to receive interference.

Evaluating interference between SFNs is a composite of the methods for evaluation of the interference in both directions between an SFN and a single transmitter operation. For each cell to be studied, the desired-signal RSL should be the one from the transmitter producing the highest field strength at the reference point for the study cell, and the receiving antenna should be assumed to be aimed toward the corresponding SFN transmitter. The undesired signal RSL used for determining interference to each study cell should be the signal aggregation of the RSLs from all the transmitters in the undesired SFN, after inclusion of the effects of antenna discrimination. The result of the analysis described will be a determination of which cells are and are not predicted to receive interference. The total population served and the population predicted to receive interference then can be calculated for the pair of SFNs, and a determination can be made whether the amount of interference is acceptable based on the *de minimis* criterion.

Software Design Tools

Designing SFNs in the modern spectrum environment would not be possible without software design tools running on reasonably fast computers. A number of different software tools can be used in designing SFNs, but certain features and characteristics of those tools are required. A tool used must be able to evaluate interference both to and from other stations when multiple transmitters are involved as the desired and/or the undesired signals. A tool is needed to compute the coverage that can be achieved with repeated iteration of the network design. A tool is needed to examine internal network interference, based on the characteristics of receiver adaptive equalizers and including consideration of the timing offsets of the transmitter emissions.

All of the enumerated software tools may be included in one software package, or they may involve separate programs. In any event, they have certain features in common. They all require a terrain-sensitive propagation model. Typically, in the United States, the Longley–Rice terrain-based propagation model is used because of its adoption by the FCC for interference prediction. Other models (*e.g.*, the Terrain-Integrated Rough-Earth Model, or TIREM) also could be applied. The tool generally used for interference analyses between stations is the *TV_Process* program written for and published by the FCC. It includes the Longley–Rice methodology, as described in FCC Office of Engineering and Technology Bulletin Number 69, and can be straightforwardly modified to treat multiple transmitters in a network on the desired, the undesired, or both sides of the D/U ratio calculation.

The TV_Process program is not conducive to interactive network design modifications with immediate display of the results; consequently, at least one other tool is needed to deal with coverage, internal network interference, transmitter timing adjustments, and stresses placed on adaptive equalizers, ideally producing a graphic display of the results of system parameter changes. Many such tools exist in the marketplace. Unless and until they are modified to meet the specific needs of SFN design as described herein, it usually will be required to extract data produced by propagation modeling tools into a standard mathematical processing and data manipulation program for completion of the necessary algorithms. The manipulated data often then can be returned to one of the original software tools for display.

Making Trade-Offs

The art of SFN design is the art of making trade-offs. It is highly unlikely, in any given case, that a design will be achieved that can perfectly meet all of the goals set out for the network. Consequently, it is important at the outset to establish goals for the SFN design that are given relative priorities. Then, when it becomes impossible during the design process to meet all of the goals perfectly, guidance will already be available on how to balance the potential choices against one another. This, of course, is the process of making trade-offs.

Trade-offs in the design of an SFN typically involve matters of coverage *versus* internal network interference, service *versus* interference to neighboring stations, and the like. There also will be aspects of cost included in the necessary trade-offs; for example, the number of transmitters included in the network, the use of one type of transmitter or another, the type of STL, and the power levels of the individual transmitters all affect both network performance and cost.

Part of the process of making trade-offs is determining where to locate the INI that is inherent in the use of multiple transmitters. Typically, the choice will be made to create regions near population centers where INI will be minimized. This result can be achieved either through the use of small cells, so as to avoid the effects of INI, or by placing larger cells so one transmitter is dominant in the population center and any INI is located in low-population regions. Depending on the specific terrain, sometimes advantage can be taken of terrain features such as ridges and mountains as dividing lines between the service areas from different transmitters, thereby avoiding the effect of INI.

When considering treatment of interference to geographically neighboring stations, trade-offs often will be between protection of the other stations, service within the service areas of the SFNs, and cost. Frequently, the best method for managing interference to other stations is through control of antenna elevation patterns. For example, it is possible to use greater than normal beam tilt in the direction of a distant station, thereby maintaining the power level in that direction while providing a reduction toward the distant service area. It also is possible to create an elevation pattern that has a notch on top of the main beam, further to reduce the signal toward the distant station. The trade-off for achieving such elevation directivity is that the antenna becomes longer than would otherwise be required, especially at low transmitter power levels, and the added length contributes both to the cost of the antenna itself and to the cost of tower space on which to install it.

In situations in which there is an adjacent channel station with a service area that overlaps that of an SFN, it similarly is possible to mitigate interference to the neighboring station. The interference reduction can be achieved again by use of elevation pattern control—in this case, through the use of a pattern having relatively uniform field strength from the transmitter site to some substantial distance from the transmitter, at which point the field strength begins to fall off. Such a pattern is shown in Figure 6.5-7 and is termed an *inverted cosecant squared* pattern. Placing SFN transmitters that use the inverted cosecant squared pattern within the service area of the adjacent channel station allows setting the D/U ratios to the neighboring station in such a way that the values selected are not likely to be exceeded within the coverage areas of the respective SFN transmitters. The trade-off to achieve the elevation pattern necessary for protection of an adjacent channel station with a service area overlapping that of an SFN is, once again, the length of the antenna and its attendant costs of acquisition and of tower space.

Design Examples

As discussed above, two design examples are presented herein. The examples given are of a four-transmitter network in which a high-power, single-transmitter facility is extended with three added transmitters to overcome terrain obstacles and of an eight-transmitter network designed from the beginning for SFN operation in an environment having substantial interference protection requirements. In both instances, the technique used is that of distributed transmitters, requiring STLs to tie the studios to the

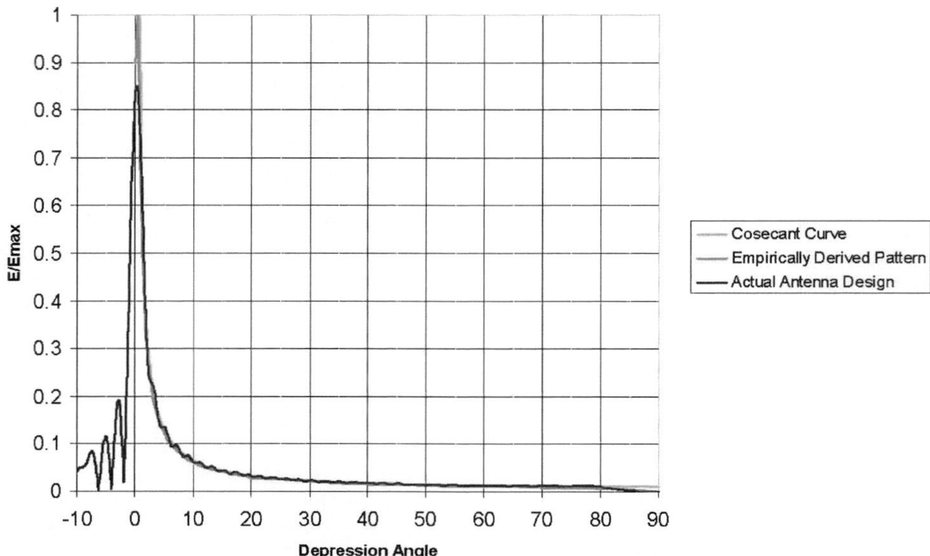

FIGURE 6.5-7 Actual inverted cosecant squared antenna design compared to empirically derived ideal elevation pattern and cosecant curve (cosecant curve applies to field values, cosecant squared to power values). The three curves essentially overlap each other. To show what small difference there is, see the color version of the image on the CD.

various transmitters, but other geographic configurations could have allowed for use of DOCRs, and greater spectrum availability could have allowed for use of DTxRs.

Example 1. A Main Transmitter and Three Gap Fillers

The first example is of a system designed to overcome the terrain blockage of a major geographic feature within the service area: a ridge that blocks service from the main transmitter (a megawatt-class facility) to the three principal population centers along the southern tier of the main transmitter service contour. The effect of the obstruction can be seen in Figure 6.5-8, which shows the noise-limited service contour of the station together with the Longley–Rice predicted field strength at locations throughout the allotted service area of the station. The population centers all are located in isolated valleys, so relocating the main transmitter near any one of them would not have provided service to all of them. Moreover, such relocation from the existing main transmitter site would have eliminated service to the rural area in the northern half of the station's service contour, where it often provides the only over-the-air television service.

The result of adding three gap filler transmitters is shown in Figure 6.5-9, which adds to the map of Figure 6.5-8 the contours of the additional transmitters and their Longley–Rice predicted field strengths. As indicated by the yellow and orange areas on the maps, the regions receiving field strengths sufficient to permit indoor antenna reception encompass the population centers in Figure 6.5-9 but not in Figure 6.5-8. Because of the size of the valleys involved and the relative containment of service from each transmitter caused by the terrain, the added DTx transmitters have power levels of 25 kW ERP to the south and southwest, while the DTxT to the southeast has a power level of 50 kW ERP. In each case, the gap filler antennas have higher than normal beam tilt values, ranging from 3 to 5 degrees' depression, down into their respective valleys, to provide signal coverage as widespread as possible within the valleys. The

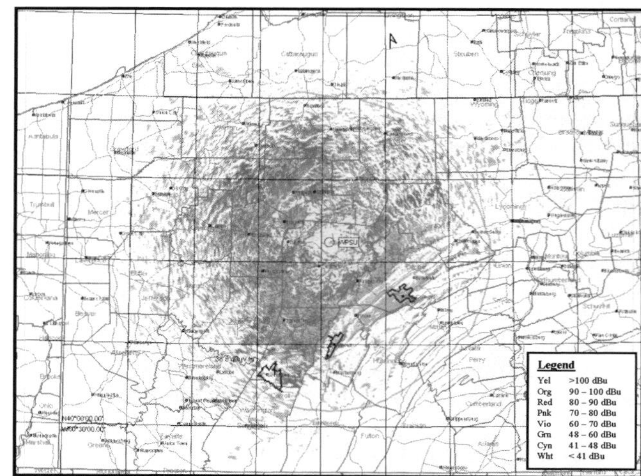

FIGURE 6.5-8 810 kW Main Transmitter at 413 m Height Above Average Terrain (HAAT), with major terrain obstructions to southeast and south-southwest cutting off service to all three major population centers

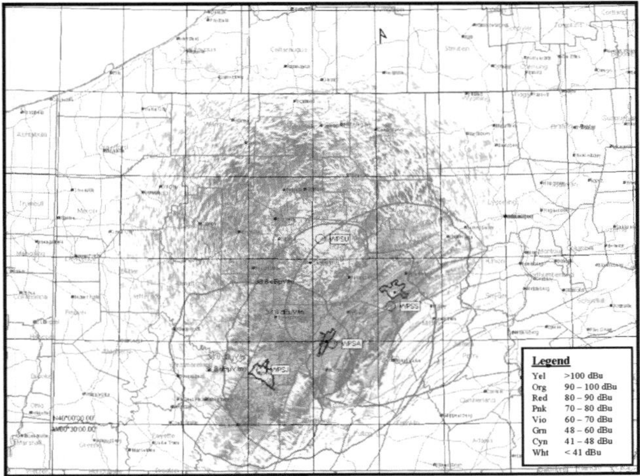

FIGURE 6.5-9 Distributed Transmission Network (DTxN) comprising Main Transmitter of Figure 8 and three Distributed Transmitter (DTxT) gap fillers, operating at power levels of 25 kW, 25 kW, and 50 kW from left to right (west to east).

antennas used are ordinary, low-power types, with transmitter power output levels on the order of 2 kW each.

Even though the terrain provides a degree of isolation between the service areas of the several DTxTs in Example 1, it still is necessary to pay attention to the INI that is caused between the transmitters. The areas of concern are shown in Figure 6.5-10, where the D/U ratios based on Longley–Rice analyses are presented. In particular, areas that are not yellow or white are places where INI must be managed. These are the areas where the signal level is above the reception threshold and where the D/U ratio is below the 16 dB

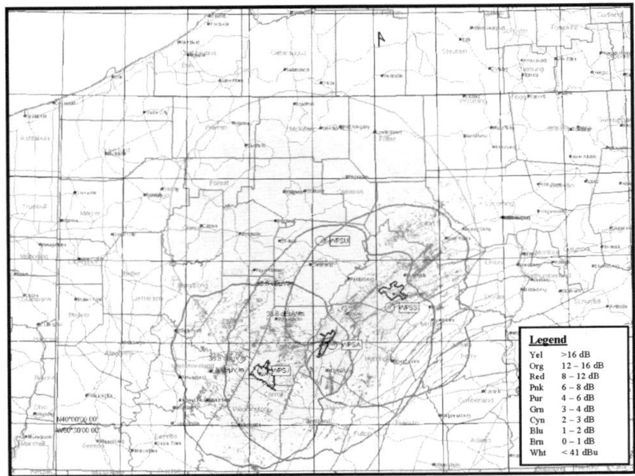

FIGURE 6.5-10 Distributed Transmission Network (DTxN) of Figure 9, showing Internal Network Interference (INI) in areas of concern.

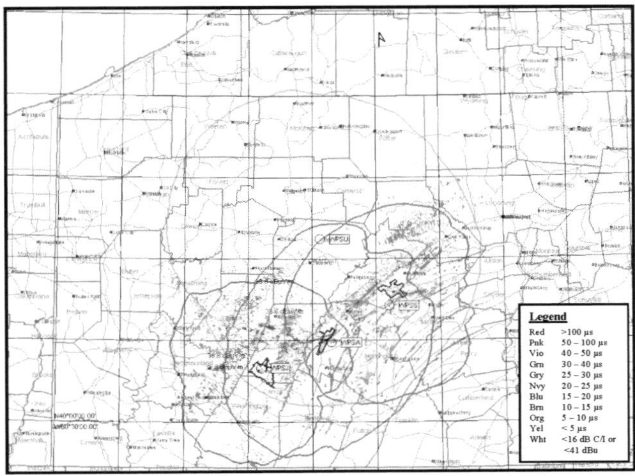

FIGURE 6.5-11 Distributed Transmission Network (DTxN) of Figure 9, showing relative timing of signals in areas having Internal Network Interference (INI) exceeding 16 dB threshold.

adaptive equalizer operation threshold shown in the mask of Figure 6.5-4. It is in these regions that the timing of the signals from the respective transmitters must be controlled so as to minimize the effect of the INI and to allow receiver adaptive equalizers to treat the multiple signals as echoes of one another. The timing of the signals, in the regions where timing matters, is shown in Figure 6.5-11. To obtain the results shown, the transmitters were set with offsets from equal emission times of 0 µsec, 115 µsec, 135 µsec, and 150 µsec, respectively, for the main, southeast, south, and southwest transmitters.

Example 2. SFN with Numerous Interference Constraints

The second example system is so constrained by interference that it takes advantage of many of the possible techniques for interference mitigation described previously. The primary constraint is an in-market, full-power station on a first adjacent channel. There also are co-channel full-service stations in adjoining markets to the northeast and southwest and a low-power station entitled to protection (*i.e.*, a class A station) also to the northeast. The protected contours of the several neighboring stations and the service contour authorized to the subject station all are shown on the map in Figure 6.5-12. On that map, the subject station's authorized contour is shown in orange, the adjacent channel protected contour is shown in red, and the co-channel protected contours are shown in maroon. Protection is required in both directions (to and from) the neighboring stations, but, because the other stations already are built, control of interference in both directions must be part of the SFN design considerations.

The best way to achieve interference protection with respect to an adjacent-channel, in-market neighbor is through collocation of the transmitters. That is

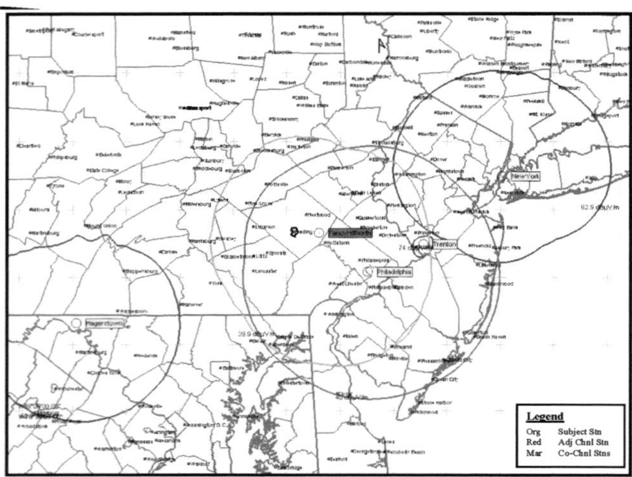

FIGURE 6.5-12 Authorized contours of station in Example 2; one in-market, adjacent-channel station; two co-channel, full service stations in adjoining markets; and one low-power, co-channel, in-market station.

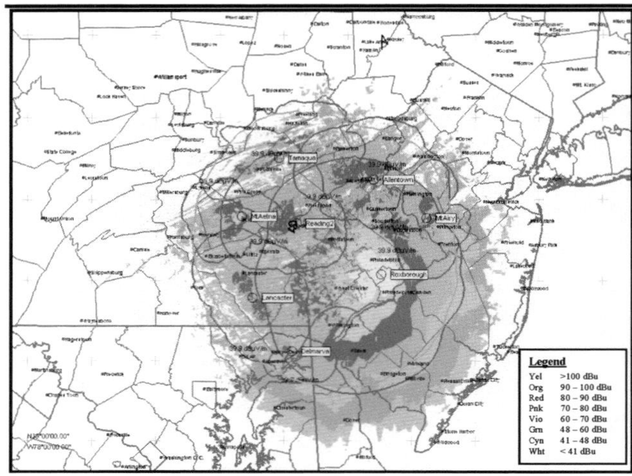

FIGURE 6.5-13 Authorized contour of station in Example 2, contours of eight Distributed Transmitters (DTxTs), and Longley-Rice predicted field strengths from Distributed Transmission Network (DTxN).

the approach taken for the highest power transmitter in the SFN, with its location in the same antenna farm as the adjacent channel neighbor. The adjacent channel is a megawatt-class station, and the power of the collocated transmitter was set about 8 dB lower, at about the same antenna height, to obtain a measure of protection from the neighboring station. At the same time, the collocated transmitter must provide protection to the two stations (class A and full service) to the northeast, and its contour must be constrained so as not to project excessively beyond the authorized service contour of the station. These goals were achieved using a panel antenna with a carefully sculpted elevation pattern that varies in the different directions. In particular, it puts a null toward the radio horizon in the direction of the stations to the northeast, resulting in a reduction of interference to both stations, as measured by population predicted to receive interference with respect to the full service station and by contour overlap with respect to the class A station.

All of the other transmitters in the SFN use the inverted cosecant squared antenna elevation pattern described previously, in conjunction with azimuth patterns chosen to match the authorized contour to the extent possible. The ERP of the smaller transmitters in the network ranges from 110 W to 4.25 kW, and antenna heights are between 30 and 76 m above ground level at the center of radiation. The overall result of the design in terms of service provided is shown on the map in Figure 6.5-13, which includes the noise-limited contours of the SFN transmitters in purple, the authorized contour of the station in orange, and the Longley–Rice predicted field strengths as shown in the legend. The areas in which INI must be mitigated are shown in Figure 6.5-14 in colors other than yellow (or white, where signal levels are predicted to be too low for service). The relative timing of arrival of the signals in the areas where their timing matters is shown in Figure 6.5-15. The emission timing of the various transmitters was adjusted to achieve the results shown on the map. With the combination of design parameters used, interference to other stations generally was reduced by a substantial amount, and the population served within the authorized contour was increased by about 3 million people over what could have been achieved with a single transmitter at the authorized location, largely as a result of overcoming adjacent channel interference from the in-market, adjacent-channel operation.

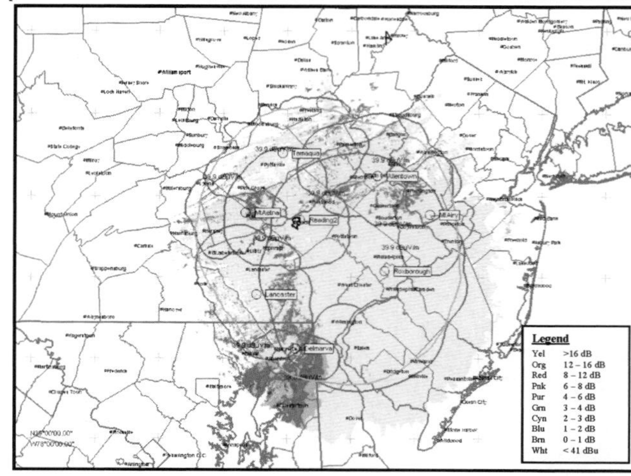

FIGURE 6.5-14 Distributed Transmission Network (DTxN) of Figure 13, showing Internal Network Interference (INI) in areas of concern.

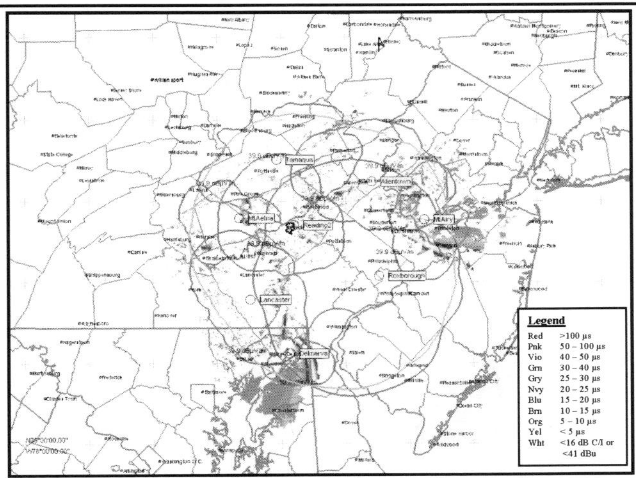

FIGURE 6.5-15 Distributed Transmission Network (DTxN) of Figure 13, showing relative timing of signals in areas having Internal Network Interference (INI) exceeding 16 dB threshold.

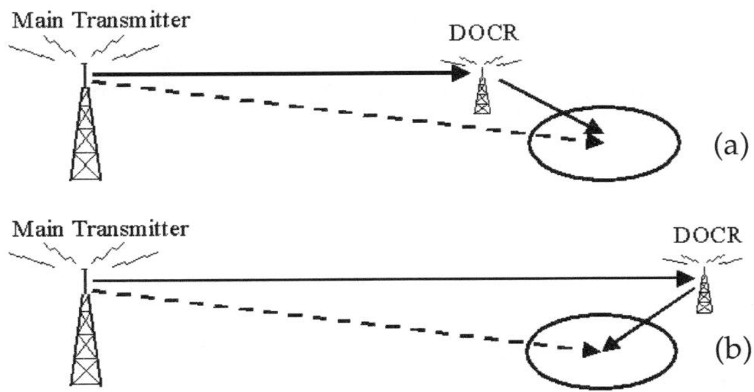

FIGURE 6.5-16 Geometric relationship between main transmitter, DOCR, and DOCR service area and its effect on leading echo time displacement. (a) Main transmitter, DOCR, and DOCR service area in approximate straight line, yielding minimum leading echo time displacement. (b) DOCR service area between main transmitter and DOCR, yielding increased leading echo time displacement. (Figure courtesy of Y. Wu and K. Salehian, Communications Research Centre Canada, Ottawa, Ontario.)

DIGITAL ON-CHANNEL REPEATERS

From an implementation standpoint, digital on-channel repeaters may be the simplest of the SFN transmitters to install, but they are not the most flexible in their application and may not be the easiest to design into a network. Fundamentally, DOCRs receive signals over the air from an earlier transmitter in the network, process those signals, and retransmit them on the same channels on which they were received. Because they receive and transmit on the same channels, an important aspect of their designs, with respect to both the equipment and the installation, is the coupling that occurs from the transmitting antenna back into the receiving antenna. Installations should be designed to minimize the antenna coupling, while the equipment may be designed to minimize the effect on signal quality of any coupling that does occur.

DOCRs are characterized by a fixed time delay through them, with the actual time delay dependent on the equipment design. Generally, the more the processing that is done in a DOCR to clean up the signal, the longer the time delay through the DOCR becomes. Some clever designs have been developed to minimize the time delay while producing the highest signal quality possible from a DOCR. Coincidentally, the same designs are those that permit the highest power outputs to be obtained from DOCRs.

Because the laws of physics dictate that DOCRs always have a positive time delay from arrival of the signal at the receiving antenna to emission of the signal from the transmitting antenna, there are certain limitations that should be applied to the geometric arrangement of the transmitter that is the source of signals to the DOCR, the location of the DOCR itself, and the location of the area to be served by the DOCR. The possible arrangements are shown in Figure 6.5-16. The limitations derive from an objective of avoiding causing, or at least minimizing, leading echoes in the area to be served by the DOCR.

As can be seen in Figure 6.5-16a, if the source of signals to the DOCR (labeled "Main Transmitter" in the figure), the DOCR itself, and the area to be served are all in roughly a straight line, with the area to be served beyond the DOCR when looking from the signal source, then the arrival time of signals in the DOCR service area from the DOCR will be approximately the arrival time of the signals from the signal source plus the delay of the DOCR. Thus, the DOCR signal will arrive in the service area later than the signal from the signal source. Assuming the DOCR signal is stronger in the DOCR service area than that of the signal source, the DOCR signal will provide the timing reference to receivers in the DOCR service area, and the signal from the signal source will appear to receivers as a leading echo. Because the two transmitters and the DOCR service area are in a straight line and in the order shown, the leading echo from the signal source will have the minimum possible time offset from the reference signal provided by the DOCR.

If the geometric relationship between the signal source (*i.e.*, main transmitter), the DOCR, and the DOCR service area places the service area to the side of the line between the two transmitters or, worse, between the two transmitters, the time delay to receivers of signals from the signal source will become relatively shorter while the time delay of signals from the DOCR will become relatively longer, as can be seen in Figure 6.5-16b. The result will be a leading echo displaced much further in time from the reference signal from the DOCR than in the case shown in Figure 6.5-16a and described above. The longer leading echo will place more stress on receiver adaptive equalizers, leaving less margin for them to correct echoes occurring in the natural environment.

As was discussed above with respect to Figure 6.5-4, receiver adaptive equalizers generally can be expected to have less margin for handling leading echoes than for trailing echoes. Because DOCRs always convert source signals into leading echoes when receivers in the DOCR service area are able to receive both the source signal and the DOCR output, whenever possible, steps should be taken in designing the DOCR installation to minimize the leading echoes. These steps include using DOCRs primarily to serve areas that have significant terrain blockage of the signal from the signal source, so the D/U ratio between the DOCR signal and the source signal will be as great as possible; placing the DOCR between the signal source and the intended DOCR service area, to the extent possible; and applying advanced DOCR signal processing techniques in those cases in which high DOCR power output is required, isolation cannot be achieved between the signals from the source and those from the DOCR, or the geometry of the transmitters and service area must be other than the desirable arrangement.

Turning to the technology of DOCRs themselves, several configurations are shown in conceptual block diagram form in Figure 6.5-17. In Figure 6.5-17a, the simplest arrangement of a receiving antenna, channel filter, amplifier, and transmitting antenna is shown. Because there is no frequency conversion, the prerequisite that the output frequency must match the input frequency will be met inherently. Because there is no demodulation, the prerequisite that the output data symbols must match the input symbols also inherently will be met. Limitations of the simple DOCR configuration are in adjacent channel selectivity, as only on-channel filters are used; in the signal-to-noise ratio that can be achieved, as there is no signal processing to recover the original signal; and in the power level that can be transmitted, as a result of the other two limitations. An advantage of the simple design is that it generally has the shortest time delay of any of the designs (measured in nanoseconds). Typically, the very simple design of Figure 6.5-17a only would be used to serve a small, isolated area and where good separation can be achieved between receiving and transmitting antennas of the DOCR.

Moving up the complexity scale, Figure 6.5-17b shows a DOCR design that uses intermediate frequency (IF) signal processing. The input signal is converted to an IF, filtered and amplified, then converted back to the same frequency for more amplification and delivery to the transmitting antenna. The IF filtering can be done with conventional tuned circuits or with a surface acoustic wave (SAW) filter, which can achieve greatly improved filtering characteristics as compared to ordinary tuned circuits. With IF signal processing, a DOCR can achieve better performance than the simple design with respect to adjacent channel emissions and increased power output because of the narrower bandwidth of the signals that it must amplify. The trade-off for these improvements is a longer delay through the DOCR, which can be up to several microseconds when a SAW filter is used. Note that, for

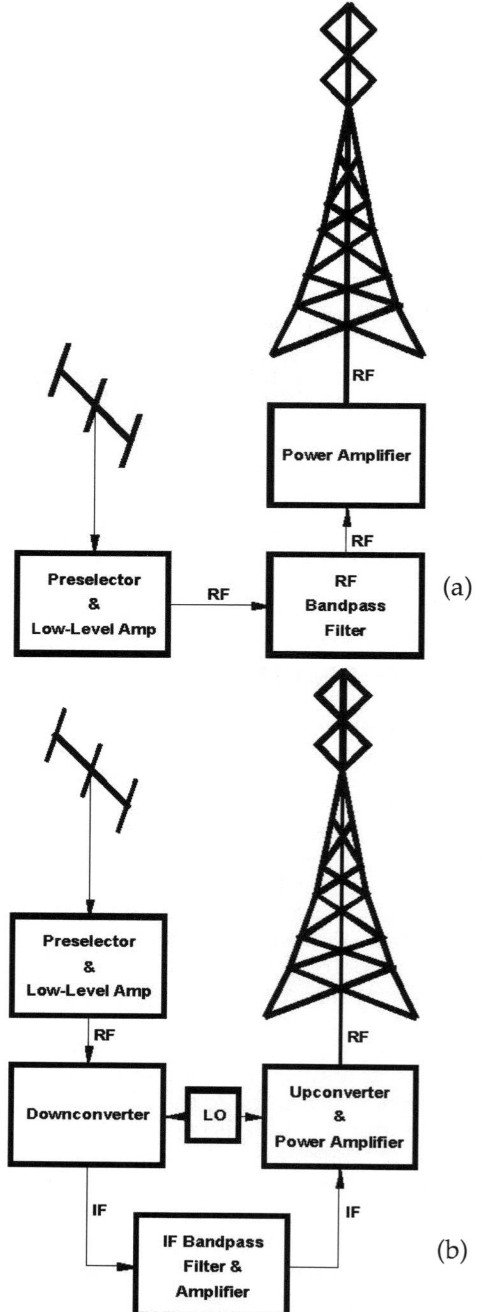

FIGURE 6.5-17 Digital on-channel repeater (DOCR) configurations: (a) RF processing DOCR, (b) IF processing DOCR, (c) Baseband equalization DOCR, (d) Baseband decoding DOCR.

DOCR designs including IF processing, use of the same local oscillator (LO) frequency for both downconversion and upconversion helps to restore the output signal to the same frequency as the input signal, to meet the prerequisite for SFN operation.

Further up the complexity scale is a design that demodulates the IF signal to baseband symbols,

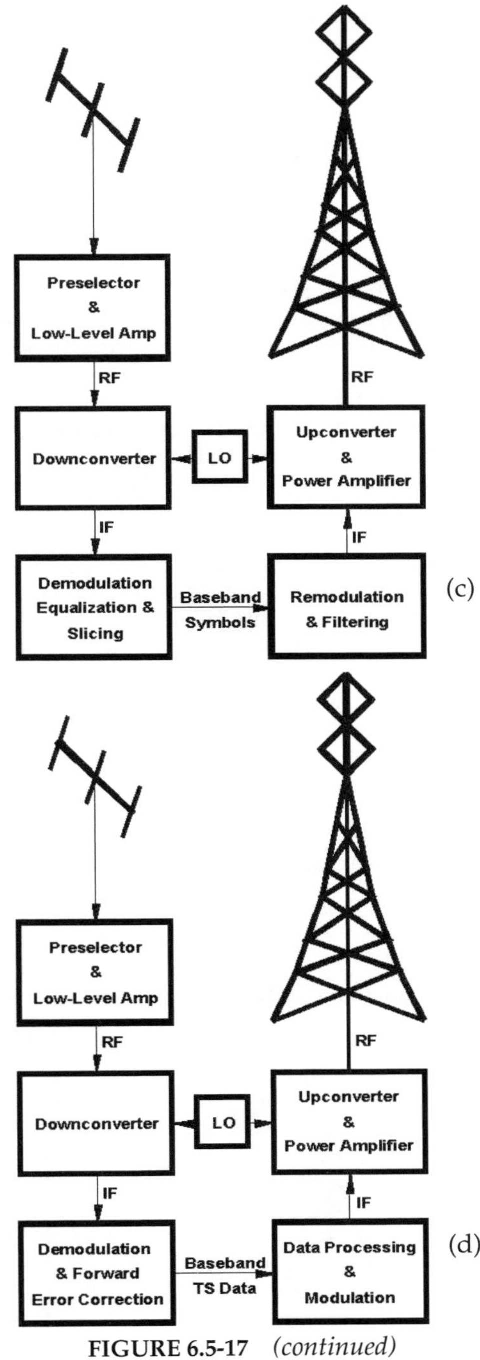

FIGURE 6.5-17 *(continued)*

applies adaptive equalization to the baseband, slices the baseband signal to reduce the effects of noise on the received signal, and remodulates the signal for further amplification and transmission. This design concept is shown in Figure 6.5-17c and is termed an equalization DOCR (E-DOCR). The advantages of this design are that the effects of signal impairments in the natural transmission channel from the signal source to the DOCR and the impairments that result from coupling of the transmitted signal into the receiving antenna both can be minimized by the combination of adaptive equalization and symbol level slicing. The result is that significantly higher gain can be applied in the DOCR, and consequently higher power can be transmitted relative to the received signal level than is possible with the simpler designs. The trade-offs are that the delay through the DOCR becomes even longer than in the simpler designs, and any errors that occur in the slicing of the received signal to restore the levels of the symbols will be built into the transmitted signal, causing potentially unrecoverable errors in consumer receivers. To partially overcome these trade-offs, careful equipment design can hold the E-DOCR delay to something on the order of 5 microseconds, and a network design producing a reasonably high signal-to-noise (S/N) ratio on the E-DOCR input can help reduce retransmitted errors. If a high enough input S/N cannot be obtained, then use of one of the simpler designs may be advisable.

The most complex DOCR design involves completely decoding the received signal to transport stream data, as shown in Figure 6.5-17d. In this case, both adaptive equalization and forward error correction can be applied to the received signal, resulting in a fully reconstructed, noise- and error-free signal for retransmission. To make the transmitted symbols match the received symbols, the location of the data frame sync and the states of the trellis coding in the received signal must be recovered and used in the remodulation process. Although theoretically possible to build, the design of Figure 6.5-17d is likely to be impractical to implement because of the very long delay that will result from all of the signal and data processing. Such a design could only be used where there is nearly complete isolation of the DOCR service area from the source signal, and, in such a case, a simpler solution would be to use retransmission of the signal, on channel, without implicating SFN techniques.

DISTRIBUTED TRANSMITTERS

Distributed transmission (DTx) is a technique that uses multiple distributed transmitters (DTxTs) that operate in a more or less conventional manner to provide service to a region. Typically, the DTxTs are fed the data stream for transmission over STLs; they individually modulate the data onto carriers at an IF frequency; the carriers are upconverted to the broadcast channel, amplified, and fed to an antenna. To operate in a DTx network (DTxN), as dictated by the prerequisites for SFN operation discussed earlier, the DTxTs must be synchronized with one another in their modulation processes, and they must produce output signals on the same frequency. A basic DTx system has the configuration shown above in Figure 6.5-2.

Although there is one basic concept for distributed transmission, there are a number of ways in which it can be implemented. The differences between methods largely involve the form in which the data stream is delivered to the several transmitters in a network and the resulting techniques required to synchronize transmitter operation and adjust the relative timing of emissions from the DTxTs. In all cases to be covered, the data is delivered to the DTxTs in digital form for local modulation and upconversion to the output channel. Not discussed in detail herein is the alternative of modulating the signal in one place and delivering an analog IF signal to all of the transmitters for upconversion. Such a system would be totally dependent on the stability of the time delays of the STLs to the several transmitters for the stability of the entire network, and experience has shown that adequate time delay stability cannot be achieved economically in the STLs for such a technique to provide the required network performance.

Synchronization Methods

There are three basic ways in which to deliver signals to the transmitters in a DTxN. The first involves modulating the data onto a carrier at a central location and distributing the resulting RF signal to the various DTxTs. At each DTxT, a receiver must recover from the RF signal both the data and the various states of the modulation system that carried the data. The data then can be remodulated and upconverted onto a carrier at the assigned frequency of the DTx network. In the course of processing the data prior to remodulation, each DTxT can extract information from it to control adjustment of the delay through that DTxT, thereby attaining the network timing objectives. A system using the RF distribution method for delivery of the data to the DTxTs is shown conceptually in Figure 6.5-18. Although it has the ability to control the transmitter emission timing, like the completely analog distribution method mentioned at the start of this section, the RF distribution method requires an analog STL with a reasonable noise margin to reach each of the DTxTs. Such a requirement can be very limiting in the types of STLs that can be used and in their availability. Moreover, the data processing at each transmitter would be much more complex than with other methods, so this approach is believed not to have been implemented as of this writing.

A second scheme for delivering the data to the several DTxTs in a network is to perform at a central location all of the data processing that normally occurs in a transmitter, delivering to all of the network transmitters the symbol data that must be input to the modulator. This approach is shown conceptually in Figure 6.5-19 and uses a digital STL to each transmitter. Because the data delivered to the DTxTs is already formatted for transmission, if any data for control of the transmitters is to be embedded in the data stream at the transmitters, the symbol data must be converted back to its original form so the needed information can be extracted from the data stream, or some form of multiplexing of the entire stream could be used. More significant, the normal process of formatting the data for transmission adds a significant amount of error correction coding data necessary to allow the payload data to survive the broadcast RF transmission channel. This additional data increases the data bandwidth required in the STL. In the case of the ATSC 8VSB system, the increase is from 19.39 Mbps to over 32.25 Mbps. This bandwidth increase in the STL makes the

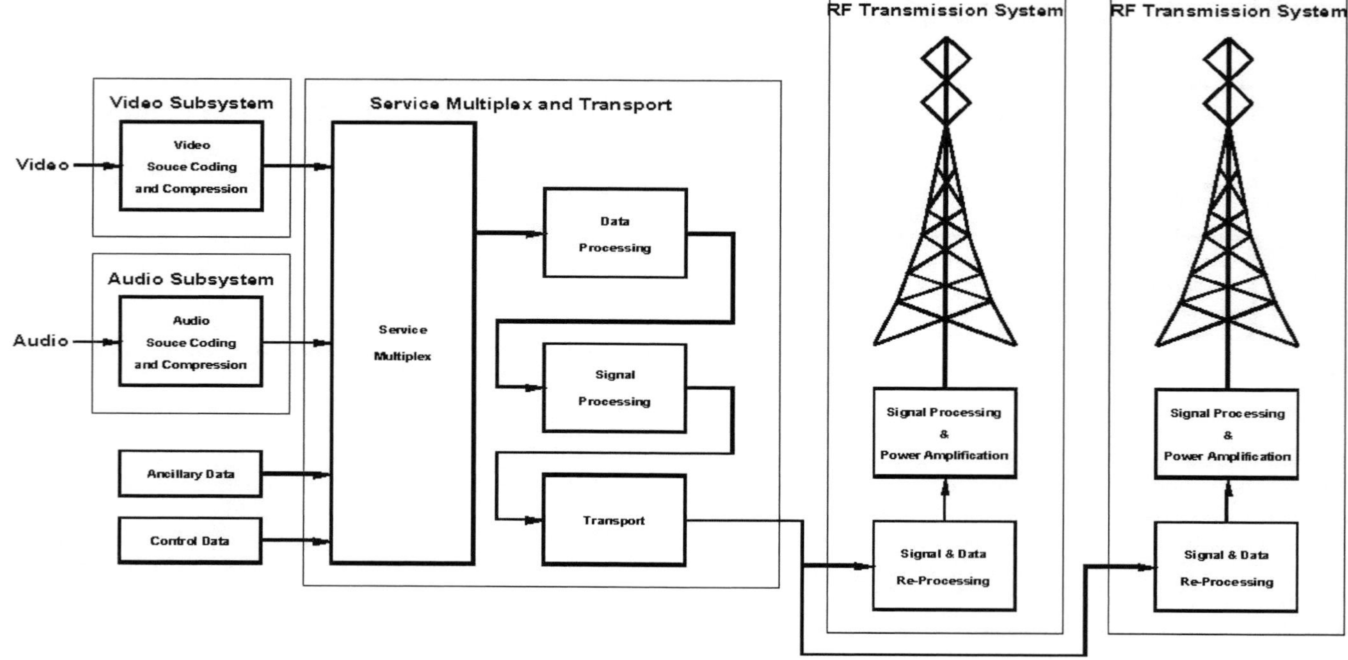

FIGURE 6.5-18 DTx system using RF distribution.

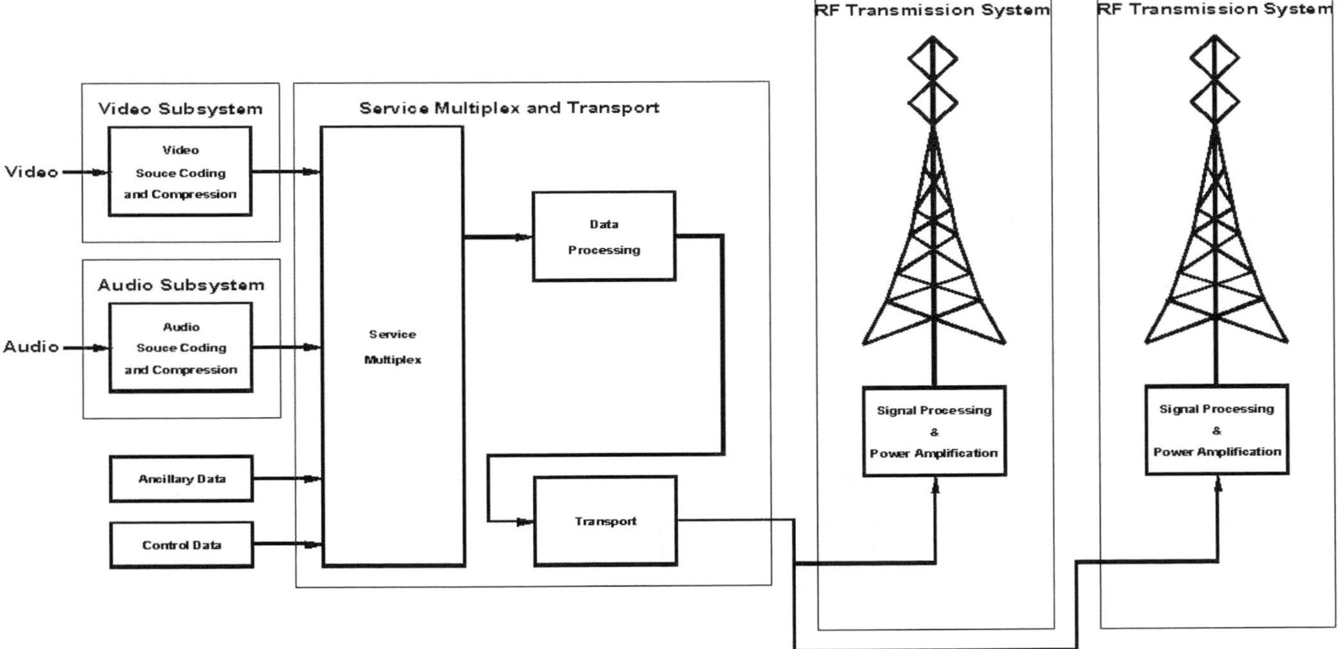

FIGURE 6.5-19 DTx system using distribution of symbol data.

centralized data processing approach uneconomical, and this method is believed not to have been implemented as of this writing.

The third method uses data processing at a central location to derive data that can be embedded in the data stream sent to the network DTxTs for synchronizing them and controlling their operation. The data stream sent over the STL operates at exactly the same data rate and in the same format as is used in a conventional, single-transmitter system (*i.e.*, for the ATSC 8VSB system, 19.39 Mbps of MPEG-2 transport stream packets). A device called a *distributed transmission adapter* (DTxA) is inserted into the transport stream signal path at the input to the STL. The DTxA develops information that is sent to all of the transmitters to permit them to tightly control the processing of the data they receive and emit, thereby synchronizing them with one another. The data sent from the DTxA also permits adjustment of the network timing by individually controlling the relative emission time of each of the DTxTs. To achieve this result, a small amount of data capacity from the transport stream (typically, 1 packet per second) is required to carry the synchronization and control information from the DTxA to the DTxTs. This method has been adopted by the ATSC for use with the 8VSB transmission system and is documented in the ATSC Synchronization Standard for Distributed Transmission (A/110). A conceptual block diagram of the scheme is shown in Figure 6.5-20. All of the remaining discussion of DTx will focus on this method, in the form adopted by the ATSC.

In the ATSC 8VSB transmission system, there are a number of data processing functions that operate either completely asynchronously or with a loosely defined relationship to the data in the stream that is to be transmitted. For several of these processes (*i.e.*, data randomization, Reed–Solomon error correction coding, and convolutional interleaving), operation is synchronous between them and with the packet structure of the MPEG-2 transport stream, but there is no association of their starting points with particular packets in the stream. Other processes (*i.e.*, precoding and trellis encoding) are completely asynchronous with one another and with the data in the stream. The result of this lack of defined relationships between the data in the transport stream and the operation of the various transmitter data processing functions is that the data processing in the 8VSB system can take any of 42,880,953,483,264 (almost 43 trillion) states for a given data input, and no two transmitters in a network are likely to produce the same output symbols at the same time, let alone a larger number of transmitters doing so.

To facilitate the transmission adapter approach to transmitter synchronization, the DTxA incorporates a model of the data processing that conventionally takes place in the transmitters. The model includes all portions of the data processing functionality through the formation of symbols. The model processes the incoming transport stream data after randomly starting its operation with respect to that data. In doing so, it establishes a relationship between the transport stream data and the output symbols that are produced. Then, it is necessary only to communicate from the DTxA to the DTxTs information about the relationship between the transport stream data and the symbols that are to be produced from them. The

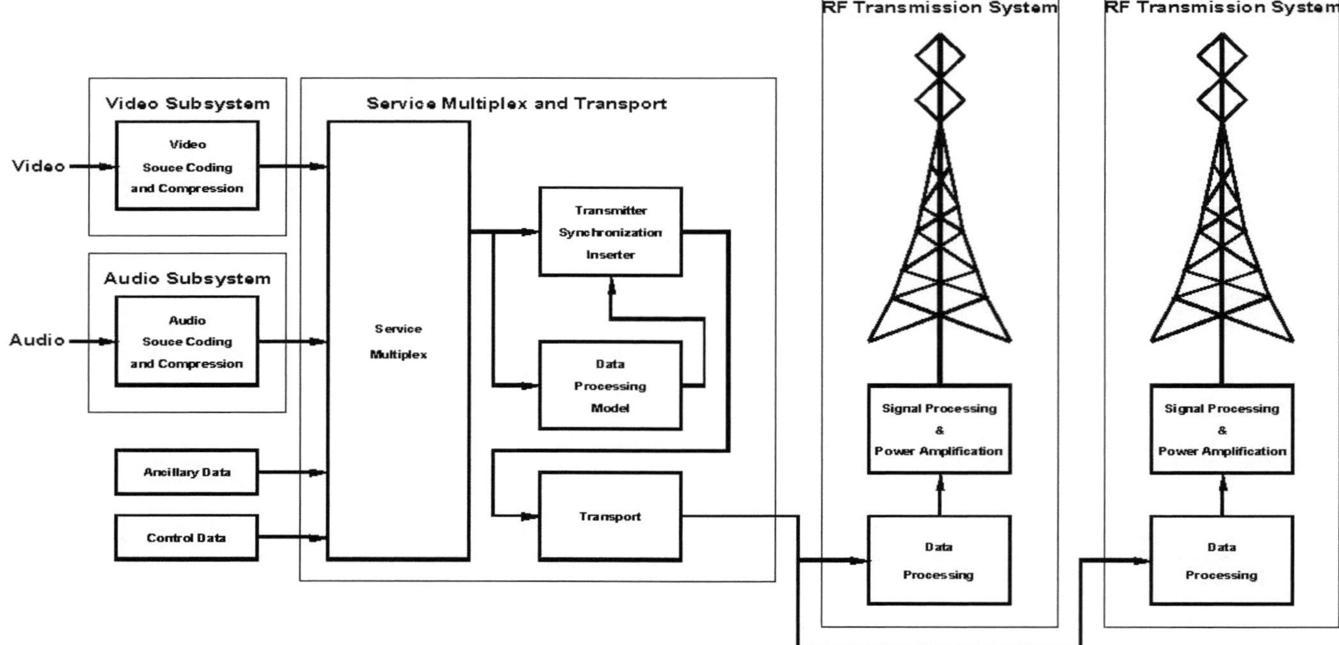

FIGURE 6.5-20 DTx system using transport stream distribution.

communication takes two forms: information about the relative timing of the processes that are synchronous with one another and with the transport steam packet structure (called *cadence synchronization*) and information about the states of the precoders and trellis encoders at particular times (used to *jam sync* the equivalent functions in the transmitters). The information is inserted by the DTxA into the transport stream so it can be extracted and acted upon identically by each transmitter in the network.

The information sent from the DTxA to the DTxTs is carried in a special packet type designated as an operations and maintenance packet (OMP), which has a dedicated packet identifier (PID) assigned. Each OMP begins with a further identifier for the particular application, in this case a distributed transmission packet (DTxP). In fact, 16 values of OMP identifier are assigned for DTxP use to support multiple layers of distributed translators, as described below. In addition to the synchronization information for the DTxTs, the DTxP carries information used to set the emission times and to control the power levels of the several transmitters. To facilitate the inclusion of DTxPs in the data stream, *precursor packets*, which can be converted to DTxPs by the DTxA, are inserted into the transport stream by the service multiplexer or a remultiplexer upstream of the DTxA. The general scheme of the emission time control system is depicted in Figure 6.5-21, which shows a timeline of the various elements of the system and their relationships.

The time relationships illustrated in Figure 6.5-21 are all based on an underlying reference time clock, available in common at all sites in the network and shown as 1-second clock ticks. The reference time can be derived from GPS or LORAN-C, for example. Certain information is carried in the DTxP, and other information is calculated at each transmitter from the data received in the DTxP plus certain locally derived values. Starting on the left side of the figure, a synchronization time stamp (STS) is sent in the DTxP to identify the time at which a particular reference bit within the DTxP is released from the DTxA. Also sent in the DTxP is a value for maximum delay (MD) that is used by all of the transmitters in the network to determine when the reference emission time is for the reference bit within the DTxP. MD is set as an input parameter in the DTxA and depends for its value on the length of time delay through the STL system to the farthest point (in time) in the network. (MD can be up to 1 second, so the system has sufficient capacity to treat STL time delays equal to almost 4 satellite hops.) All of the DTxTs in the network use the combination of the STS plus MD values to calculate the same reference emission time at each transmitter, as identified toward the right side of the timeline.

In addition to calculating the reference emission time for the network, each DTxT also calculates a number of other values to allow it emit the signal at the intended time with respect to the other transmitters in the network. Each transmitter is sent individually, in the DTxP, a value of offset delay (OD), which instructs it when to emit its signal relative to the *reference emission time* for the network. It also has an input parameter that sets the length of time between a specific point in the transmitter data processing and the antenna output—the transmitter and antenna delay (TAD). By adding the OD (which may have a positive or negative value) to the reference emission time and

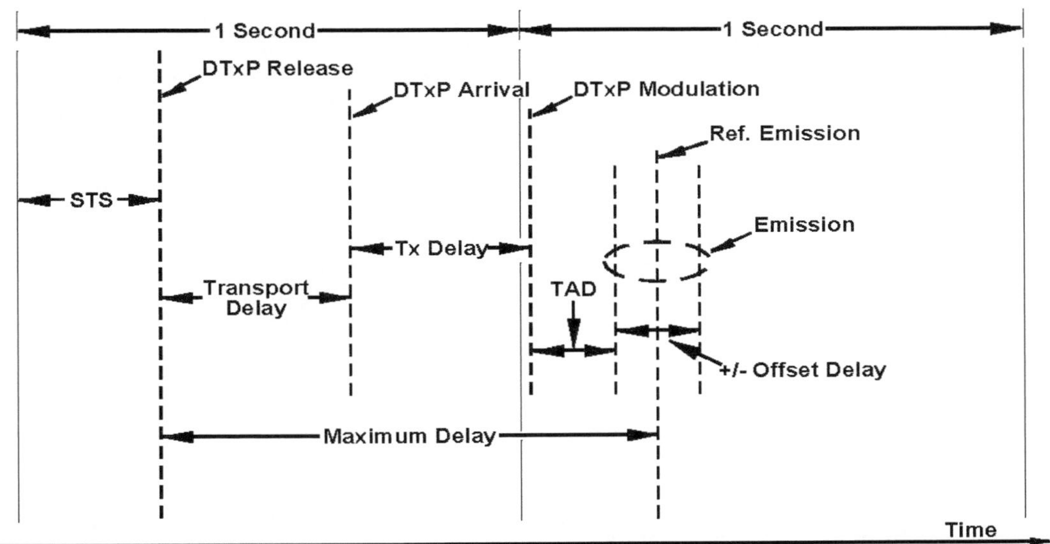

FIGURE 6.5-21 Transmitter emission timing control information.

subtracting the TAD value, the transmitter can determine when it is supposed to release the reference point in the data stream into the modulator. By determining the DTxP arrival time relative to the 1 second clock ticks, it can calculate the difference between that DTxP arrival time and the intended DTxP modulation time. That time difference is the time delay required in the particular transmitter (Tx delay) for proper operation of the network. The Tx delay is calculated whenever a DTxP is received, and its value is maintained at least until the next DTxP arrives. Because shifting the transmitter time delay can cause interruptions in the emitted signal, unless relatively complex steps are taken in the transmitter design to overcome them, some amount of hysteresis generally is applied to the Tx delay value, so it will not be changed unless a significant error has accumulated.

STL Considerations

It is a fact that the time delay of signals transiting an STL typically varies. For conventional, single-transmitter facilities, this variation is immaterial. When an STL is embedded in a DTx network, the variation, which most likely will be different on each STL path, may become quite significant. Time delay change accumulates from slight errors in the output frequency of the link, following a buffer typically used for de-jittering the signal data. The accumulation of time delay variation can occur despite (indeed, because of) the use of long-time-constant frequency control loops to recover accurately the input frequency at the output of the link.

Consider a link carrying an MPEG-2 transport stream with a data rate appropriate for an ATSC 8VSB transmission. The transport stream will operate with a data rate of 19.392658 Mbps. If the error in the STL between input and output is 2.8 ppm, half the error band permitted by ATSC standards, over a period of an hour, the error will accumulate to about half a microsecond. Over a day, it will become an error of about 12 microseconds. Depending on the amount of time that the frequency control loop runs high or low in frequency before shifting in the opposite direction to maintain the average output frequency to match the input, even greater time delay changes can accumulate in the STL. Such delay changes are likely to cause the transmitter to reset its internal delay, thereby interrupting the signal while it does. Below the hysteresis threshold at which a delay reset occurs, the delay changes will cause the relative timing of the transmitters to wander back and forth, causing the locations of equal signal arrival times in the service area from the several transmitters to wander geographically. The hysteresis threshold generally is set to a moderately low value (e.g., 0.5 microsecond) to prevent too much geographic wander, thereby reducing the time between delay resets when there is much time delay variation.

To overcome the potential for STL time delay variations, a method is provided in the ATSC A/110A standard to lock the data stream frequency to an external frequency reference at the DTxA and at the DTxTs. Because an external time reference, from a source such as GPS or LORAN-C, already is required at each node in the network, using the frequency reference available from the same external sources involves little increase in complexity of the system. There is a change required, however, in the data processing of the DTxA, which must adjust the data rate of its input to be precisely the specified value on its output. To accomplish that potential frequency change, it is required periodically to insert packets into or delete packets from the data stream. It makes the insertions and deletions using null packets whenever possible. If there are no null packets in the stream for too long a time when a deletion is needed, the DTxA can delete the precursor packet that it would have converted into a DTxP when the next one appears in the data stream.

DISTRIBUTED TRANSLATORS

Combining the concept of a distributed transmitter with that of a conventional translator results in a distributed translator (DTxR). A DTxR has its input on one channel and its output on another. The benefit of DTxRs is that they can receive their input signals over the air from earlier transmitters in the network, and they can share an output channel while mitigating interference between themselves in the same way as do DTxTs. Because of the different input and output channels, DTxRs can operate with more power than can DOCRs, and they do not require STLs. The trade-off for the additional flexibility of DTxRs is that they require at least one additional broadcast channel on which to operate, but additional spectrum may not be available in many places in which it would be desirable to utilize DTxRs.

Networks using DTxRs can be designed in layers, in which transmitters sharing the same channel in the same layer are treated as part of a separate subnetwork for purposes of interference mitigation. Indeed, it is possible to design a very extensive network using DTxRs in which only two channels are required, the channels alternating with one another from layer to layer and moving away from the initial transmitter in the network. To support the synchronization and timing requirements of the transmitters in each layer, the DTxA must produce a separate DTxP for each layer, for which purpose a number of OMP identifiers are available to indicate to which layer a particular DTxP applies. Because the DTxP carries information that depends on the data processing of the signal in the layer prior to the one to which it applies, there is implied in the DTxA data processing model a cascading of corresponding layers of data processing to develop the symbols to which the transmitters in the respective layer are to be synchronized. The basic layout of a DTxR network with one central transmitter, a single layer of translators, and using two channels appears in Figure 6.5-3.

In the DTxR itself is a receiver that demodulates and decodes the received signal to a baseband MPEG-2 transport stream. The receiver is followed by a modulator that is virtually identical to the one in a DTxT, as described previously. The only differences are that the data processing in the modulator examines the DTxP for the OMP address associated with the layer in which the DTxR resides, and provision must be made in the modulator for identification of that layer as an input parameter. The overall configuration of a DTxR nearly matches that of a baseband decoding DOCR, which decodes the received signal to a baseband transport stream, as shown above in Figure 6.5-17d. The primary difference is that the local oscillators used to downconvert and upconvert the received and transmitted signals, respectively, will be on different frequencies, as the input and output channels of the DTxR will be different.

When setting up a DTxR network, it is important to set the maximum delay for each layer, which will be an independent parameter, to a somewhat higher value for each successive layer in the network. The MD value must be high enough to allow for the signal passing through all of the preceding layers, with the attendant decoding and remodulation at each DTxR in the chain. At the same time, the MD value should be kept only a little higher than necessary to account for all of the preceding layers so as to allow for the addition of following layers. The total time available is just under 1 second, so there is adequate capacity in the system to accommodate as many layers as might be needed in the most complex of network designs.

NETWORK ADJUSTMENTS

As noted previously, among the prerequisites for SFN operation are that the symbols emitted by the transmitters in a network must be identical to one another and that the transmitters must operate on essentially the same frequency. The purpose, of course, is to make the signals from the several transmitters appear to receivers as if they all were emitted by one transmitter and followed multiple paths to the receivers, thereby creating multipath that can be treated by the receiver adaptive equalizers. Thus, for purposes of reception by consumer receivers, the transmitted signals must appear to be identical to one another.

When it is necessary to adjust the SFN, it is desirable to be able to differentiate one transmitter from another. The differentiation is needed to permit, at any given receiving location, the relative field strength and arrival time of the signal from each transmitter to be determined with respect those of the other transmitters. This determination would enable adjustment of the emission characteristics of each transmitter. Unfortunately, with identical signals being transmitted and received, such a differentiation is not possible. It is possible, however, to create a hidden difference between the transmitted signals that is not detectable by consumer receivers but that can be detected by a special receiver built for the purpose. Using such a technique allows network characteristics to be measured without interfering with the operation of the network from the standpoint of delivery of signals to consumer receivers. The technique developed for the purpose of differentiating transmitters is termed an *RF watermark.*

RF Watermark

An RF watermark is a signal that is added to a host signal (*e.g.*, an 8VSB broadcast signal) to permit the identification of that host signal and to obtain information from and about the combination of the host signal and the RF watermark without interfering with the normal operation of the host signal for broadcasting purposes. Specific RF watermark technology has been developed for use with 8VSB transmissions as part of the process of standardizing the synchronization methods for distributed transmission. The RF watermark is defined in ATSC A/110A and explained in ATSC A/111. There are several purposes for the RF watermark, which are achieved with very high performance potential by the particular technology adopted. There also are several additional applications for the technology that have become apparent because of its high performance under difficult conditions.

Purposes

As applied in an SFN environment, the RF watermark serves three main purposes: identification of transmitters, measurement of the channel impulse response from each transmitter, and carriage of information with great robustness when compared to that of the host signal. As discussed above, when the prerequisites for SFN operation are met with respect to the signals delivered to consumer receivers, it becomes impossible to tell the signals from the several transmitters apart, if nothing additional is done. Each of the transmitters would carry, as part of its program content, the station identification of the station operating the network, but individual transmitters would be indistinguishable from one another. Thus, one of the principal purposes of the RF watermark is to carry identification information that allows differentiating one transmitter from another, both when mitigating INI and when evaluating interference to other stations. For this objective, the RF watermark emitted by each network transmitter includes a code that can be recovered by a specialized receiver. The code has two portions: one part associated with the network and one part associated with the individual transmitter. The transmitter identification (TxID) codes are maintained in a publicly accessible database. The specialized receiver thereby is enabled to completely determine the source of any signals that it is receiving that are carrying RF watermarks.

The second goal of including an RF watermark in the emissions of each transmitter in an SFN is to support measurements that are necessary to adjust the network for minimum INI. Ideally, both the relative strengths and the relative arrival times of the signals from various transmitters could be ascertained at any receiving location. The channel impulse response (CIR) is designed just for the purpose of indicating the strength of a signal and its echoes over time; that is, it characterizes the propagation channel traversed by a signal. If the CIRs could be taken independently on the signals from each of the transmitters in an SFN, then presenting them overlaid on one another (*e.g.*, in different colors, one for each transmitter) on a common display and using a common time base would show exactly the information needed to adjust the transmitter amplitudes and timing in an SFN. The RF watermark completely supports such functionality.

The third priority for the RF watermark is the carriage, individually from the various transmitters in an SFN, of information that is independent of the host signal transmitted in common by all of the network transmitters. The data carried can be anything that will fit into a relatively low data rate channel and that originates from or can be delivered to each specific transmitter. The data can be carried in addition to the TxID information and can be delivered very robustly. Among the many uses proposed for the RF watermark data channel are carriage of telemetry information from the associated transmitter, carriage of time data to support use of the digital television signal for radiolocation purposes, and carriage of emergency alert data. Because the RF watermark is extremely robust in comparison to virtually all other broadcast signals, it offers the opportunity for reliable delivery of the data it carries to indoor locations that are not penetrated by other signals.

Technology

The RF watermark consists of a spread spectrum signal that is added at a very low level to the host signal carried by each transmitter. Because it operates in the amplitude region that normally would be considered to be noise in a transmitter output, the RF watermark is termed a buried spread spectrum (BSS) signal. In the case of an 8VSB host, the BSS signal is configured with 2VSB modulation that is synchronized to the modulation of the host signal and has the same spectrum occupancy as the host. It generally operates at a level 30 dB or more below that of the host, although higher levels can be used for out-of-service network testing and adjustment. The relationship between the signals, within the spectrum of the operating channel, is portrayed in Figure 6.5-22, where the host signal is shown in blue and the RF watermark is shown in green. Given the synchronization between host and watermark signals, the RF watermark has the effect of opening up an additional set of "eyes" in the amplitude *versus* time representation of the combined signal. The set of small additional eyes is apparent in Figure 6.5-23.

The 2VSB signal is modulated with a code called a *Kasami sequence*, a code that has a number of properties that make it particularly useful for the RF watermark application. The important properties relate to the auto-correlation and cross-correlation of different values of the code. The particular coding method selected permits over 16 million different patterns to be generated, thus carrying 24 bits of information. The sequences generated are sufficiently short that they can be repeated almost four times during one 8VSB data field. The code sequences are synchronized with the 8VSB data framing to make finding the start of a code sequence easier and, hence, faster. The particular Kasami sequence transmitted by each transmitter serves as its transmitter identifier and is unique to that transmitter, with part of the data represented by the code value serving as an identifier for the network in which the transmitter operates and the other part of the data serving as an identifier for the individual transmitter within the network. With 12 bits allocated to identifying networks and 12 bits allocated to identifying transmitters within networks, up to 4096 networks and up to 4096 transmitters within each network can be identified.

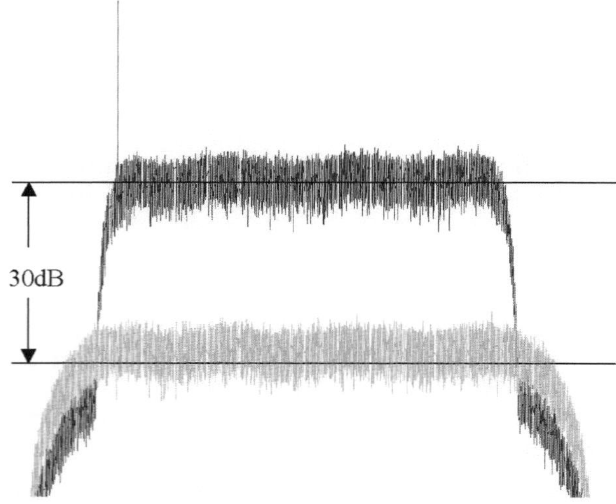

FIGURE 6.5-22 Superimposition in spectrum of host 8VSB signal and RF watermark. (Figure courtesy of X. Wang, Y. Wu, & J.-Y. Chouinard, Communications Research Centre Canada, Ottawa, Ontario.) [4].

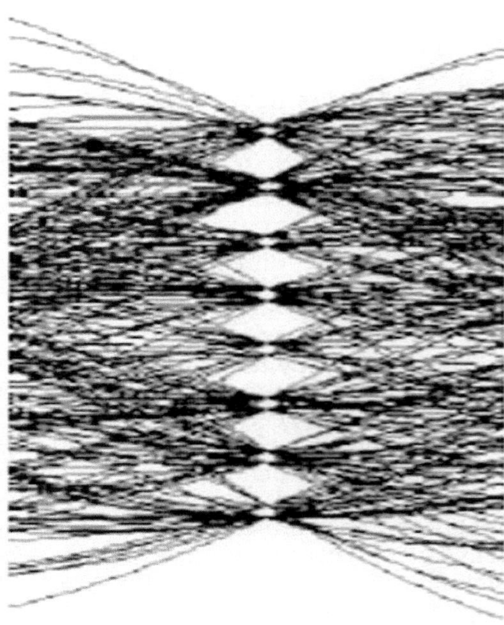

FIGURE 6.5-23 Eye openings of 8VSB host signal with added eyes of RF watermark in between. (Figure courtesy of Advanced Television Systems Committee, Washington, D.C.) [2]

Carrying additional information on the RF watermark is accomplished by inverting the phase of the data in the Kasami sequence from one repetition of the code to the next. In a receiver for the RF watermark, finding a code or its inverse is a relatively efficient operation using standard correlation techniques. Thus, the phase of the Kasami sequence data relative to the assigned code value (*i.e.,* non-inverted or inverted) can indicate binary ones and zeroes. This capability allows the transmission of a one or a zero for each repetition of the code sequence. Because there are four usable repetitions of the code sequence in each 8VSB data field and because there are approximately 40 data fields per second, a data rate of about 160 bps results. Although this data rate is quite slow when compared with the rates provided by modern data communications technology, it is carried on a highly robust channel, making it quite useful.

Performance

There is a tremendous challenge that must be met by an RF watermark if it is to be effective. One of its main purposes is to allow measurement of the relative levels of signals from multiple transmitters that can affect the operation of adaptive equalizers. As shown previously, adaptive equalizers become operational when the amplitude difference between multipath signals is on the order of 20 dB. Thus, the RF watermark must enable measurement of the level of an echo that is 20 dB below the level of the signal serving as the reference to an adaptive equalizer, but the RF watermark itself must be inserted on the order of 30 dB below the level of the host signal that carries it, so as not to interfere with its host. This combination of factors means that the RF watermark must be recoverable when it is 50 dB below the strongest signal in an ensemble of multipath signals. Not only must it be recoverable, but it must also be retrieved with a sufficient signal-to-noise ratio (*i.e.,* about 10 dB) to permit the necessary amplitude and timing relationships to be determined. It has been demonstrated that the RF watermark is capable of just such performance.

Because the RF watermark can be received with a usable margin when it is 50 dB below the amplitude of any host signal and because it is transmitted about 30 dB below the level of its own host signal, it stands to reason that the RF watermark can be recovered at a received signal level about 20 dB below that at which an 8VSB signal can be correctly demodulated. The 20 dB advantage permits reception in places where the host signal cannot be received. Moreover, when combined with the fact that the frequencies on which digital television signals operate can penetrate buildings much better than can higher frequencies, the RF watermark should be able to provide reliable reception in places not reachable by many other techniques. Furthermore, because of its 2-VSB modulation, the RF watermark should be received readily by receivers that are in motion, such as those carried by pedestrians or those in vehicles. This combination of characteristics makes possible a range of potential applications for the RF watermark beyond those originally intended for it.

Other Applications

Because of its highly robust nature, the RF watermark is well suited to a number of applications beyond the transmitter identification and channel characterization for which it was originally designed. A number of suggestions have been made to date, and they are enumerated here without evaluation. Nevertheless, the proposals for application of the RF watermark have been put forward largely by well-known and well-respected organizations, so it stands to reason that they have a relatively high probability of success. Most are tied to the ability of the RF watermark to carry additional information with very high robustness. In addition, there are potential applications of the transmitter identification capability of the RF watermark beyond the needs of SFNs.

The original proposed use of the RF watermark data channel was support of telemetry from the DTxTs in a network back to the network control point. Control data can be sent to the transmitters through the DTxPs, and, in fact, provision is made in their protocol for carriage of the basic functionality required. The DTxPs, of course, flow to the DTxTs with the data to be transmitted. To close the control loop around the transmitters, then, it is necessary to provide a means to get information from them back to the control point. The RF watermark data channel can provide a path through which to send telemetry from each transmitter to one or more central receiving locations from which the data can be relayed back to the control point. Thus, it would be possible to obtain complete remote control of the DTxTs without requiring the installation of separate control links to or from each site beyond the STL needed to deliver the data for transmission.

Another proposed application of the RF watermark data channel is carriage of the time information that could enable use of DTV signals for radiolocation purposes. The best-known radiolocation system—the global positioning system (GPS)—is well understood not to work in most indoor locations. Digital television signals usually reach those locations with sufficient strength to permit location determination, even when television reception is not possible. This is true because radiolocation based on DTV would use only the synchronization elements of the DTV signal, which are far more robust than the data that carries the television content. Thus, a radiolocation receiver could obtain the needed timing data from a DTV signal in many places where program content carried by the same DTV signal cannot be received. A radiolocation receiver would obtain the synchronization information from a number of DTV transmitters simultaneously, using the relative arrival times of those signals to determine its own position by triangulation. What is needed, in addition to the synchronization information from the multiple DTV signals, is relative time data that would allow interpretation of the DTV synchronization timing from the group of transmitters, thereby permitting determination of the location of the receiver. The RF watermark data channel could provide an ideal way to distribute the necessary relative time data to radiolocation receivers.

Because of the robustness of the RF watermark data channel and the potential for receiving it in very difficult locations, such as indoors, in basements, and in moving vehicles, it has been suggested as a means for distribution of emergency alert warnings and related data. Although the data rate is very slow and could not be used to carry, for example, audible tones or voice messages, it could carry triggers for location-based alarms, text messages directed to specific types of users, and similar information that must be delivered reliably to locations that are difficult to reach. The triggers could sound warnings and direct the public to turn to other types of services for more detailed information. The text messages could carry instructions and information to first responders using fixed, portable, and/or mobile receivers. Proposals have been made for country-wide emergency alert systems among administrations that have adopted the ATSC 8VSB transmission system.

Beyond the need to identify individual transmitters in SFNs, the TxID function could prove beneficial for use by single-transmitter operations as well. In the past, with analog transmission, interference between stations (*e.g.*, from "E-skip" propagation) resulted in distortion of the received image, typically with a "Venetian blind" effect appearing on the display and sometimes with the undesired signal capturing the receiver. The interfering signal could be identified by watching the image on the receiver and observing the station identification of the interferer, either in the foreground or as a ghost image in the background. When the time comes that analog transmissions end, there will still be such interference between digital stations, but the interference simply will result in receivers in the region receiving the interference stopping working. It no longer will be possible to observe the displayed image to determine the source of the interference. Transmission of an RF watermark by all stations, whether using single or multiple transmitters, would enable use of special RF watermark receivers to determine the origin of such interfering signals.

Field Measurements

When setting up and maintaining an SFN, it is desirable to confirm through field measurements the actual operation of the network in comparison with its design parameters. Two basic characteristics need to be measured, and there are two ways in which they can be measured. The two characteristics are:

- The field strengths of the signals from the several network transmitters that reach any particular location at which testing is performed
- The relative arrival times of the signals at that location

The two ways in which to make the measurements are through short-term visits to the test locations, such as with a mobile measuring system, and with fixed installations of monitoring equipment at critical locations within the network.

Making measurements with a mobile measurement system, if done prior to system operation or during out-of-service periods, can be facilitated by turning transmitters on and off sequentially to permit field strength data to be obtained. Determining signal arrival timing, however, requires some form of instrumentation that can respond to the presence of multiple signals and indicate their timing relationship. One way to acquire this data is with a receiver equipped to display adaptive equalizer tap weight information that can be interpreted by a human observer. Tap weights can show the presence of echoes from multiple transmitters but can make determination of the source of the various echoes very difficult, especially in natural environments having a great deal of multipath and consequently many active taps in an equalizer. Frequently, determining the sources of echoes also requires turning transmitters on and off. The difficulty of interpreting tap weights becomes even greater when the levels of the signals being received are radically different from one another, making one signal appear at a very low level on the tap weight display relative to the stronger signal that serves as a reference.

Routine monitoring of SFN operations while they are in service cannot take advantage of the turning on and off of transmitters to determine signal levels and for the same reason cannot benefit from the interpretation of tap weights by inspection of a human observer. Instead, equipment is needed at monitoring sites that autonomously can collect and analyze the necessary data from the signals of multiple, continuously operating transmitters. Multiple monitoring sites may be necessary in order to collect information about the operation of all of the transmitters in a given network. The results of the analyses at the monitoring sites then must be forwarded to the station for collection and aggregation with the data from the other sites to create a complete picture of network operation.

Virtually all of the needs for measurement and monitoring in SFNs are addressed by the RF watermark. It provides the mechanism through which to determine both the amplitude and the timing of each signal received at a location. With specialized receivers to recover the channel impulse responses of the signals from multiple transmitters, the data can be collected while the transmitters are in regular operation, the absolute amplitudes of the signals can be determined using information on the RF watermark insertion levels, and the timing of arrival of the signals can be measured both relative to one another and with respect to an external time reference. Such receivers can provide fully automatic analysis of the various signals even when those signals are received at significantly different levels from one another, thereby facilitating aggregation of the information and analysis of the overall operation of the network.

References

[1] *Synchronization Standard for Distributed Transmission*, Revision A (A/110A), Advanced Television Systems Committee, Washington, D.C., July 19, 2005.

[2] *Recommended Practice: Design of Synchronized Multiple Transmitter Networks* (A/111), Advanced Television Systems Committee, Washington, D.C., September 3, 2004.

[3] *Recommended Practice: Receiver Performance Guidelines* (A/74), Advanced Television Systems Committee, Washington, D.C., June 18, 2004.

[4] Wang, Xianbin, Yiyan Wu, and Jean-Yves Chouinard. "Robust Data Transmission Using the Transmitter Identification Sequences in ATSC DTV Signals." *IEEE Transactions on Consumer Electronics*, vol.51, no.1, Feb. 2005, pp.41-47.

Bibliography

Angueira, P., M.M. Velez, D. De La Vega, A. Arrinda, I. Landa, J.L. Ordiales, and G. Prieto. "DTV (COFDM) SFN Signal Variation Field Tests in Urban Environments for Portable Outdoor Reception." *IEEE Transactions on Broadcasting*, vol. 49, no. 1, March 2003, pp. 81–86.

Angueira, P., M.M. Velez, D. de la Vega, G. Prieto, D. Guerra, J. M. Matias, and J.L. Ordiales. "DTV Reception Quality Field Tests for Portable Outdoor Reception in a Single Frequency Network." *IEEE Transactions on Broadcasting*, vol. 50, no. 1, March 2004, pp. 42–48.

Bank, M. "On Increasing OFDM Method Frequency Efficiency Opportunity." *IEEE Transactions on Broadcasting*, vol. 50, no. 2, June 2004, pp. 165–171.

Bretl, W., W.R. Meintel, G. Sgrignoli, X. Wang, S. Merrill Weiss, and K. Salehian. "ATSC RF, Modulation, and Transmission." *Proceedings of the IEEE*, vol. 94, no. 1, Jan. 2006, pp. 44–59.

Guerra, G., P. Angueira, M.M. Velez, D. Guerra, G. Prieto, J.L. Ordiales, and A. Arrinda. "Field Measurement Based Characterization of the Wideband Urban Multipath Channel for Portable DTV Reception in Single Frequency Networks." *IEEE Transactions on Broadcasting*, vol. 51, no. 2, June 2005, pp. 171–179.

Hershberger, D.L. "Implementation of the ATSC Distributed Transmission System." *2003 NAB Broadcast Engineering Conference Proceedings*, pp. 280–289. Washington: National Association of Broadcasters, 2003.

Hershberger, D.L. "Lessons Learned from DTx Implementation and Applications," *2005 NAB Broadcast Engineering Conference Proceedings*, pp. 493–500. Washington: National Association of Broadcasters, 2005.

Kim, S.W., Y.-T. Lee, S.I. Park, H.M. Eum, J.H. Seo, and H.M. Kim. "Equalization Digital On-Channel Repeater in the Single Frequency Networks." *IEEE Transactions on Broadcasting*, vol. 52, no. 2, June 2006, pp. 137–146.

Lee, Y.-T., S.I. Park, H.-M. Eum, H.M. Kim, J.-H. Seo, S.W. Kim, B. Ledoux, S. Lafleche, and Y. Wu. "Laboratory and Field Test Results of Equalization Digital On-Channel Repeater (EDOCR)." *2005 NAB Broadcast Engineering Conference Proceedings*, pp. 485–492. Washington: National Association of Broadcasters, 2005.

Ligeti, A., and J. Zander. "Minimal Cost Coverage Planning for Single Frequency Networks." *IEEE Transactions on Broadcasting*, vol. 45, no. 1, March 1999, pp. 78–87.

Linfoot, S.L. "A Comparison of 64-QAM and 16-QAM DVB-T Under Long Echo Delay Multipath Conditions." *IEEE Transactions on Consumer Electronics*, vol. 49, no. 4, Nov. 2003, pp. 978–982.

Linfoot, S.L. and Lin-Peng Gao. "A Soft Decision 16-QAM Demodulation Algorithm for Multipath Affected DVB-T Systems." *IEEE Transactions on Consumer Electronics*, vol. 51, no. 4, Nov. 2005, pp. 1121–1128.

Malmgren, G. "On the Performance of Single Frequency Networks in Correlated Shadow Fading." *IEEE Transactions on Broadcasting*, vol. 43, no. 2, June 1997, pp. 155–165.

Mattsson, A. "Single Frequency Networks in DTV." *IEEE Transactions on Broadcasting*, vol. 51, no. 4, Dec. 2005, pp. 413–422.

Nakahara, S., S. Moriyama, T. Kuroda, M. Sasaki, S. Yamazaki, and O. Yamada. "Efficient Use of Frequencies in Terrestrial ISDB System." *IEEE Transactions on Broadcasting*, vol. 42, no. 3, Sept. 1996, pp. 173–178.

O'Leary, S. "Field Trials of an MPEG2 Distributed Single Frequency Network." *IEEE Transactions on Broadcasting*, vol. 44, no. 2, June 1998, pp. 194–205.

O'Leary, S. "Digital/Analogue Co-Channel Protection Ratio Field Measurements." *IEEE Transactions on Broadcasting*, vol. 44, no. 4, Dec. 1998, pp. 540–546.

Oziewicz, M. "The Phasor Representation of the OFDM Signal in the SFN Networks." *IEEE Transactions on Broadcasting*, vol. 50, no. 1, March 2004, pp. 63–70.

Rebhan, R., and J. Zander. "On the Outage Probability in Single Frequency Networks for Digital Broadcasting." *IEEE Transactions on Broadcasting*, vol. 39, no. 4, Dec. 1993, pp. 395–401.

Rinne, J., A. Hazmi, and M. Renfors. "Impulse Burst Position Detection and Channel Estimation Schemes for OFDM Systems." *IEEE Transactions on Consumer Electronics*, vol. 49, no. 3, Aug. 2003, pp. 539–545.

Salehian, K., M. Guillet, B. Caron, and A. Kennedy. "On-Channel Repeater for Digital Television Broadcasting Service." *IEEE Transactions on Broadcasting*, vol. 8, no. 2, June 2002, pp. 97–102.

Salehian, K., B., and M. Guillet. "Using On-Channel Repeater To Improve Reception in DTV Broadcasting Service Area" *IEEE Transactions on Broadcasting*, vol. 49, no. 3, Sept. 2003, pp. 309–313.

Salehian, K., Y. Wu, and B. Caron. "Design Procedures and Field Test Results of a Distributed-Translator Network, and a Case Study for an Application of Distributed-Transmission." *2005 NAB Broadcast Engineering Conference Proceedings*, pp. 501-510. Washington: National Association of Broadcasters, 2005.

Salehian, K., Y. Wu, and B. Caron. "Design Procedures and Field Test Results of a Distributed-Translator Network, and a Case Study for an Application of Distributed-Transmission" *IEEE Transactions on Broadcasting*, vol. 52, no. 3, Sept. 2006, pp. 281–289.

Sgrignoli, G. "DTV Repeater Emission Mask Analysis." *IEEE Transactions on Broadcasting*, vol. 49, no. 1, March 2003, pp. 32–80.

Tanyer, S.G., T. Yucel, and S. Seker. "Topography Based Design of the T-DAB SFN for a Mountainous Area." *IEEE Transactions on Broadcasting*, vol. 43, no. 3, Sept. 1997, pp. 309–319.

Wang, J.-T., J. Song, J. Wang, C.-Y. Pan, Z.-X. Yang, and L. Yang. "A General SFN Structure With Transmit Diversity for TDS-OFDM System." *IEEE Transactions on Broadcasting*, vol. 52, no. 2, June 2006, pp. 245–251.

Wang, Jian-Tao, Jian Song, Jun Wang, Chang-Yong Pan, Zhi-Xing Yang, and Lin Yang. "Corrections to 'A General SFN Structure With Transmit Diversity for TDS-OFDM System.'" *IEEE Transactions on Broadcasting*, vol. 52, no. 3, Sept. 2006, p. 412.

Wang, J., Z.-X. Yang, C.-Y. Pan, J. Song, and L. Yang. "Iterative Padding Subtraction of the PN Sequence for the TDS-OFDM Over Broadcast Channels." *IEEE Transactions on Consumer Electronics*, vol. 51, no. 4, Nov. 2005, pp. 1148–1152.

Wang, X., Y. Wu, and B. Caron. "Transmitter Identification Using Embedded Pseudo Random Sequences." *IEEE Transactions on Broadcasting*, vol. 50, no.3, Sept. 2004, pp. 244-252.

Wang, X., Y. Wu, and J.-Y. Chouinard. "Transmitter Identification in Distributed Transmission Network and its Application in Position Location and a New Data Transmission Scheme." *2005 NAB Broadcast Engineering Conference Proceedings*, pp. 511-520. Washington: National Association of Broadcasters, 2005.

Weiss, S.M. "Distributed Transmission Systems—Overcoming the Limitations of DTV Transmission." *2003 NAB Broadcast Engineering Conference Proceedings*, pp. 263–279. Washington: National Association of Broadcasters, 2003.

Weiss, S.M. "Designing Distributed Transmission Systems to Meet FCC Requirements." *2006 NAB Broadcast Engineering Conference Proceedings*, pp. 152–161. Washington: National Association of Broadcasters, 2006.

Wu, Y., X. Wang, R. Citta, B. Ledoux, S. Laflèche, and B. Caron. "An ATSC DTV Receiver with Improved Robustness to Multipath and Distributed Transmission Environments." *IEEE Transactions on Broadcasting*, vol. 50, no. 1, March 2004, pp. 32–41.

Zhang, G.L., L.Y. Qiao, and W. Zhang. "Obtaining Diversity Gain for DTV by Using MIMO Structure in SFN." *IEEE Transactions on Broadcasting*, vol. 50, no. 1, March 2004, pp. 83–90.

CHAPTER

6.6

Coaxial and Waveguide Transmission Lines

KERRY W. COZAD
Dielectric Communications
Raymond, Maine

INTRODUCTION

Transmission lines are one of the main components in the radiofrequency (RF) transmission plant of a broadcast station. Acting as the connecting link between the transmitter and the antenna, the transmission line plays a critical role in both the quality and reliability of the broadcast signal; therefore, the proper choice of a transmission line type to be used can have a significant impact on the success of the station. The choice of transmission line is typically decided based on the following criteria:

- Frequency of operation
- Power handling capacity
- Attenuation (or efficiency)
- Characteristic impedance
- Tower loading (size and weight)

With the implementation of additional digital broadcast channels, other criteria such as installation costs, connector design, reliability, and the effectiveness of transmitting multiple channels through the same transmission line are receiving renewed attention from design and broadcast engineers. There are a wide variety of transmission line types and designs to choose from. This chapter reviews the attributes of rigid coaxial and waveguide transmission lines to assist the broadcast engineer in the selection of the type best suited for a specific need.

COAXIAL TRANSMISSION LINE TYPES

A coaxial transmission line consists of two concentric conductors, the inner conductor being supported within the outer conductor through the use of a dielectric material, as illustrated in Figure 6.6-1. The dielectric material may be continuous throughout the line or, as in the case of rigid coaxial lines, located at distinct points along the line in the shapes of "pegs" or cylindrical "beads." Some general characteristics from electromagnetic field theory are:

- An infinite number of electromagnetic field configurations (modes) is possible.
- Propagating modes will be in a general form of waves traveling along the axis of the line.
- The propagation constant is different for each mode.
- There is a frequency called the *cutoff frequency* where the propagation constant is 0. Below this frequency, there is no propagation of that mode.
- There is one mode, the *transverse electromagnetic mode* (TEM), for which the electromagnetic fields are transverse to the transmission line axis. The TEM has a cutoff frequency of 0 Hz. This is the mode that is primarily used for propagating signals within coaxial transmission lines.

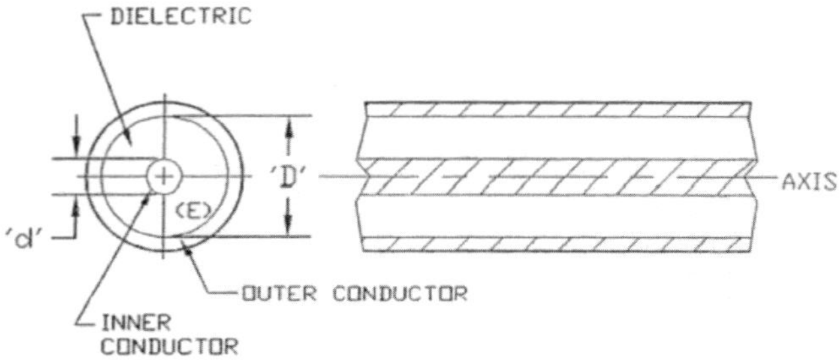

FIGURE 6.6-1 Drawing of coaxial transmission line showing cross-sections.

Based on these common characteristics, coaxial transmission lines are usually divided into three primary groups:

- Flexible RF cables
- Semi-flexible cables
- Rigid coaxial lines

RF Cables

Flexible RF cables are typically used for short interconnections between equipment. They consist of a solid or stranded inner conductor, solid plastic dielectric insulating material, and a braided outer conductor sheath. The cable is coated with a plastic outer jacket to resist moisture and abrasion. The outer conductors of most RF cables are less than 0.5 inch, and the attenuation characteristics restrict this type of cable to short runs. Because it is highly susceptible to deterioration due to aging, it is normally only used indoors.

Semi-Flexible Cables

Semi-flexible cables are designed with soft tempered copper inner and outer conductors. The dielectric material may be either solid (foam filled) or a spiral that has been wrapped in helical fashion around the inner conductor (air dielectric). A plastic jacket is applied to the outer conductor to resist abrasion. One advantage of semi-flexible cables is that they can be fabricated in diameters up to 9 inches and in continuous lengths of hundreds and even thousands of feet. This cable type is used extensively for radio and low-power television broadcasting, as well as inter-element feeders for some antenna types.

Rigid Coaxial Lines

Rigid lines are designed with hard, tempered inner and outer conductors. Dielectric insulators are used to support the inner conductor at various intervals within the outer conductor. Because the coaxial line is rigid, it must be fabricated in defined lengths, typically no longer than 20 feet. The individual lengths are then attached to each other through the use of flanges and the inner conductors are typically spliced together. Rigid coaxial lines can have diameters up to 14 inches. They have high power handling capabilities and low attenuation values. Figure 6.6-2 is a cut-away view of the end of a section of rigid transmission line showing the flange and insulator for the center conductor. Figure 6.6-3 shows the weld holding the flange to the outer conductor. The focus in this chapter is on rigid coaxial lines. See Chapter 4.11 for information on RF cable and semi-flexible types of transmission lines.

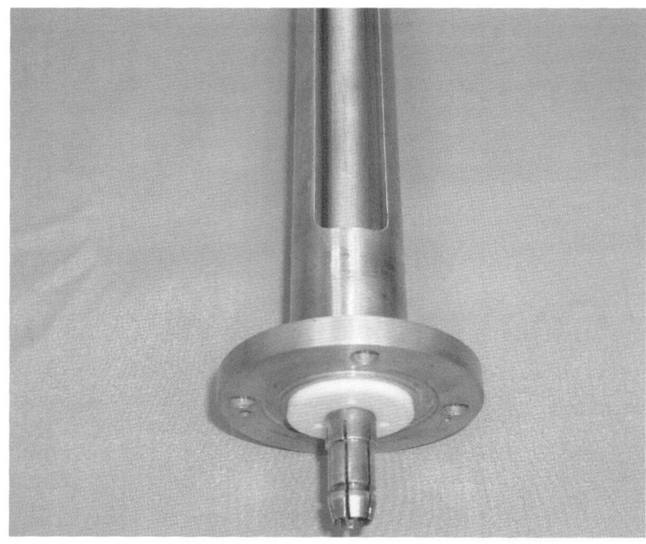

FIGURE 6.6-2 Cut-away view of the end of a section of rigid transmission line showing the flange and insulator for the center conductor.

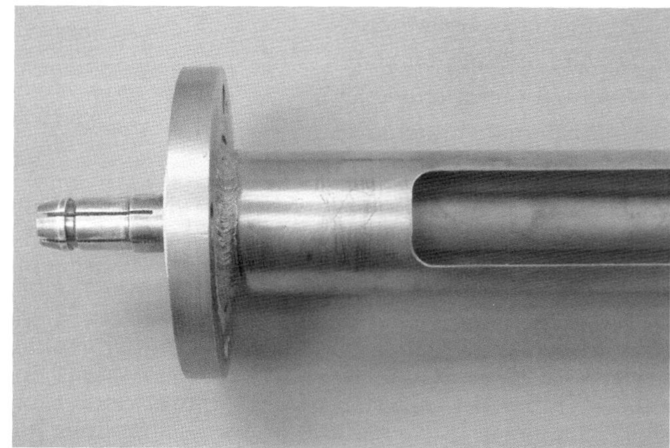

FIGURE 6.6-3 Cut-away view of end of section of rigid transmission line showing the weld holding the flange to the outer conductor.

DESIGN CRITERIA FOR COAXIAL TRANSMISSION LINES

Electrical Parameter Optimization

A basic principle in the design of most transmission lines is to find the optimum configuration of inner conductor to outer conductor dimensions where the power handling is maximized and the attenuation minimized. Each configuration will then determine a characteristic impedance for the line.

Characteristic Impedance

The characteristic impedance is determined by the relative dimensions of the inner and outer conductors and the dielectric material between them. It can be expressed by the following equation:

$$Z_o = \frac{60}{\sqrt{\varepsilon'}} * \ln\left(\frac{D}{d}\right) \tag{1}$$

where:

Z_o = Characteristic impedance.

ε' = Dielectric constant or relative permittivity of dielectric to air.

D = Inside electrical diameter of outer conductor.

d = Outside electrical diameter of inner conductor.

ln = Natural logarithm function.

Attenuation

The attenuation of the line is normally expressed in terms of loss per unit length (dB per 100 ft or dB per 100 meters). The attenuation is due to dielectric losses and conductor losses. The dielectric material loss is directly proportional to frequency. For air-dielectric lines, where the majority of the volume of space between the inner and outer conductors is air, the most commonly used dielectric materials are tetrafluoroethylene (TFE) and polyethylene. These materials produce very small losses; in the case of rigid coaxial lines, this loss is usually negligible relative to the conductor losses. Conductor losses are related to the dimension, permeability, and conductivity of the material used. The conductor loss varies with the square root of the frequency and for copper conductors:

$$\alpha = \frac{0.433}{Z_o} * \left(\frac{1}{D} + \frac{1}{d}\right) * \sqrt{f} \tag{2}$$

where α is the attenuation constant (dB per 100 ft), and f is the frequency (MHz). Attenuation is minimized when D/d is equal to 3.59, which results in an impedance of 77 ohms.

Power Handling Capacity

The power handling capabilities of coaxial lines are based on two factors: the maximum peak power (or maximum voltage that can be safely present) and the maximum average power, which is determined by the allowable temperature rise on the inner conductor.

Peak Power

The maximum electric field strength between two coaxial conductors can be calculated from:

$$E_{max} = \frac{0.278}{d} * \sqrt{\frac{P}{\ln\left(\frac{D}{d}\right)}} \tag{3}$$

where E_{max} is the maximum electrical field strength (volts per inch), and P is the power level of the signal (watts). E_{max} is at a minimum when the ratio D/d is equal to 1.65. This results in a characteristic impedance for an air-dielectric line of 30 ohms. Determining the average power capability of a line requires extensive testing or the use of complex thermal models. It has been determined that the optimum ratio of D/d is approximately 2.72, resulting in a characteristic impedance of 60 ohms.

Summary of Characteristics

From the previous analysis it is apparent that there are trade-offs between optimum configurations for attenuation, peak power, and average power. Broadcast transmission lines have standardized on 50 and 75 ohms. The 50 ohm impedance is a balance between optimum peak power and average power. Typically, if power is the controlling factor, a 50 ohm line is used. If attenuation (efficiency) is more important, then a 75 ohm line should be used.

Operational Parameters

With the characteristic impedances determined, the choice of line now is based on the desired frequency of operation (or limited by the cut-off frequency) and the actual power handling and attenuation ratings.

Cut-Off Frequency

The cut-off frequency (f_c) is the frequency above which undesirable modes of propagation can be generated. The generation of these modes results in degraded efficiency, higher reflections (voltage standing wave ratio [VSWR]), and lowered power handling. Because larger coaxial transmission lines can handle more power, f_c typically is the limiting factor where frequency of operation *versus* power handling is determined. The cut-off frequency is inversely proportional to the conductor dimensions and is determined by the following equation:

$$f_c(GHz) = \frac{7.52}{\sqrt{\varepsilon'} * (D + d)} \tag{4}$$

where ε' is defined as in Eq. 1, above.

Differences in the maximum operating frequency of specific line sizes are sometimes evident when reviewing various manufacturers' specifications. This is typically a result of a different safety factor used when deciding on the actual maximum frequency of operation. A 5% to 10% reduction in the calculated f_c is a typical safety factor and will account for manufacturing tolerances and the effects of connections and elbows; however, where numerous elbows are assembled back-to-back within a system, additional safety factors may be necessary to prevent the generation of higher order modes.

Voltage Standing Wave Ratio (VSWR)

It is important that the transfer of energy from the source (transmitter) to a load (antenna or receiver) be as efficient as possible. The attenuation constant is one factor in determining the efficiency as it represents a direct loss of energy. Another factor is the impedance mismatch between the line and the load. The greater the mismatch, the higher the reflected level of energy at the mismatched connection. This reflected energy reduces the amount of signal transmitted to the load and results in lower efficiency. The incident and reflected waves will combine to produce an uneven voltage distribution. Voltage and current maxima and minima occur, resulting in a standing wave along the line. The VSWR is defined as the ratio of maximum to minimum voltage:

$$\text{VSWR} = V_{\max}/V_{\min}$$

The effects of VSWR may be significant depending on the VSWR level and the specific conditions of operation. In general, as the VSWR increases, the maximum voltage levels increase and voltage breakdown may occur. Excessive heating along the line may also occur due to the current maxima, and softening or melting of the dielectric insulating material may cause the inner conductor to shift in position that can result in a worsening condition of VSWR.

Attenuation

Attenuation is defined as the loss created by the imperfect conductivity of the conductors and the imperfect insulating dielectric medium. In coaxial lines, losses come from both the inner and outer conductor materials and the material used to support the inner conductor. In solid dielectric cables, the dielectric loss can be appreciable and at higher frequencies actually exceed the conductor losses. For air-dielectric lines, the insulating supports are a small percentage of the total dielectric space and their losses are generally negligible. Attenuation and efficiency are directly related. Efficiency is defined as the ratio of power delivered to the load relative to the power input into the transmission line:

$$\text{Efficiency} = \text{Power out}/\text{Power in}$$

The efficiency is determined by calculating the total attenuation (in dB) of the line (α_{total}) based on its overall length. The total attenuation is then converted to efficiency:

$$Efficiency\ \% = 10^{-\frac{\alpha_{total}}{10}} * 100 \tag{5}$$

Actual attenuation can be influenced by the VSWR of the line and the operating temperature. Except in cases of extreme VSWR or temperatures, these effects are insignificant to the performance of the system. For VSWR, the loss is increased by the factor

$$\frac{1+VSWR^2}{2*VSWR}$$

The greatest effect on attenuation is a change in the temperature of the inner conductor, as the conductivity of the material is affected by temperature. The adjustment factor for attenuation (M_α) is given by:

$$M_a = \sqrt{1+\sigma_o\left(T_t - T_o\right)} \tag{6}$$

where:

T_t = Inner connector temperature (°C).

T_o = Inner conductor temperature at standard rating (°C).

σ_o = Temperature coefficient of resistance at standard rating.

The standard temperature rating is based on typical conditions for measuring the attenuation. Often this is done using test equipment under ambient temperature conditions (not under high power transmission). In that case, a normal ambient temperature (inner conductor and outer conductor at the same temperature) can be taken as 20°C (68°F). For a standard temperature rating of 20°C, σ_o = 0.00393/°C. Then:

$$M_\alpha = \sqrt{1+0.00393\left(T_t - 20\right)} \tag{7}$$

For an inner conductor temperature of 100°C, the attenuation will increase by a factor of 1.146.

POWER RATINGS

Two rating factors of concern when discussing the power handling capabilities of coaxial transmission lines are average and peak power. The *average* power rating is based on the maximum internal heating the line construction can withstand. This is normally limited by the maximum safe operating temperature of the dielectric spacing material. The *peak* power rating is based on voltage breakdown characteristics.

Peak Power

Peak power is defined as the maximum RF power that can be reached in any interval, such as an RF modulation cycle. "Peak" refers to the peak amplitude of modulation and is not the instantaneous power when the RF voltage is at a maximum. In a continuous wave (CW) carrier such as FM, the peak power equals the average power. In 100% modulation of an AM signal,

the power rises to four times the carrier power at the peaks of the modulation envelope. So, in this case, the peak power is four times the carrier power.

The peak power rating is dependent on voltage breakdown considerations, which are relatively frequency insensitive; therefore, this rating is constant with frequency. It is, however, sensitive to line size, physical conditions, pressure, and the dielectric medium. It is important that the rating be based on a predetermined set of conditions that are clearly stated. Once a rating is determined for these conditions, it can be adjusted for other conditions.

The procedure for determining a peak power rating is to establish a peak voltage the line will withstand every time under normal manufacturing processes. Because voltage breakdown levels are sensitive to dust, insulator condition, and surface irregularities, the theoretical breakdown gradient cannot be used in practice. It has become standard procedure to de-rate the theoretical breakdown to 35% of its value to determine a production DC test voltage. This test voltage (E_p) is calculated using the following equation, which is derived from the maximum voltage gradient in a coaxial line and includes the derating of 35%:

$$E_p = 3.17 * 10^4 (d\delta) * \left[\log\left(\frac{D}{d}\right)\right] * \left(1 + \frac{0.273}{\sqrt{d\delta}}\right) \tag{8}$$

where:

E_p = Production test voltage.

d = Inner conductor OD, inches.

D = Outer conductor ID, inches.

δ = Air density factor = 3.92B/T, where B is the absolute pressure (cm of mercury), and T is the temperature (°K) (δ = 1 when B = 76 cm and T = 23°C = 296°K).

Rounded off values for E_p for 50 ohm transmission lines are as follows:

E_p (DC Volts)	Nominal Line OD (in.)
6000	7/8
11,000	1-5/8
13,000	2-1/4
16,000	3
19,000	3-1/8
21,000	4
27,500	5
35,000	6-1/8

The production test voltage must now be converted to realistic RF root mean square (RMS) operating voltage,

$$E_{rf} = 0.7 * E_p * 0.707 * 1/SF \tag{9}$$

where:

E_{rf} = Maximum RF RMS operating voltage with no de-rating for VSWR or modulation, but includes a safety factor.

0.7 = DC to RF factor.

E_p = Production test voltage.

0.707 = RMS factor.

SF = Safety factor for voltage (typically 1.4 or 2).

This voltage then determines the peak power rating, P_{pk}:

$$P_{pk} = \left(E_{rf}\right)^2 / Z_o \text{ watts} \tag{10}$$

or

$$P_{pk} = \frac{\left(\frac{E_p * 0.707 * 0.7}{SF}\right)^2}{Z_o} \tag{11}$$

An adequate safety factor on peak power is necessary to safeguard against voltage breakdown that can result in permanent damage to the line. Many manufacturers have settled on a power safety factor of 2×, which is equivalent to a voltage safety factor (SF) of 1.4. When good conditions for installation or operation cannot be maintained, it is good practice to apply a higher safety factor to prevent damage on initial startup of a system. Once the peak power is determined, it must be further de-rated for the effects of modulation and VSWR. These de-ratings are calculated as follows:

$$\text{AM: } P_{\max} < \frac{P_{pk}}{(1+M)^2 * VSWR} \tag{12}$$

$$\text{FM: } P_{\max} < \frac{P_{pk}}{VSWR} \tag{13}$$

$$\text{TV: } P_{\max} < \frac{P_{pk}}{\left(1 + AU + 2\sqrt{AU}\right) * VSWR} \tag{14}$$

$$\text{DTV: } P_{\max} < \frac{P_{pk}}{VSWR} \tag{15}$$

where:

$P_{\max}$ = De-rated maximum power.

P_{pk} = Peak power rating.

M = Amplitude modulation index (100% = 1).

AU = Aural-to-visual ratio (20% aural: AU = 0.2).

Much of the technical literature written on this topic discusses methods to increase the peak power rating through the use of increased pressure and high dielectric strength gases. Although in theory a higher rating is possible, the system requires significant monitoring to prevent extensive damage due to breakdown if a failure in the pressurization integrity occurs. The decision to utilize these procedures requires extensive analysis of the proposed system and should not be used if other more stable configurations are available.

Average Power

With the implementation of digital television transmissions, power rating calculations have moved from "peak of sync + aural" for analog to "average DTV power" for digital. The reality is that power ratings for broadcast transmission lines are primarily based on the average power handling capability of the line type. That average power rating is then converted to the terms normally used in system discussions. When multichannel systems are involved, the average powers of each station are added numerically to determine the maximum average power for which the system must be designed.

The average power rating is limited by the amount of heat created due to line losses. The amount of heat allowable is primarily determined by the safe, long-term performance of the dielectric material used. Because the loss and temperature rise of the inner conductor are greater than the outer conductor, the maximum allowable temperature of the inner conductor is normally used to determine the average power rating.

A typical calculation for determining average power levels is based on allowing the inner conductor to reach a temperature of 100°C with an ambient temperature of 40°C. This means the inner conductor temperature is allowed to rise 60°C above the ambient. Based on this standard condition, the average power can be calculated from the following:

$$P_{avg} < \frac{16,380 * \sigma * D}{M_{\alpha} * \alpha} \text{ watts} \tag{16}$$

where:

P_{avg} = Average power rating for 60°C rise of inner conductor temperature.

σ = Heat emissivity coefficient of outer conductor (watts/in.^{2}).

D = Outer conductor OD (in.).

M = Correction factor for attenuation (relative to 20°C).

α = Attenuation constant (dB/100 ft at 20°C).

It should be noted that the heat emissivity coefficient is derived from experimental data and there are no industry standards as to its value for various line sizes. Also, specifications for allowable inner conductor temperatures vary between manufacturers. Therefore, it is possible to have noticeable differences between published ratings due to slight differences in these values. Typical heat emissivity values for 50 ohm rigid transmission lines are as follows:

Line Size (in.)	Emissivity (watts/in.^{2})
7/8	0.120
3-1/8	0.107
4-1/16	0.104
6-1/8	0.097

The average power is also frequency sensitive, as the attenuation constant (α) is frequency dependent. At higher frequencies, the attenuation is greater; therefore, the average power rating will be reduced. If average power is the most significant factor in choosing a transmission line, it is suggested to review data sheets from several manufacturers and discuss any differences in average power ratings. Ambient temperature and VSWR can affect the average power handling of a transmission line (as these factors directly affect the attenuation losses), but these factors are typically only significant for extreme conditions. Other factors that should be reviewed are flange connection types, inner connector design, and method of thermal compensation for the inner conductor. In this way, all factors that may affect the long-term performance of the transmission line are addressed and a sound engineering decision can be made.

For broadcasting, the average power is dependent on the nominal CW power of the transmitter and the type of modulation. For the primary broadcast modulation schemes, the average power can be calculated from:

$$AM \quad P_{avg} = P_c\left(1 + \frac{M^2}{2}\right) \tag{17}$$

$$FM \quad P_{avg} = P_T \tag{18}$$

$$TV \quad P_{avg} = (0.6 + AU)\,P_{TV} \tag{19}$$

$$DTV \quad P_{avg} = P_{DTV} \tag{20}$$

where:

P_{avg} = Average transmitter power.

P_c = Carrier power.

M = Amplitude modulation index.

P_T = FM transmitter power.

AU = Aural-to-visual ratio.

P_{TV} = TV peak sync power.

P_{DTV} = DTV transmitter power.

Increases in average power ratings can also be theoretically accomplished in a similar fashion as peak powers. The same cautions apply for average power as for peak power.

Velocity of Propagation

A final performance characteristic that has primary importance when attempting to match the phases between two or more lines is the velocity of propagation or phase velocity (V_p). For coaxial transmission lines, it is expressed as a fraction of the speed of light in a vacuum and is determined by the dielectric constant of the insulating material:

$$V_p = \frac{c}{\sqrt{\varepsilon'}}$$

$$\frac{1}{\sqrt{\varepsilon'}} = \text{relative phase velocity} \qquad (21)$$

where c = speed of light

where ε' is defined as in Eq. 1, above.

The result of adding a dielectric material other than air is to slow down the TEM wave. For two coaxial lines having a different percentage of dielectric material, there will be a difference in electrical phase between them even though they are physically the same length. This can be important when using cables for sampling the relative phases of an antenna array, for example. Most manufacturers supply this factor with the other specifications for a transmission line. Note that the phase velocity is independent of frequency and size for coaxial lines

RIGID COAXIAL TRANSMISSION LINE

Rigid transmission lines have inherently high power handling and low attenuation and VSWR which make them ideal for high power broadcast applications. The line is typically fabricated from high-conductivity, oxygen-free, hard copper tubing in nominal lengths of up to 20 ft. The inner conductor is supported inside the outer conductor by peg or disk insulators. A PTFE compound is normally used for the insulating materials due to its extremely good electrical and mechanical properties.

The inner conductors of adjacent line sections are joined together by inner connector *bullets*. These act as splices between the inner conductors and are designed with tension spring fingers to provide high insertion forces that produce low resistance electrical contacts. The high insertion force is also needed for good thermal conductivity. For added conductivity, the bullets are normally silver plated. An insulator is also attached to the middle of the bullet to anchor it between the outer conductor flanges, as shown previously in Figure 6.6-2. This provides mechanical stability when installing the line in a vertical position.

The outer conductors are normally attached by bolting the flanges at the end of each section together. The RF contact is accomplished through a raised surface near the inside diameter of the flange. A pressure seal is obtained through the use of an O-ring between the flanges. Flange types are typically made according to Electronic Industries Alliance (EIA) standards RS-225 and RS-259.[1] However, proprietary designs of various suppliers are also available. Important parameters for rigid coaxial line are given in Tables 6.6-1 and 6.6-2.

Rigid Line Section Lengths

The connections between rigid line sections present an imperfect impedance transition, because each flange produces a small reflection. At some frequency, the distance between flanges will become periodic and a VSWR spike will be generated. This spike will

[1]For more information, see http://www.tiaonline.org/standards/catalog/search.cfm, and enter EIA-225 or EIA-259 in the search field. Alternatively, see: http://www.jampro.com/tl/flangesize.pdf.

TABLE 6.6-1
Rigid Coaxial Transmission Line Characteristics

Nominal OD Outer Conductor (in.)	Characteristic Impedance (Z_o) (ohms)	Maximum Frequency (MHz)	Velocity of Propagation (%)	Nominal OD Inner Conductor (in.)	Nominal ID Outer Conductor (in.)	Net Weight (lb/ft)
7/8	50	6000	99.8	0.341	0.785	0.6
1-5/8	50	3000	99.8	0.664	1.527	1.3
3-1/8	50	1588	99.8	1.315	3.027	3.0
4-1/16	50	1197	99.8	1.711	3.935	5.6
6-1/8	50	788	99.8	2.600	5.981	7.3
6-1/8	75	900	99.7	1.711	5.981	6.75
7-3/16	75	830	99.7	2.000	7.000	9
8-3/16	75	709	99.7	2.293	8.000	11.0
9-3/16	50	530	99.7	3.910	9.000	13
9-3/16	75	600	99.7	2.580	9.000	11.45

TABLE 6.6-2
Rigid Coaxial Transmission Line Flange Parameters

EIA Flange Size (in.)	Bolt Size (in.)	No. of Bolts	Recommended Torque Value
7/8	1/4	3	80 lb-in. (9.0 N-m)
1-5/8	5/16	4	140 lb-in. (15.8 N-m)
3-1/8	3/8	6	20 lb-ft (27.1 N-m)
6-1/8	3/8	12	20 lb-ft (27.1 N-m)
7-3/8	3/8	14	20 lb-ft (27.1 N-m)
8-3/16	3/8	18	20 lb-ft (27.1 N-m)
9	3/8	20	25 lb-ft (34.0 N-m)
9-3/16	3/8	20	20 lb-ft (27.1 N-m)

typically have a narrow frequency response but can be significant based on the number of discontinuities and the magnitude of the reflections. The frequency of the VSWR spike is determined by the following equation:

$$F_{spike}(MHz) = \frac{492.15 * V_p * N}{L} \tag{22}$$

where:

F_{spike} = Spike frequency.

V_p = Relative velocity of propagation.

N = Any integer.

L = Equal distance between discontinuities (ft).

The response curve for this equation has a bandwidth of approximately ±2 MHz; therefore, the line type must be chosen so that a critical frequency of operation is a minimum of 2 MHz from the spike frequency. Because many rigid line systems are several hundred feet in length (involving many 19 or 20 ft line sections), the amplitude of the VSWR spike can be quite large; some manufacturers allow greater safety margins from the critical frequency when determining

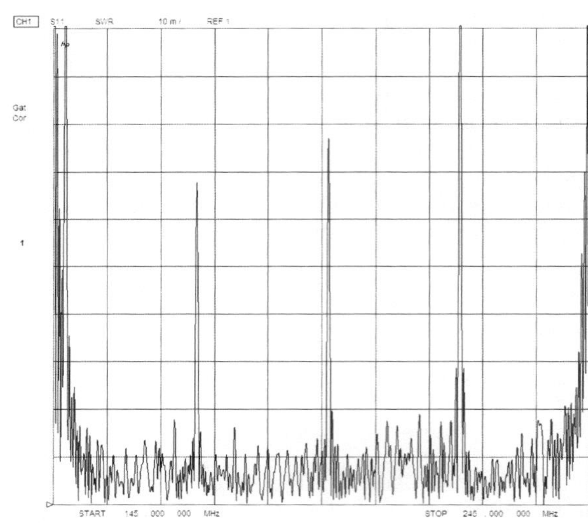

FIGURE 6.6-4 Display of the relative magnitudes of the VSWR spikes versus frequency. Line lengths are chosen to avoid spikes occurring on frequencies within the operating channel.

rigid line section lengths. Figure 6.6-4 shows the relative magnitudes of VSWR spikes *versus* frequency. Table 6.6-3 provides the line lengths required for specific TV and FM channels.

Broadband Rigid Coaxial Transmission Lines

Broadband rigid coaxial transmission line systems have been developed to accommodate multichannel antenna installations. They were also developed to minimize the need for replacing newly installed systems during the final phases of the digital television (DTV) transition due to the channel changes taking place. Each of these systems is proprietary in nature, and most utilize a variation in the length of the line sections to minimize flange reflection additions. The result is minimizing VSWR spikes, as described in the previous section, but it also raises the average VSWR across all the design channels.

TABLE 6.6-3
Rigid Coaxial Transmission Line Lengths for TV and FM Channels

20' Section	19'- 9" Section	19'-6" Section	19' Section
TV Channels			
2, 3, 5, 6, 7, 8, 11, 12, 14, 15, 18, 19, 23, 27, 31, 35, 39, 40, 43, 44, 47, 48, 52, 56, 60, 64, 68	16, 20, 24, 28, 32, 33, 36, 37, 41, 45, 49, 53, 57, 58, 61, 62, 65, 66, 69	4, 9, 10, 13, 17, 21, 22, 25, 26, 29, 30, 34, 38, 42, 46, 50, 51, 54, 55, 59, 63, 67	—
FM Radio			
88.1–95.9 MHz	—	96.1–98.3 MHz	98.5–100.1 MHz
100.3–107.9 MHz		—	—

Internal Thermal Expansion

In a high power broadcast application, the inner conductor will run substantially hotter than the outer conductor. A typical difference in operating temperatures between the inner and outer is 70°F. Because of this temperature difference, the inner conductor will expand to a longer length relative to the outer conductor. This difference can be calculated based on the thermal expansion of copper being 9.8×10^{-6} in./in./°F. With an overall length of 20 ft, the difference in lengths will be 0.164 in. In semi-flexible line, this difference would be absorbed by the corrugations in the conductors. For rigid lines, however, compensation must occur at the inner conductor and inner connector (bullet) interface.

Without the thermal expansion, the inner conductor would be cut to exactly mate up to a shoulder on the bullet. This provides the optimum electrical contact and impedance match. To account for thermal expansion, the inner is cut back the distance needed to allow for the expansion under the maximum allowable operating temperature differential between the outer and inner conductors. A typical guideline is illustrated in Figure 6.6-5 for EIA style flanged line. The actual cutback dimension is determined by the length from the center of the bullet insulator (this is where the contact surfaces of the flanges meet) to the shoulder of the bullet plus the expansion allowance (typically between 0.125 and 0.20 in.). If cut in the field, the edges of the inner conductor must be carefully deburred so as not to deform the inner conductor. Additionally, the cutback dimension should not exceed 0.20 in., or the pressure applied by the bullet fingers may not be great enough to provide good electrical and thermal contact. As the temperature varies between input power levels and ambient temperatures, the cutback allows the inner conductor to move back and forth on the bullet. For long section lengths operating at maximum power and temperature levels, this will cause material deterioration from friction and ultimately will result in either mechanical or electrical failure at the contact points.

To accommodate this movement but minimize the degradation of the contact surface, compensation devices have been developed for rigid lines. They come in two commonly used designs:

- Watchband spring attachment to the bullet
- Bellows attachment to the inner conductor

The watchband spring design provides for a spring-like contact device that is attached circumferentially around one side of the bullet as shown in Figure 6.6-6. The spring side is then inserted at the top of the lower inner conductor (the line section closer to the transmitter) such that the spring is below the insertion point. The insertion force of the upper side of the bullet is such that, during thermal changes, the inner conductor movement is primarily on the lower spring section. This allows any shavings that are caused by the movement to fall inside the inner conductor and not out into the main portion of the line. The spring is

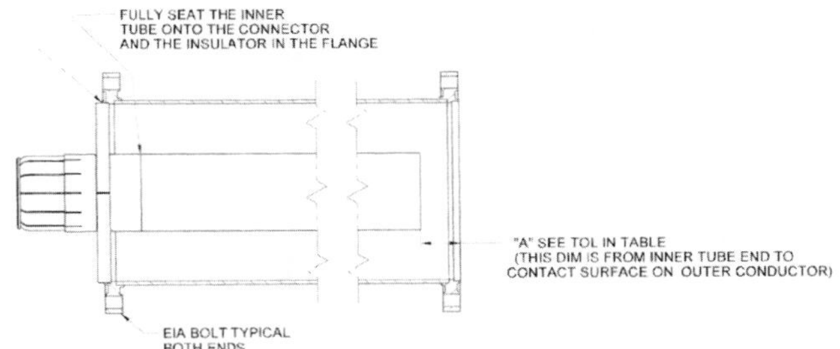

FIGURE 6.6-5 Illustration showing cut-back center conductor to accommodate uneven heating and to prevent buckling.

designed to be self-compensating in this process, providing a constant contact pressure on the inner conductor; however, the mechanical wear does limit the reliable life of the spring, based on the conditions of operation for that particular line system.

In some rigid line, a bellows device is built directly into the inner conductor, as shown in Figure 6.6-7. Thermal compensation is achieved by the expansion and contraction of the corrugated section similar to the principal used in the construction of semi-flexible lines. The benefit of this design is that the bellows absorb all of the inner conductor movement and eliminate wear at the bullet/inner conductor interface. The reliable life of the line is determined by the number of expansion/contraction cycles the bellows is designed to withstand. The different compensation methods must be evaluated by the user relative to the specific environmental and operating conditions present at the broadcast site. Factors such as installation, maintenance requirements, extreme ambient temperatures, and high operating power levels can be significant in determination of the proper concept needed for long-term reliability at minimum costs.

Rigid Line Systems

A rigid coaxial transmission line system is composed of a vertical exterior run and a horizontal run that have some components inside the transmitter building and some attached to an exterior horizontal support bridge between the building and the tower. Because rigid lines are air-dielectric types, all components subject to contamination by outside air must be under constant pressurization.

Differential Thermal Expansion

The *most important factor* to consider in an overall rigid line installation is the difference in thermal expansion coefficients between the copper line and the steel support tower. For a 100°F change in ambient temperature, the differential expansion between the line and tower will be about 0.4 in. per 100 ft of length. For a 1000 ft tower and a temperature change from 0°F to 100°F, this will result in the rigid line growing up to 4

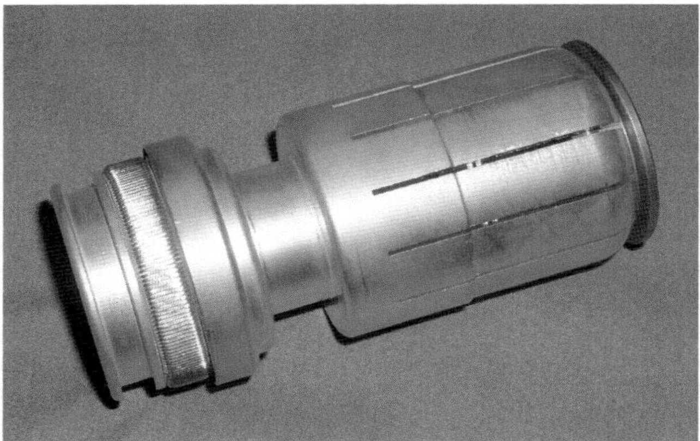

FIGURE 6.6-6 Watchband spring attachment to the bullet to accommodate thermal expansion and contraction of the center conductor.

FIGURE 6.6-7 A bellows attachment to the inner conductor is designed to accommodate thermal expansion and contraction of the center conductor.

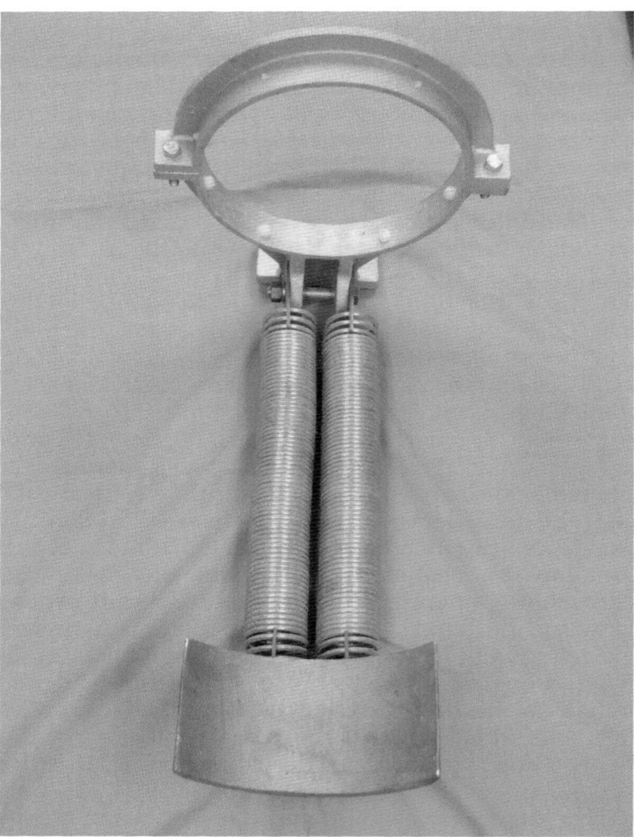

FIGURE 6.6-8 Combination rigid line hanger with spring assembly to accommodate thermal expansion and contraction of the overall transmission line. Spring hangers are used at all support points except at the top.

in. relative to the tower. Provisions must be made to support the line within the tower and at the same time allow it to float to prevent buckling strains on the tower and line. This is accomplished through the use of vertical spring hangers that attach rigidly to the tower but support the weight of the line by springs and clamps, as shown in Figure 6.6-8.

To account for ice loads and possible failures of the spring hangers, a rigid hanger without a spring is attached near the top of the vertical run. It is important that it be placed as close to the antenna input as possible to prevent thermal expansion from causing excessive force on the antenna input. Because the rigid connection is at the top of the run, the cumulative effects of the thermal expansion or contraction occur at the bottom of the vertical run. This is where it is important to allow sufficient clearance for the line to enter the tower from the horizontal run and not be damaged by tower members as the line moves up and down relative to the tower. And, because the horizontal run is anchored at the point of entry to the building, horizontal movement in and out of the tower must also be accounted for during installation. Tables 6.6-4 and 6.6-5 provide the amount of extra space needed at the bottom of the tower to accommodate thermal expansion. Figure 6.6-9 illustrates how the transmission line is configured at the bottom of the tower to accommodate thermal expansion.

Movement up and down in the horizontal run is typically limited to several feet from the elbow at the base of the tower. From that point toward the tower, either horizontal spring hanger assemblies or rollers can be used to accommodate the linear expansion of the line. Typically one support for every 10 ft of line is adequate. Once inside the building, supports for the line are usually threaded rods attached to supports in the ceiling and saddles around the line. For long runs inside, a support every ten feet is still adequate; however, if it is anticipated that the line may be taken apart for maintenance (for example, to attach a test transition), then additional supports may be desired near the area where the line will be separated.

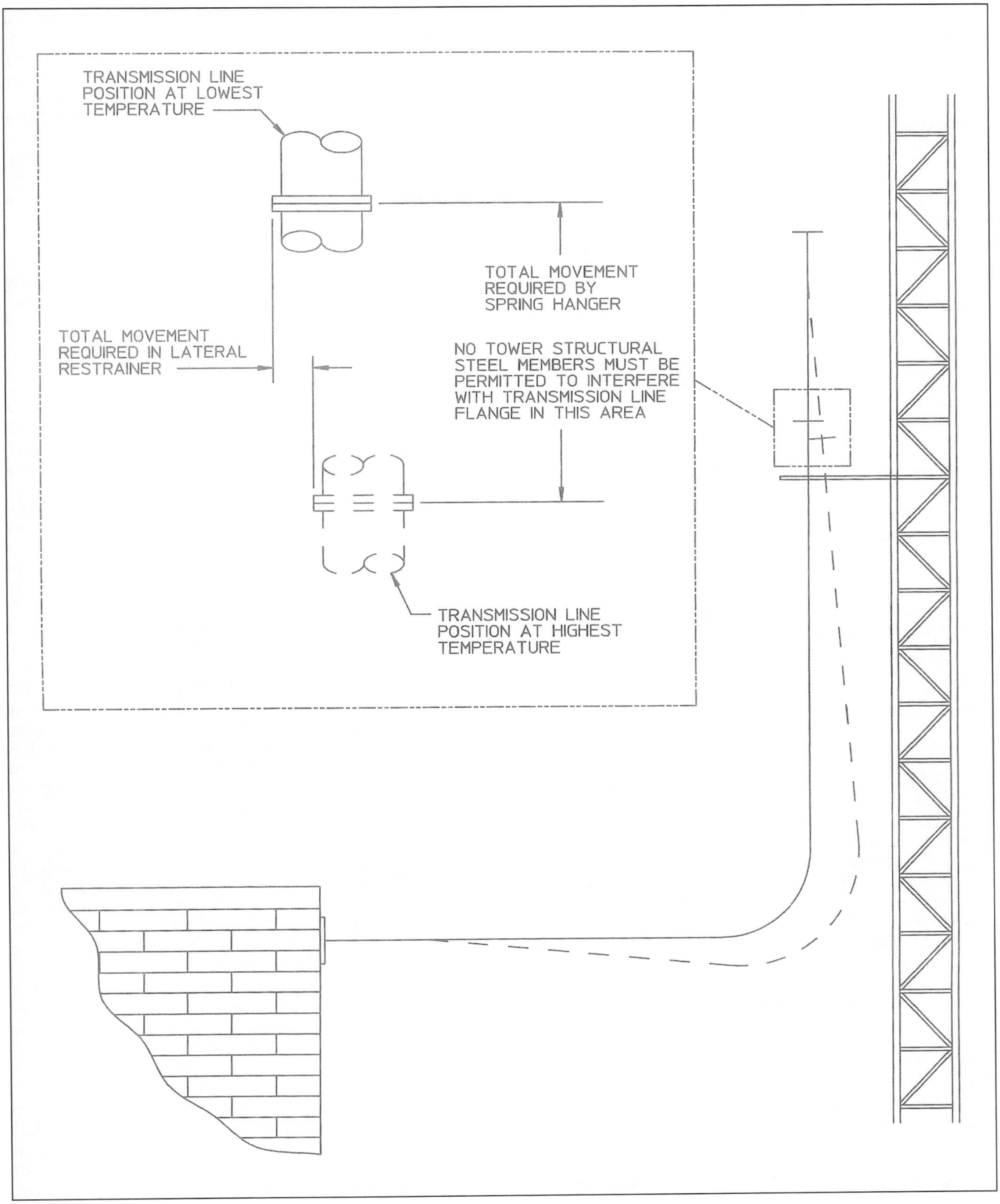

TRANSMISSION LINE POSITION AT LOWEST TEMPERATURE

TOTAL MOVEMENT REQUIRED BY SPRING HANGER

TOTAL MOVEMENT REQUIRED IN LATERAL RESTRAINER

NO TOWER STRUCTURAL STEEL MEMBERS MUST BE PERMITTED TO INTERFERE WITH TRANSMISSION LINE FLANGE IN THIS AREA

TRANSMISSION LINE POSITION AT HIGHEST TEMPERATURE

FIGURE 6.6-9 Schematic of rigid transmission line installation showing how thermal expansion is accommodated.

TABLE 6.6-4
Distance from Lowest Hanger to Location of Horizontal Run Entry into Tower

Horizontal Run (ft)	Minimum Distance to Lowest Hanger (ft)				
	3-1/8"	4-1/16"	6-1/8"	8-3/16"	9-3/16"
20	5	6	9	12	13
40	6	7	11	15	17
60	7	8	13	17	20
80	8	9	14	19	22
100	9	10	16	21	23

TABLE 6.6-5
Recommended Minimum Length of Rigid Coaxial Line Horizontal Run

Vertical Run (ft)	Recommended Length of Horizontal Run (ft)				
	3-1/8"	4-1/16"	6-1/8"	8-3/16"	9-3/16"
100	15	15	15	20	20
500	25	30	35	40	40
1000	35	40	50	60	60
1500	40	50	60	70	70
2000	45	60	70	80	80

Line Interfaces

Interface standards for rigid transmission line, established by the EIA are RS-225 for 50 ohm line and RS-259 for 75 ohm line. The user should be aware, however, that some proprietary designs used by suppliers are not completely compatible with the EIA standards or other manufacturers' components. When purchasing components from different suppliers or making repairs to a system, it is important that all suppliers are aware of the system design and individual component types so the appropriate adapters can be supplied if needed.

RIGID WAVEGUIDE TRANSMISSION LINE

Although the term *waveguide* can mean any structure that guides electromagnetic waves, for the purposes of this section it will refer to a hollow, closed metal tube of constant cross-section. Because waveguide does not use a center conductor like coaxial transmission lines, it has the advantages of higher power handling capabilities and lower attenuation. At the same time, it requires larger cross-sectional sizes than rigid coaxial line to allow propagation of the electromagnetic wave which increases the wind-load forces applied to the tower *versus* coaxial lines. For practical purposes, waveguide use is limited to UHF channels in broadcasting and typically for those special applications that require higher power levels or higher efficiencies than can be found using a coaxial transmission line.

As a result of its high power handling capabilities and high wind loading characteristics, there have been minimal applications for digital television. Lower ERP requirements for DTV have reduced the necessary transmitter power levels. An exception has been in cases where a TV broadcast station has chosen to combine the NTSC channel and an adjacent DTV channel into a single UHF broadband antenna. The combined NTSC and DTV power levels may then require the higher power capacity of waveguide instead of coax line.

Propagation Modes

The physical size of the waveguide is driven primarily by the electrical dimensions necessary to allow the propagation of the principal mode of propagation. Unlike coaxial line, which has no lower limit on the frequency of the signal that will propagate, waveguides have a lower cutoff frequency, below which propagation will not occur. This cutoff frequency is dependent on the physical dimensions of the waveguide. Table 6.6-6 provides the operating frequencies for standard types of high power broadcast waveguides.

The principal mode of propagation for rectangular waveguides is the TE_{01} mode. The electric field vector is perpendicular to the direction of propagation and is oriented from one broad wall to the other. The electric field intensity is greater at the center of the guide and least at the narrow wall. In elliptical and circular waveguides, the principal mode is TE_{11}. Again, the electric field vector is perpendicular to the direction of propagation but its specific orientation in the circular waveguide is not as easily determined because the circular cross-section has no single axis of symmetry. This results in a more complicated system to launch the signal into the waveguide and then be able to effectively retrieve the signal at the antenna end. A method used to control the electric field orientation in circular waveguide has been to install polarizing pins within the waveguide sections which act to maintain a specific orientation of the electric field.

For UHF broadcasting, the electrical operating parameters of waveguide are typically much better than coaxial lines: lower attenuation and higher peak and average power handling. For installation purposes, the lack of an inner conductor makes it much simpler to make the flange-to-flange connection. Its larger physical diameter, however, can make it more difficult to handle during installations, as it now takes up much more room inside the tower than a coaxial line will. Once it is determined that the use of waveguide *versus* coaxial line is necessary, the final determination of waveguide type will typically come down to wind loading concerns and overall system costs. Because waveguide systems used for UHF

TABLE 6.6-6
Typical Characteristics of Waveguides for UHF Television Channels

Waveguide Designation	Waveguide Type	Major Axis (a) or Diameter (in.)	TV Channels	Frequency (MHz)
WR1800	Rectangular	18	14–35	470–602
WR1500	Rectangular	15	20–56	506–728
WR1150	Rectangular	11.5	46–69	662–806
DTW1750	Doubly Truncated	17.5	14–47	470–674
DTW1500	Doubly Truncated	15	25–59	536–746
DTW1350	Doubly Truncated	13.5	37–69	608–806
GLW1750	Circular	17.5	14–19, 24–41	470–506, 530–638
GLW1700	Circular	17	20–23	506–530
GLW1500	Circular	15	39–55	620–722
GLW1350	Circular	13.5	56–69	722–806
EWG1800	Elliptical	18	14–39	470–626
EWG1600	Elliptical	16	32–52	578–704

broadcast applications are typically not limited based on peak or average power ratings, the details of the theory behind these parameters are not presented here. Performance parameters that do differ from coaxial transmission lines are presented below.

Cutoff Frequency

The lower cutoff frequencies (in MHz) for the primary modes found in broadcast waveguide types are:

- Rectangular waveguide (TE_{01}), $f_{Cl} = c/2a$.
- Circular waveguide (TE_{11}), $f_{Cl} = 6917/\text{diameter}$.

Velocity of Propagation

Due to the lower cutoff frequency, the velocity of propagation for waveguides is dependent on the frequency of operation. For waveguide:

$$V_p = c * \sqrt{\left[1-\left(f_{c1}/f\right)^2\right]} \tag{23}$$

This equation shows that, unlike coaxial lines, the velocity of propagation of a wideband signal (such as a 6 MHz television channel) changes for each frequency component because the cutoff frequency is a constant. And, because the cutoff frequency is dependent on the waveguide physical dimensions, the electrical performance of the signal as it is transmitted through the waveguide is directly affected by the size of the waveguide. This is one of the reasons why different types of waveguides may use different sizes to optimize performance for the same frequency range.

Differential Thermal Expansion

As is the case with rigid coaxial lines, the most important factor to consider in planning the installation of waveguide is the difference in thermal expansion coefficients between the aluminum waveguide and the steel support tower. For a 100°F change in ambient temperature, the differential expansion between the line and tower will be about 0.75 in. per 100 ft of length. Note that this is approximately twice the rate of change as for copper rigid coaxial line. For a 1000 ft tower and a temperature range of 0 to 100°F, this will result in the aluminum waveguide growing about 7.5 in. relative to the tower. Provisions must be made to support the line within the tower and at the same time allow it to float to prevent buckling strains on the tower and line. This is accomplished through the use of vertical spring hangers that attach rigidly to the tower but support the weight of the line by springs and clamps, as discussed for coaxial rigid line. The different shapes and weights of waveguides require the use of different hanger designs for each waveguide type. Consult the specific manufacturer's information for details regarding hanger mounting, spring tension settings, maintenance, and repair.

Another point to consider is where the waveguide enters the tower near its base, as this is where the greatest amount of change will occur because the top of the vertical run is held rigid. The larger cross-section dimensions of the waveguide will be more susceptible to interference with tower members as the line expands and contracts relative to the tower. The line must be located so tower members (horizontal and diagonal components) will not restrict this movement (both upward and downward); otherwise, the tower member will cause the line section to bend or dent, resulting in a significant change in the VSWR of the line and possibly resulting in a failure under power. Also consider the expansion and contraction into and out of the tower due to the horizontal run. Because horizontal runs for waveguide systems are typically longer than rigid coaxial systems to allow more flexibility of the straight sections near the base of the tower, the waveguide will have more relative movement into and out of the tower as diagrammed earlier in Figure 6.6-9. Therefore, tower members or other objects that are horizontally close to the waveguide at the base of the tower may also cause interference.

Special Considerations for Waveguide

From a maintenance standpoint, there are two considerations that require annual inspections (or more often in highly corrosive environments) to ensure that the system will maintain long-term reliability. The first is the connection between the aluminum waveguide and brass or copper components on waveguide to coaxial transitions. These transitions are typically used inside the transmitter building and near the top of the vertical waveguide run. They are used to interconnect with the coaxial components of the transmitter RF system output or the coaxial input of the antenna. The galvanic properties of the two dissimilar metals can result in significant corrosion at the contact joint. This corrosion, if left untreated, can eventually lead to material deterioration so severe that the gasket or O-ring can no longer hold air pressure and a pressure leak occurs. Subsequently, if the situation continues to be unaddressed, deterioration of the electrical contact surface can occur which will result in overheating and ultimate failure of the flange connection. Several actions can be taken to minimize this type of corrosion. Plating of the brass or copper flange with a material that is closer in the galvanic series to aluminum has been successful in areas of low to moderate corrosion. In areas of high corrosion such as Houston, Texas, or the coast of Florida, more aggressive methods may become necessary. These include using a dielectric spacer between the flanges to prevent prolonged moisture contact between the materials (the water acts as a conductor in the process of corrosion so by breaking the electrical path the majority of the corrosion mechanism is stopped), or filling the gap and coating the materials with a moisture-inhibiting grease.

The second condition involves corrosion that can occur between the flanges of the waveguide itself. A natural condition of aluminum is to form a coating of aluminum oxide on its surface quickly when exposed to air and moisture. In most cases, if this coating is left undisturbed, it forms a protective coating that resists further deterioration of the material's surface; however, it has been found that in areas where a corrosive solution can be trapped between two closely spaced aluminum surfaces, the buildup of corrosion can be significant. In areas with highly corrosive environments, inspection of the waveguide flanges and hardware should be performed regularly, areas of corrosion buildup cleaned, and a moisture-inhibiting grease applied for future protection.

An additional consideration for waveguide systems is pressurization. Rectangular waveguides (WR-series) are typically limited to less than 0.5 psig of internal pressurization. Higher pressure levels can result in bowing of the broad walls and deterioration of the VSWR performance of the system. Other waveguide types may allow pressurization up to 2 psig, depending on the manufacturer's recommended practice for the specific situation. In all cases, pressure relief valves are necessary to prevent overpressurization and permanent damage to the waveguide. A relief valve also provides the ability to properly purge a system or apply a vacuum to eliminate moisture that might be in the waveguide after installation.

Because of the typical low pressure levels, careful monitoring of the waveguide pressure is necessary to prevent moisture intrusion due to a leak. One of the benefits of waveguide is that small amounts of moisture do not normally result in performance deterioration, unlike coaxial lines where moisture may build up on the insulators and result in an arc; however, if a leak is left uncorrected, moisture may begin to accumulate at the lowest point, typically the elbow at the bottom of the vertical run. Instances of water in an amount sufficient to change the tuning of the waveguide system have been known to occur. In at least one case known to the author, the water did not have an effect until cold weather caused the water to freeze at night and the change in dielectric constant caused such high reflected power that the transmitter could not run at much more than 20% power. The station waited until the next day when the sun had raised the temperature sufficiently to melt the ice and they then drained the water using a plug located on the bottom side of the elbow.

TRANSMISSION LINE ATTENUATION AND POWER RATING

Tables 6.6-7 and 6.6-8 show the relative wind load on a tower, attenuation, and power ratings for popular semi-flexible and rigid coaxial as well as waveguide transmission line for the range of TV channels.

MAINTENANCE AND TESTING

Although there are some specific differences between the maintenance and testing of coaxial lines and waveguide, the primary concerns are the same. The following guidelines will apply for both except where a specific difference is noted.

Pressurization

Air dielectric transmission lines located in an uncontrolled environment must be kept under positive pressure using a dry gas to prevent moisture or other contaminants from entering the system. The moisture

TABLE 6.6-7
Windloading and Attenuation Factors for Popular Sizes of Semi-Flexible and Rigid 50 Ohm Transmission Line

	SEMI-FLEXIBLE 50 OHM						RIGID 50 OHM		
TYPE	1-5/8	2-1/4	3	4	5	5 HP	1-5/8	3-1/8	6-1/8
Relative Size for Windload	1.0	1.2	1.5	2.0	2.6	2.6	0.8	1.6	3.1
Attenuation (dB/100 feet)									
Channel 14	0.462	0.387	0.351	0.277	0.184	0.192	0.442	0.226	0.120
Channel 69	0.620	0.520	0.489	0.379	0.250	0.267	0.592	0.296	0.178
Average Power (kW)									
Channel 14	6.40	8.80	15.10	23.40	31.80	68.80	6.63	26.40	79.40
Channel 69	4.80	6.50	10.80	17.10	23.00	49.10	4.98	20.20	58.50
Peak Power (kW) (Voltage Breakdown)	305	425	640	1100	1890	1690	300	900	3000

and contaminants can accumulate on the internal surfaces and connection joints of the transmission line causing oxidation and corrosion resulting in:

- Increase in line attenuation
- Increase in VSWR
- Localized heating due to resistive losses
- Voltage flashover at dielectric surfaces

Transmission lines pressurized with dry gas maintain the initial performance of the system and reduce the risk of damage to the line and subsequent off-air situations. Dry gas is typically obtained from nitrogen bottles or an air dehydrator system. It is important not to exceed the pressure rating of any of the system components.

Prior to applying power to the transmission line system, it should be sufficiently purged of environmental air acquired during installation. This normally can be accomplished by exchanging at least three volumes of dry gas through the system by opening a purge valve or joint at the far end of the system and allowing the dry gas to pass through the system for a predetermined length of time based on the size of the system. Purging should also be performed anytime the system is opened to outside air or a positive pressure condition has not been maintained.

Evacuation of the transmission line system has become more common in the past several years. This is accomplished by attaching a vacuum pump to the line and reducing the internal pressure to cause any

TABLE 6.6-8
Windloading and Attenuation Factors for Popular Sizes of Rigid 75 Ohm Transmission Line and Several Sizes of Rectangular and Cylindrical Waveguide

	RIGID 75 OHM			RECTANGULAR WAVEGUIDE			CYLYNDRICAL WAVEGUIDE		
TYPE	6-1/8	8-3/16	9-3/16	1150	1500	1800	1350	1500	1750
Relative Size for Windload	1.0	1.3	1.5	2.8	3.7	4.5	2.2	2.4	2.9
Usable Channels	14 to 69	14 to 54	14 to 35	43 to 69	30 to 56	14 to 42	56 to 69	39 to 59	14 to 41
Attenuation (dB/100 feet)									
Channel 14	0.112	0.077	0.071	0.113	0.062	0.057	0.053	0.049	0.052
Channel 69	1.154	0.097	0.080	0.086	0.053	0.047	0.046	0.038	0.030
Average Power (kW)									
Channel 14	68.8	119.2	146.7	X	X	X	X	X	X
Channel 69	49.5	96.8	130.2	X	X	X	X	X	X
Peak Power (kW) (Voltage Breakdown)	2000	2700	3000	X	X	X	X	X	X

X Not a limiting factor for typical system. Power rating greater than 5000 kW.

moisture present to become a vapor and then removing the vapor from the line by evacuation. It is important to know all the components that are attached to the line that may be subject to this reverse internal pressure, and care must be taken not to exceed any pressure restrictions. Examples would be rectangular waveguide, which may have its broad walls deformed, and a pressurized antenna radome that may suffer cracking. A significant advantage of this process is that when nitrogen is used to replace the evacuated air, a minimal amount of oxygen is then present within the line, and natural oxidation of contact surfaces over time is minimized. This process should not be undertaken without careful review of the options by a consulting engineer or manufacturer and a discussion about any new information that may be available on the optimal pressurization system for your installation.

Electrical Testing

Electrical testing of the transmission line system should be performed for any of the following conditions:

- During initial installation
- After VSWR trips to determine the cause and location
- Periodically to document performance
- At any time the performance of the system is in doubt

Most conditions that result in less than optimal performance of the transmission line system can be detected immediately after installation. The most useful measurements are the VSWR/return loss sweep test and RF pulse testing. There are many types of equipment that can perform these tests. Over the past few years, the network analyzer with time-domain transform capabilities has become the standard type of equipment for these tests.

The VSWR/return loss test is used to determine the composite input VSWR of the transmission line system terminated by a load or antenna as seen by the output of the transmitter. Normally, measurement of the VSWR is performed as a function of the frequency within the broadcast channel of interest; however, wider frequency sweeps are recommended to determine possible performance over wide temperature variations due to line length changes from thermal expansion/contraction or antenna VSWR changes due to environmental conditions such as icing. This data may also be helpful when looking at the use of an existing transmission line system for a new channel, as may occur during the DTV transition. Significant cost savings from being able to reuse components as well as the avoidance of unanticipated anomalies in performance due to tuning processes used during the initial installation can be realized.

Because the VSWR test is a composite, it presents an overall view of the system performance. System specifications for VSWR must take into account the many individual components; therefore, long complicated systems will typically have higher VSWR responses than short simple systems. Improvements can be made through the use of fine tuners installed near known contributors of VSWR, such as elbow complexes. Additionally, known component configurations such as elbow complexes can be assembled in the factory and optimized prior to shipment to minimize the need for additional fine tuners. Exceptional care must be taken during installation of factory assembled and tuned components and complexes. They are normally match marked and must be installed exactly as indicated or significant degradation of the VSWR performance will result.

The in-channel RF pulse test displays reflections in the channel response as a function of time rather than frequency. This can be of great help in locating the cause of excessive VSWR or faults within the system. It is also used to identify the reflections from the antenna and other far-end components to determine if ghosting conditions may be present. A typical far-end reflection performance that will prevent most problems with the transmitted signal is a 32 dB return loss (1.05:1 VSWR) within the broadcast channel of interest.

Where the cause of high VSWR may be an improperly assembled flange or damaged conductor, the in-channel RF pulse does not have the resolution to accurately locate the fault. Because the width of the pulse is inversely proportional to the frequency bandwidth, a narrower pulse for resolution requires wider bandwidth. With network analyzers presently on the market, the feature to change the bandwidth is readily available. These can be used in similar fashion to the time-domain reflectometer (TDR) test. Bandwidths of 50 to 150 MHz centered on the channel will usually provide sufficient resolution to identify the above types of faults. For testing waveguide systems, note the operating frequency range of the waveguide and be sure not to exceed that range in setting up for a wider bandwidth pulse measurement. The introduction of additional modes of propagation will affect the accuracy of the measurements. Note that this test is not to determine the in-channel characteristics of the system but rather to provide a more detailed look at the transmission line. In particular, components that are frequency sensitive, such as an antenna, may present a significant reflection when using wider bandwidths. These types of components should be identified ahead of time to prevent incorrect analysis of the data.

Insulation Resistance Test

An insulation resistance test can be performed on a coaxial transmission line system after purging with dry gas to determine if it is sufficiently dry before applying power. The insulation resistance between the inner and outer conductors should be greater than 100,000 megohms. There are many inexpensive test units available to perform this test. This is especially important when installing semi-flexible cable that has

been stored outdoors for an extended length of time, and it is not known whether positive pressure has been maintained.

Operation and Maintenance

The following guidelines should be followed to ensure long-term and reliable operation of a coaxial or waveguide transmission line system:

- Establish a baseline VSWR and time-domain response of the system during installation.
- Set and maintain all VSWR or reflected power protection devices at nominal levels. If a fault occurs, do not override these devices to continue transmitting at high power.
- Always maintain positive pressure on the transmission line system.
- Do not exceed the pressure rating of the transmission line or antenna. Damage to components and personnel may result.
- Perform periodic maintenance checks on the pressurization system. Deviations from baseline levels are a first indication of potential problems with the line.
- Check for hot spots along the line during routine inspections. The outer conductor temperature is proportional to the inner temperature, and excessive heating may indicate the beginning of a failure of the inner conductor connections.
- Check the horizontal and vertical line sections for wear caused by thermal expansion.
- Measure the system VSWR and time-domain response on a regular basis and compare to previous results for changes. This is one way to avoid catastrophic failures of the transmission line system. Unless the line is damaged by environmental conditions (such as falling ice or high winds) or vandalism, most failures occur due to a deteriorating condition that may last several months or even years. Monitoring and testing on a regular basis are like regular physical check-ups for people. If detected early, a deteriorating condition can be addressed and corrected under normal maintenance conditions; otherwise, they can result in emergency actions to keep the station on the air which will likely result in significantly higher costs.
- If a repair does become necessary, it is important that all contamination inside the transmission line has been removed. Typical contamination is carbon dust that results from the dielectric insulators overheating and burning. Components that have been severely damaged due to arcing or a stubborn adherence of the molten dielectric materials should be replaced.
- Outer conductors can often be cleaned adequately by first running clean, lint-free rags through the line section followed up with rags using a quick drying cleaning solution such as denatured alcohol. Do not use soaps or other agents that may leave a film or require flushing with water unless it is possible to completely clean and dry the inside surface of the outer conductor.
- Always check for local guidelines regarding the use of potentially hazardous cleaning materials and take proper precautions regarding protective clothing and eyewear.
- Inner conductors can be wiped off and cleaned in a similar manner. Although insulators can also be cleaned, it is good practice to replace them, as any contamination that is not completely removed can become the start of a failure in the future. Note that any contaminant not removed during the cleaning process most likely would migrate to an insulator at some time in the future. This can result in the possibility of an arc occurring across the insulator. One method for checking the line cleanliness is to insert the inner conductor assembly and then remove it. The edges of the spacer insulators may reveal if contaminants are still present. Finally, when cleaning bullets, note that they are fabricated from brass and then silver-plated. Be careful not to damage the plating surface as this will result in a high resistance contact point and excessive heating under RF power. If a watchband spring is part of the connector, replace any spring that appears damaged or over-heated.

Conditions to Look for and Avoid

Over the years, the author has identified specific conditions that are common to many installations of rigid transmission lines and will result in less than optimal performance or even failure of the systems. Below are examples of some of those conditions:

- Pinched O-rings or gaskets
- Broken spring hangers due to wind vibration
- Improper lengths of line used to connect horizontal and vertical runs
- Highly corrosive environments
- Loose flange hardware
- Missing hanger components
- Improper installation of rigid hanger

Regular maintenance and inspections will help identify these conditions early. If one of these conditions is found, immediate corrective action will prevent further deterioration and probably catastrophic failure.

SUMMARY

There are several options when choosing transmission line. Careful consideration of the electrical and mechanical properties of the line types will help the broadcast engineer in recommending the optimum transmission line for the stations. For longer life and to minimize conditions that can result in catastrophic failure and off-air time, regular maintenance is time and effort well spent.

Bibliography

Cozad, K.W., A technical review of transmission line designs and specifications for transmitting television signals, in *NAB Broadcast Engineering Conference Proceedings*, 1998, pp. 16–24.

Cozad, K.W., Coaxial transmission lines, in *NAB Engineering Handbook*, Ninth ed., Whitaker, J., Ed., National Association of Broadcasters, Washington, D.C., 1999.

Cozad, K.W., Inner conductor replacement for rigid transmission lines, *Broadcast Eng.*, Dec., 54, 1991.

Cozad, K.W., *Transmission Line Considerations for HDTV*, Professional Report PR-42-05, Andrew Corp., Westchester, IL, April 2, 1993.

Dienes, G., Circular waveguide for UHF-TV: operational field experience, in *NAB Broadcast Engineering Conference Proceedings*, 1987, pp. 269–274.

Dienes, G., The antenna/transmission line system and HDTV, in *NAB Broadcast Engineering Conference Proceedings*, 1990, pp. 124–131.

Mamak, W., Save time and money: optimizing your broadcast RF system, in *NAB Broadcast Engineering Conference Proceedings*, 2003, pp. 61–65.

Margala, A.J. and Johnson, E.H., Transmission lines for broadcast use, in *NAB Engineering Handbook*, Fifth ed., Prose Walker, A., Ed., National Association of Broadcasters, Washington, D.C., 1960, pp. 2-184–2-210.

Packard, K.S. and Lowman, R.V., Transmission lines and waveguides, in *Antenna Engineering Handbook*, Jasik, H., Ed., McGraw-Hill, New York, 1961, chap. 30.

Paris, D.T. and Hurd, K.F., *Basic Electromagnetic Theory*, McGraw-Hill, New York, 1969, chap. 9.

Schmitz, A.N., Average power ratings of coaxial transmission lines, in *NAB Broadcast Engineering Conference Proceedings*, 1991, pp. 111–115.

Shumate, S.E., Methods and costs of installing initial and interim DTV transmission facilities on existing towers, in *NAB Broadcast Engineering Conference Proceedings*, 1998, pp. 195–201.

T. Vaughn Associates, *Advanced Television Transmission*, Public Broadcasting Service, National Associations of Broadcasters, 1995, pp. VI-18–VI-23.

Manufacturer Web Sites

Andrew Corporation, www.andrew.com
Belden, www.belden.com
Dielectric Communications, www.dielectric.com
Electronics Research Inc., www.eriinc.com
Jampro Antennas/RF Systems, www.jampro.com
Myatt, Inc., www.myat.com
Radio Frequency Systems, www.rfsworld.com

C H A P T E R

6.7

Filters, Combiners, and RF Components

DEREK SMALL AND STEPHEN KOLVEK

MYAT, Inc.
Mahwah, New Jersey

INTRODUCTION

This chapter examines the components and systems that comprise broadcast radio frequency (RF) systems. These components route, switch, combine, and filter the RF power once it exits the transmitter. The way that these components are arranged and coordinated will determine the versatility, reliability, and overall performance of the system. Just as the situations and challenges for each transmission facility are unique, so too can be the solutions that are engineered to meet their demands.

FILTERS

The stringent requirements of digital television, and the need to place an increased number of channels into a reduced allotment of spectrum, make RF filters an important part of the television broadcast transmission facility. Whether it is simply meeting the FCC "mask" specification, or combining multiple channels into a single antenna feed, the properly designed filter is the device that makes it possible. There are many filter types to choose from, each with their own set of attributes and limitations. Selecting the proper one for the required application will yield the desired result.

Low Pass Filters

Low pass, or harmonic filters, reduce the amplifier's second and third harmonic intermodulation distortion products at the RF system output. Low pass filter design for broadcast applications is based on two slightly different mathematical functions: Chebyshev and Achieser-Zolotarev. Realization of these functions gives similar appearing mechanical designs as illustrated in Figure 6.7-1. Design is achieved by alternately cascading high impedance sections with low-impedance sections. These sections simulate lumped element shunt capacitors and series inductors for the realization of distributed low pass filters. Elliptic function polynomials can be used for the design, but are seldom used in broadcast applications.

A low pass filter is designed around a desired cutoff frequency, which is defined as the point at which the insertion loss of the filter begins to increase due to ever-increasing reflection. The rate at which the filter's attenuation increases with frequency is determined by the number of high or low impedance sections that are cascaded in series. Each capacitive and inductive element constitutes a section. Referring again to Figure 6.7-1, there are five low impedance sections and four high impedance sections, yielding a nine-section filter.

FIGURE 6.7-1 Cascaded low and high impedance sections of a typical coaxial low pass filter inner conductor.

Chebyshev Low Pass Filters

The Chebyshev low pass filter is a true low pass design characterized by a wide passband at low frequencies, right up to the cutoff. At cutoff, the attenuation then increases rapidly with frequency to a maximum value. This design is widely applied in broadcast low pass filters, as it presents low loss in the passband and reasonable attenuations in the reject band. The primary limitation of this filter, other than power handling, is that when the high impedance sections approach a half wavelength of a frequency in the reject band, the filter will begin to exhibit random points of spurious pass frequencies in the reject band. The length of the high impedance section is determined by the cutoff frequency and the diameter of that section. Smaller diameters result in shorter inductive sections and a greater spurious free stop band, but at the cost of decreased power handling. A Chebyshev low pass filter designed for VHF will typically be free of spurious responses through the fifth harmonic of the cutoff frequency, while in a UHF filter it will be clear of spurious responses up through the third harmonic.

Achieser-Zolotarev Low Pass Filters

The Achieser-Zolotarev design uses a unique polynomial function, but its physical configuration is similar to the Chebyshev in shape and in method of construction. At first glance, the two designs appear similar, but the Achieser-Zolotarev polynomial has several advantages over the Chebyshev polynomial. The major advantage is an overall reduction of impedances in the Achieser-Zolotarev filter. This allows for the use of larger diameters in the high impedance sections resulting in greater power handling, and for the use of shorter sections, which extend the reject band of the filter. The Achieser-Zolotarev design also gives increased attenuation in the reject band. There is a cost, however, to this improvement. The Achieser-Zolotarev polynomial has a reject band appearing below the passband. The position of this reject band is controllable by the design of the filter. The general rule is to set the upper edge of the lower reject band at 0.5 of the cutoff frequency. Since most broadcast low pass filters will meet the "0.5 of cutoff" criteria, this design is desirable. A typical UHF coaxial low pass filter for broadcast television is shown in Figure 6.7-2.

Band Pass Filters

RF band pass filters are common to most broadcast facilities to absorb or block unwanted intermodulation products generated by the power amplifiers. An RF band pass filter is a two-port network configured to provide transmission of power at frequencies within the desired operating channel, or the passband, and attenuation at the undesired frequencies, or stop band. Filters were designed initially using principles of image theory, which makes use of simple constant-k and m-derived networks cascaded to achieve the desired filter response. This technique tends to be time consuming due to its iterative nature in achieving the desired filter response. Modern design techniques make use of insertion loss theory, where filter responses are synthesized based on rejection specifications. Research performed under contract by The Signal Corps, supported by the Stanford Research Institute, generated many new design techniques, papers, and what's commonly known as the bible of filter design, *Microwave Filters, Impedance-Matching Networks, and Coupling Structures* [1]. Key papers by Orchard, Temes, Levy, and Cameron furthered the work of those original treatises by simplifying the design of cross-coupled filters to realize sharper elliptic and pseudo-elliptic function responses.

Band pass filters are generally used to clean up unwanted intermodulation distortion products nearby the channel passband. Low pass filters are used to block amplifier second and third harmonic products that are missed by the band pass filter due to its size and the inability to attenuate these signals. Band stop filters tend to be used for the rejection of narrower signals, such as in the FM band. All filter responses find uses in channel-combining applications.

FIGURE 6.7-2 Typical coaxial low pass filter.

Design Concepts

All filter designs, regardless of the type, begin with the low pass prototype. The low pass prototype response is one in which the low pass response is normalized to 1 ohm, and has a passband cutoff of 1 radian. Coupling coefficients between the "n" elements are synthesized based on this response and transformed to the desired filter response using appropriate impedance and frequency mapping functions. Using the synthesized coupling coefficients, n-section band pass filters of any shape or size can be constructed with the appropriate resonant cavity design to handle desired levels of RF power.

Bandpass filters use "n" resonant cavities that store energy to obtain frequency discrimination. The storage elements for high-power RF filters may take the form of coaxial transverse electromagnetic (TEM) mode, evanescent mode, or waveguide TE/TM mode cavities. Cavity efficiency, and therefore its power handling capability, is given by the unloaded quality factor "Q_U." The quality factor associated with a resonant cavity coupled (or loaded) by an external source is expressed as

$$Q_U = Q_L\,(1+\kappa) \tag{1}$$

where Q_L is the loaded Q, and κ is the input-coupling coefficient. The coupling coefficient and loaded Q can be measured using a network analyzer, and the cavity Q_U calculated to predict the mid-band insertion loss of

a filter. Mid-band insertion loss, L_o, of an n-section filter with a Chebyshev response is approximated using

$$L_o = (4.34/Q_U)\,(F_o / BW) \sum g_k \text{ (dB)} \qquad (2)$$

where F_o is the filter center frequency, and BW is the ripple bandwidth. The normalized Chebyshev g_k values are summed from k = 1 to n. The mid-band insertion loss, as stated in equation 2, is a function of cavity Q_U, the percentage bandwidth, and the Chebyshev g_k values. Given the number of sections and filter VSWR, the g_k values are fixed, and cavity Q_U, bandwidth, and center frequency are the remaining factors that influence insertion loss. Larger cavity Q_U results in lower insertion loss and flatter passband responses. In general, as cavity volume increases, so too will the Q_U of that cavity. As cavity volume is increased, however, it is important to avoid introducing unwanted higher-order modes that tend to propagate unwanted energy to the antenna instead of reflecting it back to the source or to a reject load. Very high Q_U cavities tend to introduce undesired modes relatively close to the filter passband. Cavity geometry must be judiciously chosen to keep the higher-order modes as far as possible from the desired reject band. Low Q_U cavities tend to be smaller and decrease the roll-off rate in the reject band. For that reason, sharp rejections become difficult to achieve and attenuation peaks in cross-coupled filters become finite. Passband roll-off for low Q_U cavities increases and may make the design of narrow or sharp-tuned filters prohibitive.

Percentage bandwidth of a filter, BW/F_o, inversely affects insertion loss, particularly when it becomes excessively small. Insertion loss as a function of the percentage bandwidth, for a six-section Chebyshev filter with a passband VSWR of 1.10, is shown in Figure 6.7-3. In this example, cavities with a Q_U of 7,000 and of 30,000 are shown. In practice, the knees of these curves are slightly sharper. Filters designed for television typically exhibit a percentage bandwidth of approximately 0.75–1.7%. Percentage bandwidth for DAB (digital audio broadcasting) filters at L-band typically range from 0.11–0.14%.

Given the bandwidth and center frequency of a filter, the appropriate cavity Q_U must be chosen to meet insertion loss and power-handling specifications without contributing to excessive heat buildup. Obviously, the average power-handling capability of a low Q cavity is reduced compared to a higher Q cavity, due to heating considerations that tend to make the cavities drift with frequency. Typical 8-VSB and NTSC filter responses are illustrated in Figures 6.7-4, 6.7-5, and 6.7-6 for filters with different unloaded Q. The sharper eight-section response in Figure 6.7-5 is used where greater attenuation is required in the first upper/lower adjacent 500 kHz slot. The lower loss responses in each figure represent unloaded Q of aluminum TE111 cavities, approximately 30,000. The higher loss responses reflect smaller cavities with heavy evanescent mode propagation, approximately 14,000. When designing high-power filters, it is desirable to have the bandwidth of the passband be as wide as possible. As mentioned above, a wider bandwidth decreases insertion loss and lowers group delay variation, two factors that increase the average and peak power-handling capacity.

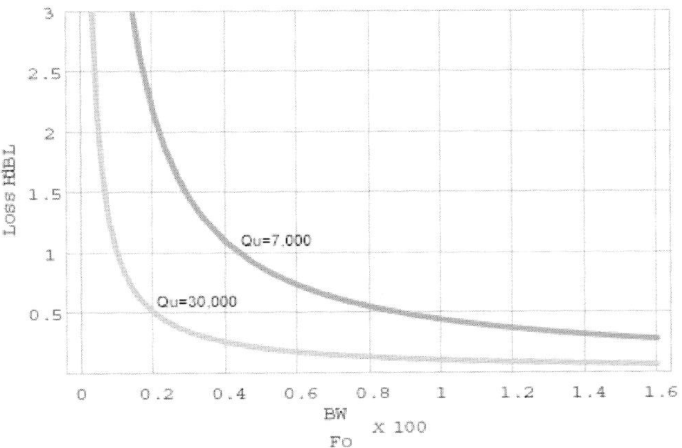

FIGURE 6.7-3 Percentage bandwidth versus mid-band loss for Q_U = 7,000 and 30,000.

An increase in the energy storage of a cavity is proportional to an increase in the group delay characteristics of the cavity. The peak power handling of a filter is estimated by calculating the voltage breakdown at the

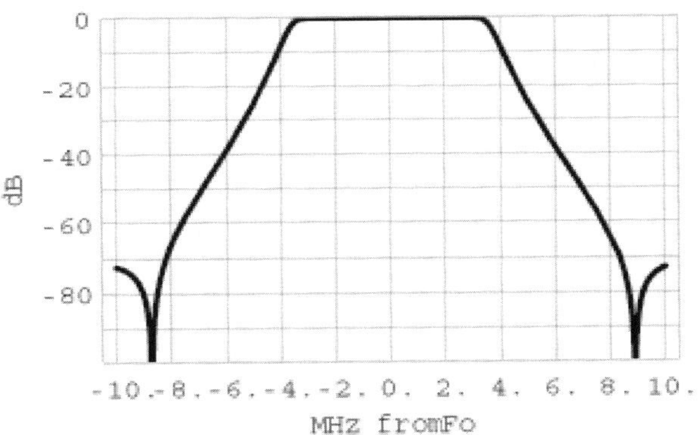

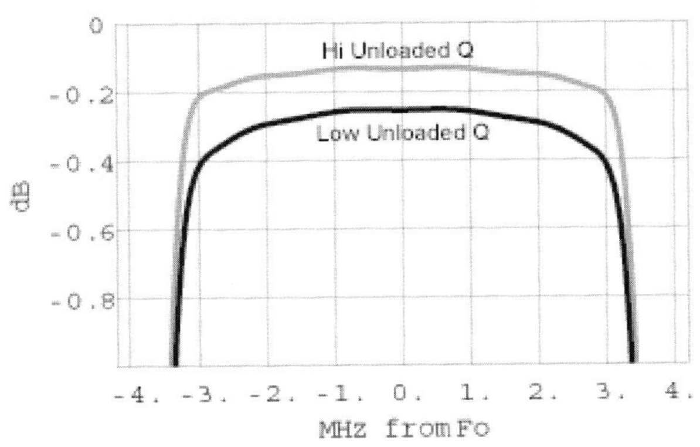

FIGURE 6.7-4 Typical six-section "mild-tuned" filter response for an 8-VSB transmitter.

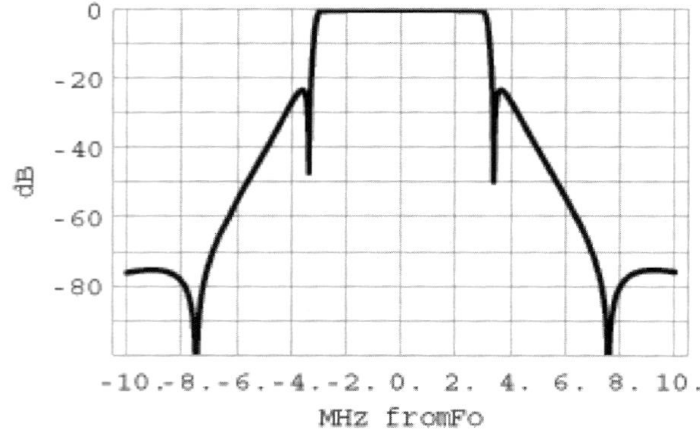

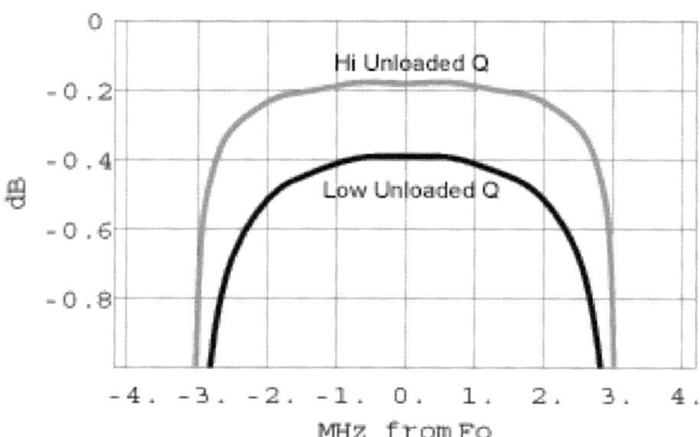

FIGURE 6.7-5 Typical eight-section "sharp-tuned" filter response for an 8-VSB transmitter.

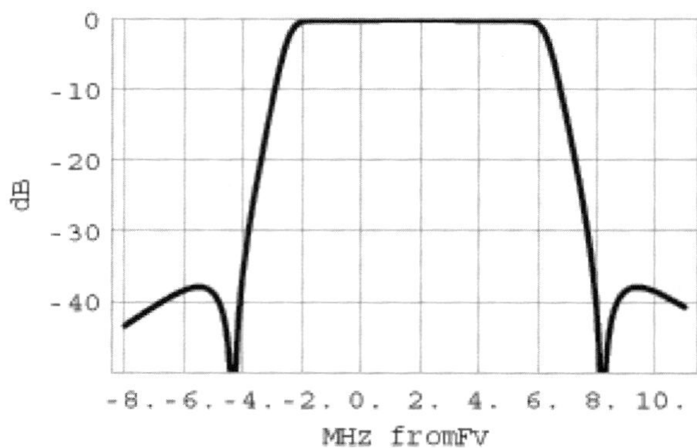

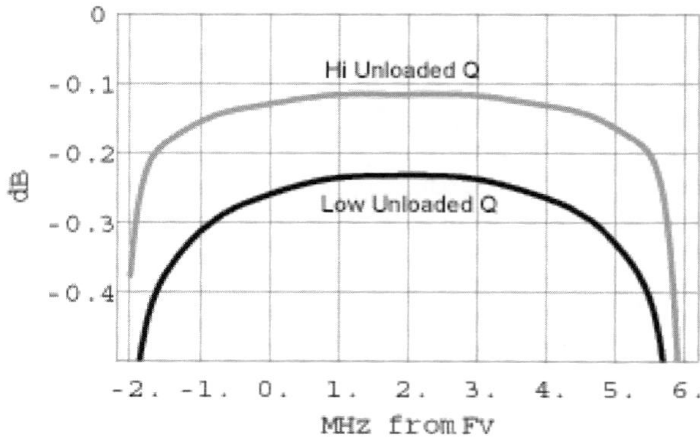

FIGURE 6.7-6 Typical six-section filter response for an NTSC transmitter.

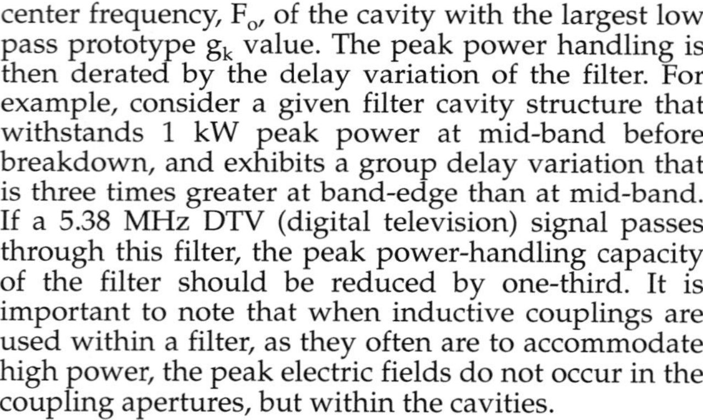

center frequency, F_o, of the cavity with the largest low pass prototype g_k value. The peak power handling is then derated by the delay variation of the filter. For example, consider a given filter cavity structure that withstands 1 kW peak power at mid-band before breakdown, and exhibits a group delay variation that is three times greater at band-edge than at mid-band. If a 5.38 MHz DTV (digital television) signal passes through this filter, the peak power-handling capacity of the filter should be reduced by one-third. It is important to note that when inductive couplings are used within a filter, as they often are to accommodate high power, the peak electric fields do not occur in the coupling apertures, but within the cavities.

Average power handling of a filter is estimated (for digital signals) by integrating the passband insertion loss so that dissipated power can be calculated. This power is then distributed evenly over inductive surfaces to find the watts/cm dissipated on the cavity surface. A thermal rise is then calculated based on cavity material and construction.

TEM Coaxial Mode Band Pass Filters

A TEM filter is one in which the components of the E and H fields lie transverse to the direction of propagation. Stated differently, the field components do not lay in the direction of propagation. A structure of this nature will have an inner conductor, as is found in coaxial line, strip-line, slab-line, and microstrip. Cavity design is typically cylindrical, square, or rectangular in cross section. The resonator inner conductor is usually cylindrical and a quarter wave long at the design frequency, differing slightly due to cavity coupling and capacitive end loading. Temperature compensation is generally achieved by using Invar rods to keep the resonator at a fixed length relative to temperature. Coaxial mode filters tend to be smaller and have higher loss compared to their waveguide counterparts. Filters in the FM and VHF bands are almost exclusively of TEM design. For a strict coaxial mode VHF filter design (Channels 7–3), the unloaded Q may only be as high as 7000. Because of the large percentage bandwidth for television filters at VHF, the mid-band loss of the eight-section sharp-tuned DTV filter shown in Figure 6.7-7 is 0.2 dB.

Due to the reduction in percent bandwidth, and the fact that a strict TEM cavity has such a low unloaded Q at UHF (e.g., 3 inch square cavity), a six-section UHF DTV filter would have a mid-band loss of approximately 0.6 dB. This filter will handle very little power due to excessive insertion loss and minimal size. The insertion loss of this strict TEM UHF filter is illustrated in Figure 6.7-8. Note the average, or integrated, loss is nearly 0.7 dB.

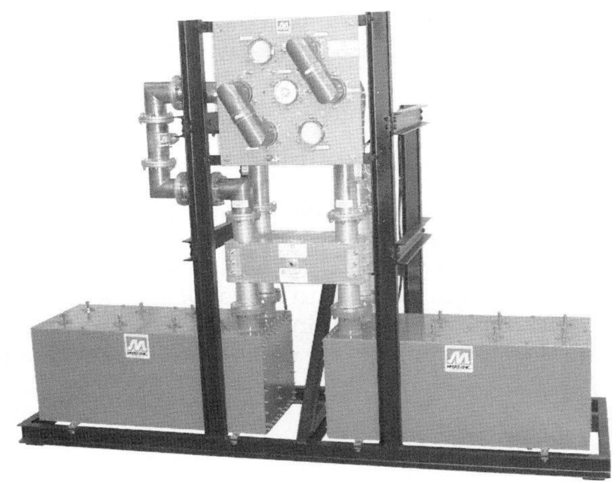

FIGURE 6.7-7 VHF eight-section pseudo-elliptic filter.

Interestingly, high-power FM filters operate in a TEM coaxial mode with percentage bandwidths less than those of the UHF filter shown in Figure 6.7-8. Cavity size, however, plays a significant role in filter design in the FM band. FM cavities can be constructed to allow evanescent modes to propagate. As a cavity design for a given frequency becomes larger, waveguide modes below cutoff begin to propagate more freely. If allowed, these modes can significantly reduce the insertion loss of a filter. For example, an FM band pass filter might have a passband of 0.45 MHz giving a percent bandwidth of approximately one-half. A strict TEM mode cavity could have an unloaded Q of approximately 8,600, giving a mid-band loss of approximately 0.33 dB. A larger cavity will naturally achieve a greater unloaded Q, however, enhancing evanescent modes can result in a third less loss when compared to a strict TEM design.

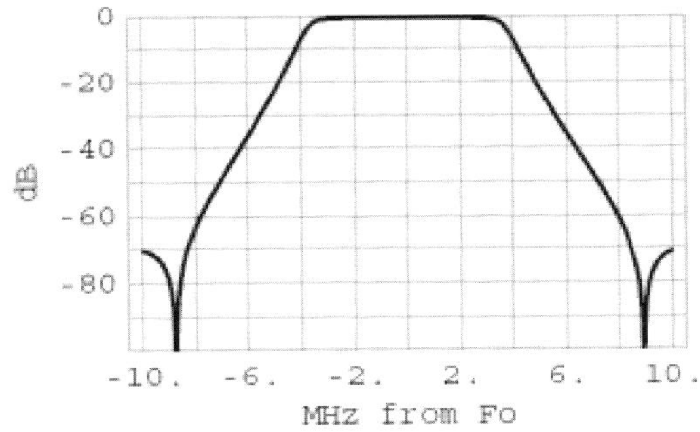

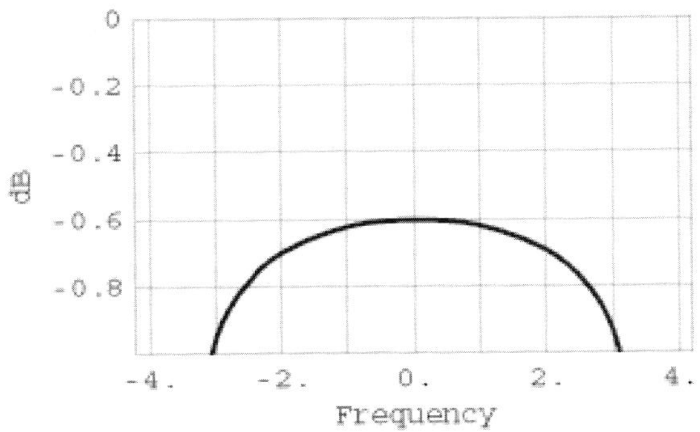

FIGURE 6.7-8 Coaxial mode UHF six-section pseudo-elliptic filter response.

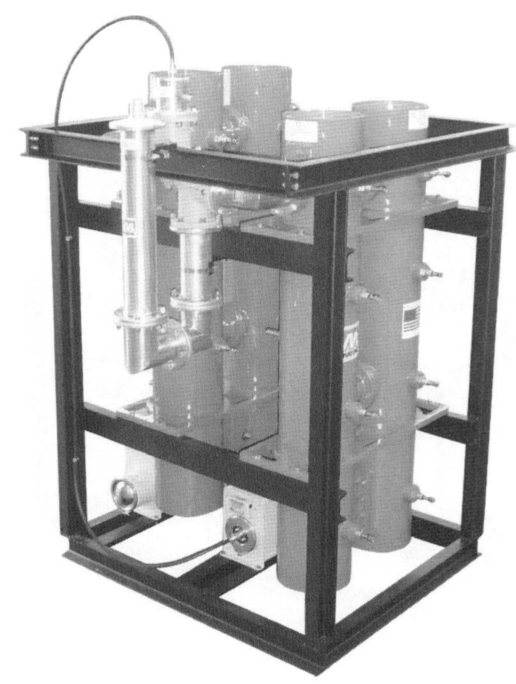

FIGURE 6.7-9 Evanescent mode UHF filter.

Techniques used to enhance evanescent mode propagation in UHF filter design allow several kilowatts of power to pass without excessive rise in operating temperature. An example of a TEM mode filter that significantly enhances evanescent mode propagation is illustrated in Figure 6.7-9. This filter, despite its relatively small size, has an insertion loss of 0.3 dB at mid-band.

Waveguide Mode Band Pass Filters

Rectangular waveguide TE10 mode band pass filters were initially used in terrestrial broadcast facilities for combining multiple stations into one antenna. With the advent of common amplification transmitters, rectangular waveguide filters were used as mask filters. Initial common amplification mask filters were designed using Chebyshev responses of seven sections. Improved designs cascaded notch filters, or traps, with six-section band pass filters for reduced insertion loss. Further improvements were made using cylindrical dual TE11n mode cavities. Dual-mode cavity filters were first utilized for terrestrial TV broadcast applications in 1993, primarily for NTSC common amplification transmitters. Dual

FIGURE 6.7-10 Typical waveguide UHF dual-mode band pass filter.

TE11n cavities allow for the design of narrow, low-loss filters with steep rejections to clean up visual/aural IMD products. Today, the output of nearly all high-power NTSC and DTV transmitters use these filters because of their high Q, low loss, and ability to realize complex filter functions. Digital applications utilize six-section, or sharper tuned eight-section, pseudo-elliptic function responses with pass bandwidths designed to straddle a specific channel (refer back to Figures 6.7-4 and 6.7-5). A typical filter with waveguide I/O is shown in Figure 6.7-10.

COMPONENTS

The components used as building blocks for RF passive systems include

- Power divider
- Power combiner
- Transmission line switches
- Directional couplers
- Loads
- Diplexers
- Channel combiners

The basic concepts for of each these items will be covered in this section. It is important to specify electrical performance required for each component. Performance requirements of each component must include frequency bandwidth, maximum peak and average power levels, the impedance of each port, and required port match for the intended application. Transmission line is another component, which is treated separately in this handbook. See Chapters 4.11 and 6.6 for information on transmission lines for FM and TV, respectively.

Power Dividers and Combiners

Power dividers are designed to split power equally between two or more output ports for a given frequency and bandwidth. Unequal power division is also possible. Both equal and unequal N-way power division can be designed in coaxial or waveguide transmission line. For power combining the reverse is true.

Unmatched Power Divider

A typical unmatched power divider is characterized by having three ports each dimensionally meeting standard port dimensions for a coaxial or waveguide transmission line. Selecting one port as an input and the remaining two as outputs, some analysis can be made. For example, in a coaxial 3-1/8 inch 50 ohm unmatched tee with ports each having an impedance of 50 ohms as shown in Figure 6.7-11, the input port would see a tee junction impedance of 25 ohms due to the two 50-ohm output ports being connected in parallel. The impedance mismatch as seen at any port results in a VSWR of 2.0:1. The return loss of the input port would be –9.5 dB with 11.1% of transmission power being reflected back to the source. The source output port would need to be protected from the high power level being reflected back. The remaining 88.9% of transmission power would be evenly split between the two output ports yielding a power division of –3.5 dB. In addition only –3.5 dB isolation exists between output ports. Additional output ports could be added, but this would result in severe degradation of all performance parameters with the severity increasing with each additional port.

The usefulness as a power divider is obviously poor but the unmatched tee becomes quite useful

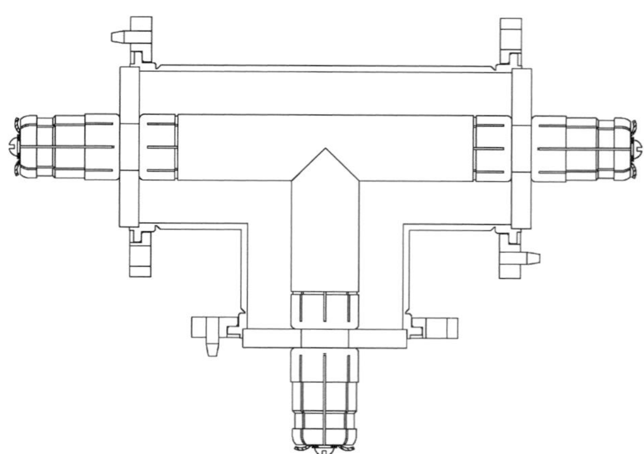

FIGURE 6.7-11 Typical coaxial unmatched tee.

when a short or open circuit is connected to one port, and the remaining two ports are used as the input and output for the device. A simple notch filter is constructed by placing an adjustable short circuit stub to one port of the tee. The electrical phase length of the short from the tee junction impedance is chosen so that it appears as a short circuit to a frequency to be blocked from passing through the device. The blocked signal is reflected back to the source. In addition, the notch response will occur at harmonics of the design frequency. Between the notch responses the short appears as high impedance and is a low-loss path to frequencies going from the input to the output ports. Shorted stub on the unmatched tee theory becomes useful when designing a junction type channel combiner (discussed in a later section). In theory, an open circuit on an unmatched tee would operate in the same manner with the proper phasing considerations. The open port represents an RF hazard to personnel and is therefore not normally used.

Matched Power Divider

A matched power divider is achieved by adding a frequency-sensitive impedance-matching network to the input port of a basic unmatched tee as shown in Figure 6.7-12. The matching network takes the form of a quarter wave transformer, which inherently has a bandwidth limitation. Increased bandwidth is achieved by adding transformer sections in the matching network. The impedance matching network must match the 50 ohm input port to 25 ohm junction impedance resulting from the two 50 ohm output ports. The transformer impedance is equal to the square root of the product between junction impedance and input impedance. The impedance of the transformer is then found to be 35.35 ohms. If the two outputs are selected to be 75 ohms, then the junction impedance becomes 37.5 ohms requiring a different matching network.

The input port exhibits excellent return loss and hence a VSWR approaching unity, provided that the output ports are properly terminated. Equal –3 dB power division occurs, sending half power to each output port. A drawback to the matched tee is that poor isolation still exists between the two output ports. Output port-to-port isolation is –6 dB, only a minor improvement to the case of an unmatched tee. Any type of change in one output port load will disturb both power division and degrade input port match. N-way power dividers are designed by calculating the junction impedance and providing an appropriate impedance matching network.

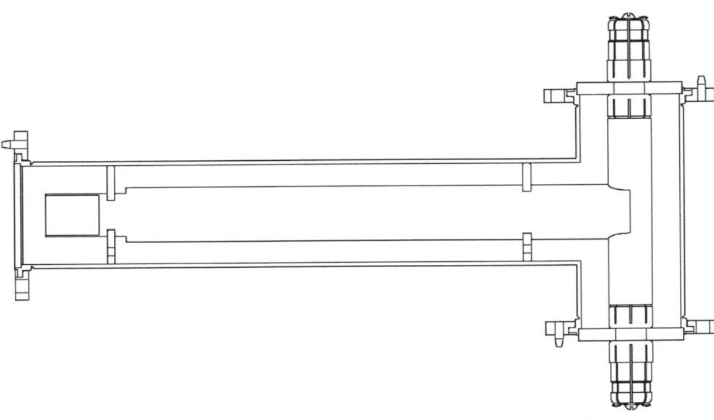

FIGURE 6.7-12 Typical matched tee power divider, 50 ohm ports.

Hybrid Coupler, π/2 Type

The hybrid coupler, $\pi/2$ type, is a four-port device that can be used to combine two inputs to one output or to divide one input into two outputs. A single hybrid design can be used for power division or for power combining. The power levels to be divided or combined, either equal or unequal, will influence the port size requirement and coupling ratio of the hybrid. The intended center frequency and bandwidth must also be considered when designing the quarter wave coupling structure. Best coupling performance occurs over a narrowband. Wide bandwidth can be achieved by permitting a small variation in coupling value versus frequency. An important benefit of the hybrid circuit is that 35 dB isolation exists between inputs when used as a combiner or between outputs when used as a divider.

Hybrid combiners are the preferred method of RF power combining for broadcast facilities. Whether coaxial or waveguide, the hybrid exhibits the same basic characteristics. The most common is the 3 dB hybrid for equal power division or combining. For example, two single-frequency transmitters, each producing 1 kW, are to be combined as shown in Figure 6.7-13. Inside this hybrid, port A is directly connected to port D with ports C and B coupled. The energy from transmitter Tx1 enters the hybrid at port A. Energy from transmitter Tx2 enters the hybrid at port B in phase quadrature, lagging 90 degrees in phase from Tx1. In a properly functioning hybrid, the sum of the hybrid inputs (A + B) will be a combined 2 kW signal delivered to port C. A load is placed at port D to dissipate the small amount of energy routed there by sight mismatches inherent in the hybrid. When both transmitters are operating properly, the hybrid provides an efficient and extremely simple solution to power combining requirements. However, when there is a fault with one of the inputs, the hybrid becomes very inefficient. Returning to Figure 6.7-13, if Tx1 fails or is taken off the air for maintenance, only one half of Tx2's power reaches port C, the output. The other half of the power (–3 dB or 500 W) from Tx2 will be directed to port D, the load; thus, the hybrid performs as a power divider. For this reason, it is important to properly size the load to meet a worst-case operating condition for port D, knowing that it could become responsible for dissipating half the power from either of the input transmitters while the other transmitter is off-line for any reason.

By introducing an intentional departure from phase quadrature at the hybrid inputs, the output power can

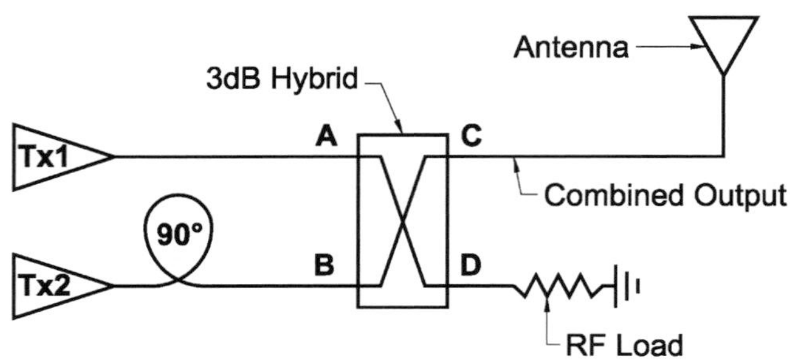

FIGURE 6.7-13 Typical cross-over hybrid 3 dB power combiner.

be redirected. When the hybrid inputs (ports A and B) are fed in phase, 90 degree phase error from quadrature, power will be split equally between the load (port D) and the antenna (port C). A 180 degrees phase error from quadrature results in a complete reversal delivering the sum of port A and port B to the load at port D. In this case the load must be sized to handle the full combined power.

Hybrids can also be used as power dividers rather than combiners. The same energy routing methodology is applied, only in reverse as shown in Figure 6.7-14. In this example, again using a 3 dB hybrid, power enters at port A. The power is then divided with half the power going to port D with a lagging 90 degree phase shift and the other half to port C. Again, a load is required at port B.

Special hybrid designs are required to combine inputs of unequal power levels. For instance, if a 2 kW transmitter is to be combined with a 1 kW transmitter, a 4.77 dB hybrid would be chosen, which is designed for combining inputs with 3 dB power differences.

Coaxial hybrids commonly take two forms. In the first form the direct connected port is on the same side of the hybrid. In the second form, the direct connected port crosses over to the other side of the hybrid. This second form is a most friendly configuration for system layout. Each approach accomplishes the same thing, just with a different physical configuration. This "cross-over" type hybrid shown in Figure 6.7-15 is for a high VHF channel application. The relatively low frequency requires the coupling structure to be physically long.

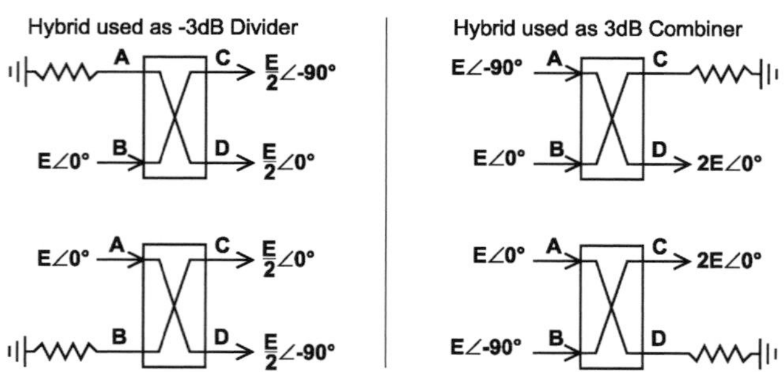

FIGURE 6.7-14 Typical hybrids: –3 dB power divider and 3 dB power combiners.

FIGURE 6.7-15 Coaxial high VHF 3 dB hybrid, cross-over type.

FIGURE 6.7-16 Coaxial UHF 3 dB hybrid, non-cross-over type. (Courtesy of SIRA, Italy.)

A direct connected port on the same side of a non-cross over-type hybrid with ports in a Y configuration, as shown in Figure 6.7-16, is commonly used in UHF applications, as the higher frequency allows for a more compact design.

Waveguide Hybrid Coupler, π/2 type

By far the most common type of waveguide hybrid is the Riblet hybrid as shown in Figure 6.7-17. Developed in 1950 by Henry Riblet, this design is simple and exhibits a relatively wide bandwidth of operation. Because of these characteristics, discussed later, the Riblet hybrid is the backbone of many UHF constant impedance channel combiners.

Another type of waveguide hybrid is the branch hybrid as shown in Figure 6.7-18. The branch hybrid is less common in broadcasting than the Riblet hybrid,

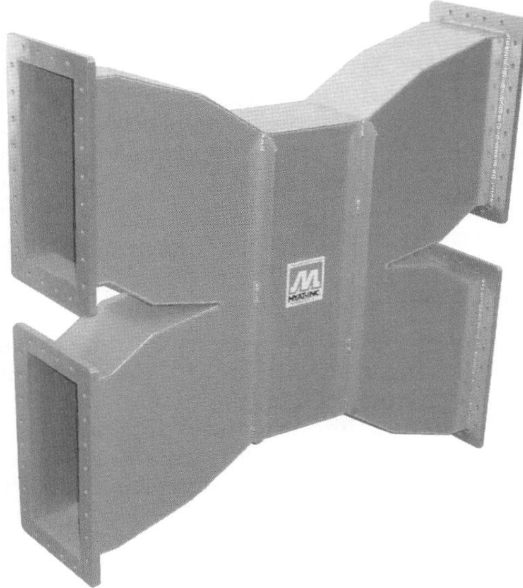

FIGURE 6.7-17 Waveguide UHF Riblet 3 dB hybrid.

primarily because it is more expensive to produce. It's advantage, however, is that it exhibits greater bandwidth. This characteristic is especially useful when designing constant impedance channel combiners with wide channel spreads.

FIGURE 6.7-18 Waveguide UHF branch 3 dB hybrid.

Power Combiners with Enhanced Functionality

Other types of power combiners are available with enhanced functionality. For instance, the Gysel combiner offers fault tolerance. A switchless combiner permits rerouting signal paths while under RF power.

Gysel Combiner/Divider

This Gysel design has many improvements over other combining techniques. The Gysel combiner is based on a circuit design by Ulrich H. Gysel [2] of the Stanford Research Institute and was presented to the IEEE (Institute of Electrical and Electronics Engineers) in 1975. The Gysel design, referred to as a Gysel network, is an improvement to the N-way combiner/divider devised by Wilkenson. The design is characterized by low insertion loss, high isolation between input ports, matched conditions at all ports, and external high-power load resistors. The main improvements are the external isolation loads and easily realizable geometry. The original design was only applied to strip-line amplifiers for power levels well below broadcast requirements. Recently the Gysel network has found its way into broadcast equipment. It is now common to find this circuit built into FM and television solid-state transmitters combining hot swappable modules of a transmitter power amplifier with output power levels around 10 kW. The Gysel network is also used for 3-way to 7-way combining of power amplifier cabinets achieving output power levels from 30 kW to 70 kW for both VHF and UHF television. Both coaxial and strip-line versions are available normally with rigid coaxial ports sized for the application.

The operation of the Gysel network is simple to understand when a 2-way Gysel network schematic is studied as shown in Figure 6.7-19. All of the transmission line sections of the schematic are one quarter-wave in length. This simple 2-way combiner schematic is a symmetrical 180 degree hybrid (π type). The inputs and output ports are 50 ohms for this example. The first requirement is to have a 50 ohm impedance at the output of the combiner Z1 and at the input ports. The impedance Z2 is a quarter-wave transformer that must match the 50 ohm inputs to the 50

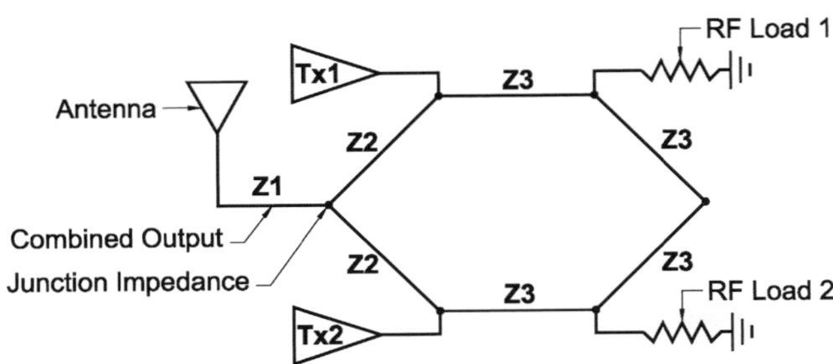

FIGURE 6.7-19 2-way Gysel combiner schematic.

ohm output impedance. The remaining quarter-wave line sections (Z3) all have 50 ohms impedance.

When equal voltages are supplied by the transmitters and arrive at the input port of this Gysel circuit in phase, the sum of both signals will be delivered to the output. There are three explanations for this. The electrical distance from each input port to the output port is equal, whether the power flows through the short path along Z2 (one quarter-wave length, 90°) or through the longer path (five quarter-wave lengths, 450°). The signal from one input arrives at a load port out of phase from the other input. The signal from one input arrives at the other input out of phase regardless of the path it takes: 180° or 360° and vice versa, hence the inputs are isolated from each other.

Under normal operating conditions everything to the right of the inputs, all Z3 lines and the RF loads, are effectively out of the circuit. Only during a fault condition on the input port, or when a transmitter is taken off-line, does this out-of-the-circuit portion of this network come into play. The input ports and the output port are under a matched condition when all transmitters are operating normally. When a transmitter is taken off-line or a fault condition is developed at an input port, the remaining input is still matched because of the inherent isolation of the network (inputs are isolated from each other). The power from the normal operating transmitter is approximately evenly split between the output and the isolation loads. The operation of this 2-way Gysel network is similar to the $\pi/2$ hybrid circuit discussed earlier.

This 2-way Gysel network can easily be expanded to 3-, 4-, 5-, 6-, or 7-way by adding additional arms and adjusting the impedance Z2 to match the inputs to the output as shown in Figure 6.7-20. It should be noted that an odd number of inputs does not require special attention as is required in $\pi/2$ hybrid combining. It is also apparent that all of the arms are identical in design and the power-handling requirement is the same for each.

A fault tolerance system can be achieved with this design. Suppose a 7-way Gysel network combiner is designed to combine seven solid-state transmitter building blocks capable of 10 kW each. If each transmitter is set for 5 kW output with an automatic gain control, an output level of 35 kW is achieved on the Gysel output port. When one transmitter is taken off-line for maintenance, or develops a fault, the remaining six transmitting outputs automatically increase to maintain the needed 35 kW power level at the output of the Gysel. In this condition, approximately 3 kW is dissipated in the isolation loads. The system can be run indefinitely with up to two transmitters faulted while maintaining the desired 35 kW output. The output power level will decrease only after three inputs are faulted or taken off-line at the same time.

Switchless Combiner

The purpose of a basic switchless combiner system is to combine two transmitters of equal amplitude, phase, and frequency. As the name suggests, the switching is done without manual patch panels, manual switches, or motorized switches. Phase shifters are used to switch modes. The basic system has the following three modes of operation:

- Transmitter 1 and transmitter 2 combined to the antenna.
- Transmitter 1 to the antenna port and transmitter 2 to the load port.
- Transmitter 2 to the antenna port and transmitter 1 to the load port.

Switchless combiners for FM and VHF can be constructed using coaxial components. UHF systems exceeding the power-handling or frequency limitations of coaxial line would be constructed using waveguide components. Two 3 dB hybrids and two phase shifters with interconnecting transmission line are all that is required to make a switchless combiner system. One hybrid is used as a divider and the other is used as a combiner as shown in Figure 6.7-21.

The input hybrid is being used as a divider. In this case Tx1 is isolated from Tx2. Each transmitter must have equal amplitude and phase delivered to the input hybrid ports, A and B. The power from Tx1 at input A is divided evenly in amplitude between the output ports C (Tx1/2 at 0 degrees) and D (Tx1/2 at –90 degrees) in phase quadrature. The power from Tx2 at input B is divided evenly in amplitude between the output ports C (Tx2/2 at –90 degrees) and D (Tx2/2 at

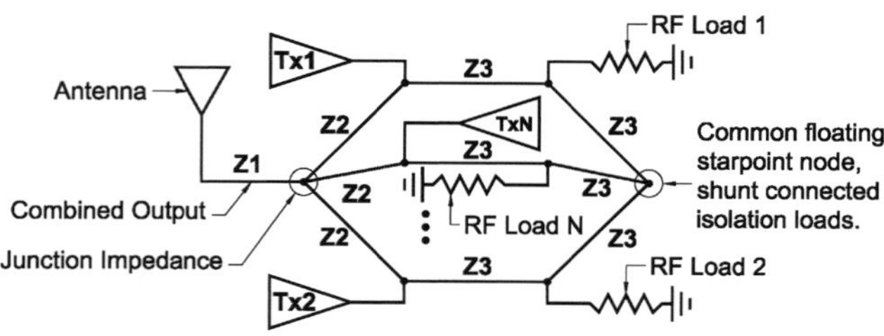

FIGURE 6.7-20 N-way Gysel combiner schematic.

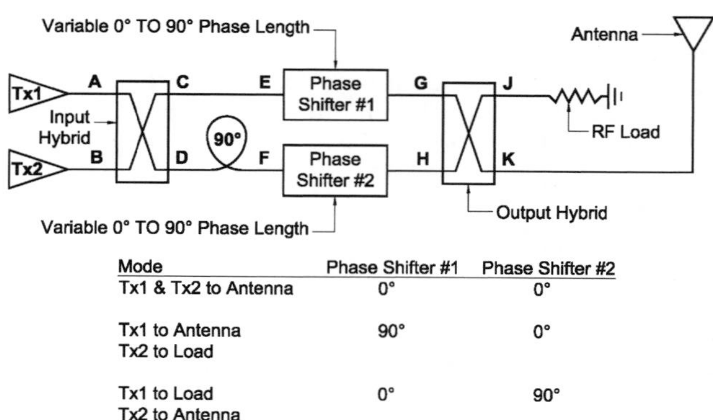

Mode	Phase Shifter #1	Phase Shifter #2
Tx1 & Tx2 to Antenna	0°	0°
Tx1 to Antenna Tx2 to Load	90°	0°
Tx1 to Load Tx2 to Antenna	0°	90°

FIGURE 6.7-21 Switchless combiner schematic.

0 degrees) in phase quadrature but opposite to the power from Tx1. The transmission path from point D to point F is one quarter-wave longer than the transmission path from point C to point E. The power present at port D will be delayed an additional 90 degrees at point F than the power is from point C arriving at point E. For analysis purposes consider that point E is the same phase as point C and apply a 90 degree delay to the signal at point D to find the signal at point F. This results in point E having Tx1/2 at 0 degrees + Tx2/2 at –90 degrees and with point F having Tx1/2 at –180 degrees + Tx2/2 at –90 degrees.

For the combining mode Tx1 and Tx2 to antenna, both phase shifters 1 and 2 are set to 0 degree phase shift. This results in the same phase relationship that was seen at the inputs to the phase shifters. At point G we find Tx1/2 at 0 degrees + Tx2/2 at –90 degrees. At point H we find Tx1/2 at –180 degrees + Tx2/2 at –90 degrees. These signals are at the inputs of the second hybrid acting as a power combiner. The sum of the power at point J (the load) is found to be opposite in phase resulting in no power delivered to the load. The sum of the power at point K is found to be in phase resulting in all of Tx1 and Tx2 power being delivered to the antenna.

For the second mode, Tx1 to antenna and Tx2 to load, phase shifter 1 is set to a 90 degree phase shift and phase shifter 2 is set to a 0 degree phase shift. This changes the signal phase at point G to Tx1/2 at –90 degrees + Tx2/2 at –180 degrees. Point H remains unchanged containing the signals Tx1/2 at –180 degrees + Tx2/2 at –90 degrees. The output hybrid acting as a power combiner is presented with new signals at the inputs. At point J (the load) we find Tx1's signal to be opposite in phase canceling out while Tx2's signal is in phase adding resulting in full Tx2 power being delivered to the load. At point K the reverse is found: Tx1's signal is in phase adding resulting in full Tx1 power being delivered to the antenna while Tx2's signal to be opposite in phase canceling out.

For the third mode, Tx1 to load and Tx2 to antenna, phase shifter 1 is returned to a 0 degree phase shift and phase shifter 2 is set to a 90 degree phase shift. Reanalyzing the effects of the phase shifters, the signal at point G becomes Tx1/2 at 0 degrees + Tx2/2 at –90 degrees. The signal at point H becomes Tx1/2 at –270 degrees + Tx2/2 at –180 degrees. At point J (the load) we find Tx2's signal to be opposite in phase canceling out while Tx1's signal is in phase adding resulting in full Tx1 power being delivered to the load. At point K, again the reverse is found: Tx2's signal is in phase adding resulting in full Tx2 power being delivered to the antenna while Tx1's signal to be opposite in phase canceling out.

The schematic diagram of this switchless combiner is using identical phase shifters. The quarter-wave delay portion of transmission path point D to F could just as easily been incorporated into the phase shifter 2 design. This can be done by adding one eighth-wave long transmission line on each end or placing a full quarter-wave long transmission line on one end of the phase shifter.

If it is required to route both transmitters to the load, Tx1 + Tx2 to load, then additional equipment is required. The load must be sized to handle the sum of the transmitters. An additional 180 degree phase shifter needs to be added to the input for Tx2 and would be kept in the 0 degree position for the three standard modes. For the new mode, Tx1 + Tx2 to load, Tx2 input phase shifter would be set to 180 degrees and phase shifters 1 and 2 need to be set for 90 degrees. A new controller for the three phase shifters and the required four modes of operation must be supplied.

The ability to change modes of operation is automated by using motors with logical position controllers to position each of the phase shifters. Typically the position commands can be done locally to the phase shifter or by remote control. Power requirements are normally specified at the time of order to be compatible with the power source available at the intended installation site.

Control voltage requirements are normally specified at the time of order to be compatible with the equipment to be used to command the controller.

A high-power UHF switchless combiner is constructed in a similar manner to that of the style previously discussed. Waveguide components are used for power-handling reasons. The main difference is that the output hybrid coupler is a π type, 180 degrees. In waveguide this coupler is also known as a *magic tee*. When the magic tee is used as a combiner, as is the case for this application, signals that are in phase at the two inputs will be combined at the sum output port and canceled at the isolated difference port as shown in Figure 6.7-22. Signals that are 180 degrees out of phase at the two inputs will be combined at the difference output port and canceled at the sum isolated port. When used as a divider, if a signal is applied to the sum input port the signal will be divided equally to the two outputs each with the same

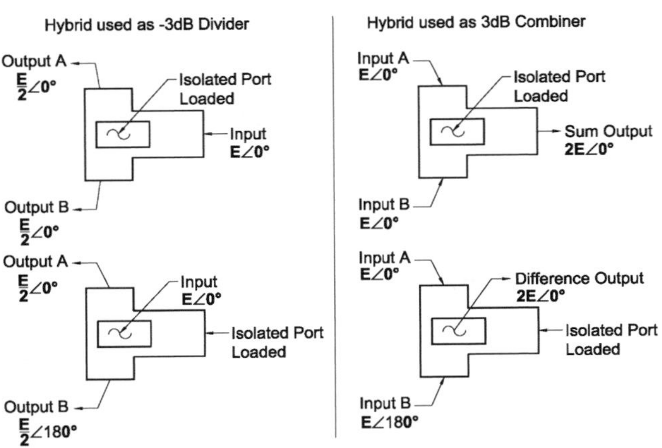

FIGURE 6.7-22 Magic tee hybrid coupler, π type.

phase. When a signal is applied to the difference input port, the power will be equally divided between the two outputs with a 180 degree phase difference.

The modes of operation are the same as before: Tx1 and Tx2 combined to the antenna, Tx1 to the antenna port, and Tx2 to the load port, and Tx2 to the antenna port and Tx1 to the load port. In order to provide proper phased signals to the input of the magic tee supporting these modes, the electrical path from point C to point E must be the same as from point D to point F when the phase shifters 1 and 2 are set to equal phase lengths as shown in Figure 6.7-23. The input hybrid evenly divides the power to two outputs in phase quadrature. Like before, point C has two signals, Tx1/2 at 0 degrees + Tx2/2 at –90 degrees, and point D has two signals, Tx1/2 at –90 degrees + Tx2/2 at 0 degrees.

In mode Tx1 and Tx2 combined to the antenna, both phase shifters are set for 0 degrees and the phase relationship at points E and F is unchanged from points C and D. The magic tee would then combine Tx1/2 at 0 degrees from point E with Tx2/2 at 0 degrees from point F, both in phase, to the main output. In addition, the magic tee would also combine Tx2/2 at –90 degrees from point E with Tx1/2 at –90 degrees from point F, both in phase, to the main output. The two-combined signal present at the main output is not in phase so the resultant signal is the vector sum Tx1 + Tx2 at –45 degrees.

In mode Tx1 to the antenna port and Tx2 to the load port, phase shifter 1 is set to 90 degrees and phase shifter 2 is set for 0 degrees. This changes the signals presented at point E to Tx1/2 at –90 degrees + Tx2/2 at –180 degrees. Point F remains unchanged with signals Tx1/2 at –90 degrees + Tx2/2 at 0 degrees. Tx1 signals are in phase resulting in combining to the sum port point H. Tx2 signals are 180 degrees out of phase resulting in combining to the difference port point G.

In the last mode, Tx2 to the antenna port and Tx1 to the load port, phase shifter 2 is set to 90 degrees and phase shifter 1 is set for 0 degrees. This results in Tx1's signals to be 180 degrees out of phase at the inputs that combine to the difference port at point G. The signals from Tx2 are in phase at the inputs to the magic tee; hence, they combine to the sum port point H.

When maintenance is to be performed on a transmitter that is part of a switchless combiner system, the maintenance personnel must be aware that the isolation between transmitters is a function of the hybrid combiner. Isolation is normally around 35 dB, which will result in some low-power RF signal on the output port for the transmitter being serviced.

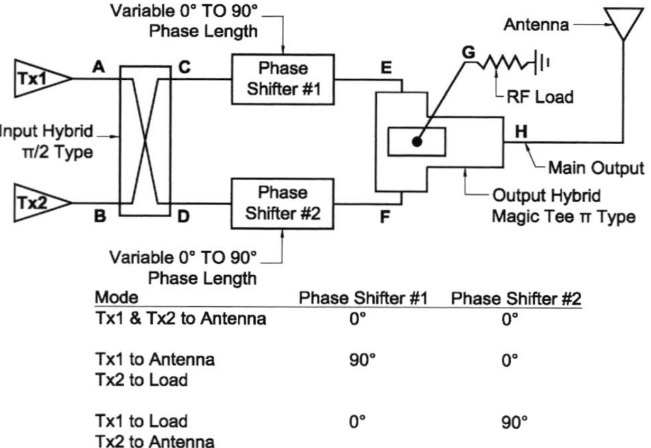

Mode	Phase Shifter #1	Phase Shifter #2
Tx1 & Tx2 to Antenna	0°	0°
Tx1 to Antenna Tx2 to Load	90°	0°
Tx1 to Load Tx2 to Antenna	0°	90°

FIGURE 6.7-23 Waveguide switchless combiner schematic using magic tee hybrid coupler.

Transmission Line Switches

A transmission line switch provides an efficient means of routing RF power to an alternate transmission path. Every broadcast transmission facility requires a level of switching that can range from simple to complex based on the needs of the station. The basic switch can be manual or automated with three or four ports and may be in the form of patch panels, either manual switch or automated switch. Switching cannot occur under RF power without catastrophic failure to the switch or patch and possibly exposing personnel to hazardous RF power levels.

Manual Patch Panels

The most basic form of transmission line switch is the patch panel, which provides a simple and convenient method to reroute transmission line inputs and outputs. Being manual their use is normally for a maintenance purpose, as they require personnel to physically make the switch. Whether constructed from coaxial or rectangular waveguide transmission line, the patching device takes the form of a U-link. The U-link is connected with some mechanical locking or fastening method that ensures a good RF connection. To prevent accidental patching while RF power is present, an interlock switch is installed for each patching position. The interlock switch is connected to transmitter control circuitry so RF power can be applied only when a U-link is installed completing a safe path for RF to propagate through. Accidental removal of a U-link under power would then disable the transmitter.

The relative inflexibility of rectangular waveguide limits the number of patching options to three in-line ports without requiring an extended length U-link. The design can be along the broad wall or narrow wall of the waveguide. Optionally, a waveguide U-link designed along the opposite wall would permit an additional port or more to one side or the other. This U-link would need to be in storage until its use is required.

Coaxial patch panels offer far more patching options. The ports of a patch panel are normally arranged to facilitate multiple patching options with a single U-link. The most common patching system has either three, four, or seven ports. The patch panel plates are most often mounted to a support frame either fabricated in the field or factory supplied.

Depending on the panel size, it may be ordered to fit a standard 19 inch, 24 inch, or 30 inch rack mount. Custom patches with many more ports can be supplied as required for any special patching options. Additionally, custom length U-links are easily supplied for reaching a nonadjacent port and in some cases capable of jumping over an existing installed U-link. The U-links are normally an unflanged design that cannot be pressurized. The power handling of a coaxial patch panel is essentially the same as the interconnecting line size without pressurization.

Coaxial Switch

The manual coaxial switch is commonly offered as a three-port SPDT transfer switch or a four-port transfer configuration. The three-port design is constructed out of the same parts as the four-port design except that the fourth port is covered over and without inner conductor parts. Only two positions are possible, position A connects port 1 to port 2 and port 3 to port 4, while position B connects port 1 to port 4 and port 2 to port 3. The RF path is constructed into a single body separated by a vane. Isolation between these two RF paths is typically better than 50 dB and often approaches 60 dB. Operating a mechanical lever or knob changes the position with a mechanism that locks the switch in position. Like the patch panel, coaxial switches cannot be switched under RF power, and interlock switches are part of the design to disable transmitter RF power during a switching operation. Power ratings for coaxial switches are derated to approximately 80% of a given line size. The derating varies between designs and suppliers although some designs have made improvement in power handling. The complexity of design requires many more parts and increased assembly time that add significantly to the cost over that of a four-port patch panel.

Automated or motorized coaxial switches are based on the manual switch. The manual operating mechanism is replaced with a motor driven positioning system as shown in Figure 6.7-24. Normally a mechanical override is supplied in case the motor drive or controller fails. Switching time varies between designs but is generally longer for larger coaxial sizes. Timing between shutting down RF power and switching should be checked so that switching under RF power cannot occur. Switch controllers are also available to locally or remotely operate the motorized switch.

FIGURE 6.7-24 Typical 3-1/8 inch four-port coaxial motorized switch. (Courtesy of Spinner, North America.)

Waveguide Switch

The waveguide switch is similar to the coaxial switch. For broadcast frequencies three rectangular waveguide sizes are normally used: WR1800 for low UHF, WR1500 for mid-band UHF, and WR1150 for high UHF. They are available as three- and four-port switches with manual or motorized control for positioning. One difference is that there are two types, the E-plane switch and the H-plane switch. The E-Plane switch is typically more compact, switching about the narrow wall of the waveguide. The RF portion is constructed into a main body with a back plane vane that rotates performing the switching. When in position the vane makes contact through finger stock. The RF path including the vane resembles a waveguide elbow. Designs vary between vendors, as does performance. Some types require tuning to a channel or over a specific bandwidth.

Directional Coupler

A directional coupler is a four-port passive device that is used to take a fractional sample of power present in a transmission line at a specified frequency. A low-loss primary transmission path is used for high RF power levels with ports sized accordingly and is referred to as the main line. The second transmission path is loosely coupled with a fractional power ratio to the primary line. The first port of the secondary line is designed to discern and measure a fractional coupled sample of the incident traveling wave in the main line. The coupling value varies at a rate of –6 dB per octave with respect to frequency. The second port of the secondary line measures the reflected traveling wave in the main line with the same fractional coupling ratio. The forward and reverse measurements can be used to calculate VSWR or return loss in the main line. Directivity is a measure of how well each secondary port can be isolated from the unintended signal. Measured signal accuracy improves as isolation from the unintended signal increases. Multiple couplings are often placed onto one main transmission line with either incident coupled port or reflected port used while the opposite port is properly terminated. The terminated port may be built in and therefore not accessible to be a reverse measuring port. The termination must be precision and sized for worst-case maximum amount coupled energy.

For broadcast applications coupling values are normally selected for less than 1 watt of power coupled into the secondary line. The coupling range is typically –40 dB to –60 dB (10,000:1 to 1,000,000:1 power ratio) depending on maximum main line power.

Directivity of a coupled port is typically –35 dB or better. Directional couplers often have from one to four or more couplers with a main line VSWR of 1.05:1 or better at a specified frequency. The main line can be coaxial or waveguide and the secondary line most often is a type N 50 ohm coupled port.

Line Stretcher Phase Shifter

This type of phase shifter permits the electrical length of a transmission path to be adjustable. The most basic form is a telescopic line section that has little use when working with rigid transmission lines.

Coaxial Phase Shifter

A coaxial telescopic line stretcher can be constructed by using a U-link and two telescopic line sections. It functions quite like a trombone with each telescopic line accounting for half of the desired phase shift. It is designed for high power and to be adjusted while under RF power.

An alternate coaxial phase shifter uses two adjustable length short circuits and a 3 dB hybrid being used both as a divider and as a combiner at the same time. The adjustable electric length to each short circuit must be kept identical. Considering the hybrid as a divider, an input signal is divided into two equal amplitude signals in phase quadrature. The shorts are connected to the divided ports of the hybrid power divider. These signals are each reflected back to the hybrid with a phase reversal, and then the hybrid combines the signal to the remaining fourth port, the output.

Waveguide Phase Shifter

The line stretcher and hybrid method are not preferred in waveguide due to difficulties in designing telescopic electrical contacts that can be adjusted under power. For broadcasting, a rectangular waveguide phase shifter takes a different approach. Phase shifting is accomplished by changing the dielectric constant of the waveguide. A movable piece of dielectric material is placed along the inside sidewall of a line section where it has a minimal effect. The phase shift occurs as the dielectric material is moved toward the center of the waveguide where the signal is greatest and has the most effect influencing waveguide dielectric constant. The velocity of propagation through the dielectric material is longer hence providing a delay. The size of the dielectric material is selected to cause the desired difference in phase at a given frequency.

Loads

High power RF loads for broadcast are generally a 50 ohm impedance capable of dissipating the energy in the form of heat. Typically the heat dissipater is air, oil, or water depending on the type.

Air-cooled loads have fans blowing across the resistive elements exhausting to the surrounding environment. Usually multiple tubular ceramic elements are connected in parallel to handle the power. Because safe operation is dependent on forced air heat dissipation, the fans should be monitored for proper operation and a thermal switch is used as an interlock for normal operation. Air-cooled loads are characterized by being a uniform relatively reflectionless termination, large in size, and a low-maintenance reliable design.

Oil-cooled loads have the resistor element submerged in oil. Thermally conductive oil is used to transfer the heat to a heat sink. An exponential tapered housing surrounds the resistor. The housing contains the RF permitting oil to flow through the housing while providing 50 ohm impedance. Large oil-cooled loads may also have forced air fans blowing across the heat sink. In some cases a heat exchanger is part of the design. Although considered a good termination, the actual impedance varies slightly with frequency and with temperature. It is good practice to use a fine matcher with oil-cooled loads.

Water-cooled loads are constructed in one of two methods. The first type uses a resistor with water flowing across and sometimes through the load. They are efficient, compact, and require a heat exchanger to dissipate the heat. The heat exchanger can normally be remotely located where heat dissipation is not a problem. The second type is constructed using a stainless-steel tube as an inner conductor resistive element with water as the dielectric. The water should be pure with a minimum of solids dissolved. When cooling systems need to operate at freezing temperatures, then ethylene glycol and water mixture is used. The mixture is most often 50% to 60% water. Chemical additives and impurities will affect the thermal and electrical characteristics of the load. The length of the element must be sufficiently long enough to dissipate all energy at a given frequency. Generally this type is designed for UHF frequencies because of the length requirement. Calorimeter loads are often used in the precise determination of transmitter power output at the manufacturing plant.

Circulators

Circulators are beginning to be found more often in high-power broadcast applications. Relatively recent advancements in circulator design for broadcast applications have made them capable of higher power levels. The basic circulator is a three-port nonreciprocal ferromagnetic device commonly referred to as a Y-junction circulator. When a signal is applied to port 1 of a circulator the power is delivered to port 2 with port 3 isolated. Any reflections from port 2 are delivered to port 3 with port 1 isolated and any signal from port 3 would be delivered to port 1 with port 2 isolated. For broadcast applications, a reject load is placed on port 3 that dissipates reflections from port 2 effectively isolating the input port 1 from the output port 2 reflections. An isolator is a circulator with a reject load connected to one port. Isolators are used because the input port impedance remains constant regardless of output port conditions, input VSWR is

independent of load VSWR, and the input port is isolated from output effects. Circulator design must consider both forward and reverse power. Average power handling is dependent on the ability to remove heat from the circulator. Peak power capacity is limited by the spacing between internal components. Two main purposes for using a circulator as an isolator in broadcast are absorption of reflections and improved isolation between transmitters.

CHANNEL COMBINERS

Channel combiners, or multiplexers, combine two or more channels to a common feed line. Inversely, multiplexers separate two or more channels on a common line to individual isolated lines designated to the channels. Diplexers separate or combine two channels, while multiplexers work on three or more channels. Combiner design must consider total output power, isolation requirements, number of channels, channel spacing, available space, future add-on capabilities, and cost. Total output power will determine whether coax or waveguide design is used. Isolation requirements become important when combining channels with large differences in power levels. The number of channels combined along with frequency separation and available space will define what types of filter functions are used, method of combining, and component definition. Incorporating mask filters within the combining system reduces system cost by eliminating the need for separate RF systems.

Terrestrial broadcasters commonly use junction combiners or directional filter/combiners. Different combining systems lie within these two general techniques and each has its own attributes.

Junction Combiners

Junction combiners consist of filters (band pass, low pass, high pass, notch, or any combination thereof) tuned to the channel center frequency and tied to a common junction using tuned line lengths. Two common techniques of junction combining are the common junction, or star point, and a manifold multiplexer as shown in Figure 6.7-25. Line sections from the filters to the junctions can be waveguide or coax.

Star Point Combiners

Star point combiners use filters tuned at the channel center frequency and are spaced approximately half-wave from the junction. Spacing can vary significantly from half-wave and depends on the type (capacitive or inductive) and structure (length to open or short) of the input coupling. A signal presented at the tuned input passes through the filter and splits equally at the junction. The divided signal reflects off filters tuned to other channels and is delivered back to the junction again where they are in proper phase to recombine and pass to the output. Wide filter passbands minimize additional system loss and delay variations when the design does not incorporate mask filters into the system. Isolation requirements are achieved from the filter response, with slight gains due to the junction. Adjacent channel combining is possible with star point combiners, but filter order can become excessive depending on the required isolation between channels. Conversely, widely separated channels (e.g., UHF and VHF) combine easily. Star point combiners are low-cost solutions requiring little space for installation but do not offer practical future expansion possibilities.

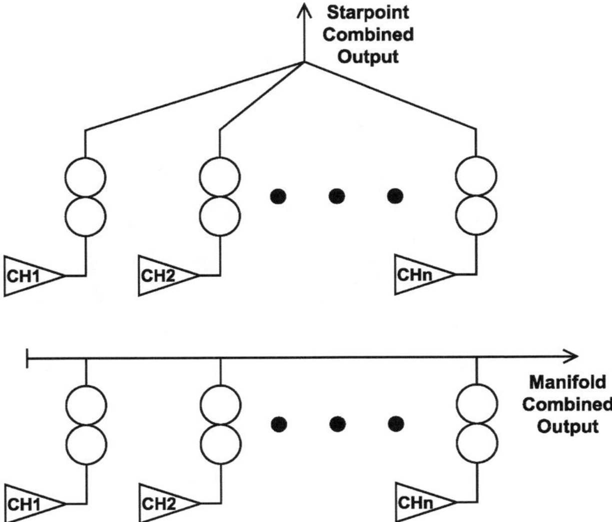

FIGURE 6.7-25 Star point and manifold combiner schematic diagram.

Manifold Combiners

Manifold combiners are similar to the star point in that tuned line lengths are required for proper function. Filters tuned to the channel center frequency are tied to the manifold using T-junctions; a short circuit is at one end and the output at the other. Channels are sequenced in ascending or descending order from the short circuit. The manifold technique provides better performance at a lower cost due to its simplicity when the number of channels to combine becomes greater than three and confined to a total transmission bandwidth of less than approximately 30% (a function of whether waveguide or coaxial manifold is used). As with star point combiners, adjacent channel combining is possible, but, filter order can become excessive depending on the required isolation between channels. Manifold combiners also tend to require little space (see Figure 6.7-26) and do not offer add-on capabilities.

Directional Filter Combiners

Directional filters used as combining modules consist of two approaches—hybrid and waveguide. A third

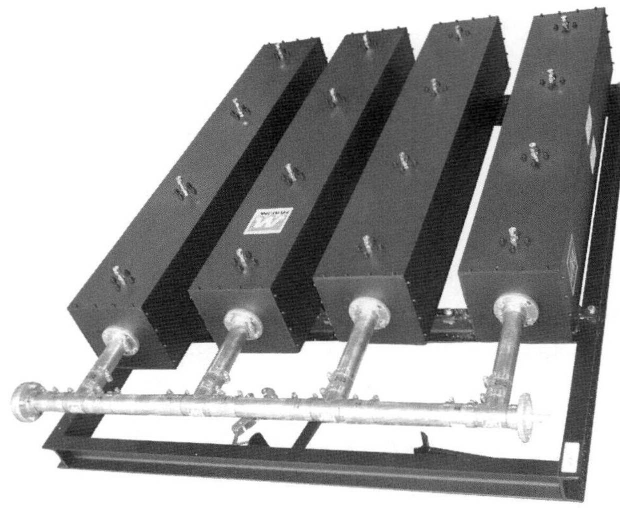

FIGURE 6.7-26 UHF manifold combiner.

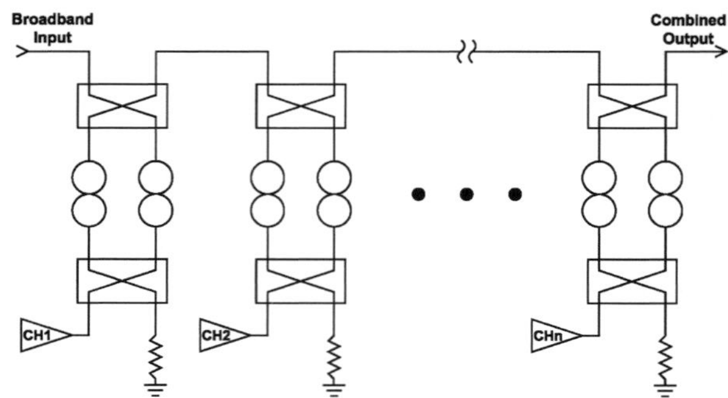

FIGURE 6.7-28 Multiple-channel hybrid directional filter/combiner schematic diagram.

approach using circulators exists but is not used for broadcast applications due to the added losses of the circulator. Directional filters are ideal for broadcast applications where future add-on capability is required. The hybrid directional filter, as shown in Figure 6.7-27, is a four-port device consisting of two identically tuned filters and two $\pi/2$ hybrids. Signals within the passband of the filter pass from ports 1 to 4 and are isolated from ports 2 and 3. Signals outside the passband of the filter pass from ports 3 to 4 and are isolated from ports 1 and 2. When band pass filters are used, the module then consists of a narrowband input, port 1, a broadband input, port 3, and an output, port 4. An appropriately sized load is normally placed at port 2. These modules are commonly used as "constant impedance" filters for tube transmitters. Constant impedance is used because unwanted emissions from the transmitter reflect off the filters and are delivered to the loaded port 2 instead of looking like a short

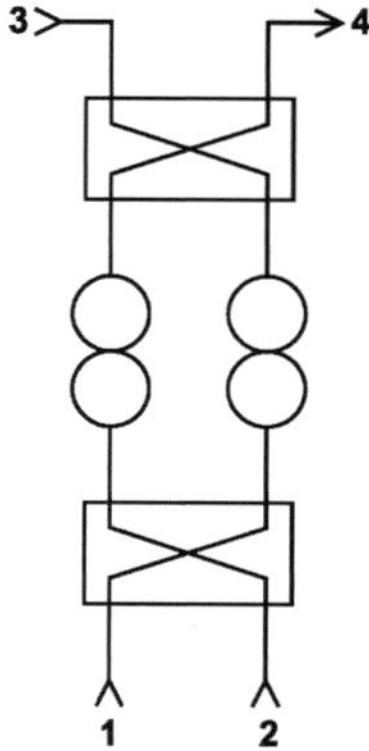

FIGURE 6.7-27 Hybrid directional filter/combiner module schematic diagram.

and being reflected back at the tube. Conversely, unwanted signals from the antenna that travel down the feed line and into this module at port 4 are reflected off the filters and delivered to port 3, which is terminated in these instances.

Directional filters can be cascaded to combine multiple channels as shown in Figure 6.7-28. Each channel has its own narrowband input and a broadband input for the addition of future channels. When not used, the broadband input is terminated with an appropriately sized load. The hybrid directional filter/combiner is limited in power-handling capability only by the size of filter and output side hybrid. As multiple high-power channels are combined, cavities must be sized accordingly. Special attention is given to spurious modes generated in oversized cavities that tend to pass signals and not reflect them as desired. Dual-mode filters and their ability to accommodate pseudo-elliptic function responses make them ideal for use in hybrid directional filter/combining modules for adjacent channel-combining applications. Output hybrids can be coaxial or waveguide depending on power requirements. Branch style hybrids are required for combining high-power UHF channels with large frequency separations.

Waveguide directional filters were initially developed to multiplex narrowband microwave channels. Extensive literature on the subject can be found in Microwave Filters, Impedance-Matching Networks, and Coupling Structures. The waveguide directional filter, as shown in Figure 6.7-29, performs a similar function as the hybrid directional filter. Instead of using two separate filters and $\pi/2$ hybrids to obtain the directional properties, TE11n cavities with dual orthogonal modes are used along with waveguide E and H coupling. Referring to the cavities in the figure, each polarity of the dual orthogonal modes is a separately tuned filter (on the same center frequency of course); therefore, the number of cavities represents the filter order. The individual filters are coupled with $\pi/2$ difference (one is E coupled, the other H coupled) and replaces the operation of the hybrids. A signal enters the narrowband port 1, where half is E-coupled to one mode polarity and the other half H-coupled to the other mode polarity. The signals pass through the

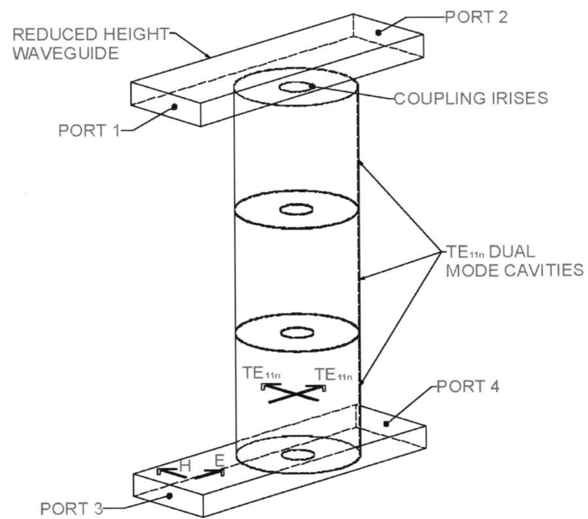

FIGURE 6.7-29 Waveguide directional filter/combiner schematic diagram.

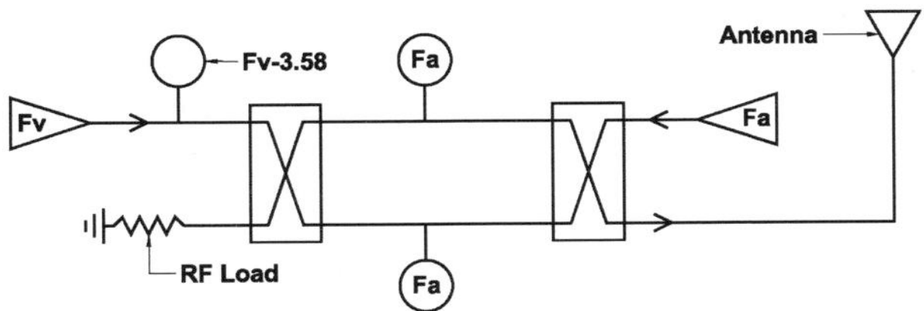

FIGURE 6.7-30 Visual/aural notch diplexer schematic.

filters and recombine through separate E and H coupling irises and are delivered to the output port 4; port 2 is terminated. Signals outside the filter passband enter port 3 and pass straight through to the output port 4.

A reduced height waveguide is required for the strong input and output couplings. Waveguide size (broad wall dimension) is chosen to operate over the band of channels to be combined. Mismatches that occur in the line due to input/output coupling are easily dealt with due to short distances between modules.

Waveguide or hybrid directional filter modules give more isolation than the junction combiners described above due to the hybrids. For example, the narrowband port is isolated from the broadband port signals by approximately 60 dB due to the well-matched hybrids, and filter rejection is usually 30 dB. The broadband port is isolated from narrowband port signals by approximately 30–36 dB; however, channels entering the broadband port are typically associated with their own filter or module and therefore can easily give 60 dB isolation total. In general, greater isolation is achieved at the expense of additional hardware.

Visual/Aural Notch Diplexer

NTSC visual/aural diplexers are used to combine visual and aural transmitters to a common feed line. Junction-type diplexers could be used for this application, however, the directional filter style shown in Figure 6.7-30 tends to be the most common. The diplexer schematic shows notch cavities (tuned to reject aural frequencies) placed between 3 dB hybrids. The directional function of the module is a complement to the above-described modules by using a notch filter as opposed to a bandpass filter (bandpass filters can be used but tend to be more costly). All signals not centered on aural frequencies (the visual signal in this case) pass directly from the Fv input to the output. A color notch filter is typically provided at the visual input to clean up Fv-3.58 MHz products at the output. The aural signal is input at the Fa port, where it is divided with $\pi/2$ phase difference, reflected off the notch cavities, and sent back to the hybrid where it recombines and is delivered to the output. The reject load supplied is of adequate power rating to permit full power operation and is supplied with sample ports for system fine-tuning of aural cavities. Isolation between visual and aural frequencies is primarily due to the hybrids and is typically 36 dB.

Visual/aural diplexers are manufactured in coax and waveguide construction depending on power and frequency band of operation. Aural cavities are temperature compensated to provide essentially constant frequency response over a specified temperature range and power levels. Aural losses are a function of the cavity unloaded Q and notch depth. Waveguide notch cavities require minimal tuning screw penetrations for efficient operation and optimal operating temperatures. Waveguide designs are chosen for high-power-handling capabilities and for high-frequency applications. Coaxial diplexers are normally used for low-frequency applications in addition to limited power handling.

SUMMARY

Each broadcast facility has it own unique system requirements. In some instances the requirements are solely due to power-handling needs affecting the size of equipment and in other applications could include special sharp-tuned filter responses protecting adjacent channels or when combining with adjacent channels adding to the design requirements. When channel combining is needed for a particular application, maximum channel spread, total number of channels, and channel spacing will greatly affect which type of combiner is required and to some extent control theoretical performance that is achievable. Basic components, such as a coaxial hybrid, may need to be optimized for a specific channel or band. Although some off-the-shelf items may work in a given system, most components will be designed and built to order or in some cases specially optimized or tuned for a specific channel or band of operation.

Advanced planning for system requirements including any future needs should be compiled and verified before requests for proposals are generated. Unnecessarily tight specifications will add to design requirements and total cost. Although future needs may be hard to predict, they can simplify adding components that otherwise may be prohibitively expensive or in some cases not possible resulting in replacing and or reworking a great deal of equipment.

References

[1] Matthaei, George L., Young, Leo, and Jones, E. M. T. *Microwave Filters, Impedance-Matching Networks, and Coupling Structures.* Dedham, MA: Artech House, Inc., 1980. Reprint of edition printed by McGraw-Hill Book Company, 1964.

[2] Gysel, Ulrich H. *A New N-way Power Divider/Combiner Suitable for High-Power Applications*, IEEE-MTTS-5, International Symposium Digest, 1975, p. 116.

CHAPTER

6.8

Television Antenna Systems

RONALD E. LILE AND TOM SILLIMAN

Electronics Research, Inc.
Chandler, Indiana

INTRODUCTION

The broadcast antenna is the single most important part of the entire broadcast transmission chain in that it is the device that actually delivers the signal to the audience. If the antenna is not well designed, built, and installed, a station may not obtain the calculated coverage, or if directional, the antenna can cause interference with neighboring stations. The antenna requirements for digital television (DTV) broadcasting are essentially the same as for National Television System Committee (NTSC) television broadcasting.

The antenna should:

- Have structural integrity.
- Be capable of being mounted on tall towers.
- Be efficient.
- Have maximum gain for minimum length.
- Have minimum loss and reflection.
- Have differential gain less than or equal to 3 dB.
- Have accurate directional characteristics if required to do so.
- Be able to handle power consistent with the effective radiated power (ERP) and feedline power requirements.

In addition, the antenna manufacturer must be able to shape the radiation pattern to satisfy coverage and interference requirements, and this pattern should be consistent over the required channel or channels. During the transition from analog to digital transmission formats, the antenna may be required to operate on a second channel so digital and analog transmissions occur simultaneously. The antennas available today and described in this chapter satisfy all DTV requirements including the multichannel requirement. This chapter reviews and discusses technical terms and FCC requirements for TV broadcast antennas, as well as antenna design considerations related to these requirements.

TERMS AND DEFINITIONS

Azimuth Pattern

The *azimuthal pattern* is a plot of the freespace radiated field strength *versus* azimuth angle measured in the far field at a specified vertical angle with respect to a horizontal plane (relative to smooth earth) passing through the center of radiation of the antenna. The *horizontal plane pattern* is an azimuthal pattern when the specified vertical angle is zero relative to the horizontal plane. For antennas with gain, where *beam tilt* is employed, the azimuthal pattern at the specified beam tilt angle is significant. In general, it has been customary to determine television broadcast antenna radiation by an azimuthal pattern at the specified beam tilt and a sufficient number of vertical plane patterns taken at various frequencies in the channel. An *omnidirectional antenna* is defined as one that radiates the same signal strength in all azimuthal directions. Antennas with variations up to ±3 dB in their freespace azimuthal patterns have provided satisfactory service and are considered to be omnidirectional. A *directional antenna* is designed to radiate greater signal

in one or more azimuthal directions than in other azimuthal directions.

Elevation Pattern

The *vertical plane pattern* is a plot of freespace radiated field strength, measured in the far field *versus* vertical angle in any specified azimuth plane containing the center of the antenna and the center of the earth.

Effective Radiated Power

The product of the antenna input power and the antenna power gain relative to a standard one-half wavelength dipole is the effective radiated power (ERP). If the antenna uses circular polarization, the term ERP will apply to the horizontal and vertical components separately. Effective isotropic radiated power (EIRP) is the product of the antenna input power and antenna power gain referenced to an isotropic source. An isotropic source radiates equally well in all directions of the radiation sphere.

Far Field

The far field or Fraunhofer region extends beyond the point where the distance between the transmitting and receiving location is:

$$\frac{2D^2}{\lambda}$$

where D is the length of the radiating portion of the antenna and λ is the wavelength at the operating frequency of the antenna. As an example, use the above formula to determine the defined far field distance for a UHF antenna for which the radiating aperture is 34.57 feet (antenna gain of 20, channel 30, 1 wavelength element spacing). The distance from the transmitting antenna to the point in space defined as the far field at the receiving antenna will be 1382.88 feet and beyond.

Polarization

The polarization (orientation of the electric field) of a broadcast television antenna can be linear horizontal or elliptical. Elliptical polarization contains both horizontal and vertical field components. Circular polarization is a special case of elliptical polarization where the E field in both the horizontal plane and the vertical planes are equal in magnitude and the phase between them is in quadrature.

Antenna Directivity/Gain

Gain, as defined in the television broadcast service in the United States, is the ratio of the maximum power output at any angle from the subject antenna to the maximum power from a thin, lossless, half wave, horizontally polarized dipole having the same power input. Antenna gain depends on several factors, including:

- The amount of power concentrated in the maximum direction
- Losses in the antenna including ohmic and other losses, such as energy radiated at a polarization other than the desired one.

The amount of power concentrated in the maximum direction can be determined by comparison with a reference antenna or by integrating the total power flow through a sphere, which is done by taking a sufficient number of vertical and azimuthal patterns. Both methods are capable of giving accurate results when the proper precautions are taken. Ohmic losses are taken into account in the comparison method or can be calculated when using the power integration method. Cross-polarized radiated energy can be measured. The measurement of gain must be carefully done with full knowledge of all the problems that are involved. The measurement of gain used in the calculation of ERP for a circular polarized antenna must be made relative to a horizontal dipole. The directive gain of a half-wavelength dipole antenna with respect to an isotropic antenna is 1.64. The field strength from a circular polarized transmitting antenna when received at a circular polarized receiving antenna of the same circular polarization sense will be 3 dB higher than when received by a linearly polarized receiving antenna. Gain requirements for a television broadcast antenna depend on the available transmitter power, economics, and field strength requirements, as determined by the terrain and the FCC population distribution.

Impedance/VSWR

Input impedance is the complex impedance consisting of resistance and reactance as measured at the antenna terminals throughout the television channel. The voltage standing wave ratio (VSWR) is the ratio of the maximum to minimum voltage resulting from the mismatch of the transmission line impedance and the antenna input impedance. When the transmission line impedance and the antenna impedance are exactly the same, the resulting ratio of voltages would be 1, resulting in a VSWR of 1:1. Most antennas are designed for the same input impedance as the transmission line used in the system. Impedance matching requirements for television antennas are generally more stringent than for other types of service because reflected energy, which would occur when the antenna does not terminate the line properly, can cause degradation of the system by introducing echoes and voltage/current peaks on the transmission line which could damage the antenna, couplings, and transmission line.

Radiation Characteristics

The television broadcast transmitting antenna most commonly has an omnidirectional pattern in the

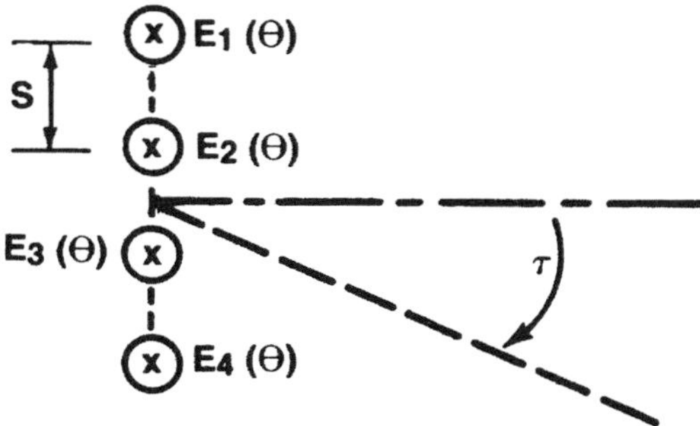

FIGURE 6.8-1 Elevation pattern calculation.

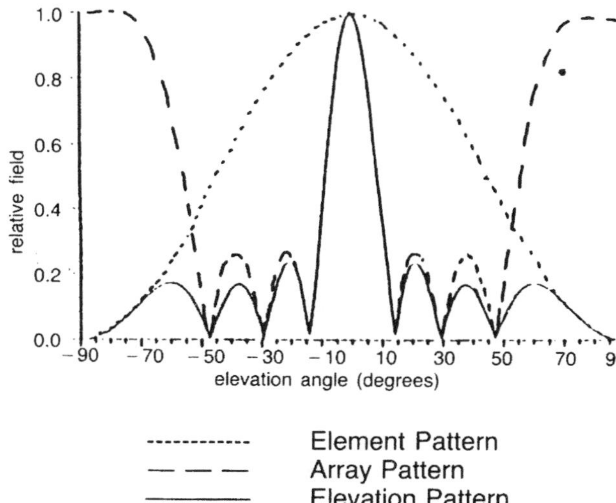

FIGURE 6.8-2 Elevation pattern for a four-element array ($\cos^2(\theta)$ element pattern, 1.02λ spacing).

azimuth plane and a narrow beam in the elevation plane. For an omnidirectional antenna, the gain is approximately one per aperture wavelength per polarization. Most broadcast antenna elements are spaced one wavelength apart; therefore, the gain of a linear polarized antenna is approximately equal to the number of elements. Directional antennas are available when specific coverage conditions warrant their use.

Elevation Pattern

The elevation pattern is the product of the element elevation pattern times the elevation array pattern (see Figure 6.8-1).

$$F(\theta) = E_n(\theta) \sum_{n=1}^{N} A_n e^{j\left[(2\pi n s)\sin\theta + \delta_n\right]}$$

where:

$E_n(\theta)$ = single element elevation pattern $\approx \cos^m(\theta) m$ = 1 or 2, typically.

s = element spacing.

θ = elevation angle.

δ_n = phase of nth element.

A_n = amplitude of nth element.

N = total number of elements.

The resultant antenna elevation pattern for a four element array pattern with 1.02λ spacing and $\cos^2(\theta)$ element pattern is shown in Figure 6.8-2.

As the spacing between the elements in the array pattern is increased, grating lobes begin to move closer to the main beam. This results in increasing the magnitude of the grating lobe in the resulting pattern. Figure 6.8-3 shows the patterns for a four element array with 1.30λ spacing. Although the main beam will narrow as the element spacing increases, energy will be lost in the high side lobe, resulting in an overall loss of gain. The optimum spacing is 1 wavelength for a $\cos^2(\theta)$ element pattern. Figure 6.8-4 shows power gain *versus* element spacing for multiple bay antennas.

Azimuth Pattern

The ripple content of the azimuth pattern is dependent on the element pattern and the distance to the phase center, as illustrated in Figure 6.8-5:

$$F(\phi) = \sum_{n=1}^{N} E_n(\phi) A_n e^{j\left[(2\pi\rho)\cos(\phi-\phi_n)+\delta_n\right]}$$

where:

$E_n(\phi)$ = element pattern

ϕ = azimuth angle

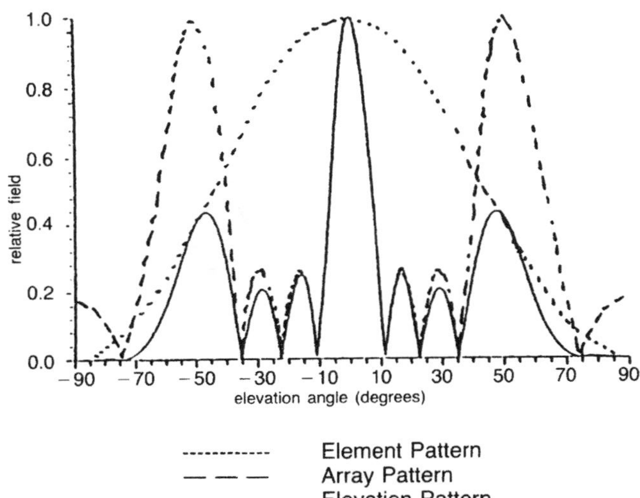

FIGURE 6.8-3 Elevation pattern for four element array ($\cos^2(\theta)$ element pattern, 1.30λ spacing).

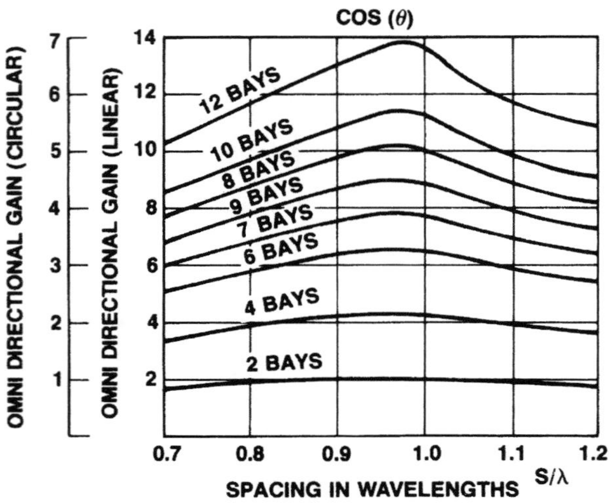

FIGURE 6.8-4 Power gain *versus* element spacing.

δ_n = phase of nth element

A_n = amplitude of nth element

ρ = distance to phase center (in wavelengths)

N = total number of elements

The value of ρ for a panel antenna is dependent on the panel or mast width. The value for a coaxial slotted antenna is dependent on the size of the coax line which in turn is dependent on the power-handling capability of the line. The value of ρ for waveguide slotted lines is dependent on the mode of operation and is a minimum of a half wavelength. This makes waveguide-fed slot antennas practical only at UHF frequencies. For some antennas, such as crossed dipoles or turnstiles, the value of ρ is zero. Others, such as panel, slotted coax, or waveguide, have a

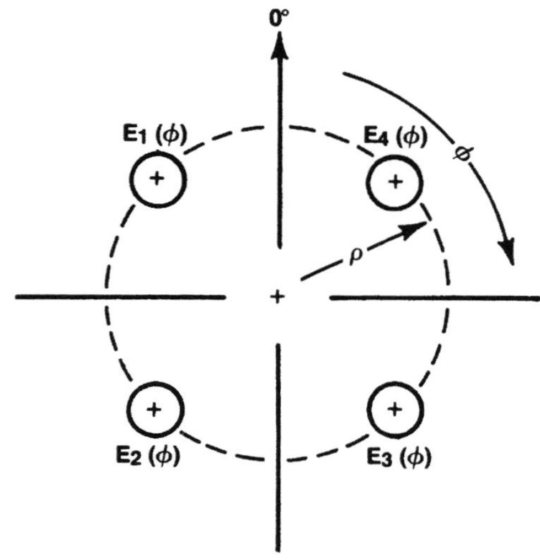

FIGURE 6.8-5 Azimuth pattern calculation.

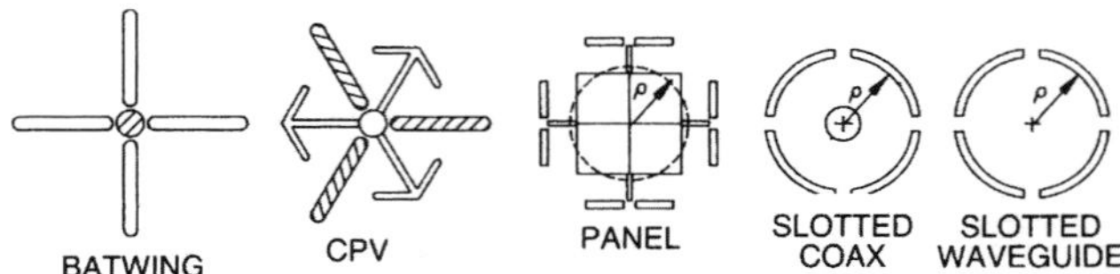

FIGURE 6.8-6 Arrangements of radiating elements (ρ).

value of ρ from 0.25λ to 2.0λ as shown in Figure 6.8-6. The ideal element pattern for a three-sided tower is $E = \cos(\phi)$; for a four-sided tower, $E = \cos^2(\phi)$. The ripple content of four panels with a $\cos^2(\phi)$ element pattern and phase center spacing of ρ = 0.5, 1.5 are shown in Figure 6.8-7.

Radiated Field Considerations

NTSC coverage is based on power (ERP) and height (height above average terrain, or HAAT) using the FCC F(50,50) field strength prediction curves. From this simple curve (one for each band), Grade A and B signal levels are determined and are defined as the coverage contours. The Grade B contour identifies the maximum reach of the station signal from which the demographics of the station are determined. DTV prediction requirements specify a different signal level for determining the coverage contour and use a different statistical model. If the station is to match the DTV and NTSC coverage, it is necessary to first determine a new NTSC contour using the new statistical model and terrain criteria. The NTSC contour obtained using the new procedure will result in a contour that is not the same as the current NTSC contour, and the antenna pattern requirement will very likely be different from the original NTSC pattern. For example, if the original antenna had an omnidirectional pattern, the recalculated pattern for DTV may not be omnidirectional. It may result in a pattern that is not physically realizable. To produce a DTV geographical contour that matches the revised NTSC contour in every case will not be possible as the rules are now written in FCC Bulletin OET-69 [11]. Stations with

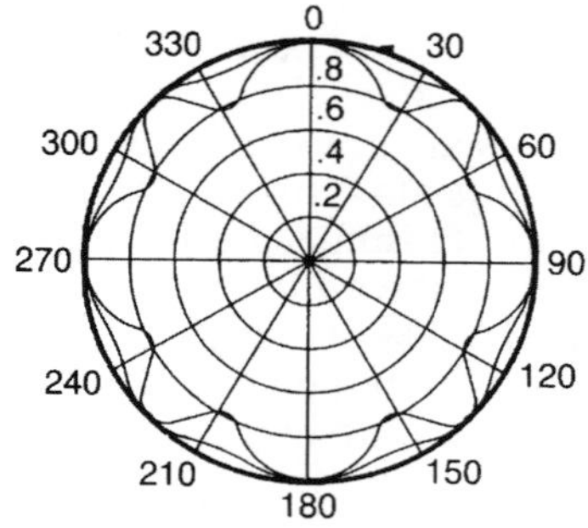

FIGURE 6.8-7 Azimuth pattern of four-sided antenna for which ρ = 0.5, 1.5.

both NTSC and DTV operations can request that their current NTSC channel be retained for their future DTV service unless the NTSC channel is out of the core band. To do this, the station must file with the FCC for use of that NTSC channel. Because the DTV antenna pattern provided by the FCC will likely differ from the current NTSC pattern, a new antenna may be required.

ERP

The maximum ERP levels currently permitted by the FCC are as follows:

	NTSC (kW)	DTV (kW)
Channels 2 to 6	100	45
Channels 7 to 13	316	160
Channels 14 to 69	5000	1000

Note that the NTSC power is peak while the DTV power is average, with peaks about 6 dB higher than average power.

Economics is a factor in antenna choice. As a general rule, for a given ERP, the combined cost of a transmitter and antenna is less when a higher gain antenna is used. In determining the ERP required to serve the area under consideration, a tradeoff must be made between using a low power transmitter and high gain antenna or high transmitter power and low antenna gain. For VHF antennas, the transmitter power to antenna gain ratio is fairly well established. For channels 2 to 6, antennas usually have gain values from about 4 to 6, depending on the length of the transmission line run. For channels 7 to 13, gain values vary from 12 to 18. For UHF antennas, higher antenna gains are required. Because of the shorter wavelengths at UHF frequencies, antennas with gains on the order of 20 to 30 can be built economically; however, the higher gain results in a narrow main beam. For a given situation, the high gain antenna may sacrifice local coverage for more distant coverage. As the gain of the antenna increases, the beam width narrows. The positions of the nulls and side lobes also move farther from the tower.

The distance in miles (R) from the tower, for a given depression angle, to the point where the beam will touch the earth (neglecting earth curvature) is:

$$R = \frac{H}{5280 \tan\theta}$$

where H is the height in feet and θ is the elevation angle.

Field Strength

The elevation pattern and relative field strength *versus* distance are shown in Figure 6.8-8 for a UHF antenna on a 1000 foot tower with constant ERP but different antenna gain. The ERP shown (1000 kW) is for a DTV

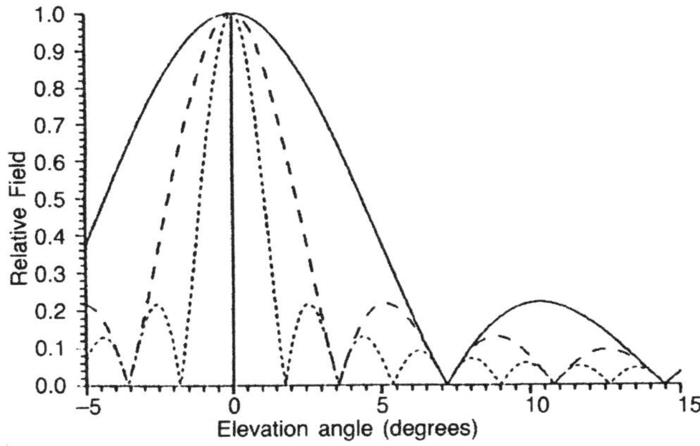

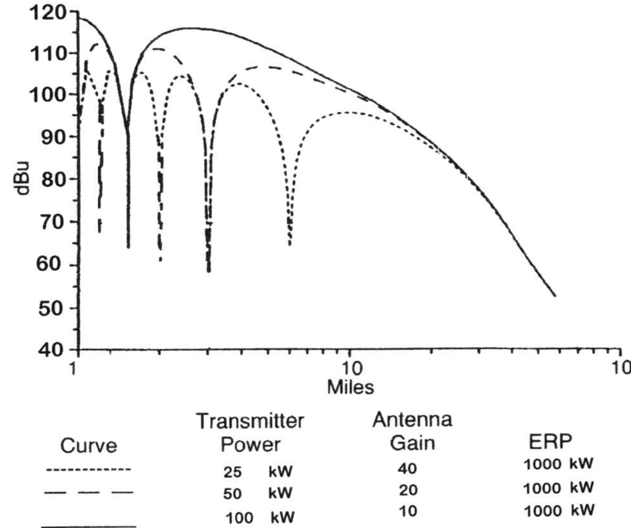

FIGURE 6.8-8 Signal *versus* antenna gain at UHF.

station. The elevation pattern and antenna gain are the same for DTV and NTSC. Note that, if the ERP and antenna height are held constant, there is no difference in the distant field signal strength. Figure 6.8-8 shows there is no change in distant relative field strength for the three normalized antenna patterns from 0° to 0.5°.

As the gain of the antenna is reduced, the main beam will widen from 1.6° (gain 40) to 7.2° (gain 10). The first null, the point where it impinges on the earth, will move from 6 miles to 1.6 miles, and the signal strength near the antenna will increase by 10 to 20 dB. It should be noted that this field increase makes a very strong signal (95 dBμV/m) much stronger (116 dBμV/m); therefore, if a higher gain antenna is contemplated, the local field strengths should be calculated using the FCC F(50,90) propagation curves for DTV. Required DTV coverage should be judged with an ERP of 1000 kW at 1000 feet and an antenna gain selected to meet this requirement. In hilly terrain, it may be desirable to increase the field strength by 10 dB or more and in heavily populated cities, with large buildings, by 6 dB or more. The field strengths specified by the FCC are

TABLE 6.8-1
Field Intensities (dBμV/m) for Specified Contours

	Low VHF	High VHF	UHF
Channel	**2–6**	**7–13**	**14–69**
DTV	28	36	41
NTSC, Grade A	68	71	74
NTSC, Grade B	47	56	64

shown in Table 6.8-1. Measurement of the field strength should follow FCC guidelines.

As an aid in reviewing coverage requirements, a brief review of the use of the charts mentioned is provided at the end of this chapter in Appendix A. All text applies to the F(50,90) chart; however, the methods are applied equally to the F(50,50) for NTSC or other charts as appropriate.

The digital contour is the noise limited contour as described in FCC Rule Section 73.622(e), "DTV Service Areas." If fields of this order cannot be achieved with a high gain antenna, the transmitter power should be increased provided maximum permitted ERP is not exceeded. Relative field and field strength *versus* distance for DTV are shown in Figure 6.8-9 for a fixed antenna gain at antenna heights of 500, 1000, and 2000 ft.

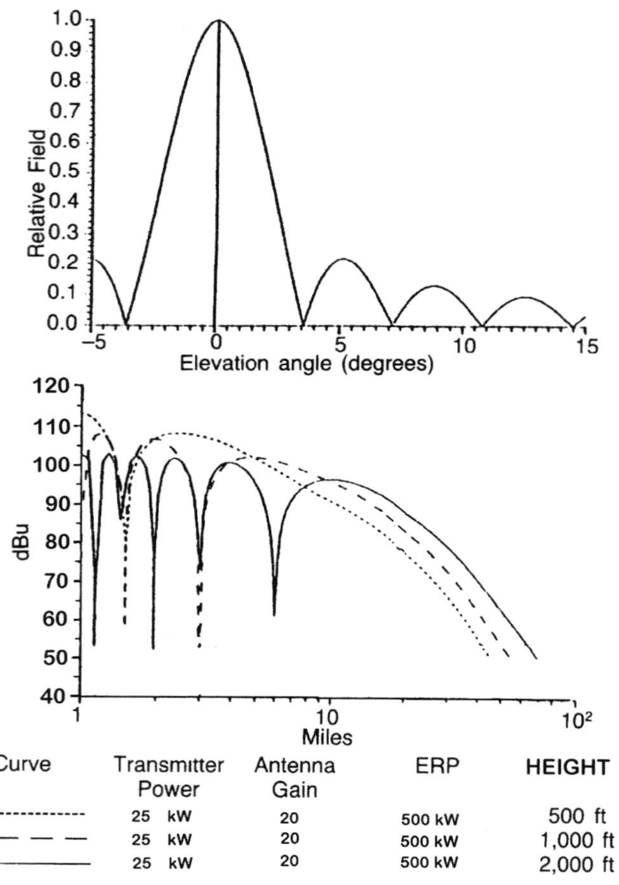

FIGURE 6.8-9 Field strength *versus* antenna height.

ANTENNA CHARACTERISTICS

Antenna Impedance/VSWR

The antenna impedance consists of the resistance and reactance that the antenna presents to the transmission line. The primary purpose of an input VSWR specification is to ensure a good match to the transmission line. If a mismatch occurs, the reflected power travels back to the transmitter, where it will be reflected again back to the antenna and radiate as a delayed but weaker signal that will be perceived as a ghost for NTSC signals. The delayed, weak signal will appear as a multipath signal for DTV signals. This reflected energy is treated as multipath in the digital receiver. The presence of this additional reflection reduces the data recovery margin of the system. The acceptable VSWR for digital service is considered to be 1.1:1 (see Figure 6.8-10). Typically, the VSWR at mid-band (middle of the channel) should be less than or equal to 1.06. The reflection coupled with the length of the transmission line that creates this multipath-like signal will be treated in more detail later in this chapter in the transmission components section.

Polarization

Polarization, illustrated in Figure 6.8-11, is defined by the plane of the electric vector (E). Elliptical polarization is the most general form. In general, an elliptically polarized wave may be expressed in terms of x and y components, given by:

$$E_x = E_1 \sin(\omega t - \beta z)$$

$$E_y = E_2 \cos(\omega t - \beta z + \delta)$$

The condition for linear polarization is when either E_1 or E_2 is zero. Horizontal polarization results when E_2 is zero. Vertical polarization results when E_1 is zero. The condition for circular polarization is when $E_1 = E_2$ and

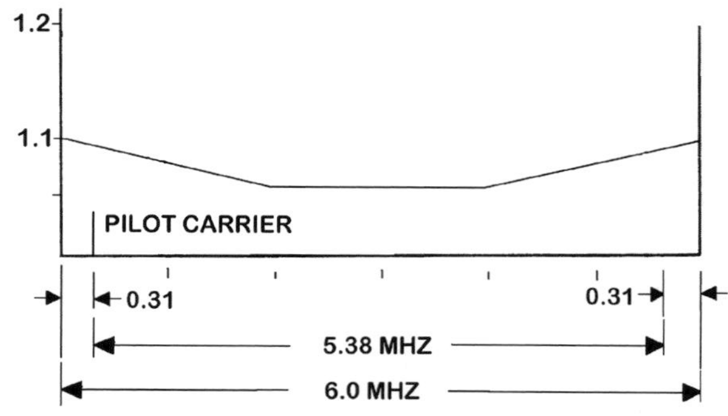

FIGURE 6.8-10 Acceptable DTV VSWR level.

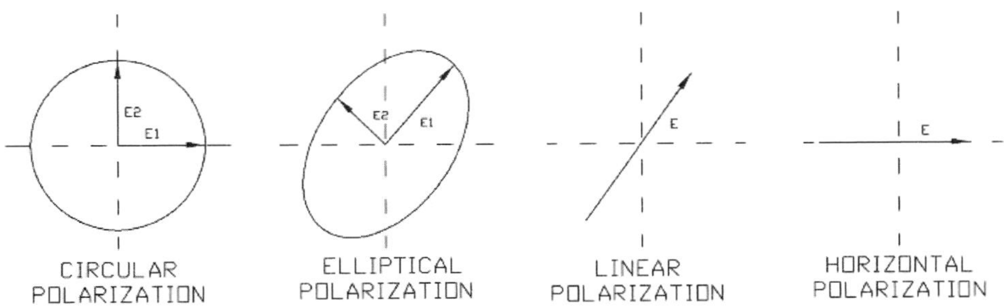

FIGURE 6.8-11 Polarizations.

δ equals 90° (the two components must be quadrature in both space and time). The transmitted wave of TV broadcasting signals is specified by the FCC to be linear and in the horizontal plane; however, FCC regulations permit the radiation of circular polarization (CP) provided the vertical component does not exceed the licensed horizontal component. The possible advantages of using circular polarization for television include:

- Less critical receive antenna azimuth orientation permits good reception on all types of indoor antennas, including rabbit ears, whips, and rings.
- Improved coverage in the fringe area is due to twice the power density; improvements on the order of 3 dB can be expected, thereby reducing the cliff edge effect of the terrestrial DTV signal.

Although mixed results were achieved in the use of CP for NTSC broadcasts, DTV applications may prove more amenable to improvement through the use of this technique. The transmission of CP has obvious drawbacks in the form of a 2× increase in required transmitter power and transmission line rating and a more complex antenna. The 2× increase comes from the fact that ERP is measured in just one radiation plane. For the DTV signal, polarization diversity can be achieved if the vertically polarized signal is transmitted through CP. A polarization diversity system at the receive antenna can provide missing signal levels when one of the horizontal or vertical signal components experiences a deep fade; thus, the inherent diversity attributes of CP operation could be put to good use.

Beam Tilt

Beam tilt is sometimes used to aim the main vertical beam tangential to the earth toward the radio horizon if the relative height of the antenna is such that substantial energy is being radiated above the horizon. The distance to the radio and optical horizon can be determined from Figure 6.8-12. Note that the antenna height over the service area may not necessarily be the height over average terrain, especially in mountainous areas. Mechanical beam tilt is useful where excess beam tilt is required to improve coverage, particularly where water or mountainous terrain limits population density. Electrical beam tilt is often used, but it reduces the power gain, especially for a higher gain antenna; however, the increase in local coverage is generally a more important consideration than the slight loss of power gain. Electrical beam tilt is accomplished by adjusting the phase of the energy fed to each element across the array or each section of a split feed antenna. Beam tilt *versus* element phase change is shown in Figure 6.8-13. Most often beam tilt is introduced by adjusting the feedline delay (length) and element spacing of the array so the desired amount of beam tilt is obtained.

Null Fill

The amount of null fill, and the number of nulls that must be filled, depends on how close the populated area is to the transmitter site. Allowance should be made for population movement toward the site in the future. If the transmitter site is in the center of the population area or on the edge of it, consideration should

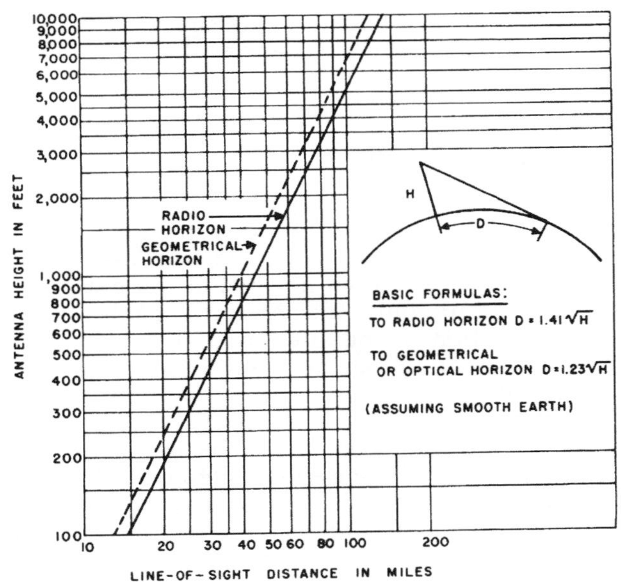

FIGURE 6.8-12 Distance to radio horizon *versus* height.

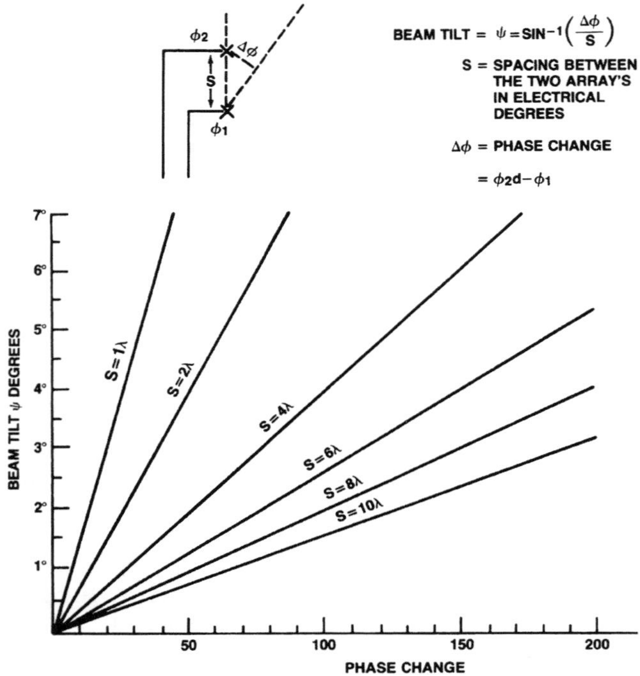

FIGURE 6.8-13 Beam tilt *versus* element phase change (Δφ).

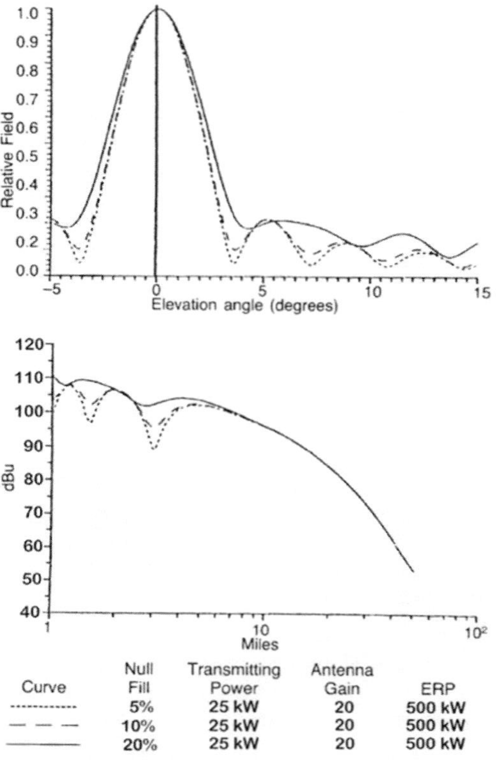

FIGURE 6.8-14 Null fill.

be given to having null fill. The exact amount of first and second null fill is not critical. Anything greater than 5% will usually result in an adequate signal level. The null depth (dB) in signal strength will correspond to the difference between the first null and first side lobe (dB). The effects of null fill on signal strength are shown in Figure 6.8-14. A combination of null fill and beam tilt is shown in Figure 6.8-15. A beam tilt of 1° will usually result in a 10 dB improvement in near-in coverage. Note that null fill and beam tilt have little effect on the magnitude of the signal at the limits of the coverage.

Pattern Considerations

For a side-mounted antenna (directional or omnidirectional), the tower will modify the as-designed patterns. For optimum coverage, the as-installed patterns must be known not just at the center frequency but throughout the entire channel before the relative position of the antenna and its azimuthal pattern orientation can be fixed. There is usually one position that will provide the optimum coverage without exceeding the structural limitations of the tower. Coverage considerations are particularly important to DTV because undesired energies, such as reflections, translate into a loss of coverage, whereas in NTSC the undesired energies translate primarily into a loss of picture quality.

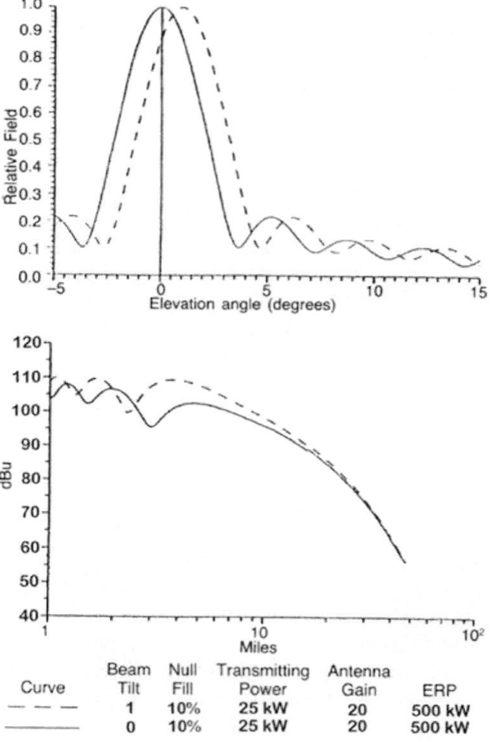

FIGURE 6.8-15 Null fill and beam tilt.

Power Handling Capability

Traditionally, power in NTSC television broadcast transmission systems has been regarded in terms of peak power, which is the instantaneous power developing during the peak of the synchronizing pulse in the visual transmitter. Because the black-level signal is 75% of the total voltage value of the pulse, the average power for a totally black picture is 56% of the peak sync power. The duty cycle of the synchronizing pulses, both horizontal and vertical, adds about 4% to this power so the average black-level power is 60% of the peak TV power. Because the aural transmitter is usually 10% of the peak TV power, the total heating or continuous wave (CW) power of the TV signal is 70% of the peak TV power. The average power level (APL) of the video signal with a typical program measured over a long period of time is 4.32 dB (37%) below the peak TV power. With digital modulation and, specifically, eight-level vestigial sideband modulation (8VSB), power for these systems will be expressed in terms of average power. Peak-to-average ratios of 6 to 7 dB exist in these systems. Typically, breakdown voltage calculations should be made assuming a peak-to-average ratio of at least 7 dB. The design of all TV antennas will allow for a sufficient safety margin to handle the power to be applied (average and peak digital or an additional channel where required), plus imperfections in the transmission line or VSWR and does not depend on pressurization.

Unwanted Antenna Radiation

The antenna pattern is the product of the array pattern times the element pattern. The array pattern for in-phase elements having one wavelength vertical spacing has a downward and upward lobe as large as the desired main lobe. If the element pattern is not a cos(θ) function, the vertical plane element pattern will not drive the array downwardlobe to zero, resulting in a very large signal in the vicinity of the tower base. Figure 6.8-16 shows the resultant antenna pattern with a $\cos^N(\theta)$, where $N = 0.0, 0.5$, and 1.0. Care should be exercised in making sure the element pattern has a distribution in the elevation plane of $\cos^1(\theta)$ or $\cos^2(\theta)$. A well designed antenna system should not create a non-ionizing radiofrequency radiation (RFR) problem. The tower height is usually selected for maximum signal coverage and should therefore be high enough above the ground so the downward radiation is well below the safe American National Standards Institute (ANSI) exposure levels.

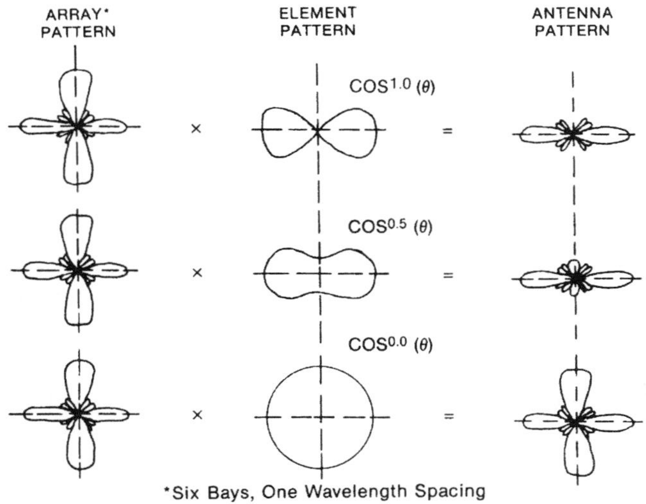

FIGURE 6.8-16 Pattern multiplication.

Equally important would be the power density level on nearby towers, which could result in tower maintenance personnel being exposed to high energy levels when in the main beam of the antenna on a nearby tower. The distance in feet along the main beam to the 1 mW/cm² contour for maximum ERP and the height above ground to the 1 mW/cm² contour (occupational) [13] are shown in Table 6.8-2. The information provided in Table 6.8-2 is for guidance; each situation should be evaluated individually with regard to requirements for meeting exposure limits.

TABLE 6.8-2
Distance for 1 mW/cm² Level

Service	Low VHF	High VHF	UHF
ERP	45 kW	160 kW	1000 kW
EIRP	73.8 kW	262.4 kW	1640 kW
Distance to nearby tower	79 ft	150 ft	607 ft
Minimum height above ground	44 ft	84 ft	340 ft

ANTENNA RADIATING STRUCTURE DESIGNS

Many possible radiating structure designs exist; the most common are discussed in this section.

Slot Antenna

The slot antenna is similar to a dipole (see Figure 6.8-17). Currents in the slot antenna spread out over the entire sheet (in which the slot is cut) and radiation take place from both sides of the sheet. The resemblance between the two becomes more pronounced when it is recognized that the field patterns of the two will be equivalent if the physical dimensions of the slot and the cross section of the dipole are the same. Furthermore, the impedance of a slot is proportional to the impedance of a dipole of the same dimensions by the relationship:

$$Z_{slot} = \frac{35,476}{Z_{Dipole}} \text{ ohms}$$

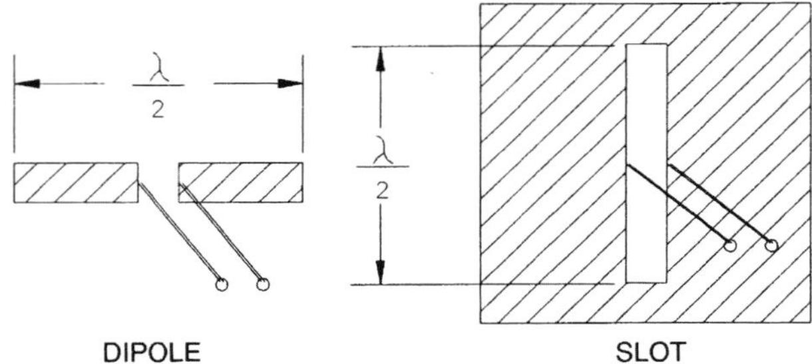

FIGURE 6.8-17 Dipole/slot equivalence.

and the bandwidth characteristics are essentially the same for both. Actually, the previous discussion is rigorously accurate only if the sheet is of infinite extent, but it is substantially correct if the edge of the sheet is half a wavelength from the slot. Bending the sheet into a cylinder results in another form of slot antenna that also takes on characteristics of a stack of coaxial rings, as shown in Figure 6.8-18. The slotted cylindrical tube can have an inner conductor and be coaxial in form or have no inner conductor and be a cylindrical waveguide.

Panel Antenna

A panel antenna has dipoles or formed wire structures mounted $\lambda/4$ in front of a reflecting sheet. These antennas will have a directional ($\cos(\theta)$ or $\cos^2(\theta)$) pattern in the azimuth and elevation planes, depending on how the antenna is mounted. Figure 6.8-19 shows typical panel antenna configurations. The pattern of a panel antenna is directional, as the energy radiated from the element sees its image in the ground plane, resulting in a $\lambda/2$ spaced end-fire radiator. The dipoles in the doublet panel are spaced $\lambda/2$ and are effective in eliminating downward radiation.

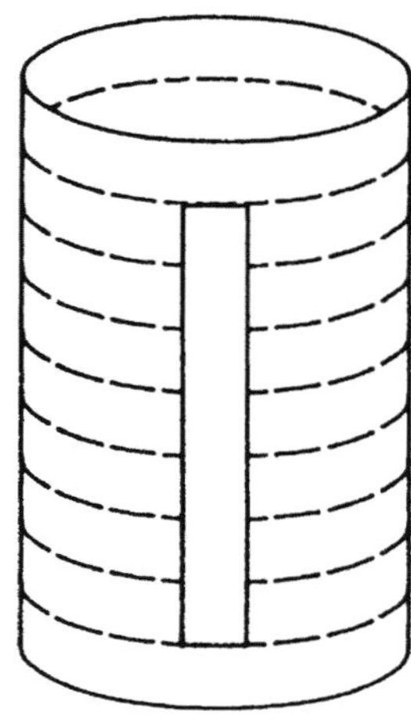

FIGURE 6.8-18 Slot in cylindrical sheet.

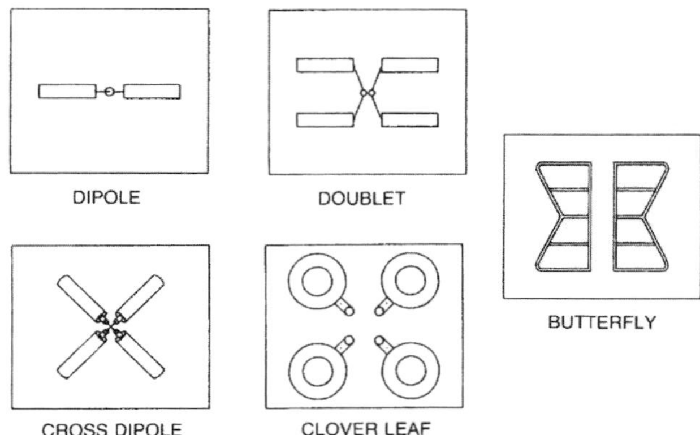

FIGURE 6.8-19 Panel configuration.

Traveling Wave

Traveling wave slot antennas can be either resonant (standing wave) or nonresonant (traveling wave). Most are bottom-end fed. When the far end is shorted, an infinite standing wave is set up in the transmission line (either coax or waveguide). When the slots are appropriately sized and spaced, energy will be coupled out. When the amplitude and phase of each slot are suitably controlled across the aperture, the resulting antenna will have high gain. The bandwidth requirement for broadcast limits this technique to single channel or adjacent channel applications. Several traveling wave antennas are shown in Figure 6.8-20.

The most common bottom fed traveling wave array is the nonresonant slot array. The number of half wavelength slots arranged about the periphery is dependent on the shape of the pattern desired. The slots, circumferentially arranged around the antenna structure, are usually displaced by $\lambda/4$ and spaced vertically one wavelength apart. The slot coupling from the bottom to the top increases because the power radiated progressively reduces the power on the line. This method produces an amplitude taper across the aperture that provides null fill. With all nonresonant arrays, a small amount of energy is leftover. This is radiated in specially designed end-loaded slots or in the opposite polarization.

Circular polarization can be obtained by orienting successive slots 90° and exciting them in phase quadrature or by exciting a vertically polarized radiator located near the slot. The linear polarized spiral (helix) is essentially a strip transmission line mounted over a ground plane; the spacing controls the radiation. Each element is a half wave radiator, and, because of the orientation, the vertical components cancel. The choice of pitch angle and length along the circumference of the spiral will result in very good horizontally polarized broadside radiation. The circular polarized spiral or multiarm helix consists of multiple wires mounted around a large cylinder, forming a current sheet. The optimum parameters are a complex function of the number of arms, the length of one coil, and the spacing to the support pole. Excellent circularity can be

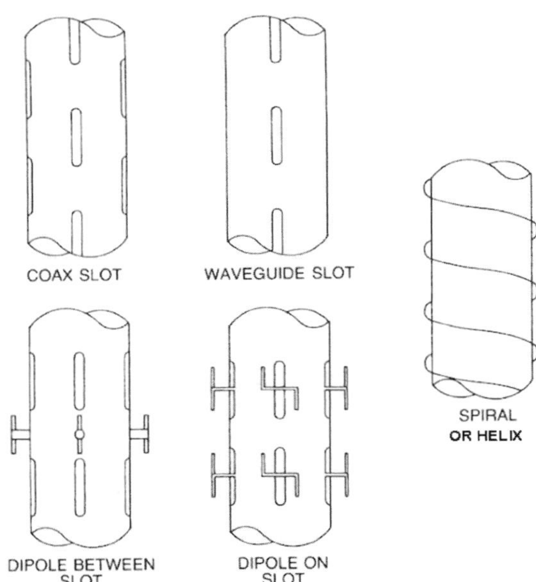

FIGURE 6.8-20 Traveling wave antennas.

obtained because in any heading the radiation is essentially that from a current ring.

ANTENNA SUPPORT STRUCTURES

The current structural standard for antenna supporting structures and antennas is ANSI/TIA/EIA-222-G, "Structural Standards for Steel Antenna Towers and Supporting Structures."

Deflection and Wind Load

The broadcast tower structural standards allow a rather liberal maximum horizontal displacement and rotation limitation on broadcast towers for service load conditions. The definition of service load is a wind load on the tower resulting from a 60 mph, 3 second gust ground level wind speed, which is equivalent to a 50 mph fastest mile wind speed. At this service load condition, the allowable rotation for a structure is up to 4° about the vertical axis, and the allowable maximum horizontal displacement is 5% of the height of the structure. The tower standard's limitation for cantilever tubular or latticed spines, poles, or similar structures mounted on the top-latticed structures is a relative horizontal displacement of 1% of the cantilever height measured between the tip of the cantilever and its base. For example, a 500 foot structure could move horizontally at its top as much as 25 feet under the 60 mph, 3 second gust ground-level wind speed, with any point on the mast rotating as much as 4°. If there were a 50 foot pole on top of this base structure it could deflect laterally at its top an additional 6 inches over its height. These deflections represent an unacceptable extreme for most broadcast

TABLE 6.8-3
Antenna and Tower Deflection

	Wind Speed, 3 Second Peak Gust † (Fastest Mile ‡)			
	28 (20)	40 (30)	51 (40)	60 (50)
Tower Height AGL*	Typical Angular Deflection of Top Mounted UHF TV Antenna Resulting from above Wind Loading			
1192′	0.07 °	0.16 °	0.33 °	0.57 °
992′	0.07 °	0.15 °	0.28 °	0.47 °
1650′	0.10 °	0.17 °	0.30 °	0.45 °
1680′	0.10 °	0.21 °	0.46 °	0.85 °

Typical Angular Deflection of Top Mounted UHF TV Antennas on Guyed Towers at Various Wind Speeds.

The variables used in this exhibit are defined below:

*Above Ground Level (AGL)

† Measured 3 second peak gust in mph at 30 ft. AGL (G)

‡ Calculated fastest mile wind speed in mph at 30 ft. AGL. Durst Method of conversion from 3 second peak gust to theoretical fastest mile. (U_{td})

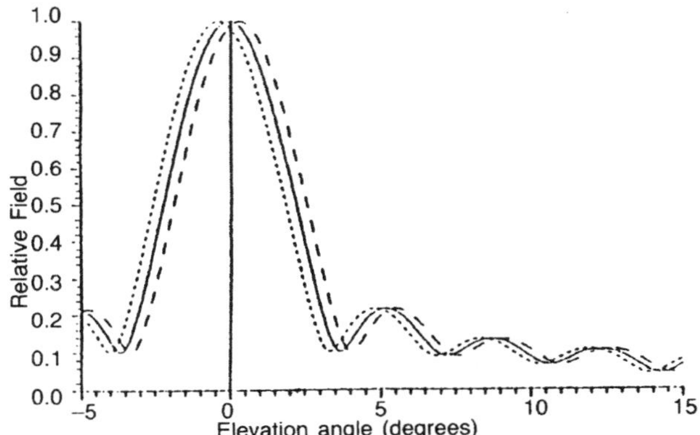

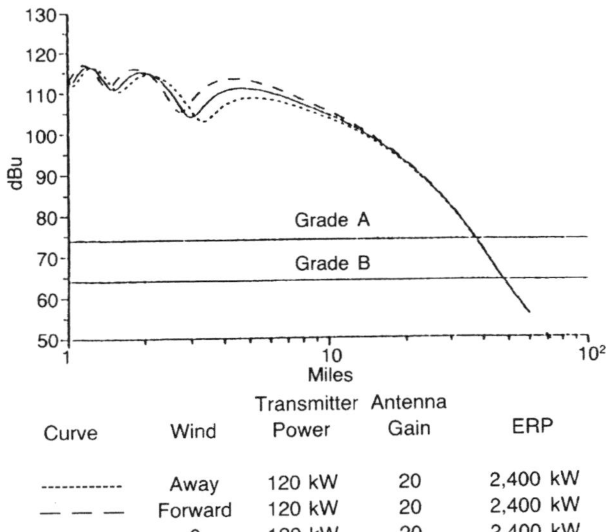

FIGURE 6.8-21 Field strength changes with a 50 mph wind.

applications. Structurally, a free standing antenna can be considered as a cantilever beam in which the deflection increases toward the end. Antenna deflection is stated as the angle from the vertical of the chord that connects the base to the top of the antenna.

The performance of a broadcast tower at service wind speed conditions varies depending on where the tower is located, to what standard it was originally designed, how tall the tower is, the design of the tower cross section, the tower guy wire size/tension, and what appurtenances are mounted on the tower. Some typical examples of the movement of the support tower top plate under various wind conditions are shown in Table 6.8-3. In these four examples, the top-mounted antenna alone would sway approximately 0.5°, which would result in antenna sway ranging from 0.65° to as much as 1.3°.

When considering a new tower, tower sway should be specified in the tower design. Top mounted antennas for broadcast service are typically designed for a maximum deflection of <0.5°. For top mounted TV antennas, the sway of the antenna mounting plate located at the top of the tower should be limited to a reasonable number such as 0.4° at service wind speed to avoid excessive antenna sway. Examples of the effect of antenna and tower movement on predicted coverage for a typical top mounted TV antenna are shown in Figures 6.8-21 and 6.8-22. Figure 6.8-21 shows the effect on TV coverage of a ±0.35° antenna sway, and Figure 6.8-22 shows the effect on TV coverage of a ±1.4° antenna sway.

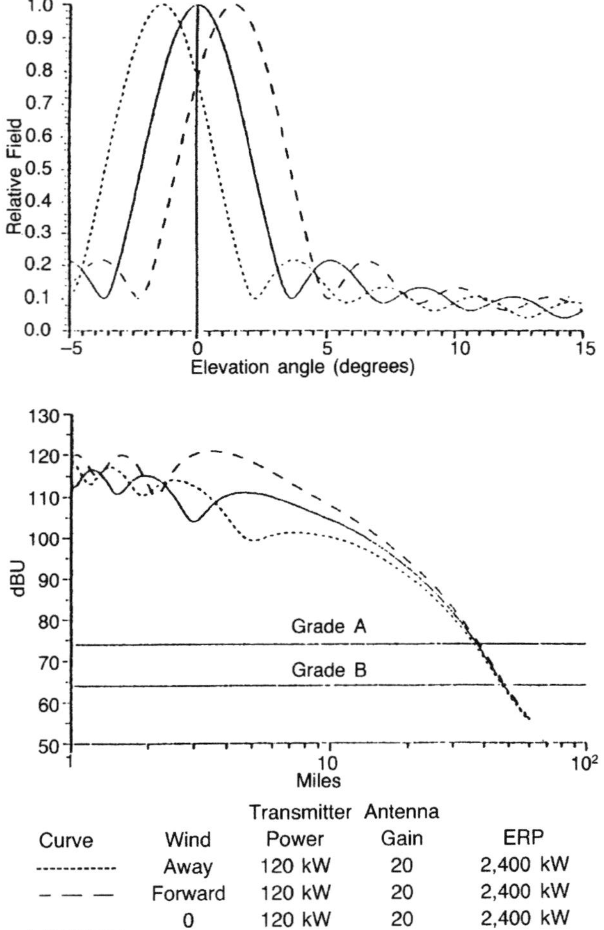

FIGURE 6.8-22 Field strength changes with a 100 mph wind.

ANTENNA FEED SYSTEMS

The feed system of a television broadcast antenna is that portion of the transmission system having its input at the antenna terminal, at the top of the vertical run of coaxial transmission line in the tower, and its output at the radiating elements. Most antenna gain figures specified by the manufacturer take the losses of the feed system into account; therefore, when system gains are calculated, the feed system loss can be excluded. In the television broadcasting field, three types of feed systems are in wide use: branch feed, standing-wave, and traveling wave.

Branch Feed Systems

Branch feed systems are necessary for wideband antennas (an antenna that covers the full VHF or UHF band). Being center fed, an antenna with branch feed eliminates beam steering or beam tilt variations with frequency. This is due to the feed system delivering the RF with the same phase relation (same path length) at each frequency. This feed system progressively divides the power as shown in Figure 6.8-23. It is used when the radiators are individual elements, each with its own terminal, such as a dipole or panel. The system shown in Figure 6.8-23a will normally have a narrower impedance bandwidth than the system shown in Figure 6.8-23b, because, for economy, one eight-way power divider is used. The junction impedance is $Z_j/8$. If the element impedance is 50

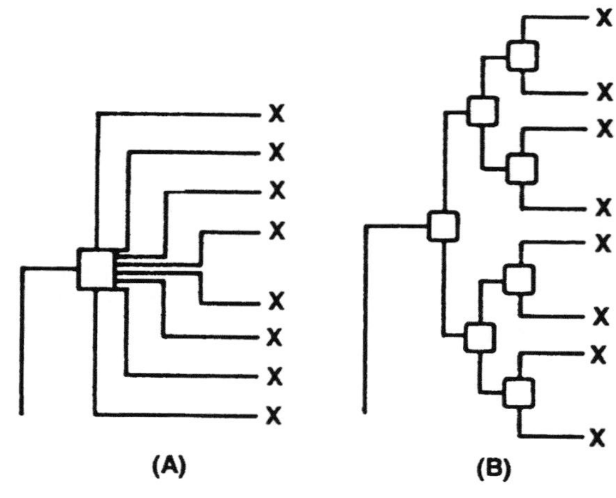

FIGURE 6.8-23 Branch feed systems.

ohms, the power divider must transform 6.25 ohms to 50 ohms. The system shown in Figure 6.8-23b uses two-way power dividers and is sometimes called a *corporate feed*. Although it has a broader bandwidth, it is less economical because it has seven power dividers with additional interconnecting cables. Null fill and beam tilt are accomplished by changing the length of the feed cables or using unequal power dividers, or both. A problem with branch feed systems is the presence of the feed line in the aperture of the lower elements. The feed lines can cause reradiation and distort the azimuth pattern. The branch feed system can be more effectively used with panel antennas that require a center support tower or mast where the transmission lines are behind the antenna.

Series Feed Systems

Standing Wave

A coaxial or waveguide transmission line can be shorted at the far end, resulting in standing waves along the length of the line. If slots or coupling probes are appropriately sized and positioned, the RF energy can be radiated and a desired amplitude and phase distribution across the aperture obtained. This resonant array structure has a desirable feature; all coupling parameters are the same and equally spaced. Its disadvantages are the narrow bandwidth and that it can only be used at high VHF and at UHF frequencies (Figure 6.8-24).

Traveling Wave

The traveling wave feed system operates on the principle of a gradual attenuation (by radiation) of the input signal as it progresses from the input along the aperture of the antenna. An application of this principle is the slot antenna or spiral antenna. Figure 6.8-25 shows the principle of this feed system using short rod radiators to illustrate the theory. A number of uniformly spaced radiators per wavelength are loosely coupled to a coaxial line. Because of the number of radiators and the relatively slight reflection between them, the effect is essentially that of a uniform loading. The result is a uniformly attenuated traveling wave in the line. Because a traveling wave has a linear-phase characteristic, the excitation of each successive radiator will be lagging from the previous one by an amount that depends on the spacing between the radiators and the velocity of propagation in the line. If the radiators are alike, their currents will have the same phase relationship as the excitation; thus, the radiating currents will be successively lagging, and repetition of phase occurs after every wavelength.

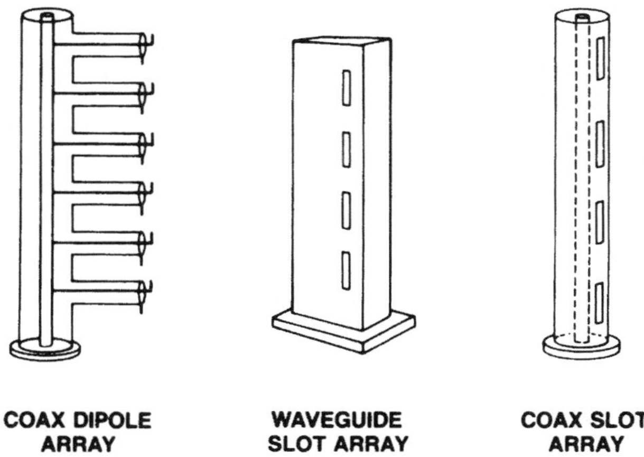

FIGURE 6.8-24 Slotted arrays.

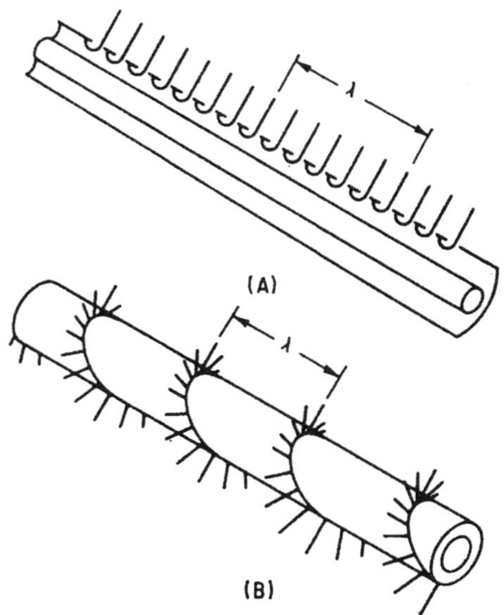

FIGURE 6.8-25 Traveling wave excitation.

To obtain an omnidirectional pattern, the radiators, instead of being in line, can be moved around the periphery to form a spiral, as shown in Figure 6.8-25b. For a horizontal main beam, the pitch of the spiral has to be equal to the wavelength in the transmission line. In this arrangement, all the radiators in any one vertical plane on one side are in phase, and the phase difference between radiators in different planes equals the azimuth angle difference between the planes. That is, the phase rotates around the periphery. The rotating phase produces a rotating field. Because of the relatively small amount of current change from layer to layer, an omnidirectional pattern is produced.

ANTENNA TYPES

The types of antennas available can be grouped into the following categories (see Table 6.8-4):

- Top or side mounted
- Linear or circular polarization
- High power, low power

TABLE 6.8-4
Commonly Used Television Transmitting Antennas

	Horizontal Polarization		Circular Polarization	
	Top Mounted	Side or Tower Mounted	Top Mounted	Side or Tower Mounted
VHF low band	Batwing	Butterfly Dual dipole	TDM	—
VHF high band	Batwing Traveling wave	Butterfly Dual dipole H-panel	Spiral Slot Dipole	TCP CBR Ring
UHF	Slot coax	Dual dipole	Slot with Z dipole Slot with dipole	Slot with Z dipole Slot with dipole

The prime purpose of the tower is to support the antenna. The antenna should have the required radiation characteristics to deliver a satisfactory signal to the viewer. Because it is desired to cover as large a viewing area as possible, tall towers are used. It is not uncommon for antennas to be mounted on 1000, 1500, or 2000 foot towers. These tall towers are usually guyed triangular towers with face dimensions of 5 to 10 feet.

The cost of a tower is heavily dependent on the wind load presented by the tower, the transmission line, and the antenna. For VHF antennas, the antenna wind load is the most significant parameter. For UHF antennas, the transmission line or waveguide is the more significant factor. The length of the antenna is related to the channel or wavelength and the gain requirements (see Table 6.8-5); therefore, the ideal antenna would be of minimum length to meet the required gain and would be physically small in surface area to minimize the loading on the tower. However, the antenna size is also a function of frequency and its cross section approximately a minimum of one-half wavelength.

At low VHF frequencies the resonant half wave element cross section of 8 feet is sufficiently large so a center support pole can be incorporated with the antenna elements (such as a batwing antenna). On the other hand, at UHF frequencies, the resonant half wavelength of 1 foot is so small that the resonant elements must mount outside the support pole or the support pole itself must be a radiator, such as a slotted coax or waveguide array.

TABLE 6.8-5
Dimensions of Typical Antenna

	Wavelength (ft)	Length for Maximum Gain (ft)	Maximum Gain
Low VHF	16	100	6
High VHF	5	90	18
UHF	2	120	60

Ideally, omnidirectional antennas should be fed from the bottom, so there are no feed lines in the aperture of the antenna to distort the pattern. If feed lines are required for the upper elements, they should be on the inside of the antenna or support structure. A conflicting requirement is that the diameter of the feed line must be large enough to satisfy the power handling requirements yet small enough not to create pattern distortion. Slotted arrays overcome this shortcoming of other antenna styles because, typically, they have no external feed lines. To minimize pattern distortion by the tower, the antennas should be mounted on top of the tower rather than on the side. Where an antenna must be side mounted, significant reradiation from tower members can occur. Tower and appurtenance effects can be calculated using simulation software as an aid to determining the optimum position for the antenna on the tower.

The radiating element for many UHF antennas is either the slot or dipole excited from an internal source such as coaxial or waveguide. Most slotted arrays are traveling wave structures that are bottom fed and contain many slots. The greater the number of slots, the narrower the beamwidth and the greater the beam steering or tilt differential across the 6 MHz DTV channel. Careful array design can limit differential gain to less than 3 dB in a slotted antenna. Some slotted arrays use the outer conductor (coax) or the waveguide itself as the structural member.

Panel antennas are fed from the back of the panel and require a secondary structural member to support the panels. Three or four panels must be mounted around the tower to produce an omnidirectional pattern. Additional panels may be required as the face size of the tower increases to maintain the desired omnidirectional azimuth pattern. Panel antennas, when fed with a branch feed harness, can preserve the wide bandwidth of the dipole and cover the full high VHF band (174–216 MHz) or UHF band (470–806 MHz).

VHF ANTENNA TYPES

Linear Polarized

Superturnstile/Batwing

The first antenna developed for commercial service was the Superturnstile. It consists of a central sectionalized steel pole on which is mounted the individual radiators, or batwings. These radiators are mounted in groups of four around the pole in north–south and east–west planes to form a section. The sections are stacked one above the other to obtain the desired gain. Figure 6.8-26 illustrates this construction. Each of the radiators of the batwing antenna is fed by its own feed line. The impedance of each is carefully matched. The feed lines, in turn, are combined at junction boxes, which perform the dual function of feeding power simultaneously to all feed lines and transforming the combined impedance of these lines to that of the main transmission line. This latter function is achieved by the use of multistage transformers immediately below the junction box.

At the base of the antenna, a combining network is used when there are more than two junction boxes. These networks accomplish power division between portions of the antenna, if desired. Batwing antennas are manufactured in various gains from 3 to 12 (12 bays are unusual) for channels 2 to 6 and gains of 6 to 18 (18 bays are unusual) for channels 7 to 13. They can also be designed for various types of null fill, and they have been used in stacked and candelabra installations. The batwing antenna is relatively wideband and can be used for two channels—for example, channels 4 and 5 or channel 6 and an FM channel, as well as in various combinations in the channel 7 to channel 13 high VHF range.

FIGURE 6.8-26 Superturnstile/batwing antenna. (Figure courtesy of Electronics Research, Inc.)

Traveling Wave

The traveling wave (TW) antenna is a slot antenna with a traveling wave feeding the slots (see Figure 6.8-27). The TW antenna is a coaxial line with pairs of slots in the outer conductor spaced at intervals of a quarter wavelength throughout its length. Probes at the center of each slot distort the field within the line to place voltages across the slots. These, in turn, drive currents on the periphery, setting up a radiated field. Attenuation of the signal by withdrawal of a portion of the power at each slot reduces it to a very low value at the upper end of the antenna. There, a special pair of slots, designed to match the line, radiates the remaining power.

Operation of the TW antenna can be better understood if the section of the aperture having pairs of slots is recognized as being, in effect, a dipole. Successive pairs of slots are alternately in one plane and in another at 90° to it, so the antenna can be simulated by stacked dipoles with a 90° angle between successive layers. In a given plane, reversal of the direction of feed every half wavelength (by placing the probes on opposite sides of the slots), together

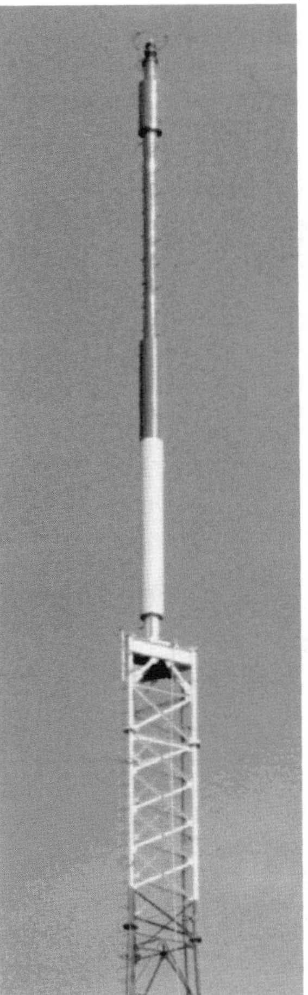

FIGURE 6.8-27 Traveling wave antenna. (Figure courtesy of Dielectric Communications.)

with the half-wave change in phase of the signal as it passes along the aperture through this distance, results in all the dipoles in that plane being fed in phase. The same action takes place in the other plane, except that they are fed 90° out of phase with the first plane, due to their 90° displacement along the antenna. Each plane of dipoles radiates essentially a figure-eight pattern. Because the planes are fed in quadrature, addition of the patterns results in an omnidirectional pattern. Because of the circular cross section and lack of obstructing radiators, the resulting horizontal pattern is nearly circular.

Panels

The panel antenna is designed to wrap around the tower. Four panels are needed for a square tower where each panel must radiate a $\cos^2(\theta)$ element pattern. For a triangular tower, each element must radiate a $\cos(\theta)$ element pattern for an omnidirectional pattern. A wide variety of azimuth and elevation patterns can be obtained by using fewer panels or changing the phase and power division to the panels. The panels are typically 0.7λ to 0.9λ in vertical length and spaced vertically approximately one wavelength. The radiating elements may be either single dipoles, such as the H-panel rhombus in Figure 6.8-28a, or a delta dipole (butterfly), which is essentially a folded-back batwing (Figure 6.8-28b). The dual dipole shown in Figure 6.8-28c is designed to minimize downward radiation and consists of four dipoles spaced one-half wavelength apart. The impedance bandwidth at the panel antenna is very wide and capable of handling more than one TV channel. The bandwidth can be further improved with the four-side configuration by using panel offset and phase rotation (90°) between each panel in the bay. The impedance at the four-way power divider will be conjugate for every two panels, resulting in wider bay impedance bandwidth than the element itself.

Circular Polarized

Transmission Dual Mode

The transmission dual mode (TDM) shown in Figure 6.8-29 is a circularly polarized antenna for channels 2 to 6 and is designed for tower top mounting. It is capable of replacing an existing six-bay Superturnstile without any increase in tower wind loading. The TDM antenna shown utilizes seven layers of radiators in a slanted dipole configuration with three radiators mounted symmetrically around the pole per layer. Each of the three radiators is fed in phase by a single feed line. Only 21 feed lines are required for the entire antenna. A branch-type feed system is used to achieve excellent vertical pattern stability. One junction box feeds the upper four layers and another feeds the lower three layers, with each box being fed by a 3-1/8 inch line. This feature allows for standby capability in the event of weather related or other damage. The TDM can be supplied with deicers or radomes,

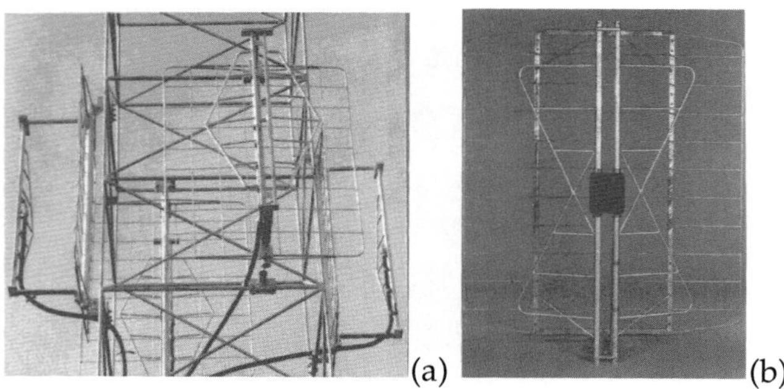

FIGURE 6.8-28 Panel antennas: (a) H-panel rhombus (Dielectric), (b) delta dipole butterfly (Dielectric), (c) dual dipole (MCI).

FIGURE 6.8-29 Transmission dual mode antenna. (Figure courtesy of Dielectric Communications.)

depending on environmental requirements. The unique design of the TDM provides for excellent pattern circularity and axial ratio. Axial ratio measurements are performed on a complete, full-scale, as-built antenna standing vertically at the factory. Because all elements are excited, the mutual effects of adjacent elements are considered.

Circularly Polarized "V"

The circularly polarized "V" (CPV) antenna shown in Figure 6.8-30 is a circular, polarized top-mounted antenna consisting of three cross "V" dipoles mounted at 120° intervals around a vertical mast. The dipoles are segmented by three vertical grids like a corner reflector, used both for isolation and to shape the element for good circularity. The cross dipoles are fed in phase quadrature and radiate circular polarization from each element. A branch feed system is used with the lines fed up the mast. Null fill and beam tilt are accomplished by changing the electrical length of the feed cables.

Panel

The cross dipole panel antenna consists of dual dipole, fed in space and time quadrature, a necessary condition for circular polarization. The transmission circularly polarized (TCP) series uses dipoles, as shown in Figure 6.8-31a, which are in the form of a clover leaf mounted on the front of a ground plane or panel. The four elements generate the required cos(θ)

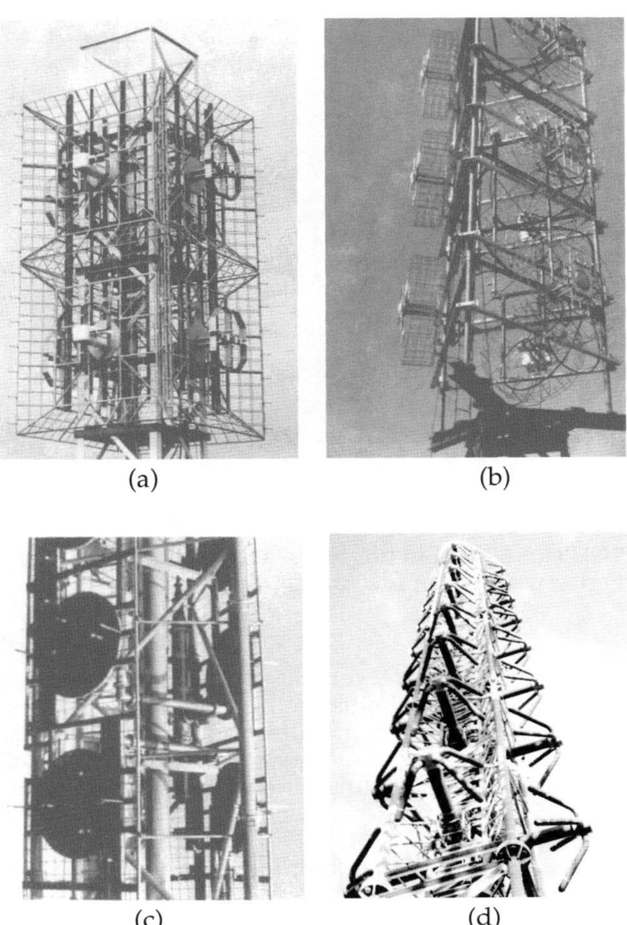

FIGURE 6.8-31 Panel antennas: (a) TCP (Dielectric), (b) DCBR (Dielectric), (c) JRP (Jampro), (d) COGWHEEL™. (Figure courtesy of Electronics Research, Inc.)

element pattern. By tilting the ground plane edges forward, as a partial corner reflector, further shaping can be obtained. The circularly polarized basket reflector (DCBR) series in Figure 6.8-31b consists of fat dipoles mounted in a basket. The element pattern is controlled by the diameter and depth of the basket. The Jampro resonant loop panel (JRP) antenna in Figure 6.8-31c is a large resonant loop mounted against a ground plane. The ring is approximately one wavelength in circumference and hence radiates a circular polarized wave in all planes perpendicular to the panel. The COGWHEEL™ dipole array in Figure 6.8-31d is similar to the other cross dipole arrays except that each set of cross dipoles is fed with a hybrid. This design offers excellent impedance bandwidth and pattern circularity in both three-around and four-around configurations.

FIGURE 6.8-30 Circularly polarized "V" antenna. (Figure courtesy of Dielectric Communications.)

Slotted Coax

The circularly polarized slotted antenna in Figure 6.8-32 is similar to the linear polarized slotted antenna

FIGURE 6.8-32 Circularly polarized slotted antenna. (Figure courtesy of Electronics Research, Inc.)

with the addition of vertical radiators. The vertical radiators are located outside the slotted cylinder, and the energy coupled to them from the slot is in phase quadrature. Any axial ratio is adjustable from elliptical to circular by adjusting the length of the radiators.

FIGURE 6.8-33 UHF pylon antenna. (Figure courtesy of Electronics Research, Inc.)

UHF ANTENNAS

Linear Polarized

Slotted Coaxial

The UHF pylon antenna in Figure 6.8-33 is a coaxial transmission line with radiating slots in the outer conductor. The number of slots (per layer) around the circumference is determined by the desired horizontal pattern—for example, one slot for a skull-shaped pattern, two for a peanut-shaped pattern, three for a tri-lobe pattern, and four or more slots, depending on the outer cylinder diameter, for an omnidirectional pattern. The layers are located at approximately one wavelength spacing along the antenna, with the number of layers determined by the vertical gain and pattern. The radiation parameters of phase and amplitude are determined by a combination of slot length and coupler bar size. This allows discrete control of the illumination along the antenna aperture at every wavelength, resulting in vertical pattern control and shaping. It also allows for maximum aperture efficiency and, in conjunction with the extremely low cross-polarized radiation component of a slot, produces the highest vertical gain for a given antenna length. The antenna shown is a bottom fed traveling-wave resonant antenna. Some antennas launch the energy into the coax radiating section at the center. Others feed the coax radiating section at the center. The bottom coax feed is located inside the radiating coax feed. The pylon uses a radome to cover the entire antenna. Designs are also available where the radome covers only the radiating slots. One version is the omnidirectional bottom fed traveling wave waveguide slot antenna. Because it is waveguide, no inner conductor is required. The signal propagates in the transverse magnetic (a waveguide propagation) mode. The resultant current rings on the inside wall are interrupted by slots cut in the wall. Another version is designed to radiate a cardioid pattern and has a single row of slots. This is also a bottom fed traveling wave waveguide array operating in the TE01 mode. It is built by exciting the fields in a rectangular waveguide and transitioning the waveguide into a cylinder.

Panel Array

The panel array antenna shown in Figure 6.8-34 consists of four dipoles, each spaced slightly greater than one-half wavelength apart. The one-half wavelength spacing reduces the downward radiation and increases the gain. The array has a VSWR less than 1.10:1 from 470 to 800 MHz. Arrays composed of panels can be arranged to produce omnidirectional or directional patterns and can handle four or more high-power channels.

Circularly Polarized

Slotted Coax

The UHF circularly polarized slot antennas are similar to the linear polarized slot antenna. The slot cut in the wall will radiate a horizontally polarized signal. If a

FIGURE 6.8-34 Panel array antenna. (Figure courtesy of IRTE.)

vertical dipole is placed above or near the slot, energy will be coupled to the dipole and reradiate as a vertically polarized component. The number of slots about the periphery of the cylinder can be varied to obtain omnidirectional or directional patterns. The circular polarized pylon antenna in Figure 6.8-35a uses a Z dipole located directly above the slot and radiates a vertical component in phase quadrature with the same elevation pattern as the horizontally polarized slot.

FIGURE 6.8-35 Circular polarized antennas: (a) Z dipole antenna (dielectric), (b) TRASAR™ vertical dipole configuration. (Figure courtesy of Electronics Research, Inc.)

The size and spacing of the graduated dipole can be used to control the amount of vertical component radiation. Another configuration consists of a series of slotted arrays that uses a vertical dipole located between the slots, as shown in Figure 6.8-35b. The pylon circularly polarized type of antennas can be either end fed or center fed. The center fed array can be fed internally for a top mounted location or externally fed for side-mounted installations. All of the traveling-wave slotted arrays have a cylindrical radome covering the full array.

TRANSMISSION LINE COMPONENT TESTS

Testing the transmission line system and components will determine that:

- The transmission line and components are properly assembled.
- The reflections from the antenna and other components at or near the tower top are sufficiently low so no visible ghost or intersymbol interference occurs.
- The impedance presented to the transmitter will result in the maximum transfer of energy.

For an extremely broadband device, such as a coaxial transmission line (which is usually designed to cover the entire, or at least a large portion of, the TV band), time-domain reflectometry (TDR) is the most effective test to determine if the line and components have been properly assembled. This is most often done using a network analyzer with time domain transformation capability.

Return loss or VSWR measurements made at the input of the transmission line measure the combined input impedance at that point only, not the discrete reflections that can cause transmission irregularities (ghosts or intersymbol interference). The TDR, on the other hand, will measure the reflection on a time rather than frequency basis so the exact position and magnitude of the reflection can be determined. Low levels of reflection are generally not in and of themselves a serious problem; however, because the source (a transmitter) is not a good match to the transmission line, the re-reflected power is then radiated by the antenna. This added energy, delayed in time, can be a source of intersymbol interference to digital signals and a multipath-like signal that produce ghosts in NTSC signals.

Generally, the two sources of reflection, assuming that the line itself is reflectionless, that will cause system impairments are:

- Lower elbow complex (V1)
- Upper elbow complex and antennas (V2)

The upper elbow complex and the antenna reflection would be more objectionable than the lower elbow complex because of its distance and time displacement rather than its magnitude. A 5% voltage reflection (VSWR 1.10) echo would not be as objectionable at 0.2 μsec as it would be at 2.0 μsec, as shown in Table 6.8-6.

TABLE 6.8-6
Reflection Comparison

Location of Reflection	Lower Elbow Complex	Antenna
Distance	100 ft	1000 ft
VSWR	1.08	1.04
Reflection	4%	2%
Δf MHz	4.96	0.49
Displacement (μsec)	0.2	2.0

In either case, the digital receiver would be able to compensate for the reflection.

When measured at the input to the system, these can be thought of as two vectors: V1 and V2. Each vector individually will rotate about the preceding vector at a frequency corresponding to the distance from the point of measurement (see Figure 6.8-36). The frequency of rotation can be calculated as follows:

$$\Delta f = \frac{496}{L} \text{ coax}$$

$$\Delta f = \frac{496}{L}\left(\lambda_o/\lambda_g\right) \text{ waveguide}$$

where 496 is half the velocity of light, and L is in feet.

Close reflections (approximately 100 feet) will have a period equivalent to 5 MHz; distant reflections (approximately 1000 feet) will have a period equivalent to 0.5 MHz. The individual and combined reflections are shown in Figure 6.8-37. The use of a network analyzer with TDR is effective in determining the location and magnitude of the antenna system VSWR and sources of reflection. Measurements of VSWR in both frequency and time domains for an antenna and 1700 feet of transmission line (waveguide) are shown in Figure 6.8-38. The near and far reflections are quite apparent in the time domain.

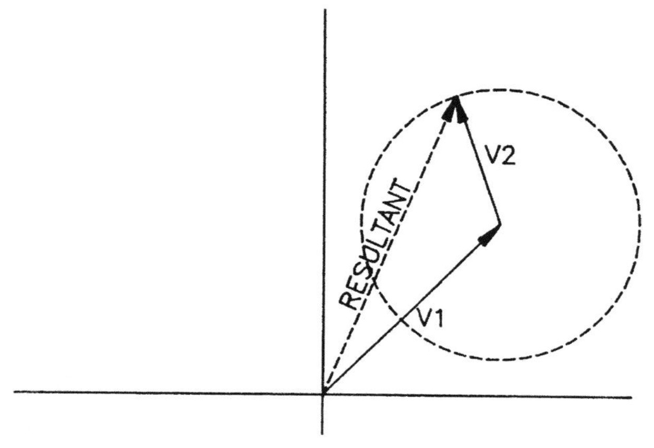

FIGURE 6.8-37 Vector relationship.

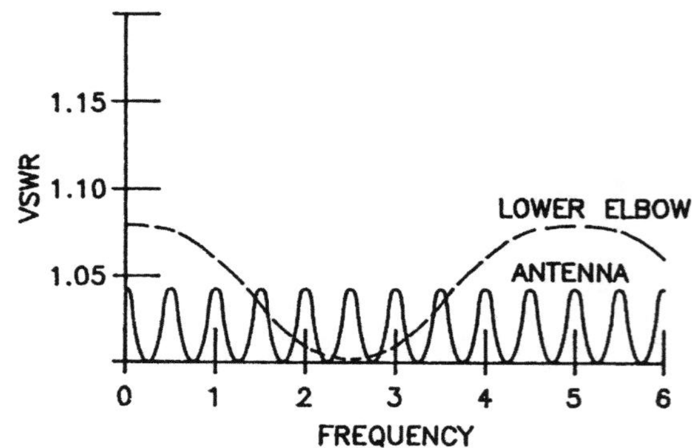

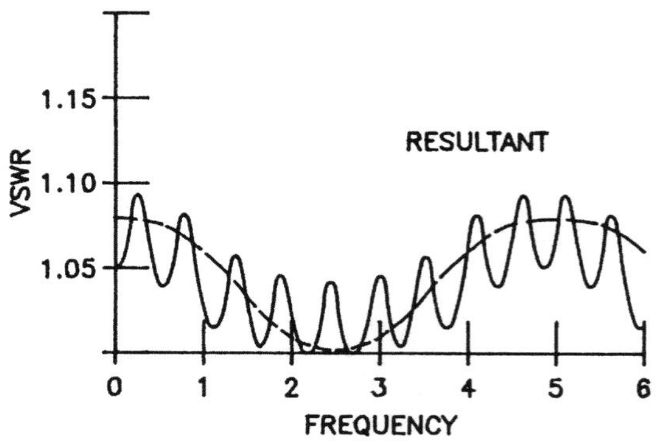

FIGURE 6.8-36 Example of combined reflections.

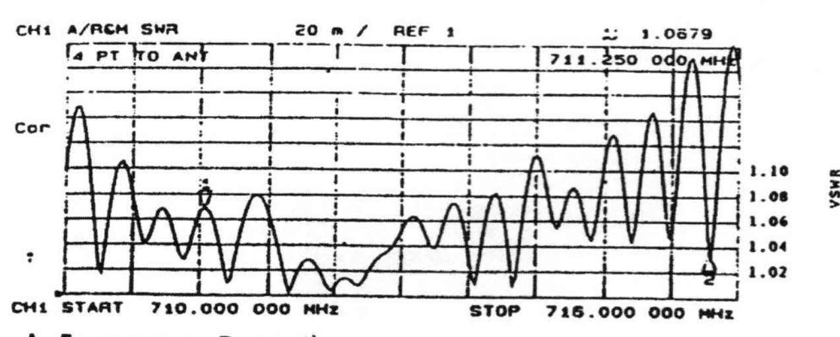

A. Frequency Domain.

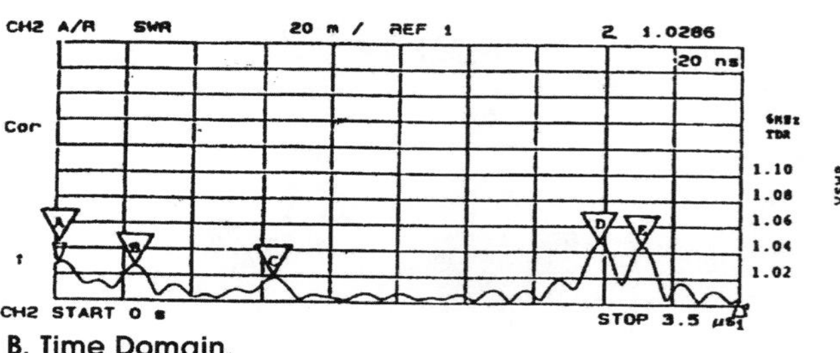

B. Time Domain.

FIGURE 6.8-38 VSWR of antenna and waveguide (1700 feet) in the frequency domain and time domain.

ANTENNA TESTS

Antennas must be tested to determine if the requirements for impedance and pattern are met. Impedance tests are usually run on all production antennas. Pattern tests are normally run on prototype and custom antennas. Antennas should always be impedance tested before shipment. These measurements should be made with the antenna completely assembled and in an area free of reflections. Using a suitable test instrument, such as a network analyzer, a determination of the antenna return loss characteristic can be made. Once the antenna arrives at the site and is installed, sweeping of the antenna and transmission system can determine proper performance. It is valuable to have, for the system sweep, both an input return loss measurement and a TDR measurement. Having these available for the next regularly scheduled inspection of the system allows a basis of comparison. This comparison will aid in the identification of irregularities which, when found and corrected, may prevent a catastrophic failure.

A pattern measurement is conducted for two reasons: (1) to determine the gain as compared with a dipole for which a substitution method could be used, and (2) to determine the amount of radiation at all vertical and horizontal angles that have an influence on the coverage. Pattern tests can be conducted on full- or partial-scale models of the final antenna. Scale models have the advantage of reduced size, which permits high gain antennas to be tested in an anechoic chamber (Figure 6.8-39), free of reflections from the ground or nearby objects that can occur when conducting full-scale tests on a range. Full-scale, single-layer models are practical at UHF channels. The use of near-field techniques for pattern measurements is also practical for production testing.

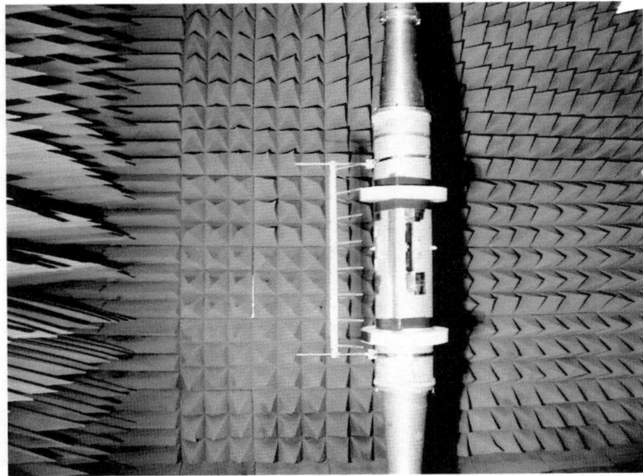

FIGURE 6.8-39 Anechoic chamber. (Figure courtesy of Electronics Research, Inc.)

References

[1] Kraus, J. D., *Antennas*, McGraw-Hill, New York, 1950, pp. 57–126.

[2] Eilers, C. and Sgrignoli, G., Reradiation (echo) analysis of a tapered tower section supporting a side-mounted DTV broadcast antenna and the corresponding azimuth pattern, *IEEE Transactions on Broadcasting*, 47(3), 249–258, 2001.

[3] Giardina, J., Vaughan, T., and Neuhaus, J., True APL levels, in *Proc. of the IEEE Broadcast Technology Society Broadcast Symposium*, 1984.

[4] Masters, R. W., The Superturnstile antenna, *Broadcast News*, January, 1946.

[5] Siukola, M., Predicting performance of Candelabra antenna by mathematical analysis, *Broadcast News*, October, 1957.

[6] Siukola, M., *Evaluation of Circularly Polarized TV Antenna Systems*, BC-24, Institute of Electrical and Electronics Engineers, New York, 1978.

[7] Stenberg, J. and Pries, W., Advances in RF system measurement techniques, in *NAB Broadcast Engineering Conference Proceedings*, National Association of Broadcasters, Washington, D.C., 1989.

[8] Vaughan, T. and Windle, J., Tall towers for super power TV, in *NAB Broadcast Engineering Conference Proceedings*, National Association of Broadcasters, Washington, D.C., 1988.

[9] Wright, R. H. and Hyde, J. V., The Hill-Tower antenna system, *RCA Engineer*, August–September, 1955.

[10] Silliman, T., Jones, E., and Davies, D., Joint rotation on tall towers and the impact on coverage for broadcast transmission systems, in *Proc. of the IEEE 54th Annual Broadcast Symp.*, Washington, D.C., October 13–15, 2004.

[11] FCC, OET-69: *Longley–Rice Methodology for Evaluating TV Coverage and Interference*, Federal Communications Commission, Washington, D.C., 1997.

[12] FCC, OET-65, Evaluating Compliance with FCC Guidelines for Human Exposure to Radiofrequency Electromagnetic Fields, Federal Communications Commission, Washington, D.C., 1997.

[13] OSHA, *Nonionizing Radiation*, 29 CFR 1910.97, Occupational Safety and Health Administration, U.S. Department of Labor, Washington, D.C.

APPENDIX A: USE OF FCC PROPAGATION CHARTS—DTV F(50,90)

To find the value of field strength given the distance and transmitting antenna height, calculate the distance of interest at the depression angle of the elevation pattern:

1. Calculate the ERP at the depression angle relative to the maximum ERP and the associated beam tilt. For example, if the maximum ERP occurs at 0.5° beam tilt and the distance occurs at 1° depression angle, then the ERP at 1° is the relative field squared times the maximum ERP at that point. The calculation is made easier by subtracting the required field (dB) from the maximum ERP (dBk).
2. Determine the desired transmitting antenna height.
3. Locate the point on the chart where the distance curve crosses the tower height line. Read the dBu above 1 kW ERP on the left chart scale.
4. Add to the field strength (dBu) the difference (dB) between the maximum ERP (dBk) and the ERP (dBk) at the depression angle of the elevation pattern.

This calculation assumes a nondirectional azimuth pattern for the antenna under consideration. The calculation can continue in the same manner described for any of the distance/beam tilts as necessary.

As an example, at the maximum 1000 kW ERP for UHF DTV and with an antenna having an elevation gain of 10, at one mile the ERP is 16 dBk for 10° beam tilt. From the F(50,90) curve at one mile (1.61 km) and 1000 feet center of radiation (305 m), the value of predicted field strength is 102 dBu. Adding 16 dB for the ERP at 10° of beam tilt to the base 102 dBu gives a field strength of 118 dBu for a digital signal at the receiving location one mile from the transmitting antenna. See Figure 6.8-40.

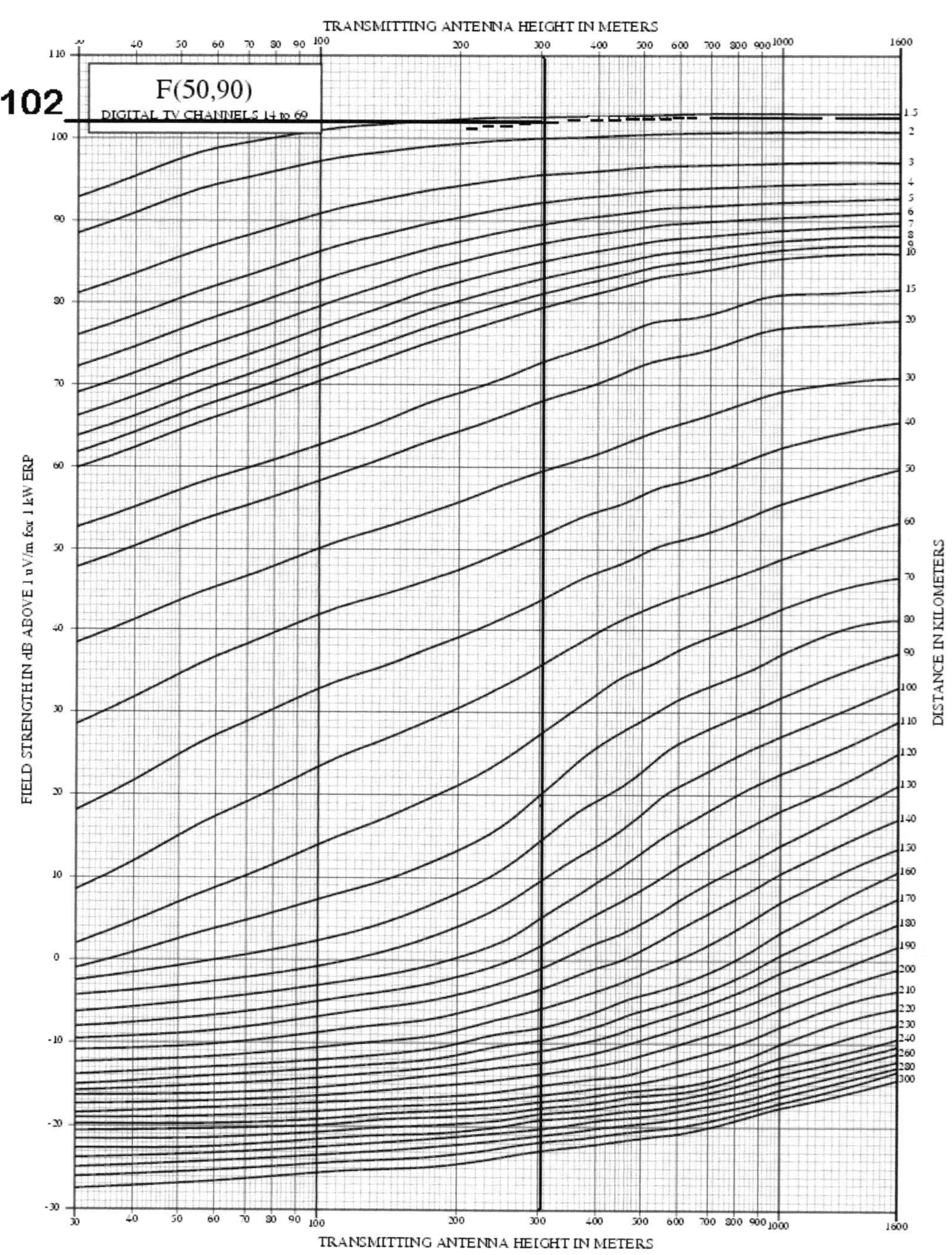

FIGURE 6.8-40 FCC coverage prediction chart, F(50,90).

CHAPTER

6.9

Television Field Strength Measurements

DONALD G. EVERIST
Cohen, Dippell and Everist, P.C.
Washington, D.C.

INTRODUCTION

The Federal Communications Commission (FCC) has always encouraged performing field strength measurements in the VHF and UHF portion of the television band. The design of television systems, whether National Television System Committee (NTSC) or digital television (DTV), depends on the field strength at the receive location. The FCC, through its rulemaking process, determines the signal strength required for the receiver input, incorporates the losses and gain from the transmission line and antenna, and specifies a minimum field strength that the receiver requires for a predictable viewer result. The value required for the receiver input is based on a number of viewing tests performed under controlled laboratory conditions. As stated in the FCC Rules, the signal strength is determined with the receive antenna at 9.1 m (30 ft) in height above ground. The ability to predict the signal level is dependent on many variables. These variables are factors that are present from the transmitter antenna to the receive antenna on what can be described as the receive path. For a comprehensive review of the characteristics and propagation of signals, see Chapter 1.8, "Propagation Characteristics of Radio Waves." Further, when performing field strength measurements using a vehicle with an elevated mast, it is important for field personnel to be familiar with all necessary safety practices (see Chapter 2.6, "Electrical Shock").

PREDICTION MODELS

The FCC uses two general types of prediction models to determine the presence or absence of coverage. The first and historically older prediction model was selected for the purpose of permitting authorization of over-the-air NTSC television stations. The service area contours show the approximate extent of coverage based on average terrain in the absence of interference from other television stations. More recently, the FCC has introduced the Longley–Rice methodology, as contained in the FCC Bulletin OET-69. This version of Longley–Rice point-to-point is terrain dependent. In the analog calculation, the outer Grade B contour is determined by the signal level that is predicted to be exceeded based on F(50,50) (*i.e.*, 50% of the locations 50% of the time) [1]. The statistical representation in predicting the field strength provides a picture that is acceptable to the median observer for 90% of the time at the best of 50% of the locations. The following text provides the definitions used in various FCC publications on VHF, such as FCC/OCE R5 77-01 [2].

PLANNING FACTORS

Grade A Planning Factors

Refer to Table 6.9-1. (In Tables 6.9-1, 6.9-2, and 6.9-3, the unit dBμV represents decibels above 1 microvolt; the unit dBμV/m represents decibels above 1 microvolt per meter.)

TABLE 6.9-1
Grade A NTSC Planning Factors

Planning Factors		Units	Channels 2–6	Channels 7–13	Channels 14–69
1	Thermal noise (at 300 ohms)	dB/μV	7	7	7
2	Receiver noise figure	dB	12	12	15
3	Peak carrier to RMS noise	dB	30	30	30
4	Transmission line loss	dB	1	2	5
5	Receive antenna gain	dB	0	0	–8
6	Dipole factor	dB	–3	6	16
7	Local field F(70,90)	dBμV/m	47	57	65
8	Terrain factor (70%)	dB	4	4	6
9	Time fading factor (90%)	dB	3	3	3
10	Median field F(50,50)	dBμV/m	54	64	74
11	To overcome urban noise	dB	14	7	0
12	Required median field	dBμV/m	68	71	74

Note: In a table of losses, a gain appears as a negative value.

NTSC Grade B Planning Factors

Refer to Table 6.9-2.

DTV Planning Factors

Refer to Table 6.9-3. For digital television, the service area is defined as the geographic area within the noise-limited contour that is predicted to be exceeded based on F(50,90) (*i.e.*, 50% of locations 90% of the time). The following provides the definitions supplied in FCC Bulletin OET-69 [3].

Minimum Height

The FCC has specified, for conducting field measurements, a receive antenna height above ground of 9.1 m (30 ft). The F(50,50) propagation curves are based on

TABLE 6.9-2
NTSC Grade B Planning Factors

Planning Factors		Units	Channels 2–6	Channels 7–13	Channels 14–69
1	Thermal noise (at 300 ohms)	dB/μV	7	7	7
2	Receiver noise figure	dB	12	12	15
3	Peak vis. car./RMS noises	dB	30	30	30
4	Transmission line loss	dB	1	2	5
5	Receive antenna gain	dB	–6	–6	–13
6	Dipole factor	dB	–3	6	16
7	Local field F(70,90)	dBμV/m	41	51	60
8	Terrain factor (70%)	dB	0	0	0
9	Time fading factor (90%)	dB	6	5	4
10	Median field F(50,50)	dBμV/m	47	56	64
11	To overcome urban noise	dB	0	0	0
12	Required median field	dBμV/m	47	56	64

Note: In a table of losses, a gain appears as a negative value.

TABLE 6.9-3
DTV Planning Factors

Planning Factor	Symbol	Low VHF	High VHF	UHF
Geometric mean frequency (MHz)	F	69	194	615
Dipole factor (dBm–dBμV/m)	K	–111.8	–120.8	–130.8
Dipole factor adjustment	K	None	None	Frequency dependent
Thermal noise (dBm)	N	–106.2	–106.2	–106.2
Antenna gain (dBd)	G	4	6	10
Downlead line loss (dB)	L	1	2	4
System noise figure (dB)	N	10	10	7
Required carrier-to-noise ratio (dB)	C/N	15	15	15

this height as well as the NTSC coverage contours. Also, digital received signal strength is based on the receive antenna height at 9.1 m (30 ft). Although other receive antenna heights can be used for specific applications, it has been found that performing measurements at 9.1 m (30 ft) yields more reliable data.

SAFETY

Performing measurements requires addressing all safety considerations, including every facet of the measurement program, such as:

- Properly mounting all measurement equipment
- Raising and lowering the mast with the calibrated antenna
- Choosing the point selection methodology
- Ensuring that there are no overhead obstructions
- Navigating the measurement vehicle in traffic

Measurements of this type must never be taken at nighttime. In addition, if field strength measurements are going to be performed for extended periods, then appropriate rest intervals should be implemented so operator vigilance can be maintained.

FIELD STRENGTH MEASUREMENTS

Field strength measurements are used in a standard proceeding to assess the transmitting antenna performance or to evaluate a station's coverage in a specific or general area. The FCC Rules provide an excellent basis for developing a measurement program. Field strength measurements are typically performed sequentially using either radials spaced at appropriate intervals or by using the grid method, explained in detail later. If the NTSC station's coverage is under scrutiny, subjective observations are useful [4]. These observations can be accomplished by either recording the signal or by viewing on an appropriate receiver. This additional step of subjective observations can assist in determining whether multipath or other propagation factors are present.

NTSC TV Field Strength Measurements

The FCC has very specific criteria for the submission of NTSC field strength measurements to the FCC. The field strength measurement data can normally be permitted in an appropriate rulemaking with reference to technical standards or if specifically requested by the FCC such as when an issue arises with a neighboring administration. The FCC has two general techniques outlined. One technique is to perform field strength measurements along a radial and the second is to determine the measured signal level over a specific community.

The FCC requires that large-scale maps similar to the U.S. Geological Survey (USGS) 7-1/2 minute scale maps be used. The radials are to be drawn to a point beyond the area of interest. Radials can be shifted to incorporate roads where field strength measurements can be taken and traverse representative types of terrain. A sufficient number of radials should be considered to accomplish the measurement goal. Radial marks at convenient intervals from the transmitter site can be shown on these large-scale maps. Alternatively, coordinates from global positioning system (GPS) devices of sufficient accuracy can also be used. The FCC desires where possible that the measurements be taken at uniform distance increments. When convenient measurement point locations cannot be found at a particular uniform increment, then a location having a similar environment nearest the designated location should be selected.

The NTSC measurement procedure should be made with a professional instrument of known accuracy utilizing a horizontally polarized antenna and shielded transmission line recommended by the instrument manufacturer. The antenna is to be raised to a height of 9.1 m (30 ft). Briefly, the following steps are specified in FCC Rule Section 73.686:

- Instrument calibration is verified.

- The antenna is raised 9.1 m (30 ft) in an area where there are no overhead obstructions or wires.
- The receive antenna is oriented for maximum signal in the direction of the station.
- A mobile run of a minimum of 30.5 m (100 ft) is performed by continuously recording the signal over the measurement path via computer or chart recorder.
- The measurement point is to be located on the previously prepared topographic map. A written record is to be made of each point describing the topography, type of surroundings with obstacles, and other relevant features. The use of digital cameras can assist in that documentation.
- If it is not possible to perform a mobile run, five spot measurements can be performed in a cluster. The spot measurements (cluster measurements) should be performed where possible but within 61 m (200 ft) of the first measurement.
- Measurement reports to the FCC must be in affidavit form and provide the observed information.
- Tables of field strength measurements should be included with the distance from the antenna, ground elevation, and geographic measurements at the midpoint of the measurement path.
- Date, time, and weather should be reported.
- The median field for the measurement locations should be shown.
- Measurement point locations should be indicated on USGS topographic maps.
- A tabulation of all related technical information for the transmitting site should be provided.

DTV Measurements

The FCC has not revised its Rules to specify the methodology to perform DTV field strength observations. In the absence of such guidance, the following general procedure is provided as outlined in the FCC Report Project TRB-00-1 Interim Report [5]. In any field measurement project, the approach can be modified to define a specific objective. Generally speaking, measurements are performed at specific intervals along radials extending out from the transmitter site. A minimum of eight radials is recommended, and the measurement intervals can be at fixed distance increments. Additional measurements can be created if a particular area is to be subject to scrutiny. This can be in the form of a square, the sides of which are determined by the size of the area to be investigated (also known as the grid method). Equidistant grids can be drawn and

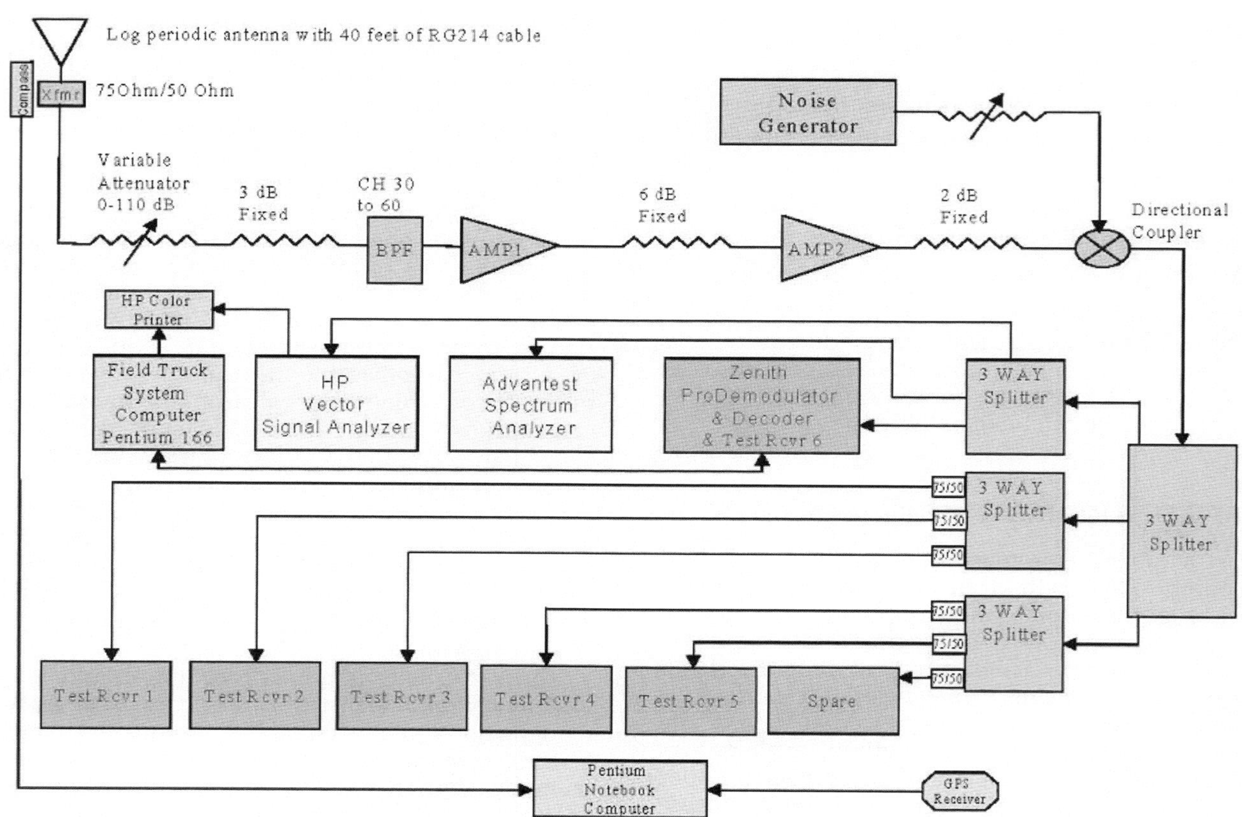

FIGURE 6.9-1 Block diagram for equipment setup for conducting 9.1 m (30 ft) antenna height DTV field tests.

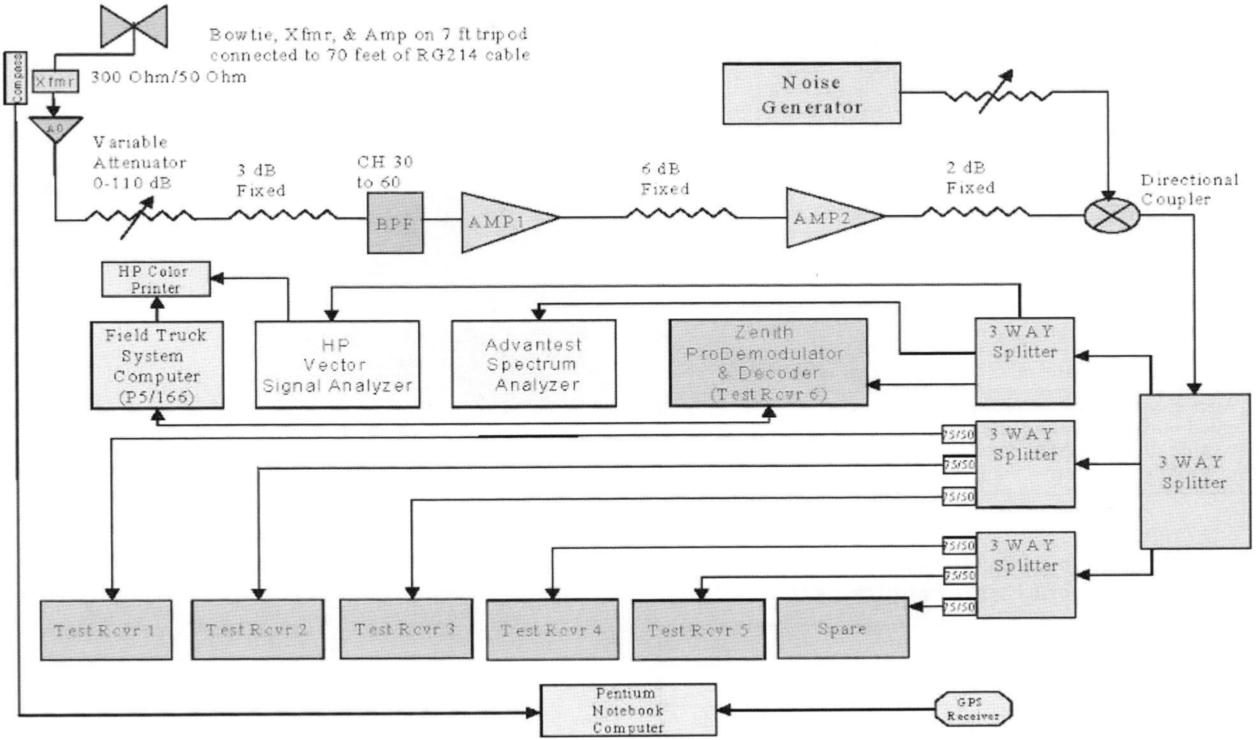

FIGURE 6.9-2 Block diagram for equipment setup for conducting 2.1 m (7 ft) antenna height DTV field tests.

measurements taken at the intersection of the grid lines within the square.

For DTV measurements, in the Interim Report the FCC recommends two heights for the receiving antennas. One system uses a 9.1 m (30 ft) mast and a log-periodic antenna with known electrical performance characteristics. The second system uses a 2.1 m (7 ft) tripod. Two antennas that can be considered are the *bow-tie* and *silver sensor* (set-top, log-periodic-like) directional antennas. This second method uses antennas that may be typically used in an indoor environment. Appendix B in the FCC Report Project TRB-00-1 Interim Report provides specific procedures it used at each test site. These procedures may be used as a guide to developing a DTV measurement program.

A block diagram used by the FCC's staff for the 9.1 m (30 ft) system in this project is shown in Figure 6.9-1 [6]. Similarly, Figure 6.9-2 provides the measurement equipment configuration for the 2.1 m (7 ft) tripod-type measurements [7]. Tables 6.9-4 and 6.9-5 provide lists of the equipment used in Figures 6.9-1 and 6.9-2, respectively.

Figures 6.9-3, 6.9-4, and 6.9-5 are photographs of a vehicle with a mast for 9.1 m (30 ft) height measurements and its interior [8].

TABLE 6.9-4
Equipment List for Figure 6.9-1

2 dB fixed attenuator
3 dB fixed attenuator
Three-way splitters, 50 ohm; 50 to 75 ohm transformers on same outputs
6 dB fixed attenuators
Advantest spectrum analyzer or equivalent
AMP1 (amplifier 1)
AMP2 (amplifier 2)
BPF (bandpass filter; will vary by desired signal)
Compass
Directional coupler
Field truck system computer
GPS receiver
HP color printer or equivalent
HP vector signal analyzer or equivalent
Log periodic antenna with 40 ft of RG214 cable
Noise generator
Notebook computer
Spare receiver

TABLE 6.9-4 *(continued)*
Equipment List for Figure 6.9-1

Test receiver 1
Test receiver 2
Test receiver 3
Test receiver 4
Test receiver 5
Variable attenuator, 0–110 db
75 to 50 ohm transformer (xfmr)
Zenith ProDemodulator and Decoder (test receiver 6)
Note: An accurate GPS receiver is recommended and should be incorporated into the system documentation. A method of overall system calibration should be developed and performed as often as necessary, with a minimum being daily verification.

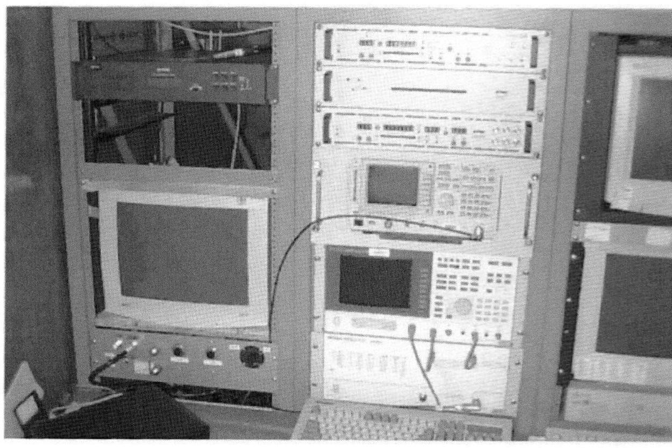

FIGURE 6.9-3 Interior of FCC laboratory field measurement vehicle showing test instrumentation.

TABLE 6.9-5
Equipment List for Figure 6.9-2

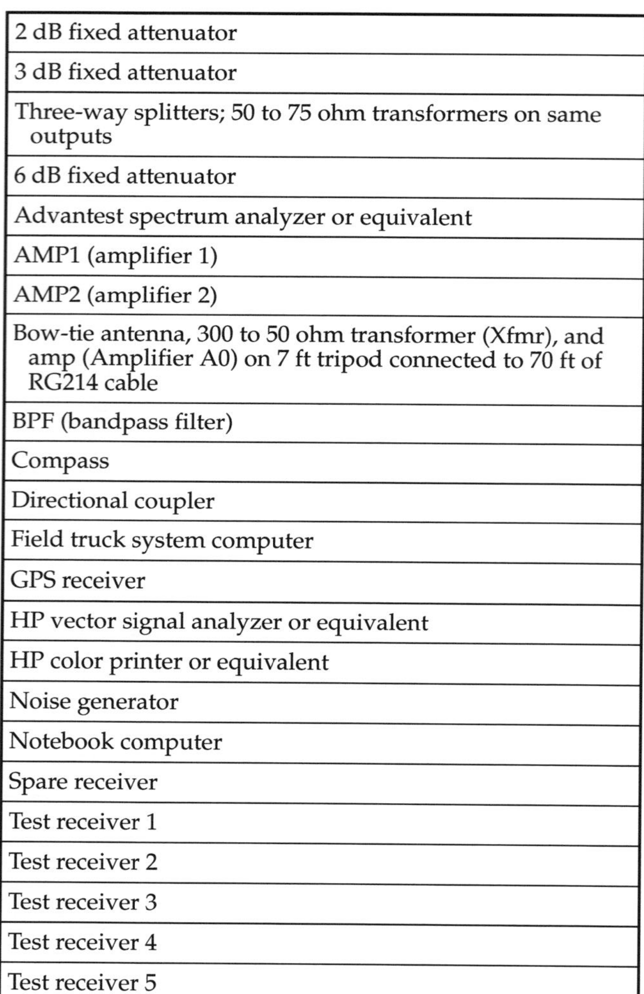

2 dB fixed attenuator
3 dB fixed attenuator
Three-way splitters; 50 to 75 ohm transformers on same outputs
6 dB fixed attenuator
Advantest spectrum analyzer or equivalent
AMP1 (amplifier 1)
AMP2 (amplifier 2)
Bow-tie antenna, 300 to 50 ohm transformer (Xfmr), and amp (Amplifier A0) on 7 ft tripod connected to 70 ft of RG214 cable
BPF (bandpass filter)
Compass
Directional coupler
Field truck system computer
GPS receiver
HP vector signal analyzer or equivalent
HP color printer or equivalent
Noise generator
Notebook computer
Spare receiver
Test receiver 1
Test receiver 2
Test receiver 3
Test receiver 4
Test receiver 5
Variable attenuator, 0–110 dB
Zenith ProDemodulator and Decoder (test receiver 6)

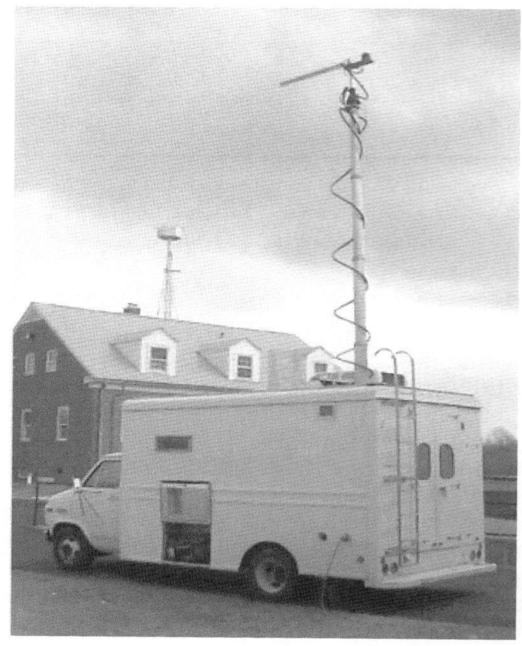

FIGURE 6.9-4 FCC laboratory field measurement vehicle with 9.1 m (30 ft) height antenna.

SATELLITE DELIVERY OF NETWORK SIGNALS

The FCC issued a Report and Order in 1999 with reference to satellite delivery of network of NTSC signals. Several subsequent determinations have been made since 1999, and each is discussed briefly here. The 1999 Report and Order, commonly referred to as the *Satellite Home Viewer Act* (SHVA), established procedures for measuring and predicting television signal intensity at individual household locations. The Report and Order was wide ranging in scope and defined Grade B intensity for determining whether or not a household

FIGURE 6.9-5 Interior of FCC laboratory field measurements vehicle showing screens of multiple receivers.

is served by a local television station. By using the objective Grade B signal as defined by the FCC, SHVA set the standard as to whether or not a household can receive local stations through the use of a conventional outdoor rooftop receiving antenna [9].

An Over-the-Air Signal of a Network Station

SHVA defined whether a household is served or unserved. Most issues addressed in the Report and Order are nontechnical. There are other technical aspects which, while important, are not necessarily ingredients to developing a measurement program. The following provides some of the technical aspects of developing a measurement program that are the basis of the current rules. FCC Rule Section 73.686(d) specifies the method for collection of field strength data to determine television signal intensity at an individual location (*i.e.*, cluster measurements):

(1) Preparation for measurements.

(i) *Testing antenna*. The test antenna shall be a standard half-wave dipole tuned to the visual carrier frequency of the channel being measured.

(ii) *Testing locations*. At the location, choose a minimum of five locations as close as possible to the specific site where the site's receiving antenna is located. If there is no receiving antenna at the site, choose the minimum of five locations as close as possible to a reasonable and likely spot for the antenna. The locations shall be at least three meters apart, enough so that the testing is practical. If possible, the first testing point should be chosen as the center point of a square whose corners are the four other locations. Calculate the median of the five measurements (in units of dBμ) and report it as the measurement result.

(iii) *Multiple signals*. If more than one signal is being measured (*i.e.*, signals from different transmitters), use the same locations to measure each signal.

(2) Measurement Procedure. Measurements shall be made in accordance with good engineering practice and in accordance with this section of the Rules. At each measuring location, the following procedure shall be employed:

(i) *Testing equipment*. Measure the field strength of the visual carrier with a calibrated instrument with a bandwidth of at least 450 kHz, but no greater than one megahertz. Perform an on-site calibration of the instrument in accordance with the manufacturer's specifications. The instrument must accurately indicate the peak amplitude of the synchronizing signal. Take all measurements with a horizontally polarized dipole antenna. Use a shielded transmission line between the testing antenna and the field strength meter. Match the antenna impedance to the transmission line, and, if using an unbalanced line, employ a suitable balun. Take account of the transmission line loss for each frequency being measured.

(ii) *Weather*. Do not take measurements in inclement weather or when major weather fronts are moving through the measurement area.

(iii) *Antenna elevation*. When field strength is being measured for a one-story building, elevate the testing antenna to 6.1 meters (20 feet) above the ground. In situations where the field strength is being measured for a building taller than one story, elevate the testing antenna 9.1 meters (30 feet) above the ground.

(iv) *Antenna orientation*. Orient the testing antenna in the direction that maximizes the value of field strength for the signal being measured. If more than one station's signal is being measured, orient the testing antenna separately for each station.

(3) Written record shall be made and shall include at least the following:

(i) A list of calibrated equipment used in the field strength survey, which, for each instrument, specifies the manufacturer, type, serial number and rated accuracy, and the date of the most recent calibration by the manufacturer or by a laboratory. Include complete details of any instrument not of standard manufacture.

(ii) A detailed description of the calibration of the measuring equipment, including field strength meters, measuring antenna, and connecting cable.

(iii) For each spot at the measuring site, all factors that may affect the recorded field, such as topography, height and types of vegetation, buildings, obstacles, weather, and other local features.

(iv) A description of where the cluster measurements were made.

(v) Time and date of the measurements and signature of the person making the measurements.

(vi) For each channel being measured, a list of the measured value of field strength (in units of dBμ and after adjustment for line loss and antenna factor) of the five readings made during the cluster measurement process, with the median value highlighted.

Prediction Model

SHVA also adopted a refined approach for the determination of signal intensity at discrete locations in lieu of the traditional contour method (Section 73.684 of the FCC Rules) for the prediction of the extent of the Grade B contour. For the prediction model, SHVA selected Longley–Rice Version 1.2.2, also known as *Individual Location Longley–Rice* (ILLR). It is similar to the point-to-point prediction method used for the DTV allocations. Some of the characteristics of the model include the following:

- The time variability factor is 50% (when the time variability factor for the predicted field strength is 50%, an acceptable quality picture should be available 90% of the time) and the confidence variability factor is 50%.
- The model is run in individual mode.
- Terrain elevation is considered every 0.1 km.
- Receiving antenna height is assumed to be 20 ft above ground for one-story buildings and 30 ft above ground for buildings taller than one story.
- Land use and land cover (*e.g.*, vegetation and buildings) are to be included when an accurate method for doing so is developed.
- Where error codes appear, they should be ignored and the predicted value accepted or the result should be tested with an on-site measurement.
- Locations both within and beyond a station's Grade B contour should be examined.
- Predicted interference is included.

Subsequently, the FCC revisited the satellite eligibility issue in ET Docket No. 00-90 [10]. That rulemaking was in response to the enacted legislation Satellite Home Viewer Improvement Act of 1999 (SHVIA) [11].

FCC Notice of Inquiry

The FCC outlined in its Notice of Inquiry (NOI), ET-Docket No. 00-90 (released May 26, 2000), the historical and current significance regarding a signal of Grade B intensity. The NOI describes the signal of Grade B intensity as a discrete value with units of dBμV/m. The NOI further notes that the absolute intensity of any broadcast signal in the VHF and UHF bands cannot be absolutely determined through any current known prediction method so the signal must have a statistical basis. The NOI reiterates the time variability planning factor used in deriving the Grade B standard. The NOI provides insight in Footnote 15 of FCC Docket No. 00-184 which in part states:

> … In the TV Allocations Sixth Report and Order, 41 FCC at 177, the Commission adopted the initial television station allocation rules and stated, "In the case of Grade B service the figures are 90 percent of the time and 50 percent of the locations." See also TV Allocations Third Notice, 16 Fed. Reg. 3072, Appendices A and B. In CS Docket No. 98- 201, supra note 4, both the broadcast and satellite parties stated the time variability factor differently than above. They described the field strength at the Grade B contour as being available to at least 50% of the locations at least 50% of the time. This apparent inconsistency arises from an adjustment the Commission adopted for the Grade B signal strength values when it originally established them. This adjustment results in a Grade B value that predicts reception of an acceptable picture 90% of the time. For example, on channels 2–6, a signal strength of 41 dBμV/m is needed for an acceptable picture. In order for this signal strength to be available 90% of the time, the median or F(50,50) field strength is set at 47 dBμV/m, which includes the addition of a time variability planning factor of 6 dB.

Further, in the NOI the term *median observer* is defined as the observer who ranks as a middle value from all of the observers' data. In the NOI, the FCC sought comment on all the previously recognized planning factors. In addition, it sought comment to incorporate the degradation of the picture due to multipath as a criteria for eligibility. Further, in a separate proceeding, a Notice of Proposed Rulemaking (NPRM) was released on January 20, 2000, which considered signal propagation factors for the ILLR [12].

The FCC adopted the First Report and Order, which presented an improved point-to-point prediction method [13]. Appendix A of this chapter provides the details of Longley–Rice Version 1.2.2, which is to be employed as the revised individual Longley–Rice for FCC Rule Section 73.683(d). The FCC was required to conduct an inquiry to foster technical standards to serve as a basis to determine if a household is unserved within the provisions of the Satellite Home Viewer Extension and Reauthorization Act of 2004 (SHVERA) [14]. Many of the issues in the Notice of Inquiry specified consideration of six specific factors, all of which had previously been addressed by the FCC to some degree [15].

The FCC prepared a report in response to SHVERA [16]. Because a rulemaking is required, insufficient time has elapsed to adopt modifications to the FCC Rules. Certain salient portions of the report are provided here that provide the essence of the FCC inquiry:

- *Test antenna*—The test antenna shall be either a standard half-wave dipole tuned to the center frequency of the channel being tested or a gain antenna, provided its antenna factor for the channel(s) under test has been determined. Use the antenna factor supplied by the antenna manufacturer as determined on an antenna range.
- *Testing locations*—At the test site, choose a minimum of five locations as close as possible to the specific site where the site's receiving antenna is located. If there is no receiving antenna at the site, choose a minimum of five locations as close as possible to a reasonable and likely spot for the antenna. The locations shall be at least three meters apart, enough so that the testing is practical. If possible, the first testing point should be chosen as the center point of a square whose corners are the four other locations. Calculate the median of the five measurements (in units of dBμ) and report it as the measurement result.
- *Multiple signals*—If more than one signal is being measured (*i.e.*, signals from different transmitters), use the same locations to measure each signal.
- *Measurement procedure*—Measurements shall be made in accordance with good engineering practice.
- *Testing equipment setup*—Perform an on-site calibration of the test instrument in accordance with the manufacturer's specifications. Tune a calibrated instrument to the center of the channel being tested. Measure the integrated average power over the full 6 megahertz bandwidth of the television signal. The i.f. of the instrument must be less than 6 megahertz, and the instrument must be capable of integrating over the selected i.f. Take all measurements with a horizontally polarized antenna. Use a shielded transmission line between the testing antenna and the field strength meter. Match the antenna impedance to the transmission line at all frequencies measured, and, if using an unbalanced line, employ a suitable balun. Take account of the transmission line loss for each frequency being measured.
- *Weather*—Do not take measurements in inclement weather or when major weather fronts are moving through the measurement area.
- *Antenna elevation*—When field strength is being measured for a one-story building, elevate the testing antenna to 6.1 meters (20 feet) above the ground. In situations where the field strength is being measured for a building taller than one story, elevate the testing antenna 9.1 meters (30 feet) above the ground.
- *Antenna orientation*—Orient the testing antenna in the direction that maximizes the value of field strength for the signal being measured. If more than one station's signal is being measured, orient the testing antenna separately for each station.
- *Test records*—Written record shall be made and shall include at least the following:
  - — A list of calibrated equipment used
  - — Detailed description of the calibration of the measuring equipment, including field strength meters, measuring antenna, and connecting cable
  - — All factors that may affect the recorded field, such as topography, height and types of vegetation, buildings, obstacles, weather, and other local features for each spot at the measuring site
  - — A description of where the cluster measurements were made
  - — The time and date of the measurements and signature of the person making the measurements
  - — A list of the measured value of field strength (in units of dBμ and after adjustment for line loss and antenna factor) of the five readings made during the cluster measurement process, with the median value highlighted for each channel being measured

SUMMARY

NTSC and DTV field strength measurements are a valuable tool for broadcasters who want to determine exactly how their transmission systems are performing and how their coverage is being affected by such factors as their towers, antennas, the actual propagation path, and other nearby signals. In the case of DTV, these field strength measurements are an optional tool, which, as DTV becomes the primary off-the-air signal to consumer homes, will help resolve many issues that are expected to arise, such as coverage and interference, and which can be satisfied by an appropriately designed field measurement plan.

ACKNOWLEDGMENTS

The author acknowledges Ross Heide, PE; Ryan Felmlee, EIT; and Martin Doczkat, EIT, for their valuable contributions and review.

References

[1] O'Connor, R. A., Understanding television's grade A and grade B service contours, *IEEE Transactions on Broadcasting*, December, 1968, p. 139.

[2] FCC, *A Review of the Technical Planning Factors for VHF Television Service*, Research Standards Division, Office of Engineering, Federal Communications Commission, Washington, D.C., March 1, 1977.

[3] FCC, *OET Bulletin No. 69: Longley–Rice Methodology for Evaluating TV Coverage and Interference*, Federal Communications Commission, Washington, D.C., February 6, 2004.

[4] See *Engineering Aspects of Television Allocation*, Report of the Television Allocations Study Organization, March 16, 1959; *Staff Report on Comparability for UHF Television: Final Report*, UHF Comparability Task Force, Office of Plans and Policy, Federal Communications Commission, Washington, D.C.; Recommendation 654, *Subjective Quality of Television Pictures in Relation to the Main Impairments of Analogue Composite Television Signal*, ITU, CCIR XI-I, Broadcasting Service (Television), Dubrovnik, 1986; Report 478, *Ghost Images in Television*, Questions 6/11 Study Programme 6A/11, ITU, CCIR XI-I, Broadcasting Service (Television), Dubrovnik, 1986.

[5] *A Study of ATSC (8-VSB) DTV Coverage in Washington, D.C., and Generational Changes in DTV Receiver Performance*, April 9, 2001,

Technical Research Branch Laboratory Division, Office of Engineering and Technology, Federal Communications Commission, Washington, D.C.

[6] See page 5 of Project TRB-00-1 Interim Report.

[7] See page 6 of Project TRB-00-1 Interim Report.

[8] *Ibid.*

[9] *In the Matter of Satellite Delivery of Network Signals to Unserved Households for Purposes of the Satellite Home Viewer Act*, Part 73, *Definition and Measurement of Signals of Grade B Intensity*, CS Docket No. 98-201, RM No. 9335, RM No. 9345, Report and Order, Adopted February 1, 1999, released February 2, 1999.

[10] *In the Matter of Technical Standards for Determining Eligibility for Satellite-Delivered Network Signals Pursuant to the Satellite Home Viewer Improvement Act*, ET Docket No. 00-90. Engineering and Technology Office, Federal Communications Commission, Washington, D.C.

[11] See Consolidated Appropriations Act for 2000, Pub. L. 106-113, §1000(9), 113 Stat. 1501 (enacting S. 1948, including the Satellite Home Viewer Information Act of 1999, Title I of the Intellectual Property and Communications Omnibus Reform Act of 1999, relating to copyright licensing and carriage of broadcast signals by satellite carriers, codified in scattered sections of 17 and 47 USC); Section 1008(a) of SHVIA added, *inter alia*, new Section 339 (Carriage of Distant Television Stations by Satellite Carriers) to the Commission's statutory charter, the Communications Act of 1934, 47 USC §151 *et seq.*

[12] *Establishment of an Improved Model for Predicting the Broadcast Television Field Strength Received at Individual Locations*, ET Docket No. 00-11, *Notice of Proposed Rulemaking*, FCC 00-17, released January 20, 2000, at Paragraphs 6–12. Engineering and Technology Office, Federal Communications Commission, Washington, D.C.

[13] First Report and Order, *In the Matter of Establishment of an Improved Model for Predicting the Broadcast Television Field Strength Received at Individual Locations*, ET Docket No. 00-11.

[14] See The Satellite Home Viewer Extension and Reauthorization Act of 2004, Pub. L. No. 108-447, Section 207, 118 Stat 2809, 3393 (2004) (codified at 47 USC Section 339c). The SHVERA was enacted as Title IX of the Consolidated Appropriations Act, 2005. Hereinafter Section 204(b) is cited as codified in 47 USC 339(c).

[15] Notice of Inquiry, *In the Matter of Technical Standards for Determining Eligibility for Satellite-Delivered Network Signals Pursuant to the Satellite Home Viewer Extension and Reauthorization Act*, ET Docket No. 05-182, released May 3, 2005.

[16] Report to Congress, *The Satellite Home Viewer Extension and Reauthorization Act of 2004: Study of Digital Television Field Strength Standards and Testing Procedures*, ET Docket No. 05-182, adopted December 6, 2005; released December 9, 2005.

APPENDIX A: TECHNICAL DATA

This appendix specifies technical details and input parameters that are to be used with Longley–Rice Version 1.2.2 to qualify the latter as the Individual Location Longley–Rice (ILLR) propagation prediction model per Section 73.683(d) of the FCC Rules. The method for including Land Use and Land Clutter (LULC) classifications of locations with attributed clutter loss values is defined here. This appendix will be republished as OET Bulletin No. 72 and included in FCC rules by reference.

Computer code for the Longley–Rice radio propagation prediction model is published in an appendix of NTIA Report 82-100, *A Guide to the Use of the ITS Irregular Terrain Model in the Area Prediction Mode* (G.A. Hufford, A.G. Longley, and W.A. Kissick, U.S. Department of Commerce, April, 1982). The report may be obtained from the U.S. Department of Commerce, National Technical Information Service, Springfield, Virginia, by requesting Accession No. PB 82-217977. Some modifications to the code were described by G.A. Hufford in a memorandum to users of the model dated January 30, 1985. With these modifications, the code is referred to as Version 1.2.2 of the Longley–Rice model. It is available for downloading at the U.S. Department of Commerce website: http://elbert.its.bldrdoc.gov/itm.html.

When run under the conditions given in Table 6.9-A1, the Longley–Rice model is the ILLR prescribed by Section 73.683(d) of the FCC rules. Note especially the following unique features of the ILLR prediction procedure (these distinguish the ILLR model from, for instance, the use of Longley–Rice for digital television coverage and interference calculations as detailed in OET Bulletin No. 69):

- The time variability factor is 50%, presuming that the ILLR field strength prediction is to be compared with a required field (the grade B field intensity defined in Section 73.683(d) of the FCC Rules) that already includes an allowance for long-term (daily and seasonal) time fading.
- The confidence variability factor is 50%, indicating median situations.
- The model is run in individual mode.
- Terrain elevation is considered every 0.1 km.
- Receiving antenna height is assumed to be 6 m (20 ft) above ground for one-story buildings and 9 m (30 ft) above ground for buildings taller than one story.
- In the rare cases that error code 3 occurs (KWX = 3), the predicted field strength is nevertheless accepted as indicative of whether a grade B field strength is available at that location.

- Land use and land cover (*e.g.*, vegetation and buildings) considerations are included through a look-up table of clutter losses additional to those inherent in the basic Longley–Rice 1.2.2 model and keyed to the Land Use and Land Cover categories defined by the U.S. Geological Survey (USGS).

The field strength of a network TV station at an individual location is predicted as follows:

1. Find the engineering facilities data for the network affiliate station of interest by, for example, consulting the FCC website at http://www.fcc.gov/mmb/vsd/. The most accurate source of these data should be used. Necessary data are station latitude and longitude, height above mean sea level of the radiation center, and the effective radiated power (ERP) in the direction of the individual location under study.
2. Run Longley–Rice 1.2.2 in the point-to-point mode with the parameters specified in Table 6.9-A1 to find the propagation path loss relative to free space propagation.
3. Find the USGS Land Use and Land Cover classification of the individual location under study by consulting the LULC database, available from the USGS webpage at http://edcwww.cr.usgs.gov/glis/hyper/guide/1_250_lulc.
4. Convert the USGS Land Use and Land Cover classification to the corresponding ILLR clutter category using Table 6.9-A2, and find the associated clutter loss from Table 6.9-A3.
5. Finally, calculate the ILLR field strength prediction from the formula:

   Field = (Free space field) – (Longley–Rice 1.2.2 path loss) – (ILLR clutter loss)

   where the free space field (in dB) = 106.92 + $10\log_{10}(\text{ERP}) - 20\log_{10}(\text{distance})$, and distance is the path length in kilometers from transmitter to the individual location under study.

HG(1) in Table 6.9-A1 is the height of the radiation center above ground. It is determined by subtracting the ground elevation above mean sea level (AMSL) at the transmitter location from the height of the radiation center AMSL. The latter may be found in the FCC's TV Engineering Database, while the former is retrieved from the terrain elevation database as a function of the transmitter site coordinates also found in the TV Engineering Database. Terrain elevation data at uniformly spaced points between the transmitter and receiver must be provided. The ILLR computer program must be linked to a terrain elevation database with values every 3 arc-seconds of latitude and longitude or closer. The program should retrieve elevations from this database at regular intervals with a spacing increment of 0.1 km (parameter XI in Table 6.9-A1). The elevation of a point of interest is determined by linear interpolation of the values retrieved for the corners of the coordinate rectangle in which the point of interest lies.

TABLE 6.9-A1
Parameter Values for ILLR Implementation of the Longley–Rice Fortran Code

Parameter	Value	Meaning/Comment
EPS	15.0	Relative permittivity of ground
SGM	0.005	Ground conductivity (siemens per meter)
ZSYS	0.0	Coordinated with setting of EN0; see p. 72 of NTIA report
EN0	301.0	Surface refractivity in N units (parts per million)
IPOL	0	Denotes horizontal polarization
MDVAR	1	Individual mode of variability calculations set by code 1
KLIM	5	Climate code 5 for continental temperate
XI	0.1 km	Distance between successive points along the radial from transmitter to individual reception point
HG(1)	See text	Height of the radiation center above ground
HG(2)	6 m or 9 m	Height of TV receiving antenna above ground; use 6 m for one-story building, otherwise 9 m
KWX	Numeric error marker	Output indicating the severity of a possible error due to parameters being out of range; accept the field strength prediction when KWX is 3

TABLE 6.9-A2
Regrouping of LULC Categories for ILLR Applications*

LULC Classification Number	LULC Classification Description		ILLR Clutter Category Number	ILLR Clutter Category Description
11	Residential		7	Residential
12	Commercial and services		9	Commercial/industrial
13	Industrial		9	Commercial/industrial
14	Transportation, communications, and utilities		1	Open land
15	Industrial and commercial complexes		9	Commercial/industrial
16	Mixed urban and built-up lands		8	Mixed urban/buildings
17	Other urban and built-up land		8	Mixed urban/buildings
21	Cropland and pasture		2	Agricultural
22	Orchards, groves, vineyards, nurseries, and horticultural		2	Agricultural
23	Confined feeding operations		2	Agricultural
24	Other agricultural land		2	Agricultural
31	Herbaceous rangeland		3	Rangeland
32	Shrub and brush rangeland		3	Rangeland
33	Mixed rangeland		3	Rangeland
41	Deciduous forest land		5	Forest land
42	Evergreen forest land		5	Forest land
43	Mixed forest land		5	Forest land
51	Streams and canals		4	Water
52	Lakes		4	Water
53	Reservoirs		4	Water
54	Bays and estuaries		4	Water
61	Forested wetland		5	Forest land
62	Non-forest wetland		6	Wetland
71	Dry salt flats		1	Open land
72	Beaches		1	Open land
73	Sandy areas other than beaches		1	Open land
74	Bare exposed rock		1	Open land
75	Strip mines, quarries, and gravel pits		1	Open land
76	Transitional areas		1	Open land
77	Mixed barren land		1	Open land
81	Shrub and brush tundra		1	Open land
82	Herbaceous tundra		1	Open land
83	Bare ground		1	Open land
84	Wet tundra		1	Open land
85	Mixed tundra		1	Open land
91	Perennial snowfields		10	Snow and ice
92	Glaciers		10	Snow and ice

Note: The U.S. Geological Survey (USGS) maintains a database on land use and land cover indicating features such as vegetation and man-made structures. It is often referred to as the LULC database and is available from the USGS web page at http://edcwww.cr.usgs.gov/glis/hyper/guide/1_250_lulc.

*This regrouping into 10 categories for use with the ILLR model was designed by EDX Engineering, Inc.

TABLE 6.9-A3
Clutter Loss as a Function of ILLR LULC Clutter Category and TV Channel

ILLR Clutter Category Number	ILLR Clutter Category Description	Clutter Loss (dB) to Be Added to Longley–Rice Prediction of Path Loss			
		Low-Band VHF Channels 2–5	High-Band VHF Channels 7–13	UHF Band	
				Channels 14–36	Channels 38–69
1	Open land	0	0	4	5
2	Agricultural	0	0	5	6
3	Rangeland	0	0	3	6
4	Water	0	0	0	0
5	Forest land	0	0	5	8
6	Wetland	0	0	0	0
7	Residential	0	0	5	7
8	Mixed urban/buildings	0	0	6	6
9	Commercial/industrial	0	0	5	6
10	Snow and ice	0	0	0	0

CHAPTER

6.10

Fiber-Optic Transmission Systems

JIM JACHETTA

Multidyne Video & Fiber Optic Systems
Locust Valley, New York

INTRODUCTION TO FIBER OPTICS

Fiber-Optic Medium

Fiber optics is a method of carrying information using optical fibers. An optical fiber is a thin strand of glass or plastic that serves as the transmission medium over which information is sent. It thus fills the same basic function as a copper cable carrying a telephone conversation, computer data, or video. Unlike the copper cable, however, the optical fiber carries light instead of electrons. In so doing, it offers many distinct advantages that make it the transmission medium of choice for applications ranging from telephone calls, television, and machine control.

The basic fiber-optic system is a link connecting two electronic circuits. Figure 6.10-1 shows a simple fiber-optic link.

There are three basic parts to a fiber-optic system:

- *Transmitter:* The transmitter unit converts an electrical signal to an optical signal. The light source is typically a light-emitting diode, LED, or a laser diode. The light source performs the actual conversion from an electrical signal to an optical signal. The driving circuit for the light source changes the electrical signal into the driving current.
- *Fiber-optic cable:* The fiber-optic cable is the transmission medium for carrying the light. The cable includes the optical fibers in their protective jacket.
- *Receiver:* The receiver accepts the light or photons and converts them back into an electrical signal. In most cases, the resulting electrical signal is identical to the original signal fed into the transmitter. There are two basic sections of a receiver. First is the detector that converts the optical signal back into an electrical signal. The second section is the output circuit, which reshapes and rebuilds the original signal before passing it to the output.

Depending on the application, the transmitter and receiver circuitry can be very simple or quite complex. Other components that make up a fiber-optic transmission system, such as couplers, multiplexers, optical amplifiers, and optical switches, provide the means for building more complex links and communications networks. The transmitter, fiber, and receiver, however, are the basic elements in every fiber-optic system.

Beyond the simple link, the fiber-optic medium is the fundamental building block for optical communications. Most electrical signals can be transported optically. Many optical components have been invented to permit signals to be processed optically without electrical conversion. Indeed, one goal of optical communications is to be able to operate entirely in the optical domain from system end to end.

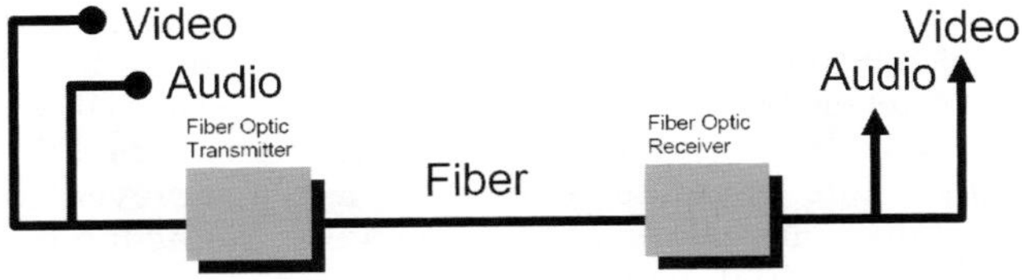

FIGURE 6.10-1 Basic building blocks of a fiber-optic system.

(a) Refraction through the boundary

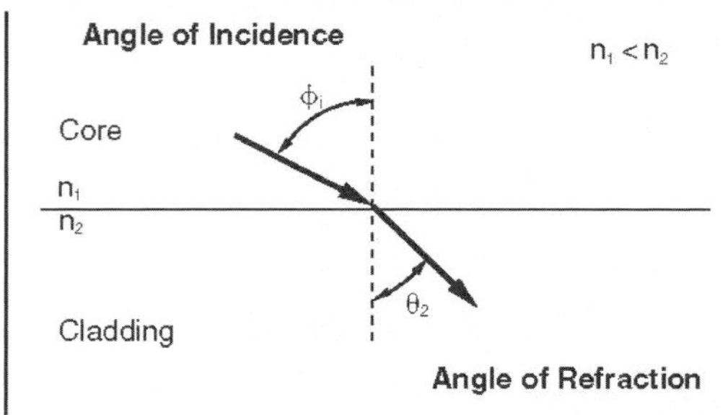

(b) Absorption along the boundary

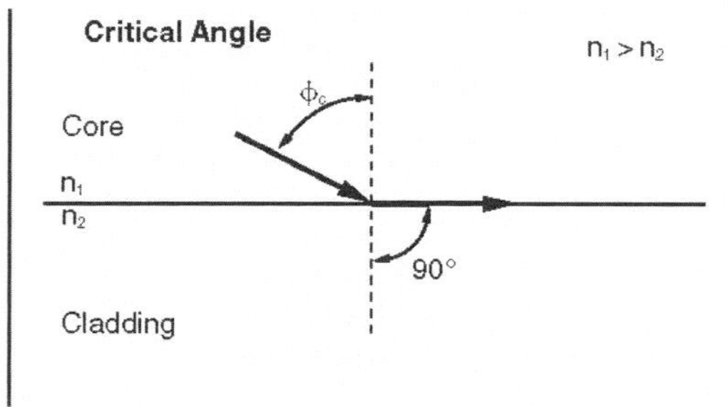

(c) Reflection back into the core

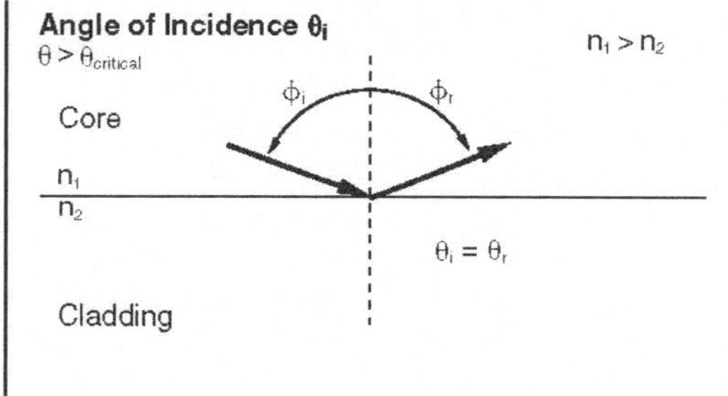

FIGURE 6.10-2 Light wave refraction principles. The refraction index of the core, n_1, is always less than that of the cladding, n_2. Light incident on the boundary at less than the critical angle, φ_1, propagates through the boundary, but is refracted away from the normal to the boundary (a) at the critical angle, φ_C, along the boundary (b). Light incident on the boundary at angles φ_1 above the critical angle is totally internally reflected (c). (Adapted from Force, Inc., illustration used with permission.)

Snell's Law

Early fiber optics exhibited high loss that limited transmission distances. To correct this, glass fibers were developed that included a separate glass coating. The innermost region of the fiber, the core, carried the light, while the glass coating or cladding prevented the light from leaking out of the core by refracting the light back into the inner boundaries of the core. *Snell's Law* explained this concept. It states that the angle at which a light reflects as it passes from one material to another depends on the refractive indices of the two materials.

In the case of fiber optics, this is the refractive index between the core and the cladding. Figure 6.10-2 illustrates the equations for Snell's Law. In this figure, the upper region of the frame, n_1, indicates a higher refractive index than the lower region n_2. The refractive index of the upper region is designated as n_1 while the lower region refractive index is n_2. The figure on the top shows the case with the angle of the indices less than the critical angle. Note that the angle of the light changes at the interface between the higher refractive index, in region 1, and the lower refractive index, in region 2. In the center figure, the angle of indices has increased to the critical angle. At this point all the refracted light rays travel parallel to the interface region. In the figure on the bottom, the angle of indices has increased to a value greater than the critical angle. In this case 100% of the light refracts at the interface region.

Advancements in laser technology next elevated the fiber-optics industry. Only the light-emitting diode or its higher powered counterpart, the laser diode, had the potential to generate large amounts of light in a focused beam small enough to be useful for fiber-optic transport.

Communications engineers quickly noticed the importance of lasers and their higher modulation frequency capabilities. Light has the capacity to carry 10,000 times more information than radio frequencies. Because environmental conditions, such as rain, snow, and fog, disrupt laser light, a transmission scheme other than free space was needed. In 1966, Charles Kao and Charles Hockham, working at the Standard Telecommunications Laboratory, presented optical fibers as an ideal transmission medium, assuming fiber-optic attenuation could be kept under 20 dB per kilometer. Optical fibers of the day exhibited losses of 1,000 dB/km or more. At a loss of 20 dB/km, 99% of the light would be lost over only 1000 meters (3300 ft).

Scientists theorized that the high levels of loss were due to impurities in the glass and not the glass itself. At the time in 1970, an optical loss of 20 dB/km was within the capabilities of electronics and opto-electronic components for short distances (less than 1 km) but not for longer distances (greater than 1 km). Dr. Robert Maurer, Donald Keck, and Peter Schultz of Corning succeeded in developing a glass fiber that exhibited attenuation at less than 20 dB/km, the limit for making fiber optics a usable technology. Other advances of the day, such as semiconductor chips, optical detectors, and optical connectors, initiated the true beginnings of the fiber-optic communications industry.

Optical Windows and Spectrum

Wavelength remains a significant factor in fiber-optic developments. Figure 6.10-3 illustrates the wavelength "windows." Table 6.10-1 shows the wavelength of each optical window and the typical application for multimode (MM) or single-mode (SM) operation.

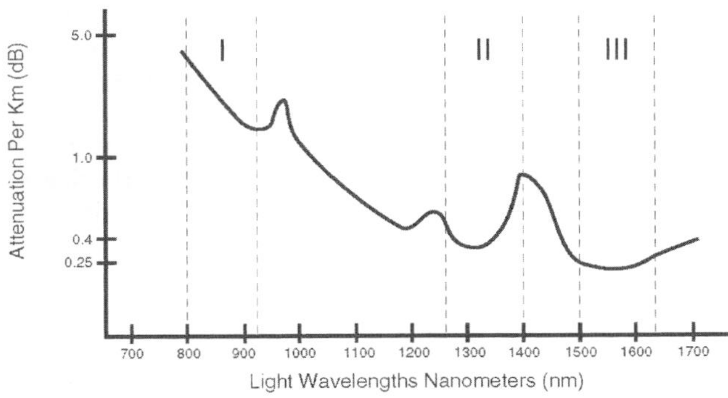

FIGURE 6.10-3 Fiber attenuation versus light wavelength characteristics.

TABLE 6.10-1
De Facto Standard Light Wavelengths

Nominal spectrum (nm)	Fiber Types		
	Window	Multimode (MM)	Single Mode (SM)
850 ± 30 (short wavelength)	I	X	
1300 ± 30 (long wavelength)	II	X	X
1550 ± 30 (extra-long wavelength)	III		X

The earliest fiber-optic systems were developed at an operating wavelength of about 850 nm. This wavelength corresponded to the so called "first window" in a silica-based optical fiber, as shown in Figure 6.10-3. This window refers to the wavelength region that will offer a low optical loss that sits between several large absorption peaks. The absorption peaks are caused primarily by moisture in the fiber and Rayleigh scattering, which is the scattering of light due to random variations in the index of refraction caused by irregularities in the structure of the glass.

The attraction to the 850 nm region came from its ability to use low-cost infrared LEDs and low-cost silicon detectors. As technology progressed, the first window lost its appeal due to its relatively high 3 dB/km losses. Most companies began to exploit the "second window" at 1310 nm with a lower attenuation of about 0.5 dB/km. In late 1977, Nippon Telegraph and Telephone developed the "third window" at 1550 nm. The third window offers an optical loss of about 0.2 dB/km.

The three optical windows—850 nm, 1310 nm, and 1550 nm—are used in many fiber-optic installations today. The visible wavelength near 660 nm is used in low-end, short-distance systems. Each wavelength has its advantages. Longer wavelengths offer higher performance, but always come with higher cost.

Table 6.10-2 provides the typical optic attenuation for each of the common wavelengths versus the fiber-optic cable diameter. A narrower core fiber has less optical attenuation.

The International Telecommunication Union (ITU), an international organization that promotes worldwide telecommunications standards, has specified six transmission bands for fiber-optic transmission. The first is the O band ("original band"), which is from 1260–1310 nm. The second band is the E band ("extended band"), which is 1360–1460 nm. The third band is the S band ("short band"), which is 1460–1530 nm. The fourth band in the spectrum is the C band ("conventional band"), which is 1530–1565 nm. The fifth band is the L band ("longer band"), which is 1560–1625 nm. The sixth band is the U band ("ultra band"), which is 1625–1675 nm. There is a seventh band that has not been defined by the ITU that is in the 850 nm region. It is mostly used in private networks. The seventh band is widely used in high-speed computer networking, video distribution, and corporate applications.

Researchers have attempted to develop new fiber optics that could reduce costs or improve performance. Some alternative fiber materials have found specialized usage. Plastic fiber is ideal for short transmission distances that are ideal for home theater installations. Lower cost glass fiber reduces the need

TABLE 6.10-2
Typical Optical Fiber Loss

Fiber		Optical Loss (dB/km)			
Size (μm)	Type	780 nm	850 nm	1300 nm	1550 nm
9/125	SM	3.0	2.5	0.5–0.8	0.2–0.4
50/125	MM	3.5–7.0	2.5–6.0	0.7–4.0	0.6–3.5
62.5/125		4.0–8.0	3.0–7.0	1.0–4.0	1.0–4.0
100/140		4.5–8.0	3.5–7.0	1.5–5.0	1.5–5.0
110/125			15		
200/230			12		

to develop longer distance plastic fiber and the higher cost of copper wire has expanded glass fiber-optic cable applications.

Types of Fiber-optic Material

There are two distinct parts of a fiber optic cable—the optical fiber that carries the signal and the protective covering that keeps the fiber safe from environmental and mechanical damage. This section deals specifically with the optical fiber.

An optical fiber has two concentric layers called *core* and *cladding*. The core (inner part) is the light-carrying part. The surrounding cladding provides the difference in refractive index that allows total internal reflection of light through the core. The index of refraction of the cladding is less than 1% lower than that of the core. Typical values, for example, are a core index of 1.47 and cladding index of 1.46. Fiber manufacturers must carefully control this difference to obtain the desired fiber characteristics.

Fibers have an additional coating around the cladding. This coating, which is usually one or more layers of polymer, protects the core and cladding from shocks that might affect their optical or physical properties. The coating has no optical properties affecting the propagation of light within the fiber. This coating is just a shock absorber.

Figure 6.10-4 shows the light traveling through a fiber. Light injected into the fiber and striking the core-to-cladding interface at a critical angle reflects back into the core. Since the angles of incident and reflection are equal, the light will again be reflected. The light will continue as expected down the length of the fiber.

Light, however, striking the interface at less than the critical angle passes into the cladding, where it is lost over distance. The cladding is usually inefficient as a light carrier, and light in the cladding becomes attenuated fairly rapidly. The propagation of light is governed by the indices of the core and cladding and by Snell's Law.

Such total internal reflection forms the basis of light propagation through a simple optical fiber. This analysis considers only *meridional rays*, the rays that pass through the fiber center axis each time they are reflected. Other rays, called *skew rays*, travel down the fiber without passing through the axis. The path of the skew ray is typically helical, wrapping around and around the center axis. To simply analyze, skewer rays are ignored in most fiber-optics analysis.

A cone known as the *acceptance cone*, shown in Figure 6.10-5, defines which light will be accepted and propagated by a total internal reflection. Light that enters the core from within this acceptance cone refracts down the fiber. Light outside the cone will not strike the core-to-cladding interface at the proper angle that allows total internal reflection. This light will not propagate.

The specific characteristics of light propagation through fiber depend on many factors. The factors include the size and composition of the fiber as well as the light source injected into the fiber. An understanding of the interplay between these properties will clarify many aspects of fiber optics.

Fiber itself has a very small diameter. Table 6.10-3 provides the core and cladding diameters of four commonly used fibers.

To realize how small these fibers are, note that human hair has a diameter of about 100 μm. Fiber sizes are usually expressed by first giving the core size, followed by the cladding size. Thus, 50/125 means a core diameter of 50 microns (μm) and a cladding diameter of 125 microns (μm).

Optical fibers are classified in two ways. One way is by the material makeup:

- *Glass fiber:* Glass fibers have a glass core and glass cladding. They are the most widely used type of fiber. The glass used in an optical fiber is an ultra pure and transparent silicon dioxide or fused quartz. If ocean water was as clear as fiber, one could see to the bottom of the Marianas Trench in the Pacific Ocean, a depth of 36,000 feet. Impurities are purposely added to the pure class to achieve the desired index of refraction. The elements germanium and phosphorus are added to increase the refractive index of the glass. Boron or fluorine is used to decrease the index. There are other impurities that are not removed when the class is purified. These additional impurities also affect the fiber properties by increasing attenuation from scattering or by the absorbing light.

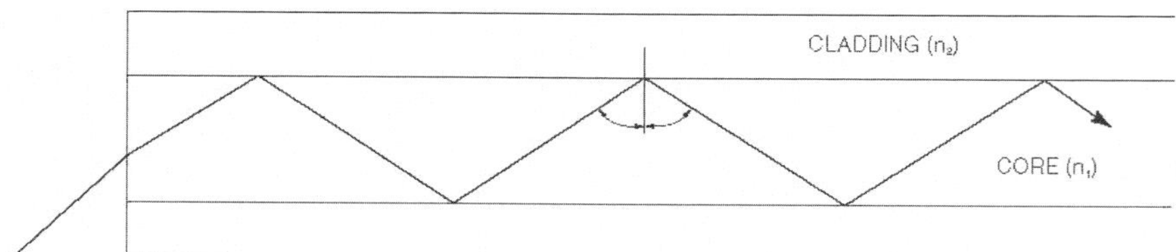

FIGURE 6.10-4 Total internal reflection in an optical fiber. Rays of light incident on the core/cladding boundary at greater than the critical angle, determined by the quotient n_1/n_2, propagate down the fiber's core at a velocity determined by that fiber's value. One ray is shown to keep the diagram simple. (From AMP, Inc., copyright illustration, used with permission.)

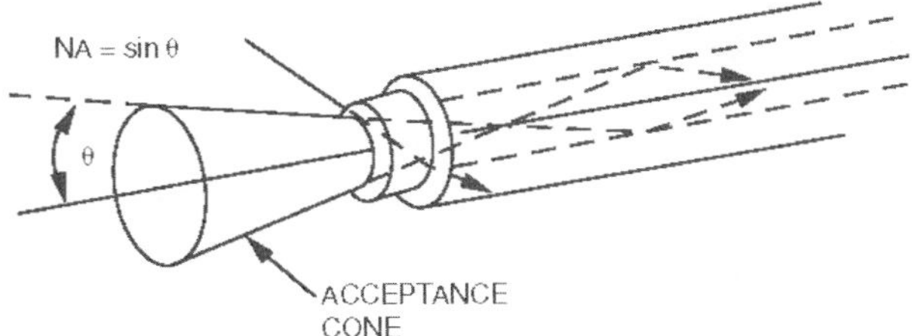

FIGURE 6.10-5 Light ray acceptance cone geometry. The acceptance cone is an imaginary right angle cone extending outward coaxially from the fiber's core. It is a measure of the light-gathering capability of a fiber. Its ray acceptance angle, called the numerical aperture (NA) of the fiber, is uniquely determined by the refractive indices of that fiber's core and cladding. (From AMP, Inc., copyright illustration, used with permission.)

- *Plastic-clad silica (PCS):* PCS fibers have a glass core and plastic cladding. The performance of PCS fiber is limited compared to a fiber made of all glass.
- *Plastic:* Plastic fibers have a plastic core and plastic cladding. Plastic fibers are limited by high optical loss and low bandwidth. The very low cost and ease of use make them attractive for applications where low bandwidth or high losses are acceptable. Plastic and PCS fibers do not have the buffer coating surrounding the cladding.

The second way to classify fibers is by the refractive index of the core and the modes that the fiber propagates. Fiber can be categorized into three general types; Figure 6.10-6 shows the three general fiber types and their basic characteristics.

Figure 6.10-6 shows the difference between the input pulse injected into a fiber and the output pulses exiting the fiber. The decrease in the height of the pulse shows the loss of optical signal power. The broadening of the pulse shows the bandwidth limiting effects of the fibers. It also shows the different paths of rays of light traveling down the fiber. And, it shows the relative index of refraction of the core and cladding for each type of fiber.

TABLE 6.10-3
Core and Cladding Diameters of Four Commonly Used Fibers

Core (μm)	Cladding (μm)
8	125
50	125
62.5	125
100	140

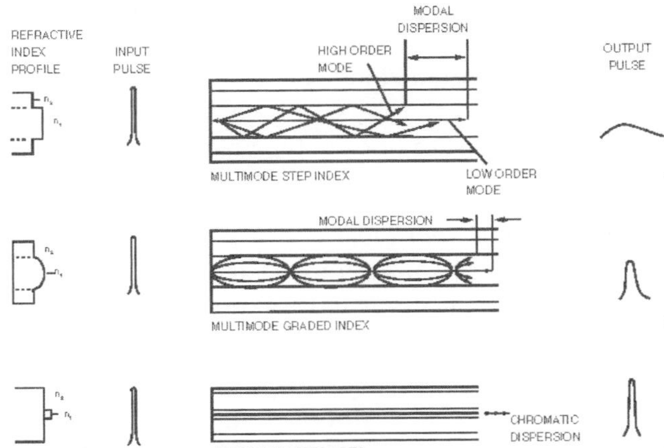

FIGURE 6.10-6 Optical fiber types. The core diameter and its refractive index characteristics determine the light propagation path or paths within the fiber's core. (From AMP, Inc., copyright illustration, used with permission.)

Modes

Mode is a mathematical or physical concept describing the propagation of an electromagnetic wave through any media. In its mathematical form, mode theory derives from Maxwell's equations. James Maxwell first developed mathematical expressions to the relationship between electric and magnetic energy. He proved that they were both a single form of electromagnetic energy, not two different forms as was then commonly believed. His equations also showed that the propagation of electromagnetic energy follows strict rules. Maxwell's equations form the basis of electromagnetic theory.

A mode is a solution to Maxwell's equations. For purposes of this chapter, a mode is simply a path that a ray of light travels down a fiber. The number of modes that a given fiber will support ranges from 1 to over 100,000 individual rays of light. This depends on the physical properties of the fiber and fiber diameter.

Refractive Index Profile

The refractive index profile describes the relationship between the indices of the core and cladding. Two main relationships exist: step index and graded index. The step index fiber has a core with a uniform index throughout. The profile shows a sharp step at the junction of the core and cladding. In contrast, graded index has a nonuniform core. The index is highest at the center of the core and gradually decreases until it matches that of the cladding. Therefore, there is no sharp transition between the core and the cladding. By this classification, there are three types of fibers:

- Multimode step index fiber, commonly called *step index fiber.*

- Single-mode step index fiber, called *single-mode fiber.*
- Multimode degraded index fiber, called *graded index fiber.*

The characteristics of each type have an important bearing on its suitability for particular applications.

Step Index Multimode Fiber

The multimode step index fiber is the simplest type. It has a core diameter from 100–970 microns. This fiber type includes glass, PCS, and plastic fibers. The step index fiber is the most widely used fiber type. This is despite relatively low bandwidth and high losses.

Since light reflects at different angles for different paths, the different rays of light take a shorter or longer time to propagate down the fiber. The ray of light that travels straight down the center of the core arrives at the other end first. Other rays of light arrive later, since they refract back and forth in a zigzag path. Therefore, rays of light that enter the fiber at the same time exit the fiber at different times. The effect is that the light has spread out in time.

This spreading of an optical pulse is called *modal dispersion*. A pulse of light that began as a tight and precisely defined shape has dispersed or spread over time. Dispersion describes the spreading of light by various mechanisms. Modal dispersion is that type of dispersion that results from the varying path lengths of each mode of light as it propagates through the fiber.

The typical modal dispersion for a stepped index fiber ranges from 15–30 ns per kilometer. This means that when rays of light enter a 1 km long fiber at the same time, the ray of light that takes the longest path will arrive 15–30 ns after the ray of light that took the shortest path.

The modal dispersion of 15–30 billionths of a second does not seem to be very much, but dispersion is a fiber's main limiting factor to bandwidth. Pulse spreading results in the overlapping of adjacent pulses, as shown in Figure 6.10-7. Eventually the pulses will merge so that one pulse cannot be distinguished from another. This results in the loss of information. Reducing the modal dispersion in a fiber will increase a fiber's bandwidth.

Graded Index Multimode Fiber

One way to reduce modal dispersion is to use graded index fiber. Here the core has numerous concentric layers of glass, somewhat like the annular rings of a tree. Each successive layer outward from the central axis of the core has a lower index of refraction.

Light travels faster in a lower index of refraction, so the further the light is from the center axis, the greater the speed. Each layer of the core refracts the light. Instead of being sharply refracted as it is in a step index fiber, the light is now bent or continually refracted in almost a sinusoidal pattern. Those rays that follow the longest path by traveling in the outside of the core have a faster average velocity. The light traveling near the center of the core has the slowest average velocity. As a result, all rays tend to reach the end of the fiber at the same time. The graded index reduces modal dispersion to 1 ns per kilometer or less.

Popular graded index fibers have a core diameter of 50 or 62.5 microns and a cladding diameter of 125 microns. The fiber is popular in applications requiring high bandwidth, especially telecommunications, local area networks, computers, and video applications.

Single-Mode Fiber

Another way to reduce modal dispersion is to reduce the core's diameter until the fiber propagates only one mode efficiently. The single-mode fiber has a very small core diameter of only 5–12 microns. The standard cladding diameter is 125 microns. The cladding diameter was chosen for three reasons:

- The cladding must be about 10 times thicker than the core in a single-mode fiber. For a fiber with an 8 or 9 μm core, the cladding should be at least 80 μm.
- It is the same size as a graded index fiber that promotes size standardization.
- It promotes easy handling because it makes the fiber less fragile and because the diameter is reasonably large so that it can be handled by technicians.

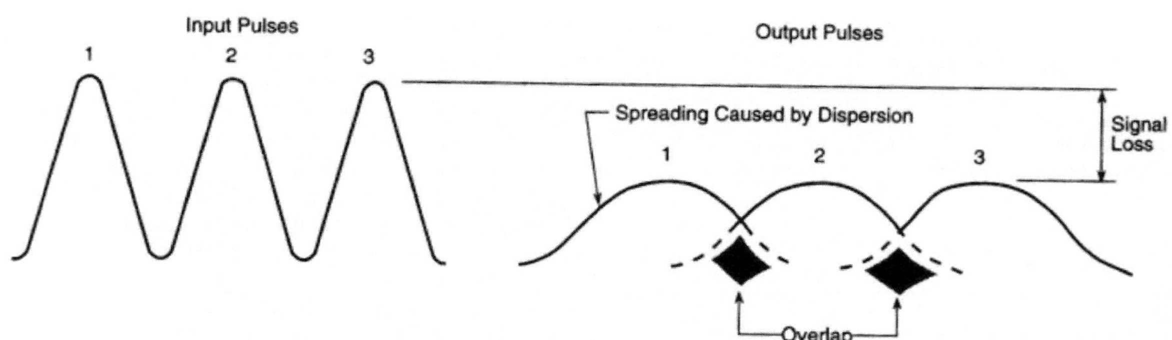

FIGURE 6.10-7 Pulse spreading due to modal dispersion.

Since the single-mode fiber only carries one mode, modal dispersion does not exist. Single-mode fibers have a potential bandwidth of 50–100 GHz-kilometers. Present fiber has a bandwidth of several GHz and allows transmissions of tens of kilometers.

Dispersion-Shifted Single-Mode Fibers

There are three types of single-mode optical fibers usually found in typical applications for telecommunications and data networks. Beyond standard single-mode fibers, there are also dispersion-shifted (DS) fibers and nonzero-dispersion-shifted (NZ-DS) fibers. The purpose of these fibers is to reduce dispersion in the transmission window having the lowest attenuation. Normally, attenuation is lowest in the 1550 nm window and dispersion is lowest in the 1310 nm window. Dispersion shifting creates a fiber that shifts the lowest dispersion to the 1550 nm region. This shifting of dispersion results in a fiber suited for highest data rates and longest transmission distances. In a standard single-mode fiber, the points of lowest loss and highest bandwidth do not coincide. Dispersion shifting brings them closer together.

EARLY APPLICATIONS FOR FIBER OPTICS

The U.S. armed services immediately took advantage of fiber optics to improve its communications and tactical systems. In the early 1970s, the U.S. navy installed a fiber-optic telephone system aboard the *USS Little Rock*. In 1976, the air force followed suit by developing its Airborne Light Optical Fiber Technology (ALOFT) program. The early successes of these applications spawned a number of military research and development programs to create stronger fibers, tactical cables, and ruggedized, high-performance components for applications ranging from aircraft to undersea.

Soon after, commercial applications began to appear. The broadcast television industry was always interested in systems that offered superior video transmission quality. In 1980, broadcasters of the winter Olympics, in Lake Placid, New York, requested a fiber-optic video transmission system for backup video feeds. The fiber-optic feed, because of its quality and reliability, soon became the primary video feed, making the 1980 winter Olympics the first use of fiber optics for a live television production in history.

The telecommunications industries took advantage of this new technology. In 1977, both AT&T and GTE created fiber-optic telephone systems in Chicago and Boston, respectively. Soon after, fiber-optic telephone networks increased in number and reach. Network designers originally specified multimode-grated index fiber, but by the early 1980s, single-mode fiber operating in the 1310 nm and later in the 1550 nm wavelength windows became the standard. In 1983, British Telecom's entire phone system used single-mode fiber exclusively. Computer and information networks slowly moved to fiber. Today fiber is favored over copper due to lighter-weight cables, lightning strike immunity, and the increased bandwidth over longer distances.

In the mid-1980s, the U.S. government deregulated telephone service, allowing small telephone companies to compete with the giant, AT&T. MCI and Sprint led the way by installing regional fiber-optic telecommunications networks throughout the world. Existing natural rights of way, such as railroad lines and gas pipes, allowed these and other companies to install thousands of miles of fiber-optic cable. With this boom, the fiber manufacturer's output capacity struggled to keep up with the demand of the optical fiber needed to increase bandwidth over greater distances.

In 1990, Bell Labs sent a 2.5 gigabit-per-second signal over 7,500 km without regeneration. With the use of soliton lasers and an erbium-doped fiber amplifier (EDFA), the light pulses maintained their shape and intensity. In 1998, Bell Labs researchers transmitted 100 simultaneous optical signals. Each optical signal was at a data rate of 10 Gbps and was transported for a distance of nearly 250 miles. The bandwidth on one fiber was increased to 1 terabit per second. This was achieved using dense wavelength-division multiplexing (DWDM) technology, which allows multiple wavelengths to be combined into one optical signal. Figure 6.10-8 illustrates a basic DWDM system.

DWDM technologies continue to develop as the need for bandwidth increases. The potential bandwidth of fiber is 50 terahertz or better. DWDM technology has decreased greatly in cost and power consumption over the years. The DWDM laser technology requires strict temperature control and compensation. This makes the device draw high amounts of power and adds to the system costs. Today it is still a rather expensive form of optical multiplexing.

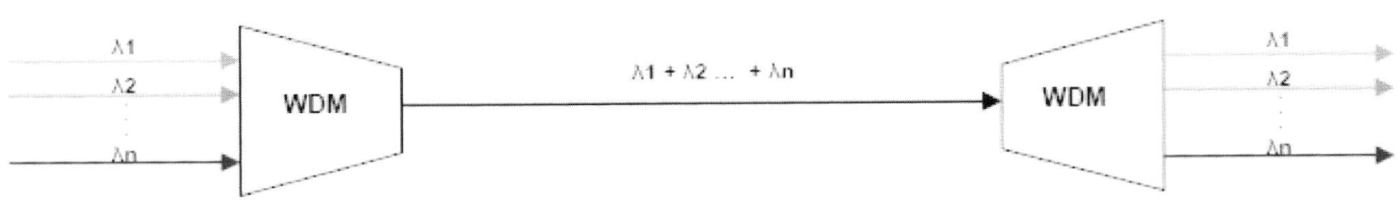

FIGURE 6.10-8 Dense wave-division multiplexing.

More commonly used in the broadcast television industry is coarse wave-division multiplexing (CWDM). CWDM technology gives the ability for up to 18 simultaneous optical signals on one fiber. This gives a usable bandwidth of more than 70 Gbps. CWDM optics are relatively common that operate at 4 Gbps.

The Federal Communications Commission (FCC) mandated that all broadcasters switch from analog to digital transmission, which provides the capacity for high definition (HDTV). This presented researchers with the challenge to provide high bandwidth fiber-optic transport for HDTV. Beyond broadcast television, however, consumers are requesting to have broadband services, including data, audio, and video, delivered to the home.

INFORMATION TRANSMISSION OVER FIBER OPTICS

A fiber-optic cable provides a pipeline that can carry large amounts of information. Copper wires or copper coaxial cable carry modulated electrical signals but only a limited amount of information, due to the inherent characteristics of copper cable.

Free-space transmission, such as radio and TV signals, provides information transmission to many people, but this transmissions scheme cannot offer private channels. Also, the free-space spectrum is becoming a costly commodity with access governed by the FCC. Fiber-optic transmission offers high bandwidth and data rates, but it does not add to the crowded free-space spectrum.

Information Modulation Schemes

The modulation scheme is the manner in which the information to be transported is encoded. Encoding information can improve the integrity of the transmission, allow more information to be sent per unit time, and in some cases, take advantage of some strength of the communication medium or overcome some weakness.

Three basic techniques exist for transmitting information such as video signals over fiber optics:

- Amplitude modulation (AM) includes baseband AM, radio frequency (RF) carrier AM, and vestigial sideband AM.
- Frequency modulation (FM) includes sine wave FM, square wave FM, pulse FM, and FM-encoded vestigial sideband.
- Digital modulation of the optical light source with the ones and zeros of a digital data stream. A simplified explanation is that the light or laser source is off for a digital zero and on for a digital one. In actual practice, the light source never completely shuts off. The light source modulates darker and lighter for digital zero and one information.

ADVANTAGES OF FIBER-OPTIC TRANSMISSION

In addition to fiber optics technical advantages, the cost of materials for Fiber optics is becoming more attractive because the cost of copper wire has risen substantially in recent years.

Longer Distances

A significant benefit of fiber-optic transmission is the capability to transport signals long distances. Basic systems are capable of sending signals up to 5 km over multimode fiber and up to 80 km over single mode without repeaters. Most modern fiber-optic systems transport information digitally. A digital fiber-optic system can be repeated or regenerated virtually indefinitely. An electro-optical repeater or an erbium doped fiber amplifier (EDFA) can be used to regenerate or amplify the optical signal.

Multiple Signals

As discussed in previous sections, fiber has a bandwidth of more than 70 GHz using typical off-the-shelf fiber-optic transport equipment. Theoretically, hundreds, even thousands, of video and audio signals can be transported over a single fiber. This is achieved by using a combination of time-division multiplexing (TDM) and optical multiplexing. Fiber-optic transport equipment is readily available to transport more than 8 video and 32 audio channels per wavelength. Off-the-shelf coarse wave-division multiplexing CWDM equipment easily provides up to 18 wavelengths. This combination of equipment provides up to 144 video and 576 audio channels, as shown in Figure 6.10-9.

Size

Fiber-optic cable is very small in diameter and size when compared to copper. A single strand of fiber-optic cable is about 3 mm. A video coaxial cable is typically much larger. Fiber cable facilitates higher capacity in building conduits. There is often limited space in existing building conduits for infrastructure expansion. In mobile and field productions for sports and news events, fiber is often the cable of choice due to space limitations in a mobile and electronic news-gathering vehicle.

Weight

A fiber-optic cable is substantially lighter in weight than copper cable. A single core PVC-jacketed fiber weighs about 25 pounds per kilometer; RG-6 copper coaxial cable may be three to four times as much.

Noise Immunity

A signal traveling on a copper cable is susceptible to electromagnetic interference. In many applications it

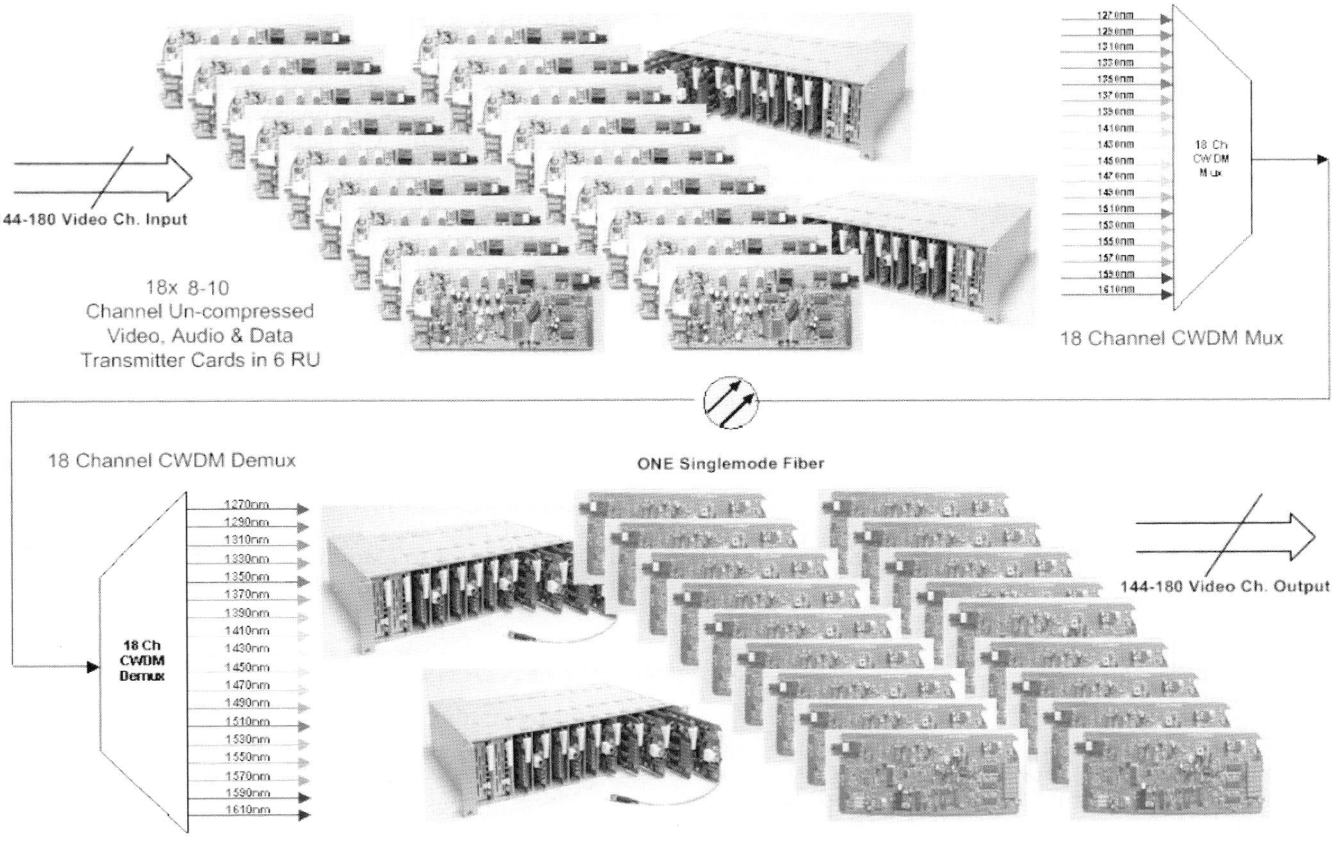

FIGURE 6.10-9 Time-division and optical multiplexing equipment offers substantial capacity for carrying video and audio signals. (Courtesy of Multidyne Video & Fiber Optic Systems.)

is unavoidable to have to route cabling near power substations; heating, ventilating, and air-conditioning (HVAC) equipment; and other industrial sources of interference. A signal traveling as photons in an optical fiber is immune to such interference. The photons traveling down a fiber cable are immune to the effects of electromagnetic interference. In military applications, fiber systems are immune to an electromagnetic pulse (EMP) generated by a nuclear explosion in the Earth's atmosphere. Fiber-optic equipment is used in command and control bunkers to isolate facilities and systems from EMP interference. A fiber-optic signal does not radiate any interference or noise.

Ease of Installation

One of the myths regarding fiber is that it is difficult to install and maintain. This may have been true in the early days, but now it is as simple to terminate an optical fiber with a connector as it is to install a BNC connector on coax. Fiber-optic termination kits are now available that require no epoxy and special polishing. Simple cable stripping tools are used, similar to those used for copper coax, to prepare the fiber for termination. Epoxy-free connectors are available to terminate both multimode and single-mode fiber-optic cable. The connectors are already prepolished. No polishing equipment is needed.

Connector Types

Over the years as fiber-optic communications have grown and changed, there have been many different types of connectors. Today there are four common connector types that are used in most fiber-optic applications (Figure 6.10-10).

The first is the ST connector (Figure 6.10-10(a)). It is a bayonet-style connector similar to a coaxial BNC connector, and is available for single-mode and multimode applications.

The next style is the FC connector (Figure 6.10-10(b)). This connector has a threaded screw–type receptacle. It is similar to an RF-type connector, and is only used for single-mode applications.

The telecommunications industry standardized on the SC connector (Figure 6.10-10(c)). It is a square snap-in–type connector and has gained popularity in the video and computer networking industries.

Telecommunications and networking applications typically require two fibers: one for transmitted data

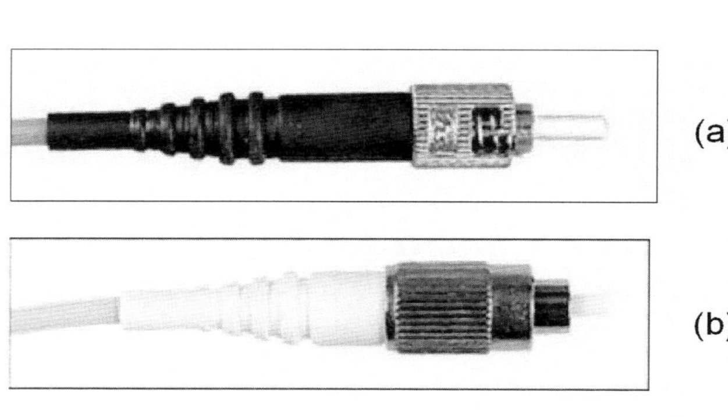

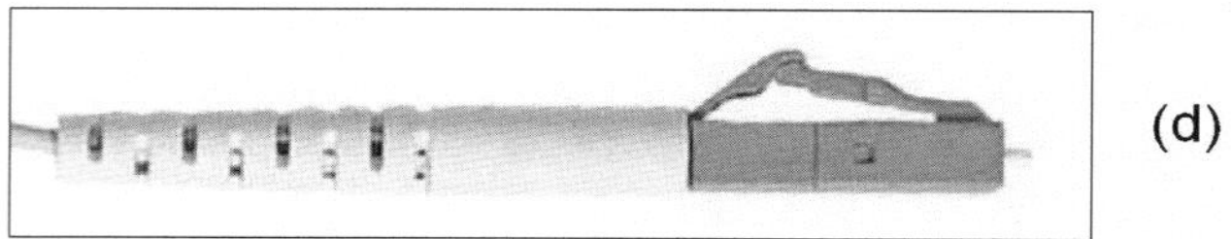

FIGURE 6.10-10 Fiber-optic communications connector types.

and one for received data traffic. Since SC-type connectors were popular in these types of applications, two SC connectors were required. As the size of fiber equipment reduced and the density of fiber-optic input/outputs (I/Os) increased, a small alternative to the SC connector was required. This led to the LC connector as shown in Figure 6.10-10(d). An LC is approximately half the size of an SC connector. It is rectangular in shape and has a locking clip.

Ease of Splicing

Another myth is the repair or maintenance of a broken or cut fiber. The cost of fusion splicing equipment has come down significantly. The fusion splicer is a small portable device that is easily carried in the field.

A fusion splice is easy to perform. First, the fiber is stripped and prepared using simple tools. The fiber is then placed in the fusion-splicing machine. An LCD screen shows the device automatically aligning the fibers. With the press of a button a fusion arc is generated to splice the fibers together. The fusion splicer even tests the connection when complete.

There is now an even simpler way to splice a fiber in the field—mechanical splicing. A mechanical splice consists of a small device that is used to splice a fiber. It is about 2 inches long by 1/2 inch wide. The process involves first stripping the fiber-optic cable and then inserting the ends into the splicing unit with mating gel. A key is used to close and clamp the unit shut. The mechanical splice gives fiber installers the ability to splice and repair with inexpensive equipment in areas where no electrical power is available.

Radiation and Security

Fiber-optic transport is a secure means of communications. Since a fiber-optic cable emits or radiates no RF energy, it is impossible to passively listen or to tap into a fiber-optic circuit. The only way to tap into a fiber-optic cable is to physically cut the cable. An eavesdropper would have to cut the fiber and install a splitter to tap into the fiber-optic link. The cut in the fiber and the inserted splitter can be detected by fiber-optic test equipment.

Environmental Conditions

Fiber-optic cable is immune to most environmental conditions. Fiber-optic cable is capable of tolerating temperature extremes. Unlike copper cable, fiber is immune to moisture. Fiber is available with jacketing that is resistant to nuclear radiation. Many fiber-optic systems are used for the inspection of nuclear reactors. Many military applications require fiber-optic equipment and cable to have resistance to radiation.

END-TO-END SYSTEM DESIGN

A common misconception is that it is difficult to design a fiber-optic system. There are simple calculations to be made using information from the fiber-optic product datasheet. When designing a fiber optic system, it is necessary to know the number and nature of the signals to be carried on the fiber as well as the transmission distance. In other words, an optical budget is to be developed.

Transmitter Launch Power

The datasheet of any fiber-optic transport system will provide the transmitter unit's output optical power. There may be different models with varying levels of output power. A more powerful transmitter can be chosen to reach a further transmission distance. A typical fiber-optic transmitter has an output optical power of –8 dBm or 0.158 mW.

Receiver Sensitivity

The receiver sensitivity is another parameter found on any fiber-optic equipment datasheet. The receiver sensitivity is the minimum optical signal or power required for the receiver unit to operate properly. Many systems have a minimum receiver sensitivity of –28 dBm or 0.00158 mW. The –28 dBm value represents an optical power that is 28 decibels below the 0 dBm or 1 mW reference point.

Optical Power Budget

The optical budget of a fiber-optic transport system takes into account the optical power of the transmitter, loss in the fiber for a given distance, receiver sensitivity, and signal-to-noise required. Optical power, like

electrical power, is measured in watts or milliwatts. Fiber-optic systems are typically designed using decibels referenced to 1 milliwatt or 0 dBm. The following formula shows the conversion from watts to decibels:

$$\text{dBm} = 10 \times \log(\text{laser power in mW}).$$

The output power of an optical laser may be 1 milliwatt. The equivalent power in dBm would be 10 * log (1 mW) = 0 dBm. For 0.5 mW laser, output power would be 10 * log(0.5 mW) = –3 dBm.

The optical attenuation of a multimode fiber at the 850 nm wavelength is about 3 dB/km. The attenuation on single-mode fiber at 1310 and 1550 nm is 0.5 and 0.2 dB/km, respectively. Using these numbers we can calculate how much optical power is required to reach a certain transmission distance. For example, a 10 km run over single-mode fiber at 1310 nm would incur a loss of 5 dB (10 km × 0.5 dB/km).

The optical budget that a fiber-optic system provides is the difference between the fiber-optic transmitter optical output power and the receiver sensitivity. For example, if the transmitter power is –8 dBm and the minimum receiver level is –28 dBm, then the maximum loss the system can withstand is 20 dBm.

In many cases it may seem that a multimode or single-mode fiber run has optical power to reach 40–60 km. When transmissions exceed about 5 km in multimode systems and about 15 km in single-mode systems, other factors due to dispersion come into play and limit the transmission distance.

Bandwidth

The optical losses and usable bandwidth of a fiber-optic system have to be taken into account. As mentioned previously, multimode fibers have greater losses and less bandwidth compared to single mode. Single mode has lower losses and very high bandwidth compared to multimode.

Most manufacturers of multimode fiber-optic cable do not specify dispersion. They will provide a figure of merit known as the bandwidth-length product or just bandwidth with units of MHz-kilometer. For example, 500 MHz-km translates to a 500 MHz signal that can be transported 1 km. The product of the required bandwidth and transmission distance cannot exceed 500:

$$BW \times L \leq 500$$

A lower bandwidth signal can be sent a longer distance. A 100 MHz signal can be sent

$$\begin{aligned} L &= BW - \text{product}/BW \\ &= 500 \text{ MHz-km}/100 \text{ MHz} \\ &= 5 \text{ km} \end{aligned}$$

Single-mode fiber typically has a dispersion specification provided by the manufacturer. The dispersion is specified in picoseconds per kilometer per nanometer of light source spectral width or ps/km/nm. This loosely translates to the wider the spectral bandwidth of the laser light source, the more dispersion. The analysis of dispersion of a single-mode fiber is very complex. An approximate calculation can be made with the following formula:

$$BW = 0.187/(\text{disp} \times SW \times L),$$

where:

disp is the dispersion of the fiber at the operating wavelength with units seconds per nanometer per kilometer.

SW is the spectral width (rms) of the light source in nanometers.

L is the length of fiber cable in kilometers.

For example, with a dispersion equal to 4 ps/nm/km, spectral width of 3 nm, and a transmission length of 20 km, then:

$$BW = 0.187/(4 \times 10^{-12} \text{ s/nm/km}) \times (3 \text{ nm}) \times (20 \text{ km})$$

$$BW = 779{,}166{,}667 \text{ Hz or about 800 MHz.}$$

If the spectral width of the laser light source is doubled to 6 nm the bandwidth will drop to about 390 MHz. This shows how significant the spectral width of the laser source is on the usable bandwidth of a fiber. If a laser light source with a narrow optical spectral width is used, or a fiber with a lower dispersion figure, the bandwidth and transmission distance will increase.

In single-mode fiber communications, there are two basic types of laser light sources. The first type is the less expensive laser that uses Fabre-Perot laser diode (FP-LD) technology. The FP-LD is an inexpensive choice for digital fiber-optic communication. With a spectral width of typically 4 nm or more, it is primarily used for lower bandwidth or short-distance applications. The second is the distributed feedback laser diode (DFB-LD) technology. These light sources are more expensive and are widely used for long-distance fiber-optic communications. The typical spectral width for a DFB laser is about 1 nm. When a DBF laser is used in combination with a low dispersion fiber, the transmission bandwidth and distance can be significantly higher.

See Table 6.10-2, which shows the typical fiber-optic cable losses, and Table 6.10-4, which shows the bandwidth for different types of fiber cable.

Optical Losses

Optical loss or attenuation can vary from 300 to 0.2 dBm/km for plastic or single-mode fibers, respectively. Optical fiber has different loss characteristics at different wavelengths. The optical windows, as mentioned earlier, are regions within the optical fiber spectrum with low loss.

The earliest fiber-optic systems operated in the first optical window in the 850 nm range. The second window is the 1310 nm range, which has zero dispersion. The third window is the 1550 nm window. A multimode fiber has an attenuation of about 4 dB/km at 850 nm and about 2.5 dB/km at 1310 nm. The multimode

TABLE 6.10-4
Typical Fiber-Optic Bandwidth

Fiber		Bandwidth-Distance Product (MHz × km)		
Size (μm)	**Type**	**850 nm**	**1300 nm**	**1550 nm**
9/125	SM	2000	20,000+	4000–20,000+
50/125	MM	200–800	400–1500	300–1500
62.5/125		100–400	200–1000	150–500

fiber spectrum attenuation curve is shown in Figure 6.10-3. Note the high loss regions at 700, 1250, and 1380 nm. The single-mode fiber attenuation curve is shown in Figure 6.10-11. There are high-loss regions at 800, 1100, and 1490 nm regions. The high-loss region at about 1100 nm is called the *mode transition region*. This is where the fiber changes from multimode to single-mode characteristics.

In order to make use of the low-loss properties of a given region in the fiber, the optic light source must generate light at that wavelength. For multimode fiber, light sources are used in the 850 and 1310 nm wavelengths. In single-mode fiber, light sources are typically at 1310 and 1550 nm. CWDM lasers are in the 1470–1610 nm range. The curve in Figure 6.10-11 shows that the fiber has low loss and a flat spectrum at these wavelengths. Corning introduced a CWDM metro fiber that eliminated the high water peak or the high-loss region centered at about 1380 nm. Most single-mode fibers, on new installation, use this flat-spectrum fiber with a usable spectrum from about 1270–1610 nm. The new fiber gives the ability to have up to 18 CWDM wavelengths on one single-mode fiber.

Most video fiber-optic systems take advantage of the 18 usable wavelengths. CWDM is far less expensive than its 42 wavelength counterpart, DWDM. With the fiber-optic systems available with up to 8 channels of video per wavelength, when combined with the

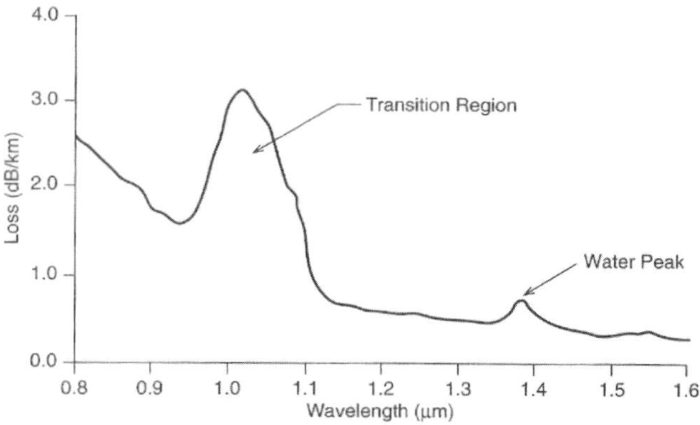

FIGURE 6.10-11 Single-mode fiber attenuation curve. (Courtesy of Corning Glass Works.)

capabilities of CWDM optical multiplexing, more than 144 channels of video can be transported over one fiber.

Plastic fiber is used over short distances due to high attenuation. The visible light region at around 650 nm is used over plastic fiber. Optical attenuation is constant at all bit rates and modulation frequencies. The attenuation in copper cable increases at higher bit rates and modulation frequencies. In a copper cable, a 100 MHz signal will be attenuated more per foot than a 50 MHz signal. This results in distances and bandwidth limitation. In a fiber cable, the 100 Mhz and 50 MHz signals are attenuated the same.

SYSTEM TESTING, TROUBLESHOOTING, AND MAINTENANCE

There are simple procedures to test, troubleshoot, and maintain a fiber-optic system. For basic procedures only simpler inexpensive equipment is required. More sophisticated equipment can be used for advanced analysis.

Measuring Optical Power

The optical power output from a fiber-optic transmitter or fiber cable can be measured with a simple and inexpensive light meter. The light meter is calibrated for each of the three optical windows—850, 1310, and 1550 nm. The meters are available with interchangeable connectors so that systems with any fiber connector type can be tested. The meter gives a reading in milliwatts or dBm.

When troubleshooting a fiber-optic system, the first step is to see if the transmitter unit is sending any optical power. The technician will attach the meter to the transmitter with a fiber patch cord. The output optical power of the transmitter can then be confirmed against the manufacturer's datasheet. If the transmitter is within specification, the next step is the see if light is making it through the fiber to the receiver side.

If the light level output of the transmitter does not meet specifications, this indicates the source of a possible failure. After reconnecting the transmitter back to the fiber, the optical meter is connected to the receiver side of the fiber. Measure the output fiber from the end of the fiber. The theoretical attenuation for the fiber length can be calculated. Using the theoretical attenuation of the fiber and subtracting it from the transmitter optical output power, the power level that should be present at the fiber end near the receiver can be calculated. As long as the optical power level at the received side is higher than the receiver optical sensitivity, the fiber link should operate. If there is a low or no optical signal at the end of the fiber on the receiver side, the fiber may be damaged or have faulty or dirty connectors. Lint-free optical wipes, isopropyl alcohol, and a can of compressed air can be used to clean all optical connectors.

A test of a fiber-optic cable can be performed prior to the purchase of fiber equipment. If a calibrated optical source is not available, handheld calibrated light sources are available as a companion device to the optical power meter. The calibrated light source can be attached to one side of the fiber and the optical meter to the other end.

Optical Time Domain Reflectometer

More extensive tests can be performed with an optical time domain reflectometer (OTDR). An OTDR is a sophisticated device that sends a calibrated light source at a specific wavelength down one end of a fiber. The unit is extremely sensitive and measures the extremely low levels of light reflected back through the fiber.

OTDR works very much like sonar. In sonar an audio tone is bounced off objects. The size of the reflection and the delay determine the size and distance of the objects. As the downstream light beam from the OTDR hits connectors, splices, and other defects in the fiber, it reflects small amounts of light back to the OTDR. Based on the size of the reflection and the time it takes for the reflection to return to the OTDR, the system will provide a calibrated representation of the attenuation and flaws in an optical fiber. The OTDR analysis can be performed at different wavelengths with various modulation schemes. An OTDR analysis is typically only necessary on very long fiber runs with many optical connectors, patch panels, and splices. It is easier to predict optical losses and bandwidth on the majority of fiber-optic cable runs since there will be a minimal number of connectors and splices. Most applications will not require an OTDR.

Cleaning and Maintaining Optical Connectors

Fiber-optic connectors should be cleaned with lint-free optical wipes, and 100% pure isopropyl alcohol should be used with the wipes. Compressed air is also useful to clean any dirt and debris from connectors or receptacles. There are cassette-type cleaning devices that have an advanceable cleaning ribbon. The tips of a male fiber-optic connector are typically ceramic. A protective cap should always be applied to the connector on the fiber cable as well as to the connector on the fiber equipment. This prevents damage and dirt build up.

FIBER-OPTIC TRANSMISSION SYSTEMS

Digital Modulation

The digital bit is the basic unit of digital information. This unit has two values: one or zero. The bit represents the electronic equivalent of the circuit being on or off where a zero equals off and a one equals on. One bit of information is limited to these two values. The digital information is transmitted through the fiber serially one bit at a time.

A digital pulse train represents the ones and zeros of digital information. The pulse train can also depict high and low electrical voltage levels or the presence and absence of a voltage.

Digital in the Television and Video Industries

A digital signal can mean different things to video and cable TV system engineers, causing much confusion. The most common types of digital video and audio are:

- Uncompressed digital video and audio
- Lossless compression of digital video and audio
- Lossy compression of digital video and audio
- Complex digital modulations schemes such as 64 QAM, 256 QAM, 16 VSB, 64 QPSK, etc.
- SONET, ATM, or other telecom base standards
- Serial digital interface (SDI)
- High-definition serial digital interface (HD-SDI)
- Digital audio or AES/EBU

The process of digitizing a standard NTSC video signal is straightforward. The typical bandwidth of a video signal is 4.5 MHz. Typically a sample rate of four times the video bandwidth is used or about 18 megasamples per second. The analog-to-digital (A/D) converter typically has a sampling resolution of 8, 10, or 12 bits.

This process generates a serial digital data stream of about 144–270 Mbps. The video signal is typically encoded in a digital format at the video source or in the video camera. Depending on the digital video format, the analog video will be encoded in one of several standard formats such as 4:2:2, 4:1:1, or 4:2:0.

While these encoding schemes are not referred to as compression, they omit or remove certain information to reduce the systems bandwidth requirement. In the encoding schemes above, the three digits refer to the three common components of video. The first component is luminance (Y) or the light intensity of the video signal. The second is the color signal of red minus luminance (R–Y). The third component is the color signal of blue minus luminance (B–Y). These three components are referred to as YUV. The numbers 4:2:2 have to do with the fact that twice the bandwidth is used for the Y channel than the two color channels. This technique is a form of compression that will be addressed later in the chapter. HDTV or high-definition video requires a data rate of 1.485 Gbps for one uncompressed signal.

The most efficient means of analog video transport utilizes analog to digital conversion. Once video and audio signals are converted to digital information, many channels can be combined into one high-speed data stream using TDM. The high-speed serial digital data stream is then converted to light via a laser or LED.

The receiver unit performs the reverse function. The light or optical signal is received by a PIN photo detector. The optical signal is converted back into a serial data stream. The data stream is demultiplexed using TDM. The digital data is then converted back to video and audio via digital-to-analog (D/A) converters.

Digital video transmission has many advantages over analog transmission. An analog fiber-optic system requires high-linearity optical components that are expensive and require fine tuning and complex calibration procedures. Once a video or audio signal has been digitized, it can be transported via fiber using readily available digital telecom optical components for both multimode and single-mode applications. A digital system has a higher immunity to noise and superior performance characteristics compared to an analog system. A digital signal can be regenerated and repeated virtually indefinitely without signal or performance degradation.

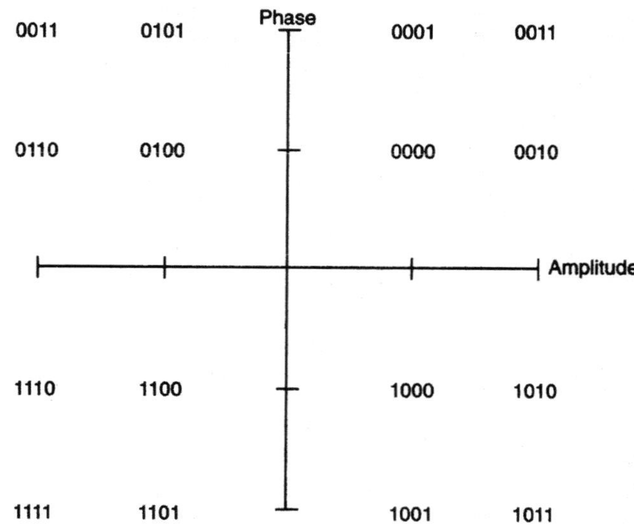

FIGURE 6.10-12 16-QAM encoding phase constellation.

Compressed Digital Video

When compression is introduced into a video transport system, a substantial reduction in bandwidth can be implemented. A digital composite signal requires 144 Mbps and an HD-SDI signal requires 1.485 Gbps. When considering a system that will transport many channels of digital video, an enormous amount of bandwidth is required. A compression system removes redundant or repetitive information from the digital data stream. A compression or transmission encoding scheme will take advantage of limitations in the human eye. The human eye has lower sensitivity or resolution to color detail. Many compression or encoding schemes take this into account and compress or omit certain color details.

There are two basic types of compression systems: lossless and lossy. A lossless compression system does not degrade the video or audio quality. The receiver unit recovers the original uncompressed information. A lossless compression system strictly removes repetitive information from the data stream. Most video content has repetitive information from one video frame to the next. For example, the background image may not change from frame to frame. Therefore, there is no need to send this information repetitive times. Unfortunately, a lossless compression scheme does not offer significant bandwidth savings. A compression rate of three to four times can be expected.

A lossy compression scheme can achieve very high levels of compression but at the cost of image or signal quality. A lossy compression algorithm removes detail from the original image. Once the information has been removed, it cannot be reconstructed. There are many compression and encoding schemes used in video transport. The 4:2:2, 4:1:1, and 4:1:0 encoding schemes mentioned earlier are a technique used to reduce bandwidth. Since the human eye has less sensitivity or less resolution for color, these encoding schemes have less bandwidth for the color information. The human eye has higher resolution horizontally than vertically. When taking this into account, most video formats have a higher horizontal resolution than vertical resolution.

QAM Digital Encoding

Quadrature amplitude modulation (QAM) is a widely used modulation technique for video transport applications, particularly in digital cable TV systems. In a serial digital modulation scheme there are only two informational states: 1 and 0, or on and off. With 256-QAM there are 256 states. The information is encoded by a varying 360 degree quadrature phase and amplitude. This modulation scheme can provide an enormous amount of data throughput in a limited amount of bandwidth, but a higher signal-to-noise band ratio is required. Figure 6.10-12 is the phase constellation for a 16-QAM signal.

MULTIPLEXING

In communication, there are many techniques to transport multiple signals over one transmission medium. These techniques apply to fiber-optic transport.

Time-division Multiplexing

Time-division multiplexing (TDM) is an encoding technique that combines many data streams into one high-speed serial digital stream by combining each data stream in turn on a time basis and converting it into a single data stream. As a result, the single data stream is a sum of the total data from the streams to be multiplexed, plus some overhead bits to organize the data stream.

To produce a serial data stream from a parallel data source, a parallel to serial converter or serializer may be employed that takes, for example, an 8-bit parallel data word and converts it into a 1-bit serial digital signal. If a system with eight 150 Mbps serializers is to be multiplexed together in a TDM system, the output serial data stream will be 8 times 150, or 1200 Mbps.

Optical Multiplexing

Optical multiplexing techniques can be used in addition to the TDM techniques mentioned above. If both optical multiplexing and TDM are combined, very large bandwidths of information can be transported by one fiber.

Wave-Division Multiplexing

Wave-division multiplexing (WDM) is the technique of taking two or more wavelengths or colors of light and combining them onto one fiber. On one end of the fiber the two wavelengths are combined, and then on the other end separated. Basic WDM uses two wavelengths. In multimode the 850 and 1310 nm wavelengths are used. In single mode, 1310 and 1550 nm are used. The two wavelengths can travel in the same direction or in opposite directions.

Coarse Wave-Division Multiplexing

Coarse wave-division multiplexing (CWDM) gives the ability to combine up to 18 wavelengths onto one fiber. The 18 wavelengths are evenly spaced from 1270–1610 nm in 20 nm increments. Each laser source is precisely tuned to a given wavelength to within ±1 nm. What makes CWDM technology possible is extreme temperature stability in laser light sources from 0–70°C without active cooling. CWDM lasers are relatively inexpensive and provide very high and scalable bandwidth. A system could be initially designed and installed using only one wavelength. At any point in the future, up to 17 more wavelengths can be added to increase the system capacity. Components are available to provide both multimode and single-mode CWDM systems.

Dense Wave-Division Multiplexing

Dense wave-division multiplexing (DWDM) takes optical bandwidth and throughput to a higher level. DWDM permits up to 80 wavelengths to share one fiber. The DWDM spectrum is spaced very tightly over a narrow range. The laser systems are complex in order to provide precise wavelength accuracy and stability over temperature. If the wavelength of a laser drifts, it will interfere with an adjacent channel.

The typical DWDM channel spacing is 0.8 nm. The tight spectrum of DWDM permits optical amplification using EDFA. DWDM technology is expensive and is seldom used in video applications. It is rare that an application will require the use of 80 wavelengths. DWDM technology is typically used for fiber-optic systems with long fiber-optic cables between continents and on the ocean floor. DWDM technology is only available for single-mode fiber.

APPLICATIONS FOR VIDEO FIBER-OPTIC TRANSPORT

There are many applications for fiber-optic communications. Any application that requires high bandwidth or high bit rate communications is ideally suited for fiber-optic transport. The television and video industries are a perfect application for fiber-optic transport. Analog television is a relatively high bandwidth signal of more than 5 MHz. Digital television (in particular HDTV) has bit rates of more than 1.5 Gbps. High-resolution computer graphics can have a bandwidth exceeding 500 MHz. All of these television and video applications are ideal for fiber.

Broadcast Television Transmission

As mentioned earlier, television production and broadcast engineers have always sought out the best technology for media events such as the Olympics. In the mid-1980s, fiber-optic transport was introduced into the television industry. Fiber optics are used in all aspects of production and distribution of video and audio signals. The state of the art for the transport of analog video is to use 12-bit video digital encoding. The serial digital bit rate can vary from about 144–300 Mbps.

With the introduction of digital video in the 1990s, fiber-optic transport continued to enjoy growth in the broadcast industry. Digital video was encoded into data rates ranging from 144–360 Mbps. These high bit rate video signals could only travel over copper up to about 300 meters. Transport distance beyond 300 meters with a coax required a repeater (which needs power) or a fiber optic system.

The transition to 100% DTV/HDTV has created a need to transport signals with a bit rate as high as 1.5 Gbps. HDTV using an SDI interface (HD-SDI), in its native or uncompressed form, is 1.485 Gbps. HD-SDI can reach about 150 m using coax cable. Fiber optics are more suitable for transmission distances greater than 150 m.

Systems can be designed using many of the technologies described above. Analog and digital signal transport can be mixed. Time-division and optical multiplexing can be combined.

A broadcast television station may typically reside in a downtown metropolitan area. The television transmitter and satellite up and down links may be on a distant mountaintop outside the city. This situation is a perfect application for fiber transport. The system may require both analog video and digital video since the station may be in the midst of its conversion from analog to digital broadcast. Signals in both directions will be required to support downlink satellite video.

Another typical application is that of back-haul feeds, where many channels of video and audio are trunked together over one fiber. Such a system can use TDM to combine groups of eight channels of video with audio into single wavelengths. The optical multiplexing or CWDM technology is used to combine the wavelengths with groups of eight videos onto one fiber. The combined technique of TDM and CWDM provides a fiber transport capacity of more than 144 video channels on one fiber.

High-Resolution Graphics and Video Transmission

The quality and fidelity of an analog signal over long distances are difficult to maintain over copper. As signals increase in bandwidth and bit rate, it becomes more and more difficult for systems to transport these high bandwidth signals even a short distance over copper. This becomes very apparent when working with high-resolution video and graphics.

A computer-generating RGB-HV or UXGA signal at 1,600 × 1,200 pixels requires an analog bandwidth of close to 500 MHz. If the signal is digitized, it requires a data transport bit rate of 3–4 Gbps. There are many copper-based products that will transport these signals but at a cost in performance and video quality. Figure 6.10-13 shows a typical RGB/UXGA high-resolution video and audio fiber-optic link.

Many applications today require the same video or graphical signal to displayed on a series of monitors. An example may be an airport terminal where arrival and departure information is displayed every 100 ft. This application requires a long daisy-chain of units that can drop and repeat the same signals to each monitor every 100 ft.

Systems are available with the drop-and-repeat or daisy-chain feature. As shown in Figure 6.10-14, one transmitter can send the video signal to the first receiver. The first receiver decodes the optical signal and generates an output for the local monitor. The receiver also repeats and regenerates the optical signal to send to the next receiver in the chain. This technique saves on installation and equipment costs. The alternative would be to run a fiber from each monitor back to the control room. Instead, one fiber can feed many monitors.

FIGURE 6.10-13 Typical RGB/UXGA high-resolution video and audio fiber-optic link. (Photo courtesy of Multidyne Video & Fiber Optic Systems.)

Optical Repeaters and Distribution Amplifiers

There are applications in fiber-optic communications where a signal requires regeneration and replication. The function required is similar to that of a distribution amplifier or digital signal reclocker. A passive splitter can be used to split an optical signal, but each signal is significantly weaker after the split. A device called an optical repeater or distribution amplifier can be used to repeat or regenerate a weak optical signal. This is helpful on long fiber-optic runs where a fiber signal is reaching its limit. The repeater can be used to regenerate the signal for further distribution.

The same device can be used to replicate an optical signal. One optical signal can be replicated up to eight times with one device. Unlike the passive split were the optical output is diminished, the output optical signals are regenerated to full optical power.

The device can also be used as a mode converter or wavelength remapper. The device can be configured with a single-mode input and multimode outputs. This gives the ability to convert from multimode to single mode or from one wavelength to another wavelength. The devise can convert optical signals to CWDM wavelengths.

Broadband Cable Television Transport

Broadband cable television signals are traditionally transported over a hybrid system of fiber and coax. Hundreds of video sources are modulated onto individual carriers and combined into one broadband RF signal. These RF signals can have a frequency bandwidth of 48–870 MHz. This is an enormous bandwidth to transport strictly over coax. Cable systems are designed to transport the high bandwidth signals over fiber to each residential community. The last mile of distribution is then accomplished over coax for delivery to each home. Multiple line amplifiers are used to transport the high bandwidth signals over coax the last mile.

Another application is in a campus environment such as a corporation, university, or military base, where the buildings are spread over some distance. Many of these facilities have an internal cable television system. These facilities will receive a commercial

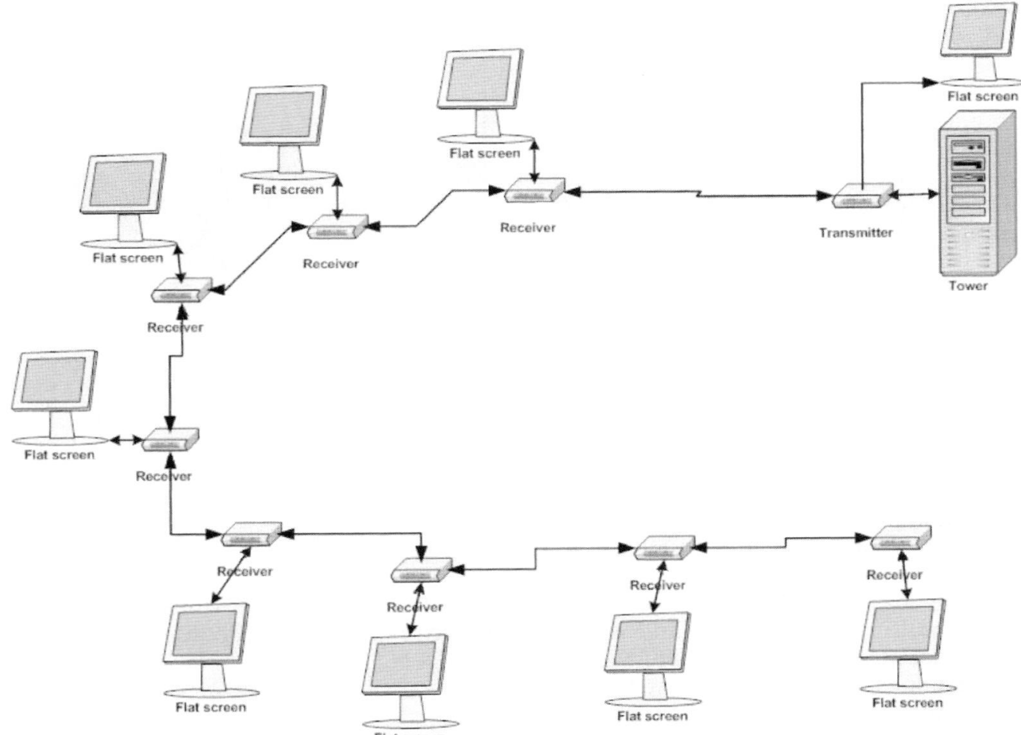

FIGURE 6.10-14 Daisy-chain or drop-and-repeat video fiber transport.

cable TV signal from a local provider. They will strip out the channels of interest, such as news and educational programming. They will then combine these channels with their own internal programming, such as human resource and training channels. The channels are then modulated and combined into one broadband RF signal.

Many of the buildings on the campus may be several miles away. Broadband fiber-optic links can be used to get the cable TV signal to each building. The cable TV is then distributed in each building via coax the last several 100 ft.

Broadband RF and Satellite Link Transport

There are many broadband RF applications for fiber. One important application is satellite uplinks and downlinks. Commercial satellites, for long wavelength applications, typically use an intermediate frequency (IF) signal as a means of communication. The IF signal is typically 70 or 140 MHz. The signal has a limited transport distance over copper coax. The noise level and sensitivity of a satellite system will suffer with the use of copper coax. The relatively high bandwidth and noise sensitivity issues make IF signal transport ideal for fiber.

Consumer and residential satellite service is finding its way into more and more commercial, corporate, and military applications. Consumer satellite equipment is used in many applications as a source of information and news. A large corporation may use consumer satellite instead of cable for news and informational content. In a corporate building or military bunker, the satellite dish may be on the roof many floors above the control or conference room. The L-band signal has a bandwidth from 950–2,150 MHz. A coax run will not transport the L-band satellite signal very far. A consumer satellite dish is small due to a shorter wavelength. It typically has a device called a low-noise block downconverter (LNB). The LNB is an active device that receives the satellite signal and translates and amplifies it to be sent down the coax to the receiver. The LNB requires a DC voltage, which is typically generated by the receiver and sent up the coax. If a fiber link is used, the DC voltage needs to be generated on the dish side of the fiber link. When designing an L-band fiber link, the system needs to provide the appropriate LNB DC power.

FIBER-OPTIC ROUTING SWITCHERS

Just about every broadcast and audio visual system today has some sort of a video and audio routing switcher. The switcher gives the user the ability to control the source and destination of a given video and audio signal. As more and more video and communications migrate from copper to fiber, it makes sense that the need for an optical routing switcher arose. The optical routing switcher is a new concept for the video market, but it has been used for many years in the telecommunications industry. It has been used to route and control telephone traffic.

As video and broadcast television industries become more and more complex with dozens of different video and encoding formats, optical switching starts to make more sense. In broadcast or video application there may be analog video, component video, SDI, and HD-SDI. An optical switch can switch most signals in the optical domain. There are two basic types of optical switching.

Photonic Fiber-Optic Switcher

The first is 100% optical switching using 3D microelectromechanical mirror (MEMS) technology. It uses electronically controlled mirrors to route optical signals. This type of switch has an optical input, an optical cross-point, and an optical output. The abbreviation for this technology is OOO. An OOO switch provides only point-to-point switching. One input cannot be multicast to many outputs. The mirrors cannot point to more than one output at a time. The use of mirrors does permit multiple wavelengths and wavelengths in both directions. Switches are available is sizes from 8 × 8 to 256 × 256. Pure optical switching is available for multimode and single-mode applications. Optical switching supports both analog and digital optical signals.

Pure optical switching is performed using 3D MEMS arrays. Tiny mirrors are fabricated out of silicon. The mirrors are positioned and controlled with electrostatic charges. The core of the optical switch is a 1 inch square cube. The cube has an array of up to 256 input fibers on the left side as shown in Figure 6.10-15. Each fiber has a lens that focuses the optical light onto a MEMS mirror. Each input has its own mirror. On the right side is an array of output fibers. Each output has a MEMS mirror. An optical connection is made when one input mirror aligns with one of the output mirrors.

Fiber-optic switching is ideal for video broadcast, production, security, and other video applications requiring transmission, switching, and replication of high-quality optical signals. The fiber-optic switcher revolutionizes how video is distributed and managed. It is based on state-of-the-art field proven photonic switching technology. Laser light is switched in a pure optical format, without electrical conversion, allowing it to support transparent connections compatible with any video or data format including uncompressed HD video at 1.5 Gbps. Also, since the switching is done optically, the switch eliminates video degradation. With a traditional electrical switcher, electrical-to-optical (EO) and optical-to-electrical (OE) conversions are required that cause signal degradation and jitter.

An optical switch supports a wide range of formats from 19.4 Mbps ATSC through 1.5 Gbps HDTV as well as NTSC, PAL, SECAM, SMPTE 259M serial digital (SDI) video, broadband analog, L-band, IF, and many more. The optical switcher will also transparently switch CWDM and DWDM signals.

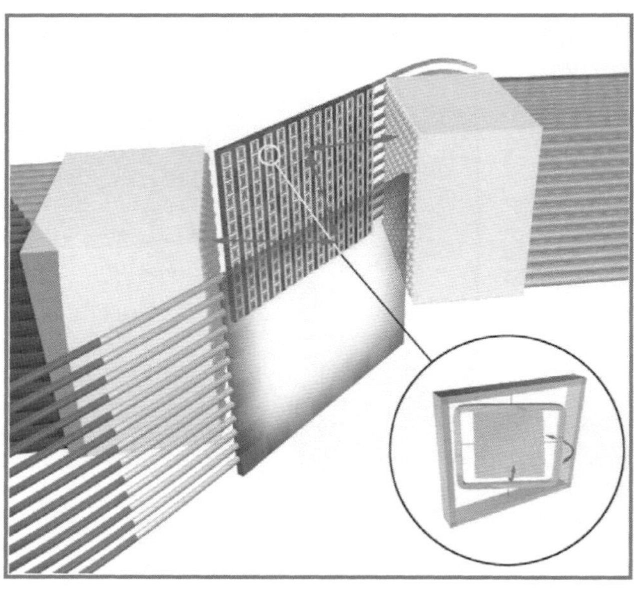

FIGURE 6.10-15 Three dimensional MEMS pure optic switching element. (Photo provided by Calient Networks.)

Optical switcher technology can be used in the field to support applications requiring reliable, high-quality video distribution such as mobile production trucks, sports venues, and professional video facilities; campus video and surveillance networks; remote video monitoring; as well as government and military. Optical layer protection and fault tolerant switching can be configured for mission critical, nonstop applications.

Optical switching is cost effective for any applications requiring 32 or more switched optical ports. It eliminates the need for expensive video transceivers to convert signals between electrical and optical formats. Switching the signals in optical format can substantially reduce the cost per port in fiber-optic transport equipment costs.

Electro-Optical Switch

The second type is the electro-optical switch. The electro-optical switch uses a hybrid approach. The input is optical, the cross-point is electrical, and the output is optical. The abbreviation for this technology is OEO. An OEO switch supports point-to-multipoint or multicast switching. Any input can be switched to every output if necessary. Since the optical signal is converted to electrical, only one wavelength can be switched at a time. Also an electrical cross-point only operates in one direction. Therefore, only one wavelength in one direction is supported.

THE FUTURE OF VIDEO FIBER-OPTIC TRANSPORT

Systems are currently in development for the transport of high-resolution video at bit rates exceeding 10 Gbps. Digital cinema and the proliferation of HDTV television will demand fiber-optic transport systems with high bandwidth capabilities. Fiber transport to the home of video, telephone, and Internet traffic is slowly becoming a reality in many North American communities. This will fuel the demand for high-speed content delivery and distribution throughout the globe.

Bibliography

A Brief History of Fiber Optic Technology, at http://www.fiber-optics.info/fiber-history.htm.

Corning website, at http://www.corning.com/.

Fiber Optics website, at http://www.wetenhall.com/Physics/History.html.

Goff, David R. *Fiber Optic Video Transmission: The Complete Guide*, 1st ed. Boston: Focal Press, 2003.

Goff, David R. *Fiber Optic Reference Guide*, 3rd ed. Boston: Focal Press, 2002.

A Short History of Fiber Optics, at http://www.sff.net/people/Jeff.Hecht/history.html.

Multidyne website, http://www.multidyne.com.

Optical Cable Corporation website, at http://www.occfiber.com/.

Sterling, Donald J. *Technicians' Guide to Fiber Optics*, 4th ed. Delmar Learning, 2004.

Fiber Optics Communications Handbook, 2nd ed. Blue Ridge Summit, PA: TAB Books, April 1991.

CHAPTER

6.11

Satellite Earth Stations and Systems

SIDNEY M. SKJEI
Skjei Telecom, Inc.
Falls Church, Virginia

JAMES H. COOK, JR.
Scientific-Atlanta, Inc.
Atlanta, Georgia

INTRODUCTION

The use of satellite technology by broadcasters is highly developed and in widespread use throughout the world. Satellites are routinely used for broadcast television and radio backhaul and network distribution services, and they provide low cost, reliable operation. These and many other services are made possible due to the unique characteristics of satellites placed in the geosynchronous satellite orbit. The capability of a single quasi-stationary repeater in the sky, visible to large, contiguous regions, offers a wide variety of distribution capabilities for terrestrial broadcast service providers as well as the public in general. Satellite communications are also particularly useful for mobile services, long-distance communication services, and for services across difficult terrain. Analog and digital satellite transmission systems are considered in detail in this chapter.

Satellites in the 22,300 mile high geosynchronous orbit rotate from west to east. They appear fixed in space to earth stations on the ground because they orbit in synchronism with the earth's rotation. A satellite that is closer to the earth orbits faster and one that is beyond synchronous orbit rotates slower than the earth. Satellites located in the geosynchronous orbit have direct lines of sight to almost half of the surface of the earth, as shown in Figure 6.11-1.

Communication by satellite was made possible by parallel advances in space technology and electronics. In 1945, Sir Arthur C. Clarke, the noted British scientist and science fiction writer, proposed relay stations in geostationary orbit for satellite communications. By 1963, due to advances in technology, solid-state electronics, and the thrust capability of rockets, the ability to place a satellite into a stationary orbit was achieved.

Frequency Bands

Communication satellites today operate at many different microwave frequencies. Excluding satellites intended primarily for mobile communications or direct to home satellites (discussed in Chapter 6.15 of this handbook), the frequencies primarily used by U.S. broadcasters are shown in Figure 6.11-2. In the U.S., the domestic commercial communications satellite networks operate in the fixed satellite services (FSS) frequency bands as defined by the Federal Communications Commission (FCC). Most of the domestic systems operate in either the C-band (6 and 4 GHz) or Ku-band (14 and 12 GHz) frequency ranges. The Ka-band frequencies (30 and 20 GHz) are also now beginning to be used for FSS operation, and the use of these frequencies will increase in the future. For completeness, it should also be noted that L-band (approximately 1.6 GHz) frequencies for the Mobile Satellite Service and S-band (approximately 2.5 GHz) frequencies for the Digital Audio Radio Service and some mobile service are also used by broadcasters. This chapter of the handbook deals primarily with the C- and Ku-band frequency FSS satellites, as those are most commonly used by broadcasters today.

International satellite systems provide services on a global basis (global beams) to all countries visible from a single orbit location and on a regional basis (spot beams). INTELSAT and Intersputnik are examples of this type of system, which uses both the C-band and Ku-band. The international satellite communication

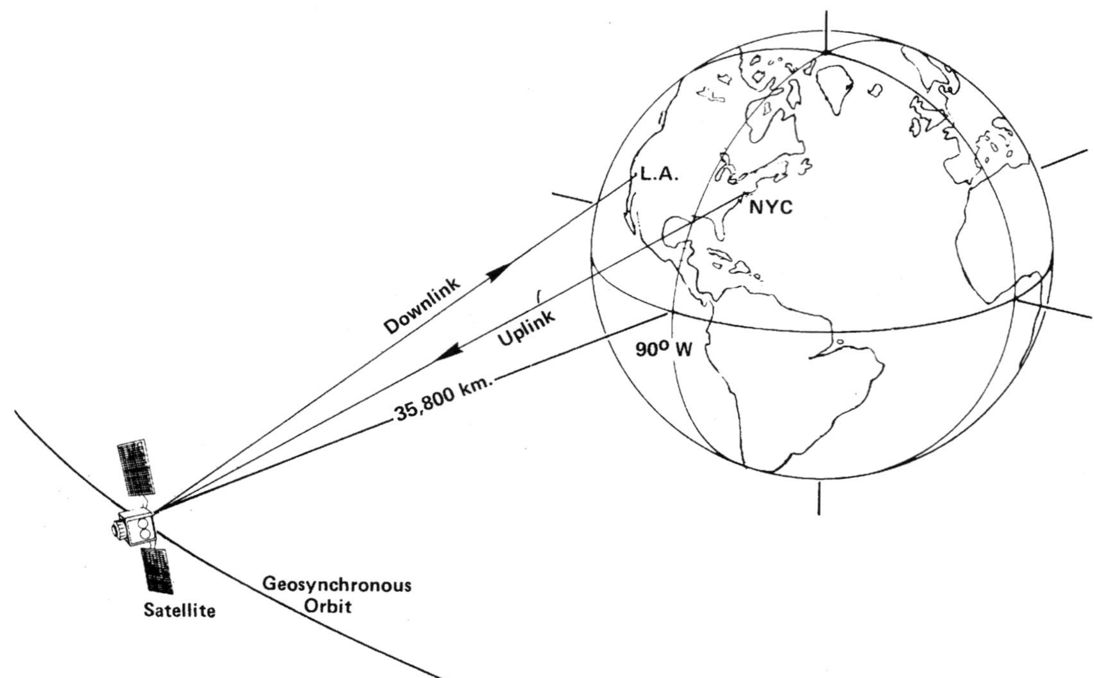

FIGURE 6.11-1 Satellite in geosynchronous orbit.

frequency bands are similar to the U.S. frequencies at C-band (except that some locations have extended this band) but somewhat different at Ku-band, which can vary on a geographical region basis. The frequency assignments are determined in joint negotiations by the countries of the world through the auspices of the Radio Communications Sector of the International Telecommunication Union (ITU-R).

Satellite Stationkeeping

It was stated above that the synchronous satellite appears stationary in space. Actually, a synchronous satellite is never perfectly stationary, because a number of forces, including the pull of the sun and the moon, perturb its orbit. If left alone, the satellite would eventually drift out of orbit. To overcome this, the position of the satellite is continuously monitored by an earth station, called a *telemetry, tracking, and control* (TT&C) station, and small jets of propellant such as hydrazine are used to keep it in position within a *station-keeping box*. The station-keeping box is typically a square ±0.05° or less on each side and oriented with the sides parallel and perpendicular to the orbital plane. Sufficient rocket propellant must be carried on board to last for the satellite's predicted life, usually from 12 to 16 years. This life can and sometimes is extended by cessation of "north–south" stationkeeping and allowing the satellite's orbital inclination to increasingly exceed the station-keeping box, normally

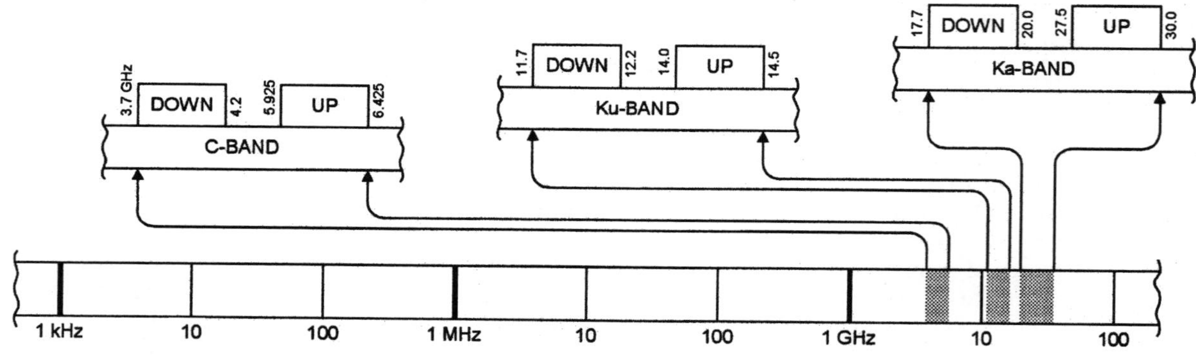

FIGURE 6.11-2 Frequencies in the microwave range of communications satellites.

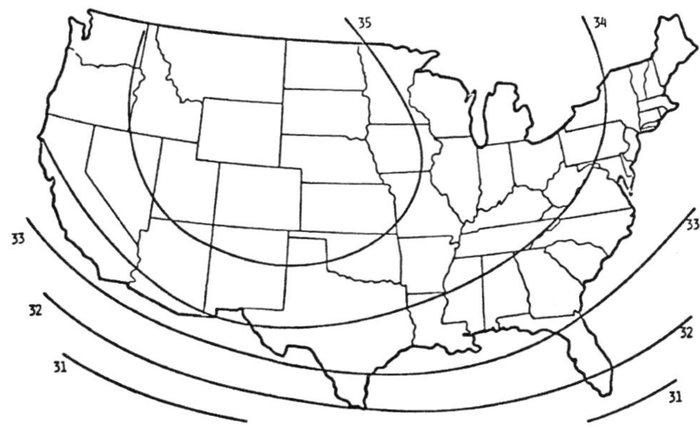

FIGURE 6.11-3 Satellite footprint. Lines with numbers indicate varying antenna gain and consequently varying levels of downlink power (EIRP) or uplink receive system performance (SFD or G/T).

at a rate of 0.9° per year. This type of operation requires FCC approval, is termed *inclined orbit operation*, and can extend the remaining life of a satellite by a factor of 12.

Satellite Footprint

The transmitting and receiving antennas on a satellite are designed to cover only specific desired regions of the earth's surface. This has several purposes. It concentrates the power radiated from the satellite into the desired direction, increases the sensitivity of its receiving antennas, and helps prevent interference with signals from other satellites. The part of the earth's surface covered by a satellite is called the satellite's *footprint*. The footprint may cover one or more relatively localized regions of the earth or almost a complete hemisphere. A typical footprint is shown in Figure 6.11-3.

Polarization

Electromagnetic waves and antennas are always *polarized* in some manner. The polarization may be linear or (approximately) circular. Linear polarizations and circular polarizations are aligned in space as shown in Figure 6.11-4. Most domestic FSS satellites are linearly polarized. A linearly polarized antenna receives maximum power from an incident linearly polarized wave if the *tilt angles* of the wave and the antenna polarizations are aligned in space as shown in Figure 6.11-4a. The wave is then said to be *co-polarized*. As the tilt angle of the wave or antenna rotates from co-polarization, the received power decreases. When the tilt angles are 90° apart, as shown in Figure 6.11-4b, the antenna is *cross-polarized* to the wave and receives (ideally) no power from it. The antenna and the wave then have *orthogonal* polarizations. A given satellite can employ two orthogonal polarizations that exist simultaneously and carry different information without interference. This principle, *frequency reuse*, is used to increase the information capacity of satellites and the geosynchronous orbit. *Circular polarizations* have either right hand (RHC) or left hand (LHC) senses. RHC and LHC polarizations are orthogonal. A circularly polarized satellite and a circularly polarized earth station are co-polarized if they have the same senses and are cross-polarized if they have the opposite senses. The relative tilt angles of circular polarized antennas and waves are of no consequence and are not defined. This represents an advantage of circular polarization over linear polarization, because the tilt angle of the earth station does have to be adjusted for a particular satellite.

SATELLITE SYSTEM CHARACTERISTICS

The design of a satellite communication system is an intricate process involving trade-offs among many variables to obtain maximum performance at a reasonable cost. The major cost and complexity trade-off occur between satellite and earth stations or, more generically, between the space segment and ground segment. The dominating design factors in both segments for systems using geostationary satellites include the following.

Space Segment

- Weight and size of satellite
- DC power generated on board
- Dimensions and complexity of satellite antennas
- Requirements of the communications payload

Ground Segment

- Allocated frequency bands
- Earth stations' antenna size and radiofrequency (RF) capabilities
- Earth stations' multiple access and signal processing techniques

The weight of the satellite is limited by the high cost of launching a spacecraft into geostationary orbit, a cost that generally increases with the weight of the satellite. For a satellite of limited weight and size, a limited number of solar cells can be deployed, which defines an upper limit on the DC power available for the communication transponders. The size and power limitations translate into the fact that the spacecraft has limited RF output power, which then must be transmitted onto particular areas of the earth (*e.g.*, the continental U.S.). Furthermore, power densities over the earth's surface are limited, depending on operating frequency bands, to allow interference free coexistence with other communications systems operating at the same frequencies.

Multiple access and multiple destinations are distinctive virtues of satellite communications. The

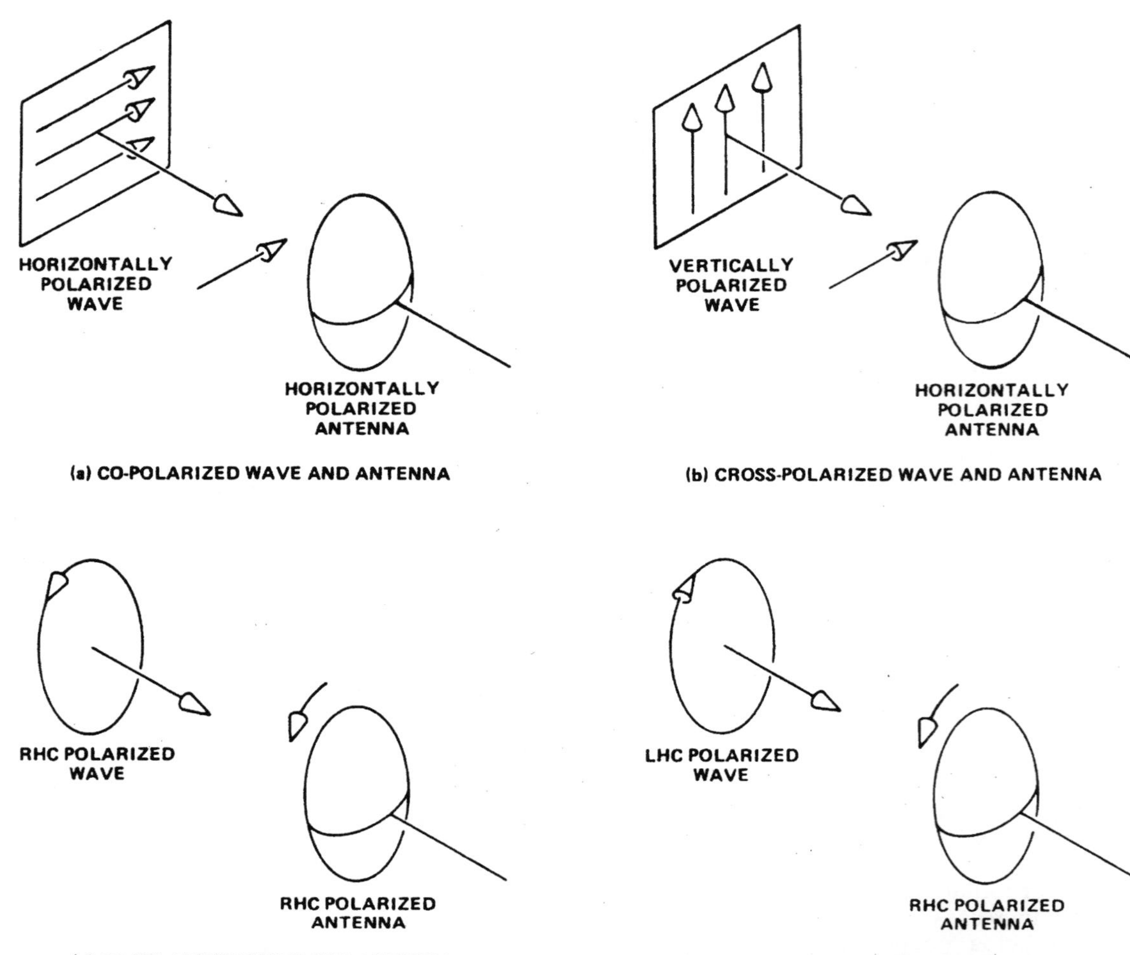

FIGURE 6.11-4 Linear and circular polarizations: (a) co-polarized wave and antenna; (b) cross-polarized wave and antenna; (c) co-polarized wave and antenna; (d) cross-polarized wave and antenna.

methods by which a large number of earth stations share one satellite or one transponder providing the required connectivity (multiple access techniques) also have a significant impact on system design. The multiple access can be achieved by sharing the transponder bandwidth in separate *frequency* slots—frequency-division multiple access (FDMA)—or the transponder availability in discrete *time* slots—time-division multiple access (TDMA). A third but less commonly used technique, code-division multiple access (CDMA) or spread spectrum, shares the transponders by allowing *coded* signals to overlap in time and frequency.

A satellite communication system must be designed to meet certain minimum performance standards, within limitations of the transmitted power, RF bandwidth, and antenna sizes. The most important performance criterion for analog systems is the signal-to-noise ratio (SNR or S/N) in the information channel or baseband. In digital systems, the performance measurement criterion is bit error rate (BER).

The SNR and BER depend on a number of factors, such as the predetection (incident upon the receive antenna) carrier-to-noise density ratio (C/N_o) and the carrier-to-noise ratio (C/N) in the receiver, the type of modulation, and the RF and baseband bandwidths. In the following section, the design and analysis of satellite communications links in terms of C/N_o are discussed; therefore, the carrier power received in an earth station receiver and the noise power density in the receiver must be calculated to establish the operating link C/N_o.

Satellite Transmission Modes for Television

Analog and digital modulation methods are both used for the transmission of television signals via satellite. Historically, analog techniques were predominant until technological developments in digital compression and modulation systems resulted in their greater efficiency and cost effectiveness in many applications.

Today, digital formats and modulation are on the rise, and the use of analog techniques is decreasing.

With regard to analog modulation methods, frequency modulation (FM) has been the prevalent analog technique because:

- It minimizes the effects of nonlinearities in the transmission channel.
- It is immune to AM noise.
- Power-limited systems can take advantage of the wider bandwidth to increase the C/N.
- Various processing techniques can be employed to optimize video transmission (multiplexing, preemphasis, and threshold extension).

The choice of the optimum modulation index **m**, the ratio of the FM deviation to the highest modulating frequency (f_m), is critical. Bandwidth and deviation are related by bandwidth = $2(\mathbf{m} + 1)f_m$. The spectral distribution of an FM signal as a function of **m** is shown in Figure 6.11-5. The selection of the optimum deviation must be based on the number of channels to be transmitted, the type of baseband signal (component or composite), the signal quality requirement, the power received, and the available bandwidth. A typical value of peak frequency deviation (Δf) for a C-band satellite transponder with a nominal bandwidth of 36 MHz is 10.75 MHz.

For satellite transmission, in addition to the deviation of the carrier by the signal it is usually necessary to subject the main carrier to a low-frequency deviation. This spreads the high concentration of carrier and sideband energy over a larger range of the spectrum and permits higher satellite effective radiated power (ERP) without exceeding the FCC's limit on watts/meter^2/kHz downlink power density.

Preemphasis/deemphasis is employed in FM systems for the transmission of video to compensate for the increase in thermal noise with increasing frequency. ITU-T J.61 (formerly CCIR Recommendation 567) specifies a standard 75 μsec preemphasis, for example.

A characteristic of FM is that the detected S/N for the video signal is higher than its C/N ratio. This difference is the FM improvement factor; satellite transmission takes advantage of this improvement, provided the received C/N is greater than the receiver operating threshold. Threshold extension demodulation (TED) is a common technique used in FM receivers to reduce video impulse noise when the C/N drops below the receiver's operating threshold. Above threshold, the receiver acts like a standard discriminator; when the C/N drops below threshold, TED circuitry automatically switches to a narrow bandwidth.

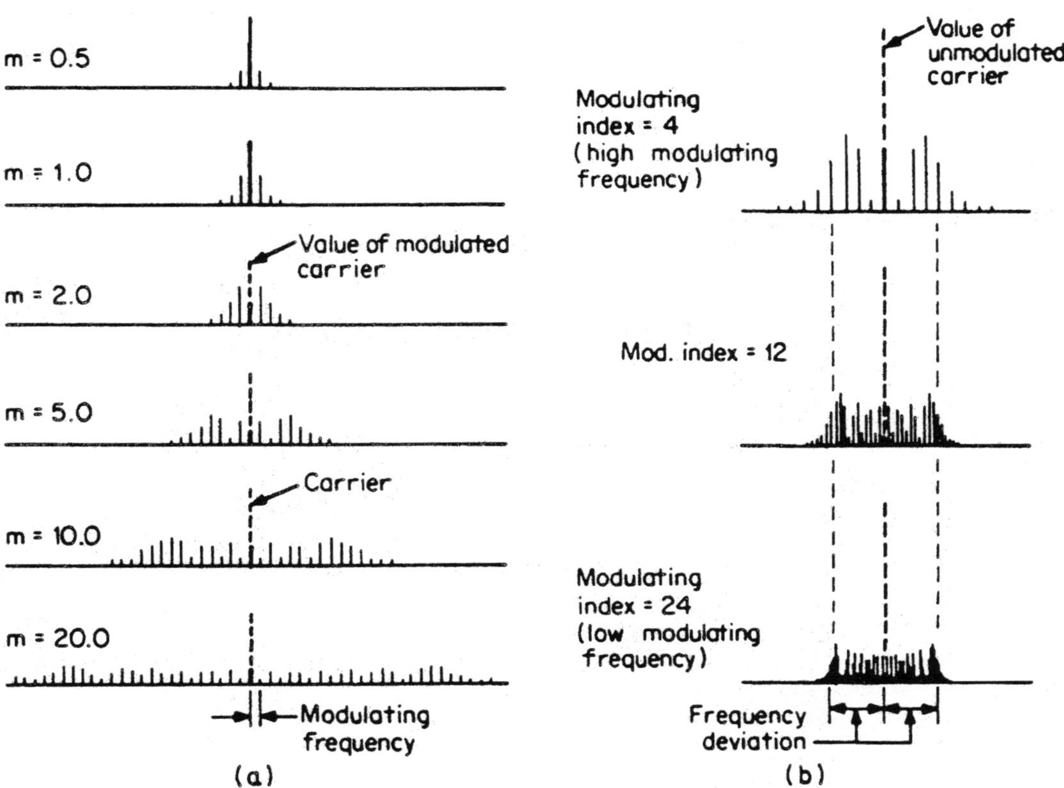

FIGURE 6.11-5 Spectra of frequency-modulated signals: (a) frequency spectra with increasing frequency deviation and constant modulating frequency; (b) frequency spectra with constant frequency deviation.

Digital Transmission

As noted above, satellite systems are increasingly being used for the transmission of digital information. For audio and video services, this information will usually be perceptually encoded (*e.g.*, by using MPEG coding algorithms) before transmission so as to make optimal use of the satellite channel. A typical digital transmission uplink consists of a modem (modulator–demodulator), upconverter, and power amplifier. The modem converts digital information to and from a modulated carrier at an intermediate frequency (IF), typically 70 MHz or 140 MHz. The upconverter converts the modulated carrier at IF to a satellite frequency and thus selects the transponder of the satellite.

Transmitted digital data is first applied to the encoder section of the modem for forward error correction (FEC) encoding. This process adds redundancy to the bitstream by appending additional bits to the original information to provide error detection and correction capability. Increasingly, two separate FEC systems (*e.g.*, a block FEC and a convolutional FEC) are utilized, with the applications separated by an interleaver that acts to break up error bursts to permit the FEC to be more effective. The signal is then scrambled using a standard algorithm to produce random-like data for a signal spectrum that will be noiselike in nature.

The aggregate data (original data plus error correction bits) is applied to the modulator for modulation onto an IF carrier. The IF carrier is selectable, typically in the range of 50 to 90 MHz (for 70 MHz operation) or 100 to 180 MHz (for 140 MHz operation). The center frequency of the modem modulator is tuned to position the signal within the satellite transponder.

A satellite digital transmission system is characterized by:

- Data rate
- Data interface
- Code rate
- Modulation scheme

Data rate refers either to the number of output bits per second transmitted by the modem (transmitted data rate) or to the data rate of the digital input stream provided to the modem for transmission (information data rate). The information data rate is typically front panel selectable. Modems typically support a number of data interfaces. The data interface refers to the connector and signal levels. Code rate refers to the FEC-encoding scheme. In most modems, the code rate is selectable. The code rate configuration is referred to as m/n, where m is the number of original bits per block of transmitted bits and n is the number of original bits plus error correction bits per block of transmitted bits. Thus, a code rate of 3/4 means that for every three data bits input four data bits are transmitted; thus, a 1024 kbps modem operating with a code rate of 3/4 would transmit 1365 kbps over the satellite channel.

The most widespread FEC coding technique currently in use is *concatenated coding*, in which two different coding methods are successively applied to the information to be transmitted, separated by an interleaver that acts to distribute burst errors (on the receive side) so the outer coding system (the one applied first) can correct errors not corrected by the inner system (the one applied second). Frequently, convolutional codes are used for the outer code and block codes are used for the inner code.

The modulation scheme refers to the method of carrying the data bits on the RF carrier. Two common digital modulation schemes employed in satellite transmission systems are quadrature phase-shift keying (QPSK) and eight-phase-shift keying (8PSK). These modulation schemes generate periodic phase shifts in the RF carrier referred to as *symbols*. The desired symbol rate (the number of symbols per second) and data transmission rate (which depends upon the information source itself as well as the amount of FEC coding employed) determine the amount of bandwidth required in the channel.

In QPSK, four phase shifts are used to represent two unique states, so each transmitted symbol represents two transmission bits. For this case, the symbol rate is equal to half the data transmission rate. The 8PSK scheme uses eight phase shifts, thus transmitting three bits per symbol. For 8PSK, the symbol rate is equal to one-third the transmission rate. For a given data transmission rate, 8PSK requires less bandwidth than QPSK, but it requires increased performance (C/N) from the channel to achieve an equivalent BER. Neither QPSK nor 8PSK employs an amplitude modulation component (all modulation is phase modulation). This permits them to be employed in a saturated (maximum power) satellite transponder when utilized in single carrier per transponder mode.

Advantages of Digital Television

The advantages of digital technology for television processing, production, storage, and distribution are:

- Perfect multigenerational reproducibility
- Precise time and level controls
- Digital storage and signal processing
- Data compression
- Easy manipulation by computers and generation of multimedia content
- Amenability to future technology changes and cost reductions
- Incorporation in digital transport packets for packet switching, broadband digital communication, and error control

The 1982 adoption of ITU-R Recommendation 601 (CCIR601) for component television and the adoption of MPEG-2 by the International Standards Organization (ISO) and the International Electrotechnical Commission (IEC) in 1993 were major milestones for the industry. The use of digital compression techniques, removing statistical redundancy (entropy coding), and taking advantage of psychophysics to remove irrelevant information (perceptual coding) have allowed the data rate of television to be reduced by significant factors, making its use very attractive in conservation of bandwidth.

C-Band Satellites

C-band was initially favored for communications satellites because of the favorable propagation characteristics at these frequencies. The specific bands in most common use are the 5925 to 6425 MHz (uplink) and the 3700 to 4200 MHz (downlink) band pair. U.S. domestic FSS requires the use of 36 MHz bandwidth channels placed on 40 MHz centers. A satellite using a single polarization can provide 12 such transponders, although all satellites are mandated to employ frequency reuse and provide 24 such transponders. Frequency reuse is implemented by the use of orthogonal polarizations and by staggering the center frequency of opposing transponders. As an example of a typical satellite, the transmit and receive frequency plans of a C-band satellite are shown in Figure 6.11-6. The numbered brackets represent each channel. The bandwidth of the channel is represented by the width of the bracket. The carrier frequency, shown above the channel number, is centered on each channel. The signals of alternate transponders in the frequency plan of Figure 6.11-6 are nominally orthogonal. If they were exactly orthogonal and the associated earth stations were ideal (with respect to polarization), there would be no interference caused by the overlapping sideband energy of adjacent transponders. In practice, the polarizations of the antennas of the satellite and earth stations are not ideal. Some small amount of interference occurs, but the combination of nearly orthogonal polarizations and the use of the staggered frequency plan provides for high-quality transmission under almost all weather conditions.

Ku-Band Satellites

The first systems using the 14.0 to 14.5 GHz (uplink) band and the 11.7 to 12.2 GHz (downlink) band were launched in 1976 by Satellite Business Systems (SBS). The higher propagation loss characteristics at these frequencies require higher spacecraft equivalent isotropic radiated power (EIRP) to achieve the same transmission performance as C-band frequencies, and this is obtained by a variety of methods, including the use of greater spacecraft antenna gains, readily achievable at the higher frequencies. Because the Ku-band frequencies are not shared with terrestrial systems (as is the case of C-band), the power flux density (PFD) limitation is much less stringent, and there is no requirement for coordination with terrestrial microwave systems; consequently, Ku-band satellites employ higher power satellite amplifiers than do C-band satellites—as much as an order of magnitude higher. The high powers permit the use of very small earth station antennas at or near the user's premises. This provides an important economic advantage for many services and makes the use of this frequency band very attractive; however, a good part of the higher satellite power achievable is necessary to offset the additional attenuation that is experienced at these frequencies during heavy rain conditions.

There is no mandated frequency plan for transponders in this frequency band, although typical

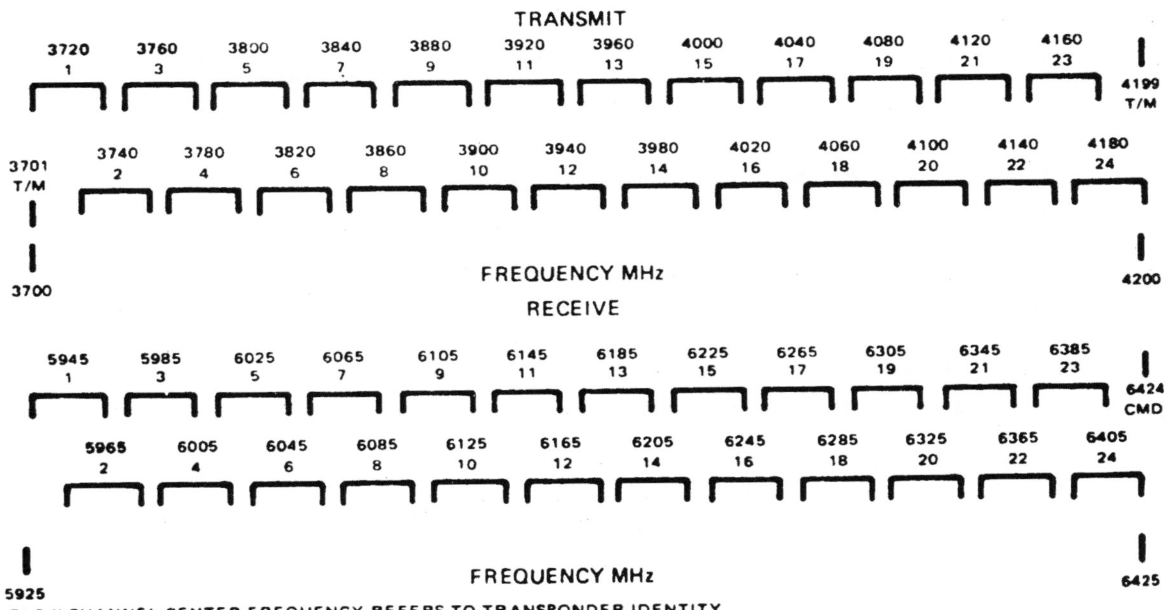

FIGURE 6.11-6 GE/RCA Satcom satellite frequency plans.

transponder bandwidths today are 36 MHz. Because the bandwidth is the same as C-band, it is possible to have a similar 24 transponder, 36 MHz frequency plan with 40 MHz channel spacing when frequency reuse is utilized.

Two important differences between C-band and Ku-band are:

- C-band FSS share frequencies with terrestrial microwave systems. This places constraints on the location of C-band earth stations and limits the permissible downlink power density. Prior to licensing a C-band transmit antenna, a frequency coordination process must be performed (such frequency coordination is not required at the Ku-band).
- Ku-band signals are subject to significant attenuation in heavy rainfall.

The advantages and disadvantages of C-band and Ku-band are summarized in Table 6.11-1. For completeness, Ka-band is also included and is discussed subsequently.

TABLE 6.11-1
Merits of C-, Ku-, and Ka-Band for Satellite Communications

C-Band Advantages
C-band is usually the most reliable due to less susceptibility to rain outages.
The C-band space segment is normally less expensive than other FSS space segments.
C-Band Disadvantages
The frequency band is congested because it is shared with terrestrial microwave, making frequency coordination a requirement.
C-band requires relatively large antennas because of low satellite EIRP levels and the necessity of narrow half-power beamwidth to allow 2° spaced satellites.
Avoiding terrestrial interference can make site selection a difficult process.
The use of artificial shielding to block interference can increase total system cost.
Faraday rotation of polarization can affect system performance.
Ku-Band Advantages
The frequency band is used only for satellite communication.
Smaller antennas may be used because of higher gain and higher satellite EIRP.
Site selection is easier because of the smaller size of antenna and reduced terrestrial interference.
Ku-band is suitable for direct-to-home application.
Channel plan is flexible.
Ku-band is not affected by Faraday rotation.
Ku-Band Disadvantages
Ku-band is affected by rain attenuation and depolarization.
Waveguide and coaxial transmission line losses are quite high.
Interference can occur from radar detectors located in passing automobiles; site surveys should assess.
Ka-Band Advantages
Space segment cost per bit transmitted is lowest because satellites have much greater capacity due to increased frequency reuse via spot beams.
Smaller antennas (than Ku-band) can be used in many areas where atmospheric attenuation is low.
Higher frequency is more amenable to spot beam use and facilitates small dish uplinking and point-to-point applications.
Smaller downlink beams can be tailored to the coverage area and exclude reception elsewhere.
Ka-Band Disadvantages
Increased atmospheric attenuation requires a greater link margin and can reduce reliability.
All-spot beam operation results in significant disadvantage for broadcast (point-to-multipoint) service where broad geographical distribution of the same signal is required.

Ka-Band Satellites

The commercial application of Ka-band satellites was first demonstrated by NASA's experimental Advanced Communications Technology Satellite (ACTS), which demonstrated successful use of the 17.5 to 22.5 GHz downlink and 29.5 to 34.5 GHz bands for various services. Use of this band for direct broadcast satellite (DBS) video and Internet service is a direct result of those experiments. It is likely that these bands will be increasingly used for video applications, with considerable interest in narrow casting and point-to-point applications.

Due to the higher frequency range, Ka-band satellites generally include many small spot beams in lieu of one large CONUS or geographically large beam. This significantly improves frequency reuse and overall satellite capacity but at the expense of connectivity, such as the ability of an uplink to simultaneously broadcast to many geographically dispersed downlinks. Ka-band signals suffer from greater attenuation due to the presence of rain and atmospheric oxygen than do C-band and Ku-band services. This is illustrated in Figure 6.11-7, where the approximate center of each band is marked by a dashed line. This attenuation problem has historically made Ka-band rather unattractive for satellite communications, but the scarcity of spectrum in other bands and desirability of small spot beams for some applications, as well as advances in satellite communications technology, override this disadvantage.

In addition to core satellite voice, data, and video offerings, Ka-band licensees will provide many low cost, broadband interactive services such as:

- Direct-to-home video
- Internet access

At the time of this writing, DirecTV is using Ka-band on the SPACEWAY 1 satellite to provide DBS video service. Although this satellite was originally designed for on-board switching, it is being used in a "bent pipe" configuration for local-into-local video, including HDTV. Another current use of Ka-band is for home and small office Internet service, which is being implemented by the Wild Blue satellite system (for Internet applications). At the time of writing, little or no Ka-band trunking or backhaul is taking place, but this type of point-to-point application would seem a natural application of Ka-band systems, particularly in drier regions of the country.

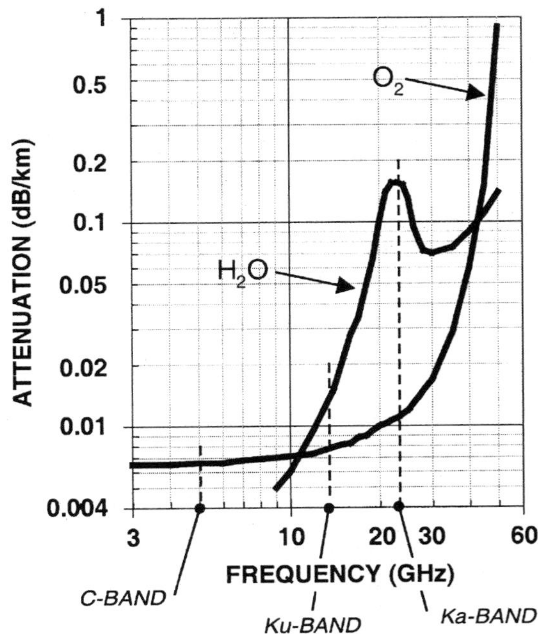

FIGURE 6.11-7 Signal attenuation *versus* frequency due to atmospheric oxygen and water vapor.

Regulatory Issues

Satellite communication systems are governed by the FCC in the U.S. and by the ITU-R on the international level. The governing agencies assign frequency bands of operation, satellite performance characteristics, and orbit location and provide technical specifications of radiated power density and radiation gain patterns for the earth stations. The FCC is the licensing body for all transmit earth stations in the U.S. and licenses C-band receive-only earth stations at the owner's request. Such receive antenna licensing ensures that the antenna technical characteristics exist in the master FCC database and must therefore be considered and analyzed from a potential interference standpoint by anyone considering licensing a transmit antenna in its vicinity. The FCC Rules and Regulations, Part 25, form the basis of the applicable documents that must be followed for the planning and implementation of any FSS band satellite communication system.

The FCC amends and interprets the rules as the technology and the requirements of satellites change through amendments, decisions, and declaratory orders; therefore, it is necessary to review the current FCC Rules prior to planning a new satellite system. The FCC established precedents for the minimum diameter apertures and sidelobe gain envelopes for earth station antennas operating in the FSS bands in the early 1970s to minimize interference between terrestrial systems and satellite systems and between satellite systems. These precedents have been modified through the years as the use of satellite services has increased. The more significant recent rulings pertaining to earth station antenna performance have resulted in improved antenna radiation patterns in the close-in sidelobe region and have established maximum radiated power densities.

The FCC Rules and Regulations (Title 47) Part 25.209 pertaining to antenna gain envelopes are mandatory for all transmit antennas. Excerpts from this standard follow (refer to the current rules publication for the entire text):

a) The gain of any antenna to be employed in transmission from an earth station in the fixed satellite service shall lie below the envelope defined below:

1. In the plane of the geostationary satellite orbit as it appears at the particular earth station location:

[29 – 25 log(θ)] dBi	$1° \le \theta \le 7°$
+ 8 dBi	$7° < \theta \le 9.2°$
[32 – 25 log(θ)] dBi	$9.2° < \theta \le 48°$
10 dBi	$48° < \theta \le 180°$

where θ is the angle in degrees from the axis of the main lobe, and dBi refers to the dB relative to an isotropic radiator. For the purposes of this section, the peak gain of an individual sidelobe may not exceed the envelope defined above for θ between 1° and 7°. For θ greater than 7°, the envelope may be exceeded by 10% of the sidelobes, but no individual sidelobe may exceed the envelope by more than 3 dB.

2. In all other directions:

Outside the main beam, the gain of the antenna shall lie below the envelope defined by:

[32 – 25 log(θ)] dBi	$1° \le \theta \le 48°$
–10 dBi	$48° < \theta \le 180°$

where θ is the angle in degrees from the axis of the main beam, and dBi refers to dB relative to an isotropic radiator. For the purposes of this section, the envelope may be exceeded by no more than 10% of the sidelobes, provided the level of no individual sidelobe exceeds the gain envelope given above by more than 6 dB.

b) The off-axis cross-polarization isolation of any antenna from an earth station to a space station in the domestic fixed-satellite service shall be defined by:

[19 – 25 log(θ)] dBi	$1.8° < \theta \le 7°$
–2 dBi	$7° < \theta \le 9.2°$

c) Any antenna licensed for reception of radio transmission from a space station in the fixed-satellite service shall be protected from radio interference caused by other space stations only to the degree to which harmful interference would not be expected to be caused to an earth station employing an antenna conforming to the standards defined in paragraphs (a) and (b) of this section.

d) The operations of any earth station with an antenna not conforming to the standards of paragraph (a) and (b) of this section shall impose no limitations upon the operation, location, and design of any terrestrial station, any other earth station, or any space station, above what would be imposed by an antenna conforming to the standards in paragraphs (a) and (b) above.

e) An earth station with an antenna not conforming to the standards of paragraphs (a) and (b) of this section will be routinely authorized upon a finding by the Commission that unacceptable levels of interference will not be caused under conditions of uniform 2° orbital spacings.

f) The antenna performance standards of small antennas operating in the 12/14 GHz band with diameters as small as 1.2 meters start at 1.25° instead of 1° as stipulated in paragraph (a) of this section.

The FCC further acknowledged that the envelope defined above is only a reference envelope in the receive band. Receiving antennas do not have to conform to this envelope to be eligible for licensing. Facilities with performance worse than the reference envelope must, however, accept higher interference levels.

FCC License

The FCC requires licensing of transmitting earth stations and permits licensing of receive-only (RO) earth stations. It is desirable for a broadcaster to license a C-band RO earth station, as licensing protects the station from future interference from domestic microwave systems. This is because a licensed antenna is included in the FCC earth station database and anyone wanting to license a new transmitter needs to frequency coordinate with all licensed and identified antennas, including receive antennas. Because the FCC Rules and Regulations are ever evolving, it is necessary to review the latest rules before filing satellite applications. Licensing information is available on the Internet at www.fcc.gov.

SYSTEM PERFORMANCE ANALYSIS TECHNIQUES

Consider an RF link as illustrated in Figure 6.11-8 with transmit power P_t and transmit gain G_t. The effective isotropic radiated power (EIRP) for the station along the main beam of the antenna is the product $G_t * P_t$. At distance R (meters) from the transmitter, the radiated flux density (S) becomes:

$$S = (G_t P_t)\frac{1}{4\pi R^2}k_a \qquad \text{Watts/m}^2 \qquad (1)$$

where k_a is the atmospheric attenuation factor < 1.

If an antenna with effective area A_e (in square meters) is receiving this flux density, the received carrier level (C) at the antenna output is:

$$C = SA_e = (P_t G_t A_e)\frac{1}{4\pi R^2}k_a \qquad \text{Watts} \qquad (2)$$

At the same antenna output point, the effective noise power density (N_o) is given by:

$$N_o = k * T_s \qquad \text{Watts/Hz} \qquad (3)$$

where k is Boltzmann's constant = $1.38 * 10^{-23}$ Joules/K or –228.6 dB, and T_s is the system noise temperature. Consequently, received carrier-to-noise density becomes:

$$\frac{C}{N_o} = (P_t G_t A_e)\frac{1}{4\pi R^2}\frac{1}{kT_s}k_a \qquad (4)$$

A fundamental relationship in antenna theory is that the gain (G_r) and the effective area of an antenna (A_e) are related by:

$$A_e = G_r \frac{\lambda^2}{4\pi} \text{m}^2 \qquad (5)$$

Substituting this relation into the expression for C/N_o,

$$\frac{C}{N_o} = (P_t G_t G_r)(\lambda/4\pi R)^2 \frac{1}{kT_s}k_a \qquad (6)$$

or

$$\frac{C}{N_o} = EIRP \frac{G_r}{T_s}(\lambda/4\pi R)^2 \frac{1}{k}k_a \qquad (7)$$

The factor $(\lambda/4\pi R)^2$ is often inverted and defined as the spreading loss or space loss factor. This spreading loss can also be expressed as:

$$L_s = (4\pi Rf/c)^2 \qquad (8)$$

where c is the speed of light = $3 * 10^{+8}$ m/sec, and f = frequency (in hertz).

Link calculations are usually carried out in decibels rather than directly from the above relations because of the ease of working in common logarithms. For clear sky conditions, C/N_o in decibels can be calculated by:

$$(C/N_o)\text{ dB} = 10\log(C/N_o)$$

$$(C/N_o)\text{ dB} = G_t P_t - L_s + (G/T) + 228.6 - k_a \qquad (9)$$

where $EIRP = 10\log(G_t P_t)$ dBW, and:

$$L_s = 20\log(4\pi Rf/c) \text{ dB} = 92.45 + 20\log R \ (R \text{ in km}) + 20\log f \ (f \text{ in GHz}) \qquad (10)$$

$$(G/T) = 10\log(G_r/T_s) \text{ dB/K} \qquad (11)$$

Alternatively, C/No can be expressed in terms of flux density S as:

$$(C/N_o) = S + (G/T) - A_i + 228.6 - k_a \text{ dBHz} \qquad (12)$$

$$S = EIRP - L_s + A_i \text{ dBW/m}^2 \qquad (13)$$

where A_i is the effective aperture of an isotropic radiator (in dB):

$$A_i = 10\log(4\pi/\lambda^2) \qquad (14)$$

Equation (9) is a fundamental tool for characterizing space link performance. It will be utilized later when calculating overall satellite link performance.

Earth Station Receive Figure of Merit: G/T

G/T is the figure of merit of a receive system. It is primarily a function of the gain of the antenna along with the antenna noise temperature, first amplifier noise temperature, and losses located between the antenna and the first amplifier. The importance of the term G/T in Equations (9) and (12) cannot be overstated. Examination of the C/N_o expression shows that, for a given available transmitting power and information format (and thus bandwidth), the only available method of controlling the received signal quality that can be used by the downlink operator is through the system G/T. Note that G/T provides a direct dB relationship with C/N_o.

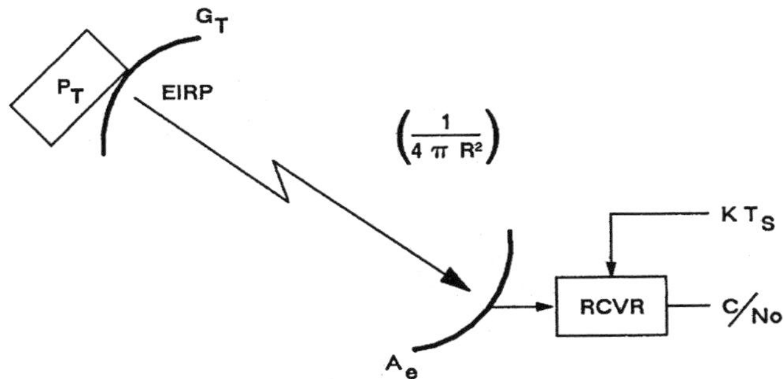

FIGURE 6.11-8 RF link diagram.

Figure 6.11-9 shows a block diagram of a typical receive system. Each device in the RF path has an associated gain or loss and a noise temperature. These contributions are combined to reflect the noise power weighted by the gain distribution through the chain. The earth station G/T is given by:

$$(G/T) = G_a - 10 * \log(T_s) \text{ dB/K} \qquad (15)$$

where G_a is the antenna gain referenced to the low-noise amplifier (LNA) or low-noise block converter (LNB) input (dBi), and T_s is the system noise temperature referenced to LNA or LNB input (K).

The system noise temperature (T_s) referenced to the LNA or LNB input can be calculated by adding as noise powers the equivalent noise temperatures of all noise contributors, weighted by the net gain between the point in which that noise has been added and the LNA or LNB input; that is:

$$T_s = (T_a/L_w) + T_0(L_w - 1)/L_w + T_{vswr} + T_{lna} + [(L_t - 1) + L_t(F_r - 1)]\,T_1/G) \qquad (16)$$

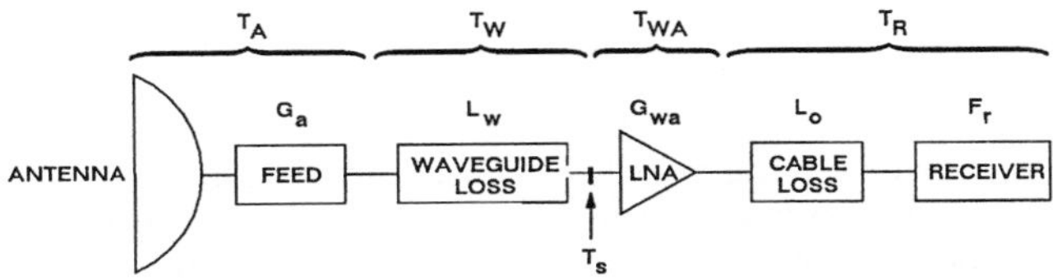

FIGURE 6.11-9 G/T system diagram: receive-only earth station.

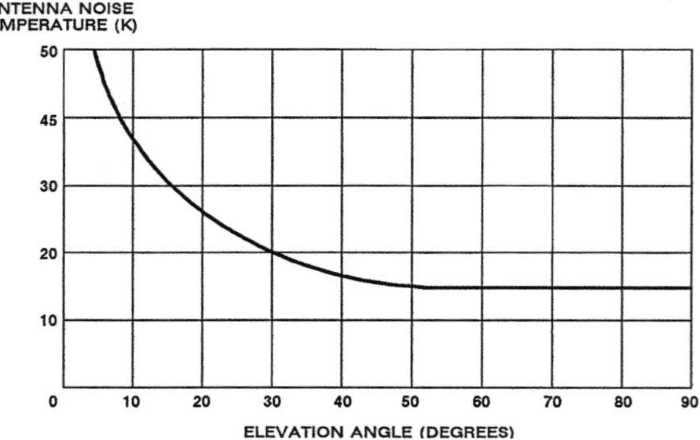

FIGURE 6.11-10 Typical antenna noise temperature variations with elevation angle.

where:

T_a = antenna noise temperature (K).

L_w = waveguide loss between antenna and LNA or LNB (linear power ratio).

T_0 = ambient temperature (K).

T_{vswr} = LNA or LNB antenna impedance mismatch noise temperature (K).

T_{lna} = LNA or LNB noise temperature (K).

L_t = transmission loss between LNA or LNB and receiver (line power ratio).

F_r = receiver noise figure (linear power ratio).

T_1 = 290 K.

G = net gain between LNA or LNB input and receiver input (linear, includes interconnect cable loss).

The antenna temperature is usually minimum at zenith, typically 15 to 25° for a low loss, C-band antenna with low wide angle sidelobes. As the elevation angle decreases, the antenna temperature increases because more of the higher level sidelobes look at the earth which has a temperature of about 290 K. A typical curve of the variation of noise temperature with elevation angle is illustrated in Figure 6.11-10. Similarly, Figure 6.11-11 shows typical G/T system performance for different C-band and Ku-band antenna diameters as a function of elevation angle.

Satellite Transponder

The orbiting spacecraft provides a one-hop carrier relay over a wide geographic area. In C-band systems, the uplink signal is transmitted near 6 GHz, received by the satellite, amplified, translated in frequency, filtered, and retransmitted near 4 GHz. Likewise, in Ku-band systems, the uplink occurs in the 14 GHz range and the downlink in the 12 GHz range; in Ka-band systems, the uplink occurs in the 30 GHz range, the downlink in the 20 GHz range (see Figure 6.11-2).

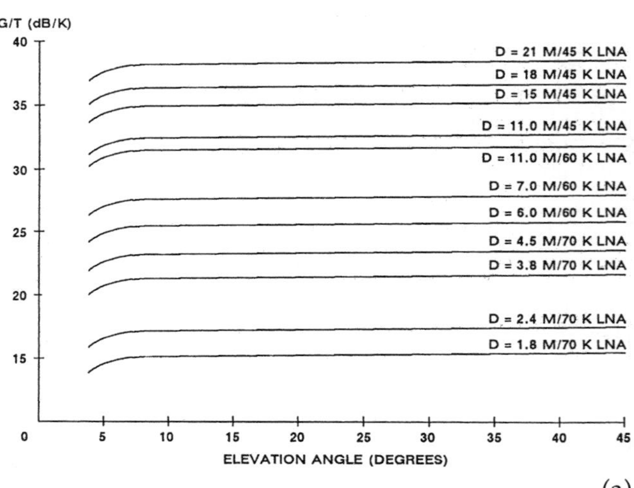

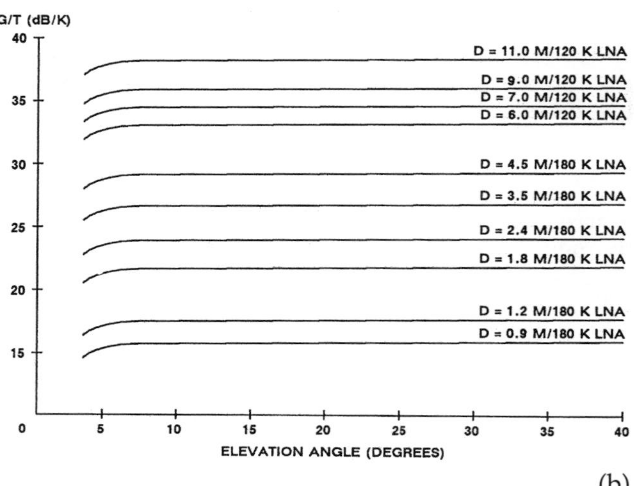

FIGURE 6.11-11 (a) Typical C-band G/T system performance *versus* elevation angle for different commonly used antenna diameters; (b) typical Ku-band G/T system performance *versus* elevation angle for different commonly used antenna diameters.

Because the satellite serves as a transmit/receive station, it must be characterized by a G/T for the uplink side and by saturated EIRP for the downlink side. To couple the uplink and downlink signal strengths and as a definition of the transponder sensitivity, the uplink RF spectral flux density required at the satellite to saturate the transponder is also specified (saturation flux density [SFD]). These three satellite parameters vary with geographic location. Contour maps, called *footprints*, are usually available for assessing these variations. Typical footprints for C-band and Ku-band satellites are shown in Figures 6.11-12 and 6.11-13, respectively.

Another important parameter that characterizes the transponder performance is the input/output power transfer and the intermodulation response. Both performance parameters are normally specified in terms

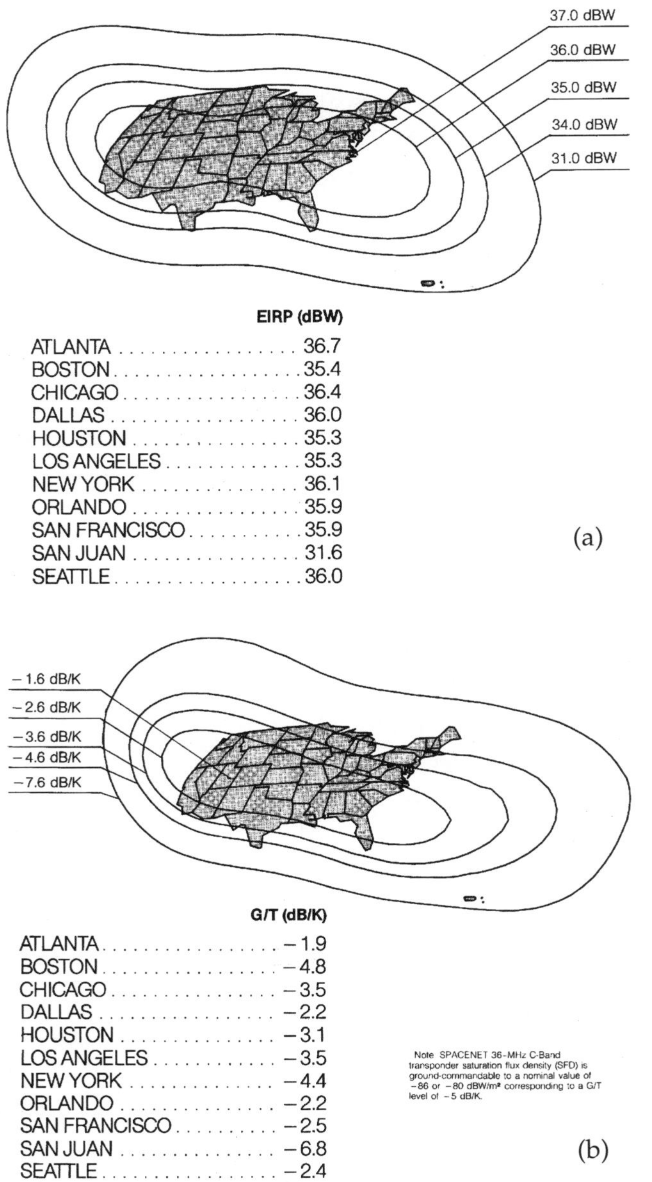

	EIRP (dBW)
ATLANTA	36.7
BOSTON	35.4
CHICAGO	36.4
DALLAS	36.0
HOUSTON	35.3
LOS ANGELES	35.3
NEW YORK	36.1
ORLANDO	35.9
SAN FRANCISCO	35.9
SAN JUAN	31.6
SEATTLE	36.0

(a)

	G/T (dB/K)
ATLANTA	−1.9
BOSTON	−4.8
CHICAGO	−3.5
DALLAS	−2.2
HOUSTON	−3.1
LOS ANGELES	−3.5
NEW YORK	−4.4
ORLANDO	−2.2
SAN FRANCISCO	−2.5
SAN JUAN	−6.8
SEATTLE	−2.4

Note SPACENET 36-MHz C-Band transponder saturation flux density (SFD) is ground-commandable to a nominal value of −86 or −80 dBW/m² corresponding to a G/T level of −5 dB/K.

(b)

FIGURE 6.11-12 (a) C-band satellite EIRP footprint; (b) C-band satellite G/T footprint.

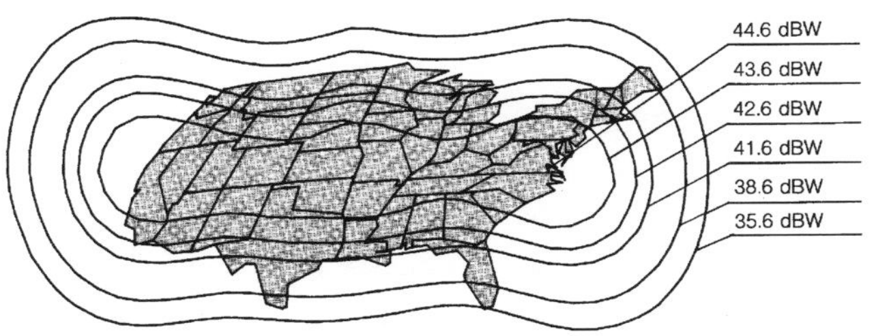

	EIRP (dBW)
ATLANTA	43.2
BOSTON	41.7
CHICAGO	42.9
DALLAS	43.2
HOUSTON	41.3
LOS ANGELES	43.0
NEW YORK	43.4
ORLANDO	39.8
SAN FRANCISCO	44.1
SEATTLE	42.0

(a)

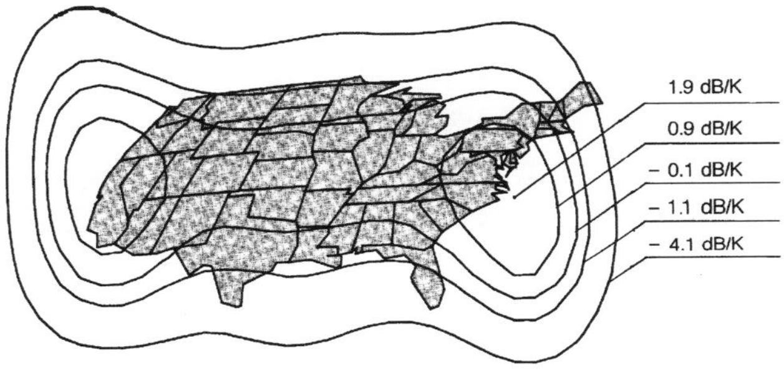

	G/T (dB/K)
ATLANTA	− .1
BOSTON	− 1.1
CHICAGO	.2
DALLAS	− .2
HOUSTON	− 1.0
LOS ANGELES	1.2
NEW YORK	.7
ORLANDO	− 1.4
SAN FRANCISCO	1.5
SEATTLE	− .3

Note: SPACENET Ku-Band transponder saturation flux density (SFD) is ground-commandable to a nominal value of −86, −80 or −74 dBW/m² corresponding to a G/T level of −2 dB/K.

(b)

FIGURE 6.11-13 (a) Ku-band satellite EIRP footprint; (b) Ku-band satellite G/T footprint.

of input back-off (BO_i) and output back-off (BO_o), as a function of the power reduction (expressed in dB) with respect to saturation. Figures 6.11-14 and 6.11-15 show typical transponder responses for a satellite equipped with a traveling wave tube (TWT) type of power amplifier. Solid-state power amplifiers (SSPAs) are also used as satellite transponder amplifiers, particularly as driver amplifiers and at C-band. Their improved linearity and increased reliability (longer device life) recommend them highly, but they cannot be used in all applications, particularly where higher powers are required.

Satellite Link Analysis

With the preliminary procedures and formulations described previously, link calculations can be conducted. First, the distance or slant range from the satellite to the earth station must be determined so the space loss may be calculated. From orbit geometry and Equation (10) above, the space loss (expressed in dB) is found to be:

$$L_s = 185.05 + 10\log[1 - 0.295\cos(H)\cos(\Delta L)] + 20\log f \quad (17)$$

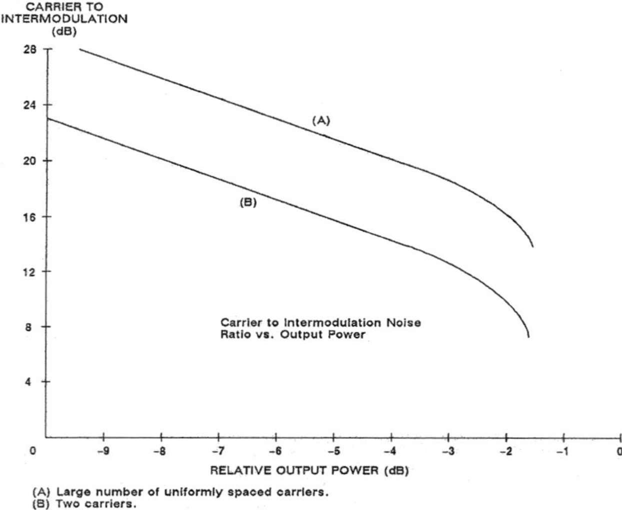

FIGURE 6.11-14 Output power normalized to single carrier saturation point.

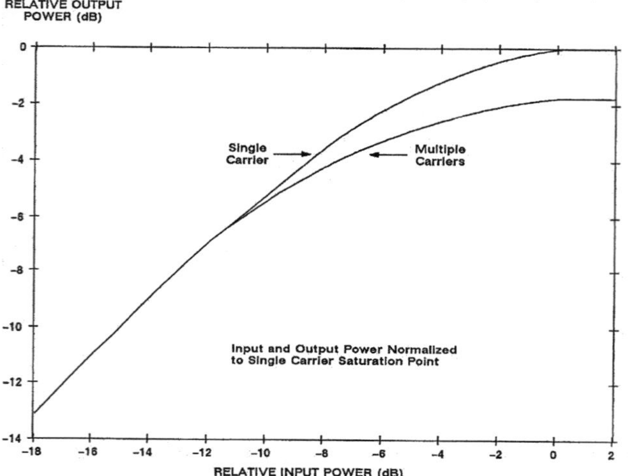

FIGURE 6.11-15 TWT power transfer characteristics.

where H is the latitude of the earth station, ΔL is the difference in longitude for earth station and satellite, and f is the frequency (in GHz). The overall satellite link can now be calculated.

Uplink C/N

From Equation (9), the uplink $(C/N_o)_u$ becomes:

$$(C/N_o)_u = EIRP_u - L_u + (G/T)_s + 228.6 - k_a \qquad (18)$$

or

$$(C/N_o)_u = S - A_i + (G/T)_s + 228.6 \qquad (19)$$

and

$$S = SFD - BO_i \qquad (20)$$

where:

$EIRP_u$ = uplink EIRP (dBW).

L_u = uplink space loss (dB).

$(G/T)_s$ = satellite G/T (dB/K).

k_a = atmospheric attenuation.

S = flux density (dBW/m^2).

$A_i = 21.5 + 20\log f\,(\text{GHz})$ (dB/m^2).

SFD = saturation flux density (dBW/m^2).

BO_i = transponder input back-off (dB).

Downlink C/N

Likewise, the downlink $(C/N_o)_d$ can be calculated by:

$$(C/N_o)_d = EIRP_d - L_d + (G/T)_{e.s} + 228.6 - k_a \qquad (21)$$

and

$$EIRP_d = EIRP_s - BO_o \qquad (22)$$

where:

$EIRP_d$ = downlink EIRP (dBW).

L_d = downlink space loss (dB).

$(G/T)_{e.s}$ = earth station G/T (dB/K).

$EIRP_s$ = saturated EIRP (dBW).

BO_o = transponder output back-off (dB).

It is important to note that Equations (20) and (22) are related by the nonlinear power transfer function of the transponder; therefore, for transponder operation below saturation, the input and output relationship must be resolved graphically with the aid of Figure 6.11-15 or its equivalent.

Once the uplink and downlink noise contributions are determined, the composite link performance in terms of total carrier-to-noise density ratio $(C/N_o)_t$ can be readily obtained by simple noise power addition, as the uplink and downlink contributions are incoherent. This yields:

$$(C/N_o)_t = \{(C/N_o)_u^{-1} + (C/N_o)_d^{-1}\}^{-1} \qquad (23)$$

This equation represents a simplified situation in that only thermal noise is added to the carriers. In practice, there are other sources of perturbations and interference, some of the more important ones of which transponder nonlinearity is the cause. As shown in Figure 6.11-15, operating the transponder near maximum power for better efficiency implies that compression, due to the instantaneous nonlinear transfer characteristic of the amplifier relative to the signal amplitude, becomes more significant. Under this condition, when more than one frequency is amplified, interaction between the signals occurs and consequently a spectrum of spurious frequencies or intermodulation products is generated. Particularly, the so-called third-order intermodulation product of the form $(2f_1 - f_2)$, a consequence of the third-order nonlinearity of the transponder, constitutes a significant interfering signal because it is the largest product and it falls in the same

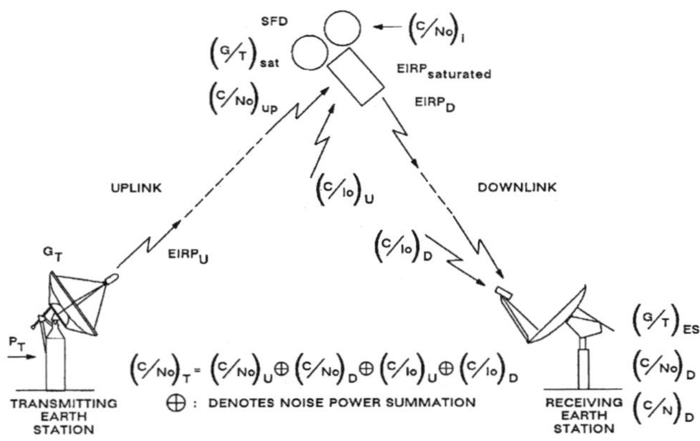

FIGURE 6.11-16 Satellite link model.

operating bandwidth of the information signal. Figure 6.11-14 shows how carrier-to-intermodulation ratio varies as a very sensitive function of the transponder operating output back-off.

System C/N

Figure 6.11-16 depicts a complete satellite link. Other sources of interference have been added, such as uplink interference due to off-beam radiation from other earth stations and uplink cross-polarization isolation combined, represented by the quantity $(C/I_o)_u$. Similarly, in the downlink the quantity $(C/I_o)_d$ represents the combined effects of downlink cross-polarization isolation and adjacent satellite interference. When all of these terms are considered, the total link $(C/N_o)_t$ can be calculated by:

$$(C/N_o)_t = \{(C/N_o)_u^{-1} + (C/I_o)_u^{-1} + (C/N_o)_d^{-1} + (C/I_o)_d^{-1}\}^{-1} \qquad (24)$$

Figure 6.11-17 shows the typical interaction of the different terms in Equation (24) as a function of transponder input back-off and in the presence of thermal and intermodulation noise. The total $(C/N_o)_t$ can be maximized by reducing the transponder input drive and adjusting the transponder gain. Backing off the TWT reduces $(C/N_o)_u$ and $(C/N_o)_d$ (through the input/output relationship of the transponder), but as $(C/N_o)_I$ increases rapidly when the input drive is reduced, an optimum value of $(C/N_o)_t$ is obtained at a specific back-off level. Interference noise can be kept down by proper antenna design, transponder sensitivity, and frequency coordination of satellite services.

Rain Effects on System Performance

Rain is the dominant factor in satellite propagation for frequencies above 10 GHz. Rain propagation has been studied intensively since the late 1960s and only a brief discussion is presented here. Due to the basic interaction of electromagnetic waves with water in liquid form, raindrops cause absorption, scattering, and depolarization phenomena. Absorption and scattering

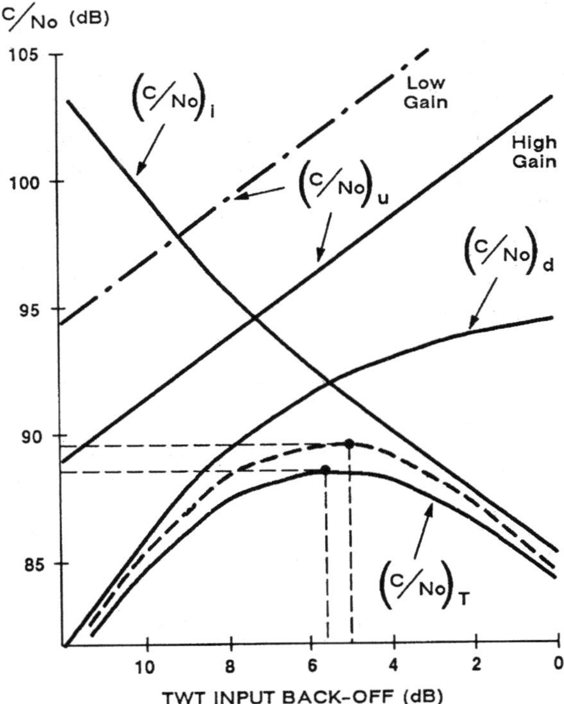

FIGURE 6.11-17 Optimum TWT operation.

result in signal attenuation and an increase in sky noise temperature, with the consequent degradation of the received C/N_o. Depolarization has an effect on dual polarization systems and creates interference between cross-polarized signals.

Signal Attenuation

The amount of attenuation depends fundamentally on the rain intensity or rain rate and the signal path length in rain. Rainfall data is available for most parts of the world; different types of climates have been defined and boundaries of their regions identified. Figures 6.11-18 through 6.11-20 show the NASA rain rate climate regions. The long-term behavior of rain is described by the cumulative probability distribution or exceedance curve. This gives the percentage of time that the rain rate exceeds a given value. Table 6.11-2 gives the rain-rate distribution values *versus* percent of year for the various rain climate regions of Figures 6.11-18 through 6.11-20. Figure 6.11-21 plots the rain rate cumulative probability distributions for the regions presented on the previous maps.

The calculation of the rain attenuation involves two basic steps. The first step is to determine the rain rate in mm/hr as a function of the cumulative probability of occurrence. This probability will be defined by the grade of service or availability of the link to be provided. The second step consists of the calculation of the actual rain attenuation associated with the rain rate that was exceeded with such probability.

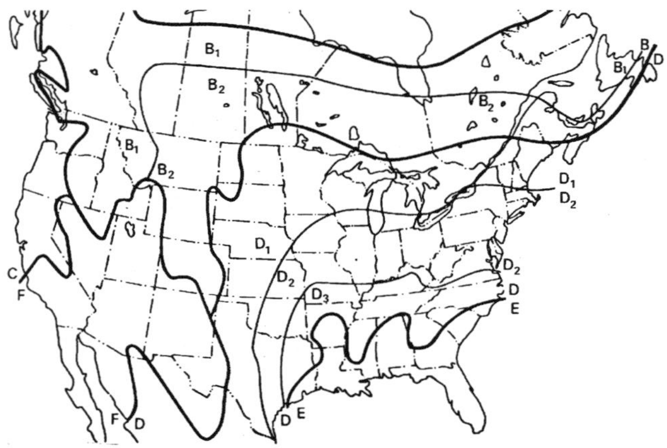

FIGURE 6.11-18 Rain rate climate regions for the continental U.S. showing the subdivision of Region D. (Reprinted with permission from *NASA Propagation Effects Handbook for Satellite Systems Design*, ORI TR 1679.)

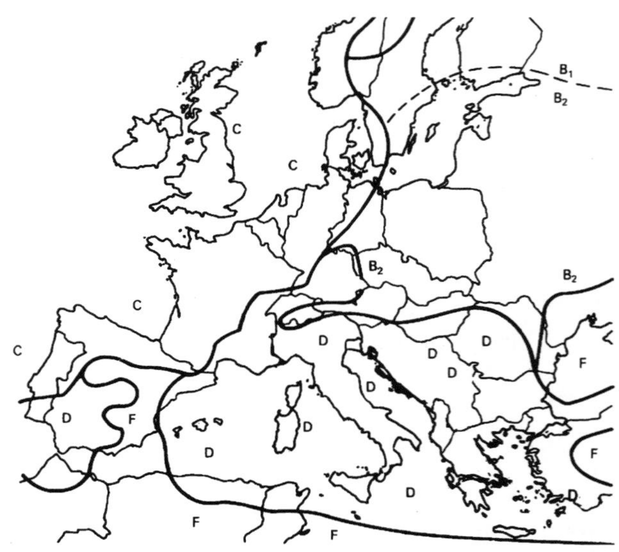

FIGURE 6.11-19 Rain rate climate regions for Europe. (Reprinted with permission from *NASA Propagation Effects Handbook for Satellite Systems Design*, ORI TR 1679.)

The attenuation per unit of length (specific attenuation), λ_r (dB/km), is tied to the rain rate R (mm/hr), by the empirically derived relationship:

$$\lambda_r = a(f)R^{b(f)} \quad \text{dB/km} \tag{25}$$

where $a(f)$ and $b(f)$ are frequency dependent coefficients. For the frequency range between 8.5 and 25 GHz, Equation (25) becomes:

$$\lambda_r = 4.21 \times 10^{-5} f^{2.42} \times 1.41 \times f^{-0.0779} R \quad \text{dB/km} \tag{26}$$

The attenuation per unit length is heavily frequency dependent; Figure 6.11-22 shows the frequency dependence of λ_r for various rain rates.

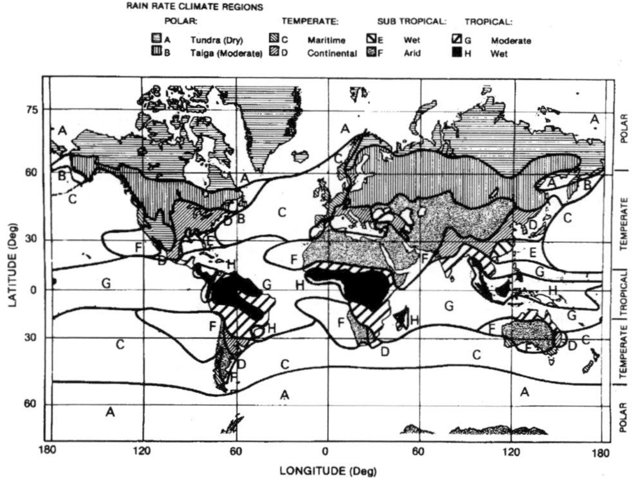

FIGURE 6.11-20 Global rain rate climate regions, including the ocean areas. (Reprinted with permission from *NASA Propagation Effects Handbook for Satellite Systems Design*, ORI TR 1679.)

Introducing the concept of equivalent path length, $L_e(R)$, the total rain attenuation in decibels is simply:

$$Ar = \lambda_r \times L_e(R) \quad \text{dB} \tag{27}$$

Equivalent path length is primarily determined by the height of the freezing level or 0° isotherm, which depends on latitude, season, and rain rate; the cosecant of the elevation angle; and site altitude. For latitudes within ±30°, the freezing level is at 4.8 km. Curves of equivalent path lengths *versus* elevation angle and for different rain rates are shown in Figure 6.11-23. The rain attenuation is required to be added to the satellite link as a margin to allow the specified availability under fading conditions. Figure 6.11-24 shows typical rain attenuations *versus* rainfall rate in the receive and transmit Ku-bands for different elevation angles.

Noise Contribution

In addition to the attenuation, rain also degrades the performance of a satellite link by increasing the earth station antenna noise temperature. In clear weather, the antenna sees the cold background of space, but in rain it receives thermal radiation from the raindrops. The increase in antenna noise temperature due to rain (T_r) may be estimated by:

$$T_r = 280(1 - 10^{-A/10}) \quad \text{K} \tag{28}$$

where A is the rain attenuation in decibels. Figure 6.11-25 shows the impact of the rain contribution of noise temperature on the normal clear sky G/T for different clear sky system temperatures. The G/T degradation corresponding to the rain attenuation for the stipulated link availability also must be added to the

TABLE 6.11-2
Point Rain Rate Distribution Values (Millimeters per Hour) *versus* Percent of the Year Rain Rate Is Exceeded

Percent of Year	Rain Climate Region										Minutes per Year	Hours per Year
	A	B	C	D_1	D_2	D_3	E	F	G	H		
0.001	28.0	54.0	80.0	90.01	102.0	127.0	164.0	66.01	129.0	251.0	5.3	0.09
0.002	24.0	40.0	62.0	72.0	86.0	107.0	144.0	51.0	109.0	220.0	10.5	0.18
0.005	19.0	26.0	41.0	50.0	64.0	81.0	117.0	34.0	85.0	178.0	26.0	0.44
0.01	15.0	19.0	28.0	37.0	49.0	63.0	98.0	23.0	67.0	147.0	53.0	0.88
0.02	12.0	14.0	18.0	27.0	35.0	48.0	77.0	14.0	51.0	115.0	105.0	1.75
0.05	8.0	9.5	11.0	16.0	22.0	31.0	52.0	8.0	33.0	77.0	263.0	4.38
0.1	6.5	6.8	7.2	11.0	15.0	22.0	35.0	5.5	22.0	51.0	526.0	8.77
0.2	4.0	4.8	4.8	7.5	9.5	14.0	21.0	3.8	14.0	31.0	1052.0	17.50
0.5	2.5	2.7	2.8	4.0	5.2	7.0	8.5	2.4	7.0	13.0	2630.0	43.80
1.0	1.7	1.8	1.9	2.2	3.0	4.0	4.0	1.7	3.7	6.4	5260.0	87.66
2.0	1.1	0.2	1.2	1.3	1.8	2.5	2.0	1.1	0.6	2.8	10,520.0	175.30

Source: NASA Propagation Effects Handbook for Satellite Systems Design, ORI TR 1679.

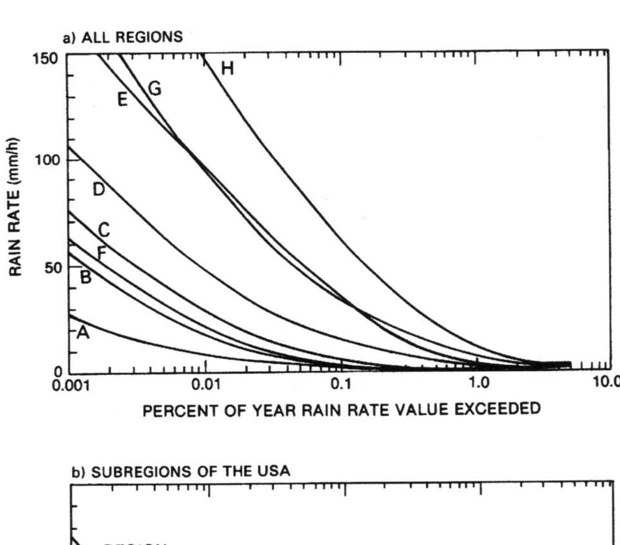

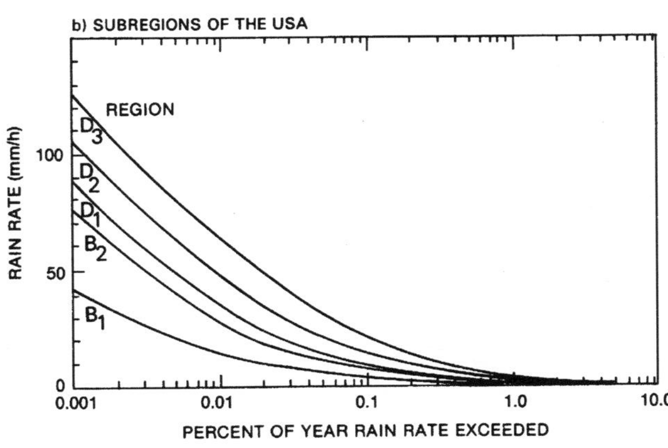

FIGURE 6.11-21 Rain rate cumulative probability distributions for the regions presented on the previous maps. (Reprinted with permission from Ippolito, L.J. et al., *Propagation Effects Handbook for Satellite Systems Design*, NASA Reference Publication 1082(03), National Aeronautics and Space Administration, Washington, D.C., June, 1983.)

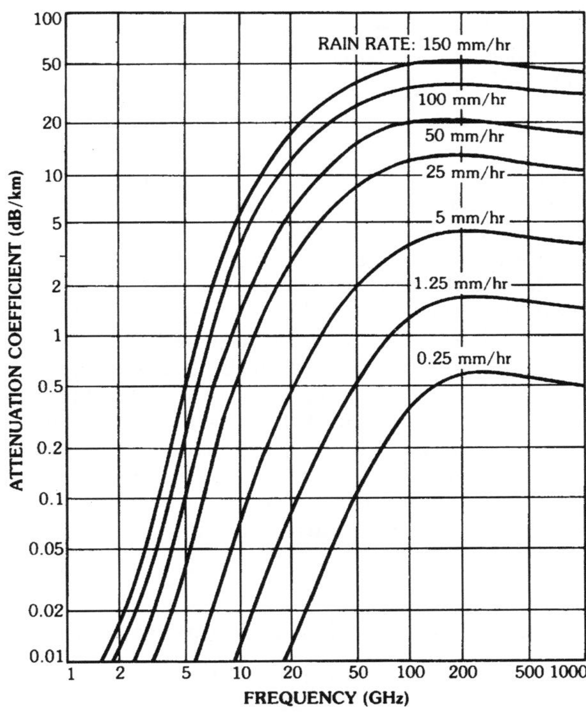

FIGURE 6.11-22 Attenuation per unit length *versus* frequency and rain rate. (Reprinted with permission from Miya, K., Ed., *Satellite Communications Technology*, KDD Engineering and Consulting, Inc., Tokyo, 1982.)

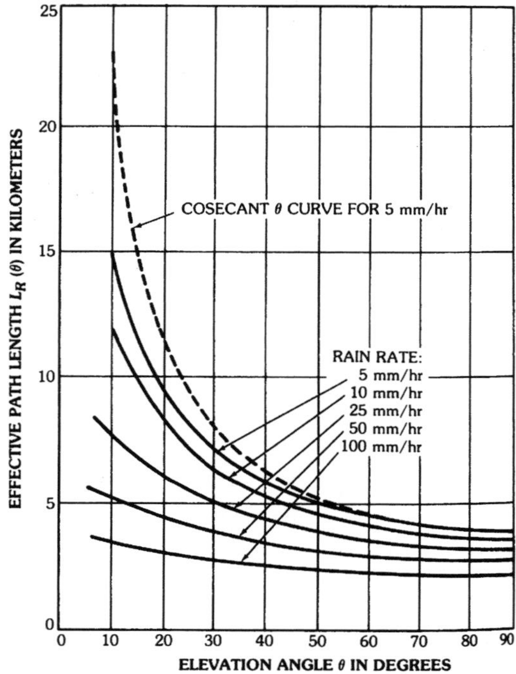

FIGURE 6.11-23 Equivalent path length *versus* rain rate and elevation angle. (Reprinted with permission from Miya, K., Ed., *Satellite Communications Technology*, KDD Engineering and Consulting, Inc., Tokyo, 1982.)

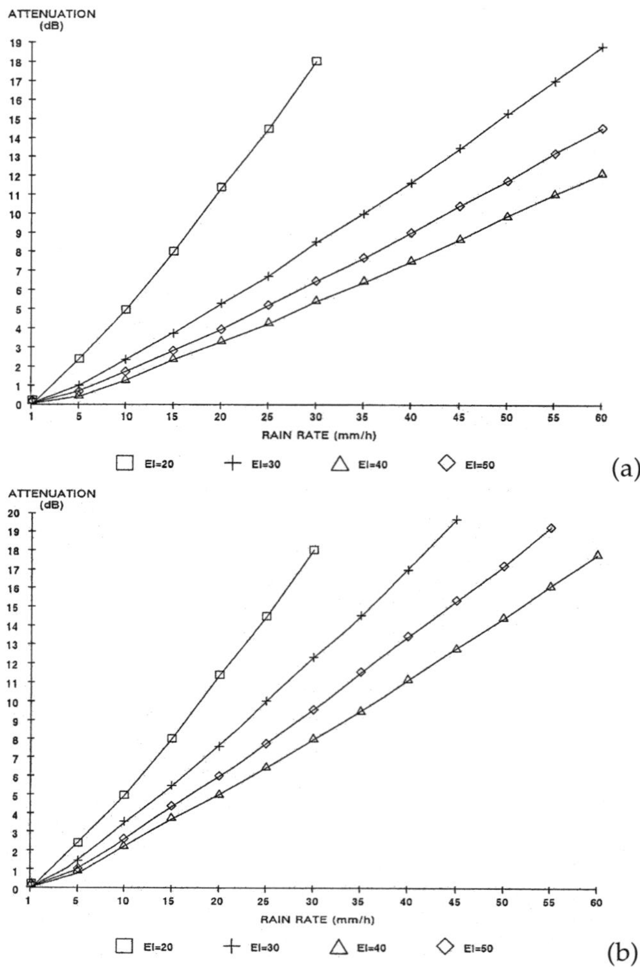

FIGURE 6.11-24 (a) Rain attenuation (11.95 GHz) 4.8 km 0° isotherm; (b) rain attenuation (14.25 GHz) 4.8 km 0° isotherm.

satellite downlink. This is to provide sufficient margin to compensate for the combined rain effect of signal attenuation and noise increase.

The allocation of rain fade margins in the uplink and downlink can be done independently, corresponding to specific availability requirements of the uplink or downlink and consistent with the availability requirement of the total link. The assumption is that, due to the localized nature of the rain fades, the uplink fade and downlink fade can be considered as two statistically independent processes; therefore, total link availability can be obtained as the reciprocal of the summation of the uplink and downlink outages calculated as if they occurred independently and one at a time.

Example of System Link Calculation: Analog

Table 6.11-3 shows a typical satellite link budget for an FM video uplink and downlink for Ku-band operation where a 5 dB uplink power control has been applied to mitigate the effects of rain fade in the uplink. A C-band link budget would contain similar terms but the uplink power control to mitigate the effects of rain would not be necessary.

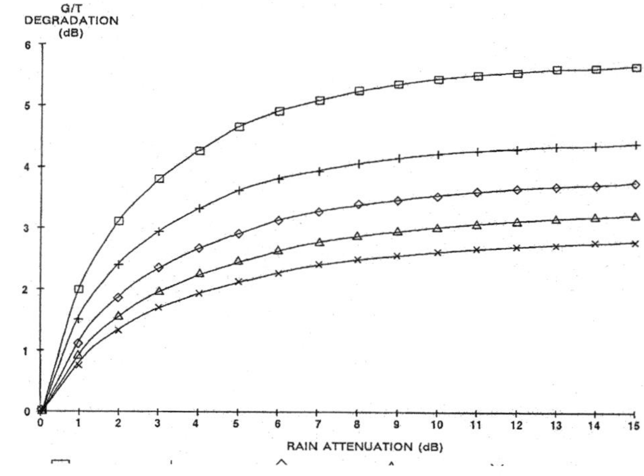

FIGURE 6.11-25 G/T rain degradation.

Baseband Performance Link Analysis

As stated previously, the overall quality of the delivered baseband signal can be expressed in analog systems by the S/N. The C/N_o *versus* S/N performance of different modulation schemes can be characterized by rather simple mathematical equations. The most common technique in satellite transmission is frequency modulation (FM). Because of its simplicity and the low cost of the receivers and demodulators, FM has historically been widely utilized in the transmission of television signals and is still in considerable use today. Equation (29) allows the computation of the S/N as a function of C/N_o and modulation parameters:

$$(S/N)_w = C/N_o + 10\log(12\Delta F^2/B_v^3) + W \qquad (29)$$

or in terms of carrier-to-total noise power ratio in bandwidth B:

$$(S/N)_w = C/N + 10\log(12AF^2/B_v^3) + 10\log(B/B_v) + W \qquad (30)$$

TABLE 6.11-3
Link Budget for Analog Ku-Band Video Satellite Link

Satellite	Spacenet-II				
Beam type	Conus (Continental U.S.)				
Type of service	FM/video				
Transmit/receive connectivity	7.0/4.5 m				
Occupied bandwidth per carrier	30.0 MHz				
Available bandwidth per carrier	36.0 MHz				
Transponder bandwidth	72.0 MHz				
Parameter		**Values**			
I. UPLINK NOISE		Clear Sky	Uplink Fade	Downlink Fade	Units
Earth station EIRP per carrier		75.0	80.0	75.0	dBW
Pointing losses		0.5	0.5	0.5	dB
Path loss		207.0	207.0	207.0	dB
Isotropic antenna area		44.5	44.5	44.5	dBW/m^2
Saturation flux density		–81.0	–81.0	–81.0	dBW/m^2
Rain attenuation		0.0	6.0	0.0	dB
G/T including footprint advantage		–1.1	–1.1	–1.1	dB/K
Input back-off per carrier		7.0	8.0	7.0	dB
Uplink thermal C/N		20.2	19.2	20.2	dB
Co-channel interference		27.0	24.0	27.0	dB
Off-beam emissions interference		26.0	25.0	26.0	dB
Total uplink C/(N+I)		18.5	17.2	18.5	dB
UPLINK AVAILABILITY	99.99%				
II. INTERMODULATION NOISE		20.0	19.0	20.0	dB
III. DOWNLINK NOISE					
Satellite saturation EIRP		43.0	43.0	43.0	dBW
Transponder output back-off/carrier		4.5	6.5	4.5	dB
EIRP per carrier		38.5	36.5	38.5	dBW
Path loss		206.0	206.0	206.0	dB
Rain attenuation		0.0	0.0	3.0	dB
Pointing losses		0.5	0.5	0.5	dB
Earth station G/T		29.5	29.5	29.5	dB/K
G/T degradation		0.0	0.0	2.2	dB
Downlink thermal C/N		15.3	13.3	10.1	dB
Co-channel interference		27.0	25.0	19.8	dB
Total downlink C/(N + I)		15.0	13.0	9.7	dB
DOWNLINK AVAILABILITY	99.85%				
IV. TOTAL C/(N + I) NOISE					
Total C/(N + I)		12.6	10.9	8.8	dB
Occupied bandwidth per carrier		74.8	74.8	74.8	dB–Hz
$C/(N_o + I_o)$ total		87.4	85.7	83.6	dB–Hz
Required C/N_o		83.5	83.5	83.5	dB–Hz
Margin		3.9	2.2	0.1	dB
TOTAL LINK AVAILABILITY	99.75%				

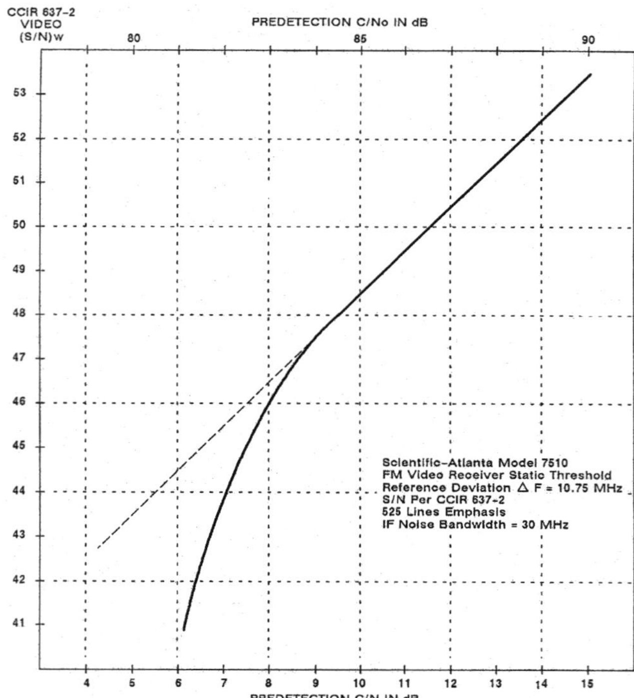

FIGURE 6.11-26 Signal-to-noise ratio performance of FM video demodulator *versus* C/N and C/N_o.

where:

$(S/N)_w$ = weighted signal-to-noise ratio (dB).

C/N_o = carrier-to-noise density ratio (dB).

ΔF = peak composite video deviation (MHz).

B = IF predetection noise bandwidth (MHz).

B_v = video filter bandwidth (MHz).

W = deemphasis and weighting improvement (dB).

For NTSC and 30 MHz bandwidth these parameters are typically ΔF = 10.75 MHz, B = 28 MHz, B_v = 4.2 MHz, and W = 13.8 dB. Figure 6.11-26 shows the result of Equation (29) with the previous transmission parameters at high levels of C/N. The departure from a linear relationship at low values of C/N is not predicted by Equation (29) but represents the actual performance of a typical FM demodulator. This phenomenon is known as the *threshold effect*.

Example of Digital System Link Calculation

In digital satellite systems, the most common modulation technique is phase-shift keying (PSK). Variations of this technique are known as binary PSK (BPSK) when logic symbol zeros and ones are mapped into RF signals 180° apart in phase and as quadrature PSK (QPSK) when the phases are 90° apart. In recent years, 8PSK, using 45° phases, has also come into widespread use. The BER performance of these systems is evaluated as a function of

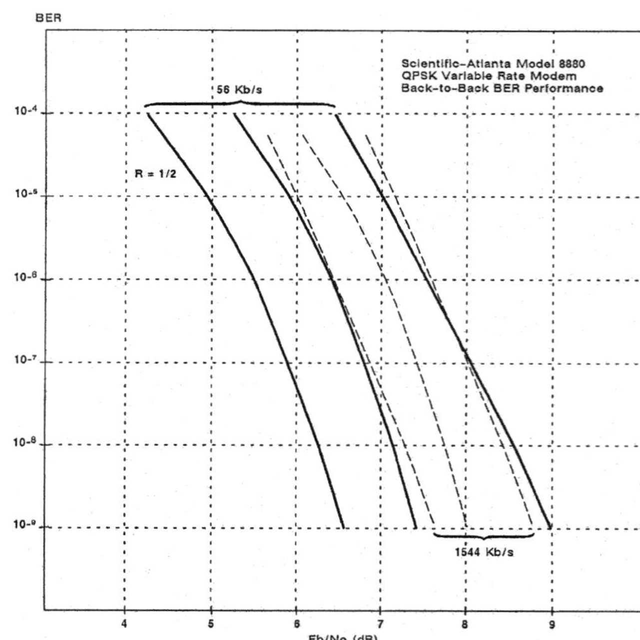

FIGURE 6.11-27 QPSK modem BER performance *versus* bit energy over noise density ratio.

the energy per bit of information transmitted (E_b) *versus* noise density (E_b/N_o). Sophisticated digital coding and decoding techniques exist that, by adding error control bits to the information data stream, allow substantial improvements in BER that can translate into transmit power reductions of up to 5 dB. The ratio between the uncoded data rate and the coded one is the *coding rate* (R). Figure 6.11-27 shows the performance of a typical QPSK modulator and demodulator for different data and coding rates.

Table 6.11-4 shows a typical link budget for a digital video signal without uplink power control. Rate 3/4 FEC and 188/204 Reed-Solomon block coding are applied. Performance is in accordance with the DVB-S specification.

Earth Station Block Diagram

The block diagram of Figure 6.11-28 depicts an earth station capable of providing uplink services for analog and digital video and digital data in the vertical and horizontal polarizations. All subsystems are redundant for maximum reliability. A computer-based monitor and control system, by means of a serial control bus, offers centralized operation of the complete earth station with the ability of monitoring all status and controlling all variable parameters of every subsystem from the local or remote terminals.

Figure 6.11-29 shows the block diagram of the corresponding dual polarization receive-only terminal. This low-cost earth station, with an L-band (950 to 1450 MHz) interfacility link (IFL), can provide simultaneous reception of analog and digital video and audio. Data capability (not shown) can be readily

TABLE 6.11-4
Link Budget for Digital Ku-Band Video Satellite Link

Satellite	SES Americom AMC-3			
Beam type	Conus (Continental U.S.)			
Type of service	Digital SCPC			
Transmit/receive connectivity	5.0 m New York to 3.0 m Dallas			
Information rate	3.0 Mbps			
Combined forward error correction (FEC)	0.69, using rate 3/4 convolutional coding and 204/188 Reed-Solomon block coding			
Transponder bandwidth	36.0 MHz			
Parameter	**Values**			
I. UPLINK NOISE	Clear Sky	Uplink Fade	Downlink Fade	Units
Earth station EIRP per carrier	55.3	55.3	55.3	dBW
Pointing losses	0.7	0.7	0.7	dB
Path loss	207.2	207.2	207.2	dB
Isotropic antenna area	44.5	44.5	44.5	dBW/m^2
Saturation flux density	–80.8	–80.8	–80.8	dBW/m^2
Rain attenuation	0.0	5.3	0.0	dB
G/T including footprint advantage	4.3	4.3	4.3	dB/K
Input back-off per carrier	18.7	24.0	18.7	dB
Uplink thermal C/N	15.5	10.2	15.5	dB
Co-channel interference	26.1	20.8	26.1	dB
Off-beam emissions interference	18.6	13.4	18.6	dB
Total uplink C/(N + I)	13.6	8.2	13.6	dB
UPLINK AVAILABILITY 99.93%				
II. INTERMODULATION NOISE	21.7	16.4	21.7	dB
III. DOWNLINK NOISE				
Satellite saturation EIRP	46.0	46.0	46.0	dBW
Transponder output back-off	4.0	4.3	4.0	dB
EIRP per carrier	31.6	26.3	31.6	dBW
Path loss	205.5	205.5	205.5	dB
Rain attenuation	0.0	0.0	6.0	dB
Pointing losses	0.6	0.6	0.6	dB
Earth station G/T	26.7	26.7	26.7	dB/K
G/T degradation	0.0	0.0	3.5	dB
Downlink thermal C/N	16.1	10.8	6.6	dB
Co-channel interference	21.1	15.8	18.1	dB
Total downlink C/(N + I)	14.4	9.6	6.4	dB
DOWNLINK AVAILABILITY 99.93%				
IV. TOTAL C/(N + I) NOISE				
Total C/(N + I)	10.8	5.5	5.5	dB
Occupied bandwidth per carrier	64.8	64.8	64.8	dB–Hz
$C/(N_o + I_o)$ total	75.6	70.3	70.3	dB–Hz
Required E_b/N_o	5.5	5.5	5.5	dB
Margin	5.3	00.0	0.0	dB
TOTAL LINK AVAILABILITY 99.86%				

added. For simplicity, analog reception is shown on one polarity and digital on another, but in an actual terminal either polarity can receive and process either or both types of video, as well as multiple signals.

Interference Considerations

The consideration of interference in a satellite communication system is important, not only from the standpoint of interference to the desired satellite signal but also as it applies to the satellite's signals generating interference into other systems. The FCC requires a proposal for a satellite transmit system in the U.S. to include a coordination filing with an interference analysis. This analysis must show the impact of the proposed system on existing operational systems and must satisfy the allowable interference requirements of the FCC.

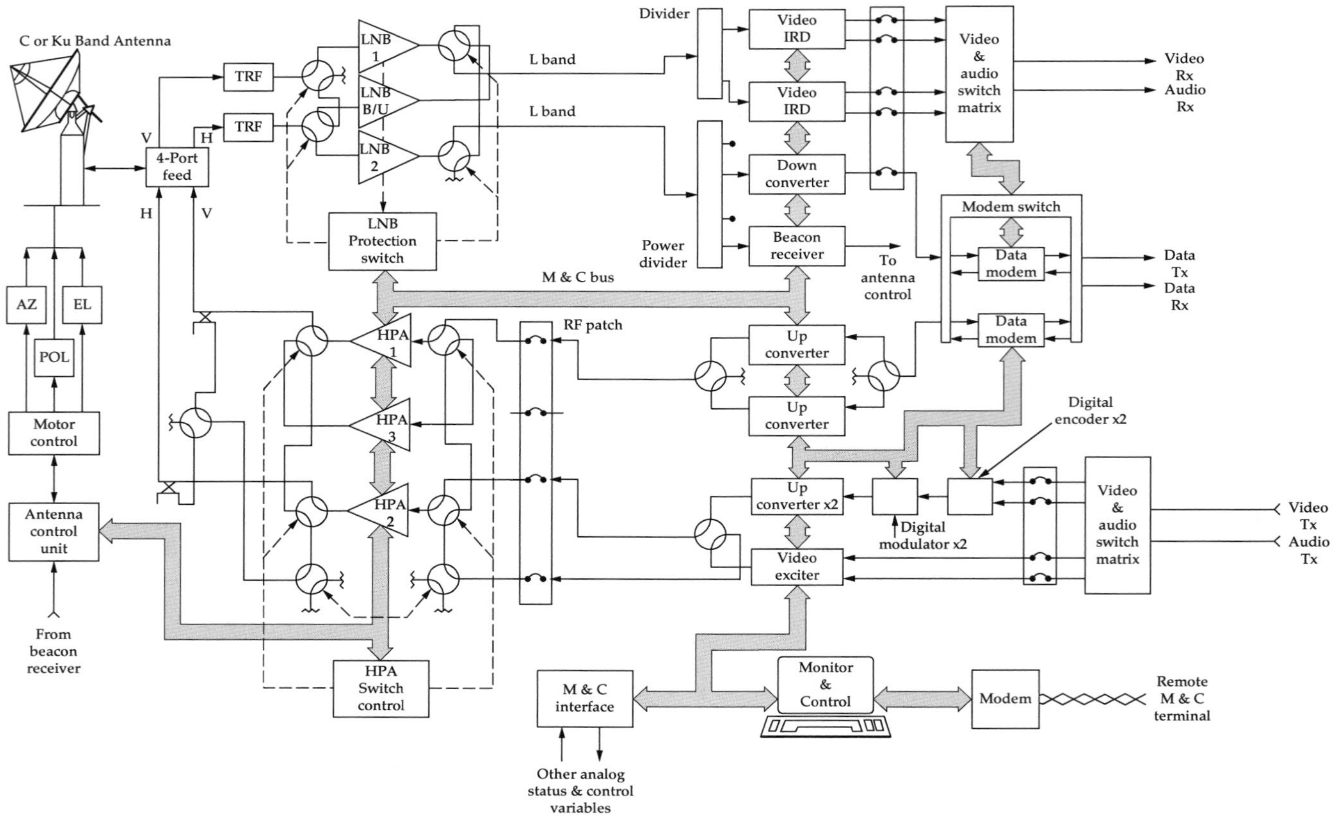

FIGURE 6.11-28 Digital video and data broadcast earth station block diagram.

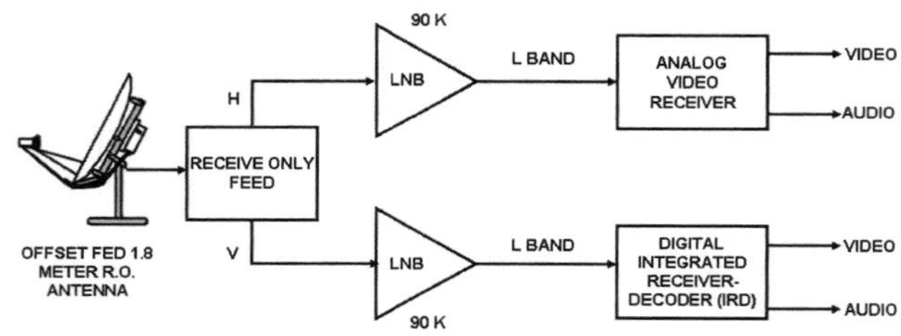

FIGURE 6.11-29 Video/audio receive-only earth station.

Antenna Characteristics

The primary characteristic of the antenna that affects the interference analysis is the angular discrimination, defined as the gain differential between the on-axis gain and the gain of an off-axis angle for the interfering source. Both receive and transmit antenna characteristics should be considered. Off-beam cross-polarization isolation of the antenna should also be considered, and in this regard it should be noted that during periods of heavy rain the polarization of the incoming and outgoing signals may be affected such that the full cross-polarization isolation is not realized.

Sources of Interference

Interference into a geostationary satellite communication system can originate from several sources, including:

- Adjacent satellite uplink or downlink signals
- Internal cross-polarization signals (half transponder frequency offsets)
- Terrestrial microwave or radar detector signals

Adjacent Satellite Interference

Interference from adjacent satellites occurs in two ways: (1) uplink interference from earth stations transmitting to adjacent satellites and (2) downlink interference from adjacent satellite transmission into the desired earth station. The interference in both the uplink and downlink consists of many possible sources, but it is primarily caused by the co-frequency channels/transponders and the two half transponder bandwidth offset-channels/transponders in a frequency reuse system. The particular interferers for a 40 MHz transponder bandwidth system as shown in Figure 6.11-30 are the following:

1. The two 20 MHz offset-frequency, co-polarized transponders on the first adjacent satellite on either side

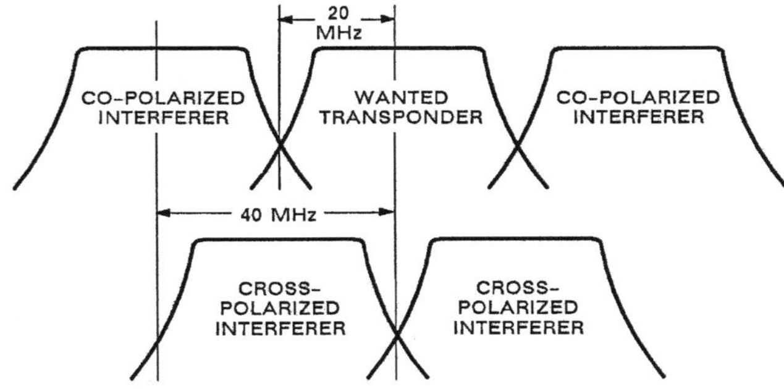

FIGURE 6.11-30 40 MHz transponder frequency reuse transponder plan.

2. The co-frequency, cross-polarized transponders on the first adjacent satellite on each side (*Note:* In Figure 6.11-31, even though the cross-polarized antenna characteristic is nominally 10 dB less than the co-polarized antenna characteristic, this specification is not always met in practice, particularly on receive.)‘
3. The co-frequency, co-polarized transponders on the second adjacent satellite on either side
4. The two 20 MHz offset-frequency, cross-polarized transponders on the second adjacent satellite on either side

The contribution to interference from satellites at orbital positions greater than 4° from the desired satellite tends to be noiselike in that it is the result of a number of small, relatively noncoherent signals.

The equations for calculation (in dB) of the adjacent satellite interference are given below:

$$(C/I)_u = (EIRP)_{es} - \sum_{i=1}^{N} \oplus \{(EIRP)_i - (G_i - G(\theta_i)) + F_i + P_i\} \tag{31}$$

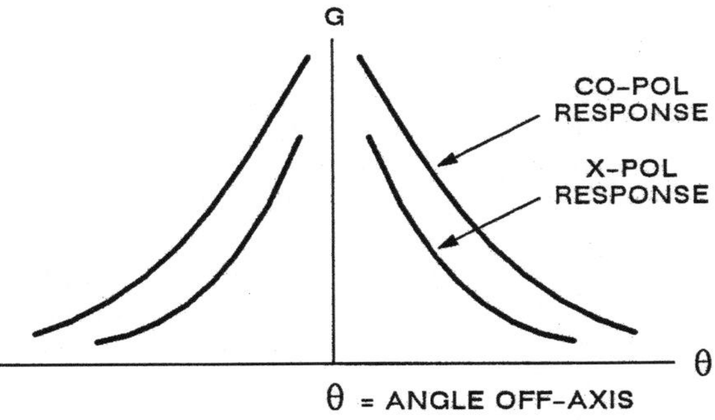

FIGURE 6.11-31 Earth station antenna radiation characteristics.

where:

$\sum_{i=1}^{N} \oplus$ = Series power summation.

$(EIRP)_{es}$ = earth station radiated power in dBW.

$(EIRP)_i$ = effective radiated power of interfering earth station in dBW.

G_i = peak gain of the interfering earth station in dBi.

$G(\theta)_i$ = gain of the interfering earth station in direction θ in dBi.

F_i = frequency discrimination factor for the *ith* earth station.

P_i = polarization discrimination factor for the *ith* earth station.

N = number of interfering earth stations

$$(C/I)_d = (EIRP)_{sat} + G_{es} - \sum_{j=1}^{M} \oplus \{(EIRP)_j - G_{es}(\theta_j) - F_j - P_j\} \tag{32}$$

where:

$(EIRP)_{sat}$ = effective radiated power of satellite in the direction of the receive earth station in dBW.

$(EIRP)_j$ = effective radiated power of interfering transponder j

G_{es} = gain of the receive earth station in dBi (main beam receive gain).

$G_{es}(\theta_j)$ = gain of the receive earth station in direction θ_j in dBi.

F_j = frequency discrimination factor for the jth transponder.

P_j = polarization discrimination factor for the jth transponder.

M = number of interfering transponders on adjacent satellites considered ($M \geq 3$).

The total adjacent satellite interference is then calculated by combining the uplink and downlink contributions in a power summation manner:

$$(C/I)_{adj.sat} = (C/I)_u \oplus (C/I)_d \tag{33}$$

The polarization discrimination factor in the previous equations is the system discrimination rather than that of the receive or transmit antenna alone. A well-designed dual linearly polarized antenna can achieve excellent cross-polarization discrimination on or near the main beam axis (greater than 35 dB or 40 dB relative to the co-polarized energy) and reasonable rejection of the cross-polarized signals in the close-in sidelobe regions. The adjacent satellite signals are received through the sidelobes of the earth station, and the [19 – 25log(θ)] envelope is normally assumed. This assumption, rather than being conservative, may be optimistic when one considers various factors such as the interactions of the ionosphere and atmosphere on the transmitted signals (from the earth station and/or the satellites) and actual off-axis cross-polarization antenna performance. A more conservative analysis may assume a slightly reduced discrimination of perhaps [21 – 25log(θ)] for the cross-polarized sidelobe energy in the off-axis regions. The frequency discrimination factor is related to the spectra of the desired and undesired signals. This factor can range from 0 to 30 dB depending on the interfering power from the different services.

Internal Interference

The internal interference in a satellite system utilizing 40 MHz transponders is primarily due to the cross-polarized 20 MHz offset transponders. The interfering power from different services is calculated by convolving the power spectra of the individual services or by addition on a power basis. This data, taken together with the appropriate polarization discrimination term, determines the amount of interference.

Terrestrial Interference

Terrestrial microwave carriers are centered on frequencies offset by 10 MHz from the satellite carriers. To analyze the effects of terrestrial carriers, it is necessary to determine the power level of the interfering signal and the spillover of the terrestrial carrier spectra into the passband of the receiver. The first factor involves site details, such as angular discrimination and distance to the interfering transmitter. The second factor can be computed from the spectral distribution projected for the terrestrial carrier and the filter characteristic of the receiver. Other sources of terrestrial interference to satellite circuits can include:

- L-band mobile radio
- Radio altimeters (C-band)
- Military radars (C-band)
- Vehicle-borne radar detectors (Ku-band)

Sun Transits and Eclipses

Communications satellite systems experience predictable service interruptions involving the sun. A sun transit outage occurs when the pointing angles from a receiving earth station to a satellite and to the sun so nearly coincide that the additional noise power presented by the sun renders transmission unusable. A solar eclipse occurs when the earth shadows the sun from the satellite. The eclipse event is not as serious as the sun transit event because the satellite has battery backup systems to augment the solar primary power.

Daily sun transits of all geostationary satellites serving an earth station occur during one week in the spring and again in the fall. The exact dates depend primarily on the latitude of the receiving earth station. The geometry and duration associated with a sun transit are controlled by the off-axis gain of the earth station antenna, the receiving system noise temperature, the solar noise power profile, and the minimum acceptable S/N ratio. In late February or early March, short daily outages affect earth station systems situated near the U.S.–Canadian border. Two or three days later, these systems experience maximum outages lasting 5 minutes or more, depending on transmission parameters and permissible S/N. Outages at these earth station locations end after an additional 2 to 3 days, and the sun transit outage paths progress southward at a rate of about 3° latitude per day. All outages affecting U.S. earth station antenna systems above north latitude 26° cease prior to mid-March. Conversely, in the fall, the daily outages progress from south to north, affecting southern U.S. earth stations beginning October 1 and ending in the north about mid-October.

Eclipses of geostationary satellites can be expected for a total of about 90 evenings per year in the spring and fall. Eclipses occur near apparent midnight of the time zone at each satellite's longitude, beginning in late February or early March and ending mid-April. Fall events begin about September 1 and end mid-October. Eclipses of about 70 minutes' duration occur on the dates of the spring and fall equinoxes. Communication satellites are provided with batteries to prevent circuit outages and to maintain pointing, attitude control, stationkeeping, telemetry, and command capabilities during eclipses.

EQUIPMENT CHARACTERISTICS

An earth station system is made up of four major subsystems:

- Antenna subsystem
- Transmitting subsystem
- Receiving subsystem
- Monitor and control subsystem

Antenna

The antenna provides the means of transmitting signals to the satellite and collecting the signal transmitted by the satellite. The antenna must not only provide the gain necessary to allow proper transmission and reception but must also have radiation characteristics that discriminate against unwanted signals and minimize interference into other satellite or terrestrial systems. A further function of the antenna is to provide the means of polarization discrimination of unwanted signals. The individual communication system operational parameters dictate to the antenna designer the necessary electromagnetic, structural, and environmental specifications necessary for the antenna. Antenna requirements can be grouped into several major categories: electrical or RF, control systems, pointing and tracking accuracies, and environmental and miscellaneous requirements such as radiation hazard and primary power distribution. Table 6.11-5 summarizes many of the more important parameters of an earth station antenna.

TABLE 6.11-5
General Considerations for Earth Station Antenna Design

Electrical Performance	Mechanical Performance	System Considerations
Frequency (bandwidth)	Angular travel	Operational function
Gain	Drive speed and acceleration	Local and/or remote operation
Noise temperature	Pointing and tracking accuracies	Availability and maintainability
Radiation pattern	Compatibility and environmental conditions	Design lifetime
Polarization	Reflector surface accuracy	Interface conditions with other subsystems
Axial ratio	Physical dimensions	Space needed in antenna hub for electronics
Voltage standing wave ratio (VSWR)	Weight	
Power handling capability	Ability to survive in strong winds	
Port-to-port isolation		
Out-of-band emissions		

Electrical Performance

The primary electrical specifications of an earth station antenna are gain, noise temperature, voltage standing wave ratio (VSWR), power rating, receive/transmit group delay, radiation pattern, polarization, axial ratio, isolation, and G/T. All of the parameters except the radiation pattern are determined by the system requirements. The radiation pattern should meet the minimum requirements set by the FCC and/or the ITU-R. Earth stations that operate in a regulated environment in the U.S. domestic system must meet the requirements set forth in the FCC regulations for earth station antennas pertaining to antenna aperture diameter, sidelobes, and/or radiated power density (see Part 25, Paragraph 25.209, of the FCC Rules and Regulations). The desired radiation properties to satisfy the communication system design dictate the choice of the type of antenna to be employed as an earth station. The three most important radiation properties are gain, sidelobe performance, and noise temperature. Most earth station antennas are designed to maximize gain and minimize noise, thereby maximizing G/T. These two criteria have led to the predominance of reflector-type antennas for earth station applications, although other types of antennas such as arrays and horns have been used.

Types of Earth Station Antennas

Several types of earth station antennas are in use in the U.S. and abroad. These antennas can be grouped into two broad categories: *single beam* and *multiple beam*. A single beam earth station antenna is defined as an antenna that generates a single beam pointed toward a satellite by means of a positioning system. A multiple beam earth station antenna is defined as an antenna that generates multiple beams by employing a common reflector aperture with multiple feeds illuminating that aperture. The axes of the beams are determined by the location of the feeds. The individual beam identified with a feed is pointed toward a satellite by positioning the feed without moving the reflector. The dual-frequency antennas may be considered another class of antennas as they produce two coincident simultaneous beams and as such are categorized as single beam antennas.

Single Beam Antennas

The majority of the earth station antennas in use is single beam antennas. Single beam antenna types used as earth stations are paraboloidal reflectors with focal point feeds (prime focus antenna), dual reflector antennas such as the Cassegrain and Gregorian configurations, horn reflector antennas, offset-fed paraboloidal antennas, and offset-fed multiple-reflector antennas. Each of these antenna types has its own unique characteristics, advantages, and disadvantages to be considered when choosing them for a particular application.

Axisymmetric Dual-Reflector Antennas

The predominant choice of many system operators has been the dual-reflector Cassegrain antenna. Cassegrain antennas can be divided into three primary types. The classical Cassegrain geometry employing a paraboloidal contour for the main reflector and a hyperboloidal contour for the subreflector (Figure 6.11-32). The paraboloidal reflector is a point focus device with diameter D_p and focal length f_p. The hyperboloidal subreflector has two foci. For proper operation, one of the two foci is the real focal point of the system and is located coincident with the phase center of the feed; the other focus, the virtual focal point, is located coincident with the focal point of the main reflector.

A geometry consisting of a paraboloidal main reflector and special-shaped, quasi-hyperboloidal subreflector shown in Figure 6.11-33, is appropriate for describing this antenna. The main difference between the classical Cassegrain and this antenna is that the subreflector has been designed such that the overall efficiency of the antenna has been enhanced, thereby yielding improved gain performance. This technique is especially useful with antenna diameters of approximately 30 to 100 wavelengths—for example, a 5 m antenna in the 6/4 GHz frequency band.

A generalization of the Cassegrain geometry consists of a specially-shaped, quasi-paraboloidal main

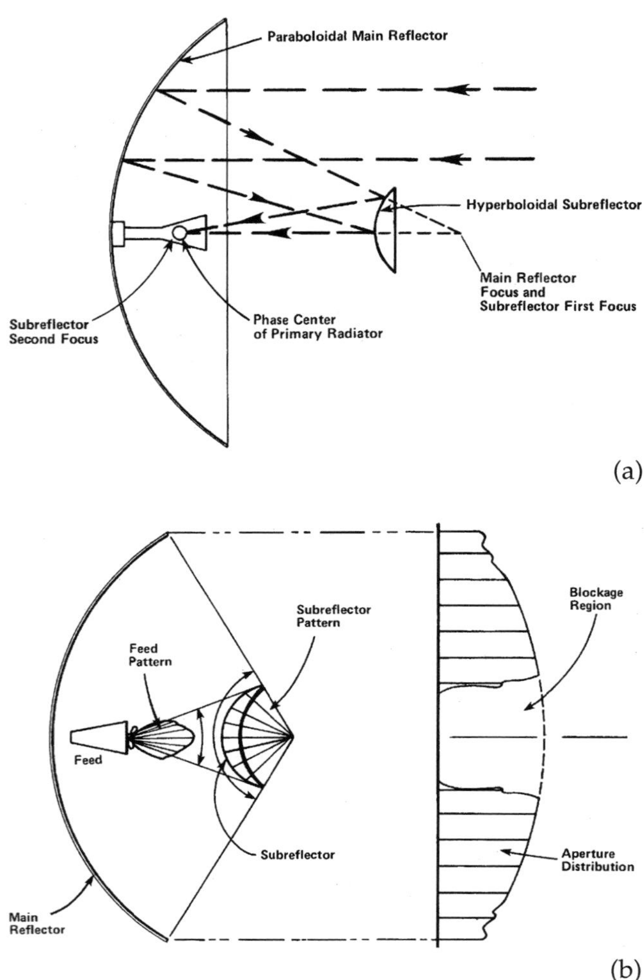

FIGURE 6.11-32 (a) Cassegrain antenna geometry; (b) aperture distribution of a Cassegrain antenna.

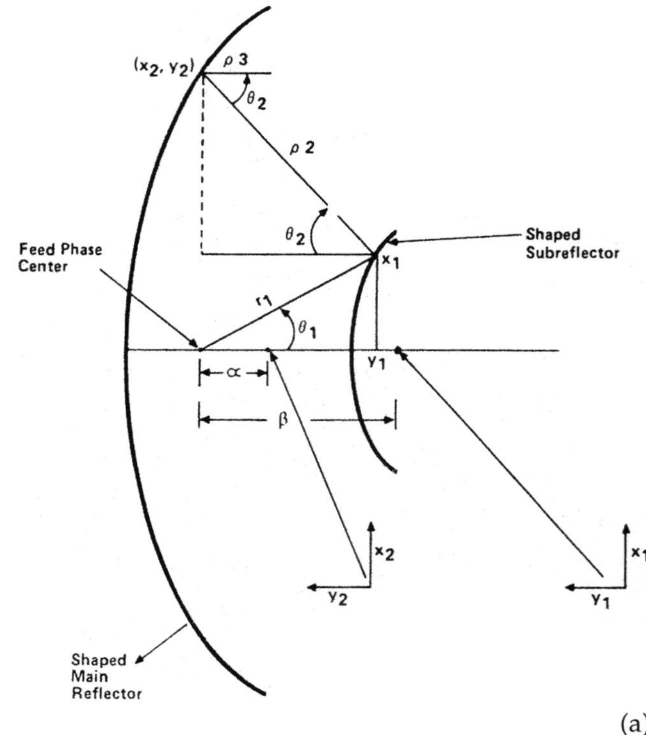

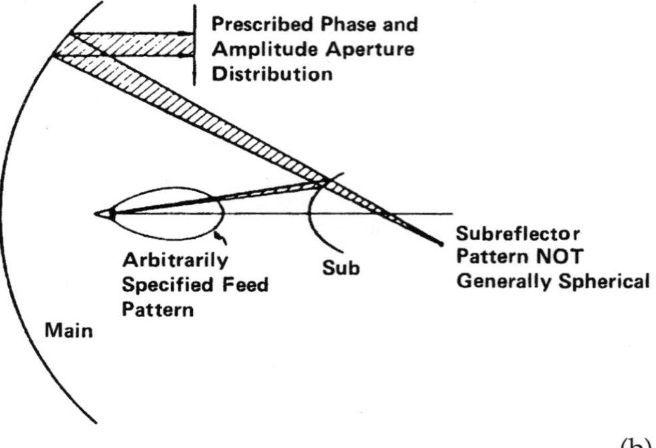

FIGURE 6.11-33 (a) Dual-shaped reflector geometry; (b) circularly symmetric dual-shaped reflectors.

reflector and a specially-shaped, quasi-hyperboloidal subreflector. The subreflector is shaped to redistribute its incident energy such that the illumination of the main reflector is optimized for high gain and desired radiation pattern. The main reflector is then shaped to correct the phase of the aperture field such that it is in phase. The feed must have a high beam efficiency, and its radiation pattern should be circular symmetric. This technique allows the antenna designer to synthesize the surfaces to achieve an arbitrary aperture distribution. The dual reflector antenna offers excellent gain performance, and for aperture sizes larger than approximately 75 wavelengths the sidelobe performance can meet the FCC pattern requirements. Dual reflector designs are employed for earth station antennas for apertures as small as 50 wavelengths to as large as 500 wavelengths.

Prime Focus-Fed Paraboloidal Antennas

The prime focus-fed paraboloidal (PFFP) antenna is another of the most often employed antennas for earth stations. This type of antenna can have excellent sidelobe performance in all angular regions except the spillover region around the edge of the reflector, but even in this region the pattern requirements of the FCC can be met. This antenna configuration has a lower cost than dual reflector antennas and offers a good compromise choice between gain and sidelobes. Its basic limitations are its location of the feed for transmit applications and, for aperture sizes less than approximately 30 wavelengths, the blockage of the feed and the feed support structure raises the sidelobes with respect to the main beam such that it becomes exceedingly difficult to meet the FCC sidelobe requirements. The PFFP antenna is used for many receive-only earth station antennas as well as for transmit/receive applications when only one transmit polarization is required.

Offset-Fed Reflector Antennas

The offset-fed reflector antenna (Figure 6.11-34) was originally used primarily in small-aperture antennas for VSAT applications but is now used for larger antennas as well. The offset-fed reflector antenna can employ a single main reflector or multiple reflectors, with two reflectors being the more prevalent of the multiple reflector designs. The offset, front-fed reflector, consisting of a section of a paraboloidal surface, eliminates the direct aperture blockage from the feed and feed supports and minimizes diffraction scattering by removing the feed and feed support structure from direct illumination of the aperture current distribution. Limitations of the offset-fed single-reflector antenna can include its polarization performance, reduced cross-polarization performance off-axis for linear polarizations, and beam squints in opposite directions for two orthogonal circular polarizations. The actual offset feed can be located above or below the antenna.

These antennas are sometimes selected for use to avoid the need for deicing systems in some locations where the need for deicing is not great. This is because an offset-fed reflector is oriented more toward the vertical than a prime focus-fed antenna. Use of an offset-fed antenna and either a cover or a hydrophobic coating can obviate the initial and recurring cost of antenna deicing systems.

The offset dual-reflector antenna (Figure 6.11-35) can be designed to have all the desirable characteristics of an axisymmetric antenna with increased gain and lower sidelobes. The polarization problems sometimes associated with the offset single-reflector design can also be compensated for with a two-reflector antenna design. The only disadvantages of the offset dual-reflector antenna are its cost of manufacturing for large apertures consisting of multiple sections and the complexity of its mount geometry and associated cost.

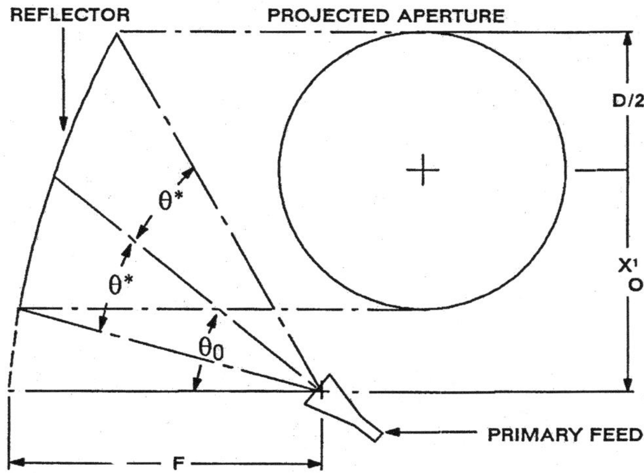

FIGURE 6.11-34 Offset single-reflector antenna.

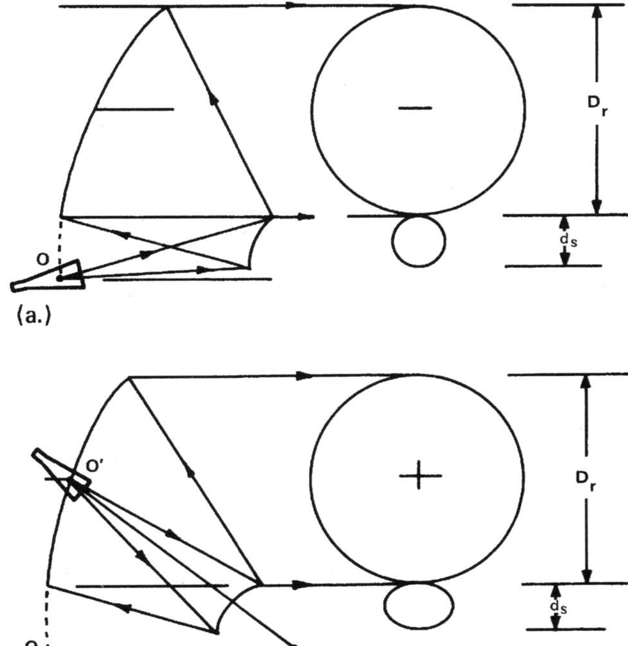

FIGURE 6.11-35 Offset dual-reflector geometries: (a) double-offset geometry (feed phase center and paraboloidal vertex at 0); (b) open Cassegrainian geometry (feed phase center located at 0′; paraboloidal vertex at 0).

Multiple Beam Antennas

Several multiple-beam antenna (MBA) configurations are used for earth station applications (see Figure 6.11-36). These include the spherical reflector, the torus antenna, and a class of offset-fed Cassegrain antennas. All of these configurations employ multiple feeds to generate the multiple beams. The multiple feeds must be physically small such that the individual beams may be pointed at desired satellites. When the desired satellites are spaced as close as 2° apart, the MBA antenna may not be practical. The obvious advantage of the MBA antenna is that a single antenna installation can transmit or receive signals to several satellites simultaneously, such as the case in DBS receive applications where three LNBs are used to receive three satellites with one antenna. Multiple-beam antennas can be particularly attractive where space is limited and many simultaneous feeds are required or when high reliability is required and antennas cannot easily be repointed in the event of a satellite failure. The disadvantages are the complexity of the feed arrangements for maintaining pointing at several satellites at the same time when the primary antenna aperture, the main reflector, remains fixed with respect to the earth's coordinates and the stringent requirements for the initial installation of the antenna system. Another disadvantage is that the MBA is not easily steerable in the dimension perpendicular to the orbital arc and can therefore be difficult to use with inclined orbit satellites.

Reflector Feed Configurations

There are many different feed configurations used in earth station antennas. The feed configurations are typically classified by the number of transmit and

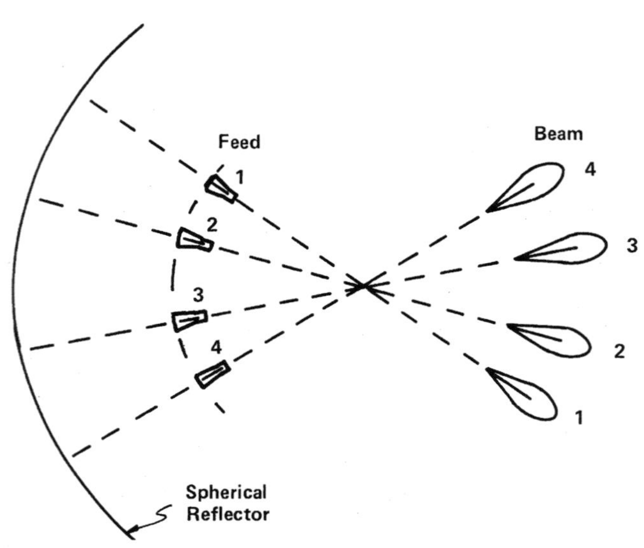

Conventional spherical multibeam antenna using extended reflector and multiple feeds

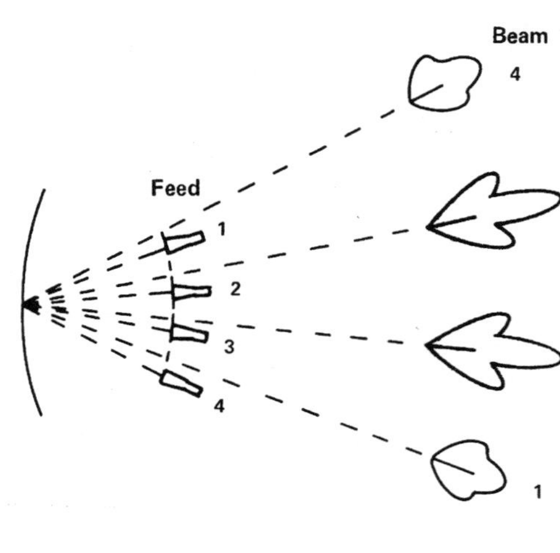

Alternative spherical multibeam antenna using minimum reflector aperture with scanned beam feeds

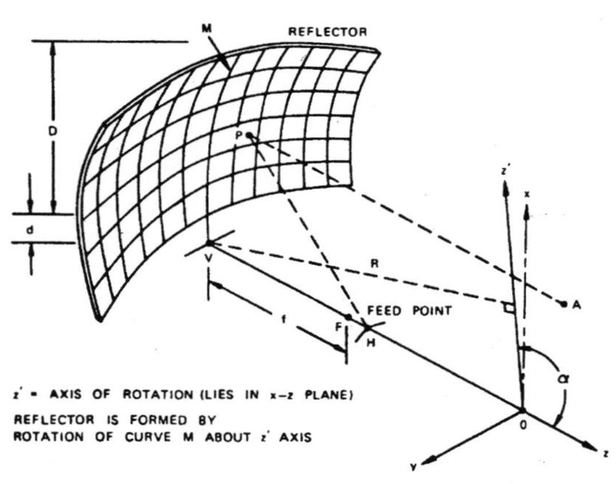

Torus-antenna geometry
(Copyright 1974, *COMSAT Technical Review*. Reprinted by permission.)

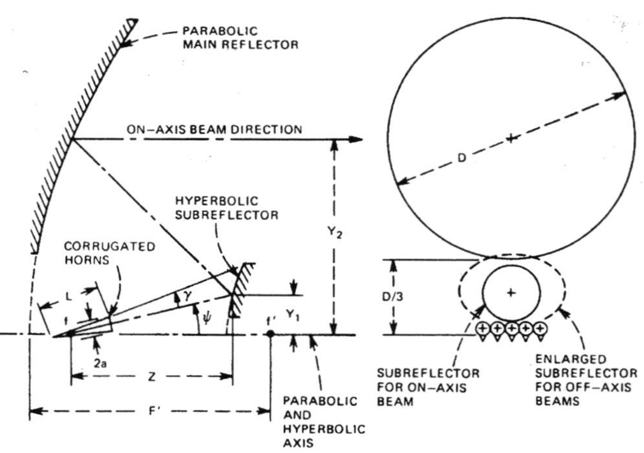

Geometry of the offset-fed multibeam Cassegrain antenna.
(Copyright, 1974, *American Telephone and Telegraph Company*. Reprinted by permission.)

FIGURE 6.11-36 Multiple-beam configurations.

receive ports available. The frequency bands of operation are those specified above for FSS operation in the U.S. or the appropriate FSS bands for international services. Note that C-band or Ku-band refers to the frequency segments for both transmit and receive bands. The more popular feed systems are classified as follows:

- The two-port feed configuration may have two orthogonally polarized receive ports or a single transmit port and a single receive port. The transmit and receive ports may be either co-polarized or cross-polarized with respect to each other. The two-port feeds are available in either C-band or Ku-band.
- Three-port feed has two receive ports and a single transmit port. The receive ports provide for two orthogonal polarizations. The three-port feeds are available in either C-band or Ku-band.
- Four-port feed provides dual polarization capability for both transmit and receive applications. This feed configuration is also referred to as a *frequency reuse feed*. The four-port feeds are available for either C-band or Ku-band.
- Dual-band feed provides for the simultaneous reception of C-band signals and Ku-band signals from a hybrid satellite. The dual-band feeds are typically a single aperture; that is, the C-band and Ku-band radiating apertures occupy the same space and usually sacrifice gain and sidelobe performance to provide the dual frequency operation. This is true for the single-reflector and dual-reflector designs.

An alternate configuration that does not sacrifice radiation performance utilizes a dual-reflector geometry where the subreflector is a frequency-selective surface. This configuration typically uses a prime focus C-band feed, a frequency-selective surface subreflector that is transparent to C-band and reflective for Ku-band, and a Ku-band dual-reflector feed. The Ku-band feed may be as simple as a single-port feed to a full frequency reuse, four-port feed whereas C-band is a prime focus, receive-only feed.

Mechanical Performance

The mechanical design of an earth station antenna must provide the structural integrity to accurately point the antenna beam toward the desired satellite and to maintain the pointing accuracy within the environmental conditions for the locale. Further, the mechanical design of the antenna must ensure the required tolerance of the radiating surface such that the radiation performance of the antenna is not compromised. The antenna pedestal must also provide the means to steer the antenna beam to the satellites of interest.

The location and size of an earth station antenna system (antenna, pedestal or mount, electronics, and control housing) usually make it subject to local building codes. The code that is almost universally accepted is *Minimum Design Loads in Buildings and Other Structures* (ASCE 7, formerly ANSI A58.1), which requires that buildings or other structures and all parts thereof be designed and constructed to support safely all loads, including dead loads, without exceeding the allowable stresses (or ultimate strengths when appropriate load factors are applied) for the materials of construction in the structural members and connections. When both wind and earthquake loads are present, only that one which produces the greater stresses needs to be considered, and both need not be assumed to act simultaneously. The loads that must be safely supported by an earth station antenna system are the weight of the antenna and the attached equipment, the expected ice and snow load, earthquake load, and the wind load. Of these, the wind load is usually the largest single contributor to the stress and deflection of the structure.

Earth station antennas have a specification that is variously called *maximum wind*, *survival wind*, or *withstand wind*. These terms should be considered synonymous. At the manufacturer's specified survival wind, the system must be safely supported without exceeding the allowable stresses for the materials. Survival wind, as defined herein, when combined with ice and dead weight results in the *design load* as defined in Standard EIA-222C. In addition to survival wind, two other sets of wind conditions are usually specified: the operational wind velocity and the drive-to-stow wind velocity. The operational wind velocity is the maximum value at which the antenna system fully meets the performance specifications. The drive-to-stow wind velocity is the maximum value the antenna may be driven through the azimuth and elevation actuators to the prescribed stow position (usually zenith).

Positioning Systems

There are two broad classes of positioning systems used for satellite earth station antennas. One class consists of orthogonal two-axis configurations; the other is the one-axis or single-axis configuration. The two-axis systems are characterized by the orientation of the lower most axis with respect to the earth. A two-axis system having its lower axis perpendicular to the ground is an *elevation-over-azimuth mount* (Figure 6.11-37a). One that has its lower axis parallel to the ground (Figure 6.11-37b) is an *X–Y mount*. One that has its lower axis parallel to the earth's axis of rotation is an *hour angle-declination* or *polar mount* (Figure 6.11-37c). Each of the three positioning systems has the beam axis or pointing direction perpendicular to the upper axis, as is illustrated in Figure 6.11-37d. Providing there are no physical limitations, all three types can theoretically point in any direction.

Development of single-axis antenna mounts was brought about by efforts to reduce the costs and to simplify the positioning of the antenna beam with respect to the geostationary orbit. This geometry has restricted applications due to its lack of capability to follow satellites in inclined orbits and its inherent error as the pointing transverses the geostationary arc.

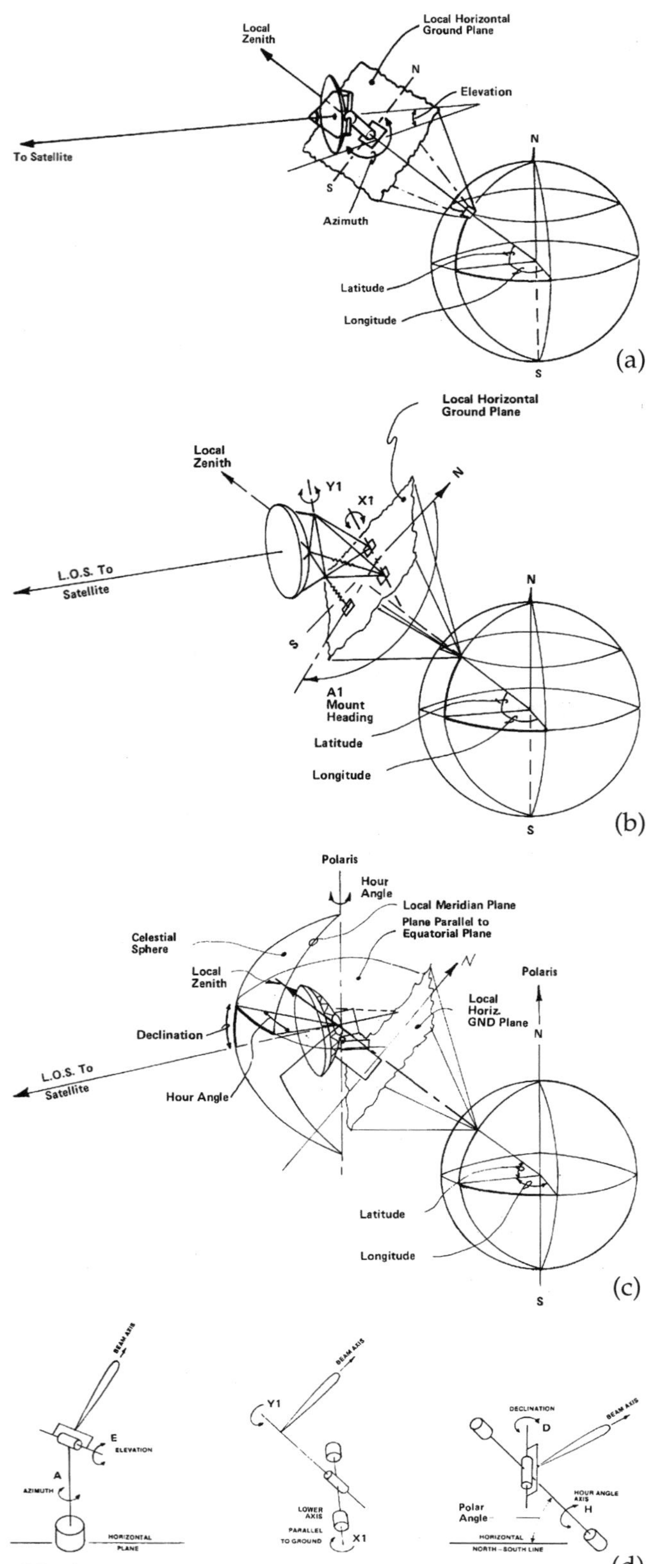

FIGURE 6.11-37 (a) Elevation-over-azimuth mount geometry; (b) X–Y mount geometry; (c) polar mount geometry; (d) schematic illustration of two-axis earth station positioning systems.

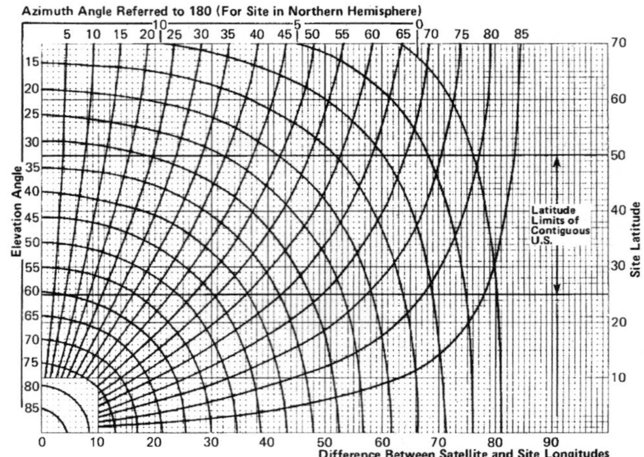

FIGURE 6.11-38 Universal azimuth–elevation look angles.

The elevation-over-azimuth positioner has become the choice for most systems. Figure 6.11-38 is a graph of azimuth and elevation angles *versus* a particular site latitude and longitudinal difference between the satellite and the site. The horizontal and vertical rectangular coordinates are site latitude and difference longitude, respectively. The curved lines running toward and labeled at the top of the graph are the required azimuth angles (add 180° if the satellite is west of the site; subtract from 180° if the satellite is east of the site). The curved lines running toward and labeled at the left margin (down to 15°) are the required elevation angles. The elevation lines of 10° and below are labeled at the top of the graph. To determine the required azimuth and elevation pointing angles, find the satellite site longitudinal difference and move vertically on this line until it intersects the horizontal latitude line. At this intersection, interpolate between bounding azimuth and elevation curves for the required angles.

The azimuth and elevation angles to a particular geostationary satellite can be calculated using the satellite longitude (Z); the site longitude (Y); the site latitude (X); and the following equations:

$$C = Z - Y \text{ degrees} \quad (34)$$

$$A \text{ (azimuth)} = 180 + \tan^{-1}[\tan(C)/\sin(X)] \text{ degrees} \quad (35)$$

$$E \text{ (elevation)} = \tan^{-1}\{[\cos(C) \times \cos(X) - 0.15126)]/[\sin^2(C) + \cos^2(C) \times \sin^2(X)]^{1/2}\} \text{ degrees} \quad (36)$$

All of the other environmental conditions at a particular site should be addressed, such as effects of solar radiation, lightning strikes, damage by salt water, acid rain, and pollution gases, in the planning and implementation of the earth station system.

Pointing and Tracking

The pointing and tracking accuracy are two very important considerations for an earth station antenna system. Pointing accuracy is defined as the precision with which an antenna can be held (for a fixed-position antenna) or steered under the specified operating conditions. The pointing error is a measure of pointing accuracy and is defined as the space angle difference between the command vector and the actual position of the antenna communication RF axis. Pointing error is usually specified to less than 0.2 of the half-power beamwidth (HPBW) of the antenna in the transmit frequency band. Tracking accuracy is the precision with which an antenna can track a source under specified operating conditions. The tracking error is a measure of tracking accuracy and is defined as the space angle difference between the communication RF axis of the antenna and the vector to the RF source. Tracking error is usually specified to be less than 0.1 of the HPBW. Table 6.11-6 lists the sources of error that should be considered in an overall budget or calculation of pointing accuracy and tracking accuracy.

TABLE 6.11-6
Pointing and Tracking Error Budget Terms

Pointing Error Budget	
Velocity lag	Breakaway friction
Wind up of gear train	Secant potentiometer
Angle encoder	Tachometer
Angle encoder coupling	Amplifier drift
Level	Amplifier bias
North alignment, initial zeroing of encoders	Motor cogging
AZ–EL axis orthogonality	Backlash
RF–EL axis orthogonality	Servo dead zone
Reflector alignment	Servo noise
Structural distortion	Axis wobble
Gravity	Radome diffraction
Ice	Boresight shift *versus* polarization
Wind	Boresight shift *versus* frequency
Thermal	
Acceleration	
Foundation displacement	
Acceleration lag	
Tracking Error Budget	
Velocity lag	Wind torque (servo)
Tracking receiver	Null axis–beam axis alignment
Acceleration lag	Breakaway friction
Tachometer	Motor cogging
Amplifier drift	Backlash
Servo noise	

Many earth station systems operate in the point mode; that is, there is no requirement for automatic tracking of the satellite. This condition exists when the satellite orbital location is maintained within a small fraction of a degree (<0.1°) and when the earth station antenna HPBW is sufficiently broad (>0.5°). Automatic tracking may become necessary as the antenna becomes large in terms of wavelengths (very narrow RF beam) or if the satellite is allowed to transverse an inclined orbit. The complexity of the tracking system is determined by the overall system accuracy requirements and the allowance in EIRP and G/T that is budgeted for impaired operation.

A hierarchy of pointing and tracking systems is as follows:

1. Initial fixed pointing is satisfactory (receive-only).
2. Repointing of the antenna is required to switch between various satellites or to correct for satellite motion.
3. Tracking is required to correct for satellite drift. Satellite position *versus* time is known and program track is satisfactory.
4. Automatic tracking is necessary but can be satisfied by a simple step-track system.
5. Full automatic tracking is necessary (extended inclined orbits).

The simple step-track system is satisfactory for most satellite communication applications when automatic tracking is required. The step-track systems generate tracking information by moving the RF beam in several steps, comparing the signal level, deciding the proper direction to move for the next step, and then continuing this process until the RF signal is maximized. The step-track system uses a very low frequency servo loop and therefore will not track out such disturbances as wind. Step-track can be susceptible to fade conditions unless the sampling circuitry is preset to cut off when a large signal loss is evident. Step-track may also be augmented with a program track mode whereby the satellite movement is memorized and then followed by a memory command circuit.

Fully automatic tracking systems are typically used for TT&C earth stations or for those earth stations operating under extreme conditions with very narrow RF beamwidths. The automatic tracking configurations include conical scan, electronic beam scanning, single-channel monopulse, and three-channel monopulse. The electronic scanning and three-channel monopulse techniques offer the advantage of providing a data channel and a transmit channel without tracking modulation superimposed on the signals. This is not possible with the conical scanning technique.

Transmit Electronics

The transmit subsystem for analog or digital transmission consists of equipment from baseband to the high-power RF amplifier. Depending on the application digital encoders, FM baseband processors for combined video and audio signals, modulators, upconverters, and high-power RF amplifiers may be employed.

Digital Video Encoders

A system used for the transmission of digital video uses a perceptual video codec (coder/decoder) such as MPEG to sample the applied video and audio waveforms and generate a digital representation of the input signals. The encoder performs compression to reduce the bit rate of the signal. Digital video systems are described elsewhere in this handbook, but in general the encoder will include an integrated (for Single Channel Per Carrier service) or external (for Multiple Channel Per Carrier service) MPEG transport stream multiplexer. This combination of encoder/multiplexer can accommodate a wide variety of input signals and quality levels, including not only video and audio but also encapsulated data signals. As video varies widely in its encoding requirements, the systems should be capable of different data rates. MPEG-2 4:2:0 or 4:2:2 video encoding may be used or newer encoding systems such as MPEG-4 part 10 AVC, SMPTE-VC-1 (Windows Media 9), or JPEG-2000, depending on the application.

FM Baseband Processor

For analog frequency-modulated signals, the incoming analog video signal is first processed by the baseband processing module. The video signal is preemphasized for either 525 line (NTSC) or 625 line (PAL/SECAM) operation and passed through a low-pass roofing filter. Preemphasis acts to improve the output video signal-to-noise ratio by compensating for the increase in noise density with frequency (triangular noise) which is characteristic of the receiver's FM demodulator. (The preemphasis is removed by a deemphasis network after the receiver discriminator.) Energy dispersal modulation is applied to the incoming video signal. Satellite transmissions of video signals are processed in this manner to disperse the RF spectrum, thus preventing concentrations of energy. This reduces interference with terrestrial microwave and other satellite links and also reduces intermodulation among the multiple carriers that exist in a real satellite. Energy dispersal modulation is applied using a triangular waveform with apexes located at the vertical intervals of the video signal.

FM Audio Subcarrier Modulator

Each subcarrier modulator modulates an audio signal onto a carrier between 5.0 and 8.5 MHz. Typical subcarrier frequencies are 6.2 MHz and 6.8 MHz. Most video exciters synthesize the center frequency of the subcarrier with a resolution (step size) of 10 kHz. Generally, a subcarrier modulator will allow for selection of preemphasis. Frequency deviation (the amount of frequency modulation applied) is also typically adjustable between 50 kHz and 500 kHz peak.

Automatic Transmit Identification System

The FCC requires that analog video uplinks must incorporate an automatic transmit identification system (ATIS), to identify the source of the transmission. The ATIS signal is an FM subcarrier positioned at 7.1 MHz that contains a message composed of international Morse code characters to identify the source of the signal and provide a telephone number for communication with its operator. The message is repeated every 30 seconds and includes a unique 10 digit ID code that is unchangeable by the operator. The subcarrier frequency of 7.1 MHz was chosen because it is very close to the second harmonic of the color subcarrier and therefore not usable for any possible revenue source. The subcarrier injection level of –26 dB referenced to the unmodulated main carrier represents a reasonable compromise between ATIS system sensitivity, resistance to interference, and power taken from the main carrier. This injection level is approximately 0.05 of the normal level of a monaural TV associated audio subcarrier.

Audio Electronics Service

Audio (*e.g.*, radio) signals are transmitted by satellite in both analog and digital form. Most of the domestic U.S. nationally distributed audio material is delivered in digital format. The satellite distribution service encodes program material in digital form at the source and distributes the information in that form. The digital audio system typically supports four types of signals: voice-grade, 7.5 kHz audio, 15 kHz audio, and data. The 7.5 kHz and 15 kHz audio signals are sampled at 16 kHz and 32 kHz, respectively.

Digital Modem

The compressed digital audio or video signal is modulated onto the carrier using a digital modem or modulator/demodulator. The output frequency is typically 70 MHz. The modem will normally accommodate a broad selection of input digital data rates and will offer a wide variety of FEC techniques such as block codes, convolutional codes, turbo codes, and low-density parity check (LDPC) codes. Often a combination of two of these FEC techniques separated by an interleaver is used to permit a greater range of error performance. Although QPSK predominates at the time of this writing, use of 8PSK is increasing due to its increased bandwidth efficiency. Other modulation systems such as 16 QAM (or APSK) or 32 QAM/APSK or even forms of minimum-shift keying (MSK) will likely see increased use in the future because of their bandwidth efficiency.

Upconverter

The upconverter converts the modulated 70 MHz IF signal up to C-band or Ku-band. Normally, dual-frequency conversion techniques are used and the transmitted spectrum is not inverted. Note that the upconverter contains built-in amplifiers and it is important that these are operated in the linear region.

Normally, 10 dB backoff from the upconverter's 1 dB compression point is adequate.

Video Exciter

For many applications it is desirable to combine the modulation and upconversion functions in a device termed an *exciter*. The advantage of using an exciter is that it is more economical of space and cost for uplinks such as satellite news gathering (SNG) vehicles and flyaway (checked baggage air transportable) earth terminals. Block diagrams of typical analog and digital video exciters are shown in Figure 6.11-39. In the analog exciter, audio signals are separately modulated and then combined with the video into a composite baseband which in turn is frequency modulated. After appropriate filtering, the modulated IF signal is upconverted to RF for transmission to the satellite. Digital exciters are not as common as analog exciters but are often used where cost and space are at a premium. Input video and audio signals not already in digital format are converted to digital using an analog-to-digital (A/D) converter. They are then typically MPEG encoded and combined into an MPEG transport stream. This transport stream is usually digitally modulated as QPSK or 8PSK and then upconverted to RF for transmission to the satellite.

High-Power Amplifier

The high-power amplifier (HPA) amplifies the RF output signal from the upconverter to the required power level for transmission to the satellite. Amplifiers for satellite video applications are typically sized in the range from 1 W to 3 kW. Amplifiers in the 1 to 200 W range are available in solid-state power amplifier (SSPA) configurations. Traveling wave tube amplifiers (TWTAs) are available in configurations up to approximately 3000 W, although lower power TWTAs (approximately 200 to 400 watts) are more common.

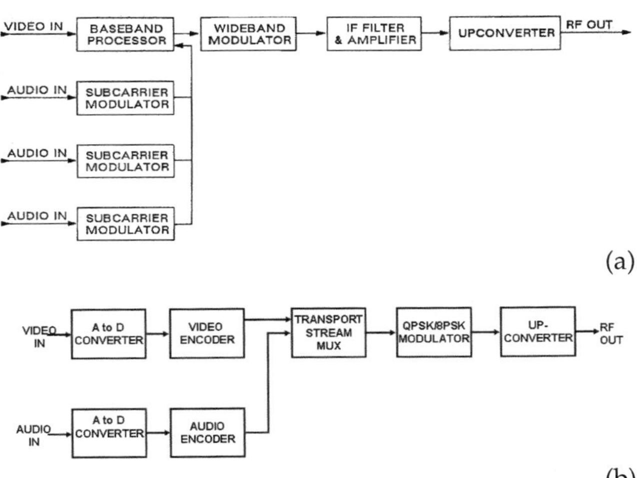

FIGURE 6.11-39 (a) Typical analog video exciter block diagram; (b) typical digital video exciter.

For power levels above 750 W, klystron tube amplifiers are frequently used, but these are typically limited to an output frequency range of less than 140 MHz without retuning, whereas SSPAs and TWTAs have a 500 MHz output range. The lower power SSPAs restrict their application, particularly for multicarrier operation where the amplifier must be operated in the linear region to prevent the generation of intermodulation product interference; however, for applications where SSPAs can be used, they are generally preferred because of greater reliability and linearity.

To improve the linearity for multicarrier operation, a device known as a linearizer, is sometimes used with TWT or klystron amplifiers to achieve greater output power levels without generating intermodulation products. Solid-state amplifiers normally do not require the use of linearizers. The HPA usually contains bandpass filters to reject harmonics and power sampling circuits for monitoring the output transmit power and the reflected power from the antenna. Often, protection circuitry is added to turn off the HPA when the reflected power exceeds some predetermined level.

Receive Electronics

The receive electronics are similar in scope to the transmit subsystem but operate in the reverse order. The incoming RF signal is filtered, amplified, downconverted (optional, depending on the frequency band of operation), and passed to the receiver, where the signal is further downconverted, amplified, and demodulated and decoded to baseband video and audio.

Low-Noise Amplifiers/Low-Noise Block Converters

The first active signal processing of a downlinked satellite signal occurs at the low-noise amplifier (LNA) or low-noise block converter (LNB). Traditional C-band broadcast applications use an LNA or LNB mounted at the antenna and connected to the indoor electronics through a length of coaxial cable. Typical Ku-band systems use an LNB at the antenna that amplifies and downconverts the signal to L-band (950 MHz to 1450 MHz). The LNA or LNB:

- Provides high gain and low noise to establish a high system G/T.
- Provides transition from antenna waveguide to coaxial cable to eliminate long expensive waveguide runs.
- Provides adequate mechanical strength to be mounted directly to the antenna waveguide and to connect a coaxial cable to go to the receiver.
- Provides radiofrequency interference (RFI)/electromagnetic interference (EMI) tight weatherproof housing for the sensitive amplifier circuitry.

A block diagram of a typical LNA is shown in Figure 6.11-40.

Low-noise block converters convert the RF signal block of frequencies to a standard L-band intermedi-

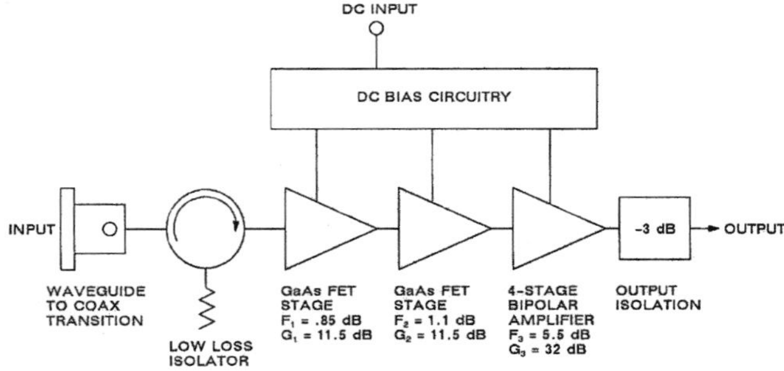

FIGURE 6.11-40 LNA block diagram.

ate frequency of 950 to 1450 MHz (Figure 6.11-41). This conversion may be accomplished with or without the use of an external reference signal and with or without a phase-locked loop stage in the LNB. Due to the normal selection of the oscillator frequencies (5150 MHz at C-band and 10,750 MHz at the U.S. FSS Ku-band), spectral inversion occurs for the C-band signal but not for the Ku-band signal.

Downconverter

The downconverter converts the RF-input signal to an intermediate frequency prior to demodulation. This intermediate frequency is typically 70 MHz. The downconverter may provide either a single input or multiple inputs for multiple-antenna, multiple-polarization operation. Single-input downconverters may employ an external relay to select the appropriate polarization input. Like upconverters, downconverters contain internal amplifiers which should be operated in the linear region to avoid generation of intermodulation products or spectrum regrowth of digital signals. The downconverter input signal is 3700 to 4200 MHz for C-band operation or 950 to 1450 MHz (called L-band) for C-band and Ku-band operation. For Ku-band, the LNB downconverts the input RF signal to L-band frequencies. The need for a separate downconverter can be avoided by using LNBs with L-band outputs and video receivers or integrated receiver–decoders (IRDs) with L-band inputs.

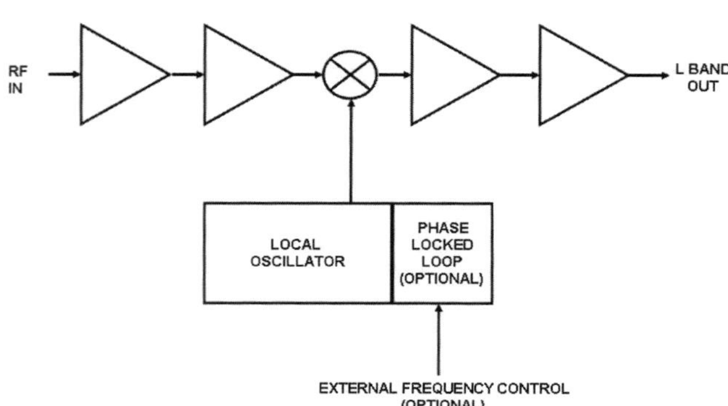

FIGURE 6.11-41 LNB block diagram.

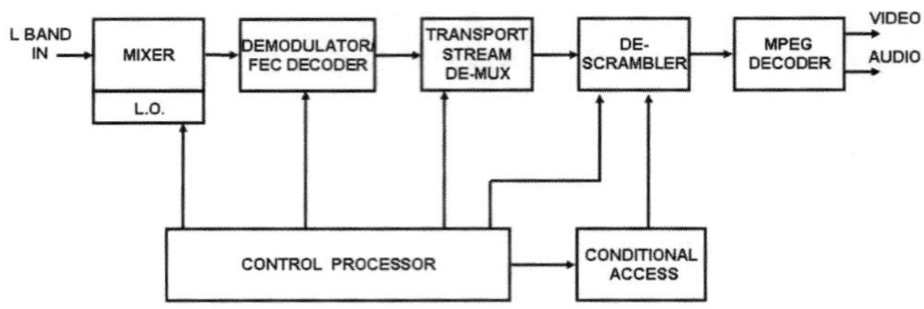

FIGURE 6.11-42 Integrated receiver–decoder (IRD).

FM (Analog) Video Receivers

The video receiver takes the received satellite signal and produces baseband video and one or more audio signals. Although C-band video receivers still exist, the increased use of Ku-band and decreasing costs for LNBs permit most receivers to use L-band inputs. A block diagram of a typical analog video receiver is shown in Figure 6.11-42. The downconverter converts the input RF signal to 70 MHz IF.

IF Filter/Amplifier

For FM analog video, the output of the downconverter is routed to the 70 MHz IF filter/amplifier. The signal is first bandpass filtered. In the early days of satellite video services, the video receivers provided a single IF filter bandwidth. More recently, most video receivers provide multiple IF filter bandwidths to provide half transponder as well as full transponder operation. Some video receivers provide up to six IF filter bandwidths, made economical because of the availability of surface acoustic wave (SAW) filters for this application.

Demodulator

The filtered, amplified IF signal is fed to the FM demodulator from which the baseband video and multiple audio subcarriers are obtained.

Analog Video Processing

After demodulation, the analog video baseband contains the 30 Hz triangular energy dispersal waveform, which is removed by a "clamp" circuit. A low-pass filter separates the audio subcarriers from the video baseband signal.

Subcarrier Demodulator

Video receivers generally provide up to four audio subcarrier demodulators and provide baseband audio (generally 600 Ω balanced) outputs.

Integrated Receiver–Decoder

For digital video signals, an IRD is typically used. The IRD takes an L-band input from the LNB, demodulates, de-interleaves, FEC decodes, and decrypts (as needed) the compressed video and audio signals and provides them to the MPEG decoder, which digitally decodes the signals and provides component serial

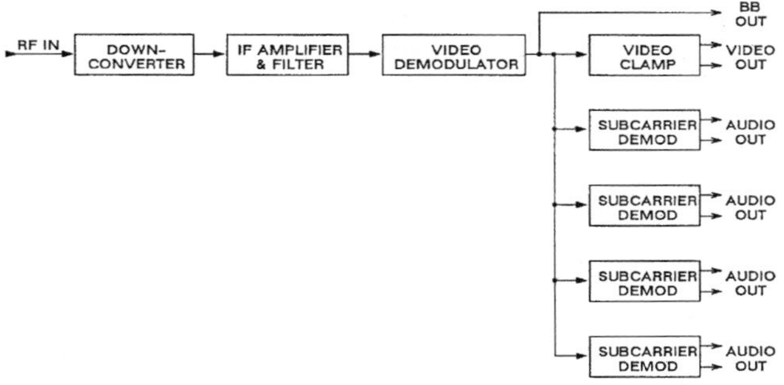

FIGURE 6.11-43 Typical analog video receiver block diagram.

digital interface (SDI) or analog video outputs, as well as associated data and closed-captioning signals if included. Figure 6.11-43 shows a basic block diagram for an IRD that includes a control processor, as a wide variety of data rates, FEC, and digital video compression options must be accommodated.

Protection Switching

With the exception of receive-only systems, most satellite transmission systems contain redundant subsystems to meet high availability specifications. Satellite teleports advertise availability specifications as high as 99.995%. Protection switching is used to implement automatic subsystem redundancy. The protection switch monitors one or more online subsystems for failure. Upon detecting a failure in the subsystem, the protection switch switches to the backup unit and configures the backup to the configuration of the failed online system.

The configuration of the subsystems protected by a protection switch is referred to as *m*:*n*, where *m* is the number of backups available and *n* is the number of online units monitored and protected by the protection switch. The simplest configuration is 1:1—a single backup is available to replace a single online unit. A typical 1:1 protection system for a Ku-band uplink in shown in Figure 6.11-44. A typical 1:1 C-band downlink is shown in Figure 6.11-45. Operation of the 1:1 configuration is simplified in that the backup may be tuned to the same operating parameters as the online unit; switching to the backup merely requires switching the source of the input and output signals from the

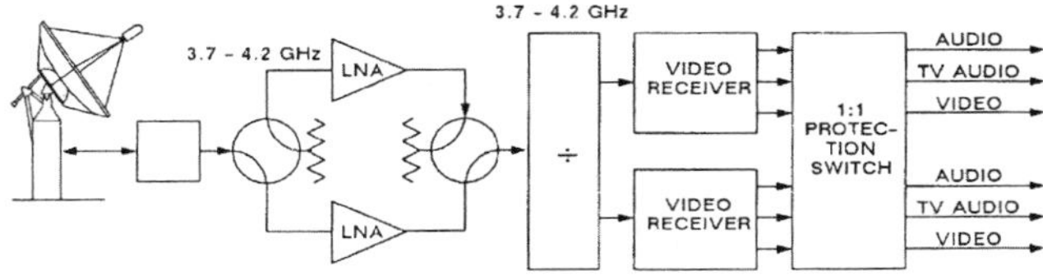

FIGURE 6.11-45 Redundant C-band downlink protection system.

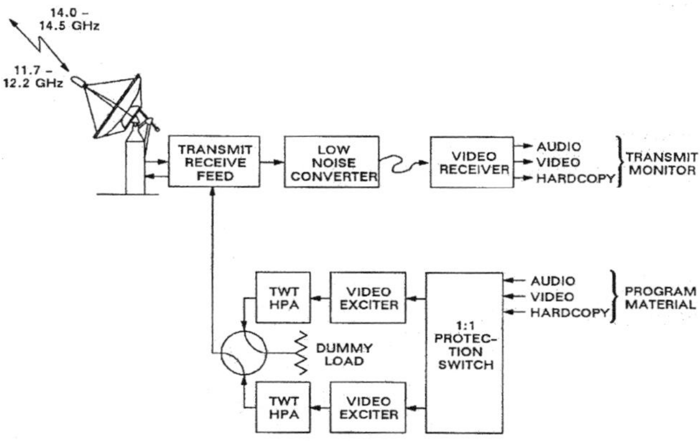

FIGURE 6.11-44 Redundant Ku-band uplink protection system.

online unit to the backup. An example of a larger configuration is 2:6, where two backup units protect six online units; this configuration is often used with large multiple-channel uplinks. The larger configurations become more cost effective as the number of online channels increases. Earth station monitoring and control systems, described below, are increasingly being used to implement protection switching, often eliminating the need for dedicated (separate) redundancy switchover logic devices.

Monitor and Control Systems

Monitor and control refers to systems used to monitor earth station components for failures and provide manual and automatic control of the components. These systems are widely used for a variety of reasons. Although most earth station components provide front panel monitor and control functions, there are generally too many earth station components to monitor from the front panels of the respective components. A monitor and control system provides a single point of monitoring and control for the operator, thus easing the operator workload and allowing the operator to handle more transmissions. Because many earth stations are located away from the studio or master control room, a monitor and control system permits remote operation of the earth station. Remote operation can employ the Internet or a low-speed data circuit between the earth station equipment and the monitor and control computer in the studio or master control. This circuit can be a fiberoptic or wireless channel, a satellite data circuit, a data subcarrier on a microwave radio channel or another data carrier, a dial-up phone circuit with computer modems, or a dedicated EIA-422 hardwire connection (for distances less than 1000 m).

Earth Station Control Computer

Monitor and control systems are typically based on the use of a general-purpose computer executing applications software specifically designed for communications systems. This computer is referred to as the *earth station controller*. In addition to the centralized method for earth station monitoring and control, redundancy switching for equipment failure or for diversity uplink or downlink switching for rain fade protection, automatic uplink power control, and even remote spectrum analyzer control and display can also be incorporated.

Operator Interface

Earth station control computers provide a video monitor for display and a keyboard for operator interaction. The system typically displays earth station status in the form of a hierarchical, graphical display and a high-level block diagram of the earth station. The operator may see detailed displays of subsystems by selecting the subsystem symbol from the screen. Earth station components and subsystems are coded with color to indicate state (red, failure; amber, ready or standby; green, normal). Audible alarms are also provided.

Earth Station Interface

The computer interfaces to the earth station components through a number of interfaces, including serial, contact closures, and other customer supplied interfaces:

- *Serial interfaces*. The most common interface is a serial ASCII-protocol-based interface. The interface is usually an asynchronous-character-oriented scheme utilizing EIA-232C or EIA-422 signal levels. EIA-232C interfaces are used for short cable distances (computer to device) less than 5 m or for connection to a modem. EIA-422 interfaces may be used for cable lengths of up to 1000 m and are used in multi-drop mode, thus allowing many devices to share a single interface port. The protocols are usually ASCII based because there is no satellite communication standard and most vendors use their own version of an interface protocol. Thus, earth station computers must offer a number of serial equipment interface ports and also support a variety of protocols on those ports.
- *Contact closure interfaces*. Older components typically provided contact closure interfaces and offered no serial interface functions. Many of these components are still in use today. In addition, components such as waveguide switches and shelter alarms (intrusion, air conditioner, emergency generator, etc.) offer only a contact closure interface; for example, a waveguide switch may provide two status points and two control points. Thus, earth station computers must provide some method of accommodating contact closure controls. Some vendors offer systems that connect the contact closure directly to the computer. Others offer general-purpose interfaces (GPIs) that reside in the earth station and interface to the earth station computer through a serial interface. Status inputs to the earth station computer are usually optically isolated. The earth station computer supplies the optical isolator. The monitored device sinks current through the isolator to indicate one of two states. The alternate state is no current flow. Control outputs are of two types. The most flexible interface is the Form C output. The Form C output provides a common connection and a normally opened (NO) and normally closed (NC) connection. The second type of control output is the open collector output. This interface provides a connection to the collector of a transistor to sink current. One control state is with the transistor on, thus sinking current from the controlled device. The alternate state is with the transistor off.
- *Vendor-supplied computer interfaces*. A vendor-supplied, device-specific contact closure adaptor. Most TWT amplifiers provide contact-closure-based remote monitor and control interfaces. Additionally, forward and reflected power indications are provided by a signal with voltage level proportional to power. Some TWT amplifiers require an analog current signal to control the power level attenuator; however, most TWT amplifier vendors offer an interface adaptor that converts the contact closures to a serial-protocol-based EIA-232C or EIA-485 interface. These adaptors reside in the rack with the TWT amplifiers and connect to the earth station computer through a serial connection.

SUMMARY

There are many aspects of the design, installation, operation, and maintenance of an earth station antenna system that have not been discussed in this chapter. Site selection and preparation are, in particular, critical to the successful operation of the system as well as the foundation design. Details of this aspect of the earth station design should be accomplished with the assistance of experienced engineers and frequency coordination experts. Although the operations building and equipment houses should be in close proximity to the earth station, remote operation is both possible and practical. The power requirements for the earth station should also be carefully planned to provide adequate, reliable, and conditioned power for the electronics, including the transmitter equipment and any power required for antenna deicing where applicable.

Bibliography

Cook, Jr., J.H. and Hollis, S., Eds., *Communications Symposium '83 Notebook*, Scientific-Atlanta, Atlanta, GA, 1983.

DVB-EN 300 421, *Framing Structure, Channel Coding and Modulation for 11/12 GHz.*

DVB-EN 302 307, *Second-Generation Framing Structure, Channel Coding and Modulation Systems for Broadcasting, Interactive Services, News Gathering and Other Broadband Satellite Applications* (DVB-S2).

DVB-TR 102 154, *Implementation Guidelines for the Use of MPEG-2 Systems, Video and Audio in Contribution Applications.*

DVB-TSI-TR 101 154 V1.4.1, *Digital Video Broadcasting (DVB): Implementation Guidelines for the Use of MPEG-2 Systems, Video and Audio in Satellite, Cable and Terrestrial Broadcasting Applications*, Annex B, July, 2000.

Inglis, A. F., Ed., *Electronic Communications Handbook*, McGraw-Hill, New York, chapters 2, 5, 6, 17.

Ippolito, L. J., Kaul, R. D., and Wallace, R. G., *Propagations Effects Handbook for Satellite Systems Design*, NASA Ref. Publ. 1082, National Aeronautics and Space Administration, Washington, D.C., 1981.

ISO/IEC 13818-1, International Standard, *Information Technology: Generic Coding of Moving Pictures and Associated Audio Information: Systems.*

ISO/IEC IS 13818-2, International Standard, *MPEG-2 Video*, 1996.

Johnson, R. C. and Jasik, H., Earth station antennas, in *Antenna Engineering Handbook*, Cook, Jr., J.H., Ed., McGraw-Hill, New York, chapter 36.

Jordan, E. C., *Reference Data for Engineers: Radio Electronics, Computers and Communications*, Howard W. Sams & Co., Indianapolis, IN.

Martin, J., *Communications Satellite Systems*, Prentice-Hall, Englewood Cliffs, NJ.

Morgan, W. L. and Gordon, G. D., *Communications Satellite Handbook*, Wiley-Interscience, New York.

Pratt, T. and Bostian, C. W., *Satellite Communications*, John Wiley & Sons, New York.

Pritchard, W. L. and Sciulli, J. A., *Satellite Communication Systems Engineering*, Prentice-Hall, Englewood Cliffs, NJ.

CHAPTER

6.12

Low Power TV and TV Translators

GREG BEST

Greg Best Consulting
Kansas City, Missouri

INTRODUCTION

This chapter focuses on current technology and accepted policies and procedures that support Low Power TV (LPTV) and translator transmission systems.

HISTORY

LPTV and translator transmission systems have been authorized by the FCC for use since 1956 to provide TV service for areas where service had either not been provided or could not be provided by full-service TV stations for various economic, geographic, or technical reasons. Both LPTV and translator stations are used to extend the coverage of a full-power TV station. According to a *DecisionMark* study prepared for the National Translator Association in September 2003, it is estimated that about 7–10 million households receive their broadcast coverage through the use of these translators and LPTV stations. In 2004, the FCC added digital low power televisions as a new service to increase the capability for the general viewing audience to receive over-the-air television service; see FCC Docket 03-185.

The term "LPTV" refers to a class of service and does not necessarily refer to a specific range of output power from the authorized transmission systems. The transmission systems themselves are as diverse as the communities they serve. There are currently (as of September 2006) about 6,700 LPTV and translator stations on the air and more being added each day. With the ongoing conversion to digital TV (DTV) service, these LPTV entities will eventually have to convert their operational facilities to digital in order to continue to serve their constituents. The conversion of LPTV transmission systems to digital is presently not required to occur at the same date as is the conversion of full-service TV facilities.

FCC RULES AND TYPES OF OPERATION

FCC Rules that govern the LPTV service are provided in 47 CFR Part 74 Subpart G, whereas full-service TV FCC Rules are in Part 73. The current LPTV range of permissible operation includes LPTV stations that originate program material and *TV translators* that convert an incoming signal to a different channel and rebroadcast the same signal. Within those basic classes of operation, there are other categories such as *booster stations*, which take the input signal of a full-service station and rebroadcast it on the same channel as the input signal, and other stations (such as a studio-to-transmitter link or STL) that relay a program to a different location for rebroadcast. Translator networks of multiple transmission systems have been created that feed programming from one area to the next to deliver television over large land areas and are an efficient method of providing TV service. For more details on permissible services for both analog and digital LPTV, see the appropriate FCC Rules in CFR 47 Part 74 Subpart G.

The licensed category of operation of the LPTV transmission system—digital or analog—is based solely on the transmitted signal. So it is possible, for example, for an analog LPTV system to receive either a

DTV or analog signal and then retransmit an analog signal (and similarly for a digital LPTV system).

Application Process

The application process for obtaining an LPTV license is also described in the FCC Rules. The FCC currently permits applications to be filed for such licenses only during certain "windows" of time that the FCC announces in advance. The FCC expects to have future windows associated with only digital LPTV service and does not expect any further windows for analog service.

In contrast to full-service TV transmission systems, where license applications are accepted only for channels listed in the national allotment table adopted by FCC rulemaking, an LPTV station or translator may be built almost anywhere, on any channel, subject to the FCC Rules that ensure it will not cause interference to other stations, other authorized services, or any pending LPTV applications.

Available Channels for LPTV Service

Previously, the FCC authorized LPTV operation on TV channels 2 to 69. All new licenses are targeted for the channel range from 2 to 51. However, existing LPTV licensees or permittees may still continue to operate on channels 52–69 as long as no interference exists to other broadcast services or other land mobile or public services which have been granted licenses to operate on those channels. New licenses using channels 52–59 will be granted only if there are no available channels between 2 and 51 for the coverage area in question.

UHF Channels for STL or Relay Purposes

In less populated regions, UHF spectrum is less congested, and translators may be used as relay stations to provide a signal to a main transmitter site from a studio, or from one translator site to another site. Typically, UHF transmission equipment is more economical than microwave frequency equipment for the same level of equivalent radiated power (ERP) and distances to be covered. Broadcasters interested in constructing these types of facilities should consult the FCC Rules since there are restrictions regarding the ERP and the antenna polarity for the signals transmitted from these stations; see FCC broadcast auxiliary service (BAS) Rule 74.602 (h).

Interference Limits to or from Other Services

The LPTV service is a secondary service and cannot cause interference to primary service (i.e., full-service) TV stations, analog or digital. In addition, an LPTV station may not cause interference to Class A and other LPTV service stations, permittees, and applications already filed and accepted by the FCC. The interference limits to other broadcast entities are different depending on whether the station receiving interference is a DTV or analog full-service station, LPTV, or a Class A station.

If a full-service TV station moves or changes its transmission site and/or facilities, it does not have to provide interference protection to LPTV stations. If predicted or actual interference is caused to an LPTV station by such a move or new allocation of a TV channel, the LPTV station can be "displaced"; i.e., it may be allowed to move its site and change channel subject to the provisions of no interference to other stations and maintaining some overlap of its existing coverage contour after proper application and construction permit grant from the FCC. Applications for displacement are handled with higher priority than other standard minor LPTV change applications. For a definition of minor and major changes, refer to FCC Rule 1.929.

The previous method of determining interference to other broadcast entities involved a projected contour-overlap using FCC statistical contours at certain field strengths. The method used for all future DTV LPTV applications is a procedure outlined in FCC OET Bulletin 69. It uses estimated signal strengths in a small geographic cell based on a signal strength prediction algorithm known as *Longley-Rice*.

The process of determining interference starts with first defining a "protected" service area, which for LPTV stations uses a higher field strength in terms of dBu (dB above 1 μV/m) than that used for full-service stations. The defined "protected area" for each LPTV/Class A station is based on its ERP, antenna height, and channel of operation. For DTV LPTV service, the protected service area is defined by the area where 50% of the locations will receive a signal at least 90% of the time for the defined field strength. For analog service, this area is where 50% of the locations will receive the defined signal strength at least 50% of the time. This protected area is the area where other LPTV or Class A entities cannot create interference as identified in Part 74 of the FCC Rules.

As shown in Figure 6.12-1, the coverage area within the contour protected from interference is divided into square cells. When one determines whether proposed facilities will not cause interference to other broadcast

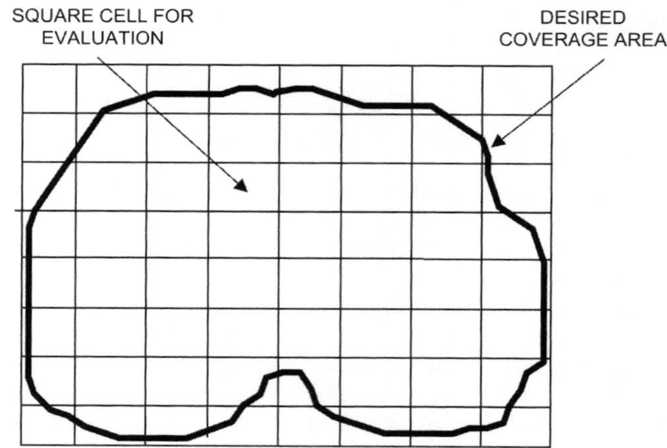

FIGURE 6.12-1 Representation of coverage area using square cells.

TABLE 6.12-1
Typical Coverage Areas for Various NTSC Antenna Heights and ERPs (from FCC 50-50 Curves)

ERP/Freq Band	Coverage Distance for Antenna Height, km (mi)		
	HAAT = 30 Meters	HAAT = 50 Meters	HAAT = 100 meters
100 W low band VHF*	12 (7.5)	15.4 (9.6)	22 (13.8)
100 W high band VHF†	9.5 (5.9)	12.2 (7.6)	17.6 (11.0)
100 W UHF‡	4.5 (2.8)	5.8 (3.6)	8.2 (5.1)
1000 W low band VHF	22 (13.8)	27 (16.9)	37 (23.1)
1000 W high band VHF	17 (10.6)	22 (13.8)	30.5 (19.1)
1000 W UHF	7.9 (4.9)	10.3 (6.4)	14.4 (9.0)
10 kW UHF	14 (8.8)	18.8 (11.8)	26.3 (16.4)
150 kW UHF**	27.7 (17.3)	33.8 (21.1)	42.3 (26.4)

*47 dBu signal strength for 50% of locations for 50% of time.
†54 dBu signal strength for 50% of locations for 50% of time.
‡64 dBu signal strength for 50% of locations for 50% of time.
**Maximum ERP allowed.

facilities, the amount of interference is calculated by adding the population affected in each cell based on certain channel relationships (i.e., co-channel, adjacent channel, etc.) and desired-to-undesired (D/U) signal strength ratios. The population affected is then compared to the associated limit for the class of station receiving interference. Typically, LPTV service will be evaluated using a square cell size of 1 km on a side.

ERP limits for analog LPTV service (for either transmitters or translators) are 3 kW for VHF and 150 kW for UHF, whereas ERP limits for digital LPTV service are 300 W for VHF and 15 kW for UHF. Some types of digital translators called *heterodyne translators* are limited to 3 W (VHF) or 30 W (UHF); see FCC Rule 74.795 (c). Heterodyne translators are a subset of LPTV transmission systems and are described later in the Transmitters subsection below.

LPTV COVERAGE AREAS

LPTV transmission systems generally cover smaller areas and smaller populations than full-service TV stations do, but not always. This depends on the TV market where the service is to be provided. Tables 6.12-1 and 6.12-2 describe typical coverage area (not

TABLE 6.12-2
Typical Coverage Areas for Various ATSC DTV Antenna Heights and ERPs (using propagation simulation software)

ERP / Freq Band	Coverage distance for antenna height, km (mi)		
	HAAT = 30 meters	HAAT = 50 meters	HAAT = 100 meters
10 W low band VHF*	19.9 (12.4)	24.7 (15.4)	32.7 (20.4)
10 W high band VHF†	14.7 (9.2)	19.0 (11.9)	26.3 (16.4)
10 W UHF‡	9.4 (5.9)	12.0 (7.5)	16.8 (10.5)
100 W low band VHF	31.3 (19.6)	38.0 (23.8)	48.4 (30.3)
100 W high band VHF	25.4 (15.9)	31.2 (19.5)	41.4 (25.9)
100 W UHF	16.6 (10.4)	21.7 (13.6)	29.0 (18.1)
1000 W UHF	27.3 (17.1)	32.9 (20.6)	40.7 (25.4)
15 kW UHF**	39.1 (24.4)	45.5 (28.4)	53.7 (33.6)
Note: These distances assume the frequency is the middle of the applicable band.			

*28 dBu signal strength for 50% of locations for 90% of time.
†36 dBu signal strength for 50% of locations for 90% of time.
‡41 dBu signal strength for 50% of locations for 90% of time.
**Maximum ERP allowed.

"protected" area described earlier) based on given ERP and antenna heights over average terrain for NTSC analog TV and ATSC DTV, respectively. These are typical values only, based on FCC planning factors; when planning an actual service, broadcasters should conduct a more careful estimate of coverage using the techniques described in the OET 69 bulletin previously mentioned (or some other, similar computer-based estimation tool).

OPERATIONAL AND OVERALL SYSTEM CONSIDERATIONS

Design considerations for LPTV stations are basically the same as for full-service stations but smaller in scope. The number and variety of programming sources for LPTV systems are generally much smaller than a full-service station and permit a greater amount of flexibility in determining the size and location of the studio for the LPTV station. In many cases, the studio and transmitter locations are the same for LPTV systems. Of course, translators do not use studios by their very nature; however, it is permissible for them to insert messages of public interest.

Operating requirements for LPTV service are generally more lenient than for full service, but LPTV stations that are eligible for "must-carry" status on cable television systems are subject to all FCC Part 73 Rule programming requirements. LPTV stations are also subject to certain Emergency Alert System (EAS) requirements, whereas translators may rely on EAS messages transmitted by the station they rebroadcast; see FCC Rule 11.11. As with all broadcast stations, unattended operation is permitted, subject to certain monitoring requirements for alerting an operator when equipment malfunctions. In certain circumstances, such as a loss in programming or loss of the signal provided to the transmitter site, equipment is configured so as to place the transmitter in a nonradiating state when this type of malfunction occurs.

Transmission Facilities

Transmission facility considerations for LPTV stations and translators are similar to those for full-power stations, except that the costs are generally more modest. In general, redundancy in the equipment facilities is not as prevalent as in full-service stations; equipment is smaller and less expensive and requires less power. A transmitter site situated at the highest possible elevation is desirable, unless a lower elevation is needed to avoid causing prohibited interference to another station or because of terrain limitations. Antenna selection is important since use of directional patterns in LPTV and translator facilities is common. In many cases, "composite" antennas are used, made from combinations of standard antennas with appropriate power dividers to route the correct power to each antenna to yield the desired pattern and coverage area while not causing objectionable interference to other stations.

FIGURE 6.12-2 Typical antennas for LPTV. Top mounted panel (a) and directional Yagi (b). (Photos courtesy of Kathrein-Scala, Inc.)

Studio equipment varies widely from station to station but may often be of high-grade consumer type. Where possible, LPTV operators may co-locate their transmitter and antenna facilities with those of other LPTV operators in order to minimize operational expenses such as tower space usage and installation or leasing/rental, back-up power, antennas and transmission lines, basic site maintenance, and engineering personnel costs. Site location often involves trade-offs between available support facilities (such as road and building access and power), distance from the community of coverage, and height of antenna placement.

Antennas are available from a wide variety of manufacturers. The antennas often mimic full-service designs but are more modest in cost and power-handling capabilities. Examples are shown in Figure 6.12-2. LPTV antennas are typically horizontally polarized since the signal is normally intended to be broadcast to the general public (in cases where a UHF channel is used as an STL or relay station, vertical polarization is required to minimize interference unless sufficient justification can be provided to the FCC to approve horizontal polarization). Examples of different azimuth patterns are shown in Figure 6.12-3. These types of patterns are achievable through use of slotted coaxial antennas and also by using log periodic antennas or other highly directional antennas. In many cases in which an LPTV or translator signal is focused on a narrow area such as a valley or a small community, a highly directional azimuth pattern is used. The pattern directivity reduces the transmitter power required while still providing excellent coverage. Some azimuth patterns are created by "composite" or custom antennas made up of more than one type of antenna. An example is shown in Figure 6.12-4. Elevation pattern gains depend entirely on the

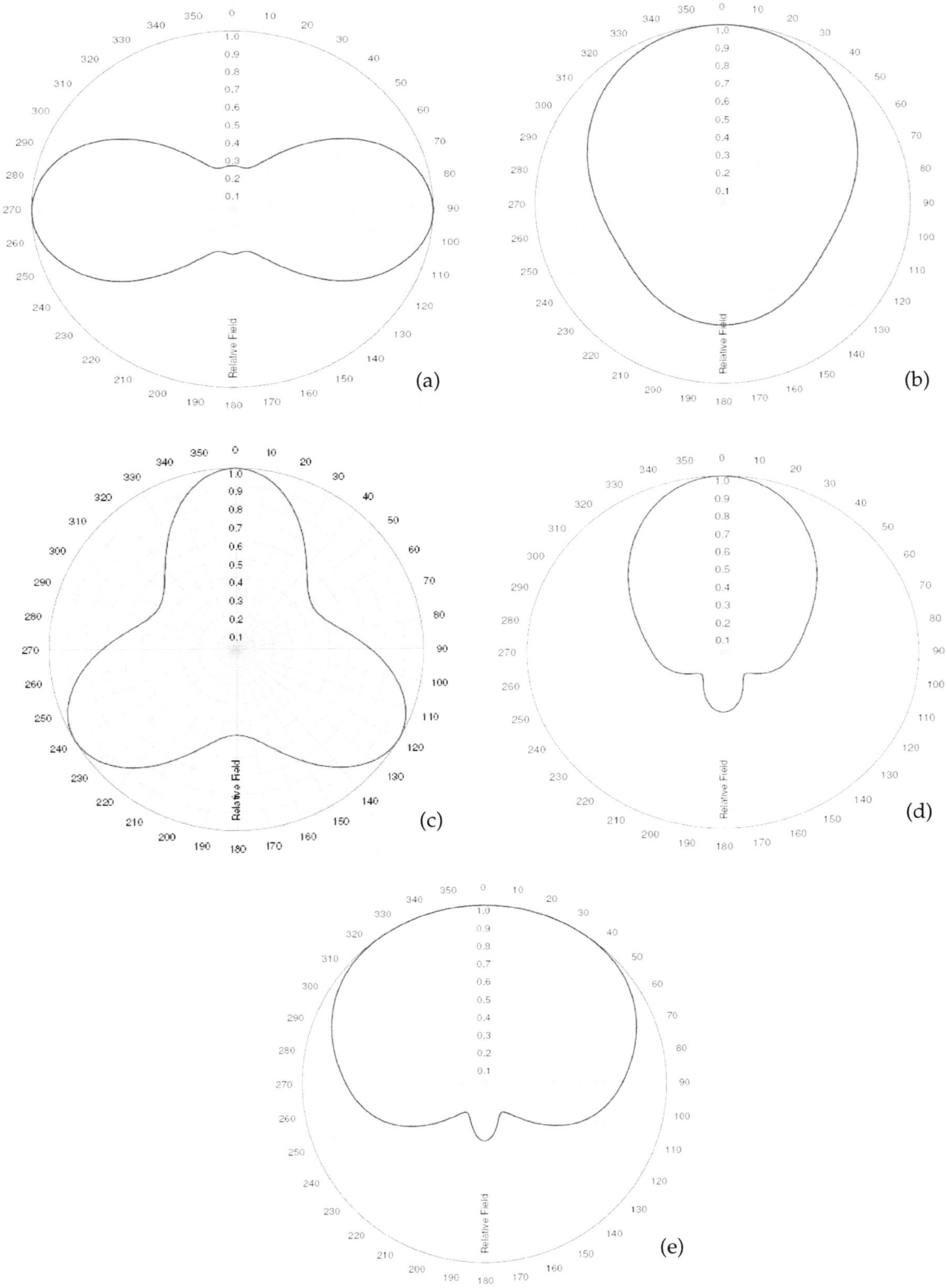

FIGURE 6.12-3 Peanut azimuth antenna pattern (a), omnioid or "skull" azimuth antenna pattern (b), trilobe azimuth antenna pattern (c), narrow cardioid azimuth antenna pattern (d), and extended cardioid azimuth antenna pattern (e). (Courtesy of ERI.)

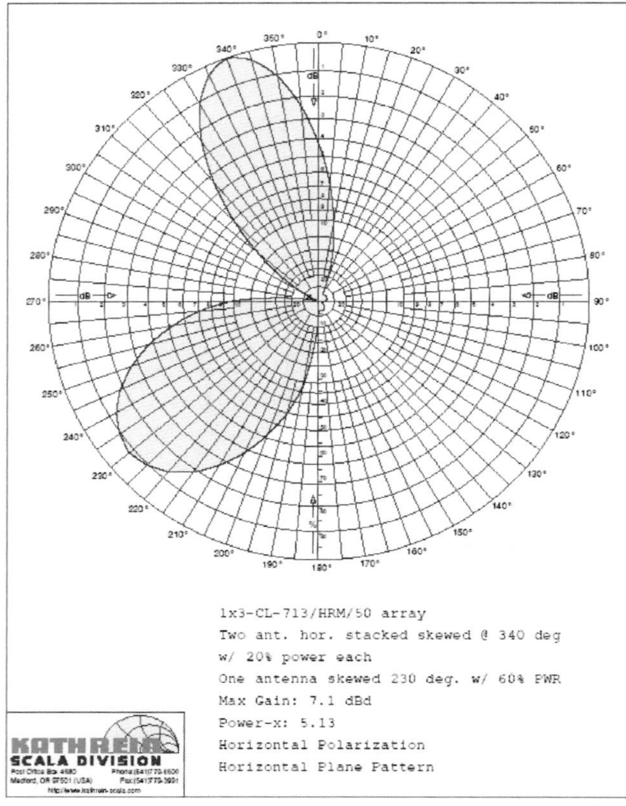

FIGURE 6.12-4 Composite azimuth pattern made by combining two different antennas. (Courtesy of Kathrein-Scala, Inc.)

construction but generally range from 1 to 16 with some gains as high as 32 where close-in coverage is not required.

In situations where a desired coverage area is shielded from full-service stations by terrain, translators may be utilized. The receiving antenna (which receives the signal to be translated) is usually placed at a high location and if, for example, the desired coverage area is located at the bottom of a valley, the transmitting antenna beam can be tilted toward the coverage area rather than mounting the antenna so the main beam is horizontal. This can be done using mechanical and/or electrical (i.e., the antenna is electrically designed to accomplish this) means to "tilt" the beam. The beam tilt results in larger values of signal hitting the desired spot. For more information, see Chapter 6.8, "Television Antenna Systems."

Antennas designed for analog service should be analyzed before being used for DTV service. The manufacturer can be contacted to determine if the performance of the antenna for DTV is suitable. For some antennas, the performance of the antenna has been optimized at the (analog) visual carrier frequency and ignored for the rest of the channel. If the antenna (on the receive side of the LPTV/translator station) has an amplitude versus frequency response that is not flat enough over the desired channel, then the equalizer designed to compensate for linear distortion in the transmitter portion of the station (if so equipped) will have to work harder to optimize the signal-to-noise ratio of the DTV signal. Removing this linear distortion in the receiver increases noise in the receiver, and this may result in a small reduction of the desired service area at the fringe reception area.

In general, it is good engineering practice to "sweep" the antenna (i.e., measure the amplitude and delay characteristics versus frequency) to ensure that the amplitude response and group delay response are nearly constant across the desired channel. Antennas that have good amplitude versus frequency response (< 1 dB variation) and good group delay response (< 25 nanosecond variation) across the desired channel should perform well for DTV.

If LPTV antennas are mounted at lower heights with respect to ground level, RF exposure levels to people must not be exceeded. This is especially true for antennas mounted on buildings or on towers with other radiators because the RF exposure levels must be below the required levels anywhere the general public has access. Even if only occupational personnel have access to the site, the allowable RF exposure levels must be established, and the personnel must be adequately trained regarding RF exposure. See Chapter 2.4, "Human Exposure to Radio Frequency Energy," for more information on this subject.

Transmission Lines

Transmission lines connecting the antenna to the transmitters are generally smaller in size than their full-service counterparts because the transmitters generally produce less power. Flexible or rigid coaxial transmission lines are commonly used. Both air dielectric and foam dielectric cables are utilized. The common enemy of all transmission lines is moisture or water because it will degrade the performance of the transmission lines.

One method to keep a transmission line dry is to pressurize it with either dry air or nitrogen. Foam dielectric lines may be used in place of air dielectric cables. The foam cables have greater loss but are favored in some instances where pressurization with nitrogen or dry air may not be practical. Smaller coax will have greater attenuation than larger coax at the same frequency so it is important to always factor the cable attenuation into the transmission system design. No matter what type of transmission line is used, it is worthwhile to inspect it at least annually to ensure that its original parameters have been maintained and that the characteristics have not degraded.

Transmitters

Output power of transmitters for LPTV/translator service may run from one watt to tens of kilowatts. The design of both analog and digital transmitters is basically the same. An analog TV transmitter consists of a modulator that takes the video and audio signals and converts them to a modulated IF signal followed by an upconverter that places the modulated IF signal on the desired RF channel, a power amplifier that increases the signal power to a level suitable for broadcasting, and a filter that attenuates any undesired energy outside the desired channel that is created in the upconverter or the power amplifier before the signal reaches the antenna.

A DTV transmitter is composed of the same items except, instead of receiving a video and an audio signal, a DTV transmitter receives an MPEG transport bitstream. In most cases, the bitstream format will be according to Advanced Television Systems Committee (ATSC) digital television standards, and the steps used to create a modulated RF signal are quite different from those followed for the analog case. The digital modulation technique utilized for terrestrial broadcast is called 8-VSB, and is described in ATSC A/53 Standard [1]. The ATSC has developed some recommended standards for performance of 8-VSB transmitters for full service in ATSC A/64 Standard [2]. No such standards exist for DTV LPTV transmitters, but minimum performance standards do exist within the FCC Rules (see FCC Rules section 74).

Because both analog and digital TV transmitters share common elements, it is possible and it may be advantageous to convert an analog LPTV or translator transmitter to DTV operation. Another option is to keep the basic amplifier of the existing analog system, exchange the downconverter (if dealing with a translator) and IF upconverter for those which accommodate DTV, and add a DTV emission mask filter at the output. A third option is to replace the entire transmitter, which becomes a more attractive option as the price of 8-VSB equipment drops. The FCC does permit individual LPTV stations to convert their existing analog equipment to DTV transmission provided certain requirements are met; see FCC Rule 74.796 (c). This also includes translators.

Translators are simply a combination of a receiver and transmitter. A block diagram of a translator with optional configurations for analog or digital operation is shown in Figure 6.12-5.

There are two types of translators. One is called a *heterodyne translator,* which downconverts the received signal to a convenient intermediate frequency (IF) for filtering and processing and then upconverts the signal to another authorized TV channel for retransmission. In contrast, *baseband translators* for analog service and *regenerative translators* for digital service demodulate the signal to video and audio (for analog service) or MPEG transport bitstreams (for DTV service) and then remodulate the signal to meet the FCC requirements for retransmission on another channel. The dashed line in Figure 6.12-5 indicates the signal path for a regenerative translator, and the solid line indicates the signal path for a heterodyne translator.

Emission Mask Filters

Most transmitters have some sort of filter following the power amplifier. In NTSC transmitters this filter is used primarily to attenuate specific intermodulation products located at +9 MHz, +13.5 MHz, –4.5 MHz, –9 MHz (all relative to the visual carrier frequency), and harmonics. In ATSC DTV transmitters, the filter will provide attenuation to energy in the adjacent channel produced by intermodulation in the power amp as well as attenuate harmonics. These filters are often called emission mask filters because they ensure that the transmitter systems will meet the emission masks designated by the FCC.

Two types of DTV emission masks are specified by the FCC depending on the degree of frequency

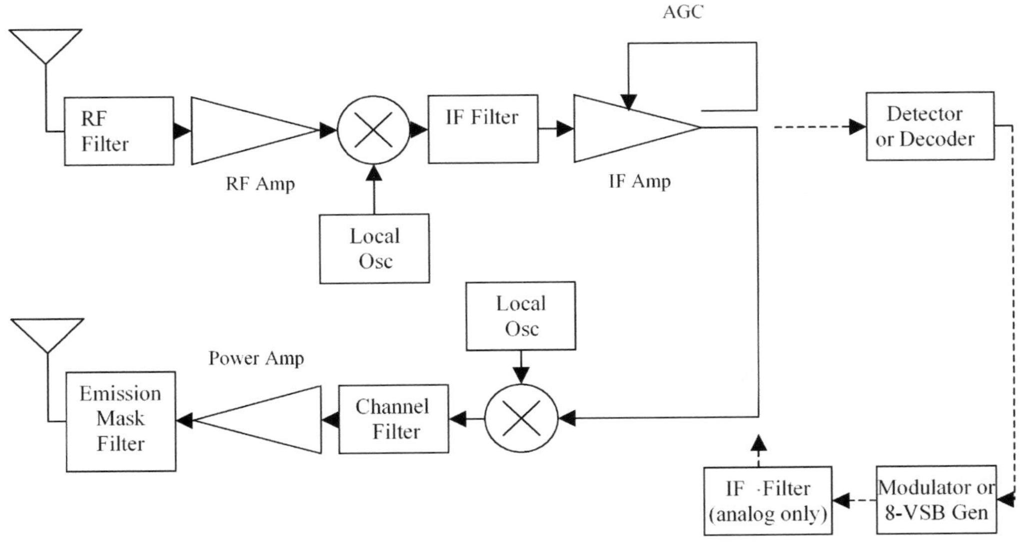

FIGURE 6.12-5 Block diagram of typical analog or digital TV translator with options for remodulator type translators.

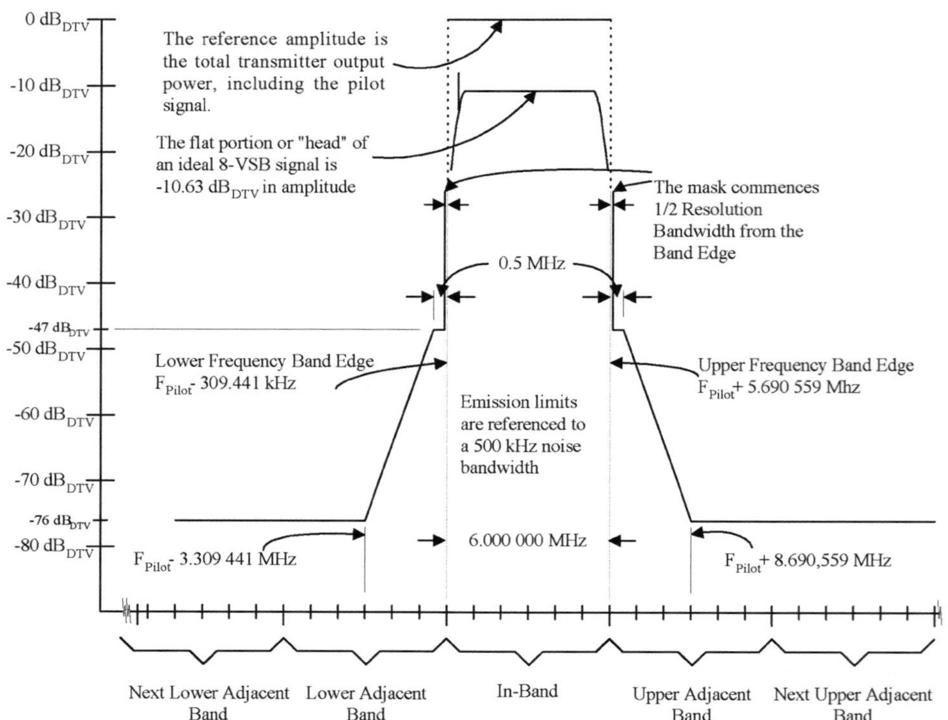

FIGURE 6.12-6 8-VSB stringent emission mask. (Courtesy of IEEE Broadcast Technology Society.)

congestion and interference conditions for the station under design. The "simple" emission mask is used where congestion is typically not a problem. The "stringent" emission mask is used for areas of greater congestion and where the simple emission mask is not sufficient to eliminate the interference. The stringent and simple emission masks for DTV LPTV/translator systems are shown in Figures 6.12-6 and 6.12-7, respectively.

Because the 8-VSB signal has a nearly uniform spectral density (save for the pilot tone) and looks very similar to noise, there are no obvious carriers visible as there are with NTSC. With LPTV DTV transmitters, the FCC has introduced the requirement for additional attenuation in the GPS band. Specifically, transmission systems operating on channels 22–24, 32–36, 38, and channels 65–69 must provide 85 dB of attenuation to frequencies in the three GPS bands (1164 MHz to 1215 MHz, 1215 MHz to 1240 MHz, and 1559 MHz to 1610 MHz); see FCC Rule 74.794 (b).

SITE ISSUES, MAINTENANCE, AND TEST EQUIPMENT

The same factors that influence full-service station reliability also determine the reliability of LPTV stations. A stable temperature environment within the operational specifications of the equipment is important if operations are to be reliable. The size of LPTV transmitter facilities can range from about 20 square feet to as large as the environment permits. Communications site facilities, including tower space, are often shared among multiple LPTV stations. Tower space for LPTV antennas is often shared among tenants. Where situations require that many communications services share common facilities, it is important that the communications services do not interfere with one another. Therefore, it is good practice for an intermodulation or comprehensive interference study to be performed for an entire communications site before initial operation begins, when new licensees are added, or when changes in equipment are contemplated.

Test equipment for modern LPTV facilities need not be exorbitant in price nor extensive in scope. A spectrum analyzer, power meter, DTV signal analyzer, and/or radio service monitor are almost necessities. Note that some power meters used for analog NTSC signals may not yield accurate results for 8-VSB signals. It is important to check with the power meter manufacturer to see if this is the case. Important features to have in a spectrum analyzer for LPTV/translator use are a frequency counter mode and band power measurement capability (especially for measuring 8-VSB signals). A receiver for the type of service under measurement (NTSC or ATSC) is also required. With NTSC it is often possible to detect problems and to determine the source of the problem by simply looking at the various waveforms (RF and baseband) which describe the service. On the other hand, ATSC signal analysis is much more complicated. There are

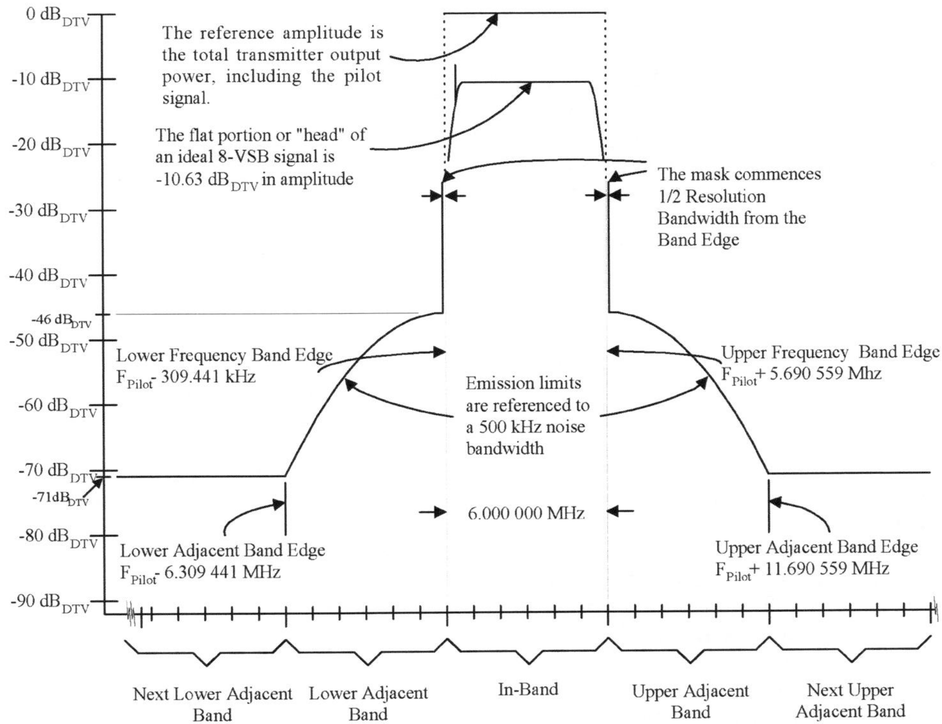

FIGURE 6.12-7 8-VSB simple emission mask. (Courtesy of IEEE Broadcast Technology Society.)

modest cost 8-VSB signal analyzers available for digital LPTV transmitter analysis. Even in very low budget circumstances, it may be possible to obtain more expensive equipment through the pooling of resources to purchase this equipment, which can then be shared among the group which made the purchase.

DESIGN EXAMPLE

The sample station design discussed here (an actual LPTV facility) will illustrate some of the considerations for a typical DTV LPTV or DTV translator.

Station WQXT plans to build an LPTV DTV station at its current transmitter coordinates and use the same tower as the existing analog station . WQXT transmits on analog channel 22 and uses an ALP16 narrow cardioid pattern antenna. The transmitter building has sufficient room to handle an extra (LPTV) DTV transmitter.

In analyzing other channels for a possible (LPTV) DTV transmission system, channel 21 is first considered. This would be a very convenient choice for WQXT because the analog signal for channel 22 and the DTV signal for channel 21 could be combined using a constant impedance combiner, and both signals could be fed to the antenna, which happens to be wide enough to cover both channel 21 and channel 22. This would eliminate the need to purchase additional antenna and transmission line. However, an engineering analysis indicated that channel 21 is not a good candidate due to interference to another station.

In evaluating other channels, channel 14 appears to provide the best compromise of most population covered and the least amount of interference to other existing stations, permittees, and applications on file. Since channel 14 is at the low end of the TV channel range, it is possible that adjacent channel interference may occur to public service transmission equipment (also true for the other end of the TV band at channel 69). In this case, no other interference was predicted to viewers of WQXT or from the planned station.

The next stage is to determine the maximum radiated power and antenna pattern choice. Since WQXT is located on the east side of the state of Florida, radiating a large signal to the east (into the Atlantic Ocean) would be a waste of transmitter power. Most of the desired population to be served is located north and south of the station transmission coordinates, so a peanut pattern, such as that shown in Figure 6.12-3a (rotated to line up with the areas of desired population), is chosen. The required ERP is determined to be at the maximum power level of the LPTV DTV license class of 15 kW.

The next item to determine is the antenna gain. The gain of the antenna and the transmitter power minus the transmission line loss will equal the ERP. In this case, it is not desirable to have maximum antenna gain and minimum transmitter power because doing this would result in a very narrow transmitted elevation pattern from the antenna. The consequence of this would be that the population close to the antenna (1 mile or less) would likely suffer from a lack of signal due to the antenna elevation pattern nulls. A wider elevation pattern is desired to cover the close-in population. So in this case, the goal is to achieve a good balance between moderate antenna gain (so as not to suffer close-in coverage) and moderate transmitter power.

This part of the process is usually iterative. If the antenna gain of the combined azimuth and elevation patterns is approximately 22.3 (or 13.5 dBd), and there is a transmission line loss of 1.6 dB (using Andrew HJ8-50B 3" heliax), then a required transmitter output power of 970 W is the result. This can be accomplished by using a 1 kW solid state transmitter without pushing the transmitter beyond its limits.

Now that the basic system has been designed, there are other details to be addressed. First, the need for antenna beam tilt needs to be determined. In this case, the closest population is approximately 1 km from the tower, but a large part of the desired population is 48 km away. A beam tilt of only 0.5 degrees is chosen to ensure the desired population will be reached with the major portion of the antenna beam.

For this example, the antenna is a side-mount design, so the proximity of the tower may have an effect on the radiation pattern. If the tower face width is 3 feet or larger, then a "scattering analysis" should be conducted. This will indicate the impact of the tower on the radiation pattern and on other antennas located effectively at the same height as the antenna for channel 14.

The next step is to determine how the antenna is connected to the transmission line and how the transmission line is routed up through the tower. Once tentative plans are made for the antenna mounting location and the location of all transmission lines, the design should be reviewed by a structural engineer to ensure that the tower, including guy wires if so equipped, is not overstressed.

In this case, the transmitter power consumption is estimated to be about 7 kW, so the main AC power and the air conditioning (or cooling fan) requirements can be determined knowing the transmitter power consumption and other AC loads in the building. Additionally, the transmitter must be tied into the transmitter building ground system. This is best accomplished by placing a ground strap (e.g., 3" wide) from the transmitter equipment rack to the transmitter building ground system. Likewise, the legs of the tower should also be grounded by using heavy wire or strap from the tower legs to the facility ground system.

In order to keep moisture out of the transmission line and antenna, it is necessary to pressurize the transmission line (and antenna if equipped) with nitrogen or dry air. The volume and pressure can be determined from the antenna and transmission line specifications so that moisture and condensation problems are eliminated.

To determine if there are any concerns about RF exposure at the ground level, the equation given in OET Bulletin 65 can be used to evaluate the RF at ground level. In this case, the calculated value of power density, F, of 0.39 μwatts/cm^2 is determined from the equation

$$F = \left(33.4 \times 0.1^2\right) \times \frac{A}{\left((B \times 0.3048) - 2\right)^2}$$

where A is the maximum ERP of 15 kW, B is the antenna center of radiation in feet above ground at 377 feet, and F is the predicted radiation exposure in μwatts/cm^2 at the ground level for a UHF transmitter. At this frequency, the limit of RF exposure is 320 μwatts/cm^2. The calculated value is only 0.12% of limit so there is no cause for concern in this instance. Once installation of the transmitter is complete, some method to determine power independent of the transmitter itself such as a directional coupler and power meter should be used to verify that the operating power of the transmitter matches the power indicated on the FCC license.

SUMMARY

Transmission equipment designed for LPTV/translator applications exhibits much of the same characteristics as does the equipment used for full-service stations. While LPTV and translator transmission equipment serve smaller communities in general, the operation of these RF transmission sites must meet many of the same requirements, either by FCC Rule, or by general operating practices, as do their full-service counterparts. In many cases, the environments in which LPTV and translator equipment operate present more difficulties than those typically encountered by full-service stations.

LPTV system designs are in general more cost-sensitive than full-service transmission system designs, and maintenance is generally performed less often. Consequently, the reliability of an LPTV system is a key factor when evaluating equipment.

The basic transmitter and receiver block diagrams for LPTV and translator systems have not changed substantially with the transition to DTV. Replacement of the normal modulator and upconverter by the DTV exciter and the addition of a filter at the transmitter output constitute the major changes. Both heterodyne and regenerative translators are suitable for analog and DTV usage. The main advantage of a regenerative translator is that the signal at the output of the LPTV/translator system can be improved over that which is received at the input of the system.

References

[1] Advanced Television Systems Committee, *A/53E: ATSC Digital Television Standard with Amendment No. 1*, 27 December 2005 (Amendment No. 1 dated 18 April 2006), available online at www.atsc.org

[2] Advanced Television Systems Committee, *A/64A: Transmission Measurement and Compliance for Digital Television*, 30 May 2000, available online at www.atsc.org

CHAPTER

6.13

Audio and Video over IP Networks and Internet Broadcasting*

WES SIMPSON
Telecom Product Consulting

INTRODUCTION

Internet Protocol (IP) is the basic communication method that is used by over a billion people through the public Internet. It is also a popular choice for a wide variety of video transport applications, including:

- Local communication within the confines of a television production facility;
- Wide area communication between venues, production facilities, and broadcasters;
- Delivering video signals to consumers over dedicated networks in the form of Internet Protocol TV (IPTV), or over shared networks such as the Internet;
- Streaming live and prerecorded video and audio content to audiences around the globe.

IP is popular for a variety of reasons. Three major motivations for using IP networks for video and audio transport are:

- *Flexibility:* IP networks can be used to provide a wide variety of different services, including e-mail, file transfer, instant messaging, web surfing, voice communication, and video/audio for conferencing and entertainment. This flexibility allows IP to serve many different types of users with a single converged data transport system.
- *Cost:* Local area networks that transport IP data at speeds up to a gigabit per second can be constructed for a few hundred dollars per user, significantly less than most other networking technologies. Connections to wide area IP networks covering metropolitan or national areas are also economical compared to other communication technologies.
- *Ubiquity:* IP networks reach every corner of the globe and over half of the households in the U.S. have access to broadband Internet service, along with a substantial majority of business Internet users. This means that any broadcaster wishing to send video over the Internet has a large potential audience.

There are many different applications in which IP networks have been used successfully for video applications. Here are a few examples:

- Television networks and local affiliates often host public Internet websites containing video content that can be downloaded or streamed to viewers on-demand.
- Local exchange carriers (telephone companies and other network providers) are increasingly turning to high-capacity IP networks (such as Digital Subscriber Lines or DSLs) to deliver multichannel video content to large groups of households.
- A wide variety of devices are available for transmitting live video feeds over IP networks, ranging from uncompressed high-definition (HD) video signals to highly compressed, low-resolution video

*This chapter is based on material extracted and adapted from the book *Video Over IP, A Practical Guide to Technology and Applications* by Wes Simpson and published by Focal Press (ISBN 0-240-80557-7) by permission of the publisher.

signals via dial-up networks and satellite telephone links. These feeds have been successfully used for sports, entertainment, and particularly news coverage from around the world.

- Video signals that have been stored in the form of computer data files have been transported over IP networks for many different applications, including television and cinema post-production, news clips, advertisement distribution, content logging/ archiving, and distribution of on-demand video to users (such as movie trailers via the Internet).

IP BASICS

IP is a useful mechanism to enable communications between computers. It provides a uniform addressing scheme so that computers on one network can communicate with computers on a distant network. IP also provides a set of functions that make it easy for different types of applications (such as e-mail, web browsing, or video streaming) to work alongside each other on a single computer or share a common network. Plus, IP allows different types of computers (mainframes, PCs, Macs, Linux machines, etc.) to communicate with each other.

IP is flexible because it is not tied to a specific physical communication method; in fact, IP links have been successfully established over a wide variety of different network technologies. The most common technology for IP transport is Ethernet (also known as 10BaseT or IEEE 802.3). Many other technologies can support IP, including dial-up modems, wireless links (such as WiFi), and SONET and ATM telecom links. IP will also work across connections where several network technologies are combined, such as a wireless home access link that connects to a cable TV system offering cable modem services, which in turn sends customer data to the Internet by means of a fiber-optic backbone.

IP depends on other software and hardware, and other software in turn depends on IP. Figure 6.13-1 illustrates how IP fits in between the functions of data transport performed by physical networks and the software applications that use IP to communicate with applications running on other servers.

IP is neither a user application nor an application protocol. However, many user applications employ IP to accomplish their tasks, such as sending e-mail, playing a video, or browsing the web. These applications use application protocols such as the HyperText Transfer Protocol (HTTP) or Simple Mail Transfer Protocol (SMTP). These protocols provide services to applications. For example, one of the services provided by HTTP is a uniform method for giving the location of resources on the Internet, which goes by the name Uniform Resource Locator (URL).

IP by itself is not a reliable means of communications; it does not provide a mechanism to resend data that might be lost or corrupted in transmission. Other protocols that employ IP are responsible for that (such as Transmission Control Protocol or TCP).

IP Addresses

One aspect of IP that is easy to recognize is the special format used to give an IP address. This format is called "dotted decimal" and consists of a series of four numbers separated by periods (or "dots"). For example, "209.116.240.200" is the IP address for www.nab.org. Individuals who have had to configure their own home network or laptop connection have probably seen information in this form. A dotted decimal number represents a 32-bit number, which is broken up into four 8-bit numbers.

Most people have a hard time remembering and typing all of those digits correctly. So, to make life

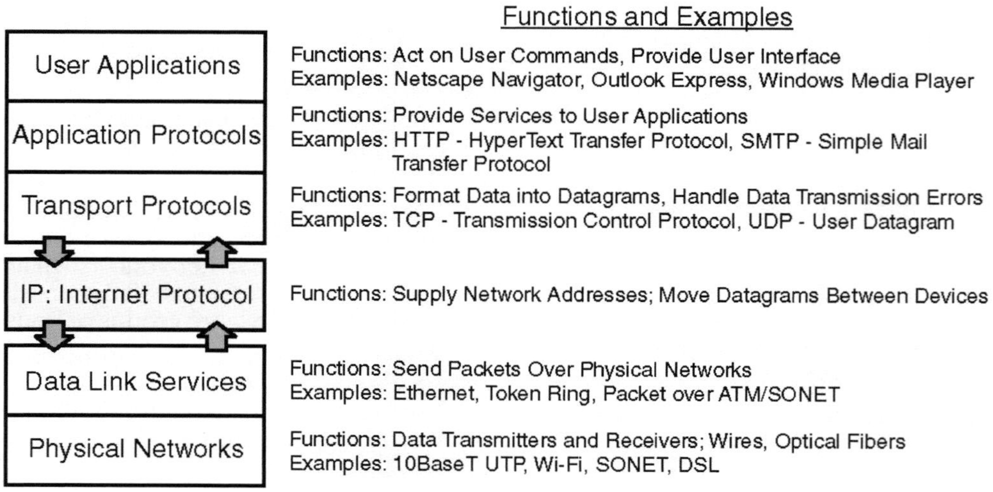

FIGURE 6.13-1 How IP fits into the data communications hierarchy.

easy, the Domain Name System (DNS) was invented. DNS provides a translation service for web browsers and other software applications that takes easy-to-remember domain names (such as "nab.org") and translates them into IP addresses (such as 209.116.240.200).

IP Datagrams and Packets

IP works as a delivery service for "datagrams." They can also (somewhat confusingly) be called "IP packets." A datagram is a single message unit that can be sent over IP, with very specific format and content rules. The key header elements of an IP datagram are listed here and illustrated in Figure 6.13-2:

- *Destination Address:* The IP address indicating where the datagram is being sent.
- *Source Address:* The IP address indicating where the datagram came from.
- *Header Checksum:* A basic error-checking mechanism for the header information in a datagram. Any datagram with an error in its header will be destroyed.
- *Protocol Identifier:* A byte that indicates the type of protocol used to format the datagram. Common examples include TCP (value of 6) and UDP (value of 17).
- *Time-to-Live:* A counter that is decremented each time a datagram is forwarded in an IP network. When this value reaches zero, the datagram is destroyed. This mechanism prevents packets from endlessly looping around in a network.
- *Fragmentation Control:* These bits and bytes are used when a datagram needs to be broken up into smaller packets to travel across a specific network segment.
- *Total Length:* A count, in bytes, of the length of the entire datagram, including the header.

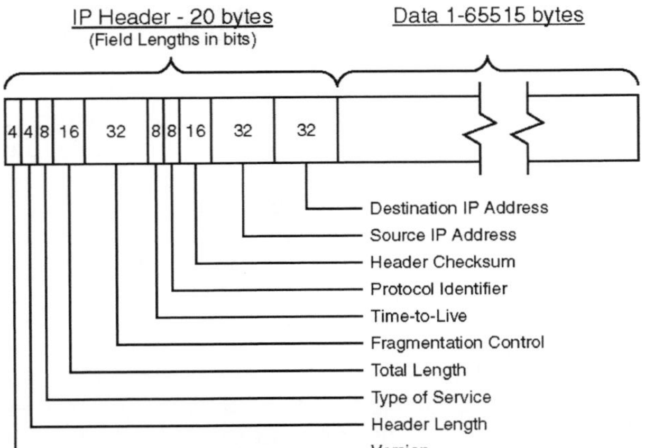

FIGURE 6.13-2 IP datagram format.

- *Type of Service:* A group of bits that can be used to indicate the relative importance of one datagram as compared to another. Note that these are often ignored by many networks, including the public Internet.
- *Header Length:* The length of the packet header, in 32-bit words. Since IP headers are normally 20 bytes long, the length is normally 5 32-bit words ($5 \times 32 = 20 \times 8$ bits).
- *Version:* Two versions of IP are in widespread use today: IPv4, which is used throughout the public Internet and in most private networks, and IPv6, which is starting to be used in new networks.

Ethernet and IP

Ethernet and IP are not synonymous, but they are closely linked in many people's minds. This has happened because Ethernet is one of the most widely used physical networking systems for IP traffic. The following sections describe how Ethernet addressing works and how IP and Ethernet interoperate.

Ethernet Addressing

Ethernet equipment uses Media Access Control (MAC) addresses for each piece of equipment. MAC addresses can be recognized because they use the numbers 0–9 and the letters A–F for their addresses, which are then separated into six fields of two characters each. For example, a MAC address on an Ethernet card inside a PC might be 00:01:02:6A:F3:81. These numbers are assigned to each piece of equipment, and the first six digits can be used to identify the manufacturer of the hardware. MAC addresses are uniquely assigned to each piece of hardware by the manufacturer and do not change.[1]

The difference between MAC addresses and IP addresses can be illustrated with a simple analogy. When an automobile is manufactured, it is assigned a serial number that is unchanging and stays with that car permanently. This is similar to a MAC address for a piece of hardware. When an owner goes to register the auto, a license plate (a.k.a. number plate or marker tag) is received. The number of the license plate is controlled by the rules of the local jurisdiction, such as a state in the USA, a province in Canada, or a country in Europe. During its lifetime, one auto may have several different license plate numbers, as it changes owners or if the owner registers the car in another jurisdiction.

Similarly, one piece of physical hardware may have different IP addresses assigned at different times in its life, as the hardware moves from network to network. Stretching the analogy a bit further, an auto's serial number is somewhat private information, of interest only to the owner of the car and the agency that issues the license plates. In contrast, the number of the license plate is public information, and emblazoned

[1]Some pieces of equipment, including many small routers made for sharing home broadband connections, have the ability to "clone" a MAC address of a different device. This allows a user to set up the router to match a MAC address stored at the service provider.

on the auto for all to see. Similarly, a MAC address is important only to the local Ethernet connection, commonly on a private network. The IP address is public and is used by other computers all over the Internet to communicate with a particular machine.

Ethernet Interfaces

Unshielded Twisted Pair (UTP) cabling is the most common physical interface used today for Ethernet cabling, also known as Category 5 or 6 UTP. The category indicates the rules established in ANSI/TIA/EIA-568-B that were followed for constructing the cables: Category 6 supports higher data transmission speeds than Category 5. Each cable contains four pairs of wires (a total of eight wires in all). Each pair of wires is twisted in a very precise pattern, and the pairs are not shielded from each other (hence the name UTP). Other forms of Ethernet wiring, including coaxial cable and fiber, are also widely used.

Ethernet interfaces that use UTP cabling are a strict physical star. That is, each end device (such as a PC, printer, or server) must have a direct, uninterrupted connection to a port on a network device. This network device can be a hub, bridge, switch, or router. The network device accepts the signals from the end device and then retransmits the data to other end devices. In the case of a hub, this retransmission is done without any processing; any incoming signal is retransmitted out to every other port of the hub. With bridges, switches, and routers, processing is done on each incoming Ethernet data frame to send it out only on the proper destination ports.

Common UTP interfaces include:

- 10BaseT, which operates at a nominal bit rate of 10 Mbps;
- 100BaseT, which operates at a nominal bit rate of 100 Mbps;
- 1000BaseT, also known as Gigabit Ethernet, which operates at a nominal bit rate of 1000 Mbps.

Note that all of the preceding speeds are described as "nominal." This is a reflection of reality because, due to overhead, required dead time, and turnaround time, actual sustained throughput will be much less than the nominal rate, depending on the application. For system implementers, it is important to realize that it isn't possible to send a 10 Mbps MPEG video stream over a 10BaseT link.

Wireless Ethernet

Wireless Ethernet standards, such as 802.11a, 802.11b, and 802.11g, offer unprecedented flexibility for users to move devices from one location to another within a network's coverage area. Low-cost wireless access devices are available for desktop PCs, laptops, and for connecting to (and sharing) high-speed network connections, such as DSL and cable modems.

Video users must be careful about using wireless links for high-quality, streaming video connections. The speed and quality of a wireless link depend on a number of factors, including the distance between the wireless transmitter and receiver, the configuration of antennas, and any local sources of signal interference or attenuation. If the local environment changes (say, a laptop is moved, or a metal door is closed), the error rate on the wireless link can change rapidly. If the packet loss rate increases significantly, 802.11 links are designed to drop down to a lower operating speed, which helps lower the error rate. Conversely, if the packet loss rate goes down, the transmission speed can be increased.

IP Encapsulation

Encapsulation is the process of taking a data stream, formatting it into datagrams, and adding the headers and other data required to comply with Internet Protocol standards. This is not simply a matter of taking a well-established formula and applying it. Rather, the process of encapsulation can be adjusted to meet the performance requirements of different applications and networks.

As discussed previously, any data that is to flow over an IP network must be encapsulated into IP datagrams, commonly called IP packets. This is true whether the data is a prerecorded file or a live digital video stream. Also, as shown in Figure 6.13-2, a variety of data must be inserted into each packet's IP header, including both the source and the destination IP addresses, the Time-to-Live of the packet, the packet's priority indication, and the protocol that was used to format the data contained in the packet (such as TCP or UDP, which are explained later in this chapter). Software tools to perform encapsulation are included in a wide variety of devices, including desktop PCs and file servers. Packet encapsulation is done in real time, just before the packets are sent out over the network because much of the data going into the packet headers changes for each data stream (such as the destination IP address).

Packet Size

Performance of IP video signals will be affected by the choices that are made for the video packet size. The length of the packets must meet the minimum and maximum sizes that are specified in the standards for IP. However, within those constraints, there are advantages to using long packets, just as there are advantages to using short packets. Long packet advantages include less overhead as a percentage of the total data stream and reduced load (due to fewer packet headers to process) on routers and other devices. Short packet advantages include lower latency for low bit rate streams, less sensitivity to lost packets, and less need for fragmentation in networks.

Typically, video signals tend to use the longest possible packet sizes that won't result in fragmentation on the network. Fine-tuning the packet sizes can help make a network run better. A mistake in setting these values will generally not prevent the video from flowing, but it can create extra work for devices all along the path of a connection.

TRANSPORT PROTOCOLS

Transport protocols are used to control the transmission of data packets in conjunction with IP. Three major protocols that are commonly used in transporting real-time video are:

- *UDP* or User Datagram Protocol: This is one of the simplest and earliest of the IP protocols. UDP is often used for video and other data that is time sensitive.
- *TCP* or Transmission Control Protocol: This is a well-established Internet protocol that is widely used for data transport. Most of the devices that connect to the Internet are capable of supporting TCP over IP (or simply TCP/IP).
- *RTP* or Real-time Transport Protocol (also known as Real Time Protocol): This protocol has been specifically developed to support real-time data transport, such as video streaming.

In the networking hierarchy, all three protocols are considered to operate above the IP protocol because they rely on IP's datagram transport services to actually move data from one computer to another computer. These protocols also provide service to other functions inside each computer (such as user applications) and so are considered to sit "below" those functions in the networking protocol hierarchy. Figure 6.13-3 shows how UDP, TCP, and RTP fit into the networking hierarchy. Note that RTP actually uses some of the functions of UDP; it operates on top of UDP.

Ports

A common feature of UDP and TCP is that they use logical *ports* for data communications. Ports serve as logical addresses within a device for high-level protocols and user applications.[2] Packets that are part of an ongoing file transfer will be addressed to a different port number than packets that contain network management information, even though they may be going to a single IP address on a single device. When a remote device wishes to access a specific application in another device, it must send data to the correct IP address and then indicate to UDP or TCP which port the data is intended for. A basic set of ports that support well-known services such as e-mail or web browsing are defined by the Internet Assigned Number Authority (IANA).

By standardizing the numbering of ports for well-known services, the IANA has done a tremendous service. For example, a web server that is supporting an HTTP application always listens for remote devices to request connections on port 80. This makes it easy for a remote user with a standard browser to begin an HTTP session simply by knowing the IP address of the server and sending a request packet to port 80 to begin an HTTP session.

UDP

User Datagram Protocol (UDP) is a connectionless transport mechanism that can support high-speed information flows such as digital video. It can support many other types of data transport and is frequently used when the overhead of setting up a connection (as is done by TCP) is not needed. For example, UDP is often used for broadcasting messages from one device to all the other devices on a network segment (say, a print server alerting all the users that the printer is out of paper).

UDP is a connectionless protocol, which means that there is no mechanism to set up or control a connection

[2]It is important not to confuse the concept of logical ports used by protocols with the reality of physical ports on devices, such as a 10BaseT "port" on an Ethernet switch. Logical ports are used to control how datagrams are passed from the IP layer and to make sure they end up in the proper application program.

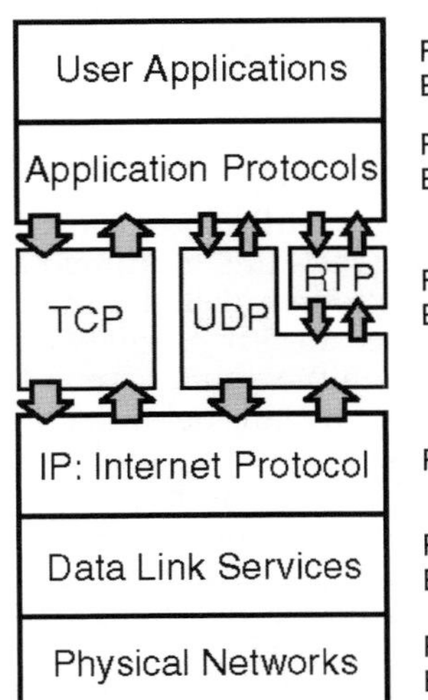

Functions: Act on User Commands, Provide User Interface
Examples: Netscape Navigator, Outlook Express, Windows Media Player

Functions: Provide Services to User Applications
Examples: HTTP - HyperText Transfer Protocol, SMTP - Simple Mail Transfer Protocol

Functions: Format Data into Datagrams, Handle Data Transmission Errors
Examples: TCP - Transmission Control Protocol, UDP - User Datagram Protocol, RTP - Real-time Transport Protocol

Functions: Supply Network Addresses; Move Datagrams Between Devices

Functions: Send Packets Over Physical Networks
Examples: Ethernet, Token Ring, Packet over ATM/SONET

Functions: Data Transmitters and Receivers; Wires, Optical Fibers
Examples: 10BaseT UTP, Wi-Fi, SONET, DSL

FIGURE 6.13-3 Transport protocol hierarchy.

between a sending device and a receiving device. The UDP sender simply formats datagrams with the correct destination IP address and port number and passes them to IP for transport. There is no coordination between a UDP data transmitter and a UDP data receiver to ensure that the data is transferred completely and correctly from the sender to the receiver.

This lack of coordination could seem to make UDP unsuitable for video data transfer, since missing video data can interfere with the receiver's capability to display a correct sequence of complete images. However, it is important to remember that each image in an NTSC video stream is displayed for only 33 milliseconds (40 msec for PAL video). If part of the data for an image is missing, the receiver would ideally need to:

- Recognize that the data was missing;
- Send a message back to the sender to tell it which data needed to be retransmitted;
- Receive and process the retransmitted data;
- Make sure that all this happened before the image was due to be displayed.

Since completing all of these steps within a 33 msec window could be impractical, it may be wise not to even try retransmitting the missing data. Instead, many video stream formats include a mechanism for detecting and correcting errors. For example, MPEG transport streams can include bytes for Reed-Solomon forward error correction. When these bytes are included, the MPEG decoder can correct for bit errors and can sometimes re-create lost packets. In UDP, the originating device can control how rapidly data flows into the network. For many types of video streams, UDP is a logical choice for the transport protocol, since it does not add unneeded overhead to streams that already have built-in error correction functions.

TCP

Transmission Control Protocol (TCP) is a connection-oriented protocol that provides highly reliable data communications services. TCP is easily the most widely used protocol on the Internet.

TCP is "connection oriented," meaning that TCP requires a connection be set up between the data sender and the data receiver before any data transmission can take place. Either the sender or the receiver of the data can initiate a connection, but acknowledgment messages must be sent between both ends of the circuit before it is ready for use. A standard handshake sequence has been defined for TCP that handles essentially all of the different things that can and do go wrong in this process.

One of the important features of TCP is its capability to handle transmission errors, particularly lost packets. TCP counts and keeps track of each byte of data that flows across a connection, using a field in the header of each packet called the Sequence Identifier. In each packet, the Sequence Identifier indicates the sequence number of the first byte of that packet. If a packet is lost or arrives out of order, the Sequence Identifier will be incorrect, indicating that an error has occurred. Through a fairly elaborate system of control flags, the receiver in a TCP circuit tells the sender that some data is missing and needs to be retransmitted. This is how TCP ensures that the receiver gets every byte from the sender. This system is appropriate for transmitting files that cannot withstand corruption, such as e-mails or executable program code.

TCP also has the capability to control the flow rate of data across a connection. This feature operates by using some of the same control flags that are used to identify lost packets and through the use of a data buffer in the receiver. The receiver tells the sender how big a buffer it is using and the sender must not send more data than will fit into the buffer. More data can be sent only after the receiver acknowledges that it has processed at least some of the data in its buffer, and there is room for more.

Unfortunately, some of these very mechanisms that make TCP valuable for data transmission can interfere with video transport. Remember, the data for a video signal needs to arrive not only intact, but also on time. So, a mechanism that retransmits lost packets can be harmful in two ways:

- If a packet is retransmitted but arrives too late for display, it can tie up the receiver while the packet is examined and then discarded.
- When packets are retransmitted, they can occupy network bandwidth that is needed to send new data.

Also, the flow control mechanism that is built into TCP can interfere with video transport. If packets are lost before they are delivered to the receiver, TCP can go into a mode where the transmission speed is automatically reduced. This is a sensible policy, except when a real-time video signal needs to be sent. Reducing the transmit speed to less than the minimum needed for the video stream can prevent enough data from getting through to form any video image at all. With lost packets on a real-time video stream, the better policy is to ignore the losses and keep sending the data as fast as necessary to keep up with the video (and audio) content.

RTP

The Real-time Transport Protocol (or Real Time Protocol) is intended for real-time multimedia applications, such as voice and video over the Internet. RTP was specifically designed to carry signals where time is of the essence. For example, in many real-time signals, if the packet delivery rate falls below a critical threshold, it becomes impossible to form a useful output signal at the receiver. For these signals, packet loss can be tolerated better than late packet delivery. RTP was created for these kinds of signals—to provide a set of functions that are useful for real-time video and audio transport over the Internet.

One example of a non-video real-time signal that is well suited to RTP is a voice conversation. As most early users of mobile telephones can attest, an occasional noise artifact (such as an audible "click") is not enough to grossly interfere with an ongoing conversation. In contrast, if a mobile phone were designed to stop transmission and rebroadcast the voice data packets each time an error occurred, then the system would become constantly interrupted and virtually useless. Video is similar: A short, transient disruption is better than a "perfect" signal that is continuously stopped and restarted to allow missing bits to be rebroadcast. RTP is built on this same philosophy: Occasional data errors or lost packets are not automatically retransmitted. Similarly, RTP does not try to control the bit rate used by the sending application; it does not have the automatic rate reduction functions that are built into TCP to handle congested networks.

RTP provides a time-stamping function that allows multiple streams from the same source to be synchronized. Each form of payload (video, audio, voice) has a specific way of being mapped into RTP. Each payload type is carried in separately by RTP, which allows a receiver to, for example, receive only the audio portion of a videoconference call. (This capability can be very useful for a conference participant who has access only via a slow dial-up link.) Each source inserts time stamps into the outgoing packet headers, which can be processed by the receiver to recover each stream's clock signal that is required to synchronize audio and video clips.

RTP is not strictly a transport protocol, like UDP or TCP, as illustrated in Figure 6.13-3. In fact, RTP is designed to use UDP as a packet transport mechanism. That is, RTP adds a header to each packet, which is then passed to UDP for further processing (and another header).

Overall, RTP adds a lot of functionality on top of UDP, without adding a lot of the unwanted functions of TCP. For example, RTP does not automatically throttle down transmission bandwidth if packet loss occurs. Instead, RTP provides information to the sending application to let it know that congestion is happening. The application can then determine what it wants to do to compensate for the congestion. The application could, for example, lower the video encoding bit rate (sacrificing some quality), or it could simply ignore the report if, say, only a few of the receivers were affected by the congestion. RTP also supports multicasting (discussed later), which can be a very efficient way to transport video over a network.

CALCULATING NETWORK BANDWIDTH

Calculating the amount of actual bandwidth consumed by an MPEG stream is important. Here is how one manufacturer, HaiVision Systems of Montreal, Quebec, does this for one MPEG–2 product: the hai500 series Multi-stream encoder/decoder/multiplexer.

As in most MPEG devices, the hai500 user is given control over the rate of the raw MPEG stream. For this device (which is considered typical), the video bit rate can be set anywhere from 800 kbps to 15 Mbps, in steps of 100 kbps. The user can also set the audio stream rates over a range from 32 kbps to 448 kbps.

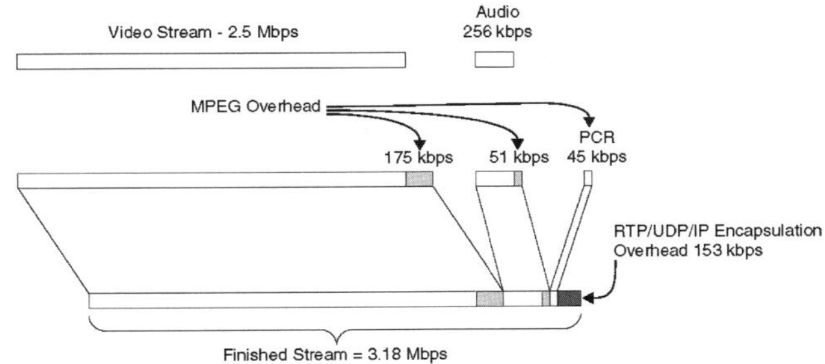

FIGURE 6.13-4 Sample total bandwidth calculation for an MPEG 2 stream.

Consider an example using a video bandwidth of 2.5 Mbps and audio bandwidth of 256 kbps.

In order to transport these raw streams over a network, the first step is to convert the raw MPEG into a transport stream (TS). This will add about 20% to the raw audio bandwidth and 7% to the raw video bandwidth. The original audio stream becomes a 307 kbps TS, and the video TS occupies 2.675 Mbps. A 45 kbps program clock reference is also added. Figure 6.13-4 illustrates this example.

The next step is to calculate the IP and Ethernet packet overhead. The hai500 uses RTP over UDP, requiring a 12-byte RTP header and an 8-byte UDP header. A 20-byte IP header and a 26-byte Ethernet header must also be added, bringing the total of all the headers to 66 bytes. One 1500-byte Ethernet frame can accommodate seven MPEG standard TS packets of 188 bytes each, for a total of 1,316 bytes. With 66 bytes of header for 1,316 bytes of data, the overhead is approximately 5%. The resulting total bandwidth for both the audio and video streams (with rounding because partial packets are not permitted) comes to 3.18 Mbps. This calculates to 15.4% overhead on the original raw streams (2.5 Mbps plus 256 kbps). In terms of packets, this equates to roughly 288 packets per second.

FIREWALLS

A firewall is used to control the flow of information between two networks. In many cases, firewalls are used between user computers and the Internet, to keep unauthorized users from gaining access to private information or network resources.

Firewalls vary greatly in their level of sophistication and in their methods for providing security. Most firewalls provide at least a packet filtering function, which examines each packet passing through the firewall on a packet-by-packet basis. More sophisticated firewalls provide *stateful inspection*, which not only examines the contents of each packet, but also tracks the connection status of each of the TCP or UDP ports that are used by the packet. A firewall could, for example, block any packets from the public Internet that have not been requested by a user behind the firewall.

Firewalls can interfere with video traffic in two ways. First, the firewall can block outside users from being able to access a video source or server. Since many video servers will not send video data to a user until requested to do so, it can be impossible for a user to make a request if all the well-known ports (such as port 80 used by HTTP) are blocked by the firewall. Placing the video server outside the firewall, at an Internet hosting service, for example, can solve this problem. Alternatively, a special subnetwork can be created that is only partially protected by the firewall, commonly called a DMZ (after the military term "Demilitarized Zone").

The second way in which firewalls can interfere with video traffic is by blocking all UDP traffic, which is a very common setup. UDP traffic is harder to police than TCP traffic, because it is possible for an intruder to insert unauthorized packets into a UDP stream. UDP has a number of benefits for streaming video, and many video sources use the UDP protocol. So, if UDP is to be used for streaming, any firewalls that are along the path from source to destination must be configured to allow this type of traffic.

VIDEO STREAMING AND MEDIA PLAYERS

Video streaming is the process of delivering video content to a viewing device for immediate display. Streaming is often the first application that springs to mind when people think about transporting video over a network. "Video streaming" is a generic term that covers a few different technologies. The most common ones are:

- *True streaming,* where the video signal arrives in real time and is displayed to the viewer immediately. With true streaming, a 2-minute video sequence takes 2 minutes to deliver to the viewer—not more, and not less. Live content is transmitted through the use of true streaming.
- *Download and play,* where a file that contains the compressed video/audio data is downloaded onto the user's device before playback begins. With download and play, a 2-minute sequence video could take 10 seconds to download on a fast network, or 10 minutes to download on a slow network. Podcasting uses download and play technology.
- *Progressive download and play,* which is a hybrid of the two preceding technologies that tries to capture the benefits of both. For this technique, the video program is broken up into small files, each of which is downloaded to the user's device during playback. With progressive download and play, a 2-minute video sequence might be broken up into 20 files, each 6 seconds long, that would be successively downloaded to the viewer's device before each file is scheduled to play.

True Streaming

True streaming over an IP network starts by taking a digital video signal and breaking it up into IP packets. The video signal can be uncompressed, but generally, when discussing streaming, the video content has been compressed using MPEG or other compression method. These packets are then "streamed" out over the network, which means that the packets are sent at a data rate that matches the rate of the video. In other words, a video program that lasts 10 minutes, 6 seconds will take 10 minutes, 6 seconds to stream. Special software, called "player software," accepts the incoming packets and creates an image on the viewing device.

For streaming to work properly, video content must arrive at the display right when it is needed. This is not as simple as it sounds, because many factors can affect the timely delivery of the video packet data. Servers running specialized software (and often containing specialized hardware) are used to send video content in a smooth, constant speed stream of packets over a network. The network must be capable of delivering the stream to the viewer intact without losing or changing the timing of the packets. The player software must accept the incoming packets and deal with any imperfections in the data flow that were caused by the network. Typically, this requires a small amount of buffering in the player device.

A number of software packages available commercially support video streaming. Apple's QuickTime Streaming Server supports several different video and audio formats, and can support live streaming as well as prerecorded content. Users with Macintosh or Windows PCs can download and install a free copy of the QuickTime Player to view the media. Similar products are available from Microsoft, RealNetworks, and other suppliers.

Download and Play: Podcasting

Download and play takes a video content file and transfers a copy to a viewing device, where it is temporarily stored and then immediately shown to the viewer. *Podcasting* is a variation wherein the content file is stored permanently on the viewer's device (for example, an Apple iPod, the device from which "Podcasting" got its name), where it can be played back at any time. Both technologies are very similar to the process used by websites, where user browsers request web pages and files from a server. In fact, download and play frequently uses the same protocols as normal web surfing: HTTP and TCP.

Whether web pages or video clips are being requested, operation of a download and play network is fairly simple: Each time content is requested, a file containing the content is sent to the requesting device. If the content is a 2 kB web page, then the download can occur fairly rapidly, even if a slow network connection is used. If, however, the content is a 5-minute video clip, the time to download the clip to the playing device can be quite long, even on a fast network connection.

Video and audio content can be hosted on standard web servers when download and play is used. Although the files can be very large, the protocols and procedures needed to send the content to a viewer are the same for video and audio content as for simple pages of HTML text. However, just like any web server, system performance may need to be tuned to handle high volumes of requests for large blocks of content. This is no different than what would be required of a server designed to handle a number of large documents or images.

One advantage of download and play technology is that it can work over any speed network connection. The reason is that there is no requirement for the packets containing the video data to arrive at any specific time, since all of the content is delivered before playback begins.

Progressive Download

Progressive download takes a video content file and breaks it up into smaller segments, each one of which can be sent in turn to the player software. Progressive download is a variation on download and play. It is used to simulate streaming for applications in which streaming won't work properly, such as when true streaming behavior is blocked by a firewall. As soon as a segment is completely downloaded, the player can begin to process it and display it, while the next segment is downloaded. As long as each new segment arrives before its designated time to play, the playback device will be able to create a smooth, unbroken video image. Progressive download can best be thought of as a compromise between streaming and download and play that takes some of the benefits of each.

Download and play was prevalent in the early days of multimedia on the web, and typically was the only way for audio and video content to be distributed over low-speed links. Progressive download evolved from basic download and play as server and viewer software became more sophisticated.

STREAMING SYSTEM ARCHITECTURE

When implemented on an IP (or any other technology) network, streaming requires a fair amount of infrastructure. The key item is the streaming server, which has the responsibility to deliver the video just when it is needed. Another important piece is the viewer software that actually receives the incoming video signal from the IP network and generates an image on the user's display. The final pieces in this system are the content preparation station and the transport network between the server and the viewing device. Figure 6.13-5 shows how these several pieces work together.

Content Preparation

Raw video content, such as a live image generated by a camera or video that has been recorded to a tape, normally needs to be prepared to make it ready for streaming. Preparation can include capture/format conversion, video compression, labeling and indexing, and publishing for streaming. The content preparation station may also support video editing and preprocessing. These six functions are listed in the order that they are normally performed:

- *Capture:* Gathering video content and placing it into the preparation system in a common format.
- *Editing (optional):* Organizing the content into the form that viewers will see, complete with synchronized audio, music, overlaid text, and other visual and audio effects.
- *Preprocessing (optional):* Conditioning the edited video prior to compression, and can include color correction, noise removal, image resizing, and other processing needed to get the best results possible from compression.
- *Compression:* Converting the video and audio streams into the formats that will actually be streamed out to viewers. If different stream rates, video resolutions, or types of player software are to be supported, then the compression process is normally repeated for each combination.
- *Labeling and Indexing:* Making the content accessible to viewers, by providing descriptions of the content

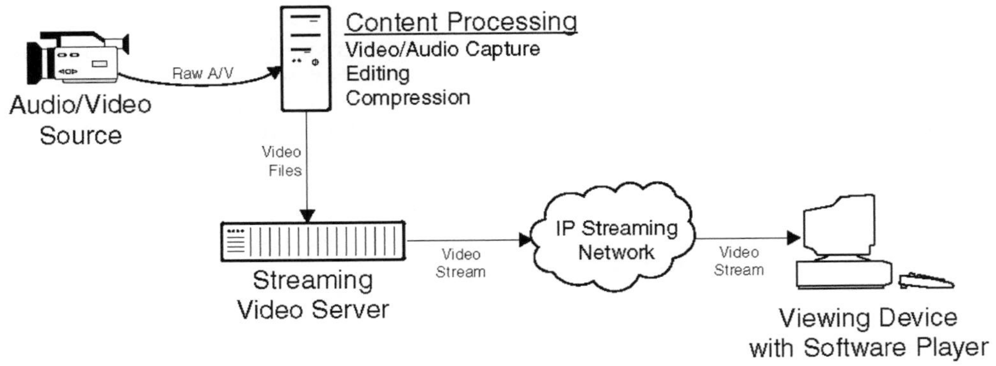

FIGURE 6.13-5 Architecture of a typical streaming system.

and organizing it so that viewers can locate the form of the content that is most suitable to their needs.

- *Publishing:* Transferring the content to one or more streaming servers and creating the web pages and other materials that will contain links to the media streams.

Streaming Server

Streaming servers are responsible for distributing media streams to viewers. They take media content that has been stored internally and create a stream for each viewer request. These streams can be either unicast or multicast, and can be controlled by a variety of mechanisms.

One of the main functions of a streaming server is content storage and retrieval. When content is prepared, it is normally produced in a variety of compression ratios so that users with different network connection speeds can select the video rate that they require. For example, on many websites the user is given a choice between playback speeds that are suitable for a dial-up connection (56 kbps, sometimes called "low speed"), a medium-speed connection (such as 100 or 128 kbps, sometimes called "ISDN" speed), and a high-speed connection (300 kbps or more, sometimes called "broadband" or "ADSL/cable modem" speed). Each of these different playback speeds requires a different version of the content file to be created during the compression process. This means that one piece of content may be contained in three different files inside the server—one for each of the playback speed choices.

The most processing intensive job of the streaming server is to create packets for each outbound stream. As discussed earlier, each IP packet must have a source and a destination IP address. In order for the video packets to reach the correct destination, the server must create headers for these packets with the correct IP destination address. If the server is sending out standard RTP packets, the video and audio streams must be sent as separate packet streams, each to a different port on the receiving device.

Another responsibility of the server is to encrypt the packets in the outgoing stream, if required by the content owner. Each stream will then be encrypted while it is being sent out of the streaming server with a unique key for each user.

Sometimes, streaming servers are given the responsibility to change the rate of the stream being sent to a viewer based on changing network conditions. For example, a stream could start out at 300 kbps and then drop to 100 kbps if the network became congested. This process, also know as "scaling" a stream, is a feature of many advanced streaming systems.

Streaming servers supplying a number of simultaneous viewers require high-speed storage and network connections. A 2 GHz Pentium server could easily process enough simultaneous video streams to overwhelm a 1.5 Mbps T1 line several times over, and seriously load a 10BaseT Ethernet link. For large servers handling broadband streams, use of Gigabit Ethernet network interfaces is becoming common. Also, in order to support hundreds or thousands of users, the video content can be copied to multiple servers at different physical locations around the Internet.

IP Streaming Network

Although it would be useful if any IP network could be used for streaming, in reality, a streaming system will function better if some of the more important network performance variables are controlled. Network parameters that can affect streaming performance include:

- *Packet Loss Ratio:* If too many packets are lost, then streaming video performance will suffer. (An occasional packet lost every now and then can generally be hidden.) Packet loss ratios should be kept below one out of every thousand, if possible.
- *Packet Delay Variation:* Because streaming is a real-time activity, if a packet arrives too late, it can't be used in the playback. Conversely, if packets arrive too early, they need to be stored, which can overload the incoming signal buffer and cause errors. Ideally, if the delay variation can be kept below 50 milliseconds, the player can usually accommodate the variation.
- *End-to-End Delay:* This parameter is not terribly important, unless the link is being used for a two-way conversation. In this latter case, total one-way delay must be kept below 400 milliseconds.

Player Software

Player software is responsible for accepting the incoming stream and converting it into an image that can be shown on the viewer's display. A number of functions need to be performed in this software, and the performance of this software can have an effect on how satisfied the user is with the streaming system overall.

Before streaming can begin, the user must select the content. For most private streaming applications, the content supplier must maintain this list. Typically, the list is presented to the user inside a web page that is hosted on the streaming server, and the user simply selects the appropriate hot link to begin content playback.

If the streaming server has encrypted the content, the player software must decrypt the incoming packets. This is a fairly simple process once the key is known. The player can obtain the keys by communicating with the server or by connecting to a third-party authentication service.

The player software is responsible for managing a buffer that receives the incoming packets. This buffer is necessary to absorb any timing variations in the incoming packets and to place packets that arrive out of order back into the correct order. Buffer underflows (too little data) and overflows (too much data) can severely affect the display of the stream, so the buffers must be sized appropriately for the application. On

the downside, large buffers can take awhile to fill when a new stream is starting, or if the user decides to select a new stream. Overall, careful buffer design is an important factor in player success.

One of the most intensive jobs of the player software is to decompress the incoming signal and create an image for display. The amount of processing required varies depending on the size of the image and on the compression method. Older compression systems (such as MPEG–1) tend to be easier to decode than newer systems (such as MPEG–4) and therefore place less of a burden on the decoding device. Smaller images, with fewer pixels to process, are also easier to decode. Stand-alone devices such as set top boxes usually have hardware-based decoders.

Most recent-vintage personal computers are also capable of running player software. This category includes desktop and laptop machines, as well as Windows-, Macintosh-, and Linux-based systems. Some hand-held devices also are capable of running player software. For high-quality video decoding, a number of hardware decoder cards can also be added to a personal computer to boost performance.

TECHNOLOGIES FOR STREAMING

As streaming software vendors continue to innovate, the quality of streamed video continues to go up even as bit rates drop. Today's compression algorithms are 20–30% more bandwidth efficient than those introduced a few years ago. Progress continues, as more powerful compression systems allow even more sophisticated techniques to be used on the encoder, and more powerful PCs are available to users for running player software. In addition, research continues on new algorithms that can make lower bit rate pictures even more appealing to the human eye.

Real-Time Streaming Protocol (RTSP)

RTSP provides a means for users to control video, audio, and multimedia sessions. RTSP does not actually provide for the transport of video signals; it allows these signals to be controlled by a user. Like a dispatcher for a delivery service, RTSP does not go out and actually deliver packages; it controls when and how packages are delivered by other protocols such as RTP.

RTSP is like HTTP for real-time files. A command such as "rtsp://content.com/mymovie.rm" begins playback of the video file named "mymovie" on the server named "content." This is very similar to the command that would fetch the page named "webpage" from the same site: "http://content.com/webpage.htm." In the first case, the actual video would be transported by RTP over UDP. In the second case, the web page would be transported by TCP.

RTSP is specifically intended for use with time-oriented content, including streaming audio and video. It has the capability to move around inside content based on time stamps contained in the video, allowing a user to, say, skip ahead exactly 10 seconds in a video clip while maintaining audio synchronization. Contrast this with HTTP, which doesn't have this rich set of features for managing timed content.

RTSP uses URLs to identify content on servers, just like HTTP. When a server receives an RTSP command, it will begin to play the stream to a specific destination, which is generally to the client that sent the RTSP command. RTSP requests can go from a client to a server, or from a server to a client. Both the server and the client must keep track of the current status of the stream so that both will know when playback has begun.

Synchronized Multimedia Integration Language (SMIL)

SMIL was developed to allow the design of websites that combined many different types of media, including audio, video, text, and still images. With SMIL, the web page author can control the timing of when objects appear or play, and can make the behavior of objects depend on the behavior of other objects.

SMIL is like HTML for multimedia. SMIL scripts can be written and embedded into standard web pages to cause actions or reactions to user inputs. For example, a SMIL script could be written as a portal into a library of multimedia content. SMIL could allow the viewer to select several pieces of content, play them back in any order, and provide a smooth video transition (such as a dissolve or a wipe) between each piece.

A number of commonly used software packages include support for SMIL, including Microsoft's Internet Explorer (versions 5.5 and 6.0) and RealNetworks' RealOne platform.

IP MULTICASTING

IP *multicasting* is the process of simultaneously sending a single video signal to multiple users over an IP network. All viewers get the same signal at the same time, just as in traditional radio or TV broadcasting.[3] However, when IP networks are used, multicasting is the exception rather than the rule.

Most IP networks, including the public Internet, are configured to support only *unicasting*. In unicasting, each video stream is sent to exactly one recipient. If multiple recipients want the same video, the source must create a separate unicast stream made up of packets that are specifically addressed to each recipient. These streams then flow all the way from the source to each destination over the IP network.

In IP multicasting, a single video stream is sent simultaneously to multiple users. Through the use of

[3]With the introduction of digital terrestrial broadcasting, it became possible for a broadcaster to transmit multiple simultaneous video streams on a single digital channel. Confusingly, this technique has also been dubbed "multicasting," as shorthand for "multiplexed broadcasting." It is important to distinguish between these two terms, as they describe two completely different concepts.

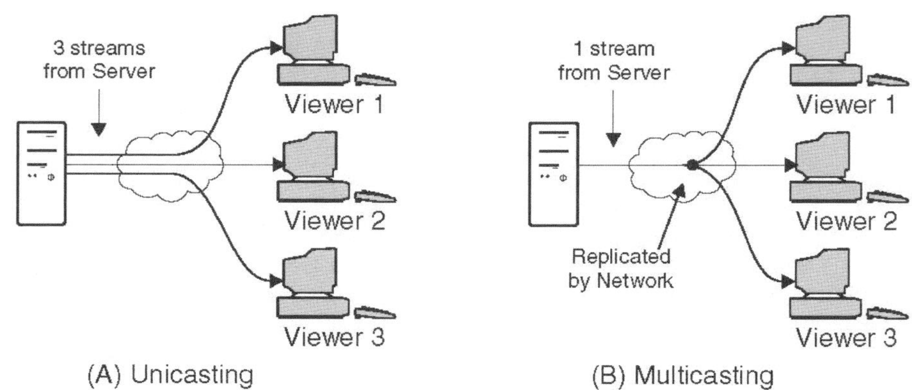

FIGURE 6.13-6 IP unicasting (a) and IP multicasting (b).

special protocols, the network is directed to make copies of the video stream for every recipient. This process of copying occurs inside the network, rather than at the video source. In true IP multicasting, the packets are sent only to the devices that specifically request to receive them, by "joining" the multicast. Copies are made at each point in the network only where they are needed. Figure 6.13-6 shows the difference in the way data flows under unicasting and multicasting.

Unicasting

Unicasting is the traditional way that packets are sent from a source to a single destination. The source formats each packet with the destination IP address, and the packets travel across the network. When the same data must be sent to multiple destinations, the source prepares separate packets for each destination. As the number of simultaneous viewers increases, the load on the source increases, since it must continuously create individual packets for each viewer. This can require a significant amount of processing power, and also requires a network connection that is big enough to carry all the outbound packets. For example, if a video source were equipped to send 20 different users a video stream of 2.5 Megabits per second (Mbps), it would require a network connection of at least 50 Mbps.

An important benefit of unicasting is that each viewer receives a custom-tailored video stream. This allows the video source to offer specialized features such as pause, rewind, and fast-forward video. This is normally practical only with prerecorded content but can be a popular feature with users.

Multicasting

In multicasting, the burden of creating copies of streams for each user shifts from the video source to the network. Inside the network, specialized protocols allow the network to recognize packets that are being multicast and send them to multiple destinations. This is accomplished by giving the multicast packets special addresses that are reserved for multicasting. There is also a special protocol for users that allows them to inform the network that they wish to join the multicast.

Many existing network devices (such as IP routers) have the capability of performing multicasting, but this capability is not enabled in many networks because multicasting can place a tremendous burden on the processing resources of the equipment. Consider an IP router that has users on 12 different ports that want to watch the same multicast stream. (These ports could be connected directly to end users or could be connected to another router farther down the network.) Multicasting requires the router to make 12 copies of each multicast packet and forward one copy out on each port. In addition, the router must be able to listen for and to process requests for adding new ports to the multicast, and to drop ports that are no longer requesting the multicast. These functions are typically handled in the router's control software, so the processor burden can be significant.

Multicasts operate in one direction only, just like an over-the-air broadcast. This means that any interactivity between the endpoints and the video source must be handled by some other mechanism.

Session Announcement Protocol (SAP) is used to periodically inform multicast-enabled receivers about programs (such as video or audio streams) that are currently being multicast on a network. SAP is used to send out a description of the multicast streams that are available on a network, along with the multicast address of the stream. The user device needs this address in order to send a request into the network to join that multicast. A single multicast source may generate multiple streams; each one of these can be announced via SAP. For example, a video encoder could provide two versions of a video stream (one for high-quality and one for low-bandwidth users) and three different audio streams (at varying bandwidths and quality levels). An end user equipped with a low-speed dial-up connection might choose to join only the low-bandwidth audio multicast, whereas a cable modem user might want to join a high-quality video multicast and a surround-sound audio multicast.

The Internet Group Management Protocol (IGMP) is designed to allow devices to become members of multicast groups in order to receive the multicast content. In addition, IGMP messages can be used to periodically poll each of the devices in the group to gather statistics about the multicast performance.

"Joining" and "leaving" a multicast are key concepts for multicasting. When a user device joins a multicast, the network must be reconfigured to deliver a copy of the required packets to the user's port. Similarly, when a user device leaves a multicast, the network must stop delivering those packets to the user, so as to make the network bandwidth available for other uses.

Joining a multicast can be a fairly complex process. If the user device is the first one to request a particular multicast stream, then all the network devices that lie along the path between the multicast source and the user device must be configured to transport the stream. This process operates on a hop-by-hop basis, beginning with the requesting user's device. Requests are passed upstream to each router in succession, forming a continuous chain from the requesting device to the multicast source. Each router along the path must process the IGMP messages and prepare to forward the multicast stream packets when they arrive. Routers must also listen for IGMP "leave" messages and stop forwarding the multicast packets after they are received.

Simulated Multicasting: Content Delivery Networks

Simulated multicasting occurs when the result of multicasting is achieved without using actual multicasting technology. In other words, a stream is sent to multiple recipients simultaneously, but the protocols discussed previously (SAP and IGMP) are not used. Instead, specialized servers are used to receive a broadcast stream and then retransmit it to a number of users on a unicast basis.

The term "content delivery network" (CDN) is generally taken to mean a system of servers distributed around the edges of the Internet that assists companies in delivering content to end users. One major application of a CDN is to provide simulated multicasts of video and audio streams to users.

Currently (spring 2007), the Internet is not multicast enabled. If a single video stream is to be sent to a number of users simultaneously, a unicast stream must be set up to each user from a server. Each user will receive a unique stream of packets, targeted to a specific IP address. This type of broadcast is difficult to scale up, because the load on the server increases as each additional viewer is added. In large broadcast applications, with hundreds or thousands of viewers, the burden on a central server can be very heavy. A CDN can be used to provide what is called a "reflected multicast" wherein the content is "reflected" by the remote servers. As shown in Figure 6.13-7, a central server delivers a unicast copy of the video stream to each remote server. The remote servers take that incoming stream and make a copy of it for each viewer. Software is needed to establish the connections between the viewers and the remote servers, but once this happens, the streams can be sent continuously.

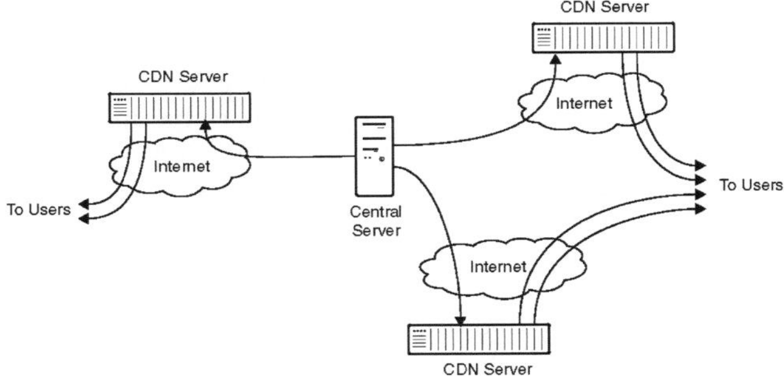

FIGURE 6.13-7 Content delivery network servers being used for reflected multicast.

IPTV: IP VIDEO TO THE VIEWER

The term "IPTV" is often used to denote a multichannel television programming service that is delivered over an IP network, primarily to home viewers, but sometimes to business viewers. For home viewers, purpose-built networks are often used to deliver services in competition with cable television, satellite television, and over-the-air broadcasts. For business viewers, existing data communications networks are typically shared between IPTV and other business applications, such as e-mail, client-server networking, and many others.

IPTV to the Home

Three delivery mechanisms are widely used to deliver IPTV to the home: DSL, Passive Optical Networks (PONs), and traditional cable TV lines.

Current generation DSL technologies, such as Asymmetric Digital Subscriber Line (ADSL), provide limited amounts of bandwidth from the service provider to the consumer and lower bandwidth links from the consumer back to the provider (hence the "Asymmetrical" element of the acronym ADSL). Newer VDSL technology (Very high-speed Digital Subscriber Line), where available, supports significantly more bandwidth on each subscriber line. More video channels can be transmitted to each VDSL subscriber, and HD video is also possible. A drawback to VDSL is that the distance of operation is less than that of ADSL, so subscribers must be closer to the service provider facilities than for ADSL service.

Each subscriber requires a set-top box to decode the incoming video. These boxes can also act as the residential gateway in the home and provide connections for other voice and data communications equipment.

Because of DSL speed limitations, each time a viewer changes channels on the DSL IPTV system, a command must be sent back to the service provider to indicate that a new video stream is to be delivered. Also, when a user orders a "Video on Demand" service, a stream must be delivered the entire distance from the server providing the content to the subscriber.

Passive Optical Networks, or PONs, are all-optical networks with no active components between the service provider and the customer. IPTV systems can be implemented over the high-speed data circuits that are commonly found on PONs.

Cable TV providers have begun to offer triple-play video, voice, and data services over a standard CATV plant. Often, to make this operate, providers will transmit different IP streams, each containing one of the services over the cable network. Modulation schemes such as QAM (Quadrature Amplitude Modulation), OFDM (Orthogonal Frequency Division Multiplexing), and VSB (Vestigial Side Band) are used to transmit the digital data, at capacities on the order of 35–40 Mbps in place of a 6 MHz analog video channel. These modulation schemes are quite useful if an average digital video stream operates at 2.5 Mbps (including overhead); the system can carry 15 digital video channels (for a total of 37.5 Mbps) in place of the single analog video channel. Since CATV systems typically provide in excess of 100 analog video channels, use of digital technology could permit 1,500 digital video channels to be transmitted.

Business IPTV

In order to deliver IPTV services to widely dispersed viewers for narrowcasting and corporate video applications, it is not normally economical to construct special-purpose networks, such as DSL or CATV. Instead, it is common practice to share existing broadband IP networks to reach these viewers. These networks can take many forms, such as corporate networks or the public Internet or as part of a standard ISP's offerings. To ensure success, several factors must be considered, including bandwidth requirements from source to destination, multicasting capability of the network, and supported interactivity levels.

Adequate bandwidth is the key requirement for any type of video delivery system. This bandwidth must be available over the entire path from the video source to each destination. The amount of bandwidth required at each point in the network depends on the number and size of the video streams that pass through that point. This calculation can be difficult, particularly for corporate networks where large clusters of potential viewers are interconnected by limited bandwidth backbone links. Public Internet connections can also be used but have unpredictable performance. For an IPTV system to work properly, each segment of the network must have enough capacity to handle the total bandwidth of all the simultaneous video streams passing over that segment.

IP multicasting is one way to reduce the amount of bandwidth needed to deliver IPTV services on a private network. If the network is multicast enabled, users who are viewing the identical content can share one video stream from the source to the point where the paths to the viewers diverge. With IP multicasting, all of the users who share one stream will need to view the same content at the same time, thereby preventing the use of VCR-like interactive playback controls.

Use of VCR-like interactivity in IPTV systems implemented over shared networks must be carefully planned. Interactivity places two burdens on a network. First, a path must be provided for the interactive control signals to travel from the viewer back to the IPTV source. Second, the amount of two-way data transit delay between the video sources and the viewers must be controlled so that user commands (say to play, pause, or rewind) can be processed promptly. This can be an issue for Internet-based systems.

Overall, it is quite feasible to implement IPTV services on a variety of different shared networks. A number of active implementations are in service today. The key to a successful IPTV deployment is ensuring that adequate capacity is available to service the peak number of simultaneous viewers, both in terms of video server capability and network capacity.

SUMMARY

Many different types of data can flow over modern IP networks, including digital video and audio in a variety of forms. Special protocols have been developed to handle the timing and synchronization issues inherent in media transport; if properly used, the results can be quite pleasing to a viewer. IP networks do not have to be specially designed to handle video and audio signals, as long as they have adequate capacity and reasonable end-to-end delay performance. There are numerous examples of companies and organizations that are profitably delivering video and audio content over IP networks today.

Bibliography

Text

Austerberry, David. *The Technology of Video & Audio Streaming, Second Edition.* Focal Press, 2004. http://www.focalpress.com

Hall, Eric A. *Internet Core Protocols—The Definitive Guide.* O'Reilly, 2000. http://www.oreilly.com/catalog/coreprot/index.html

Simpson, Wes. *Video Over IP, A Practical Guide to Technology and Applications.* Focal Press, 2006. http://www.focalpress.com

Internet Resources

Basic tutorial on desktop video streaming:

http://www.iec.org/online/tutorials/desk_stream/index.html

Cisco's IP Multicast Planning and Deployment Guide:

International Engineering Consortium, Desktop Streaming Media Production. http://www.cisco.com/warp/public/cc/techno/tity/ipmu/tech/ipcas_dg.htm

For information on streaming media and players:

http://www.apple.com/quicktime/win.html
http://www.microsoft.com/windows/windowsmedia/default.mspx
http://www.realnetworks.com

Good article on TCP/IP and IP in general:

Moss, Julian. Understanding TCP/IP, *PC Advisor*, Issue 87, September 1997. http://www.pcsupportadvisor.com/c04100.htm

CHAPTER

6.14

Cable Television Systems

PAUL HEARTY
Ryerson University
Toronto, Ontario, Canada

INTRODUCTION

This chapter is provided as guidance for engineering professionals in terrestrial broadcasting concerning technology and practice in cable television.[1] The intent of the chapter is to provide an overview of cable systems sufficient to convey an overall understanding but, in particular, to focus on those aspects of cable most likely to be of operational and practical interest to the broadcaster. The information given in this chapter reflects cable in North America and should not be assumed to accurately portray cable elsewhere.[2]

Given the depth and breadth of the topic and the limited space available, it is not possible to present this material in any detail. Accordingly, this chapter identifies key standards and other sources that will allow those interested to pursue topics in greater depth. For a more detailed picture of digital video/audio in cable, readers might examine Ciciora *et al.* [1], Thomas and Edgington [2], and Ovadia [3].

[1]This chapter is based in part on "Carriage of Digital Video and Other Services by Cable in North America," by Paul J. Hearty, which appeared in *Proceedings of the IEEE*, vol. 94, no. 1, 2006, 148–157, ©2006 IEEE. Those portions of the paper reflected here are provided with the permission of the IEEE. Other acknowledgments are provided at the end of the chapter.

[2]In this respect, it is important to note that, although there are many commonalities among cable systems worldwide, there are differences in channel plans and in many aspects of technology and operation. In particular, there are significant differences among systems in digital services; there, North American systems conform to Society of Cable and Telecommunications Engineers (SCTE) standards, while systems in Europe, for example, conform to Digital Video Broadcasting (DVB) standards.

CABLE IN THE UNITED STATES

At its beginnings in the late 1940s, cable was conceived, first as an extension to terrestrial broadcast which "filled in" coverage in difficult-to-reach areas, then as a community antenna television (CATV) system, which provided an alternative to the proliferation of terrestrial receive antennas. Since those early days, however, cable has developed to be a full-scale communications medium that offers a broad array of one-way and two-way services over vast geographical areas.

The next stage in the development of cable began in the late 1950s, when operators began to use CATV systems to deliver "distant" television signals (signals originated outside the local broadcast coverage area). This development, although seminal to the concept of cable as we understand it today, provoked competitive concerns among local broadcasters and led to regulatory actions that slowed development of the cable service concept.

The 1960s could be characterized as a period of subscriber growth, with little change in the cable service concept. For example, the National Cable and Telecommunications Association (NCTA) estimates that cable penetration grew from about 70 systems with about 14,000 subscribers in 1952 to about 800 systems with about 850,000 subscribers.[3]

The 1970s saw several developments that enabled much of the concept of cable as we see it today. First, relaxation of regulations allowed operators to enter-

[3]*Source:* National Cable and Telecommunications Association (NCTA), http://www.ncta.com.

tain new service concepts. Second, the launch of national television service delivery over satellite by Home Box Office in 1972 established a new model for the delivery of cable programming. And, third, the emergence of new program providers, such as the "Superstations," not only added program offerings, but also cemented the concept of the cable system as a platform for diverse program delivery.

The 1980s were a period of considerable growth in the cable industry. Subscriptions grew from about 16M at the end of the 1970s to about 53M at the end of the 1980s. Subscriber growth and revenue, as well as the availability of new program sources, fueled upgrades to systems to provide greater capacity, which, in turn, allowed and stimulated the emergence of new program sources. By the end of the 1980s, there were 79 cable television networks, up from 28 a decade previously.

The pattern of growth in subscribers and program offerings continued through the 1990s. Upgrades continued, not only to provide for greater capacity, but also to enable radically new service delivery options: digital television (including high-definition television, HDTV), broadband (data services using a cable modem), and voice (telephony). By the middle of the decade, many cable systems were transitioning to a new architecture, hybrid fiber-coax, which allowed high bandwidth capacity at the core of the network using fiber and cost-effective improvements to network elements closer to customer premises through upgrades to the existing coaxial infrastructure.

The first digital offerings in the United States, comprising both digital cable and broadband (cable modem) services, were launched in 1996. From initial launch in 1996 through September 2004, digital cable achieved a penetration of 32.1% of cable subscribers, while broadband achieved a penetration of 26.9% of cable subscribers.[4]

By the end of the 1990s, the number of cable subscribers had grown to more than 65M, and number of national cable programmers stood at more than 170. By early 2001, subscribers to digital television and broadband services had grown to 12.2M and 5.5M, respectively.[5]

The 2000s saw further growth in digital and broadband deployments, as well as the introduction of new services and concepts. In the early 2000s, operators began pilot testing of new services, such as interactive TV. In 2003, operators accelerated deployments of HDTV, launched Video on Demand (content access with VCR-like capabilities), and introduced cable telephony service (Voice over Internet Protocol, VoIP).

As of early-to-mid 2006, there were 65.5M cable subscribers in the United States, representing a penetration of 59.1% of television households. Penetration of digital cable had reached 26.9M. That for broadband had exceeded 27.6M, while that for cable telephony had exceeded 6.6M. Of nearly 119M homes passed by cable, 96M had access to HDTV service.[6]

Currently, operators are exploring even more advanced service concepts. For example, work currently is underway to add wireless extensions to cable systems, allowing cable service to extend beyond the wired cable infrastructure to mobile, personal devices.

ECONOMICS AND CONTEXT

The U.S. cable industry is expected to give a strong financial showing in 2006, with subscriber revenues of $69.5B and advertising revenues of $24.6B.[7] Despite this strong showing, however, the cable industry faces a number of challenges.

First, there is the ever-increasing demand for program bandwidth. Cable operators must sustain the viability of their legacy analog services, while at the same time responding to demand for new digital and high-definition television services. For reference, support for 80 legacy analog services in a 550 MHz plant would leave only 13% of capacity available for digital television, HDTV, and other services, while support for the same would leave only 45% of capacity in an 870 MHz plant.[8]

Second, there is competition from other multichannel video program distributors (MVPDs—primarily digital satellite and xDSL sources—that have a subscriber base of about 28M, or approximately 30% of the multichannel market[9]). In this respect, it is important to note that other MVPDs do not have legacy analog service to maintain and can concentrate all of their bandwidth resources on digital programming and other services.

Third, the competitive landscape has evolved beyond the unidirectional delivery of television and audio programming. High-definition television, which requires approximately five times the bandwidth of standard-definition television, is a key competitive offering. Similarly, broadband Internet continues to pose both opportunity and challenge; subscribers and revenues continue to be significant, but competitive pressures have driven user bandwidth requirements from their early-day limits of 1.5–3.0 Mbps to values as high as 8 Mbps. Additional services, such as Video on Demand (VOD) and Voice over Internet Protocol (VoIP), round out the portfolio but add to the bandwidth challenge.

Recognizing the need to extract the maximum performance from their plant, operators expect to expend $11.1B in plant upgrades in 2006, approximately 16% of cable subscriber revenues.[10]

[4] *Source:* NCTA, http://www.ncta.com.

[5] *Source:* NCTA, http://www.ncta.com.

[6] *Source:* NCTA, http://www.ncta.com.

[7] *Source:* NCTA, http://www.ncta.com.

[8] When technical constraints on the use of frequencies below about 54 MHz are considered, these percentages become more like 3% and 39%.

[9] Some viewers subscribe to both cable and another multiprogram service, so this percentage is approximate.

[10] *Source:* NCTA, http://www.ncta.com.

CABLE SYSTEM ARCHITECTURE

Cable systems are closed communications networks designed to meet particular service objectives over specific geographical topologies. Systems vary in size, both in geographical area and in subscriber population, in bandwidth capacity, and in degree of technical sophistication.

Accordingly, it is difficult to describe a typical plant with any degree of detail, as it is in the details that they differ. However, at a generic level, the system is composed of four main structural elements:

- The head-end, which accommodates signal ingest and origination, as well as network control, provisioning, and monitoring.
- The trunk, which conveys signals from the head-end to area distribution points. Trunks typically transport signals up to 10 miles or so.
- The distribution plant, which delivers signals from the trunk to the neighborhood. Distribution plants typically transport signals about 1–2 miles.
- The drop, which delivers signals from the distribution plant to customer premises. Drops typically transport signals 100–300 feet.

Originally, all three stages in the signal delivery chain were mediated by coaxial cable—rigid cable in trunk and distribution and flexible cable in the drop. However, recent years have seen increasing use of fiber in trunk and distribution. Modern hybrid fiber/coaxial systems deliver not only improvements in capacity, but also increased quality due to a reduction in the number of amplifiers needed for service delivery and due to its inherent indifference to ingress from Radio Frequency transmissions.

PLANT AND SERVICE TOPOLOGY

A large cable network may consist of a Master Head-end, a number of Regional Head-ends, a number of Local Head-ends (often called *hubs*), and a large number of District/Neighborhood *nodes*. The interchange of content and other data between Master and Regional Head-ends may be by satellite or fiber (telecommunications link). That between Regional Head-ends and Local Head-ends typically is by fiber. That between Local Head-ends and District/Neighborhood nodes almost always is by fiber. The final link, that between the District/Neighborhood node and customer premises, typically is by coaxial cable.

Bandwidth

Plant bandwidths typically are in the range of 550–870 MHz, although systems with higher and lower bandwidths exist.

Frequency and Channel Plans

Frequency utilization typically is 8-40 MHz for signals originating in subscriber terminals and 50–870 MHz, or higher, for signals directed toward subscribers.

For historical reasons, the television channel plan is based on the 6 MHz channel bandwidths established by the FCC for terrestrial VHF/VSB broadcast. Historically, three channel assignment plans have been used:

- The standard plan, in which carriers are spaced at 6 MHz increments, beginning at 55.25 MHz.
- The incrementally related carriers (IRC) plan. Here, carriers are phase locked and are located at multiples of 6 MHz beginning at 55.25 MHz.
- The harmonically related carriers (HRC) plan. Here, carriers are phase locked and are located at multiples of 6 MHz beginning at 54 MHz.

In the analog domain, noise is the limiting factor for television service quality in systems carrying small numbers of channels (say, fewer than 30). However, as the number of channels increases, the limiting factor becomes distortions introduced by cascaded amplifiers in the distribution plant. The IRC and, in particular, the HRC assignment plans were developed to minimize the visible impact of these distortions.

Due to the existence of legacy analog television services, digital television, radio/audio, and broadband (cable modem) services use the existing 6 MHz assignment plan. Analog FM radio/audio is carried between 88 MHz (90 MHz for IRC) and 108 MHz with 200 kHz carrier spacing.

Service Topology

Currently, available bandwidth typically is partitioned by frequency among five types of services:

- Analog television and audio services (legacy);
- Digital television and audio services;
- Video on Demand (VOD) services;
- Cable modem (broadband) services; and
- Voice over IP (VoIP) services.

In the near future, it is to be expected that the digital bandwidth will be shared between conventional (broadcast) services and switched (scheduled unicast or multicast) services. In the less immediate future, additional services, such as those supporting mobile devices, also will occupy spectrum. At some point, when digital-capable devices become ubiquitous, legacy analog services may be terminated.

Standards

The primary body for standardization in North American cable television is the Society of Cable Telecommunications Engineers (SCTE), an American National Standards Institute (ANSI) accredited Standards Development Organization.[11] All standards developed by SCTE are available at http://www.scte.org. In addition to the ANSI-SCTE standards, many elements of cable technology are recognized in International Telecommunications Union standards.

ANALOG TELEVISION AND AUDIO SERVICES

The first analog cable services were launched in the late 1940s, and analog services support more than half of cable subscribers today, nearly 60 years later. The technology, of course, has developed considerably during that period, providing better quality, greater capacity, and more robust and reliable service.

Operative subcommittees for standards in analog cable plants in North America include the SCTE's Interface Practices and In-home Cabling and Hybrid Management Sub-layer Subcommittees.

Analog Network Architecture

An analog cable network, although in some respects simpler than its digital companion described later, nevertheless is a complex assembly of many components. An exhaustive description of these components and their interrelation would be beyond the scope of this chapter. Moreover, because analog cable plants vary widely, such a detailed description could give rise to potentially misleading impressions of specific plant architectures.

At a conceptual level, however, the key elements of the analog video/audio delivery architecture are as given in Figure 6.14-1. As the figure shows, a variety of analog source signals are made available to the cable plant. Some of these are received off-air from broadcasters in analog NTSC and are simply passed on unencrypted, at delivery frequency, through *Heterodyne Processors*. Signals from other sources are demodulated and decoded (if needed) by *Receiver/Decoders*, and modulated in AM-VSB in accordance with the NTSC standard by *NTSC Modulators*, and placed on-frequency for delivery.

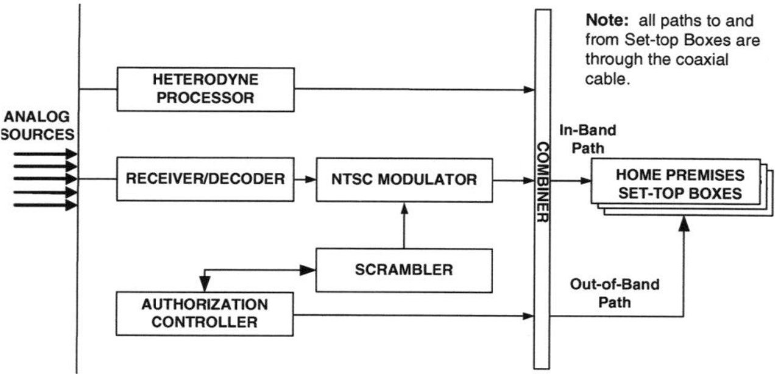

FIGURE 6.14-1 Conceptual representation of key architectural elements in an analog cable network.

[11]Standards from other bodies, such as the Consumer Electronics Association, also inform certain elements of cable practice.

After they have been modulated and placed on-frequency for delivery, the signals comprising the analog cable service are directed to a *Combiner* or set of Combiners, which interface to the delivery plant.

Most signals originated from non-broadcast sources are encrypted for secure delivery. In plants that provide addressable Access Control, an *Authorization Controller* is used to deliver authorizations either in-band or through an out-of-band channel. In plants that do not use addressable control, scrambling is applied to provide authorized access for recognized customer premises equipment.

Services and Components

Television

Television services consist of standard-definition video (525 lines, with a 4.2 MHz bandwidth), with associated audio on an FM subcarrier 4.5 MHz above the video carrier that contains monaural (L+R) audio with a bandwidth of 15 kHz, a stereo pilot at 15.734 kHz, and a subcarrier at 31.468 kHz for L–R difference audio for stereo services. The signal may also contain a 10 kHz audio bandwidth Secondary Audio Program (SAP) channel on a subcarrier at 78.7 kHz and closed captions and/or extended data services on line 21 of the video vertical interval per CEA-608-C [4].

Additional data, such as SID/AMOL (for which standards are under development in SCTE and CEA) and Content Advisory information, may be conveyed in the Vertical Blanking Interval (lines 21 and 22). Additional data services, such as *teletext*, which may or may not be related to the program, also may be present.

Audio/Radio

Cable systems may carry analog radio services in the FM band (88–108 MHz). Main audio channel bandwidth is 15 kHz for monaural services (L+R). A stereo pilot is placed at 19 kHz with a subcarrier at 38 kHz for L–R difference audio for stereo services.

Program Guide

To assist viewers in accessing programming, operators typically devote an analog channel to delivery of a noninteractive program guide. The guide presents, in grid format, channel offerings as a function of time. The time window is fixed at any given time, and the video scrolls down through the channels at a fixed pace.

Service Delivery Paradigms

Linear Programming

In analog cable, all program delivery is linear, in that it is delivered according to a fixed program schedule. Access to such programming is through basic and

tiered subscription, as well as Pay Per View (PPV); access is secured by a variety of security mechanisms (see following sections).

Near Video on Demand

Pay Per View (PPV) content can be staggered in time to allow greater choice in program access times. However, such offerings still are linear, in that they are delivered according to a fixed delivery schedule. In analog cable, staggered delivery is very costly in terms of bandwidth utilization.

Technical and Service Issues

In the analog domain, the challenge is to provide the maximum number of television (and audio) services while maintaining quality and reliability of service. Unlike digital services, which maintain their initial quality until error protection is overtaxed, analog services exhibit impairments reflecting imperfections in the plant and interference from other signals.

The level of the signal delivered to the subscriber's television (after any intervening decoding/descrambling devices) has to fall within a constrained range. Measured in dB relative to 1 mV (0 dBmV) across 75 ohms, signals should be in the range of 0–3 dBmV. Delivered signal levels below this range present snowy pictures, while those above this range (for example, at 10 dBmV) may begin to overload television receiver tuners, introducing cross-modulation, which presents as horizontal and vertical synchronizing bars moving through the picture.[12]

The operator has to deal with two classes of impairments in television pictures: noncoherent impairments, such as noise, which present as "snow" when visible; and coherent impairments, which present as pattern when visible. The primary source of noncoherent impairment is noise; sources of coherent impairments include cross-modulation of video signals, cross-modulation of carriers within signal, signal reflections resulting from discontinuities in impedance along the transmission line (also called microreflections, these are similar in aspect to multipath in terrestrial broadcast), and ingress of RF signals originating outside the cable system. Of the cross-modulation impairments, which arise primarily from nonlinear amplifiers, the primary concerns are composite second-order (CSO), distortion from second-order carrier inter-modulation products (which presents as a herringbone pattern), and composite triple beat (CTB), distortion from third-order carrier inter-modulation products (which presents as noise at low levels and as streaks at high levels).

Analog cable services are implicitly noise limited. Noise will become apparent in pictures as the carrier-to-noise ratio (CNR) approaches 43–44 dB and objectionable as CNR approaches 40–41 dB; a good design target is 48–50 dB.[13]

[12]*Source:* Ciciora, 9th ed. *NAB Engineering Handbook*, Chapter 6.13.
[13]*Source:* Ciciora, 9th ed.

As the number of channels in a cable system is increased, inter-modulation products become increasingly determinative of service quality, as the occupied frequency range increases to the point at which higher order products fall within occupied spectrum. For systems with 12 channels or fewer (appropriately placed), inter-modulation is not an issue. As the number of channels is increased to 30 or so, cross-modulation may become the limiting factor. For systems with 60 or more channels, CSO and CTB become limiting factors; a good design target for each of CSO and CTB is –53 dB relative to carrier.[14]

Ingest

In the early days, when cable content consisted entirely of local broadcast content, ingest primarily was accomplished by direct, over-the-air reception of the broadcast signal. Later, as more distant signals were added, radio and microwave relays were employed. With the advent of analog satellite distribution in the 1970s, cable operators began to acquire content from satellite, initially from C-Band (4 GHz) and, later, from a mix of C-Band and Ku-Band (12 GHz) sources. In the 1990s, cable operators increasingly began to acquire program content, including local broadcast content, by telecommunication links, such as DS3 and fiber. By the mid-1990s, the analog satellite feeds began to be replaced by feeds using digital compression and transmission.

Terrestrial Broadcast Sources

Analog Broadcast Sources

Originally, cable operators captured the terrestrial broadcast signal off-air, subjected the incoming signal to heterodyne processing to render it to intermediate frequency, and then up-converted the signal to the frequency used for transmission through the cable plant. However, capture of analog broadcast signals off-air is a less than desirable approach in that the signal transmitted through the cable plant will include, not only any impairments that might originate in the plant, but also any impairments introduced in the terrestrial broadcast path.

At present, most local broadcaster signals are received by fiber or satellite. In the case of fiber, analog-to-fiber equipment is installed at the broadcast facility for the analog feed and receive equipment is installed in the cable head-end. In the case of satellite, the local broadcast signals are up-linked by the broadcaster or its service provider; the down-linked signal is captured by satellite receive equipment in the head-end.

In principle, the entire signal is passed through the plant, although certain non-program-related data services may be removed, depending on particular operator-broadcaster business arrangements.

[14]*Source:* Ciciora, 9th ed. CSO and CTB are measured by a standardized procedure with 35 channels.

Digital Broadcast Sources

Currently, broadcasters provide both analog and digital signals so, with rare exceptions, there is no need to ingest the digital signal for distribution in analog cable.

Non-Broadcast Sources

Broadcaster-originated programming is, of course, only part of the content delivered by the cable operator. Most of the content delivered by the cable operator derives from non-broadcast sources. For the most part, this content is delivered by satellite transmission, although some content is delivered by fiber-optic link.

For signals delivered by fiber-optic link, the incoming signal must be demodulated, converted to electronic form, and rendered to baseband analog. For signals delivered by digital satellite, two modulation schemes are currently in use: Quaternary Phase-Shift Keying (QPSK), the more common; and 8-state Phase-Shift Keying (8-PSK). In both cases, the signal must be demodulated, decrypted, and decoded to analog baseband. Originally, the device used for such processing was an Integrated Receiver Decoder (IRD); now, other head-end devices can provide the same functionality, but with a denser configuration.

Regulations and Requirements for Carriage

Operators are subject to a number of technical and other regulatory requirements, as well as a number of voluntary standards established by industry consensus. Some of these requirements are specific to broadcaster-originated signals, whereas others relate to compatibility with consumer electronics equipment.

The cable operator is subject to the FCC's Code of Federal Regulations (CFR) 47, Section 76 (basic technical) [5]. An excellent summary of regulatory technical requirements established by the FCC is given in Ciciora *et al.* [1, page 621]. The operator also operates in cognizance of CFR 47, Section 15.118 (Cable Ready) [6] and is bound by parts of the Telecommunications Act of 1996, notably Section 304 (Competitive Availability of Navigation Devices) [7].

For all signals, the cable operator is required to pass through closed caption data per CEA-608-C [4], when present. With regard to broadcast-originated signals, the operator is additionally required to pass through Second Audio Program (SAP), Content Advisory information, and other program-related data, if present.

Delivery

If they are not already so modulated, the analog sources for services are modulated in AM-VSB in accordance with NTSC standard and up-converted to the channel frequencies assigned by the cable operator.

Security

In analog cable, there are two general approaches to secured program delivery: trapping and scrambling.

In trapping, a filter is applied to the drop to the premises. Trapping can be done in one of two ways.

- In negative trapping, the filter removes signals to which the household is not entitled. This works well for homes that are not subscribed to cable service. However, homes that are subscribed for cable services may experience reduced quality for services delivered on frequencies adjacent to those of trapped services due to limited selectivity of the filters, particularly at higher frequencies.
- In positive trapping, an interfering (or jamming) carrier is introduced to the protected signal in the cable plant, and the trapping filter is used to remove the interfering carrier at the drop to an authorized viewing household. This method, particularly with modern filter technology, offers a fair level of security, but the filter may reduce the resolution of the protected video.

Both negative and positive trapping can be defeated, the former by physical removal of the trap and the latter by introduction of a "pirate" filter.

In scrambling, the protected signal is perturbed such that it is not usable without the application of a restorative (descrambling) process. There are two scrambling methods in use:

- In RF Synchronization Suppression, various methods are used to prevent proper synchronization to the protected signal. In the most common methods, horizontal or vertical synchronizing pulses in the protected content may be attenuated (gated or pulsed suppression), or the video carrier may be modulated with a sine wave to prevent synchronization (sine-wave suppression). Authorized descramblers restore the signal to synchronizable form by restoring the synchronizing pulses to proper level or by modulating the signal with the inverse to the sine wave used in scrambling.
- In Baseband Scrambling, protected content is subjected to pseudorandom synchronization suppression, video inversion, or both. In some cases, encrypted digital audio is introduced into the horizontal and vertical synchronizing intervals of the protected video. Authorized descramblers perform the inverse process to the initial scrambling operation.

Both scrambling approaches offer a higher level of security than trapping. However, inasmuch as they physically perturb the protected signal, special effort must be made to minimize impacts on the quality of the protected video.

Trapping typically is done outside the customer premises. Descrambling typically is done inside customer premises with a set-top converter but can also be done with specialized equipment outside customer premises. Security devices may be addressable (controlled by a data signal from the cable head-end delivered in the VBI or by separate carrier) or nonaddressable (programmed prior to arrival at customer premises).

DIGITAL TELEVISION AND AUDIO SERVICES

The North American cable industry began the transition to digital cable delivery in the early 1990s, with the development, and subsequent approval by the International Telecommunications Union (ITU) in 1995, of its specification for 64-QAM and 256-QAM modulation formats (ITU Recommendation J.83, Annex B) [8]. The initial draft specification provided for 64-QAM, with a payload of 26.97 Mbps; a later development added 256-QAM, with a payload of 38.8 Mbps.

The operative subcommittee for standards in digital cable television in North America is the SCTE's Digital Video Subcommittee.

Digital Network Architecture

A digital cable network is a complex assembly of many interrelated elements, and an exhaustive description of these elements would be beyond the scope of this chapter. At a conceptual level, however, the key elements of the digital video/audio delivery architecture are as given in Figure 6.14-2. It should be noted, however, that the actual embodiments of these elements vary. For example, in some implementations, encryption, modulation, and up-conversion may be carried out in the same device.

Referring to the figure, the *Authorization Controller* deals with all aspects of subscriber authorization, whether for individual programs or for tiers (packages) of programs. Alone, or in conjunction with a separate key server, it is responsible for the generation and secure handling of authorization keys, which are propagated through the network to the customer's set-top box. The Authorization Controller also interfaces with the operator's business system to confirm subscription entitlements and to report program and event purchases.

The *MPEG-2 Multiplexers* accept compressed MPEG-2 single-program transport streams and assemble them into MPEG-2 multiple-program transport streams. They also insert the transport infrastructure needed to facilitate program access and assembly at the set-top box. Note: In practice, Multiplexer topology is varied. The Multiplexer may be interfaced to an assembly of analog-to-digital converters and MPEG-2 encoders that process a group of analog program sources intended to constitute the line-up for a given cable channel. Alternatively, the Multiplexer may transcode (and pass through) an entire incoming digital multiplex (as delivered by satellite, for example) that is intended to constitute the line-up for a channel. And, in still other cases, the Multiplexer may process a number of in-bound single-program and/or multiple-program digital sources, selectively dropping individual sources to assemble a multiple-program transport stream for a channel.[15]

[15]The most sophisticated Multiplexer of this sort can adjust the bit rates of the individual single-program transport streams.

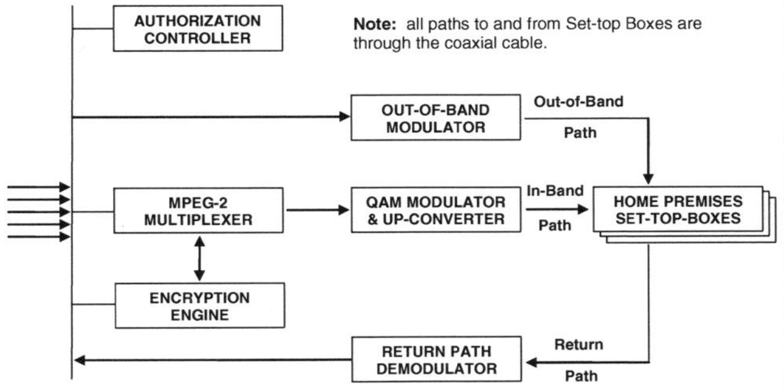

FIGURE 6.14-2 Conceptual representation of key architectural elements in a digital cable network.

The *Encryption Engines* interface with the Authorization Controller and the MPEG-2 Multiplexers to encrypt outbound content using the appropriate keys. The *Quadrature Amplitude Modulation (QAM) Modulators* apply Forward Error Correction and QAM modulation to the outbound MPEG-2 transport streams, while the *Up-converters* shift the QAM-modulated Intermediate Frequency (IF) signals to the frequencies used for transmission to the home.

The *Out-of-Band Modulators* deliver System Information, authorization data (for example, decryption keys, etc.), messages, and application data to the set-top through a separate channel that is processed by a second demodulator in the set-top. The *Return-Path Demodulators* collect messages transmitted to the network by the set-tops using a modulator/up-converter integrated in the set-top. Messages issued to the network using this path include Instant (or Impulse) Pay Per View (IPPV) and VOD requests, program purchase/credit status reports, service access requests, and status and health information.

Services and Components

Television

Digital cable television services consist of ANSI/SCTE 43 2004 [9] video, one or more ATSC A/53C Annex B (Dolby™ AC-3) [10] audio program streams, and closed captioning per CEA-608-C [4] (ANSI/SCTE 20 2004 [11]) and per CEA-708-B [12] (ATSC A/53C [10]), as appropriate, as well as other data, as identified in ANSI/SCTE 43 2004 [9], ANSI/SCTE 54 2004 [13], ANSI/SCTE 19 2006 [14], and ANSI/SCTE 27 2003 [15].

The basic video compression in digital cable conforms to ISO/IEC 13818-2 Main Profile [16]. Specifically, it is Main Profile with constraints, spanning the range from Main Level through High 1440 to High Level. The video coding also incorporates compatible extensions to support captions and other data services. The video specification is given in ANSI/SCTE 43 2004 [9].

The video formats used in digital cable also are specified in ANSI/SCTE 43 2004 [9]. For convenience, they are shown in Table 6.14-1.

TABLE 6.14-1
Compression Format Constraints

Vertical_size_value	Horizontal_size_value	aspect_ratio_information	frame_rate_code	Progressive_sequence
1080	1920	1, 3	1, 2, 4, 5	1
			4, 5	0
	1440	3	1, 2, 4, 5	1
			4, 5	0
720	1280	1, 3	1, 2, 4, 5, 7, 8	1
480	720	2, 3	1, 2, 4, 5, 7, 8	1
			4, 5	0
	704	2, 3	1, 2, 4, 5, 7, 8	1
			4, 5	0
	640	1, 2	1, 2, 4, 5, 7, 8	1
			4, 5	0
	544	2	1	1
			4	0
	528	2	1	1
			4	0
	352	2	1	1
			4	0

Legend:
aspect_ratio_information: 1 = square pixels; 2 = 4:3 aspect ratio; 3 = 16:9 aspect ratio.
frame_rate_code: 1 = 23.976 Hz; 2 = 24 Hz; 4 = 29.97 Hz; 5 = 30 Hz; 7 = 59.94 Hz; 8 = 60 Hz.
Progressive_sequence: 0 = interlaced; 1 = progressive.

ANSI/SCTE 43 2004 [9] supports a number of extensions for delivery of program-related services through video_user_data. Specifically, it supports delivery of advanced DTV captions (encoded per CEA-708-B [12], with transport per ATSC A/53C [10]), NTSC closed captions (encoded per CEA-608-C [4], with transport according to ANSI/SCTE 20 2004 [11], ATSC A/53C [10], or both), and other NTSC Vertical Blanking Interval data (encoded per ANSI/SCTE 20 2004 [11] and ANSI/SCTE 21 2001 [17]). In addition, it supports delivery of Bar Data and Active Format Descriptor data per ATSC A/53C [10].

Although ANSI/SCTE 43 2004 [9] is very similar to ATSC A/53C [10], there are a few differences:

1. In compression format constraints, digital cable allows more formats within the Main-Profile/Main-Level to Main-Profile/High-Level envelope. In addition to those formats supported by ATSC (shaded cells in the table), digital cable supports other 1080-line and 480-line formats. The additional formats provide options for different horizontal resolutions; these reflect prior practice in the digital cable environment, mostly to allow for more bandwidth-efficient transmission.
2. In sequence_header constraints, digital cable allows a bit rate of 26.97 Mbps for 64-QAM, while ATSC allows a value of 19.39 Mbps for 8-VSB. This difference reflects differences in payload between 64-QAM and 8-VSB. Values for 256-QAM and 16-VSB are identical at 38.8 Mbps.
3. In sequence_display_extension constraints, digital cable requires encoding of 480-line formats in accordance with SMPTE 170M [18] colorimetry, unless otherwise indicated; for such formats, ATSC acknowledges, but does not require, SMPTE 170M [18]. This difference reflects prior practice in digital cable.
4. In transport of NTSC captions (CEA 608-C [4]), digital cable allows carriage in accordance with ANSI/SCTE 20 2004 [11], A/53C [10], or both, while ATSC requires carriage in A/53C [10] only. This difference reflects a balance between prior practice in digital cable and the need to ensure harmonization between digital cable and digital terrestrial broadcast.
5. In transport of other "NTSC" Vertical Blanking Interval (VBI) data, digital cable employs ANSI/SCTE 20 2004 [11] and ANSI/SCTE 21 2001 [17]. This difference reflects prior practice in digital cable.

The basic audio compression in digital cable is as given in ATSC A/53C, Annex B (Dolby™ AC-3) [10].

Audio can be offered in monaural, stereo, or 5.1-channel surround. Monaural audio, which rarely is used, typically would be offered at 64 or 128 kbps. Stereo audio, which currently is most common, typically would be offered at 160 or 192 kbps. The 5.1-channel audio typically is offered at 384 kbps, although 448 kbps is not precluded.

Audio service configurations and constraints are the same as those given in ATSC A/53C, Annex B [10].

Audio/Radio

Audio services conform to ATSC A/53C, Annex B (Dolby™ AC-3) [10] and may be associated with other data, as with television services.

Electronic Program Guide

In the multichannel universe of digital cable, the Electronic Program Guide is fundamental to providing the viewer with a satisfactory experience. Currently deployed set-tops host proprietary guide applications, with "look and feel" typically customized to the cable operator's needs. However, the North American cable industry is considering the need for standardized protocols for guide applications.

Service Delivery Paradigms

Linear Programming

At launch, all programming on digital cable in North America was linear, that is, available according to a pre-established program schedule. Access to such programming was through basic and tiered delivery or through Call-Ahead or Instant (Impulse) Pay Per View (CAPPV or IPPV). The former Pay Per View option involves a telephone call to the cable operator's support center to arrange authorization for a future program offering, whereas the latter involves direct access through the set-top box, which is provisioned with an amount of purchase credit and which periodically reports purchases and seeks refreshment of the credit amount. However, both PPV options are linear programming, in that the program availability schedule is fixed by the operator.

Near Video on Demand

Soon after the launch of digital cable in North America, operators recognized viewer demand for more choice in program delivery times and introduced Near Video on Demand (NVOD). In NVOD, PPV offerings are staggered in time to allow more frequent and convenient access times from the viewer's perspective, but this is done at a cost in system bandwidth. Such offerings also are linear, in that offerings are by pre-established schedule.

Video on Demand

Initial launches of Video on Demand (VOD) began in 2000. In VOD, the digital cable portion of the network is subdivided into service groups comprising a group of nodes, each of which serves a number of homes. Delivery of VOD programming to a service group is conditional upon a viewer request, either for access or for adaptation of delivery in accordance with VCR-like control requests such as Pause, Fast Forward, Rewind, and so on. VOD represents the first instance of the delivery of nonlinear programming services.

Essentially, VOD involves a high-speed interactive application enacted through a "thin client" application hosted by the set-top box, a "thick" application hosted by the digital cable network, and a set of server and processing resources made available through the network. A schematic of a VOD infrastructure is provided in Figure 6.14-3.

The VOD architecture makes use of the existing downstream and upstream out-of-band communication resources of the digital cable plant. The viewer's request, interpreted by the client application, is directed to the cable plant through the QPSK return channel. This request is processed by the *Network VOD Controller,* which accesses programming resources from the VOD Servers and directs this content to the *Multiplexer/Encryptor/Modulator/Up-converter* assemblies that serve the viewer's node. Authorization data, as well as content access information, is directed to the viewer's set-top through the out-of-band QPSK downstream channel. The VOD application in the viewer's set-top processes this information and accesses the "new" program content supplied for the viewer.

The VOD architecture scales well with demand. As more VOD-active subscribers come on line, service groups are subdivided (potentially down to the level of a single node serving 500 or fewer homes), and additional Multiplexer/Encryptor/Modulator/Up-converter assemblies are introduced to support the newly created service groups. Quality of Service is maintained by ensuring that nodes are sized to ensure that contention for VOD resources among VOD-active

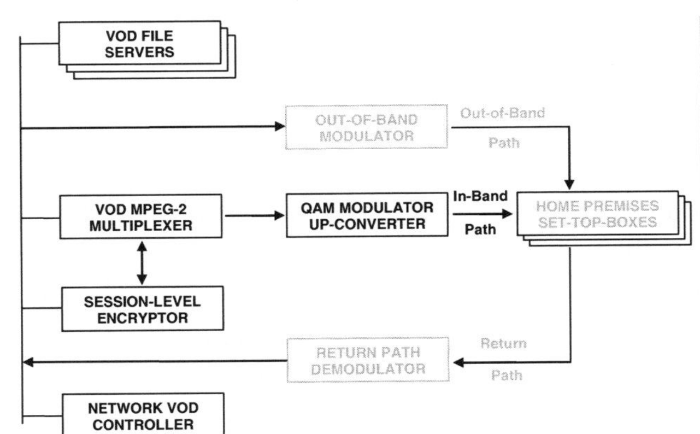

FIGURE 6.14-3 Conceptual representation of incremental architectural elements necessary to support VOD in a digital cable network (basic digital infrastructure that already is in place, but is leveraged by VOD, is shown in gray).

viewers cannot exceed a given service threshold (such as 4% contention) at peak VOD viewing times.

At this point, cable networks in North America have nearly complete VOD coverage, defined in terms of the availability of VOD programming to subscribers with digital set-top boxes. The limiting factor in VOD penetration is the number of subscribers with suitably equipped set-tops. At present, business models for VOD range from free access (either as an introductory offering or as an advertising-supported service) through subscription to Pay Per View.

Digital Simulcast

Some operators have begun service delivery in what is called digital simulcast. In digital simulcast, analog services are encoded in MPEG-2 and replicated in digital service tiers. This is a strategic move to facilitate the transition to all-digital service and eventual reclamation of spectrum dedicated to analog services. This allows the deployment of lower cost set-tops but, more importantly, allows the operator to offer higher quality pictures and to enhance competition in the Multichannel Video Program Delivery (MVPD) environment.

Switched Program Delivery

Building on its success in launching VOD, the North American digital cable industry recently has begun experimentation with switched program delivery. This future potential offering could leverage the infrastructure developed for VOD but use this infrastructure to deliver programming that previously would have been available network-wide as basic subscription or subscription-tier programming. The key element of this approach is that programming would be delivered linearly (by schedule) but would be provided to a given service group only when requested by one or more viewers in that group. It should be noted, however, that this flexibility can be achieved only at the expense of increased infrastructure with more Multiplexer/Encryptor/Modulator/Up-converter assemblies.

Interactive Programming

Electronic Program Guide and VOD, of course, are interactive services. The cable industry has developed specifications for common middleware to support other interactive applications, and set-tops that support these specifications are being deployed worldwide. The key specifications are given in Open Cable Applications Platform (OCAP) and are reflected in SCTE Cable Applications Platform (CAP) standards and in ITU standards J.200 [19], J.201 [20], and J.202 [21].

Technical and Service Issues

As noted previously, analog services reveal impairments arising from imperfections in plant and from ingress by external RF signals. Digitally transmitted services, however, retain their initial picture and sound quality up to the point at which their error protection and concealment are overtaxed.

That said, however, noise, cross-modulation, intermodulation, ingress, and the like do challenge the digital signal. If the strength of such sources is too great, visible and audible impairments will become apparent and, in cases at the limit, the signal may be lost entirely. Moreover, because the material, and in particular the video, is highly compressed, the visible or audible impact of uncorrected transmission errors can be very substantial.

Service targets for digital cable are given in ANSI/SCTE 40 2004 [22] and are incorporated in the FCC's Code of Federal Regulations CFR 47, section 76.640 [5].

Ingest

Terrestrial Broadcast Sources

Currently, most broadcasters generate two signals: a conventional analog NTSC (standard-definition) signal and a digital 8-VSB signal, which may host a digital high-definition television program or a multiplex of multiple digital standard-definition programs.

Analog Broadcast Sources

Ingest of the analog NTSC signal is as described in the section on analog cable services. However, when this is the only source available for the digital cable service, the operator is obliged to convert the analog baseband signal to digital baseband and execute MPEG-2 and AC-3 compression encoding. The ensuing MPEG-2 single-program multiplex then is inserted into an MPEG-2 Multiplexer.

Digital Broadcast Sources

Ingest of the digital signal typically is by fiber, although over-the-air and satellite receptions are not uncommon.

Interestingly, because off-air transmission is error protected (from digital channel coding), it is now possible to take the signal off-air at the same quality it had at source, provided the received signal is reliably above the 8-VSB threshold. In the instance of off-air reception, an 8-VSB receiver delivers the MPEG-2 transport in DVB-ASI [23] or SMPTE-310M [24], which is inserted into an MPEG-2 Multiplexer. The multiplexing device combines individual program streams and processes PSIP data to eliminate potential conflicts and to aggregate data; see CFR47, Section 76.640(b)(1)(iv) [5]. Typically, two DTV signals are combined into a single 256-QAM channel, although other aggregation arrangements are possible.

At the time of writing, analog standard-definition and digital high-definition signals are available to the operator. In rare cases, digital standard-definition signals also are available.

Non-Broadcast Sources

Receipt of signals from non-broadcast sources is as described in the section on analog cable service.

However, instead of being decoded to analog and inserted into the plant on frequency in NTSC, the digital signals are directed to the appropriate resources for transmission through the plant. In the particular case of digital satellite reception (for example, *Head in the Sky*, HITS), two scenarios exist. In some cases, the entire incoming transport multiplex is QAM-modulated, up-converted, and passed through the plant. In others, the incoming programs are multiplexed together with other nonbroadcast or locally originated programs for greater bandwidth efficiency in transmission.

Regulations and Requirements

In all cases, the operator is obliged to pass through CEA-608-C [4] and CEA-708-B [12] captions, as well as Content Advisory data, if present, in program streams delivered through the cable system.

In the case of broadcaster-originated content, there are no specific requirements for carriage of content.

The operator also is bound by FCC CFR 47, Section 76.640 (Support for Unidirectional Cable product) [5].

Delivery

Preparation for Delivery

Forming a transport multiplex for delivery through the plant is a complex exercise involving the selection and combination of programs (video and/or audio and associated data). In general, there are three possible operating scenarios.

In the first, an entire group of services may simply be passed through. This typically occurs with incoming satellite signals, for example, HITS feeds, which have been assembled for maximum bandwidth efficiency through the cable plant.

In the second, the transport payload simply is assembled, without further processing, from single-program transport streams (SPTS) drawn from multiple sources. This is likely to be inefficient in that it is unlikely that the sum of all payload components will reliably fully occupy the transport capacity of the transmission channel. This inefficiency is worsened when constituent signals are encoded in constant bit rate (CBR), as is the case with a 19.39 Mbps broadcast HDTV source. Because the bit rate actually required at a given point in time varies quite widely (as a function of individual field or frame requirements), the output from CBR encoding tends to include null or zero-value data packets and, sometimes, a considerable quantity of such packets.

In the third, the transport payload is groomed from a group of single-program transport streams. This is the most efficient scenario in terms of system bandwidth and other resources such as multiplexers, modulators, and up-converters.

At its simplest, grooming supports the assembly of multiplexes with minimal null-packet payload. This kind of grooming has absolutely no impact on quality and demonstrates that throughput bit rate should not be a point for discussion or contention.

In its more complex form, sometimes called *Statistical Multiplexing* (StatMUX) or *rate-shaping*, grooming can achieve enhancements to efficiency yet, if done with due care, will have little or no impact on quality. New developments in Statistical Multiplexing, which employ a different approach from that taken in previous generations of StatMUX products, offer the promise of very significant improvements in efficiency with no loss in quality whatsoever.

Transport

The syntax and semantics of the transport for digital cable conform to ISO/IEC 13818-1 [25]. More specifically, transport for digital cable is a compatible subset of ISO/IEC 13818-1 [25]. Constraints and extensions to ISO/IEC 13818-1 are as specified in ANSI/SCTE 54 2004 [13].

The key elements of the digital video transport are the Program Association Table (PAT), the Conditional Access Table (CAT), and the Transport Stream Program Map Tables (PMTs). The PAT and CAT are identified according to Packet Identifiers (PIDs) 0x00 and 0x01; the PIDs for the PMTs corresponding to individual programs are identified in the PAT.

Programs compliant with SCTE or ATSC standards may be identified with MPEG-2 Registration Descriptors of 0x5343 5445 ("SCTE" in ASCII) or 0x4741 3934 ("GA94"). These descriptors may be provided as part of the program_map_section describing the program in its PMT.

Under the standard, MPEG-2 programs are constrained to carry no more than one MPEG-2 video stream, which is identified by a stream_type code of 0x02 or 0x80. If the program contains audio components, at least one such is to be a complete main audio service, as specified in A/53C [10]. Audio programs complying with A/53C [10] are identified by a stream_type code of 0x81. Transport stream packets identified by a particular PMT_PID value are constrained to carry only one program definition, as described in the PMT.

The standard imposes a number of restrictions on the timing and total bandwidths of the PAT, CAT, and PMT sections. Additional constraints are imposed on the use of adaptation headers and on Packetized Elementary Stream packet headers and extensions.

The standard also provides for delivery of Emergency Alert Messaging per ANSI-J-STD-042-2002, a joint SCTE-CEA standard [26]. Such messages are carried in transport packets identified with PID 0x1FFB.

In comparison with ATSC A/53C, digital cable transport imposes fewer restrictions on ISO/IEC 13818-1 [25]. The more notable of these are:

1. Digital cable does not constrain PES packets always to begin with a video access unit (alignment_type value of 0x02) aligned with the packet header.
2. Digital cable does not constrain PES packets to contain only one video frame.
3. Digital cable permits use of MPEG-2 still pictures.

Both ATSC A/53C [10] and ANSI/SCTE 54 2004 [13] permit specification of private data services. However, ANSI/SCTE 54 2004 [13] permits not only the ATSC services, but also additional SCTE-specific services.

Because throughput rates differ between 64-QAM and 8-VSB, the transport encoder bit rates differ between the SCTE and ATSC standards. Those for 256-QAM and 16-VSB are essentially identical.

System/Service Information

Out-of-Band

The primary method in digital cable for the delivery of System/Service Information is through the Out-of-Band Channel (Downstream). System/Service Information (SI) delivered out-of-band is as described in ANSI/SCTE 65 2002 [27]. This specification presents six profiles for potential use in SI. They range from baseline, which corresponds to current practice in digital cable, through a series of intermediate profiles that add features and/or elements of ATSC A/65B (PSIP) [28], to a further profile that consists solely of PSIP. The SI tables are formatted in accordance with the data structures defined for Program Specific Information in ISO/IEC 13818-1 [25].

The key tables in the Baseline Profile are the Network Information Table (PID 0xC2, which comprises the Carrier Definition Sub-table and the Modulation Mode Sub-table), the optional Network Text Table (PID 0xC3, which comprises the Source [Program Provider] Name Sub-table), and the Short-form Virtual Channel Table (0xC4, which comprises the Virtual Channel Map, the Defined Channels Map, and, optionally, the Inverse Channel Map).

Descriptions of other profiles are given in ANSI/SCTE 65 2002, Annex A [27].

In-Band

ANSI/SCTE 54 2004 [13] recognizes the potential presence in the digital cable network of devices that are unable to process out-of-band SI data (that is, digital cable-ready television receivers or set-top boxes not equipped with CableCARD™[16]). Accordingly, it allows for provision of in-band SI data in PID 0x1FFB, in accordance with ATSC A/65B [28]. Specifically, it provides for the inclusion of, and the repetition rates for, the PSIP Master Guide Table, System Time Table, and either a Cable Virtual Channel Table or a Terrestrial Virtual Channel Table. If the PMT or an Event Information Table in the Master Guide Table includes a content_advisory_descriptor referring to a Rating_Region_Table other than the United States or its possessions, the standard also requires inclusion of the Rating Region Table. Details are available in ANSI/SCTE 54 2004 [13].

[16]A CableCARD™ is a PCMCIA card that incorporates the security and other network-specific elements of the digital cable network.

Transmission

In-Band Modulation

A conceptual diagram of the in-band Forward Error Correction (FEC) encoding and decoding process, as well as the modulation and demodulation processes, is given in Figure 6.14-4.

Forward Error Correction uses a concatenated coding approach for high performance with modest complexity. The outer stage involves a (128, 122) Reed-Solomon Block Encoding. This allows correction of up to three symbols per Reed-Solomon block. Inner coding involves fixed (I=128; J=1) or variable (I=128,64,32,16,8; J=1,2,3,4,8,16) convolutional interleaving. This allows the operator to explicitly trade off the level of burst-error protection against decoding latency. Inner coding continues with randomization to facilitate demodulator synchronization and then with trellis coding to further mitigate random channel errors. The trellis coding uses rate 1/2 convolutional encoding with a 4/5 puncture rate.

Modulation in digital cable may be in either 64-QAM or 256-QAM, as specified in ANSI/SCTE 07 2006 [29] (also, ITU Recommendation J.83, Annex B [8]), although many systems currently employ only the 256-QAM mode. FEC framing in 64-QAM consists of 80 Reed-Solomon blocks, followed by a 42-bit sync trailer. That for 256-QAM consists of 80 Reed-Solomon blocks, followed by a 40-bit sync trailer.

The principal parameters of the downstream modulation approach are given in Table 6.14-2.

As was indicated previously, the 64-QAM mode was standardized and development begun before the FCC Advisory Committee process had reached conclusion. The 256-QAM mode was added in 1995.

Out-of-Band Signaling and Modulation

There are two types of out-of-band signaling: that from the network to the set-top and that from the set-top to the network. In both cases, modulation is Quaternary Phase-Shift Keying (QPSK), which provides high robustness with good payload. For each case, there are

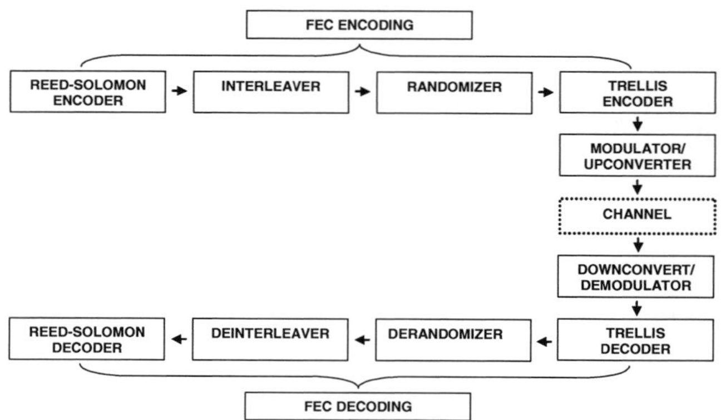

FIGURE 6.14-4 Block diagram for coding and transmission in in-band channel.

TABLE 6.14-2
Principal Parameters of Downstream In-Band Modulation

Parameter	64-QAM	256-QAM
Modulation	64-QAM, rotationally invariant	256-QAM, rotationally invariant
Symbol Size	3 bits for I and 3 bits for Q	4 bits for I and 4 bits for Q
Transmission Band	54–860 MHz	54–860 MHz
Channel Spacing	6 MHz	6 MHz
Symbol Rate	5.056941 Msps +/− 5 ppm	5.360537 Msps +/− 5 ppm
Information Bit Rate*	26.97035 Mbps +/− 5 ppm	38.81070 Mbps +/− 5 ppm
Frequency Response (Nyquist pulse shaping filters)	Square Root Raised Cosine (band-edge roll-off ≈ 0.18)	Square Root Raised Cosine (band-edge roll-off ≈ 0.12)
FEC Framing	42-bit sync trailer following 60 Reed-Solomon blocks	40-bit sync trailer following 88 Reed-Solomon blocks
QAM Constellation Mapping	6 bits per symbol	8 bits per symbol

*The information bit rate is the effective payload once forward error correction overhead is applied.

two protocols: that in systems provided by Motorola and that in systems provided by Scientific-Atlanta. These are given in ANSI/SCTE 55-1 2002 (Motorola) [30] and ANSI/SCTE 55-2 2002 [31] (Scientific-Atlanta).

With the advent of Open Cable Applications Platform (OCAP) middleware, operators recognized the need for a more flexible, higher capacity mechanism for out-of-band signaling to support interactive and other high-demand applications. Leveraging the robust and mature data infrastructure of Data over Cable Service Interface Specification (DOCSIS), which already was in place, the operators developed the DOCSIS Signaling Gateway (DSG). DSG uses the DOCSIS channel on an on-demand basis to support faster downstream and upstream data transfer and smoother running interactive applications. DSG can efficiently support many applications, including software download. It also allows for simplification in the design of set-top terminal devices by elimination of the out-of-band downstream tuner and upstream modulator. For the foreseeable future, in recognition of the deployed base of set-tops and Consumer Electronics devices, DSG will be used in parallel with the conventional signaling methods described in forthcoming sections of this chapter. Specifications for DSG are given in ANSI/SCTE 106 2005 [32] and ITU J.128 [33].

Downstream

At the Physical Layer, both downstream encoding approaches involve Reed-Solomon Block Encoding, followed by Convolutional Interleaving. The principal parameters of the two downstream modulation approaches are given in Table 6.14-3.

At the Data Link Layer, both downstream approaches employ roughly similar architectures. At source, Network-related control messages are formatted as Media Access Control (MAC) messages, complete with MAC headers and trailers (including Cyclic Recovery Clock, CRC). These are encapsulated in MPEG-2 transport stream packets and delivered as MPEG-2 private streams. On receipt, the set-top accesses the MPEG-2 private stream, de-encapsulates to retrieve the MAC packets, and performs error-correction to recover the MAC packet payload. The payload then is processed by the terminal.

Return Path (Upstream)

At the Physical Layer, both upstream encoding approaches involve Reed-Solomon Block Encoding, followed by Convolutional Interleaving. The principal parameters of the two upstream modulation approaches are given in Table 6.14-4.

At the terminal, both upstream approaches employ roughly similar architectures. The higher-level protocols hand off a Service Data Unit to the Link Layer, which forms a Protocol Data Unit (PDU), complete with upstream link header and trailer. The PDU is parsed into MAC-packet-payload-sized data fields, which are encapsulated into MAC packets, complete with headers and trailers (FEC). These, then, are transmitted upstream to the head-end, using one or another form of contention-based traffic management schemes.

Security

Digital cable systems in North America employ very robust security systems. Encryption is done using Data Encryption Standard (DES) Cipher Block Chaining, as described in ANSI/SCTE 52 2002 [34]. The Access Control systems, which manage the generation and secure distribution of decryption keys, as well as general authorization, provisioning, and service configuration, are proprietary systems supplied by the North American cable industry's primary OEM developers, Motorola and Scientific-Atlanta.

TABLE 6.14-3
Principal Parameters of Out-of-Band Downstream Modulation Approaches

Parameter	Motorola	Scientific-Atlanta	
		Grade A	Grade B
Modulation	Differentially encoded Quaternary Phase-Shift Keying (QPSK)		
Transmission Rate	2.048 Mbps	1.544 Mbps	3.088 Mbps
Channel Spacing	1.8 MHz	1.0 MHz	2.0 MHz
Frequency Band	70–130 MHz		
Carrier Center Frequency	75.25 MHz ± 0.01%*	70–130 MHz in 250 kHz steps	
Frequency Response (Nyquist pulse shaping filters at receiver)	Raised Cosine (band-edge roll-off = 0.5)	Square Root Raised Cosine (band-edge roll-off = 0.3)	
R-S Forward Error Correction	(96,94)	(59,53)	
Convolutional Interleaving	(96,8)	(55,5)	

*72.75 MHz and 104.2 MHz are optionally supported.

Embedded Security

Digital cable set-top boxes supplied by North American cable operators are manufactured primarily by Motorola and Scientific-Atlanta, although other manufacturers are entitled to build under license. The majority of set-tops deployed to date incorporates embedded security; that is, the security element, which accomplishes decryption and other Access Control functions, is entirely internal to the set-top, rather than partially externalized, as is the case with "Smart Card" approaches, such as the CableCARD™ described in the following section. This embedded security approach allows high levels of physical protection for the secure element and eliminates opportunities for wrongful interception of secure messages such as Control Words.

CableCARD™

Responding to regulatory and business requirements for the availability of digital cable-ready devices at retail and portability of devices between different networks, the North American cable industry has collaborated with the Consumer Electronics industry and other interested parties to develop specifications for CableCARD™. Originally called the Point of Deployment (POD) module, the CableCARD™ is a PCMCIA card that incorporates the security and other network-specific elements of North American digital cable

TABLE 6.14-4
Principal Parameters of Out-of-Band Upstream Modulation Approaches

Parameter	Motorola	Scientific-Atlanta		
		Grade A	Grade B	Grade C
Modulation	Differentially encoded Quaternary Phase-Shift Keying (QPSK)			
Transmission Rate	256 kbps ± 50 ppm*	256 kbps	1.544 Mbps	3.088 Mbps
Channel Spacing	192 kHz	200 kHz	1.0 MHz	2.0 MHz
Frequency Band	8–40.16 MHz	8–26.5 MHz		
Carrier Center Frequency	8–40.16 MHz (in 8 kHz steps)	8–26.5 MHz in 50 kHz steps		
Frequency Response (Nyquist pulse shaping filters at receiver)	Square Root Raised Cosine (band-edge roll-off = 0.5)	Square Root Raised Cosine (band-edge roll-off =0.3)		
R-S Forward Error Correction	(62,54)	(59,53)		
Convolutional Interleaving	(96,8)	(53,6)		

* Higher rates are available with Extended Practice.

systems, allowing appropriately manufactured Consumer Electronics devices to support one-way linear digital cable services and allowing suitably equipped Consumer Electronics devices and set-tops to be moved across system boundaries.

The basic specification for the Host-CableCARD™ interface is given in ANSI/SCTE 28 2004 [35]. A supplementary specification for Host-CableCARD™ Copy Protection is given in ANSI/SCTE 41 2004 [36]. A specification for a one-way (receive-only) digital cable receiver is given in ANSI/SCTE 105 2005 [37], while discussions concerning a specification for two-way capability are in process.

Copy Protection

The North American cable community recognizes that providers of high-value content have the right to be assured that their content, when transmitted through the digital cable network, is protected against unauthorized copying. To that end, the content community, in collaboration with the Consumer Electronics industry and other interested parties, has developed a standard for Copy Protection.

Copy Protection technology provides methods for authentication and revocation of devices, for securely binding devices, for generating copy protection keys, and for rescrambling high-value content locally to protect against unauthorized copying. A full description of such systems would be beyond the scope of this chapter. However, at the heart of the system is an agreement between the content provider and the program distributor concerning copying permissions for program content. Information concerning these permissions is delivered securely to a device, which uses this information to enforce restrictions on copying. Sample permissions are shown in Table 6.14-5.

Similar capabilities are available to set-tops with embedded security.

New Security Initiatives

Recently, operators have begun exploring new Conditional Access technologies to replace the CableCARD™. Under current FCC rules for separable security, operators must use CableCARD™ technology in leased-out set-top boxes after July 1, 2007. The Downloadable Conditional Access System (DCAS) was created to replace the older CableCARD™ technology, which is expensive and cumbersome to use, with downloadable Conditional Access technology. DCAS uses a secure microprocessor in the subscriber's terminal device to host the downloaded Conditional Access system software. DCAS allows operators to use multiple Conditional Access systems within or across cable systems and to change CA systems as necessary or appropriate. DCAS, which has been successfully demonstrated to the FCC and across the United States, accomplishes separability objectives, while allowing considerable savings in device design and license fees. PolyCipher is the name of the official licensing authority and keeper of the DCAS specifications.

TABLE 6.14-5
Sample Digital Copy Permission Values

Encryption Mode Indicator Value	Digital Copy Permission
00	Copying not restricted
01	No further copying is permitted
10	One generation copy is permitted
11	Copying is prohibited

New Technologies

The cable industry continues to explore ways of improving existing services and of leveraging the digital plant to offer new or expanded services. In video/audio services, operators are considering advanced video and audio compression systems (AVC/MPEG-4 Part 10, SMPTE VC-1, Dolby™ E-AC-3, and MPEG-4 AAC).

BROADBAND (CABLE MODEM) AND IP TELEPHONY SERVICES

Coverage of broadband and IP telephone services would be well beyond the scope of this chapter. However, it is important to note that these are key services in the cable operator's portfolio and represent, not only increasing revenues to the operator, but also increasing challenges for bandwidth. Both services have exhibited remarkable uptake by subscribers and are making increasing demands for bandwidth to support Quality of Service objectives. Broadband, in particular, shows this trend. At launch, the Broadband downstream path was limited to a maximum of 3 Mbps, with actual user allocations often much lower. However, user demand and competition from xDSL have put pressure on operators to raise downstream limits, such that many operators now allow downstream rates of 6–8 Mbps.

The operative subcommittee for standards in Broadband and IP Telephony services in North America is the SCTE's Data Standards Subcommittee.

OTHER SERVICES

The industry is seriously investigating wireless service extension. In this offering, the cable plant would be "extended" using wireless transmission to support access to "cable" services with mobile devices.

The industry also is considering CableHome™, an initiative to leverage the two-way capabilities of the digital cable plant to offer convenient support for the management of communications, appliances, and security in the home. Further details are available at http://www.cablelabs.com/.

It is not as yet clear what impacts such new services might have on analog and digital cable television services.

CONCLUSION

This chapter provides a brief overview of architectures, technologies, and services in cable in North America. Given the depth and breadth of the topic and the limited space available, it was not possible to present this material in any detail. Instead, the chapter identifies key standards and other sources that will allow those interested to pursue topics in greater depth.

ACKNOWLEDGMENTS

The author would like to acknowledge with gratitude the contributions of Bill Warga and Andy Scott, who gave freely of their time to improve the quality and accuracy of this chapter. The author also would like to acknowledge the *NAB Engineering Handbook*, 9th edition, predecessor to this chapter, an excellent overview of cable by Dr. Walter Ciciora. Dr. Ciciora's chapter helped me ascertain what would be of interest to this audience and provided many important insights.

References

[1] Ciciora, W., Farmer, J., Large, D., and Adams, M. Modern cable television technology (2nd ed.). Morgan Kaufmann, 2004.

[2] Thomas, J., and Edgington, F. Digital basics for cable TV systems. Prentice Hall, 1998.

[3] Ovadia, S. Broadband cable TV access networks: From technologies to applications. Prentice Hall, 2001.

[4] Consumer Electronics Association, CEA-608-C, "Line 21 Data Services," 2005.

[5] FCC Code of Federal Regulations (CFR) 47, Section 76, "Multichannel video and cable television service."

[6] FCC Code of Federal Regulations (CFR) 47, Section 15, Part 118, "Cable ready consumer electronics equipment."

[7] The Telecommunications Act of 1996, 47 U.S.C., Section 304.

[8] International Telecommunications Union (ITU), Recommendation J.83, "Digital multi-programme systems for television, sound and data services for cable distribution."

[9] Society of Cable Telecommunications Engineers, ANSI/SCTE 43 2004, "Digital video systems characteristics for cable television," 2004. Available: http://www.scte.org/standards.

[10] Advanced Television Systems Committee, ATSC A/53C, "Digital television standard," 2004. Available: http://www.atsc.org/standards.html.

[11] Society of Cable Telecommunications Engineers, ANSI/SCTE 20 2004, "Method for carriage of closed captions and non-real-time sampled video," 2004. Available: http://www.scte.org/standards.

[12] Consumer Electronics Association, CEA-708-B, "Digital television (DTV) closed captioning," 1999.

[13] Society of Cable Telecommunications Engineers, ANSI/SCTE 54 2004, "Digital video service multiplex and transport system for digital cable," 2004. Available: http://www.scte.org/standards.

[14] Society of Cable Telecommunications Engineers, ANSI/SCTE 19 2006, "Methods for isochronous data services transport," 2006. Available: http://www.scte.org/standards.

[15] Society of Cable Telecommunications Engineers, ANSI/SCTE 27 2003, "Subtitling methods for broadcast and cable," 2002. Available: http://www.scte.org/standards.

[16] [MPEG-2] International Standards Organization/International Electrotechnical Commission, Standard ISO/IEC 13818-2, "Information technology—generic coding of moving pictures and associated audio—Part 2: Video coding," 1994.

[17] Society of Cable Telecommunications Engineers, ANSI/SCTE 21 2001, "Standard for carriage of NTSC VBI data in cable digital transport streams," 2001. Available: http://www.scte.org/standards.

[18] Society of Motion Picture and Television Engineers, SMPTE 170M-2004, "Television—Composite analog video signal—NTSC for studio applications," 2004.

[19] International Telecommunications Union (ITU), Recommendation J.200, "Worldwide common core—Application environment for digital interactive television services."

[20] International Telecommunications Union (ITU), Recommendation J.201, "Harmonization of declarative content format for interactive television applications."

[21] International Telecommunications Union (ITU), Recommendation J.202, "Harmonization of procedural content formats for interactive TV applications."

[22] Society of Cable Telecommunications Engineers, ANSI/SCTE 40 2004, "Digital cable network interface standard," 2004. Available: http://www.scte.org/standards.

[23] European Telecommunications Standards Institute, EN 50083-9, "Cabled distribution systems for television, sound and interactive multimedia signals; Part 9: Interfaces for CATV/SMATV headends and similar professional equipment for DVB/MPEG-2 transport streams," (DVB Blue Book, A010), Annex B, Asynchronous Serial Interface.

[24] Society of Motion Picture and Television Engineers, SMPTE 310M-2004, "Television—Synchronous serial interface for MPEG-2 digital transport streams," 2004.

[25] [MPEG-2] International Standards Organization/International Electrotechnical Commission, Standard ISO/IEC 13818-1, "Information technology—generic coding of moving pictures and associated audio—Part 1: Systems," 1994.

[26] American National Standards Institute, ANSI-J-STD-042-2002, "Emergency alert message for cable," 2002.

[27] Society of Cable Telecommunications Engineers, ANSI/SCTE 65 2002, "Service information delivered out-of-band for digital cable television," 2002. Available: http://www.scte.org/standards.

[28] Advanced Television Systems Committee, ATSC A/65B, "Program and system information protocol for terrestrial broadcast and cable, 2003. Available: http://www.atsc.org/standards.html.

[29] Society of Cable Telecommunications Engineers, ANSI/SCTE 07 2006, "Digital video transmission standard for television," 2006. Available: http://www.scte.org/standards.

[30] Society of Cable Telecommunications Engineers, ANSI/SCTE 55-1 2002, "Digital broadband delivery system part 1: Mode A," 2002. Available: http://www.scte.org/standards.

[31] Society of Cable Telecommunications Engineers, ANSI/SCTE 55-2 2002, "Digital broadband delivery system part 2: Mode B," 2002. Available: http://www.scte.org/standards.

[32] Society of Cable Telecommunications Engineers, ANSI/SCTE 106 2005, "DOCSIS Set-Top Gateway (DSG) specification," 2005. Available: http://www.scte.org/standards.

[33] International Telecommunications Union (ITU), Recommendation J.128, "Set top Gateway specification for transmission systems for interactive cable television services."

[34] Society of Cable Telecommunications Engineers, ANSI/SCTE 52 2002, "Data encryption standard cipher block chaining packet encryption," 2002. Available: http://www.scte.org/standards.

[35] Society of Cable Telecommunications Engineers, ANSI/SCTE 28 2004, "Host-POD interface standard," 2004. Available: http://www.scte.org/standards.

[36] Society of Cable Telecommunications Engineers, ANSI/SCTE 41 2004, "POD copy protection system," 2004. Available: http://www.scte.org/standards.

[37] Society of Cable Telecommunications Engineers, ANSI/SCTE 105 2005, "Uni-directional receiving device standard for digital cable," 2005. Available: http//www.scte.org/standards.

CHAPTER

6.15

Direct-to-Home Satellite Systems

LEON STANGER
Consulting Engineer
Farmington, Utah

JOHN P. GODWIN
Gretna Green Associates
Los Angeles, California

STEPHEN P. DULAC
DIRECTV
El Segundo, California

INTRODUCTION

Over the past decade, direct-to-home (DTH) satellite systems, sometimes called direct broadcast satellite (DBS) systems, have emerged as one of the primary delivery methods of digital television (DTV) to consumers. Approximately one-fourth of U.S. households depend on subscription-based satellite systems to meet their entertainment, news, educational, and interactive television needs.[1] This trend may continue as more channels and services become available from satellite providers. This chapter provides an introduction to the system arrangements and technologies used for modern digital DTH satellite broadcasting and mentions some of the recent changes in the regulations governing satellite broadcasting in the United States.[2]

DTH SYSTEM ARCHITECTURE

Figure 6.15-1 provides a simplified diagram of an all-digital multichannel direct-to-home satellite system, which comprises four main subsystems:

- Program contribution
- DTH broadcasting facility and uplink
- Broadcasting satellite(s)
- Customer equipment

[1]FCC Twelfth Annual Report (2005) on video programming competition.

[2]This chapter is based in part on: "Satellite Direct to Home," by Stephen P. Dulac and John P. Godwin, which appeared in the *Proceedings of the IEEE*, vol. 94, no. 1, 2006, 158–172, © 2006 IEEE. Those portions of the paper reflected here are provided with the permission of the IEEE.

Program Contribution

DTH program channels originate at a variety of remote locations and most are also available to other program distributors. These channels arrive at the DTH broadcasting or uplink facility via existing "backhaul" satellites or fiber-optic links. Some programming, such as theatrical films for pay-per-view, arrive at the facility as prerecorded digital tapes.

The delivery of local television channels to consumers via satellite has necessitated the use of in-market facilities to preprocess and backhaul the signals to the DTH broadcasting facility via leased terrestrial transmission facilities or satellite links. Figure 6.15-2 summarizes the components of a typical solution for backhauling local channels. Each DTH service provider has a local receive facility (LRF) in each market served. These local, unmanned facilities provide signal collection and encoding as illustrated in the figure. A later section provides more information on quality considerations and methods for delivery of program content to the DTH service provider.

DTH Broadcasting Facility

Baseband Systems

The DTH broadcasting facility shown in Figure 6.15-1 provides several functions common to any broadcasting facility, such as incoming signal monitoring, adjustment, resynchronization, and signal routing within the facility. For prerecorded content, quality control, cloning, and playback are common functions. The program content for most channels is unchanged

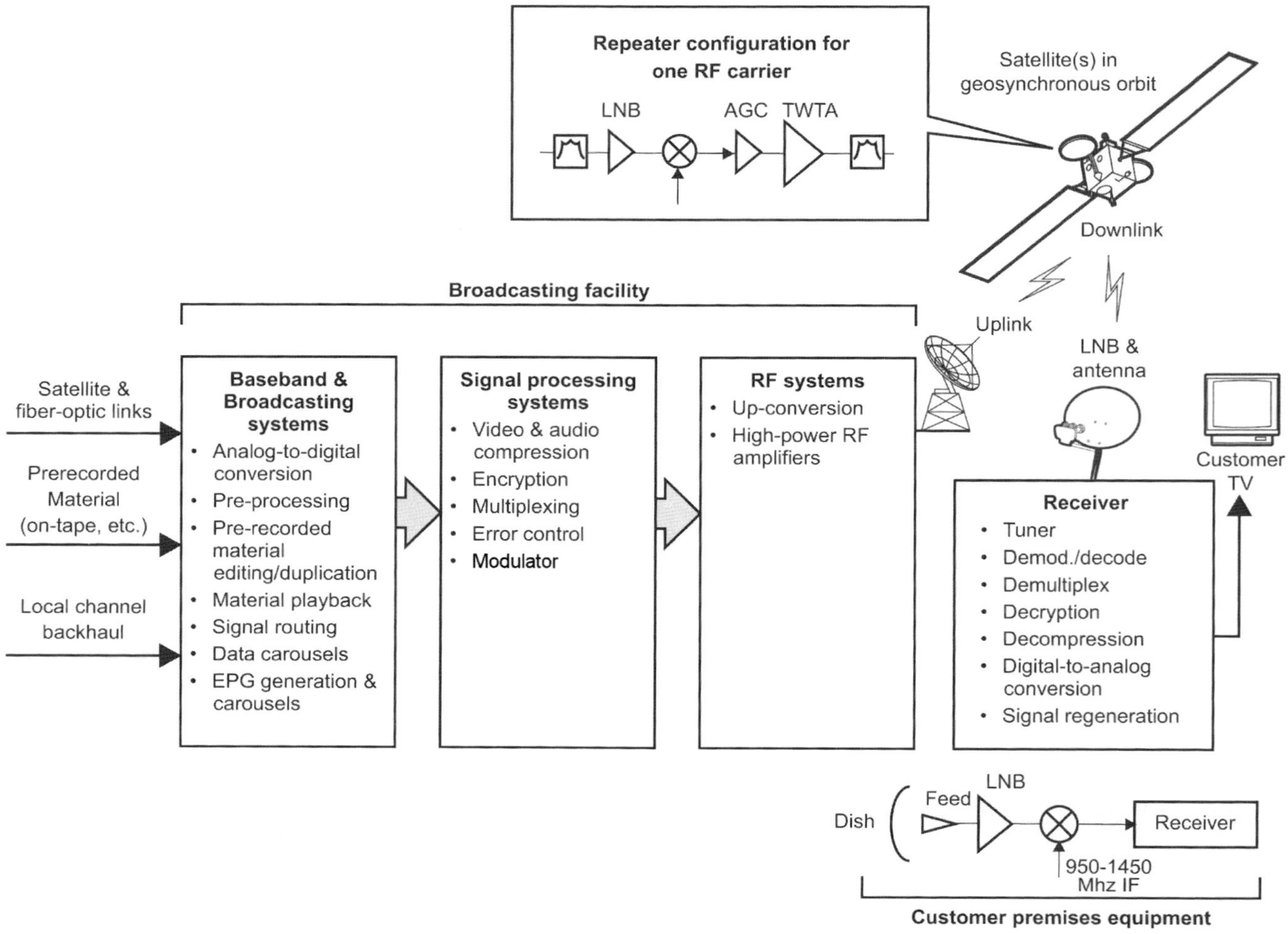

Repeater configuration for one RF carrier
LNB
AGC
TWTA
Satellite(s) in geosynchronous orbit
Downlink
Broadcasting facility
Uplink
LNB & antenna
Satellite & fiber-optic links
Prerecorded Material (on-tape, etc.)
Local channel backhaul
Baseband & Broadcasting systems
• Analog-to-digital conversion
• Pre-processing
• Pre-recorded material editing/duplication
• Material playback
• Signal routing
• Data carousels
• EPG generation & carousels
Signal processing systems
• Video & audio compression
• Encryption
• Multiplexing
• Error control
• Modulator
RF systems
• Up-conversion
• High-power RF amplifiers
Customer TV
Receiver
• Tuner
• Demod./decode
• Demultiplex
• Decryption
• Decompression
• Digital-to-analog conversion
• Signal regeneration
Dish
Feed
LNB
Receiver
950-1450 Mhz IF
Customer premises equipment

FIGURE 6.15-1 All-digital multichannel satellite DTH system.

VHF/UHF
Local Station Studio
Local Station Studio
Local Station Studio
TV-1
(analog fiber)
Satellite service provider Local Receive Facility in a broadcast TV local market
MPEG Video/ Audio Encoding
Transport Layer Mux
Digital Telco I/F
Telco I/F
Telco digital (e.g., DS-3)
or Satellite backhaul
DTH Broadcast Facility
Local station signals
Local Receive Facility in a neigboring market(s)

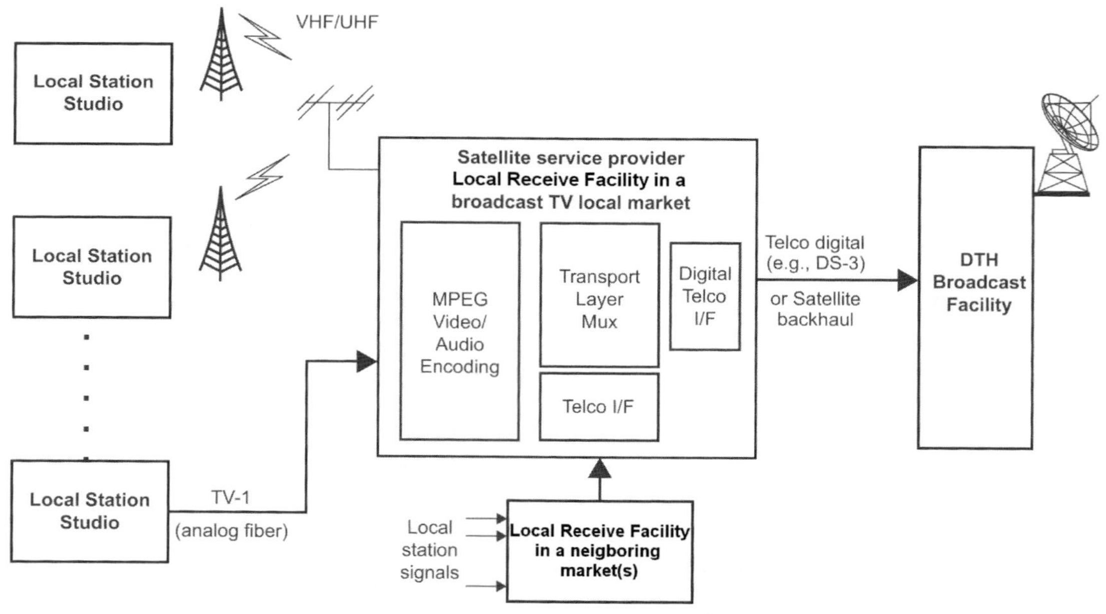

FIGURE 6.15-2 In-market local television backhaul solution.

by the facility. Certain channels, by agreement with the originator, may have commercials or promotional spots inserted at points identified (e.g., by in-band tones) by the originator. Prerecorded material is copied from digital tape masters to video file servers. The video servers use redundant arrays of independent disk (RAID) technology and play back the content on a digital satellite channel at a time established by the daily broadcast schedule.

The "pay" business model used for DTH in the United States requires that the DTH broadcast site provide conditional access equipment. The conditional access system, which includes equipment within the home, permits customer access to programming services only when certain conditions are met. Each audio and video program stream may be independently encrypted at the discretion of the DTH provider.

Other signal processing equipment includes the service information/electronic program guide (SI/EPG) generators, compression encoders, and multiplexing, error control, and modulation equipment. The SI/EPG equipment creates data streams that are used by the in-home electronics to display information about the programming channels and the individual programs. The EPG data typically include program title, start and end times, synopsis, program rating for parental control, alternate languages, and other descriptive information.

Compression

Standard definition (SD) video/audio is typically routed within the DTH broadcasting facility in serial digital component format [1] at 270 Mbps, but is reduced to average rates of 2–4 Mbps with compression encoding prior to transmission. High definition (HD) program content is routed in plant at 1.5 Gbps using the standard SMPTE 292M serial interface [2] and is compressed to average rates of 7–15 Mbps for transmission. Signal compression dramatically reduces the transmission path investment—in satellites, for example—and, conversely, also increases the entertainment channels available for a given amount of transmission bandwidth and investment. Most operational digital DTH systems in the Americas use the Moving Picture Experts Group encoding standard MPEG-2 [3] or MPEG-4 advanced video and audio compression (AVC, AAC) [4, 5] for newer systems. Each of the constituent streams may be encoded using a constant bitrate (CBR) although variable bitrate (VBR) is more commonly used. With VBR, the bitrate is determined by the instantaneous image complexity. Simple scenes with little detail and no motion will consume a very low bitrate, while scenes with large amounts of action and complex detail, such as sports, require much higher bitrates and will typically peak at the maximum bitrate allowed. VBR streams from a group of channels are statistically multiplexed together to achieve more efficient bandwidth utilization. Audio channels are generally encoded as CBR streams using the compression standard chosen by the DTH provider and multiplexed together with the video streams.

Error Control and Modulation

The composite high-speed stream containing multiple program channels, SI/EPG information, and other data is processed by forward error correction (FEC) logic. The FEC method, often concatenating convolutional and block codes, provides excellent delivered quality at transmission thresholds below those available with previous analog methods. For systems deployed in the 1990s, quadrature phase shift keying (QPSK) modulation was used in virtually every instance using the DVB-S standard [6]. This modulation is more bandwidth-efficient than binary PSK, but has a constant envelope modulation and therefore is appropriate for satellite transmission systems with repeater output stages driven into the limiting region. More recent systems use the DVB-S2 standard [7] for modulation and coding. This standard provides 8-PSK modulation for highly nonlinear channels and also 16-ary and 32-ary amplitude and phase shift keying modes for more linear applications. The standard's FEC uses a Bose-Chaudhuri-Hoequenghem (BCH) code concatenated with a low-density parity check (LDPC) inner code yielding performance within 0.7 dB of the Shannon limit. Figure 6.15-3 shows the ~30% bandwidth efficiency of DVB-S2 versus the DVB-S modulation method.

After modulation, the signal is upconverted to the radio frequency (RF) channel used for transmission to the satellite and fed to a high-gain dish antenna for the uplink.

Broadcasting Satellites

Each uplink signal from a DTH broadcasting facility or facilities is received and rebroadcast by an RF "transponder" of a frequency-translating repeater on board a geosynchronous communications satellite. For

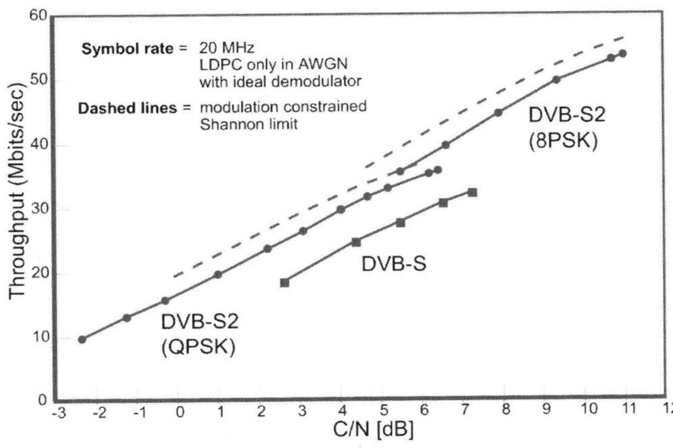

FIGURE 6.15-3 DVB-S2 satellite modulation and coding provide increased throughput per unit bandwidth.

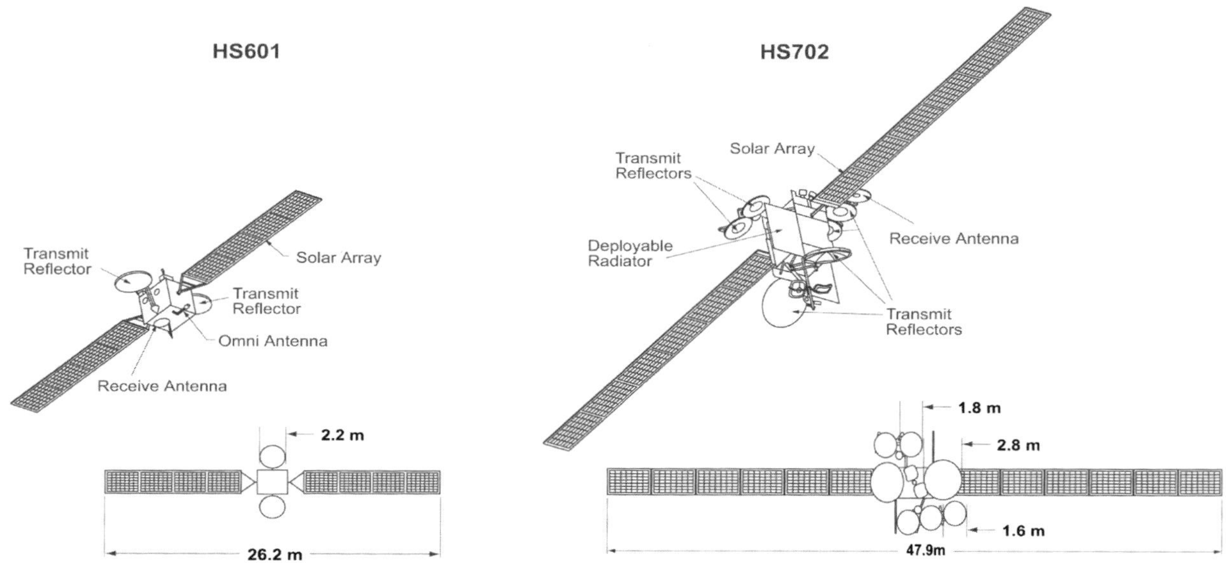

FIGURE 6.15-4 Progress in satellite platforms provided more delivered bandwidth per spacecraft and bandwidth reuse via spot-beam technology: (a) HS601, and (b) HS702.

Broadcasting Satellite Service (BSS) band operation, the satellite receives signals in the range 17.3–17.8 GHz [8], downconverts each signal by 5.1 GHz, and retransmits each signal in the range 12.2–12.7 GHz [9]. The satellites used in DTH systems are similar in architecture to geosynchronous communications satellites that have been deployed for international and domestic telecommunications since the mid-1960s. For DTH systems, the satellites' greatly increased physical size and weight permit relatively high levels of received solar energy, and hence DC power, and relatively large onboard antennas enabling downlink beam shaping. Figure 6.15-4(a) illustrates a typical DTH satellite as deployed in the 1990s: the HS601 from Hughes Space and Communications Company (now Boeing Satellite Systems) [10,11]. The configuration is dominated by the sun-oriented solar panels for DC power generation and the parabolic reflectors used to create the downlink beams. Spot beams are used for delivery of channels intended for a local market. They allow downlink frequencies to be reused but require more complex antenna systems, as shown in Figure 6.15-4(b), which illustrates a more recent DTH satellite, the HS702 [12]. Each DTH provider may shape these beams differently. For example, from one provider a given spot beam may cover only major metropolitan areas in a given market while the beam from the competing satellite provider may cover essentially the entire market.

Each satellite's communications "payload" is a microwave frequency-translating repeater. A broadband front-end receiver, one per polarization, downconverts to the downlink frequency and drives multiple RF chains, one per carrier, with each RF chain or "transponder" having a high-power traveling wave tube (TWT) transmitter [13]. Typically, each TWT amplifier has a saturated power rating of 240 watts [10]. The primary repeater functional elements are shown in Figure 6.15-1.

In the United States, to increase the total available capacity, a single system operator often uses multiple satellites at a given orbital location and, additionally, multiple satellites at adjacent orbital locations. Multiple satellites at a single orbital location—actually, separated in longitude by at least 0.1 degree—give full use of the available spectrum by effectively pooling the capabilities of several satellites. This implements the futuristic visions from past decades for massive "earth-facing communications relay platforms" without the necessity for a single physical vehicle. Use of adjacent orbital locations permits spectrum reuse by a single system operator. This is illustrated in Figure 6.15-5.

Customer Electronics

The DTH customer electronics consists of a small aperture antenna and low-noise block downconverter, an integrated receiver/decoder (IRD) unit (or simply "receiver"), and a handheld remote control. The antenna is typically an offset parabolic reflector in the range of 45–60 cm in diameter. This size is chosen as a tradeoff of several factors, including aesthetics for customer locations, link margin including rain attenuation, and beam width for easy alignment. Using reflectors larger than 90 cm to improve reception is generally not recommended since the beam width may become too narrow for easy alignment and may have difficulty receiving from multiple satellites in a closely spaced constellation. The RF signal collected by the feed horn at the focus is coupled with a low-noise amplifier (LNA) and then block downconverted to an L-band IF of 950–1450 MHz, or as wide as 250–2150

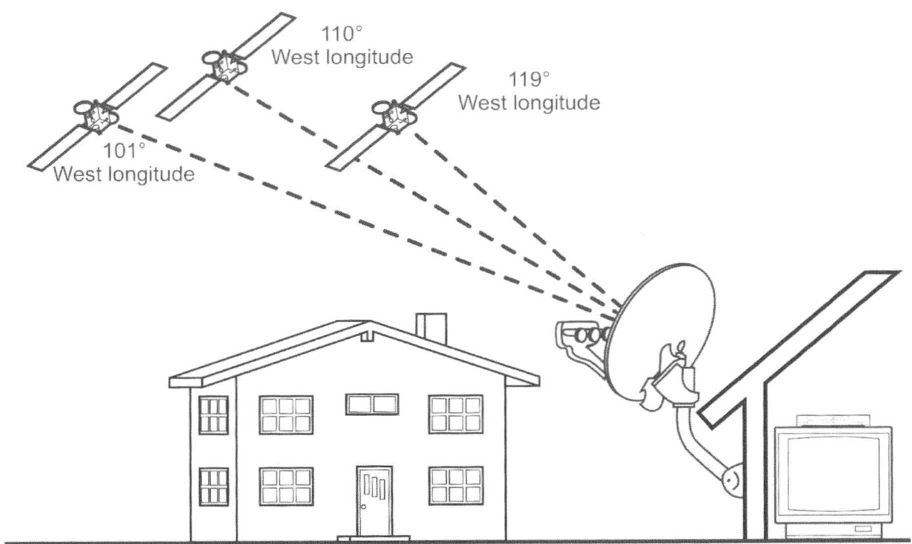

FIGURE 6.15-5 Use of adjacent orbital locations permits spectrum reuse.

MHz for recent models. An elliptical dish with multiple feed horns and LNA/block converters is used for receiving satellites at adjacent orbital locations.

The outdoor electronics, shown diagrammatically in Figure 6.15-1, receives low-voltage DC power via the coaxial cable used to deliver the downconverted signal into the customer's home and, specifically, to the receiver. The indoor receiver provides the many functions listed on the far right in Figure 6.15-1. The unit's circuitry includes an IF tuner, QPSK demodulator, stream demultiplexer (to capture a single programming channel), decryption under conditional access control, MPEG video/audio decoder, and TV signal regeneration.

In the western hemisphere, most DTH receivers utilize a replaceable "smart card" with an embedded secure microprocessor used to generate cryptographic keys for decryption of the individual services. In the event that security is compromised, the system operator may only need to replace the smart card to allow economic upgrade of a portion of the conditional access logic instead of the far more costly replacement of entire receivers. The receiver provides output signals to various home-entertainment devices, such as standard and high definition televisions and audio amplifier systems. The receiver may have front panel controls but it is routinely controlled via signals from a handheld remote control using IR, and in many cases RF, transmission.

ITU Recommendation ITU-R BO.1516 [15], published in 2001, presents a generic reference model for a digital DTH receiver. This model presents the common functions required in a satellite IRD and an adaptation of Figure 1 of that recommendation is reproduced here as Figure 6.15-6. The reference model is arranged in layers, with the physical layer located at the lowest level of abstraction, and the services layer located at the highest level.

Terrestrial DTV receivers share these reference model functional elements, with notable differences that reflect both business and technical differences in these services, as follows:

- The physical and link layers of the terrestrial receiver are designed to support the antennas and modulations required for off-air (terrestrial) signal reception.

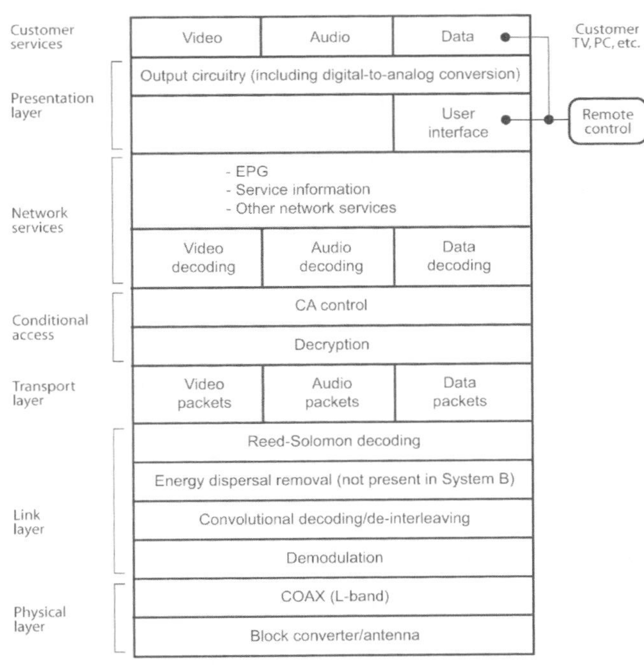

FIGURE 6.15-6 Generic reference model for a digital DTH receiver. (Reproduced with the permission of the ITU.)

- The conditional access layer of the terrestrial receiver is optional, whereas in satellite systems all services, even local channel rebroadcasts, tend to be encrypted. While there are "digital cable-ready" digital televisions having decrypt capabilities, there are no "satellite-ready" digital televisions.
- The EPG and interactive service capabilities tend to be highly customized in a satellite receiver to meet the competitive needs of the service operator.

COMMON FUNCTIONAL ELEMENTS OF DIGITAL DTH SYSYTEMS IN THE AMERICAS

Beginning in 1994 and driven by business imperative, the first digital DTH satellite systems were launched prior to the creation of industry standards for either modulation and coding or transport and multiplexing or for video and audio source encoding. Nevertheless, standards for the digital DTH application did follow, and there are four now in use for the Americas: ITU System A/DVB (used by Dish Network, Sky Brasil, Sky Mexico, and Bell ExpressVu); ITU System B (used by DIRECTV and DIRECTV Latin America); ITU System C (used by Star Choice); and the more recent ATSC A/81 standard for direct-to-home satellite broadcast (adopted in 2003, but not yet in use).

A high-level description of the first three standards is presented in Table 6.15-1. This shows excerpts of summary characteristics from Table 1 of the ITU document [14], which concludes that common receiver designs supporting all of these systems are possible. Note that as their service offerings evolve, the architectures of service providers in the Americas are subject to changes not reflected in these ITU standards.

The ATSC A/81 specification [15] defines extensions to audio, video, transport, and PSIP subsystems as defined in ATSC standards A/53B and A/65A. It also includes carriage of data broadcasting as defined in ATSC standard A/90 without requiring extensions. Transmission and conditional access subsystems are not defined in A/81, allowing service providers to use existing subsystems.

DELIVERING PROGRAMMING CONTENT TO DTH SERVICE PROVIDERS

Program contribution was discussed in the DTH System Architecture section. System arrangements should provide high standards of quality and reliability.

Quality Considerations

Ideally, program content delivered to DTH systems should be of studio or contribution quality. This will allow the DTH provider to compress the signal for satellite delivery and still meet the quality expectations of high-end customers. If the source quality is degraded due to heavy compression, noise, or other distortions, the quality delivered to DTH customers will be compromised. It is difficult to quantify the effects of multiple compression cycles, but the rule of thumb is to retain contribution quality until the final compression segment for customer delivery. Contribution quality signals, even if compressed, will utilize the maximum resolution (720 × 480 for SD, 1920 × 1080i or 1280 × 720p for HD), 4:2:2 chroma profile, and a sufficiently high bitrate to be essentially free of compression artifacts.

Where it is not feasible to provide contribution quality, for example, when DTV signals are taken off-air, the content provider should still maintain the highest quality possible, recognizing that the program content will be reencoded for satellite delivery. The resolution and bitrate utilized should be as high as the delivery system will allow.

Audio can be delivered in analog, uncompressed digital, or compressed digital form. For DTV off-air and other surround-sound applications, the AC-3 format defined in ATSC A/52 [16] is often passed through the satellite system directly to customers. The fixed latency induced by reencoding the video must be added to the audio channel to maintain audio/video timing or lip-sync. The content provider should assure tight control of audio/video timing and loudness variations. The audio metadata indicating the correct Dialog Normalization Value should be applied to each program segment to maintain comfortable listening levels and consistency with other channels.

The information content used in electronic program guides distributed over DTH satellite systems is typically derived from one or more of the media service companies. The media service companies derive their databases from TV networks, stations, and other content providers. It is essential that program providers maintain accuracy in program descriptions, air times, and ratings for parental controls. Tuning information for local DTV stations is also derived from the same database.

Delivery Methods

Methods commonly used for program content delivery to DTH providers include:

- *Digital tape.* Digital tape provides excellent source quality for SD and HD material to be aired at a later time. Digital Betacam® [17], HDCAM™ [17], D-5 [18], or HD-D5 [18] are commonly used. The DTH provider will generally transfer the content onto a video server for playback at a later time. Other tape formats may also be accepted by the DTH provider.
- *Analog satellite.* Analog satellite transmission [19] of SD content using frequency modulation (FM) has been the mainstay for delivery to cable systems for many years. The quality is also generally suitable for DTH applications.
- *Digital satellite and digital fiber.* Operating costs can be reduced by utilizing MPEG digital compression [3,4,5] for backhaul and content delivery to DTH providers. Where possible, contribution quality

should be maintained. Digital compression will allow multiple channels to be carried over satellite links that formerly carried a single analog channel. Digital compression over fiber circuits may also be used to reduce the bitrate required and makes fiber a cost-effective alternative to satellite for program backhaul and distribution.

TABLE 6.15-1
Summary of ITU Direct-to-Home Formats as of 2001*

	System A	System B	System C
Modulation Scheme	QPSK	QPSK	QPSK
Symbol Rate	Not specified	Fixed 20Mbaud	Variable 19.5 and 29.3 Mbaud
Necessary bandwidth (-3 dB)	Not specified	24 MHz	19.5 and 29.3 MHz
Roll-off rate	0.35 (raised cosine)	0.2 (raised cosine)	0.55 and 0.33 (4th order Butterworth filter)
Reed-Solomon outer code	(204,188,T=8)	(146,130,T=8)	(204,188,T=8)
Interleaving	Convolutional, I=12, M=17 (Forney)	Convolutional, N1=13, N2=146 (Ramsey II)	Convolutional, I=12, M=19 (Forney)
Inner coding	Convolutional	Convolutional	Convolutional
Constraint length	K=7	K=7	K=7
Basic code	1/2	1/2	1/3
Generator polynomial	171, 133 (octal)	171, 133 (octal)	117, 135, 161 (octal)
Inner coding rate	1/2,2/3,3/4,5/6,7/8	1/2,2/3,6/7	1/2,2/3,3/4,3/5,4/5,5/6,5/11,7/8
Net data rate	23.754 to 41.570 Mbits/s given symbol rate of 27.776 Mbaud	17.69 to 30.32 Mbits/s at fixed 20 Mbaud symbol rate	16.4 to 31.5 Mbits/s given symbol rate of 19.5 Mbaud
Packet Size	188 bytes	130 bytes	188 bytes
Transport layer	MPEG-2	Non-MPEG	MPEG-2
Commonality with other media (i.e., terrestrial, cable, etc)	MPEG transport stream basis	MPEG elementary stream basis	MPEG transport stream basis
Video source coding	MPEG-2 at least main level/main profile	MPEG-2 at least main level/main profile	MPEG-2 at least main level/main profile
Aspect ratios	4:3 16:9 (2.12:1 optionally)	4:3 16:9	4:3 16:9
Image supported formats	Not restricted, Recommended: 720x576, 704x576 544x576, 480x576 352x576, 352x288	720x480, 704x480 544x480, 480x480 352x480, 352x240 720x1280, 1280x1024 1920x1080	720(704)x576 720(704)x480 528x480, 528x576 352x480, 352x576 352x288, 352x240
Frame rates at monitor (per s)	25	29.97	25 or 29.97
Audio source decoding	MPEG-2, Layers I and II	MPEG-1, Layer II; ATSC A/53 (AC3)	ATSC A/53 or MPEG-2 Layers I and II
Service information	ETS 300 468	System B	ATSC A/56 SCTE DVS/011
EPG	ETS 300 707	System B	User selectable
Teletext	Supported	Not specified	Not specified
Subtitling	Supported	Supported	Supported
Closed caption	Not specified	Yes	Yes

*Reproduced with the permission of the ITU.

- *Uncompressed digital fiber.* High-bandwidth digital circuits with 270 Mbps for SD or 1.5 Gbps for HD are offered by some local telephone companies at competitive rates. Where available, these circuits provide very high quality for short-haul applications.
- *"TV1" fiber.* TV1 circuits [20] from local telephone companies are suitable for short-haul SD links if off-air reception is not acceptable at the local receive facility. Under SHVERA rules (see next section), it is the TV station's responsibility to provide a good quality signal either by good off-air reception or an alternate means of delivery.
- *Over-the-air analog TV.* Conventional over-the-air TV is commonly used for receiving local SD channels as long as the reception quality is good. This option will not be available when NTSC analog broadcasting ceases in 2009. It is recommended that stations should utilize the ghost canceling reference (GCR) signal [21] to allow the receiver to correct ghosting, smearing, or frequency-response errors caused by multipath distortion at the receive location.
- *Over-the-air digital TV.* Where a digital fiber feed is not available, over-the-air digital TV broadcast is used as the source for HD programming from local TV stations. DTV broadcast is also increasingly used as an alternate means of delivering an equivalent to the analog SD channel. When this is done, the programming and aspect ratio must be identical to the analog channel (until discontinued in 2009). The compression bitrate for the SD service must be sufficiently high to preserve equal or better quality than the analog channel.

U.S. LOCAL CHANNELS REGULATORY EVOLUTION

In the late 1980s, local U.S. broadcasters were concerned about nationwide satellite delivery of national network broadcast programming, as it meant that viewers could watch network programming without watching their local network affiliate. Moreover, copyright holders (movie studios, sports leagues, etc.) argued that satellite carriers lacked copyright authority to engage in such retransmissions. In 1988 the U.S. Congress passed the Satellite Home Viewer Act (SHVA), which granted satellite operators copyright authority to deliver this programming, but allowed such delivery only to "unserved" households (e.g., approximately 10% of the U.S. population not well covered by local broadcasters' over-the-air signals).

For many years, satellite operators used this license to deliver signals originating from New York and Los Angeles to unserved households throughout the United States. By way of comparison, cable operators generally cannot deliver "distant signals," that is, signals originating in a market other than that in which the subscriber resides. Thus, satellite operators typically delivered distant signals, but only to some subscribers, while cable operators typically delivered only "local signals," that is, signals originating in the subscriber's local market, but to all subscribers.

By 1999, it was recognized that consumers would be better served, and the competitive playing field between satellite and cable operators would be more level, if satellite operators could, like cable, also redistribute local broadcast signals to all subscribers. As a result, the Satellite Home Improvement Viewing Act (SHIVA) was passed and both U.S. operators (DIRECTV and Dish Network) began offering "local-into-local" services to the largest markets. While the more popular local stations will negotiate for carriage, SHIVA's "carry one, carry all" requirement obligates a satellite carrier to carry all qualified local stations in each market where it carries any such station.

Renewed for another five years in 2004 as SHVERA (Satellite Home Viewer Extension and Reauthorization Act), satellite DTH operators' rights to carry local broadcast stations are now more comparable to the rights enjoyed by cable operators. Satellite operators, like cable operators, now have the option to retransmit nearby out-of-market stations that are "significantly viewed" into certain communities in a market. But satellite operators, like cable operators, may generally no longer sign up new subscribers for distant signals in markets where they provide local signals. SHVERA also banned a two-dish practice employed by one of the satellite operators, in which reception of stations that elected mandatory carriage required a second consumer antenna that was not installed unless requested by the consumer.

These laws are codified in Title 17 of the United States Code, administered through the U.S. Copyright Office, as well as Title 47, for which the U.S. Federal Communications Commission (FCC) is responsible. At the time of this writing, the FCC is finalizing rules that define cable and satellite operators' rights and obligations for carriage of local broadcasters' signals as they transition from analog to digital ATSC broadcasts. For more information, see http://www.copyright.gov and http://www.fcc.gov.

FURTHER INFORMATION

This chapter has provided an introduction to a complex subject with sophisticated engineering, particularly for the space segment of the system. For more information on the standards and technologies involved, the reader is referred to the references provided below.

References

[1] Society of Motion Picture and Television Engineers. SMPTE 259M Television—10-bit 4:2:2 Component and 4 fsc Composite Digital Signals: Serial Digital Interface (1997). Available at http://www.smpte.org.

[2] Society of Motion Picture and Television Engineers. SMPTE 292M Television—Bit Serial Digital Interface for High-definition Systems. Available at http://www.smpte.org.

[3] [MPEG-2] ISO/IEC 13818-1:2000. Information Technology—Generic Coding of Moving Pictures and Audio Information: Systems.

[4] [MPEG-4 AVC] ISO/IEC 14496-10. Information Technology—Coding of Audio-Visual Objects, Part 10: Advanced Video Coding.
[5] [MPEG-4 AAC] ISO/IEC 14496-3. Information Technology—Coding of Audio-Visual Objects, Part 3: Audio.
[6] ETSI EN 300 421 (V1.1.2). Digital Video Broadcasting (DVB): Framing Structure, Channel Coding, and Modulation for 11/12 GHz Satellite Services. Available at http://www.dvb.org.
[7] ETSI EN 302 307. Digital Video Broadcasting (DVB): Second Generation Framing Structure, Channel Coding, and Modulation Systems for Broadcasting, Interactive Services, News Gathering, and Other Broadband Satellite Applications. Available at http://www.dvb.org.
[8] International Telecommunication Union (ITU). *Radio Regulations*, 1982, rev. 1985, 1986, 1988, vol. 2, Appendix 30A.
[9] International Telecommunication Union (ITU). *Radio Regulations*, 1982, rev. 1985, 1986, 1988, vol. 2, Appendix 30.
[10] The Boeing Company. DIRECTV. Available at http://www.boeing.com/defense-space/space/bss/factsheets/601/dbs/dbs.html.
[11] The Boeing Company. Boeing 601 Fleet. Available at http://www.boeing.com/defense-space/space/bss/factsheets/601/601fleet.html.
[12] The Boeing Company. Boeing 702 Fleet. Available at http://www.boeing.com/defense-space/space/bss/factsheets/702/702fleet.html.
[13] Godwin, J. P. Direct Satellite Television Broadcasting. In J. G. Webster, ed., *Wiley Encyclopedia of Electrical and Electronics Engineering*. New York: Wiley, 1999, pp. 590–602.
[14] Recommendation ITU-R BO.1516. Digital Multiprogramme Television Systems for Use By Satellites Operating in the 11/12 GHz Frequency Range, 2001.
[15] Advanced Television Systems Committee. ATSC Standard: Direct-to-Home Satellite Broadcast Standard, Doc. A/81, 2003. Available at http://www.atsc.org.
[16] Advanced Television Systems Committee. ATSC Standard: Digital Audio Compression (AC-3, E-AC-3), Rev. B Doc. A/52B, 2003. Available at http://www.atsc.org.
[17] Registered trademark of Sony Corporation.
[18] Society of Motion Pictures and Television Engineers. SMPTE 279M-2001 Digital Video Recording: Half-Inch Type D-5 Standard-definition Component Video and Type HD-D5 High-definition Video Compressed Data.
[19] Code of Federal Regulations, Title 47—Telecommunication, Part 25, Satellite Communications.
[20] Telecordia. GR-338-CORE, "Television Special Access and Local Services—Transmission Parameter Limits and Interface Combinations."
[21] Advanced Television Systems Committee. ATSC Standard: Ghost Canceling Reference Signal for NTSC, Doc. A/49, 1993. Available at http://www.atsc.org.

SECTION

7

BROADCAST TOWERS

CHAPTER

7.1

Tower Safety and OSHA Requirements

PHIL RAYSON
Radian Communication Services Corp.
Oakville, Ontario, Canada

JOHN WAHBA
Turris Corp
Mississauga, Ontario, Canada

INTRODUCTION

The information in this chapter is only intended as a guide. Before using any fall protection or tower rigging equipment the proper training, knowledge, and experience must be obtained. Nothing contained in this chapter precludes the implementation and application of OSHA (Occupational Safety and Health Administration) guidelines [1]. This chapter discusses the fall protection and climbing safety aspects for accessing or construction of broadcast towers. The information below will help tower owners and contractors to be compliant with legislation and maintain a safe and healthy work site resulting in reduced incidents, lower exposure to liability, and reduced insurance costs.

A number of serious injuries, accidents, and fatal falls have occurred as a result of climbing towers that could have been reduced, if not eliminated, by:

- Identifying and controlling hazards.
- Using proper equipment and conducting site inspections.
- Providing formal training programs specific to the tower industry.

Training programs such as OSHA Outreach 10 and 30 Hour, Rigging, Fall Protection, Job Planning, RF Awareness, and Electrical Awareness are some of the safety programs that will help provide the knowledge necessary to achieve a good skill level and employee competency required in the industry and by OSHA.

FALL PROTECTION REQUIREMENTS

The following fall protection information provides the key elements and fundamentals to help a company ensure a safe working environment when working aloft:

- A company fall protection standard that includes rules, practices, and consequences of noncompliance with company standards.
- Identification of the types of fall protection safety equipment and components approved for use by the company.
- Specific legislated requirements pertaining to fall protection requirements of the company.

Fall protection systems are required any time a worker is 6 ft or 1.8 m above the ground. As a safe practice, a company should require 100% fall protection at all times when a worker is 6 ft or more above ground and should not allow free climbing at any time. If an employee is observed climbing without required fall protection he or she may be immediately dismissed from employment.

FALL PROTECTION SYSTEM BASICS

There are some key elements of all fall protection systems that must be understood, as shown in Figure 7.1-1.

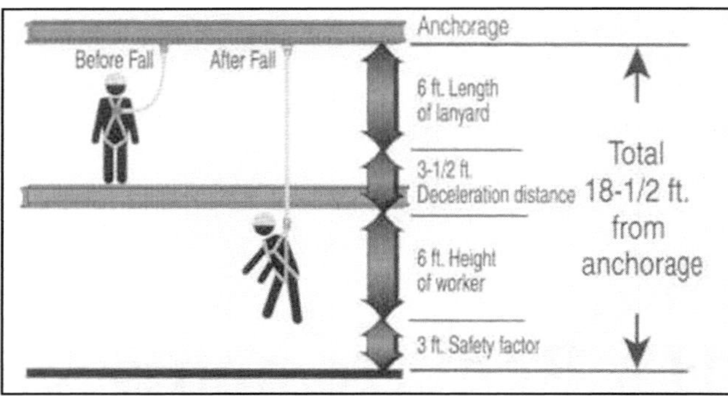

FIGURE 7.1-1 Example of how to calculate fall protection.

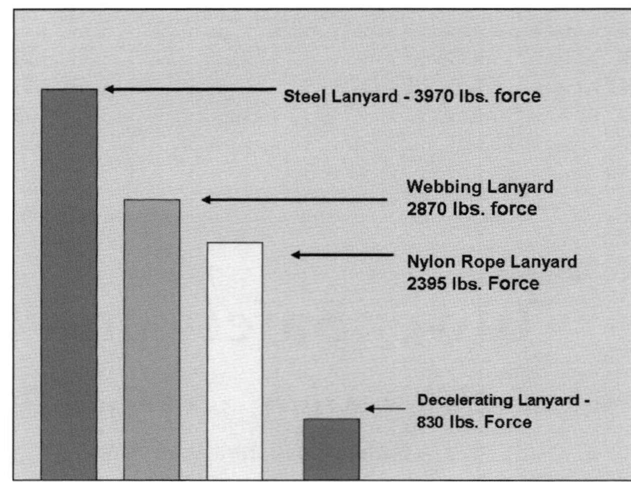

FIGURE 7.1-2 Relative strength of different materials used in fall protection systems.

Total Fall Distance

The total fall distance must be known to ensure no one will come in contact with a substructure, equipment, or machinery after a fall occurs. The total fall distance also is used to determine how much force can be generated in a fall. The total fall distance is calculated from the position of a person before a fall to the position of a person after a fall. The calculations include the length of lanyard, deceleration device length, anchor position, height of worker, and distance to nearest contact point after a fall.

The example in Figure 7.1-1 demonstrates that if a 6 ft decelerating lanyard is used, the total distance of travel may be 9.5 ft due to the 3.5 ft deceleration device after full deceleration or shock absorption takes place. To decrease the fall distance the system anchor should be as high as possible. Then the fall forces will not be high enough to generate large amounts of energy. A deceleration device takes 900 lbs or 4 kN of energy to begin a deployment. As the fall energy is absorbed, the fall slows down so that if the end of the deceleration device is reached there should be very little energy left to impact the body. However, if there is machinery or equipment at the end of the deceleration distance, then the worker could be injured from the machine or equipment. Note that the higher the anchor the better and safer the system. A 3 ft safety factor should be included in a total fall distance calculation.

Maximum Arresting Force or Peak Arresting Force

Figure 7.1-2 shows a graph of tests completed during fall arrest testing using a 6 ft fall distance and a 220 lb load. The testing was done using load cells that measured the maximum arresting force generated with the different connections. Note that with a steel lanyard, which has no shock absorption, the arresting force was similar to the weight of a car, almost 2 tons in just 6 ft of fall. This energy would be the total force acting on the body during a fall with a 6 ft steel lanyard. However the deceleration device in the same test only generated 830 lbs of arresting force. This testing was used to determine legislation and why deceleration devices or decelerators are required in all fall arrest systems.

A company should have approved fall protection systems in place that are documented, staff that are trained in their use, and their use should be ensured on the job. The following are four approved fall protection systems that a company may use. Each system has variations but the approved systems all ensure legislated standards are met or exceeded. Any additional changes or modifications to these systems must be approved by the company safety department before any application in the field. The four systems are:

- Fall Arrest System (F.A.S.)
- Center Climb (safety climb) System (C.C.S.)
- Work Positioning System (W.P.S.)
- Travel Restrict System (T.R.S.)

FALL PROTECTION SAFETY EQUIPMENT

Only safety equipment that has been approved by American National Standards Institute (ANSI) and/or Canadian Standards Association (CSA) shall be used in a fall protection system. Approved fall protection components listed in this section are identified by type, standard, and legislated requirements for each piece of the approved equipment.

The standards may refer to pounds of force or kN. The term kN stands for kilo-newtons and is the same as foot-pounds. A kN is equivalent to 225 lbs (e.g., 22.2 kN = 225 × 22.2 = 4,995 lbs; OSHA would round this number to 5,000 lbs).

The following are key fall protection equipment items that are used in proper fall protection systems and required by OSHA. For greater detail, 29 CFR 1926.502 Subpart M provides additional criteria for proper fall protection systems and equipment.

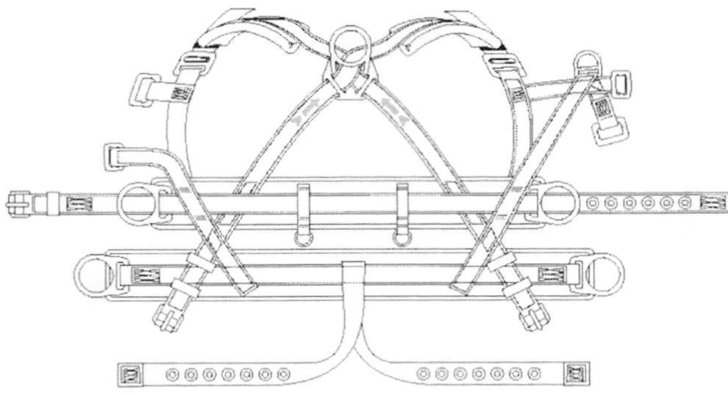

FIGURE 7.1-3 Example of approved full-body harness.

Body Harness

The purpose of a full-body harness, shown in Figure 7.1-3, is to distribute the arresting fall forces proportionately throughout the thighs, pelvis, waist, and upper body. This distribution of forces prevents one body area from taking the full impact of the arresting fall force.

All approved harnesses are designed to meet ANSI and CSA criteria for fall protection. OSHA, and State OSHA, cite-specific ANSI and CSA standards for manufacturing and testing of fall protection equipment. The maximum arresting force allowed by legislation on a harness system is 1,800 lbs or 8 kN, and with a total fall distance of 6 ft or 1.8 m.

Each harness has various benefits and should be selected for climbing preference and protection. All harnesses used in the tower industry should have a D-ring in the back and a chest D-ring in the front.

Body Belt

The body belt or lineman's belt, shown in Figure 7.1-4, has been used by the tower industry for years. As of January 1, 1998, body belts were *not* acceptable as fall arrest equipment. A body belt is used with a positioning strap and is approved for use in a work positioning system (W.P.S.). It does not have fall arrest capability on its own and only allows a worker to fall a maximum of 2 ft when the positioning strap is properly attached with each D-ring connected to the connecting strap. The fall forces generated at 2 ft are distributed across the waist and back. Lanyards or a positioning strap must never be attached to only one D-ring as this will cause all the fall forces to impact one point at the waist and could break the user's back or cause major internal damage.

If a fall arrest system is needed when climbing towers, rails, cables, or double hooking, a light-body harness is required to be used, along with the body belt. The light-body harness is used to properly connect a lanyard so a fall arrest can be made safely and distributed over the body with little potential for injury. At no time should a cable, rail, or rope grab be attached to the body belt D-rings.

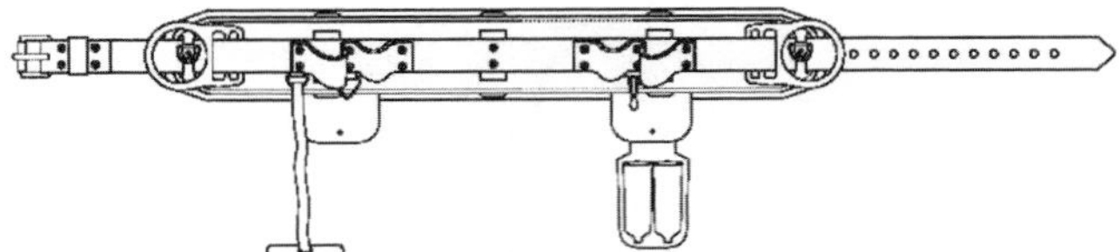

FIGURE 7.1-4 Example of a body belt that is not acceptable for fall arrest.

Approved belts have at least a 3-1/2 inch cushion for comfort in the back and are 5/32 inch thick. The maximum force allowed by legislation on a belt is 900 lbs or 4 kN with a total fall distance of 2 ft. This can be maintained by adjusting the strap or by wrapping it around an approved anchor support to the proper length.

Vertical Lifeline

The vertical lifeline is used as an anchorage for an approved rope grab attachment. The proper size and type of rope grab is free to travel on the lifeline so that should a fall occur, the vertical lifeline will anchor the fall. A low stretch ½ inch or 12 mm Kernmantle rope is recommended. This lifeline exceeds legislated standards for fall protection and exceeds the minimum breaking strength of 5000 lbs or 22.2 kN. Other ropes can be used as long as the rope meets the 5000 lbs or 22.2 kN breaking strength requirement.

Ropes should be stored in a closed container for protection from weather or adverse environments. Wet ropes should be dried before storing.

Rope Grab

An approved rope grab is a device that travels along a lifeline and will automatically engage within 6 inches of movement to arrest a fall, as shown in Figure 7.1-5. Only approved rope grabs designed for a specific type of lifeline should be used.

The new CSA standard requires a deceleration device to be used whenever the climber has the rope grab attached at the back D-ring of the harness. The deceleration device is integrally attached to the rope grab by the manufacturer. If an older version single rope grab is used, it is recommended that a rope grab with deceleration device is used to reduce the fall forces placed on a body during a fall.

Decelerating Lanyard

Approved decelerating lanyards have a rating of 5,000 lbs or 22.2 kN and have an integral deceleration device. They meet or exceed legislated standards. They come in single, double, or linked configurations, as shown in Figure 7.1-6. They should be attached to the D-ring on the back of the harness with the deceleration device component always attached to the person.

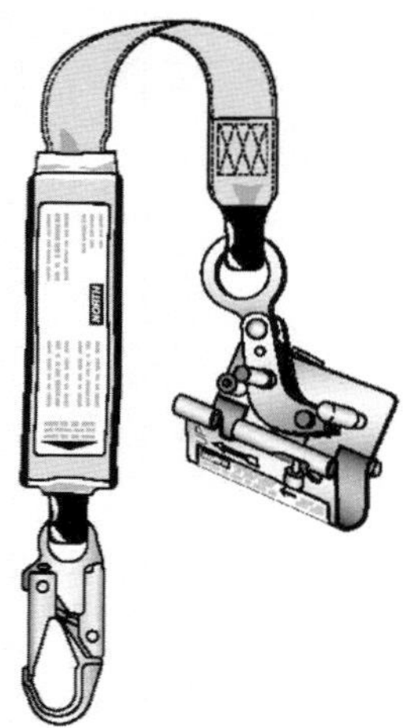

FIGURE 7.1-5 Example of approved rope grab device.

The pelican hooks are designed to go over a wide variety of steel for a secure anchorage. When the steel is too large, the anchor sling is to be used. An approved deceleration lanyard limits the fall forces acting on a person's body to 900 lbs or 4 kN when the deceleration device is engaged.

A deceleration lanyard of no more than 36 inches or 0.9 m shall be attached to the rope grab on a vertical rope. No lanyard longer than 6 ft or 1.8 m may be used in fall protection systems, which would allow a total fall distance of greater than 6 ft. A standard lanyard must never be attached back to itself. It must always be attached in a straight configuration so the fall forces act in line with the hook design and ensure the webbing is not cut by the snap hook in the event of a fall.

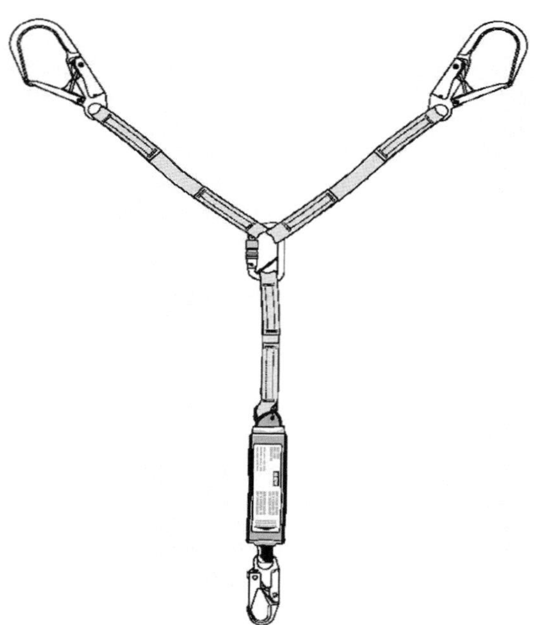

FIGURE 7.1-6 Example of decelerating lanyard.

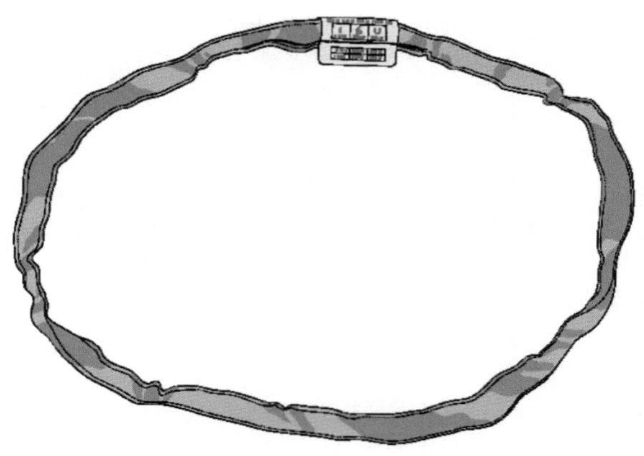

FIGURE 7.1-7 Example of anchorage sling.

Anchorage Sling

An approved anchor sling is an endless, polyester sling that exceeds the legislated 5000 lbs or 22.2 kN standards for anchorage points. Vertical lifelines, lanyards, retractables, and other fall protection devices may be attached to the sling to anchor the fall protection system. The sling should be used in a basket configuration, to ensure the best working load limit (WLL) application. Only one lifeline per anchored sling is permitted. There are a number of anchor slings and tools on the market that are acceptable for fall protection applications.

When hooking to the tower, be sure the steel or connection point is capable of withstanding a 5000 lb or 22.2 kN force. This sling is not to be used for any other purpose than fall protection. It must never be used for rigging as other slings are provided for that purpose. A well-designed anchorage point is located such that it covers all elements of reduced free-fall distance, deceleration distance is calculated, individual's body length, existing substructure below, and reduced pendulum or swing fall.

Karabiners

Approved karabiners, such as the one shown in Figure 7.1-8, are dropped forged, pressed, or formed steel, or made of equivalent materials.

Ensure that all karabiners used in fall protection systems have been rated for fall protection use. Karabiners have a corrosion-resistant finish and the surfaces are smooth to prevent damage to other parts. They have an autoclose or double-locking mechanism, which requires two distinct motions to close or open the connector (e.g., twist then push or pull to open). The connectors have 5,000 lb or 22.2 kN capacities stamped on them. The 22.2 kN requirement is across the length of the karabiner not the width. If there is no

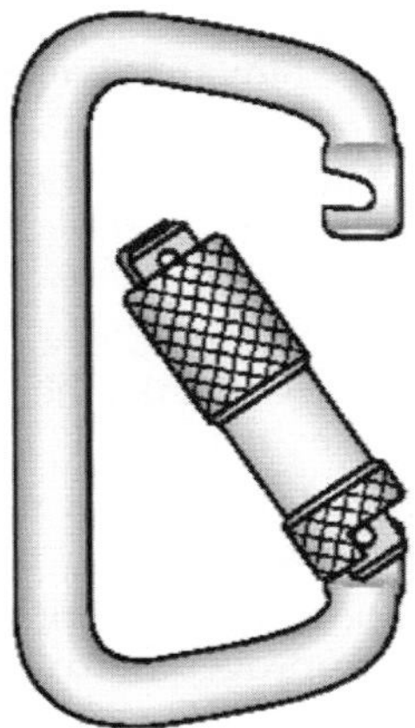

FIGURE 7.1-8 Example of a karabiner.

FIGURE 7.1-9 Example of cable trolley with attached karabiner or snap ring.

stamping, or it does not meet the legislated standard, it shall not be used in fall protection systems.

Snap Hooks

Approved snap hooks are dropped forged, pressed, or formed steel, or made of equivalent materials and have a minimum tensile strength of 5,000 lbs or 22.2 kN, which may be indicated with a 5M designation on the body of the snap hook. Double-locking-type snap hooks are designed to prevent disengagement of the snap hook by the contact of the snap hook keeper at the connected member. Single-locking snap hooks are not to be used in any fall protection systems and must be removed from service. A single-locking snap hook can "roll out" of connection and cause a loss of connection in the event of a fall. Double-locking snap hooks can prevent "roll out" as the keeper takes two motions to open the gate.

Cable Grabs

Cable grab trolleys, as shown in Figure 7.1-9, are used in two applications and are only to be used on specific cable types such as 3/8 inch cable. One is for aircraft or stainless steel cable and the other is galvanized cable. There are two types of cable trolleys for these systems and the correct grab must be used with the appropriate cable.

Approved cable trolleys meet or exceed legislated standards. They operate similar to a rope grab so that, when properly attached, they will activate within 6 inches to prevent a long fall distance. They are connected to a full-body harness in the chest D-ring. Ensure the appropriate distance of 9 inches is maintained between the chest D-ring and the cable by directly connecting the device to the chest D-ring using a small karabiner. The proper distance ensures a total fall distance of less than 18 inches is maintained and reduces the fall arrest loading on the person.

Never have a deceleration device in between the chest D-ring and the device. The cable grab and cable must be inspected daily before use. Ensure all moving parts operate freely. Light oil may be applied to the moving parts only. There should be no oil, galvanizing buildup, or dirt on the grabbing surface of the trolley. Field repairs or modifications are not permitted. Connection to a waist D-ring is not allowed in systems due to the inappropriate fall forces placed on the lower back in the event of a slip or fall.

> *Important note:* Fall protection equipment actually involved in a fall prevention action must be taken out of service and returned to appropriate safety personnel. It must never be used after a fall due to potential tearing and component breakdowns. Legislation and manufacturers also advise that any fall renders the equipment defective until inspected by the manufacturer.

FALL PROTECTION EQUIPMENT USE AND INSPECTION

All components of a fall protection system must be inspected each day before use. This inspection must be recorded. Following are the fall protection and all protective equipment requirements:

- Inspect the equipment daily.
- Replace defective equipment. If there is any doubt about the safety of the equipment, do not use it. Send it to the safety personnel for inspection or return to the manufacturer.
- Replace any equipment, including ropes, that is involved in a fall immediately and return it to safety personnel or the manufacturer for inspection.
- Inspect tool loops and belt sewing for broken or stretched loops.
- Check bag rings and knife snaps to determine that they are secure and working properly. Check tool

loop rivets. Check for thread separation or rotting, both inside and outside the body pad belt.

- Inspect snaps for hook and eye distortions, cracks, corrosion, or pitted surfaces. The keeper (latch) should be seated into the snap nose without binding and should not be distorted or obstructed. The keeper spring should exert sufficient force to close the keeper firmly.

Basic care to prolong the life of fall protection equipment and contribute to proper performance includes:

- Wipe off all surface dirt with a sponge dampened in plain water. Rinse the sponge and squeeze it dry. Dip the sponge in a mild solution of water and commercial soap or detergent. Work up a thick lather with a vigorous back-and-forth motion.
- Rinse the webbing in clean water.
- Wipe the belt dry with a clean cloth. Hang freely to dry.
- Dry the belt and other equipment away from direct heat and out of long periods of sunlight.
- Store in a clean, dry area, free of fumes, sunlight, or corrosive materials, and in such a way that it does not warp or distort the belt.

SAFE CLIMBING PRACTICES SUMMARY

Do not use any fall protection equipment without proper-use training, rescue training, and emergency planning.

- Always inspect the existing structure before and during a climb.
- All fall arrest anchorages must be able to withstand a maximum force of 5,000 lbs or 22.2 kN.
- Anchor slings must be used to wrap around an adequate anchorage point when connecting a lanyard or lifeline. Do not wrap the lifeline or lanyard around the anchorage point and hook it back into itself. This practice will reduce the strength of the hook and there is also the danger of a cutting action on the rope or lanyard.
- On the body harness, the front D-ring is designed for the use of a safety cable or rail system only. Note: Do not use the safety climb as a body-positioning device. Do not exceed 9 inches between the safety climb carrier and the front D-ring of the harness.
- The decelerating lanyard is to be attached to the back D-ring of the harness at all times. Never attach a lanyard to the front D-ring.
- When climbing to a work position with one vertical rope, only one person can be climbing at one time. If there are two people on one rope, then one person must be tied off to the structure while the other person is in the act of climbing. Maintain a distance of not less than 15 ft from the other climber.
- When in the work position it is recommended to have a W.P.S. and a lanyard attached as a second connection.
- If there is any situation where there is the possibility of falling more than 6 ft, then a fall protection system must be used.
- All lifelines should be stored in a dry container out of the weather, this will make them easier to work with and also make the rope last longer.
- Use a deceleration device or decelerating lanyard for all fall arrest systems.
- Connect all components of fall protection equipment using only approved hardware.
- Attach fall protection equipment to a suitable anchor.
- Keep potential fall distances to a minimum.
- Select the most appropriate fall systems and equipment for the job.
- Always show proper care for fall arrest equipment and inspect it daily before use as though a life depended on it, because it does.
- Always ensure a rescue plan is in place in case a fall should occur.

The above fall protection information meets or exceeds OSHA's Fall Protection requirements as prescribed in 29 CFR 1926 Subpart M—Fall Protection, and 29 CFR 1910.268—Telecommunications.

Refer to ANSI A10.14-1991 standards, CSA Z259 standards, and other state and local regulations to ensure that the fall arrest system to be employed complies with applicable requirements.

RIGGING

Rigging is often overlooked as a key component in maintaining a safe work environment on a tower site. Rigging failures can lead to a catastrophic tower failure resulting in fatalities, costly equipment/material damage, and potential OSHA citations. It should not be assumed that rigging equipment is in good condition and has adequate capacity to perform the lift. Conduct an inspection of equipment prior to each use. Guidelines to ensure safety and adequate capacity include:

- Training employees in rigging basics should include, but not be limited to, rigging selection and capacity; rigging inspection (synthetic slings, wire rope slings, synthetic rope, and rigging hardware); sling capacity; sling angle factors; block angle factors; and knots.
- Critical lifts should be defined, engineered, and a documented procedure provided to employees before work commences and for use as a prejob plan.
- Load weights identified for the heaviest items to be lifted.

- Cable, slings, and rigging hardware capacity identified and adequate for load weights to be lifted.
- Maintain a 5:1 factor of safety for all rigging components and 10:1 factor of safety for personnel lifts.
- Cables, slings, and rigging hardware inspected, and defective equipment removed from service and destroyed or tagged out.
- Identify sling angle as this directly affects the capacity of the hoisting sling.
- Use tag lines for load control when required.
- Lifting points on the load and tower itself must be determined prior to hoisting; verification can be provided by a qualified engineer.
- Hoisting equipment must be set up properly, and the load must be within the manufacturer's recommended lifting capacity (hoists, ginpoles, and cranes).
- Lifting equipment must be inspected according to manufacturer specifications.
- Communication between qualified equipment operators and rigging personnel must be established prior to hoisting operations.
- A documented lift plan identifying all of the above should be in place prior to any hoisting operation. The lift plan must be communicated to all personnel involved in the rigging and hoisting operations.
- 29 CFR 1926.251 Subpart H—Rigging Equipment for Material Handling is an excellent source of information for developing a rigging program.
- 29 CFR 1926.550 Subpart N—Cranes and Derricks provides information that pertains to rigging and safe hoisting operations.

ROLE OF OSHA

The mission of OSHA is to assure the safety and health of America's workers by setting and enforcing standards; providing training, outreach, and education; establishing partnerships; and encouraging continual improvement in workplace safety and health.

OSHA provides the minimum standards for general industry (29 CFR 1910) and construction (29 CFR 1926) to help reduce workplace accidents and injuries. OSHA has both federal and state programs. There are 26 state programs in which many have adopted the federal standards.

Worker Injuries, Illnesses, and Fatalities Statistics for 2003

In 2003, occupational injury and illness rates declined again to 5.0 cases per 100 workers, with 4.4 million injuries and illnesses among private sector firms. About 32% of work-related injuries occurred in goods-producing industries and 68% in services.

There were 5,559 worker deaths in 2003, a slight increase from 2002, accounted for by 114 additional deaths among self-employed workers and 61 more through workplace violence. The fatality rate of 4.0 deaths per 100,000 workers remained the same. Fatalities related to highway incidents, falls, and electrocutions declined while homicides and deaths related to fires and explosions and contact with objects or equipment increased.

- Federal inspection statistics for the fiscal year 2004: 39,167 inspections resulting in 86,708 violations and $85,192,940 in penalties.
- State inspection statistics for the fiscal year 2004: 57,866 inspections resulting in 133,873 violations and $74,058,959 in penalties.

OSHA Multiemployer Citation Policy

Under Multiemployer Worksite Legislation, in addition to citing the contractor violating the OSHA regulations and creating the hazard, the owner and other contractors on the worksite can also be issued the same citation or more citations than the creating employer.

Multiemployer worksites hold the owner equally responsible for legislated compliance and safety, as well as the contractor violating the laws. If the owner or general contractors have a contractor approval process in place to manage subcontractors, by verifying that the subcontractor works in compliance with OSHA legislation, the owner will probably not be cited. An owner is not required to be an expert, but must exercise due diligence when hiring and managing subcontractors.

The lowest-bidding contractor may not always be the right choice when it comes to working in compliance with OSHA. It is important that the OSHA CPL 2-0.124 Multiemployer Citation Policy be reviewed and understood by all owners and subcontractors to reduce liability. OSHA defines four types of employers:

- Creating employer
- Exposing employer
- Correcting employer
- Controlling employer

The *creating employer* is the employer that caused the hazardous condition that violates an OSHA standard. An example would be an employer who removes a guard from a piece of equipment and doesn't replace it. Or, a contractor that digs a foundation hole and doesn't cover or place a barrier around it.

The *exposing employer* is an employer whose own employees are exposed to the hazard. An example would be subcontractor A performing brickwork over a building entrance without protective barriers in place. Subcontractor B instructs his employees to move equipment into the building under the bricklayers. Subcontractor B employees were exposed to possible injuries due to material that could fall.

The *correcting employer* is responsible for correcting the hazard. The correcting employer is the employer who has the ability to correct the hazard.

The *controlling employer* has supervisory authority over the work site, and can correct hazards or instruct others to do so.

Example of Enforcement by OSHA

On August 25, 2005, OSHA cited a Colorado cellular contractor for unsafe working conditions, following a fatal accident at a communications tower. Proposed penalties totaled $115,500.

One worker fell more than 180 ft and was killed on February 25, 2005, during the installation of cellular phone antennas on a tower located west of Yuma, CO. "This accident could have been avoided by following recognized safe work practices for climbing communication towers," said Herb Gibson, the OSHA area director in Denver (OSHA Regional News Release, August 25, 2005).

Citations issued by OSHA's Denver area office alleged one serious and two willful violations of OSHA standards. The willful violations, with proposed penalties of $112,000, involved improper use of a hoist for lifting personnel up the tower, lack of fall protection, and failing to remove a defective snap hook from service. Additional penalties of $3,500 were proposed for lack of fall protection training.

Willful violations are those committed with intentional disregard of, or plain indifference to, the requirements of the Occupational Safety and Health Act. A serious citation is issued when there is substantial probability that death or serious physical harm could result from a hazard about which the employer knew or should have known.

The employer has 15 working days from receipt of the citations to request an informal conference with the OSHA area director, or to contest the citations and proposed penalties before the independent Occupational Safety and Health Review Commission.

Employers are responsible for providing a safe and healthful workplace for their employees. OSHA's role is to assure the safety and health of America's workers by setting and enforcing standards; providing training, outreach, and education; establishing partnerships; and encouraging continual improvement in workplace safety and health. For more information, visit news releases at http://www.osha.gov/html/a-z-index.html#E.

Examples of Fines for Fall Protection Violations

On June 6, 2005, a construction company in Augusta, ME, was fined $20,930 for fall protection violations (Region 1 News Release: 05-970-BOS/BOS 2005-141, Report I.D. #0111100). On April 8, 2005, three contractors faced a total of $67,600 in fines for a variety of fall hazards found at a construction site of five townhouses in Hampton, NH (Region 1 News Release: 05-537-BOS/BOS 2005-074).

On February 2, 2005, a construction company was fined $5,000 for fall protection violations for residential construction projects in Brooklyn, NY (Region 2 News Release: 05-163-NEW/BOS 2005-018, Report I.D. #0216000).

ROLE OF THE FCC

The Federal Communications Commission (FCC) is charged with regulating interstate and international communications by radio, television, wire, satellite, and cable. The FCC issued radio frequency (RF) radiation exposure regulations on August 1, 1996, that took effect October 15, 1997. These regulations point out several problems that must be solved by each organization/tower owner in the wireless industry. The regulations include resolving complaints, investigate, and take or recommend enforcement action for violations. See Chapter 2.4 for additional information on human exposure to radio frequency fields [2].

The FCC's Enforcement Bureau (EB) is responsible for resolution of complaints involving public safety and technical issues such as tower registration, marking and lighting, and equipment requirements. EB is also responsible for taking enforcement actions regarding such violations as unauthorized construction or operation of radio stations [3].

Telecommunication employees who may be exposed to RF energy during the course of their work duties must be made aware of and recognize possible RF exposure to nonionizing radiation. This understanding can be achieved through enacting an RF safety program (consisting of administrative, engineering, and work practice controls); proper training in the use of RF monitors; OSHA required safety training; and a basic knowledge of antennas and their operating characteristics.

These activities should incorporate RF energy in regard to sources, effects on the human body, how to recognize potential overexposure, and ways to minimize exposure. The RF safety program must also address "controlled" and "uncontrolled" areas.

For more information on RF energy in the workplace, see Chapter 2.4, Human Exposure to Radio Frequency Energy, in this book.

TOWER-MOUNTED ELEVATORS

Tower-mounted elevators are installed on many of the taller towers in the industry, which range from a few hundred feet to two thousand feet. Tower-mounted elevators eliminate hours of climbing during routine maintenance, troubleshooting, and equipment repair operations. A tower-mounted elevator is designed to transport personnel up and down the tower and must not be used to hoist material unless specified by the elevator manufacturer. Before operation, hazards such as wind, lightning, ice, and RF should be identified and addressed for the safety of personnel required to operate the elevator. Always follow the elevator manufacturer operating instructions and manual.

Safe Operation

To ensure safe operation the following checks can be used as a guide:

- Ensure that the state elevator inspection certificate is current.
- Ensure that all regular maintenance has been completed according to manufacturer requirements.
- Check the capacity rating plate; ensure that the capacity is clearly identified.

Visual Inspection

Visually inspect the following components before and during operation:

- Condition of the car, gate, and controls.
- Condition of the elevator cables.
- Condition of the elevator guide rails.
- Ensure the counterweights are secured properly and the counterweight rails are clear of obstructions.
- Check the condition of the traction sheaves for dirt, debris, and paint.
- Check limit switches, interlocks, and safety devices; ensure that each has not been overridden or its operation altered by devices, tape, etc.
- Check the condition of the elevator rails during ascent.

Operating Instruction

It is extremely important to always follow the manufacturer's operating instructions.

- Review manufacturer operating instructions provided by the station owner or a qualified person.
- Identify power source for emergency shutdown.
- Identify personnel and total weight to be elevated. Ensure weight is within the elevator's capacity. (Limited by weight and number of persons.)
- Ensure that all passengers are within the car and the gate is closed before ascending or descending.
- Ensure good communication is maintained at all times during operation.

Operational Checks

Prior to use, ensure that all operational checks have been completed in accordance to the manufacturer.

- Check all limit switches, interlocks, and safety devices for proper operation.
- Check brakes to ensure the car is stopping and holding according to manufacturer's instructions.
- Check sheaves, cable, and counterweight for proper operation.
- Check the emergency operation feature to ensure proper operation.

Note: If a defect is found at any time that affects the safe operation of the elevator, do not operate the elevator. Lock and tag the elevator out-of-service and contact the person in charge of having the elevator inspected and repaired.

SUMMARY

An owner or an employer can ensure a safe work environment and reduce the potential for an accident if he or she provides or hires contractors with proper fall protection systems, equipment, and training. The risk of incidents can be further reduced by ensuring good rigging practices and that proper RF management is used by staff and contractors. These efforts will result in reduced liability and exposure to OSHA and FCC citations.

References

[1] See http://www.osha.gov/SLTC/fallprotection/index.html.
[2] See http://www.fcc.gov/Bureaus/Engineering_Technology/Documents/bulletins/oet65/oet65a.pdf.
[3] See http://www.fcc.gov/eb/.

CHAPTER

7.2

Tower Design, Erection, and Maintenance

THOMAS J. HOENNINGER
Stainless LLC
North Wales, Pennsylvania

INTRODUCTION

The purpose of this chapter is to provide broadcast engineers information concerning the design, erection, and maintenance of antenna structures. While fundamental principles of the design and behavior of these structures will be discussed, this chapter is not intended to enable readers to design and build their own tower, but instead provides a basic understanding of these unique structures to facilitate planning, modifying, and maintaining broadcast facilities.

TOWER CHARACTERISTICS

All towers may be classified in one of the two basic groups, guyed or self-supporting. As their names imply, guyed towers depend on cables extending from the tower to anchors located some distances from the tower base for their structural integrity, while self-supporting towers rely solely on their own construction as a cantilevered space truss.

With only a few exceptions, the cost of the actual tower structure and foundations is considerably less for a guyed tower than for one that is self-supporting. The advantage of the self-supporting tower is the relatively small land area required. Therefore, the choice between guyed or self-supporting depends to a large degree on the availability and cost of real estate.

A self-supporting tower requires a nearly square plot of land with equal sides that are 8% to 20% of the tower's height, provided local zoning rules do not require tower height radius of land.

The amount of land required for a guyed tower depends on the distance between the tower base and the guy anchors. This distance is preferably between 70% and 80% of the height, which would require a rectangular plot having sides equal to 125% and 145% of the height. Because of the great flexibility in guyed tower design, it is possible to reduce the anchor distance to as little as 35% of the height, thereby requiring a much smaller land area. However, the cost of the tower increases as the anchor distance decreases. The approximate relationship of cost to anchor distance for a representative 1,200 ft television broadcast tower is shown in Figure 7.2-1. It is often possible to position a guyed tower on an irregularly shaped plot or to obtain long-term lease agreements or easements for guy paths and anchor locations in order to minimize the tower cost without obtaining large, rectangular land areas.

Self-Supporting Towers

Self-supporting towers may be either square or triangular in cross section. While it is usually more economical to use a triangular cross section, there are situations where a square cross section is a better choice. The principle structural elements are the legs, the web bracing in each face, and if required for stability, horizontal diaphragm bracing. The legs are usually sloped (tapered) to provide adequate strength and stability as the height increases. The degree of slope is an option of the designer to suit the equipment supported, the required rigidity, and the available land area. The slope is sometimes varied within a tower to maintain a desirable balance between the costs of leg

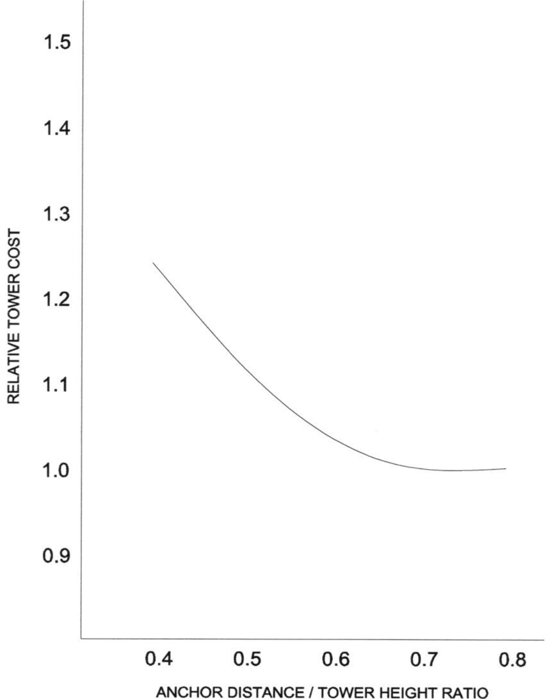

FIGURE 7.2-1 Effects of anchor distance on the cost of a 1,200 ft guyed TV broadcast tower using ANSI/TIA Standard 222-G-2005.

members and bracing, or to reduce the foundation loads. Frequently, the legs in the top section of the tower will be parallel to simplify the mounting of equipment (see Figure 7.2-2).

There are several different configurations of bracing members for the individual truss panels. The choice is influenced by the width of the panel, magnitude of the wind and ice loads imposed, location of equipment, and required stability. Continuity in transferring the applied loads through the structure without significant eccentricity is essential regardless of the configuration used.

Guyed Towers

Guyed towers are almost always of triangular cross section although there are a few unique conditions for microwave and panel type FM and TV antenna supports where a square cross section is advantageous. The principle structural elements are the legs, the web bracing in each face, and the guy support systems (see Figure 7.2-3). Except for sections at the tower base and locations where the width changes, the legs are parallel. The width of the tower is usually constant throughout the height of the tower with the exception of sections supporting antennas requiring a specific width of support structure. The base section is often tapered to a single point to provide a pivot support to eliminate large bending and torsional moments.

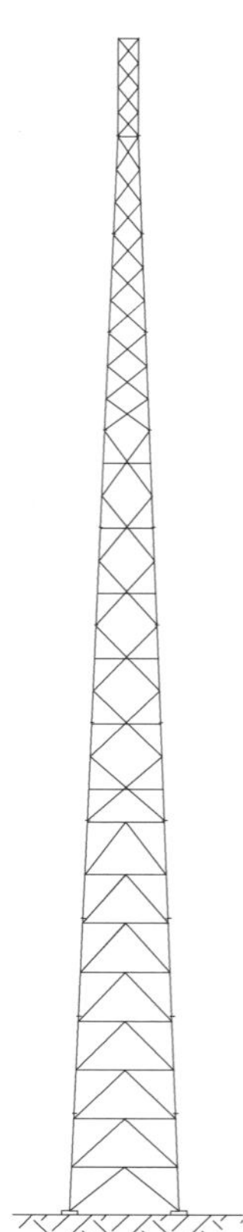

FIGURE 7.2-2 Elevation view of a typical self-supporting tower.

Theoretically, there are an infinite number of arrangements of guy cables to support a tower. The most common arrangement is three cables spaced at 120 degrees at each guy level with one cable attached to each leg as shown in Figure 7.2-4(a). This is the min-

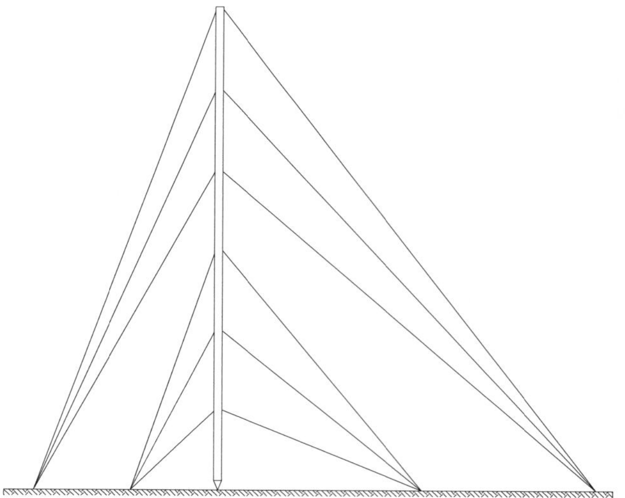

FIGURE 7.2-3 Elevation view of a typical guyed tower.

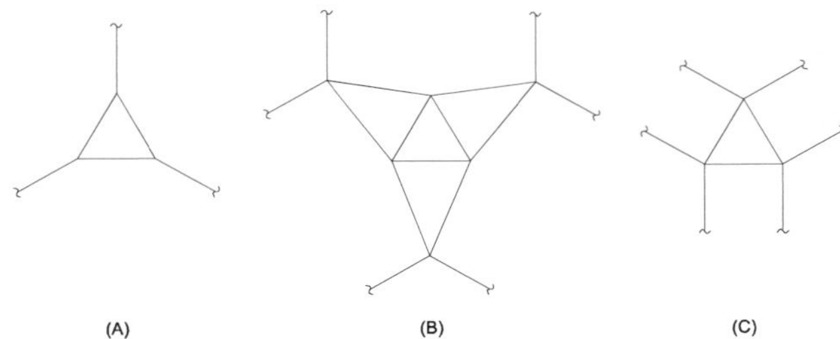

FIGURE 7.2-4 Typical guy arrangements.

imum number of cables that can be used. When the tower supports equipment, which imposes large twisting moments (torque), it is necessary to provide six cables at a guy level to maintain torsional stability. If the torque is localized, the guys at this location may be attached to triangular frames as shown in Figure 7.2-4(b). If the torque occurs throughout the height, it may be necessary to double-guy the tower at every level as shown in Figure 7.2-4(c).

The number of guy cable levels required to support the tower is dependent on a number of factors including the height of the tower, width, location of equipment, and environmental loading conditions. Because the tower is an axially compressed column, its strength is a function of its slenderness. While design codes permit slenderness ratios resulting in triangular towers having a span-to-width ratio as great as 49, it is usually economical to limit the ratio to a maximum of 30. While there is no upper limit to the number of guy levels imposed by any code, a practical limit for economical design is 10.

The position of equipment on the tower is an important factor in determining the location of guy levels. Preferably, guy attachments should not be located within the apertures of side-mounted TV and FM broadcast antennas. Equipment producing large localized wind loads, such as microwave antennas or clusters of two-way radio cabinets and antennas, should not be positioned near the center of a span between guys.

If the tower will be subjected to ice loading, it is desirable to reduce the number of guy levels to minimize loads imposed on the tower by ice accumulation on the guy cables.

The number of anchors in each guy direction is dependent on several factors including the number of guy levels, soil conditions, topography, and obstacles. As a general guideline, it is desirable to limit the number of guy levels attached to a single anchor to five. However, there is nothing absolute about this number, and other conditions may dictate using an anchor for a greater number. There are some soil conditions where it may be economical to provide two or more smaller anchors, while in another instance the use of one large anchor might be desirable. If minimizing the area within which the tower would fall in the event of collapse is a consideration, a minimum of two anchors should be used in each direction. Where the elevations of the anchors differ from the tower base, it is desirable to vary the distance of the anchors from the tower base to maintain nearly equal initial tensions in the guy cables. Anchors higher than the tower base should be moved toward the tower, and anchors that are lower, away from the tower. The designer should specify the amount of movement.

Materials

Nearly all broadcast towers are made from steel because it provides the most economical structure. The selection of the grade and shape of steel is obviously an important design consideration.

Steels used for towers commonly have low carbon content with yield strengths in the range of 36,000 psi to 60,000 psi. These materials have good ductility and are suitable for welding. Some towers have been built using higher grade materials with yield strengths up to 100,000 psi, but the savings in weight are more than offset by higher material prices and increased fabrication costs. Regardless of the grade of material, the mechanical and chemical properties of the steel should be certified by the producing mill to ensure that the properties conform to the design requirements.

The shape of the material as well as its size and strength affect the tower's load-carrying capacity. The shape also has a significant effect on the magnitude of loads produced by wind. Design standards permit a reduced wind load on round members as little as 57% of the wind load for flat or angular members of the same width. For this reason solid round bars and round hollow structural sections are often used. This advantage in wind load is offset somewhat by increased fabrication costs, due to the necessity of welding plates to connect the various members. There is no one grade or shape of materials that is best. The choice depends to a large degree on the preference of the designer and the type of fabricating facilities available.

A factor equally as important as the selection of the grade and shape of the structural steel is the design of the connections. For shop-welded connections, the compatibility of the base and filler metals and required preheat temperatures must be considered. The procedures used must be qualified and the welders certified to those procedures. Inspection procedures should be compatible with the weld design. The American Institute of Steel Construction (AISC) and the American Welding Society (AWS) have quality fabricator certification programs that fabricators can follow. Having this type of certification confirms that quality fabricated steel is being produced.

Bolts for field assembly may be of various types. Usually those for the main load-carrying members are high strength. If positive resistance to slippage of the connections is required, they should be designed as slip-critical connections.

Guys

The most common material for tower guys is galvanized strand. This material has excellent strength and durability. Its "structural" elongation due to seating of the individual wires in the strand is small and can be almost entirely eliminated by prestressing the strand to 45% of its breaking strength at the factory. This should be performed for guys on tall towers with factory-connected end fittings.

For guys on AM towers, and those close to FM and TV antenna apertures, a nonconductive material is sometimes desirable. Two such materials that have been used are Kevlar rope and fiberglass rods. When using these materials, careful attention must be given to protection against corona effects, fatigue, and deterioration from exposure to ultraviolet light. Also, their elongation characteristics under load must be evaluated. They require delicate handling at all times.

Just as for the tower structure, the connections for the guys are as important as the guy material itself. Some of the most common connections are as follows and are shown in Figure 7.2-5:

- *Sockets* of forged or cast steel attached with molten zinc or epoxy resins develop the full strength of the guy. They are normally installed at the factory and proof loaded to 55% of the guy breaking strength. This type of fitting is most common for the larger guys used on tall towers.
- *Dead-end grips* are preformed spiral wire loops in the shape of large hairpins. The two legs of the hairpin are wrapped around the guy with its closed end forming an eye. These grips are used for guys up to

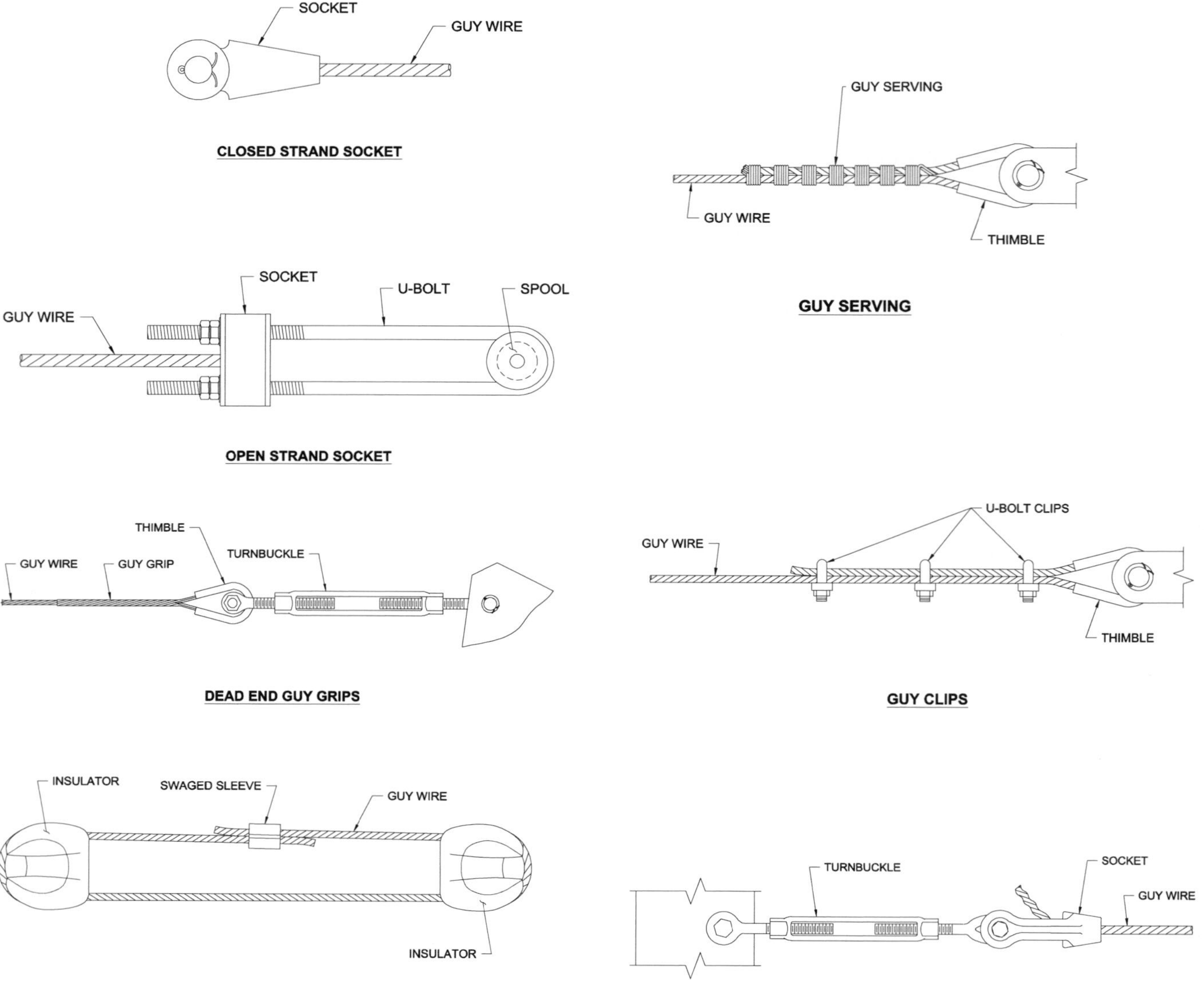

FIGURE 7.2-5 Examples of guy connections.

1 inch in diameter and usually develop their full strength. They are easily installed in the field, but the ends must be completely snapped into place and a protective device installed to prevent ice from sliding down the guy and loosening the grip.

- *Clips* used to clamp the ends of guys (when properly applied and tightened) develop 90% of the guy's strength for sizes up to and including 7/8 inch and 80% for larger sizes. To install them it is necessary to bend the strand back on itself to form a loop; thus, the use of clips on large cables is difficult. The saddle of U-bolt type clips must be installed on the load side and not the dead-end side, which provides potential error in their installation.
- *Swaged sleeves* develop between 85% to 100% of a guy's strength depending on the size of the guy and equipment used to squeeze the sleeve. These fittings are usually installed at the factory and can be proof loaded. They are advantageous for connecting closely spaced insulators where the length of dead-end grips is unacceptable.
- *Wedge-type sockets* are available for guys up to 1-7/16 inch diameter and develop 100% of their strength. They are most advantageous for guys larger than those for which dead-end grips are available.
- A *serving* is a connection made by rolling the individual wires of a strand back on the strand itself. This method has for the most part been replaced with dead-end grips, but it is advantageous for small guys with closely spaced insulators.

Insulators

Insulators in radio frequency applications must withstand mechanical and electrical stress in a varied, changing exterior environment. Selection of insulators should be made with these factors in mind. Insulators primarily designed for 60 Hz applications are unsuitable, particularly at high RF powers.

The most common insulating material is wet-process porcelain, which has excellent compressive strength and good insulating capabilities for frequencies up to 2 MHz. Synthetic materials are also used.

- *Base insulators* for AM towers are made from porcelain with appropriate steel end plates or ferrous/nonferrous castings. For guyed towers, a rocking arrangement is provided in the form of a convex plate and pin at the top, or a pivot pin at the bottom of the assembly to hold the tower in place and relieve the porcelain from bending loads that could cause cracking. For self-supporting towers, the insulators are bolted between the tower leg and base pier, and are designed to sustain both uplift and download while keeping the porcelain in compression.
- *Sectionalizing insulators* are sometimes required to isolate a section of a guyed tower. Where only a compression load is applied, a guyed tower base insulator with minor modifications can be used. If a tension load is anticipated, a push-pull insulator similar to the type used for self-supporting towers is required. Under no circumstances should the porcelain be put in tension.
- *Guy insulators* are available for primary insulation and for break-up purposes. Primary insulation (insulators next to the tower) should be selected to withstand the full voltage appearing at the guy attachment point. This is to ensure that sufficient insulation remains if all the break-up insulators in the guy line flash over. Break-up insulators, used to reduce reradiation, are selected to withstand the transmitter-induced voltage and static voltage. Break-up insulators are usually low-voltage types, sometimes protected from flashover and subsequent power arc by a static dissipation device. Guy insulators are available in many styles classified as either compression or tension types. Compression insulators are designed such that the porcelain element is in compression. Simple low-voltage types are a single piece of porcelain placed between interlocking loops of the guy. Such insulators are available for mechanical working loads up to 40,000 lbs. For higher loads, oil-filled and open types are used. The most common uses of compression insulators in broadcasting are for break-ups and as primary insulators (in groups of three or four) on low-power antennas. Tension insulators come in many forms, including porcelain rods (not permitted in structural applications), fiberglass rods, synthetic ropes, and oil-filled safety core types. Tension insulators are used as primary insulators with corona rings to reduce the electrical field stress at the end fittings. One insulator is required at each guy attachment point. Since the voltage level is different at each point on the tower, different voltage ratings may be required for some insulators at certain guy attachment points. Tension insulators are available in a wide range of electrical and mechanical ratings to meet most needs.

Finishes

Steel is susceptible to deterioration from atmospheric corrosion. To prevent deterioration, the tower members and hardware must be given a protective coating. This coating is usually zinc, which has excellent resistance to corrosion, and, because it is higher in the electrochemical series of the periodic table of elements, it provides cathodic protection to exposed steel surfaces adjacent to it. Even though the zinc coating may be scraped or otherwise damaged, it continues to inhibit corrosion of these exposed areas, and rust will not develop beneath adjacent zinc coats.

There are several methods for applying the zinc including hot-dip galvanizing, flame spraying, electroplating, and painting. All must be applied to clean surfaces.

- *Hot-dip galvanizing* consists of dipping the steel into a bath of molten zinc. A metallurgical bond develops between the steel and the zinc, which adheres to it. When galvanizing tubular members, it is necessary to provide holes in both ends to ensure that the inside surfaces are coated. Careful attention must be given to the type of base and weld metals used, as well as to the welding and forming procedures used in fabrication, to safeguard against possible embrittlement of the steel when galvanized. When properly applied, this process provides the most durable coating.

- *Flame spraying* consists of spraying molten zinc at high pressure onto the steel surfaces. The bond in this process is mechanical rather than metallurgical. The coating produced is more porous and has less resistance to abrasion than the hot-dip galvanized coating. It cannot be used for the inside of hollow sections or other cavities where access is difficult.
- *Electroplating*, while suitable for small objects, does not produce a coating thick enough to withstand a hostile environment. This method is not recommended for tower parts or hardware.
- *Zinc-rich paint* consists of extremely finely divided zinc in an inorganic or organic vehicle. It is not a metal coating method, but rather a painting procedure. Its resistance to abrasion and durability are less than hot-dip galvanizing. This procedure is, however, useful for maintenance.

Ice Prevention

Coatings are available to reduce the adherence of water to surfaces and subsequently the formation of ice on them. However, no reliable means exists to completely remove the risk of severe ice accretion.

Access Facilities

A tower must have some access facilities in order to maintain it and the equipment it supports. For small towers, the bracing members of the tower itself often serve as steps, or step bolts are attached to one leg or face. For taller broadcast towers, a fixed ladder inside the tower is desirable. The Occupational Safety and Health Administration (OSHA) standards for these ladders require a minimum clear width between side rails of 16 inches and a maximum rung spacing of 12 inches. OSHA also requires that any continuous ladder more than 20 ft in height be equipped with a safety device. This device consists of a continuous rail, either rigid or cable, running up the center of the ladder. A clamping device attached to the climber's safety belt rides along this rail. As long as the climber is in a normal position, the clamp slides freely; if the climber begins to fall, a cam-actuated mechanism freezes the clamp to the rail and prevents the person from falling.

The ANSI/TIA 222-G-2005 (222-G) standard specifies towers to be built as either a class A or class B climbing and working facility. A class A facility can be accessed by both basic and skilled climbers where class B can only be accessed by skilled climbers. Class A facilities will be more expensive due to designing and installing additional safety items in the tower to allow for basic climbers.

Elevators

For tall towers supporting multiple antennas, it is often desirable to install an elevator. Most tower elevators are of the electric power, cable-driven type with a capacity of 500–750 lbs and a speed between 80–100 ft per minute. They consist of a drive mechanism, car, guide rails, hoist cable with supporting sheaves, tension weights, electronic controls, and two-way communications system.

Considerable attention must be given to elevator safety features, including, but not limited to:

- Limit switches to prevent travel beyond the upper and lower landings on the tower.
- An automatic brake on the driving mechanism that is activated by an interruption in power.
- A mechanism to automatically clamp the car to the rails in the event of a broken hoist cable.
- Interlocks to prevent operation with the car gate open.

It is important to determine the applicable state or municipal government regulations that may apply and whether permits, tests, and inspections are required before the tower and elevator system are designed.

The added wind and dead loads from an elevator system are substantial and must be considered in the tower design. Also, careful attention must be given to the positioning of the ladder, RF transmission lines, and electrical conduits in relationship to the elevator. The ladder must be positioned so it is accessible from the elevator car and can be used for an emergency descent. While the elevator hoist cables can be restrained in guides on the return side, they are free to move about under wind load on the lifting side. Therefore, the conduits and transmission lines must be protected from hoist cables striking and damaging them. If a side-mounted TV or FM antenna produces a high RF field within the hoistway, protection must be provided to prevent arcing between the hoist cables, tower structure, and other appurtenances.

Transmission Line Bridges

To allow for thermal expansion and contraction of transmission lines, it is necessary to locate broadcast towers some distance from the transmitter building. Unless the transmission line is placed underground, it is necessary to provide a structural support for it at a height compatible with the transmitter location in the building. The top of the support can be covered with steel grating or plate to protect the line from falling ice. The details of this structure can become quite involved for sites with multiple antennas, uneven terrain and roadways, or obstacles between the tower and building.

Stairways

The lower landing for a tall, guyed broadcast tower with an elevator is often 30 ft or more above ground level. A stairway may be desirable to permit easier access to the landing. This structure can be combined with the transmission line support bridge, or it may be completely separate. It may also be desirable to install a small capacity boom above the lower landing to lift radio cabinets or other equipment onto the landing.

Marking and Lighting

Marking and lighting antenna structures to meet FAA/FCC requirements are addressed in Chapter 7.4, "Tower Lighting and Monitoring."

Other Electrical Systems

During the planning and design stages for an antenna structure there are other electrical circuits, in addition to lights, that are necessary or should be considered to make its operation more efficient and provide a source for future income. Most tall antenna structures require deicing circuits with their associated control devices. An AC utility circuit can provide access to 120 volt power at selected elevations on the structure and will reduce costly maintenance time. If the structure's height justifies an elevator, control circuits for the elevator are necessary. Circuits to provide power and multipair cables to various platforms are needed for rental communication system customers.

DESIGN STANDARDS

The vast majority of towers in the United States has been designed in accordance with the Electronic Industries Association EIA-222, Structural Standard for Antenna Supporting Structures and Antennas.

Industry Standard 222

This standard has been used since 1959 when it replaced the Radio-Electronic-Television Manufacturers Association (RETMA) Standard TR-116. The following is a chronology of the standards:

RETMA TR-116 (1949)	ANSI/EIA 222-D (1987)
EIA RS-222 (1959)	ANSI/EIA/TIA 222-E (1991)
EIA RS-222-A (1966)	ANSI/TIA/EIA 222-F (1996)
EIA RS-222-B (1972)	ANSI/TIA 222-G (2005)
EIA RS-222-C (1976)	

The current revision "G" of TIA-222 became effective January 1, 2006. It is an approved American National Standard and carries the designation ANSI/TIA-222-G-2005 (termed "222-G" in this chapter). This standard is intended to provide minimum criteria for specifying and designing antenna supporting structures and antennas. Unlike general specifications and building codes, it is applicable only to antenna supporting structures and antennas. As such it contains criteria specific to these structures that are not readily available elsewhere. Therefore, it is important to specify the tower must conform to this standard.

Annex A: Procurement and User Guidelines of this standard is intended to assist in the procurement of towers and for informational purposes. Designing new towers and analyzing existing towers in accordance with 222-G requires much more information and decisions from the customer than in previous 222 standards. The major items that the customer must specify are:

- *Structure classification I, II, or III.* The structure classification is based on reliability criteria. Basically, Class I represents a low hazard to human life, Class II represents a substantial hazard to human life, and Class III represents a high hazard to human life. The default classification is Class II.
- *Basic wind speed and design ice thickness.* The minimum basic wind speed without ice, the minimum basic wind speed with ice, and the minimum design ice thickness are provide in Annex B of 222-G. Annex B should be used unless the local climatic conditions dictate otherwise.
- *Exposure categories B, C, and D.* The exposure category is based on the characteristics of the ground surface irregularities at the site. Category B is used for urban and suburban areas. Category C is used for open terrain with scattered obstructions having heights less than 30 ft and hurricane coastlines. Category D is used for flat terrain with wind flowing over open water but does not include hurricane coastlines. The default exposure category is category C. 222-F used only one exposure category, which was category C.
- *Topographic categories 1, 2, 3, 4, and 5.* Wind speed-up effects at isolated hills, ridges, and unobstructed escarpments increase wind load on towers. The topographic categories determine the increase in wind load that needs to be included in the tower design. The default category is category 1, which means wind speed-up effects do not occur.
- *Earthquake loading site classes A, B, C, D, E, and F.* There are many factors that need to be considered when evaluating earthquake loading. Many of these factors are dependent on the earthquake loading site class. The site classes are based on the type of soil at the tower site. The default is site class D, which is a stiff soil type.
- *Climbing and working facilities classes A and B.* Class A climbing facilities have been designed and installed to allow for both authorized (basic) and competent (skilled) climbers. This means additional safety precautions and clearances have been installed to allow for easier access. Class B climbing facilities are for skilled climbers only. The default is class B.

Refer to Annex A of 222-G for a detailed explanation of each item.

Statutory

Most municipal and state governments have statutory codes regulating the design of structures. Since it is necessary to comply with the applicable statutory requirements, it is important to determine what these requirements are and include them in the purchase specifications for the tower. Many of these are patterned after or include the international building code (IBC). The IBC covers all types of structures and is directed primarily toward conventional types of buildings. However, Section 3108 of IBC 2006 states that towers shall be designed in accordance with ANSI/TIA/EIA 222-F. This IBC reference standard for towers will more than likely be changed to 222-G in the IBC 2009.

LOADS, ANALYSIS, AND RESISTANCE FACTORS

In addition to a tower's own dead weight and the dead weight of the appurtenances and equipment it supports, the tower must withstand the forces of nature, wind, ice, temperature changes, and earthquakes.

Wind Load

Wind produces the principal load on tower structures. For design purposes it is represented as a horizontal static force.

Wind load is specified in terms of a three second gust basic wind speed at 10 meters (33 ft) above ground level. 222-G provides a tabulation of recommended minimum values for this speed for each county in the United States. This standard also provides specific procedures and factors for calculating wind loads considering:

- Wind pressure is proportional to the square of wind speed.
- Wind speed and, consequently, wind pressure vary with respect to the height above ground.
- Ice thickness varies with respect to the height above ground.
- The effects of gusts of brief duration.
- Topographic wind speed-up effects.
- The effects of the configuration, size, proportions, shape, and orientation with respect to the wind direction of the structural components of the tower and its appurtenances.

Since the wind may act from any direction it is necessary to apply the calculated wind loads in any horizontal direction to determine the maximum stresses produced in the structure. For a triangular tower, a minimum of three directions must be considered; while for a square tower, a minimum of two directions must be considered. These are shown in Figure 7.2-6.

In addition to this direct load in the direction of the wind (drag), there may also be a component of load

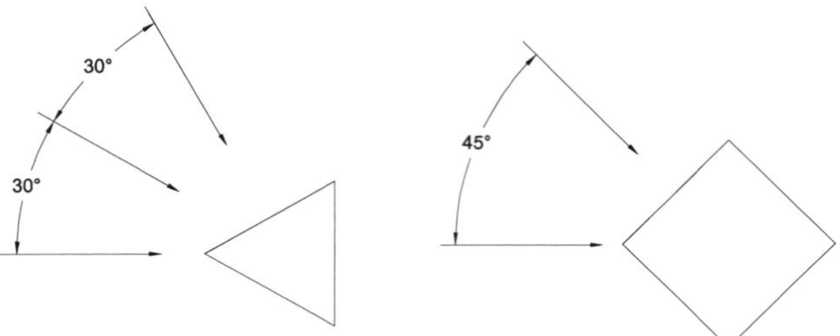

FIGURE 7.2-6 Wind directions to be considered.

perpendicular to the wind direction (lift). These lift components are calculated in a manner similar to that for drag forces using different shape coefficients that vary with respect to the angle of attack between the member's geometric axis and the wind direction. They are most significant for wind acting on guy cables, microwave antennas, and rectangular waveguides.

Ice Loads

Ice accumulations have two effects on a tower. The weight of the ice acts directly on the structure in the same manner as the dead weight. The ice accumulation also increases the area exposed to the wind and consequently the load produced by the wind. This increase is substantial on small components such as guy cables, tension rods, ladders, small diameter transmission lines, and reflector screens for antennas. It is also possible for the ice accumulation to alter the aerodynamic shape of members, thereby requiring the use of a different coefficient in calculating the wind load. An example is a set of closely spaced parallel coaxial lines. Without ice each would be considered a round cylindrical member. With accumulated ice, they would present a large flat area to the wind requiring a different coefficient.

Ice produces an entirely different stress distribution in a tower than wind, so it is not reasonable to merely increase the design wind load to provide for ice accumulation. It is also a misconception that ice will break up and blow off the tower and, therefore, ice and wind need not be considered simultaneously. 222-G provides minimum basic wind speeds in combination with ice thickness in the county listing in Annex B of the standard. This is the first 222 standard that includes a mandatory ice loading. As with wind speed the ice thickness increases with respect to height above ground.

Temperature Changes

Changes in temperature have no significant load-producing effects on self-supporting towers, but they can on a guyed tower. Because of their differences in length, the guy cables expand and contract different amounts than the tower itself and thereby require elastic deformations from stress changes. The effects are greatest for those cables having the flattest angle with

the ground. While the stresses produced are considerably less than those produced by wind and ice loads, they should be considered in the design of a guyed tower.

Topographic Effects

It has been determined that wind speed-up occurs at isolated hills, ridges, and escarpments. This wind speed-up effect increases the wind load on towers and must be considered in tower design.

Earthquake Loads

Loads due to earthquakes are considered to act horizontally and are dependent on the mass and stiffness of the tower. They are usually less than those produced by wind but are distributed in a different manner. Procedures for calculating these loads are provided in 222-G. While a tower properly designed for wind loads is usually adequate for earthquake loads, it cannot be neglected in areas with frequent and intense earthquake occurrences.

STRUCTURAL MODELS AND ANALYSIS

A self-supporting tower may be described structurally as a cantilevered space frame or truss. Although it may have many different members, it is a relatively simple structure, and the determination of the forces in the individual members due to the applied static loads is easily done using fundamental principles of structural mechanics. The potential modes of failure are buckling of individual leg or bracing members under compressive loads, and shear or tension failures of the connections.

A guyed tower is a much more complex structure than a self-supporting tower. Whereas there is only one basic path through a self-supporting tower for the loads to be transferred to the ground, there are several for a guyed tower. The distribution of the loads among these paths is dependent on the relative stiffness of guy systems and the tower shaft.

Each span of the tower has stiffness with respect to relative deflections from axial and shear forces and bending and torsional moments. This stiffness is a function of several variables, including the geometric configuration, the mechanical properties, and the sizes of the individual members.

Each guy cable also has stiffness with respect to movement of its attachment point to the tower that is a function of the amount of initial tension, the magnitude of ice load, and the magnitude and direction of wind load on the cables. By evaluating all of these, it is possible to simulate all the guys at a given level as a spring having a specific stiffness. Because of the nonlinearity of some of the relationships involved, the spring constant derived is only valid for a specific set of conditions and for a finite range of translation. Similarly, a torsional spring constant can be derived. It is interdependent with the translation stiffness and is also valid for only a finite range of translation.

Another difference between a guyed and self-supporting tower is the magnitude and significance of the axial load. For a self-supporting tower this is composed only of the gravity loads from the tower, its appurtenances, and any ice load. It is independent of wind load, and its effects on individual member loads are relatively small. The axial load for a guyed tower includes, in addition to the gravity loads, the vertical components of the tensions in the various guys. Since these tensions are directly affected by the wind load, the axial load is now dependent on wind load, and its effects on the individual leg members are relatively large. Tension in the guy wires also produces an additional bending moment on the tower equal to the product of the axial load and the deflection of the tower.

Despite the complexity of the relationships involved, the availability and widespread use of computer systems permit accurate structural analysis of guyed towers. There are several different structural models that may be used:

- Models the tower shaft as a continuous beam-column on nonlinear elastic supports or cable elements (the guys) subjected to simultaneous transverse (wind and/or seismic) and axial (dead, ice, and vertical components of guy tensions) loads.
- Models the tower shaft as elastic three-dimensional truss members, producing only axial forces in the members, on nonlinear elastic supports or cable elements (the guys) subjected to simultaneous transverse (wind and/or seismic) and axial (dead, ice, and vertical components of guy tensions) loads.
- Models the tower shaft leg members as continuous elastic three-dimensional frame members and the tower shaft bracing members as three-dimensional truss members on nonlinear elastic supports or cable elements (the guys) subjected to simultaneous transverse (wind and/or seismic) and axial (dead, ice, and vertical components of guy tensions) loads.

All of the above models result in similar results if the input data is accurate.

Dynamic Considerations

As previously mentioned, even though wind and earthquakes involve kinetic energy, their effects are simulated by equivalent static loads determined in accordance with the design standards. In recent years there have been more sophisticated efforts to investigate the actual response of tower structures to the dynamic aspects of wind gusts. A conclusion drawn from these studies is that the bending moments in the upper portions of tall, guyed towers are considerably higher than those determined by the usual static analysis. Consequently, 222-G has addressed this issue by including pattern loading in the static analysis of towers, which emulates the dynamic load and the dynamic response of the tower and provides for a more reliable design.

There have been dynamic issues with numerous top-mounted, cylindrical-shaped antennas. Excessive antenna movement due to vortex shedding can occur. Increasing the stiffness of the antenna and/or disrupting the wind flow around the antenna have been successful methods in minimizing this dynamic occurrence. There are two other phenomena related to the dynamics of wind that are important in guyed tower design—aeolian vibrations and galloping, both of which involve periodic loading.

Aeolian vibrations are low amplitude, high-frequency movements, which occur in the tower guy cables due to a phenomenon known as vortex shedding. If they are not suppressed through the use of dampers, they can result in destruction of the filaments in the tower lights at the least, or fatigue failure of guy cable and collapse of the tower at the worst. Dampers attached at one or both ends of the guy cables have proven effective in controlling these vibrations and should be considered for all tall, guyed towers.

Galloping is a condition of instability involving large amplitude, low-frequency movements. It is caused by the perpetual amplifications of periodic loads due to the motion of the body itself. The most dramatic and well-known example of galloping is the collapse of the Tacoma Narrows suspension bridge in 1940.

For tower structures, galloping is usually associated with guy cables on tall towers, but in at least one instance it was related to a large rectangular waveguide. There have been several different methods involving detuning and energy dissipation used for preventing galloping in guy cables that appear to be successful. In the case of the rectangular waveguide, galloping was controlled by moving the waveguide inside the tower, along the centroidal axis, from its original position on the outside of one face. This reduced the torsional rotation of the structure, which was the source of the perpetuating force. Based on this experience, it would appear prudent to always install this type of waveguide inside the tower unless adequate torsional rigidity is provided throughout the height of the tower.

Load and Resistance Factor Design

One of the major changes included in 222-G was to convert from an allowable stress design (ASD) to a load and resistance factor design (LRFD). Basically an LRFD design means the strength provided in the design must be at least equal to the factored load acting on it. The load factors were derived based on the probability of an overload. An ASD design means the strength provided in the design must be at least greater by a certain margin to the load acting on it. This "certain margin" is the safety factor. In ASD the load is a service load and therefore is not factored. The load factors in LRFD were determined by probabilistic methods and the safety factors in ASD were determined by experience and judgment. An LRFD design should result in a more reliable design.

222-G refers to the American Institute of Steel Construction (AISC) Load and Resistance Factor Design Specifications for Buildings for the design of the structure's members and to the American Concrete Institute (ACI) Building Code Requirement for Reinforced Concrete Structures for the design of the reinforced concrete foundations and guy anchors.

EFFECTS OF ANTENNAS AND TRANSMISSION LINES

Except for AM radiators, the tower supports broadcast antennas and transmission lines at a suitable height above ground. Thus, the effects of this equipment are of paramount importance.

Loads

Every antenna imposes a wind load and a dead load on the tower. If the antenna is mounted atop the tower, it also imposes an overturning moment. If it is mounted on a side of the tower, the antenna imposes a torsional moment. For TV and FM broadcast and microwave antennas these loads are relatively large and their location has a significant effect on the placement of guy cables.

Transmission lines feeding the various antennas also impose wind and dead loads on the tower. These loads are distributed uniformly between the antenna and their entry point near the base of the tower. The total produced by a coaxial line or waveguide is frequently greater than that produced by the antenna itself. The shape of the transmission line influences the magnitude of the wind load, with circular or elliptical lines having loads that are 60% of those for rectangular lines with the same projected area.

It is important not to overlook the support system required for transmission lines. Some large waveguides have support systems that require nearly continuous vertical structural members that add substantial wind and dead loads. Small, flexible lines require supports at a maximum interval of 3–4 ft, which is often less than the vertical spacing of horizontal members in the tower. Thus, it may be necessary to provide an additional support structure for these lines, again adding to the total load.

An important consideration when locating transmission lines is that 222-G permits a reduction in wind load based on the size, shape, and location of the transmission line in the tower.

Width Restrictions

Some antennas impose restrictions on the width of the supporting tower. One common example is a side-mounted FM antenna requiring a maximum width of 18–24 inches. For antennas with more than eight bays, this results in a very slender structure. When placed at the top of a tall, guyed tower, the design of the guy system for this structure becomes extremely critical.

Use of a cantilevered pole structure above the top of the main tower should be considered for these cases.

Another example of width restriction is a panel-type TV or FM antenna mounted on the faces of the tower. Here, too, it is often better to support these antennas on a cantilevered structure above the main tower rather than placing guys within the aperture of the antenna.

As previously mentioned, it is desirable to place large transmission lines inside the tower near the vertical centroidal axis to prevent large torsional loads. This requires a tower having a large face width to accommodate the transmission line and its supports and climbing facility.

Initial and Future Loading Considerations

Because the antennas and transmission lines have such a significant effect on the tower design, it is important to consider all possible uses for a tower before it is designed. It is better to have unused capacity than to undergo expensive modifications or replacement in several years to obtain additional height or accommodate another antenna. This has become apparent in recent years with the proliferation of microwave, two-way communications, cellular telephone, and personal communication systems.

In a new tower design the initially installed equipment loading should be analyzed in addition to designing for full loading. In guyed towers a reduction in load in one area can cause tower member overstresses in another.

When providing for multiple antennas, it is important to determine not only the number and type of antennas and lines, but also their location on the tower. The distribution of load is equally important as magnitude.

Triangular-top platforms ("candelabras") to support broadcast antennas on each corner have been successfully used for many years. They have the advantage of placing all antennas at the same height. A variation of this platform to support only two antennas ("tee-bars") has also been used. Both of these systems require multiple guy cables at the top platform to provide adequate torsional stability. It is possible to design the tower for a multiple antenna support platform without installing all antennas at the same time.

Another arrangement of multiple antennas is stacking (i.e., installing one antenna atop the tower and arranging others along the tower) one below the other. This arrangement can also be combined with a multiple antenna support platform.

If capacity for microwave antennas is required, it should be provided near guy levels and preferably above to minimize interference with the guy cables. The guy system and web bracing at these levels must be designed to provide adequate torsional rigidity.

Capacity for small antennas may be provided at various locations throughout the height of the tower. One arrangement for a large number of antennas is to provide a platform around the outside of the tower that is large enough to support the radio equipment for these antennas. The antennas can be mounted on the outside railing of the platform, thereby requiring only a short run of coax. Electrical power must be provided to the platform. This arrangement imposes a large concentrated load at the platform location with a relatively small uniform load between the tower base and the platform. If the same number of antennas were mounted along the tower and each fed by an individual coax line from the base, there would be only small concentrated loads at the antenna locations, but a relatively large uniform load due to the lines. This is an entirely different distribution of load, and would have a pronounced effect on the design.

Another important consideration for future antennas or height extension is the electrical system. If an extension in height is planned, the wiring for the aircraft warning light system should be designed so that any additional lights can be connected to the system without adding or replacing wires in the existing conduit. The same holds true for any circuits required for future antennas. If the necessary wiring cannot be provided during the initial installations, capacity should be provided for additional conduits to hold the future circuits.

Serviceability Requirements

Antenna performance from an operational standpoint must be considered. Some antennas require limits on structure movement at operational wind speeds to obtain acceptable performance. The 222-G operational wind speed is 60 mph three second gust basic wind speed. In lieu of limits provided by antenna manufacturers, 222-G specifies deformation limits on towers with regards to rotation and displacements. 222-G also has displacement limits on cantilevered tubular or latticed spines poles, antennas, or similar structures.

Replacement, Relocation, or Additions to Existing Towers

Since every tower has been designed for a specified arrangement of equipment, changes should not be made without considering their effects on the structural adequacy of the tower.

Two common misconceptions related to changes in equipment are "lower is better," and "smaller is better." Neither is necessarily correct, especially for guyed towers. Decisions based on these premises can have serious consequences.

It is much better to have a structural analysis of the tower made by a structural engineer experienced in tower design. Because of the significant changes that have been made in the methods of specifying loads in the various revisions of the design standard, the analysis should be made using the same criteria for wind and ice loads used for the original design and also for the current revision of the standard. This analysis will determine if any overstresses would occur in the tower or its foundations, and what modifications and reinforcing would be required to retain the structural integrity. To make this analysis, it is necessary for the engineer to have complete data on the tower and its foundation including configuration, member sizes, and material strengths. It is also necessary for the customer to provide the engineer with a complete inventory of existing antennas and transmission lines on the tower along with the proposed equipment. A plan view drawing of the tower, showing where the transmissions lines, conduits, ladder, etc. are located, is also required. The use of presumptive values can result in an analysis with little value.

The 222-G standard has broken down structural analyses into two types. They are:

- *Feasibility structural analysis.* A feasibility structural analysis is used as a preliminary review to identify the impact of proposed changed conditions. The type of analysis determines overall stability and adequacy of the main tower members. It does not include an evaluation of connections and foundations. Acceptance of a changed condition cannot be based on a feasibility structural analysis. It must be based on a rigorous structural analysis.
- *Rigorous structural analysis.* A rigorous structural analysis is used to determine the final acceptance of proposed changed conditions and/or required modifications. This type of analysis determines the overall stability and adequacy of structural members, foundations, and connection details.

The structural analysis report must state the type of analysis (feasibility or rigorous). A feasibility report must state that final acceptance of changed conditions must be based on a rigorous analysis.

FOUNDATIONS AND ANCHORS

It is most difficult to predict the cost of the foundation system of a tower installation. This is due to the nonhomogeneous nature of soils and the uncertainty of the conditions that may exist below grade. Therefore, it is necessary to have an investigation made of the subsurface soil conditions.

It is important to note that even though for class I and II structures as defined in 222-G, site-specific geotechnical design parameters are not required, however, it is recommended a site-specific geotechnical investigation be performed. The soil design parameters given in the 222-G should be viewed as a basis for preliminary design and estimating of foundation cost prior to obtaining specific soil data. They should not be used for the final design without verification by geotechnical investigation.

Soil Investigation

The soil investigation should be made by an engineering firm that specializes in soil investigation and evaluation and is familiar with the general area of the tower site. It should consist of making a test boring at each foundation and guy anchor location, analysis of soil samples taken from the borings, determination of groundwater levels, recommendations of parameters for designing the foundations, identification of any special construction procedures required, and recommended backfill specifications. Other soil characteristics or properties may be required because of local conditions, such as the soil resistivity, to determine if any special corrosion control methods should be implemented. If piles or rock anchors are necessary, recommendations related to these should be provided. It should also address requirements for frost protection and buoyancy effects.

Because the loads imposed on tower foundations are unique from those for conventional buildings (tower foundations have large uplift and horizontal components), it is important to provide the soil engineer with the loading conditions before he or she makes the investigation. This will enable the work to be planned in a manner suitable for obtaining and reporting the characteristics relevant to designing for the projected foundation loads.

Self-Supporting Tower Foundations

Except for relatively small towers with narrow base spreads, isolated foundations at each leg are usually more economical than a single mat for all legs. These foundations may be spread footings, drilled caissons, or driven piles. If sound rock is present at shallow depths it is often economical to anchor the footing to the rock. These anchors should be proof-loaded to ensure their holding capacity in uplift.

Since these foundations are subjected to large uplift forces, it is important to consider buoyancy effects if groundwater is present. Also, if driven or cast-in-place piles are used, they must be adequately anchored to the reinforced concrete cap.

Guyed Tower Base Foundations

Foundations may be spread footings, drilled caissons, or driven piles. Since they are subject only to downloads with relatively small horizontal forces, they require no special anchorage details for uplift, unless they are placed above expansive soils. Buoyancy is usually not a problem.

Guy Anchors

Buried reinforced concrete blocks (deadmen), drilled caissons, or driven piles may be used for these foundations. If solid rock is present at shallow depths it is often economical to anchor the foundation to the rock.

These foundations are subject to large horizontal forces as well as vertical uplift. Therefore, deadmen must have a large enough frontal area bearing against the soil to resist sliding; drilled caissons must have sufficient diameter and depth to prevent excessive lateral deflections as well as pull out from uplift; and driven piles must be sloped to prevent large lateral loads being imposed on them. Rock anchors may be installed along the slope of the resultant horizontal and vertical loads, or they may be installed vertically and posttensioned to clamp the concrete cap to the rock to prevent sliding. Because of the uplift forces, it is important to consider buoyancy due to groundwater and to provide adequate anchorage for driven or cast-in-place piles.

Construction

Because nearly the entire foundation system will be below finished grade and not subject to later inspection, it is important to carefully monitor its construction. The following items should be verified:

- Location and alignment of anchors in plan and elevation.
- Condition of excavation surfaces on which concrete will be placed.
- Position, size, and grade of reinforcement steel.
- Placement of concrete to prevent voids and air pockets.
- Strength of concrete using test cylinders for 7- and 28-day break tests.
- Protection of concrete against freezing during the curing period.
- Placement and compaction of backfill.
- Driving records and/or load tests of piles.
- Proof loading and posttensioning of rock anchors.

For towers with extensive foundation systems, it is important to retain an independent inspection service for this work. Often the firm making the subsurface soil investigation can also provide this service.

ERECTION

The erection of towers is a highly specialized field and should be performed only by firms having the proper equipment and experienced rigging personnel. It is important that the firm have adequate insurance coverage including workers' compensation, general and automobile liability, and builder's all-risk for direct damage to the tower and antennas being erected. It is also important that erectors have an adequate safety/training plan that includes such items as fall protection training, personal protection equipment training, equipment that meets OSHA requirements, a drug program, and man-rated hoists (if personnel will be riding the line). The erection firm should prepare and submit site-specific technical procedures and a rigging plan that define all the critical steps that are involved prior to starting the project. The firm should also ensure the technical procedures and rigging plan are being followed as the work is being performed.

Owner's Preparation

Prior to the arrival on site of the erection crew, the site should be made ready for work to begin. These preparations include:

- *Access.* Suitable access from public roads for delivery of the tower materials and erection equipment is required. While a paved roadway is not necessary, the access must be able to handle heavy trucks and construction equipment.
- *Permits.* All necessary building and construction permits should be obtained and posted as required. Any inspections required during construction should be noted.
- *Clearing.* A work area must be cleared to permit unloading, sorting, and assembling the tower. Paths from the tower base to the guy anchors must be cleared for a width adequate to permit hauling the guy cables to the anchors and pulling them to the tower. Paths must also be cleared for the hoist line from the tower base to the hoist location, and for the tag line used to stabilize the loads as they are lifted. The sizes and locations of these cleared areas should be agreed upon beforehand with the erector. A typical layout is shown in Figure 7.2-7.
- *Electrical power.* Power for operating temporary aircraft warning lights must be available before erection begins.

Assembly

The usual procedure for erecting a guyed tower is to assemble the individual sections on the ground and then lift them one at a time as an assembled unit. For a self-supporting tower, the wider sections near the bottom of the tower are often assembled in the air as the tower is constructed.

Assembly of the tower sections should be done on a level bed to ensure that they will be straight and not racked or twisted. Bolts must be properly tightened and have a locking device.

Stacking

For a guyed tower, the first group of three to six sections is often joined together on the ground and then

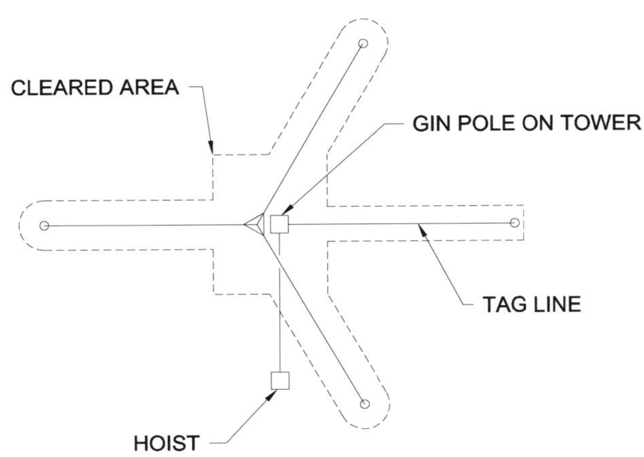

FIGURE 7.2-7 Typical layout for guyed tower.

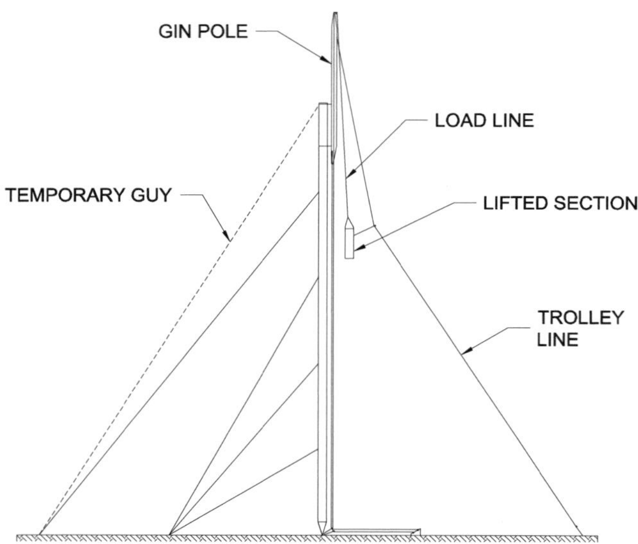

FIGURE 7.2-8 Typical erection setup for guyed tower.

lifted into place using a crane. This portion of the tower is then guyed with temporary cables, and the remaining sections are erected one at a time using a vertical boom or "gin pole." This boom is moved or "jumped" up the tower as each section is installed. This arrangement is shown in Figure 7.2-8. Temporary guys to stabilize the tower should be used when instructed by the designer.

For a self-supporting tower, a crane is often used to lift as many of the tower sections as possible, after which a gin pole is installed and used for the upper sections beyond the crane's reach. Temporary aircraft warning lights must be installed at the top of the construction at the end of each day. The tower should be grounded as soon as the first section is in place.

Guy Installation

When the tower reaches a guy attachment level, the cables at that level are installed. The guys in all three directions should be pulled out simultaneously to prevent any large unbalanced loads on the tower.

The tower should be checked for plumb as each set of guys is installed and tensioned. Maintaining a plumb tower during erection eliminates the need for time-consuming adjustments later. Final tensioning of the guy cables and a plumb check are performed after the entire tower is erected.

Rigging and Temporary Supports

It is the responsibility of the erector to ensure the rigging and temporary supports have been properly designed and installed. It is also the erector's responsibility to ensure the rigging loads imposed on the tower do not adversely affect the structure. This may require the erector to retain the services of a rigging engineer to review the erector's rigging plan.

REINFORCEMENT AND MODIFICATIONS OF EXISTING TOWERS

When equipment is replaced, relocated, or added to a tower it is often necessary to reinforce or replace existing structural components. The details and specifications for this work must be developed by the structural engineer who analyzed the tower.

Leg members may be strengthened by installing additional bracing or by field-welding additional material to them. Specific procedures in accordance with The American Welding Society's Structural Welding Code must be provided and followed if field welding is required.

Bracing members and guy cables may be replaced with stronger components. Careful attention must be given to the connections for the new components to ensure their compatibility with the existing tower, as well as providing the required strength for the new components. When replacing components it is essential that temporary bracing or guy cables be installed before removing any existing component, and that they remain in place until the new component has been installed.

It is necessary that damage to the protective finish on existing members due to field welding or reaming be repaired. If required, the affected areas must be painted for aviation obstruction marking.

Foundations and guy anchors are the most difficult components to strengthen, and they may prove to be the limiting factor in determining a tower's capacity. The nature and feasibility of strengthening these components depend on the specific soil conditions.

As with new tower erection, only qualified erectors should install reinforcement and modification material. It is again important that the erection firm has adequate insurance and a safety/training plan in place. Also that technical procedures and a rigging plan are prepared, submitted, and followed.

INSPECTION AND MAINTENANCE PROCEDURES

To ensure trouble-free performance of a tower and its appurtenances, it is desirable to have a regular inspection and maintenance program. Portions of the program can be performed by station personnel while others require experienced tower personnel.

Safety precautions should be observed at all times when working on or around the tower. If the tower itself is energized or if a high-intensity RF field exists from antennas mounted on the tower, no work should be done on the tower without consulting with the station engineer and RF power has been reduced to the appropriate levels or the RF source turned off and locked out/tagged out until such time that workers are clear. When climbing the tower, safety belts and climbing devices should always be used. Automatic safety features on elevators should never be bypassed to save time. It is a good idea to never work alone. Failure to observe proper safety measures can result in serious injury or death.

Tower Structure

A visual inspection should be made of the entire tower structure to determine if any of the members have been deformed or damaged. Any bowed or kinked member should be noted as to type, location in tower, and nature and magnitude of deformation or damage. This information should be reported to the tower designer for evaluation and recommended action.

Condition of Paint

A visual inspection should be made of the entire tower structure to determine the condition of the paint. If the painting of the tower is for aircraft observation marking only, and not for corrosion protection, it is necessary only to note any general deterioration rather than small blemishes and scratches. If repainting is necessary, it is important to properly prepare all surfaces and select paints that are compatible with the existing finish.

Corrosion

Small scratches in the galvanized surface are not detrimental, as the exposed surfaces will be protected by cathodic action of the adjacent zinc. If corrosion is observed, the source should be determined and noted. The affected areas should be wire-brushed clean to bare metal then painted with a zinc-rich prime coat and, if necessary, a finish enamel coat of the appropriate color.

Connections

All bolts should be checked for tightness. Any loose bolts should be tightened in accordance with the original installation instructions.

Alignment

The tower structure should be checked for alignment using an engineer's transit. This check should be done only on a calm day (i.e., with wind velocity less than 10 mph) and in conjunction with measuring the guy tensions (described later).

Both plumb and twist of a tower can be calculated from the measured horizontal deviations of each tower leg member from true vertical. Thus, three transit setups (one on each leg azimuth) are required for a triangular tower, and four for a square tower. When the transit has been properly leveled, set the vertical cross hair on the centerline of the vertical leg at the tower base and lock the instrument in this position. By moving the telescope upward, it is then possible to observe the straightness of the leg over its entire height. The magnitude of misalignment can be accurately estimated by comparison with the tower leg diameter. A record should be made of the observations of each leg at each guy level.

Tolerances for plumb and straightness should be provided by the designer. 222-G provides a plumb tolerance that limits the horizontal distance between the vertical centerlines at any two elevations to 0.25% of the vertical distance between the two elevations. This should never be exceeded. A good rule of thumb in the absence of other data is to keep the tower plumb and straight within the diameter of the leg members. 222-G provides a twist tolerance of 0.5 degrees in any 10 ft and total twist limit of 5 degrees.

If straightening of the tower is required, it should be performed by adjusting the guy wires as described later.

When checking the plumb of top-mounted poles and pylon antennas, the effects of direct sunlight on them must be considered. It is best to make these checks early in the morning or on a cloudy day.

Guys and Guy Insulators

Inspection of the guys can be done visually only for those portions adjacent to the anchors and tower. The range of this visual inspection can be extended by using binoculars, but its reliability is limited. If experienced riggers are available, it is possible to ride down the guy on a boson's chair, but this method should be used only under the supervision of qualified personnel.

Other maintenance requirements include the following: A visual inspection should be made of the guy cables, insulators, and hardware. Cables and dead-end grips should be checked for nicks or cuts in the individual strands. All porcelain insulators should be checked for chips, cracks, and oil leaks where appropriate. Fiberglass rods should be checked for surface tracking (black carbon track marks on the surface of the rod); breakdown of the epoxy surface; and exposure of the individual glass strands. The manufacturers should be consulted with regard to corrective action.

- *Corrosion.* If the guy cables show signs of corrosion, consideration should be given to coating or replacing them. The cost of cleaning and coating the cables should be considered along with the life expectancy of the coating when comparing it to the cost of replacement. All guy hardware should be checked using the same procedures for inspection and corrective actions as previously described for the tower structure.
- *Connections.* All pins should be checked for tightness and the condition of the cotter keys. Dead-end grips should be checked to ensure that their ends are completely snapped close, preventing any ice from forming inside. The surface appearance of the guy strand immediately next to the connections should be noted for evidence of slippage. Threads should be given a light petroleum coating.
- *Tensions.* Guy tensions should be checked in conjunction with the tower alignment. These tensions should be measured at the anchor end and compared to the specified values. It is important to

remember that they are dependent on the ambient temperature.

For the usual guy arrangement with cables in three directions, it is necessary to measure the tensions in only one direction while keeping the tower plumb in all directions. For guy arrangements with cables in four or more directions, it is necessary to measure the tensions in only one of the two guys in the same vertical plane while keeping the tower plumb in that plane.

There are several methods of measuring guy tension with varying degrees of accuracy. For small guys up to 3/4 inch, a shunt dynamometer calibrated for the size and type of strand is often used.

For larger guys, a series dynamometer may be placed in a temporary line between the anchor and a clamp on the cable. This line is then tightened until the permanent connection is relieved and the tension is indicated on the dynamometer. Hydraulic jacks with a calibrated pressure gauge or load cells can be used in place of the temporary line and dynamometer. These are particularly effective for large guys attached with bridge sockets.

There are two indirect methods of measuring tensions in guys that do not have any large insulators or other loads in them. The intercept method consists of sighting along a straight bar attached at the bottom of the guy and measuring the vertical distance between the point when the line of sight intercepts the tower and the point where the guy is attached. This distance can be accurately estimated by counting the number of bracing panels. The tension in the guy is directly related to this intercept distance, and the weight, length, and slope of the guy.

The tension in a guy cable is also directly related to its length, weight, and natural frequency of free vibration. The natural frequency can be determined by putting the guy in motion manually and measuring the fundamental period with a stopwatch. It should be noted that because a guy slopes, the tension on it varies along its length, and this method will only provide the average tension and not the tension at the anchor point. For long cables, this difference can be significant.

All tension measurements should be recorded along with temperature, wind speed, and direction. If any substantial changes are noted from the values previously measured, careful checks for slippage of all connections should be made.

Tolerances for guy tensions should be as provided by the designer. In the absence of any other tolerance, tensions should be within plus or minus 5% of the specified values.

Any necessary adjustments in tensions can be made by adjusting the turnbuckle or bridge socket at the anchor. Make such adjustments slowly and carefully. Never leave less than three threads sticking through the turnbuckle body or nut on the socket U-bolt. Remember that the tower must be kept plumb.

Base Insulator

The porcelain surface should be wiped clean with a soft cloth to remove accumulated dirt. A check should be made for cracks or chips on the porcelain surface. Scratches are often mistaken for cracks. Oil-filled insulators will display a wet surface or leak if cracked. If an oil stain or leak appears at the bottom of the porcelain on an oil-filled insulator and a crack cannot be found, incorrect loading possibly due to settlement of the pier should be suspected. A cracked base insulator should be replaced as soon as practical. Any sign of corrosion in the upper and lower bearing plates, rain shield, or lightning gap should be noted and corrected in a manner similar to that described for the tower structure. The lightning gap should be adjusted in accordance with instructions from the station engineer.

Tower Base and Guy Anchors

The tower base and guy anchors above grade should be visually inspected for spalling and cracking of the concrete. The soil surrounding the tower base foundation should be inspected for evidence of settlement. The anchor arms and surrounding soil should be examined for evidence of movement of the anchor. Any such settlement or movement should be noted.

Steel anchor shafts exposed directly to the soil should be inspected below grade for evidence of galvanic or electrolytic corrosion, especially in areas of high ground conductivity. Extreme caution should be exercised when excavating and backfilling during this inspection to ensure that the anchor's effectiveness is maintained.

Appurtenances

The ladder and its connections should be checked for corrosion and tightness along with the tower. The sleeve and belt of the safety device should be visually examined and tested near the ground level before each use.

Inspection and maintenance of the elevator system should be in accordance with the manufacturer's instructions. It is a good practice to operate the elevator at least once a month.

Inspection and maintenance of the lighting system should be in accordance with the manufacturer's instructions. Checks for corrosion in the conduit, junction boxes, and light fixtures should be made along with the tower inspection. Any obstructions in the breather or drain in the conduit should be removed. Broken or cracked glass and any leaking gaskets should be replaced.

If the tower is equipped with an isolation transformer, its surface should be inspected for cracking and splitting. The surface should be painted with a good quality alkyd varnish. Badly cracked surfaces should be filled with a mixture of varnish and microscopic glass ball powder and the area cotton taped over and varnished.

Frequency of Inspection and Maintenance

A suggested schedule for inspections and maintenance performance is shown in Table 7.2-1.

REPORTS

A written report of each maintenance and inspection procedure performed should be made and filed with the station engineer.

Additional Resources

American Concrete Institute (ACI): www.aci-int.org.
American Institute of Steel Construction (AISC): www.aisc.org.
ANSI/TIA-222-G-2005: www.tiaonline.org.
American Welding Society (AWS): www.aws.org.
International Building Code (IBC) 2006: www.iccsafe.org.
Stainless LLC: www.stainlessllc.com.

TABLE 7.2-1
Suggested Inspection and Maintenance Schedule

SUGGESTED INSPECTION AND MAINTENANCE SCHEDULE						
ITEM	DAILY	MONTHLY	BEFORE EACH USE	ANNUALLY	AFTER A MAJOR WIND OR ICE STORM	MANUFACTURER'S RECOMMENDATION
TOWER STRUCTURES:						
DAMAGED OR DEFORMED MEMBERS				X	X	
CONDITION OF PAINT				X		
CORROSION				X		
CONNECTIONS				X		
ALIGNMENT				X	X	
GUYS AND INSULATORS:						
DAMAGED COMPONENTS				X	X	
CORROSION				X		
CONNECTIONS				X	X	
TENSIONS				X	X	
BASE INSULATORS				X		
TOWER BASE AND GUY ANCHORS				X	X	
LADDER SAFETY DEVICE			X			
ELEVATOR SYSTEM						X
OPERATE		X				
LIGHTING SYSTEM						X
LAMP FAILURE	X					
CONDUIT SYSTEM, FIXTURES				X		

CHAPTER

7.3

Lightning Protection for Tower Structures

EDWARD A. LOBNITZ

TLC Engineering for Architecture
Orlando, Florida

INTRODUCTION

Understanding how lightning works as well as its effects on tower structures, grounding, and antenna systems is very helpful when trying to apply protection techniques that are practical. In fact, lightning protection is never 100% perfect and understanding the limitations or cost-effective application of protection is an important part of the design process. Also, since no two installations are ever exactly alike, repetitive siting of towers and locating antennas on existing towers will always require a review of standard protection to assure that expected protection has not been compromised.

Lightning is potentially one of the most hazardous natural occurrences on our planet. It can kill, maim, start fires, cause explosions, damage equipment, interrupt critical data transmissions, stop a shuttle from flying, and cause many other disasters. It is also beautiful, awe inspiring, exciting, mysterious, and challenging to all who are enchanted by its technical fascination. It is both predictable and unpredictable. Statistics such as the following are both surprising and sobering when attempting to protect against such a power:

- From 1959–2003, the total injuries in the United States were 3,696.
- Fatalities by state from 1959–2003:
  - —Florida: 425
  - —Texas: 195
  - —North Carolina: 181
  - —New York: 134
  - —Tennessee: 133
- A typical Florida thunderstorm can unleash 10,000 lightning strikes.
- In Florida, lightning kills approximately 10 people and injures 35 people every year. It is more deadly, in recent years, than hurricanes, tornadoes, or floods.
- Nationally, about 60 people a year die from lightning and 300 are injured.
- Lightning strikes the earth 100 times each second. Over the continental 48 states, an average of 20,000,000 cloud-to-ground flashes have been detected every year since 1989. About half of all flashes have more than one strike point, so at least 30 million points on the ground are struck each year.
- There are about 45,000 thunderstorms annually across the United States.
- Property and crop losses totaled $25.7 million in 2003.

Methods of protection from lightning have been changing ever since Benjamin Franklin's original studies of lightning and the creation of his lightning rod system, today called the Franklin Rod System. Although the exact physics of structure lightning rod protection originally was not fully understood, extensive research has greatly enhanced our understanding of lightning physics while also creating great controversies regarding theory and protection techniques. Also, electronic equipment has entered the picture in one of the fastest growing technologies of our century, but with inherent weaknesses to the effects of lighting. This has created even another huge industry known as surge protection, with its own confusing mix of technologies, theories, manufacturing methods, and jargon to help further the protection methodologies and challenge protection designers with almost unlimited technical choices.

Fortunately, there are some standards that have been and are being developed to help protection designers through the maze of technologies available in order to evaluate effectiveness, technology comparisons, test results, performance claims, and other criteria. Many of these standards are listed in the references at the end of this chapter.

These standards and publications are considered *consensus* standards since their committee members or testing agencies are made up of members from across the industry or third-party testing agencies. Only equipment that is verified to be in compliance with these standards and testing agencies should be used by protection designers. Claims by other equipment manufacturers regarding performance or construction should not be considered unless full verification and proof of claims by third-party testing are provided. Lightning can cause great losses to life and property, and assumed protection must be equal to the task without question. More cannot be expected from lightning protection than it can be proven to provide, regardless of guarantee or guarantee claims.

The purpose of this section, therefore, is not only to examine protection techniques for the present generation of towers, antenna system equipment, and protection methods and equipment, but to aid in preparing the reader to analyze new and everchanging environments. Additional images of grounding techniques for towers and buildings are contained in the CD-ROM for this book.

LIGHTNING AND STORM DEVELOPMENT

The intensity of lightning storm activity throughout the world varies by location. Until recently, thunderstorm activity was measured by the number of *thunderstorm days* observed and reported. In the United States, data is reported through the National Weather Service and displayed on an isokeraunic map (see Figure 7.3-1). A thunderstorm day is defined as any day during which thunder is heard at a specific observation point. Since these observations merely confirm the presence of lightning and do not provide information regarding severity, number of ground strokes, and stroke location, more complete data was needed to determine exposure risk. Atmospheric measurements throughout the United States [1] and other countries have led to new maps depicting ground stroke intensity by location (see Figure 7.3-2). Data is now available in the United States and increasingly throughout the world to determine not only whether lightning occurred, but also stroke coordinates, discharge characteristics, flash density, direction of storm movement, and other data that can be manipulated into very accurate charts, graphs, trend analyses, histographs, etc. Access to this extensive database can be secured through private and public sources via the internet and may be useful in selection of broadcast antenna sites and level of protection.

Thunderstorms are of two general types: convection storms and frontal storms.

Convection Storms

Convection storms are usually local in extent, of relatively short duration, and account for the majority of storms in the United States. They are caused by local heating of the air near the earth and, therefore, occur mainly during summer months and in warmer climates where moisture is present, although many convective storms are generated over mountainous areas. These storms can generate to great heights of over 40,000 ft where vast amounts of moisture and high temperatures are present, producing intense lightning activity. However, they dissipate quickly as accompanying cloud shade and rain cool the earth and dissipate the clouds' source of energy.

Frontal Storms

Frontal storms extend over greater areas, may continue for several hours, and are usually more dangerous, causing greater damage than convection storms. These storms develop from the meeting of a warm, moist weather front with a cold front that may at times extend for several hundred miles, exposing large areas to severe lightning discharges.

Figures 7.3-3 and 7.3-4 show typical convection (summer) discharges and frontal (winter) discharges, respectively.

The formation of lightning in all thunderstorms is generally believed [1] to be caused by ice/dust particles colliding with moisture particles in vertically generated air currents. The collisions create positive and negative charges within the cloud that eventually separate out so that negative charge concentration occurs near the bottom of the cloud and positive charges concentrate at the top. The negative charge at the bottom of the cloud causes a corresponding positive charge on the earth below the cloud. This charge difference between cloud and earth will commonly develop electrostatic field strengths of 100 KV per meter and higher. The field strength above the top of a 50 m, well-grounded, broadcast tower, therefore, could approach several million volts and result in upward charge streamers and corona discharge. A lightning strike to the tower is then a very real possibility.

The lightning process actually begins as a faintly visible stepped leader originating normally from the base of the storm cell. This leader carries the negative cloud charge toward the positive ground charge, forking and branching due to wind and random charge pockets in the air. As the leader nears the earth, one or more upward streamers will be initiated, usually from the tallest grounded bodies in near proximity to the downward leader. At some point, the downward leader will connect to an upward leader and return stroke currents flow in an effort to equalize the charge difference between the storm cell and the earth. The distance

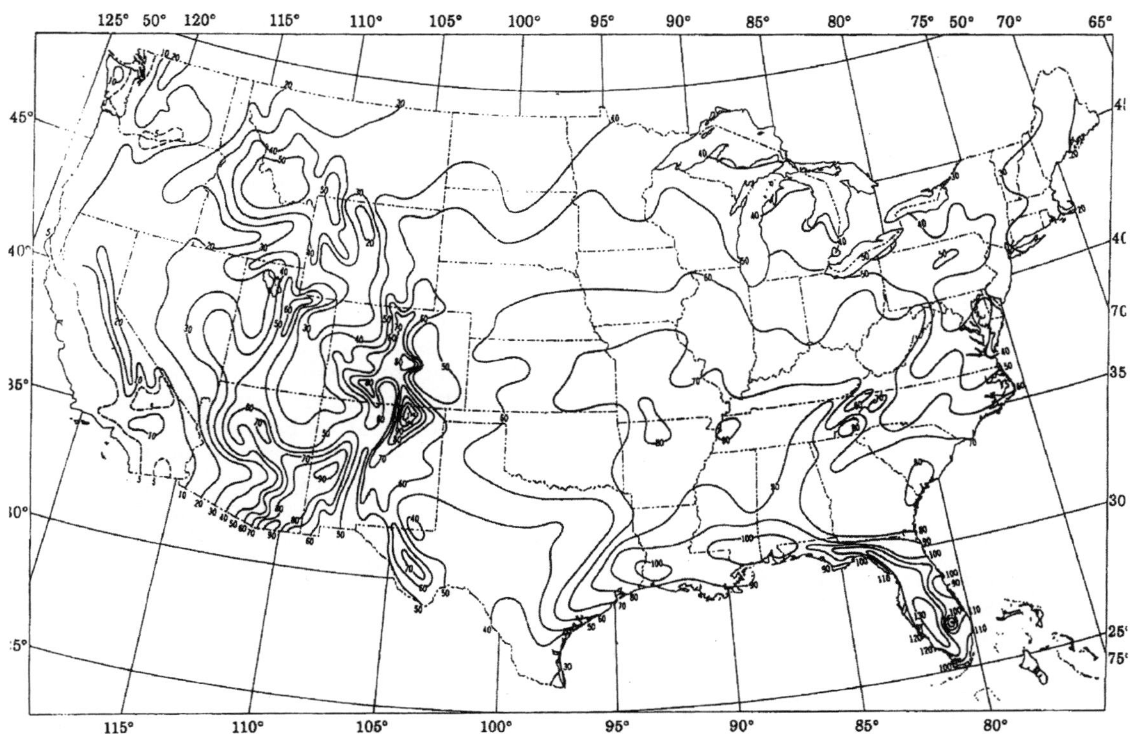

FIGURE 7.3-1 Average number of thunderstorm days annually. (Reprinted with permission from *NFPA 780, Installation of Lightning Protection Systems*, © 1997, National Fire Protection Association, Quincy, MA 02269. This reprinted material is not the complete and official position of the National Fire Protection Association, on the reference subject that is represented only by the standard in its entirety.)

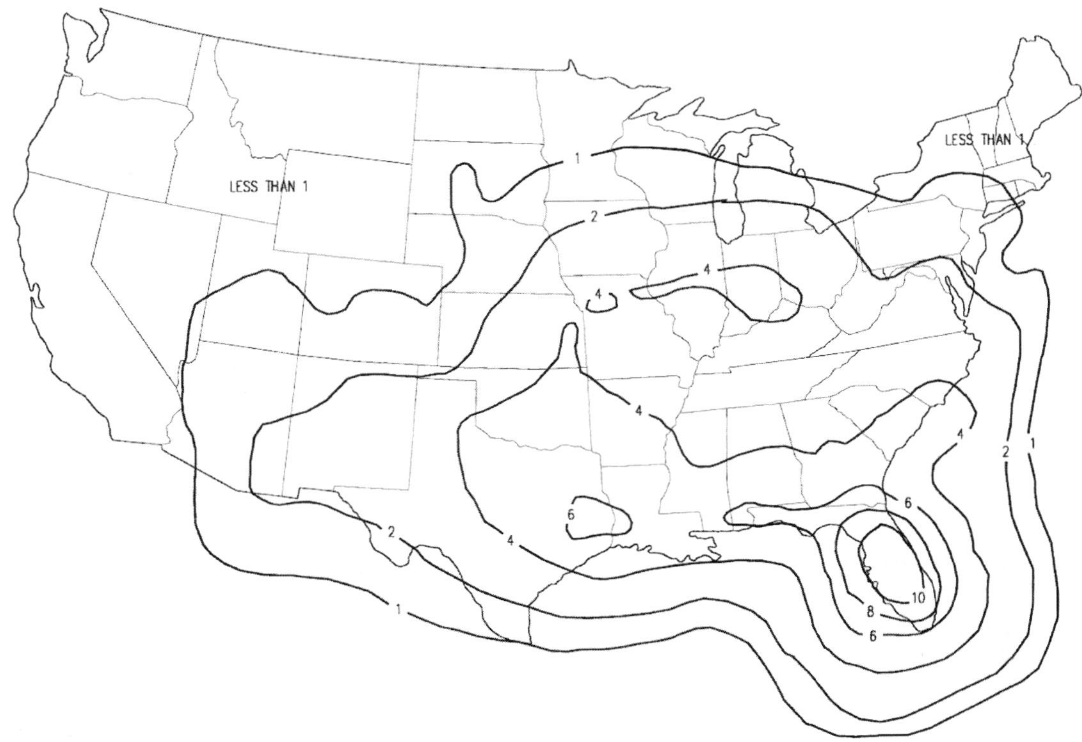

FIGURE 7.3-2 Composite of measured lightning flash density per year, 1989–1992; flashes per square kilometer. (Data supplied by the National Lightning Detection Network.)

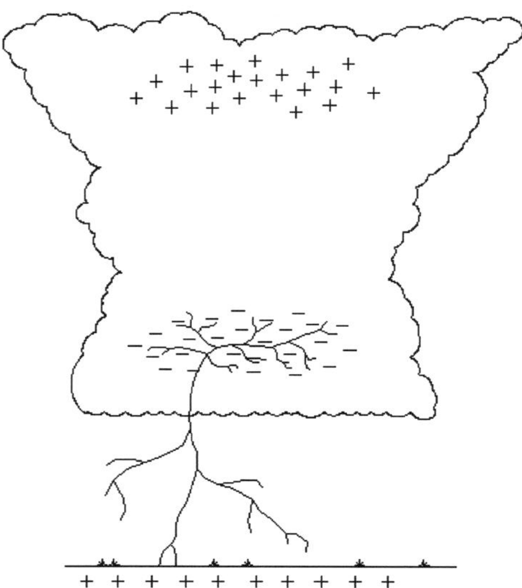

FIGURE 7.3-3 Typical mode of discharge in summer storms (negative stroke predominance).

between the connection of the downward leader and the object producing the upward leader is called the *striking distance* and is the basis for the rolling ball theory to be discussed later. The striking distance is most greatly influenced by the amount of charge in the downward leader such that the greater the charge, the greater the striking distance. Basically, the amount of leader charge and the surge impedance of the object being struck determine the value of lightning current. Lightning, however, does not always perform as would be expected from the above discussion.

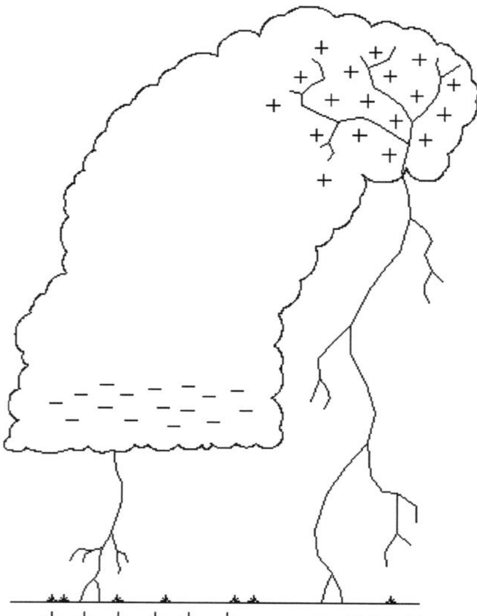

FIGURE 7.3-4 Typical discharge in winter storms (positive stroke predominance).

In addition to bottom of cloud strikes, severe, long-traveling lightning strokes can emanate from the positively charged top-of-storm cells to negative earth charges. Also grounded objects above 150 ft tall can initiate upward-leading charges, reversing the above process. Grounded objects taller than 150 ft, such as broadcast towers, can be struck on the side of the tower in lieu of the top and can subject side-mounted antennas to direct strikes.

Lightning currents range from an average value of 20,000 amperes to a high recorded value of over 400,000 amperes. Many strikes contain up to 30 strokes just milliseconds apart. These strokes appear to flicker or strobe; are usually wider, brighter, and hotter than single strikes; and can cause serious damage if not adequately dissipated. Figure 7.3-5 shows typical lightning propagation development. Figure 7.3-6 provides probability values of lightning stroke peak values.

Lightning can affect broadcast tower structures and associated equipment or studio buildings in basically two ways. These generally take the form of direct strikes and remote strikes. Direct strikes usually attach to the tower structure, building air terminals, tower-mounted antennas, and tower warning lighting. Remote lightning strikes can inductively couple the surge field into the tower structure, coaxial cable shields, warning lighting conduit systems, building rod systems, and antennas. Both types of strikes can cause considerable damage if proper protection is not provided in the design of the tower/building system and in applied surge/lightning protection equipment. Design techniques include proper grounding and bonding, selection of tower/building location, coaxial cable routing, and bulkhead design. Protection techniques include proper selection and placement of surge protection devices for coaxial cables, lighting warning circuit, equipment building power service, tower mounted preamp equipment, and emergency power equipment, as well as air terminal protection for side-mounted antennas mounted above 150 ft and for the equipment building located outside the tower-protected zone. Locating new antennas on existing towers requires a special analysis of existing conditions as will be discussed later.

As mentioned above, the radiated impulse field from a remote lightning strike can pose a significant hazard to equipment and systems, particularly those that are interconnected by long lengths of cable. Figure 7.3-7 is a composite made by a number of researchers of lightning electromagnetic field measurements that have been normalized to a distance of 10 kilometers. The figure shows a frequency domain distribution that peaks at about 10 kilohertz at an intensity of slightly more than one volt per meter. It is important to realize that nearer strikes can create field strengths many orders of magnitude higher than those shown. The predominant low frequency component is also very effective in coupling energy into systems of wiring, producing continuous frequency, and ringing waveforms due to inductance, capacitance, and resonant conditions, even if wiring is buried in the ground.

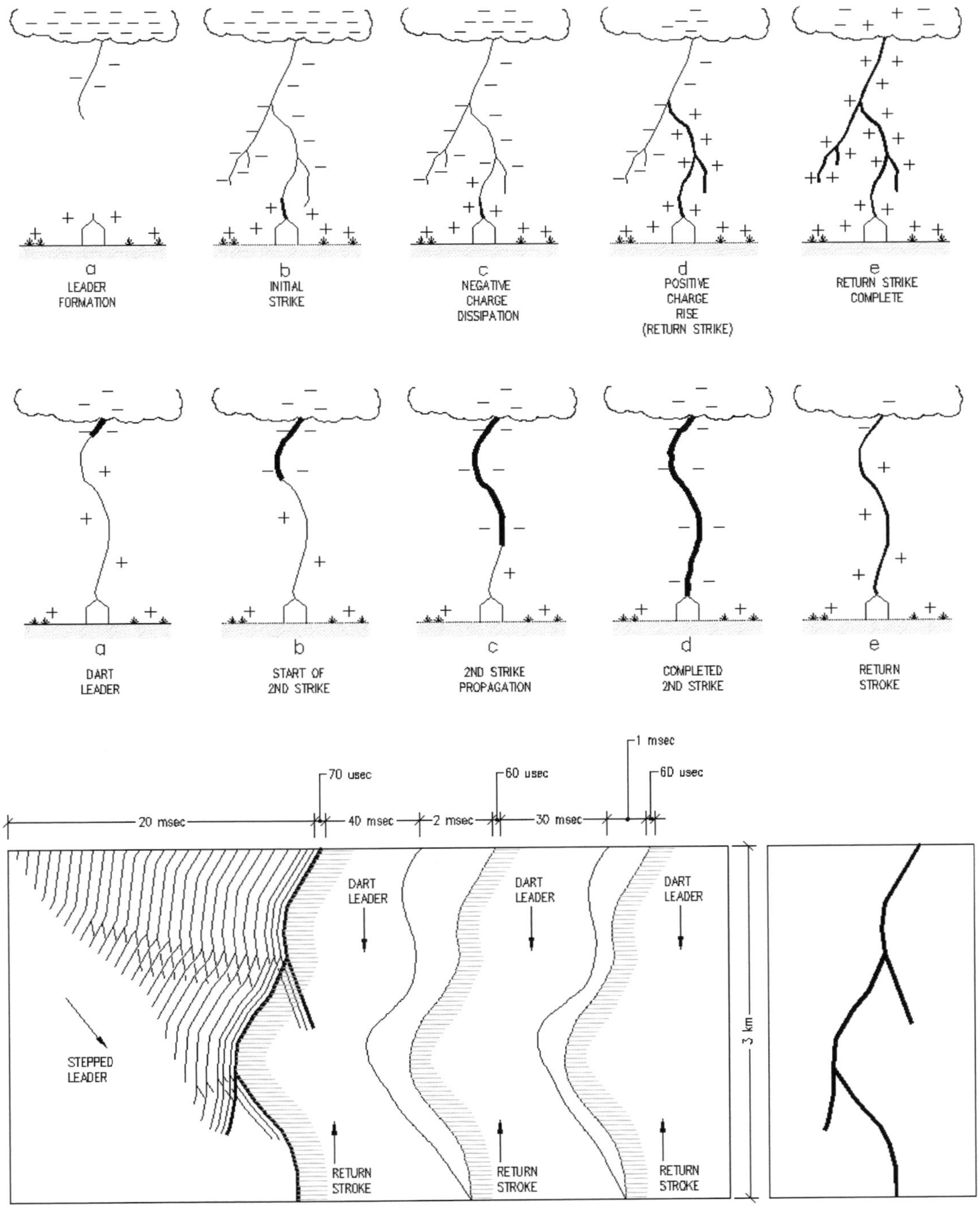

FIGURE 7.3-5 Typical lightning propagation development.

The time domain current waveform associated with a typical lightning strike is characterized by a very fast leading edge or rise time, followed by a more gradual decay. Technically, rise time is the period of time required for the wave to increase from 10% to 90% of its crest value. Decay time is normally expressed as the time measured between the wave crest and 50% of the crest value. A description of a waveform such as 1.5 × 50 microseconds for a 10–20 KA strike would indicate a single impulse waveform with a rise time of 1.5 microseconds and a decay time of 50 microseconds, to half peak value.

There are many variations on actual lightning waveforms seen in real-world circuits. Waveforms such as those shown in Figure 7.3-8 may be found with rise times measured in a fraction of a micro-

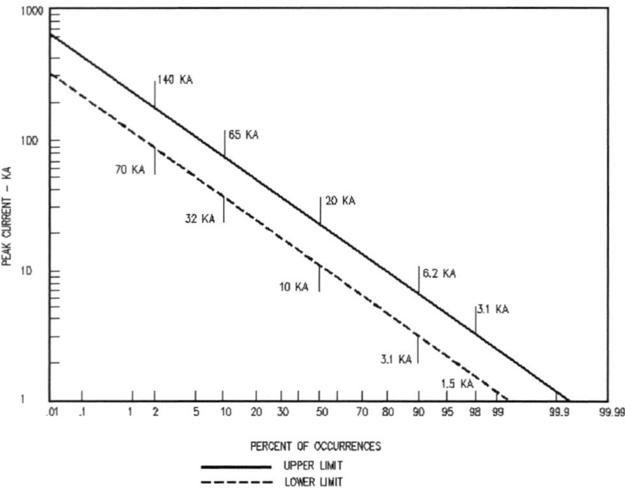

FIGURE 7.3-6 Lightning stroke intensity.

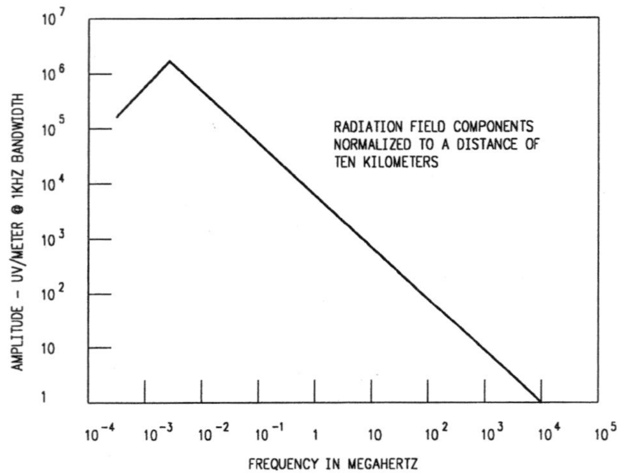

FIGURE 7.3-7 Lightning signal amplitude versus frequency.

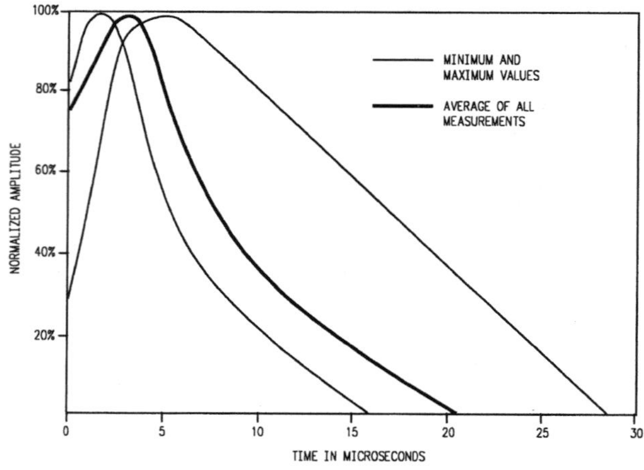

FIGURE 7.3-8 Typical lightning waveforms.

second near the point of lightning entry to a circuit. Normally, however, the higher the lightning current, the slower the rise time and decay period such that a 250,000 amp strike will have a much slower rise time (30 μs) and 250 μs decay compared to a 10–20 KA strike. As a wave propagates through a wiring system, the rise time and the decay time will lengthen. The polarity of the impulse may be either positive or negative. As in remote strokes, inductive and capacitive properties of the wiring system may cause the circuit to act as a resonant tuned circuit, producing a ringing wave that alternates in polarity. References that categorize the waveforms and current levels for several types of circuits are listed at the end of this chapter.

PROTECTION OBJECTIVES

Primary lightning protection objectives, specifically for broadcast towers and associated building equipment wiring, may be grouped into the following two basic categories: equipment and personnel protection.

Equipment Protection

Control of small potential differences that are deadly to electronic equipment and antenna cable, as measured between active circuitry and grounded media, is key to protecting equipment. Controlling the potential differences to a value below the equipment damage threshold will ensure the equipment survives. Providing tighter voltage control to a value below the equipment upset threshold will help to ensure the system rides through the lightning event without any noticeable effect. All protection, however, is predicated on the assumption of average or standard lightning strike parameters, such as energy waveform, and can, therefore, never be 100% effective.

Personnel Protection

Protecting service personnel from the threat of a direct lightning strike, secondary flashing (side flash), and controlling differences in potential (step and touch potentials) between different parts of their bodies during a lightning event must be integrated into any comprehensive tower protection scheme. Step potentials are voltage gradients seen along the surface or near surface of the earth as lightning current radiates hemispherically from its point of entry into the soil or grounding system. Touch potentials are voltage differences developed in horizontal and vertical elements of a structure, natural object, or system during the passage of lightning current. Both step and touch potentials can be hazardous and must be minimized.

Secondarily, the broadcast tower and associated building should also be reviewed for protection against lightning's effects. In most cases, however, towers and equipment buildings will be self-protected, as will be discussed later, and provide a means to intercept lightning strikes, conduct the lightning current safely through or around the structure, and dissipate the current into the earth. These characteristics, however, are critical to personnel and equipment protection. Lightning current passing uncontrolled through a structure may result in deterioration of tower joints, ignition of combustible materials, generation of explosive forces in masonry and other moisture-bearing materials, and burning or tearing of roofing systems. Secondary flashing between the primary current path and nearby unbonded grounded objects may also pose a threat to persons in or near elements of the structure.

PROTECTION STRATEGIES

A typical broadcast tower installation will consist of an antenna tower in association with a studio building or an equipment hut for repeater locations. The preferred building type would consist of structural steel framing to maximize equipotential grounding and bonding characteristics. Where the buildings are not fully within the zone of protection provided by the tower or other adjacent structure, they must be provided with a Franklin Rod–type system. Design requirements for this system are adequately documented and described in NFPA 780, Standard for the Installation of Lighting Protection Systems [2] and as supported by UL 96A [3].

Broadcast towers are normally three- or four-legged structures constructed of galvanized tubular or structural steel with sectional, vertical truss elements. They are either self-supporting or guyed and can be of various heights from a satellite dish near the ground to a 1600 ft or higher antenna-support structure. Lately, however, some architecturally creative–type towers of both steel and concrete have been introduced that create some interesting challenges to lightning protection methodologies. Also, some towers are ground mounted, some are mounted on top of concrete or steel buildings of various heights, and some are large enough to span multilegged over the broadcast building. It is, therefore, impossible to describe detailed protection techniques for all type of towers. Basic requirements, however, will be presented so that application to any type tower can be easily extrapolated. Where new antenna are mounted on existing leased towers, an analysis of existing conditions and protective techniques must be made and improvements or corrections made as necessary to maximize protection. Leasors may, however, object to making changes.

The tower structure itself is the basic lightning protection element that helps dissipate direct lighting strikes; minimize voltage rise on associated antennas, coaxial cable, and equipment; and protects adjacent buildings. Where new towers are located in populated areas, increased lightning activity and dissipation of ground currents can cause local controversy, increased risks, and influence tower design complexity beyond the basic requirements presented herein.

Tower protection elements consist of the following:

- Grounding
- Coaxial cable shield bonding
- Enhanced strike attachment
- Warning light bonding
- Maintenance of joints
- Protection for side- and top-mounted antenna
- Grounded versus isolated guys

Proper tower grounding is essential for the protection of all other systems and facilities interconnected with the tower. It should consist of exothermic weld connections between the tower base or legs and foundation-reinforcing steel, anchor bolts, grounding radials, counterpoise rings, guy anchors, and adjacent building counterpoise and fences.

Coaxial cable shields from antennas must be bonded to the tower at the top, near the antenna, at each guy attachment point, at the midpoint of any 150–200 ft interval between established bonding points, and at the bottom where the cable leaves the tower. The cable should exit the tower as low as possible to minimize voltage gradient and current dissipation at the bulkhead panel on the building.

Air terminals mounted on top of the tower will enhance the probability that the rod will be struck in lieu of the tower structure or any side-mounted antennas, thereby minimizing maintenance from damaged antennas and pitting of tower steel. Any antenna or other tower mounted components should not be mounted within 12 inches of the top of the air termainal or beyond a 45-degree cone of protection from the top of the air terminals. If the tower is over 150 ft high, side-mounted antennas above this level are vulnerable to direct hits and should be protected by mounting horizontal air terminals above and below the antennas, protruding at least 6 inches beyond the antenna. Since they are horizontal and located in the end nulls of the antenna pattern, the air terminals should not affect system performance. See Figure 7.3-9.

The warning light(s) on the tower must be circuited in rigid metal galvanized conduit and the fixture itself should be constructed of metal to minimize surge currents on the wiring that would be difficult and expensive to eliminate from the source panel board in the adjacent building. Surge suppression on the lighting circuit would still be required at the panel board but it could be selected to handle minimal surge current. Also, any low voltage power or control wiring serving tower mounted amplifiers, heaters, and other equipment must also be installed in rigid metal conduit.

A separate lightning protection down conductor is not necessary for adequate dissipation of lightning currents. The tower itself will always conduct the bulk of any lightning strike although a portion will travel on the coax shield and warning light conduit. An additional down conductor that is attached to the tower at intervals will have a negligible effect on the division of current paths due to the inductance distribution of all components.

Lightning currents traveling down towers can, over time, cause arcing at tower joints, stripping away galvanizing and resulting in rusting. Maintenance, therefore, of the tower system is important to long lasting,

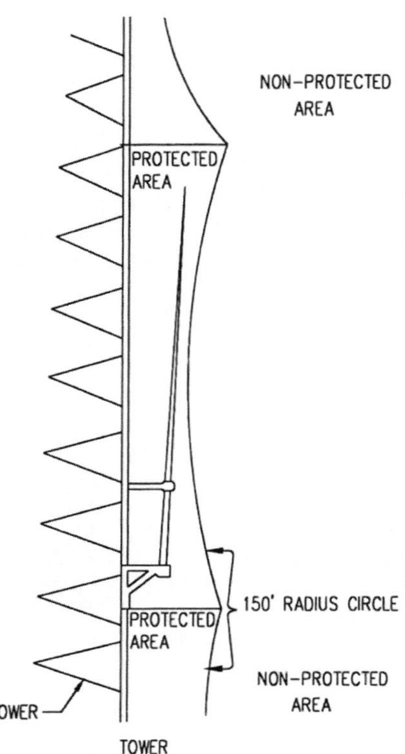

FIGURE 7.3-9 Side antenna protection method.

well performing components. Air terminals subject to heavy lightning currents of long duration can cause pitting and melting of points as well as severe mechanical stresses in mounting hardware. Bonding straps, cable fasteners, and other miscellaneous hardware can loosen and allow movement or "whipping" of conductors from electromagnetic forces associated with unequal downcurrents in various components of the tower system. Therefore, at least yearly maintenance is recommended.

For guyed towers, guy wires can be very effective in dividing the lightning currents into additional paths to ground, therefore lowering the current and voltage stress to tower components and the service building bulkhead. Where guy wires contain insulators, special precautions must be observed. The insulators will be subject to voltage differences during lightning strikes occurring between guy-to-tower attachment points and the ground potential at the anchors due to ground currents and induced guy wire voltages. These voltage differences can reach hundreds of thousands of volts and can create arcing or "tracking" over insulators, resulting in ineffective isolation over time. Therefore, maintenance inspection of the insulators and selection of maximum insulative values for the insulators are very important.

The rolling sphere concept is a protection strategy that is fully described in NFPA 780 and effectively applies to towers up to 150 ft high above level ground. Figure 7.3-10 shows traditional lightning protection for a studio facility with an adjacent microwave relay tower. Protective zones are based on an imaginary "rolling sphere" that is 150 ft in radius being passed over the structure. The sphere is also rolled around the structure tangent to earth. Air terminals are placed in such a way that the ball never contacts the structure or other objects requiring protection. Properly protected masts, light poles, and adjacent buildings may be taken into account when establishing zones of protection.

The protective zones created by the 150 ft radius sphere have proven statistically adequate for most facilities and are the basis behind most codes and standards [4,5,6]. An interesting way to think about structural lightning protection design is to imagine rolling an inked sphere of 150 ft radius around and over a structure. The ball should roll only on the earth, air terminals, and other suitable metallic components that have been connected to the lightning protection system. Any other area that receives ink should be considered unprotected. Where grounded guy wires are present, the sphere rolls over the wire in similar fashion to protect items under the projected zone. This is similar to overhead ground wire protection described fully in NFPA 780.

Most broadcast towers, and much of the equipment they support, have sufficient mass and conductivity to resist damage from direct lightning contact. Localized burning, pitting, and mechanical forces may occur in the immediate area of contact with the lightning channel. However, the brief duration of a lightning stroke creates little heating in areas away from the point of lightning contact. Air terminals, shown in both Figures 7.3-10 and 7.3-11, when properly bonded to the tower, may be used to protect more sensitive objects. Whenever practical, coaxial transmission lines and other wiring should be extended up the inside face of the tower to minimize the possibility of direct lightning contact.

Personnel Protection

The threat to personnel during a lightning strike ranges from the obvious danger of direct contact with a lightning strike to the more obscure effects of step and touch voltages. Protection from a direct strike, when near or within structures, is accomplished with traditional rolling sphere concept methods. However, to ensure that an adequate protective zone is provided in areas frequented by personnel, the rolling sphere should be reduced to 100 ft radius.

Step and touch potentials are created as a lightning current passes through resistive soil and other available paths as it dissipates into the earth. A person in contact with only one point of the gradient will simply rise and fall in potential with the gradient without injury. A person in contact with multiple points on the earth or objects at different potentials along the gradient will become part of the current path and may sustain injury or death.

Figure 7.3-11 [7,8] indicates a number of methods for protecting personnel from the direct and secondary

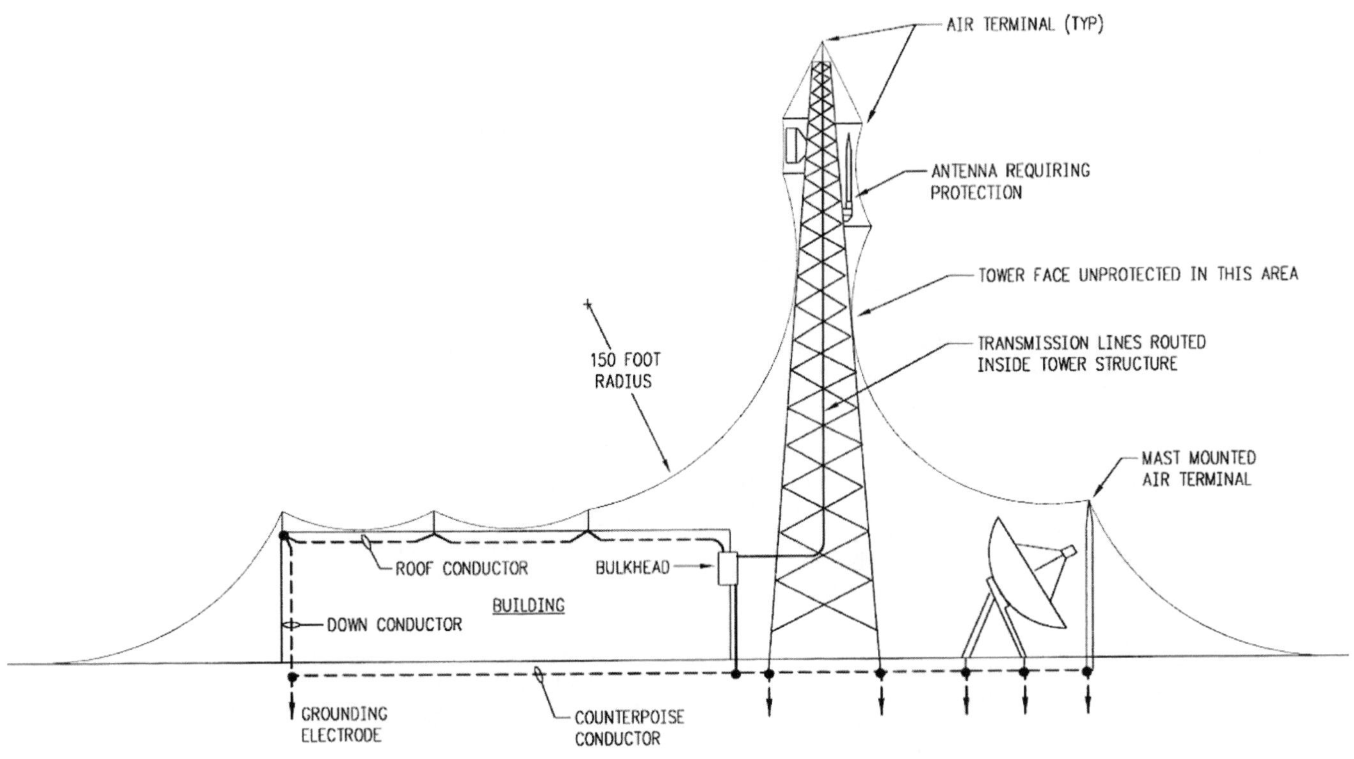

FIGURE 7.3-10 Example of rolling ball theory application.

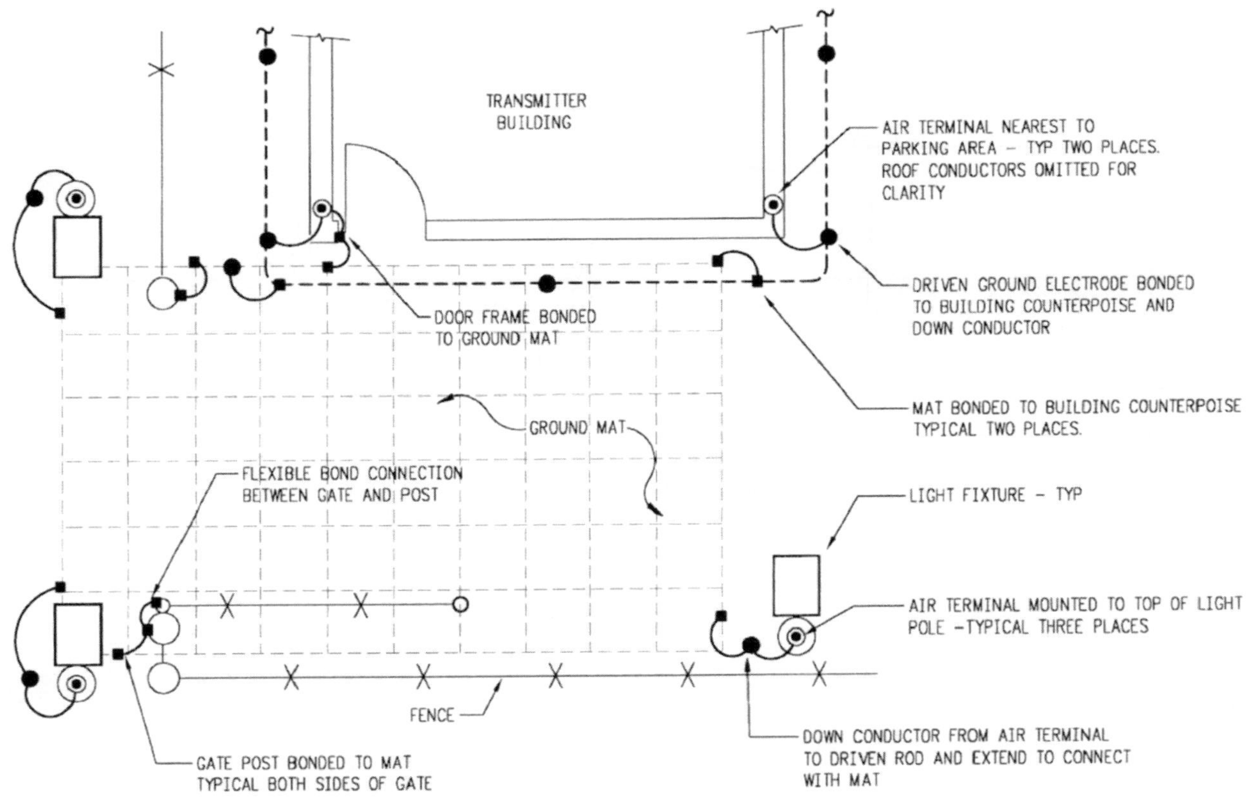

FIGURE 7.3-11 Personnel protection methods.

effects of lightning. A typical tower/transmitter site is used as an example. A technician responding to a service problem during a thunderstorm would likely exit his or her vehicle outside the gate, unlock and open the gate, and move his or her vehicle into the inside yard. The technician would then leave the vehicle and enter the building.

The threat of a direct lightning strike to the technician has been minimized by establishing a protective zone over the areas to be traversed. This zone is created by the tower and air terminals mounted atop light poles.

Step potentials are minimized through the use of a ground mat buried just below the surface of the area where the technician is expected to be outside the vehicle. Ground mats are commercially available, fabricated in a 6 × 6 inch square pattern using #8 AWG bare copper wire. Each intersection is welded creating, for all practical purposes, an equipotential plane that short-circuits the step potential gradient in the area above the mat. The mat, as a whole, will rise and fall in potential due to lightning current discharges, however, there will be little difference in potential between the technician's feet. Mats should be covered with 6 inches of crushed stone or pavement.

The threat of dangerous touch potentials is minimized by bonding the ground mat to the building perimeter counterpoise, the fence at each side of the gate opening, bonding to the door frame of the transmitter building door, and providing a flexible bonding connection between the swing gate and its terminal post. Such bonding ensures that the object being touched by the technician is at or near the same potential as his or her feet.

Bonding both sides of the gate opening to the mat helps to ensure that the technician and both sides of the gate are at approximately the same potential while the gate is being handled. The flexible bond between the gate and its support post may be accomplished using a commercially available kit or by exothermically welding a short length of flexible 2/0 AWG welding cable between the two elements.

External Ground System

The effectiveness of a grounding system is a function of the type and extent of the electrode system used and the resistivity of the surrounding soil. Soil resistivity is dependent on the quantity of free ions (chemical salts) in the soil, temperature, and moisture content. The character of the soil below a particular site may also vary significantly with depth and location due to layering of different types of soil, the presence of hardpan layers, and subsurface rock.

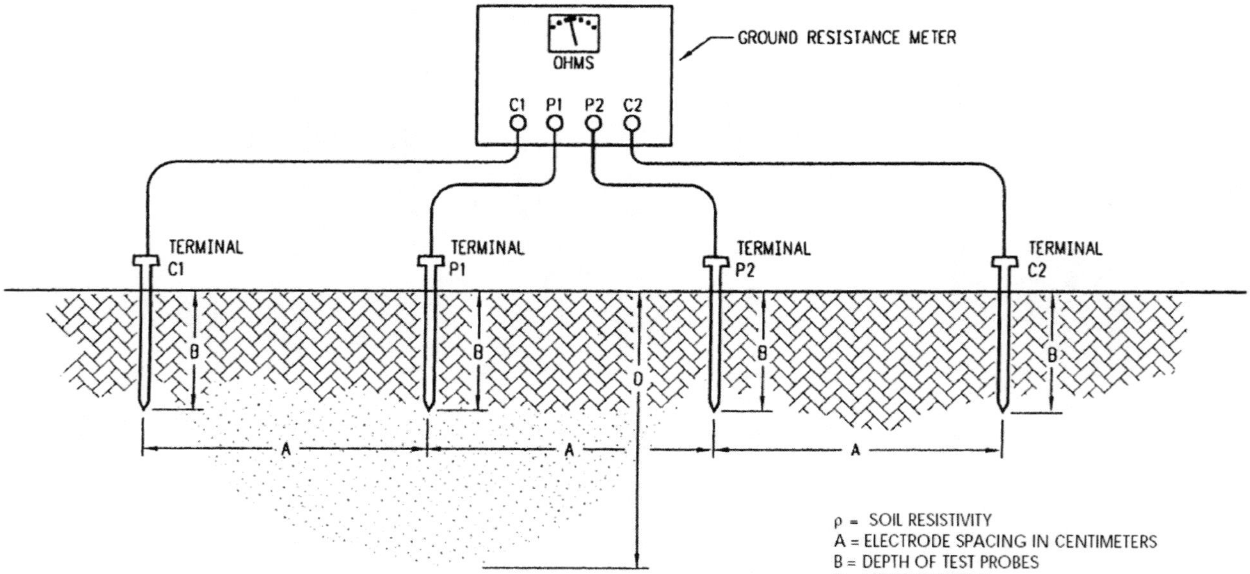

GROUND RESISTANCE METER

OHMS

C1 P1 P2 C2

TERMINAL C1 · TERMINAL P1 · TERMINAL P2 · TERMINAL C2

ρ = SOIL RESISTIVITY
A = ELECTRODE SPACING IN CENTIMETERS
B = DEPTH OF TEST PROBES
R = METER READING

NOTES:
1) DEPTH OF TEST PROBES (B) MUST BE LESS THAN A/20
2) DIMENSION D IS THE DEPTH AT WHICH RESISTIVITY IS DETERMINED. THIS DEPTH CHANGES WITH PROBE SPACING
3) PROBE SPACING (A) MUST BE EQUAL
4) ONE FOOT EQUALS 30.5 CENTIMETERS

$$\rho = 2\Pi AR$$

FIGURE 7.3-12 Soil resistivity test method.

Temperature is a major concern in shallow grounding systems as it has a major effect on soil resistivity. During winter months, the grounding system resistance may rise to unacceptable levels due to freezing of liquid water in the soil. The same shallow grounding system may also suffer from high resistance in the summer as moisture is evaporated from soil. It is wise to determine the natural frost line and moisture profile for an area before attempting design of a grounding system.

Figure 7.3-12 describes a four-point method for in-place measurement of soil resistivity. Four uniformly spaced probes are placed in a linear arrangement and connected to a ground resistance test meter. An alternating current (at a frequency other than 60 hertz) is passed between the two most distant probes resulting in a potential difference between the center potential probes. The meter display in ohms of resistance may then be applied to the formula to determine the average soil resistivity in ohm-centimeters for the hemispherical area between the C1 and P2 probes.

Soil resistivity measurements should be repeated at a number of locations to establish a resistivity profile for the site. The depth of measurement may be controlled by varying the spacing between the probes. In no case should the probe length exceed 20% of the spacing between probes.

Once the soil resistivity for a site is known, calculations can be made to determine the effectiveness of a variety of grounding system configurations. Figure 7.3-13 presents equations for several driven rod and radial cable configurations that, after the soil resistivity is known, may be used for the purpose of estimating total system resistance. Generally, driven rod systems are appropriate where soil resistivity continues to improve with depth or where temperature extremes indicate seasonal frozen or dry soil conditions. See Figure 7.3-14 for a typical U.S. soil resistivity map.

Radials are also quite effective if placed below the frost line. They are often the only practical solution in areas with shallow subsurface rock. There have been instances at bald rock mountain-top sites where radials were either grouted into saw cuts in the rock or simply pinned against the face of the rock.

The performance of a grounding system in high-resistivity soil can often be improved through the addition of chemical salts or conductive concrete. Salts leach into the soil, increasing the number of free ions with a proportionate decrease of soil resistivity in the area of the rod. Magnesium sulphate (epsom salts), copper sulphate (blue virol), calcium chloride, sodium chloride (table salt), and potassium nitrate have been used for this purpose. Conductive concrete increases the volumetric area for surge dissipation and is very effective in poor soil conditions or rocky soil.

Figure 7.3-15 describes the trench and well methods for applying chemical treatment. A typical precharged chemical ground rod installation is also shown. In a precharged rod, moisture from the air enters the rod through breather holes at the top of the rod and leaches through chemicals inside, gradually exiting the rod through weep holes. As one might expect, chemically enriched grounds require recharging after a number of years to maintain their effectiveness. It is also wise to check with governing environmental agencies before introducing any foreign chemical into the soil.

The bentonite method of grounding is also shown in Figure 7.3-14. Instead of driving the rod, it is placed in the center of a 6–12 inch augered hole. A slurry consisting of bentonite clay and water (well drillers mud) is then poured around the rod. As the water settles out, the resulting clay remains moist through absorption of moisture from the surrounding soil. Popular additives to the slurry include up to 75% powdered gypsum (calcium sulphate) and up to 5% sodium sulphate (galvanic anode backfill). Conductive concrete is installed in a similar manner and can be utilized in horizontal trenches or mounded on top of rocky soil to enclose grounding conductors.

Ground Electrode Testing

Testing of all grounding electrodes before they are connected to form a complex network is a fairly simple process that is well described in the documentation included with all ground electrode meters and therefore will not be described here. Also, the system as a whole should be tested after all interconnections are made, providing a benchmark for future tests.

On a new site, it is often possible to perform ground system tests before the power company ground/neutral conductor is attached to the system. It is worthwhile to conduct a before and after test with probes in the same position to determine the influence of the power company attachment during future tests. It is also worthwhile to install permanent electrodes and marker monuments at the original P2 and C2 probe positions to ensure the repeatability of future tests. At existing sites, ground testing may be impossible without shutting down existing power systems so strict adherence to recommended grounding practices must be followed. Also, clamp-on ground test meters are now available to test existing ground terminals for comparative analysis.

Tower-Building System

In typical tower-building arrangements, the tower is normally subject to more frequent and larger lightning currents than the station building. It is therefore reasonable to place emphasis on the tower grounding system with less emphasis on that for the station building. Improved grounding at the tower will result in less current flowing between the tower and station building grounding systems, reducing potential differences between the two systems. The potential rise on the tower due to a lightning strike is a result of fast-rising lightning current, tower inductance, and ground resistance such that $E = IR + L(di/dt)$, where E is the potential rise from top of tower to ground, I is the magnitude of lightning current, R is the grounding resistance, L is the tower inductance, and di/dt is the

	Hemisphere radius a	$R = \frac{\rho}{2\pi a}$
•	One ground rod length L, radius a	$R = \frac{\rho}{2\pi L}\left(\ln\frac{4L}{a} - 1\right)$
• •	Two ground rods $s > L$; spacing s	$R = \frac{\rho}{4\pi L}\left(\ln\frac{4L}{a} - 1\right) + \frac{\rho}{4\pi s}\left(1 - \frac{L^2}{3s^2} + \frac{2L^4}{5s^4}\right)$
••	Two ground rods $s < L$; spacing s	$R = \frac{\rho}{4\pi L}\left(\ln\frac{4L}{a} + \ln\frac{4L}{s} - 2 + \frac{s}{2L} - \frac{s^2}{16L^2} + \frac{s^4}{512L^4}\cdots\right)$
—	Buried horizontal wire length $2L$, depth $s/2$	$R = \frac{\rho}{4\pi L}\left(\ln\frac{4L}{a} + \ln\frac{4L}{s} - 2 + \frac{s}{2L} - \frac{s^2}{16L^2} + \frac{s^4}{512L^4}\cdots\right)$
L	Right-angle turn of wire length of arm L, depth $s/2$	$R = \frac{\rho}{4\pi L}\left(\ln\frac{2L}{s} + \ln\frac{2L}{s} - 0.2373 + 0.2146\frac{s}{L} + 0.1035\frac{s^2}{L^2} - 0.0424\frac{s^4}{L^4}\cdots\right)$
⅄	Three-point star length of arm L, depth $s/2$	$R = \frac{\rho}{16\pi L}\left(\ln\frac{2L}{s} + \ln\frac{2L}{s} + 1.071 - 0.209\frac{s}{L} + 0.238\frac{s^2}{L^2} - 0.054\frac{2}{L^4}\cdots\right)$
+	Four-point star length of arm L, depth $s/2$	$R = \frac{\rho}{8\pi L}\left(\ln\frac{2L}{a} + \ln\frac{2L}{s} + 2.912 - 1.071\frac{s}{L} + 0.645\frac{s^2}{L^2} - 0.145\frac{s^4}{L^4}\cdots\right)$
✳	Six-point star length of arm L, depth $s/2$	$R = \frac{\rho}{12\pi L}\left(\ln\frac{2L}{a} + \ln\frac{2L}{s} + 6.851 - 3.128\frac{s}{L} + 1.758\frac{s^2}{L^2} - 0.490\frac{s^4}{L^4}\cdots\right)$
✳	Eight-point star length of arm L, depth $s/2$	$R = \frac{\rho}{16\pi L}\left(\ln\frac{2L}{a} + \ln\frac{2L}{s} + 10.98 - 5.51\frac{s}{L} + 3.26\frac{s^2}{L^2} - 1.17\frac{s^4}{L^4}\cdots\right)$
○	Ring of Wire-diameter of ring D, diameter of wire d, depth $s/2$	$R = \frac{\rho}{2\pi^2 D}\left(\ln\frac{8D}{d} + \ln\frac{4D}{s}\right)$
—	Buried horizontal strip length $2/L$, section $\propto$ by b, depth $s/2$, $b < a/8$	$R = \frac{\rho}{4\pi L}\left(\ln\frac{4L}{a} + \frac{a^2 - \pi ab}{2(a-b)^2} + \ln\frac{4L}{S} - 1 + \frac{s}{2L} - \frac{s^2}{16L^2} + \frac{s^4}{512L^4}\cdots\right)$
⊘	Buried horizontal round plate, radius a, depth $s/2$	$R = \frac{\rho}{8a} + \frac{\rho}{4\pi s}\left(1 - \frac{7}{12}\frac{a^2}{s^2} + \frac{33}{40}\frac{a^4}{s^4}\cdots\right)$
	Buried vertical round plate, radius a, depth $s/2$	$R = \frac{\rho}{8a} + \frac{\rho}{4\pi s}\left(1 + \frac{7}{24}\frac{a^2}{s^2} + \frac{99}{320}\frac{99}{320}\frac{a^4}{s^4}\cdots\right)$

Notes:

1. Approximate formulas, including the effect of images
2. Dimensions must be in centimeters to return result in ohms
3. ρ = resistivity of earth in ohm-centimeters
4. For 10 ft. (3 m) rods of $\frac{1}{2}''$ (12.7 mm), $\frac{5}{8}''$ (15.88 mm) and $\frac{3}{4}''$ (19.05 mm) diameters, the grounding resistance may be quickly determined by dividing the soil resistivity ρ in ohm-centimeters by 292, 302 and 311 respectively.
5. Data source IEE Green Book (Std. 142-1982)

FIGURE 7.3-13 Formulas for calculation of resistances to ground.

lightning current rise with respect to time. Tower structures have a certain amount of inductance per foot. The amount of this inductance is dependent on the geometric configuration as well as the width of the tower. This width-to-height ratio will determine the total inductance of a tower. A 150 ft tower, for instance, with 35 inch side widths can have an inductance of 40 µH. This value of inductance can be approximated (W/H $\leq$ 1%) by treating the tower as a quarter-wave antenna using:

$$f = \frac{468 \times 10^6}{2(\text{H in feet})} \quad \text{then inductance } L = \frac{377}{2\pi f} \qquad (1)$$

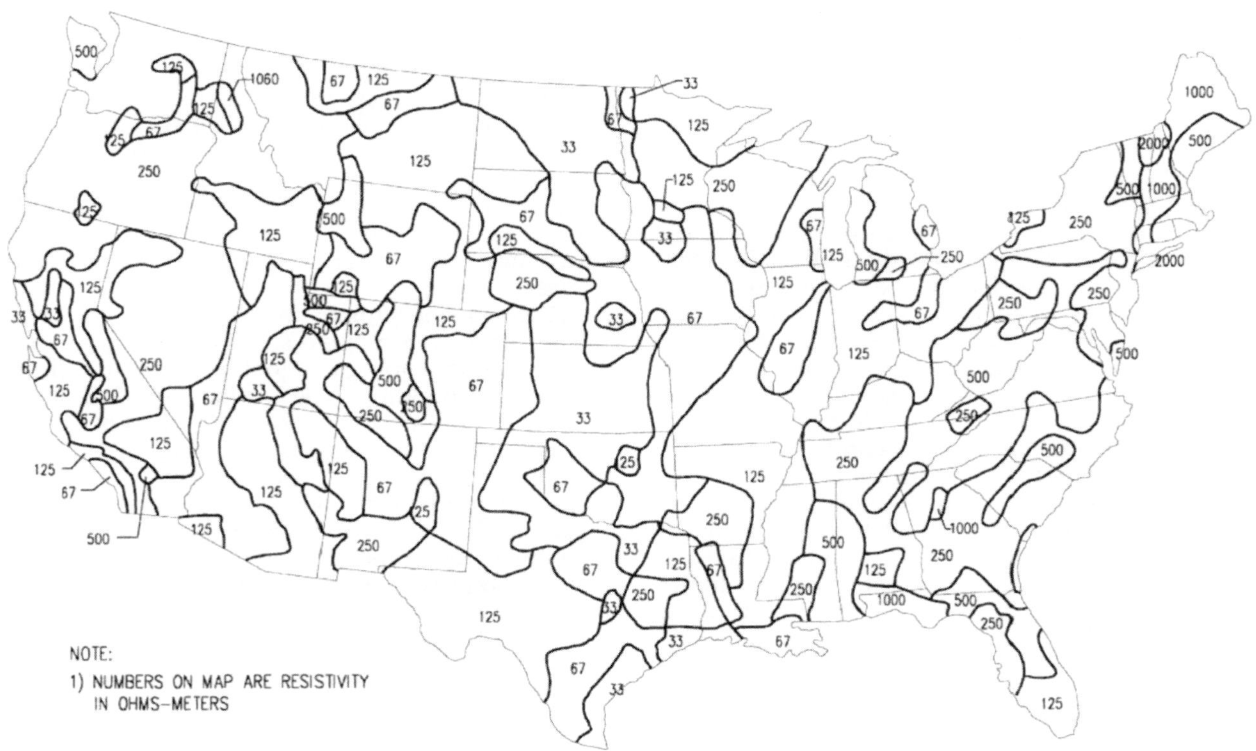

FIGURE 7.3-14 Estimated average earth resistivity in the United States.

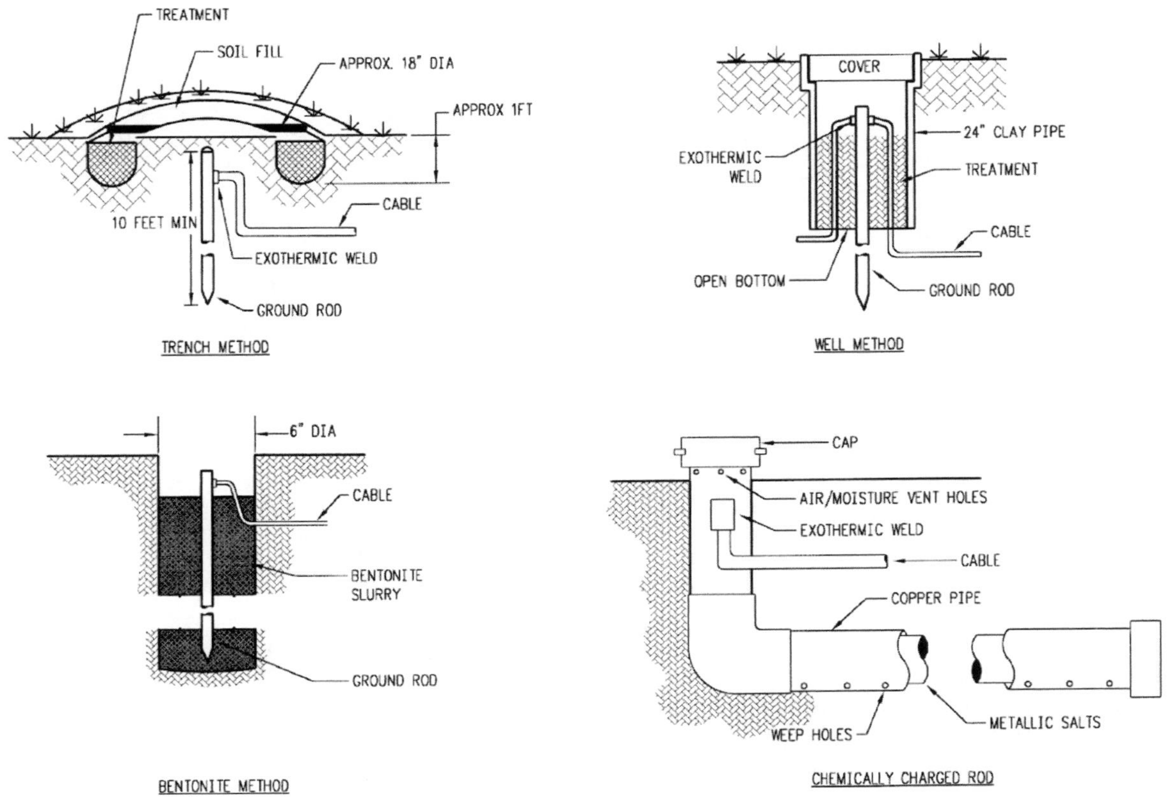

FIGURE 7.3-15 Common chemically treated ground systems.

Therefore, there is not much that can be done about the tower inductance once the tower structure is known. Grounding, however, can be minimized with proper grounding techniques.

Figure 7.3-16 describes a typical grounding configuration for a guyed tower and associated transmitter building. Tower grounding may be accomplished either by a system of interconnected driven electrodes between the tower base and guy anchor points or by radial counterpoise conductors without rods. The chosen method should be determined from previously described electrode calculations and earth resistivity testing. In the driven electrode configuration, one radial counterpoise conductor is extended from the base of the tower to each guy anchor point. This conductor interconnects driven or ring electrodes near the tower base, at guy anchor locations and at intermediate points to assist in dissipating wire currents.

The rings shown feeding the system of radial conductors should stop within a few feet of the tower base. The complex mesh created by multiple bonds between the rings and radials will help to feed lightning current efficiently from the tower legs into each of the radial conductors. Apart from providing more copper in contact with the earth, there is no advantage in adding additional rings in the area between the tower base and guy anchor points. Current flowing from the tower base out on the radials will produce approximately equal potentials between adjacent radials. With nearly equal potentials at both ends, additional bonding conductors between the radials will carry little or no current. There is only a slight advantage in bonding between radials for guy wire currents introduced into the grounding system at guy anchor points.

A perimeter ground ring (actually shown as a square) or counterpoise is shown that encircles the station building helping to equalize potential differences within the building. The station building ring also serves as a connection point for driven electrodes, fencing, and other objects that must be bonded. A bonding conductor is shown between the station building ground ring and the tower base or grounding system. This conductor will equalize potentials and minimize the level of current carried between the tower and station building by the coaxial lines.

A commercially available bulkhead plate is shown on the side wall of the station building and is bonded

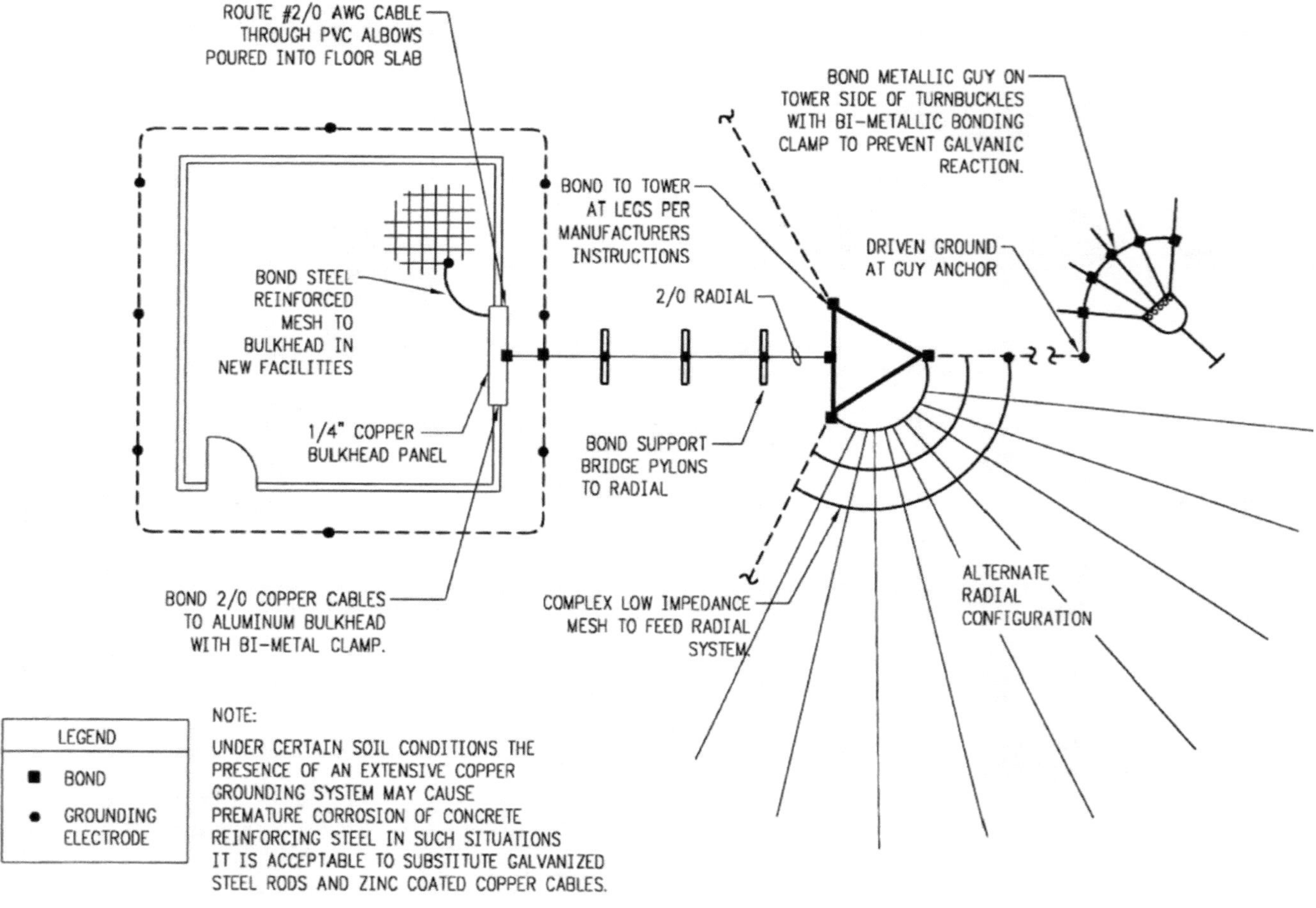

FIGURE 7.3-16 Typical tower ground system.

to the station grounding ring. This plate serves as a single point ground for all equipment within the station building and the coax lines. In new construction, the steel reinforcing mesh in the station building floor should be bonded together to the bulkhead panel to minimize potential differences between the equipment and floor during a lightning strike.

EQUIPMENT DAMAGE PROCESS

Most lightning damage to equipment occurs as the result of potential differences that exceed the tolerance level of the equipment. These potential differences may be presented to the equipment or system through external metallic circuits as a conducted current or induced transient voltage surge. They may also occur as the result of differences in ground potential at various items of equipment that are connected together to form a system. Figure 7.3-17 describes these situations in greater detail.

Common mode surges, which are also referred to as "longitudinal modes" in many documents, arrive at the equipment with approximately equal potential on both sides of a balanced pair or on a number of circuits simultaneously. These surges may be induced into a wiring system by nearby lightning, directly coupled into the circuit or even created by the action of an upstream suppression device as it clamps two or more conductors together. Common mode surges may enter equipment on power phases, signal, and other circuits. Damage is normally sustained due to potential differences between the affected circuit(s) and equipment chassis or other uninvolved circuits.

Differential or normal mode surges are often more damaging than their common mode counterpart as most equipment is designed to operate in a differential fashion. In power circuits, a differential mode surge may appear on one or more phases relative to the neutral and ground conductors. In signal circuits, especially those operating on a balanced differential basis, the tolerance to differential mode surges is lower than for common mode. Common mode surges on electrical power systems are routinely converted to differential mode at electrical services where one side of the

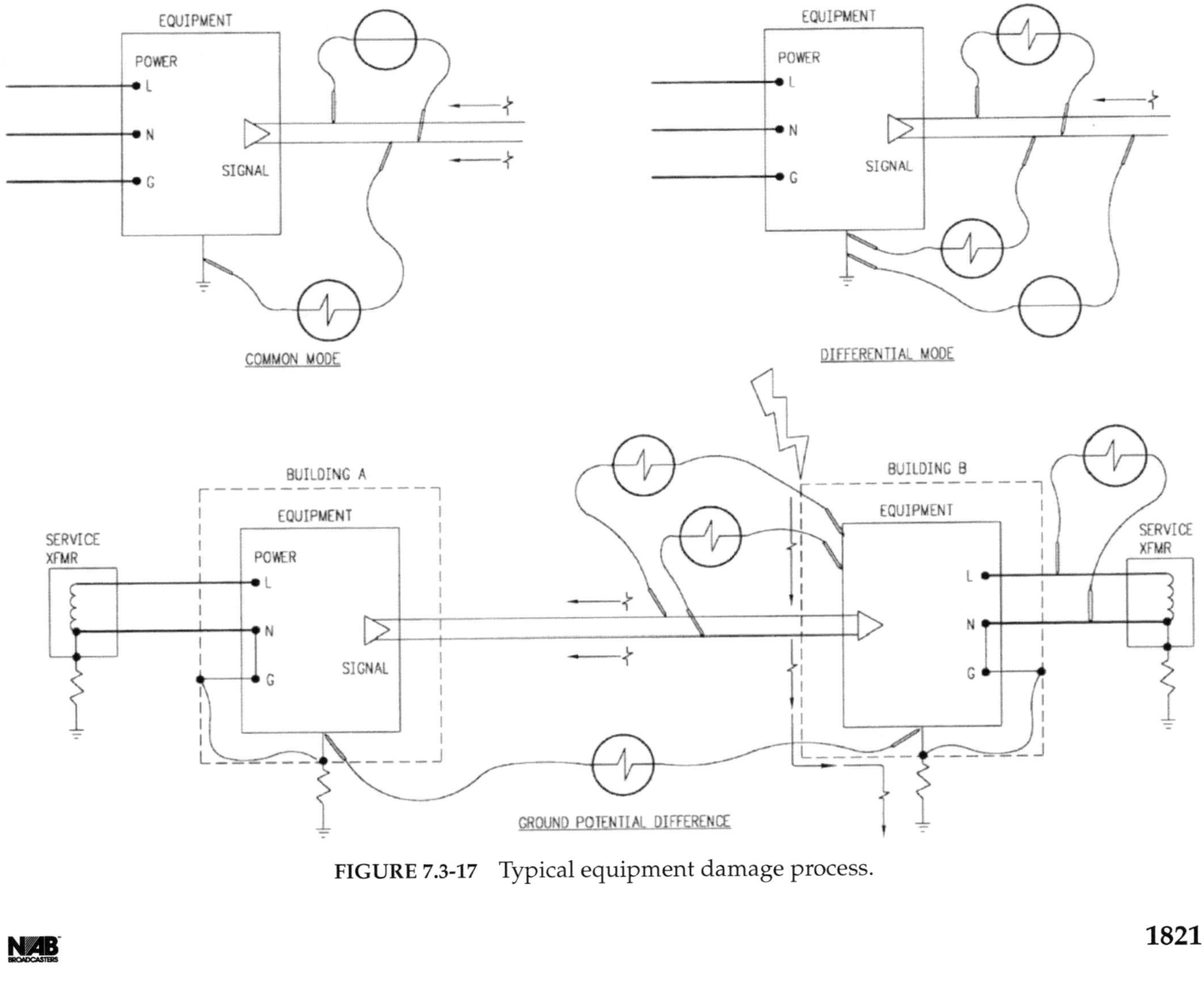

FIGURE 7.3-17 Typical equipment damage process.

service is referenced to ground. The same conversion process can occur on a balanced circuit when an upstream surge protective device (SPD) clamps one side of a pair before the other.

Ground differential damage is a bit more obscure than either of the other mechanisms. However, it is responsible for a great deal of damage to systems with equipment in multiple locations. Equipment in different buildings, or even equipment within different areas of the same building, can be damaged through ground differentials.

In the simple example of Figure 7.3-17, assume that building B receives a lightning strike of 20,000 amperes. Also assume the grounding system resistance is 2 ohms. As the lightning current flows into the earth through the 2 ohm grounding resistance, a 40,000 volt potential rise will be produced in building B's grounding system. Since the equipment in building B references the local building through its power cord and bonding conductor, its chassis will rise to about the same potential.

Circuitry within building B's equipment will attempt to track the building ground potential rise, except for the components that attach to wiring from building A. These components see a large difference in potential between the balance of their circuitry and the wiring to building A. Building A has not been involved in the ground potential rise so these circuits are still near ground potential.

Component breakdown occurs within the equipment in building B and a small fraction of the total lightning current attempts to find a path to ground through the wiring leaving for building A. Upon reaching the equipment at the remote building A, this current presents itself as a common mode surge causing damage to the equipment.

EQUIPMENT PROTECTION STRATEGIES

Protecting equipment from the effects of lightning involves a combination of grounding, bonding, and surge suppression. Grounding provides a path to introduce lightning currents into the earth. Bonding serves to equalize lightning potential differences between various elements of equipment. Surge suppression limits differences in potential on active circuits that cannot be directly bonded.

Bonding is a means of equalizing potentials during a lightning strike. Figure 7.3-18 repeats the example in Figure 7.3-17 regarding damage due to ground potential differences. This time, however, a bonding conductor is provided between the two grounding systems in an attempt to keep both at the same potential.

Kirchhoff's law tells us that current will divide itself among all of the available parallel paths through a circuit in proportion to the impedance of each path. Lightning currents behave in the same way, flowing through all available paths to ground. In direct current

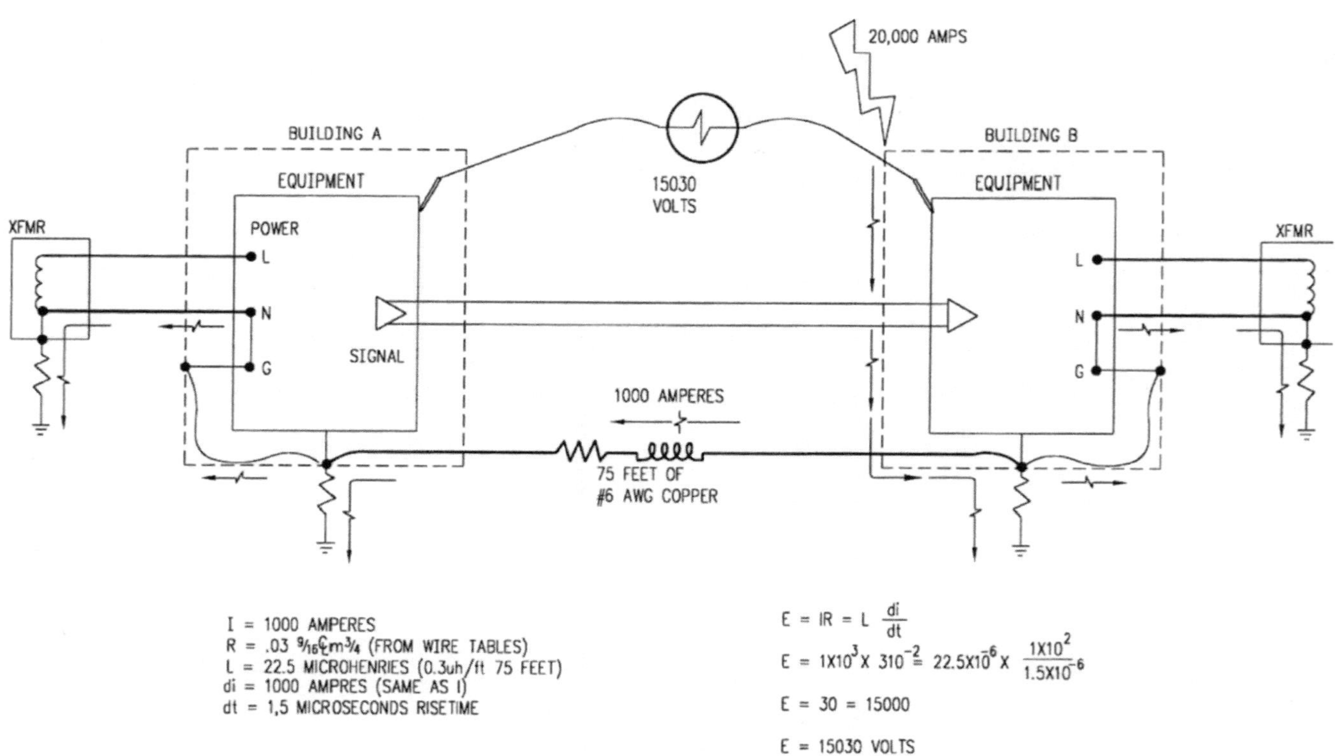

FIGURE 7.3-18 Example of ground potential difference.

circuits, the voltage produced across any circuit component is the product of current and resistance. When dealing with rapidly changing lightning current, inductance of the circuit plays a far larger role than simple resistance. Recall that an inductor tends to oppose any change in current until it has stabilized its magnetic field.

The bonding conductor shown in Figure 7.3-18 by virtue of its connection to the ground at remote building A, serves as one of several paths for lightning current to follow on its way to ground. In this case, only 5% or 1000 amperes of the lightning current flows through the 75 ft conductor with the remainder of the 20,000 amperes flowing into the grounding system of building B and its electrical service. Assuming a 1.5 microsecond rise time for the lightning current, the peak end-to-end voltage on the conductor is 15,030 volts, 15,000 volts of which is the result of inductance in the conductor. The remaining 30 volts are the result of the conductor's resistance.

The normal reaction to lowering the potential difference between buildings in this example is to suggest a larger cable. After all, larger cables have less resistance and lower voltage drop. Changing to a larger cable, however, has little effect on the circuit inductance, affecting primarily the 30 volt portion of the total.

Figure 7.3-19 provides a comparison of inductance values for a 1 ft length of various sizes of round conductors, strip materials, and coaxial cables. Strip materials are considerably more effective for the same cross-sectional areas as a round conductor, making them more attractive as a bonding medium for lightning protection purposes and bulkhead grounding. Another

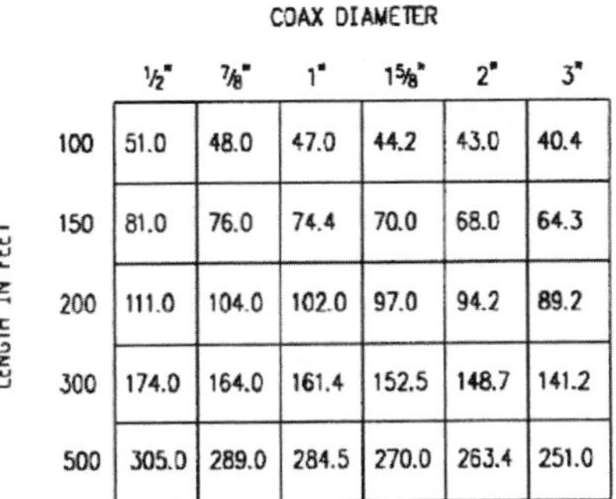

LENGTH IN FEET	COAX DIAMETER ½"	⅞"	1"	1⅝"	2"	3"
100	51.0	48.0	47.0	44.2	43.0	40.4
150	81.0	76.0	74.4	70.0	68.0	64.3
200	111.0	104.0	102.0	97.0	94.2	89.2
300	174.0	164.0	161.4	152.5	148.7	141.2
500	305.0	289.0	284.5	270.0	263.4	251.0

COAXIAL CABLE

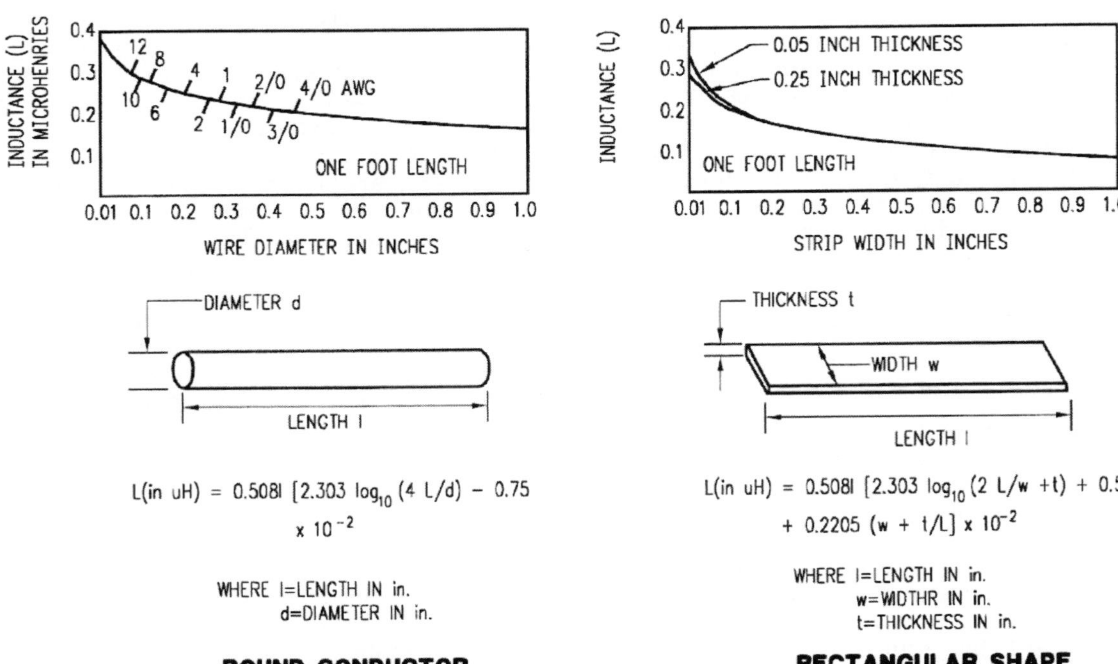

FIGURE 7.3-19 Inductance values for common engineering materials.

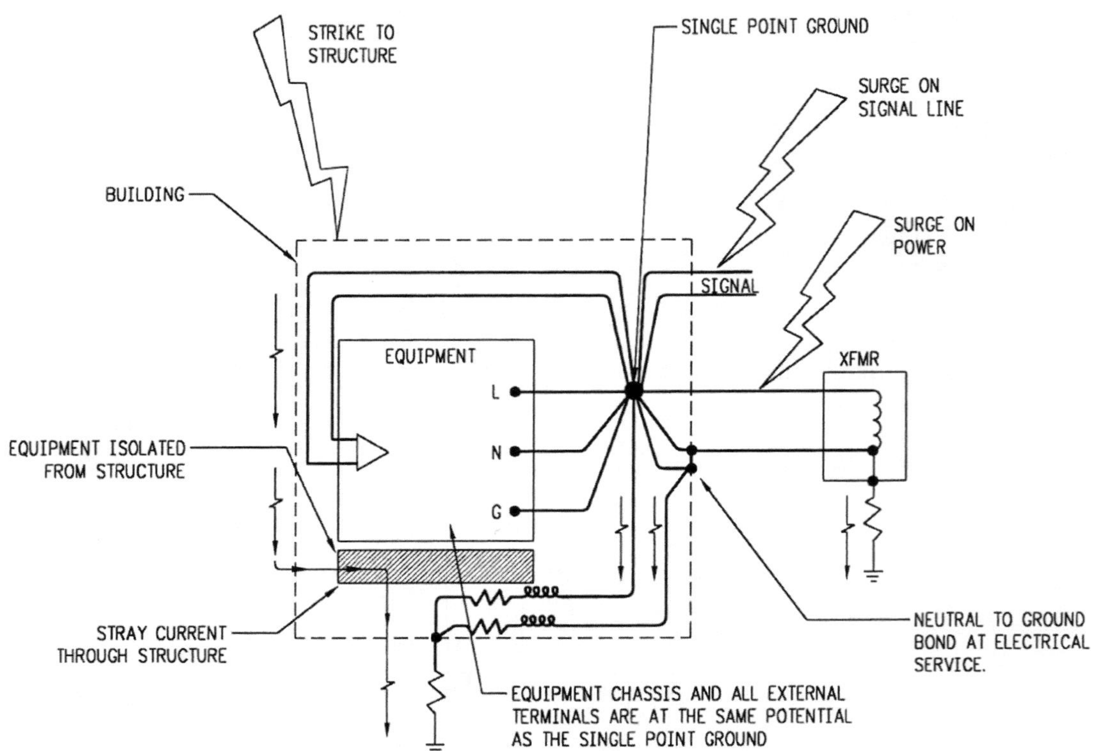

FIGURE 7.3-20 Single-point grounding concept.

interesting property of strip material is that it is not proportionally more effective once a width of about 4 inches is reached. The major reason for thickness is mechanical strength and mounting convenience.

As a final note on the inductance of bonding materials, *never expect an insulated conductor in steel conduit to carry lightning current effectively.* Ferrite beads make a reasonable effective low pass filter when placed around an insulated conductor. Steel conduit around an insulated cable creates the same effect, increasing the inductance of the cable within the conduit at least an order of magnitude similar to a choke coil. Where this condition exists (and there are many locations) a marginal compromise is to bond both ends of the cable to the conduit, permitting the conduit to serve as part of the circuit.

Single-Point Grounding

If, after the exercise in bonding, it is concluded there is no way of preventing potential differences in a conductor carrying lightning current, that is correct. There is, however, a method of preventing lightning current flow through a bonding circuit. With no current flow there can be no potential difference between the bonded items. This method is called *single-point grounding*.

In Figure 7.3-20, the equipment chassis and all metallic circuits leaving the equipment for the outside world have been bonded together and to a ground conductor at a single point. There is no possibility of a difference in potential between the circuits entering the equipment or between these circuits and the chassis as they are all bonded together. There is also no possibility of current flow from the single-point ground into the equipment through any of the circuits as the equipment is isolated from the structure.

A surge entering on the power or signal lines cannot present itself to the equipment in differential mode as the lines are all connected together. A common mode surge arriving at the single-point ground will pass harmlessly to ground through the grounding conductor and ground electrode resistance. There will be potential rise at the single-point ground due to inductance and resistance in the grounding circuit, however, no current can flow through the equipment as it remains isolated from other points of ground reference. The equipment will simply rise and fall in potential, tracking the potential of the single-point ground.

A lightning strike to the building or other structures connected to its grounding system will also cause the single-point ground to rise and fall in potential. The equipment, however, sees no potential difference as its chassis and all external metallic circuits are tied together at the single-point ground. The only difficulty with the example in Figure 7.3-20 is that *nothing works!*

Figure 7.3-21 provides a more realistic approach to single-point grounding. The only difference between

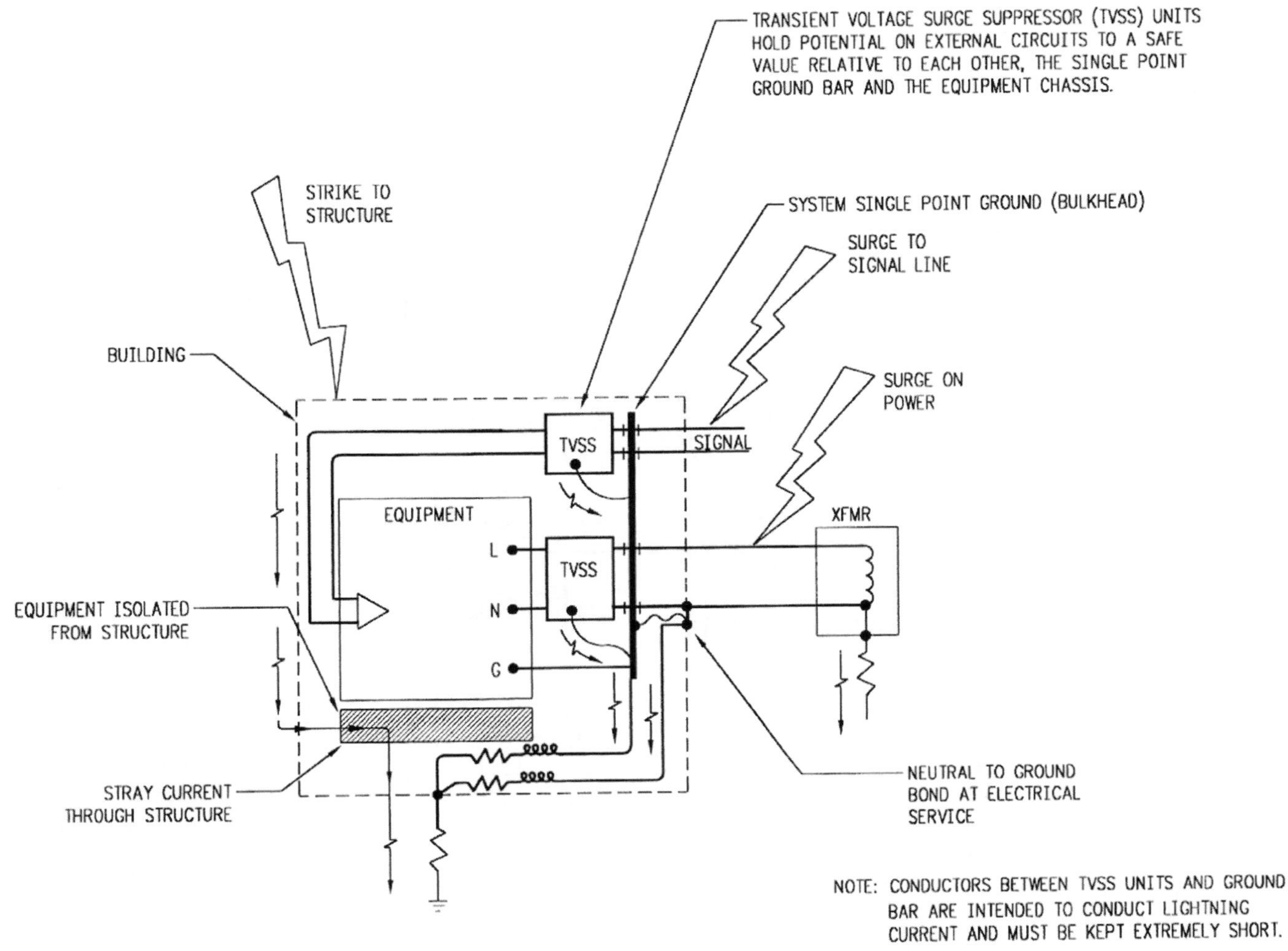

FIGURE 7.3-21 Practical application of single-point grounding with TVSS.

Figure 7.3-21 and the preceding example is that transient voltage surge suppression devices (TVSS) are used on the active circuits, which, for obvious reasons, cannot be directly bonded to the single-point ground.

For the purpose of this discussion, it is helpful to think of surge protectors as a conditional bond, clamping or limiting the excursion of voltage on active circuits to a safe level relative to each other and to the single-point ground. The single-point ground may rise and fall in potential as the protectors discharge current into it or during a strike to the building, but the difference in potential presented to the protected equipment is always held within safe limits.

It is worthwhile noting that SPDs will clamp in response to a rise in potential on their ground terminal as well as for legitimate transients on their active conductors. A strike to the building or nearby structure will cause a significant elevation in ground potential. The single-point ground will rise in potential by virtue of its connection(s) to the building grounding system. The SPD units, seeing their building ground terminals rise in potential above their remotely connected active circuits, will clamp, forcing the active circuits to track the potential of the single-point ground and the chassis of the protected equipment. Again, the voltage excursion seen by the equipment is held to a safe level and no damage is sustained.

Applications of Single-Point Grounding

The application of single-point grounding is normally limited to equipment within a room or a group of rooms. While it is possible to design larger configurations, the need to bring circuits in at different locations soon dictates the need for multiple locations, each treated as an island of equipment with its own protection devices and single point of ground reference. Larger single-point grounding systems are also more susceptible to induced voltages from nearby lightning by virtue of their increased cable lengths.

As an example of two extremes in scale, a computer room that serves terminals throughout a station complex may be engineered with a single-point grounding system and proper surge protection on its

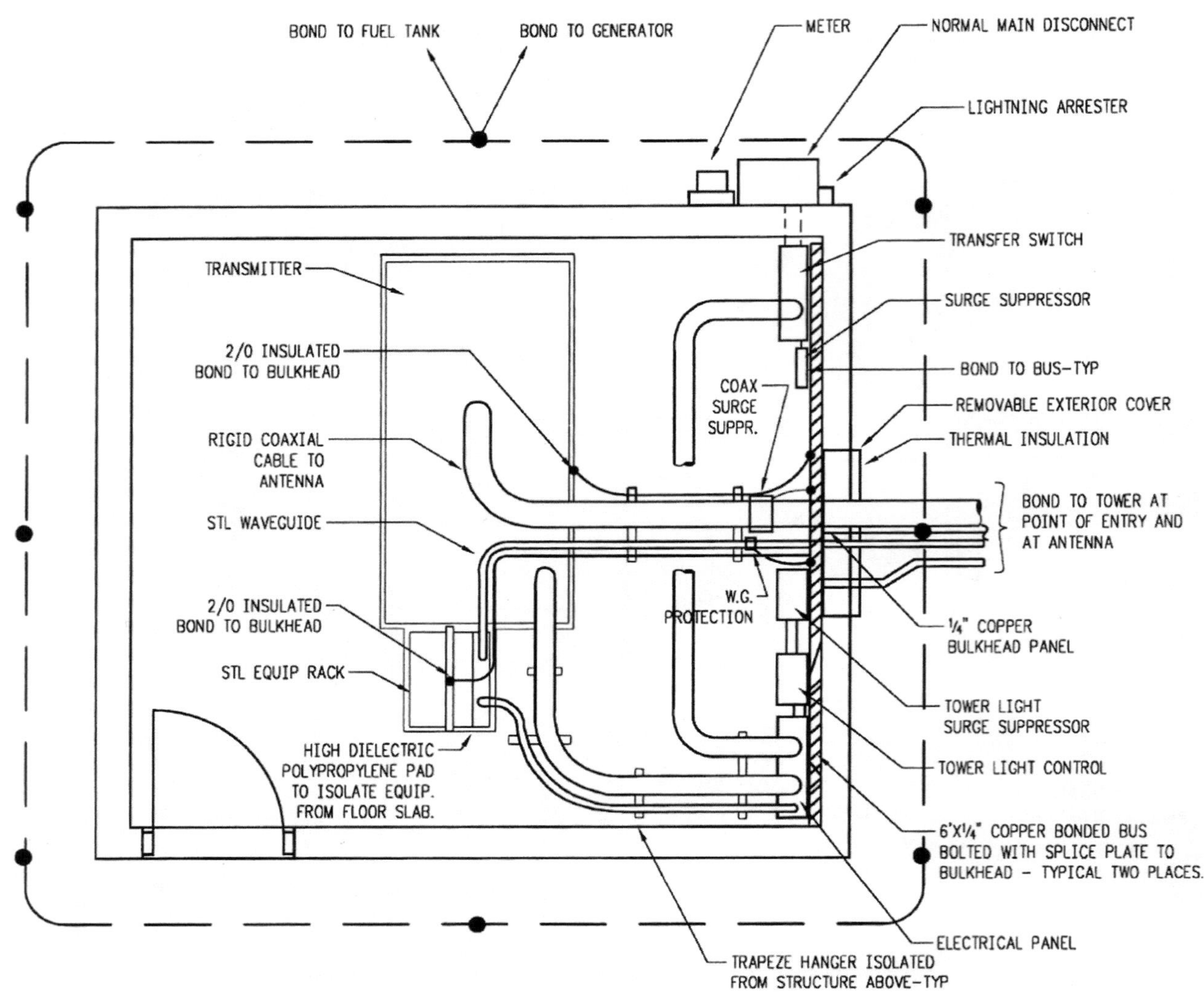

FIGURE 7.3-22 Single-point transmitter building grounding.

external circuits. The terminals and their printers, however, are scattered throughout the building, referencing ground at each location through their power cords. It is possible to designate the ground pin on the receptacle for each terminal/printer combination as the single-point ground for the equipment at that location. A combination power and data protector may be provided for each location that ensures that these conductors are held within safe limits of the receptacle ground pin and chassis of the equipment. The equipment is isolated from stray grounds by placement on a desktop.

Figures 7.3-22 and 7.3-23 are examples of how single-point grounding applies to a typical broadcast tower-building installation. A bulkhead panel in the wall of the building serves as a single-point ground reference for all equipment within the small facility. All coaxial cables, waveguides, and raceways from the tower are bonded to the bulkhead as they pass into the building.

The physical size of the electrical equipment dictates that it cannot be located directly at the bulkhead panel. To minimize the effect of bonding system inductance, a 6 inch wide bonding bus is extended to each side of the bulkhead, or for large installations, an overhead ground ring is provided. Width of the strip provides the necessary low inductance. Its 1/4 inch thickness, while not necessary for electrical reasons, provides the installer with a bus that may be drilled and tapped to accept short bonding pigtails to the equipment.

Support hangers for cables and raceways serving the protected equipment are isolated from the roof structure to prevent inadvertent current flow through the raceways. Isolation is provided between the equipment feet and floor slab by a high-dielectric polypropylene pad. Such isolation may not be necessary if adequate isolation is provided by the equipment feet. Nylon bolts may be used with conventional expansion anchors to secure equipment to the floor

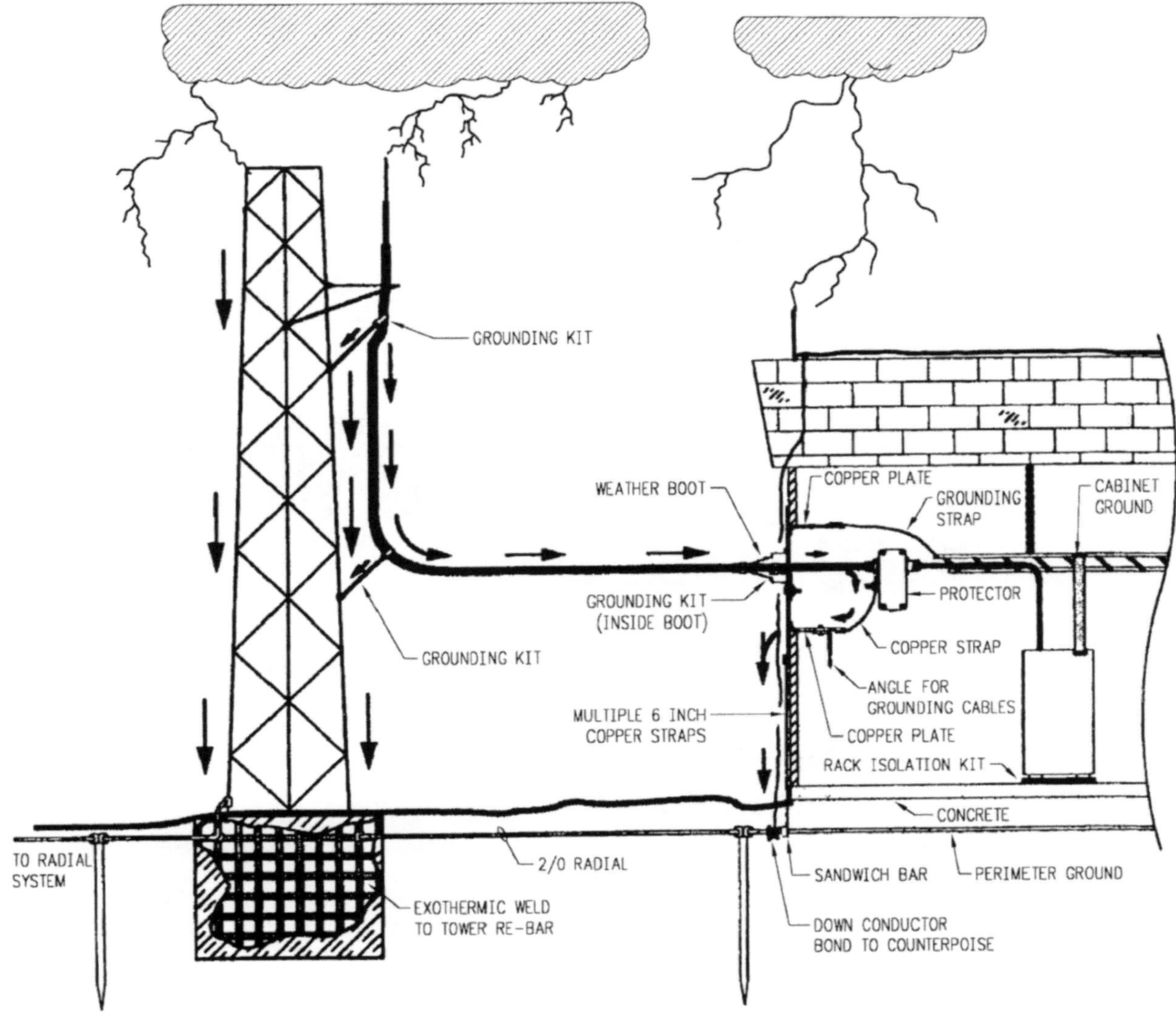

FIGURE 7.3-23 Typical bulkhead installation with single-point grounding.

without violating the integrity of the single-point grounding system.

SURGE PROTECTION

In Figure 7.3-24, surge protectors are shown at the transfer switch emergency feed to protect the transfer switch from surges on the emergency power feeder whenever the feeder and generator starting controls are not run in rigid metal conduit. Where metal conduit is used, it must be bonded to the perimeter ground ring. Main service surge protectors should be connected through individual or integral fused switches or breakers with short-circuit interrupting ratings equal to or higher than the available service fault current. Also, a protector is installed on the circuit for tower lights. Both protectors should be bonded to the grounding bus, keeping the length of their bonding lead as short as possible. Surge protection should also be provided on all metallic, control, monitoring, and communications lines and circuits serving lighting, winches, and other electrical items outside the building. Again, locate these suppressors at the point of entry for the circuits and bond their ground leads to the bus with the shortest possible lead length.

The lightning arrester shown at the main disconnect outside the building serves an important purpose. Under normal operation, the main disconnect is closed and the transfer switch is connected to utility power. During operation on emergency power the normal transfer switch position is open. Potentially, a lightning strike to the utility line will propagate along the line as a traveling wave in both directions from the point of lightning contact. Upon reaching the open circuit input of the transfer switch, the wave will reflect back on itself, potentially doubling its initial

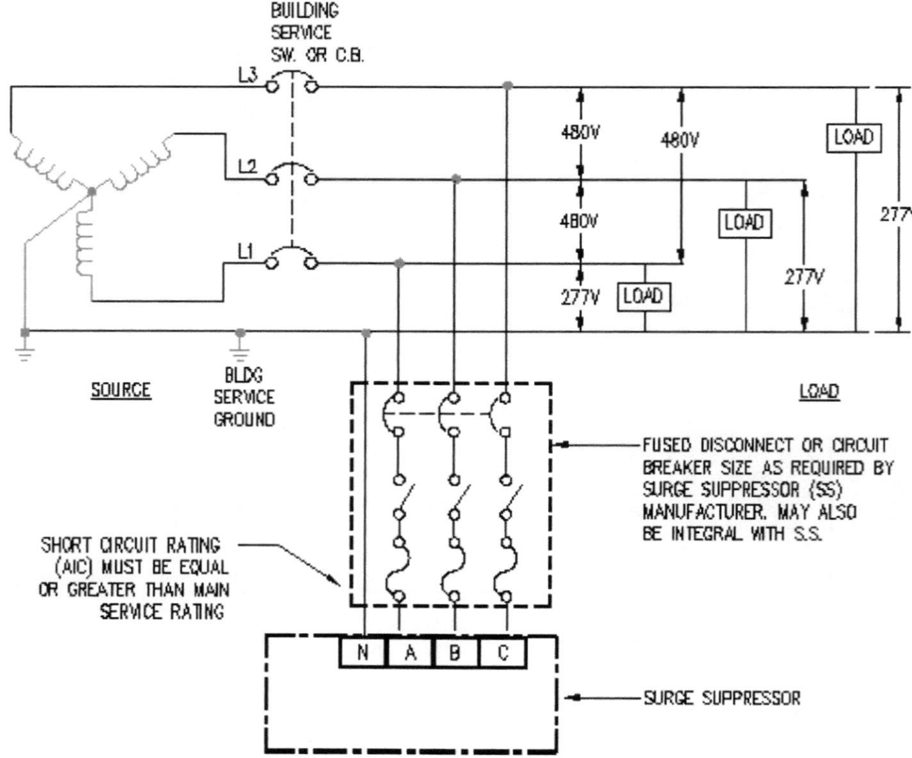

FIGURE 7.3-24 Main service surge-suppression installation.

crest value. The same condition can occur at the main utility company disconnect if it should be open during servicing and each condition can cause equipment flashover.

Flashover within electrical equipment is serious, however, the problem is compounded when operating voltage is also present. The flashover arc provides a low-impedance path for 60 Hz follow current and significant damage to the equipment may occur. The arrester will prevent this condition from occurring by limiting the traveling wave voltage.

It is of great importance to minimize the length of the suppressor leads and that of the grounding conductor as their inductive voltage drop is additive with the initial clamping voltage of the suppressor.

Signal Line Suppressors

Figure 7.3-25 shows a common surge suppressor configuration for signal line and coax applications. Most suppressors of this type are both shunt type and multistage hybrid devices utilizing a high-energy first stage, a fast-acting second stage, and impedance in series between the two stages to coordinate their clamping behavior. Because of the multistage design, these devices must be installed in series with the protected circuits.

The treatment of shields is often an issue when dealing with signal line surge suppressors. Figure 7.3-25 also shows shields being bonded to the suppressor ground bus to force them to track the single-point ground. If ground loop or other technical restrictions prevent direct bonding of the shields, they should be protected with suppressors as any other active circuit.

The bonding lead distance for signal line suppressors is often more critical than for power devices. Tolerance levels of signal circuits are normally lower than for power-supply inputs, and the relatively small voltages developed in the suppressor ground leads can become significant.

Signal line and coaxial surge suppressors are sold in a variety of shapes and sizes for different applications. The suppressors should generally clamp transient voltage on a circuit to within 150% of normal peak-operating voltages and even lower in some applications. Since most signal line suppressors are inserted in series with the circuit being protected, it is wise to evaluate the effect of their series impedance and capacitance on the insertion loss of a circuit.

The effect of suppressor capacitance can be important in many high speed data, RF, and video applications. One simple way of evaluating the effect of this capacitance is to equate it to equivalent cable feet. For example, if the desired suppressor exhibits capacitance of 100 picofarads, and the cable used in the circuit is rated at 10 picofarads per foot, will the circuit tolerate an additional 10 ft of cable? If so, the suppressor capacitance should produce no noticeable effect on the circuit.

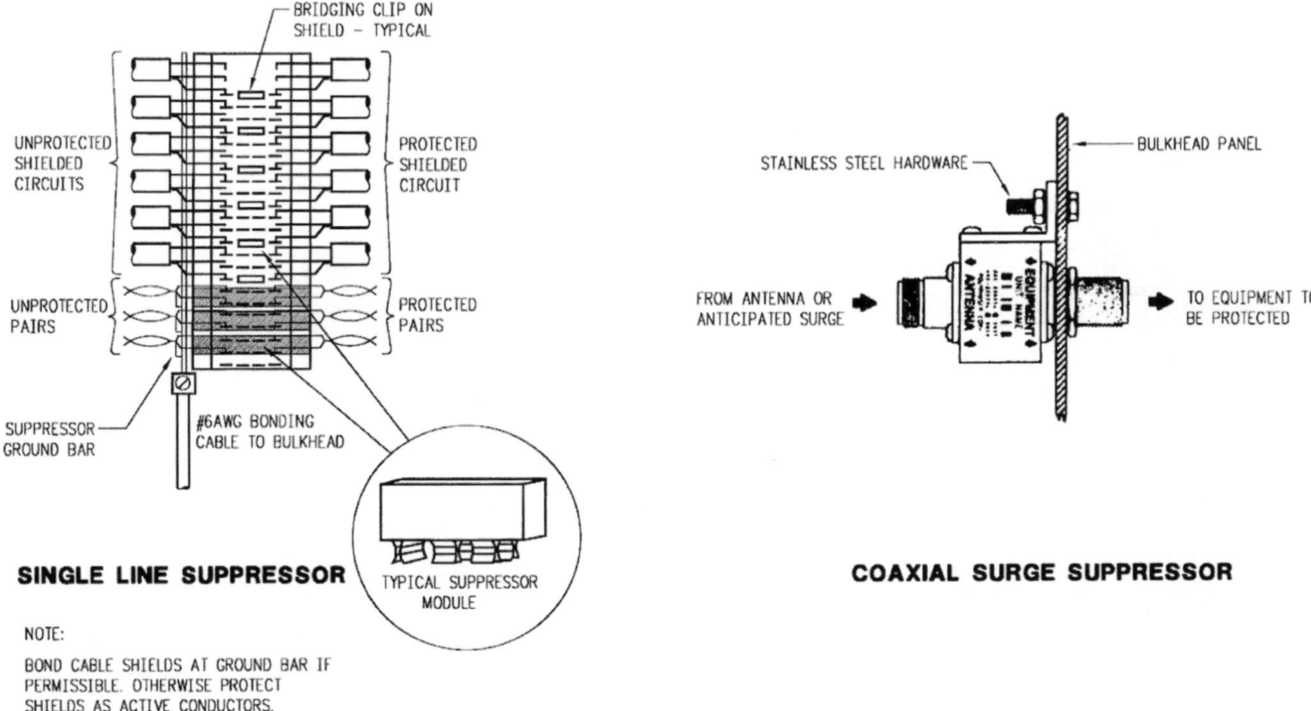

FIGURE 7.3-25 Common signal and coaxial suppressors.

Surge Protection Selection

Surge protectors and lightning arresters come in all shapes, sizes, types, and accessories, and a complete discussion of application methods and evaluation is not possible in this document. Contacting surge protector manufacturers, however, may be extremely confusing due to the myriad of choices available, but eventually a choice must be made. Contacting the chairmen of the referenced IEEE, UL and IEC standard committees may also be helpful in understanding application issues. The following guidelines therefore may be helpful:

- Select protectors and arresters based on IEEE C62.41.1 (2002).
- All surge protectors should be UL 1449, 2nd edition (or future 3rd edition) tested, approved, and labeled.
- Ask for published proof of testing to back up claims, especially clamping levels, energy handling capabilities (joules, watts, or current), and speed of response.
- Look for a minimum of 3–5 year unconditional warranties. Many companies will offer 10 year warranties, but only *unconditional* warranties are of any value.
- Review the six major performance characteristics: response time, voltage protection level, power dissipation, disturbance-free operation, reliability, and operating life. Make sure maximum power dissipation level and voltage protection level using specific waveforms are stated at the same point to avoid misinterpretation.
- Talk to other similar users for advice and recommendations.
- Consult industry standards for performance requirements. At present, there is no testing or application standard for coaxial surge suppressors, so a comparative review of reputable manufacturer data will be necessary. Compare: impedance (50 or 75 ohm); speed; frequency range; number of transmit, transceive- or receive-only signals; transmit power; presence of AC or DC power with the RF signals; mounting; and connector type and sex.

ISOLATED GROUND RECEPTACLE

Figure 7.3-26 shows an isolated ground receptacle circuit commonly used in computer room grounding applications. The receptacles used in this type of circuit differ from the norm in that their ground pins are electrically isolated from their mounting tabs. They are, therefore, isolated from their outlet box and structural ground at each receptacle location. A dual system of grounding conductors ensures that equipment plugged into an isolated ground (IG) receptacle references ground first at the single-point ground.

The use of isolated ground receptacles helps to ensure that plug-in terminals, printers, diagnostic, and

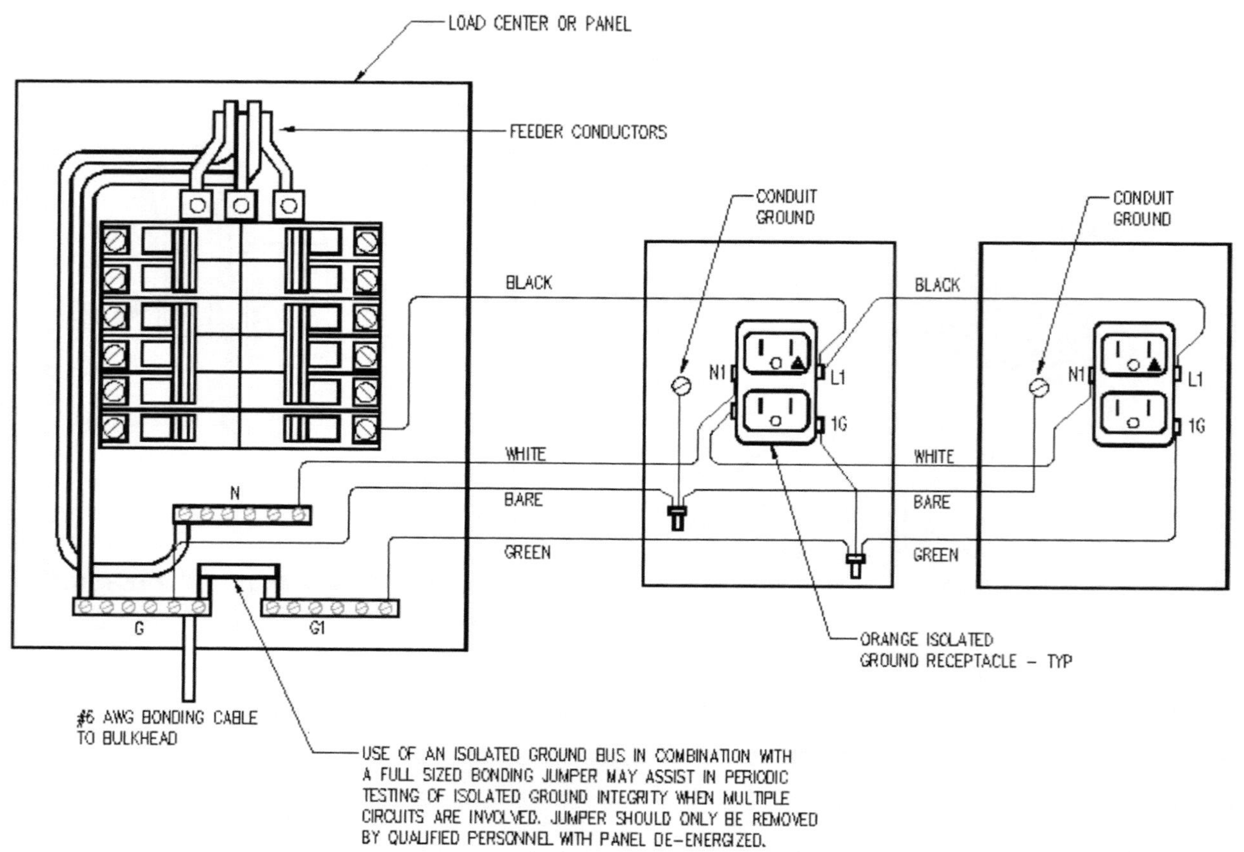

FIGURE 7.3-26 Isolated ground receptacle.

other ancillary equipment are properly referenced to the single-point ground and not the local structure. It only takes one item of equipment connected between the protected equipment and a remotely grounded receptacle to compromise the integrity of the grounding system.

Figure 7.3-27 is a composite of the bonding and grounding recommendations for the typical broadcast site. While complicated in appearance, each component has its purpose as part of asimple-to-understand subsystem. Finally, Figure 7.3-28 illustrates a comprehensive single-point grounding system that ties all elements of the grounding system together.

NEW VERSUS EXISTING SITES

The basic suggestions enumerated in this article should apply equally to both new and existing installations, however, sites having existing towers, buildings, grounding systems, and site appurtenances offer a more challenging design effort to establish truly complying grounding and bonding solutions. For instance, a new tower installation can take advantage of tower base reinforcing-steel grounding whereas grounding for an existing tower must be enhanced to account for the loss of such a massive and low impedance grounding component. Also of critical concern in existing sites is that the existing grounding systems cannot be isolated sufficiently without shutting down the entire broadcast station to permit accurate grounding measurements. The challenge for the designer, therefore, is to approximate the ideal conditions with creative technical modifications that do not violate the basic tenets of grounding and bonding presented herein and in associated publications. In most cases, calculating exact grounding and impedance values to compare to some ideal value is not as important as *relative* values between various portions of the grounding and bonding components.

Site Audit Report

Performing a site audit report for an existing site requires more time and attention to detail than at a new site, obviously, but the audit report will form the basis of the grounding and bonding desired. Detailed sketches of all grounding systems, pictures of all conditions, and uncovering buried systems to minimize guesswork are critical to establishing a basic understanding of the existing systems on which to build or improve.

Designing and investigating an existing site with maintenance in mind will enhance the ability of those

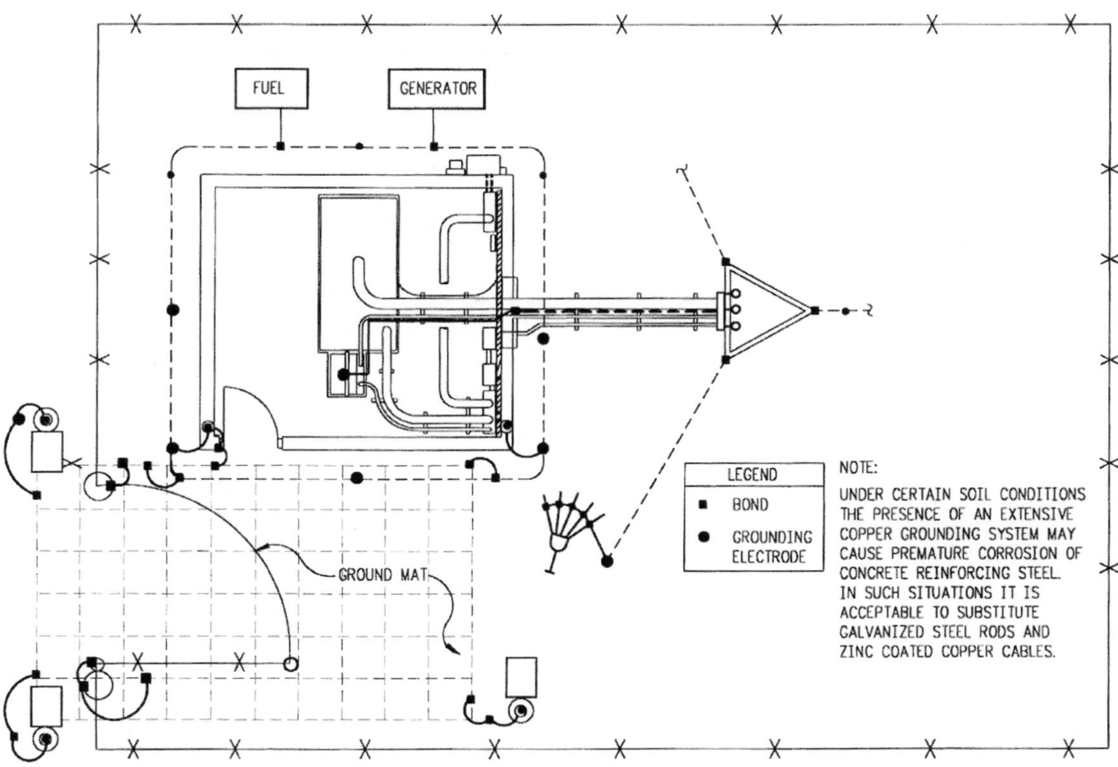

FIGURE 7.3-27 Typical composite bonding and grounding installation.

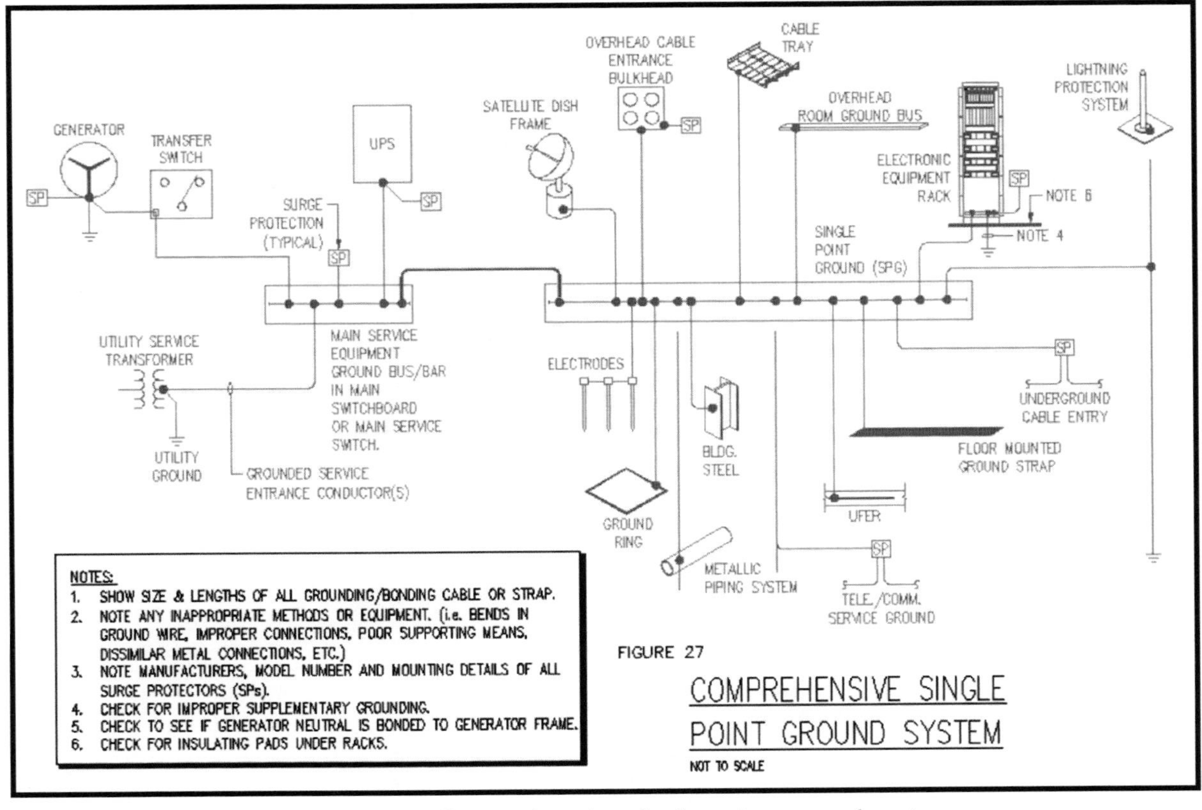

FIGURE 7.3-28 Comprehensive single-point ground system.

who will be asked to maintain a secure, quality system to provide the tools necessary to do their job. Documentation of existing equipment types and model numbers, wire/cable types and sizes, conductor routings, connector and splice types and locations, sketches, pictures, plans, etc. should all be cataloged and properly tabbed for ease of future reference.

SUMMARY

This chapter has been written in tutorial form as every site is different and no single set of recommendations will apply to every situation. The principles set forth, while tailored to a broadcast environment, apply equally to other systems. There are still a few mysteries to be solved in completely understanding lightning, but once it enters a wiring system, it becomes an electrical current that is both predictable and understandable. A list of publications that deal with the subject material in greater depth is included at the end of this chapter.

Images of Typical Grounding Situations Found at Broadcasting Stations

A series of images of typical grounding situations at broadcast stations can be found on the CD-ROM for this book.

References

[1] Earle R. Williams. The Electrification of Thunderstorms, *Scientific American*, November 1988.

[2] NFPA 780, Lightning Protection Code, National Fire Protection Association, Quincy, MA.

[3] UL-96A, Standard for Lightning Protection Systems, Underwriters Laboratories, Northbrook, IL.

[4] See [2] above.

[5] See [3] above.

[6] LPI-175, Standard of Practice, Lightning Protection Institute, Woodstock, IL.

[7] Military Handbook 419A, Grounding, Bonding, and Shielding for Electronic Systems, December, 1987. Available from the U.S. Government Printing Office, Philadelphia, PA.

[8] IEEE 142, Grounding Practices for Electrical Systems.

Additional Resources

ANSI/IEEE C62.33 (1989), Standard Test Specifications for Varistor Surge Protective Devices.

ANSI/IEEE C62.35 (1987), Standard Test Specification for Avalanche Junction Semiconductor Surge Protective Devices.

ANSI/IEEE C62.36 (2000), Standard Test Methods for Surge Protectors Used in Low-voltage AC Power Circuits.

ANSI/IEEE C62.41.1 (2002), IEEE Guide on the Surge Environment in Low-voltage (1000 V and less) AC Power Circuits.

ANSI/IEEE C62.41.2 (2002), IEEE Recommended Practice on Characterization of Surges in Low-voltage (1000 V and less) AC Power Circuits.

ANSI/IEEE C62.34 (R2001) (1996), IEEE Guide for Performance Low-voltage Surge-protective Devices (Secondary Arresters).

ANSI/IEEE C62.45 (1992), Guide on Surge Testing for Equipment Connected to Low-voltage AC Power Circuits.

Federal Information Processing Standards (FIPS) Publication 94, CCITT Rec. K-17, Waveform Specification for Electronic Systems.

IEEE 518, Recommended Guide on Electrical Noise.

IEEE 142 (1991), Grounding Practices for Electrical Systems ("green book").

Military Standard (MIL-STD-220A) (1952).

NEMA LS-1 (1992), Low-voltage Surge Protective Devices.

NFPA 70 (2005), National Electrical Code.

NFPA 75 (2003), Standard for Protection of Electronic Computer Systems.

NFPA 780 (2004), Standard for the Installation of Lightning Protection Systems.

UL-1283 (2nd edition), Standard for Safety—Electromagnetic Interference Filters.

UL-1449 (1998), Standard for Safety—Transient Voltage Surge Suppressors.

CHAPTER

7.4

Tower Lighting and Monitoring

RICHARD G. HICKEY

Flash Technology
Franklin, Tennessee

INTRODUCTION

The Federal Communications Commission (FCC) and Federal Aviation Administration (FAA) take tower lighting and marking seriously. Broadcasters and other owners and operators of towers that are required to be lighted and marked according to specific rules and regulations have received substantial fines and penalties when those requirements are not met. Solutions include strict adherence to the rules and regularly reviewing current tower lighting and marking conditions to ensure compliance. If a broadcaster is uncertain about whether a tower meets FCC and FAA requirements a conversation with a broadcast consultant, tower manufacturer, or FAA airspace specialist would be in order. In addition to keeping a current version of the FCC rules on hand for review of applicable tower lighting regulations, the latest version of FAA Advisory Circular AC 70/7460-1K is available from the FAA website at http://www.faa.gov/ATS/ata/ai/index.html.

In this chapter the following important issues concerning tower lighting and marking are reviewed: general information, types of lighting systems available, lighting versus painting, maintenance, and methods of tower light monitoring.

GENERAL INFORMATION

The FAA is the agency designated by the FCC to author the regulations for lighting and marking structures, yet the FCC levies the fines. Simply put, the FAA makes the rules and regulations on marking, lighting, and monitoring and the FCC enforces them. The following documents represent the essential elements of information regarding tower lighting:

- The FAA Advisory Circular AC 70/7460-1K. This document, which can be found at http://www.faa.gov/ats/ata/ai/AC70_7460_1K.pdf, describes the requirements for marking, lighting, and monitoring.
- FAA Advisory Circular 150/5345-43F. This document, which can be found at http://www.faa.gov, describes the requirements for paint standards and the parameters by which aviation obstruction lighting (AOL) must perform to be tested and certified.
- FAA Advisory Circular AC 50/5345-53C. This document, which can be found at http://www.faa.gov, lists all approved manufacturers, equipment, and testing labs.
- Form 7460-1. This document, which is a request for alteration of a structure, can be found at http://forms.faa.gov/forms/faa7460-1.pdf.

Before starting a review of the various regulations, it is important to note that if your tower lighting and marking scheme was performed in accordance with the FCC/FAA rules at the time the determination was given to you, the tower is probably "grandfathered." As long as no changes are made to the structure and the devices attached to it, it is likely that it is only necessary to maintain the lighting and marking as it is. However, if the tower has been or will be modified in any manner, such as changing the overall height or adding or replacing an antenna, it is likely that a new determination will be returned bearing the statement,

"Must be lighted and marked in accordance with AC 70/7460-1K," naming the specific chapters that apply to your structure. In this case the tower may be required to meet current standards for lighting and/or marking.

TOWER LIGHTING BASICS

Although there are some rare exceptions, marking and lighting a broadcast tower is straightforward, standardized, and the rules are rarely altered.

Structure height is measured and recorded in two different ways: above mean sea level (AMSL), and above ground level (AGL). Lighting standards are generally determined by the structure's height AGL.

The term *candela* is used to describe light output from lighting fixtures. In laymen's terms, the viewable output of a lighting source is measured in candlepower. The directed beam of the lighting source is measured in candela. One candela has a directional radiant intensity of 1/683 watt per steradian, equal to one "directed" candlepower. The candela is abbreviated cd and its standard symbol is I_V.

Lighting types common to the broadcast structures and listed as approved on the FAA publication 53B include the following.

L-810: Side Lights and Marker Lights

These lights, as shown in Figure 7.4-1, are most often used as the nonflashing intermediate lights used between tiers of L-864 red lighting. Their use is for night only, and they are not acceptable for daytime marking. General requirements include a light output minimum of 32.5 candelas and may use incandescent filament, LED, halogen, and neon technologies, and a 360 degree omnidirectional light output. For "red and paint" combination or dual-lighting applications, one fixture per tower leg is needed on each tier level for structures exceeding 350 ft in height. Two or more are required per level on designated structures under 350 ft in height AGL.

L-864: Medium-Intensity Red Beacon

These traditional flashing red beacons are acceptable for nighttime marking only and operate at 2000 candela ± 25%. Various technologies include incandescent filament, LED, and strobe. These lights must provide a 360 degree omnidirectional unobstructed view, necessitating the use of two beacons per tier below the tower apex. Figure 7.4-2 shows an example of a medium-intensity beacon.

L-865: Medium-Intensity White Strobe

These lights, as shown in Figure 7.4-2, are used primarily on structures under 500 ft AGL and as an AOL (in this case, appurtenance or antenna obstruction light) beacon in a high-intensity configuration. They may be used, with FAA and zoning approval, instead of aviation orange and white paint for daytime, and instead of red beacons for night marking. The approved technology for this application at the time of this writing is for strobe lights only. Output is 20,000 candelas for day and twilight, 2000 candelas for night, ± 25%. These lights must provide a 360 degree omnidirectional unobstructed view, necessitating the use of two beacons per tier below the tower apex. Note that the medium-intensity strobe does not change intensity for twilight mode.

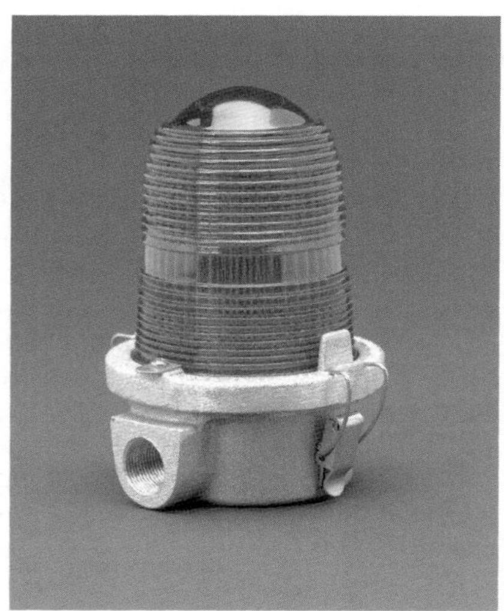

FIGURE 7.4-1 Example of typical marker light for broadcast towers.

FIGURE 7.4-2 Example of dual-technology medium-intensity tower light beacon.

FIGURE 7.4-3 Example of typical high-intensity white strobe tower light.

L-856: High-Intensity White Strobe

These are normally used, with the proper approvals, instead of FAA orange and white aviation paint on structures exceeding 500 ft AGL in height including appurtenances. Since no option for omnidirectional design currently exists, a minimum of three unidirectional (120° output) fixtures is required per tier level, as shown in Figure 7.4-4. Light output must be within ± 25% of 270,000 candelas for day, 20,000 candelas for twilight, and 2000 candelas for night.

FIGURE 7.4-4 Example of dual-lighting system on each leg of tower. Note the small ice shields above each lighting fixture. (Photo courtesy of Flash Technology.)

CONFIGURATION

Configuration and number of tier-level requirements may be found in the FAA Advisory Circular AC 70/7460-1K (see Figure 7.4-5). The configuration specifications vary between lighting types. With the exception of "red and paint" or "A" designation towers, the 500 ft AGL mark is very significant.

White strobe lighting may be used for day marking under the following conditions:

- If the tower including all appurtenances (note that the lightning protection over the uppermost beacon is considered an appurtenance) is less than 500 ft AGL, the strobe lighting will normally be medium intensity, or L-865. The number and placement of light levels are determined by the overall height of the structure including all appurtenances.
- If any part of the structure including all appurtenances exceeds 500 ft AGL, the overall height of the main (normally guyed, but not always) structure will determine the number and placement of the

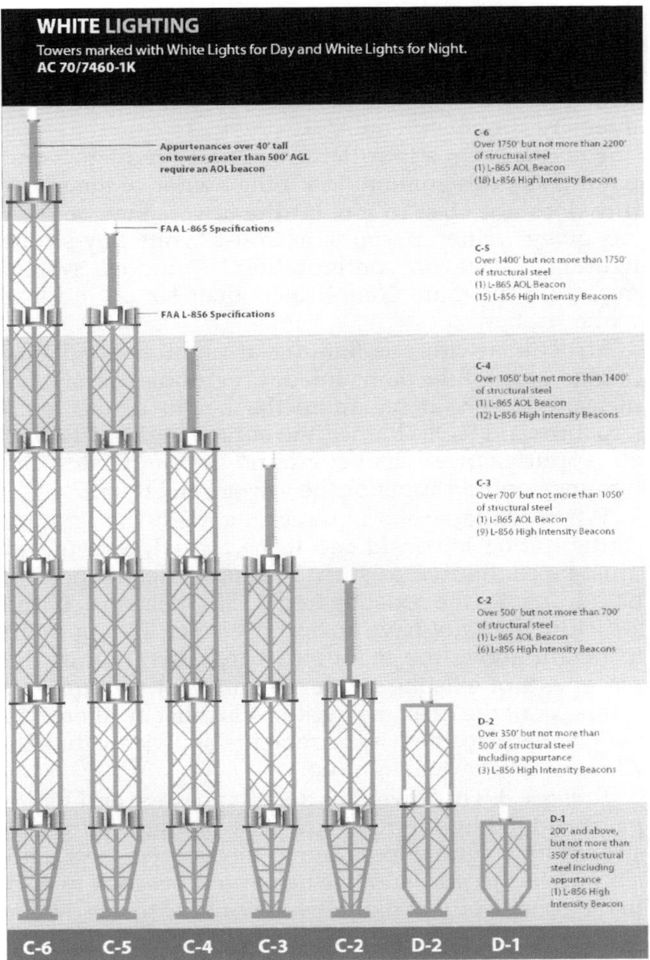

FIGURE 7.4-5 Tower light intervals for white lighting per FAA AC 70/7460-1K.

lighting tier levels. If an appurtenance or antenna exceeds 40 ft above the top of the main structure, the appurtenance must have an AOL light mounted at the top. If the appurtenance is incapable of supporting the AOL beacon, the AOL beacon must be mounted on a separate supporting pole or mast within 40 ft of the apex of the appurtenance. Two beacons may be necessary to provide an unobstructed view and mark the appurtenance in this manner. The FAA regards anything over 7/8 inches in diameter as an obstruction. If the tower has more than one mast, as in the case of a T-bar or candelabra with two or three masts, the AOL beacon(s) should be mounted on the tallest appurtenance. If all appurtenances are of the same height, only one is required to have an AOL on top.

The stipulations above apply whether the tower only has white strobe lighting or has a combination of red lighting for night and white strobe for day and twilight, commonly referred to as "dual" lighting.

TOWER LIGHTING SYSTEM EXAMPLES

Example 1: White strobe lighting is desired on a tower that is 490 ft AGL in height and has a 15 ft top mast making the overall height 505 ft. In this case the FAA will normally designate the lighting system to be high-intensity strobe with no AOL.

Example 2: On a four-tier tower using red lights and paint (FAA designation "A4") the owner of the structure does not wish to repaint the tower, and the FAA has given written permission to use white day-strobe lighting. However, configuration as a dual system may be a bit more complicated than simply adding white strobes.

Structures with red lighting at night and aviation orange and white paint for day marking are always measured without regard to the height of the main structure and AOL. Rather, the entire height including all appurtenances above ground level dictates the lighting configuration for the tower. See Figure 7.4-6.

When white strobe lighting is added, the lighting configuration is considered to be a dual system and must be configured as such. This may dictate moving or adding to the existing red L-864 beacons. If the structure does not have an appurtenance 40 ft or more in height above the main structure, alteration of the placement of existing tier levels may not be required. If this is not the case, the light levels must be altered to match the placement requirements shown in the AC 70/7460-1K for dual lighting.

Example 3: The tower has aviation orange and white paint and red lighting. The owner wishes to add white strobes for daytime to augment the safety aspect, and also wishes to add another tier of red beacons to add to the conspicuity at night. Even with the tower owner exceeding the requirements for lighting on his tower, it is necessary to notify the FAA. The FAA requires prior notification for voluntary marking and lighting. Failure to do so is a direct violation of the regulations and could result in FCC fines. Do not assume "more is better."

Example 4: The tower in this example is relatively old and is susceptible to corrosion. While the tower must be painted for maintenance purposes anyway, the owner is considering using just the aviation orange and white paint and saving the cost of the strobes.

This configuration may be acceptable, but there are other considerations. To do this the tower owner must apply for and receive authority from the FAA to use only the aviation orange and white paint. In the AC 70/7460-1K Chapter 3, Section 36, Subsection (a), the FAA states, "The high-intensity lighting systems are more effective than aviation orange and white paint and therefore can be recommended instead of marking."

Should an unfortunate event occur with this tower, the tower owner should be prepared to defend in a legal proceeding the decision to use aviation orange and white paint instead of high-intensity strobe lighting. Also, aviation orange and white paint

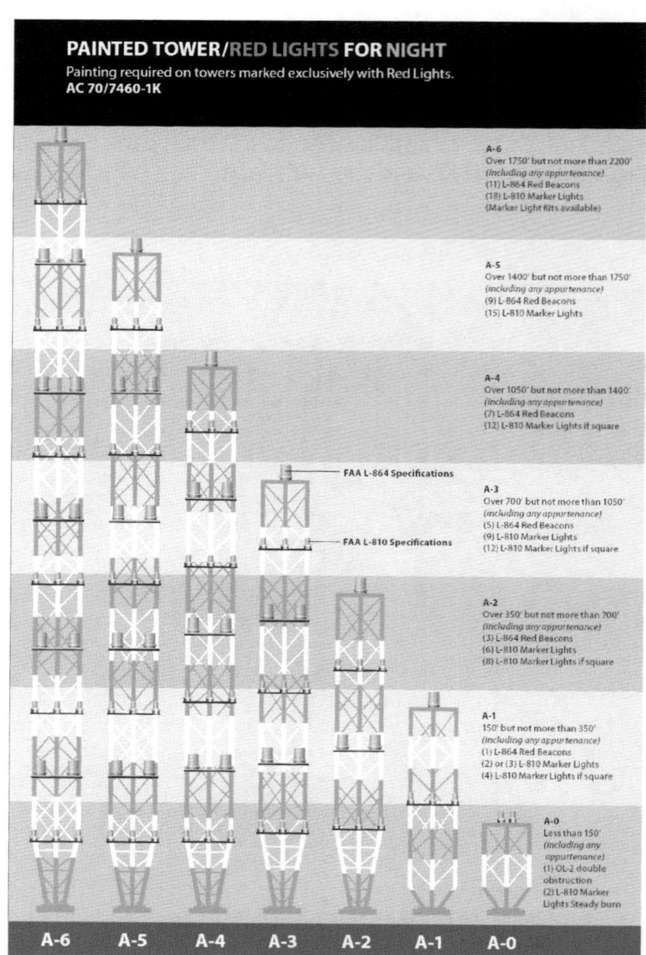

FIGURE 7.4-6 Tower light intervals for painted tower with red lighting per FAA AC 70/7460-1K.

may not be as durable as some of the rust-inhibiting paints available on the market. The aviation orange and white requires attention when fading, chipping, or peeling occurs or the tower owner risks the chance of incurring an FCC fine. Protective paint used for maintenance can be reapplied by a maintenance timetable and may provide better shielding from the elements.

EVALUATING LIGHTING TECHNOLOGIES

While only one technology (the white strobe) has met the FAA requirements for medium- or high-intensity day lighting, several choices exist for red lighting. An analysis of each FAA recognized technology is provided:

- *Incandescent filament.* This is the old standard lighting beacon, using two 620 watt lamps.

  Advantages include simplicity and reliability as these beacons normally experience very few problems. Lamp life expectancy is usually around 1 year, and the lamps are comparatively inexpensive. Beacon replacement cost is also very attractive when compared with other technologies.

  The disadvantage is the total electrical consumption of one beacon is 1240 watts. This makes the incandescent beacon the most expensive L-864 option in terms of energy consumption.

  Also, the loss of either of the two lamps requires a NOTAM (notice to air men) be filed with the FAA.

- *LED.* The newest technology for obstruction lighting.

  Advantages include very low electrical consumption and long life. These beacons normally operate at 240 watts or less. Most manufacturers warrant the LED beacons for 5 years and low maintenance make them especially attractive to broadcasters, who may lose significant revenue due to shutdowns for maintenance that could occur during rating sweeps periods.

  The disadvantage is that the purchase price is comparatively high and maintenance includes replacement of expensive components.

- *Strobe.* Middle ground between the other options.

  Advantages include low power consumption that is only slightly higher than LED and the expected life of flash tubes is roughly 2–3 years. While these lamps are more expensive than the incandescent version, they are less expensive than the lighting modules in the LED systems.

  The disadvantage is that a separate power converter or power unit is required for each flash head, adding more wind load to tall towers.

DETERMINING WHICH LIGHTING CONFIGURATION TO USE

The first consideration is the FAA determination and the zoning requirements. For this example, it is assumed that no extenuating circumstances exist and the tower may use any lighting or marking combination that are allowed within the parameters outlined in the circular 7460. The tower will be a three-leg, 12 ft face, guyed, 1200 ft tall with an 80 ft antenna mast on top. The FAA designation for this tower would be one of three configurations: A4 with four tiers of flashing red L-864 beacons and four intermediate tiers of three L-810 side markers; C4 with four tiers of L-856 high-intensity strobe beacons plus an AOL beacon at the top of the mast; or F4 with four tiers of both L-856 and L-864 beacons with four intermediate tiers of L-810 side markers, plus an L-864 and L-865 (may be separate units or a combined unit) beacon at the top of the antenna mast.

See Figures 13–18 in FAA Advisory Circular AC 70/7460-1K for complete configuration illustrations.

Lighting System Start-Up Costs

The initial investment of a lighting system would be least with the A4 system, provided the focus is on the cost of the lighting system alone and not the ongoing chore of maintaining the paint (keeping the paint colors within FAA specification usually requires repainting the structure every 4–5 years). The system would consist of 7 L-864 beacons (three tiers of 2 beacons plus 1 at the apex), 12 L-810 side markers (four tiers of 3), and a controller and photocell for a total of 19 lighting fixtures. With fewer lights less maintenance should be required. The system should be relamped and given a thorough preventative maintenance (PM) routine annually. The long-term basis in electrical consumption is substantial—approximately $6,200 annually in most cases (assumes $0.14/kW hour for 12 average hours of operation per day). Note that with the cost and maintenance of the paint, short-term savings may be overshadowed by the long-term operating expenses.

Relamping costs are approximately $1,800/year, and prorated painting cost averages $8,000 to $12,000 per year.

C4 White Strobe Lighting System

The C4 white strobe system would cost more initially but would produce a substantial savings in the long run due to the savings in electrical operating expenses. This system would consist of 12 high-intensity beacons (four tiers of 3 beacons) and 1 AOL beacon atop the mast, for a total of 13 beacons. Scheduled relamps (about $500 per strobe plus installation labor) should be accompanied by minor PMs, with major PMs factored for every third relamp. Strobe lights are considered to be a reliable method of lighting and will last for many years provided they are properly maintained. Considering the cost of maintaining paint and the higher consumption of electricity by the incandescent system, the white strobe system may be a more attractive economic option.

F4 Dual-Lighting System

The most expensive initial option is the F4 dual system, consisting of white strobes and either red option listed above. The higher purchase price is primarily due to the number of flash heads required at each tier level, normally five per level. Flash tube replacement costs therefore increase substantially. Operating costs vary with regard to the power consumption and duty cycle of the red options. Again, the PM schedule should be identical to the C4 system above. However, a dual-lighting system may be necessary to meet local concerns about strobe lighting systems illuminating the sky at night. See Figure 7.4-7.

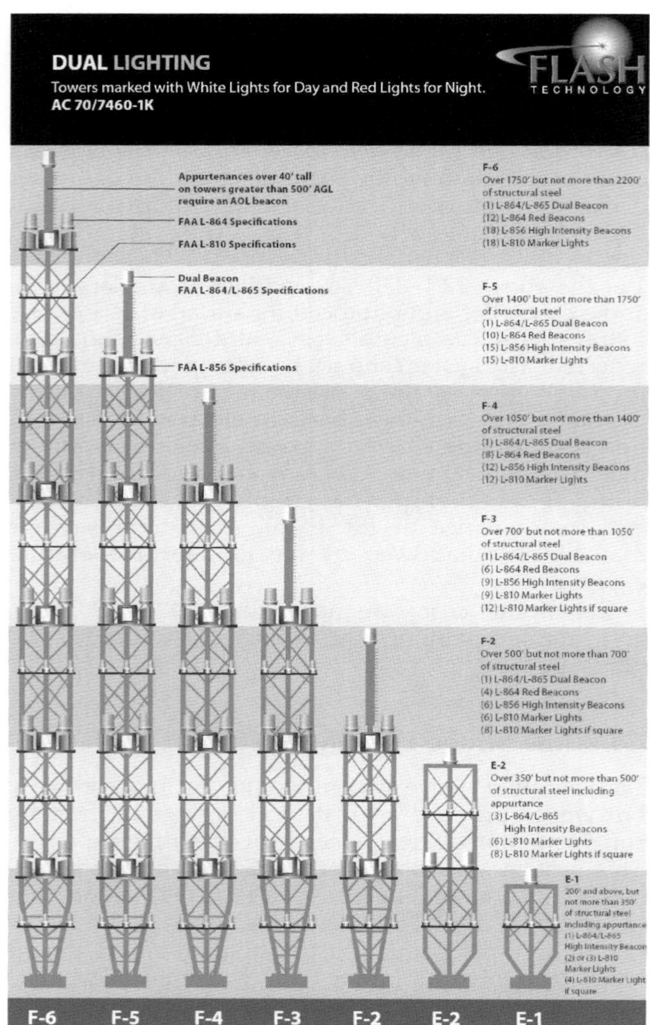

FIGURE 7.4-7 Tower light intervals for dual lighting per FAA AC 70/7460-1K.

MAINTENANCE REQUIREMENTS

Because of the seriousness with which both the FAA and FCC take with regard to tower lighting and marking and the fact that improper lighting, marking, and maintenance may be a safety-of-life issue, it is important that service and maintenance be taken equally seriously by the tower owner and operator. Therefore, it is necessary to inform the FAA of any outage or abnormal operation tower light condition whether it is because of malfunction or scheduled maintenance.

Special Maintenance Note

Licensees of broadcast stations must adhere to certain regulations for "human exposure to RFR" while workers are present on towers, as shown in Figure 7.4-8, and in close proximity to antennas. These regulations apply to personnel involved in tower lighting maintenance as well as to antenna and transmission line. Refer to Chapter 2.4, "Human Exposure to Radio Frequency Energy," for a detailed discussion of this topic.

FIGURE 7.4-8 Worker installing tower lighting system. Note the use of safety devices. (Photo courtesy of Flash Technology.)

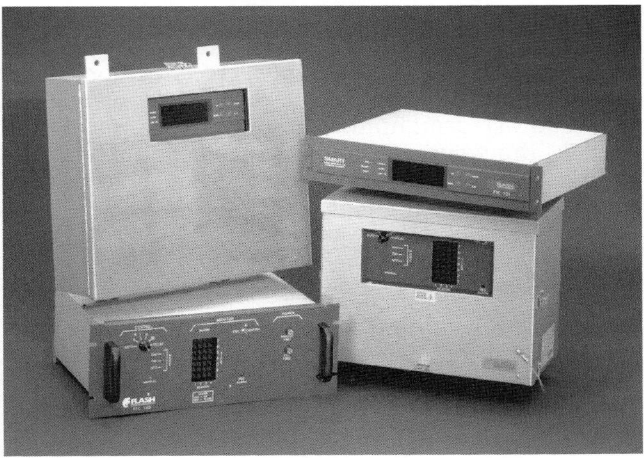

FIGURE 7.4-9 Examples of tower lighting system controllers—indoor and outdoor configurations. (Photo courtesy of Flash Technology.)

Tower lighting controllers are available in both indoor and outdoor configurations, as shown in Figure 7.4-9. Examples of poor maintenance of conduit and a beacon controller are shown in Figures 7.4-10 and 7.4-11, respectively. Lack of adequate maintenance will lead to premature failure of tower lighting systems and the potential of citations by the FCC or FAA.

Selecting a Tower Lighting Service Company

The same crew that services the antennas on a tower many not be qualified to maintain the tower lighting system. While some crews are highly trained strobe

FIGURE 7.4-10 Example of poor maintenance on an electrical conduit. (Photo courtesy of Flash Technology.)

FIGURE 7.4-11 Example of poor maintenance of a beacon lighting system controller. Note the evidence of corrosion from water leakage. (Photo courtesy of Flash Technology.)

technicians, others may know little about strobe theory and repair.

- It is important to check credentials, obtain proof of training, and find third-party references. As an option, the tower owner or operator could contact the manufacturer of the lighting equipment or ask other broadcasters in the area for recommendations.
- Request the potential service company candidates to provide proof of insurance and compliance with OSHA regulations. Check questionable claims with the governing agencies.
- Ask for a guaranteed lead time, and inquire how the lead time is backed. A guarantee is worthless without compensation of some sort in the event of failure to perform. *Note:* Resist the urge to force a repair in dangerous or marginal conditions. It is better to wait a few days than to cause an injury or death. Of course the FAA must be notified of the existing lighting outage and again upon restoration of normal operation.
- Ask to see the parts the service company has replaced. Inspect, but do not demand to keep the parts. Some parts may have some trade-in value to the service company, and demanding these parts could result in higher maintenance costs.
- Have an up-to-date wiring diagram available for use by service personnel that may not be familiar with a specific tower lighting system. Figure 7.4-12 shows an example of the wiring complexity of a dual-lighting system.
- Explore service options before it is time to call for service. Waiting until service is needed may result in a hasty, uninformed, and potentially costly decision.

FIGURE 7.4-12 Example of a wiring junction box for dual tower lighting system that illustrates the complexity of the system and the necessity of having up-to-date wiring diagrams for servicing.

MONITORING

An important decision involved in tower lighting is how the system will be monitored. While the purchase of the latest technology, top-of-the-line certified lighting system may provide some degree of comfort, the ongoing status of operation must be known in order to avoid service and regulatory problems.

FCC rule Section 17.47 addresses various requirements for monitoring, including an automatic alarm system, the requirement to visually or automatically observe the tower lighting system at least once every 24 hours, and the requirement to physically inspect the system at all intervals (each mode) at least every 3 months (quarterly) for proper operation.

FCC rule Section 17.48 requires reporting to the FAA any "observed or otherwise known extinguishment or improper functioning of any flashing obstruction light, regardless of its position on the antenna structure, not corrected within 30 minutes." This would include a power failure in addition to a light or control system malfunction.

Refer to AC 70/7460 Chapter 2, Section 23, Light Failure Notification, for more specific information concerning notification. Users of red incandescent systems should be aware that "when one of the lamps in an incandescent L-864 flashing red beacon fails, it should be reported" (p. 3, footnote). Section 24 also addresses the requirement to notify the FAA when the *normal operation is restored*. Of special interest is the statement, "The FCC advises that noncompliance with notification procedures could subject its sponsor to penalties or monetary forfeitures" and posts recent enforcement actions at the following website: http://www.fcc.gov/eb/broadcast/asml.html. A quick review of the names, infractions, and fines incurred should be informative. Suffice to say the FCC is serious about safety and cooperating with the FAA.

Monitoring Methods

Because tower lighting may range from a single non-flashing lamp on a short AM or microwave tower to an elaborate dual-lighting system on a 2000 ft tower, the method of monitoring can take many forms (see Figure 7.4-13). Examples include:

- Contract the tower light monitoring to a professional monitoring center.
- An in-depth detailed device that instantly reports operational data of each beacon through the transmitter remote control system.
- Use a dry contact lighting system monitor circuit through a dial-out device to call the maintenance center of the tower owner or operator.
- Use an employee of the station to perform monitoring service.
- Pay someone who lives near the tower to observe the operation of the lighting system on a daily basis and maintain a log of this activity (visual check).
- Some other method that complies with the requirements set forth by the FAA.

Automated monitoring is recommended for towers in rural locations. Records of daily monitoring activities are often kept in best form by professional monitoring services, and may save the tower owner substantial grief in the event of an impact by airborne transportation or a fine by the FCC.

The decision on what method to use for tower lighting monitoring may require having answers to some of the following questions:

- How much risk of financial exposure is acceptable in case of an unfortunate lighting system event?
- Is the system in compliance with FAA/FCC regulations?
- Will the "visual check" contractor be reliable, keep the required logs, and contact the owner (or the FAA) according to the requirements?
- Can a station employee responsibly check the operating status of the lighting system on a daily basis? What if the employee calls in sick, quits, or is on vacation?
- This is a 24 hour, 7 days per week, 365 days per year responsibility. To what extent is the employee liable in the event of an FCC notification of liability for improper tower lighting system operation?
- How much does a monitoring service cost, and will the service comply with applicable FAA/FCC rules and local requirements?

FIGURE 7.4-13 Example of a commercial tower lighting monitoring facility.

- If an outage occurs, can the monitoring method offer some insight to the problem, possibly expediting the repair?
- How does a monitoring system or service communicate with the tower light controller (e.g., satellite, microwave, Ethernet, LAN, IP, dial-up or dedicated line, or some wireless method)?

The FAA requires a daily monitoring log to be kept for each tower site. The FAA holds the tower owner as the responsible party in all scenarios. The tenants of the tower may also be implicated in the case of an event on the tower.

OTHER TOWER LIGHTING SYSTEM ISSUES

Power Outage

Most outages are a result of power loss, so a backup generator may be used. However, since all tower lighting systems are either capacitance discharge or on/off designs, the generator should be of ample size to prevent "loping," which is a tendency to lag and thrust due to an effort to catch up to the power draw. A load balance resistor may not always be necessary, but will almost always facilitate proper operation of the generator. Also, an FAA NOTAM should be opened if communication is lost to the tower (called "no comm."). The NOTAM is good for 15 days after opening. This provides ample time in most situations to restore power or communications to the site. In any event, it is important to provide backup power to any remote monitoring device (communications system) as well. Many site-monitoring devices will interface with sensors that will indicate fuel level, building temperature, and gate access, as well as many other site functions.

Spare Parts

While spare parts can be kept at the tower site to permit lighting system repair crews to have the right fixture on hand for immediate installation, some spare parts have a shelf life, and most have a warranty that begins the day of shipment. In the event of an outage, a NOTAM must be filed within 30 minutes or fix the system, not much time to call the service company, drive to the site, locate the part, climb the tower, and fix the light. That is assuming power is still on at the site. Since the NOTAM is probably necessary anyway and most manufacturers can ship all current parts overnight, at a minimum keep only fuses and MOVs on hand.

Neighbor Complaints

When malfunctions occur some lighting systems default to maximum brightness so that at night the bright lights may cause neighbors to be unhappy and complain. Or, people in the immediate vicinity of the tower may be unhappy with the lighting in general. In this event most manufacturers of the tower lighting system may provide documentation and assistance in the event of legal action. However, the most common requests from zoning boards are to alter the system with lower output, shielding, or aiming that would take the system out of compliance. Normally all the manufacturer can provide is the test data stating that the system has been tested and approved to be within the parameters set forth by the FAA and FCC. For example, Flash Technology keeps this test data for each individual beacon by serial number and will appear before the various zoning boards to help explain these requirements and offer proof that the systems provided meet or exceed all the regulatory standards.

On a side note, some aiming is allowed on high-intensity systems according to FAA Advisory Circular AC 70/7460-1K, Chapter 7, paragraph 74.

SUMMARY

Tower lighting is a requirement that must be carefully considered throughout the lifespan of a tower. Rely on the FAA/FCC and the documents they provide. Follow the guidelines. Get to know your FAA airspace specialists. While FCC fines can represent a financial hardship, the liability and publicity over an incident caused by a faulty tower lighting system can be potentially fatal to a business. Take studied precautions and seek professional engineering and legal guidance to ensure the tower lighting and marking are compliant, and, importantly, document all relevant actions taken with respect to tower lighting and marking. Tower owners must ensure that their towers are in full compliance with all the specifications set forth in the FCC issued Antenna Structure Registration (FCC Form 854R).

SECTION

8

SIGNAL MEASUREMENT AND ANALYSIS

CHAPTER

8.1

Audio Signal Analysis

STANLEY SALEK, P.E.
Hammett & Edison, Inc., Consulting Engineers
San Francisco, California

THOMAS KITE and DAVID MATHEW
Audio Precision, Inc.
Beaverton, Oregon

INTRODUCTION

Audio signal analysis has been one of the fundamental chapters in the *NAB Engineering Handbook* because audio is an essential part of all broadcasting. With the widespread use of digital audio systems, this chapter has nearly doubled in size. Analog measurement definitions and techniques are maintained, and a new section devoted to digital audio test and measurements has been added.

The ability to quantify audio signals in terms of characteristics and qualities is paramount in audio engineering. Comparisons can be made with reference to established standards and requirements, including measurements relating to amplitude, frequency content, distortion, noise, and phase. The observation of such attributes allows a virtually complete characterization of an analog electrical audio system.

In the broadcast environment, audio measurements are used to gauge the overall quality of equipment such as amplifiers, recording systems, mixing consoles, digital audio devices, and other networks throughout the broadcast signal path. These networks are increasingly being replaced with digital audio distribution systems, which require special consideration in measuring performance characteristics. While some digital audio measurements can be straightforward, quantitative performance measurements of nonlinear audio systems employing various types of bit-rate compression can be challenging or virtually impossible.

AMPLITUDE MEASUREMENT

The most basic of needs in audio measurement is to determine a value relating to the size, or amplitude, of an analog audio signal. Since an audio waveform is rapidly changing, methods have been developed to convert peak, root mean square (RMS), and average values of the changing waveform into corresponding proportional DC voltages that can be more easily observed.

There are specific cases in which the peak value is the most direct measure of magnitude. It gives an indication of the largest excursions (either positive or negative) of an audio waveform. As shown in Figure 8.1-1, the audio signal is applied to an absolute value circuit, which rectifies the waveform such that the output is all positive. A diode is then used to couple the signal into C and R. These serve as memory and decay time elements, respectively, that can be adjusted in value to conform to the desired ballistics. Although the output is still changing with time, along with the input, the excursions corresponding to the peak values of the original waveform are much slower and more easily observed on metering devices. As the value of resistor R is increased, the decay time of the output is proportionally increased as well. If the resistor is completely removed, a peak-hold circuit results.

Peak (actually peak-to-peak) functions can also be observed on an oscilloscope, although this technique is often impractical because of the difficulty in reading the random waveforms typical in most audio material. Storage oscilloscopes can perform a peak-to-peak hold function.

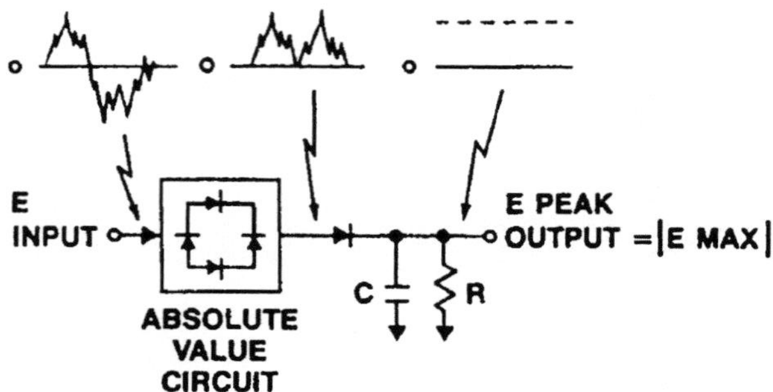

FIGURE 8.1-1 Peak value detection.

While there are many cases in which the peak value is of considerable use, the RMS value of a signal is generally most meaningful since it gives indication of the energy content of the signal without regard to its waveform. In analog audio measurement, however, it is usually simpler to detect the average amplitude of the given waveform and relate it to an associated RMS value, with reference to a sine wave. The RMS level can be defined as follows:

$$\mathrm{E_{rms}} = \sqrt{\frac{E_1(t)^2 + E_2(t)^2 + \ldots + E_n(t)^2}{n}} \tag{1}$$

where E_1 through E_n are successive measurements over a total of n samples. As can be seen from its name, the value is computed by taking the average of n samples of E squared. Performing the square root function completes the calculation. This function is also commonly referred to as "true RMS" [1]. Figure 8.1-2 shows how this technique is accomplished electrically.

Through the use of the absolute value circuit of Figure 8.1-1 and the R-C configuration found in the RMS detector of Figure 8.1-2, an average detector can be made, as shown electrically in Figure 8.1-3 and mathematically:

$$\mathrm{E_{average}} = \frac{E_1(t) + E_2(t) + \ldots + E_n(t)}{n} \tag{2}$$

In terms of audio perception, the average value of an audio signal is related to program material density,

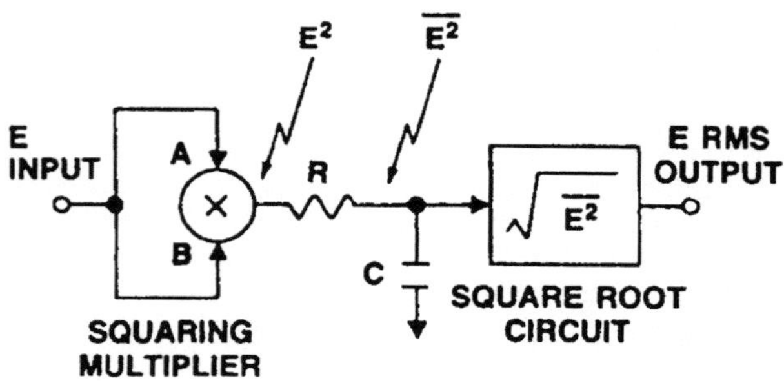

FIGURE 8.1-2 True RMS detection.

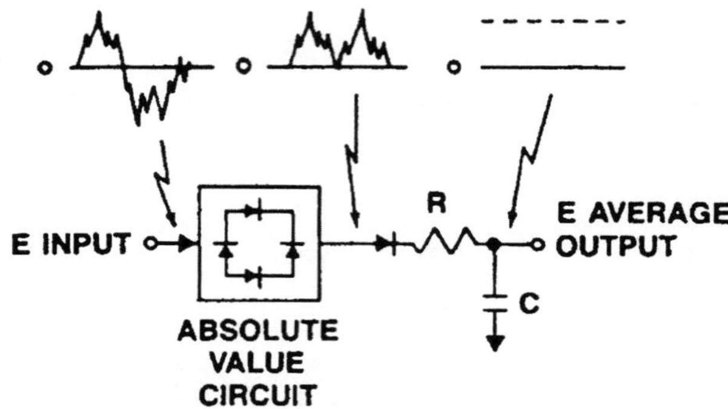

FIGURE 8.1-3 Average value detection.

where the peak value described earlier relates to a maximum. Since the peak value defines the upper limit of allowable modulation in a transmission system, it is often technically desirable that the peak-to-average ratio be as low as possible to attain highest perceived loudness and signal-to-noise ratio (SNR). Achieving this effect may require compromising aesthetic goals and may not always be appropriate, depending on the type of program material.

The decibel (dB) is a unit for comparing relative levels of voltage or power in transmission systems. In broadcast audio systems, the most common representation of decibels is dBm. This is the value of a signal with reference to 1 mW into a 600 ohm load. The level in dBm of a signal can be found using the following relation:

$$\mathrm{dBm} = 20\log\left(\frac{\mathrm{E}}{0.775}\right) \tag{3}$$

where E is in volts. The number 0.775 represents the voltage level reference of 0 dBm. Note that, strictly speaking, this formula is true only when the circuit impedance is 600 ohms. In practice, the formula is used typically without regard to the impedance level, although such application can lead to significant error. Voltage levels obtained from the peak, RMS, and average circuits described previously can be used for possible values of E. When this is done, some common types of metering can be synthesized to observe the activity of audio material.

Metering

Two popular types of metering for the characterization of program audio are the standard volume unit indicator, commonly known as the volume unit (VU) meter, and the peak program meter (PPM). Although VU metering has been more common in U.S. broadcast equipment, the standard PPM indicator is often found as well, especially in audio mixing consoles.

The VU meter was introduced in 1939 to serve as a standard program-level indicating device—see Figures 8.1-4(a) and 8.1-4(b) [2]. Its original purpose was to be the reference between broadcasters (as well as other programming suppliers) and the telephone

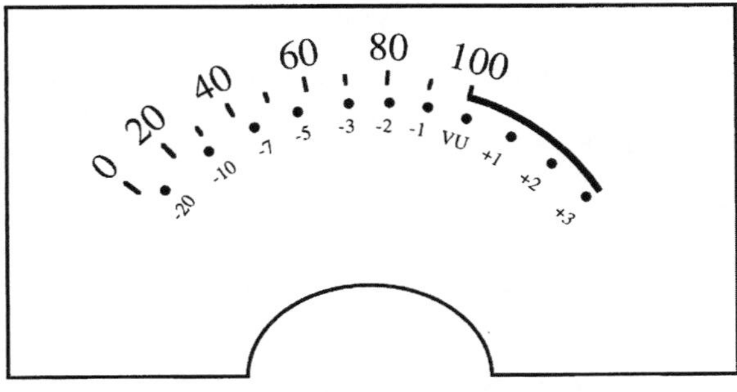

FIGURE 8.1-4(a) VU meter.

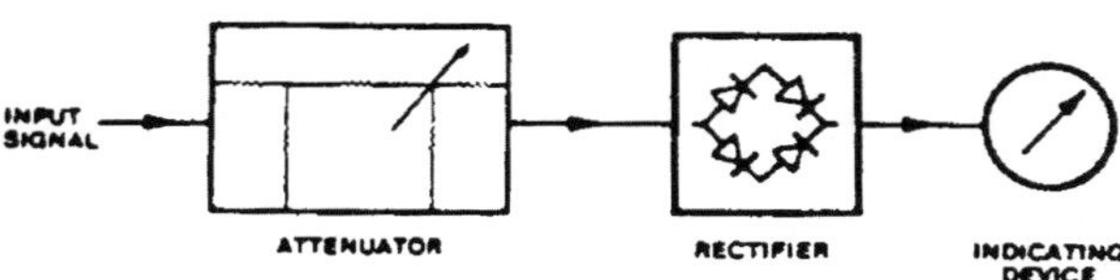

FIGURE 8.1-4(b) Block diagram of the stages of a typical VU indicator.

company. A VU meter is the combination of a bridge rectifier, a resistive attenuator, and an ammeter with an approximately linear voltage scale to produce an average responding AC voltmeter. The VU meter is calibrated to read 0 VU across a circuit in which a sinusoid develops 1 mW in a resistance equal to the circuit impedance (0.775 volts RMS across 600 ohms). This configuration allows the meter to be powered directly by a 600 ohm program line, with the attenuator typically set to read 0 VU at +8 dBm.

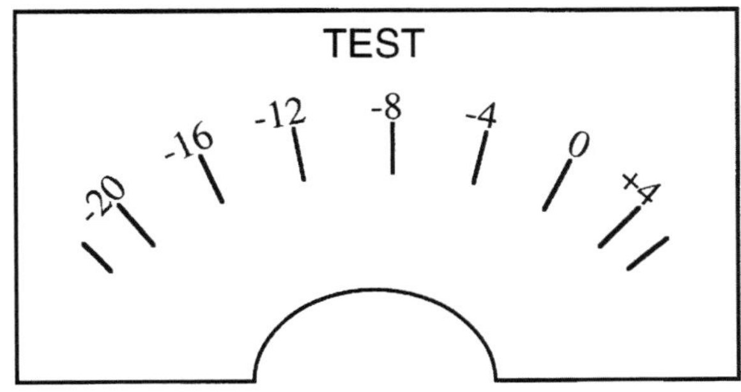

FIGURE 8.1-4(c) Arrangement of a typical PPM scale.

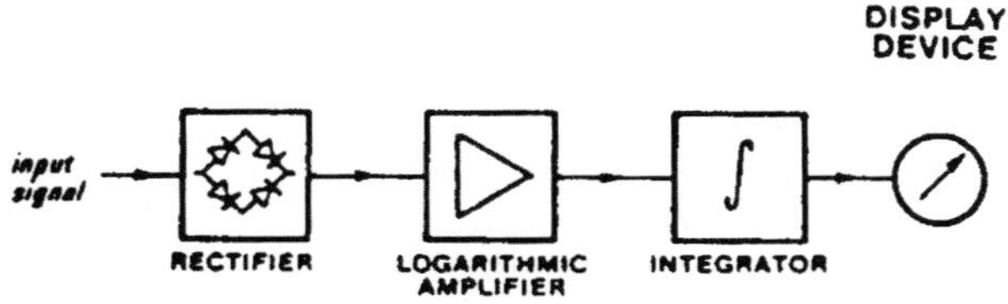

FIGURE 8.1-4(d) Block diagram of the stages of a typical PPM.

FIGURE 8.1-4(e) LED-based meters are available which combine aspects of both PPM and VU meters. (Courtesy Dorrough Electronics.)

Beyond reading continuous tones, VU meter dynamic characteristics are set so that it will display 99% of its steady-state reading on a sine wave tone burst 300 ms long, with a fall to 5% of the reading in 300 ms. Essentially an average responding device, the VU meter will not respond to short-duration program peaks. Therefore, levels normally should be set with a 10 dB margin (headroom) before the point of clipping [3].

The PPM is designed to read nearly the full peak value of the audio signal—see Figures 8.1-4(c) and 8.1-4(d). It uses a rectifier and an integrator, producing a fast rise and slow fall effect on the display device. Typical standards require the PPM to read –2±0.5 dB of the steady-state value for a tone burst of 10 ms, and take 2.8 seconds for the pointer to fall 20 dB [4].

Typically, a PPM exhibits flat frequency response and is calibrated such that the nominal peak program level corresponds to a 0 dB meter reading near full scale (generally +16 dBm).

Other types of metering devices have been developed to read wideband audio in a simultaneous mode. One such system consists of an LED bar graph display exhibiting peak program content. Riding on this is a brighter display (utilizing the same display elements) corresponding to VU standards. Such an indicator allows continuous monitoring of program material compression and dynamic characteristics. Another system combines a VU movement meter with peak indicating LED flashers, as shown in Figure 8.1-4(e). A hold function is sometimes associated with these flashers to allow an operator sufficient time to observe the approximate peak content.

FREQUENCY ANALYSIS

Amplitude analysis methods, as described in the previous section, are generally used to provide an indication of signal levels simultaneously over the entire audio range. It is sometimes more desirable, however, to be able to measure discrete frequencies in an audio system, allowing frequency response measurement as well as dynamic measurement of energy content throughout the audio spectrum.

Simply stated, frequency response is the capability of a device or system to pass or amplify equally all frequencies within a specified range. As far as audio in the broadcast environment is concerned, the range of interest is generally 50 Hz to 15 kHz or 20 kHz. Although few musical instruments produce fundamental frequencies greater than 4 kHz and the human voice does not contain fundamental frequencies much above 1 kHz, the reproducing device or system must be able to pass the harmonics that accompany the fundamental frequencies. Without adequate bandwidth or with uneven frequency response, an unnatural coloration of the perceived sound becomes evident. To overcome this potential problem, amplifiers with flat frequency response over the specified range are employed. Since the responses of series-connected amplifiers are additive, it is important to verify the flatness of each device in a system.

Several methods are available to measure audio frequency response. They include discrete measurement and swept frequency methods. Parallel analysis and Fast Fourier Transform (FFT) techniques also can be used.

Discrete Frequency Analysis

The discrete frequency measurement method is uncomplicated and inexpensive. A simple measurement system consists of a low-distortion audio frequency oscillator and a wideband AC voltmeter. The oscillator output is connected to the input of the device or system to be characterized. The voltmeter is used to observe the level at the output of the device or at a desired intermediate point in a system.

The measurement is made by first setting the generator output level to the nominal input operating level of the device. Generally, a 400 Hz or 1 kHz frequency is chosen initially for most audio systems. The device output level is read on the AC voltmeter, and this quantity is noted as a zero dB relative reference. Provided the generator itself has a flat frequency response, measurements at frequencies through the audio band can be made while recording the corresponding dB output levels with respect to the reference. A convenient and commonly used technique is to increment the frequency in a 1, 2, 5 sequence (100 Hz, 200 Hz, 500 Hz, 1 kHz, 2 kHz, 5 kHz, etc.). This method permits plotting the final response data on 4-cycle "LOG/LIN" graph paper, providing regularly spaced frequency increments horizontally. The logarithmic amplitude data are plotted along the horizontal axis, with the zero dB relative reference placed in a convenient position on the linear vertical axis.

Although the discrete frequency measurement technique is straightforward, it is also often tedious and time-consuming. Numerous frequency measurements must be made to ensure adequate testing. This method is most usable in the response measurement of single-ended devices that do not have a suitable input port for connection to an audio generator. Examples of these types of devices include transcription equipment, such as compact disc players. Test recordings supplied by the equipment manufacturer and other sources are used to provide the tones necessary for discrete frequency response characterization [5,6].

Swept Frequency Analysis

A faster and more efficient means of measuring frequency response is the swept frequency method. This process employs a sweep frequency generator as a signal source and measures response over the entire range of interest in one sweep. The detector for these measurements is most often a tracking type that follows the signal source and measures a narrow band of frequencies centered around the source frequency. Use of a tracking detector provides a better assurance that the amplitude measured is that of the tone generator and is not influenced by spurious noise or harmonics.

Devices specifically designed to conduct swept frequency measurements include spectrum analyzers and dedicated automated audio test systems. Spectrum analyzers directly produce response images on a built-in display, while the dedicated systems generally use a connected personal computer (with appropriate software and hardware) for data display and storage.

Network analyzers are also useful in audio frequency domain measurements. They are used to characterize two-port networks (devices having an input and output) as to frequency, phase, and delay responses. They are employed where substantial accuracy in the measurement of these parameters is required. RF subsystem and semiconductor device design have been the major application for network analyzers, although some newer generation equipment includes audio-frequency coverage. In the case of audio systems, network analyzers allow precise response measurement of amplifier and filter designs.

Figure 8.1-5 shows a typical setup for measuring amplifier frequency response using the swept frequency method. The signal source used to drive the test device is the tracking oscillator output of the analyzer. The device output is terminated with an appropriate characteristic load impedance and connected to the analyzer input. Measurement of the frequency response is made by manually or automatically sweeping the analyzer across the frequency range of interest. A plotter or personal computer (with appropriate software and hardware) can be used to record a permanent record of the test device response characteristics.

Real-Time Analysis

A real-time audio analyzer (RTA) consists of a sequential collection of one octave or one-third octave filters having individual detectors and indicators at each output. The program audio is simultaneously fed to the inputs of all the filters. The output signal of each filter is proportional to the amount of energy occurring in that particular frequency band. This technique is also referred to as parallel analysis.

A simplified version of an RTA is presented in Figure 8.1-6. As shown, it is intended to break the audio band into three sections using lowpass, bandpass, and highpass filtering. Signal detectors are then used to condition the audio for display on a suitable indicator, one set for each of the three bands. The detectors can

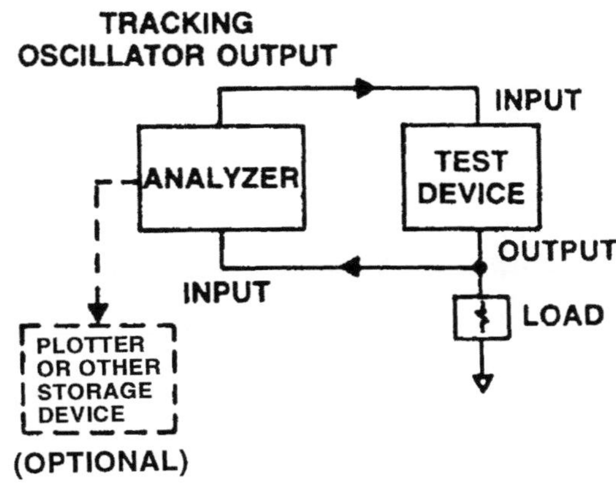

FIGURE 8.1-5 Frequency response measurement using swept method.

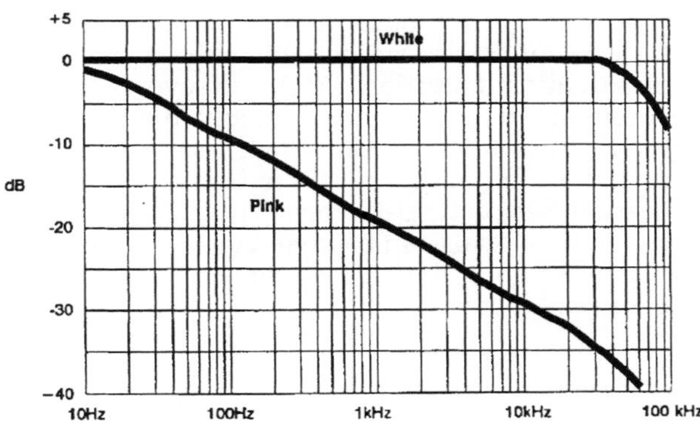

FIGURE 8.1-7 Response versus frequency for white noise and pink noise.

be (and often are) the same peak, RMS, or average circuits described earlier. Typical readout indicators are bar graph displays with dB-calibrated scales. When arranged side by side, the readouts provide a graphical presentation of amplitude versus frequency, similar to a spectrum analyzer display. Unlike the spectrum analyzer, however, an RTA does not rely on a fixed sweep speed.

Parallel techniques using the RTA are often used for dynamic program material and room acoustics analysis. This type of analyzer is also useful for measuring frequency response of audio devices when used in conjunction with a "pink" noise source. Pink noise has a constant mean squared voltage per octave of frequency. This characteristic makes it popular in audio work, since it allows correlation between successive octaves by ensuring the same voltage amplitude is available as a reference. By connecting the pink noise source to the input of a device to be characterized, and the RTA to its output, a response curve can be displayed almost instantly. The characteristic response of pink noise is shown in Figure 8.1-7, as compared with white noise. White noise is unweighted, since its response is flat with frequency, except that it is attenuated for frequencies above the audio spectrum.

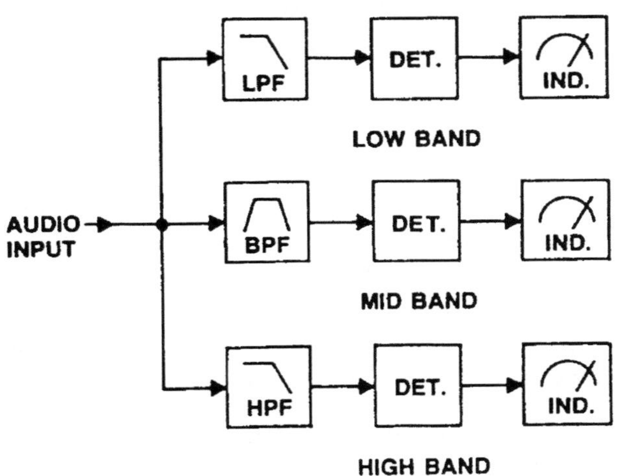

FIGURE 8.1-6 Basic real-time analyzer.

FFT analyzers have the capability to convert a snapshot sampling of an audio or other time-varying source and mathematically transform the result into a display of the frequency components present. Because the conversion process is done by a specialized digital signal processing (DSP) microcomputer, an FFT analyzer often can produce a complete spectrum display as much as an order of magnitude faster than conventional swept spectrum analyzers. While FFT analyzers have been most helpful in low-frequency measurement (where a swept analyzer would require a very slow sweep time to resolve closely spaced components), their costs have been decreasing, so they are becoming increasingly popular for audio system analysis and measurement.

DISTORTION MEASUREMENT

When a two-port device is driven beyond its range of linear operation or through areas of discontinuity, signal distortion occurs. As a result, additional frequencies appear at the device output that were not present at its input. In cases in which distortion becomes extreme, it can be identified through listening. Odd-order distortion (such as clipping distortion) can become audible at around 1.25% (distortion is expressed in a percentage of the amplitude of the undistorted waveform). Even-order distortion, characterized by a coloration of the program material, becomes audible at about 5%. Generally, systems with a wider frequency response capability need to maintain lower distortion levels to be acceptable. Since distortion is not always obvious to many people, techniques are available to measure its various types.

Classic audio distortion can be characterized in two basic ways: harmonic distortion and intermodulation distortion. While the two associated methods produce uncorrelated measurement values, each gives a quantitative result of device quality in terms of a single number. Although total harmonic distortion (THD) content is determined by only one method, intermodulation distortion (IMD) has several accepted measurement practices, most notably the Society of Motion Picture and Television Engineers (SMPTE) and International Telecommunication Union—Radiocommunication Sector (ITU-R, formerly CCIR) methods. Digital audio signals can be characterized using THD and IMD methods, but they also are susceptible to other types of distortion.

Total Harmonic Distortion

Total harmonic distortion is a measure of individual harmonic amplitudes with respect to the amplitude of the fundamental frequency. In practice, harmonics greater than third order often add little to the resultant value because of their negligible amplitude. THD is defined as

$$\text{THD\%} = 100 \times \frac{\sqrt{A_2^{\,2} + A_3^{\,2} + A_4^{\,2} + \ldots + A_n^{\,2}}}{A_1} \qquad (4)$$

where A_2 through A_n are the amplitudes of the individual harmonics and A_1 is the amplitude of the fundamental.

As shown in Figure 8.1-8, a 1 kHz sine wave with harmonic distortion shows only minor differences when it is overlaid with an undistorted signal (as viewed on an oscilloscope). The amplitude and slope errors do not lead directly to a numeric result. But when a spectrum photo of the same waveform is observed (see Figure 8.1-9), the preceding relation can be applied. With the fundamental at 0 dBm (0.775V), the second harmonic at –26 dBm (38.8 mV), and the third harmonic at –50 dBm (2.5 mV), the harmonic distortion can be calculated as follows:

$$\text{THD\%} = 100 \times \frac{\sqrt{(0.0388)^2 + (0.0025)^2}}{0.775} = 5.0\% \qquad (5)$$

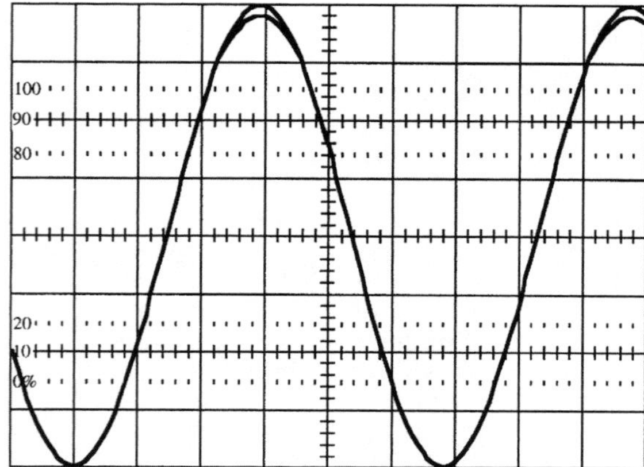

FIGURE 8.1-8 Comparing the distorted output of an amplifier with its undistorted 1 kHz input component.

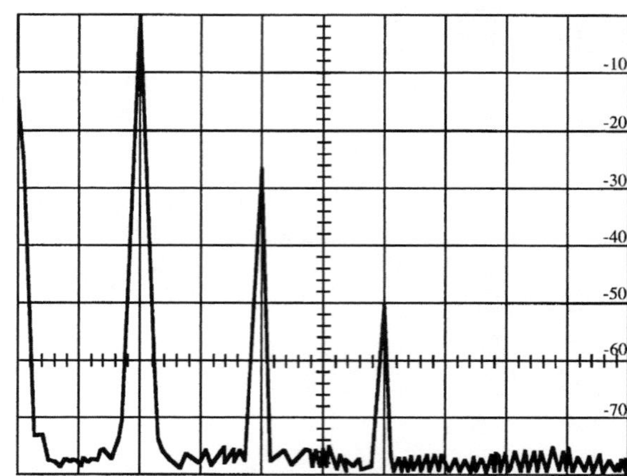

FIGURE 8.1-9 Measured THD using a spectrum analyzer (V: 10 dB/div.; H: 500 Hz/div.).

Although spectrum analysis can produce accurate THD measurement results, a simpler and more cost-effective procedure that produces a direct numeric quantity is more popular. Figure 8.1-10 shows the block diagram of a typical THD analyzer. An oscillator (with much less harmonic distortion than the device or system to be measured) is connected to the test device input. The distorted output signal of the device is filtered to remove A_1, the fundamental component. This produces a signal that, when RMS detected, is proportional to the THD produced by the device being tested.

THD measurement is often conducted using the same 1, 2, 5 sequence of frequencies mentioned for discrete response measurement. The THD results can be plotted on the same graph to characterize the device under test on a single figure. THD measurements may be taken over various input levels, but as the level is reduced, noise characteristics may affect the readings. In such cases, the spectrum analyzer method could produce more meaningful results.

Intermodulation Distortion

The intermodulation method of measuring distortion uses a test signal composed of two (generally sinusoidal) signals of different frequencies. After summation, they produce the effect of an amplitude modulated carrier when applied to a circuit having IMD. The intermodulation method is useful because the harmonic distortion of the signal sources does not affect the measurement.

The SMPTE method uses a low-frequency (f_1) and a relatively high frequency (f_2) signal (usually 60 Hz and 7 kHz, respectively) that are mixed at a four-to-one amplitude ratio (see Figures 8.1-11 and 8.1-12). This method involves the measurement of the relative amplitude of the modulation sidebands added to the higher frequency signal. For diagnostic purposes, it is often useful to determine even-order and odd-order distortions separately, although this is best done by

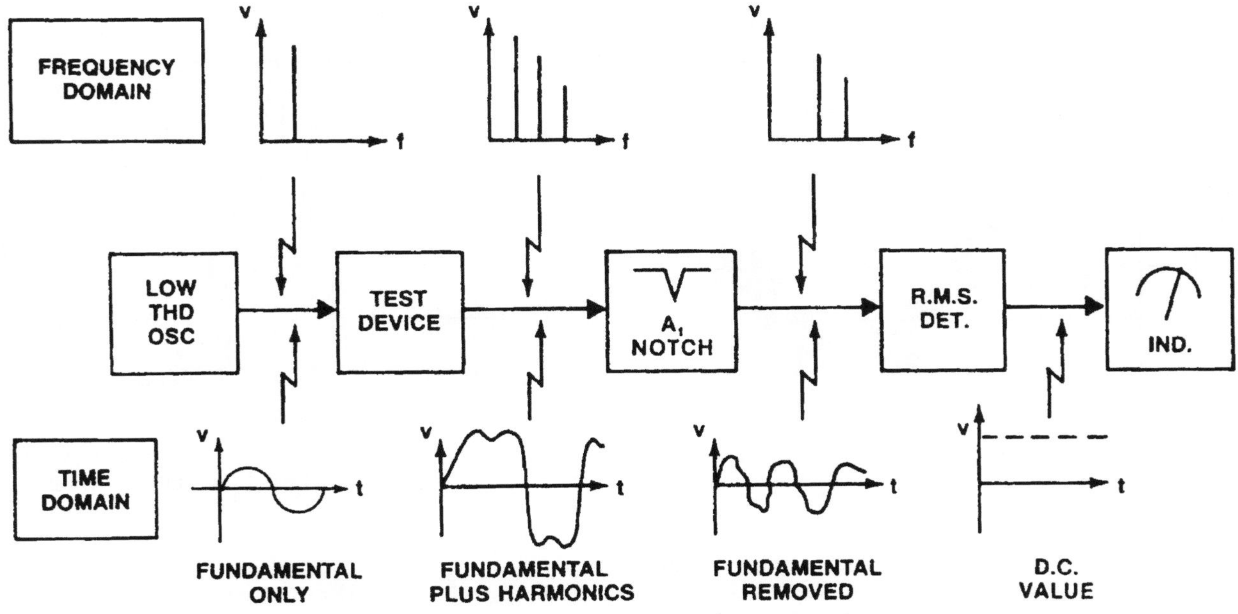

FIGURE 8.1-10 THD analyzer.

spectrum measurement techniques. Even-order distortion usually can be characterized by the ratio of the sum of the amplitudes of only the two second-order spurious frequencies, $f_2 - f_1$ and $f_1 + f_2$, to the amplitude of the carrier signal, f_2:

$$\text{SMPTE IMD\%} = \left[\frac{A_{(f_2-f_1)} + A_{(f_1+f_2)}}{A_{f_2}}\right] \times 100 \qquad (6)$$

(second order)

In a similar manner, odd-order distortion can be characterized by the ratio of the sum of the amplitudes of the two third-order spurious frequencies, $f_2 - 2f_1$ and $2f_1 + f_2$ to the amplitude of f_2:

$$\text{SMPTE IMD\%} = \left[\frac{A_{(f_2-2f_1)} + A_{(2f_1+f_2)}}{A_{f_2}}\right] \times 100 \qquad (7)$$

(third order)

Figure 8.1-13 shows the output signal of an amplifier with IMD, as viewed on an oscilloscope. Note the elongated trough as compared to Figure 8.1-12. As with THD, spectrum analysis can be used to determine the numerical amount of distortion present.

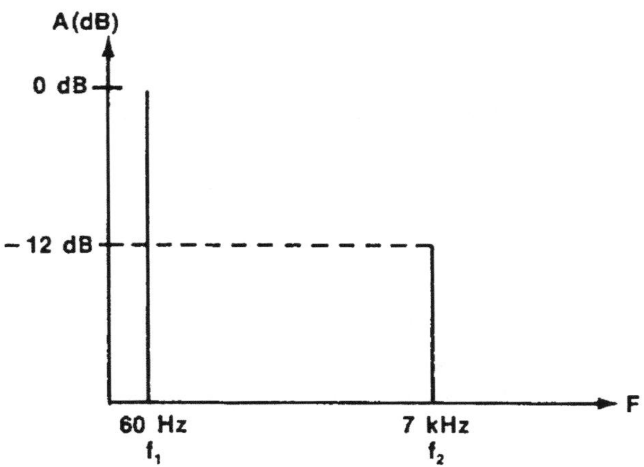

FIGURE 8.1-11 Spectrum of SMPTE IMD test signal ratios.

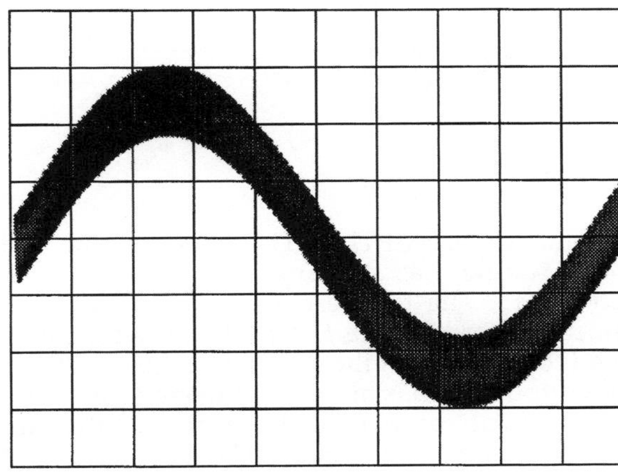

FIGURE 8.1-12 SMPTE IMD test signal as viewed on an oscilloscope.

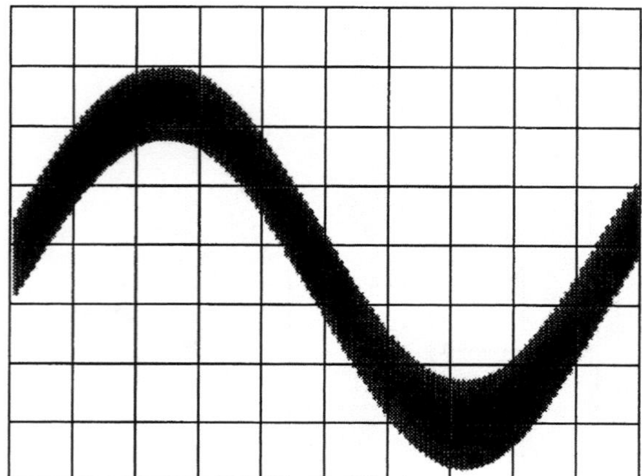

FIGURE 8.1-13 Output of an amplifier exhibiting significant SMPTE IMD.

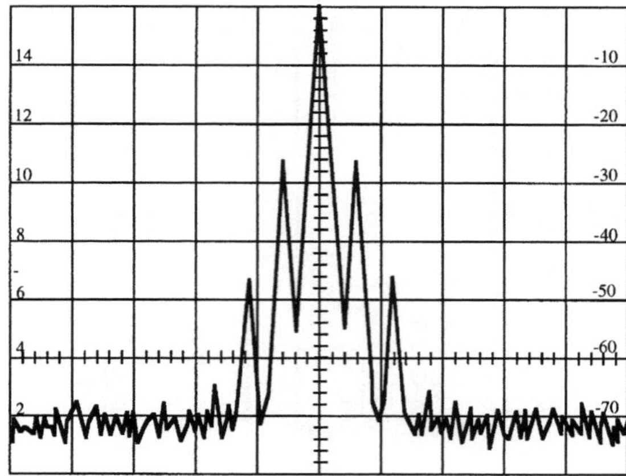

FIGURE 8.1-14 SMPTE IMD measurement (V: 10 dB/div., top of screen –12 dBm; H: 100 Hz/div., center frequency 7.0 kHz).

Intermodulation sidebands can be seen around f_2 in the spectrum photo of Figure 8.1-14. Second- and third-order distortion percentages for this example are calculated as follows:

$$A_{f_2} = -12\,\text{dBm} = 195\,\text{mV}$$

$$A_{(f_2-f_1)} = A_{(f_1+f_2)} = -38\,\text{dBm} = 9.75\,\text{mV}$$

$$\text{SMPTE IMD\%} = \left[\frac{9.76+9.76}{195}\right] \times 100 = 10.0\%$$

(second order)

$$A_{(f_2-2f_1)} = A_{(2f_1+f_2)} = -58\,\text{dBm} = 0.98\,\text{mV}$$

$$\text{SMPTE IMD\%} = \left[\frac{0.98+0.98}{195}\right] \times 100 = 1.0\% \tag{8}$$

(third order)

As shown, the contribution of even-order distortion products is usually greater than that of the odd-order. To express the result as a single quantity, the vector sum of the two quantities is taken:

$$\text{SMPTE IMD\%} = \sqrt{(IMD\%Even)^2 + (IMD\%Odd)^2} \tag{9}$$

(total)

$$= \sqrt{10^2 + 1^2} = 10.05\%$$

As with THD, SMPTE IMD has a direct method of numeric solution, as shown in the block diagram of Figure 8.1-15. The two test frequency oscillators are summed to produce the $f_1 + f_2$ signal, which is then applied to the input of the device to be tested. The distorted output signal is high-pass filtered to remove the f_1 fundamental component, leaving only the amplitude modulated f_2 component. Using a standard AM demodulator and low-pass filter, the residual f_1 component is obtained. After RMS detection, a DC level proportional to the distortion is produced that can be viewed on a direct-reading indicator.

Of some historical note, *wow and flutter* is a term that describes a special case of IMD normally associated with analog audio tape recorders. It is caused by variations in tape velocity across the recording and/or reproducing heads, due to imperfections in the mechanical drive system. These variations result in frequency modulation of the recorded and reproduced signal. The frequency spectrum obtained is similar to that of the SMPTE IMD measurement method, except the f_1 low-frequency signal is generated by fluctuations in tape speed and is not of any set amplitude.

The ITU-R intermodulation method uses a combination of two higher frequency sinusoidal signals (f_3, f_4) of equal amplitude. They are typically 1 kHz apart and found at 5/6 kHz, 14/15 kHz, or 19/20 kHz in many applications. One of the spurious frequencies generated is low in frequency, whereas others are gathered around the two driving frequencies. Figures 8.1-16 and 8.1-17 spectrally show the driving frequencies before and after passing through a test amplifier. As with the SMPTE IMD measurement, the generated spurious products can be classified as even-order or odd-order. Even-order distortion is expressed as the ratio of the amplitude of the difference component ($f_4 - f_3$) to the sum of the two driving frequencies (f_3, f_4):

$$\text{ITU-R IMD\%} = \left[\frac{A_{(f_4-f_3)}}{A_{f_3} + A_{f_4}}\right] \times 100 \tag{10}$$

(second order)

Odd-order distortion is determined by calculating the ratio of the sum of the amplitude of the two third-order products, $2f_3 - f_4$ and $2f_4 - f_3$, to the sum of the amplitudes of the two driving frequencies, f_3 and f_4:

$$\text{ITU-R IMD\%} = \left[\frac{A_{(2f_3-f_4)} + A_{(2f_4-f_3)}}{A_{f_3} + A_{f_4}}\right] \times 100 \tag{11}$$

(third order)

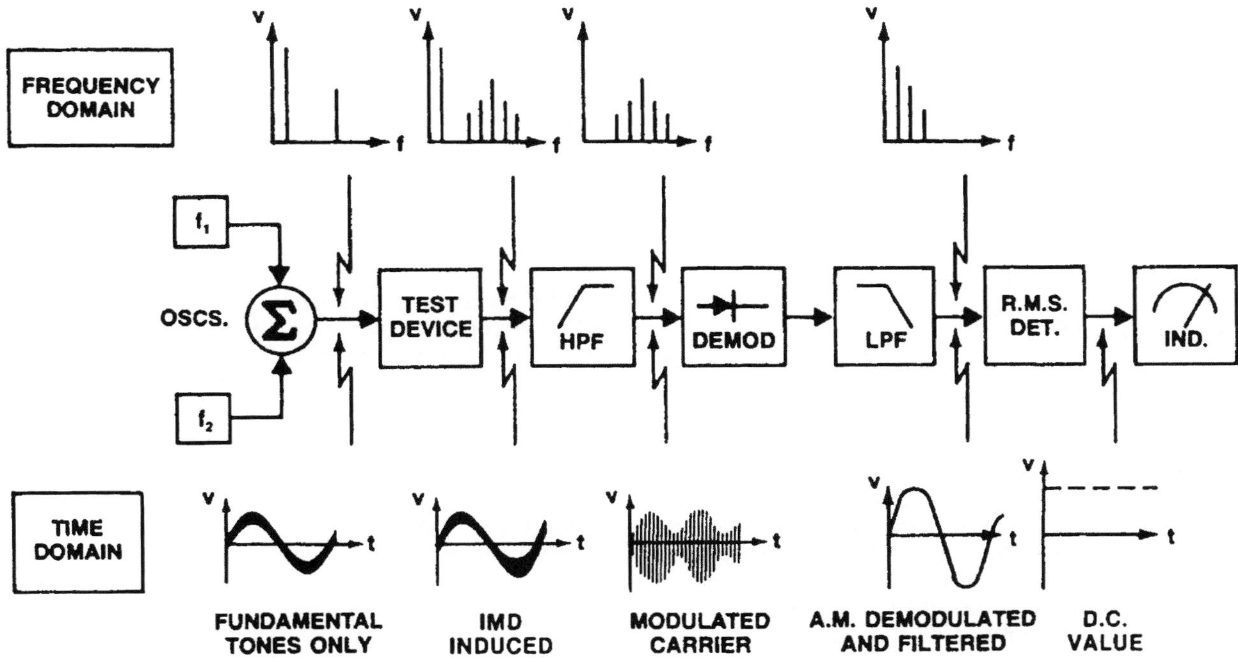

FIGURE 8.1-15 SMPTE IMD analyzer.

In the case of Figure 8.1-17, the driving frequencies f_3 and f_4 are at 5 kHz and 6 kHz, even-order products at 1 kHz, and third-order products at 4 kHz and 7 kHz, respectively. Distortion percentages for this example are calculated as follows:

$$A_{f_3} = A_{f_4} = 0 \text{ dBm} = 775 \text{ mV}$$

$$A_{(f_4-f_3)} = -64 \text{ dBm} = 0.49 \text{ mV}$$

$$\text{ITU-R IMD\%} = \left[\frac{0.49}{775+775}\right] \times 100 = 0.032\%$$

(second order)

$$A_{(2f_3-f_4)} = -62 \text{ dBm} = 0.62 \text{ mV}$$

$$A_{(2f_4-f_3)} = -61 \text{ dBm} = 0.69 \text{ mV}$$

$$\text{ITU-R IMD\%} = \left[\frac{0.62+0.69}{775+775}\right] \times 100 = 0.085\%$$

(third order)

It is a common practice for direct-reading metered analyzers to measure only the amplitude of the difference product ($f_4 - f_3$) with respect to the driving signal amplitudes. A device that performs this task is called an ITU-R second-order difference frequency distortion

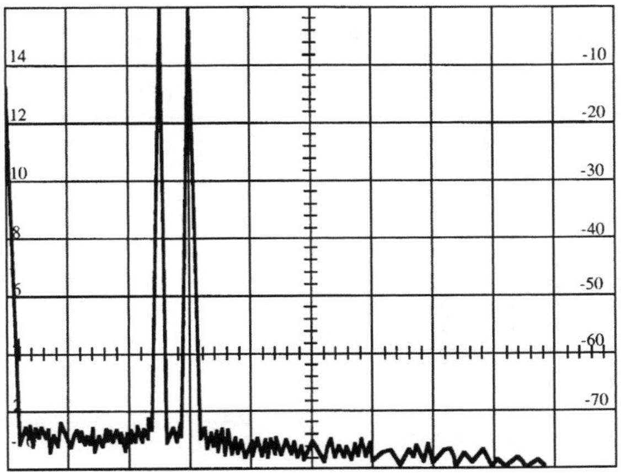

FIGURE 8.1-16 Spectrum of ITU-R IMD test signal (V: 10 dB/div.; H: 2 kHz/div.).

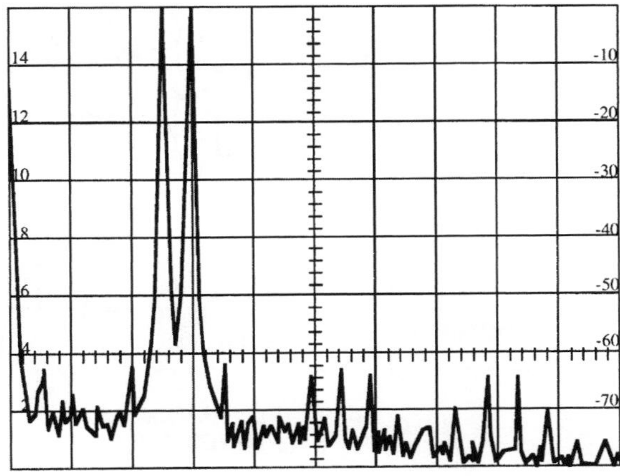

FIGURE 8.1-17 ITU-R IMD measurement example (V: 10 dB/div.; H: 2 kHz/div.).

analyzer. Figure 8.1-18 illustrates how the measurement is made.

Additional forms of intermodulation distortion measurement have been devised to uncover potential issues with particular audio systems. Some use multiple audio frequencies, whereas others employ nonsinusoidal audio waveforms. One such example is transient intermodulation (TIM) distortion, found only in amplifiers that utilize negative feedback. When this feedback is excessive, a fast-rising transient signal applied to the input of the amplifier can produce an internal overshoot that saturates the circuits in the amplifier.

The most popular procedure used to measure TIM distortion is called the sine-square wave method. The test signal employed uses a square wave to induce nonlinearity in the test device by saturating the amplifier's internal current, caused by its alternate rises and falls. Mixed with this square wave is a low-level, high-frequency sine wave, which is unrelated harmonically. In one definition, the frequency of the square wave is 3.18 kHz and that of the sine wave 15 kHz, where the peak-to-peak amplitude ratio of the former to the latter is four to one [7]. Similar to the other IMD spectrum analyzer evaluation methods, the components present at the output can be combined to express a TIM distortion figure as a single percentage.

Other TIM measurement methods include a sawtooth wave method that takes amplifier slew rate into account and a noise-square wave method, where the sine wave of the sine-square wave method is replaced by a narrow-band noise spectrum [8].

Added Filtering in Distortion Measurement

Filtering as part of distortion measurement is often useful to remove components that are of little interest and as a diagnostic aid, especially with THD measurement. A 20 kHz or 30 kHz high-pass filter placed in series with the output of the device being measured is useful for testing broadcast equipment. This practice is often acceptable, since the harmonics produced outside the transmission bandwidth can be eliminated, producing a more realistic result.

A high-pass filter also can serve as an important analytical aid. With a cutoff frequency in the 400 Hz range, for example, it can be placed in series with the test device output and used to determine the contribution of AC line frequency hum (50 or 60 Hz) and related harmonics (100/120 and 150/180 Hz) to a THD measurement utilizing a fundamental frequency of 1 kHz or greater. Verification of adequate grounding used in the test setup also can be evaluated using this filter.

When filtering is used in IMD measurement, it is important to verify that in-band distortion components are not inadvertently removed. Also, sharp cutoff analog filter designs may produce overshoot components that could affect measurement results.

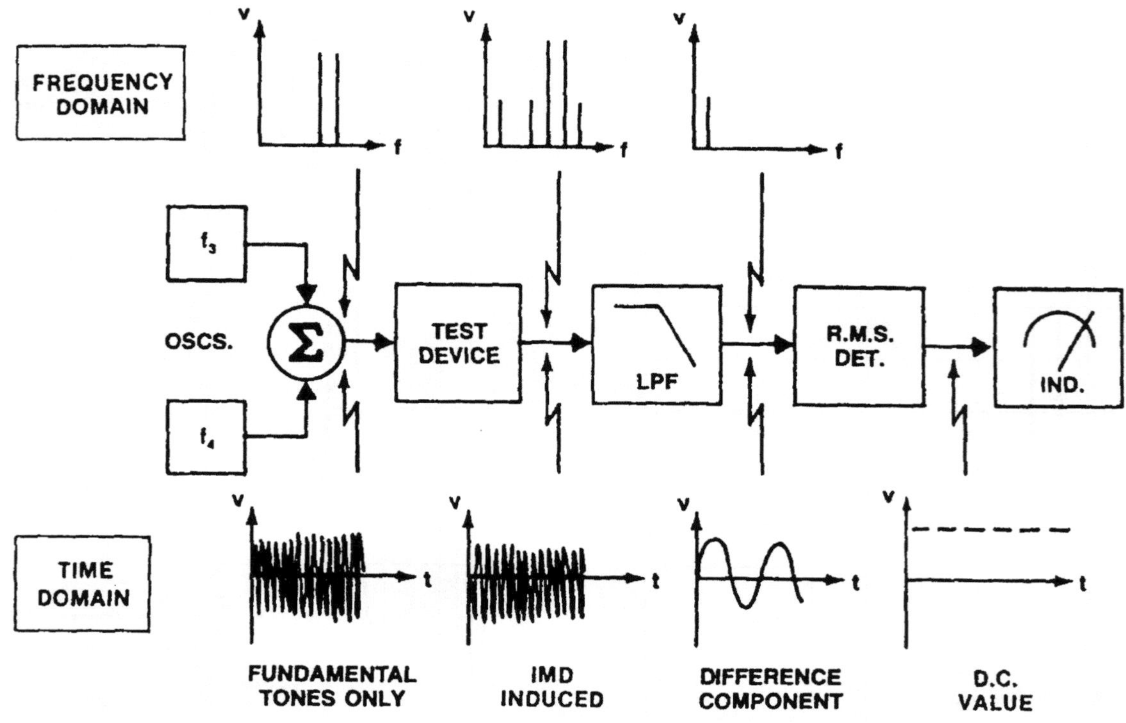

FIGURE 8.1-18 ITU-R second-order IMD analyzer.

NOISE MEASUREMENT

In audio engineering, noise is a random energy distribution in which individual spectral components are not clearly resolved. Primary sources of noise in analog circuits and amplifiers are in the resistive circuit elements [9]. It is important to control noise in amplifiers as their gain increases to preserve a high signal-to-noise ratio, which is the ratio of the operating signal level to the noise level inherent in the amplifier itself.

To understand the origin of the noise, we can model a passive resistive element as a noiseless resistor in series with a noise voltage generator, E_r:

$$E_r = \sqrt{4KTBR}(volts) \quad (12)$$

where:

K = Boltzman's constant (1.38×10^{-23} W-sec/°K)

T = Temperature in degrees Kelvin

B = Noise bandwidth (Hz)

R = Resistance in ohms

In the equation, noise voltage is a physical phenomenon that can be worsened by an increase in any of the variable factors. Therefore, noise cannot be eliminated, but it can be reduced. This optimization process is often accomplished by proper selection of the resistive components, because of an additional factor known as excess noise, which is proportional to the voltage drop across the resistor and related to the material from which it is made. Of the different available types, carbon composition resistors are prone to the most excess noise contribution, whereas metal-film devices show the least.

At times, the actual spectral distribution of noise is of less importance than the noise voltage within a given bandwidth. For audio frequencies, a 15 or 20 kHz bandwidth is of interest. With a low-pass filter in this range connected in series with an amplifier output, and the input of the amplifier grounded, an unweighted but band-limited noise measurement can be made. When the noise output level is obtained, it can be expressed as a ratio with a standard operating level and reference frequency. This produces an indication of the amplifier's SNR.

When the gain of the amplifier is known, this same technique can be used to determine equivalent input noise voltage, that being the voltage of the noise that would be found at the input of the amplifier if the amplifier were completely noiseless [10].

A practical goal in the measurement of audio noise is to obtain data that correlates well with the subjective perception of noise [11]. For example, two amplifiers with identical SNRs can sound very different because one may have a uniform noise spectrum and the other may have most of the noise concentrated over a limited frequency range. Hence, the latter amplifier would sound noisier than the former, having to do with the way the ear perceives the loudness of a signal that is uniform in amplitude across the audio band. To make comparative noise measurements more meaningful, several weighting filters have been used to alter noise spectra over the frequency band of interest.

"A" weighting is based on the inverse of early measurements by Fletcher and Munson of the ear's sensitivity at low sound pressure levels [12]. A later weighting curve utilizes the CCIR/ARM method, an updated scheme that places the zero dB reference at 2 kHz instead of 1 kHz [13]. It is believed that this method, which is based on the obtrusiveness as well as the levels of different kinds of noise, provides a more commercially acceptable result when used to characterize modern, wide-range audio equipment. Figure 8.1-19 compares the two curves.

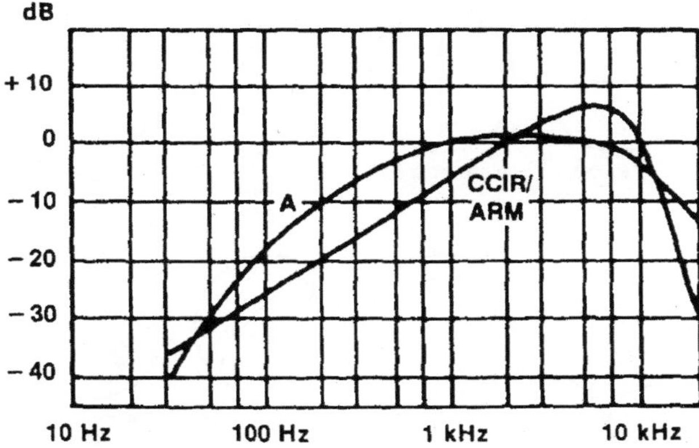

FIGURE 8.1-19 Noise measurement weighting curves.

PHASE MONITORING AND MEASUREMENT

An (L+R) summation is the monophonic compatible signal for AM, FM, and TV stereo broadcasting. In analog systems, separation information is usually transmitted via an (L-R) signal. Since these two signals are created through a summation and difference process of the original left and right channel stereophonic source, it is important that they can be recombined properly at the receiver [14]. Amplitude and phase errors must be minimized in the transmission system to accomplish this goal. Phase measurement is important in accomplishing this task.

In a stereo program system, if left and right audio information is correlated but delayed in phase, the error would not be evident on a stereophonic receiver. A monophonic signal, however, would be degraded because of inexact summation. To monitor audio systems for the presence of this problem, a phase meter can be used. A simple version would take phase and amplitude variances into account simultaneously by functioning as a two-input subtractor. When both characteristics are identical in each of the channels, the output becomes zero. A meter that measures only phase information compares the zero-crossing times of the two input signals, and the resulting time difference is used to generate a DC voltage proportional to the

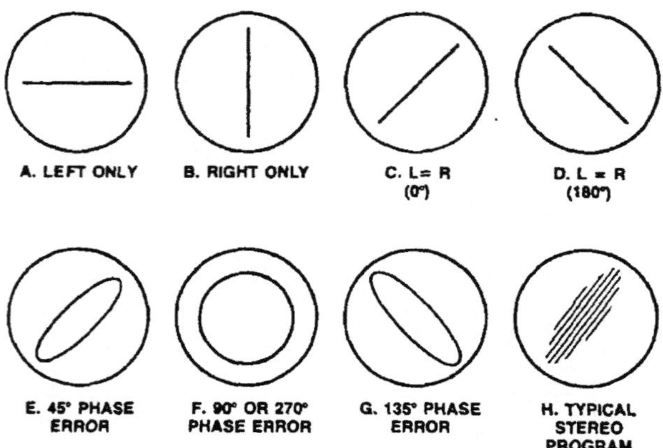

FIGURE 8.1-20 Interpretation of Lissajous patterns.

phase difference [15]. Phase detectors operating in this manner often limit the input signals in order to remove all amplitude information.

More popular, however, is the Lissajous figure method, involving the use of an oscilloscope in the X-Y mode. The patterns produced are shown in Figure 8.1-20. An oscilloscope is connected such that the left channel audio causes an X-axis deflection and right channel audio produces a Y-axis deflection, as shown in Figures 8.1-20A and 8.1-20B, respectively. When each channel contains the same program material, the pattern of Figure 8.1-20C is produced, along the L+R axis. If one of the channels is inverted, the pattern of Figure 8.1-20D becomes evident. This is often called the L-R axis. Monophonic program material that follows this axis is said to be inverted in polarity because no sum or L+R information is present.

The patterns of Figures 8.1-20E, 8.1-20F, and 8.1-20G are commonly seen when phase errors exist between two channels carrying the same discrete audio tone signal. Stereo program material, in unprocessed form, generally modulates the L+R axis while simultaneously deviating in the L-R direction to a lesser amount, as shown in Figure 8.1-20H.

DIGITAL AUDIO

Increasingly, audio is recorded, mixed, edited, stored, and transmitted digitally. *Digital audio* refers to the parts of the audio chain in which the audio signal is represented by a sequence of discrete numerical samples, rather than by a continuous variation in a physical parameter, such as electrical voltage. The primary benefit of a digital representation is that it is robust—a digital signal can in principle be , copied, and transmitted without loss, with much greater immunity to the degrading effects (noise, interference, crosstalk, etc.) commonly associated with analog systems. In addition, a numerical representation enables signal processing that would be more expensive or impossible using analog techniques.

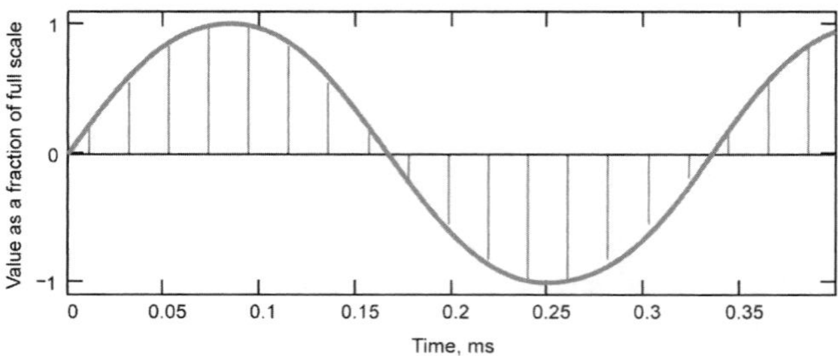

FIGURE 8.1-21 3 kHz sine wave with 48 kHz sampling instants shown.

Sampling

Analog audio can be transformed into a digital audio signal using specialized quantization hardware to convert the continuous time-varying amplitude information into a discrete series of numbers, with each related to the instantaneous value of the signal. A digital audio sample represents the instantaneous level of an audio signal at a specified time as a discrete numerical value. A digital audio stream comprises a sequence of such samples taken at regular intervals, a representation known as *discrete-time*.

The *sampling theorem* states that a bandwidth-limited signal can be perfectly represented by a discrete-time sequence of samples, provided that the rate at which the samples are taken is at least twice the signal bandwidth. An audio signal with a bandwidth of DC to 20 kHz therefore requires a sampling rate of at least 40 kHz to be perfectly represented by a sequence of samples. In practice, a somewhat higher sampling rate is used, to allow spectral room for physically realizable filters to operate. Figure 8.1-21 shows a 3 kHz sine wave with 48 kHz sampling instants.

Real-world signals are not truly band-limited, and therefore filtering is needed to remove signal energy above half the sampling frequency before sampling takes place. If filtering is not performed, or is inadequate, tones with frequencies above half the sampling rate will *alias* into the passband, leading to potentially objectionable products in the sampled signal that are not tonally related to the original signal.

For a true digital representation of audio, it is also necessary to ascribe a discrete amplitude to the instantaneous level of the signal and to represent this amplitude as a numeric value. This process is known as *quantization*. The combined processes of sampling and quantization are performed by an analog-to-digital converter (ADC).

Sampling Rate and Dynamic Range

Professional and broadcast audio typically uses a sampling rate of 48 kHz. This covers the standard 20 Hz to 20 kHz audio bandwidth, with some margin above 20 kHz for filtering. Other sampling rates commonly found in broadcast systems are 32 kHz and 44.1 kHz, with some professional applications using 96 kHz and even 192 kHz. The 32 kHz sampling rate is adequate for 15 kHz audio often used in broadcast systems, while 44.1 kHz is the consumer sampling rate used for compact discs and other media. Newer consumer technologies such as DVD-Audio use higher sampling rates. In a typical broadcast environment, sample rate converters (SRCs) are sometimes needed to interface digital equipment operating at different sampling rates.

A digital audio system has a theoretical dynamic range that is determined by the granularity of the quantization used to represent the audio samples. In a 16-bit system (the minimum generally accepted as providing high quality audio), each audio sample is represented as a 16-bit binary word, giving a maximum dynamic range of approximately 98 dB. In practice, the dynamic range is somewhat less, because of the combined effects of noise in the analog front end, inherent converter noise, the need to avoid clipping, and the noise contribution of *dither*, low-level noise added to linearize the digital system. Professional systems often employ 20-bit or 24-bit quantization to reduce these effects, and to allow multiple digital streams to be combined without degrading noise performance to unacceptable levels.

Analog-to-Digital Conversion

An ADC must perform the following functions, most or all of which are performed by modern ADCs in a single, integrated circuit:

- Bandwidth limit the input signal to satisfy the sampling theorem;
- Sample the signal at regular intervals;
- Quantize each sample to the nearest of a predefined set of discrete levels;
- Represent the discrete level as a *digital word* and make it available externally.

Typically, the digital words representing the audio signal are made available in real time on the output pins of the ADC, for consumption by subsequent devices. These words are represented as pulse code modulation (PCM), a binary format with a fixed word length, such as 16 bits or 24 bits. Stand-alone ADCs incorporate a *digital transmitter* to transmit the digital output words over a standard transport (see "Transporting Digital Audio" later in the chapter).

A *sample clock* is provided to the converter by an external clock circuit. In broadcast and professional audio, the sampling rate is typically 48 kHz, but other rates are sometimes used, such as 96 kHz and 32 kHz. In professional systems, it is often required to lock the sampling rate of the ADC to an external reference, or *house sync*, so that all converters in the system run at the same rate. Internally, clock synchronization circuitry locks the local clock to the external clock. This process can lead to audible distortion or even total signal loss if there is a problem locking to the house sync. See "Interface Jitter" later in the chapter.

The block diagram of a typical stand-alone stereo ADC is shown in Figure 8.1-22. Most ADCs employ *oversampling*, which uses a high sample rate at the front end, allowing the use of a simple passive analog lowpass filter for band limiting. The ADC samples and digitizes the analog signal at the oversampled rate, and then reduces the sample rate to the desired rate using an internal digital filter and decimator. The output of this block is typically 16 to 24 bits wide. Typically the two channels are multiplexed and serialized onto a single data line, with an accompanying bit clock and frame clock on separate lines. The I^2S standard is often used.

A stand-alone converter must make the audio data available externally for consumption by subsequent devices. Figure 8.1-22 shows one method: the I^2S data and clock signals are combined with auxiliary data and bi-phase mark encoded for transmission over a cable.

Transporting Digital Audio

A digital audio signal can be transported as a serial bitstream, referred to as the *digital interface signal*. The AES3 standard specifies the physical, electrical, and data characteristics of a transport for stereo digital

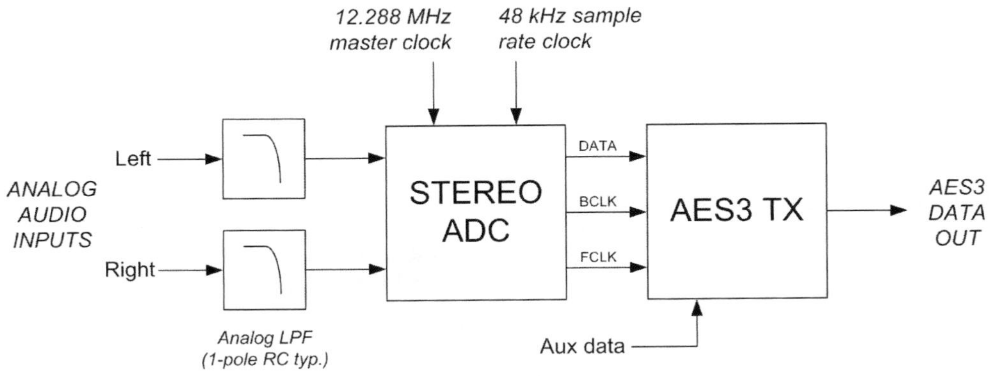

FIGURE 8.1-22 Stereo digital audio encoder (AES/EBU or S/PDIF).

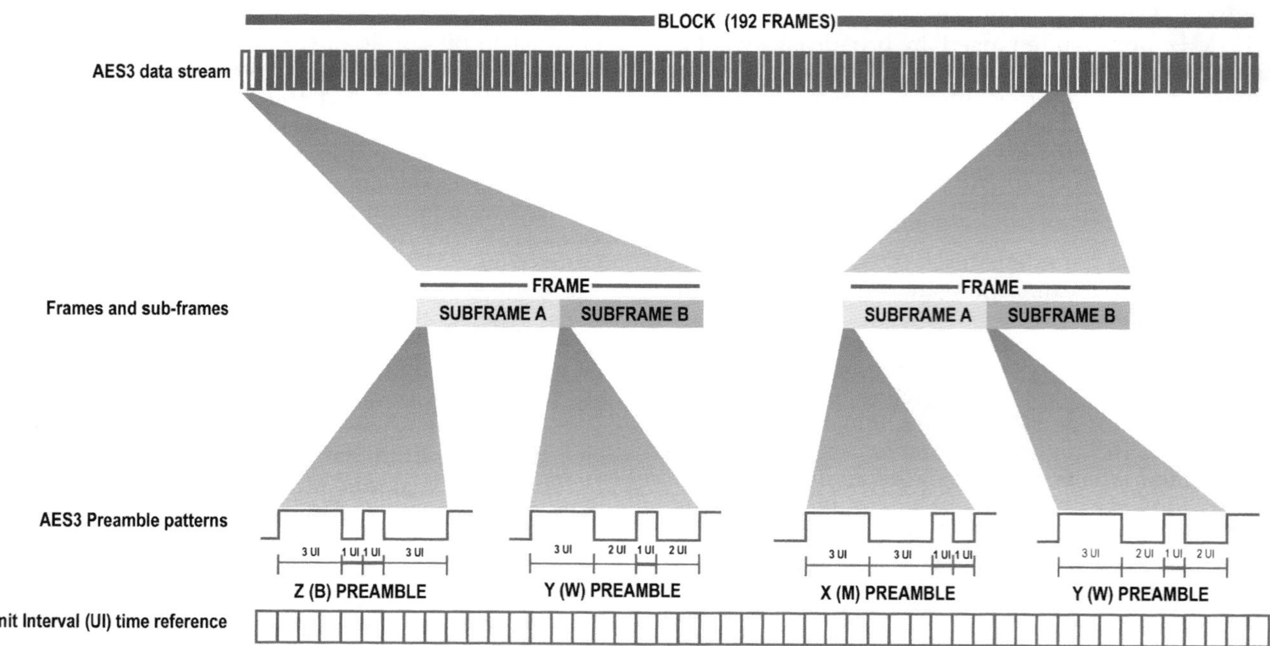

FIGURE 8.1-23 AES3 bitstream.

audio using balanced connectors and coaxial cable (see Figure 8.1-23). The AES3id information document and SMPTE276M extend AES3 to unbalanced connectors. The consumer S/PDIF standard is similar to AES3id, but uses lower voltages, and allows the use of optical (known as "Toslink") connectors. These standards share an interface format, with only minor differences in the status data carried alongside the audio information, and therefore are largely compatible with one another at the data layer level.

The digital interface signal, or *carrier*, consists of a sequence of *frames*. Each frame consists of two *subframes*, one for each channel. Each subframe consists of the following:

- A unique *preamble* signifying the start of a sample;
- Up to 24 bits of embedded audio data;
- 4 bits of *metadata*: parity, validity, channel status, and user bits.

The audio data and metadata are biphase-mark encoded. This makes the polarity of the signal unimportant, ensures that there is no DC, and allows the sample clock to be recovered at the receiver by maintaining a steady stream of signal transitions on the line. The shortest time between these transitions is known as a *unit interval*, or UI. For the AES3 interface format, there are 128 UIs per sample, so for a sample frequency of 48 kHz, 1 UI is about 163 ns.

The AES/EBU interface (discussed later) is intended to drive cables of up to 100 (AES-3) or 1000 (AES3id) meters in length, whereas the consumer interface is intended for equipment interconnection within 10 meters.

The channel status bits for each channel form a 192-bit block, delimited by a special *block preamble* that occurs every 192 samples. The first bit of the channel status block specifies whether the format is professional or consumer; the meaning of each bit in the block is different in the professional and consumer standards. It is important that the receiver of the digital bitstream react appropriately to the information in the channel status bits. For example, the audio/non-audio bit is set when compressed audio is being carried; a DAC should recognize this and mute its output. Further information on the meaning of these bits, and the expected response to them, can be found in the references.

Digital-to-Analog Conversion

A digital audio stream must be converted to analog form for presentation to the user. This *reconstruction* is performed by a digital-to-analog converter (DAC). A DAC must perform the following functions, most or all of which are performed by modern DACs in a single integrated circuit:

- Accept a sequence of digital words from an external source;
- Convert each word to an instantaneous analog level (typically an electrical current);
- Bandwidth limit the sequence of analog levels to half the sampling rate.

The last step is necessary because the sampling process creates replicas, or *images*, of the original signal at multiples of the sampling frequency. These images must be removed during reconstruction so that high

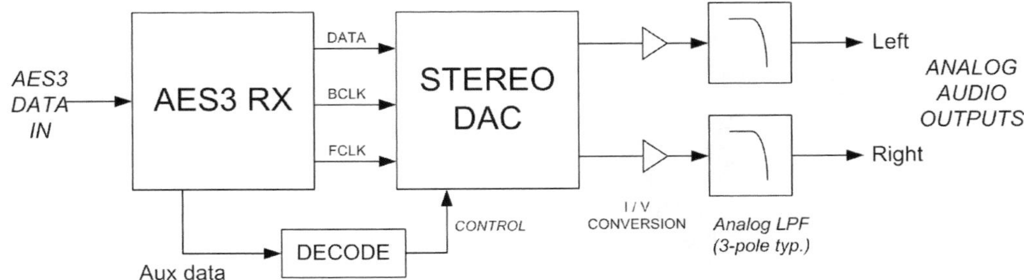

FIGURE 8.1-24 Stereo digital audio decoder (AES/EBU or S/PDIF).

energy levels above the system bandwidth do not interfere with the operation of subsequent devices in the audio chain.

As with ADCs, the clock is provided to the DAC externally. Stand-alone DACs incorporate a digital receiver to receive the digital input words over a standard transport (see "Transporting Digital Audio," earlier). Standard transports transmit a clock signal with the audio itself; the digital receiver extracts this clock signal from the transmitted signal and feeds it to the DAC. Clock extraction can be a problem if the signal at the DAC is poor. See "Interface Jitter," later in the chapter.

The block diagram of a typical stand-alone stereo DAC is shown in Figure 8.1-24. A standard interface, in this case AES3, is used to convey the digital audio signal. An AES3 receiver conditions the signal to remove any noise incurred during transport, and extracts the word clock, sampling clock, and stereo audio data. These are typically made available on three lines in an I^2S format. The auxiliary data is also extracted and made available on a separate interface.

A DAC IC receives the stereo I^2S data and converts it to analog form. Most DACs employ *oversampling*, in which a digital interpolator first increases the sample rate, and then a digital filter removes images below half the oversampled rate. Only a simple analog filter is then needed to remove the oversampled images from the final analog output.

MONITORING VERSUS MEASUREMENT

Monitoring and *metering* refer to measurements made on the audio chain when it is carrying a live signal, such as program material, over which the test instrument has no control. Typically, these measurements are limited to level and a graphical representation of the stereo or surround image. Monitoring is an important part of the broadcasting chain, and is discussed elsewhere.

For proper *measurement* of an audio device or chain, it is necessary to insert a known signal at one point and measure a signal at a downstream point. The known signal is provided by a *generator*. An *analyzer* acquires the downstream signal and characterizes it. The term *audio analyzer* is also commonly used to describe a complete measurement system consisting of an audio generator and an audio analyzer. The generator and analyzer work together to make audio measurements.

DIGITAL AUDIO MEASUREMENT

The measurement of audio characteristics in the analog domain traditionally makes use of an analog generator, such as a sine wave oscillator; analog signal conditioning, such as ranging and lowpass filtering; and analog measurement, such as RMS detection and metering. A dedicated piece of circuitry is typically required for each analog function.

In contrast, the measurement of audio characteristics in the digital domain requires a dedicated *software algorithm* for each function. Algorithms designed specifically for generating, manipulating, and analyzing signals in the digital domain are known as *digital signal processing* (DSP) algorithms. The continual decline in the cost of computing power and memory has led to DSP implementations of audio analyzers that are cheaper than their analog counterparts.

In addition, DSP makes possible signal generation and analysis methods which cannot be performed using analog techniques, such as the Fast Fourier Transform (FFT), and multitone analysis.

DACs and ADCs extend the reach of DSP into the analog domain. A DSP-based audio analyzer can therefore be used to measure both analog and digital audio systems. The analyzer interfaces to analog devices through DACs and ADCs, and to digital devices through digital interface transmitters and receivers. The analyzer can also measure cross-domain devices by using the analog and digital interfaces in combination.

When measuring digital devices, or when using DSP and converters to measure analog or mixed-domain devices, there are two broadly defined approaches:

- *Time domain methods:* Typically, these are DSP implementations of analog techniques, using digital sine wave generators, digital filters, and digital RMS detectors. The speed and accuracy of these methods are similar to the analog versions, though usually with a larger dynamic range.

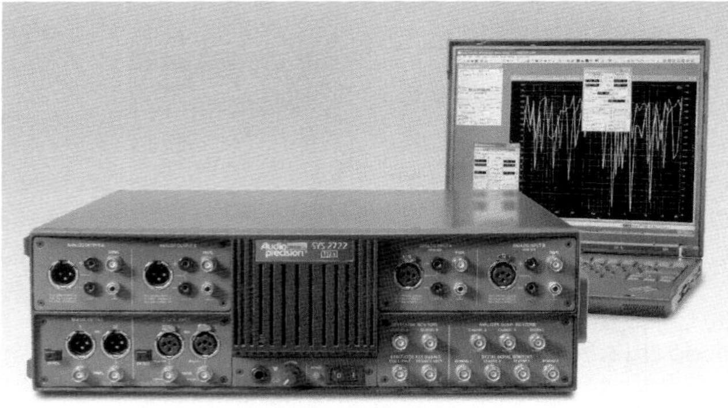

FIGURE 8.1-25 Dual-domain analog and digital audio analyzer. (Photo courtesy of Audio Precision.)

- *Frequency domain methods:* The FFT allows signals to be transformed efficiently into the frequency domain. Frequency domain processing facilitates certain operations that are impractical in the time domain, such as deconvolution, and can make some tasks easier, such as filtering. This opens up new possibilities for audio measurement, such as very fast methods of measuring transfer functions.

DSP-based audio test equipment (see Figure 8.1-25) may use time or frequency domain methods internally when making traditional measurements. Usually, the user does not need to be aware of the method being used.

The types of measurements performed by a DSP-based audio analyzer can be categorized as follows:

- *Traditional measurements*, such as level, frequency, and total harmonic distortion plus noise (THD+N) ratio. These measurements can be performed on analog, digital, and mixed-domain devices.
- *Nontraditional measurements*, such as power spectrum, continuous exponential sweep, and maximum length sequence (MLS). These measurements can be performed on analog, digital, and mixed-domain devices.
- *Converter-specific measurements*, such as stopband attenuation and jitter sensitivity. These are performed on ADCs, DACs, and sample-rate converters (SRCs, devices designed to convert digital bitstreams from one sample rate to another).
- *Digital interface measurements,* such as bit exactness, jitter, and eye patterns. Although these characterize the digital interface itself, rather than the audio signal carried on it, interface problems can be a major cause of audio signal degradation.

Traditional Measurements Using DSP

The following traditional measurements provide an overview of the performance of an audio device under test (DUT):

- *Level and gain*: The basic level measurement shows the signal level at the output of an audio device. Typically, the level is measured with a root mean square (RMS) detector. Depending on the device, the result might be displayed in volts (analog), decibels full-scale (dBFS, digital), or related units. Gain is the ratio of the DUT output level to its input level when the DUT is driven with a sine wave.
- *Total harmonic distortion plus noise (THD+N)*: THD+N measurements have been used for many years as a comprehensive single-value statement of an audio device's performance. The DUT is driven with a sine wave. The output of the DUT contains the sine wave fundamental (possibly changed in level because of DUT gain), together with distortion and noise generated inside the DUT. The fundamental is removed and the level of the residual is measured. The result is usually expressed as a ratio in dB of the residual level to the total level.
- *Frequency response*: This test shows how the gain of a device varies across the audible spectrum when driven with a sine wave. For most devices, a "flat" response curve, in which all frequencies over a specified bandwidth are passed with the same gain, is desirable.
- *Signal-to-noise ratio (SNR)*: A single figure that indicates how "noisy" an audio device is. Two measurements are required: The first measures the DUT output level when driven by a sine wave (usually at either maximum level or at a nominal operating level); the second measures the residual noise in the device. The result is expressed as a ratio in dB of the signal level to the noise level.
- *Crosstalk*: In multichannel DUTs, crosstalk is a measure of signal leakage between channels. All practical devices have some crosstalk. Crosstalk is measured by applying a high-frequency sine wave to at least one channel and measuring the level of the tone in a nondriven channel. The result is expressed as a ratio in dB of the level in the nondriven channel to the level in the driven channel (or channels). Crosstalk typically increases with increasing frequency because of capacitive coupling, so high-frequency tones provide a more sensitive test.
- *Phase*: Phase is a measure of lag or lead of a single tone with regard to a reference tone. Interchannel phase measures the phase relationships between channels in a multichannel device; one channel is chosen as the reference. Absolute phase measures the phase at the output of a device relative to its input. Phase is expressed in degrees.

Nontraditional Measurements Using DSP

DSP makes available measurement methods which cannot be performed using analog circuitry. These methods have gained acceptance in the audio industry because of their accuracy, speed, and resolution. They include

- *Power spectrum:* Also known as the *FFT spectrum,* this is a display of the spectral content of a signal. When the DUT is driven with a sine wave, for example, the power spectrum of the DUT output shows the harmonic distortion, noise floor, and hum level of the DUT on the same graph (see Figure 8.1-26).
- *Real-time analyzer (RTA):* An RTA can be implemented with analog circuitry, but DSP-based systems are much more common. The RTA uses a bank of bandpass filters of constant *Q* (fractional width) to divide the input signal into bands and then measures the level in each band with an RMS detector. The result is typically displayed as a vertical bar chart, with one bar for each band. Third-octave bands are often used, giving 31 bands over the full audio spectrum (DC-15 kHz). In conjunction with a *pink-noise generator,* an RTA can be used to quickly determine the frequency response of a DUT.
- *Multitone:* A periodic collection of sine tones is applied to the DUT by the generator (see Figure 8.1-27). Simultaneously, the DUT output is acquired by the analyzer. An FFT is performed on the acquired signal. The FFT is post-processed in multiple ways to compute the frequency and phase responses (using the FFT bins which contain the generator tones), the distortion performance (using the FFT bins which do not contain tones), and the noise performance (by using an acquisition of twice the generator length, leading to FFT bins which contain only noise generated in the DUT). A multitone has a dissonant but not unpleasant quality, and typically lasts for less than 250 ms.
- *Continuous exponential sweep:* A method for measuring the linear and harmonic nonlinear impulse responses of a system. A sine tone whose frequency increases exponentially in time is applied to the DUT, while the DUT output is acquired (see Figure 8.1-28). A mathematical deconvolution operation recovers the linear impulse response of the DUT and its harmonic responses independently. The linear impulse response can be transformed into the frequency domain to recover the frequency and phase responses. The harmonic responses can also be transformed to recover the harmonic frequency responses, or power summed to recover the THD response. If desired, the impulse response can be truncated before frequency transformation to reject room reflections in acoustic measurements.

  Continuous sweep techniques are very fast; a commercially available analyzer system has demonstrated the ability to make 14 different measurements using data acquired in a 1-second sweep. The sweep has a chirping sound, with a typical duration of 1000 ms or less.
- *Time delay spectrometry (TDS):* Like RTA, TDS can be implemented with analog circuitry, but is more commonly done with DSP. A sine tone whose frequency increases linearly in time is applied to the DUT. The output of the DUT is applied to a bandpass filter whose center frequency follows the generator frequency. An RMS detector measures the output level of the bandpass filter. A plot of the RMS level against generator frequency reveals the frequency response of the DUT. TDS is popular in acoustic measurement, where the tracking bandpass filter is set to lag the generator by a fixed time; this allows the direct radiation from the acoustic source to be measured, while rejecting room reflections, which arrive later and are rejected by the filter. The TDS signal has a chirping sound and typically lasts for several seconds.
- *Maximum length sequence (MLS):* A method for measuring the impulse response of a system (and therefore its frequency and phase response). A pseudo-

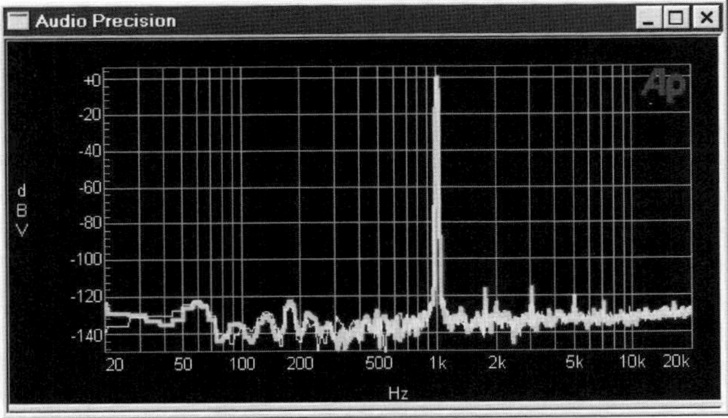

FIGURE 8.1-26 Display of power spectrum for two channels. Stimulus is a 1 kHz sine wave—note visibility of noise (at 120 dB below the stimulus), hum (small peaks at 60, 120, 180 Hz), and harmonic distortion (small peaks at 3, 5, 7 kHz).

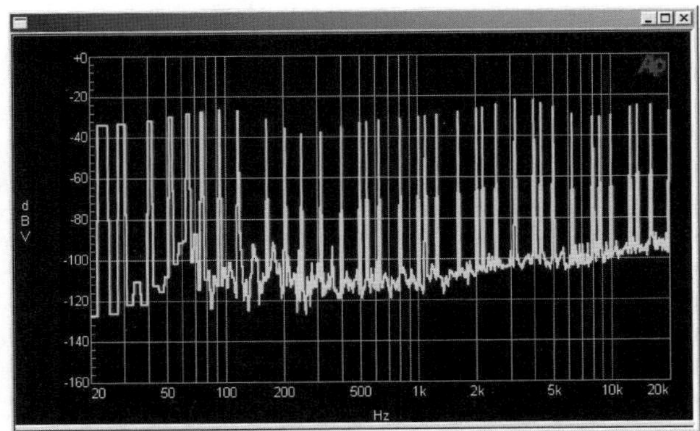

FIGURE 8.1-27 Power spectrum of a DUT output using a 36-tone multitone stimulus. The level variation across the tones shown here reflects the DUT frequency response. Phase, distortion, noise, and interchannel phase results can also be calculated from this one brief acquisition. (Photo courtesy of Audio Precision.)

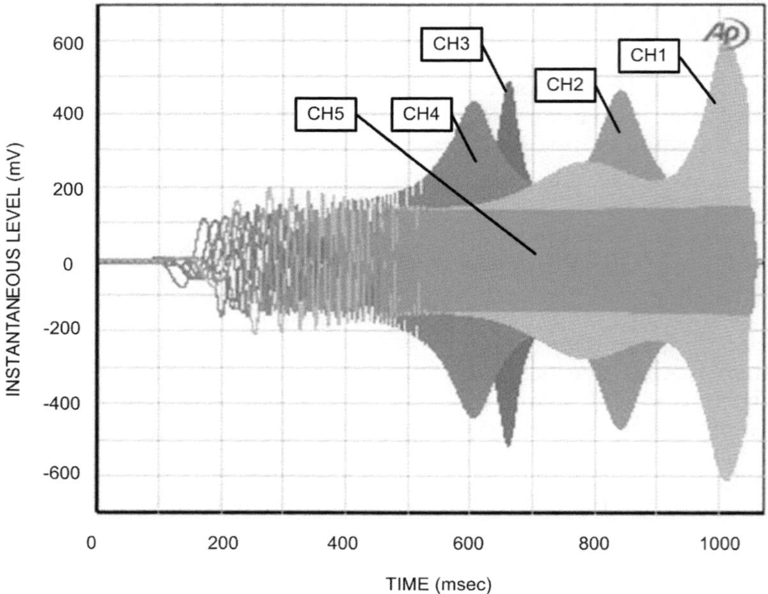

FIGURE 8.1-28 A multichannel continuous sweep acquisition, unprocessed. Each channel was subjected to different equalization, resulting in the modulation envelopes shown. Further processing of this 1-second acquisition can provide impulse response, level, frequency response, interchannel phase, distortion, group delay, and other results.

random signal is applied to the DUT, while the DUT output is acquired. A mathematical deconvolution operation is performed between the acquired and generated signals. This extracts the impulse response of the DUT. The impulse response can be transformed into the frequency domain to recover the frequency and phase responses. MLS is popular in acoustic measurement, since the recovered impulse response can be truncated before it is frequency transformed, thus preventing room reflections from corrupting the measurement. The MLS sounds noise-like and typically lasts for around 500 ms.

Converter-Specific Measurements

ADCs and DACs, being hybrid analog/digital devices, exhibit distortions and anomalies not seen in purely analog equipment. To a large extent, their performance can be measured using the traditional and nontraditional measurements described previously. However, further measurements may be needed to characterize them fully. For ADCs, these measurements include

- *Anti-alias filter stopband attenuation*: Tones at frequencies above half the sampling rate must be filtered out before sampling; otherwise, aliasing will occur. In this test, a full-scale analog sine wave is presented to the ADC, and the digital output level is measured as the sine wave frequency is varied. Tones in the stopband should exhibit very high attenuation.
- *Idle channel noise spectrum*: Some ADCs produce audible tones at low level when the input signal is small or nonexistent. A high-resolution FFT of the output spectrum of the ADC with the input grounded can show these tones.
- *Jitter modulation*: *Jitter* is variation in timing of clock edges from their nominal positions. If there is jitter on the sample clock, distortion is introduced in the sampling process, and this distortion is measurable in the digital output. Sample clock jitter is most likely to occur if the ADC is synchronized to an external clock. The external clock itself may have jitter, or the clock synchronization circuitry inside the ADC may introduce jitter of its own. An FFT of the output spectrum of the ADC when the input is driven with a sine wave can show jitter sidebands around the fundamental tone.

Measurements specific to DACs include

- *Reconstruction filter ripple and image rejection*: The images of the original signal are typically removed with a digital filter. This filter has a nonideal response that exhibits ripples in the passband and noninfinite attenuation in the stopband. To measure passband ripple, a full-scale digital sine wave is presented to the DAC, and the analog output level is measured as the sine wave frequency is varied. To measure stopband attenuation, the analog output level is measured at the known image frequencies. The passband ripple should be very low, and the stopband attenuation should be very high.

Digital Interface Measurements

The measurements described so far are concerned with the quality of the audio signal itself. It is also possible to measure characteristics of the embedded data (bit exactness) and of the digital bitstream or interface carrier signal. Errors in the data or degradation of the carrier can both cause distortion or failure in recovery of the embedded audio.

In particular, the carrier is subject to influences (primarily noise and high-frequency line loss) that can result in jitter at the receiver. A moderately jittered carrier can cause distortion in the decoded audio; in the worst situations, it may not be possible to convert the digital signal back to analog form.

Bit Exactness

A digital storage or transmission medium must be able to store or convey embedded audio without introducing errors. Digital storage and broadcast systems make use of *error correction* to detect and correct the inevitable errors that arise from dropouts, transmission interference, and the like. Digital transport mechanisms such as AES3, however, use a simple parity scheme to protect the embedded audio signal. The scheme used in AES3 is not capable of correcting errors, or of detecting an even number of errors in a single subframe. It is therefore important to verify that a digital link of this type is performing properly.

In a test for bit exactness, a prescribed sequence of digital samples is transported over the medium to be tested. At the receiving end, the recovered sequence is compared to the known transmitted sequence, and any errors reported. The sequence is often pseudo-random to allow it to be generated easily and to ensure that all of the bits in the audio word are exercised. The transmitter and receiver must share the sequence generation function but do not need to be synchronized.

Interface Jitter

A digital audio receiver must recover a clock signal from the bitstream it receives. In AES3, the clock signal is recovered from the transitions in the bitstream; typically, the zero-crossing of a transition is designated as the transition time. Although the transitions have a short rise time at the transmitter, the rise time at the receiver is longer because of high frequency roll-off in the transmission line. This nonzero rise time introduces the possibility that the time at which the digital receiver detects the transition differs from the true transition time (see Figure 8.1-29). This can occur in several ways:

- Noise on the transmission line causes the transitions to cross through zero earlier or later than the true transition time. This is known as *noise-induced jitter* and is illustrated in Figure 8.1-30. In this figure, the bottommost trace represents the amalgamation of the bit sequences above, where each of these sequences has been "smeared" by the effect of the rise time of the transmission line; that is, the transition from 1 to 0 and 0 to 1 is no longer sharp but is now an exponential curve. The circled zero-crossing illustrates the jitter which results from this smearing.

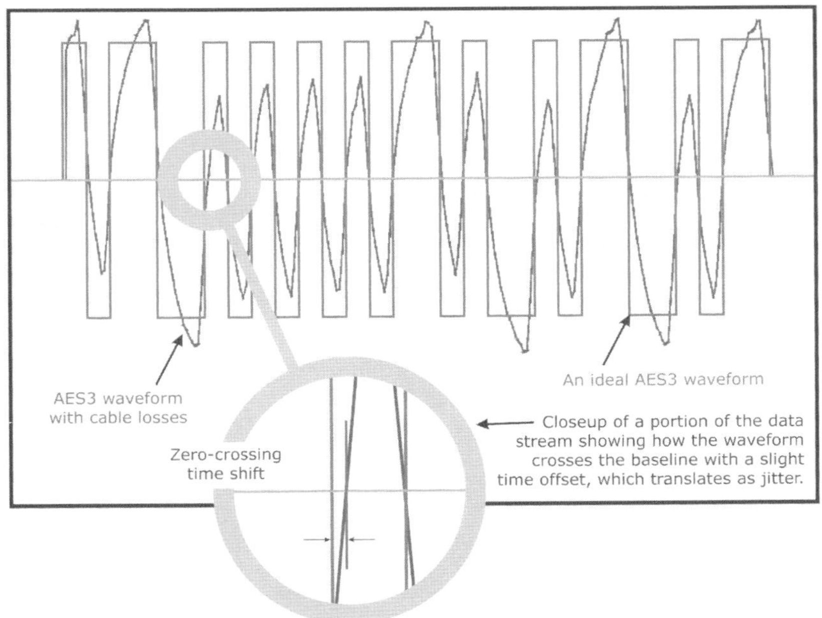

FIGURE 8.1-29 The transitions of the ideal and received waveforms differ because of cable losses, resulting in jitter.

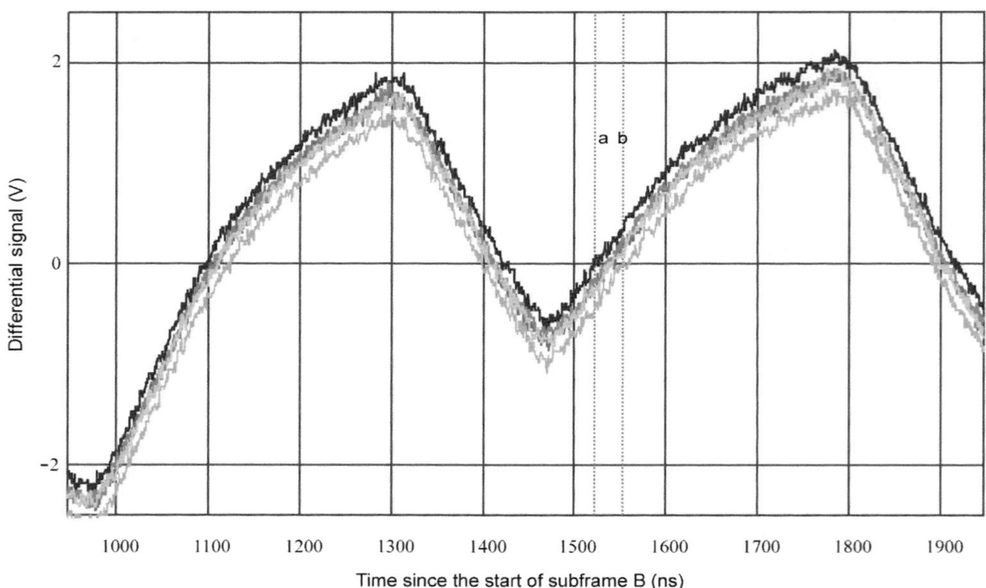

FIGURE 8.1-30 Noise-induced jitter.

- In transmission lines where the rise time is comparable to the unit interval (UI), the position of the zero-crossing is affected by the previously transmitted bits. This is known as *inter-symbol interference,* and affects even noise-free lines (see Figure 8.1-31). In the figure, a jittery clock signal has been captured using a signal monitor that has available a stable, jitter-free reference clock. Each individual waveform in the figure represents a successive trace of the (jittery) signal under test—the fact that the zero-crossings do not line up is evidence of jitter. Note that the zero-crossing extremes of this group of waveforms are indicated by the "a" and "b" markers; jitter is often characterized by the differences in the zero-crossing extremes and, for this case, is about 300 nsec.

Therefore, because of impairments in the transmission line, the clock recovered at the receiver will contain some jitter. Digital receivers are designed to pass jitter below a certain frequency and attenuate jitter above that frequency. This characteristic allows the receiver to follow slow variations in sample rate that occur intentionally at the transmitter, while rejecting high-frequency variations characteristic of line-induced jitter. The AES17 standard specifies the amount of jitter that a digital receiver must tolerate.

Eye Pattern

If noise or roll-off in a transmission line is severe, the clock (and therefore the embedded audio) may be partially or completely unrecoverable, leading to signal errors and dropouts. The *eye pattern* test allows this problem to be diagnosed quickly (see Figure 8.1-32). The signal at the receiving end is acquired for many UI (typically tens of thousands) and plotted on a 1 UI-wide graph, superimposing the many UI traces to show the worst case for pulse width, timing, and voltage swing.

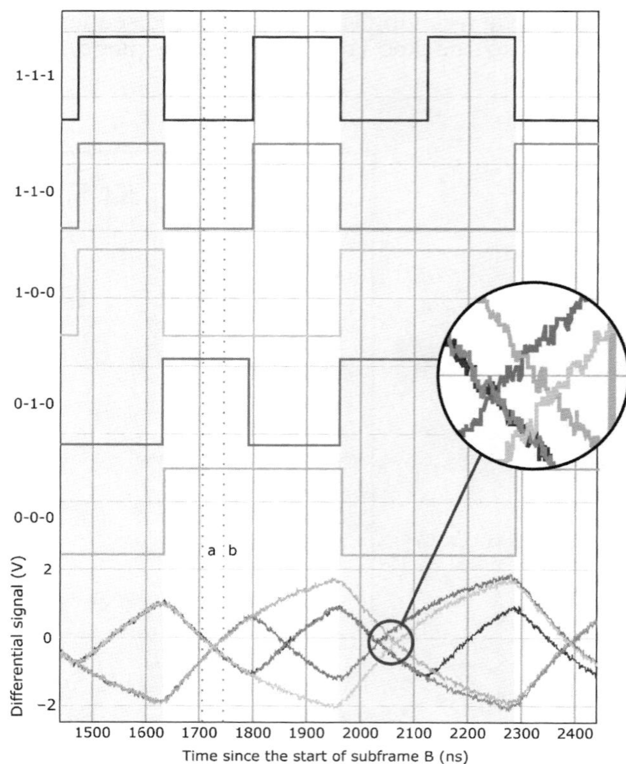

FIGURE 8.1-31 Inter-symbol interference and the resultant jitter.

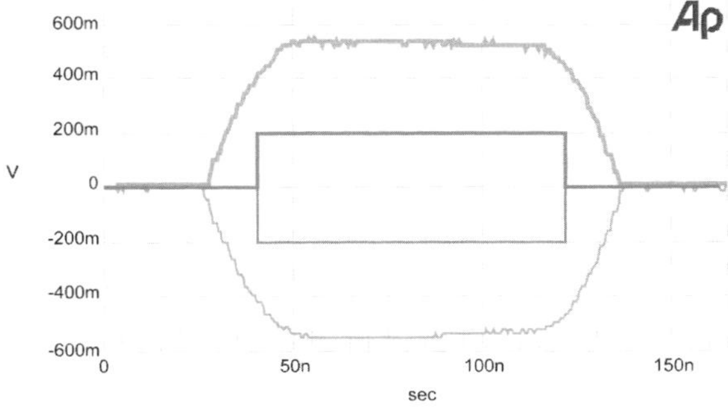

FIGURE 8.1-32 Eye pattern. The small box in the center is the AES17 minimum eye specification; the trace surrounding it is the eye pattern created from many acquisitions of a measured interface signal.

Jitter causes the eye to close horizontally, whereas noise causes the eye to close vertically. The AES17 standard specifies the minimum size the eye must have for the transmission system to be deemed acceptable. If the eye closes beyond this minimum size, it is likely that data errors will occur.

ADDITIONAL TOPICS

Some additional information relevant to digital audio signal analysis not covered previously is given in the following subsections.

Multichannel Audio

Throughout the first half-century of sound technology, recorded and transmitted audio was almost exclusively monaural (single channel). Audio generators, meters, and analyzers were also monaural. Two-channel stereo arrived in recording and film sound in the 1950s, spread to FM radio in the 1960s, and was added to television in the 1980s. Audio analyzers took longer to adopt stereo; in many cases, a single-channel tester switched between stereo channels was adequate. However, some important measurements, such as crosstalk and inter-channel phase, required both channels to be stimulated or analyzed simultaneously. By the 1980s, cutting-edge analyzers were stereo, and the best test and measurement systems could perform almost all the tests needed on two channels simultaneously. This brought the added benefit of reducing test time.

Alternative languages and other associated audio services are driving both television and radio broadcasters to support the transport of multichannel audio within their facilities. In addition, multichannel surround sound is established in film sound and DVDs, and has been adopted by digital television standards. These so-called 5.1 and 7.1 systems offer six and eight channels of audio, respectively, with the ".1" referring to a low bandwidth channel intended for low-frequency effects in film soundtracks. The large channel

FIGURE 8.1-33 Multichannel (8 input, 8 output) audio analyzer. (Photo courtesy Audio Precision.)

counts of these systems present a monitoring and testing challenge.

When these systems are tested with stereo analyzers, it is necessary to be able to choose which pair of channels to test at any one time. It is also desirable to automatically test all the channels without moving cables. An *output switcher* is a demultiplexing device which allows the generator signal to be routed to any of the DUT channels. An *input switcher* is a multiplexing device which allows any of the DUT channels to be routed to the analyzer. Under software control, these devices allow automated testing of multichannel systems on conventional analyzers.

However, measurement of crosstalk requires stimulation or analysis of more than two channels simultaneously. In addition, the time required to test eight channels two at a time may be prohibitive. Multichannel audio analyzers are now available with the capability to generate and analyze eight or more audio signal channels simultaneously (see Figure 8.1-33). Such analyzers do not require input or output switchers.

Psychoacoustic Data Compression

In audio delivery channels such as DVD media and digital television and radio broadcast, it is desirable to maintain audio quality while minimizing bandwidth, to allow more content or more channels. An AES bitstream with a 48 kHz sampling rate has a bit rate of 6.144 Mbps and a special content that extends beyond 20 MHz [16], which is too great to be practical for direct delivery over most wireline and radio links. Psychoacoustic data compression can be exploited to avoid transmitting data that is inaudible or of low perceptual importance (see Chapter 3.7 for additional information on digital audio data compression). A *perceptual audio encoder* converts the input signal into a data-reduced form for transmission, and a *perceptual audio decoder* at the receiver reconstructs a facsimile of the original signal from the data-reduced bitstream.

The encoder uses a model of the human auditory system to determine the perceptually important content of the signal. This content is then represented in a compact way, typically making use of a frequency transform. Since the transmitted data rate (or *bit rate*) is typically fixed, the encoder allocates the available bits to minimize the audible difference between the reconstructed signal and the original. The compact representation of the audio signal is then transmitted or stored. At the receiving end, the decoder converts the compact representation back into a conventional (or *linear*) audio stream. The encoder and decoder must agree on the meaning of the compact representation so that decoding can proceed correctly.

The *compression ratio* is the ratio of the input bit rate to the compressor to the output bit rate. Compression systems for delivery to the consumer, such as Dolby® Digital, maximize compression ratio, whereas compression systems for transport within a facility, such as Dolby® E, maximize robustness to multiple encode/decode cycles.

A compressed digital audio signal may be carried on an AES3 connection. An extra layer of protocol, defined by the IEC61937 standard, is used to identify the signal to the receiver as compressed rather than linear.

Most audio test equipment generates and analyzes linear (uncompressed) signals. It may therefore be necessary to convert between uncompressed and compressed formats for testing. When testing a transport designed to carry Dolby E audio, for instance, the stimulus signal must be encoded with a Dolby E encoder before applying it to the transport. At the receiving end, the signal must be decoded with a Dolby E decoder before it can be accepted by the audio analyzer.

Dolby E

Digital audio can be encoded for transport using Dolby E compression, which uses perceptual coding techniques to allow one AES3 connection to carry up to eight channels of audio information in the IEC61937 layer. Dolby E sacrifices compression ratio to allow multiple encode/decode cycles without audible degradation.

Dolby Digital

Digital audio can be encoded for distribution to end users using Dolby Digital (AC-3) compression. Dolby Digital allows one AES3 interconnection to carry 5.1 or 7.1 channels of surround sound in the IEC61937 layer. The ".1" refers to a low bandwidth channel, typically used for low-frequency effects in movie soundtracks.

Embedding

An AES3 transport carries two channels of linear PCM; multiple AES3 connections must be used to carry more than two channels. Alternatively, multichannel digital audio may be transported over a single connection using *embedding* techniques. In embedding, multiple linear digital audio signals are placed in a high-bandwidth bitstream to be conveyed over a single connection. The connection may also carry other signals simultaneously, such as digital video and metadata.

Most audio test equipment makes use of standard audio interfaces such as S/PDIF. To interface to an embedded system, it is necessary to use an embedder at the generator end and a de-embedder at the analyzer end. If the embedded system is carrying audio in a compressed format, it may be necessary to daisy-chain equipment. For example, when the audio quality of an SDI transport carrying Dolby E compressed audio is tested, it is necessary to first compress the digital output of the audio analyzer using an outboard Dolby E encoder and then use an SDI embedder to embed the Dolby E data into the SDI bitstream. At the receiving end, an SDI de-embedder extracts the Dolby E data, and a Dolby E decoder converts it to linear digital. This digital bitstream can then be accepted by the audio analyzer.

SDI

In television studios, digital audio is often embedded with video for transport within a Serial Digital Interface (SDI) signal. For standard definition (SD) video, the SDI format is specified in SMPTE 259M; SMPTE 292M specifies HD-SDI for high-definition (HD) video. The serial interfaces provide for eight AES3-compatible 48 kHz 20- or 24-bit audio data streams (each with two channels of audio), for a total of 16 linear digital channels of audio information.

SUMMARY

Radio and TV broadcast engineers alike need to be aware of the principles and techniques involved in the analysis and characterization of audio signals. This chapter has provided an introduction into these topics for both analog and digital audio systems. DSP-based audio analyzers are a powerful tool for characterizing both analog and digital audio, accommodating

- *Traditional measurements*, such as level, frequency, and total harmonic distortion plus noise (THD+N) ratio;
- *Nontraditional measurements*, such as power spectrum, continuous exponential sweep, and maximum length sequence (MLS);
- *Converter-specific measurements*, such as stopband attenuation and jitter sensitivity;
- *Digital interface measurements*, such as bit exactness, jitter, and eye patterns.

Digital audio does not degrade gradually as is the case with analog, but suddenly and annoyingly. Having and regularly using proper test equipment will identify performance that is marginal before it fails and becomes difficult to find among the many components in a digital audio facility.

References

[1] Graeme, J., *Applications of Operational Amplifiers*, McGraw-Hill, 1973, pp. 132–139, 202–203.

[2] "A New Standard Volume Indicator and Reference Level," *Proceedings of the IRE*, January 1940, pp. 1–17.

[3] Harry, D., "Audio Program Analysis," as presented to the National Association of Broadcasters, April 1985.

[4] IEEE Standard: Recommended Practice for Audio Program Level Measurement, DOC. G-2.1.2/13, 1988.
[5] Salek, S., "The NAB Test CD—Use and Applications," as presented to the National Association of Broadcasters, April 1989.
[6] National Association of Broadcasters, "The NAB Broadcast and Audio System Test CD" instruction and applications booklet, 1988 (www.nab.org).
[7] Skritek, P., and H. Pichler, "Extended Application of T.I.M. Test Procedures," preprint no. 1557, 64th AES Convention, November 1979.
[8] Takahashi, S., and S. Tanaka, "A New Method of Measuring Transient Intermodulation Distortion: A Comparison with the Conventional Method," preprint no. 1539, 64th AES Convention, November 1979.
[9] National Semiconductor Corporation, *Audio/Radio Handbook,* 1980, pp. 2.3–2.10.
[10] See note 11.
[11] Rane Corporation, Pro Audio Reference, on the web at http://www.rane.com/par-w.html, "weighting filters" entry.
[12] National Association of Broadcasters, 1965, *Magnetic Tape Recording and Reproducing* (NAB-2/65), Figure 6, pg. 23.
[13] Dolby, R., D. Robinson, and K. Gundry, "CCIR/ARM: A Practical Noise-Measurement Method," *Journal of the Audio Engineering Society,* March 1979, vol. 27, pp. 149–157.
[14] Mendenhall, G., "The Composite Signal-Key to Quality FM Broadcasting," *Broadcast Electronics.*
[15] Graeme, J., *Designing with Operational Amplifiers*, McGraw-Hill, 1977, pp. 76–77, 251–253.
[16] Robin, M., "The AES/EBU Digital Audio Signal Distribution Standard," Technical Notes – 1, Miranda Technologies.
[17] Salek, S., and A. Clegg, "Digital Audio Broadcasting," *The Electronics Handbook, Second Edition*, Taylor and Francis Group, 2005, pp. 1687–1689.

Bibliography

AES3-1992, "Recommended Practice for Digital Audio Engineering—Serial Transmission Format for Two-Channel Linearly Represented Digital Audio Data," *Journal of the Audio Engineering Society,* vol. 40, no. 3, pp. 147–165 (June 1992) www.aes.org

AES-3id-1995, "AES Information Document for Digital Audio Engineering—Transmission of AES3 Formatted Data by Unbalanced Coaxial Cable," *Journal of the Audio Engineering Society,* vol. 43, no. 10, pp. 827–844 (October 1995)

AES17, "AES Standard Method for Digital Audio Engineering—Measurement of Digital Audio Equipment," *Journal of the Audio Engineering Society,* vol. 46, no. 5, pp. 428–447 (May 1998)

Dunn, Julian, *Measurement Techniques for Digital Audio,* Audio Precision, Beaverton, Oregon (2002)

IEC-60958-1:1999, "Digital Audio Interface—Part 1: General," International Electrotechnical Commission, Geneva (1999)

IEC-60958-3:1999, "Digital Audio Interface—Part 3: Consumer Applications," International Electrotechnical Commission, Geneva (1999)

IEC-60958-4:1999, "Digital Audio Interface—Part 4: Professional Applications," International Electrotechnical Commission, Geneva (1999)

IEC-61937, "Digital Audio—Interface for Non-linear PCM Encoded Audio Bitstreams Applying IEC 60958," First Edition, International Electrotechnical Commission, Geneva (2000)

Metzler, Robert, *Audio Measurement Handbook,* Audio Precision, Beaverton, Oregon (1993) www.ap.com

SMPTE 276M-1995: for Television, "Transmission of AES/EBU Digital Audio Signals Over Coaxial Cable," Society of Motion Picture and Television Engineers, White Plains, NY, USA (1995) www.smpte.org

CHAPTER

8.2

Digital Video Signal and Bitstream Analysis

DAVE GUERRERO*

Harris Corporation
Pottstown, Pennsylvania

DANNY WILSON**

Pixelmetrix Corporation
Singapore, Indonesia

INTRODUCTION

The migration to digital video for studio production and distribution has required major changes in the way that video signals are monitored, measured, and analyzed, both for quality control and troubleshooting problems when they arise. The initial change from composite NTSC analog to 601 standard definition (SD) digital video introduced station engineers to a new generation of test and measurement equipment. It also made clear that while digital video has many advantages and quality improvements compared to analog, various constraints also need to be considered in working with the new signal formats. The more recent addition of high definition (HD) video has further expanded the requirements for test and measurement, while the much wider bandwidths and higher data rates have imposed more stringent requirements on cabling and station infrastructure.

With the introduction of digital television (DTV), broadcasters can now add a multiplicity of services on top of conventional audio and video programming. As well as the improved quality of HDTV pictures, broadcasters today may offer 5.1 surround sound audio, multiple audio services, sophisticated captioning, and data broadcasting and interactive services. In addition, thanks to compression technology, more channels can be squeezed into the transmission pipe, increasing the revenue potential. These are significant ways in which DTV is superior to traditional analog television.

However, the increased complexity of digital video means that more things can go wrong. Problems can happen anywhere along the transmission chain as the data is processed, manipulated, and re-sent. Although in some respects digital video is more robust than analog, it is an "all or nothing" technology and does not degrade gracefully. For instance, a compressed transport stream for distribution or transmission of compressed bitstreams has little tolerance for error. If a table that maps an audio stream to a video stream contains a typo, the correct audio will not be delivered with the video.

With the current mandate to implement DTV and new video technologies, many new challenges are being presented to engineers of all disciplines. In the past, most broadcast facilities supported a large technical staff that consisted of specialists in maintaining and operating audio, video, RF, videotape, routing, and control systems, including an expert well versed for each respective field. Especially at the local station level, many of these specialist positions are gone. Today's engineer is a member of an engineering team or group in which every engineer and technician is often responsible for many facets of the broadcast operation. The modern engineer is really in the Quality Assurance business. His job is to ensure that every technical detail related to program content meets industry specifications. Fortunately, as digital video has become more complex, the equipment available to test, monitor, and measure the signals has increased in

*Dave Guerrero wrote the sections related to baseband digital video signals.
**Danny Wilson wrote the sections related to compressed bitstreams.

power and sophistication, greatly simplifying what would otherwise be an impossible task of monitoring and maintaining digital video facilities.

This chapter looks at some of the issues relating to digital video and the main areas that need to be addressed for analysis and quality control. The first part of the chapter discusses the characteristics and analysis of baseband digital video signals, particularly HD, and the later part covers compressed bitstreams for distribution and transmission. Many of the figures for this chapter are reproduced in color on the accompanying CD; when enlarged, they provide greater clarity than the black and white print reproductions.

HD VERSUS SD SIGNALS INTERFACE CONSIDERATIONS

Traditionally, broadcast engineers have used copper coaxial interconnections to transport video and audio signals within the facility. The physical design of the cable defines its characteristics and capability to transfer a signal from source to destination, with minimum impairments.

Serial digital interfaces for standard definition video apply to video data rates of 270 Mbps. The standards defining serial digital high definition video describe high digital data rates including 3.0 Gbps (more than ten times standard definition data rates). The HD serial digital signals, much like analog UHF signals, require special handling in order to properly transfer video data. The cable carrying the HD signal is essentially an RF transmission line. The cable's critical specification as a transmission line is its characteristic impedance. In order to transfer maximum energy (signal) from a source to destination through a coaxial cable, the cable's characteristic impedance must precisely match that of the source and destination.

Mechanical construction is the principal factor that determines the capability of a transmission line to carry a signal without loss. For example, in composite analog video systems, the losses attributed to the impedance of the transmission line directly influence the high frequency characteristics of the image. The high frequency chroma information is attenuated with increased cable length; the phase of the video changes approximately 2° per foot of cable, chroma and luma will have slightly different phase angles, and so on.

These analog signal aberrations do not occur in images using serial digital interfaces. However, digital systems present a new set of requirements for coaxial cable manufacturers. In particular, the frequencies of high-definition serial digital signals are much greater and require more precise methods of handling and testing when engineers design systems using cable interfaces.

Copper interfaces (standard coaxial cables) are the currently accepted method to transport high definition serial digital interface (HD-SDI) video when working with distances up to about 100 meters (330 feet). Depending on the cable, however, the general rule of thumb is to transport HD-SDI signals no more than about 50 meters (165 feet) on a coaxial cable before requiring a reclocking distribution amplifier. Fiber-optic transports may be used to extend the operating distance between high definition transceivers, with capability for miles of useful transmission.

High Definition Frequencies and Data Rates

The digital video signal is a serial data bitstream that includes video information, and (optionally) associated metadata and audio data encoded and embedded into the video data. The SMPTE HD-SDI standard 292M specifies that *single link* high definition serial digital video shall be formatted in 10-bit words at a data rate of approximately 1.5 Gbps (actually 1.485 Gbps), while some formats actually use 1.485/1.001 Gbps (1.483516 Gbps).

A direct correlation exists between the serial digital signal's data rate and its equivalent analog frequency. The equivalent analog frequency is half the data rate:

$$\frac{\text{Data rate}}{2} = \text{Frequency}$$

For HD serial digital video, the data rates are 1.485 Gbps and 1.4835 Gbps, so the equivalent analog frequencies are

$$\frac{1.485\text{ Gbps}}{2} = 742.5\text{ MHz} \quad \frac{1.4835\text{ Gbps}}{2} = 741.7\text{ MHz}$$

These high definition equivalent data rate frequencies are 742.5 MHz and 741.7 MHz. However, since the digital video signal is constructed in the form of square waves, the coaxial cables must be able to minimally pass the third and fifth harmonics of the signals as well as the fundamental frequencies.

Just as an RF system's bandwidth is defined by its fundamental and sideband frequencies, a properly designed high definition video transmission line must be able to pass the video data as a carrier and its odd harmonics. In Figure 8.2-1, waveform A is the fundamental frequency (approximately 741 MHz). Adding the third harmonic (three times the fundamental), B, results in a waveform approximating a square wave

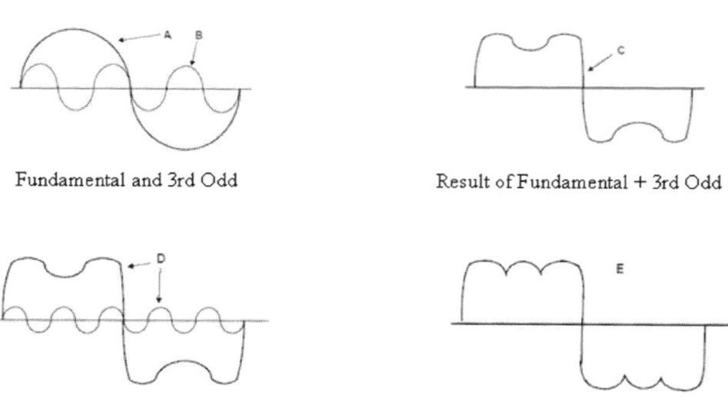

FIGURE 8.2-1 Creating a square wave using the fundamental and the third and fifth harmonics.

similar to C. Next, the addition of the fifth harmonic (five times the fundamental) to C, as in D, results in nearly a square wave as in E.

The third and fifth odd harmonics of 1.485 Gbps high-definition video are 2.2275 GHz and 3.7125 GHz, respectively. To put these frequencies into perspective, consider that TV channels 14–69 in the United States are located in the frequency band from 470 MHz to 806 MHz. In the world of radio frequency (RF) communications, these frequencies require special handling. Some operators refer to this handling as "black magic." Transmission lines resemble plumbing, and transformers are produced by wrapping a few turns of wire around a second coil of wire previously wrapped around a pencil! High definition video data circuits do not require the same transmission elements; however, they do require careful consideration when designing critical signal paths.

At the time of writing (Fall 2006), there is a movement for manufacturers to provide camera and recording equipment using 1080p60 video formats over a single link at a rate of 3 Gbps. Working at these rates will add more challenges for the systems engineer; resolving the fifth harmonic will require circuitry capable of passing a bandwidth of approximately 7 GHz from the output of one item of equipment to the input of the next.

TESTING THE HD PHYSICAL INTERFACE

The importance of testing the physical interface (transmission line) between a high definition source and destination is to ensure that it is achieving the best possible bit error rate. Losses of video data may result in artifacts visible in the picture. All coaxial cables suffer from *attenuation*, in which the video signal's amplitude is reduced due to the resistance of the center conductor. High-frequency information is even more susceptible to loss due to the capacitance of the cable. These losses are a function of the cable length.

In addition to attenuation, an important measurement of the performance of cable as it transfers a signal from point A to point B is its *return loss*. Signal degradation due to these characteristics means that a precision coaxial cable able to carry an analog video signal 333 m (about 1000 ft) may be able to carry a high-definition signal only 80 m (260 ft) or less.

Testing Return Loss

Coaxial cables are not perfect, in that the impedance may vary along the length of the cable creating standing waves.

Return loss is a measure of the mismatch of impedance along the cable or at a termination at the end of the cable. It is similar to VSWR in a transmission line. Any change of impedance can cause a signal reflection. Reflected or "return" signals interfere with the forward signal and result in variations in amplitude and phase of the combined signal. Return loss (RL) is determined using the following formula:

$$RL = -20\log\left[\frac{Z_1 - Z_{2A}}{Z_1 + Z_{2B}}\right]$$

where

Z_1 = nominal impedance

$Z_{2A} = Z_1 - \Delta Z$

$Z_{2B} = Z_1 + \Delta Z$

ΔZ = variation (tolerance)

A typical HD video cable has a nominal impedance of 75 Ω; the manufacturer specifies this impedance with a tolerance of +/− 1.5 Ω. The return loss of this cable can be calculated as

$$RL = -20\log\left[\frac{75-73.5}{75+76.5}\right] = -20\log(0.0099)$$
$$= -40.08 \text{ dB}$$

The preceding return loss calculation demonstrates that the coaxial cable's performance is directly related to its impedance (Z). The impedance is determined by the physical characteristics of the cable:

$$Z = 138\left\{\frac{V_{prop}}{100}\log\left|\frac{D_i}{D_C}\right|\right\}$$

where

Z = cable impedance

V_{prop} = cable velocity of propagation

D_i = diameter of insulation

D_c = diameter of conductor

A typical HD video cable has a nominal propagation velocity of 66, and insulation and conductor diameters of 0.198" and 0.031", respectively. The impedance of this cable can be calculated as

$$Z = 138\left\{\frac{66}{100}\log\left|\frac{0.198}{0.031}\right|\right\} = 73.34\ \Omega$$

Engineers use equations similar to those shown here to calculate the nominal impedance for the manufacturers' published specifications, in this case 73.34 Ω.

The next equation accounts for the impedance change within the same cable having a 10% reduction (0.0198") in the diameter of the cable's insulation at periodic intervals along its length. This minor periodic difference of less than 0.02" in diameter can easily be caused by overly tightened cable ties commonly used for installation:

$$Z = 138\left\{\frac{66}{100}\log\left|\frac{0.1782}{0.031}\right|\right\} = 69.179\ \Omega$$

Calculating the return loss for this cable

$$RL = -20\log\left[\frac{75-69.179}{75+69.179}\right] = -20\log(0.0040)$$
$$= -27.8786 \text{ dB}$$

Installation Measurements

The preceding calculations show that return loss of the installed cable may be reduced from –40.08 dB to –27.88 dB due to a series of overly tightened cable ties used to install the cable. This example clearly demonstrates the need to measure the physical system characteristics following any installation, or following a change to an installation.

Similarly, in a studio facility, manual video patching is usually employed to provide system flexibility. Patch cords and connections in the patch-bay all need to be measured for return loss. The quality of the electrical connection begins with the physical connection.

Data Recovery and Measuring HD-SDI Performance

As an HD serial digital video signal (see Figure 8.2-2) travels from a source device through the coaxial cable to a destination device, its signal will be attenuated and distorted. The cable's attenuation and return loss characteristic will determine the signal amplitude and waveform delivered to the destination device.

Clock and Data Recovery

Based on the amount that the high-definition signal has been attenuated, the receiving device has to recover the digital video signal from the background noise that is present. First, the signal is generally returned it to its nominal voltage level of 800 mV by automatically adding an appropriate amount of equalization. The problem is that this also amplifies noise with the signal, as illustrated in Figures 8.2-3 through 8.2-5. The data recovery process begins by extracting the clock information from the data, which is then used to decode the video data. Video data can be extracted perfectly in the presence of large levels of

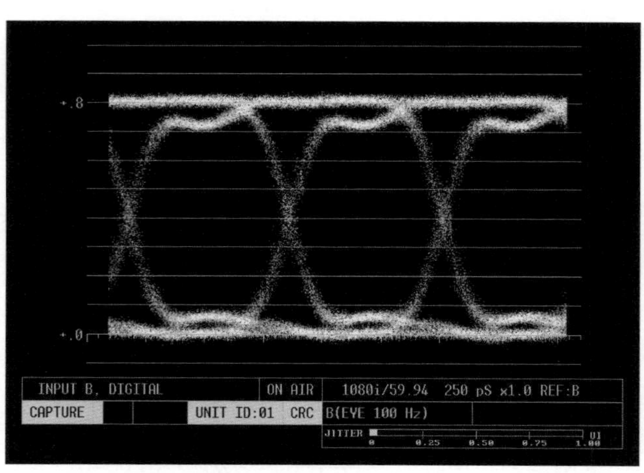

FIGURE 8.2-2 Eye pattern of "clean" serial data signal.

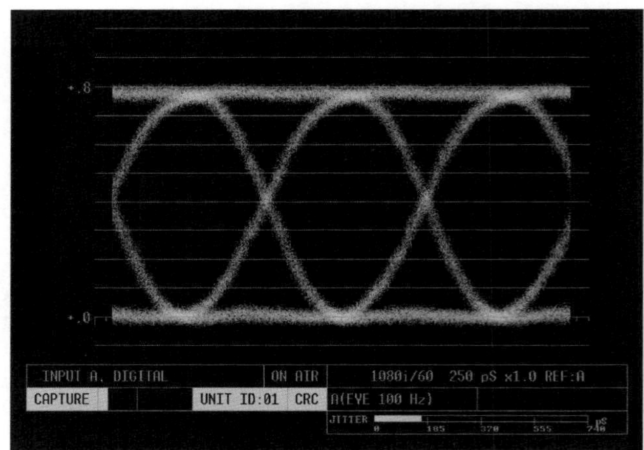

FIGURE 8.2-3 Eye pattern of serial data signal with 10% signal loss.

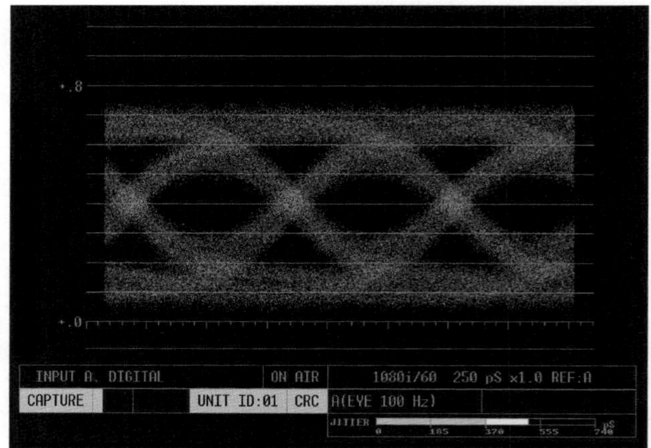

FIGURE 8.2-4 Eye pattern of serial data signal with 50% signal loss.

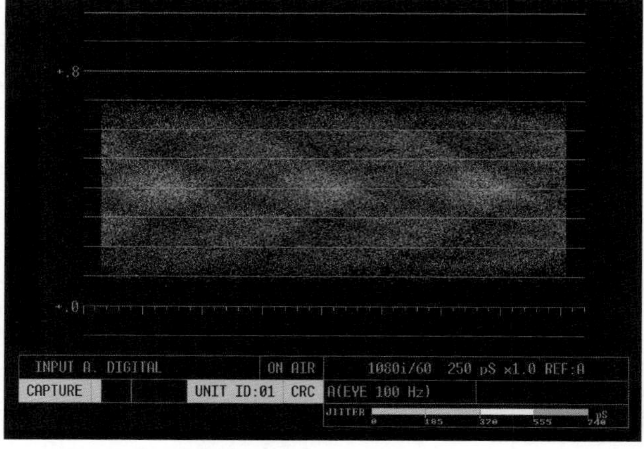

FIGURE 8.2-5 Eye pattern of serial data signal with 100% signal loss.

noise, but if the signal is attenuated to a point that it is little more than the noise floor of the interface circuitry, the data recovery process becomes increasingly difficult. The receiver will attempt to equalize the level of the incoming signal, but if the level is not much greater than that of the noise, the chances are that the clock cannot be recovered perfectly. The spikes produced by random noise become difficult to differentiate from signal edges after the input receiver's level equalization, resulting in signal jitter.

Measuring the Serial Digital Interface Using an Eye Pattern

Cable manufacturers use impedance bridges and other very sophisticated equipment to measure performance characteristics. However, the television engineer typically uses a different tool, a waveform monitor, capable of displaying an *eye diagram* of the applied signal. Instead of measuring the impedance of the interfaces, the eye diagram presents the voltage waveform of the video data signal arriving at the test instrument. Examples of such signals received with varying levels of degradation are illustrated in the following figures.

Figure 8.2-2 shows an example of a "clean" serial data signal, from which video data can be perfectly recovered. Figures 8.2-3 through 8.2-5 depict 10%, 50%, and 100% signal loss, respectively. The signal loss is simply due to interconnecting cables and their connectors.

Eye pattern and jitter measurements can be performed on the output of any cable run to ensure that it is producing serial digital data that can be recovered by receiving equipment. Figure 8.2-6 shows the effect of approximately 90% signal loss on a standard test signal. Persistence has been added to the display to demonstrate the accumulation of jitter in the recovered signal.

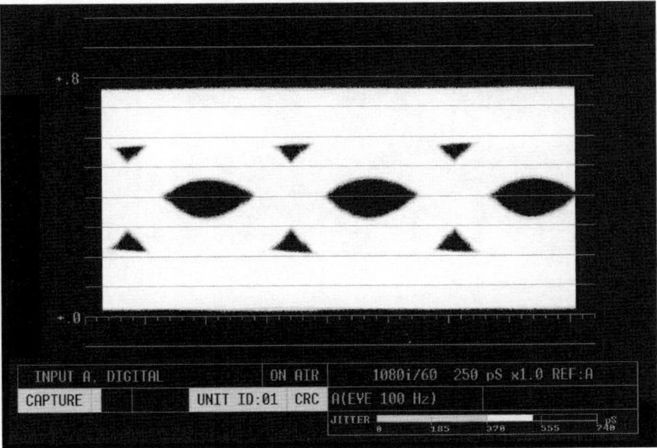

FIGURE 8.2-6 Eye pattern of serial data signal with 90% signal loss, with persistence added to demonstrate the accumulation of jitter in the recovered signal.

MONITORING PROTECTED AND ANCILLARY DATA IN HD DIGITAL VIDEO

Serial digital video is composed of active video data and protected data describing the start of active video (SAV), end of active video (EAV), timing reference, and other parameters of the video data. This information is used to identify the video in the serial digital data stream. The protected information uses unique "headers" that describe the beginning and end of each video line. The number $3FF_h$ (all ones) followed by two occurrences of 000_h (all zeros) signal the beginning of a new line (SAV). Subsequent information describes the line number, field number, parity, and cyclic redundancy check (CRC) values of the video data.

TABLE 8.2-1
Common Ancillary Data Identification (DID) Words

DID	Definition
200	Undefined format
180	Marks packets for deletion
260	Ancillary Time code
284	Data End Marker File
288	Data Start Marker Packet
1E3	299M, HDTV, control, Group 1
2E2	299M, HDTV, control, Group 2
2E1	299M, HDTV, control, Group 3
1E0	299M, HDTV, control, Group 4
2E7	299M, HDTV, audio, Group 1
1E6	299M, HDTV, audio, Group 2
1E5	299M, HDTV, audio, Group 3
2E4	299M, HDTV, audio, Group 4
2FF	AES audio data, Group 1
1FD	AES audio data, Group 2
1FB	AES audio data, Group 3
2F9	AES audio data, Group 4
1EF	AES control packet, Group 1
2EE	AES control packet, Group 2
2ED	AES control packet, Group 3
1EC	AES control packet, Group 4
1FE	AES extended packet, Group 1
2FC	AES extended packet, Group 2
2FA	AES extended packet, Group 3
FF8	AES extended packet, Group 4
2F0	Metadata packets

TABLE 8.2-1 *(continued)*
Common Ancillary Data Identification (DID) Words

DID	Definition
1F4	Error detection (EDH)
2F5	Time code
161/101	Closed Caption (CEA-708-B)
161/102	Closed caption (CEA-608 data)
162/101	Program Description DTV
162/102	Data Broadcast (DTV)
162/203	162/203 VBI Data

Ancillary data is information carried as packets within the serial digital stream that are completely unrelated to the recovery of the video image. This information may include (but is not limited to) embedded AES/EBU audio,[1] audio validity, AC-3 audio, Dolby E, closed captions, teletext, subtitles, copy protection bit, AFD and Bar Data, Pan-scan information, wide screen signaling, extended data system (XDS) data, ancillary time code, AMOL, test signals, and other data.

Ancillary data includes a "header" similar to active video; however, its flag begins with a sequence of 000_h, two occurrences of $3FF_h$, and followed by a data identification (DID) word. Table 8.2-1 contains a list of DIDs for common ancillary data. Most data analyzers provide a method to specifically search for these DIDs and other custom values as well.

Testing of protected and ancillary data is performed using one of two methods. The packetized data found in serial digital video is formatted with a unique DID. The simplest method is to test for presence of the DID header information. A test instrument checking the video data produces an indication based on the detection of the DID. This method works well when the operator needs only to know presence.

In the second method, a test instrument capable of displaying the data hex values for serial digital video signals, as shown in Figure 8.2-7, can be extremely helpful. Using automatic or manual data analysis, identification of all ancillary and video data can be verified. Manually analyzing the data pixel by pixel and line by line, with an associated numeric readout, verifies all serial digital data and its relative position within the video frame. This procedure can, however, be rather time consuming, as it requires manually moving a cursor through the ancillary and active video data.

EVALUATING PICTURE DATA

The data displayed on a serial digital video analyzer can provide crucial information when diagnosing a problem. In Figure 8.2-7, (B) is representative of the magnified view of the area surrounding the analyzer target, (A), and (C) displays the actual hexadecimal numeric value of the pixel data at the target position.

In evaluating the digital video data shown in Figure 8.2-8, a user has moved the eight samples wide, pixel cursor (6) onto line 163 from the field with odd-numbered lines (1). The magnified view of the picture displays the surrounding pixels (7), providing a visual presentation. If the surrounding pixels were different

[1]See Chapter 1.15, "Audio Standards and Practices."

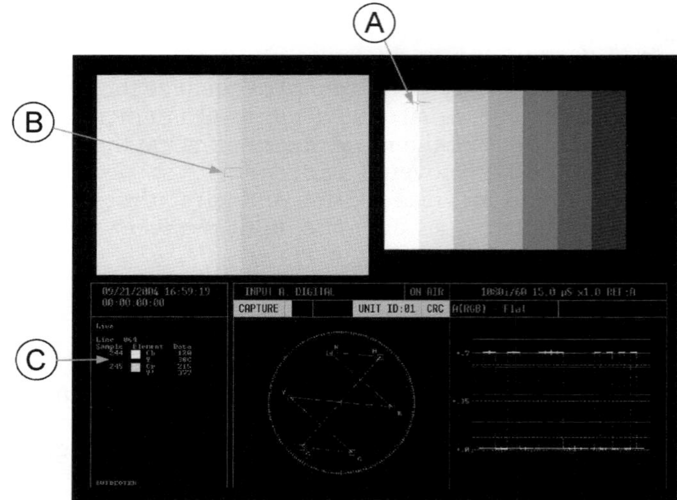

FIGURE 8.2-7 Advanced digital video signal analyzer with multiple display panels, including picture monitor with target pixel selector, magnified pixel display, waveform monitor, vectorscope, audio, pixel, and data analysis. (See CD for detailed color version of this and other figures.)

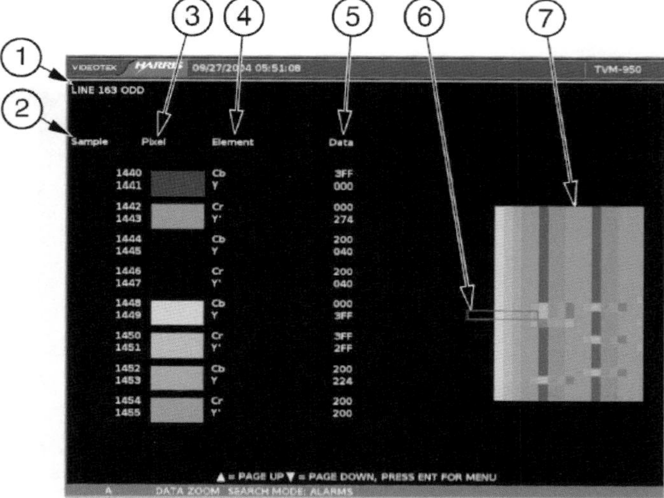

FIGURE 8.2-8 Advanced digital video analyzer with capability for showing individual screens for waveform, vectorscope, color gamut, timing, audio, and data analysis. Figure shows pixel data analysis.

in color, chances are that the pixels selected by the cursor would be incorrect in value.

The display shown in Figure 8.2-8 also demonstrates the 4:2:2 sequence of broadcast video. Samples 1440 and 1441 (2) of line 163 define a pixel (3) containing components C_B and Y (4); the values of these samples are 3FF and 000 (5), respectively. The next samples horizontally, 1442 and 1443, contain C_R and Y′ having values of 000 and 274. Note there is one Y sample for each chroma component; thus, the Y sample rate is two times the chroma sample rate.

When an engineer is looking for problem pixels or samples of a high-definition picture containing millions of pixels, the task may be seem difficult. However, modern video cameras provide a means to mask bad pixels of the imaging "chip block." The location information is critical in repairing cameras to this level of detail. Some test instruments have the capability to automatically place the cursor on values that are detected as faulty, therefore greatly reducing the time required to analyze a discrepancy.

For day-to-day operations, test instruments are required that simply indicate presence of pertinent information and decode the data for verification. For example, the operator may be required to monitor closed caption presence; simply decoding the data and displaying it in a picture will meet needs for this evaluation. The monitoring example shown in Figure 8.2-9 indicates that closed captions are present and contain valid information.

Another common and simple test for ancillary data is picture delay. In Figure 8.2-10 the image is displaced horizontally and vertically to display the ancillary data. The presence of embedded audio can be verified without any special test equipment. AES/EBU audio is embedded evenly in the H_{ANC} (horizontal ancillary) data, as seen at (A). Each vertical column represents one group of audio. If the column is not continuous or not present, it cannot be decoded.

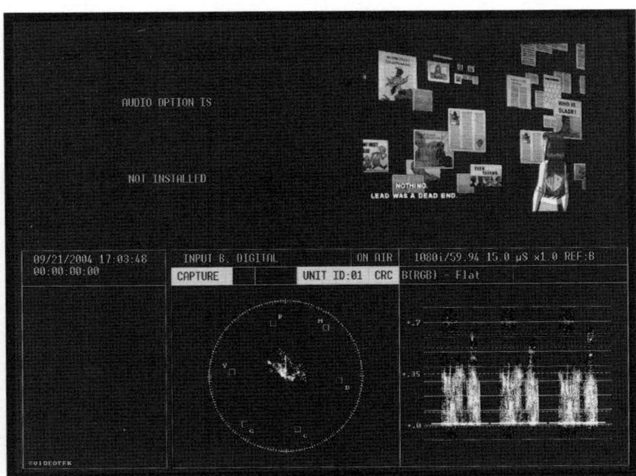

FIGURE 8.2-9 Monitoring of closed captions with a digital video analyzer.

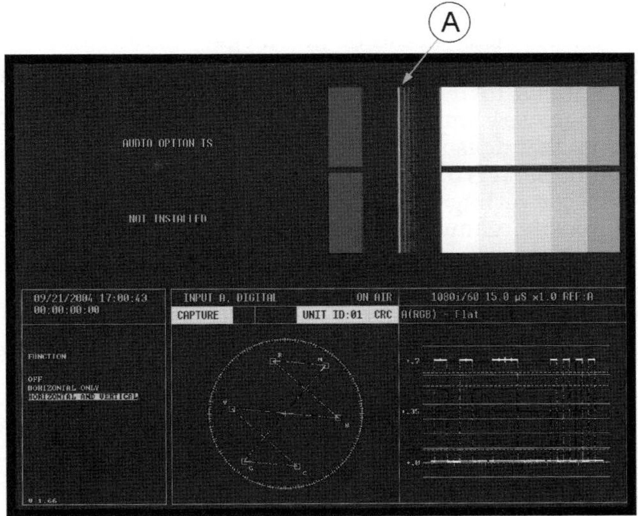

FIGURE 8.2-10 Displaying ancillary data on a digital video analyzer.

VERIFYING HD COLOR GAMUT

The issues of color gamut that have arisen with HD video began in the early stages of color television. In the United States, the NTSC color system was developed with a particular set of color parameters that matched the color characteristics of the phosphors that were available for commercial television receiver cathode ray tube displays. The limited available technology of the time did not allow for the manufacture of phosphors that could replicate the full gamut of colors perceptible to human vision (per CIE 15.2 specification, 1931, see Chapter 5.2). This limited set of colors also helped with limiting the amount of information that had to be sent (if the display systems cannot reproduce specific colors, there is no need to transmit them).

Because of improvement in both camera and display technology, digital television standards were designed to include a wider range of color values (that is a larger gamut) than the original NTSC color gamut as defined in the SMPTE 170M standard.

Conversion Issues

When an engineer is working solely with SD digital video and hardware that conforms to the ITU-R BT.601 standard, equipment that is designed properly will not have gamut-related problems, and users should not encounter a need to compensate digital video for color space limitations. However, as soon as the engineer needs to prepare digital SD video for conversion to HD or for broadcast by an analog transmission channel, gamut measurements and possible adjustments may be necessary.

A program produced in analog video (composite or RGB) will have no gamut problems when converted to SD or HD digital video. The digital format allows a broader gamut of colors than the analog counterpart; therefore, there should be no "illegal" colors resulting from this conversion. Of course, there are always exceptions to the rule, and it is technically possible (though very improbable) to have a composite video signal that doesn't conform to the SMPTE 170M specifications and is out of gamut when converted to digital.

In converting a digital SD or HD signal to analog composite, gamut errors are easily produced. Traditionally, *legalizers* have been used to modify the illegal digital values prior to conversion to analog composite or RGB. They are used when program material produced in a digital format needs to broadcast in an analog format. A legalizer alters the serial digital video data on a pixel-by-pixel basis rather than a field-by-field basis (as in a processing amplifier) and does so using a lookup table (LUT), replacing "illegal sample values" with those that are legal for broadcast. Legalizers can be configured as "set and forget" devices and should be used in every digital signal path that is converted to analog for broadcast.

Monitoring Color Gamut Errors

Monitoring gamut errors in a serial digital video signal is a little more difficult than measuring other parameters. The traditional instruments used to measure and monitor video are the vectorscopes and waveform monitors with which most engineers are very familiar. Serial digital test and measurement instruments are not capable of displaying a true composite representation of the serial digital signal, as all pertinent blanking information is not available in SDI signals. Using a waveform monitor and/or vectorscope to monitor or measure gamut is nearly impossible, as gamut is a measure of the relationship of chroma versus luma.

The problem of measuring gamut using these instruments is limited by the composition of their displays. The vectorscope plots a polar diagram of the phase and amplitude of the chrominance content of the video signal, while the waveform monitor is designed to show the amplitude of the video signal as a voltage per unit of time. Neither scope was designed to monitor gamut. Gamut errors are produced when the combination (mathematical relationship) of a video signal's luminance and chrominance creates a color that is not legal (not defined) for a display or transmitting device. An engineer could try to analyze a vectorscope and waveform monitor (composite display) and then guess when the gamut parameters may or may not be proper. This "measurement" would be very impractical and inefficient. A specialized serial digital video analyzer with color gamut capabilities is required.

Composite Gamut Compliance for Serial Digital Systems

The gamut display is designed specifically to display video gamut and is used to verify inter-format gamut errors. Similar in appearance to a vectorscope display, the composite/digital gamut displays serial digital video and maps the color space information into a circular diagram using vector position calculations. Since the composite/digital gamut display is produced digitally, the plot of the gamut information is extremely accurate and consistent. The gamut indications are electronically generated traces placed into a polar display using their corresponding luminance and chrominance sample values. All active video samples are mathematically converted to a representative composite value prior to plotting into the display. Each composite color sample is mapped into the composite/digital gamut display as a gamut dot-pair. The mapped dot-pair is derived from two vector calculations:

Luminance sample value + 1/2 Color Saturation sample value

Luminance sample value – 1/2 Color Saturation sample value

Processing the color samples in this manner calculates and plots a maximum and minimum indication for each positive and negative chrominance packet peak versus luminance. As shown in Figure 8.2-11, the outer mark (plotted farthest from center) represents the sum of the luminance and half the color saturation. The inner mark (plotted nearest the center) is the difference of the luminance and half the color saturation value. A sample that has no color saturation is plotted on the 0° vector (x-axis), and a bar depicting the magnitude of the samples containing luma information (monochrome) is shown next to the gamut chroma polar plot.

Use of the Display for Composite Gamut Compliance

A color bar test signal is shown in Figure 8.2-11. Although the test instrument's input signal is HD- or SD-SDI, the display presents the digital video as if it was transcoded to composite. The composite/digital gamut graticule includes marks indicating the minimum and maximum gamut for native color bar signals. The gamut reference indicators assist the engineer in understanding the relationship between a display of gamut color versus a color value created by

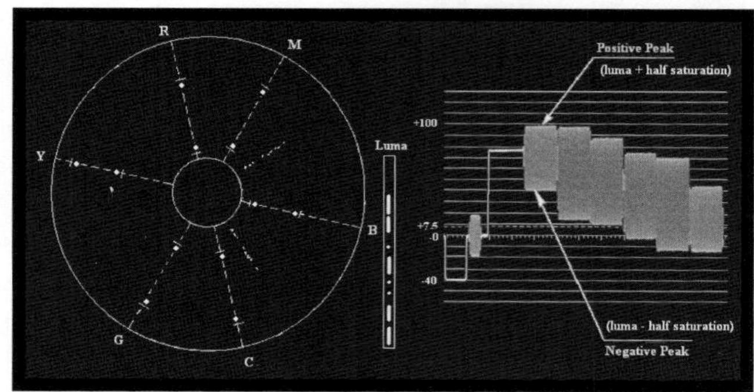

FIGURE 8.2-11 The gamut iris display plots two values depicting the composite gamut of each color.

a standard color bar generator. For example, the composite gamut dot-pair indicators for yellow are near the outer edge of the yellow vector; any color composed of more saturation or luminance than the yellow packet of the color bar signal will be out of gamut. A color with the same vector phase but less luminance (or chrominance) will be within the gamut range (brown) for composite video. Conversely, the color of the blue bar is at the low end of the gamut scale. There is plenty of room for "blue colors" with more saturation or increased luminance in the range of composite gamut and none for those of lower values.

Checking RGB Gamut Compliance in Serial Digital Systems

A serial digital video signal (HD or SD) monitored for RGB color space uses a similar display as it plots RGB sample information. The RGB/digital gamut displays in Figures 8.2-12 and 8.2-13 display video amplitudes (in mV) via a polar map representing color information. The amplitude of each R, G, and B pixel is plotted at a radial representing the derived RGB color sample values. Unlike the composite gamut plot of two points per pixel, the RGB gamut vector plots three points: the YCbCr sample values decoded into RGB values. For a purely red, 100% saturated signal, a mark is plotted at a magnitude of 700mV (red vector), and two marks each representing the green and blue vectors are plotted at 0mV. Again, monochrome signals, those of equal RGB value, are displayed in a separate bar of the display. Video sample values greater than the maximum gamut range setting are displayed outside the outer circle; values less than the minimum range setting are plotted inside the inner circle.

The RGB/digital gamut's graticule scales appear as inner and outer boundary circles representing the minimum and maximum gamut excursion for the system under test. The boundary circles are adjustable via user settings, if the maximum parameter is increased; the outer circle increases in diameter. This increase in diameter creates the illusion that the gamut is larger. If the negative limit of the gamut range is made more negative, the inner circle decreases in diameter, again appearing that the gamut range has increased. As the upper boundary of gamut decreases, the gamut range is made smaller. If a color sample is out of gamut, the RGB/digital gamut display will plot it outside (or inside) the appropriate boundary circle. When any sample is calculated as violating the gamut range, the color of the gamut graticule changes to red, adding to the usefulness of the display. This display presents an instant indication of the color of out of gamut video.

The fact that the scales of the RGB/digital gamut display change color upon a chroma/luma range violation enables its use by inexperienced operators. As a go/no-go indication, operators can easily be trained to monitor color gamut violations. Also, the experienced engineer will find that determining gamut parameters is simple: The display indicates the actual hue of the violating color space, so there is no guesswork. The colors are read just like those in a traditional vectorscope.

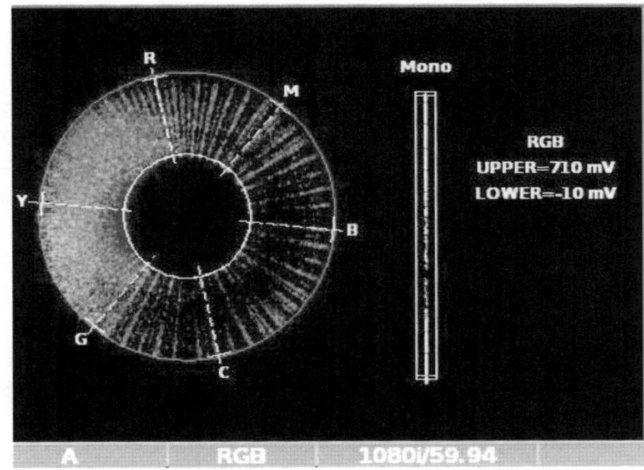

FIGURE 8.2-13 Digital gamut iris display of a live video signal.

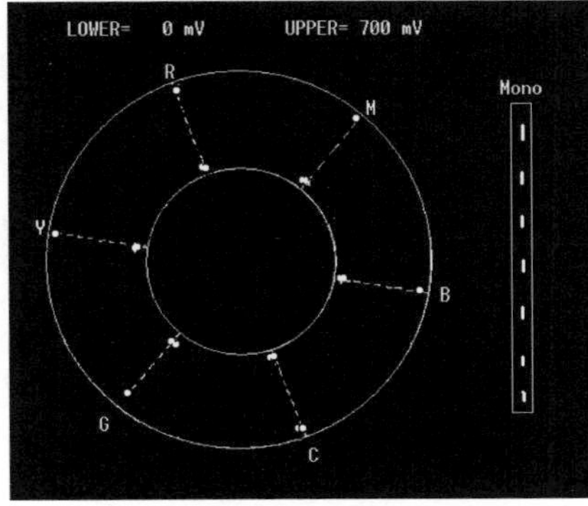

FIGURE 8.2-12 RGB gamut plots three polar coordinates based on the RGB values.

COLORIMETRY AND COLOR SPACE

HD versus SD Color Space

For HD video, the ITU-R BT.709-4 recommendation colorimetry is defined as

$$E'_Y = 0.2126\ E'_R + 0.7152\ E'_G + 0.0722\ E'_B$$

$$E'_{C_B} = 0.5389\ (E'_B - E'_Y)$$

$$E'_{C_R} = 0.6350\ (E'_R - E'_Y)$$

For NTSC (SMPTE170M) and SD-SDI (ITU-R BT.601-5 or SMPTE-259M) the colorimetry matrix is defined as

$$E'_Y = 0.299\ E'_R + 0.587\ E'_G + 0.114\ E'_B$$

$$E'_{C_B} = 0.564\ (E'_B - E'_Y)$$

$$E'_{C_R} = 0.713\ (E'_R - E'_Y)$$

These equations demonstrate that the HD luma will contain a different mixture of color than the SD equivalents; the result is a slightly different overall image colorimetry. This change was made to compensate HD video for the inevitable display on flat-panel monitors instead of CRTs. SD video was designed to be viewed on monitors using CRT technology with phosphor dots splattered on the screen.

The color matrix was created to make monochrome video (luma only) appear black and white—void of any color—when displayed on a monitor. If HD video is simply down-converted and not color compensated, the HD signal will appear on a CRT to have a higher green content and not be purely monochromatic. This also becomes problematic when viewing color. As the C_B and C_R components are added to the luma, the resulting display will not be of the correct hue.

Linear RGB video systems allow for as many as 16 million colors. Accounting for system headroom (the limiting factor is the interface, the electrical output of the camera), linear RGB video is reduced to approximately 10 million colors. The native YCbCr system is capable of nearly 100 times this value; however, current technology of standard definition production equipment cannot take advantage of its full range. RGB transcoded into YCbCr (nonlinear is limited to 700mV – 8 bits) and composite video are capable of about 2.5 million colors. Color limiting is always necessary when down-converting from HD.

The design of any video processing equipment should not allow operators to adjust video levels beyond the protected sampling range of native digital video formats. With the introduction of digital video to the "prosumer" market, hardware designed for nonbroadcast applications is finding its way into some broadcast facilities, and it may be possible that some products produce invalid digital video data. If video sample values infringe on reserved values, the signal may become incompatible with recording, transmission, or switching equipment and therefore technically useless.

Figure 8.2-14 demonstrates that the color space of RGB transcoded to YCbCr is a subset of the native SDI format. Due to its limitation of 700mV per color component (effectively 8-bit chroma resolution), RGB gamut is limited to approximately 2.5 million colors.

Limiting high definition video to RGB specifications requires the application of an algorithm that reduces the minimum and maximum sample levels to a range of 0–700 mV. The high definition active video sample range is 1019 samples, which equates to 746.1 mV of video. The conversion of the YCbCr component into RGB is the simplest form of legalization.

In this form of legalization, the algorithm transcodes sample numeric values of one format to another (HD native YCbCr to nonlinear RGB). Any value seen above the equivalent of 700 mV is set to 700 mV; any value seen under 0 mV is set to 0 mV. In peak component limiting, limiting an individual color component may affect other colors. For example, if the amount of blue is reduced, other colors may be affected, such as magenta and cyan.

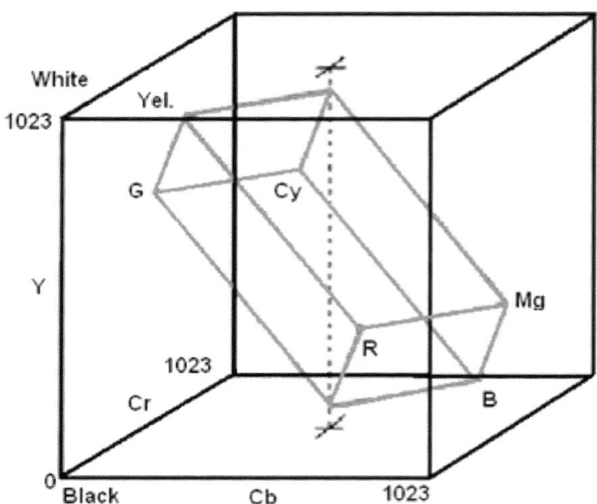

FIGURE 8.2-14 RGB versus HD-SDI color space nonlinear RGB transcoded to 700 mV limits (8 bits, 255 samples), approximately 4 million colors.

Conforming to Composite Color Space

The color space of composite video is representative of broadcast standards and specifications for the analog equipment passing or processing composite signals. Some operators clip luma at 100 IRE[2] (714 mV) and overall composite at 105 IRE (749 mV), whereas others may elect to clip at 110 IRE (785 mV). The composite color space as it relates to HD-SDI is plotted in Figure 8.2-15.

Figure 8.2-15 demonstrates that the color space of composite video is a subset of the native HD-SDI format. Due to its broadcast limitations, composite gamut is limited to approximately 2.5 million colors. Notice that the limitations of composite video are a function of luma versus chroma saturation. The higher or lower the luma, the less allowable chroma saturation. Colors with medium luma are allowed the highest color saturation.

Conforming HD video to composite (NTSC and PAL) standards becomes more complex due to the nature of composite video. In composite video a matrix determines the proportion of each color component. All composite components contain some proportion of red, green, and blue (Y, I, Q, or YUV).

Adjusting the overall peak excursion of the signal can modify a single color's saturation. If a color with a high luma value approaches the maximum value for peak limiting, the peak limiter will compress the luma and chroma, resulting in a lower saturation of the "bright" color. Vector limiting maintains the original hue of the image, while reducing overall chroma saturation. Peak limiting may desaturate an individual color, leaving all others with normal saturation.

[2]IRE stands for "Institute of Radio Engineers," and is a unit of video level with the blanking level equaling 0 IRE and the white level equaling +100 IRE.

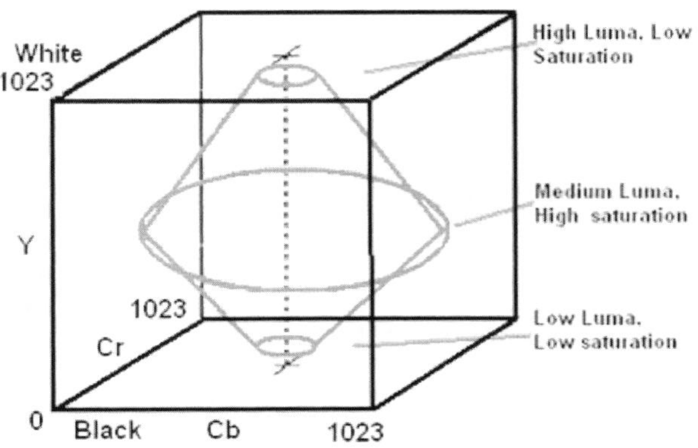

FIGURE 8.2-15 Composite versus HD-SDI color space. Composite color space is a subset of SDI color space with approximately 2.5 million colors. Not all colors allow equal saturation.

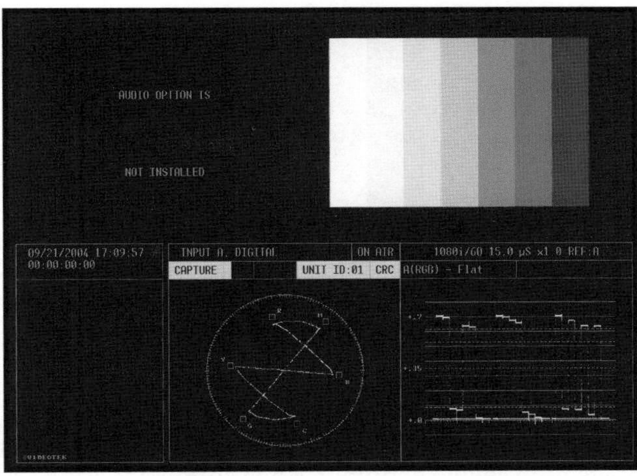

FIGURE 8.2-17 Test signal with approximately 7% vector limiting.

Vector Limiting

Figure 8.2-16 shows a picture of a normal test signal. Figure 8.2-17 depicts the same signal with approximately 7% vector limiting. Notice that all colors are reduced by the same amount. Vector limiting sets the maximum excursion of the chroma information to an outer diameter. Any color in excess of the maximum vector setting will be limited. With the application of vector limiting, as in Figure 8.2-17, the operator can ensure that colors will not exceed a set maximum and that their hue will not change.

Severe vector limiting, nearly 50%, is shown in Figure 8.2-18; degradation of the image hue is not perceivable. This sort of legalization, in real time, is much more efficient than color correcting an image. Changing any individual color component will change other color values of the image. Vector limiting ensures that there is no change to the image's color balance, while ensuring it is legal for broadcast.

Using knee (or softness) adjustments to the composite settings allows the engineer/operator to make artistic decisions in low light or high light areas of the image, while keeping the pixel values within legal range. A hard clip, for instance, will cause all sample values above a threshold to be the same. Therefore, when an image is viewed, a bright area of the scene, such as billowy clouds, will become splotches of white in the sky and the image will lose any highlight latitude. For HD systems, the means of altering the sample values will be more critical due to the nature of the receivers and viewers. The home audience will be more aware of subjective quality issues in HD systems than SD systems in that the content's quality is fundamental to the HD experience.

In the three images of Figures 8.2-19, the sky and rock formation contain areas brighter than other portions of the picture. In Figure 8.2-19, (A) is a normally

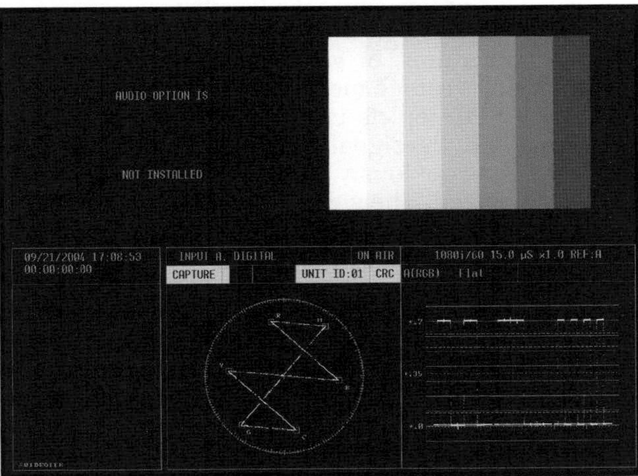

FIGURE 8.2-16 Normal test signal.

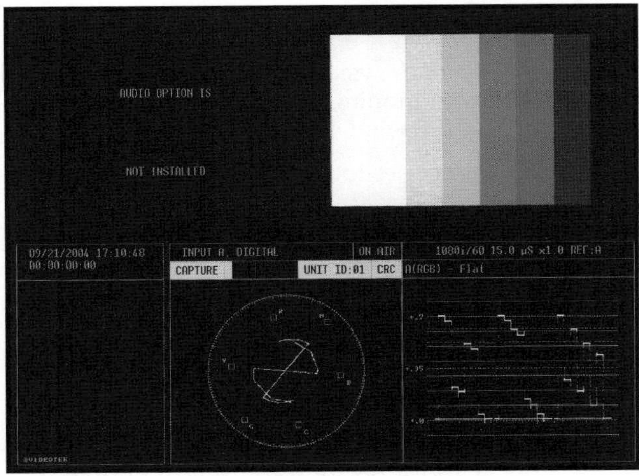

FIGURE 8.2-18 Severe vector limiting of nearly 50%.

(A)

(B)

(C)

FIGURE 8.2-19 Adding softness to the clipping process can improve an overexposed image with severe clipping.

exposed image. Figure 8.2-19 (B) is an overexposed image with severe clipping. When softness is added to the clipping process, the image can regain its highlight latitude, as shown in Figure 8.2-19 (C) and still be broadcast legal.

VIDEO SYSTEM TIMING

Many current video systems are capable of accepting both composite analog video as well as HD/SD serial digital video sources for switching. In master control and production systems, these signals must arrive to the switcher's input connections in "time," within a given tolerance, in order to produce proper transitions and effects while mixing input sources. Switching severely mistimed video may also introduce signal faults in the data being recorded on a server or videotape machine.

Modern master control and production switchers have a large, though not indefinite, tolerance for input timing differences; however, switcher throughput delay time can be as much as a frame. These system timing tolerances are part of the installation instructions usually accompanying the equipment and must be adhered to in order to get optimum performance from the switcher.

Analog System Timing

In the past, analog system timing was performed using a waveform monitor and vectorscope combination. The procedure went something like this:

1. Connect the scopes to an external blackburst reference.
2. Connect a signal that is correctly timed to the video input of the scopes.
3. Turn on DC restore.
4. Select the input for the device under test on both the waveform monitor and vectorscope.
5. Press the Ext. Reference button on both the waveform monitor and vectorscope.
6. Turn the H and V position controls of both the waveform monitor and vectorscope to a graticule mark, denoting the position of sync for a properly timed signal.
7. Magnify the scope's traces in order to view the position of leading edge of sync.
8. Connect the signal with questionable timing to the video inputs of both the waveform monitor and vectorscope.
9. Adjust the H and V timing controls of the video source until the same edge of sync lies on the same "tick" mark of the graticule as the correctly timed source.
10. Adjust the subcarrier controls of the video source until the color vector information matches that of the signal with correct timing.
11. Now connect the next signal to be checked for system timing accuracy.

An alternative method used a waveform monitor and vectorscope with A/B overlay. This method compared the timing of a displayed input (it is the timing reference) to the unknown timing of a signal under test, instead of an applied blackburst reference, with steps as follows:

1. Connect a signal that is correctly timed to the A video input of the waveform monitor and vectorscope.
2. Connect the signal under test to the B input, on both the waveform monitor and vectorscope.
3. Turn the H and V position controls of both the waveform monitor and vectorscope to a position convenient to view both the overlaid signals.
4. Magnify the scope's traces in order to view the leading edge of sync.
5. Adjust the H and V timing controls of the source video until the same edge of sync of the signal under test lies directly over the signal with the reference timing.
6. Adjust the subcarrier controls of the video source until the color vector information matches that of the signal with correct timing.
7. Now connect the next signal to be checked for system timing accuracy to the B input of both scopes.

These procedures, or variants of them, have been used for many years to measure system timing.

Figure 8.2-20 depicts two analog video inputs (A and B) overlaid; the signal designated as timing reference has been previously positioned to a convenient area of the screen. The signal under test is displayed by enabling the AB overlay feature of the waveform monitor. The difference in position between the leading edge of sync is approximately 1.2 microseconds, indicating that the signal under test is 1.2 microseconds later (lagging) than the timing reference signal. If the signal under test were positioned to the left of the timing reference, it would be leading the timing reference.

This method of using the A/B display for timing is popular in the analog world, and most professional test instruments provide this as a standard feature. A/

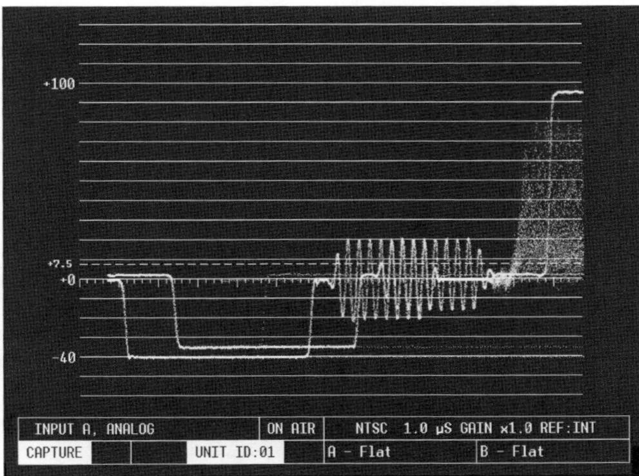

FIGURE 8.2-20 Displaying two analog video inputs (A and B) overlaid.

B comparisons work well when signals are of the same format; however, when analog signals are timed to digital signals, this method is unusable, and more complicated procedures to measure system timing are required.

HD, SD, and Analog Mixed Format Video Systems Timing

In the HD world, timing adjustments may need to be a few microseconds, lines, or fields in difference. When the signals being switched are time-aligned, this interruption is not visually noticeable as long as the switch is performed during vertical blanking and, of course, the signals are synchronous; that is, their fundamental clock rates have been generated or locked together. The brief interruption that does occur is unseen since it occurs during video blanking. Further masking any visible effect is the fact that the video levels are guaranteed to be identical at the video switch point.

Aligning two video signals was at least conceptually quite simple in the days of a single video format. A combination waveform monitor and vectorscope was all that was needed to align composite analog signals mentioned earlier; if the signals were aligned at a line rate, field rate, and the burst vectors aligned, those signals were perfectly in time.

Even when component digital signals needed to be aligned to an analog blackburst reference (a very common situation), this could be readily accomplished because commonly used test equipment supported this capability and analog-to-digital timing was clearly specified (see ITU-R BT.601 for details). The old analog waveform monitor technique—using blackburst as an external reference for the waveform monitor that was used to align the component digital input—worked satisfactorily for mixed analog/digital standard definition formats. The reason is that (for a given frame rate) both composite analog and component digital video have an identical number of lines per field, are interlaced, and have identical line and field timing. This is by design; the component digital formats were meant to duplicate the composite analog timing to make format conversions simple.

The situation is completely different when using analog 525-line 29.97 frames per second blackburst to reference HD video. There no longer exists a one-to-one relationship between the signal and reference numbers of lines, line rates, and even frame rate, since some progressive formats in HD (such as 24p and 60p) have lower or higher frame rates than does SD.

Table 8.2-2 compares signal timing information for commonly used HD and SD video standards.

This table shows that in using any HD format there is no simple timing relationship between it and composite video, blackburst, or an SD reference. Even though many of the HD formats use the same frame rate as blackburst, there is still no way to account for the widely varying line timing. Furthermore, even if the HD signal has the same frame rate as the blackburst reference, the question still exists of how to align it from a frame standpoint. How is this signal to be synchronized when almost everything about it is different?

The SMPTE Recommended Practice RP-168 (2002) deals with the switching point in vertical interval, which is the point where one video source may be switched to a subsequent video source with minimum disruption to the output video. Because of the common use of blackburst as a reference signal, the most recent version of this RP now explicitly defines the timing relationships between different formats. In addition to explaining how an analog signal relates to a high-definition signal from a timing standpoint, this RP also defines reference timing points for different HD signals, even though a 750-line system will have a frame rate twice as fast as an interlaced 1125-line system. As long as the field or frame rates are the same, the signals can be aligned.

Knowing how the signals are supposed to be aligned is one matter; the question of how to actually do it still remains. A waveform monitor is the classic waveform timing apparatus. When a reference that is exactly the same format as the video signal is used, this method works fairly well, but when conventional test equipment is used, it is difficult (if not impossible) to determine if the reference lines in vertical interval are precisely aligned as specified in SMPTE RP-168. A waveform monitor is *extremely* limited in what it can do with mixed format inputs, where an HD input can be externally referenced to blackburst, but an engineer cannot make any real quantitative judgments about the timing alignment without seeing both the reference and the HD signal. A specialized device for monitoring these signals is required.

A specific display designed for HD/SD relative timing alignment is shown in Figure 8.2-21. The relative timing display is deceptively simple, as it employs a simple number line, representing timing differences between a video signal under test and a video reference video source. The basic operation of

TABLE 8.2-2
SD and HD Signal Timing Information

Format	Total Number of Lines	Interlaced or Progressive	Frame Rate (Hz)	Duration of One Video Line
NTSC	525	Interlaced	29.97	63.56
PAL	625	Interlaced	25	64.0
525/59.94	525	Interlaced	29.97	63.56
625/50	625	Interlaced	25	64.0
1280 × 720p	750	Progressive	60	22.22
1280 × 720p	750	Progressive	59.94	22.24
1280 × 720p	750	Progressive	50	26.67
1280 × 720p	750	Progressive	30	44.44
1280 × 720p	750	Progressive	29.97	44.49
1280 × 720p	750	Progressive	25	53.33
1280 × 720p	750	Progressive	24	55.56
1280 × 720p	750	Progressive	23.98	55.6
1920 × 1080i	1125	Interlaced	30	29.63
1920 × 1080i	1125	Interlaced	29.97	29.66
1920 × 1080i	1125	Interlaced	25	35.56
1920 × 1080p	1125	Progressive	30	29.63
1920 × 1080p	1125	Progressive	29.97	29.66
1920 × 1080p	1125	Progressive	25	35.56
1920 × 1080p	1125	Progressive	24	37.04
1920 × 1080p	1125	Progressive	23.98	37.07

the relative timing display is quite simple: use a graph of timing differences between two signals, one called the "reference" and the other the "test" input. Two pointers move along the graph to show the relative timing between these signals. The first pointer is drawn on the top of the graph and represents timing differences in full line increments ("line" referring to the "test" signal when mixed formats are being displayed). The line scale represents the coarse timing differences between the signals. The second pointer is drawn on the bottom of the graph and represents finer timing differences; this pointer's units are microseconds.

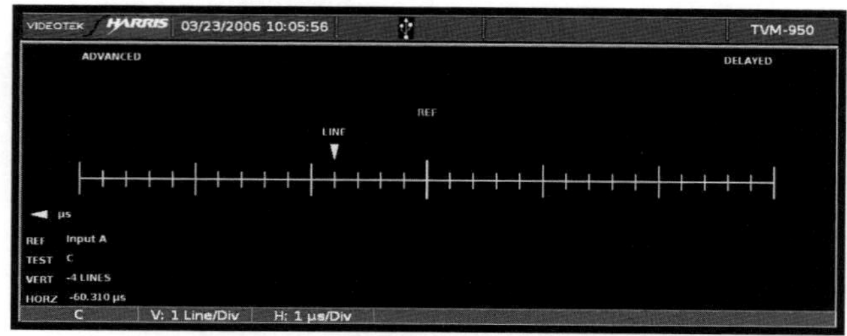

FIGURE 8.2-21 HD/SD relative timing display.

Reading timing differences is straightforward. When the test signal is perfectly aligned with the reference signal, regardless of their formats, the two pointers will perfectly align in the middle of the display. As a further visual indicator the "REF" text above the reference point changes from red to green, making it easy to tell even from across the room when doing timing alignments.

Any timing offset will cause one or both of the pointers to deviate from the center position. Note that the direction in which they deviate is significant. Like an oscilloscope or waveform monitor, any deviation to the left of the center point represents being advanced relative to the reference, whereas being to the right of center represents delay.

When the relative timing display pointers are both centered, a test signal that is the same format as the reference will be aligned with it from both a line and field standpoint. When the test signal is a different format than the reference, the alignment points specified by SMPTE RP-168 will be used.

Offset Timing

The timing display can easily show timing relationship errors when two video signals are intended to be

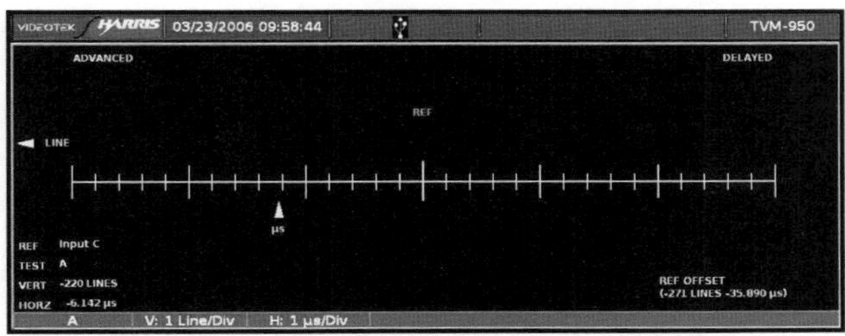

FIGURE 8.2-22 Timing display with offset.

aligned. The goal in system setup, however, is often not to actually align the timing of two signals but to create a precise timing offset, such as a 2-video line advance at the input of a switcher.

An obvious method for accomplishing this would be to use the relative timing display by monitoring the desired offset rather than aligning the inputs exactly. A timing offset can be introduced from a previously aligned source, so that once the correct timing is determined, this condition may be declared "aligned." The timing display then functions exactly as before; only now the additional timing offset information is displayed as well to indicate a nonzero timing alignment. Figure 8.2-22 demonstrates the timing offset feature of the relative timing display; the offset for this source is –271 lines plus –35.890 microseconds.

COMPRESSED VIDEO FOR DISTRIBUTION AND TRANSMISSION

The following sections look at digital video and how various problems can occur as the video is compressed and then combined with other data and transmitted out to the viewer at home. The discussion begins with an overview of the topology for digital television broadcasting before examining the basics of video compression. It then looks at the transmission chain and all the things that can go wrong, examining the encoding process and problems with System Information/Program and System Information Protocol (SI/PSIP) servers, conditional access (CA) servers, and bandwidth.

OVERVIEW OF THE BROADCAST TOPOLOGY AND ITS COMPONENTS

To understand the potential issues faced by compressed video, it is necessary to understand the digital broadcast topology (see Figure 8.2-23). Programs are delivered from various sources, such as a video server, a live feed, or a satellite feed. Typically, for a broadcast station, these sources are switched through a master control switcher. The selected source is then sent to encoders that compress the video and audio in real time. The compressed video and audio are then sent from the encoders to the multiplexer.

At this point, the SI/PSIP server will send descriptive data to the multiplexer to group the video together with the appropriate audio and with any additional services needed, such as closed captioning and data. It also adds other relevant information such as start and end time, duration, and rating as part of an electronic program guide (EPG). The multiplexer combines the data stream from all the different encoders, with the additional information, into a transport stream. The transport stream is distributed via the different transmission platforms, which could be terrestrial broadcast, cable television, satellite, or

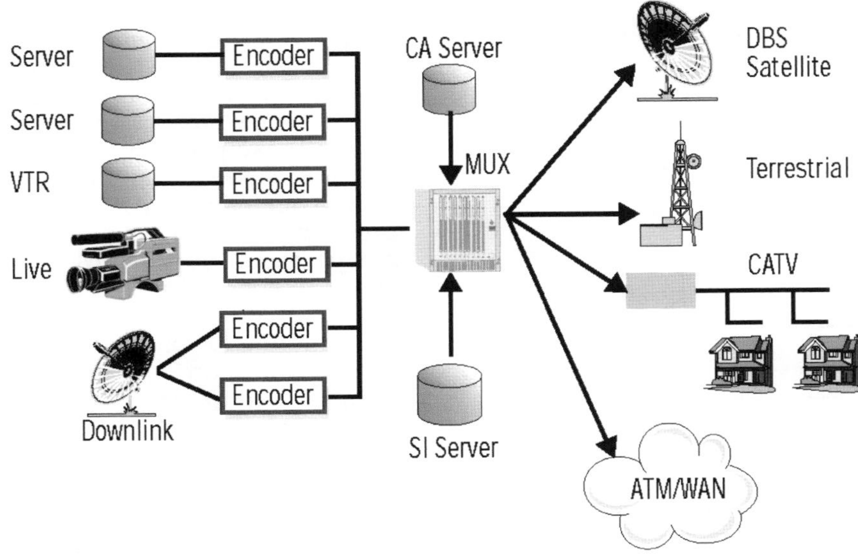

FIGURE 8.2-23 Digital broadcast topology.

asynchronous transfer mode/wide area network (ATM/WAN). These different platforms deliver the video to the viewer in the home. The common transport stream is a key advantage of compressed digital video broadcasting, as it means that generally the same equipment can be used for different transmission platforms.

The Realities of Video Compression

Compressed video utilizes a layered architecture for transmission (see Figure 8.2-24). The data that is carried on the higher layers (video or audio) is independent of the lower layers that carry it—wire, radio frequency (RF), ATM, Internet Protocol (IP). As a result, a video stream will be the same whether it is carried by an RF carrier, an IP link, or an Asynchronous Serial Interface/Program and System Information Protocol (ASI/PSIP) interface.

A problem in a lower level of the stack might, though not necessarily, lead to a problem in a higher level (see Figure 8.2-25). For instance, a problem in the physical layer, such as a lightning strike, could affect the transport layer, which could lead to errors in the coding that would affect the quality of the final product.

There are numerous Information Technology (IT)-based components throughout the entire transmission chain, all of which need to be programmed correctly, and all of which might need a reboot, as computers occasionally do. These components include the video servers, the conditional access (CA) server, and the various SI/PSIP servers. This means numerous points in the chain where human error or software failure could result in the failure to deliver a promised service.

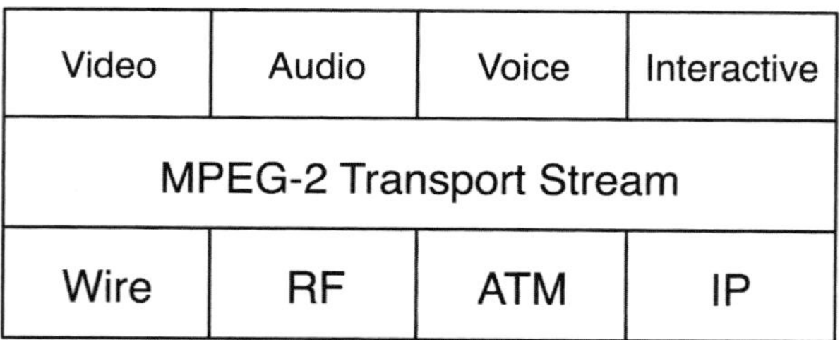

FIGURE 8.2-24 Multimedia protocol stack.

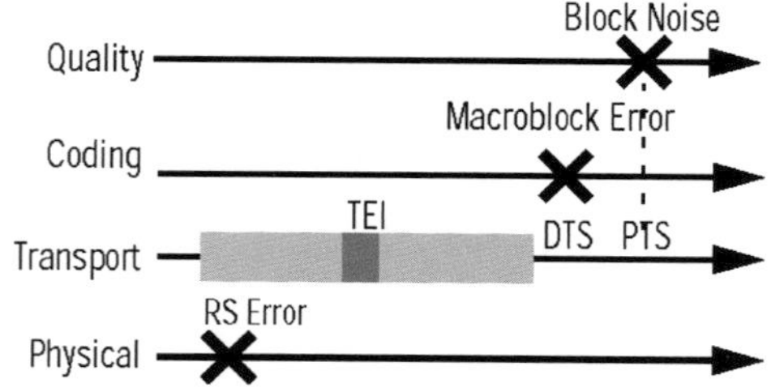

FIGURE 8.2-25 Error propagation for compressed systems.

How and Why We Monitor

The dictionary defines the verb "monitor" as meaning to "observe and check the progress or quality (of something) over a period of time." In the world of analog video, this meant an operator would visually inspect different TV screens, waveform monitors, and vectorscopes looking for problems. For analog video, this made sense, as there was usually a strong correlation between a problem and a potential cause. For example, if there were line noise, the picture would show interference. The solution would be to look into the physical connections to find the source of the line noise. However, in the digital world, there are numerous possible causes of a visual defect. A blocky picture could be caused by line noise, by bandwidth starvation, or by timing jitter. Even if the picture shows no defect, something could still be wrong. For example, the audio might be in the wrong language, the closed captioning might be missing, or the parental rating code may be set incorrectly. None of these problems may be directly visible to the operator.

In the analog world, adjusting a setting for a device—say a voltage or some other parameter within the system—might fix a problem. In the digital world, a service is either present, or it is not. With only a few exceptions, testing compressed video is much more focused on confirming the existence of a parameter rather than with the measurement of the continuous value against the parameter.

What We Monitor

Complete monitoring means verifying the signal and content attributes at each step along the transmission chain. The answer to the question "Does my broadcast meet expectations?" is that it depends on what the expectations are. The end result should be that the correct number of television services with the correct number of attributes has been delivered to viewers. To ensure that the broadcast meets expectations, monitoring must take place along three dimensions: geography, protocol layer, and time (see Figure 8.2-26).

By geography, we mean that the monitoring needs to measure the service quality throughout the transmission chain, from ingress all the way through to viewer reception. Distortion or failure can take place anywhere along this chain.

Monitoring the protocol layer means looking at the physical, transport, and service layers. At the physical layer, we need to check the electrical or radio modulation characteristics of the signal. If there is no inbound RF signal at the monitoring site, for example, then there is no need to troubleshoot MPEG parameters or SI/PSIP tables.

At the transport layer, we need to look at the MPEG-2 transport stream that carries the multiple programs in a single serial data stream. The transport stream has

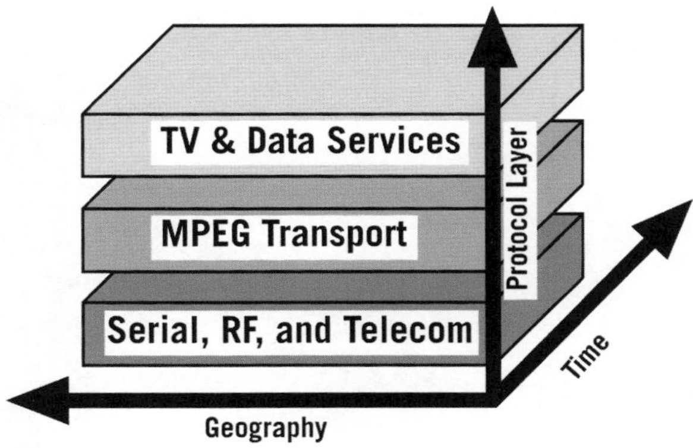

FIGURE 8.2-26 Monitoring in three dimensions.

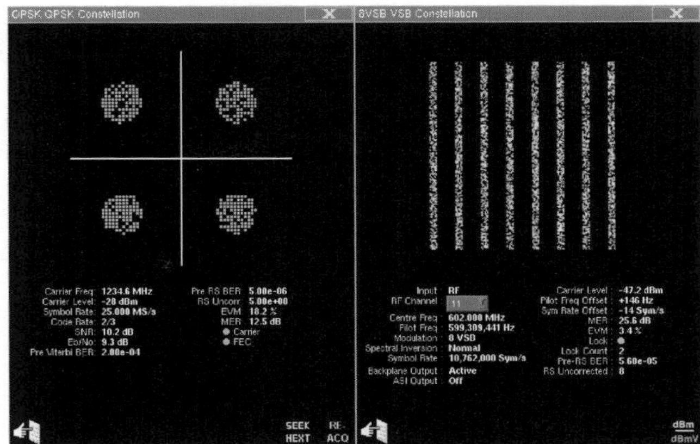

FIGURE 8.2-27 Monitoring 8-PSK and 8-VSB.

to ensure proper timing. In addition, the System Information (SI/PSIP) should also be accurate. SI/PSIP provides the additional descriptive information regarding the transport stream, including service names, channel branding, and the electronic program guide.

At the service layer, monitoring needs to ensure that the correct services and attributes are being delivered. Is there the correct number of programs? Are they the right programs? Are there missing components, such as audio?

Finally, the monitoring must have the time element. It needs to look beyond the current signal and service quality and include the signal and service quality of the past. The ability to study log files and conduct in-depth analysis of measurement results is important in helping to pin down problems.

Monitoring Points

Monitoring the transmission chain starts with monitoring any RF and/or transport stream (TS) incoming signals at the ingress point, for instance, from a network distribution satellite or terrestrial links. Monitoring here will depend on the particular systems. For RF systems, it will include checking the signal or carrier power and modulation quality (see Figure 8.2-27). Standard techniques such as Error Vector Magnitude and/or Modulation Error Ratio are useful in this area.

Incoming transport streams should have all parameters verified, as for the outgoing signal. The monitoring platform should be configured to watch for any changes to the nominal signal quality and sound the alarm in the event of any degradation.

The next point in the chain to monitor will be the Compression Center, which consists of the encoders, the multiplexer, the SI/PSIP servers, and the conditional access (CA) server (if used). Monitoring of the signal generated by the compression center, usually at the point when it is ready to leave the building, requires in-depth analysis of the MPEG-2 TS. This includes checks that the assembled TS has all the services expected of it (video, audio, captioning, program guide, etc.); it requires multiple levels of bandwidth measurement, timing, and jitter analysis, as well as confirmation of all System Information, PSIP, and program guide components.

Finally, it is necessary to confirm that the signal and content have been transmitted correctly and are reaching the viewer. In the case of a local broadcast station, it is almost always possible to monitor an off-air signal received from the transmitter. This is an essential part of the monitoring process in order to validate that the bitstream has not been degraded by the transmission process.

Where local off-reception is not possible, or to monitor remote transmitters, translators, or repeaters in a network, it is possible to access telemetry and measurements from a receiver at a remote location using a local area network (LAN) connection, to validate that the viewers are receiving everything they are meant to.

TRANSPORT STREAM MONITORING

Transport stream monitoring involves monitoring basic things such as channel lineup verification and bandwidth, as well as Program Clock Reference (PCR) jitter, SI/PSIP table verification, content validation, and conditional access.

Channel Lineup Verification and Program Display

The MPEG-2 transport stream multiplexes a number of video, audio, and other data signals. Each component (audio, video, captioning, etc.) is carried within a separate data stream that has a packet ID (PID) number. The first step in assessing broadcast integrity is to determine whether the expected "channels" are actually there or not (see Figure 8.2-28). Once the expected channels are verified, the next step is to ensure that all the attributes (main audio, second language audio, captioning, etc.) match what is expected.

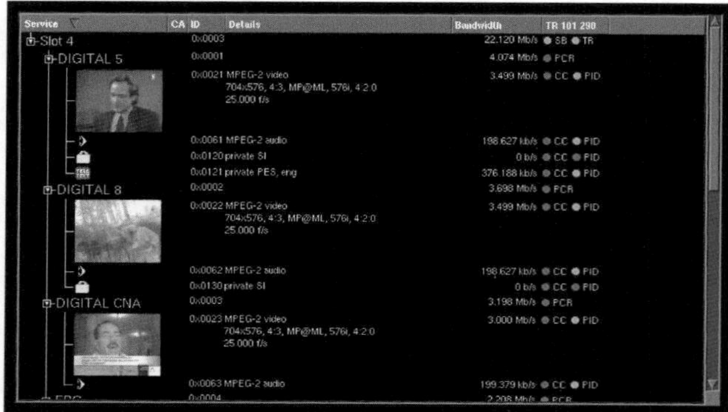

FIGURE 8.2-28 High-level service content and critical parameter warnings screen. This determines whether expected channels exist in a video stream and what their status is.

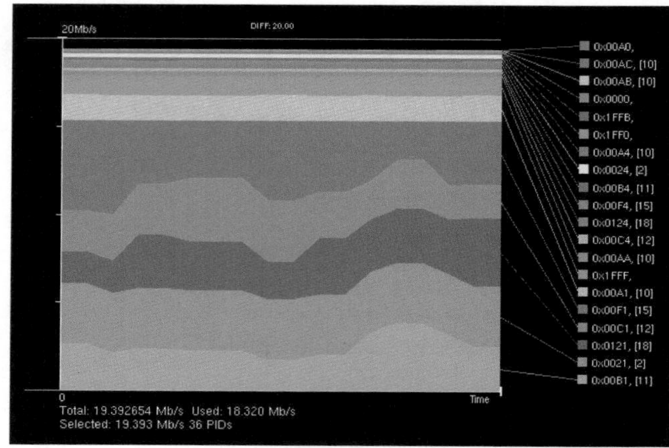

FIGURE 8.2-29 Graph illustrating how use of variable bit rate coding makes for very efficient use of the transport stream. Each color band (see CD) represents an element of the stream; time is represented along the x-axis and bits per second along the y-axis.

Bandwidth

The overall multiplexed MPEG transport stream is of a fixed bandwidth. The commonly used DVB asynchronous serial interface is fixed at a maximum line rate of 216 Mbps. However, it utilizes stuffing bytes to allow a lower (fixed) transport stream bit rate. Other physical interfaces support other maximum bit rates; for example, SMPTE 310M is fixed at 19.39 Mbps, while Gigabit Ethernet, which is becoming increasingly common, supports transport streams close to 1000 Mbps. Whatever the maximum might be, it is important to note that there is a maximum bit rate and that the transport stream cannot exceed this amount.

Within the transport stream multiplexer, individual PIDs may be set by the encoder and/or multiplexer to a fixed bit rate. The use of a transport stream analyzer allows verification of those settings. That is relatively straightforward. However, since video compression standards require more bits for encoding moving pictures compared with still pictures, a fixed bit rate assignment is not very efficient. To increase efficiency, therefore, variable bit rate (VBR) encoding evolved. Figure 8.2-29 shows an example of the makeup of a VBR stream.

Testing VBR streams requires validating the average, minimum, and maximum bit rates against their expected values as set in the multiplexer. Again, a transport stream analyzer can measure those bit rates, and alarms can be set against the low and high bit rates to warn of a misconfiguration or other problems.

However, VBR encoding can create a situation in which the instantaneous bit rate of all the VBR flows is greater than the total available bit rate. This happens when the overall bit rate is set slightly higher than the maximum bandwidth, based on the assumption that it is rare that the different data that is being compressed will all need to be compressed at the maximum bit rate at the same time. Multiplexers handle this problem with different levels of finesse. Some MUX equipment will proportionately lower the bit rate of all input streams, whereas other multiplexers will starve traffic on one PID in favor of another PID. Bandwidth starvation in the latter situation creates picture distortions but is very difficult to detect through measurements. The reason is that simple bandwidth measurements smooth out variations to provide a view of the situation over time. The instantaneous drop in bit rate of one PID thus becomes impossible to detect because it is smoothed out in the analysis.

The only way to identify bandwidth starvation is to measure the statistical distribution of packet arrival time of null packets. An inbound data stream to the MUX that is very bursty creates bandwidth allocation difficulties and subsequent data loss in the multiplexer. Identifying this requires a detailed analysis of the null packet distribution, looking for short periods when no null packets were sent at all as meaning the MUX runs out of bandwidth for a short period of time. However, commercial transport analyzers cannot display bandwidth over such a short integration period and the short periods of zero bandwidth cannot be detected.

In this situation, real-time packet interval measurement is needed because it allows the measurement of how frequently a specific packet (in this case, null packets) actually arrives. If no null packets leave the MUX for a certain amount of time, this can indicate a stressed situation for a multiplexer.

Setting the transport analyzer to plot a real-time histogram of the inter-arrival times of the specific PIDs within the transport stream allows the identification of bandwidth starvation in the multiplexer. The resultant graph (see Figure 8.2-30) shows the statistical distribution of packet arrivals, with the arrival time on the horizontal axis and the number of packets on the vertical axis.

A single, concentrated grouping, as shown in Figure 8.2-30 (A), indicates constant arrival rate (CBR), whereas other patterns indicate more random arrival.

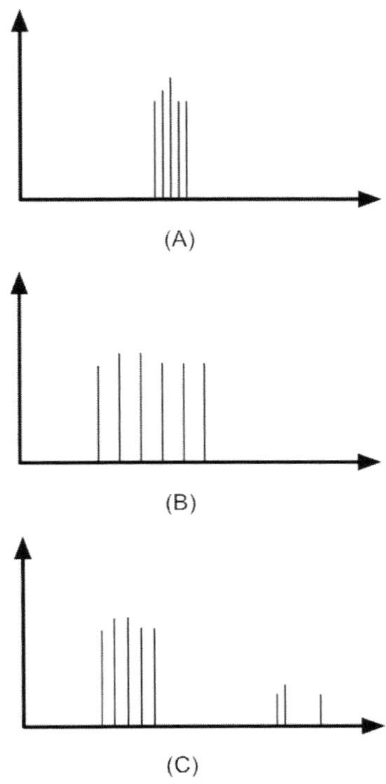

FIGURE 8.2-30 Histograms of inter-arrival times of specific PIDs can reveal "bandwidth starvation" in the multiplexer. (A) CBR, (B) VBR, (C) Interval problem, e.g., buffer overflow.

A bimodal distribution as shown in Figure 8.2-30 (C) could point to the periodic overflow of an internal buffer within the MUX.

DVB TR101-290 and ATSC Table Testing

The proper operation of all of the features of a broadcast system based on MPEG-2 transport stream requires not only that the expected PIDs and administrative tables be present, but that they occur with the correct frequency. To simplify testing of DVB transport streams, Digital Video Broadcasting (DVB) defined a basic set of health checks for MPEG-2 that assess the overall condition of the transport stream. This test is called the TR101-290. The tests are not an exhaustive analysis of transport stream integrity but are at least a good place to start when assessing overall broadcast quality of service. For example, the TR101-290 evaluates whether the Service Description Table (SDT) is being sent often enough, but the test does not check whether it contains the correct information, for example, what channels should be present in the multiplex.

Since DVB transmission systems are widely used for program distribution, TR101-290 testing is applicable worldwide. However, for terrestrial broadcasters using the ATSC standard, TR101-290 is not fully suitable, since ATSC utilizes different tables to represent the channel guide, program guide, etc. Until recently there has been no equivalent ATSC standard.[3] However, some transport stream analyzers support ATSC-specific tests that mimic the spirit of the DVB tests with acceptable test coverage.

TR101-290 divides the checks into three sections sorted by their importance. Priority 1 contains a basic set of parameters required to receive and decode a transport stream. Priority 2 contains additional parameters that are recommended by DVB for continuous monitoring. Priority 3 parameters are optional but, if applicable to the transmission system, would be useful for further monitoring. Table 8.2-3 shows the types of errors detected by TR101-290.

Priority 1 Checks

- *TS sync loss:* Loss of synchronization with the physical layer is one of the first things to look out for. The actual synchronization of the TS depends on the number of correct sync bytes necessary for the device to synchronize and also on the number of distorted sync bytes that the device cannot cope

[3]In September 2006, the ATSC published "Recommended Practice: Transport Stream Verification (A/78)." It is anticipated that bitstream analyzers reflecting the recommendations in that document will be introduced.

TABLE 8.2-3
Types of Errors Detected by TR101-290

Priority 1	Priority 2	Priority 3
TS sync loss	Transport errors	NIT errors
Sync byte errors	CRC errors	SI/PSIP repetition errors
PAT errors	PCR errors	Buffer errors
Continuity count errors	PTS errors	Unreferenced PIDs
PMT errors	CAT errors	SDT errors
PID errors		EIT errors
		RST errors
		TDT errors
		Empty buffer errors
		Data delay errors

with. Five consecutive correct sync bytes (ISO/IEC 138181, annex G.01) should be sufficient for sync acquisition, and two or more consecutive corrupted sync bytes would indicate sync loss. After synchronization has been achieved, evaluation of the other parameters can be carried out.

- *Sync byte error:* This error occurs when the correct sync byte (0x47) does not appear after 188 or 204 bytes. This error is fundamental because this structure is used throughout the channel encoder and decoder chains for synchronization. It is also important that every sync byte is checked, since the encoders may not necessarily check the sync byte. Some encoders use the sync byte flag signal on the parallel interface to control randomizer reseeding and byte inversion without checking that the corresponding byte is a valid sync byte.
- *PAT errors:* The Program Association Table (PAT), which appears only in PID 0x0000 packets, tells the decoder what programs are in the TS and points to the Program Map Tables (PMTs), which in turn point to the component video, audio, and data streams that make up the program. If the PAT is missing, then the decoder can do nothing and no program is decodable. Nothing other than a PAT should be contained in a PID 0x0000. This error also occurs when the PAT consists of several (consecutive) sections with the same table ID, 0x00.
- *Continuity count error:* This error occurs when there is a loss of continuity and/or sequence according to the continuity count field within the MPEG-2 transport packet. This error indicates incorrect packet order, a duplicated packet, or that a packet has been lost. Incorrect packet order and lost packets could cause problems for integrated receiver-decoders (IRDs) that are not equipped with additional buffer storage and intelligence. A duplicated packet may be symptomatic of a deeper problem that the service provider should keep under observation.
- *PMT error:* The PMTs contain the information where the parts for any given event can be found. Parts in this context are the video stream (normally one) and the audio streams and the data stream (e.g., Teletext). Without a PMT, the corresponding program is not decodable.

  A PMT error occurs if sections with table ID 0x02 (i.e., a PMT) do not occur at least every 0.5 second on the PID which is referred to in the PAT, or if the scrambling control field is not 00 for all PIDs containing sections with table ID 0x02 (i.e., a PMT). This error also occurs when sections with table ID 0x02 (i.e., a PMT) do not occur at least every 0.5 second on each program map PID which is referred to in the PAT, or when the scrambling control field is not 00 for all packets containing information of sections with table ID 0x02 (i.e., a PMT) on each program map PID which is referred to in the PAT.
- *PID error:* This error occurs when a referred PID does not occur for a user-specified period. TR101-290 checks whether a data stream for each PID that occurs does indeed exist. This error might occur where the TS is multiplexed, or demultiplexed and again remultiplexed. The user-specified period should not exceed 5 seconds for video or audio PIDs. Data services and audio services with ISO 639 language descriptors with a type greater than "0" should be excluded from this 5-second limit. In principle, a different user-specified period can be defined for each PID.

Priority 2 Checks

- *Transport error:* The transport error indicator in the TS-Header is set to "1." This bit should be set by equipment earlier in the transmission chain to indicate an error has occurred earlier. The primary transport error indicator is Boolean, but there should also be a resettable binary counter which counts the erroneous TS packets. This counter is intended for statistical evaluation of the errors. If an error occurs, no further error indication should be derived from the erroneous packet. There may be value in providing a more detailed breakdown of the erroneous packets, such as providing a separate transport error counter for each program stream or including the PID of each erroneous packet in a log of transport error events. Such extra analysis is regarded as optional.
- *CRC error:* The CRC checks the CAT, PAT, PMT, NIT, EIT, BAT, SDT, and TOT to see if the content of the corresponding table is corrupted. If it is corrupted, a CRC error will occur.
- *PCR error:* The PCRs are used to regenerate the local 27 MHz system clock. If the PCR does not arrive with sufficient regularity, then this clock may jitter or drift. The receiver-decoder may even go out of lock. In DVB, a repetition period of not more than 40 ms is recommended.

  PCR errors happen for a variety of reasons. One event that triggers this error is when the difference between two consecutive PCR values is outside the range of 0–100 ms without the discontinuity indicator set. A different PCR error can occur when the PCR accuracy of a selected program is not within ±500 ns. This test should be performed only on a constant bit rate TS as defined in ISO/IEC 13818-1 Section 2.1.7.
- *PTS error:* The Presentation Time Stamps (PTS) should occur at least every 700 ms. This error occurs when the PTS does not occur within the time set. Note that the PTS is accessible only if the TS is not scrambled.
- *CAT error:* The Conditional Access Table (CAT) contains pointers that enable the IRD to find the Entitlement Management Messages (EMMs) associated with the CA system or systems that it uses. If the CAT is not present, the receiver is not able to receive management messages and consequently will be unable to decrypt the content.

  This error occurs when packets with transport scrambling control not 00 are present, but no sec-

tion with table ID = 0x01 (i.e., a CAT) is present. It also presents itself when a section with table ID other than 0x01 (i.e., not a CAT) is found on PID 0x0001.

Priority 3 Checks

- *NIT errors:* Network Information Tables (NITs), as defined by DVB, contain information on frequency, code rates, modulation, polarization, etc., of various programs that the decoder can use.

  NIT errors occur when a table other than a Network Information Table is found on PID 0x0010, or there has been no NIT found on PID 0x0010 for more than 10 seconds. These errors also occur when the NIT related to the respective TS is not present in the TS, or it has the wrong PID. Further NITs can be present under a separate PID and refer to other TSs to provide more information on programs available on other channels. Their distribution is not mandatory, and the checks should be performed only if they are present.
- *SI repetition error:* For SI/PSIP tables, maximum and minimum periodicities are specified in EN 300 468 and ETR 211. They are checked for this indicator. This indicator should be set in addition to other indicators of repetition errors for specific tables.
- *Buffer error:* This error occurs when the MPEG-2 reference decoder has an underflow or an overflow. The individual checks defined within TR101-290 are
  - *TB buffering error:* Overflow of transport buffer (TB . TBn)
  - *TBsys buffering error:* Overflow of transport buffer for System Information (Tb . Tbsys sys)
  - *MB buffering error:* Overflow of multiplexing buffer (MB MBn), or if the *vbv delay method* is used, underflow of multiplexing buffer (Mb Mbn)
  - *EB buffering error:* Overflow of elementary stream buffer (EB . EBn), or if the *leak method* is used, underflow of elementary stream buffer (EB EBn) though low delay flag and DSM trick mode flag are set to 0; else (*vbv delay method*) underflow of elementary stream buffer (EB) Ebn
  - *B buffering error:* Overflow or underflow of main buffer (B . Bn)
  - *Bsys buffering error:* Overflow of PSI/PSIP input buffer (B . Bsys sys)
- *Unreferenced PID:* Each nonprivate program data stream should have its PID listed in the PMTs. Unreferenced PID checks that a PID not explicitly defined to be present (e.g., a PID for one of the standard tables such as NIT) or that a PID not referred to by a PMT or CAT occurs within 0.5 second. This check is useful for tracking down multiplexer misconfigurations or identifying *phantom PIDs.*
- *SDT error:* The SDT describes the services available to the viewer. It is split into subtables containing details of the contents of the current TS (mandatory) and other TS (optional). Without the SDT, the IRD is unable to give the viewer a list of what services are available. It is also possible to transmit a BAT on the same PID, which groups services into "bouquets."

  SDT errors occur when an SDT actual table is not present on PID 0x0011 for more than 2 seconds, when other types of tables besides SDT are found on that same PID, when two sections of an SDT actually occur within 25 ms or less, or when identical sections of an SDT other table (i.e., sections of an SDT other table having the same section number) occur greater than 10s apart.
- *EIT error:* The Event Information Table (EIT) describes what is on now and next on each service, and optionally details the complete programming schedule. The EIT is divided into several subtables, with only the "present and following" information for the current TS being mandatory. The EIT schedule information is accessible only if the TS is not scrambled.

  EIT errors occur when there is no EIT present on PID 0x0012 for more than 2 seconds, when tables other than EIT are found on PID 0x0012, when two sections of an EIT table occur within 25 ms or less, or when one subtable of an EIT-P/F is present and the other subtable is missing.
- *RST error:* The Running Status Table (RST) is a quick updating mechanism for the status information carried in the EIT. An RST error occurs when a table other than the Running Status Table is present on PID 0x0013 or any two sections of the RST within 25 ms or less.
- *TDT error:* The Time and Date Table (TDT) carries the current UTC time and date information. In addition to the TDT, a TOT can be transmitted which gives information about a local time offset in a given area.

  TDT error occurs when a Time and Date Table is not present on PID 0x0014 for more than 30 seconds; tables other than a TDT, ST, or TOT are found on PID 0x0014; or any two sections of a TDT occur within 25 ms or less.

  The carriage of the *NIT other, SDT other, EIT P/F other, EIT schedule other,* and *EIT schedule actual* tables is optional; therefore, these tests should be performed only when the respective table is present.

  When these tables are present, this should be done automatically by measuring the interval rather than the occurrence of the first section. As a further extension of the checks and measurements mentioned here, an additional test concerning the SI/PSIP is recommended; all mandatory descriptors in the SI/PSIP tables should be present, and the information in the tables should be consistent.

- *Empty buffer error:* This error occurs if the transport buffer or the transport buffer for System Information does not empty at least once per second, or if the *leak method* is used, the multiplexing buffer does not empty at least once per second.
- *Data delay error:* This error occurs if the delay of nonvideo data traffic through the TSTD buffers is greater than 1 second or the delay of still picture video data through the TSTD buffers is greater than 60 seconds.

ATSC-Specific Tests

Since the ATSC table structures used to represent the channel guide, program guide, etc., are quite different from their DVB counterparts, much of TR101-290 cannot be used to completely check the health of ATSC-compliant bitstreams. However, as supported by some transport stream analyzers, some common ATSC table tests are described here and can provide acceptable test coverage against operational objectives.[4]

Since the priority 1 and 2 tests identified by DVB TR101-290 are common with the tables used within ATSC systems, the TR101-290 tests can be directly used. However, the priority 3 tests are quite different, as they involve tables specific to ATSC (see Figure 8.2-31).

Priority 3 Checks for ATSC

- *SI/PSIP Repetition Rate:* Checks whether the repetition rate of any SI/PSIP table is outside its specified limits.
- *Unreferenced PID:* Checks that a PID not explicitly defined to be present (e.g., a PID for one of the standard tables such as NIT) or that a PID not referred to by a PMT or CAT occurs within 0.5 second. This check is useful for tracking down multiplexer misconfigurations or identifying *phantom PIDs.*
- *EIT:* Occurs if an EIT-0 table is not present for more than 500 ms or that all of the EIT-0, EIT-1, EIT-2, and EIT-3 tables are not present.
- *STT:* Occurs if a System Time Table (STT) is not received within 1 second.
- *MGT:* Occurs if the Master Guide Table (MGT) is not present for more than 150 ms.
- *TVCT:* Occurs if the Terrestrial Virtual Channel Table (TVCT) is not present for more than 400 ms.
- *CVCT:* Occurs if the Cable Virtual Channel Table (CVCT) is not present for more than 400 ms.
- *RRT:* Occurs if the Regional Ratings Table (RRT) is not present for more than 60s.

[4]The new ATSC RP A/78 explicitly describes the elements and parameters of A/53 and A/65 that should be verified in an ATSC Transport Stream for it to be considered a proper emission. It does not cover RF, captioning, or elementary streams.

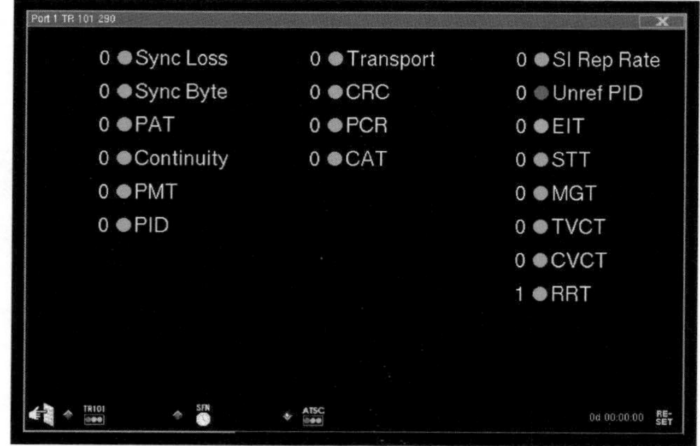

FIGURE 8.2-31 Summary status screens for monitoring the health of an ATSC transport stream.

PCR Jitter

Moving video images must be delivered in real time and with a consistent rate of presentation in order to preserve the illusion of motion. However, delays introduced by coding, multiplexing, and transmission can cause variable delay for video packets arriving at the decoder. This delay wreaks havoc in the decoding process, mandating buffers in the decoder. The MPEG-2 standard provides an additional mechanism to ensure video frames can be decoded and delivered to the viewer with a consistent rate of display. That mechanism is called the Program Clock Reference (PCR).

PCR is fundamental to the timing recovery mechanism for MPEG-2 transport streams. PCR values are embedded into the adaptation field within the transport packets of defined PIDs. The PCR consists of two parts totaling 42 bits. The PCR values increment with a standard clock rate of 27 MHz. PCR values roll over roughly once in a day. As PCR is used by the IRDs to derive the clock reference, any jitter or drift in the PCR clock can have a damaging effect on the IRD's performance.

The irregularities in the PCR can be broadly classified into two types: jitter or offset. Jitter in the PCR is mainly attributed to two sources: as systematic jitter (or PCR accuracy error, PCR AC) and network jitter. Systematic jitter and network jitter combined create overall jitter (PCR OJ).

Systematic jitter arises primarily because the repetition rate of the transport stream packets on the PID containing PCR is not a multiple of the PCR clock time. Apart from this, other systematic malfunctions along the transport chain may also result in PCR jitter. Improper management of input and output data rates on the transport buffers is another possible cause of PCR jitter.

Network jitter is a result of variations in the propagation delay in the transmission path. Network jitter is also referred to as PCR Arrival time jitter. PCR offset is the difference of the PCR clock from the required clock rate of 27 MHz.

Watching the SI/PSIP

MPEG, DVB-based satellite transmission systems, and the terrestrial ATSC system all use a variety of tables to describe the overall structure of the broadcast multiplex and the attributes of its components.

The Program Association Table is the master index for all the subsequent tables. It is linked to the Program Mapping Tables that indicate for each channel where a particular channel's audio, video, and additional data can be found in the data stream. The Virtual Channel Table, which has the channel name and other channel information, points to the Program Mapping Table to locate the audio and video PIDs for that channel. Sometimes, the VCT also contains the audio and video PIDs themselves. In that case, you must confirm that both the PMT and TVCT are consistent. Some parts of the System Information require conversion, such as from an inbound satellite signal (based on DVB) to the ATSC standard.

The most basic MPEG level tables are generated by the multiplexer. The MPEG PSI/PSIP is usually created by the MUX, while the ATSC-specific PSIP components are created by a separate PC server running PSIP generation software. Sometimes, this is even divided further with the "static PSIP" and the "dynamic PSIP" (really the electronic program guide) split between two different machines.

All the different equipment has to be configured. Sometimes, the configuration is done using an Excel spreadsheet, where someone has slowly typed in all the information. Sometimes the configuration has to be done via a green screen. All these different configurations might be put in by different people or different departments. Making sure that everything links correctly to everything else is thus another problem.

Syncing the information on all the different servers and tables is challenging and manually tedious. Given the complex interconnection of the various tables, manual inspection and fault isolation are extremely difficult. However, a number of monitoring platforms provide tools to automatically validate linkages and detect conflicts within tables and alert the operator in the event of a misconfiguration (see Figures 8.2-32 and 8.2-33). These platforms ensure that everything defined is used properly, and everything used is defined properly.

However, even with the right tools to validate the linkages, detailed troubleshooting may be needed. Close inspection of the decoded table data may sometimes be required. In these instances, it will be necessary to have a monitoring tool that can clearly reveal the bit level detail in each table (see Figure 8.2-34). Through such close analysis, each field, parameter, and descriptor is displayed, allowing the engineer to track down problems.

It is worth noting that table standards are continually evolving with the introduction of new features

TVCT								
transport_stream_id	0x5000							
version_number	1							
num_channels_in_section	4							
short_name	CH1		CH2		CH3		CH4	
Major/minor_channel_number	10-1		10-2		10-3		10-4	
channel_TSID	0x5000		0x5000		0x5000		0x5000	
program_number	0x101		0x102		0x103		0x104	
ETM_location	1		1		1		1	
source_id	100		100		100		100	
descriptors	service_location		service_location		service_location		service_location	
	PCR	0x200	PCR	0x300	PCR	0x400	PCR	0x500
	Video	0x200	Video	0x300	Video	0x400	Video	0x500
	Audio	0x201	Audio	0x301	Audio	0x401	Audio	0x501
			Audio	0x302			Audio	0x502

PMT								
PID	0x1001		0x1002		0x1003		0x1004	
program_number	0x101		0x102		0x103		0x104	
version_number	1		1		1		1	
	PCR	0x200	PCR	0x300	PCR	0x400	PCR	0x500
	Video	0x200	Video	0x300	Video	0x400	Video	0x500
	Audio	0x201	Audio	0x301	Audio	0x401	Audio	0x501
			Audio	0x302			Audio	0x502

FIGURE 8.2-32 Example of automatic validation of linkages between tables (a check for VCT-PMT consistency).

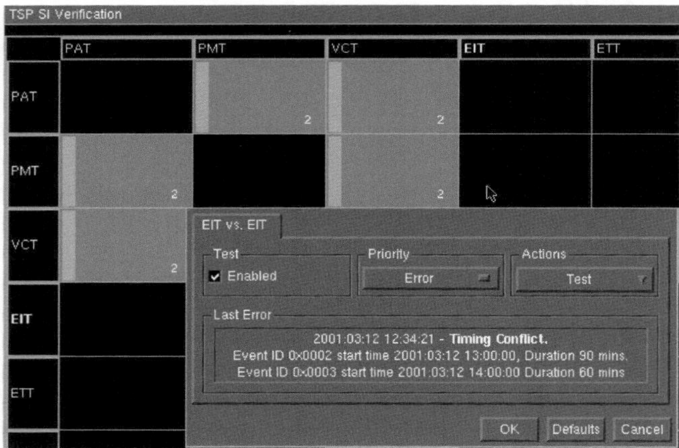

FIGURE 8.2-33 Example of automatic detection of timing conflict within a table—overlap of EIT.

and capabilities. In addition, some equipment (especially encryption related) generates proprietary "private SI" which cannot be decoded by most analyzers without modifications or additional software.

Content Validation

Delivering to expectations is the ultimate goal. Beyond whether simply the bits and bytes are correct, validation is required that the correct channels are present within the multiplex and that those channels have the correct attributes, audio channels, languages, interactive components, etc.

In the old days of single-stream analog broadcasting, the signal path was straight and simple. Taking voltage, timing, and spectrum measurements along that chain could quickly identify which components were in need of adjustment or replacement.

Now, digital broadcasting is a world of IT, Ethernet, servers, and multiplexers—each with numerous parameters that must be individually configured. Not only that, but the settings of each piece of equipment must be in agreement with each other piece of equipment (for instance, if a program is carried on PID 0x100, both the MUX and PSIP generator must be set to reflect that). In essence, it is an information management and synchronization problem, not a technical engineering problem.

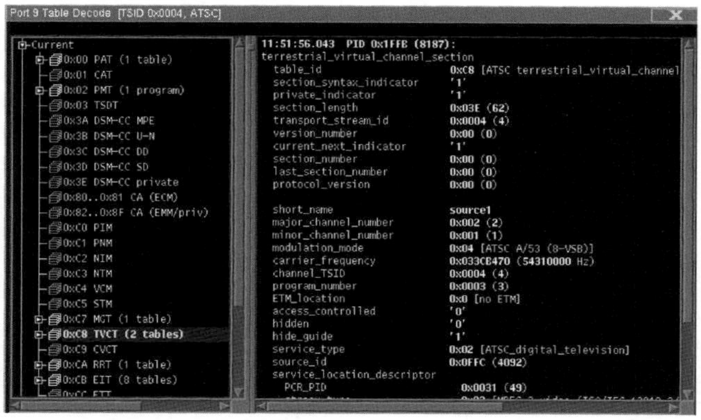

FIGURE 8.2-34 Bit-level detail table analysis—ATSC terristrial virtual channel.

Conditional Access

Conditional access systems utilize encryption techniques in order to protect content or restrict viewers from watching particular programs without authorization or payment.

A detailed explanation of the cryptographic techniques used within conditional access systems is out of the scope of this chapter. What is relevant to note here is that the actual scrambling itself takes place usually at the transport packet layer (sometimes, though rarely, it takes place at the PES layer).

For transport-layer scrambling, the transport scrambling control bits in the header of each transport stream packet indicate either that the packet is unscrambled or that it is scrambled with an even or an odd key.

Naturally, scrambled video services cannot be viewed on an analyzer without first being decrypted with the appropriate hardware—in theory. In practice, simple human error could cause a particular channel to be delivered unencrypted. This is costly, as it means lost revenue. It could also lead to adult channels being available to children because the adult content is no longer encrypted.

The solution is to use a transport stream analyzer to indicate in either the service list or bandwidth display that a particular PID and/or service is encrypted.

Streams employing transport-level scrambling will be scrambled with either an even or odd encryption key. Furthermore, to increase the level of protection against potential code breakers, the transport scrambling control is usually alternated between even and odd anywhere from once per second to once per month.

A transport stream analyzer can be used to validate whether the correct and expected services are, in fact, scrambled, as well as indicate the status of the scrambling even/odd keys. In a monitoring environment, an alarm can be set to warn operators in the event of a failure of the key generation server.

REMOTE CONTROL AND MONITORING

The increasing use of IT and LAN technologies in the broadcast environment has enabled encoding, transmission, and final broadcast operations to be separated, in some cases by large distances. The "*central casting*" philosophy allows a centralized master control or network operations center, with the possibility of centralized compression, to feed a network of remote stations and transmitter sites. The remote sites may even insert local commercials automatically using digital program insertion "cue tones" embedded within the transport stream.

It is now possible to monitor and troubleshoot these remote locations without skilled staff onsite, as an increasing amount of video and audio and transport stream analysis equipment comes with remote control features and interfaces for network management systems. This allows fault and performance telemetry to be accessed quickly by the people needing the information at a centralized network operations center (NOC).

Alarms and Notifications

While, of course, it is possible to measure various parameters and manually assess whether or not they are correct, a monitoring system can be used to *continuously* validate content, video, audio, and transport streams for compliance to expectations. Modern test and monitoring equipment allows alarm thresholds to be set for each parameter, which will be triggered once the parameter goes out of range. The alarms could be local in nature (such as a beep or log file entry), or alternatively configured to send an SNMP TRAP to a remote network management system.

SNMP Network Management Integration

The Simple Network Management Protocol (SNMP) was devised by the IT community as a simple and standard way for a network management system to communicate with, monitor, and control the behavior of various devices—whether as simple as a printer or as complex as an IP router. For a single engineer responsible for multiple locations, SNMP-based network management systems provide an excellent way to consolidate and concentrate technical information and highlight faults and alarms in a clear manner.

Technically, the monitoring system will send a TRAP message to the Network Management System (NMS) when a user-specified parameter goes out of range. At that point the NMS could display a log file, highlight an icon on the GUI, or respond in some other way.

Alternatively, the NMS might poll the remote test instrument to read the status of various alarms and/or measurements. In this way, the NMS can reconstruct an image of the working state of the remote site and present the information in a form suitable for the user's environment.

For all of this to work, of course, the remote monitoring system must support the SNMP protocol but also must have a comprehensive Management Information Base (MIB).

The MIB is a data structure on the remote device, which, in a very structured way, holds the information being made available to the NMS. Note that while a test instrument may be able to make many measurements, if those measurements are not made available in the MIB, they will not be available to a central NMS even though they can be accessed directly by the local GUI. Consequently, when remote control and management using SNMP are considered, it is essential to closely study the MIBs of the equipment and validate that the measurements of interest are, in fact, provided for within the MIB.

SUMMARY

Modern video systems are quite different from those of only a few years ago and require new methods of quality control and signal analysis to ensure that high technical standards are maintained for program making and for transmission. Compressed digital technology has been a boon for companies in the broadcasting business. It allows broadcasters to offer more features, use bandwidth more efficiently, and lower the costs of operations. Systems and signals have become more diverse, but the sheer number of channels brings its own challenges.

Digital video, especially compressed digital video, is inherently more complex than analog. Many more things can go wrong and, unlike in the analog world, it is not always obvious what is the cause of the problem. In addition, the digital world is not a tolerant one. With analog video, failing equipment causes a slow decline in quality. In the digital world, failing equipment simply causes failure. Nevertheless, sophisticated, automated, and computer-aided test and monitoring equipment enables the station engineer to have a higher level of confidence than ever before in the signals leaving the station. All companies need to ensure that they deliver to their expectations and the expectations of their customers. To ensure that expectations are met, monitoring every aspect of the transmission chain is critical.

Bibliography

Advanced Television Systems Committee. Recommended Practice: Transport Stream Verification, Doc A/78, 2006. Available at: http://www.atsc.org

ETSI, European Telecommunications Standards Institute, DVB. Measurement Guidelines for DVB Systems, Doc. TR 101-290

Harris test and measurement application notes and white papers. Available at: http://www.harris.com.videotek/support/default.asp

Tektronix application notes and technical documents. Available at: http://www.tek.com/service/

Pixelmetrix application notes. Available at http://www.pixelmetrix.com/datasheets/fdvstation.html

CHAPTER

8.3

Radio Frequency Signal Analysis

DONALD MARKLEY

Markley & Associates
Peoria, Illinois

INTRODUCTION

The analysis of radio frequency (RF) signals has benefited greatly from the newer generation of digital test equipment. In particular, the digital spectrum analyzers and vector network analyzers have changed the way in which transmitted signals and RF networks can be analyzed and adjusted.

In its simplest form, the spectrum analyzer can be visualized as a self-tuning receiver with adjustable bandwidth. The operator selects the center frequency around which measurements are desired. The receiver bandwidth and the width of the band of frequencies to be measured can then be set. The spectrum analyzer then scans across those frequencies and displays all received signals. With analog systems, the display was sometimes difficult to interpret as certain combinations of sweep characteristics resulted in a display that was not constantly visible. Experienced operators learned to cope with the display and, through experience, could evaluate even the weakest of signals. However, digital spectrum analyzers have made it easier for the occasional or new user to benefit from the device.

The digital spectrum analyzer functions in a similar fashion to a traditional analog device as far as general measurement principles are concerned. In fact, it is often possible to switch to a totally analog display. However, in the digital display mode, the unit will display each measured signal without fading between signal traces. The display is simply updated each time the system measures the desired frequencies. For transient measurements, the system holds the display as long as desired to show a single trace. Most modern units will also store one or more sets of measurements for plotting or printing using a computer interface.

An additional feature of the digital spectrum analyzer is the ability to store the maximum value of the signal noted on any frequency within the selected band or frequencies over a period of time. This feature is particularly useful in the analysis of spurious signals or the observation of a modulation envelope.

AM RADIO SIGNALS

Standard broadcast stations (AM radio) are required to perform annual measurements of spurious radiation and the occupied bandwidth of their signals. Section 73.44 of the Federal Communications Commission (FCC) rules specifically identifies the instrument to be used as a suitable swept-frequency RF spectrum analyzer. The instrument must use a 300 Hz resolution bandwidth and perform the measurement over a 10 minute period. The peak signal values at each frequency during that period are to be displayed. See Chapter 1.9 for specific details of the NRSC analog and digital radio standards [1]. Prior to June 30, 1994, AM stations were assumed to comply with FCC rule 73.44 if they employed an audio processor with a National Radio Systems Committee (NRSC) compliant filter. After that date, AM stations are required to perform annual measurements with no more than 14 months between measurements. FCC rules 73.1590(a)(1), 73.1590(a)(2), and 73.1590(a)(3) specify additional circumstances where these measurements must be performed.

Other specialized receivers such as the Delta Electronics SM-1 Splatter Monitor may be employed [2]. However, in the case of disagreement over the results, the peak-storing spectrum analyzer is to be used to determine the correct value. The FCC rules also specify the manner in which the measurements are to be performed. In particular, the station must be operated normally and the measurements taken at approximately 1 km from the center of the antenna system. The rules further explain how this is to be done for directional antenna systems.

It is often difficult to obtain a sufficient signal to perform the measurement at the specified 1 km distance, especially for lower-power stations. Loop antennas are available that provide good response and gain characteristics. In addition, the use of a loop antenna can eliminate the signals of other stations through proper orientation. If the loop does not yield sufficient signal to allow for the dynamic range necessary to show compliance with the limits in the rules, a preamplifier may be added to the system as long as its response does not significantly alter the overall frequency response of the system.

Figure 8.3-1 illustrates the results of an AM modulation envelope measurement. The limits specified by the FCC are shown on the plot. In addition to the basic envelope, the rules require measurement of the harmonics of the station's assigned frequency. All harmonics are to be either 43 + $\log_{10}$ (power in watts) or 80 dB below the carrier, whichever is the lesser attenuation. Stations operating at less then 158 watts are only required to show harmonic attenuation of at least 65 dB.

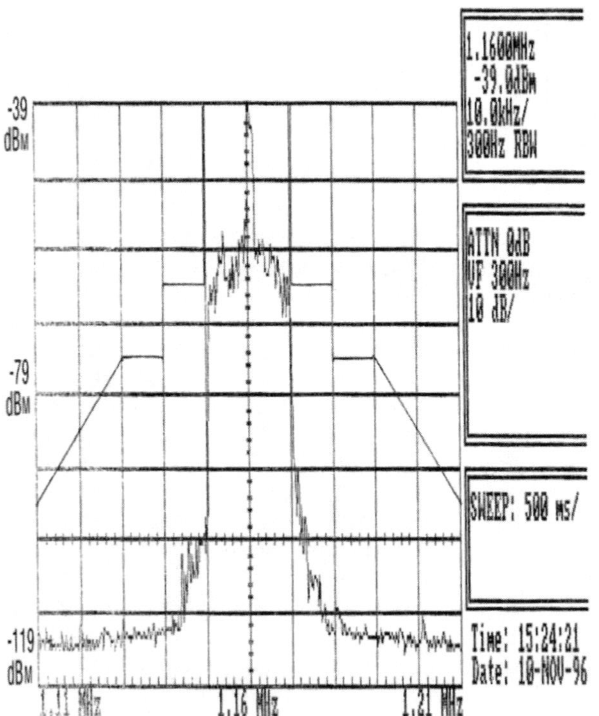

FIGURE 8.3-1 Measurement of AM transmitter with FCC limits.

While measurement of the station's harmonics can be performed with the spectrum analyzer, the results are often masked by noise. More accurate results can often be obtained by using a field strength meter that has a calibrated antenna. One significant source of error in measuring harmonics with a spectrum analyzer is the gain of the antenna used. The gain will vary considerably over the range of frequencies from the carrier to the third harmonic. Accurate measurements can only be obtained when the entire system is calibrated.

DIGITAL RADIO SIGNALS

With the adoption of digital transmissions employing carriers within the FCC-assigned bandwidth simultaneously with conventional analog signals, equipment must be available to monitor both analog and digital modes of transmission.

AM and FM digital in-band-on-channel (IBOC) transmissions (also called HD Radio™) are made up of individual carriers on both sides of the main carrier. The injection level of these carriers referenced to the main carrier is one of the measurements performed by HD modulation monitors. In addition to the RF spectrum, these devices also monitor the audio spectrum, HD PAD data, time synchronization alignment of the digital and analog audio for AM IBOC, multiple streams (HD2, etc.), 5.1 Surround, RDS/RBDS, SCA, and Tomorrow Radio for FM IBOC.

The analog and digital audio left (L), right (R), left plus right (L + R), and left minus right (L – R) are also displayed. It is important to monitor the RF spectrum for system nonlinearities that could result in spectral regrowth and sideband energy of the main analog carrier that could interfere with the IBOC carriers.

Manufacturers such as Audemat-Aztec, Belar, Inovonics, and others produce HD modulation monitors capable of the above functions. Many are frequency agile and accept low-level antenna as well as high-level direct from transmitter RF inputs. Most spectral and audio level displays are graphic with a separate screen for showing tabular information for data.

Different models of analog modulation monitors respond with varying degrees of accuracy in the presence of IBOC carriers. Additional information and definitions on AM and FM IBOC equipment and systems can be found in Chapter 4.13.

TELEVISION SIDEBAND MEASUREMENTS

The FCC has rigorous requirements for the sideband response of analog television transmitters. The measurements to confirm proper operation, as well as the tuning of the transmitters themselves, require both a spectrum analyzer and a suitable signal generator. The more common method of measurement has involved

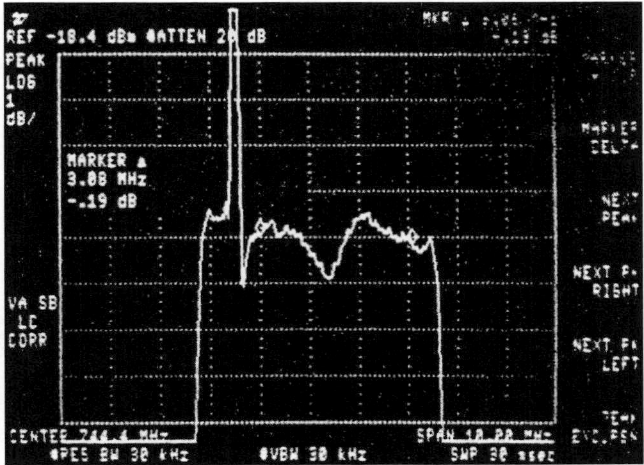

FIGURE 8.3-2 Television transmitter sideband response. Vertical scale divisions are 1 dB.

the use of a sideband adapter, such as the Tektronix 1405 or Aerodynex NTSC Sideband Adapter, which, together with the spectrum analyzer, provides a stable and easily analyzed waveform. The adapter and spectrum analyzer are interconnected so that their frequencies track. The sideband adapter generates a sweep test signal that includes markers to aid in determining the system frequency response. Figure 8.3-2 shows the response of a UHF television transmitter using a sideband adapter. In that figure, the vertical scale is 1 dB/div.

Another method of sweeping the TV transmitter uses a (sin x)/x signal from a digital television test signal generator. The wide-frequency response of that signal, together with the accuracy inherent in modern spectrum analyzers, provides sufficient accuracy to meet the FCC's measurement requirements. An advantage of this method is that the cost of a separate sideband adapter is avoided.

Digital Television Signal Measurements

For digital television, the FCC has established an emission mask to ensure that the digital signal will not interfere with adjacent channel stations. The mask is described in Section 73.622(h) of the FCC rules for full service stations and in Section 74.794 for low-power stations.

The out-of-channel emission requirements are specified for a measurement resolution bandwidth of 500 kHz and are defined as attenuations relative to the total in-channel power. Many older spectrum analyzers may find making that measurement difficult as they may not have that resolution bandwidth available. In addition, the response curve of the analyzers may result in leakage of power into the skirts of the resolution filter that may cause false indications of out-of-band emissions. The FCC has established various methods of performing that measurement and published them in May 10, 2005, as DA 05-1321. That document, which clarifies the measurement procedure, can be obtained from http://hraunfoss.fcc.gov/edocs_public/attachmatch/DA-05-1321A1.doc.

A more convenient way of measuring the system response to ensure compliance with the mask is to use one of the several measurement instruments designed for DTV. An example would be the Tektronix RFA300 series that has the ability to measure the mask without other test equipment. Other manufacturers have similar testing abilities, such as the Rhode & Schwarz FSH3-TV. It is essentially impossible to set up or monitor a DTV transmission system without some integrated test equipment systems and the majority of those systems will monitor the mask compliance.

Note that compliance with the DTV mask is more critical than the sideband response measurements previously specified for analog transmitters. DTV transmitters are permitted to operate on first adjacent channels where the protection between stations is primarily provided by compliance with the mask requirements. However, that compliance is a function of the band pass filters that are on the output of the DTV transmitter.

TRACKING GENERATOR MEASUREMENTS

A general problem exists when a signal generator and spectrum analyzer are used to measure the frequency response of systems. In simplest terms, the signal generator only generates one frequency at any given instant. The generator then sweeps continuously over a desired band of frequencies or, in the case of modern, digitally swept generators, moves in increments across that band. Since the spectrum analyzer can only measure one frequency at a given instant, some means must be found to cause the analyzer to perform measurements on the same frequency that is being generated at that time. This is accomplished with a device known as a *tracking generator.*

The tracking generator is either installed in a spectrum analyzer or mounted externally and connected directly to an analyzer. The two units are synchronized so that the generator frequency and the frequency measured by the analyzer are the same at all times. For analog equipment, the output of the tracking generator, when connected directly to the input of the spectrum analyzer, will appear as a straight line at the generator output magnitude. For digital equipment, the output of the tracking generator appears as a row of pulses.

Figure 8.3-3 illustrates one way this equipment can be used. The device under test is connected between the tracking generator output and the input to the spectrum analyzer. Since the magnitude of the tracking generator signal output is known, this test setup allows the frequency response of the device to be measured accurately over the range of interest. It must be noted that this measurement only shows the magnitude of the response. The spectrum analyzer is a scalar device and does not provide vector (magnitude and phase) data.

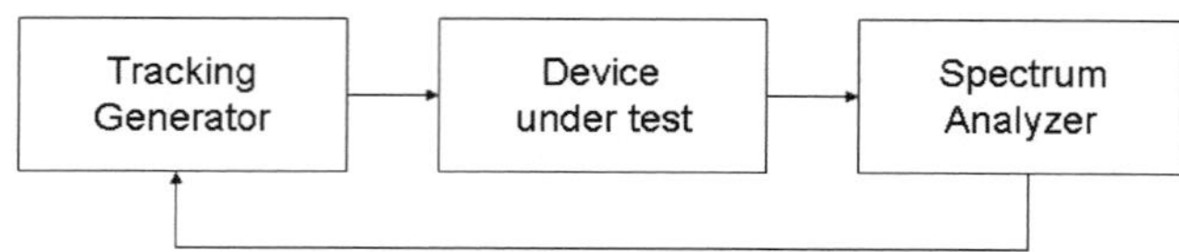

FIGURE 8.3-3 Block diagram of tracking generator and spectrum analyzer test setup.

Using a return loss bridge, the tracking generator and spectrum analyzer combination can measure the return loss from antenna systems. The return loss can in turn be used to calculate voltage standing wave ratio (VSWR) and the reflection coefficient of the load as shown in equation 1.

$$|\Gamma| = 10^{(RL/20)}$$

$$\text{VSWR} = \frac{1+|\Gamma|}{1-|\Gamma|} \qquad (1)$$

where:

Γ = reflection coefficient; and

RL = return loss in dB.

However, measurements made on a system in this fashion do not provide sufficient information to determine the location of any problems in the system—only the frequency response of the overall system is measured. Determining the location of faults requires either vector measurement information or the use of a time domain reflectometer.

TIME DOMAIN REFLECTOMETERS

The time domain reflectometer (TDR) applies a pulse or step of voltage to a transmission line or antenna system. An oscilloscope is normally used as the display device and is supplied signals from the TDR. The oscilloscope is usually operated in the x-y mode with the horizontal and vertical signal provided directly from the TDR. The oscilloscope displays the measured signal on the transmission line input as a function of time. Some TDRs incorporate an LCD display or a chart recorder or provide an output for a printer.

The applied pulse will travel down the cable or transmission line at a known rate that is the velocity of propagation of the cable under test (usually available from the cable manufacturer). When the pulse passes an irregularity or discontinuity in the cable, a portion of the pulse will be reflected back to the source. The strength of the reflection will be determined by the impedance change at the point of the discontinuity. Equation 2 can be used to determine the coefficient of reflection at the impedance change. The reflected signal is simply equal to the incident signal, or the pulse, multiplied by the coefficient of reflection:

$$\Gamma = (Z - Z_0)/(Z + Z_0) \qquad (2)$$

where:

Γ = reflection coefficient;

Z = impedance at discontinuity; and

Z_0 = characteristic impedance of line.

The reflected signal travels back to the input of the transmission line where the TDR is connected. The distance from the TDR to the irregularity is determined by equation 3. The factor of 1/2 in equation 3 is needed to account for the fact that the signal travels over the distance from the TDR to the fault twice, first in the forward direction and then in the reverse direction. An antenna on the end of the transmission line will normally look like either an open or a short circuit to the TDR pulse antenna that has DC continuity will appear to be a short with a reflection coefficient of –1; if the antenna is an open at DC, it will appear to be an open circuit with a reflection coefficient of +1.

$$d = 3 \times 10^2 \times V_p \times t/2 \qquad (3)$$

where:

d = distance to fault in meters;

V_p = velocity of propagation in cable; and

t = time between incident and reflected pulse.

A TDR is usually equipped with an adjustable delay calibrated in distance to fault. The initial adjustment of zero distance establishes a reference on the oscilloscope. The delay can then be adjusted to place any reflected signal at the same reference point. The distance to the fault is then read directly from the TDR if the instrument has been calibrated for the correct velocity of propagation. This is relatively easy to do if the exact length of the cable is known or if the velocity is provided by the manufacturer. The velocity of propagation setting can be adjusted until the reflection from the end is at the correct distance. All points between the TDR and the end of the cable will then also be at the correct distance.

Newer TDR equipment may allow for keypad entry of the velocity of propagation value. In some cases, the type of transmission line is selected by the user from the display and the velocity is then automatically selected by the unit. The calibration can be cross-checked if the cable length is known.

The sensitivity of TDR measurements is usually limited by the noise level present on the line. With an antenna present at the end of the line, the RF noise level present may mask minor reflections. Two solutions exist to that problem: one is to use a higher-powered TDR, and the other, easier method is to terminate the transmission line in either an open or a short circuit. The more complete the open or short is made, the smaller the discontinuity that can be seen. An open termination should be shielded to prevent some radiation from the end or the reception of unwanted signals. A good TDR on a tightly sealed rigid transmission line should be able to display every connection between sections as well as the insulators along the center conductor.

A transmission line that has experienced an arc across one of the insulators will leave a carbon path although that path can be difficult to find. Obviously, the TDR does not use a sufficiently high voltage to cause the arc to reoccur. If the line can be tightly sealed to lower the noise floor as much as possible, the location of the carbon path may become discernable. Faults of this type are difficult to find when operating at test instrument power levels. Unfortunately, their location will probably become quite evident after the normal transmitter power level recreates the arc a few more times.

Situations do exist that will cause highly misleading indications on a TDR. For example, a gradual change in the line impedance (versus distance) will not cause a distinct reflection leading the operator to conclude that no problem exists. This situation can occur when an air dielectric line with a slight slope in the horizontal run accumulates water from being improperly pressurized. The impedance change along the line will be gradual as the level of the water approaches the center conductor. To the TDR, the gradual change does not present a clean point of reflection and the overall change in the line may be missed. To avoid this pitfall, the operator should always try to first identify (using the TDR) the end of the line. If it appears to be missing, a problem obviously exists that the TDR method cannot identify.

NETWORK ANALYZER MEASUREMENTS

The digital network analyzer is essentially an incorporation of several pieces of test equipment in one instrument. This equipment includes a synthesized signal generator, up to three separate receivers, a controller/analyzer system, and a display. Depending on the model, the system may also include a reflection or transmission test set and, in some cases, an s-parameter test set. Figure 8.3-4 is a block diagram of a typical network analyzer.

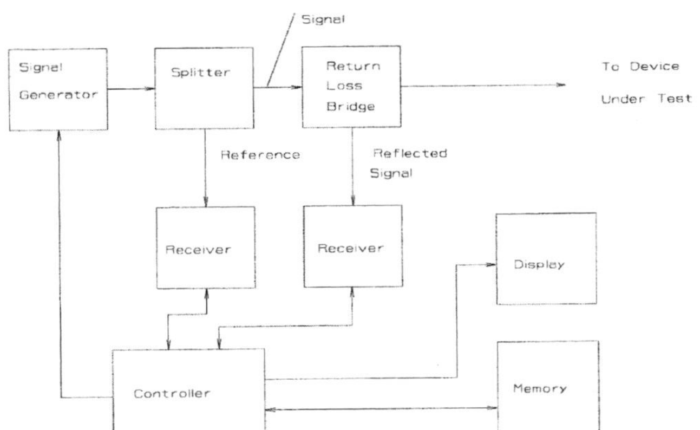

FIGURE 8.3-4 Generic block diagram of a network analyzer.

Traditionally, analog network analyzers were swept-frequency devices and were not as capable of compensating for system errors, such as test setup amplitude distortions and internal noise, as are the newer digital systems. The newer digital analyzers have eliminated that problem.

Digital Network Analyzer

The digital network analyzer does not do swept-frequency measurements. Rather, the band of interest is analyzed on a point-by-point basis. The initial instructions to the system controller include the number of points to be measured. That number of measurement points is then uniformly (or in some cases logarithmically) distributed over the range of interest. For each point, the generator and receivers are set to a selected frequency. Once the generator has locked on frequency, the output is directed to a transmission test set or s-parameter test set. For illustrative purposes, assume that the more common transmission test set is in use. Within the transmission test set, the signal from the generator is divided into two equal signals: one is returned to one receiver channel as a reference while the other signal is applied to a return loss bridge or directional coupler. This second signal is then routed to the system under test and then to another network analyzer receiver channel for comparison to the reference signal. The difference between those signals represents the reflection coefficient.

Scalar Network Analyzer

In a scalar network analyzer, the system performs the required calculations and the results are displayed as the simple magnitude of the reflection, the return loss in dB or the VSWR. Such a VSWR plot is shown in Figure 8.3-5. Traditionally, the display has been a CRT, although some of the newest models have LCD displays. That change has resulted in a significant reduction in system weight without any reduction in the quality of the presentation.

Vector Network Analyzer

A vector network analyzer goes further in determining both the magnitude and the phase of the reflected signal. That allows a Smith Chart presentation of the measured impedance, as shown in Figure 8.3-6. Some systems also permit a display of phase only as well as a polar plot of the system impedance. Markers can be used to identify the critical frequencies, such as visual carrier, aural carrier, or color carrier frequencies.

Field Calibration

A major feature of digital vector network analyzers is the ability to calibrate the unit in the field. To do this, three accurate terminations are required. The normal procedure is to use a short circuit, an open circuit, and a calibrated termination. The open and short circuits

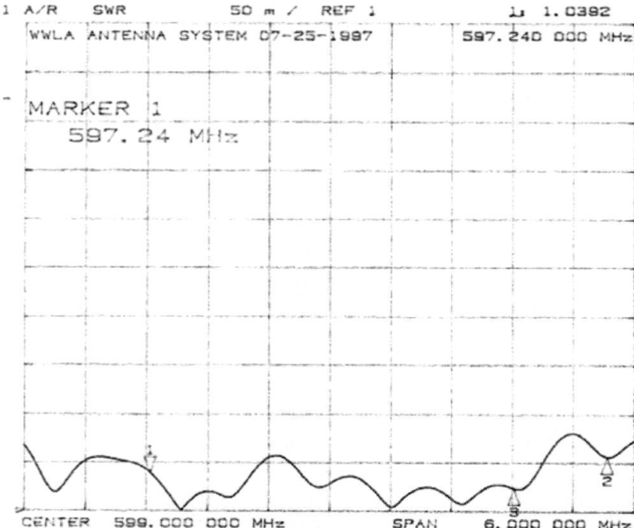

FIGURE 8.3-5 VSWR plot of TV antenna and transmission line.

are placed in their respective connectors at the same location as the load in the resistive termination. The open is shielded to eliminate any errors due to stray RF signals. For each of the three terminations, the system measures the impedance of the termination at each of the frequencies across the selected range. After the three sets of measurements are completed, a set of correction coefficients are calculated to correct the measured data to accurately represent an open, a short, and a perfect 50 Ω resistive load. That correction matrix is used on all subsequent measurements to

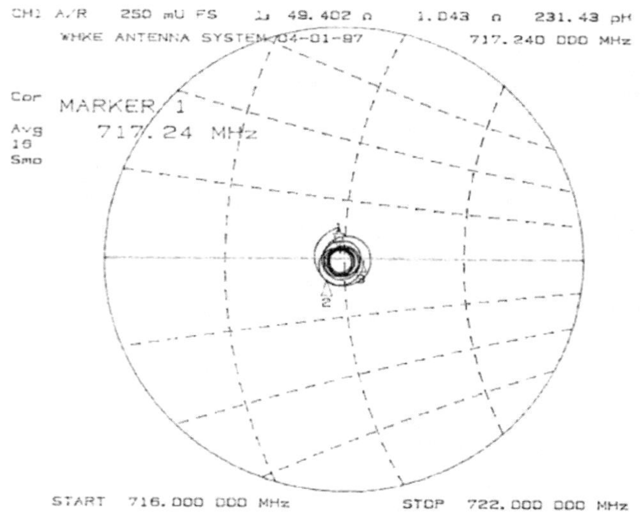

FIGURE 8.3-6 Smith Chart plot of TV antenna and transmission line with good impedance match at the antenna.

compensate for system errors. This corrects for cable leakage or resistance problems, errors in the power divider and return loss bridge, and other minor system variations. However, the final result is only as accurate as the calibrating devices.

If the network analyzer, after undergoing calibration, determines the system VSWR at some frequency to be 1.05, this result is actually based on the value of the resistive termination being 50 Ω with zero reactance. Any error in the value of this termination causes a corresponding system error in all subsequent measurements.

The ability to calibrate the system in the field has opened an additional level of measurement ability. In the past, it was necessary to lift the measurement equipment to the top of the tower if measuring the actual antenna was desired (assuming that this discussion refers to a large antenna). With the ability to calibrate the system using the three precision terminations, it is possible to perform the calibration at the load end of the transmission line rather than at the output of the test equipment, correcting for the presence of the transmission line. Once calibrated, the display will now show the actual VSWR of the antenna itself irrespective of the connecting hardware. This is useful when elements of the antenna need to be tuned in the field.

A major problem occurs when performing measurements on antenna systems, particularly in the UHF television band. The adapters used to convert from type N connectors to large transmission lines must be carefully tuned to eliminate errors. Without tuning, it is not unusual for adapters (such as a type N to 6-1/8 inch flange, or to a larger cable) themselves to introduce a VSWR of 1.1:1 or more. When measuring VSWR system values of less than 1.05:1, such errors in the adapters are unacceptable. When manufactured, these adapters are often tuned using a slotted line, an approach not normally available in the field. In summary, the purpose of all of the initial tuning and calibration is to eliminate the errors caused by the instrumentation itself.

Some network analyzers permit the "gating out" of reflections between specified distances along the line. The display then is the result of the measurements less those reflections. This can lead to significant errors if the input mismatch is significant and has simply been gated away. Significant mismatches at or near the input result in distortion or loss of reflections occurring further down the line. The instrument can only respond to actual measured signals. If they are disturbed or eliminated due to input mismatches, they cannot be recreated by the test equipment. To avoid this problem, it is often necessary to eliminate the mismatches at the input end of the system before continuing with problems at a greater distance.

Measuring TV Antennas

Modern analog television transmitting antennas are normally tuned for a VSWR of 1.05:1 or less at visual carrier, 1.08:1 or less at aural and color carriers, and a maximum value of 1.1:1 across the 6 MHz channel. Some antenna manufacturers request the field tuning of DTV antenna systems for a minimum return loss of 30 dB across the channel. That corresponds to a VSWR of 1.065:1. Clearly, good instrument calibration will be important when making measurements of these types.

As no critical frequency for DTV systems exists, such as visual carrier in analog systems, it is widely accepted that an average over the channel of 1.05:1 is satisfactory even though individual frequencies may vary as high as 1.1:1.

SIMULATED TDR MEASUREMENTS

When tuning transmitting antennas, a TDR does not provide any information concerning the match to the antenna itself. In fact, since the antenna normally appears to be either an open or a short at DC, the antenna will present a very large reflection. On the other hand, measurement of the VSWR with a network analyzer or a spectrum analyzer with a tracking generator will not provide any information concerning the location of a mismatch in the band of frequencies of interest. It is here that another feature of the vector network analyzer becomes of interest. It is possible to determine the distance to faults at RF frequencies and the magnitude of those faults.

When the network analyzer performs the measurement of the reflection coefficient at the selected number of points, it is building a series of values that represent the system response as a function of frequency. This can be said to be a function in the frequency domain. When the inverse Fourier transform (equation 4) [3] is applied to that measured data, the result is the response expressed in the time domain. That is, the response demonstrates the distance to faults along the line and the magnitude of those faults. For the analysis of TV antenna systems, that response is calculated across the 6 MHz channel and shown as VSWR versus distance.

$$f(t) = \frac{1}{2\Pi} \int_{-\infty}^{+\infty} \mathrm{F}(w) c^{iwt} dw \qquad (4)$$

where:

$f(t)$ = function of time;

F(w) = function of frequency;

d = distance;

$w = 2\pi f$; and

f = frequency in Hertz.

MEASUREMENT PRECAUTIONS AND PROCEDURES

There are some precautions to be taken in determining the distance to faults by this method. Due to the limitation on the amount of data used in the calculation, some erroneous indications can occur due to a phenomenon called *aliasing*. The maximum distance range to avoid aliasing is determined by equation 5 and is a function of the frequency span over which the measurements are performed, the number of data points used, and the relative propagation velocity of the transmission line:

$$d_{\max} = \left(1.5 \times 10^{8}\right)\left(N_{p}\right)\left(V_{p}\right) / \left(F_{2} - F_{1}\right) \qquad (5)$$

where:

d_{max} = maximum measurement distance without aliasing;

N_p = number of measurement points;

V_p = velocity of propagation;

F_1 = starting frequency; and

F_2 = stopping frequency.

The most common problem in the analysis of a TV-transmitting antenna system is the mismatch between the transmission line and the antenna at the input to the antenna. Figure 8.3-7 shows the time domain response of a UHF TV antenna system with a minor mismatch at the input of the antenna. Normally, such a mismatch is eliminated through the use of a fine-matching network or a variable transmission line transformer at that point in the system. Mismatches are also shown at the input and at the elbow at the tower base. Figure 8.3-7(a) is the measured plot before any tuning with Figure 8.3-7(b) showing the response after all mismatches have been eliminated by tuning or other corrective measures.

Input mismatches are common and are usually due to mistuned input adapters and can be corrected by retuning or replacement. Any input mismatch must be eliminated as it may mask reflections from the line and antenna. It is important to tune the input connector off of the transmission line to avoid masking such problems as bad gas blocks or irregularities very near the input.

Elbows can be tuned by slugs on the inner conductor or by probes through the outer conductor. The preferred method for new systems is to simply return the elbow displaying a mismatch to the manufacturer. For other systems and as a field solution, a short length of threaded rod can be used as a probe with a nut welded to the outside of the outer conductor. The rod can be inserted as needed to counter the irregularity causing the mismatch. This approach to the problem should only be undertaken with good test equipment and by someone experienced in such adjustments. Otherwise, the attempted repair can cause more problems than it cures.

As a casual standard, UHF antennas should be tuned for a maximum time domain VSWR of 1.03 for the 6 MHz span of frequencies in the desired channel. To find problems with the transmission line system, the analysis should be made over a much broader range of frequencies. The greater the range of frequencies and the larger the number of points measured, the more detailed the information will be concerning the

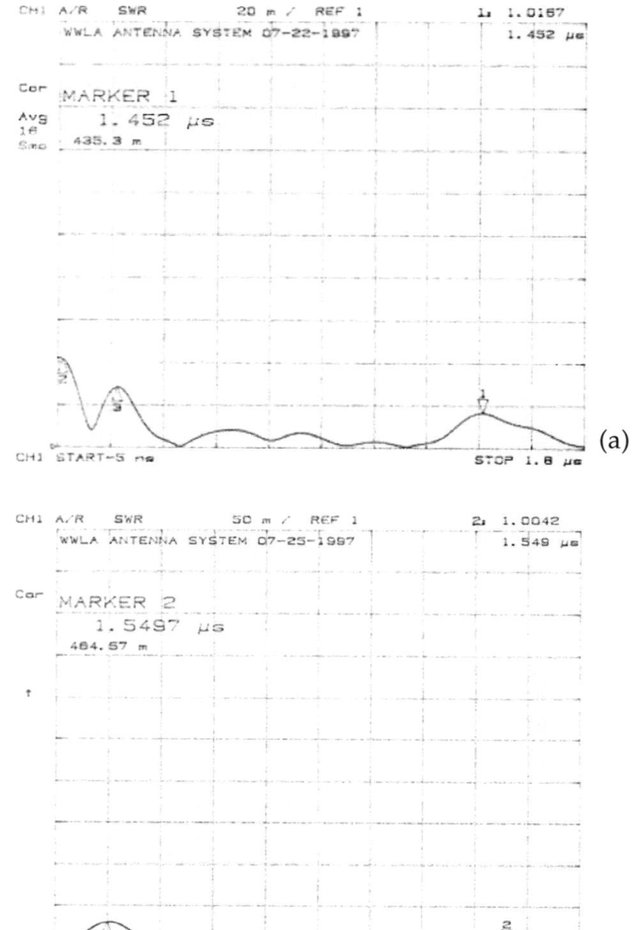

FIGURE 8.3-7 (a) Time domain of TV antenna system with marker 1 at elbow complex at top of transmission line before tuning. (b) Time domain of TV antenna system with marker 1 on antenna input after system tuning.

distance to faults. For some measurement systems, measurements are taken at 1,601 points permitting details as small as the connections at each flange to be seen on a rigid coaxial line. Systems are commercially available that measure over 20,000 points across the selected frequency range. With such equipment, small irregularities, such as a bent spacing pin on the center conductor, can be found.

There are two simple rules that can help the measurement process when using the network analyzer. First, the definition of the response plots is best if the maximum number of measurement points is used. However, that causes the slowest response to change in the system. When tuning the system in any manner a smaller number of points, such as 201 or 401, are preferred to increase the response speed of the analyzer. So, use fewer points for tuning and the maximum number of points for the final results. Note that recalibration is normally necessary when the number of points is changed.

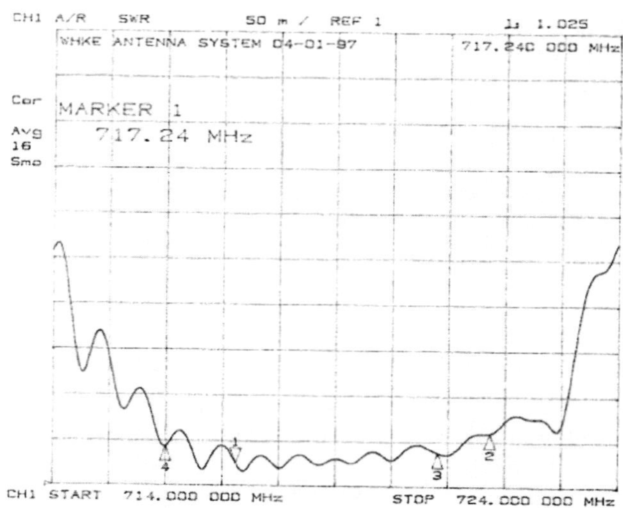

FIGURE 8.3-8 VSWR plot of a tuned system: TV antenna with waveguide.

Second, when performing network analyzer measurements the range of frequencies in use must be within the limits of the cable or waveguide. If measurements extend above the usable values for a transmission line or below the cut-off limits for waveguide, highly incorrect results will be obtained.

Waveguide Analysis

Due to its basic nature, a TDR will not work on a waveguide system because a waveguide does function at low frequencies. Therefore, a TDR will always see a waveguide as an open circuit occurring at the transition from coaxial cable to the waveguide. However, in its simulated TDR mode, the network analyzer can determine the distance to any mismatch along a rectangular or truncated elliptical waveguide. Tuning sections can then be installed (or tuning straps in the case of standard rectangular waveguide). Circular waveguide at UHF frequencies is normally not adjustable in the field other than the transitions from rectangular waveguide or coaxial cable.

Figure 8.3-8 shows the results of a properly tuned antenna system with waveguide. The network analyzer normally will store such information on a disk that permits the data to be displayed or printed out.

AURAL MODULATION MONITORS

Aural modulation monitors are specialized, accurate, and low-distortion instruments. They replace several more complicated, general-purpose test instruments of similar accuracy as well as provide a convenient means of ensuring compliance with FCC rules.

Broadcast aural modulation monitors are also classified according to the type of modulation to be monitored:

- AM monaural
- AM stereo
- FM monaural (and subsidiary communication services [SCA] as needed)
- FM stereo (and SCA as needed)
- TV monaural
- TV stereo (and multichannel sound, SAP, and PRO)

Monaural and stereophonic functions are usually combined in one instrument.

The major reasons for using frequency and aural modulation monitors at a broadcast station are described in the following sections.

Coverage

Because coverage of the broadcast service area is enhanced by a high modulation level, it is desirable to maintain the modulation level at the maximum legal limit in order to maintain good coverage and improve signal-to-noise ratio (SNR) for the audience. Modulation monitors determine whether overmodulation occurs.

Proof of Performance

By using the monitor as a precision test instrument, the station engineer can perform proof-of-performance measurements to ensure that the transmitter is within the modulation and frequency technical specifications of the manufacturer and the operating requirements of the FCC. The FCC no longer requires a regular proof of performance. However, stations are still required to meet certain technical requirements.

Most aural modulation monitors have built-in facilities to measure baseband audio frequency response and SNR. Outputs are provided for total harmonic distortion measurements by an external distortion analyzer. Modulation monitors with stereo and SCA functions can measure stereo separation, subcarrier injection levels, crosstalk between service channels, and provide demodulated outputs for stereo and SCA channels. TV aural broadcast modulation monitors are often comprised of several integrated pieces of test equipment to test many functions of TV aural transmitters.

Compliance with FCC Rules

The monitor enables the station engineers to operate the transmitter in accordance with FCC rules regarding aural modulation levels and carrier frequency tolerances. In addition, Part 73.1590(a) requires proof-of-performance measurements of all main transmitters, except class D noncommercial educational FM stations operating under 10 watts, to be taken as follows:

- Upon initial installations.
- Upon modifications of transmission facilities.
- Installation of AM stereo.
- Installation of FM stereo or subcarrier.
- Installation of TV stereo or subcarrier.
- Annually on AM stations as specified in FCC rule 73.44.
- When required by other special provisions of the station license.

Aural Modulation Limits

Part 73.1570 of the FCC rules states that modulation percentage is to be maintained at the highest level consistent with good transmission quality and broadcast service, not to exceed the following limits:

AM stations. Modulation must not exceed 100% on negative peaks of frequent recurrence, or 125% on positive peaks at any time. There are additional regulations for AM stereo and telemetry transmissions.

FM stations. Total modulation must not exceed 100% on peaks of frequent recurrence referenced to 75 kHz deviation. However, stereo stations simultaneously providing subsidiary communication services (SCA) on subcarriers may increase the total peak modulation 0.5% for each 1% of subcarrier injection modulation; but the total carrier modulation must not exceed 110%. If two or more SCA subcarriers are used in conjunction with the stereo channel, the maximum allowable peak deviation is ±82.5 kHz or 110%.

TV stations. In general, the modulation of the aural carrier must not exceed ±25 kHz deviation (monaural) on peaks of frequent recurrence. Stations transmitting multiplexed subcarrier signals on the aural carrier must limit the modulation of the aural carrier by the arithmetic sum of the subcarrier(s) allowable deviation and the total modulation must not exceed ±75 kHz deviation.

Modulation requirements for the Broadcast Television Systems Committee (BTSC) stereo sound are subject to the criteria set forth in FCC rule 73.682(c) and the FCC Office of Engineering and Technology (OET) Bulletin 60A. Also see Chapter 6.3, "Multichannel Television Sound."

FREQUENCY MONITORING AND TOLERANCES

Part 73.1540(a) of the FCC rules requires that center frequencies of AM, FM, and TV stations must be measured or determined as often as necessary to ensure that they are maintained within the required tolerances as follows:

- *AM stations:* ±20 Hz of the assigned frequency.
- *FM stations:* ±2000 Hz of the assigned frequency (±3,000 Hz for stations of 10 watts or less).

- *TV stations:* Visual carrier ±1000 Hz of the assigned frequency. Aural carrier must be 4.5 MHz ±1000 Hz above the actual visual carrier frequency.

FCC RULES AFFECTING MONITORING

The FCC no longer specifies the type of aural modulation monitor or measuring equipment a broadcast station must use. It is the responsibility of the station licensee to decide what monitoring equipment is needed to ensure that the station transmitter emissions comply with FCC frequency and modulation requirements.

FREQUENCY MONITORS, METERS, AND COUNTERS

While most broadcast stations employ frequency counters to measure carrier and subcarrier frequencies as well as other signal-generating circuits, frequency monitors or meters still used in some stations must be capable of accurately measuring and displaying the carrier frequency requirements. The recommended resolution for AM monitoring is 1 Hz while FM and TV monitors should provide a resolution of 10 Hz or better. These tolerances call for an accurate and stable internal frequency standard with an aging rate of 1 part per million (ppm) per year or better that is traceable to NIST. Many frequency counters have an input for an external 10 MHz GPS reference.

If the transmitter is to be monitored at some distance from the transmitter site, a built-in or external preselector is generally required to raise the input level for receiving the signal off-the-air. The following features are highly desirable in a frequency monitor:

- Digitally tuned RF preselector for multiple off-air applications.
- Digitally tuned RF preselector for multiple off-air applications combined with the aural modulation monitor.
- An output to operate an alarm when preset frequency limits are exceeded.
- Provision for calibrating the internal frequency standard against an NIST station or other highly accurate standard such as GPS.
- An output for automatic logging.

Peak Modulation Duration

Although the FCC's current rules contain no precise definition of the maximum allowable peak modulation duration, some manufacturers of modulation monitors are using the pre-1983 de facto rules as their design guidelines. The de facto rules allowed a 1 millisecond response time for the peak modulation indicators and the permissible overmodulation limit was 10 counts per minute.

AURAL MODULATION MONITORS

At a minimum, all aural modulation monitors should have a quasipeak reading modulation meter to give a direct indication of modulation percentage. Meter accuracy should be ±4% or better. Other desirable features common to all modulation monitors are:

- Park indicators, accurate to ±2% or better, to indicate when maximum positive and negative modulation peaks are occurring.
- Adjustable peak indicator trigger points to indicate when modulation peaks exceed preset levels.
- Adjustable peak modulation duration detecting circuit in the event the FCC and the broadcast community agree on the maximum allowable peak duration.
- An internal modulation level calibrator to check the accuracy of the modulation meter and peak flashers and a means for recalibrating the meter and peak flasher circuits.
- An output to operate an overmodulation alarm or a built-in alarm.
- An output to operate an external alarm or a built-in alarm for when the modulation drops below a certain level (10%) for a specified period of time.
- Outputs for a remote meter and peak flasher.

AM Radio Modulation Monitors

AM radio monaural modulation monitors should have the features described in the preceding paragraph. If an RF preselector is used for off-air monitoring with the monitor, its sensitivity should be approximately 100 V for a 35 dB SNR and 1 mV for a 50 dB SNR. Selectivity should be at least –40 dB at ±40 kHz, and image rejection should be at least 50 dB. The RF preselector must be linear to avoid causing erroneous readings on the modulation monitor.

It is also desirable that the AM modulation monitor be equipped with NRSC deemphasis circuits so that the audio output of the monitor matches that of the AM transmitter modified with the NRSC audio response characteristics. AM stereo monitors should have the following additional features:

- L + R and L – R channel decoding and outputs.
- L and R channel separation measurement capability.
- L and R channel SNR measurement capability.
- L and R channel frequency response measurement capability.
- Pilot carrier injection level measurement capability.
- Channel crosstalk measurement capability.
- Signal output for distortion measurements.

See Chapter 4.2, "AM Transmitters and AM Stereo," for additional information.

FM Modulation Monitors

FM modulation monitors should have at least a 70 dB signal-to-noise ratio. The discriminator must have a distortion figure of 0.1% or better and a baseband frequency response of at least 25 Hz to 100 kHz, so that it can pass and accurately measure an SCA channel up to 92 kHz. FM monitors often consist of both frequency and modulation monitors in one package.

If an RF preselector is used for off-air monitoring, its intermediate frequency (IF) amplifier should have a linear phase response curve, yet be narrow enough to reject adjacent channels. A built-in multipath detector is highly desirable to help minimize multipath interference. The FM monitor should be equipped to measure synchronous and nonsynchronous AM noise of the FM carrier. FM stereo monitors should be able to measure:

- The L + R channel level (30 Hz to 15 kHz).
- The L – R channel level (23 kHz to 53 kHz).
- The 19 kHz pilot injection level.
- The 38 kHz subcarrier level.
- Crosstalk between the main channel and subcarriers.
- Separation of left and right channels (up to 60 dB).

If one or more SCA subcarriers are transmitted, an SCA monitor should be used in addition to the FM stereo modulation monitor. The SCA monitor is usually an add-on to the main unit, which takes a composite feed from the demodulated output of the baseband monitor. Some manufacturers offer an optional RF and baseband demodulator, so that the SCA monitor can be used independently from the main monitor. The capability for user selection of SCA frequencies is important for future expansion of SCA service. The SCA monitor should be able to measure:

- Modulation percentage.
- SCA injection level on the composite signal.
- Signal-to-noise ratio.
- Crosstalk.

The SCA modulation measurement should be selectable for ±4 kHz or ±6 kHz as the deviation level for a meter indication of 100%.

Analog TV Monitors

The features for analog TV monaural and stereo monitors are similar to those for FM monitors. When transmitting BTSC multichannel sound, the operator should be able to monitor the main channel and stereo channel as well as SAP and PRO channels (if they are utilized). Monitoring the modulation level of the BTSC signal is important for achieving good stereo separation.

BTSC Stereo Separation and Modulation Accuracy

Stereo separation in the BTSC format is sensitive to gain and phase errors in the transmission path. This is because the L + R and L – R signals are treated differently. In particular, L – R is companded while L + R is simply preemphasized and deemphasized. The L – R and L + R signals must arrive at the receiver's decoder matrix, which yields L and R, with very small errors in gain and phase across the aural baseband from 50 Hz to 50 kHz. Figure 8.3-9 shows how stereophonic separation is affected by gain and phase errors in the L – R signal relative to the L + R signal at the input of the final matrix.

Subjective tests have shown that an average listener begins to "perceive" a loss in the spatial character of stereophonic music material when the separation drops below 18 dB. A separation of 15 dB ±3 dB is considered "adequate" by the average listener.

Although the subjective effects of separation depend on the spectral distribution and other aspects of the audio material, it appears that a good engineering objective for the entire system is for the separation to exceed 20 dB in the midrange, decreasing somewhat at frequencies above 8 kHz. Figure 8.3-9 shows that a separation of 20 dB requires a gain error smaller than 1 dB, and a phase error of less than 10°. The BTSC standards require that the separation of the radiated signal exceeds 30 dB in the midband from 100 Hz to 8 kHz, but that it may decrease at low frequencies to 26 dB at 50 Hz, and to 20 dB at 14 kHz. This requires that

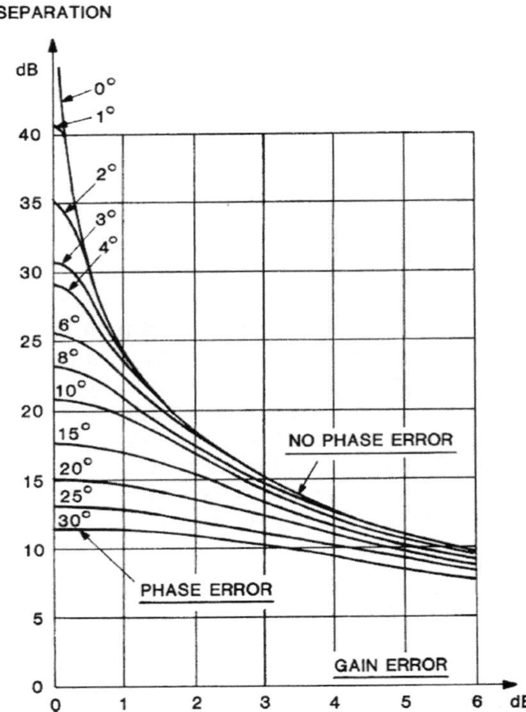

FIGURE 8.3-9 Stereo separation as a function of gain and phase errors in the L – R versus the L + R paths. (Reprinted with permission from TFT, TV Aural Proof of Performance Guide.)

the gain and phase errors in the midband be less than 0.3 dB and 3.0 degrees, respectively.

The total modulation level accuracy in BTSC stereo is more critical than in FM stereo radio broadcasting to produce acceptable stereo separation. Because the L + R and L – R signal paths in BTSC are processed differently, a small change of modulation level in the BTSC system will affect the stereo separation. This is because the amplitude and phase relationship between the L + R and L – R channel is altered. If the total modulation level of the BTSC system is not maintained accurately, the dbx decoder in the receiver will see an incorrect RMS level and reproduce an L – R signal with altered amplitude and phase. That is, if an incorrect L – R signal is fed to the decoding matrix, the consequence will be poor stereo separation. Because the RMS level to the input of the decoder is directly proportional to the total modulation level, the total modulation level in the BTSC transmitter must, therefore, be accurately monitored in order to maintain good stereo separation and high-quality audio performance.

ON-SITE MONITORING TECHNIQUES

In a studio-transmitter collocated operation, the monitor is normally connected directly, or through an attenuator, to an RF sampling point of the transmission line feeding the antenna via a directional coupler. It is important to know the RF voltage level of this sampling point so that it meets the input requirement of the monitor.

AM Monitoring

The modulation percentage of a monaural AM carrier is normally displayed directly on a front-panel meter.

Peak flashers, used on some monitors, are intended to catch fast transients and peaks that the meter cannot respond to. There may be one flasher to indicate maximum allowable negative peaks (100%) and another to indicate maximum allowable positive peaks (125%). The monitor may also have an adjustable peak flasher that can be preset by means of digital switches so that it flashes when the modulation percentage exceeds the preset switch value.

For monitoring modulation of an AM stereo transmitter, an AM stereo monitor or a stereo monitor plus a compatible AM modulation monitor are required. The equipment should permit the operator to simultaneously read the modulation percentage on both left and right channels and to measure separation between channels and crosstalk between the main channel and subchannels.

FM Monitoring

On a typical monaural FM monitor, the modulation percentage is normally displayed on a front-panel meter. Generally, either positive, negative, or combined modulation peaks can be selected for monitoring. Some monitors also provide peak flashers to indicate modulation peaks that exceed a preset percentage. The RF input must be adjusted to the correct level, as described in the instruction manual, before accurate readings can be taken.

For monitoring a stereo FM transmitter, a modulation monitor with a compatible stereo monitor must be used. Left channel, right channel, and total modulation are generally read on front-panel meters. With the typical monitor, these meters can also be used to measure separation between the left and right channels as well as crosstalk between the main channel and subchannels as well as pilot level injection.

SCA monitors may be accessories to the baseband monitor and are either fed from the FM composite signal or from the RF carrier through a separate FM demodulator.

Analog TV Aural Monitoring

The typical TV frequency monitor will provide separate displays of the visual carrier, aural carrier, and intercarrier frequency errors (rather than actual carrier frequency). The RF input to the monitor should be adjusted to the correct level as described in the instruction manual before measurements are made.

Aural modulation percentage can be read from a front-panel meter. If the TV transmitter is also transmitting BTSC stereo sound, a monitor with a stereo decoder for left and right channels must be used. If an SAP or PRO channel is employed, the monitor should also be used to provide modulation level and frequency.

Due to the critical relationship in BTSC stereo separation between the recovery of the companded L – R channel and the modulation level of the transmitter described previously, it is essential to maintain accurate modulation levels when broadcasting multichannel television sound. One convenient method of achieving this objective is to monitor the pilot carrier injection level, which has a constant level. After calibration of modulation levels for maximum stereo separation, the pilot carrier injection level is the best reference for maintaining maximum stereo performance.

OFF-SITE MONITORING TECHNIQUES

For remote, off-air monitoring, the monitor should incorporate a built-in preselector (RF amplifier) or an external RF preselector connected to an outdoor antenna.

Off-site monitoring techniques are generally the same as on-site techniques. RF input level to the preselector or monitor must be carefully adjusted, as described in the manufacturers' instruction manuals. It is even more important to know the RF level so that it does not overload the RF preselector and create intermodulation products. Some monitors are equipped with a multipath detector that enables the user to rotate the receiving antenna for minimum multipath interference.

MODULATION MONITOR CALIBRATION

The aural modulation meter and peak flashers should be calibrated regularly. Most monitors have built-in calibrators, so that meter and flasher accuracy can be checked by pressing front-panel switches and observing the peak flashers and the meter reading. If the reading is in error, an adjustment usually corrects the error.

If the built-in modulation calibrator is of the frequency marker type for establishing the ±100% peak modulation level, and the frequency markers are generated from single crystal sources, the accuracy can usually be maintained within ±1%. If this type of monitor is being used, it is not necessary to calibrate the FM modulator using a Bessel null method.

If an FM modulation monitor has no internal calibrator, the monitor must be calibrated against a laboratory standard or by means of a Bessel null measurement using a spectrum analyzer and a precision audio frequency generator. See Chapter 4.7, "FM and Digital Radio Transmitters," for details on using this method.

If an AM modulation monitor has no internal calibrator, the monitor can be calibrated using an RF signal generator. The generator must be capable of very low-distortion amplitude modulation, and the level of this modulation should be accurately observed by using a good-quality oscilloscope.

An AM monitor is usually calibrated by comparing the peak amplitude of the waveform against an amplitude reference established by an oscilloscope with good linearity. A digitally generated reference is frequently used as a built-in calibration standard as in an FM modulation monitor.

MONITOR MAINTENANCE

A broadcast station monitor should be maintained in the same way as other precision laboratory instruments. If should be calibrated regularly as described in the preceding paragraphs. Manufacturers and specialty services offer maintenance and calibration to broadcasters.

DIGITAL TELEVISION MEASUREMENT ISSUES

For analog signals, some transmission impairments are tolerable because the effect at the receiver is often negligible, even for some fairly significant faults. With digital television (DTV), however, an improperly adjusted transmitter could mean the loss of viewers in the edge of the grade B coverage area (or worse, intermittent reception). DTV reception does not degrade gracefully, it simply disappears as the receiver fails to correct errors or the signal drops below the decoding threshold level. Attention to several parameters is required for satisfactory operation of the ATSC 8-VSB modulation system used by DTV. First, there is the basic FCC requirement against creating interference to other over-the-air services. To verify that there is no spill-over into adjacent channels, out-of-band emission testing is required. Second, the composite signal to noise, which may consist of many transmission system impairments, is measured to ensure it is not contributing to errors in the receiver in addition to impairments encountered between the transmitter and receiver [4]. For a detailed explanation of DTV signal measurement and analysis, see Chapter 8.4.

In DTV, flat frequency response across the channel passband is required comparable to that required by analog transmitters. Where analog TV is concerned, group delay problems result in chrominance/luminance delay, which can degrade the displayed picture but still leave it viewable. Group delay problems in DTV transmitters, however, result in intersymbol interference (ISI) and a rise in the bit error rate (BER), causing the receiver to drop in and out of lock. Even low levels of ISI may cause receivers operating near the threshold level to lose the picture and sound completely. Amplitude and phase errors may cause similar problems, again resulting in reduced viewer coverage.

Eye patterns and BER have become well-known parameters of digital signal measurement, although they may not always be the best parameters to monitor for 8-VSB transmission. The constellation diagram and modulation error ratio, on the other hand, provide insight into the overall system health. RF constellations are displayed on the (*I*) in phase and (*Q*) quadrature components of the received RF signal, commonly displayed on the x and y axes of the constellation diagram, respectively. Constellations of tight vertical dot patterns with no slanting or bending indicate proper operation, as illustrated in Figure 8.3-10(a). Since 8-VSB levels are the in-phase signal they are displayed left to right.

An 8-VSB signal is a single sideband signal with a pilot carrier added. In a single sideband signal, phase does not remain constant. Therefore, the constellation points (dots) occur in a vertical pattern. As long as the dot pattern is vertical and the points form narrow lines of equal height, the signal is considered good and can be decoded.

Figure 8.3-10(b) shows an 8-VSB signal that has noise and phase shift. Noise is indicated by the spreading of the dot pattern. Phase problems are indicated by the slant along the *Q*-axis.

While BER is a valid measurement for 8-VSB, a more accurate method involves monitoring the modulation error ratio (MER), which usually will reveal problems before the BER is affected. In many cases, MER can provide enough warning time to correct problems that would result in an increase in the BER. MER provides an indication of how far the points in the constellation have migrated from the ideal. There can be considerable migration before boundary limits are exceeded. Degradation in the BER is only apparent when those limits have been exceeded. MER is compa-

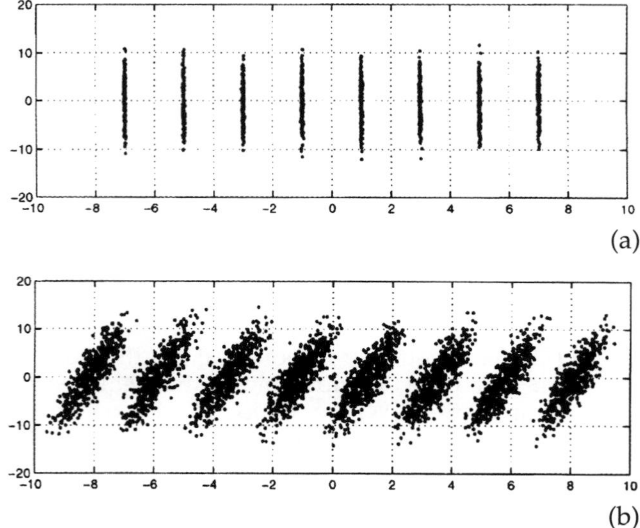

FIGURE 8.3-10 8-VSB constellation diagram: (a) a near-perfect condition, and (b) constellation diagram with noise and phase shift (the spreading of the pattern is caused by noise; the slant is caused by phase shift).

rable to signal-to-noise ratio in that impairments manifest as noise that can be summed into a composite noiselike signal.

Principle DTV System Parameters

The FCC specifies the out-of-band emissions mask for DTV terrestrial transmission according to Section 73.622. That section requires that transmitter out-of-band emissions be attenuated consistent with the following:

- At the channel edge, emissions must be attenuated no less than 47 dB below the average transmitted power.
- At more than 6 MHz from the channel edge, emissions must be attenuated no less than 110 dB below the average transmitted power.
- At any frequency between 0 and 6 MHz from the channel edge, emissions must be attenuated no less than the value determined by the following formula, which is based on a measurement bandwidth of 500 kHz: attenuation in dB = $11.5 + (\Delta f + 3.6)$, where Δf = the frequency difference in MHz from the edge of the channel.

More attenuation may be required if interference is created. Other values of attenuation apply to DTV translators and LPTV stations.

Power Specification and Measurement

NTSC broadcast transmissions are allowed a power variation ranging between 80% and 110% of authorized power. These values correspond to –0.97 dB and +0.41 dB, respectively. Because of the cliff effect at the fringes of the service coverage area for a DTV signal, a more than 1 dB variation in power level can affect DTV reception if the signal level is already near the reception threshold. One dB reduction represents approximately a one mile (1.6 km) reduction in coverage distance from the transmitter (for an average high-power TV facility). Therefore, maintaining the average power of the DTV transmitted signal at rated power level can be important. In contrast the ATSC recommends a more stringent range of operating power with a lower limit of 95% and an upper limit of 105% of authorized power [5].

SUMMARY

Not all measurement requirements for broadcast transmission have been covered in this chapter. Transmitter manufacturers usually provide extensive documentation on the tests and measurements recommended and/or required for their products. In-factory or on-site instruction is also available from many manufacturers. Because the purchase of a transmitter or antenna system is a substantial investment, it is essential that the proper transmission measurement instrumentation be available along with the knowledge of how it is used in the routine business of broadcasting that will ensure long-term reliability and an acceptable level of quality performance.

References

[1] See http://www.nrscstandards.org.
[2] See http://www.deltaelectronics.com/data/sm1data.htm
[3] See http://mwt.e-technik.uni-ulm.de/world/lehre/basic_mathematics/fourier/node6.php3, Example 7 for details on the inverse Fourier transform.
[4] Portions of the DTV Measurements section extracted from the ninth edition of the *NAB Engineering Handbook* were contributed by Joe Wu, TFT, Inc., Santa Clara, CA.
[5] ATSC, Transmission Measurement and Compliance for Digital Television, Advanced Television Systems Committee, Washington, DC, Doc. A/64A, May 30, 2000.

CHAPTER

8.4

Digital Television Transmitter Measurements

LINLEY GUMM
Consultant
Beaverton, Oregon

GARY SGRIGNOLI
Meintel, Sgrignoli, & Wallace
Mount Prospect, Illinois

INTRODUCTION

The transmitter is the last major processing block before the digital television (DTV) signal leaves the control of the broadcaster. In the transmitter, the signal is randomized and interleaved, forward error-correction coding and synchronization information is added, and critical band-shaping filtering is performed. It is important that the broadcaster verifies that all these processes have been performed correctly and that the transmitted signal meets FCC requirements.

There are three modes of DTV transmitter measurements: monitoring, diagnostics, and maintenance and verification.

Monitoring Mode

In monitoring mode, the operator or automatic equipment is able to quickly monitor basic parameters for acceptable transmitter performance. If performance is out of tolerance, the process shifts to the diagnostic mode.

Diagnostic Mode

In diagnostic mode, the operator or automatic equipment tracks down the source of the problem and decides if it necessary to shift to maintenance and verification mode.

Maintenance and Verification Mode

In maintenance and verification mode, maintenance is performed and the system is verified (tested) to determine that it meets operational and regulatory requirements once again.

The rest of this chapter provides information to support all three of these measurement modes as they relate to Advanced Television Systems Committee (ATSC) 8-VSB transmitters. The section on signal quality measurements provides the background and the techniques required for the monitoring and diagnostic modes. The sections on transmitter power measurement, frequency measurement, and emissions measurement provide most of what is required to support verification measurements.

EQUIPMENT REQUIRED

Several pieces of test equipment (some specialized) must be available to perform 8-VSB transmitter verification and adjustments. It is important to understand that conventional test equipment used on analog TV transmitters may not be suitable for DTV transmitter measurements.

- *Power meter.* This is used to measure the total average transmitter power (in the 6 MHz channel).
- *VSB analyzer.* This is sometimes called a vector signal analyzer (VSA). This instrument measures packet error rate (PER), signal-to-noise ratio (SNR), constellation diagram, phase noise, peak-to-average power ratio, and pilot parameters.
- *Spectrum analyzer.* This is used to measure DTV signal emissions and frequency. A suitable spectrum analyzer should have band power markers and at least 110 dB of dynamic range between its two-tone,

third-order intercept (TOI) amplitude and its displayed average noise level (DANL) in a 10 kHz resolution bandwidth. An analyzer with a 5 dB step internal attenuator will simplify measurement procedures. An analyzer that corrects low-amplitude readings for the proximity of its own noise floor is also desirable. To maximize its usefulness the analyzer should also have a built-in tracking generator for sweeping filters and other components.

- *Band stop filter.* When tuned to the channel to be tested, the band stop filter is used to decrease the 8-VSB signal's amplitude to allow emissions to be measured by the spectrum analyzer. The filter must lower all emissions in the transmitter's 6 MHz channel by at least 46 dB, yet pass emissions 2 MHz beyond the channel's edges with no more than 3 dB of loss. The filter should be capable of handling 1 watt.
- *Calibrated variable attenuator.* This is required only if the spectrum analyzer does *not* have an internal 5 dB step attenuator. It allows exact adjustment of the spectrum analyzer's input amplitude for optimum dynamic range. It must have 5 dB step size, at least a 1 watt rating (i.e., +30 dBm), and 10 dB range, and the overall accuracy must be ±1 dB.
- *Low-loss, shielded cable, coaxial adapters, etc.* These are important for interconnecting equipment with minimal degradation.

These items are often useful but are not strictly necessary:

- *Calibrated 10 dB attenuator (pad).* This is required only when the calibrated directional coupler signal level is greater than the safe operating range of the test equipment.
- *Calibrated radio frequency (RF) preamplifier.* This is helpful for use with low-level signal devices or in the field (may be part of a spectrum analyzer).
- *Frequency reference.* A GPS or other source for a precision frequency reference. This is required only when a high-accuracy internal frequency reference is not incorporated within the spectrum analyzer itself.
- *CW RF signal source.* An RF source of some type covering the TV channel to be measured is required to characterize the band stop filter. It is required only if the spectrum analyzer does not have a built-in tracking generator.

SIGNAL QUALITY MEASUREMENTS

When transmitting a digital signal using an RF carrier, the carrier's amplitude and phase are often simultaneously modulated by the digital signal. In the 8-VSB system, the waveforms used to create the 8-VSB signal have been specifically designed to minimize the amount of bandwidth required to carry the digital MPEG signal. The MPEG signal is sent by adding error correction bits and converting the resulting digital data into multilevel symbols. Each symbol carries three bits of the combined data by encoding the signal into one of eight amplitude levels; thus, the "8" in 8-VSB.

A further goal when designing the 8-VSB waveform is to create instants separated by the period of the symbol clock when the waveform's amplitude can be sampled without being affected by the value of previous and subsequent symbols. This is referred to as having no *intersymbol interference* (ISI).

Figure 8.4-1 shows the instantaneous amplitude and phase of the RF 8-VSB signal plotted like a vectorscope display of the chroma signal in analog NTSC color television. The RF signal's instantaneous amplitude is proportional to its distance from the origin, and the RF signal's instantaneous phase with respect to the 8-VSB signal's pilot signal is equal to the angle from the positive real axis.

At first glance, all that is seen is a swirling mass. However, when the trace is sampled at the edges of the symbol clock (at the dots in the figure), eight vertical lines of data points can be made out (along the dotted lines) representing the eight real-axis symbol values of the transmitted 8-VSB signal. Unlike the chroma signal, the imaginary-axis signal is used to suppress the 8-VSB signal's unwanted sideband, and thus carries no information.

If the amplitudes of the 8-VSB waveform at only the symbol instants are plotted, a *constellation* pattern is obtained. This can be seen schematically in the line drawing in Figure 8.4-2. The constellation consists of eight discrete real-axis values spread out continuously along the imaginary-axis. The imaginary-axis values of the symbols may be essentially any random value along the constellation line.

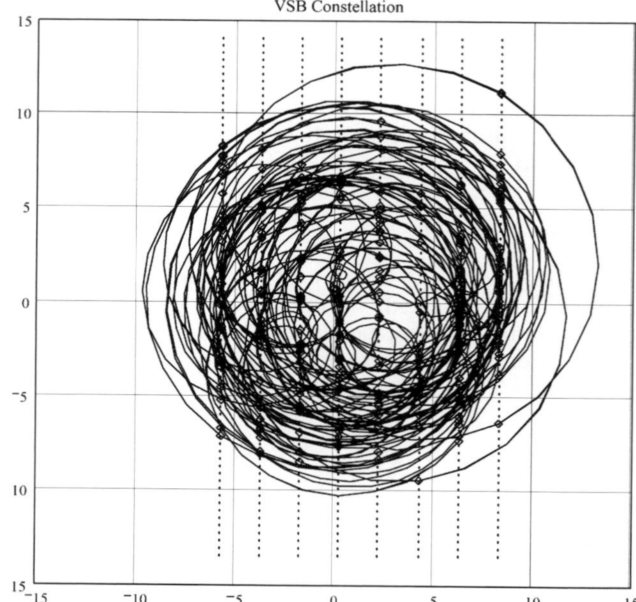

FIGURE 8.4-1 Continuous vector display of instantaneous amplitude and phase of an 8-VSB signal.

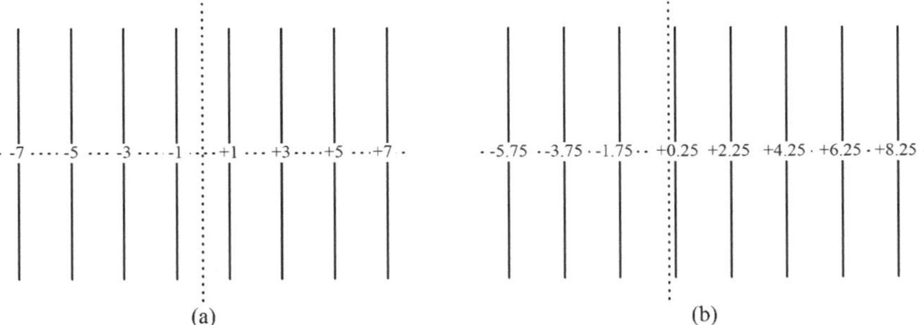

FIGURE 8.4-2 (a) Basic 8-VSB constellation diagram, and (b) 8-VSB constellation as transmitted with a +1.25 CU offset to create the pilot signal.

In principle, the constellation is symmetric as shown in Figure 8.4-2(a). The eight displayed states are at –7, –5, –3, –1, +1, +3, +5, and +7 constellation units (CU). However, the 8-VSB system features a small amount of in-phase carrier, or pilot signal, to aid receiver locking. To obtain this pilot, the baseband constellation is offset in amplitude by +1.25 CU resulting in the transmitted eight constellation states of –5.75, –3.75, –1.75, +0.25, +2.25, +4.25, +6.25, and +8.25. The pilot carrier contributes only 0.3 dB to the total average power. These are the values shown in Figure 8.4-2(b) and in Figure 8.4-1. *Note:* When measuring the 8-VSB signal quality, the pilot offset is removed before the measurements are made.

Figure 8.4-1 deserves further discussion. Note that (1) the RF signal amplitude spends a good deal of its time well beyond the maximum 8-VSB constellation symbol value (i.e., 8.25), and (2) that the maximum carrier amplitude always occurs when the carrier is roughly in quadrature with the real axis in the period *between* symbols (i.e., when both the in-phase and quadrature components are at their maximum values). These two facts are important later when considering causes of signal degradation.

Another way to look at the 8-VSB signal is to plot the amplitude of the demodulated real-axis signal versus time as in Figure 8.4-3. Since the imaginary axis carries no data information, it's not plotted. Because the waveforms are designed so that their value at the symbol clock edge represents one of the eight symbol values transmitted in the 8-VSB system, we see seven "eyes" at the instant of the symbols where the waveform has only eight distinct values. In between the symbol instants, the waveform looks like noise as it

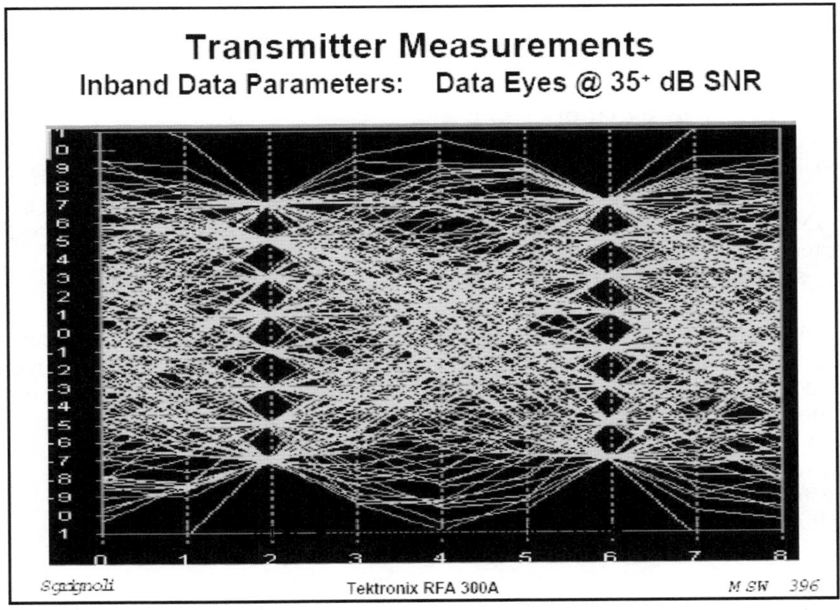

FIGURE 8.4-3 Display of the real-axis amplitude of the RF signal versus time. The data "eyes" occur at the symbol instants where the signal may be measured to determine what value is being sent.

takes on many different values depending on the values of many of the past and future symbols.

Effect of Signal Degradations

The eight symbol values are easily decoded unless something happens to the signal to cause the waveforms at the symbol instant to stray away from the desired value and near another symbol's value. This is said to "close the data eyes," and, if the data eyes close enough, errors will be made when the signal is sampled (or "sliced") at the symbol instant. If only a few errors are made per unit time, the forward error-correction schemes incorporated into the 8-VSB signal's encoding will correct them. However, if too many errors are made per unit time, the correction processes will fail, the data become garbled, and the picture and sound are "lost." The difference between an error-free, perfect picture-sound combination and an all-error, frozen picture with no sound can occur with an amplitude change of less than 1 dB. This is the so-called digital "cliff" effect.

To the digital signal, any unwanted disturbance that causes the data eyes to close appears to be noise. The receiver, using sophisticated digital signal-processing techniques in its linear equalizer, will remove linear distortion such as multipath and frequency response distortion of the transmitted signal. It does this *before* the signal reaches the data slicer (where the decision of which of the eight levels was transmitted is made), but at the cost of some increase in noise (i.e., a lower SNR). However, at some point near the edge of the transmitter's coverage area, or at a low signal level location, the 8-VSB signal will become weak enough so that the SNR drops below the 15 dB transmission system threshold; the signal can no longer be decoded and the picture and sound are lost.

Unlike analog TV, the signal distortions within the DTV transmitter can cause coverage problems. That is, if the transmitter distorts the 8-VSB signal prior to its transmission, the receiver sees the distortion as just another source of noise. Thus, as the signal leaves the transmitter, even though it is at very high amplitude, it exhibits a limited SNR based on how much the transmitter has distorted the 8-VSB signal. At the receiver, the transmitter's "noise" is added to the environment's noise. The various noise sources tend to act independently of each other and add, just as uncorrelated noise power adds in a chain of video or analog processing steps. If the transmitter adds too much noise to the signal, as the signal's amplitude falls with distance with respect to the environment's noise, the decoder's signal-to-noise threshold will be reached sooner than was expected. Thus, unlike analog TV, misadjustment can limit a DTV transmitter's coverage area.

While this effect is known in principle, exactly predicting how the transmitter's SNR will affect the coverage area is virtually impossible. For instance, the transmitter's typical linear distortion (more below) will be easily corrected by the receiver's linear equalizer with a fairly small noise penalty while the transmitter's nonlinear distortion will not. Further, all of the reflections, signal path losses, etc. found in the propagation environment are virtually impossible to predict yet may cause receiver problems at any particular site. Therefore, a transmitter's output signal, if registering at least 27 dB SNR or higher, and it is meeting the FCC emission mask, will degrade the 15 dB system noise threshold in the DTV receivers by less than 0.1 dB, and is therefore why 27 dB SNR is recommended by the ATSC.

Proper maintenance of the transmitter's SNR is, therefore, very important. Viewers deep in the fringe coverage areas may have little idea that their DTV receiver is operating near its SNR threshold. For those viewers, a slight decrease (just a few dB) of the transmitter's SNR could cause the picture and sound to freeze or be lost, and from the viewers' viewpoint, for no apparent reason.

Since "noise" is a major concern in digital transmitters, it is desirable to have a simple, all-inclusive measurement of the transmitter's *signal quality*; one that is a simple scalar metric that can be readily interpreted as being "good" or "bad." In fact, there are three such quality measures in use in the DTV industry today: signal-to-noise ratio, error vector magnitude, and modulation error ratio. While at first glance they appear different, they are all based on a very similar concept.

Signal-to-Noise Ratio

The first measure, the one used in ATSC documentation, is signal-to-noise ratio (SNR). Mathematically, it is defined as:

$$\mathrm{SNR} = 10\log\left(\frac{\sum_n \mathrm{I}_n^2}{\sum_n \delta\mathrm{I}_n^2}\right)\mathrm{dB} \tag{1}$$

where:

I_n is the amplitude of the nth *ideal* real-axis symbol value of a very long record of ideal real-axis values (with the pilot offset removed); and

$\delta\mathrm{I}_n$ is the amount of real-axis *error* exhibited by the nth symbol value.

Since all the values are proportional to the signal's *voltage*, their *power* is then proportional to the value squared. The Σ sign indicates a summation of the values of I_n^2 and δI_n^2 over the entire record length. Thus, SNR is the ratio, expressed in dB, of the summation of the ideal in-phase signal power divided by the summation of the actual in-phase noise power associated with that signal.

SNR is useful in that it measures the 8-VSB signal in the same way that the DTV receiver will decode the signal; along the real (in-phase) axis. Its weakness is that it does not directly measure the effects of signal compression or other nonlinearities that occur on the imaginary (quadrature) axis. Therefore, while SNR is the final arbiter of what the receiver sees, it may miss quadrature axis effects when diagnosing transmitter impairments.

Error Vector Magnitude

Prior to the development of 8-VSB, there were two other signal versus noise measures in current use for other digital systems that use both the real (I) and imaginary (Q) axis to transmit information. These measures were subsequently applied to 8-VSB signals. The first of these is error vector magnitude (EVM).

EVM is similar to SNR except it includes quadrature axis errors, scales the result differently, and presents the results in percent. Note that while a *larger* SNR value is better, a *smaller* EVM value is better. EVM is defined as:

$$\mathrm{EVM} = \sqrt{\frac{\frac{1}{N}\sum_{n=1}^{N}\left(\delta I_n^2 + \delta Q_n^2\right)}{S_{\max}^2}} * 100\% \quad (2)$$

where:

δI_n is the amount of error that the nth real-axis symbol value exhibited;

δQ_n is the amount of error that the nth imaginary-axis symbol value exhibited; and

$S_{\max}$ is the maximum value or state along the real axis (i.e., 7, since the pilot signal offset is removed before the EVM calculation is made).

It should be noted that $\sqrt{\delta I_n^2 + \delta Q_n^2}$ is the magnitude of a vector drawn from the ideal signal value to the actual signal value; thus the name error vector magnitude. EVM is therefore the RMS value of the error vector expressed as a percentage of the constellation's largest value (i.e., its outermost state).

EVM is similar to SNR but, because the errors are normalized to a different value, a 1% EVM does *not* correspond to a 40 dB SNR (see below).

Modulation Error Ratio

Simply put, modulation error ratio (MER) is a complex version of SNR. It is computed using the same equation as SNR, but includes the quadrature axis signal and the quadrature noise power. It is defined as:

$$\mathrm{MER} = 10\log\left(\frac{\sum_{n}\left(I_n^2 + Q_n^2\right)}{\sum_{n}\left(\delta I_n^2 + \delta Q_n^2\right)}\right)\mathrm{dB} \quad (3)$$

where:

I_n is the nth ideal real-axis symbol value of a very long record of ideal real-axis values (with pilot offset removed);

Q_n is the nth ideal imaginary-axis symbol value of a very long record of ideal imaginary-axis values;

δI_n is the amount of real-axis error exhibited by the nth value; and

δQ_n is the amount of imaginary-axis error exhibited by the nth value.

As in the SNR calculation, all values are *voltages*. Their *power* is proportional to the voltage squared. The Σ sign indicates a summation so the signal's complex power is, in effect, being divided by the power of the signal's complex error (noise) with the resulting ratio expressed in dB.

SNR, EVM, and MER are often used in quantifying digital RF transmission errors. Because each scalar metric defines the error signal in a different way, it is not possible to write a rigorous definition of the relationship between them. However, *because MER is just a complex version of SNR, its value will normally be about the same as SNR* (i.e., assuming the amount of distortion in the quadrature-phase channel is the same as in the in-phase channel).

The relationship between EVM and SNR can be *approximated*. In the SNR definition, since each of the eight states has an equal probability of occurring, the value of the signal term in the equation can be calculated as:

$$\mathrm{Signal_{SNR}} = \sqrt{\frac{1}{N}\sum_{j=1}^{N} I_j^2} = \sqrt{\frac{1}{8}\left[2\left(7^2+5^2+3^2+1^2\right)\right]}$$

$$\mathrm{Signal_{SNR}} = \sqrt{21} \quad (4)$$

In the EVM definition, the signal term is defined as:

$$\mathrm{Signal_{EVM}} = \sqrt{S_{\max}^2} = \sqrt{49} = 7 \quad (5)$$

In the EVM noise term, assume that, summed over the large record length, the quadrature error power is roughly equal to the real-axis error power, or:

$$\sum_{j=1}^{N}\left(\delta I_j^2 + \delta Q_j^2\right) \approx 2\sum_{j=1}^{N}\left(\delta I_j^2\right) \quad (6)$$

Using the signal relationships above, EVM can be expressed in terms of SNR as:

$$\mathrm{EVM} \approx \left[10^{\frac{39.3-\mathrm{SNR(dB)}}{20}}\right]\% \quad (7)$$

and SNR can be expressed in terms of EVM as:

$$\mathrm{SNR} \approx 39.3 - 20\log\left[\mathrm{EVM}(\%)\right]\mathrm{dB} \quad (8)$$

These relationships are shown graphically in Figure 8.4-4. For example, a 0.9% EVM is roughly the same as 40 dB SNR (or MER).

DTV SIGNAL IMPAIRMENTS

There are eight known impairments that can cause SNR reduction in 8-VSB transmitters. Only six of these

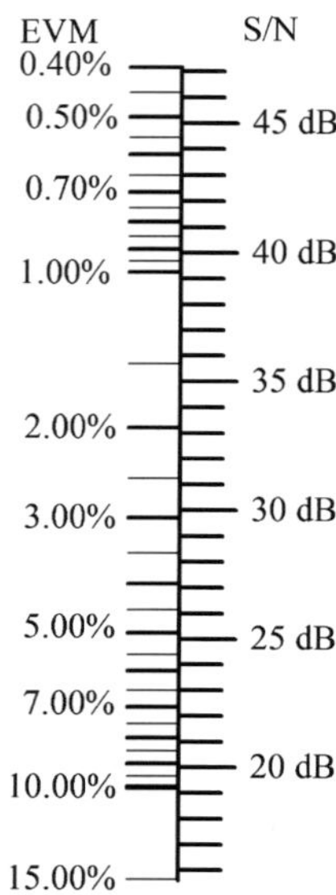

FIGURE 8.4-4 Approximate relationship between EVM and SNR.

are of concern to the transmitter operator. The impairments are grouped here into three broad categories: linear errors, nonlinear errors, and miscellaneous errors.

Linear Errors

Linear errors are those caused by abnormal filter response or transmitter mistuning and include:

- Frequency response error: A magnitude response that differs from the ideal root-raised cosine frequency response as specified in the ATSC standard [1].
- Group delay error: A group delay characteristic that differs from flat or zero across the TV channel.

Nonlinear Errors

Nonlinear errors typically are caused by amplifier nonlinearity and include:

- Amplitude error: Caused by amplifier gain variation as a function of signal amplitude. Sometimes called AM/AM conversion.
- Phase error: Caused by phase variation as a function of signal amplitude. Sometimes called AM/PM conversion and also known as *incidental carrier phase modulation*, or ICPM, in analog TV transmitters.

Miscellaneous Errors

The following signal impairment errors if excessive produce the equivalent of increased noise, thus lowering the SNR of the system:

- Phase noise: Random variation of the RF carrier's phase.
- Broadband noise: Introduction of excessive wideband or white noise into the 8-VSB-signal channel—a rare condition within a transmitter.
- DSP noise: This is noise created by the transmitter's DSP (digital signal processor) that creates the 8-VSB signal—a rare condition with today's advanced technology.
- Reflected signal noise: Noise introduced by inclusion of a delayed and attenuated version of the transmitter's RF signal. Typically caused by the 8-VSB signal being partially reflected from the antenna, traveling back down the feed line to the transmitter, where it is re-reflected and sent back toward the antenna with a time delay.

It has been determined that, to a first-order approximation, since the VSB signal consists of random noiselike symbols, each of the above sources acts independently of the others, and the results, therefore, can be added as *noise* power. For example, the noise power caused by the transmitter being unflat can be added directly to the noise power caused by an intermodulation source to determine the combined effect of the two sources.

To maintain predicted coverage, ATSC recommends that transmitters exhibit an SNR of no less than 27 dB [2]. That is, the transmitter should be designed and operated so that the noise from each of the above sources, when added together, creates a signal with at least a 27 dB SNR or better value.

Since the effects of each impairment results are measured in terms of SNR, EVM, or MER in order to facilitate troubleshooting poor transmitter performance, it is useful to have an idea of the sensitivity of the various parameters. The data given below was created using state-of-the-art laboratory equipment to generate an 8-VSB signal with a single known impairment. By measuring these single impairment signals, the relationship between the degree of impairment and resulting SNR was determined. The assumption that each impairment acts approximately independently was also roughly confirmed [3].

Frequency Response

Frequency response is a linear frequency domain distortion error. Yet, if incorrect, it causes digital noise because it creates intersymbol interference. Frequency response error is assumed here to be the difference between the ideal spectral response of the 8-VSB signal (i.e., a flat central spectrum with root-raised cosine transition regions) and that of the transmitted signal's actual spectral response. The two types of frequency response errors modeled are shown in Figure 8.4-5. The *linear model* is a ramp function across the TV channel, similar to what could be encountered if something were only slightly mistuned. The *square law model* approximates the crowned response one could encounter if, for instance, an IOT power amplifier has been mistuned (input and/or output tuned cavities).

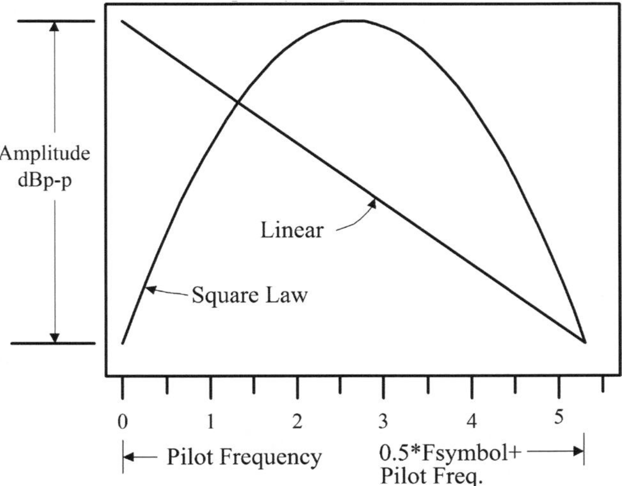

FIGURE 8.4-5 Frequency response flatness error models.

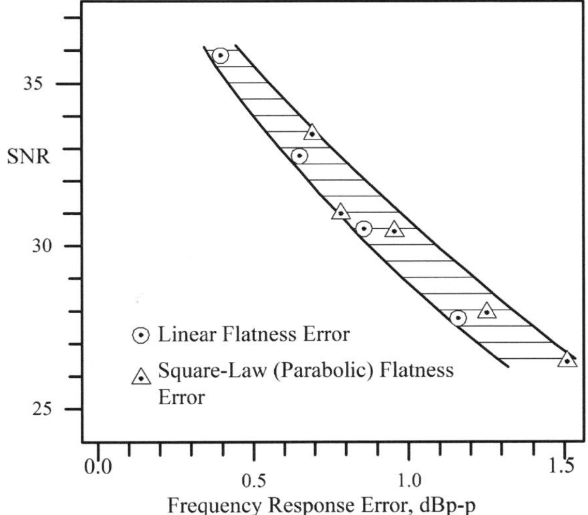

FIGURE 8.4-6 Measured SNR versus frequency response error.

Data is given for both models, varying only in the number of dBp-p of unflatness across the 6 MHz channel.

The measured SNR as a function of frequency response error for the two different types of response errors is shown in Figure 8.4-6. The hatched band is an approximation of the SNR resulting from flatness errors given that the flatness errors of real transmitters will differ from the model. Note that the *amount* of unflatness is more important than its exact shape.

Group or Envelope Delay Error

Group delay error is also a linear frequency domain distortion. Group delay can be thought of as the time delay (expressed in nanoseconds or ns) that each portion of the spectrum experiences as it goes through a circuit. Mathematically, group delay is:

$$\tau(f) = \left| \frac{1}{360} \frac{d\Phi^{\circ}(f)}{df} \right| - T_0 \tag{9}$$

where:

$\tau(f)$ is the delay at frequency f;

$\Phi(f)$ is the phase shift experienced by the signal in degrees as a function of frequency;

f is in Hz; and

T_0 is the group delay offset constant, typically taken to be the value of group delay exhibited at either the channel's center or at the pilot frequency.

An ideal group delay curve is a constant (i.e., flat) across the channel. To model group delay error, two functions were used. The first is a linear ramp function and the second is a square law function (see Figure 8.4-7). The ramp is an approximation of a mistuned broadband network and the parabolic function is approximately the shape that one will encounter when the signal is passed through a band pass filter.

The resulting SNR versus $ns_{p\text{-}p}$ of group delay error is shown in Figure 8.4-8. Unlike the frequency response case, the *shape* of the group delay curve makes a large difference. One of the authors' (Gumm) previous experience in extensive simulation of 8-VSB signal characteristics indicates that as the group delay curve becomes more complicated, with high-order curvature or ripple, more p-p delay variation can be tolerated for a given SNR. It is difficult to establish a relationship between group delay and SNR if the shape of the curve is unknown.

The effects of linear distortions can be seen in the schematic of the constellation diagram in Figure 8.4-9.

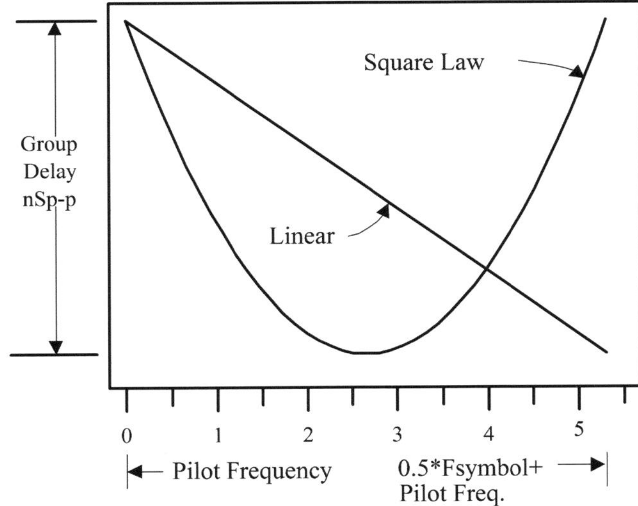

FIGURE 8.4-7 Group delay models.

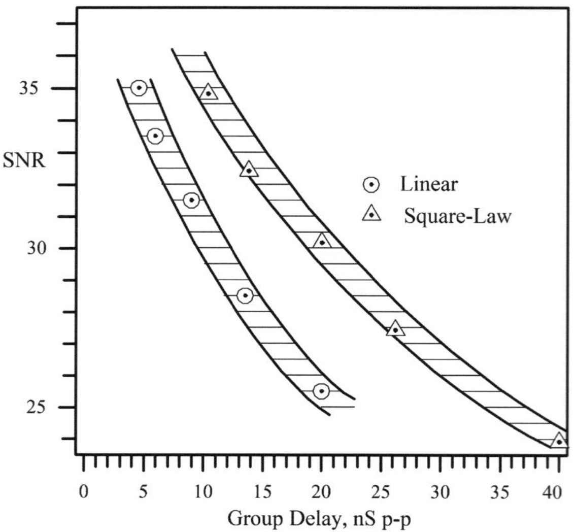

FIGURE 8.4-8 Measured SNR as a function of group delay error.

As the SNR declines, the eight vertical lines become "fat" or "fuzzy" but remain relatively straight. The amount of error (i.e., line "thickness") is essentially constant along the length of each line and gets wider as the SNR decreases.

Frequency response and envelope delay errors cause lower SNR but do *not* cause out-of-channel emissions.

Nonlinear Distortions or Intermodulation

Intermodulation is caused when a device is nonlinear. There are two types of nonlinear distortions in transmitter power amplifiers. The first is when the amplifier's gain is not constant as the signal's amplitude changes. Typically, the gain decreases somewhat as the amplifier's output power increases (an extreme example would be "clipping"). The second is when the signal's phase shift changes as a function of signal amplitude. In analog TV transmitters, this is a well-known phenomenon known as incidental carrier phase modulation.

Both forms of nonlinear response create new signal components. The well-known example is the two-tone, third-order intermodulation test. Here, two RF "CW" tones (F1 and F2) are passed through a nonlinear amplifier (or device) with the result that the output contains the original two tones plus two additional tones (2 F1-F2 and 2 F2-F1). The broadband 8-VSB signal acts, to a first-order approximation, like thousands of independent tones. Nonlinearities in the transmitter cause multiple new signal or intermodulation components to be generated, some of which are at frequencies *inside* the transmitter's assigned channel and some that fall *outside* the channel. The new signals that fall inside the transmitter's channel are a source of noise power that reduces the transmitted signal quality or SNR. The ones that fall outside the channel are the cause of the out-of-channel emissions that are considered below.

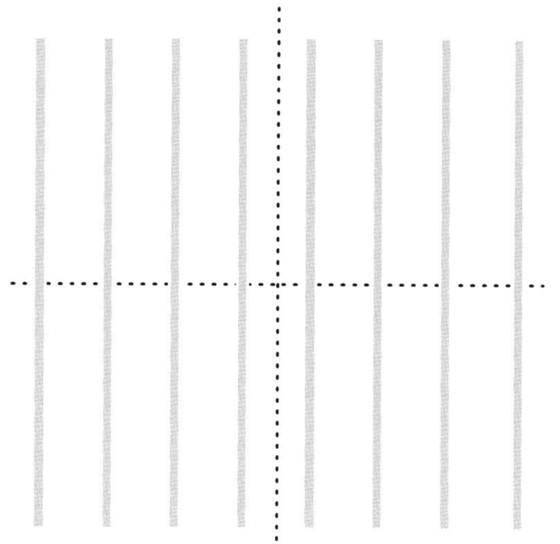

FIGURE 8.4-9 Linear errors scatter the points in the 8-VSB constellation making the lines "fat" or "fuzzy."

Intermodulation: Amplitude Error

Amplitude error can be visualized as gain change as a function of instantaneous variations in the signal's magnitude, as shown in Figure 8.4-10. It is sometimes called AM-to-AM conversion. The models used assume that the 8-VSB source is essentially linear but experiences small negative gain changes with increasing signal magnitude with a second- or third-order curvature.

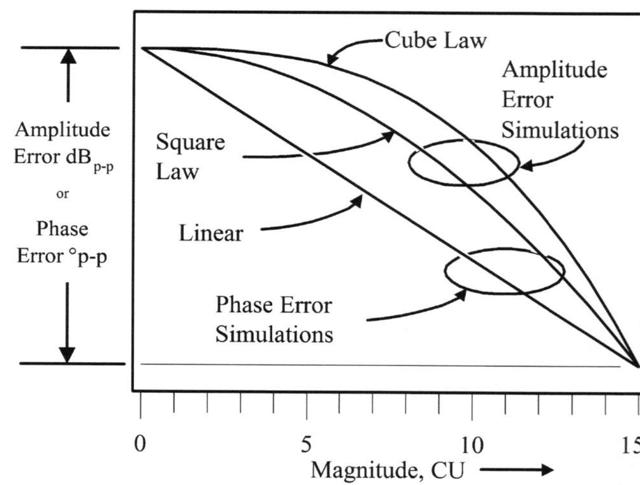

FIGURE 8.4-10 Amplitude and phase error models.

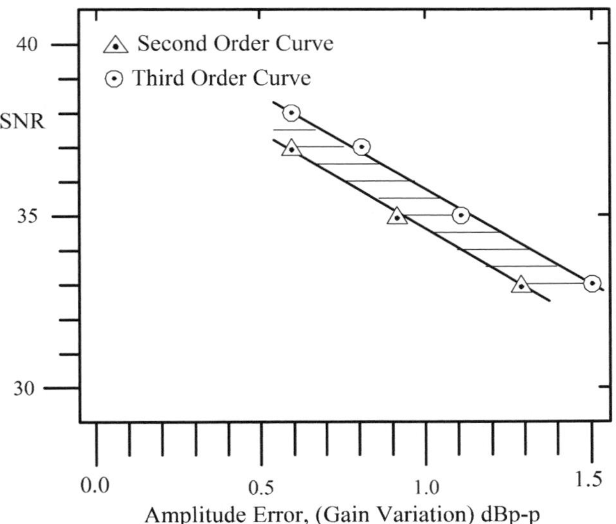

FIGURE 8.4-11 SNR impairment caused by amplitude error.

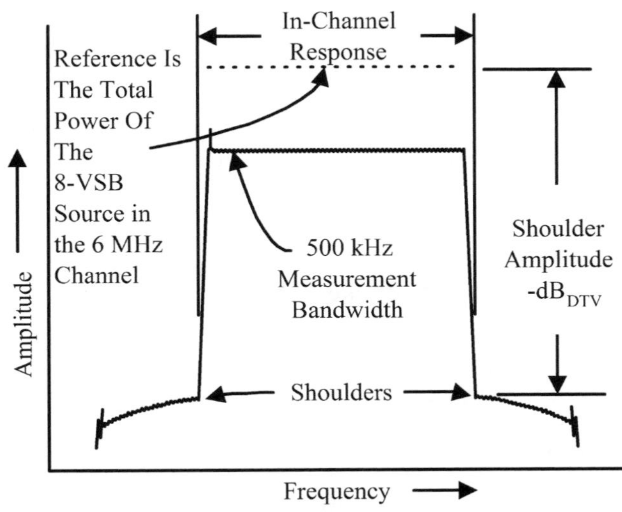

FIGURE 8.4-13 Definition of shoulder amplitude.

The measured results are shown in Figure 8.4-11. To a good approximation, the SNR impairment caused by amplitude error is dependent only on the p-p gain variation and not the type of curvature.

As noted, some of the signals caused by the nonlinearity fall outside the assigned channel. This is sometimes called spectral regrowth or splatter, and is regulated by the FCC. Figure 8.4-12 shows the relationship between the amplitude of the intermodulation components at the channel's edge (i.e., the shoulder amplitude) versus the SNR impairment caused by amplitude nonlinearity. The definition of shoulder amplitude is shown in Figure 8.4-13.

The FCC requires that the shoulder be at least –46 dB$_{DTV}$ or –47 dB$_{DTV}$ (depending on the transmitter type such as full service or low power, respectively). (The term dB$_{DTV}$ designates average splatter power in a 500 kHz band relative to the total average in-band power in the 6 MHz channel.) Any transmitter meeting this emissions requirement will not have low SNR caused by non-linearities.

The typical effects of amplitude error can also be seen in the constellation diagram in Figure 8.4-14. The result is that the corners of the high-amplitude lines will be pulled toward the center of the constellation by the amplifier's compression (the amount of compression shown is exaggerated to demonstrate the effect). Note that the compression (curved lines) of the right

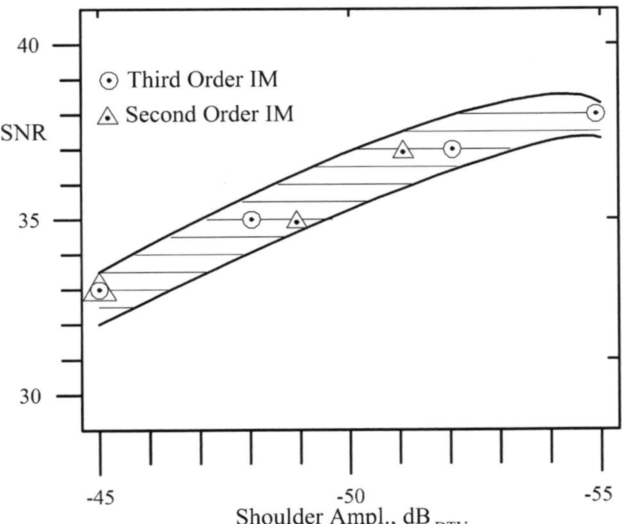

FIGURE 8.4-12 Shoulder amplitude caused by amplitude error.

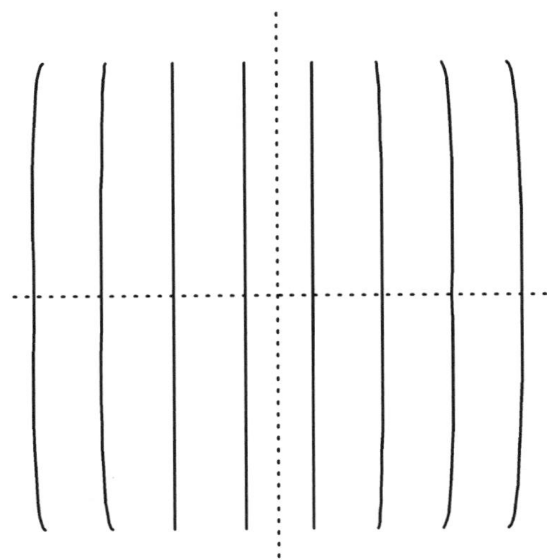

FIGURE 8.4-14 Constellation resulting from amplifier compression.

or positive states is greater than that of the left or negative states. This is due to the in-phase pilot offset, which shifts the constellation in a positive direction by 1.25 CU. This offset is typically removed from the constellation pattern before display. However, as transmitted, the offset causes the signal amplitude of the positive states (and thus the compression evidenced) to be noticeably greater than the negative states.

Intermodulation: Phase Error

A second type of nonlinear behavior is called phase error. This is phase variation imparted on a signal passing through a system as a function of the signal's magnitude. This is also called AM-to-PM conversion (or, in the case of analog TV transmitters, ICPM). The models used for this simulation are also shown in Figure 8.4-10. They are similar to the amplitude error model used above.

The resulting relationship between SNR and phase error is shown in Figure 8.4-15. Again, the two different curvatures cause approximately the same SNR degradation. Also, because this is a nonlinear process, the shoulder amplitude versus SNR caused by the phase error is important. This is shown in Figure 8.4-16.

The effects of phase nonlinearities can also be seen in the constellation diagram. In this case, as shown in Figure 8.4-17, the phase error will cause each point to twist about the center of the constellation by an amount proportional to the signal's amplitude. This is evident by the eight lines of the constellation tilting or twisting. In principle, the baseline or center of the pattern is also twisted. But its twist is at least partially removed by the measurement instrument's software as it finds the optimum demodulation angle and often cannot be seen.

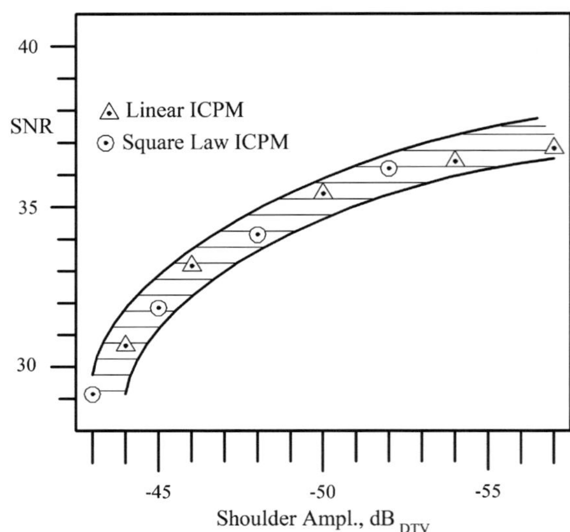

FIGURE 8.4-16 Shoulder amplitude versus SNR caused by phase error.

Phase Noise

Simulation and measurements by one of the authors (Gumm) have determined that the effect of RF carrier phase noise on SNR is primarily determined not by its amplitude at any given offset from the carrier but by the *integral* of the entire phase noise curve. This integral represents the carrier's RMS phase jitter. The integral is computed over the range of offset frequencies starting at the cutoff frequency of the receiver's pilot carrier tracking loop and ending at some high frequency. There is very little information available about what tracking loop bandwidths are used in 8-VSB receivers. Virtually the only data point available is the

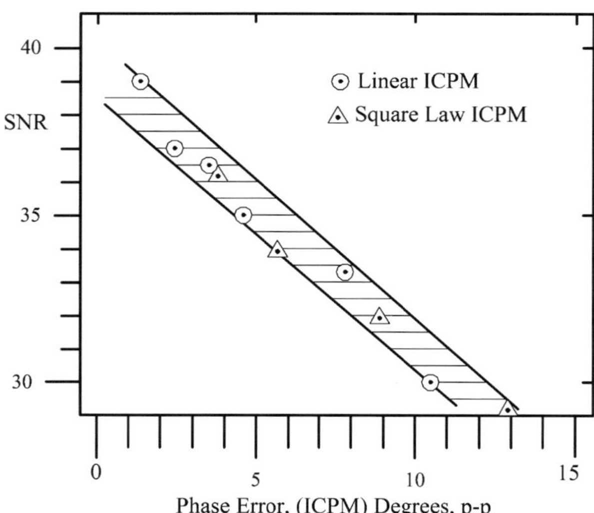

FIGURE 8.4-15 SNR versus phase error.

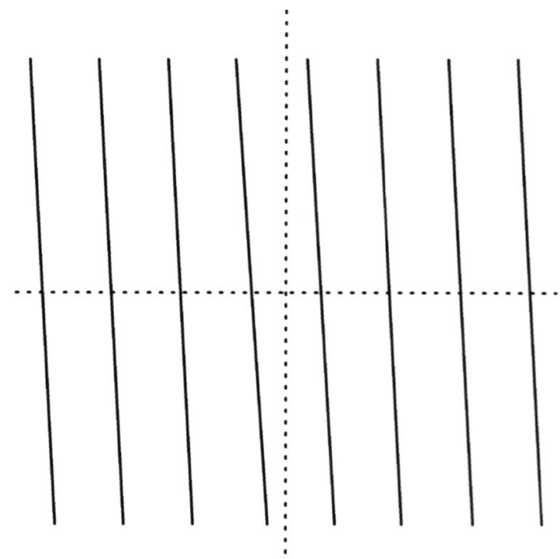

FIGURE 8.4-17 The effect of phase error on the constellation diagram.

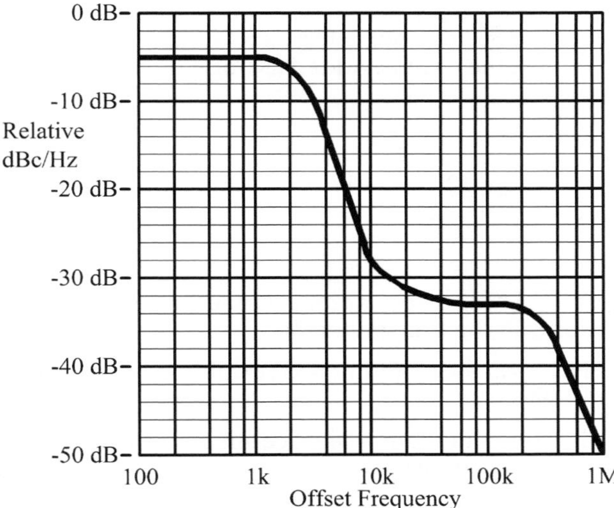

FIGURE 8.4-18 Phase noise spectral model.

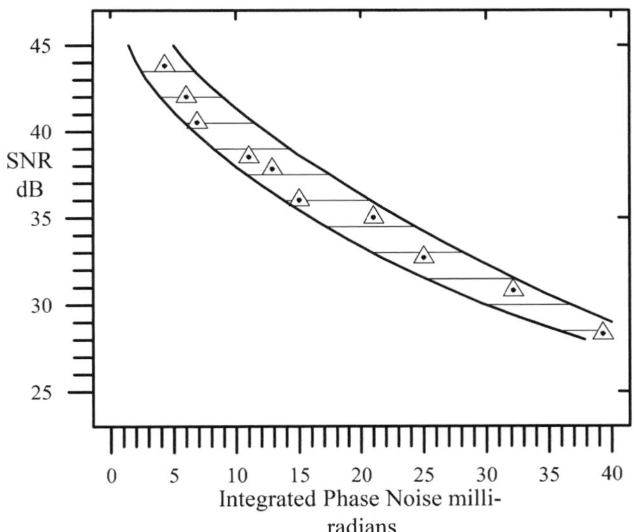

FIGURE 8.4-19 SNR as a function of integrated phase jitter.

2 kHz value designed by one of the authors (Sgrignoli) into the Grand Alliance prototype, and described in the ATSC documents [4]. The spectral shape of a UHF oscillator's phase noise or jitter is such that the phase noise at low-frequency offsets dominates. Therefore, the upper frequency bound of the integral is relatively immaterial and was arbitrarily set at 300 kHz in early test equipment by one of the authors (Gumm).

The procedure of developing 8-VSB signal waveforms with a known impairment was *not* used to determine the relationship between phase noise and SNR. Instead, a shaped audio noise spectrum was used to phase modulate an otherwise ideal 8-VSB source's up-conversion oscillator. The spectrum used had the spectral shape shown in Figure 8.4-18, which is similar to many RF sources in the VHF/UHF frequency range. The overall amount of phase noise caused by this spectrum was varied to produce the information shown in Figure 8.4-19.

Phase noise affects the constellation pattern by causing the entire constellation to randomly twist a few degrees. This error does not cause much error along the real axis. However, at points with a large quadrature component, the random phase will cause the tips of the constellation lines to scatter as shown in Figure 8.4-20. Fortunately, in a VSB signal, while the 8-VSB in-phase data levels are all equally probable, the quadrature-channel values have a Gaussian distribution centered about the real axis (with zero-value mean). This means that the quadrature channel spends more time proportionally near the real axis where the phase noise has less effect.

Broadband Noise

Broadband noise is almost never an issue in properly designed transmitters where the signal amplitudes are large. It is typically an issue in receiver designs.

DSP Noise

DSP is the digital signal processor in a digital transmitter. It accurately creates the 8-VSB signal and literally makes DTV possible. DSP noise may be created in a number of ways. One cause is that the DSP unit fails to include enough information about the *previous* and *subsequent* 8-VSB symbols.

Another way DSP may cause noise is if its arithmetic fails to perform the calculations to a sufficient precision. Once designed, neither of these sources of noise can be changed. Further, in a well-designed

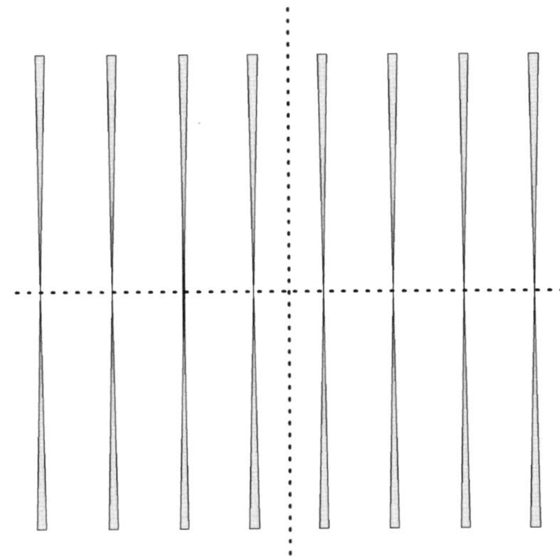

FIGURE 8.4-20 Effects of phase noise on the constellation pattern.

transmitter, neither will be detectable considering the other sources of noise.

Reflected Signal Noise

No hardware simulation is available for this noise source. The process that creates reflected "noise" is shown in Figure 8.4-21. The 8-VSB signal leaves the transmitter and experiences a loss of α dB in the transmission feedline on its way to the antenna. Upon reaching the antenna, a small portion of it, ρ_{ant} dB (i.e., the value of the antenna's return loss) is reflected back down the transmission line where it again experiences α dB of loss. Arriving at the transmitter, a portion of it, ρ_{tx} dB, similarly reflects off of the transmitter due to an imperfect back termination and again travels toward the antenna. Since the 8-VSB signal is essentially random from one symbol to the next, the reflected signal will be seen as noiselike interference as long as the round trip delay is greater than about one symbol duration (i.e., about 93 nsec or greater).

The reflected signal is thought to be an SNR source equal to the combined losses it has experienced in one round trip. That is:

$$SNR_{Reflected} = 2\alpha\,dB + \rho_{ant}\,dB + \rho_{tx}\,dB \qquad (10)$$

For broadcast systems, α is often about 1.5 dB and ρ_{ant}, for an antenna with a VSWR of 1.10:1 is 26 dB. ρ_{tx} is uncertain, probably a value between 3 dB and 10 dB. Therefore, the SNR from a reflected signal alone, in a typical system is perhaps 32 dB to 39 dB; a large enough value that it normally does not contribute significantly to the total transmission system SNR. Many transmitters detect and correct this noise source as part of their closed loop correction system.

Combining Noise Sources

As was mentioned previously, simulation and experimental results indicate that, at least to a good approximation, the presence of the various sources of SNR impairments does not affect each other; acting independently, they can be added as noise power. That is,

$$SNR_{Total} = -10\log\left[10^{-\frac{SNR_a}{10}} + 10^{-\frac{SNR_b}{10}} + 10^{-\frac{SNR_c}{10}} \ldots\right] dB \qquad (11)$$

where:

SNR_a, SNR_b, ... etc. are the various individual SNR causes in the transmitter; and

SNR_{Total} is the resulting SNR exhibited by the transmitter.

There are only four major requirements for an operational 8-VSB transmitter:

- Its output has a nominal SNR ≥ 27 dB.
- Out-of-channel emissions stay within the FCC mask (see below).
- The effective radiated power (ERP) stays within +10/–20% of its FCC-authorized value.
- The DTV pilot carrier is within 1 kHz of its allocated frequency (unless it is an upper adjacent channel to an analog signal, which then requires a precision offset that is accurate to within 3 Hz).

The critical issue about the mask that needs to be considered here is that the out-of-channel emissions must be ≤ –47 dB_{DTV} at the channel edge where the emissions are typically unattenuated by the channel band pass filter. The amplitude of emissions at the channel edge is thus essentially controlled by the transmitter's AM–AM and AM–PM performance. This fact provides a starting place to apportion the SNR impairments within a transmitter.

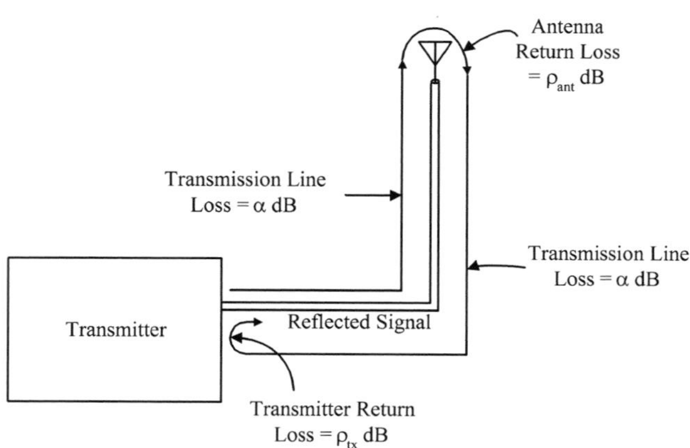

FIGURE 8.4-21 Reflected noise is created by the 8-VSB signal being reflected off of the antenna then reflected off of the transmitter.

TABLE 8.4-1
Typical Transmitter SNR Budget Values

SNR Impairment Cause	SNR Impairment	Specification
Amplitude Unflatness	32 dB	0.7 dB p-p
Group Delay	32 dB	10 nsec p-p
Phase Noise	40 dB	10 milli-radians RMS
Amplitude Error	36 dB	0.7 dB p-p, 50 dB shoulder
Phase Error	35 dB	6.5° p-p, 50 dB shoulder
Reflected Noise	37 dB	Combined antenna and transmitter return loss ≥ 37 dB, round trip delay ≥ 90 nsec
Total SNR	26.7 dB	
Total Shoulder Amplitude		–47 dB

A transmitter's SNR budget can be arranged in an infinite number of ways, as long as the total SNR adds up to being at least 27 dB. An example budget leading to an overall SNR of 27 dB is shown in Table 8.4-1.

The exact details of a given transmitter's noise budget are unimportant. *What is important is to understand that the transmitter's SNR is the result of several different factors, all of which contribute to the final value.* The goal for a transmitter's operator is to learn the normal operating range of the various SNR-determining parameters when the transmitter is operating well. With that knowledge in hand, when the SNR falls below 27 dB, it will be much easier to locate and correct the problem.

PRACTICAL SIGNAL QUALITY MEASUREMENTS

Signal quality measurements must be performed by test equipment designed for the 8-VSB signal (see previous list of equipment required). To perform the measurement, the VSB analyzer accurately demodulates the 8-VSB signal to at least the symbol level, where the I and Q signal amplitudes for each symbol are carefully measured. At this point, the symbol value actually sent may be determined allowing the ideal values of I and Q to be calculated. The actual and ideal values are then compared at the symbol times in the process of determining the value of the SNR, EVM, or MER.

TROUBLESHOOTING

When the SNR or MER falls below 27 dB or the EVM exceeds 4%, the cause must be determined and remedied. Many VSB analyzers allow their internal equalizer to be turned on or off. It is best to routinely make measurements with the equalizer OFF, measuring the transmitter's output as sent. If the SNR becomes too low, the first step in the diagnosis process is to turn its (linear) equalizer ON. If turning the equalizer on, which only removes the transmitter's linear distortions, improves the signal's quality to essentially its initial level, then the most likely problem is that the transmitter's frequency response is in error. The amplitude response may have become unflat or perhaps the envelope delay has changed, or both. At this point, the power amplifier tuning or perhaps the channel emission mask filter's frequency response should be checked until the problem is found. Note that this error can also be caused by excessive antenna reflections as described below.

If the equalizer does not improve the SNR, then the problem is one of the nonlinear or miscellaneous problems listed above. An easy test is to measure the shoulder amplitude of the transmitter's out-of-band emissions. Beyond that, some VSB transmitters also have a built-in signal quality function to diagnose transmitter AM-to-AM and AM-to-PM intermodulation problems. Also, the peak-to-average value read from the *complementary cumulative distribution function* (CCDF) can be checked for excessive amplifier compression. If the 8-VSB signal's shoulder amplitude has increased, it may be that the transmitter's output power has been inadvertently increased. This would cause the intermodulation problems to increase, lowering signal quality and increasing the emissions at the shoulders of the 8-VSB spectrum. If the power output is correct, but the shoulder amplitude is high, some amplifier in the system may be beginning to fail and automatic gain controls are attempting to correct the output power level but, in turn, are causing a nonlinear condition to be magnified.

The next layer in the problem is to consider if any one of the miscellaneous causes is the problem, such as a change in the antenna system or increasing the energy reflected from the antenna. An antenna change that is enough to cause signal quality problems should be clearly evident by an increase of the reflected power (VSWR) by a factor of two or three. This can be anything from ice on the radiating elements to burned transmission line components. An antenna problem normally will also manifest itself as unflatness and excessive envelope delay variations.

Another factor may be a failure or partial failure in the transmitter's frequency control system that causes an increase in its phase noise. If the VSB analyzer has the ability to perform an in-service phase noise test, this parameter may be immediately determined. If not, out-of-service tests of the local oscillators in the transmitter using the spectrum analyzer may be in order. The probability of this problem occurring is much smaller than the others, so it should be investigated last after other possible causes of an SNR decline have been ruled out.

Many transmitters use closed loop feedback techniques to dynamically correct both linear and nonlinear transmitter errors. This is typically achieved by predistorting the signal in the transmitter's exciter before applying it to the power amplifiers. If the errors become too extreme, the exciter may run out of adjustment range or other undesirable effects may occur. It is best if the closed-loop system is turned off from time to time and the uncorrected transmitter's SNR noted. Likewise, the various operating parameters can be measured and compared against previously recorded values. Any change should be tracked to its source and corrected.

TRANSMITTER POWER OUTPUT

Unlike analog TV, which specifies the transmitter's peak sync-tip (envelope) output power, DTV transmitters are specified in terms of their total *average* output power (in 6 MHz). This actually makes power measurements of DTV transmitters somewhat easier since many power-measuring instruments measure average power directly.

The transmitter's total average output power should be measured at a point in the system after all filtering is performed and the transmitter's signal is ready for delivery to the antenna feedline (i.e., at the output of the coupler in Figure 8.4-24). Transmitter power output

(TPO) may be measured by several methods. The typical methods use a calorimeter, or a calibrated transmission line coupler and a power meter, or utilize a specially designed in-line power meter.

Calorimeter Measurements

The calorimeter will be familiar to many broadcasters who have used it as their primary determination of output power. Since the calorimeter is essentially an instrumented dummy load, the transmitter must be taken off air to measure output power and operated into the calorimeter dummy load. By knowing the rate of the cooling fluid's flow, the temperature rise of that fluid as it passes over the load, and the fluid's specific heat, the total average power can readily be calculated.

To achieve sufficient accuracy (≈ 5%), great care must be taken. Since this procedure is well known, further detail will be omitted here except to emphasize that unlike measuring analog TV output power, where the sync tip power (i.e., peak envelope power, or PEP) must be calculated from the total average output power, in DTV the measured value is the desired value.

Calibrated Coupler and Power Meter Measurements

Another common method of determining a transmitter's TPO is to insert a carefully calibrated directional coupler in the transmission line at the transmitter's output, and then read the average power using a broadband RF power meter connected directly to the coupler's output. The power meter reading, combined with the knowledge of the coupling factor, allows the transmitter's TPO to be easily determined.

For best accuracy, the transmitter should be operating into a dummy load since the broadband power meter measures the power of *all* the signals at its terminals. A dummy load excludes any signals that might arrive by ingress via the antenna and transmission line. However, accurate off-air measurements may be made if a spectrum analyzer is used to determine that any other signals present are at least 25 dB below the transmitter's signal (this limits the error to 0.3% per signal present at that amplitude).

Before measurements are made, the power meter should be carefully calibrated using the power meter's internal calibrator (typically a 1 mW or 0 dBm, 50 MHz signal source) and its auto-zero feature activated to remove any effects due to temperature drift. A thermocouple-type of power meter sensor is recommended. This type of sensor provides sufficient accuracy and linearity over its dynamic range and is not affected by the noise-like 8-VSB signal's high-peak-to-average ratio.

In-Line Power Meters

Several vendors provide an inline power meter that accurately measures total average output power of the DTV signal. In reality, this type of meter actually consists of a calibrated directional coupler combined with a built-in power meter so, except for convenience, it is very similar in concept to the calibrated coupler system above. It is important that the unit is designed for the correct transmission line, the correct frequency, and the correct power range of the 8-VSB transmitter.

TRANSMITTER FREQUENCY MEASUREMENTS

Most transmitters do not provide facilities to directly measure the carrier frequency using a standard frequency counter. When this is the case, its operating frequency must be then determined from the 8-VSB signal itself.

Fortunately, many spectrum analyzers have the ability to accurately count the frequency of a signal positioned at its marker. Inside the analyzer, this is performed by phase-locking all of its local oscillators, and then measuring the frequency of the signal within its IF passband. The input frequency is then calculated from these known factors. The signal is thus measured to the accuracy of the spectrum analyzer's frequency reference.

To make frequency measurements with the spectrum analyzer, it is essential that a spectrum analyzer with a high-accuracy internal frequency reference be employed or that an accurate external frequency reference for the spectrum analyzer is used. Both approaches have advantages. If an accurate reference is built-in, the instrument can be easily moved from place to place without the complication of multiple boxes and cables. However, some means must be provided to periodically check and perhaps calibrate the analyzer's frequency reference. This requires having access to a high-accuracy frequency reference or the instrument must be periodically sent to a calibration service.

If an external reference is used, then the use of a GPS disciplined oscillator or a cesium or rubidium source is recommended. With these sources, periodic returns to a calibration laboratory for frequency calibration can be avoided.

Using the spectrum analyzer's built-in counter supported with an accurate internal or external frequency reference makes transmitter frequency measurements straightforward. Use a relatively narrow span (10 kHz) and a relatively narrow-resolution bandwidth (e.g., 100 Hz) and center the 8-VSB signal's pilot, as shown in Figure 8.4-22. Place the marker on the pilot signal, and engage the analyzer's internal frequency counter mode. For normal measurements, be certain to adjust the counter to obtain at least 100 Hz resolution.

If the transmitter is *not* frequency-locked to a nearby lower adjacent channel analog TV transmitter (to minimize interference to the analog signal), the pilot's frequency should be 309.441 kHz above the nominal frequency of the channel's lower edge to an accuracy of ±1 kHz. The analyzer's possible frequency

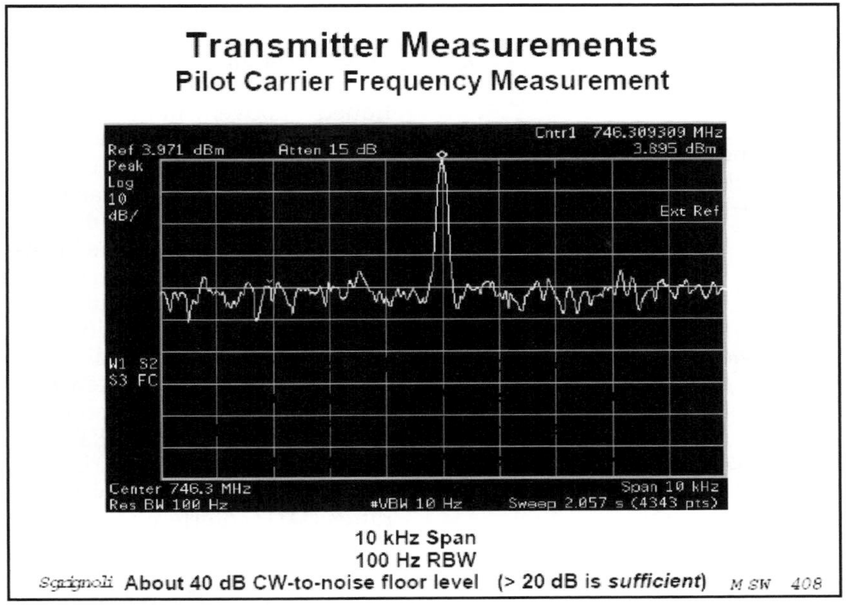

FIGURE 8.4-22 Measuring pilot frequency with a spectrum analyzer's counter.

measurement error must be taken into account to ensure that the transmitter's frequency is within the appropriate range. For example, consider the case of a channel 53 transmitter measured with a spectrum analyzer with a frequency reference that is specified to be within 0.2 ppm, and the analyzer's counter measures to only 100 Hz accuracy.

The nominal pilot frequency of a channel 53 transmitter is 704 MHz + 309.441 kHz = 704.309441 MHz ± 1000 Hz. The effect of the analyzer's reference error is to insert the same relative error into the frequencies it measures. An error of 0.2 ppm will result in an uncertainty of ±141 Hz in the measurement at 709.309 MHz. Further, the readout will be the nearest 100 Hz so the 141 Hz value is rounded up to ±200 Hz for this measurement. To be sure that the transmitter's frequency is correct and within the FCC requirement, the measured value must be within ±(1000 Hz − 200 Hz) = ±800 Hz of the transmitter's specified frequency; that is 704.309441 ± 800 Hz.

If the transmitter is frequency-locked to the picture carrier of an adjacent channel analog station, its pilot frequency must be 5.082138 MHz ±3 Hz above the analog picture carrier's frequency. For this measurement, a 1 Hz counter resolution (or better) is required. The carrier frequency of each transmitter is measured to at least a 1 Hz resolution. Then the difference frequency between the two transmitters is calculated.

Again, the frequency tolerance of the spectrum analyzer's reference must be factored in. Since the 5.082 MHz frequency *difference* is the important number, if a 0.2 ppm reference is again used, when measuring this difference, the counter could have as much as 5.082 MHz × 0.2 ppm = 1 Hz of error. Since the counter resolution is also 1 Hz in this example, no rounding is used. Therefore, the frequency difference between the two transmitters must be 5.082138 MHz ± 2 Hz.

EMISSION MEASUREMENTS

The theory as well as the measurement of DTV out-of-channel emissions is often complex. This section begins with a brief review of the theory behind the measurement, and then gives a procedure to measure the emissions near the transmitter's frequency. A measurement standard, "IEEE Standard for 8-VSB Terrestrial Transmission Mask Compliance for the USA," is, as of this writing, still under development.

Background

The 8-VSB signal is, by design, noise-like and nearly 6 MHz wide. When measured, its amplitude varies directly with the instrument measurement (i.e., resolution) bandwidth. The exception is the 8-VSB's small, in-phase pilot signal. Because the pilot is a coherent or CW-like signal, its amplitude is constant with changing resolution bandwidth.

Units of Measurement

All of the FCC's emission mask requirements are given in terms of the *ratio* of power measured in a 500 kHz band divided by the total 8-VSB signal power in 6 MHz, as expressed in dB; this quantity is defined as amplitude in dB_{DTV}. The FCC allows the procedure described here, where the power in a 500 kHz bandwidth is determined by using the spectrum analyzer's band power markers to actually measure the power in a series of 500 kHz sub-band frequency ranges. The total average power in the DTV signal is likewise determined by using the analyzer's band power markers to measure the power within the channel's 6 MHz frequency range. The two readings are used to determine the value expressed in dB_{DTV}. In practice, the power in the 500 kHz sub-band and the 6 MHz channel is measured in dBm, and the division (i.e., the ratio) is accomplished by subtraction of their logarithmic values:

$$PdB_{DTV} = 10\log\left(\frac{\text{Power: 500 kHz Sub-band}}{\text{Power: Total in Channel}}\right) = P500kHz_{dBm} - PTotal_{dBm} \quad (12)$$

where:

Power: 500 kHz Sub-band, is the average power, in mW, measured within a 500 kHz range using band power markers;

Power: Total in Channel, is the average power, in mW, measured across the DTV signal's 6 MHz channel using band power markers;

$P500\ kHz_{dBm}$, is the average power, in dBm, measured within a 500 kHz sub-band using band power markers;

$PTotal_{dBm}$, is the average power, in dBm, measured across the DTV signal's 6 MHz channel using band power markers; and

PdB_{DTV}, is the unit of measurement used to define the FCC's masks.

8-VSB Signal Power Relationships

If the pilot signal is removed, the total average power of the 8-VSB signal will be reduced by 0.31 dB. Further, the *equivalent* noise bandwidth of the 8-VSB signal without the pilot is 5.38 MHz. Therefore, as displayed in the FCC's mask (i.e., in a 500 kHz bandwidth), the *head* or central flat portion of the 8-VSB signal will be:

$$P_{\text{Head}} = P_{\text{total}}(dB_{DTV}) - P_{\text{pilot}}(dB) = 10\log\left(\frac{500\ \text{KHz}}{5.38\ \text{MHz}}\right) - 0.31 = -10.63\ dB_{DTV} \quad (13)$$

If the 8-VSB signal is decreased in amplitude by 0.31 dB when its pilot is removed, the pilot's amplitude must then be

$$10\log\left(1 - 10^{\frac{-0.31}{10}}\right) = -11.6\ \text{dB} \quad (14)$$

with respect to the amplitude of the entire 8-VSB signal. That is, for an ideal 8-VSB signal, the pilot is 11.6 dB below the total average in-channel power. These are useful facts to remember when measuring the signal's emissions.

FCC Emission Requirements

Out-of-channel emissions, regardless of the source, may cause problems to other spectrum users. As mentioned above, nonlinear amplitude and phase response within the DTV transmitter cause its output spectrum to spread into the adjacent channels. Therefore, the FCC requires each DTV transmitter to meet one of three emissions masks, depending upon the details of the transmitter's license. Figure 8.4-23 shows the three masks: Full Service, Stringent, and Simple. All emissions must fit within (i.e., see below) the mask specified for that transmitter.

Performance to the Full Service mask is required for all DTV transmitters licensed under part 73 of the FCC's rules. A DTV transmitter licensed under part 74 of the FCC's rules (i.e., LPTV, Class A, and Translators) must meet either the Stringent or the Simple mask, depending on the desired to undesired interference ratios (D/U interference) assumed in its license application.

As noted above, all measurements are expressed in terms of dB_{DTV}. The reference amplitude, or 0 dB_{DTV}, is the amplitude of the total 8-VSB signal *including* the pilot. Because the DTV signal is measured in an *equivalent* bandwidth of 500 kHz, the head or flat portion of the 8-VSB signal is –10.6 dB_{DTV} as shown.

The Full Service and stringent masks have nearly the same shape. They both require that any emissions from the channel edge out to 500 kHz beyond the channel edge be less than –47 dB_{DTV}. Both these two masks then slope down at the rate of –11.5 dB per MHz. The Stringent and Full Service masks differ only in the amount of ultimate attenuation required. The stringent mask requires that emissions be –76 dB_{DTV} or less at all frequencies greater than 3 MHz from either channel edge. The Full Service mask continues at the same –11.5 dB per MHz slope until it reaches –110 dB_{DTV} for all frequencies 6 MHz beyond either channel edge.

The Simple mask starts at –46 dB_{DTV} at the channel edge and decreases in amplitude by ($\Delta F^2/1.44$) dB from the channel edge (where ΔF is the frequency in MHz from the channel edge) until an ultimate attenuation of –71 dB_{DTV} is reached at all frequencies greater than 6 MHz away from the channel edge.

The ultimate attenuation requirement shown in each mask extends to all other frequencies beyond those shown in Figure 8.4-23. That is, all emissions further than 6 MHz from the channel edge must be below the minimum value given at the mask's edge.

Further, transmitters using either the Stringent or the Simple masks, operated on channels 22–34, channels 36–38, or channels 65–69, are also required to have a minimum of 85 dB of attenuation between the power amplifier's output terminals and the antenna output in the frequency ranges of 1164–1188 MHz, 1215–1240 MHz, and 1559–1610 MHz. This filtering, often accomplished in part by the harmonic low pass filter, is intended to ensure sufficient protection for GPS signals from the transmitter's harmonics.

All measurements are made with respect to the allocated channel edges unless the DTV transmitter's

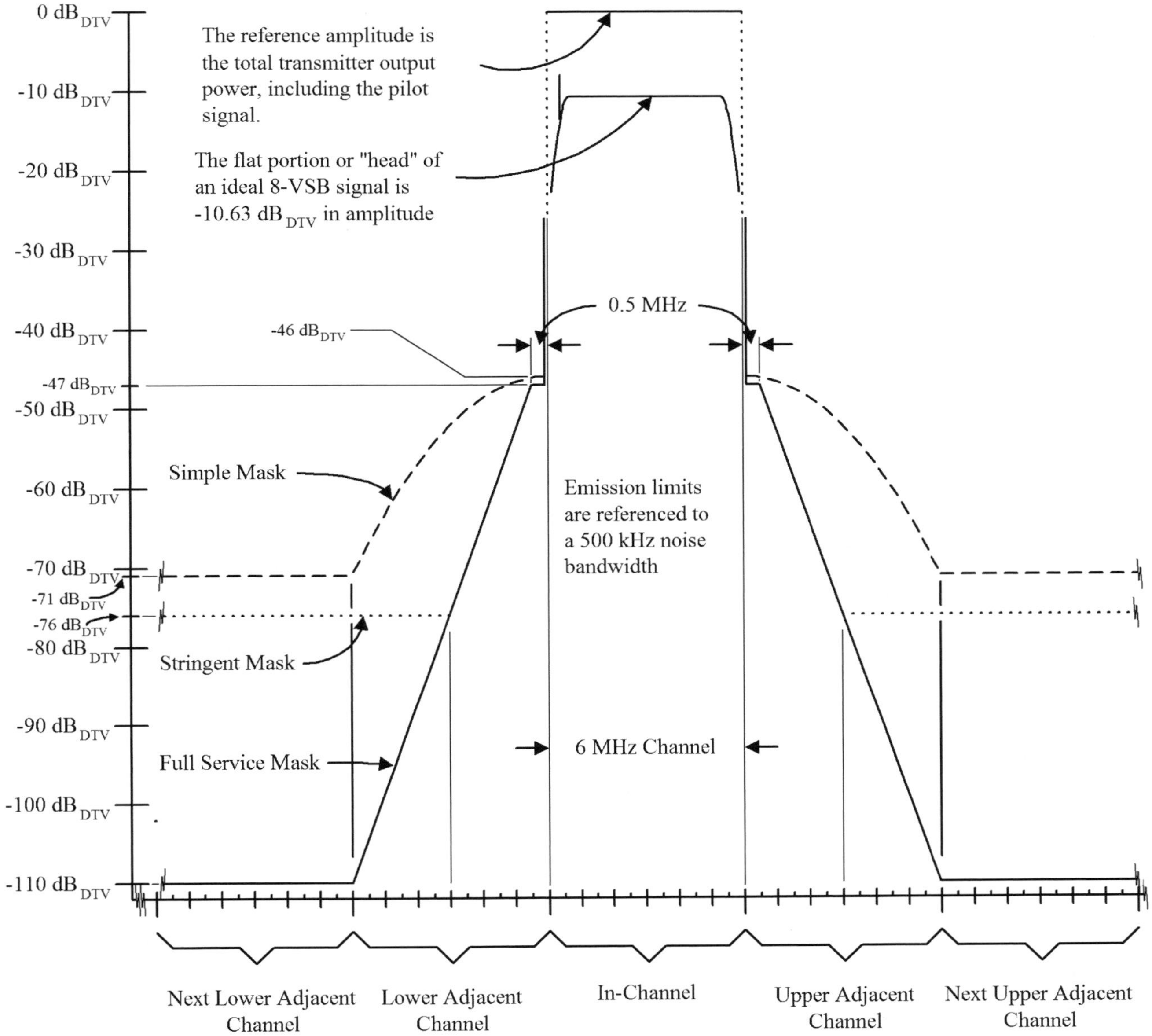

FIGURE 8.4-23 The Full Service, Stringent, and Simple FCC emissions masks.

frequency is referenced to the frequency of a lower adjacent channel analog TV transmitter's visual carrier. In that case, all measurement frequencies are offset from the normally used frequencies by the same amount that the pilot is removed from its nominal frequency.

Measuring the VSB Signal

The procedure given here is primarily designed for the Full Service mask. Testing to the Stringent and Simple masks is similar in concept but the dynamic range measurement requirements are less critical.

The FCC specifies that all measurements be made at the transmitter's output terminals beyond any filters that may be employed. If available, it is *recommended* that the transmitter be operated into a dummy load for testing emissions. Its normal antenna may be used as a load if necessary, but care must be taken to exclude from the emissions tests any signal ingress from nearby stations.

As shown in Figure 8.4-24, for Full Service transmitters, the signal is often obtained from a directional coupler inserted in the antenna line *after* the channel filter. For accurate measurements the directional coupler must have less than 0.5 dB peak-to-peak unflatness over an 18 MHz range (centered on the channel)

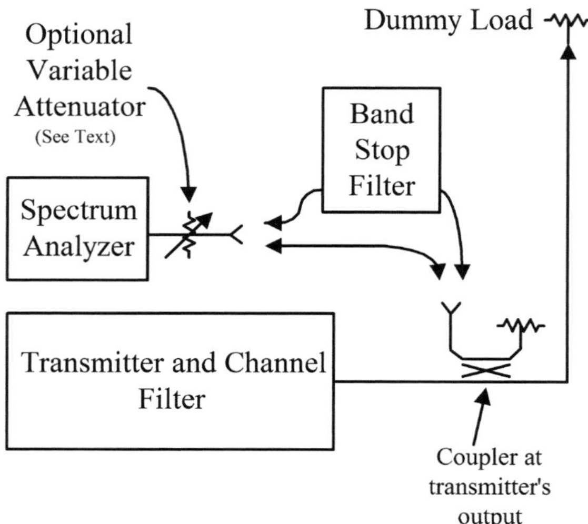

FIGURE 8.4-24 Test signal locations for full-service transmitters. Low-power transmitters may be able to use a power attenuator instead of the coupler.

and have a return loss of 18 dB or better. For testing full-service transmitters, the signal amplitude from the directional coupler must be at least 127 dB greater than the spectrum analyzer's sensitivity in a 10 kHz resolution bandwidth (greater amplitudes make the measurements easier). The typical spectrum analyzer's sensitivity is usually about –100 dBm with a 10 kHz bandwidth. Therefore, the total average amplitude of the 8-VSB signal should be about +27 dBm (less than the +30 dBm typical spectrum analyzer input power limit).

Measurements for the Stringent and Simple masks have a reduced dynamic range, so less test signal is required. The Stringent mask requires a signal only 93 dB greater than the analyzer's noise floor (or about –7 dBm). The Simple mask requires even less, only 88 dB greater than the analyzer's noise floor (or about –12 dBm). For *all* emissions measurements, the exact value of coupling is unimportant (i.e., a *calibrated* directional coupler is unnecessary).

When measuring low-power DTV transmitters, a power attenuator to simultaneously provide a dummy load for the transmitter and to provide a reasonable and safe signal amplitude for the test instrument may be used. Ensure that the attenuator's power rating is adequate and that the test signal amplitude obtained from it is appropriate for the test equipment.

Avoid damaging the test instrument due to signal overload when connecting it to the 8-VSB signal. Because the signal is noise-like and is nearly always viewed in a relatively narrow bandwidth, and because the noise-like data spectrum scales with the resolution bandwidth, there is a possibility that a signal that *appears* to be relatively low amplitude on the display can physically damage the instrument (e.g., input mixer or attenuator). Before connecting to the signal, adjust the instrument to its maximum reference level, and select a 10 MHz span and a center frequency that will position the 8-VSB channel's pilot carrier at mid-screen. As the connection is made, observe the amplitude of the 8-VSB signal's pilot signal. If the pilot is at least 12 dB below the instrument's maximum reference level, then the 8-VSB signal will probably not damage the instrument. *If the pilot is above that amplitude, the 8-VSB signal should be immediately disconnected.*

Emission Measurement Procedure

Some instruments, specifically designed for FCC-specified 8-VSB measurements, have built-in algorithms that will simplify emissions measurements. Otherwise measurements can be made using a general-purpose spectrum analyzer with band power markers.

The measurement method, specified by the FCC in its notice that allows 500 kHz sub-band measurements [5], is described here.

After an initial setup, the spectrum analyzer's band power markers are first used to measure the total average power of the 8-VSB signal within the 6 MHz channel. Then the band markers are used to measure the emissions in a series of twelve 500 kHz sub-bands on each side of the channel, as shown in Figure 8.4-25. The amplitude of the emissions in each sub-band, measured in dBm, is converted to dB_{DTV} by subtracting the amplitude of the total 8-VSB signal, also measured in dBm. The resulting dB_{DTV} values are then compared against the appropriate values obtained at the *midpoints* of each of the sub-bands from the mask in Figure 8.4-23. These values are given numerically in Table 8.4-2.

The dynamic range of spectrum analyzers is insufficient to directly measure the Full Service mask. Some spectrum analyzers can directly measure the Stringent mask and many, but not all, can measure the Simple mask.

For Full Service transmitters, the 8-VSB signal in-channel amplitude must be quite large to ensure the amplitude of the emissions that are well away from the carrier is large enough for the analyzer to measure accurately. Such a large in-channel signal will overload the analyzer's input mixer causing it to create intermodulation products that are much larger than the transmitter's out-of-channel emissions. This problem is solved by inserting a band stop filter, with steep skirts, that reduces the amplitude of the in-channel 8-VSB signal by at least 46 dB, while passing, with little attenuation, emissions greater than 2 MHz into the adjacent channels. For a typical analyzer, only 13 dB of rejection would be required to measure to a Stringent mask and 7 dB for a simple mask. This reduced in-channel signal power decreases the amplitude of the internally-created intermodulation products below the analyzer's noise floor and allows accurate measurements to be made.

Initial Instrument Adjustment

Using the precautions given above (e.g., setting the reference level to its maximum value before making a connection and watching for overload as the connection is

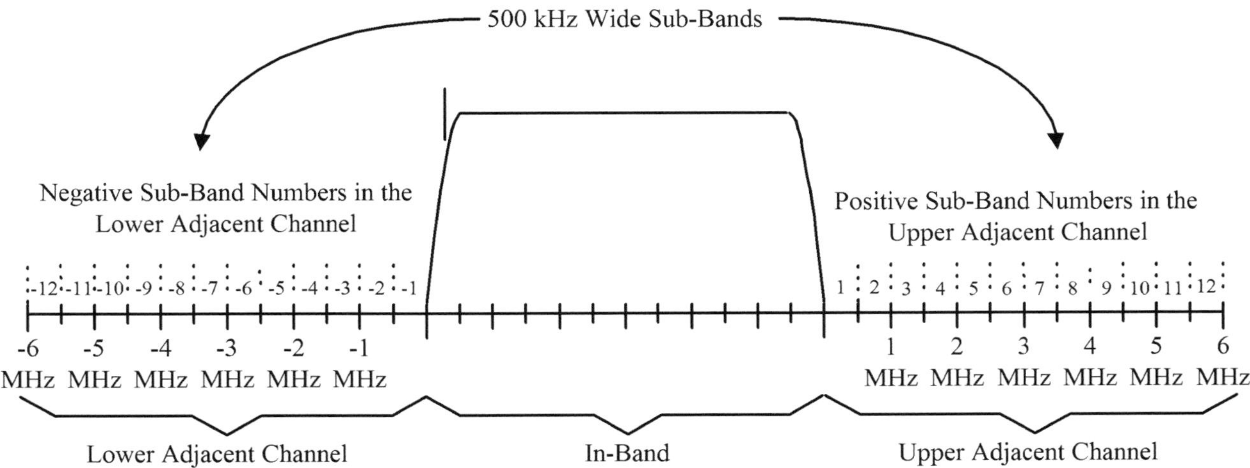

FIGURE 8.4-25 Twelve 500 kHz sub-bands near the channel frequency.

made), connect the instrument to the directional coupler at the transmitter's output. Select a 10 MHz span, a 10 kHz resolution bandwidth, and a center frequency that centers the 8-VSB signal in the display. Select a 10 dB/div. vertical scale factor. (Note: This in-channel power measurement assumes that the instrument has at least 1001 horizontal points in its display. If it has less than 1001 points but more than 334 points, use a 30 kHz resolution bandwidth.)

Preferably, select a video detection mode of *power* (some analyzers refer to this mode as RMS) or *sampling* mode. Otherwise select an *averaging* mode. The point is to use a mode that will give a true picture of the 8-VSB signal's power. Avoid peak-to-peak or rise-and-fall video detector modes of display. If available, engage an (ensemble) averaging mode, selecting 100 or more acquisitions to be averaged into the final reading. If such an averaging mode is not available, select a narrow video filter bandwidth such as 300 Hz or 100 Hz to obtain an averaged amplitude reading.

The optimum signal amplitude at the input mixer of a spectrum analyzer is often difficult to determine.

TABLE 8.4-2
Center Frequencies for the 12 500 kHz Wide Sub-Bands

Sub-Band	Lower Sub-Band Center Frequencies		Upper Sub-Band Center Frequencies	
	Sub-Band Center, with Respect to Lower Channel Edge	Sub-Band Center, with Respect to Pilot Frequency	Sub-Band Center, with Respect to Upper Channel Edge	Sub-Band Center, with Respect to Pilot Frequency
1	–0.250 MHz	–0.559 MHz	0.250 MHz	5.941 MHz
2	–0.750 MHz	–1.059 MHz	0.750 MHz	6.441 MHz
3	–1.250 MHz	–1.559 MHz	1.250 MHz	6.941 MHz
4	–1.750 MHz	–2.059 MHz	1.750 MHz	7.441 MHz
5	–2.250 MHz	–2.559 MHz	2.250 MHz	7.941 MHz
6	–2.750 MHz	–3.059 MHz	2.750 MHz	8.441 MHz
7	–3.250 MHz	–3.559 MHz	3.250 MHz	8.941 MHz
8	–3.750 MHz	–4.059 MHz	3.750 MHz	9.441 MHz
9	–4.250 MHz	–4.559 MHz	4.250 MHz	9.941 MHz
10	–4.750 MHz	–5.059 MHz	4.750 MHz	10.441 MHz
11	–5.250 MHz	–5.559 MHz	5.250 MHz	10.941 MHz
12	–5.750 MHz	–6.059 MHz	5.750 MHz	11.441 MHz

Note: All measurements are 500 kHz wide.

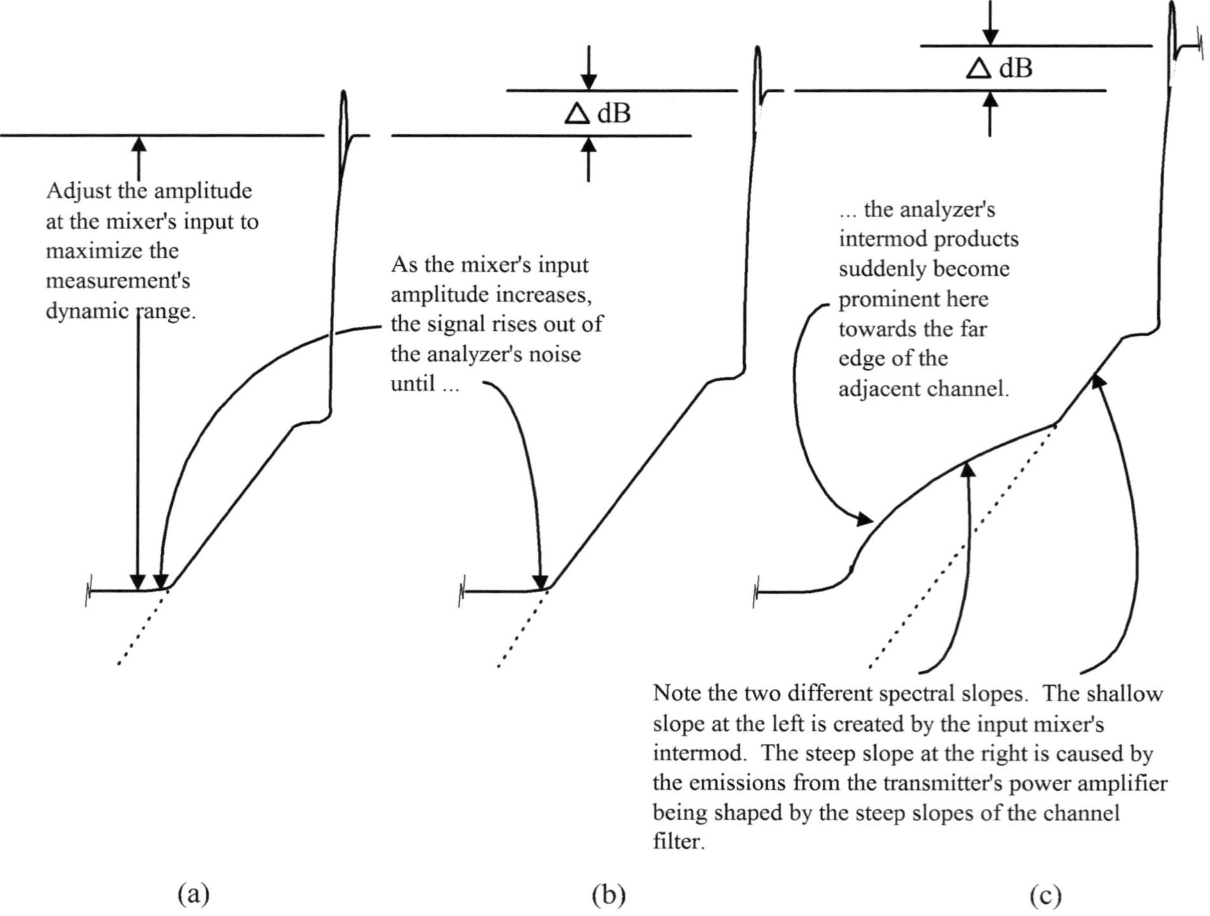

Adjust the amplitude at the mixer's input to maximize the measurement's dynamic range.
Δ dB
As the mixer's input amplitude increases, the signal rises out of the analyzer's noise until ...
Δ dB
... the analyzer's intermod products suddenly become prominent here towards the far edge of the adjacent channel.
Note the two different spectral slopes. The shallow slope at the left is created by the input mixer's intermod. The steep slope at the right is caused by the emissions from the transmitter's power amplifier being shaped by the steep slopes of the channel filter.
(a)
(b)
(c)

FIGURE 8.4-26 Adjusting the mixer input amplitude with the internal or external variable attenuator to achieve maximum measurement dynamic range. The mixer's input amplitude is too low at (a), optimum at (b), and too large at (c).

If too little, a major portion of the signal will be below the analyzer's noise floor. If too much, the instrument itself will produce distortion products that look very similar to those created by the transmitter (i.e., *before* the mask filter). Figure 8.4-26 shows part of the analyzer's display with three different amplitudes of the 8-VSB signal applied to the analyzer's input mixer. The amplitude variation is performed by changing the analyzer's input attenuator or by changing an external step attenuator. At (a), the mixer's input signal is too low, "submerging" parts of the 8-VSB signal below the *flat* analyzer noise floor. At (b) the analyzer's mixer input signal has been increased by Δ dB bringing *more* of the 8-VSB signal's emissions out of the noise. When the mixer's input amplitude is increased again Δ dB at (c), the intermodulation products created in the instrument increase at a three-times rate (that is, if the 8-VSB signal input is increased 2 dB in amplitude, the instrument's intermodulation products will increase by 6 dB). Note that these intermodulation products are created within the instrument and have a different, shallower slope, than the transmitter's emissions, which are shaped by the channel filter. The analyzer's products obscure the 8-VSB signal being measured at the far end of the adjacent channels.

Put another way, when the VSB signal at the mixer input is too small (a), the VSB splatter falls below the *flat* analyzer noise floor. When the VSB signal at the mixer input is too large (c), the *shallow* spectrum analyzer mixer splatter adds to the much steeper splatter caused by the shallow transmitter power amplifier splatter passing through the very steep mask filter. This is what causes the dual-slope sidebands.

A straightforward way to adjust the signal amplitude at the analyzer input mixer for an optimum level (or "sweet spot") is now described. When the analyzer has an internal 5 dB step attenuator, the 8-VSB in-band signal power remains constant on the screen with small changes in the analyzer's internal attenuator, and only the noise floor or intermod splatter will change. Therefore, by carefully observing the splatter around 2 MHz to 3 MHz into the adjacent channel, the internal attenuator can be adjusted for minimum splatter levels by visual inspection. The "sweet spot" will be the best compromise (lowest observable energy at these frequencies) between the noise floor and the analyzer intermod.

When an external attenuator is used ahead of the analyzer, the signal amplitude varies while the noise floor is constant as the attenuator is varied. If a small external attenuator (e.g., 10 dB) is combined with large internal attenuator steps, both the positions of the signal and the noise floor will vary on screen. In these cases, carefully observe the *shape* of the response as the mixer's amplitude is increased. The signal will first rise out of the analyzer's noise floor with the steep slope of the channel filter reaching down to the noise as shown in Figure 8.4-26(a) and (b). As the amplitude is increased further, there will be a slope change as shown in Figure 8.4-26(c) before the steep slope reaches the analyzer's noise floor. The "sweet spot" is found when the attenuation is increased to just move the changed slope region down to the instrument's noise floor.

Finding the "sweet spot" when measuring transmitters with the Simple mask is more difficult because the shape of the transmitter's emissions is the same as the analyzer's internally generated intermod products. In this case, the amplitude at the input mixer should be adjusted to maximize the amount the entire signal is above the noise while observing the signal head-to-shoulder amplitude difference. When the mixer signal becomes too large, the head-to-shoulder amplitude difference will begin to shrink. When that occurs, lower the mixer's input amplitude two or three dB to ensure that the analyzer's intermod products are well below the transmitter's emissions.

In either case, it is generally better to have the input amplitude a little low rather than a bit too high. This keeps the amplitude of any internal intermod products created in the analyzer well below the noise floor and prevents errors when the instrument corrects for the presence of its own noise floor.

Because the 110 dB dynamic range requirement (500 kHz splatter bandwidth to 6 MHz in-band bandwidth) is substantial, the required VSB signal level is required to be large (e.g., +25 to +30 dBm). Therefore, for optimum performance, there will typically be between 40 and 50 dB of attenuation ahead of the analyzer's input mixer, whether internal or external to the instrument.

Measure the Total Amplitude

Using the input attenuator (internal or external) and the instrument as just optimized, set the band power markers to measure the 6 MHz range of the transmitter's assigned channel, from the lower to the upper edge of the channel. If the transmitter is offset from the normal channel frequency due to a requirement of tracking the frequency of a lower adjacent channel analog transmitter, adjust the markers to measure the frequency range between 309.44 kHz below to 5.69056 MHz above the pilot. Measure and record the 8-VSB transmitter total average power in the 6 MHz channel (in dBm), being cautious to use the same detector mode, video filter, and averaging selected in the initial adjustments.

Sub-Band Measurements

The sub-bands are measured using the same adjustments and input amplitude except that the span may be narrowed to 1 or 2 MHz, if desired. (If a 30 kHz resolution bandwidth was used for step 1, it should now be readjusted for a 10 kHz bandwidth.) Set the band power markers to measure a 500 kHz frequency range (sub-band) beyond the channel edge. Adjust the center of that frequency range to each of the sub-bands, and use the band power markers to measure and record the total power (in dBm) within that band. If the minimum edge of the sub-band is within 8 dB of the instrument's noise floor, corrections should be made. Remove the signal from the analyzer and measure the power of the instrument's noise floor in the sub-band using the same band power settings. Then correct the amplitude of the transmitter's measured sub-band emissions by:

$$\text{Corrected Reading} = 10\log\left[10^{\frac{\text{Sub-Band Measured Emissions Power, dBm}}{10}} - 10^{\frac{\text{Sub-Band Measured Instrument Noise Power, dBm}}{10}}\right]\text{dBm} \tag{15}$$

Insert the Band Stop Filter

After the total in-channel power and the first four sub-bands have been measured, the band stop filter must be inserted in the circuit in order to measure the remaining sub-bands that are otherwise buried in the analyzer noise floor.

Before the band stop filter can be inserted at the input of the spectrum analyzer, its attenuation must be carefully measured and documented across an 18 MHz band by either a network analyzer or a spectrum analyzer. Within the center 6 MHz channel, the filter should have more than 46 dB stop band attenuation. In its passband, defined as frequencies beyond 2 MHz into each adjacent channel, the attenuation should be less than 3 dB. Specifically, the filter's passband insertion loss at the *center* of sub-bands ±4 to ±12 should be carefully documented, as these will be used to correct these 500 kHz sub-band measurements.

With the band stop filter in the signal path the total in-channel average power (in 6 MHz) should drop by *more* than 46 dB (perhaps as much as 50 dB). Of course, the analyzer's Reference Level can now be reduced by *approximately* the same amount, which has the same effect as lowering the analyzer noise floor by this same value. If the amplitude of signal available from the coupler is close to the value required to make the measurement, it will be necessary to adjust the Reference Level to remove essentially all the attenuation. Thus, the remaining transmitter emissions that were previously buried in the noise are now "uncovered," allowing 500 kHz sub-band measurements down to the 110 dB (in a 500 kHz sub-band) below the total average in-channel power (in 6 MHz).

Upon reduction of the input attenuator, observe the DTV splatter near the *edges* of each adjacent channel (4–5 MHz) to make sure that no excessive analyzer mixer splatter is present (e.g., presence of a dual slope). If significant splatter is observed, it will be necessary to find another "sweet spot" for mixer input level.

If an external attenuator is used, note the value of attenuation before and after this readjustment. The *difference* of these two attenuation values must be accounted for to align the measurements taken before and after the filter was inserted.

Measure the Remaining Sub-Bands

Measure the emissions in the remaining sub-bands, with the exception of the new Reference Level setting, using the same technique and instrument adjustments as before. After obtaining the sub-band amplitude, add the previously measured filter insertion loss as measured in the *center* of that sub-band. This procedure accounts for the fact that the stop band filter does not have zero passband insertion loss, and, therefore, causes artificially low sub-band power measurements. Thus, the filter's passband insertion loss must be *added* to the sub-band power measurement to raise the value back up to where it actually exists. If an external attenuator was used, the *difference* of attenuation (original external attenuation value minus the new external attenuation value) observed when the filter was inserted must be *subtracted* from the measurements.

TABLE 8.4-3
Maximum Amplitudes dDBTV for Each 500 kHz Sub-Band

Sub-Band	Full Service Mask	Stringent Mask	Simple Mask
1	–47.0	–47.9	–46.0
2	–49.9	–49.9	–46.4
3	–55.6	–55.6	–47.1
4	–61.4	–61.4	–48.1
5	–67.1	–67.1	–49.5
6	–71.9	–71.9	–51.3
7	–78.6	–76.0	–53.3
8	–84.4	–76.0	–55.8
9	–90.1	–76.0	–58.5
10	–95.9	–76.0	–61.7
11	–101.6	–76.0	–65.1
12	–107.4	–76.0	–69.0
>12	–110.0	–76.0	–71.0

Convert Sub-Band Measurements to dB_{DTV}

Convert the additional sub-band power measurements made with the presence of the stop band filter into units of dB_{DTV} as required by the FCC. As before, this is performed by subtracting the total signal power measured in the 6 MHz channel from each of the *corrected* amplitudes measured in each sub-band.

For example, if the total carrier power measured in step 2 is –2 dBm and the emissions in 500 kHz frequency sub-band 1 measured in step 3 is –51 dBm, the emissions in dB_{DTV} for sub-band 1 is then:

$$\text{Emissions}_{dB_{DTV}} = -51\,\text{dBm} - (-2\,\text{dBm}) = -49\,\text{dB}_{DTV}. \tag{16}$$

Determine if the Transmitter Is in Compliance

Using the corrected sub-band emissions in dB_{DTV}, consult Table 8.4-3. If the measured emissions in every sub-band is less than that listed in the table for the appropriate mask, the transmitter is in FCC compliance with the mask.

SUMMARY

Measuring DTV transmitters is a very broad topic but basic measurement techniques along with descriptions of operating parameters and transmission impairments have been provided in some depth. DTV measurement technology has many differences from traditional analog systems but parameters and procedures will become clearer as more experience is gained. The station engineer should continue to learn more about DTV transmitters, transmission systems in general, and measurement techniques in particular. It is important to realize that without adequate test and measurement equipment, some impairments at the transmit end that could cause failures to decode in the viewer's receivers, may not be apparent. That in turn may result in substantial service outages even though the transmitter itself is on the air. The authors believe that the information presented in this chapter will provide a firm foundation to enable the user to operate and maintain a DTV transmitter while the learning process continues.

References

[1] Advanced Television Systems Committee, ATSC A/53E, "Digital Television Standard," 2006, p. 48. Available at http://www.atsc.org/standards.

[2] Advanced Television Systems Committee, ATSC A/64A, "Transmission Measurement and Compliance for Digital Television," 2000. Available at http://www.atsc.org/standards.

[3] Gumm, L. Signal-to-Noise Relationships in 8-VSB, at http://www.tektronix.com/Measurement/App_Notes/Technical_Briefs/25W_13224_0.pdf..

[4] Advanced Television Systems Committee, ATSC A/54A, "Guide to the Use of the ATSC Digital Television Standard," 2003, p. 107. Available at http://www.atsc.org/standards.

[5] FCC Public Notice DA 05-1321, May 10, 2005.

SECTION

9

ANCILLARY BROADCAST SYSTEMS

C H A P T E R

9.1

STL Systems for AM–FM–TV

ERNEST M. HICKIN
Myton Associates, Inc.
Winchester, Massachusetts

JAMES H. ROONEY III
Microwave Radio Communications
Chelmsford, Massachusetts

Updated for the 10th Edition by

RICHARD MILLER
Microwave Radio Communications
Chelmsford, Massachusetts

GEORGE MAIER
Orion Broadcast Solutions
Sudbury, Massachusetts

INTRODUCTION

For many years, and for both radio and television broadcasters, the typical studio-to-transmitter link (STL) has been a privately owned microwave system, with leased telephone lines being a popular option for radio stations. The last decade of the twentieth century saw a marked increase in alternative choices, with fiber-optic and digital carrier circuits gaining steadily in popularity. The dawn of digital broadcasting has spawned the need for even greater diversity in connecting the studio to the transmitter. In some cases, radio and TV broadcasters have turned to unlicensed, spread spectrum microwave for STLs or data links, while more than a few radio stations have deployed Internet protocol (IP) over digital subscriber lines (DSLs) as an STL.

The conversion by TV broadcasters in the U.S. to the Advanced Television Systems Committee (ATSC) digital television (DTV) system created a need to support multiple video streams and increased ancillary data. In AM/FM radio broadcasting, the development of the NRSC-5-A Standard for in-band/on-channel (IBOC) digital radio, implemented using the iBiquity Digital Corporation's HD radio system, has created a similar situation in that additional STL capacity is needed to support both multicasting (in digital) as well as simulcast digital and analog programming.

Frequency modulation (FM) techniques have been the mainstay in microwave STL communications for decades, but are clearly on the decline as the digital transition increases the complexity of STLs and transmitter-to-studio links (TSLs). While microwave radio still accounts for the largest proportion of STL systems in use, this chapter will also examine the alternatives so as to provide information that will help an engineer decide which is best for a particular situation.

MICROWAVE PROPAGATION

With certain exceptions, microwave radio communication is a line-of-site method, and is well suited to broadcast STLs and intercity relay applications. However, microwave paths are subject to numerous atmospheric anomalies, most of which can adversely affect reliability. These effects may be random in nature, or they may be seasonal. They may be very short in duration, or they may last for hours, even days.

Free Space Loss

Free space path loss for any electromagnetic wave arises from the spreading of the wavefront as it propagates from its source. A simple way to visualize this effect is to observe a flashlight beam, which closely resembles a microwave signal in most respects. At its origin, the light beam is strong and narrowly concentrated. As the distance from the light source increases, the beam spreads out and grows weaker.

After passing beyond the *near field* radius of its transmitting antenna, a radio signal will loose 6 dB (a 4:1 power ratio) every time the distance from the transmitter doubles (at microwave frequencies where the wavelengths are short, the near field is usually a fraction of a mile). Since a 2:1 increase in range leads to a 2^2:1 reduction in power, this relationship is referred to as the *inverse square law.*

The formula for free space path attenuation (A) at microwave frequencies, which is independent of ground or atmospheric effects, is given by:

$$A\ (\text{dB}) = 96.6 + 20\log F(\text{GHz}) + 20\log D(\text{miles}),$$

or

$$A\ (\text{dB}) = 92.45 + 20\log F(\text{GHz}) + 20\log D(\text{km}).$$

This is the loss between isotropic antennas (theoretical antennas that radiate or receive equally in all directions).

The gain (G) of a microwave antenna is then expressed in dBi (gain relative to an isotropic antenna). For a parabolic reflector antenna, diameter d (ft), with an efficiency of 55% (which is typical of all but the smallest antennas), gain is given by:

$$\text{G(dBi)} = 20\log d(\text{ft}) + 20\log F(\text{MHz}) - 52.6.$$

Strictly speaking, an antenna being a passive device cannot possess power gain. The value G is the amount by which the radiation in a desired direction has been increased by redirecting energy, which would have been radiated in unwanted directions by an isotropic antenna. It should be noted that below 1 GHz it is usual to express gain relative to a dipole where 0 dBd = 2.2 dBi; thus, the gain of a VHF or UHF TV transmitting antenna is most likely to be quoted in dBd.

The ratio of received power to transmitted power between two correctly aligned antennas of gains G1 and G2, when D miles apart, will be given by:

$$\frac{\text{Pr}}{\text{Pt}} = G1 - 36.6 - 20\log F - 20\log D + G2 \text{ (units of dB)}.$$

This number will be negative; the numerical value (i.e., absolute value) is referred to as *path loss*.

As examples, if the antennas are 6 ft diameter and the path is 30 miles long, then at 2 GHz:

$$\frac{\text{Pr}}{\text{Pt}} = 28.5 - 36.6 - 66.0 - 29.5 + 28.5$$

$$= -75.1 \text{ or path loss} = 75.1 \text{ dB}.$$

At 7 GHz:

$$\frac{\text{Pr}}{Pt} = 39.8 - 36.6 - 76.9 - 29.5 + 39.8$$

$$= -63.4 \text{ or path loss} = 63.4 \text{ dB}.$$

Note that in these equations doubling the frequency increases the total antenna gain by 12 dB (2 × 6) while the path loss increases by only 6 dB. This means that lower-frequency systems require larger antennas or more power, or both, for a given received carrier level.

Atmospheric Bending and *k* Factor

The pressure and hence the density of the atmosphere surrounding the earth vary with height, getting less as the height increases and the weight of the air above decreases. As a result the dielectric constant also decreases with height and this has a prismatic effect causing microwaves (and light waves) to bend toward the earth. Under normal conditions, the bending is less than the curvature of the earth but nonetheless microwaves will go farther than simple geometry would suggest. A convenient way to allow for this when drawing profiles is to increase the radius of the earth until the microwaves appear to be traveling in straight lines.

The ratio of this effective earth's radius to the true earth's radius is called *k* and its value is approximately 4/3 or 1.33 for over 90% of the time in most parts of the world. However, there are times when *k* can be anything from infinity to as low as 0.45. When *k* is equal to infinity, the earth appears to be flat; it is a condition where mirages are seen and radar echoes are received from hundreds of miles away. Values of *k* between 1 and 0.45 can occur for a few percent of the hours in a year and it is necessary to allow for this if a reliable link is to be established. Figure 9.1-1 is a map of the continental United States showing contours of equal minimum *k* factor; this is based on refractive index measurements made by the Central Radio Propagation Laboratory.

Path Clearance Requirements

One of the major tasks required to engineer an STL system is the path analysis between the STL transmitter at the studio and the STL receiver location. To determine what constitutes a clear path, the concept of *Fresnel zones* for optical theory is applied to radio waves. Most of the electromagnetic energy at a receiving point is concentrated in an elliptical volume that is a function of the distance between the transmit and receive points and the wavelength. The energy outside this volume either cancels or reinforces the energy within the volume, depending on whether the distance that the energy travels to the receive point is longer by an even or odd number of one-quarter wavelengths. Even distances result in

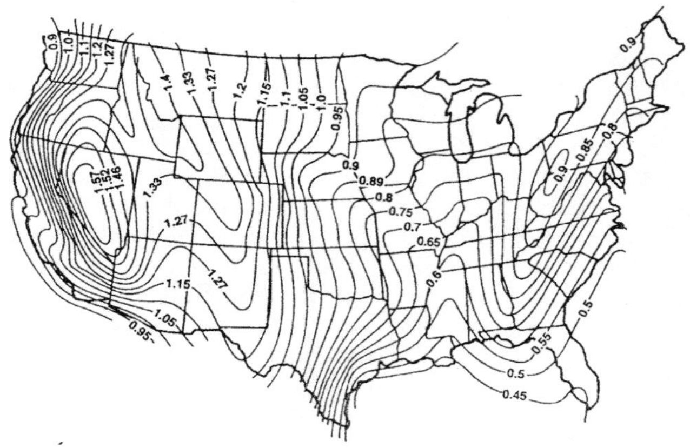

FIGURE 9.1-1 Map of the continental United States showing contours of equal estimated minimum *k* factor. [2]

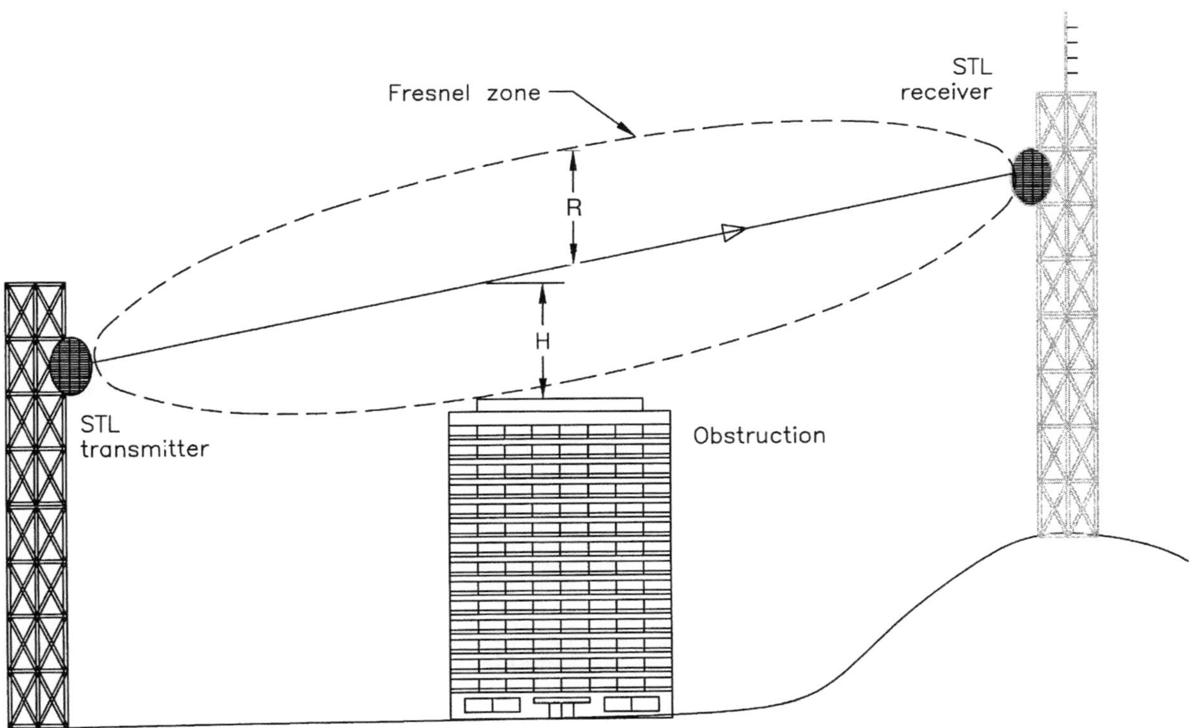

FIGURE 9.1-2 Fresnel zone clearance for an STL path. [2]

radio wave cancellations; odd distances result in radio wave reinforcement (see Figure 9.1-2).

The radius of the first Fresnel zone, which defines the boundary of the elliptical volume, is given by the following:

$$F_1 = 72.1\sqrt{\frac{d1 \times d2}{f \times D}},$$

where:

F_1 is the first Fresnel zone radius in feet.

$d1$ is the distance from the transmitting antenna to the obstruction in miles.

D is the total path length in miles.

$d2$ is $D - d1$ in miles.

f is the frequency in GHz.

H is the distance from the top of the obstruction to the radio path (as shown in Figure 9.1-2).

The significance of the first Fresnel zone is that 96% of the transmitted power is contained within it, thus requiring adequate first-zone clearance to be maintained over the entire path. For reliable operation, obstructions should not project into the area thus defined. Empirical studies, however, have shown that performance is substantially the same for H greater than 0.6 F_1.

Figure 9.1-3 shows the extent to which a microwave signal is attenuated when it passes close to an obstruction. Clearance is stated as a fraction or multiple of the first Fresnel zone, F_1. This figure illustrates that the attenuation with a Fresnel Zone Clearance (FZC) clearance of 0.6 is equal to the free space attenuation. However, as noted earlier, *k*-factor variations will mean that more than this clearance will have to be built in to allow for values less than 4/3, and a typical design parameter is to plan for 0.3 FZC for the lowest value of *k* expected on the path, as derived from Figure 9.1-1. While such a clearance will introduce 2–8 dB of loss, this is well within the fade margin of a well-designed link.

Fade Margin

The excess of signal over the minimum required for satisfactory service is called the *fade margin.* Systems are typically designed to have fade margins in the range 26–46 dB and (ideally) larger for higher frequencies, longer paths, and over water or similar difficult situations.

The choice of fade margin may involve compromises, for example, when the need to use existing towers or masts may limit the size of antenna. It will also be influenced by the environment, requiring a higher than normal margin in humid, flat country, whereas in dry, mountainous regions, a lower than normal margin may be used.

Often times, a microwave signal will reach its destination not just by the direct path, but by one or more indirect paths as well. Depending on its phase

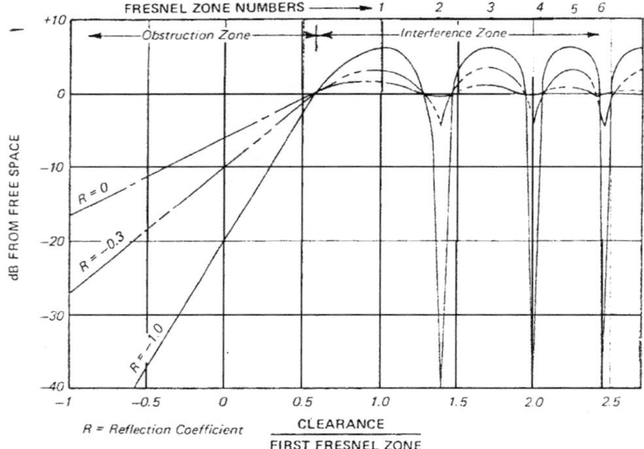

FIGURE 9.1-3 Attenuation versus path clearance for various types of obstruction. [2]

relationship to the direct signal, a signal reflection can add to or cancel the direct signal, causing increases of up to 6 dB or reductions of more than 50 dB. *Multipath fading* is defined as the fading caused by the desired signal arriving by multiple paths, caused by reflections from water, hills, buildings, or atmospheric discontinuities. Although multipath reflections are typically atmospheric, or ground based (including water), reflections may also be caused by internal system discontinuities. Radio frequency (RF) channel filters, band filters, circulators, power dividers, transmission line, connectors, and the antenna itself are all potential sources of reflections, and should be carefully evaluated when problems arise.

There are multiple factors that contribute to a system's fade margin, including:

- *Thermal fade margin (TFM):* A reduction in the signal level reaching the receiver; not frequency selective. Virtually all path performance and fade margin calculations are based on thermal margin, unless otherwise specified.
- *Dispersive fade margin (DFM):* A frequency-selective fade, usually caused by a multipath reflection that manifests itself as a notch moving through the receiver passband. Multipath can alter the ratio of the carrier to the sidebands in an FM system, which modifies the effective deviation and therefore the level of the demodulated signal. In a digital system, multipath can introduce intersymbol interference to a level that the demodulator cannot tolerate, causing temporary path failure.
- *External interference fade margin (EIFM):* The receiver threshold degradation due to external system interference (independent of thermal noise).
- *Adjacent channel interference fade margin (AIFM):* The receiver threshold degradation due to interference from adjacent channel transmitters in one's own system, or a neighboring system. This is caused by closely spaced systems at the same site, on an opposite polarization, or a different antenna.

The ability of a receiver to perform in a fading environment is determined by the type of modulation and the effectiveness of the adaptive equalizer (if present). When digital systems are involved, manufacturers typically characterize receiver fading performance by moving a phase delayed 6.3 ns notch across the intermediate frequency (IF) passband. As the notch frequency is changed, the depth of the notch is adjusted to produce a 10E-6 bit error rate (BER, a measure of digital system performance). The result of a typical radio response is shown in Figure 9.1-4. The response shown is referred to as the *equipment signature*, *M* or *W curves*.

These four fade margins are power added to derive a composite fade margin (CFM) as follows:

$$\text{CFM} = 10\log\left(10^{-\text{TFM}/10} + 10^{-\text{DFM}/10} + 10^{-\text{EIFM}/10} + 10^{-\text{AIFM}/10}\right).$$

The outage time due to multipath fading is calculated by:

$$T = \left(rT_o \times 10^{-(\text{CFM}/10)}\right)/I_o$$

where:

T is outage time in seconds.

r is fade occurrence.

T_o is $(t/50)(8 \times 10^6)$ = length of fade season in seconds.

t is the average annual temperature in degrees Fahrenheit.

CFM is composite fade margin.

I_o is the space diversity improvement factor; factor = 1 for no diversity; ≥1 for space diversity.

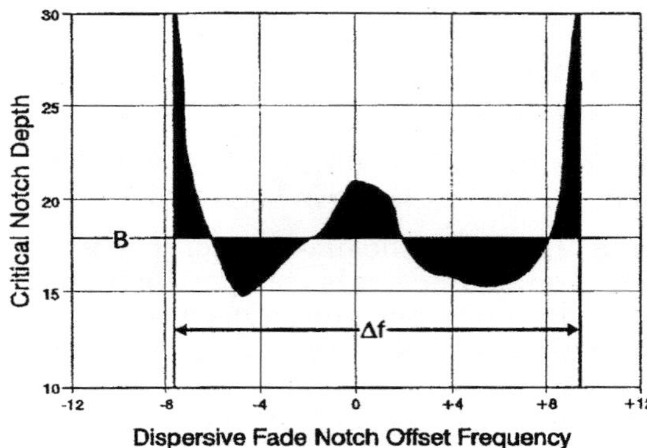

FIGURE 9.1-4 Dispersive fade margin measurement—W curve. [2]

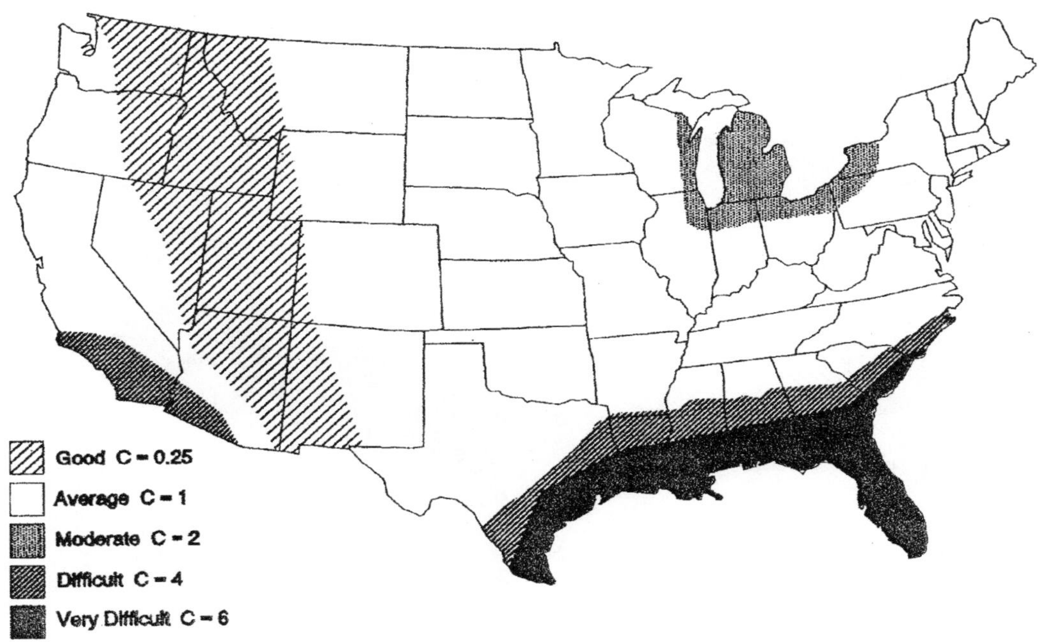

FIGURE 9.1-5 Values of climate factor for the continental United States. [2]

The fade occurrence factor, *r*, is calculated from the basic outage equation for atmospheric multipath fading:

$$r = c(f/4)(D/1.6)^3 \times 10^{-5} \text{ (English)}$$

$$r = c(f/4)D^3 \times 10^{-5} \text{ (metric)}$$

where:

c is climate or C factor (see Figures 9.1-5 and 9.1-6).

f is frequency (GHz).

D is path length (miles or km).

A close look at the formula reveals that path length has an exponential effect on outage, which better explains the need for higher power or space diversity antennas on longer paths.

System Availability

The outage time due to propagation effects is usually given in seconds or minutes per month or per year. System availability, which is the reciprocal of unavailability, is typically expressed as a percentage of a year or a month; based on the fact that there are 525,600 minutes in a 365-day year.

For example, if the annual outage time has been calculated as 9.5 minutes, express this as a percentage:

Unavailability is 9.5/525,600 × 100 = 0.00181% of a year.

If there were no outages ever, the availability would be 100%, so availability may be calculated subtracting unavailability from the ideal, or 100% minus 0.00181% = 99.99819%. There is a strong tendency among engineers to strive for availability numbers that are close to 99.99999%, however, this may not be an economically achievable goal.

Digital STL systems require a significantly higher degree of RF linearity, which results in lower system gain figures, and a lower fade margin than comparative analog systems. The inherent lower system gain of a digital system may require larger antennas or space diversity to maintain the same reliability. The spacing for diversity antennas is based largely on empirical data, but there is broad agreement among microwave engineers that 30–40 ft of vertical separation is a good rule of thumb, while some point to 40 wavelengths as being the proper spacing. Field experience has shown that spacing as low as 15–20 ft can provide a valuable improvement factor.

The improvement factor of space-diversity receiving antennas may be more accurately predicted by using a formula developed by Arvid Vigants of Bell Laboratories in the 1970s:

$$\text{Isd} = \left(7.0 \times 10^{-5} \times f \times s^2 \times 10^{F2/10}\right)/D,$$

where:

Isd is improvement factor.

f is frequency in GHz.

s is the vertical antenna spacing in feet, center to center.

D is path length in miles.

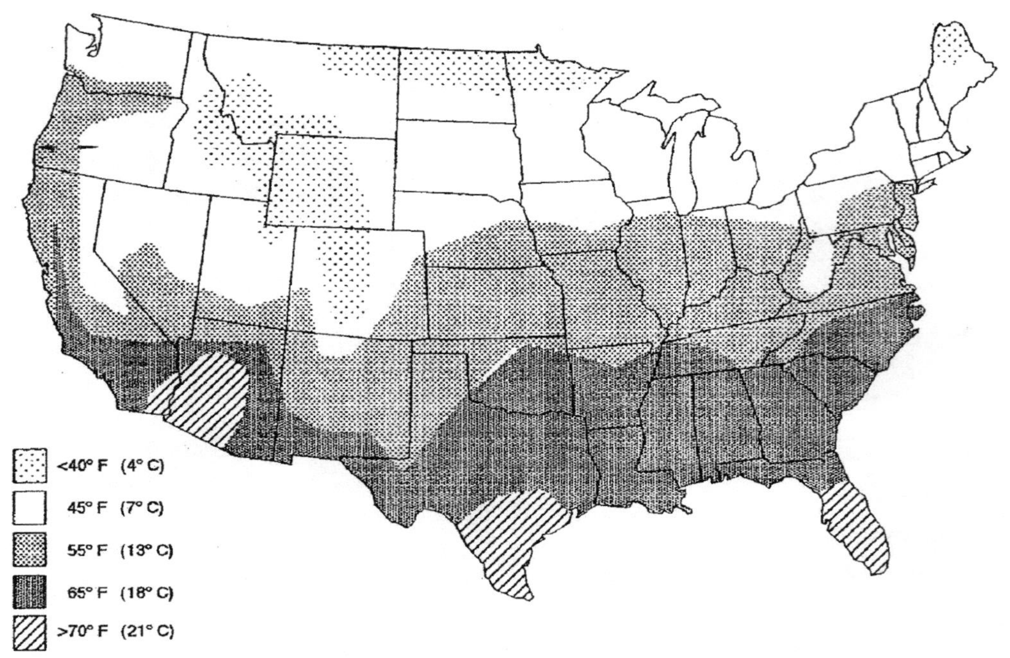

FIGURE 9.1-6 Average annual temperature for the United States. [2]

F2 is the fade margin in dB of the antenna with the lower fade margin (the antenna with the higher fade margin should be used to calculate U).

Isd as calculated above represents a ratio, for example, if Isd is calculated to be equal to 43, the improvement factor is 43 to 1. To calculate the total unavailability for a space diversity system:

$$\text{Usd} = U/\text{Isd},$$

where:

Usd is unavailability with space diversity.

U is unavailability, nondiversity [1].

Isd is improvement factor.

Moisture Absorption and Rainfall

The impact of rain on microwave systems has been studied increasingly over the last two decades. The principal finding is that rain causes increasing attenuation at frequencies above 10 GHz, and should always be taken into account in path calculations at these frequencies. The peak rain rate rather than the annual average rainfall is the most important parameter to analyze. A day-long drizzle and a short thundershower may precipitate the same volume of water, however, a dense cell in a thunderstorm will cause the most severe outage.

On longer paths at 13 or 15 GHz, and for all paths above 18 GHz, the rainfall outage dominates and multipath effects can usually be ignored.

Path Profiles

Microwave terminals are likely to be located at studios, earth stations, or TV transmitters, all of which are known, fixed sites. Should a new ENG site or a microwave repeater site be required, the choice may be constrained by availability and access. A new access road and utility service connections are often the most costly part of building a new site. As such, it may prove more economical to avoid the highest point of ground and use a taller tower or other support structure located closer to an existing road and utility connections. In some cases, an existing tower may be close enough to the desired path or coverage area to warrant approaching the owner about a leasing arrangement.

Assessing the suitability of two or more sites, and determining the required antenna heights for establishing point-to-point microwave paths, requires an accurate knowledge of the intervening terrain and any obstructions along the path. The standard method of gathering this information is to record the vertical elevation data between the sites, from which a vertical earth profile may be drawn. This can be done quickly and easily with the aid of economical computer programs that draw profiles based on USGS topographical survey data. Some programs now include aerial photos and three-dimensional mapping that allow the user to "see" in all directions from ground level at any spot on the map. Still, others provide a complete set of path performance and reliability calculations, including rain fade, based on inputs from the user. Manually drawn profiles may be done by taking data from USGS topographical maps, if available.

Most programs draw a flat, linear profile on a rectangular format, much like the manual method of drawing on rectangular graph paper. At one time, special *k*-factor paper was available from a few vendors, but was limited to one *k* factor only. Flat profiles have the distinct advantage of allowing the engineer to analyze multiple *k* factors, and plot reflection points, all on the same page with allowances included for average tree heights and known manmade obstructions at obvious critical points along the way. A critical point is defined as one that has the potential for blocking the microwave beam, if sufficient antenna height cannot be achieved at one or both ends of the path.

Once a decision has been made to pursue a particular path, an on-site examination of key path elements is recommended to ensure that tree heights are correct and no new buildings have been added in the path since the data was compiled, or since the maps were printed. The typical criteria used in programs and in manual calculations are as follows:

$$1.0F_1 + k = 4/3 \text{ for "normal" clearance, and}$$

$$0.3F_1 + k = 2/3 \text{ for "worst case" clearance}$$

where F_1 is the Fresnel zone radius in feet. Although a clearance of $0.6F_1$ is adequate, it is viewed as the absolute minimum, and leaves no room for future tree growth or topographical survey errors.

FIGURE 9.1-7 Standard performance parabolic antenna. (Photo courtesy of Andrew Corporation.)

ANTENNA SYSTEMS

For point-to-point microwave systems, the usual antenna choice is a parabolic reflector type. The parabolic antenna consists of a frequency-sensitive feed assembly at the focus of a reflector; which may be placed in the aperture for minimum sidelobes, or in front of the aperture for maximum gain.

In the 950 MHz aural broadcast auxiliary service (BAS) band, the sectional parabolic is favored in a majority of STL systems, followed by the grid parabolic, and finally the solid parabolic. Yagi arrays are still being used in some AM and FM STLs but are not recommended for new installations, and should be replaced in existing installations due to their poor sidelobe patterns and low front-to-back ratios.

At frequencies up to 3.7 GHz, the feed is typically a dipole with a subreflector and a solid main reflector. At frequencies above 3.7 GHz, waveguide feeds and solid reflectors are required to maintain the same level of efficiency. Grid-style main reflectors may be used up to 2.7 GHz to reduce wind load, but are limited to a single polarization.

The gain ratings of microwave antennas are referenced to an isotropic radiator, as noted by the reference to gain in dBi. An isotropic radiator is a hypothetical antenna with a radiation pattern that is perfectly spherical. Directional antennas are designed to enhance the radiation in a given direction at the expense of other directions, which has the effect of concentrating the signal in the desired direction and reducing the signal in other directions.

Antenna Types and Applications

The range of typical antenna types in use today for aural and television service include the following.

Standard Parabolic

For many years, the workhorse antenna of microwave communications has been the standard parabolic, which offers a good compromise between gain, sidelobe performance, and front-to-back ratio (Figure 9.1-7). Standard antennas may be single or dual polarized, and include both solid and grid reflectors, although grid antennas cannot be dual polarized. Low VSWR versions are available in most models, and in some cases may be the only type available.

Sectional Parabolic

Although based on a parabolic design, the truncated sectional parabolic does not employ the typical round reflector, but only a portion of the reflector. In practice, these antennas employ grid reflectors (Figure 9.1-8), and have elliptical radiation patterns as shown in Figure 9.1-9. A sectional antenna may be successfully employed in uncontested areas, where potential interference is minimal. The only drawback to a sectional antenna stems from its elliptical radiation pattern and the fact that the optimum sidelobe

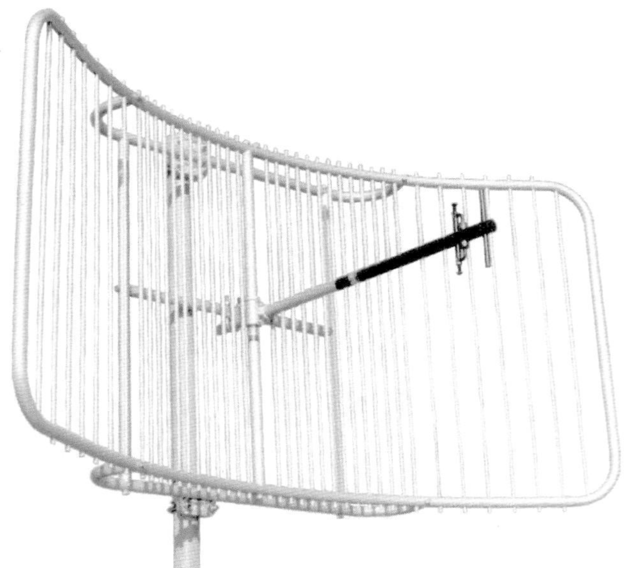

FIGURE 9.1-8 Kathrein—Scala PR-950 Paraflector; sectional parabolic antenna. (Photo courtesy of Kathrein-Scala.)

suppression is realized only when the antenna is vertically polarized. While the Federal Communications Commission (FCC) rules do not currently specify a minimum beam width for the 950 MHz aural band, they do enforce a de facto standard that may preclude use of a paraflector. In addition, the SBE has asked the FCC to set antenna standards for the 950 MHz band, and future action may be taken.

Grid Antennas

Grid-style reflectors were developed to reduce wind load, and are built with grid elements spaced close enough to act like a solid reflector in the specified band of interest. The grids must be aligned in parallel with the polarization of the feed, or the gain will be reduced at least 20 dB. Over the years, they have been very popular for the 950 MHz aural band, and the 2 GHz TV STL band. Grids are effective to around 2.7 GHz, and are restricted to one polarization only. Above 2.7 GHz, the grid spacing is reduced to the point that it no longer provides a wind load advantage, and the feed also becomes less efficient.

Enhanced Performance Parabolic

In the last decade, a number of enhanced designs have come on the market to fill the gap between standard-performance and high-performance antennas. The enhanced-performance parabolic usually includes a deeper than normal reflector, and is sometimes called a "deep dish" antenna (Figure 9.1-10). The deep dish designs may also include special feed assemblies, and selective shield attachments at the edge of the reflector. The main advantage of enhanced designs is lower sidelobe radiation that may meet FCC category A compliance where a standard antenna of the same size might only meet category B compliance. Some enhanced designs may sacrifice a small amount of gain as part of the tradeoff for cleaner sidelobe performance. The significance of categories A and B is based on sidelobe suppression, where category A is more stringent than category B. Table 9.1-1 compares category A to category B in the 7 GHz television broadcast auxiliary services band. The sidelobe suppression values are in dB, and relative to maximum gain, with reference to azimuth in degrees from peak.

High-Performance Antennas

The term *high performance* with regard to a parabolic antenna refers to an antenna that has been purpose-built to suppress the sidelobes to a much higher degree than a standard antenna of equal size (Figure 9.1-11). It is not unusual to find, for example, that a 6 ft high-performance antenna will provide better adjacent channel protection than does an 8 ft standard-performance antenna, and with less wind load. Vendors typically offer several models of high-performance antennas, with increasingly tighter patterns or higher polarization discrimination in order to support difficult frequency coordination requirements in a crowded environment.

Angle Diversity Antennas

The angle diversity antenna is a highly specialized design, intended to help reduce multipath fading on paths that require space diversity, but are limited to one antenna at each site. The antenna is built with two separate feed assemblies that propagate identical beams with a small angular displacement. Theoretically, the multipath fade mechanism will affect each beam differently, and effectively reduce the fading.

Radiation Patterns

In order to comply with FCC and European Telecommunications Standards Institute (ETSI) standards, the radiation pattern for every variation of microwave antenna that is offered for sale in fixed link applications must be measured, recorded, and made available to customers and frequency coordinators.

A typical radiation pattern is shown in Figure 9.1-12. The pattern shown provides information about the typical gain and sidelobe performance of an 8 ft dual-polarized antenna for the 7 GHz BAS STL band.

Radomes and Loss

Depending on the manufacturer and antenna type, a radome may or may not be included with a specific antenna. Whether included or optional, the addition of a radome to any microwave antenna provides several benefits. Protection from weather extremes and wind-loading reduction are at the top of the list. A radome-equipped antenna is also better protected against falling ice and the potential for feed horn damage as a result.

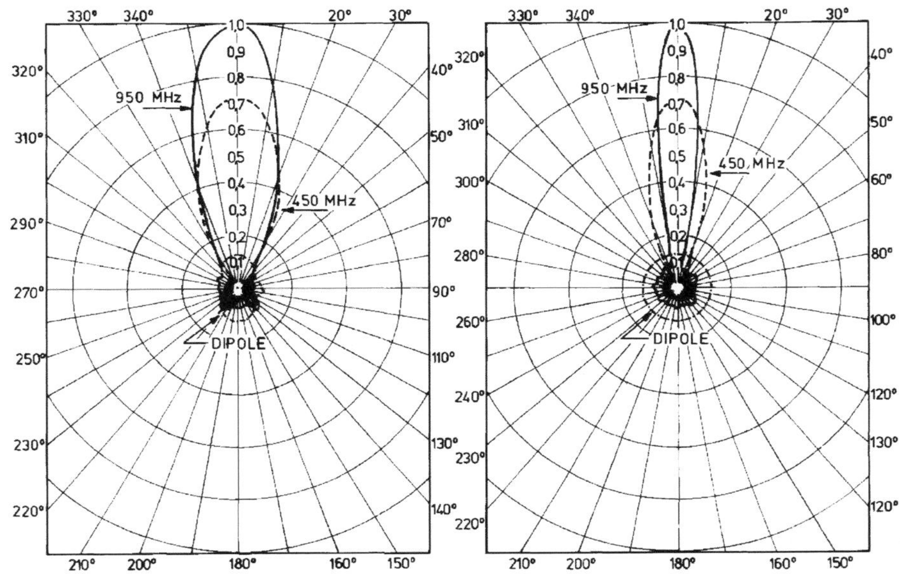

FIGURE 9.1-9 Radiation plots sectional antennas at 450 and 950 MHz; (a) is the pattern for a horizontally polarized antenna, and (b) is the pattern for a vertically-polarized pattern. [2]

TABLE 9.1-1
Comparison of FCC Category A and Category B Compliance Requirements in the 7 GHz TV BAS Band

Category	Maximum 3 dB Beam Width	Minimum Suppression 5–10°	Minimum Suppression 10–15°	Minimum Suppression 15–20°	Minimum Suppression 20–30°	Minimum Suppression 30–100°	Minimum Suppression 100–140°
A	1.5	26	29	32	34	38	41
B	2.0	21	25	29	32	35	39

FIGURE 9.1-10 Enhanced performance "deep dish" parabolic antenna. (Photo courtesy of Andrew Corporation.)

FIGURE 9.1-11 Typical high-performance, shrouded antenna. (Photo courtesy of Andrew Corporation.)

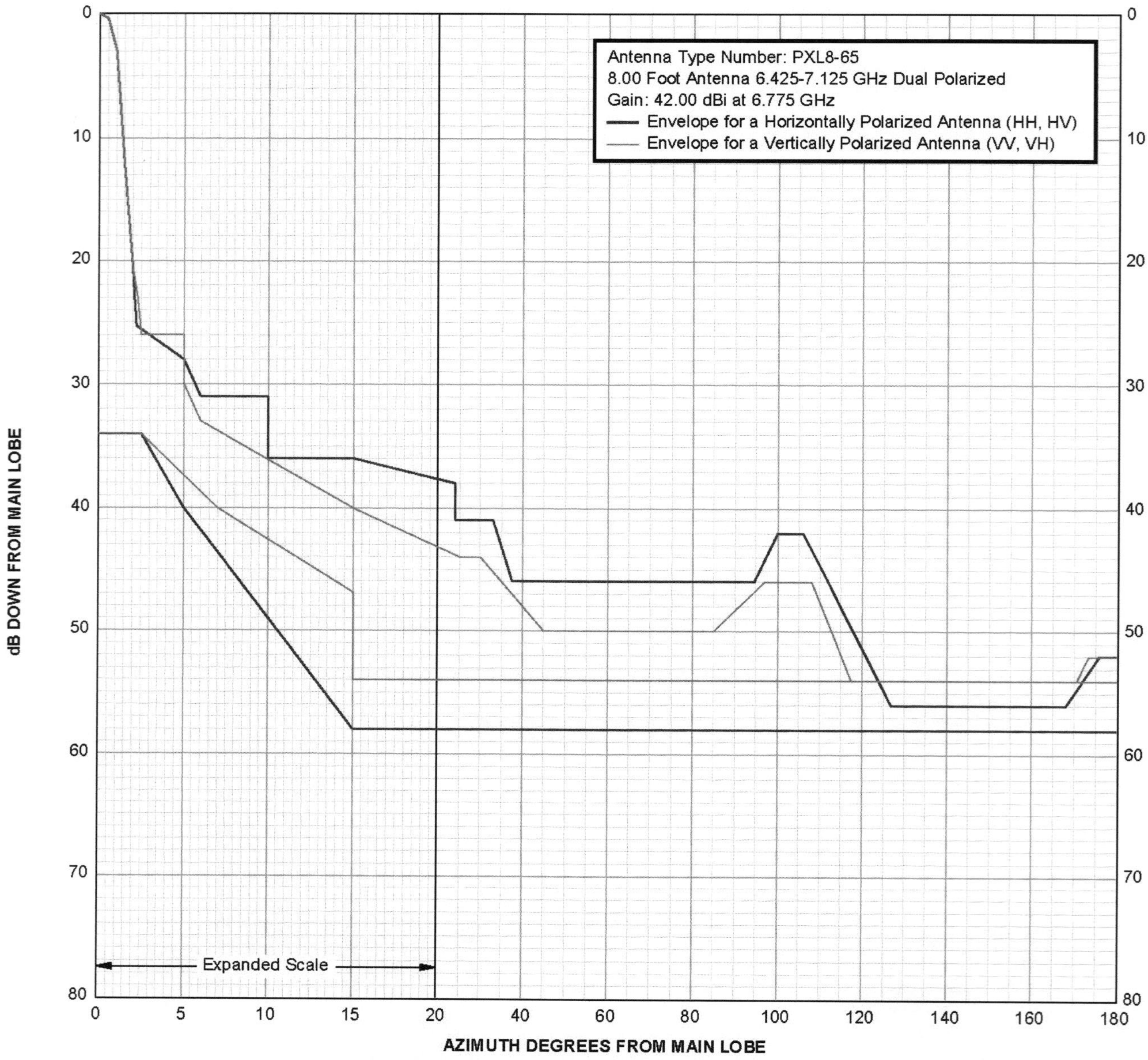

FIGURE 9.1-12 Radiation pattern envelope for an 8 ft standard parabolic antenna at 7 GHz, with a dual-polarized feed. (Courtesy of Andrew Corporation.)

For standard antennas, the radome is usually convex in shape and made of molded fiberglass or ABS plastic. Radome options include embedded heating elements, extra high-strength versions, and custom colors to blend better with the environment.

High-performance shrouded antennas always include flexible unheated radomes made with rubber-coated nylon or polymer-coated fiberglass sheeting. The flexibility is an advantage in windy conditions, as the movement tends to shed any ice buildup.

While the loss of a radome is included in all high-performance antenna gain specifications, it is generally not included in standard antenna gain figures, and must be accounted for when performing path calculations. Table 9.1-2 provides average radome loss for typical standard antennas, interpolated from catalog values for the broadcast auxiliary service bands. Exact values can be obtained from the specific manufacturer.

It can be seen that the loss is almost negligible in the 2 GHz band, but rather substantial in the 13 GHz band. When making path calculations, the manufacturers' stated loss figures should always be used. Losses have been omitted at 950 MHz, as they are negligible, and most aural STL installations do not include radomes. In the 18 and 23 GHz bands, most

TABLE 9.1-2
Typical Radome Loss for Standard Parabolic Antennas [2]

Frequency Band	Antenna Size		
	6 ft	8 ft	10 ft
2 GHz	0.1dB	0.1 dB	0.2 dB
7 GHz	0.7 dB	0.8 dB	1.1 dB
13 GHz	1.7 dB	1.8 dB	2.1 dB

antennas are of the high-performance type, and the losses are included in the gain figures.

Transmission Lines, Tuning, and Pressurization

Below 3.7 GHz, antennas are normally fed with coaxial cable that is rated for microwave use. The cable may use a foam dielectric between the inner and outer conductors, or an air dielectric construction that includes a spiral wound center-to-outer spacer. Air dielectric cables exhibit the lowest transmission loss, however, foam dielectric cables do not require dehydration equipment and are often completely adequate in spite of higher losses.

Above 3.7 GHz (and also below if the cable loss is excessive), elliptical waveguide is the typical choice for transmission line. In cases where the loss of elliptical waveguide is a problem, circular waveguide may be deployed. In all cases, microwave transmission line is delicate in nature, and *must* be installed to manufacturers' specifications for bending radius, support points, and pressurization.

Connectors for both cable and waveguide need careful attention during installation to avoid mismatch, air leaks, and the ensuing possibility of moisture ingress over time. Special flanging and attachment tools can be purchased or rented from manufacturers to aid in the correct and consistent assembly of connectors.

As the broadcast world converts to digital, and as more radio channels are combined on a single path, tuning the transmission line becomes much more important. RF reflections can reduce the effective range of adaptive equalizers in digital radios by stressing them needlessly, or, they can cause excessive distortion at spot frequencies in analog radios. Sweep tuning the waveguide during installation, with occasional sweep tests at periodic intervals thereafter, will assure that system performance is at its best when fades do occur, or when new RF channels are added.

Installations that employ waveguide or air dielectric cables must be maintained at a positive pressure (typically 0.5–5.0 lb/in^2) above atmospheric with dry air or nitrogen. A wide range of dehydrators and accessories are available to facilitate proper pressurization and control.

Dehydrators can be fully automatic (where the desiccant is automatically dried out by the unit) or semi-automatic (where the desiccant must be periodically dried out or replaced). For remote sites, automatic dehydrators are strongly recommended to avoid the possibility of moisture ingress and damage associated with desiccant exhaustion. Proper sizing of the dehydrator is also important for reliability. An undersized dehydrator may operate at a high duty cycle and shorten the compressor life. To avoid rapid cycling of the compressor, the volume of air in the system should not be less than 1.5 cubic feet; this can be achieved in small systems by adding a regulating tank.

In systems with multiple transmission lines, it is good practice to install a pressurization manifold with shutoff valves and metering for each transmission line. Should one of the lines develop a leak, it can be isolated for repairs, while pressure is maintained on the others.

The Andrew Corporation, Radio Frequency Systems (RFS), and Kathrein-Scala each provide a wealth of information on their websites and in their catalogs relative to antennas, transmission lines, connectors, and all of the accessories needed for installation.

Antenna Alignment Techniques

The highly directive nature of a parabolic antenna can present quite a challenge when attempting to align a microwave path. At lower frequencies, such as the 950 MHz band, the job is made easier by a comparatively broad main lobe, however at 7 GHz and above, the narrower beam widths can make it difficult to find the signal when performing an initial alignment. Knowing a few tricks of the trade can be very helpful when the tower crew arrives and needs direction in the alignment process. More often than not, the crew has an idea of the basics, but lacks technical savvy. Although this section focuses on the alignment of a new installation, a realignment is essentially the same after the point that a signal has been acquired.

Prealign the Antenna

First and foremost, it is vitally important to know the exact pointing azimuth for each antenna, and to prealign the antenna as closely as possible to its final position. If antennas or their mounting structures can be seen with the naked eye or with a pair of binoculars from one end of the path to the other, the job should be straightforward. If the other end of the path is not visible, another way must be found to prealign the antennas. One source of determining the azimuth and elevation data can be found on the FCC website at http://www.fcc.gov/mb/audio/bickel/distance.html.

The first step in prealigning the elevation is usually a matter of setting the peak of the main lobe on the horizon (zero degrees elevation referenced to the horizon), unless there is a great difference in antenna height at opposite ends of the path. A short path will exaggerate the effect, while a long path reduces it. For extremely narrow beam widths, in most cases an "eyeball" elevation alignment is not sufficient, and a leveling tool must be used. The simplest tool is an inexpensive angle finder held against the structural mounting ring on the back of the antenna. An angle finder is a handheld device that uses a weighted indicator to measure the angle of any surface with respect to gravity, and can be found in most hardware stores. The scale is usually calibrated in 0.5 degree increments, making it an easy matter to adjust the antenna bore sight elevation to the specific path requirements.

Prealigning the azimuth is a bit more complicated. USGS topographical maps and virtually all path programs use true north as a reference, however, the most common direction-finding tool is a magnetic compass. Let's say, for example, that a crew in downtown San Francisco must align an antenna to an azimuth of 115 degrees. In San Francisco, the magnetic declination, or the difference between true north and magnetic north, is +14.7 degrees. In order to find the correct direction referenced to true north, the declination must be subtracted from the compass reading, which we now know is 14.7 degrees beyond where the antenna should point. So, 115 degrees minus 14.7 degrees yields a magnetic azimuth of 100.3 degrees. On the east coast, the declinations are negative, and must be added to the compass reading.

Path performance calculations should always be done in advance, to provide an accurate estimate of the expected receive carrier level.

It is extremely important to verify that the antennas were assembled according to the manufacturers' instructions, especially the positioning of the feed assembly. Failure to do so can degrade the performance and distort the pattern, causing lower than expected signal levels and increased susceptibility to interference. For example, if the antennas are cross-polarized at the start, the signal will be at least 20 dB down from where it should be.

Also, the transmission lines should be checked carefully for any kinks, dents, or extreme bends, and the connectors should be checked that they have been correctly installed. It is a good idea to tune adjustable connectors and sweep the transmission lines connected to the antenna using a microwave network analyzer and a return loss bridge, to benchmark the performance immediately after installation.

Alignment Hints

First and foremost, it is very important that the polarization is identical at both ends of the path, and that it matches the polarization that the path is licensed for. Coaxial and waveguide feeds usually have arrows near the input connector showing the polarization of the feed that can be checked for verification purposes.

When the tower crew is in place, and the antennas are prealigned, check to see if there is any signal at all in the receiver. If not, have the crew on one end of the path start to pan the azimuth slowly about 10 degrees to each side of the preset center. If the first attempt at receiving a signal is not successful, return the antenna to its preset azimuth, and try the same procedure at the other end of the path.

Once a signal is acquired, there are numerous ways to coach the tower crew into a perfect alignment. To begin with, everyone involved in the antenna alignment should be in voice communications. VHF and UHF handy talkies are useful for such communications.

One successful method of directing an alignment is to call out automatic gain control (AGC) readings to the person that his panning the antenna, after explaining the significance of the changes. The tower crew should be instructed in advance on the plan, in particular what to expect with regard to the feedback that they will be receiving while listening to the AGC readouts. The antennas should be moved very slowly during this process, or the signal can be missed completely. Many crews have done this type of work before, but it helps to set ground rules that everyone can understand.

AURAL STL RADIO SYSTEMS

With the ongoing move toward digital well under way, radio broadcasting is now going through the most radical change since it began in the early 20th century. While a substantial number of analog STL systems are still feeding analog AM and FM exciters, the landscape is changing, and the rate of change is accelerating. The majority of new and upgraded AM and FM STL systems employs digital modulation formats that support the AES/EBU audio interface and an Ethernet stream feeding new digital exciters. The most current exciters are capable of supporting standard analog and the NRSC-5-A digital format, also known as IBOC or by the trade name of *HD Radio*. The obvious benefit of an all-digital STL and exciter is that they can still continue to produce analog transmission, however, they are ready to add IBOC/HD Radio when the station wishes to do so.

The principal frequency range used by aural STL systems in the U.S. is 944.5–951.5 MHz. Some stations with very short paths have chosen 18 or 23 GHz, while some have ventured into the unlicensed spread spectrum bands at 900 MHz, 2.4 GHz, and 5.8 GHz.

Legacy Systems

An FM-modulated monaural STL has traditionally been used to feed AM or monaural FM transmitters, with an audio bandwidth capability up to 15 kHz for the program channel, and a single subcarrier above the audio at approximately 39 kHz, as shown in Figure 9.1-13(a). Prior to the development of composite systems, stations often used dual-parallel systems for stereo or dual-mono applications with a common antenna at the transmitter and receiver. At the transmit site, a combiner/isolator adds the signals and prevents the generation of intermodulation products. At the receive site, a low-noise antenna splitter feeds the individual receivers.

The composite STL for FM radio was developed as an improvement over discrete systems because it has

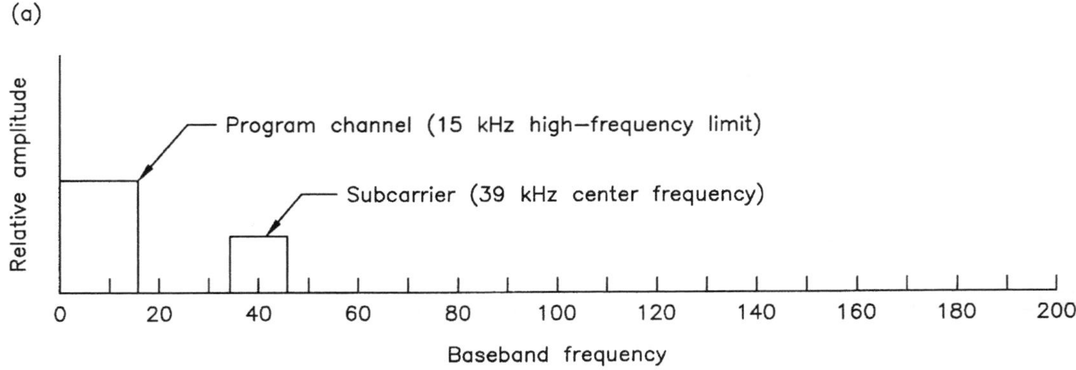

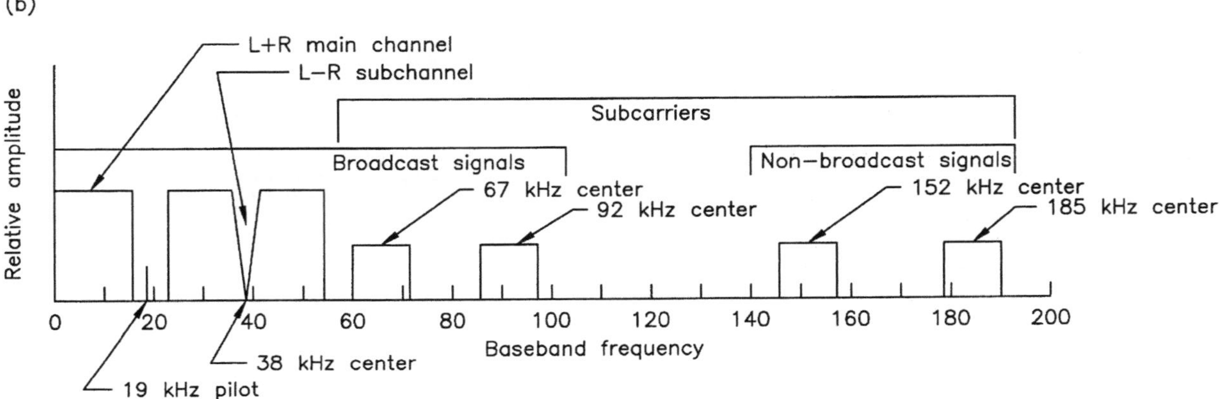

FIGURE 9.1-13 Baseband spectrum of STL systems: (a) monaural, and (b) composite. [2]

sufficient baseband bandwidth (220 kHz) to accommodate the output of an FM stereo generator and several subcarriers. A low pass filter is used at the output of the receiver to separate the signals to be broadcast over the air from the nonbroadcast (or closed-circuit) signals. A typical configuration is illustrated in Figure 9.1-13(b). The composite STL provides superior stereo performance compared to dual-monaural radio links in several respects, including:

- Elimination of interchannel phase and amplitude errors that are found in a dual-channel system.
- Elimination of audio headroom considerations because the STL input signal has already been passed through the station's audio processing system and the stereo generator, which are located at the studio in this configuration.

For FM applications, the composite signal from the STL receiver is fed directly into the FM exciter. An alternate method involves the use of a *reciter* configuration, which functions as a combined STL receiver and FM exciter but with an IF interface rather than a baseband interface. The elimination of one stage of demodulation and remodulation reduces noise and distortion in the transmitted FM signals.

Should a composite STL be used to feed an AM stereo exciter, or an FM exciter with discrete audio inputs, a complementary stereo decoder is used to recover the left and right channel audio to provide the required discrete right and left channel audio inputs.

Composite aural STLs may be found in a number of hardware variations, all designed to accomplish the same goal: a low-distortion relay of a baseband signal from the studio exciter to the transmitter. Figure 9.1-14 illustrates the basic STL configurations for an AM stereo and FM stereo application. A block diagram of a representative composite STL is given in Figure 9.1-15.

The RF carrier is generated by a *voltage-controlled oscillator* (VCO) that is phase locked to a modulating VCO operating at an intermediate frequency of approximately 70 MHz. Direct composite FM modulation takes place at the IF VCO, which is also phase locked to an oven-controlled crystal oscillator that serves as the master time base for the transmitter. Because frequency multiplication is not used, the RF carrier is an exact reproduction of the IF VCO. Therefore, a number of potential compromises resulting from frequency multiplication and post-heterodyne filtering are eliminated, including:

- Degradation of S/N
- Generation of spurious signals
- Degraded stereo performance

IF modulation provides excellent overall stereo performance, as documented in Table 9.1-3.

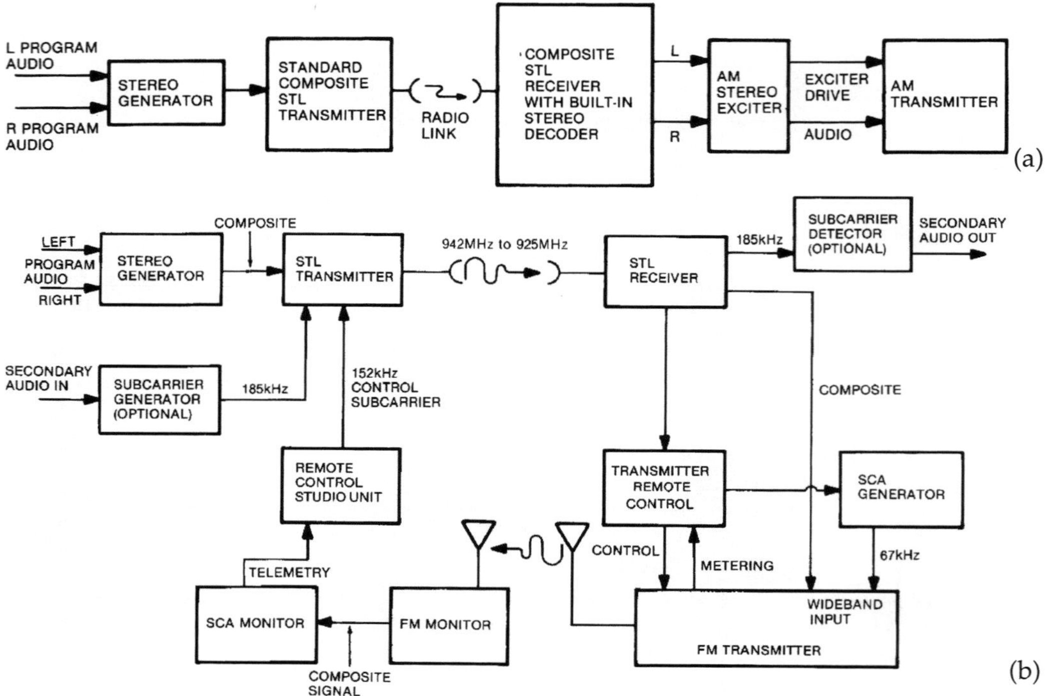

FIGURE 9.1-14 Analog composite STL links: (a) AM radio application, and (b) FM radio application. [2]

In this design, the operating frequency of the transmitter is typically programmable by internal dual inline package (DIP) switches in 12.5 kHz steps, so that frequency changes may be made in the field. The composite baseband signal accommodates the full stereo baseband, including two SCA and/or multiplex (MUX) channels. These features permit all processing equipment for analog operation to be located at the studio (an illustration of the composite baseband of such a system is shown in Figure 9.1-13(b)).

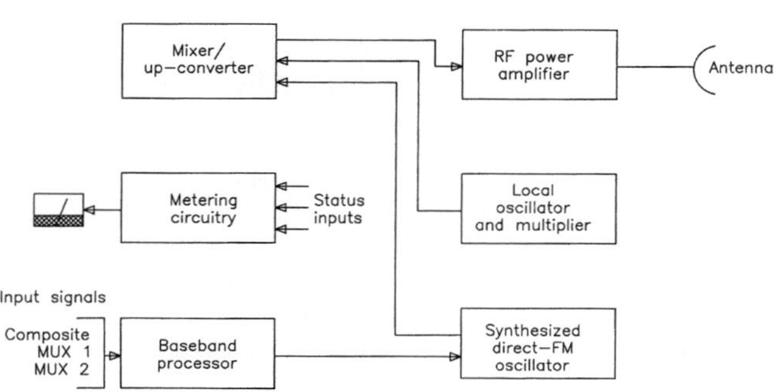

FIGURE 9.1-15 Block diagram of a composite STL transmitter. [2]

A block diagram of the companion composite STL receiver is shown in Figure 9.1-16. Like the transmitter, the receiver is user-programmable in 12.5 kHz steps typically through the use of internal DIP switches. The front end uses cascaded high-Q cavity filters and *surface acoustic wave* (SAW) IF filters to provide high selectivity and phase linearity. Triple conversion IF is used

TABLE 9.1-3
Specifications for a Typical Composite STL System

Parameter	Specification
Power output	6–8 W
Frequency stability	±0.0002%, 0–50°C
Spurious emissions	60 dB below maximum carrier power
Baseband frequency response	±0.1 dB or less, 30 Hz to 75 kHz
Stereo separation	Greater than 55 dB at 1 kHz
Total harmonic distortion	0.02%, 75 μs deemphasis
S/N	85 dB below ±75 kHz deviation, 75 μs deemphasis
Nonlinear crosstalk	50 dB or less
Subchannel-to-main crosstalk	60 dB or less

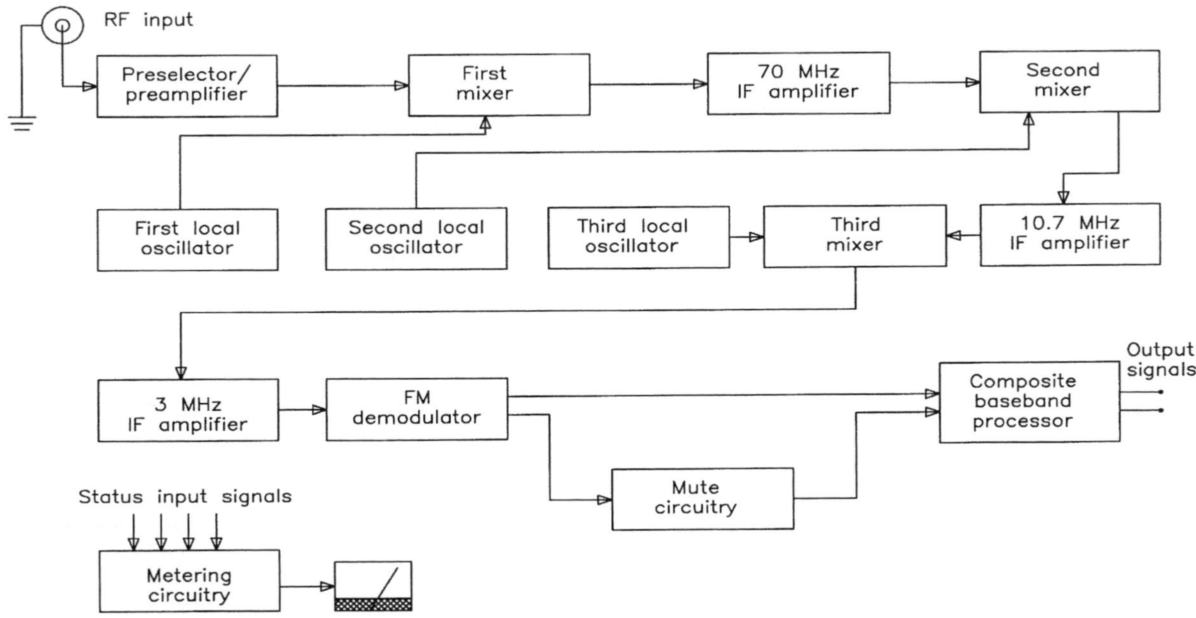

FIGURE 9.1-16 Block diagram of a composite STL receiver. [2]

to feed a pulse-counting discriminator for linear baseband demodulation.

Digital STLs for Radio Broadcast

As the conversion to all-digital radio broadcasting moves forward, so does the sophistication of aural digital STLs. By its nature, digital transmission offers more robust path performance during fades, up to the point of reaching a threshold at which point the digital system performance drops significantly. As long as the received signal strength and carrier-to-noise exceed a defined minimum value, the *bit error rate* remains relatively constant, and the decoded baseband signal is an exact duplicate of the input signal at the studio (Figure 9.1-17). With adequate error correction, signals can be received error free. Another advantage can be found where multiple microwaves are required. Digital STLs regenerate the digital stream at each repeater, eliminating additive signal degradations found in analog systems.

It is fair to point out that digitization of the audio signal always brings with it degradation in the form of quantization errors. However, the high sampling rates typically used for professional audio applications reduce such degradation to extremely low levels that are virtually inaudible.

The process of quantization is illustrated in Figure 9.1-18. The sampling rate and quality of the sampling circuit determine, in large part, the overall quality of the digital system. A properly operating transmission channel can be assumed to provide error-free throughput. This being the case, the digital signal can be regenerated at the receiving point as an exact duplicate of the input waveform, quantization errors excepted. Figure 9.1-19 shows a general representation of a digital communications channel. In the case of a radio link, such as an STL, the transmission medium is analog in nature (FM). The circuits used to excite the modulator, however, are essentially identical to those used for an all-digital link, such as fiber-optic cable.

The functions of the encoder and decoder, shown in Figure 9.1-19(a), usually are formed into a single device, or set of devices (a chip set), known as a *codec* (*coding and decoding* device). At the transmission end, the codec provides the necessary filtering to band limit the analog signal to avoid aliasing, thereby preventing *analog-to-digital* (A/D) conversion errors. At the receiver, the codec performs the reciprocal *digital-to-analog* (D/A) conversion and interpolates (smoothes) the resulting analog waveform.

The benefits of a digital STL for radio broadcasters can best be appreciated by comparing the performance of a digital system and an analog system. A digital STL typically permits broadcasters to extend the fade margin of an existing analog link by 20 dB or more. Furthermore, audio *signal-to-noise* (S/N) improvements of at least 10 dB can be expected for a given RF signal strength. Alternatively, for the same S/N, the maximum possible path distance of a given composite STL transmitter and receiver can be extended. These features could, in some cases, mean the difference between a one-hop system and a two-hop system.

The spectrum-efficiency of a digital STL is of great importance today in highly congested markets. The system may, for example, be capable of relaying four program channels and two voice-grade channels. The use of digital coding also makes the signals more tolerant of cochannel interference than a comparable analog STL.

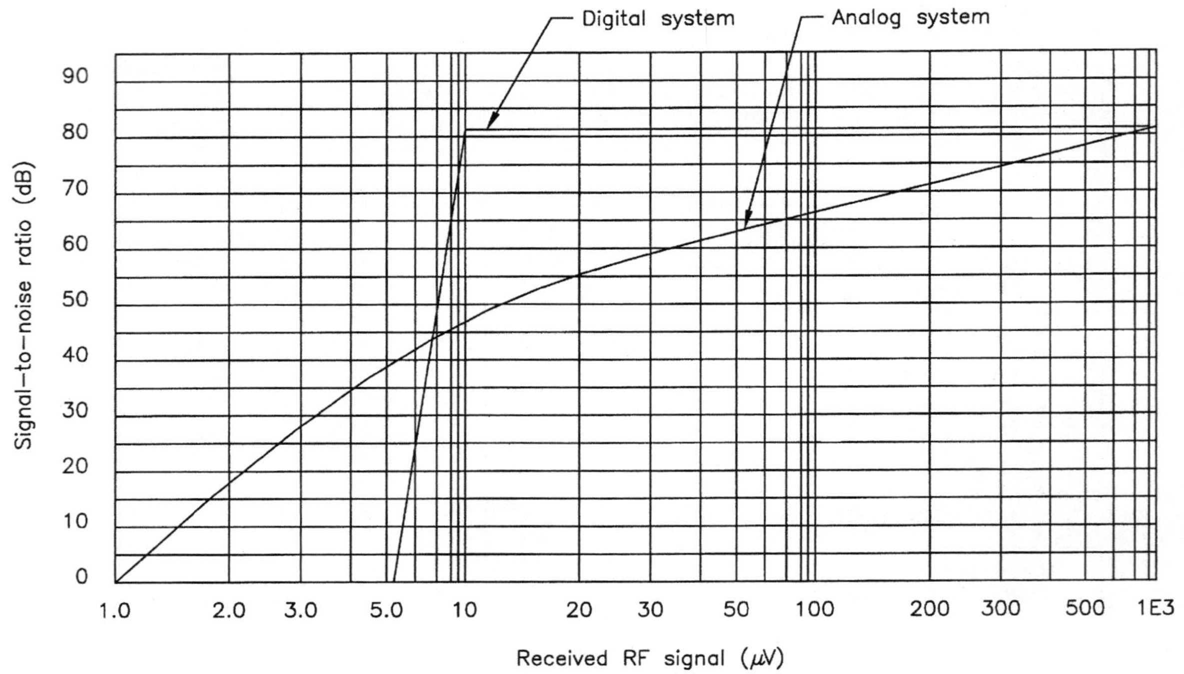

FIGURE 9.1-17 The benefits of digital versus analog STL systems in terms of S/N and received RF level. [2]

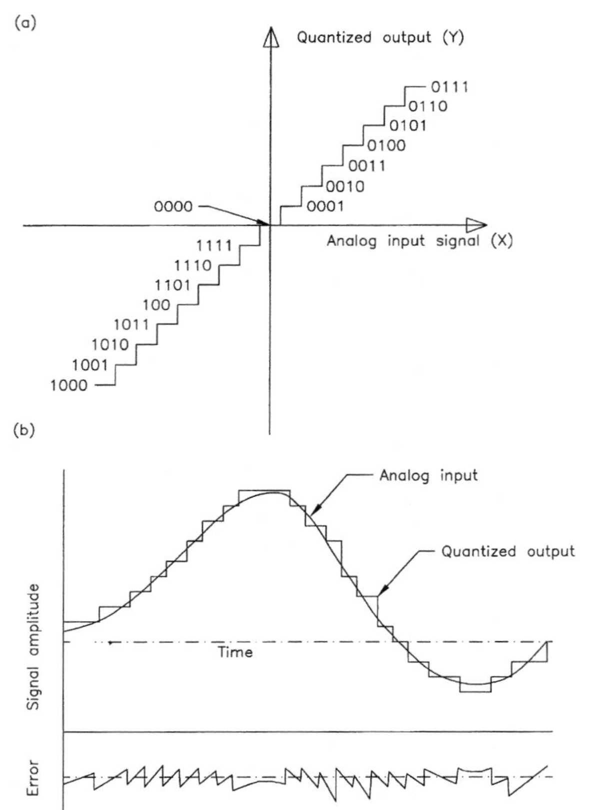

FIGURE 9.1-18 Quantization of an input signal: (a) quantization steps, and (b) quantization error signal. [2]

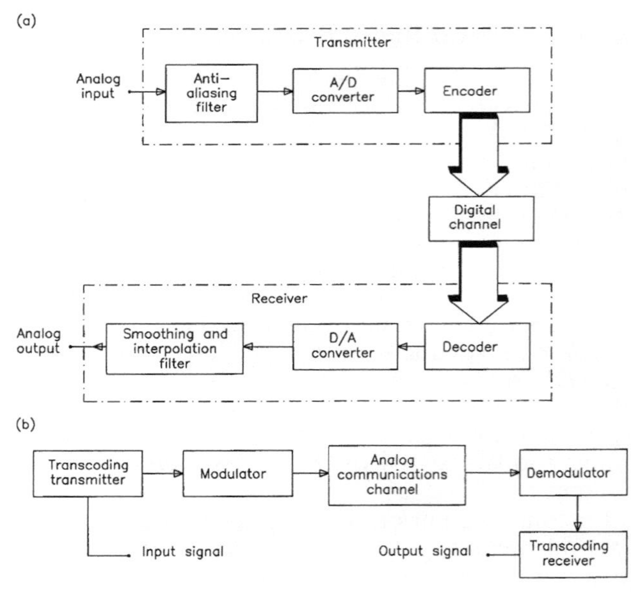

FIGURE 9.1-19 Digital transmission system: (a) coding/decoding functions, and (b) overall communications link. [2]

Coding System

Several approaches may be used to digitize or encode the input audio signals. The complexity of the method used is a function of the availability of processing power and encoder memory and determines the resulting delay incurred due to the encoding/decoding process. For a real-time function such as an STL, significant encoding/decoding delay is unacceptable. *Pulse code modulation* (PCM) is a common scheme that meets the requirements for speed and accuracy. In the PCM process, the sampled analog values of the input waveform are coded into unique and discrete values. This quantization may be uniform, as illustrated in Figure 9.1-18, or nonuniform. With nonuniform quantization, compression at the coder and subsequent expansion at the decoder is performed. By using larger quantization steps for high-energy signals and smaller steps for low-energy signals, efficient use is made of the data bits, while maintaining a specified signal-to-quantization noise level. This process is known as *companding* (compression and expansion).

PCM encoding, in a simple real-time system, provides a high-speed string of discrete digital values that represent the input audio waveform. Each value is independent of all previous samples. No encoder memory is required. This approach, while simple and fast, is not particularly efficient insofar as the transmission channel is concerned. There are many redundancies in any given input signal. By eliminating the redundancies, and taking advantage of the *masking* effects of human hearing, greater transmission efficiency can be realized. Viewed from another perspective, for a given radio transmission bandwidth, more information can be transferred by using a compression system that removes nonessential data bits.

NRSC-5-A—IBOC STL Requirements

The requirements for an in-band-on-channel (IBOC) STL using the HD Radio implementation include delivery of an Ethernet stream to the transmitter. There are a number of Ethernet delivery options open to the station engineer, but microwave is likely to be the most dominant for several reasons:

1. Many stations already have a microwave in operation.
2. Microwave is private pipe and not subject to public network problems.
3. Many transmitter sites do have access to high-speed public networks.

There are challenges involved with microwave, however. The current generation HD Radio system requires an AES digital audio channel sampled at 44.1 kHz and a 400 bps Ethernet stream. The 44.1 kHz AES channel digital is transported by the STL to the transmitter site where the audio processing and IBOC exciter are located. The IBOC exciter splits the audio into two streams. The audio signal destined for the legacy analog transmitter is time aligned with the digital, sent to the analog audio chain processor and to the legacy analog transmitter. The signal destined for the IBOC exciter is sent to the digital audio chain processing then returned to the IBOC exciter where it is encoded into the main program service (MPS) digital signal for over-the-air broadcast.

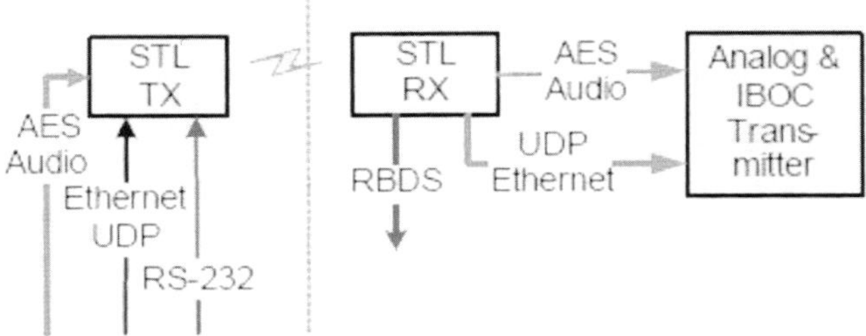

FIGURE 9.1-20 950 MHz STL with Ethernet capability. (Courtesy of iBiquity Digital Corporation.)

The program service data (PSD) is delivered separately to the IBOC exciter on a 400 bps Ethernet data stream. The supplementary program service (SPS) audio is also delivered separately on additional STL channels. Data for advanced application services, (AAS) when used, would require a separate delivery path as well. Transmitting of all the required IBOC components individually via microwave or T1E1 represents a serious challenge, however, solutions are available.

Figure 9.1-20 shows a basic block diagram of an STL system, like the Moseley Starlink SL9003Q, which is an example of a system that can be equipped for data transport in addition to AES digital program audio. Moseley can provide audio source encoder and decoder modules with a built-in RS-232 data channel that does not require extra bandwidth, and can support RBDS or the command side of a remote control system. Optional multiplexer and Ethernet modules can provide the one-way Ethernet UPD data path to the transmitter site for IBOC.

While a UDP path will provide the needed connectivity, it also poses a problem in that there is no bidirectional handshake with the far end. To solve this problem, Moseley offers Lanlink, a duplex local area

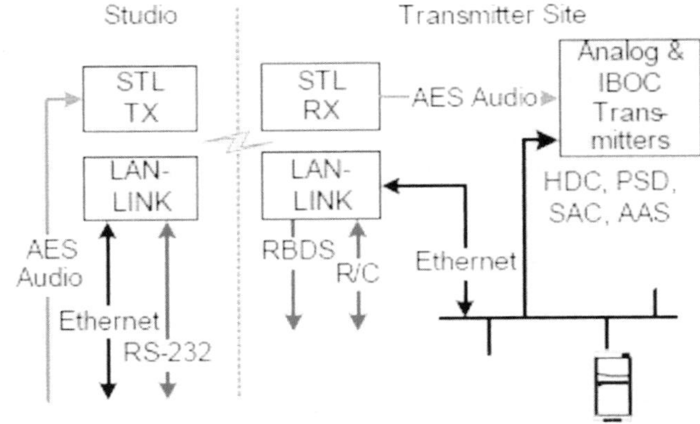

FIGURE 9.1-21 950 MHz STL with AES capability, and 900 MHz LAN extender. (Courtesy of iBiquity Digital Corporation.)

network (LAN) extension radio 902-928 MHz ISM band.

Using the existing 950 MHz STL antenna system, the LAN extension radio can transport bidirectional Ethernet and RS-232 serial data using digital frequency hopping spread spectrum technology producing signals that can still be recoverable even with a very low signal-to-noise ratio. The Ethernet data transport capabilities are up to 512 kbps of throughput for 10baseT IP/Ethernet connections. A representative block diagram is shown in Figure 9.1-21.

RADIO EQUIPMENT FOR TV STL

Digital video microwave systems are rapidly becoming a dominant methodology for TV STLs. With the advent of digital broadcasting, the requirement to transport an ATSC transport stream to a new DTV exciter was the primary impetus in the development of STL total digital solutions for the television broadcaster.

The earliest requirement was to transport 19.392 Mbps from the studio to the transmitter site. This demanded a new definition of transmitter and receiver architecture, and new modulation techniques that were completely different than the prior analog industry standard. Although the digital transmitter and receiver architectures are similar, to those used in FM designs, linearity improvements, oscillator phase noise, and frequency stability all had to be addressed.

In the analog world, one set of ITU-R recommendations had the effect of harmonizing performance parameters and interoperability standards among the manufacturers of broadcast-centric FM microwave equipment. Interfaces at baseband, IF, and RF were common across the industry, making it an easy matter for standard configurations and universal data sets.

For decades, the telecom industry has employed a set of interface standards that are utilized to support defined data rates, such as DS-1, DS-3, OC-3, and OC-12, including some with variability, like G.703. In addition, bandwidth allocations have been standardized dependent on the aggregate data rate and modulation type, which is typically QPSK, QAM, and TCM. This situation, much like analog video, makes life easier for engineers assigned to perform frequency, link budget, and interference calculations. The broadcast migration to digital transmission systems has complicated this effort.

Broadcast engineers have enjoined digital radio and digital modem manufacturers to efficiently utilize the bandwidth allocated to them—typically a 17 or 25 MHz channel. A derivative of MPEG-2 design efforts was the introduction of the Asynchronous Serial Interface, or ASI, to the video industry. The fact that MPEG encoders can produce an asynchronous data transmission format has encouraged modem manufacturers to take advantage of it in their designs. Instead of being restricted to a rigid telecom hierarchy of 45 Mbps, or some multiple thereof, there now exists the possibility of supporting any desired rate in a given bandwidth. For example, if there is a need to transmit one or more sources of program material at more than 45 Mbps but less than 90 Mbps, a single ASI stream could transport all of the MPEG data and elementary streams, without the need to bit-stuff the remaining unused bandwidth. Where the benefit as well as the complication reside is in the ability to optimize the bandwidth as a function of the required transport data rate by adjustment of the constellation symbol rate, and thus allowing more efficient use of the spectrum.

Heterodyne Radio

The heterodyne microwave process generates new frequencies by mixing two (or more) signals in a nonlinear device, such as a transistor or diode, resulting in the creation of two (or more) new frequencies at the sum and the difference of the original frequencies. The new mixed or sum frequency is typically that of a signal reference or local oscillator, and that of a lower frequency, such as a modulated 70 MHz source. The lower applied frequency in this discussion is a signal originating from a digital modulator. The second frequency (known as a beat frequency) is rejected in the image reject mixer.

The heterodyne principle, as used in receivers, allows certain obstacles in high-performance radio design to be overcome. Tuned RF receivers would suffer from poor selectivity and frequency stability issues due to high bandwidths, even with high-Q filters at microwave radio frequencies.

In a heterodyne receiver design, all microwave RF signal frequencies that the receiver can tune to are converted to a fixed lower frequency before detection. This fixed lower frequency is called the intermediate frequency, or IF.

A heterodyne receiver mixes a signal from its local oscillator, which is tuned above or below the RF frequency by the IF difference frequency, in this case 70 MHz. When the user tunes the radio to its desired RF frequency, it is actually being tuned by adjusting the local oscillator to the desired 70 MHz difference frequency. In the low-noise converter or mixer, a signal output is produced at both the sum and the difference of the two input frequencies, and either the higher or lower frequency is chosen as the IF, while all others are filtered out. The IF is then equalized and amplified (and may have AGC) for delivery to the appropriate demodulator.

The key advantage to this approach is that the IF signal path is required to pass only a narrow bandwidth, however the RF amplifier stage must be sensitive to a wide range of frequencies without overloading in the presence of strong multiple inputs.

The input of a receiver in the 6 GHz BAS band is typically capable of supporting RF bandwidths of more than 300 MHz (6.8–7.1 GHz), requiring the front end to maintain the capability of receiving and linearly delivering a selected signal within this band to the low-noise converter/mixer. Prefiltering at RF helps eliminate undesired out-of-band interference. Once the signal is processed and downconverted to the receiver IF, the IF filtering sets the final bandwidth according to the modulation scheme and provides

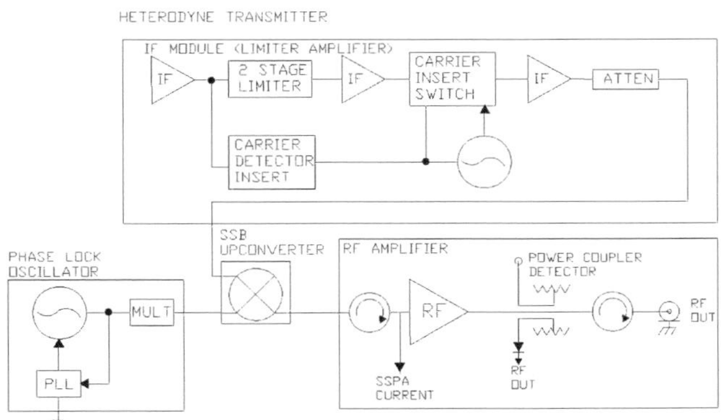

FIGURE 9.1-22 Heterodyne transmitter—functional block diagram.

further discrimination from unwanted mixer products as well as undesired adjacent channel interference.

A typical heterodyne transmitter block diagram is shown in Figure 9.1-22. It consists of upconverting the mixer with the output of an IF amplifier on one port and a microwave source feeding a second port to produce the desired RF frequency at the output port. One sideband, usually the upper, is selected by a filter or an image-reject mixer, or both. The mixer output, which is typically at a level of around 0 dBm, is then amplified by 30–40 dB to provide the final transmitter output power.

The same basic receiver design is used in directly modulated and heterodyne systems. Figure 9.1-23 is a block diagram of a typical receiver. In fixed link (STL) equipment, the RF input normally includes a channel filter to facilitate RF multiplexing with other receivers or transmitters, suppress local oscillator radiation, and eliminate any response at the image frequency, which is ±140 MHz from the received signal with a 70 MHz IF. In Figure 9.1-23, the input is amplified at the RF frequency by a low-noise amplifier (LNA) before the image-rejecting mixer. Following the mixer an IF amplifier, filter, and equalizer provide a flat passband with high adjacent channel rejection followed by a high gain amplifier with AGC. This latter feature ensures a constant level of 70 MHz at the receiver output with input level changes of 50–60 dB in the received carrier level (RCL).

The local oscillator may be a dielectric resonance (DRO)-type oscillator in the remodulating case, where a stability of 0.005% is sufficient as any errors are not passed on to the following transmitter. In the heterodyne case, frequency errors will add up requiring a stability of 0.001% or better, or a phase-locked AFC circuit, which will ensure a stable 70 MHz output.

Optimum receiver noise performance is achieved when the minimum necessary bandwidth is used, typically 20–25 MHz between the 3 dB points for 7 and 13 GHz radios. The licensed RF channel bandwidth will dictate IF filter bandwidth, however, the characteristics of a filter-equalizer must be carefully controlled for phase versus amplitude linearity.

The main IF amplifier, which follows the filter, typically provides 50–65 dB of gain. The RF input level at the front end of the receiver (referred to as the received signal level, or RSL) will vary between –20 and –88 dBm, and the output must remain substantially constant at –10 dBm in digital mode, or +5 dBm in analog mode, requiring a wide range IF AGC to be incorporated. When very large signal variations are expected, as in ENG and portable applications where the path length can be very short, AGC action may be extended to the LNA to help prevent overloading of the mixer.

Hybrid Radio Systems

The broadcast industry faced a challenge in the late 1990s and into the twenty-first century. The FCC mandated the delivery of digital television (DTV) to the broadcast audience beginning on November 1, 1998, and to cease analog transmission by February 17, 2009. The ATSC developed digital system guidelines to enable uniform delivery of the digital transport and all of its associated data services. The challenge to microwave radio manufacturers was to transport this signal over the STL at a data rate of 19.392 Mbps, while continuing to support analog until the 2009 cutoff date.

During the early implementation of this standard, many hurdles were presented. Most of the current MPEG-2 broadcast encoders offer an ASI interface as a standard output. Most exciters now offer the choice of converting an ASI to SMPTE 310M internally, or taking SMPTE 310M directly. In addition, MPEG-2 encoder and digital modem manufacturers have improved their products to meet more stringent transmission parameters. Frequency and clock drift translate into "jitter," and jitter tolerance measured in the tens of nanoseconds is required. Figures 9.1-24 and 9.1-25 show the jitter transfer tolerance masks.

The unit of digital radio performance is BER rather than S/N noise accumulation, which was the measure for analog radio link design. The BER is the ratio of the number of *errored* bits received to the *total* number of bits received over a given time interval. In digital telephony systems, the BER threshold is approximately 1×10^{-3}; this was established around the distribution of voice data. In digital video radio systems, the BER threshold is approximately 1×10^{-6}; this criterion is established around the video encoding and decoding equipment.

Digital television transmitter designs are necessarily more sophisticated than their analog counterparts. Network management, control, and monitoring may require wider data channels to transport the data needed to support Ethernet and web browser interfaces. Graphical representations of transmitter control points can be highly sophisticated applications, with some employing their own MAC layers to communicate. A customized graphical user interface (GUI) is typically employed in most microprocessor-controlled devices today.

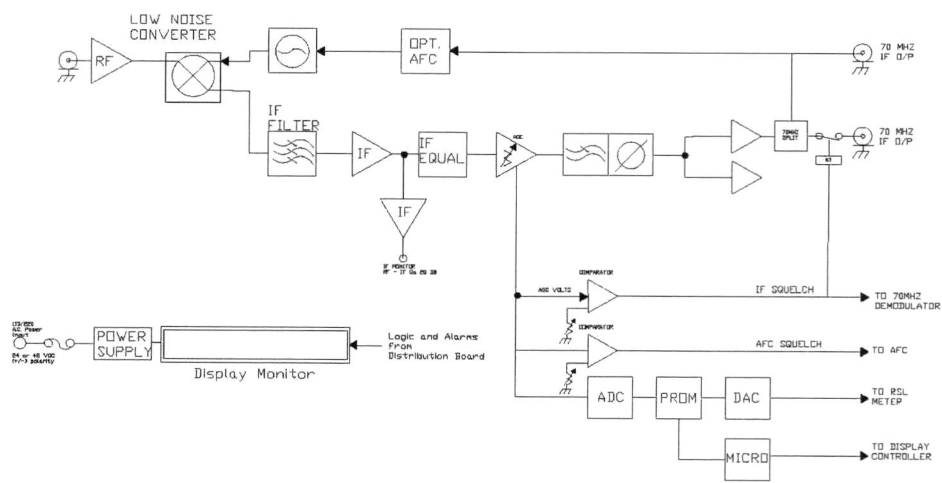

FIGURE 9.1-23 Heterodyne receiver—functional block diagram.

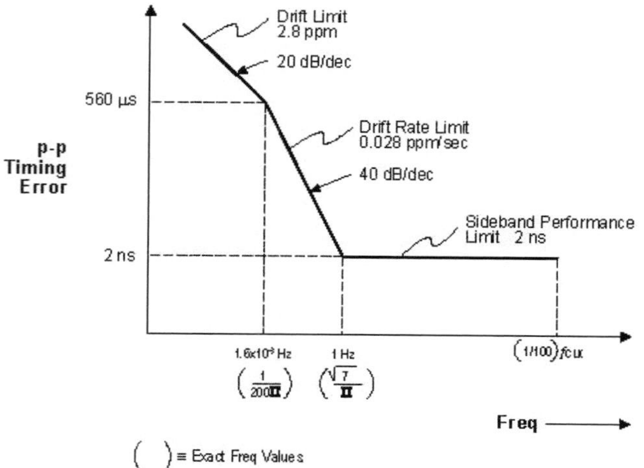

FIGURE 9.1-24 SMPTE 310M jitter template.

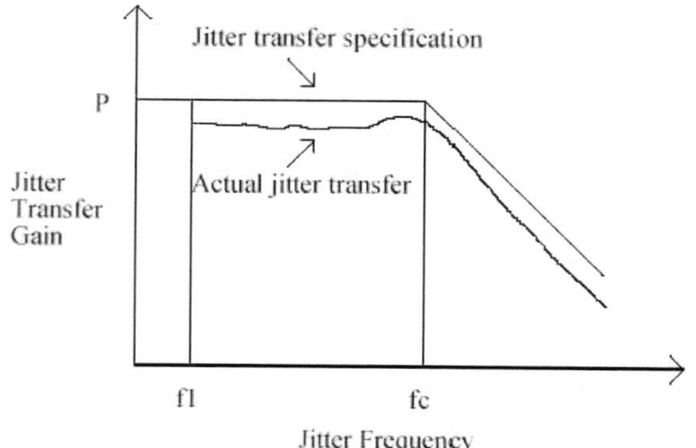

FIGURE 9.1-25 SMPTE RP-184 jitter transfer.

During the ATSC transition, broadcast stations and networks must continue to provide their analog transmission services while viewer demand for digital reception continues to grow. The desire to maintain the traditional "purity" of analog transmission, while supporting a new ATSC digital stream, created an interesting challenge, especially in crowded metro areas where additional frequencies are in short supply.

What developed was a new patented design known as the "hybrid" radio. The FCC adopted this design and instituted new emission designators. This new concept addressed a technological solution to all of the new and existing requirements, and delivered a range of products to the broadcast industry.

The hybrid radio is efficient in its design in that it takes full advantage of the available 25 MHz BAS channels at 7 and 13 GHz to deliver a combined payload. Dubbed *TwinStream* by MRC, and *DualStream* by Nucomm, the hybrid design contains two discrete transmission technologies that enable continuation of the NTSC transport at full deviation with up to four standard audio subcarriers, and delivers the full ATSC transport stream at its required 19.34 Mbps data rate, plus an added "wayside" data channel with up to an E1 rate capability.

The transmitter and receiver are essentially of heterodyne design incorporating both analog and digital modulation.

The Hybrid (Heterodyne) Transmitter

The hybrid transmitter enables the simultaneous transport of the legacy analog and ATSC transport stream television signals. Traditional FM video with multiple audio subcarriers at full deviation allows preservation of the signal quality to the analog TV transmitter, while the ATSC signal employs digital modulation techniques (typically 16 QAM) to advance the industry capability of providing both STL services without the need for added RF channel capacity.

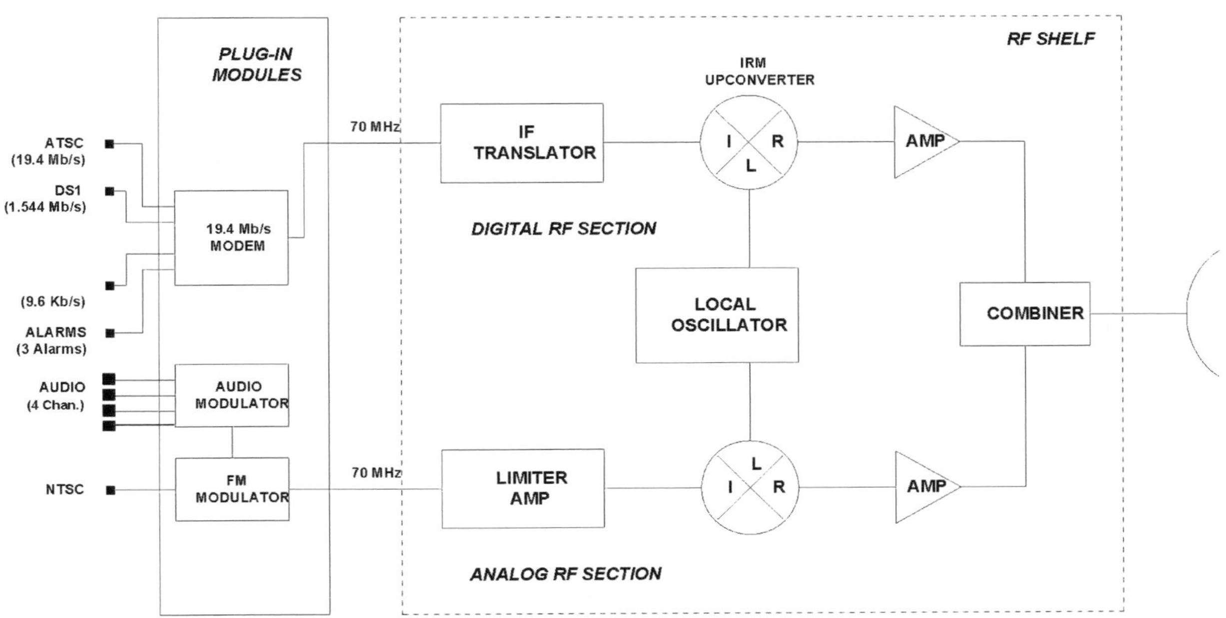

FIGURE 9.1-26 Hybrid transmitter single oscillator—functional block diagram.

The transmitter is a heterodyne design effectively employing several discrete signals in parallel within the same RF channel. The final amplification stage of each signal takes place prior to summing or combining the discrete RF content into a single RF channel filter for delivery to the antenna. Alarm and control functions effectively monitor both paths.

Figures 9.1-26 and 9.1-27 show two designs of the transmitter. One employs a single local oscillator with frequency translation to satisfy both discrete RF frequencies. The second and more robust design employs dedicated oscillators for each signal path, offering a form of layered, fail-soft protection ensuring that both RF paths are not reliant on a single point of failure.

The spectrum is managed by strategically setting the LO frequencies for plus and minus offsets from the center channel frequency. A full set of compliance measurements were taken to ensure that FCC Part 74 spectral mask requirements are met. Intermodulation or interference between the two carriers is negligible due to C/I and frequency separation along with the spectral efficiency of the digital modulator. Proper selection of the power ratios and the C/I frequency separation allows the transmit PA to operate in their most linear regions to maintain acceptable digital signal quality. The power reduction required to avoid compression in the digital mode is approximately 6 dB below saturated power at 16 QAM modulation.

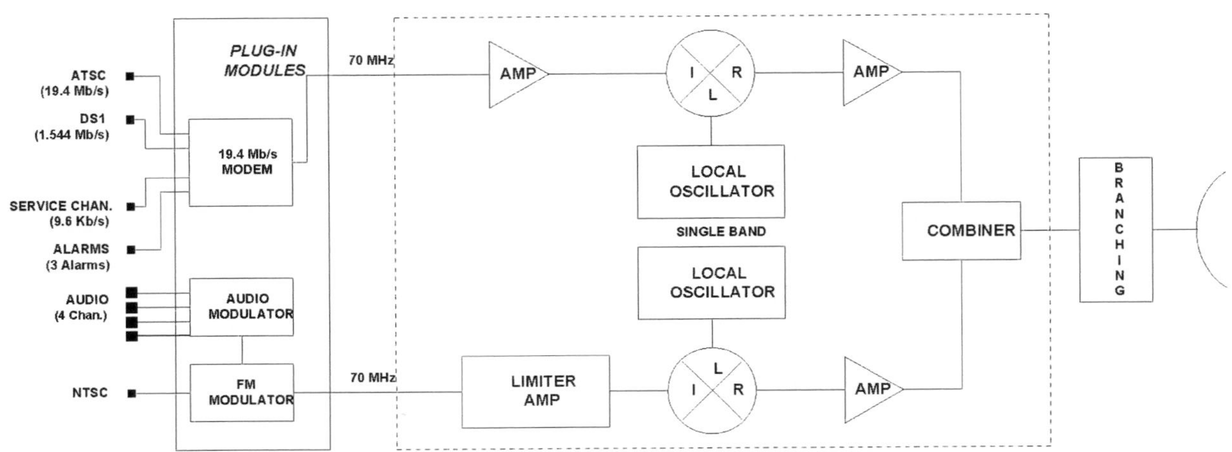

FIGURE 9.1-27 Hybrid transmitter dual oscillator—functional block diagram.

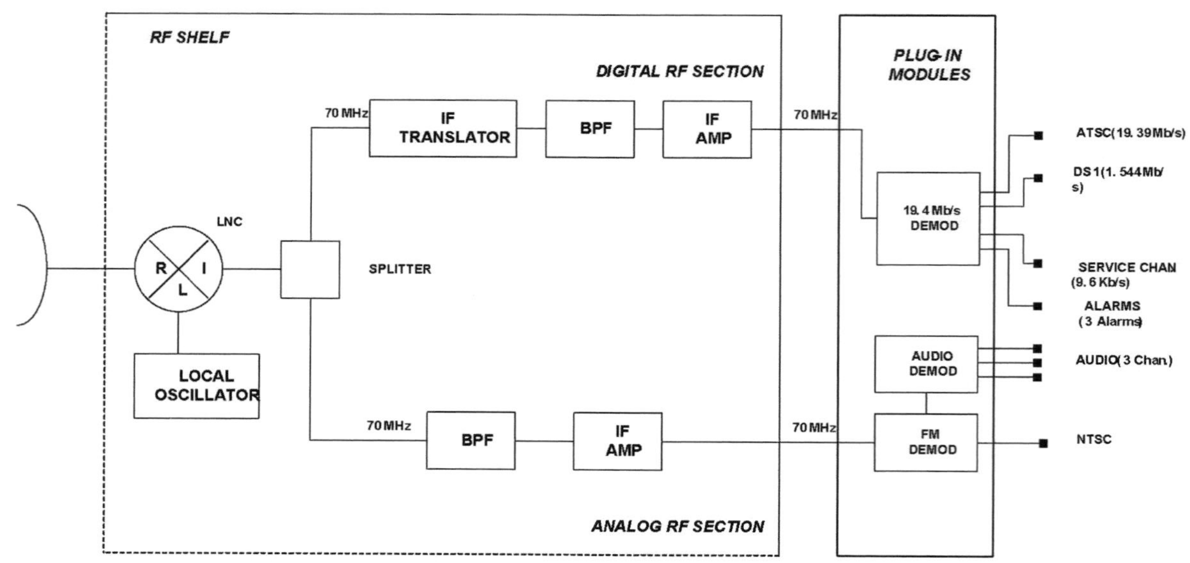

FIGURE 9.1-28 Hybrid receiver single oscillator—functional block diagram.

The Hybrid (Heterodyne) Receiver

The hybrid receiver is a complement to the transmitter nearly identical in its simplicity (Figures 9.1-28 and 9.1-29). Once again, as in the transmitter single- or dual-oscillator design, all baseband and RF components allow for two discrete signal paths. The high stability oscillators (.0005%) allow for virtually no frequency drift to ensure the discrete RF paths are protected from internal intermodulation errors due to intolerable frequency errors. Maintenance of these systems is relatively simple—typical analog practice for video and audio verification should be implemented on a periodic basis.

Digital verification of the transmission path may be observed through several internal diagnostic applications embedded into the modem design. Internal BER threshold monitoring is typical as well as internal generation of a bit stream. The internal generator will typically provide either an all ones (A1S) or "pseudo-random" bit sequence (PRBS) at the transmitter modulator. The demodulator will decode this test data stream and analyze this for bit error rate (BER) performance. The demodulator may have an actual BER display or indication that bit errors are being accumulated. Under normal operation, these capabilities can be used to isolate the microwave link from the transport stream input to verify the integrity of the data channel.

Shown in Figure 9.1-30 is a simple STL using a DS3 solution. This solution does not preserve the "purity" of the analog signal delivery to the TV transmitter. This method uses a digital compression scheme and

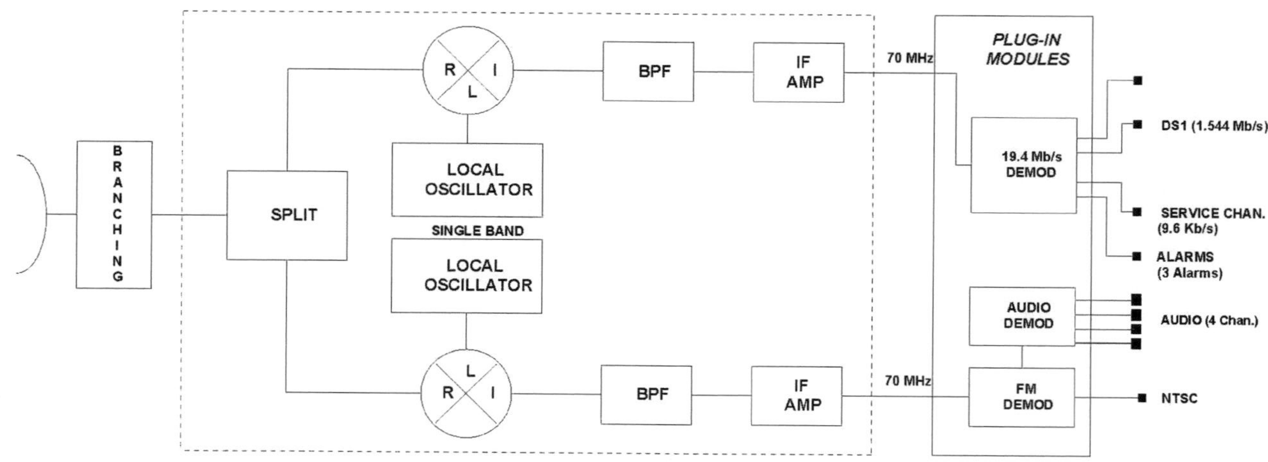

FIGURE 9.1-29 Hybrid receiver dual oscillator—functional block diagram.

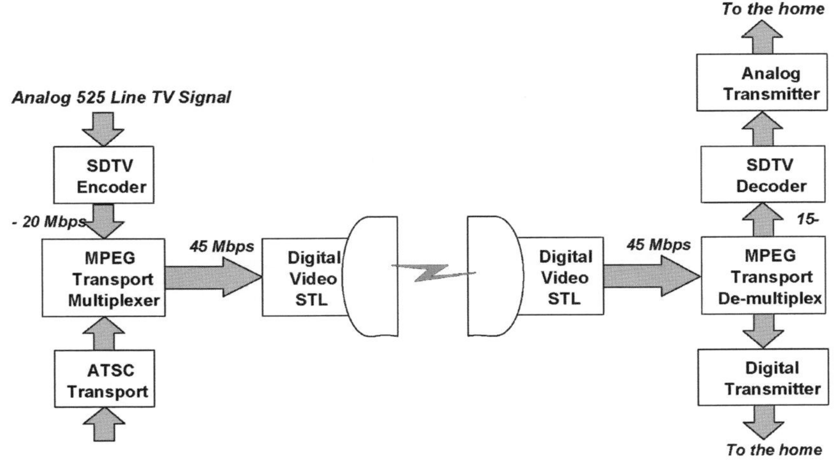

FIGURE 9.1-30 DS3 STL solution.

multiplex solution to combine the two elementary (analog and ATSC) streams into a single primary transport stream.

Digital Radio Systems

The future of digital transport solutions in the broadcast industry remains very realistic. Bandwidth today is being reallocated to provide more users the spectrum they require to service a very large wireless infrastructure. On August 8, 2004, the FCC released the relocation framework in 04-168, Report and Order, Fifth Report and Order, Fourth Memorandum Opinion and Order, and Order. This document covered:

- WT Docket 02-55
- ET Docket 00-258
- RM-9498
- RM-10024
- ET Docket 95-18

In essence, the FCC had taken a segment of the 2 GHz broadcast auxiliary services (BAS) band and reallocated it for advanced wireless services. In doing so, the FCC granted Nextel the right to operate on a portion of reallocated frequencies, provided that Nextel, at its expense, relocated all existing broadcast licensees as a condition of the grant, and do so within 30 months of Nextel's acceptance of the FCC rulemaking.

The relocation plan would move all 2 GHz fixed systems to their same channel designation under the "new" 12 MHz frequency plan with digital equipment, to help minimize interference by improving spectrum utilization.

FCC records confirm that there are still a substantial number of 2 GHz fixed links in service, which are allowed as a perfectly legitimate use of the band, in spite of its popularity for ENG and remote pickup. Estimates are that a substantial number of STLs will remain in 2 GHz for a variety of reasons, with long paths at the top of the list.

Reallocation of spectrum is a real possibility in other portions of the BAS bands as well. Today, analog microwave is viewed as a bandwidth hog. Analog STLs are, by today's standards, archaic and nonefficient users of valuable spectrum.

Digital compression along with microwave transport via modern digital modulation techniques mitigate this perspective. It is not only possible but very much in practice today to transport multiple video, audio, and data services in a single efficient manner.

In a properly designed digital video microwave link, the major advantage to be had over the analog link is that the video S/N ratio is constant over the microwave receiver's operating range. This inherent digital feature provides constant picture quality over the radio receiver's dynamic range. Figure 9.1-31 shows the S/N comparison between an FM analog and 16-QAM DS3 radio system. The digital S/N is limited by the A/D converters employed in the video encoding system and not the received signal level (as in the analog case), as illustrated by Figure 9.1-31. The rule of thumb is 6 dB of S/N for every bit used in the A/D. For the digital system shown in Figure 9.1-31, the video encoder employs a 10 bit A/D.

In the coming years, there will be a great temptation to replace analog radios with digital radios in many of the existing links around the country. However, simply replacing an analog radio with a digital radio will not always provide the performance shown in Figure 9.1-31. It is the responsibility of the broadcast engineer to ensure the path will meet acceptable availability criteria. Fortunately, the broadcast engineer will not be alone during this transition period. Radio manufacturers and consultants are available to analyze existing links and to determine what changes are necessary to provide reliable performance. If the broadcast engineer is so inclined, there are many software analysis programs available to aid in link design.

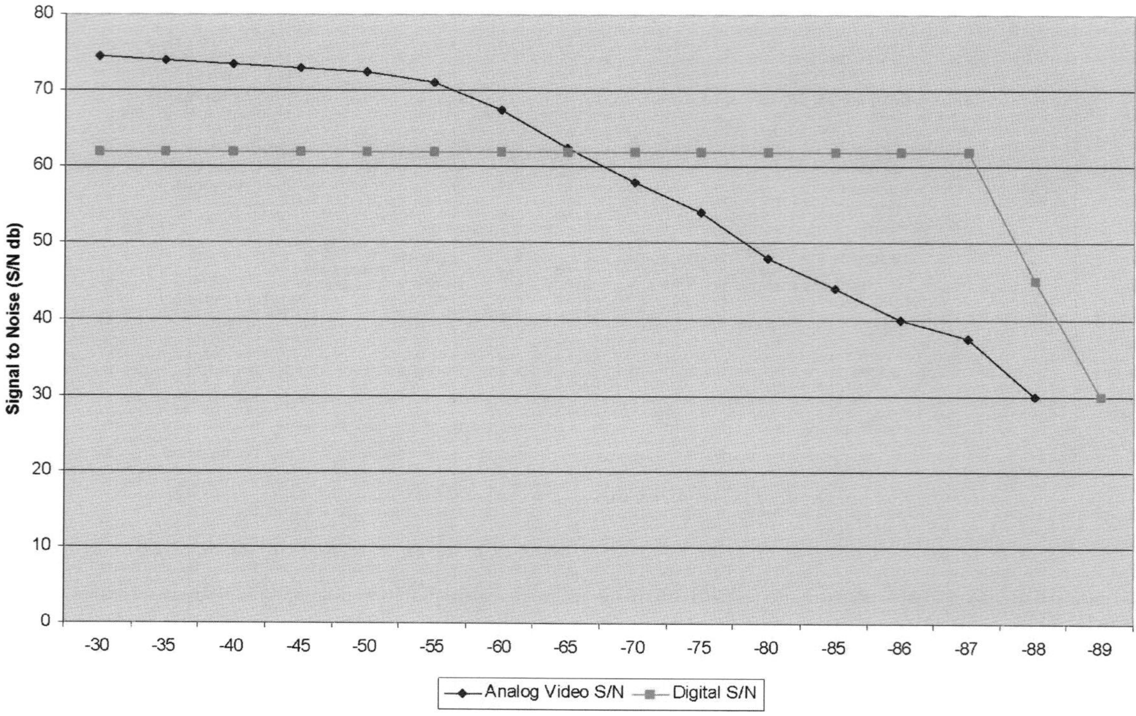

FIGURE 9.1-31 Video signal-to-noise ratio versus received signal level in dBm for analog and digital video signals.

Modulation Techniques

The FM system is one in which the frequency of the carrier is caused to vary in accordance with some specified information-carrying signal. For the simple case of a sinusoidal modulating signal at frequency f_m, the corresponding frequency modulated sign is given by $f(t)$ where:

$$f(t) = a \cos \omega_m t,$$

where, a is the fixed amplitude of the modulated signal and ω_m ($\omega_m = 2\pi f_m$) is the rate of modulation of the modulating signal.

Simply stated, there is a fixed amplitude sine wave of varying angular frequency ω_m. In the FM system all of the information is encoded as a function of the continuously varying angular frequency ω_m.

In digital systems, the amplitude, frequency, and phase properties of the RF are quantized by the modulating signal. Digital implies a fixed set of discrete values. A digital radio waveform, then, can assume one of a discrete set of amplitude levels, frequencies, or phases as a result of the modulating signal.

Frequency Shift Keying

Frequency shift keying (FSK) consists of shifting the frequency of a sinusoidal carrier from a mark frequency (corresponding to a binary 1) to a space frequency (corresponding to a binary 0). This can be expressed as:

$$f_{c1}(t) = a \cos w_1 t$$
$$f_{c0}(t) = a \cos w_2 t,$$

where a is the fixed amplitude of the modulated signal, a binary one corresponds to frequency f_1, and a binary zero to frequency f_2. An alternative representation of an FSK waveform consists of letting $f_1 = f_c - \Delta f$ and $f_2 = f_c + \Delta f$. The two frequencies then differ by $2\Delta f$ hertz and thus we can write:

$$f_c(t) = a \cos (w_c \pm \Delta w)t,$$

where $\Delta\omega = 2\pi\Delta f$. The frequency then deviates $\pm\Delta f$ about f_c. Δf is commonly called the *frequency deviation*.

Phase Shift Keying

Phase shift keying (PSK) consists of shifting the phase of a sinusoidal carrier 180 degrees with a unipolar binary signal. This can be expressed as:

$$f_c(t) = \pm \cos w_c t.$$

Here a binary 1 in the baseband binary stream corresponds to positive polarity and a binary 0 to negative polarity. This modulation is commonly called *binary phase shift keying* (BPSK).

For BPSK, the values of 0 degrees and 180 degrees are typically used. These values are used to maximize (or to separate) the decision distance. There is no reason why 45 degrees cannot be assigned as the binary 1

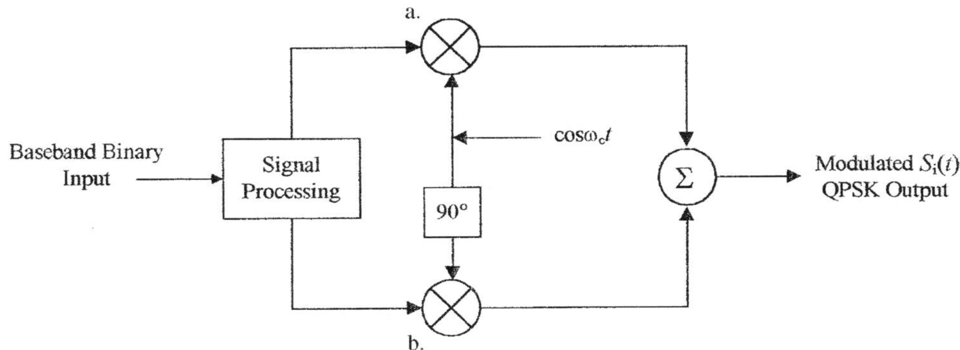

FIGURE 9.1-32 Diagram of QPSK modulator. [2]

and 225 degrees as the binary 0, as long as the distance between the two states is maximized.

Continuing along this line, instead of using two phase states, four can be used, each separated by 90 degrees. The binary values to each of the four phase states can now be assigned as: 0° = 0,1; 90° = 0,0; 180° = 1,1; and 270° = 1,0. This type of multisymbol signaling is commonly called *quaternary phase shift keying* (QPSK) modulation or 4-PSK modulation. It can be expressed as:

$$S_i(t) = a_i \cos w_c t + b_i \sin w_c t.$$

A simplified block diagram of a QPSK modulator is shown in Figure 9.1-32.

It is useful to represent the signal of $S_i(t)$ in a two-dimensional diagram locating the various points (a_i, b_i). The horizontal axis corresponding to the location of a_i is called the *in-phase axis*. The vertical axis, along which b_i is located, is called the *quadrature axis*. The four signals of $S_i(t)$ assigned above then appear as shown in Figure 9.1-33(a). The points are said to represent a signal constellation, or as it is sometimes called, a *signal state diagram*.

More general types of multisymbol signaling schemes may be generated by letting a_i and b_i take on multiple values themselves. The resultant signals are called *quadrature amplitude modulation* (QAM) signals. These signals may be interpreted as having multilevel amplitude modulation applied independently on each of the quadrature carriers. A simplified block diagram of a QAM modulator is shown in Figure 9.1-34. The output of such a modulator can be expressed as:

$$S_i(t) = r_i \cos\left(w_c t + \theta_i\right),$$

where the amplitude r_i and the phase angle θ_i are given by the appropriate combinations of (a_i, b_i). The signal constellations for 4-, 16-, and 64-QAM modulation are shown in Figure 9.1-33.

If the constellations shown in Figure 9.1-33 were measured with a vector signal analyzer (VSA) they would demonstrate nearly perfect vector alignment. This would be representative of maximum signal-to-noise, low-phase noise, good error vector magnitude (EVM), and modulation error ratio (MER).

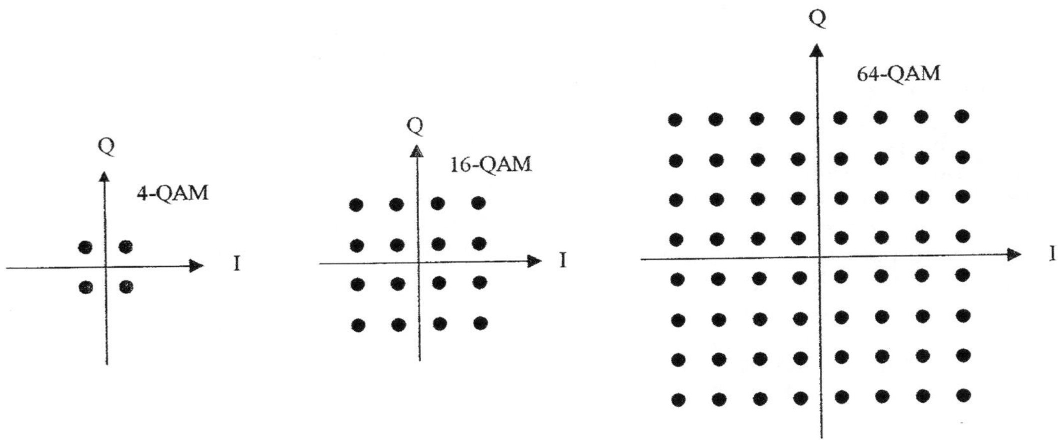

FIGURE 9.1-33 Signal constellations for (a) 4QAM (identical to QPSK), (b) 16 QAM, and (c) 64 QAM. [2]

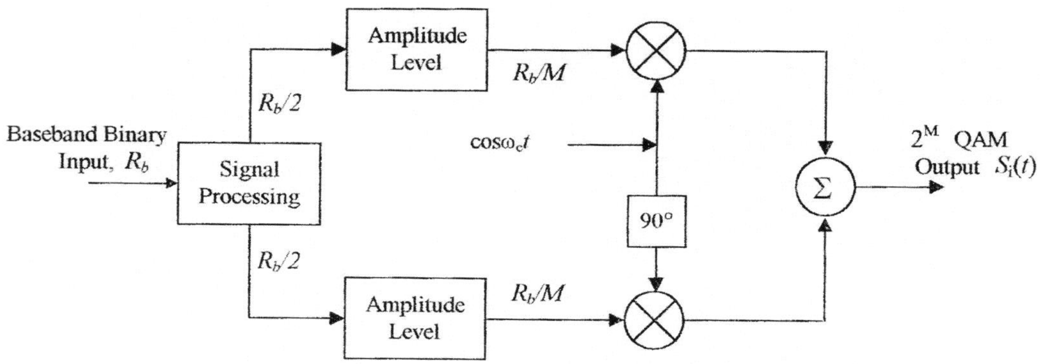

FIGURE 9.1-34 Diagram of QAM modulator. [2]

Additional constellations are shown in Figure 9.1-35. The tool used for these complex measurements is a VSA. This is a tool that is rapidly replacing the swept-tuned spectrum analyzer as a design and diagnostics tool by RF engineers. The ability of the VSA to time-capture and record these complex waveforms allows for the in-depth time domain and frequency spectrum analysis. These analyzers combine high-speed digital signal and analog-to-digital conversion processes, super heterodyne frequency conversion, and advanced time domain analysis.

In each part of Figure 9.1-35, the upper left quadrant is a vector display of a QAM modulated signal's I (in-phase) and Q (quadrature) magnitude components. This is also referred to as the constellation display. The display in the lower left quadrant is representative of the swept-tuned frequency spectrum of the modulated input signal under test. The upper right quadrant represents the time component of the QAM signal. Digital modulation is three-dimensional; the IQ constellation represents two of these dimensions, the time domain being the third dimension. Last, at the lower right, is the symbol table. This represents the error summary data for all of the analyzed bits represented below the data table.

EVM analysis is a useful tool for transmitter and receiver signal chain optimization and allows prediction of dynamic system performance.

Notice that in all cases shown in Figure 9.1-35 a large percentage of the symbols are falling outside the appropriate vector alignment. This is representative of noise due to poor path performance and radio distortions (such as third-order intermodulation distortion, also called spectral regrowth) at the amplifier stage. This analysis shows poor pedestal definition in the spectrum display due to poor C/N; note the blurred constellation points, and the pedestal being only a few dB above the noise floor. Depending on where these measurements were taken this could represent poor amplifier performance due to oversaturation at the amplifier input (nonlinear operation) or, if at the receiver IF, this could represent poor path performance (low signal level or multipath distortion) or signal gain error in the down-conversion stage. Frequency errors could amount to signal distortion, however, they would not account for the system noise floor as viewed in this spectrum display.

After careful examination of the PSK and QAM signal constellations, one begins to appreciate the complexity required in a modem that can resolve the amplitude and phase information in multisymbol signaling schemes. One has to be more impressed with the modem that can resolve this information when the digital radio system is subject to external noise factors. To get a feel of how different digital modulations compare in the presence of thermal noise we can look at the bit energy-to-noise density ratio, E_b/N_o, where E_b is the energy per bit and N_o is the single-sided noise spectral density. E_b/N_o can be expressed as the ratio of receive signal level (RSL) to the bit rate; in terms of log arithmetic, this can be written as:

$$E_b = \text{RSL}_{\text{dBm}} - 10\log(\text{bit rate}).$$

The single-sided noise spectral density of a perfect receiver in a 1 Hz bandwidth with a noise figure (NF) can be expressed as:

$$N_o = -174\text{dBm} + \text{NF}_{\text{dB}}.$$

The expression for the bit energy-to-noise density ratio is:

$$E_b/N_o = \left(\text{RSL}_{\text{dBm}} - 10\log(\text{bit rate})\right) - \left(-174\,\text{dBm} + \text{NF}_{\text{dB}}\right)$$
$$E_b/N_o = \text{RSL}_{\text{dBm}} - 10\log(\text{bit rate}) + 174\,\text{dBm} - \text{NF}_{\text{dB}}$$

This expression shows that as RSL increases, so does the E_b/N_o. In other words, the greater that E_b/N_o needs to be, the less sensitive the receiver will be. This will be an important factor when determining the type of modulation to employ when converting existing analog links to digital.

Table 9.1-4 shows the E_b/N_o ratios based on a BER of 10^{-6} for some common digital modulation formats and their bandwidth efficiencies based on the Nyquist bandwidth[1] (bandwidth numerically equal to the bit

[1]The Nyquist bandwidth, in this context, represents the minimum theoretical bandwidth required to transmit the modulated signal if a brick wall filter were used in the modulation process. As a practical matter, the actual bandwidth required is always greater than this.

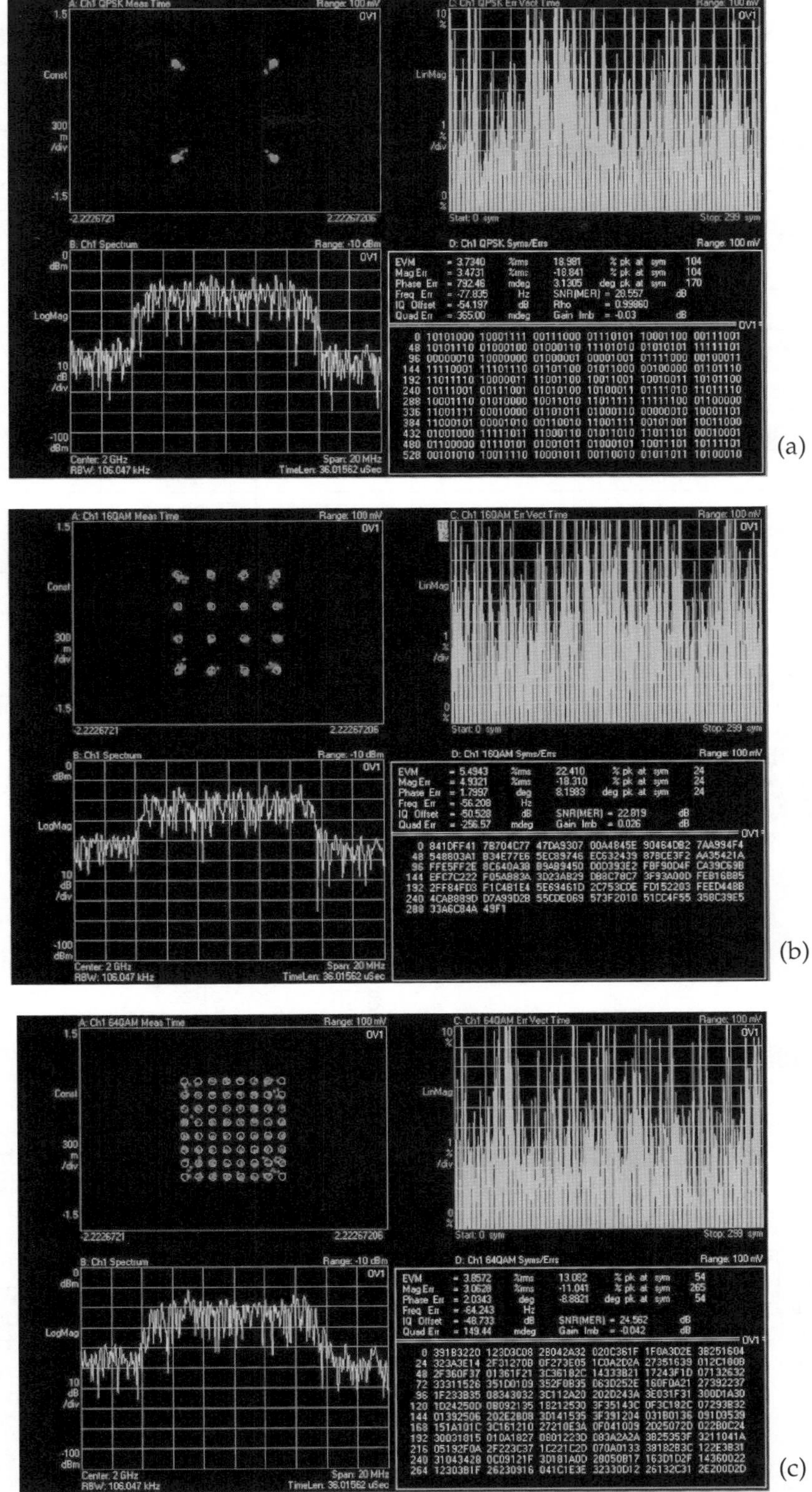

FIGURE 9.1-35 (a) Diagram of 4 QAM (QPSK) signal showing signal distortion, (b) diagram of 16 QAM signal showing signal distortion, and (c) diagram of 64 QAM signal showing signal distortion.

TABLE 9.1-4
E_b/N_o at a BER of 10^{-6} for Various Modulation Schemes (without FEC)

Modulation	E_b/N_o (dB)	S/N (dB)	Nyquist Bandwidth
2-state FSK with discriminator detection	13.4	13.4	B
3-state FSK (duo-binary)	15.9	15.9	B
4-state FSK	20.1	23.1	B/2
2-state PSK with coherent detection	10.5	10.5	B
4-state PSK with coherent detection	10.5	13.5	B/2
8-state PSK with coherent detection	14.0	18.8	B/3
16-state PSK with coherent detection	18.4	24.4	B/4
16-QAM with coherent detection	10.96	12.8	B/4
32-QAM with coherent detection	12.76	15.24	B/5
64-QAM with coherent detection	17.79	20.46	B/6
128-QAM with coherent detection	20.96	24.04	B/7
256-QAM with coherent detection	26.71	30.63	B/8

rate). The parameters listed in Table 9.1-4 are idealistic because the only source of errors is due to thermal noise in the receiver. No modulation implementation loss is considered. With the information given in Table 9.1-4 and from the E_b/N_o equation above, the receiver threshold for a T3 (45 Mbps) 16-QAM radio system with a radio noise figure of 4 dB can be calculated.

If we solve the E_b/N_o equation for $RSL_{(min\ dBm)}$ we have the expression:

$$RSL_{(min\ dBm)} = E_b/N_o + 10\log(\text{bit rate}) - 174\text{ dBm} + NF_{dB}$$

$$RSL_{(min\ dBm)} = 17.0\text{ dB} + 10\log(44.736\times 10^6) - 174\text{ dBm} + 4\text{ dB}$$

$$RSL_{(min\ dBm)} = -76.5\text{ dB}.$$

For a 64-QAM system the $RSL_{(min\ dBm)} = -71.0$ dBm.

For a 256-QAM system the $RSL_{(min\ dBm)} = -65.7$ dBm.

The example clearly shows that for higher-order modulation schemes a greater E_b/N_o is required to enable the demodulator to extract the phase and amplitude information from the modulated signal. This inherent digital radio feature also causes the system gain to decrease as the modulation complexity increases. The previous example limits the distance between the transmitter and receiver as a function of the modulation scheme.

As a means to increase system gain, modem designers employ error-correcting codes (or forward error correction, FEC) to help improve performance. Error-correcting coding usually requires redundancy (increasing the bit rate) and, therefore, poses a contradictory requirement to obtaining maximum spectral efficiency. There is no set standard for the amount of FEC a manufacturer will employ in a given modem. During the modem design phase, tradeoffs are made concerning BER performance, dispersive fade margin, spectral efficiency, and cost.

For the applications being discussed in this chapter, satellite modems require more FEC than terrestrial modems to improve the system performance at threshold. Satellite links cover great distances and thermal noise becomes the dominant impairment. The composite bit rate at the output of a satellite modulator can be as much as 50% higher than the data rate into the modulator and as much as 10 dB of coding gain can be realized.

A well-designed terrestrial digital link is not fighting thermal noise as its major impairment. A modem designed for terrestrial use has to operate in a more hostile dispersive environment than the vacuum of space. The terrestrial modem designer has to balance coding gain, dispersive fade margin, spectral efficiency, and cost. Terrestrial modems typically employ 5–15% of error-correction redundancy. Table 9.1-5 shows the effect that 6.7% of error-correction redundancy has on QAM modulation (compare these values with the corresponding values in Table 9.1-4).

Again, FEC is not a cure-all for digital radio performance. The amount of error-correction redundancy is up to the discretion of the modem designer and is just one of many factors that need to be optimized during the modem design.

Transmitter

The transmitter in the digital radio link is the device that upconverts the modulated signal, amplifies it, and then delivers it to the antenna system for transmission into free space. A simplified block diagram of a digital radio transmitter is shown in Figure 9.1-36.

TABLE 9.1-5
E_b/N_o at a BER of 10^{-6} for Various Modulation Schemes (with FEC Added)

Modulation	E_b/N_o (dB)	S/N (dB)	Nyquist Bandwidth
16-QAM with coherent detection	7.75	9.62	B/4 × (1 + .067)
32-QAM with coherent detection	10.83	13.32	B/5 × (1 + .067)
64-QAM with coherent detection	16.37	19.03	B/6 × (1 + .067)
128-QAM with coherent detection	19.83	22.97	B/7 × (1 + .067)
256-QAM with coherent detection	25.90	29.83	B/8 × (1 + .067)

As shown in Figure 9.1-36, an IF signal is first generated by the modulation portion of the system. The IF frequency is typically 70 or 140 MHz. This signal is then upconverted by an image reject mixer. This translation of the modulated IF signal up to the RF carrier frequency is made by a mixing operation or multiplication between two frequencies. The result of this mixing operation is a summation of frequency components that can be expressed as:

$$nF_{\mathrm{LO}} \pm mF_{\mathrm{IF}},$$

where $n = 1, 2, 3, \ldots$ and $m = 1, 2, 3, \ldots$.

The frequencies that are of most concern for the case when $n = m = 1$, or:

$$F_{\mathrm{LO}} + F_{(\mathrm{IF})}, \text{ and}$$
$$F_{\mathrm{LO}} - F_{(\mathrm{IF})}.$$

The desired frequency for transmission is:

$$F_c = F_{\mathrm{co}} + F_{(\mathrm{IF})}.$$

$F_{\mathrm{LO}} - F_{(\mathrm{IF})}$ is the difference frequency, which is commonly called the *image frequency*. The image frequency is typically suppressed 20–25 dB by the image rejection properties of the mixer.

In addition to the sum and difference frequencies, there will be some LO leakage present at the output port of the mixer. The LO leakage is attenuated approximately 20 dB by the mixer's LO to RF isolation. Some digital transmitter designs may employ a band pass filter to remove the image and LO leakage signals.

The local oscillator that drives the mixer in a digital radio is a high-stability, low-phase noise frequency source. The low-phase noise LO is required when implementing multisymbol modulation. Phase noise (or short-term instability, or jitter) is seen near the carrier and causes degradation of BER. As can be seen in the signal constellation diagrams shown in Figure 9.1-35, LO phase noise causes the constellation points to jitter about. If the LO in the transmitter and receiver were an ideal frequency source the constellation points would not move at all. The more phase noise the LO induces onto the modulated IF signal, the more difficult signal recovery becomes. Figure 9.1-37 shows the empirically derived phase noise requirements for QPSK, 16-QAM, and 64-QAM modulation. Notice that as the modulation complexity increases the LO phase noise needs to decrease.

Continuing the journey through the transmitter, the signal out of the frequency translating mixer is input into a linear operating amplifier (typically solid state). A linear operating amplifier is one that is operating below its 1 dB gain compression point. The 1 dB gain compression (called $G_{1\mathrm{dB}}$) is defined as the power gain where the amplifier's nonlinearity reduces the power gain by 1 dB over the small signal linear power gain. That is,

$$G_{1\mathrm{dB}}(\mathrm{dB}) = G_o(\mathrm{dB}) - 1,$$

where G_o(dB) is the small signal linear power gain in decibels. Since the power gain is defined as

$$G_p = P_{\mathrm{out}}/P_{\mathrm{in}}, \text{ or}$$

$$P_{\mathrm{out}}(\mathrm{dBm}) = G_p(\mathrm{dB}) + P_{in}(\mathrm{dBm}).$$

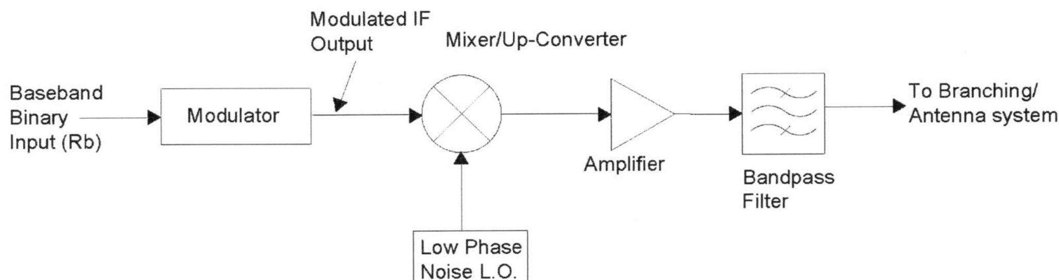

FIGURE 9.1-36 Simplified digital radio transmitter block diagram.

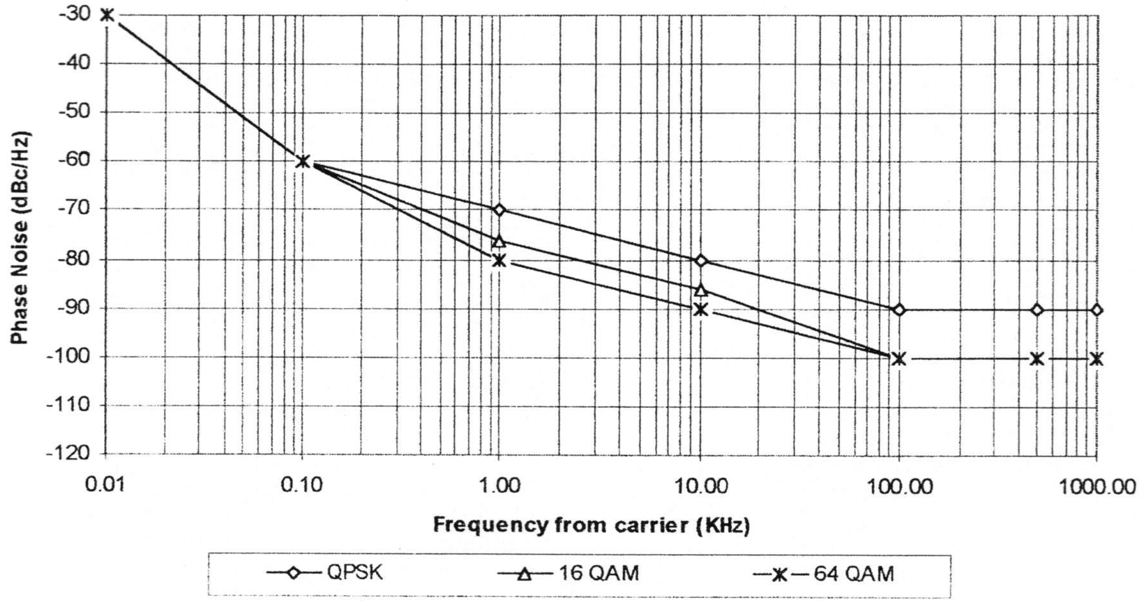

FIGURE 9.1-37 LO phase noise required for QPSK, 16-QAM, and 64-QAM. [2]

The output power at the 1 dB gain compression point, called P_{1dB}, is

$$P_{1dB}(\text{dBm}) = G_{1dB}(\text{dB}) + P_{in}(\text{dBm}).$$

Substituting $G_{1dB}(\text{dB}) = G_o(\text{dB}) - 1$ into the above equation gives

$$P_{1dB}(\text{dBm}) - P_{in}(\text{dBm}) = G_o(\text{dB}) - 1.$$

The equation shows that the 1 dB gain compression point is that point at which the output power minus the input power in dBm is equal to the small signal power gain minus 1 dB. A typical plot of P_{out} versus P_{in} illustrating the 1 dB gain compression point is shown in Figure 9.1-38.

In the digital radio transmitter that employs linear modulation, distortion can be caused by the power amplifier operating near or beyond G_{1dB}(dB). This distortion is caused by intermodulation (IM) products that arise when two or more sinusoidal frequencies are applied to a nonlinear amplifier. The output of the amplifier will contain additional frequencies called *intermodulation products*. For example, if two sinusoidal signals

$$v(t) = a \cos 2\pi f_1 t + b \cos 2\pi f_2 t$$

are applied to a nonlinear amplifier whose output voltage can be represented by the power series

$$v_o(t) = \alpha_1 v(t) + \alpha_2 v^2(t) + \alpha_3 v^3(t)$$

the output signal will contain frequency components at dc, $f_1, f_2, 2f_1, 2f_2, 3f_1, 3f_2, f_1 \pm f_2, 2f_1 \pm f_2$, and $2f_2 \pm f_1$. The frequencies $2f_1$ and $2f_2$ are the second harmonics, $3f_1$ and $3f_2$ are the third harmonics, $f_1 \pm f_2$ are the second-order intermodulation products (since the sum of the f_1 and f_2 coefficients is 2), $2f_1 \pm f_2$, and $2f_2 \pm f_1$ are the third-order intermodulation products (since the sum the f_1 and f_2 coefficients is 3), and so on. The input and output power spectra for a typical solid-state amplifier are shown in Figure. 9.1-39.

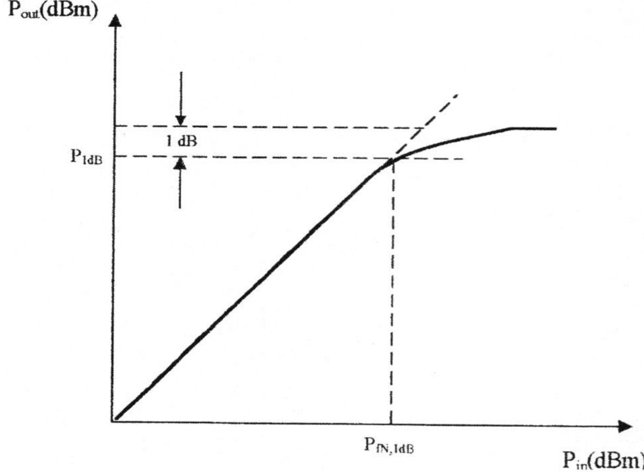

FIGURE 9.1-38 One dB amplifier gain compression point. [2]

Figure 9.1-39 shows that the third-order intermodulation products at $2f_1 - f_2$ and $2f_2 - f_1$ are close to the fundamental frequencies f_1 and f_2 and (typically) fall within the amplifier bandwidth, producing distortion in the output.

The digital radio system that employs linear modulation and operates the power amplifier near or into saturation (beyond G_{1dB}(dB)) will see distortions that are caused by the third-order IM products. The distortions can be seen as a spreading of the spectrum that

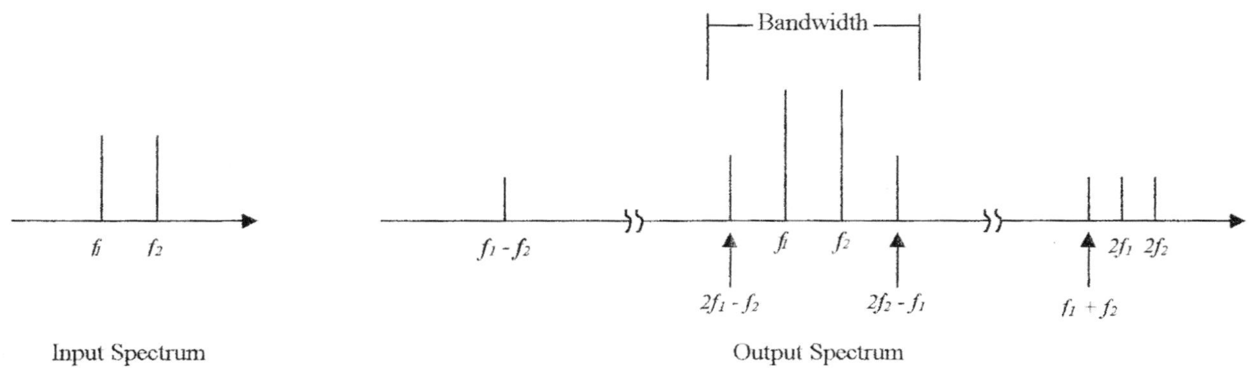

FIGURE 9.1-39 Input and output power spectrum for a typical solid-state amplifier. [2]

can occupy up to three times the bandwidth of the original spectrum. Third-order IM distortion will degrade BER performance and can cause interference to adjacent channels.

The severity of third-order IM distortion is a function of where the amplifier output level is operating in relation to its P_{1dB} point. In addition, the more complex the modulation scheme, the larger the required P_{1dB} back off to minimize IM distortion. Table 9.1-6 shows the typical power amplifier back offs for FSK, QPSK, and QAM modulation.

Figure 9.1-40 shows the spectrum plots of a 16-QAM 70 MHz modulated signal at the modem output and the same signal upconverted to 7,037.8 MHz at the transmitter output. In this case the transmitter power amplifier was backed off by 6 dB from P_{1dB}. It is important to note that even at this operating point, some spreading still occurs.

TABLE 9.1-6
Typical Power Amplifier P_{1dB} Back Off

System	Variants	Typical Back Off from P_{1dB}
FSK	2-state FSK with discriminator detection	0
	3-state FSK (duo-binary)	0
	4-state FSK	0
PSK	2-state PSK with coherent detection	1
	4-state PSK with coherent detection	2
	8-state PSK with coherent detection	4
QAM	16-QAM with coherent detection	6
	64-QAM with coherent detection	10
	256-QAM with coherent detection	12

Finally, the amplified signal is band pass filtered and directed to the antenna system for free-space transmission. At the far end of the free-space path the modulated microwave signal is captured by the receive antenna system.

Receiver

The receiver in the digital radio link is the device that receives the modulated signal from the antenna system; filters, amplifies, and downconverts it to IF; filters and amplifies it again; and then delivers it to the demodulator for signal processing back to baseband. A simplified block diagram of a digital radio receiver is shown in Figure 9.1-41.

In the digital link, the receiver performs the lion's share of the work. In Figure 9.1-33, the 4-, 16-, and 64-QAM signal constellations were shown. The figure shows that as the value of M (M = 4, 16, 64) increases, the space between the constellation points decreases. For the digital link to operate, the digital receiver must be capable of resolving which signal point was transmitted. For 64-QAM modulation it has to resolve 64 points, for 512-QAM modulation, it has to resolve 512 points!

All of the transmitter issues previously discussed concerning LO phase noise and amplifier linearity hold true for the digital receiver. In addition to these internal concerns, the receiver is also susceptible to outside disturbances that tend to deteriorate system performance. It is an understatement to say that the receiver must be capable of resolving which signal point was transmitted. The receiver must be capable of resolving which signal point was transmitted in the presence of transmitter distortions, receiver distortions, white Gaussian noise, cochannel and/or adjacent channel interference, and multipath distortions.

The transmitter and receiver internal distortions can be reduced by employing low-phase noise local oscillators and operating internal amplifiers sufficiently below their P_{1dB} point.

Additive white Gaussian noise is unavoidable. The bandwidth efficient modem requires a higher E_b/N_o for a given symbol rate as the number of bits per second per Hertz is increased.

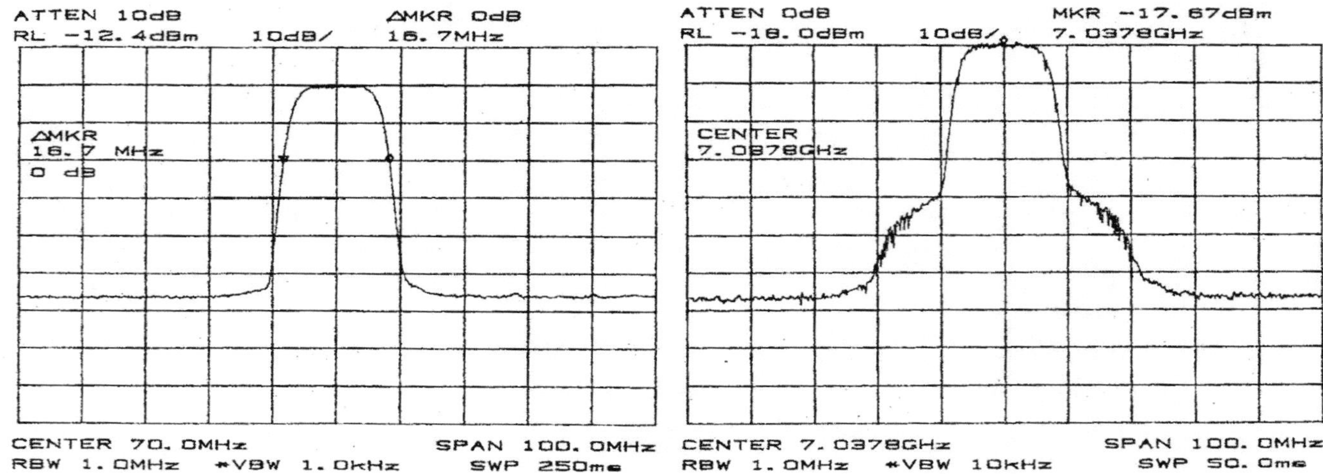

FIGURE 9.1-40 The RF spectrum for a 16-QAM signal at the modulator output and transmitter output. [2]

Cochannel and/or adjacent channel interference in most cases can be avoided by frequency allocation management, channel filtering, and meeting spectral masks.

Multipath distortions are caused by the transmission channel itself. To minimize dispersive distortions the broadcast engineer must design the link to avoid ground reflections, implement space and/or frequency diversity, and employ adaptive equalization.

System Considerations

The implementation of digital microwave links for both terrestrial LOS and ENG mobile radio links is accelerating in the broadcast industry. This is due primarily the emergence of MPEG-2 technology. The transmission of the ATSC suite of content (over the STL) typically contains one or more MPEG-2 compressed digital video signals as primary or elementary content on the transport stream. These digital video streams take the form of standard and or high-definition content.

MPEG-2 compression in the ENG environment has primarily been standard definition (SD) utilizing coded orthogonal frequency division multiplex (COFDM) transmission as the de facto industry standard for modulation.

Today, the industry sees the migration to high-definition (HD) field contribution and ENG. This approach to news and content delivery complicates the ENG effort all the more. The broadcast engineer must make decisions concerning the type of ENG infrastructure necessary to meet the station's requirements. Cameras, compression schemes, and interfaces all need to be considered relative to the available bandwidth. Current COFDM modulation will support specific bandwidth and bit rate requirements with restriction per the DVB modulation format for COFDM.

Current RF channel plans allow bandwidths to 17 MHz. This also is under change; the available bandwidth is soon to be narrowed to 12 MHz per the FCC 2 GHz band reallocation. With the convergence of varying MPEG and interface formats, the need for broader bit rates is at hand. Various types of modulation schemes are available to allow the broadcast engineer this capability.

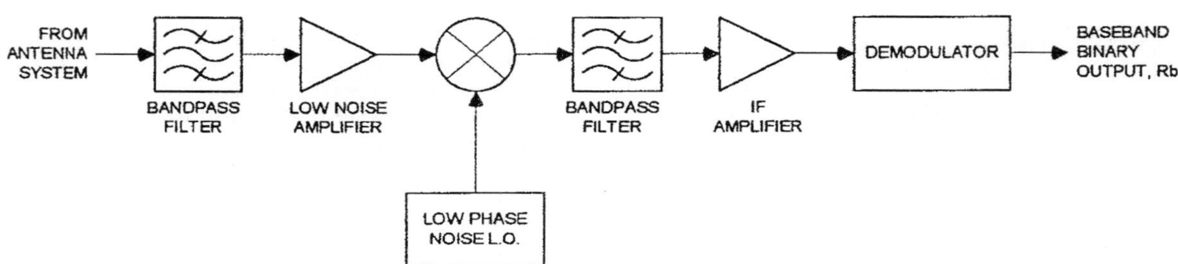

FIGURE 9.1-41 Simplified block diagram of a digital receiver from the 9th edition. [2]

High-definition transport from ENG vehicle or airborne platform is all in operation today. There has been unheralded success in transporting respectable bit rate, approximately 20 Mbps, from aircraft with COFDM modulation in QAM operating modes. Airborne challenges have been overcome by specifically concentrating on improved MER performance of the transmitter, minimizing ground reflections by incorporating auto elevation control of the antenna element, and minimizing the use of omnidirectional antennas. The next obvious step is to focus on the delivery of higher bit rates. With the improvements in the single-carrier core QAM modulator, and the demodulator equalizers, along with focus on improvements in the areas of multipath performance and frequency errors due to Doppler shift, it has been proven that single-carrier QAM modulation is adequate to meet these demands.

There continues to be many discussions on the appropriate modulation platform to deliver this bandwidth efficiently to the broadcast environment. Single-carrier modulation has been in use in Europe and Asia for many years. There, the infrastructure has been developed with this implementation in mind. Here in the United States, an analog infrastructure has been primarily in use for 40 years.

To help with this discussion there are several charts given here that compare the different modulation formats. The user needs to define what the goals and challenges for the broadcast station are. The chart in Figure 9.1-42 depicts COFDM and single-carrier modulation as applied to 12 MHz and 25 MHz channel bandwidths. Recently, tests have proven that delivering 36 Mbps from aircraft utilizing single-carrier modulation operates as effectively as COFDM at half the bit rate.

Strictly from an ENG BAS perspective it is important to understand exactly what is indeed possible. With the pending migration to 12 MHz channels, the broadcast environment has to operate in a much tighter spectrum. The telecom industry has been operating this way for many decades. The difference is that telecom traditionally operates in the terrestrial domain with focus on spectrum planning to minimize interference from adjacent or cochannel operators. In the mobile BAS services this is not a luxury that can be afforded, therefore, it would be necessary to know what the system parameters are before deployment. Quick link analysis can be performed to allow confidence in the field deployment. Interference criteria are well published by the various equipment manufacturers. One advantage the broadcast operator has is a flexible equipment configuration. For example, channel interference by adjacent operators can be mitigated by the ability to adjust polarization of the antenna, IF

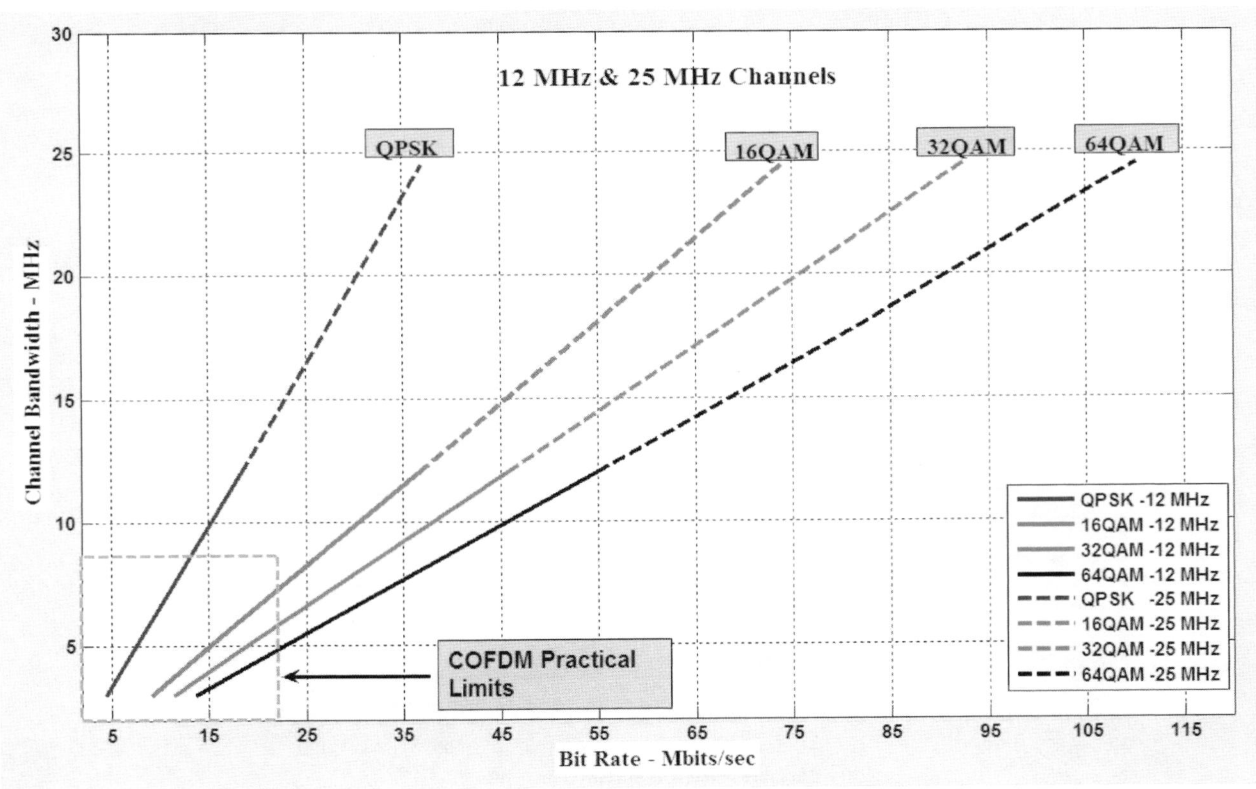

FIGURE 9.1-42 RF bandwidth versus bit rate.

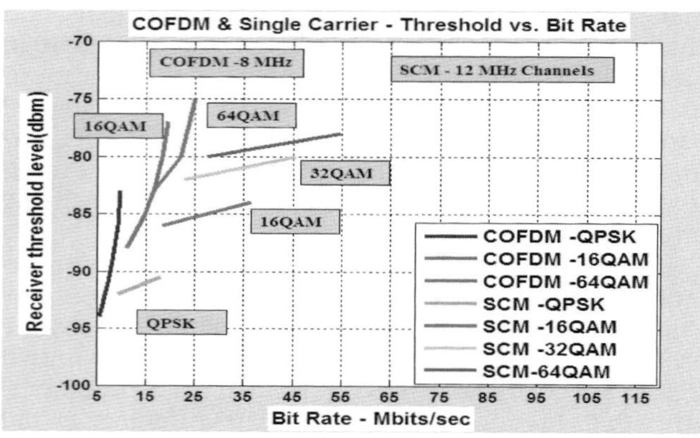

FIGURE 9.1-43 12 MHz data rate versus modulation.

filter selection, and dynamic gain control of amplifiers and LNAs.

The following charts (Figures 9.1-43 through 9.1-45) are shown as an aid to extract some of this common data on two predominant types of modulation: COFDM multicarrier modulation and single-carrier QPSK and QAM modulation.

Primary Power and Backup

With improvements in equipment and systems, engineering the reliability of the primary power source becomes a major factor in overall system availability. Where the terminals are in studios or TV transmitter sites the AC supplies may well be protected with standby generators; in this case, AC sourcing is the obvious choice. At repeaters, the AC may be supplied to a remote site by overhead lines and these may be subject to interruption under adverse weather conditions. Since solid-state microwave equipment has essentially low power requirements, batteries are the most popular form of standby power. Where long outages are to be expected (12 hours or more) an AC generator will be needed to recharge the batteries; this can be fuelled by gasoline, diesel, or liquified natural gas (propane).

Industrial lead-acid batteries are the most common type. They can be sealed to prevent evaporation of liquid, reducing maintenance and the need to ventilate the explosive gas given off when a battery is charged at too high a rate. The cells are floated across the charger, which supplies the station load until the AC fails. Since different types of lead-acid cell have different float voltages (2.15, 2.2, and 2.23 VDC for lead-antimony, lead-calcium, and lead-selenium, respectively) it is important that the charger and battery are matched—for this reason it is recommended that both are ordered from the same supplier.

It must be remembered that a battery of 12 cells with a nominal voltage of 24 VDC will float at 26.8 VDC, dropping to 24 when the charging fails and finally falling to 21 VDC at the end of the standby

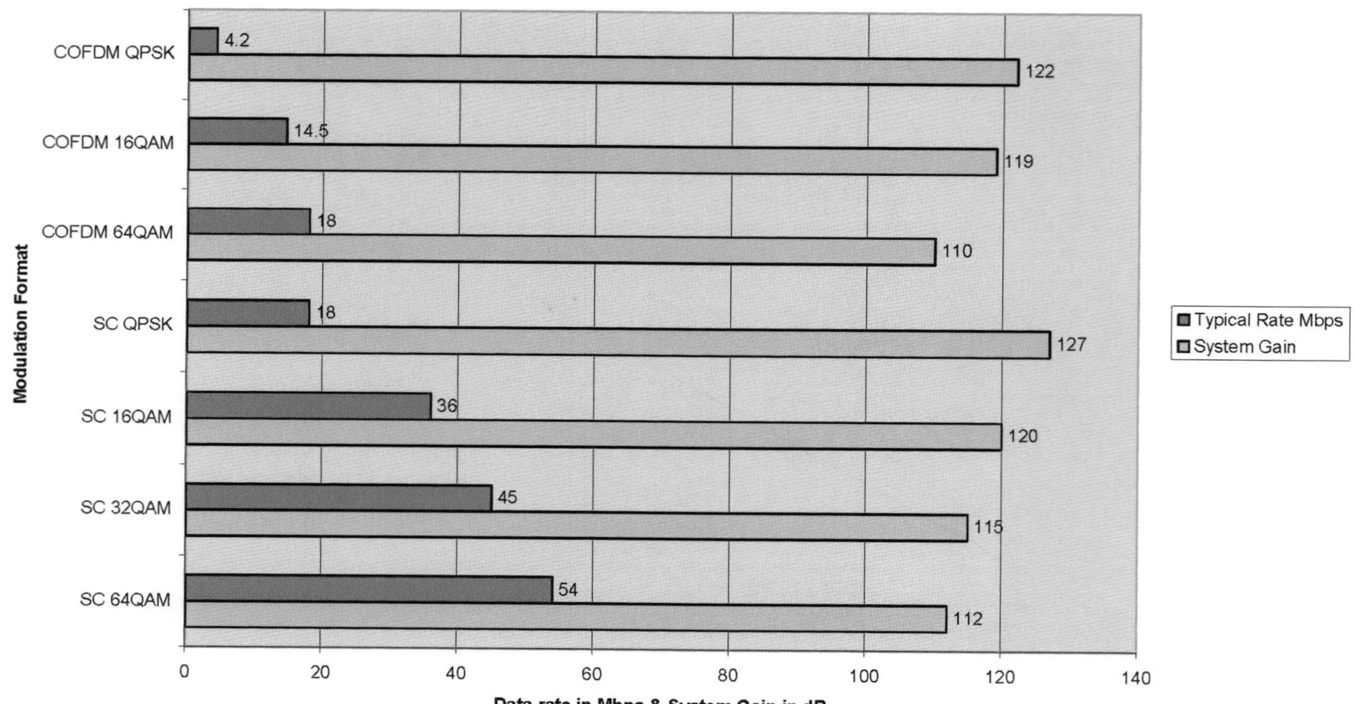

FIGURE 9.1-44 System gain comparison—12 MHz data rate versus modulation (referenced to 2 GHz BAS with 312 MHz channel).

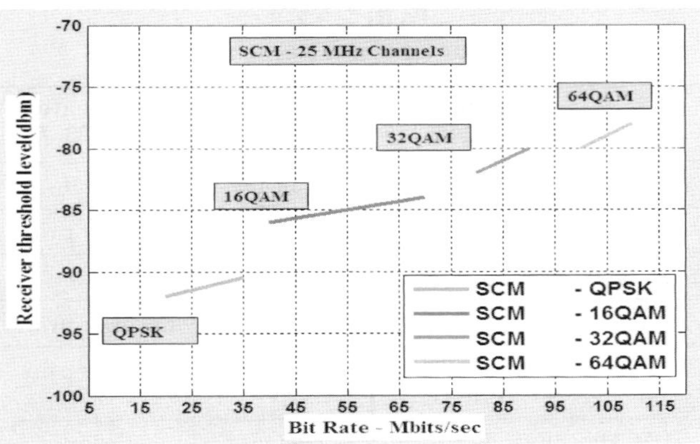

FIGURE 9.1-45 System threshold performances for single-carrier modulation.

time. At a minimum, the electronic equipment must accept this range. To determine the capacity of each cell required to give a particular standby time (*H* hours) the steady current drain must be determined (*A* amperes). Depending on the type of regulation used in the electronic equipment this may be higher at 26.8 VDC or 21 VDC (usually the latter)—take the higher value.

To arrive at the required cell capacity, multiply *A* by *H* to get the capacity in ampere-hours (*AH*). The minimum size of charger (*C* amperes) to meet the station load and at the same time recharge a discharged battery in *R* hours is given by

$$C = (A + AH) \times 1.1/R.$$

Since charger failure would lead to station failure after *H* hours it is common practice to use a duplicated charger for full protection at key sites; in that case each charger need only have the capacity *A* amperes rather than *C* as calculated above. The chargers must be designed to share the load.

The battery acts as a large capacitor, reducing the ripple voltage generated by the charger and protecting against power line surges. A good charger will have low-voltage protection to isolate the battery when discharged and over-voltage protection and charge rate limiting to protect against surges and gassing by the cells. When diesel generators are used as the primary source of power, they should be run at 75% or more of their rating (after allowing for any derating for altitude).

Solar power using photovoltaic cells is attractive as there are no moving parts requiring maintenance. However, the batteries used to maintain the supply during hours of darkness or heavy overcast (which could exist for several days in some parts of the country) can be the most expensive part of the installation. It may pay to have a small AC generator and charger to back up the batteries if the solar input fails for more than 3 days, rather than batteries to maintain the supply for 6 days, for example. Such a generator could also power lights and test equipment during routine site visits.

FREQUENCY COORDINATION AND LICENSING

Late in 2003, the rules for frequency coordination in broadcast auxiliary services were changed by FCC Docket 01-75. Until that time, coordination had been done locally, on a case-by-case basis through the cooperation of the respective chief engineers, and under the guidance of a regional frequency coordinator appointed by the Society of Broadcast Engineers (SBE). The SBE regional coordinators consist of a voluntary group that maintained an up-to-date log of existing licenses and proposed applications.

The antenna standards in Part 74 did not change, but the newly imposed frequency coordination rules require a more careful study of the potential interference that may be caused to other licensees by a new or modified microwave link. Except for the 950 MHz aural band, the only antennas that meet existing standards are parabolic reflectors.

In the 950 MHz band, the only stated FCC requirement is found in 74.536, which states: "Aural broadcast STL and ICR stations are required to use a directional antenna with the minimum beam width necessary, consistent with good engineering practice, to establish the link." Experience has shown that the FCC considers 22 degrees to be the maximum half-power beam width that they will license in this band. Based on this de facto standard, it is reasonable to assume that actual standards may be imposed some time in the future.

For TV STLs and intercity relay systems, the FCC has minimum rules that must be met with regard to the radiation pattern, as spelled out in Part 74.641. Table 9.1-7 summarizes those standards. The sidelobe suppression specifications have been omitted, and an example of the minimum acceptable antenna size has been added. Antenna sizes were based on data taken from vendor catalogs, however, they may vary plus or minus one size depending on feed structure and reflector shaping. It is good practice to review a vendor's specifications for compliance in the different categories.

TABLE 9.1-7
Typical Minimum Antenna Sizes per FCC 74.641

Band	Category	3 dB B/W	Minimum Gain	Minimum Size (ft)
2.02–2.11	A	5	n/a	8
	B	8	n/a	6
6.8–7.1	A	1.5	n/a	8
	B	2	n/a	6
12.7–13.2	A	1	n/a	6
	B	2	n/a	4
17.7–19.7	A	2.2	38	2
	B	2.2	38	2

While these standards apply to transmitting antennas, it is important to note that the FCC also states the following in 74.641: "The choice of receiving antennas is left to the discretion of the licensee. However, licensees will not be protected from interference which results from the use of antennas with poorer performance than identified in the table of this section." Station engineers are urged to strongly consider applying the FCC antenna standards equally at both ends of the path. For example, if a path has been coordinated using an 8 ft transmit antenna, it is good practice to use a matching, compliant antenna at the other end, even if it is only used for receiving.

Part 101 references TIA Telecommunications Systems Bulletin TSB 10-F, *Interference Criteria for Microwave Systems,* which specifies a 60 dB cochannel carrier/interferer ratio for analog FM video systems. This is an ideal, but in many cases may not prove possible. Lowering the C/I ratio will raise the effective noise level and hence reduce the fade margin. Given a threshold of –85 dBm the noise level will be approximately –95 dBm and an interferer at –95 dBm will double the noise level and raise the threshold to –82 dBm. This, or an even higher level, may be acceptable in a difficult situation. The interferer must at all times be below the threshold of the receiver or capture will occur if the wanted signal fails.

For digital systems, the rules change significantly due to the profound differences in their tolerance for cochannel and adjacent channel interference, which varies as a function of the modulation type and data rate employed. Digital systems are also characterized by their threshold to interference, or T/I ratios, which may be positive or negative, depending on the circumstances.

With regard to where category A or category B antennas should be deployed, there is some ambiguity in the rules in that regard; no clear reference is given other than, without a commitment to the contrary from the FCC, one must anticipate that category A antennas will be required, and that high-performance antennas would be recommended to all concerned.

Frequency Coordination Requirements and PCNs

Under the new Part 74 coordination rules, frequency assignments for fixed stations above 944 MHz will apply the interference protection criteria specified in FCC Part 101.105(a), (b), and (c), and the frequency usage coordination procedures of FCC Part 101.103(d) for each frequency authorized. Exceptions are made for mobile/ENG operations in most bands, and for fixed operation in the 2 GHz band, all of which are still coordinated locally as they always have been, and the SBE continues to provide local coordinators. Table 9.1-8 shows Part 74 frequency coordination requirements before and after the rules changes in FCC Docket 01-75.

To license a new frequency, or to make a major change to an existing license, as defined by the FCC, the licensee must enlist the services of a coordinator who is capable of providing an interference analysis meeting the FCC rules. Once the analysis is complete, a prior coordination notice is sent to all other cochannel and adjacent channel licensees within a specific set of boundaries known as the "keyhole." The keyhole refers to the shape of the area that must be analyzed for potential interference. For fixed, point-to-point service below 15 GHz, the area is essentially a circle with a diameter of 200 km (125 miles), except within ±5 degrees of the main beam, where the area extends to 400 km (248.6 miles). The size of the keyhole also varies with frequency and class of service. At higher frequencies, and when mobile stations are involved, the area is smaller.

TABLE 9.1-8
BAS Coordination Requirements

Band	All before 12/03	Fixed after 12/03	Mobile after 12/03
2 Ghz	Local	Local	Local
2.5 GHz	Local	Prior	Local or prior
6.5 GHz	Prior	Prior	Prior
7 GHz	Local	Prior	Local or prior
13 GHz	Interference criteria	Prior	Local or prior
18 GHz	Local	Prior	Prior

BAS Bands and POFS Bands

Most broadcast microwave links are operated in the traditional broadcast auxiliary services bands, however, it is important to recognize that other options are available and should be employed when required. Most specifically, broadcasters may obtain licenses to operate in the private operational fixed service (POFS) bands, provided that they are never used as the final link feeding a broadcast transmitter. Table 9.1-9 summarizes the various microwave bands available to the broadcaster, by FCC part number and by application.

Broadcasters may legally hold licenses for video transmission in FCC Part 101 bands between 5.9 and 6.8 GHz, 10.7–11.7 GHz, provided that the system is not used as the final link to a TV transmitter.

TABLE 9.1-9
FCC BAS Bands

Band	Limits	Channels	FCC Rules
2 GHz	2.025–2.110	12 MHz	Part 74
2.5 GHz	2.45–2.835	17 MHz	Part 74
6.5 GHz	6.425–6.525	25 MHz	Parts 74 and 101
7 GHz	6.875–7.125	25 MHz	Part 74
13 GHz	12.7–13.25	25 MHz	Parts 74 and 78

Prior Coordination Process

As noted above, Part 74 licensees must now employ the same frequency coordination procedures that Part 101 licensees use. With the exception of ENG/mobile operation and 2 GHz fixed links, which are still coordinated locally, all other fixed links must use the Part 101 procedure.

In essence, if a new path is to be constructed, or a major change is made to an existing path, coordination is required. While it is possible to manually coordinate by following the procedure outlined in the FCC rules, it would take an inordinate amount of time to research every licensee within the area of interest and calculate the potential for interference. A much easier solution is working with one of the commercial coordinators, such as those listed on the FCC website at http://wireless.fcc.gov/microwave/coordinators.html.

Once a coordinator has been chosen, an interference analysis will be conducted in accordance with FCC rules, and the licensee will be presented with a list of frequency and polarization options, assuming that some are available. If a frequency is not available on the band of choice, another band may be an option. If the study shows that a frequency is available to support a new or modified path, the coordinator will send out written prior coordination notices on behalf of the licensee, by email or direct mail, to all cochannel and adjacent channel licensees advising them of the proposed operation, and providing a statement that the FCC's rules have been followed in allowing 30 days for a response. If the 30-day waiting period ends without a protest, the licensee is free to file for a new or modified license.

Major versus Minor Changes

The criteria for minor versus major changes may be found in Part 1 of FCC rules. The most current rules may be found on the FCC website, in the Wireless Telecommunications Bureau pages, at http://wireless.fcc.gov/rules.html. In summary, if a microwave station licensee makes any of the following changes, it will be considered major, and will require coordination:

1. Any change in transmit antenna location by more than 5 seconds in latitude or longitude for fixed point-to-point facilities (e.g., a 5 second change in latitude, longitude, or both would be minor).
2. Any increase in frequency tolerance.
3. Any increase in bandwidth.
4. Any change in emission type.
5. Any increase in EIRP greater than 3 dB.
6. Any increase in transmit antenna height (above mean sea level) more than 3 meters.
7. Any increase in transmit antenna beam width.
8. Any change in transmit antenna polarization.
9. Any change in transmit antenna azimuth greater than 1 degree.

Also, if a number of minor changes are made, they may add up to a major change. Refer to the FCC website for the complete rules.

FCC Data Resources and Online License Applications

The FCC website includes many informational resources as noted in previous sections, and also supports online filing of licenses and amendments. Any transaction with the Commission requires the individual or company to obtain an FCC registration number, or FRN. This may be obtained free of charge on the FCC website by clicking the FRN registration link on the Wireless Telecommunication page at http://wireless.fcc.gov/.

Once an FRN is obtained, it is a relatively easy matter to apply for or modify licenses online. Provisions are made for payment of associated fees online, as well as the ability to upload required supporting documentation, such as a supplemental showing certifying that frequency coordination procedures have been completed in accordance with FCC rules.

ALTERNATIVE STL SYSTEMS

Over the years, the largest percentage of STL systems for both radio and television has been private microwave, however, alternatives have always been available and in use. Leased telephone lines have been the dominant alternate in radio, while leased fiber circuits have become the standard in television.

The growth of digital broadcasting and IP connectivity has spawned a whole new generation of alternatives.

Carrier-Based Fiber-Optic Systems

Incumbent local exchange carriers (ILECs) and competitive local exchange carriers (CLECs) have become increasingly active in offering a greater variety of broadcast-oriented services. The types of circuits that are now available include:

- DS1 for transport of AES audio in radio STL and remote pick-up applications.
- DS3 for compressed video.
- SDI at 270 Mbps for remote SDTV applications.
- ATSC at 19.39 Mbps for STL applications.
- OC-3/ STS-1 for multiple videos or lightly-compressed HDTV.
- OC-48 for multichannel video.

One thing to remember is that LECs are generally bound by tariffs, which amount to nonnegotiable prices. The only way to achieve better terms in most cases is a longer agreement. The pricing structure is normally based on an initial installation fee, plus monthly charge for the basic service and an additional charge for total mileage from point to point and any local access and transport area (LATA) crossings.

Interexchange carriers (IXCs) can provide long-distance connectivity, such as might be needed between New York and Chicago, for example. The fees are generally unregulated and negotiable except for the "last mile," which may be supplied by an LEC at one or both ends of a circuit.

In carrier circuits, the connection is virtual. The carrier provides a demarcation point at the customer premises that terminates in the protocol or connection that is being leased. For example, a carrier would provide a 75 ohm unbalanced BNC interface at the correct level for an SDI circuit, or 110 ohms balanced for a DS1.

Private IP and DSL Systems

In some cases, radio stations have chosen to deploy a DSL or similar IP service using an Internet service provider (ISP) as the carrier. Some ISPs can guarantee a minimum bit rate through a negotiated service level agreement (SLA), contracted in advance. This has an element of risk involved due to reliance on the public network as a transmission medium, but the circuit charges are quite low and some feel the risk is worth it. The most simplistic protection in such cases might be a dial-up analog or ISDN line that could be used in an emergency to replace a failed DSL line.

It is not unusual, however, to find ISPs that will offer business-class services with a guaranteed quality of service (QoS). The prices will be higher, but the danger of lost packets will be reduced considerably.

Unlicensed Microwave Band Systems

Unlicensed microwave spectrum in what is known as the industrial, scientific, and medical (ISM) bands has been available for over 20 years, and is popular with many communications users. The good news is that a web search on unlicensed microwaves will turn up quite a few companies that can provide radio equipment; the bad news is that an operator has absolutely no protection against interference. If another user causes interference, there is no recourse.

Still, some have chosen the unlicensed route as a way to establish a low-cost link for an STL or TSL application. While this may work in suburban applications, the closer a path gets to a metro area, the more likely the chances for interference. One problem is power and EIRP limitations as shown in Table 9.1-10.

TABLE 9.1-10
Unlicensed ISM Bands Available for Point-to-Point Applications

Band	Maximum Power Output	EIRP Limits
902–928 MHz	+ 30 dBm (1 watt)	+36 dBm
2.400–2.485 GHz	+ 30 dBm (1 watt)	+36 dBm
5.725–5.850 GHz	+ 30 dBm (1 watt)	+53 dBm

Two modulation types are allowed: frequency hopping spread spectrum (FHSS) and direct sequence spread spectrum (DSSS). FHSS tends to be more robust, is highly resistant to multipath, and has a higher immunity to interference, while DSSS can support much higher packet sizes and therefore greater throughput. The FCC allows FHSS systems a top-end emission bandwidth of 500 kHz in the 900 and 2.4 GHz bands, and 1.0 MHz in the 5.8 GHz band. Direct sequence is the opposite, where the FCC allows a minimum emission bandwidth of 500 kHz.

The 900 MHz and 2.4 GHz FHSS bands are mostly used for T1 and IP LAN traffic, while the 5.8 GHz radios may support a DS-3 or equivalent. The EIRP limits imposed on these bands do not lend themselves to long paths, and most radios automatically throttle down the throughput on longer paths to optimize path reliability versus data capacity.

SUMMARY

Frequency modulation (FM) techniques have been the mainstay in microwave studio-to-transmitter (STL) communications for decades, but are clearly on the decline as the digital transition increases the complexity of STLs and transmitter-to-studio links (TSLs). While microwave radio still accounts for the largest proportion of STL systems in use, this chapter examines the alternatives so as to provide information that will help an engineer decided which is best for a particular situation.

Reference

[1] White, Robert. *Engineering Considerations for Microwave Communications Systems*, 2nd ed. GTE Lenkurt, 1983, pp. 61–62.

[2] Whitaker, Jerry C. *A Primer: Digital Aural Studio to Transmitter Links*, TFT, Santa Clara, CA, 1994.

Bibliography

Anderson T., Detweiler, J., Gopal, M., and Iannuzzelli, R. *HD Radio™ Data Network Requirements*. Columbia, MD: iBiquity Digital Corporation, 2006.

Andrew Catalog 38. Orland Park, IL: Andrew Corporation, 1997, at www.andrew.com.

Cablewave Systems Catalogs 720C and 800. North Haven, CT: Radio Frequency Systems, Inc., Cablewave Systems Division, 1992.

C.C.I.R. Documents of the Seventeenth Plenary Assembly, Düsseldorf, Germany, and Geneva, Switzerland: ITU, 1990.

Effects of Multipath Propagation on the Design and Operation of Line-of-Sight Digital Relay Systems, ITU-R Rec., F.1093, 1994 F Series Vol., Part 1, Geneva, Switzerland: ITU, 1994.

Electrical Performance Standards for Television Relay Facilities, Standard EIA\TIA 250-C. Washington, DC: Telecommunications Industries Association, 1990.

Feher, K. *Digital Communications Microwave Applications*. Englewood Cliffs, NJ: Prentice-Hall, 1981.

Freeman, Roger L. *Radio System Design for Telecommunications*, 2nd ed. New York: John Wiley & Sons, Inc., 1997.

Freeman, Roger L. *Telecommunications System Engineering*, 3rd ed. New York: John Wiley & Sons, Inc., 1996.

Hickin, E. M. Microwave Engineering for the Broadcaster, in *NAB Engineering Handbook*, 8th ed. **[[AU: Editors?]]** Washington, DC: National Association of Broadcasters, 1992.

Hogg, D. C. Statistics on Attenuation of Microwaves by Intense Rain, *Bell System Technical Journal*, November, 1969.

Interference Criteria for Microwave Systems, TIA Telecommunications Systems Bulletin TSB 10-F. Washington, DC: Telecommunications Industries Association, 1994.

Ivanek, Ferdo. *Terrestrial Digital Microwave Communications.* Norwood, MA; Artech House, 1989.

Kolberg, Erik L. *Microwave and Millimeter-Wave Mixers.* New York: IEEE Press, 1984.

National Archives and Records Administration. Code of Federal Regulations, Title 47 Telecommunications, chapter 1, parts 0–101, October 1, 2005.

Oster, J., and Bachner, E. *Angle Diversity: A Practical Technique for Reducing Fades*, SP20-45, Orland Park, IL: Andrew Corporation, 1989.

Pathloss, Contract Telecommunication Engineering Ltd., Coquitlam, BC, Canada.

Schwartz, M. *Information Transmission, Modulation, and Noise*, 3rd ed. New York: McGraw-Hill, 1984.

Serafin, R. *LO Phase Noise Requirements for QSPK, 16-QAM, and 64-QAM Modulation*, Microwave Radio Communications Memo, Microwave Communications, Chelmsford, MA, September 3, 1996.

Additional information may be found at the following websites:

iBiquity Digital Corporation, at http://www.ibiquity.com/.
Marti Electronics, at http://www.martielectronics.com/.
Moseley Associates, at http://www.moseleysb.com/.
Nucomm, Inc., at http://www.nucomm.com/.
TFT Inc., at http://www.tftinc.com/.

C H A P T E R

9.2

AC Power Conditioning

JERRY C. WHITAKER
Advanced Television Systems Committee
Washington, D.C.

JOSÉ R. ALVAREZ*
Wavetech Associates
Hackensack, New Jersey

INTRODUCTION

Utility companies make a good-faith attempt to deliver clean, well-regulated power to their customers. Unfortunately, most disturbances on the AC line are beyond their control. Large load changes imposed by customers on a random basis, power factor (PF) correction switching, lightning, and accident-related system faults all combine to produce an environment in which tight control over AC power quality is difficult to maintain. Therefore, the responsibility for ensuring AC power quality must rest with the users of sensitive equipment—in this case, the broadcaster.

The selection of a protection method for a given facility is as much an economic question as it is a technical one. A wide range of power line conditioning and isolation equipment is available. A logical decision about how to proceed can be made only with accurate, documented data on the types of disturbances typically found on the AC power service to the facility. The protection equipment chosen must be matched to the problems that exist on the line. Using inexpensive basic protectors may not be much better than operating equipment directly from the AC line. On the other hand, the use of sophisticated protectors designed to shield the plant from every conceivable power disturbance may not be economically justifiable.

Purchasing transient suppression equipment is only one element in the selection process. In addition to the capital costs, it is necessary to consider the costs associated with site preparation, installation, and maintenance. Also, protection units that are placed in series with the load will consume a certain amount of power and therefore generate heat. Thus, it is necessary to consider the operating efficiency of the system. It is useful to prepare a complete life-cycle cost analysis of the protection methods proposed. The study may reveal that the long-term operating expense of one system outweighs the lower purchase price of another.

The amount of money a facility manager is willing to spend on protection from utility company disturbances generally depends on the engineering budget and how much the plant has at stake. For example, spending $250,000 on system-wide protection for a major market television station is easily justified. At smaller operations, the justification of these amounts may not be as straightforward.

THE KEY TOLERANCE ENVELOPE

The susceptibility of electronic equipment to failure because of disturbances on the AC power line has been studied by many organizations. A benchmark study was conducted by the Naval Facilities Engineering Command from 1968–1978. The program, directed by Lt. Thomas Key, identified three distinct categories of recurring disturbances on utility company power systems [1]. As shown in Table 9.2-1, it is

*This chapter is an update of the ninth edition version with major new material by José R. Alvarez.

TABLE 9.2-1
Types of Voltage Disturbances Identified in Key's Report

Parameter	Type 1	Type 2	Type 3
Definition	Transient and oscillatory overvoltage	Momentary undervoltage or overvoltage	Power outage
Causes	Lightning, power network switching, operation of other loads	Power system faults, large load changes, utility company equipment malfunctions	Power system faults, unacceptable load changes, utility equipment malfunctions
Threshold*	200–400% of rated rms voltage or higher (peak instantaneous above or below rated rms)	Below 80-85% and above110% of rated rms voltage	Below 80–85% of rated rms voltage
Duration	Transients 0.5–200 (ms wide and oscillatory up to 16.7 ms at frequencies of 200 Hz to 5 kHz and higher	From 4–6 cycles, depending on the type of power system distribution equipment	From 2–60 s if correction is automatic; from15 min to 4 hr if manual

* The approximate limits beyond which the disturbance is considered to be harmful to the load equipment

not the magnitude of the voltage, but the duration of the disturbance that determines the classification.

In the study, Key found that most data processing (DP) equipment failures caused by AC line disturbances occurred during bad weather, as shown in Table 9.2-2. According to a report on the findings, the incidence of thunderstorms in an area may be used to predict the number of failures. The type of power transmission system used by the utility company also was found to affect the number of disturbances observed on power company lines (see Table 9.2-3). For example, an analysis of utility system problems in Washington, DC, Norfolk, VA, and Charleston, SC, demonstrated that underground power distribution systems experienced one-third fewer failures than overhead lines in the same areas. Based on his research, Key developed the recommended voltage tolerance envelope shown in Figure 9.2-1. The design goals illustrated are recommendations to computer manufacturers for implementation in new equipment. In the table, cycles refer to cycles per second (Hz) of the AC power system.

ASSESSING THE LIGHTNING HAZARD

As identified by Key in his Naval Facilities study, the extent of lightning activity in an area significantly affects the probability of equipment failure caused by transient activity. The threat of a lightning flash to a facility is determined, in large part, by the nature of the installation and its geographic location. The type and character of the lightning flash are also important factors.

TABLE 9.2-2
Causes of Power-Related Computer Failures, Northern Virginia, 1976

Recorded Cause	No. of Disturbances		Number of Computer Failures
	Undervoltage	Outage	
Wind and lightning	37	14	51
Utility equipment failure	8	0	8
Construction or traffic accident	8	2	10
Animals	5	1	6
Tree limbs	1	1	2
Unknown	21	2	23
Totals	80	20	100

TABLE 9.2-3
Effects of Power System Configuration on Incidence of Computer Failures

Configuration	No. of Disturbances		Recorded Failures
	Undervoltage	Outage	
Overhead radial	12	6	18
Overhead spot network	22	1	23
Combined overhead (weighted*)	16	4	20
Underground radial	6	4	10
Underground network	5	0	5
Combined underground (weighted*)	5	2	7

*The combined averages weighted based on the length of time monitored (30 to 53 months).

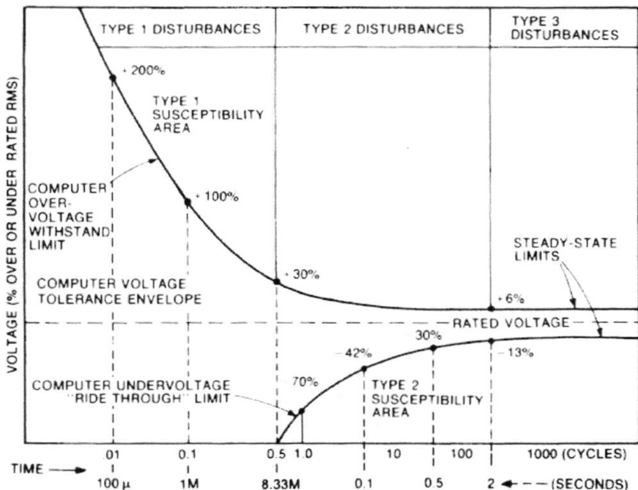

FIGURE 9.2-1 The recommended voltage tolerance envelope for computer equipment. This chart is based on pioneering work done by the Naval Facilities Engineering Command. The study identified how the magnitude and duration of a transient pulse must be considered in determining the damaging potential of a spike. The design goals illustrated in the chart are recommendations to computer manufacturers for implementation of new equipment. In the chart, cycles refer to cycles per second (Hz) of the AC power system [1].

The Keraunic number of a geographic location describes the likelihood of lightning activity in that area. Figure 9.2-2 shows the isokeraunic map of the United States, which estimates the number of lightning days per year across the country. On average, 30 storm days occur per year across the continental United States. This number does not fully describe the lightning threat because many individual lightning flashes occur during a single storm.

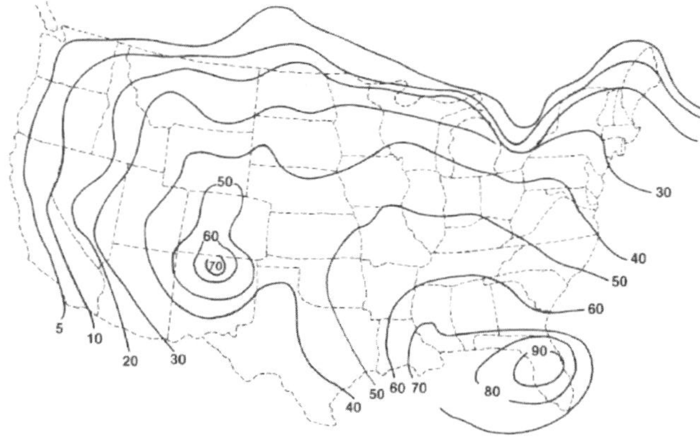

FIGURE 9.2-2 The isokeraunic map of the United States, showing the approximate number of lightning days per year.

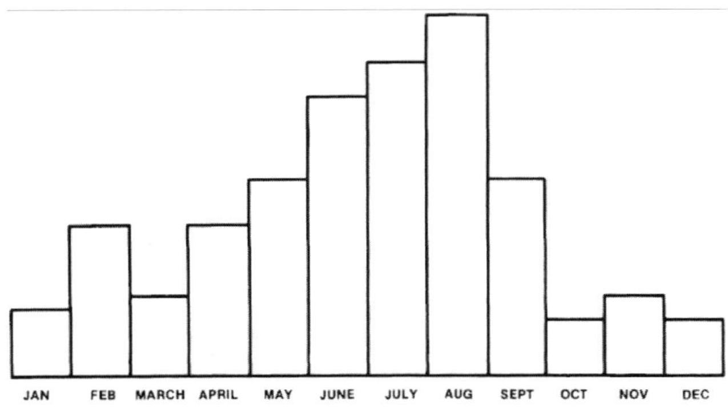

FIGURE 9.2-3 The relative frequency of power problems in the United States, classified by month.

The structure of a facility has a significant effect on the exposure of equipment to potential lightning damage. Higher structures tend to collect and even trigger localized lightning flashes. Because storm clouds tend to travel at specific heights above the earth, conductive structures in mountainous areas more readily attract lightning activity. The plant exposure factor is a function of the size of the facility and the isokeraunic rating of the area. The larger the physical size of an installation, the more likely it is to be hit by lightning during a storm. The longer a transmission line (AC or RF), the more lightning flashes it is likely to receive.

The relative frequency of power problems is seasonal in nature. As shown in Figure 9.2-3, most problems are noted during June, July, and August. These high problem rates can be traced primarily to increased thunderstorm activity.

FIPS PUBLICATION 94

In 1983, the U.S. Department of Commerce published a guideline summarizing the fundamentals of powering, grounding, and protecting sensitive electronic devices. The document, known as Federal Information Processing Standards Publication 94 (FIPS Pub. 94), was first reviewed by governmental agencies and sent to the Computer Business Equipment Manufacturers Association (CBEMA) for review [2]. When CBEMA approved the document, the data-processing industry was provided with an important guideline for power quality.

FIPS Pub. 94 was written to cover automatic data-processing (ADP) equipment, which at that time constituted the principal equipment that was experiencing difficulty running on normal utility supplied power. Since then, IEEE Standard P1100 was issued [3], which applies to all sensitive electronic equipment, including computer based broadcast systems. FIPS Pub. 94 is a guideline intended to provide a cost/benefit course of action. As a result, it can be relied on to give the best solution for the least amount of money to typical problems that will be encountered.

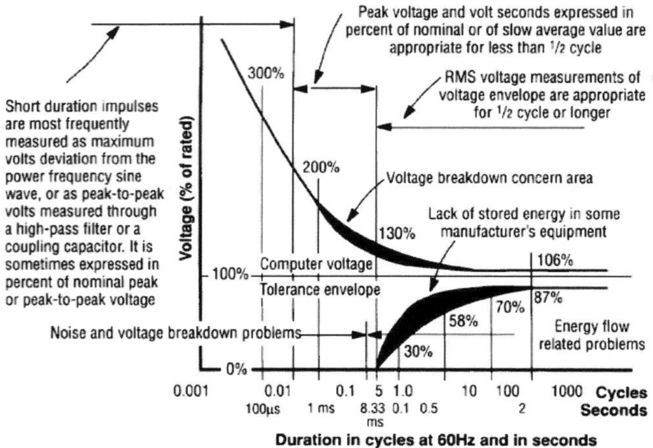

FIGURE 9.2-4 The CBEMA curve from FIPS Pub. 94.

In addition to approving FIPS Pub. 94, CBEMA provided a curve that had been used as a guideline for their members in designing power supplies for modern electronic equipment. The CBEMA curve from the FIPS document is shown in Figure 9.2-4. (Note the similarity to the Key tolerance envelope shown in Figure 9.2-1.)

The curve is a susceptibility profile. In order to better explain its meaning, the curve has been simplified and redrawn in Figure 9.2-5. The vertical axis of the graph is the percent of voltage that is applied to the power circuit, and the horizontal axis is the time factor involved (in ms to sec). In the center of the chart is the acceptable operating area, and on the outside is a danger area on top and bottom. The danger zone at the top is a function of the tolerance of equipment to excessive voltage levels. The danger zone on the bottom sets the tolerance of equipment to a loss or reduction in applied power. The CBEMA guideline states that if the voltage supply stays within the acceptable area given by the curve, the sensitive load equipment will operate as intended.

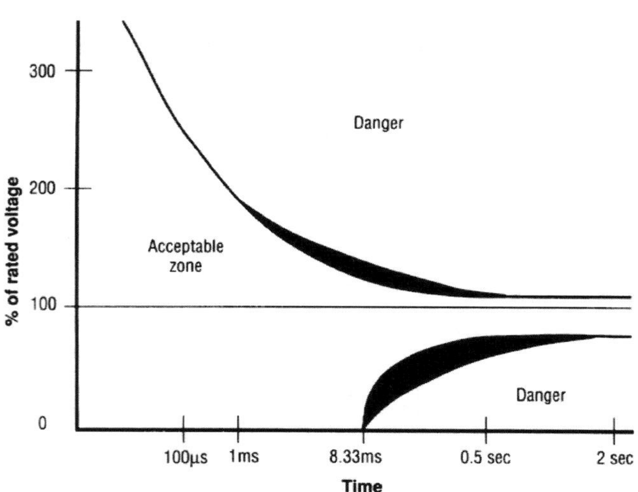

FIGURE 9.2-5 A simplified version of the CBEMA curve. Voltage levels outside the acceptable zone result in potential system shutdown and hardware and software loss [2].

ITIC CURVE

More recently the Information Technology Industry Council (ITIC) developed the curve shown in Figure 9.2-6 to more accurately reflect the operating parameters of single-phase loads that operate in an office environment. Since these devices are electronic in nature, this curve can also apply to other electronic loads, single or three phase. [4]

TRANSIENT PROTECTION ALTERNATIVES

A facility can be protected from transient disturbances in two basic ways: the systems approach or the discrete device approach. Table 9.2-4 outlines the major alternatives available:

- Uninterruptible power supply (UPS) and standby generator.
- UPS stand-alone system.
- Secondary AC spot network where power is delivered to a facility via separate paths from the utility company.

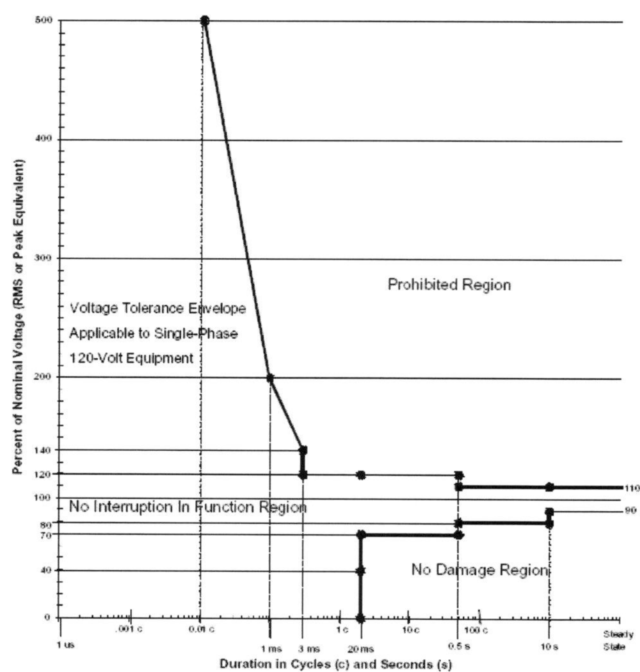

FIGURE 9.2-6 ITIC curve (revised in 2000) applies to 120, 120/208, and 120/240 volt systems.

- Secondary selective AC network that is identical to the secondary AC spot network, except that a static transfer switch is included to permit load switching without interruption in service.
- Rotary systems.
- Shielded isolation transformer.
- Suppressors, filters, and lightning arrestors.
- Solid-state line voltage regulator/filter.

TABLE 9.2-4
Types of System-Wide Protection Equipment Available to Facility Managers and the AC Line Abnormalities That Each Approach Can Handle

System	Disturbance* Type 1	Disturbance Type 2	Disturbance Type 3
UPS system and standby generator	All source transients; no load transients	All	All
UPS system	All source transients; no load transients	All	All outages shorter than the battery supply discharge time
Secondary spot network†	None	None	Most, depending on the type of outage
Secondary selective network‡	None	Most	Most, depending on the type of outage
Motor-generator set	All source transients; no load transients	Most	Only brown-out conditions
Shielded isolation transformer	Most source transients; no load transients	None	None
Suppressors, filters, lightning arrestors	Most transients	None	None
Solid state line voltage regulator/filter	Most source transients; no load transients	Some, depending on the response time of the system	Only brown-out conditions

*Dual power feeder network.
†Dual power feeder network using a static (solidstate) transfer switch.
‡Disturbance types illustrated graphically in Figure 9.2-1.

Table 9.2-5 lists the relative benefits of each protection method. Because each installation is unique, a thorough investigation of facility needs should be conducted before purchasing any equipment. The systems approach offers the advantages of protection engineered to a particular application and need, and (usually) high level factory support during equipment design and installation. The systems approach also means higher costs for the end user.

SPECIFYING SYSTEM-PROTECTION HARDWARE

Developing specifications for system-wide power-conditioning/backup hardware requires careful analysis of various factors before a particular technology or a specific vendor is selected. Key factors in this process relate to the load hardware and load application. The power required by sensitive loads may vary widely depending on the configuration of the system. The principle factors that apply to system specifications include:

- Power requirements, including voltage, current, power factor, harmonic content, and transformer configuration.
- Voltage regulation requirements of the load.
- Frequency stability required by the load, and the maximum permissible slew rate (the rate of change of frequency per second).
- Effects of unbalanced loading.
- Overload and inrush current capacity.
- Bypass capability.
- Primary/standby path transfer time.
- Maximum standby power reserve time.
- System reliability and maintainability.
- Operating efficiency.

An accurate definition of critical applications will aid in the specification process for a given site. The potential for future expansion also must be considered in all plans.

Power requirements can be determined either by measuring the actual installed hardware or by checking the nameplate ratings. Most nameplate ratings include significant safety margins. Moreover, the load normally will include a diversity factor; all individual elements of the load will not necessarily be operating at the same time.

Every load has a limited tolerance to noise and harmonic distortion. Total harmonic distortion (THD) is a measure of the quality of the waveform applied to the load. It is calculated by taking the geometric sum of the harmonic voltages present in the waveform, and expressing that value as a percentage of the fundamental voltage. Critical loads typically can withstand 5% THD, where no single harmonic exceeds 3%. The power conditioning system must provide this high-quality output waveform to the load, regardless of the

TABLE 9.2-5
Relative Merits of System-Wide Protection Equipment

System	Strong Points	Weak Points	Technical Profile
UPS system and standby generator	Full protection from power outage failures and transient disturbances; ideal for critical technical and life-safety loads	Hardware is expensive and may require special construction; electrically and mechanically complex; noise may be a problem; high annual maintenance costs	Efficiency 85–95%; typical high impedance presented to the load may be a consideration; frequency stability good; harmonic distortion determined by UPS system design
UPS system	Completely eliminates transient disturbances; eliminates surge and sag conditions; provides power outage protection up to the limits of the battery supply; ideal for critical load applications	Hardware is expensive; depending on battery supply requirements, special construction may be required; noise may be a problem; periodic maintenance required	Efficiency 85–95%; typical high impedance presented to the load may be a consideration; frequency stability good; harmonic content determined by inverter type
Secondary spot network*	Simple; inexpensive when available in a given area; protects against local power interruptions; no maintenance required by user	Not available in all locations; provides no protection from area-wide utility failures; provides no protection against transient disturbances or surge/sag conditions	Virtually no loss, 100% efficient; presents low impedance to the load; no effect on frequency or harmonic content
Secondary selective network†	Same as above; provides faster transfer from one utility line to the other	Same as above	Same as above
Motor-generator set	Electrically simple; reliable power source; provides up to 0.5 s power-fail ride-through in basic form; completely eliminates transient and surge/sag conditions	Mechanical system requires regular maintenance; noise may be a consideration; hardware is expensive; depending on m-g set design, power-fail ride-through may be less than typically quoted by manufacturer	Efficiency 80–90%; typical high impedance presented to the load may be a consideration; frequency stability may be a consideration, especially during momentary power-fail conditions; low harmonic content
Shielded isolation transformer	Electrically simple; provides protection against most types of transients and noise; moderate hardware cost; no maintenance required	Provides no protection from brown-out or outage conditions	No significant loss, essentially 100% efficient; presents low impedance to the load; no effect on frequency stability; usually low harmonic content
Suppressors, filters, lightning arrestors	Components inexpensive; units can be staged to provide transient protection exactly where needed in a plant; no periodic maintenance required	No protection from Type 2 or 3 disturbances; transient protection only as good as the installation job	No loss, 100% efficient; some units subject to power-fail conditions; no effect on impedance presented to the load; no effect on frequency or harmonic content
Solid state line voltage regulator/ filter	Moderate hardware cost; uses a combination of technologies to provide transient suppression and voltage regulation; no periodic maintenance required	No protection against power outage conditions; slow response time may be experienced with some designs	Efficiency 92–98%; most units present low impedance to the load; usually no effect on frequency; harmonic distortion content may be a consideration

*Dual power feeder network.
†Dual power feeder network using a static (solidstate) transfer switch.

level of noise and/or distortion present at the AC input terminals.

If a power conditioning/standby system does not operate with high reliability, the results often can be disastrous. In addition to threats to health and safety, there is a danger of lost revenue and hardware damage. Reliability must be considered from three different viewpoints:

- Reliability of utility AC power in the area.
- Impact of line voltage disturbances on computer-based loads.
- Ability of the protection system to maintain reliable operation when subjected to expected and unexpected external disturbances.

The environment in which the power conditioning system operates will have a significant effect on reliability. Extremes of temperature, altitude, humidity, and vibration can be encountered in various applications. Extreme conditions can precipitate premature component failure and unexpected system shutdown. Most power protection equipment is rated for operation from 0°C to 40°C. During an HVAC system failure, however, the ambient temperature of the equipment room can easily exceed either value, depending on the exterior temperature. Operating temperature derating typically is required for altitudes in excess of 1000 ft.

Additional Considerations

While this chapter cannot address all instances of the need for specialized power conditioning requirements there are some broadcaster-specific events that can be taken into account.

Crowbar Events

Broadcast transmitters using inductive output tubes (IOT) have a protection circuit in the event of an arc-over inside the tube. This crowbar circuit grounds out the high-voltage DC power supply within a few milliseconds, thereby protecting the tube. Unfortunately this essentially looks like a short circuit to any source that powers this type of load. This event must to be taken into account when sizing power protection equipment. Usually oversizing the source to prevent frequent transfers to bypass can accomplish the desired result. Some systems can be set up to delay this transfer to bypass thereby keeping the load protected during these events. This testing should be performed during a factory witness test where factory personnel can be on hand to tune the product for the desired performance.

Power Problem Types

IEEE Std. 1100-1999 [4] defines power problems that can be organized into the following nine categories, many of which can adversely affect broadcast equipment:

- Power failure
- Sag
- Surge
- Undervoltage
- Overvoltage
- Line noise
- Transient
- Frequency variation
- Harmonic distortion

It is important to note that many power protection technologies only address some of these problems. A product that may be labeled a UPS, or uninterruptible power supply, however, does not ensure all power-related problems are addressed. Some standby UPS systems only address three of the problems listed above whereas a double conversion UPS system addresses all nine.

Environmental Concerns

Some of the solutions listed below use batteries as the stored energy source in the event of a power failure. There can be special requirements for temperature and ventilation that must to be taken into account. Usually these systems are placed within the facility, near the equipment being protected. Some of the rotary and hybrid technologies do not have these limitations and are often placed in containers outside the facility, freeing up space that can be used for the core business.

UPS

Uninterruptible power supply (UPS) systems, such as the one shown in Figure 9.2-7, have become a virtual necessity for powering large or small computer systems where the application serves a critical need and continuity of service is essential, such as broadcasting. Computers and data communications systems are no more reliable than the power from which they operate. Below are various currently available technologies.

Solid-State (Static) UPS Technologies

Recent designs in static UPS employ insolated gate bipolar transistors (IGBT) to create the pulse width modulated (PWM) conversion not only from DC to AC in the inverter but also to chop the AC down to DC in the rectifier section. The main reason for the change to this technology is the speed at which the transistors can be switched. This improves the output performance of the UPS—its ability to handle overloads, step loads, and harmonics generated by these loads.

Designs employing IGBT technology also reduce the parts count that both improves reliability and

FIGURE 9.2-7 Typical parallel UPS installation. (Courtesy of Eaton Powerware.)

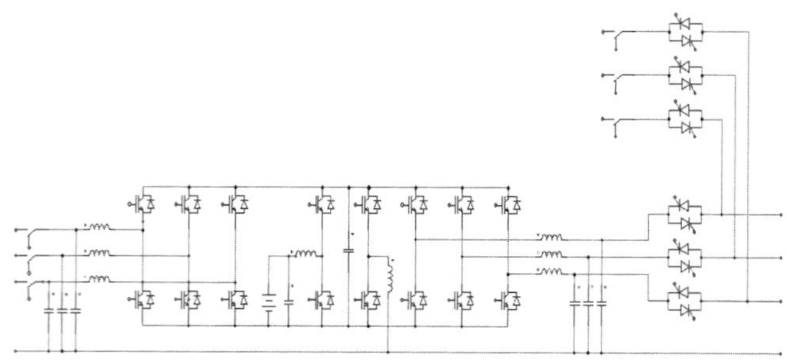

FIGURE 9.2-8 Double-conversion UPS with IGBT rectifier and inverter (three phase detail shown).

reduces cost. This creates further benefits in the way of reduction of footprint (physical size) and less heat emission (greater efficiency). The high speed conversion in the rectifier removes the requirement for an input filter to reduce reflected harmonics back to the service entrance. This again results in greater efficiency, lower component count, and higher reliability of the system especially as it relates to engine generators and the possible interaction with the input filter on the UPS.

Double-Conversion UPS

This design is by far the most popular approach for large critical applications. Utility power is converted to DC, where the batteries are charged, and subsequently reconverted to AC to power the load, as shown in Figure 9.2-8. The two conversions isolate the load from any and all power-quality problems on the primary side. During normal operation the load is fed from the inverter. In the event of a failure or overload condition, the static bypass would activate to keep the load running. This approach provides the highest reliability in a single module UPS. Higher reliability can be achieved through paralleling for redundancy or additional capacity.

Delta-Conversion UPS

This approach has been used for years in smaller applications and has recently become available for larger three phase systems. Incoming AC power is conditioned and powers the critical load directly as shown in Figure 9.2-9. Input power also feeds the rectifier/inverter that charges the battery and provides energy to help condition power to the load. During a utility outage the system switches to the battery system to power the load. Due to the nature of this design, the incoming frequency will pass on through to the load requiring the use of the battery system if frequency shifts outside a predetermined window. Efficiency is usually higher than comparable double-conversion designs.

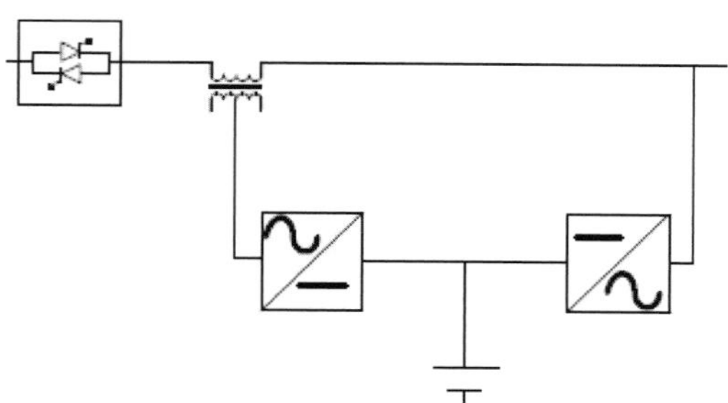

FIGURE 9.2-9 Simplified block diagram of delta-conversion UPS.

Line-Interactive and Standby UPS

Primarily used in single-phase applications, these systems provide good point-of-use backup in case of AC input failure. These simple technologies provide little or no conditioning of the utility power and only some minor surge protection and filtering. Incoming AC power serves the load directly and in the event of a power failure the system switches over to the battery system to supply the load. This system provides lower availability of the load than double-conversion UPS. Since the load is normally powered by the utility and switches to battery only during an anomaly, if a problem arises with the inverter there is a potential for load drop (the failure mode for double conversion, since normal operation runs off the inverter, is to go to bypass keeping the load operational).

Battery Supply

UPS systems typically are supplied with sufficient battery capacity to carry a critical load for periods ranging from 5 minutes to 1 hour or more. Longer backup time periods are usually handled by a standby diesel generator. Batteries require special precautions. For large installations, they almost always are placed in a room dedicated to that purpose. Proper temperature control is important for long life and maximum discharge capacity.

Four battery types are found in UPS systems, the first two being more common:

- *Sealed lead-calcium.* A gel-type electrolyte is used that does not require the addition of water. There is no outgassing or corrosion. This type of battery is used when the devices are integral to small UPS units, or when the batteries must be placed in occupied areas. The lifespan of a sealed lead-calcium battery, under ideal conditions, is about 5 years.
- *Conventional lead-calcium (wet cell).* The most common battery type for large UPS installations, these units require watering and terminal cleaning about every 3–6 months. Expected lifetime ranges up to 20 years. Conventional lead-calcium batteries outgas hydrogen under charge conditions and must be located in a secure, ventilated area.
- *Lead-antimony.* Traditional lead-acid batteries, these devices are equivalent in performance to lead-calcium batteries. Maintenance is required every 3

months. Expected lifetime is about 10 years. To retain their capacity, lead-antimony batteries require a monthly equalizing charge.

- *Nickel-cadmium.* Advantages of the nickel-cadmium battery include small size and low weight for a given capacity. These devices offer excellent high- and low-temperature properties. Life expectancy is nearly that of a conventional lead-calcium battery. Nickel-cadmium batteries require a monthly equalizing charge, as well as periodic discharge cycles to retain their capacity. Nickel-cadmium batteries are the most expensive of the devices typically used for UPS applications.

ROTARY AND HYBRID UPS TECHNOLOGIES

In addition to standard motor-generator sets, new developments in rotary and hybrid technologies have increased the variety, flexibility, and reliability of these systems.

Motor-Generator UPS

Motor-generator sets have been used successfully in power-quality applications for decades. As the name implies, a motor-generator (MG) set consists of a motor powered by the AC utility supply that is mechanically connected to a generator, which feeds the load. This provides complete electrical isolation from any transients on the utility feed. There are several iterations converting this technology into a UPS, one of which is shown in Figure 9.2-10. The design shown employs the use of a simple solid-state UPS ahead of the MG set. Utility AC energy is rectified and used to drive an inverter, which provides a regulated frequency source to power the synchronous motor. The output from the DC-to-AC inverter need not be a well-formed sine wave, nor a well-regulated source. The output from the generator will provide a well-regulated sine wave for the load. The maintenance requirements and increasing cost of copper have minimized the use of this design.

Dynamic Diesel UPS

This design consists of a diesel engine, induction coupling, clutch, and generator, as shown in Figure 9.2-11. The diesel engine sits idle until the utility power fails. The induction coupling contains an inner rotor that stores sufficient energy to bridge the diesel start time. The generator provides electrical power to the load in diesel mode. In normal operating mode, the generator provides filtering and power factor correction. Due to cost, these systems are impractical below 500 KVA and are typically used to power entire buildings at over 1 MW.

DC Flywheel Applications

There have been significant advances in the use of flywheels to provide stored energy in the form of DC to complement or supplant batteries in a UPS system. The main drawback in applying flywheels in the past has been the fact that frequency decay occurred quickly and provided limited ride-through. By utilizing this technology on the DC side, longer outage protection can be achieved, up to 15 seconds, which provides enough time to transfer over to a generator in the case of a sustained utility outage.

DC flywheels can also be used to increase the life and reliability of a battery system in a UPS. The DC flywheel can be adjusted to take the shorter power "hits" leaving the longer outages to be handled by the

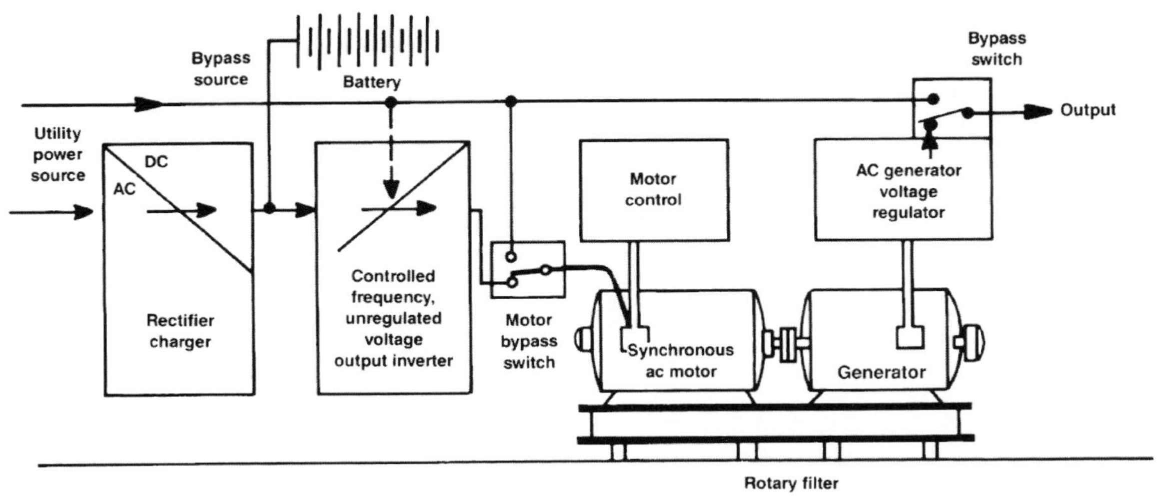

FIGURE 9.2-10 Uninterruptible MG set using a synchronous AC motor.

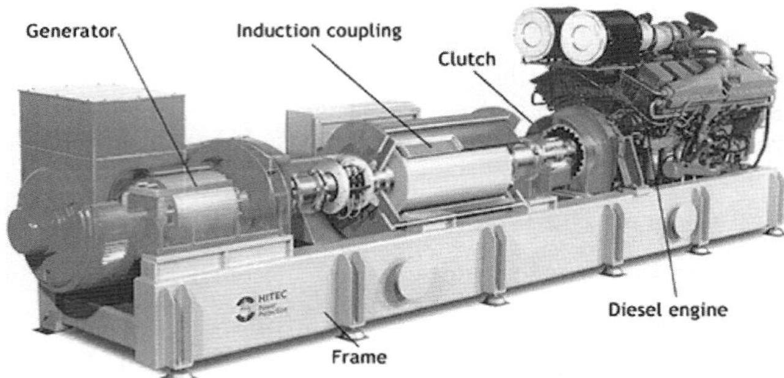

FIGURE 9.2-11 Dynamic diesel UPS. (Courtesy of Hitec Power Protection.)

battery, as shown in Figure 9.2-12. This prevents the batteries from aging prematurely in remote applications that may have frequent outages. It can also prevent the loss of load in case a cell opens during battery operation. This is the weakest link in a UPS and is the reason most critical installations use redundant battery strings. This battery-hardening technique provides another alternative in improving reliability.

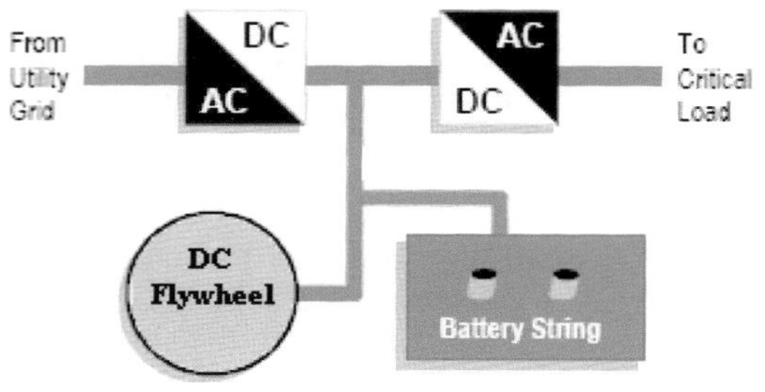

FIGURE 9.2-12 Double-conversion UPS with DC flywheel and batteries.

Thermal and Compressed-Air Storage DC

A new development in energy storage uses compressed air to provide run times in excess of 15 minutes at 80 KW. During a power outage, compressed air is sent through a thermal storage unit and greatly expands. This heated air spins a turbine that generates the power to support the DC bus on a double-conversion UPS. Bridging the gap until the turbine spins up is the flywheel effect of a small MG set, as shown in the one-line diagram in Figure 9.2-13. A useful byproduct of this technology is the creation of cold air, which can help in keeping the space cool during an extended outage. Initial high hardware costs and the lack of codes for high-pressure piping could limit the initial application of this technology.

CONDITIONING TECHNOLOGIES

While UPS and MG sets inherently provide for power conditioning because they isolate the source from the load, power conditioning is still necessary for those times when the load is connected directly to the source, such as during maintenance and bypass situations.

Motor Generator Sets

As previously stated, MG sets are a mature technology that provide complete electrical isolation from utility power and tight voltage regulation within a wide range of input power variations. They can also provide useful ride-through in the case of short utility outages. Maintenance and high hardware costs, however, have limited their use.

Electronic Tap-Changing High-Isolation Transformer

This system is built around a high-attenuation isolation transformer with a number of primary winding

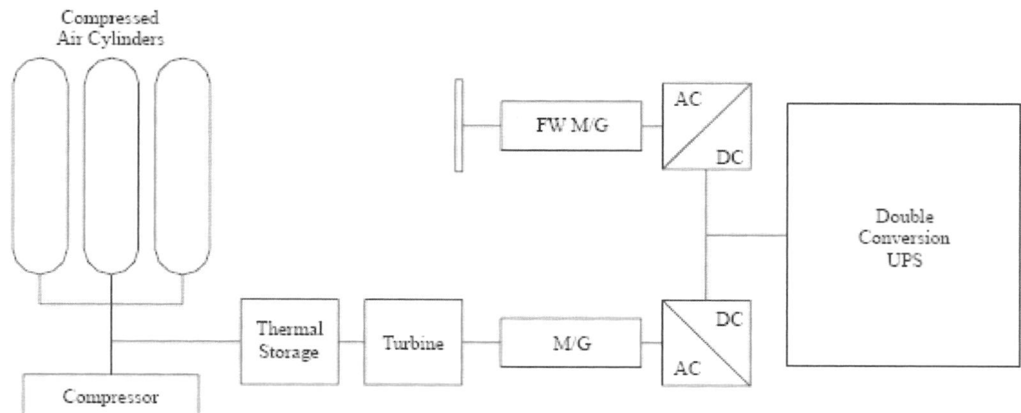

FIGURE 9.2-13 Basic one-line diagram of compressed-air and thermal storage UPS system.

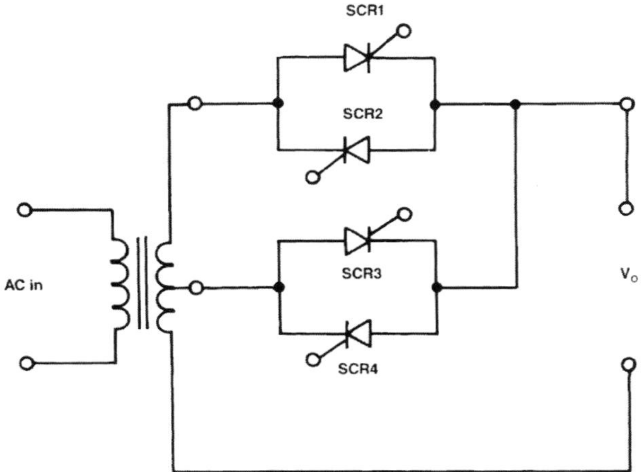

FIGURE 9.2-14 Secondary side-synchronous, tap-changing transformer.

taps. Silicon controlled rectifier (SCR) pairs control voltage input to each tap, as in a normal tap-changing regulator. Tap changing also can be applied to the secondary, as shown in Figure 9.2-14. The electronic tap changer is an efficient design that effectively regulates voltage output and prevents noise propagation to the load.

Output Transfer Switch Technology

Fault conditions, maintenance operations, and system reconfiguration require the load to be switched from one power source to another. This work is accomplished with an output transfer switch. Electromechanical or motor-driven relays operate too slowly for most loads. As discussed previously, most UPS systems use electronic (static) switching. Static transfer switches can be configured as:

- *Break-before-make*. Power output is interrupted before transfer is made to the new source.
- *High speed break-before-make*. Static transfer switch, STS, with less than 4 ms transfer time.
- *Make-before-break*. The two power sources are overlapped briefly so as to prevent any interruption in AC power to the load.

Figure 9.2-15 illustrates each approach to load switching.

For critical load applications, a seamless transfer of the load is necessary. This will necessitate either a make-before-break transfer or a high-speed break-before-make. To prevent faults from taking down both sides, most applications call for static transfer switches. For the switchover to be accomplished with minimum disturbance to the load, both power sources must be synchronized. The UPS system must, therefore, be capable of synchronizing to the utility AC power line (or other appropriate power source).

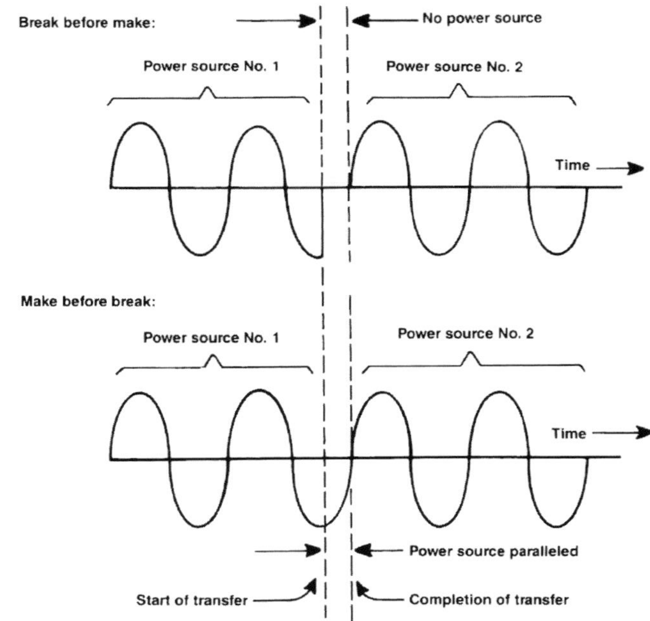

FIGURE 9.2-15 Static transfer switch modes.

SUMMARY

Due to the large selection of power-conditioning technologies available and great variations in the site requirements, a reliable site is dependent on a tailored approach that takes these and other factors into account. A successful application should be determined only after defining the needs as well as understanding the loads to be conditioned. It is rarely practical to shield a facility from every conceivable disturbance that can occur on an AC power line. On the other hand it is possible and feasible to fashion a practical and cost effective protection system that meets the needs of the broadcast facility for maximum reliability and uninterrupted operation.

References

[1] Key, Lt. Thomas. The Effects of Power Disturbances on Computer Operation, IEEE Industrial and Commercial Power Systems Conference, Cincinnati, OH, June 7, 1978.

[2] Federal Information Processing Standards Publication No. 94, Guideline on Electrical Power for ADP Installations, U.S. Department of Commerce, National Bureau of Standards, Washington, DC, 1983.

[3] IEEE Recommended Practice for Powering and Grounding Sensitive Electronic Equipment (IEEE Green Book), IEEE Std. 1100-1992.

[4] Information Technology Industry Council (ITI), Washington, DC. Website: http//: www.itic.org

[5] IEEE Recommended Practice for Powering and Grounding Sensitive Electronic Equipment (IEEE Emerald Book), IEEE Std. 1100-1999.

Bibliography

How to Correct Power Line Disturbances, Edison, NJ: Dranetz Technologies, 1985.

Martzloff, F. D. The Development of a Guide on Surge Voltages in Low-voltage AC Power Circuits, 14th Electrical/Electronics Insulation Conference, IEEE, Boston, October 1979.

Newman, Paul. UPS Monitoring: Key to an Orderly Shutdown, in *Microservice Management*. Overland Park, KS: Intertec Publishing, March 1990.

Noise Suppression Reference Manual, San Diego, CA: Topaz Electronics.

Pettinger, Wesley. The Procedure of Power Conditioning, in *Microservice Management*. Overland Park, KS: Intertec Publishing, March 1990.

Practical Guide to Quality Power for Sensitive Electronic Equipment, 2nd ed., EC&M, Overland Park, KS: Intertec Publishing, August 1997.

Smeltzer, Dennis. Getting Organized about Power, in *Microservice Management*. Overland Park, KS: Intertec Publishing, March 1990.

Whitaker, Jerry C. *AC Power Systems Handbook,* 2nd ed., Boca Raton, FL: CRC Press, 1998.

CHAPTER

9.3

Facility Grounding Practice

RICHARD B. BERNHARDT, P.E.

John-Winston Engineers & Consultants, Inc.
*Loch Arbour, New Jersey**

INTRODUCTION

A facility can be defined as something that is built, installed, or established to serve a particular purpose [1]. A facility is usually thought of as a single building or group of buildings. The National Electrical Code (NEC) uses the term *premises* to refer to a facility when it defines premises wiring as the interior and exterior (facility) wiring, such as power, lighting, control, and signal systems. Premises wiring includes the service and all permanent and temporary wiring between the service and the load equipment. Premises wiring does not include wiring internal to any load equipment. This chapter will:

- Define grounding and list the reasons for grounding.
- Discuss equipment grounding.
- Discuss system grounding and earth connections.
- Apply recommended facility grounding practices to the safe design and construction of electronics installations.
- Present grounding terminology.

In addition, applicable codes and industry standards will be emphasized.

DEFINITION AND REASONS FOR GROUNDING

The Institute of Electrical and Electronic Engineers (IEEE) defines *grounding* as a conducting connection, whether intentional or accidental, by which an electric circuit or equipment is connected to the earth, or to some conducting body of relatively large extent that serves in place of the earth. It is used for establishing and maintaining the potential of the earth (or of the conducting body), or approximately that potential, on conductors connected to it, and for conducting ground current to and from the earth (or the conducting body) [2].

Based on the IEEE definition, the reasons for grounding can be identified as:

- Personnel safety by limiting potentials between all noncurrent-carrying metal parts of an electrical distribution system and all noncurrent-carrying metal parts of an electrical distribution system and the earth.
- Personal safety and control of electrostatic discharge (ESD) by limiting potentials between all noncurrent-carrying metal parts of an electrical distribution system and earth.
- Fault isolation and equipment safety by providing a low-impedance fault return path to the power source to facilitate the operation of overcurrent devices during a ground fault.

*This chapter contains material from the ninth edition authored by W.E. DeWitt, Purdue University, Sweetwater, TN.

The IEEE definition makes an important distinction between *ground* and *earth*. *Earth* refers to planet Earth and *ground* refers to the equipment grounding system, which includes equipment grounding conductors, metallic raceways, cable armor, enclosures, cabinets, frames, building steel, and all other noncurrent-carrying metal parts of the electrical distribution system.

There are other reasons for grounding not implicit in the IEEE definition. Overvoltage control has long been a benefit of power-system grounding, and is described in IEEE Standard 142, *The Green Book* [3]. With the increasing use of electronic computer systems, noise control has become associated with the subject of grounding, and is described in IEEE Standard 1100, *The Emerald Book* [4].

EQUIPMENT GROUNDING

Personnel safety is achieved by interconnecting all noncurrent-carrying metal parts of an electrical distribution system, and then connecting the interconnected metal parts to the earth. This process of interconnecting metal parts is called *equipment grounding* and is illustrated in Figure 9.3-1, where the equipment grounding conductor is used to interconnect the metal enclosures. Equipment grounding ensures that there is no difference of potential, and thus no shock hazard between noncurrent-carrying metal parts anywhere in the electrical distribution system. Connecting the equipment grounding system to earth ensures that there is no difference of potential between the earth and the equipment grounding system. It also prevents static charge buildup.

SYSTEM GROUNDING

System grounding, which is also illustrated in Figure 9.3-1, is the process of intentionally connecting one of the current-carrying conductors of the electrical distribution system to ground. Figure 9.3-1 shows the neutral conductor intentionally connected to ground and earth. This conductor is called the *grounded* conductor because it is intentionally grounded. The purpose of system grounding is overvoltage control and equipment safety through fault isolation. An ungrounded system is subject to serious overvoltages under conditions, such as intermittent ground faults, resonant conditions, and contact with higher voltage systems. Fault isolation is achieved by providing a low-impedance return path from the load back to the source that will ensure operation of overcurrent devices in the event of a ground fault. The system ground connection makes this possible by connecting the equipment grounding system to the low side of the voltage source. Methods of system grounding include solidly grounded, ungrounded, and impedance grounded.

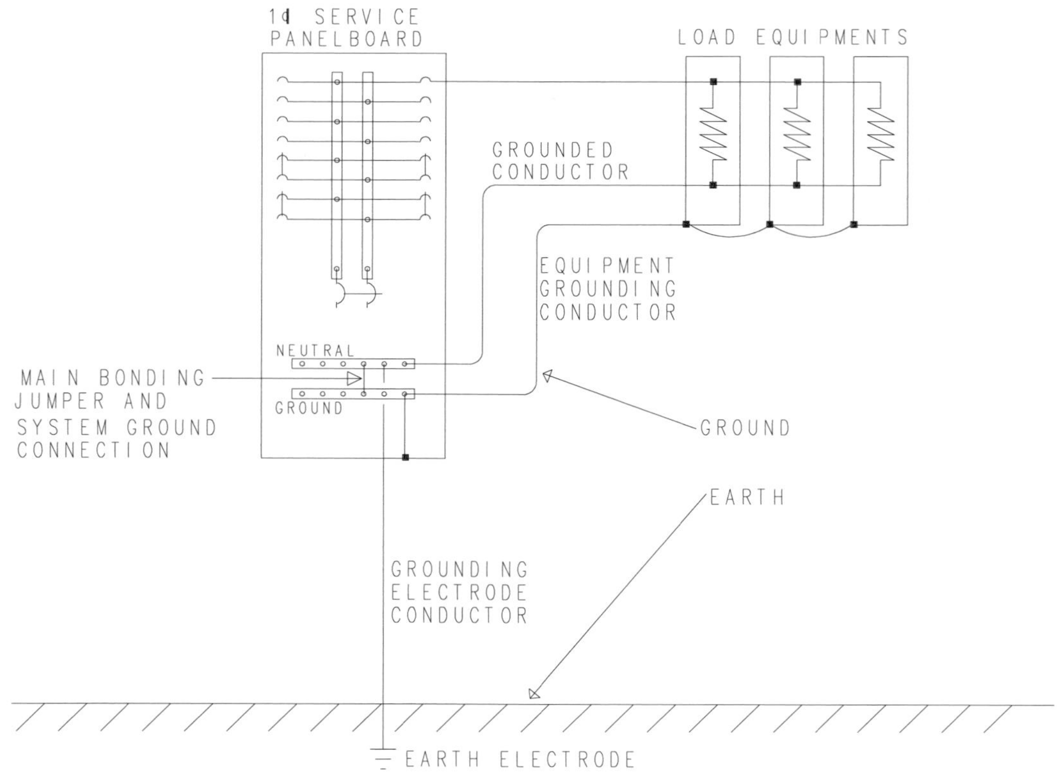

FIGURE 9.3-1 Equipment grounding and system grounding illustrated.

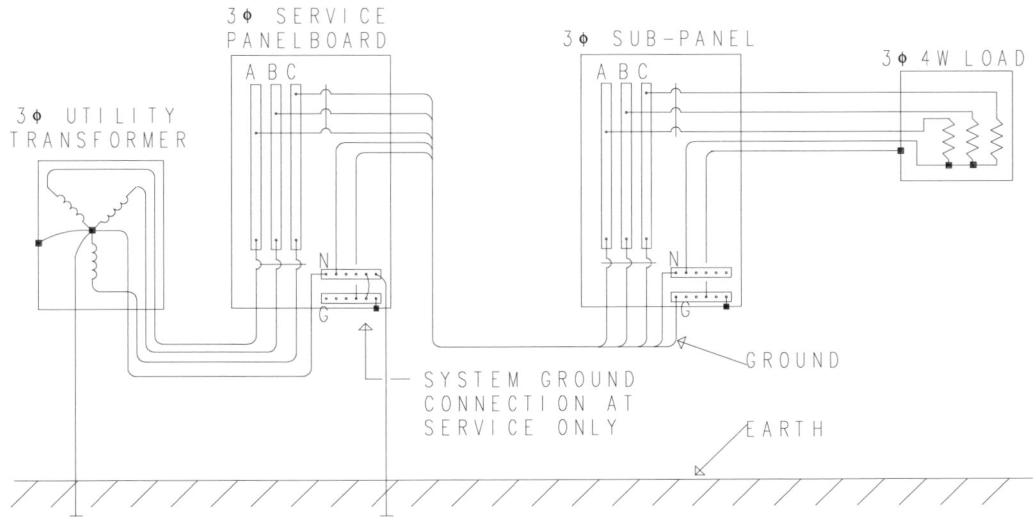

FIGURE 9.3-2 Solidly grounded wye system.

Solidly Grounded

Solidly grounded means that an intentional zero impedance connection is made between a current-carrying conductor and ground. The single-phase (1) system shown in Figure 9.3-1 is solidly grounded. A solidly grounded, three-phase, four-wire, wye system is illustrated in Figure 9.3-2. The neutral is connected directly to ground with no impedance installed in the neutral circuit. The National Electric Code (NFPA 70) permits this connection to be made at the service only. The advantages of a solidly grounded wye system include reduced magnitude of transient overvoltages, improved fault protection, and faster location of ground faults. There is one disadvantage of the solidly grounded wye system. For low-level arcing ground faults, the application of sensitive, properly coordinated, ground fault protection (GFP) devices is necessary to prevent equipment damage from arcing ground faults. The NEC requires arcing ground fault protection at 480Y/277V services, and a maximum sensitivity limit of 1200 A is permitted. Severe damage is less frequent at the lower voltage 208 V systems, where the arc may be self-extinguishing.

Ungrounded

Ungrounded means that there is no intentional connection between a current-carrying conductor and ground. However, charging capacitance will create unintentional capacitive coupling from each phase to ground making the system essentially a capacitance-grounded system. A three-phase, three-wire system from an ungrounded delta source is illustrated in Figure 9.3-3. The most important advantage of an ungrounded system is that an accidental ground fault in one phase does not require immediate removal. This allows for continuity of service, which made the ungrounded delta system very popular in the past. However, ungrounded systems have serious disadvantages.

Since there is no fixed system ground point, it is difficult to locate the first ground fault and to sense the magnitude of fault current. As a result, the fault is often permitted to remain on the system for a long time. If a second fault should occur before the first one is removed, and the second fault is on a different phase, the result will be a double line-to-ground fault causing serious arcing damage. Another problem with the ungrounded delta system is the occurrence of high transient overvoltages from phase-to-ground. Transient overvoltages are caused by intermittent ground faults, with overvoltages capable of reaching a phase-to-ground voltage of from six to eight times the phase-to-neutral voltage. Sustained overvoltages may ultimately result in insulation failure and thus more ground faults.

Impedance Grounded

Impedance grounded means that an intentional impedance connection is made between a current-carrying conductor and ground. The *high resistance grounded* wye system, illustrated in Figure 9.3-4, is an alternative to solidly grounded and ungrounded systems. High resistance grounding will limit ground fault current to a few amperes, thus removing the potential for arcing damage inherent in solidly grounded systems. The ground reference point is fixed, and relaying methods can locate first faults before damages from second faults occur. Internally generated transient overvoltages are reduced since the neutral-to-ground resistor dissipates any charge that may build up on system-charging capacitance.

Table 9.3-1 compares the three most common methods of system grounding. There is no one *best* system-grounding method for all applications. In choosing among the various options, the designer must consider

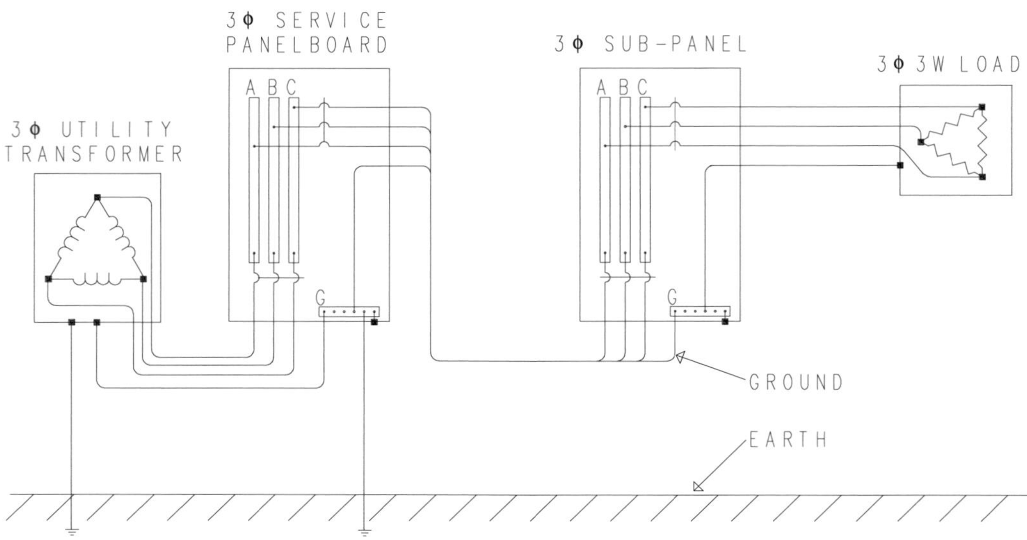

FIGURE 9.3-3 Ungrounded delta system.

the requirements for safety, continuity of service, and cost. Generally, low-voltage systems should be operated solidly grounded. For applications involving continuous processes in industrial plants or where shutdown might create a hazard, a high resistance grounded wye system or a solidly grounded wye system with an alternate power supply may be used. The high resistance grounded wye system combines many of the advantages of the ungrounded delta system and the solidly grounded wye system. IEEE Standard 142 suggests that medium voltage systems less than 15 kV be low-resistance grounded to limit ground fault damage, yet permit sufficient current for detection and isolation of ground faults. IEEE Standard 142 also suggests that medium voltage systems over 15 kV be solidly grounded. Solid grounding should include sensitive ground fault relaying in accordance with the NEC.

Earth Connections

The process of connecting the grounding system to the earth is called *earthing*, and consists of immersing a metal electrode or system of electrodes into the earth. The conductor that connects the grounding system to earth is called the *grounding electrode conductor* (illustrated in Figure 9.3-1). Its function is to keep the entire grounding system at earth potential (voltage equalization during lightning and other transients) rather than

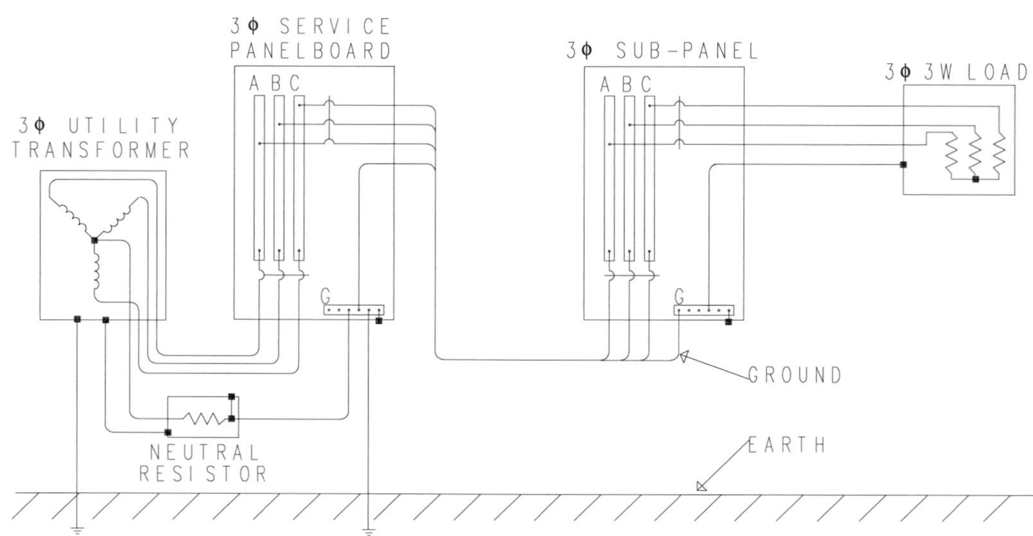

FIGURE 9.3-4 High resistance grounded wye system.

TABLE 9.3-1
Comparison of System-Grounding Methods

Characteristic Assuming No Fault Escalation	System Grounding Method		
	Solidly Grounded	Un-Grounded	High Resistance
Operation of Overcurrent Device on First Ground Fault	Yes	No	No
Control of Internally Generated Transient Overvoltages	Yes	No	Yes
Control of Steady-State Overvoltages	Yes	No	Yes
Flash Hazard	Yes	No	No
Equipment Damage from Arcing Ground-Faults	Yes	No	No
Overvoltage (on Unfaulted Phases) from Ground-Fault	L-N Voltage	>> L-L Voltage	L-L Voltage
Can Serve Line-to-Neutral Loads	Yes	No	No

for conducting ground fault current. Therefore, the NEC allows reduced sizing requirements for the grounding electrode conductor when connected to *made* electrodes (see NEC Section 250 for exact requirements) as described in the next section of this chapter.

The basic measure of effectiveness of an earth electrode system is called *earth electrode resistance*. Earth electrode resistance is the resistance, in ohms, between the point of connection and a distant point on the earth called *remote earth*. Remote earth, about 25 ft from the driven electrode, is the point where earth electrode resistance does not increase appreciably when this distance is increased. Earth electrode resistance consists of the sum of the resistance of the metal electrode (negligible) plus the contact resistance between the electrode and the soil (negligible) plus the soil resistance itself. Thus, for all practical purposes, earth electrode resistance equals the soil resistance. The soil resistance is nonlinear, with most of the earth resistance contained within several feet of the electrode. Furthermore, current flows only through the electrolyte portion of the soil, not the soil itself. Thus, soil resistance varies as the electrolyte content (moisture and salts) of the soil varies. Without electrolyte, soil resistance would be very large.

Soil resistance is a function of soil resistivity. A one cubic meter sample of soil with a resistivity (ρ) of 1 Ω-meter will present a resistance (R) of 1 Ω between opposite faces. A broad variation of soil resistivity occurs as a function of soil types, and soil resistivity can be estimated or measured directly. Soil resistivity is usually measured by injecting a known current into a given volume of soil and measuring the resulting voltage drop. When soil resistivity is known, the earth electrode resistance of any given configuration (single rod, multiple rods, or ground ring) may be determined by using standard equations developed by Sunde [6], Schwarz [7], and others. Three standard grounding equations are listed here:

$$R_1 = \frac{0.52\rho}{lr}\left(\ln\frac{96lr}{Dr} - 1\right), \tag{1}$$

where:

R_1 is the resistance in ohms of a single driven ground rod;

ρ is the soil resistivity in ohm-meters;

lr is the length of ground rod in feet; and

Dr is the diameter of ground rod in inches.

$$R_N = \frac{1}{N}\left(R_1 + \frac{1.05\rho}{S}\left(\frac{1}{2}+\frac{1}{3}+\ldots+\frac{1}{N}\right)\right), \tag{2}$$

where:

R_N is the resistance in ohms of N driven ground rods;

N is the number of equally spaced ground rods;

R_1 is the resistance in ohms of a single driven ground rod; and

S is the uniform spacing of ground rods in feet.

$$R_{GR} = \frac{0.52\rho}{(N)(lr)}\left(\ln\left(\frac{96lr}{Dr}\right) - 1 + \frac{2K_1 lr}{\sqrt{A}}\left(\sqrt{N}-1\right)^2\right),$$

where:

R_{GR} is the resistance in ohms of ground ring configuration;

A is the area of coverage at farthest dimensions in square feet; and

K_1 is the length-to-width ratio coefficient.

Length-to-Width Ratio	$\approx K_1$	Length-to-Width Ratio	$\approx K_1$
1	1.37	5	1.18
2	1.31	6	1.15
3	1.26	7	1.12
4	1.22	8	1.10

Earth resistance values should be as low as practicable, but are a function of the application. The NEC approves the use of a single made electrode if the earth resistance does not exceed 25 Ω. IEEE Standard 1100 reports that the very low earth resistance values specified for computer systems in the past are not necessary. Methods of reducing earth resistance values include the use of multiple electrodes in parallel, ground rings, increased ground rod lengths installation of ground rods to the permanent water level, increased area of coverage of ground rings, and use of concrete-encased electrodes, ground wells, and electrolytic electrodes.

Earth Electrodes

Earth electrodes may be *made* electrodes, *natural* electrodes, or *special purpose* electrodes. Made electrodes include driven rods, buried conductors, ground mats, buried plates, and ground rings. The electrode selected is a function of the type of soil and the available depth. Driven electrodes are used where bedrock is 10 ft or more below the surface. Mats or buried conductors are used for lesser depths. Buried plates are not widely used because of the higher cost when compared to rods. Ground rings employ equally spaced driven electrodes interconnected with buried conductors. Ground rings are used around large buildings, around small unit substations, and in areas having high soil resistivity.

Natural electrodes include buried water pipe electrodes and concrete-encased electrodes. The NEC lists underground metal water piping, available on the premises and not less than 10 ft in length, as part of a preferred grounding electrode system. Since the use of plastic pipes in new water systems will impair the effectiveness of water pipe electrodes, the NEC requires that metal underground water piping be supplemented by an additional approved electrode. Concrete below ground level is a good electrical conductor. Thus, metal electrodes encased in such concrete will function as excellent grounding electrodes. The application of concrete-encased electrodes is covered in IEEE Standard 142.

A special purpose electrode called the *electrolytic* or *chemically charged electrode*, consists of a ground rod installed in a canister containing an electrolytic salt. The salt absorbs moisture from the atmosphere through *breather* holes at the top of the canister. As the salt dissolves in the moisture, it forms a homogeneous electrolytic solution. Gravity and changes in atmospheric pressure cause the electrolytic solution to leak out through *weep* holes at the bottom of the canister. As the electrolytic solution soaks into the surrounding soil, it reduces the soil resistivity and also helps to hold it relatively constant. The result is a consistently lowered and uniform earth electrode resistance.

FACILITY GROUND SYSTEM TESTING

Considering the importance of the grounding system to the safe and reliable operation of the electrical power system, verifying the ground system's condition is critical for the responsible operation of facilities. The justification for periodic testing verification of the grounding system is clear when, in response to system malfunctions, it is often found that the source of the trouble is traced to grounding problems.

The methods for ground system testing are long established procedures that follow basic electrical concepts and are straightforward procedurally. The primary goals of ground testing are to verify the following conditions:

- The resistance of earth electrodes meets the design requirements.
- The physical connection bonding the electrical grounding system to equipment and buildings is mechanically adequate.

The basic earth electrode test is commonly known as the "fall of potential method" test. The test will measure the resistance of the soil to electrode interface along with the ground system conductors. An excellent outline of the specific tasks involved is outlined in the International Electrical Testing Association's Maintenance Testing Specification (http://www.netaworld.org).

Figure 9.3-5 illustrates the basic "fall of potential method" testing concept.

The basic test places the most remote electrode 100 ft away from the ground system being tested. The test unit supplies a voltage and current at the ground system being tested. The test set supplies current to the remote probe that is injected into the ground returning to the current probe connected to the equipment being tested. The movable potential probe is used by the test set to measure the voltage drop at various points along the current return path. Through experience it has been accepted that the ohmic value calculated at the 62 ft mark is the ground resistance of the test equipment.

The testing frequency of grounding systems is determined by several factors that include how thoroughly the existing facility was originally commissioned, the level of maintenance that has been applied, local history of grounding problems, and criticality of

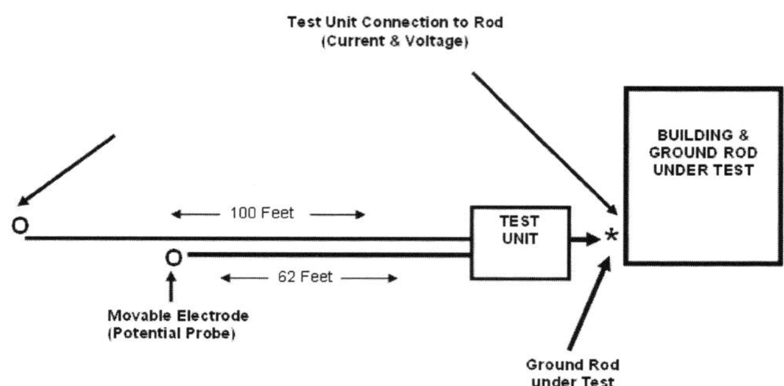

FIGURE 9.3-5 Fall of potential method ground-resistance test.

the facility to provide a high-reliability performance level. A guideline for general testing is contained in the International Electrical Testing Association's MTS-2001, Appendix B, Maintenance Testing Specification, which outlines a risk-weighted frequency of inspection, maintenance, and testing tasks.

FACILITIES GROUNDING FOR COMPUTER AND ELECTRONICS EQUIPMENT

Noise control is an important aspect of computer and electronic systems. The process of noise control through proper grounding techniques is called *referencing*. For this discussion, electronic systems will be viewed as a multiplicity of signal sources transmitting signals to a multiplicity of loads. The ideal electronic circuit consists of a simple signal source supplying a load via a pair of leads as shown in Figure 9.3-6. This source load electronic pair is ideal and free of interference because the impedance of the signal return path is zero and the signal return is a dedicated path.

Practically speaking, however, the impedance of the signal return path is not zero and dedicated return paths for each source load pair are not practical. Packaged electronics systems typically incorporate a common signal reference plane that serves as a common return path for numerous source load pairs (see Figure 9.3-7(a)). The signal reference plane may be a large dedicated area on a printed circuit board, the metal chassis or enclosure of the electronic equipment, or the metal frame or mounting rack that houses several different units. Ideally, the signal reference plane offers zero impedance to the signal current. Practically, however, the signal reference plane has a finite impedance. The practical result is called *common impedance* or *conductive coupling* (see Figure 9.3-7(b)). Since a practical signal reference plane has a finite impedance, current flow in the plane will produce potential differences between various points on the plane. Source load pairs referenced to the plane will, therefore, experience interference as a result. Z_R is common to both circuits referenced to the plane in Figure 9.3-7(b). Thus, I_1 and I_2 returning to their respective sources will produce interference voltages by flowing through Z_R. The total interference voltage drop felt across Z_R causes the source reference A to be at a different potential than the load reference B. This difference in potential is often called *ground voltage shift* (even though ground may not even be involved), and is a major source of noise and interference in electronic circuits.

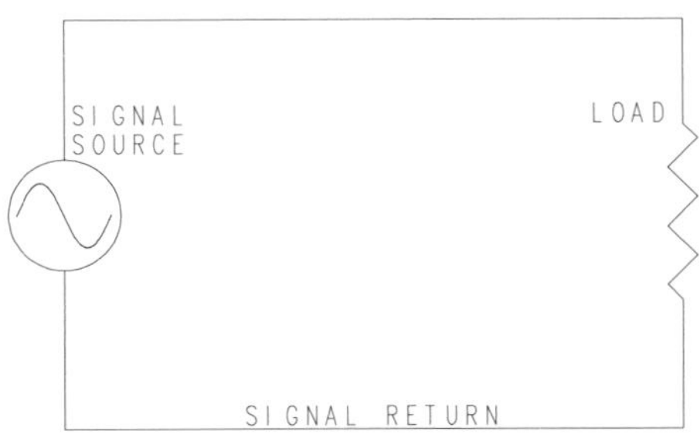

FIGURE 9.3-6 Ideal electronic circuit.

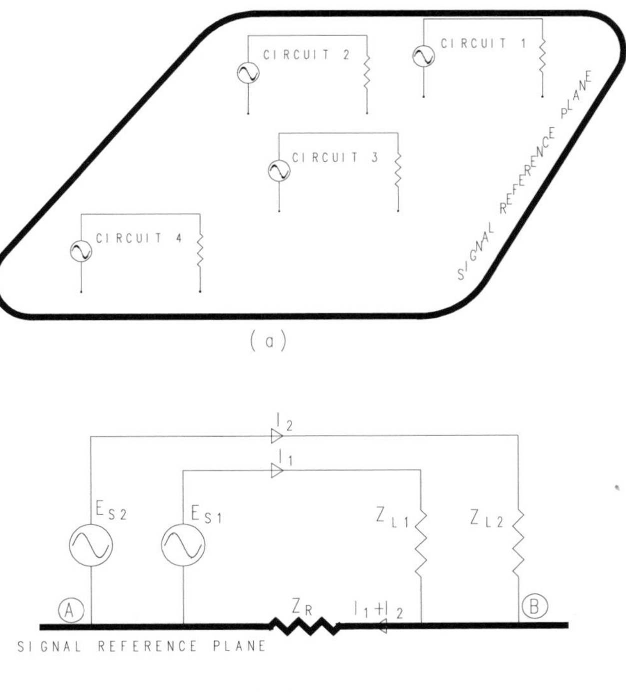

FIGURE 9.3-7 Equipotential plane.

Ground voltage shifts can also be caused by electromagnetic or electrostatic fields in close proximity to the source load pairs. The interference source induces interference voltages into any closed loop by antenna action. This loop is called a *ground loop* (even though ground may not be involved). Interference voltages can be minimized by reducing the loop area as much as possible. This may be very difficult if the loop includes an entire room. The interference voltage can be eliminated entirely by breaking the loop.

Within individual electronic equipments, the signal reference plane consists of a metal plate or the metal enclosure or a rack assembly as previously discussed. Between units of equipment that are located in different rooms on different floors or even in different buildings, the signal reference planes of each unit must be connected together via interconnected wiring such as coax shields or separate conductors. This action, of course, increases the impedance between signal reference planes and makes noise control more difficult. Reducing noise caused by common impedance coupling is a matter of reducing the impedance of the interconnected signal reference planes. Regardless of the configuration encountered (circuit board, individual electronic equipment, or equipment remotely located within a facility or in separate buildings), the next question to be answered is should the signal reference be connected to ground? Floating signal grounding, single-point grounding, and multipoint grounding are methods of accomplishing this signal reference-to-ground connection.

IEEE Standard 1100 recommends multipoint grounding and a signal reference plane for most computer and electronics installations. An ideal equipotential signal reference plane is one that has 0 V (thus zero impedance) between any two points on the plane. Since an *ideal* equipotential plane is not attainable, a *nominal* equipotential plane is accepted. Multipoint grounding connections are made to the plane, which ensures minimum ground voltage shift between signal reference systems connected to the plane. Collectively, the current flow in an equipotential plane may be quite large. Between any two equipments, however, the current flow should be low due to the many current paths available.

Practical versions of an equipotential plane include a bolted stringer system of a raised computer floor, flat copper strips bonded together at 2 ft centers, copper conductors bonded together at 2 ft centers, and single or multiple flat copper strips connected between equipment.

ISOLATED GROUNDING

Isolated grounding schemes, where the signal reference plane is isolated from equipment ground but connected to an *isolated* electrode in the earth, *do not work, are unsafe,* and *violate the NEC.* It is thought by some people that the isolated earth connection is *clean* since there is no connection between it and the *dirty* system ground connection at the service entrance. The clean, isolated earth connection is also viewed, incorrectly, as a point where noise currents can flow into the earth and be dissipated. *Kirchhoff's Current Law* teaches that any current flowing into the isolated ground must return to the source through another earth connection. Current cannot be dissipated. It must always return to its source. Even lightning current is not dissipated into the earth. It must have a return path that is the electrostatic and electromagnetic fields that created the charge buildup and the lightning strike in the first place.

Consider what might happen if the previously described system above is subjected to a lightning strike. Assume that a transient current of 2000 A flows into the earth and through an earth resistance of 5 Ω between the system ground electrode and the isolated electrode. A more realistic resistance might be even higher, such as 25 Ω. Two thousand amperes flowing through 5 Ω results in a transient potential of 10,000 V between the two electrodes. Since this potential is impressed between the equipment frame (system ground electrode) and the signal reference plane (isolated electrode), it could result in equipment damage and personnel hazard. Dangerous potential differences between grounding subsystems can be reduced by bonding together all earth electrodes on a facility. NEC Article 250 states that the bonding together of separate grounding electrodes will limit potential differences between them and their associated wiring systems [5].

A facility ground system, then, can be defined as an electrically interconnected system of multiple conducting paths to the earth electrode or system of electrodes. The facility grounding system includes all electrically interconnected grounding subsystems such as the equipment grounding subsystem, signal reference subsystem, fault protection subsystem, and lightning protection subsystem.

Isolated ground (IG) receptacles, which are a version of single-point grounding, are permitted by the NEC. Proper application of IG receptacles is very important. They must be used with an insulated equipment grounding conductor, not a bare conductor. Also, only metallic conduit should be used.

SEPARATELY DERIVED SYSTEMS

A *separately derived system* is a premises wiring system whose power is derived from generator, transformer, or converter windings and that has no direct electrical connection, including a solidly connected grounded circuit conductor, to supply conductors originating in another system [5]. Solidly grounded, wye connected, isolation transformers used to supply power to computer room equipment are examples of separately derived systems. Figure 9.3-8 illustrates the bonding and grounding requirements of separately derived systems. NEC Article 250 permits the bonding and grounding connections to be made at the source of the separately derived system or at the first disconnecting means, but not both locations.

If the ground neutral connection is made in two locations, as shown in Figure 9.3-8, normal neutral current would be normally carried in both the neutral and the ground conductors. This is unsafe and pre-

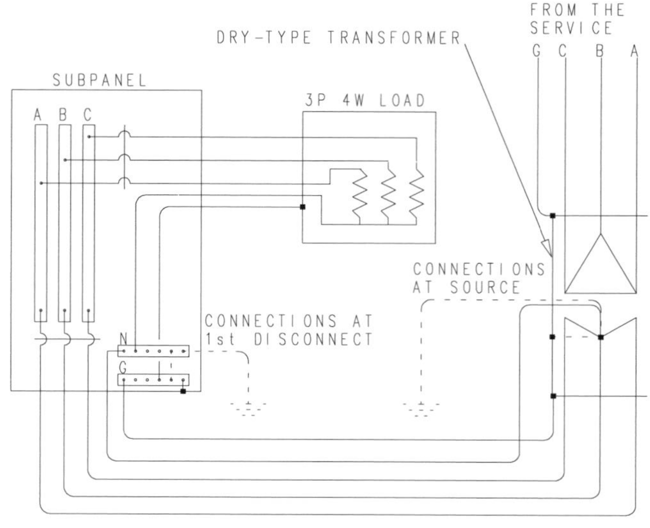

FIGURE 9.3-8 Separately derived system.

sents a hazard to equipment and personnel. In practice a separately derived system shall have neutral and earth ground bonded only once at one location.

Other examples of separately derived systems include service transformers, generators, distributed generation systems, and UPS systems. Note that all earth electrodes are bonded together via the equipment grounding conductor system. This is consistent with the recommendations listed in NEC Article 250 and described in the previous paragraph.

FACILITY GROUNDING OF LIGHTNING PROTECTION SYSTEMS

The application details and performance requirements of lightning protection to buildings are not specifically addressed in the NEC. The application of lightning protection systems can be certified by independent testing laboratories. Underwriter Laboratories (UL) has a field inspection program where it will inspect, list deficiencies, and certify the system when properly completed.

The key concept in applying a lightning protection system, as it relates to the facility grounding system, is that all the lightning protection is bonded to the facility earth electrode. An example of a typical building lightning protection system includes rooftop air terminals directly routed to the ground floor of the building and attached to earth electrodes. These earth electrodes are bonded to the facility grounding system and building steel to form an integrated, single ground reference, earth grid to create the lowest impedance path for the lightning's energy to be dissipated to earth. Oftentimes in steel buildings additional bond connections are added to lower the impedance of the original structural steel-bolted connections. See Chapter 7.3, "Lightning Protection," for more information.

Key Terms

Equipment grounding conductor: The conductor used to connect the noncurrent-carrying metal parts of equipment, raceways, and other enclosures to the system grounded conductor, the grounding electrode conductor, or both, at the service equipment or at the source of a separately derived system.

Equipotential plane: A mass of conducting material that offers a negligible impedance to current flow, thus producing zero volts (equipotential) between points on the plane.

Grounded conductor: A system or circuit conductor that is intentionally grounded.

Floating signal grounding: A nongrounding system in which all electronic signal references are isolated from ground.

Grounding electrode conductor: The conductor used to connect the grounding electrode to the equipment grounding conductor, the grounded conductor, or both, of the circuit at the service equipment or at the source of a separately derived system.

Main bonding jumper: The connection between the grounded circuit conductor and the equipment grounding conductor at the service.

Multipoint grounding: A grounding system in which all electronic signal references are grounded at multiple points.

Service: The conductors and equipment for delivering energy from the electricity supply system to the wiring system of the premises served.

Service conductors: The supply conductors that extend from the street main or from transformers to the service equipment of the premises supplied.

Service equipment: The necessary equipment, usually consisting of a circuit breaker or switch and fuses, and their accessories, located near the point of entrance of supply conductors to a building or other structure, or an otherwise defined area, and intended to constitute the main control and means of cutoff of the supply.

Single-point grounding: A grounding system in which all electronic signal references are bonded together and grounded at a single point.

References

[1] *Webster's New Collegiate Dictionary.* G. & C. Merriam Company, Springfield, MA.

[2] IEEE Standard 100, Definitions of Electrical and Electronic Terms, Institute of Electrical and Electronics Engineers, Inc., New York, NY.

[3] IEEE Standard 142, Recommended Practice for Grounding Industrial and Commercial Power Systems, Institute of Electrical and Electronics Engineers, Inc., New York, NY, 1982.

[4] IEEE Standard 1100, Recommended Practice for Powering and Grounding Sensitive Electronics Equipment, Institute of Electrical and Electronics Engineers, Inc., New York, NY, 1992.

[5] NFPA Standard 70, The National Electrical Code, The National Fire Protection Association, Inc., Quincy, MA, 1996.

[6] Sunde, E. D. *Earth Conduction Effects in Transmission Systems.* Von Nonstrand Co., 1949.

[7] Schwarz, S. J. Analytical Expression for Resistance of Grounding Systems, *AIEE Transactions*, 73, Part III-B, 1954, pp. 1011–1016.

Further Reading

"Ground Fault Protection for Solidly-Grounded Low-Voltage Systems," General Electric Bulletin EESG II-AP-7, December 1974.

Kaufmann, R. H. "Some Fundamentals of Equipment Grounding Circuit Design," GEB #957, Industrial Power Systems Applications, General Electric Corporation.

Lazar, Irwin. "System Grounding in Industrial Power Systems," *Specifying Engineer*, May 1978–Jan. 1979.

NEMA PB2.2, "Application Guide for Ground Fault Protective Devices for Equipment," National Electrical Manufacturers Association, Washington, DC, current edition.

Techniques of Electrical Design and Construction, Volume 4: Protecting against Ground Faults. Boston: McGraw-Hill, 1979.

CHAPTER

9.4

Standby Power Systems

GEOFFREY A. KRENKEL, P.E.
John-Winston Engineers & Consultants, Inc.
*Loch Arbour, New Jersey**

INTRODUCTION

When utility company power problems are discussed, most people immediately think of blackouts. The lights go out, and everything stops. With the facility down and in the dark, there is nothing to do but sit and wait until the utility company finds the problem and corrects it. This process generally takes only a few minutes. There are times, however, when it can take hours. In some remote locations, it can even take days.

Because there is no way of predicting where or when a power outage will occur, therefore, an alternative power source should be available at all times. Today's broadcast facility engineers are typically equipped with electrical distribution systems that integrate commercial power with on-site standby power generation in the event that a catastrophic power failure occurs.

Standby power sources may come in the form of natural gas or diesel generators and/or turbines. Less-resilient fuel cells and photovoltaic solar cells are beginning to make their presence and present great assurance that an environmentally friendly approach to producing power may someday be possible on a grand scale. However, the key to understanding a particular power source is to understand its characteristics and the application in which it is used. It is for this reason that several different standby power systems will be examined and explored to help provide insight as to why choosing the correct standby power system for a distinct application is important.

Power Blackouts

Blackouts are, without a doubt, the most troublesome utility company problem that a broadcast facility will have to face. Statistics show that power failures are, generally speaking, a rare occurrence in most areas of the country and are short in duration. Studies have shown that 50% of blackouts last 6 seconds or less, and 35% are less than 11 minutes long. These failure rates usually are not cause for concern to commercial users, except where computer-based operations, transportation control systems, medical facilities, and communications sites (including radio and television stations) are concerned.

A broadcast facility that is down for even a few minutes can suffer a significant loss of audience that may take hours or days to rebuild. A blackout affecting a transportation or medical center could be life threatening. Coupled with this threat is the possibility of extended power service loss due to severe storm conditions. Many broadcast and communications relay sites are located in remote, rural areas or on mountaintops. Neither of these kinds of locations is known for reliable power. It is not uncommon in mountainous areas for utility company service to be out for extended periods after a major storm. Few operators are willing to take such risks with their business. Most choose to install standby power systems at appropriate points in the equipment chain. A distinction must be made between *emergency* and *standby* power sources. Strictly speaking, emergency systems

*This chapter contains material from the ninth edition authored by Jerry Whitaker, Technical Press, Morgan Hill, CA.

supply circuits legally designated as being essential for safety to life and property. Standby power systems are used to protect a facility against the loss of productivity resulting from a utility company power outage.

STANDBY POWER OPTIONS

The cost of standby power for a facility can be substantial, and an examination of the possible alternatives should be conducted before any decision on equipment is made. Management must clearly define the direct and indirect costs and weigh them appropriately. Include the following items in the cost versus risk analysis:

- Standby power system equipment purchase and installation cost.
- Exposure of the system to utility company power failure.
- Alternative operating methods available to the facility.
- Direct and indirect costs of lost uptime because of blackout conditions.

Several types of standby power supplies are currently available to broadcasts for transmission and studio operations. These include:

- Dual-power utility feeders
- Classic power generators
- Fuel cells
- UPS systems

To ensure the continuity of AC power, many commercial/industrial facilities depend on either two separate utility services or one utility service plus on-site generation. Because of the growing complexity of electrical systems, attention must be given to power supply reliability.

Dual-Power Utility Feeder System

In some areas, usually metropolitan centers, two utility company power drops can be brought into a facility as a means of providing a source of standby power. As shown in Figure 9.4-1, two separate utility service drops—from separate power distribution systems—are brought into the plant, and an automatic transfer switch changes the load to the backup line in the event of a main line failure. The dual feeder system provides an advantage over the diesel generator arrangement in that power transfer from main to standby can be made in a fraction of a second if a static transfer switch is used. Time delays in the diesel generator system limit its usefulness to power failures lasting more than several minutes.

The dual feeder system of protection is based on the assumption that each of the service drops brought into the facility is routed via different paths. This being the case, the likelihood of a simultaneous failure on both power lines is remote. The dual feeder system will not, however, protect against area-wide power failures that may occur from time to time.

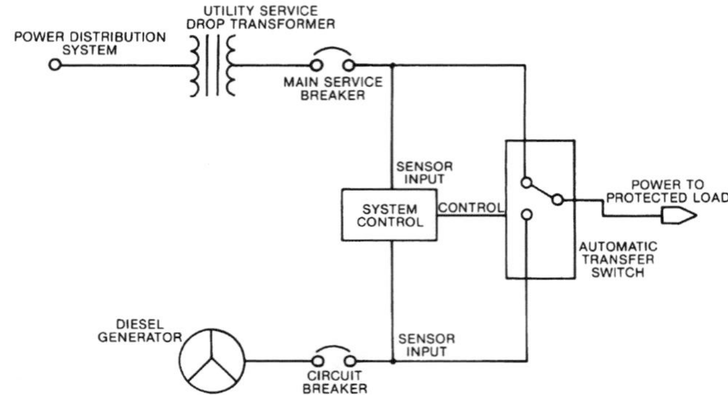

FIGURE 9.4-1 The dual utility feeder system of AC power loss protection. An automatic transfer switch changes the load from the main utility line to the standby line in the event of a power interruption.

The dual feeder system is limited primarily to urban areas. Rural or mountainous regions generally are not equipped for dual redundant utility company operation. Even in urban areas, the cost of bringing a second power line into a facility can be high, particularly if special lines must be installed for the feed. If two separate utility services are available at or near the site, these redundant feeds generally will be less expensive than engine driven generators of equivalent capacity.

Figure 9.4-2 illustrates a dual feeder system that employs both utility inputs *simultaneously* at the facility. During normal operation, both AC lines feed facility loads, and the *tie* circuit breaker is open. In the event of a loss of either line, the circuit-breaker switches reconfigure the load to place the entire facility on the single remaining AC feed. Switching is normally performed automatically while manual control

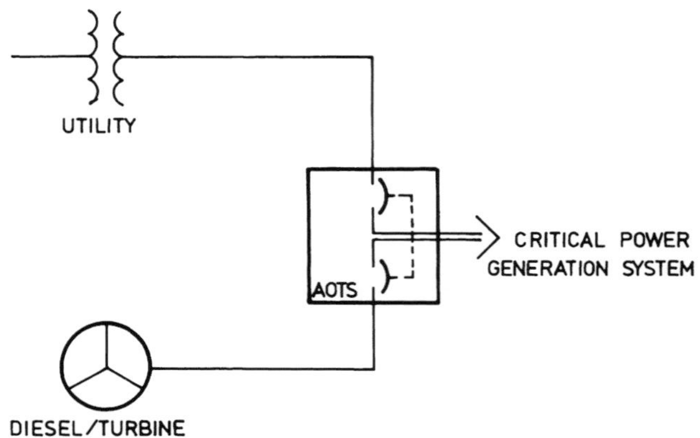

FIGURE 9.4-2 A dual utility feeder system with interlocked circuit breakers.

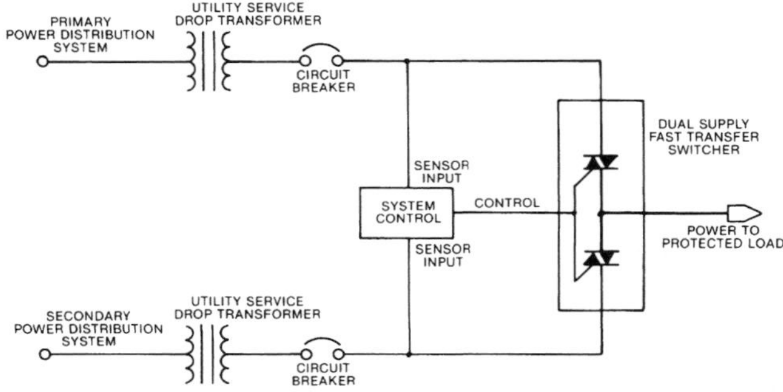

FIGURE 9.4-3 The classic standby power system using an engine-generator set. This system protects a facility from prolonged utility company power failures.

is provided in the event of a planned shutdown on one of the lines or for routine testing.

Classic Power Generators

The engine-generator shown in Figure 9.4-3 is the classic standby power system. An automatic transfer switch monitors the AC voltage coming from the utility company line for power failure conditions. Upon detection of an outage for a predetermined period of time (generally 1–10 seconds), the standby generator is started; after the generator is up to speed, the load is transferred from the utility to the local generator. Upon return of the utility feed, the load is switched back, and the generator is stopped. This basic type of system is used widely in industry and provides economical protection against prolonged power outages (5 minutes or more).

The transfer device shown in Figure 9.4-3 is a contactor-type, break-before-make unit. By replacing the simple transfer device shown with an automatic overlap (*static*) transfer switch, as shown in Figure 9.4-4, additional functionality can be gained. The overlap transfer switch permits the on-site generator to be synchronized with the load, making a clean switch from one energy source to another. This functionality offers the following benefits:

- Switching back to the utility feed from the generator can be accomplished without interruption in service.
- The load can be cleanly switched from the utility to the generator *in anticipation* of utility line problems (such as an approaching severe storm) or for routine testing.
- The load can be switched to and from the generator to accomplish *load* shedding or peak power shaving objectives.

Peak Power Shaving

Figure 9.4-5 illustrates the use of a backup diesel generator for both standby power and *peak power shaving* applications. Commercial power customers often can

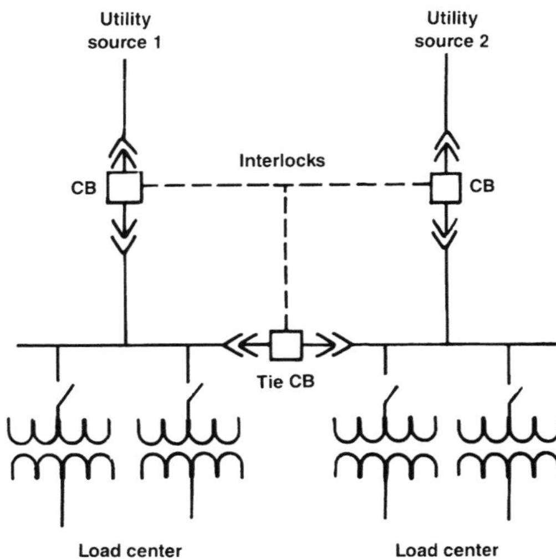

FIGURE 9.4-4 The use of a static transfer switch to transfer the load from the utility company to the on-site generator.

realize substantial savings on utility company bills by reducing their energy demand during certain hours of the day. An automatic overlap transfer switch is used to change the load from the utility company system to the local diesel generator. The changeover is accomplished by a static transfer switch that does not disturb the operation of load equipment. This application of a standby generator can provide financial return to the facility, whether or not the unit is ever needed to carry the load through a commercial power failure.

Choosing a Generator

Engine-generator sets are available for power levels ranging from less than 1 kVA to several thousand kVA

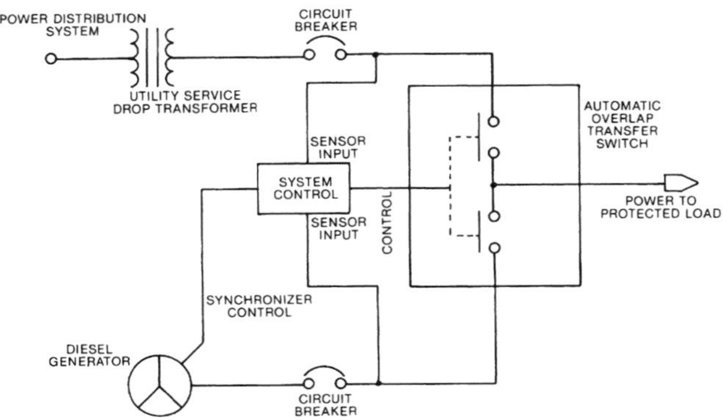

FIGURE 9.4-5 The use of a diesel generator for standby power and peak power shaving applications. The automatic overlap (static) transfer switch changes the load from the utility feed to the generator instantly so that no disruption of normal operation is encountered.

or more. Machines also may be paralleled to provide greater capacity or redundancy when high load capacities are involved. Engine-generator sets typically are classified by the type of power plant used:

- *Diesel.* Advantages: rugged and dependable, low fuel costs, low fire and/or explosion hazard. Disadvantages: somewhat more costly than other engines, heavier in smaller sizes, longer startup times, fuel requires special attention when not used for long periods and for long-term storage.
- *Natural and liquefied petroleum gas.* Advantages: quick starting after long shutdown periods, long life, low maintenance. Disadvantage: availability of natural gas during area-wide power failure subject to question.
- *Gasoline.* Advantages: rapid starting, low initial cost. Disadvantages: greater hazard associated with storing and handling gasoline, generally shorter mean time between overhaul.
- *Gas turbine.* Advantages: smaller and lighter than piston engines of comparable horsepower, rooftop installations practical, rapid response to load changes. Disadvantages: longer time required to start and reach operating speed, sensitive to high input air temperature.

The type of power plant chosen is usually determined primarily by the environment in which the system will be operated and by the cost of ownership. For example, a standby generator located in an urban area office complex may be best suited to the use of an engine powered by natural gas, because of the problems inherent in storing large amounts of fuel. State or local building codes may place expensive restrictions on fuel storage tanks and make the use of a gasoline or diesel-powered engine impractical. The use of propane usually is restricted to rural areas. The availability of propane during periods of bad weather (when most power failures occur) also must be considered.

The generator rating for a standby power system should be chosen carefully and should take into consideration the anticipated future growth of the plant. It is good practice for a standby power system to be rated for at least 25% greater output than the current peak facility load. This headroom gives a margin of safety for the standby equipment and allows for future expansion of the facility without overloading the system.

An engine driven standby generator typically incorporates automatic starting controls and an automatic transfer switch, as shown in Figure 9.4-6. Control circuits monitor the utility supply and start the engine when there is a failure or a sustained predetermined voltage drop. The switch transfers the load as soon as the generator reaches operating voltage and frequency. Upon restoration of the utility supply for a predetermined period, the switch returns the load and initiates engine shutdown. The automatic transfer switch must meet demanding requirements, including:

- Carrying the full-rated current continuously.

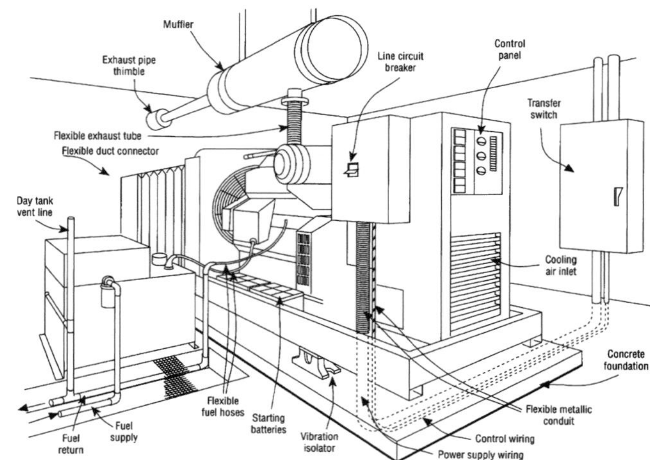

FIGURE 9.4-6 Typical configuration of an engine-generator set. (From [1], used with permission.)

- Withstanding fault currents without contact separation.
- Handling high inrush currents.
- Withstanding many interruptions at full load without damage.

The nature of most power outages requires a sophisticated monitoring system for the engine-generator set. Most power failures occur during periods of bad weather. Most standby generators are unattended. More often than not, the standby system will start, run, and shut down without any human intervention or supervision. For reliable operation, the monitoring system must check the status of the machine continually to ensure that all parameters are within normal limits. Time-delay periods usually provided by the controller require an outage to last from 5–10 seconds before the generator is started and the load is transferred. This prevents false starts that needlessly exercise the system. A time delay of 5–30 minutes usually is allowed between the restoration of utility power and the return of the load. This delay permits the utility power lines to stabilize before the load is reapplied.

The transfer of electric motor loads, such as air conditioning systems and air handlers, may require special consideration depending on the size and type of motors used at a plant. If the residual voltage of the motor is out of phase with the power source to which the motor is being transferred, serious damage may result to the motor. Excessive current draw also may trip overcurrent protective devices. Motors above 50 HP with relatively high load inertia in relation to torque requirements, such as flywheels and fans, may require special controls. Restart time delays are a common solution.

Automatic starting and synchronizing controls are used for multiple engine-generator installations. The output of two or three smaller units can be combined to feed the load. This capability offers additional protection for the facility in the event of a failure in any one machine. As the load at the facility increases, additional engine-generator systems can be installed on the standby power bus.

Generator Types

Generators for standby power applications may be *induction* or *synchronous* machines. Most engine-generator systems in use today are synchronous because of the versatility, reliability, and capability of operating independently that this approach provides. Most modern synchronous generators are of the *revolving field alternator* design. Essentially, this means that the armature windings are held stationary and the field is rotated. Therefore, generated power can be taken directly from the stationary armature windings. Revolving armature alternators are less popular because the generated output power must be derived via slip rings and brushes.

The exact value of the AC voltage produced by a synchronous machine is controlled by varying the current in the DC field windings, while frequency is controlled by the speed of rotation. Power output is controlled by the torque applied to the generator shaft by the driving engine. In this manner, the synchronous generator offers precise control over the power it can produce.

Practically all modern synchronous generators use a *brushless exciter*. The exciter is a small AC generator on the main shaft and the AC voltage produced is rectified by a three-phase rotating rectifier assembly also on the shaft. The DC voltage thus obtained is applied to the main generator field, which is also on the main shaft. A voltage regulator is provided to control the exciter field current, and in this manner, the field voltage can be precisely controlled, resulting in a stable output voltage.

The frequency of the AC power produced is dependent on two factors: the number of poles built into the machine and the speed of rotation (rpm). Because the output frequency must normally be maintained within strict limits (whether 60 Hz or 50 Hz), control of the generator speed is essential. This is accomplished by providing precise rpm control of the *prime mover* (the engine), which is performed by a governor.

There are many types of governors; however, for auxiliary power applications, the *isochronous governor* is normally selected. The isochronous governor controls the speed of the engine so that it remains constant from no load to full load, assuring a constant AC power output frequency from the generator. A modern system consists of two primary components: an electronic speed control and an actuator that adjusts the speed of the engine. The electronic speed control senses the speed of the machine and provides a feedback signal to the mechanical/hydraulic actuator, which in turn positions the engine throttle or fuel control to maintain accurate engine rpm.

The National Electrical Code (NEC) provides guidance for safe and proper installation of on-site engine-generator systems. Local codes may vary and must be reviewed during early design stages of standby power systems.

Location of Generators

Engine-generators are often located outside the building for which standby power is supplied. While most safety concerns are easily met with outside generators, the downside is that the generator requires space that may be difficult to obtain at reasonable cost, produces noise that may be objectionable to neighbors (solved by expensive "hospital grade" insulation and noise baffling), and subject to extreme variations in temperature and humidity that could impair starting reliability (manifold heaters help solve this problem). Generators located inside of a building such as a basement or special addition are less susceptible to environmental variations but in turn are subject to rigorous safety, code, fuel, and operational requirements, and expensive ventilation systems for exhaust and cooling. In some cases when neither a suitable outside or inside location is possible, generators can be placed on rooftops. This is an extreme situation but may be necessary when no other location seems either possible or practical. Rooftop generators are expensive to install, operate, and maintain. Installation may require substantial enhancements of the support structure, rental of a crane to hoist the generator to the roof, and special fuel systems, such as an extension of natural gas to the roof or expensive storage and management of liquid fuel. In addition, special variants in local building codes may be required and the permission of neighbors who may object to both the visibility of the generator and its operating noises. Regular testing of the rooftop generator will likely produce vibration and noise and concern by the building occupants. Maintenance may be difficult if major components of the generator system must be replaced or overhauled.

Standby Power System Noise

Acoustic noise produced by backup power systems can be a serious problem if not addressed properly. Standby generators and UPS systems produce noise that can disturb building occupants and irritate neighbors and/or landlords.

The acoustic noise associated with electrical generation usually is related to the drive mechanism, most commonly an internal combustion engine. The amplitude of the noise produced is directly related to the size of the engine-generator set. First, consider whether noise reduction is a necessity. Many building owners have elected to tolerate the noise produced by a standby power generator because its use is limited to short-term outage situations. During an outage, when the normal source of power is unavailable, most people will tolerate noise associated with a standby generator.

If the decision is made that building occupants can live with the noise of the generator, care must be taken in scheduling the required testing and exercising of the unit. Whether testing occurs monthly or weekly, it should be done on a regular schedule.

If it has been determined that the noise should be controlled, or at least minimized, the easiest way to achieve this objective is to physically separate the machine from occupied areas—this may be easier said than done. Because engine noise is predominantly low frequency in character, walls and floor/ceiling construction used to contain the noise must be massive. Lightweight construction, even though it may involve several layers of resiliently mounted drywall, is ineffective in reducing low-frequency noise. Exhaust noise is a major component of engine noise but, fortunately, it is easier to control. When selecting an engine-generator set, select the highest quality exhaust muffler available, such as hospital grade mufflers.

Engine-generator sets also produce significant vibration. The machine should be mounted securely to a slab-on grade or an isolated basement floor, or it should be installed on vibration-isolation mounts. Such mounts usually are specified by the manufacturer.

Because a UPS system is a source of continuous power, it must run continuously. Noise must be adequately controlled. Physical separation is the easiest and most effective method of shielding occupied areas from noise. Enclosure and environmental control of UPS equipment usually is required, but noise control is significantly easier than for an engine-generator because of the lower noise levels involved. Nevertheless, the low-frequency 120 Hz fundamental of a UPS system is difficult to contain adequately and extensive vibration control may be necessary.

FUEL CELL SYSTEMS

Mostly recognized as a power source that can deliver clean and efficient power, fuel cells are becoming the promising future for alternate power-source options. The underlying principle of a fuel cell is to convert chemical energy into electrical energy. In theory, two electric plates (an anode and cathode) are sandwiched together between an electrolyte (an ion conductor) to form a fuel cell stack that produces electricity, water, and heat using fuel and oxygen in the air, as shown in Figure 9.4-7. When numerous stacks are combined, a substantial amount of energy can be obtained.

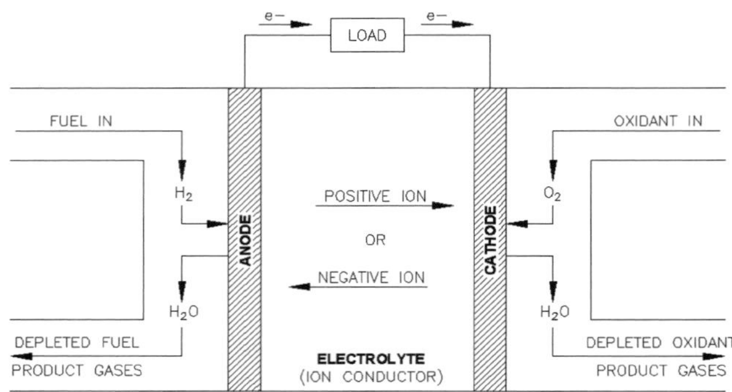

FIGURE 9.4-7 Schematic of an individual fuel cell. (From Hirschenhofer, et al. [2], 1998.)

There are several different types of fuel cells (see Table 9.4-1) with pros and cons that may support or undermine a particular application. Typically fuel cells are categorized by their electrolyte. Although there are presently several types of fuel cells gaining noticeable recognition, those described herein have been selected based on their successful commercialization and present dedication to research and development.

In addition to a fuel stack, a fuel cell power system requires several other components, including:

- *Fuel reformer.* For the processing of hydrogen from fossil fuels.
- *Power inverter.* To change direct current into alternating current.
- *Common coupling point.* Point at which power production and the distribution network (typically the load side of the power network) interconnect.

From a performance perspective, although a fuel cell can achieve up to 60% efficiency, it does not handle surge currents very well. Surges typically occur when large motor loads, such as pumps, chiller, large mechanical equipment loads, and transmitter power

TABLE 9.4-1
Comparison of Fuel Cell Types

Fuel Cell Type	Electrolyte	Anode Gas	Cathode Gas	Temperature
Proton exchange membrane (PEM)	Solid polymer membrane	Hydrogen	Pure or atmospheric oxygen	80°C
Phosphoric acid (PAFC)	Phosphorus	Hydrogen	Atmospheric oxygen	200°C
Molton carbonate (MCFC)	Alkali-carbonates	Hydrogen, methane	Atmospheric oxygen	650°C
Solid oxide (SOFC)	Ceramic oxide	Hydrogen, methane	Atmospheric oxygen	800–1000°C

amplifiers, are first energized. Surges can also be intermittently experienced in inductive output tubes (IOT) that are sometimes found in large broadcast transmitters. However, recent interest in hybrid fuel cells, consisting of two or more generation technologies that make the best use of their operating characteristics to obtain efficiencies higher than those that could be obtained by a single power source, holds promise that someday a commercially available system may be capable of handling large inrush currents.

Although independently not ideal for inrush currents, today's fuel cells are finding a useful purpose by replacing standard rechargeable batteries. Their storage efficiency and size make them ideal for replacing bulky batteries that can be found, for example, on professional broadcast cameras.

UPS SYSTEMS

An uninterruptible power supply (UPS) is used to bridge the gap of time when no available power sources are present, such as between the time the commercial power fails and the generator picks up the load. This bridge time is usually long enough to switch to an alternate power source. Although a UPS has other purposes, such as voltage and frequency regulation, its main emphasis is often realized when commercial power is no longer available. A typical commercial UPS (double on-line conversion) is made up of a static switch, rectifier (converts AC power to DC power), inverter (converts DC power back to AC power), and batteries as means of storing DC power until needed. Its fast-acting transfer switch allows for an almost seamless transfer between batteries and normal commercial power.

From a broadcast transmitter application perspective, designing a reliable uninterruptible power supply has been a challenge for most UPS manufacturers. This in part is due to the fact that many broadcasting transmitters today can exhibit high inrush currents that are intermittently caused by, among other kinds of loads, high-power amplifiers such as an inductive output tube (IOT). IOTs typically make up the last output amplifier in certain types of transmitters and are being successfully incorporated into analogue UHF and digital transmitters [3]. Their large inrush capabilities are characteristic of a short circuit current and must be dealt with cautiously when choosing a UPS.

When planning for a UPS system, consider the following:

- Reserve power capacity for future growth of the facility.
- Inverter current surge capability (if the system will be driving inductive loads, such as electric motors and switching power supplies).
- Output voltage and frequency stability over time and with varying loads.
- Required battery supply voltage and current. Battery costs vary greatly, depending on the type of units needed.
- Type of UPS system (forward-transfer type or reverse-transfer type) required by the particular application. Some sensitive loads may not tolerate even brief interruptions of the AC power source.
- Inverter efficiency at typical load levels. Some inverters have good efficiency ratings when loaded at 90% capacity, but poor efficiency when lightly loaded.
- Size and environmental requirements of the UPS system. High-power UPS equipment requires a large amount of space for the inverter/control equipment and batteries. Battery banks often require special ventilation and ambient temperature control.
- Maintenance of the UPS system cabinet and controls.

In order to perform UPS system maintenance a three-circuit-breaker wraparound bypass configuration can be designed so that complete removal of the UPS can be achieved without interrupting the load. A typical 225 kVA UPS with maintenance bypass is shown in Figure 9.4-8. The output transformer allows for stepping the voltage down to 208/120 volt.

Sometimes a transformer is applied on the input of a UPS (not shown) to filter the supply source or dilute high-input available fault currents. There are many different system configurations that can be considered when employing a UPS. Some classical design methods include:

- Stand alone (single UPS)
- Isolated redundant (two UPS in series)
- Independent redundant (two UPS in parallel)
- Paralleling systems (typically three or more UPS modules with a shared output bus)

While arguments can be made for or against a particular method, it is important to realize that the complexity of a UPS system design can sometimes be inversely related to its reliability.

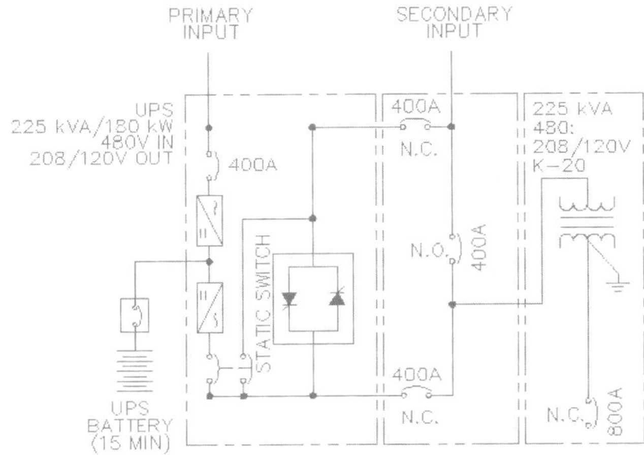

FIGURE 9.4-8 UPS with maintenance bypass options.

Batteries

Batteries are part of most UPS systems. Important characteristics include:

- Charge capacity—how long the battery will support the UPS load
- Weight
- Charging characteristics
- Durability/ruggedness

Additional features that add to the utility of the battery include:

- Built-in status/temperature/charge indicator and/or data output port
- Built-in overtemperature and overcurrent protection with auto-reset capabilities
- Environmental friendliness

The last point deserves some attention. Many battery types must be recycled or disposed through some prescribed process. Proper disposal of a battery at the end of its useful life is, therefore, an important consideration. Battery packaging may contain disposal instructions but local environmental regulations will likely supercede any battery manufacturer's recommendations. Failure to follow the proper procedures could have serious health and cost consequences.

Research has brought about a number of different battery chemistries, each offering distinct advantages. The most common and promising rechargeable chemistries include the following:

- *Nickel cadmium (NiCd).* Used for portable radios, cellular phones, video cameras, laptop computers, and power tools. NiCds have good load characteristics, are economically priced, and are simple to use.
- *Lithium ion (Li-Ion).* Now commonly available and typically used for video cameras. This battery promises to replace some NiCds for high-energy/density applications.
- *Sealed lead acid (SLA).* Used for UPS, video cameras, and other demanding applications where the energy-to-weight ratio is not critical and low battery cost is desirable.
- *Nickel metal hydride (NiMH).* Used for cellular phones, video cameras, and laptop computers where high energy is of importance and cost is secondary.
- *Lithium polymer (Li-Polymer).* This battery has the highest energy density and lowest self-discharge of common battery types, but its load characteristics will likely only suit low-current applications.
- *Reusable alkaline.* Used for light-duty applications. Because of its low self-discharge, this battery is suitable for portable entertainment devices and other noncritical appliances that are used only occasionally.

No single battery offers all the answers; rather each chemistry is based on a number of compromises.

A battery, of course, is only as good as its charger. Common attributes for the current generation of charging systems include quick-charge capability, automatic battery condition analysis, and subsequent *intelligent* charging.

Sealed Lead-Acid Battery

The lead-acid battery is a commonly used chemistry. The *flooded* version is found in automobiles and large UPS battery banks. Most smaller, portable systems use the *sealed* version, also referred to as *gelcell* or SLA (sealed lead acid).

The lead-acid chemistry is commonly used when high power is required, weight is not critical, and cost must be kept low. The typical current range of a medium-sized SLA device is 2–50 Ah. Because of its minimal maintenance requirements and predictable storage characteristics, the SLA has found wide acceptance in the UPS industry, especially for *point-of-application* systems. These are often found in smaller UPS units for personal computers and related data-handling devices.

The SLA is not subject to memory. No harm is done by leaving the battery on float charge for a prolonged time. On the negative side, the SLA does not lend itself well to fast charging. Typical charge times are 8–16 hours. The SLA must always be stored in a charged state because a discharged SLA will *sulphate* (a chemical process that ruins the battery). If left discharged, a recharge may be difficult or even impossible.

Unlike the common NiCd, the SLA prefers a shallow discharge. A full discharge reduces the number of times the battery can be recharged, similar to a mechanical device that wears down when placed under stress. In fact, each discharge-charge cycle reduces (slightly) the storage capacity of the battery. This wear-down characteristic also applies to other chemistries, including the NiMH.

The charge algorithm of the SLA differs from that of other batteries in that a *voltage-limit* rather than *current-limit* is used. Typically, a multistage charger applies three charge stages consisting of a *constant-current charge*, *topping charge*, and *float charge* (see Figure

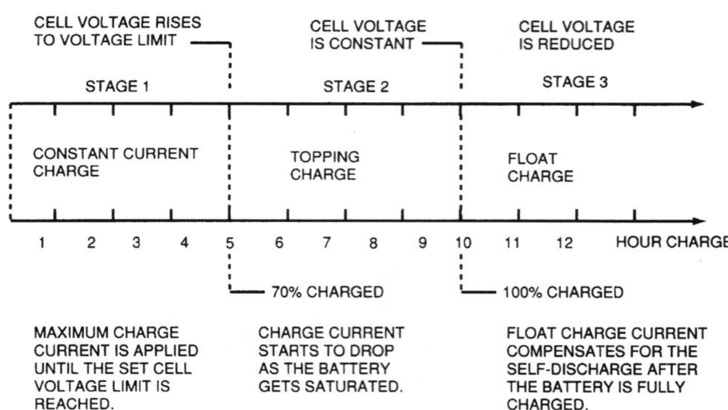

FIGURE 9.4-9 The charge states of an SLA battery. (From [4], used with permission.)

TABLE 9.4-2
Recommended Charge Voltage Limit for the SLA Battery [4]

	2.30 V to 2.35 V/cell	2.40 V to 2.45 V/cell
Advantage	Maximum service life; battery remains cool on charge; battery may be charged at ambient temperature exceeding 30°C (86°F).	Faster charge times; higher and more consistent capacity readings; less subject to damage because of under charge condition.
Disadvantage	Slow charge time; capacity readings may be low and inconsistent. Produces under charge condition that may cause sulphation and capacity loss if the battery is not periodically cycled.	Battery life may be reduced because of elevated battery temperature while charging. A hot battery may fail to reach the cell voltage limit, causing harmful over charge.

9.4-9). During the constant-current stage, the battery charges to 70% in about 5 hours; the remaining 30% is completed by the topping charge. The slow topping charge, lasting another 5 hours, is essential for the performance of the battery. If not provided, the SLA eventually loses the ability to accept a full charge and the storage capacity of the battery is reduced. The third stage is the float charge that compensates for self-discharge after the battery has been fully charged.

During the constant-current charge, the SLA battery is charged at a high current, limited by the charger itself or temperature of the battery. After the voltage limit is reached, the topping charge begins and the current starts to gradually decrease. Full charge is reached when the current drops to a preset level or reaches a low-end plateau.

The proper setting of the cell voltage limit is critical and is related to the conditions under which the battery is charged. A typical voltage limit range is from 2.30–2.45 V. If a slow charge is acceptable, or if the room temperature may exceed 30°C (86°F), the recommended voltage limit is 2.35 V/cell. If a faster charge is required and the room temperature remains below 30°C, 2.40 or 2.45 V/cell may be used. Table 9.4-2 compares the advantages and disadvantages of the different voltage settings.

Key Terms

The following terms are commonly used to specify and characterize batteries:

Cell voltage: The output voltage of the basic battery element. The cell voltage multiplied by the number of cells provides the battery terminal voltage.

Current rate: The *C*-rate is a unit by which charge and discharge times are scaled. If discharged at 1 *C*, a 100 Ah battery provides a current of 100 A; if discharged at 0.5 *C*, the available current is 50 A.

Cycle life: The typical number of charge-discharge cycles for a given battery before the capacity decreases from the nominal 100% to approximately 80%, depending on the application.

Energy density: The storage capacity of a battery measured in *watt-hours per kilogram* (Wh/kg).

Exercise requirement: This parameter indicates the frequency that the battery needs to be exercised to achieve maximum service life.

Fast charge time: The time required to fully charge an empty battery.

Load current: The maximum recommended current the battery can provide.

Self-discharge: The discharge rate when the battery is not in use.

References

[1] DeDad, John A. *Auxiliary Power, in Practical Guide to Power Distribution for Information Technology Equipment.* Overland Park, KS: Intertec Publishing, 1997, pp. 31–39.

[2] Hirschenhofer, J.H., et al. *Fuel Cell Handbook*, 4th ed. U.S. Department of Energy, Office of Fossil Energy, Federal Energy Technology Center, Morgantown, VA, 1998.

[3] Hepinstall, R., and Clayworth, G.T. The Inductive Output Tube: The Latest Generation of Amplifier for Digital Terrestrial Television Transmission, EBU Technical Review, Autumn 1997.

[4] Buchmann, Isidor. *Batteries, in The Electronics Handbook*, ed., J.C. Whitaker. Boca Raton, FL: CRC Press, 1996, p. 1058.

Bibliography

Angevine, Eric. *Controlling Generator and UPS Noise, in Broadcast Engineering*. Overland Park, KS: Intertec Publishing, March 1989.

Baietto, Ron. *How to Calculate the Proper Size of UPS Devices, in Microservice Management*. Overland Park, KS: Intertec Publishing, March 1989.

Federal Information Processing Standards Publication No. 94, Guideline on Electrical Power for ADP Installations, U.S. Department of Commerce, National Bureau of Standards, Washington, DC, 1983.

Highnote, Ronnie L. *The IFM Handbook of Practical Energy Management*. Old Saybrook, CT: Institute for Management, 1979.

Rajashekara, Kaushik. Hybrid Fuel Cell Stategies for Clean Power Generation, *IEEE Transactions on Industry Applications*, 41(3), May/June 2005.

Smith, Morgan. *Planning for Standby AC Power, Broadcast Engineering*. Overland Park, KS: Intertec Publishing, March 1989.

C H A P T E R

9.5

Transmitter Remote Control and Monitoring Systems

HAROLD HALLIKAINEN

hallikainen.com
Santa Maria, California

INTRODUCTION

This chapter presents an overview of the technology in current use for the control and monitoring of broadcast transmission systems. Since the majority of stations operates without a transmitter operator at the transmitter itself, it concentrates on the operation of remote transmitters and other transmission facilities. These techniques can also be applied to locally controlled equipment.

The chapter covers:

- A review of the regulatory and technical history of transmission control systems
- Current control system technology
- Transmission and monitoring equipment interface
- Communications systems
- User interface
- Automation

REGULATORY HISTORY

In 1910, Congress passed Public Law 262, which required licensed operators be in charge of all radio transmitting apparatus. For a summary of the history of broadcast operator requirements, see the *NAB Guide for Unattended Transmitter Operation*.[1]

Through the first 30 or 40 years of broadcasting, the Federal Communications Commission (FCC) and its predecessor, the Federal Radio Commission, required a licensed operator be at the transmitter to observe its operation and make necessary adjustments.

In 1950, the FCC authorized class D noncommercial educational FM stations (stations with an authorized transmitter power output of 10 watts or less) to operate by remote control. Stations were required to have "positive on-off control" of the transmitter such that faults in the communications circuit would not turn the transmitter on or prevent the licensed operator at the control point from turning the transmitter off. This "fail-safe" provision generally required a transmitter shutdown on loss of control of the transmitter. The fail-safe requirement remained in the rules until 1984. A remnant of this requirement remains in FCC Rules Section 73.1350(b)(2), which requires the transmitter control personnel have the capability of turning the transmitter off at all times.

In 1953, the FCC authorized remote control of FM stations and nondirectional AM stations operating with 10 kW or less. Though no telemetry requirements were specified, operators at remote control points were required to comply with the same rules as operators at the transmitter site. Since those rules required the logging of specified transmitter parameters every 30 minutes, it was necessary for those parameters to be available to the operator at the remote control site.

In 1957, the FCC authorized remote control of high power and directional stations. Prior to being authorized remote control operation, stations had to submit logs demonstrating the transmission system had sufficient stability to operate properly without the constant supervision of an on-site operator.

In 1963, the FCC authorized automatic logging for AM, FM, and NCE-FM stations. Logging was done

[1]An excerpt of which can be viewed at www.hallikainen.org/nab/unattended/RegHistory.pdf.

with chart recorders that were required to have a scale of at least two inches. Automatic alarms were to check parameters at least every 10 minutes. Also in 1963, the FCC authorized remote control of UHF television stations. Remote control of VHF television stations was proposed in 1965, but was rejected due to the heavily used spectrum surrounding VHF stations.

In 1967, the use of subaudible tones on AM stations was proposed for the return of metering data. It was approved in 1969. Also in 1969, the FCC proposed remote reading of directional AM station antenna monitors. Prior to this, stations did not remote the antenna monitor, but, instead, had to inspect the antenna system within 2 hours of starting directional operation.

After several proposals, the FCC finally authorized the remote control of VHF television stations in 1971. As with previous remote control rules, VHF stations were required to have a control fail-safe system that would shut down the transmitter on loss of the control circuit. In addition, these stations were required to have a telemetry fail-safe that would automatically shut down the transmitter on loss of required telemetry. Manufacturers generally provided a shutdown on loss of the transmitter to studio telemetry circuit or loss of a sample voltage from the transmitter. The new rules also required VHF stations have various monitors at the control point, including an aural modulation monitor, a waveform monitor, and a vectorscope (if transmitting color). UHF stations that were already operating by remote control had a year to comply with the new rules. Remote controlled television stations were required to insert vertical interval test signals (VITS) on the video leaving the control point. The VITS requirement was deleted in 1982.

In 1974, the FCC adopted rules permitting the use of extension meters. As the name implies, extension meters extend the transmitter meters a limited distance. The rules permitted the meters to be extended 100 feet and did not require the transmitter controls to be extended since the operator could quickly get to the transmitter to make adjustments due to the limited distance.

Automatic control of transmitters was first proposed in 1957. In 1975, the FCC started an inquiry into the possibility of automatic transmission systems (ATS). ATS would monitor critical parameters and shut the transmitter down if any of these parameters went outside limits. FM and nondirectional AM stations were authorized to use ATS in 1977. The requirements were complex and did not reduce operator requirements. Thus, ATS was not widely adopted.

In 1984, the FCC dropped the control fail-safe requirement. Loss of telemetry for 3 hours required a transmitter shutdown. This rule change permitted the use of non-permanent circuits for transmitter control and telemetry. This permitted the use of dial-up (Public Switched Telephone Network) circuits for transmitter control and telemetry. The low cost of equipment, the low cost of dial-up circuits, and the lack of a requirement for special studio equipment have made dial-up remote control very popular. These rules were further clarified by the FCC in 1988.

In 1986, the FCC deleted the how-to details of the ATS rules, vastly simplifying the rules. Stations utilizing ATS still did not get any relief from operator requirements, so, again, ATS was not widely adopted.

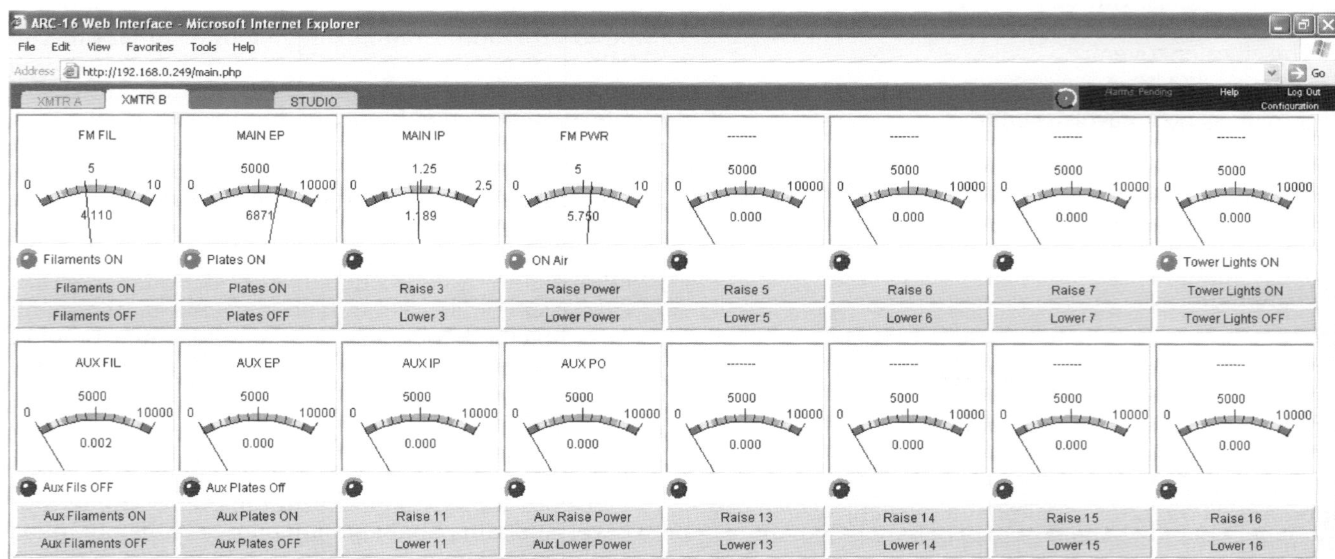

FIGURE 9.5-1 The channel select—raise or lower—user interface remains popular today. Note that each analog channel has two associated control buttons (raise and lower) that can be labeled by the user. Each channel also has an associated status channel. See the CD for image details. (Photo courtesy of Burk Technology.)

By 1995, Congress had removed the requirement that broadcast stations have a licensed operator in attendance. The FCC simplified the remote control and ATS rules (though the simplification tended to make them vague) and permitted unattended transmitter operation. The FCC has interpreted the main studio location rule, FCC Rule Section 73.1125, to require minimum staffing levels at the main studio during "normal" business hours. While a transmitter may be operated unattended, the main studio cannot (during business hours).

In recent years, the FCC has recognized that broadcast transmitters need less supervision than they have in the past. The FCC largely leaves decisions regarding transmitter control and telemetry to licensees. However, should the FCC find a station operating outside its licensed parameters, it may question the adequacy of the station's monitoring equipment and procedures. Station personnel should be familiar with all the FCC rules affecting the station's operation. The discussion in the FCC Report and Order establishing a particular rule can provide guidance as to the intent of the rule. Further guidance is available by reviewing FCC enforcement actions on a particular rule. Finally, stations should seek qualified legal counsel.

CURRENT TRANSMITTER CONTROL REGULATION

A brief discussion of some of the rules regarding transmission system control is included here to provide some background.

FCC Rule Section 73.1125 specifies the allowed locations for the main studio of a broadcast station. In 1988, in response to petitions that, among other things, asked for a definition of the main studio, the FCC said that to fulfill the function of serving the needs and interests of the station's community of license, "a station must . . . maintain a meaningful management and staff presence." In 1991, in Jones Eastern of the Outer Banks, the FCC found that the main studio must have a full-time management and a full-time staff presence. Successive FCC decisions have relied on this decision to prohibit fully unattended broadcast stations. Further, FCC Rule Sections 73.3526 and 73.3527 require the public inspection file be available to members of the public during normal business hours. While a transmitter may be unattended, a station may not be unattended during normal office hours.[2]

FCC Rule Section 73.1300 permits stations to operate transmitters attended (supervised by a person) or unattended (relying on either high-stability equipment or automatic controls to maintain proper operation in the absence of a person).

FCC Rule Section 73.1350 reminds station licensees of their responsibility to ensure the station operates within the requirements of the rules and the station license. This section also requires the station licensee to designate a chief operator and allows the licensee to designate one or more technically competent persons (operators) to adjust the transmitter as required. Operators can make adjustments at the transmitter itself or from an off-site location. Operators must be able to turn the transmitter off at all times. If these operators are not at the transmitter site, the shut down capability must be continuous or there must be an alternate means of acquiring control so the transmitter can be shutdown. Most stations using temporary circuits (generally dial-up) provide an alternate transmitter shutdown method through the studio-to-transmitter link (STL) that carries programming to the transmitter. The transmitter may be shut down on loss of STL signal, loss of audio, loss of a subcarrier, or some other control signal carried on the STL. This "turn off the transmitter capability" is the current version of the fail-safe requirement that was introduced when remote transmitter control was first authorized in 1950.

Licensees are to establish monitoring procedures and schedules to ensure specified parameters are within limits. The specified parameters are power (common point or base current power for AM stations, transmission line input power for FM or TV stations), AM mode, modulation level, and directional array parameters. While the FCC no longer specifies a schedule, which could, perhaps, be a weekly inspection of equipment at the transmitter site, tradition, carried over from previous rules, calls for these parameters to be available to transmitter operators. Operation with any of these parameters outside their permitted limits must be suspended within 3 hours, though operation of an AM station at a power or mode other than authorized (such as daytime pattern or power operation at night) must be discontinued within 3 minutes. Alternate methods of continued operation (low-power operation, etc.) are also listed in this section. Finally, the FCC must be notified of any transmitter control point other than the studio or transmitter within 3 days.

FCC Rule Section 73.1400 states, "Any method of complying with applicable tolerances is permissible." It describes typical methods of transmission system operation. A typical attended station could operate by direct control where the operator is at the transmitter site. Remote control makes critical parameters and control available to the operator at a control point other than the transmitter (recall the requirement that the FCC be notified of control points other than the transmitter or main studio). Finally, an attended station may utilize an ATS to automatically adjust transmitter parameters as required. The operator receives an alarm from the ATS should it fail to bring a critical parameter back within limits.

The question of which parameters deserve to be remotely monitored is often asked. Tradition, as mentioned above, dictates that indication of power, modulation, and AM directional array parameters are available on the remote control. FCC Rule Section 73.1400, however, states, "In the case of remote control or ATS operation, not every station parameter need be monitored or controlled if the licensee has

[2]For a history of the main studio rule, see www.hallikainen.org/rw/insite/insite95.html.

good reason to believe that its stability is so great that its monitoring and control are unnecessary."

FCC Rule Section 73.1400 continues to describe unattended operation as operation without human supervision. This can be accomplished in a couple ways. One way is to use an ATS that monitors and adjusts appropriate transmission system parameters. Failure to keep parameters within allowed limits results in the ATS shutting down the transmitter. This is in contrast with attended operation with ATS where the system notifies the operator instead of shutting down the transmitter. With unattended operation, there is no operator to notify, so the ATS shuts down the transmitter when an out-of-tolerance condition that cannot be corrected is detected. Finally, unattended operation does not necessarily imply the use of ATS. The rules permit unattended operation if the system is stable enough to operate within licensed parameters for extended periods of time.

TECHNICAL HISTORY

A basic transmitter remote control system is merely extended meters and extended controls bringing these meters and controls to the operator's location. Extension metering (discussed above) brought metering from the transmitter and monitors to the operator, generally with each meter utilizing another pair of wires. Though the extension metering rules did not require extension control, another pair of wires could have been brought to the operator for each control switch.

In the 1950s, remote control circuits were generally DC loops (metallic loops that were DC coupled, thus did not contain transformers or amplifiers) rented from the local telephone company. To reduce the number of circuits required, simple multiplexing was used.

Early remote control equipment utilized electromechanical stepper switches at the transmitter site to select which meter sample was to be sent to the control point. A telephone dial at the control point sent pulses to the stepper. The stepper also selected which control circuits were to be actuated. Two control functions were generally associated with each meter position, allowing the operator to raise or lower the selected parameter. On/off controls, such as transmitter tube filament voltage and plate voltage, were normally on the same metering channels that displayed filament voltage and plate voltage so the operator had feedback that the command had actually been executed. An AM transmitter plate current sample might be paired with day/night power select, and transmitter power trim paired with antenna or common point current. This association of raise and lower control functions to metering channels continues today in many transmitter control systems.

These early control systems utilized two DC loops, one for control and one for metering. Various control functions (stepper step, stepper reset, raise, and lower) were transmitted by sending different voltages and different polarities on the control pair, or as differing voltages to ground instead of just voltage across the pair. The second pair was used for metering. Sample voltages from the transmitter site equipment were placed on the metering pair by the stepper switch. These voltages were read by an analog meter at the control point. The control system had a calibration pot (potentiometer) for each metering channel that would scale the sample voltage for appropriate display at the control point. The meter at the control point generally had several scales. The station engineer would adjust the calibration pots at least once each week so the operators could get an accurate reading of the transmitter parameter on one of the scales on the analog meter at the control point. Finally, since telephone lines vary in resistance, the control point unit would include a pot in series with the phone line. One position on the stepper would feed a known constant voltage down the metering pair. The operator would adjust this control point calibration pot for some specified level on the local meter (generally half or full scale).

The next major step in remote control systems was to eliminate one of the telephone lines. Using vacuum tube circuitry, manufacturers converted the control signals to audio tones that were carried on the DC loop to the transmitter site. The selected metering sample was still returned as a DC voltage on the line.

With the introduction of integrated circuits, remote control manufacturers eliminated the telephone dial and, later, the stepper. Digital-integrated circuits emulated the telephone dial, generating control pulses, while analog-integrated circuits replaced the vacuum tube circuitry that had generated and detected tones. Early solid-state remote controls continued to use the stepper. Later ones emulated the stepper using integrated circuit counters and numerous relays.

At this point (about 1970), telephone circuits with DC continuity started to become more difficult to obtain. In addition, stations were using 950 MHz studio-to-transmitter links to carry programming to the transmitter instead of equalized wire line circuits. Manufacturers started providing systems that did not need DC continuity on the telephone line. A typical system of this time used 300–400 Hz for control (frequency shift keying where the duration of the shift determined the control function) and 800–1200 Hz for metering (where 800 Hz was zero and 1200 Hz was full scale). These audible control and metering signals could be carried on telephone lines, STL subcarriers, FM subcarriers, and telemetry return links (UHF links authorized for operational communications in 74.402(e)(9)). AM stations returned their metering as subaudible tones as authorized in 1969. A 20 Hz tone represented zero and 30 Hz represented full scale.

In the late 1970s, digital metering appeared. Systems used a dual-slope analog-to-digital converter at the transmitter site. Electromechanical relays continued to be used to select the analog sample to be measured and to route the raise and lower control signals to the appropriate control circuitry. Both metering and control were transmitted as frequency shift keyed audio tones. Instead of using a continuously variable tone or a variety of tones for various functions, systems now used standard modem techniques to transmit serial data from the control point to the transmitter site and the responses back. Data formatting was handled by standard UART chips along with a few other chips. At this time "status" indicators also started to appear in systems. Continuous contact closures at the transmitter site would cause continuous lighting of individual LEDs at the control point. Status indicators would mimic the indicators on the transmitter and other equipment showing overloads and various on/off conditions (such as tower lights, doors open, temperature high/low).

In the 1980s, microprocessors started to appear in remote control systems. At first, these systems largely emulated the existing digital controls, replacing the data-formatting logic with software. Later, circuitry and software to drive standard EIA-232 video display terminals was added, presenting operators with a full-screen text display showing all transmitter parameters on one screen. Some of these systems offered automatic logging, alarms, and automatic control. Also in the 1980s, dial-up remote control was introduced. These systems use TouchTone® (DTMF or dual-tone multifrequency) tone bursts for control and synthesized or recorded voice to return metering values.

In the 1990s and later, manufacturers of transmitter control systems developed techniques to deal with the monitoring of multiple sites, added more automation capabilities, improved the user interface, and adapted systems to digital communications (especially the use of Internet protocol over Ethernet). Some of the data analysis and presentation duties were handed off to standard computers. The remainder of this chapter takes a closer look at the current state of these control systems.

TRANSMISSION EQUIPMENT INTERFACE

Remote control interface to transmitter site equipment started as analog voltages for metering and momentary relay contact closures for control. Later, status inputs that accepted contact closures were added to the transmitter site end of remote control equipment. Most recently, various serial digital interfaces and network interfaces have been added.

Analog Metering Interface

Analog metering interfaces remain the most universal metering interface between remote transmitters and remote control equipment. Low-voltage (typically 1–10 VDC full scale) samples of transmitter parameters are available on the remote control interface terminals of the transmitter. These samples are developed by resistive voltage dividers for sampled voltage and current sense resistors (shunts) for sampled current. The sample is directly proportional to the parameter being measured. The actual ratio of the sample to the original parameter is unimportant as long as the sample is within the acceptable range of the remote control equipment. The remote control equipment will scale the sample using a calibration factor to display the actual parameter to the operator. What is important, however, is the stability of the sample ratio, both with time and temperature. Changes in the sample scaling ratio adversely affect the accuracy of the value of the parameter displayed to the operator.

The calibration factor used to convert the scaled sample back to a display of the original value is, today, a user-defined constant held in nonvolatile memory. The control system multiplies the measured sample by this calibration factor to yield the displayed parameter value. Early systems used a potentiometer (calibration pot) for each metering channel to multiply the sample voltage by a user-set constant for display at the control point. Analog control systems required the technician at the transmitter site to adjust the calibration pot for each channel until the operator at the control point indicated the calibration was correct. In digital systems, the display at the control point duplicates one at the transmitter site. This allows one person calibration. The calibration factor (whether a value in nonvolatile memory or set by a pot) is adjusted until the local display indicates the proper value. The control point display shows the same value.

Note that analog systems and early digital systems generally did not include a decimal point in the displayed value. It was up to the operator to properly place the decimal point. A display of 512 might indicate 5.12 kV or 512 mA, depending on which metering channel was displayed. More recent systems properly place the decimal point, either through the use of floating point calibration factors or through programmed decimal point positioning.

As discussed in the technical history section, early systems transmitted the sample voltage on a DC pair provided by the local telephone company. The selection of the metering sample was done with a stepper switch that had separate contacts for each side of the metering sample. The metering circuitry at the control point was "floating," having no connection to ground. The system, therefore, could tolerate common mode voltage:

$$V_{cm} = (V_a + V_b)/2,$$

where V_a and V_b are the voltages to ground on each leg of the sample. The common mode voltage was limited only by safety concerns (telephone companies did not like high voltages to ground being placed on their circuits).

Systems using audible and subaudible metering systems continued to use relay contacts to select the metering sample. The metering circuitry (voltage-controlled oscillator) operated from a floating power supply, and the metering output was transformer coupled to the remaining circuitry. These systems also tolerated relatively high common mode voltage (in the 50 V range) on the metering samples. Early digital systems also used relays and floating power supplies, then optically coupled the data to the modem circuitry. Thus, these systems also tolerated common mode voltage on the samples.

Current microprocessor-based systems use integrated circuit-based analog multiplexers to switch sample voltages to the input of the analog-to-digital converter. These multiplexers have longer life than relays (though relays are more rugged) and are faster. They also are smaller than relays (saving space on the printed circuit board) and cost less. Most analog multiplexer integrated circuits, however, will not tolerate any voltage higher than the power supply voltage going to the chip (generally in the +/–15 V range). Further, many current remote control systems have "single-ended" metering inputs where one side of the sample is grounded. If the transmitter has a differential sample (where one side is not grounded), some sort of interface is required. This may take the form of a differential amplifier, an instrumentation amplifier, or an isolation amplifier. Each of these provides a single-ended output that is proportional to the differential input voltage ($V_a - V_b$) and is not affected by the common mode voltage. The actual output is A_d (V_{diff}) + A_{cm}(V_{cm}). Ideally, the common mode rejection ratio (A_d/A_{cm}) is high making V_{cm} have little effect.

Differential and instrumentation amplifiers also have limits to the common mode input voltage they will accept. Most such circuits require the input voltages be between the supply rails, but some recent chips allow a couple hundred volts of common mode voltage.

An isolation amplifier can handle larger common mode voltages, but is more complex. An isolated power supply powers the input circuitry that encodes the floating differential input voltage, then transmits the voltage value across an insulating barrier optically, inductively, or capacitively.

A final approach to handling differential samples is to determine the differential voltage in software by having the A/D converter measure the voltage on each of the lines, then subtract those voltages. Some

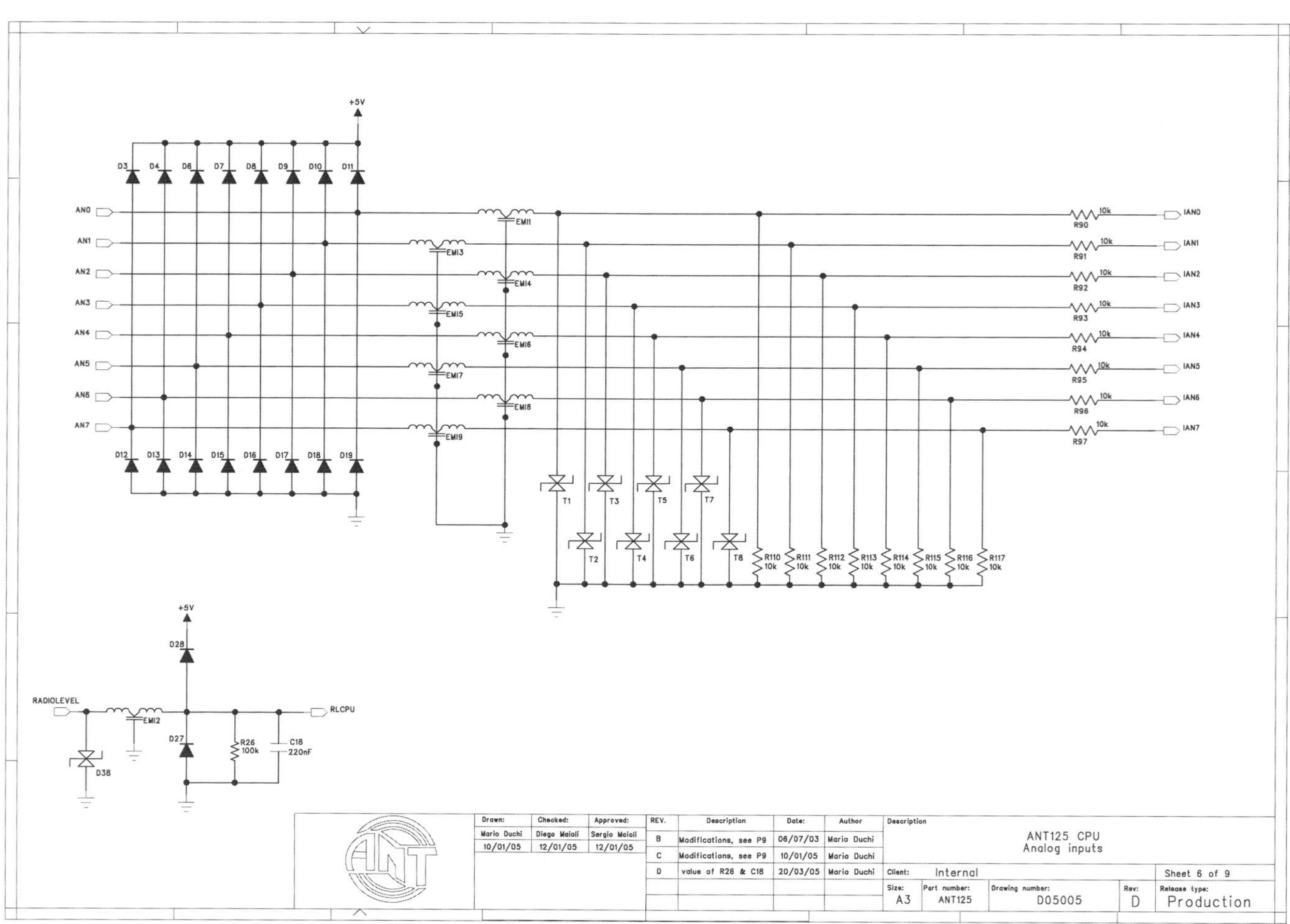

FIGURE 9.5-2 Typical analog input-conditioning circuitry. This circuitry provides radio frequency (RF) rejection, transient voltage protection, and divides input sample voltages by two. Note that the inputs are single ended and that outputs (on the left) are clamped to between 0 and +5 V. See the CD for image details. (Drawing courtesy of Antenna Nord Telecomunicazioni.)

systems support various calculations based on the actual analog inputs, often dropping the results in "virtual channels," memory locations that can be accessed in a manner similar to actual analog inputs.

Many remote controls with single-ended inputs will only accept positive sample voltages, while some transmitter samples may be negative voltages. The sample can be converted to a positive voltage using a differential amplifier, negating the sample. Another approach to the problem is to bias the sample up to a positive voltage. If, for example, the sample voltage varies between 0 and –5 V, a pair of equal-value series resistors can be placed between the sample and a +5 V supply. The junction of the two resistors will be 0 V when the sample is –5 V, and +2.5 V when the sample is 0 V. This bias can be subtracted out in software (again, perhaps through virtual channels). Resistor values need to be high enough to avoid loading the sample.

Some monitors (e.g., AM directional antenna monitors) provide a positive sample voltage even if the parameter being sampled is negative (e.g., the phase indication of a particular tower). Some of these monitors provide an additional output to indicate the sign of the phase. Others indicate the phase by a change in the indication when a delay is inserted in the radio frequency (RF) sample. If the phase indication is typically far from 0 or 180 degrees, most stations assume the sign of the phase angle will not change, so they do not provide a remote indication of the sign. Instead, the sign of the phase is "hard wired" in the remote indication. If the remote indication is to be positive, calibration is done the same as for any other linear sample. If the sign is negative, the operator may do a mental sign correction, or, if the system supports it, a negative calibration factor can be applied, yielding a negative display with a positive sample voltage.

Single-ended samples are, of course, grounded at both the transmitter sample output and the remote control input. This is a classic recipe for a ground loop. Ground loops, however, typically result in power line frequency or RF being added to the DC sample. These are filtered out in the remote control through a combination of hardware filters and choice of A/D sampling time (so power line frequency interference cancels). It is important, however, to make sure the equipment is well grounded. In one installation, where the sampling lines formed a lower impedance RF path to ground than other paths available to the transmitter, the sample lines melted due to the high RF current.

Most samples are DC. Occasionally an AC sample needs to be monitored, such as filament voltage or tower light current. If the required accuracy is not high, a simple rectifier and filter can be used to derive a DC sample from an AC sample. Such a circuit suffers nonlinearity due to diode drops and further inaccuracy if the wave shape changes (especially from sinusoidal). With AC voltage samples, nonlinearity can be minimized by using a transformer to step the sample voltage to a voltage considerably higher than the diode drop, then rectifying the voltage and dividing it down to an appropriate DC sample. Rectifying at a high voltage makes the nonlinearity due to the diode drop less significant.

If the AC sample is a current, such as tower light current, nonlinearity caused by diode voltage drop can be reduced by putting the current sense transformer terminating resistor (the "burden" resistor) after a full-wave bridge rectifier, then sampling the voltage across the terminating resistor. The transformer will develop enough voltage to force the desired current through the terminating resistor, ignoring the diode drops. For the most accuracy, an RMS-to-DC converter circuit can be used between the AC sample and the remote control, or the remote control equipment may directly support AC samples through an internal hardware or software RMS to DC conversion.

Installers should carefully review the analog samples available from the transmitting equipment to make sure they are compatible with the inputs on the remote control. Careful attention should be paid to whether the system accepts AC samples, whether sample inputs are single ended or differential, and what the minimum and maximum sample voltages can be.

The discussion thus far has assumed the parameter being displayed is proportional to the sample voltage. Some samples (especially AM, FM, and TV transmitter output forward and reflected power) are proportional to the square of the sample voltage (directly proportional to the transmission line voltage or current). Correction for this nonlinearity has been handled in a variety of ways. The differences are where the correction is applied and whether the correction is done through analog circuitry, software, or directly by the operator. Analog remote controls included a logarithmic scale duplicating the scale on the transmitter meter. Correction for sample nonlinearity was handled by this scale through proper interpretation of the indication by the operator. In a digital system, the operator could read an uncorrected value, then correct it using a chart or calculator. Prior to microprocessor-based digital systems, the required squaring of the sample was done in hardware at the transmitter site through an analog "power to linear converter." This circuit produced an output voltage proportional to the square of the input voltage, thus proportional to the transmitter power. The linear A/D conversion and display in the control system then tracked the transmitter power. Microprocessor-based control systems do the squaring in software. The sample voltage is measured by the A/D converter, the result squared, then multiplied by a calibration factor to directly yield the transmitter power.

Broadcast sample voltages tend to be either linear or "square law," so most systems allow the user to select one of these curves during system calibration. Other systems have more complex functions available, allowing the user to set various coefficients in the function. Other systems allow the user complete freedom in providing a mathematical expression to describe the relationship between the sample voltage and the parameter being measured. The "output" of such a user-defined function is often dropped into a "virtual channel" that is interpreted the same as actual analog input channels. Just as with normal channels, these virtual channels can be displayed, subjected to limit checks, etc. These virtual channels are similar to cells in a spreadsheet program that contain functions based on the contents of other cells.

Some monitors (most notably directional antenna monitors) provide an analog sample that switches function under operator control. While some antenna monitors provide separate analog samples for each tower, some monitors provide a single analog output for phase and another for current or current ratio. With local operation, the operator presses front panel switches to determine which tower is being monitored. With remote control operation, it is common to use the raise and lower control outputs (described below) to drive the tower select inputs of the antenna monitor. For example, one might select channel 10R (channel 10, then send the "raise" command) to read tower 1 phase, 10L (10 lower) to read tower 2 phase, 11R to read tower 3 phase, and 11L to read tower 4 phase. Similarly, channel 12R could read tower 1 ratio, channel 12L would read tower 2 ratio, channel 13R would read tower 3 ratio, and channel 14L would read tower 4 ratio. In this example, the phase sample output of the monitor would drive the analog metering inputs of channels 10 and 11 on the remote control, while the ratio output would drive channels 13 and 14. Of course, operators (or automatic systems) must allow sufficient time for the newly selected reading to stabilize.

Some remote control manufacturers also provide "probes" that sample various parameters and interface directly to the remote control equipment. Temperature probes have been common for many years. Probes that measure various characteristics of an RF signal, including modulation analysis for AM, FM, and analog and digital television, are now also available for direct interface to some systems.

Parallel Control Interface

As described previously, early stepper-based remote control systems provided momentary normally open floating-relay contact control outputs. Two sets of contacts were provided for each metering channel, one called "raise," the other called "lower." Different transmitter parameters could be adjusted or turned on and off using the raise and lower functions on each channel. The channel raise-lower system is carried over to most of today's control systems.

Once control systems converted to digital circuitry, they tended to provide open collector control outputs, though still a raise and a lower for each metering channel. The open collector outputs are less expensive to provide since several outputs can be provided by one integrated circuit instead of using one relay for each output. These outputs can directly drive transmitter controls that have "active-low" (control line pulled to ground to activate the control) relay, logic, or opto-coupled control inputs. Some transmitters have opto-coupled control inputs that can be configured to have active-high or active-low control inputs. Many transmitters, however, have only active-high control inputs (that is, the control line must be driven to some positive voltage to activate the control). Transmitter manufacturers use active-high controls for safety. A control line is less likely to develop a short to a positive supply than to ground. Use of active-high control inputs on the transmitter makes it less likely a control will be activated by accident. Remote control manufacturers often provide an optional relay panel that can be placed between the open collector control outputs and the transmitter control inputs to solve the active-high versus active-low problem.

Most transmitters and control systems utilize control pulses rather than latched or holding contact closures (or their open collector equivalent). Use of momentary closures allows paralleling of control sources. No control source or control state has priority over other sources or states. Should a transmitter require a latched contact, some control systems provide a latched output as a user-configured option. As an alternative, external latching relays can be used to convert momentary control pulses to latched contacts.

Parallel Status Interface

Status inputs of a transmitter-control system are designed to relay system status (on or off, true or false) information to the operator. They generally accept contact closures to ground or open collector outputs. These inputs often run through RF filters and input protection circuitry directly to logic circuit inputs. Some systems may provide opto-coupled inputs. These (more costly) inputs accept inputs that are not ground referenced. Even inputs that drive logic inputs can be designed to handle a wide range of (ground-referenced) input voltages through the use of current limit and voltage clamping circuitry on the status inputs.

Traditionally, status inputs have been distinct from analog inputs. In some systems, however, a status input is merely an analog input with a logic threshold set in software. Any voltage above some level is one logic level. Below that threshold (or perhaps another threshold if there is hysteresis) is the other logic level. Setting an input to be a status input sets this threshold and may also enable a pull-up resistor so the input can be driven by a contact closure or open collector. Some A/D converters use the resistor bias method described earlier to handle negative inputs (biasing them to positive). This bias network can be enabled and disabled under software control as the need for metering of negative voltage is required on some channels and not on others. If this bias network is enabled and the input is an open circuit (open relay contacts or an open collector), the bias network pulls the input up, just as a pull-up resistor on a typical status input would. Grounding the input (again, with a relay contact or an open collector transistor) would pull the A/D input half way to ground, which is easily detected in software. Systems using such an A/D converter have an especially easy time making analog inputs status inputs since no additional components (pull-up resistors) are needed. The pull-up is already there and under software control.

Status inputs can be active high or active low. An active high input is "true" when the status line has a positive voltage on it, while an active low input is "false" when a positive voltage is present. Early systems generally assumed status inputs were active low. Grounding the input at the transmitter site caused an LED at the control point to light. Later systems made the active-high or active-low decision under program control. The user could configure the system as required.

Some systems would latch the true state on a status indicator. A momentary pulse of a status line to its active state would light the control point LED until a reset button was pressed. The idea was to latch alarm indications until they were acknowledged by the operator. This is generally not the case on current LED status displays. Instead, alarm latching is done by the equipment driving the remote control system. A transmitter generally continues to display its alarm LEDs until an alarm reset switch on the transmitter is pressed. This switch is generally "remoted" so an operator at the remote control point can reset the alarm LED at both locations. Once a full-screen display (typically a text or graphics CRT) is utilized, however, alarms may be latched by the software and reset by the user.

EIA-232 Equipment Interface

Some transmitters and monitoring equipment include an EIA-232 port. These generally are intended to drive a remote video terminal dedicated to this particular piece of equipment. Transmitter or monitor parameters are formatted for a person to read the data. Keystrokes on the terminal can control the equipment. This data can be extended long distances using standard modems. Some remote control equipment can extract transmitter or monitor readings from the stream of EIA-232 data based on the position of the data (line number and column number as it would be displayed on a terminal) or through the use of regular expressions that identify a particular parameter by the surrounding text. The extracted parameters can be dropped into system variables (or "virtual channels") that can be displayed and evaluated by the remainder of the system. In addition, commands can be assembled that simulate operator keystrokes on the EIA-232 terminal.

EIA-232 equipment interfaces are desirable because a large amount of data can be sent on few wires. However, a separate port is required for each EIA-232 piece of equipment, since EIA-232 does not support a multidrop mode (like EIA-485, discussed below). Low-cost interfaces can drop multiple pieces of equipment with EIA-232 ports onto an EIA-485 network or Ethernet network. Some remote control equipment includes an EIA-485 interface for transmitter site equipment interface and provides "interface boxes" for equipment that does not support EIA-485 or support the particular protocol used by that manufacturer. Once the EIA-485 data is inside the remote control equipment, the extracted data and commands can be handled the same as other data and commands (such as analog metering channels, control outputs, etc.). Once on an Ethernet network, terminal emulators or other applications can use "virtual comm ports" to access the data from each piece of equipment. Software can extract data from the particular piece of equipment (similar to pulling data from websites using "web scraping") and present it to another application directly or through a database.

EIA-485 Equipment Interface

EIA-485 is a two-wire (plus ground) balanced data-transmission system. It is normally arranged as a data bus with several devices on the bus (and the ends of the bus terminated to avoid reflections). Each device on the bus listens for data addressed to it and is able to transmit data to any other device(s) on the bus. EIA-485 sets the electrical standard for the bus, but does not set a data protocol. Manufacturers are free to design their own data formats (packet structures), contention avoidance, collision resolution, addressing methods, etc. At this writing, no industry-wide standard exists. Some remote control manufacturers have defined their own standards and make adapters to various pieces of equipment. EIA-485 has several advantages over EIA-232. These include the use of a balanced transmission line, giving more noise immunity and faster data speed, and the multidrop mode allowing one port to communicate with several devices. The lack of standard protocol, however, makes it more difficult for a user to build a system with equipment from several manufacturers. As previously mentioned, remote control manufacturers often provide interfaces for a variety of different pieces of equipment that convert data to the particular manufacturer's EIA-485 data format.

Ethernet Equipment Interface

Some transmitters and other equipment have moved much of the remote monitoring and control responsibilities to within the particular piece of equipment and provided an Ethernet interface to the outside world. At this writing, most equipment uses the Internet protocol (IP) to provide a web browser interface to the system or requires the use of a custom application running on the remote computer to view parameters and control the system. The web interface, while universal, generally prevents integration of information from various pieces of equipment into an overall view of the transmitter site or multiple sites. Clever users, however, can use "web-scraping" techniques to extract data from the web pages generated by this equipment and drop the data into a database for further display and analysis.

The use of custom applications to view and control a piece of equipment generally presents a useful view of that particular piece of equipment, but, again, prohibits integration of the data and control with other equipment. These systems could provide a second "machine interface" to go along with their existing "human interface" to allow such system integration. A couple of possibilities in this area would be to include an HTTP CGI (common gateway interface) that could be polled for parameters and accept commands, or include an SNMP (simple network management protocol) interface that could similarly be polled and could also provide unsolicited status messages in the form of SNMP "trap" messages. These will be discussed further in the sections on user interface and automatic control.

STUDIO CONTROL AND TELEMETRY

Transmission control systems have traditionally been used to monitor and control equipment at unstaffed remote sites. There is, however, often transmission equipment at the studio that can benefit from being monitored and controlled by the same system. STL transmitters, audio processors, stereo generators, and various other pieces of equipment located at the studio can be connected to the system through interfaces similar to those used at transmitter sites. System users then have a universal user interface to all the station equipment. Such a system allows a user to view STL transmitter parameters and perhaps bring up an alternate STL transmitter.

INTERSITE COMMUNICATIONS

While all the techniques discussed here can be applied to local transmitters, they are more commonly applied to remote transmitters, where the operator and the transmitter are separated by considerable distance. This raises the requirement for a communications circuit between the operator and the transmission equipment.

When remote transmitter control was first authorized by the FCC, DC loops (twisted pair lines leased from the local telephone company) were the standard communications circuit for transmitter control and telemetry. As discussed previously, early systems used one pair for control and another pair for metering, using DC pulses and levels for both control and telemetry. Later, to save on phone line costs, equipment moved to the use of audio tones for control and telemetry.

The audible tone interface became the standard communications interface for many years (and is still the most common) due to the wide availability of circuits designed to carry voice. As stations moved from leased telephone lines to 950 MHz STL links to carry their programming, a subcarrier was set aside to carry audible control to the transmitter site. FM and TV stations could return audible metering on a subcarrier above the audio. AM stations could return subaudible metering as a low-frequency subcarrier below the audio. Subaudible metering could carry either analog or digital data at a relatively slow data rate (up to about 10 bits per second). On AM stations, the subaudible spectrum found several proposed uses in later years, including utility load management and a pilot signal for AM stereo.

The use of the broadcast carrier for return of telemetry information worked well when the transmitter was working. Should the broadcast transmitter fail, however, the operator had no way of remotely diagnosing the problem. The most common solution to this problem is the use of a telemetry return link, a one-way radio voice grade link using one of the UHF frequencies set aside for operational communications in 74.402(e)(9) of the FCC rules. These links commonly transmit the telemetry data continuously using frequency shift-keyed audio at low data speeds (300–1,200 bits per second). In many areas, these frequencies are fully occupied by these one-way, low-speed data circuits.

PACKET COMMUNICATIONS SYSTEMS

A more spectrum-efficient method of utilizing the UHF channels set aside for operational communications is the use of packet communications. Some remote control systems directly support packet communications. Packet capabilities can be added to other systems through standard terminal node controllers (TNC) used by amateur radio operators. These systems transmit data only when needed instead of continuously transmitting the same data over and over. Further, the RF carrier is brought up only when data is to be transmitted, making the RF channel available for other users. Packet systems also set up a two-way data circuit so control data can be moved from the STL subcarrier to the packet network. This frees the STL subcarrier for other uses and allows remote diagnostics of the system on failure of the STL.

The key to the increased spectrum efficiency of packet systems is mostly due to the reduction of transmission of redundant data. Systems that include an internal packet data system provide immediate response to user data requests and update that data as time goes on. The update rate backs off as the time since the last user request increases, making the channel available for other data. Systems not designed with packet communications in mind tend to repeatedly send the same current telemetry data at the same update rate, even though the operator has likely walked away from the display. These systems can still benefit from the RF packet communications systems if the data rate on the RF channel is substantially higher than the data rate being delivered by the remote control system. The TNC gathers the low-speed data for a period of time, forms a packet that includes the data, addressing information, and error-checking information, then sends the packet at high speed over the RF channel. Between transmissions, the RF channel is available for other users.

Though packet communications where the RF carrier is dropped between packets allows other users to use the channel, it does suffer some spectrum inefficiency because of the time it takes to detect the RF carrier before transmission of data can begin. A more spectrum-efficient approach would be to leave the RF carrier up continuously and packetize data from multiple sites onto the same carrier. Using two RF channels, two-way data circuits are set up between the multiple sites in the system. Data is routed through these RF links as needed based on the address headers in the packets and routing tables in the equipment at the sites. This is difficult to accomplish, however, due to the very limited number of UHF channels available for this purpose. The use of highly directional antennas may allow sufficient frequency reuse to make the approach work in some areas. Currently, most stations using UHF data links are using the traditional one-way TRL approach. Once the channels are all occupied, other stations have to find another way to get their telemetry data.

DIGITAL LINKS

Most of the previously described communications circuits were "voice-grade" circuits that carried data represented as modem tones, though the packet radio system based on TNCs, discussed above, presents an EIA-232 interface to the station control equipment. Other communications circuits are available that provide data as EIA-232 or Ethernet. In particular, digital RF STLs may provide a low-speed data circuit or a voice-grade circuit for transmitter control. Since an STL is one way, the station gets to figure out a way to get data back from the transmitter site. This may, again, take the form of a UHF radio telemetry return link.

OTHER DATA COMMUNICATIONS TECHNIQUES

The FCC has set aside various bands for license-free spread-spectrum data communications. One of these bands is close enough to the 950 MHz band used for studio-to-transmitter links that a spread-spectrum link transceiver can be combined with the STL signal at both ends of the path to provide a high-speed digital circuit (about 10 Mbps). The highly directional antennas used in STL systems minimize interference and make up for a portion of the path loss. To the user, the link appears to be a long Ethernet cable. The RF link can extend the control point local area network (LAN) to the transmitter site to provide high-speed control, telemetry, and access to other data on the LAN. With appropriate routing equipment, several of these links could be combined to form a network that includes several transmitter sites and control points.

This Ethernet link between sites can also be provided a variety of different ways. Some methods include other RF links; use of a portion of a leased data line DS1, DS3, etc. that is carrying program material to the site; or use of the public Internet. With appropriate security (encryption, password protection, etc.), the various sites in the system can connect to the Internet through telephone company digital subscriber loop, CATV system cable modems, satellite-based Internet, cellular telephone digital networks, or wireless Internet service providers. The extremely low cost of Internet connectivity makes this quite attractive. Appropriate security measures are extremely important, however, to avoid someone anywhere on earth breaking into the system. Encrypted protocols (SSH, HTTPS, etc.) are suggested. Careful consideration of port-forwarding and port-blocking settings in routers can limit exposure of the system to the outside world. Equipment and router logs can be reviewed to reveal attempts at intrusion into the system.

DIAL-UP COMMUNICATIONS

Dial-up communications with transmitter sites over the public switched telephone network is also cost effective. The circuit between the control point and a transmitter site is only established when there is a need to send information. The rest of the time telephone company facilities are used to handle communications for other purposes. The control point initiates a connection when a control command needs to be sent or polling of telemetry data is desired. The transmitter site initiates a connection when it has an alarm condition to report.

Dial-up circuits are especially easy to deal with when there are several sites that must be monitored by a control point. The telephone company does all the data routing. One phone line at the control point serves all the sites. In very large systems, more lines may be used, but any line can reach any site.

Dial-up circuits most often carry dual-tone multi-frequency (DTMF) control data generated by a telephone keyboard. Telemetry data is sent back as voice. Most systems also include a data modem so a video display terminal (or a computer emulating one) can provide the user with a full-screen display of the status of the site and present a simple menu for control of the site.

In locations where no telephone lines are available, the cellular telephone network can often be used. Cellular transceivers are available that emulate the wired public switched telephone network (PSTN). They provide a two-wire interface, dial tone, ring voltage, and accept DTMF dialing. To the dial-up control system, it's a normal dial-up phone line.

Many consumer cellular telephones include an EIA-232 port that can be used to dial in to modems on the PSTN. These phones emulate a modem connected to the PSTN and can be used for remote control communications if the remote control system provides an EIA-232 interface for the intersite communications interface.

LIGHTNING AND RF PROTECTION

Broadcast transmitter sites tend to attract lightning. Lightning can be conducted into the transmitter from the tower and from the transmitter to monitoring equipment, including remote control equipment. Since many transmitter sites are located at high elevations, power and telephone lines are also likely to be struck by lightning or have high transient voltages induced into them by nearby lightning strikes. Every conductor entering the transmitter building requires adequate lightning protection devices connected to ground connections that can handle very high surge currents with low voltage rise. Further, transmitter control equipment generally includes transient suppressors on every conductor "entering the box" to further limit lightning damage (note the transient suppressors on each line in Figure 9.5-2). Even with all this effort, lightning damage is probably the most frequent problem with transmitter remote control equipment, so improvements are always possible.

Transmitter sites, obviously, are high RF environments. RF can be radiated or conducted into equipment at the site. Transmitter site equipment, including transmitter control equipment, is generally in shielded cabinets and includes extensive RF filtering on all conductors entering and leaving the box. These conductors include the power input, telephone and other telecommunications circuits, metering, and control circuits. Excessive RF fields may keep the equipment from operating at all, may cause it to operate erratically, or may cause it to give erroneous indications. If the unit operates properly with sample, control, and telephone lines disconnected, but does not operate properly when these lines are connected, additional RF filtering should be added as close to the unit as possible. This filtering may be as simple as running cables through ferrite beads (especially effective at VHF and UHF frequencies). More difficult cases may require the addition of RC or LC filters on each line.

USER INTERFACE

As the number of stations under the control of a single licensee grows, the importance of a unified user interface to all the transmission equipment increases. The current trend is to have a single color computer screen that shows the overall status of all the transmission equipment being monitored. This screen, as shown in Figure 9.5-3, may organize the sites geographically or in a signal flowchart. Alarm conditions at a particular site are shown as a flashing color along with an audible alarm. The user then clicks the mouse on the problem site to "drill down" toward the particular problem. Clicking on a site brings up a screen showing all the equipment at that site with the problem equipment identified. Clicking on the problem piece of equipment brings up another screen showing the

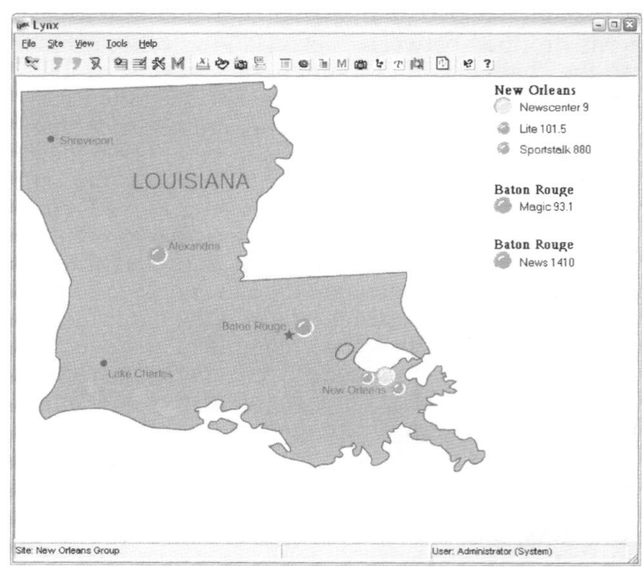

FIGURE 9.5-3 A user screen with a geographic system overview. Users view overall system status and can "drill down" to see detailed data from a particular site. See the CD for image details. (Image courtesy of Burk Technology.)

details of that particular piece of equipment. There are various approaches to building such a system.

The approach that appears most common for a "unified-view" system, as described above, is to have a server that is building a database (this may indeed be a true SQL or similar database, or may be a simpler "flat file" text log file) of information from the remote sites. The server may poll these sites for current information, or it may send unsolicited data on an alarm condition or other significant change in the status of the site. The user interface, which may be generated on the server or another computer, pulls the required data from the database to build the screens. The database provides a consistent interface for the user-interface software. All system data is available in one location and in a consistent format. The software that generates user screens may use historic data from the database to generate trend graphs (see Figure 9.5-4). Finally, data can be pulled from the database to build printed logs, as required. The user may also develop custom applications for analysis of transmission system data.

In such a system, interface hardware and software is required to convert the various data formats and hardware interfaces to a consistent format for storage in the database.

Different suppliers use a variety of methods to allow users to build custom user-interface screens. Some use scripting languages and others use drag-and-drop interfaces where a user builds the screen graphically (resulting in a script). These systems generally require that a program, devoted to transmitter control, be running on the computer where the opera-

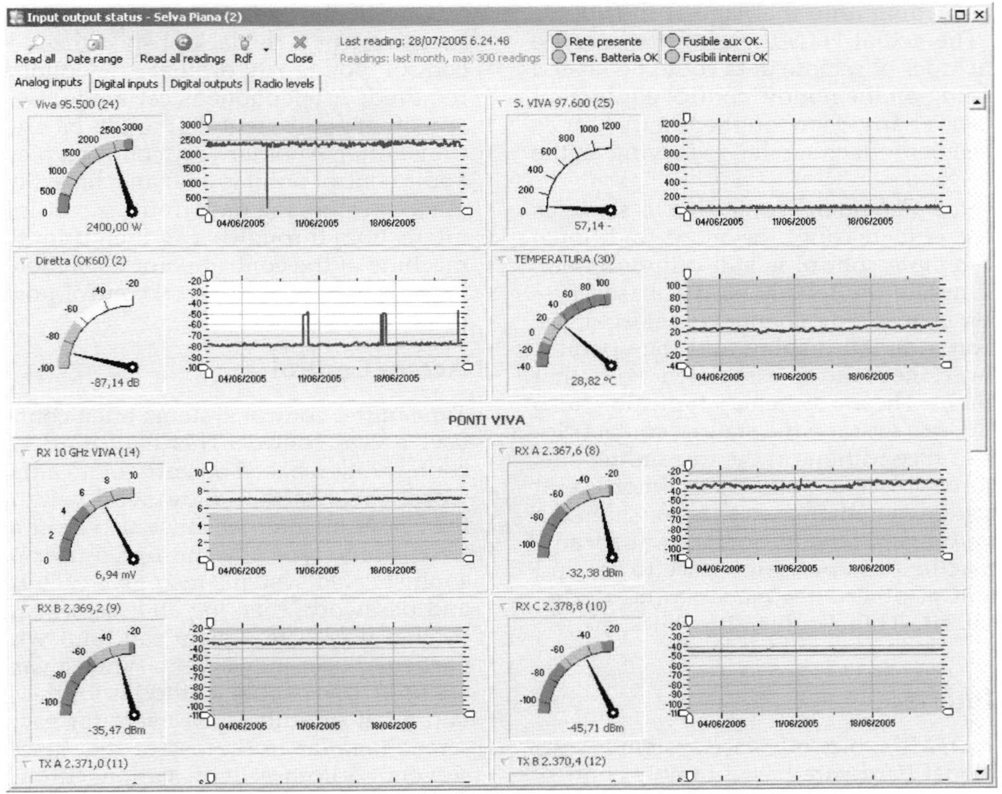

FIGURE 9.5-4 This user screen shows current analog parameters as simulated analog meters with color indicating limits. Values are also shown in digital form as numeric values with proper decimal point positioning and units. This view also has a "strip chart" view of each parameter, showing trends and easily identifying out-of-tolerance operation. See the CD for image details. (Image courtesy of Antenna Nord Telecomunicazioni.)

tor interacts with the system. This application pulls data from the database and builds user screens.

A technique that does not require a special application running on the user-interface computer is to use a standard web browser for the user interface. A couple of approaches are commonly used:

The browser loads and runs a Java application that provides the user interface and interacts with the database or directly interacts with the control equipment. This is the same as running a custom application, but the application need not be installed on the computer. Instead, it is downloaded and run as needed by the web browser.

Use the web browser interface to include an embedded web server in the control equipment. The user can usually configure the screens presented by the equipment to show appropriate labels and units for analog channels, and labels for status and control channels. Each time the page is loaded, updated data is presented to the user. The web server may include a metatag that forces a refresh of the screen at some interval so the operator has relatively current data. Systems with an embedded server similar to this are very easy to set up (as is the Java-based system described above) in that there is no central equipment. A browser directly addresses a site and displays the data.

Most systems with an embedded server (whether HTML or Java) are limited to displaying data from that particular site or control system on one screen (or in one window). HTML-based systems could be expanded to allow the display of data from multiple sites on user-defined screens by including a simple CGI script that would return data from an analog or status channel specified in the URL. HTML does not directly support "client-side includes" that would allow a page to be built with data from several different servers (image tags are a form of client-side include, but only allow the insertion of an image, not text that would represent a meter reading). However, HTML pages can include tags that load and execute a short piece of JavaScript. If a CGI on a remote site were to return a JavaScript document write statement containing a meter reading, a standard web browser would drop that reading onto the page at the appropriate location. The actual HTML pages that include the client side includes of remote data could be hosted anywhere, including on the remote control equipment itself, on a web server anywhere on the Internet, or as a local file on the client machine that is serving as the user interface.

While the use of a standard computer to serve as the user interface is increasing, the more traditional hardware-based remote control is still quite popular. These systems emulate the telephone dial-based systems of the 1950s. A metering channel is selected, the value is shown on a digital display, and the operator can raise or lower the control associated with that metering channel. These systems typically also include status LEDs to indicate the state of contact closures or other on/off conditions at the transmitter site. Some manufacturers place a similar user interface at all sites in the system, whether the site is considered a control point or a transmitter site. Someone at a transmitter site can use the front panel interface to see and control equipment at other sites (such as the studio, intermediate relay sites, other transmitter sites, etc.).

DTMF/Voice Interface

Operation of the DTMF/voice interface system is similar to the traditional hardware-based remote control. The operator dials the telephone number for the transmitter site, enters a user identification and/or password, and is given the status of the site using recorded or synthesized voice. The user can select various metering or status channels using the DTMF keyboard on the phone. The system responds with voice. The user can raise or lower the control associated with a particular channel using additional keys. These systems typically also have alarm reporting capabilities. Should an alarm condition occur (typically an analog value outside limits), the system will call a sequence of phone numbers to report the alarm. These dial-up systems may also report data through additional methods. If a DTMF response to the voice request for password is not received, the system may put a standard modem answer tone on the line to attempt to establish modem communications. If modem communications are established, the username and password are collected, and then access to the system is granted using ASCII text. A full screen of site status collected by the system is presented to the user. The screen also includes a menu for the various control operations. Log entries not yet reported to the system (routine meter readings, alarm conditions, system logins, and adjustments) can be logged to an attached printer or, if a computer is being used to emulate a terminal, to a hard drive. If logged to a hard drive, which can generally be accomplished by choosing "text file" as the printer, the log format might be tab-delimited ASCII to make machine analysis of the log easier. These systems might also call in daily at a specified time to automatically verify the system is working properly and to update log data to that time.

One of the great advantages of a simple DTMF/voice system is the lack of specialized hardware at the control point. An ordinary telephone is all that is required. A telephone is certainly less expensive than any other equipment that might be used to control the transmitter. Another very common piece of equipment at the control point is a fax machine. Some dial-up systems can store log data (routine readings, alarms, user interaction) through a day, then print the log to the fax machine at the control point. A station gets automatic logging with no additional control-point hardware.

Access Control

Transmitter control systems often contain some sort of user access control. This may limit access to system configuration based on entry of an administrator password. Systems that allow worldwide access through the PSTN or Internet must also limit access to authorized users, typically through a user password. More sophisticated systems may give each user a username and password, then log all accesses and other actions by that user. The system can limit which functions a particular user may access. While a variety of schemes are used, one simple method is for the system to maintain a list of authorized users for each user-interface screen (similar to the use of the .htaccess file in web servers). A licensee with stations throughout the country may give its corporate engineers access to every screen in the system. District engineers would only be able to access screens holding information on sites in their district. Station engineers would only see screens for their station. Station operators would have access to a limited subset of those station screens.

AUTOMATIC CONTROL

A control system can evaluate the parameters it has collected and make adjustments to the equipment in the system. At this time, most transmitters and their associated equipment are quite stable, so very limited automatic control is typically required. Automatic control is typically limited to minor power adjustments, AM day/night pattern and power switch, and automatic switching to standby equipment on the failure of the main transmitter. Transmitter output power is generally quite stable, or the transmitter may include its own automatic power control. FM stations relying on the indirect method of power determination (final amplifier voltage × current × efficiency) may want to have the external control equipment adjust power based on the calculated indirect power, as most transmitter automatic power control systems use the direct method of determining power (wattmeter on the transmitter output). As transmitter efficiency varies, it would be possible for the output power to be within limits if determined by the direct method and out of limits if determined by the indirect method. Whichever method the station licensee chooses, the power is to remain within limits.

Automatic pattern and power changes for AM stations are easily accomplished by current control systems. Some systems calculate sunrise and sunset based on the location of the system and the date. Others rely on a table of sunrise and sunset times. FCC licenses for AM stations that require a power or pattern change specify the times to change for each month of the year. These times are the average sunrise and sunset times for the month rounded to the nearest quarter hour.

Pattern change requires the control system to drive several pieces of equipment: the transmitter(s), the antenna system, and the antenna monitor. The interaction of the various components in the system could be handled by the remote control system, but is more commonly handled by custom electromechanical relay systems that interlock the equipment, ensuring that RF switches do not switch while energized and that all the RF switches are in the proper position before the RF is applied again. Some stations have utilized programmable logic controllers (PLCs) to implement the required interlocking and sequencing.

A similar arrangement can be used to switch transmitters should the main transmitter fail. In most cases, a transmitter swap is similar to a pattern change since a high-power RF switch is involved. An automatic control system could watch the output power of the transmitter that is on the air. If it is unable to keep that power within limits, the main transmitter is shut down, the antenna switched to the auxiliary transmitter, and that transmitter brought up. As with a pattern change, this sequence could be done by the control system, but is commonly done through custom relay circuitry or a PLC.

DIVISION OF RESPONSIBILITY

A wide variety of control systems is available to a particular station. A transmitter may include a sophisticated control system within the transmitter cabinet and provide an Ethernet, EIA-232, or EIA-485 interface to the outside world. An Ethernet link or EIA-232 link to a remote computer may be all that is required. Such a system, without a central controller, makes it difficult to have various pieces of equipment interact. If, however, the system provides machine-readable data, a control system or computer can integrate the data from the various pieces of equipment and issue appropriate commands and alarms. Devices with an Ethernet interface can provide a variety of data on the same connector. It may present data to a custom user-interface application or web browser while simultaneously responding to requests from another computer (perhaps SNMP commands or requests for data). EIA-232 links tend to generate data in human-friendly (and machine-unfriendly) formats, though it's possible to parse usable data out of almost anything. EIA-485 data tends to be more machine oriented, making it easy for a control system to send commands and get data from the transmitter and other equipment. The EIA-485 interface tends to duplicate the traditional analog metering and contact-closure control, but with only two wires, which may be shared among several pieces of equipment.

How much "intelligence" should be in the transmitter, how much in the transmitter site control equipment, and how much at the control point, and should there perhaps be intelligence somewhere else? Another consideration in the placement of intelligence is the available communications circuits and existing hardware in a system. For example, some systems add a web interface with hardware and software at the control point while maintaining voice-grade communications with the transmitter site(s) and more traditional control equipment at the transmitter site(s).

In large multisite systems, the intelligence (sorting out data, archiving it, generating alarms, making adjustments, etc.) is tending to move to a server that is typically at one of the control points (there may be several points that have access to the data). Where there is no central server and the control point equipment is of limited intelligence (e.g., a DTMF telephone), the system intelligence is put back at the transmitter site, typically in a separate control system, and possibly within the transmitter itself.

SUMMARY

The combination of equipment and communications circuits that define each broadcast facility is unique. Station personnel have the opportunity to combine a variety of monitoring equipment in a system that uniquely serves the needs of the station. Small facilities do fine with a DTMF/voice system and a telephone (and large facilities may include such a system as a backup system). Larger facilities will likely combine control and telemetry data from a variety of sources into an Ethernet link. Data can be directly dropped into a database server (or scraped from HTML or text pages, then dropped into the database). Other applications present the user with a variety of views of the data pulled from the database. Another application may handle automatic control of the system. Station engineers are making use of equipment designed for other industries. Programmable logic controllers can handle automatic sequencing of control functions and may be adapted to various other uses. Implementation of IP technology is a logical approach for future remote control of broadcast transmission systems.

Index

NAB BROADCASTERS